Errorless

49 YEARS JEE ADVANCED (1978 - 2026)

&

25 YEARS JEE MAIN (2002 - 2026)

CHEMISTRY

CHAPTER-WISE & TOPIC-WISE PYQs

DISHA™
Publication Inc

DISHA Publication Inc.

A-23 FIEE Complex, Okhla Phase II

New Delhi-110020

Tel: 49842349/ 49842350

We have tried and made our best efforts to provide accurate up-to-date information in this book.

Edited By

Kalpana Bhargav

Rawindar Singh

Diksha Jhanwar

Typeset By

DISHA DTP Team

Contents

CLASS XI A1 - A208

1. **Some Basic Concepts of Chemistry** A1 – A14
   - **Topic 1:** Laws of Chemical Combinations and Mole Concept
   - **Topic 2:** Percent Composition and Empirical Formula
   - **Topic 3:** Stoichiometric Calculations
2. **Structure of Atom** A15 – A34
   - **Topic 1:** Developments Leading to the Bohr's Model of Atom
   - **Topic 2:** Bohr's Model for Hydrogen Atom (Emission and Absorption Spectra)
   - **Topic 3:** Towards Quantum Mechanical Model of the Atom (Dual Behaviour and Uncertainty Principle)
   - **Topic 4 :** Quantum Mechanical Model of Atom
3. **Classification of Elements and Periodicity in Properties** A35 – A46
   - **Topic 1:** Modern Periodic Table
   - **Topic 2:** Periodic Trends in Properties of Elements
4. **Chemical Bonding and Molecular Structure** A47 – A66
   - **Topic 1:** Electrovalent and Covalent Bonding
   - **Topic 2:** Octet Rule and Resonance Structures
   - **Topic 3:** Dipole Moment and Fajan's Rule
   - **Topic 4:** VSEPR Theory, Valence Bond Theory and Concept of Hybridisation
   - **Topic 5:** Molecular Orbital Theory and Hydrogen Bonding
5. **States of Matter** A67 – A77
   - **Topic 1 :** Intermolecular Forces, , Gas Laws and Ideal Gas Equation
   - **Topic 2 :** Kinetic Theory of Gases and Molecular Speeds
   - **Topic 3 :** Deviation from Ideal Gas Behaviour, Liquefaction of Gases and Liquid State
6. **Thermodynamics** A78 – A100
   - **Topic 1:** First Law and Fundamentals of Thermodynamics
   - **Topic 2:** Laws of Thermochemistry
   - **Topic 3:** Entropy and Second Law of Thermodynamics
   - **Topic 4:** Spontaneity and Gibb's Free Energy
7. **Equilibrium** A101 – A127
   - **Topic 1:** Law of Chemical Equilibrium, Equilibrium Constant (K_c and K_p) and its Application)
   - **Topic 2:** Relation between K, Q and G and Factors Effecting Equilibrium
   - **Topic 3:** Theories of Acids and Bases, Ionic Product of Water and pH Scale
   - **Topic 4:** Ionisation of Weak Acids and Bases and Relations between K_a and K_b
   - **Topic 5:** Common Ion Effect, Salt Hydrolysis, Buffer Solutions and Solubility Product
8. **Redox Reaction** A128 – A132
   - **Topic 1:** Oxidation and Reduction Reactions
   - **Topic 2:** Oxidation Number
   - **Topic 3:** Disproportionation and Balancing of Redox Reactions
9. **Hydrogen** A133 – A137
   - **Topic 1 :** Preparation and Properties of Hydrogen and Hydrides
   - **Topic 2 :** Preparation and Properties of H_2O and D_2O
   - **Topic 3 :** Preparation and Properties of H_2O_2
10. **The *s*-Block Elements** A138– A143
    - **Topic 1 :** Group -1 Elements (Alkali Metals)
    - **Topic 2 :** Group -2 Elements (Alkaline Earth Metals)
    - **Topic 3 :** General Characteristics of Alkaline Earth Metals and their Compounds
    - **Topic 4 :** Some Importance Compounds of Alkaline Earth Metals
11. **The *p*-Block Elements (Group-13 and 14)** A144 – A151
    - **Topic 1:** Boron Family
    - **Topic 2:** Carbon Family
12. **Organic Chemistry-Some Basic Principles and Techniques** A152 – A175
    - **Topic 1:** Classification and Nomenclature of Organic Compounds
    - **Topic 2:** Isomerism in Organic Compounds
    - **Topic 3:** Concept of Reaction Mechanism in Organic Compounds, Purification and Quantitative Analysis
13. **Hydrocarbons** A176 – A204
    - **Topic 1:** Alkanes
    - **Topic 2:** Alkenes
    - **Topic 3:** Alkynes
    - **Topic 4:** Aromatic Hydrocarbons
14. **Environmental Chemistry** A205 – A208
    - **Topic 1 :** Air Pollution
    - **Topic 2 :** Water and Soil Pollution

Hints & Solutions (Class XI) A209 - A356

1. Some Basic Concepts of Chemistry A209 – A225
2. Structure of Atom A226 – A238
3. Classification of Elements and Periodicity in Properties A239 – A244
4. Chemical Bonding and Molecular Structure A245 – A260
5. States of Matter A261 – A268
6. Thermodynamics A269 – A284
7. Equilibrium A285 – A309

Reaction A310 – A312

9. Hydrogen A313 – A314
10. The *s*-Block Elements A315 – A317
11. The *p*-Block Elements (Group-13 and 14) A318 – A321
12. Organic Chemistry-Some Basic Principles and Techniques A322 – A335
13. Hydrocarbons A336 – A354
14. Environmental Chemistry A355 – A356

CLASS XII

B1 - B340

1. The Solid State B1 – B6
Topic 1 : Classification of Solids, Crystal Structure and Properties of Solids
Topic 2 : Close Packed Structure, Packing Efficiency and Imperfection in Solids
2. Solutions B7 – B26
Topic 1: Concentration of Solutions and Solubility
Topic 2: Vapour Pressure, Laws of Solutions, Ideal and Non-ideal Solutions
Topic 3: Colligative Properties and Abnormal Molecular Masses
3. Electrochemistry B27 – B50
Topic 1: Cells and Electrode Potential, Nernst Equation
Topic 2: Conductance and Conductivity
Topic 3: Electrolytic Cells and Electrolysis
Topic 4: Commercial Cells and Corrosion
4. Chemical Kinetics and Nuclear Chemistry B51 – B75
Topic 1: Rate of Reaction, Rate Laws and Rate Constant
Topic 2: Order of Reaction and Half Life Period
Topic 3: Theories of Rate of Reaction
Topic 4: Nuclear Chemistry
5. Surface Chemistry B76 – B84
Topic 1: Adsorption
Topic 2: Catalysis and Theories of Catalysis
Topic 3: Colloids and Emulsions
6. General Principles and Process of Isolation of Elements B85 – B93
Topic 1: Occurrence of Metals and Metallurgical Processes
Topic 2: Purification and Uses of Metals
7. The *p*-Block Elements (Group 15, 16, 17 & 18) B94– B107
Topic 1: Nitrogen Family
Topic 2: Oxygen Family
Topic 3: Halogen Family
Topic 4: Noble Gases
8. The *d*-and *f*-Block Elements B108 – B123
Topic 1: Electronic Configuration and Properties of Transition Elements
Topic 2: Compounds of Transition Elements
Topic 3: Lanthanoids and Actinoids
9. Co-ordination Compounds B124 – B147
Topic 1: Werner's Theory, Important terms, Nomenclature and Isomerism of Coordination Compounds Coordination Compounds
Topic 2: Bonding in Coordination Compounds, (Valence Bond Theory and Crystal Field Theory)
Topic 3: Importance of Coordination Compounds
10. Haloalkanes and Haloarenes B148 – B175
Topic 1: Preparation and Properties of Haloalkanes
Topic 2: Preparation and Properties of Haloarenes
Topic 3: Some Important Polyhalogen Compounds
11. Alcohols, Phenols and Ethers B176– B201
Topic 1: Preparation and Properties of Alcohols
Topic 2: Preparation and Properties of Phenols
Topic 3: Preparation and Properties of Ethers
12. Aldehydes, Ketones and Carboxylic Acids B202 – B254
Topic 1: Methods of Preparation of Carbonyl Compounds
Topic 2: Properties of Carbonyl Compounds
Topic 3 : Preparation and Properties of Carboxylic Acids and their Derivatives
13. Amines B255 – B290
Topic 1: Aliphatic and Aromatic Amines
Topic 2: Preparation, Chemical Reactions and Importance of Diazonium Salts
14. Biomolecules B291 – B309
Topic 1: Carbohydrates
Topic 2: Proteins and Amino Acids
Topic 3: Enzymes and Nucleic Acid
Topic 4: Vitamins and Hormones
15. Polymers B310 – B316
Topic 1: Classification of Polymers
Topic 2: Preparation and Properties of Polymers
Topic 3: Uses of Polymers
16. Chemistry in Every Day Life B317 – B321
Topic 1: Drugs Classification, Therapeutic Action of Drugs and Chemicals in Food
17. Principles Related to Practical Chemistry B322 – B340
Topic 1: Detection of Extra Elements, Detection of Organic Functional Groups and Preparation of Organic Compounds
Topic 2: Qualitative Salt Analysis and Preparation of Inorganic Compounds
Topic 3: Principles Involved in Titrimetric Exercises, Calorimetry, Preparation of Colloidal Sols and Kinetic Study of Reaction Between I^- and H_2O_2

Hints & Solutions (Class XII)

B341 - B556

1. The Solid State B341 – B344
2. Solutions B345 – B361
3. Electrochemistry B362 – B380
4. Chemical Kinetics and Nuclear Chemistry B381 – B399
5. Surface Chemistry B400 – B403
6. General Principles and Process of Isolation of Elements B404 – B407
7. The *p*-Block Elements (Group 15, 16, 17 & 18) B408 – B417
8. The *d*-and *f*-Block Elements B418 – B427
9. Co-ordination Compounds B428 – B449
10. Haloalkanes and Haloarenes B450 – B462
11. Alcohols, Phenols and Ethers B463 – B475
12. Aldehydes, Ketones and Carboxylic Acids B476 – B512
13. Amines B513 – B534
14. Biomolecules B535 – B542
15. Polymers B543 – B545
16. Chemistry in Every Day Life B546 – B547
17. Principles Related to Practical Chemistry B548 – B556

Disha's **SMART PREP EdGe**

CRACK JEE

Deep JEE Analysis

JEE advanced number of appeared and seats status- last 6 years

Stats	2026	2025	2024	2023	2022	2021
Eligible	250000	250000	250000	250000	262175	260010
Registered	190000	193000	186584	189744	160038	151209
Appeared		180422	180200	155538	141699	150838
Qualified		54378	48248	43773	40712	41862
Seats		17740	17760	17385	16598	16296

JEE Advanced Cutoff

Category	2025 (%)	2024 (%)	2023 (%)	2022 (%)
General (CRL)	20.56	30.34	23.89	15.28
GEN-EWS	18.50	27.30	21.50	13.89
SC	10.28	15.17	11.95	7.78
ST	10.28	15.17	11.95	7.78
PwD	10.28	15.17	11.95	7.78

2025 Qualifying Candidates according to Gender

Gender	Registered	Appeared in both Papers	Qualified
Male	143810	139085	44974
Female	43413	41337	9404
Total	187223	180422	54378

In 2025,
Total Female candidates who qualified: 9404
Seats available only for female candidates: 3566

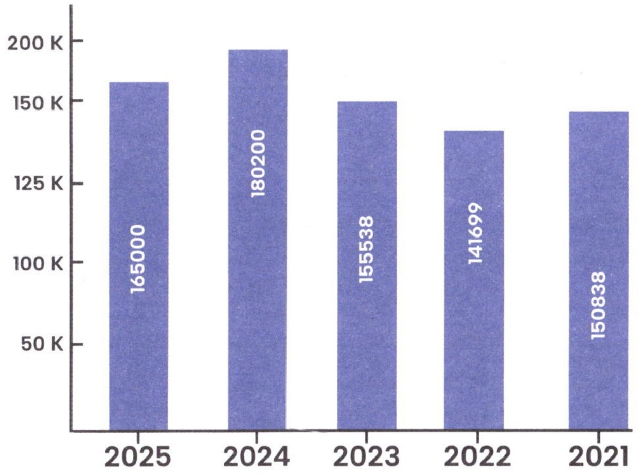

This Reduction of **15,000** in the No. of Candidates directly affect candidates' **Rank & Percentile**.

2025 JEE Advanced Qualifying Marks

Rank List	Minimum Marks in Each Subject	Minimum Aggregate Marks
Common Rank List (CRL)	7	74
OBC-NCL Rank List	6	66
GEN-EWS Rank List	6	66
SC Rank List	3	37
ST Rank List	3	37
Common-PwD Rank List (CRL-PwD)	3	37
OBC-NCL-PwD Rank List	3	37
GEN-EWS-PwD Rank List	3	37
SC-PwD Rank List	3	37
ST-PwD Rank List	3	37
Preparatory Course (PC) Rank List	1	18

JEE Advanced RANK vs MARKS (in percentage) : last 10 years

Rank	2025	2024	2023	2022	2021	2020	2019	2018	2017	2016
1	92.2%	98%	70%	87%	86%	87%	93%	74%	72%	86%
100	77.2%	83%	67%	52%	73%	75%	80%	60%	60%	66%
500	65%	72%	62%	47%	67%	70%	75%	57%	56%	55%
1000	57.8%	65%	55%	43%	63%	65%	70%	54%	52%	50%
2000	50.2%	57%	50%	38%	58%	60%	65%	50%	48%	45%
3000	45.8%	54%	46%	35%	54%	56%	61%	47%	45%	42%
4000	42.8%	52%	44%	33%	52%	54%	59%	45%	43%	40%
5000	40.3%	47%	43%	32%	49%	52%	57%	43%	41%	39%
6000	38.3%	45%	41%	30%	47%	50%	55%	42%	40%	37%
7000	36.9%	43%	39%	29%	45%	48%	53%	40%	38%	36%
8000	35.5%	41%	38%	28%	44%	46%	51%	38%	36%	35%
9000	34.4%	39%	36%	27%	42%	44%	49%	36%	35%	34%
10000	33.3%	37%	34%	26%	40%	43%	47%	35%	33%	33%
12000	31.3%	34%	31%	24%	37%	40%	44%	32%	31%	30%
15000	28.8%	30%	28%	21%	34%	36%	40%	30%	28%	28%

Computer Engineering @ Top 10 IIT's

Institute	2025 Rank	2024 Rank	2023 Rank	2022 Rank
IIT Bombay	66	68	67	61
IIT Delhi	125	42	43	46
IIT Madras	171	159	148	175
IIT Kanpur	270	252	238	238
IIT Kharagpur	450	414	430	443
IIT Roorkee	535	481	500	521
IIT Guwahati	699	607	650	670
IIT Hyderabad	667	656	685	710
IIT Varanasi	1350	1071	1100	1120
IIT Indore	1628	1389	1385	1400

CHEMISTRY

Topic-wise weightage

Topic/Year	Organic Chemistry	Inorganic Chemistry	Physical Chemistry
2026	38%	30%	32%
2025	37%	28%	34%
2024	34%	33%	33%
2023	30%	30%	30%
2022	33%	33%	33%
2021	26%	24%	50%
2020	28%	22%	50%
2019	33%	33%	33%
2018	31%	22%	47%
2017	35%	28%	39%
2016	33%	31%	36%
2015	35%	30%	35%

Paper 1 & 2 - JEE Advanced 2026 CHEMISTRY Chapter HEAT MAP

Class XI

Ch 1	Ch 2: 2 (E1, A0, D1)	Ch 3: 1 (E1, A0, D0)	Ch 4: 2 (E2, A0, D0)	Ch 5: 1 (E0, A1, D0)
Ch 6: 2 (E0, A1, D1)	Ch 7: 1 (E1, A0, D0)	Ch 8	Ch 9	Ch 10
Ch 11	Ch 12	Ch 13: 3 (E1, A0, D2)		

Class XII

Ch 1	Ch 2: 4 (E1, A3, D0)	Ch 3: 1 (E0, A1, D0)	Ch 4: 1 (E1, A0, D0)	Ch 5: 1 (E0, A1, D0)
Ch 6	Ch 7: 3 (E3, A1, D0)	Ch 8	Ch 9: 2 (E1, A0, D1)	Ch 10: 1 (E0, A0, D1)
Ch 11	Ch 12: 4 (E1, A1, D2)	Ch 13: 1 (E0, A1, D0)	Ch 14: 2 (E0, A1, D1)	Ch 15: 2 (E0, A2, D0)
Ch 16	Ch 17			

Color Codes indicating the number of Questions that appeared in JEE Adv. 2026 (from a particular Chapter)

0 Qns 1 Qns 2 Qns 3 Qns 4 More Qns

Question distribution as per Class, Chapter & Section 2026

Class	Chapter	Paper 1	Paper 2	Section	Question Type	Difficulty
11	Structure of atom	Present	Present	Section 2 & 3	One or more than one correct, Numerical Value	Easy, Difficult
11	Chemical Bonding & Molecular Structure	Present		Section 1 & 4	MCQ (Single Correct) Matching	Easy
11	Thermodynamics	Present		Section 1 & 4	MCQ (Single Correct) Matching	Difficult, Moderate
11	Equilibrium	Present		Section 1	MCQ (Single Correct)	Easy
11	Hydrocarbon	Present	Present	Section 3 & 4	Numerical Value	Easy, Difficult
12	Solutions		Present	Section 3 & 4	Numerical Value	Moderate
12	Electrochemistry		Present	Section 1	MCQ (Single Correct)	Moderate
12	Chemical Kinetics		Present	Section 2	One or More than One Correct	Easy
12	Surface Chemistry		Present	Section 3	Numerical Value	Moderate
12	p-Block Elements (Groups 15–18)	Present	Present	Section 2, 1 & 2	One or more than one correct, MCQ (SIngle correct, One or more than)	Easy
12	Coordination Compounds	Present	Present	Section 3	Numerical Value	Difficult, Easy
12	Aldehydes, Ketones and Carboxylic Acids	Present		Section 1, 2, 3 & 4	MCQ, One or more than one correct, Numerical Value, Matching	Difficult, Moderate, Easy, Moderate
12	Amines		Present	Section 2	One or More than one correct	Difficult
12	Biomolecules		Present	Section 1 & 2	MCQ, One or more than one correct	Moderate, Difficult

Chapter-1

Some Basic Concepts of Chemistry

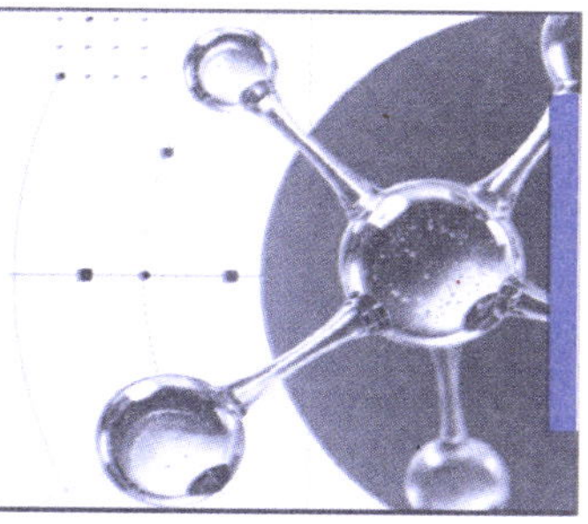

Part-A (JEE Main)

Topic 1- Laws of Chemical Combinations and Mole Concept

MULTIPLE CHOICE QUESTIONS (MCQs)

1. Match List-I with List-II. [April 08, 2026 (II)]

List-I Mass of substance	List-II Number of atoms
A. 1.8 mg water	I. $2 \times 10^{-4} \times N_A$
B. 9.8 mg sulphuric acid	II. $1.5 \times 10^{-4} \times N_A$
C. 1.8 mg carbon	III. $3 \times 10^{-4} \times N_A$
D. 5.85 mg salt (NaCl)	IV. $7 \times 10^{-4} \times N_A$

Choose the correct answer from the options given below:
(1) A-IV, B-III, C-I, D-II (2) A-III, B-II, C-IV, D-I
(3) A-III, B-IV, C-II, D-I (4) A-III, B-IV, C-I, D-II

2. Which of the following contain the same number of atoms? (Given : Molar mass in g mol^{-1} of H, He, O and S are 1,4,16 and 32 respectively) [April 06, 2026 (II)]
A. 2g of O_2 gas B. 4g of SO_2 gas
C. 1400 mL of O_2 at STP D. 0.05 L of He at STP
E. 0.0625 mol of H_2 gas
Choose the correct answer from the options given below:
(1) A and B only (2) B and C only
(3) C and D only (4) A, C and E only

3. The correct order of total number of atoms of present in
(A) 2 moles of cyclohexane
(B) 684 g of sucrose
(C) 90.8 L of dihydrogen at STP [April 04, 2026 (II)]
(1) C > A > B (2) C > B > A
(3) B > C > A (4) B > A > C

4. Number of moles and number of molecules in 1.4187 L of SO_2 at STP respectively are [April 04, 2026 (I)]
(1) 0.1266; 3.812×10^{22} (2) 0.0633; 3.812×10^{22}
(3) 0.1266; 7.6238×10^{22} (4) 0.0633; 7.6238×10^{22}

5. Among 10^{-9} g (each) of the following elements, which one will have the highest number of atoms? Element : Pb, Po, Pr and Pt [April 3, 2025 (I)]
(1) Po (2) Pr (3) Pb (4) Pt

6. Choose the correct statements. [Jan. 29, 2025 (I)]
(A) Weight of a substance is the amount of matter present in it.
(B) Mass is the force exerted by gravity on an object.
(C) Volume is the amount of space occupied by a substance.
(D) Temperatures below 0°C are possible in Celsius scale, but in Kelvin scale negative temperature is not possible.
(E) Precision refers to the closeness of various measurements for the same quantity.
(1) (B), (C) and (D) Only (2) (A), (B) and (C) Only
(3) (A), (D) and (E) Only (4) (C), (D) and (E) Only

7. 2.8×10^{-3} mol of CO_2 is left after removing 10^{21} molecules from its 'x' mg sample. The mass of CO_2 taken initially is Given : $N_A = 6.02 \times 10^{23}\ mol^{-1}$. [Jan. 23, 2025 (I)]
(1) 196.2 mg (2) 98.3 mg
(3) 150.4 mg (4) 48.2 mg

8. The candela is the luminous intensity, in a given direction, of a source that emits monochromatic radiation of frequency 'A' $\times 10^{12}$ hertz and that has a radiant intensity in that direction of $\frac{1}{'B'}$ watt per steradian. 'A' and 'B' are respectively
(1) 540 and $\frac{1}{683}$ (2) 540 and 683 [April 9, 2024 (II)]
(3) 450 and $\frac{1}{683}$ (4) 450 and 683

9. The incorrect postulates of the Dalton's atomic theory are :
(A) Atoms of different elements differ in mass. [April 5, 2024 (I)]
(B) Matter consists of divisible atoms.
(C) Compounds are formed when atoms of different element combine in a fixed ratio.
(D) All the atoms of given element have different properties including mass.
(E) Chemical reactions involve reorganisation of atoms.
Choose the correct answer from the options given below:
(1) (B), (D), (E) only (2) (A), (B), (D) only
(3) (C), (D), (E) only (4) (B), (D) only

10. Choose the Incorrect Statement about Dalton's Atomic Theory
(1) Compounds are formed when atoms of different elements combine in any ratio [April 4, 2024 (II)]
(2) All the atoms of a given element have identical properties including identical mass
(3) Matter consists of indivisible atoms
(4) Chemical reactions involve reorganization of atoms

11. The number of molecules and moles in 2.8375 litres of O_2 at STP are respectively [April 10, 2023 (I)]
(1) 7.527×10^{22} and 0.250 mol
(2) 1.505×10^{23} and 0.250 mol
(3) 7.527×10^{23} and 0.125 mol
(4) 7.527×10^{22} and 0.125 mol

12. Match List I with List II [April 10, 2023 (II)]

List I	List II
A. 16g of $CH_4(g)$	I. Weighs 28g
B. 1g of $H_2(g)$	II. 60.2×10^{23} electrons
C. 1 mole of $N_2(g)$	III. Weighs 32g
D. 0.5 mol of $SO_2(g)$	IV. Occupies 11.4 L volume at STP

Choose the correct answer from the options given below:
(1) A-I, B-III, C-II, D-IV (2) A-II, B-III, C-IV, D-I
(3) A-II, B-IV, C-III, D-I (4) A-II, B-IV, C-I, D-III

13. Hemoglobin contains 0.34% of iron by mass. The number of Fe atoms in 3.3 g of hemoglobin is : (Given : Atomic mass of Fe is 56 u, N_A is 6.022×10^{23} mol^{-1}) [July 26, 2022 (II)]
(1) 1.21×10^5 (2) 12.0×10^{16}
(3) 1.21×10^{20} (4) 3.4×10^{22}

14. A commercially sold conc. HCl is 35% HCl by mass. If the density of this commercial acid is 1.46 g/mL, the molarity of this solution is : [June 26, 2022 (I)]
(Atomic mass : Cl = 35.5 amu, H = 1 amu)
(1) 10.2 M (2) 12.5 M (3) 14.0 M (4) 18.2 M

15. A solution of two components containing n_1 moles of the 1st component and n_2 moles of the 2nd component is prepared. M_1 and M_2 are the molecular weights of component 1 and 2 respectively. If d is the density of the solution in g mL^{-1}, C_2 is the molarity and x_2 is the mole fraction of the 2nd component, then C_2 can be expressed as: [Sep. 06, 2020 (I)]
(1) $C_2 = \dfrac{1000\, x_2}{M_1 + x_2(M_2 - M_1)}$ (2) $C_2 = \dfrac{d\, x_2}{M_2 + x_2(M_2 - M_1)}$
(3) $C_2 = \dfrac{1000\, d\, x_2}{M_1 + x_2(M_2 - M_1)}$ (4) $C_2 = \dfrac{d\, x_1}{M_2 + x_2(M_2 - M_1)}$

16. Amongst the following statements, that which was not proposed by Dalton was: [Jan. 07, 2020 (I)]
(1) Chemical reactions involve reorganization of atoms. These are neither created nor destroyed in a chemical reaction.
(2) All the atoms of a given element have identical properties including identical mass. Atoms of different elements differ in mass.
(3) When gases combine or reproduced in a chemical reaction they do so in a simple ratio by volume, provided all gases are at the same T & P.
(4) Matter consists of indivisible atoms.

17. 5 moles of AB_2 weigh 125×10^{-3} kg and 10 moles of A_2B_2 weigh 300×10^{-3} kg. The molar mass of A (M_A) and molar mass of B (M_B) in kg mol^{-1} are: [April 12, 2019 (I)]
(1) $M_A = 10 \times 10^{-3}$ and $M_B = 5 \times 10^{-3}$
(2) $M_A = 50 \times 10^{-3}$ and $M_B = 25 \times 10^{-3}$
(3) $M_A = 25 \times 10^{-3}$ and $M_B = 50 \times 10^{-3}$
(4) $M_A = 5 \times 10^{-3}$ and $M_B = 10 \times 10^{-3}$

18. The percentage composition of carbon by mole in methane is : [April 8, 2019 (II)]
(1) 75% (2) 80% (3) 25% (4) 20%

19. 8 g of NaOH is dissolved in 18 g of H_2O. Mole fraction of NaOH in solution and molality (in mol kg^{-1}) of the solution respectively are : [Jan. 12, 2019 (II)]
(1) 0.2, 22.20 (2) 0.2, 11.11
(3) 0.167, 11.11 (4) 0.167, 22.20

20. The amount of sugar ($C_{12}H_{22}O_{11}$) required to prepare 2 L of its 0.1 M aqueous solution is: [Jan. 10, 2019 (II)]
(1) 136.8 g (2) 17.1 g (3) 68.4 g (4) 34.2 g

21. An unknown chlorohydrocarbon has 3.55% of chlorine. If each molecule of the hydrocarbon has one chlorine atom only, chlorine atoms present in 1g of chlorohydrocarbon are:
(Atomic wt. of Cl = 35.5 u; Avogadro constant = 6.023×10^{23} mol^{-1}) [Online April 16, 2018]
(1) 6.023×10^9 (2) 6.023×10^{23}
(3) 6.023×10^{21} (4) 6.023×10^{20}

22. Excess of NaOH (aq) was added to 100 mL of $FeCl_3$ (aq) resulting into 2.14 g of $Fe(OH)_3$. The molarity of $FeCl_3$ (aq) is : (Given molar mass of Fe=56 g mol^{-1} and molar mass of Cl = 35.5 g mol^{-1}) [Online April 8, 2017]
(1) 0.2 M (2) 0.3 M (3) 0.6 M (4) 1.8 M

23. 5 L of an alkane requires 25 L of oxygen for its complete combustion. If all volumes are measured at constant temperature and pressure, the alkane is : [Online April 9, 2016]
(1) Isobutane (2) Ethane
(3) Butane (4) Propane

24. The molecular formula of a commercial resin used for exchanging ions in water softening is $C_8H_7SO_3^-\,Na^+$ (Mol. wt. 206. What would be the maximum uptake of Ca^{2+} ions by the resin when expressed in mole per gram resin ? [2015]
(1) $\dfrac{2}{309}$ (2) $\dfrac{1}{412}$ (3) $\dfrac{1}{103}$ (4) $\dfrac{1}{206}$

25. Dissolving 120 g of a compound of mol. wt. 60 in 1000 g of water gave a solution of density 1.12 g/mL. The molarity of the solution is: [Online April 9, 2014]
(1) 1.00 M (2) 2.00 M (3) 2.50 M (4) 4.00 M

26. The amount of oxygen in 3.6 moles of water is: [Online April 9, 2014]
(1) 115.2 g (2) 57.6 g (3) 28.8 g (4) 18.4 g

27. The density of 3M solution of sodium chloride is 1.252 g mL^{-1}. The molality of the solution will be :
(molar mass, NaCl = 58.5 g mol^{-1}) [Online April 22, 2013]
(1) 260 m (2) 2.18 m (3) 2.79 m (4) 3.00 m

28. Number of atoms in the following samples of substances is the largest in : [Online April 23, 2013]
(1) 4.0 g of hydrogen (2) 71.0 g of chlorine
(3) 127.0 g of iodine (4) 48.0 g of magnesium

29. 10 mL of 2M NaOH solution is added to 200 mL of 0.5 M of NaOH solution. What is the final concentration ? [Online April 25, 2013]
(1) 0.57 M (2) 5.7 M
(3) 11.4 M (4) 1.14 M

30. The concentrated sulphuric acid that is peddled commercial is 95% H_2SO_4 by weight. If the density of this commercial acid is 1.834 g cm^{-3}, the molarity of this solution is [Online May 7, 2012]
(1) 17.8 M (2) 12.0 M (3) 10.5 M (4) 15.7 M

31. The ratio of number of oxygen atoms (O) in 16.0 g ozone (O_3), 28.0 g carbon monoxide (CO) and 16.0 oxygen (O_2) is (Atomic mass : C = 12, O = 16 and Avogadro's constant $N_A = 6.0 \times 10^{23}$ mol^{-1}) [Online May 7, 2012]
(1) 3 : 1 : 2 (2) 1 : 1 : 2 (3) 3 : 1 : 1 (4) 1 : 1 : 1

32. The ppm level of F^- in a 500 g sample of a tooth paste containing 0.2 g F^- is [Online May 12, 2012]
(1) 400 (2) 1000 (3) 250 (4) 200

33. 5 g of benzene on nitration gave 6.6 g of nitrobenzene. The theoretical yield of the nitrobenzene will be [Online May 12, 2012]
(1) 4.5 g (2) 5.6 g (3) 8.09 g (4) 6.6 g

34. How many moles of magnesium phosphate, $Mg_3(PO_4)_2$ will contain 0.25 mole of oxygen atoms? [2006]
(1) 1.25×10^{-2} (2) 2.5×10^{-2}
(3) 0.02 (4) 3.125×10^{-2}

35. If we consider that 1/6, in place of 1/12, mass of carbon atom as the relative atomic mass unit, the mass of one mole of the substance will [2005]
(1) be a function of the molecular mass of the substance
(2) remain unchanged
(3) increase two fold
(4) decrease twice

36. 6.02×10^{20} molecules of urea are present in 100 mL of its solution. The concentration of urea solution is [2004]
(1) 0.02 M (2) 0.01 M (3) 0.001 M (4) 0.1 M
(Avogadro constant, $N_A = 6.02 \times 10^{23}$ mol^{-1})

37. 25 mL of a solution of barium hydroxide on titration with a 0.1 molar solution of hydrochloric acid gave a titre value of 35 mL. The molarity of barium hydroxide solution was [2003]
(1) 0.14 (2) 0.28 (3) 0.35 (4) 0.07

38. With increase of temperature, which of these changes? [2002]
(1) molality (2) weight fraction of solute
(3) molarity (4) mole fraction

39. Number of atoms in 558.5 gram Fe (at. wt. of Fe = 55.85 g mol^{-1}) is
(1) twice that in 60 g carbon [2002]
(2) 6.023×10^{22}
(3) half that in 8 g He
(4) $558.5 \times 6.023 \times 10^{23}$

NUMERICAL VALUE QUESTIONS (NVQs)

40. 0.1 mol of the following given antiviral compound (P) will weigh _______ $\times 10^{-1}$ g [April 2, 2025 (I)]

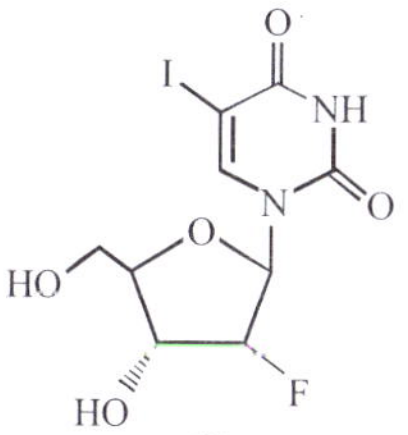

(P)

(Given : molar mass in g mol^{-1} H: 1, C : 12, N : 14, O : 16, F : 19, I : 127)

41. 0.05 cm thick coating of silver is deposited on a plate of 0.05 m^2 area. The number of silver atoms deposited on plate are ____ $\times 10^{23}$. (At mass Ag = 108, d = 7.9 g cm^{-3}) [Jan. 30, 2024 (I)]

42. In the given reaction, $X + Y + 3Z \rightleftharpoons XYZ_3$

if one mole of each of X and Y with 0.05 mol of Z gives compound XYZ_3. (Given : Atomic masses of X, Y and Z are 10, 20 and 30 amu, respectively). The yield of XYZ_3 is __________ g. (Nearest integer) [July 28, 2022 (I)]

43. Chlorophyll extracted from the crushed green leaves was dissolved in water to make 2L solution of Mg of concentration 48 ppm. The number of atoms of Mg in this solution is $x \times 10^{20}$ atoms. The value of x is ____. (Nearest Integer)
(Given: Atomic mass of Mg is 24 g mol^{-1};
$N_A = 6.02 \times 10^{23}$ mol^{-1}) [July 26, 2022 (I)]

44. 56.0 L of nitrogen gas is mixed with excess of hydrogen gas and it is found that 20 L of ammonia gas is produced. The volume of unused nitrogen gas is found to be____ L. [July 25, 2022 (II)]

45. Two elements A and B which form 0.15 moles of A_2B and AB_3 type compounds. If both A_2B and AB_3 weigh equally, then the atomic weight of A is _____ times of atomic weight of B. [June 27, 2022 (I)]

46. The number of N atoms in 681 g of $C_7H_5N_3O_6$ is $x \times 10^{21}$. The value of x is __________ ($N_A = 6.02 \times 10^{23}$ mol^{-1}) (Nearest Integer) [June 25, 2022 (I)]

47. The number of atoms in 8 g of sodium is $x \times 10^{23}$. The value of x is________. (Nearest integer)
[Given : $N_A = 6.02 \times 10^{23}$ mol^{-1}
Atomic mass of Na = 23.0 u] [Sep. 1, 2021 (II)]

48. The molarity of HNO_3 in a sample which has density 1.4 g/mL and mass percentage of 63% is ______. (Molecular Weight of HNO_3 = 63) [Jan. 09, 2020 (I)]

Topic 2 - Percent Composition and Empirical Formula

MULTIPLE CHOICE QUESTIONS (MCQs)

49. An oxide of iron contains 69.9% iron, its empirical formula, is : (Given : Molar mass of Fe and O are 56 and 16 g mol^{-1} respectively.) [April 06, 2026 (I)]
(1) FeO (2) Fe_2O_3 (3) Fe_3O_4 (4) FeO_3

50. On combustion 0.210 g of an organic compound containing C, H and O gave 0.127 g H_2O and 0.307 g CO_2. The percentages of hydrogen and oxygen in the given organic compound respectively are: [April 8, 2025 (II)]
(1) 53.41, 39.6 (2) 6.72, 53.41
(3) 7.55, 43.85 (4) 6.72, 39.87

51. On complete combustion 1.0 g of an organic compound (X) gave 1.46 g of CO_2 and 0.567 g of H_2O. The empirical formula mass of compound (X) is _________ g.
(Given molar mass in g mol^{-1} C : 12, H : 1, O : 16) [April 2, 2025 (I)]
(1) 30 (2) 45 (3) 60 (4) 15

52. The elemental composition of a compound is 54.2%, C, 9.2% H and 36.6% O. If the molar mass of the compound is 132 g mol^{-1}, the molecular formula of the compound is :
[Given : The relative atomic mass of C : H : O = 12 : 1 : 16]
(1) $C_4H_9O_3$ (2) $C_6H_{12}O_6$ [Jan. 24, 2025 (II)]
(3) $C_6H_{12}O_3$ (4) $C_4H_8O_2$

53. An organic compound has 42.1% carbon, 6.4% hydrogen and remainder is oxygen. If its molecular weight is 342, then its molecular formula is : [April 5, 2024 (I)]
(1) $C_{11}H_{18}O_{12}$ (2) $C_{12}H_{20}O_{12}$
(3) $C_{14}H_{20}O_{10}$ (4) $C_{12}H_{22}O_{11}$

54. A sample of $CaCO_3$ and $MgCO_3$ weighed 2.21 g is ignited to constant weight of 1.152 g. The composition of mixture is :
(Given molar mass in g mol^{-1} $CaCO_3$:100, $MgCO_3$: 84)
(1) 1.187 g $CaCO_3$ + 1.023 g $MgCO_3$ [Jan. 31, 2024 (II)]
(2) 1.023 g $CaCO_3$ + 1.023 g $MgCO_3$
(3) 1.187 g $CaCO_3$ + 1.187 g $MgCO_3$
(4) 1.023 g $CaCO_3$ + 1.187 g $MgCO_3$

55. A metal chloride contains 55.0% of chlorine by weight. 100 mL vapours of the metal chloride at STP weight 0.57 g. The molecular formula of the metal chloride is
(Given : Atomic mass of chlorine is 35.5u) [April 12, 2023 (I)]
(1) MCl_2 (2) MCl_4 (3) MCl_3 (4) MCl

56. Compound A contains 8.7% Hydrogen, 74% Carbon and 17.3% Nitrogen. The molecular formula of the compound is,
Given : Atomic masses of C, H and N are 12, 1 and 14 amu respectively. [June 28, 2022 (II)]
The molar mass of the compound A is 162 g mol^{-1}.
(1) $C_4H_6N_2$ (2) C_2H_3N (3) C_5H_7N (4) $C_{10}H_{14}N_2$

57. 120 of an organic compound that contains only carbon and hydrogen gives 330 g of CO_2 and 270 g of water on complete combustion. The percentage of carbon and hydrogen, respectively are. [June 24, 2022 (II)]
(1) 25 and 75 (2) 40 and 60
(3) 60 and 40 (4) 75 and 25

58. Complete combustion of 750 g of an organic compound provides 420 g of CO_2 and 210 g of H_2O. The percentage composition of carbon and hydrogen in organic compound is 15.3 and ____ respectively. (Round off to the Nearest Integer). [March 16, 2021 (I)]

59. Complete combustion of 1.80 g of an oxygen containing compound ($C_xH_yO_z$) gave 2.64 g of CO_2 and 1.08 g of H_2O. The percentage of oxygen in the organic compound is: [Feb. 25, 2021 (I)]
(1) 50.33 (2) 53.33 (3) 51.63 (4) 63.53

60. The average molar mass of chlorine is 35.5 g mol^{-1}. The ratio of ^{35}Cl to ^{37}Cl in naturally occurring chlorine is close to: [Sep. 06, 2020 (II)]
(1) 4 : 1 (2) 3 : 1 (3) 2 : 1 (4) 1 : 1

61. The ratio of mass percent of C and H of an organic compound ($C_XH_YO_Z$) is 6 : 1. If one molecule of the above compound ($C_XH_YO_Z$) contains half as much oxygen as required to burn one molecule of compound C_XH_Y completely to CO_2 and H_2O. The empirical formula of compound $C_XH_YO_Z$ is : [2018]
(1) $C_3H_6O_3$ (2) C_2H_4O
(3) $C_3H_4O_2$ (4) $C_2H_4O_3$

62. The most abundant elements by mass in the body of a healthy human adult are : [2017]
Oxygen (61.4%) ; Carbon (22.9%), Hydrogen (10.0%) ; and Nitrogen (2.6%). The weight which a 75 kg person would gain if all ^1H atoms are replaced by ^2H atoms is
(1) 15 kg (2) 37.5 kg (3) 7.5 kg (4) 10 kg

63. What quantity (in mL) of a 45% acid solution of a mono-protic strong acid must be mixed with a 20% solution of the same acid to produce 800 mL of a 29.875% acid solution? [Online April 9, 2017]
(1) 320 (2) 325 (3) 316 (4) 330

64. At 300 K and 1 atm, 15 mL of a gaseous hydrocarbon requires 375 mL air containing 20% O_2 by volume for complete combustion. After combustion the gases occupy 330 mL. Assuming that the water formed is in liquid form and the volumes were measured at the same temperature and pressure, the formula of the hydrocarbon is: [2016]
(1) C_4H_8 (2) C_4H_{10} (3) C_3H_6 (4) C_3H_8

65. Choose the incorrect formula out of the four compounds for an element X below : [Online April 11, 2015]
(1) X_2O_3 (2) X_2Cl_3
(3) $X_2(SO_4)_3$ (4) XPO_4

66. A gaseous compound of nitrogen and hydrogen contains 12.5% (by mass) of hydrogen. The density of the compound relative to hydrogen is 16. The molecular formula of the compound is: [Online April 11, 2014]
(1) NH_2 (2) N_3H (3) NH_3 (4) N_2H_4

67. The amount of $BaSO_4$ formed upon mixing 100 mL of 20.8% $BaCl_2$ solution with 50 mL of 9.8% H_2SO_4 solution will be: [Online April 12, 2014]
(Ba = 137, Cl = 35.5, S = 32, H = 1 and O = 16)
(1) 23.3 g (2) 11.65 g (3) 30.6 g (4) 33.2 g

68. A gaseous hydrocarbon gives upon combustion 0.72 g of water and 3.08 g of CO_2. The empirical formula of the hydrocarbon is : [2013]
(1) C_2H_4 (2) C_3H_4 (3) C_6H_5 (4) C_7H_8

69. A transition metal M forms a volatile chloride which has a vapour density of 94.8. If it contains 74.75% of chlorine, the formula of the metal chloride will be [Online May 26, 2012]
(1) MCl_3 (2) MCl_2 (3) MCl_4 (4) MCl_5

70. In a compound C, H and N atoms are present in 9 : 1 : 3.5 by weight. Molecular weight of compound is 108. Molecular formula of compound is [2002]
(1) $C_2H_6N_2$ (2) C_3H_4N (3) $C_6H_8N_2$ (4) $C_9H_{12}N_3$.

NUMERICAL VALUE QUESTIONS (NVQs)

71. The cycloalkane (X) on bromination consumes one mole of bromine per mole of (X) and gives the product (Y) in which C : Br ratio is 3 : 1. The percentage of bromine in the product (Y) is _______ %. (Nearest integer)
(Given : Molar mass in g mol^{-1} H : 1, C : 12, O : 16, Br : 80) [Jan. 22, 2026 (I)]

72. Thyroxine, the hormone has given below structure

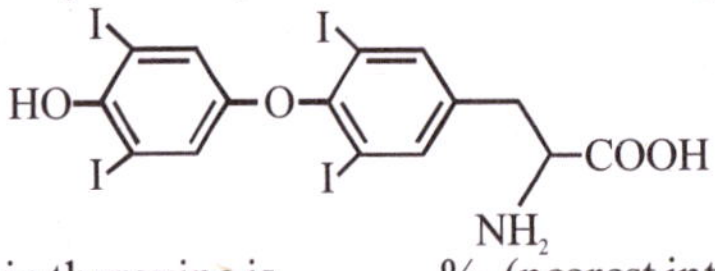

The percentage of iodine in thyroxine is .... %. (nearest integer)
(Given molar mass in g mol–1 *C*:12. H:1, *O*:16, *N*:14, I:127) [April 7, 2025 (I)]

73. An organic compound weighing 500 mg, produced 220 mg of CO_2. on complete combustion. The percentage composition of carbon in the compound is .... %. (nearest integer) [April 7, 2025 (I)]
(Given molar mass in g mol^{-1} of C : 12, O : 16)

74. The hydrocarbon (X) with molar mass 80 g mol^{-1} and 90% carbon has ________ degree of unsaturation. [Jan. 24, 2025 (II)]

75. An organic compound gives 0.220 g of CO_2 and 0.126 g of H_2O on complete combustion If the % of carbon is 24 then the % hydrogen is ______ × 10^{-1}. (Nearest integer) [April 13, 2023 (I)]

76. A solution of sugar is obtained by mixing 200 g of its 25% solution and 500 g of its 40% solution (both by mass). The mass percentage of the resulting sugar solution is ______. (Nearest integer) [April 11, 2023 (I)]

77. A sample of a metal oxide has formula $M_{0.83}O_{1.00}$. The metal M can exist in two oxidation states +2 and +3. In the sample of $M_{0.83}O_{1.00}$, the percentage of metal ions existing in +2 oxidation state is _____ % (nearest integer) [Jan. 31, 2023 (II)]

78. Uracil is base present in RNA with the following structure. % of N in uracil is _______. [Jan. 24, 2023 (I)]

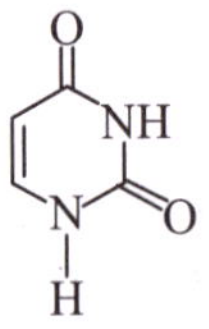

Given :
Molar mass N = 14 g mol^{-1}; O = 16 g mol^{-1};
C = 12 g mol^{-1}; H = 1 g mol^{-1}

79. Number of hydrogen atoms per molecule of a hydrocarbon A having 85.8% carbon is _________ (Given : Molar mass of A = 84 g mol^{-1}) [Jan. 25, 2023 (II)]

80. The complete combustion of 0.492 g of an organic compound containing 'C', 'H' and 'O' gives 0.793g of CO_2 and 0.442 g of H_2O. The percentage of oxygen composition in the organic compound is ____________. (nearest integer) [June 28, 2022 (II); Same July 28, 2022 (I)]

81. 116 g of a substance upon dissociation reaction, yields 7.5 g of hydrogen, 60g of oxygen and 48.5 g of carbon. Given that the atomic masses of H, O and C are 1, 16 and 12 respectively. The data agrees with how many formulae of the following______
(1) CH_3COOH (2) HCHO [June 27, 2022 (II)]
(3) CH_3OOCH_3 (4) CH_3CHO

82. On complete combustion 0.30 g of an organic compound gave 0.20 g of carbon dioxide and 0.10 g of water. The percentage of carbon in the given organic compound is ________ (Nearest Integer) [June 26, 2022 (I)]

83. The ratio of the mass percentages of 'C & H' and 'C & O' of a saturated acyclic organic compound 'X' are 4 : 1 and 3 : 4 respectively. Then, the moles of oxygen gas required for complete combustion of two moles of organic compound 'X' is _________. [Sep. 02, 2020 (II)]

Topic 3 - Stoichiometric Calculations

MULTIPLE CHOICE QUESTIONS (MCQs)

84. What volume of hydrogen gas at STP would be liberated by action of 50 mL of H_2SO_4 of 50% purity (density = 1.3 g mL^{-1}) on 20 g of zinc ?
Given : Molar mass of H, O, S, Zn are 1, 16, 32, 65 g mol^{-1} respectively. [April 05, 2026 (II)]
(1) 5.824 L (2) 7.428 L (3) 6.892 L (4) 8.375 L

85. How many grams of residue is obtained by heating 2.76 g of silver carbonate ?
(Given : Molar mass of C, O and Ag are 12, 16 and 108 g mol^{-1} respectively) [April 05, 2026 (I)]
(1) 1.08 g (2) 2.16 g (3) 3.24 g (4) 4.32 g

86. Complete combustion of X g of an organic compound gave 0.25 g of CO_2 and 0.12 g of H_2O. If the % of carbon is 25% and of hydrogen is 4.89%, then X = ______ × 10^{-3} g. (Nearest integer) [April 02, 2026 (II)]
(Molar mass of C, H and O are 12, 1 and 16 g mol^{-1} respectively.)
(1) 273 (2) 27 (3) 2730 (4) 227

87. The ratio of mass percentage (w/w) of C : H in a hydrocarbon is 12 : 1. It has two carbon atoms. The weight (in g) of CO_2(g) formed when 3.38 g of this hydrocarbon is completely burnt in oxygen is :
(Given : Molar mass in g mol^{-1} C : 12, H : 1, O : 16) [April 02, 2026 (II)]
(1) 5.68 (2) 11.44 (3) 22.74 (4) 17.05

88. The mass of iron converted into Fe_3O_4 by the action of 18 g of steam is : [April 02, 2026 (I)]
(Given : Molar mass of H, O and Fe are 1, 16 and 56 g mol^{-1} respectively)
Assume iron is present in excess :
(1) 2.1 g (2) 4.2 g (3) 21 g (4) 42 g

89. For the given reaction;
$CaCO_3 + 2HCl \rightarrow CaCl_2 + H_2O + CO_2$
If 90 g $CaCO_3$ is added to 300 mL of HCl which contains 38.55% HCl by mass and has density 1.13 g mL^{-1}, then which of the following option is **correct**?
Given molar mass of H, Cl, Ca and O are 1, 35.5, 40 and 16 g mol^{-1} respectively. [Jan. 28, 2026 (II)]
(1) 64.97 g of HCl remains unreacted
(2) 32.85 g of $CaCO_3$ remains unreacted
(3) 97.30 g of HCl reacted
(4) 60.32 g of HCl reamains unreacted

90. One mole of Cl_2 (g) was passed into 2 L of cold 2M KOH solution. After the reaction, the concentrations of Cl^-, ClO^- and OH^- are respectively (assume volume remains constant) [Jan. 24, 2026 (II)]
(1) 0.75 M, 0.75 M, 1 M (2) 0.5 M, 0.5 M, 0.5 M
(3) 0.5 M, 0.5 M, 1 M (4) 1 M, 1M, 1 M

91. $A + 2B \rightarrow AB_2$
36.0 g of 'A' (Molar mass : 60 g mol^{-1}) and 56.0 g of 'B' (Molar mass : 80 g mol^{-1}) are allowed to react. Which of the following statements are **correct**? [Jan. 22, 2026 (II)]
(1) 'A' is the limting reagent
(2) 77.0 g of AB_2 is formed
(3) Molar mass of AB_2 is 140 g mol^{-1}
(4) 15.0 g of A is left unreacted after the completion of reaction.
Choose the **correct** answer from the options given below:
(1) C and D only (2) A and C only
(3) B and D only (4) A and B only

92. In the reaction, [Jan. 22, 2026 (I)]
$2Al(s) + 6HCl(aq) \rightarrow 2Al^{3+}(aq) + 6Cl^-(aq) + 3H_2(g)$
(1) 11.2 L H_2 (g) at STP is produced for every mole of HCl consumed.
(2) 67.2 L H_2 (g) at STP is produced for every mole of Al that reacts.
(3) 12 L HCl (aq) is consumed for every 6L H_2(g) produced.
(4) 33.6 L H_2 (g) is produced regardless of temperature and pressure for every mole of Al that reacts.

93. Aqueous HCl reacts with MnO_2(s) to form $MnCl_2$(aq), Cl_2(g) and $H_2O(l)$. What is the weight (in g) of Cl_2 liberated when 8.7 g of MnO_2(s) is reacted with excess aqueous HCl solution? (Given Molar mass in g mol^{-1} Mn = 55, Cl = 35.5, O = 16, H = 1) [Jan. 21, 2026 (II)]
(1) 7.1 (2) 71 (3) 21.3 (4) 14.2

94. 14.0 g of calcium metal is allowed to react with excess HCl at 1.0 atm pressure and 273 K. [Jan. 21, 2026 (I)]
Which of the following statements is **incorrect**?
[**Given :** Molar mass in g mol^{-1} of Ca=40, Cl=35.5, H=1]
(1) 0.35 mol of H_2 gas is evolved.
(2) 7.84 L of H_2 gas is evolved.
(3) 33.3 of $CaCl_2$ is produced.
(4) The limiting reagent is calcium metal.

95. 80 mL of a hydrocarbon on mixing with 264 mL of oxygen in a closed U-tube undergoes complete combustion. The residual gases after cooling to 273 K occupy 224 mL. When the system is treated with KOH solution, the volume decreases to 64 mL. The formula of the hydrocarbon is: [Jan. 21, 2026 (I)]
(1) C_2H_4 (2) C_4H_{10} (3) C_2H_2 (4) C_2H_6

96. Mass of magnesium required to produce 220 mL of hydrogen gas at STP on reaction with excess of dil. HCl is
Given : Molar mass of Mg is 24 g mol^{-1}. [April 3, 2025 (II)]
(1) 235.7 g (2) 0.24 mg (3) 236 mg (4) 2.444 g

97. $CaCO_3(s) + 2HCl(aq) \rightarrow CaCl_2(aq) + CO_2(g) H_2O(l)$ Consider the above reaction, what mass of $CaCl_2$ will be formed if 250 mL of 0.76 M HCl reacts with 1000 g of $CaCO_3$?
(Given : Molar mass of Ca, C, O, H and Cl are 40, 12, 16, 1 and 35.5 g mol^{-1}, respectively) [April 2, 2025 (I)]
(1) 3.908 g (2) 2.636 g (3) 10.545 g (4) 5.272 g

98. 0.1 M solution of KI reacts with excess of H_2SO_4 and KIO_3 solution. According to equation

$$5I^- + IO_3^- + 6H^+ \rightarrow 3I_2 + 3H_2O$$

Identify the correct statements : **[Jan. 29, 2025 (II)]**
(A) 200 mL of KI solution reacts with 0.004 mol of KIO_3
(B) 200 mL of KI solution reacts with 0.006 mol of H_2SO_4
(C) 0.5 L of KI solution produced 0.005 mol of I_2
(D) Equivalent weight of KIO_3 is equal to $\left(\frac{\text{Molecular weight}}{5}\right)$

Choose the correct answer from the options given below :
(1) (A) and (D) only (2) (B) and (C) only
(3) (A) and (B) only (4) (C) and (D) only

99. Concentrated nitric acid is labelled as 75% by mass. The volume in mL of the solution which contains 30 g of nitric acid is ________. **[Jan. 28, 2025 (II)]**
Given : Density of nitric acid solution is 1.25 g/mL
(1) 45 (2) 55 (3) 32 (4) 40

100. What amount of bromine will be required to convert 2 g of phenol into 2, 4, 6-tribromophenol?
(Given molar mass in g mol^{-1} of C, H, O, Br are 12, 1, 16, 80 respectively) **[Jan. 23, 2025 (I)]**
(1) 10.22 g (2) 6.0 g (3) 4.0 g (4) 20.44 g

101. Density of 3 M NaCl solution is 1.25 g/mL. The molality of the solution is : **[Jan. 22, 2025 (II)]**
(1) 1.79m (2) 2m (3) 3m (4) 2.79m

102. Given below are two statements: **[Jan. 22, 2025 (I)]**
Statement I: One mole of propyne reacts with excess of sodium to liberate half a mole of H_2 gas.
Statement II : Four g of propyne reacts with $NaNH_2$ to liberate NH_3 gas which occupies 224 mL at STP.
In the light of the above statements, choose the most appropriate answer from the options given below:
(1) Statement I is correct but Statement II is incorrect.
(2) Both Statement I and Statement II are incorrect
(3) Statement I is incorrect but Statement II is correct
(4) Both Statement I and Statement II are correct.

103. The number of moles of methane required to produce 11 g $CO_2(g)$ after complete combustion is :
(Given molar mass of methane in g mol^{-1} : 16) **[April 5, 2024 (II)]**
(1) 0.75 (2) 0.25 (3) 0.35 (4) 0.5

104. What is the mass ratio of ethylene glycol ($C_2H_6O_2$, molar mass = 62 g/mol) required for making 500 g of 0.25 molal aqueous solution and 250 mL of 0.25 molar aqueous solution ?
[Jan. 25, 2023 (II)]
(1) 1 : 1 (2) 3 : 1 (3) 2 : 1 (4) 1 : 2

105. $\underset{20\text{ g}}{N_2(g)} + \underset{5\text{ g}}{3H_2(g)} \rightleftharpoons 2NH_3(g)$
Consider the above reaction, the limiting reagent of the reaction and number of moles of NH_3 formed respectively are:
(1) H_2, 1.42 moles (2) H_2, 0.71 moles **[July 29, 2022 (I)]**
(3) N_2, 1.42 moles (4) N_2, 0.71 moles

106. Consider the reaction

$$4HNO_3(l) + 3KCl(s) \longrightarrow Cl_2(g) + NOCl(g) + 2H_2O(g) + 3KNO_3(s)$$

The amount of HNO_3 required to produce 110.0 g of KNO_3 is:
(Given : Atomic masses of H, O, N and K are 1, 16, 14 and 39, respectively.) **[July 29, 2022 (II)]**
(1) 32.2 g (2) 69.4 g (3) 91.5 g (4) 162.5 g

107. $C(s) + O_2(g) \rightarrow CO_2(g) + 400 \text{ kJ}$

$C(s) + \frac{1}{2}O_2(g) \rightarrow CO(g) + 100 \text{ kJ}$ **[July 29, 2022 (II)]**

When coal of purity 60% is allowed to burn in presence of insufficient oxygen, 60% of carbon is converted into 'CO' and the remaining is converted into 'CO_2'.
The heat generated when 0.6 kg of coal is burnt is ______.
(1) 1600 kJ (2) 3200 kJ (3) 4400 kJ (4) 6600 kJ

108. 250 g solution of D-glucose in water contains 10.8% of carbon by weight. The molality of the solution is nearest to
(Given: Atomic Weights are H, 1u ; C, 12u ; O, 16u)
[July 27, 2022 (I)]
(1) 1.03 (2) 2.06 (3) 3.09 (4) 5.40

109. SO_2Cl_2 on reaction with excess of water results into acidic mixture
$SO_2Cl_2 + 2H_2O \rightarrow H_2SO_4 + 2HCl$
16 moles of NaOH is required for the complete neutralisation of the resultant acidic mixture.
The number of moles of SO_2Cl_2 used is: **[July 25, 2022 (I)]**
(1) 16 (2) 8 (3) 4 (4) 2

110. Production of iron in blast furnace follows the following equation
$Fe_3O_4(s) + 4CO(g) \rightarrow 3Fe(l) + 4CO_2(g)$
when 4.640 kg of Fe_3O_4 and 2.520 kg of CO are allowed to react then the amount of iron (in g) produced is :
[Given : Molar Atomic mass (g mol^{-1}): Fe = 56
Molar Atomic mass (g mol^{-1}) : O = 16
Molar Atomic mass (g mol^{-1}): C = 12 **[June 29, 2022 (I)]**
(1) 1400 (2) 2200 (3) 3360 (4) 4200

111. If a rocket runs on a fuel ($C_{15}H_{30}$) and liquid oxygen, the weight of oxygen required and CO_2 released for every litre of fuel respectively are:
(Given: density of the fuel is 0.756 g/mL) **[June 24, 2022 (I)]**
(1) 1188 g and 1296 g (2) 2376 g and 2592 g
(3) 2592g and 2376 g (4) 3429 g and 3142 g

112. The ammonia (NH_3) released on quantitative reaction of 0.6 g urea (NH_2CONH_2) with sodium hydroxide (NaOH) can be neutralized by: **[Jan. 07, 2020 (II)]**
(1) 200 mL of 0.4 N HCl (2) 200 mL of 0.2 N HCl
(3) 100 mL of 0.2 N HCl (4) 100 mL of 0.1 N HCl

113. The minimum amount of $O_2(g)$ consumed per gram of reactant is for the reaction : (Given atomic mass : Fe = 56, O = 16, Mg = 24, P = 31, C = 12, H = 1) **[April 10, 2019 (II)]**
(1) $4\,Fe(s) + 3\,O_2(g) \rightarrow 2\,Fe_2O_3(s)$
(2) $P_4(s) + 5\,O_2(g) \rightarrow P_4O_{10}(s)$
(3) $C_3H_8(g) + 5\,O_2(g) \rightarrow 3\,CO_2(g) + 4\,H_2O\,(l)$
(4) $2\,Mg(s) + O_2(g) \rightarrow 2\,MgO(s)$

114. For a reaction, $N_2(g) + 3\,H_2(g) \rightarrow 2\,NH_3(g)$; identify dihydrogen ($H_2$) as a limiting reagent in the following reaction mixtures.
[April 9, 2019 (I)]
(1) 56 g of N_2 + 10 g of H_2 (2) 35 g of N_2 + 8 g of H_2
(3) 28 g of N_2 + 6 g of H_2 (4) 14 g of N_2 + 4 g of H_2

115. 0.27 g of a long chain fatty acid was dissolved in 100 cm^3 of hexane. 10 mL of this solution was added dropwise to the surface of water in a round watch glass. Hexane evaporates and a monolayer is formed. The distance from edge to centre of the watch glass is 10 cm. What is the height of the monolayer?
[Density of fatty acid = 0.9 g cm^{-3} ; π = 3] **[April 8, 2019 (II)]**
(1) 10^{-6} m (2) 10^{-8} m
(3) 10^{-2} m (4) 10^{-4} m

116. 50 mL of 0.5 M oxalic acid is needed to neutralize 25 mL of sodium hydroxide solution. The amount of NaOH in 50 mL of the given sodium hydroxide solution is: **[Jan. 12, 2019 (I)]**
(1) 40 g (2) 10 g (3) 20 g (4) 80 g

117. A 10 mg effervescent tablet containing sodium bicarbonate and oxalic acid releases 0.25 mL of CO_2 at $T = 298.15$ K and $P =$ 1 bar. If molar volume of CO_2 is 25.0 L under such condition, what is the percentage of sodium bicarbonate in each tablet? [Molar mass of $NaHCO_3 = 84\ g\ mol^{-1}$] **[Jan. 11, 2019 (I)]**
(1) 0.84 (2) 33.6 (3) 16.8 (4) 8.4

118. 25 mL of the given HCl solution requires 30 mL of 0.1 M sodium carbonate solution. What is the volume of this HCl solution required to titrate 30 mL of 0.2 M aqueous NaOH solution? **[Jan. 11, 2019 (II)]**
(1) 25 mL (2) 75 mL (3) 50 mL (4) 12.5 mL

119. For the following reaction, the mass of water produced from 445 g of $C_{57}H_{110}O_6$ is: **[Jan. 9, 2019 (II)]**
$$2C_{57}H_{110}O_6(s) + 163O_2(g) \rightarrow 114CO_2(g) + 110H_2O(l)$$
(1) 490 g (2) 445 g (3) 495 g (4) 890 g

120. A sample of $NaClO_3$ is converted by heat to NaCl with a loss of 0.16 g of oxygen. The residue is dissolved in water and precipitated as AgCl. The mass of AgCl (in g) obtained will be: (Given: Molar mass of AgCl = 143.5 g mol^{-1}) **[Online April 15, 2018 (I)]**
(1) 0.35 (2) 0.54 (3) 0.41 (4) 0.48

121. 1 gram of a carbonate (M_2CO_3) on treatment with excess HCl produces 0.01186 mole of CO_2. The molar mass of M_2CO_3 in g mol^{-1} is: **[2017]**
(1) 1186 (2) 84.3 (3) 118.6 (4) 11.86

122. The amount of arsenic pentasulphide that can be obtained when 35.5 g arsenic acid is treated with excess H_2S in the presence of conc. HCl (assuming 100% conversion) is: **[Online April 9, 2016]**
(1) 0.25 mol (2) 0.50 mol (3) 0.333 mol (4) 0.125 mol

123. The volume of 0.1 N dibasic acid sufficient to neutralize 1 g of a base that furnishes 0.04 mole of OH^- in aqueous solution is: **[Online April 10, 2016]**
(1) 400 mL (2) 600 mL (3) 200 mL (4) 800 mL

124. A sample of a hydrate of barium chloride weighing 61 g was heated until all the water of hydration is removed. The dried sample weighed 52 g. The formula of the hydrated salt is: (atomic mass, Ba = 137 amu, Cl = 35.5 amu) **[Online April 10, 2015]**
(1) $BaCl_2 . 4H_2O$ (2) $BaCl_2 . 3H_2O$
(3) $BaCl_2 . H_2O$ (4) $BaCl_2 . 2H_2O$

125. $A + 2B + 3C \rightleftharpoons AB_2C_3$
Reaction of 6.0 g of A, 6.0×10^{23} atoms of B, and 0.036 mol of C yields 4.8 g of compound AB_2C_3. If the atomic mass of A and C are 60 and 80 amu, respectively, the atomic mass of B is (Avogadro no. = 6×10^{23}): **[Online April 11, 2015]**
(1) 50 amu (2) 60 amu (3) 70 amu (4) 40 amu

126. 6 litres of an alkene require 27 litres of oxygen at constant temperature and pressure for complete combustion. The alkene is: **[Online April 25, 2013]**
(1) Ethene (2) Propene (3) 1-Butene (4) 2-Butene

127. The molality of a urea solution in which 0.0100 g of urea, $[(NH_2)_2CO]$ is added to 0.3000 dm^3 of water at STP is: **[2011RS]**
(1) 5.55×10^{-4} m (2) 33.3 m
(3) 3.33×10^{-2} m (4) 0.555 m

128. The density (in g mL^{-1}) of a 3.60 M sulphuric acid solution that is 29% H_2SO_4 (molar mass = 98 g mol^{-1}) by mass will be **[2007]**
(1) 1.45 (2) 1.64 (3) 1.88 (4) 1.22

129. In the reaction, **[2007]**
$$2Al(s) + 6HCl(aq) \rightarrow 2Al^{3+}(aq) + 6Cl^-(aq) + 3H_2(g)$$
(1) 11.2 L H_2(g) at STP is produced for every mole of HCl(aq) consumed
(2) 6 L HCl(aq) is consumed for every 3 L of H_2(g) produced
(3) 33.6 L H_2(g) is produced regardless of temperature and pressure for every mole of Al that reacts
(4) 67.2 H_2(g) at STP is produced for every mole of Al that reacts.

130. Density of a 2.05M solution of acetic acid in water is 1.02 g/mL. The molality of the solution is **[2006]**
(1) 2.28 mol kg^{-1} (2) 0.44 mol kg^{-1}
(3) 1.14 mol kg^{-1} (4) 3.28 mol kg^{-1}

131. Two solutions of a substance (non electrolyte) are mixed in the following manner: 480 mL of 1.5 M first solution + 520 mL of 1.2 M second solution. What is the molarity of the final mixture? **[2005]**
(1) 2.70 M (2) 1.344 M (3) 1.50 M (4) 1.20 M

132. To neutralise completely 20 mL of 0.1 M aqueous solution of phosphorous acid (H_3PO_3), the valume of 0.1 M aqueous KOH solution required is **[2004]**
(1) 40 mL (2) 20 mL (3) 10 mL (4) 60 mL

133. What volume of hydrogen gas, at 273 K and 1 atm. pressure will be consumed in obtaining 21.6 g of elemental boron (atomic mass = 10.8) from the reduction of boron trichloride by hydrogen? **[2003]**
(1) 67.2 L (2) 44.8 L (3) 22.4 L (4) 89.6 L

NUMERICAL VALUE QUESTIONS (NVQs)

134. One mole of an alkane (x) requires 8 mole oxygen for complete combustion. Sum of number of carbon and hydrogen atoms in the alkane (x) is ______. **[April 02, 2026 (I)]**

135. 0.25 g of an organic compound "A" containing carbon, hydrogen and oxygen was analysed using the combustion method. There was an increase in mass of $CaCl_2$ tube and potash tube at the end of the experiment. The amount was found to be 0.15 g and 0.1837 g, respectively. The percentage of oxygen in compound A is ________%. (Nearest integer) **[Jan. 24, 2026 (II)]**
(Given : molar mass in g mol^{-1} H : 1, C : 12, O : 16)

136. The mass of benzanilide obtained from the benzoylation reaction of 5.8 g of aniline, if yield of product is 82%, is _________ g (nearest integer).
(Given molar mass in g mol^{-1} H:1, C:12, N:14, O:16) **[Jan. 22, 2026 (II)]**

137. 20 mL of sodium iodide solution gave 4.74 g silver iodide when treated with excess of silver nitrate solution. The molarity of the sodium iodide solution is ______ M. (Nearest Integer value)
(Given : Na = 23, I = 127, Ag = 108, N = 14, O = 16 g mol^{-1}) **[April 8, 2025 (II)]**

138. Butane reacts with oxygen to produce carbon dioxide and water following the equation given below **[April 7, 2025 (II)]**
$$C_4H_{10}(g) + \frac{13}{2}O_2(g) \rightarrow 4CO_2(g) + 5H_2O(l)$$
If 174.0 kg of butane is mixed with 320.0 kg of O_2, the volume of water formed in litres is _____. (Nearest integer)
[Given : (a) Molar mass of C, H, O are 12, 1, 16 g mol^{-1} respectively, (b) Density of water = 1 g mL^{-1}]

139. The amount of calcium oxide produced on heating 150 kg limestone (75% pure) is ______kg. (Nearest integer)
Given : Molar mass (in g mol^{-1}) of Ca-40, O-16, C-12. **[April 4, 2025 (II)]**

140. 0.5 g of an organic compound on combustion gave 1.46 g of CO_2 and 0.9 g of H_2O. The percentage of carbon in the compound is _____. (Nearest integer)
[Given : Molar mass (in g mol^{-1}) C : 12, H : 1, O : 16] **[April 3, 2025 (I)]**

141. The molarity of a 70% (mass/mass) aqueous solution of a monobasic acid (X) is ________ $\times 10^{-1}$ M(Nearest integer)
[Given : Density of aqueous solution of (X) is 1.25 g mL^{-1} Molar mass of the acid is 70 g mol^{-1}] **[Jan. 28, 2025 (I)]**

142. Consider the following reaction occurring in the blast furnace.
$Fe_3O_4(s) + 4CO(g) \rightarrow 3Fe(l) + 4CO_2(g)$ **[Jan. 24, 2025 (I)]**
'x' kg of iron is produced when 2.32×10^3 kg Fe_3O_4 and 2.8×10^2 kg CO are brought together in the furnace. The value of 'x' is ______. (nearest integer)
{Given:
Molar mass of $Fe_3O_4 = 232$ g mol^{-1}
Molar mass of CO = 28 g mol^{-1}
Molar mass of Fe = 56 g mol^{-1}}

143. Xg of benzoic acid on reaction with aq. $NaHCO_3$ release CO_2 that occupied 11.2 L volume at STP.
X is ______ g. **[Jan. 24, 2025 (I)]**

144. When 81.0 g of aluminium is allowed to react with 128.0 g of oxygen gas, the mass of aluminium oxide produced in grams is ______. (Nearest integer)
Given:
Molar mass of Al is 27.0 g mol^{-1}
Molar mass of O is 16.0 g mol^{-1} **[Jan. 23, 2025 (II)]**

145. 0.01 mole of an organic compound (X) containing 10% hydrogen, on complete combustion produced 0.9 g H_2O. Molar mass of (X) is ______ g mol^{-1}. **[Jan. 23, 2025 (II)]**

146. 20 mL of 2M NaOH solution is added to 400 mL of 0.5 M NaOH solution. The final concentration of the solution is _______ $\times 10^{-2}$ M.(Nearest integer). **[Jan. 22, 2025 (II)]**

147. Some CO_2 gas was kept in a sealed container at a pressure of 1 atm and at 273 K. This entire amount of CO_2 gas was later passed through an aqueous solution of $Ca(OH)_2$. The excess unreacted $Ca(OH)_2$ was later neutralized with 0.1M of 40 mL HCl. If the volume of the sealed container of CO_2 was x, then x is ______ cm^3 (nearest integer). **[Jan. 22, 2025 (I)]**
[Given: The entire amount of $CO_2(g)$ reacted with exactly half the initial amount of $Ca(OH)_2$ present in this aqueous solution.]

148. Consider the following reaction :
$3PbCl_2 + 2(NH_4)_3PO_4 \rightarrow Pb_3(PO_4)_2 + 6NH_4Cl$
If 72 mmol of $PbCl_2$ is mixed with 50 mmol of $(NH_4)_3PO_4$, then the amount of $Pb_3(PO_4)_2$ formed is ________ mmol (nearest integer). **[Feb. 1, 2024 (I)]**

149. 10 mL of gaseous hydrocarbon on combustion gives 40 mL of $CO_2(g)$ and 50 mL of water vapour. Total number of carbon and hydrogen atoms in the hydrocarbon is ________. **[Feb. 1, 2024 (II)]**

150. Number of moles of methane required to produce 22g $CO_2(g)$ after combustion is $x \times 10^{-2}$ moles. The value of x is ______. **[Jan. 31, 2024 (I)]**

151. If 50 mL of 0.5M oxalic acid is required to neutralise 25 mL of NaOH solution, the amount of NaOH in 50 mL of given NaOH solution is _____ g. **[Jan. 29, 2024 (II)]**

152. Mass of methane required to produce 22 g of CO_2 after complete combustion is ______g.
(Given Molar mass in g mol^{-1} C = 12.0, H = 1.0, O = 16.0) **[Jan. 27, 2024 (I)]**

153. Volume of 3 M NaOH (formula weight 40 g mol^{-1}) which can be prepared from 84 g of NaOH is ________ $\times 10^{-1}$ dm^3. **[Jan. 27, 2024 (II)]**

154. 9.3 g of aniline is subjected to reaction with excess of acetic anhydride to prepare acetanilide. The mass of acetanilide produced if the reaction is 100% completed is _________ $\times 10^{-1}$ g.
(Given molar mass in g mol^{-1} N : 14, O : 16, C : 12, H : 1) **[Jan. 27, 2024 (II)]**

155. 1 g of a carbonate (M_2CO_3) on treatment with excess HCl produces 0.01 mol of CO_2. The molar mass of M_2CO_3 is ______ g mol^{-1}. (Nearest integer) **[April 13, 2023 (II)]**

156. The volume of hydrogen liberated at STP by treating 2.4 g magnesium with excess of hydrochloric acid is _____ $\times 10^{-2}$ L. Given: Molar volume of gas is 22.4 L at STP. Molar mass of magnesium is 24 g mol^{-1}. **[April 11, 2023 (II)]**

157. 0.5 g of an organic compound (X) with 60% carbon will produce _______ $\times 10^{-1}$ g of CO_2 on complete combustion. **[April 08, 2023 (I)]**

158. If 5 moles of $BaCl_2$ is mixed with 2 moles of Na_3PO_4, the maximum number of moles of $Ba_3(PO_4)_2$ formed is.......... (Nearest integer) **[April 06, 2023 (I)]**

159. On complete combustion, 0.492 g of an organic compound gave 0.792 g of CO_2.
The % of carbon in the organic compound is _____ (Nearest integer) **[Jan. 31, 2023 (I)]**

160. Zinc reacts with hydrochloric acid to give hydrogen and zinc chloride. The volume of hydrogen gas produced at STP from the reaction of 11.5 g of zinc with excess HCl is ____ L (Nearest integer)
(Given : Molar mass of Zn is 65.4g mol^{-1} and Molar volume of H_2 at STP = 22.7L) **[Jan. 31, 2023 (I)]**

161. Assume carbon burns according to following equation :
$2C(s) + O_2(g) \rightarrow 2CO(g)$
When 12 g carbon is burnt in 48 g of oxygen, the volume of carbon monoxide produced is ______ $\times 10^{-1}$ L at STP [nearest integer]
[Given : Assume CO as ideal gas, Mass of C is 12 g mol^{-1}, Mass of O is 16 g mol^{-1} and molar volume of an ideal gas at STP is 22.7 L mol^{-1}] **[Jan. 31, 2023 (II)]**

162. A 300 mL bottle of soft drink has 0.2 M CO_2 dissolved in it. Assuming CO_2 behaves as an ideal gas, the volume of the dissolved CO_2 at STP is _________ mL. (Nearest integer)
Given: At STP, molar volume of an ideal gas is 22.7L mol^{-1}. **[Jan. 30, 2023 (I)]**

163. The strength of 50 volume solution of hydrogen peroxide is ______ g/L (Nearest integer).
Given: Molar mass of H_2O_2 is 34 g mol^{-1}
Molar volume of gas at STP = 22.7 L. **[Jan. 30, 2023 (II)]**

164. When 0.01 mol of an organic compound containing 60% carbon was burnt completely, 4.4 g of CO_2 was produced. The molar mass of compound is __________ g mol^{-1} (Nearest integer) **[Jan. 29, 2023 (II)]**

165. The volume of HCl, containing 73 g L^{-1}, required to completely neutralise NaOH obtained by reacting 0.69 g of metallic sodium with water, is _______ mL. (Nearest Integer)
(Given : Molar masses of Na, Cl, O, H are 23, 35.5, 16 and 1 g mol^{-1} respectively) [Jan. 29, 2023 (II)]

166. When $Fe_{0.93}O$ is heated in presence of oxygen, it converts to Fe_2O_3. The number of correct statement/s from the following is _______. [Jan. 24, 2023 (I)]

A. The equivalent weight of $Fe_{0.93}O$ is $\frac{\text{Molecular weight}}{0.79}$

B. The number of moles of Fe^{2+} and Fe^{3+} in 1 mole of $Fe_{0.93}O$ is 0.79 and 0.14 respectively.

C. $Fe_{0.93}O$ is metal deficient with lattice comprising of cubic closed packed arrangement of O^{2-} ions.

D. The % composition of Fe^{2+} and Fe^{3+} in $Fe_{0.93}O$ is 85% and 15% respectively.

167. Blister copper is produced by reaction of copper oxide with copper sulphide.
$2Cu_2O + Cu_2S \rightarrow 6Cu + SO_2$
When 2.86×10^3 g of Cu_2O and 4.77×10^3 g of Cu_2S are used for reaction, the mass of copper produced is ____ g. (nearest integer) [June 30, 2022 (I)]
(Atomic mass of Cu = 63.5 a.m. u)
S = 32.0 a.m. u
O = 16.0 a.m. u)

168. CNG is an important transportation fuel. When 100 g CNG is mixed with 208g oxygen in vehicles, it leads to the formation of CO_2 and H_2O and produces large quantity of heat during this combustion, then the amount of carbon dioxide, produced in grams is________. [nearest integer]
[Assume CNG to be methane] [June 26, 2022 (II)]

169. The moles of methane required to produce 81 g of water after complete combustion is ______ $\times 10^{-2}$ mol. [nearest integer] [June 26, 2022 (II)]

170. Number of grams of bromine that will completely react with 5.0g of pent-1-ene is __________ $\times$ 10^{-2}g. (Atomic mass of Br = 80 g/mol) (Nearest Integer) [June 25, 2022 (I)]

171. 1 L aqueous solution of H_2SO_4 contains 0.02 m mol H_2SO_4. 50% of this solution is diluted with deionized water to give 1 L solution (A). In solution (A), 0.01 m mol of H_2SO_4 are added. Total m mols of H_2SO_4 in the final solution is _______ $\times 10^3$ m mols. [June 25, 2022 (I)]

172. A protein 'A' contains 0.30% of glycine (molecular weight 75). The minimum molar mass of the protein 'A' is _______ $\times 10^3$ g mol^{-1} [nearest integer] [June 25, 2022 (II)]

173. The neutralization occurs when 10 mL of 0.1 M acid 'A' is allowed to react with 30 mL of 0.05 M base $M(OH)_2$. The basicity of the acid 'A' is _______. [M is a metal] [June 25, 2022 (II)]

174. If a rocket runs on a fuel ($C_{15}H_{30}$) and liquid oxygen, the weight of oxygen required and CO_2 released for every litre of fuel respectively are:
(Given: density of the fuel is 0.756 g/mL)

175. When 35 mL of 0.15 M lead nitrate solution is mixed with 20 mL of 0.12 M chromic sulphate solution, ________ $\times 10^{-5}$ moles of lead sulphate precipitate out.
(Round off to the Nearest Integer). [March 16, 2021 (II)]

176. Sodium oxide reacts with water to produce sodium hydroxide. 20.0 g of sodium oxide is dissolved in 500 mL of water. Neglecting the change in volume, the concentration of the resulting NaOH solution is _____ $\times 10^{-1}$ M. (Nearest integer)
[Atomic mass : Na = 23.0, O = 16.0, H = 1.0] [Aug. 31, 2021 (II)]

177. 100 g of propane is completely reacted with 1000 g of oxygen. The mole fraction of carbon dioxide in the resulting mixture is $x \times 10^{-2}$. The value of x is _______. (Nearest integer)
[Atomic weight : H = 1.008; C = 12.00; O = 16.00] [Aug. 27, 2021 (II)]

178. The density of NaOH solution is 1.2 g cm^{-3}. The molality of this solution is ________ m.
(Round off to the Nearest Integer)
[Use : Atomic masses : Na : 23.0 u O : 16.0 u H : 1.0 u
Density of H_2O : 1.0 g cm^{-3}] [July 27, 2021 (I)]

179. When 10 mL of an aqueous solution of Fe^{2+} ions was titrated in the presence of dil H_2SO_4 using diphenylamine indicator, 15 mL of 0.02 M solution of $K_2Cr_2O_7$ was required to get the end point. The molarity of the solution containing Fe^{2+} ions is $x \times 10^{-2}$ M. The value of x is _____. (Nearest integer) [July 25, 2021 (I)]

180. Consider the complete combustion of butane, the amount of butane utilized to produce 72.0 g of water is ________ $\times 10^{-1}$ g. (in nearest integer) [July 25, 2021 (I); Same March 18, 2021 (I)]

181. If the concentration of glucose ($C_6H_{12}O_6$) in blood is 0.72 g L^{-1}, the molarity of glucose in blood is _______ $\times 10^{-3}$ M. (Nearest integer)
[Given : Atomic mass of C = 12, H = 1, O = 16 u] [July 22, 2021 (II)]

182. 250 mL of 0.5 M NaOH was added to 500 mL of 1 M HCl. The number of unreacted HCl molecules in the solution after complete reaction is _______ $\times 10^{21}$. (Nearest integer)
($N_A = 6.022 \times 10^{23}$) [July 20, 2021 (I)]

183. 4 g equimolar mixture of NaOH and Na_2CO_3 contains x g of NaOH and y g of Na_2CO_3. The value of x is _______ g. (Nearest integer) [July 20, 2021 (II)]

184. ________ grams of 3-Hydroxypropanal (MW = 74) must be dehydrated to produce 7.8 g of acrolein (MW = 56) (C_3H_4O) if the percentage yield is 64.
(Rounded off to the Nearest Integer). [March 18, 2021 (I)]

185. A reaction of 0.1 mole of Benzylamine with bromomethane gave 23 g of Benzyl trimethyl ammonium bromide. The number of moles of bromomethane consumed in this reaction are $n \times 10^{-1}$, when n = _____ (Rounded off to the Nearest Integer).
[Given : Atomic masses are C : 12.0 u, H : 1.0 u, N : 14.0 u, Br : 80.0 u] [March 18, 2021 (I)]

186. Complete combustion of 3 g of ethane gives x $\times 10^{22}$ molecules of water. The value of x is ________.
(Rounded off to the Nearest Integer).
[Use : $N_A = 6.023 \times 10^{23}$;
Atomic masses in u : C : 12.0 ; O : 16.0 ; H : 1.0] [March 18, 2021 (I)]

187. The number of chlorine atoms in 20 mL of chlorine gas at STP is ________ $\times 10^{21}$. (Round off to the Nearest integer)
[Assume chlorine is an ideal gas at STP]
$R = 0.083$ L bar $mol^{-1} K^{-1}$, $N_A = 6.023 \times 10^{23}$ [March 17, 2021 (II)]

188. 15 mL of aqueous solution of Fe^{2+} in acidic medium completely reacted with 20 mL of 0.03 M aqueous $Cr_2O_7^{2-}$. The molarity of the Fe^{2+} solution is ___________ $\times 10^{-2}$ M.
(Round off to the Nearest Integer). [March 17, 2021 (I)]

189. The $NaNO_3$ weighed out to make 50 mL of an aqueous solution containing 70.0 mg Na^+ per mL is __________ g. (Rounded off to the nearest integer)
[Given : Atomic weight in g mol^{-1} - Na : 23; N : 14; O : 16] [Feb. 26, 2021 (II)]

190. 4.5 g of compound A (MW = 90) was used to make 250 mL of its aqueous solution. The molarity of the solution is $x \times 10^{-1}$. The value of x is ________ (Rounded off to the nearest integer) [Feb. 24, 2021 (I)]

191. The formula of a gaseous hydrocarbon which requires 6 times of its own volume of O_2 for complete oxidation and produces 4 times its own volume of CO_2 is C_xH_y. The value of y is ______. [Feb. 24, 2021 (II)]

192. The minimum number of moles of O_2 required for complete combustion of 1 mole of propane and 2 moles of butane is ______. [Sep. 05, 2020 (I)]

193. The mass of ammonia in grams produced when 2.8 kg of dinitrogen quantitatively reacts with 1 kg of dihydrogen is __________. [Sep. 04, 2020 (I)]

194. A 100 mL solution was made by adding 1.43 g of $Na_2CO_3 \cdot xH_2O$. The normality of the solution is 0.1 N. The value of x is __________. [Sep. 04, 2020 (II)]
(The atomic mass of Na is 23 g/ mol)

195. The mole fraction of glucose ($C_6H_{12}O_6$) in an aqueous binary solution is 0.1. The mass percentage of water in it, to the nearest integer, is ________. [Sep. 03, 2020 (I)]

196. 6.023×10^{22} molecules are present in 10 g of a substance 'x'. The molarity of a solution containing 5 g of substance 'x' in 2 L solution is __________ $\times 10^{-3}$. [Sep. 03, 2020 (II)]

197. The volume (in mL) of 0.1 N NaOH required to neutralise 10 mL of 0.1 N phosphinic acid is _________. [Sep. 03, 2020 (II)]

198. 10.30 mg of O_2 is dissolved into a liter of sea water of density 1.03 g/mL. The concentration of O_2 in ppm is _____. [Jan. 09, 2020 (II)]

199. Ferrous sulphate heptahydrate is used to fortify foods with iron. The amount (in grams) of the salt required to achieve 10 ppm of iron in 100 kg of wheat is ______.Atomic weight: Fe = 55.85; S = 32.00; O = 16.00 [Jan. 08, 2020 (I)]

200. $NaClO_3$ is used, even in spacecrafts, to produce O_2. The daily consumption of pure O_2 by a person is 492 L at 1 atm, 300 K. How much amount of $NaClO_3$, in grams, is required to produce O_2 for the daily consumption of a person at 1 atm, 300 K? ________.

$NaClO_3(s) + Fe(s) \rightarrow O_2(g) + NaCl(s) + FeO(s)$
$R = 0.082$ L atm mol^{-1} K^{-1} [Jan. 08, 2020 (II)]

Part-B (JEE Advanced)

Topic 1 - Laws of Chemical Combinations and Mole Concept

1 MCQs with One Correct Answer

1. Which has maximum number of atoms? [2003S]
(a) 24 g of C (12) (b) 56 g of Fe (56)
(c) 27 g of Al (27) (d) 108 g of Ag (108)

2. How many moles of electron weigh one kilogram? [2002S]
(a) 6.023×10^{23} (b) $\frac{1}{9.108} \times 10^{31}$
(c) $\frac{6.023}{9.108} \times 10^{54}$ (d) $\frac{1}{9.108 \times 6.023} \times 10^{8}$

3. The largest number of molecules is in [1979]
(a) 36 g of water (b) 28 g of carbon monoxide
(c) 46 g of ethyl alcohol (d) 54 g of nitrogen pentoxide

4. The total number of electrons in one molecule of carbon dioxide is [1979]
(a) 22 (b) 44 (c) 66 (d) 88

5. 27 g of Al will react completely with how many grams of oxygen?
(a) 8 g (b) 16 g (c) 32 g (d) 24 g [1978]

3 Numeric / New Stem Based Questions

6. If the value of Avogadro number is 6.023×10^{23} mol^{-1} and the value of Boltzmann constant is 1.380×10^{-23} J K^{-1}, then the number of significant digits in the calculated value of the universal gas constant is [Adv. 2014]

7. Calculate the molarity of water if its density is 1000 kg/m^3. [2003 - 2 Marks]

4 Fill in the Blanks

8. The weight of 1×10^{22} molecules of $CuSO_4 \cdot 5H_2O$ is .... [1991 - 1 Mark]

9. The modern atomic mass unit is based on .... [1980]

10. The total number of electrons present in 18 mL of water is .... [1980]

10 Subjective Problems

11. A plant virus is found to consist of uniform cylindrical particles of 150 Å in diameter and 5000 Å long. The specific volume of the virus is 0.75 cm^3/g. If the virus is considered to be a single particle, find its molar mass. [1999 - 3 Marks]

12. (a) One litre of a sample of hard water contains 1 mg of $CaCl_2$ and 1 mg of $MgCl_2$. Find the total hardness in terms of parts of $CaCO_3$ per 10^6 parts of water by weight.
(b) A sample of hard water contains 20 mg of Ca^{++} ions per litre. How many milli-equivalent of Na_2CO_3 would be required to soften 1 litre of the sample? [1980]
(c) 1 g of Mg is burnt in a closed vessel which contains 0.5 g of O_2.
(i) Which reactant is left in excess?
(ii) Find the weight of the excess reactants?
(iii) How may milliliters of 0.5 N H_2SO_4 will dissolve the residue in the vessel.

13. Find [1980]
(i) The total number of neutrons and
(ii) The total mass of neutron in 7 mg of ^{14}C.
(Assume that mass of neutron = mass of hydrogen atom)

Topic 2 - Percent Composition and Empirical Formula

1 MCQs with One Correct Answer

14. If two compounds have the same empirical formula but different molecular fomulae they must have [1987 - 1 Mark]
(a) different percentage composition (b) different molecular weight
(c) same viscosity (d) same vapour density

15. A gaseous mixture contains oxygen and nitrogen in the ratio of 1 : 4 by weight. Therefore, the ratio of their number of molecules is [1979]
(a) 1 : 4 (b) 1 : 8 (c) 7 : 32 (d) 3 : 16

16. A compound was found to contain nitrogen and oxygen in the ratio 28 g and 80 g respectively. The formula of compound is
(a) NO (b) N_2O_3 (c) N_2O_5 (d) N_2O_4 [1978]

3 Numeric / New Stem Based Questions

17. The composition of a sample of Wurtzite is $Fe_{0.93}O_{1.00}$. What percentage of the iron is present in the form of Fe (III)? [1994-2Marks]

18. A compound contains 28 percent of nitrogen and 72 percent of metal by weight. 3 atoms of metal combine with 2 atoms of N. Find the atomic weight of metal. [1980]

6 MCQs with One or More than One Correct Answer

19. To check the principle of multiple proportions, a series of pure binary compounds (P_mQ_n) were analyzed and their composition is tabulated below. The correct option(s) is (are) [Adv. 2022]

Compound	Weight % of P	Weight % of Q
1	50	50
2	44.4	55.6
3	40	60

(a) If empirical formula of compound 3 is P_3Q_4, then the empirical formula of compound 2 is P_3Q_5.
(b) If empirical formula of compound 3 is P_3Q_2 and atomic weight of element P is 20, then the atomic weight of Q is 45.
(c) If empirical formula of compound 2 is PQ, then the empirical formula of the compound 1 is P_5Q_4.
(d) If atomic weight of P and Q are 70 and 35, respectively, then the empirical formula of compound 1 is P_2Q.

10 Subjective Problems

20. A hydrocarbon contains 10.5 g of carbon per gram of hydrogen. 1 litre of the vapour of the hydrocarbon at 127 °C and 1 atmosphere pressure weighs 2.8 g. Find the molecular formula. [1980]

Topic 3 - Stoichiometric Calculations

1 MCQs with One Correct Answer

21. In a metal deficient oxide sample, $M_XY_2O_4$ (M and Y are metals), M is present in both +2 and +3 oxidation states and Y is in +3 oxidation state. If the fraction of M^{2+} ions present in M is $\frac{1}{3}$, the value of X is _____. [Adv. 2024]
(a) 0.25 (b) 0.33 (c) 0.67 (d) 0.75

22. Mixture X = 0.02 mol of $[Co(NH_3)_5SO_4]Br$ and 0.02 mol of $[Co(NH_3)_5Br]SO_4$ was prepared in 2 litre of solution.
1 litre of mixture X + excess $AgNO_3 \longrightarrow Y$.
1 litre of mixture X + excess $BaCl_2 \longrightarrow Z$
No. of moles of Y and Z are [2003S]
(a) 0.01, 0.01 (b) 0.02, 0.01
(c) 0.01, 0.02 (d) 0.02, 0.02

23. An aqueous solution of 6.3 g oxalic acid dihydrate is made up to 250 mL. The volume of 0.1 N NaOH required to completely neutralize 10 mL of this solution is [2001S]
(a) 40 mL (b) 20 mL (c) 10 mL (d) 4 mL

24. In the standardization of $Na_2S_2O_3$ using $K_2Cr_2O_7$ by iodometry, the equivalent weight of $K_2Cr_2O_7$ is [2001S]
(a) (molecular weight)/2 (b) (molecular weight)/6
(c) (molecular weight)/3 (d) same as molecular weight

25. The normality of 0.3 M phosphorous acid (H_3PO_3) is, [1999 - 2 Marks]
(a) 0.1 (b) 0.9 (c) 0.3 (d) 0.6

26. The equivalent weight of $MnSO_4$ is half of its molecular weight when it is converted to : [1988 - 1 Mark]
(a) Mn_2O_3 (b) MnO_2 (c) MnO_4^- (d) MnO_4^{2-}

27. In which mode of expression, the concentration of a solution remains independent of temperature? [1988 - 1 Mark]
(a) Molarity (b) Normality
(c) Formality (d) Molality

28. A molal solution is one that contains one mole of a solute in: [1986 - 1 Mark]
(a) 1000 g of the solvent (b) one litre of the solvent
(c) one litre of the solution (d) 22.4 litres of the solution

29. If 0.50 mole of $BaCl_2$ is mixed with 0.20 mol of Na_3PO_4, the maximum number of moles of $Ba_3(PO_4)_2$ that can be formed is [1981 - 1 Mark]
(a) 0.70 (b) 0.50 (c) 0.20 (d) 0.10

30. M is molecular weight of $KMnO_4$. The equivalent weight of $KMnO_4$ when it is converted into K_2MnO_4 is [1980]
(a) M (b) $M/3$ (c) $M/5$ (d) $M/7$

31. 2.76 g of silver carbonate on being strongly heated yields a residue weighing [1979]
(a) 2.16 g (b) 2.48 g (c) 2.32 g (d) 2.64 g

2 Integer Value Answer

32. The stoichiometric reaction of 516 g of dimethyldichlorosilane with water results in a tetrameric cyclic product X in 75% yield. The weight (in g) of X obtained is______. [Adv. 2023]
[Use, molar mass (g mol^{-1}): H = 1, C = 12, O = 16, Si = 28, Cl = 35.5]

33. H_2S (5 moles) reacts completely with acidified aqueous potassium permanganate solution. In this reaction, the number of moles of water produced is x, and the number of moles of electrons involved is y. The value of (x + y) is _______. [Adv. 2023]

3 Numeric / New Stem Based Questions

Question Stem for Question Nos. 34 and 35

A sample (5.6 g) containing iron is completely dissolved in cold dilute HCl to prepare a 250 mL of solution. Titration of 25.0 mL of this solution requires 12.5 mL of 0.03 M $KMnO_4$ solution to reach the end point. Number of moles of Fe^{2+} present in 250 mL solution is $x \times 10^{-2}$ (consider complete dissolution of $FeCl_2$). The amount of iron present in the sample of y% by weight.
(Assume : $KMnO_4$ reacts only with Fe^{2+} in the solution
Use : Molar mass of iron as 56 g mol^{-1})

34. The value of x is ____. [Adv. 2021]
35. The value of y is ____. [Adv. 2021]

36. The ammonia prepared by treating ammonium sulphate with calcium hydroxide is completely used by $NiCl_2.6H_2O$ to form a stable coordination compound. Assume that both the reactions are 100% complete. If 1584 g of ammonium sulphate and 952 g of $NiCl_2.6H_2O$ are used in the preparation, the combined weight (in grams) of gypsum and the nickel-ammonia coordination compound thus produced is________.
(Atomic weights in g mol^{-1}: H = 1, N = 14, O = 16, S = 32, Cl = 35.5, Ca = 40, Ni = 59) **[Adv. 2018]**

37. Galena (an ore) is partially oxidized by passing air through it at high temperature. After some time, the passage of air is stopped, but the heating is continued in a closed furnace such that the contents undergo self-reduction. The weight (in kg) of Pb produced per kg of O_2 consumed is______.
(Atomic weights in g mol^{-1}: O = 16, S = 32, Pb = 207) **[Adv. 2018]**

38. How many millilitres of 0.5 M H_2SO_4 are needed to dissolve 0.5 g of copper(II) carbonate? **[1999 - 3 Marks]**

39. One gram of commercial $AgNO_3$ is dissolved in 50 mL. of water. It is treated with 50 mL. of a KI solution. The silver iodide thus precipitated is filtered off. Excess of KI in the filterate is titrated with (M/10) KIO_3 solution in presence of 6M HCl till all I^- ions are converted into ICl. It requires 50 mL of (M/10) KIO_3 solution. 20 mL. of the same stock solution of KI requires 30 mL. of (M/10)KIO_3 under similar conditions. Calculate the percentage of $AgNO_3$ in the sample.
(Reaction : $KIO_3 + 2KI + 6HCl \rightarrow 3ICl + 3KCl + 3H_2O$) **[1992 - 4 Marks]**

40. A 1.0 g sample of Fe_2O_3 solid of 55.2% purity is dissolved in acid and reduced by heating the solution with zinc dust. The resultant solution is cooled and made upto 100.0 mL. An aliquot of 25.0 mL of this solution requires 17.0 mL of 0.0167 M solution of an oxidant for titration. Calculate the number of electrons taken up by the oxidant in the reaction of the above titration. **[1991 - 4 Marks]**

41. Calculate the molality of 1 litre solution of 93% H_2SO_4 (weight/volume). The density of the solution is 1.84 g/mL. **[1990 - 1 Marks]**

42. A sample of hydrazine sulphate ($N_2H_6SO_4$) was dissolved in 100 mL of water, 10 mL of this solution was reacted with excess of ferric chloride solution and warmed to complete the reaction. Ferrous ion formed was estimated and it required 20 mL of M/50 potassium permanganate solution. Estimate the amount of hydrazine sulphate in one litre of the solution. **[1988 - 3 Marks]**

Reaction : $4Fe^{3+} + N_2H_4 \rightarrow N_2 + 4Fe^{2+} + 4H^+$

$MnO_4^- + 5Fe^{2+} + 8H^+ \rightarrow Mn^{2+} + 5Fe^{3+} + 4H_2O.$

43. Hydroxylamine reduces iron (III) according to the equation:
$2NH_2OH + 4Fe^{3+} \rightarrow N_2O(g) \uparrow + H_2O + 4Fe^{2+} + 4H^+$
Iron (II) thus produced is estimated by titration with a standard permanganate solution. The reaction is :

$MnO_4^- + 5Fe^{2+} + 8H^+ \rightarrow Mn^{2+} + 5Fe^{3+} + 4H_2O$

A 10 mL. sample of hydroxylamine solution was diluted to 1 litre. 50 mL. of this diluted solution was boiled with an excess of iron (III) solution. The resulting solution required 12 mL. of 0.02 M $KMnO_4$ solution for complete oxidation of iron (II). Calculate the weight of hydroxylamine in one litre of the original solution. (H = 1, N = 14, O = 16, K = 39, Mn = 55, Fe = 56) **[1982 - 4 Marks]**

44. A 1.00 g sample of H_2O_2 solution containing X per cent H_2O_2 by weight requires X mL of a $KMnO_4$ solution for complete oxidation under acidic conditions. Calculate the normality of the $KMnO_4$ solution. **[1981 - 3 Marks]**

45. 4.215 g of a metallic carbonate was heated in a hard glass tube and the CO_2 evolved was found to measure 1336 mL at 27°C and 700 mm pressure. What is the equivalent weight of the metal? **[1979]**

46. What weight of AgCl will be precipitated when a solution containing 4.77 g of NaCl is added to a solution of 5.77 g of $AgNO_3$? **[1978]**

47. Igniting MnO_2 converts it quantitatively to Mn_3O_4. A sample of pyrolusite is of the following composition : MnO_2 80%, SiO_2 and other inert constituents 15%, rest being water. The sample is ignited in air to constant weight. What is the percentage of Mn in the ignited sample? **[1978]**
[O = 16, Mn = 54.9]

9 Assertion and Reason/Statement Type Questions

Each question contains **STATEMENT-1 (Assertion)** and **STATEMENT-2 (Reason)**. Each question has 4 choices (a), (b), (c) and (d) out of which ONLY ONE is correct. Mark your answer as

(a) If both Statement -1 and Statement -2 are correct, and Statement -2 is the correct explanation of the Statement -2.
(b) If both Statement -1 and Statement -2 are correct, but Statement - 2 is not the correct explanation of the Statement -1.
(c) If Statement -1 is correct but Statement -2 is incorrect.
(d) If Statement -1 is incorrect but Statement -2 is correct.

48. Read the following statement and explanation and answer as per the options given below :
Statement-1 : In the titration of Na_2CO_3 with HCl using methyl orange indicator, the volume required at the equivalence point is twice that of the acid required using phenolphthalein indicator.
Statement-2 : Two moles of HCl are required for the complete neutralization of one mole of Na_2CO_3. **[1991 - 2 Marks]**

10 Subjective Problems

49. 3 g of a salt of molecular weight 30 is dissolved in 250 g of water. The molality of the solution is .... **[1983 - 1 Mark]**

50. Hydrogen peroxide solution (20 mL) reacts quantitatively with a solution of $KMnO_4$ (20 mL) acidified with dilute H_2SO_4. The same volume of the $KMnO_4$ solution is just decolourised by 10 mL of $MnSO_4$ in neutral medium simultaneously forming a dark brown precipitate of hydrated MnO_2. The brown precipitate is dissolved in 10 mL of 0.2 M sodium oxalate under boiling condition in the presence of dilute H_2SO_4. Write the balanced equations involved in the reactions and calculate the molarity of H_2O_2. **[2001 - 5 Marks]**

51. An aqueous solution containing 0.10 g KIO_3 (formula weight = 214.0) was treated with an excess of KI solution. The solution was acidified with HCl. The liberated I_2 consumed 45.0 mL of thiosulphate solution to decolourise the blue starch-iodine complex. Calculate the molarity of the sodium thiosulphate solution. **[1998 - 5 Marks]**

52. A 3.00 g sample containing Fe_3O_4, Fe_2O_3 and an inert impure substance, is treated with excess of KI solution in presence of dilute H_2SO_4. The entire iron is converted into Fe^{2+} along with the liberation of iodine. The resulting solution is diluted to 100 mL. A 20 mL of the diluted solution requires 11.0 mL of 0.5 M $Na_2S_2O_3$ solution to reduce the iodine present. A 50 mL of the diluted solution, after complete extraction of the iodine requires 12.80 mL of 0.25 M $KMnO_4$ solution in dilute H_2SO_4 medium for the oxidation of Fe^{2+}. Calculate the percentages of Fe_2O_3 and Fe_3O_4 in the original sample. **[1996 - 5 Marks]**

53. 8.0575×10^{-2} kg of Glauber's salt is dissolved in water to obtain 1 dm^3 of a solution of density 1077.2 kg m^{-3}. Calculate the molarity, molality and mole fraction of Na_2SO_4 in the solution. **[1994 - 3 Marks]**

54. Upon mixing 45.0 mL. of 0.25 M lead nitrate solution with 25.0 mL of 0.10 M chromic sulphate solution, precipitation of lead sulphate takes place. How many moles of lead sulphate are formed? Also, calculate the molar concentrations of the species left behind in the final solution. Assume that lead sulphate is completely insoluble. **[1993 - 3 Marks]**

55. A 2.0 g sample of a mixture containing sodium carbonate, sodium bicarbonate and sodium sulphate is gently heated till the evolution of CO_2 ceases. The volume of CO_2 at 750 mm Hg pressure and at 298 K is measured to be 123.9 mL. A 1.5g of the same sample requires 150 mL. of (M/10) HCl for complete neutralisation. Calculate the % composition of the components of the mixture. **[1992 - 5 Marks]**

56. A solution of 0.2 g of a compound containing Cu^{2+} and $C_2O_4^{2-}$ ions on titration with 0.02 M $KMnO_4$ in presence of H_2SO_4 consumes 22.6 mL. of the oxidant. The resultant solution is neutralized with Na_2CO_3, acidified with dil. acetic acid and treated with excess KI. The liberated iodine requires 11.3 mL of 0.05 M $Na_2S_2O_3$ solution for complete reduction. Find out the molar ratio of Cu^{2+} to $C_2O_4^{2-}$ in the compound. Write down the balanced redox reactions involved in the above titrations. **[1991 - 5 Marks]**

57. A mixture of $H_2C_2O_4$ (oxalic acid) and $NaHC_2O_4$ weighing 2.02 g was dissolved in water and solution made upto one litre. Ten millilitres of the solution required 3.0 mL. of 0.1 N sodium hydroxide solution for complete neutralization. In another experiment, 10.0 mL. of the same solution, in hot dilute sulphuric acid medium. require 4.0 mL. of 0.1 N potassium permanganate solution for complete reaction. Calculate the amount of $H_2C_2O_4$ and $NaHC_2O_4$ in the mixture. **[1990 - 5 Marks]**

58. A solid mixture (5.0 g) consisting of lead nitrate and sodium nitrate was heated below 600 °C until the weight of the residue was constant. If the loss in weight is 28.0 per cent, find the amount of lead nitrate and sodium nitrate in the mixture. **[1990 - 4 Marks]**

59. An equal volume of a reducing agent is titrated separately with 1M $KMnO_4$ in acid neutral and alkaline media. The volumes of $KMnO_4$ required are 20 mL. in acid, 33.4 mL in neutral and 100 mL. in alkaline media. Find out the oxidation state of manganese in each reduction product. Give the balanced equations for all the three half reactions. Find out the volume of 1M $K_2Cr_2O_7$ consumed; if the same volume of the reducing agent is titrated in acid medium. **[1989 - 5 Marks]**

60. A sugar syrup of weight 214.2 g contains 34.2 g of sugar ($C_{12}H_{22}O_{11}$). Calculate : (i) molal concentration and (ii) mole fraction of sugar in the syrup. **[1988 - 2 Marks]**

61. (*i*) What is the weight of sodium bromate and molarity of solution necessary to prepare 85.5 mL of 0.672 N solution when the half-cell reaction is

$$BrO_3^- + 6H^+ + 6e^- \rightarrow Br^- + 3H_2O$$

(*ii*) What would be the weight as well as molarity if the half-cell reaction is :

$$2BrO_3^- + 12H^+ + 10e^- \rightarrow Br_2 + 6H_2O$$ **[1987 - 5 Marks]**

62. Five mL of 8N nitric acid, 4.8 mL of 5N hydrochloric acid and a certain volume of 17M sulphuric acid are mixed together and made upto 2litre. Thirty mL. of this acid mixture exactly neutralise 42.9 mL of sodium carbonate solution containing one gram of $Na_2CO_3.10H_2O$ in 100 mL. of water. Calculate the amount in gram of the sulphate ions in solution. **[1985 - 4 Marks]**

63. 2.68×10^{-3} moles of a solution containing an ion A^{n+} require 1.61×10^{-3} moles of MnO_4^- for the oxidation of A^{n+} to AO_3^- in acid medium. What is the value of *n*? **[1984 - 2 Marks]**

64. The density of a 3 M sodium thiosulphate solution ($Na_2S_2O_3$) is 1.25 g per mL. Calculate (i) the percentage by weight of sodium thiosulphate, (ii) the mole fraction of sodium thiosulphate and (iii) the molalities of Na^+ and $S_2O_3^{2-}$ ions. **[1983 - 5 Marks]**

65. 4.08 g of a mixture of BaO and an unknown carbonate MCO_3 was heated strongly. The residue weighed 3.64 g. This was dissolved in 100 mL of 1 N HCl. The excess acid required 16 mL of 2.5 N NaOH solution for complete neutralization. Identify the metal *M*. **[1983 - 4 Marks]**
(At. wt. H = 1, C = 12, O = 16, Cl = 35.5, Ba = 138)

66. (i) A sample of $MnSO_4 \cdot 4H_2O$ is strongly heated in air. The residue is Mn_3O_4.
(ii) The residue is dissolved in 100 mL of 0.1 N $FeSO_4$ containing dilute H_2SO_4.
(iii) The solution reacts completely with 50 mL of $KMnO_4$ solution.
(iv) 25 mL of the $KMnO_4$ solution used in step (iii) requires 30 mL of 0.1 N $FeSO_4$ solution for complete reaction.
Find the amount of $MnSO_4 \cdot 4H_2O$ present in the sample. **[1980]**

67. A mixture contains NaCl and unknown chloride MCl.
(i) 1 g of this is dissolved in water. Excess of acidified $AgNO_3$ solution is added to it. 2.567 g of white ppt. is formed.
(ii) 1 g of original mixture is heated to 300 °C. Some vapours come out which are absorbed in acidified $AgNO_3$ solution, 1.341 g of white precipitate was obtained.
Find the molecular weight of unknown chloride. **[1980]**

68. 5 mL of a gas containing only carbon and hydrogen were mixed with an excess of oxygen (30 mL) and the mixture exploded by means of an electric spark. After the explosion, the volume of the mixed gases remaining was 25 mL. On adding a concentrated solution of potassium hydroxide, the volume further diminished to 15 mL of the residual gas being pure oxygen. All volumes have been reduced to N.T.P. Calculate the molecular formula of the hydrocarbon gas. **[1979]**

69. One gram of an alloy of aluminium and magnesium when treated with excess of dil. HCl forms magnesium chloride, aluminium chloride and hydrogen. The evolved hydrogen, collected over mercury at 0 °C has a volume of 1.20 litres at 0.92 atm. pressure. Calculate the composition of the alloy.
[H = 1, Mg = 24, Al = 27] **[1978]**

Answer Key

Part-A (JEE Main)																			
1	(3)	21	(4)	41	(11.00)	61	(4)	81	(2)	101	(4)	121	(2)	141	(125)	161	(227)	181	(4)
2	(4)	22	(1)	42	(2)	62	(3)	82	(18)	102	(1)	122	(4)	142	(420)	162	(1362)	182	(226)
3	(4)	23	(4)	43	(24)	63	(3)	83	(5)	103	(2)	123	(1)	143	(61)	163	(150)	183	(1)
4	(2)	24	(2)	44	(46)	64	(4)	84	(3)	104	(3)	124	(4)	144	(153)	164	(200)	184	(16)
5	(2)	25	(2)	45	(2)	65	(2)	85	(2)	105	(3)	125	(1)	145	(100)	165	(15)	185	(3)
6	(4)	26	(2)	46	(5418)	66	(4)	86	(1)	106	(3)	126	(2)	146	(57)	166	(4)	186	(18)
7	(1)	27	(3)	47	(2)	67	(2)	87	(2)	107	(4)	127	(1)	147	(45)	167	(3810)	187	(1)
8	(2)	28	(1)	48	(14.00)	68	(4)	88	(4)	108	(2)	128	(4)	148	(24)	168	(143)	188	(24)
9	(4)	29	(1)	49	(2)	69	(3)	89	(1)	109	(3)	129	(1)	149	(14)	169	(225)	189	(13)
10	(1)	30	(1)	50	(2)	70	(3)	90	(3)	110	(3)	130	(1)	150	(50)	170	(1143)	190	(2)
11	(4)	31	(4)	51	(1)	71	(66)	91	(3)	111	(3)	131	(2)	151	(4)	171	(0)	191	(8)
12	(4)	32	(1)	52	(3)	72	(65)	92	(1)	112	(3)	132	(1)	152	(8)	172	(25)	192	(18)
13	(3)	33	(3)	53	(4)	73	(12)	93	(1)	113	(1)	133	(1)	153	(7)	173	(3)	193	(3400)
14	(3)	34	(4)	54	(1)	74	(3)	94	(3)	114	(1)	134	(17)	154	(135)	174	(3)	194	(10)
15	(3)	35	(2)	55	(1)	75	(56)	95	(3)	115	(1)	135	(73)	155	(100)	175	(525)	195	(47)
16	(3)	36	(2)	56	(4)	76	(36)	96	(3)	116	(N)	136	(10)	156	(224)	176	(13)	196	(25)
17	(4)	37	(4)	57	(4)	77	(59)	97	(3)	117	(4)	137	(1)	157	(11)	177	(19)	197	(10)
18	(4)	38	(3)	58	(3)	78	(25)	98	(1)	118	(1)	138	(138)	158	(1)	178	(5)	198	(10.00)
19	(3)	39	(1)	59	(2)	79	(12)	99	(3)	119	(3)	139	(63)	159	(44)	179	(18)	199	(4.96)
20	(3)	40	(372)	60	(2)	80	(46)	100	(1)	120	(4)	140	(80)	160	(4)	180	(464)	200	(2130)
Part-B (JEE Advanced)																			
1	(a)	6	(4)	14	(b)	19	(b, c)	25	(d)	30	(a)	35	(18.75)	40	(6.0)	45	(12.15)		
2	(d)	7	(55.55)	15	(c)	21	(d)	26	(b)	31	(a)	36	(2992)	41	(10.43)	46	(4.87)		
3	(a)	8	(4.14 g)	16	(c)	22	(a)	27	(d)	32	(222)	37	(6.47)	42	(6.5)	47	(59.33)		
4	(a)	9	(carbon (c-12)	17	(15.05)	23	(a)	28	(a)	33	(18)	38	(8.09)	43	(39.6)	48	(b)		
5	(d)	10	(6.02 × 1024)	18	(24)	24	(b)	29	(d)	34	(1.875)	39	(85)	44	(0.58)				

Chapter-2

Structure of Atom

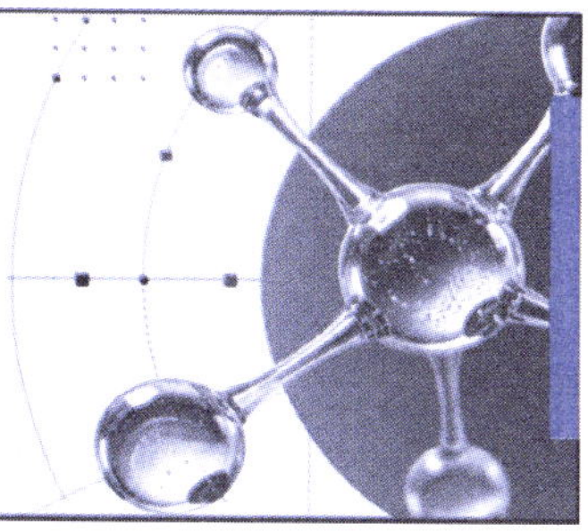

Part-A (JEE Main)

Topic 1 - Developments Leading to the Bohr's Model of Atom

MULTIPLE CHOICE QUESTIONS (MCQs)

1. The wavelength of photon 'A' is 400 nm. The frequency of p4esrfvhoton 'B' is $10^{16} s^{-1}$. The wave number of photon 'C' is $10^4 cm^{-1}$. The correct order of energy of these photons is : **[Jan. 28, 2026 (II)]**

(1) C > B > A (2) B > A > C
(3) A > B > C (4) A > C > B

2. The work functions of two metals (M_A and M_B) are in the 1 : 2 ratio. When these metals are exposed to photons of energy 6 eV, the kinetic energy of liberated electrons of $M_A : M_B$ is in the ratio of 2.642 : 1. The work function (in eV) of M_A and M_B are respectively. **[Jan. 23, 2026 (II)]**

(1) 3.1, 6.2 (2) 2.3, 4.6
(3) 1.4, 2.8 (4) 1.5, 3.0

3. Identify the **INCORRECT** statements from the following: **[Jan. 23, 2026 (II)]**

A. Notation $^{24}_{12}Mg$ represents 24 protons and 12 neutrons.
B. Wavelength of a radiation of frequency $4.5 \times 10^{15} s^{-1}$ is 6.7×10^{-8} m.
C. One radiation has wavelength = λ_1(900 nm) and energy = E_1. Other radiation has wavelength = λ_2(300 nm) and energy = E_2. $E_1 : E_2 = 3 : 1$.
D. Number of photons of light of wavelength 2000 pm that provides 1 J of energy is 1.006×10^{16}.

Choose the **correct** answer from the options given below:

(1) A and D only (2) A and C only
(3) A and B only (4) B and C only

4. Visible spectrum

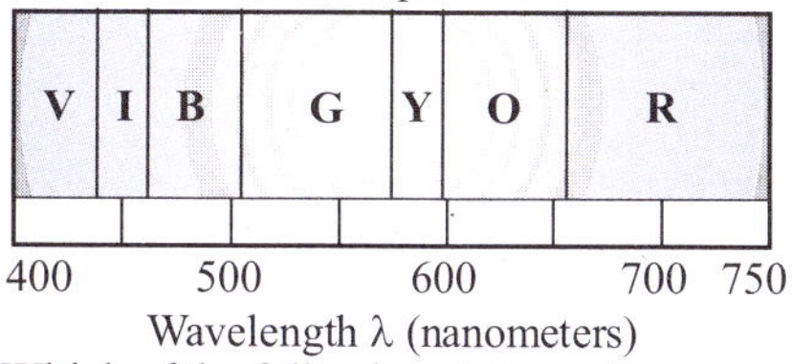

Wavelength λ (nanometers)

Which of the following statements are correct, if the threshold frequency of cesium is 5.16×10^{14} Hz ? **[April 7, 2025 (I)]**

A. When *Cs* is placed inside a vacuum chamber with an ammeter connected to it and yellow light is focused on *Cs* the ammeter shows the presence of current.
B. When the brightness of the yellow light is dimmed, the value of the current in the ammeter is reduced.
C. When a red light is used instead to the yellow light, the current produced is higher with respect to the yellow light.
D. When a blue light is used, the ammeter shows the formation of current.
E. When a white light is used, the ammeter shows formation of current.

Choose the **correct** answer from the options given below :

(1) A, D and E Only (2) B, C and D Only
(3) A, C, D and E Only (4) A, B, D and E Only

5. Given below are two statements : one is labelled as Assertion A and the other is labelled as Reason R :

Assertion A : In the photoelectric effect, the electrons are ejected from the metal surface as soon as the beam of light of frequency greater than threshold frequency strikes the surface.

Reason R : When the photon of any energy strikes an electron in the atom, transfer of energy from the photon to the electron takes place.

In the light of the above statements, choose the most appropriate answer from the options given below : **[April 11, 2023 (I)]**

(1) Both A and R are correct but R is NOT the correct explanation of A.
(2) A is correct but R is not correct.
(3) Both A and R are correct and R is the correct explanation of A.
(4) A is not correct but R is correct.

6. Given below are two statements: one is labelled as assertion and the other is labelled as reason. **[April 06, 2023 (I)]**

Assertion: Loss of electron from hydrogen atom results in nucleus of ~1.5×10^{-3} pm size.

Reason: Proton (H^+) always exists in combined form

In the light of the above statements, choose the most appropriate answer from the options given below:

(1) Both A and R are correct R is the correct explanation of A.
(2) A is correct but R is not correct.
(3) A is not correct but R is correct.
(4) Both A and R are correct but R is NOT the correct explanation of A.

7. The energy of one mole of photons of radiation of wavelength 300 nm is **[June 24, 2022 (II)]**
(Given : $h = 6.63 \times 10^{-34}$ Js, $N_A = 6.02 \times 10^{23} mol^{-1}$, $c = 3 \times 10^8 ms^{-1}$)

(1) 235 kJ mol^{-1} (2) 325 kJ mol^{-1}
(3) 399 kJ mol^{-1} (4) 435 kJ mol^{-1}

8. The minimum energy that must be possessed by photons in order to produce the photoelectric effect with platinum metal is:
[**Given:** The threshold frequency of platinum is $1.3 \times 10^{15} s^{-1}$ and $h = 6.6 \times 10^{-34}$ J s.] **[June 25, 2022 (II)]**

(1) 3.21×10^{-14} J (2) 6.24×10^{-16} J
(3) 8.58×10^{-19} J (4) 9.76×10^{-20} J

9. If p is the momentum of the fastest electron ejected from a metal surface after the irradiation of light having wavelength λ, then for $1.5p$ momentum of the photoelectron, the wavelength of the light should be:
(Assume kinetic energy of ejected photoelectron to be very high in comparison to work function): [April 8, 2019 (II)]
(1) $\frac{3}{4}\lambda$ (2) $\frac{1}{2}\lambda$ (3) $\frac{2}{3}\lambda$ (4) $\frac{4}{9}\lambda$

10. What is the work function of the metal if the light of wavelength 4000 Å generates photoelectrons of velocity 6×10^5 ms^{-1} from it ?
(Mass of electron $=9\times10^{-31}$ kg [Jan. 12, 2019 (I)]
Velocity of light $=3\times10^8$ ms^{-1}
Planck's constant $=6.626\times10^{-34}$ Js
Charge of electron $=1.6\times10^{-19}$ JeV^{-1})
(1) 0.9 eV (2) 3.1 eV (3) 2.1 eV (4) 4.0 eV

11. Which of the graphs shown below does not represent the relationship between incident light and the electron ejected from metal surface? [Jan. 10, 2019 (I)]

(1)
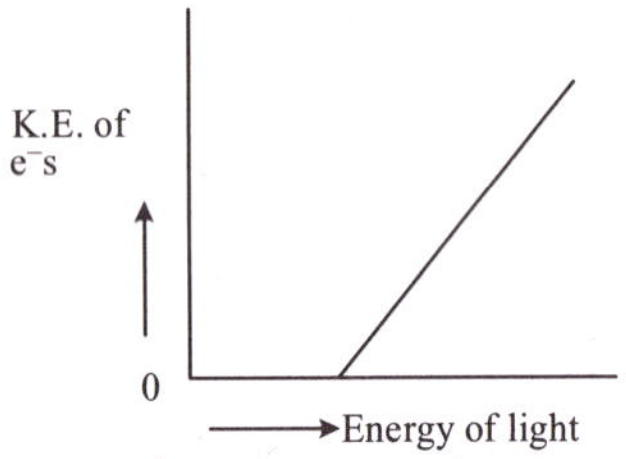

(2)
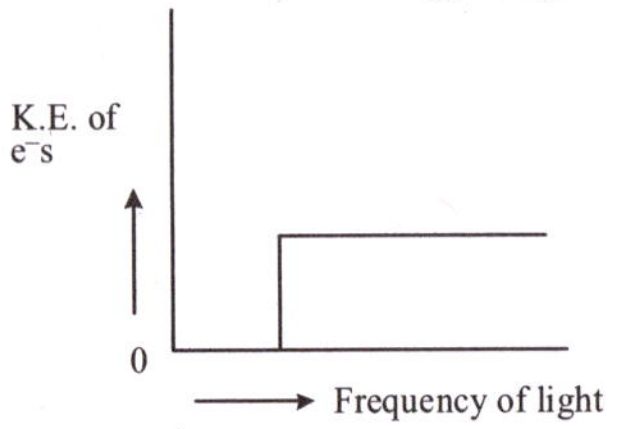

(3)
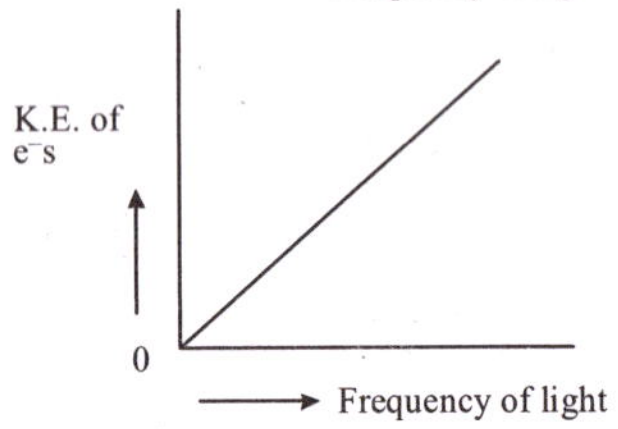

(4)
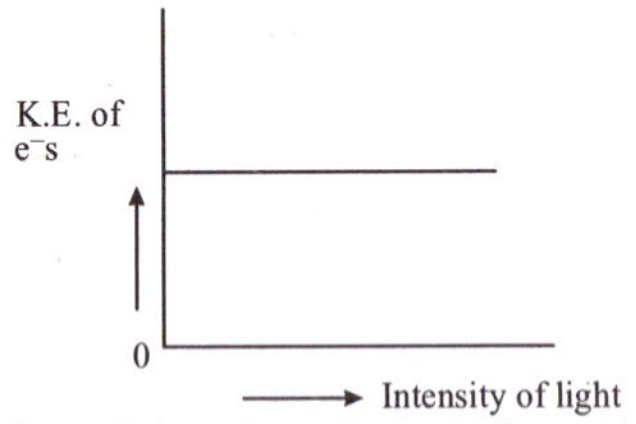

12. Ejection of the photoelectron from metal in the photoelectric effect experiment can be stopped by applying 0.5V when the radiation of 250 nm is used. The work function of the metal is : [Online April 15, 2018 (I)]
(1) 4 eV (2) 5.5 eV (3) 4.5 eV (4) 5 eV

13. Excited hydrogen atom emits light in the ultraviolet region at 2.47×10^{15} Hz. With this frequency, the energy of a single photon is: ($h = 6.63\times10^{-34}$ Js) [Online April 12, 2014]
(1) 8.041×10^{-40} J (2) 2.680×10^{-19} J
(3) 1.640×10^{-18} J (4) 6.111×10^{-17} J

14. Ionization energy of gaseous Na atoms is 495.5 kJ mol^{-1}. The lowest possible frequency of light that ionizes a sodium atom is ($h = 6.626\times10^{-34}$ Js, $N_A = 6.022\times10^{23}$ mol^{-1}) [Online April 19, 2014]
(1) 7.50×10^4 s^{-1} (2) 4.76×10^{14} s^{-1}
(3) 3.15×10^{15} s^{-1} (4) 1.24×10^{15} s^{-1}

15. If λ_0 and λ be threshold wavelength and wavelength of incident light, the velocity of photoelectron ejected from the metal surface is: [Online April 11, 2014]
(1) $\sqrt{\frac{2h}{m}(\lambda_0-\lambda)}$ (2) $\sqrt{\frac{2hc}{m}(\lambda_0-\lambda)}$
(3) $\sqrt{\frac{2hc}{m}\left(\frac{\lambda_0-\lambda}{\lambda\lambda_0}\right)}$ (4) $\sqrt{\frac{2h}{m}\left(\frac{1}{\lambda_0}-\frac{1}{\lambda}\right)}$

16. The energy required to break one mole of Cl – Cl bonds in Cl_2 is 242 kJ mol^{-1}. The longest wavelength of light capable of breaking a single Cl – Cl bond is [2010]
($c = 3\times10^8$ ms^{-1} and $N_A = 6.02\times10^{23}$ mol^{-1}).
(1) 594 nm (2) 640 nm
(3) 700 nm (4) 494 nm

17. Match List I with List II [Jan, 29, 2024 (II)]

	LIST I (Spectral Series for Hydrogen)		LIST II (Spectral Region / Higher Energy State)
A.	Lyman	I.	Infrared region
B.	Balmer	II.	UV region
C.	Paschen	III.	Infrared region
D.	Pfund	IV.	Visible region

Choose the correct answer from the options given below:
(1) A-I, B-III, C-II, D-IV (2) A-II, B-III C-I, D-IV
(3) A-II, B-IV, C-III, D-I (4) A-I, B-II, C-III, D-IV

18. The region in the electromagnetic spectrum where the Balmer series lines appear is : [Sep. 04, 2020 (I)]
(1) Visible (2) Microwave
(3) Infrared (4) Ultraviolet

19. In Bohr series of lines of hydrogen spectrum, the third line from the red end corresponds to which one of the following inter-orbit jumps of the electron for Bohr orbits in an atom of hydrogen [2003]
(1) $5\rightarrow2$ (2) $4\rightarrow1$
(3) $2\rightarrow5$ (4) $3\rightarrow2$

Numerical Value Questions (NVQs)

20. Consider two radiations of wavelengths :
1. $\lambda_1 = 2000$ Å
2. $\lambda_2 = 6000$ Å

The ratio of the energies of these two radiations $\left(\frac{E_1}{E_2}\right)$ is ________. (Nearest integer) [April 08, 2026 (II)]

21. The surface of sodium metal is irradiated with radiation of wavelength x nm. The kinetic energy of ejected electrons is 2.8×10^{-20} J. The work function of sodium is 2.3 eV. The value of x is ________ $\times10^2$ nm. (Nearest integer)
(Given : $h = 6.6\times10^{-34}$ J s; 1eV $= 1.60\times10^{-19}$ J; $c = 3.0\times10^8$ ms^{-1}) [April 02, 2026 (II)]

22. The ionization energy of sodium in kJ mol^{-1}, if electromagnetic radiation of wavelength 242 nm is just sufficient to ionize sodium atom is _____. [Jan. 31, 2024 (I)]

23. Values of work function (W_0) for a few metals are given below

Metal	Li	Na	K	Mg	Cu	Ag
W_0 / eV	2.42	2.3	2.25	3.7	4.8	4.3

The number of metals which will show photoelectric effect when light of wavelength 400 nm falls on it is ______
Given : $h = 6.6 \times 10^{-34}$ J s
$c = 3 \times 10^8$ m s^{-1}
$e = 1.6 \times 10^{-19}$ C [April 12, 2023 (I)]

24. The number of incorrect statement/s about the black body from the following is _______ [April 10, 2023 (I)]
(A) Emit or absorb energy in the form of electromagnetic radiation
(B) Frequency distribution of the emitted radiation depends on temperature
(C) At a given temperature, intensity vs frequency curve passes through a maximum value
(D) The maximum of the intensity vs frequency curve is at a higher frequency at higher temperature compared to that at lower temperature activation energy.

25. The number of following statement/s which is/are incorrect is_______ [April 08, 2023 (I)]
(A) Line emission spectra are used to study the electronic structure
(B) The emission spectra of atoms in the gas phase show a continuous spread of wavelength from red to violet
(C) An absorption spectrum is like the photographic negative of an emission spectrum
(D) The element helium was discovered in the sun by spectroscopic method

26. The energy of one mole of photons of radiation of frequency 2×10^{12} Hz in Jmol^{-1} is _______. [Jan. 30, 2023 (I)]
(Nearest integer) (Given: $h = 6.626 \times 10^{-34}$ Js; $N_A = 6.022 \times 10^{23}$ mol^{-1})

27. Following figure shows spectrum of an ideal black body at four different temperatures. The number of correct statement/s from the following is______. [Jan. 24, 2023 (II)]

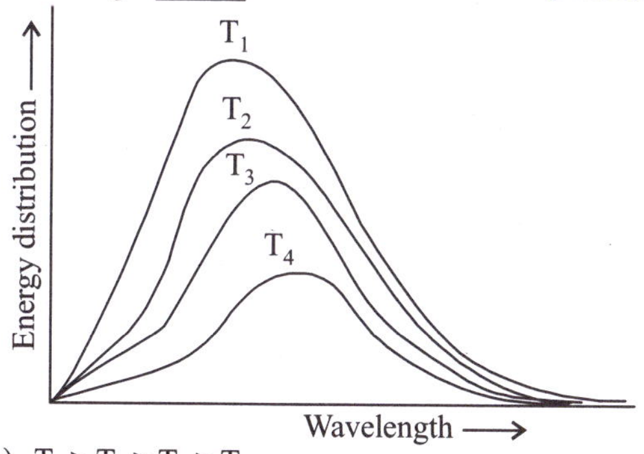

(A) $T_4 > T_3 > T_2 > T_1$
(B) The black body consists of particles performing simple harmonic motion.
(C) The peak of the spectrum shifts to shorter wavelength as temperature increases.
(D) $\frac{T_1}{\nu_1} = \frac{T_2}{\nu_2} = \frac{T_3}{\nu_3} \neq$ constant
(E) The given spectrum could be explained using quantisation of energy.

28. If the wavelength for an electron emitted from H-atom is 3.3×10^{-10} m, then energy absorbed by the electron in its ground state compared to minimum energy required for its escape from the atom, is _____ times. (Nearest integer) [Given : $h = 6.626 \times 10^{-34}$ J s]
Mass of electron = 9.1×10^{-31} kg [July 28, 2022 (II)]

29. When the excited electron of a H atom from n = 5 drops to the ground state, the maximum number of emission lines observed are __________. [July 25, 2022 (II)]

30. If the work function of a metal is 6.63×10^{-19} J, the maximum wavelength of the photon required to remove a photoelectron from the metal is _______ nm. (Nearest integer)
[Given : $h = 6.63 \times 10^{-34}$ J s, and $c = 3 \times 10^8$ m s^{-1}] [June 28, 2022 (I)]

31. A 50 watt bulb emits monochromatic red light of wavelength of 795 nm. The number of photons emitted per second by the bulb is $x \times 10^{20}$. The value of x is _________.
[Given : $h = 6.63 \times 10^{-34}$ Js and $c = 3.0 \times 10^8$ ms^{-1}] [Sep. 1, 2021 (II)]

32. The number of photons emitted by a monochromatic (single frequency) infrared range finder of power 1 mW and wavelength of 1000 nm, in 0.1 second is $x \times 10^{13}$. The value of x is _____. (Nearest integer) [Aug. 27, 2021 (II)]
($h = 6.63 \times 10^{-34}$ Js, $c = 3.00 \times 10^8$ ms^{-1})

33. A metal surface is exposed to 500 nm radiation. The threshold frequency of the metal for photoelectric current is 4.3×10^{14} Hz. The velocity of ejected electron is ________ $\times 10^5$ ms^{-1} (Nearest integer) [Aug. 26, 2021 (II)]
[Use : $h = 6.63 \times 10^{-34}$ Js, $m_e = 9.0 \times 10^{-31}$ kg]

34. A source of monochromatic radiation of wavelength 400 nm provides 1000 J of energy in 10 seconds. When this radiation falls on the surface of sodium, $x \times 10^{20}$ electrons are ejected per second. Assume that wavelength 400 nm is sufficient for ejection of electron from the surface of sodium metal. The value of x is _______. (Nearest integer) ($h = 6.626 \times 10^{-34}$ Js) [July 25, 2021 (I)]

35. Electromagnetic radiation of wavelength 663 nm is just sufficient to ionise the atom of metal A. The ionization energy of metal A in kJ mol^{-1} is ____. (Rounded-off to the nearest integer)
[$h = 6.63 \times 10^{-34}$ Js, $c = 3.00 \times 10^8$ ms–1, $N_A = 6.02 \times 10^{23}$ mol^{-1}] [Feb. 25, 2021 (II)]

36. The work function of sodium metal is 4.41×10^{-19} J. If photons of wavelength 300 nm are incident on the metal, the kinetic energy of the ejected electrons will be ($h = 6.63 \times 10^{-34}$ J s; $c = 3 \times 10^8$ m/s) _____ $\times 10^{-21}$ J. [Sep. 02, 2020 (II)]

Topic 2 - Bohr's Model for Hydrogen Atom (Emission and Absorption Spectra)

MULTIPLE CHOICE QUESTIONS (MCQs)

37. The Bohr radius of a hydrogen like species is 70.53 pm. The species and the stationary state(n) are respectively
(Given: Hydrogen atom Bohr radius is 52.9 pm) [April 06, 2026 (II)]
(1) Li^{2+}, 3 (2) He^+, 3 (3) He^+, 2 (4) Li^{2+}, 2

38. If shortest wavelength of hydrogen atom in Lyman series is x, then longest wavelength in Balmer series of He^+ is : [April 06, 2026 (I)]
(1) $\frac{9x}{5}$ (2) $\frac{36x}{5}$ (3) $\frac{x}{4}$ (4) $\frac{5x}{9}$

39. The species having identical radii according to the Bohr's theory are : [April 04, 2026 (II)]
A. H (first orbit)
B. He^+ (first orbit)
C. He^+ (second orbit)
D. Li^{2+} (first orbit)
E. Be^{3+} (Second orbit)
Choose the correct answer from the options given below:
(1) A and C Only (2) A and E Only
(3) B and E Only (4) C and D Only

40. What is the ratio of wave number of first line (lowest energy line) of Balmer series of H atomic spectrum to first line of its Brackett series? [April 04, 2026 (I)]
(1) 5 : 1 (2) 5 : 0.81 (3) 5 : 1.75 (4) 5 : 27

41. What is the energy (in J $atom^{-1}$) required for the following process ? [April 02, 2026 (I)]
$Li^{2+}(g) \rightarrow Li^{3+}(g) + e^-$
(Take the ionization energy for the H atom in the ground state as 2.18×10^{-18} J $atom^{-1}$)
(1) 8.72×10^{-18} (2) 1.962×10^{-18}
(3) 1.962×10^{-17} (4) 6.54×10^{-17}

42. The wave numbers of three spectral lines of H atom are considered. Identify the set of spectral lines belonging to Balmer series. [Jan. 28, 2026 (I)]
(R = Rydberg constant)
(1) $\frac{5R}{36}, \frac{3R}{16}, \frac{21R}{100}$ (2) $\frac{5R}{36}, \frac{8R}{9}, \frac{15R}{16}$
(3) $\frac{7R}{144}, \frac{3R}{16}, \frac{16R}{255}$ (4) $\frac{3R}{4}, \frac{3R}{16}, \frac{7R}{144}$

43. The wavelength of spectral line obtained in the spectrum of LI^{2+} ion, when the transition takes place between two levels whose sum is 4 and difference is 2, is [Jan. 24, 2026 (II)]
(1) 2.28×10^{-7} cm (2) 2.28×10^{-6} cm
(3) 1.14×10^{-7} cm (4) 1.14×10^{-6} cm

44. Which of the following statements regarding the energy of the stationary state is **true** in the following one-electron system? [Jan. 23, 2026 (I)]
(1) -1.09×10^{-18} J for second orbit of H atom.
(2) $+2.18 \times 10^{-18}$ J for second orbit of He^+ ion
(3) $+8.72 \times 10^{-18}$ J for first orbit of He^+ ion
(4) -2.18×10^{-18} J for third orbit of Li^{2+} ion

45. The energy of first (lowest) Balmer line of H atom is x J. The energy (in J) of second Balmer line of H atom is : [Jan. 22, 2026 (II)]
(1) x^2 (2) $\frac{x}{1.35}$ (3) $2x$ (4) $1.35x$

46. The energy required by electrons, present in the first Bohr orbit of hydrogen atom to be excited to second Bohr orbit is ______ J mol^{-1}. (**Given :** $R_H = 2.18 \times 10^{-11}$ ergs.) [Jan. 22, 2026 (I)]
(1) 1.635×10^{-18} (2) 9.835×10^{5}
(3) 9.835×10^{12} (4) 1.635×10^{-11}

47. Consider the following spectral lines for atomic hydrogen:
A. First line of Paschen series
B. Second line of Balmer series
C. Third line of Paschen series
D. Fourth line of Bracket sereis
The correct arrangement of the above lines in ascending order of energy is: [Jan. 21, 2026 (II)]
(1) D < C < A < B (2) A < B < C < D
(3) C < D < B < A (4) D < A < C < B

48. Given below are two statements: [Jan. 21, 2026 (I)]
Statement-I: When an electric discharge is passed through gaseous hydrogen, the hydrogen molecules dissociate and the energetically excited hydrogen atoms produce electromagnetic radiation of discrete frequencies.
Statement-II: The frequency of second line of Balmer series obtained from He^+ is equal to that of first line of Lyman series obtained from hydrogen atom.
In the light of the above statements, choose the correct answer from the options given below:
(1) Both Statement-I and Statement-II are true.
(2) Both Statement-I and Statement-II are false.
(3) Statement-I is false but Statement-II is true.
(4) Statement-I is true but Statement II is false.

49. Which of the following postulate of Bohr's model of hydrogen atom in not in agreement with quantum mechanical model of an atom ? [April 3, 2025 (I)]
(1) An atom in a stationary state does not emit electromagnetic radiation as long as it stays in the same state
(2) An atom can take only certain distinct energies E_1, E_2, E_3, etc. These allowed states of constant energy are called the stationary states of atom
(3) When an electron makes a transition from a higher energy stationary state to a lower energy stationary state, then it emits a photon of light
(4) The electron in a H atom's stationary state moves in a circle around the nucleus

50. According to Bohr's model of hydrogen atom, which of the following statement is **incorrect**? [April 2, 2025 (I)]
(1) Radius of 3rd orbit is nine times larger than that of 1st orbit.
(2) Radius of 8th orbit is four times larger than that of 4th orbit.
(3) Radius of 6th orbit is three time larger than that of 4th orbit.
(4) Radius of 4th orbit is four times larger than that of 2nd orbit.

51. For hydrogen like species, which of the following graphs provides the most appropriate representation of E vs Z plot for a constant n ? [Jan. 29, 2025 (II)]
[E : energy of the stationary state, Z : atomic number, n = principal quantum number]
(1) 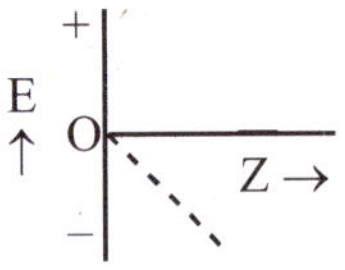

(2) 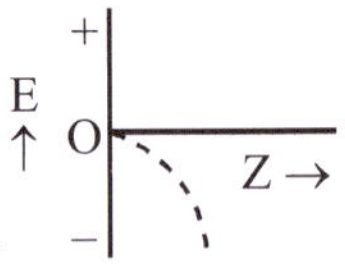

(3) 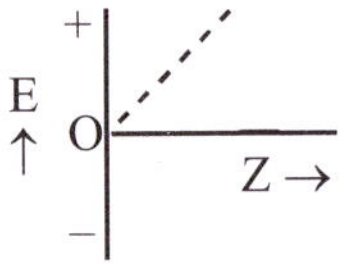

(4) 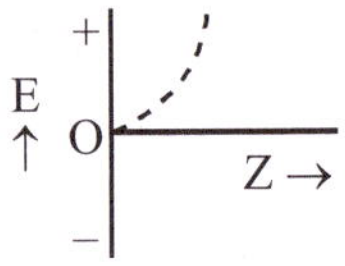

52. Heat treatment of muscular pain involves radiation of wavelength of about 900 nm. Which spectral line of H atom is suitable for this? Given: Rydberg constant [Jan. 23, 2025 (I)]
$R_H = 10^5$ cm^{-1}, $h = 6.6 \times 10^{-34}$ J s, $c = 3 \times 10^8$ m/s)
(1) Paschen series, $\infty \rightarrow 3$ (2) Lyman series, $\infty \rightarrow 1$
(3) Balmer series, $\infty \rightarrow 2$ (4) Paschen series, $5 \rightarrow 3$

53. Radius of the first excited state of Helium ion is given as:
$a_0 \rightarrow$ radius of first stationary state of hydrogen atom. [Jan. 22, 2025 (I)]
(1) $r = \frac{a_0}{2}$ (2) $r = \frac{a_0}{4}$ (3) $r = 4a_0$ (4) $r = 2a_0$

54. Given below are two statements: [April 15, 2023 (I)]
Statement I : According to Bohr's model of hydrogen atom, the angular momentum of an electron in a given stationary state is quantised.
Statement II : The concept of electron in Bohr's orbit, violates the Heisenberg uncertainty principle.
In the light of the above statements, choose the most appropriate answer from the options given below:

(1) Both Statement I and Statement II are correct.
(2) Statement I is correct but Statement II is incorrect.
(3) Statement I is incorrect but Statement II is correct.
(4) Both Statement I and Statement II are incorrect.

55. The energy of an electron in the first Bohr orbit of hydrogen atom is -2.18×10^{-18} J. Its energy in the third Bohr orbit is ________. **[April 13, 2023 (I)]**

(1) $\frac{1}{27}$ of this value (2) One third of this value
(3) Three times of this value (4) $\frac{1}{9}$ th of this value

56. Which transition in the hydrogen spectrum would have the same wavelength as the Balmer type transition from n = 4 to n = 2 of He^+ spectrum **[Jan. 31, 2023 (I)]**

(1) n = 2 to n = 1 (2) n = 1 to n = 3
(3) n = 1 to n = 2 (4) n = 3 to n = 4

57. The shortest wavelength of hydrogen atom in Lyman series is λ. The longest wavelength in Balmer series of He^+ is **[Jan. 29, 2023 (I)]**

(1) $\frac{5}{9\lambda}$ (2) $\frac{9\lambda}{5}$ (3) $\frac{36\lambda}{5}$ (4) $\frac{5\lambda}{9}$

58. The radius of the 2nd orbit of Li^{2+} is x. The expected radius of the 3rd orbit of Be^{3+} is **[Jan. 25, 2023 (I)]**

(1) $\frac{9}{4}x$ (2) $\frac{4}{9}x$ (3) $\frac{27}{16}x$ (4) $\frac{16}{27}x$

59. If the radius of the 3rd Bohr's orbit of hydrogen atom is r_3 and the radius of 4th Bohr's orbit is r_4.
Then : **[June 26, 2022 (I)]**

(1) $r_4 = \frac{9}{16} r_3$ (2) $r_4 = \frac{16}{9} r_3$
(3) $r_4 = \frac{3}{4} r_3$ (4) $r_4 = \frac{4}{3} r_3$

60. Given below are two statements.
Statement I : According to Bohr's model of an atom, qualitatively the magnitude of velocity of electron increases with decrease in positive charges on the nucleus as there is no strong hold on the electron by the nucleus.
Statement II : According to Bohr's model of an atom, qualitatively the magnitude of velocity of electron increases with decrease in principal quantum number.
In the light of the above statements, choose the most appropriate answer from the options given below **[Aug. 26, 2021 (I)]**

(1) Both Statement I and Statement II are false.
(2) Both Statement I and Statement II are true.
(3) Statement I is false but Statement II is true.
(4) Statement I is true but Statement II is false.

61. According to Bohr's atomic theory: **[Feb. 24, 2021 (II)]**

(A) Kinetic energy of electron is $\propto \frac{Z^2}{n^2}$.
(B) The product of velocity (v) of electron and principal quantum number (n), 'vn' $\propto Z^2$.
(C) Frequency of revolution of electron in an orbit is $\propto \frac{Z^3}{n^3}$.
(D) Coulombic force of attraction on the electron is $\propto \frac{Z^3}{n^4}$.

Choose the most appropriate answer from the options given below :

(1) (C) only (2) (A) only
(3) (A), (C) and (D) only (4) (A) and (D) only

62. The difference between the radii of 3rd and 4th orbits of Li^{2+} is ΔR_1. The difference between the radii of 3rd and 4th orbits of He^+ is ΔR_2. Ratio $\Delta R_1 : \Delta R_2$ is : **[Sep. 05, 2020 (I)]**

(1) 8 : 3 (2) 3 : 8 (3) 2 : 3 (4) 3 : 2

63. The shortest wavelength of H atom in the Lyman series is λ_1. The longest wavelength in the Balmer series of He^+ is: **[Sep. 04, 2020 (II); Same Online, April 8, 2017]**

(1) $\frac{36\lambda_1}{5}$ (2) $\frac{5\lambda_1}{9}$ (3) $\frac{9\lambda_1}{5}$ (4) $\frac{27\lambda_1}{5}$

64. The radius of the second Bohr orbit, in terms of the Bohr radius, a_0, in Li^{2+} is: **[Jan. 08, 2020 (II)]**

(1) $\frac{2a_0}{3}$ (2) $\frac{4a_0}{9}$ (3) $\frac{4a_0}{3}$ (4) $\frac{2a_0}{9}$

65. For the Balmer series in the spectrum of H atom, $\bar{v} = R_H \left\{ \frac{1}{n_1^2} - \frac{1}{n_2^2} \right\}$, the correct statements among (I) to (IV) are: **[Jan. 08, 2020 (I)]**

(I) As wavelength decreases, the lines in the series converge
(II) The integer n_1 is equal to 2
(III) The lines of longest wavelength corresponds to $n_2 = 3$
(IV) The ionization energy of hydrogen can be calculated from wave number of these lines

(1) (I), (III), (IV) (2) (I), (II), (III)
(3) (I), (II), (IV) (4) (II), (III), (IV)

66. The radius of the second Bohr orbit in terms of the Bohr radius, a_0, in Li^{2+} is: **[Jan. 08, 2020 (II)]**

(1) $\frac{2a_0}{3}$ (2) $\frac{4a_0}{9}$ (3) $\frac{4a_0}{3}$ (4) $\frac{2a_0}{9}$

67. The ratio of the shortest wavelength of two spectral series of hydrogen spectrum is found to be about 9. The spectral series are : **[April 10, 2019 (II)]**

(1) Lyman and Paschen (2) Balmer and Brackett
(3) Brackett and Pfund (4) Paschen and Pfund

68. For any given series of spectral lines of atomic hydrogen, let $\Delta\bar{v} = \bar{v}_{max} - \bar{v}_{min}$ be the difference in maximum and minimum frequencies in cm^{-1}. The ratio $\Delta\bar{v}_{Lyman}/\Delta\bar{v}_{Balmer}$ is : **[April 9, 2019 (I)]**

(1) 4 : 1 (2) 9 : 4 (3) 5 : 4 (4) 27 : 5

69. Which one of the following about an electron occupying the 1s orbital in a hydrogen atom is incorrect? (The Bohr radius is represented by a_0). **[April 9, 2019 (II)]**

(1) The probability density of finding the electron is maximum at the nucleus.
(2) The electron can be found at a distance $2a_0$ from the nucleus.
(3) The magnitude of the potential energy is double that of its kinetic energy on an average.
(4) The total energy of the electron is maximum when it is at a distance a_0 from the nucleus.

70. The ground state energy of hydrogen atom is –13.6 eV. The energy of second excited state of He^+ ion in eV is: **[Jan. 10, 2019 (II), Same 2002]**

(1) –54.4 (2) –3.4
(3) –6.04 (4) –27.2

71. For emission line of atomic hydrogen from $n_i = 8$ to $n_f = n$, the plot of wave number ($\bar{v}$) against $\left(\frac{1}{n^2}\right)$ will be (The Rydberg constant, R_H is in wave number unit) [Jan. 9, 2019 (I)]
(1) Linear with intercept $-R_H$ (2) Non linear
(3) Linear with slope R_H (4) Linear with slope $-R_H$

72. Which of the following statements is false? [Online April 16, 2018]
(1) Splitting of spectral lines in electrical field is called Stark effect
(2) Frequency of emitted radiation from a black body goes from a lower wavelength to higher wavelength as the temperature increases
(3) Photon has momentum as well as wavelength
(4) Rydberg constant has unit of energy

73. The electron in the hydrogen atom undergoes transition from higher orbitals to orbital of radius 211.6 pm. This transition is associated with : [Online April 9, 2017]
(1) Lyman series (2) Balmer series
(3) Paschen series (4) Brackett series

74. If the shortest wavelength in Lyman series of hydrogen atom is A, then the longest wavelength in Paschen series of He^+ is : [Online April 8, 2017]
(1) $\frac{5A}{9}$ (2) $\frac{9A}{5}$ (3) $\frac{36A}{5}$ (4) $\frac{36A}{7}$

75. The radius of the second Bohr orbit for hydrogen atom is : (Plank's const. $h = 6.6262 \times 10^{-34}$ Js ; mass of electron = 9.1091×10^{-31} kg ; charge of electron e = 1.60210×10^{-19} C ; permittivity of vaccum $\epsilon_0 = 8.854185 \times 10^{-12}\ kg^{-1}\ m^{-3}\ A^2$) [2017]
(1) 1.65Å (2) 4.76Å (3) 0.529Å (4) 2.12Å

76. Which of the following is the energy of a possible excited state of hydrogen ? [2015]
(1) –3.4 eV (2) +6.8 eV (3) +13.6eV (4) –6.8 eV

77. The energy of an electron in first Bohr orbit of H-atom is – 13.6 eV. The energy value of electron in the excited state of Li^{2+} is: [Online April 9, 2014]
(1) –27.2 eV (2) 30.6 eV
(3) –30.6 eV (4) 27.2 eV

78. Based on the equation: $\Delta E = -2.0 \times 10^{-18} J\left(\frac{1}{n_2^2} - \frac{1}{n_1^2}\right)$
the wavelength of the light that must be absorbed to excite hydrogen electron from level $n = 1$ to level $n = 2$ will be: ($h = 6.625 \times 10^{-34}$ Js, $c = 3 \times 10^8\ ms^{-1}$) [Online April 11, 2014]
(1) 1.325×10^{-7} m (2) 1.325×10^{-10} m
(3) 2.650×10^{-7} m (4) 5.300×10^{-10} m

79. If m and e are the mass and charge of the revolving electron in the orbit of radius r for hydrogen atom, the total energy of the revolving electron will be: [Online April 12, 2014]
(1) $\frac{1}{2}\frac{e^2}{r}$ (2) $-\frac{e^2}{r}$ (3) $\frac{me^2}{r}$ (4) $-\frac{1}{2}\frac{e^2}{r}$

80. Energy of an electron is given by $E = -2.178 \times 10^{-18} J\left(\frac{Z^2}{n^2}\right)$. Wavelength of light required to excite an electron in an hydrogen atom from level $n = 1$ to $n = 2$ will be : [2013]
($h = 6.62 \times 10^{-34}$ Js and $c = 3.0 \times 10^8\ ms^{-1}$)
(1) 1.214×10^{-7} m (2) 2.816×10^{-7} m
(3) 6.500×10^{-7} m (4) 8.500×10^{-7} m

81. The wave number of the first emission line in the Balmer series of H-spectrum is : [Online April 22, 2013]
(R = Rydberg constant) :
(1) $\frac{5}{36}R$ (2) $\frac{9}{400}R$ (3) $\frac{7}{6}R$ (4) $\frac{3}{4}R$

82. The limiting line in Balmer series will have a frequency of (Rydberg constant, $R_\infty = 3.29 \times 10^{15}$ cycles/s) [Online May 7, 2012]
(1) $8.22 \times 10^{14}\ s^{-1}$ (2) $3.29 \times 10^{15}\ s^{-1}$
(3) $3.65 \times 10^{14}\ s^{-1}$ (4) $5.26 \times 10^{13}\ s^{-1}$

83. The frequency of light emitted for the transition $n = 4$ to $n = 2$ of the He^+ is equal to the transition in H atom corresponding to which of the following ? [2011RS]
(1) $n = 2$ to $n = 1$ (2) $n = 3$ to $n = 2$
(3) $n = 4$ to $n = 3$ (4) $n = 3$ to $n = 1$

84. Ionisation energy of He^+ is 19.6×10^{-18} J atom^{-1}. The energy of the first stationary state ($n = 1$) of Li^{2+} is [2010]
(1) 4.41×10^{-16} J atom^{-1} (2) -4.41×10^{-17} J atom^{-1}
(3) -2.2×10^{-15} J atom^{-1} (4) 8.82×10^{-17} J atom^{-1}

85. The ionization enthalpy of hydrogen atom is 1.312×10^6 J mol^{-1}. The energy required to excite the electron in the atom from $n = 1$ to $n = 2$ is [2008]
(1) 8.51×10^5 J mol^{-1} (2) 6.56×10^5 J mol^{-1}
(3) 7.56×10^5 J mol^{-1} (4) 9.84×10^5 J mol^{-1}

86. According to Bohr's theory, the angular momentum of an electron in 5th orbit is [2006]
(1) $10\ h/\pi$ (2) $2.5\ h/\pi$ (3) $25\ h/\pi$ (4) $1.0\ h/\pi$

87. The wavelength of the radiation emitted, when in a hydrogen atom electron falls from infinity to stationary state 1, would be (Rydberg constant = $1.097 \times 10^7\ m^{-1}$) [2004]
(1) 406 nm (2) 192 nm
(3) 91 nm (4) 9.1×10^{-8} nm

88. In a hydrogen atom, if energy of an electron in ground state is 13.6. eV, then that in the 2nd excited state is [2002]
(1) 1.51 eV (2) 3.4 eV (3) 6.04 eV (4) 13.6eV.

NUMERICAL VALUE QUESTIONS (NVQs)

89. The hydrogen spectrum consists of several spectral lines in Lyman series (L_1, L_2, L_3 ....; L_1 has lowest energy among Lyman series). Similarly it consists of several spectral lines in Balmer series (B_1, B_2, B_3 ......; B_1 has lowest energy among Balmer lines). The energy of L_1 is x times the energy of B_1. The value of x is ____ $\times 10^{-1}$ (Nearest integer) [Jan. 24, 2026 (I)]

90. The energy of an electron in first Bohr orbit of H-atom is –13.6 eV. The magnitude of energy value of electron in the first excited state of Be^{3+} is ______ eV. (nearest integer value) [April 8, 2025 (II)]

91. A hypothetical electromagnetic wave is show below.

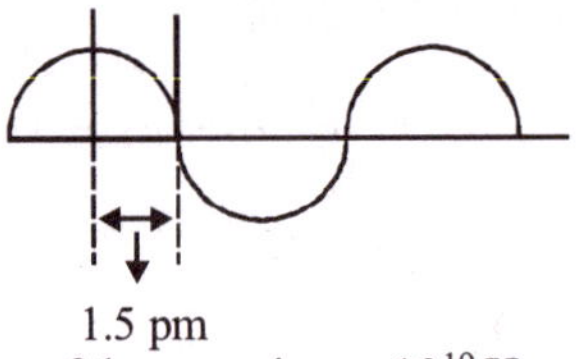

1.5 pm

The frequency of the wave is x $\times 10^{19}$ Hz.
x = ______ (nearest integer) [April 8, 2024 (I)]

92. Wavenumber for a radiation having 5800 Å wavelength is x $\times$ 10 cm^{-1}. The value of x is _____. [April 8, 2024 (II)]

93. For hydrogen atom, energy of an electron in first excited state is – 3.4 eV, K.E. of the same electron of hydrogen atom is x eV. Value of x is_____ $\times 10^{-1}$ eV. (Nearest integer) [April 6, 2024 (II)]

94. The value of Rydberg constant (R_H) is 2.18×10^{-18} J. The velocity of electron having mass 9.1×10^{-31} kg in Bohr's first orbit of hydrogen atom = ... $\times 10^5 ms^{-1}$ (nearest integer) [April 5, 2024 (I)]

95. The de-Broglie's wavelength of an electron in the 4th orbit is ______ πa_0. (a_0 = Bohr's radius) [April 4, 2024 (I)]

96. Number of spectral lines obtained in He^+ spectra, when an electron makes transition from fifth excited state to first excited state will be [Jan. 30, 2024 (II)]

97.

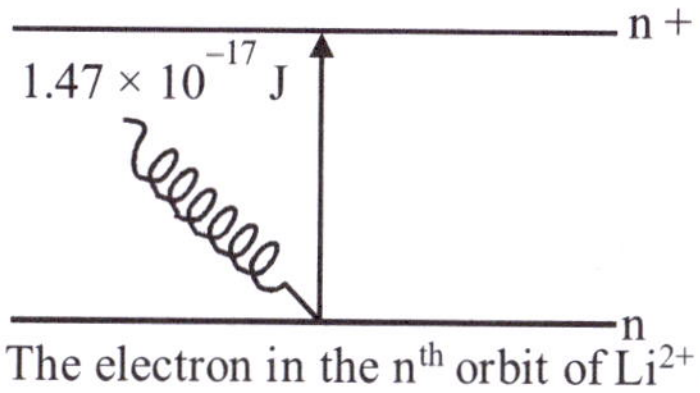

The electron in the nth orbit of Li^{2+} is excited to (n + 1) orbit using the radiation of energy 1.47×10^{-17}J (as shown in the diagram). The value of n is _______ . Given $R_H = 2.18 \times 10^{-18}$J [April 10, 2023 (II)]

98. Assume that the radius of the first Bohr orbit of hydrogen atom is 0.6 Å. The radius of the third Bohr orbit of He^+ is ________ picometer. (Nearest Integer) [Jan. 29, 2023 (II)]

99. If wavelength of the first line of the Paschen series of hydrogen atom is 720 nm, then the wavelength of the second line of this series is _______ nm.
(Nearest integer) [Jan. 24, 2023 (I)]

100. The longest wavelength of light that can be used for the ionisation of lithium atom (Li) in its ground state is $x \times 10^{-8}$ m. The value of x is __________. (Nearest Integer)
(Given : Energy of the electron in the first shell of the hydrogen atom is -2.2×10^{-18} J; h = 6.63×10^{-34} Js and c = $3 \times 10^8 ms^{-1}$) [June 25, 2022 (I)]

101. The kinetic energy of an electron in the second Bohr orbit of a hydrogen atom is equal to $\frac{h^2}{xma_0^2}$. The value of $10x$ is ________ . (a_0 is radius of Bohr's orbit) (Nearest integer) [Given : $\pi = 3.14$] [Aug. 27, 2021 (I)]

Topic 3 - Towards Quantum Mechanical Model of the Atom (Dual Behaviour and Uncertainty Principle)

MULTIPLE CHOICE QUESTIONS (MCQs)

102. Identify the correct statements from the following : [April 05, 2026 (I)]
A. Heisenberg uncertainty principle is applicable to electron.
B. The size of $2p_x$ orbital is less than the size of $3p_x$ orbital.
C. The energy of 2s orbital of H atom is equal to the energy of 2s orbital of Li.
D. The electronic configuration of Cr is $[Ar]\ 3d^5 4s^1$
Choose the correct answer from the options given below:
(1) A, B and C only (2) A, B and D only
(3) B, C and D only (4) A, C and D only

103. Given below are two statements :
Statement I: It is impossible to specify simultaneously with arbitrary precision, both the linear momentum and the position of a particle. [Jan. 29, 2025 (II)]
Statement II : If the uncertainty in the measurement of position and uncertainty in measurement of momentum are equal for an electron, then the uncertainty in the measurement of velocity is $\geq \sqrt{\frac{h}{\pi}} \times \frac{1}{2m}$.
In the light of the above statements, choose the correct answer from the options given below :
(1) Statement I is true but Statement II is false.
(2) Both Statement I and Statement II are true.
(3) Statement I is false but Statement II is true.
(4) Both Statement I and Statement II are false.

104. If a_0 is denoted as the Bohr radius of hydrogen atom, then what is the de-Broglie wavelength (λ) of the electron present in the second orbit of hydrogen atom ? [n : any integer] [Jan. 29, 2025 (I)]
(1) $\frac{2a_0}{n\pi}$ (2) $\frac{8\pi a_0}{n}$ (3) $\frac{4\pi a_0}{n}$ (4) $\frac{4n}{\pi a_0}$

105. According to the wave particle duality of matter by de - Brogile, which of the following graph plot presents ,most appropriate relationship between wavelength of electron (λ) and momentum of electron (p)? [Feb. 1, 2024 (I)]

(1)
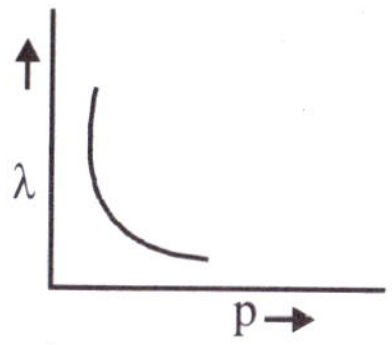

(2)
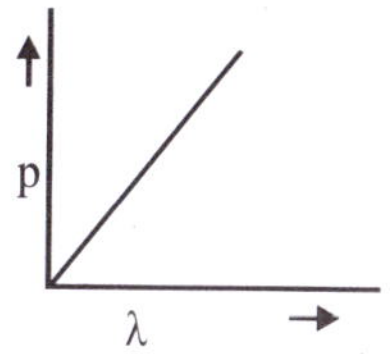

(3)
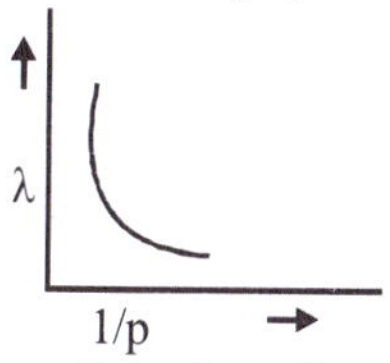

(4)
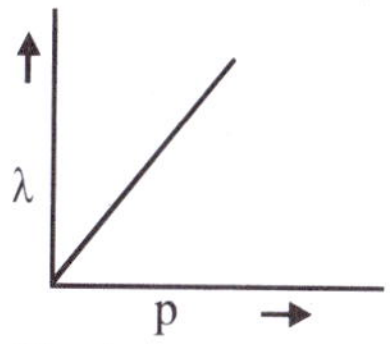

106. If the radius of the first orbit of hydrogen atom a_0, then de-Broglie's wavelength of electron in 3rd orbit is [April 06, 2023 (II)]
(1) $\frac{\pi a_0}{6}$ (2) $\frac{\pi a_0}{3}$ (3) $6\pi a_0$ (4) $3\pi a_0$

107. The figure that is **not** a direct manifestation of the quantum nature of atoms is : [Sep. 02, 2020 (I)]

(1)
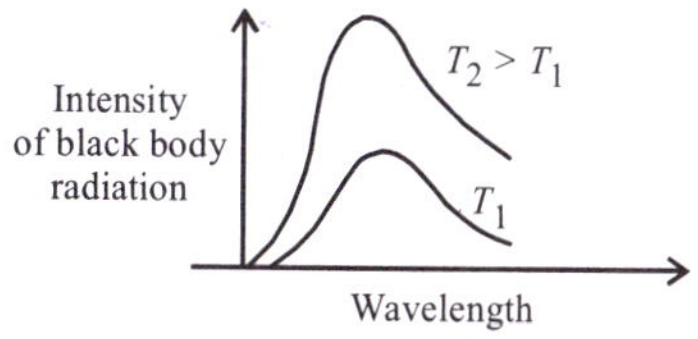

(2)
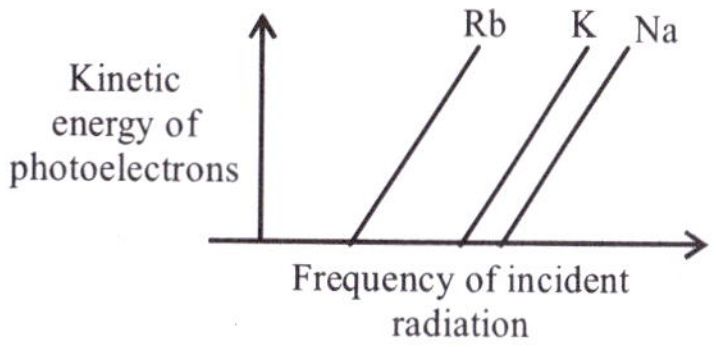

(3) 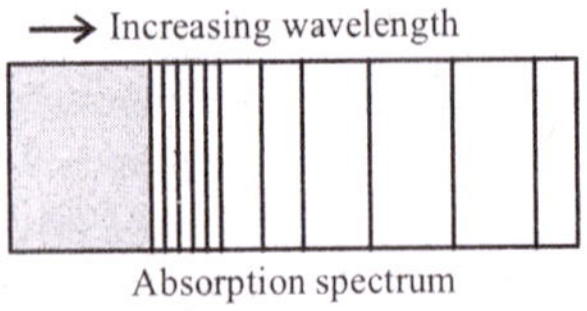

(4) 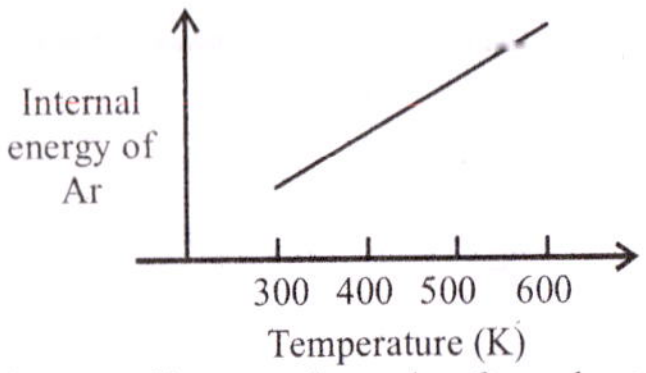

108. The de Broglie wavelength of an electron in the 4th Bohr orbit is: **[Jan. 09, 2020 (I)]**
(1) $2\pi a_0$ (2) $4\pi a_0$ (3) $6\pi a_0$ (4) $8\pi a_0$

109. If the de Broglie wavelength of the electron in n^{th} Bohr orbit in a hydrogenic atom is equal to $1.5\ \pi a_0$ (a_0 is Bohr radius), then the value of n/Z is : **[Jan. 12, 2019 (II)]**
(1) 0.40 (2) 1.50 (3) 1.0 (4) 0.75

110. The de Broglie wavelength (λ) associated with a photoelectron varies with the frequency (v) of the incident radiation as, [v_0 is threshold frequency]: **[Jan. 11, 2019 (II)]**

(1) $\lambda \propto \frac{1}{(v - v_0)}$ (2) $\lambda \propto \frac{1}{(v - v_0)^{\frac{1}{4}}}$

(3) $\lambda \propto \frac{1}{(v - v_0)^{\frac{3}{2}}}$ (4) $\lambda \propto \frac{1}{(v - v_0)^{\frac{1}{2}}}$

111. The de-Broglie's wavelength of electron present in first Bohr orbit of 'H' atom is: **[Online April 15, 2018 (II)]**
(1) 4×0.529Å (2) $2\pi \times 0.529$Å

(3) $\frac{0.529}{2\pi}$Å (4) 0.529Å

112. A stream of electrons from a heated filament was passed between two charged plates kept at a potential difference V volt. If 'e' and m are charge and mass of an electron, respectively, then the value of h/λ (where λ is wavelength associated with electron wave) is given by: **[2016]**
(1) $\sqrt{meV}$ (2) $\sqrt{2meV}$ (3) meV (4) $2meV$

113. At temperature T, the average kinetic energy of any particle is $\frac{3}{2}kT$. The de Broglie wavelength follows the order : **[2003 (S), 2013, Online April 11, 2015]**
(1) Visible photon > Thermal neutron > Thermal electron
(2) Thermal proton > Thermal electron > Visible photon
(3) Thermal proton > Visible photon > Thermal electron
(4) Visible photon > Thermal electron > Thermal neutron

114. The de-Broglie wavelength of a particle of mass 6.63 g moving with a velocity of 100 ms^{-1} is: **[Online April 12, 2014]**
(1) 10^{-33} m (2) 10^{-35} m (3) 10^{-31} m (4) 10^{-25} m

115. The de Broglie wavelength of a car of mass 1000 kg and velocity 36 km/hr is: **[Online April 23, 2013]**
(1) 6.626×10^{-34} m (2) 6.626×10^{-38} m
(3) 6.626×10^{-31} m (4) 6.626×10^{-30} m

116. If the kinetic energy of an electron is increased four times, the wavelength of the de-Broglie wave associated with it would become **[Online May 19, 2012]**
(1) one fourth (2) half
(3) four times (4) two times

117. If the radius of first orbit of H atom is a_0, the de-Broglie wavelength of an electron in the third orbit is **[Online May 12, 2012]**
(1) $4\pi a_0$ (2) $8\pi a_0$ (3) $6\pi a_0$ (4) $2\pi a_0$

118. Calculate the wavelength (in nanometer) associated with a proton moving at 1.0×10^3 m-s^{-1}. (Mass of proton = 1.67×10^{-27} kg and $h = 6.63 \times 10^{-34}$ Js) **[2009]**
(1) 0.40 nm (2) 2.5 nm
(3) 14.0 nm (4) 0.32 nm

119. In an atom, an electron is moving with a speed of 600 m/s with an accuracy of 0.005%. Certainity with which the position of the electron can be located is ($h = 6.6 \times 10^{-34}$ kg m^{2}s^{-1}, mass of electron, $e_m = 9.1 \times 10^{-31}$ kg) : **[2002 (S), 2006(S), 2009]**
(1) 5.10×10^{-3} m (2) 1.92×10^{-3} m
(3) 3.84×10^{-3} m (4) 1.52×10^{-4} m

120. Uncertainty in the position of an electron (mass = 9.1×10^{-31} kg) moving with a velocity 300 ms^{-1}, accurate upto 0.001% will be ($h = 6.63 \times 10^{-34}$ Js) **[2006]**
(1) 1.92×10^{-2} m (2) 3.84×10^{-2} m
(3) 19.2×10^{-2} m (4) 5.76×10^{-2} m

121. The de Broglie wavelength of a tennis ball of mass 60 g moving with a velocity of 10 metres per second is approximately **[2003]**
(1) 10^{-31} metres (2) 10^{-16} metres
(3) 10^{-25} metres (4) 10^{-33} metres
Planck's constant, $h = 6.63 \times 10^{-34}$ Js

122. Uncertainty in position of a minute particle of mass 25 g in space is 10^{-5} m. What is the uncertainty in its velocity (in ms^{-1})? ($h = 6.6 \times 10^{-34}$ Js) **[2002]**
(1) 2.1×10^{-34} (2) 0.5×10^{-34}
(3) 2.1×10^{-28} (4) 0.5×10^{-23}.

NUMERICAL VALUE QUESTIONS (NVQs)

123. Two positively charged particles m_1 and m_2 have been accelerated across the same potential difference of 200 keV as shown below. **[Jan. 28, 2026 (II)]**

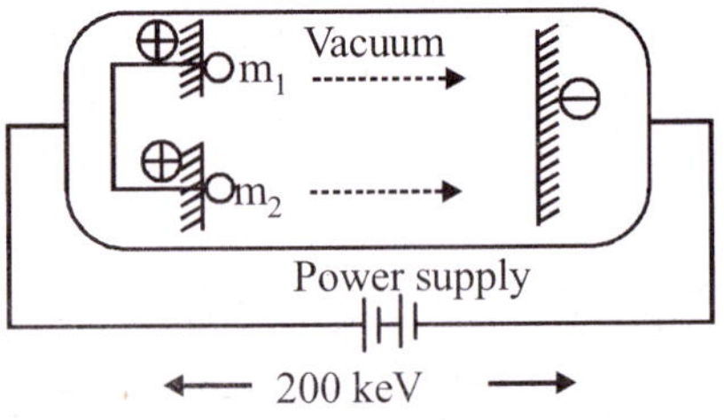

[Given mass of m_1 = 1 amu and m_2 = 4 amu]
The de-Broglie wavelength of m_1 will be x times of m_2. The value of x is ________. (nearest integer)

124. Based on Heisenberg's uncertainty principle, the uncertainty in the velocity of the electron to be found within an atomic nucleus of diameter 10^{-15} m is ________ $\times 10^9$ ms^{-1} (nearest integer)
[Given : mass of electron = 9.1×10^{-31} kg, Plank's constant (h) = 6.626×10^{-34} Js]
(Value of π = 3.14) **[April 9, 2024 (II)]**

125. Frequency of the de-Broglie wave of electron in Bohr's first orbit of hydrogen atom is ____________ $\times 10^{13}$ Hz (nearest integer).
[Given : R_H (Rydberg constant) = 2.18×10^{-18} J. h (Plank's constant) = 6.6×10^{-34} J.s.] **[April 6, 2024 (I)]**

126. The wavelength of an electron of kinetic energy 4.50×10^{-29} J is .... $\times 10^{-5}$ m. (Nearest integer) Given : mass of electron is 9×10^{-31} kg, $h = 6.6 \times 10^{-34}$ J s [April 06, 2023 (I)]

127. Electrons in a cathode ray tube have been emitted with a velocity of 1000 m s^{-1}. The number of following statements which is/are true about the emitted radiation is ______.
Given : $h = 6 \times 10^{-34}$ J s, $m_e = 9 \times 10^{-31}$ kg. [Feb. 01, 2023 (I)]
(A) The deBroglie wavelength of the electron emitted is 666.67 nm.
(B) The characteristic of electrons emitted depend upon the material of the electrodes of the cathode ray tube.
(C) The cathode rays start from cathode and move towards anode.
(D) The nature of the emitted electrons depends on the nature of the gas present in cathode ray tube.

128. The minimum uncertainty in the speed of an electron in an one-dimensional region of length $2a_0$ (Where a_0 = Bohr radius 52.9 pm) is _____ km s^{-1}.
(Given : Mass of electron = 9.1×10^{-31} kg, Planck's constant h = 6.63×10^{-34} Js) [July 29, 2022 (I)]

129. The wavelength of an electron and a neutron will become equal when the velocity of the electron is x times the velocity of neutron. The value of x is _______. (Nearest Integer) (Mass of electron is 9.1×10^{-31} kg and mass of neutron is 1.6×10^{-27} kg) [July 26, 2022 (I)]

130. If the uncertainty in velocity and position of a minute particle in space are, 2.4×10^{-26} (ms^{-1}) and 10^{-7} (m) respectively. The mass of the particle in g is _________ (Nearest integer)
(Given : $h = 6.626 \times 10^{-34}$ Js) [June 27, 2022 (I)]

131. An accelerated electron has speed of 5×10^6 ms^{-1} with an uncertainty of 0.02%. The uncertainty in finding its location while in motion is $x \times 10^{-9}$m. The value of x is ______. (Nearest integer) [Use mass of electron = 9.1×10^{-31} kg, $h = 6.63 \times 10^{-34}$Js, $\pi = 3.14$] [July 25, 2021 (II)]

132. The wavelength of electrons accelerated from rest through a potential difference of 40 kV is $x \times 10^{-12}$ m. The value of x is _______. (Nearest integer)
Given : Mass of electron = 9.1×10^{-31} kg
Charge on an electron = 1.6×10^{-19} C
Planck's constant = 6.63×10^{-34} Js [July 20, 2021 (II)]

133. When light of wavelength 248 nm falls on a metal of threshold energy 3.0 eV, the de-Broglie wavelength of emitted electrons is ____ Å. (Round off to the Nearest Integer).
[Use : $\sqrt{3} = 1.73, h = 6.63 \times 10^{-34}$ Js,
$m_e = 9.1 \times 10^{-31}$ kg; $c = 3.0 \times 10^8$ ms^{-1}; 1 eV = 1.6×10^{-19} J]
[March 16, 2021 (I)]

134. A ball weighing 10 g is moving with a velocity of 90 ms^{-1}. If the uncertainty in its velocity is 5%, then the uncertainty in its position is ___________ $\times 10^{-33}$ m. (Rounded off to the nearest integer)
[Given : $h = 6.63 \times 10^{-34}$ Js] [Feb. 26, 2021 (II)]

135. A proton and a Li^{3+} nucleus are accelerated by the same potential. If λ_{Li} and λ_p denote the de Broglie wavelengths of Li^{3+} and proton respectively, then the value of $\frac{\lambda_{Li}}{\lambda_p}$ is $x \times 10^{-1}$. The value of x is ______
(Rounded off to the nearest integer)
(Mass of Li^{3+} = 8.3 times mass of proton) [Feb. 24, 2021 (I)]

Topic 4 - Quantum Mechanical Model of Atom

MULTIPLE CHOICE QUESTIONS (MCQs)

136. Match the List-I with List-II : [April 06, 2026 (I)]

List-I Orbital		List-II Radial nodes & nodal plane	
A.	2s	I.	1 Radial node + two nodal planes
B.	3s	II.	1 Radial node + one nodal plane
C.	3p	III.	3 Radial node + No nodal plane
D.	4d	IV.	1 Radial node + No nodal plane

Choose the correct answer from the options given below:
(1) A-IV, B-I, C-III, D-II (2) A-IV, B-II, C-III, D-I
(3) A-III, B-I, C-IV, D-II (4) A-IV, B-III, C-II, D-I

137. Which of the following statement(s) is/are true? [April 05, 2026 (II)]
(A) If two orbitals have the same value of (n + ℓ), the orbital with lower value of n will have lower energy.
(B) Energies of the orbitals in the same subshell increase with increase in atomic number.
(C) The size of 2px, orbital is less than the size of 3px, orbital.
(D) Among 5f, 6s, 4d, 5p and 5d orbitals, none of the orbitals have 2 radial nodes.
Choose the correct answer from the options given below:
(1) A, B and C only (2) A and C only
(3) C and D only (4) A only

138. Arrange the following atomic orbitals of multi electron atoms in order of increasing energy.
(A) n = 3, ℓ = 2, m = +1 (B) n = 4, ℓ = 0, m = 0
(C) n = 6, ℓ = 1, m = 0 (D) n = 5, ℓ = 1, m = +1
(E) n = 2, ℓ = 1, m = +1 [April 05, 2026 (I)]
Choose the correct answer from the options given below:
(1) C < D < B < A < E (2) B < A < E < C < D
(3) E < C < D < B < A (4) E < B < A < D < C

139. Which of the following is correct set of 4 quantum number of 19th electron in Chromium (Atomic number = 24) in accordance with Aufbau principle? [April 04, 2026 (I)]
(1) n = 3, ℓ = 2, m = +2, s = $+\frac{1}{2}$
(2) n = 3, ℓ = 2, m = –2, s = $+\frac{1}{2}$
(3) n = 4, ℓ = 1, m = 0, s = $+\frac{1}{2}$
(4) n = 4, ℓ = 0, m = 0, s = $+\frac{1}{2}$

140. 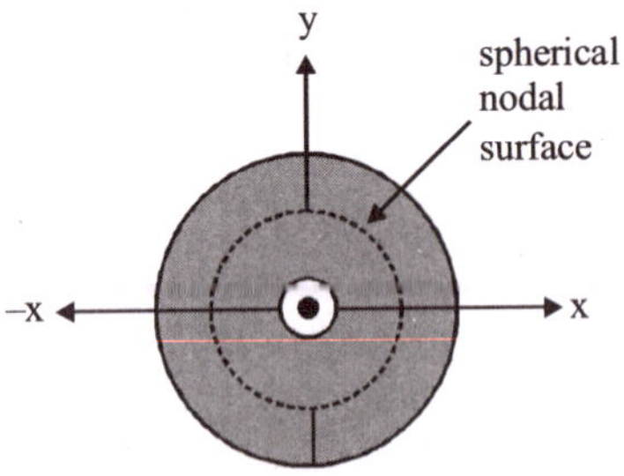

Figure 1. electron probability density for 2s orbital

[Jan. 28, 2026 (I)]

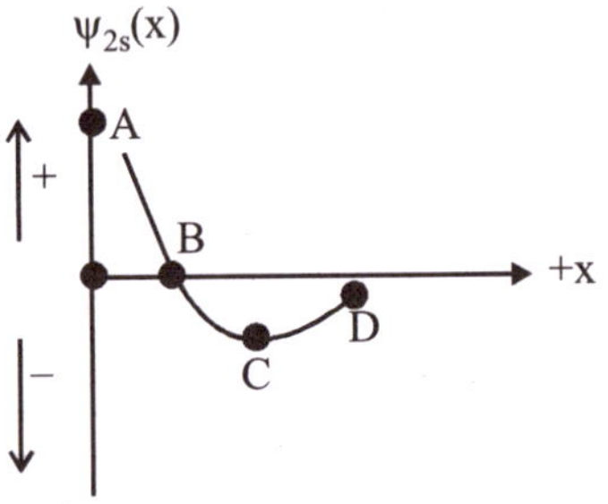

Figure 2. wave function for 2s orbital

Which of the following point in Figure 2 most accurately represents the nodal surface as shown in Figure 1 ?

(1) B (2) D (3) C (4) A

141. Given, [Jan. 23, 2026 (I)]

(A) $n = 5, m_l = -1$

(B) $n = 3, l = 2, m_l = -1, m_s = +\frac{1}{2}$

The maximum number of electron(s) in an atom that can have the quantum numbers as given in (A) and (B) respectively are:

(1) 26 and 1 (2) 4 and 1
(3) 2 and 4 (4) 8 and 1

142. Correct statements for an element with atomic number 9 are

A. There can be 5 electrons for which $m_s = +\frac{1}{2}$ and 4 electrons for which $m_s = -\frac{1}{2}$ [April 8, 2025 (II)]

B. There is only one electron in p_z orbital

C. The last electron goes to orbital with n = 2 and $l = 1$

D. The sum of angular nodes of all the atomic orbitals is 1.

Choose the **correct** answer from the options given below:

(1) C and D only (2) A and C only
(3) A, C and D only (4) A and B only

143. The extra stability of half-filled subshell is due to [April 7, 2025 (II)]

(A) Symmetrical distribution of electrons
(B) Smaller coulombic repulsion energy
(C) The presence of electrons with the same spin in non-degenerate orbitals
(D) Larger exchange energy
(E) Relatively smaller shielding of electrons by one another

Identify the **correct** statements

(1) (B), (D) and (E) only (2) (A), (B), (D) and (E) only
(3) (B), (C) and (D) only (4) (A), (B) and (D) only

144. Consider the ground state of chromium atom (Z = 24). How many electrons are with Azimuthal quantum number $l = 1$ and $l = 2$ respectively ? [April 4, 2025 (II)]

(1) 12 and 4 (2) 16 and 4
(3) 12 and 5 (4) 16 and 5

145. Which one of the following about an electron occupying the 1s orbital in a hydrogen atom is **incorrect** ? (Bohr's radius is represented by a_0) [April 4, 2025 (I)]

(1) The probability density of finding the electron is maximum at the nucleus
(2) The electron can be found at a distance $2a_0$ from the nucleus
(3) The 1s orbital is spherically symmetrical
(4) The total energy of the electron is maximum when it is at a distance a_0 from the nucleus

146. For electron in '2s' and '2p' orbitals, the orbital angular momentum values, respectively are : [April 3, 2025 (II)]

(1) $\sqrt{2}\frac{h}{2\pi}$ and 0 (2) $\frac{h}{2\pi}$ and $\sqrt{2}\frac{h}{2\pi}$
(3) 0 and $\sqrt{6}\frac{h}{2\pi}$ (4) 0 and $\sqrt{2}\frac{h}{2\pi}$

147. Which of the following statements are **true** ? [April 2, 2025 (II)]

(A) The subsidiary quantum number l describes the shape of the orbital occupied by the electron.

(B) 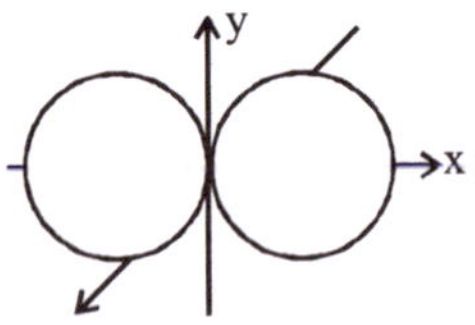

is the boundary surface diagram of the $2p_x$ orbital.

(C) The + and – signs in the wave function of the $2p_x$ orbital refer to charge.

(D) The wave function of $2p_x$ orbital is zero everywhere in the xy plane.

(1) (B) and (D) only (2) (A), (B) and (C) only
(3) (C) and (D) only (4) (A) and (B) only

148. Which of the following is/are not correct with respect to energy of atomic orbitals of hydrogen atom? [Jan. 28, 2025 (II)]

(A) 1s < 2p < 3d < 4s (B) 1s < 2s = 2p < 3s = 3p
(C) 1s < 2s < 2p < 3s < 3p (D) 1s < 2s < 4s < 3d

Choose the correct answer from the options given below :

(1) (B) and (D) only (2) (A) and (C) only
(3) (C) and (D) only (4) (A) and (B) only

149. In a multielectron atom, which of the following orbitals described by three quantum numbers will have same energy in absence of electric and magnetic fields? [Jan. 28, 2025 (I)]

(A) $n = 1, l = 0, m_l = 0$ (B) $n = 2, l = 0, m_l = 0$
(C) $n = 2, l = 1, m_l = 1$ (D) $n = 3, l = 2, m_l = 1$
(E) $n = 3, l = 2, m_l = 0$

Choose the correct answer from the options given below :

(1) A and B only (2) B and C only
(3) C and D only (4) D and E only

150. For hydrogen atom, the orbital/s with lowest energy is/are: [Jan. 24, 2025 (II)]

(A) 4s (B) $3p_x$ (C) $3d_{x^2-y^2}$ (D) $3d_{z^2}$
(E) $4p_z$

Choose the correct answer from the options given below :

(1) (A) and (E) only (2) (B) only
(3) (A) only (4) (B), (C) and (D) only

151. Given below are two statements: [Jan. 23, 2025 (II)]

Statement (I): For a given shell, the total number of allowed orbitals is given by n^2.

Statement (II): For any subshell, the spatial orientation of the orbitals is given by $-l$ to $+l$ values including zero.

In the light of the above statements, choose the correct answer from the options given below

(1) Statement I is true but Statement II is false
(2) Statement I is false but Statement II is true
(3) Both Statement I and Statement II are true
(4) Both Statement I and Statement II are false

152. Given below are two statements: [Jan. 22, 2025 (II)]

Statement (I): A spectral line will be observed for a $2p_x \rightarrow 2p_y$ transition.

Statement (II): $2p_x$ and $2p_y$ are degenerate orbitals.

In the light of the above statements, choose the correct answer from the options given below:

(1) Both Statement I and Statement II are true.
(2) Both Statement I and Statement II are false.
(3) Statement I is true but Statement II is false.
(4) Statement I is false but Statement II is true.

153. Compare the energies of following sets of quantum numbers for multielectron system. [April 9, 2024 (I)]

(A) $n = 4, l = 1$ (B) $n = 4, l = 2$
(C) $n = 3, l = 1$ (D) $n = 3, l = 2$
(E) $n = 4, l = 0$

Choose the correct answer from the options given below:

(1) (B) > (A) > (C) > (E) > (D)
(2) (E) > (C) < (D) < (A) < (B)
(3) (E) > (C) > (A) > (D) > (B)
(4) (C) < (E) < (D) < (A) < (B)

154. The number of radial nodes for 3p orbital is: [Feb. 1, 2024 (II)]

(1) 3 (2) 2 (3) 1 (4) 4

155. The four quantum numbers for the electron in the outer most orbital of potassium (atomic no. 19) are [Jan. 31, 2024 (II)]

(1) $n = 4, l = 2, m = -1, s = +\frac{1}{2}$

(2) $n = 4, l = 0, m = 0, s = +\frac{1}{2}$

(3) $n = 3, l = 0, m = 1, s = +\frac{1}{2}$

(4) $n = 2, l = 0, m = 0, s = +\frac{1}{2}$

156. Given below are two statements:

Statement I : The orbitals having same energy are called as degenerate orbitals. [Jan. 30, 2024 (I)]

Statement II : In hydrogen atom, 3p and 3d orbitals are not degenerate orbitals.

In the light of the above statements, choose the **most appropriate** answer from the options given below:

(1) Statement I is true but Statement II is false.
(2) Both Statement I and Statement II are true.
(3) Both Statement I and Statement II are false.
(4) Statement I is false but Statement II is true.

157. The correct set of four quantum numbers for the valence electron of rubidium atom ($Z = 37$) is : [Jan. 29, 2024 (I)]

(1) $5, 1, 0, +\frac{1}{2}$ (2) $5, 1, 1, +\frac{1}{2}$

(3) $5, 0, 1, +\frac{1}{2}$ (4) $5, 0, 0, +\frac{1}{2}$

158. Given below are two statement : one is labelled as **Assertion A** and the other is labelled as **Reason R**

Assertion A : 5f electrons can participate in bonding to a far greater extent than 4f electrons

Reason R : 5f orbitals are not as buried as 4f orbitals

In the light of the above statements, choose the *correct* answer from the options given below. [April 12, 2023 (I)]

(1) Both A and R are true and R is NOT the correct explanation of A.
(2) Both A and R are true but R is the correct explanation of A.
(3) A is false but R is true.
(4) A is true but R is false.

159. Arrange the following orbitals in decreasing order of energy? [Jan. 31, 2023 (II)]

(A) $n = 3, l = 0, m = 0$ (B) $n = 4, l = 0, m = 0$
(C) $n = 3, l = 1, m = 0$ (D) $n = 3, l = 2, m = 1$

The correct option for the order is :

(1) B > D > C > A (2) D > B > C > A
(3) A > C > B > D (4) D > B > A > C

160. Maximum number of electrons that can be accommodated in shell with $n = 4$ are: [Jan. 30, 2023 (II)]

(1) 16 (2) 32 (3) 50 (4) 72

161. The wave function (Ψ) of 2s is given by

$$\Psi_{2s} = \frac{1}{2\sqrt{2\pi}}\left(\frac{1}{a_0}\right)^{1/2}\left(2 - \frac{r}{a_0}\right)e^{-r/2a_0}$$

At $r = r_0$, radial node is formed. Thus, r_0 in terms of a_0 [Jan. 30, 2023 (II)]

(1) $r_0 = a_0$ (2) $r_0 = 4a_0$

(3) $r_0 = \frac{a_0}{2}$ (4) $r_0 = 2a_0$.

162. Given below are the quantum numbers for 4 electrons.

A. $n = 3, l = 2, m_l = 1, m_s = +1/2$
B. $n = 4, l = 1, m_l = 0, m_s = +1/2$
C. $n = 4, l = 2, m_l = -2, m_s = -1/2$
D. $n = 3, l = 1, m_l = -1, m_s = +1/2$

The correct order of increasing energy is : [July 29, 2022 (II)]

(1) D < B < A < C (2) D < A < B < C
(3) B < D < A < C (4) B < D < C < A

163. Identify the incorrect statement from the following. [July 28, 2022 (I)]

(1) A circular path around the nucleus in which an electron moves is proposed as Bohr's orbit.
(2) An orbital is the one electron wave function (ψ) in an atom.
(3) The existence of Bohr's orbits is supported by hydrogen spectrum.
(4) Atomic orbital is characterised by the quantum numbers n and l only

164. Given below are two statements. One is labelled as **Assertion A** and the other is labelled as **Reason R.**

Assertion A : Energy of 2s orbital of hydrogen atom is greater than that of 2s orbital of lithium.

Reason R : Energies of the orbitals in the same subshell decrease with increase in the atomic number.

In the light of the above statements, choose the **correct** answer from the options given below. [July 27, 2022 (I)]

(1) Both A and R are true and R is the correct explanation of A.
(2) Both A and R are true but R is NOT the correct explanation of A.
(3) A is true but R is false.
(4) A is false but R is true.

165. The correct decreasing order of energy, for the orbitals having, following set of quantum numbers: [July 27, 2022 (II)]

(A) $n=3, l=0, m=0$ (B) $n=4, l=0, m=0$
(C) $n=3, l=1, m=0$ (D) $n=3, l=2, m=1$
(1) (D) > (B) > (C) > (A) (2) (B) > (D) > (C) > (A)
(3) (C) > (B) > (D) > (A) (4) (B) > (C) > (D) > (A)

166. Which of the following sets of quantum numbers is not allowed? [July 25, 2022 (I)]

(1) $n = 3, 1 = 2, m_1 = 0, s = +\frac{1}{2}$

(2) $n = 3, 1 = 2, m_1 = -2, s = +\frac{1}{2}$

(3) $n = 3, 1 = 3, m_1 = -3, s = -\frac{1}{2}$

(4) $n = 3, 1 = 0, m_1 = 0, s = -\frac{1}{2}$

167. The number of radial nodes and total number of nodes in 4p orbital respectively are: [June 30, 2022 (I)]

(1) 2 and 3 (2) 2 and 2 (3) 3 and 4 (4) 4 and 4

168. Which of the following statements are **correct** ?

(A) The electronic configuration of Cr is [Ar] $3d^5 4s^1$.
(B) The magnetic quantum number may have a negative value.
(C) In the ground state of an atom, the orbitals are filled in order of their increasing energies.
(D) The total number of nodes are given by n – 2 .

Choose the **most appropriate** answer from the options given below : [June 29, 2022 (I)]

(1) (A), (C) and (D) only (2) (A) and (B) only
(3) (A) and (C) only (4) (A), (B) and (C) only

169. Which of the following is the correct plot for the probability density $\psi^2(r)$ as a function of distance 'r' of the electron form the nucleus for 2s orbital? [June 29, 2022 (II)]

(1) 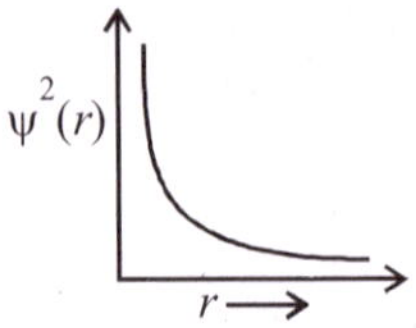

(2) 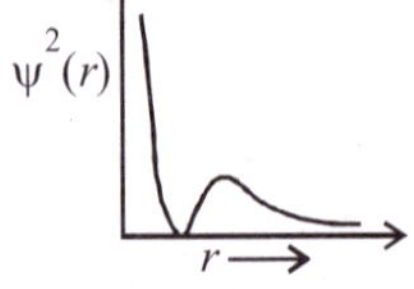

(3) 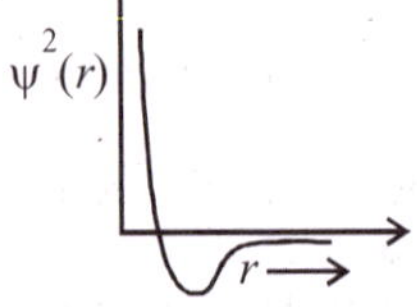

(4) 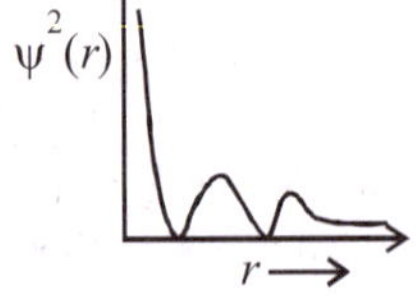

170. Consider the following statements : [June 28, 2022 (II)]

(A) The principal quantum number 'n' is a positive integer with values of 'n' = 1, 2, 3, ...
(B) The azimuthal quantum number 'l' for a given 'n' (principal quantum number) can have values as 'l' = 0, 1, 2, ... n
(C) Magnetic orbital quantum number 'm_l' for a particular 'l' (azimuthal quantum number) has $(2l + 1)$ values.
(D) ±1/2 are the two possible orientations of electron spin.
(E) For $l = 5$, there will be a total of 9 orbital.

Which of the above statements correct?

(1) (A), (B) and (C) (2) (A), (C), (D) and (E)
(3) (A), (C) and (D) (4) (A), (B), (C) and (D)

171. The number of radial and angular nodes in 4d orbital are respectively [June 26, 2022 (II)]

(1) 1 and 2 (2) 3 and 2 (3) 1 and 0 (4) 2 and 1

172. Consider the following pairs of electrons

(A) (a) $n = 3, 1 = 1, m_1 = 1, m_s = +\frac{1}{2}$

(b) $n = 3, 1 = 2, m_1 = 1, m_s = +\frac{1}{2}$

(B) (a) $n = 3, 1 = 2, m_1 = -2, m_s = -\frac{1}{2}$

(b) $n = 3, 1 = 2, m_1 = -1, m_s = -\frac{1}{2}$

(C) (a) $n = 4, 1 = 2, m_1 = 2, m_s = +\frac{1}{2}$

(b) $n = 3, 1 = 2, m_1 = 2, m_s = +\frac{1}{2}$

The pairs of electron present in degenerate orbitals is/are: [June 24, 2022 (I)]

(1) Only A (2) Only B
(3) Only C (4) (B) and (C)

173. A certain orbital has no angular nodes and two radial nodes. The orbital is [March 18, 2021 (I)]

(1) 2p (2) 3p (3) 3s (4) 2s

174. The orbital having two radial as well as two angular nodes is : [Feb. 26, 2021 (I)]

(1) 3p (2) 4d (3) 5d (4) 4f

175. In the sixth period, the orbitals that are filled are : [Sep. 05, 2020 (I)]

(1) 6s, 4f, 5d, 6p (2) 6s, 5d, 5f, 6p
(3) 6s, 5f, 6d, 6p (4) 6s, 6p, 6d, 6f

176. The correct statement about probability density (except at infinite distance from nucleus) is : [Sep. 05, 2020 (II)]

(1) It can be zero for 1s orbital.
(2) It can be negative for 2p orbital.
(3) It can be zero for 3p orbital.
(4) It can never be zero for 2s orbital.

177. Consider the hypothetical situation where the azimuthal quantum number, l, takes values 0, 1, 2, ... $n + 1$, where n is the principal quantum number. Then, the element with atomic number : [Sep. 03, 2020 (II)]

(1) 9 is the first alkali metal
(2) 13 has a half-filled valence subshell
(3) 8 is the first noble gas
(4) 6 has a 2p-valence subshell

178. The number of subshells associated with $n = 4$ and $m = -2$ quantum numbers is : **[Sep. 02, 2020 (II)]**

(1) 8 (2) 2 (3) 16 (4) 4

179. The number of orbitals associated with quantum numbers $n = 5, m_s = +\frac{1}{2}$ is: **[Jan. 07, 2020 (I)]**

(1) 11 (2) 25 (3) 50 (4) 15

180. Among the following, the energy of 2s orbital is lowest in: **[2017 (S); April 12, 2019 (II)]**

(1) K (2) H (3) Li (4) Na

181. The electrons are more likely to be found : **[April 12, 2019 (I)]**

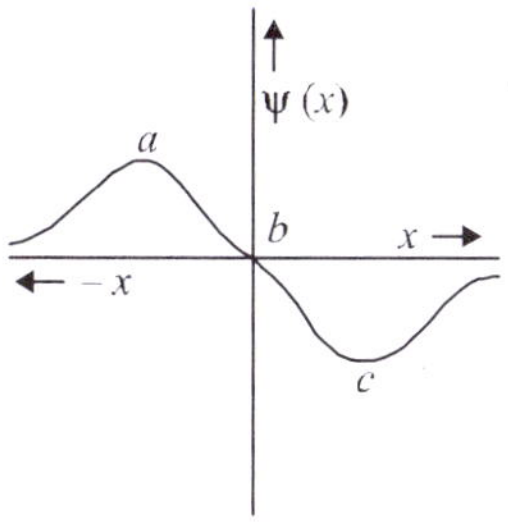

(1) in the region a and c (2) in the region a and b
(3) only in the region a (4) only in the region c

182. The graph between $|\psi|^2$ and r (radial distance) is shown below. This represents : **[April 10, 2019 (I)]**

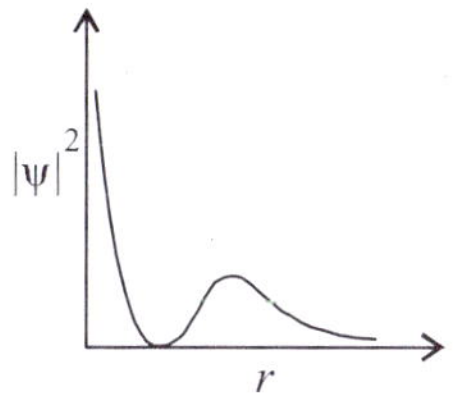

(1) 3s orbital (2) 2s orbital
(3) 1s orbital (4) 2p orbital

183. The size of the iso-electronic species Cl^-, Ar and Ca^{2+} is affected by : **[April 8, 2019 (I)]**

(1) azimuthal quantum number of valence shell
(2) electron-electron interaction in the outer orbitals
(3) principal quantum number of valence shell
(4) nuclear charge

184. The quantum number of four electrons are given below :

I. $n = 4, l = 2, m_l = -2, m_s = -1/2$
II. $n = 3, l = 2, m_l = 1, m_s = +1/2$
III. $n = 4, l = 1, m_l = 0, m_s = +1/2$
IV. $n = 3, l = 1, m_l = 1, m_s = -1/2$

The correct order of their increasing energies will be : **[2007(S), April 8, 2019 (I)]**

(1) IV < III < II < I (2) I < II < III < IV
(3) IV < II < III < I (4) I < III < II < IV

185. Which of the following combination of statements is true regarding the interpretation of the atomic orbitals? **[Jan. 9, 2019 (II)]**

(A) An electron in an orbital of high angular momentum stays away from the nucleus than an electron in the orbital of lower angular momentum.
(B) For a given value of the principal quantum number, the size of the orbit is inversely proportional to the azimuthal quantum number.
(C) According to wave mechanics, the ground state angular momentum is equal to $\frac{h}{2\pi}$
(D) The plot of ψ vs r for various azimuthal quantum numbers, shows peak shifting towards higher r value.

(1) (A), (D) (2) (A), (B)
(3) (A), (C) (4) (B), (C)

186. The total number of orbitals associated with the principal quantum number 5 is : **[Online April 9, 2016]**

(1) 20 (2) 25 (3) 10 (4) 5

187. If the principal quantum number $n = 6$, the correct sequence of filling of electrons will be : **[Online April 10, 2015]**

(1) $ns \to (n-2)f \to np \to (n-1)d$
(2) $ns \to (n-2)f \to (n-1)d \to np$
(3) $ns \to np \to (n-1)d \to (n-2)f$
(4) $ns \to (n-1)d \to (n-2)f \to np$

188. The correct set of four quantum numbers for the valence electrons of rubidium atom ($Z = 37$) is: **[2014]**

(1) $5, 0, 0, +\frac{1}{2}$ (2) $5, 1, 0, +\frac{1}{2}$
(3) $5, 1, 1, +\frac{1}{2}$ (4) $5, 0, 1, +\frac{1}{2}$

189. In an atom how many orbital(s) will have the quantum numbers; $n = 3, l = 2$ and $m_l = +2$? **[Online April 9, 2013]**

(1) 5 (2) 3 (3) 1 (4) 7

190. Given **[Online April 25, 2013]**

(A) $n = 5, m_\ell = +1$
(B) $n = 2, \ell = 1, m_\ell = -1, m_s = -1/2$

The maximum number of electron(s) in an atom that can have the quantum numbers as given in (A) and (B) are respectively:

(1) 25 and 1 (2) 8 and 1
(3) 2 and 4 (4) 4 and 1

191. The increasing order of the ionic radii of the given isoelectronic species is : **[2012]**

(1) $Cl^-, Ca^{2+}, K^+, S^{2-}$ (2) $S^{2-}, Cl^-, Ca^{2+}, K^+$
(3) $Ca^{2+}, K^+, Cl^-, S^{2-}$ (4) $K^+, S^{2-}, Ca^{2+}, Cl^-$

192. The following sets of quantum numbers represent four electrons in an atom. **[Online May 26, 2012]**

(i) $n = 4, l = 1$ (ii) $n = 4, l = 0$
(iii) $n = 3, l = 2$ (iv) $n = 3, l = 1$

The sequence representing increasing order of energy, is

(1) (iii) < (i) < (iv) < (ii) (2) (iv) < (ii) < (iii) < (i)
(3) (i) < (iii) < (ii) < (iv) (4) (ii) < (iv) < (i) < (iii)

193. Which one of the following constitutes a group of the isoelectronic species? **[2008]**

(1) C_2^{2-}, O_2^-, CO, NO (2) $NO^+, C_2^{2-}, CN^-, N_2$
(3) $CN^-, N_2, O_2^{2-}, C_2^{2-}$ (4) N_2, O_2^-, NO^+, CO

194. Which of the following sets of quantum numbers represents the highest energy of an atom? **[2007]**

(1) $n = 3, l = 0, m = 0, s = +1/2$
(2) $n = 3, l = 1, m = 1, s = +1/2$
(3) $n = 3, l = 2, m = 1, s = +1/2$
(4) $n = 4, l = 0, m = 0, s = +1/2$.

195. Which one of the following sets of ions represents a collection of isoelectronic species? **[2006]**
(1) $N^{3-}, O^{2-}, F^-, S^{2-}$ (2) $Li^+, Na^+, Mg^{2+}, Ca^{2+}$
(3) $K^+, Cl^-, Ca^{2+}, Sc^{3+}$ (4) $Ba^{2+}, Sr^{2+}, K^+, Ca^{2+}$

196. Of the following sets which one does NOT contain isoelectronic species? **[2005]**
(1) $BO_3^{3-}, CO_3^{2-}, NO_3^-$
(2) $SO_3^{2-}, CO_3^{2-}, NO_3^-$
(3) CN^-, N_2, C_2^{2-}
(4) $PO_4^{3-}, SO_4^{2-}, ClO_4^-$

197. The electrons identified by quantum numbers n and ℓ : **[2005]**
(A) $n=4, l=1$ (B) $n=4, l=0$
(C) $n=3, l=2$ (D) $n=3, l=1$
can be placed in order of increasing energy as :
(1) (C) < (D) < (B) < (A)
(2) (D) < (B) < (C) < (A)
(3) (B) < (D) < (A) < (C)
(4) (A) < (C) < (B) < (D)

198. In a multi-electron atom, which of the following orbitals described by the three quantum members will have the same energy in the absence of magnetic and electric fields?
(A) $n=1, l=0, m=0$ (B) $n=2, l=0, m=0$
(C) $n=2, l=1, m=1$ (D) $n=3, l=2, m=1$
(E) $n=3, l=2, m=0$ **[2005]**
(1) (D) and (E) (2) (C) and (D)
(3) (B) and (C) (4) (A) and (B)

199. Which of the following sets of quantum numbers is correct for an electron in 4f orbital ? **[2004]**
(1) $n=4, l=3, m=+1, s=+½$
(2) $n=4, l=4, m=-4, s=-½$
(3) $n=4, l=3, m=+4, s=+½$
(4) $n=3, l=2, m=-2, s=+½$

200. Which one of the following sets of ions represents the collection of isoelectronic species? **[2004]**
(1) $K^+, Cl^-, Mg^{2+}, Sc^{3+}$ (2) $Na^+, Ca^{2+}, Sc^{3+}, F^-$
(3) $K^+, Ca^{2+}, Sc^{3+}, Cl^-$ (4) $Na^+, Mg^{2+}, Al^{3+}, Cl^-$
(Atomic nos. : F = 9, Cl = 17, Na = 11, Mg = 12, Al = 13, K = 19, Ca = 20, Sc = 21)

201. Consider the ground state of Cr atom (X = 24). The number of electrons with the azimuthal quantum numbers, $l = 1$ and 2 are, respectively **[2004]**
(1) 16 and 4 (2) 12 and 5
(3) 12 and 4 (4) 16 and 5

202. The number of d-electrons retained in Fe^{2+} (At. no. of Fe = 26) ion is **[2003]**
(1) 4 (2) 5 (3) 6 (4) 3

203. The orbital angular momentum for an electron revolving in an orbit is given by $\sqrt{l(l+1)}.\frac{h}{2\pi}$. This momentum for an s-electron will be given by **[2003]**
(1) zero (2) $\frac{h}{2\pi}$
(3) $\sqrt{2}.\frac{h}{2\pi}$ (4) $+\frac{1}{2}.\frac{h}{2\pi}$

204. Which one of the following groupings represents a collection of isoelectronic species ?(At. nos. : Cs : 55, Br : 35) **[2003]**
(1) N^{3-}, F^-, Na^+ (2) Be, Al^{3+}, Cl^-
(3) Ca^{2+}, Cs^+, Br (4) Na^+, Ca^{2+}, Mg^{2+}

NUMERICAL VALUE QUESTIONS (NVQs)

205. In an atom, total number of electrons having quantum numbers $n = 4, |m_l| = 1$ and $m_s = -\frac{1}{2}$ is **[April 5, 2024 (II)]**

206. The maximum number of orbitals which can be identified with n = 4 and $m_l = 0$ is_____ **[April 4, 2024 (II)]**

207. The number of electrons present in all the completely filled sub shells having n=4 and $s = +\frac{1}{2}$ is _____.
(Where n = principal quantum number and s = spin quantum number) **[Jan. 27, 2024 (I)]**

208. The orbital angular momentum of an electron in 3s orbital is $\frac{xh}{2\pi}$. The value of x is _______. **[April 13, 2023 (II)]**

209. The number of correct statements from the following _______. **[April 11, 2023 (II)]**
A. For 1s orbital, the probability density is maximum at the nucleus.
B. For 2s orbital, the probability density first increases to maximum and then decreases sharply to zero.
C. Boundary surface diagrams of the orbitals encloses a region of 100% probability of finding the electron.
D. p and d-orbitals have 1 and 2 angular nodes respectively.
E. Probability density of p-orbital is zero at the nucleus.

210. The number of atomic orbitals from the following having 5 radial nodes is ________.
7s, 7p, 6s, 8p, 8d **[April 08, 2023 (II)]**

211. The number of given orbitals which have electron density along the axis is _______
$p_x, p_y, p_z, d_{xy}, d_{yz}, d_{xz}, d_{z^2}, d_{x^2-y^2}$ **[Jan. 25, 2023 (II)]**

212. Consider the following set of quantum numbers:

	n	l	m_l
A.	3	3	–3
B.	3	2	–2
C.	2	1	+1
D.	2	2	+2

The number of correct sets of quantum numbers is _______. **[June 27, 2022 (II)]**

213. The value of magnetic quantum number of the outermost electron of Zn^+ ion is _________. **[Aug. 31, 2021 (II)]**

214. The Azimuthal quantum number for the valence electrons of Ga^+ ion is ______. **[July 20, 2021 (I)]**
(Atomic number of Ga = 31)

215. A certain orbital has $n = 4$ and $m_l = -3$. The number of radial nodes in this orbital is __________.
(Round off to the Nearest Integer). **[March 17, 2021 (I)]**

216. The number of orbitals with $n = 5, m_l = +2$ is _______.
(Round off to the Nearest Integer). **[March 16, 2021 (II)]**

Part-B (JEE Advanced)

Topic 1 – Developments Leading to the Bohr's Model of Atom

1 MCQs with One Correct Answer

1. Given that the abundances of isotopes ^{54}Fe, ^{56}Fe and ^{57}Fe are 5%, 90% and 5%, respectively, the atomic mass of Fe is [2009S]
(a) 55.85 (b) 55.95 (c) 55.75 (d) 56.05

2. Rurtherford's experiment, which established the nuclear model of the atom, used a beam of [2002S]
(a) β-particles, which impinged on a metal foil and got absorbed
(b) γ-rays, which impinged on a metal foil and ejected electrons
(c) helium atoms, which impinged on a metal foil and got scattered
(d) helium nuclei, which impinged on a metal foil and got scattered

3. Which of the following does not characterise X-rays?
(a) The radiation can ionise gases [1992 - 1 Mark]
(b) It causes ZnS to fluorescence
(c) Deflected by electric and magnetic fields
(d) Have wavelengths shorter than ultraviolet rays

4. The wavelength of a spectral line for an electronic transition is inversely related to : [1988 - 1 Mark]
(a) the number of electrons undergoing the transition
(b) the nuclear charge of the atom
(c) the difference in the energy of the energy levels involved in the transition
(d) the velocity of the electron undergoing the transition.

5. The triad of nuclei that is isotonic is [1988 - 1 Mark]
(a) $^{14}_{6}C$, $^{15}_{7}N$, $^{17}_{9}F$ (b) $^{12}_{6}C$, $^{14}_{7}N$, $^{19}_{9}F$
(c) $^{14}_{6}C$, $^{14}_{7}N$, $^{17}_{9}F$ (d) $^{14}_{6}C$, $^{14}_{7}N$, $^{19}_{9}F$

6. The ratio of the energy of a photon of 2000 Å wavelength radiation to that of 4000 Å radiation is : [1986 - 1 Mark]
(a) 1/4 (b) 4 (c) 1/2 (d) 2

7. Rutherford's alpha particle scattering experiment eventually led to the conclusion that : [1986 - 1 Mark]
(a) mass and energy are related
(b) electrons occupy space around the nucleus
(c) neutrons are buried deep in the nucleus
(d) the point of impact with matter can be precisely determined.

8. Electromagnetic radiation with maximum wavelength is :
(a) ultraviolet (b) radiowave [1985 - 1 Mark]
(c) *X*-ray (d) infrared

9. The radius of an atomic nucleus is of the order of :
[1985 - 1 Mark]
(a) 10^{-10} cm (b) 10^{-13} cm (c) 10^{-15} cm (d) 10^{-8} cm

10. The increasing order (lowest first) for the values of e/m (charge/mass) for electron (e), proton (p), neutron (n) and alpha particle (α) is : [1984 - 1 Mark]
(a) e, p, n, α (b) n, p, e, α
(c) n, p, α, e (d) n, α, p, e

11. Rutherford's scattering experiment is related to the size of the [1983 - 1 Mark]
(a) nucleus (b) atom (c) electron (d) neutron

12. Rutherford's experiment on scattering of α-particles showed for the first time that the atom has [1981 - 1 Mark]
(a) electrons (b) protons
(c) nucleus (d) neutrons

13. The number of neutrons in dipositive zinc ion with mass number 70 is [1979]
(a) 34 (b) 36 (c) 38 (d) 40

3 Numeric / New Stem Based Questions

14. Wavelength of high energy transition of H–atoms is 91.2nm. Calculate the corresponding wavelength of He atoms.
[2003 - 2 Marks]

4 Fill in the Blanks

15. The light radiations with discrete quantities of energy are called .... [1993 - 1 Mark]

16. Elements of the same mass number but of different atomic numbers are known as .... [1983 - 1 Mark]

17. Isotopes of an element differ in the number of .... in their nuclei. [1982 - 1 Mark]

18. The mass of a hydrogen atom is .... kg. [1982 - 1 Mark]

5 True / False

19. In a given electric field, β-particles are deflected more than α-particles in spite of α-particles having larger charge.
[1993 - 1 Mark]

20. Gamma rays are electromagnetic radiations of wavelengths of 10^{-6} cm to 10^{-5} cm. [1983 - 1 Mark]

6 MCQs with One or More than One Correct Answer

21. The sum of the number of neutrons and proton in the isotope of hydrogen is : [1986 - 1 Mark]
(a) 6 (b) 2 (c) 4 (d) 3

22. When alpha particles are sent through a thin metal foil, most of them go straight through the foil because : [1984 - 1 Mark]
(a) alpha particles are much heavier than electrons
(b) alpha particles are positively charged
(c) most part of the atom is empty space
(d) alpha particle move with high velocity

23. Many elements have non-integral atomic masses because:
(a) they have isotopes [1984 - 1 Mark]
(b) their isotopes have non-integral masses
(c) their isotopes have different masses
(d) the constitutents, neutrons, protons and electrons, combine to give fractional masses

24. An isotone of $^{76}_{32}Ge$ is : [1984 - 1 Mark]
(a) $^{77}_{32}Ge$ (b) $^{77}_{33}As$ (c) $^{77}_{34}Se$ (d) $^{78}_{34}Se$

10 Subjective Problems

25. Naturally occurring boron consists of two isotopes whose atomic weights are 10.01 and 11.01. The atomic weight of natural boron is 10.81. Calculate the percentage of each isotope in natural boron. **[1978]**

Topic 2 - Bohr's Model for Hydrogen Atom (Emission and Absorption Spectra)

1 MCQs with One Correct Answer

26. According to Bohr's model, the highest kinetic energy is associated with the electron in the **[Adv. 2024]**
(a) First orbit of H atom (b) First orbit of He^+
(c) Second orbit of He^+ (d) Second orbit of Li^{2+}

27. The kinetic energy of an electron in the second Bohr orbit of a hydrogen atom is [a_0 is Bohr radius] : **[2012]**
(a) $\frac{h^2}{4\pi^2 ma_0^2}$ (b) $\frac{h^2}{16\pi^2 ma_0^2}$
(c) $\frac{h^2}{32\pi^2 ma_0^2}$ (d) $\frac{h^2}{64\pi^2 ma_0^2}$

28. The radius of which of the following orbit is same as that of the first Bohr's orbit of hydrogen atom? **[2004S]**
(a) $He^+ (n=2)$ (b) $Li^{2+} (n=2)$
(c) $Li^{2+} (n=3)$ (d) $Be^{3+} (n=2)$

29. Bohr model can explain : **[1985 - 1 Mark]**
(a) the spectrum of hydrogen atom only
(b) spectrum of an atom or ion containing one electron only
(c) the spectrum of hydrogen molecule
(d) the solar spectrum

30. Which electronic level would allow the hydrogen atom to absorb a photon but not to emit a photon? **[1984 - 1 Mark]**
(a) 3*s* (b) 2*p* (c) 2*s* (d) 1*s*

2 Integer Value Answer

31. For He^+, a transition takes place from the orbit of radius 105.8 pm to the orbit of radius 26.45 pm.
The wavelength (in nm) of the emitted photon during the transition is ____. **[Adv. 2023]**
[**Use:** Bohr radius, a = 52.9 pm; Rydberg constant, $R_H = 2.2 \times 10^{-18}$ J; Planck's constant, $h = 6.6 \times 10^{-34}$ J s; Speed of light, $c = 3 \times 10^8$ m s^{-1}]

32. The work function (ϕ) of some metals is listed below. The number of metals which will show photoelectric effect when light of 300 nm wavelength falls on the metal is **[2011]**

Metal	Li	Na	K	Mg	Cu	Ag	Fe	Pt	W
ϕ (eV)	2.4	2.3	2.2	3.7	4.8	4.3	4.7	6.3	4.75

3 Numeric / New Stem Based Questions

33. X^{a+} and Y^{b+} are hydrogen-like species. The wavelength of light absorbed during the transition between the states with principal quantum numbers n = 1 and n = 2 of X^{a+} is λ. The wavelength of light absorbed during the transition between the states with principal quantum numbers n = 2 and n = 4 of Y^{b+} is 9λ. The lowest possible value of (a + b) is ____. **[Adv. 2026]**

34. Calculate the wave number for the shortest wavelength transition in the Balmer series of atomic hydrogen. **[1996 - 1 Mark]**

35. According to Bohr's theory, the electronic energy of hydrogen atom in the nth Bohr's orbit is given by $E_n = \frac{-21.76 \times 10^{-19}}{n^2}$ J.
Calculate the longest wavelength of light that will be needed to remove an electron from the third Bohr orbit of the He^+ ion. **[1990 - 3 Marks]**

36. Calculate the wavelength in Angstrom of the photon that is emitted when an electron in the Bohr orbit, *n* = 2 returns to the orbit, *n* = 1 in the hydrogen atom. The ionization potential of the ground state hydrogen atom is 2.17×10^{-11} erg per atom. **[1982 - 4 Marks]**

37. The energy of the electron in the second and the third Bohr's orbits of the hydrogen atom is -5.42×10^{-12} erg and -2.41×10^{-12} erg respectively. Calculate the wavelength of the emitted radiation when the electron drops from the third to the second orbit. **[1981 - 3 Marks]**

6 MCQs with One or More than One Correct Answer

38. The 2s and the 2p orbital energies of hydrogen atom are $E_{2s}(H)$ and $E_{2p}(H)$, respectively. The 2s and the 2p orbital energies of lithium atom are $E_{2s}(Li)$ and $E_{2p}(Li)$, respectively. The correct option(s) about the orbital energies is(are) **[Adv. 2026]**
(a) $E_{2s}(Li) < E_{2p}(Li)$ (b) $E_{2s}(H) = E_{2p}(H)$
(c) $E_{2p}(H) < E_{2s}(Li)$ (d) $E_{2s}(H) > E_{2s}(Li)$

39. The energy of an electron in the first Bohr orbit of H atom is –13.6 eV. The possible energy value(s) of the excited state(s) for electrons in Bohr orbits of hydrogen is (are) **[1998 - 2 Marks]**
(a) –3.4 eV (b) –4.2 eV (c) –6.8 eV (d) –1.5 eV

7 Match the Following

40. Consider the Bohr's model of a one – electron atom where the electron moves around the nucleus. In the following List–I contains some quantities for the nth orbit of the atom and List–II contains options showing how they depend on *n* **[Adv. 2019]**

List–I	List–II
(I) Radius of the nth orbit	(P) $\propto n^{-2}$
(II) Angular momentum of the electron in the nth orbit	(Q) $\propto n^{-1}$
(III) Kinetic energy of the electron in the nth orbit	(R) $\propto n^0$
(IV) Potential energy of the electron in the nth orbit	(S) $\propto n^1$
	(T) $\propto n^2$
	(U) $\propto n^{1/2}$

Which of the following options has the correct Combination considering List–I and List–II?
(a) (II), (R) (b) (II), (Q)
(c) (I), (P) (d) (I), (T)

10 Subjective Problems

41. Calculate the energy required to excite one litre of hydrogen gas at 1 atm and 298 K to the first excited state of atomic hydrogen. The energy for the dissociation of H–H bond is 436 kJ mol^{-1}. **[2000 - 4 Marks]**

42. Consider the hydrogen atom to be a proton embedded in a cavity of radius a_0 (Bohr radius) whose charge is neutralised by the addition of an electron to the cavity in vacuum, infinitely slowly. Estimate the average total energy of an electron in its ground state in a hydrogen atom as the work done in the above neutralisation process. Also, if the magnitude of the average kinetic energy is half the magnitude of the average potential energy, find the average potential energy. **[1996 - 2 Marks]**

43. Iodine molecule dissociates into atoms after absorbing light of 4500Å. If one quantum of radiation is absorbed by each molecule, calculate the kinetic energy of iodine atoms. (Bond energy of $I_2 = 240$ kJ mol^{-1}) **[1995 - 2 Marks]**

44. Find out the number of waves made by a Bohr electron in one complete revolution in its 3rd orbit. **[1994 - 3 Marks]**

45. What transition in the hydrogen spectrum would have the same wavelength as the Balmer transition $n = 4$ to $n = 2$ of He^+ spectrum? **[1993 - 3 Marks]**

46. Estimate the difference in energy between 1st and 2nd Bohr orbit for a hydrogen atom. At what minimum atomic number, a transition from $n = 2$ to $n = 1$ energy level would result in the emission of X-rays with $\lambda = 3.0 \times 10^{-8}$m? Which hydrogen atom-like species does this atomic number correspond to ? **[1993 - 5 Marks]**

47. The electron energy in hydrogen atom is given by $E = (-21.7 \times 10^{-12})/n^2$ ergs. Calculate the energy required to remove an electron completely from the $n = 2$ orbit. What is the longest wavelength (in cm) of light that can be used to cause this transition? **[1984 - 3 Marks]**

Topic 3 - Towards Quantum Mechanical Model of the Atom (Dual Behaviour and Uncertainty Principle)

1 MCQs with One Correct Answer

48. The wavelength associated with a golf ball weighing 200 g and moving at a speed of 5 m/h is of the order **[2001S]**
(a) 10^{-10} m (b) 10^{-20} m (c) 10^{-30} m (d) 10^{-40} m

49. Which of the following relates to photons both as wave motion and as a stream of particles? **[1992 - 1 Mark]**
(a) Inference (b) $E = mc^2$
(c) Diffraction (d) $E = h\nu$

2 Integer Value Answer

50. Consider a helium (He) atom that absorbs a photon of wavelength 330 nm. The change in the velocity (in cm s^{-1}) of He atom after the photon absorption is ___.
(Assume : Momentum is conserved when photon is absorbed.
Use : Planck constant = 6.6×10^{-34} J s, Avogadro number = 6×10^{23} mol^{-1}, Molar mass of He = 4 g mol^{-1}) **[Adv. 2021]**

51. The atomic masses of 'He' and 'Ne' are 4 and 20 a.m.u., respectively. The value of the de Broglie wavelength of 'He' gas at —73 °C is "M" times that of the de Broglie wavelength of 'Ne' at 727 °C. 'M' is **[Adv. 2013]**

4 Fill in the Blanks

52. Wave functions of electrons in atoms and molecules are called .... **[1993 - 1 Mark]**

53. The uncertainty principle and the concept of wave nature of matter were proposed by .... and .... respectively. (Heisenberg, Schrodinger, Maxwell, de Broglie) **[1988 - 1 Mark]**

6 MCQs with One or More than One Correct Answer

54. Among the following, the correct statement(s) for electrons in an atom is(are) **[Adv. 2024]**
(a) Uncertainty principle rules out the existence of definite paths for electrons.
(b) The energy of an electron in 2s orbital of an atom is lower than the energy of an electron that is infinitely far away from the nucleus.
(c) According to Bohr's model, the most negative energy value for an electron is given by $n = 1$, which corresponds to the most stable orbit.
(d) According to Bohr's model, the magnitude of velocity of electrons increases with increase in values of n.

10 Subjective Problems

55. Find the velocity (ms^{-1}) of electron in first Bohr's orbit of radius a_0. Also find the de Broglie's wavelength (in m). Find the orbital angular momentum of 2p orbital of hydrogen atom in units of $h/2\pi$. **[2005 - 2 Marks]**

56. A ball of mass 100 g is moving with 100 ms^{-1}. Find its wavelength. **[2004 - 1 Mark]**

57. The Schrodinger wave equation for hydrogen atom is

$$\Psi_{2s} = \frac{1}{4\sqrt{2\pi}}\left(\frac{1}{a_0}\right)^{3/2}\left(2 - \frac{r_0}{a_0}\right)e^{-r_0/a_0}$$

Where a_0 is Bohr's radius. If the radial node in 2s be at r_0, then find r_0 in terms of a_0. **[2004 - 2 Marks]**

Topic 4 - Quantum Mechanical Model of Atom

1 MCQs with One Correct Answer

58. The number of radial nodes of 3s and 2p orbitals are respectively **[2005S]**
(a) 2, 0 (b) 0, 2 (c) 1, 2 (d) 2, 1

59. If the nitrogen atom has electronic configuration $1s^7$, it would have energy lower than that of the normal ground state configuration $1s^2 2s^2 2p^3$, because the electrons would be closer to the nucleus. Yet $1s^7$ is not observed because it violates. **[2002S]**
(a) Heisenberg uncertainty principle
(b) Hund's rule
(c) Pauli exclusion principle
(d) Bohr postulate of stationary orbits

60. The quantum numbers +1/2 and –1/2 for the electron spin represent **[2001S]**
(a) rotation of the electron in clockwise and anticlockwise direction respectively
(b) rotation of the electron in anticlockwise and clockwise direction respectively
(c) magnetic moment of the electron pointing up and down respectively
(d) two quantum mechanical spin states which have no classical analogue

61. The electronic configuration of an element is $1s^2, 2s^2 2p^6, 3s^2 3p^6 3d^5, 4s^1$. This represents its [2000S]
(a) excited state (b) ground state
(c) cationic form (d) anionic form

62. The number of nodal planes in a p_x orbital is [2000S]
(a) one (b) two (c) three (d) zero

63. The electrons, identified by quantum numbers n and l, (i) $n=4, l=1$, (ii) $n=4, l=0$, (iii) $n=3, l=2$, and (iv) $n=3, l=1$ can be placed in order of increasing energy, from the lowest to highest, as [1999 - 2 Marks]
(a) (iv) < (ii) < (iii) < (i) (b) (ii) < (iv) < (i) < (iii)
(c) (i) < (iii) < (ii) < (iv) (d) (iii) < (i) < (iv) < (ii)

64. For a d-electron, the orbital angular momentum is [1997 - 1 Mark]
(a) $\sqrt{6}(h/2\pi)$ (b) $\sqrt{2}(h/2\pi)$
(c) $(h/2\pi)$ (d) $2(h/2\pi)$

65. The orbital angular momentum of an electron in 2s orbital is: [1996 - 1 Mark]
(a) $+\frac{1}{2}.\frac{h}{2\pi}$ (b) Zero (c) $\frac{h}{2\pi}$ (d) $\sqrt{2}.\frac{h}{2\pi}$

66. A 3p orbital has : [1995S]
(a) two non spherical nodes
(b) two spherical nodes
(c) one spherical and one non spherical node
(d) one spherical and two non spherical nodes

67. The correct set of quantum numbers for the unpaired electron of chlorine atom is : [1989 - 1 Mark]

	n	l	m		n	l	m
(a)	2	1	0	(b)	2	1	1
(c)	3	1	1	(d)	3	0	0

68. The correct ground state electronic configuration of chromium atom is : [1989 - 1 Mark]
(a) [Ar] $3d^5 4s^1$ (b) [Ar] $3d^4 4s^2$
(c) [Ar] $3d^6 4s^0$ (d) [Ar] $4d^5 4s^1$

69. The outermost electronic configuration of the most electronegative element is [1988 - 1 Mark]
(a) $ns^2 np^3$ (b) $ns^2 np^4$ (c) $ns^2 np^5$ (d) $ns^2 np^6$

70. The orbital diagram in which the Aufbau principle is violated is : [1988 - 1 Mark]

	2s	2p
(a)	↑↓	↑↓ | ↑ |
(b)	↑	↑↓ | ↑ | ↑
(c)	↑↓	↑ | ↑ | ↑
(d)	↑↓	↑↓ | ↑ | ↑

71. Which one of the following sets of quantum numbers represents an impossible arrangement? [1986 - 1 Mark]

	n	l	m_l	m_s		n	l	m_l	m_s
(a)	3	2	–2	½	(b)	4	0	0	½
(c)	3	2	–3	½	(d)	5	3	0	–½

72. Correct set of four quantum numbers for the valence (outermost) electron of rubidium ($Z=37$) is : [1984 - 1 Mark]
(a) 5, 0, 0, +½ (b) 5, 1, 0, +½
(c) 5, 1, 1, +½ (d) 6, 0, 0, +½

73. Any p-orbital can accommodate upto [1983 - 1 Mark]
(a) four electrons
(b) six electrons
(c) two electrons with parallel spins
(d) two electrons with opposite spins

74. The principal quantum number of an atom is related to the
(a) size of the orbital [1983 - 1 Mark]
(b) spin angular momentum
(c) orbital angular momentum
(d) orientation of the orbital in space

2 Integer Value Answer

75. Not considering the electronic spin, the degeneracy of the second excited state ($n=3$) of H atom is 9, while the degeneracy of the second excited state of H^- is [Adv. 2015]

76. In an atom, the total number of electrons having quantum numbers $n=4, |m_l|=1$ and $m_s=-\frac{1}{2}$ is [Adv. 2014]

77. The maximum number of electrons that can have principal quantum number, $n=3$, and spin quantum $m_s=-\frac{1}{2}$, is [2011]

3 Numeric / New Stem Based Questions

78. What is the maximum number of electrons that may be present in all the atomic orbitals with principal quantum number 3 and azimuthal quantum number 2? [1985 - 2 Marks]

4 Fill in the Blanks

79. The outermost electronic configuration of Cr is .... [1994 - 1 Mark]

80. The $2p_x$, $2p_y$ and $2p_z$ orbitals of atom have identical shapes but differ in their .... [1993 - 1 Mark]

81. When there are two electrons in the same orbital, they have .... spins. [1982 - 1 Mark]

5 True / False

82. The electron density in the XY plane in $3d_{x^2-y^2}$ orbital is zero. [1986 - 1 Mark]

83. The energy of the electron in the 3d-orbital is less than that in the 4s-orbital in the hydrogen atom. [1983 - 1 Mark]

84. The outer electronic configuration of the ground state chromium atom is $3d^4 4s^2$. [1982 - 1 Mark]

6 MCQs with One or More than One Correct Answer

85. The ground state energy of hydrogen atom is –13.6 eV. Consider an electronic state ψ of He^+ whose energy, azimuthal quantum number and magnetic quantum number are –3.4 eV, 2 and 0, respectively. Which of the following statement(s) is (are) true from the state ψ? [Adv. 2019]
(a) It is a 4d state
(b) It has 3 radial nodes
(c) It has 2 angular nodes
(d) The nuclear charge experienced by the electron in this state is less than 2e, where e is the magnitude of the electronic charge

86. Ground state electronic configuration of nitrogen atom can be represented by [1999 - 3 Marks]
(a) ↑↓ ↑↓ ↑ ↑ ↑ (b) ↑↓ ↑↓ ↑ ↓ ↑
(c) ↑↓ ↑↓ ↑ ↓ ↓ (d) ↑↓ ↑↓ ↓ ↓ ↓

87. Which of the following satement(s) is (are) correct? [1998 - 2 Marks]

(a) The electronic configuration of Cr is [Ar] $3d^5 4s^1$. (Atomic Number of Cr = 24)
(b) The magnetic quantum number may have a negative value.
(c) In silver atom, 23 electrons have a spin of one type and 24 of the opposite type. (Atomic Number of Ag = 47)
(d) The oxidation state of nitrogen in HN_3 is – 3.

7 Match the Following

(Qs. 88-92) are based on the table, having 3 columns and 4 rows. Each question has four options (A), (B), (C) and (D). Only one of these four options is correct. By appropriately matching the information given in the three columns of the following table.

The wave function, ψ_{n,l,m_l} is a mathematical function whose value depends upon spherical polar coordinates (r, θ, ϕ) of the electron and characterized by the quantum numbers n, l and m_l. Here r is distance from nucleus, θ is colatitude and ϕ is azimuth. In the mathematical functions given in the table, Z is atomic number and a_0 is Bohr radius. [Adv. 2017]

Column-1	Column-2	Column-3
(I) 1s orbital	(i) $\psi_{n,l,m_l} \propto \left(\frac{Z}{a_0}\right)^{\frac{3}{2}} e^{-\left(\frac{Zr}{a_0}\right)}$	(P) 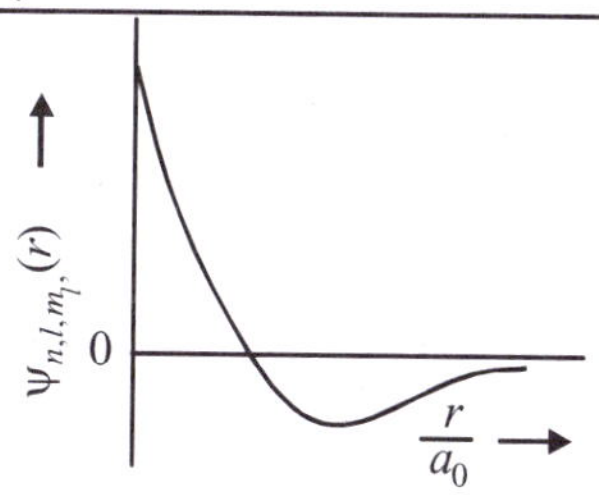
(II) 2s orbital	(ii) One radial node	(Q) Probability density at nucleus $\propto \frac{1}{a_0^3}$
(III) $2p_z$ orbital	(iii) $\psi_{n,l,m_l} \propto \left(\frac{Z}{a_0}\right)^{\frac{5}{2}} re^{-\left(\frac{Zr}{2a_0}\right)} \cos\theta$	(R) Probability density is maximum at nucleus
(IV) $3d_{z^2}$ orbital	(iv) xy-plane is a nodal plane	(S) Energy needed to excite electron from $n = 2$ state to $n = 4$ state is $\frac{27}{32}$ times the energy needed to excite electron from $n = 2$ state to $n = 6$ state

88. For the given orbital in Column 1, the only CORRECT combination for any hydrogen-like species is
(a) (I) (ii) (S) (b) (IV) (iv) (R) (c) (II) (ii) (P) (d) (III) (iii) (P)

89. For hydrogen atom, the only CORRECT combination is
(a) (I) (i) (S) (b) (II) (i) (Q) (c) (I) (i) (P) (d) (I) (iv) (R)

90. For He^+ ion, the only INCORRECT combination is
(a) (I) (i) (R) (b) (II) (ii) (Q) (c) (I) (iii) (R) (d) (I) (i) (S)

91. Match the entries in Column I with the correctly related quantum number(s) in Column II. [2008 - 6M]

Column-I	Column-II
(A) Orbital angular momentum of the electron in a hydrogen-like atomic orbital	(p) Principal quantum number
(B) A hydrogen-like one-electron wave function obeying Pauli principle	(q) Azimuthal quantum number
(C) Shape, size and orientation of hydrogen- like atomic orbitals	(r) Magnetic quantum number
(D) Probability density of electron at the nucleus in hydrogen-like atom	(s) Electron spin quantum number

92. E_n = Total energy, K_n = Kinetic energy, V_n = Potential energy, r_n = Radius of nth orbit Match the following :

Column-I	Column-II
(A) $V_n / K_n = ?$	(p) 0
(B) If radius of nth orbit $\propto E_n^x$, $x = ?$	(q) –1
(C) Angular momentum in lowest orbital	(r) –2
(D) $\frac{1}{r_n} \propto Z^y$, $y = ?$	(s) 1

8 Comprehension Based Questions

The hydrogen-like species Li^{2+} is in a spherically symmetric state S_1 with one radial node. Upon absorbing light the ion undergoes transition to a state S_2. The state S_2 has one radial node and its energy is equal to the ground state energy of the hydrogen atom. [2010]

93. The state S_1 is :
(a) 1s (b) 2s
(c) 2p (d) 3s

94. Energy of the state S_1 in units of the hydrogen atom ground state energy is :
(a) 0.75 (b) 1.50 (c) 2.25 (d) 4.50

95. The orbital angular momentum quantum number of the state S_2 is :
(a) 0 (b) 1 (c) 2 (d) 3

10 Subjective Problems

96. Give reasons why the ground state outermost electronic configuration of silicon is : [1985 - 2 Marks]

3s	3p		3s	3p
↑↓	↑ ↑ ☐	and not	↑ ↓	↑ ↓ ☐

Answer Key

Part-A (JEE Main)

1	(2)	23	(3)	45	(4)	67	(1)	89	(54)	111	(2)	133	(9)	155	(2)	177	(3)	199	(1)
2	(2)	24	(0)	46	(2)	68	(2)	90	(54)	112	(2)	134	(1)	156	(1)	178	(2)	200	(3)
3	(2)	25	(1)	47	(4)	69	(4)	91	(5)	113	(4)	135	(2)	157	(4)	179	(2)	201	(2)
4	(4)	26	(798)	48	(1)	70	(3)	92	(1724)	114	(1)	136	(4)	158	(2)	180	(1)	202	(3)
5	(1)	27	(2)	49	(4)	71	(3)	93	(34)	115	(2)	137	(2)	159	(2)	181	(1)	203	(1)
6	(4)	28	(2)	50	(3)	72	(2)	94	(22)	116	(2)	138	(4)	160	(2)	182	(2)	204	(1)
7	(3)	29	(10)	51	(2)	73	(2)	95	(8)	117	(3)	139	(4)	161	(4)	183	(4)	205	(6)
8	(3)	30	(300)	52	(1)	74	(4)	96	(10)	118	(1)	140	(1)	162	(2)	184	(3)	206	(4)
9	(4)	31	(2)	53	(4)	75	(4)	97	(1)	119	(2)	141	(4)	163	(4)	185	(1)	207	(16)
10	(3)	32	(50)	54	(1)	76	(1)	98	(270)	120	(1)	142	(2)	164	(1)	186	(2)	208	(0)
11	(2, 3)	33	(5)	55	(4)	77	(3)	99	(492)	121	(4)	143	(2)	165	(1)	187	(2)	209	(3)
12	(3)	34	(2)	56	(1)	78	(1)	100	(N)	122	(3)	144	(3)	166	(3)	188	(1)	210	(3)
13	(3)	35	(181)	57	(2)	79	(4)	101	(3155)	123	(2)	145	(4)	167	(1)	189	(3)	211	(5)
14	(4)	36	(222)	58	(3)	80	(1)	102	(2)	124	(58)	146	(4)	168	(4)	190	(2)	212	(2)
15	(3)	37	(4)	59	(2)	81	(1)	103	(2)	125	(661)	147	(4)	169	(2)	191	(3)	213	(0)
16	(4)	38	(1)	60	(3)	82	(1)	104	(2)	126	(7)	148	(3)	170	(3)	192	(2)	214	(0)
17	(3)	39	(2)	61	(4)	83	(1)	105	(1)	127	(2)	149	(4)	171	(1)	193	(2)	215	(0)
18	(1)	40	(2)	62	(3)	84	(2)	106	(3)	128	(548)	150	(4)	172	(2)	194	(3)	216	(3)
19	(1)	41	(3)	63	(3)	85	(4)	107	(4)	129	(1758)	151	(3)	173	(3)	195	(3)		
20	(3)	42	(1)	64	(3)	86	(2)	108	(4)	130	(22)	152	(4)	174	(3)	196	(2)		
21	(5)	43	(4)	65	(2)	87	(3)	109	(4)	131	(58)	153	(4)	175	(1)	197	(2)		
22	(494)	44	(4)	66	(3)	88	(1)	110	(4)	132	(6)	154	(3)	176	(3)	198	(1)		

Part-B (JEE Advanced)

1	(b)	9	(b)	21	(b, d)	30	(d)	38	(a, b, d)	58	(a)	66	(c)	74	(a)	85	(a, c)	93	(b)
2	(d)	10	(d)	22	(a, c)	31	(30)	39	(a, d)	59	(c)	67	(c)	75	(3)	86	(a, d)	94	(c)
3	(c)	11	(a)	23	(a, c)	32	(4)	40	(d)	60	(d)	68	(a)	76	(6)	87	(a, b, c)	95	(b)
4	(c)	12	(c)	24	(b, d)	33	(3.00)	48	(a)	61	(b)	69	(c)	77	(9)	88	(c)		
5	(a)	13	(d)	26	(b)	34	(27419)	49	(d)	62	(a)	70	(b)	78	(10)	89	(a)		
6	(d)	14	(22.8)	27	(c)	35	(2055)	50	(30)	63	(a)	71	(c)	82	(False)	90	(c)		
7	(b)	19	(True)	28	(d)	36	(1220)	51	(5)	64	(a)	72	(a)	83	(True)	91	(A)-(q); (B)-(p, q, r, s); (C)-(p, q, r); D-(p, q, r)		
8	(b)	20	(Falsc)	29	(b)	37	(660)	54	(a, b, c)	65	(b)	73	(d)	84	(False)	92	(A)-(r); (B)-(q); (C)-(p); (D)-(s)		

Chapter-3

Classification of Elements and Periodicity in Properties

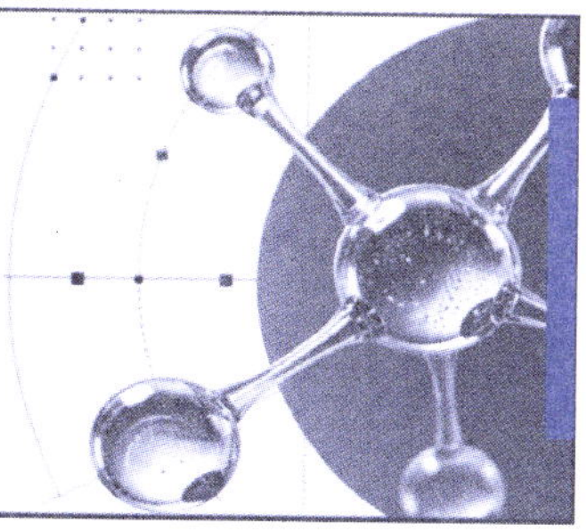

Part-A (JEE Main)

Topic 1 - Modern Periodic Table

MULTIPLE CHOICE QUESTIONS (MCQs)

1. A monoatomic anion (A^-) has 45 neutrons and 36 electrons. Atomic mass, group in the periodic table and physical state at room temperature of the element (A) respectively are : [April 04, 2026 (I)]
(1) 80, 17, liquid (2) 81, 16, solid
(3) 80, 16, gas (4) 81, 15, gas

2. Match the **LIST-I** with **LIST II**. [April 3, 2025 (II)]

LIST-I (Family)		**LIST-II (Symbol of Element)**	
A.	Pnictogen (group 15)	I.	Ts
B.	Chalcogen	II.	Og
C.	Halogen	III.	Lv
D.	Noble gas	IV.	Mc

Choose the **correct** answer from the options given below:
(1) A-IV, B-I, C-II, D-III (2) A-IV, B-III, C-I, D-II
(3) A-III, B-I, C-IV, D-II (4) A-II, B-III, C-IV, D-I

3. Given below are two statements : [Jan. 28, 2025 (II)]
Statement (I) : According to the Law of Octaves, the elements were arranged in the increasing order of their atomic number.
Statement (II) : Meyer observed a periodically repeated pattern upon plotting physical properties of certain elements against their respective atomic numbers.
In the light of the above statements, choose the correct answer from the options given below :
(1) Statement I is false but Statement II is true
(2) Both Statement I and Statement II are true
(3) Statement I is true but Statement II is false
(4) Both Statement I and Statement II are false

4. Which of the following Statements are NOT true about the periodic table? [Jan. 24, 2025 (I)]
A. The properties of elements are function of atomic weights.
B. The properties of elements are function of atomic numbers.
C. Elements having similar outer electronic configuration are arranged in same period.
D. An element's location reflects the quantum numbers of the last filled orbital.
E. The number of elements in a period is same as the number of atomic orbitals available in energy level that is being filled.
Choose the correct answer from the options given below:
(1) A, C and E Only (2) D and E Only
(3) A and E Only (4) B, C and E Only

5. Given below are two statements about X-ray spectra of elements:
Statement (I) : A plot of $\sqrt{\nu}$ (ν = frequency of X-rays emitted) vs atomic mass is a straight line.
Statement (II) : A plot of ν (ν = frequency of X-rays emitted) vs atomic number is a straight line.
In the light of the above statements choose the correct answer from the options given below: [Jan. 23, 2025 (II)]
(1) Statement I is true but Statement II is false
(2) Both Statement I and Statement II are true
(3) Both Statement I and Statement II are false
(4) Statement I is false but Statement II is true

6. The element that does not belong to the same period of the remaining elements (modern periodic table) is: [Jan. 23, 2025 (I)]
(1) Palladium (2) Iridium
(3) Osmium (4) Platinum

7. Match **List-I** with **List-II**. [April 9, 2024 (II)]

List-I (Element)	**List-II (Electronic Configuration)**
(A) N	(I) [Ar] $3d^{10}4s^2 4p^5$
(B) S	(II) [Ne] $3s^2 3p^4$
(C) Br	(III) [He] $2s^2 2p^3$
(D) Kr	(IV) [Ar] $3d^{10} 4s^2 4p^6$

Choose the correct answer from the options given below:
(1) A-(IV), B-(III), C-(II), D-(I)
(2) A-(III), B-(II), C-(I), D-(IV)
(3) A-(I), B-(IV), C-(III), D-(II)
(4) A-(II), B-(I), C-(IV), D-(III)

8. Henry Moseley studied characteristic X-ray spectra of elements. The graph which represents his observation correctly is : [April 08, 2023 (II)]
(Given ν = frequency of X-ray emitted; Z = atomic number)

(1) 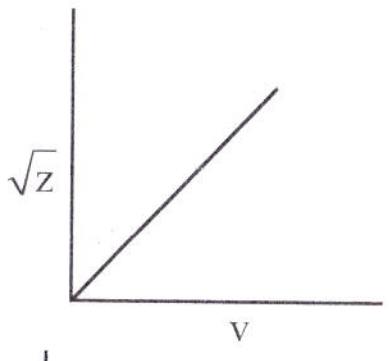

(2) 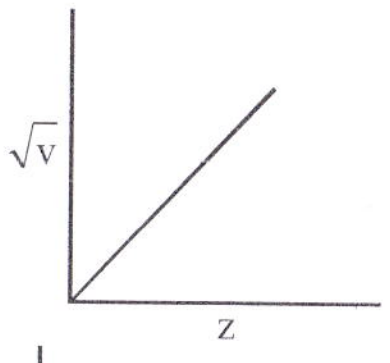

(3) 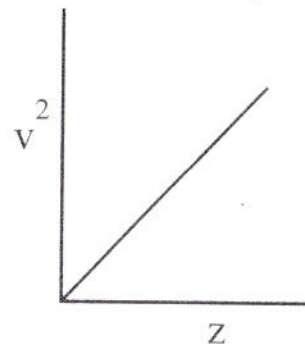

(4) 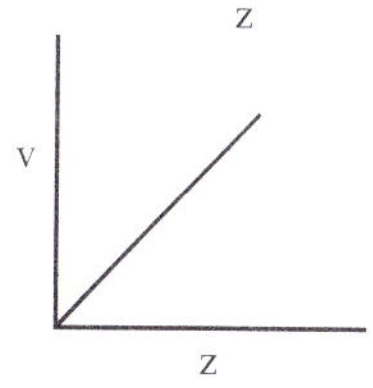

9. It is observed that characteristic X-ray spectra of elements show regularity. When frequency to the power 'n' i.e. v^n of X-rays emitted is plotted against atomic number 'Z', following graph is obtained. [Jan. 24, 2023 (I)]

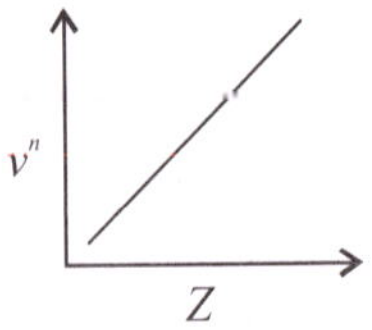

The value of 'n' is

(1) 1 (2) 2 (3) $\frac{1}{2}$ (4) 3

10. Match **List I** with **List II** [Jan. 30, 2023 (I)]

List-I (Atomic number)		List-II (Block of periodic table)
(A) 37	I.	p-block
(B) 78	II.	d-block
(C) 52	III.	f-block
(D) 65	IV.	s-block

Choose the correct answer from the options given below:
(1) A – II, B – IV, C – I, D – III
(2) A – I, B – III, C – IV, D – II
(3) A – IV, B – III, C – II, D – I
(4) A – IV, B – II, C – I, D – III

11. The characteristics of elements X, Y and Z with atomic numbers, respectively, 33, 53 and 83 are [March 16, 2021 (II)]
(1) X and Y are metalloids and Z is a metal.
(2) X is a metalloid, Y is a non-metal and Z is a metal.
(3) X and Z are non-metals and Y is a metalloid.
(4) X, Y and Z are metals.

12. Similarity in chemical properties of the atoms of elements in a group of the periodic table is most closely related to: [Online April 12, 2014; Same 2003]
(1) atomic numbers
(2) atomic masses
(3) number of principal energy levels
(4) number of valence electrons

13. According to the periodic law of elements, the variation in properties of elements is related to their [2003]
(1) nuclear masses
(2) atomic numbers
(3) nuclear neutron-proton number ratios
(4) atomic masses

NUMERICAL VALUE QUESTIONS (NVQs)

14. If IUPAC name of an element is "Unununnium" then the element belongs to n^th^ group of Periodic table. The value of n is ____. [Jan. 30, 2024 (I)]

Topic 2 - Periodic Trends in Properties of Elements

MULTIPLE CHOICE QUESTIONS (MCQs)

15. Match **List-I** with **List-II**. [April 08, 2026 (II)]

List-I Electronic configuration of neutral atom (where n = 2)		List-II 1st Ionization Energy (kJ mol^{-1})
A. ns^2	I.	2080
B. ns^2np^1	II.	899
C. ns^2np^3	III.	800
D. ns^2np^6	IV.	1402

Choose the correct answer from the options given below:
(1) A-II, B-III, C-IV, D-I (2) A-IV, B-III, C-II, D-I
(3) A-III, B-II, C-IV, D-I (4) A-III, B-II, C-I, D-IV

16. In a period, the first ionisation enthalpy of the element at extreme left and the negative electron gain enthalpy of the extreme right element, except noble gases, are respectively. [April 06, 2026 (II)]
(1) lowest and lowest (2) highest and lowest
(3) lowest and highest (4) highest and highest

17. Given below are two statements: [April 05, 2026 (II)]
Statement-I: The number of pairs among [Al_2O_3, Cr_2O_3], [Cl_2O_7, Mn_2O_7], [Na_2O, V_2O_3] and [CO, N_2O] that contain oxides of same nature (acidic, basic, neutral or amphoteric) is 4.
Statement-II: Among Na_2O, Al_2O_3, CO and Cl_2O_7, the most basic and acidic oxides are Na_2O and Cl_2O_7, respectively.
In the light of the above statements, choose the correct answer from the options given below :
(1) Both Statement I and Statement II are true
(2) Both Statement I and Statement II are false
(3) Statement I is true but Statement II is false
(4) Statement I is false but Statement II is true

18. The 1st ionization enthalpy for Mg is +737 kJ/mol. The most probable estimated value of the 2nd ionization enthalpy of Mg is____. [April 04, 2026 (II)]
(1) –906 kJ/mol (2) –856 kJ/mol
(3) +1450 kJ/mol (4) +590 kJ/mol

19. The correct set that contains all kinds (basic, acidic, amphoteric and neutral) of oxides is : [April 02, 2026 (II)]
(1) Na_2O, K_2O, Al_2O_3 and As_2O_3
(2) Al_2O_3, As_2O_3, CO and NO
(3) K_2O, Cl_2O_7, As_2O_3 and NO
(4) Na_2O, N_2O, Al_2O_3 and CO

20. Given below are two statements : [April 02, 2026 (I)]
Statement (I) : The first ionisation enthalpy of the elements Na, Mg, Cl and Ar follows the order Na > Mg > Cl > Ar
Statement (II) : Among Ca, Al, Fe and B, the third ionisation enthalpy is very high for Ca.
In the light of the above statements, choose the **correct answer** from the options given below :
(1) Both Statement I and Statement II are true.
(2) Both Statement I and Statement II are false.
(3) Statement I is true but Statement II is false.
(4) Statement I is false but Statement II is true.

21. Consider the elements N, P, O, S, Cl and F. The number of valence electrons present in the elements with most and least metallic character from the above list is respectively. [Jan. 28, 2026 (II)]
(1) 7 and 5 (2) 5 and 6
(3) 5 and 7 (4) 6 and 7

22. In period 4 of the periodic table, the elements with highest and lowest atomic radii are respectively. [Jan. 28, 2026 (I)]
(1) Na & Cl (2) K & Se (3) K & Br (4) Rb & Br

23. The correct order of C, N, O and F in terms of second ionisation potential is [Jan. 24, 2026 (II)]
(1) F < N < C < O (2) C < O < N < F
(3) C < N < F < O (4) C < F < N < O

24. Given below are two statements:
Statement I: K > Mg > Al > B is the correct order in terms of metallic character. [Jan. 24, 2026 (I)]
Statement II: Atomic radius is always greater than the ionic radius for any element.
In the light of the above statements, choose the correct answer from the options given below.
(1) Both Statement I and Statement II are true.
(2) Both Statement I and Statement II are false.
(3) Statement I is false but Statement II is true.
(4) Statement I is true but Statement II is false.

25. Given below are two statements: [Jan. 23, 2026 (II)]
Statement-I: The second ionisation enthalpy of Na is larger than the corresponding ionisation enthalpy of Mg.
Statement-II: The ionic radius of O^{2-} is larger than that of F^-
In the light of the above statements, choose the correct answer from the options given below.
(1) Both Statement-I and Statement-II are true.
(2) Both Statement-I and Statement-II are false.
(3) Statement-I is false but Statement-II is true.
(4) Statement-I is true but Statement-II is false.

26. The correct trend in the first ionization enthalpies of the elements in the 3rd period of periodic table is : [Jan. 23, 2026 (I)]
(1) Al < Si < S < P < Cl (2) Al < S < P < Si < Cl
(3) Si < S < Al < P < Cl (4) S < Si < Al < P < Cl

27. Given below are two statements : [Jan. 22, 2026 (II)]
Statement-I : C < O < N < F is the correct order in terms of first ionization enthalpy values.
Statement-II : S > Se > Te > Po > O is the correct order in terms of the magnitude of electron gain enthalpy values.
In the light of the above statements, choose the **correct answer** from the options given below :
(1) Statement-I is false but Statement-II is true.
(2) Both Statement-I and Statement-II are true.
(3) Both Statement-I and Statement-II are false.
(4) Statement-I is true but Statement-II is false.

28. Given below are two statements : [Jan. 22, 2026 (II)]
Statement-I : Element 'X' and 'Y' are the most and least electronegative elements, respectively among N, As, Sb and P. The nature of the oxides X_2O_3 and Y_2O_3 is acidic and amphoteric, respectively.
Statement-II : BCl_3 is covalent in nature and gets hydrolysed in water. It produces $[B(OH)_4]^-$ and $[B(H_2O)_6]^{3+}$ in aqueous medium.
In the light of the above statements, choose the **correct answer** from the options given below :
(1) Both Statement-I and Statement-II are false
(2) Statement-I is true but Statement-II is false
(3) Statement-I is false but Statement-II is false.
(4) Both Statement-I and Statement-II are true

29. A '*p*'-block element (E) and hydrogen form a binary cation $(EH_x)^+$, while EH_3 on treatment with K_2HgI_4 in alkaline medium gives a precipitate of basic mercury (II) amido-iodine. Given below are first ionisation enthalpy values ($kJ\ mol^{-1}$) for first element each from group 13, 14, 15 and 16. Identify the correct first ionisation enthalpy value for element E. [Jan. 22, 2026 (I)]
(1) 1312 (2) 1086 (3) 1402 (4) 801

30. Given below are two statements: [Jan. 21, 2026 (II)]
Statement-I: The correct order in terms of atomic/ionic radii is $Al > Mg > Mg^{2+} > Al^{3+}$.
Statement-II: The correct order in terms of the magnitude of electron gain enthalpy is Cl > Br > S > O.
In the light of the above statements, choose the **correct** answer from the options given below:
(1) Both Statement I and Statement II are false.
(2) Statement I is false but Statement II is true.
(3) Statement I is true but Statement II is false.
(4) Both Statement I and Statement II are true.

31. Which of the following represents the correct trend for the mentioned property? [Jan. 21, 2026 (I)]
A. F > P > S > B — First Ionization Energy
B. Cl > F > S > P — Electron Affinity
C. K > Al > Mg > B — Metallic Character
D. $K_2O > Na_2O > MgO > Al_2O_3$ — Basic Character
Choose the correct answer from the option given below.
(1) A, B and D only (2) A, B, C and D
(3) A and B only (4) B and C only

32. The atomic number of the element from the following with lowest 1st ionisation enthalpy is : [April 8, 2025 (II)]
(1) 32 (2) 35 (3) 87 (4) 19

33. Choose the incorrect trend in the atomic radii (r) of the elements : [April 7, 2025 (II)]
(1) $r_{Br} < r_K$ (2) $r_{Mg} < r_{Al}$
(3) $r_{Rb} < r_{Cs}$ (4) $r_{At} < r_{Cs}$

34. Which of the following statements are correct? [April 3, 2025 (I)]
A. The process of the addition an electron to a neutral gaseous atom is always exothermic
B. The process of removing an electron from an isolated gaseous atom is always endothermic
C. The 1st ionization energy of the boron is less than that of the beryllium
D. The electronegativity of C is 2.5 in CH_4 and CCl_4
E. Li is the most electropositive among elements of group I
Choose the **correct** answer from the options gives below
(1) B and C only (2) A, C and D only
(3) B and D only (4) B, C and E only

35. The nature of oxide (TeO_2) and hydride (TeH_2) formed by Te, respectively are : [April 2, 2025 (II)]
(1) Oxidising and acidic (2) Reducing and basic
(3) Reducing and acidic (4) Oxidising and basic

36. Electronic configuration of four elements A, B, C and D are given below : [April 2, 2025 (II)]
(A) $1s^22s^22p^3$ (B) $1s^22s^22p^4$
(C) $1s^22s^22p^5$ (D) $1s^22s^22p^2$
Which of the following is the correct order of increasing electronegativity (Pauling's scale) ?
(1) A < D < B < C (2) A < C < B < D
(3) A < B < C < D (4) D < A < B < C

37. The type of oxide formed by the element among Li, Na, Be, Mg, B and Al that has the least atomic radius is : [Jan. 29, 2025 (II)]
(1) A_2O_3 (2) AO_2 (3) AO (4) A_2O

38. Given below are two statements : [Jan. 29, 2025 (I)]
Statement (I) : The radii of isoelectronic species increases in the order. $Mg^{2+} < Na^+ < F^- < O^{2-}$
Statement (II) : The magnitude of electron gain enthalpy of halogen decreases in the order. Cl > F > Br > I
In the light of the above statements, choose the most appropriate answer from the options given below :
(1) Statement I is incorrect but Statement II is correct
(2) Both Statement I and Statement II are incorrect
(3) Statement I is correct but Statement II is incorrect
(4) Both Statement I and Statement II are correct

39. The incorrect decreasing order of atomic radii is : [Jan. 28, 2025 (I)]
(1) $Mg > Al > C > O$ (2) $Al > B > N > F$
(3) $Be > Mg > Al > Si$ (4) $Si > P > Cl > F$

40. The successive 5 ionisation energies of an element are 800, 2427, 3658, 25024 and 32824 kJ/mol, respectively. By using the above values predict the group in which the above element is present : [Jan. 24, 2025 (II)]
(1) Group 2 (b) Group 13
(3) Group 4 (4) Group 14

41. Given below are two statements : [Jan. 22, 2025 (II)]
Statement (I) : An element in the extreme left of the periodic table forms acidic oxides.
Statement (II) : Acid is formed during the reaction between water and oxide of a reactive element present in the extreme right of the periodic table.
In the light of the above statements, choose the correct answer from the options given below :
(1) Statement I is false but Statement II is true.
(2) Both Statement I and Statement II are false.
(3) Statement I is true but Statement II is false.
(4) Both Statement I and Statement II are true.

42. Match List-I with List-II [Jan. 22, 2025 (I)]

	List-I		**List-II**
(A)	$Al^{3+} < Mg^{2+} < Na^+ < F^-$	(I)	Ionisation Enthalpy
(B)	$B < C < O < N$	(II)	Metallic character
(C)	$B < Al < Mg < K$	(III)	Electronegativity
(D)	$Si < P < S < Cl$	(IV)	Ionic radii

Choose the correct answer from the options given below:
(1) A-IV, B-I, C-III, D-II (2) A-II, B-III, C-IV, D-I
(3) A-IV, B-I, C-II, D-III (4) A-III, B-IV, C-II, D-I

43. Which of the following electronegativity order is incorrect? [Jan. 22, 2025 (I)]
(1) $Al < Mg < B < N$ (2) $Al < Si < C < N$
(3) $Mg < Be < B < N$ (4) $S < Cl < O < F$

44. Match **List-I** with **List-II** : [April 8, 2024 (I)]

	List-I (Elements)		**List-II (Properties in their respective groups)**
(A)	Cl,S	(I)	Elements with highest electronegativity
(B)	Ge, As	(II)	Elements with largest atomic size
(C)	Fr, Ra	(III)	Elements which show properties of both metals and non metal
(D)	F, O	(IV)	Elements with highest negative electron gain enthalpy

Choose the **correct** answer from the options given below:
(1) A-(II), B-(III), C-(IV), D-(I)
(2) A-(III), B-(II), C-(I), D-(IV)
(3) A-(IV), B-(III), C-(II), D-(I)
(4) A-(II), B-(I), C-(IV), D-(III)

45. Identify the correct statements about p-block elements and their compounds. [April 8, 2024 (II)]
(A) Non metals have higher electronegativity than metals.
(B) Non metals have lower ionisation enthalpy than metals.
(C) Compounds formed between highly reactive nonmetals and highly reactive metals are generally ionic.
(D) The non-metal oxides are generally basic in nature.
(E) The metal oxides are generally acidic or neutral in nature.
(1) (D) and (E) only (2) (A) and (C) only
(3) (B) and (E) only (4) (B) and (D) only

46. The electron affinity value are negative for :
A. $Be \rightarrow Be^-$ B. $N \rightarrow N^-$
C. $O \rightarrow O^{2-}$ D. $Na \rightarrow Na^-$
E. $Al \rightarrow Al^-$
Choose the most appropriate answer from the options given below : [April 6, 2024 (I)]
(1) D and E only (2) A, B, D and E only
(3) A and D only (4) A, B and C only

47. Match **List-I** with **List-II**. [April 6, 2024 (II)]

	List-I (Alkali Metal)		**List-II (Emission Wavelength in nm)**
(A)	Li	(I)	589.2
(B)	Na	(II)	455.5
(C)	Rb	(III)	670.8
(D)	Cs	(IV)	780.0

Choose the **correct** answer from the options given below:
(1) (A)-(I), (B)-(IV), (C)-(III), (D)-(II)
(2) (A)-(III), (B)-(I), (C)-(IV), (D)-(II)
(3) (A)-(IV), (B)-(II), (C)-(I), (D)-(III)
(4) (A)-(II), (B)-(IV), (C)-(III), (D)-(I)

48. Evaluate the following statements related to group 14 elements for their correctness. [April 6, 2024 (II)]
(A) Covalent radius decreases down the group from C to Pb in a regular manner.
(B) Electronegativity decreases from C to Pb down the group gradually.
(C) Maximum covalence of C is 4 whereas other elements can expand their covalence due to presence of d orbitals.
(D) Heavier elements do not form $p\pi$-$p\pi$ bonds.
(E) Carbon can exhibit negative oxidation states.
Choose the **correct** answer from the options given below:
(1) (C), (D) and (E) Only (2) (A) and (B) Only
(3) (A), (B) and (C) Only (4) (C) and (D) Only

49. The statement(s) that are **correct** about the species O^{2-}, F^-, Na^+ and Mg^{2+}. [April 5, 2024 (I)]
(A) All are isoelectronic
(B) All have the same nuclear charge
(C) O^{2-} has the largest ionic radii
(D) Mg^{2+} has the smallest ionic radii
Choose the **most appropriate** answer from the options given below :
(1) (B), (C) and (D) only (2) (A), (B), (C) and (D)
(3) (C) and (D) only (4) (A), (C) and (D) only

50. Given below are two statements: [April 5, 2024 (II)]
Statement I : The metallic radius of Na is 1.86 Å and the ionic radius of Na^+ is lesser than 1.86 Å.
Statement II : Ions are always smaller in size than the corresponding elements.
In the light of the above statements, choose the **correct** answer from the options given below :
(1) **Statement I** is correct but **Statement II** is false
(2) Both **Statement I** and **Statement II** are true
(3) Both **Statement I** and **Statement II** are false
(4) **Statement I** is incorrect but **Statement II** is true

51. The correct order of the first ionization enthalpy is
(1) Al > Ga > Tl (2) Ga > Al > B [April 4, 2024 (II)]
(3) B > Al > Ga (4) Tl > Ga > Al

52. Given below are two statements: [April 4, 2024 (II)]
Statement I : The correct order of first ionization enthalpy values of Li, Na, F and Cl is Na < Li < Cl < F.
Statement II : The correct order of negative electron gain enthalpy values of Li, Na, F and Cl is Na < Li < F < Cl
In the light of the above statements, choose the **correct** answer from the options given below :
(1) Both Statement I and Statement II are true.
(2) Both Statement I and Statement II are false.
(3) Statement I is false but Statement II is true.
(4) Statement I is true but Statement II is false.

53. The correct order of first ionization enthalpy values of the following elements is : [April 4, 2024 (I)]
(A) O (B) N (C) Be (D) F (E) B
Choose the correct answer from the options given below:
(1) B < D < C < E < A (2) E < C < A < B < D
(3) C < E < A < B < D (4) A < B < D < C < E

54. In case of isoelectronic species the size of F^- Ne and Na^+ is affected by: [Feb. 1, 2024 (I)]
(1) Principal quantum number (n)
(2) None of the factors because their size is the same
(3) Electron - electron interaction in the outer orbitals
(4) Nuclear charge (z)

55. Given below are two statements: [Feb. 1, 2024 (II)]
Statement (I) : Both metals and non - metals exist in *p* and *d* - block elements.
Statement (II) : Non - metals have higher ionisation enthalpy and higher electronegativity than the metals.
In the light of the above statements, choose the most appropriate answer from the options given below:
(1) Both Statement I and Statement II are false.
(2) Both Statement I and Statement II are true.
(3) Statement I is false but Statement II is true.
(4) Statement I is true but Statement II is false.

56. The correct sequence of electron gain enthalpy of the elements listed below is [Jan. 31, 2024 (I)]
A. Ar B. Br C. F D. S
Choose the **most appropriate** from the options given below:
(1) A > D > B > C (2) D > C > B > A
(3) A > D > C > B (4) C > B > D > A

57. Consider the following elements.
Group ↓ A' B' ⟶ Period
C' D'
Which of the following is/are true about A', B', C' and D' ? [Jan. 31, 2024 (II)]
A. Order of atomic radii: B' < A' < D' < C'
B. Order of metallic character : B' < A' < D' < C'
C. Size of the element : D' < C' < B' < A'
D. Order of ionic radii : $B'^{+} < A'^{+} < D'^{+} < C'^{+}$
Choose the correct answer from the options given below:
(1) A only (2) A, B and D only
(3) A and B only (4) B, C and D only

58. Given below are two statements : One is labelled as Assertion **(A)** and the other is labelled as Reason **(R)**. [Jan. 30, 2024 (I)]
Assertion (A) : There is a considerable increase in covalent radius from N to P. However from As to Bi only a small increase in covalent radius is observed.
Reason (R) : Covalent and ionic radii in a particular oxidation state increases down the group.
In the light of the above statements, choose the **most appropriate** answer from the options given below:
(1) **(A)** is false but **(R)** is true.
(2) Both **(A)** and **(R)** are true but **(R)** is not the correct explanation of **(A)**.
(3) **(A)** is true but **(R)** is false.
(4) Both **(A)** and **(R)** are true and **(R)** is the correct explanation of **(A)**.

59. Given below are two statements: [Jan. 30, 2024 (II)]
Statements I: Along the period, the chemical reactivity of the elements gradually increases from group 1 to group 18.
Statement II: The nature of oxides formed by group 1 elements is basic while that of group 17 elements is acidic.
In the light of the above statements, choose the most appropriate from the options given below:
(1) Both Statement I and Statement II are true.
(2) Statement I is true but Statement II is false.
(3) Statement I is false but Statement II is true.
(4) Both Statement I and Statement II are false.

60. Given below are two statements : one is labelled as **Assertion A** and the other is labelled as **Reason R :** [Jan. 29, 2024 (I)]
Assertion A : The first ionisation enthalpy decreases across a period.
Reason R : The increasing nuclear charge outweighs the shielding across the period.
In the light of the above statements, choose the **most appropriate** from the options given below:
(1) **A** is false but **R** is true.
(2) **A** is true but **R** is false.
(3) Both **A** and **R** are true and **R** is the correct explanation of **A**.
(4) Both **A** and **R** are true but **R** is NOT the correct explanation of **A**.

61. Given below are two statements [Jan. 29, 2024 (I)]
Statement I : The electronegativity of group 14 elements from Si to Pb, gradually decreases.
Statement II : Group 14 contains non-metallic, metallic, as well as metalloid elements.
In the light of the above statements, choose the **most appropriate** from the options given below:
(1) Statement I is true but Statement II is false.
(2) Statement I is false but Statement II is true.
(3) Both Statement I and Statement II are false.
(4) Both Statement I and Statement II are true.

62. The element having the highest first ionization enthalpy is [Jan. 29, 2024 (II)]
(1) C (2) Al (3) Si (4) N

63. Given below are two statements :
Statement I : The 4f and 5f - series of elements are placed separately in the Periodic table to reserve the principle of classification. [Jan. 27, 2024 (I)]
Statement II : s-block elements can be found in pure form in nature.
In the light of the above statements, choose the **most appropriate** answer from the options given below:
(1) Statement I is false but Statement II is true.
(2) Both Statement I and Statement II are true.
(3) Statement I is true but Statement II is false.
(4) Both Statement I and Statement II are false.

64. Which one of the following sets of ions represents a collection of isoelectronic species ?
(Given : Atomic number : F : 9, Cl : 17, Na = 11, Mg = 12, Al = 13, K = 19, Ca = 20, Sc = 21) [Feb. 01, 2023 (II)]
(1) $N^{3-}, O^{2-}, F^{-}, S^{2-}$ (2) $Ba^{2+}, Sr^{2+}, K^{+}, Ca^{2+}$
(3) $K^{+}, Cl^{-}, Ca^{2+}, Sc^{3+}$ (4) $Li^{+}, Na^{+}, Mg^{2+}, Ca^{2+}$

65. Which of the following statements are **not** correct?
A. The electron gain enthalpy of F is more negative than that of Cl
B. Ionization enthalpy decreases in a group of periodic table
C. The electronegativity of an atom depends upon the atoms bonded to it.
D. Al_2O_3 and NO are examples of amphoteric oxides.
Choose the most appropriate answer from the options given below: [April 13, 2023 (I)]
(1) A, B, C and D (2) A, C and D Only
(3) B and D Only (4) A, B and D Only

66. Given below are two statements : [April 12, 2023 (I)]
Statement I : $SbCl_5$ is more covalent than $SbCl_3$.
Statement II : The higher oxides of halogens also tend to be more stable than the lower ones.
In the light of the above statements, choose the most appropriate answer from the options given below
(1) Both Statement I and Statement II are correct
(2) Both Statement I and Statement II are incorrect
(3) Statement I is correct but Statement II is incorrect
(4) Statement I is incorrect but Statement II is correct

67. For elements B, C, N, Li, Be, O and F the correct order of first ionization enthalpy is [April 11, 2023 (I)]
(1) $Li < Be < B < C < N < O < F$
(2) $B > Li > Be > C > N > O > F$
(3) $Li < B < Be < C < O < N < F$
(4) $Li < Be < B < C < O < N < F$

68. Given below are two statements: one is labelled as **Assertion A** and the other is labelled as **Reason R.**
Assertion A: The energy required to form Mg^{2+} from Mg is much higher than that required to produce Mg^{+}.
Reason R: Mg^{2+} is small ion and carry more charge than Mg^{+}.
In the light of the above statements, choose the **correct** answer from the options given below: [April 10, 2023 (II)]
(1) Both A and R are true but R is NOT the correct explanation of A.
(2) A is true but R is false.
(3) A is false but R is true.
(4) Both A and R are true and R is the correct explanation of A.

69. The correct order of metallic character is: [April 10, 2023 (II)]
(1) $Be > Ca > K$ (2) $Ca > K > Be$
(3) $K > Ca > Be$ (4) $K > Be > Ca$

70. The correct order of electronegativity for given elements is: [April 08, 2023 (I)]
(1) $C > P > At > Br$ (2) $Br > P > At > C$
(3) $P > Br > C > At$ (4) $Br > C > At > P$

71. The difference between electron gain enthalpies will be maximum between: [April 06, 2023 (I)]
(1) Ne and F (2) Ne and Cl
(3) Ar and Cl (4) Ar and F

72. Group-13 elements react with O_2 in amorphous form to form oxides of type M_2O_3 (M = element). Which among the following is the most basic oxide? [April 06, 2023 (II)]
(1) Al_2O_3 (2) Ga_2O_3 (3) Tl_2O_3 (4) B_2O_3

73. The correct increasing order of the ionic radii is [Jan. 31, 2023 (I)]
(1) $Cl^{-} < Ca^{2+} < K^{+} < S^{2-}$ (2) $K^{+} < S^{2-} < Ca^{2+} < Cl^{-}$
(3) $S^{2-} < Cl^{-} < Ca^{2+} < K^{+}$ (4) $Ca^{2+} < K^{+} < Cl^{-} < S^{2-}$

74. Given below are two statements: one is labelled as Assertion (A) and the other is labelled as Reason (R)
Assertion (A) : The first ionization enthalpy of 3d series elements is more than that of group 2 metals
Reason (R) : In 3d series of elements successive filling of d-orbitals takes place.
In the light of the above statements, choose the correct answer from the options given below : [Jan. 31, 2023 (II)]
(1) Both (A) and (R) are true and (R) is the correct explanation of (A).
(2) Both (A) and (R) are true but (R) is not the correct explanation of (A).
(3) (A) is false but (R) is true.
(4) (A) is true but (R) is false.

75. Which of the following represents the correct order of metallic character of the given elements ? [Jan. 25, 2023 (II)]
(1) $Si < Be < Mg < K$ (2) $Be < Si < Mg < K$
(3) $K < Mg < Be < Si$ (4) $Be < Si < K < Mg$

76. The first ionization enthalpy of Na, Mg and Si, respectively, are: 496, 737 and 786 kJ mol^{-1}. The first ionization enthalpy (kJ mol^{-1}) of Al is: [July 29, 2022 (I)]
(1) 487 (2) 768 (3) 577 (4) 856

77. In which of the following pairs, electron gain enthalpies of constituent elements are nearly the same or identical ?
(A) Rb and Cs (B) Na and K [July 28, 2022 (I)]
(C) Ar and Kr (D) I and At
Choose the correct answer from the options given below :
(1) (A) and (B) only (2) (B) and (C) only
(3) (A) and (C) only (4) (C) and (D) only

78. The correct decreasing order for metallic character is [July 28, 2022 (II)]
(1) $Na > Mg > Be > Si > P$ (2) $P > Si > Be > Mg > Na$
(3) $Si > P > Be > Na > Mg$ (4) $Be > Na > Mg > Si > P$

79. The **incorrect** statement is [July 27, 2022 (I)]
(1) The first ionization enthalpy of K is less than that of Na and Li
(2) Xe does not have the lowest first ionization enthalpy in its group
(3) The first ionization enthalpy of element with atomic number 37 is lower than that of the element with atomic number 38.
(4) The first ionization enthalpy of Ga is higher than that of the d-block element with atomic number 30.

80. Outermost electronic configurations of four elements A, B, C, D are given below: [July 27, 2022 (II)]
(A) $3s^2$ (B) $3s^2 3p^1$ (C) $3s^2 3p^3$ (D) $3s^2 3p^4$
The **correct** order of first ionization enthalpy for them is:
(1) (A) < (B) < (C) < (D) (2) (B) < (A) < (D) < (C)
(3) (B) < (D) < (A) < (C) (4) (B) < (A) < (C) < (D)

81. An element A of group 1 shows similarity to an element B belonging to group 2. If A has maximum hydration enthalpy in group 1 then B is: [July 27, 2022 (II)]
(1) Mg (2) Be (3) Ca (4) Sr

82. Given two statements below: [July 26, 2022 (I)]
Statement I: In Cl_2 molecule the covalent radius is double of the atomic radius of chlorine.
Statement II: Radius of anionic species is always greater than their parent atomic radius.
Choose the **most appropriate** answer from options given below:
(1) Both statement I and statement II are correct.
(2) Both statement I and statement II are incorrect.
(3) Statement I is correct but statement II is incorrect.
(4) Statement I is incorrect but statement II is correct.

83. The first ionization enthalpies of Be, B, N and O follow the order [July 25, 2022 (II)]
(1) $O < N < B < Be$ (2) $Be < B < N < O$
(3) $B < Be < N < O$ (4) $B < Be < O < N$

84. The correct order of electron gain enthalpy (– ve value) is: [June 30, 2022 (I)]
(1) $O > S > Se > Te$ (2) $O < S < Se < Te$
(3) $O < S > Se > Te$ (4) $O < S > Se < Te$

85. Given below are two statements. One is labelled as **Assertion A** and the other is labelled as **Reason R**.
Assertion A : The first ionization enthalpy for oxygen is lower than that of nitrogen.
Reason R : The four electrons in $2p$ orbitals of oxygen experience more electron-electron repulsion.
In the light of the above statements, choose the correct answer from the options given below. [June 29, 2022 (II)]
(1) Both A and R are correct and R is the correct explanation of A.
(2) Both A and R are correct but R is NOT the correct explanation of A.
(3) A is correct but R is not correct.
(4) A is not correct but R is correct.

86. Among the following basic oxide is : [June 28, 2022 (II)]
(1) SO_3 (2) SiO_2 (3) CaO (4) Al_2O_3

87. Match List-I with List-II. [June 28, 2022 (II)]

List-I (Oxide)	List-II (Nature)
(A) Cl_2O_7	(I) Amphoteric
(B) Na_2O	(II) Basic
(C) Al_2O_3	(III) Neutral
(D) N_2O	(IV) Acidic

Choose the correct answer from the options given below:
(1) (A) – (IV), (B) – (III), (C) – (I), (D) – (II)
(2) (A) – (IV), (B) – (II), (C) – (I), (D) – (III)
(3) (A) – (II), (B) – (IV), (C) – (III), (D) – (I)
(4) (A) – (I), (B) – (II), (C) – (III), (D) – (IV)

88. Given below are two statements: one is labelled as **Assertion (A)** and the other is labelled as **Reason(R)**
Assertion (A) : The ionic radii of O^{2-} and Mg^{2+} are same.
Reason (R) : Both O^{2-} and Mg^{2+} are isoelectronic species
In the light of the above statements, choose the correct answer from the options given below : [June 27, 2022 (I)]
(1) Both (A) and (R) are true and (R) is the correct explanation of (A).
(2) Both (A) and (R) are true but (R) is not the correct explanation of (A).
(3) (A) is true but (R) is false.
(4) (A) is false but (R) is true.

89. The correct order of increasing ionic radii is [June 27, 2022 (II)]
(1) $Mg^{2+} < Na^+ < F^- < O^{2-} < N^{3-}$
(2) $N^{3-} < O^{2-} < F^- < Na^+ < Mg^{2+}$
(3) $F^- < Na^+ < O^{2-} < Mg^{2+} < N^{3-}$
(4) $Na^+ < F^- < Mg^{2+} < O^{2-} < N^{3-}$

90. The correct order of melting point is : [June 26, 2022 (I)]
(1) $Be > Mg > Ca > Sr$ (2) $Sr > Ca > Mg > Be$
(3) $Be > Ca > Mg > Sr$ (4) $Be > Ca > Sr > Mg$

91. Which of the following elements is considered as a metalloid? [June 26, 2022 (II)]
(1) Sc (2) Pb (3) Bi (4) Te

92. The pair, in which ions are isoelectronic with Al^{3+} is : [June 25, 2022 (I)]
(1) Br^- and Be^{2+} (2) Cl^- and Li^+
(3) S^{2-} and K^+ (4) O^{2-} and Mg^{2+}

93. The correct order of electron gain enthalpies of Cl, F, Te and Po is [June 25, 2022 (II)]
(1) $F < Cl < Te < Po$ (2) $Po < Te < F < Cl$
(3) $Te < Po < Cl < F$ (4) $Cl < F < Te < Po$

94. Metals generally melt at very high temperature. Amongst the following, the metal with the highest melting point will be [June 24, 2022 (II)]
(1) Hg (2) Ag (3) Ga (4) Cs

95. The ionic radii of K^+, Na^+, Al^{3+} and Mg^{2+} are in the order: [July 25, 2021 (I)]
(1) $Na^+ < K^+ < Mg^{2+} < Al^{3+}$ (2) $Al^{3+} < Mg^{2+} < K^+ < Na^+$
(3) $Al^{3+} < Mg^{2+} < Na^+ < K^+$ (4) $K^+ < Al^{3+} < Mg^{2+} < Na^+$

96. The set in which compounds have different nature is : [July 20, 2021 (I)]
(1) $B(OH)_3$ and H_3PO_3 (2) $B(OH)_3$ and $Al(OH)_3$
(3) NaOH and $Ca(OH)_2$ (4) $Be(OH)_2$ and $Al(OH)_3$

97. The set of elements that differ in mutual relationship from those of the other sets is [March 17, 2021 (II)]
(1) B – Si (2) Li – Na
(3) Be – Al (4) Li – Mg

98. The first ionization energy of magnesium is smaller as compared to that of elements X and Y, but higher than that of Z. The elements X, Y and Z, respectively, are [March 18, 2021 (II)]
(1) Chlorine, lithium and sodium
(2) argon, lithium and sodium
(3) argon, chlorine and sodium
(4) neon, sodium and chlorine

99. Which pair of oxides is acidic in nature? [Feb. 26, 2021 (II)]
(1) CaO, SiO_2 (2) B_2O_3, CaO
(3) B_2O_3, SiO_2 (4) N_2O, BaO

100. The correct order of electron gain enthalpy is: [Feb. 26, 2021 (II)]
(1) $O > S > Se > Te$ (2) $Te > Se > S > O$
(3) $S > O > Se > Te$ (4) $S > Se > Te > O$

101. Consider the elements Mg, Al, S, P and Si, the correct increasing order of their first ionization enthalpy is : [February 24, 2021 (I)]
(1) $Mg < Al < Si < S < P$ (2) $Al < Mg < Si < S < P$
(3) $Mg < Al < Si < P < S$ (4) $Al < Mg < S < Si < P$

102. The set that contains atomic numbers of only transition elements, is : [Sep. 06, 2020 (I)]
(1) 37, 42, 50, 64 (2) 21, 25, 42, 72
(3) 9, 17, 34, 38 (4) 21, 32, 53, 64

103. The correct order of the ionic radii of O^{2-}, N^{3-}, F^-, Mg^{2+}, Na^+ and Al^{3+} is : [Sep. 04, 2020(I) (S)]
(1) $N^{3-} < O^{2-} < F^- < Na^+ < Mg^{2+} < Al^{3+}$
(2) $Al^{3+} < Na^+ < Mg^{2+} < O^{2-} < F^- < N^{3-}$
(3) $Al^{3+} < Mg^{2+} < Na^+ < F^- < O^{2-} < N^{3-}$
(4) $N^{3-} < F^- < O^{2-} < Mg^{2+} < Na^+ < Al^{3+}$

104. The process that is NOT endothermic in nature is : [Sep. 04, 2020 (II)]
(1) $Ar(g)+e^- \to Ar^-(g)$
(2) $H(g)+e^- \to H^-(g)$
(3) $O^-(g)+e^- \to O^{2-}(g)$
(4) $Na(g) \to Na^+(g)+e^-$

105. The ionic radii of O^{2-}, F^-, Na^+ and Mg^{2+} are in the order: [Sep. 04, 2020 (I)]
(1) $F^- > O^{2-} > Na^+ > Mg^{2+}$ (2) $O^{2-} > F^- > Na^+ > Mg^{2+}$
(3) $Mg^{2+} > Na^+ > F^- > O^{2-}$ (4) $O^{2-} > F^- > Mg^{2+} > Na^+$

106. Among the statements (I - IV), the **correct** ones are : [Sep. 03, 2020 (II)]
(I) Be has smaller atomic radius compared to Mg.
(II) Be has higher ionization enthalpy than Al.
(III) Charge/radius ratio of Be is greater than that of Al.
(IV) Both Be and Al form mainly covalent compounds.
(1) (I), (II) and (IV) (2) (I), (III) and (IV)
(3) (II), (III) and (IV) (4) (I), (II) and (III)

107. The five successive ionization enthalpies of an element are 800, 2427, 3658, 25024 and 32824 kJ mol^{-1}. The number of valence electrons in the element is : [Sep. 03, 2020 (II)]
(1) 5 (2) 4 (3) 3 (4) 2

108. In general, the property (magnitudes only) that shows an opposite trend in comparison to other properties across a period is : [Sep. 02, 2020 (I); Same Jan. 09, 2019 (I)]
(1) Ionization enthalpy (2) Electronegativity
(3) Electron gain enthalpy (4) Atomic radius

109. Three elements X, Y and Z are in the 3rd period of the periodic table. The oxides of X, Y and Z, respectively, are basic, amphoteric and acidic. The correct order of the atomic numbers of X, Y and Z is : [Sep. 02, 2020 (II)]
(1) $Z < Y < X$ (2) $X < Y < Z$
(3) $X < Z < Y$ (4) $Y < X < Z$

110. B has a smaller first ionization enthalpy than Be. Consider the following statements:
(I) it is easier to remove 2p electron than 2s electron
(II) 2p electron of B is more shielded from the nucleus by the inner core of electrons than the 2s electrons of Be
(III) 2s electron has more penetration power than 2p electron
(IV) atomic radius of B is more than Be
(atomic number B = 5, Be = 4)
The correct statements are:
[April 12, 2019 (II) (S), Jan. 09, 2020 (I)]
(1) (I), (II) and (IV) (2) (II), (III) and (IV)
(3) (I), (II) and (III) (4) (I), (III) and (IV)

111. The acidic, basic and amphoteric oxides, respectively, are: [Jan. 09, 2020 (I)]
(1) Na_2O, SO_3, Al_2O_3 (2) Cl_2O, CaO, P_4O_{10}
(3) N_2O_3, Li_2O, Al_2O_3 (4) MgO, Cl_2O, Al_2O_3

112. The first and second ionisation enthalpies of a metal are 496 and 4560 kJ mol^{-1}, respectively. How many moles of HCl and H_2SO_4, respectively, will be needed to react completely with 1 mole of the metal hydroxide? [Jan. 09, 2020 (II)]
(1) 1 and 1 (2) 2 and 0.5 (3) 1 and 2 (4) 1 and 0.5

113. In general, the properties that decrease and increase down a group in the periodic table, respectively, are: [Jan. 9, 2019 (I)]
(1) atomic radius and electronegativity.
(2) electron gain enthalpy and electronegativity.
(3) electronegativity and atomic radius.
(4) electronegativity and electron gain enthalpy.

114. The first ionization energy (in kJ/mol) of Na, Mg, Al and Si respectively, are: [Jan. 08, 2020 (I)]
(1) 496, 737, 577, 786 (2) 496, 577, 737, 786
(3) 786, 737, 577, 496 (4) 496, 577, 786, 737

115. The increasing order of the atomic radii of the following elements is: [Jan. 08, 2020 (II)]
(i) C (ii) O (iii) F (iv) Cl (v) Br
(1) (ii) < (iii) < (iv) < (i) < (v)
(2) (iv) < (iii) < (ii) < (i) < (v)
(3) (iii) < (ii) < (i) < (iv) < (v)
(4) (i) < (ii) < (iii) < (iv) < (v)

116. The electron gain enthalpy (in kJ/mol) of fluorine, chlorine, bromine and iodine, respectively, are:
(1) –296, –325, –333 and –349 [Jan. 07, 2020 (I); Same 2011]
(2) –349, –333, –325 and –296
(3) –333, –349, –325 and –296
(4) –333, –325, –349 and –296

117. Within each pair of elements F & Cl, S & Se, and Li & Na, respectively, the elements that release more energy upon an electron gain are: [Jan. 07, 2020 (II)]
(1) Cl, Se and Na (2) Cl, S and Li
(3) F, S and Li (4) F, Se and Na

118. In comparison to boron, berylium has: [April 12, 2019 (II)]
(1) lesser nuclear charge and lesser first ionisation enthalpy.
(2) greater nuclear charge and lesser first ionisation enthalpy.
(3) greater nulear charge and greater first ionisation enthalpy.
(4) lesser nuclear charge and greater first ionisation enthalpy.

119. The isoelectronic set of ions is
[2003 (S), 2004 (S), 2005 (S), 2006(S), 2008 (S), April 10, 2019 (I)]
(1) N^{3-}, O^{2-}, F^- and Na^+ (2) N^{3-}, Li^+, Mg^{2+} and O^{2-}
(3) F^-, Li^+, Na^+ and Mg^{2+} (4) Li^+, Na^+, O^{2-} and F^-

120. The element having greatest difference between its first and second ionization energies, is: [April 9, 2019 (I)]
(1) Ca (2) Sc (3) Ba (4) K

121. The correct order of the atomic radii of C, Cs, Al, and S is : [Jan. 11, 2019 (I)]
(1) $C < S < Al < Cs$ (2) $S < C < Cs < Al$
(3) $S < C < Al < Cs$ (4) $C < S < Cs < Al$

122. The correct option with respect to the Pauling electronegativity values of the elements is: [Jan. 11, 2019 (II)]
(1) $Te > Se$ (2) $Ga < Ge$ (3) $Si < Al$ (4) $P > S$

123. When the first electron gain enthalpy($\Delta_{eg}H$) of oxygen is – 141 kJ/mol, its second electron gain enthalpy is:
(1) a more negative value than the first [Jan. 09, 2019 (II)]
(2) almost the same as that of the first
(3) negative, but less negative than the first
(4) a positive value

124. For Na^+, Mg^{2+}, F^- and O^{2-}; the correct order of increasing Ionic radii is : [Online April 15, 2018 (I)]
(1) $O^{2-} < F^- < Na^+ < Mg^{2+}$ (2) $Na^+ < Mg^{2+} < F^- < O^{2-}$
(3) $Mg^{2+} < Na^+ < F^- < O^{2-}$ (4) $Mg^{2+} < O^{2-} < Na^+ < F^-$

125. The correct order of electron affinity is: [Online April 15, 2018 (II)]
(1) $O > F > Cl$ (2) $F > O > Cl$
(3) $F > Cl > O$ (4) $Cl > F > O$

126. Both lithium and magnesium display several similar properties due to the diagonal relationship; however, the one which is incorrect is :
(1) Both form basic carbonates [2017]
(2) Both form soluble bicarbonates
(3) Both form nitrides
(4) Nitrates of both Li and Mg yield NO_2 and O_2 on heating

127. The group having isoelectronic species is : [2017]
(1) O^{2-}, F^-, Na^+, Mg^{2+} (2) O^-, F^-, Na, Mg^+
(3) O^{2-}, F^-, Na, Mg^{2+} (4) O^-, F^-, Na^+, Mg^{2+}

128. The electronic configuration with the highest ionization enthalpy is : [Online April 9, 2017]
(1) [Ne] $3s^2 3p^1$ (2) [Ne] $3s^2 3p^2$
(3) [Ne] $3s^2 3p^3$ (4) [Ar] $3d^{10} 4s^2 4p^3$

129. Consider the following ionization enthalpies of two elements 'A' and 'B'. [Online April 8, 2017]

Element	Ionization	enthalpy	(kJ/mol)
	1st	2nd	3rd
A	899	1757	14847
B	737	1450	7731

Which of the following statements is correct ?
(1) Both 'A' and 'B' belong to group–1 where 'B' comes below 'A'.
(2) Both 'A' and 'B' belong to group–1 where 'A' comes below 'B'.
(3) Both 'A' and 'B' belong to group–2 where 'B' comes below 'A'.
(4) Both 'A' and 'B' belong to group–2 where 'A' comes below 'B'.

130. Which of the following atoms has the highest first ionization energy? [2016]
(1) K (2) Sc (3) Rb (4) Na

131. The following statements concern elements in the periodic table. Which of the following is true ? [Online April 10, 2016]
(1) For Group 15 elements, the stability of +5 oxidation state increases down the group
(2) Elements of Group 16 have lower ionization enthalpy values compared to those of Group 15 in the corresponding periods.
(3) The Group 13 elements are all metals
(4) All the elements in Group 17 are gases.

132. The ionic radii (in Å) of N^{3-}, O^{2-} and F^- are respectively :
(1) 1.71, 1.40 and 1.36 (2) 1.71, 1.36 and 1.40 [2015]
(3) 1.36, 1.40 and 1.71 (4) 1.36, 1.71 and 1.40

133. Which of the following arrangements represents the increasing order (smallest to largest) of ionic radii of the given species O^{2-}, S^{2-}, N^{3-}, P^{3-}? [Online April 12, 2014(s), 2020]
(1) $O^{2-} < N^{3-} < S^{2-} < P^{3-}$ (2) $O^{2-} < P^{3-} < N^{3-} < S^{2-}$
(3) $N^{3-} < O^{2-} < P^{3-} < S^{2-}$ (4) $N^{3-} < S^{2-} < O^{2-} < P^{3-}$

134. Which of the following series correctly represents relations between the elements from X to Y?
$X \rightarrow Y$ [Online April 11, 2014]
(1) ${}_3Li \rightarrow {}_{19}K$ Ionization enthalpy increases
(2) ${}_9F \rightarrow {}_{35}Br$ Electron gain enthalpy (negative sign) increases
(3) ${}_6C \rightarrow {}_{32}Ge$ Atomic radii increases
(4) ${}_{18}Ar \rightarrow {}_{54}Xe$ Noble character increases

135. Which one of the following has largest ionic radius?
[2004 (S), Online April 19, 2014]
(1) Li^+ (2) O_2^{2-} (3) B^{3+} (4) F^-

136. Which of the following represents the correct order of increasing first ionization enthalpy for Ca, Ba, S, Se and Ar ?
(1) Ca < S < Ba < Se < Ar (2) S < Se < Ca < Ba < Ar [2013]
(3) Ba < Ca < Se < S < Ar (4) Ca < Ba < S < Se < Ar

137. The order of increasing sizes of atomic radii among the elements O, S, Se and As is : [Online April 22, 2013]
(1) As < S < O < Se (2) Se < S < As < O
(3) O < S < As < Se (4) O < S < Se < As

138. Which is the correct order of second ionization potential of C, N, O and F in the following ?
[2006 (S), Online May 26, 2012; April 23, 2013]
(1) O > N > F > C (2) O > F > N > C
(3) F > O > N > C (4) C > N > O > F

139. Which among the following elements has the highest first ionization enthalpy? [Online May 12, 2012]
(1) Nitrogen (2) Boron
(3) Carbon (4) Oxygen

140. The increasing order of the ionic radii of the given isoelectronic species is : [2012]
(1) Cl^-, Ca^{2+}, K^+, S^{2-} (2) S^{2-}, Cl^-, Ca^{2+}, K^+
(3) Ca^{2+}, K^+, Cl^-, S^{2-} (4) K^+, S^{2-}, Ca^{2+}, Cl^-

141. The electron affinity of chlorine is 3.7 eV. 1 gram of chlorine is completely converted to Cl^- ion in a gaseous state. (1 eV = 23.06 kcal mol^{-1}).
Energy released in the process is [Online May 7, 2012]
(1) 4.8 kcal (2) 7.2 kcal (3) 8.2 kcal (4) 2.4 kcal

142. In which of the following arrangements, the sequence is not strictly according to the property written against it?
[2005 (S), 2008, Online May 7, 2012]
(1) $CO_2 < SiO_2 < SnO_2 < PbO_2$: increasing oxidising power
(2) $NH_3 < PH_3 < AsH_3 < SbH_3$: increasing basic strength
(3) HF < HCl < HBr < HI: increasing acid strength
(4) B < C < O < N: increasing first ionisation enthalpy.

143. The correct order of electron gain enthalpy with negative sign of F, Cl, Br and I, having atomic numbers 9, 17, 35 and 53 respectively, is : [2011RS]
(1) F > Cl > Br > I (2) Cl > F > Br > I
(3) Br > Cl > I > F (4) I > Br > Cl > F

144. The correct sequence which shows decreasing order of the ionic radii of the elements is [2010]
(1) $Al^{3+} > Mg^{2+} > Na^+ > F^- > O^{2-}$
(2) $Na^+ > Mg^{2+} > Al^{3+} > O^{2-} > F^-$
(3) $Na^+ > F^- > Mg^{2+} > O^{2-} > Al^{3+}$
(4) $O^{2-} > F^- > Na^+ > Mg^{2+} > Al^{3+}$

145. The increasing order of the first ionization enthalpies of the elements B, P, S and F (Lowest first) is [2006]
(1) B < P < S < F (2) B < S < P < F
(3) F < S < P < B (4) P < S < B < F

146. Following statements regarding the periodic trends of chemical reactivity of alkali metals and halogens are given. Which of these statements gives the correct picture? [2006]
(1) Chemical reactivity increases with increase in atomic number down the group in both the alkali metals and halogens
(2) In alkali metals, the reactivity increases but in the halogens, it decreases with increase in atomic number down the group
(3) The reactivity decreases in the alkali metals but increases in the halogens with increase in atomic number down the group
(4) In both, alkali metals and halogens, chemical reactivity decreases with increase in atomic number down the group

(4) In both, alkali metals and halogens, chemical reactivity decreases with increase in atomic number down the group

147. In which of the following arrangements, the order is NOT according to the property indicated against it?

(1) $Li < Na < K < Rb$: Increasing metallic radius [2005]

(2) $I < Br < F < Cl$: Increasing electron gain enthalpy (with negative sign)

(3) $B < C < N < O$ Increasing first ionization enthalpy

(4) $Al^{3+} < Mg^{2+} < Na^{+} < F^{-}$ Increasing ionic size

148. Which of the following oxides is amphoteric in character? [2003 (S), 2005]

(1) SnO_2 (2) SiO_2 (3) CO_2 (4) CaO

149. Among Al_2O_3, SiO_2, P_2O_3 and SO_2, the correct order of acid strength is [2004]

(1) $Al_2O_3 < SiO_2 < SO_2 < P_2O_3$

(2) $SiO_2 < SO_2 < Al_2O_3 < P_2O_3$

(3) $SO_2 < P_2O_3 < SiO_2 < Al_2O_3$

(4) $Al_2O_3 < SiO_2 < P_2O_3 < SO_2$

150. The formation of the oxide ion $O^{2-}(g)$ requires first an exothermic and then an endothermic step as shown below

$$O(g) + e^{-} = O^{-}(g)\ \Delta H^\circ = -142\ \text{kJmol}^{-1}$$

$$O^{-}(g) + e^{-} = O^{2-}(g)\ \Delta H^\circ = +844\ \text{kJmol}^{-1}$$

This is because [2004]

(1) O^- ion will tend to resist the addition of another electron

(2) Oxygen has high electron affinity

(3) Oxygen is more elecronegative

(4) O^- ion has comparatively larger size than oxygen atom

151. Which one of the following ions has the highest value of ionic radius ? [2004]

(1) O^{2-} (2) B^{3+} (3) Li^+ (4) F^-

152. Which one of the following is an amphoteric oxide ? [2003]

(1) Na_2O (2) SO_2 (3) B_2O_3 (4) ZnO

Numerical Value Questions (NVQs)

153. The lowest oxidation number of an atom in a compound A_2B is –2. The number of electrons in its valence shell is _______. [Feb. 1, 2024 (I)]

154. The total number of isoelectronic species from the given set is _______. [April 15, 2023 (I)]

O^{2-}, F^-, Al, Mg^{2+}, Na^+, O^+, Mg, Al^{3+}, F

155. Among the following oxides of p - block elements, number of oxides having amphoteric nature is _________.

Cl_2O_7, CO, PbO_2, N_2O, NO, Al_2O_3, SiO_2, N_2O_5, SnO_2 [Feb. 1, 2024 (I)]

156. In the following reactions, the total number of oxygen atoms in X and Y is _______. [April 10, 2023 (I)]

$Na_2O + H_2O \rightarrow 2X$ $\quad$ $Cl_2O_7 + H_2O \rightarrow 2Y$

157. Total number of acidic oxides among

N_2O_3, NO_2, N_2O, Cl_2O_7, SO_2, CO, CaO, Na_2O and NO is _________ . [Jan. 29, 2023 (II)]

Part-B (JEE Advanced)

Topic 1 - Modern Periodic Table

1 MCQs with One Correct Answer

1. Which of the following has the maximum number of unpaired electrons? [1996 - 1 Mark]

(a) Mg^{2+} (b) Ti^{3+} (c) V^{3+} (d) Fe^{2+}

2. The statement that is not correct for the periodic classification of element is [1992 - 1 Mark]

(a) The properties of elements are the periodic functions of their atomic numbers

(b) Non-metallic elements are lesser in number than metallic elements

(c) The first ionisation energies of elements along a period do not vary in a regular manner with increase in atomic number

(d) For transition elements the *d*-subshells are filled with electrons monotonically with increase in atomic number.

6 MCQs with One or More than One Correct Answer

3. The statements that are true for the long form of the periodic table are : [1988 - 1 Mark]

(a) it reflects the sequence of filling the electrons in the order of sub-energy level s, p, d and f.

(b) it helps to predict the stable valency states of the elements

(c) it reflects trends in physical and chemical properties of the elements

(d) it helps to predict the relative ionicity of the bond between any two elements.

Topic 2 - Periodic Trends in Properties of Elements

1 MCQs with One Correct Answer

4. Identify the correct order of acidic strengths of CO_2, CuO, CaO, H_2O [2002S]

(a) $CaO < CuO < H_2O < CO_2$ (b) $H_2O < CuO < CaO < CO_2$

(c) $CaO < H_2O < CuO < CO_2$ (d) $H_2O < CO_2 < CaO < CuO$

5. The correct order of radii is [2000S]

(a) $N < Be < B$ (b) $F^- < O^{2-} < N^{3-}$

(c) $Na < Li < K$ (d) $Fe^{3+} < Fe^{2+} < Fe^{4+}$

6. The correct order of acidic strength is [2000S]

(a) $Cl_2O_7 > SO_2 > P_4O_{10}$ (b) $CO_2 > N_2O_5 > SO_3$

(c) $Na_2O > MgO > Al_2O_3$ (d) $K_2O > CaO > MgO$

7. Amongst H_2O, H_2S, H_2Se and H_2Te, the one with the highest boiling point is [2000S]

(a) H_2O because of hydrogen bonding

(b) H_2Te because of higher molecular weight

(c) H_2S because of hydrogen bonding

(d) H_2Se because of lower molecular weight

8. Which has most stable +2 oxidation state : [1995S]

(a) Sn (b) Pb

(c) Fe (d) Ag

9. Amongst the following elements (whose electronic configurations are given below), the one having the highest ionization energy is : [1990 - 1 Mark]
(a) [Ne] $3s^2 3p^1$ (b) [Ne] $3s^2 3p^3$
(c) [Ne] $3s^2 3p^2$ (d) [Ne] $3d^{10} 4s^2 4p^3$

10. Which one of the following is the strongest base? [1989 - 1 Mark]
(a) AsH_3 (b) NH_3
(c) PH_3 (d) SbH_3

11. Which one of the following is the smallest in size? [1989 - 1 Mark]
(a) N^{3-} (b) O^{2-}
(c) F^- (d) Na^+

12. The first ionisation potential of Na, Mg, Al and Si are in the order [1988 - 1 Mark]
(a) Na < Mg > Al < Si (b) Na > Mg > Al > Si
(c) Na < Mg < Al < Si (d) Na > Mg > Al < Si

13. The electronegativity of the following elements increases in the order [1987 - 1 Mark]
(a) C, N, Si, P (b) N, Si, C, P
(c) Si, P, C, N (d) P, Si, N, C

14. The first ionisation potential in electron volts of nitrogen and oxygen atoms are respectively given by [1987 - 1 Mark]
(a) 14.6, 13.6 (b) 13.6, 14.6
(c) 13.6, 13.6 (d) 14.6, 14.6

15. Atomic radii of fluorine and neon in Ångstorm units are respectively given by [1987 - 1 Mark]
(a) 0.72, 1.60 (b) 1.60, 1.60
(c) 0.72, 0.72 (d) None of these values

16. The element with the highest first ionization potential is [1982 - 1 Mark]
(a) boron (b) carbon
(c) nitrogen (d) oxygen

17. The correct order of second ionisation potential of carbon, nitrogen, oxygen and fluorine is [1981 - 1 Mark]
(a) C > N > O > F (b) O > N > F > C
(c) O > F > N > C (d) F > O > N > C

2 Integer Value Answer

18. The 1st, 2nd, and the 3rd ionization enthalpies, I_1, I_2, and I_3, of four atoms with atomic numbers n, $n+1$, $n+2$, and $n+3$, where $n < 10$, are tabulated below. What is the value of n? [Adv. 2020]

Atomic number	Ionization Enthalpy (kJ/mol)		
	I_1	I_2	I_3
n	1681	3374	6050
$n+1$	2081	3952	6122
$n+2$	496	4562	6910
$n+3$	738	1451	7733

19. Among the following, the number of elements showing only one non-zero oxidation state is :
O, Cl, F, N, P, Sn, Tl, Na, Ti [2010]

4 Fill in the Blanks

20. On Mulliken scale, the average of ionization potential and electron affinity is known as .... [1985 - 1 Mark]

21. The energy released when an electron is added to a neutral gaseous atom is called .... of the atom. [1982 - 1 Mark]

5 True / False

22. The basic nature of the hydroxides of group 13 (Gr. III B) decreases progressively down the group. [1993 - 1 Mark]

23. The decreasing order of electron affinity of F, Cl, Br is F > Cl > Br. [1993 - 1 Mark]

24. In group IA, of alkali metals, the ionisation potential decreases on moving down the group. Therefore, lithium is a poor reducing agent. [1987 - 1 Mark]

6 MCQs with One or More than One Correct Answer

25. The correct statement(s) regarding the periodic properties of elements is(are) [Adv. 2026]
(a) Second ionization enthalpy of carbon atom is less than that of boron atom.
(b) Increasing order of ionic radii: $Al^{3+} < Mg^{2+} < Na^+$
(c) Under identical conditions, in solid state, the density of potassium metal is more than density of sodium metal.
(d) The H–H bond is weaker than F–F bond.

26. The option(s) with only amphoteric oxides is(are) [Adv. 2017]
(a) Cr_2O_3, BeO, SnO, SnO_2 (b) Cr_2O_3, CrO, SnO, PbO
(c) NO, B_2O_3, PbO, SnO_2 (d) ZnO, Al_2O_3, PbO, PbO_2

27. Ionic radii of [1999 - 3 Marks]
(a) $Ti^{4+} < Mn^{7+}$ (b) $^{35}Cl^- < {}^{37}Cl^-$
(c) $K^+ > Cl^-$ (d) $P^{3+} > P^{5+}$

28. Sodium sulphate is soluble in water whereas barium sulphate is sparingly soluble because : [1989 - 1 Mark]
(a) the hydration energy of sodium sulphate is more than its lattice energy
(b) the lattice energy of barium sulphate is more than its hydration energy
(c) the lattice energy has no role to play in solubility
(d) the hydration energy of sodium sulphate is less than its lattice energy.

9 Assertion and Reason/Statement Type Questions

Each question contains STATEMENT-1 (Assertion) and STATEMENT-2 (Reason). Each question has 4 choices (a), (b), (c) and (d) out of which ONLY ONE is correct. Mark your answer as
(a) If both Statement -1 and Statement -2 are correct, and Statement -2 is the correct explanation of the Statement -2.
(b) If both Statement -1 and Statement -2 are correct, but Statement - 2 is not the correct explanation of the Statement -1.
(c) If Statement -1 is correct but Statement -2 is incorrect.
(d) If Statement -1 is incorrect but Statement -2 is correct.

29. Read the following statement and explanation and answer as per the options given below :
Assertion : The first ionization energy of Be is greater than that of B. [2000S]
Reason : $2p$ orbital is lower in energy than $2s$.

10 Subjective Problems

30. Arrange the following in :
(*i*) Arrange the following ions in order of their increasing radii : Li^+, Mg^{2+}, K^+, Al^{3+}. [1997 - 1 Mark]
(*ii*) Increasing order of basic character : MgO, SrO, K_2O, NiO, Cs_2O [1991 - 1 Mark]

(*iii*) Increasing order of ionic size : N^{3-}, Na^+, F^-, O^{2-}, Mg^{2+} [1991 - 1 Mark]

(*iv*) Increasing size : Cl^-, S^{2-}, Ca^{2+}, Ar [1986 - 1 Mark]

(*v*) Increasing first ionization potential : Mg, Al, Si, Na [1985 - 1 Mark]

(*vi*) Increasing acidic property : ZnO, Na_2O_2, P_2O_5, MgO [1985 - 1 Mark]

(*vii*) Decreasing ionic size : Mg^{2+}, O^{2-}, Na^+, F^- [1985 - 1 Mark]

31. The first ionization energy of carbon atom is greater than that of boron atom whereas, the reverse is true for the second ionization energy. [1989 - 2 Marks]

Answer Key

Part-A (JEE Main)

1	(1)	17	(1)	33	(2)	49	(4)	65	(2)	81	(1)	97	(2)	113	(3)	129	(3)	145	(2)
2	(2)	18	(3)	34	(1)	50	(1)	66	(1)	82	(4)	98	(3)	114	(1)	130	(2)	146	(2)
3	(4)	19	(3)	35	(1)	51	(4)	67	(3)	83	(4)	99	(3)	115	(3)	131	(2)	147	(3)
4	(1)	20	(4)	36	(4)	52	(1)	68	(4)	84	(3)	100	(4)	116	(3)	132	(1)	148	(1)
5	(3)	21	(3)	37	(1)	53	(2)	69	(3)	85	(1)	101	(2)	117	(2)	133	(1)	149	(4)
6	(1)	22	(3)	38	(4)	54	(4)	70	(4)	86	(3)	102	(2)	118	(4)	134	(3)	150	(1)
7	(2)	23	(3)	39	(3)	55	(3)	71	(2)	87	(2)	103	(3)	119	(1)	135	(2)	151	(1)
8	(2)	24	(4)	40	(2)	56	(1)	72	(3)	88	(4)	104	(2)	120	(4)	136	(3)	152	(4)
9	(3)	25	(1)	41	(1)	57	(2)	73	(4)	89	(1)	105	(2)	121	(1)	137	(4)	153	(6)
10	(4)	26	(1)	42	(3)	58	(2)	74	(2)	90	(4)	106	(1)	122	(2)	138	(2)	154	(5)
11	(2)	27	(2)	43	(1)	59	(3)	75	(1)	91	(4)	107	(3)	123	(4)	139	(1)	155	(3)
12	(4)	28	(2)	44	(3)	60	(1)	76	(3)	92	(4)	108	(4)	124	(3)	140	(3)	156	(5)
13	(2)	29	(3)	45	(2)	61	(2)	77	(3)	93	(2)	109	(2)	125	(4)	141	(4)	157	(4)
14	(11)	30	(2)	46	(4)	62	(4)	78	(1)	94	(2)	110	(3)	126	(N)	142	(2)		
15	(1)	31	(1)	47	(2)	63	(3)	79	(4)	95	(3)	111	(3)	127	(1)	143	(2)		
16	(3)	32	(3)	48	(1)	64	(3)	80	(2)	96	(2)	112	(4)	128	(3)	144	(4)		

Part-B (JEE Advanced)

1	(d)	4	(a)	7	(a)	10	(b)	13	(c)	16	(c)	19	(2)	24	False	27	(d)		
2	(d)	5	(b)	8	(b)	11	(d)	14	(a)	17	(c)	22	False	25	(a, b)	28	(a, b)		
3	(a, c, d)	6	(a)	9	(b)	12	(a)	15	(a)	18	(9)	23	False	26	(a, d)	29	(c)		

Chapter-4

Chemical Bonding and Molecular Structure

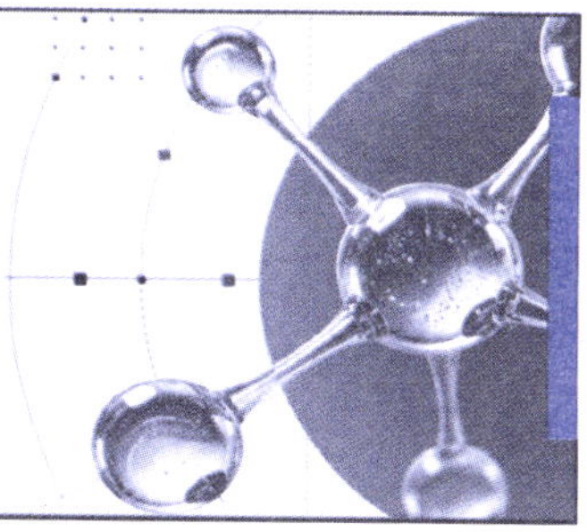

Part-A (JEE Main)

Topic 1- Electrovalent and Covalent Bonding

MULTIPLE CHOICE QUESTIONS (MCQs)

1. The covalent radii of atoms A and B are r_A and r_B, respectively. The covalent bond length and total length of AB molecule are respectively: [April 05, 2026 (II)]

(1) $(r_A+r_B), 2(r_A+r_B)$ (2) $\frac{1}{2}(r_A+r_B),(r_A+r_B)$

(3) $(r_A+r_B),(r_A+r_B)$ (4) $2(r_A+r_B),\frac{1}{2}(r_A+r_B)$

2. The correct increasing order of C – H(A), C–O(B), C=O(C) and C≡N(D) bonds in terms of covalent bond length is: [Jan. 21, 2026 (II)]

(1) A < B < C < D (2) A < D < C < B
(3) D < C < B < A (4) D < C < A < B

3. Which of the following compounds contain(s) no covalent bond(s)? KCl, PH_3, O_2, B_2H_6, H_2SO_4 [2018]

(1) KCl, B_2H_6, PH_3 (2) KCl, H_2SO_4
(3) KCl (4) KCl, B_2H_6

4. Which of these statements is not true? [Online April 19, 2014]

(1) NO^+ is not isoelectronic with O_2
(2) B is always covalent in its compounds
(3) In aqueous solution, the Tl^+ ion is much more stable than Tl (III)
(4) $LiAlH_4$ is a versatile reducing agent in organic synthesis.

5. Lattice energy of an ionic compound depends upon [2005]

(1) Charge on the ion and size of the ion
(2) Packing of ions only
(3) Size of the ion only
(4) Charge on the ion only

Topic 2 - Octet Rule and Resonance Structures

MULTIPLE CHOICE QUESTIONS (MCQs)

6. The pairs among A = $[SO_3^{2-}, CO_3^{2-}]$, B = $[O_2^{2-}, F_2]$, C = $[CN^-, CO]$, D = $[NH_3, H_3O^+]$ and E = $[MnO_4^{2-}, CrO_4^{2-}]$ that do not have similar Lewis dot structure are : [April 06, 2026 (I)]

(1) A, B and E (2) A and E
(3) B, C and D (4) C and D

7. Identify the molecule (X) with maximum number of lone pairs of electrons (obtained using Lewis dot structure) among HNO_3, H_2SO_4, NF_3 and O_3.
Choose the **correct** bond angle made by the central atom of the molecule (X). [Jan. 23, 2026 (I)]

(1) 120° (2) 107° (3) 102° (4) 116°

8. The formal charges on the atoms marked as (1) to (4) in the Lewis representation of HNO_3 molecule respectively are [Jan. 22, 2026 (I)]

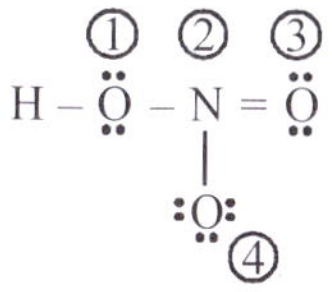

(1) +1, 0, 0, –1 (2) 0, –1, 0, +1
(3) 0, +1, 0, –1 (4) 0, 0, –1, +1

9. From the following, the least stable structure is: [Jan. 21, 2026 (I)]

(1) 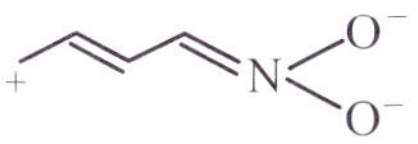 (2) 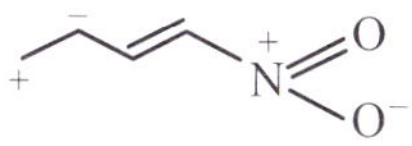

(3) 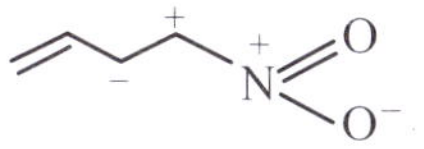 (4) 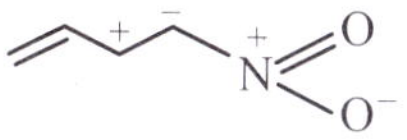

10. Given below are two statements : [Jan. 24, 2025 (II)]

Statement I: Experimentally determined oxygen-oxygen bond lengths in the O_3 are found to be same and the bond length is greater than that of a O = O (double bond) but less than that of a single (O – O) bond.

Statement II: The strong lone pair-lone pair repulsion between oxygen atoms is solely responsible for the fact that the bond length in ozone is smaller than that of a double bond (O=O) but more than that of a single bond (O – O).

In the light of the above statements, choose the correct answer from the options given below:

(1) Statement I is true but Statement II is false
(2) Both Statement I and Statement II are true
(3) Both Statement I and Statement II are false
(4) Statement I is false but Statement II is true

11. Match the List-I with List-II [Jan. 23, 2025 (I)]

List-I (Classification of molecules based on octet rule)		List-II (Example)	
A.	Molecules obeying octet rule	I.	NO, NO_2
B.	Molecules with incomplete octet	II.	$BCl_3, AlCl_3$
C.	Molecules with incomplete octet with odd electron	III.	H_2SO_4, PCl_5
D.	Molecules with expanded octet	IV.	CCl_4, CO_2

Choose the correct answer from the options given below:
(1) A-IV, B-II, C-I, D-III (2) A-III, B-II, C-I, D-IV
(3) A-IV, B-I, C-III, D-II (4) A-II, B-IV, C-III, D-I

12. The difference in energy between the actual structure and the lowest energy resonance structure for the given compound is [Jan. 29, 2024 (I)]
(1) electromeric energy (2) ionization energy
(3) hyperconjugation energy (4) resonance energy

13. Resonance in carbonate ion (CO_3^{2-}) is

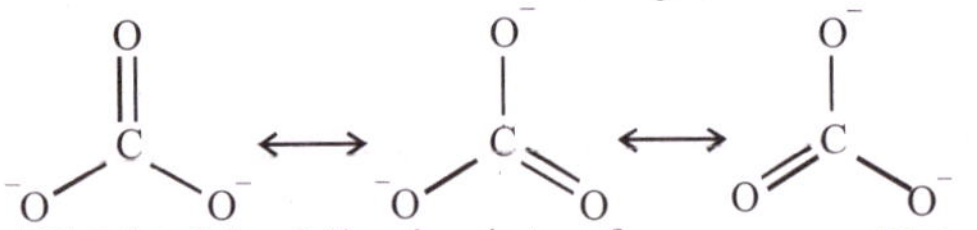

Which of the following is true? [Feb. 01, 2023 (I)]
(1) It is possible to identify each structure individually by some physical or chemical method.
(2) All these structures are in dynamic equilibrium with each other.
(3) CO_3^{2-} has a single structure i.e., resonance hybrid of the above three structures.
(4) Each structure exists for equal amount of time.

14. Increasing order of stability of the resonance structure is:

A. 

B. 

C. 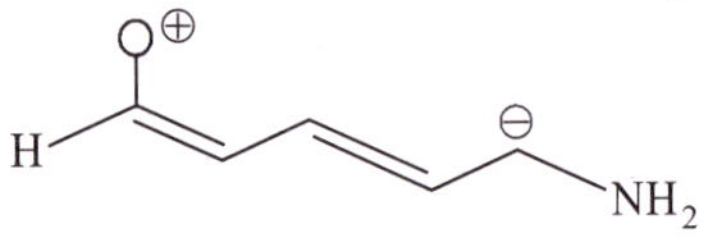

D. 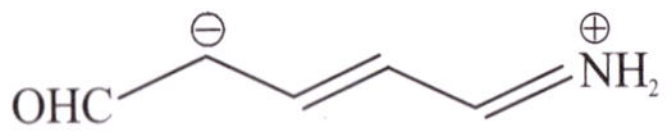

[Jan. 24, 2023 (I)]
(1) C, D, B, A (2) C, D, A, B
(3) D, C, A, B (4) D, C, B, A

15. Which of the following pair of molecules contain odd electron molecule and an expanded octet molecule? [July 29, 2022 (I)]
(1) BCl_3 and SF_6 (2) NO and H_2SO_4
(3) SF_6 and H_2SO_4 (4) BCl_3 and NO

16. Number of electron deficient molecules among the following: PH_3, B_2H_6, CCl_4, NH_3, LiH and BCl_3 is [June 25, 2022 (I)]
(1) 0 (2) 1 (3) 2 (4) 3

17. Identify the species having one π-bond and maximum number of canonical forms from the following: [July 25, 2021 (II)]
(1) SO_3 (2) O_2 (3) SO_2 (4) CO_3^{2-}

NUMERICAL VALUE QUESTIONS (NVQs)

18. Resonance in X_2Y can be represented as

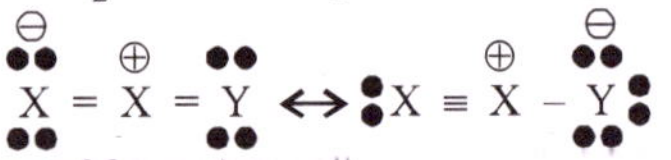

The enthalpy of formation of

$$X_2Y\left(X \equiv X(g) + \frac{1}{2}Y = Y(g) \rightarrow X_2Y(g)\right) \text{ is } 80\ kJ\ mol^{-1}.$$

The magnitude of resonance energy of X_2Y is ________ kJ mol^{-1} (nearest integer value)
Given : Bond energies of $X \equiv X, X=X, Y=Y$ and $X=Y$ are 940, 410, 500 and 602 kJ mol^{-1} respectively.
valence X : 3, Y : 2 [April 8, 2025 (II)]

19. Total number of non-bonded electrons present in NO_2^- ion based on Lewis theory is _____ . [Jan. 29, 2025 (II)]

20. Number of molecules from the following which are exceptions to octet rule is _____.
CO_2, NO_2, H_2SO_4, BF_3, CH_4, SiF_4, ClO_2, PCl_5, BeF_2, C_2H_6, $CHCl_3$, CBr_4. [April 8, 2024 (I)]

21. Consider the following reactions

$NiS + HNO_3 + HCl \rightarrow A + NO + S + H_2O$

$$A + NH_4OH + \begin{array}{c} H_3C - C = N - OH \\ | \\ H_3C - C = N - OH \end{array} \rightarrow B + NH_4Cl + H_2O$$

The number of protons that do not involve in hydrogen bonding in the product B is______. [April 6, 2024 (II)]

22. In the lewis dot structure for NO_2^-, total number of valence electrons around nitrogen is .... [April 5, 2024 (I)]

23. The number of molecules or ions from the following, which do not have odd number of electrons are ________ . [Jan. 29, 2023 (I)]
(A) NO_2 (B) ICl_4^- (C) BrF_3 (D) ClO_2
(E) NO_2^+ (F) NO

24. The total number of lone pairs of electrons on oxygen atoms of ozone is_____ [Jan. 25, 2023 (I)]

Topic 3 - Dipole Moment and Fajan's Rule

MULTIPLE CHOICE QUESTIONS (MCQs)

25. Consider three metal chlorides x, y and z, where x is water soluble at room temperature, y is sparingly soluble in water at room temperature and z is soluble in hot water. x, y and z respectively. [Jan. 24, 2026 (I)]
(1) $MgCl_2$, AgCl and $AlCl_3$
(2) AgCl, Hg_2Cl_2 and $PbCl_2$
(3) $AlCl_3$, $PbCl_2$ and $BaCl_2$
(4) $CuCl_2$, AgCl and $PbCl_2$

26. Given below are statements about some molecules/ions. Identify the CORRECT statements. [Jan. 24, 2026 (I)]
A. The dipole moment value of NF_3 is higher than that of NH_3.
B. The dipole moment value of BeH_2 is zero.
C. The bond order of O_2^{2-} and F_2 is same.
D. The formal charge on the central oxygen atom of ozone is –1.
E. In NO_2 all the three atoms satisfy the octet rule, hence it is very stable.

Choose the correct answer from the options given below:
(1) A, B, C, D & E (2) B & C only
(3) B, C & D only (4) A, C & D only

27. Among H_2S, H_2O, NF_3, NH_3 and $CHCl_3$, identify the molecule (X) with lowest dipole moment value. [Jan. 22, 2026 (II)]
The number of lone pairs of electrons present on the central atom of the molecule (X) is :
(1) 2 (2) 0 (3) 1 (4) 3

28. Given below are two statements : [Jan. 21, 2026 (II)]
Statement I: The correct order in terms of bond dissociation enthalpy is $Cl_2 > Br_2 > F_2 > I_2$
Statement II: The correct trend in the covalent character of the metal halides is $[SnCl_4 > SnCl_2]$, $[PbCl_4 > PbCl_2]$ and $[UF_4 > UF_6]$
In the light of the above statements, choose the correct answer from the options given below:
(1) Statement I is true but Statement II is false.
(2) Both Statement I and Statement II are true.
(3) Statement I is false but Statement II is true.
(4) Both Statement I and Statement II are false.

29. Given below are two statements.
Statement I : The dipole moment of
$\overset{4}{C}H_3-\overset{3}{C}H=\overset{2}{C}H-\overset{1}{C}H=O$ is greater than
$\overset{4}{C}H_3-\overset{3}{C}H_2-\overset{2}{C}H_2-\overset{1}{C}H=O$
Statement II : C1–C2 bond length of
$\underset{4}{C}H_3-\underset{3}{C}H-\underset{2}{C}H-\underset{1}{C}H=O$ is greater than C_1-C_2 bond length
of $\underset{4}{C}H_3-\underset{3}{C}H_2-\underset{2}{C}H_2-\underset{1}{C}H=O$
In the light of the above statements, choose the correct answer from the options given below: [April 4, 2025 (I)]
(1) Statement I is false but Statement II is true
(2) Both Statement I and Statement II are false
(3) Statement I is true but Statement II is false
(4) Both Statement I and Statement II are true

30. An element 'E' has the ionisation enthalpy value of 374 kJ mol^{-1}. 'E' reacts with elements A, B, C and D with electron gain enthalpy values of –328, –349, –325 and –295 kJ mol^{-1}, respectively. [Jan. 29, 2025 (I)]
The correct order of the products EA, EB, EC and ED in terms of ionic character is
(1) EB > EA > EC > ED (2) ED > EC > EA > EB
(3) EA > EB > EC > ED (4) ED > EC > EB > EA

31. Arrange the following compounds in increasing order of their dipole moment : [Jan. 22, 2025 (II)]
HBr, H_2S, NF_3 and $CHCl_3$
(1) $NF_3 < HBr < H_2S < CHCl_3$
(2) $HBr < H_2S < NF_3 < CHCl_3$
(3) $H_2S < HBr < NF_3 < CHCl_3$
(4) $CHCl_3 < NF_3 < HBr < H_2S$

32. The correct increasing order for bond angles among BF_3, PF_3 and ClF_3 is : [April 9, 2024 (II)]
(1) $PF_3 < BF_3 < ClF_3$ (2) $BF_3 < PF_3 < ClF_3$
(3) $ClF_3 < PF_3 < BF_3$ (4) $BF_3 = PF_3 < ClF_3$

33. Given below are two statements : one is labelled as Assertion (A) and the other is labelled as Reason (R). [April 5, 2024 (II)]
Assertion (A) : NH_3 and NF_3 molecule have pyramidal shape with a lone pair of electrons on nitrogen atom. The resultant dipole moment of NH_3 is greater than that of NF_3.
Reason (R) : In NH_3, the orbital dipole due to lone pair is in the same direction as the resultant dipole moment of the N–H bonds. F is the most electronegative element.
In the light of the above statements, choose the correct answer from the options given below:
(1) Both (A) and (R) are true and (R) is the correct explanation of (A)
(2) (A) is false but (R) is true
(3) (A) is true but (R) is false
(4) Both (A) and (R) are true but (R) is NOT the correct explanation of (A)

34. Which one of the following molecules has maximum dipole moment ? [April 4, 2024 (I)]
(1) NF_3 (2) CH_4 (3) NH_3 (4) PF_5

35. Arrange the bonds in order of increasing ionic character in the molecules. LiF, K_2O, N_2, SO_2 and ClF_3: [Feb. 1, 2024 (I)]
(1) $ClF_3 < N_2 < SO_2 < K_2O < LiF$
(2) $LiF < K_2O < ClF_3 < SO_2 < N_2$
(3) $N_2 < SO_2 < ClF_3 < K_2O < LiF$
(4) $N_2 < ClF_3 < SO_2 < K_2O < LiF$

36. Which of the following is least ionic ? [Jan. 31, 2024 (II)]
(1) $BaCl_2$ (2) AgCl (3) KCl (4) $CoCl_2$

37. Given below are two statements: [Jan. 30, 2024 (II)]
Statement I: Since Fluorine is more electronegative than nitrogen, the net dipole moment of NF_3 is greater than NH_3.
Statement II: In NH_3, the orbital dipole due to lone pair and the dipole moment of NH bonds are in opposite direction, but in NF_3, the orbital dipole due to lone pair and dipole moments of N- F bonds are in same direction.
In the light of the above statements, choose the most appropriate from the options given below:
(1) Statement I is true but Statement II is false.
(2) Both Statement I and Statement II are false.
(3) Both Statement I and Statement II are true.
(4) Statement I is false but Statement II is true.

38. Choose the polar molecule from the following : [Jan. 27, 2024 (I)]
(1) CCl_4 (2) CO_2
(3) $CH_2=CH_2$ (4) $CHCl_3$

39. Among the following compounds, the one which shows highest dipole moment is [April 13, 2023 (I)]
(1) 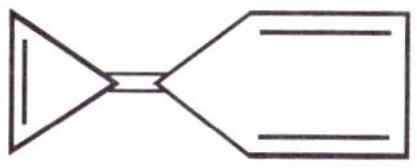
(2) 
(3) 
(4) 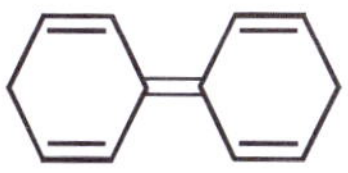

40. Which one of the following pairs is an example of polar molecular solids? [April 11, 2023 (II)]
(1) $SO_2(s)$, $NH_3(s)$ (2) $SO_2(s)$, $CO_2(s)$
(3) HCl(s), AlN(s) (4) MgO(s), $SO_2(s)$

41. The pair from the following pairs having both compounds with net non-zero dipole moment is [April 10, 2023 (I)]
(1) Benzene, anisidine
(2) 1,4-Dichlorobenzene, 1,3-Dichlorobenzene
(3) CH_2Cl_2, $CHCl_3$
(4) cis-butene, trans-butene

42. **Statement I:** Dipole moment is a vector quantity and by convention it is depicted by a small arrow with tail on the negative centre and head pointing towards the positive centre. [Jan. 25, 2023 (II)]
Statement II: The crossed arrow of the dipole moment symbolizes the direction of the shift of charges in the molecules.
In the light of the above statements, choose the most appropriate answer from the options given below :-
(1) Both Statement I and Statement II are correct.
(2) Statement I is incorrect but Statement II is correct.
(3) Both Statement I and Statement II are incorrect.
(4) Statement I is correct but Statement II is incorrect.

43. Order of Covalent bond; [Jan. 24, 2023 (I)]
A. KF > KI; LiF > KF
B. KF < KI; LiF > KF
C. $SnCl_4 > SnCl_2$; CuCl > NaCl
D. LiF > KF; CuCl < NaCl
E. KF < KI; CuCl > NaCl
(1) C, E only (2) B, C only
(3) B, C, E only (4) A, B only

44. Arrange the following in increasing order of their covalent character. [July 26, 2022 (II)]
(A) CaF_2 (B) $CaCl_2$
(C) $CaBr_2$ (D) CaI_2
Choose the correct answer from the options given below.
(1) B < A < C < D (2) A < B < C < D
(3) A < B < D < C (4) A < C < B < D

45. Arrange the following in the decreasing order of their covalent character : [June 29, 2022 (I)]
(A) LiCl (B) NaCl
(C) KCl (D) CsCl
Choose the **most appropriate** answer from the options given below :
(1) (A) > (C) > (B) > (D) (2) (B) > (A) > (C) > (D)
(3) (A) > (B) > (C) > (D) (4) (A) > (B) > (D) > (C)

46. The dipole moments of CCl_4, $CHCl_3$ and CH_4 are in the order: [Jan. 07, 2020 (I)]
(1) $CHCl_3 < CH_4 = CCl_4$ (2) $CCl_4 < CH_4 < CHCl_3$
(3) $CH_4 < CCl_4 < CHCl_3$ (4) $CH_4 = CCl_4 < CHCl_3$

47. Which compound exhibits maximum dipole moment among the following ? [Online April 11, 2015]
(1) 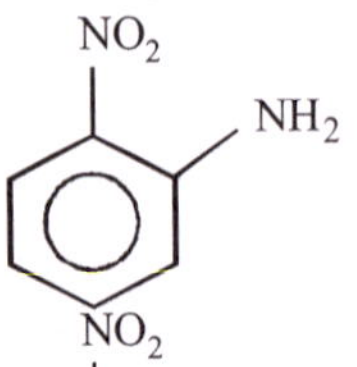
(2) 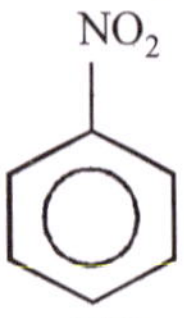
(3) 
(4) 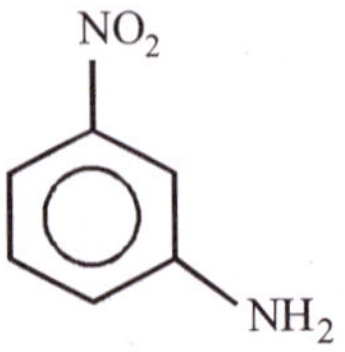

48. Molecule AB has a bond length of 1.617Å and a dipole moment of 0.38 D. The fractional charge on each atom (absolute magnitude) is : ($e_0 = 4.802 \times 10^{-10}$ esu) [Online April 11, 2015]
(1) 0.5 (2) 0.05 (3) 0 (4) 1.0

49. Amongst LiCl, RbCl, $BeCl_2$ and $MgCl_2$ the compounds with the greatest and the least ionic character, respectively are: [Online April 19, 2014]
(1) LiCl and RbCl (2) RbCl and $BeCl_2$
(3) $MgCl_2$ and $BeCl_2$ (4) RbCl and $MgCl_2$

50. Bond distance in HF is 9.17×10^{-11} m. Dipole moment of HF is 6.104×10^{-30} Cm. The percentage ionic character in HF will be : (electron charge = 1.60×10^{-19} C) [Online April 23, 2013]
(1) 61.0% (2) 38.0% (3) 35.5% (4) 41.5%

51. Which one of the following molecules is polar ? [Online April 9, 2013]
(1) XeF_4 (2) IF_5 (3) SbF_5 (4) CF_4

52. Dipole moment is shown by [Online May 26, 2012]
(1) 1, 2-dichlorobenzene (2) *trans*-2, 3-dichloro-2-butene
(3) 1, 4-dichlorobenzene (4) *trans*-1, 2-dinitroethene

53. Among the following, the molecule with the lowest dipole moment is [Online May 19, 2012]
(1) $CHCl_3$ (2) CH_3Cl (3) CH_2Cl_2 (4) CCl_4

54. Among the following chloro-compound having the lowest dipole moment is [Online May 12, 2012]
(1) CH_3Cl (2) 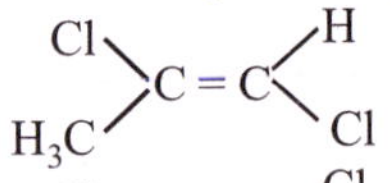
(3) CH_2Cl_2 (4) 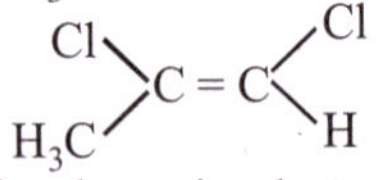

55. Although CN^- ion and N_2 molecule are isoelectronic, yet N_2 molecule is chemically inert because of [Online May 12, 2012]
(1) presence of more number of electrons in bonding orbitals
(2) low bond energy
(3) absence of bond polarity
(4) uneven electron distribution.

56. The charge/size ratio of a cation determines its polarizing power. Which one of the following sequences represents the increasing order of the polarizing power of the cationic species, K^+, Ca^{2+}, Mg^{2+}, Be^{2+}? [2007]
(1) $Ca^{2+} < Mg^{2+} < Be^{+} < K^{+}$ (2) $Mg^{2+} < Be^{2+} < K^{+} < Ca^{2+}$
(3) $Be^{2+} < K^{+} < Ca^{2+} < Mg^{2+}$ (4) $K^{+} < Ca^{2+} < Mg^{2+} < Be^{2+}$.

57. Which one of the following pairs of molecules will have permanent dipole moments for both members ? [2003]
(1) NO_2 and CO_2 (2) NO_2 and O_3
(3) SiF_4 and CO_2 (4) SiF_4 and NO_2

Numerical Value Questions (NVQs)

58. Number of compounds from the following with zero dipole moment is _____. [April 5, 2024 (II)]
HF, H_2, H_2S, CO_2, NH_3, BF_3, CH_4, $CHCl_3$, SiF_4, H_2O, BeF_2

59. Number of compounds/species from the following with non-zero dipole moment is _____
$BeCl_2$, BCl_3, NF_3, XeF_4, CCl_4, H_2O H_2S, HBr, CO_2, H_2, HCl [April 4, 2024 (II)]

60. A diatomic molecule has a dipole moment of 1.2 D. If the bond distance is 1Å, then fractional charge on each atom is _____ $\times 10^{-1}$ esu .
(Given 1D = 10^{-18} esu cm) [Jan. 31, 2024 (II)]

61. The total number of molecules with zero dipole moment among CH_4, BF_3, H_2O, HF, NH_3, CO_2 and SO_2 is _____. [Jan. 29, 2024 (II)]

62. The number of non - polar molecules from the following is________ [Jan. 27, 2024 (II)]
HF, H_2O, SO_2, H_2, CO_2, CH_4, NH_3, HCl, $CHCl_3$, BF_3

63. The number of following factors which affect the percent covalent character of the ionic bond is____ [April 08, 2023 (I)]
(A) Polarising power of cation
(B) Extent of distortion of anion
(C) Polarisability of the anion
(D) Polarising power of anion

64. Amongst the following, the number of molecule/(s) having net resultant dipole moment is _________.
NF_3, BF_3, BeF_2, $CHCl_3$, H_2S, SiF_4, CCl_4, PF_5 [June 30, 2022 (I)]

65. Amongst BeF_2, BF_3, H_2O, NH_3, CCl_4 and HCl, the number of molecules with non-zero net dipole moment is ____. [June 25, 2022 (II)]

Topic 4 - VSEPR Theory, Valence Bond Theory and Concept of Hybridisation

MULTIPLE CHOICE QUESTIONS (MCQs)

66. Bromine trifluoride autoionizes to form $BrF_2^{\oplus}$ and $BrF_4^{\ominus}$. The shapes of the cation and anion are respectively _____, and _____. [April 08, 2026 (II)]
(1) bent, square planar (2) linear, square planar
(3) bent, see-saw (4) linear, tetrahedral

67. Given below are two statements : [April 06, 2026 (II)]
Statement I: $F_2O < H_2O < Cl_2O$ is the correct trend in terms of bond angle.
Statement II: SiF_4, SnF_4 and PbF_4 are ionic in nature.
In the light of the above statements, choose the correct answer from the options given below:
(1) Both Statement I and Statement II are true
(2) Both Statement I and Statement II are false
(3) Statement I is true but Statement II is false
(4) Statement I is false but Statement II is true

68. Given below are two statements: [April 06, 2026 (II)]
Statement I: The number of compounds among SO_2, SO_3, SF_4, SF_6 and H_2S in which sulphur does not obey the Octet rule is 3.
Statement II: Among [H_2O, ClF_3, SF_4], [NH_3, BrF_5, SF_4], [BrF_5, ClF_3, XeF_4] and [XeF_4, ClF_3, H_2O], the number of sets in which all the molecules have one lone pair of electrons on the central atom is 1.
In the light of the above statements, choose the correct answer from the options given below :
(1) Both Statement I and Statement II are true
(2) Both Statement I and Statement II are false
(3) Statement I is true Statement II is false
(4) Statement I is false Statement II is true

69. Given below are two statements: [April 05, 2026 (II)]
Statement I: Aluminium upon reaction with NaOH forms $[Al(OH)_6]^{3-}$ ion.
Statement II: The geometry of ICl_4^-, ClO_3^- and IBr_2^- is square planar, pyramidal and linear respectively.
In the light of the above statements, choose the correct answer from the options given below :
(1) Both Statement I and Statement II are true
(2) Both Statement I and Statement II are false
(3) Statement I is true but Statement II is false
(4) Statement I is false but Statement II is true

70. SF_4 is isostructural with : [April 02, 2026 (II)]
A. $BrF_4^{\ominus}$ B. CH_4
C. $IF_4^{\oplus}$ D. XeF_4
E. XeO_2F_2
Choose the correct answer from the options given below:
(1) C only (2) C and E only
(3) A and D only (4) B and E only

71. Match List-I with List-II according to shape. [Jan. 28, 2026 (II)]

List-I		List-II	
A.	XeO_3	I.	BrF_5
B.	XeF_2	II.	NH_3
C.	XeO_2F_2	III.	$[I_3]^-$
D.	$XeOF_4$	IV.	SF_4

Choose the correct answer from the options given below:
(1) A-II, B-I, C-III, D-IV (2) A-II, B-III, C-IV, D-I
(3) A-II, B-III, C-I, D-IV (4) A-III, B-II, C-IV, D-I

72. Among the following, the CORRECT combinations are: [Jan. 24, 2026 (I)]
A. $IF_3 \rightarrow$ T-shaped (sp^3d)
B. $IF_5 \rightarrow$ Square pyramidal (sp^3d^2)
C. $IF_7 \rightarrow$ Pentagonal bipyramidal (sp^3d^3)
D. $ClO_4^- \rightarrow$ Square planar (sp^2d)
Choose the correct answer from the options given below:
(1) A, B and C only (2) A and B only
(3) A, B, C and D (4) B, C and D only

73. Two *p*-block elements X and Y form fluroides of the type EF_3. The fluoride compound XF_3 is a Lewis acid and YF_3 is a Lewis base. The hybridization of the central atoms of XF_3 and YF_3 respectively are [Jan. 22, 2026 (I)]
(1) Both sp^3 (2) sp^2 and sp^3
(3) sp^3 and sp^2 (4) Both sp^2

74. Given below are two statements: [Jan. 21, 2026 (I)]
Statement I: The number of species among SF_4, NH_4^+, $[NiCl_4]$, XeF_4, $[PtCl_4]^{2-}$, SeF_4 and $[Ni(CN)_4]^{2-}$, that have tetrahedral geometry is 3.
Statement II: In the set [NO_2, BeH_2, BF_3, $AlCl_3$], all the molecules have incomplete octet around central atom. In the light of the above statements, choose the correct answer from the options given below:
(1) Statement I is true but Statement II is false.
(2) Both Statement I and Statement II are false.
(3) Statement I is false but Statement II is true.
(4) Both Statement I and Statement II are true.

75. In SO_2, NO_2^- and N_3^- the hybridizations at the central atom are respectively : [April 7, 2025 (II)]
(1) sp^2, sp^2 and sp (2) sp^2, sp and sp
(3) sp^2, sp^2 and sp^2 (4) sp, sp^2 and sp

76. Match the LIST-I with LIST-II. [April 7, 2025 (I)]

LIST-I Molecule/ion		LIST-II Bond pair : lone pair (on the central atom)	
A.	ICl_2^-	I.	4 : 2
B.	H_2O	II.	4 : 1
C.	SO_2	III.	2 : 3
D.	XeF_4	IV.	2 : 2

Choose the **correct** answer from the options given below :

(1) A-IV, B-III, C-II, D-I (2) A-III, B-IV, C-II, D-I

(3) A-III, B-IV, C-I, D-II (4) A-II, B-I, C-IV, D-III.

77. Given below are two statements: [April 4, 2025 (II)]

Statement I: for $\ddot{\underset{..}{Cl}}F_3$, all three possible structures may be drawn as follows.

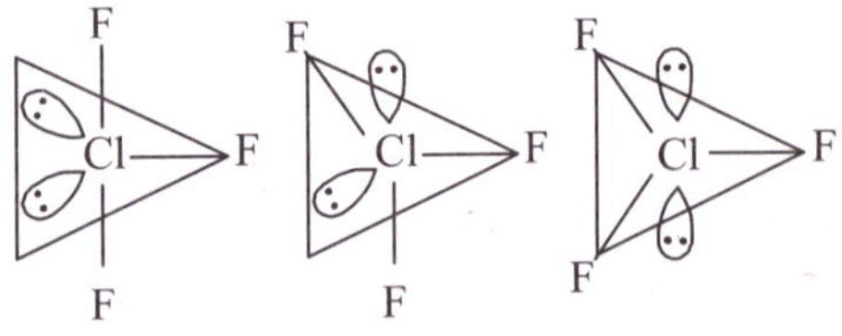

I II III

Statement II: Structure III is most stable, as the orbitals having the lone pairs are axial, where the lp – bp repulsion is minimum.

In the light of the above statements, choose the **most appropriate answer** from the options given below:

(1) Statement I is incorrect but statement II is correct.

(2) Statement I is correct but statement II is incorrect.

(3) Both Statement I and statement II are correct.

(4) Both Statement I and statement II are incorrect.

78. Match the **LIST-I** with **LIST-II** [April 3, 2025 (I)]

LIST-I (Molecules/ion)	LIST-II (Hybridisation of central atom)
A. PF_5	I dsp^2
B. SF_6	II. sp^3d
C. $Ni(CO)_4$	III. sp^3d^2
D. $[PtCl_4]^{2-}$	IV. sp^3

Choose the **correct** answer from the options given below:

(1) A-II, B-III, C-IV, D-I (2) A-IV, B-I, C-II, D-III

(3) A-I, B-II, C-III, D-IV (4) A-III, B-I, C-IV, D-II

79. Which among the following molecules is (a) involved in sp^3d hybridization, (b) has different bond lengths and (c) has lone pair of electrons on the central atom ? [April 2, 2025 (II)]

(1) PF_5 (2) XeF_4 (3) SF_4 (4) XeF_2

80. In 3,3-dimethylhex-1-ene-4-yne, there are _____ sp^3, _____ sp^2 and _____ sp hybridised carbon atoms respectively : [April 2, 2025 (II)]

(1) 4, 2, 2 (2) 3, 3, 2

(3) 2, 4, 2 (4) 2, 2, 4

81. A molecule with the formula AX_4Y has all it's elements from p-block. Element A is rarest, monoatomic, non-radioactive from its group and has the lowest ionization enthalpy value among A, X and Y. Elements X and Y have first and second highest electronegativity values respectively among all the known elements. The shape of the molecule is : [April 2, 2025 (I)]

(1) Square pyramidal (2) Octahedral

(3) Pentagonal planar (4) Trigonal bipyramidal

82. Among SO_2, NF_3, NH_3, XeF_2, ClF_3 and SF_4, the hybridization of the molecule with non-zero dipole moment and highest number of lone-pairs of electrons on the central atom is [April 2, 2025 (I)]

(1) sp^3 (2) dsp^2 (3) sp^3d^2 (4) sp^3d

83. The molecules having square pyramidal geometry are [Jan. 28, 2025 (I)]

(1) BrF_5 & $XeOF_4$ (2) SbF_5 & $XeOF_4$

(3) SbF_5 & PCl_5 (4) BrF_5 & PCl_5

84. In the given structure, number of sp and sp^2 hybridized carbon atoms present respectively are : [Jan. 24, 2025 (II)]

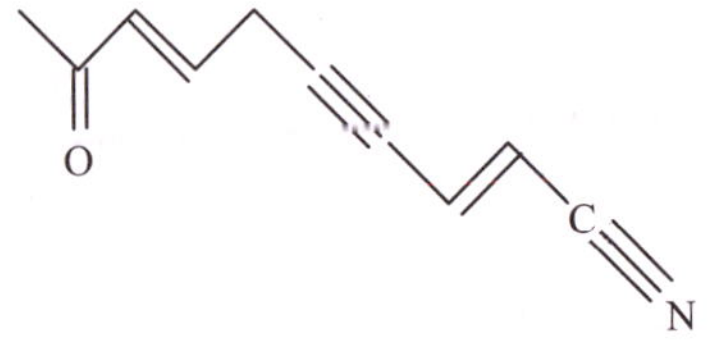

(1) 3 and 6 (2) 3 and 5

(3) 4 and 6 (4) 4 and 5

85. Which of the following statement is true with respect to H_2O, NH_3 and CH_4? [Jan. 24, 2025 (I)]

A. The central atoms of all the molecules are sp^3 hybridized.

B. The H-O-H, H-N-H and H-C-H angles in the above molecules are 104.5°, 107.5° and 109.5° respectively.

C. The increasing order of dipole moment is $CH_4 < NH_3 < H_2O$.

D. Both H_2O and NH_3 are Lewis acids and CH_4 is a Lewis base

E. A solution of NH_3 in H_2O is basic. In this solution NH_3 and H_2O act as Lowry-Bronsted acid and base respectively.

Choose the correct answer from the options given below:

(1) A, B and C only (2) C, D and E only

(3) A, D and E only (4) A, B, C and E only

86. In which one of the following pairs the central atoms exhibit sp^2 hybridization ? [April 9, 2024 (I)]

(1) BF_3 and NO_2^- (2) NH_2^- and H_2O

(3) H_2O and NO_2 (4) NH_2^- and BF_3

87. Given below are two statements : [April 9, 2024 (I)]

Statement I: The oxidation state of an element in a particular compound is the charge acquired by its atom on the basis of electron gain enthalpy consideration from other atoms in the molecule.

Statement II: $p\pi$-$p\pi$ bond formation is more prevalent in second period elements over other periods.

In the light of the above statements, choose the **most appropriate** answer from the options given below :

(1) Both Statement I and Statement II are incorrect

(2) Statement I is correct but Statement II is incorrect

(3) Both Statement I and Statement II are correct

(4) Statement I is incorrect but Statement II is correct

88. When Ψ_A and Ψ_B are the wave functions of atomic orbitals, then σ^* is represented by : [April 8, 2024 (II)]

(1) $\Psi_A - 2\Psi_B$ (2) $\Psi_A - \Psi_B$

(3) $\Psi_A + 2\Psi_B$ (4) $\Psi_A + \Psi_B$

89. Match **List-I** with **List-II**: [April 8, 2024 (I)]

List-I (Molecule)	List-II (Shape)
(A) NH_3	(I) Square pyramid
(B) BrF_5	(II) Tetrahedral
(C) PCl_5	(III) Trigonal pyramidal
(D) CH_4	(IV) Trigonal bipyramidal

Choose the **correct** answer from the options below:

(1) A-(IV), B-(III), C-(I), D-(II)

(2) A-(II), B-(IV), C-(I), D-(III)

(3) A-(III), B-(I), C-(IV), D-(II)

(4) A-(III), B-(IV), C-(I), D-(II)

90. Given below are two statements:
Statement I: PF_5 and BrF_5 both exhibit sp^3d hybridisation.
Statement II: Both SF_6 and $[Co(NH_3)_6]^{3+}$ exhibit sp^3d^2 hybridisation. [April 6, 2024 (II)]
In the light of the above statements, choose the **correct** answer from the options given below:
(1) Statement I is true but Statement II is false
(2) Both Statement I and Statement II are true
(3) Both Statement I and Statement II are false
(4) Statement I is false but Statement II is true

91. Match **List-I** with **List-II**: [April 6, 2024 (I)]

List-I (Compound/ Species)	List - II (Shape/ Geometry)
(A) SF_4	(I) Tetrahedral
(B) BrF_3	(II) Pyramidal
(C) BrO_3^-	(III) See saw
(D) NH_4^+	(IV) Bent T-shape

Choose the **correct** answer from the options given below:
(1) A-(II), B-(III), C-(I), D-(IV)
(2) A-(III), B-(IV), C-(II), D-(I)
(3) A-(II), B-(IV), C-(III), D-(I)
(4) A-(III), B-(II), C-(IV), D-(I)

92. Match **List-I** with **List-II**: [April 6, 2024 (I)]

List-I (Molecule / Species)	List-II (Property / Shape)
(A) SO_2Cl_2	(I) Paramagnetic
(B) NO	(II) Diamagnetic
(C) NO_2^-	(III) Tetrahedral
(D) I_3^-	(IV) Linear

Choose the **correct** answer from the options given below:
(1) A-(IV), B-(I), C-(III), D-(II)
(2) A-(III), B-(I), C-(II), D-(IV)
(3) A-(II), B-(III), C-(I), D-(IV)
(4) A-(III), B-(IV), C-(II), D-(I)

93. Match **List-I** with **List-II**: [April 6, 2024 (I)]

List-I (Hybridization)	List-II (Orientation in Space)
(A) sp^3	(I) Trigonal bipyramidal
(B) dsp^2	(II) Octahedral
(C) sp^3d	(III) Tetrahedral
(D) sp^3d^2	(IV) Square planar

Choose the **correct** answer from the options given below:
(1) A-(III), B-(I), C-(IV), D-(II)
(2) A-(II), B-(I), C-(IV), D-(III)
(3) A-(IV), B-(III), C-(I), D-(II)
(4) A-(III), B-(IV), C-(I), D-(II)

94. Match **List-I** with **List-II** [April 5, 2024 (II)]

List- I	List-II
(A) ICl	(I) T-Shape
(B) ICl_3	(II) Square pyramidal
(C) ClF_5	(III) Pentagonal bipyramidal
(D) IF_7	(IV) Linear

Choose the **correct** answer from the options given below:
(1) A–(I), B–(IV), C–III, D–(II)
(2) A–(I), B–(III), C–(II), D–(IV)
(3) A–(IV), B–(I), C–(II), D–(III)
(4) A–(IV), B–(III), C–(II), D–(I)

95. The number of species from the following that have pyramidal geometry around the central atom is ________
$S_2O_3^{2-}, SO_4^{2-}, SO_3^{2-}, S_2O_7^{2-}$ [April 4, 2024 (II)]
(1) 4 (2) 3 (3) 1 (4) 2

96. Number of molecules/ions from the following in which the central atom is involved in sp^3 hybridization is ______.
NO_3^-, BCl_3, ClO_2^-, ClO_3^- [April 4, 2024 (I)]
(1) 2 (2) 4 (3) 3 (4) 1

97. The molecule /ion with square pyramidal shape is [Jan. 30, 2024 (II)]
(1) $[Ni(CN)_4]^{2-}$ (2) PCl_5
(3) BrF_5 (4) PF_5

98. Match **List - I** with **List - II** [Jan. 30, 2024 (I)]

List - I (Molecule)	List - II (Shape)
(A) BrF_5	(I) T - shape
(B) H_2O	(II) See saw
(C) ClF_3	(III) Bent
(D) SF_4	(IV) Square pyramidal

Choose the correct answer from the options given below:
(1) (A) - (I), (B) - (II), (C) - (IV), (D) - (III)
(2) (A) - (II), (B) - (I), (C) - (III), (D) - (IV)
(3) (A) - (III), (B) - (IV), (C) - (I), (D) - (II)
(4) (A) - (IV), (B) - (III), (C) - (I), (D) - (II)

99. Given below are two statements: [April 13, 2023 (II)]
Statement I: SO_2 and H_2O both possess V-shaped structure.
Statement II: The bond angle of SO_2 is less than that of H_2O.
In the light of the above statements, choose the **most appropriate** answer from the options given below:
(1) Both Statement I and Statement II are correct.
(2) Statement I is correct but Statement II is incorrect.
(3) Both Statement I and Statement II are incorrect.
(4) Statement I is incorrect but Statement II is correct.

100. Match List-I with List-II : [April 11, 2023 (I)]

List - I Species	List - II Geometry / Shape
A. H_3O^+	I. Tetrahedral
B. Acetylide anion	II. Linear
C. NH_4^+	III. Pyramidal
D. ClO_2^-	IV. Bent

Choose the correct answer from the options given below:
(1) A-III, B-II, C-I, D-IV (2) A-III, B-I, C-II, D-IV
(3) A-III, B-IV, C-I, D-II (4) A-III, B-IV, C-II, D-I

101. Match List I with List II. [Jan. 31, 2023 (I)]

List I	List II
A. XeF_4	I. See-saw
B. SF_4	II. Square planar
C. NH_4^+	III. Bent T-shaped
D. BrF_3	IV. Tetrahedral

Choose the correct answer from the options given below:
(1) A-IV, B-III, C-II, D-I (2) A-II, B-I, C-III, D-IV
(3) A-IV, B-I, C-II, D-III (4) A-II, B-I, C-IV, D-III

102. Match **List I** with **List II.** [Jan. 30, 2023 (I)]

List-I (molecules/ions)	List-II (No. of lone pairs of e^- on central atom)
(A) IF_7	I. Three
(B) ICl_4^-	II. One
(C) XeF_6	III. Two
(D) XeF_2	IV. Zero

Choose the **correct** answer from the options given below:
(1) A – II, B – III, C – IV, D – I
(2) A – IV, B – III, C – II, D – I
(3) A – II, B – I, C – IV, D – III
(4) A – IV, B – I, C – II, D – III

103. For OF_2 molecule consider the following: [Jan. 30, 2023 (I)]
(A) Number of lone pairs on oxygen is 2.
(B) FOF angle is less than 104.5°.
(C) Oxidation state of O is –2.
(D) Molecule is bent 'V' shaped.
(E) Molecular geometry is linear.
Correct options are:
(1) C, D, E only (2) B, E, A only
(3) A, C, D only (4) A, B, D only

104. Number of lone pairs of electrons in the central atom of SCl_2, O_3, ClF_3 and SF_6, respectively, are: [July 29, 2022 (I)]
(1) 0, 1, 2 and 2 (2) 2, 1, 2 and 0
(3) 1, 2, 2 and 0 (4) 2, 1, 0 and 2

105. Match **List - I** with **List - II** [July 26, 2022 (I)]

List - I (Compound)	List - II (Shape)
(A) BrF_5	(I) bent
(B) $[CrF_6]^{3-}$	(II) square pyramidal
(C) O_3	(III) trigonal bipyramidal
(D) PCl_5	(IV) octahedral

Choose the **correct** answer from the options given below:
(1) (A) – (I), (B) – (II), (C) – (III), (D) – (IV)
(2) (A) – (IV), (B) – (III), (C) – (II), (D) – (I)
(3) (A) – (II), (B) – (IV), (C) – (I), (D) – (III)
(4) (A) – (III), (B) – (IV), (C) – (II), (D) – (I)

106. Consider the species CH_4, NH_4^+ and BH_4^-. Choose the correct option with respect to the three species:
(1) They are isoelectronic and only two have tetrahedral structures [June 29, 2022 (II)]
(2) They are isoelectronic and all have tetrahedral structures
(3) Only two are isoelectronic and all have tetrahedral structures
(4) Only two are isoelectronic and only two have tetrahedral structures

107. Number of lone pair(s) of electrons on central atom and the shape of BrF_3 molecule respectively, are : [June 29, 2022 (II)]
(1) 0, triangular planar. (2) 1, pyramidal.
(3) 2, bent T-shape. (4) 1, bent T-shape.

108. In the structure of SF_4, the lone pair of electrons on S is in
(1) equatorial position and there are two lone pair-bond pair repulsions at 90° [June 28, 2022 (II)]
(2) equatorial position and there are three lone pair-bond pair repulsions at 90°
(3) axial position and there are three lone pair-bond pair repulsion at 90°
(4) axial position and there are two lone pair-bond pair repulsion at 90°

109. Identify the **incorrect** statement for PCl_5 from the following. [June 27, 2022 (II)]
(1) In this molecule, orbitals of phosphorous are assumed to undergo sp^3d hybridization.
(2) The geometry of PCl_5 is trigonal bipyramidal.
(3) PCl_5 has two axial bonds stronger than three equatorial bonds.
(4) The three equatorial bonds of PCl_5 lie in a plane.

110. Based upon VSEPR theory, match the shape (geometry) of the molecules in **List-I** with the molecules in **List-II** and select the most appropriate option. [June 27, 2022 (I)]

List-I (Shape)	List-II (Molecules)
(A) T-shaped	(I) XeF_4
(B) Trigonal planar	(II) SF_4
(C) Square planar	(III) ClF_3
(D) See-saw	(IV) BF_3

(1) (A) – (I), (B) – (II), (C) – (III), (D) – (IV)
(2) (A) – (III), (B) – (IV), (C) – (I), (D) – (II)
(3) (A) – (III), (B) – (IV), (C) – (II), (D) – (I)
(4) (A) – (IV), (B) – (III), (C) – (I), (D) – (II)

111. Match List-I with List-II [June 24, 2022 (I)]

List-I	List-II
(A) $[PtCl_4]^{2-}$	(I) sp^3 d
(B) BrF_5	(II) d^2sp^3
(C) PCl_5	(III) dsp^2
(D) $[Co(NH_3)_6]^{3+}$	(IV) sp^3d^2

(1) (A) → (II), (B) → (IV), (C) → (I), (D) → (III)
(2) (A) → (III), (B) → (IV), (C) → (I), (D) → (II)
(3) (A) → (III), (B) → (I), (C) → (IV), (D) → (II)
(4) (A) → (II), (B) → (I), (C) → (IV), (D) → (III)

112. The hybridisations of the atomic orbitals of nitrogen in NO_2^-, NO_2^+ and NH_4^+ respectively are: [July 20, 2021 (II)]
(1) sp^3, sp^2 and sp (2) sp, sp^2 and sp^3
(3) sp^3, sp and sp^2 (4) sp^2, sp and sp^3

113. A central atom in a molecule has two lone pairs of electrons and forms three single bonds. The shape of this molecule is [March 17, 2021 (I)]
(1) trigonal pyramidal (2) see-saw
(3) T-shaped (4) planar triangular

114. Which among the following species has unequal bond lengths? [Feb. 25, 2021 (II); Same 2006]
(1) XeF_4 (2) BF_4^- (3) SF_4 (4) SiF_4

115. The compound that has the largest H – M – H bond angle (M = N, S, C), is : [Sep. 05, 2020 (II)]
(1) H_2O (2) NH_3 (3) H_2S (4) CH_4

116. If AB_4 molecule is a polar molecule, a possible geometry of AB_4 is : [Sep. 02, 2020 (I)]
(1) Square pyramidal (2) Tetrahedral
(3) Rectangular planar (4) Square planar

117. The shape/ structure of $[XeF_5]^-$ and XeO_3F_2, respectively, are : [Sep. 02, 2020 (II)]
(1) pentagonal planar and trigonal bipyramidal
(2) octahedral and square pyramidal
(3) trigonal bipyramidal and pentagonal planar
(4) trigonal bipyramidal and trigonal bipyramidal

118. The molecular geometry of SF_6 is octahedral. What is the geometry of SF_4 (including lone pair(s) of electrons, if any)? [Sep. 02, 2020 (II)]
(1) Tetrahedral (2) Trigonal bipyramidal
(3) Pyramidal (4) Square planar

119. The correct statement about ICl_5 and ICl_4^- is : [April 8, 2019 (II)]
(1) both are isostructural.
(2) ICl_5 is trigonal bipyramidal and ICl_4^- is tetrahedral.
(3) ICl_5 is square pyramidal and ICl_4^- is tetrahedral.
(4) ICl_5 is square pyramidal and ICl_4^- is square planar.

120. The ion that has sp^3d^2 hydridisation for the central atom, is: [April 8, 2019 (II)]
(1) $[ICl_4]^-$ (2) $[ICl_2]^-$ (3) $[IF_6]^-$ (4) $[BrF_2]^-$

121. In XeO_3F_2, the number of bond pair(s), π-bond(s) and lone pair(s) on Xe atom respectively are: [Online April 15, 2018 (II)]
(1) 5, 3, 0 (2) 5, 2, 0 (3) 4, 2, 2 (4) 4, 4, 0

122. The decreasing order of bond angles in BF_3, NH_3, PF_3 and I_3^- is : [Online April 15, 2018 (I)]
(1) $I_3^- > BF_3 > NH_3 > PF_3$ (2) $BF_3 > I_3^- > PF_3 > NH_3$
(3) $BF_3 > NH_3 > PF_3 > I_3^-$ (4) $I_3^- > NH_3 > PF_3 > BF_3$

123. Which of the following conversions involves change in both shape and hybridisation? [Online April 16, 2018]
(1) $H_2O \to H_3O^+$ (2) $BF_3 \to BF_4^-$
(3) $CH_4 \to C_2H_6$ (4) $NH_3 \to NH_4^+$

124. The incorrect geometry is represented by __________. [Online April 16, 2018]
(1) NF_3- trigonal planar
(2) BF_3- trigonal planar
(3) AsF_5- trigonal bipyramidal
(4) H_2O- bent

125. Identify the pair in which the geometry of the species is T-shape and square pyramidal, respectively [Online April 15, 2018 (I)]
(1) ICl_2^- and ICl_5 (2) IO_3^- and $IO_2F_2^-$
(3) ClF_3 and IO_4^- (4) $XeOF_2$ and $XeOF_4$

126. Total number of lone pair of electrons in I_3^- ion is : [2018]
(1) 3 (2) 6
(3) 9 (4) 12

127. The group having triangular planar structures is : [Online April 9, 2017]
(1) BF_3, NF_3, CO_3^{2-} (2) CO_3^{2-}, NO_3, SO_3
(3) NH_3, SO_3, CO_3^{2-} (4) NCl_3, BCl_3, SO_3

128. sp^3d^2 hybridisation is not displayed by : [Online April 8, 2017]
(1) BrF_5 (2) SF_6 (3) $[CrF_6]^{3-}$ (4) PF_5

129. The bond angle H–X–H is the greatest in the compound : [Online April 10, 2016]
(1) PH_3 (2) CH_4 (3) NH_3 (4) H_2O

130. The group of molecules having identical shape is : [Online April 9, 2016]
(1) PCl_5, IF_5, XeO_2F_2 (2) BF_3, PCl_3, XeO_3
(3) SF_4, XeF_4, CCl_4 (4) $ClF_3, XeOF_2, XeF_3^+$

131. The species in which the N atom is in a state of *sp* hybridisation is : [2016]
(1) NO_3^- (2) NO_2 (3) NO_2^+ (4) NO_2^-

132. The type of hybridisation and number of lone pair(s) of electrons of Xe in $XeOF_4$, respectively, are: [Online April 10, 2015 (S)]
(1) sp^3d^2 and 1 (2) sp^3d and 2
(3) sp^3d^2 and 2 (4) sp^3d and 1

133. The number and type of bonds in C_2^{2-} ion in CaC_2 are: [Online April 9, 2014; Same 2011 RS]
(1) One σ bond and one π-bond
(2) One σ bond and two π-bond
(3) Two σ bond and two π-bond
(4) Two σ bond and one π-bond

134. Which of the following molecules has two sigma (σ) and two pi (π) bonds? [Online April 12, 2014]
(1) C_2H_4 (2) N_2F_2
(3) $C_2H_2Cl_2$ (4) HCN

135. In which of the following sets, all the given species are isostructural ? [2012 Online April 25, 2013]
(1) CO_2, NO_2, ClO_2, SiO_2 (2) $PCl_3, AlCl_3, BCl_3, SbCl_3$
(3) BF_3, NF_3, PF_3, AlF_3 (4) $BF_4^-, CCl_4, NH_4^+, PCl_4^+$

136. The shape of IF_6^- is : [Online April 23, 2013]
(1) Trigonally distorted octahedron
(2) Pyramidal
(3) Octahedral
(4) Square antiprism

137. The formation of molecular complex $BF_3 \leftarrow NH_3$ results in a change in hybridisation of boron [Online May 12, 2012]
(1) from sp^2 to dsp^2 (2) from sp^2 to sp^3
(3) from sp^3 to sp^2 (4) from sp^3 to sp^3d

138. Which of the following has the square planar structure? [Online May 19, 2012]
(1) XeF_4 (2) NH_4^+
(3) BF_4^- (4) CCl_4

139. Among the following species which two have trigonal pyramidal shape? [Online May 26, 2012]
(I) NI_3 (II) I_3^- (III) SO_3^{2-} (IV) NO_3^-
(1) I and III (2) III and IV
(3) I and IV (4) II and III

140. In which of the following pairs, the two species are not isostructural ? [2012]
(1) CO_3^{2-} and NO_3^- (2) PCl_4^+ and $SiCl_4$
(3) PF_5 and BrF_5 (4) AlF_6^{3-} and SF_6

141. The bond dissociation energy of B – F in BF_3 is 646 kJ mol^{-1}, whereas that of C – F in CF_4 is 515 kJ mol^{-1}. The correct reason for higher B – F bond dissociation energy as compared to that of C – F bond is [2008]
(1) stronger σ bond between B and F in BF_3 as compared to that between C and F in CF_4.
(2) significant pπ – pπ interaction between B and F in BF_3, whereas there is no possibility of such interaction between C and F in CF_4.
(3) lower degree of pπ – pπ interaction between B and F in BF_3 than that between C and F in CF_4.
(4) smaller size of B– atom as compared to that of C– atom.

142. The decreasing values of bond angles from NH_3 (106°) to SbH_3 (101°) down group-15 of the periodic table is due to [2006]
(1) decreasing *lp-bp* repulsion
(2) decreasing electronegativity
(3) increasing *bp-bp* repulsion
(4) increasing *p*-orbital character in sp^3

143. The states of hybridization of boron and oxygen atoms in boric acid (H_3BO_3) are respectively [2004]
(1) sp^3 and sp^2 (2) sp^2 and sp^3
(3) sp^2 and sp^2 (4) sp^3 and sp^3

144. The maximum number of 90° angles between bond pair-bond pair of electrons is observed in [2004]
(1) dsp^2 hybridization (2) sp^3d hybridization
(3) dsp^3 hybridization (4) sp^3d^2 hybridization

145. Which one of the following has the regular tetrahedral structure ? [2004]

(1) BF_4^- (2) SF_4

(3) XeF_4 (4) $[Ni(CN)_4]^{2-}$

(Atomic nos. : B = 5, S = 16, Ni = 28, Xe − 54)

146. The pair of species having identical shapes for molecules of both species is [2003]

(1) XeF_2, CO_2 (2) BF_3, PCl_3

(3) PF_5, IF_5 (4) CF_4, SF_4

147. Hybridisation of the underline atom changes in: [2002]

(1) $\underline{Al}H_3$ changes to AlH_4^-

(2) $H_2\underline{O}$ changes to H_3O^+

(3) $\underline{N}H_3$ changes to NH_4^+

(4) in all cases

148. In which of the following species the interatomic bond angle is 109° 28'? [2002]

(1) NH_3, BF_4^- (2) NH_4^+, BF_3

(3) NH_3, BF_4 (4) NH_2^-, BF_3.

Numerical Value Questions (NVQs)

149. Consider the following species

$BrF_5, XeF_5^-, BF_4^-, ICl_4^-, XeF_4, SF_4, NH_4^+, ClF_3, XeF_2, ICl_2^-$

Number of species having sp^3d hybridized central atom is ______. [April 05, 2026 (I)]

150. According to Lewis theory, the total number of σ bond-pairs and lone pair of electrons around the central atom of XeO_6^{4-} ion is _____. [April 04, 2026 (I)]

151. The number of molecules/ions that show linear geometry among the following is _____. [Jan. 22, 2025 (I)]

$SO_2, BeCl_2, CO_2, N_3^-, NO_2, F_2O, XeF_2, NO_2^+, I_3^-, O_3$

152. Total number of species from the following with central atom utilising sp^2 hybrid orbitals for bonding is..........

$NH_3, SO_2, SiO_2, BeCl_2, C_2H_2, C_2H_4, BCl_3, HCHO, C_6H_6, BF_3, C_2H_4Cl_2$ [April 6, 2024 (II)]

153. The number of molecules/ions/ having trigonal bipyramidal shape is _____. [Feb. 1, 2024 (I)]

$PF_5, BrF_5, PCl_5, [PtCl_4]^{2-}, BF_3, Fe(CO)_5$

154. The number of species from the following in which the central atom uses sp^3 hybrid orbitals in its bonding is _____.

$NH_3, SO_2, SiO_2, BeCl_2, CO_2, H_2O, CH_4, BF_3$ [Jan. 31, 2024 (I)]

155. The total number of molecular orbitals formed from 2s and 2p atomic orbitals of a diatomic molecule is __________. [Jan. 30, 2024 (I)]

156. Number of compounds with one lone pair of electrons on central atom amongst following is ________

$O_3, H_2O, SF_4, ClF_3, NH_3, BrF_5, XeF_4$ [Jan. 29, 2024 (I)]

157. The maximum number of lone pairs of electrons on the central atom from the following species is ______

ClO_3^-, XeF_4, SF_4, and I_3^- [April 11, 2023 (II)]

158. The number of molecules from the following which contain only two lone pair of electrons is ______

$H_2O, N_2, CO, XeF_4, NH_3, NO, CO_2, F_2$ [April 10, 2023 (II)]

159. The number of bent-shaped molecule/s from the following is ______ [April 10, 2023 (I)]

$N_3^-, NO_2^-, I_3^-, O_3, SO_2$

160. The number of species having a square planar shape from the following is ________ [April 06, 2023 (II)]

$XeF_4, SF_4, SiF_4, BF_4^-, BrF_4^-, \left[Cu(NH_3)_4\right]^{2+}, \left[FeCl_4\right]^{2-}, \left[PtCl_4\right]^{2-}$

161. The number of species from the following which have square pyramidal structure is

$PF_5, BrF_4^-, IF_5? BrF_5, XeOF_4, ICl_4^-$ [April 06, 2023 (I)]

162. Amongst the following, the number of species having the linear shape is __. [Jan. 31, 2023 (II)]

$XeF_2, I_3^+, C_3O_2, I_3^-, CO_2, SO_2, BeCl_2$ and BCl_2^-

163. The number of molecule(s) or ion(s) from the following having non-planar structure is______.

$NO_3^-, H_2O_2, BF_3, PCl_3, XeF_4, SF_4, XeO_3, PH_4^+, SO_3, [Al(OH)_4]^-$ [July 27, 2022 (II)]

164. Consider, $PF_5, BrF_5, PCl_3, SF_6, [ICl_4]^-, ClF_3$ and IF_5. Amongst the above molecule(s)/ion(s), the number of molecule(s)/ion(s) having sp^3d^2 hybridisation is____. [July 29, 2022 (II)]

165. The hybridization of P exhibited in PF_5 is sp^xd^y. The value of y is _________. [June 28, 2022 (I)]

166. Amongst SF_4, XeF_4, CF_4 and H_2O, the number of species with two lone pairs of electrons ______ . [June 26, 2022 (II)]

167. The number of species having non–pyramidal shape among the following is_______. [Aug. 27, 2021 (II)]

(A) SO_3 (B) NO_3^- (C) PCl_3 (D) CO_3^{2-}

168. AB_3 is an interhalogen T-shaped molecule. The number of lone pairs of electrons on A is _______. [Aug. 26, 2021 (I)]

169. The number of lone pairs of electrons on the central I atom in I_3^- is __________. [July 20, 2021 (I)]

170. The number of species below that have two lone pairs of electrons in their central atom is ______.

$SF_4, BF_4^-, ClF_3, AsF_3, PCl_5, BrF_5, XeF_4, SF_6$ [March 18, 2021 (II)]

Topic 5 - Molecular Orbital Theory and Hydrogen Bonding

Multiple Choice Questions (MCQs)

171. Which of the following pictorial diagram most correctly represents the π* (π – antibonding) molecular orbital between two atoms if the internuclear axis is taken to be in the z-direction ($\xrightarrow{z-axis}$)? [April 04, 2026 (II)]

(1)  (2) 

(3)  (4) 

172. Given below are two statements : [April 02, 2026 (I)]

Statement I: The correct sequence of bond lengths in the following species is :

$O_2^+ < O_2 < O_2^- < O_2^{2-}$

Statement II: The correct sequence of number of unpaired electrons in the following species is :

$O_2 > O_2^+ > O_2^- > O_2^{2-}$

In the light of the above statements, choose the correct answer from the options given below :
(1) Both Statement I and Statement II are true
(2) Both Statement I and Statement II are false
(3) Statement I is true but Statement II is false
(4) Statement I is false but Statement II is true

173. Given below are two statements:

Statement I: The number of species among BF_4^-, SiF_4, XeF_4 and SF_4, that have unequal E–F bond lengths is two. Here, E is the central atom.

Statement II: Among O_2^-, O_2^{2-}, F_2 and O_2^+, O_2^- has the highest bond order. [Jan. 28, 2026 (I)]

In the light of the above statements, choose the correct answer from the options given below
(1) Both Statement I and Statement II are false
(2) Both Statement I and Statement II are true
(3) Statement I is true but Statement II is false
(4) Statement I is false but Statement II is true

174. Pair of species among the following having same bond order as well as paramagnetic character will be- [Jan. 24, 2026 (II)]
(1) O_2^+, N_2^{2-} (2) O_2^-, N_2^+
(3) O_2^+, N_2 (4) O_2^-, N_2^-

175. Which of the following molecules(s) show/s paramagnetic behavior ? [April 4, 2025 (I)]
(A) O_2 (B) N_2 (C) F_2 (D) S_2
(E) Cl_2
Choose the correct answer from the options given below :
(1) B only (2) A & C only
(3) A & E only (4) A & D only

176. The steam volatile compounds among the following are : [Jan. 29, 2025 (I)]

(A) 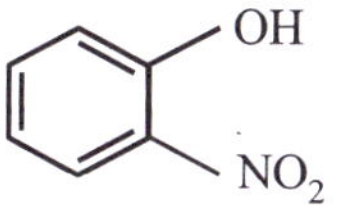
(B) 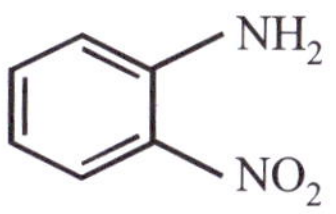
(C) 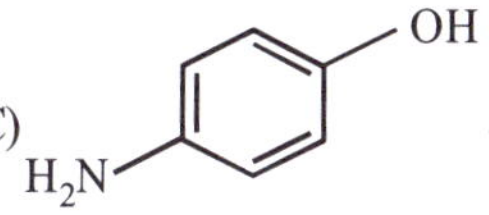
(D) 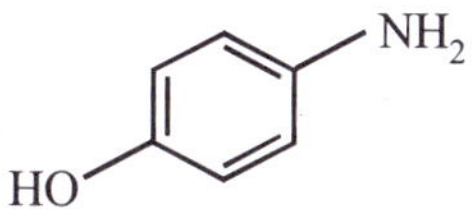

Choose the correct answer from the options given below :
(1) (B) and (D) only (2) (A) and (C) only
(3) (A) and (B) only (4) (A),(B) and (C) only

177. Which of the following linear combination of atomic orbitals will lead to formation of molecular orbitals in homonuclear diatomic molecules [internuclear axis in z-direction]? [Jan. 24, 2025 (I)]
A. $2p_z$ and $2p_x$ B. $2s$ and $2p_x$
C. $3d_{xy}$ and $3d_{x^2-y^2}$ D. $2s$ and $2p_z$
E. $2p_z$ and $3d_{x^2-y^2}$
(1) E Only (2) A and B Only
(3) D Only (4) C and D Only

178. The correct statement(s) about Hydrogen bonding is/are : [April 4, 2024 (II)]
(A) Hydrogen bonding exists when H is covalently bonded to the highly electronegative atom.
(B) Intermolecular H bonding is present in *o*-nitro phenol
(C) Intramolecular H bonding is present in HF.
(D) The magnitude of H bonding depends on the physical state of the compound.
(E) H-bonding has powerful effect on the structure and properties of compounds.
Choose the correct answer from the options given below:
(1) A only (2) A, D, E only
(3) A, B, D only (4) A, B, C only

179. Select the compound from the following that will show intramolecular hydrogen bonding. [Feb. 1, 2024 (II)]
(1) 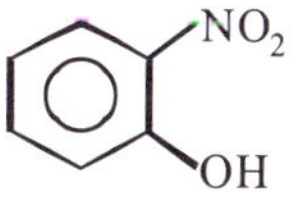 (2) H_2O
(3) C_2H_5OH (4) NH_3

180. Given below are two statements: [Feb. 1, 2024 (II)]
Statement I: A π bonding MO has lower electron density above and below the inter - nuclear axis.
Statement II: The π^* antibonding MO has a node between the nuclei.
In the light of the above statements, choose the most appropriate answer from the options given below:
(1) Both Statement I and Statement II are true
(2) Both Statement I and Statement II are false
(3) Statement I is true but Statement II is false
(4) Statement I is false but Statement II is true

181. The linear combination of atomic orbitals to form molecular orbitals takes place only when the combining atomic orbitals [Jan. 31, 2024 (I)]
A. have the same energy
B. have the minimum overlap
C. have same symmetry about the molecular axis
D. have different symmetry about the molecular axis
Choose the most appropriate from the options given below:
(1) A, B, C only (2) A and C only
(3) B and D only (4) B, C, D only

182. Consider the following statements: [April 15, 2023 (I)]
(A) NF_3 molecule has a trigonal planar structure.
(B) Bond length of N_2 is shorter than O_2.
(C) Isoelectronic molecules or ions have identical bond order.
(D) Dipole moment of H_2S is higher than that of water molecule.
Choose the correct answer from the option below:
(1) (A) and (D) are correct (2) (C) and (D) are correct
(3) (A) and (B) are correct (4) (B) and (C) are correct

183. In which of the following processes, the bond order increases and paramagnetic character changes to diamagnetic one? [April 13, 2023 (I); Same 2019 (II)]
(1) $O_2 \rightarrow O_2^{2-}$ (2) $NO \rightarrow NO^+$
(3) $N_2 \rightarrow N_2^+$ (4) $O_2 \rightarrow O_2^+$

184. The bond order and magnetic property of acetylide ion are same as that of [April 12, 2023 (I)]
(1) NO^+ (2) O_2^+ (3) O_2^- (4) N_2^+

185. According to MO theory the bond orders for O_2^{2-}, CO and NO^+ respectively, are [Jan. 29, 2023 (II)]
(1) 1, 3 and 3 (2) 1, 3 and 2
(3) 1, 2 and 3 (4) 2, 3 and 3

186. What is the number of unpaired electron(s) in the highest occupied molecular orbital of the following species :

$N_2; N_2^+; O_2; O_2^+$? **[Jan. 24, 2023 (II)]**

(1) 0, 1, 2, 1 (2) 2, 1, 2, 1
(3) 0, 1, 0, 1 (4) 2, 1, 0, 1

187. All structures given below are of vitamin C. Most stable of them is: **[Feb. 01, 2023 (II)]**

(1) 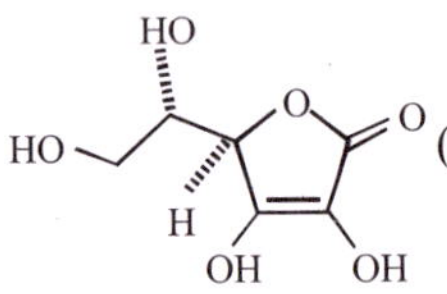

(2) 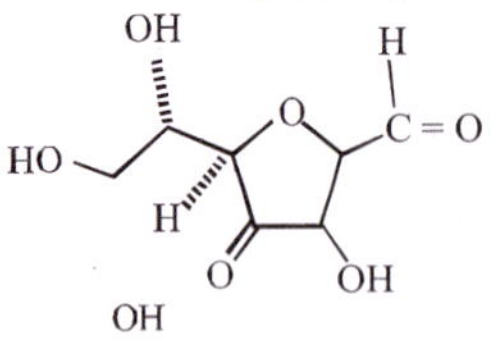

(3) 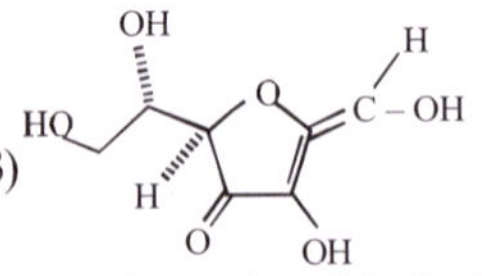

(4) 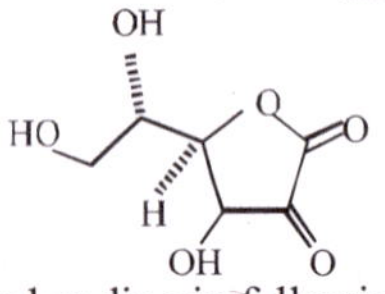

188. Decreasing order of the hydrogen bonding in following forms of water is correctly represented by **[Jan. 24, 2023 (I)]**

A. Liquid water B. Ice
C. Impure water

(1) A>B>C (2) B>A>C
(3) C>B>A (4) A>B>C

189. Given below are two statements: One is labelled as **Assertion A** and the other is labelled as **Reason R**.

Assertion A: Zero orbital overlap is an out of phase overlap.

Reason R: It results due to different orientation/direction of approach of orbitals.

In the light of the above statements, choose the correct answer from the options given below: **[July 28, 2022 (II)]**

(1) Both A and R are true and R is the correct explanation of A.
(2) Both A and R are true but R is NOT the correct explanation of A.
(3) A is true but R is false.
(4) A is false but R is true.

190. Match **List-I** with **List-II** **[July 27, 2022 (II)]**

List-I	List-II
(A) $\Psi_{MO} = \Psi_A - \Psi_B$	(I) Dipole moment
(B) $\mu = Q \times r$	(II) Bonding molecular orbital
(C) $\dfrac{N_b - N_a}{2}$	(III) Anti-bonding molecualr orbital
(D) $\Psi_{MO} = \Psi_A + \Psi_B$	(IV) Bond order

(1) (A)-(II), (B)-(I), (C)-(IV), (D)-(III)
(2) (A)-(III), (B)-(IV), (C)-(I), (D)-(II)
(3) (A)-(III), (B)-(I), (C)-(IV), (D)-(II)
(4) (A)-(III), (B)-(IV), (C)-(II), (D)-(I)

191. The correct order of increasing intermolecular hydrogen bond strength is **[June 27, 2022 (II)]**

(1) $HCN < H_2O < NH_3$ (2) $HCN < CH_4 < NH_3$
(3) $CH_4 < HCN < NH_3$ (4) $CH_4 < NH_3 < HCN$

192. Consider the ions/molecule O_2^+, O_2, O_2^-, O_2^{2-}

For increasing bond order the correct option is :
[June 26, 2022 (I); Same July 25, 2021 (II)]

(1) $O_2^{2-} < O_2^- < O_2 < O_2^+$ (2) $O_2^- < O_2^{2-} < O_2 < O_2^+$
(3) $O_2^- < O_2^{2-} < O_2^+ < O_2$ (4) $O_2^- < O_2^+ < O_2^{2-} < O_2$

193. Bonding in which of the following diatomic molecule(s) become(s) stronger, on the basis of MO Theory, by removal of an electron ? **[June 25, 2022 (I)]**

(A) NO (B) N_2 (C) O_2 (D) C_2 (E) B_2

Choose the most appropriate answer from the options given below :

(1) (A), (B), (C) only (2) (B), (C), (E) only
(3) (A), (C) only (4) (D) only

194. The correct order of bond orders of C_2^{2-}, N_2^{2-} and O_2^{2-} is, respectively. **[June 24, 2022 (II)]**

(1) $C_2^{2-} < N_2^{2-} < O_2^{2-}$ (2) $O_2^{2-} < N_2^{2-} < C_2^{2-}$
(3) $C_2^{2-} < O_2^{2-} < N_2^{2-}$ (4) $N_2^{2-} < C_2^{2-} < O_2^{2-}$

195. The bond order and magnetic behaviour of O_2^- ion are, respectively : **[Aug. 26, 2021 (II)]**

(1) 1.5 and paramagnetic (2) 1.5 and diamagnetic
(3) 2 and diamagnetic (4) 1 and paramagnetic

196. Match **List-I** with **List-II**

List-I (Molecule)	List-II (Bond order)
(A) Ne_2	(I) 1
(B) N_2	(II) 2
(C) F_2	(III) 0
(D) O_2	(IV) 3

Choose the correct answer from the options given below **[Feb. 26, 2021 (II)]**

(1) (A)-(IV); (B)-(III); (C)-(II); (D)-(I)
(2) (A)-(II); (B)-(I); (C)-(IV); (D)-(III)
(3) (A)-(I); (B)-(II); (C)-(III); (D)-(IV)
(4) (A)-(III); (B)-(IV); (C)-(I); (D)-(II)

197. According to molecular orbital theory, the species among the following that does not exist is **[Feb. 25, 2021 (I)]**

(1) He_2^- (2) Be_2 (3) He_2^+ (4) O_2^{2-}

198. The potential energy curve for the H_2 molecule as a function of internuclear distance is : **[Sep. 05, 2020 (I)]**

(1) 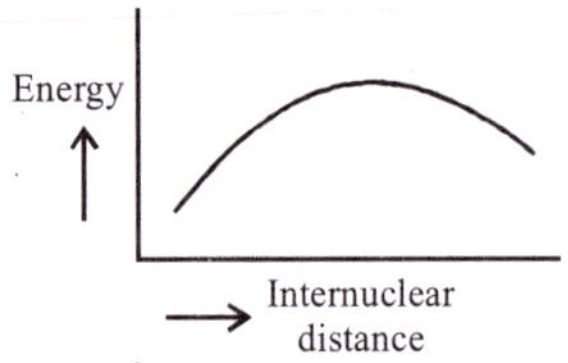

(2) 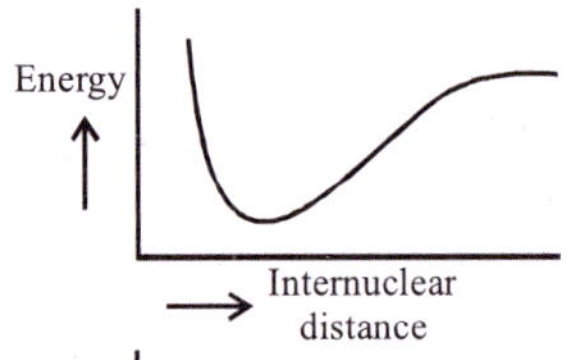

(3) 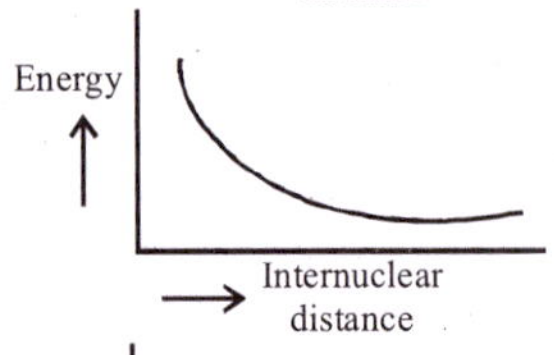

(4) 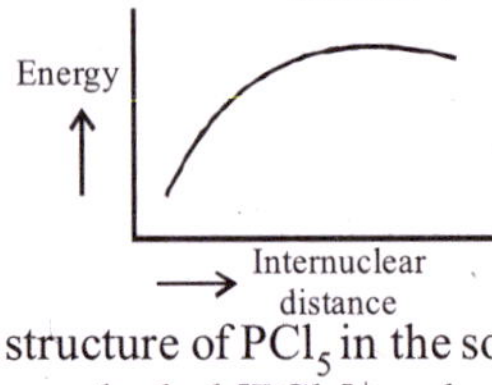

199. The structure of PCl_5 in the solid state is: **[Sep. 05, 2020 (I)]**

(1) tetrahedral $[PCl_4]^+$ and octahedral $[PCl_6]^-$
(2) square planar $[PCl_4]^+$ and octahedral $[PCl_6]^-$
(3) square pyramidal
(4) trigonal bipyramidal

200. The molecule in which hybrid MOs involve only one *d*-orbital of the central atom is : **[Sep. 04, 2020 (II)]**
(1) $[Ni(CN)_4]^{2-}$ (2) BrF_5
(3) XeF_4 (4) $[CrF_6]^{3-}$

201. Of the species, NO, NO^+, NO^{2+} and NO^-, the one with minimum bond strength is : **[Sep. 03, 2020 (I)]**
(1) NO^+ (2) NO
(3) NO^{2+} (4) NO^-

202. If the magnetic moment of a dioxygen species is 1.73 B.M, it may be: **[Jan. 09, 2020 (I)]**
(1) O_2^- or O_2^+ (2) O_2 or O_2^+
(3) O_2 or O_2^- (4) O_2, O_2^- or O_2^+

203. The bond order and the magnetic characteristics of CN^- are: **[Jan. 07, 2020 (II)]**
(1) $2\frac{1}{2}$, diamagnetic (2) 3, diamagnetic
(3) 3, paramagnetic (4) $2\frac{1}{2}$, paramagnetic

204. During the change of O_2 to O_2^-, the incoming electron goes to the orbital : **[April 10, 2019 (I)]**
(1) $\pi 2p_y$ (2) $\sigma^* 2p_z$ (3) $\pi^* 2p_x$ (4) $\pi 2p_x$

205. Among the following species, the diamagnetic molecule is: **[April 9, 2019 (II)]**
(1) NO (2) CO (3) B_2 (4) O_2

206. Among the following, the molecule expected to be stabilised by anion formation is:
C_2, O_2, NO, F_2 **[April 9, 2019 (I)]**
(1) C_2 (2) F_2 (3) NO (4) O_2

207. Among the following molecules/ions, C_2^{2-}, N_2^{2-}, O_2^{2-}, O_2
Which one is diamagnetic and has the shortest bond length? **[April 8, 2019 (II)]**
(1) O_2 (2) N_2^{2-} (3) O_2^{2-} (4) C_2^{2-}

208. Two pi and half sigma bonds are present in: **[Jan. 10, 2019 (I)]**
(1) O_2^+ (2) N_2 (3) O_2 (4) N_2^+

209. According to molecular orbital theory, which of the following is true with respect to Li_2^+ and Li_2^-? **[Jan. 9, 2019 (I)]**
(1) Li_2^+ is unstable and Li_2^- is stable
(2) Li_2^+ is stable and Li_2^- is unstable
(3) Both are stable
(4) Both are unstable

210. (I) (II)
H – N - - - N - - - N
In hydrogen azide, the bond orders of bonds (I) and (II) are ________. **[Online April 15, 2018 (I)]**
(1) I < 2, II > 2 (2) I > 2, II > 2
(3) I > 2, II < 2 (4) I < 2, II < 2

211. Which of the following best describes the diagram of molecular orbital? **[Online April 15, 2018 (II)]**

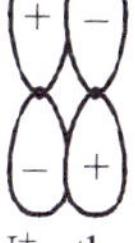

(1) A bonding π orbital
(2) A non-bonding orbital
(3) An antibonding σ orbital
(4) An antibonding π orbital

212. In the molecular orbital diagram for the molecular ion, N_2^+, the number of electrons in the σ_{2p} molecular orbital is: **[Online April 15, 2018 (I)]**
(1) 0 (2) 2 (3) 3 (4) 1

213. According to molecular orbital theory, which of the following will not be a viable molecule? **[2018]**
(1) He_2^{2+} (2) He_2^+ (3) H_2^- (4) H_2^{2-}

214. Which of the following is paramagnetic ? **[Online April 9, 2014(s), Online April 8, 2017(s)]**
(1) NO^+ (2) CO (3) O_2^{2-} (4) B_2

215. Which of the following species is not paramagnetic ? **[2017]**
(1) NO (2) CO (3) O_2 (4) B_2

216. After understanding the assertion and reason, choose the correct option. **[Online April 10, 2015]**
Assertion : In the bonding molecular orbital (MO) of H_2, electron density is increased between the nuclei.
Reason : The bonding MO is $\Psi_A + \Psi_B$, which shows destructive interference of the combining electron waves.
(1) Assertion is incorrect, reason is correct.
(2) Assertion is correct, reason is incorrect.
(3) Assertion and reason are correct and reason is the correct explanation for the assertion.
(4) Assertion and reason are correct, but reason is not the correct explanation for the assertion.

217. The correct order of bond dissociation energy among N_2, O_2, O_2^- is shown in which of the following arrangements? **[Online April 11, 2014]**
(1) $N_2 > O_2^- > O_2$ (2) $O_2^- > O_2 > N_2$
(3) $N_2 > O_2 > O_2^-$ (4) $O_2 > O_2^- > N_2$

218. Which one of the following molecules is paramagnetic? **[Online April 19, 2014]**
(1) N_2 (2) NO (3) CO (4) O_3

219. Which of the following has unpaired electron(s)? **[Online April 9, 2014]**
(1) N_2 (2) O_2^- (3) N_2^{2+} (4) O_2^{2-}

220. Which one of the following properties is **not** shown by NO? **[2014]**
(1) It is diamagnetic in gaseous state
(2) It is neutral oxide
(3) It combines with oxygen to form nitrogen dioxide
(4) Its bond order is 2.5

221. In which of the following ionization processes the bond energy has increased and also the magnetic behaviour has changed from paramagnetic to diamagnetic ? **[Online April 9, 2013; Same 2007]**
(1) $NO \rightarrow NO^+$ (2) $N_2 \rightarrow N_2^+$
(3) $C_2 \rightarrow C_2^+$ (4) $O_2 \rightarrow O_2^+$

222. Which one of the following molecules is expected to exhibit diamagnetic behaviour ? **[2013]**
(1) C_2 (2) N_2 (3) O_2 (4) S_2

223. Which of the following is the wrong statement **[2013]**
(1) ONCl and ONO^- are isoelectronic.
(2) O_3 molecule is bent
(3) Ozone is violet-black in solid state
(4) Ozone is diamagnetic gas.

224. In which of the following pairs of molecules/ions, both the species are not likely to exist ? **[2013]**
(1) H_2^+, He_2^{2-} (2) H_2^-, He_2^{2-}
(3) H_2^{2+}, He_2 (4) H_2^-, He_2^{2+}

225. Stability of the species Li_2, Li_2^- and Li_2^+ increases in the order of : **[2013]**
(1) $Li_2 < Li_2^+ < Li_2^-$ (2) $Li_2^- < Li_2^+ < Li_2$
(3) $Li_2 < Li_2^- < Li_2^+$ (4) $Li_2^- < Li_2 < Li_2^+$

226. The internuclear distances in O – O bonds for O_2^+, O_2, O_2^- and O_2^{2-} respectively are : [Online April 25, 2013]
(1) 1.30 Å, 1.49 Å, 1.12 Å, 1.21 Å
(2) 1.49 Å, 1.21 Å, 1.12 Å, 1.30 Å
(3) 1.21 Å, 1.12 Å, 1.49 Å, 1.30 Å
(4) 1.12 Å, 1.21 Å, 1.30 Å, 1.49 Å

227. Bond order normally gives idea of stability of a molecular species. All the molecules viz. H_2, Li_2 and B_2 have the same bond order yet they are not equally stable. Their stability order is [Online April 22, 2013]
(1) $H_2 > B_2 > Li_2$ (2) $Li_2 > H_2 > B_2$
(3) $Li_2 > B_2 > H_2$ (4) $B_2 > H_2 > Li_2$

228. Among the following, the species having the smallest bond length is [Online May 7, 2012]
(1) NO^- (2) NO^+ (3) O_2 (4) NO

229. *ortho*-Nitrophenol is less soluble in water than *p*- and *m*- nitrophenols because : [2012]
(1) *o*-nitrophenol is more volatile steam than those of *m*- and *p*-isomers.
(2) *o*-nitrophenol shows intramolecular H-bonding
(3) *o*-nitrophenol shows intermolecular H-bonding
(4) Melting point of *o*-nitrophenol is lower than those of *m*- and *p*-isomers.

230. Using MO theory, predict which of the following species has the shortest bond length? [2008]
(1) O_2^+ (2) O_2^- (3) O_2^{2-} (4) O_2^{2+}

231. Which one of the following pairs of species have the same bond order? [2008]
(1) CN^- and NO^+ (2) CN^- and CN^+
(3) O_2^- and CN^- (4) NO^+ and CN^+

232. Which of the following species exhibits the diamagnetic behaviour ? [2007]
(1) NO (2) O_2^{2-} (3) O_2^+ (4) O_2

233. Which of the following hydrogen bonds is strongest? [2007]
(1) O – H - - - F (2) O – H - - - H
(3) F – H - - - F (4) O – H - - - O

234. Which of the following molecules/ions does not contain unpaired electrons? [2006]
(1) N_2^+ (2) O_2 (3) O_2^{2-} (4) B_2

235. Which of the following species is diamagnetic in nature? [2005]
(1) H_2^- (2) H_2^+ (3) H_2 (4) He_2^+

236. The bond order in NO is 2.5 while that in NO^+ is 3. Which of the following statements is true for these two species ? [2004]
(1) Bond length in NO^+ is equal to that in NO
(2) Bond length in NO is greater than in NO^+
(3) Bond length in NO^+ is greater than in NO
(4) Bond length is unpredictable

237. An ether is more volatile than an alcohol having the same molecular formula. This is due to [2003]
(1) alcohols having resonance structures
(2) intermolecular hydrogen bonding in ethers
(3) intermolecular hydrogen bonding in alcohols
(4) dipolar character of ethers

238. Which of the following are arranged in an increasing order of their bond strength? [2002]
(1) $O_2^- < O_2 < O_2^+ < O_2^{2-}$ (2) $O_2^{2} < O_2^- < O_2 < O_2^+$
(3) $O_2^- < O_2^{2-} < O_2 < O_2^+$ (4) $O_2^+ < O_2 < O_2^- < O_2^{2-}$

NUMERICAL VALUE QUESTIONS (NVQs)

239. Total number of molecules/species from following which will be paramagnetic is ______.

O_2, O_2^+, O_2^-, NO, NO_2, CO, $K_2[NiCl_4]$,
$[Co(NH_3)_6]Cl_3$, $K_2[Ni(CN)_4]$ [Jan. 28, 2025 (II)]

240. Total number of electron present in (π^*) molecular orbitals of O_2, O_2^+ and O_2^- is ______. [April 9, 2024 (II)]

241. The total number of species from the following in which one unpaired electron is present, is _______.

$N_2, O_2, C_2^-, O_2^-, O_2^{2-}, H_2^+, CN^-, He_2^+$ [April 9, 2024 (I)]

242. Number of molecules having bond order 2 from the following molecule is _______.
C_2, O_2, Be_2, Li_2, Ne_2, N_2, He_2 [April 8, 2024 (II)]

243. Number of molecules from the following which can exhibit hydrogen bonding is ______. (nearest integer) [April 6, 2024 (I)]

CH_3OH, H_2O, C_2H_6, C_6H_6, 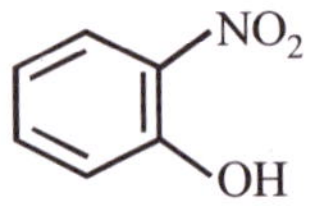

HF, NH_3

244. Number of molecules/species from the following having one unpaired electron is ________.
O_2, O_2^-, NO, CN^-, O_2^{2-} [April 4, 2024 (I)]

245. The number of species from the following which are paramagnetic and with bond order equal to one is _____.
H_2, He_2^+, O_2^+, N_2^{2-}, O_2^{2-}, F_2, Ne_2^+, B_2 [Jan. 29, 2024 (I)]

246. The total number of anti bonding molecular orbitals, formed from 2s and 2p atomic orbitals in a diatomic molecule is _____. [Jan. 29, 2024 (II)]

247. The total number of 'Sigma' and 'Pi' bonds in 2-formylhex-4-enoic acid is ____. [Jan. 29, 2024 (II)]

248. Sum of bond order of CO and NO^+ is ________. [Jan. 27, 2024 (I)]

249. The number of paramagnetic species among the following is ________. [July 28, 2022 (I)]
B_2, Li_2, C_2, C_2^-, O_2^{2-}, O_2^+ and He_2^+

250. According to MO theory, number of species/ions from the following having identical bond order is _____:
$CN^-, NO^+, O_2, O_2^+, O_2^{2+}$ [July 27, 2022 (I)]

251. Among the following species [July 25, 2022 (I)]
N_2, N_2^+, N_2^-, N_2^{2-}, O_2, O_2^+, O_2^-, O_2^{2-} the number of species showing diamagnetism is ________.

252. According to molecular orbital theory, the number of unpaired electron(s) in O_2^{2-} is : [Aug. 31, 2021 (II); Same July 27, 2021 (II)]

253. The difference between bond orders of CO and $NO^{\oplus}$ is $\frac{x}{2}$ where x = _______.
(Round off to the Nearest Integer) [July 27, 2021 (I)]

254. The total number of electrons in all bonding molecular orbitals of O_2^{2-} is _______.
(Round of to the nearest integer) [July 27, 2021 (II)]

255. AX is a covalent diatomic molecule where A and X are second row elements of periodic table. Based on Molecular orbital theory, the bond order of AX is 2.5.
The total number of electrons in AX is __________. [March 18, 2021 (I)]

Part-B (JEE Advanced)

Topic 1- Electrovalent and Covalent Bonding

1 MCQs with One Correct Answer

1. Which one is most ionic : [1995S]
(a) P_2O_5 (b) CrO_3 (c) MnO (d) Mn_2O_7

2. Pick out the isoelectronic structures from the following; [1993 - 1 Mark]
I. CH_3^+ II. H_3O^+ III. NH_3 IV CH_3^-
(a) I and II (b) III and IV
(c) I and III (d) II, III and IV

3. The bond between two identical non-metal atoms has a pair of electrons : [1986 - 1 Mark]
(a) unequally shared between the two
(b) transferred fully from one atom to another
(c) with identical spins
(d) equally shared between them

4. The types of bonds present in $CuSO_4 \cdot 5H_2O$ are only
(a) electrovalent and covalent [1983 - 1 Mark]
(b) electrovalent and coordinate covalent
(c) electrovalent, covalent and coordinate covalent
(d) covalent and coordinate covalent

5. Which of the following is soluble in water [1980]
(a) CS_2 (b) C_2H_5OH (c) CCl_4 (d) $CHCl_3$

6. Which of the following compounds are covalent? [1980]
(a) H_2 (b) CaO (c) KCl (d) Na_2S

7. Element X is strongly electropositive and element Y is strongly electronegative. Both are univalent. The compound formed would be [1980]
(a) X^+Y^- (b) X^-X^+ (c) $X-Y$ (d) $X \rightarrow Y$

8. The compound which contains both ionic and covalent bonds is [1979]
(a) CH_4 (b) H_2 (c) KCN (d) KCl

2 Integer Value Answer

9. Among B_2H_6, $B_3N_3H_6$, N_2O, N_2O_4, $H_2S_2O_3$, $H_2S_2O_8$, the total number of molecules containing covalent bond between two atoms of the same kind is ________ [Adv. 2019]

4 Fill in the Blanks

10. The two types of bonds present in B_2H_6 are covalent and .... [1994 - 1 Mark]

11. There are .... π bonds in a nitrogen molecule. [1982 - 1 Mark]

5 True / False

12. Linear overlap of two atomic p-orbitals leads to a sigma bond. [1983 - 1 Mark]

6 MCQs with One or More than One Correct Answer

13. Read the following Assertion and Reason and answer as per the options given below : [1998 - 2 Marks]
Assertion : LiCl is predominantly a covalent compound.
Reason : Electronegativity difference between Li and Cl is too small.

Topic 2 - Octet Rule and Resonance

1 MCQs with One Correct Answer

14. The cyanide ion, CN^- and N_2 are isoelectronic. But in contrast to CN^-, N_2 is chemically inert, because of
(a) low bond energy [1992 - 1 Mark]
(b) absence of bond polarity
(c) unsymmetrical electron distribution
(d) presence of more number of electrons in bonding orbitals

15. The ion that is isoelectronic with CO is [1982 - 1 Mark]
(a) CN^- (b) O_2^+ (c) O_2^- (d) N_2^+

16. The total number of electrons that take part in forming the bond in N_2 is [1980]
(a) 2 (b) 4 (c) 6 (d) 10

17. The octet rule is not valid for the molecule [1979]
(a) CO_2 (b) H_2O (c) O_2 (d) CO

2 Integer Value Answer

18. Consider the following compounds in the liquid form:
O_2, HF, H_2O, NH_3, H_2O_2, CCl_4, $CHCl_3$, C_6H_6, C_6H_5Cl.
When a charged comb is brought near their flowing stream, how many of them show deflection as per the following figure? [Adv. 2020]

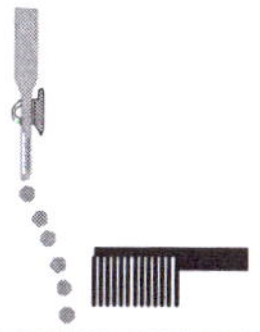

6 MCQs with One or More than One Correct Answer

19. The option(s) in which at least three molecules follow Octet Rule is (are) [Adv. 2024]
(a) CO_2, C_2H_4, NO and HCl (b) NO_2, O_3, HCl and H_2SO_4
(c) BCl_3, NO, NO_2 and H_2SO_4 (d) CO_2, BCl_3, O_3 and C_2H_4

9 Assertion and Reason/Statement Type Questions

Each question contains STATEMENT-1 (Assertion) and STATEMENT-2 (Reason). Each question has 4 choices (a), (b), (c) and (d) out of which ONLY ONE is correct. Mark your answer as
(a) If both Statement -1 and Statement -2 are correct, and Statement -2 is the correct explanation of the Statement -2.
(b) If both Statement-1 and Statement-2 are correct, but Statement-2 is not the correct explanation of the Statement-1.
(c) If Statement -1 is correct but Statement -2 is incorrect.
(d) If Statement -1 is incorrect but Statement -2 is correct.

20. Read the following Assertion and Reason and answer as per the options given below : [1998 - 2 Marks]

Assertion : The electronic structure of O_3 is 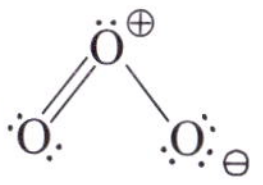

Reason : 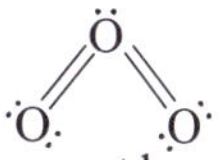 structure is not allowed because octet around O cannot be expanded.

10 Subjective Problems

21. Write the Lewis dot structure of the following :

$O_3, COCl_2$ [1986 - 1 Mark]

22. Write the Lewis dot structural formula for each of the following. Give, also, the formula of a neutral molecule, which has the same geometry and the same arrangement of the bonding electrons as in each of the following. An example is given below in the case of H_3O^+ :

$\left[\begin{array}{c} H \\ H:\ddot{O}:H \end{array}\right]^+$ $\left[\begin{array}{c} H \\ H:\ddot{N}:H \end{array}\right]$

Lewis dot structure | Neutral molecule

(i) O_2^{2-}; (ii) CO_3^{2-}; (iii) CN^-; (iv) NCS^- [1983 - 1 × 4 = 4 Marks]

23. Water is liquid while H_2S is a gas at room temperature. Why? [1978]

Topic 3 - Dipole Moment and Fajan's Rule

1 MCQs with One Correct Answer

24. The correct order of dipole moments for the given species is [Adv. 2026]
(a) $BF_3 = NH_4^+ < NF_3 < NH_3$
(b) $BF_3 < NH_4^+ < NF_3 < NH_3$
(c) $NH_4^+ < BF_3 < NH_3 < NF_3$
(d) $BF_3 < NH_4^+ < NH_3 < NF_3$

25. The geometry of H_2S and its dipole moment are [1999 - 2 Marks]
(a) angular and non-zero (b) angular and zero
(c) linear and non-zero (d) linear and zero

26. The critical temperature of water is higher than that of O_2 because the H_2O molecule has [1997 - 1 Mark]
(a) fewer electrons than O_2 (b) two covalent bonds
(c) V-shape (d) dipole moment.

27. Which contains both polar and non-polar bonds? [1997 - 1 Mark]
(a) NH_4Cl (b) HCN (c) H_2O_2 (d) CH_4

28. The molecule which has zero dipole moment is : [1989 - 1 Mark]
(a) CH_2Cl_2 (b) BF_3 (c) NF_3 (d) ClO_2

29. Carbon tetrachloride has no net dipole moment because of [1983 - 1 Mark]
(a) its planar structure
(b) its regular tetrahedral structure
(c) similar sizes of carbon and chlorine
(d) similar electron affinities of carbon and chlorine

30. The compound with no dipole moment is [1982 - 1 Mark]
(a) methyl chloride (b) carbon tetrachloride
(c) methylene chloride (d) chloroform

31. If a molecule MX_3 has zero dipole moment, the sigma bonding orbitals used by M (atomic number < 21) are [1981 - 1 Mark]
(a) pure *p* (b) *sp* hybrid
(c) sp^2 hybrid (d) sp^3 hybrid

3 Numeric / New Stem Based Questions

32. The dipole moment of KCl is 3.336×10^{-29} Coulomb meters which indicates that it is a highly polar molecule. The interatomic distance between K^+ and Cl^- in this molecule is 2.6×10^{-10} m. Calculate the dipole moment of KCl molecule if there were opposite charges of one fundamental unit located at each nucleus. Calculate the percentage ionic character of KCl. [1993 - 2 Marks]

5 True / False

33. The dipole moment of CH_3F is greater than that of CH_3Cl. [1993 - 1 Mark]

34. The presence of polar bonds in a poly-atomic molecule suggests that the molecule has non-zero dipole moment. [1990 - 1 Mark]

35. All molecules with polar bonds have dipole moment. [1985 - ½ Mark]

6 MCQs with One or More than One Correct Answer

36. Each of the following options contains a set of four molecules. Identify the option(s) were all four molecules posses permanent dipole moment at room temperature. [Adv. 2019]
(a) BF_3, O_3, SF_6, XeF_6
(b) $NO_2, NH_3, POCl_3, CH_3Cl$
(c) $SO_2, C_6H_5Cl, H_2Se, BrF_5$
(d) $BeCl_2, CO_2, BCl_3, CHCl_3$

37. The molecules that will have dipole moment are [1992 - 1 Mark]
(a) 2, 2-dimethylpropane
(b) *trans*-2-pentene
(c) *cis*-3-hexene
(d) 2,2,3,3-tetramethylbutane

Topic 4 - VSEPR Theory, Valence Bond Theory and Concept of Hybridisation

1 MCQs with One Correct Answer

38. The species having pyramidal shape is : [2010]
(a) SO_3 (b) BrF_3 (c) SiO_3^{2-} (d) OSF_2

39. Which species has the maximum number of lone pair of electrons on the central atom? [2005S]
(a) $[ClO_3]^-$ (b) XeF_4 (c) SF_4 (d) $[I_3]^-$

40. Which of the following are isoelectronic and isostructural? $NO_3^-, CO_3^{2-}, ClO_3^-, SO_3$ [2003S]
(a) NO_3^-, CO_3^{2-} (b) SO_3, NO_3^-
(c) ClO_3^-, CO_3^{2-} (d) CO_3^{2-}, SO_3

41. Specify the coordination geometry around and hybridisation of N and B atoms in a 1 : 1 complex of BF_3 and NH_3 [2002S]
(a) N : tetrahedral, sp^3; B : tetrahedral, sp^3
(b) N : pyramidal, sp^3; B : pyramidal, sp^3
(c) N : pyramidal, sp^3; B : planar, sp^2
(d) N : pyramidal, sp^3; B : tetrahedral, sp^3

42. The correct order of hybridization of the central atom in the following species NH_3, $[PtCl_4]^{2-}$, PCl_5 and BCl_3 is [2001S]
(a) dsp^2, dsp^3, sp^2, sp^3 (b) sp^3, dsp^2, sp^3d, sp^2
(c) dsp^2, sp^2, sp^3, dsp^3 (d) dsp^2, sp^3, sp^2, dsp^3

43. The hybridisation of atomic orbitals of nitrogen in NO_2^+, NO_3^- and NH_4^+ are [2000S]
(a) sp, sp^3 and sp^2 respectively
(b) sp, sp^2 and sp^3 respectively
(c) sp^2, sp and sp^3 respectively
(d) sp^2, sp^3 and sp respectively

44. Molecular shapes of SF_4, CF_4 and XeF_4 are [2000S]
(a) the same, with 2, 0 and 1 lone pairs of electrons respectively

(b) the same, with 1, 1 and 1 lone pairs of electrons respectively
(c) different, with 0, 1 and 2 lone pairs of electrons respectively
(d) different, with 1, 0 and 2 lone pairs of electrons respectively

45. The geometry and the type of hybrid orbital present about the central atom in BF_3 is [1998 - 2 Marks]
(a) linear, sp (b) trigonal planar, sp^2
(c) tetrahedral, sp^3 (d) pyramidal, sp^3.

46. Which one of the following compounds has sp^2 hydridization? [1997 - 1 Mark]
(a) CO_2 (b) SO_2 (c) N_2O (d) CO

47. Among the following species, identify the isostructural pairs.
NF_3, NO_3^-, BF_3, H_3O^+, HN_3 [1996 - 1 Mark]
(a) $[NF_3, NO_3^-]$ and $[BF_3, H_3O^+]$
(b) $[NF_3, HN_3]$ and $[NO_3^-, BF_3]$
(c) $[NF_3, H_3O^+]$ and $[NO_3^-, BF_3]$
(d) $[NF_3, H_3O^+]$ and $[HN_3, BF_3]$

48. The type of hybrid orbitals used by the chlorine atom in ClO_2^- is [1992 - 1 Mark]
(a) sp^3 (b) sp^2
(c) sp (d) none of these

49. The compound in which $\overset{*}{C}$ uses its sp^3 hybrid orbitals for bond formation is : [1989 - 1 Mark]
(a) $H\overset{*}{C}OOH$ (b) $(H_2N)_2\overset{*}{C}O$
(c) $(CH_3)_3\overset{*}{C}OH$ (d) $CH_3\overset{*}{C}HO$

50. The molecule which has pyramidal shape is : [1989 - 1 Mark]
(a) PCl_3 (b) SO_3 (c) CO_3^{2-} (d) NO_3^-

51. The species in which the central atom uses sp^2 hybrid orbitals in its bonding is [1988 - 1 Mark]
(a) PH_3 (b) NH_3 (c) CH_3^+ (d) SbH_3

52. The molecule that has linear structure is [1988 - 1 Mark]
(a) CO_2 (b) NO_2 (c) SO_2 (d) SiO_2

53. The hybridisation of sulphur in sulphur dioxide is : [1986 - 1 Mark]
(a) sp (b) sp^3 (c) sp^2 (d) dsp^2

54. On hybridization of one s and one p orbitals we get : [1984 - 1 Mark]
(a) two mutually perpendicular orbitals
(b) two orbitals at 180°
(c) four orbitals directed tetrahedrally
(d) three orbitals in a plane

2 Integer Value Answer

55. A mong $[I_3]^+$, $[SiO_4]^{4-}$, SO_2Cl_2, XeF_2, SF_4, ClF_3, $Ni(CO)_4$, XeO_2F_2, $[PtCl_4]^{2-}$, XeF_4, and $SOCl_2$, the total number of species having sp^3 hybridised central atom is ______. [Adv. 2023]

56. The sum of the number of lone pairs of electrons on each central atom in the following species is
$[TeBr_6]^{2-}$, $[BrF_2]^+$, SNF_3 and $[XeF_3]^-$
(Atomic numbers: N = 7, F = 9, S = 16, Br = 35, Te = 52, Xe = 54) [Adv. 2017]

57. Among the triatomic molecules/ions, $BeCl_2$, N_3^-, N_2O, NO_2^+, O_3, SCl_2, ICl_2^-, I_3^- and XeF_2, the total number of linear molecule(s)/ion(s) where the hybridization of the central atom does not have contribution from the d-orbital(s) is
[Atomic number : S = 16, Cl = 17, I = 53 and Xe = 54] [Adv. 2015]

58. A list of species having the formula XZ_4 is given below.
XeF_4, SF_4, SiF_4, BF_4^-, BrF_4^-, $[Cu(NH_3)_4]^{2+}$, $[FeCl_4]^{2-}$, $[CoCl_4]^{2-}$ and $[PtCl_4]^{2-}$.
Defining shape on the basis of the location of X and Z atoms, the total number of species having a square planar shape is [Adv. 2014]

59. Based on VSEPR theory, the number of 90 degree F–Br–F angles in BrF_5 is [2010]

4 Fill in the Blanks

60. The shape of $[CH_3]^+$ is .... [1990 - 1 Mark]

61. The angle between two covalent bonds is maximum in .... (CH_4, H_2O, CO_2) [1981 - 1 Mark]

62. .... hybrid orbitals of nitrogen atom are involved in the formation of ammonium ion. [1982 - 1 Mark]

5 True / False

63. sp^2 hybrid orbitals have equal s and p character. [1987 - 1 Mark]

64. In benzene, carbon uses all the three p-orbitals for hybridisation. [1987 - 1 Mark]

65. $SnCl_2$ is a non-linear molecule. [1985 - ½ Mark]

6 MCQs with One or More than One Correct Answer

66. The compound(s) with TWO lone pairs of electrons on the central atom is(are) [Adv. 2016]
(a) BrF_5 (b) ClF_3 (c) XeF_4 (d) SF_4

67. The linear structure is assumed by : [1991 - 1 Mark]
(a) $SnCl_2$ (b) NCO^-
(c) CS_2 (d) NO_2^+

68. CO_2 is isostructural with : [1986 - 1 Mark]
(a) $HgCl_2$ (b) $SnCl_2$ (c) C_2H_2 (d) NO_2

7 Match the Following

69. Consider the following species: [Adv. 2026]
$SOCl_2$, $XeOF_4$, ClF_3, ClF_5, XeF_5^+, SO_3^{2-}, XeF_3^+, SF_4
List-I contains different molecular shapes and **List-II** contains total number of species with the same molecular shapes from the given species. Match each entry in **List-I** with the appropriate entry in **List-II** and choose the correct option.

List-I	List-II
(P) See-saw	(1) one
(Q) T-Shaped	(2) two
(R) Trigonal Planar	(3) three
(S) Square Pyramidal	(4) four
	(5) zero

(a) P → 1, Q → 2, R → 5, S → 3
(b) P → 5, Q → 4, R → 2, S → 3
(c) P → 3, Q → 2, R → 1, S → 4
(d) P → 1, Q → 3, R → 5, S → 4

70. Based on **VSEPR** model, match the xenon compounds given in **List-I** with the corresponding geometries and the number of lone pairs on xenon given in **List-II** and choose the correct option. [Adv. 2024)]

List-I	List-II
(P) XeF_2	(1) Trigonal bipyramidal and two lone pair of electrons
(Q) XeF_4	(2) Tetrahedral and one lone pair of electrons

(R) XeO_3 (3) Octahedral and two lone pair of electrons
(S) XeO_3F_2 (4) Trigonal bipyramidal and no lone pair of electrons
(5) Trigonal bipyramidal and three lone pair of electrons

(a) P-5, Q-2, R-3, S-1 (b) P-5, Q-3, R-2, S-4
(c) P-4, Q-3, R-2, S-1 (d) P-4, Q-2, R-5, S-3

10 Subjective Problems

71. Draw the structure of XeF_4 and OSF_4 according to VSEPR theory, clearly indicating the state of hybridisation of the central atom and lone pair of electrons (if any) on the central atom. [2004 - 2 Marks]

72. Using VSEPR theory, draw the shape of PCl_5 and BrF_5. [2003 - 2 Marks]

73. Interpret the non-linear shape of H_2S molecule and non-planar shape of PCl_3 using valence shell electron pair repulsion (VSEPR) theory. (Atomic numbers : H = 1, P = 15, S = 16, Cl = 17.) [1998 - 4 Marks]

74. Using the VSEPR theory, identify the type of hybridization and draw the structure of OF_2. What are the oxidation states of O and F ? [1994 - 3 Marks]

Topic 5 - Molecular Orbital Theory and Hydrogen Bonding

1 MCQs with One Correct Answer

75. The correct molecular orbital diagram for F_2 molecule in the ground state is [Adv. 2023]

(a)
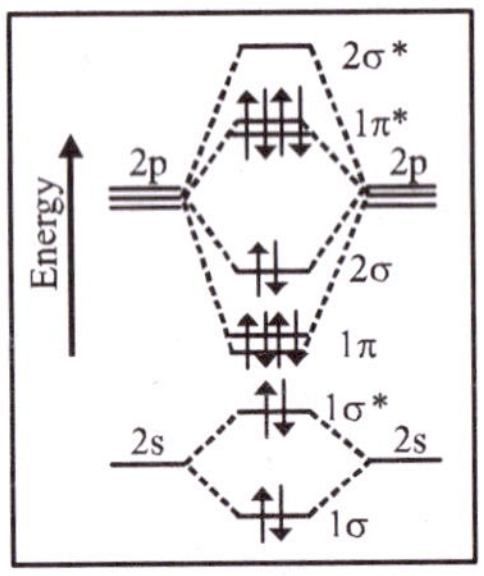

(b)
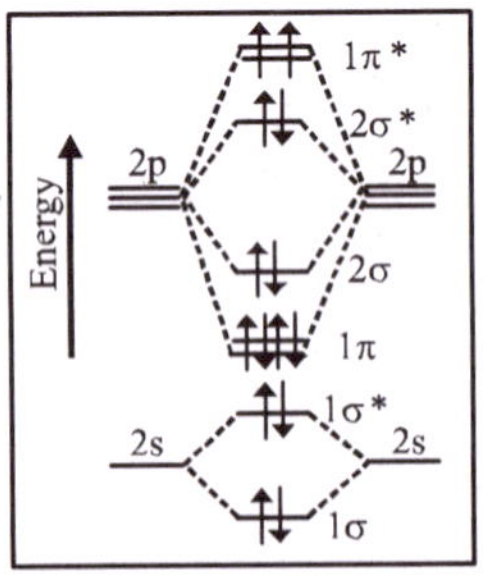

(c)
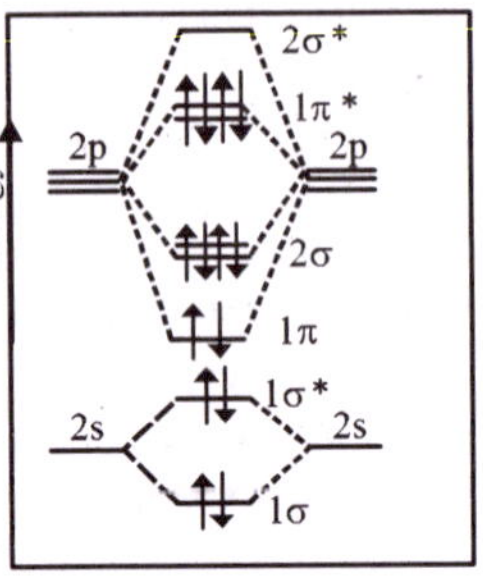

(d)
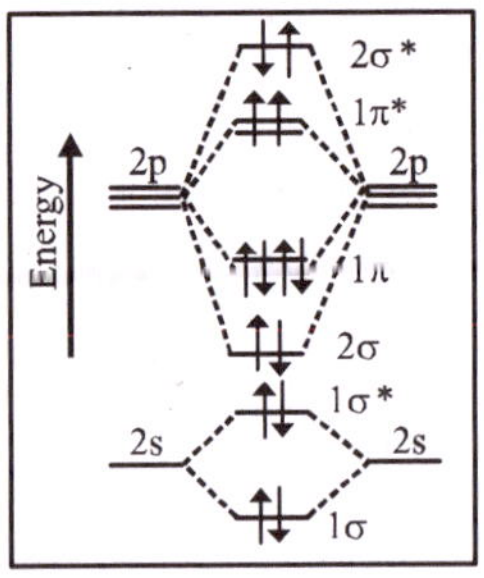

76. Assuming 2*s*-2*p* mixing is **NOT** operative, the paramagnetic species among the following is [Adv. 2014]
(a) Be_2 (b) B_2 (c) C_2 (d) N_2

77. Assuming that Hund's rule is violated, the bond order and magnetic nature of the diatomic molecule B_2 is [2010]
(a) 1 and diamagnetic (b) 0 and dimagnetic
(c) 1 and paramagnetic (d) 0 and paramagnetic

78. The species having bond order different from that in CO is [2007]
(a) NO^- (b) NO^+ (c) CN^- (d) N_2

79. Among the following, the paramagnetic compound is [2007]
(a) Na_2O_2 (b) O_3 (c) N_2O (d) KO_2

80. According to molecular orbital theory, which of the following statement about the magnetic character and bond order is correct regarding O_2^+ ? [2004S]
(a) Paramagnetic and Bond order < O_2
(b) Paramagnetic and Bond order > O_2
(c) Diamagnetic and Bond order < O_2
(d) Diamagnetic and Bond order > O_2

81. Which of the following molecular species has unpaired electron(s) ? [2002S]
(a) N_2 (b) F_2 (c) O_2^- (d) O_2^{2-}

82. Identify the least stable ion amongst the following : [2002S]
(a) Li^- (b) Be^- (c) B^- (d) C^-

83. The common features among the species CN^-, CO and NO^+ are [2001S]
(a) bond order three and isoelectronic
(b) bond order three and weak field ligands
(c) bond order two and π–acceptors
(d) isoelectronic and weak field ligands

84. The correct order of increasing C — O bond length of CO, CO_3^{2-}, CO_2, is [1999 - 2 Marks]
(a) $CO_3^{2-} < CO_2 < CO$ (b) $CO_2 < CO_3^{2-} < CO$
(c) $CO < CO_3^{2-} < CO_2$ (d) $CO < CO_2 < CO_3^{2-}$

85. The number and type of bonds between two carbon atoms in CaC_2 are : [1996 - 1 Mark]
(a) one sigma (σ) and one pi (π) bonds
(b) one sigma (σ) and two pi (π) bonds
(c) one sigma (σ) and one and a half pi (π) bonds
(d) one sigma (σ) bond.

86. Number of paired electrons in O_2 molecule is : [1995S]
(a) 7 (b) 8 (c) 16 (d) 14

87. The maximum possible number of hydrogen bonds a water molecule can form is [1992 - 1 Mark]
(a) 2 (b) 4 (c) 3 (d) 1

88. Which of the following is paramagnetic? **[1989 - 1 Mark]**
(a) O_2^- (b) CN^- (c) CO (d) NO^+

89. Hydrogen bonding is maximum in **[1987 - 1 Mark]**
(a) Ethanol (b) Diethyl ether
(c) Ethyl chloride (d) Triethylamine

90. The molecule having one unpaired electron is: **[1985 - 1 Mark]**
(a) NO (b) CO (c) CN^- (d) O_2

91. The hydrogen bond is strongest in : **[1986 - 1 Mark]**
(a) O–H..........S (b) S–H..........O
(c) F–H..........F (d) F–H..........O

92. Which one among the following does not have the hydrogen bond? **[1983 - 1 Mark]**
(a) phenol (b) liquid NH_3
(c) water (d) liquid HCl

2 Integer Value Answer

93. Thermal decomposition of $AgNO_3$ produces two paramagnetic gases. The total number of electrons present in the antibonding molecular orbitals of the gas that has the higher number of unpaired electrons is _______. **[Adv. 2022]**

3 Numeric / New Stem Based Questions

94. Among H_2, He_2^+, Li_2, Be_2, B_2, C_2, N_2, O_2^- and F_2, the number of diamagnetic species is
(Atomic numbers: H = 1, He = 2, Li = 3, Be = 4, B = 5, C = 6, N = 7, O = 8, F = 9) **[Adv. 2017]**

4 Fill in the Blanks

95. When N_2 goes to N_2^+, the N–N bond distance ..., and when O_2 goes to O_2^+ the O–O bond distance.... **[1996 - 1 Mark]**

96. Pair of molecules which forms strongest intermolecular hydrogen bond is ....
(SiH_4 and SiF_4, $CH_3-\underset{\underset{O}{||}}{C}-CH_3$ and $CHCl_3$, $H-\underset{\underset{O}{||}}{C}-OH$ and $CH_3-\underset{\underset{O}{||}}{C}-OH$) **[1981 - 1 Mark]**

6 MCQs with One or More than One Correct Answer

97. Regarding the molecular orbital (MO) energy levels for homonuclear diatomic molecules, the **INCORRECT** statement(s) is(are) **[Adv. 2025]**
(a) Bond order of Ne_2 is zero.
(b) The highest occupied molecular orbital (HOMO) of F_2 is σ-type.
(c) Bond energy of O_2^+ is smaller than the bond energy of O_2.
(d) Bond length of Li_2 is larger than the bond length of B_2.

98. For diatomic molecules, the correct statement(s) about the molecular orbitals formed by the overlap of two $2p_z$ orbitals is (are) **[Adv. 2022]**
(a) σ orbital has a total of two nodal planes.
(b) σ* orbital has one node in the *xz*-plane containing the molecular axis.
(c) π orbital has one node in the plane which is perpendicular to the molecular axis and goes through the center of the molecule.
(d) π* orbital has one node in the *xy*-plane containing the molecular axis.

99. According to Molecular Orbital Theory, **[Adv. 2016]**
(a) C_2^{2-} is expected to be diamagnetic
(b) O_2^{2+} is expected to have a longer bond length than O_2
(c) N_2^+ and N_2^- have the same bond order
(d) He_2^+ has the same energy as two isolated He atoms

7 Match the Following

100. Match the orbital overlap figures shown in List-I with the description given in List-II and select the correct answer using the code given below the lists. **[Adv. 2014]**

List-I	List-II
P. 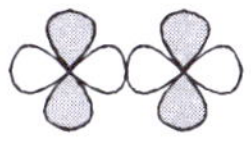	1. $p-d\ \pi$ antibonding
Q. 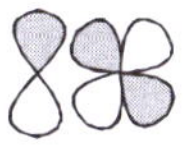	2. $d-d\ \sigma$ bonding
R. 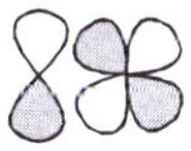	3. $p-d\ \pi$ bonding
S. 	4. $d-d\ \sigma$ antibonding

Code:

	P	Q	R	S
(a)	2	1	3	4
(b)	4	3	1	2
(c)	2	3	1	4
(d)	4	1	3	2

101. Match each of the diatomic molecules in **Column I** with its property/properties in **Column II**. **[2009]**

Column I	Column II
(A) B_2	(p) Paramagnetic
(B) N_2	(q) Undergoes oxidation
(C) O_2^-	(r) Undergoes reduction
(D) O_2	(s) Bond order ≥ 2
	(t) Mixing of 's' and 'p' orbital

10 Subjective Problems

102. Arrange the following :
(i) N_2, O_2, F_2, Cl_2 in increasing order of bond dissociation energy. **[1988 - 1 Mark]**
(ii) Increasing strength of hydrogen bonding (X–H–X) : O, S, F, Cl, N **[1991 - 1 Mark]**
(iii) In the decreasing order of the O – O bond length present in them **[2004 - 4 Marks]**
O_2, KO_2 and $O_2[AsF_4]$

103. How many sigma bonds and how many pi-bonds are present in a benzene molecule? **[1985 - 1 Mark]**

Answer Key

Part-A (JEE Main)

1	(1)	27	(3)	53	(4)	79	(3)	105	(3)	131	(3)	157	(3)	183	(2)	209	(2)	235	(3)
2	(2)	28	(1)	54	(2)	80	(1)	106	(2)	132	(1)	158	(4)	184	(1)	210	(1)	236	(2)
3	(3)	29	(3)	55	(3)	81	(1)	107	(3)	133	(2)	159	(3)	185	(1)	211	(4)	237	(3)
4	(N)	30	(1)	56	(4)	82	(4)	108	(1)	134	(4)	160	(4)	186	(1)	212	(4)	238	(2)
5	(1)	31	(1)	57	(2)	83	(1)	109	(3)	135	(4)	161	(3)	187	(1)	213	(4)	239	(6)
6	(2)	32	(3)	58	(6)	84	(2)	110	(2)	136	(1)	162	(5)	188	(2)	214	(4)	240	(6)
7	(3)	33	(1)	59	(5)	85	(1)	111	(2)	137	(2)	163	(6)	189	(1)	215	(2)	241	(4)
8	(3)	34	(3)	60	(2.5)	86	(1)	112	(4)	138	(1)	164	(4)	190	(3)	216	(2)	242	(2)
9	(3)	35	(3)	61	(3)	87	(4)	113	(3)	139	(1)	165	(1)	191	(3)	217	(3)	243	(5)
10	(1)	36	(2)	62	(4)	88	(2)	114	(3)	140	(3)	166	(2)	192	(1)	218	(2)	244	(2)
11	(1)	37	(2)	63	(3)	89	(3)	115	(4)	141	(2)	167	(3)	193	(3)	219	(2)	245	(1)
12	(4)	38	(4)	64	(3)	90	(3)	116	(1)	142	(2)	168	(2)	194	(2)	220	(1)	246	(4)
13	(3)	39	(1)	65	(3)	91	(2)	117	(1)	143	(2)	169	(3)	195	(1)	221	(1)	247	(22)
14	(N)	40	(1)	66	(1)	92	(2)	118	(2)	144	(4)	170	(2)	196	(4)	222	(1, 2)	248	(6)
15	(2)	41	(3)	67	(3)	93	(4)	119	(4)	145	(1)	171	(3)	197	(2)	223	(1)	249	(4)
16	(3)	42	(2)	68	(4)	94	(3)	120	(1)	146	(1)	172	(3)	198	(2)	224	(3)	250	(3)
17	(4)	43	(3)	69	(4)	95	(3)	121	(1)	147	(1)	173	(1)	199	(1)	225	(2)	251	(2)
18	(98)	44	(2)	70	(2)	96	(1)	122	(1)	148	(1)	174	(3)	200	(1)	226	(4)	252	(0)
19	(12)	45	(3)	71	(2)	97	(3)	123	(2)	149	(4)	175	(4)	201	(4)	227	(N)	253	(0)
20	(6)	46	(4)	72	(1)	98	(4)	124	(1)	150	(6)	176	(3)	202	(1)	228	(2)	254	(10)
21	(12)	47	(3)	73	(2)	99	(2)	125	(4)	151	(6)	177	(N)	203	(2)	229	(2)	255	(15)
22	(8)	48	(2)	74	(3)	100	(1)	126	(3)	152	(6)	178	(2)	204	(3)	230	(4)		
23	(3)	49	(2)	75	(1)	101	(4)	127	(2)	153	(3)	179	(1)	205	(2)	231	(1)		
24	(6)	50	(4)	76	(2)	102	(2)	128	(4)	154	(3)	180	(4)	206	(1)	232	(2)		
25	(4)	51	(2)	77	(2)	103	(4)	129	(2)	155	(8)	181	(2)	207	(4)	233	(3)		
26	(2)	52	(1)	78	(1)	104	(2)	130	(4)	156	(4)	182	(4)	208	(4)	234	(3)		

Part-B (JEE Advanced)

1	(c)	12	True	26	(d)	37	(b, c)	48	(a)	59	(4)	70	(6)	85	(b)	97	(b, c)		
2	(d)	13	(c)	27	(c)	38	(d)	49	(c)	60	(Planar)	75	(c)	86	(d)	98	(a, d)		
3	(d)	14	(b)	28	(b)	39	(d)	50	(a)	61	CO_2	76	(c)	87	(b)	99	(a, c)		
4	(c)	15	(a)	29	(b)	40	(a)	51	(c)	62	sp^3	77	(a)	88	(a)	100	(c)		
5	(b)	16	(c)	30	(b)	41	(a)	52	(a)	63	False	78	(a)	89	(a)	101	(A) –p, r, t ; (B) – s,t ; (C) – p, q ; (D) – p, q, s		
6	(a)	17	(d)	31	(c)	42	(b)	53	(c)	64	False	79	(d)	90	(a)				
7	(a)	18	(6)	32	(80.09)	43	(b)	54	(b)	65	True	80	(b)	91	(c)				
8	(c)	19	(a, d)	33	False	44	(d)	55	(5)	66	(b, c)	81	(c)	92	(d)				
9	(4)	20	(a)	34	False	45	(b)	56	(6)	67	(b,c,d)	82	(b)	93	(6)				
10	(Banana bond)	24	(a)	35	False	46	(b)	57	(4)	68	(a, c)	83	(a)	94	(6)				
11	2	25	(a)	36	(b, c)	47	(c)	58	(4)	69	(a)	84	(d)	95	(Increases, decreases)				

Chapter-5

States of Matter

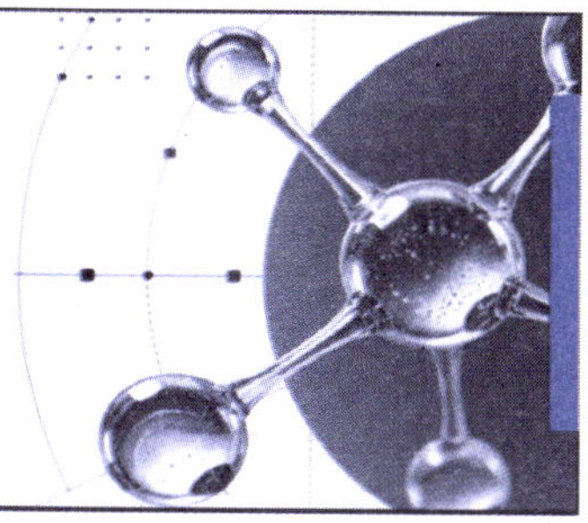

Part-A (JEE Main)

Topic 1 - Intermolecular Forces, Gas Laws and Ideal Gas Equation

MULTIPLE CHOICE QUESTIONS (MCQs)

1. Which amongst the given plots is the correct plot for pressure (p) vs density (d) for an ideal gas? [June 27, 2022 (II)]

(1) $T_3 > T_2 > T_1$

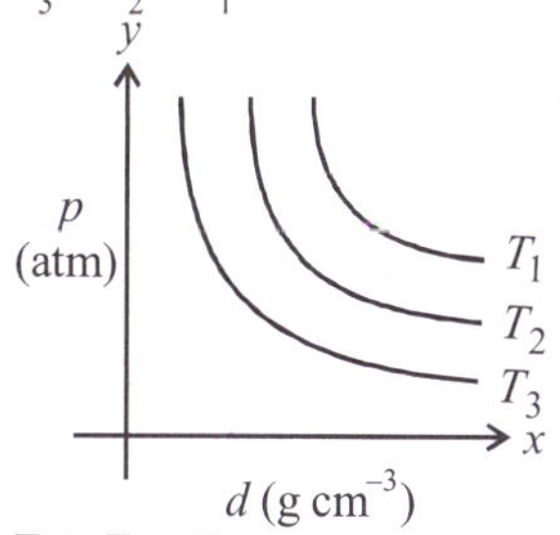

(2) $T_3 > T_2 > T_1$

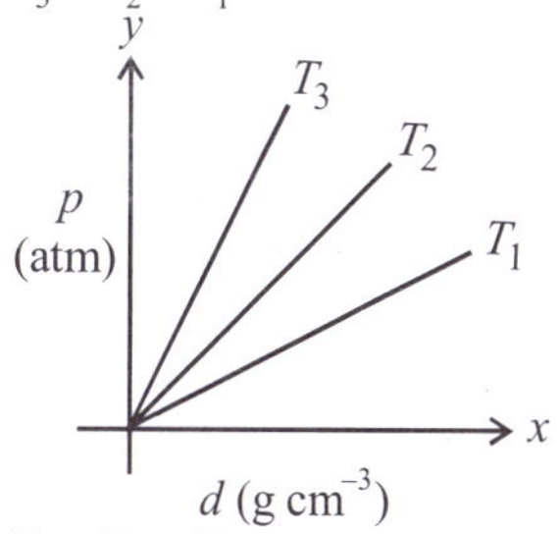

(3) $T_3 > T_2 > T_1$

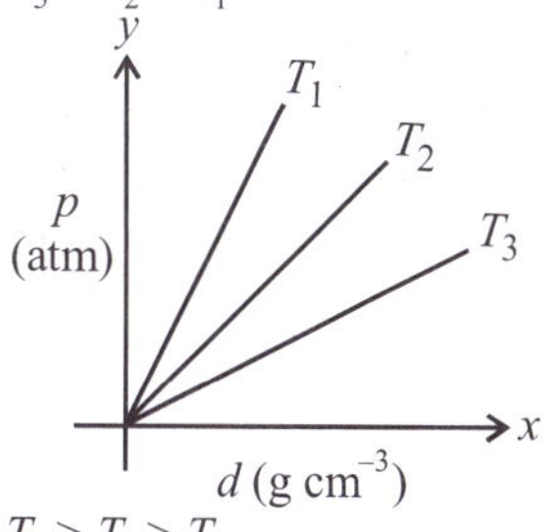

(4) $T_3 > T_2 > T_1$

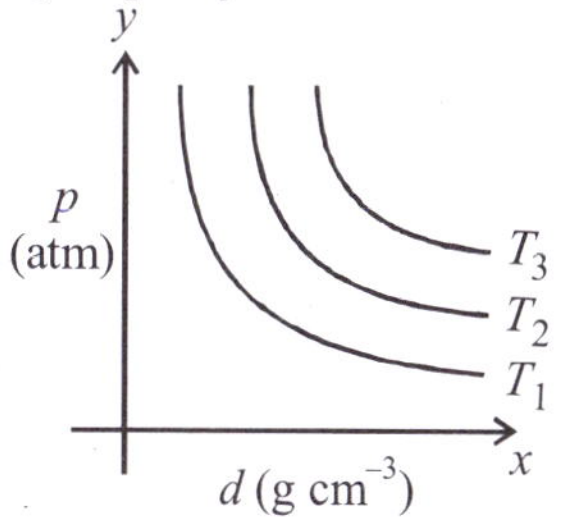

2. An evacuated glass vessel weighs 40.0 g when empty, 135.0 g when filled with a liquid of density 0.95 g mL^{-1} and 40.5 g when filled with an ideal gas at 0.82 atm at 250 K. The molar mass of the gas in g mol^{-1} is :
(Given : R = 0.082 L atm K^{-1} mol^{-1}) [June 26, 2022 (I)]
(1) 35 (2) 50 (3) 75 (4) 125

3. The interaction energy of London forces between two particles is proportional to r^x, where r is the distance between the particles. The value of x is : [Aug. 26, 2021 (II)]
(1) 3 (2) –3 (3) –6 (4) 6

4. A mixture of one mole each of H_2, He and O_2 each are enclosed in a cylinder of volume V at temperature T. If the partial pressure of H_2 is 2 atm, the total pressure of the gases in the cylinder is : [Sep. 03, 2020 (II)]
(1) 6 atm (2) 38 atm (3) 14 atm (4) 22 atm

5. Which one of the following graphs is **not** correct for ideal gas ? [Sep. 02, 2020 (I)]

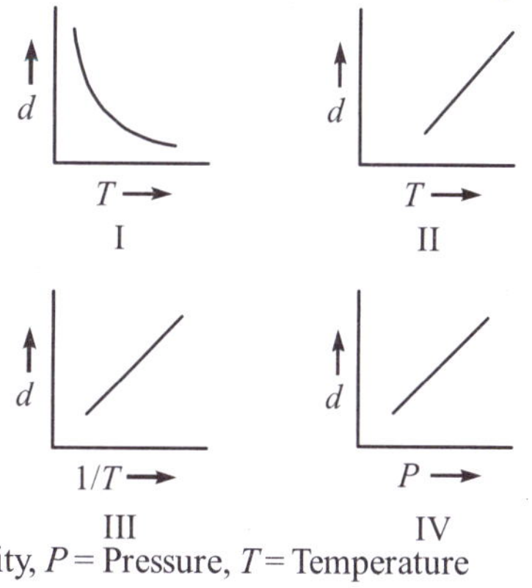

d = Density, P = Pressure, T = Temperature
(1) I (2) II (3) IV (4) III

6. Match the type of interaction in column A with the distance dependence of their interaction energy in column B :

A	B
(I) ion-ion	(A) $1/r$
(II) dipole-dipole	(B) $1/r^2$
(III) London dispersion	(C) $1/r^3$
	(D) $1/r^6$

[Sep. 02, 2020 (II)]
(1) (I)-(B), (II)-(D), (III)-(C) (2) (I)-(A), (II)-(B), (III)-(D)
(3) (I)-(A), (II)-(B), (III)-(C) (4) (I)-(A), (II)-(C), (III)-(D)

7. The predominant intermolecular forces present in ethyl acetate, a liquid, are: [Jan. 08, 2020 (I)]
(1) London dispersion and dipole-dipole
(2) hydrogen bonding and London dispersion
(3) Dipole-dipole and hydrogen bonding
(4) London dispersion, dipole-dipole and hydrogen bonding

8. The relative strength of interionic/ intermolecular forces in decreasing order is: [Jan. 07, 2020 (I)]
(1) dipole-dipole > ion-dipole > ion-ion
(2) ion-dipole > ion-ion > dipole-dipole
(3) ion-dipole > dipole-dipole > ion-ion
(4) ion-ion > ion-dipole > dipole-dipole

9. 0.5 moles of gas A and x moles of gas B exert a pressure of 200 Pa in a container of volume 10 m^3 at 1000 K. Given, R is the gas constant in JK^{-1} mol^{-1}, x is: [Jan. 9, 2019 (I)]
(1) $\frac{2R}{4+R}$ (2) $\frac{2R}{4-R}$ (3) $\frac{4+R}{2R}$ (4) $\frac{4-R}{2R}$

10. An open vessel at 27 °C is heated until two fifth of the air (assumed as an ideal gas) in it has escaped from the vessel. Assuming that the volume of the vessel remains constant, the temperature at which the vessel has been heated is : [Online May 19, 2012, Jan. 12, 2019 (II)]
(1) 500 °C (2) 500 K (3) 750 °C (4) 750 K

11. Assuming ideal gas behaviour, the ratio of density of ammonia to that of hydrogen chloride at same temperature and pressure is: (Atomic wt. of Cl = 35.5 u) [Online April 16, 2018]
(1) 1.46 (2) 1.64 (3) 0.46 (4) 0.64

12. Among the following, the incorrect statement is : [Online April 11, 2015, Online April 8, 2017]
(1) At low pressure, real gases show ideal behaviour.
(2) At very low temperature, real gases show ideal behaviour.
(3) At very large volume, real gases show ideal behaviour.
(4) At Boyle's temperature, real gases show ideal behaviour.

13. At 300 K, the density of a certain gaseous molecule at 2 bar is double to that of dinitrogen (N_2) at 4 bar. The molar mass of gaseous molecule is : [Online April 9, 2017]
(1) 28 g mol^{-1} (2) 56 g mol^{-1}
(3) 112 g mol^{-1} (4) 224 g mol^{-1}

14. Two closed bulbs of equal volume (V) containing an ideal gas initially at pressure p_i and temperature T_1 are connected through a narrow tube of negligible volume. The temperature of one of the bulbs is then raised to T_2. The final pressure p_f is : [2016]

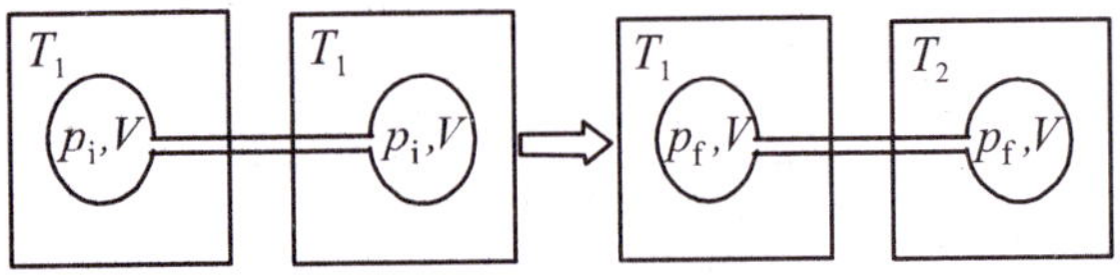

(1) $2p_i\left(\frac{T_2}{T_1+T_2}\right)$ (2) $2p_i\left(\frac{T_1T_2}{T_1+T_2}\right)$
(3) $p_i\left(\frac{T_1T_2}{T_1+T_2}\right)$ (4) $2p_i\left(\frac{T_1}{T_1+T_2}\right)$

15. The initial volume of a gas cylinder is 750.0 mL. If the pressure of gas inside the cylinder changes from 840.0 mm Hg to 360.0 mm Hg, the final volume the gas will be: [Online April 11, 2014]
(1) 1.750 L (2) 3.60 L (3) 4.032 L (4) 7.50 L

16. Sulphur dioxide and oxygen were allowed to diffuse through a porous partition. 20 dm^3 of SO_2 diffuses through the porous partition in 60 seconds. The volume of O_2 in dm^3 which diffuses under the similar condition in 30 seconds will be (atomic mass of sulphur = 32 u): [Online April 19, 2014]
(1) 7.09 (2) 14.1 (3) 10.0 (4) 28.2

17. For 1 mol of an ideal gas at a constant temperature T, the plot of (log P) against (log V) is a (P : Pressure, V : Volume)
(1) Straight line parallel to x-axis. [Online May 7, 2012]
(2) Straight line with a negative slope.
(3) Curve starting at origin.
(4) Straight line passing through origin.

18. When CO_2(g) is passed over red hot coke, it partially gets reduced to CO(g). Upon passing 0.5 L of CO_2(g) over red hot coke, the total volume of the gases increased to 700 mL. The composition of the gaseous mixture at STP is
(1) CO_2 = 300 mL; CO = 400 mL [Online May 19, 2012]
(2) CO_2 = 0.0 mL; CO = 700 mL
(3) CO_2 = 200 mL; CO = 500 mL
(4) CO_2 = 350 mL; CO = 350 mL

19. The intermolecular interaction that is dependent on the inverse cube of distance between the molecules is: [2015]
(1) London force (2) hydrogen bond
(3) ion - ion interaction (4) ion - dipole interaction

20. If 10^{-4} dm^3 of water is introduced into a 1.0 dm^3 flask at 300 K, how many moles of water are in the vapour phase when equilibrium is established ? [2010]
(Given : Vapour pressure of H_2O at 300 K is 3170 Pa; R = 8.314 J K^{-1} mol^{-1})
(1) 5.56×10^{-3} mol (2) 1.53×10^{-2} mol
(3) 4.46×10^{-2} mol (4) 1.27×10^{-3} mol

21. Value of gas constant R is [2002]
(1) 0.082 litre atm (2) 0.987 cal mol^{-1} K^{-1}
(3) 8.3 J mol^{-1} K^{-1} (4) 83 erg mol^{-1} K^{-1}.

NUMERICAL VALUE QUESTIONS (NVQs)

22. 'x' g of molecular oxygen (O_2) is mixed with 200 g of neon (Ne). The total pressure of the non-reactive mixture of O_2 and Ne in the cylinder is 25 bar. The partial pressure of Ne is 20 bar at the same temperature and volume. The value of 'x' is_____.
[Given: Molar mass of O_2 = 32 g mol^{-1}.
Molar mass of Ne = 20 g mol^{-1}]. [July 29, 2022 (II)]

23. A 10 g mixture of hydrogen and helium is contained in a vessel of capacity 0.0125 m^3 at 6 bar and 27 °C. The mass of helium in the mixture is _______ g. (nearest integer) Given : R = 8.3 JK^{-1} mol^{-1} (Atomic masses of H and He are 1u and 4u, respectively) [July 26, 2022 (II)]

24. The pressure of a moist gas at 27°C is 4 atm. The volume of the container is doubled at the same temperature. The new pressure of the moist gas is _________ $\times 10^{-1}$ atm. (Nearest integer)
(Given : The vapour pressure of water at 27°C is 0.4 atm.) [July 25, 2022 (I)]

25. A sealed flask with a capacity of 2 dm^3 contains 11 g of propane gas. The flask is so weak that it will burst if the pressure becomes 2 MPa. The minimum temperature at which the flask will burst is _____ °C. [Nearest integer]
(Given: R = 8.3 JK^{-1} mol^{-1}. Atomic masses of C and H are 12u and 1u respectively). (Assume that propane behaves as an ideal gas.) [July 25, 2022 (II)]

26. Geraniol, a volatile organic compound, is a component of rose oil. The density of the vapour is 0.46 gL^{-1} at 257 °C and 100 mm Hg. The molar mass of geraniol is ___________ (Nearest Integer)
[Given R = 0.082 L atm K^{-1} mol^{-1}] [June 29, 2022 (I)]

27. 100 g of an ideal gas is kept in a cylinder of 416 L volume at 27°C under 1.5 bar pressure. The molar mass of the gas is _______ g mol^{-1}. (Nearest integer)
(Given : R = 0.083 L bar K^{-1} mol^{-1}) [June 28, 2022 (II)]

28. A rigid nitrogen tank stored inside a laboratory has a pressure of 30 atm at 06:00 am when the temperature is 27 °C. At 03:00 pm, when the temperature is 45°C, the pressure in the tank will be __________ atm. [nearest integer] [June 25, 2022 (II)]

29. At 300 K, a sample of 3.0 g of gas A occupies the same volume as 0.2 g of hydrogen at 200 K at the same pressure. The molar mass of gas A is____ g mol^{-1} (nearest integer) Assume that the behaviour of gases as ideal. (Given: The molar mass of hydrogen (H_2) gas is 2.0 g mol^{-1}) [June 24, 2022 (II)]

30. An empty LPG cylinder weighs 14.8 kg. When full, it weighs 29.0 kg and shows a pressure of 3.47 atm. In the course of use at ambient temperature, the mass of the cylinder is reduced to 23.0 kg. The final pressure inside of the cylinder is ______ atm. (Nearest integer)
(Assume LPG of be an ideal gas) [Sep. 1, 2021 (II)]

31. A home owner uses 4.00×10^3 m^3 of methane (CH_4) gas, (assume CH_4 is an ideal gas) in a year to heat his home. Under the pressure of 1.0 atm and 300 K, mass of gas used is $x \times 10^5$ g. The value of x is ________ . (Nearest integer)
(Given $R = 0.083$ L atm K^{-1} mol^{-1}) [July 25, 2021 (I)]

32. An LPG cylinder contains gas at a pressure of 300 kPa at 27 °C. The cylinder can withstand the pressure of 1.2×10^6 Pa. The room in which the cylinder is kept catches fire. The minimum temperature at which the bursting of cylinder will take place is ________ °C. (Nearest integer) [July 25, 2021 (II)]

33. The pressure exerted by a non-reactive gaseous mixture of 6.4 g of methane and 8.8 g of carbon dioxide in a 10 L vessel at 27 °C is .... kPa. [March 17, 2021 (I)]
(Round off to the Nearest Integer).
[Assume gases are ideal, $R = 8.314$ J mol^{-1} K^{-1}
Atomic masses are C : 12.0 u, H : 1.0 u, O : 16.0 u]

34. A car tyre is filled with nitrogen gas at 35 psi at 27 °C. It will burst if pressure exceeds 40 psi. The temperature in °C at which the car tyre will burst is _______ . (Rounded-off to the nearest integer) [Feb. 25, 2021 (I)]

35. The volume occupied by 4.75 g of acetylene gas at 50 °C and 740 mmHg pressure is ________ L.
(Rounded off to the nearest integer)
[Given $R = 0.0826$ L atm K^{-1} mol^{-1}] [Feb. 24, 2021 (II)]

36. A spherical balloon of radius 3 cm containing helium gas has a pressure of 48×10^{-3} bar. At the same temperature, the pressure of a spherical balloon of radius 12 cm containing the same amount of gas will be ______ $\times 10^{-6}$ bar. [Sep. 06, 2020 (I)]

Topic 2 - Kinetic Theory of Gases and Molecular Speeds

MULTIPLE CHOICE QUESTIONS (MCQs)

37. Identify the correct labels of A, B and C in the following graph from the options given below:

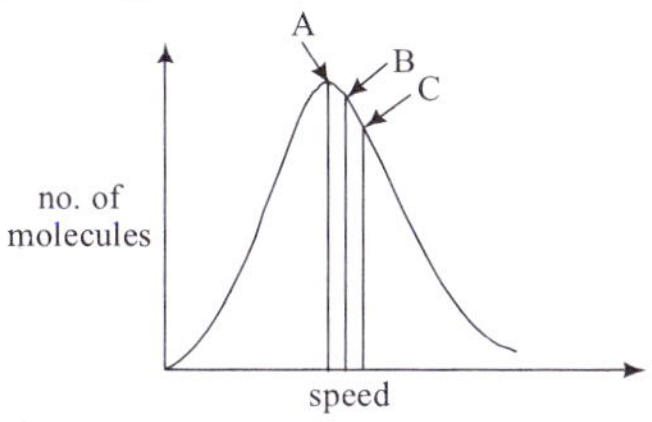

Root mean square speed (V_{rms}); most probable speed (V_{mp}); average speed (V_{av}) [Online May 12, 2012(S), Jan. 07, 2020 (II)]
(1) A – V_{mp}; B – V_{rms}; C – V_{av}
(2) A – V_{av}; B – V_{rms}; C – V_{mp}
(3) A – V_{rms}; B – V_{mp}; C – V_{av}
(4) A – V_{mp}; B – V_{av}; C – V_{rms}

38. Points I, II and III in the following plot respectively correspond to (V_{mp} : most probable velocity) [April 10, 2019 (II)]

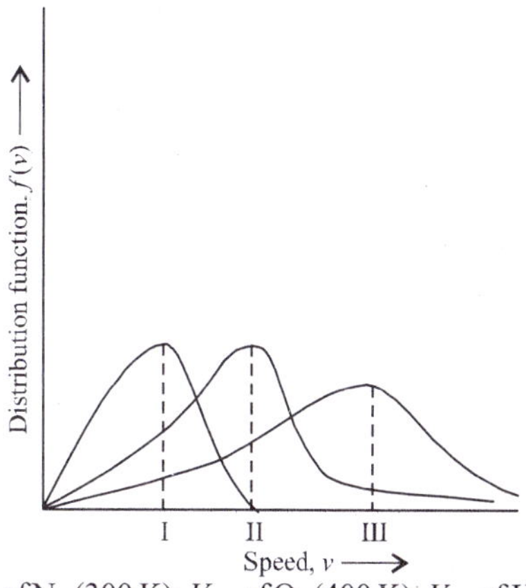

(1) V_{mp} of N_2 (300 K); V_{mp} of O_2 (400 K); V_{mp} of H_2 (300 K)
(2) V_{mp} of O_2 (400 K); V_{mp} of N_2 (300 K); V_{mp} of H_2 (300 K)
(3) V_{mp} of N_2 (300 K); V_{mp} of H_2 (300 K); V_{mp} of O_2 (400 K)
(4) V_{mp} of H_2 (300 K); V_{mp} of N_2 (300 K); V_{mp} of O_2 (400 K)

39. Initially, the root mean square (rms) velocity of N_2 molecules at certain temperature is u. If this temperature is doubled and all the nitrogen molecules dissociate into nitrogen atoms, then the *rms* velocity will be : [Online April 10, 2016]
(1) $2u$ (2) $14u$ (3) $4u$ (4) $u/2$

40. Which of the following is not an assumption of the kinetic theory of gases? [Online April 10, 2015]
(1) Gas particles have negligible volume.
(2) A gas consists of many identical particles which are in continual motion.
(3) At high pressure, gas particles are difficult to compress.
(4) Collisions of gas particles are perfectly elastic.

41. The ratio of masses of oxygen and nitrogen in a particular gaseous mixture is 1 : 4. The ratio of number of their molecules is: [2014]
(1) 1 : 4 (2) 7 : 32 (3) 1 : 8 (4) 3 : 16

42. The temperature at which oxygen molecules have the same root mean square speed as helium atoms have at 300 K is: Atomic masses: He = 4 u, O = 16 u) [Online April 9, 2014]
(1) 300 K (2) 600 K (3) 1200 K (4) 2400 K

43. For gaseous state, if most probable speed is denoted by C*, average speed by $\overline{C}$ and mean square speed by C, then for a large number of molecules, the ratios of these speeds are : [Online May 26, 2012, 2013]
(1) $C^* : \overline{C} : C = 1.225 : 1.128 : 1$
(2) $C^* : \overline{C} : C = 1.128 : 1.225 : 1$
(3) $C^* : \overline{C} : C = 1 : 1.128 : 1.225$
(4) $C^* : \overline{C} : C = 1 : 1.225 : 1.128$

44. By how many folds the temperature of a gas would increase when the root mean square velocity of the gas molecules in a container of fixed volume is increased from 5×10^4 cm/s to 10×10^4 cm/s ? [Online April 9, 2013]
(1) Two (2) Three (3) Six (4) Four

45. Which one of the following is the wrong assumption of kinetic theory of gases ? [Online April 25, 2013]
(1) Momentum and energy always remain conserved.
(2) Pressure is the result of elastic collision of molecules with the container's wall.
(3) Molecules are separated by great distances compared to their sizes.
(4) All the molecules move in straight line between collision and with same velocity.

46. When r, P and M represent rate of diffusion, pressure and molecular mass, respectively, then the ratio of the rates of diffusion (r_A/r_B) of two gases A and B, is given as [2011RS]
(1) $(P_A/P_B)(M_B/M_A)^{1/2}$
(2) $(P_A/P_B)^{1/2}(M_B/M_A)$
(3) $(P_A/P_B)(M_A/M_B)^{1/2}$
(4) $(P_A/P_B)^{1/2}(M_A/M_B)$

47. The molecular velocity of any gas is : [2011RS]
(1) inversely proportional to absolute temperature.
(2) directly proportional to square of temperature.
(3) directly proportional to square root of temperature.
(4) inversely proportional to the square root of temperature.

48. Which one of the following statements is NOT true about the effect of an increase in temperature on the distribution of molecular speeds in a gas? [2005]
(1) The area under the distribution curve remains the same as under the lower temperature
(2) The distribution becomes broader
(3) The fraction of the molecules with the most probable speed increases
(4) The most probable speed increases

49. As the temperature is raised from 20 °C to 40 °C, the average kinetic energy of neon atoms changes by which factor [2004]
(1) $313/293$ (2) $\sqrt{(313/293)}$
(3) $1/2$ (4) 2

50. According to the kinetic theory of gases, in an ideal gas between two successive collisions, a gas molecule travels [2003]
(1) in a wavy path
(2) in a straight line path
(3) with an accelerated velocity
(4) in a circular path

Topic 3 - Deviation from Ideal Gas Behaviour, Liquefaction of Gases and Liquid State

Multiple Choice Questions (MCQs)

51. Given below are the critical temperatures of some of the gases : [July 29, 2022 (II)]

Gas	Critical temperature (K)
He	5.2
CH_4	190
CO_2	304.2
NH_3	405.5

The gas showing least adsorption on a definite amount of charcoal is :
(1) He (2) CH_4 (3) CO_2 (4) NH_3

52. Consider the following table :

Gas	$a/(\text{k Pa dm}^6\text{mol}^{-1})$	$b/(\text{dm}^3\text{mol}^{-1})$
A	642.32	0.05196
B	155.21	0.04136
C	431.91	0.05196
D	155.21	0.4382

a and b are van der Waals constants. The correct statement about the gases is : [April 10, 2019 (I)]
(1) Gas C will occupy more volume than gas A; gas B will be more compressible than gas D
(2) Gas C will occupy lesser volume than gas A; gas B will be lesser compressible than gas D
(3) Gas C will occupy more volume than gas A; gas B will be lesser compressible than gas D
(4) Gas C will occupy lesser volume than gas A; gas B will be more compressible than gas D

53. Consider the van der Waals constants, a and b, for the following gases,

Gas	Ar	Ne	Kr	Xe
$a/(\text{atm dm}^6\text{ mol}^{-2})$	1.3	0.2	5.1	4.1
$b/(10^{-2}\text{ dm}^3\text{ mol}^{-1})$	3.2	1.7	1.0	5.0

Which gas is expected to have the highest critical temperature? [April 9, 2019 (I)]
(1) Kr (2) Ne (3) Xe (4) Ar

54. At a given temperature T, gases Ne, Ar, Xe and Kr are found to deviate from ideal gas behaviour. Their equation of state is given as [April 9, 2019 (II)]

$$P = \frac{RT}{V-b} \text{ at } T.$$

Here, b is the van der Waals constant. Which gas will exhibit steepest increase in the plot of Z (compression factor) vs P?
(1) Xe (2) Kr (3) Ne (4) Ar

55. The volume of gas A is twice than that of gas B. The compressibility factor of gas A is thrice than that of gas B at same temperature. The pressures of the gases for equal number of moles are: [Jan. 12, 2019 (I)]
(1) $3P_A=2P_B$ (2) $2P_A=3P_B$
(3) $P_A=3P_B$ (4) $P_A=2P_B$

56. At very high pressures, the compressibility factor of one mole of a gas is given by : [2012, Online April 9, 2016]
(1) $1+\frac{Pb}{RT}$ (2) $\frac{Pb}{RT}$ (3) $1-\frac{Pb}{RT}$ (4) $1-\frac{b}{(VRT)}$

57. If Z is a compressibility factor, van der Waals equation at low pressure can be written as: [2014]
(1) $Z=1+\frac{RT}{Pb}$ (2) $Z=1-\frac{a}{VRT}$
(3) $Z=1-\frac{Pb}{RT}$ (4) $Z=1+\frac{Pb}{RT}$

58. van der Waals equation for a gas is stated as,

$$P=\frac{nRT}{V-nb}-a\left(\frac{n}{V}\right)^2.$$ [Online April 9, 2014]

This equation reduces to the perfect gas equation, $P=\frac{nRT}{V}$ when,

(1) temperature is sufficient high and pressure is low.
(2) temperature is sufficient low and pressure is high.
(3) both temperature and pressure are very high.
(4) both temperature and pressure are very low.

59. In van der Waals equation of state of the gas law, the constant 'b' is a measure of [2004]
(1) volume occupied by the molecules
(2) intermolecular attractions
(3) intermolecular repulsions
(4) intermolecular collisions per unit volume

NUMERICAL VALUE QUESTIONS (NVQs)

60. For a real gas at 25°C temperature and high pressure (99 bar) the value of compressibility factor is 2, so the value of Van der Waals constant 'b' should be ________ $\times 10^{-2}$ L mol^{-1} (Nearest integer) (Given R = 0.083 L bar K^{-1} mol^{-1}) [July 27, 2022 (II)]

61. A certain gas obeys $P(V_m - b) = RT$. The value of $\left(\frac{\partial Z}{\partial P}\right)_T$ is $\frac{xb}{RT}$. The value of x is ________. (Integer answer) (Z : compressibility factor) [Feb. 26, 2021 (I)]

Part-B (JEE Advanced)

Topic 1 – Intermolecular Forces, Gas Laws and Ideal Gas Equation

1 MCQs with One Correct Answer

1. The ratio of the rate of diffusion of helium and methane under identical condition of pressure and temperature will be [2005S]
(a) 4 (b) 2 (c) 1 (d) 0.5

2. Positive deviation from ideal behaviour takes place because of [2003S]
(a) Molecular interaction between atoms and $PV/nRT > 1$
(b) Molecular interaction between atoms and $PV/nRT < 1$
(c) Finite size of atoms and $PV/nRT > 1$
(d) Finite size of atoms and $PV/nRT < 1$

3. Which of the following volume (V) - temperature (T) plots represents the behaviour of one mole of an ideal gas at one atmospheric pressure ? [2002S]

(a) 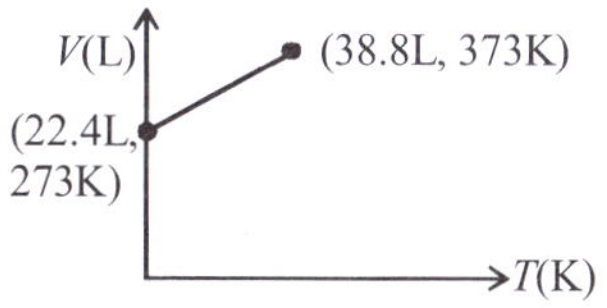

(b) 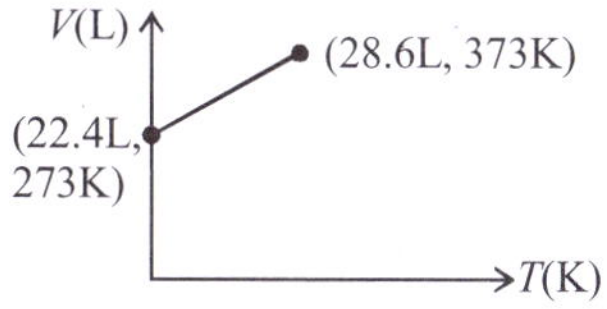

(c) 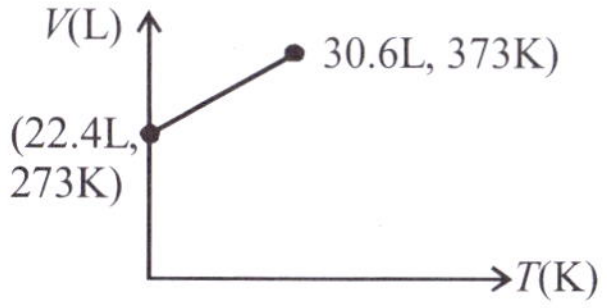

(d) 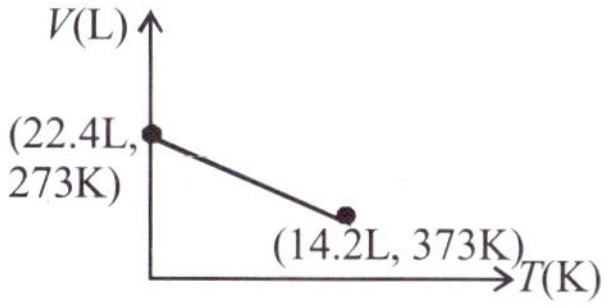

4. At 100 °C and 1 atm, if the density of liquid water is 1.0 g cm^{-3} and that of water vapour is 0.0006 g cm^{-3}, then the volume occupied by water molecules in 1 litre of steam at that temperature is [2000S]
(a) 6 cm^3 (b) 60 cm^3 (c) 0.6 cm^3 (d) 0.06 cm^3

5. A gas will approach ideal behaviour at [1999 - 2 Marks]
(a) low temperature and low pressure.
(b) low temperature and high pressure.
(c) high temperature and low pressure.
(d) high temperature and high pressure.

6. X mL of H_2 gas effuses through a hole in a container in 5 seconds. The time taken for the effusion of the same volume of the gas specified below under identical conditions is : [1996 - 1 Mark]
(a) 10 seconds : He (b) 20 seconds : O_2
(c) 25 seconds : CO (d) 55 seconds : CO_2

7. One mole of N_2O_4(g) at 300 K is kept in a closed container under one atmosphere. It is heated to 600 K when 20% by mass of N_2O_4(g) decomposes to NO_2(g). The resultant pressure is : [1996 - 1 Mark]
(a) 1.2 atm (b) 2.4 atm (c) 2.0 atm (d) 1.0 atm

8. At constant volume, for a fixed number of moles of a gas the pressure of the gas increases with rise in temperature due to [1992 - 1 Mark]
(a) Increase in average molecular speed
(b) Increased rate of collisions amongst molecules
(c) Increase in molecular attraction
(d) Decrease in mean free path

9. The density of neon will be highest at [1990 - 1 Mark]
(a) S.T.P. (b) 0°C, 2 atm
(c) 273°C, 1 atm. (d) 273°C, 2 atm.

10. The rate of diffusion of methane at a given temperature is twice that of a gas X. The molecular weight of X is [1990 - 1 Mark]
(a) 64.0 (b) 32.0 (c) 4.0 (d) 8.0

11. A bottle of dry ammonia and a bottle of dry hydrogen chloride connected through a long tube are opened simultaneously at both ends. The white ammonium chloride ring first formed will be [1988 - 1 Mark]
(a) at the centre of the tube.
(b) near the hydrogen chloride bottle.
(c) near the ammonia bottle.
(d) throughout the length of the tube.

12. Rate of diffusion of a gas is : [1985 - 1 Mark]
(a) directly proportional to its density.
(b) directly proportional to its molecular weight.
(c) directly proportional to the square root of its molecular weight.
(d) inversely proportional to the square root of its molecular weight.

13. Equal weights of methane and hydrogen are mixed in an empty container at 25 °C. The fraction of the total pressure exerted by hydrogen is : [1984 - 1 Mark]

(a) $\frac{1}{2}$ (b) $\frac{8}{9}$ (c) $\frac{1}{9}$ (d) $\frac{16}{17}$

14. The temperature at which a real gas obeys the ideal gas laws over a wide range of pressure is [1981 - 1 Mark]

(a) Critical temperature (b) Boyle temperature
(c) Inversion temperature (d) Reduced temperature

2 Integer Value Answer

15. A closed tank has two compartments **A** and **B**, both filled with oxygen (assumed to be ideal gas). The partition separating the two compartments is fixed and is a perfect heat insulator (Figure 1). If the old partition is replaced by a new partition which can slide and conduct heat but does **NOT** allow the gas to leak across (Figure 2), the volume (in m^3) of the compartment **A** after the system attains equilibrium is_______. [Adv. 2018]

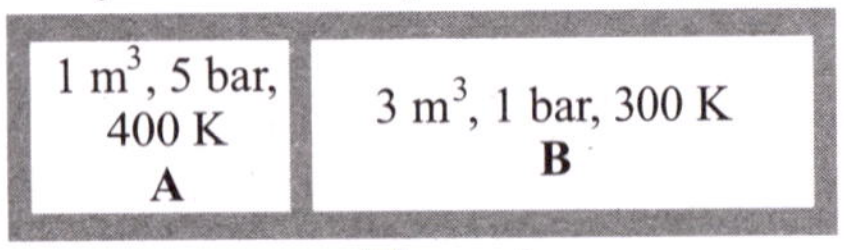

Figure 1

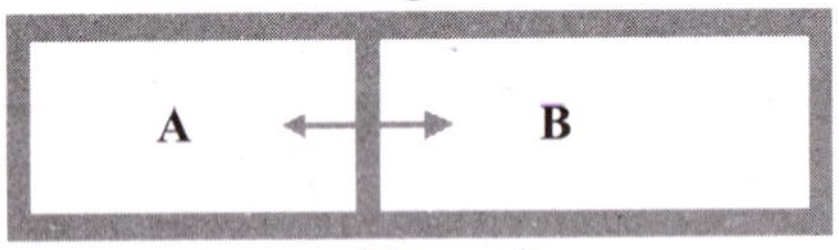

Figure 2

16. The diffusion coefficient of an ideal gas is proportional to its mean free path and mean speed. The absolute temperature of an ideal gas is increased 4 times and its pressure is increased 2 times. As a result, the diffusion coefficient of this gas increases x times. The value of x is [Adv. 2016]

17. To an evacuated vessel with movable piston under external pressure of 1 atm, 0.1 mol of He and 1.0 mol of an unknown compound (vapour pressure 0.68 atm. at 0°C) are introduced. Considering the ideal gas behaviour, the total volume (in litre) of the gases at 0°C is close to [2011]

3 Numeric / New Stem Based Questions

18. Two cylinders, both fitted with frictionless pistons, are filled with mixtures of He and Ar gases. In the first cylinder, the masses of He and Ar are m_1 and m_2, respectively. In the second cylinder, the masses of He and Ar are m_2 and m_1, respectively. The molar mass of Ar is 10 times the molar mass of He. The external pressure applied by the piston on the first cylinder needs to be 5 times that on the second cylinder so that the volume of the gas mixtures in both the cylinders are equal at the same temperature. Assuming He and Ar behave like ideal gases, the value of (m_1/m_2) is ______. [Adv. 2026]

19. The figure below is the plot of potential energy versus internuclear distance (d) of H_2 molecule in the electronic ground state. What is the value of the net potential energy E_0 (as indicated in the figure) in kJ mol^{-1}, for $d = d_0$ at which the electron-electron repulsion and the nucleus-nucleus repulsion energies are absent? As reference, the potential energy of H atom is taken as zero when its electron and the nucleus are infinitely far apart. [Adv. 2020]
Use Avogadro constant as 6.023×10^{23} mol^{-1}.

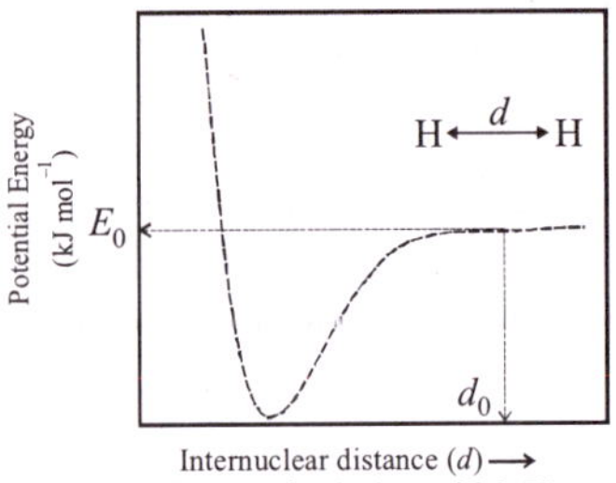

20. The degree of dissociation is 0.4 at 400 K and 1.0 atm for the gaseous reaction $PCl_5 \rightleftharpoons PCl_3 + Cl_2$. Assuming ideal behaviour of all gases, calculate the density of equilibrium mixture at 400 K and 1.0 atmosphere. (Relative atomic mass of P = 31.0 and Cl = 35.5) [1998 - 3 Marks]

21. For the reaction, $N_2O_5(g) \rightarrow 2NO_2(g) + 0.5\,O_2(g)$, calculate the mole fraction of $N_2O_5(g)$ decomposed at a constant volume and temperature, if the initial pressure is 600 mm Hg and the pressure at any time is 960 mm Hg. Assume ideal gas behaviour. [1998 - 3 Marks]

22. At room temperature, ammonia gas at 1 atm pressure and hydrogen chloride gas at P atm pressure are allowed to effuse through identical pin holes from opposite ends of a glass tube of one metre length and of uniform cross-section. Ammonium chloride is first formed at a distance of 60 cm from the end through which HCl gas is sent in. What is the value of P? [1982 - 4 Marks]

23. A straight glass tube has two inlets X and Y at two ends. The length of tube is 200 cm. HCl gas through inlets X and NH_3 gas through inlet Y are allowed to enter the tube at the same time. White fumes appear at point P inside the tube. Find distance of P from X. [1980]

24. Calculate density of NH_3 at 30°C and 5 atm pressure. [1978]

4 Fill in the Blanks

25. The value of PV for 5.6 litres of an ideal gas is .... RT, at N.T.P. [1987 - 1 Mark]

26. The rate of diffusion of gas is .... proportional to both .... and square root of molecular mass. [1986 - 1 Mark]

27. $C_p - C_v$ for an ideal gas is .... [1984 - 1 Mark]

5 True / False

28. A mixture of ideal gases is cooled upto liquid helium temperature (4.22 K) to form an ideal solution. [1996 - 1 Mark]

6 MCQs with One or More than One Correct Answer

29. The correct statement(s) about intermolecular forces is(are)

(a) The potential energy between two point charges approaches zero more rapidly than the potential energy between a point dipole and a point charge as the distance between them approaches infinity. [Adv. 2025]

(b) The average potential energy of two rotating polar molecules that are separated by a distance r has $\frac{1}{r^3}$ dependence.

(c) The dipole-induced dipole average interaction energy is independent of temperature.

(d) Nonpolar molecules attract one another even though neither has a permanent dipole moment.

30. Refer to the figure given : **[2006 - 5M; –1]**
Which of the following statements is wrong?

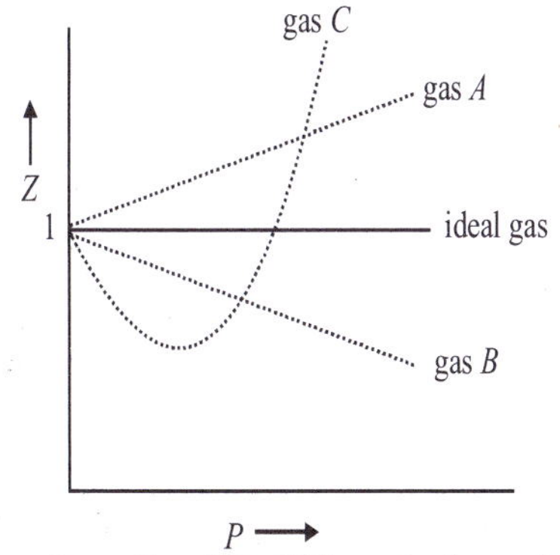

(a) For gas A, $a = 0$ and Z will linearly depend on pressure
(b) For gas B, $b = 0$ and Z will linearly depend on pressure
(c) Gas C is a real gas and we can find 'a' and 'b' if intersection data is given
(d) All van der Waal gases will behave like gas C and give positive slope at high pressure

31. According to Graham's law, at a given temperature the ratio of the rates of diffusion r_A/r_B of gases A and B is given by **[1998 - 2 Marks]**
(a) $(P_A/P_B)(M_A/M_B)^{1/2}$ (b) $(M_A/M_B)(P_A/P_B)^{1/2}$
(c) $(P_A/P_B)(M_B/M_A)^{1/2}$ (d) $(M_A/M_B)(P_B/P_A)^{1/2}$
(Where P and M are pressures and molecular weights of gases A and B respectively.)

32. Equal weights of ethane and hydrogen are mixed in an empty container at 25 °C. The fraction of the total pressure exerted by hydrogen is **[1993 - 1 Mark]**
(a) 1 : 2 (b) 1 : 1 (c) 1 : 16 (d) 15 : 16

33. If a gas is expanded at constant temperature : **[1986 - 1 Mark]**
(a) the pressure decreases
(b) the kinetic energy of the molecules remains the same
(c) the kinetic energy of the molecules decreases
(d) the number of molecules of the gas increases

34. When an ideal gas undergoes unrestrained expansion, no cooling occurs because the molecules : **[1984 - 1 Mark]**
(a) are above the inversion temperature
(b) exert no attractive forces on each other
(c) do work equal to loss in kinetic energy
(d) collide without loss of energy

8 Comprehension Based Questions

X and Y are two volatile liquids with molar weights of 10 g mol^{-1} and 40 g mol^{-1} respectively. Two cotton plugs, one soaked in X and the other soaked in Y, are simultaneously placed at the ends of a tube of length L = 24 cm, as shown in the figure. The tube is filled with an inert gas at 1 atmosphere pressure and a temperature of 300 K. Vapours of X and Y react to form a product which is first observed at a distance d cm from the plug soaked in X. Take X and Y to have equal molecular diameters and assume ideal behaviour for the inert gas and the two vapours.

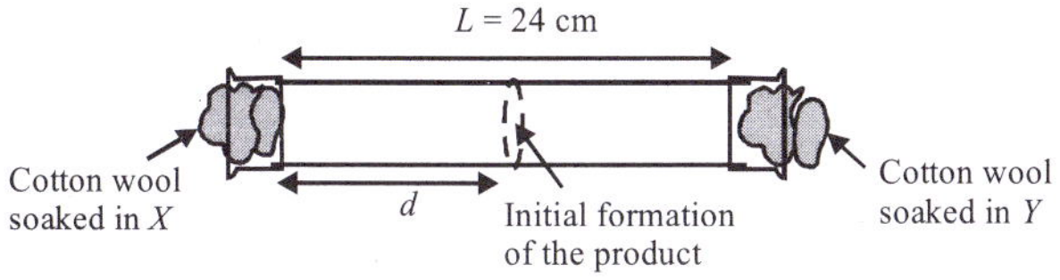

35. The value of d in cm (shown in the figure), as estimated from Graham's law, is **[Adv. 2014]**
(a) 8 (b) 12 (c) 16 (d) 20

36. The experimental value of d is found to be smaller than the estimate obtained using Graham's law. This is due to **[Adv. 2014]**
(a) Larger mean free path for X as compared to that of Y
(b) Larger mean free path for Y as compared to that of X
(c) Increased collision frequency of Y with the inert gas as compared to that of X with the inert gas
(d) Increased collision frequency of X with the inert gas as compared to that of Y with the inert gas

10 Subjective Problems

37. The density of the vapour of a substance at 1 atm pressure and 500 K is 0.36 kg m^{-3}. The vapour effuses through a small hole at a rate of 1.33 times faster than oxygen under the same condition. **[2002 - 5 Marks]**
(a) Determine (i) molecular weight,(ii) molar volume, (iii) compression factor (Z) of the vapour and (iv) which forces among the gas molecules are dominating, the attractive or the repulsive?
(b) If the vapour behaves ideally at 1000 K, determine the average translational kinetic energy of a molecule.

38. One mole of nitrogen gas at 0.8 atm takes 38 s to diffuse through a pinhole, whereas one mole of an unknown compound of xenon with flourine at 1.6 atm takes 57 s to diffuse through the same hole. Calculate the molecular formula of the compound. **[1999 - 5 Marks]**

39. The pressure exerted by 12 g of an ideal gas at temperature t°C in a vessel of volume V litre is one atm. When the temperature is increased by 10 degrees at the same volume, the pressure increases by 10%. Calculate the temperature t and volume V. (Molecular weight of the gas = 120.) **[1999 - 5 Marks]**

40. An evacuated glass vessel weighs 50.0 g when empty, 148.0 g when filled with a liquid of density 0.98 g mL^{-1} and 50.5 g when filled with an ideal gas at 760 mm Hg at 300K. Determine the molar mass of the gas. **[1998 - 3 Marks]**

41. A mixture of ethane (C_2H_6) and ethene (C_2H_4) occupies 40 litres at 1.00 atm and at 400 K. The mixture reacts completely with 130 g of O_2 to produce CO_2 and H_2O. Assuming ideal gas behaviour, calculate the mole fractions of C_2H_4 and C_2H_6 in the mixture. **[1995 - 4 Marks]**

42. The composition of the equilibrium mixture ($Cl_2 \rightleftharpoons 2Cl$), which is attained at 1200°C, is determined by measuring the rate of effusion through a pin–hole. It is observed that at 1.80 mmHg pressure, the mixture effuses 1.16 times as fast as krypton effuses under the same conditions. Calculate the fraction of the chlorine molecules dissociated into atoms. (Relative atomic mass of Kr = 84.) **[1995 - 4 Marks]**

43. A 20.0 cm^3 mixture of CO, CH_4 and He gases is exploded by an electric discharge at room temperature with excess of oxygen. The volume contraction is found to be 13.0 cm^3. A further contraction of 14.0 cm^3 occurs when the residual gas is treated with KOH solution. Find out the composition of the gaseous mixture in terms of volume percentage. **[1995 - 4 Marks]**

44. A 4 : 1 molar mixture of He and CH_4 is contained in a vessel at 20 bar pressure . Due to a hole in the vessel, the gas mixture leaks out. What is the composition of the mixture effusing out initially? **[1994 - 2 Marks]**

45. An LPG (liquefied petroleum gas) cylinder weighs 14.8 kg when empty. When full, it weighs 29.0 kg and shows a pressure of 2.5 atm. In the course of use at 27 °C, the weight of the full cylinder reduces to 23.2 kg. Find out the volume of the gas in cubic meters used up at the normal usage conditions, find the final pressure inside the cylinder. Assume LPG to be *n*-butane with normal boiling point of 0 °C. [1994 - 3 Marks]

46. At 27 °C, hydrogen is leaked through a tiny hole into a vessel for 20 minutes. Another unknown gas at the same temperature and pressure as that of H_2 is leaked through the same hole for 20 minutes. After the effusion of the gases the mixture exerts a pressure of 6 atmosphere. The hydrogen content of the mixture is 0.7 mole. If the volume of the container is 3 litres, what is the molecular weight of the unknown gas? [1992 - 3 Marks]

47. At room temperature the following reactions proceed nearly to completion : [1992 - 4 Marks]

$$2NO + O_2 \rightarrow 2NO_2 \rightarrow N_2O_4$$

The dimer, N_2O_4, solidifies at 262 K. A 250 mL flask and a 100 mL. flask are separated by a stop-cock. At 300 K, the nitric oxide in the larger flask exerts a pressure of 1.053 atm. and the smaller one contains oxygen at 0.789 atm. The gases are mixed by opening the stopcock and after the end of the reaction the flasks are cooled at 220 K. Neglecting the vapour pressure of the dimer, find out the pressure and composition of the gas remaining at 220 K. (Assume the gases to behave ideally).

48. Calculate the volume occupied by 5.0 g of acetylene gas at 50 °C and 740 mm pressure. [1991 - 2 Marks]

49. A spherical balloon of 21 cm diameter is to be filled up with hydrogen at N.T.P. from a cylinder containing the gas at 20 atmospheres at 27 °C. If the cylinder can hold 2.82 litres of water, calculate the number of balloons that can be filled up. [1987 - 5 Marks]

50. Oxygen is present in 1 litre flask at a pressure of 7.6×10^{-10} mm of Hg. Calculate the number of oxygen molecules in the flask at 0 °C. [1983 - 2 Marks]

51. When 2 g of a gas *A* is introduced into an evacuated flask kept at 25 °C, the pressure is found to be one atmosphere. If 3 g of another gas *B* is then added to the same flask, the total pressure becomes 1.5 atm. Assuming ideal gas behaviour, calculate the ratio of the molecular weights $M_A : M_B$. [1983 - 2 Marks]

52. 1 litre of mixture of CO and CO_2 is taken. The mixture is passed through a tube containing red hot charcoal. The volume now becomes 1.6 litre. The volumes are measured under the same conditions. Find the composition of mixture by volume. [1980]

53. 3.7 g of a gas at 25 °C occupied the same volume as 0.184 g of hydrogen at 17 °C and at the same pressure. What is the molecular weight of the gas? [1979]

Topic 2 - Kinetic Theory of Gases and Molecular Speeds

1 MCQs with One Correct Answer

54. A closed vessel contains 10 g of an ideal gas **X** at 300 K, which exerts 2 atm pressure. At the same temperature, 80 g of another ideal gas **Y** is added to it and the pressure becomes 6 atm. The ratio of root mean square velocities of **X** and **Y** at 300 K is [Adv. 2024]

(a) $2\sqrt{2} : \sqrt{3}$ (b) $2\sqrt{2} : 1$
(c) 1 : 2 (d) 2 : 1

55. If the distribution of molecular speeds of a gas is as per the figure shown below, then the ratio of the most probable, the average, and the root mean square speeds, respectively, is [Adv. 2020]

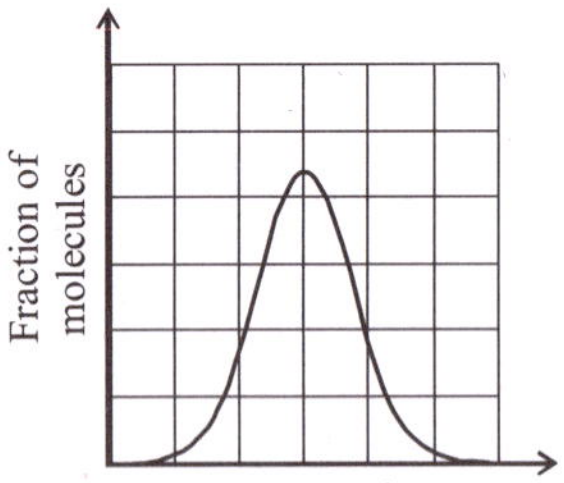

(a) 1 : 1 : 1 (b) 1 : 1 : 1.224
(c) 1 : 1.128 : 1.224 (d) 1 : 1.128 : 1

56. The root mean square velocity of one mole of a monoatomic gas having molar mass *M* is $u_{r.m.s.}$. The relation between the average kinetic energy (*E*) of the gas and $u_{r.m.s.}$ is [2004S]

(a) $u_{r.m.s.} = \sqrt{\frac{3E}{2M}}$ (b) $u_{r.m.s.} = \sqrt{\frac{2E}{3M}}$

(c) $u_{r.m.s.} = \sqrt{\frac{2E}{M}}$ (d) $u_{r.m.s.} = \sqrt{\frac{E}{3M}}$

57. The root mean square velocity of an ideal gas at constant pressure varies with density (*d*) as [2001S]

(a) d^2 (b) d (c) $\sqrt{d}$ (d) $1/\sqrt{d}$

58. The rms velocity of hydrogen is $\sqrt{7}$ times the rms velocity of nitrogen. If *T* is the temperature of the gas, then [2000S]

(a) $T(H_2) = T(N_2)$ (b) $T(H_2) > T(N_2)$
(c) $T(H_2) < T(N_2)$ (d) $T(H_2) = \sqrt{7}\, T(N_2)$

59. The ratio between the root mean square speed of H_2 at 50 K and that of O_2 at 800 K is, [1996 - 1 Mark]

(a) 4 (b) 2 (c) 1 (d) 1/4

60. Longest mean free path stands for : [1995S]

(a) H_2 (b) N_2 (c) O_2 (d) Cl_2

61. According to kinetic theory of gases, for a diatomic molecule [1991 - 1 Mark]

(a) the pressure exerted by the gas is proportional to mean velocity of the molecule
(b) the pressure exerted by the gas is proportional to the root mean velocity of the molecule
(c) the root mean square velocity of the molecule is inversely proportional to the temperature
(d) the mean translational kinetic energy of the molecule is proportional to the absolute temperature.

62. The average velocity of an ideal gas molecule at 27 °C is 0.3 m/sec. The average velocity at 927 °C will be: [1986 - 1 Mark]

(a) 0.6 m/sec (b) 0.3 m/sec
(c) 0.9 m/sec (d) 3.0 m/sec

63. Helium atom is two times heavier than a hydrogen molecule. At 298 K, the average kinetic energy of a helium atom is [1982 - 1 Mark]

(a) two times that of a hydrogen molecule.
(b) same as that of a hydrogen molecule.
(c) four times that of a hydrogen molecule.
(d) half that of a hydrogen molecule.

64. The ratio of root mean square velocity to average velocity of a gas molecule at a particular temperature is **[1981 - 1 Mark]**
(a) 1.086 : 1 (b) 1 : 1.086 (c) 2 : 1.086 (d) 1.086 : 2

2 Integer Value Answer

65. At 400 K, the root mean square (rms) speed of a gas **X** (molecular weight = 40) is equal to the most probable speed of gas **Y** at 60 K. The molecular weight of the gas **Y** is **[2009]**

3 Numeric / New Stem Based Questions

66. The average velocity of gas molecules is 400 m/sec. Calculate its rms velocity at the same temperature. **[2003 - 2 Marks]**

4 Fill in the Blanks

67. Eight gram each of oxygen and hydrogen at 27 °C will have the total kinetic energy in the ratio of .... **[1989 - 1 Mark]**

68. The total energy of one mole of an ideal monoatomic gas at 27 °C is .... calories. **[1984 - 1 Mark]**

5 True / False

69. Kinetic energy of a molecule is zero at 0 °C. **[1985 - ½ Mark]**

70. A gas in a closed container will exert much higher pressure due to gravity at the bottom than at the top. **[1985 - ½ Mark]**

6 MCQs with One or More than One Correct Answer

71. Which of the following statement(s) is (are) correct regarding the root mean square speed (U_{rms}) of a molecule in a gas at equilibrium? **[Adv. 2019]**
(a) E_{av} at a given temperature does not depend on its molecular mass
(b) U_{rms} is inversely proportional to the square root of its molecular mass
(c) U_{rms} is doubled when its temperature is increased four times
(d) E_{av} is doubled when its temperature is increased four times

72. According to kinetic theory of gases **[2011]**
(a) collisions are always elastic
(b) heavier molecules transfer more momentum to the wall of the container
(c) only a small number of molecules have very high velocity
(d) between collisions, the molecules move in straight lines with constant velocities

9 Assertion and Reason/Statement Type Questions

Each question contains **STATEMENT-1 (Assertion)** and **STATEMENT-2 (Reason)**. Each question has 4 choices (a), (b), (c) and (d) out of which ONLY ONE is correct. Mark your answer as
(a) If both Statement -1 and Statement -2 are correct, and Statement -2 is the correct explanation of the Statement -2.
(b) If both Statement -1 and Statement -2 are correct, but Statement -2 is not the correct explanation of the Statement -1.
(c) If Statement -1 is correct but Statement -2 is incorrect.
(d) If Statement -1 is incorrect but Statement -2 is correct.

73. Read the following statement and explanation and answer as per the options given below :
Assertion : The pressure of a fixed amount of an ideal gas is proportional to its temperature
Reason : Frequency of collisions and their impact both increase in proportion to the square root of temperature. **[2000S]**

10 Subjective Problems

74. A gas bulb of 1 litre capacity contains 2.0×10^{21} molecules of nitrogen exerting a pressure of 7.57×10^3 Nm^{-2}. Calculate the root mean square (r.m.s) speed and the temperature of the gas molecules. If the ratio of the most probable speed to the root mean square speed is 0.82, calculate the most probable speed for these molecules at this temperature. **[1993 - 4 Marks]**

75. The average velocity at T_1K, and the most probable velocity at T_2K of CO_2 gas is 9.0×10^4 cm sec^{-1}. Calculate the value of T_1 and T_2. **[1990 - 4 Marks]**

76. Calculate the root mean square velocity of ozone kept in a closed vessel at 20 °C and 82 cm mercury pressure. **[1985 - 2 Marks]**

77. Calculate the average of kinetic energy, in Joules of the molecules in 8.0 g of methane at 27 °C. **[1982 - 2 Marks]**

Topic 3 - Deviation from Ideal Gas Behaviour, Liquefaction of Gases and Liquid State

1 MCQs with One Correct Answer

78. The qualitative sketches I, II and III given below show the variation of surface tension with molar concentration of three different aqueous solutions of KCl, CH_3OH and $CH_3(CH_2)_{11}OSO_3^-\ Na^+$ at room temperature. The correct assignment of the sketches is **[Adv. 2016]**

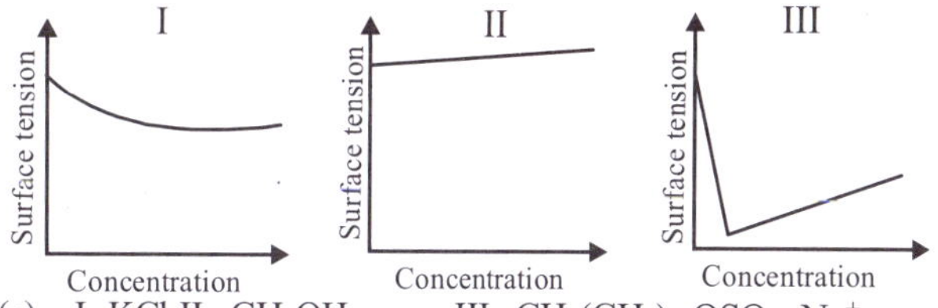

(a) I : KCl II : CH_3OH III : $CH_3(CH_2)_{11}OSO_3^-\ Na^+$
(b) I : $CH_3(CH_2)_{11}OSO_3^-\ Na^+$ II : CH_3OH III : KCl
(c) I : KCl II : $CH_3(CH_2)_{11}OSO_3^-\ Na^+$ III : CH_3OH
(d) I : CH_3OH II : KCl III : $CH_3(CH_2)_{11}OSO_3^-\ Na^+$

79. For one mole of a van der Waals gas when $b = 0$ and $T = 300$ K, the PV vs, $1/V$ plot is shown below. The value of the van der Waals constant a (atm. liter^2 mol^{-2}) is : **[2012]**

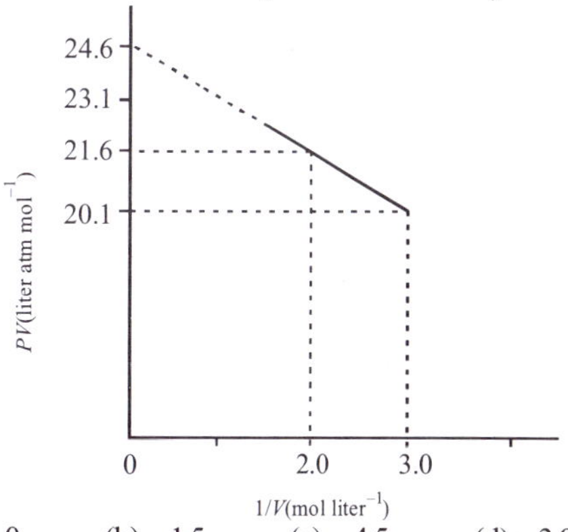

(a) 1.0 (b) 1.5 (c) 4.5 (d) 3.0

80. The term that corrects for the attractive forces present in a real gas in the van der Waals equation is **[2009 - 3M; –1]**

(a) nb (b) $\frac{an^2}{V^2}$ (c) $-\frac{an^2}{V^2}$ (d) $-nb$

81. When the temperature is increased, surface tension of water [2002S]
(a) increases
(b) decreases
(c) remains constant
(d) shows irregular behaviour

82. The compressibility of a gas is less than unity at STP. Therefore, [2000S]
(a) $V_m > 22.4$ litres (b) $V_m < 22.4$ litres
(c) $V_m = 22.4$ litres (d) $V_m = 44.8$ litres

83. The compressibility factor for an ideal gas is [1997 - 1 Mark]
(a) 1.5 (b) 1.0 (c) 2.0 (d) ∞

84. The values of van der Waals constant 'a' for the gases O_2, N_2, NH_3 and CH_4 are 1.360, 1.390, 4.170 and 2.253 L^2 atm mol^{-2} respectively. The gas which can most easily be liquified is: [1989 - 1 Mark]
(a) O_2 (b) N_2 (c) NH_3 (d) CH_4

85. In van der Waals equation of state for a non-ideal gas, the term that accounts for intermolecular forces is [1988 - 1 Mark]
(a) $(V-b)$ (b) RT
(c) $\left(P+\frac{a}{V^2}\right)$ (d) $(RT)^{-1}$

2 Integer Value Answer

86. A gas has a compressibility factor of 0.5 and a molar volume of 0.4 $dm^3\,mol^{-1}$ at a temperature of 800 K and pressure x atm. If it shows ideal gas behaviour at the same temperature and pressure, the molar volume will be y $dm^3\,mol^{-1}$. The value of x/y is ______. [Adv. 2023]
[**Use:** Gas constant, $R = 8 \times 10^{-2}$ L atm K^{-1} mol^{-1}]

3 Numeric / New Stem Based Questions

87. Molar volume (V_m) of a van der Waals gas can be calculated by expressing the van der Waals equation as a cubic equation with V_m as the variable. The ratio (in mol dm^{-3}) of the coefficient of V_m^2 to the coefficient of V_m for a gas having van der Waals constants a = 6.0 dm^6 atm mol^{-2} and $b = 0.060$ dm^3 mol^{-1} at 300 K and 300 atm is ______.
Use: Universal gas constant (R) = 0.082 dm^3 atm mol^{-1} K^{-1} [Adv. 2025]

88. Calculate the pressure exerted by one mole of CO_2 gas at 273 K if the van der Waals constant a = 3.592 dm^6 atm mol^{-2}. Assume that the volume occupied by CO_2 molecules is negligible. [2000 - 2 Marks]

5 True / False

89. In the van der Waals equation $\left(P+\frac{n^2a}{V^2}\right)(V-nb)=nRT$ the constant 'a' reflects the actual volume of the gas molecules. [1993 - 1 Mark]

6 MCQs with One or More than One Correct Answer

90. One mole of a monoatomic real gas satisfies the equation $p(V - b) = RT$ where b is a constant. The relationship of interatomic potential $V(r)$ and interatomic distance r for the gas is given by [Adv. 2015]

(a)
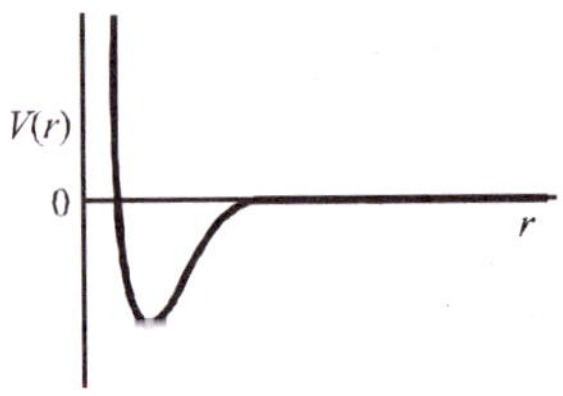

(b)
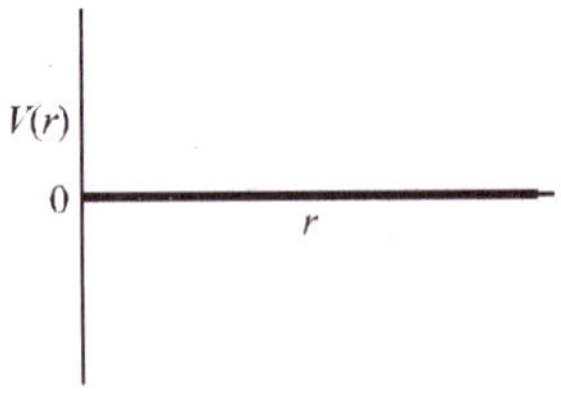

(c)
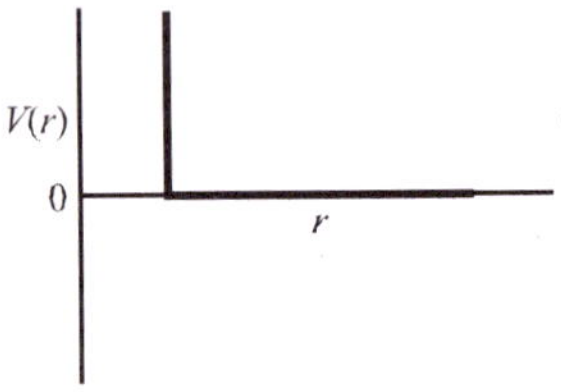

(d)
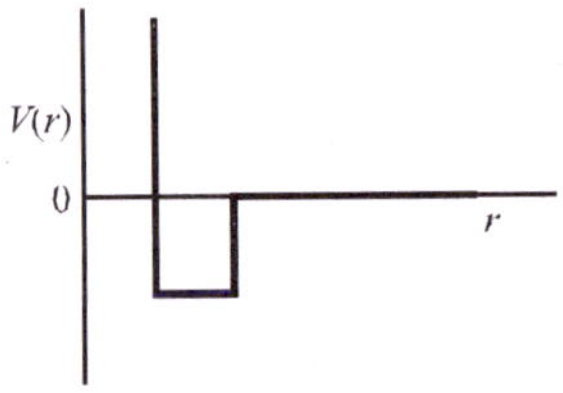

91. A gas described by van der Waals equation – [2008- 1 Mark]
(a) behave similar to an ideal gas in the limit of large molar volumes
(b) behaves similar to an ideal gas is in limit of large pressures
(c) is characterised by van der Waals coefficients that are dependent on the identity of the gas but are independent of the temperature.
(d) has the pressure that is lower than the pressure exerted by the same gas behaving ideally

7 Match the Following

92. Match gases under specified conditions listed in **Column I** with their properties/laws in **Column II**. Indicate your answer by darkening the appropriate bubbles of the 4 × 4 matrix given in the ORS. [2007]

Column I	Column II
(A) Hydrogen gas ($P = 200$ atm, $T = 273$ K)	(p) Compressibility factor $\neq 1$
(B) Hydrogen gas ($P \sim 0$, $T = 273$ K)	(q) Attractive forces are dominant
(C) CO_2 ($P = 1$ atm, $T = 273$ K)	(r) $PV = nRT$
(D) Real gas with very large molar volume	(s) $P(V - nb) = nRT$

93. Arrange the van der Waals constant (a) for the gases: [1995S]

I	$C_6H_6(g)$	A.	0.217
II	$C_6H_5.CH_3(g)$	B.	5.464
III	Ne(g)	C.	18.000
IV.	$H_2O(g)$	D.	24.060

(a) I-A, II-D, III-C, IV-B (b) I-D, II-A, III-B, IV-C
(c) I-C, II-D, III-A, IV-B (d) I-B, II-C, III-A, IV-D

9 Assertion and Reason/Statement Type Questions

Each question contains **STATEMENT-1 (Assertion)** and **STATEMENT-2 (Reason)**. Each question has 4 choices (a), (b), (c) and (d) out of which ONLY ONE is correct. Mark your answer as

(a) If both Statement -1 and Statement -2 are correct, and Statement -2 is the correct explanation of the Statement -2.

(b) If both Statement -1 and Statement -2 are correct, but Statement -2 is not the correct explanation of the Statement -1.

(c) If Statement -1 is correct but Statement -2 is incorrect.

(d) If Statement -1 is incorrect but Statement -2 is correct.

94. Read the following statement and explanation and answer as per the options given below :

Assertion : The value of van der Waals' constant 'a' is larger for ammonia than for nitrogen. **[1998 - 2 Marks]**

Reason : Hydrogen bonding is present in ammonia.

10 Subjective Problems

95. A graph is plotted between PV_m along Y-axis and P along X-axis, where V_m is the molar volume of a real gas. Find the intercept along Y-axis. **[2004 - 2 Marks]**

96. The compression factor (compressibility factor) for one mole of a van der Waals gas at 0 °C and 100 atmospheric pressure is found to be 0.5. Assuming that the volume of a gas molecule is negligible, calculate the van der Waals constant a. **[2001 - 5 Marks]**

97. Using van der Waals equation, calculate the constant, 'a' when two moles of a gas confined in a four litre flask exerts a pressure of 11.0 atmospheres at a temperature of 300 K. The value of 'b' is 0.05 L mol^{-1}. **[1998 - 4 Marks]**

Answer Key

Part-A (JEE Main)																	
1	(2)	**8**	(4)	**15**	(1)	**22**	(80)	**29**	(45)	**36**	(750)	**43**	(3)	**50**	(2)	**57**	(2)
2	(4)	**9**	(4)	**16**	(2)	**23**	(8)	**30**	(2)	**37**	(4)	**44**	(4)	**51**	(1)	**58**	(1)
3	(3)	**10**	(2)	**17**	(2)	**24**	(22)	**31**	(26)	**38**	(1)	**45**	(4)	**52**	(1)	**59**	(1)
4	(1)	**11**	(3)	**18**	(1)	**25**	(1655)	**32**	(927)	**39**	(1)	**46**	(1)	**53**	(1)	**60**	(25)
5	(2)	**12**	(2)	**19**	(2)	**26**	(152)	**33**	(150)	**40**	(3)	**47**	(3)	**54**	(1)	**61**	(1)
6	(4)	**13**	(3)	**20**	(4)	**27**	(4)	**34**	(70)	**41**	(2)	**48**	(3)	**55**	(2)		
7	(1)	**14**	(1)	**21**	(3)	**28**	(32)	**35**	(5)	**42**	(4)	**49**	(1)	**56**	(1)		
Part-B (JEE Advanced)																	
1	(b)	**9**	(b)	**17**	(7)	**25**	(0.25)	**35**	(c)	**60**	(a)	**68**	(900)	**80**	(b)	**88**	(0.99)
2	(c)	**10**	(a)	**18**	(9.80)	**28**	False	**36**	(d)	**61**	(d)	**69**	False	**81**	(b)	**89**	False
3	(c)	**11**	(b)	**19**	(–5242.42)	**29**	(c, d)	**54**	(d)	**62**	(a)	**70**	False	**82**	(b)	**90**	(c)
4	(c)	**12**	(d)	**20**	(4.53)	**30**	(b)	**55**	(b)	**63**	(b)	**71**	(a, b, c)	**83**	(b)	**91**	(a, c)
5	(c)	**13**	(b)	**21**	(0.4)	**31**	(c)	**56**	(c)	**64**	(a)	**72**	(a, b, c, d)	**84**	(c)	**93**	(c)
6	(b)	**14**	(b)	**22**	(2.197)	**32**	(d)	**57**	(d)	**65**	(4)	**73**	(a)	**85**	(c)	**94**	(a)
7	(b)	**15**	(2.22)	**23**	(85.2)	**33**	(a, b)	**58**	(c)	**66**	(434)	**78**	(d)	**86**	(100)		
8	(a)	**16**	(4)	**24**	(3.42)	**34**	(b)	**59**	(c)	**67**	(1:16)	**79**	(b)	**87**	(–7.11)		

Chapter-6

Thermodynamics

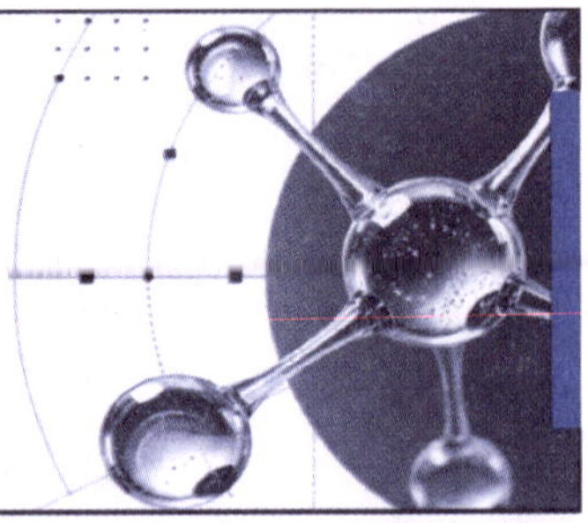

Part-A (JEE Main)

Topic 1- First Law and Fundamentals of Thermodynamics

Multiple Choice Questions (MCQs)

1. Match **List-I** with **List-II** [April 06, 2026 (II)]

Given V_1 and V_2 are initial and final volumes respectively:

List-I (Isothermal process)		List-II (Expression)	
A.	Reversible expansion	I.	$q = 0$
B.	Free expansion	II.	$q = nRT \ln \frac{V_2}{V_1}$
C.	Irreversible Compression	III.	$w = -p_{ext}(V_1 - V_2)$
D.	Cyclic reversible	IV.	$\frac{q_{rev}}{T} = 0$

Choose the correct answer from the options given below :

(1) A-II, B-III, C-I, D-IV (2) A-II, B-I, C-IV, D-III

(3) A-II, B-I, C-III, D-IV (4) A-I, B-II, C-III, D-IV

2. Arrange the following isothermal processes in order of the magnitude of the work (p – V) involved between states 1 and 2. [April 06, 2026 (I)]

A. Expansion in single stage w_A

B. Expansion in multi stages w_B

C. Compression in single stage w_C

D. Compression in multi stages w_D

(1) $|w_B| > |w_A| > |w_C| > |w_D|$ (2) $|w_C| > |w_D| > |w_A| > |w_B|$

(3) $|w_C| > |w_D| > |w_B| > |w_A|$ (4) $|w_B| > |w_A| > |w_D| > |w_C|$

3. The correct order of molar heat capacities measured at 298 K and 1 bar is : [April 05, 2026 (II)]

(1) Copper (s) > Bromine (l) > Helium (g)

(2) Bromine (l) > Copper (s) > Helium (g)

(3) Helium (g) > Bromine (l) > Copper (s)

(4) Helium (g) > Bromine (l) = Copper (s)

4. Given below are two statements: [April 04, 2026 (I)]

Statement I : For an ideal gas, heat capacity at constant volume is always greater than the heat capacity at constant pressure.

Statement II : In a constant volume process, no work is produced and all the heat withdrawn goes into the chaotic motion and is reflected by a temperature increase of the ideal gas.

In the light of the above statements, choose the **correct** answer from the options given below

(1) Both Statement I and Statement II are true

(2) Both Statement I and Statement II are false

(3) Statement I is true but Statement II is false

(4) Statement I is false but Statement II is true

5. Gas 'A' undergoes change from state 'X' to state 'Y'. In this process, the heat absorbed and work done by the gas is 10 J and 18 J respectively. Now gas is brought back to state 'X' by another process during which 6 J of heat is evolved. In the reverse process of 'Y' to 'X', [April 02, 2026 (II)]

(1) 18 J of the work is done by the gas 'A'.

(2) 2 J of the work is done by the gas 'A'.

(3) 12 J of the work is done on the gas 'A' by the surrounding.

(4) 14 J of the work is done on the gas 'A' by the surrounding.

6. 20.0 dm^3 of an ideal gas 'X' at 600 K and 0.5 MPa undergoes isothermal reversible expansion until pressure of the gas is 0.2 MPa. Which of the following option is correct ?

(Given: log 2 = 0.3010 and log 5 = 0.6989) [Jan. 28, 2026 (I)]

(1) $w = -9.1$ kJ, $\Delta U = 0$, $\Delta H = 0$, $q = 9.1$ kJ

(2) $w = 9.1$ J, $\Delta U = 9.1$ J, $\Delta H = 0$; $q = 0$

(3) $w = +4.1$ kJ, $\Delta U = 0$, $\Delta H = 0$; $q = -4.1$ kJ

(4) $w = -3.9$ kJ, $\Delta U = 0$, $\Delta H = 0$; $q = 3.9$ kJ

7. Match the List-I with List-II. [Jan. 24, 2026 (I)]

List-I (Isothermal process for ideal gas system)		List-II Work done (Vf > Vi)	
A.	Reversible expansion	I.	$w = 0$
B.	Free expansion	II.	$w = -nRT \ln \frac{V_f}{V_i}$
C.	Irreversible expansion	III.	$w = -p_{ex}(V_f - V_i)$
D.	Irreversible compression	IV.	$w = -p_{ex}(V_i - V_f)$

Choose the **correct** answer from the options given below:

(1) A-IV, B-I, C-III, D-II (2) A-IV, B-II, C-III, D-I

(3) A-I, B-III, C-II, D-IV (4) A-II, B-I, C-III, D-IV

8. A cup of water at 5°C (system) is placed in a microwave oven and the oven is turned on for one minute during which, the water begins to boil. Which of the following option is **true**? [Jan. 23, 2026 (I)]

(1) $q = +ve$, $w = 0$, $\Delta U = -ve$

(2) $q = +ve$, $w = -ve$, $\Delta U = +ve$

(3) $q = -ve$, $w = -ve$, $\Delta U = -ve$

(4) $q = +ve$, $w = -ve$, $\Delta U = -ve$

9. Match the **LIST-I** with **LIST-II** [Jan. 22, 2026 (I)]

List-I (Thermodynamic Process)		List–II (Magnitude in kJ)	
A.	Work done in reversible, isothermal expansion of 2 mol of ideal gas from 2 dm^3 to 20 dm^3 at 300 K.	I.	4

B.	Work done in irreversible isothermal expansion of 1 mol ideal gas from 1 m^3 to 3 m^3 at 300 K against A constant pressure of 3kPa.	II.	11.5
C.	Change in internal energy for adiabatic expansion of 1 mol ideal gas with change of temperature $= 320\text{ K}$ and $\overline{C}_V = \frac{3}{2}R$	III.	6
D.	Change in enthalpy at constant pressure of 1 mole ideal gas with change of temperature 337 K and $\overline{C}_P = \frac{5}{2}R$	IV.	7

Choose the **correct** answer from the option given below :

(1) A-III, B-II, C-IV, D-I (2) A-II, B-III, C-I, D-IV

(3) A-I, B-II, C-III, D-IV (4) A-II, B-I, C-III, D-IV

10. Which of the following graphs between pressure 'P' versus volume 'V' represent the maximum work done ? [Jan. 21, 2026 (I)]

(1) 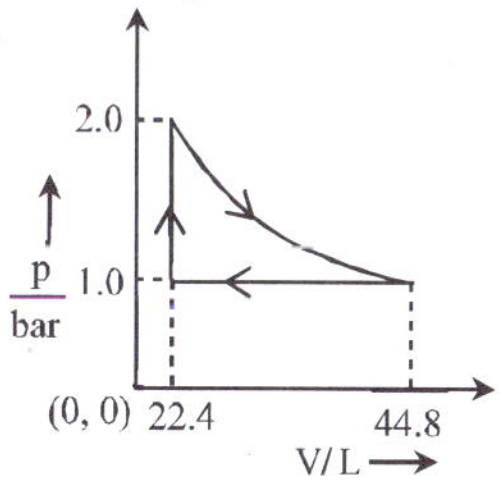

(2) 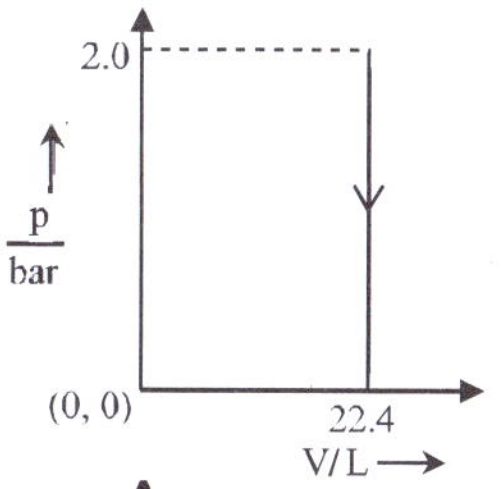

(3) 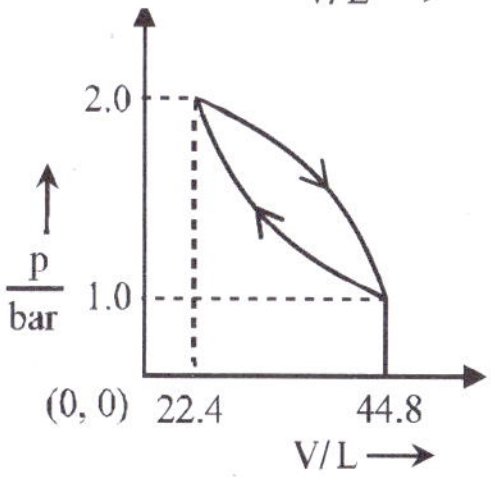

(4) 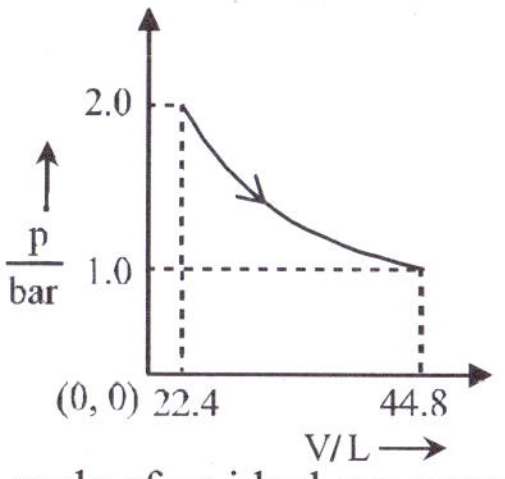

11. One mole of an ideal gas expands isothermally and reversibly from 10 dm^3 to 20 dm^3 at 300 K. ΔU, q and work done in the process respectively are :

Given : $R = 8.3\ JK^{-1}$ and mol^{-1}

$\ln 10 = 2.3$

$\log 2 = 0.30$

$\log 3 = 0.48$ [April 4, 2025 (I)]

(1) 0, 21.84 kJ, –1.26 kJ (2) 0, –17.18 kJ, 1.718 J

(3) 0, 21.84 kJ, 21,84 kJ (4) 0, 1.78 kJ, –1.718 kJ

12. Which of the following properties will change when system containing solution 1 will become solution 2 ? [April 3, 2025 (I)]

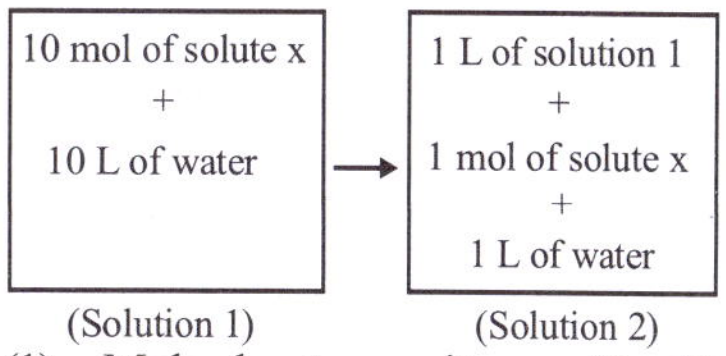

(1) Molar heat capacity (2) Density

(3) Concentration (4) Gibbs free energy

13. Arrange the following in order of magnitude of work done by the system / on the system at constant temperature :

(a) $|w_{reversible}|$ for expansion in infinite stage. [April 2, 2025 (II)]

(b) $|w_{irreversible}|$ for expansion in single stage.

(c) $|w_{reversible}|$ for compression in infinite stage.

(d) $|w_{irreversible}|$ for compression in single stage.

Choose the **correct** answer from the options given below :

(1) $a > b > c > d$ (2) $d > c = a > b$

(3) $c = a > d > b$ (4) $a > c > b > d$

14. 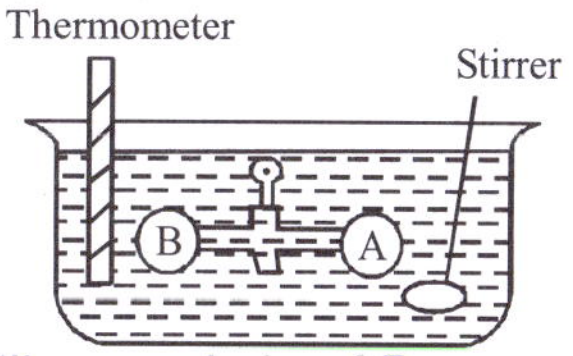

Two vessels A and B are connected via stopcock. The vessel A is filled with a gas at a certain pressure. The entire assembly is immersed in water and is allowed to come to thermal equilibrium with water. After opening the stopcock the gas from vessel A expands into vessel B and no change in temperature is observed in the thermometer. Which of the following statement is **true**? [April 2, 2025 (I)]

(1) $dw \neq 0$

(2) $dq \neq 0$

(3) $dU \neq 0$

(4) The pressure in the vessel B before opening the stopcock is zero.

15. 500 J of energy is transferred as heat to 0.5 mol of Argon gas at 298 K and 1.00 atm. The final temperature and the change in internal energy respectively are :

Given : $R = 8.3\ JK^{-1}\ mol^{-1}$ [Jan. 29, 2025 (I)]

(1) 348 K and 300 J (2) 378 K and 300 J

(3) 368 K and 500 J (4) 378 K and 500 J

16. [Jan. 28, 2025 (II)]

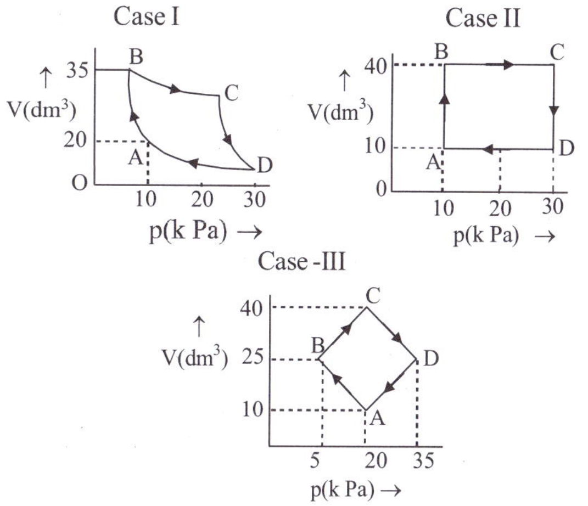

An ideal gas undergoes a cyclic transformation starting from the point A and coming back to the same point by tracing the path A → B → C → D → A as shown in the three cases above. Choose the correct option regarding ΔU.

(1) ΔU (Case-III) > ΔU (Case-II) > ΔU (Case-I)
(2) ΔU (Case-I) > ΔU (Case-II) > ΔU (Case-III)
(3) ΔU (Case-I) > ΔU (Case-III) > ΔU (Case-II)
(4) ΔU (Case-I) = ΔU (Case-II) = ΔU (Case-III)

17. A liquid when kept inside a thermally insulated closed vessel at 25°C was mechanically stirred from outside. What will be the correct option for the following thermodynamic parameters? **[Jan. 22, 2025 (I)]**

(1) $\Delta U > 0, q = 0, w > 0$ (2) $\Delta U = 0, q = 0, w = 0$
(3) $\Delta U < 0, q = 0, w > 0$ (4) $\Delta U = 0, q < 0, w > 0$

18. Choose the correct option for free expansion of an ideal gas under adiabatic condition from the following: **[Feb. 1, 2024 (I)]**

(1) $q = 0, \Delta T \neq 0, w = 0$ (2) $q = 0, \Delta T < 0, w \neq 0$
(3) $q = \neq 0, \Delta T = 0, w = 0$ (4) $q = 0, \Delta T = 0, w = 0$

19. What happens when methane undergoes combustion in systems A and B respectively? **[April 13, 2023 (II)]**

Adiabatic system	Diathermic container
System A	System B

(1)

System A	System B
Temperature rises	Temperature remains same

(2)

System A	System B
Temperature falls	Temperature rises

(3)

System A	System B
Temperature falls	Temperature remains same

(4)

System A	System B
Temperature remains same	Temperature rises

20. Which of the following relation is not correct? **[July 28, 2022 (I)]**

(1) $\Delta H = \Delta U - P\Delta V$ (2) $\Delta U = q + W$
(3) $\Delta S_{sys} + \Delta S_{surr} \geq 0$ (4) $\Delta G = \Delta H - T\Delta S$

21. Five moles of an ideal gas at 1 bar and 298 K is expanded into vacuum to double the volume. The work done is : **[Sep. 04, 2020 (II)]**

(1) $C_V(T_2 - T_1)$ (2) $-RT(V_2 - V_1)$
(3) $-RT \ln V_2/V_1$ (4) zero

22. For one mole of an ideal gas, which of these statements must be **true**? **[Sep. 04, 2020 (I)]**

(A) U and H each depends only on temperature
(B) Compressibility factor Z is not equal to 1
(C) $C_{P,m} - C_{V,m} = R$
(D) $dU = C_V dT$ for any process

(1) (A) and (C) (2) (B), (C) and (D)
(3) (C) and (D) (4) (A), (C) and (D)

23. An ideal gas is allowed to expand from 1 L to 10 L against a constant external pressure of 1 bar. The work done in kJ is : **[April 12, 2019 (I)]**

(1) −9.0 (2) +10.0 (3) −0.9 (4) −2.0

24. Among the following, the set of parameters that represents path functions, is: **[April 9, 2019 (I)]**

(A) $q + w$ (B) q (C) w (D) $H - TS$

(1) (B) and (C) (2) (B), (C) and (D)
(3) (A) and (D) (4) (A), (B) and (C)

25. During compression of a spring the work done is 10 kJ and 2 kJ escaped to the surroundings as heat. The change in internal energy, U (in kJ) is: **[April 9, 2019 (II)]**

(1) −12 (2) −8 (3) 8 (4) 12

26. Which one of the following equations does not correctly represent the first law of thermodynamics for the given processes involving an ideal gas? (Assume non-expansion work is zero) **[April 8, 2019 (I)]**

(1) Cyclic process : $q = -w$
(2) Adiabatic process : $\Delta U = -w$
(3) Isochoric process: $\Delta U = q$
(4) Isothermal process: $q = -w$

27. 5 moles of an ideal gas at 100 K are allowed to undergo reversible compression till its temperature becomes 200 K. If $C_V = 28$ J K^{-1} mol^{-1}, calculate ΔU and ΔPV for this process. ($R = 8.0$ J K^{-1} mol^{-1}) **[April 8, 2019 (II)]**

(1) $\Delta U = 14$ kJ; $\Delta(PV) = 18$ kJ
(2) $\Delta U = 14$ kJ; $\Delta(PV) = 0.8$ kJ
(3) $\Delta U = 14$ kJ; $\Delta(PV) = 4$ kJ
(4) $\Delta U = 14$ kJ; $\Delta(PV) = 8.0$ kJ

28. For a diatomic ideal gas in a closed system, which of the following plots does not correctly describe the relation between various thermodynamic quantities? **[Jan. 12, 2019 (I)]**

(1)
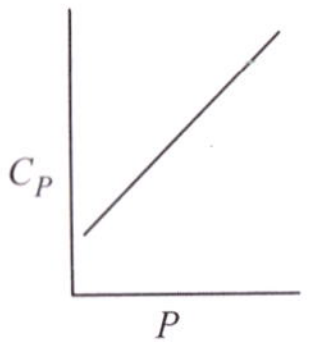

(2)

(3)
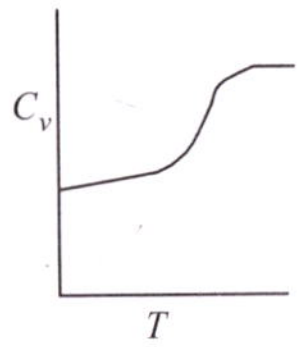

(4)
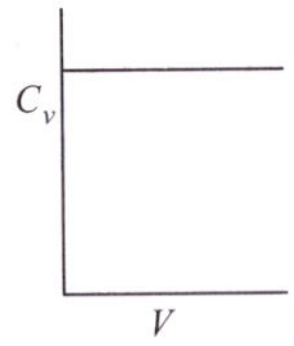

29. The combination of plots which does not represent isothermal expansion of an ideal gas is: **[Jan. 12, 2019 (II)]**

(A)
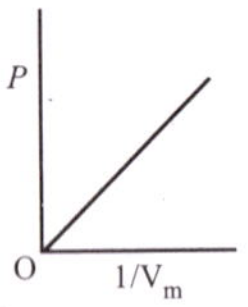

(B)
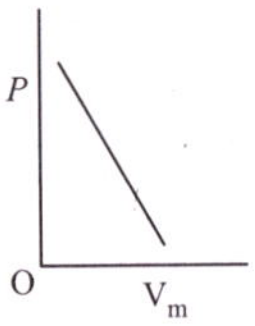

(C)
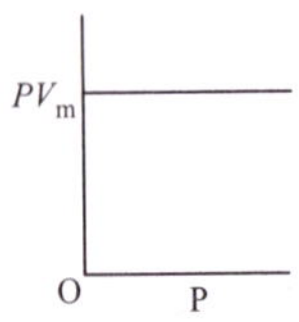

(D)
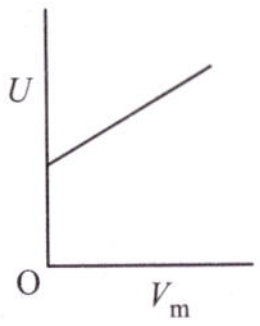

(1) (B) and (D) (2) (A) and (C)
(3) (B) and (C) (4) (A) and (D)

30. An ideal gas undergoes isothermal compression from 5 m^3 to 1 m^3 against a constant external pressure of 4 Nm^{-2}. Heat released in this process is used to increase the temperature of 1 mole of Al. If molar heat capacity of Al is 24 J mol^{-1} K^{-1}, the temperature of Al increases by: **[Jan. 10, 2019 (II)]**

(1) $\frac{3}{2}$ K (2) 2 K (3) $\frac{2}{3}$ K (4) 1 K

31. Consider the reversible isothermal expansion of an ideal gas in a closed system at two different temperatures, T_1 and T_2 ($T_1 < T_2$). The correct graphical depiction of the dependence of work done (w) on the final volume (V) is: [Jan. 9, 2019 (I)]

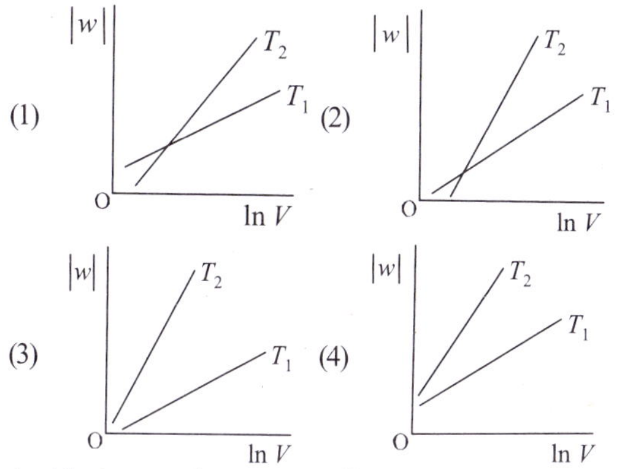

32. An ideal gas undergoes a cyclic process as shown in figure.

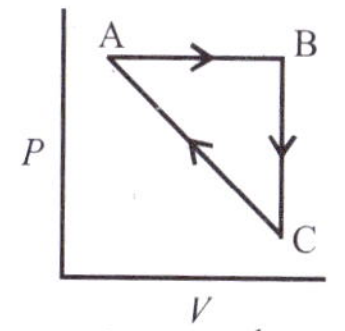

$\Delta U_{BC} = -5$ kJ mol^{-1}, $q_{AB} = 2$ kJ mol^{-1}

$W_{AB} = -5$ kJ mol^{-1}, $W_{CA} = 3$ kJ mol^{-1}

Heat absorbed by the system during process CA is: [Online April 15, 2018 (I)]

(1) -5 kJ mol^{-1} (2) $+5$ kJ mol^{-1}
(3) 18 kJ mol^{-1} (4) -18 kJ mol^{-1}

33. ΔU is equal to [2017]

(1) Isochoric work (2) Isobaric work
(3) Adiabatic work (4) Isothermal work

34. A gas undergoes change from state A to state B. In this process, the heat absorbed and work done by the gas is 5 J and 8 J, respectively. Now gas is brought back to A by another process during which 3 J of heat is evolved. In this reverse process of B to A: [Online April 9, 2017]

(1) 10 J of the work will be done by the gas.
(2) 6 J of the work will be done by the gas.
(3) 10 J of the work will be done by the surrounding on gas.
(4) 6 J of the work will be done by the surrounding on gas.

35. If 100 mole of H_2O_2 decomposes at 1 bar and 300 K, the work done (kJ) by one mole of O_2(g) as it expands against 1 bar pressure is : [Online April 10, 2016]

$$2H_2O_2(l) \rightleftharpoons 2H_2O(l) + O_2(g)$$

($R = 8.3$ JK^{-1} mol^{-1})

(1) 124.50 (2) 249.00
(3) 498.00 (4) 62.25

36. A piston filled with 0.04 mol of an ideal gas expands reversibly from 50.0 mL to 375 mL at a constant temperature of 37.0 °C. As it does so, it absorbs 208 J of heat. The values of q and w for the process will be:

($R = 8.314$ J/mol K) (ln 7.5 = 2.01) [2013]

(1) $q = +208$ J, $w = -208$ J
(2) $q = -208$ J, $w = -208$ J
(3) $q = -208$ J, $w = +208$ J
(4) $q = +208$ J, $w = +208$ J

37. Which of the following statements/relationships is **not** correct in thermodynamic changes? [Online April 23, 2013]

(1) $\Delta U = 0$ (isothermal reversible expansion of a gas)

(2) $w = -nRT \ln \frac{V_2}{V_1}$ (isothermal reversible expansion of an ideal gas)

(3) $w = nRT \ln \frac{V_2}{V_1}$ (isothermal reversible expansion of an ideal gas)

(4) For a system of constant volume, heat involved directly changes to internal energy.

38. Assuming that water vapour is an ideal gas, the internal energy change (ΔU) when 1 mol of water is vapourised at 1 bar pressure and 100 °C, (given : molar enthalpy of vapourisation of water at 1 bar and 373 K = 41 kJ mol^{-1} and R = 8.3 J mol^{-1} K^{-1}) will be [2007]

(1) 41.00 kJ mol^{-1} (2) 4.100 kJ mol^{-1}
(3) 3.7904 kJ mol^{-1} (4) 37.904 kJ mol^{-1}

39. An ideal gas is allowed to expand both reversibly and irreversibly in an isolated system. If T_i is the initial temperature and T_f is the final temperature, which of the following statements is correct?

(1) $(T_f)_{rev} = (T_f)_{irrev}$ [2006]
(2) $T_f = T_i$ for both reversible and irreversible processes
(3) $(T_f)_{irrev} > (T_f)_{rev}$
(4) $T_f > T_i$ for reversible process but $T_f = T_i$ for irreversible process

40. An ideal gas expands in volume from 1×10^{-3} to 1×10^{-2} m^3 at 300 K against a constant pressure of 1×10^5 Nm^{-2}. The work done is

(1) 270 kJ (2) -900 kJ (3) -900 J (4) 900 kJ [2004]

41. The internal energy change when a system goes from state A to B is 40 kJ/mole. If the system goes from A to B by a reversible path and returns to state A by an irreversible path what would be the net change in internal energy ? [2003]

(1) >40 kJ (2) <40 kJ (3) Zero (4) 40 kJ

42. A heat engine abosrbs heat Q_1 at temperature T_1 and heat Q_2 at temperature T_2. Work done by the engine is $(Q_1 + Q_2)$. This data [2002]

(1) violates 1st law of thermodynamics
(2) violates 1st law of themodynamics if Q_1 is –ve
(3) violates 1st law of thermodynamics of Q_2 is –ve
(4) does not violate 1st law of themodynamics.

43. The heat required to raise the temperature of body by 1 K is called [2002]

(1) specific heat (2) thermal capacity
(3) water equivalent (4) none of these.

NUMERICAL VALUE QUESTIONS (NVQs)

44.

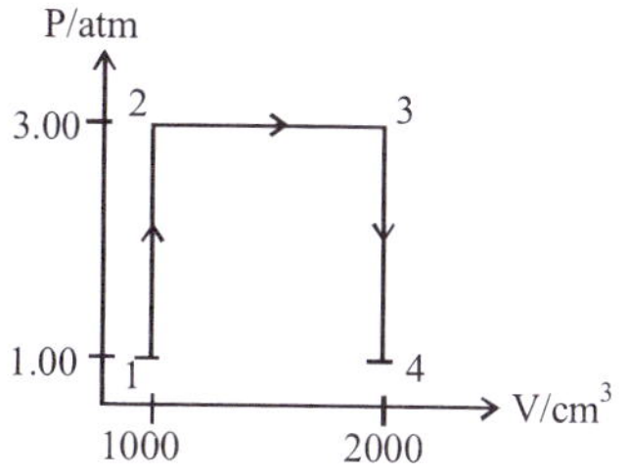

A perfect gas (0.1 mol) having $\bar{C}_V = 1.50$ R (independent of temperature) undergoes the above transformation from point 1 to point 4. If each step is reversible, the total work done (w) while going from point 1 to point 4 is (–)___ J (nearest integer) [Given : R = 0.082 L atm K^{-1} mol^{-1}] [April 3, 2025 (II)]

45. 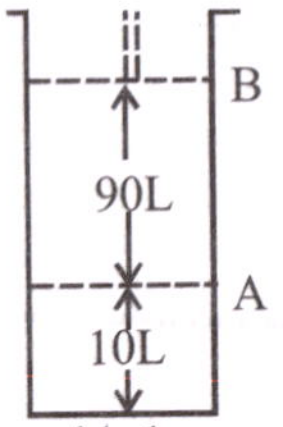

Consider the figure provided. [April 8, 2024 (I)]
1 mol of an ideal gas is kept in a cylinder, fitted with a piston, at the position A, at 18°C. If the piston is moved to position B, keeping the temperature unchanged, then 'x' L atm work is done in this reversible process.
x = ______ L atm. (nearest integer)
[Given : Absolute temperature = °C + 273.15, R = 0.08206 L atm mol^{-1} K^{-1}]

46. $\Delta_{vap}H^{\ominus}$ for water is +40.79 kJ mol^{-1} at 1 bar and 100°C. Change in internal energy for this vapourisation under same condition is ______ kJ mol^{-1}. (Integer answer)
(Given R = 8.3 JK^{-1} mol^{-1}) [April 8, 2024 (II)]

47. An ideal gas, $\bar{C}v = \frac{5}{2}R$, is expanded adiabatically against a constant pressure of 1 atm untill it doubles in volume. If the initial temperature and pressure is 298 K and 5 atm, respectively then the final temperature is ________ K (nearest integer). [April 6, 2024 (I)]
[$\bar{C}v$ is the molar heat capacity at constant volume]

48. Three moles of an ideal gas are compressed isothermally from 60 L to 20 L using constant pressure of 5 atm. Heat exchange Q for the compression is ______ Lit. atm. [April 4, 2024 (II)]

49. If 5 moles of an ideal gas expands from 10 L to a volume of 100 L at 300 K under isothermal and reversible condition then work, w, is –x J. The value of x is ______.
(Given R = 8.314 J K^{-1} mol^{-1}) [Jan. 31, 2024 (II)]

50. An ideal gas undergoes a cyclic transformation starting from the point A and coming back to the same point by tracing the path A → B → C → A as shown in the diagram above. The total work done in the process is ________ J. [Jan. 30, 2024 (I)]

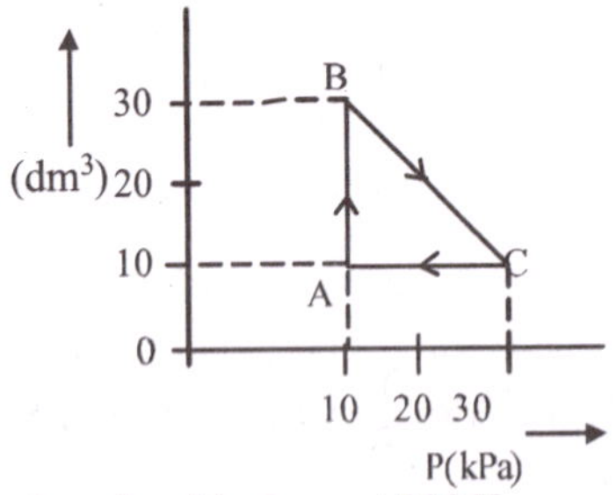

51. If three moles of an ideal gas at 300 K expand isothermally from 30 dm^3 to 45 dm^3 against a constant opposing pressure of 80 kPa, then the amount of heat transferred is ________ J. [Jan. 27, 2024 (I)]

52. The total number of intensive properties from the following is ______
Volume, Molar heat capacity, Molarity, E° cell, Gibbs free energy change, Molar mass, Mole. [April 11, 2023 (II)]

53. The number of endothermic process/es from the following is ______ [April 10, 2023 (II)]
A. $I_2(g) \rightarrow 2I(g)$ B. $HCl(g) \rightarrow H(g) + Cl(g)$
C. $H_2O(l) \rightarrow H_2O(g)$ D. $C(s) + O_2(g) \rightarrow CO_2(g)$
E. Dissolution of ammonium chloride in water

54. When a 60 W electric heater is immersed in a gas for 100s in a constant volume container with adiabatic walls, the temperature of the gas rises by 5°C. The heat capacity of the given gas is_____J K^{-1} (Nearest integer) [April 08, 2023 (I)]

55. When 2 litre of ideal gas expands isothermally into vacuum to a total volume of 6 litre, the change in internal energy is ______ J. (Nearest integer) [Jan. 30, 2023 (I)]

56. 1 mole of ideal gas is allowed to expand reversibly and adiabatically from a temperature of 27°C. The work done is 3 kJ mol^{-1}. The final temperature of the gas is _______K (Nearest integer). Given C_v = 20 J $mol^{-1}K^{-1}$. [Jan. 30, 2023 (II)]

57. One mole of an ideal monoatomic gas is subjected to changes as shown in the graph. The magnitude of the work done (by the system or on the system) is _______ J (nearest integer). [Jan. 24, 2023 (II)]

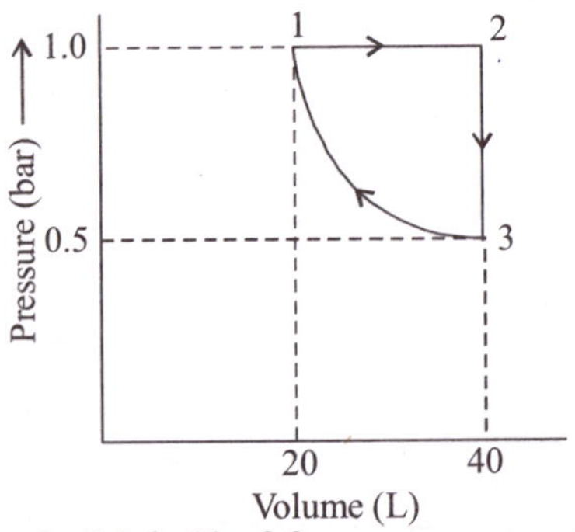

Given : log 2 = 0.3, ln 10 = 2.3

58. Among the following the number of state variables is ________. [July 28, 2022 (II)]
Internal energy (U)
Volume (V)
Heat (q)
Enthalpy (H)

59. The molar heat capacity for an ideal gas at constant pressure is 20.785 J K^{-1} mol^{-1}. The change in internal energy is 5000 J upon heating it from 300K to 500K. The number of moles of the gas at constant volume is ___ [Nearest integer]
(Given: R = 8.314 J K^{-1} mol^{-1}) [July 27, 2022 (I)]

60. 1.0 mol of monoatomic ideal gas is expanded from state 1 to state 2 as shown in the figure. The magnitude of the work done for the expansion of gas from state 1 to state 2 at 300 K is _______ J. (Nearest integer)
(Given : R = 8.3 J K^{-1} mol^{-1}, ln 10 = 2.3, log 2 = 0.30) [June 30, 2022 (I)]

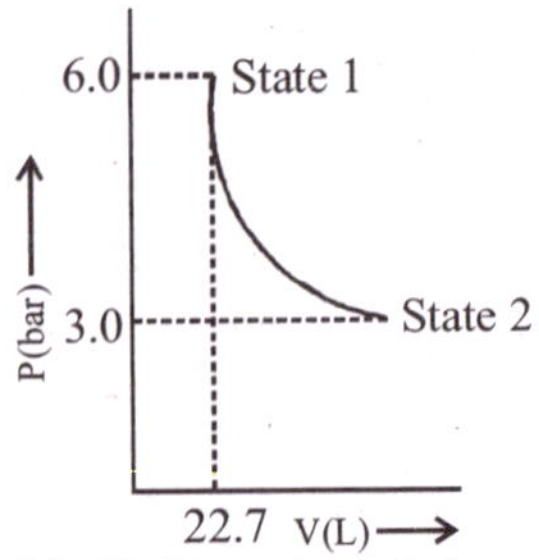

61. 2.2 g of nitrous oxide (N_2O) gas is cooled at a constant pressure of 1 atm from 310 K to 270 K causing the compression of the gas from 217.1 mL to 167.75 mL. The change in internal energy of the process, ΔU is '–x' J. The value of 'x' is ___. [nearest integer]
(Given: atomic mass of N = 14 g mol^{-1} and of O = 16 g mol^{-1}. Molar heat capacity of N_2O is 100 JK^{-1} mol^{-1}) [June 29, 2022 (II)]

62. 4.0 L of an ideal gas is allowed to expand isothermally into vacuum until the total volume is 20 L. The amount of heat absorbed in this expansion is ___________ L atm. [June 28, 2022 (I)]

63. When 5 moles of He gas expand isothermally and reversibly at 300 K from 10 litre to 20 litre, the magnitude of the maximum work obtained is ____ J. [nearest integer] (Given: R = 8.3 J $K^{-1}mol^{-1}$ and log 2 = 0.3010) [June 27, 2022 (II)]

64. Two flasks I and II shown below are connected by a valve of negligible volume. [Aug. 27, 2021 (II)]

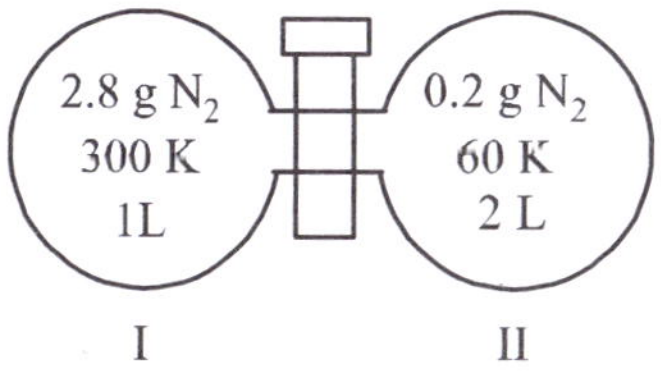

When the valve is opened, the final pressure of the system in bar is $x \times 10^{-2}$. The value of x is ______. (Integer answer)
[Assume–Ideal gas; 1 bar = 10^5 Pa; Molar mass of $N_2 = 28.0\ g\ mol^{-1}$; $R = 8.31\ J\ mol^{-1}K^{-1}$]

65. For water $\Delta_{vap}H = 41$ kJ mol^{-1} at 373 K and 1 bar pressure. Assuming that water vapour is an ideal gas that occupies a much larger volume than liquid water, the internal energy change during evaporation of water is_____kJ mol^{-1}
[Use : $R = 8.3\ J\ mol^{-1}\ K^{-1}$] [Aug. 26, 2021 (II)]

66. For water at 100°C and 1 bar,
$\Delta_{vap}H - \Delta_{vap}U$ = ______ $\times 10^2\ J\ mol^{-1}$.
(Round off to the Nearest Integer)
[Use : $R = 8.31\ J\ mol^{-1}\ K^{-1}$]
[Assume volume of $H_2O(l)$ is much smaller than volume of $H_2O(g)$. Assume $H_2O(g)$ treated as an ideal gas] [July 27, 2021 (I)]

67. A system does 200 J of work and at the same time absorbs 150 J of heat. The magnitude of the change in internal energy is ______ J. (Nearest integer) [July 25, 2021 (II)]

68. At 25 °C, 50 g of iron reacts with HCl to form $FeCl_2$. The evolved hydrogen gas expands against a constant pressure of 1 bar. The work done by the gas during this expansion is _______ J. (Round off to the Nearest Integer).
[Given : $R = 8.314\ J\ mol^{-1}\ K^{-1}$. Assume, hydrogen is an ideal gas] [Atomic mass of Fe is 55.85 u] [March 16, 2021 (II)]

69. The magnitude of work done by a gas that undergoes a reversible expansion along the path ABC shown in the figure is ________ [Jan. 08, 2020 (I)]

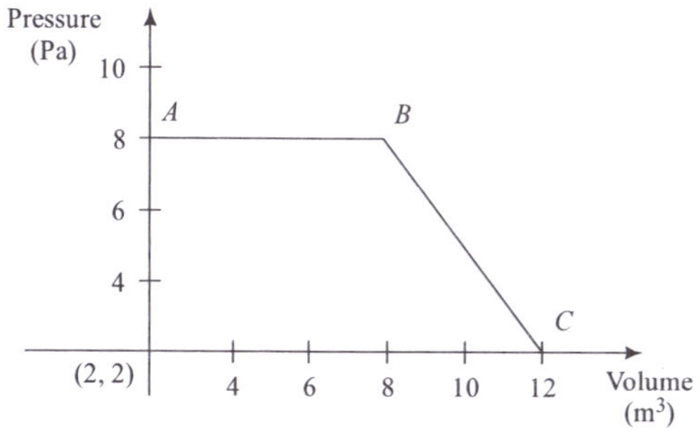

70. At constant volume, 4 mol of an ideal gas when heated from 300 K to 500 K changes its internal energy by 5000 J. The molar heat capacity at constant volume is ______. [Jan. 08, 2020 (II)]

Topic 2 - Laws of Thermochemistry

MULTIPLE CHOICE QUESTIONS (MCQs)

71. Consider the following data [April 02, 2026 (I)]
(i) $2Al(s) + 6HCl(aq) \rightarrow Al_2Cl_6(aq) + 3H_2(g) + 1200$ kJ/mol.
(ii) $H_2(g) + Cl_2(g) \rightarrow 2HCl(g) + 164$ kJ/mol.
(iii) $HCl(g) + aq \rightarrow HCl(aq) + 83$ kJ/mol.
(iv) $Al_2Cl_6(s) + aq \rightarrow Al_2Cl_6(aq) + 663$ kJ/mol.
The enthalpy of formation of anhydrous solid Al_2Cl_6 is :
(1) $-648\ kJ\ mol^{-1}$ (2) $-1350\ kJ\ mol^{-1}$
(3) $-2002\ kJ\ mol^{-1}$ (4) $-1527\ kJ\ mol^{-1}$

72. The heat of atomisation of methane and ethane are 'x' kJ mol^{-1} and 'y' kJ mol^{-1} respectively. The longest wavelength (λ) of light capable of breaking the C-C bond can be expressed in SI unit as: [Jan. 24, 2026 (II)]
(1) $\frac{hc}{1000}\left(\frac{y-6x}{4}\right)^{-1}$ (2) $\frac{N_A hc}{250(4y-6x)}$
(3) $\frac{N_A hc}{250(y-6x)}$ (4) $N_A hc\left(y-\frac{6x}{4}\right)^{-1}$

73. Consider the following data: [Jan. 21, 2026 (II)]
$\Delta_f H^\ominus$ (methane, g) = – X kJ mol^{-1}
Enthalpy of sublimation of graphite – Y kJ mol^{-1}
Dissociation enthalpy of H_2 = Z kJ mol^{-1}
The bond enthalpy of C – H bond is given by:
(1) $\frac{X+Y+2Z}{4}$ (2) $\frac{X+Y+4Z}{2}$
(3) X + Y + Z (4) $\frac{-X+Y+Z}{4}$

74. The correct statement amongst the following is : [April 7, 2025 (II)]
(1) The term 'standard state' implies that the temperature is 0°C
(2) The standard state of pure gas is the pure gas at a pressure of 1 bar and temperature 273 K
(3) $\Delta_f H^0_{298}$ is zero for O(g)
(4) $\Delta_f H^0_{500}$ is zero for $O_2(g)$

75. The hydration energies of K^+ and Cl^- are $-x$ and $-y$ kJ/mol respectively. If lattice energy of KCl is $-z$ kJ/mol, then the heat of solution of KCl is : [April 7, 2025 (II)]
(1) $+x - y - z$ (2) $x + y + z$
(3) $z - (x + y)$ (4) $-z - (x + y)$

76. Total enthalpy change for freezing of 1 mol of water at 10°C to ice at – 10°C is ______ [April 7, 2025 (I)]
(Given : $\Delta_{fus}H = x$ kJ/mol
$C_p[H_2O(l)] = y\ J\ mol^{-1}\ K^{-1}$
$C_p[H_2O(s)] = z\ J\ mol^{-1}\ K^{-1}$
(1) $-x - 10y - 10z$ (2) $-10(100x + y + z)$
(3) $10(100x + y + z)$ (4) $x - 10y - 10z$

77. Consider the given data : [April 4, 2025 (II)]
(A) $HCl(g) + 10H_2O(l) \rightarrow HCl \cdot 10H_2O$ $\Delta H = -69.01\ kJ\ mol^{-1}$
(B) $HCl(g) + 40H_2O(l) \rightarrow HCl \cdot 40H_2O$ $\Delta H = -72.79\ kJ\ mol^{-1}$

Choose the **correct** statement :
(1) Dissolution of gas in water is an endothermic process
(2) The heat of solution depends on the amount of solvent.
(3) The heat of dilution for the HCl ($HCl \cdot 10H_2O$ to $HCl \cdot 40H_2O$) is 3.78 kJ mol^{-1}.
(4) The heat of formation of HCl solution is represented by both (A) and (B)

78. If $C(diamond) \rightarrow C(graphite) + X\ kJ\ mol^{-1}$
$C(diamond) + O_2(g) \rightarrow CO_2(g) + Y\ kJ\ mol^{-1}$
$C(graphite) + O_2(g) \rightarrow CO_2(g) + Z\ kJ\ mol^{-1}$
At constant temperature. Then **[Jan. 29, 2025 (II)]**
(1) $X = Y + Z$ (2) $-X = Y + Z$
(3) $X = -Y + Z$ (4) $X = Y - Z$

79. $S(g) + \frac{3}{2}O_2(g) \rightarrow SO_3(g) + 2x\,kcal$
$SO_2(g) + \frac{1}{2}O_2(g) \rightarrow SO_3(g) + y\,kcal$
The heat of formation of $SO_2(g)$ is given by : **[Jan. 24, 2025 (II)]**
(1) $\frac{2x}{y}$ kcal (2) $y - 2x$ kcal
(3) $2x + y$ kcal (4) $x + y$ kcal

80. Combustion of glucose ($C_6H_{12}O_6$) produces CO_2 and water. The amount of oxygen (in g) required for the complete combustion of 900 g of glucose is: [Molar mass of glucose in g mol^{-1} = 180] **[April 8, 2024 (I)]**
(1) 480 (2) 960 (3) 800 (4) 32

81. Given **[April 10, 2023 (I)]**
(A) $2CO(g) + O_2(g) \rightarrow 2CO_2(g)\ \Delta H_1^\circ = -x\ kJ\ mol^{-1}$
(B) $C(graphite) + O_2(g) \rightarrow CO_2(g)\ \Delta H_2^\circ = -y\ kJ\ mol^{-1}$
The ΔH° for the reaction
$C(graphite) + \frac{1}{2}O_2(g) \rightarrow CO(g)$ is
(1) $\frac{x-2y}{2}$ (2) $\frac{x+2y}{2}$ (3) $\frac{2x-y}{2}$ (4) $2y - x$

82. At 25°C and 1 atm pressure, the enthalpy of combustion of benzene (*l*) and acetylene (g) are –3268 kJ mol^{-1} and –1300 kJ mol^{-1}, respectively. The change in enthalpy for the reaction $3\,C_2H_2(g) \rightarrow C_6H_6(l)$, is **[June 25, 2022 (II)]**
(1) +324 kJ mol^{-1} (2) +632 kJ mol^{-1}
(3) –632 kJ mol^{-1} (4) –732 kJ mol^{-1}

83. At 25°C and 1 atm pressure, the enthalpies of combustion are as given below:

Substance	H_2	C (graphite)	$C_2H_6(g)$
$\frac{\Delta_c H^\circ}{kJmol^{-1}}$	–286.0	–394.0	–1560.0

The enthalpy of formation of ethane is **[June 24, 2022 (II)]**
(1) +54.0 kJ mol^{-1} (2) –68.0 kJ mol^{-1}
(3) –86.0 kJ mol^{-1} (4) +97.0 kJ mol^{-1}

84. Lattice enthalpy and enthalpy of solution of NaCl are 788 kJ mol^{-1} and 4 kJ mol^{-1}, respectively. The hydration enthalpy of NaCl is : **[Sep. 05, 2020 (II)]**
(1) –780 kJ mol^{-1} (2) 780 kJ mol^{-1}
(3) –784 kJ mol^{-1} (4) 784 kJ mol^{-1}

85. The intermolecular potential energy for the molecules *A, B, C* and *D* given below suggests that : **[Sep. 04, 2020 (I)]**

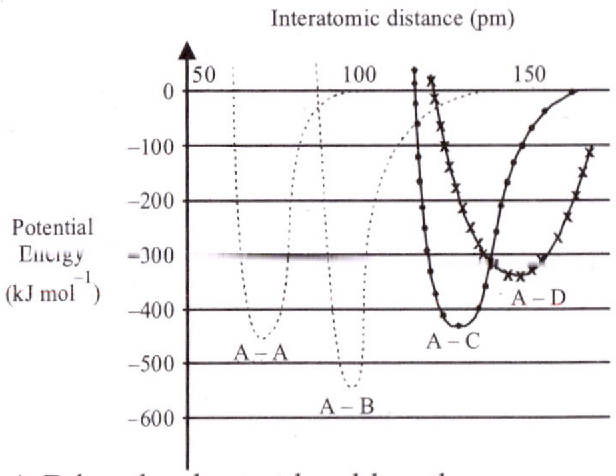

(1) A-D has the shortest bond length
(2) A-A has the largest bond enthalpy
(3) D is more electronegative than other atoms
(4) A-B has the stiffest bond

86 If enthalpy of atomisation for $Br_2(l)$ is *x* kJ/mol and bond enthalpy for Br_2 is *y* kJ/mol, the relation between them: **[Jan. 09, 2020 (I)]**
(1) is $x = y$ (2) does not exist
(3) is $x > y$ (4) is $x < y$

87. Enthalpy of sublimation of iodine is 24 cal g^{-1} at 200 °C. If specific heat of $I_2(s)$ and $I_2(vap)$ are 0.055 and 0.031 cal g^{-1} K^{-1} respectively, then enthalpy of sublimation of iodine at 250 °C in cal g^{-1} is : **[April 12, 2019 (I)]**
(1) 2.85 (2) 5.7 (3) 22.8 (4) 11.4

88. The difference between ΔH and ΔU ($\Delta H - \Delta U$), when the combustion of one mole of heptane (l) is carried out at a temperature *T*, is equal to : **[April 10, 2019 (II)]**
(1) $-4RT$ (2) $-3RT$ (3) $4RT$ (4) $3RT$

89. For silver, $C_p(JK^{-1}mol^{-1}) = 23 + 0.01T$. If the temperature (*T*) of 3 moles of silver is raised from 300 K to 1000 K at 1 atm pressure, the value of ΔH will be close to: **[April 8, 2019 (I)]**
(1) 62 kJ (2) 16 kJ (3) 21 kJ (4) 13 kJ

90. (i) $C(graphite) + O_2(g) \rightarrow CO_2(g);\ \Delta_r H^0 = x\ kJ\ mol^{-1}$
(ii) $C(graphite) + \frac{1}{2}O_2(g) \rightarrow CO(g);\ \Delta_r H^0 = y\ kJ\ mol^{-1}$
(iii) $CO(g) + \frac{1}{2}O_2(g) \rightarrow CO_2(g);\ \Delta_r H^0 = z\ kJ\ mol^{-1}$
Based on the above thermochemical equations, find out which one of the following algebraic relationships is correct? **[Jan. 12, 2019 (II)]**
(1) $x = y + z$ (2) $z = x + y$
(3) $y = 2z - x$ (4) $x = y - z$

91. The combustion of benzene (l) gives CO_2 (g) and H_2O (l). Given that heat of combustion of benzene at constant volume is –3263.9 kJ mol^{-1} at 25 °C; heat of combustion (in kJ mol^{-1}) of benzene at constant pressure will be :
(R= 8.314 JK^{-1} mol^{-1}) **[2018]**
(1) 4152.6 (2) –452.46
(3) 3260 (4) –3267.6

92. For which of the following reactions, ΔH is equal to ΔU? **[Online April 15, 2018 (I)]**
(1) $N_2(g) + 3H_2(g) \rightarrow 2NH_3(g)$
(2) $2HI(g) \rightarrow H_2(g) + I_2(g)$
(3) $2SO_2(g) + O_2(g) \rightarrow 2SO_3(g)$
(4) $2NO_2(g) \rightarrow N_2O_4(g)$

93. Given

$C(graphite) + O_2(g) \rightarrow CO_2(g)$; $\Delta_r H° = -393.5$ kJ mol^{-1}

$H_2(g) + \frac{1}{2}O_2(g) \rightarrow H_2O(l)$; $\Delta_r H° = -285.8$ kJ mol^{-1}

$CO_2(g) + 2H_2O(l) \rightarrow CH_4(g) + 2O_2(g)$;
$\Delta_r H° = +890.3$ kJ mol^{-1}

Based on the above thermochemical equations, the value of $\Delta_r H°$ at 298 K for the reaction,

$C(graphite) + 2H_2(g) \rightarrow CH_4(g)$ will be : [2017]

(1) +74.8 kJ mol^{-1} (2) +144.0 kJ mol^{-1}
(3) −74.8 kJ mol^{-1} (4) −144.0 kJ mol^{-1}

94. For a reaction, $A(g) \rightarrow A(l)$; $\Delta H = -3RT$. The correct statement for the reaction is : [Online April 8, 2017]

(1) $\Delta H = \Delta U \neq O$ (2) $\Delta H = \Delta U = O$
(3) $|\Delta H| < |\Delta U|$ (4) $|\Delta H| > |\Delta U|$

95. The enthalpy change on freezing of 1 mol of water at 5 °C to ice at −5 °C is : [Online April 8, 2017]

(Given $\Delta_{fus}H = 6$ kJ mol^{-1} at 0 °C,
$C_p(H_2O, l) = 75.3$ J mol^{-1} K^{-1}, $C_p(H_2O, s) = 36.8$ J mol^{-1} K^{-1})

(1) 5.44 kJ mol^{-1} (2) 5.81 kJ mol^{-1}
(3) 6.56 kJ mol^{-1} (4) 6.00 kJ mol^{-1}

96. The heats of combustion of carbon and carbon monoxide are −393.5 and −283.5 kJ mol^{-1}, respectively. The heat of formation (in kJ) of carbon monoxide per mole is: [2016; same 2004]

(1) −676.5 (2) −110.5 (3) 110.5 (4) 676.5

97. The heat of atomization of methane and ethane are 360 kJ/mol and 620 kJ/mol, respectively. The longest wavelength of light capable of breaking the C – C bond is (Avogadro number = 6.02×10^{23}, $h = 6.62 \times 10^{-34}$ J s) : [Online April 10, 2015]

(1) 2.48×10^4 nm (2) 1.49×10^3 nm
(3) 2.48×10^3 nm (4) 1.49×10^4 nm

98. The standard enthalpy of formation of NH_3 is – 46.0 kJ/mol. If the enthalpy of formation of H_2 from its atoms is – 436 kJ/mol and that of N_2 is – 712 kJ/mol, the average bond enthalpy of N - H bond in NH_3 is: [Online April 9, 2014]

(1) −1102 kJ/mol (2) −964 kJ/mol
(3) +352 kJ/mol (4) +1056 kJ/mol

99. The standard enthalpy of formation ($\Delta_f H°_{298}$) for methane, CH_4 is – 74.9 kJ mol^{-1}. In order to calculate the average energy given out in the formation of a C – H bond from this, it is necessary to know which one of the following? [Online April 12, 2014; same 2006]

(1) The dissociation energy of the hydrogen molecule, H_2.
(2) The first four ionisation energies of carbon.
(3) The dissociation energy of H_2 and enthalpy of sublimation of carbon (graphite).
(4) The first four ionisation energies of carbon and electron affinity of hydrogen.

100. Given : [Online April 9, 2013]

(I) $H_2(g) + \frac{1}{2}O_2(g) \rightarrow H_2O(l)$;

$\Delta H°_{298K} = -285.9$ kJ mol^{-1}

(II) $H_2(g) + \frac{1}{2}O_2(g) \rightarrow H_2O(g)$;

$\Delta H°_{298K} = -241.8$ kJ mol^{-1}

The molar enthalpy of vapourisation of water will be :

(1) 241.8 kJ mol^{-1} (2) 22.0 kJ mol^{-1}
(3) 44.1 kJ mol^{-1} (4) 527.7 kJ mol^{-1}

101. Given

Reaction	Energy Change (in kJ)
$Li(s) \rightarrow Li(g)$	161
$Li(g) \rightarrow Li^+(g)$	520
$\frac{1}{2}F_2(g) \rightarrow F(g)$	77
$F(g) + e^- \rightarrow F^-(g)$	(Electron gain enthalpy)
$Li^+(g) + F^-(g) \rightarrow LiF(s)$	−1047
$Li(s) + \frac{1}{2}F_2(g) \rightarrow LiF(s)$	−617

Based on data provided, the value of electron gain enthalpy of fluorine would be : [Online April 22, 2013]

(1) −300 kJ mol^{-1} (2) −350 kJ mol^{-1}
(3) −328 kJ mol^{-1} (4) −228 kJ mol^{-1}

102. In which of the following exothermic reactions, the heat liberated per mole is the highest ? [Online April 25, 2013]

(1) $CaO + H_2O \rightarrow Ca(OH)_2$
(2) $SrO + H_2O \rightarrow Sr(OH)_2$
(3) $BaO + H_2O \rightarrow Ba(OH)_2$
(4) $MgO + H_2O \rightarrow Mg(OH)_2$

103. Given that: [Online April 25, 2013]

(i) $\Delta_f H°$ of N_2O is 82 kJ mol^{-1}
(ii) Bond energies of $N\equiv N$, $N=N$, $O=O$ and $N=O$ are 946, 418, 498 and 607 kJ mol^{-1} respectively,

The resonance energy of N_2O is :

(1) −88kJ (2) 66kJ (3) −62kJ (4) −44kJ

104. The difference between the reaction enthalpy change ($\Delta_r H$) and reaction internal energy change ($\Delta_r U$) for the reaction:

$2C_6H_6(l) + 15O_2(g) \longrightarrow 12CO_2(g) + 6H_2O(l)$

at 300 K is ($R = 8.314$ J mol^{-1}K^{-1}) [Online May 12, 2012]

(1) 0 J mol^{-1} (2) 2490 J mol^{-1}
(3) −2490 J mol^{-1} (4) −7482 J mol^{-1}

105. The enthalpy of neutralisation of NH_4OH with HCl is −51.46 kJ mol^{-1} and the enthalpy of neutralisation of NaOH with HCl is −55.90 kJ mol^{-1}. The enthalpy of ionisation of NH_4OH is [Online May 19, 2012]

(1) −107.36 kJ mol^{-1} (2) −4.44 kJ mol^{-1}
(3) +107.36 kJ mol^{-1} (4) +4.44 kJ mol^{-1}

106. Consider the reaction :

$4NO_2(g) + O_2(g) \rightarrow 2N_2O_5(g)$, $\Delta_r H = -111$ kJ.

If $N_2O_5(s)$ is formed instead of $N_2O_5(g)$ in the above reaction, the $\Delta_r H$ value will be :

(given, ΔH of sublimation for N_2O_5 is 54 kJ mol^{-1}) [2011RS]

(1) +54 kJ (2) +219 kJ
(3) −219 J (4) −165 kJ

107. The value of enthalpy change (ΔH) for the reaction

$C_2H_5OH(l) + 3O_2(g) \rightarrow 2CO_2(g) + 3H_2O(l)$

at 27 °C is – 1366.5 k J mol^{-1}. The value of internal energy change for the above reaction at this temperature will be : [2011RS]

(1) −1369.0 kJ (2) −1364.0 kJ
(3) −1361.5 kJ (4) −1371.5 kJ

108. The standard enthalpy of formation of NH_3 is – 46.0 kJ mol^{-1}. If the enthalpy of formation of H_2 from its atoms is – 436 kJ mol^{-1} and that of N_2 is – 712 kJ mol^{-1}, the average bond enthalpy of N – H bond in NH_3 is [2010]

(1) −964 kJ mol^{-1} (2) +352 kJ mol^{-1}
(3) +1056 kJ mol^{-1} (4) −1102 kJ mol^{-1}

109. On the basis of the following thermochemical data : $(\Delta_f H^\circ H^+(aq) = 0)$ **[2009]**

$H_2O(l) \rightarrow H^+(aq) + OH^-(aq); \Delta H = 57.32$ kJ

$H_2(g) + \frac{1}{2}O_2(g) \longrightarrow H_2O(l);\ \Delta H = -286.20$ kJ

The value of enthalpy of formation of OH^- ion at 25° C is:

(1) –228.88 kJ (2) +228.88 kJ
(3) –343.52 kJ (4) –22.88 kJ

110. Oxidising power of chlorine in aqueous solution can be determined by the parameters indicated below:

$$\frac{1}{2}Cl_2(g) \xrightarrow{\frac{1}{2}\Delta_{diss}H^o} Cl(g) \xrightarrow{\Delta_{eg}H^o} Cl^-(g) \xrightarrow{\Delta_{hyd}H^o} Cl^-(aq)$$

The energy envolved in the conversion of

$\frac{1}{2}Cl_2(g)$ to $Cl^{-1}(aq)$

(using the data,

$\Delta_{diss}H^o_{Cl_2} = 240$ kJ mol^{-1}, $\Delta_{eg}H^o_{Cl} = -349$ kJ mol^{-1},

$\Delta_{hyd}H^o_{Cl^-} = -381$ kJ mol^{-1}), will be **[2008]**

(1) + 152 kJ mol^{-1} (2) – 610 kJ mol^{-1}
(3) – 850 kJ mol^{-1} (4) + 120 kJ mol^{-1}

111. The enthalpy changes for the following processes are listed below : **[2006]**

$Cl_2(g) = 2Cl(g)$, 242.3 kJ mol^{-1}
$I_2(g) = 2I(g)$, 151.0 kJ mol^{-1}
$ICl(g) = I(g) + Cl(g)$, 211.3 kJ mol^{-1}
$I_2(s) = I_2(g)$, 62.76 kJ mol^{-1}

Given that the standard states for iodine and chlorine are $I_2(s)$ and $Cl_2(g)$, the standard enthalpy of formation for ICl(g) is :

(1) +16.8 kJ mol^{-1} (2) +244.8 kJ mol^{-1} **[2006]**
(3) –14.6 kJ mol^{-1} (4) –16.8 kJ mol^{-1}

112. $(\Delta H - \Delta U)$ for the formation of carbon monoxide (CO) from its elements at 298 K is **[2006]**

(R = 8.314 J K^{-1} mol^{-1})

(1) –2477.57 J mol^{-1} (2) 2477.57 J mol^{-1}
(3) –1238.78 J mol^{-1} (4) 1238.78 J mol^{-1}

113. Consider the reaction : $N_2 + 3H_2 \rightarrow 2NH_3$ carried out at constant temperature and pressure. If ΔH and ΔU are the enthalpy and internal energy changes for the reaction, which of the following expressions is true ? **[2005]**

(1) $\Delta H > \Delta U$ (2) $\Delta H < \Delta U$
(3) $\Delta H = \Delta U$ (4) $\Delta H > 0$

114. If the bond dissociation energies of XY, X_2 and Y_2 (all diatomic molecules) are in the ratio of 1 : 1 : 0.5 and ΔH_f for the formation of XY is – 200 kJ mole^{-1}. The bond dissociation energy of X_2 will be **[2005]**

(1) 400 kJ mol^{-1} (2) 300 kJ mol^{-1}
(3) 200 kJ mol^{-1} (4) 100 kJ mol^{-1}

115. If at 298 K, the bond energies of C—H, C—C, C=C and H—H bonds are respectively 414, 347, 615 and 435 kJ mol^{-1}, the value of enthalpy change for the reaction **[2003]**

$H_2C = CH_2(g) + H_2(g) \rightarrow H_3C—CH_3(g)$ at 298 K will be

(1) – 250 kJ (2) + 125 kJ
(3) – 125 kJ (4) + 250 kJ

116. The enthalpy change for a reaction does **not** depend upon

(1) use of different reactants for the same product **[2003]**
(2) the nature of intermediate reaction steps
(3) the differences in initial or final temperatures of involved substances
(4) the physical states of reactants and products

NUMERICAL VALUE QUESTIONS (NVQs)

117. Consider the reaction :

$2H_2S(g) + 3O_2(g) \longrightarrow 2H_2O(l) + 2SO_2(g)$

The magnitude of enthalpy change for the reaction in kJ mol^{-1} is ______. (Nearest integer)

Given: $\Delta_f H^\ominus(H_2S) = -20.1$ kJ mol^{-1}
$\Delta_f H^\ominus(H_2O) = -286.0$ kJ mol^{-1}
$\Delta_f H^\ominus(SO_2) = -297.0$ kJ mol^{-1} **[April 08, 2026 (II)]**

118. First and second ionization enthalpies of lithium are 520 kJ mol^{-1} and 7297 kJ mol^{-1} respectively.

Energy required to convert 3.5 mg lithium (g) into $Li^{2+}(g)$ [$Li(g) \rightarrow Li^{2+}(g)$] is _____ kJ mol^{-1}. (nearest integer)

[Molar mass of Li = 7 g mol^{-1}] **[April 06, 2026 (I)]**

119. If 3.365 g of ethanol (*l*) burnt completely in a bomb calorimeter at 298.15K, the heat produced is 99.472 kJ. The $|\Delta H_f^\circ|$ of ethanol at 298.15 K is __________ $\times 10^2$ kJ mol^{-1}. (Nearest integer)

Given : **[April 04, 2026 (II)]**

Standard enthalpy of combustion of graphite = – 393.5 kJ mol^{-1}

Standard enthalpy of formation of water (*l*) = – 285.8 kJ mol^{-1}

Molar mass in g mol^{-1} of C, H, O are 12, 1 and 16 respectively.

120. If the enthalpy of sublimation of Li is 155 kJ mol^{-1}, enthalpy of dissociation of F_2 is 150 kJ mol^{-1}, ionization enthalpy of Li is 520 kJ mol^{-1}, electron gain enthalpy of F is –313 kJ mol^{-1}, standard enthalpy of formation of LiF is –594 kJ mol^{-1}. The magnitude of lattice enthalpy of LiF is _________ kJ mol^{-1} (Nearest integer). **[Jan. 22, 2026 (II)]**

121. Given : **[April 3, 2025 (I)]**

$\Delta H^\ominus_{sub}$ [C(graphite)] = 710 kJ mol^{-1}

$\Delta_{C-H}H^\ominus = 414$ kJ mol^{-1}
$\Delta_{H-H}H^\ominus = 436$ kJ mol^{-1}
$\Delta_{C=H}H^\ominus = 611$ kJ mol^{-1}

The $\Delta H_f^\ominus$ for $CH_2 = CH_2$ is ______ kJ mol^{-1} (nearest integer value)

122. Consider the following data :

Heat of formation of $CO_2(g) = -393.5$ kJ mol^{-1}
Heat of formation of $H_2O(l) = -286.0$ kJ mol^{-1}
Heat of combustion of benzene = –3267.0 kJ mol^{-1}

The heat of formation of benzene is _____ kJ mol^{-1}. (Nearest integer) **[Jan. 28, 2025 (II)]**

123. The formation enthalpies, $\Delta H_f^\ominus$ for $H_{(g)}$ and $O_{(g)}$ are 220.0 and 250.0 kJ mol^{-1}, respectively, at 298.15 K, and $\Delta H_f^\ominus$ for $H_2O_{(g)}$ is –242.0 kJ mol^{-1} at the same temperature. The average bond enthalpy of the O–H bond in water at 298.15 K is _______ kJ mol^{-1} (nearest integer). **[Jan. 28, 2025 (I)]**

124. The bond dissociation enthalpy of X_2 ΔH^o_{bond} calculated from the given data is ______ kJ mol^{-1}. (Nearest integer)

$M^+X^-(s) \rightarrow M^+(g) + X^-(g)\ \Delta H^o_{lattice} = 800$ kJ mol^{-1}

$M(s) \rightarrow M(g)\ \Delta H^\circ_{sub} = 100\ kJ\ mol^{-1}$
$M(g) \rightarrow M^+(g) + e^-(g)\ \Delta H^\circ_i = 500\ kJ\ mol^{-1}$
$X(g) + e^-(g) \rightarrow X^-(g)\ \Delta H^\circ_{eg} = -300\ kJ\ mol^{-1}$

$M(s) + \frac{1}{2}X_2(g) \rightarrow M^+X^-(s)\ \Delta H^\circ_f = -400\ kJ\ mol^{-1}$

[Given : M^+X^- is a pure ionic compound and X forms a diatomic molecule X_2 is gaseous state] **[Jan. 23, 2025 (II)]**

125. Consider the following cases of standard enthalpy of reaction

[ΔH_r^o in kJ mol^{-1}]

$C_2H_6(g) + \frac{7}{2}O_2(g) \rightarrow 2CO_2(g) + 3H_2O(\ell)\ \Delta H_1^0 = -1550$

$C(graphite) + O_2(g) \rightarrow CO_2(g)\ \Delta H_2^0 = -393.5$

$H_2(g) + \frac{1}{2}O_2(g) \rightarrow H_2O(\ell)\ \Delta H_3^0 = -286$

The magnitude of $\Delta H^0_{fC_2H_6(g)}$ is ______ kJ mol^{-1} (Nearest integer). **[Jan. 22, 2025 (II)]**

126. The heat of combustion of solid benzoic acid at constant volume is –321.30 kJ at 27°C. The heat of combustion at constant pressure is (–321.30 – xR) kJ, the value of x is .... **[April 5, 2024 (I)]**

127. Combustion of 1 mole of benzene is expressed at

$C_6H_6(1) + \frac{15}{2}O_2(g) \rightarrow CO_2(g) + 3H_2O(1).$

The standard enthalpy of combustion of 2 mol of benzene is 'x' kJ.

x = ______. **[April 5, 2024 (II)]**

(A) standard Enthalpy of formation of 1 mol of $C_6H_6(1)$, for the reaction
$6C(graphite) + 3H_2(g) \rightarrow C_6H_6(1)$ is 48.5 kJ mol^{-1}.

(B) Standard Enthalpy of formation of 1 mol of $CO_2(g)$, for the reaction
$C(graphite) + O_{2(g)} \rightarrow CO_2(g)$ is –393.5 kJ mol^{-1}.

(C) Standard and Enthalpy of formation of 1 mol of $H_2O(1)$, for the reaction

$H_2(g) + \frac{1}{2}O_2(g) \rightarrow H_2O(1)$ is –286 kJ mol^{-1}.

128. The enthalpy of formation of ethane (C_2H_6) from ethylene by addition of hydrogen where the bond-energies of C – H, C – C, C = C, H – H are 414 kJ, 347 kJ, 615 kJ and 435 kJ respectively is – ______ kJ. **[April 4, 2024 (I)]**

129. Two reactions are given below: **[Jan. 30, 2024 (II)]**

$2Fe(s) + \frac{3}{2}O_2(g) \rightarrow Fe_2O_3(s), \Delta H^\circ = -822 kJ/mol$

$C(s) + \frac{1}{2}O_2(g) \rightarrow CO(g), \Delta H^\circ = -110 kJ/mol$

Then enthalpy change for following reaction $3C(s) + Fe_2O_3(s) \rightarrow 2Fe(s) + 3CO(g)$ is __________ kJ/mol.

130. Standard enthalpy of vapourisation for CCl_4 is 30.5 kJ mol^{-1}. Heat required for vapourisation of 284g of CCl_4 at constant temperature is ______ kJ. **[Jan. 29, 2024 (II)]**
(Given molar mass in g mol^{-1}; C = 12, Cl = 35.5)

131. $A_2 + B_2 \rightarrow 2AB, \Delta H_f^\circ = -200\ kJmol^{-1}$

AB, A_2 and B_2 are diatomic molecule. If the bond enthalpies of A_2, B_2 and AB are in the ratio 1 : 0.5 : 1, then the bond enthalpy of A_2 is ______ kJmol^{-1} (Nearest integer) **[April 13, 2023 (I)]**

132. Solid fuel used in rocket is a mixture of Fe_2O_3 and Al (in ratio 1 : 2). The heat evolved (kJ) per gram of the mixture is ________ (Nearest integer) **[April 11, 2023 (I)]**

Given : $\Delta H_f^\circ(Al_2O_3) = -1700\ kJ\ mol^{-1}$

$\Delta H_f^\circ(Fe_2O_3) = -840\ kJ\ mol^{-1}$

Molar mass of Fe, Al and O are 56, 27 and 16 g mol^{-1} respectively.

133. Consider the following data **[April 06, 2023 (II)]**
Heat of combustion of $H_2(g)$ = –241.8 kJ mol^{-1}
Heat of combustion of C(s) = –393.5 kJ mol^{-1}
Heat of combustion of $C_2H_5OH(l)$ = –1234.7 kJ mol^{-1}
________ kJ mol^{-1} (Nearest integer)
The heat of formation of $C_2H_5OH(l)$ is (–)

134. At 25°C, the enthalpy of the following processes are given:
$H_2(g) + O_2(g) \rightarrow 2OH(g)\quad \Delta H^\circ = 78\ kJ\ mol^{-1}$
$H_2(g) + 1/2\ O_2(g) \rightarrow H_2O(g)\quad \Delta H^\circ = -242\ kJ\ mol^{-1}$
$H_2(g) \rightarrow 2H(g)\ \Delta H^\circ = 436\ kJ\ mol^{-1}$
$1/2\ O_2(g) \rightarrow O(g)\ \Delta H^\circ = 249\ kJ\ mol^{-1}$
What would be the value of X for the following reaction? ________ (Nearest integer)
$H_2O(g) \rightarrow H(g) + OH(g)\ \Delta H^\circ = X\ kJ\ mol^{-1}$ **[Feb. 01, 2023 (I)]**

135. The enthalpy change for the conversion of $\frac{1}{2}Cl_2(g)$ to $Cl^-(aq)$ is (-) ___________ kJ mol^{-1} (Nearest integer)

Given : $\Delta_{dis}H^\circ_{Cl_{2(g)}} = 240 kJmol^{-1}$. **[Jan. 31, 2023 (I)]**

$\Delta_{eg}H^\circ_{Cl_{(g)}} = -350 kJmol^{-1}$, $\Delta_{hyd}H^\circ_{Cl^-_{(g)}} = -380 kJmol^{-1}$

136. Enthalpies of formation of $CCl_4(g)$, $H_2O(g)$, $CO_2(g)$ and HCl(g) are –105, –242, –394 and –92 KJ mol^{-1} respectively. The magnitude of enthalpy of the reaction given below is ______ kJ mol^{-1} (nearest integer)
$CCl_4(g) + 2H_2O(g) \rightarrow CO_2(g) + 4HCl(g)$ **[Jan. 31, 2023 (II)]**

137. 28.0 L of CO_2 is produced on complete combustion of 16.8 L gaseous mixture of ethene and methane at 25°C and 1 atm. Heat evolved during the combustion process is ________ kJ. Given : $\Delta H_C(CH_4) = -900\ kJ\ mol^{-1}$
$\Delta H_C(C_2H_4) = -1400\ kJ\ mol^{-1}$ **[Jan. 25, 2023 (II)]**

138. An athlete is given 100 g of glucose ($C_6H_{12}O_6$) for energy. This is equivalent to 1800 kJ of energy. The 50% of this energy gained is utilized by the athlete for sports activities at the event. In order to avoid storage of energy, the weight of extra water he would need to perspire is_______g (Nearest integer)
Assume that there is no other way of consuming stored energy.
Given : The enthalpy of evaporation of water is 45 kJ mol^{-1}
Molar mass of C, H & O are 12. 1 and 16 g mol^{-1}. **[Jan. 25, 2023 (I)]**

139. When 600 mL of 0.2 M HNO_3 is mixed with 400 mL of 0.1M NaOH solution in a flask, the rise in temperature of the flask is _______ $\times 10^{-2}$ °C.
(Enthalpy of neutralisation = 57 kJ mol^{-1} and Specific heat of water = 4.2 JK^{-1} g^{-1}) (Neglect heat capacity of flask)
Density of water = 1.00 g cm^{-3}]
(Assume no volume change on mixing) **[July 29, 2022 (I)]**

140. For the reaction
$H_2F_2(g) \rightarrow H_2(g) + F_2(g)$
$\Delta U = -59.6\ kJ\ mol^{-1}$ at 27 °C.
The enthalpy change for the above reaction is (–) ______ kJ mol^{-1} [nearest integer]
Given : R = 8.314 JK^{-1} mol^{-1}. **[July 26, 2022 (II)]**

141. 17.0 g of NH_3 completely vapourises at – 33.42 °C and 1 bar pressure and the enthalpy change in the process is 23.4 kJ mol^{-1}. The enthalpy change for the vapourisation of 85 g of NH_3 under the same conditions is ________ kJ. [June 29, 2022 (I)]

142. The enthalpy of combustion of propane, graphite and dihydrogen at 298 K are –2220.0 kJ mol^{-1}, –393.5 kJ mol^{-1} and – 285.8 kJ mol^{-1} respectively. The magnitude of enthalpy of formation of propane (C_3H_8) is ____ kJ mol^{-1}. (Nearest integer) [July 25, 2022 (I)]

143. While performing a thermodynamics experiment, a student made the following observations,

$HCl + NaOH \rightarrow NaCl + H_2O, \Delta H = -57.3\ kJ\ mol^{-1}$

$CH_3COOH + NaOH \rightarrow CH_3COONa + H_2O$

$\Delta H = -55.3\ kJ\ mol^{-1}$

The enthalpy of ionization of CH_3COOH as calculated by the student is _____ kJ mol^{-1}, (nearest integer) [July 25, 2022 (II)]

144. For combustion of one mole of magnesium in an open container at 300 K and 1 bar pressure, $\Delta_C H^\circ = -601.70\ kJ\ mol^{-1}$, the magnitude of change in internal energy for the reaction is _____ kJ. (Nearest integer)

(Given : $R = 8.3\ J\ K^{-1}\ mol^{-1}$) [June 28, 2022 (II)]

145. A fish swimming in water body when taken out from the water body is covered with a film of water of weight 36 g. When it is subjected to cooking at 100 °C, then the internal energy for vaporization in kJ mol^{-1} is ________. [nearest integer]

[Assume steam to be an ideal gas. Given $\Delta_{vap} H^{\ominus}$ for water at 373 K and 1 bar is 41.1 kJ mol^{-1} ; $R = 8.31\ JK^{-1}mol^{-1}$] [June 26, 2022 (II)]

146. 200 mL of 0.2 M HCl is mixed with 300 mL of 0.1 M NaOH. The molar heat of neutralization of this reaction is –57.1 kJ. The increase in temperature in °C of the system on mixing is $x \times 10^{-2}$. The value of x is ________. (Nearest integer)

[Given : Specific heat of water = 4.18 J g^{-1} K^{-1}

Density of water = 1.00 g cm^{-3}]

(Assume no volume change on mixing) [Aug. 27, 2021 (I)]

147. These are physical properties of an element

(A) Sublimation enthalpy

(B) Ionisation enthalpy

(C) Hydration enthalpy

(D) Electron gain enthalpy

The total number of above properties that affect the reduction potential is _________ (Integer answer) [Aug. 26, 2021 (I)]

148. The Born-Haber cycle for KCl is evaluated with the following data :

$\Delta_f H^\circ$ for KCl = –436.7 kJ mol^{-1}; $\Delta_{sub} H^\circ$ for K = 89.2 kJ mol^{-1};

$\Delta_{ionization} H^\circ$ for K = 419.0 kJ mol^{-1}; $\Delta_{electron\ gain} H^\circ$ for $Cl_{(g)}$ = –348.6 kJ mol^{-1}; $\Delta_{bond} H^\circ$ for Cl_2 = 243.0 kJ mol^{-1}

The magnitude of lattice enthalpy of KCl in J mol^{-1} is _______ (Nearest integer) [Aug. 26, 2021 (I)]

149. When 400 mL of 0.2 M H_2SO_4 solution is mixed with 600 mL of 0.1 M NaOH solution, the increase in temperature of the final solution is ______ $\times 10^{-2}$ K. (Round off to the nearest integer).

[Use : $H^+(aq) + OH^-(aq) \rightarrow H_2O : \Delta H = -57.1\ kJ\ mol^{-1}$]

Specific heat of H_2O = 4.18 J K^{-1} g^{-1}.

Density of H_2O = 1.0 g cm^{-3} [July 27, 2021 (II)]

Assume no change in volume of solution on mixing.

150. At 298 K, the enthalpy of fusion of a solid (X) is 2.8 kJ mol^{-1} and the enthalpy of vaporisation of the liquid (X) is 98.2 kJ mol^{-1}. The enthalpy of sublimation of the substance (X) in kJ mol^{-1} is ________. (in nearest integer) [July 25, 2021 (I)]

151. If the standard molar enthalpy change for combustion of graphite powder is -2.48×10^2 kJ mol^{-1}, the amount of heat generated on combustion of 1 g of graphite powder is ______ kJ. (Nearest integer) [July 22, 2021 (II)]

152. For the reaction

$C_2H_6 \rightarrow C_2H_4 + H_2$

the reaction enthalpy $\Delta_r H$ = ________ kJ mol^{-1}.

(Rounded off to the Nearest Integer).

[Given : Bond enthalpies in kJ mol^{-1} :

C–C : 347, C=C : 611;

C–H : 414, H–H : 436] [March 18, 2021 (I)]

153. The average S-F bond energy in kJ mol^{-1} of SF_6 is ___________. (Rounded off to the nearest integer)

[Given : The values of standard enthalpy of formation of $SF_6(g)$, S(g) and F(g) are – 1100, 275 and 80 kJ mol^{-1} respectively.] [Feb. 26, 2021 (II)]

154. The ionization enthalpy of Na^+ formation from Na(g) is 495.8 kJ mol^{-1}, while the electron gain enthalpy of Br is –325.0 kJ mol^{-1}. Given the lattice enthalpy of NaBr is –728.4 kJ mol^{-1}. The energy for the formation of NaBr ionic solid is (–) _______ $\times 10^{-1}$ kJ mol^{-1}. [Feb. 25, 2021 (I)]

155. Assuming ideal behaviour, the magnitude of log *K* for the following reaction at 25°C is $x \times 10^{-1}$. The value of *x* is ________. (Integer answer).

$3HC \equiv CH(g) \rightleftharpoons C_6H_6\ (l)$

[Given : $\Delta_f G^0 (HC \equiv CH) = -2.04 \times 10^5\ J\ mol^{-1}$;

$\Delta_f G^0 (C_6H_6) = -1.24 \times 10^5\ J\ mol^{-1}$;

$R = 8.314\ J K^{-1}\ mol^{-1}$] [Feb. 24, 2021 (II)]

156. The internal energy change (in J) when 90 g of water undergoes complete evaporation at 100 °C is ________. (Given : ΔH_{vap} for water at 373 K = 41 kJ/ mol, $R = 8.314\ JK^{-1}\ mol^{-1}$) [Sep. 02, 2020 (I)]

157. The heat of combustion of ethanol into carbon dioxide and water is – 327 kcal at constant pressure. The heat evolved (in cal) at constant volume and 27 °C (if all gases behave ideally) is (R = 2 cal mol^{-1} K^{-1}) __________. [Sep. 02, 2020 (II)]

158. The standard heat of formation ($\Delta_f H^\circ_{298}$) of ethane (in kJ/mol), if the heat of combustion of ethane, hydrogen and graphite are –1560, –393.5 and –286 kJ/mol, respectively is _______. [Jan. 07, 2020 (II)]

Topic 3 - Entropy and Second Law of Thermodynamics

MULTIPLE CHOICE QUESTIONS (MCQs)

159. Ice at –5°C is heated to become vapor with temperature of 110°C at atmospheric pressure. The entropy change associated with this process can be obtained from : [Jan. 23, 2025 (I)]

(1) $\int_{268K}^{383K} C_p\, dT + \frac{\Delta H_{melting}}{273} + \frac{\Delta H_{boiling}}{373}$

(2) $\int_{268K}^{273K} \frac{C_{p.m}}{T} dT + \frac{\Delta H_{m}, fusion}{T_f} + \frac{\Delta H_{m,\ vaporisation}}{T_b} + \int_{273K}^{373K} \frac{C_{p.m} dT}{T} + \int_{373K}^{383K} \frac{C_{p.m} dT}{T}$

(3) $\int_{268K}^{383K} C_p dT + \frac{q_{rev}}{T}$

(4) $\int_{268K}^{373K} C_{p.m} dT + \frac{\Delta H_m, \text{fusion}}{T_f} + \frac{\Delta H_{m.\text{vaporisation}}}{T_b}$

$+ \int_{273K}^{373K} C_{p.m} dT + \int_{373K}^{383K} C_{p.m} dT$

160. Match List-I with List-II. **[Jan. 22, 2025 (II)]**

List-I (Partial Derivatives)	List-II (Thermodynamic Quantity)
(A) $\left(\frac{\partial G}{\partial T}\right)_P$	(I) C_P
(B) $\left(\frac{\partial H}{\partial T}\right)_P$	(II) –S
(C) $\left(\frac{\partial G}{\partial P}\right)_T$	(III) C_v
(D) $\left(\frac{\partial U}{\partial T}\right)_V$	(IV) V

Choose the correct answer from the option given below:
(1) (A)-(II), (B)-(I), (C)-(III), (D)-(IV)
(2) (A)-(II), (B)-(I), (C)-(IV), (D)-(III)
(3) (A)-(I), (B)-(II), (C)-(IV), (D)-(III)
(4) (A)-(II), (B)-(III), (C)-(I), (D)-(IV)

161. The true statement amongst the following is : **[Jan. 09, 2020 (II)]**
(1) Both ΔS and S are functions of temperature.
(2) Both S and ΔS are not functions of temperature.
(3) S is not a function of temperature but ΔS is a function of temperature.
(4) S is a function of temperature but ΔS is not a function of temperature.

162. Two blocks of the same metal having same mass and at temperature T_1, and T_2, respectively, are brought in contact with each other and allowed to attain thermal equilibrium at constant pressure. The change in entropy, ΔS, for this process is : **[Jan. 11, 2019 (I)]**

(1) $C_P \ln\left[\frac{(T_1+T_2)^2}{4T_1T_2}\right]$ (2) $2C_P \ln\left[\frac{(T_1+T_2)^{\frac{1}{2}}}{T_1T_2}\right]$

(3) $2C_P \ln\left(\frac{T_1+T_2}{4T_1T_2}\right)$ (4) $2C_P \ln\left[\frac{T_1+T_2}{2T_1T_2}\right]$

163. The process with negative entropy change is: **[Jan. 10, 2019 (II)]**
(1) Dissociation of $CaSO_4(s)$ to $CaO(s)$ and $SO_3(g)$
(2) Sublimation of dry ice
(3) Dissolution of iodine in water
(4) Synthesis of ammonia from N_2 and H_2

164. The entropy change associated with the conversion of 1 kg of ice at 273 K to water vapours at 383 K is:
(Specific heat of water liquid and water vapour are 4.2 kJ K^{-1} kg^{-1} and 2.0 kJ $K^{-1} kg^{-1}$; heat of liquid fusion and vapourisation of water are 334 kJ kg^{-1} and 2491 kJ kg^{-1}, respectively). **[Jan. 9, 2019 (II)]**
(log 273 = 2.436, log 373 = 2.572, log 383 = 2.583)
(1) 7.90 kJ $kg^{-1} K^{-1}$ (2) 2.64 kJ $kg^{-1} K^{-1}$
(3) 8.49 kJ $kg^{-1} K^{-1}$ (4) 9.26 kJ $kg^{-1} K^{-1}$

165. For which of the following processes, ΔS is negative? **[Online April 16, 2018]**
(1) C(diamond) $\rightarrow$ C(graphite)
(2) $N_2(g, 1\text{ atm}) \rightarrow N_2(g, 5\text{ atm})$
(3) $N_2(g, 273\text{ K}) \rightarrow N_2(g, 300\text{ K})$
(4) $H_2(g) \rightarrow 2H(g)$

166. The molar heat capacity (C_p) of CD_2O is 10 cals at 1000 K. The change in entropy associated with cooling of 32 g of CD_2O vapour from 1000 K to 100 K at constant pressure will be: (D = deuterium, atomic mass = 2 u) **[Online April 11, 2014]**
(1) 23.03 cal deg^{-1} (2) – 23.03 cal deg^{-1}
(3) 2.303 cal deg^{-1} (4) – 2.303 cal deg^{-1}

167. The (S°) of the following substances are:
CH_4 (g) 186.2 $JK^{-1} mol^{-1}$; O_2 (g) 205.2 $JK^{-1} mol^{-1}$
CO_2 (g) 213.6 $JK^{-1} mol^{-1}$; H_2O (l) 69.9 $JK^{-1} mol^{-1}$
The entropy change (ΔS°) for the reaction
$CH_4(g) + 2O_2(g) \rightarrow CO_2(g) + 2H_2O$ (l) is: **[Online April 12, 2014]**
(1) –312.5 J $K^{-1} mol^{-1}$ (2) –242.8 J $K^{-1} mol^{-1}$
(3) –108.1 J $K^{-1} mol^{-1}$ (4) –37.6 J $K^{-1} mol^{-1}$

168. The entropy of a sample of a certain substance increases by 0.836 J K^{-1} on adding reversibly 0.3344 J of heat at constant temperature. The temperature of the sample is: **[Online May 7, 2012]**
(1) 2.5 K (2) 0.3 K (3) 0.016 K (4) 0.4 K

169. One mole of an ideal gas is expanded isothermally and reversibly to half of its initial pressure. ΔS for the process in J $K^{-1} mol^{-1}$ is [ln 2 = 0.693 and R = 8.314, J/(mol K)] **[Online May 26, 2012]**
(1) 6.76 (2) 5.76 (3) 10.76 (4) 8.03

170. Standard entropy of X_2, Y_2 and XY_3 are 60, 40 and 50 J $K^{-1} mol^{-1}$, respectively. For the reaction,
$\frac{1}{2}X_2 + \frac{3}{2}Y_2 \rightarrow XY_3, \Delta H = -30$ kJ, to be at equilibrium,
the temperature will be **[2008]**
(1) 1250 K (2) 500 K (3) 750 K (4) 1000 K

NUMERICAL VALUE QUESTIONS (NVQs)

171. When ΔH_{vap} = 30 kJ/mol and ΔS_{vap} = 75 J $mol^{-1} K^{-1}$, then the temperature of vapour, at one atmosphere is ______ K. **[April 9, 2024 (II)]**

172. For the reaction at 298 K, $2A + B \rightarrow C$. ΔH = 400 kJ mol^{-1} and ΔS = 0.2 kJ $mol^{-1} K^{-1}$. The reaction will become spontaneous above______ K. **[April 6, 2024 (II)]**

173. 30.4 kJ of heat is required to melt one mole of sodium chloride and the entropy change at the melting point is 28.4 J $K^{-1} mol^{-1}$ at 1 atm. The melting point of sodium chloride is _______ K (Nearest Integer) **[April 15, 2023 (I)]**

174. One mole of an ideal gas at 350K is in a 2.0 L vessel of thermally conducting walls, which are in contact with the surroundings. It undergoes isothermal reversible expansion from 2.0 L to 3.0 L against a constant pressure of 4 atm. The change in entropy of the surroundings (ΔS) is _________ J K^{-1} (Nearest integer) Given : R = 8.314 J $K^{-1} Mol^{-1}$. **[April 12, 2023 (I)]**

175. For independent process at 300 K. **[Jan. 24, 2023 (I)]**

Process	ΔH/kJ mol^{-1}	ΔS/J K^{-1}
A	–25	–80
B	–22	40
C	25	–50
D	22	20

The number of non-spontaneous process from the following is _______.

176. For a given chemical reaction $A \rightarrow B$ at 300 K the free energy change is -49.4 kJ mol^{-1} and the enthalpy of reaction is 51.4 kJ mol^{-1}. The entropy change of the reaction is ______ J K–1 mol^{-1}. **[July 20, 2021 (II)]**

177. For a chemical reaction $A + B \rightleftharpoons C + D$ ($\Delta_r H^\circ = 80$ kJ mol^{-1}) the entropy change $\Delta_r S^\circ$ depends on the temperature T (in K) as $\Delta_r S^\circ = 2T\,(J\,K^{-1}\,mol^{-1})$.
Minimum temperature at which it will become spontaneous is ____________ K. (Integer) **[Feb. 26, 2021 (I)]**

Topic 4 - Spontaneity and Gibb's Free Energy

MULTIPLE CHOICE QUESTIONS (MCQs)

178. Consider the following data for the reaction **[April 05, 2026 (II)]**

$$X_2(g) + Y_2(g) \rightleftharpoons 2XY(g)$$

at 600 K. The $\Delta_r G^\ominus$ (in kJ mol^{-1}) for the reaction is:

Compound	$\Delta_f H^\ominus_{600K}$ (kJ mol^{-1})	$S^\ominus_{600K}$ (J $mol^{-1}K^{-1}$)
$XY(g)$	42	200
$X_2(g)$	8	140
$Y_2(g)$	80	250

(1) –21000 (2) –10
(3) –1000 (4) –9.012

179. It is noticed that Pb^{2+} is more stable than Pb^{4+} but Sn^{2+} is less stable than Sn^{4+}. Observe the following reactions

$PbO_2 + Pb \rightarrow 2PbO;\ \Delta_r G^\circ$ ...(1)

$SnO_2 + Sn \rightarrow 2SnO;\ \Delta_r G^\circ$ ...(2)

Identify the correct set from the following. **[Jan. 23, 2026 (II)]**
(1) $\Delta_r G^\circ(1) > 0; \Delta_r G^\circ(2) < 0$ (2) $\Delta_r G^\circ(1) < 0; \Delta_r G^\circ(2) < 0$
(3) $\Delta_r G^\circ(1) < 0; \Delta_r G^\circ(2) > 0$ (4) $\Delta_r G^\circ(1) > 0; \Delta_r G^\circ(2) > 0$

180. Let us consider a reversible reaction at temperature, T.
In this reaction, both ΔH and ΔS were observed to have positive values. If the equilibrium temperature is Te, then the reaction becomes spontaneous at : **[April 4, 2025 (I)]**
(1) $T = Te$ (2) $Te > T$
(3) $T > Te$ (4) $Te = 5T$

181. Which of the following graphs correctly represents the variation of thermodynamic properties of Haber's process ? **[April 2, 2025 (II)]**

(1)
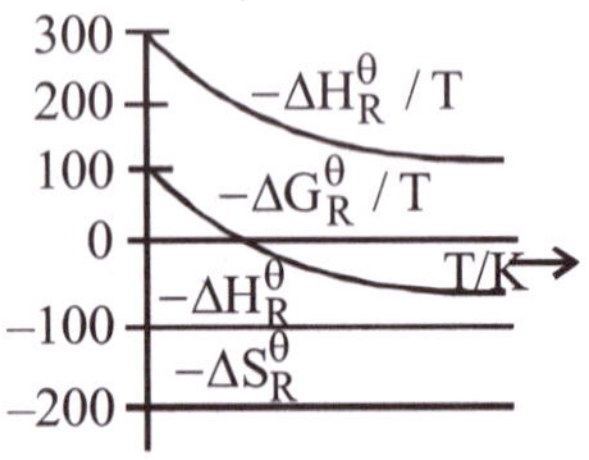

(2)
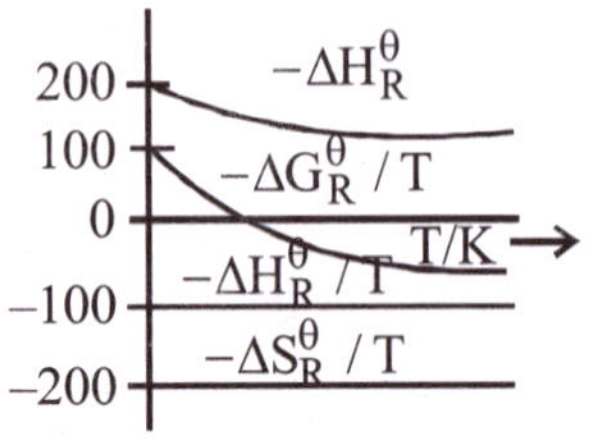

(3)
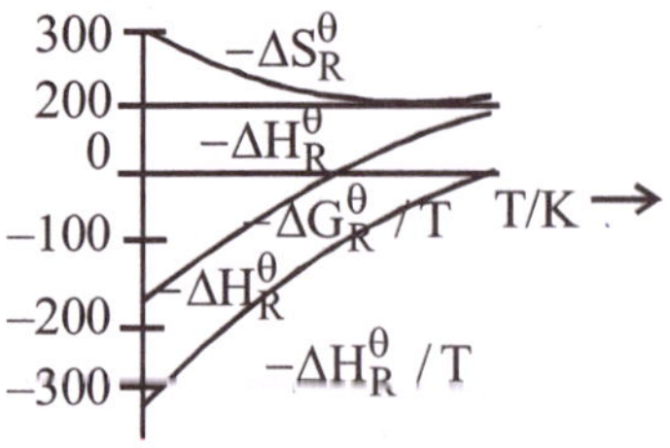

(4)
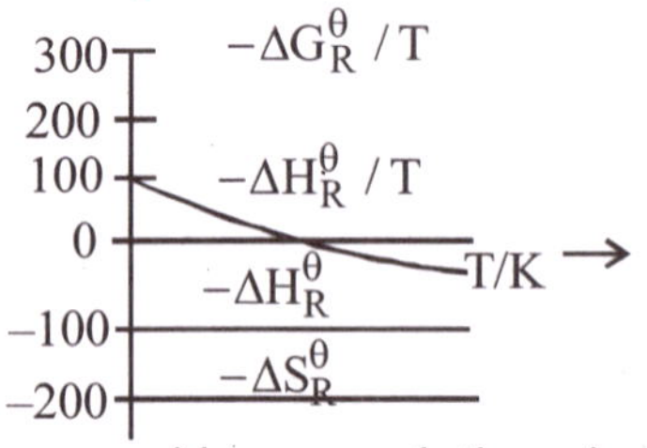

182. Let us consider an endothermic reaction which is non-spontaneous at the freezing point of water. However, the reaction is spontaneous at boiling point of water. Choose the correct option. **[Jan. 24, 2025 (I)]**
(1) Both ΔH and ΔS are (+ve) (2) ΔH is (–ve) but ΔS is (+ve)
(3) ΔH is (+ve) but ΔS is (–ve) (4) Both ΔH and ΔS are (–ve)

183. The effect of temperature on spontaneity of reactions are represented as: **[Jan. 23, 2025 (II)]**

	ΔH	ΔS	Temperature	Spontaneity
(A)	+	–	any T	Non spontaneous
(B)	+	+	low T	spontaneous
(C)	–	–	low T	Non spontaneous
(D)	–	+	any T	spontaneous

The incorrect combinations are :
(1) (B) and (D) only (2) (A) and (D) only
(3) (B) and (C) only (4) (A) and (C) only

184. Which of the following is **not** correct? **[Jan. 29, 2024 (I)]**
(1) ΔG is zero for a reversible reaction
(2) ΔG is negative for a spontaneous reaction
(3) ΔG is positive for a spontaneous reaction
(4) ΔG is positive for a non-spontaneous reaction

185. Which of the following relations are correct?
(A) $\Delta U = q + p\Delta V$ (B) $\Delta G = \Delta H - T\Delta S$
(C) $\Delta S = \dfrac{q_{rev}}{T}$ (D) $\Delta H = \Delta U - \Delta nRT$

Choose the most appropriate answer from the options given below : **[Jan. 29, 2023 (II)]**
(1) C and D only (2) B and C only
(3) A and B only (4) B and D only

186. Given below are two statements: One is labelled as **Assertion A** and the other is labelled as **Reason R**.
Assertion A: The reduction of a metal oxide is easier if the metal formed is in liquid state than solid state.
Reason R: The value of ΔG° becomes more on negative side as entropy is higher liquid state than solid state.
In the light of the above statements, choose the correct answer from the options given below: **[July 28, 2022 (II)]**
(1) Both A and R are correct and R is the correct explanation of A.
(2) Both A and R are correct but R is NOT the correct explanation of A.
(3) A is correct but R is not correct.
(4) A is not correct but R is correct.

187. Match List-I with List-II **[June 27, 2022 (I)]**

List-I	List-II
(A) Spontaneous process	(I) $\Delta H < 0$
(B) Process with $\Delta P = 0$, $\Delta T = 0$	(II) $\Delta G_{T,P} < 0$
(C) $\Delta H_{reaction}$	(III) Isothermal and isobaric process
(D) Exothermic process	(IV) [Bond energies of molecules in reactants] - [Bond energies of product molecules]

Choose the correct answer from the options given below:
(1) (A) – (III), (B) – (II), (C) – (IV), (D) – (I)
(2) (A) – (II), (B) – (III), (C) – (IV), (D) – (I)
(3) (A) – (II), (B) – (III), (C) – (I), (D) – (IV)
(4) (A) – (II), (B) – (I), (C) – (III), (D) – (IV)

188. A process will be spontaneous at all temperatures if : **[April 10, 2019 (I)]**
(1) $\Delta H < 0$ and $\Delta S < 0$ (2) $\Delta H > 0$ and $\Delta S < 0$
(3) $\Delta H < 0$ and $\Delta S > 0$ (4) $\Delta H > 0$ and $\Delta S > 0$

189. For the chemical reaction $X \rightleftharpoons Y$, the standard reaction Gibbs energy depends on temperature T (in K) as $\Delta_r G^\circ$ (in kJ mol^{-1}) $= 120 - \frac{3}{8} T$

The major component of the reaction mixture at T is : **[Jan. 11, 2019 (I)]**
(1) Y if $T = 300$ K (2) Y if $T = 280$ K
(3) X if $T = 350$ K (4) X if $T = 315$ K

190. The reaction
$MgO(s) + C(s) \rightarrow Mg(s) + CO(g)$, for which
$\Delta H^\circ = +491.1$ kJ mol^{-1} and $\Delta S^\circ = 198.0$ JK^{-1}mol^{-1},
is not feasible at 298 K. Temperature above which reaction will be feasible is **[Jan. 11, 2019 (II)]**
(1) 2040.5 K (2) 1890.0 K
(3) 2480.3 K (4) 2380.5 K

191. The standard reaction Gibbs energy for a chemical reaction at an absolute temperature T is given by $\Delta G^\circ = A - BT$ where A and B are non-zero constants. Which of the following is true about this reaction? **[Jan. 11, 2019 (II)]**
(1) Exothermic if $B < 0$
(2) Endothermic if $A > 0$
(3) Endothermic if $A < 0$ and $B > 0$
(4) Exothermic if $A > 0$ and $B < 0$

192. A process has $\Delta H = 200$ J mol^{-1} and $\Delta S = 40$ JK^{-1}mol^{-1}. Out of the values given below, choose the minimum temperature above which the process will be spontaneous: **[Jan. 10, 2019 (I)]**
(1) 20 K (2) 12 K (3) 5 K (4) 4 K

193. Which of the following lines correctly show the temperature dependence of equilibrium constant, K, for an exothermic reaction? **[2018]**

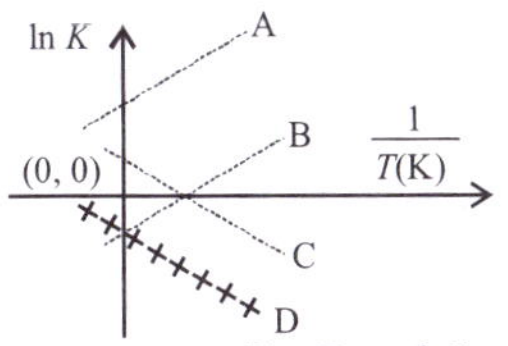

(1) A and B (2) B and C
(3) C and D (4) A and D

194. At 320 K, a gas A_2 is 20% dissociated to A(g). The standard free energy change at 320 K and 1 atm in J mol^{-1} is approximately: ($R = 8.314$ JK^{-1} mol^{-1}; ln 2 = 0.693; ln 3 = 1.098) **[Online April 16, 2018]**
(1) 1844 (2) 2068 (3) 4281 (4) 4763

195. $\Delta_f G^\circ$ at 500 K for substance 'S' in liquid state and gaseous state are + 100.7 kcal mol^{-1} and + 103 kcal mol^{-1}, respectively. Vapour pressure of liquid 'S' at 500 K is approximately equal to: ($R = 2$ cal K^{-1}mol^{-1}). **[Online April 15, 2018 (II)]**
(1) 100 atm (2) 1 atm (3) 10 atm (4) 0.1 atm

196. Given **[Online April 15, 2018 (II)]**
(i) $2Fe_2O_3(s) \rightarrow 4Fe(s) + 3O_2(g)$;
$\Delta_r G^\circ = +1487.0$ kJ mol^{-1}
(ii) $2CO(g) + O_2(g) \rightarrow 2CO_2(g)$;
$\Delta_r G^\circ = -514.4$ kJ mol^{-1}
Free energy change, $\Delta_r G^\circ$ for the reaction
$2Fe_2O_3(s) + 6CO(g) \rightarrow 4Fe(s) + 6CO_2(g)$ will be:
(1) –112.4 kJ mol^{-1} (2) –56.2 kJ mol^{-1}
(3) –208.0 kJ mol^{-1} (4) –168.2 kJ mol^{-1}

197. An ideal gas undergoes isothermal expansion at constant pressure. During the process : **[Online April 9, 2017]**
(1) enthalpy increases but entropy decreases.
(2) enthalpy remains constant but entropy increases.
(3) enthalpy decreases but entropy increases.
(4) Both enthalpy and entropy remain constant.

198. A reaction at 1 bar is non–spontaneous at low temperature but becomes spontaneous at high temperature. Identify the correct statement about the reaction among the following : **[Online April 9, 2016]**
(1) ΔH is negative while ΔS is positive
(2) Both ΔH and ΔS are negative
(3) ΔH is positive while ΔS is negative
(4) Both ΔH and ΔS are positive.

199. The incorrect expression among the following is: **[2012]**
(1) $\frac{\Delta G_{system}}{\Delta S_{total}} = -T$
(2) In isothermal process, $w_{reversible} = -nRT \ell n \frac{V_f}{V_i}$
(3) $\ln K = \frac{\Delta H^\circ - T\Delta S^\circ}{RT}$
(4) $K = e^{-\Delta G^\circ/RT}$

200. For a particular reversible reaction at temperature T, ΔH and ΔS were found to be both +ve. If T_e is the temperature at equilibrium, the reaction would be spontaneous when **[2010]**
(1) $T_e > T$ (2) $T > T_e$
(3) T_e is 5 times T (4) $T = T_e$

201. In conversion of lime-stone to lime,
$CaCO_3(s) \rightarrow CaO(s) + CO_2(g)$ the values of ΔH° and ΔS° are + 179.1 kJ mol^{-1} and 160.2 J/K respectively at 298 K and 1 bar. Assuming that ΔH° and ΔS° do not change with temperature, temperature above which conversion of limestone to lime will be spontaneous is **[2007]**
(1) 1118 K (2) 1008 K
(3) 1200 K (4) 845 K.

202. Identify the correct statement regarding a spontaneous process: [2007]
(1) Lowering of energy in the process is the only criterion for spontaneity.
(2) For a spontaneous process in an isolated system, the change in entropy is positive.
(3) Endothermic processes are never spontaneous.
(4) Exothermic processes are always spontaneous.

203. In an irreversible process taking place at constant T and P and in which only pressure-volume work is being done, the change in Gibbs free energy (dG) and change in entropy (dS), satisfy the criteria [2003]
(1) $(dS)_{V,E} > 0, (dG)_{T,P} < 0$ (2) $(dS)_{V,E} = 0, (dG)_{T,P} = 0$
(3) $(dS)_{V,E} = 0, (dG)_{T,P} > 0$ (4) $(dS)_{V,E} < 0, (dG)_{T,P} < 0$

204. The correct relationship between free energy change in a reaction and the corresponding equilibrium constant K_c is [2003]
(1) $-\Delta G = RT \ln K_c$ (2) $\Delta G^o = RT \ln K_c$
(3) $-\Delta G^o = RT \ln K_c$ (4) $\Delta G = RT \ln K_c$

205. If an endothermic reaction is non-spontaneous at freezing point of water and becomes feasible at its boiling point, then [2002]
(1) ΔH is –ve, ΔS is +ve (2) ΔH and ΔS both are +ve
(3) ΔH and ΔS both are –ve (4) ΔH is +ve, ΔS is –ve

206. For the reactions, [2002]

$C + O_2 \rightarrow CO_2$; $\Delta H = -393$ J

$2Zn + O_2 \rightarrow 2ZnO$; $\Delta H = -412$ J

(1) carbon can oxidise Zn
(2) oxidation of carbon is not feasible
(3) oxidation of Zn is not feasible
(4) Zn can oxidise carbon

NUMERICAL VALUE QUESTIONS (NVQs)

207. Consider the reaction $X \rightleftharpoons Y$ at 300 K. If ΔH^θ and K are 28.40 kJ mol^{-1} and 1.8×10^{-7} at the same temperature, then the magnitude of ΔS^θ for the reaction in $J K^{-1} mol^{-1}$ is ______.(Nearest Integer). (Given : R = 8.3 JK^{-1} mol^{-1}, ln10 = 2.3, log 3 = 0.48, log 2 = 0.30) [April 06, 2026 (I)]

208. The values of pressure equilibrium constant recorded at different temperatures for the following equilibrium reaction have been given below: $A(g) \rightleftharpoons B(g) + C(g)$

$\frac{1}{T}(K^{-1})$	$Log_{10}K_P$
0.05	3.5
0.06	2.5
0.07	1.5

The magnitude of $\frac{\Delta H^o}{R}$ calculated from the above data is ______. (Nearest integer) [April 05, 2026 (I)]

209. At the transition temperature T, $A \rightleftharpoons B$ and $\Delta G^0 = 105 - 35 \log T$ where A and B are two states of substance X. The transition temperature in °C when pressure is 1 atm is _____. (Nearest integer) [April 02, 2026 (I)]

210. The standard enthalpy and standard entropy of decomposition of N_2O_4 to NO_2 are 55.0 kJ mol^{-1} and 175.0 J/K/mol respectively. The standard free energy change for this reaction at 25°C in J mol^{-1} is _______ (Nearest integer) [Jan. 23, 2025 (I)]

211. Consider the graph of Gibbs free energy G vs Extent of reaction. The number of statement/s from the following which are true with respect to points (a), (b) and (c) is............. [April 06, 2023 (I)]

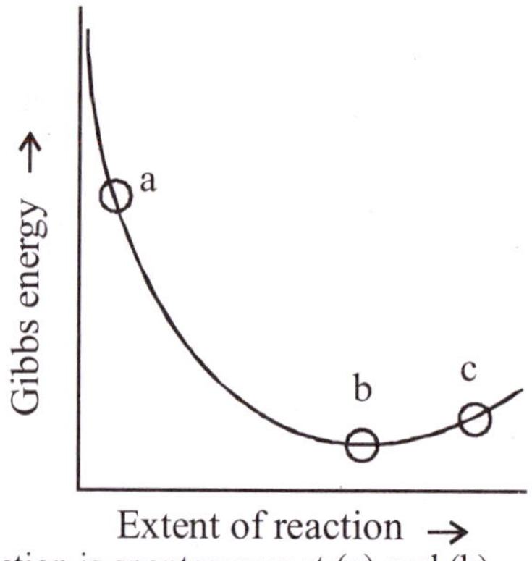

A. Reaction is spontaneous at (a) and (b)
B. Reaction is at equilibrium at point (b) and non–spontaneous at point (c)
C. Reaction is spontaneous at (a) and non–spontaneous at (c)
D. Reaction is non–spontaneous at (a) and (b)

212. The standard entropy change for the reaction $4Fe(s) + 3O_2(g) \rightarrow 2Fe_2O_3(s)$ is $-550\ JK^{-1}$ at 298 K.
[Given : The standard enthalpy change for the reaction is –165 kJ mol^{-1}]. The temperature in K at which the reaction attains equilibrium is __________. (Nearest Integer) [June 25, 2022 (I)]

213. The standard free energy change (ΔG^o) for 50% dissociation of N_2O_4 into NO_2 at 27°C and 1 atm pressure is – x J mol^{-1}. The value of x is __________. (Nearest Integer)
[Given : R = 8.31 J K^{-1} mol^{-1}, log 1.33 = 0.1239 ln 10 = 2.3] [June 25, 2022 (I)]

214. For the reaction $2NO_2(g) \rightleftharpoons N_2O_4(g)$, when $\Delta S = -176.0\ JK^{-1}$ and $\Delta H = -57.8$ kJ mol^{-1}, the magnitude of ΔG at 298 K for the reaction is_____ kJ mol^{-1}. (Nearest integer) [Sep. 1, 2021 (II)]

215. Data given for the following reaction is as follows :
$FeO(s) + C(graphite) \rightarrow Fe(s) + CO(g)$

Substance	ΔH^o_f (kJ mol^{-1})	ΔS^o (J $mol^{-1}K^{-1}$)
FeO(s)	–266.3	57.49
C(graphite)	0	5.74
Fe(s)	0	27.28
CO(g)	–110.5	197.6

The minimum temperature in K at which the reaction becomes spontaneous is _________ . (Integer answer) [Aug. 27, 2021 (II)]

216. For a dimerization reaction, $2A(g) \rightarrow A_2(g)$, at 298 K, $\Delta U^o = -20$ kJ mol^{-1}, $\Delta S^o = -30$ J K^{-1} mol^{-1}, then the ΔG^o will be ___________ J. [Sep. 05, 2020 (II)]

217. For the reaction ; $A(l) \rightarrow 2B(g)$
$\Delta U = 2.1$ kcal, $\Delta S = 20$ cal K^{-1} at 300 K.
Hence ΔG in kcal is _______. [Jan. 07, 2020 (I)]

Part-B (JEE Advanced)

Topic 1- First Law and Fundamentals of Thermodynamics

1 MCQs with One Correct Answer

1. An ideal gas (0.5 mol), initially at 2 bar pressure, is compressed at a constant temperature of 600 K in two steps: first, against a constant external pressure of P bar (2 < P < 8), and then against constant external pressure of 8 bar. At each step, the compression is stopped only when the pressure of the gas becomes equal to the external pressure. The total work done on the gas in these steps is W. Considering all possible values of P (2 < P < 8) and taking the gas constant as R (in $J K^{-1} mol^{-1}$), the minimum value of |W| (in J) is [Adv. 2026]
(a) 207 R (b) 600 R
(c) 630 R (d) 900 R

2. One mole of an ideal gas at 300 K in thermal contact with surroundings expands isothermally from 1.0 L to 2.0 L against a constant pressure of 3.0 atm. In this process, the change in entropy of surroundings (ΔS_{surr}) in JK^{-1} is
(1 L atm = 101.3 J) [Adv. 2016]
(a) 5.763 (b) 1.013
(c) –1.013 (d) –5.763

3. A mono-atomic ideal gas undergoes a process in which the ratio of P to V at any instant is constant and equals to 1. What is the molar heat capacity of the gas [2006 - 3M; –1]
(a) $\frac{3R}{2}$ (b) $2R$ (c) 0 (d) $\frac{5R}{2}$

4. When one mole of monoatomic ideal gas at T K undergoes adiabatic change under a constant external pressure of 1 atm, volume changes from 1 litre to 2 litre. The final temperature in Kelvin would be [2005S]
(a) $\frac{T}{2^{(2/3)}}$ (b) $T+\frac{2}{3}\times 0.082$
(c) T (d) $T-\frac{2}{3}\times 0.082$

5. Two moles of an ideal gas is expanded isothermally and reversibly from 1 litre to 10 litres at 300 K. The enthalpy change (in kJ) for the process is [2004S]
(a) 11.4 kJ (b) – 11.4 kJ (c) 0 kJ (d) 4.8 kJ

6. One mole of a non-ideal gas undergoes a change of state (2.0 atm, 3.0L, 95 K) → (4.0 atm, 5.0 L, 245 K) with a change in internal energy, ΔU = 30.0 L atm. The change in enthalpy (ΔH) of the process in L atm is [2002S]
(a) 40.0
(b) 42.3
(c) 44.0
(d) not defined, because pressure is not constant

7. Which one of the following statements is false? [2001S]
(a) Work is a state function.
(b) Temperature is a state function.
(c) Change in the state is completely defined when the initial and final states are specified.
(d) Work appears at the boundary of the system.

8. In thermodynamics, a process is called reversible when [2001S]
(a) surroundings and system change into each other.
(b) there is no boundary between system and surroundings.
(c) the surroundings are always in equilibrium with the system.
(d) the system changes into the surroundings spontaneously.

2 Integer Value Answer

9. Consider the following volume–temperature (V–T) diagram for the expansion of 5 moles of an ideal monoatomic gas.

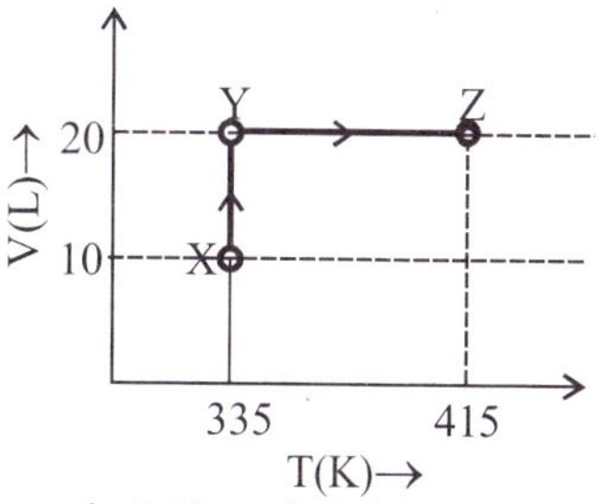

Considering only P-V work is involved, the total change in enthalpy (in Joule) for the transformation of state in the sequence **X→Y→Z** is ______. [Adv. 2024]
[Use the given data: Molar heat capacity of the gas for the given temperature range, $C_{V,m} = 12\ J K^{-1} mol^{-1}$ and gas constant, $R = 8.3\ J K^{-1} mol^{-1}$].

10. One mole of an ideal monoatomic gas undergoes two reversible processes (A → B and B → C) as shown in the given figure:

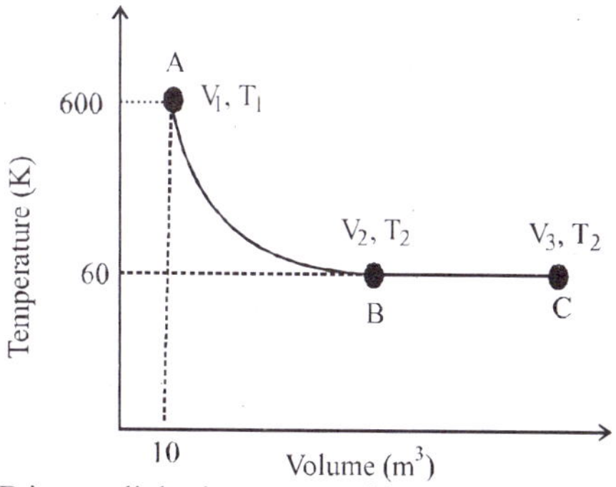

A → B is an adiabatic process. If the total heat absorbed in the entire process (A → B and B → C) is $RT_2 \ln 10$, the value of 2 log V_3 is ______. [Adv. 2023]
[Use, molar heat capacity of the gas at constant pressure, $C_{p,m} = \frac{5}{2}R$]

11. In a one–litre flask, 6 moles of A undergoes the reaction A (g) ⇌ P (g). The progress of product formation at two temperatures (in Kelvin), T_1 and T_2, is shown in the figure:

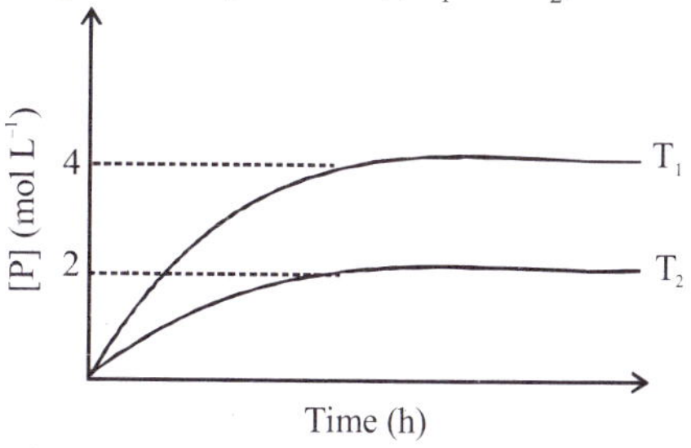

If $T_1 = 2T_2$ and $\left(\Delta G_2^\circ - \Delta G_1^\circ\right) = RT_2 \ln x$, then the value of x is ______. **[Adv. 2023]**

[ΔG_1° and ΔG_2° are standard Gibb's free energy change for the reaction at temperatures T_1 and T_2, respectively.]

12. One mole of an ideal gas at 900 K, undergoes two reversible processes, I followed by II, as shown below. If the work done by the gas in the two processes are same, the value of $\ln \frac{V_3}{V_2}$ is ______. **[Adv. 2021]**

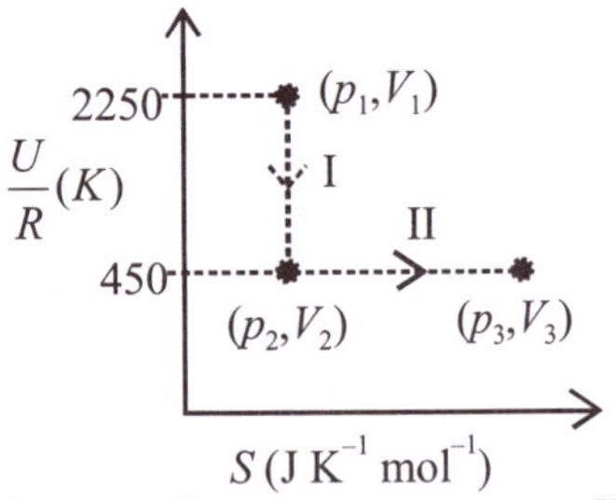

(U: internal energy, S: entropy, p: pressure, V: volume, R: gas constant)

(Given: molar heat capacity at constant volume, $C_{V,m}$ of the gas is $\frac{5}{2}R$)

13. One mole of an ideal gas is taken from a to b along two paths denoted by the solid and the dashed lines as shown in the graphs below. If the work done along the solid line path w_s and that along the dotted line path is w_d, then the integer closest to the ratio w_d / w_s is : **[2010]**

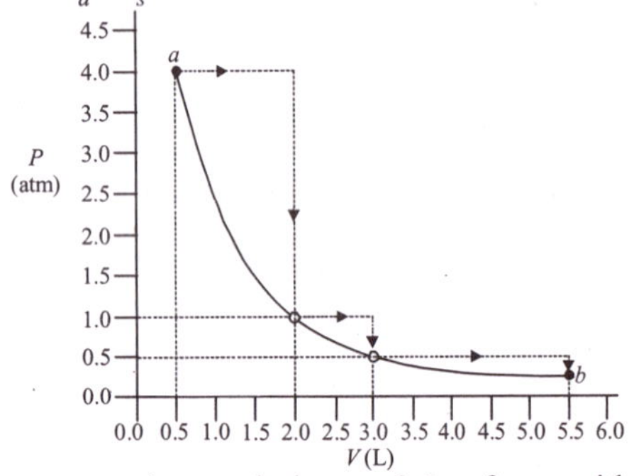

14. In a constant volume calorimeter, 3.5 g of a gas with molecular weight 28 was burnt in excess oxygen at 298.0 K. The temperature of the calorimeter was found to increase from 298.0 K to 298.45 K due to the combustion process. Given that the heat capacity of the calorimeter is 2.5 kJ K^{-1}, the numerical value for the enthalpy of combustion of the gas in kJ mol^{-1} is **[2009 - 6M]**

3 Numeric / New Stem Based Questions

15. Considering ideal gas behavior, the expansion work done (in kJ) when 144 g of water is electrolyzed completely under constant pressure at 300 K is ______.
Use: Universal gas constant (R) = 8.3 J K^{-1} mol^{-1}; Atomic mass (in amu): H = 1, O = 16 **[Adv. 2025]**

Qusetion Stem for Question no. 15 & 16

The entropy versus temperature plot for phases α and β at 1 bar pressure is given. S_T and S_0 are entropies of the phases at temperatures T and 0 K, respectively.

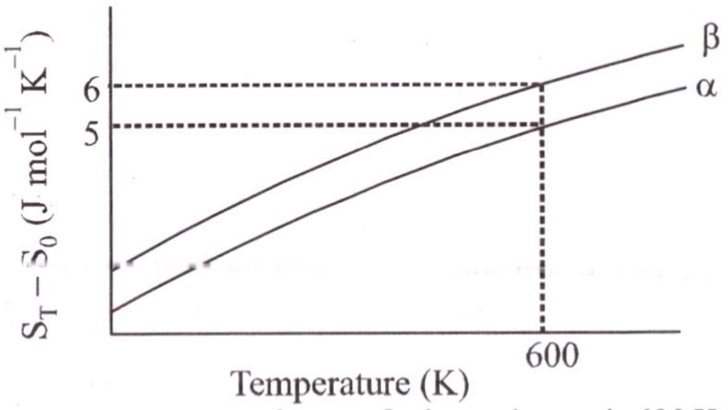

The transition temperature for α to β phase change is 600 K and $C_{p,\beta} - C_{p,\alpha} = 1$ J mol^{-1} K^{-1}. Assume $(C_{p,\beta} - C_{p,\alpha})$ is independent of temperature in the range of 200 to 700 K. $C_{p,\alpha}$ and $C_{p,\beta}$ are heat capacities of α and β phases, respectively.

16. The value of entropy change, $S_\beta - S_\alpha$ (in J mol^{-1} K^{-1}), at 300 K is __________. **[Adv. 2023]**
[Use: ln 2 = 0.69
Given: $S_\beta - S_\alpha = 0$ at 0 K]

17. The value of enthalpy change, $H_\beta - H_\alpha$ (in J mol^{-1}), at 300 K is __________. **[Adv. 2023]**

18. The value of standard enthalpy, ΔH^ϕ (in kJ mol^{-1}) for the given reaction is______. **[Adv. 2021]**

19. The value of ΔS^ϕ (in J K^{-1} mol^{-1}) for the given reaction, at 1000 K is________. **[Adv. 2021]**

20. Tin is obtained from cassiterite by reduction with coke. Use the data given below to determine the minimum temperature (in K) at which the reduction of cassiterite by coke would take place.

At 298 K: $\Delta_f H^0(SnO_2(s)) = -581.0$ kJ mol^{-1}, **[Adv. 2020]**

$\Delta_f H^0(CO_2(g)) = -394.0$ kJ mol^{-1},

$S^0(SnO_2(s)) = 56.0$ J K^{-1} mol^{-1},

$S^0(Sn(s)) = 52.0$ J K^{-1} mol^{-1},

$S^0(C(s)) = 6.0$ J K^{-1} mol^{-1},

$S^0(CO_2(g)) = 210.0$ J K^{-1} mol^{-1}.

Assume that the enthalpies and the entropies are temperature independent.

21. For the reaction, $2CO + O_2 \longrightarrow 2CO_2$; $\Delta H = -560$ kJ. Two moles of CO and one mole of O_2 are taken in a container of volume 1 L. They completely form two moles of CO_2, the gases deviate appreciably from ideal behaviour. If the pressure in the vessel changes from 70 to 40 atm, find the magnitude (absolute value) of ΔU at 500 K. (1 L atm = 0.1 kJ) **[2006 - 6M]**

22. A sample of argon gas at 1 atm pressure and 27 °C expands reversibly and adiabatically from 1.25 dm^3 to 2.50 dm^3. Calculate the enthalpy change in this process. $C_{V,m}$ for argon is 12.48 JK^{-1} mol^{-1}. **[2000 - 4 Marks]**

23. An athlete is given 100 g of glcuose ($C_6H_{12}O_6$) of energy equivalent to 1560 kJ. He utilizes 50 percent of this gained energy in the event. In order to avoids storage of energy in the body, calculate the weight of water he would need to perspire. The enthalpy of evaporation of water is 44 kJ/mole. **[1989 - 2 Marks]**

4 Fill in the Blanks

24. Enthalpy is an .... property. **[1997 - 1 Mark]**

25. A system is said to be .... if it can neither exchange matter nor energy with the surroundings. **[1993 - 1 Mark]**

5 True / False

26. Heat capacity of a diatomic gas is higher than that of a monoatomic gas. **[1985 - ½ Mark]**

27. First law of thermodynamics is not adequate in predicting the direction of a process. **[1982 - 1 Mark]**

6 MCQs with One or More than One Correct Answer

28. An ideal gas undergoes a reversible isothermal expansion from state I to state II followed by a reversible adiabatic expansion from state II to state III. The correct plot(s) representing the changes from state I to state III is (are)
(p : pressure, V : volume, T : temperature, H : enthalpy, S : entropy) **[Adv. 2021]**

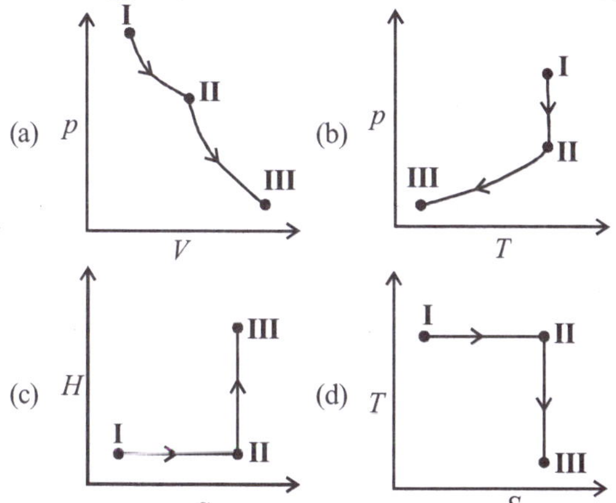

29. In thermodynamics, the $P-V$ work done is given by

$$w=-\int dV\,P_{ext}.$$

For a system undergoing a particular process, the work done is,

$$w=-\int dV\left(\frac{RT}{V-b}-\frac{a}{V^2}\right).$$

This equation is applicable to a **[Adv. 2020]**

(a) system that satisfies the van der Waals equation of state.
(b) process that is reversible and isothermal.
(c) process that is reversible and adiabatic.
(d) process that is irreversible and at constant pressure.

30. A reversible cyclic process for an ideal gas is shown below. Here, P, V, and T are pressure, volume and temperature, respectively. The thermodynamic parameters q, w, H and U are heat, work, enthalpy and internal energy, respectively. **[Adv. 2018]**

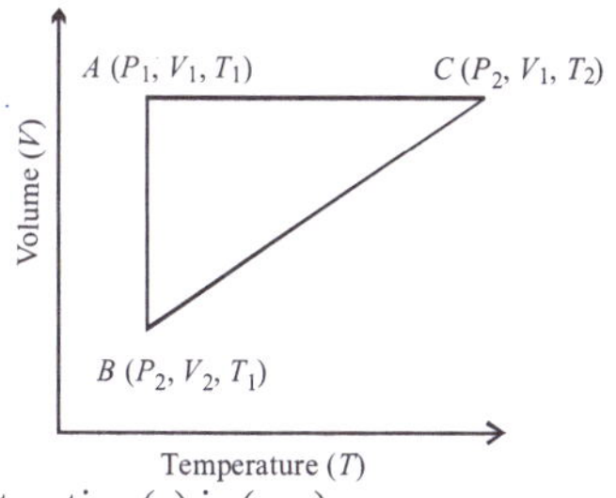

The correct option(s) is (are)

(a) $q_{AC}=\Delta U_{BC}$ and $w_{AB}=P_2(V_2-V_1)$
(b) $w_{BC}=P_2(V_2-V_1)$ and $q_{BC}=\Delta H_{AC}$
(c) $\Delta H_{CA}<\Delta U_{CA}$ and $q_{AC}=\Delta U_{BC}$
(d) $q_{BC}=\Delta H_{AC}$ and $\Delta H_{CA}>\Delta U_{CA}$

31. An ideal gas is expanded from (P_1, V_1, T_1) to (P_2, V_2, T_2) under different conditions. The correct statement(s) among the following is (are) **[Adv. 2017]**

(a) The work done on the gas is maximum when it is compressed irreversibly from (P_2, V_2) to (P_1, V_1) against constant pressure P_1
(b) If the expansion is carried out freely, it is simultaneously both isothermal as well as adiabatic
(c) The work done by the gas is less when it is expanded reversibly from V_1 to V_2 under adiabatic conditions as compared to that when expanded reversibly from V_1 to V_2 under isothermal conditions
(d) The change in internal energy of the gas is (i) zero, if it is expanded reversibly with $T_1=T_2$, and (ii) positive, if it is expanded reversibly under adiabatic conditions with $T_1\neq T_2$

32. An ideal gas in a thermally insulated vessel at internal pressure $=P_1$, volume $=V_1$ and absolute temperature $=T_1$ expands irreversibly against zero external pressure, as shown in the diagram. The final internal pressure, volume and absolute temperature of the gas are P_2, V_2 and T_2, respectively. For this expansion, **[Adv. 2014]**

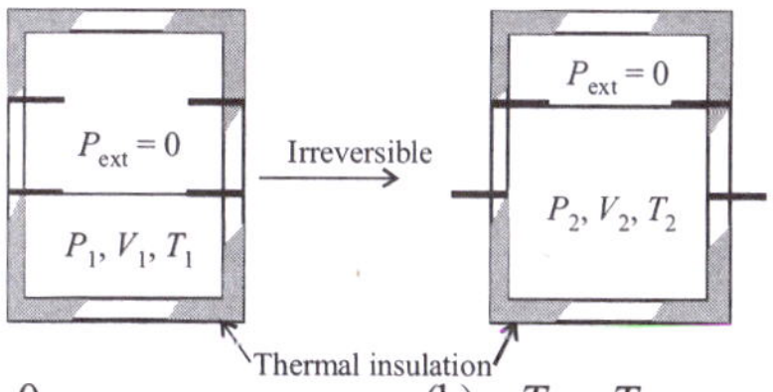

(a) $q=0$
(b) $T_2=T_1$
(c) $P_2V_2=P_1V_1$
(d) $P_2V_2^{\gamma}=P_1V_1^{\gamma}$

33. The reversible expansion of an ideal gas under adiabatic and isothermal conditions is shown in the figure. Which of the following statement(s) is (are) correct ? **[2012 - II]**

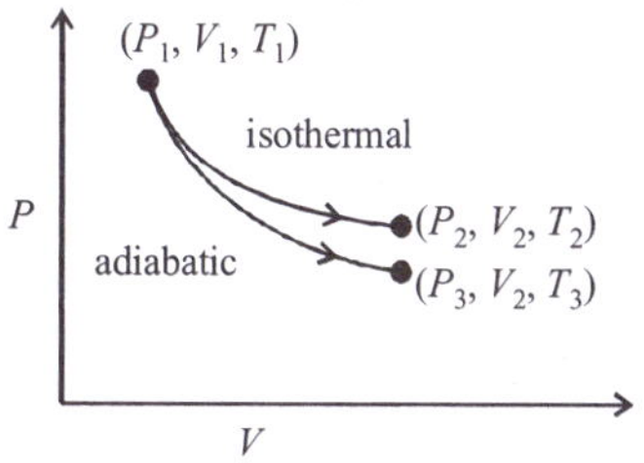

(a) $T_1=T_2$
(b) $T_3>T_1$
(c) $w_{isothermal}>w_{adiabatic}$
(d) $\Delta U_{isothermal}>\Delta U_{adiabatic}$

34. For an ideal gas, consider only $P-V$ work in going from an initial state X to the final state Z. The final state Z can be reached by either of the two paths shown in the figure. Which of the following choice(s) is (are) correct?
[Take ΔS as change in entropy and w as work done]. **[2012]**

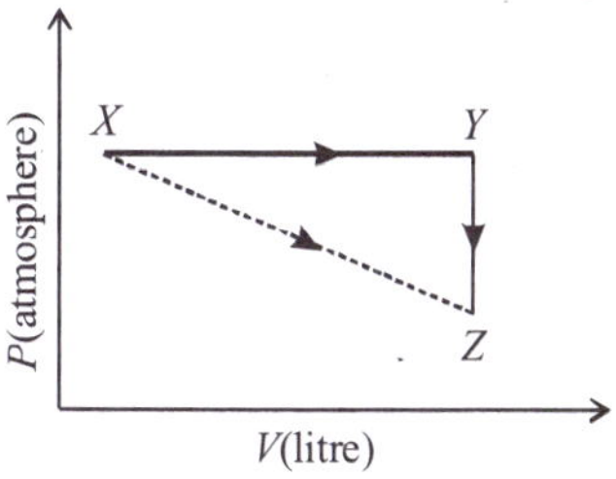

(a) $\Delta S_{X\to Z}=\Delta S_{X\to Y}+\Delta S_{Y\to Z}$
(b) $w_{X\to Z}=w_{X\to Y}+w_{Y\to Z}$
(c) $w_{X\to Y\to Z}=w_{X\to Y}$
(d) $\Delta S_{X\to Y\to Z}=\Delta S_{X\to Y}$

35. Among the following, the intensive property is (properties are) **[2010]**

(a) molar conductivity (b) electromotive force
(c) resistance (d) heat capacity

36. Among the following the state function(s) is (are) **[2009]**

(a) Internal energy
(b) Irreversible expansion work
(c) Reversible expansion work
(d) Molar enthalpy

37. Identify the intensive quantities from the following: **[1993 - 1 Mark]**

(a) Enthalpy (b) Temperature
(c) Volume (d) Refractive Index

7 Match the Following

38. Match the thermodynamic processes given under Column-I with the expressions given under Column-II. **[Adv. 2015]**

Column-I	Column-II
(A) Freezing of water at 273 K and 1 atm	(p) $q = 0$
(B) Expansion of 1 mol of an ideal gas into a vacuum under isolated conditions	(q) $w = 0$
(C) Mixing of equal volumes of two ideal gases at constant temperature and pressure in an isolated container	(r) $\Delta S_{sys} < 0$
(D) Reversible heating of $H_2(g)$ at 1 atm from 300 K to 600 K, followed by reversible cooling to 300 K at 1 atm	(s) $\Delta U = 0$
	(t) $\Delta G = 0$

8 Comprehension Based Questions

Passage-I

The amount of energy required to break a bond is same as the amount of energy released when the same bond is formed. In gaseous state, the energy required for *homolytic cleavage* of a bond is called Bond Dissociation Energy (BDE) or Bond Strength. BDE is affected by *s*-character of the bond and the stability of the radicals formed. Shorter bonds are typically stronger bonds. BDEs for some bonds are given below :

$$H_3C-H(g) \longrightarrow H_3C^{\bullet}(g) + H^{\bullet}(g) \qquad \Delta H^{\circ} = 105 \text{ kcal mol}^{-1}$$

$$Cl-Cl(g) \longrightarrow Cl^{\bullet}(g) + Cl^{\bullet}(g) \qquad \Delta H^{\circ} = 58 \text{ kcal mol}^{-1}$$

$$H_3C-Cl(g) \longrightarrow H_3C^{\bullet}(g) + Cl^{\bullet}(g) \qquad \Delta H^{\circ} = 85 \text{ kcal mol}^{-1}$$

$$H-Cl(g) \longrightarrow H^{\bullet}(g) + Cl^{\bullet}(g) \qquad \Delta H^{\circ} = 103 \text{ kcal mol}^{-1}$$

39. Correct match of the C–H bonds (shown in bold) in Column J with their BDE in Column K is

Column J Molecule	Column K BDE (kcal mol–1)
(P) $H-CH(CH_3)_2$	(i) 132
(Q) $H-CH_2Ph$	(ii) 110
(R) $H-CH=CH_2$	(iii) 95
(S) $H-C\equiv CH$	(iv) 88

[Adv. 2021]

(a) P – iii, Q – iv, R – ii, S – i (b) P – i, Q – ii, R – iii, S – iv
(c) P – iii, Q – ii, R – i, S – iv (d) P – ii, Q – i, R – iv, S – iii

40. For the following reaction

$$CH_4(g) + Cl_2(g) \xrightarrow{\text{light}} CH_3Cl(g) + HCl(g)$$

the correct statement is **[Adv. 2021]**

(a) Initiation step is exothermic with $\Delta H^{\circ} = -58$ kcal mol^{-1}
(b) Propagation step involving $^{\bullet}CH_3$ formation is exothermic with $\Delta H^{\circ} = -2$ kcal mol^{-1}.
(c) Propagation step involving CH_3Cl formation is endothermic with $\Delta H^{\circ} = +27$ kcal mol^{-1}.
(d) The reaction is exothermic with $\Delta H^{\circ} = -25$ kcal mol^{-1}.

Passage-II

A fixed mass '*m*' of a gas is subjected to transformation of states from *K* to *L* to *M* to *N* and back to *K* as shown in the figure **[Adv. 2013]**

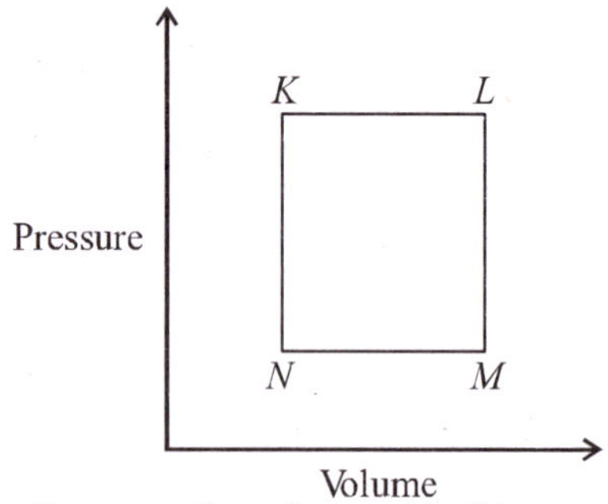

41. The succeeding operations that enable this transformation of states are

(a) Heating, cooling, heating, cooling
(b) Cooling, heating, cooling, heating
(c) Heating, cooling, cooling, heating
(d) Cooling, heating, heating, cooling

42. The pair of isochoric processes among the transformation of states is

(a) *K* to *L* and *L* to *M* (b) *L* to *M* and *N* to *K*
(c) *L* to *M* and *M* to *N* (d) *M* to *N* and *N* to *K*

9 Assertion and Reason/Statement Type Questions

Each question contains **STATEMENT-1 (Assertion)** and **STATEMENT-2 (Reason)**. Each question has 4 choices (a), (b), (c) and (d) out of which ONLY ONE is correct. Mark your answer as

(a) If both Statement -1 and Statement -2 are correct, and Statement -2 is the correct explanation of the Statement -1.
(b) If both Statement-1 and Statement-2 are correct, but Statement-2 is not the correct explanation of the Statement-1.
(c) If Statement -1 is correct but Statement -2 is incorrect.
(d) If Statement -1 is incorrect but Statement -2 is correct.

43. **Statement - 1 :** There is a natural asymmetry between converting work to heat and converting heat to work.
Statement - 2 : No process is possible in which the sole result is the absorption of heat form a reservoir and its complete conversion into work. **[2008S]**

44. Read the following statement and explanation and answer as per the options given below :
Assertion : The heat absorbed during the isothermal expansion of an ideal gas against vacuum is zero.
Reason : The volume occupied by the molecules of an ideal gas is zero. **[2000S]**

10 Subjective Problems

45. An insulated container contains 1 mol of a liquid, molar volume 100 mL, at 1 bar. When liquid is steeply pressed to 100 bar, volume decreases to 99 mL. Find. ΔH and ΔU for the process. [2004 - 2 Marks]

46. C_v value of He is always $3R/2$ but C_v value of H_2 is $3R/2$ at low temperature and $5R/2$ at moderate temperature and more than $5R/2$ at higher temperature. Explain in two to three lines. [2003 - 2 Marks]

47. Two moles of a perfect gas undergo the following processes: [2002 - 5 Marks]
    (a) a reversible isobaric expansion from (1.0 atm, 20.0 L) to (1.0 atm, 40.0 L);
    (b) a reversible isochoric change of state from (1.0 atm, 40.0 L) to (0.5 atm, 40.0 L);
    (c) a reversible isothermal compression from (0.5 atm, 40.0 L) to (1.0 atm, 20.0 L).
    (i) Sketch with labels each of the processes on the same *P-V* diagram.
    (ii) Calculate the total work (W) and the total heat change (q) involved in the above processes.
    (iii) What will be the values of ΔU, ΔH and ΔS for the overall process?

48. "The heat energy q, absorbed by a gas is ΔH", is true at what condition(s). [1984 - 1 Mark]

Topic 2 - Laws of Thermochemistry

1 MCQs with One Correct Answer

49. The standard enthalpies of formation of $CO_2(g)$, $H_2O(l)$ and glucose(s) at 25 °C are –400 kJ/mol, –300 kJ/mol and –1300 kJ/mol, respectively. The standard enthalpy of combustion per gram of glucose at 25 °C is [Adv. 2013-I]
    (a) +2900 kJ (b) –2900 kJ
    (c) –16.11 kJ (d) +16.11 kJ

50. The species which by definition has **ZERO** standard molar enthalpy of formation at 298 K is [2010]
    (a) $Br_2(g)$ (b) $Cl_2(g)$ (c) $H_2O(g)$ (d) $CH_4(g)$

51. Which of the reaction defines ΔH_f°? [2003S]
    (a) $C(\text{diamond}) + O_2(g) \longrightarrow CO_2(g)$
    (b) $\frac{1}{2}H_2(g) + \frac{1}{2}F_2(g) \longrightarrow HF(g)$
    (c) $N_2(g) + 3H_2(g) \longrightarrow 2NH_3(g)$
    (d) $CO(g) + \frac{1}{2}O_2(g) \longrightarrow CO_2(g)$

52. The ΔH_f^0 for $CO_2(g)$, $CO(g)$ and $H_2O(g)$ are –393.5, –110.5 and –241.8 kJ mol^{-1} respectively. The standard enthalpy change (in kJ) for the reaction
    $CO_2(g) + H_2(g) \rightarrow CO(g) + H_2O(g)$ is [2000S]
    (a) 524.1 (b) 41.2 (c) –262.5 (d) –41.2

53. For which change $\Delta H \neq \Delta E$: [1995S]
    (a) $H_2(g) + I_2(g) \rightarrow 2HI(g)$
    (b) $HCl + NaOH \rightarrow NaCl$
    (c) $C(s) + O_2(g) \rightarrow CO_2(g)$
    (d) $N_2(g) + 3H_2(g) \rightarrow 2NH_3(g)$

54. The difference between heats of reaction at constant pressure and constant volume for the reaction: [1991 - 1 Mark]
    $2C_6H_6(l) + 15O_2(g) \rightarrow 12CO_2(g) + 6H_2O(l)$ at 25 °C in kJ is
    (a) –7.43 (b) +3.72 (c) –3.72 (d) +7.43

3 Numeric / New Stem Based Questions

55. 2 mol of Hg(g) is combusted in a fixed volume bomb calorimeter with excess of O_2 at 298 K and 1 atm into HgO(s). During the reaction, temperature increases from 298.0 K to 312.8 K. If heat capacity of the bomb calorimeter and enthalpy of formation of Hg(g) are 20.00 kJ K^{-1} and 61.32 kJ mol^{-1} at 298 K, respectively, the calculated standard molar enthalpy of formation of HgO(s) at 298 K is X kJ mol^{-1}. The value of |X| is ________.
    [Given: Gas constant R = 8.3 J K^{-1} mol^{-1}] [Adv. 2022]

56. Diborane is a potential rocket fuel which undergoes combustion according to the reaction. [2000 - 2 Marks]
    $B_2H_6(g) + 3O_2(g) \rightarrow B_2O_3(s) + 3H_2O(g)$
    From the following data, calculate the enthalpy change for the combustion of diborane.
    $2B(s) + \frac{3}{2}O_2(g) \longrightarrow B_2O_3(s) \quad \Delta H = -1273 \text{ kJ mol}^{-1}$
    $H_2(g) + \frac{1}{2}O_2(g) \longrightarrow H_2O(\ell) \quad \Delta H = -286 \text{ kJ mol}^{-1}$
    $H_2O(l) \longrightarrow H_2O(g) \quad \Delta H = 44 \text{ kJ mol}^{-1}$
    $2B(s) + 3H_2(g) \longrightarrow B_2H_6(g) \quad \Delta H = 36 \text{ kJ mol}^{-1}$

57. Estimate the average S–F bond energy in SF_6. The values of standard enthalpy of formation of $SF_6(g)$, S(g) and F(g) are: –1100, 275 and 80 kJ mol^{-1} respectively. [1999 - 3 Marks]

58. From the following data, calculate the enthalpy change for the combustion of cyclopropane at 298 K. The enthalpy of formation of $CO_2(g)$, $H_2O(l)$ and propene(g) are –393.5, –285.8 and 20.42 kJ mol^{-1} respectively. The enthalpy of isomerisation of cyclopropane to propene is – 33.0 kJ mol^{-1}. [1998 - 5 Marks]

59. Compute the heat of formation of liquid methyl alcohol in kilojoules per mole, using the following data. Heat of vaporization of liquid methyl alcohol = 38 kJ/mol. Heat of formation of gaseous atoms from the elements in their standard states; H, 218 kJ/mol; C, 715 kJ/mol; O, 249kJ / mol. Average bond energies: [1997 - 5 Marks]
    C – H = 415kJ/mol, C – O = 365 kJ/mol, O – H = 463 kJ/mol

60. The standard molar enthalpies of formation of cyclohexane (l) and benzene (l) at 25° C are – 156 and + 49 kJ mol^{-1} respectively. The standard enthalpy of hydrogenation of cyclohexene (l) at 25° C is –119 kJ mol^{-1}. Use these data to estimate the magnitude of the resonance energy of benzene. [1996 - 2 Marks]

61. The polymerisation of ethylene to linear polyethylene is represented by the reaction [1994 - 2 Marks]
    $nCH_2 = CH_2 \longrightarrow -\!\!\!\!\!\!\left[CH_2 - CH_2 \right]\!\!\!\!\!\!-_n$
    where n has a large integral value. Given that the average enthalpies of bond dissociation for C = C and C – C at 298 K are + 590 and + 331 kJ mol^{-1}, respectively, calculate the enthalpy of polymerisation per mole of ethylene at 298 K.

62. Determine the enthalpy change of the reaction. [1992 - 3 Marks]
    $C_3H_8(g) + H_2(g) \rightarrow C_2H_6(g) + CH_4(g)$, at 25 °C, using the given heat of combustion values under standard conditions:

Compound	$H_2(g)$	$CH_4(g)$	$C_2H_6(g)$	C(graphite)
ΔH° (kJ/mol)	–285.8	–890.0	–1560.0	–393.5

The standard heat of formation of $C_3H_8(g)$ is –103.8 kJ/mol.

63. A gas mixture of 3.67 litres of ethylene and methane on complete combustion at 25 °C produces 6.11 litres of CO_2. Find out the amount of heat evolved on burning one litre of the gas mixture. The heats of combustion of ethylene and methane are –1423 and –891 kJ mol^{-1} at 25 °C. **[1991 - 5 Marks]**

64. The standard enthalpy of combustion at 25 °C of hydrogen, cyclohexene (C_6H_{10}) and cyclohexane (C_6H_{12}) are –241, –3800 and –3920 kJ/mole respectively. Calculate the heat of hydrogenation of cyclohexene. **[1989 - 2 Marks]**

65. An intimate mixture of ferric oxide, Fe_2O_3, and aluminium, Al, is used in solid fuel rockets. Calculate the fuel value per gram and fuel value per cc of the mixture. Heats of formation and densities are as follows : **[1988 - 2 Marks]**
$H_f(Al_2O_3) = 399$ kcal/mole; $H_f(Fe_2O_3) = 199$ kcal/mole;
Density of Fe_2O_3 = 5.2 g/cc; Density of Al = 2.7 g/cc.

66. The standard molar heats of formation of ethane, carbon dioxide and liquid water are –21.1, –94.1 and –68.3 kcal respectively. Calculate the standard molar heat of combustion of ethane. **[1986 - 2 Marks]**

67. The bond dissociation energies of gaseous H_2, Cl_2 and HCl are 104, 58 and 103 kcal/mole respectively. Calculate the enthalpy of formation of HCl gas. **[1985 - 2 Marks]**

68. Given the following standard heats of reactions :
(i) heat of formation of water = –68.3 kcal;
(ii) heat of combustion of acetylene = –310.6 kcal;
(iii) heat of combustion of ethylene = –337.2 kcal;
Calculate the heat of reaction for the hydrogenation of acetylene at constant volume (25 °C). **[1984 - 4 Marks]**

69. The molar heats of combustion of $C_2H_2(g)$, C(graphite) and $H_2(g)$ are 310.62 kcal, 94.05 kcal and 68.32 kcal, respectively. Calculate the standard heat of formation of $C_2H_2(g)$. **[1983 - 2 Marks]**

70. The enthalpy for the following reaction (ΔH^o) at 25 °C are given below : **[1981 - 2 Marks]**

(i) $\frac{1}{2}H_2(g) + \frac{1}{2}O_2(g) \rightarrow OH(g)$ 10.06 kcal
(ii) $H_2(g) \rightarrow 2H(g)$ 104.18 kcal
(iii) $O_2(g) \rightarrow 2O(g)$ 118.32 kcal
Calculate the O–H bond energy in the hydroxyl radical.

4 Fill in the Blanks

71. When Fe(s) is dissolved in aqueous hydrochloric acid in a closed vessel, the work done is .... **[1997 - 1 Mark]**

72. The heat content of the products is more than that of the reactants in an .... reaction. **[1993 - 1 Mark]**

6 MCQs with One or More than One Correct Answer

73. Choose the reaction(s) from the following options, for which the standard enthalpy of reaction is equal to the standard enthalpy of formation **[Adv. 2019]**

(a) $\frac{1}{8}S_8(s) + O_2(g) \rightarrow SO_2(g)$
(b) $2H_2(g) + O_2(g) \rightarrow 2H_2O(l)$
(c) $\frac{3}{2}O_2(g) \rightarrow O_3(g)$
(d) $2C(g) + 3H_2(g) \rightarrow C_2H_6(g)$

74. The following is (are) endothermic reaction(s): **[1999 - 3 Marks]**
(a) Combustion of methane
(b) Decomposition of water
(c) Dehydrogenation of ethane to ethylene
(d) Conversion of graphite to diamond

7 Match the Following

75. **List-I** contains various physical/chemical processes, and **List-II** contains combinations of changes in enthalpy (ΔH) and entropy (ΔS). Match each entry in **List-I** to the appropriate entry in **List-II**, and choose the correct option. **[Adv. 2026]**

List-I	List-II
(P) Physisorption	(1) $\Delta H > 0$ and $\Delta S > 0$
(Q) Diamond $\rightarrow$ Graphite	(2) $\Delta H < 0$ and $\Delta S < 0$
(R) Denaturation of protein	(3) $\Delta H < 0$ and $\Delta S = 0$
(S) Propene $\rightarrow$ Cyclopropane	(4) $\Delta H > 0$ and $\Delta S < 0$
	(5) $\Delta H < 0$ and $\Delta S > 0$

(a) $P \rightarrow 2, Q \rightarrow 3, R \rightarrow 5, S \rightarrow 4$
(b) $P \rightarrow 4, Q \rightarrow 3, R \rightarrow 5, S \rightarrow 1$
(c) $P \rightarrow 2, Q \rightarrow 5, R \rightarrow 1, S \rightarrow 4$
(d) $P \rightarrow 2, Q \rightarrow 5, R \rightarrow 1, S \rightarrow 3$

10 Subjective Problems

76. Anhydrous $AlCl_3$ is covalent. From the data given below, predict whether it would remain covalent or become ionic in aqueous solution. (Ionisation energy for Al = 5137 kJ mol^{-1}; $\Delta H_{hydration}$ for Al^{3+} = –4665 kJ mol^{-1}; $\Delta H_{hydration}$ for Cl^- = – 381 kJ mol^{-1}.) **[1997 - 2 Marks]**

77. In order to get maximum calorific output, a burner should have an optimum fuel to oxygen ratio which corresponds to 3 times as much oxygen as is required theoretically for complete combustion of the fuel. A burner which has been adjusted for methane as fuel (with *x* litre/hour of CH_4 and 6*x* litre/hour of O_2) is to be readjusted for butane, C_4H_{10}. In order to get the same calorific output, what should be the rate of supply of butane and oxygen ? Assume that losses due to incomplete combustion, *etc*, are the same for both the fuels and the gases behave ideally. Heats of combustion :
$CH_4 = 809$ kJ/mol; $C_4H_{10} = 2878$ kJ/mol **[1993 - 3 Marks]**

78. Using the data (all values are in kcal mol^{-1} at 25 °C) given below, calculate the bond energy of C–C and C–H bonds. **[1990 - 5 Marks]**
$\Delta H^o_{combustion}$(ethane) = –372.0
$\Delta H^o_{combustion}$(propane) = –530.0
$\Delta H^o_{C(s) \rightarrow C(g)}$ = 172.0
Bond energy of H–H = 104.0
ΔH^o_f of $H_2O(l)$ = –68.0
ΔH^o_f of $CO_2(g)$ = –94.0

Topic 3 - Entropy and Second Law of Thermodynamics

1 MCQs with One Correct Answer

79. For the process **[Adv. 2014]**
$H_2O(l) \rightarrow H_2O(g)$
at *T* = 100°C and 1 atmosphere pressure, the correct choice is
(a) $\Delta S_{system} > 0$ and $\Delta S_{surroundings} > 0$
(b) $\Delta S_{system} > 0$ and $\Delta S_{surroundings} < 0$
(c) $\Delta S_{system} < 0$ and $\Delta S_{surroundings} > 0$
(d) $\Delta S_{system} < 0$ and $\Delta S_{surroundings} < 0$

80. The enthalpy of vapourization of liquid is 30 kJ mol^{-1} and entropy of vapourization is 75 J mol^{-1} K. The boiling point of the liquid at 1 atm is [2004S]
(a) 250 K (b) 400 K (c) 450 K (d) 600 K

Topic 4 – Spontaneity and Gibb's Free Energy

1 MCQs with One Correct Answer

81. The standard state Gibbs free energies of formation of C(graphite) and C(diamond) at $T = 298$ K are
$\Delta_f G^0$ [C(graphite)] = 0 kJ mol^{-1}
$\Delta_f G^0$ [C(diamond)] = 2.9 kJ mol^{-1}
The standard state means that the pressure should be 1 bar, and substance should be pure at a given temperature. The conversion of graphite [C(graphite)] to diamond [C(diamond)] reduces its volume by 2×10^{-6} m^3 mol^{-1}. If C(graphite) is converted to C(diamond) isothermally at $T = 298$ K, the pressure at which C(graphite) is in equilibrium with C(diamond), is
[Useful information : 1 J = 1 kg m^2s^{-2}; 1 Pa = 1 kg m^{-1} s^{-2}; 1 bar = 10^5 Pa] [Adv. 2017]
(a) 14501 bar (b) 58001 bar
(c) 1450 bar (d) 29001 bar

82. For the process $H_2O(l)$ (1 bar, 373 K) $\rightarrow H_2O(g)$ (1 bar, 373 K), the correct set of thermodynamic parameters is [2007]
(a) $\Delta G = 0, \Delta S = +ve$ (b) $\Delta G = 0, \Delta S = -ve$
(c) $\Delta G = +ve, \Delta S = 0$ (d) $\Delta G = -ve, \Delta S = +ve$

83. The value of $\log_{10} K$ for a reaction $A \rightleftharpoons B$ is
(Given : $\Delta_r H^\circ_{298K} = -54.07$ kJ mol^{-1}, $\Delta_r S^\circ_{298K}$ = 10 JK^{-1} mol^{-1} and $R = 8.314$ JK^{-1} mol^{-1};
$2.303 \times 8.314 \times 298 = 5705$) [2007]
(a) 5 (b) 10 (c) 95 (d) 100

3 Numeric / New Stem Based Questions

84. The surface of copper gets tarnished by the formation of copper oxide. N_2 gas was passed to prevent the oxide formation during heating of copper at 1250 K. However, the N_2 gas contains 1 mole % of water vapour as impurity. The water vapour oxidises copper as per the reaction given below: $2Cu(s) + H_2O(g) \rightarrow Cu_2O(s) + H_2(g)$
p_{H2} is the minimum partial pressure of H_2 (in bar) needed to prevent the oxidation at 1250 K. The value of $\ln(p_{H2})$ is______.
(Given: total pressure = 1 bar, R (universal gas constant) = 8 J K^{-1} mol^{-1}, ln (10) = 2.3. Cu (s) and Cu_2O (s) are mutually immiscible.
At 1250 K : $2\,Cu(s) + \frac{1}{2}O_2(g) \rightarrow Cu_2O(s)$; $\Delta G^\circ = -78{,}000$ J mol^{-1}
$H_2(g) + \frac{1}{2}O_2(g) \rightarrow H_2O(g)$; $\Delta G^\circ = -1{,}78{,}000$ J mol^{-1};
(G is the Gibbs energy) [Adv. 2018]

6 MCQs with One or More than One Correct Answer

85. For a reaction taking place in a container in equilibrium with its surroundings, the effect of temperature on its equilibrium constant K in terms of change in entropy is described by [Adv. 2017]
(a) With increase in temperature, the value of K for exothermic reaction decreases because the entropy change of the system is positive
(b) With increase in temperature, the value of K for endothermic reaction increases because unfavourable change in entropy of the surroundings decreases
(c) With increase in temperature, the value of K for endothermic reaction increases because the entropy change of the system is negative
(d) With increase in temperature, the value of K for exothermic reaction decreases because favourable change in entropy of the surroundings decreases

7 Match the Following

86. Match the transformations in **column I** with appropriate options in **column II** [2011]

Column-I	Column-II
(A) $CO_2(s) \rightarrow CO_2(g)$	(p) phase transition
(B) $CaCO_3(s) \rightarrow CaO(s) + CO_2(g)$	(q) allotropic change
(C) $2H^\bullet \rightarrow H_2(g)$	(r) ΔH is positive
(D) $P_{(white,\ solid)} \rightarrow P_{(red,\ solid)}$	(s) ΔS is positive
	(t) ΔS is negative

10 Subjective Problems

87. In the following equilibrium $N_2O_4(g) \rightleftharpoons 2NO_2(g)$ [2004 - 2 Marks]
When 5 moles of each is taken and the temperature is kept at 298 K, the total pressure was found to be 20 bar.
Given : $\Delta G_f^\circ(N_2O_4) = 100$kJ; $\Delta G_f^\circ(NO_2) = 50$ kJ
(i) Find ΔG of the reaction at 298 K.
(ii) Find the direction of the reaction

88. When 1-pentyne (A) is treated with 4 N alcoholic KOH at 175 °C, it is converted slowly into an equilibrium mixture of 1.3% 1-pentyne (A), 95.2% 2-pentyne (B) and 3.5% of 1, 2-pentadiene (C). The equilibrium was maintained at 175 °C. Calculate ΔG° for the following equilibria :

$$B \rightleftharpoons A \quad \Delta G_1^\circ = ? \qquad B \rightleftharpoons C \quad \Delta G_2^\circ = ?$$

From the calculated value of ΔG_1° and ΔG_2° indicate the order of stability of (A), (B) and (C). Write a reasonable reaction mechanism showing all intermediates leading to (A), (B) and (C). [2001 - 10 Marks]

89. Show that the reaction $CO(g) + \frac{1}{2}O_2(g) \longrightarrow CO_2(g)$ at 300 K, is spontaneous and exothermic, when the standard entropy change is -0.094 kJ mol^{-1} K^{-1}. The standard Gibbs free energies of formation for CO_2 and CO are -394.4 and -137.2 kJ mol^{-1}, respectively. [2000 - 3 Marks]

Answer Key

Part-A (JEE Main)																			
1	(3)	**23**	(3)	**45**	(55)	**67**	(50)	**89**	(1)	**111**	(1)	**133**	(278)	**155**	(855)	**177**	(200)	**199**	(3)
2	(3)	**24**	(1)	**46**	(38)	**68**	(2218)	**90**	(1)	**112**	(4)	**134**	(499)	**156**	(189494)	**178**	(2)	**200**	(2)
3	(2)	**25**	(3)	**47**	(274)	**69**	(48.00)	**91**	(4)	**113**	(2)	**135**	(610)	**157**	(– 326400)	**179**	(3)	**201**	(1)
4	(4)	**26**	(2)	**48**	(200)	**70**	(6.25)	**92**	(2)	**114**	(N)	**136**	(173)	**158**	(–192.5)	**180**	(3)	**202**	(2)
5	(4)	**27**	(3)	**49**	(28721)	**71**	(4)	**93**	(3)	**115**	(3)	**137**	(925)	**159**	(2)	**181**	(1)	**203**	(1)
6	(1)	**28**	(1)	**50**	(200)	**72**	(2)	**94**	(4)	**116**	(2)	**138**	(360)	**160**	(2)	**182**	(1)	**204**	(3)
7	(4)	**29**	(1)	**51**	(1200)	**73**	(1)	**95**	(3)	**117**	(1126)	**139**	(54)	**161**	(1)	**183**	(3)	**205**	(2)
8	(2)	**30**	(3)	**52**	(4)	**74**	(4)	**96**	(2)	**118**	(4)	**140**	(57)	**162**	(1)	**184**	(3)	**206**	(1)
9	(2)	**31**	(3)	**53**	(4)	**75**	(3)	**97**	(2)	**119**	(3)	**141**	(117)	**163**	(4)	**185**	(2)	**207**	(34)
10	(4)	**32**	(2)	**54**	(1200)	**76**	(2)	**98**	(3)	**120**	(1031)	**142**	(–103.7)	**164**	(4)	**186**	(1)	**208**	(230)
11	(4)	**33**	(3)	**55**	(0)	**77**	(2)	**99**	(3)	**121**	(25)	**143**	(2)	**165**	(2)	**187**	(2)	**209**	(727)
12	(4)	**34**	(4)	**56**	(150)	**78**	(4)	**100**	(3)	**122**	(48)	**144**	(600)	**166**	(2)	**188**	(3)	**210**	(2850)
13	(2)	**35**	(1)	**57**	(620)	**79**	(2)	**101**	(3)	**123**	(466)	**145**	(38)	**167**	(2)	**189**	(4)	**211**	(2)
14	(4)	**36**	(1)	**58**	(3)	**80**	(2)	**102**	(3)	**124**	(200)	**146**	(82)	**168**	(4)	**190**	(3)	**212**	(300)
15	(1)	**37**	(3)	**59**	(2)	**81**	(1)	**103**	(1)	**125**	(95)	**147**	(3)	**169**	(2)	**191**	(2)	**213**	(710)
16	(4)	**38**	(4)	**60**	(1718)	**82**	(3)	**104**	(4)	**126**	(150)	**148**	(718)	**170**	(3)	**192**	(3)	**214**	(5)
17	(1)	**39**	(3)	**61**	(195)	**83**	(3)	**105**	(4)	**127**	(6535)	**149**	(82)	**171**	(400)	**193**	(1)	**215**	(964)
18	(4)	**40**	(3)	**62**	(0)	**84**	(3)	**106**	(3)	**128**	(125)	**150**	(101)	**172**	(2000)	**194**	(3)	**216**	(– 13538)
19	(1)	**41**	(3)	**63**	(8630)	**85**	(4)	**107**	(2)	**129**	(492)	**151**	(21)	**173**	(1070)	**195**	(4)	**217**	(–2.70)
20	(1)	**42**	(4)	**64**	(84)	**86**	(3)	**108**	(2)	**130**	(56)	**152**	(128)	**174**	(3)	**196**	(2)		
21	(4)	**43**	(2)	**65**	(38)	**87**	(3)	**109**	(1)	**131**	(800)	**153**	(309)	**175**	(2)	**197**	(2)		
22	(4)	**44**	(304)	**66**	(31)	**88**	(1)	**110**	(2)	**132**	(4)	**154**	(5576)	**176**	(336)	**198**	(4)		

Part-B (JEE Advanced)																			
1	(b)	**10**	(7)	**19**	(141.34)	**30**	(b, c)	**39**	(a)	**51**	(b)	**59**	(–266)	**67**	(–22)	**79**	(b)		
2	(c)	**11**	(8)	**20**	(935)	**31**	(a, b, c)	**40**	(d)	**52**	(b)	**60**	(–152)	**68**	(–41.104)	**80**	(b)		
3	(b)	**12**	(10)	**21**	(557)	**32**	(a, b, c)	**41**	(c)	**53**	(d)	**61**	(–72)	**69**	(54.20)	**81**	(a)		
4	(d)	**13**	(2)	**22**	(115.87)	**33**	(a, c, d)	**42**	(b)	**54**	(a)	**62**	(–55.7)	**70**	(101.19)	**82**	(a)		
5	(c)	**14**	(9)	**23**	(319.1)	**34**	(a, c)	**43**	(a)	**55**	(90.39)	**63**	(50.90)	**71**	zero	**83**	(b)		
6	(c)	**15**	(–29.88)	**26**	(True)	**35**	(a, b)	**44**	(b)	**56**	(–2035)	**64**	(–121)	**72**	endothermic	**84**	(– 14.6)		
7	(a)	**16**	(0.31)	**27**	(True)	**36**	(a, d)	**49**	(c)	**57**	(309.16)	**65**	(3.94)	**73**	(a, c)	**85**	(b, d)		
8	(c)	**17**	(300)	**28**	(a, b, d)	**37**	(b, d)	**50**	(b)	**58**	(–2091.32)	**66**	(–372.0)	**74**	(b, c, d)				
9	(8120)	**18**	(166.28)	**29**	(a, b, c)	**38**	A- (r, t); B -(p, q, s); C -(p, q, s); D- (p, q, s, t)							**75**	(c)				

Chapter-7

Equilibrium

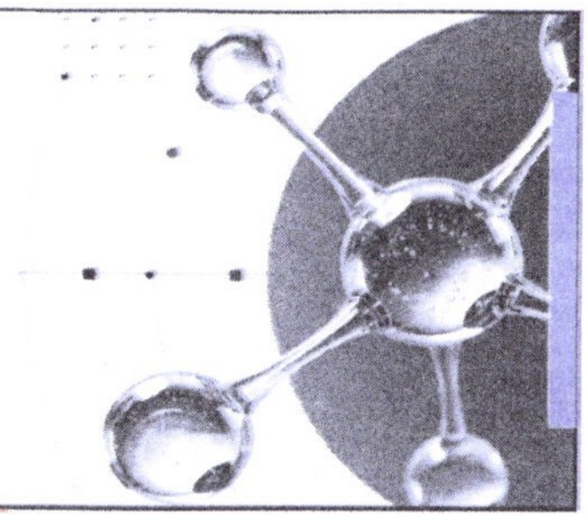

Part-A (JEE Main)

Topic 1- Law of Chemical Equilibrium, Equilibrium Constant (K_c and K_p) and its Application)

MULTIPLE CHOICE QUESTIONS (MCQs)

1. Consider the following reactions in which all the reactants and products are present in gaseous state [April 08, 2026 (II)]

$2xy \rightleftharpoons x_2 + y_2 \quad K_1 = 2.5 \times 10^5$

$xy + \frac{1}{2}z_2 \rightleftharpoons xyz \quad K_2 = 5 \times 10^{-3}$

The value of K_3 for the equilibrium

$\frac{1}{2}x_2 + \frac{1}{2}y_2 + \frac{1}{2}z_2 \rightleftharpoons xyz$ is:

(1) 2.5×10^{-3} (2) 2.5×10^3
(3) 1.0×10^{-5} (4) 5×10^{-3}

2. Given is a concentrated solution of a weak electrolyte A_xB_y of concentration 'c' and dissociation constant 'K'. the degree of dissociation is given by: [April 06, 2026 (II)]

(1) $[K \times c^{x+y-1} x^x y^y]^{x+y}$ (2) $\left(\frac{K}{c^{x+y-1} x^x y^y}\right)^{\frac{1}{x+y}}$

(3) $\left(\frac{c^{x+y-1} x^x y^y}{K}\right)^{x+y}$ (4) $\left(\frac{c^{x+y-1} x^x y^y}{K}\right)^{\frac{1}{x+y}}$

3. One mole each of He and A(g) are taken in a 10 L closed flask and heated to 400 K to establish the following equilibrium.

$A(g) \rightleftharpoons B(g)$ [April 06, 2026 (I)]

K_C for this reaction at 400 K is 4.0. The partial pressures (in atm) of He and B(g) are respectively (at equilibrium).
(Assume He, A(g) and B(g) behave as ideal gases)
(Given : $R = 0.082$ L atm K^{-1} mol^{-1})

(1) 3.28, 2.624 (2) 2.624, 3.28
(3) 3.28, 0.656 (4) 0.656, 6.56

4. The reaction $A(g) \rightleftharpoons B(g) + C(g)$ was initiated with the amount 'a' of A(g). At equilibrium it is found that the amount of A(g) remaining is $(a - x)$ at a total pressure of p.
The equilibrium constant K_p of the reaction can be calculated from the expression: [April 05, 2026 (II)]

(1) $\frac{x^2}{a^2 + x^2} \times p$ (2) $\frac{x^2}{a^2 - x^2} \times p$

$a^2 + x^2$ $\times$ p (4) $\frac{a - x^2}{x^2} \times p$

5. At T(K), the equilibrium constant of $A_2(g) + B_2(g) \rightleftharpoons C(g)$ is 2.7×10^{-5}.
What is the equilibrium constant for

$\frac{1}{3}A_2(g) + \frac{1}{3}B_2(g) \rightleftharpoons \frac{1}{3}C(g)$ at the same temperature? [April 04, 2026 (I)]

(1) $(2.7 \times 10^{-5})^3$ (2) 6×10^{-2}
(3) $\sqrt{2.7 \times 10^{-5}}$ (4) 3×10^{-2}

6. Observe the following equilibrium in a 1 L flask. [Jan. 28, 2026 (II)]

$A(g) \rightleftharpoons B(g)$

At T(K), the equilibrium concentrations of A and B are 0.5 M and 0.375 M respectively. 0.1 moles of A is added into the flask and heated to T(K) to establish the equilibrium again. The new equilibrium concentrations (in M) of A and B are respectively.

(1) 0.367, 0.275 (2) 0.53, 0.4
(3) 0.742, 0.557 (4) 0.557, 0.418

7. Consider the following gaseous equilibrium in a closed container of volume "V" at T(K).

$P_2(g) + Q_2(g) \rightleftharpoons 2PQ(g)$

2 moles each of $P_2(g)$, $Q_2(g)$ and PQ(g) are present at equilibrium. Now one mole each of 'P_2' and 'Q_2' are added to the equilibrium keeping the temperature at T(K). The number of moles of P_2, Q_2 and PQ at the new equilibrium, respectively, are- [Jan. 24, 2026 (II)]

(1) 2.67, 2.67, 2.67 (2) 1.21, 2.24, 1.56
(3) 1.66, 1.66, 1.66 (4) 2.56, 1.62, 2.24

8. Consider the general reaction given below at 400 K

$xA(g) \rightleftharpoons yB(g)$

The values of K_p and K_c are studied under the same condition of temperature but variation in x and y.
(i) $K_p = 85.87$ and $K_c = 2.586$ appropriate units
(ii) $K_p = 0.862$ and $K_c = 28.62$ appropriate units
The value of x and y in (i) and (ii) respectively are: [Jan. 23, 2026 (I)]

(1) (i) 3, 1 ; (ii) 3, 1 (2) (i) 4, 1 ; (ii) 4, 1
(3) (i) 1, 3 ; (ii) 2, 1 (4) (i) 1, 2 ; (ii) 2, 1

9. Given below are two statements :
Statement I : When a system containing ice in equilibrium with water (liquid) is heated, heat is absorbed by the system and there is no change in the temperature of the system until whole ice gets melted.
Statement II : At melting point of ice, there is absorption of heat in order to overcome intermolecular forces of attraction

within the molecules of water in ice and kinetic energy of molecules is not increased at melting point

In the light of the above statements, choose the **correct** answer from the options given below: [April 3, 2025 (II)]

(1) Statement I is true but Statement II is false
(2) Both Statement I and Statement II are false
(3) Both Statement I and Statement II are true
(4) Statement I is false but Statement II is true

10. In the following system,

$PCl_5(g) \rightleftharpoons PCl_3(g) + Cl_2(g)$ at equilibrium, upon addition of xenon gas at constant T & P, the concentration of [April 3, 2025 (I)]

(1) PCl_5 will increase
(2) Cl_2 will decrease
(3) PCl_5, PCl_3 & Cl_2 remain constant
(4) PCl_3 will increase

11. Given below are two statements [April 3, 2025 (I)]

Statement I : A catalyst cannot alter the equilibrium constant (K_C) of the reaction, temperature remaining constant

Statement II : A homogenous catalyst can change the equilibrium composition of a system temperature remaining constant

In the light of the above statements, choose the **correct** answer from the options given below

(1) Statement I is false but Statement II is true
(2) Both Statement I and Statement II are true
(3) Both Statement I and Statement II is false
(4) Statement I is true but Statement II is false

12. Consider the following chemical equilibrium of the gas phase reaction at a constant temperature : [April 2, 2025 (II)]

$A(g) \rightleftharpoons B(g) + C(g)$

If p being the total pressure, K_p is the pressure equilibrium constant and α is the degree of dissociation, then which of the following is true at equilibrium ?

(1) If p value is extremely high compared to K_p, $\alpha \approx 1$
(2) When p increases α decreases
(3) If k_p value is extremely high compared to p, α becomes much less than unity
(4) When p increases α increases

13. At temperature T, compound $AB_{2(g)}$ dissociates as $AB_{2(g)} \rightleftharpoons AB_{(g)} + \frac{1}{2}B_{2(g)}$ having degree of dissociation x (small compared to unity). The correct expression for x in terms of K_p and p is [Jan. 29, 2025 (I)]

(1) $\sqrt[3]{\frac{2K_p}{p}}$ (2) $\sqrt[4]{\frac{2K_p}{p}}$

(3) $\sqrt[3]{\frac{2K_p^2}{p}}$ (4) $\sqrt{K_p}$

14. Ice and water are placed in a closed container at a pressure of 1 atm and temperature 273.15 K. If pressure of the system is increased 2 times, keeping temperature constant, then identify correct observation from following : [Jan. 28, 2025 (I)]

(1) Volume of system increases.
(2) Liquid phase disappears completely.
(3) The amount of ice decreases.
(4) The solid phase (ice) disappears completely.

15. For the reaction, [Jan. 24, 2025 (II)]

$H_2(g) + I_2(g) \rightleftharpoons 2HI(g)$

Attainment of equilibrium is predicted correctly by:

(1)
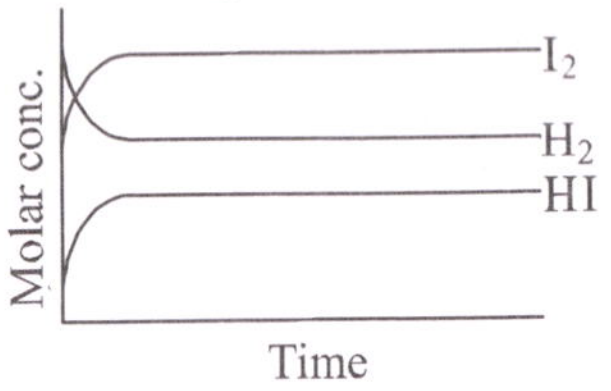

(2)
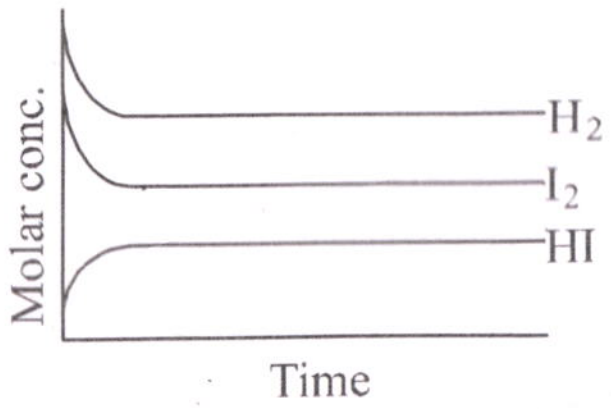

(3)
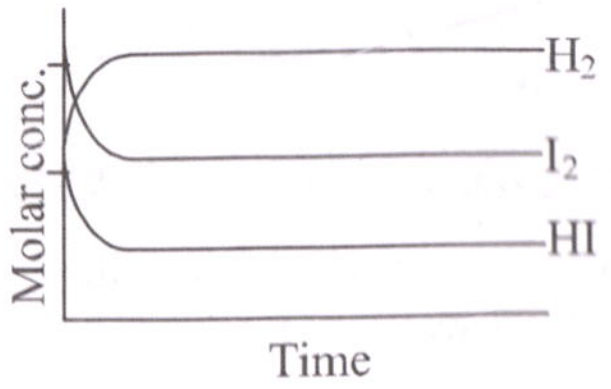

(4)
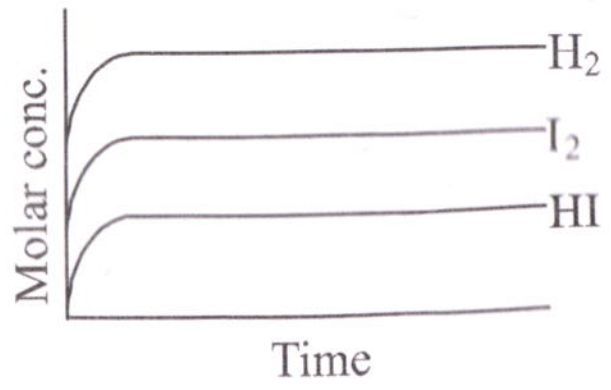

16. For a reaction, $N_2O_{5(g)} \rightarrow 2NO_{2(g)} + \frac{1}{2}O_{2(g)}$ in a constant volume container, no products were present initially. The final pressure of the system when 50% of reaction gets completed is [Jan. 24, 2025 (I)]

(1) 7/2 times of initial pressure
(2) 5 times of initial pressure
(3) 5/2 times of initial pressure
(4) 7/4 times of initial pressure

17. Consider the reaction [Jan. 23, 2025 (II)]

$X_2Y(g) \rightleftharpoons X_2(g) + \frac{1}{2}Y_2(g)$

The equation representing correct relationship between the degree of dissociation (x) of $X_2Y(g)$ with its equilibrium constant Kp is________.

Assume x to be very very small.

(1) $x = \sqrt[3]{\frac{2Kp}{p}}$ (2) $x = \sqrt[3]{\frac{2Kp^2}{p}}$

(3) $x = \sqrt[3]{\frac{Kp}{2p}}$ (4) $x = \sqrt[3]{\frac{Kp}{p}}$

18. A vessel at 1000 K contains CO_2 with a pressure of 0.5 atm. Some of CO_2 is converted into CO on addition of graphite. If total pressure at equilibrium is 0.8 atm, then K_p is: [Jan. 22, 2025 (I)]

(1) 0.18 atm (2) 1.8 atm
(3) 0.3 atm (4) 3 atm.

19. For the given hypothetical reactions, the equilibrium constants are as follows:
$X \rightleftharpoons Y ; K_1 = 1.0$
$Y \rightleftharpoons Z ; K_2 = 2.0$
$Z \rightleftharpoons W ; K_3 = 4.0$
The equilibrium constant for the reaction
$X \rightleftharpoons W$ is **[April 8, 2024 (I)]**
(1) 6.0 (2) 12.0 (3) 8.0 (4) 7.0

20. At –20 °C and 1 atm pressure, a cylinder is filled with equal number of H_2. I_2 and HI molecules for the reaction $H_2(g) + I_2(g) \rightleftharpoons 2HI(g)$, the K_P for the process is $x \times 10^{-1}$.
x = ________. **[April 6, 2024 (I)]**
[Given : $R = 0.082$ L atm K^{-1} mol^{-1}]
(1) 2 (2) 1 (3) 10 (4) 0.01

21. The ratio $\frac{K_P}{K_C}$ for the reaction
$CO_{(g)} + \frac{1}{2}O_{2(g)} \rightleftharpoons CO_{2(g)}$ is: **[April 6, 2024 (II)]**
(1) $(RT)^{1/2}$ (2) RT (3) 1 (4) $\frac{1}{\sqrt{RT}}$

22. The equilibrium constant for the reaction
$SO_3(g) \rightleftharpoons SO_2(g) + \frac{1}{2}O_2(g)$ is $K_c = 4.9 \times 10^{-2}$.
The value of K_c for the reaction given below is
$2SO_2(g) + O_2(g) \rightleftharpoons 2SO_3(g)$ is **[April 4, 2024 (II)]**
(1) 4.9 (2) 41.6 (3) 49 (4) 416

23. For the given reaction, choose the correct expression of K_C from the following :-
$Fe^{3+}(aq) + SCN^-(aq) \rightleftharpoons (FeSCN)^{2+}(aq)$ **[Jan. 31, 2024 (I)]**
(1) $K_C = \frac{[FeSCN^{2+}]^2}{[Fe^{3+}][SCN^-]}$
(2) $K_C = \frac{[FeSCN^{2+}]}{[Fe^{3+}]^{-2}[SCN^-]^2}$
(3) $K_C = \frac{[Fe^{3+}][SCN^-]}{[FeSCN^{2+}]}$
(4) $K_C = \frac{[FeSCN^{2+}]}{[Fe^{3+}][SCN^-]}$

24. $A(g) \rightleftharpoons B(g) + \frac{C}{2}(g)$ The correct relationship between K_P, α and equilibrium pressure P is **[Jan. 31, 2024 (II)]**
(1) $K_P = \frac{\alpha^{1/2}p^{1/2}}{(2+\alpha)^{1/2}}$ (2) $K_P = \frac{\alpha^{3/2}p^{1/2}}{(2+\alpha)^{1/2}(1-\alpha)}$
(3) $K_P = \frac{\alpha^{1/2}p^{3/2}}{(2+\alpha)^{3/2}}$ (4) $K_P = \frac{\alpha^{1/2}p^{1/2}}{(2+\alpha)^{3/2}}$

25. For a concentrated solution of a weak electrolyte (K_{eq} = equilibrium constant) A_2B_3 of concentration 'c', the degree of dissociation 'α' is **[April 06, 2023 (I)]**
(1) $\left(\frac{K_{eq}}{108c^4}\right)^{\frac{1}{5}}$ (2) $\left(\frac{K_{eq}}{6c^5}\right)^{\frac{1}{5}}$
(3) $\left(\frac{K_{eq}}{5c^4}\right)^{\frac{1}{5}}$ (4) $\left(\frac{K_{eq}}{25c^2}\right)^{\frac{1}{5}}$

26. The equilibrium constant for the reversible reaction
$2A(g) \rightleftharpoons 2B(g) + C(g)$ is K_1 and for the reaction
$\frac{3}{2}A(g) \rightleftharpoons \frac{3}{2}B(g) + \frac{3}{4}C(g)$ is K_2
K_1 and K_2 are related as: **[June 30, 2022 (I)]**
(1) $K_1 = \sqrt{K_2}$ (2) $K_2 = \sqrt{K_1}$
(3) $K_2 = K_1^{3/4}$ (4) $K_1 = K_2^{3/4}$

27. 4.0 moles of argon and 5.0 moles of PCl_5 are introduced into an evacuated flask of 100 litre capacity at 610 K. The system is allowed to equilibrate. At equilibrium, the total pressure of mixture was found to be 6.0 atm. The K_p for the reaction is
[Given : $R = 0.082$ L atm K^{-1} mol^{-1}] **[June 29, 2022 (II)]**
(1) 2.25 (2) 6.24 (3) 12.13 (4) 15.24

28. For a reaction at equilibrium
$A(g) \rightleftharpoons B(g) + \frac{1}{2}C(g)$
the relation between dissociation constant (K), degree of dissociation (α) and equilibrium pressure (p) is given by: **[June 24, 2022 (I)]**
(1) $K = \frac{\alpha^{\frac{1}{2}} p^{\frac{3}{2}}}{\left(1+\frac{3}{2}\alpha\right)^{\frac{1}{2}}(1-\alpha)}$
(2) $K = \frac{\alpha^{\frac{3}{2}} p^{\frac{1}{2}}}{(2+\alpha)^{\frac{1}{2}}(1-\alpha)}$
(3) $K = \frac{(\alpha p)^{\frac{3}{2}}}{\left(1+\frac{3}{2}\alpha\right)^{\frac{1}{2}}(1-\alpha)}$
(4) $K = \frac{(\alpha p)^{\frac{3}{2}}}{(1+\alpha)(1-\alpha)^{\frac{1}{2}}}$

29. For the reaction **[Sep. 06, 2020 (I)]**
$Fe_2N(s) + \frac{3}{2}H_2(g) \rightleftharpoons 2Fe(s) + NH_3(g)$
(1) $K_c = K_p(RT)$ (2) $K_c = K_p(RT)^{\frac{-1}{2}}$
(3) $K_c = K_p(RT)^{\frac{1}{2}}$ (4) $K_c = K_p(RT)^{\frac{3}{2}}$

30. The value of K_c is 64 at 800 K for the reaction
$N_2(g) + 3H_2(g) \rightleftharpoons 2NH_3(g)$. The value of K_c for the following reaction is :
$NH_3(g) \rightleftharpoons \frac{1}{2}N_2(g) + \frac{3}{2}H_2(g)$ **[Sep. 06, 2020 (II)]**
(1) 1/64 (2) 8 (3) 1/4 (4) 1/8

31. For the equilibrium $A \rightleftharpoons B$, the variation of the rate of the forward (a) and reverse (b) reaction with time is given by: [Sep. 04, 2020 (I)]

(1) 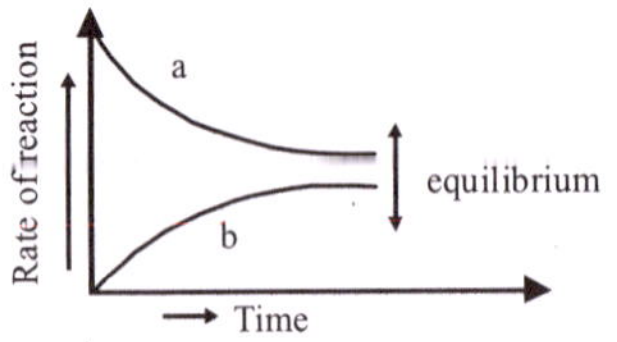

(2) 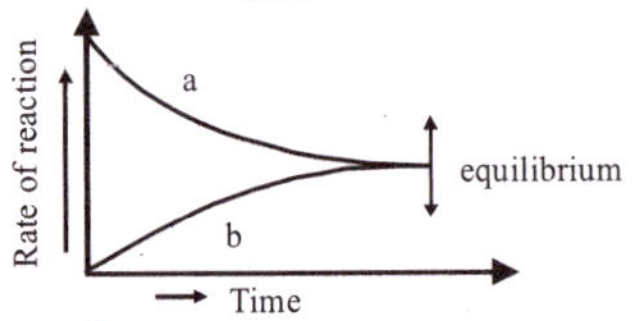

(3) 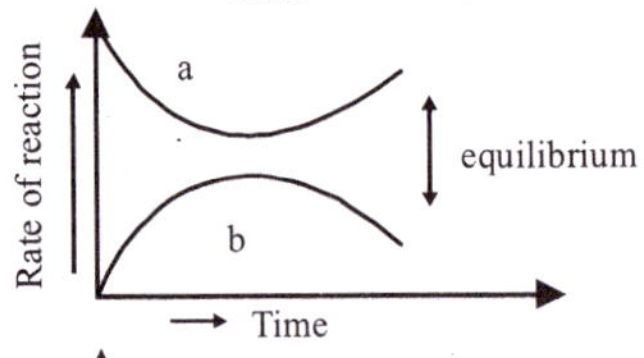

(4) 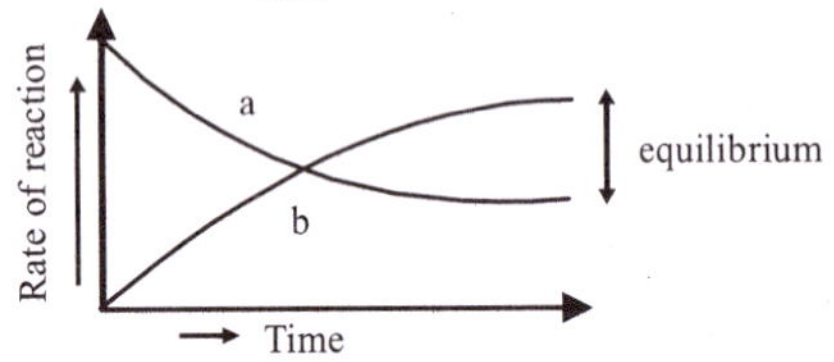

32. If the equilibrium constant for $A \rightleftharpoons B + C$ is $K_{eq}^{(1)}$ and that of $B + C \rightleftharpoons P$ is $K_{eq}^{(2)}$, the equilibrium constant for $A \rightleftharpoons P$ is: [Sep. 04, 2020 (II)]

(1) $K_{eq}^{(1)} / K_{eq}^{(2)}$ (2) $K_{eq}^{(2)} - K_{eq}^{(1)}$

(3) $K_{eq}^{(1)} + K_{eq}^{(2)}$ (4) $K_{eq}^{(1)}\ K_{eq}^{(2)}$

33. In the figure shown below, reactant A (represented by square) is in equilibrium with product B (represented by circle). The equilibrium constant is: [Jan. 09, 2020 (II)]

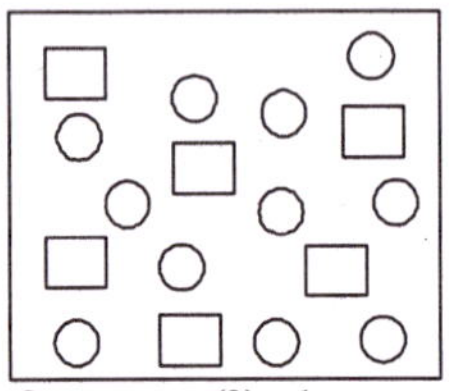

(1) 4 (2) 8 (3) 1 (4) 2

34. In which one of the following equilibria, $K_p \neq K_c$? [April 12, 2019 (II)]

(1) $2C(s) + O_2(g) \rightleftharpoons 2CO(g)$

(2) $2HI(g) \rightleftharpoons H_2(g) + I_2(g)$

(3) $NO_2(g) + SO_2(g) \rightleftharpoons NO(g) + SO_3(g)$

(4) $2NO(g) \rightleftharpoons N_2(g) + O_2(g)$

35. For the following reactions, equilibrium constants are given:

$S(s) + O_2(g) \rightleftharpoons SO_2(g);\ K_1 = 10^{52}$

$2S(s) + 3O_2(g) \rightleftharpoons 2SO_3(g);\ K_2 = 10^{129}$

The equilibrium constant for the reaction,

$2SO_2(g) + O_2(g) \rightleftharpoons 2SO_3(g)$ is: [April 8, 2019 (II)]

(1) 10^{154} (2) 10^{181} (3) 10^{25} (4) 10^{77}

36. Two solids dissociate as follows

$A(s) \rightarrow B(g) + C(g);\ K_{P_1} = x\ atm^2$

$D(s) \rightarrow C(g) + E(g);\ K_{P_2} = y\ atm^2$

The total pressure when both the solids dissociate simultaneously is: [Jan. 12, 2019 (I)]

(1) $\sqrt{x+y}$ atm (2) $2\left(\sqrt{x+y}\right)$ atm

(3) $(x+y)$ atm (4) $x^2 + y^2$ atm

37. In a chemical reaction, $A + 2B \overset{K}{\rightleftharpoons} 2C + D$, the initial concentration of B was 1.5 times of the concentration of A, but the equilibrium concentrations of A and B were found to be equal. The equilibrium constant (K) for the aforesaid chemical reaction is: [Jan. 12, 2019 (I); Same Online April 9, 2013]

(1) 4 (2) 16 (3) $\frac{1}{4}$ (4) 1

38. Consider the reaction

$N_2(g) + 3H_2(g) \rightleftharpoons 2NH_3(g)$

The equilibrium constant of the above reaction is K_P. If pure ammonia is left to dissociate, the partial pressure of ammonia at equilibrium is given by (Assume that $P_{NH_3} << P_{total}$ at equilibrium) [Jan. 11, 2019 (I)]

(1) $\frac{3^{3/2} K_P^{1/2} P^2}{16}$ (2) $\frac{K_P^{1/2} P^2}{16}$

(3) $\frac{K_P^{1/2} P^2}{4}$ (4) $\frac{3^{3/2} K_P^{1/2} P^2}{4}$

39. 5.1 g NH_4SH is introduced in 3.0 L evacuated flask at 327°C, 30% of the solid NH_4SH decomposed to NH_3 and H_2S as gases. The K_p of the reaction at 327°C is (R = 0.082 L atm mol^{-1} K^{-1}, molar mass of S = 32 g mol^{-1}, molar mass of N = 14 g mol^{-1}) [Jan. 10, 2019 (II)]

(1) $0.242 \times 10^{-4}\ atm^2$ (2) $1 \times 10^{-4}\ atm^2$

(3) $4.9 \times 10^{-3}\ atm^2$ (4) $0.242\ atm^2$

40. The values of K_p/K_c for the following reactions at 300 K are, respectively: (At 300 K, $RT = 24.62\ dm^3\ atm\ mol^{-1}$)

$N_2(g) + O_2(g) \rightleftharpoons 2NO(g)$

$N_2O_4(g) \rightleftharpoons 2NO_2(g)$

$N_2(g) + 3H_2(g) \rightleftharpoons 2NH_3(g)$ [Jan. 10, 2019 (I)]

(1) 1, 24.62 $dm^3\ atm\ mol^{-1}$, 606.0 $dm^6\ atm^2\ mol^{-2}$

(2) 1, 24.62 $dm^3\ atm\ mol^{-1}$, $1.65 \times 10^{-3}\ dm^{-6}\ atm^{-2}\ mol^2$

(3) 1, $4.1 \times 10^{-2}\ dm^{-3}\ atm^{-1}$ mol, 606 $dm^6\ atm^2\ mol^{-2}$

(4) 24.62 $dm^3\ atm\ mol^{-1}$, 606.0 $dm^6\ atm^2\ mol^{-2}$, $1.65 \times 10^{-3}\ dm^{-6}\ atm^{-2}\ mol^2$

41. Consider the following reversible chemical reactions:

$A_2(g) + B_2(g) \xrightleftharpoons{K_1} 2AB(g)$......(1)

$6AB(g) \xrightleftharpoons{K_2} 3A_2(g) + 3B_2(g)$.....(2)

The relation between K_1 and K_2 is: [Jan. 9, 2019 (II)]

(1) $K_1K_2 = \frac{1}{3}$ (2) $K_2 = K_1^3$

(3) $K_2 = K_1^{-3}$ (4) $K_1K_2 = 3$

42. At a certain temperature in a 5L vessel, 2 moles of carbon monoxide and 3 moles of chlorine were allowed to reach equilibrium according to the reaction, $CO + Cl_2 \rightleftharpoons COCl_2$. At equilibrium, if one mole of CO is present then equilibrium constant (K_c) for the reaction is: [Online April 15, 2018 (II)]

(1) 2.5 (2) 4 (3) 2 (4) 3

43. A solid XY kept in an evacuated sealed container undergoes decomposition to form a mixture of gases X and Y at temperature T. The equilibrium pressure is 10 bar in the vessel. K_p for this reaction is: [Online April 10, 2016]
(1) 25 (2) 100 (3) 10 (4) 5

44. Gaseous N_2O_4 dissociates into gaseous NO_2 according to the reaction
$[N_2O_4(g) \rightleftharpoons 2NO_2(g)]$
At 300 K and 1 atm pressure, the degree of dissociation of N_2O_4 is 0.2. If one mole of N_2O_4 gas is contained in a vessel, then the density of the equilibrium mixture is : [Online April 10, 2015]
(1) 1.56 g/L (2) 6.22 g/L
(3) 3.11 g/L (4) 4.56 g/L

45. For the reaction $SO_2(g)+\frac{1}{2}O_2(g) \rightleftharpoons SO_3(g)$, if $K_p = K_c(RT)^x$ where the symbols have usual meaning then the value of x is (assuming ideality): [2014]
(1) -1 (2) $-\frac{1}{2}$ (3) $\frac{1}{2}$ (4) 1

46. At a certain temperature, only 50% HI is dissociated into H_2 and I_2 at equilibrium. The equilibrium constant is: [Online April 9, 2014]
(1) 1.0 (2) 3.0 (3) 0.5 (4) 0.25

47. For the decomposition of the compound, represented as
$NH_2COONH_4(s) \rightleftharpoons 2NH_3(g)+CO_2(g)$
the $K_p = 2.9 \times 10^{-5}$ atm^3.
If the reaction is started with 1 mol of the compound, the total pressure at equilibrium would be: [Online April 19, 2014]
(1) 1.94×10^{-2} atm (2) 5.82×10^{-2} atm
(3) 7.66×10^{-2} atm (4) 38.8×10^{-2} atm

48. The ratio $\frac{K_p}{K_c}$ for the reaction
$CO(g)+\frac{1}{2}O_2(g) \rightleftharpoons CO_2(g)$ is: [Online April 25, 2013]
(1) $\frac{1}{\sqrt{RT}}$ (2) $(RT)^{1/2}$ (3) RT (4) 1

49. (i) $N_2(g)+3H_2(g) \rightleftharpoons 2NH_3(g), K_1$
(ii) $N_2(g)+O_2(g) \rightleftharpoons 2NO(g), K_2$
(iii) $H_2(g)+\frac{1}{2}O_2(g) \rightleftharpoons H_2O(g), K_3$
The equation for the equilibrium constant of the reaction
$2NH_3(g)+\frac{5}{2}O_2(g) \rightleftharpoons 2NO(g)+3H_2O(g), (K_4)$ in terms of K_1, K_2 and K_3 is : [Online April 23, 2013]
(1) $\frac{K_1.K_2}{K_3}$ (2) $\frac{K_1.K_3^2}{K_2}$
(3) K_1K_2K (4) $\frac{K_2.K_3^3}{K_1}$

50. The equilibrium constant (K_c) for the reaction $N_2(g)+O_2(g) \rightarrow 2NO(g)$ at temperature T is 4×10^{-4}. The value of K_c for the reaction
$NO(g) \rightarrow \frac{1}{2}N_2(g)+\frac{1}{2}O_2(g)$ at the same temperature is: [2012]
(1) 0.02 (2) 2.5×10^2 (3) 4×10^{-4} (4) 50.0

51. K_1, K_2 and K_3 are the equilibrium constants of the following reactions (I), (II) and (III) respectively: [Online May 7, 2012]
(I) $N_2+2O_2 \rightleftharpoons 2NO_2$
(II) $2NO_2 \rightleftharpoons N_2+2O_2$
(III) $NO_2 \rightleftharpoons \frac{1}{2}N_2+O_2$
The correct relation from the following is
(1) $K_1 = \frac{1}{K_2} = \frac{1}{K_3}$ (2) $K_1 = \frac{1}{K_2} = \frac{1}{(K_3)^2}$
(3) $K_1 = \sqrt{K_2} = K_3$ (4) $K_1 = \frac{1}{K_2} = K_3$

52. For the following three reactions (i), (ii) and (iii), equilibrium constants are given: [2008]
(i) $CO(g)+H_2O(g) \rightleftharpoons CO_2(g)+H_2(g); K_1$
(ii) $CH_4(g)+H_2O(g) \rightleftharpoons CO(g)+3H_2(g); K_2$
(iii) $CH_4(g)+2H_2O(g) \rightleftharpoons CO_2(g)+4H_2(g); K_3$
Which of the following is correct?
(1) $K_1\sqrt{K_2} = K_3$ (2) $K_2K_3 = K_1$
(3) $K_3 = K_1K_2$ (4) $K_3.K_2^3 = K_1^2$

53. 8 mol of AB_3(g) are introduced into a 1.0 dm^3 vessel. If it dissociates as
$2AB_3(g) \rightleftharpoons A_2(g) + 3B_2(g)$. At equilibrium, 2 mol of A_2 are found to be present. The equilibrium constant of this reaction is [Online May 12, 2012]
(1) 2 (2) 3 (3) 27 (4) 36

54. The value of K_p for the equilibrium reaction
$N_2O_4(g) \rightleftharpoons 2NO_2(g)$ is 2.
The percentage dissociation of N_2O_4(g) at a pressure of 0.5 atm is [Online May 19, 2012]
(1) 25 (2) 88 (3) 50 (4) 71

55. One mole of O_2(g) and two moles of SO_2(g) were heated in a closed vessel of one-litre capacity at 1098 K. At equilibrium 1.6 moles of SO_3(g) were found. The equilibrium constant K_c of the reaction would be [Online May 26, 2012]
(1) 30 (2) 40 (3) 80 (4) 60

56. The equilibrium constants K_{p_1} and K_{p_2} for the reactions
$X \rightleftharpoons 2Y$ and $Z \rightleftharpoons P + Q$, respectively are in the ratio of 1 : 9. If the degree of dissociation of X and Z be equal, then the ratio of total pressures at these equilibria is [2008]
(1) 1:36 (2) 1:1
(3) 1:3 (4) 1:9

57. The equilibrium constant for the reaction
$SO_3(g) \rightleftharpoons SO_2(g)+\frac{1}{2}O_2(g)$
is $K_c = 4.9 \times 10^{-2}$. The value of K_c for the reaction
$2SO_2(g)+O_2(g) \rightleftharpoons 2SO_3(g)$
will be [2006]
(1) 9.8×10^{-2} (2) 4.9×10^{-2}
(3) 416 (4) 2.40×10^{-3}

58. Phosphorus pentachloride dissociates as follows, in a closed reaction vessel [2006]
$PCl_5(g) \rightleftharpoons PCl_3(g)+Cl_2(g)$

If total pressure at equilibrium of the reaction mixture is P and degree of dissociation of PCl_5 is x, the partial pressure of PCl_3 will be

(1) $\left(\frac{x}{x-1}\right)P$ (2) $\left(\frac{x}{1-x}\right)P$

(3) $\left(\frac{x}{x+1}\right)P$ (4) $\left(\frac{2x}{1-x}\right)P$

59. For the reaction : **[2005]**

$2NO_2(g) \rightleftharpoons 2NO(g) + O_2(g)$,

($K_c = 1.8 \times 10^{-6}$ at 184 °C) (R = 0.0831 kJ/ (mol. K))

When K_p and K_c are compared at 184 °C, it is found that

(1) Whether K_p is greater than, less than or equal to K_c depends upon the total gas pressure

(2) $K_p = K_c$

(3) K_p is less than K_c

(4) K_p is greater than K_c

60. An amount of solid NH_4HS is placed in a flask already containing ammonia gas at a certain temperature and 0.50 atm pressure. Ammonium hydrogen sulphide decomposes to yield NH_3 and H_2S gases in the flask. When the decomposition reaction reaches equilibrium, the total pressure in the flask rises to 0.84 atm. The equilibrium constant for NH_4HS decomposition at this temperature is **[2005]**

(1) 0.11 (2) 0.17 (3) 0.18 (4) 0.30

61. For the reaction, $CO(g) + Cl_2(g) \rightleftharpoons COCl_2(g)$ the K_p/K_c is equal to **[2004]**

(1) $\sqrt{RT}$ (2) RT (3) $1/RT$ (4) 1.0

62. What is the equilibrium expression for the reaction $P_4(s) + 5O_2(g) \rightleftharpoons P_4O_{10}(s)$? **[2004]**

(1) $K_c = [O_2]^5$

(2) $K_c = [P_4O_{10}]/5[P_4][O_2]$

(3) $K_c = [P_4O_{10}]/[P_4][O_2]^5$

(4) $K_c = 1/[O_2]^5$

63. For the reaction equilibrium, **[2003]**

$N_2O_4(g) \rightleftharpoons 2NO_2(g)$

the concentrations of N_2O_4 and NO_2 at equilibrium are 4.8×10^{-2} and 1.2×10^{-2} mol L^{-1} respectively. The value of K_c for the reaction is

(1) 3×10^{-1} mol L^{-1} (2) 3×10^{-3} mol L^{-1}

(3) 3×10^{3} mol L^{-1} (4) 3.3×10^{2} mol L^{-1}

64. For the reaction $CO(g) + (1/2)O_2(g) \rightleftharpoons CO_2(g)$, K_p/K_c is **[2002]**

(1) RT (2) $(RT)^{-1}$ (3) $(RT)^{-1/2}$ (4) $(RT)^{1/2}$

NUMERICAL VALUE QUESTIONS (NVQs)

65. Solid carbon, CaO and $CaCO_3$ are mixed and allowed to attain equilibrium at T K.

$CaCO_3 \rightleftharpoons CaO(s) + CO_2(g)$ $Kp_1 = 0.08$ atm

$C(s) + CO_2(g) \rightleftharpoons 2CO(g)$ $Kp_2 = 2$ atm

The partial pressure of CO is _____ $\times 10^{-1}$ atm. **[April 08, 2026 (II)]**

66. Consider the following gas phase reaction being carried out in a closed vessel at 25°C.

$2A(g) \rightarrow 4B(g) + C(g)$

time (min)	total pressure of the system (mm Hg)
30	300
∞	600

The pressure of C(g) at 30 minutes time interval would be _________ mm Hg. (Nearest integer) **[April 02, 2026 (II)]**

67. $X_2(g) + Y_2(g) \rightleftharpoons 2Z(g)$

$X_2(g)$ and $Y_2(g)$ are added to a 1 L flask and it is found that the system attains the above equilibrium at T (K) with the number of moles of $X_2(g)$, $Y_2(g)$ and Z(g) being 3, 3 and 9 mol respectively (equilibrium moles). Under this conditions of equilibrium, 10 mol of Z(g) is added to the flask and the temperature is maintained at T (K). Then the number of moles of Z(g) in the flask when the new equilibrium is established is ______. (Nearest integer). **[Jan. 23, 2026 (II)]**

68. For the following gas phase equilibrium reaction at constant temperature,

$NH_3(g) \rightleftharpoons \frac{1}{2}N_2(g) + \frac{3}{2}H_2(g)$

If the total pressure is $\sqrt{3}$ atm and the pressure equilibrium constant (K_p) is 9 atm, then the degree of dissociation is given as $(x \times 10^{-2})^{-1/2}$.

The value of x is ________ (Nearest integer) **[Jan. 23, 2026 (I)]**

69. The equilibrium constant for decomposition of $H_2O(g)$

$H_2O(g) \rightleftharpoons H_2(g) + \frac{1}{2}O_2(g)$ ($\Delta G° = 92.34$ kJ mol^{-1}) is 8.0×10^{-3} at 2300 K and total pressure at equilibrium is 1 bar. Under this condition, the degree of dissociation (α) of water is _____ $\times 10^{-2}$ (nearest integer value).

[Assume α is negligible with respect to 1] **[April 8, 2025 (II)]**

70. Consider the following equilibrium, **[April 2, 2025 (I)]**

$CO(g) + 2H_2(g) \rightleftharpoons CH_3OH(g)$

0.1 mol of CO along with a catalyst is present in a 2 dm^3 flask maintained at 500 K. Hydrogen is introduced into the flask until the pressure is 5 bar and 0.04 mol of CH_3OH is formed.

The K_p^0 is ________ $\times 10^{-3}$ (nearest integer).

Given : R = 0.08 dm^3 bar K^{-1} mol^{-1}

Assume only methanol is formed as the product and the system follows ideal gas behaviour.

71. 37.8 g N_2O_5 was taken in a 1 L reaction vessel and allowed to undergo the following reaction at 500 K

$2N_2O_{5(g)} \rightarrow 2N_2O_{4(g)} + O_{2(g)}$

The total pressure at equilibrium was found to be 18.65 bar.

Then, K_p = _______ $\times 10^{-2}$ [nearest integer]

Assume N_2O_5 to behave ideally under these conditions

Given : R = 0.082 bar L mol^{-1} K^{-1}. **[Jan. 24, 2025 (I)]**

72. For the reaction $N_2O_4(g) \rightleftharpoons 2NO_2(g)$, $K_p = 0.492$ atm at 300K. K_c for the reaction at same temperature is ______ $\times 10^{-2}$.

(Given : R = 0.082 L atm mol^{-1} K^{-1}) **[Jan. 29, 2024 (I)]**

73. The following concentrations were observed at 500K for the formation of NH_3 from N_2 and H_2. At equilibrium : $[N_2] = 2 \times 10^{-2}$ M, $[H_2] = 3 \times 10^{-2}$ M and $[NH_3] = 1.5 \times 10^{-2}$ M. Equilibrium constant for the reaction is _____ . **[Jan. 29, 2024 (II)]**

74. A mixture of 1 mole of H_2O and 1 mole of CO is taken in a 10 litre container and heated to 725 K. At equilibrium 40% of water by mass reacts with carbon monoxide according to the equation :

$CO(g) + H_2O(g) \rightleftharpoons CO_2(g) + H_2(g)$.

The equilibrium constant $K_C \times 10^2$ for the reaction is ________. (Nearest integer) **[April 11, 2023 (I)]**

75. 4.5 moles each of hydrogen and iodine is heated in a sealed ten litre vessel. At equilibrium, 3 moles of HI were found. The equilibrium constant for $H_2(g) + I_2(g) \rightleftharpoons 2HI(g)$ is ______. **[April 11, 2023 (II)]**

76. The number of correct statement/s involving equilibria in physical processes from the following is **[April 10, 2023 (I)]**

(A) Equilibrium is possible only in a closed system at a given temperature.

(B) Both the opposing processes occur at the same rate.

(C) When equilibrium is attained at a given temperature, the value of all its parameters became equal.

(D) For dissolution of solids in liquids, the solubility is constant at a given temperature.

77. The equilibrium composition for the reaction

$PCl_3 + Cl_2 \rightleftharpoons PCl_5$ at 298 K is given below.

$[PCl_3]_{eq} = 0.2$ mol L^{-1} $[Cl_2]_{eq} = 0.1$ mol L^{-1},

$[PCl_5]_{eq} = 0.40$ mol L^{-1}

If 0.2 mol of Cl_2 is added at the same temperature, the equilibrium concentrations of PCl_5 is ____________ $\times 10^{-2}$ mol L^{-1}.

Given: K_c for the reaction at 298 K is 20 **[April 06, 2023 (II)]**

78. (i) $X(g) \rightleftharpoons Y(g) + Z(g)$ $K_{p1} = 3$

(ii) $A(g) \rightleftharpoons 2B(g)$ $K_{p2} = 1$

If the degree of dissociation and initial concentration of both the reactants X(g) and A(g) are equal, then the ratio of the total pressure at equilibrium $\left(\frac{P_1}{P_2}\right)$ is equal to x:1. The value of x is ______ (Nearest integer) **[Feb. 01, 2023 (I)]**

79. For reaction : $SO_2(g) + \frac{1}{2}O_2(g) \rightleftharpoons SO_3(g)$

$K_P = 2 \times 10^{12}$ at 27°C and 1 atm pressure. The K_c for the same reaction is ________ $\times 10^{13}$. (Nearest integer)

(Given R = 0.082 L atm K^{-1} mol^{-1}) **[Jan. 31, 2023 (I)]**

80. Water decomposes at 2300 K

$$H_2O(g) \rightarrow H_2(g) + \frac{1}{2}O_2(g)$$

The percent of water decomposing at 2300 K and 1 bar is ________ (Nearest integer).

Equilibrium constant for the reaction is 2×10^{-3} at 2300 K **[Jan. 29, 2023 (I)]**

81. At 298 K **[Jan. 29, 2023 (II)]**

$N_2(g) + 3H_2(g) \rightleftharpoons 2NH_3(g)$, $K_1 = 4 \times 10^5$

$N_2(g) + O_2(g) \rightleftharpoons 2NO(g)$, $K_2 = 1.6 \times 10^{12}$

$H_2(g) + \frac{1}{2}O_2(g) \rightleftharpoons H_2O(g)$, $K_3 = 1.0 \times 10^{-13}$

Based on above equilibria, the equilibrium constant of the reaction,

$2NH_3(g) + \frac{5}{2}O_2(g) \rightleftharpoons 2NO(g) + 3H_2O(g)$

is ______ $\times 10^{-33}$ (Nearest integer)

82. At 600K, 2 mol of NO are mixed with 1 mol of O_2.

$$2NO(g) + O_2(g) \rightleftharpoons 2NO_2(g)$$

The reaction occurring as above comes to equilibrium under a total pressure of 1 atm. Analysis of the system shows that 0.6 mol of oxygen are present at equilibrium. The equilibrium constant for the reaction is ________. (Nearest integer) **[July 28, 2022 (II)]**

83. At 298 K, the equilibrium constant is 2×10^{15} for the reaction:

$Cu(s) + 2Ag^+(aq) \rightleftharpoons Cu^{2+}(aq) + 2Ag(s)$

The equilibrium constant for the reaction

$\frac{1}{2}Cu^{2+}(aq) + Ag(s) \rightleftharpoons \frac{1}{2}Cu(s) + Ag^+(aq)$

is $x \times 10^{-8}$. The value of x is ________. (Nearest Integer) **[July 26, 2022 (I)]**

84. A box contains 0.90 g of liquid water in equilibrium with water vapour at 27 °C. The equilibrium vapour pressure of water at 27 °C 32.0 Torr. When the volume of the box is increased, some of the liquid water evaporates to maintain the equilibrium pressure. If all the liquid water evaporates, then the volume of the box must be____ litre. [nearest integer]

(Given: $R = 0.082$ L atm K^{-1} mol^{-1})

(Ignore the volume of the liquid water and assume water vapours behave as an ideal gas.) **[June 29, 2022 (II)]**

85. $2NOCl(g) \rightleftharpoons 2NO(g) + Cl_2(g)$

In an experiment, 2.0 moles of NOCl was placed in a one-litre flask and the concentration of NO after equilibrium established, was found to be 0.4 mol/L. The equilibrium constant at 30°C is ______ $\times 10^{-4}$. **[June 27, 2022 (I)]**

86. PCl_5 dissociates as

$PCl_5(g) \rightleftharpoons PCl_3(g) + Cl_2(g)$

5 moles of PCl_5 are placed in a 200 litre vessel which contains 2 moles of N_2 and is maintained at 600 K. The equilibrium pressure is 2.46 atm. The equilibrium constant K_p for the dissociation of PCl_5 is_____ $\times 10^{-3}$. (nearest integer)

(Given: R = 0.082 L atm K^{-1} mol^{-1} : Assume ideal gas behaviour) **[June 24, 2022 (II)]**

87. The number of moles of NH_3, that must be added to 2 L of 0.80 M $AgNO_3$ in order to reduce the concentration of Ag^+ ions to 5.0×10^{-8} M ($K_{formation}$ for $[Ag(NH_3)_2]^+ = 1.0 \times 10^8$) is ________ (Nearest integer)

[Assume no volume change on adding NH_3] **[Aug. 27, 2021 (I)]**

88. When 5.1 g of solid NH_4HS is introduced into a two litre evacuated flask at 27 °C, 20% of the solid decomposes into gaseous ammonia and hydrogen sulphide. The K_p for the reaction at 27 °C is $x \times 10^{-2}$. The value of x is _______. (Integer answer)

[Given $R = 0.082$ L atm K^{-1} mol^{-1}] **[Aug. 27, 2021 (II)]**

89. The equilibrium constant K_c at 298 K for the reaction $A + B \rightleftharpoons C + D$ is 100. Starting with an equimolar solution with concentrations of A, B, C and D all equal to 1M, the equilibrium concentration of D is_____ $\times 10^{-2}$M. **[Aug. 26, 2021 (II); July 25, 2021 (I); Same 2016]**

90. The reaction rate for the reaction

$[PtCl_4]^{2-} + H_2O \rightleftharpoons [Pt(H_2O)Cl_3]^- + Cl^-$

was measured as a function of concentrations of different species. It was observed that

$$\frac{-d\left[[PtCl_4]^{2-}\right]}{dt} = 4.8\times10^{-5}\left[[PtCl_4]^{2-}\right] - 2.4\times10^{-3}\left[[Pt(H_2O)Cl_3]^{-}\right][Cl^{-}].$$

where square brackets are used to denote molar concentrations. The equilibrium constant K_c =_____. (Nearest integer) [Aug. 26, 2021 (II)]

91. $PCl_5 \rightleftharpoons PCl_3 + Cl_2; K_c = 1.844$
3.0 moles of PCl_5 is introduced in a 1 L closed reaction vessel at 380 K. The number of moles of PCl_5 at equilibrium is ________ $\times 10^{-3}$.
(Round off to the Nearest Integer) [July 27, 2021 (I)]

92. The equilibrium constant for the reaction
$A(s) \rightleftharpoons M(s) + \frac{1}{2}O_2(g)$
is $K_p = 4$. At equilibrium, the partial pressure of O_2 is _____ atm. (Round off to the nearest integer) [July 27, 2021 (II)]

93. $2SO_2(g) + O_2(g) \rightarrow 2SO_3(g)$
The above reaction is carried out in a vessel starting with partial pressure
$P_{SO_2} = 250$ m bar, $P_{O_2} = 750$ m bar and $P_{SO_3} = 0$ m bar
When the reaction is complete, the total pressure in the reaction vessel is _____ m bar. (Round off of the nearest integer). [July 27, 2021 (II)]

94. For the reaction
$A + B \rightleftharpoons 2C$
the value of equilibrium constant is 100 at 298 K. If the initial concentration of all the three species is 1 M each, then the equilibrium concentration of C is $x \times 10^{-1}$ M. The value of x is ________. (Nearest integer) [July 25, 2021 (I)]

95. Value of K_p for the equilibrium reaction
$N_2O_4(g) \rightleftharpoons 2NO_2(g)$ at 288 K is 47.9. The K_C for this reaction at same temperature is ________. (Nearest integer)
$[R = 0.083$ L bar K^{-1} $mol^{-1}]$ [July 22, 2021 (II)]

96. $2SO_2(g) + O_2(g) \rightleftharpoons 2SO_3(g)$
In an equilibrium mixture, the partial pressures are
$P_{SO_3} = 43$ kPa; $P_{O_2} = 530$ Pa and $P_{SO_2} = 45$ kPa. The equilibrium constant K_p = ______ $\times 10^{-2}$. (Nearest integer) [July 20, 2021 (I)]

97. Consider the reaction $N_2O_4 \rightleftharpoons 2NO_2(g)$. The temperature at which $K_C = 20.4$ and $K_p = 600.1$, is ________ K. (Round off to the Nearest Integer).
[Assume all gases are ideal and $R = 0.0831$ L bar K^{-1} mol^{-1}] [March 17, 2021 (II)]

98. A homogeneous ideal gaseous reaction $AB_2(g) \rightleftharpoons A(g) + 2B(g)$ is carried out in a 25 litre flask at 27 °C. The initial amount of AB_2 was 1 mole and the equilibrium pressure was 1.9 atm. The value of K_p is $x \times 10^{-2}$. The value of x is ______. (Integer answer) $[R = 0.08206$ dm^3 atm K^{-1} $mol^{-1}]$ [Feb. 26, 2021 (I)]

99. At 1990 K and 1 atm pressure, there are equal number of Cl_2 molecules and Cl atoms in the reaction mixture. The value of K_p for the reaction $Cl_2(g) \rightleftharpoons 2Cl(g)$ under the above conditions is $x \times 10^{-1}$. The value of x is ________ (Rounded off to the nearest integer) [February 24, 2021 (I)]

100. For a reaction $X + Y \rightleftharpoons 2Z$, 1.0 mol of X, 1.5 mol of Y and 0.5 mol of Z were taken in a 1 L vessel and allowed to react. At equilibrium, the concentration of Z was 1.0 mol L^{-1}. The equilibrium constant of the reaction is $\frac{x}{15}$. The value of x is ______. [Sep. 05, 2020 (II)]

Topic 2 - Relation between K, Q and G, and Factors Effecting Equilibrium

MULTIPLE CHOICE QUESTIONS (MCQs)

101. For the reaction, $N_2O_4 \rightleftharpoons 2NO_2$, graph is plotted as shown below. Identify correct statements.
A. Standard free energy change for the reaction is -5.40 kJ mol^{-1}.
B. As $\Delta G^\ominus$ in graph is positive, N_2O_4 will not dissociate into NO_2 at all.
C. Reverse reaction will go to completion.
D. When 1 mole of N_2O_4 changes into equilibrium mixture, value of $\Delta G^\ominus = -0.84$ kJ mol^{-1}
E. When 2 mole of NO_2, changes into equilibrium mixture, $\Delta G^\ominus$ for equilibrium mixture is -6.24 kJ mol^{-1}

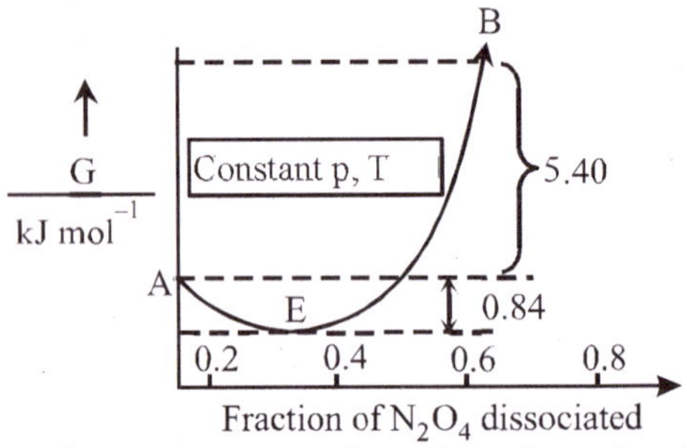

Choose the correct answer from the options given below: [Jan. 21, 2026 (I)]
(1) D and E only (2) C and E only
(3) A and D only (4) B and E only

102. Consider the equilibrium [Jan. 29, 2025 (II)]
$CO(g) + 3H_2(g) \rightleftharpoons CH_4(g) + H_2O(g)$
If the pressure applied over the system increases by two fold at constant temperature then
(A) Concentration of reactants and products increases.
(B) Equilibrium will shift in forward direction.
(C) Equilibrium constant increases since concentration of products increases.
(D) Equilibrium constant remains unchanged as concentration of reactants and products remain same.
Choose the correct answer from the options given below :
(1) (A) and (B) only (2) (A), (B) and (D) only
(3) (B) and (C) only (4) (A), (B) and (C) only

103. The following reaction occurs in the Blast furnance where iron ore is reduced to iron metal
$Fe_2O_{3(s)} + 3CO_{(g)} \rightleftharpoons Fe_{(l)} + 3CO_{2(g)}$
Using the Le-chatelier's principle, predict which one of the following will not disturb the equilibrium. [April 5, 2024 (I)]
(1) Addition of Fe_2O_3 (2) Addition of CO_2
(3) Removal of CO (4) Removal of CO_2

104. The effect of addition of helium gas to the following reaction in equilibrium state, is :
$PCl_5(g) \rightleftharpoons PCl_3(g) + Cl_2(g)$ [Feb. 01, 2023 (II)]
(1) addition of helium will not affect the equilibrium.
(2) helium will deactivate PCl_5 and reaction will stop.
(3) the equilibrium will go backward due to suppression of dissociation of PCl_5.
(4) the equilibrium will shift in the forward direction and more of Cl_2 and PCl_3 gases will be produced.

105. The variation of equilibrium constant with temperature is given below: [Sep. 06, 2020 (I)]

Temperature	Equilibrium Constant
$T_1 = 25\,°C$	$K_1 = 10$
$T_2 = 100\,°C$	$K_2 = 100$

The values of $\Delta H°$, $\Delta G°$ at T_1 and $\Delta G°$ at T_2 (in kJ mol^{-1}) respectively, are close to
[use $R = 8.314\ J K^{-1} mol^{-1}$]
(1) 28.4, – 7.14 and – 5.71
(2) 0.64, – 7.14 and – 5.71
(3) 28.4, – 5.71 and – 14.29
(4) 0.64, – 5.71 and – 14.29

106. Consider the following reaction : [Sep. 05, 2020 (I)]
$N_2O_4(g) \rightleftharpoons 2NO_2(g); \Delta H^0 = +58$ kJ
For each of the following cases ((i), (ii)), the direction in which the equilibrium shifts is :
(i) Temperature is decreases
(ii) Pressure is increased by adding N_2 at constant T.
(1) (i) towards product, (ii) towards product
(2) (i) towards reactant, (ii) towards product
(3) (i) towards reactant, (ii) no change
(4) (i) towards product, (ii) no change

107. The INCORRECT match in the following is : [April 12, 2019 (II)]
(1) $\Delta G^0 < 0, K > 1$ (2) $\Delta G^0 = 0, K = 1$
(3) $\Delta G^0 > 0, K < 1$ (4) $\Delta G^0 < 0, K < 1$

108. For the reaction,
$2SO_2(g) + O_2(g) \rightleftharpoons 2SO_3(g)$,
$\Delta H = -57.2$ kJ mol^{-1} and $K_c = 1.7 \times 10^{16}$
Which of the following statement is INCORRECT ? [April 10, 2019 (II)]
(1) The equilibrium constant is large suggestive of reaction going to completion and so no catalyst is required.
(2) The equilibrium will shift in forward direction as the pressure increases.
(3) The equilibrium constant decreases as the temperature increases.
(4) The addition of inert gas at constant volume will not affect the equilibrium constant.

109. For the equilibrium
$2H_2O \rightleftharpoons H_3O^+ + OH^-$; the value of $\Delta G°$ at 298 K is approximately: [Jan. 11, 2019 (II)]
(1) 100 kJ mol^{-1} (2) –80 kJ mol^{-1}
(3) 80 kJ mol^{-1} (4) –100 kJ mol^{-1}

110. In which of the following reactions, an increase in the volume of the container will favour the formation of products? [Online April 15, 2018 (I)]
(1) $4NH_3(g) + 5O_2(g) \rightleftharpoons 4NO(g) + 6H_2O(l)$
(2) $2NO_2(g) \rightleftharpoons 2NO(g) + O_2(g)$
(3) $3O_2 \rightleftharpoons 2O_3(g)$
(4) $H_2(g) + I_2(g) \rightleftharpoons 2HI(g)$

111. The gas phase reaction $2NO_2(g) \longrightarrow N_2O_4(g)$ is an exothermic reaction. The decomposition of N_2O_4, in equilibrium mixutre of $NO_2(g)$ and $N_2O_4(g)$, can be increased by : [Online April 16, 2018]
(1) addition of an inert gas at constant pressure
(2) lowering the temperature
(3) increasing the pressure
(4) addition of an inert gas at constant volume

112. The following reaction occurs in the Blast Furnace where iron ore is reduced to iron metal :
$Fe_2O_3(s) + 3CO(g) \rightleftharpoons 2Fe(l) + 3CO_2(g)$
Using the Le Chatelier's principle, predict which one of the following will not disturb the equilibrium ? [Online April 9, 2017]
(1) Removal of CO (2) Removal of CO_2
(3) Addition of CO_2 (4) Addition of Fe_2O_3

113. The plot shows the variation of $-\ln K_p$ versus temperature for the two reactions. [Online April 9, 2016]
$$M(s) + \frac{1}{2}O_2(g) \rightarrow MO(s) \text{ and } C(s) + \frac{1}{2}O_2(g) \rightarrow CO(s)$$
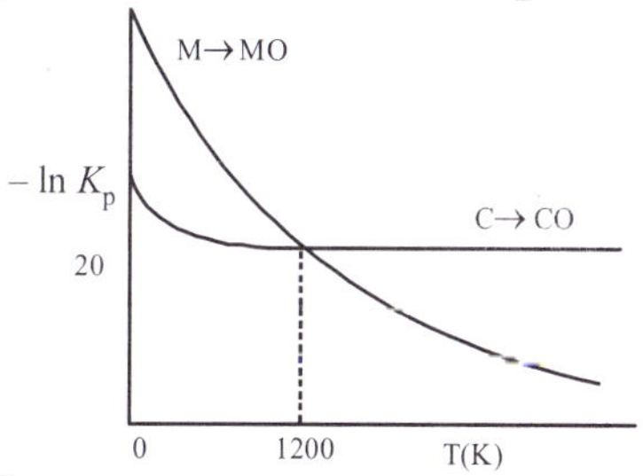

Identify the correct statement:
(1) At T < 1200 K, oxidation of carbon is unfavourable.
(2) Oxidation of carbon is favourable at all temperatures.
(3) At T < 1200 K, the reaction $MO(s) + C(s) \rightarrow M(s) + CO(g)$ is spontaneous.
(4) At T > 1200 K, carbon will reduce MO(s) to M(s).

114. For the reaction,
$A(g) + B(g) \rightarrow C(g) + D(g)$, $\Delta H°$ and $\Delta S°$ are, respectively, –29.8 kJ mol^{-1} and –0.100 kJ K^{-1} mol^{-1} at 298 K. The equilibrium constant for the reaction at 298 K is : [Online April 9, 2016]
(1) 1.0×10^{-10} (2) 10
(3) 1 (4) 1.0×10^{10}

115. The standard Gibbs energy change at 300 K for the reaction $2A \rightleftharpoons B + C$ is 2494.2 J. At a given time, the composition of the reaction mixture is $[A] = \frac{1}{2}$, $[B] = 2$ and $[C] = \frac{1}{2}$. The reaction proceeds in the : [$R = 8.314$ J/K/mol, $e = 2.718$] [2015]
(1) forward direction because $Q < K_c$
(2) reverse direction because $Q < K_c$
(3) forward direction because $Q > K_c$
(4) reverse direction because $Q > K_c$

116. The following reaction is performed at 298 K.
$$2NO(g) + O_2(g) \rightleftharpoons 2NO_2(g)$$
The standard free energy of formation of NO(g) is 86.6 kJ/mol at 298 K. What is the standard free energy of formation of $NO_2(g)$ at 298 K? ($K_p = 1.6 \times 10^{12}$) [2015]
(1) $86600 - \dfrac{\ln(1.6 \times 10^{12})}{R\,(298)}$
(2) $0.5[2 \times 86{,}600 - R(298)\ln(1.6 \times 10^{12})]$
(3) $R(298)\ln(1.6 \times 10^{12}) - 86600$
(4) $86600 + R(298)\ln(1.6 \times 10^{12})$

117. The increase of pressure on ice $\rightleftharpoons$ water system at constant temperature will lead to **[Online April 11, 2015]**
(1) a decrease in the entropy of the system
(2) an increase in the Gibb's energy of the system
(3) no effect on the equilibrium
(4) a shift of the equilibrium in the forward direction

118. What happens when an inert gas is added to an equilibrium keeping volume unchanged? **[Online April 12, 2014]**
(1) More product will form
(2) Less product will form
(3) More reactant will form
(4) Equilibrium will remain unchanged

119. The exothermic formation of ClF_3 is represented by the equation :

$Cl_2(g) + 3F_2(g) \rightleftharpoons 2ClF_3(g)$; $\Delta H = -329$ kJ

Which of the following will increase the quantity of ClF_3 in an equilibrium mixture of Cl_2, F_2 and ClF_3 ?
(1) Adding F_2 **[2005]**
(2) Increasing the volume of the container
(3) Removing Cl_2
(4) Increasing the temperature

120. Consider the reaction equilibrium

$2SO_2(g) + O_2(g) \rightleftharpoons 2SO_3(g)$; $\Delta H^o = -198$ kJ

On the basis of Le Chatelier's principle, the condition favourable for the forward reaction is **[2003]**
(1) increasing temperature as well as pressure
(2) lowering the temperature and increasing the pressure
(3) any value of temperature and pressure
(4) lowering of temperature as well as pressure

121. Change in volume of the system does not alter which of the following equilibria? **[2002]**
(1) $N_2(g) + O_2(g) \rightleftharpoons 2NO(g)$
(2) $PCl_5(g) \rightleftharpoons PCl_3(g) + Cl_2(g)$
(3) $N_2(g) + 3H_2(g) \rightleftharpoons 2NH_3(g)$
(4) $SO_2Cl_2(g) \rightleftharpoons SO_2(g) + Cl_2(g)$.

NUMERICAL VALUE QUESTIONS (NVQs)

122. In a closed flask at 600 K, one mole of $X_2Y_4(g)$ attains equilibrium as given below:

$X_2Y_4(g) \rightleftharpoons 2XY_2(g)$

At equilibrium, 75% $X_2Y_4(g)$ was dissociated and the total pressure is 1 atm. The magnitude of $\Delta_rG^{\ominus}$ (in kJ mol^{-1}) at this temperature is_____. (Nearest Integer)
(Given: $R = 8.3\ J\ mol^{-1}K^{-1}$; ln 10 = 2.3, log 2 = 0.3, log 3 = 0.48, log 5 = 0.69, log 7 = 0.84) **[April 06, 2026 (II)]**

123. For the following reaction at 50°C and 2 atm pressure,
$2N_2O_5 \rightleftharpoons (g)\ 2N_2O_4(g) + O_2(g)$
N_2O_5 is 50% dissociated
The magnitude of standard free energy change at this temperature is x.
x = _________ J mol^{-1} [Nearest integer]
Given : $R = 8.314\ mol^{-1}K^{-1}$, log 2 = 0.30,
log 3 = 0.48, ln 10 = 2.303, °C + 273 = K **[April 04, 2026 (II)]**

124. Dissociation of a gas A_2 takes place according to the following chemical reaction. At equilibrium, the total pressure is 1 bar at 300K.

$A_2(g) \rightleftharpoons 2A(g)$

The standard Gibbs energy of formation of the involved substances has been provided below:

Substances	$\Delta G°t$ / kJ mol–1
A_2	–100.00
A	–50.832

The degree of dissociation of $A_2(g)$ is given by $(x \times 10^{-2})^{1/2}$ where x = _________. (Nearest integer).
[**Given :** $R = 8.3\ J\ mol^{-1}K^{-1}$, log 2 = 0.3010, log 3 = 0.48] **[Jan. 22, 2026 (I)]**

125. Standard entropies of X_2, Y_2 and XY_5 are 70, 50 and 110 J K^{-1} mol^{-1} respectively. The temperature in Kelvin at which the reaction **[Jan. 24, 2025 (I)]**

$\frac{1}{2}X_2 + \frac{5}{2}Y_2 \rightarrow XY_5$ $\Delta H^- = -35$ kJ mol^{-1}

Will be at equilibrium is ____ (Nearest integer)

126. For a certain reaction at 300 K, K = 10, then $\Delta G°$ for the same reaction is – _______ $\times 10^{-1}$ KJ mol^{-1}. (Given R = 8.314 JK^{-1} mol^{-1}) **[Feb. 1, 2024 (II)]**

127. Consider the following reaction at 298 K.

$\frac{3}{2}O_{2(g)} \rightleftharpoons O_{3(g)}$. $K_p = 2.47 \times 10^{-29}$.

$\Delta_r G^{\ominus}$ for the reaction is _____ kJ. (Given R = 8.314 JK^{-1} mol^{-1}) **[Jan. 31, 2024 (I)]**

128. For a certain thermochemical reaction $M \rightarrow N$ at T = 400K, $\Delta H^{\ominus} = 77.2$ kJ mol^{-1}, $\Delta S = 122\ JK^{-1}$, log equilibrium constant (log K) is – _________ $\times 10^{-1}$. **[Jan. 27, 2024 (II)]**

129. The value of log K for the reaction $A \rightleftharpoons B$ at 298 K is – (Nearest integer)
Given : $\Delta H° = -54.07$ kJ mol^{-1}
$\Delta S° = 10\ JK^{-1}\ mol^{-1}$
(Take 2.303 × 8.314 × 298 = 5705) **[April 06, 2023 (I)]**

130. Consider the following equation :
$2SO_2(g) + O_2(g) \rightleftharpoons 2SO_3(g)$, $\Delta H = -190$ kJ
The number of factors which will increase the yield of SO_3 at equilibrium from the following is _______
A. Increasing temperature **[Jan. 30, 2023 (II)]**
B. Increasing pressure
C. Adding more SO_2
D. Adding more O_2
E. Addition of catalyst

131. Consider the following reaction approaching equilibrium at 27°C and 1 atm pressure

$A + B \underset{K_r = 10^2}{\overset{K_f = 10^3}{\rightleftharpoons}} C + D$

The standard Gibb's energy change ($\Delta_rG°$) at 27°C is (–) _______ kJ mol^{-1}.
(Nearest integer). (Given : R = 8.3 J K^{-1} mol^{-1} and ln 10 = 2.3) **[Jan. 29, 2023 (I)]**

132. 40% of HI undergoes decomposition to H_2 and I_2 at 300 K. $\Delta G°$ for this decompostion reaction at one atmopsphere pressure is________ J mol^{-1}. [nearest integer]
(Use R = 8.31 J K^{-1} mol^{-1}; log 2 = 0.3010.
ln 10 = 2.3, log 3 = 0.477) **[June 26, 2022 (II)]**

133. $2O_3(g) \rightleftharpoons 3O_2(g)$
At 300 K, ozone is fifty percent dissociated. The standard free energy change at this temperature and 1 atm pressure is (–) __J mol^{-1} (Nearest integer)
[Given: ln 1.35 = 0.3 and R = 8.3 J K^{-1} mol^{-1}] **[June 24, 2022 (I)]**

134. For the reaction $A(g) \rightleftharpoons B(g)$ at 495 K, $\Delta_r G° = -9.478$ kJ mol^{-1}.
If we start the reaction in a closed container at 495 K with 22 millimoles of A, the amount of B in the equilibrium mixture is _____ millimoles. (Round off to the Nearest Integer).
[$R = 8.314$ J mol^{-1} K^{-1}; ln 10 = 2.303] **[March 16, 2021 (I)]**

135. The gas phase reaction
$2A(g) \rightleftharpoons A_2(g)$ at 400 K has $dG° = +25.2$ kJ mol^{-1}
The equilibrium constant K_c for this reaction is ______ $\times 10^{-2}$. (Round off to the nearest integer).
[Use: $R = 8.3$ J mol^{-1} K^{-1}, ln 10 = 2.3 $\log_{10} 2 = 0.30$, 1 atm = 1 bar]
[antilog (–0.3) = 0.501] **[March 18, 2021 (II)]**

136. For the reaction $A(g) \rightarrow B(g)$, the value of the equilibrium constant at 300 K and 1 atm is equal to 100.0. The value of $\Delta_r G$ for the reaction at 300 K and 1 atm in J mol^{-1} is $-xR$, where x is ________ (Rounded off to the nearest integer)
($R = 8.31$ J mol^{-1} K^{-1} and ln 10 = 2.3) **[Feb. 24, 2021 (I)]**

Topic 3 - Theories of Acids and Bases, Ionic Product of Water and pH Scale

MULTIPLE CHOICE QUESTIONS (MCQs)

137. An aqueous solution of HCl with pH 1.0 is diluted by adding equal volume of water (ignoring dissociation of water). The pH of HCl solution would (Given log 2 = 0.30) **[April 7, 2025 (I)]**
(1) reduce to 0.5 (2) increase to 1.3
(3) remain same (4) increase to 2

138. A weak acid HA has degree of dissociation x. Which option gives the correct expression of ($pH - pK_a$) ? **[Jan. 28, 2025 (I)]**
(1) $\log(1+2x)$ (2) $\log\left(\frac{1-x}{x}\right)$
(3) 0 (4) $\log\left(\frac{x}{1-x}\right)$

139. pH of water is 7 at 25°C. If water is heated to 80°C, it's pH will: **[Jan. 23, 2025 (II)]**
(1) Decrease
(2) Remains the same
(3) H^+ concentration increases, OH^- concentration decreases
(4) Increase

140. What will be the decreasing order of basic strength of the following conjugate bases ?
$^{-}OH, R\bar{O}, CH_3CO\bar{O}, C\bar{l}$ **[April 4, 2024 (I)]**
(1) $C\bar{l} > OH > R\bar{O} > CH_3CO\bar{O}$
(2) $R\bar{O} > {}^{-}OH > CH_3CO\bar{O} > C\bar{l}$
(3) ${}^{-}OH > R\bar{O} > CH_3CO\bar{O} > C\bar{l}$
(4) $C\bar{l} > R\bar{O} > {}^{-}OH > CH_3CO\bar{O}$

141. Which of the following statement(s) is/are correct? **[April 15, 2023 (I)]**
(A) The pH of 1×10^{-8} M HCl solution is 8.
(B) The conjugate base of $H_2PO_4^-$ is HPO_4^{2-}.
(C) K_w increases with increase in temperature.
(D) When a solution of weak monoprotic acid is titrated against a strong base at half neutralisation point,
$$pH = \frac{1}{2} pK_a$$
Choose the correct answer from the option given below.
(1) (B), (C), (D) (2) (A), (D)
(3) (A), (B), (C) (4) (B), (C)

142. When the hydrogen ion concentration [H^+] changes by a factor of 1000, the value of pH of the solution _______. **[Jan. 25, 2023 (II)]**
(1) increases by 1000 units (2) decreases by 3 units
(3) decreases by 2 units (4) increases by 2 units

143. 200 mL of 0.01 M HCl is mixed with 400 mL of 0.01M H_2SO_4. The pH of the mixture is ____. **[July 29, 2022 (II)]**
(1) 1.14 (2) 1.78 (3) 2.34 (4) 3.02

144. A compound 'X' is a weak acid and it exhibits colour change at pH close to the equivalence point during neutralization of NaOH with CH_3COOH. Compound 'X' exists in ionized form in basic medium. The compound 'X' is : **[July 29, 2022 (II)]**
(1) methyl orange (2) methyl red
(3) phenolphthalein (4) erichrome Black T

145. Given below are two statements : one is labelled as **Assertion A** and the other is labelled as **Reason R**.
Assertion A : Phenolphthalein is a pH dependent indicator, remains colourless in acidic solution and gives pink colour in basic medium
Reason R : Phenolphthalein is a weak acid. It doesn't dissociate in basic medium. **[July 26, 2022 (II)]**
In the light of the above statements, choose the **most appropriate** answer from the options given below :
(1) Both A and R are true and R is the correct explanation of A.
(2) Both A and R are true but R is NOT the correct explanation of A.
(3) A is true but R is false.
(4) A is false but R is true.

146. Given below are two statements one is labelled as **Assertion A** and the other is labelled as **Reason R:**
Assertion A : The amphoteric nature of water is explained by using Lewis acid/base concept.
Reason R : Water acts as an acid with NH_3 and as a base with H_2S.
In the light of the above statements choose the correct answer from the options given below : **[June 25, 2022 (II)]**
(1) Both A and R are true and R is the correct explanation of A.
(2) Both A and R are true but R is NOT the correct explanation of A.
(3) A is true but R is false.
(4) A is false but R is true.

147. Which of the following compound CANNOT act as a Lewis base? **[March 17, 2021 (I)]**
(1) NF_3 (2) PCl_5 (3) ClF_3 (4) SF_4

148. Arrange the following solutions in the decreasing order of pOH : **[Sep. 06, 2020 (I)]**
(A) 0.01 M HCl (B) 0.01 M NaOH
(C) 0.01 M CH_3COONa (D) 0.01 M NaCl
(1) (A) > (C) > (D) > (B) (2) (A) > (D) > (C) > (B)
(3) (B) > (C) > (D) > (A) (4) (B) > (D) > (C) > (A)

149. 100 mL of 0.1 M HCl is taken in a beaker and to it 100 mL of 0.1 M NaOH is added in steps of 2 mL and the pH is continuously measured. Which of the following graphs **correctly** depicts the change in pH? **[Sep. 03, 2020 (II)]**

(1) 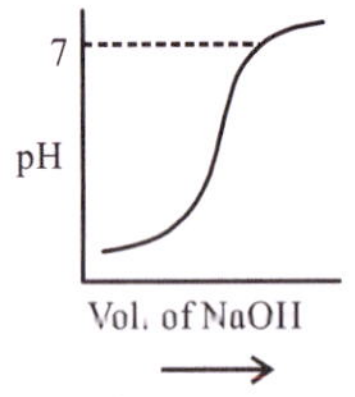

(2) 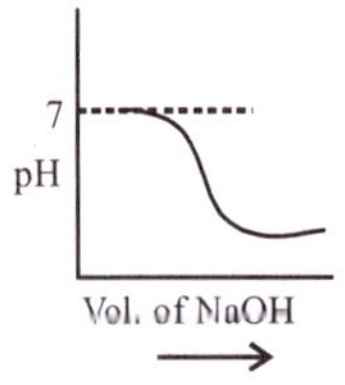

(3) 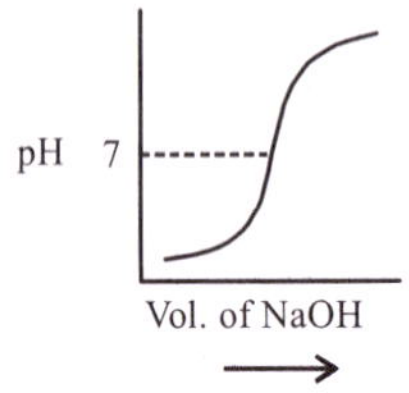

(4) 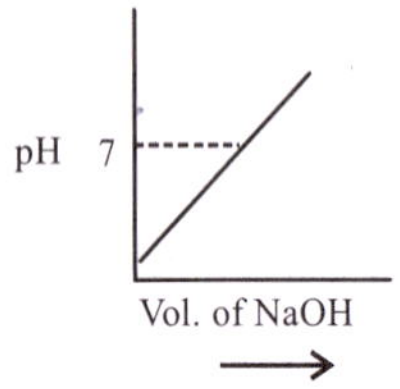

150. For the following Assertion and Reason, the correct option is:

Assertion: The pH of water increases with increase in temperature. **[Jan. 08, 2020 (II)]**

Reason: The dissociation of water into H^+ and OH^- is an exothermic reaction.

(1) Both assertion and reason are true, and the reason is the correct explanation for the assertion.
(2) Both assertion and reason are false.
(3) Both assertion and reason are true, but the reason is not the correct explanation for the assertion.
(4) Assertion is not true, but reason is true.

151. Consider the following statements **[April 10, 2019 (I)]**

(A) The pH of a mixture containing 400 mL of 0.1 M H_2SO_4 and 400 mL of 0.1 M NaOH will be approximately 1.3.
(B) Ionic product of water is temperature dependent.
(C) A monobasic acid with $K_a = 10^{-5}$ has a pH = 5. The degree of dissociation of this acid is 50%.
(D) The Le Chatelier's principle is not applicable to common-ion effect.

The correct statements are :

(1) (A), (B), and (D) (2) (A), (B) and (C)
(3) (B) and (C) (4) (A) and (B)

152. In an acid base titration, 0.1 M HCl solution was added to the NaOH solution of unknown strength. Which of the following correctly shows the change of pH of the titration mixture in this experiment? **[April 9, 2019 (II)]**

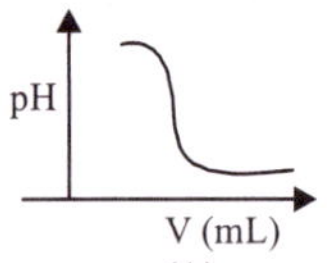

(A)

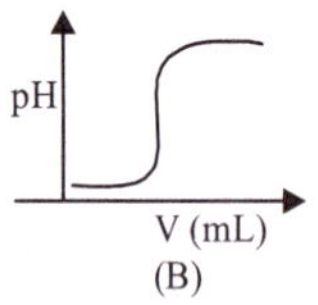

(B)

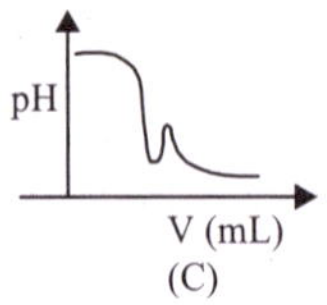

(C)

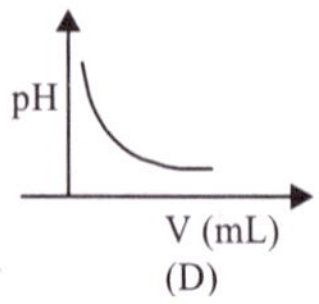

(D)

(1) (B) (2) (A) (3) (C) (4) (D)

153. Which amongst the following is the strongest acid? **[Jan. 9, 2019 (I)]**

(1) $CHBr_3$ (2) CHI_3 (3) $CH(CN)_3$ (4) $CHCl_3$

154. Following four solutions are prepared by mixing different volumes of NaOH and HCl of different concentrations, pH of which one of them will be equal to 1? **[Online April 15, 2018 (II)]**

(1) $55\ mL\ \frac{M}{10}HCl + 45\ mL\ \frac{M}{10}NaOH$

(2) $75\ mL\ \frac{M}{5}HCl + 25\ mL\ \frac{M}{5}NaOH$

(3) $100\ mL\ \frac{M}{10}HCl + 100\ mL\ \frac{M}{10}NaOH$

(4) $60\ mL\ \frac{M}{10}HCl + 40\ mL\ \frac{M}{10}NaOH$

155. Which of the following is a Lewis acid? **[Online April 15, 2018 (I)]**

(1) PH_3 (2) NF_3 (3) NaH (4) $B(CH_3)_3$

156. 50 mL of 0.2 M ammonia solution is treated with 25 mL of 0.2 M HCl. If pK_b of ammonia solution is 4.75, the pH of the mixture will be : **[Online April 9, 2017]**

(1) 3.75 (2) 4.75 (3) 8.25 (4) 9.25

157. The conjugate base of hydrazoic acid is: **[Online April 12, 2014]**

(1) N^{-3} (2) N_3^- (3) N_2^- (4) HN_3^-

158. How many litres of water must be added to 1 litre an aqueous solution of HCl with a pH of 1 to create an aqueous solution with pH of 2 ? **[2013]**

(1) 0.1 L (2) 0.9 L (3) 2.0 L (4) 9.0 L

159. NaOH is a strong base. What will be pH of 5.0×10^{-2} M NaOH solution ? (log 2 = 0.3) **[Online April 22, 2013]**

(1) 14.00 (2) 13.70 (3) 13.00 (4) 12.70

160. Equimolar solutions of the following compounds are prepared separately in water. Which will have the lowest pH value ? **[Online April 23, 2013]**

(1) $BeCl_2$ (2) $SrCl_2$ (3) $CaCl_2$ (4) $MgCl_2$

161. What is the pH of a 10^{-4} M OH^- solution at 330*K*, if K_w at 330 K is $10^{-13.6}$? **[Online April 23, 2013]**

(1) 4 (2) 9.0 (3) 10 (4) 9.6

162. Which one of the following arrangements represents the correct order of the proton affinity of the given species : **[Online April 25, 2013]**

(1) $I^- < F^- < HS^- < NH_2^-$ (2) $HS^- < NH_2^- < F^- < I^-$

(3) $F^- < I^- < NH_2^- < HS^-$ (4) $NH_2^- < HS^- < I^- < F^-$

163. Four species are listed below: **[2008]**

i. HCO_3^- ii. H_3O^+ iii. HSO_4^- iv. HSO_3F

Which one of the following is the correct sequence of their acid strength?

(1) iv < ii < iii < i (2) ii < iii < i < iv
(3) i < iii < ii < iv (4) iii < i < iv < ii

164. Hydrogen ion concentration in mol/L in a solution of pH = 5.4 will be : **[2005]**

(1) 3.98×10^{-6} (2) 3.68×10^{-6}
(3) 3.88×10^{6} (4) 3.98×10^{8}

165. What is the conjugate base of OH^- ? **[2005]**

(1) O^{2-} (2) O^- (3) H_2O (4) O_2

166. The conjugate base of $H_2PO_4^-$ is **[2004]**

(1) H_3PO_4 (2) P_2O_5
(3) PO_4^{3-} (4) HPO_4^{2-}

167. Which one of the following statements is not true? **[2003]**

(1) pH + pOH = 14 for all aqueous solutions

(2) The pH of 1×10^{-8} M HCl is 8

(3) 96,500 coulombs of electricity when passed through a $CuSO_4$ solution deposits 1 gram equivalent of copper at the cathode

(4) The conjugate base of $H_2PO_4^-$ is HPO_4^{2-}

168. When rain is accompanied by a thunderstorm, the collected rain water will have a pH value [2003]

(1) slightly higher than that when the thunderstorm is not there

(2) uninfluenced by occurrence of thunderstorm

(3) that depends on the amount of dust in air

(4) slightly lower than that of rain water without thunderstorm.

169. Species acting as both Bronsted acid and base is [2002]

(1) HSO_4^- (2) Na_2CO_3 (3) NH_3 (4) OH^-

NUMERICAL VALUE QUESTIONS (NVQs)

170. x mg of $Mg(OH)_2$ (molar mass = 58) is required to be dissolved in 1.0 L of water to produce a pH of 10.0 at 298 K. The value of x is____ mg. (Nearest integer)

(**Given :** $Mg(OH)_2$ is assumed to dissociate completely in H_2O). [April 4, 2025 (II)]

171. An analyst wants to convert 1 L HCl of pH = 1 to a solution of HCl of pH 2. The volume of water needed to do this dilution is _________ mL. (Nearest integer) [April 12, 2023 (I)]

172. 600 mL of 0.01M HCl is mixed with 400 mL of 0.01 M H_2SO_4. The pH of the mixture is _________ $\times 10^{-2}$. (Nearest integer)

[Given log 2 = 0.30, log 3 = 0.48

log 5 = 0.69 log 7 = 0.84

log 11 = 1.04] [Jan. 30, 2023 (I)]

173. Millimoles of calcium hydroxide required to produce 100 mL of the aqueous solution of pH 12 is $x \times 10^{-1}$. The value of x is _______ (Nearest integer).

Assume complete dissociation. [Jan. 29, 2023 (I)]

174. A litre of buffer solution contains 0.1 mole of each of NH_3 and NH_4Cl. On the addition of 0.02 mole of HCl by dissolving gaseous HCl, the pH of the solution is found to be _________ $\times 10^{-3}$ (Nearest integer)

[Given : $pK_b(NH_3) = 4.745$

log 2 = 0.301 log 3 = 0.477 T = 298 K] [Jan. 25, 2023 (I)]

175. The density of a monobasic strong acid (Molar mass 24.2 g/mol) is 1.21 kg/L. The volume of its solution required for the complete neutralization of 25 mL of 0.24 M NaOH is_____________ $\times 10^{-2}$ mL (Nearest integer) [Jan. 25, 2023 (I)]

176. pH value of 0.001 M NaOH solution is______. [June 27, 2022 (II)]

177. The pH of a solution obtained by mixing 50 mL of 1 M HCl and 30 mL of 1 M NaOH is $x \times 10^{-4}$. The value of x is ______. (Nearest integer)

[log 2.5 = 0.3979] [Aug. 31, 2021 (II)]

178. A soft drink was bottled with a partial pressure of CO_2 of 3 bar over the liquid at room temperature. The partial pressure of CO_2 over the solution approaches a value of 30 bar when 44 g of CO_2 is dissolved in 1 kg of water at room temperature. The approximate pH of the soft drink is ______ $\times 10^{-1}$. [Sep. 05, 2020 (I)]

(First dissociation constant of H_2CO_3 = 4.0×10^{-7}; log 2 = 0.3; density of the soft drink = 1 g mL^{-1})

179. Two solutions, A and B, each of 100 L was made by dissolving 4 g of NaOH and 9.8 g of H_2SO_4 in water, respectively. The pH of the resultant solution obtained from mixing 40 L of solution A and 10 L of solution B is ________. [Jan. 07, 2020 (I)]

180. 3 g of acetic acid is added to 250 mL of 0.1 M HCl and the solution made up to 500 mL. To 20 mL of this solution, $\frac{1}{2}$ mL of 5 M NaOH is added. The pH of the solution is ________. [Jan. 07, 2020 (II)]

[Given: pK_a of acetic acid = 4.75, molar mass of acetic acid = 60 g/mol, log 3 = 0.4771] Neglect any changes in volume.

Topic 4 - Ionisation of Weak Acids and Bases and Relation between K_a and K_b

MULTIPLE CHOICE QUESTIONS (MCQs)

181. Arrange the following resultant mixtures in increasing order of their pH values.

A. 10 mL 0.2 M $Ca(OH)_2$ + 25 mL 0.1 M HCl

B. 10 mL 0.01 M H_2SO_4 + 10 mL 0.01 M $Ca(OH)_2$

C. 10 mL 0.1 M H_2SO_4 + 10 mL 0.01 M KOH

Choose the **correct** answer from the options given below: [April 05, 2026 (I)]

(1) B < C < A (2) C < A < B

(3) C < B < A (4) A < C < B

182. The first and second ionization constants of a weak dibasic acid H_2A are 8.1×10^{-8} and 1.0×10^{-13} respectively. 0.1 mol of H_2A was dissolved in 1L of 0.1 M HCl solution. The concentration of HA^- in the resultant solution is : [April 02, 2026 (II)]

(1) 0.1 M (2) 9.53×10^{-6} M

(3) 8.1×10^{-8} M (4) 1.0×10^{-13} M

183. Given below are two statements : [Jan. 27, 2024 (I)]

Statement (I) : Aqueous solution of ammonium carbonate is basic.

Statement (II) : Acidic/basic nature of salt solution of a salt of weak acid and weak base depends on K_a and K_b value of acid and the base forming it.

In the light of the above statements, choose the most appropriate answer from the options given below :

(1) Both Statement I and Statement II are correct

(2) Statement I is correct but Statement II is incorrect

(3) Both Statement I and Statement II are incorrect

(4) Statement I is incorrect but Statement II is correct

184. 20 mL of 0.1 M NH_4OH is mixed with 40 mL of 0.05 M HCl. The pH of the mixture is nearest to:

(Given : $K_b(NH_4OH) = 1 \times 10^{-5}$, log 2 = 0.30, log 3 = 0.48, log 5 = 0.69, log 7 = 0.84, log 11 = 1.04) [July 25, 2022 (I)]

(1) 3.2 (2) 4.2 (3) 5.2 (4) 6.2

185. K_{a1}, K_{a2} and K_{a3} are the respective ionization constants for the following reactions (A), (B), and (C).

(A) $H_2C_2O_4 \rightleftharpoons H^+ + HC_2O_4^-$

(B) $HC_2O_4^- \rightleftharpoons H^+ + HC_2O_4^{2-}$

(C) $H_2C_2O_4 \rightleftharpoons 2H^+ + C_2O_4^{2-}$

The relationship between K_{a_1}, K_{a_2} and K_{a_3} is given as [July 25, 2022 (II)]

(1) $K_{a_3} = K_{a_1} + K_{a_2}$ (2) $K_{a_3} = K_{a_1} + K_{a_2}$

(3) $K_{a_3} = K_{a_1} / K_{a_2}$ (4) $K_{a_3} = K_{a_1} \times K_{a_2}$

186. Assuming that the degree of hydrolysis is small, the pH of 0.1 M solution of sodium acetate ($K_a = 1.0 \times 10^{-5}$) will be: **[Online April 11, 2014]**

(1) 5.0 (2) 6.0 (3) 8.0 (4) 9.0

187. Values of dissociation constant, K_a are given as follows :

Acid	K_a
HCN	6.2×10^{-10}
HF	7.2×10^{-4}
HNO_2	4.0×10^{-4}

Correct order of increasing base strength of the base CN^-, F^- and NO_2^- will be : **[Online April 22, 2013]**

(1) $F^- < CN^- < NO_2^-$ (2) $NO_2^- < CN^- < F^-$

(3) $F^- < NO_2^- < CN^-$ (4) $NO_2^- < F^- < CN^-$

188. What would be the pH of a solution obtained by mixing 5 g of acetic acid and 7.5 g of sodium acetate and making the volume equal to 500 mL?

($K_a = 1.75 \times 10^{-5}$, $pK_a = 4.76$) **[Online April 25, 2013]**

(1) pH=4.70

(2) pH<4.70

(3) pH of solution will be equal to pH of acetic acid

(4) 4.76<pH<5.0

189. The pH of a 0.1 molar solution of the acid HQ is 3. The value of the ionization constant, K_a of the acid is : **[2012]**

(1) 3×10^{-1} (2) 1×10^{-3}

(3) 1×10^{-5} (4) 1×10^{-7}

190. Given **[Online May 12, 2012]**

(i) $HCN(aq) + H_2O(l) \rightleftharpoons H_3O^+(aq) + CN^-(aq)$

$K_a = 6.2 \times 10^{-10}$

(ii) $CN^-(aq) + H_2O(l) \rightleftharpoons HCN(aq) + OH^-(aq)$

$K_b = 1.6 \times 10^{-5}$.

These equilibria show the following order of the relative base strength,

(1) $OH^- > H_2O > CN^-$ (2) $OH^- > CN^- > H_2O$

(3) $H_2O > CN^- > OH^-$ (4) $CN^- > H_2O > OH^-$

191. An acid HA ionises as

$HA \rightleftharpoons H^+ + A^-$

The pH of 1.0 M solution is 5. Its dissociation constant would be : **[2011RS]**

(1) 5 (2) 5×10^{-8}

(3) 1×10^{-5} (4) 1×10^{-10}

192. pK_a of a weak acid (HA) and pK_b of a weak base (BOH) are 3.2 and 3.4, respectively. The pH of their salt (AB) solution is **[2017]**

(1) 7.2 (2) 6.9 (3) 7.0 (4) 1.0

193. Three reactions involving $H_2PO_4^-$ are given below :

(i) $H_3PO_4 + H_2O \rightarrow H_3O^+ + H_2PO_4^-$

(ii) $H_2PO_4^- + H_2O \rightarrow HPO_4^{2-} + H_3O^+$

(iii) $H_2PO_4^- + OH^- \rightarrow H_3PO_4 + O^{2-}$

In which of the above does $H_2PO_4^-$ act as an acid ? **[2010]**

(1) (ii) only (2) (i) and (ii)

(3) (iii) only (4) (i) only

194. In aqueous solution the ionization constants for carbonic acid are

$K_1 = 4.2 \times 10^{-7}$ and $K_2 = 4.8 \times 10^{-11}$.

Select the correct statement for a saturated 0.034 M solution of the carbonic acid. **[2010]**

(1) The concentration of CO_3^{2-} is 0.034 M.

(2) The concentration of CO_3^{2-} is greater than that of HCO_3^-.

(3) The concentrations of H^+ and HCO_3^- are approximately equal.

(4) The concentration of H^+ is double that of CO_3^{2-}.

195. The pK_a of a weak acid, HA, is 4.80. The pK_b of a weak base, BOH, is 4.78. The pH of an aqueous solution of the correspondng salt, BA, will be **[2008]**

(1) 9.58 (2) 4.79 (3) 7.01 (4) 9.22

196. The first and second dissociation constants of an acid H_2A are 1.0×10^{-5} and 5.0×10^{-10} respectively. The overall dissociation constant of the acid will be **[2007]**

(1) 0.2×10^5 (2) 5.0×10^{-5}

(3) 5.0×10^{15} (4) 5.0×10^{-15}.

NUMERICAL VALUE QUESTIONS (NVQs)

197. Consider the dissociation equilibrium of the following weak acid HA $\rightleftharpoons H^+(aq) + A^-(aq)$ If the pKa of the acid is 4, then the pH of 10 mM HA solution is ________. (Nearest integer) **[Jan. 28, 2026 (I)]**

[Given : The degree of dissociation can be neglected with respect to unity]

198. The first and second ionization constants of H_2X are 2.5×10^{-8} and 1.0×10^{-13} respectively. The concentration of X^{2-} in 0.1 M H_2X solution is ______ $\times 10^{-15}$ M. (Nearest Integer) **[Jan. 21, 2026 (II)]**

199. Use the following data : **[Jan. 21, 2026 (I)]**

Substance	$\Delta_f H^\ominus$ (500 K) kJ mol^{-1}	$S^\ominus$ (500 K) J K^{-1} mol^{-1}
AB(g)	32	222
$A_2(g)$	6	146
$B_2(g)$	X	280

One mole each of $A_2(g)$ and $B_2(g)$ are taken in a 1 L closed flask and allowed to establish the equilibrium at 500K.

$A_2(g) + B_2(g) \rightleftharpoons 2AB(g)$

The value of x (in kJ mol^{-1}) is __________ (Nearest integer)

(**Given:** log K=2.2, R = 8.3 JK^{-1} mol^{-1})

200. The pH of a 0.01 M weak acid HX ($K_a = 4 \times 10^{-10}$) is found to be 5. Now the acid solution is diluted with excess of water so that the pH of the solution changes to 6. The new concentration of the diluted weak acid is given as $x \times 10^{-4}$ M. The value of x is ________(nearest integer) **[April 4, 2025 (I)]**

201. If 1 mM solution of ethylamine produces pH = 9, then the ionization constant (K_b) of ethylamine is 10^{-x}. The value of x is ______ (nearest integer).

[The degree of ionization of ethylamine can be neglected with respect to unity.] **[Jan. 23, 2025 (I)]**

202. K_a for CH_3COOH is 1.8×10^{-5} and K_b for NH_4OH is 1.8×10^{-5}. The pH of ammonium acetate solution will be _____. **[Feb. 1, 2024 (I)]**

203. K_a for butyric acid (C_3H_7COOH) is 2×10^{-5}. The pH of 0.2 M solution of butyric acid is ___ $\times 10^{-1}$. (Nearest integer) [Given log 2 = 0.30] **[July 28, 2022 (I)]**

204. 0.01 moles of a weak acid HA ($K_a = 2.0 \times 10^{-6}$) is dissolved in 1.0 L of 0.1 M HCl solution. The degree of dissociation of HA is ______ $\times 10^{-5}$ (Round off to the Nearest Integer).

[Neglect volume change on adding HA.

Assume degree of dissociation <<1] **[March 17, 2021 (I)]**

205. Sulphurous acid (H_2SO_3) has $k_{a1} = 1.7 \times 10^{-2}$ and $k_{a2} = 6.4 \times 10^{-8}$. The pH of 0.588 M H_2SO_3 is ________. (Round off to the Nearest Integer). [March 16, 2021 (II)]

206. The pH of ammonium phosphate solution, if pK_a of phosphoric acid and pK_b of ammonium hydroxide are 5.23 and 4.75 respectively, is ________. [Feb. 26, 2021 (II)]

Topic 5 - Common Ion Effect, Salt Hydrolysis, Buffer Solutions and Solubility Product

MULTIPLE CHOICE QUESTIONS (MCQs)

207. M_3A_2 is a sparingly soluble salt of molar mass y g mol^{-1} and solubility x g L^{-1}. The ratio of the molar concentration of the anion (A^{3-}) to the solubility product of the salt is [April 05, 2026 (I)]

(1) $\frac{1}{54} \cdot \frac{y^4}{x^4}$ (2) $\frac{y^5}{108x^4}$

(3) $108 \cdot \frac{x^5}{y^5}$ (4) $\frac{1}{108} \frac{y^4}{x^4}$

208. 20 mL of a solution of acetic acid required 28.4 mL of 0.1 M NaOH for its neutralization. A solution (X) was prepared by mixing 20 mL of the above acetic acid and 14.2 mL of 0.1 M NaOH solution. What is the pH of the solution (X) ? (pK_a value of acetic acid is 4.75). [April 04, 2026 (II)]

(1) 7.0 (2) 4.75 (3) 3.5 (4) 4.82

209. At 25°C, 20.0 mL of 0.2 M weak monoprotic acid HX is titrated against 0.2 M NaOH. The pH of the solution (a) at the start of the titration (when NaOH has not been added) and (b) when 10 mL of NaOH is added respectively, are :

Given : $K_a = 5 \times 10^{-4}$, $pK_a = 3.3$, $\alpha << 1$ [April 02, 2026 (II)]

(1) 0.7, 2.0 (2) 2.0, 3.3

(3) 1.1, 2.2 (4) 3.0, 2.2

210. The solubility product constants of Ag_2CrO_4 and AgBr are 32x and 4y respectively at 298 K. The value of $\left(\frac{\text{molarity of } Ag_2CrO_4}{\text{molarity of AgBr}}\right)$ can be expressed as : [April 02, 2026 (I)]

(1) $\frac{2\sqrt[3]{x}}{y}$ (2) $2\sqrt{\frac{x}{y}}$

(3) $\sqrt{\frac{x}{y}}$ (4) $\frac{\sqrt[3]{x}}{y}$

211. Consider a weak base 'B' of $pK_b = 5.699$. 'x' mL of 0.02 M HCl and 'y' mL of 0.02 M weak base 'B' are mixed to make 100 mL of a buffer of pH 9 at 25°C. The values of 'x' and 'y' respectively are: [Jan. 28, 2026 (I)]

[Given: log 2 = 0.3010, log 3 = 0.4771, log 5 = 0.699]

(1)

x	y
11.1	88.9

(2)

x	y
42.7	57.3

(3)

x	y
14.3	85.7

(4)

x	y
85.7	14.3

212. Which of the following mixture gives a buffer solution with pH = 9.25 ? [Jan. 22, 2026 (II)]

Given : pK_b (NH_4OH) = 4.75

(1) 0.2M NH_4OH (0.4 L) + 0.1M HCl (1L)
(2) 0.2M NH_4OH (0.5 L) + 0.1M HCl (0.5 L)
(3) 0.5M NH_4OH (0.2 L) + 0.2M HCl (0.5 L)
(4) 0.4M NH_4OH (1 L) + 0.1M HCl (1L)

213. If equal volumes of AB_2 and XY (both are salts) aqueous solutions are mixed, which of the following combination will give a precipitate of AY_2 at 300 K?

(Given K_{sp} (at 300 K) for $AY_2 = 5.2 \times 10^{-7}$) [April 2, 2025 (I)]

(1) 3.6×10^{-3} M AB_2, 5.0×10^{-4} M XY
(2) 2.0×10^{-4} M AB_2, 0.8×10^{-3} M XY
(3) 2.0×10^{-2} M AB_2, 2.0×10^{-2} M XY
(4) 1.5×10^{-4} M AB_2, 1.5×10^{-3} M XY

214. Arrange the following in increasing order of solubility product : [Jan. 28, 2025 (II)]

$Ca(OH)_2$, AgBr, PbS, HgS

(1) PbS < HgS < $Ca(OH)_2$ < AgBr
(2) HgS < PbS < AgBr < $Ca(OH)_2$
(3) $Ca(OH)_2$ < AgBr < HgS < PbS
(4) HgS < AgBr < PbS < $Ca(OH)_2$

215. Ksp for $Cr(OH)_3$ is 1.6×10^{-30}. What is the molar solubility of this salt in water? [Jan. 24, 2025 (I)]

(1) $\sqrt[4]{\frac{1.6 \times 10^{-30}}{27}}$ (2) $\frac{1.8 \times 10^{-30}}{27}$

(3) $\sqrt[5]{1.8 \times 10^{-30}}$ (4) $\sqrt[2]{1.6 \times 10^{-30}}$

216. Which of the following happens when NH_4OH is added gradually to the solution containing 1M A^{2+} and 1M B^{3+} ions ? [Jan. 23, 2025 (I)]

Given: $K_{sp}[A(OH)_2] = 9 \times 10^{-10}$ and $K_{sp}[B(OH)_3] = 27 \times 10^{-18}$ at 298 K.

(1) $B(OH)_3$ will precipitate before $A(OH)_2$
(2) $A(OH)_2$ and $B(OH)_3$ will precipitate together
(3) $A(OH)_2$ will precipitate before $B(OH)_3$
(4) Both $A(OH)_2$ and $B(OH)_3$ do not show precipitation with NH_4OH

217. The molar solubility(s) of zirconium phosphate with molecular formula $(Zr^{4+})_3 \left(PO_4^{3-}\right)_4$ is given by relation: [Jan. 22, 2025 (II)]

(1) $\left(\frac{K_{sp}}{6912}\right)^{\frac{1}{7}}$ (2) $\left(\frac{K_{sp}}{5348}\right)^{\frac{1}{6}}$

(3) $\left(\frac{K_{sp}}{8435}\right)^{\frac{1}{7}}$ (4) $\left(\frac{K_{sp}}{9612}\right)^{\frac{1}{3}}$

218. For a sparingly soluble salt AB_2, the equilibrium concentrations of A^{2+} ions and B^- ions are 1.2×10^{-4} M and 0.24×10^{-3} M, respectively. The solubility product of AB_2 is : [April 9, 2024 (II)]

(1) 0.069×10^{-12} (2) 6.91×10^{-12}

(3) 0.276×10^{-12} (4) 27.65×10^{-12}

219. Given below are two statements: [April 8, 2024 (II)]

Statement (I): A Buffer solution is the mixture of a salt and an acid or a base mixed in any particular quantities.

Statement (II): Blood is naturally occurring buffer solution whose pH is maintained by $H_2CO_3/HCO_3^{\ominus}$ concentrations.

In the light of the above statements, choose the **correct** answer from the options given below:
(1) **Statement I** is false but **Statement II** is true
(2) Both **Statement I** and **Statement II** is true
(3) Both **Statement I** and **Statement II** is false
(4) **Statement I** is true but **Statement II** is false

220. Given below are two statements: **[April 5, 2024 (II)]**
Statement I : On passing HCl(g) through a saturated solution of $BaCl_2$, at room temperature white turbidity appears.
Statement II : When HCl gas is passed through a saturated solution of NaCl, sodium chloride is precipitated due to common ion effect.
In the light of the above statements, choose the **most appropriate** answer from the options given below:
(1) **Statement I** is correct but **Statement II** is incorrect
(2) Both **Statement I** and **Statement II** are incorrect
(3) **Statement I** is incorrect but **Statement II** is correct
(4) Both **Statement I** and **Statement II** are correct

221. Solubility of calcium phosphate (molecular mass, M) in water is Wg per 100 mL at 25° C. Its solubility product at 25°C will be approximately. **[Feb. 1, 2024 (II)]**
(1) $10^7\left(\frac{W}{M}\right)^3$ (2) $10^3\left(\frac{W}{M}\right)^5$
(3) $10^7\left(\frac{W}{M}\right)^5$ (4) $10^5\left(\frac{W}{M}\right)^5$

222. Class XII students were asked to prepare one litre of buffer solution of pH 8.26 by their chemistry teacher. The amount of ammonium chloride to be dissolved by the student in 0.2 M ammonia solution to make one litre of the buffer is (Given pK_b (NH_3) = 4.74; Molar mass of NH_3 = 17 g mol^{-1}; Molar mass of NH_4Cl = 53.5 g mol^{-1}) **[July 26, 2022 (II)]**
(1) 53.5 g (2) 72.3 g (3) 107.0 g (4) 126.0 g

223. The solubility of AgCl will be maximum in which of the following ? **[June 29, 2022 (I)]**
(1) 0.01 M KCl (2) 0.01 M HCl
(3) 0.01 M $AgNO_3$ (4) Deionised water

224. A student needs to prepare a buffer solution of propanoic acid and its sodium salt with pH 4. The ratio of $\frac{[CH_3CH_2COO^-]}{[CH_3CH_2COOH]}$ required to make buffer is ....
Given : $K_a(CH_3CH_2COOH) = 1.3 \times 10^{-5}$ **[June 28, 2022 (II)]**
(1) 0.03 (2) 0.13 (3) 0.23 (4) 0.33

225. The K_{sp} for bismuth sulphide (Bi_2S_3) is 1.08×10^{-73}. The solubility of Bi_2S_3 in mol L^{-1} at 298 K is **[June 25, 2022 (II)]**
(1) 1.0×10^{-15} (2) 2.7×10^{-12}
(3) 3.2×10^{-10} (4) 4.2×10^{-8}

226. The solubility of $Ca(OH)_2$ in water is :
[Given : The solubility product of $Ca(OH)_2$ in water = 5.5×10^{-6}] **[Feb. 25, 2021 (II)]**
(1) 1.77×10^{-2} (2) 1.11×10^{-2}
(3) 1.77×10^{-6} (4) 1.11×10^{-6}

227. The solubility of AgCN in a buffer solution of pH = 3 is *x*. The value of *x* is: **[Feb. 25, 2021 (I)]**
[Assume: No cyano complex is formed;
$K_{sp}(AgCN) = 2.2 \times 10^{-16}$ and $K_a(HCN) = 6.2 \times 10^{-10}$]
(1) 1.9×10^{-5} (2) 1.6×10^{-6}
(3) 2.2×10^{-16} (4) 0.625×10^{-6}

228. An acidic buffer is obtained on mixing: **[Sep. 03, 2020 (I)]**
(1) 100 mL of 0.1 M CH_3COOH and 100 mL of 0.1 M NaOH
(2) 100 mL of 0.1 M HCl and 200 mL of 0.1 M NaCl
(3) 100 mL of 0.1 M CH_3COOH and 200 mL of 0.1 M NaOH
(4) 100 mL of 0.1 M HCl and 200 mL of 0.1 M CH_3COONa

229. For the following **Assertion** and **Reason**, the correct option is **[Sep. 02, 2020 (I)]**
Assertion (A) : When Cu (II) and sulphide ions are mixed, they react together extremely quickly to give a solid.
Reason (R) : The equilibrium constant of $Cu^{2+}(aq) + S^{2-}(aq) \rightleftharpoons CuS(s)$ is high because the solubility product is low.
(1) **(A)** is false and **(R)** is true
(2) Both **(A)** and **(R)** are false
(3) Both **(A)** and **(R)** are true but **(R)** is not the explanation for **(A)**
(4) Both **(A)** and **(R)** are true and **(R)** is the explanation for **(A)**

230. The K_{sp} for the following dissociation is 1.6×10^{-5}

$$PbCl_2(s) \rightleftharpoons Pb^{2+}(aq) + 2Cl^-(aq)$$

Which of the following choices is correct for a mixture of 300 mL 0.134 M $Pb(NO_3)_2$ and 100 mL 0.4 M NaCl? **[Jan. 09, 2020 (I)]**
(1) Not enough data provided
(2) $Q < K_{sp}$
(3) $Q > K_{sp}$
(4) $Q = K_{sp}$

231. The solubility product of $Cr(OH)_3$ at 298 K is 6.0×10^{-31}. The concentration of hydroxide ions in a saturated solution of $Cr(OH)_3$ will be: **[Jan. 09, 2020 (II)]**
(1) $(2.22 \times 10^{-31})^{1/4}$ (2) $(18 \times 10^{-31})^{1/4}$
(3) $(18 \times 10^{-31})^{1/2}$ (4) $(4.86 \times 10^{-29})^{1/4}$

232. The stoichiometry and solubility product of a salt with the solubility curve given below is, respectively: **[Jan. 08, 2020 (I)]**

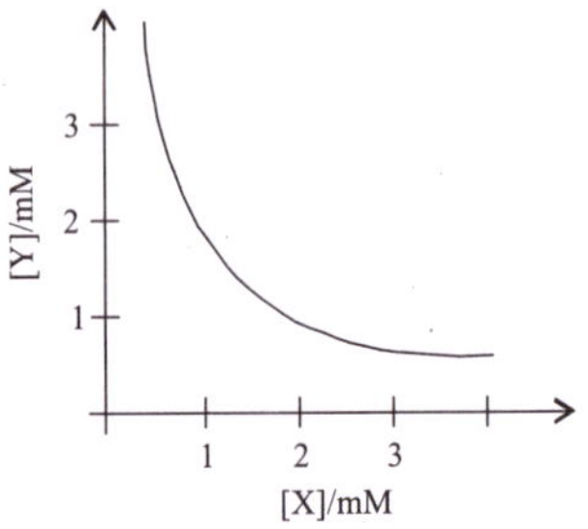

(1) $X_2Y, 2 \times 10^{-9} M^3$ (2) $XY_2, 4 \times 10^{-9} M^3$
(3) $XY_2, 1 \times 10^{-9} M^3$ (4) $XY, 2 \times 10^{-6} M^3$

233. The molar solubility of $Cd(OH)_2$ is 1.84×10^{-5} M in water. The expected solubility of $Cd(OH)_2$ in a buffer solution of pH = 12 is : **[April 12, 2019 (II)]**
(1) 1.84×10^{-9} M (2) $\frac{2.49}{1.84} \times 10^{-9}$ M
(3) 6.23×10^{-11} M (4) 2.49×10^{-10} M

234. What is the molar solubility of $Al(OH)_3$ in 0.2 M NaOH solution ? Given that, solubility product of $Al(OH)_3 = 2.4 \times 10^{-24}$: **[April 12, 2019 (I)]**
(1) 3×10^{-19} (2) 12×10^{-21}
(3) 3×10^{-22} (4) 12×10^{-23}

235. The pH of a 0.02 M NH_4Cl solution will be [given $K_b(NH_4OH) = 10^{-5}$ and log 2 = 0.301] **[April 10, 2019 (II)]**
(1) 2.65 (2) 4.35
(3) 4.65 (4) 5.35

236. If solubility product of $Zr_3(PO_4)_4$ is denoted by K_{sp} and its molar solubility is denoted by S, then which of the following relation between S and K_{sp} is correct ? [April 8, 2019 (I)]

(1) $S=\left(\frac{K_{sp}}{144}\right)^{1/6}$ (2) $S=\left(\frac{K_{sp}}{6912}\right)^{1/7}$

(3) $S=\left(\frac{K_{sp}}{929}\right)^{1/9}$ (4) $S=\left(\frac{K_{sp}}{216}\right)^{1/7}$

237. 20 mL of 0.1 M H_2SO_4 solution is added to 30 mL of 0.2 M NH_4OH solution. The pH of the resultant mixture is: [pK_b of $NH_4OH = 4.7$]. [Jan. 9, 2019 (I)]

(1) 5.2 (2) 9.0
(3) 5.0 (4) 9.4

238. If K_{sp} of Ag_2CO_3 is 8×10^{-12}, the molar solubility of Ag_2CO_3 in 0.1 M $AgNO_3$ is : [Jan. 12, 2019 (II)]

(1) 8×10^{-12} M (2) 8×10^{-11} M
(3) 8×10^{-10} M (4) 8×10^{-13} M

239. A mixture of 100 m mol of $Ca(OH)_2$ and 2 g of sodium sulphate was dissolved in water and the volume was made up to 100 mL. The mass of calcium sulphate formed and the concentration of OH^- in resulting solution, respectively, are : (Molar mass of $Ca(OH)_2$, Na_2SO_4 and $CaSO_4$ are 74, 143 and 136 g mol^{-1}, respectively; K_{sp} of $Ca(OH)_2$ is 5.5×10^{-6}) [Jan. 10, 2019 (I)]

(1) 1.9 g, 0.28 mol L^{-1} (2) 13.6 g, 0.28 mol L^{-1}
(3) 1.9 g, 0.14 mol L^{-1} (4) 13.6 g, 0.14 mol L^{-1}

240. Which of the following salts is the most basic in aqueous solution? [2018]

(1) $Al(CN)_3$ (2) CH_3COOK
(3) $FeCl_3$ (4) $Pb(CH_3COO)_2$

241. An aqueous solution contains 0.10 M H_2S and 0.20 M HCl. If the equilibrium constants for the formation of HS^- from H_2S is 1.0×10^{-7} and that of S^{2-} from HS^- ions is 1.2×10^{-13} then the concentration of S^{2-} ions in aqueous solution is : [2018]

(1) 5×10^{-8} (2) 3×10^{-20}
(3) 6×10^{-21} (4) 5×10^{-19}

242. An aqueous solution contains an unknown concentration of Ba^{2+}. When 50 mL of a 1 M solution of Na_2SO_4 is added, $BaSO_4$ just begins to precipitate. The final volume is 500 mL. The solubility product of $BaSO_4$ is 1×10^{-10}. What is the original concentration of Ba^{2+}? [2018]

(1) 5×10^{-9} M (2) 2×10^{-9} M
(3) 1.1×10^{-9} M (4) 1.0×10^{-10} M

243. The minimum volume of water required to dissolve 0.1g lead (II) chloride to get a saturated solution (K_{sp} of $PbCl_2 = 3.2 \times 10^{-8}$; atomic mass of Pb = 207 u) is : [Online April 15, 2018 (I)]

(1) 1.798 L (2) 0.36 L
(3) 17.95 L (4) 0.18 L

244. Addition of sodium hydroxide solution to a weak acid (HA) results in a buffer of pH 6. If ionisation constant of HA is 10^{-5}, the ratio of salt to acid concentration in the buffer solution will be : [Online April 8, 2017]

(1) 4 : 5 (2) 1 : 10 (3) 10 : 1 (4) 5 : 4

245. In some solutions, the concentration of H_3O^+ remains constant even when small amounts of strong acid or strong base are added to them. These solutions are known as: [Online April 11, 2014]

(1) Ideal solutions (2) Colloidal solutions
(3) True solutions (4) Buffer solutions

246. Consider the following equilibrium

$$AgCl\downarrow + 2NH_3 \rightleftharpoons \left[Ag(NH_3)_2\right]^+ + Cl^-$$

White precipitate of AgCl appears on adding which of the following? [Online April 11, 2014]

(1) NH_3 (2) aqueous NaCl
(3) aqueous HNO_3 (4) aqueous NH_4Cl

247. Zirconium phosphate $[Zr_3(PO_4)_4]$ dissociates into three zirconium cations of charge + 4 and four phosphate anions of charge – 3. If molar solubility of zirconium phosphate is denoted by S and its solubility product by K_{sp} then which of the following relationship between S and K_{sp} is correct? [Online April 19, 2014]

(1) $S = \{K_{sp}/(6912)^{1/7}\}$ (2) $S = \{K_{sp}/144\}^{1/7}$
(3) $S = \{K_{sp}/6912\}^{1/7}$ (4) $S = \{K_{sp}/6912\}^7$

248. Solid $Ba(NO_3)_2$ is gradually dissolved in a 1.0×10^{-4} M Na_2CO_3 solution. At which concentration of Ba^{2+}, precipitate of $BaCO_3$ begins to form ? (K_{sp} for $BaCO_3 = 5.1 \times 10^{-9}$) [2009, Online April 9, 2013]

(1) 5.1×10^{-5} M (2) 7.1×10^{-8} M
(3) 4.1×10^{-5} M (4) 8.1×10^{-7} M

249. Which one of the following arrangements represents the correct order of solubilities of sparingly soluble salts Hg_2Cl_2, $Cr_2(SO_4)_3$, $BaSO_4$ and $CrCl_3$ respectively ? [Online April 22, 2013]

(1) $BaSO_4 > Hg_2Cl_2 > Cr_2(SO_4)_3 > CrCl_3$
(2) $BaSO_4 > Hg_2Cl_2 > CrCl_3 > Cr_2(SO_4)_3$
(3) $BaSO_4 > CrCl_3 > Hg_2Cl_2 > Cr_2(SO_4)_3$
(4) $Hg_2Cl_2 > BaSO_4 > CrCl_3 > Cr_2(SO_4)_3$

250. If K_{sp} of CaF_2 at 25 °C is 1.7×10^{-10}, the combination amongst the following which gives a precipitate of CaF_2 is [Online May 19, 2012]

(1) 1×10^{-2} M Ca^{2+} and 1×10^{-3} M F^-
(2) 1×10^{-4} M Ca^{2+} and 1×10^{-4} M F^-
(3) 1×10^{-2} M Ca^{2+} and 1×10^{-5} M F^-
(4) 1×10^{-3} M Ca^{2+} and 1×10^{-5} M F^-

251. The solubility of PbI_2 at 25 °C is 0.7 g L^{-1}. The solubility product of PbI_2 at this temperature is (molar mass of $PbI_2 = 461.2$ g mol^{-1}) [2003, Online May 26, 2012]

(1) 1.40×10^{-9} (2) 0.14×10^{-9}
(3) 140×10^{-9} (4) 14.0×10^{-9}

252. The solubility (in mol L^{-1}) of AgCl ($K_{sp} = 1.0 \times 10^{-10}$) in a 0.1 M KCl solution will be [Online May 7, 2012]

(1) 1.0×10^{-9} (2) 1.0×10^{-10}
(3) 1.0×10^{-5} (4) 1.0×10^{-11}

253. The K_{sp} for $Cr(OH)_3$ is 1.6×10^{-30}. The solubility of this compound in water is : [2011RS]

(1) $\sqrt[4]{1.6 \times 10^{-30}}$ (2) $\sqrt[4]{1.6 \times 10^{-30}/27}$
(3) $1.6 \times 10^{-30/27}$ (4) $\sqrt[2]{1.6 \times 10^{-30}}$

254. Solubility product of silver bromide is 5.0×10^{-13}. The quantity of potassium bromide (molar mass taken as 120 g mol^{-1}) to be added to 1 litre of 0.05 M solution of silver nitrate to start the precipitation of AgBr is [2010]

(1) 1.2×10^{-10} g (2) 1.2×10^{-9} g
(3) 6.2×10^{-5} g (4) 5.0×10^{-8} g

255. At 25 °C, the solubility product of $Mg(OH)_2$ is 1.0×10^{-11}. At which pH, will Mg^{2+} ions start precipitating in the form of $Mg(OH)_2$ from a solution of 0.001 M Mg^{2+} ions? [2010]

(1) 9 (2) 10 (3) 11 (4) 8

256. Solid $Ba(NO_3)_2$ is gradually dissolved in a 1.0×10^{-4} M Na_2CO_3 solution. At what concentration of Ba^{2+} will a precipitate begin to form? (K_{sp} for for $BaCO_3 = 5.1 \times 10^{-9}$) **[2009]**
(1) 5.1×10^{-5} M (2) 8.1×10^{-8} M
(3) 8.1×10^{-7} M (4) 4.1×10^{-5} M

257. In a saturated solution of the sparingly soluble strong electrolyte $AgIO_3$ (molecular mass = 283) the equilibrium which sets in is $AgIO_3(s) \rightleftharpoons Ag^+(aq) + IO_3^-(aq)$. If the solubility product constant K_{sp} of $AgIO_3$ at a given temperature is 1.0×10^{-8}, what is the mass of $AgIO_3$ contained in 100 mL of its saturated solution? **[2007]**
(1) 1.0×10^{-4} g (2) 28.3×10^{-2} g
(3) 2.83×10^{-3} g (4) 1.0×10^{-7} g.

258. The pK_a of a weak acid (HA) is 4.5. The pOH of an aqueous buffer solution of HA in which 50% of the acid is ionized is **[2007]**
(1) 7.0 (2) 4.5 (3) 2.5 (4) 9.5

259. The solubility product of a salt having general formula MX_2, in water, is : 4×10^{-12}. The concentration of M^{2+} ions in the aqueous solution of the salt is **[2005]**
(1) 4.0×10^{-10} M (2) 1.6×10^{-4} M
(3) 1.0×10^{-4} M (4) 2.0×10^{-6} M

260. The molar solubility (in mol L^{-1}) of a sparingly soluble salt MX_4 is 's'. The corresponding solubility product is K_{sp}. 's' is given in term of K_{sp} by the relation : **[2004]**
(1) $s = (256 K_{sp})^{1/5}$ (2) $s = (128 K_{sp})^{1/4}$
(3) $s = (K_{sp}/128)^{1/4}$ (4) $s = (K_{sp}/256)^{1/5}$

261. The solubility in water of a sparingly soluble salt AB_2 is 1.0×10^{-5} mol L^{-1}. Its solubility product will be **[2003]**
(1) 4×10^{-10} (2) 1×10^{-15}
(3) 1×10^{-10} (4) 4×10^{-15}

262. 1 M NaCl and 1 M HCl are present in an aqueous solution. The solution is **[2002]**
(1) not a buffer solution with pH < 7
(2) not a buffer solution with pH > 7
(3) a buffer solution with pH < 7
(4) a buffer solution with pH > 7.

263. Let the solubility of an aqueous solution of $Mg(OH)_2$ be x, then its K_{sp} is **[2002]**
(1) $4x^3$ (2) $108x^5$ (3) $27x^4$ (4) $9x$.

NUMERICAL VALUE QUESTIONS (NVQs)

264. The pH a solution obtained by mixing 5 mL of 0.1 M NH_4OH solution with 250 mL of 0.1 M NH_4Cl solution is _____ $\times 10^{-2}$. (Nearest integer)
Given : pK_b (NH_4OH) = 4.74
log 2 = 0.30
log 3 = 0.48
log 5 = 0.70 **[April 04, 2026 (I)]**

265. Consider two Group IV metal ions X^{2+} and Y^{2+}.
A solution containing 0.01 MX^{2+} and 0.01 MY^{2+} is saturated with H_2S. The pH at which the metal sulphide YS will form as a precipitate is ______ (Nearest integer)
(Given :
K_{sp} (Xs) = 1×10^{-22} at 25°C, K_{sp} (YS) = 4×10^{-16} at 25°C, $[H_2S] = 0.1$ M in solution, $K_{a1} \times K_{a2}$ (H_2S) = 1.0×10^{-21}, log 2 = 0.30, log 3 = 0.48, log 5 = 0.70) **[Jan. 24, 2026 (I)]**

266. One litre buffer solution was prepared by adding 0.10 mol each of NH_3 and NH_4Cl in deionised water. The change in pH on addition of 0.05 mol of HCl to the above solution is ____ $\times 10^{-2}$, (Nearest integer)
(**Given :** pK_b of NH_3 = 4.745 and $\log_{10}3 = 0.477$) **[April 7, 2025 (II)]**

267. The pH at which $Mg(OH)_2$ [$K_{sp} = 1 \times 10^{-11}$] begins to precipitate from a solution containing 0.10 M Mg^{2+} ions is ___________ . **[Jan. 30, 2024 (I)]**

268. The pH of an aqueous solution containing 1M benzoic acid (pK_a = 4.20) and 1M sodium benzoate is 4.5. The volume of benzoic acid solution in 300 mL of this buffer solution is ________ mL. (given : log 2 = 0.3) **[Jan. 30, 2024 (II)]**

269. 20 mL of 0.1 M NaOH is added to 50 mL of 0.1 M acetic acid solution. The pH of the resulting solution is _________ $\times 10^{-2}$ (Nearest integer)
Given : pKa (CH_3COOH) = 4.76
log 2 = 0.30
log 3 = 0.48 **[April 13, 2023 (II)]**

270. 25.0 mL of 0.050 M ($Ba(NO_3)_2$) is mixed with 25.0 mL of 0.020 M NaF. K_{sp} of BaF_2 is 0.5×10^{-6} at 298 K. The ratio of $[Ba^{2+}][F^-]^2$ and K_{sp} is ________ (Nearest integer) **[April 13, 2023 (I)]**

271. The solubility product of $BaSO_4$ is 1×10^{-10} at 298K. The solubility of $BaSO_4$ in 0.1 M K_2SO_4(aq) solution is ______ $\times 10^{-9}$ g L^{-1} (nearest integer).
Given : Molar mass of $BaSO_4$ is 233 g mol^{-1} **[April 08, 2023 (II)]**

272. At 298 K, the solubility of silver chloride in water is 1.434×10^{-3} g L^{-1}. The value of $-\log K_{sp}$ for silver chloride is ______.
(Given mass of Ag is 107.9 g mol^{-1} and mass of Cl is 35.5 g mol^{-1}) **[Jan. 31, 2023 (II)]**

273. The dissociation constant of acetic is $x \times 10^{-5}$.
When 25 mL of 0.2 M CH_3COONa solution is mixed with 25 mL of 0.02 M CH_3COOH solution, the pH of the resultant solution is found to be equal to 5. The value of x is ______. **[Jan. 24, 2023 (I)]**

274. If the pKa of lactic acid is 5, then the pH of 0.005 M calcium lactate solution at 25° C is __________ $\times 10^{-1}$ (Nearest integer) **[Jan. 24, 2023 (II)]**

Lactic acid 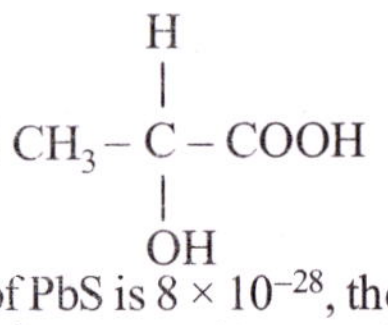

275. If the solubility product of PbS is 8×10^{-28}, then the solubility of PbS in pure water at 298 K is $x \times 10^{-16}$ mol L^{-1}. The value of x is ________. (Nearest integer)
[Given $\sqrt{2} = 1.41$] **[July 29, 2022 (I); Same Feb. 25, 2021 (II)]**

276. At 310 K, the solubility of CaF_2 in water is 2.34×10^{-3}g/100 mL. The solubility product of CaF_2 is _____ $\times 10^{-8}$ $(mol/L)^3$. (Given molar mass : CaF_2 = 78 g mol^{-1}) **[July 27, 2022 (I)]**

277. 50 mL of 0.1 M CH_3COOH is being titrated against 0.1 M NaOH. When 25 mL of NaOH has been added, the pH of the solution will be ________ $\times 10^{-2}$. (Nearest integer)

(Given : pK_a (CH_3COOH) = 4.76)
log 2 = 0.30 log 3 = 0.48
log 5 = 0.69 log 7 = 0.84 log 11 = 1.04
[June 26, 2022 (I)]

278. The molar solubility of $Zn(OH)_2$ in 0.1 M NaOH solution is $x \times 10^{-18}$ M. The value of x is _____(Nearest integer)
(Given : The solubility product of $Zn(OH)_2$ is 2×10^{-20})
[Sep. 1, 2021 (II)]

279. A_3B_2 is a sparingly soluble salt of molar mass M (g mol^{-1}) and solubility x g L^{-1}. The solubility product satisfies
$K_{sp} = a\left(\frac{x}{M}\right)^5$. The value of a is ________.
(Integer answer) [Aug. 31, 2021 (I)]

280. The OH^- concentration in a mixture of 5.0 mL of 0.0504 M NH_4Cl and 2 mL of 0.0210 M NH_3 solution is $x \times 10^{-6}$ M. The value of x is _______. (Nearest integer)
[Given $K_w = 1 \times 10^{-14}$ and $K_b = 1.8 \times 10^{-5}$] [Aug. 26, 2021 (I)]

281. In order to prepare a buffer solution of pH 5.74, sodium acetate is added to acetic acid. If the concentration of acetic acid in the buffer is 1.0 M, the concentration of sodium acetate in the buffer is _______ M.
(Rounded off to the Nearest Integer).
[Given : pK_a (acetic acid) = 4.74] [March 18, 2021 (I)]

282. The solubility of $CdSO_4$ in water is 8.0×10^{-4} mol L^{-1}. Its solubility in 0.01 M H_2SO_4 solution is ________ $\times 10^{-6}$ mol L^{-1}. (Round off to the nearest integer).
(Assume that solubility is much less than 0.01 M)
[March 18, 2021 (II)]

283. Two salts A_2X and MX have the same value of solubility product of 4.0×10^{-12}. The ratio of their molar solubilities,
i.e., $\frac{s(A_2X)}{s(MX)}$ = _____. (Round off to the Nearest Integer)
[March 16, 2021 (I)]

284. The solubility product of PbI_2 is 8.0×10^{-9}. The solubility of lead iodide in 0.1 molar solution of lead nitrate is $x \times 10^{-6}$ mol/L. The value of x is ________. (Rounded off to the nearest integer) [Given : $\sqrt{2} = 1.41$]
[Feb. 24, 2021 (II)]

285. If the solubility product of AB_2 is $3.20 \times 10^{-11} M^3$, then the solubility of AB_2 in pure water is ______ $\times 10^{-4}$ mol L^{-1}. [Assuming that neither kind of ion reacts with water]
[Sep. 06, 2020 (II)]

Part-B (JEE Advanced)

Topic 1- Law of Chemical Equilibrium, Equilibrium Constant (K_c and K_p) and its Application)

1 MCQs with One Correct Answer

1. For a reversible reaction $R \rightleftharpoons P$, at constant temperature, both the forward and the backward reactions are first order elementary reactions with rate constants k_f and k_b, respectively. At time zero, the concentration of R is $[R]_0$ and the concentration of P is zero. At any given time, [R] and [P] are the concentrations of R and P, respectively. If $k_b = 4k_f$, the correct graphical representation of the reaction is [Adv. 2026]

(a)
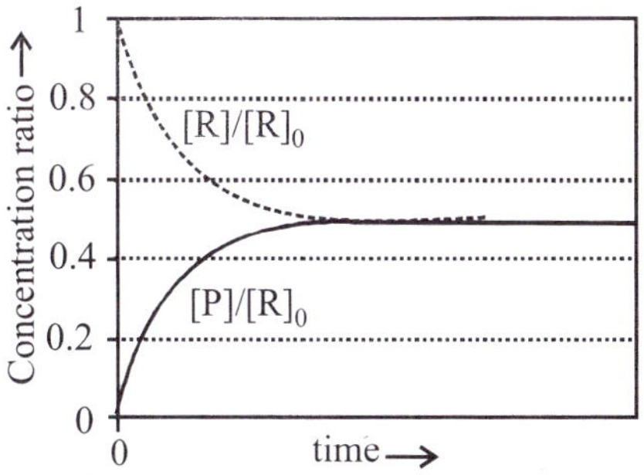

(b)
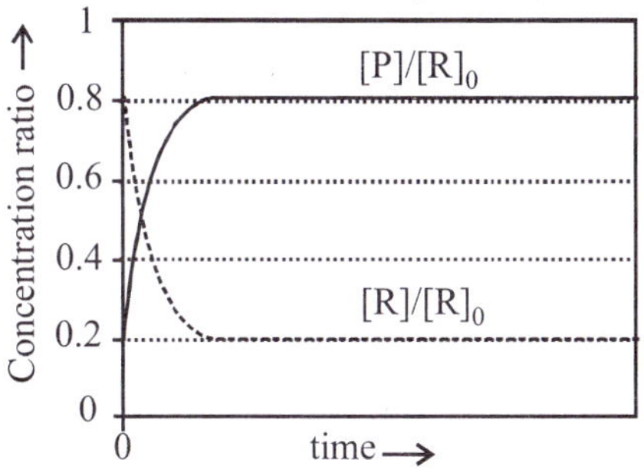

(c)
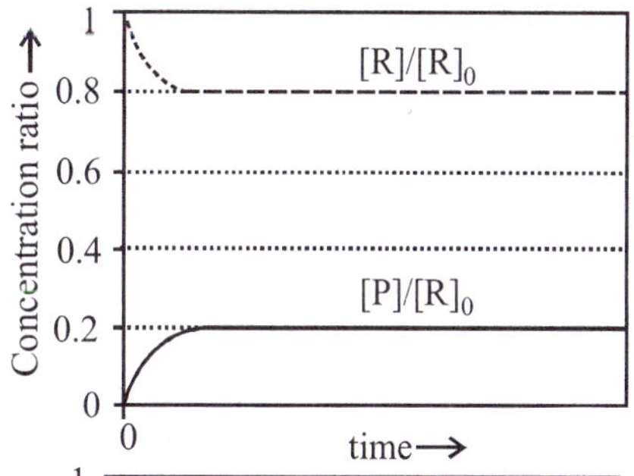

(d)
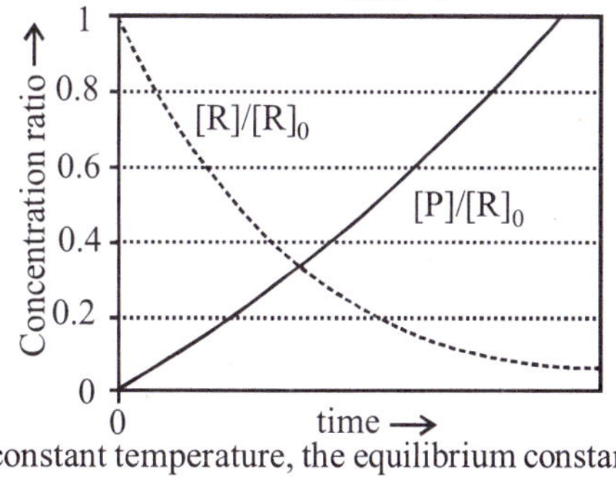

2. At constant temperature, the equilibrium constant (K_p) for the decomposition reaction $N_2O_4 \rightleftharpoons 2NO_2$ is expressed by $K_p = (4x^2P)/(1-x^2)$, where P = pressure, x = extent of decomposition. Which one of the following statements is true? [2001S]
(a) K_p increases with increase of P
(b) K_p increases with increase of x
(c) K_p increases with decrease of x
(d) K_p remains constant with change in P and x

3. When two reactants, A & B are mixed to give products C & D, the reaction quotient Q, at the initial stages of the reaction [2000S]
(a) is zero (b) decreases with time
(c) is independent of time (d) increases with time

4. For the reversible reaction, $N_2(g) + 3H_2(g) \rightleftharpoons 2NH_3(g)$ at 500 °C, the value of K_p is 1.44×10^{-5} when partial pressure is measured in atmospheres. The corresponding value of K_c, with concentration in mole litre^{-1}, is [2000S]

(a) $\dfrac{1.44\times10^{-5}}{(0.082\times500)^{-2}}$ (b) $\dfrac{1.44\times10^{-5}}{(8.314\times773)^{-2}}$

(c) $\dfrac{1.44\times10^{-5}}{(0.082\times773)^{2}}$ (d) $\dfrac{1.44\times10^{-5}}{(0.082\times773)^{-2}}$

5. An example of a reversible reaction is : [1985 - 1 Mark]
(a) $Pb(NO_3)_2(aq) + 2NaI(aq) \rightarrow PbI_2(s) + 2NaNO_3(aq)$
(b) $AgNO_3(aq) + HCl(aq) \rightarrow AgCl(s) + NaNO_3(aq)$
(c) $2Na(s) + H_2O(l) \rightarrow 2NaOH(aq) + H_2(g)$
(d) $KNO_3(aq) + NaCl(aq) \rightarrow KCl(aq) + NaNO_3(aq)$

6. Pure ammonia is placed in a vessel at a temperature where its dissociation constant (α) is appreciable. At equilibrium: [1984 - 1 Mark]
(a) K_p does not change significantly with pressure.
(b) α does not change with pressure.
(c) concentration of NH_3 does not change with pressure.
(d) concentration of hydrogen is less than that of nitrogen.

7. A liquid is in equilibrium with its vapour at its boiling point. On the average, the molecules in the two phases have equal : [1984 - 1 Mark]
(a) inter-molecular forces (b) potential energy
(c) total energy (d) kinetic energy

8. For the reaction : [1981 - 1 Mark]

$$H_2(g) + I_2(g) \rightleftharpoons 2HI(g)$$

the equilibrium constant K_p changes with
(a) total pressure
(b) catalyst
(c) the amounts of H_2 and I_2 present
(d) temperature

3 Numeric / New Stem Based Questions

9. For the following reaction, the equilibrium constant K_c at 298 K is 1.6×10^{17}. [Adv. 2019]

$$Fe^{2+}(aq) + S^{2-}(aq) \rightleftharpoons FeS(s)$$

When equal volumes of 0.06 M Fe^{2+} (aq) and 0.2 M S^{2-} (aq) solutions are mixed, the equilibrium concentration of Fe^{2+} (aq) is found to be $Y \times 10^{-17}$ M. The value of Y is

10. The equilibrium constant of the reaction
$A_2(g) + B_2(g) \rightleftharpoons 2AB(g)$
at 100 °C is 50. If a one litre flask, containing one mole of A_2 is connected to a two litre flask, containing two mole of B_2, how many mole of AB will be formed at 373 °C? [1985 - 4 Marks]

4 Fill in the Blanks

11. A ten-fold increase in pressure on the reaction, $N_2(g)+3H_2(g) \rightleftharpoons 2NH_3(g)$ at equilibrium results in .... in K_p. [1996 - 1 Mark]

12. For a given reversible reaction at a fixed temperature, equilibrium constants K_p and K_c are related by........ [1994 - 1 Mark]

5 True / False

13. If equilibrium constant for the reaction $A_2 + B_2 \rightleftharpoons 2AB$, is K, then for the backward reaction $AB \rightleftharpoons \frac{1}{2} A_2 + \frac{1}{2} B_2$, the equilibrium constant is $1/K$. [1984 - 1 Mark]

6 MCQs with One or More than One Correct Answer

14. The % yield of ammonia as a function of time in the reaction $N_2(g) + 3H_2(g) \rightleftharpoons 2NH_3(g)$, $\Delta H < 0$ at (P, T_1) is given below

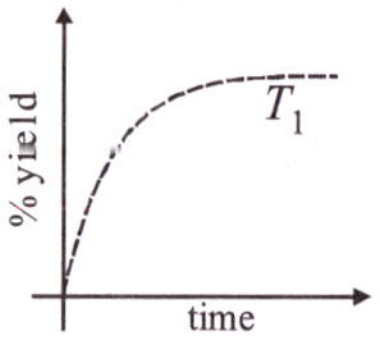

If this reaction is conducted at (P, T_2), with $T_2 > T_1$, the %yield of ammonia as a function of time is represented by [Adv. 2015]

(a)
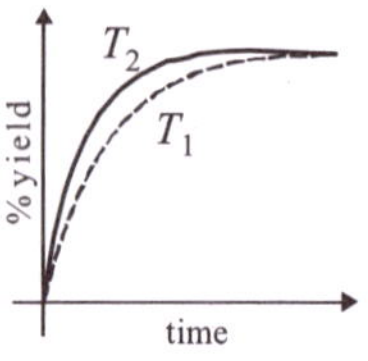

(b)
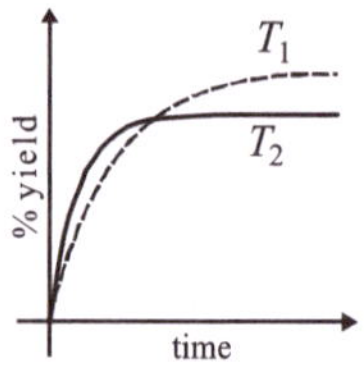

(c)
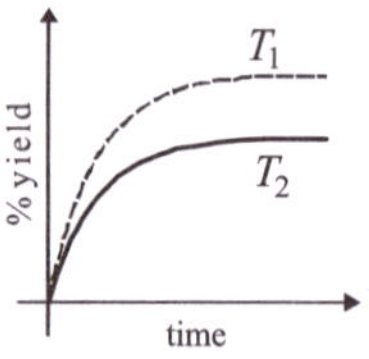

(d)
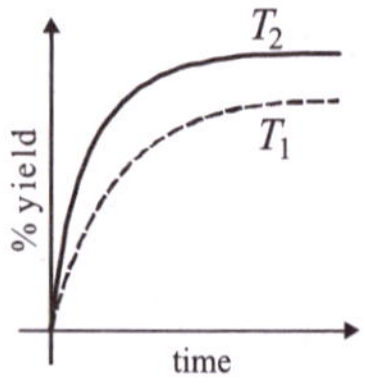

15. The thermal dissociation equilibrium of $CaCO_3(s)$ is studied under different conditions [Adv. 2013]
$CaCO_3(s) \rightleftharpoons CaO(s) + CO_2(g)$. For this equilibrium, the correct statement(s) is (are)
(a) ΔH is dependent on T
(b) K is independent of the initial amount of $CaCO_3$
(c) K is dependent on the pressure of CO_2 at a given T
(d) ΔH is independent of catalyst, if any

8 Comprehension Based Questions

PASSAGE

Thermal decomposition of gaseous X_2 to gaseous X at 298 K takes place according to the following equation:

$$X_2(g) \rightleftharpoons 2X(g)$$

The standard reaction Gibbs energy, $\Delta_r G°$, of this reaction is positive. At the start of the reaction, there is one mole of X_2 and no X. As the reaction proceeds, the number of moles of X formed is given by β. Thus, $\beta_{equilibrium}$ is the number of moles of X formed at equilibrium. The reaction is carried out at a constant total pressure of 2 bar. Consider the gases to behave ideally.
(Given $R = 0.083$ L bar K^{-1} mol^{-1}) [Adv. 2016]

16. The equilibrium constant K_p for this reaction at 298 K, in terms of $\beta_{equilibrium}$, is

(a) $\dfrac{8\beta^2_{equilibrium}}{2-\beta_{equilibrium}}$ (b) $\dfrac{8\beta^2_{equilibrium}}{4-\beta^2_{equilibrium}}$

(c) $\dfrac{4\beta^2_{equilibrium}}{2-\beta_{equilibrium}}$ (d) $\dfrac{4\beta^2_{equilibrium}}{4-\beta^2_{equilibrium}}$

17. The INCORRECT statement among the following, for this reaction, is
(a) Decrease in the total pressure will result in formation of more moles of gaseous X

(b) At the start of the reaction, dissociation of gaseous X_2 takes place spontaneously
(c) $\beta_{equilibrium} = 0.7$
(d) $K_C < 1$

10 Subjective Problems

18. When 3.06 g of solid NH_4HS is introduced into a two litre evacuated flask at 27° C, 30% of the solid decomposes into gaseous ammonia and hydrogen sulphide. (i) Calculate K_c and K_p for the reaction at 27°C. (ii) What would happen to the equilibrium when more solid NH_4HS is introduced into the flask? [1999 - 7 Marks]

19. At temperature T, a compound AB_2 (g) dissociates according to the reaction [1994 - 4 Marks]

$$2AB_2(g) \rightleftharpoons 2AB(g) + B_2(g)$$

with a degree of dissociation x which is small compared with unity. Deduce the expression for x in terms of the equilibrium constant K_p and the total pressure, P.

20. 0.15 mole of CO taken in a 2.5 L flask is maintained at 750 K along with a catalyst so that the following reaction can take place :

$$CO(g) + 2H_2(g) \rightleftharpoons CH_3OH(g)$$

Hydrogen is introduced until the total pressure of the system is 8.5 atmosphere at equilibrium and 0.08 mole of methanol is formed. Calculate (i) K_p and K_c and (ii) the final pressure if the same amount of CO and H_2 as before are used, but with no catalyst so that the reaction does not take place. [1993 - 5 Marks]

21. For the reaction : $CO(g) + 2H_2(g) \rightleftharpoons CH_3OH(g)$ hydrogen gas is introduced into a five litre flask at 327°C, containing 0.2 mole of CO(g) and a catalyst, until the pressure is 4.92 atm. At this point 0.1 mole of $CH_3OH(g)$ is formed. Calculate the equilibrium constant, K_p and K_c. [1990 - 5 Marks]

22. The equilibrium constant K_p of the reaction :

$$2SO_2(g) + O_2(g) \rightleftharpoons 2SO_3(g)$$

is 900 atm. at 800 K. A mixture containing SO_3 and O_2 having initial partial pressure of 1 and 2 atm. respectively is heated at constant volume to equilibrate. Calculate the partial pressure of each gas at 800 K. [1989 - 3 Marks]

23. N_2O_4 is 25% dissociated at 37°C and one atmosphere pressure. Calculate (i) K_p and (ii) the percentage dissociation at 0.1 atmosphere and 37°C. [1988 - 4 Marks]

24. At a certain temperature equilibrium constant (K_c) is 16 for the reaction. [1987 - 5 Marks]

$$SO_2(g) + NO_2(g) \rightleftharpoons SO_3(g) + NO(g)$$

If we take one mole each of all the four gases in a one litre container, what would be the equilibrium concentrations of NO(g) and NO_2(g)?

25. One mole of Cl_2 and 3 moles of PCl_5 are placed in a 100 litre vessel heated to 227°C. The equilibrium pressure is 2.05 atmosphere. Assuming ideal behaviour, calculate the degree of dissociation for PCl_5 and K_p for the reaction :
$PCl_5(g) \rightleftharpoons PCl_3(g) + Cl_2(g)$. [1984 - 6 Marks]

26. One mole of nitrogen is mixed with three moles of hydrogen in a 4 litre container. If 0.25 per cent of nitrogen is converted to ammonia by the following reaction [1981 - 4 Marks]

$$N_2(g) + 3H_2(g) \rightleftharpoons 2NH_3(g),$$

calculate the equilibrium constant (K_c) in concentration units. What will be the value of K_c for the following equilibrium ?

$$\frac{1}{2}N_2(g) + \frac{3}{2}H_2(g) \rightleftharpoons NH_3(ag)$$

Topic 2 - Relation between K, Q and G, and Factors Effecting Equilibrium

1 MCQs with One Correct Answer

27. The Haber's process for the formation of NH_3 at 298 K is $N_2 + 3H_2 \rightleftharpoons 2NH_3$; $\Delta H = -46.0$ kJ; Which of the following is the correct statement [2006 - 3M, –1]
(a) The condition for equilibrium is
$$G_{N_2} + 3G_{H_2} = 2G_{NH_3}$$
where G is Gibb's free energy per mole of the gaseous species measured at that partial pressure.
(b) On adding N_2, the equilibrium will shift to forward direction because according to IInd law of thermodynamics, the entropy must increase in the direction of spontaneous reaction
(c) The catalyst will increase the rate of forward reaction by 2 times and that of backward reaction by 1.5 times
(d) None of these

28. Consider the following equilibrium in a closed container [2002S]

$$N_2O_4(g) \rightleftharpoons 2NO_2(g)$$

At a fixed temperature, the volume of the reaction container is halved. For this change, which of the following statements holds true regarding the equilibrium constant (K_p) and degree of dissociation (α)?
(a) neither K_p nor α changes
(b) both K_p and α change
(c) K_p changes, but α does not change
(d) K_p does not change, but α changes

29. For the chemical reaction $3X(g) + Y(g) \rightleftharpoons X_3Y(g)$, the amount of X_3Y at equilibrium is affected by [1999 - 2 Marks]
(a) temperature and pressure
(b) temperature only
(c) pressure only
(d) temperature, pressure and catalyst

30. The oxidation of SO_2 by O_2 to SO_3 is an exothermic reaction. The yield of SO_3 will be maximum if [1981 - 1 Mark]
(a) temperature is increased and pressure is kept constant
(b) temperature is reduced and pressure is increased
(c) both temperature and pressure are increased
(d) both temperature and pressure are reduced

3 Numeric / New Stem Based Questions

31. Consider the reaction $\mathbf{A} \rightleftharpoons \mathbf{B}$ at 1000 K. At time t', the temperature of the system was increased to 2000 K and the system was allowed to reach equilibrium. Throughout this experiment the partial pressure of A was maintained at 1 bar. Given below is the plot of the partial pressure of B with time. What is the ratio of the standard Gibbs energy of the reaction at 1000 K to that at 2000 K? [Adv. 2020]

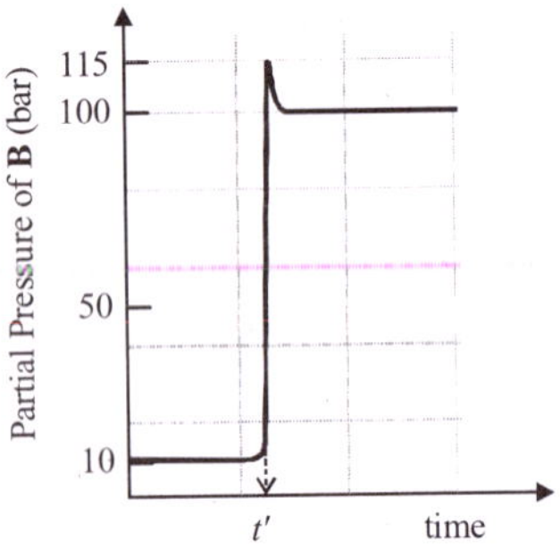

5 True / False

32. When a liquid and its vapour are at equilibrium and the pressure is suddenly decreased, cooling occurs. [1984 - 1 Mark]

6 MCQs with One or More than One Correct Answer

33. For the reaction $CO(g) + H_2O(g) \rightleftharpoons CO_2(g) + H_2(g)$ at a given temperature, the equilibrium amount of $CO_2(g)$ can be increased by [1998 - 2 Marks]
(a) adding a suitable catalyst
(b) adding an inert gas
(c) decreasing the volume of the container
(d) increasing the amount of CO(g).

34. For the reaction : [1991 - 1 Mark]
$$PCl_5(g) \rightarrow PCl_3(g) + Cl_2(g)$$
The forward reaction at constant temperature is favoured by
(a) introducing an inert gas at constant volume
(b) introducing an inert gas at constant pressure
(c) increasing the volume of the container
(d) introducing PCl_5 at constant volume

35. The equilibrium : [1989 - 1 Mark]
$$SO_2Cl_2(g) \rightleftharpoons SO_2(g) + Cl_2(g)$$
is attained at 25 °C in a closed container and an inert gas, helium is introduced. Which of the following statements are correct?
(a) Concentration of SO_2, Cl_2 and SO_2Cl_2 do not change
(b) More chlorine is formed
(c) Concentration of SO_2 is reduced
(d) More SO_2Cl_2 is formed.

36. When $NaNO_3$ is heated in a closed vessel, oxygen is liberated and $NaNO_2$ is left behind. At equilibrium. [1986 - 1 Mark]
(a) addition of $NaNO_2$ favours reverse reaction
(b) addition of $NaNO_3$ favours forward reaction
(c) increasing temperature favours forward reaction
(d) increasing pressure favours reverse reaction

37. For the gas phase reaction : [1984 - 1 Mark]
$$C_2H_4 + H_2 \rightleftharpoons C_2H_6 \ (\Delta H = -32.7 \text{ kcal})$$
carried out in a vessel, the equilibrium concentration of C_2H_4 can be increased by :
(a) increasing the temperature
(b) decreasing the pressure
(c) removing some H_2
(d) adding some C_2H_6

9 Assertion and Reason/Statement Type Questions

Each question contains **STATEMENT-1 (Assertion)** and **STATEMENT-2 (Reason)**. Each question has 4 choices (a), (b), (c) and (d) out of which ONLY ONE is correct. Mark your answer as
(a) If both Statement -1 and Statement -2 are correct, and Statement -2 is the correct explanation of the Statement -2.
(b) If both Statement -1 and Statement -2 are correct, but Statement -2 is not the correct explanation of the Statement -1.
(c) If Statement -1 is correct but Statement -2 is incorrect.
(d) If Statement -1 is incorrect but Statement -2 is correct.

38. **Statement -1** For every chemical reaction at equilibrium, standard Gibbs energy of reaction is zero.
Statement-2 At constant temperature and pressure, chemical reactions are spontaneous in the direction of decreasing Gibbs energy. [2008]

39. **Statement-1** The endothermic reactions are favoured at lower temperature and the exothermic reactions are favoured at higher temperature.
Statement-2 When a system in equilibrium is disturbed by changing the temperature, it will tend to adjust itself so as to overcome the effect of change. [1991 - 2 Marks]

Topic 3 - Theories of Acids and Bases, Ionic Product of Water and pH Scale

1 MCQs with One Correct Answer

40. On decreasing the pH from 7 to 2, the solubility of a sparingly soluble salt (MX) of a weak acid (HX) increased from 10^{-4} mol L^{-1} to 10^{-3} mol L^{-1}. The pK_a of HX is [Adv. 2023]
(a) 3 (b) 4 (c) 5 (d) 2

41. Molten sodium chloride conducts electricity due to the presence of [1981 - 1 Mark]
(a) free electrons
(b) free ions
(c) free molecules
(d) atoms of sodium and chlorine

42. The set with correct order of acidity is [2001S]
(a) $HClO < HClO_2 < HClO_3 < HClO_4$
(b) $HClO_4 < HClO_3 < HClO_2 < HClO$
(c) $HClO < HClO_4 < HClO_3 < HClO_2$
(d) $HClO_4 < HClO_2 < HClO_3 < HClO$

43. The pH of 0.1 M solution of the following salts increases in the order. [1999 - 2 Marks]
(a) $NaCl < NH_4Cl < NaCN < HCl$
(b) $HCl < NH_4Cl < NaCl < NaCN$
(c) $NaCN < NH_4Cl < NaCl < HCl$
(d) $HCl < NaCl < NaCN < NH_4Cl$

44. The following acids have been arranged in the order of decreasing acid strength. Identify the correct order.
ClOH (I), BrOH (II), IOH (III) [1996 - 1 Mark]
(a) I > II > III (b) II > I > III
(c) III > II > I (d) I > III > II

45. Which one is more acidic in aqueous solution. [1995S]
(a) $NiCl_2$ (b) $FeCl_3$ (c) $AlCl_3$ (d) $BeCl_2$

46. Which of the following solutions will have pH close to 1.0? [1992 - 1 Mark]
(a) 100 ml of (M/10) HCl + 100 ml of (M/10) NaOH
(b) 55 ml of (M/10) HCl + 45 ml of (M/10) NaOH
(c) 10 ml of (M/10) HCl + 90 ml of (M/10) NaOH
(d) 75 ml of (M/5) HCl + 25 ml of (M/5) NaOH

47. The following equilibrium is established when hydrogen chloride is dissolved in acetic acid.
$$HCl + CH_3COOH \rightleftharpoons Cl^- + CH_3COOH_2^+$$
The set that characterises the conjugate acid-base pairs is [1992 - 1 Mark]

(a) (HCl, CH_3COOH) and ($CH_2COOH_2^+$, Cl^-)
(b) (HCl, $CH_3COOH_2^+$) and (CH_3COOH, Cl^-)
(c) ($CH_3COOH_2^+$, HCl) and (Cl^-, CH_3COOH)
(d) (HCl, Cl^-) and ($CH_3COOH_2^+$, CH_3COOH)

48. The reaction which proceeds in the forward direction is [1991 - 1 Mark]
(a) $Fe_2O_3 + 6HCl \rightarrow 2FeCl_3 + 3H_2O$
(b) $NH_3 + H_2O + NaCl \rightarrow NH_4Cl + NaOH$
(c) $SnCl_4 + Hg_2Cl_2 \rightarrow SnCl_2 + 2HgCl_2$
(d) $2CuI + I_2 + 4K^+ \rightarrow 2Cu^{2+} + 4KI$

49. Which one of the following is the strongest acid? [1989 - 1 Mark]
(a) $ClO_3(OH)$ (b) $ClO_2(OH)$
(c) $SO(OH)_2$ (d) $SO_2(OH)_2$

50. The pK_a of acetylsalicyclic acid (aspirin) is 3.5. The pH of gastric juice in human stomach is about 2-3 and the pH in the small intestine is about 8. Aspirin will be [1988 - 1 Mark]
(a) unionised in the small intestine and in the stomach
(b) completely ionised in the small intestine and in the stomach
(c) ionised in the stomach and almost unionised in the small intestine
(d) ionised in the small intestine and almost unionised in the stomach.

51. The compound whose 0.1 M solution is basic is : [1986 - 1 Mark]
(a) ammonium acetate (b) ammonium chloride
(c) ammonium sulphate (d) sodium acetate

52. The compound insoluble in acetic acid is : [1986 - 1 Mark]
(a) calcium oxide (b) calcium carbonate
(c) calcium oxalate (d) calcium hydroxide

53. The compound that is not a Lewis acid is : [1985 - 1 Mark]
(a) BF_3 (b) $AlCl_3$ (c) $BeCl_2$ (d) $SnCl_4$

54. The conjugate acid of NH_2^- is : [1985 - 1 Mark]
(a) NH_3 (b) NH_2OH (c) NH_4^+ (d) N_2H_4

55. The best indicator for detection of end point in titration of a weak acid and a strong base is : [1985 - 1 Mark]
(a) methyl orange (3 to 4)
(b) methyl red (5 to 6)
(c) bromothymol blue (6 to 7.5)
(d) phenolphthalein (8 to 9.6)

56. At 90°C, pure water has $[H_3O^+]$ 10^{-6} mole litre^{-1}. What is the value of K_w at 90°C? [1981 - 1 Mark]
(a) 10^{-6} (b) 10^{-12} (c) 10^{-14} (d) 10^{-8}

57. Of the given anions, the strongest Bronsted base is [1981 - 1 Mark]
(a) ClO^- (b) ClO_2^- (c) ClO_3^- (d) ClO_4^-

58. The pH of a 10^{-8} molar solution of HCl in water is [1981 - 1 Mark]
(a) 8 (b) –8
(c) between 7 and 8 (d) between 6 and 7

2 Integer Value Answer

59. A solution of 0.1 M weak base (B) is titrated with 0.1 M of a strong acid (HA). The variation of pH of the solution with the volume of HA added is shown in the figure below. What is the pK_b of the base? The neutralization reaction is given by $B + HA \rightarrow BH^+ + A^-$ [Adv. 2020]

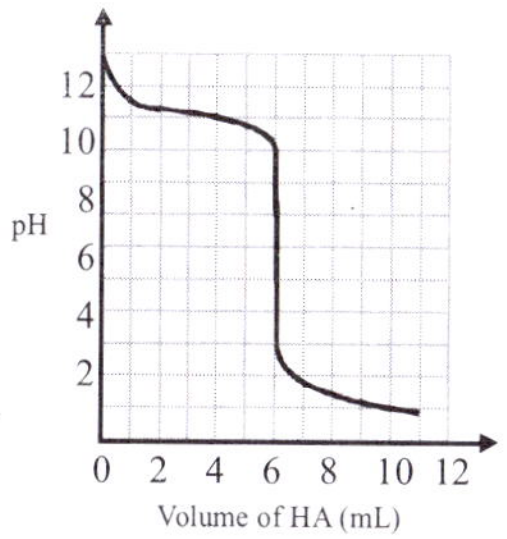

60. Amongst the following, the total number of compounds whose aqueous solution turns red litmus paper blue is
KCN, K_2SO_4, $(NH_4)_2C_2O_4$, NaCl, $Zn(NO_3)_2$, $FeCl_3$, K_2CO_3, NH_4NO_3 and LiCN [2010]

61. The total number of diprotic acids among the following is:
H_3PO_4, H_2SO_4, H_3PO_3, H_2CO_3, $H_2S_2O_7$, H_3BO_3, H_3PO_2, H_2CrO_4 and H_2SO_3. [2010]

62. The dissociation constant of a substituted benzoic acid at 25°C is 1.0×10^{-4}. The pH of a 0.01 M solution of its sodium salt is [2009 - 2 Marks]

63. 0.1 M NaOH is titrated with 0.1 M *HA* till the end point; K_a for *HA* is 5.6×10^{-6} and degree of hydrolysis is less compared to 1. Calculate pH of the resulting solution at the end point. [2004 - 2 Marks]

64. An acid type indicator, HIn differs in colour from its conjugate base (In^-). The human eye is sensitive to colour differences only when the ratio$[In^-]/[HIn]$ is greater than 10 or smaller than 0.1. What should be the minimum change in the pH of the solution to observe a complete colour change ($K_a = 1.0 \times 10^{-5}$)? [1997 - 2 Marks]

3 Numeric / New Stem Based Questions

65. A solution is prepared by mixing 0.01 mol each of H_2CO_3, $NaHCO_3$, Na_2CO_3, and NaOH in 100 mL of water. *p*H of the resulting solution is ________.
[Given: pK_{a_1} and pK_{a_2} of H_2CO_3 are 6.37 and 10.32, respectively; log 2 = 0.30] [Adv. 2022]

66. The average concentration of SO_2 in the atmosphere over a city on a certain day is 10 ppm, when the average temperature is 298 K. Given that the solubility of SO_2 in water at 298 K is 1.3653 moles litre^{-1} and the pK_a of H_2SO_3 is 1.92, estimate the pH of rain on that day. [2000 - 5 Marks]

67. What will be the resultant pH when 200mL of an aqueous solution of HCl (pH = 2.0) is mixed with 300 mL of an aqueous solution of NaOH (pH = 12.0) ? [1998 - 2 Marks]

68. An aqueous solution of a metal bromide MBr_2 (0.05M) is saturated with H_2S. What is the minimum pH at which *MS* will precipitate? [1993 - 3 Marks]

K_{sp} for $MS = 6.0 \times 10^{-21}$; concentration of saturated $H_2S = 0.1$ M

$K_1 = 10^{-7}$ and $K_2 = 1.3 \times 10^{-13}$, for H_2S.

4 Fill in the Blanks

69. In the reaction $I^- + I_2 \rightarrow I_3^-$, the Lewis acid is .... [1997 - 1 Mark]

70. An element which can exist as a positive ion in acidic solution and also as a negative ion in basic solution is said to be .... [1984 - 1 Mark]

71. The conjugate base of HSO_4^- in aqueous solution is .... [1982 - 1 Mark]

5 True / False

72. Aluminium chloride ($AlCl_3$) is a Lewis acid because it can donate electrons. [1982 - 1 Mark]

6 MCQs with One or More than One Correct Answer

73. Which of the following statements(s) is (are) correct? [1998 - 2 Marks]
(a) The pH of 1.0×10^{-8} M solution of HCl is 8
(b) The conjugate base of $H_2PO_4^-$ is HPO_4^{2-}

(c) Autoprotolysis constant of water increases with temperature

(d) When a solution of a weak monoprotic acid is titrated against a strong base, at half-neutralisation point $pH = (1/2)\, pK_a$.

8 Comprehension Based Questions

PASSAGE

When 100 mL of 1.0 M HCl was mixed with 100 mL of 1.0 M NaOH in an insulated beaker at constant pressure, a temperature increase of 5.7 °C was measured for the beaker and its contents (Expt. 1). Because the enthalpy of neutralization of a strong acid with a strong base is a constant (–57.0 kJ mol^{-1}), this experiment could be used to measure the calorimeter constant. In a second experiment (Expt. 2), 100 mL of 2.0 M acetic acid ($K_a = 2.0 \times 10^{-5}$) was mixed with 100 mL of 1.0 M NaOH (under identical conditions to Expt. 1) where a temperature rise of 5.6 °C was measured.

(Consider heat capacity of all solutions as 4.2 J g^{-1} K^{-1} and density of all solutions as 1.0 g mL^{-1}) [Adv. 2015]

74. Enthalpy of dissociation (in kJ mol^{-1}) of acetic acid obtained from the Expt.2 is
(a) 1.0 (b) 10.0 (c) 24.5 (d) 51.4

75. The pH of the solution after Expt. 2 is
(a) 2.8 (b) 4.7 (c) 5.0 (d) 7.0

9 Assertion and Reason/Statement Type Questions

Each question contains STATEMENT-1 (Assertion) and STATEMENT-2 (Reason). Each question has 4 choices (a), (b), (c) and (d) out of which ONLY ONE is correct. Mark your answer as

(a) If both Statement -1 and Statement -2 are correct, and Statement -2 is the correct explanation of the Statement -2.

(b) If both Statement -1 and Statement -2 are correct, but Statement -2 is not the correct explanation of the Statement -1.

(c) If Statement -1 is correct but Statement -2 is incorrect.

(d) If Statement -1 is incorrect but Statement -2 is correct.

76. **Statement -1** HNO_3 is a stronger acid than HNO_2
Statement -2 In HNO_3 there are two nitrogen-to-oxygen bonds whereas in HNO_2 there is only one. [1998 - 2 Marks]

10 Subjective Problems

77. Arrange the following oxides in the decreasing order of Bronsted basicity :
$BaO, SO_3, CO_2, Cl_2O_7, B_2O_3$ [2004 - 2 Marks]

78. 500 mL of 0.2 M aqueous solution of acetic acid is mixed with 500 mL of 0.2 M HCl at 25°C.
(i) Calculate the degree of dissociation of acetic acid in the resulting solution and pH of the solution.
(ii) If 6 g of NaOH is added to the above solution, determine the final pH.
[Assume there is no change in volume on mixing; K_a of acetic acid is 1.75×10^{-5} mol L^{-1}]. [2002 - 5 Marks]

79. The ionisation constant of NH_4^+ in water is 5.6×10^{-10} at 25 °C. The rate constant for the reaction of NH_4^+ and OH^- to form NH_3 and H_2O at 25 °C is 3.4×10^{10} L $mol^{-1}s^{-1}$. Calculate the rate constant for proton transfer from water to NH_3. [1996 - 3 Marks]

80. Increasing basicity order : $H_2O, OH^-, CH_3OH, CH_3O^-$ [1992 - 1 Mark]

81. Increasing acid strength : $HClO_3, HClO_4, HClO_2, HClO$ [1986 - 1 Mark]

82. Increasing bond length : F_2, N_2, Cl_2, O_2 [1985 - 1 Mark]

83. The dissociation constant of a weak acid HA is 4.9×10^{-8}. After making the necessary approximations, calculate (i) percentage ionization, (ii) pH and (iii) OH^- concentration in a decimolar solution of the acid. Water has a pH of 7. [1983 - 2 Marks]

84. A solution contains Na_2CO_3 and $NaHCO_3$. 10 mL of solution requires 2.5 mL of 0.1 M H_2SO_4 for neutralisation using phenolphthalein as an indicator. Methyl orange is then added when a further 2.5 mL of 0.2 M H_2SO_4 was required. Calculate the amount of Na_2CO_3 and $NaHCO_3$ in one litre of the solution. [1979]

Topic 4 - Ionisation of Weak Acids and Bases and Relation between K_a and K_b

1 MCQs with One Correct Answer

85. 2.5 mL of (2/5) M weak monoacidic base ($K_b = 1 \times 10^{-12}$ at 25° C) is titrated (2/15) M HCl in water at 25°C. The concentration of H^+ at equivalence point is ($K_w = 1 \times 10^{-14}$ at 25°C) [2008S]
(a) 3.7×10^{-14} M (b) 3.2×10^{-7} M
(c) 3.2×10^{-2} M (d) 2.7×10^{-2} M

86. 0.1 mole of CH_3NH_2 ($K_b = 5 \times 10^{-4}$) is mixed with 0.08 mole of HCl and diluted to one litre. What will be the H^+ concentration in the solution? [2005S]
(a) 8×10^{-2} M (b) 8×10^{-11} M
(c) 1.6×10^{-11} M (d) 8×10^{-5} M

87. A 0.004 M solution of Na_2SO_4 is isotonic with 0.010 M solution of glucose at same temperature. The percentage dissociation of Na_2SO_4 is [2004S]
(a) 25% (b) 50% (c) 75% (d) 85%

88. A weak acid HX has the dissociation constant 1×10^{-5} M. It forms a salt NaX on reaction with alkali. The percentage hydrolysis of 0.1 M solution of NaX is [2004S]
(a) 0.0001 % (b) 0.01 %
(c) 0.1% (d) 0.15 %

89. The degree of dissociation of water at 25°C is 1.9×10^{-7}% and density is 1.0 g cm^{-3}. The ionic constant for water is : [1995S]
(a) 1.0×10^{-10} (b) 1.0×10^{-14}
(c) 1.0×10^{-16} (d) 1.0×10^{-8}

90. A certain weak acid has a dissociation constant of 1.0×10^{-4}. The equilibrium constant for its reaction with a strong base is : [1984 - 1 Mark]
(a) 1.0×10^{-4} (b) 1.0×10^{-10}
(c) 1.0×10^{10} (d) 1.0×10^{14}

3 Numeric / New Stem Based Questions

91. At 25°C, the concentration of H^+ ions in 1.00×10^{-3} M aqueous solution of a weak monobasic acid having acid dissociation constant (K_a) of 4.00×10^{-11} is $X \times 10^{-7}$ M. The value of X is ______. [Adv. 2025]
Use: Ionic product of water (K_w) = 1.00×10^{-14} at 25°C

92. What is the pH of a 0.50 M aqueous NaCN solution? pK_b of CN^- is 4.70. [1996 - 2 Marks]

93. Calculate the pH of an aqueous solution of 1.0 M ammonium formate assuming complete dissociation. [1995 - 2 Marks]
(pK_a of formic acid = 3.8 and pK_b of ammonia = 4.8.)

6 MCQs with One or More than One Correct Answer

94. The initial rate of hydrolysis of methyl acetate (1M) by a weak acid (HA, 1M) is 1/100th of that of a strong acid (HX, 1M), at 25°C. The K_a of HA is [Adv. 2013]
(a) 1×10^{-4} (b) 1×10^{-5}
(c) 1×10^{-6} (d) 1×10^{-3}

7 Match the Following

95. Dilution processes of different aqueous solutions, with water, are given in LIST-I. The effects of dilution of the solutions on $[H^+]$ are given in LIST-II.
(Note: Degree of dissociation (α) of weak acid and weak base is << 1; degree of hydrolysis of salt <<1; $[H^+]$ represents the concentration of H^+ ions) [Adv. 2018]

LIST-I	LIST-II
P. (10 mL of 0.1 M NaOH + 20 mL of 0.1 M acetic acid) diluted to 60 mL	1. the value of $[H^+]$ does not change on dilution
Q. (20 mL of 0.1 M NaOH + 20 mL of 0.1 M acetic acid) diluted to 80 mL	2. the value of $[H^+]$ changes to half of its initial value on dilution
R. (20 mL of 0.1 M HCl + 20 mL of 0.1 M ammonia solution) diluted to 80 mL	3. the value of $[H^+]$ changes to two times of its initial value on dilution
S. 10 mL saturated solution of $Ni(OH)_2$ in equilibrium with excess solid $Ni(OH)_2$ is diluted to 20 mL (solid $Ni(OH)_2$ is still present after dilution).	4. the value of $[H^+]$ changes to $\frac{1}{\sqrt{2}}$ times of its initial value on dilution
	5. the value of $[H^+]$ changes to $\sqrt{2}$ times of its initial value on dilution

Match each process given in LIST-I with one or more effect(s) in LIST-II. The correct option is
(a) P-4; Q-2 ; R-3; S-1 (b) P-4 ; Q-3 ; R-2; S-3
(c) P-1 ; Q-4 ; R-5; S-3 (d) P-1 ; Q-5 ; R-4 ; S-1

10 Subjective Problems

96. A 40.0 mL solution of weak base, BOH is titrated with 0.1N HCl solution. The pH of the solution is found to be 10.04 and 9.14 after adding 5.0 mL and 20.0 mL of the acid respectively. Find out the dissociation constant of the base. [1991 - 6 Marks]

97. What is the pH of 1.0 M solution of acetic acid? To what volume must one liter of this solution be diluted so that the pH of the resulting solution will be twice the original value? Given : $K_a = 1.8 \times 10^{-5}$. [1990 - 4 Marks]

98. What is the pH of the solution when 0.20 mole of hydrochloric acid is added to one litre of a solution containing. [1987 - 5 Marks]
(i) 1 M each of acetic acid and acetate ion?
(ii) 0.1 M each of acetic acid and acetate ion?
Assume the total volume is one litre.
K_a for acetic acid = 1.8×10^{-5}.

99. The $[H^+]$ in 0.2 M solution of formic acid is 6.4×10^{-3} mole litre^{-1}. To this solution sodium formate is added so as to adjust the concentration of sodium formate to one mole litre^{-1}. What wil be pH of this solution? K_a for HCOOH is 2.4×10^{-4} and degree of dissociation of HCOONa is 0.75. [1985 - 3 Marks]

100. Twenty mL of 0.2 M sodium hydroxide is added to 50 mL of 0.2 M acetic acid to give 70 mL of the solution. What is the pH of this solution? Calculate the additional volume of 0.2 M NaOH required to make the pH of the solution 4.74. The ionization constant of acetic acid is 1.8×10^{-5}. [1982 - 3 M1arks]

101. How many moles of sodium propionate should be added to one litre of an aqueous solution containing 0.020 mole of propionic acid to obtain a buffer solution of pH 4.75? What will be pH if 0.010 mole of hydrogen chloride is dissolved in the above buffer solution. Compare the last pH value with the pH of 0.010 molar HCl solution. Dissociation constant of propionic acid, K_a at 25°C = 1.34×10^{-5}. [1981 - 4 Marks]

Topic 5 – Common Ion Effect, Salt Hydrolysis, Buffer Solutions and Solubility Product

1 MCQs with One Correct Answer

102. Solubility product constant (K_{sp}) of salts of types MX, MX_2 and M_3X at temperature T are 4.0×10^{-8}, 3.2×10^{-14} and 2.7×10^{-15}, respectively. Solubilities (mol dm^{-3}) of the salts at temperature 'T' are in the order – [2008S]
(a) $MX > MX_2 > M_3X$ (b) $M_3X > MX_2 > MX$
(c) $MX_2 > M_3X > MX$ (d) $MX > M_3X > MX_2$

103. For a sparingly soluble salt A_pB_q, the relationship of its solubility product (L_S) with its solubility (S) is [2001S]
(a) $L_S = S^{p+q}.p^p.q^q$ (b) $L_S = S^{p+q}.p^q.q^p$
(c) $L_S = S^{pq}.p^p.q^q$ (d) $L_S = S^{pq}.(pq)^{p+q}$

104. Amongst the following hydroxides, the one which has the lowest value of K_{sp} at ordinary temperature (about 25 °C) is [1990 - 1 Mark]
(a) $Mg(OH)_2$ (b) $Ca(OH)_2$
(c) $Ba(OH)_2$ (d) $Be(OH)_2$

105. When equal volumes of the following solutions are mixed, precipitation of AgCl ($K_{sp} = 1.8\times10^{-10}$) will occur only with [1988 - 1 Mark]
(a) 10^{-4} M (Ag^+) and 10^{-4} M (Cl^-)
(b) 10^{-5} M (Ag^+) and 10^{-5} M (Cl^-)
(c) 10^{-6} M (Ag^+) and 10^{-6} M (Cl^-)
(d) 10^{-10} M (Ag^+) and 10^{-10} M (Cl^-)

106. A certain buffer solution contains equal concentration of X^- and HX. The K_b for X^- is 10^{-10}. The pH of the buffer is: [1984 - 1 Mark]
(a) 4 (b) 7 (c) 10 (d) 14

107. The precipitate of [1982 - 1 Mark]
$$CaF_2(K_{sp} = 1.7 \times 10^{-10})$$
is obtained when equal volumes of the following are mixed
(a) 10^{-4}M Ca^{2+} + 10^{-4}M F^- (b) 10^{-2}M Ca^{2+} + 10^{-3}M F^-
(c) 10^{-5}M Ca^{2+} + 10^{-3}M F^- (d) 10^{-3}M Ca^{2+} + 10^{-5}M F^-

108. An acidic buffer solution can be prepared by mixing the solutions of [1981 - 1 Mark]
(a) ammonium acetate and acetic acid
(b) ammonium chloride and ammonioum hydroxide
(c) sulphuric acid and sodium sulphate
(d) sodium chloride and sodium hydroxide.

2 Integer Value Answer

109. Concentration of H_2SO_4 and Na_2SO_4 in a solution is 1 M and 1.8×10^{-2} M, respectively. Molar solubility of $PbSO_4$ in the same solution is $X \times 10^{-Y}$ M (expressed in scientific notation). The value of Y is ________.
[Given: Solubility product of $PbSO_4$ $(K_{sp}) = 1.6 \times 10^{-8}$. For H_2SO_4, K_{a1} is very large and $K_{a2} = 1.2 \times 10^{-2}$] **[Adv. 2022]**

110. In 1 L saturated solution of AgCl $[K_{sp}(AgCl) = 1.6 \times 10^{-10}]$, 0.1 mol of CuCl $[K_{sp}(CuCl) = 1.0 \times 10^{-6}]$ is added. The resultant concentration of Ag^+ in the solution is 1.6×10^{-x}. The value of "x" is **[2011]**

3 Numeric / New Stem Based Questions

111. The solubility of barium iodate in an aqueous solution prepared by mixing 200 mL of 0.010 M barium nitrate with 100 mL of 0.10 M sodium iodate is $X \times 10^{-6}$ mol dm^{-3}. The value of **X** is ______.
Use: Solubility product constant (K_{sp}) of barium iodate $= 1.58 \times 10^{-9}$ **[Adv. 2025]**

112. An acidified solution of 0.05 M Zn^{2+} is saturated with 0.1 M H_2S. What is the minimum molar concentration (M) of H^+ required to prevent the precipitation of ZnS?
Use $K_{sp}(ZnS) = 1.25 \times 10^{-22}$ and overall dissociation constant of H_2S, $K_{NET} = K_1K_2 = 1 \times 10^{-21}$. **[Adv. 2020]**

113. The solubility of a salt of weak acid (**AB**) at pH 3 is $Y \times 10^{-3}$ mol L^{-1}. The value of **Y** is________. **[Adv. 2018]**
(Given that the value of solubility product of **AB** $(K_{sp}) = 2 \times 10^{-10}$ and the value of ionization constant of **HB** $(K_a) = 1 \times 10^{-8}$)

114. The pH of blood stream is maintained by a proper balance of H_2CO_3 and $NaHCO_3$ concentrations. What volume of 5M $NaHCO_3$ solution should be mixed with a 10 mL sample of blood which is 2M in H_2CO_3 in order to maintain a pH of 7.4 ? K_a for H_2CO_3 in blood is 7.8×10^{-7}. **[1993 - 2 Marks]**

5 True / False

115. Solubility of sodium hydroxide increases with increase in temperature. **[1985 - ½ Mark]**

6 MCQs with One or More than One Correct Answer

116. The K_{sp} of Ag_2CrO_4 is 1.1×10^{-12} at 298 K. The solubility (in mol/L) of Ag_2CrO_4 in a 0.1 M $AgNO_3$ solution is **[Adv. 2013]**
(a) 1.1×10^{-11} (b) 1.1×10^{-10}
(c) 1.1×10^{-12} (d) 1.1×10^{-9}

117. Aqueous solutions of HNO_3, KOH, CH_3COOH and CH_3COONa of identical concentrations are provided. The pair(s) of solutions which form a buffer upon mixing is(are) **[2010]**
(a) HNO_3 and CH_3COOH
(b) KOH and CH_3COONa
(c) HNO_3 and CH_3COONa
(d) CH_3COOH and CH_3COONa

118. A buffer solution can be prepared from a mixture of **[1999 - 3 Marks]**
(a) sodium acetate and acetic acid in water
(b) sodium acetate and hydrochloric acid in water
(c) ammonia and ammonium chloride in water
(d) ammonia and sodium hydroxide in water

10 Subjective Problems

119. The solubility of $Pb(OH)_2$ in water is 6.7×10^{-6} M. Calculate the solubility of $Pb(OH)_2$ in a buffer solution of pH = 8. **[1999 - 4 Marks]**

120. Given : $Ag(NH_3)_2^+ \rightleftharpoons Ag^+ + 2NH_3$, $K_c = 6.2 \times 10^{-8}$ and K_{sp} of AgCl $= 1.8 \times 10^{-10}$ at 298 K. If ammonia is added to a water solution containing excess of AgCl(s) only, calculate the concentration of the complex in 1.0 M aqueous ammonia. **[1998 - 5 Marks]**

121. A sample of AgCl was treated with 5.00 mL of 1.5 M Na_2CO_3 solution to give Ag_2CO_3. The remaining solution contained 0.0026 g of Cl^- per litre. Calculate the solubility product of AgCl $(K_{sp}(Ag_2CO_3) = 8.2 \times 10^{-12})$. **[1997 - 5 Marks]**

122. For the reaction

$$[Ag(CN)_2]^- \rightleftharpoons Ag^+ + 2CN^-$$

the equilibrium costant, at 25°C, is 4.0×10^{-19}. Calculate the silver ion concentration in a solution which was originally 0.10 molar in KCN and 0.03 molar in $AgNO_3$. **[1994 - 3 Marks]**

123. The solubility product (K_{sp}) of $Ca(OH)_2$ at 25°C is 4.42×10^{-5}. A 500 mL of saturated solution of $Ca(OH)_2$ is mixed with equal volume of 0.4 M NaOH. How much $Ca(OH)_2$ in milligrams is precipitated? **[1992 - 4 Marks]**

124. The solubility product of $Ag_2C_2O_4$ at 25°C is 1.29×10^{-11} mol^3 L^{-3}. A solution of $K_2C_2O_4$ containing 0.1520 mole in 500 mL water is shaken at 25°C with excess of Ag_2CO_3 till the following equilibrium is reached : **[1991 - 4 Marks]**

$$Ag_2CO_3 + K_2C_2O_4 \rightleftharpoons Ag_2C_2O_4 + K_2CO_3$$

At equilibrium, the solution contains 0.0358 mole of K_2CO_3. Assuming the degree of dissociation of $K_2C_2O_4$ and K_2CO_3 to be equal, calculate the solubility product of Ag_2CO_3.

125. Freshly precipitated aluminium and magnesium hydroxides are stirred vigorously in a buffer solution containing 0.25 mole/L of ammonium chloride and 0.05 mole/L of ammonium hydroxide. Calculate the concentration of aluminium and magnesium ions in solution:

$K_b[NH_4OH] = 1.80 \times 10^{-5}$ **[1989- 3 Marks]**
$K_{sp}[Mg(OH)_2] = 6 \times 10^{-10}$
$K_{sp}[Al(OH)_3] = 6 \times 10^{-32}$

126. How many gram-mole of HCl will be required to prepare one litre of buffer solution (containing NaCN and HCl) of pH 8.5 using 0.01 gram formula weight of NaCN?
K dissociation (HCN) $= 4.1 \times 10^{-10}$. **[1988 - 4 Marks]**

127. The solubility of $Mg(OH)_2$ in pure water is 9.57×10^{-3} g/litre. Calculate its solubility (in g/litre) in 0.02 M $Mg(NO_3)_2$ solution. **[1986 - 5 Marks]**

128. A solution contains a mixture of Ag (0.10 M) and Hg_2^{++} (0.10 M) which are to be separated by selective precipitation. Calculate the maximum concentration of iodide ion at which one of them gets precipitated almost completely. What percentage of that metal ion is precipitated?
$[K_{sp}: AgI = 8.5 \times 10^{-17}; Hg_2I_2 = 2.5 \times 10^{-26}]$ **[1984 - 4 Marks]**

PART-A (JEE Main)

1	(3)	**31**	(2)	**61**	(3)	**91**	(1400)	**121**	(1)	**151**	(2)	**181**	(3)	**211**	(3)	**241**	(2)	**271**	(233)
2	(2)	**32**	(4)	**62**	(4)	**92**	(16)	**122**	(8)	**152**	(2)	**182**	(3)	**212**	(2)	**242**	(3)	**272**	(10)
3	(1)	**33**	(4)	**63**	(2)	**93**	(875)	**123**	(2474)	**153**	(3)	**183**	(1)	**213**	(3)	**243**	(4)	**273**	(10)
4	(2)	**34**	(1)	**64**	(3)	**94**	(25)	**124**	(33)	**154**	(2)	**184**	(3)	**214**	(2)	**244**	(3)	**274**	(85)
5	(4)	**35**	(3)	**65**	(4)	**95**	(2)	**125**	(700)	**155**	(4)	**185**	(4)	**215**	(1)	**245**	(4)	**275**	(282)
6	(4)	**36**	(2)	**66**	(20)	**96**	(172)	**126**	(57)	**156**	(4)	**186**	(4)	**216**	(1)	**246**	(3)	**276**	(0)
7	(1)	**37**	(1)	**67**	(15)	**97**	(354)	**127**	(163)	**157**	(2)	**187**	(3)	**217**	(1)	**247**	(3)	**277**	(476)
8	(4)	**38**	(1)	**68**	(125)	**98**	(72)	**128**	(37)	**158**	(4)	**188**	(4)	**218**	(2)	**248**	(1)	**278**	(2)
9	(3)	**39**	(4)	**69**	(5)	**99**	(5)	**129**	(10)	**159**	(4)	**189**	(3)	**219**	(1)	**249**	(2)	**279**	(108)
10	(4)	**40**	(2)	**70**	(74)	**100**	(16)	**130**	(3)	**160**	(1)	**190**	(2)	**220**	(4)	**250**	(1)	**280**	(3)
11	(4)	**41**	(3)	**71**	(962)	**101**	(1)	**131**	(6)	**161**	(4)	**191**	(4)	**221**	(3)	**251**	(4)	**281**	(10)
12	(2)	**42**	(1)	**72**	(2)	**102**	(N)	**132**	(2735)	**162**	(1)	**192**	(2)	**222**	(3)	**252**	(1)	**282**	(64)
13	(3)	**43**	(1)	**73**	(417)	**103**	(1)	**133**	(747)	**163**	(3)	**193**	(1)	**223**	(4)	**253**	(2)	**283**	(50)
14	(4)	**44**	(3)	**74**	(44)	**104**	(1,4)	**134**	(20)	**164**	(1)	**194**	(3)	**224**	(2)	**254**	(2)	**284**	(141)
15	(2)	**45**	(2)	**75**	(1)	**105**	(3)	**135**	(166)	**165**	(1)	**195**	(3)	**225**	(1)	**255**	(2)	**285**	(2)
16	(4)	**46**	(4)	**76**	(3)	**106**	(3)	**136**	(1380)	**166**	(4)	**196**	(4)	**226**	(2)	**256**	(1)		
17	(2)	**47**	(2)	**77**	(48.6)	**107**	(4)	**137**	(2)	**167**	(2)	**197**	(3)	**227**	(1)	**257**	(3)		
18	(2)	**48**	(1)	**78**	(12)	**108**	(1)	**138**	(4)	**168**	(4)	**198**	(100)	**228**	(4)	**258**	(4)		
19	(3)	**49**	(4)	**79**	(1)	**109**	(3)	**139**	(1)	**169**	(1)	**199**	(70)	**229**	(4)	**259**	(3)		
20	(3)	**50**	(4)	**80**	(2)	**110**	(2)	**140**	(2)	**170**	(3)	**200**	(25)	**230**	(3)	**260**	(4)		
21	(4)	**51**	(2)	**81**	(4)	**111**	(1)	**141**	(4)	**171**	(9000)	**201**	(7)	**231**	(2)	**261**	(4)		
22	(4)	**52**	(3)	**82**	(2)	**112**	(4)	**142**	(2)	**172**	(186)	**202**	(7)	**232**	(2)	**262**	(1)		
23	(4)	**53**	(3)	**83**	(2)	**113**	(3)	**143**	(2)	**173**	(5)	**203**	(27)	**233**	(4)	**263**	(1)		
24	(2)	**54**	(4)	**84**	(29)	**114**	(3)	**144**	(3)	**174**	(9079)	**204**	(2)	**234**	(3)	**264**	(756)		
25	(1)	**55**	(3)	**85**	(125)	**115**	(4)	**145**	(3)	**175**	(12)	**205**	(1)	**235**	(4)	**265**	(4)		
26	(3)	**56**	(1)	**86**	(1107)	**116**	(2)	**146**	(4)	**176**	(11)	**206**	(7)	**236**	(2)	**266**	(48)		
27	(1)	**57**	(3)	**87**	(4)	**117**	(4)	**147**	(2)	**177**	(6021)	**207**	(1)	**237**	(2)	**267**	(9)		
28	(2)	**58**	(3)	**88**	(6)	**118**	(4)	**148**	(2)	**178**	(37)	**208**	(2)	**238**	(3)	**268**	(100)		
29	(3)	**59**	(4)	**89**	(182)	**119**	(1)	**149**	(3)	**179**	(10.60)	**209**	(2)	**239**	(1)	**269**	(458)		
30	(4)	**60**	(1)	**90**	(0)	**120**	(2)	**150**	(2)	**180**	(5.23)	**210**	(4)	**240**	(2)	**270**	(5)		

PART-B (JEE Advanced)

1	(c)	**13**	False	**32**	(True)	**42**	(b)	**52**	(c)	**62**	(8)	**72**	False	**90**	(c)	**106**	(a)	**116**	(b)
2	(d)	**14**	(b)	**33**	(d)	**43**	(a)	**53**	(a)	**63**	(9)	**73**	(b, c)	**91**	(2.23)	**107**	(b)	**117**	(c, d)
3	(d)	**15**	(a,b,d)	**34**	(b, c, d)	**44**	(c)	**54**	(d)	**64**	(2)	**74**	(a)	**92**	(11.5)	**108**	(a)	**118**	(a, c)
4	(d)	**16**	(b)	**35**	(a)	**45**	(d)	**55**	(b)	**65**	(10.02)	**75**	(b)	**93**	(6.5)	**109**	(6)		
5	(d)	**17**	(c)	**36**	(c, d)	**46**	(d)	**56**	(a)	**66**	(4.86)	**76**	(c)	**94**	(a)	**110**	(7)		
6	(a)	**27**	(a)	**37**	(a,b,c,d)	**47**	(a)	**57**	(d)	**67**	(11.30)	**85**	(d)	**95**	(d)	**111**	(3.95)		
7	(d)	**28**	(d)	**38**	(d)	**48**	(a)	**58**	(b)	**68**	(0.983)	**86**	(b)	**102**	(d)	**112**	(0.20)		
8	(d)	**29**	(a)	**39**	(d)	**49**	(d)	**59**	(3)	**69**	I_2	**87**	(c)	**103**	(a)	**113**	(4.47)		
9	(8.93)	**30**	(b)	**40**	(b)	**50**	(d)	**60**	(3)	**70**	(amphoteric)	**88**	(b)	**104**	(d)	**114**	(78.36)		
10	(1.886)	**31**	(0.25)	**41**	(a)	**51**	(c)	**61**	(6)	**71**	SO_4^{2-}	**89**	(b)	**105**	(a)	**115**	True		

Chapter-8

Redox Reaction

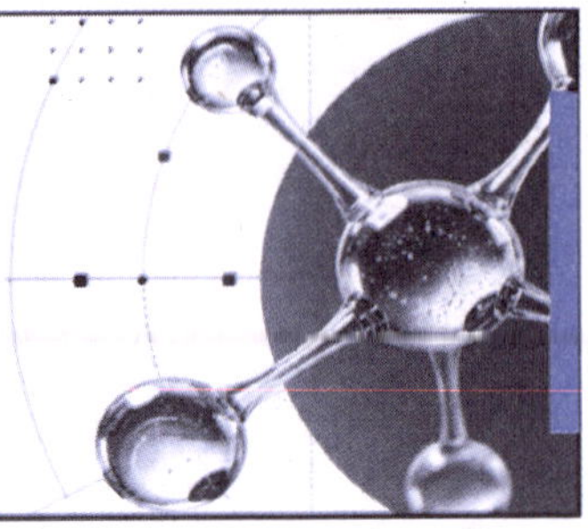

Part-A (JEE Main)

Topic 1- Oxidation and Reduction Reactions

MULTIPLE CHOICE QUESTIONS (MCQs)

1. Consider the following reduction processes : [Jan. 22, 2026 (II)]

$Al^{3+} + 3e^- \rightarrow Al(s), E^o = -1.66$ V

$Fe^{3+} + e^- \rightarrow Fe^{2+}, E^o = +0.77$ V

$Co^{3+} + e^- \rightarrow Co^{2+}, E^o = +1.81$ V

$Cr^{3+} + 3e^- \rightarrow Cr(s), E^o = -0.74$ V

The tendency to act as reducing agent decreases in the order :

(1) $Al > Cr > Fe^{2+} > Co^{2+}$ (2) $Al > Fe^{2+} > Cr > Co^{2+}$

(3) $Al > Cr > Co^{2+} > Fe^{2+}$ (4) $Cr > Fe^{2+} > Al > Co^{2+}$

2. The group 14 elements A and B have the first ionisation enthalpy values of 708 and 715 kJ mol^{-1} respectively. The above values are lowest among their group members. The nature of their ions A^{2+} B^{4+} respectively is [April 7, 2025 (I)]

(1) both reducing (2) both oxidising

(3) reducing and oxidising (4) oxidising and reducing

3. Which of the following cannot function as an oxidising agent? [Jan. 27, 2024 (II)]

(1) N^{3-} (2) SO_4^{2-} (3) BrO_3^- (4) MnO_4^-

4. The compound that cannot act both as oxidising and reducing agent is [Jan. 09, 2020 (I)]

(1) H_3PO_4 (2) HNO_2 (3) H_2SO_3 (4) H_2O_2

5. The redox reaction among the following is: [Jan. 07, 2020 (II)]

(1) formation of ozone from atmospheric oxygen in the presence of sunlight

(2) reaction of $[Co(H_2O)_6]Cl_3$ with $AgNO_3$

(3) reaction of H_2SO_4 with NaOH

(4) combination of dinitrogen with dioxygen at 2000 K

6. Which of the following reactions is an example of a redox reaction? [2017, 2002 (S)]

(1) $XeF_4 + O_2F_2 \rightarrow XeF_6 + O_2$

(2) $XeF_2 + PF_5 \rightarrow [XeF]^+ PF_6^-$

(3) $XeF_6 + H_2O \rightarrow XeOF_4 + 2\,HF$

(4) $XeF_6 + 2\,H_2O \rightarrow XeO_2F_2 + 4\,HF$

7. Copper becomes green when exposed to moist air for a long period. This is due to: [Online April 12, 2014]

(1) the formation of a layer of cupric oxide on the surface of copper.

(2) the formation of a layer of basic carbonate of copper on the surface of copper.

(3) the formation of a layer of cupric hydroxide on the surface of copper.

(4) the formation of basic copper sulphate layer on the surface of the metal.

8. Which of the following chemical reactions depict the oxidizing behaviour of H_2SO_4? [2006]

(1) $NaCl + H_2SO_4 \longrightarrow NaHSO_4 + HCl$

(2) $2\,PCl_5 + H_2SO_4 \longrightarrow 2\,POCl_3 + 2\,HCl + SO_2Cl_2$

(3) $2\,HI + H_2SO_4 \longrightarrow I_2 + SO_2 + 2\,H_2O$

(4) $Ca(OH)_2 + H_2SO_4 \longrightarrow CaSO_4 + 2\,H_2O$

NUMERICAL VALUE QUESTIONS (NVQs)

9. 1 mole of PbS is oxidised by "X" moles of O_3 to get "Y" moles of O_2. X + Y = __________. [Jan. 27, 2024 (II)]

Topic 2 - Oxidation Number

MULTIPLE CHOICE QUESTIONS (MCQs)

10. Oxidation number of potassium in K_2O, K_2O_2 and KO_2, respectively, is: [Jan. 07, 2020 (I)]

(1) +2, +1 and $+\frac{1}{2}$ (2) +1, +1 and +1

(3) +1, +4 and +2 (4) +1, +2 and +4

11. The oxidation states of Cr in $[Cr(H_2O)_6]Cl_3$, $[Cr(C_6H_6)_2]$, and $K_2[Cr(CN)_2(O)_2(O_2)(NH_3)]$ respectively are : [2018]

(1) +3, +4, and +6 (2) +3, +2, and +4

(3) +3, 0, and +6 (4) +3, 0, and +4

12. In which of the following reactions, hydrogen peroxide acts as an oxidizing agent ? [Online April 8, 2017]

(1) $HOCl + H_2O_2 \rightarrow H_3O^+ + Cl^- + O_2$

(2) $I_2 + H_2O_2 + 2OH^- \rightarrow 2I^- + 2H_2O + O_2$

(3) $2MnO_4^- + 3H_2O_2 \rightarrow 2MnO_2 + 3O_2 + 2H_2O + 2OH^-$

(4) $PbS + 4H_2O_2 \rightarrow PbSO_4 + 4H_2O$

13. Amongst the following, identify the species with an atom in + 6 oxidation state: [Online April 19, 2014]

(1) $[MnO_4]^-$ (2) $[Cr(CN)_6]^{3-}$

(3) Cr_2O_3 (4) CrO_2Cl_2

14. Oxidation state of sulphur in anions SO_3^{2-}, $S_2O_4^{2-}$ and $S_2O_6^{2-}$ increases in the orders : [Online April 22, 2013]
(1) $S_2O_6^{2-} < S_2O_4^{2-} < SO_3^{2-}$
(2) $SO_3^{2-} < S_2O_4^{2-} < S_2O_6^{2-}$
(3) $S_2O_4^{2-} < SO_3^{2-} < S_2O_6^{2-}$
(4) $S_2O_4^{2-} < S_2O_6^{2-} < SO_3^{2-}$

15. Consider the reaction:
$$H_2SO_3(aq) + Sn^{4+}(aq) + H_2O(l) \rightarrow Sn^{2+}(aq) + HSO_4^-(aq) + 3H^+(aq)$$
Which of the following statements is correct? [Online April 19, 2014]
(1) Sn^{4+} is the oxidizing agent because it undergoes oxidation
(2) Sn^{4+} is the reducing agent because it undergoes oxidation
(3) H_2SO_3 is the reducing agent because it undergoes oxidation
(4) H_2SO_3 is the reducing agent because it undergoes reduction

16. Which one of the following cannot function as an oxidising agent ? [Online April 25, 2013]
(1) I^- (2) S (s)
(3) $NO_3^-(aq)$ (4) $Cr_2O_7^{2-}$

Numerical Value Questions (NVQs)

17. From the given list, the number of compounds with + 4 oxidation state of Sulphur ________.
SO_3, H_2SO_3, $SOCl_2$, SF_4, $BaSO_4$, $H_2S_2O_7$ [Jan. 27, 2024 (I)]

18. In Chromyl chloride, the oxidation state of chromium is (+) ____________ [April 15, 2023 (I)]

19. In ammonium–phosphomolybdate, the oxidation state of Mo is .... [April 06, 2023 (I)]

20. On reaction with stronger oxidizing agent like KIO_4, hydrogen peroxide with the evolution of O_2. The oxidation number of I in KIO_4 changes to ______. [July 28, 2022 (II)]

21. The oxidation states of iron atoms in compounds (A), (B) and (C), respectively, are *x*, *y* and *z*. The sum of *x*, *y* and *z* is __________. [Sep. 02, 2020 (I)]

$Na_4[Fe(CN)_5(NOS)]$ (A) $Na_4[FeO_4]$ (B) $[Fe_2(CO)_9]$ (C)

22. The oxidation states of transition metal atoms in $K_2Cr_2O_7$, $KMnO_4$ and K_2FeO_4, respectively, are *x*, *y* and *z*. The sum of *x*, *y* and *z* is _______. [Sep. 02, 2020 (II)]

Topic 3 - Disproportionation and Balancing of Redox Reactions

Multiple Choice Questions (MCQs)

23. Match the List-I with List-II [Jan. 28, 2025 (I)]

List-I (Redox Reaction)	List-II (Type of Redox Reaction)
(A) $CH_4(g) + 2O_2(g) \xrightarrow{\Delta} CO_2(g) + 2H_2O(l)$	(I) Disproportionation reaction
(B) $2NaH(s) \xrightarrow{\Delta} 2Na(s) + H_2(g)$	(II) Combination reaction
(C) $V_2O_5(s) + 5Ca(s) \xrightarrow{\Delta} 2V(s) + 5CaO(s)$	(III) Decomposition reaction
(D) $2H_2O_2(aq) \xrightarrow{\Delta} 2H_2O(l) + O_2(g)$	(IV) Displacement reaction

Choose the correct answer from the options given below :
(1) A-II, B-III, C-IV, D-I (2) A-II, B-III, C-I, D-IV
(3) A-III, B-IV, C-I, D-II (4) A-IV, B-I, C-II, D-III

24. The species which does not undergo disproportionation reaction is : [Jan. 22, 2025 (II)]
(1) ClO_2^- (2) ClO_4^- (3) ClO^- (4) ClO_3^-

25. Thiosulphate reacts differently with iodine and bromine in the reaction given below : [April 8, 2024 (I)]
$$2S_2O_3^{2-} + I_2 \rightarrow S_4O_6^{2-} + 2I^-$$
$$S_2O_3^{2-} + 5Br_2 + 5H_2O \rightarrow 2SO_4^{2-} + 4Br^- + 10H^+$$
Which of the following statement justifies the above dual behaviour of thiosulphate?
(1) Bromine undergoes oxidation and iodine undergoes reduction by iodine in these reactions
(2) Thiosulphate undergoes oxidation by bromine and reduction by iodine in these reaction
(3) Bromine is a stronger oxidant than iodine
(4) Bromine is a weaker oxidant than iodine

26. Among the following halogens
F_2, Cl_2, Br_2 and I_2 [April 8, 2024 (I)]
Which can undergo disproportionation reaction?
(1) Only I_2 (2) Cl_2, Br_2 and I_2
(3) F_2, Cl_2 and Br_2 (4) F_2 and Cl_2

27. Match List-I with List-II. [April 6, 2024 (II)]

List-I (Reaction)	List-II (Type of redox reaction)
(A) $N_2(g) + O_2(g) \rightarrow 2NO(g)$	(I) Decomposition
(B) $2Pb(NO_3)_2(s) \rightarrow 2PbO(s) + 4NO_2(g) + O_2(g)$	(II) Displacement
(C) $2Na(s) + 2H_2O(l) \rightarrow 2NaOH(aq.) + H_2(g)$	(III) Disproportionation
(D) $2NO_2(g) + 2^-OH(aq.) \rightarrow NO_2^-(aq.) + NO_3^-(aq.) + H_2O(l)$	(IV) Combination

Choose the correct answer from the options given below:
(1) (A)-(I), (B)-(II), (C)-(III), (D)-(IV)
(2) (A)-(III), (B)-(II), (C)-(I), (D)-(IV)
(3) (A)-(II), (B)-(III), (C)-(IV), (D)-(I)
(4) (A)-(IV), (B)-(I), (C)-(II), (D)-(III)

28. In acidic medium, $K_2Cr_2O_7$ shows oxidising action as represented in the half reaction:
$$Cr_2O_7^{2-} + XH^+ + Ye^\ominus \rightarrow 2A + ZH_2O$$
X, Y, Z and A are respectively are: [Feb. 1, 2024 (I)]
(1) 8, 6, 4 and Cr_2O_3 (2) 14, 7, 6 and Cr^{3+}
(3) 8, 4, 6 and Cr_2O_3 (4) 14, 6, 7 and Cr^{3+}

29. Which of the following reactions are disproportionation reactions ? **[Feb. 1, 2024 (I)]**

(A) $Cu^+ \rightarrow Cu^{2+} + Cu$

(B) $3MnO_4^{2-} + 4H^+ \longrightarrow 2MnO_4^- + MnO_2 + 2H_2O$

(C) $2KMnO_4 \longrightarrow K_2MnO_4 + MnO_2 + O_2$

(D) $2MnO_4^- + 3Mn^{2+} + 2H_2O \longrightarrow 5MnO_2 + 4H^+$

Choose the correct answer from the options given below:

(1) (A), (B) (2) (B), (C), (D)
(3) (A), (B), (C) (4) (A), (D)

30. Given below are two statements : **[Jan. 31, 2024 (II)]**

Statement I: S_8 solid undergoes disproportionation reaction under alkaline conditions to form S^{2-} and $S_2O_3^{2-}$

Statement II: ClO_4^- can undergo disproportionation reaction under acidic condition.

In the light of the above statements, choose the **most appropriate answer** from the options given below :

(1) Statement I is correct but statement II is incorrect.
(2) Statement I is incorrect but statement II is correct.
(3) Both statement I and statement II are incorrect.
(4) Both statement I and statement II are correct.

31. Chlorine undergoes disproportionation in alkaline medium as shown below: **[Jan. 29, 2024 (I)]**

$a\,Cl_2(g) + b\,OH^-(aq) \rightarrow c\,ClO^-(aq) + d\,Cl^-(aq) + e\,H_2O(l)$

The values of a, b, c and d in a balanced redox reaction are respectively :

(1) 2, 2, 1 and 3 (2) 1, 2, 1 and 1
(3) 2, 4, 1 and 3 (4) 3, 4, 4 and 2

32. $2IO_3^- + xI^- + 12H^+ \rightarrow 6I_2 + 6H_2O$ **[April 08, 2023 (I)]**

What is the value of x?

(1) 12 (2) 2 (3) 6 (4) 10

33. Which of the following options are correct for the reaction

$2[Au(CN)_2]^-(aq) + Zn(s) \rightarrow 2Au(s) + [Zn(CN)_4]^{-2}(aq)$ **[April 06, 2023 (I)]**

A. Redox reaction
B. Displacement reaction
C. Decomposition reaction
D. Combination reaction

Choose the correct answer from the options given below:

(1) A and B only (2) A only
(3) C and D only (4) A and D only

34. Which of the given reactions is not an example of disproportionation reaction? **[July 26, 2022 (I)]**

(1) $2H_2O_2 \rightarrow 2H_2O + O_2$
(2) $2NO_2 + H_2O \rightarrow HNO_3 + HNO_2$
(3) $MnO_4^- + 4H^+ + 3e^- \rightarrow MnO_2 + 2H_2O$
(4) $3MnO_4^{2-} + 4H^+ \rightarrow 2MnO_4^- + MnO_2 + 2H_2O$

35. Which one of the following is an example of disproportionation reaction? **[June 26, 2022 (II)]**

(1) $3MnO_4^{2-} + 4H^+ \rightarrow 2MnO_4^- + MnO_2 + 2H_2O$
(2) $MnO_4^{2-} + 4H^+ + 4e^- \rightarrow MnO_2 + 2H_2O$
(3) $10I^- + 2MnO_4^- + 16H^+ \rightarrow 2Mn^{2+} + 8H_2O + 5I_2$
(4) $8MnO_4^- + 3S_2O_3^{2-} + H_2O \rightarrow 8MnO_2 + 6SO_4^{2-} + 2OH^-$

36. The species given below that does NOT show disproportionation reaction is : **[July 20, 2021 (I)]**

(1) BrO_4^- (2) BrO^- (3) BrO_2^- (4) BrO_3^-

37. An example of a disproportionation reaction is: **[April 12, 2019 (I)]**

(1) $2\,MnO_4^- + 10\,I^- + 16\,H^+ \rightarrow 2\,Mn^{2+} + 5\,I_2 + 8\,H_2O$
(2) $2\,NaBr + Cl_2 \rightarrow 2\,NaCl + Br_2$
(3) $2\,KMnO_4 \rightarrow K_2MnO_4 + MnO_2 + O_2$
(4) $2\,CuBr \rightarrow CuBr_2 + Cu$

38. In order to oxidise a mixture of one mole of each of FeC_2O_4, $Fe_2(C_2O_4)_3$, $FeSO_4$ and $Fe_2(SO_4)_3$ in acidic medium, the number of moles of $KMnO_4$ required is: **[April 8, 2019 (I)]**

(1) 2 (2) 1 (3) 3 (4) 1.5

39. In the reaction of oxalate with permanganate in acidic medium, the number of electrons involved in producing one molecule of CO_2 is: **[Jan. 10, 2019 (II)]**

(1) 1 (2) 10 (3) 2 (4) 5

40. How many electrons are involved in the following redox reaction? **[Online April 19, 2014]**

$Cr_2O_7^{2-} + Fe^{2+} + C_2O_4^{2-} \rightarrow Cr^{3+} + Fe^{3+} + CO_2$ (Unbalanced)

(1) 3 (2) 4 (3) 6 (4) 5

41. In the following balanced reaction,

$$X\,MnO_4^- + Y\,C_2O_4^{2-} + Z\,H^+ \rightleftharpoons X\,Mn^{2+} + 2Y\,CO_2 + \frac{Z}{2}H_2O$$

values of *X*, *Y* and *Z* respectively are **[Online May 12, 2012; 2013]**

(1) 2, 5, 16 (2) 8, 2, 5
(3) 5, 2, 16 (4) 5, 8, 4

42. Given :

$X\,Na_2HAsO_3 + Y\,NaBrO_3 + Z\,HCl \rightarrow NaBr + H_3AsO_4 + NaCl$

The values of X, Y and Z in the above redox reaction are respectively : **[Online April 9, 2013, 2012 (S)]**

(1) 2, 1, 2 (2) 2, 1, 3
(3) 3, 1, 6 (4) 3, 1, 4

NUMERICAL VALUE QUESTIONS (NVQs)

43. Consider the following reactions: **[Jan. 21, 2026 (I)]**

$NaCl + K_2Cr_2O_7 + H_2SO_4 \rightarrow A + KHSO_4 + NaHSO_4 + H_2O$
$A + NaOH \rightarrow B + NaCl + H_2O$
$B + H_2SO_4 + H_2O_2 \rightarrow C + Na_2SO_4 + H_2O$

In the product 'C', 'X' is the number of O_2^{2-} units, 'Y' is the total number oxygen atoms present and 'Z' is the oxidation state of Cr. The value of X + Y + Z is ______ .

44. Number of moles of H^+ ions required by 1 mole of MnO_4^- to oxidise oxalate ion to CO_2 is ____. **[Jan. 31, 2024 (II)]**

45. $2MnO_4^- + bI^- + cH_2O \rightarrow xI_2 + yMnO_2 + z\overline{O}H$

If the above equation is balanced with integer coefficients, the value of z is ________ . **[Jan. 30, 2024 (I)]**

46. Total number of species from the following which can undergo disproportionation reaction is _______ .

$H_2O_2, ClO_3^-, P_4, Cl_2, Ag, Cu^{+1}, F_2, NO_2, K^+$

[Jan. 30, 2024 (II)]

47. See the following chemical reaction: [April 13, 2023 (II)]

$Cr_2O_7^{2-} + XH^+ + 6Fe^{2+} \rightarrow YCr^{3+} + 6Fe^{3+} + ZH_2O$

The sum of X. Y and Z is ________.

48. $2\,MnO_4^- + b\,C_2O_4^{2-} + c\,H^+ \rightarrow x\,Mn^{2+} + y\,CO_2 + z\,H_2O$

If the above equation is balanced with integer coefficients, the value of c is _____. [March 16, 2021 (I)]

49. The reaction of sulphur in alkaline medium is given below:

$S_8(s) + a\,OH^-(aq) \rightarrow b\,S^{2-}(aq) + c\,S_2O_3^{2-}(aq) + d\,H_2O(l)$

The values of 'a' is ________. (Integer answer) [Feb. 24, 2021 (I)]

50. Consider the following equations :

$2\,Fe^{2+} + H_2O_2 \rightarrow x\,A + y\,B$ (in basic medium)

$2\,MnO_4^- + 6\,H^+ + 5\,H_2O_2 \rightarrow x'C + y'D + z'E$

(in acidic medium)

The sum of the stoichiometric coefficients x, y, x', y' and z' for products A, B, C, D and E, respectively, is ________. [Sep. 04, 2020 (II)]

Part-B (JEE Advanced)

Topic 1- Oxidation and Reduction Reactions

1 MCQs with One Correct Answer

1. Of the halide ions, _____ is the most powerful reducing agent. [1978]

5 True / False

2. Cu^+ disproportionates to Cu^{2+} and elemental copper in solution. [1991 - 1 Mark]

3. Copper metal reduces Fe^{2+} in an acid medium. [1982 - 1 Mark]

7 Match the Following

4. Match the reactions in Columns I with nature of the reactions/ type of the products in Column II. Indicate your answer by darkening the appropriate bubbles of the 4 × 4 matrix given in the ORS. [2007]

Column I	Column II
(A) $O_2^- \rightarrow O_2 + O_2^{2-}$	(p) redox reaction
(B) $CrO_4^{2-} + H^+ \rightarrow$	(q) one of the products has trigonal planar structure
(C) $MnO_4^- + NO_2^- + H^+ \rightarrow$	(r) dimeric bridged tetrahedral metal ion
(D) $NO_3^- + H_2SO_4 + Fe^{2+} \rightarrow$	(s) disproportionation

Topic 2 - Oxidation Number

1 MCQs with One Correct Answer

5. Consider a titration of potassium dichromate solution with acidified Mohr's salt solution using diphenylamine as indicator. The number of moles of Mohr's salt required per mole of dichromate is [2007]

(a) 3 (b) 4 (c) 5 (d) 6

6. The pair of the compounds in which both the metals are in the highest possible oxidation state is [2004S]

(a) $[Fe(CN)_6]^{3-}, [Co(CN)_6]^{3-}$ (b) CrO_2Cl_2, MnO_4^-

(c) TiO_3, MnO_2 (d) $[Co(CN)_6]^{3-}, MnO_3$

7. Amongst the following identify the species with an atom in +6 oxidation state [2000S]

(a) MnO_4^- (b) $Cr(CN)_6^{3-}$

(c) NiF_6^{2-} (d) CrO_2Cl_2

8. The oxidation number of sulphur in S_8, S_2F_2, H_2S respectively, are [1999 - 2 Marks]

(a) 0, +1 and –2 (b) +2, +1 and –2

(c) 0, +1 and +2 (d) –2, +1 and –2

9. The oxidation states of the most electronegative element in the products of the reaction, BaO_2 with dil. H_2SO_4 is [1991 - 1 Mark]

(a) 0 and –1 (b) –1 and –2

(c) –2 and 0 (d) –2 and +1

10. The oxidation number of phosphorus in $Ba(H_2PO_2)_2$ is : [1990 - 1 Mark]

(a) +3 (b) +2 (c) +1 (d) –1

11. The brown ring complex compound is formulated as $[Fe(H_2O)_5(NO)]SO_4$. The oxidation state of iron is : [1987 - 1 Mark]

(a) 1 (b) 2 (c) 3 (d) 0

12. The oxidation number of carbon in CH_2O is [1982 - 1 Mark]

(a) – 2 (b) + 2 (c) 0 (d) + 4

13. One mole of N_2H_4 loses ten moles of electrons to form a new compound Y. Assuming that all the nitrogen appears in the new compound, what is the oxidation state of nitrogen in Y? (There is no change in the oxidation state of hydrogen). [1981 - 1 Mark]

(a) –1 (b) –3 (c) +3 (d) +5

2 Integer Value Answer

14. Consider the following molecules: Br_3O_8, F_2O, $H_2S_4O_6$, $H_2S_5O_6$, and C_3O_2.

Count the number of atoms existing in their zero oxidation state in each molecule. [Adv. 2023]

Their sum is_______.

15. The difference in the oxidation numbers of the two types of sulphur atoms in $Na_2S_4O_6$ is [2011]

4 Fill in the Blanks

16. The compound $YBa_2Cu_3O_7$, which shows super-conductivity, has copper in oxidation state.................., assume that the rare earth element yttrium is in its usual + 3 oxidation state. [1994 - 1 Mark]

10 Subjective Problems

17. Arrange the following in increasing oxidation number of iodine. I_2, HI, HIO_4, ICl [1986 - 1 Mark]

Topic 3 - Disproportionation and Balancing of Redox Reactions

1 MCQs with One Correct Answer

18. The reaction, $3ClO^-(aq) \rightarrow ClO_3^-(aq) + 2Cl^-(aq)$, is an example of [2001S]
(a) oxidation reaction
(b) reduction reaction
(c) disproportionation reaction
(d) decomposition reaction

19. For the redox reaction : [1992 - 1 Mark]

$$MnO_4^- + C_2O_4^{2-} + H^+ \rightarrow Mn^{2+} + CO_2 + H_2O$$

the correct coefficients of the reactants for the balanced reaction are [1992 - 1 Mark]

	MnO_4^-	$C_2O_4^{2-}$	H^+
(a)	2	5	16
(b)	16	5	2
(c)	5	16	2
(d)	2	16	5

6 MCQs with One or More than One Correct Answer

20. For the reaction [Adv. 2014]

$$I^- + ClO_3^- + H_2SO_4 \rightarrow Cl^- + HSO_4^- + I_2$$

The correct statement(s) in the balanced equation is/are
(a) Stoichiometric coefficient of HSO_4^- is 6
(b) Iodide is oxidized
(c) Sulphur is reduced
(d) H_2O is one of the products

10 Subjective Problems

21. Complete and balance the following reactions :
(i) $Ag^+ + AsH_3 \rightarrow H_3AsO_3 + H^+$ [1986 - 1 Mark]
(ii) $ClO_3^- + I^- + H_2SO_4 \rightarrow Cl^- + HSO_4^-$ [1986 - 1 Mark]
(iii) $S + OH^- \rightarrow S^{2-} + S_2O_3^{2-}$ [1986 - 1 Mark]
(iv) $Mn^{2+} + PbO_2 \rightarrow MnO_4^- + H_2O$ [1986 - 1 Mark]
(v) $Cl_2 + OH^- \rightarrow Cl^- + ClO^-$ [1983 - 1 Mark]
(vi) $Ce^{3+} + S_2O_8^{2-} \rightarrow SO_4^{2-} + Ce^{4+}$ [1983 - 1 Mark]
(vii) $HNO_3 + HCl \rightarrow NO + Cl_2$ [1983 - 1 Mark]
(viii) $Cr_2O_7^{2-} + C_2H_4O \rightarrow C_2H_4O_2 + Cr^{3+}$ [1983 - 1 Mark]
(ix) $Zn + NO_3^- \rightarrow Zn^{2+} + NH_4^+$ [1983 - 1 Mark]

22. Balance the following equations.
(i) $Cu_2O + H^+ + NO_3^- \rightarrow Cu^{2+} + NO + H_2O$ [1981 - 1 Mark]
(ii) $K_4[Fe(CN)_6] + H_2SO_4 + H_2O \rightarrow K_2SO_4 + FeSO_4 + (NH_4)_2SO_4 + CO$ [1981 - 1 Mark]
(iii) $C_2H_5OH + I_2 + OH^- \rightarrow CHI_3 + HCO_3^- + I^- + H_2O$ [1981 - 1 Mark]

Answer Key

Part-A (JEE Main)

1	(1)	6	(1)	11	(3)	16	(1)	21	(6)	26	(2)	31	(2)	36	(1)	41	(1)	46	(6)
2	(3)	7	(2)	12	(4)	17	(3)	22	(19)	27	(4)	32	(4)	37	(4)	42	(3)	47	(23)
3	(1)	8	(3)	13	(4)	18	(6)	23	(1)	28	(4)	33	(1)	38	(1)	43	(13)	48	(16)
4	(1)	9	(8)	14	(3)	19	(6)	24	(2)	29	(1)	34	(3)	39	(1)	44	(8)	49	(12)
5	(4)	10	(2)	15	(3)	20	(5)	25	(3)	30	(1)	35	(1)	40	(3)	45	(8)	50	(19)

Part-B (JEE Advanced)

1	I^-	4	(A - p, s); (B - r); (C - p, q); (D - p)					13	(c)	16	(+7/3)	20	(a, b, d)		
2	True	5	(d)	7	(d)	9	(b)	11	(b)	14	(6)	18	(c)		
3	False	6	(b)	8	(a)	10	(c)	12	(c)	15	(5)	19	(a)		

Chapter-9

Hydrogen

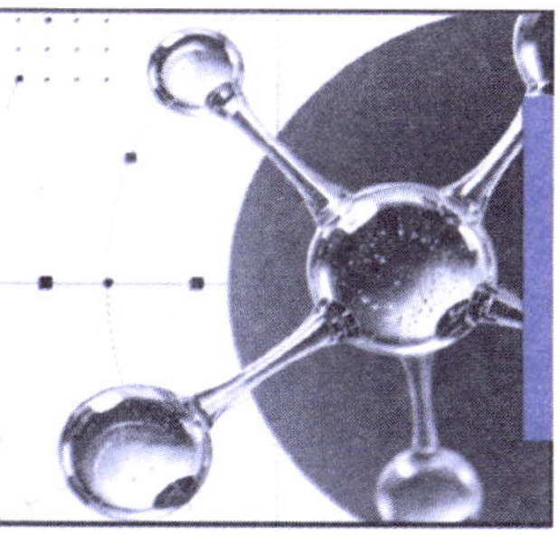

Part-A (JEE Main)

Topic 1 - Preparation and Properties of Hydrogen and Hydrides

MULTIPLE CHOICE QUESTIONS (MCQs)

1. High purity (>99.95%) dihydrogen is obtained by [July 25, 2022 (II)]
   (1) react oin of zinc with aqueous alkali.
   (2) electrolysis of acidified water using platinum electrodes.
   (3) electrolysis of warm aqeuous barium hydroxide solution between nickel electrodes.
   (4) reaction of zinc with dilute acid.

2. Match List - I with List - II: [June 30, 2022 (I)]

List - I	List - II
(A) Sodium hydride	(i) Lewis acid
(B) Silane	(ii) Saline hydride
(C) Vanadium hydride	(iii) Molecular hydride
(D) Aluminium hydride	(iv) Non-stoichiometric hydride

   (1) (A)-(ii), (B)-(iii), (C)-(iv), (D)-(i)
   (2) (A)-(ii), (B)-(iv), (C)-(iii), (D)-(i)
   (3) (A)-(iii), (B)-(ii), (C)-(i), (D)-(iv)
   (4) (A)-(i), (B)-(iii), (C)-(iv), (D)-(ii)

3. The highest industrial consumption of molecular hydrogen is to produce compounds of element: [June 24, 2022 (I)]
   (1) Carbon (2) Nitrogen
   (3) Oxygen (4) Chlorine

4. Deuterium resembles hydrogen in properties but :
   (1) reacts slower than hydrogen [Aug. 27, 2021 (I)]
   (2) reacts vigorously than hydrogen
   (3) reacts just as hydrogen
   (4) emits β^+ particles

5. The number of neutrons and electrons, respectively, present in the radioactive isotope of hydrogen is :-
   [July 27, 2021 (II); Similar Jan. 10, 2019 (I); Jan. 09, 2019 (I)]
   (1) 1 and 1 (2) 3 and 1
   (3) 2 and 1 (4) 2 and 2

6. The single largest industrial application of dihydrogen is : [July 20, 2021 (II)]
   (1) Manufacture of metal hydrides
   (2) Rocket fuel in space research
   (3) In the synthesis of ammonia
   (4) In th synthesis of nitric acid

7. Dihydrogen of high purity (> 99.95%) is obtained through:
   (1) the reaction of Zn with dilute HCl. [Sep. 06, 2020 (II)]
   (2) the electrolysis of acidified water using Pt electrodes.
   (3) the electrolysis of bringe solution.
   (4) the electrolysis of warm $Ba(OH)_2$ solution using Ni electrodes.

8. The equation that represents the water-gas shift reaction is: [2008 (S), Sep. 05, 2020 (I)]
   (1) $CH_4(g)+H_2O(g)\xrightarrow[Ni]{1270K}CO(g)+3H_2(g)$
   (2) $2C(s)+O_2(g)+4N_2(g)\xrightarrow{1273K}2CO(g)+4N_2(g)$
   (3) $C(s)+H_2O(g)\xrightarrow{1270K}CO(g)+H_2(g)$
   (4) $CO(g)+H_2O(g)\xrightarrow[Catalyst]{673K}CO_2(g)+H_2(g)$

9. 5 g of zinc is treated separately with an excess of
   (A) dilute hydrochloric acid and
   (B) aqueous sodium hydroxide.
   The ratio of the volumes of H_2 evolved in these two reactions is: [Jan. 09, 2020 (II)]
   (1) 1 : 2 (2) 1 : 1 (3) 1 : 4 (4) 2 : 1

10. Hydrogen has three isotopes (A), (B) and (C). If the number of neutron(s) in (A), (B) and (C) respectively, are (x), (y) and (z), the sum of (x), (y) and (z) is: [Jan. 08, 2020 (II)]
   (1) 3 (2) 2 (3) 4 (4) 1

11. The metal that gives hydrogen gas upon treatment with both acid as well as base is : [April 12, 2019 (I)]
   (1) magnesium
   (2) mercury
   (3) zinc
   (4) iron

12. The correct statements among (i) to (iv) are : [April 10, 2019 (II)]
   (i) saline hydrides produce H_2 gas when reacted with H_2O.
   (ii) reaction of $LiAlH_4$ with BF_3 leads to B_2H_6.
   (iii) PH_3 and CH_4 are electron – rich and electron – precise hydrides, respectively,
   (iv) HF and CH_4 are called as molecular hydrides.
   (1) (i), (ii), (iii) and (iv) (2) (iii) and (iv) only
   (3) (i), (iii) and (iv) only (4) (i), (ii) and (iii) only

13. The correct statements among (1) to (4) regarding H_2 as a fuel are : [Jan. 11, 2019 (I)]
(i) It produces less pollutants than petrol.
(ii) A cylinder of compressed dihydrogen weighs ~ 30 times more than a petrol tank producing the same amount of energy.
(iii) Dihydrogen is stored in tanks of metal alloys like $NaNi_5$.
(iv) On combustion, values of energy released per gram of liquid dihydrogen and LPG are 50 and 142 kJ, respectively.
(1) (ii) and (iv) only (2) (i) and (iii) only
(3) (ii), (iii) and (iv) only (4) (i), (ii) and (iii) only

14. NaH is an example of: [Jan. 11, 2019 (I)]
(1) Electron-rich hydride (2) Metallic hydride
(3) Saline hydride (4) Molecular hydride

15. Among the following reactions of hydrogen with halogens, the one that requires a catalyst is: [Jan. 10, 2019 (II)]
(1) $H_2 + I_2 \rightarrow 2\,HI$ (2) $H_2 + Cl_2 \rightarrow 2\,HCl$
(3) $H_2 + Br_2 \rightarrow 2\,HBr$ (4) $H_2 + F_2 \rightarrow 2\,HF$

16. Which physical property of dihydrogen is wrong ? [Online April 11, 2015]
(1) Odourless gas (2) Tasteless gas
(3) Colourless gas (4) Non–inflammable gas

17. Very pure hydrogen (99.9) can be made by which of the following processes ? [2012]
(1) Reaction of methane with steam
(2) Mixing natural hydrocarbons of high molecular weight
(3) Electrolysis of water
(4) Reaction of salts like hydrides with water

18. Which of the following statements in relation to the hydrogen atom is correct ? [2005]
(1) 3*s*, 3*p* and 3*d* orbitals all have the same energy
(2) 3*s* and 3*p* orbitals are of lower energy than 3*d* orbital
(3) 3*p* orbital is lower in energy than 3*d* orbital
(4) 3*s* orbital is lower in energy than 3*p* orbital

Topic 2 - Preparation and Properties of H_2O and D_2O

MULTIPLE CHOICE QUESTIONS (MCQs)

19. The metal salts formed during softening of hardwater using Clark's method are : [July 28, 2022 (I)] [July 28, 2022 (II)]
(1) $Ca(OH)_2$ and $Mg(OH)_2$ (2) $CaCO_3$ and $Mg(OH)_2$
(3) $Ca(OH)_2$ and $MgCO_3$ (4) $CaCO_3$ and $MgCO_3$

20. Consider the following reaction :

$$2HSO_4^-(aq) \xrightarrow[(2)\,\text{Hydrolysis}]{(1)\,\text{Electrolysis}} 2HSO_4^- + 2H^+ + A$$

The dihedral angle in product **A** in its solid phase at 110 K is : [June 26, 2022 (I)]
(1) 104° (2) 111.5° (3) 90.2° (4) 111.0°

21. Boiling of hard water is helpful in removing the temporary hardness by converting calcium hydrogen carbonate and magnesium hydrogen carbonate to [June 26, 2022 (II)]
(1) $CaCO_3$ and $Mg(OH)_2$ (2) $CaCO_3$ and M_2CO_3
(3) $Ca(OH)_2$ and $MgCO_3$ (4) $Ca(OH)_2$ and $Mg(OH)_2$

22. Which one of the following methods is most suitable for preparing deionized water? [Aug. 26, 2021 (I)]
(1) Synthetic resin method (2) Clark's method
(3) Calgon's method (4) Permutit method

23. The water having more dissolved O_2 is : [July 22, 2021 (II)]
(1) boiling water (2) water at 80 °C
(3) polluted water (4) water at 4 °C

24. The functional groups that are responsible for the ion-exchange property of cation and anion exchange resins, respectively, are [March 17, 2021 (II)]
(1) $-SO_3H$ and $-NH_2$ (2) $-NH_2$ amd $-COOH$
(3) $-SO_3H$ and $-COOH$ (4) $-NH_2$ and $-SO_3H$

25. The one that is NOT suitable for the removal of permanent hardness of water is : [Sep. 05, 2020 (II)]
(1) Clark's method
(2) Ion-exchange method
(3) Calgon's method
(4) Treatment with sodium carbonate

26. Amongst the following, the form of water with the lowest ionic conductance at 298 K is : [Jan. 09, 2020 (II)]
(1) distilled water
(2) saline water used for intravenous injection
(3) water from a well
(4) sea water

27. In comparison to the zeolite process for the removal of permanent hardness, the synthetic resins method is: [Jan. 07, 2020 (I)]
(1) less efficient as it exchanges only anions
(2) more efficient as it can exchange both cations as well as anions
(3) less efficient as the resins cannot be regenerated
(4) more efficient as it can exchange only cations

28. The temporary hardness of a water sample is due to compound X. Boiling this sample converts X to compound Y. X and Y, respectively, are : [April 12, 2019 (II)]
(1) $Mg(HCO_3)_2$ and $Mg(OH)_2$
(2) $Ca(HCO_3)_2$ and $Ca(OH)_2$
(3) $Mg(HCO_3)_2$ and $MgCO_3$
(4) $Ca(HCO_3)_2$ and CaO

29. The synonym for water gas when used in the production of methanol is : [April 10, 2019 (I)]
(1) natural gas (2) fuel gas
(3) laughing gas (4) syn gas

30. The number of water molecules(s) not coordinated to copper ion directly in $CuSO_4\cdot5H_2O$, is: [April 9, 2019 (I)]
(1) 2 (2) 3 (3) 1 (4) 4

31. 100 mL of a water sample contains 0.81 g of calcium bicarbonate and 0.73 g of magnesium bicarbonate. The hardness of this water sample expressed in terms of equivalents of $CaCO_3$ is: (molar mass of calcium bicarbonate is 162 g mol^{-1} and magnesium bicarboante is 146 g mol^{-1}) [April 8, 2019 (I)]
(1) 5,000 ppm (2) 1,000 ppm
(3) 100 ppm (4) 10,000 ppm

32. The temporary hardness of water is due to: [Jan. 9, 2019 (II)]
(1) Na_2SO_4 (2) NaCl (3) $Ca(HCO_3)_2$ (4) $CaCl_2$

33. Which one of the following statements about water is **FALSE**? [2016]
(1) There is extensive intramolecular hydrogen bonding in the condensed phase.
(2) Ice formed by heavy water sinks in normal water.

(3) Water is oxidized to oxygen during photosynthesis.
(4) Water can act both as an acid and as a base.

34. Identify the incorrect statement regarding heavy water : [Online April 9, 2016]
(1) It reacts with SO_3 to form deuterated sulphuric acid (D_2SO_4).
(2) It is used as a coolant in nuclear reactors.
(3) It reacts with CaC_2 to produce C_2D_2 and $Ca(OD)_2$.
(4) It reacts with Al_4C_3 to produce CD_4 and $Al(OD)_3$.

35. Permanent hardness in water cannot be cured by : [Online April 10, 2015]
(1) Treatment with washing soda
(2) Boiling
(3) Calgon's method
(4) Ion exchange method

36. The numbers of protons, electrons and neutrons in a molecule of heavy water are respectively : [Online April 23, 2013]
(1) 8, 10, 11 (2) 10, 10, 10
(3) 10, 11, 10 (4) 11, 10, 10

NUMERICAL VALUE QUESTIONS (NVQs)

37. The hardness of a water sample containing 10^{-3} M $MgSO_4$ expressed as $CaCO_3$ equivalents (in ppm) is ______ .
(molar mass of $MgSO_4$ is 120.37 g/mol)
[Jan. 09, 2020 (I), Jan 12, 2019 (I) (S)]

Topic 3 - Preparation and Properties of H_2O_2

MULTIPLE CHOICE QUESTIONS (MCQs)

38. Given below are two statements:
Statement I : Hydrogen peroxide can act as an oxidizing agent in both acidic and basic conditions.
Statement II: Density of hydrogen peroxide at 298 K is lower than that of D_2O.
In the light of the above statements. Choose the **correct** answer from the options given below: [July 27, 2022 (I)]
(1) Both statement I and statement II are ture
(2) Both statement I and statement II are false
(3) Statement I is true but statement II is false
(4) Statement I is false but statement II is true

39. Which of the following can be used to prevent the decomposition of H_2O_2? [July 26, 2022 (I)]
(1) Urea (2) Formaldehyde
(3) Formic acid (4) Ethanol

40. The products obtained from a reaction of hydrogen peroxide and acidified potassium permanganate are [July 26, 2022 (II)]
(1) Mn^{4+}, H_2O only
(2) Mn^{2+}, H_2O only
(3) Mn^{4+}, H_2O, O_2 only
(4) Mn^{2+}, H_2O, O_2 only

41. The reaction of H_2O_2 with potassium permanganate in acidic medium leads to the formation of mainly: [July 25, 2022 (I)]
(1) Mn^{2+} (2) Mn^{4+} (3) Mn^{3+} (4) Mn^{6+}

42. Which one of the following reactions indicates the reducing ability of hydrogen peroxide in basic medium ? [June 29, 2022 (I)]
(1) $HOCl + H_2O_2 \rightarrow H_3O^+ + Cl^- + O_2$
(2) $PbS + 4H_2O_2 \rightarrow PbSO_4 + 4H_2O$
(3) $2MnO_4^- + 3H_2O_2 \rightarrow 2MnO_2 + 3O_2 + 2H_2O + 2OH^-$
(4) $Mn^{2+} + H_2O_2 \rightarrow Mn^{4+} + 2OH^-$

43. In basic medium, H_2O_2 exhibits which of the following reactions? [March 18, 2021 (II)]
(A) $Mn^{2+} \rightarrow Mn^{4+}$ (B) $I_2 \rightarrow I^-$
(C) $PbS \rightarrow PbSO_4$
Choose the most appropriate answer the options given below.
(1) (A), (C) only (2) (A), (B) only
(3) (A) only (4) (B) only

44. The correct statements about H_2O_2 are [March 16, 2021 (II)]
(A) used in the treatment of effluents.
(B) used as both oxidising and reducing agents.
(C) the two hydroxyl groups lie in the same plane.
(D) miscible with water.
Choose the **correct** answer from the options given below
(1) (A), (B), (C) and (D)
(2) (A), (B) and (D) only
(3) (B), (C) and (D) only
(4) (A), (C) and (D) only

45. Hydrogen peroxide, in the pure state, is: [Sep. 05, 2020 (II)]
(1) non-planar and almost colorless
(2) linear and blue in color
(3) linear and almost colorless
(4) planar and blue in color

46. The strengths of 5.6 volume hydrogen peroxide (of density 1 g/mL) in terms of mass percentage and molarity (M), respectively, are : [Sep. 03, 2020 (II)]
(Take molar mass of hydrogen peroxide as 34 g/ mol)
(1) 1.7 and 0.5 (2) 0.85 and 0.25
(3) 1.7 and 0.25 (4) 0.85 and 0.5

47. Among statements (A)-(D), the correct ones are: [Jan. 07, 2020 (II)]
(A) Decomposition of hydrogen peroxide gives dioxygen.
(B) Like hydrogen peroxide, compounds, such as $KClO_3$, $Pb(NO_3)_2$ and $NaNO_3$ when heated liberate dioxygen.
(C) 2-Ethylanthraquinone is useful for the industrial preparation of hydrogen peroxide.
(D) Hydrogen peroxide is used for the manufacture of sodium perborate.
(1) (A) (B), (C) and (D) (2) (A), (B) and (C) only
(3) (A), (C) and (D) only (4) (A) and (C) only

48. The strength of 11.2 volume solution of H_2O_2 is : [Given that molar mass of H = 1 g mol^{-1} and O = 16 g mol^{-1}] [April 8, 2019 (II)]
(1) 13.6% (2) 3.4% (3) 34% (4) 1.7%

49. The chemical nature of hydrogen peroxide is: [Jan. 10, 2019 (I)]
(1) Oxidising agent in acidic medium, but not in basic medium.
(2) Reducing agent in basic medium, but not in acidic medium.
(3) Oxidising and reducing agent in acidic medium, but not in basic medium.
(4) Oxidising and reducing agent in both acidic and basic medium.

50. Hydrogen peroxide oxidises $[Fe(CN)_6]^{4-}$ to $[Fe(CN)_6]^{3-}$ in acidic medium but reduces $[Fe(CN)_6]^{3-}$ to $[Fe(CN)_6]^{4-}$ in alkaline medium. The other products formed are respectively: [2018]

(1) (H_2O+O_2) and H_2O (2) (H_2O+O_2) and (H_2O+OH^-)
(3) H_2O and (H_2O+O_2) (4) H_2O and (H_2O+OH^-)

51. From the following statements regarding H_2O_2, choose the incorrect statement : [2015]
(1) It has to be stored in plastic or wax lined glass bottles in dark
(2) It has to be kept away from dust
(3) It can act only as an oxidizing agent
(4) It decomposes on exposure to light

52. Hydrogen peroxide acts both as an oxidising and as a reducing agent depending upon the nature of the reacting species. In which of the following cases H_2O_2 acts as a reducing agent in acid medium? [Online April 12, 2014]
(1) MnO_4^- (2) $Cr_2O_7^{2-}$ (3) SO_3^{2-} (4) KI

NUMERICAL VALUE QUESTIONS (NVQs)

53. The volume strength of 8.9 M H_2O_2 solution calculated at 273 K and 1 atm is ________. ($R = 0.0821$ L atm K^{-1} mol^{-1}) (rounded off to the nearest integer) [Jan. 12, 2019 (II) (S), Sep. 03, 2020 (I)]

Part-B (JEE Advanced)

Topic 1 - Preparation and Properties of Hydrogen and Hydrides

9 Assertion and Reason/Statement Type Questions

Each question contains STATEMENT-1 (Assertion) and STATEMENT-2 (Reason). Each question has 4 choices (a), (b), (c) and (d) out of which ONLY ONE is correct. Mark your answer as
(a) If both Statement -1 and Statement -2 are correct, and Statement -2 is the correct explanation of the Statement -2.
(b) If both Statement -1 and Statement -2 are correct, but Statement -2 is not the correct explanation of the Statement -1.
(c) If Statement -1 is correct but Statement -2 is incorrect.
(d) If Statement -1 is incorrect but Statement -2 is correct.

1. Read the following statement and explanation and answer as per the options given statement below :
Statement - 1 : The alkali metals can form ionic hydrides which contain the hydride ion H^-.
Statement - 2 : The alkali metals have low electronegativity; their hydrides conduct electricity when fused and liberate hydrogen at the anode. [1994 - 2 Marks]

Topic 2 - Preparation and Properties of H_2O and D_2O

1 MCQs with One Correct Answer

2. Polyphosphates are used as water softening agents because they [2002S]
(a) form soluble complexes with anionic species
(b) precipitate anionic species
(c) form soluble complexes with cationic species
(d) precipitate cationic species

3. Heavy water is [1983 - 1 Mark]
(a) $H_2{}^{18}O$
(b) water obtained by repeated distillation
(c) D_2O
(d) water at 4°C

4. The temporary hardness of water due to calcium bicarbonate can be removed by adding – [1979]
(a) $CaCO_3$ (b) $Ca(OH)_2$ (c) $CaCl_2$ (d) HCl

6 MCQs with One or More than One Correct Answer

5. The reagent(s) used for softening the temporary hardness of water is (are) [2010]
(a) $Ca_3(PO_4)_2$ (b) $Ca(OH)_2$
(c) Na_2CO_3 (d) NaOCl

6. The species present in solution when CO_2 is dissolved in water are [2006 - 5M, –1]
(a) $CO_2, H_2CO_3, HCO_3^-, CO_3^{2-}$
(b) H_2CO_3, CO_3^{2-}
(c) CO_3^{2-}, HCO_3^-
(d) CO_2, H_2CO_3

7. When zeolite, which is hydrated sodium aluminium silicate, is treated with hard water the sodium ions are exchanged with [1990 - 1 Mark]
(a) H^+ ions (b) Ca^{++} ions
(c) SO_4^{--} ions (d) Mg^{++} ions
(e) OH^- ions

Topic 3 - Preparation and Properties of H_2O_2

1 MCQs with One Correct Answer

8. Hydrogen peroxide in its reaction with KIO_4 and NH_2OH respectively, is acting as a [Adv. 2014]
(a) Reducing agent, oxidising agent
(b) Reducing agent, reducing agent
(c) Oxidising agent, oxidising agent
(d) Oxidising agent, reducing agent

9. The volume strength of 1.5 N H_2O_2 solution is [1991 - 1 Mark]
(a) 4.8 (b) 8.4
(c) 3.0 (d) 8.0

3 Numeric / New Stem Based Questions

10. To a 25mL H_2O_2 solution, excess of acidified solution of potassium iodide was added. The iodine liberated required 20 mL of 0.3 N sodium thiosulphate solution. Calculate the volume strength of H_2O_2 solution. [1997 - 5 Marks]

11. A 5.0 cm^3 solution of H_2O_2 liberates 0.508 g of iodine from an acidified KI solution. Calculate the strength of H_2O_2 solution in terms of volume strength at STP. [1995 - 2 Marks]

10 Subjective Problems

12. Hydrogen peroxide acts both as an oxidising and as a reducing agent in alkaline solution towards certain first row transition metal ions. Illustrate both these properties of H_2O_2 using chemical equations. [1998 - 4 Marks]
13. Give reactions for the oxidation of hydrogen peroxide with potassium permangamate in acidic medium. [1997 - 1 Mark]
14. Give reasons for the following :
    (*i*) Hydrogen peroxide acts as an oxidising as well as a reducing agent. [1992 - 1 Mark]
    (*ii*) Hydrogen peroxide is a better oxidising agent that water. [1986 - 1 Mark]
15. The mixture of hydrazine and hydrogen peroxide with a copper(II) catalyst is used as a rocket propellant. [1987 - 1 Mark]
16. Write down the balanced equations for the reactions when: An alkaline solution of potassium ferricyanide is reacted with hydrogen peroxide. [1982 - 1 Mark]

Answer Key

Part-A (JEE Main)																	
1	(3)	**7**	(4)	**13**	(4)	**19**	(2)	**25**	(1)	**31**	(4)	**37**	(100)	**43**	(2)	**49**	(4)
2	(1)	**8**	(4)	**14**	(3)	**20**	(3)	**26**	(1)	**32**	(3)	**38**	(3)	**44**	(2)	**50**	(3)
3	(2)	**9**	(2)	**15**	(1)	**21**	(1)	**27**	(2)	**33**	(1)	**39**	(1)	**45**	(1)	**51**	(3)
4	(1)	**10**	(1)	**16**	(4)	**22**	(1)	**28**	(1)	**34**	(2)	**40**	(4)	**46**	(1)	**52**	(1)
5	(3)	**11**	(3)	**17**	(4)	**23**	(4)	**29**	(4)	**35**	(2)	**41**	(1)	**47**	(1)	**53**	(100)
6	(3)	**12**	(1)	**18**	(1)	**24**	(1)	**30**	(3)	**36**	(2)	**42**	(3)	**48**	(2)		
Part-B (JEE Advanced)																	
1	(a)	**3**	(c)	**5**	(b, c, d)	**7**	(b, d)	**9**	(b)	**11**	(4.48)						
2	(c)	**4**	(b)	**6**	(a)	**8**	(a)	**10**	(1.344)								

Chapter-10

The *s*-Block Elements

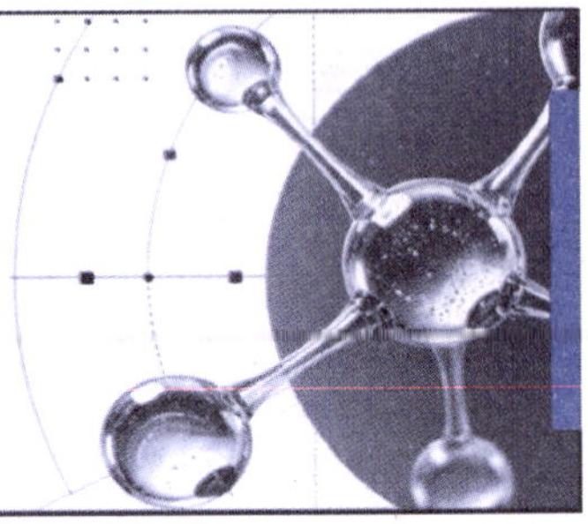

Part-A (JEE Main)

Topic 1 - Group-1 Elements (Alkali Metals)

MULTIPLE CHOICE QUESTIONS (MCQs)

1. Lithium nitrate and sodium nitrate, when heated separately, respectively, give : **[July 29, 2022 (I)]**
   (1) $LiNO_2$ and $NaNO_2$ (2) Li_2O and Na_2O
   (3) Li_2O and $NaNO_2$ (4) $LiNO_2$ and Na_2O

2. Which of the following statement is incorrect ? **[July 28, 2022 (I)]**
   (1) Low solubility of LiF in water is due to its small hydration enthalpy.
   (2) KO_2 is paramagnetic.
   (3) Solution of sodium in liquid ammonia is conducting in nature.
   (4) Sodium metal has higher density than potassium metal

3. Given below are two statements : one is labelled as **Assertion A** and the other is labelled as **Reason R**.
   Assertion A : LiF is sparingly soluble in water.
   Reason R : The ionic radius of Li^+ ion is smallest among its group members, hence it has least hydration enthalpy.
   In the light of the above statements, choose the **most appropriate** answer from the options given below : **[July 26, 2022 (II)]**
   (1) Both A and R are true and R is the correct explanation of A.
   (2) Both A and R are true but R is NOT the correct explanation of A.
   (3) A is true but R is false.
   (4) A is false but R is true.

4. Choose the **correct** order of density of the alkali metals: **[July 25, 2022 (I)]**
   (1) $Li < K < Na < Rb < Cs$ (2) $Li < Na < K < Rb < Cs$
   (3) $Cs < Rb < K < Na < Li$ (4) $Li < Na < K < Cs < Rb$

5. Correct statement about alkali metal oxides is: **[June 30, 2022 (I)]**
   (1) peroxides are coloured.
   (2) superoxides are paramagnetic.
   (3) oxides are paramagnetic.
   (4) peroxides are both coloured and paramagnetic.

6. Match the **List-I** with **List-II**. **[June 29, 2022 (I)]**

List-I (Metal)	**List-II (Emitted light wavelength (nm))**
(A) Li	(I) 670.8
(B) Na	(II) 589.2
(C) Rb	(III) 780.0
(D) Cs	(IV) 455.5

   Choose the **most appropriate** answer from the options given below:
   (1) (A)-(I), (B)-(II), (C)-(III), (D)-(IV)
   (2) (A)-(III), (B)-(II), (C)-(I), (D)-(IV)
   (3) (A)-(III), (B)-(I), (C)-(II), (D)-(IV)
   (4) (A)-(IV), (B)-(II), (C)-(I), (D)-(III)

7. Which of the following statements are correct ?
   (A) Both LiCl and $MgCl_2$ are soluble in ethanol.
   (B) The oxides Li_2O and MgO combine with excess of oxygen to give superoxide.
   (C) LiF is less soluble in water than other alkali metal fluorides.
   (D) Li_2O is more soluble in water than other alkali metal oxides.
   Choose the most appropriate answer from the options given below: **[June 24, 2022 (I)]**
   (1) (A) and (C) only (2) (A), (C) and (D) only
   (3) (B) and (C) only (4) (A) and (D) only

8. Match List-I with List-II : **[February 24, 2021 (II)]**

List-I (Salt)	**List-II (Flame colour wavelength)**
(A) LiCl	(i) 455.5 nm
(B) NaCl	(ii) 670.8 nm
(C) RbCl	(iii) 780.0 nm
(D) CsCl	(iv) 589.2 nm

   Choose the correct answer from the options given below:
   (1) (A)-(iv), (B)-(ii), (C)-(iii), (D)-(i)
   (2) (A)-(ii), (B)-(i), (C)-(iv), (D)-(iii)
   (3) (A)-(i), (B)-(iv), (C)-(ii), (D)-(iii)
   (4) (A)-(ii), (B)-(iv), (C)-(iii), (D)-(i)

9. Given below are two statements, one is labelled as **Assertion (A)** and the other is labelled as **Reason (R)**.
   Assertion (A) : Lithium salts are hydrated.
   Reason (R) : Lithium has higher polarising power than other alkali metal group members.
   In the light of the above statements, choose the **most appropriate** answer from the options given below : **[Aug. 31, 2021 (II)]**

(1) Both (A) and (R) are correct but (R) is NOT the correct explanation of (A).
(2) (A) is correct but (R) is not correct .
(3) (A) is not correct but (R) is correct.
(4) Both (A) and (R) are correct and (R) is the correct explanation of (A).

10. Choose the **correct** statement from the following : [Aug. 27, 2021 (II)]
(1) The standard enthalpy of formation for alkali metal bromides becomes less negative on descending the group.
(2) The low solubility of CsI in water is due to its high lattice enthalpy.
(3) Among the alkali metal halides, LiF is least soluble in water.
(4) LiF has least negative standard enthalpy of formation among alkali metal fluorides.

11. Match List I with List II : [July 25, 2021 (II)]

	List-I Elements		List-II Properties
(A)	Li	(i)	Poor water solubility of I^- salt
(B)	Na	(ii)	Most abundant element in cell fluid
(C)	K	(iii)	Bicarbonate salt used in fire extinguisher
(D)	Cs	(iv)	Carbonate salt decomposes easily on heating

Choose the correct answer from the options given below:
(1) (A)-(iv), (B)-(iii), (C)-(ii), (D)-(i)
(2) (A)-(i), (B)-(iii), (C)-(ii), (D)-(iv)
(3) (A)-(iv), (B)-(ii), (C)-(iii), (D)-(i)
(4) (A)-(i), (B)-(ii), (C)-(iii), (D)-(iv)

12. Metallic sodium does not react normally with : [July 20, 2021 (II)]
(1) gaseous ammonia (2) But-2-yne
(3) Ethyne (4) tert-butyl alcohol

13. Sodium metal on dissolution in liquid ammonia gives a deep blue solution due to the formation of: [Jan. 10, 2019 (II)]
(1) sodium-ammonia complex
(2) sodamide
(3) sodium ion-ammonia complex
(4) ammoniated electrons

14. The correct order of conductivity of ions in water is [March 17, 2021 (I), Similar April 8, 2019 (I)]
(1) $Na^+ > K^+ > Rb^+ > Cs^+$ (2) $Rb^+ > Na^+ > K^+ > Li^+$
(3) $Cs^+ > Rb^+ > K^+ > Na^+$ (4) $K^+ > Na^+ > Cs^+ > Rb^+$

15. On combustion of Li, Na and K in excess of air, the major oxides formed, respectively, are : [Sep. 04, 2020 (I), 2016 (S)]
(1) Li_2O_2, Na_2O_2 and K_2O_2 (2) Li_2O, Na_2O_2 and KO_2
(3) Li_2O, Na_2O and K_2O_2 (4) Li_2O, Na_2O_2 and K_2O

16. Two elements A and B have similar chemical properties. They don't form solid hydrogencarbonates, but react with nitrogen to form nitrides A and B, respectively, are : [Sep. 02, 2020 (II)]
(1) Na and Rb (2) Na and Ca
(3) Cs and Ba (4) Li and Mg

17. A metal on combution in excess of air forms X. X upon hydrolysis with water yields H_2O_2 and O_2 along with another product. The metal is: [Jan. 12, 2019 (I)]
(1) Na (2) Rb (3) Mg (4) Li

18. The correct statement(s) among I to III with respect to potassium ions that are abundant within the cell fluids is/are: [Jan. 12, 2019 (II)]
I. They activate many enzymes
II. They participate in the oxidation of glucose to produce ATP
III. Along with sodium ions, they are responsible for the transmission of nerve signals
(1) I and II only (2) I and III only
(3) I, II and III (4) III only

19. The metal that forms nitride by reacting directly with N_2 of air, is: [Jan. 9, 2019 (II)]
(1) K (2) Li (3) Rb (4) Cs

20. Which one of the following is an oxide ? [Online April 9, 2017]
(1) KO_2 (2) BaO_2 (3) SiO_2 (4) CsO_2

21. Which of the alkaline earth metal halides given below is essentially covalent in nature ? [Online April 11, 2015]
(1) $SrCl_2$ (2) $CaCl_2$ (3) $BaCl_2$ (4) $MgCl_2$

22. The first ionisation potential of Na is 5.1 eV. The value of electron gain enthalpy of Na^+ will be [2013]
(1) -2.55 eV (2) -5.1 eV (3) -10.2 eV (4) $+2.55$ eV

23. The solubility order for alkali metal fluoride in water is : [Online April 22, 2013]
(1) $LiF < RbF < KF < NaF$ (2) $RbF < KF < NaF < LiF$
(3) $LiF > NaF > KF > RbF$ (4) $LiF < NaF < KF < RbF$

24. Based on lattice energy and other considerations, which one of the following alkali metal chloride is expected to have the highest melting point? [2005, Online May 7, 2012]
(1) NaCl (2) KCl (3) LiCl (4) RbCl

25. Which one of the following will react most vigorously with water? [Online May 19, 2012]
(1) Li (2) K (3) Rb (4) Na

26. The ionic mobility of alkali metal ions in aqueous solution is maximum for [2006]
(1) Li^+ (2) Na^+ (3) K^+ (4) Rb^+

27. The metallic sodium disolves in liquid ammonia to form a deep blue coloured solution. The deep blue colour is due to formation of: [2002]
(1) solvated electron, $e(NH_3)_x^-$
(2) solvated atomic sodium, $Na(NH_3)_y$
(3) $(Na^+ + Na^-)$
(4) $NaNH_2 + H_2$

NUMERICAL VALUE QUESTIONS (NVQs)

28. Among the following, number of metal/s which can be used as electrodes in the photoelectric cell is ______. (Integer answer) [Feb. 25, 2021 (II), Similar Sep. 02, 2020 (I)]
(A) Li (B) Na (C) Rb (D) Cs

Topic 2 - Group-2 Elements (Alkaline Earth Metals)

MULTIPLE CHOICE QUESTIONS (MCQs)

29. **Statement I :** An alloy of lithium and magnesium is used to make aircraft plates.
Statement II: The magnesium ions are important for cell-membrane integrity.
In the light of the above statements, choose the correct answer from the options given below: [July 28, 2022 (II)]
(1) Both Statement I and Statement II are true.
(2) Both Statement I and Statement II are false.
(3) Statement I is true but Statement II is false.
(4) Statement I is false but Statement II is true.

30. Amongst baking soda, caustic soda and washing soda carbonate anion is present in : [June 29, 2022 (II)]

(1) washing soda only.
(2) washing soda and caustic soda only.
(3) washing soda and baking soda only.
(4) baking soda, caustic soda and washing soda.

31. In the industrial production of which of the following, molecular hydrogen is obtained as a byproduct? [June 24, 2022 (II)]
(1) NaOH (2) NaCl
(3) Na metal (4) Na_2CO_3

32. Which one of the following compounds is used as a chemical in certain type of fire extinguishers? [June 24, 2022 (II)]
(1) Baking Soda (2) Soda ash
(3) Washing Soda (4) Caustic Soda

33. The INCORRECT statement is : [April 12, 2019 (II), Similar 2011]
(1) Lithium is the strongest reducing agent among the alkali metals.
(2) Lithium is least reactive with water among the alkali metals.
(3) $LiNO_3$ decomposes on heating to give $LiNO_2$ and O_2.
(4) LiCl crystallises from aqueous solution as $LiCl \cdot 2H_2O$.

34. A hydrated solid X on heating, initially gives a monohydrated compound Y. Y upon heating above 373 K leads to an anhydrous white powder Z. X and Z, respectively, are :
(1) Washing soda and soda ash [April 10, 2019 (II)]
(2) Baking soda and dead burnt plaster.
(3) Washing soda and dead burnt plaster.
(4) Baking soda and soda ash.

35. Lithium aluminium hydride reacts with silicon tetrachloride to form: [Online April 15, 2018 (II)]
(1) LiCl, AlH_3 and SiH_4 (2) LiCl, $AlCl_3$ and SiH_4
(3) LiH, $AlCl_3$ and $SiCl_2$ (4) LiH, AlH_3 and SiH_4

36. Which of the following statements about Na_2O_2 is not correct?
(1) It is diamagnetic in nature [Online April 11, 2014]
(2) It is derivative of H_2O_2
(3) Na_2O_2 oxidises Cr^{3+} to CrO_4^{2-} in acid medium.
(4) It is the super oxide of sodium

37. Fire extinguishers contain H_2SO_4 and which one of the following? [Online May 26, 2012]
(1) $NaHCO_3$ and Na_2CO_3 (2) Na_2CO_3
(3) $NaHCO_3$ (4) $CaCO_3$

38. KO_2 (potassium super oxide) is used in oxygen cylinders in space and submarines because it [2002]
(1) absorbs CO_2 and increases O_2 content
(2) eliminates moisture
(3) absorbs CO_2
(4) produces ozone.

Topic 3 - General Characteristics of Alkaline Earth Metals and their Compounds

MULTIPLE CHOICE QUESTIONS (MCQs)

39. Given below are two statements:
Statement I : The chlorides of Be and Al have Cl-bridged structure. Both are soluble in organic solvents and act as Lewis bases.
Statement II: Hydroxides of Be and Al dissolve in excess alkali to give beryllate and aluminate ions.
In the light of the above statements. Choose the correct answer from the options given below. [July 27, 2022 (I)]
(1) Both statement I and statement II are true
(2) Both statement I and statement II are false
(3) Statement I is true but statement II is false
(4) Statement I is false but statement II is true

40. The correct order of density is [July 25, 2022 (II)]
(1) Be > Mg > Ca > Sr (2) Sr > Ca > Mg > Be
(3) Sr > Be > Mg > Ca (4) Be > Sr > Mg > Ca

41. s-block element which cannot be qualitatively confirmed by the flame test is [June 26, 2022 (II)]
(1) Li (2) Na (3) Rb (4) Be

42. Which one of the following alkaline earth metal ions has the highest ionic mobility in its equeous solution? [June 25, 2022 (I)]
(1) Be^{2+} (2) Mg^{2+} (3) Ca^{2+} (4) Sr^{2+}

43. Match List-I with List-II [July 22, 2021 (II)]

List-I (Elements)	List-II (Properties)
(A) Ba	(i) Organic solvent soluble compounds
(B) Ca	(ii) Outer electronic configuration $6s^2$
(C) Li	(iii) Oxalate insoluble in water
(D) Na	(iv) Formation of very strong monoacidic base

Choose the correct answer from the options given below:
(1) (A)-(ii), (B)-(iii), (C)-(i) and (D)-(iv)
(2) (A)-(iv), (B)-(i), (C)-(ii) and (D)-(iii)
(3) (A)-(iii), (B)-(ii), (C)-(iv) and (D)-(i)
(4) (A)-(i), (B)-(iv), (C)-(ii) and (D)-(iii)

44. Among the sulphates of alkaline earth metals, the solubilities of $BeSO_4$ and $MgSO_4$ in water, respectively, are : [Sep. 06, 2020 (I)]
(1) poor and poor (2) high and poor
(3) high and high (4) poor and high

45. An alkaline earth metal 'M' readily forms water soluble sulphate and water insoluble hydroxide. Its oxide MO is very stable to heat and does not have rock-salt structure. M is : [Sep. 04, 2020 (II)]
(1) Sr (2) Ca (3) Mg (4) Be

46. Among the statements (A)–(D), the correct ones are: [Jan. 09, 2020 (II)]
(A) Lithium has the highest hydration enthalpy among the alkali metals.
(B) Lithium chloride is insoluble in pyridine.
(C) Lithium cannot form ethynide upon its reaction with ethyne.
(D) Both lithium and magnesium react slowly with H_2O.
(1) (A), (B) and (D) only (2) (A), (C) and (D) only
(3) (B) and (D) only (4) (A) and (D) only

47. A metal (A) on heating in nitrogen gas gives compound B. B on treatment with H_2O gives a colourless gas which when passed through $CuSO_4$ solution gives a dark blue-violet coloured solution. A and B respectively, are: [Jan. 08, 2020 (II), Online May 12, 2012 (S)]
(1) Na and $NaNO_3$ (2) Na and Na_3N
(3) Mg and Mg_3N_2 (4) Mg and $Mg(NO_3)_2$

48. The correct sequence of thermal stability of the following carbonates is : [April 12, 2019 (I)]
(1) $BaCO_3 < CaCO_3 < SrCO_3 < MgCO_3$
(2) $MgCO_3 < CaCO_3 < SrCO_3 < BaCO_3$

(3) $MgCO_3 < SrCO_3 < CaCO_3 < BaCO_3$
(4) $BaCO_3 < SrCO_3 < CaCO_3 < MgCO_3$

49. The alloy used in the construction of aircrafts is : [April 10, 2019 (I)]
(1) Mg – Al (2) Mg – Zn
(3) Mg – Sn (4) Mg – Mn

50. Magnesium powder burns in air to give: [April 9, 2019 (I)]
(1) $Mg(NO_3)_2$ and Mg_3N_2 (2) MgO and Mg_3N_2
(3) MgO only (4) MgO and $Mg(NO_3)_2$

51. The structures of beryllium chloride in the solid state and vapour phase, respectively, are: [April 9, 2019 (II)]
(1) chain and chain (2) dimeric and dimeric
(3) chain and dimeric (4) dimeric and chain

52. The covalent alkaline earth metal halide (X = Cl, Br, I) is : [April 8, 2019 (II)]
(1) MgX_2 (2) CaX_2 (3) BeX_2 (4) SrX_2

53. The metal used for making X-ray tube window is: [Jan. 10, 2019 (I)]
(1) Mg (2) Na (3) Be (4) Ca

54. The alkaline earth metal nitrate that does not crystallise with water molecules, is: [Jan. 9, 2019 (I)]
(1) $Mg(NO_3)_2$ (2) $Sr(NO_3)_2$
(3) $Ca(NO_3)_2$ (4) $Ba(NO_3)_2$

55. The correct order of the solubility of alkaline–earth metal sulphates in water is : [Online April 9, 2016]
(1) Mg > Ca > Sr > Ba (2) Mg > Sr > Ca > Ba
(3) Mg < Ca < Sr < Ba (4) Mg < Sr < Ca < Ba

56. Which one of the following alkaline earth metal sulphates has its hydration enthalpy greater than its lattice enthalpy ? [2015]
(1) $BaSO_4$ (2) $SrSO_4$ (3) $CaSO_4$ (4) $BeSO_4$

57. The correct order of thermal stability of hydroxides is: [Online April 10, 2015]
(1) $Ba(OH)_2 < Ca(OH)_2 < Sr(OH)_2 < Mg(OH)_2$
(2) $Mg(OH)_2 < Sr(OH)_2 < Ca(OH)_2 < Ba(OH)_2$
(3) $Mg(OH)_2 < Ca(OH)_2 < Sr(OH)_2 < Ba(OH)_2$
(4) $Ba(OH)_2 < Sr(OH)_2 < Ca(OH)_2 < Mg(OH)_2$

58. One mole of magnesium nitride on reaction with an excess of water gives [2004]
(1) two moles of ammonia (2) one mole of nitric acid
(3) one mole of ammonia (4) two moles of nitric acid

59. The solubility of carbonates decreases down the magnesium group due to a decrease in [2003]
(1) hydration energy of cations
(2) inter-ionic attraction
(3) entropy of solution formation
(4) lattice energy of solids

60. Which one of the following processes will produce hard water ? [2003]
(1) Saturation of water with $MgCO_3$
(2) Saturation of water with $CaSO_4$
(3) Addition of Na_2SO_4 to water
(4) Saturation of water with $CaCO_3$

61. A metal M readily forms its sulphate MSO_4 which is water-soluble. It forms its oxide MO which becomes inert on heating. It forms an insoluble hydroxide $M(OH)_2$ which is soluble in NaOH solution. Then M is [2002]
(1) Mg (2) Ba (3) Ca (4) Be

Numerical Value Questions (NVQs)

62. Number of amphoteric compound among the following is ________ [Feb. 24, 2021 (I), Similar Jan. 11, 2019 (I)]
(A) BeO (B) BaO
(C) $Be(OH)_2$ (D) $Sr(OH)_2$

Topic 4 - Some Importance Compounds of Alkaline Earth Metals

Multiple Choice Questions (MCQs)

63. Portland cement contains 'X' to enhance the setting time. What is 'X'? [July 29, 2022 (II)]
(1) $CaSO_4 \cdot \frac{1}{2}H_2O$ (2) $CaSO_4 \cdot 2H_2O$
(3) $CaSO_4$ (4) $CaCO_3$

64. Reaction of $BeCl_2$ with $LiAlH_4$ gives: [July 26, 2022 (I)]
(A) $AlCl_3$ (B) BeH_2 (C) LiH (D) LiCl
(E) $BeAlH_4$
Choose the correct answer from options given below:
(1) (A), (D) and (E) (2) (A), (B) and (D)
(3) (D) and (E) (4) (B), (C) and (D)

65. Addition of H_2SO_4 to BaO_2 produces: [June 27, 2022 (I)]
(1) BaO, SO_2 and H_2O (2) $BaHSO_4$ and O_2
(3) $BaSO_4$, H_2 and O_2 (4) $BaSO_4$ and H_2O_2

66. $BeCl_2$ reacts with $LiAlH_4$ to give [June 27, 2022 (I)]
(1) $Be + Li[AlCl_4] + H_2$ (2) $Be + AlH_3 + LiCl + HCl$
(3) $BeH_2 + LiCl + AlCl_3$ (4) $BeH_2 + Li[AlCl_4]$

67. Given below are two Statements : One is labelled as Assertion A and the other is labelled as Reason R :
Assertion A : During the boiling of water having temporary hardness, $Mg(HCO_3)_2$ is converted to $MgCO_3$.
Reason R : The solubility product of $Mg(OH)_2$ is greater than that of $MgCO_3$.
In the light of the above statements, choose the most appropriate answer from the options given below: [March 18, 2021 (I)]
(1) Both A and R are true but R is NOT the correct explanation of A
(2) A is true but R is false
(3) Both A and R are true and R is the correct explanation of A
(4) A is false but R is true

68. Calgon is used for water treatment. Which of the following statement is NOT true about Calgon? [Feb. 26, 2021 (II)]
(1) It is polymeric compound and is water soluble.
(2) Calgon contains the 2nd most abundant element by weight in the Earth's crust.
(3) It is also known as Graham's salt.
(4) It does not remove Ca^{2+} ion by precipitation.

69. Match the following compounds (Column-I) with their uses (Column-II): [Sep. 06, 2020 (II)]

Column-I	Column-II
(I) $Ca(OH)_2$	(A) casts of statues
(II) NaCl	(B) white wash
(III) $CaSO_4 \cdot \frac{1}{2}H_2O$	(C) antacid
(IV) $CaCO_3$	(D) washing soda preparation

(1) (I)-(D), (II)-(A), (III)-(C), (IV)-(B)
(2) (I)-(B), (II)-(D), (III)-(A), (IV)-(C)
(3) (I)-(B), (II)-(C), (III)-(D), (IV)-(A)
(4) (I)-(C), (II)-(D), (III)-(B), (IV)-(A)

70. When gypsum is heated to 393 K, it forms: [Jan. 08, 2020 (I)]
(1) Anhydrous $CaSO_4$ (2) $CaSO_4 \cdot 5H_2O$
(3) $CaSO_4 \cdot 0.5H_2O$ (4) Dead burnt plaster

71. Match the following items in column-I with the corresponding items in column-II. [Jan. 11, 2019 (II)]

Column-I	Column II
(i) $Na_2CO_3 \cdot 10H_2O$	(A) Portland cement ingredient
(ii) $Mg(HCO_3)_2$	(B) Castner-Kellner process
(iii) NaOH	(C) Solvay process
(iv) $Ca_3Al_2O_6$	(D) Temporary hardness

(1) (i)→(B); (ii)→(C); (iii)→(A); (iv)→(D)
(2) (i)→(C); (ii)→(B); (iii)→(D); (iv)→(A)
(3) (i)→(D); (ii)→(A); (iii)→(B); (iv)→(C)
(4) (i)→(C); (ii)→(D); (iii)→(B); (iv)→(A)

72. The commercial name for calcium oxide is : [Online April 10, 2016]
(1) Quick lime (2) Milk of lime
(3) Slaked lime (4) Limestone

73. Which of the following on thermal decomposition yields a basic as well as acidic oxide ? [2012]
(1) $NaNO_3$ (2) $KClO_3$
(3) $CaCO_3$ (4) NH_4NO_3

74. The substance **not** likely to contain $CaCO_3$ is [2003]
(1) calcined gypsum (2) sea shells
(3) dolomite (4) a marble statue

75. In curing cement plasters, water is sprinkled from time to time. This helps in [2003]
(1) developing interlocking needle-like crystals of hydrated silicates
(2) hydrating sand and gravel mixed with cement
(3) converting sand into silicic acid
(4) keeping it cool

Numerical Value Questions (NVQs)

76. BeO reacts with HF in presence of ammonia to give [A] which on thermal decomposition produces [B] and ammonium fluoride. Oxidation state of Be in [A] is _______. [June 27, 2022 (II)]

Part-B (JEE Advanced)

Topic 1 - Group-1 Elements (Alkali Metals)

1 MCQs with One Correct Answer

1. The metallic lustre exhibited by sodium is explained by [1987 - 1 Mark]
(a) diffusion of sodium ions
(b) oscillation of loose electrons
(c) excitation of free protons
(d) existence of body centered cubic lattice

2. The pair of compounds which cannot exist together in solution is : [1986 - 1 Mark]
(a) $NaHCO_3$ and NaOH (b) Na_2CO_3 and $NaHCO_3$
(c) Na_2CO_3 and NaOH (d) $NaHCO_3$ and NaCl

3. The oxide that gives hydrogen peroxide on treatment with a dilute acid is : [1985 - 1 Mark]
(a) PbO_2 (b) Na_2O_2 (c) MnO_2 (d) TiO_2

4. A solution of sodium metal in liquid ammonia is strongly reducing due to the presence of [1981 - 1 Mark]
(a) sodium atoms (b) sodium hydride
(c) sodium amide (d) solvated electrons

4 Fill in the Blanks

5. The electrolysis of molten sodium hydride liberates .... gas at the .... [1989 - 1 Mark]

6. Sodium dissolved in liquid ammonia conducts electricity because .... [1985 - 1 Mark]

5 True / False

7. Sodium when burnt in excess of oxygen gives sodium oxide. [1987 - 1 Mark]

8. The softness of group I-A metals increases down the group with increasing atomic number. [1986 - 1 Mark]

6 MCQs with One or More than One Correct Answer

9. The compound(s) formed upon combustion of sodium metal in excess air is (are) [2009 - 5M, –1]
(a) Na_2O_2 (b) Na_2O (c) NaO_2 (d) NaOH

10. Highly pure dilute solution of sodium in liquid ammonia [1998 - 2 Marks]
(a) shows blue colour
(b) exhibits electrical conductivity
(c) produces sodium amide
(d) produces hydrogen gas.

9 Assertion and Reason/Statement Type Questions

Each question contains STATEMENT-1 (Assertion) and STATEMENT-2 (Reason). Each question has 4 choices (a), (b), (c) and (d) out of which ONLY ONE is correct. Mark your answer as
(a) If both Statement -1 and Statement -2 are correct, and Statement -2 is the correct explanation of the Statement -2.
(b) If both Statement -1 and Statement -2 are correct, but Statement -2 is not the correct explanation of the Statement -1.
(c) If Statement -1 is correct but Statement -2 is incorrect.
(d) If Statement -1 is incorrect but Statement -2 is correct. [2007]

11. **Statement-1 :** Alkali metals dissolve in liquid ammonia to give blue solutions.
Statement-2 : Alkali metals in liquid ammonia give solvated species of the type $[M(NH_3)_n]^+$ (M = alkali metals).

10 Subjective Problems

12. Explain the difference in the nature of bonding in LiF and LiI. [1996 - 2 Marks]

13. Write down the balanced equations for the reactions when:
(*i*) Potassium ferricyanide reacts with hydrogen peroxide in basic solution. [1989 - 1 Mark]
(*ii*) Carbon dioxide is passed through a concentrated aqueous solution of sodium chloride saturated with ammonia. [1988 - 1 Mark]

14. Give reason of the following :
Sodium carbonate is made by Solvay process but the same process is not extended to the manufacture of potassium carbonate. [1981 - 1 Mark]

Topic 2 - Group-2 Elements (Alkaline Earth Metals)

1 MCQs with One Correct Answer

15. A sodium salt on treatment with $MgCl_2$ gives white precipitate only on heating. The anion of the sodium salt is [2004S]
(a) HCO_3^- (b) CO_3^{2-} (c) NO_3^- (d) SO_4^{2-}

16. The set representing the correct order of first ionization potential is [2001S]
(a) K > Na > Li (b) Be > Mg > Ca
(c) B > C > N (d) Ge > Si > C

17. The following compounds have been arranged in order of their increasing thermal stabilities. Identify the correct order. [1996 - 1 Mark]
K_2CO_3 (I) $MgCO_3$ (II) $CaCO_3$ (III) $BeCO_3$ (IV)
(a) I < II < III < IV (b) IV < II < III < I
(c) IV < II < I < III (d) II < IV < III < I

18. Molecular formula of Glauber's salt is : [1985 - 1 Mark]
(a) $MgSO_4 \cdot 7H_2O$ (b) $CuSO_4 \cdot 5H_2O$
(c) $FeSO_4 \cdot 7H_2O$ (d) $Na_2SO_4 \cdot 10H_2O$

19. Calcium is obtained by [1980]
(a) electrolysis of molten $CaCl_2$.
(b) electrolysis of solution of $CaCl_2$ in water.
(c) Reduction of $CaCl_2$ with carbon.
(d) roasting of limestone.

20. HCl is added to following oxides. Which one would give H_2O_2? [1980]
(a) MnO_2 (b) PbO_2 (c) $BaO_2 \cdot 8H_2O$ (d) NO_2

21. A substance absorbs CO_2 and violently reacts with water. The substance is [1978]
(a) $CaCO_3$ (b) CaO (c) H_2SO_4 (d) ZnO

4 Fill in the Blanks

22. Ca^{2+} has a smaller ionic radius than K^+ because it has .... [1993 - 1 Mark]

23. Anhydrous $MgCl_2$ is obtained by heating hydrated salt with .... [1980]

5 True / False

24. $MgCl_2.6H_2O$ on heating give anhydrous $MgCl_2$. [1982 - 1 Mark]

6 MCQs with One or More than One Correct Answer

25. $MgSO_4$ on reaction with NH_4OH and Na_2HPO_4 forms a white crystalline precipitate. What is its formula? [2006 - 5M, –1]
(a) $Mg(NH_4)PO_4$ (b) $Mg_3(PO_4)_2$
(c) $MgCl_2.MgSO_4$ (d) $MgSO_4$

26. The species that do not contain peroxide ions are [1992 - 1 Mark]
(a) PbO_2 (b) H_2O_2 (c) SrO_2 (d) BaO_2

10 Subjective Problems

27. Give reason of the following:
$BeCl_2$ can be easily hydrolysed [1999 - 2 Marks]

28. Work out the following using chemical equation:
Chlorination of calcium hydroxide produces bleaching powder. [1998 - 2 Marks]

29. Arrange the following sulphates of alkaline earth metals in order of decreasing thermal stability : $BeSO_4$, $MgSO_4$, $CaSO_4$, $SrSO_4$ [1997 - 1 Mark]

30. The crystalline salts of alkaline earth metals contain more water of crystallisation than the corresponding alkali metal salts. [1997 - 2 Marks]

31. Element *A* burns in nitrogen to give an ionic compound *B*. Compound *B* reacts with water to give *C* and *D*. A solution of *C* becomes 'milky' on bubbling carbon dioxide. Identify *A*, *B*, *C* and *D*. [1997 - 3 Marks]

32. Mg_3N_2 when reacted with water gives off NH_3 but HCl is not obtained from $MgCl_2$ on reaction with water at room temperature. [1995 - 2 Marks]

33. Complete and balance the following reactions :
$$Ca_5(PO_4)_3F + H_2SO_4 + H_2O \xrightarrow{\text{Heat}} ....+ 5CaSO_4.2H_2O + ...$$ [1994-1 Mark]

34. Give briefly the isolation of magnesium from sea water by the Dow process. Give equations for the steps involved. [1993 - 3 Marks]

35. Write down the balanced equations for the reactions when:
Carbon dioxide is passed through a suspension of lime stone in water. [1991 - 1 Mark]

Answer Key

PART-A (JEE Main)																			
1	(3)	9	(1)	17	(2)	25	(3)	33	(3)	41	(4)	49	(1)	57	(3)	65	(4)	73	(3)
2	(1)	10	(3)	18	(3)	26	(4)	34	(1)	42	(4)	50	(2)	58	(1)	66	(3)	74	(1)
3	(3)	11	(1)	19	(2)	27	(1)	35	(2)	43	(1)	51	(3)	59	(1)	67	(N)	75	(1)
4	(1)	12	(2)	20	(3)	28	(1)	36	(4)	44	(3)	52	(3)	60	(2)	68	(2)	76	(2)
5	(2)	13	(4)	21	(4)	29	(2)	37	(1)	45	(4)	53	(3)	61	(4)	69	(2)		
6	(1)	14	(3)	22	(2)	30	(1)	38	(1)	46	(2)	54	(4)	62	(2)	70	(3)		
7	(1)	15	(2)	23	(4)	31	(1)	39	(4)	47	(3)	55	(1)	63	(2)	71	(4)		
8	(4)	16	(4)	24	(1)	32	(1)	40	(3)	48	(2)	56	(4)	64	(2)	72	(1)		
Part-B (JEE Advanced)																			
1	(b)	3	(b)	7	False	9	(a, b)	11	(b)	16	(b)	18	(d)	20	(c)	24	False	26	(a)
2	(a)	4	(d)	8	True	10	(a, b)	15	(a)	17	(b)	19	(a)	21	(b)	25	(a)		

Chapter-11

The *p*-Block Elements (Group-13 and 14)

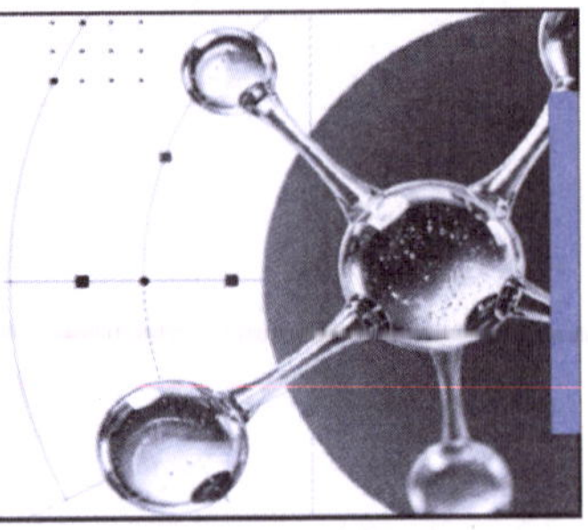

Part-A (JEE Main)

Topic 1 - Boron Family

Multiple Choice Questions (MCQs)

1. Given below are two statements : [April 06, 2026 (I)]

Statement I: Aluminium is more electropositive than thallium as the standard electrode potential value of $E^{\circ}_{Al^{3+}/Al}$ is negative and $E^{\circ}_{Tl^{3+}/Tl}$ is positive.

Statement II: The sum of first three ionization enthalpies of boron is very high when compared to that of aluminium. Due to this reason boron forms covalent compounds only and aluminium forms $A1^{3+}$ ion.

In the light of the above statements, choose the correct answer from the options given below.

(1) Both Statement I and Statement II are true.
(2) Both Statement I and Statement II are false.
(3) Statement I is true but Statement II is false.
(4) Statement I is false but Statement II is true.

2. Given below are two statements : [April 02, 2026 (II)]

Statement I : The second ionization enthalpy of B, Al and Ga is in the order of B > Al > Ga.

Statement II : The correct order in terms of first ionization enthalpy is Si < Ge < Pb < Sn.

In the light of the above statements, choose the correct answer from the options given below :

(1) Both Statement I and Statement II are true.
(2) Both Statement I and Statement II are false.
(3) Statement I is true but Statement II is false.
(4) Statement I is false but Statement II is true.

3. The correct statements from the following are: [Jan. 23, 2026 (I)]

(A) Ionic radii of trivalent cations of group 13 elements decreases down the group.
(B) Electronegativity of group 13 elements decreases down the group.
(C) Among the group 13 elements, Boron has highest first ionisation enthalpy.
(D) The trichloride and triiodide of group 13 elements are covalent in nature.

Choose the correct answer from the otptions given below:

(1) A and C only (2) A and D only
(3) C and D only (4) B and D only

4. The correct statements from the following are : [April 7, 2025 (II)]

(A) Tl^{3+} is a powerful oxidising agent
(B) Al^{3+} does not get reduced easily
(C) Both Al^{3+} and Tl^{3+} are very stable in solution
(D) Tl^{+} is more stable than Tl^{3+}
(E) Al^{3+} and Tl^{+} are highly stable

Choose the correct answer from the options given below:

(1) (A), (B), (C), (D) and (E) (2) (A), (B), (D) and (E) only
(3) (B), (D) and (E) only (4) (A), (C) and (D) only

5. The elements of Group 13 with highest and lowest first ionisation enthalpies are respectively: [April 4, 2025 (II)]

(1) B & Ga (2) B & Tl
(3) Tl & B (4) B & In

6. Given below are the pairs of group 13 elements showing their relation in terms of atomic radius.

(B < Al), (Al < Ga), (Ga < In) and (In < Tl)

Identify the elements present in the incorrect pair and in that pair find out the element (*X*) that has higher ionic radius (M^{3+}) that the other one. The atomic number of the element (*X*) is [April 4, 2025 (I)]

(1) 31 (2) 49 (3) 13 (4) 81

7. The correct orders among the following are [April 3, 2025 (II)]

(A) Atomic radius : B < Al < Ga < In < Tl
(B) Electronegativity : Al < Ga < In < Tl < B
(C) Density : Tl < In < Ga < Al < B
(D) 1st Ionisation Energy : In < Al < Ga < Tl < B

Choose the correct answer from the options given below :

(1) B and D Only (2) A and C Only
(3) C and D Only (4) A and B Only

8. Given below are two statements : [April 2, 2025 (I)]

Statement I : The metallic radius of Al is less than that of Ga.

Statement II : The ionic radius of Al^{3+} is less than that of Ga^{3+}.

In the light of the above statements, choose the most appropriate answer from the options given below :

(1) Both Statement I and Statement II are incorrect
(2) Statement I is incorrect but Statement II is correct
(3) Statement I is correct but Statement II is incorrect
(4) Both Statement I and Statement II are correct

9. Match List-I with List-II. [April 9, 2024 (II)]

List-I		List-II
(A) Melting point [K]	(I)	$Tl > In > Ga > Al > B$
(B) Ionic Radius $[M^{+3}/pm]$	(II)	$B > Tl > Al \approx Ga > In$
(C) $\Delta_{I.E}H_1$ [kJ mol^{-1}]	(III)	$Tl > In > Al > Ga > B$
(D) Atomic Radius [pm]	(IV)	$B > Al > Tl > In > Ga$

Choose the correct answer from the options given below:
(1) A-(III), B-(IV), C-(I), D-(II)
(2) A-(II), B-(III), C-(IV), D-(I)
(3) A-(IV), B-(I), C-(II), D-(III)
(4) A-(I), B-(II), C-(III), D-(IV)

10. Given below are two statements : One is labelled as Assertion A and the other is labelled as Reason R: [April 8, 2024 (I)]
Assertion A : The stability order of +1 oxidation state of Ga, In and Tl is Ga < In < Tl.
Reason R : The inert pair effect stabilizes the lower oxidation state down the group.
In the light of the above statements, choose the correct answer from the options given below :
(1) Both A and R are true and R is the correct explanation of A.
(2) A is true but R is false.
(3) Both A and R are true but R is NOT the correct explanation of A.
(4) A is false but R is true.

11. Given below are two statements :
Statement I : Gallium is used in the manufacturing of thermometers. [April 6, 2024 (I)]
Statement II : A thermometer containing gallium is useful for measuring the freezing point (256 K) of brine solution.
In the light of the above statement, choose the correct answer from the options given below :
(1) Both Statement I and Statement II are false.
(2) Statement I is false but Statement II is true.
(3) Both Statement I and Statement II are true.
(4) Statement I is true but Statement II is false.

12. Given below are two statements :
Statement I: In group 13, the stability of +1 oxidation state increases down the group.
Statement II: The atomic size of gallium is greater than that of aluminium. [April 5, 2024 (I)]
In the light of the above statements, choose the most appropriate answer from the options given below:
(1) Statement I is incorrect but Statement II is correct
(2) Both Statement I and Statement II are correct
(3) Both Statement I and Statement II are incorrect
(4) Statement I is correct but Statement II is incorrect

13. The number of neutrons present in the more abundant isotope of boron is 'x'. Amorphous boron upon heating with air forms a product, in which the oxidation state of boron is 'y'. The value of x + y is ... [April 5, 2024 (I)]
(1) 4 (2) 6 (3) 3 (4) 9

14. The correct statements from the following are : [April 5, 2024 (II)]
(A) The decreasing order of atomic radii of group 13 elements is $Tl > In > Ga > Al > B$.
(B) Down the group 13 electronegativity decreases from top to bottom.
(C) Al dissolves in dil. HCl and liberate H_2 but conc. HNO_3 renders Al passive by forming a protective oxide layer on the surface.
(D) All elements of group 13 exhibits highly stable +1 oxidation state.
(E) Hybridisation of Al in $[Al(H_2O)_6]^{3+}$ ion is sp^3d^2.
Choose the correct answer from the options given below:
(1) (C) and (E) only
(2) (A), (C) and (E) only
(3) (A), (B), (C) and (E) only
(4) (A) and (C) only

15. Given below are two statements : [Jan. 31, 2024 (II)]
Statement I : Group 13 trivalent halides get easily hydrolyzed by water due to their covalent nature.
Statement II : $AlCl_3$ upon hydrolysis in acidified aqueous solution forms octahedral $[Al(H_2O)_6]^{3+}$ ion.
In the light of the above statements, choose the correct answer from the options given below :
(1) Statement I is true but statement II is false
(2) Statement I is false but statement II is true
(3) Both statement I and statement II are false
(4) Both statement I and statement II are true

16. Aluminium chloride in acidified aqueous solution forms an ion having geometry. [Jan. 30, 2024 (I)]
(1) Octahedral (2) Square planar
(3) Tetrahedral (4) Trigonal bipyramidal

17. Given below are two statements : one is labelled as Assertion (A) and the other is labelled as Reason (R) [Jan. 27, 2024 (I)]
Assertion (A) : Melting point of Boron (2453 K) is unusually high in group 13 elements.
Reason (R) : Solid Boron has very strong crystalline lattice.
In the light of the above statements, choose the most appropriate answer from the options given below :
(1) Both (A) and (R) are correct but (R) is not the correct explanation of (A)
(2) Both (A) and (R) are correct and (R) is the correct explanation of (A)
(3) (A) is true but (R) is false
(4) (A) is false but (R) is true

18. The covalency and oxidation state respectively of boron in $[BF_4]^-$, are [April 13, 2023 (II)]
(1) 4 and 3 (2) 4 and 4
(3) 3 and 4 (4) 3 and 5

19. Given below are two statements : [April 12, 2023 (I)]
Statement I : Boron is extremely hard indicating its high lattice energy
Statement II : Boron has highest melting and boiling point compared to its other group members.
In the light of the above statements, choose the most appropriate answer from the options given below:
(1) Statement I is incorrect but Statement II is correct
(2) Both Statement I and Statement II is correct
(3) Statement I is correct but Statement II is incorrect
(4) Both Statement I and Statement II is incorrect

20. Which one of the following elements will remain as liquid inside pure boiling water? [April 06, 2023 (II)]
(1) Cs (2) Ga
(3) Li (4) Br

21. The Lewis acid character of boron tri halides follows the order: [Jan. 31, 2023 (II)]
(1) $BBr_3 > BI_3 > BCl_3 > BF_3$ (2) $BCl_3 > BF_3 > BBr_3 > BI_3$
(3) $BF_3 > BCl_3 > BBr_3 > BI_3$ (4) $BI_3 > BBr_3 > BCl_3 > BF_3$

22. Given below are two statements: [Jan. 29, 2023 (II)]
Statement I : The decrease in first ionization enthalpy from B to Al is much larger than that from Al to Ga.
Statement II : The d orbitals in Ga are completely filled.
In the light of the above statements, choose the most appropriate answer from the options given below
(1) Statement I is incorrect but statement II is correct.
(2) Both the statements I and II are correct
(3) Statement I is correct but statement II is incorrect
(4) Both the statements I and II are incorrect

23. Given below are two statements: one is labelled as Assertion (A) and the other is labelled as Reason (R).
Assertion (A) : Boron is unable to form BF_6^{3-}
Reason (R) : Size of B is very small. [July 27, 2022 (II)]
In the light of the above statements, choose the correct answer from the options given below:
(1) Both (A) and (R) are true and (R) is the correct explanation of (A).
(2) Both (A) and (R) are true but (R) is not the correct explanation of (A).
(3) (A) is true but (R) is false.
(4) (A) is false but (R) is true.

24. The metal that has very low melting point and its periodic position is closer to a metalloid is : [July 26, 2022 (II)]
(1) Al (2) Ga
(3) Se (4) In

25. Given below are two statements. [July 29, 2022 (II)]
Statement I: Stannane is an example of a molecular hydride.
Statement II: Stannane is a planar molecule.
In the light of the above statement, choose the most appropriate answer from the options given below :
(1) Both Statement I and statement II are true.
(2) Both Statement I and statement II are false.
(3) Statement I is true but statement II is false.
(4) Statement I is false but statement II is true.

26. Match the List-I with List-II. [June 29, 2022 (I)]

List-I (Metal)	List-II (Application)
(A) Cs	(I) High temperature thermometer
(B) Ga	(II) Water repellent sprays
(C) B	(III) Photoelectric cells
(D) Si	(IV) Bullet proof vest

Choose the most appropriate answer from the option given below:
(1) (A)-(III), (B)-(I), (C)-(IV), (D)-(II)
(2) (A)-(IV), (B)-(III), (C)-(II), (D)-(I)
(3) (A)-(II), (B)-(III), (C)-(IV), (D)-(I)
(4) (A)-(I), (B)-(IV), (C)-(II), (D)-(III)

27. Given below are two statements : one is labelled as Assertion (A) and the other is labelled as Reason (R).
Assertion (A) : In TlI_3, isomorphous to CsI_3, the metal is present in +1 oxidation state.
Reason (R) : Tl metal has fourteen f electrons in its electronic configuration.
In the light of the above statements, choose the most appropriate answer from the options given below : [Feb. 26, 2021 (II)]
(1) Both A and R are correct but R is NOT the correct explanation of A
(2) Both A and R are correct and R is the correct explanation of A
(3) A is correct but R is not correct
(4) A is not correct but R is correct

28. Which one of the following compounds of Group–14 elements is not known? [July 25, 2021 (I)]
(1) $[GeCl_6]^{2-}$ (2) $[Sn(OH)_6]^{2-}$
(3) $[SiCl_6]^{2-}$ (4) $[SiF_6]^{2-}$

29. Compound A is used as a strong oxidizing agent is amphoteric in nature. It is the part of lead storage batteries. Compound A is [Feb. 26, 2021 (I)]
(1) Pb_3O_4 (2) $PbSO_4$ (3) PbO (4) PbO_2

30. Choose the correct stability order of group 13 elements in their +1 oxidation state. [June 26, 2022 (I), Same Jan. 11, 2019 (II)]
(1) Al < Ga < In < Tl (2) Tl < In < Ga < Al
(3) Al < Ga < Tl < In (4) Al < Tl < Ga < In

31. The hydride that is NOT electron deficient is: [Jan. 11, 2019 (II)]
(1) SiH_4 (2) B_2H_6
(3) GaH_3 (4) AlH_3

32. The electronegativity of aluminium is similar to: [Jan. 10, 2019 (I)]
(1) Carbon (2) Beryllium
(3) Boron (4) Lithium

33. Aluminium is usually found in +3 oxidation state. In contrast, thallium exists in +1 and + 3 oxidation states. This is due to: [Jan. 9, 2019 (I)]
(1) inert pair effect (2) diagonal relationship
(3) lattice effect (4) lanthanoid contraction

34. Beryllium and aluminium exhibit many properties which are similar. But, the two elements differ in [2004]
(1) forming covalent halides
(2) forming polymeric hydrides
(3) exhibiting maximum covalency in compounds
(4) exhibiting amphoteric nature in their oxides

Topic 2 - Carbon Family

Multiple Choice Questions (MCQs)

35. The electronegativity of a group 13 element 'E' is same as that of Ge (on Pauling scale and upto one decimal point). The correct statements about E^{3+} are
A. It can act as a reducing agent.
B. It can act as an oxidizing agent.
C. E^{3+} is more stable than E^{+}.
D. The standard electrode potential value for E^{3+}/E is positive. [April 04, 2026 (II)]
Choose the correct answer from the options given below.
(1) A and C Only (2) B and C Only
(3) B and D Only (4) A and D Only

36. Choose the INCORRECT statement. [Jan. 24, 2026 (II)]
(1) Among the isotopes of carbon, ^{13}C is a radioactive isotope.
(2) Carbon exhibits negative oxidation states along with +4 and +2.
(3) Carbon cannot exceed its covalency more than four.
(4) CO_2 is the most acidic oxide among the dioxides of group of 14 elements.

37. Given below are two statements: [Jan. 21, 2026 (I)]
Statement I: The number of pairs among [SiO_2, CO_2], [SnO, SnO_2], [PbO, PbO_2] and [GeO, GeO_2], which contain oxides that are both amphoteric is 2.
Statement II: BF_3 is an electron deficient molecule can act as a Lewis acid, forms adduct with NH_3 and has a trigonal planar geometry.
In the light of the above statement, choose the correct answer from the option given below.
(1) Both Statement I and Statement II are true.
(2) Both Statement I and Statement II are false.
(3) Statement I is true but Statement II is false.
(4) Statement I is false but Statement II is true.

38. Given below are two statements : [April 4, 2025 (II)]
Statement I : The first ionisation enthalpy of group 14 elements is higher than the corresponding elements of group 13.
Statement II : Melting points and boiling points of group 13 elements are in general much higher than those the corresponding elements of group 14.
In the light of the above statements, choose the most appropriate answer from the options given below :
(1) Statement I is correct but Statement II is incorrect
(2) Statement I is incorrect but Statement II is correct
(3) Both Statement I and Statement II are incorrect
(4) Both Statement I and Statement II are correct

39. Consider the following elements In, Tl, Al, Pb, Sn and Ge.
The most stable oxidation states of elements with highest and lowest first ionisation enthalpies, respectively, are [Jan. 28, 2025 (I)]
(1) +2 and +3 (2) +4 and +3
(3) +4 and +1 (4) +1 and +4

40. Given below are two statements : [Jan. 24, 2025 (II)]
Statement I : The first ionization energy of Pb is greater than that of Sn
Statement II : The first ionization energy of Ge is greater than that of Si.
In the light of the above statements, choose the correct answer from the options given below :
(1) Statement I is true but Statement II is false
(2) Both Statement I and Statement II are false
(3) Statement I is false but Statement II is true
(4) Both Statement I and Statement II are true

41. Given below are the atomic numbers of some group 14 elements. The atomic number of the element with lowest melting point is: [Jan. 23, 2025 (II)]
(1) 14 (2) 6 (3) 82 (4) 50

42. The number of ions from the following that are expected to behave as oxidising agent is: [April 6, 2024 (II)]
Sn^{4+}, Sn^{2+}, Pb^{2+}, Tl^{3+}, Pb^{4+}, Tl^{+}
(1) 3 (2) 4 (3) 1 (4) 2

43. Which of the following material is not a semiconductor. [April 6, 2024 (I)]
(1) Germanium (2) Graphite
(3) Silicon (4) Copper oxide

44. Given below are two statements : [Feb. 1, 2024 (II)]
Statement I : SiO_2 and GeO_2 are acidic while SnO and PbO are amphoteric in nature.
Statement II : Allotrophic forms of carbon are due to property of catenation and $p\pi - d\pi$ bond formation.
In the light of the above statements, choose the most appropriate answer from the options given below:
(1) Statement I is false but Statement II is false
(2) Both Statement I and Statement II are false
(3) Both Statement I and Statement II are true.
(4) Statement I is true but Statement II is false.

45. Consider the oxides of group 14 elements
SiO_2, GeO_2, SnO_2, PbO_2, CO and GeO. The amphoteric oxides are [Jan. 31, 2024 (I)]
(1) GeO, GeO_2 (2) SiO_2, GeO_2
(3) SnO_2, CO (4) SnO_2, PbO_2

46. The correct order of bond enthalpy (kJ mol^{-1}) is: [Feb. 01, 2023 (II)]
(1) $C-C > Si-Si > Ge-Ge > Sn-Sn$
(2) $C-C > Si-Si > Sn-Sn > Ge-Ge$
(3) $Si-Si > C-C > Sn-Sn > Ge-Ge$
(4) $Si-Si > C-C > Ge-Ge > Sn-Sn$

47. The correct order of catenation is : [April 10, 2019 (I)]
(1) C > Sn > Si > Ge (2) C > Si > Ge > Sn
(3) Si > Sn > C > Ge (4) Ge > Sn > Si > C

48. The INCORRECT statement regarding the structure of C_{60} is [March 16, 2021 (II); Same April 9, 2019 (I)]
(1) It contains 12 six-membered rings and 24 five-membered rings
(2) The six-membered rings are fused to both six and five-membered rings
(3) Each carbon atom forms three sigma bonds
(4) The five-membered rings are fused only to six-membered rings

49. The element that does NOT show catenation is : [Jan. 12, 2019 (II)]
(1) Ge (2) Si (3) Sn (4) Pb

50. The C – C bond length is maximum in: [April 12, 2019 (II)]
(1) graphite (2) C_{70}
(3) C_{60} (4) diamond

51. C_{60}, an allotrope of carbon cantains: [April 9, 2019 (I)]
(1) 12 hexagons and 20 pentagons.
(2) 18 hexagons and 14 pentagons.
(3) 16 hexagons and 16 pentagons.
(4) 20 hexagons and 12 pentagons.

52. The element that shows greater ability to form $p\pi - p\pi$ multiple bonds, is : [Jan. 12, 2019 (II)]
(1) Sn (2) C (3) Ge (4) Si

53. Which of the following are Lewis acids? [2018]
(1) PH_3 and BCl_3 (2) $AlCl_3$ and $SiCl_4$
(3) PH_3 and $SiCl_4$ (4) BCl_3 and $AlCl_3$

54. In graphite and diamond, the percentage of *p*-characters of the hybrid orbitals in hybridisation are respectively: [Online April 15, 2018 (I)]
(1) 33 and 25 (2) 67 and 75
(3) 50 and 75 (4) 33 and 75

55. **Assertion :** Among the carbon allotropes, diamond is an insulator, whereas, graphite is a good conductor of electricity.
Reason : Hybridization of carbon in diamond and graphite are sp^3 and sp^2, respectively. [Online April 10, 2016]
(1) Both assertion and reason are correct, but the reason is not the correct explanation for the assertion
(2) Both assertion and reason are correct, and the reason is the correct explanation for the assertion
(3) Both assertion and reason are incorrect
(4) Assertion is incorrect statement, but the reason is correct.

56. The catenation tendency of C, Si and Ge is in the order Ge < Si < C. The bond energies (in kJ mol^{-1}) of C – C, Si – Si and Ge – Ge bonds are respectively; [Online April 25, 2013]
(1) 348, 297, 260 (2) 297, 348, 260
(3) 348, 260, 297 (4) 260, 297, 348

57. The stability of dihalides of Si, Ge, Sn and Pb increases steadily in the sequence [2007]
(1) $PbX_2 << SnX_2 << GeX_2 << SiX_2$
(2) $GeX_2 << SiX_2 << SnX_2 << PbX_2$
(3) $SiX_2 << GeX_2 << PbX_2 << SnX_2$
(4) $SiX_2 << GeX_2 << SnX_2 << PbX_2$.

58. A metal, M forms chlorides in its +2 and +4 oxidation states. Which of the following statements about these chlorides is correct? [2006]
(1) MCl_2 is more ionic than MCl_4
(2) MCl_2 is more easily hydrolysed than MCl_4
(3) MCl_2 is more volatile than MCl_4
(4) MCl_2 is more soluble in anhydrous ethanol than MCl_4

59. Graphite is a soft solid lubricant extremely difficult to melt. The reason for this anomalous behaviour is that graphite [2003]
(1) is an allotropic form of diamond
(2) has molecules of variable molecular masses like polymers
(3) has carbon atoms arranged in large plates of rings of strongly bound carbon atoms with weak interplate bonds
(4) is a non-crystalline substance

Part-B (JEE Advanced)

Topic 1 - Boron Family

1 MCQs with One Correct Answer

1. The increasing order of atomic radii of the following Group 13 elements is [Adv. 2016]
(a) Al < Ga < In < Tl (b) Ga < Al < In < Tl
(c) Al < In < Ga < Tl (d) Al < Ga < Tl < In

2. How can the following reaction be made to proceed in forward direction? [2006 - 3M, –1]
$B(OH)_3 + NaOH \rightleftharpoons NaBO_2 + Na[B(OH)_4] + H_2O$
(a) addition of borax
(b) addition of *cis* -1,2-diol
(c) addition of Na_2HPO_4
(d) addition of *trans* -1,2-diol

3. H_3BO_3 is: [2003S]
(a) Monobasic and weak Lewis acid
(b) Monobasic and weak Bronsted acid
(c) Monobasic and strong Lewis acid
(d) Tribasic and weak Bronsted acid

4. Which of the following statements about anhydrous aluminium chloride is correct? [1981 - 1 Mark]
(a) it exists as $AlCl_3$ molecules
(b) it is not easily hydrolysed
(c) it sublimes at 100 °C under vacuum
(d) it is a strong Lewis base

2 Integer Value Answer

5. Three moles of B_2H_6 are completely reacted with methanol. The number of moles of boron containing product formed is [Adv. 2015]

6. The coordination number of Al in the crystalline state of $AlCl_3$ is [2009]

4 Fill in the Blanks

7. Hydrogen gas is liberated by the action of aluminium with concentrated solution of .... [1987 - 1 Mark]

5 True / False

8. All the Al–Cl bonds in Al_2Cl_6 are equivalent. [1989 - 1 Mark]

6 MCQs with One or More than One Correct Answer

9. The compound(s) which react(s) with NH_3 to give boron nitride (BN) is(are) [Adv. 2022]
(a) B (b) B_2H_6 (c) B_2O_3 (d) HBF_4

10. Among the following, the correct statement(s) is (are) [Adv. 2017]
(a) $Al(CH_3)_3$ has the three–centre two–electron bonds in its dimeric structure
(b) BH_3 has the three–centre two–electron bonds in its dimeric structure
(c) $AlCl_3$ has the three–centre two–electron bonds in its dimeric structure
(d) The Lewis acidity of BCl_3 is greater than that of $AlCl_3$

11. The crystalline form of borax has [Adv. 2016]
(a) tetranuclear $[B_4O_5(OH)_4]^{2-}$ unit
(b) all boron atoms in the same plane
(c) equal number of sp^2 and sp^3 hybridized boron atoms
(d) one terminal hydroxide per boron atom

12. The correct statement(s) for orthoboric acid is/are [Adv. 2014]
(a) It behaves as a weak acid in water due to self ionization.
(b) Acidity of its aqueous solution increases upon addition of ethylene glycol
(c) It has a three dimensional structure due to hydrogen bonding
(d) It is a weak electrolyte in water

7 Match the Following

13. Match the following : [2006 - 6M]

Column I	Column II
(A) $Bi^{3+} \longrightarrow (BiO)^+$	(p) Heat
(B) $[AlO_2]^- \longrightarrow Al(OH)_3$	(q) Hydrolysis
(C) $[SiO_4]^{4-} \longrightarrow [Si_2O_7]^{6-}$	(r) Acidification
(D) $[B_4O_7]^{2-} \longrightarrow [B(OH)_3]$	(s) Dilution by water

9 Assertion and Reason/Statement Type Questions

Each question contains STATEMENT-1 (Assertion) and STATEMENT-2 (Reason). Each question has 4 choices (a), (b), (c) and (d) out of which ONLY ONE is correct. Mark your answer as
(a) If both Statement -1 and Statement -2 are correct, and Statement -2 is the correct explanation of the Statement -2.
(b) If both Statement -1 and Statement -2 are correct, but Statement -2 is not the correct explanation of the Statement -1.
(c) If Statement -1 is correct but Statement -2 is incorrect.
(d) If Statement -1 is incorrect but Statement -2 is correct.

14. Statement-1 : Boron always forms covalent bond.
Statement-2 : The small size of B^{3+} favours formation of covalent bond. [2007]

15. Statement-1 : In water, orthoboric acid behaves as a weak monobasic acid.
Statement-2 : In water, orthoboric acid acts as a proton donor. [2007]

16. Statement-1 : $Al(OH)_3$ is amphoteric in nature
Statement-2 : Al–O and O–H bonds can be broken with equal ease in $Al(OH)_3$. [1998 - 2 Marks]

10 Subjective Problems

17. AlF_3 is insoluble in anhydrous HF but it becomes soluble in presence of little amount of KF. Addition of boron trifluoride to the resulting solution causes reprecipitation of AlF_3. Explain with balanced chemical equations. [2004 - 2 Marks]

18. How is boron obtained from borax? Give chemical equations with reaction conditions. Write the structure of B_2H_6 and its reaction with HCl. [2002 - 5 Marks]

19. Compound (*X*) on reduction with $LiAlH_4$ gives a hydride (*Y*) containing 21.72% hydrogen along with other products. The compound (*Y*) reacts with air explosively resulting in boron trioxide. Identify (*X*) and (*Y*). Give balanced reactions involved in the formation of (*Y*) and its reaction with air. Draw the structure of (*Y*). [2001 - 5 Marks]

20. State the conditions under which the following preparation is carried out. Give the necessary equations which need not be balanced : Alumina from aluminium. [1983 - 1 Mark]

21. State with balanced equations what happens when :
(*i*) Aluminium sulphide gives a foul odour when it becomes damp. Write a balanced chemical equation for the reaction. [1997 - 2 Marks]
(*ii*) Reaction of aluminium with aqueous sodium hydroxide. [1997 - 1 Mark]

22. Give reason of the following :
The hydroxides of aluminium and iron are insoluble in water. However, NaOH is used to separate one from the other. [1991 - 1 Mark]

23. Complete the following equations (no balancing is needed)
(*i*) $HCO_3^- + Al^{3+} \longrightarrow Al(OH)_3 + ...$ [1981 - 1 Mark]
(*ii*) $AlBr_3 + K_2Cr_2O_7 + H_3PO_4 \longrightarrow K_3PO_4 + AlPO_4 + H_2O + ... + ...$ [1981 - 1 Mark

Topic 2 - Carbon Family

1 MCQs with One Correct Answer

24. Name the structure of silicates in which three oxygen atoms of $[SiO_4]^{4-}$ are shared. [2005S]
(a) Pyrosilicate
(b) Sheet silicate
(c) Linear chain silicate
(d) Three dimensional silicate

25. $(Me)_2SiCl_2$ on hydrolysis will produce [2003S]
(a) $(Me)_2Si(OH)_2$ (b) $(Me)_2Si=O$
(c) $-[-O-(Me)_2Si-O-]_n-$ (d) $Me_2SiCl(OH)$

26. Which one of the following oxides is neutral? [1996 - 1 Mark]
(a) CO (b) SnO_2
(c) ZnO (d) SiO_2

27. Which of the following halides is least stable and has doubtful existence? [1996 - 1 Mark]
(a) CI_4 (b) GeI_4 (c) SnI_4 (d) PbI_4

28. The material used in the solar cells contains [1993 - 1 Mark]
(a) Cs (b) Si (c) Sn (d) Ti

29. Moderate electrical conductivity is shown by [1982 - 1 Mark]
(a) silica (b) graphite
(c) diamond (d) carborundum

30. Lead pencil contains [1980]
(a) Pb (b) FeS
(c) Graphite (d) PbS

2 Integer Value Answer

31. The value of n in the molecular formula $Be_nAl_2Si_6O_{18}$ is [2010]

4 Fill in the Blanks

32. Compounds that formally contain Pb^{4+} are easily reduced to Pb^{2+}. The stability of the lower oxidation state is due to .... [1997 - 1 Mark]

33. A liquid which is permanently supercooled is frequently called a .... [1997 - 1 Mark]

34. One recently discovered allotrope of carbon (*e.g.*, C_{60}) is commonly known as.......... . [1994 - 1 Mark]

35. The hydrolysis of trialkylchlorosilane R_3SiCl, yields .... [1994 - 1 Mark]

36. The hydrolysis of alkyl substituted chlorosilanes gives .... [1991 - 1 Mark]

5 True / False

37. The tendency for catenation is much higher for C than for Si. [1993 - 1 Mark]

38. Diamond is harder than graphite. [1993 - 1 Mark]

39. Graphite is better lubricant on the moon than on the earth. [1987 - 1 Mark]

40. Carbon tetrachloride is inflammable. [1985 - ½ Mark]

41. Carbon tetrachloride burns in air when lighted to give phosgene. [1983 - 1 Mark]

42. When PbO_2 reacts with a dilute acid, it gives hydrogen peroxide. [1982 - 1 Mark]

6 MCQs with One or More than One Correct Answer

43. Choose the correct statement(s) among the following. [Adv. 2020]
(a) $SnCl_2 \cdot 2H_2O$ is a reducing agent.
(b) SnO_2 reacts with KOH to form $K_2[Sn(OH)_6]$.
(c) A solution of $PbCl_2$ in HCl contains Pb^{2+} and Cl^- ions.
(d) The reaction of Pb_3O_4 with hot dilute nitric acid to give PbO_2 is a redox reaction.

44. A tin chloride Q undergoes the following reactions (not balanced) [Adv. 2019]

$$Q + Cl^- \rightarrow X$$
$$Q + Me_3N \rightarrow Y$$
$$Q + CuCl_2 \rightarrow Z + CuCl$$

X is a mono anion having pyramidal geometry. Both Y and Z are neutral compounds. Choose the correct option(s).
(a) The oxidation state of the central atom in Z is +2
(b) The central atom in Z has one lone pair of electrons
(c) The central atom in X is sp^3 hybridized
(d) There is a coordinate bond in Y

45. Under hydrolytic conditions, the compounds used for preparation of linear polymer and for chain termination, respectively, are [Adv. 2015]
(a) CH_3SiCl_3 and $Si(CH_3)_4$
(b) $(CH_3)_2SiCl_2$ and $(CH_3)_3SiCl$
(c) $(CH_3)_2SiCl_2$ and CH_3SiCl_3
(d) $SiCl_4$ and $(CH_3)_3SiCl$

46. With respect to graphite and diamond, which of the statement(s) given below is (are) correct ? [2012]
(a) Graphite is harder than diamond.
(b) Graphite has higher electrical conductivity than diamond.
(c) Graphite has higher thermal conductivity than diamond.
(d) Graphite has higher C–C bond order than diamond.

7 Match the Following

47. Match gases under specified conditions listed in **Column I** with their properties/laws in **Column II**. Indicate your answer by darkening the appropriate bubbles of the 4 × 4 matrix given in the ORS. [1995S]

Column I	Column II
(A) Explosive	(p) NaN_3
(B) Artificial gem	(q) Fe_3O_4
(C) Self reduction	(r) Cu
(D) Magnetic material	(s) Al_2O_3
	(t) $Pb(N_3)_2$
	(u) Fe_2O_3
	(v) Cu
	(w) SiC

48. Each entry in column **X** is in some way related to the entries in column **Y** and **Z**. Match the appropriate entries : [1989]

X	Y	Z
A. Mica	(a) Graphite crystallite	(i) Abrasive
B. Superphosphate	(b) Cubic	(ii) Insulator
C. Carbon fibres	(c) Layer structure	(iii) Fertilizer
D. Rock salt	(d) Diamond structure	(iv) Reinforced plastics
E. Carborundum	(e) Bone ash	(v) Preservative

9 Assertion and Reason/Statement Type Questions

Each question contains **STATEMENT-1 (Assertion)** and **STATEMENT-2 (Reason)**. Each question has 4 choices (a), (b), (c) and (d) out of which ONLY ONE is correct. Mark your answer as
(a) If both Statement -1 and Statement -2 are correct, and Statement -2 is the correct explanation of the Statement -2.
(b) If both Statement -1 and Statement -2 are correct, but Statement -2 is not the correct explanation of the Statement -1.
(c) If Statement -1 is correct but Statement -2 is incorrect.
(d) If Statement -1 is incorrect but Statement -2 is correct.

49. **Statement-1 :** Pb^{+4} compounds are stronger oxidising agents than Sn^{4+} compounds [2008]
Statement-2 : The higher oxidation states for the group 14 elements are more stable for the heavier members of the group due to 'inert pair effect'.

50. **Statement-1 :** Between $SiCl_4$ and CCl_4, only $SiCl_4$ reacts with water.
Statement-2 : $SiCl_4$ is ionic and CCl_4 is covalent. 2001S]

10 Subjective Problems

51. Starting from $SiCl_4$, prepare the following in steps not exceeding the number given in parentheses (give reactions only):
   (*i*) Silicon (1)
   (*ii*) Linear silicone containing methyl groups only (4)
   (*iii*) Na_2SiO_3 (3) **[2001 - 5 Marks]**
52. Draw the structure of a cyclic silicate, $(Si_3O_9)^{6-}$ with proper labelling. **[1998 - 4 Marks]**
53. Complete and balance the following chemical reactions :
   $Sn + 2KOH + 4H_2O \longrightarrow .... + ...$ **[1994 - 1 Mark]**
54. State with balanced equations what happens when :
   (*i*) $SnCl_4 + C_2H_5Cl + Na \rightarrow$ **[1998 - 1 Mark]**
   (*ii*) Write balanced equations for the preparation of crystalline silicon from $SiCl_4$. **[1990 - 1 Mark]**
55. Give reasons for the following :
   (*i*) $(SiH_3)_3N$ is a weaker base than $(CH_3)_3N$. **[1995 - 2 Marks]**
   (*ii*) The molecule of magnesium chloride is linear whereas that of stannous chloride is angular. **[1987 - 1 Mark]**
   (*iii*) Graphite is used as a solid lubricant; **[1985 - 1 Mark]**
   (*iv*) Solid carbon dioxide is known as dry ice. **[1983 - 1 Mark]**
   (*v*) Carbon acts as an abrasive and also as a lubricant. **[1981 - 1 Mark]**
56. State with balanced equations, what happens when :
   (*i*) Tin is treated with moderately concentrated nitric acid.
   (*ii*) Aluminium is reacted with hot concentrated caustic soda solution. **[1979]**

Answer Key

Part-A (JEE Main)																			
1	(1)	7	(1)	13	(4)	19	(2)	25	(3)	31	(1)	37	(1)	43	(2)	49	(4)	55	(2)
2	(2)	8	(2)	14	(1)	20	(2)	26	(1)	32	(2)	38	(1)	44	(4)	50	(4)	56	(1)
3	(3)	9	(3)	15	(4)	21	(4)	27	(1)	33	(1)	39	(2)	45	(4)	51	(4)	57	(4)
4	(2)	10	(1)	16	(1)	22	(2)	28	(3)	34	(3)	40	(1)	46	(1)	52	(2)	58	(1)
5	(4)	11	(4)	17	(2)	23	(2)	29	(4)	35	(3)	41	(4)	47	(2)	53	(2, 4)	59	(3)
6	(1)	12	(4)	18	(1)	24	(2)	30	(1)	36	(1)	42	(4)	48	(1)	54	(2)		
Part-B (JEE Advanced)																			
1	(b)	8	False	14	(a)	26	(a)	31	(3)	40	False	46	(b, d)						
2	(b)	9	(a, b, c)	15	(c)	27	(d)	33	glass	41	False	47	A-(t); B-(s); C-(v); D-(q)						
3	(a)	10	(a, b, d)	16	(c)	28	(b)	34	Fullerene	42	False	48	A-(c)-(ii); B-(e)-(iii); C-(a)-(iv); D-(b)-(v); E-(d)-(i).						
4	(c)	11	(a, c, d)	24	(b)	29	(b)	37	True	43	(a, b)	49	(c)						
5	(6)	12	(b, d)	25	(c)	30	(c)	38	True	44	(c, d)	50	(c)						
6	(6)	13	(A)-(q), (B)-(s), (C)-(p), (D)-(r)					39	True	45	(b)								

Chapter-12

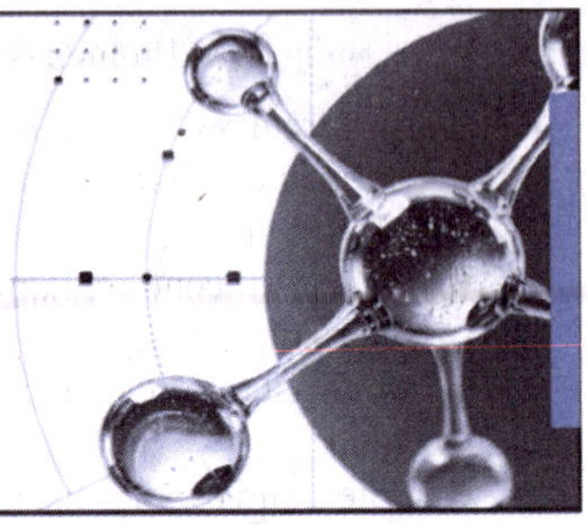

Organic Chemistry - Some Basic Principles and Techniques

Part-A (JEE Main)

Topic 1- Classification and Nomenclature of Organic Compounds

Multiple Choice Questions (MCQs)

1. Identify the correct IUPAC name of hydrocarbon (x) containing three primary carbon atoms and with molar mass 72 g mol^{-1}. [April 05, 2026 (II)]
   (1) 1, 1-Dimethylcyclopropane
   (2) 2, 2-Dimethylpropane
   (3) 2-Methylbutane
   (4) n-pentane

2. In IUPAC nomenclature, the correct order of decreasing priority of functional group is : [April 02, 2026 (I)]
   (1) $-CONH_2, >C=O, -CHO, -NH_2, -C\equiv C-$
   (2) $-CONH_2, -COOCH_3, -CHO, -NH_2, -OH$
   (3) $-CONH_2, -CHO, >C=O, -NH_2-C\equiv C-$
   (4) $-CONH_2, -CHO, -CN, -NH_2, -C\equiv C-$

3. The IUPAC name of the following comopound is : [Jan. 22, 2026 (II)]

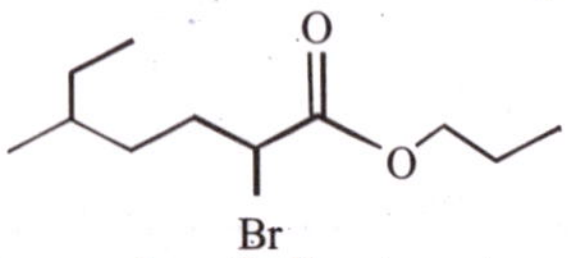

   (1) n-propyl-2-bromo-5-methylheptanoate
   (2) 2-bromo-5-methylhexylpropanoate
   (3) 2-bromo-5-methylpropanoate
   (4) n-propyl-1-bromo-4-methylhexanoate

4. What is the correct IUPAC name of [April 8, 2025 (II)]

   (1) 4-Ethyl-1-hydroxycyclopent-2-ene
   (2) 1-Ethyl-3-hydroxycyclopent-2-ene
   (3) 1-Ethylcyclopent-2-en-3-ol
   (4) 4-Ethylcyclopent-2-en-1-ol

5. Which of the following is the correct IUPAC name of given organic compound (X) ? [April 7, 2025 (I)]

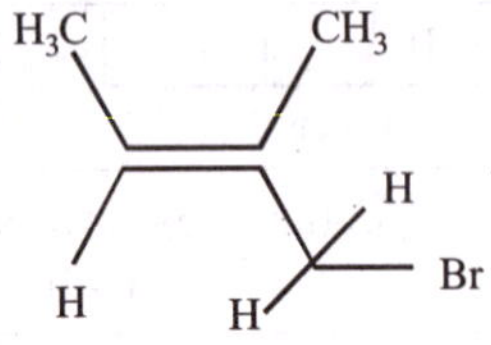

(X)

   (1) 2-Bromo-2-methylbut-2-ene
   (2) 3-Bromo-3-methylprop-2-ene
   (3) 1-Bromo-2-methylbut-2-ene
   (4) 4-Bromo-3-methylbut-2-ene

6. The IUPAC name of the following compound is – [April 4, 2025 (II)]

$$HC\equiv C-CH_2-\underset{}{\overset{OH}{\overset{|}{C}H}}-CH_2-CH=CH_2$$

   (1) 4-Hydroxyhept-1-en-6-yne
   (2) 4-Hydroxyhept-6-en-1-yne
   (3) Hept-6-en-1-yn-4-ol
   (4) Hept-1-en-6-yn-4-ol

7. What is the correct IUPAC name of [April 3, 2025 (II)]

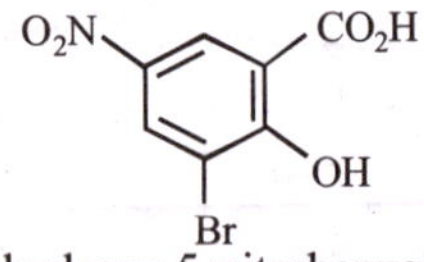

   (1) 3-Bromo-2-hydroxy-5-nitrobenzoic acid
   (2) 3-Bromo-4-hydroxy-1-nitrobenzoic acid
   (3) 2-Hydroxy-3-bromo-5-nitrobenzoic acid
   (4) 5-Nitro-3-bromo-2-hydroxybenzoic acid

8. Match List-I with List-II. [Jan. 29, 2025 (I)]

List-I (Structure)	List-II (IUPAC Name)
(A) $H_3CCH_2\underset{C_2H_5}{\underset{\mid}{C}H}CH_2\underset{CH_3}{\underset{\mid}{C}H}C_2H_5$	(I) 4-Methylpent-1-ene
(B) $(CH_3)_2C(C_3H_7)_2$	(II) 3-Ethyl-5-methylheptane
(C) 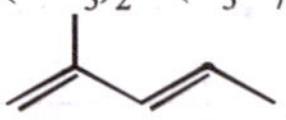	(III) 4,4-Dimethylheptane
(D) 	(IV) 2-Methyl-1,3-pentadiene

Choose the correct answer from the options given below:
   (1) (A)-(III), (B)-(II), (C)-(IV), (D)-(I)
   (2) (A)-(III), (B)-(II), (C)-(I), (D)-(IV)
   (3) (A)-(II), (B)-(III), (C)-(IV), (D)-(I)
   (4) (A)-(II), (B)-(III), (C)-(I), (D)-(IV)

9. The alkane from below having two secondary hydrogens is : [Jan. 22, 2025 (II)]
   (1) 4-Ethyl-3, 4-dimethyloctane
   (2) 2, 2, 4, 4-Tetramethylhexane
   (3) 2, 2, 3, 3-Tetramethylpentane
   (4) 2, 2, 4, 5-Tetramethylheptane

10. The IUPAC name of the following compound is: [Jan. 22, 2025 (I)]

$$CH_3-\overset{COOH}{\overset{|}{C}H}-CH_2-CH_2-\overset{COOCH_3}{\overset{|}{C}H}-CH_3$$

   (1) 2-Carboxy-5-methoxycarbonylhexane.
   (2) Methyl-6-carboxy-2, 5-dimethylhexanoate.
   (3) Methyl-5-carboxy-2-methylhexanoate.
   (4) 6-Methoxycarbonyl-2,5-dimethylhexanoic acid.

11. IUPAC name of following hydrocarbon (X) is :

$CH_3-\underset{|}{\underset{CH_3}{C}}H-CH_2-CH_2-\underset{|}{\underset{CH_3}{C}}H-\underset{|}{\underset{CH_3}{C}}H-CH_2-CH_3$ (X)

[April 8, 2024 (II)]

(1) 2-Ethyl-3,6-dimethylheptane
(2) 2-Ethyl-2,6-diethylheptane
(3) 2,5,6-Trimethyloctane
(4) 3,4,7-Trimethyloctane

12. Given below are two statements: [April 8, 2024 (I)]

Statement I : 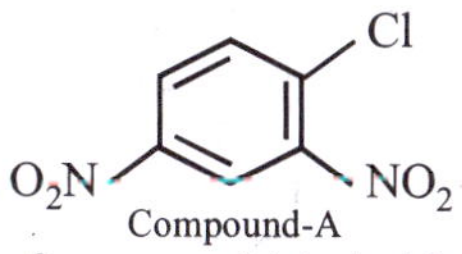
Compound-A

IUPAC name of compound A is 4-chloro-1 3-dinitrobenzene:

Statement II: 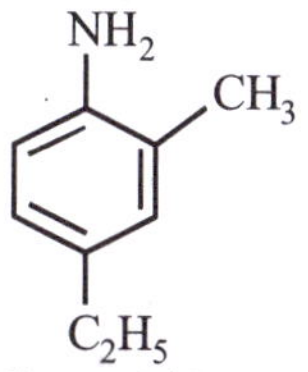
Compound-B

IUPAC name of Compound B is 4-ethyl-2-methylaniline.

In the light of the above statements, choose the **most appropriate** answer from the options given below:
(1) Both Statement I and Statement II are correct
(2) Statement I is incorrect but Statement II is correct
(3) Statement I is correct but Statement II is incorrect
(4) Both Statement I and Statement II are incorrect

13. Functional group present in sulphonic acid is : [April 6, 2024 (I)]

(1) SO_4H (2) SO_3H
(3) $-\underset{\underset{O}{\|}}{S}-OH$ (4) $-SO_2$

14. The correct nomenclature for the following compound is: [April 5, 2024 (II)]

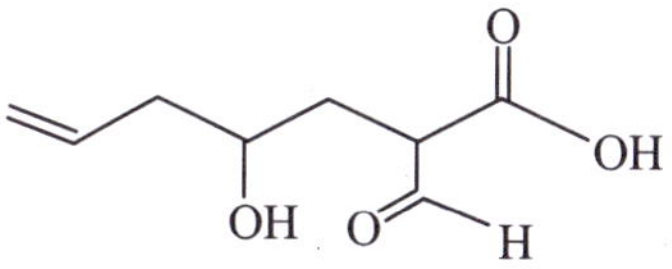

(1) 2–carboxy–4–hydroxyhept–6–enal
(2) 2–carboxy–4–hydroxyhept–7–enal
(3) 2–formyl–4–hydroxyhept–6–enoic acid
(4) 2–formyl–4–hydroxyhept–7–enoic acid

15. Identify structure of 2,3-dibromo-1-phenylpentane. [Jan. 31, 2024 (II)]

(1) 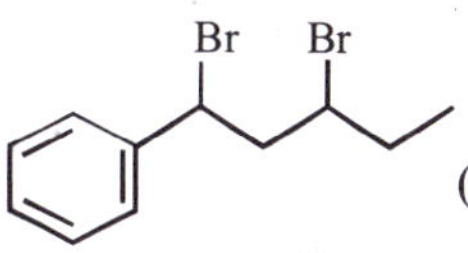 (2) 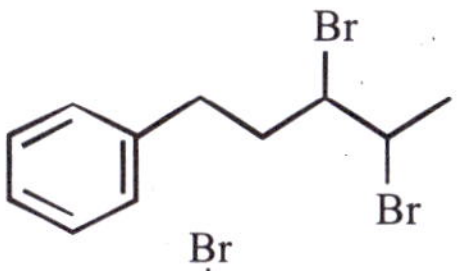

(3) 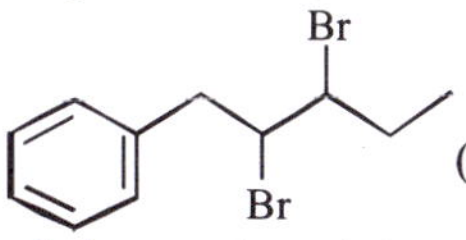 (4) 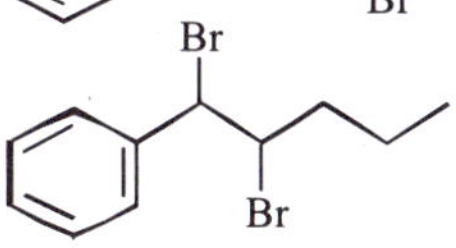

16. Given below are two statements [Jan. 31, 2024 (I)]

Statement I : IUPAC name of $HO-CH_2-(CH_2)_3-CH_2-COCH_3$ is 7-hydroxyheptan-2-one.

Statement II : 2-oxoheptan-7-ol is the correct IUPAC name for above compound.

In the light of the above statements, choose the **most appropriate answer** from the options given below:
(1) Statement I is correct but Statement II is incorrect.
(2) Statement I is incorrect but Statement II is correct.
(3) Both Statement I and Statement II are correct.
(4) Both Statement I and Statement II are incorrect.

17. IUPAC name of following compound is: [Jan. 30, 2024 (II)]

$CH_3-\underset{\underset{NH_2}{|}}{C}H-CH_2-CN$

(1) 2-Aminopentanenitrile (2) 2-Aminobutanenitrile
(3) 3-Aminobutanenitrile (4) 3-Aminopropanenitrile

18. Structure of 4- methylpent -2- enal is: [Jan. 30, 2024 (I)]

(1) $H_2C=\underset{\underset{H}{|}}{C}-\overset{\overset{CH_3}{|}}{\underset{\underset{H}{|}}{C}}-CH_2-\overset{\overset{O}{\|}}{C}-H$

(2) $CH_3-CH_2-\underset{\underset{CH_3}{|}}{C}=CH-\overset{\overset{O}{\|}}{C}-H$

(3) $CH_3-CH_2-CH=\underset{\underset{CH_3}{|}}{C}-\overset{\overset{O}{\|}}{C}-H$

(4) $CH_3-\underset{\underset{CH_3}{|}}{C}H-CH=CH-\overset{\overset{O}{\|}}{C}-H$

19. According to IUPAC system, the compound 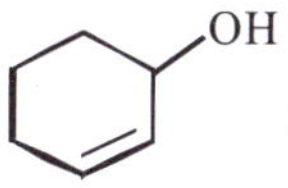 is named as [Jan. 29, 2024 (II)]

(1) Cyclohex-1-en-3-ol (2) Cyclohex-2-en-1-ol
(3) Cyclohex-1-en-2-ol (4) 1-Hydroxyhex-2-ene

20. Bond line formula of $HOCH(CN)_2$ is: [Jan. 27, 2024 (II)]

(1) 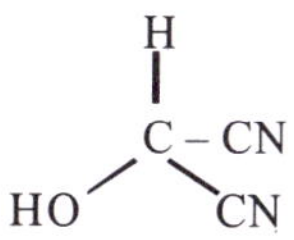 (2) 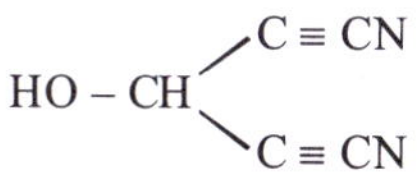

(3) 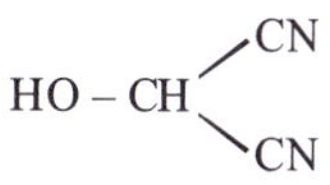 (4) 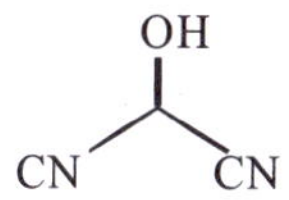

21. IUPAC name of following compound (P) is : [Jan. 27, 2024 (I)]

(1) 1-Ethyl-5, 5-dimethylcyclohexane
(2) 3-Ethyl-1,1-dimethylcyclohexane
(3) 1-Ethyl-3, 3-dimethylcyclohexane
(4) 1,1-Dimethyl-3-ethylcyclohexane

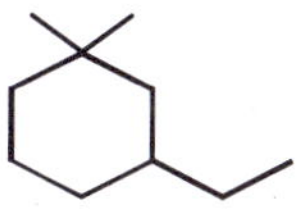
(P)

22. Cyclohexene  is _________ type of an organic compound. [Jan. 27, 2024 (I)]

(1) Benzenoid aromatic
(2) Benzenoid non-aromatic
(3) Acyclic
(4) Alicyclic

23. The correct IUPAC nomenclature for the following compound is [April 08, 2023 (II)]

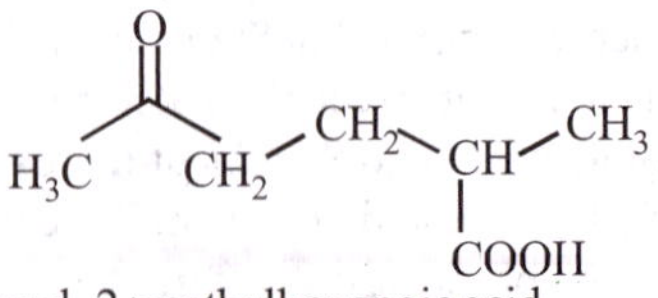

(1) 5–Formyl–2–methylhexanoic acid
(2) 2–Methyl–5–oxohexanoic acid
(3) 2–Formyl–5–methylhexan–6–oic acid
(4) 5–Methyl–2–oxohexan–6–oic acid

24. Match List-I with List-II. [July 28, 2022 (I)]

List-I	List-II
(A) 	(I) Spiro compound
(B) 	(II) Aromatic compound
(C) 	(III) Non-planar heterocyclic compound
(D) 	(IV) Bicyclo compound

Choose the correct answer from the options given below :
(1) (A) – (II), (B) – (I), (C) – (IV), (D) – (III)
(2) (A) – (IV), (B) – (III), (C) – (I), (D) – (II)
(3) (A) – (III), (B) – (IV), (C) – (I), (D) – (II)
(4) (A) – (IV), (B) – (III), (C) – (II), (D) – (I)

25. The correct decreasing order of priority of functional groups in naming an organic compound as per IUPAC system of nomenclature is : [July 26, 2022 (II)]
(1) —COOH > —$CONH_2$ > —COCl > —CHO
(2) —SO_3H > —COCl > —$CONH_2$ > —CN
(3) —COOH > —COCl > —NH_2 > $\rangle C=O$
(4) —COOH > —COOR > —$CONH_2$ > —COCl

26. The correct IUPAC name of the following compound is : [June 28, 2022 (II)]

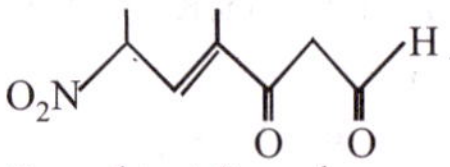

(1) 4-methyl-2-nitro-5-oxohept-3-enal
(2) 4-methyl-5-oxo-2-nitrohept-3-enal
(3) 4-methyl-6-nitro-3-oxohept-4-enal
(4) 6-formyl-4-methyl-2-nitrohex-3-enal

27. The IUPAC name of ethylidene chloride is :- [June 25, 2022 (I)]
(1) 1-Chloroethene (2) 1-Chloroethyne
(3) 1, 2-Dichloroethane (4) 1, 1-Dichloroethane

28. Choose the **correct** name for compound given below : [Aug. 31, 2021 (I)]

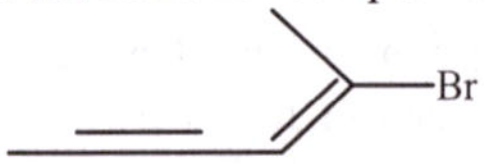

(1) (4*E*)-5-Bromohex-4-en-2-yne
(2) (2*E*)-2-Bromohex-4-yn-2-ene
(3) (2*E*)-2-Bromohex-2-en-4-yne
(4) (4*E*)-5-Bromohex-2-en-4-yne

29. Mesityl oxide is a common name of [March 17, 2021 (I)]
(1) 3-Methylcyclohexane carbaldehyde
(2) 2, 4-Dimethylpentan-3-one
(3) 2-Methylcyclohexanone
(4) 4-Methylpent-3-en-2-one

30. The IUPAC name of the following compound is: [Sep. 06, 2020 (II)]

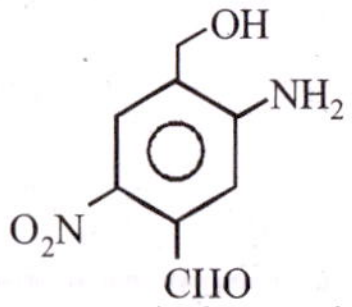

(1) 2-nitro-4-hydroxymethyl-5-amino benzaldehyde
(2) 3-amino-4-hydroxymethyl-5-nitrobenzaldehyde
(3) 5-amino-4-hydroxymethyl-2-nitrobenzaldehyde
(4) 4-amino-2-formyl-5-hydroxymethyl nitrobenzene

31. The IUPAC name of the following compound is : [Sep. 04, 2020 (I)]

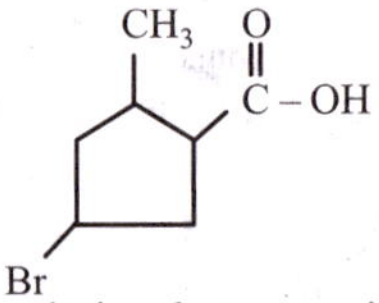

(1) 5-Bromo-3-methylcyclopentanoic acid
(2) 4-Bromo-2-methylcyclopentane carboxylic acid
(3) 3-Bromo-5-methylcyclopentanoic acid
(4) 3-Bromo-5-methylcyclopentane carboxylic acid

32. The IUPAC name for the following compound is : [Sep. 02, 2020 (I)]

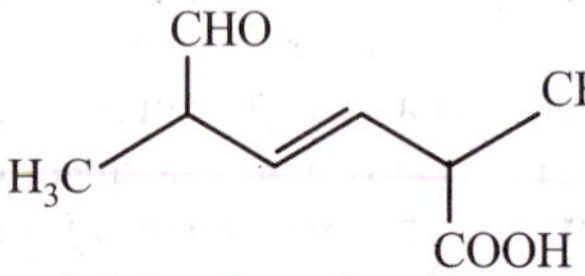

(1) 2, 5-dimethyl-5-carboxy-hex-3-enal
(2) 2, 5-dimethyl-6-carboxy-hex-3-enal
(3) 2, 5-dimethyl-6-oxo-hex-3-enoic acid
(4) 6-formyl-2-methyl-hex-3-enoic acid

33. The IUPAC name for the following compound is : [April 12, 2019 (II)]

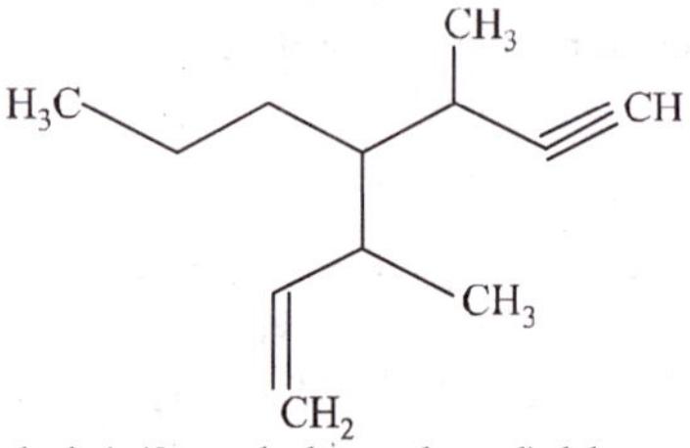

(1) 3-methyl-4-(3-methylprop-l-enyl)-l-heptyne
(2) 3,5-dimethyl-4-propylhept-6-en-l-yne
(3) 3-methyl-4-(1-methylprop-2-ynyl)-l-heptene
(4) 3,5-dimethyl-4-propylhept-1-en-6-yne

34. The correct IUPAC name of the following compound is: [April 9, 2019 (I)]

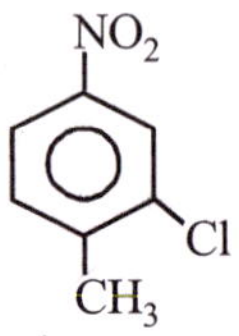

(1) 5-chloro-4-methyl-1-nitrobenzene
(2) 2-chloro-1-methyl-4-nitrobenzene
(3) 3-chloro-4-methyl-1-nitrobenzene
(4) 2-methyl-5-nitro-1-chlorobenzene

35. The IUPAC name of the following compound is :

$$\begin{array}{c} \quad\quad CH_3\ \ OH \\ \quad\quad | \quad\ \ | \\ H_3C-CH-CH-CH_2-COOH \end{array}$$

[April 8, 2019 (I)]
(1) 4, 4-Dimethyl-3-hydroxybutanoic acid
(2) 2-Methyl-3-hydroxypentan-5-oic acid
(3) 3-Hydroxy-4-methylpentanoic acid
(4) 4-Mcthyl-3-hydroxypentanoic acid

36. What is the IUPAC name of the following compound?

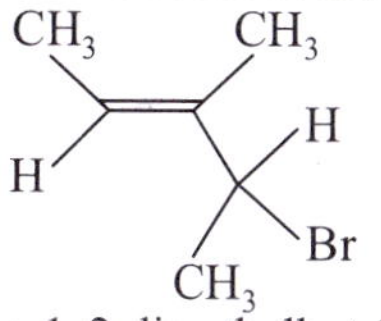

[Jan. 10, 2019 (II)]

(1) 3-Bromo-1, 2-dimethylbut-1-ene
(2) 3-Bromo-3-methyl-1, 2-dimethylprop-1-ene
(3) 2-Bromo-3-methylpent-3-ene
(4) 4-Bromo-3-methylpent-2-ene

37. The IUPAC name of the following compound is :

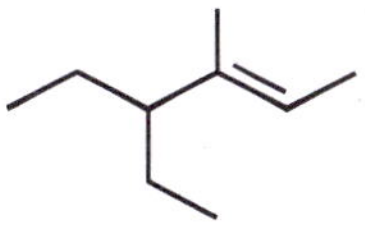

[Online April 15, 2018(I); Same Online April 19, 2014]

(1) 3-ethyl-4-methylhex-4-ene
(2) 4, 4-diethyl-3-methylbut-2-ene
(3) 4-methyl-3-ethylhex-4-ene
(4) 4-ethyl-3-methylhex-2-ene

38. The IUPAC name of the following compound is :

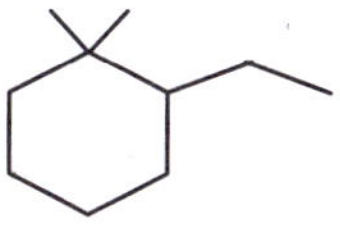

(1) 1, 1–dimethyl–2–ethylcyclohexane
(2) 2–ethyl–1,1–dimethylcyclohexane
(3) 1–ethyl–2,2–dimethylcyclohexane
(4) 2, 2–dimethyl–1–ethylcyclohexane

39. The hydrocarbon with seven carbon atoms containing a neopentyl and a vinyl group is : [Online April 9, 2016]

(1) 2, 2–dimethyl–4– pentene
(2) 4, 4– dimethylpentene
(3) isopropyl–2–butene
(4) 2, 2–dimethyl–3–pentene

40. In allene (C_3H_4), the type(s) of hybridization of the carbon atoms is (are): [Online April 11, 2014]

(1) sp and sp^3 (2) sp^2 and sp
(3) only sp^2 (4) sp^2 and sp^3

41. Dipole moment is shown by [Online May 26, 2012]

(1) 1, 2-dichlorobenzene
(2) *trans*-2, 3-dichloro-2-butene
(3) 1, 4-dichlorobenzene
(4) *trans*-1, 2-dinitroethene

42. The IUPAC name of the following compound is [Online May 19, 2012]

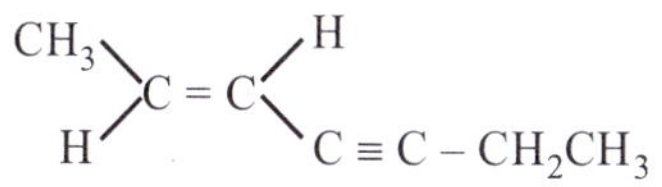

(1) (*E*)-2-hepten-4-yne (2) (*Z*)-5-hepten-3-yne
(3) (*E*)-5-hepten-3-yne (4) (*Z*)-2-hepten-4-yne

43. The IUPAC name of the compound

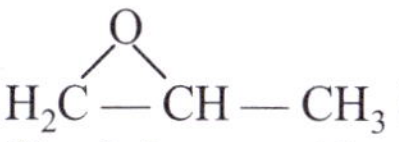 is [Online May 7, 2012]

(1) 1, 2-propoxide (2) propylene oxide
(3) 1, 2-oxo propane (4) 1, 2-epoxypropane

44. The IUPAC name of the following compound is [2009]

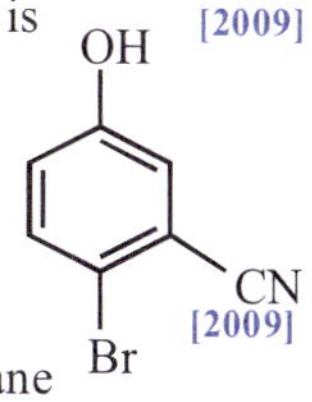

(1) 4-bromo-3-cyanophenol
(2) 2-bromo-5-hydroxybenzonitrile
(3) 2-cyano-4-hydroxybromobenzene
(4) 6-bromo-3-hydroxybenzonitrile

45. The IUPAC name of neopentane is [2009]

(1) 2, 2-dimethylpropane (2) 2-methylpropane
(3) 2, 2-dimethylbutane (4) 2- methylbutane

46. The correct decreasing order of priority for the functional groups of organic compounds in the IUPAC system of nomenclature is [2008]

(1) $-COOH, -SO_3H, -CONH_2, -CHO$
(2) $-SO_3H, -COOH, -CONH_2, -CHO$
(3) $-CHO, -COOH, -SO_3H, -CONH_2$
(4) $-CONH_2, -CHO, -SO_3H, -COOH$

47. The IUPAC name of 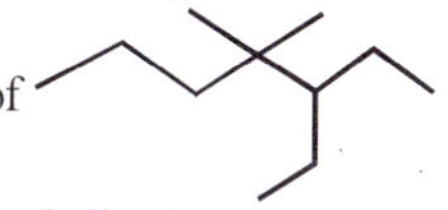 is [2007]

(1) 3-ethyl-4,4-dimethylheptane
(2) 1, 1-diethyl-2,2-dimethylpentane
(3) 4, 4 dimethyl-5,5-diethylpentane
(4) 5, 5-diethyl-4,4-dimethylpentane.

48. The IUPAC name of the compound shown below is : [2006]

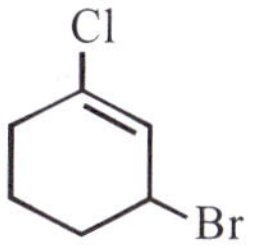

(2) 1-bromo-3-chlorocyclohexene
(3) 2-bromo-6-chlorocyclohex-1-ene
(4) 6-bromo-2-chlorocyclohexene

49. The IUPAC name of C_6H_5COCl is [2006]

(1) benzene chloro ketone
(2) chlorobenzyl ketone
(3) chlorophenyl ketone
(4) benzene carbonyl chloride

50. The IUPAC name of the compound is [2004]

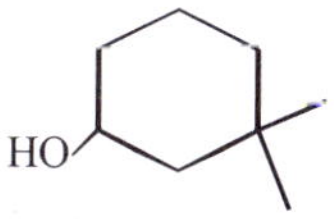

(2) 1, 1-dimethyl-3-hydroxy cyclohexane
(3) 3, 3-dimethyl-1-hydroxy cyclohexane
(4) 1, 1-dimethyl-3-cyclohexanol

51. Which one of the following does not have sp^2 hybridised carbon ? [2004]

(1) Acetonitrile (2) Acetic acid
(3) Acetone (4) Acetamide

52. The general formula $C_nH_{2n}O_2$ could be for open chain [2003]

(1) carboxylic acids (2) diols
(3) dialdehydes (4) diketones

53. The IUPAC name of $CH_3COCH(CH_3)_2$ is [2003]

(1) 2-methyl-3-butanone (2) 4-methylisopropyl ketone
(3) 3-methyl-2-butanone (4) Isopropylmethyl ketone

54. Which of the following compounds has wrong IUPAC name [2002]

(1) $CH_3–CH_2–CH_2–COO–CH_2CH_3$ → ethyl butanoate

(2) $CH_3-\underset{\underset{CH_3}{|}}{CH}-CH_2-CHO$ → 3-methyl-butanal

(3) $CH_3-\underset{\underset{OH}{|}}{CH}-\underset{\underset{CH_3}{|}}{CH}-CH_3$ → 2-methyl-3-butanol

(4) $CH_3-\underset{\underset{CH_3}{|}}{CH}-\overset{\overset{O}{||}}{C}-CH_2-CH_3$ → 2-methyl-3-pentanone

NUMERICAL VALUE QUESTIONS (NVQs)

55. Number of compounds among the following which contain sulphur as heteroatom is ______.
Furan, Thiophene, Pyridine, Pyrrole, Cysteine, Tyrosine

[Jan. 29, 2024 (I)]

Topic 2 - Isomerism in Organic Compounds

MULTIPLE CHOICE QUESTIONS (MCQs)

56. Given below are two statements : [April 06, 2026 (I)]

Statement-I: 2, 6-diethylcyclohexanone and 6-methyl-2-n-propylcyclohexanone are metamers.

Statement-II: 2, 2, 6, 6-tetramethylcyclohexanone exhibits keto-enol tautomerism.

In the light of the above statements, choose the **correct** answer from the options given below.

(1) Both Statement-I and Statement-II are true.
(2) Both Statement-I and Statement-II are false.
(3) Statement-I is true but Statement-II is false.
(4) Statement-I is false but Statement-II is true.

57. Given below are four compounds: [Jan. 21, 2026 (II)]

A. n-propyl choride B. iso-propyl chloride
C. sec-butyl chloride D. neo-pentyl chloride

Percentage of carbon in the one which exhibits optical isomerism is:

(1) 52 (2) 56 (3) 46 (4) 40

58. Match **List-I** with **List-II**. [Jan. 21, 2026 (II)]

List-I (Pair of Compounds)		List-II (Type of Isomers)	
A.	2-Methylpropene and but-1-ene	I.	Stereoisomers
B.	Cis-but-2-ene and trans-but-2-ene	II.	Position isomers
C.	2-Butanol and diethyl ether	III.	Chain isomers
D.	But-1-ene and but-2-ene	IV.	Functional group isomers

Choose the **correct** answer from the options given below:

(1) A-III, B-I, C-IV, D-II (2) A-III, B-I, C-II, D-IV
(3) A-I, B-IV, C-III, D-II (4) A-II, B-I, C-IV, D-III

59. Identify correct statement from the following: [Jan. 21, 2026 (I)]

A. Propanal and propanone are functional isomers.
B. Ethoxyethane and methoxypropane are metamers.
C. But-2-ene shows optical isomerism.
D. But-1-ene and but-2-ene are functional isomers.
E. Pentane and 2, 2-dimethyl propane are chain isomers.

Choose the **correct** answer from the options given below:

(1) B, C and D only (2) A, B and C only
(3) A, B and E only (4) C, D and E only

60. Identify the correct statements from the following [April 3, 2025 (I)]

A. 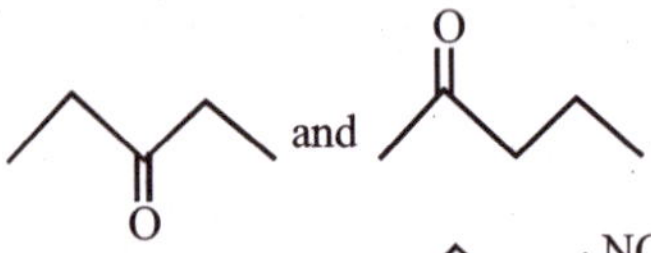 are metamers

B. 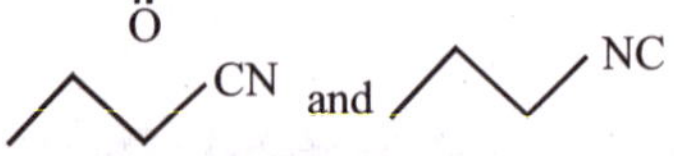 are functional isomers

C. 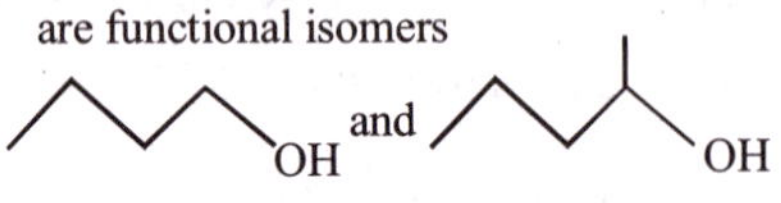 are position isomers

D. 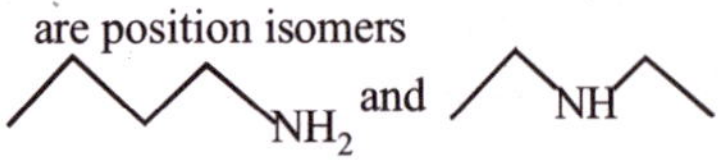 are homologous

Choose the **correct** answer from the options given below

(1) C & D only (2) B & C only
(3) A & B only (4) A, B & C only

61. Given below are two statements : [Jan. 28, 2025 (II)]

Statement (I) : 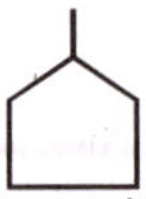 and  are isomeric compounds.

Statement (II) :  and  are functional group isomers.

In the light of the above statements, choose the correct answer from the options given below :

(1) Both Statement I and Statement II are false
(2) Both Statement I and Statement II are true
(3) Statement I is true but Statement II is false
(4) Statement I is false but Statement II is true

62. How many different stereoisomers are possible for the given molecule? [Jan. 22, 2025 (I)]

$$CH_3-\underset{\underset{OH}{|}}{CH}-CH=CH-CH_3$$

(1) 3 (2) 1 (3) 2 (4) 4

63. Total number of stereo isomers possible for the given structure: [April 9, 2024 (II)]

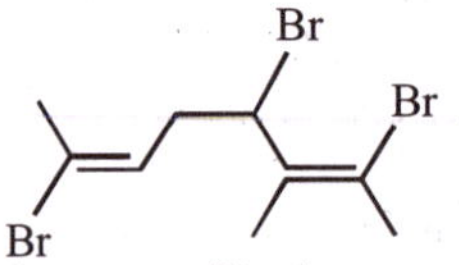

(1) 8 (2) 2 (3) 4 (4) 3

64. Which of the following is metamer of the given compound (X)? [April 6, 2024 (I)]

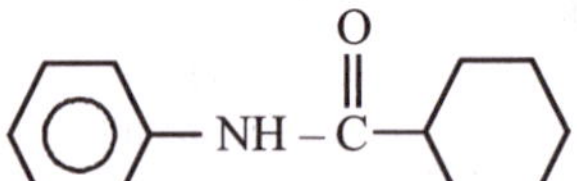

(X)

(1) 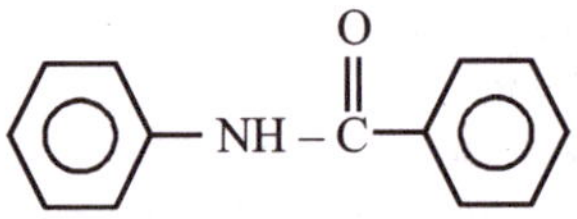

(2) 

(3) 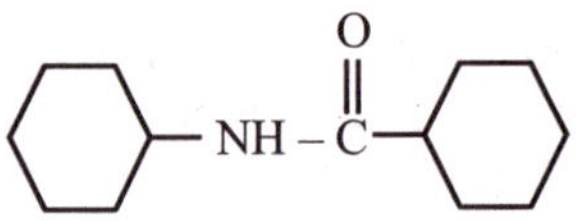

(4) 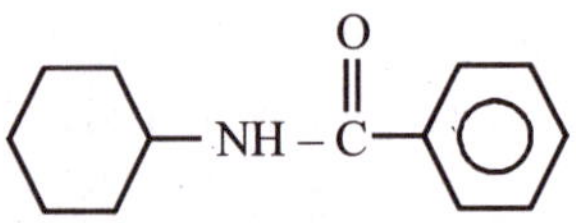

65. The incorrect statement regarding the geometrical isomers of 2-butene is: [April 6, 2024 (II)]

(1) cis-2-butene and trans-2-butene are not interconvertible at room temperature.
(2) cis-2-butene has less dipole moment than trans-2-butene.
(3) trans-2-butene is more stable than cis-2-butene.
(4) cis-2-butene and trans-2-butene are stereoisomers.

66. Match **List-I** with **List-II** [April 5, 2024 (II)]

List-I (Pair of Compounds)	List-II (Isomerism)
(A) n-propanol and Isopropanol	(I) Metamerism
(B) Methoxypropane and ethoxyethane	(II) Chain Isomerism
(C) Propanone and propanal	(III) Position Isomerism
(D) Neopentane and Isopentane	(IV) Functional Isomerism

(1) (A)–(II), (B)–(I), (C)–(IV), (D)–(III)
(2) (A)–(III), (B)–(I), (C)–(II), (D)–(IV)
(3) (A)–(I), (B)–(III), (C)–(IV), (D)–(II)
(4) (A)–(III), (B)–(I), (C)–(IV), (D)–(II)

67. Match **List I** with **List II**. [Jan. 25, 2023 (II)]

List I (Isomeric pairs)	List II (Type of isomers)
A. Propanamine and N-Methylethanamine	I. Metamers
B. Hexan-2-one and Hexan-3-one	II. Positional isomers
C. Ethanamide and Hydroxyethanimine	III. Functional isomers
D. o-nitrophenol and p-nitrophenol	IV. Tautomers

Choose the correct answer from the options given below:-
(1) A-III, B-IV, C-I, D-II (2) A-IV, B-III, C-I, D-II
(3) A-II, B-III, C-I, D-IV (4) A-III, B-I, C-IV, D-II

68. Compound with molecular formula C_3H_6O can show [March 18, 2021 (I)]
(1) Positional isomerism
(2) Both positional isomerism and metamerism
(3) Metamerism
(4) Functional group isomerism

69. The absolute configuration of

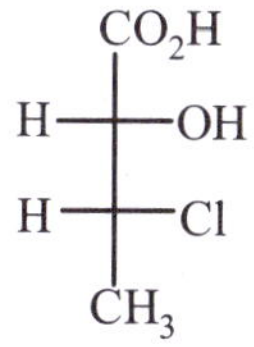

is : [2016, 2008 (S)]
(1) (2*S*, 3*S*) (2) (2*R*, 3*R*) (3) (2*R*, 3*S*) (4) (2*S*, 3*R*)

70. The number of structural isomers for C_6H_{14} is : [Online April 11, 2015]
(1) 4 (2) 3 (3) 6 (4) 5

71. Which of the following pairs of compounds are positional isomers ? [Online April 11, 2015]
(1) $CH_3-CH_2-CH_2-C(=O)-CH_3$ and $CH_3-CH_2-C(=O)-CH_2-CH_3$
(2) $CH_3-CH_2-CH_2-CH_2-CHO$ and $CH_3-CH_2-CH_2-C(=O)-CH_3$
(3) $CH_3-CH_2-CH_2-C(=O)-CH_3$ and $CH_3-CH(CH_3)-CH_2-CHO$
(4) $CH_3-CH_2-C(=O)-CH_2-CH_3$ and $(CH_3)_2CH-CH_2-CHO$

72. The optically inactive compound from the following is : [Online April 10, 2015]
(1) 2 - chloropropanal (2) 2 - chlorobutane
(3) 2 - chloropentane (4) 2 - chloro - 2- methylbutane

73. Which one of the following acids does not exhibit optical isomerism? [Online April 12, 2014]
(1) Lactic acid (2) Tartaric acid
(3) Maleic acid (4) α-amino acids

74. Maleic acid and fumaric acids are [Online May 26, 2012]
(1) chain isomers (2) functional isomers
(3) tautomers (4) geometrical isomers

75. How many cyclic structures are possible for C_4H_6? [Online May 7, 2012]
(1) 3 (2) 5 (3) 6 (4) 4

76. Out of the following, the alkene that exhibits optical isomerism is [2010]
(1) 3-methyl-2-pentene (2) 4-methyl-1-pentene
(3) 3-methyl-1-pentene (4) 2-methyl-2-pentene

77. The alkene that exhibits geometrical isomerism is: [2009]
(1) 2- methyl propene (2) 2-butene
(3) 2- methyl -2- butene (4) propene

78. The number of stereoisomers possible for a compound of the molecular formula [2009]
$CH_3-CH=CH-CH(OH)-Me$ is:
(1) 2 (2) 4 (3) 6 (4) 3

79. Which of the following molecules is expected to rotate the plane of plane-polarised light? [2007]

(1) 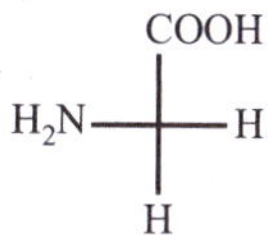
(2) 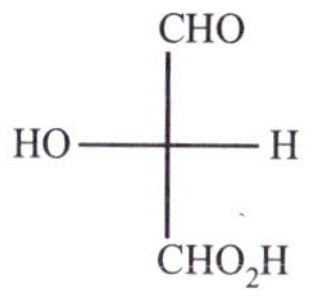
(3) 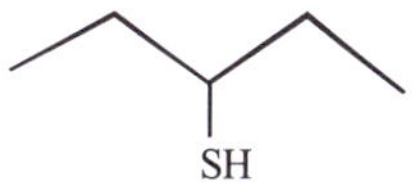
(4) 

80. Which types of isomerism is shown by 2, 3-dichlorobutane? [2005]
(1) Structural (2) Geometric
(3) Optical (4) Diastereo

81. Which of the following will have a mesoisomer also? [2004]
(1) 2, 3- Dichloropentane
(2) 2, 3-Dichlorobutane
(3) 2-Chlorobutane
(4) 2-Hydroxypropanoic acid

82. Which of the following compounds is not chiral? [2004]
(1) 1-chloro-2-methyl pentane
(2) 2-chloropentane
(3) 1-chloropentane
(4) 3-chloro-2-methyl pentane

83. Among the following four structures I to IV, [2003]

$C_2H_5-CH(CH_3)-C_3H_7$ (I), $CH_3-C(=O)-CH(CH_3)-C_2H_5$ (II),

$H-C^{\oplus}H_2$ (III), $C_2H_5-CH(CH_3)-C_2H_5$ (IV)

it is true that

(1) only I and II are chiral compounds
(2) only III is a chiral compound
(3) only II and IV are chiral compounds
(4) all four are chiral compounds

84. A similarity between optical and geometrical isomerism is that [2002]
(1) each forms equal number of isomers for a given compound
(2) if in a compound one is present then so is the other
(3) both are included in stereoisomerism
(4) they have no similarity

85. Racemic mixture is formed by mixing two [2002]
(1) isomeric compounds
(2) chiral compounds
(3) meso compounds
(4) enantiomers with chiral carbon

86. Following types of compounds (as I, II) [2002]

$CH_3CH=CHCH_3$ (I) $\quad$ $CH_3CH(OH)CH_2CH_3$ (II)

are studied in terms of isomerism in:
(1) chain isomerism (2) position isomerism
(3) conformers (4) stereoisomerism

Numerical Value Questions (NVQs)

87. The total number of structural isomers possible for the substituted benzene derivatives with the molecular formula C_9H_{12} is ______. [April 3, 2025 (II)]

88. The possible number of stereoisomers for 5-phenylpent-4-en-2-ol is ______. [Jan. 24, 2025 (II)]

89. Total number of possible stereoisomers of dimethyl cyclopentane is ______. [June 27, 2022 (I)]

90. The number of acyclic structural isomers (including geometrical isomers) for pentene are ______ [July 22, 2021 (II)]

Topic 3 - Concept of Reaction Mechanism in Organic Compounds, Purification and Quantitative Analysis

Multiple Choice Questions (MCQs)

91. Given below are two statements: [April 08, 2026 (II)]
Statement I: Vapours of the liquid with higher boiling point condense before vapours of the liquid with lower boiling points in fractional distillation.
Statement II: The vapours rising up in the fractionating column become richer in high boiling component of the mixture.
In the light of the above statements, choose the **correct** answer from the option given below:
(1) Both Statement I and Statement II are true
(2) Both Statement I and Statement II are false
(3) Statement I is true but Statement II are false
(4) Statement I is false but Statement II are true

92. Given below are two statements: [April 06, 2026 (II)]
Statement I: A mixture of $C_{12}H_{22}O_{11}$ (sugar) and NaCl can be separated by dissolving sugar in alcohol, due to differential solubility
Statement II: Rose essence from rose petals is separated by steam distillation due to its high volatility and insolubility in H_2O.
In the light of the above statements, choose the **correct** answer from the options given below:
(1) Both Statement I and Statement II are true.
(2) Both Statement I and Statement II are false.
(3) Statement I is true Statement II is false.
(4) Statement I is false Statement II is true.

93. R_f value for 2-methylpropene in a solvent system (Ethyl acetate + ether) is 0.42. 2-methylpropene is treated with dilute H_2SO_4 to give major organic product (X). R_f value for (X) in the same solvent system under identical condition will be : [April 06, 2026 (I)]
(1) 0.42 (2) 0.82 (3) 0.62 (4) 0.12

94. IUPAC name of the some alkenes are given below :
Find out the correct stability order. [April 05, 2026 (II)]
A. 2-Methylbut-2-ene B. *cis*-But-2-ene
C. 2,3-Dimethylbut-2-ene D. Prop-1-ene
Choose the correct answer from the options given below:
(1) C > A > B > D (2) C > A > D > B
(3) B > D > A > C (4) A > B > C > D

95. Match List-I with List-II. [April 05, 2026 (II)]

List-I (Purification technique)		List-II (Used to separate)	
A.	Simple distillation	I.	Steam volatile compound
B.	Fractional distillation	II.	Two liquids with large difference in boiling points
C.	Steam distillation	III.	Liquid decomposing at its boiling point
D.	Distillation under reduced pressure	IV.	Two liquids with close boiling points

Choose the correct answer from the options given below:
(1) A-II, B-III, C-I, D-IV (2) A-II, B-IV, C-I, D-III
(3) A-II, B-IV, C-III, D-I (4) A-IV, B-III, C-II, D-I

96. Given below are two statements: [April 05, 2026 (I)]
Statement I: On the basis of inductive effect, the order of stability of alkyl carbanions is $CH_3^- > CH_3-CH_2^- > (CH_3)_2CH^- > (CH_3)_3C^-$.
Statement II: Allyl and benzyl carbanions are more stabilised by inductive effect and not by resonance effect.
In the light of the above statements choose the **correct** answer from the options given below:
(1) Both Statement I and Statement II are correct
(2) Both Statement I and Statement II are incorrect
(3) Statement I is correct and Statement II is incorrect
(4) Statement I is incorrect and Statement II is correct

97. Increasing order of electron withdrawing power of following functional groups is : [April 04, 2026 (I)]
A. –CN B. –COOH
C. $-NO_2$ D. –I
(1) C < B < D < A (2) C < A < B < D
(3) D < B < A < C (4) A < B < C < D

98. Given below are two statements : [April 02, 2026 (II)]
Statement I : In

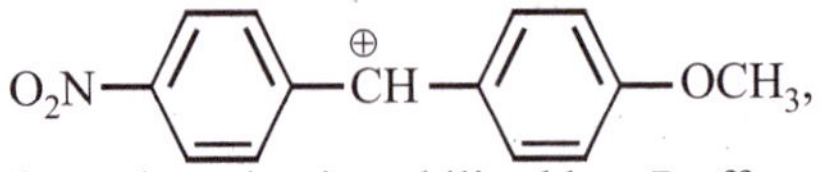

the carbocation is stabilised by +R effect of $-OCH_3$ group.
Statement II : In

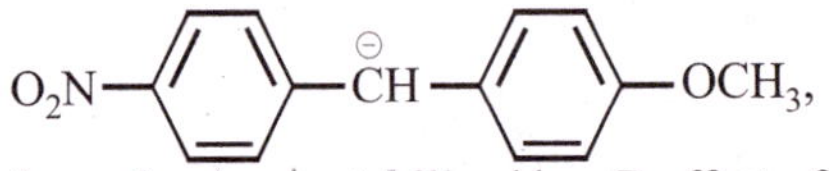

the carbanion is stabilised by –R effect of $-NO_2$ group.

In the light of the above statements, choose the **correct answer** from the options given below :
(1) Both Statement I and Statement II are true.
(2) Both Statement I and Statement II are false.
(3) Statement I is true but Statement II is false.
(4) Statement I is false but Statement II is true.

99. For the given molecule, "x", the preferred site for the attack of the electrophile is : [April 02, 2026 (I)]

x =
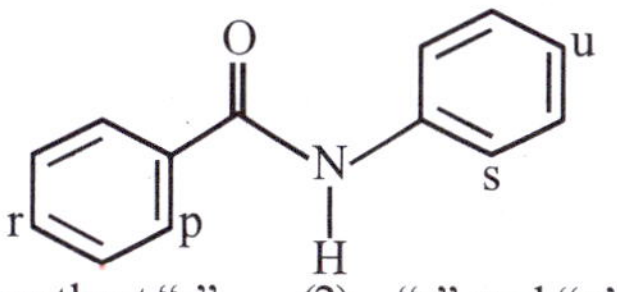

(1) Predominantly at "r" (2) "r" and "u"
(3) "p" and "s" (4) Predominantly at "u"

100. Given below are two statements : [April 02, 2026 (I)]
Statement (I) : 1, 2, 3-Trihydroxypropane can be separated from water by simple distillation.
Statement (II) : An azotropic mixture cannot be separated by fractional distillation.
In the light of the above statements, choose the **correct answer** from the options given below :
(1) Both Statement I and Statement II are true.
(2) Both Statement I and Statement II are false.
(3) Statement I is true but Statement II is false.
(4) Statement I is false but Statement II is true.

101. A student has been given 0.314 g of an organic compound and asked to estimate sulphur. During the experiment, the student has obtained 0.4813 g of barium sulphate. The percentage of sulphur present in the compound is ______ . [Jan. 28, 2026 (II)]
(Given Molar mass in g mol^{-1} S:32, $BaSO_4$: 233)
(1) 42.10% (2) 63.15% (3) 21.05% (4) 48.24%

102. The cyclic cations having the same number of hyperconjugation are : [Jan. 28, 2026 (II)]

A. 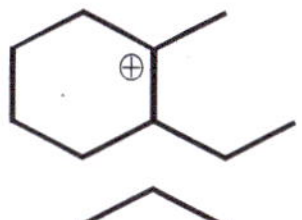
B. 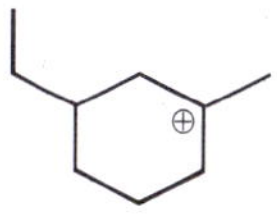
C. 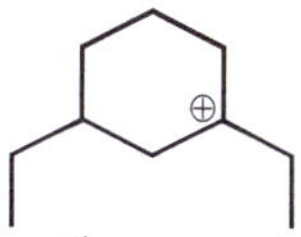
D. 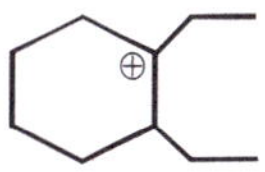

Choose the **correct** answer from the options given below:
(1) A and C Only (2) B and C Only
(3) A and B Only (4) A, C and D only

103. Method used for separation of mixture of products (B and C) obtained in the following reaction is: [Jan. 28, 2026 (I)]

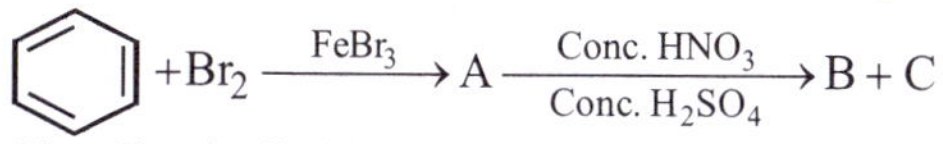

(1) Simple distillation (2) sublimation
(3) steam distillation (4) fractional distillation

104. CORRECT order of stability for the following is
$CH_2=CH^-, CH_3-CH_2^-, CH\equiv C^-$ [Jan. 28, 2026 (I)]
(1) $CH_3-CH_2^- > CH_2=CH^- > CH\equiv C^-$
(2) $CH_2=CH^- > CH\equiv C^- > CH_3-CH_2^-$
(3) $CH\equiv C^- > CH_2=CH^- > CH_3-CH_2^-$
(4) $CH\equiv C^- > CH_3-CH_2^- > CH_2=CH^-$

105. Find out the statement which are not true. [Jan. 24, 2026 (II)]
A. Reasonating structure with more number of covalent bonds and lesser charge separation are more stable.
B. In electromeric effect, an unsaturated system shows + E effect with nucleophile and –E effect with electrophile.
C. Inductive effect is responsible for high melting point, boiling point and dipole moment of polar compounds.
D. The greater the number of alkyl groups attached to the doubly bonded carbon atoms, higher is the heat of hydrogenation.
E. Stability of carbanion increases with the increase in s-character of the carbon carrying the negative charge.
Choose the correct answer from the options given below.
(1) A, D & E only (2) B, D & E only
(3) A, C & D only (4) B & D only

106. Arrange the following carbanions in the decreasing order of stability [Jan. 24, 2026 (I)]
I. $p-Br-C_6H_4-\overset{\ominus}{\ddot{C}}H_2$
II. $C_6H_5\overset{\ominus}{\ddot{C}}H_2$
III. $p-CH_3O-C_6H_4-\overset{\ominus}{\ddot{C}}H_2$
IV. $p-CHO-C_6H_4\overset{\ominus}{\ddot{C}}H_2$
V. $p-CH_3-C_6H_4\overset{\ominus}{\ddot{C}}H_2$
Choose the correct answer from the options given below:
(1) I > II > IV > V > III (2) I > IV > II > V > III
(3) IV > I > II > V > III (4) IV > II > I > III > V

107. In Carius method 0.2425 g of an organic compounds gave 0.5253 g silver chloride. The percentage of chlorine in the organic compound is [Jan. 23, 2026 (II)]
(1) 53.58% (2) 87.65% (3) 37.57% (4) 34.79%

108. Given below are two statements: [Jan. 23, 2026 (II)]
Statement-I: $(CH_3)_3\overset{\oplus}{C}$ is more stable than $\overset{\oplus}{C}H_3$ as nine hyperconjugation interactions are possible in $(CH_3)_3\overset{\oplus}{C}$.
Statement-II: $\overset{\oplus}{C}H_3$ is less stable than $(CH_3)_3\overset{\oplus}{C}$ as only three hyperconjugation interactions are possible in $\overset{\oplus}{C}H_3$. In the light of the above statements, choose the **correct** answer from the options given below.
(1) Statement-I is true but Statement-II is false.
(2) Both Statement-I and Statement-II are true.
(3) Both Statement-I and Statement-II are false.
(4) Statement-I is false but Statement-II is true.

109. Given below are two statements: [Jan. 23, 2026 (I)]
Statement-I: Sublimation is used for the separation and purification of compounds with low melting point.
Statement-II: The boiling point of a liquid increases as the external pressure is reduced.
In the light of the above statements, choose the **correct** answer from the options given below:
(1) Statement-I is false but Statement-II is true.
(2) Statement-I is true but Statement-II is false.
(3) Both Statement-I and Statement-II are true.
(4) Both Statement-I and Statement-II are false.

110. When 1 g of compound (X) is subjected to Kjeldahl's method for estimation of nitrogen, 15 mL, 1M H_2SO_4 was neutralized by ammonia evolved. The percentage of nitrogen in compound (X) is : [Jan. 22, 2026 (II)]
(1) 21 (2) 0.42 (3) 42 (4) 0.21

111. By usual analysis, 1.00g of compound (X) gave 1.79g of magnesium pyrophosphate. The percentage of phosphorus in compound (X) is: (nearest integer)
(Given, molar mass in g mol^{-1}: O = 16, Mg = 24, P = 31) [Jan. 21, 2026 (II)]
(1) 50 (2) 30 (3) 20 (4) 40

112. In Carius method, 0.75 g of an organic compound gave 1.2 g of barium sulphate, find percentage of sulphur (molar mass 32 g mol^{-1}). Molar mass of barium sulphate is 233 g mol^{-1}. [Jan. 21, 2026 (I)]
(1) 4.55% (2) 10.30%
(3) 21.97% (4) 16.48%

113. Match the LIST-I with LIST-II [April 8, 2025 (II)]

LIST-I	LIST-II
A. Carbocation	I. Species that can supply a pair of electrons.
B. C-Free radical	II. Species that can receive a pair of electrons.
C. Nucleophile	III. sp^2 hybridized carbon with empty p-orbital.
D. Electrophile	IV. sp^2/sp^3 hybridized carbon with one unpaired electron.

Choose the correct answer from the options given below:
(1) A-IV, B-II, C-III, D-I (2) A-II, B-III, C-I, D-IV
(3) A-III, B-IV, C-II, D-I (4) A-III, B-IV, C-I, D-II

114. Match List-I with List-II - [April 4, 2025 (II)]

List-I (Separation of)	List-II (Separation Technique)
(A) Aniline from aniline-water mixture	(I) Simple distillation
(B) Glycerol from spent-lye in soap industry	(II) Fractional distillation
(C) Different fractions of crude oil in petroleum industry	(III) Distillation at reduced pressure
(D) Chloroform-Aniline mixture	(IV) Steam distillation

Choose the correct answer from the options given below:
(1) (A)-(IV), (B)-(III), (C)-(II), (D)-(I)
(2) (A)-(I), (B)-(II), (C)-(III), (D)-(IV)
(3) (A)-(III), (B)-(IV), (C)-(I), (D)-(II)
(4) (A)-(II), (B)-(I), (C)-(IV), (D)-(III)

115. In which pairs, the first ion is more stable than the second? [April 4, 2025 (II)]

(A) 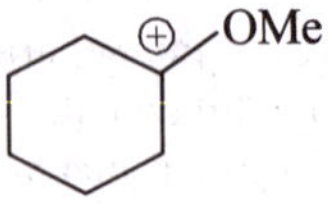 & 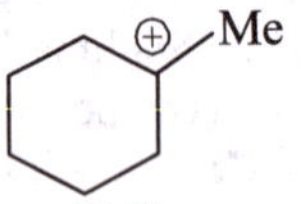

(B) 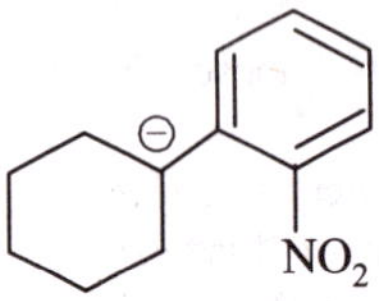 & 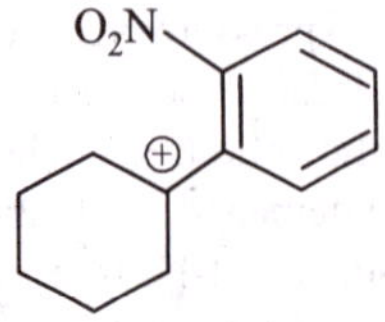

(C) 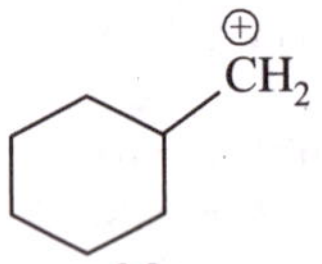 & 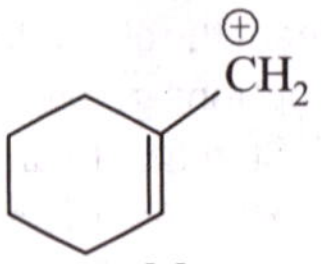

(D) 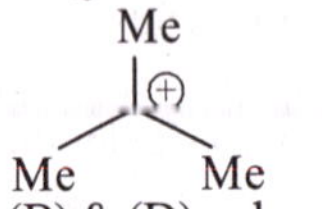 & 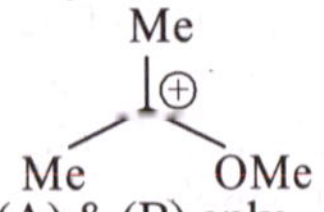

(1) (B) & (D) only (2) (A) & (B) only
(3) (B) & (C) only (4) (A) & (C) only

116. Given below are two statements :
Statement I : Hyperconjugation is not a permanent effect.
Statement II : In general, greater the number of alkyl groups attached to a positively charged C-atom, greater is the hyperconjugation interaction and stabilization of the cation.
In the light of the above statements, choose the correct answer from the options given below [April 3, 2025 (II)]
(1) Statement I is true but Statement II is false
(2) Both Statement I and Statement II are false
(3) Statement I is false but Statement II is true
(4) Both Statement I and Statement II are true

117. In Dumas' method for estimation of nitrogen 0.4 g of an organic compound gave 60 mL of nitrogen collected at 300 K temperature and 715 mm Hg pressure. The percentage composition of nitrogen in the compound is
(Given : Aqueous tension at 300 K = 15 mm Hg) [April 3, 2025 (II)]
(1) 15.71% (2) 20.95% (3) 17.46% (4) 7.85%

118. In Dumas' method for estimation of nitrogen, 0.5 gram of an organic compound gave 60 mL of nitrogen collected at 300 K temperature and 715 mm Hg pressure. The percentage composition of nitrogen in the compound (Aqueous tension at 300 K = 15 mm Hg) is [April 2, 2025 (II)]
(1) 1.257 (2) 20.87 (3) 18.67 (4) 12.57

119. Match List-I with List-II [April 2, 2025 (II)]

List-I (Purification technique)	List-II (Mixture of organic compounds)
(A) Distillation (simple)	(I) Diesel + Petrol
(B) Fractional distillation	(II) Aniline + Water
(C) Distillation under reduced pressure	(III) Chloroform + Aniline
(D) Steam distillation	(IV) Glycerol + Spent-lye

Choose the correct answer from the options given below:
(1) (A)-(II), (B)-(III), (C)-(IV), (D)-(I)
(2) (A)-(II), (B)-(IV), (C)-(I), (D)-(III)
(3) (A)-(III), (B)-(IV), (C)-(II), (D)-(I)
(4) (A)-(III), (B)-(I), (C)-(IV), (D)-(II)

120. Consider the following compound (X) [April 2, 2025 (I)]

$$\overset{\text{I}}{H-C}\equiv\overset{\text{II}}{C}-\overset{\text{III}}{CH_2}-\overset{\text{IV}}{\underset{\underset{\displaystyle CH_3}{|}}{CH}}-CH_3$$

(X)

The most stable and least stable carbon radicals, respectively, produced by homolytic cleavage of corresponding C – H bond are :
(1) II, IV (2) III, II (3) I, IV (4) II, I

121. Given below are two statements :
Statement (I) : In partition chromatography, stationary phase is thin film of liquid present in the inert support.
Statement (II) : In paper chromatography, the material of paper acts as a stationary phase.

In the light of the above statements, choose the correct answer from the options given below : [Jan. 29, 2025 (II)]
(1) Both Statement I and Statement II are false
(2) Statement I is true but Statement II is false
(3) Both Statement I and Statement II are true
(4) Statement I is false but Statement II is true

122. Total number of nucleophiles from the following is:

NH_3, PhSH, $(H_3C)_2S$, $H_2C{=}CH_2$, $\overset{\ominus}{O}H$, $H_3O^{\oplus}$, $(CH_3)_2CO$,

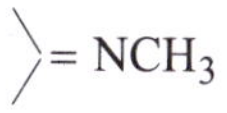 [Jan. 29, 2025 (I)]

(1) 5 (2) 4 (3) 7 (4) 6

123. The purification method based on the following physical transformation is : [Jan. 28, 2025 (II)]

$$\underset{(X)}{\text{Solid}} \xrightarrow{\text{Heat}} \underset{(X)}{\text{Vapour}} \xrightarrow{\text{Cool}} \underset{(X)}{\text{Solid}}$$

(1) Sublimation (2) Distillation
(3) Crystallization (4) Extraction

124. The correct order of stability of following carbocations is: [Jan. 28, 2025 (I)]

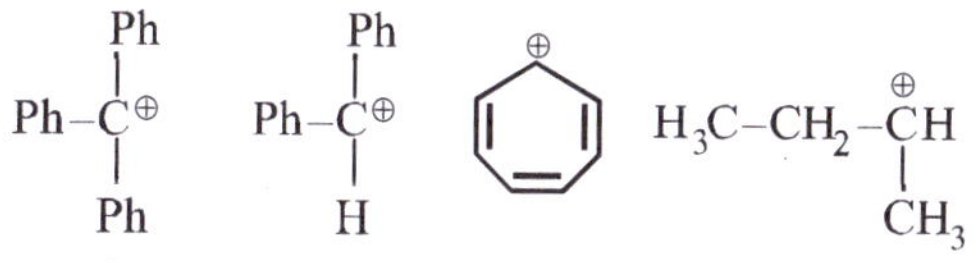

A B C D

(1) A > B > C > D (2) B > C > A > D
(3) C > B > A > D (4) C > A > B > D

125. Given below are two statements I and II. [Jan. 24, 2025 (I)]
Statement I : Dumas method is used for estimation of "Nitrogen" in an organic compound.
Statement II : Dumas method involves the formation of ammonium sulphate by heating the organic compound with conc H_2SO_4.
In the light of the above statements, choose the correct answer from the options given below.
(1) Both Statement I and Statement II are true
(2) Statement I is false but Statement II is true
(3) Both Statement I and Statement II are false
(4) Statement I is true but Statement II is false

126. Which one of the carbocations from the following is most stable? [Jan. 24, 2025 (I)]

(1) 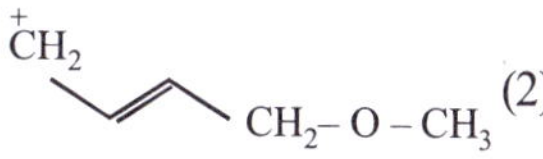 (2) 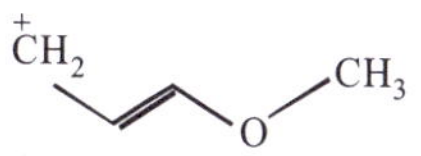

(3) 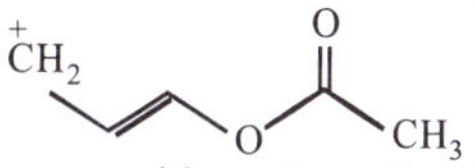 (4) 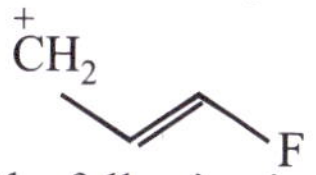

127. The most stable carbocation from the following is : [Jan. 22, 2025 (II)]

(1) 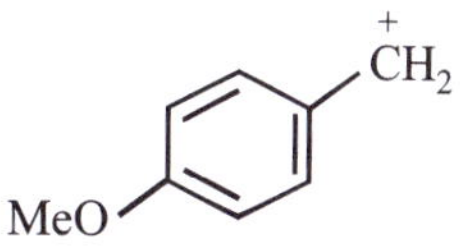 (2) 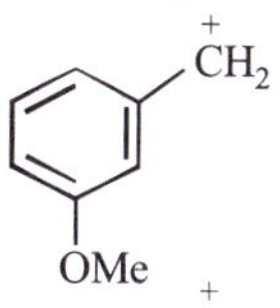

(3) 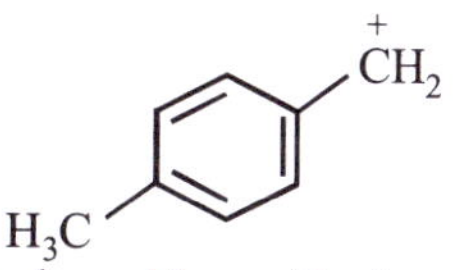 (4) 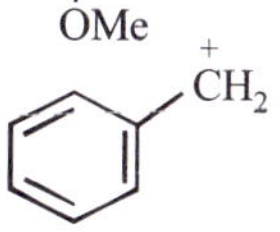

128. Methods used for purification of organic compounds are based on :
(1) neither on nature of compound nor on the impurity present.
(2) nature of compound only. [April 9, 2024 (I)]
(3) nature of compound and presence of impurity.
(4) presence of impurity only.

129. Relative stability of the contributing structures is :

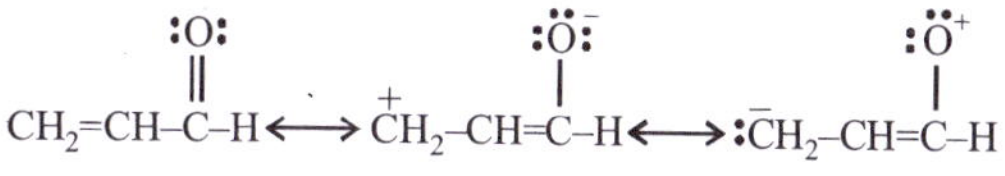

(I) (II) (III)

[April 9, 2024 (I)]

(1) (I) > (III) > (II) (2) (I) > (II) > (III)
(3) (II) > (I) > (III) (4) (III) > (II) > (I)

130. The shape of carbocation is : [April 8, 2024 (II)]
(1) trigonal planar (2) diagonal pyramidal
(3) tetrahedral (4) diagonal

131. Given below are two statements: [April 8, 2024 (II)]
Statement (I): Kjeldahl method is applicable to estimate nitrogen in pyridine.
Statement (II): The nitrogen present in pyridine can easily be converted into ammonium sulphate in Kjeldahl method.
In the light of the above statements, choose the **correct** answer from the options given below:
(1) Both **Statement I** and **Statement II** is false
(2) **Statement I** is false but **Statement II** is true
(3) Both **Statement I** and **Statement II** is true
(4) **Statement I** is true but **Statement II** is false

132. Which of the following statements are correct ? [April 6, 2024 (I)]
A. Glycerol is purified by vacuum distillation because it decomposes at its normal boiling point.
B. Aniline can be purified by steam distillation as aniline is miscible in water.
C. Ethanol can be separated from ethanol water mixture by azeotropic distillation because it forms azeotrope.
D. An organic compound is pure, if mixed M.P. is remained same.
Choose the **most appropriate** answer from the options given below:
(1) A, B, C only (2) A, C, D only
(3) B, C, D only (4) A, B, D only

133 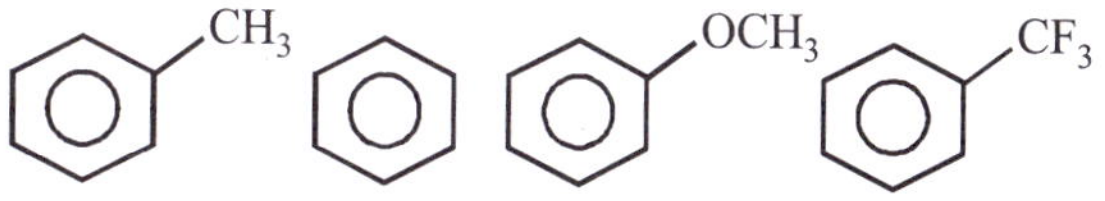

(I) (II) (III) (IV)

The **correct** arrangement for decreasing order of electrophilic substitution for above compounds [April 6, 2024 (II)]
(1) (IV) > (I) > (II) > (III) (2) (III) > (I) > (II) > (IV)
(3) (II) > (IV) > (III) > (I) (4) (III) > (IV) > (II) > (I)

134. The correct statements among the following, for a "chromatography" purification method is: [April 6, 2024 (II)]
(1) Organic compounds run faster than solvent in the thin layer chromatographic plate.
(2) Non-polar compounds are retained at top and polar compounds come down in column chromatography.
(3) R_f of a polar compound is smaller than that of a non-polar compound.
(4) R_f is an integral value.

135. Match List - I with List II : [April 4, 2024 (I)]

List-I (Mechanism steps)	List-II (Effect)
(A) 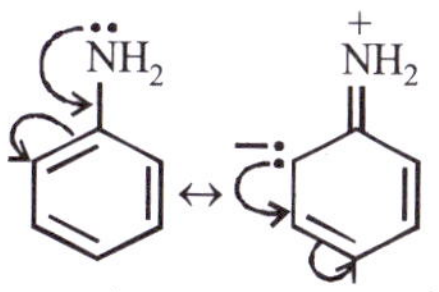	(I) – E effect
(B) 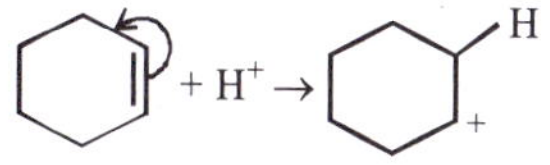	(II) – R effect

(C) 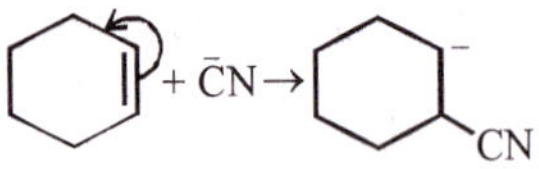 (III) + E effect

(D) 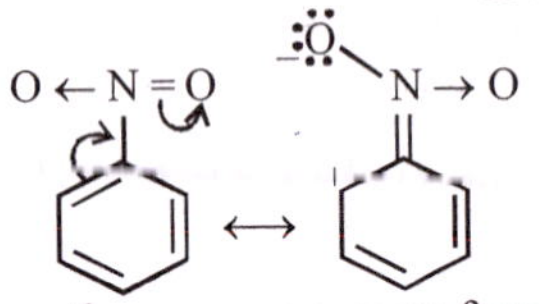 (IV) + R effect

Choose the **correct** answer from the options given below:
(1) A – (IV), B – (III), C – (I), D – (II)
(2) A – (III), B – (I), C – (II), D – (IV)
(3) A – (II), B – (IV), C) – (III), D – (I)
(4) A – (I), B – (II), C – (IV), D – (III)

136. Which among the following is **incorrect** statement? [April 4, 2024 (I)]
(1) Electromeric effect dominates over inductive effect
(2) The electromeric effect is a temporary effect
(3) The organic compound shows electromeric effect in the presence of the reagent only
(4) Hydrogen ion (H^+) shows negative electromeric effect

137. The adsorbent used in adsorption chromatography is/are
(A) silica gel (B) alumina
(C) quick lime (D) magnesia
Choose the **most appropriate** answer from the options given below:
(1) B only (2) C and D only [April 4, 2024 (II)]
(3) A and B only (4) A only

138. In Kjeldahl's method for estimation of nitrogen, $CuSO_4$ acts as: [Feb. 1, 2024 (I)]
(1) reducing agent (2) catalytic agent
(3) hydrolysis agent (4) oxidising agent

139. Ionic reactions with organic compounds proceed through: [Feb. 1, 2024 (I)]
(A) homolytic bond cleavage
(B) heterolytic bond cleavage
(C) free radical formation
(D) primary free radical
(E) secondary free radical
Choose the correct answer from the options given below:
(1) (A) only (2) (C) only
(3) (B) only (4) (D) and (E) only

140. The functional group that shows negative resonance effect is: [Feb. 1, 2024 (II)]
(1) –OH (2) –OR (3) –COOH (4) $-NH_2$

141. A species having carbon with sextet of electrons and can act as electrophile is called [Jan. 31, 2024 (I)]
(1) pentavalent carbon (2) carbon free radical
(3) carbocation (4) carbanion

142. 'Adsorption' principle is used for which of the following purification method? [Jan. 31, 2024 (I)]
(1) Chromatography (2) Sublimation
(3) Extraction (4) Distillation

143. Match List I with List II [Jan. 31, 2024 (I)]

LIST I (Technique)		LIST II (Application)	
A.	Distillation	I.	Separation of glycerol from spent-lye
B.	Fractional distillation	II.	Aniline- Water mixture
C.	Steam distillation	III.	Separation of crude oil fractions
D.	Distillation under reduced pressure	IV.	Chloroform - Aniline

Choose the correct answer from the options given below:
(1) A-IV, B-I, C-II, D-III (2) A-IV, B-III, C-II, D-I
(3) A-I, B-II, C-IV, D-III (4) A-II, B-III, C-I, D-IV

144. The fragrance of flowers is due to the presence of some steam volatile organic compounds called essential oils. These are generally insoluble in water at room temperature but are miscible with water vapour in vapour phase. A suitable method for the extraction of these oils from the flowers is - [Jan. 31, 2024 (II)]
(1) crystallisation
(2) distillation under reduced pressure
(3) distillation
(4) steam distillation

145. Which among the following purification methods is based on the principle of "Solubility" in two different solvents? [Jan. 30, 2024 (II)]
(1) Column Chromatography (2) Sublimation
(3) Distillation (4) Differential Extraction

146. The correct stability order of carbocations is [Jan. 30, 2024 (II)]
(1) $(CH_3)_3C^+ > CH_3-\overset{+}{C}H_2 > (CH_3)_2\overset{+}{C}H > \overset{+}{C}H_3$
(2) $\overset{+}{C}H_3 > (CH_3)_2\overset{+}{C}H > CH_3-\overset{+}{C}H_2 > (CH_3)_3\overset{+}{C}$
(3) $(CH_3)_3\overset{+}{C} > (CH_3)_2\overset{+}{C}H > CH_3-\overset{+}{C}H_2 > \overset{+}{C}H_3$
(4) $\overset{+}{C}H_3 > CH_3-\overset{+}{C}H_2 > CH_3-\underset{CH_3}{\overset{+}{C}H} > (CH_3)_3C^+$

147. The interaction between π bond and lone pair of electrons present on an adjacent atom is responsible for [Jan. 29, 2024 (I)]
(1) Resonance effect (2) Electrometric effect
(3) Inductive effect (4) Hyper conjugation

148. The ascending acidity order of the following H atoms is [Jan. 29, 2024 (II)]

HC ≡ C –(H)
A

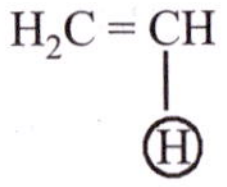
B

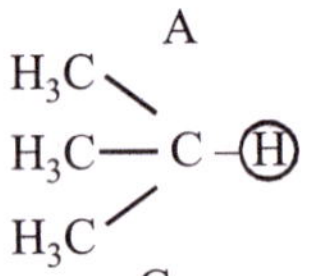
C

H_3C-CH_2 –(H)
D

(1) A < B < C < D (2) D < C < B < A
(3) A < B < D < C (4) C < D < B < A

149. Chromatographic technique/s based on the principle of differential adsorption is/are [Jan. 29, 2024 (II)]
A. Column chromatography
B. Thin layer chromatography
C. Paper chromatography
Choose the **most appropriate** answer from the options given below:
(1) B only (2) A only (3) A & B only (4) C only

150. Which of the following has highly acidic hydrogen? [Jan. 27, 2024 (I)]

(1) 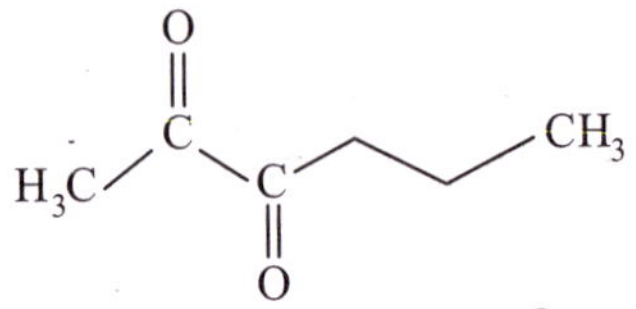

(2) 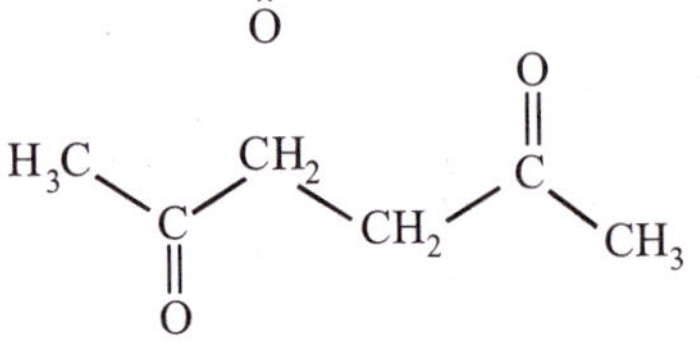

(3) 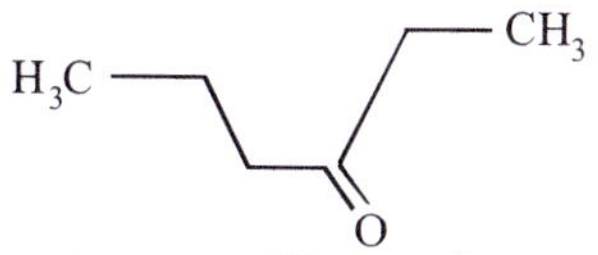

(4) 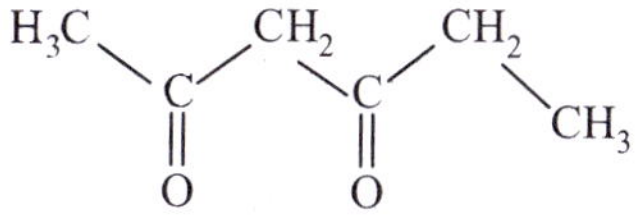

151. The order of relative stability of the contributing structure is: **[Jan. 27, 2024 (II)]**

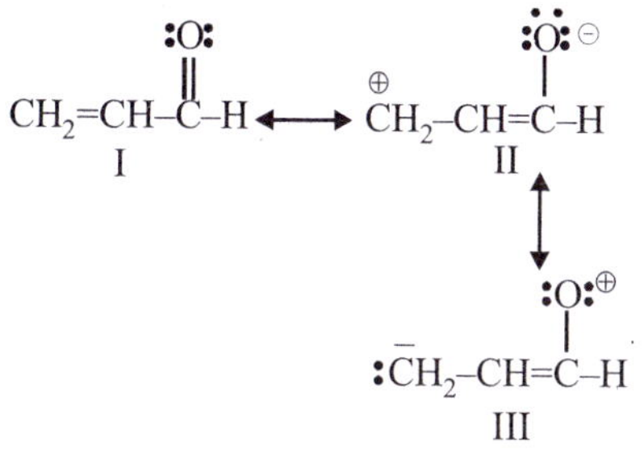

Choose the correct answer from the options given below:

(1) I > II > III (2) II > I > III
(3) I = II = III (4) III > II > I

152. The technique used for purification of steam volatile water immiscible substance is: **[Jan. 27, 2024 (II)]**

(1) Fractional distillation
(2) Fractional distillation under reduced pressure
(3) Distillation
(4) Steam distillation

153. Decreasing order of reactivity towards electrophilic substitution for the following compounds is : **[April 15, 2023 (I)]**

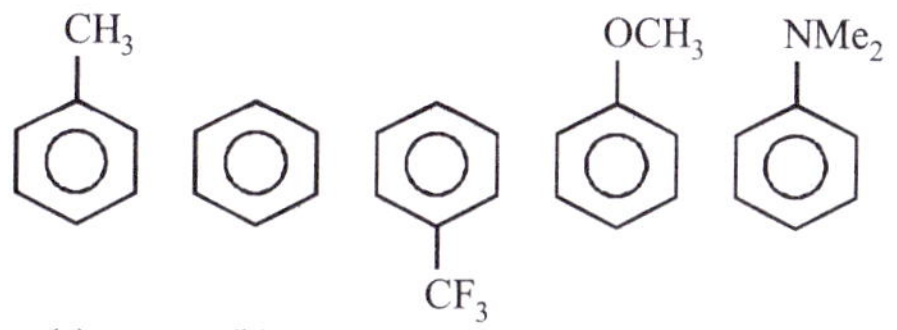

(a) (b) (c) (d) (e)

(1) c > b > a > d > e (2) e > d > a > b > c
(3) a > d > e > b > c (4) d > a > e > c > b

154. Which of the following statement is correct for paper chromatography ? **[April 15, 2023 (I)]**

(1) Water present in the mobile phase gets absorbed by the paper which then forms the stationary phase.
(2) Water present in the pores of the paper forms the stationary phase.
(3) Paper sheet forms the stationary phase.
(4) Paper and water present in its pores together form the stationary phase.

155. **[April 13, 2023 (I)]**

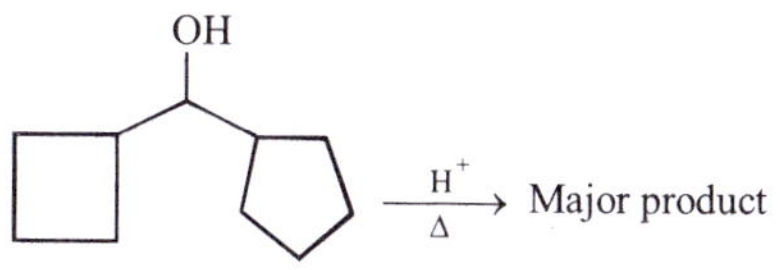

In the above reaction. Left hand side and right hand side rings are named as 'A' and 'B' respectively. The undergo ring expansion. The correct statement for this process is:

(1) Finally both rings will become six membered each.
(2) Finally both rings will become five membered each.
(3) Only 'A' will become 6 membered.
(4) Ring expansion can go upto seven membered rings.

156. Thin layer chromatography of a mixture shows the following observation : **[April 11, 2023 (I)]**

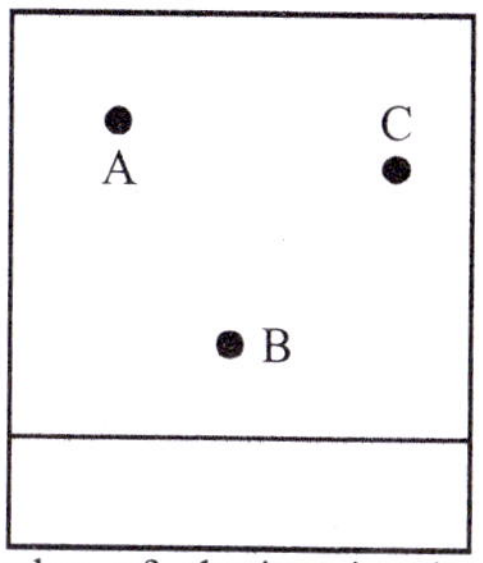

The correct order of elution in the silica gel column chromatography is

(1) A, C, B (2) B, C, A (3) C, A, B (4) B, A, C

157. Arrange the following compounds in increasing order of rate of aromatic electrophilic substitution reaction **[April 11, 2023 (I)]**

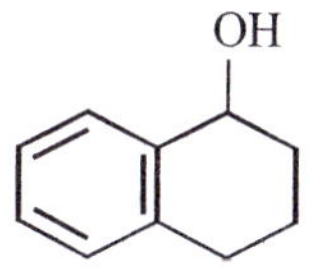
(a)

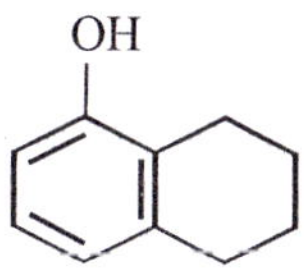
(b)

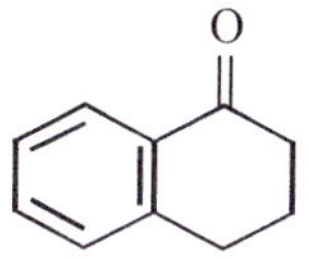
(c)

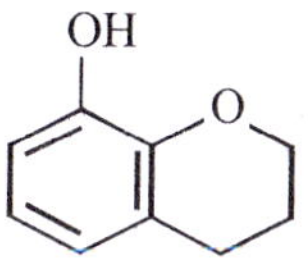
(d)

(1) d, b, c, a (2) b, c, a, d
(3) c, a, b, d (4) d, b, a, c

158. Compound from the following that will not produce precipitate on reaction with $AgNO_3$ is: **[April 11, 2023 (II)]**

(1) 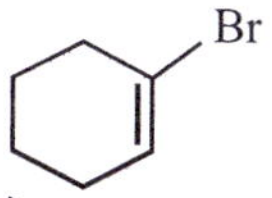

(2) 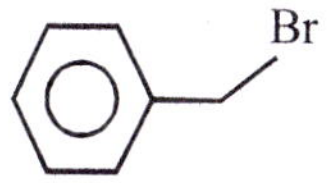

(3) 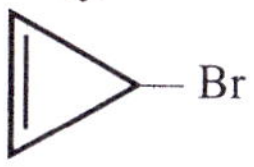

(4) 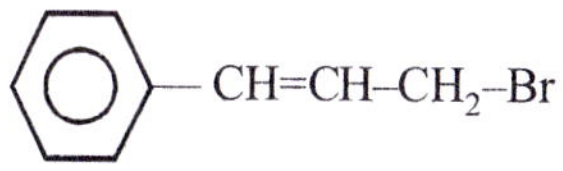

159. Using column chromatography, mixture of two compounds 'A' and 'B' was separated. 'A' eluted first, this indicates 'B' has **[April 10, 2023 (I)]**

(1) low R_f, weaker adsorption
(2) high R_f, stronger adsorption
(3) high R_f, weaker adsorption
(4) low R_f, stronger adsorption

160. In Carius tube, an organic compound 'X' is treated with sodium peroxide to form a mineral acid 'Y'. The solution of $BaCl_2$ is added to 'Y' to form a precipitate 'Z'. 'Z' is used for the quantitative estimation of an extra element. 'X' could be: **[April 10, 2023 (II)]**

(1) Cytosine (2) Chloroxylenol
(3) A nucleotide (4) Methionine

161. The decreasing order of hydride affinity for following carbocations is: **[April 10, 2023 (II)]**

A. $CH_2 = CH - \overset{+}{C}(CH_3) - CH_3$

B. $C_6H_5 - \overset{+}{C}(C_6H_5) - C_6H_5$

C. $H_3C - \overset{+}{C}(CH_3) - CH_3$

D. 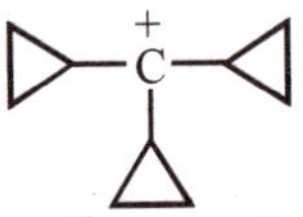

Choose the correct answer from the options given below:

(1) A, C, B, D (2) C, A, B, D

(3) C, A, D, B (4) A, C, D, B

162. Match List I with List II [April 06, 2023 (I)]

List I (Element detected)	List II (Reagent used/ Product formed)
A Nitrogen	I. $Na_2[Fe(CN)_5NO]$
B Sulphur	II. $AgNO_3$
C Phosphorus	III. $Fe_4[Fe(CN)_6]_3$
D Halogen	IV. $(NH_4)_2MoO_4$

Choose the correct answer from the options given below:

(1) A–II, B–IV, C–I, D–III (2) A–IV, B–II, C–I, D–III

(3) A–II, B–I, C–IV, D–III (4) A–III, B–I, C–IV, D-II

163. From the figure of column chromatography given below, identify incorrect statements. [April 06, 2023 (II)]

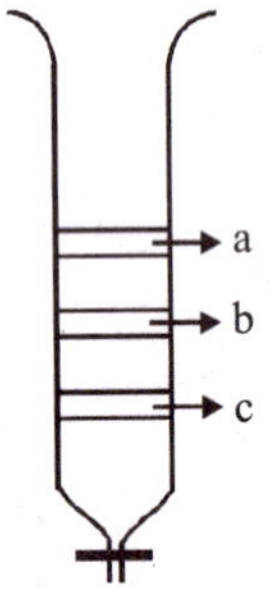

A. Compound 'c' is more polar than 'a' and 'b'

B. Compound 'a' is least polar

C. Compound 'b' comes out of the column before 'c' and after 'a'

D. Compound 'a' spends more time in the column

Choose the correct answer from the options given below:-

(1) A, B and C only (2) B, C and D only

(3) A, B and D only (4) B and D only

164. Match items of column I and II [Jan. 31, 2023 (I)]

Column I (Mixture of compounds)	Column II (Separation technique)
A. H_2O/CH_2Cl_2	(i) Crystallization
B. 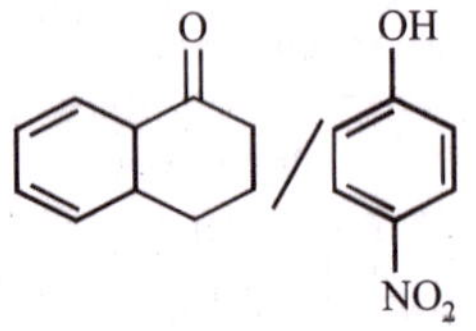	(ii) Differential solvent extraction
C. Kerosene/Naphthalene	(iii) Column chromatography
D. $C_6H_{12}O_6$/NaCl	(iv) Fractional distillation

Correct match is :

(1) A-(iii), B-(iv), C-(ii), D-(i)

(2) A-(i), B-(iii), C-(ii), D-(iv)

(3) A-(ii), B-(iii), C-(iv), D-(i)

(4) A-(ii), B-(iv), C-(i), D-(iii)

165. In Dumas method for the estimation of N_2, the sample is heated with copper oxide and the gas evolved is passed over: [Jan. 31, 2023 (II)]

(1) Ni (2) Copper gauze

(3) Pd (4) Copper oxide

166. What is the correct order of acidity of the protons marked A–D in the given compounds ? [Jan. 30, 2023 (I)]

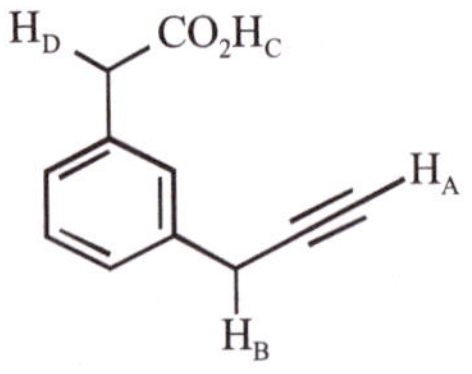

(1) $H_C > H_D > H_B > H_A$ (2) $H_C > H_D > H_A > H_B$

(3) $H_D > H_C > H_B > H_A$ (4) $H_C > H_A > H_D > H_B$

167. The most stable carbocation for the following is: [Jan. 30, 2023 (II)]

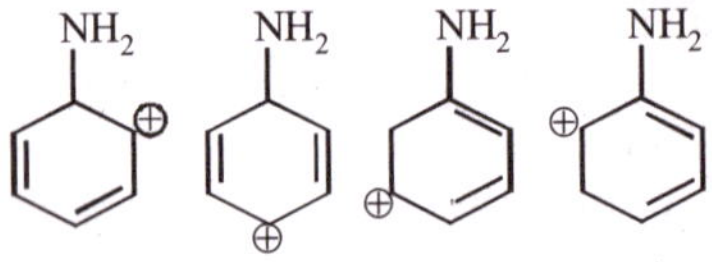

(a) (b) (c) (d)

(1) c (2) d (3) b (4) a

168. Match List I with List II: [Jan. 30, 2023 (II)]

List I (Mixture)	List II (Separation technique)
(A) $CHCl_3 + C_6H_5NH_2$	I Steam distillation
(B) $C_6H_{14} + C_5H_{12}$	II Differential extraction
(C) $C_6H_5NH_2 + H_2O$	III Distillation
(D) Organic compound in H_2O	IV Fractional distillation

(1) A-IV, B-I, C-III, D-II (2) A-III, B-IV, C-I, D-II

(3) A-II, B-I, C-III, D-IV (4) A-III, B-I, C-IV, D-II

169. Given below are two statements : One is labelled as Assertion A and the other is labelled as Reason R.

Assertion A : Thin layer chromatography is an adsorption chromatography.

Reason R : A thin layer of silica gel is spread over a glass plate of suitable size in thin layer chromatography which acts as an adsorbent.

In the light of the above statements, choose the correct answer from the options given below: [July 28, 2022 (II)]

(1) Both A and R are true and R is the correct explanation of A.

(2) Both A and R are true but R is NOT the correct explanation of A.

(3) A is true but R is false.

(4) A is false but R is true.

170. Match List-I with List-II [July 27, 2022 (II)]

List-I (Mixture)	List-II (Purification Process)
(A) Chloroform and Aniline	(I) Steam distillation
(B) Benzoic acid and Naphthalene	(II) Sublimation
(C) Water and Aniline	(III) Distillation
(D) Naphthalene and Sodium chloride	(IV) Crystallisation

(1) (A)-(IV), (B)-(III), (C)-(I), (D)-(II)
(2) (A)-(III), (B)-(I), (C)-(IV), (D)-(II)
(3) (A)-(III), (B)-(IV), (C)-(II), (D)-(I)
(4) (A)-(III), (B)-(IV), (C)-(I), (D)-(II)

171. Which technique, among the following, is most appropriate in separation of a mixture of 100 mg of *p*-nitrophenol and picric acid? [July 26, 2022 (I)]
(1) Steam distillation
(2) 2-5 ft long column of silica gel
(3) Sublimation
(4) Preparative TLC (Thin Layer Chromatography)

172. Which of the following carbocations is most stable: [June 29, 2022 (II)]

(1) 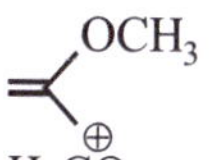
(2) 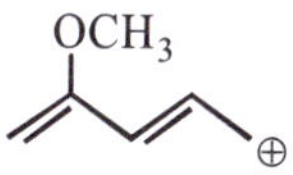
(3) 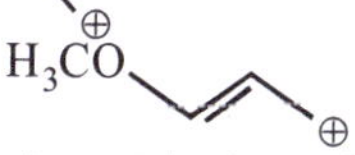
(4) 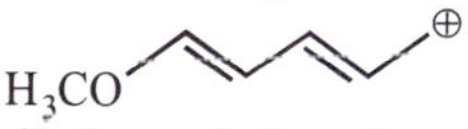

173. In Carius method of estimation of halogen, 0.45 g of an organic compound gave 0.36 g of AgBr. Find out the percentage of bromine in the compound.
(Molar masses : AgBr = 188 g mol^{-1}: Br = 80 g mol^{-1})
[July 27, 2022 (I); Same June 28, 2022 (I), Aug. 27, 2021 (I), July 20, 2021 (II)]
(1) 34.04% (2) 40.04% (3) 36.03% (4) 38.04%

174. Which one of the following techniques is not used to spot components of a mixture separated on thin layer chromatographic plate? [June 28, 2022 (I)]
(1) I_2 (Solid)
(2) U.V. Light
(3) Visualisation agent as a component of mobile phase
(4) Spraying of an appropriate reagent

175. Which of the following is structure of a separating funnel? [June 27, 2022 (I)]

(1) 
(2) 
(3) 
(4) 

176. In Carius method, halogen containing organic compound is heated with fuming nitric acid in the presence of: [July 20, 2021 (II)]
(1) HNO_3 (2) $AgNO_3$
(3) $CuSO_4$ (4) $BaSO_4$

177. 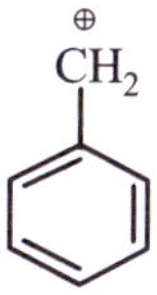 $CH_2=\overset{\oplus}{C}H$ $CH_3-\overset{\oplus}{C}H_2$ $HC\equiv\overset{\oplus}{C}$

A B C D

[July 27, 2021 (I)]
The correct order of stability of given carbocation is :
(1) A > C > B > D (2) D > B > C > A
(3) D > B > A > C (4) C > A > D > B

178. Which purification technique is used for high boiling liquid compound (decomposes near its boiling point) ? [July 22, 2021 (II)]
(1) Simple distillation
(2) Steam distillation
(3) Fractional distillation
(4) Reduced pressure distillation

179. Which of the following compounds does not exhibit resonance? [July 22, 2021 (II)]
(1) $CH_3CH_2OCH=CH_2$ (2) 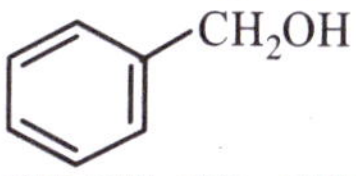
(3) $CH_3CH_2CH_2CONH_2$ (4) $CH_3CH_2CH=CHCH_2NH_2$

180. 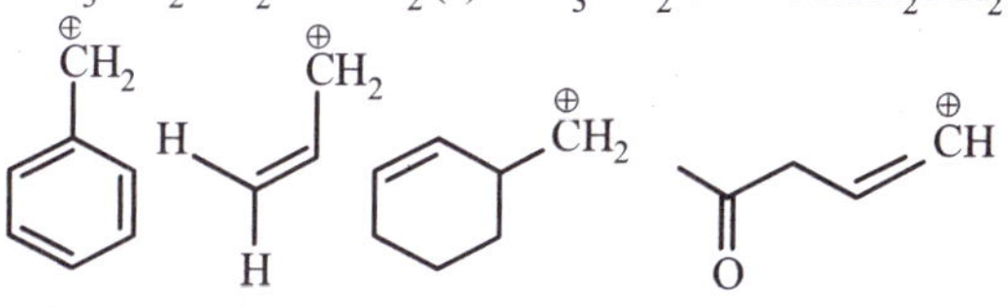

(A) (B) (C) (D)

Among the given species, the resonance stabilised carbocations are: [July 20, 2021 (I)]
(1) (C) and (D) only (2) (A), (B) and (D) only
(3) (A) and (B) only (4) (A), (B) and (C) only

181. In the following molecule,

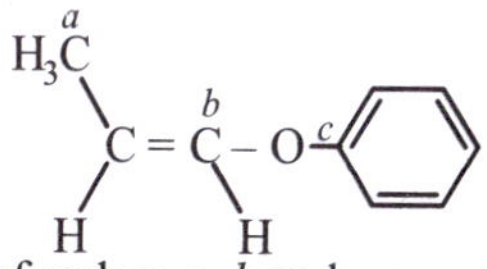

Hybridisation of carbon *a*, *b* and *c* respectively are : [March 18, 2021 (II)]
(1) sp^3, sp, sp (2) sp^3, sp^2, sp^2
(3) sp^3, sp, sp^2 (4) sp^3, sp^2, sp

182. In chromatography technique, the purification of compound is independent of [March 16, 2021 (I)]
(1) Solubility of the compound
(2) Mobility or flow of solvent system
(3) Length of the column or TLC plate
(4) Physical state of the pure compound

183. Which of the following is a 'FALSE' statement? [Feb. 26, 2021 (I)]
(1) Carius method is used for the estimation of nitrogen in an organic compound
(2) Phosphoric acid produced on oxidation of phosphorus present in an organic compound is precipitated as $Mg_2P_2O_7$ by adding magnesia mixture
(3) Kjeldahl's method is used for the estimation of nitrogen in an organic compund
(4) Carius tube is used for the estimation of sulphur in an organic compound

184. The Kjeldahl method of nitrogen estimation fails for which of the following reaction products? [Sep. 03, 2020 (I)]

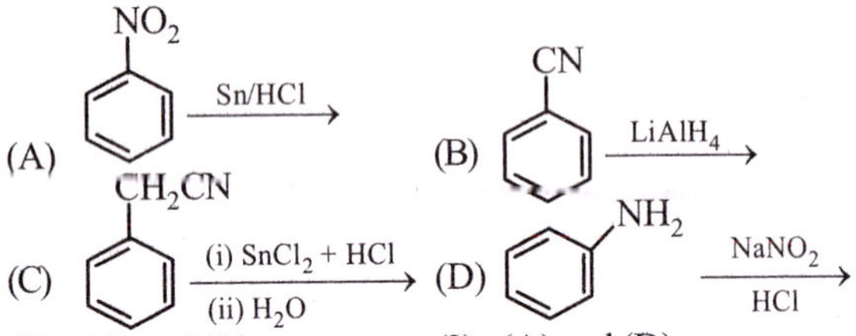

(1) (C) and (D) (2) (A) and (D)
(3) (A), (C) and (D) (4) (B) and (C)

185. Which one of the following compounds possesses the most acidic hydrogen? [Sep. 03, 2020 (I)]

(1)  (2) $H_3C-C\equiv C-H$

(3) 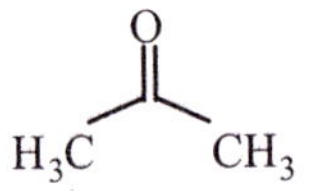 (4) 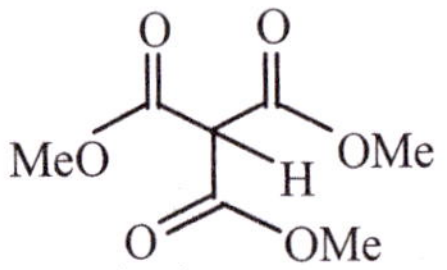

186. Glycerol is separated in soap industries by : [Sep. 03, 2020 (I); Same 2016]
(1) Fractional distillation
(2) Differential extraction
(3) Steam distillation
(4) Distillation under reduced pressure

187. The increasing order of basicity for the following intermediates is (from weak to strong) [Jan. 09, 2020 (I)]

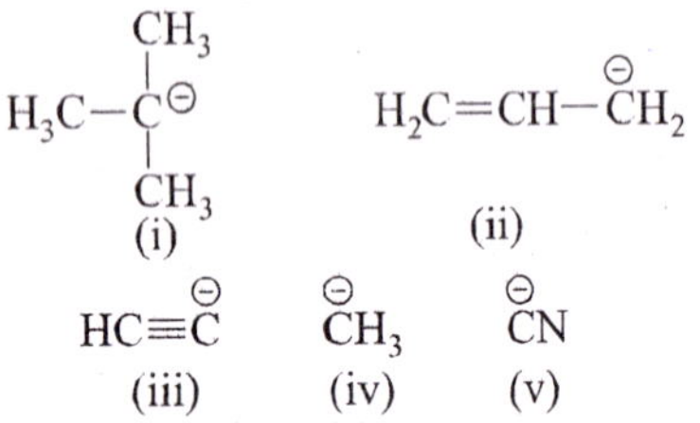

(1) (iii) < (i) < (ii) < (iv) < (v)
(2) (v) < (i) < (iv) < (ii) < (iii)
(3) (v) < (iii) < (ii) < (iv) < (i)
(4) (iii) < (iv) < (ii) < (i) < (v)

188. A flask contains a mixture of isohexane and 3-methylpentane. One of the liquids boils at 63 °C while the other boils at 60 °C. What is the best way to separate the two liquids and which one will be distilled out first? [Jan. 08, 2020 (I)]
(1) fractional distillation, isohexane
(2) simple distillation, 3-methylpentane
(3) simple distillation, isohexane
(4) fractional distillation, 3-methylpentane

189. The correct order of stability for the following alkoxides is: [Jan. 07, 2020 (II)]

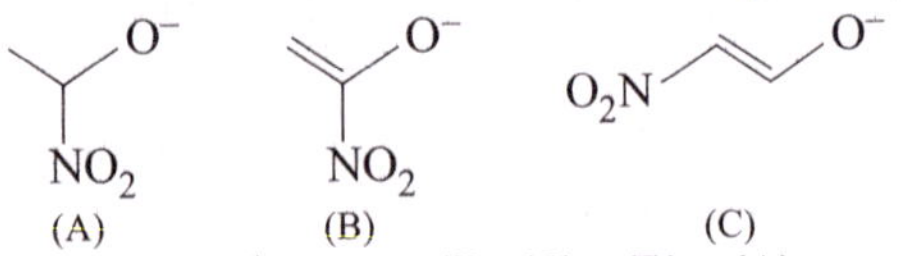

(1) (B) > (A) > (C) (2) (C) > (B) > (A)
(3) (C) > (A) > (B) (4) (B) > (C) > (A)

190. The increasing order of nucleophilicity of the following nucleophiles is : [April 10, 2019 (II)]

(i) $CH_3CO_2^-$ (ii) H_2O

(iii) $CH_3SO_3^-$ (iv) OH^-

(1) (i) < (iv) < (iii) < (ii) (2) (ii) < (iii) < (iv) < (i)
(3) (iv) < (i) < (iii) < (ii) (4) (ii) < (iii) < (i) < (iv)

191. In Carius method of estimation of halogen, 0.172 g of an organic compound showed presence of 0.08 g of bromine. Which of these is the **correct** structure of the compound? [Sep. 02, 2020 (I)]

(1) H_3C-CH_2-Br (2) 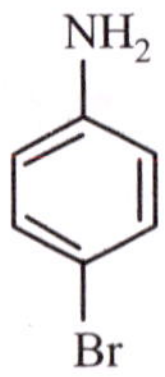

(3) 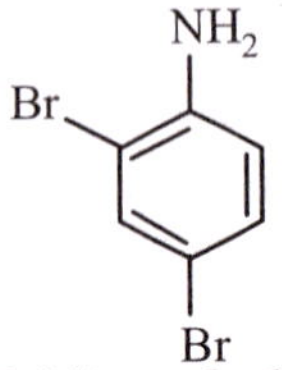 (4) H_3C-Br

192. Kjeldahl's method cannot be used to estimate nitrogen for which of the following compounds? [Jan. 08, 2020 (II)]
(1) $C_6H_5NH_2$ (2) $CH_3CH_2-C\equiv N$
(3) $C_6H_5NO_2$ (4) $NH_2-\overset{O}{\overset{\|}{C}}-NH_2$

193. A chromatography column, packed with silica gel as stationary phase, was used to separate a mixture of compounds consisting of (A) benzanilide (B) aniline and (C) acetophenone. When the column is eluted with a mixture of solvents, hexane: ethyl acetate (20:80), the sequence of obtained compounds is: [Jan. 07, 2020 (II)]
(1) (B), (C) and (A) (2) (B), (A) and (C)
(3) (C), (A) and (B) (4) (A), (B) and (C)

194. The principle of column chromatography is : [April 10, 2019 (I)]
(1) Gravitational force.
(2) Capillary action.
(3) Differential absorption of the substances on the solid phase.
(4) Differential adsorption of the substances on the solid phase.

195. In chromatography, which of the following statement is INCORRECT for R_f? [April 10, 2019 (II)]
(1) R_f value depends on the type of chromatography.
(2) The value of R_f can not be more than one.
(3) Higher R_f value means higher adsorption.
(4) R_f value is dependent on the mobile phase.

196. Which of the following compounds will show the maximum 'enol' content? [April 8, 2019 (II)]
(1) $CH_3COCH_2COOC_2H_5$ (2) $CH_3COCH_2COCH_3$
(3) CH_3COCH_3 (4) $CH_3COCH_2CONH_2$

197. The correct match between items I and II is : [Jan. 11, 2019 (I)]

Item - I (Mixture)	Item - II (Separation method)
(A) H_2O : Sugar	(P) Sublimation
(B) H_2O : Aniline	(Q) Recrystallization
(C) H_2O : Toluene	(R) Steam distillation
	(S) Differential extraction

(1) (A) → (S); (B) → (R); (C) → (P)
(2) (A) → (Q); (B) → (R); (C) → (S)
(3) (A) → (R); (B) → (P); (C) → (S)
(4) (A) → (Q); (B) → (R); (C) → (P)

198. An organic compound is estimated through Dumas method and was found to evolve 6 moles of CO_2, 4 moles of H_2O and 1 mole of nitrogen gas. The formula of the compound is: [Jan. 11, 2019 (I)]
(1) $C_{12}H_8N$ (2) $C_{12}H_8N_2$
(3) $C_6H_8N_2$ (4) C_6H_8N

199. If dichloromethane (DCM) and water (H_2O) are used for differential extraction, which one of the following statements is correct? **[Jan. 10, 2019 (I)]**

(1) DCM and H_2O would stay as lower and upper layer respectively in the separating funnel (S.F.)
(2) DCM and H_2O will make turbid/colloidal mixture
(3) DCM and H_2O would stay as upper and lower layer respectively in the separating funnel (S.F.)
(4) DCM and H_2O will be miscible clearly

200. The correct match between **Item- I** and **Item- II** is: **[Jan. 9, 2019 (II)]**

Item-I	Item-II
(A) Benzaldehyde	(P) Mobile phase
(B) Alumina	(Q) Adsorbent
(C) Acetonitrile	(R) Adsorbate

(1) (A) → (Q) ; (B) → (P) ; (C) → (R)
(2) (A) → (R) ; (B) → (Q) ; (C) → (P)
(3) (A) → (Q) ; (B) → (R) ; (C) → (P)
(4) (A) → (P) ; (B) → (R) ; (C) → (Q)

201. The correct match between items of **List-I** and **List-II** is : **[Online April 15, 2018 (I)]**

List-I	List-II
(A) Coloured impurity	(p) Steam distillation
(B) Mixture of *o*-nitrophenol and *p*-nitrophenol	(q) Fractional distillation
(C) Crude naphtha	(r) Charcoal treatment
(D) Mixture of glycerol and sugar	(s) Distillation under reduced pressure

(1) (A)-(r), (B)-(s), (C)-(p), (D)-(q)
(2) (A)-(p), (B)-(s), (C)-(r), (D)-(q)
(3) (A)-(r), (B)-(p), (C)-(q), (D)-(s)
(4) (A)-(r), (B)-(p), (C)-(s), (D)-(q)

202. TLC analysis of a mixture having 3 components (A, B, C) using silica gel as the stationary phase gave following R_f values; for A = 0.72, B = 0.48, C = 0.20.
Regarding the above observations, which one of the following statements is **not** correct for column chromatography of the mixture? **[June 30, 2022 (I)]**

(1) A is the most non-polar component
(2) C is the most soluble component in mobile phase
(3) A will be eluted first
(4) C is the most strongly absorbed component

203. The correct order of nucleophilicity is **[June 26, 2022 (II)]**

(1) $F^- > OH^-$ (2) $H_2\ddot{O} > OH^-$
(3) $R\ddot{O}H > RO^-$ (4) $NH_2^- > NH_3$

204. Arrange the following carbocations in decreasing order of stability. **[June 24, 2022 (II)]**

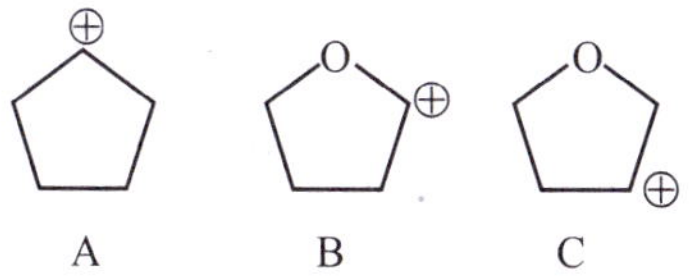

A B C

(1) A > C > B (2) A > B > C
(3) C > B > A (4) C > A > B

205. Which of the following compounds will be suitable for Kjeldahl's method for nitrogen estimation? **[2018]**

(1) 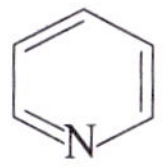 (2) 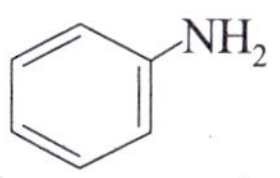
(3) 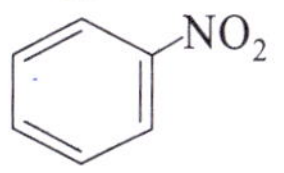 (4) 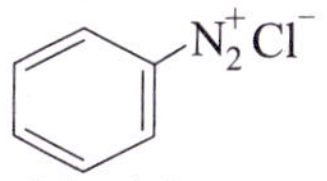

206. The increasing order of basicity of the following compounds is **[2018]**

(i)  (ii) 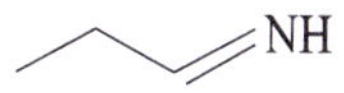
(iii) 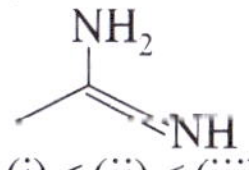 (iv) 

(1) (i) < (ii) < (iii) < (iv) (2) (ii) < (i) < (iii) < (iv)
(3) (ii) < (i) < (iv) < (iii) (4) (iv) < (ii) < (i) < (iii)

207. The most polar compound among the following is : **[Online April 16, 2018]**

(1) 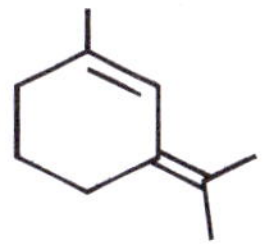 (2) 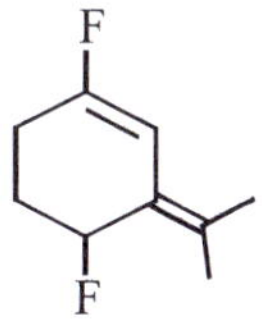
(3)  (4) 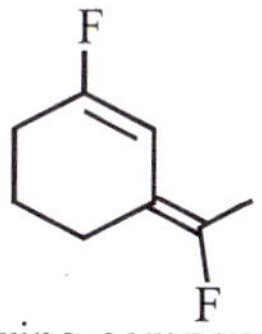

208. On the treatment of the following compound with a strong acid, the most susceptible site for bond cleavage is:

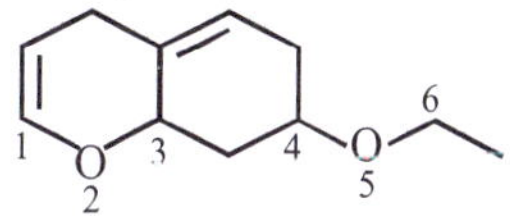

[Online April 15, 2018(II)]

(1) O2 – C3 (2) O5 – C6
(3) C4 – O5 (4) C1 – O2

209. Two compounds I and II are eluted by column chromatography (adsorption of I > II). Which one of the following is a correct statement? **[Online April 15, 2018 (II)]**

(1) II moves slower and has higher R_f value than I
(2) II moves faster and has higher R_f value than I
(3) I moves faster and has higher R_f value than II
(4) I moves slower and has higher R_f value than II

210. Which of the following statements is not true about partition chromatography? **[Online April 8, 2017]**

(1) Mobile phase can be a gas
(2) Stationary phase is a finely divided solid adsorbent
(3) Separation depends upon equilibration of solute between a mobile and a stationary phase
(4) Paper chromatography is an example of partition chromatography

211. The increasing order of the boiling points for the following compounds is : **[Online April 9, 2017]**

C_2H_5OH	C_2H_5Cl	$C_2H_5CH_3$	$C_2H_5OCH_3$
(I)	(II)	(III)	(IV)

(1) (III) < (IV) < (II) < (I) (2) (IV) < (III) < (I) < (II)
(3) (II) < (III) < (IV) < (I) (4) (III) < (II) < (I) < (IV)

212. A mixture containing the following four compounds is extracted with 1M HCl. The compound that goes to aqueous layer is : [Online April 8, 2017]

(I) 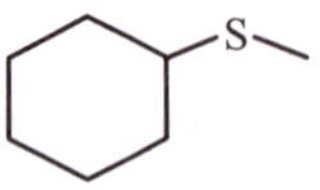 (II) 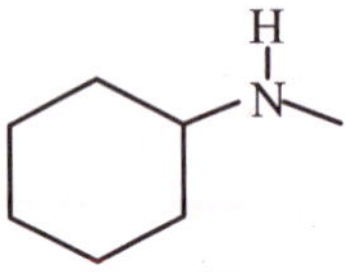

(III) 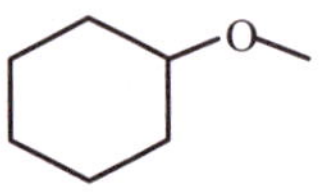 (IV) 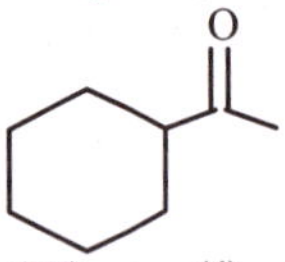

(1) (I) (2) (II) (3) (III) (4) (IV)

213. Which of the following molecules is least resonance stabilized? [2017]

(1) 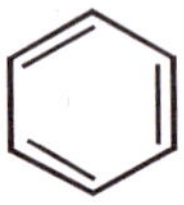 (2) 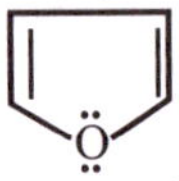

(3) 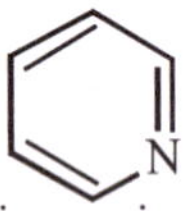 (4) 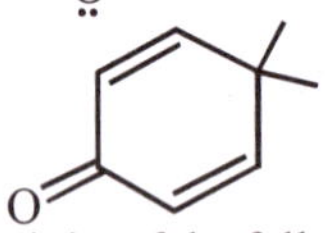

214. The increasing order of the reactivity of the following halides for the S_N1 reaction is [2017]

$CH_3CH(Cl)CH_2CH_3$ (I) $CH_3CH_2CH_2Cl$ (II)

p-H_3CO–C_6H_4–CH_2Cl (III)

(1) (III) < (II) < (I) (2) (II) < (I) < (III)
(3) (I) < (III) < (II) (4) (II) < (III) < (I)

215. For the estimation of nitrogen, 1.4 g of an organic compound was digested by Kjeldahl method and the evolved ammonia was absorbed in 60 mL of $\frac{M}{10}$ sulphuric acid. The unreacted acid required 20 mL of $\frac{M}{10}$ sodium hydroxide for complete neutralization. The percentage of nitrogen in the compound is: [Online April 10, 2014]

(1) 6% (2) 10% (3) 3% (4) 5%

216. For which of the following molecule significant $\mu \neq 0$? [2014]

(i) 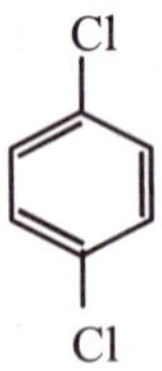 (ii) 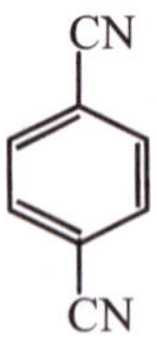

(iii) 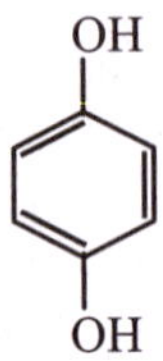 (iv) 

(1) Only (i) (2) (i) and (ii)
(3) Only (iii) (4) (iii) and (iv)

217. In which of the following pairs, A is more stable than B? [Online April 9, 2014]

	A	B
(1)	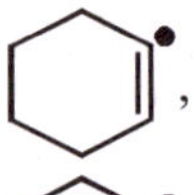,	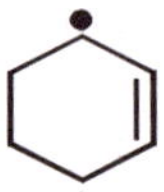
(2)	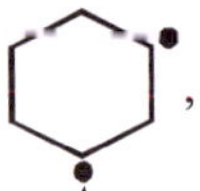,	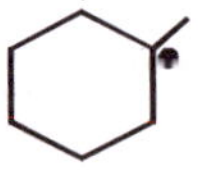
(3)	,	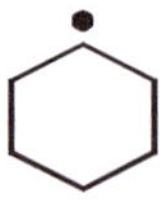
(4)	$Ph_3C^{\bullet}$,	$(CH_3)_3C^{\bullet}$

218. A solution of (–) – 1 – chloro –1– phenylethane in toluene racemises slowly in the presence of a small amount of $SbCl_5$, due to the formation of: [2013]

(1) carbanion (2) carbene
(3) carbocation (4) free radical

219. The order of stability of the following carbocations : [2013]

$CH_2=CH-\overset{\oplus}{C}H_2$ (I) ; $CH_3-CH_2-\overset{\oplus}{C}H_2$ (II) ; 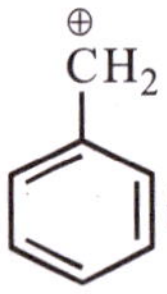 (III) is :

(1) III > II > I (2) II > III > I
(3) I > II > III (4) III > I > II

220. Which one of the following is most stable ? [Online April 9, 2013]

(1) 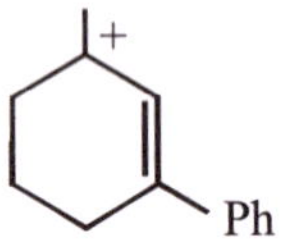 (2) 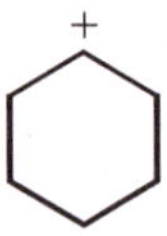

(3) 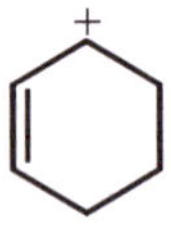 (4) 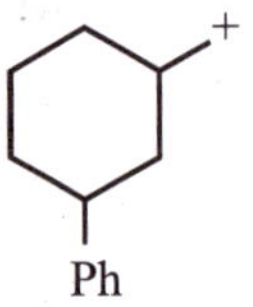

221. Given [Online April 25, 2013]

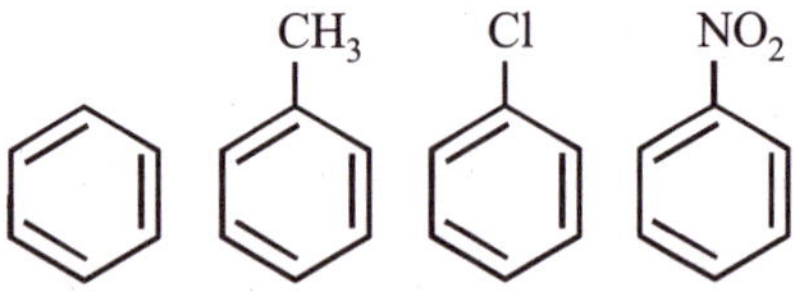

(A) (B) (C) (D)

In the above compounds, correct order of reactivity in electrophilic substitution reactions will be :

(1) B > A > C > D (2) D > C > B > A
(3) A > B > C > D (4) B > C > A > D

222. In nucleophilic substitution reaction, order of halogens as incoming (attacking) nucleophile is:

$I^- > Br^- > Cl^-$

The order of halogens as departing nucleophile should be : [Online April 25, 2013]

(1) $Br^- > I^- > Cl^-$ (2) $I^- > Br^- > Cl^-$
(3) $Cl^- > Br^- > I^-$ (4) $Cl^- > I^- > Br^-$

223. In the below mentioned compounds the decreasing order of reactivity towards electrophilic substitution is [Online May 12, 2012]

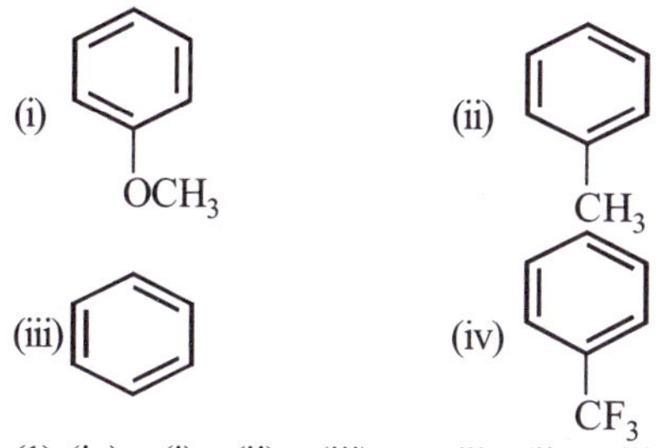

(1) (iv) > (i) > (ii) > (iii) (2) (ii) > (iii) > (i) > (iv)
(3) (iii) > (i) > (iv) > (ii) (4) (i) > (ii) > (iii) > (iv)

224. Which of the following cannot be represented by resonance structures? [Online May 7, 2012]
(1) Dimethyl ether (2) Nitrate anion
(3) Carboxylate anion (4) Toluene

225. The correct order of increasing basicity of the given conjugate bases (R = CH_3) is [2010]
(1) $RCO\bar{O} < HC \equiv \bar{C} < \bar{R} < \bar{N}H_2$
(2) $\bar{R} < HC \equiv \bar{C} < RCO\bar{O} < \bar{N}H_2$
(3) $RCO\bar{O} < \bar{N}H_2 < HC \equiv \bar{C} < \bar{R}$
(4) $RCO\bar{O} < HC \equiv \bar{C} < \bar{N}H_2 < \bar{R}$

226. Arrange the carbanions, [2009]
$(CH_3)_3\bar{C}, \bar{C}Cl_3, (CH_3)_2\bar{C}H, C_6H_5\bar{C}H_2$
in order of their decreasing stability.
(1) $(CH_3)_2\bar{C}H > \bar{C}Cl_3 > C_6H_5\bar{C}H_2 > (CH_3)_3\bar{C}$
(2) $\bar{C}Cl_3 > C_6H_5\bar{C}H_2 > (CH_3)_2\bar{C}H > (CH_3)_3\bar{C}$
(3) $(CH_3)_3\bar{C} > (CH_3)_2\bar{C}H > C_6H_5\bar{C}H_2 > \bar{C}Cl_3$
(4) $C_6H_5\bar{C}H_2 > \bar{C}Cl_3 > (CH_3)_3\bar{C} > (CH_3)_2\bar{C}H$

227. The electrophile, $E^{\oplus}$ attacks the benzene ring to generate the intermediate σ–complex. Of the following, which σ–complex has lowest energy? [2008]

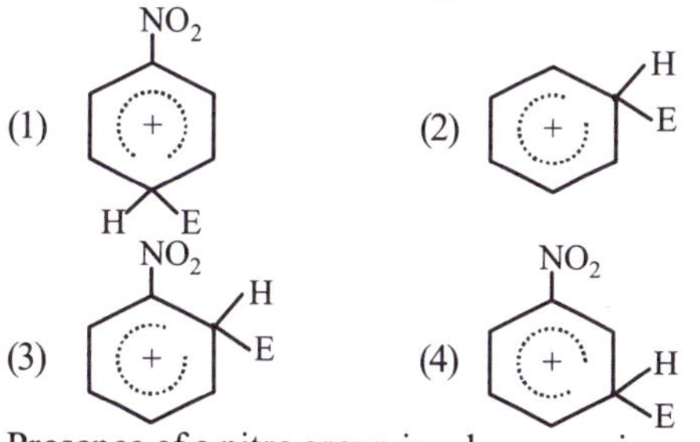

228. Presence of a nitro group in a benzene ring [2007]
(1) deactivates the ring towards electrophilic substitution
(2) activates the ring towards electrophilic substitution
(3) renders the ring basic
(4) deactivates the ring towards nucleophilic substitution.

229. The increasing order of stability of the following free radicals is [2006]
(1) $(C_6H_5)_2\dot{C}H < (C_6H_5)_3\dot{C} < (CH_3)_3\dot{C} < (CH_3)_2\dot{C}H$
(2) $(CH_3)_2\dot{C}H < (CH_3)_3\dot{C} < (C_6H_5)_2\dot{C}H < (C_6H_5)_3\dot{C}$
(3) $(CH_3)_3\dot{C} < (CH_3)_2\dot{C}H < (C_6H_5)_2\dot{C}H < (C_6H_5)_3\dot{C}$
(4) $(C_6H_5)_3\dot{C} < (C_6H_5)_2\dot{C}H < (CH_3)_3\dot{C} < (CH_3)_2\dot{C}H$

230. $CH_3Br + Nu^- \longrightarrow CH_3 - Nu + Br^-$
The decreasing order of the rate of the above reaction with nucleophiles (Nu^-) A to D is [2006]
[Nu^- = (A) PhO^-, (B) AcO^-, (C) HO^-, (D) CH_3O^-]
(1) A > B > C > D (2) B > D > C > A
(3) D > C > A > B (4) D > C > B > A

231. 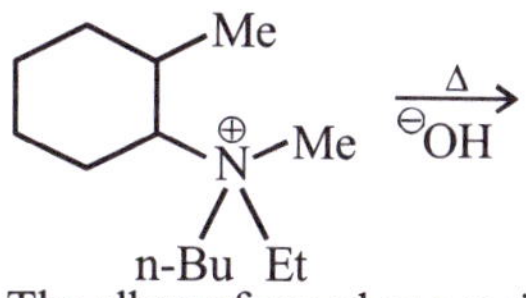

The alkene formed as a major product in the above elimination reaction is [2006]

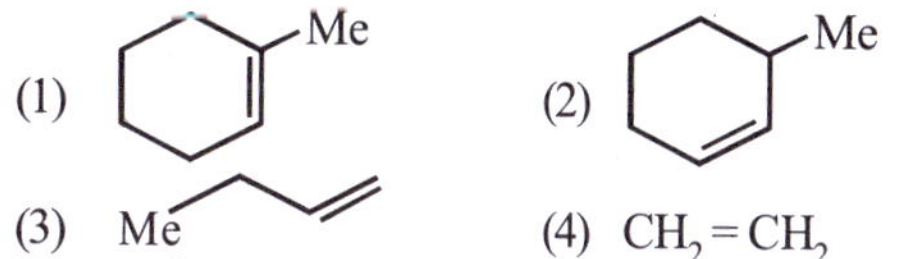

(4) $CH_2 = CH_2$

232. Due to the presence of an unpaired electron, free radicals are: [2005]
(1) cations (2) anions
(3) chemically inactive (4) chemically reactive

233. The decreasing order of nucleophilicity among the nucleophiles is [2005]
(A) $CH_3C(=O)-O^-$
(B) CH_3O^-
(C) CN^-
(D) 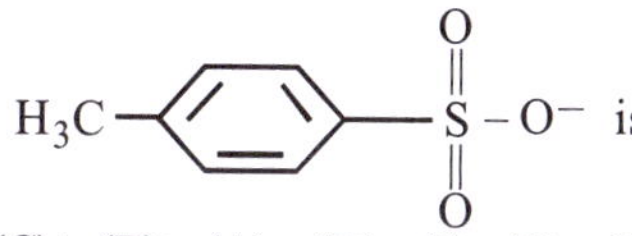

is

(1) (C) > (B) > (A) > (D) (2) (B) > (C) > (A) > (D)
(3) (D) > (C) > (B) > (A) (4) (A) > (B) > (C) > (D)

234. The reaction [2005, 2004]
$$R-C(=O)-X + \overset{\ominus}{N}u \longrightarrow R-C(=O)-Nu + X^{\ominus}$$
is fastest when X is
(1) OCOR (2) OC_2H_5 (3) NH_2 (4) Cl

235. Consider the acidity of the carboxylic acids :
(A) PhCOOH (B) *o*-$NO_2C_6H_4COOH$
(C) *p*-$NO_2C_6H_4COOH$ (D) *m*-$NO_2C_6H_4COOH$
Which of the following order is correct ? [2004]
(1) B > D > A > C (2) B > D > C > A
(3) A > B > C > D (4) B > C > D > A

236. Which of the following is the strongest base ? [2004]

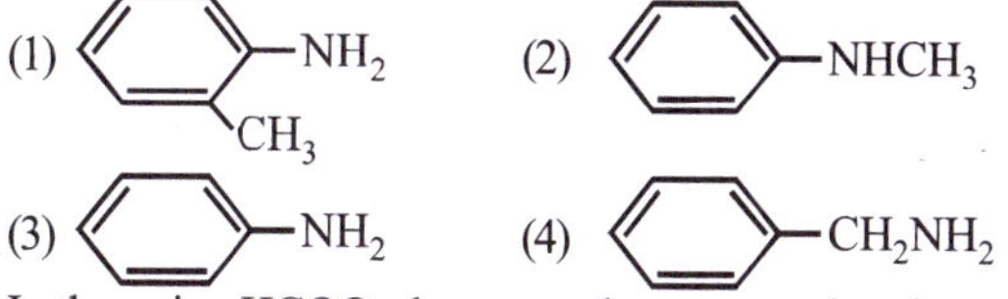

237. In the anion $HCOO^-$ the two carbon-oxygen bonds are found to be of equal length. What is the reason for it ? [2003]
(1) The C = O bond is weaker than the C — O bond
(2) The anion $HCOO^-$ has two equivalent resonating structures
(3) The anion is obtained by removal of a proton from the acid molecule
(4) Electronic orbitals of carbon atom are hybridised

238. In which of the following species is the underlined carbon having sp^3 hybridisation? **[2002]**

(1) $CH_3\underline{C}OOH$ (2) $CH_3\underline{C}H_2OH$
(3) $CH_3\underline{C}OCH_3$ (4) $CH_2=\underline{C}H-CH_3$

239. Arrangement of $(CH_3)_3C-$, $(CH_3)_2CH-$, CH_3-CH_2- when attached to benzyl or an unsaturated group in increasing order of inductive effect is **[2002]**

(1) $(CH_3)_3C- < (CH_3)_2CH- < CH_3-CH_2-$
(2) $CH_3-CH_2- < (CH_3)_2CH- < (CH_3)_3C-$
(3) $(CH_3)_2CH- < (CH_3)_3C- < CH_3-CH_2-$
(4) $(CH_3)_3C- < CH_3CH_2- < (CH_3)_2CH-$

240. The correct stability order of the following resonance structures of $CH_3-CH=CH-CHO$ is **[April 9, 2024 (II)]**

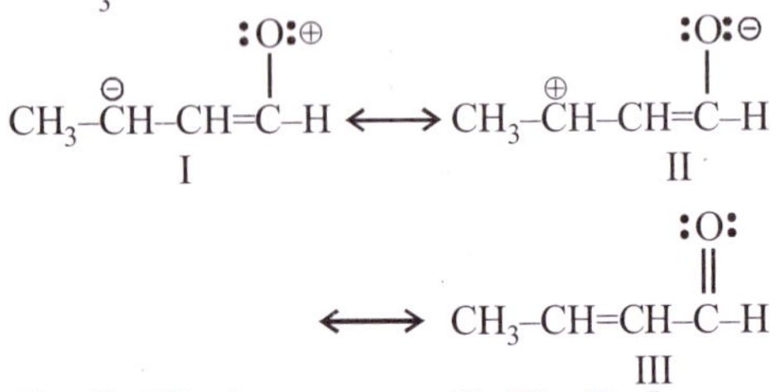

(1) II > III > I (2) III > II > I
(3) I > II > III (4) II > I > III

NUMERICAL VALUE QUESTIONS (NVQs)

241. In an estimation of sulphur by Carius method 0.2g of the substance gave 0.6 g of $BaSO_4$. The percentage of sulphur in the substance is ______%.
(Given molar mass in g mol^{-1} S : 32, $BaSO_4$: 231) **[April 05, 2026 (I)]**

242. In sulphur estimation, 2.0×10^{-3} mol of an organic compound (X) (molar mass 76 g mol^{-1}) gave 0.4813 g of barium sulphate (molar mass 233 g mol^{-1}). The percentage of sulphur in the compound (X) is ________ $\times 10^{-1}$ %
(Nearest integer) **[April 04, 2026 (II)]**

243. 2.0 g of a bromo hydrocarbon (X) was subjected to Carius analysis, gave 3.36 g of AgBr. The percentage of carbon in the compound (X) is 26.7%. Total number of carbon atoms in the empirical formula for compound (X) is ____.
(Given molar mass in g mol^{-1} H : 1, C : 12, Br : 80, Ag : 108) **[April 04, 2026 (I)]**

244. 0.53 g of an organic compound (x) when heated with excess of nitric acid (concentrated) and then with silver nitrate gave 0.75 g of silver bromide precipitate. 1.0 g of (x) gave 1.32 g of CO_2 gas on combustion. The percentage of hydrogen in the compound (x) is _______. [Nearest In teger] **[Jan. 28, 2026 (I)]**
[Given : Molar mass in g mol^{-1} H : 1, C : 12, Br : 80, Ag : 108, O : 16; Compound (x) : $C_xH_yBr_z$]

245. In Dumas method for estimation of nitrogen, 0.50 g of an organic compound gave 70 mL of nitrogen collected at 300 K and 715 mm pressure. The percentage of nitrogen in the organic compound is ______%
(Aqueous tension at 300 K is 15 mm). **[Jan. 24, 2026 (I)]**

246. Sodium fusion extract of an organic compound (Y) with $CHCl_3$ and chlorine water gives violet color to the $CHCl_3$ layer. 0.15g of (Y) gave 0.12 g of the silver halide precipitate in Carius method. Percentage of halogen in the compound (Y) is __________. (Nearest integer).
(**Given :** Molar mass g mol^{-1} C : 12, H : 1, Cl : 35.5, Br : 80, I : 127) **[Jan. 22, 2026 (I)]**

247. In Duma's method 292 mg of an organic compound released 50 mL of nitrogen gas (N_2) at 300 K temperature and 715 mm Hg pressure. The percentage composition of 'N' in the organic compound is _____ % (Nearest integer)
(Aqueous tension at 300 K = 15 mm Hg) **[April 7, 2025 (II)]**

248. In Dumas' method for estimation of nitrogen 1g of an organic compound gave 150 mL of nitrogen collected at 300K temperature and 900 mm Hg pressure. The percentage composition of nitrogen in the compound is _____ % (nearest integer). **[April 4, 2025 (I)]**
(Aqueous tension at 300 K = 15mm Hg)

249. During estimation of nitrogen by Dumas' method of compound X (0.42 g) : **[April 3, 2025 (I)]**

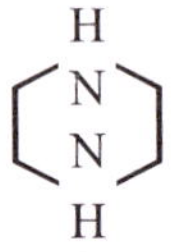

X

__________ mL of N_2 gas will be liberated at STP. (nearest integer)

250. In the sulphur estimation 0.20 g of a pure organic compound gave 0.40 g of barium sulphate.
The percentage of sulphur in the compound is _____ $\times 10^{-1}$ %. **[Jan. 29, 2025 (II)]**
(Molar mass : O = 16, S = 32, Ba = 137 in g mol^{-1})
(Given molar mass in g mol^{-1} : C : 12, H : 1, N : 14)

251. Quantitative analysis of an organic compound (X) shows following % composition. **[Jan. 28, 2025 (I)]**
C : 14.5% ; Cl : 64.46% ; H : 1.8%
(Empirical formula mass of the compound (X) is ____ $\times 10^{-1}$
(Given molar mass in g mol^{-1} of C : 12, H : 1, O : 16, Cl : 35.5)

252. In Carius method of estimation of halogen, 0.25 g of an organic compound gave 0.15 g of silver bromide (AgBr). The percentage of Bromine in the organic compound is ___________ $\times 10^{-1}$% (Nearest integer).
(Given : Molar mass of Ag is 108 and Br is 80 g mol^{-1}) **[Jan. 24, 2025 (II)]**

253. During "S" estimation, 160 mg of an organic compound gives 466 mg of barium sulphate. The percentage of Sulphur in the given compound is ______%
(Given molar mass in g mol^{-} of Ba : 137, S : 32, O : 16) **[Jan. 23, 2025 (I)]**

254. In Carius method for estimation of halogens, 180 mg of an organic compound produced 143.5 mg of AgCl. The percentage composition of chlorine in the compound is _______ %.
[Given : molar mass in g mol^{-1} of Ag : 108, Cl = 35.5] **[Jan. 22, 2025 (I)]**

255. How many compounds among the following compounds show inductive, mesomeric as well as hyperconjugation effects? **[April 9, 2024 (I)]**

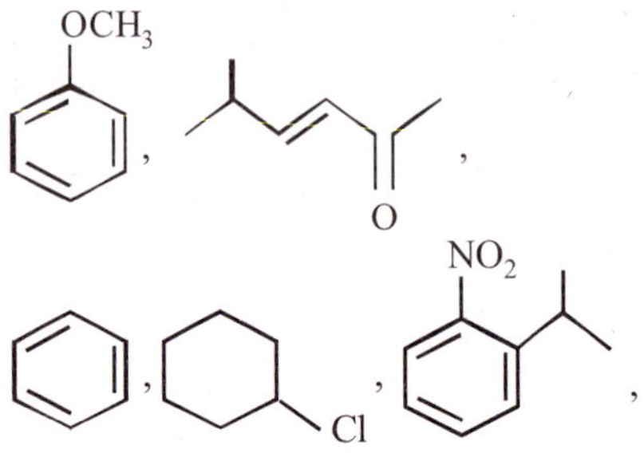

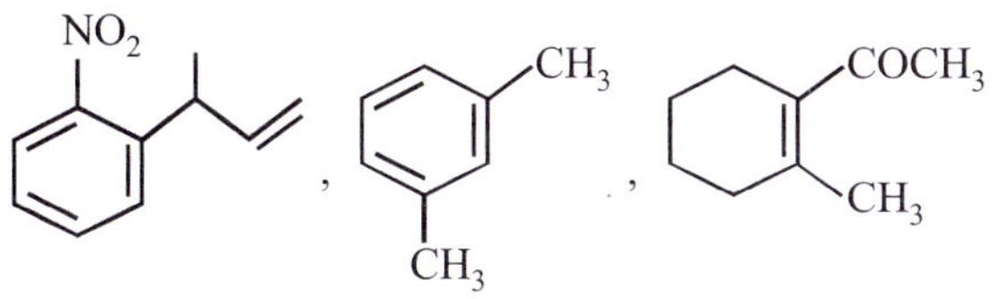

256. 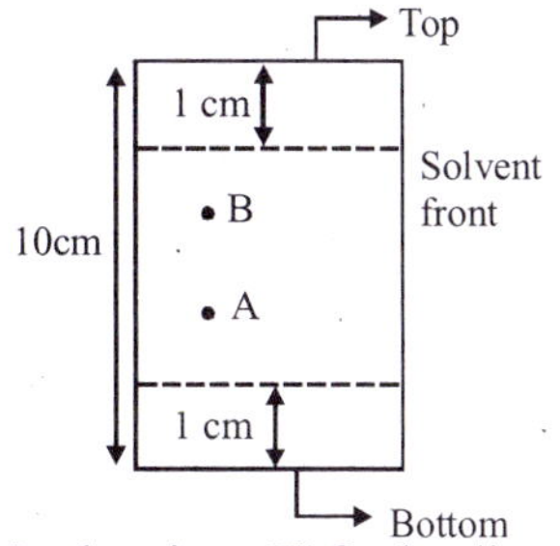

In the given TLC, the distance of spot A & B are 5 cm & 7 cm, from the bottom of TLC plate, respectively. R_f value of B is $x \times 10^{-1}$ times more than A. The value of x is_____. [April 9, 2024 (II)]

257. Number of carbocation from the following that are **not** stabilized by hyperconjugation is........... . [April 6, 2024 (II)]

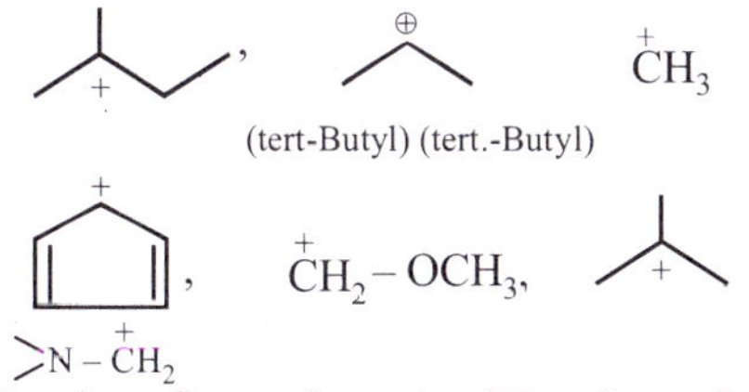

258. Using the given figure, the ratio of R_f values of sample A and sample C is $x \times 10^{-2}$. Value of x is ________. [April 5, 2024 (II)]

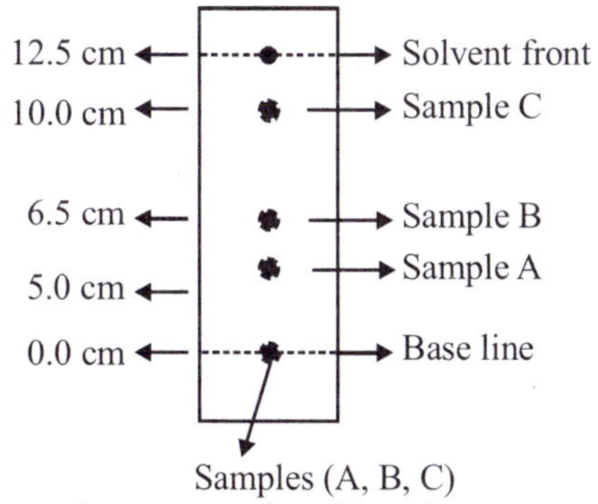

Fig : Paper chromatography of Samples

259. Following Kjeldahl's method, 1g of organic compound released ammonia, that neutralised 10 mL of 2M H_2SO_4. The percentage of nitrogen in the compound is ______ %. [Feb. 1, 2024 (II)]

260. On a thin layer chromatographic plate, an organic compound moved by 3.5 cm, while the solvent moved by 5 cm. The retardation factor of the organic compound is ______ $\times 10^{-1}$. [Jan. 30, 2024 (I)]

261. Among the following, total number of meta directing functional groups is (Integer based)
$-OCH_3$, $-NO_2$, $-CN$, $-CH_3$ $-NHCOCH_3$, $-COR$, $-OH$, $-COOH$, $-Cl$ [Jan. 27, 2024 (I)]

262. 0.400 g of an organic compound (X) gave 0.376 g of AgBr in Carius method for estimation of bromine. % of bromine in the compound (X) is ______.
(Given: Molar mass AgBr = 188 g mol^{-1} Br = 80 g mol^{-1}) [April 13, 2023 (II)]

263. Three organic compounds A, B and C were allowed to run in thin layer chromatography using hexane and gave the following result (see figure). The R_f value of the most polar compound is ______ $\times 10^{-2}$ [April 12, 2023 (I)]

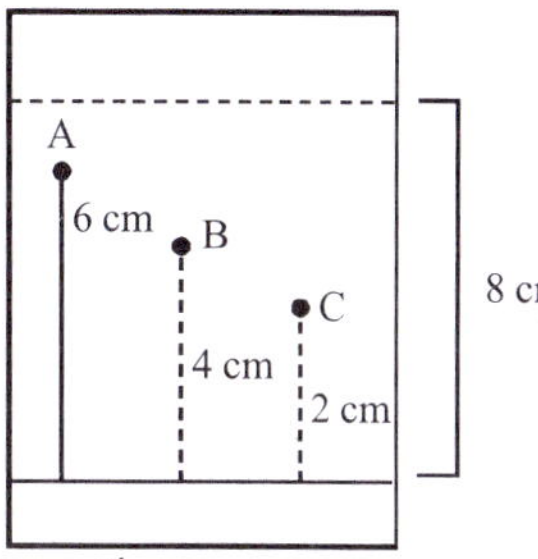

264. For the given reaction [April 13, 2023 (I)]

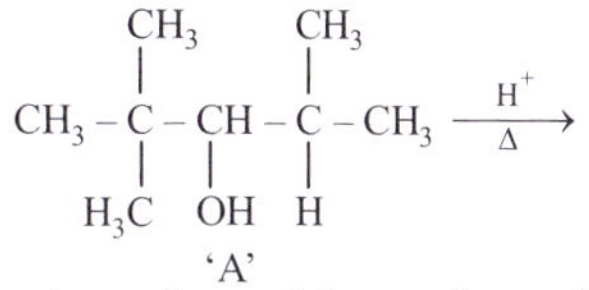

The total number of possible products formed by tertiary carbocation of A is ______.

265. 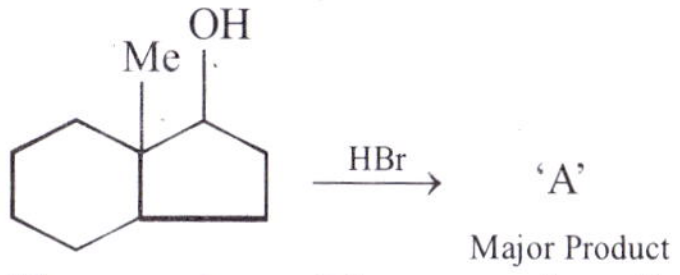

The number of hyperconjugation structures involved to stabilize carbocation formed in the above reaction is ______. [April 11, 2023 (I)]

266. Following chromatogram was developed by adsorption of compound 'A' on a 6 cm TLC glass plate. Retardation factor of the compound 'A' is ______ $\times 10^{-1}$. [Jan. 29, 2023 (I)]

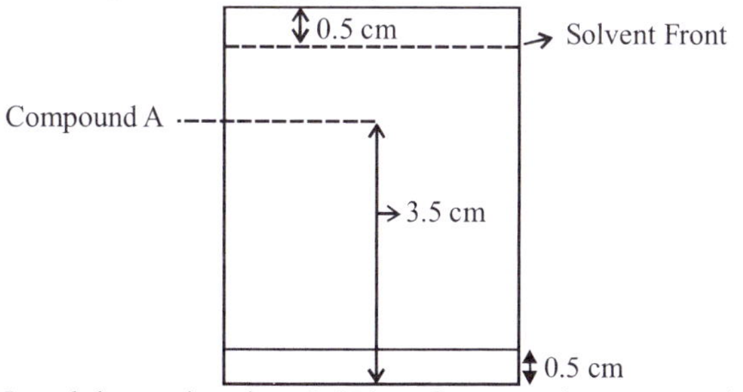

267. In sulphur estimation. 0.471 g of an organic compound gave 1.4439 g of barium sulphate.
The percentage of sulphur in the compound is ________(Nearest Integer) [Jan. 25, 2023 (I)]
(Given: Atomic mass Ba: 137 u: S: 32 u, O: 16 u)

268. Number of moles of AgCl formed in the following reaction is ________. [Jan. 24, 2023 (I)]

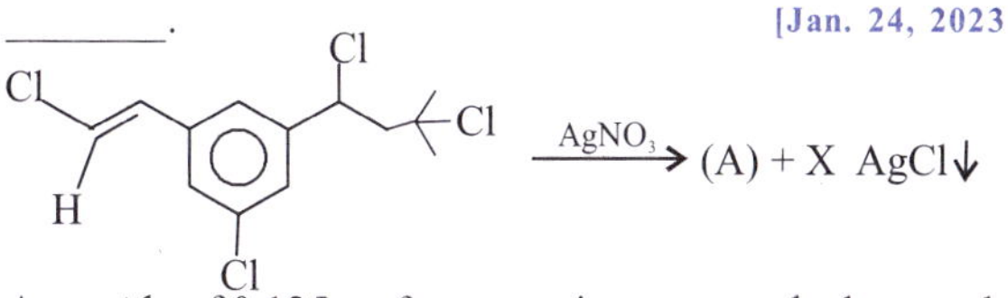

269. A sample of 0.125 g of an organic compound when analyzed by Duma's method yields 22.78 mL of nitrogen gas collected over KOH solution at 280 K and 759 mm Hg. The percentage of nitrogen in the given organic compound is _______. (Nearest integer) [July 28, 2022 (II)]
Given:
(a) The vapour pressure of water of 280 K is 14.2 mm Hg.
(b) R = 0.082 L atm K^{-1} mol^{-1}

270. The number of sp^3 hybridised carbons in an acyclic neutral compound with molecular formula C_4H_5N is _________ . [July 25, 2022 (I)]

271. The separation of two coloured substances was done by paper chromatography. The distances travelled by solvent front, substance A and substance B from the base line are 3.25 cm, 2.08 cm and 1.05 cm respectively. The ratio of R_f values of A to B is _____. [July 25, 2022 (II)]

272. In the following reaction

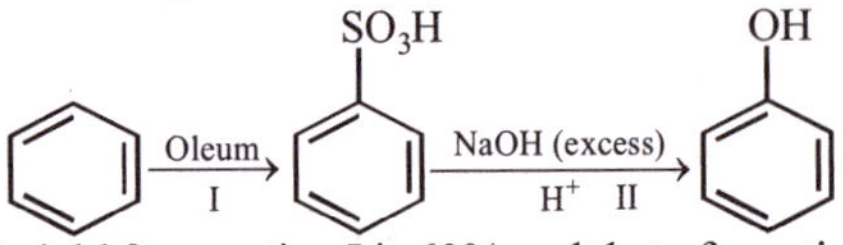

The % yield for reaction I is 60% and that of reaction II is 50%. The overall yield of the complete reaction is ___% [nearest integer] [July 27, 2022 (I)]

273. While estimating the nitrogen present in an organic compound by Kjeldahl's method, the ammonia evolved from 0.25 g of the compound neutralized 2.5 mL of 2 M H_2SO_4. The percentage of nitrogen present in organic compound is _________ . [July 25, 2022 (I)]

274. Kjeldahl's method was used for the estimation of nitrogen in an organic compound. The ammonia evolved from 0.55 g of the compound neutralised 12.5 mL of 1 M H_2SO_4 solution. The percentage of nitrogen in the compound is _________. (Nearest integer) [June 29, 2022 (I)]

275. 0.25 g of an organic compound containing chlorine gave 0.40 g of silver chloride in Carius estimation. The percentage of chlorine present in the compound is _____. [in nearest integer] (Given: Molar mass of Ag is 108 g mol^{-1} and that of Cl is 35.5 g mol^{-1}) [June 27, 2022 (II)]

276. Number of electrophilic centre in the given compound is ________ [June 24, 2022 (I)]

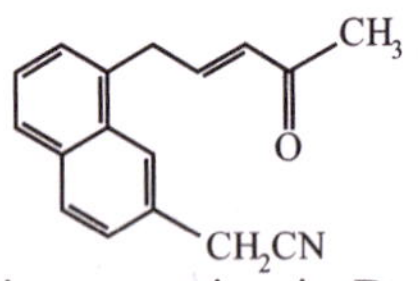

277. The transformation occurring in Duma's method is given below: [Aug. 31, 2021 (II)]

$$C_2H_7N + \left(2x + \frac{y}{2}\right)CuO \rightarrow$$

$$xCO_2 + \frac{y}{2}H_2O + \frac{z}{2}N_2 + \left(2x + \frac{y}{2}\right)Cu$$

The value of y is ________. (Integer answer)

278. The number of moles of CuO, that will be utilized in Dumas method for estimation nitrogen in a sample of 57.5g of N, N-dimethylaminopentane is ________ $\times 10^{-2}$. (Nearest integer) [Aug. 27, 2021 (I)]

279. In the sulphur estimation, 0.471 g of an organic compound gave 1.44 g of barium sulphate. The percentage of sulphur in the compound is _____%. (Nearest integer) (Atomic Mass of Ba = 137 u) [Aug. 26, 2021 (II)]

280. An organic compound is subjected to chlorination to get compound A using 5.0 g of chlorine. When 0.5 g of compound A is reacted with $AgNO_3$ [Carius Method], the percentage of chlorine in compound A is _______ when it forms 0.3849 g of AgCl. (Round off to the nearest integer) (Atomic masses of Ag and Cl are 107.87 and 35.5 respectively) [July 27, 2021 (I)]

281. Using the provided information in the following paper chromatogram:

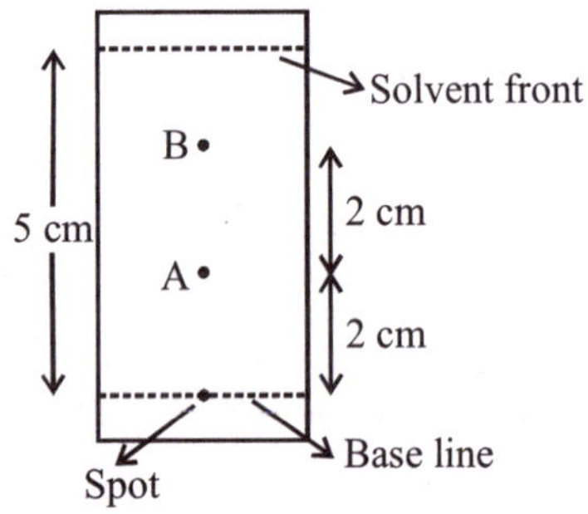

Paper chromatography for compounds A and B

The calculated R_f value of A _______ $\times 10^{-1}$. [Feb. 25, 2021 (I)]

282. In Duma's method of estimation of nitrogen, 0.1840 g of an organic compound gave 30 mL of nitrogen collected at 287 K and 758 mm of Hg pressure. The percentage composition of nitrogen in the compound is_________. (Round off to the nearest Integer). [March 16, 2021 (II)] [Given : Aqueous tension at 287 K = 14 mm of Hg]

283. In an estimation of bromine by Carius method, 1.6 g of an organic compound gave 1.88 g of AgBr. The mass percentage of bromine in the compound is ______. (Atomic mass, Ag = 108, Br = 80 g mol^{-1}) [Sep. 06, 2020 (I)]

Part-B (JEE Advanced)

Topic 1- Classification and Nomenclature of Organic Compounds

1 MCQs with One Correct Answer

1. The IUPAC name of the following compound is [2009]

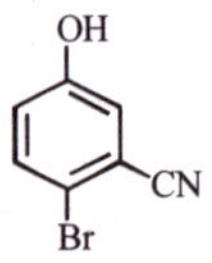

(a) 4-Bromo-3-cyanophenol
(b) 2-Bromo-5-hydroxybenzonitrile
(c) 2- Cyano-4-hydroxybromobenzene
(d) 6-Bromo-3-hydroxybenzonitrile

2. The IUPAC name of C_6H_5COCl is [2006 - 3M, –1]
(a) Benzene chloro ketone (b) Benzoyl chloride
(c) Chloro phenyl ketone (d) Benzene carbonyl chloride

3. The compound which has one isopropyl group is : [1989]
(a) 2, 2, 3, 3-tetramethylpentane
(b) 2, 2-dimethylpentane
(c) 2, 2, 3-trimethylpentane
(d) 2-methylpentane

4. The IUPAC name of the compound [1987]
$CH_2=CH-CH(CH_3)_2$ is
(a) 1, 1-dimethyl-2-propene (b) 3-methyl-1-butene
(c) 2-vinylpropane (d) 1-isopropylethylene

5. The IUPAC name of the compound having the formula is: [1984]

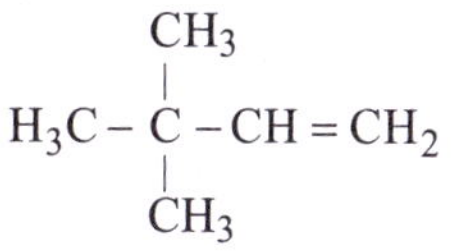

(a) 3, 3, 3-Trimethyl-1-propene
(b) 1, 1, 1-Trimethyl-2-propene
(c) 3, 3-Dimethyl-1-butene
(d) 2, 2-Dimethyl-3-butene

4 Fill in the Blanks

6. The IUPAC name of succinic acid is .... [1990]
7. The kind of delocalization involving sigma bond orbitals is called .... [1994]
8. A .... diol has two hydroxyl groups on .... carbon atoms. [1986]

6 MCQs with One or More than One Correct Answer

9. The IUPAC name(s) of the following compound is (are)

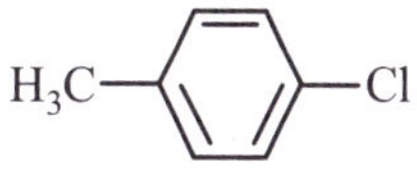

[Adv. 2017]

(a) 1-chloro-4-methylbenzene
(b) 4-chlorotoluene
(c) 4-methylchlorobenzene
(d) 1-methyl-4-chlorobenzene

10 Subjective Problems

10. (*i*) Write the IUPAC name for the following : [1991]

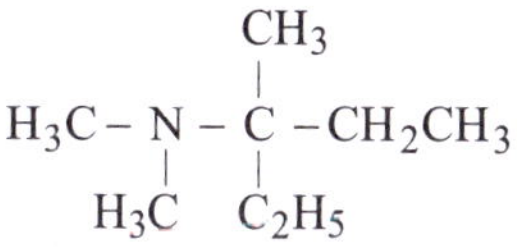

(*ii*) Give the IUPAC name of the following compound : [1990]

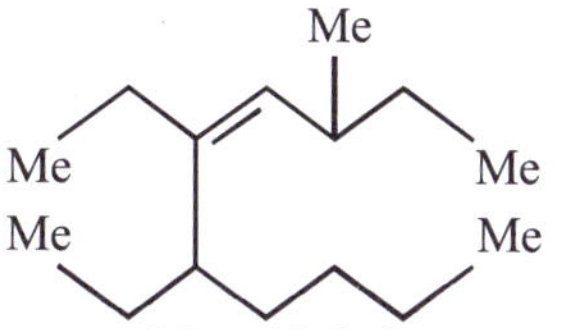

(Me = Methyl)

(*iii*) Write the IUPAC name of : [1986]
$CH_3CH_2CH=CHCOOH$

11. Write the structural formula of 4-chloro-2-pentene. [1988]

Topic 2 – Isomerism in Organic Compounds

1 MCQs with One Correct Answer

12. The number of isomers of C_6H_{14} is [1987, 2007]
(a) 4 (b) 5 (c) 6 (d) 7

13. The number of isomers for the compound with molecular formula $C_2BrClFI$ is [2001S]
(a) 3 (b) 4 (c) 5 (d) 6

14. The structure

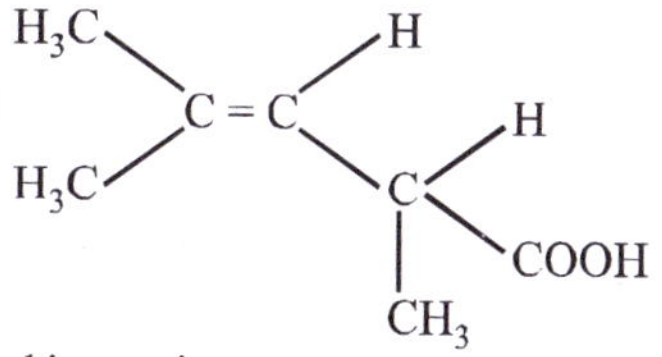

shows : [1995S]
(a) geometrical isomerism
(b) optical isomerism
(c) geometrical & optical isomerism
(d) tautomerism.

15. Isomers which can be interconverted through rotation around a single bond are [1992]
(a) Conformers (b) Diastereomers
(c) Enantiomers (d) Positional isomers

16. An isomer of ethanol is : [1986]
(a) methanol (b) diethyl ether
(c) acetone (d) dimethyl ether

17. The compound which is not isomeric with diethyl ether is [1981]
(a) *n*-propyl methyl ether (b) butan-1-ol
(c) 2-methylpropan-2-ol (d) butanone

6 MCQs with One or More than One Correct Answer

18. Amongst the given options, the compound(s) in which all the atoms are in one plane in all the possible conformations (if any), is (are) [2011]

(a) 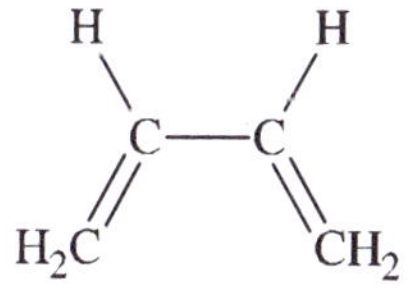
(b) 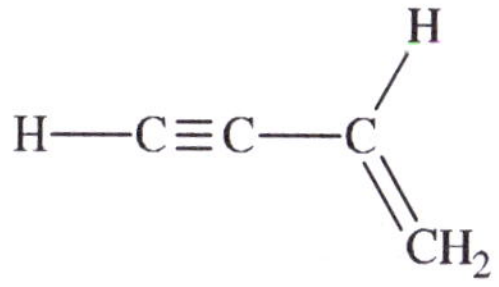

(c) $H_2C{=}C{=}O$ (d) $H_2C{=}C{=}CH_2$

19. The correct statements(s) concerning the structures E,F and G is (are) – [2008]

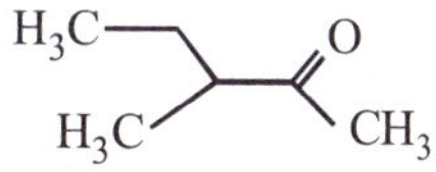

(E)

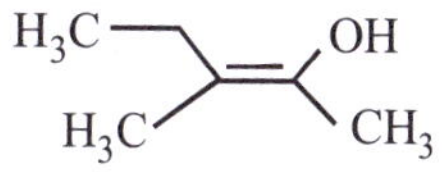

(F)

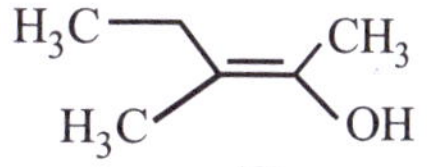

(G)

(a) E, F and G are resonance structures
(b) E, F and E, G are tautomers
(c) F and G are geometrical isomers
(d) F and G are diastereomers

20. Tautomerism is exhibited by [1998]

(a) 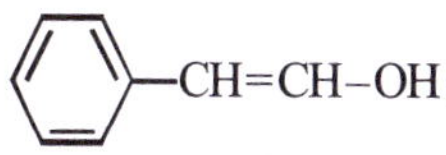
(b) 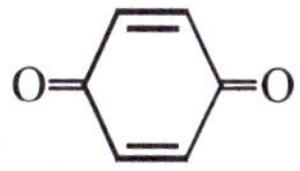
(c) 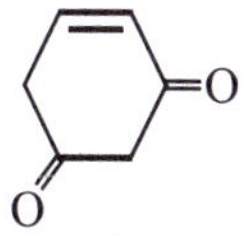
(d) 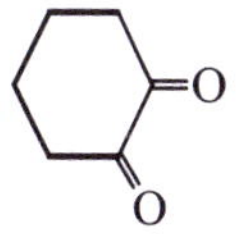

21. Only two isomeric monochloro derivatives are possible for: [1986]
(a) *n*-butane (b) 2, 4-dimethylpentane
(c) benzene (d) 2-methylpropane

10 Subjective Problems

22. Write tautomeric forms for phenol. [1992]

Topic 3 - Concept of Reaction Mechanism in Organic Compounds, Purification and Quantitative Analysis

1 MCQs with One Correct Answer

23. Consider the depicted hydrogen **(H)** in the hydrocarbons given below. The most acidic hydrogen **(H)** is [Adv. 2025]

(a) 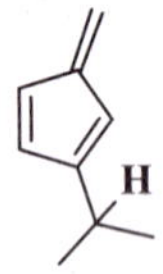

(b) 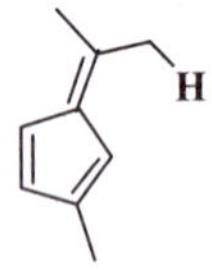

(c) 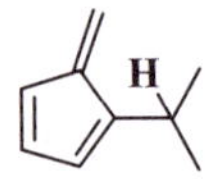

(d) 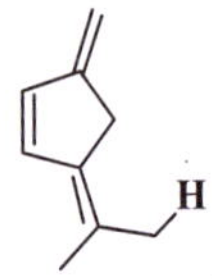

24. In CH_3CH_2OH, the bond that undergoes heterolytic cleavage most readily is [1988]
(a) C—C (b) C—O (c) C—H (d) O—H

25. The Cl—C—Cl angle in 1,1,2,2-tetrachloroethene and tetrachloromethane respectively will be about [1988]
(a) 120° and 109.5° (b) 90° and 109.5°
(c) 109.5° and 90° (d) 109.5° and 120°

26. The bond between carbon atom (1) and carbon atom (2) in compound $N \equiv \underset{1}{C} - \underset{2}{CH} = CH_2$ involves the hybrids as [1987]
(a) sp^2 and sp^2 (b) sp^3 and sp
(c) sp and sp^2 (d) sp and sp

27. Out of the following compounds, which will have a zero dipole moment? [1987]
(a) 1, 1-dichloroethylene
(b) *cis*-1, 2-dichloroethylene
(c) *trans*-1, 2-dichloroethylene
(d) None of these compounds

28. The compound 1, 2-butadiene has [1983]
(a) only sp hybridized carbon atoms
(b) only sp^2 hybridized carbon atoms
(c) both sp and sp^2 hybridized carbon atoms
(d) sp, sp^2 and sp^3 hybridized carbon atoms

29. Molecule in which the distance between the two adjacent carbon atoms is largest is [1981]
(a) Ethane (b) Ethene (c) Ethyne (d) Benzene

2 Integer Value Answer

30. The total number of contributing structures showing hyperconjugation (involving C–H bonds) for the following carbocation is [2011]

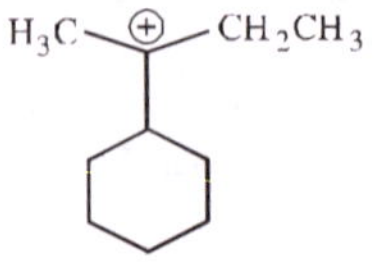

3 Numeric / New Stem Based Questions

31. The monomer (X) involved in the synthesis of Nylon 6,6 gives positive carbylamine test. If 10 moles of **X** are analyzed using Dumas method, the amount (in grams) of nitrogen gas evolved is ____.
Use: Atomic mass of N (in amu) = 14 [Adv. 2025]

4 Fill in the Blanks

32. The valence atomic orbitals on carbon in silver acetylide is .... hybridized. [1990]

33. The compound having both sp and sp^2 hybridized carbon atoms is .... [1981]
(propene, propane, propadiene)

34. Among the given cations, ... is most stable. [1981]
(*sec*-butyl carbonium ion; *tert*-butyl carbonium ion; *n* butyl carbonium ion)

6 MCQs with One or More than One Correct Answer

35. With respect to the compounds **I-V**, choose the correct statement(s). [Adv. 2020]

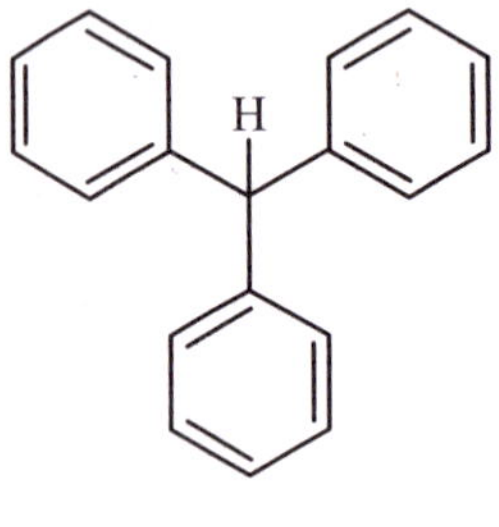

I

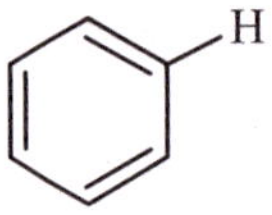

II

$H-CH_3$

III

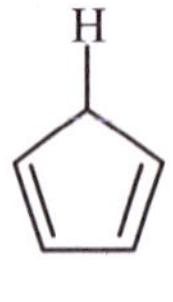

IV

V

(a) The acidity of compound **I** is due to delocalization in the conjugate base.
(b) The conjugate base of compound **IV** is aromatic.
(c) Compound **II** becomes more acidic, when it has a $-NO_2$ substituent.
(d) The acidity of compounds follows the order **I > IV > V > II > III**.

36. The hyperconjugative stabilities of tert-butyl cation and 2-butene, respectively, are due to [Adv. 2013]
(a) $\sigma \to p$ (empty) and $\sigma \to \pi^*$ electron delocalisations
(b) $\sigma \to \sigma^*$ and $\sigma \to \pi$ electron delocalisations
(c) $\sigma \to p$ (filled) and $\sigma \to \pi$ electron delocalisations
(d) p (filled) $\to \sigma^*$ and $\sigma \to \pi^*$ electron delocalisations

37. Which of the following molecules, in pure form, is (are) **unstable** at room temperature ? [2012]

(a)  (b)  (c) 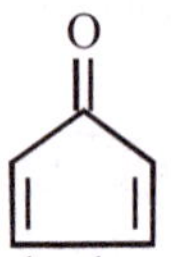

(d) 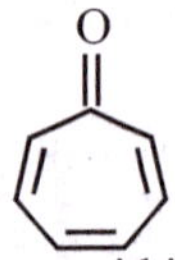

38. Among the following compounds, the strongest acid is [1998]
(a) $HC \equiv CH$ (b) C_6H_6
(c) C_2H_6 (d) CH_3OH

39. What is the decreasing order of strength of the bases OH^-, NH_2^-, $HC \equiv C^-$ and $CH_3CH_2^-$? [1993]
(a) $CH_3-CH_2^- > NH_2^- > H-C\equiv C^- > OH^-$
(b) $H-C\equiv C^- > CH_3-CH_2^- > NH_2^- > OH^-$
(c) $OH^- > NH_2^- > H-C\equiv C^- > CH_3-CH_2^-$
(d) $NH_2^- > H-C\equiv C^- > OH^- > CH_3-CH_2^-$

40. Dipole moment is shown by : [1986]
(a) 1, 4-dichlorobenzene
(b) *cis*- 1, 2-dichloroethane
(c) *trans* -1, 2-dichloroethene
(d) *trans*- 1, 2-dichloro-2-pentene

41. Resonance structures of a molecule should have : [1984]
(a) identical arrangement of atoms
(b) nearly the same energy content
(c) the same number of paired electrons
(d) identical bonding

10 Subjective Problems

42. (*i*) $\mu_{obs} = \sum_i \mu_i x_i$, where μ_i is the dipole moment of a stable conformer of the molecule, $Z-CH_2-CH_2-Z$ and x_i is the mole fraction of the stable conformer.
Given : $\mu_{obs} = 1.0$ D and $x_{(Anti)} = 0.82$
Draw all the stable conformers of $Z-CH_2-CH_2-Z$ and calculate the value of $\mu_{(Gauche)}$.
(*ii*) Draw the stable conformer of Y – CHD – CHD – Y (meso form), when Y = CH_3 (rotation about C_2-C_3) and Y = OH (rotation about C_1-C_2) in Newman projection. [2005]

43. Which of the following is more acidic and why ? [2004]

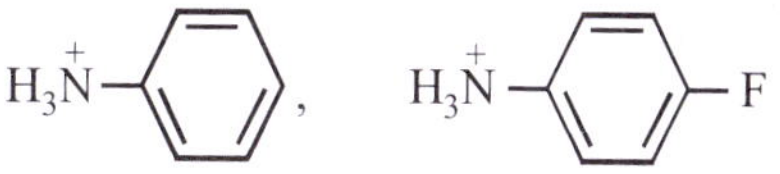

44. [2003]

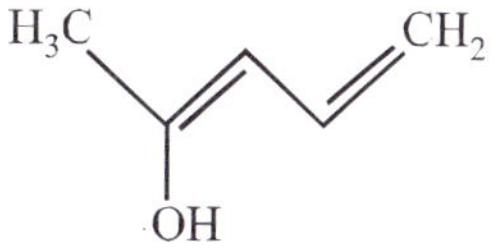

Write resonance structure of the given compound.

45. Which one is more soluble in diethyl ether - anhydrous $AlCl_3$ or hydrous $AlCl_3$? Explain in terms of bonding. [2003]

46. Give reasons for the following :
(*i*) $CH_2{=}CH^-$ is more basic than $HC \equiv C^-$. [1994]
(*ii*) Phenyl group is known to exert negative inductive effect. But each phenyl ring in biphenyl ($C_6H_5-C_6H_5$) is more reactive than benzene towards electrophilic substitution. [1992]
(*iii*) Carbon oxygen bond lengths in formic acid are 1.23Å and 1.36Å and both the carbon oxygen bonds in sodium formate have the same value i.e. 1.27Å. [1988]

47. *n*-butane, *n*-butanol, *n*-butyl chloride, isobutane in increasing order of boiling point. [1988]

48. For nitromethane molecule, write structure(s).
(*i*) showing significant resonance stabilisation. [1986]
(*ii*) indicating tautomerism. [1986]

Answer Key

Part-A (JEE Main)																			
1	(3)	**31**	(2)	**61**	(2)	**91**	(3)	**121**	(2)	**151**	(1)	**181**	(2)	**211**	(1)	**241**	(42)	**271**	(2)
2	(3)	**32**	(3)	**62**	(4)	**92**	(1)	**122**	(4)	**152**	(4)	**182**	(4)	**212**	(2)	**242**	(435)	**272**	(30)
3	(1)	**33**	(4)	**63**	(1)	**93**	(4)	**123**	(1)	**153**	(2)	**183**	(1)	**213**	(4)	**243**	(5)	**273**	(56)
4	(4)	**34**	(2)	**64**	(4)	**94**	(1)	**124**	(4)	**154**	(2)	**184**	(1)	**214**	(2)	**244**	(4)	**274**	(64)
5	(3)	**35**	(3)	**65**	(2)	**95**	(2)	**125**	(4)	**155**	(1)	**185**	(4)	**215**	(2)	**245**	(15)	**275**	(40)
6	(4)	**36**	(4)	**66**	(4)	**96**	(3)	**126**	(2)	**156**	(1)	**186**	(4)	**216**	(4)	**246**	(43)	**276**	(3)
7	(1)	**37**	(4)	**67**	(4)	**97**	(3)	**127**	(1)	**157**	(3)	**187**	(3)	**217**	(4)	**247**	(18)	**277**	(7)
8	(3)	**38**	(2)	**68**	(4)	**98**	(1)	**128**	(3)	**158**	(1)	**188**	(1)	**218**	(3)	**248**	(20)	**278**	(1125)
9	(3)	**39**	(2)	**69**	(4)	**99**	(4)	**129**	(2)	**159**	(4)	**189**	(2)	**219**	(4)	**249**	(109)	**279**	(42)
10	(4)	**40**	(2)	**70**	(4)	**100**	(4)	**130**	(1)	**160**	(4)	**190**	(4)	**220**	(1)	**250**	(275)	**280**	(19)
11	(3)	**41**	(1)	**71**	(1)	**101**	(3)	**131**	(1)	**161**	(2)	**191**	(2)	**221**	(1)	**251**	(1655)	**281**	(4)
12	(2)	**42**	(1)	**72**	(4)	**102**	(1)	**132**	(2)	**162**	(4)	**192**	(3)	**222**	(2)	**252**	(255)	**282**	(19)
13	(2)	**43**	(4)	**73**	(3)	**103**	(4)	**133**	(2)	**163**	(1)	**193**	(3)	**223**	(4)	**253**	(40)	**283**	(50)
14	(3)	**44**	(2)	**74**	(4)	**104**	(3)	**134**	(3)	**164**	(3)	**194**	(4)	**224**	(1)	**254**	(20)		
15	(3)	**45**	(1)	**75**	(2)	**105**	(4)	**135**	(1)	**165**	(2)	**195**	(3)	**225**	(4)	**255**	(4)		
16	(1)	**46**	(1)	**76**	(3)	**106**	(3)	**136**	(4)	**166**	(2)	**196**	(2)	**226**	(2)	**256**	(15)		
17	(3)	**47**	(1)	**77**	(2)	**107**	(1)	**137**	(3)	**167**	(1)	**197**	(2)	**227**	(2)	**257**	(5)		
18	(4)	**48**	(1)	**78**	(2)	**108**	(1)	**138**	(2)	**168**	(2)	**198**	(3)	**228**	(1)	**258**	(50)		
19	(2)	**49**	(4)	**79**	(2)	**109**	(4)	**139**	(3)	**169**	(1)	**199**	(1)	**229**	(2)	**259**	(56)		
20	(4)	**50**	(1)	**80**	(3)	**110**	(3)	**140**	(3)	**170**	(4)	**200**	(2)	**230**	(3)	**260**	(7)		
21	(2)	**51**	(1)	**81**	(2)	**111**	(1)	**141**	(3)	**171**	(4)	**201**	(3)	**231**	(4)	**261**	(4)		
22	(4)	**52**	(1)	**82**	(3)	**112**	(3)	**142**	(1)	**172**	(4)	**202**	(2)	**232**	(4)	**262**	(40)		
23	(2)	**53**	(3)	**83**	(1)	**113**	(4)	**143**	(2)	**173**	(1)	**203**	(4)	**233**	(2)	**263**	(25)		
24	(3)	**54**	(3)	**84**	(3)	**114**	(1)	**144**	(4)	**174**	(3)	**204**	(N)	**234**	(4)	**264**	(5)		
25	(2)	**55**	(2)	**85**	(4)	**115**	(2)	**145**	(4)	**175**	(1)	**205**	(2)	**235**	(4)	**265**	(7)		
26	(3)	**56**	(3)	**86**	(4)	**116**	(3)	**146**	(3)	**176**	(2)	**206**	(3)	**236**	(4)	**266**	(6)		
27	(4)	**57**	(1)	**87**	(8)	**117**	(1)	**147**	(1)	**177**	(1)	**207**	(3)	**237**	(2)	**267**	(42)		
28	(3)	**58**	(1)	**88**	(4)	**118**	(4)	**148**	(4)	**178**	(4)	**208**	(2)	**238**	(2)	**268**	(2)		
29	(4)	**59**	(3)	**89**	(6)	**119**	(4)	**149**	(3)	**179**	(4)	**209**	(2)	**239**	(2)	**269**	(22)		
30	(3)	**60**	(3)	**90**	(6)	**120**	(4)	**150**	(4)	**180**	(3)	**210**	(2)	**240**	(2)	**270**	(1)		
Part-B (JEE Advanced)																			
1	(b)	**5**	(c)	**14**	(b)	**18**	(b, c)	**23**	(b)	**27**	(c)	**31**	(280)	**38**	(d)				
2	(b)	**9**	(a, b)	**15**	(a)	**19**	(b, c, d)	**24**	(d)	**28**	(d)	**35**	(a, b, c)	**39**	(a)				
3	(d)	**12**	(b)	**16**	(d)	**20**	(a, c, d)	**25**	(a)	**29**	(a)	**36**	(a)	**40**	(b, d)				
4	(b)	**13**	(d)	**17**	(d)	**21**	(a, d)	**26**	(c)	**30**	(6)	**37**	(b, c)	**41**	(a, b, c)				

Chapter-13

Hydrocarbons

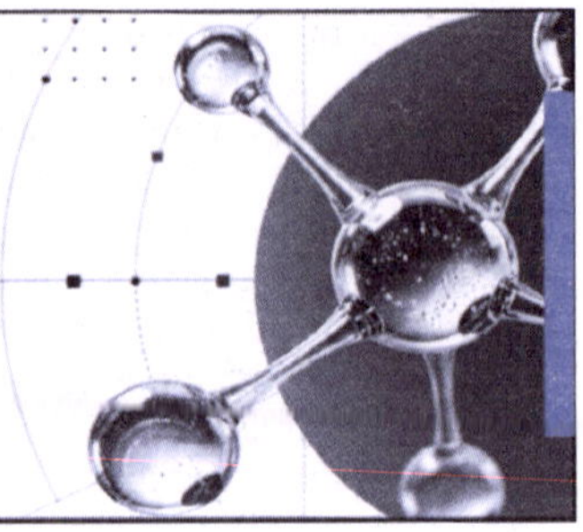

Part-A (JEE Main)

Topic 1- Alkanes

Multiple Choice Questions (MCQs)

1. n-Butane on monochlorination under photochemical conditions gives an optically active compound "P". "P" on further chlorination gives dichloro compounds.
The number of dichloro compounds obtained (ignore stereoisomers) is : [April 08, 2026 (II)]
(1) 3 (2) 4
(3) 5 (4) 6

2. Given below are two statements : [April 06, 2026 (I)]
Statement-I : Methane can be prepared by decarboxylation of sodium ethanoate, Kolbe's electrolysis of sodium acetate and reaction of CH_3MgBr with water.
Statement-II: Methane cannot be prepared from unsaturated hydrocarbons and by Wurtz reaction.
In the light of the above statements, choose the **correct** answer from the options given below.
(1) Both Statement-I and Statement-II are true.
(2) Both Statement-I and Statement-II are false.
(3) Statement-I is true but Statement-II is false.
(4) Statement-I is false but Statement-II is true.

3. The reactions which produce alcohol as the product area:
A. $CH_4 + O_2 \xrightarrow[\Delta]{Mo_2O_3}$ [Jan. 28, 2026 (II)]
B. $2CH_3CH_3 + 3O_2 \xrightarrow[\Delta]{(CH_3COO)_2Mn}$
C. $(CH_3)_3CH \xrightarrow{KMnO_4}$
D. $2CH_4 + O_2 \xrightarrow{Cu/523K/100\ atm.}$
E. $CH_3–CH=CH–CH_3 \xrightarrow{KMnO_4/H^+}$
Choose the **correct** answer from the options given below:
(1) A and D Only (2) A, C and E Only
(3) C and D Only (4) B, D and E Only

4. Given below are two statements: [Jan. 24, 2026 (II)]
Statement I: There are several conformers for n-butane. Out of those conformers,

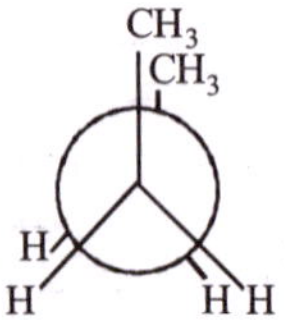

(X)

is the least stable and most stable conformer is

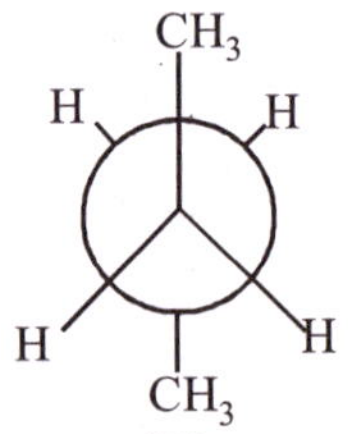

(Y)

Statement II : As the dihedral angle increases, torsional strain decreases from (X) to (Y).
In the light of the above statements, choose the correct answer from the options given below :
(1) Both Statement I and Statement II are false.
(2) Statement I is false but Statement II is true.
(3) Statement I is true but Statement II is false.
(4) Both Statement I and Statement II are true.

5. Given below are two statements : [April 2, 2025 (II)]
Statement (I) : Neopentane forms only one monosubstituted derivative.
Statement (II) : Melting point of neopentane is higher than n-pentane
In the light of the above statements, choose the **most appropriate answer** from the options given below :
(1) **Statement I** is correct but **Statement II** is incorrect
(2) Both **Statement I** and **Statement II** are correct
(3) Both **Statement I** and **Statement II** are incorrect
(4) **Statement I** is incorrect but **Statement II** is correct

6. Propane molecule on chlorination under photochemical condition gives two di-chloro products, "x" and "y". Amongst "x" and "y", "x" is an optically active molecule. How many tri-chloro products (consider only structural isomers) will be obtained from "x" when it is further treated with chlorine under the photochemical condition? [Jan. 23, 2025 (I)]
(1) 4 (2) 2 (3) 5 (4) 3

7. When sec-butylcyclohexane reacts with bromine in the presence of sunlight, the major product is: [Jan. 22, 2025 (II)]

(1) 

(2) 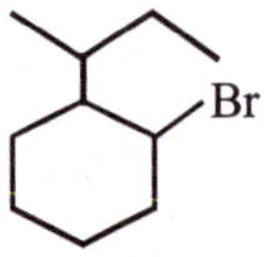

(3) 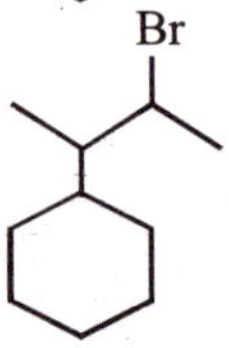

(4) 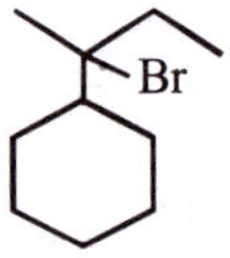

8. In the given compound, the number of 2° carbon atom/s is _____. [April 8, 2024 (I)]

$CH_3-\underset{|}{\overset{}{C}}(CH_3)-\underset{|}{C}H-\underset{|}{C}(CH_3)-CH_3$ (with H on each of the three middle carbons)

(1) Three (2) One (3) Two (4) Four

9. The incorrect statement regarding conformations of ethane is: [Jan. 27, 2024 (II)]
(1) Ethane has infinite number of conformations
(2) The dihedral angle in staggered conformation is 60°
(3) Eclipsed conformation is the most stable conformation.
(4) The conformations of ethane are inter-convertible to one-another

10. In the following reaction 'X' is [April 13, 2023 (I)]

$CH_3(CH_2)_4CH_3 \xrightarrow[HCl.\Delta]{Anhy.AlCl_3}$ 'X' Major product

(1) $CH_3(CH_2)_4CH_2Cl$ (2) $Cl-CH_2-(CH_2)_4-CH_2-Cl$
(3) $CH_3CH-(CH_2)_2CH_3$ (with CH_3 on the second carbon) (4) 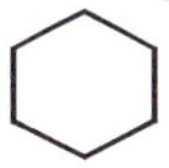

11. When a hydrocarbon A undergoes complete combustion it requires 11 equivalents of oxygen and produces 4 equivalents of water. What is the molecular formula of A? [Jan. 31, 2023 (II)]
(1) C_9H_8 (2) $C_{11}H_4$ (3) C_5H_8 (4) $C_{11}H_8$

12. When a hydrocarbon A undergoes combustion in the presence of air, it requires 9.5 equivalents of oxygen and produces 3 equivalents of water. What is the molecular formula of A? [Jan. 29, 2023 (II)]
(1) C_8H_6 (2) C_9H_9 (3) C_6H_6 (4) C_9H_6

13. Which of the following conformations will be the most stable? [Jan. 25, 2023 (I)]

(1) 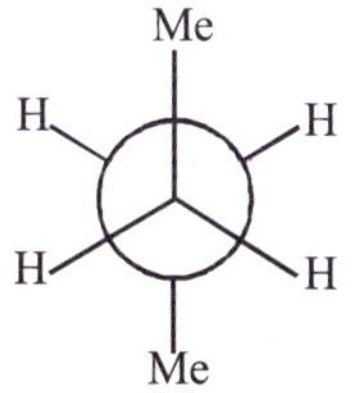

(2) 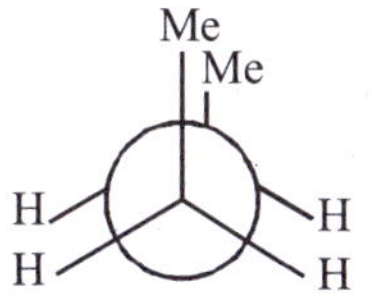

(3) 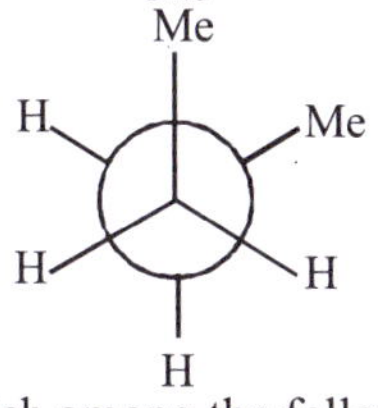

(4) 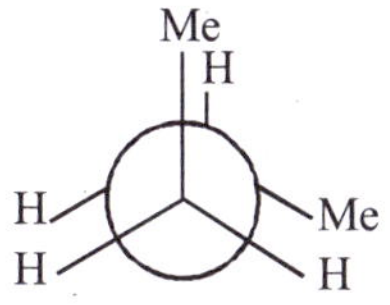

14. Which among the following pairs of the structures will give different products on ozonolysis? (Consider the double bonds in the structures are rigid and not delocalized.) [July 29, 2022 (I)]

(1) 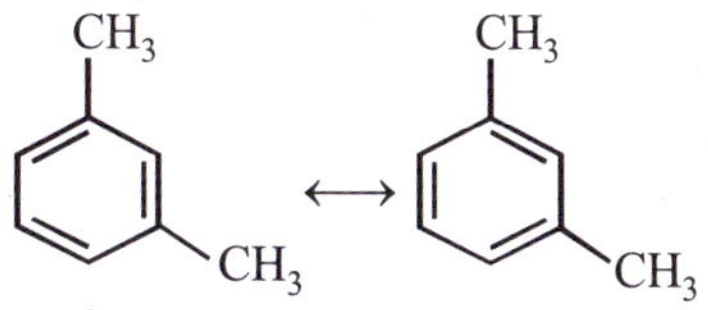

(2) 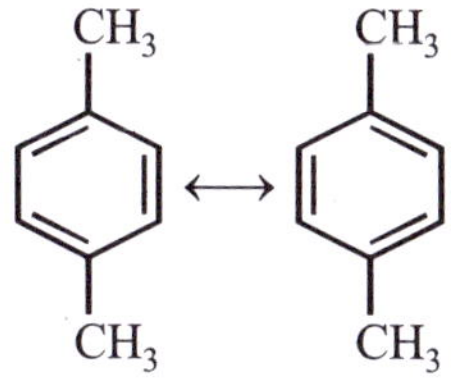

(3) 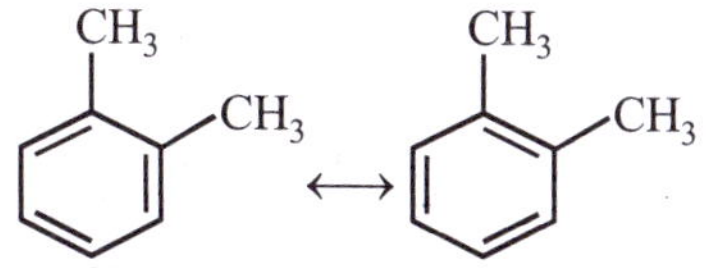

(4) 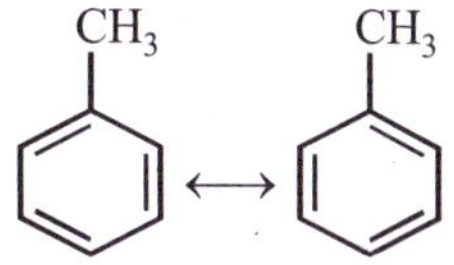

15. $(C_7H_5O_2)_2 \xrightarrow{hv} [X] + 2\dot{C}_6H_5 + 2CO_2$

Consider the above reaction and identify the intermediate 'X' [June 26, 2022 (I)]

(1) $C_6H_5-\overset{O}{\overset{\|}{C}}{}^{\oplus}$ (2) $C_6H_5-\overset{O}{\overset{\|}{C}}-O^{\ominus}$
(3) $C_6H_5-\overset{O}{\overset{\|}{C}}-\ddot{\underset{..}{O}}$ (4) $C_6H_5-\overset{O}{\overset{\|}{C}}-\ddot{\underset{..}{O}}\cdot$

16. In the following structures, which one is having staggered conformation with maximum dihedral angle? [June 25, 2022 (I)]

(1) 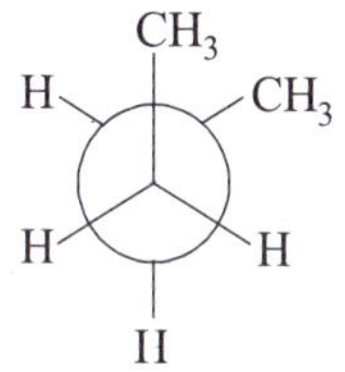

(2) 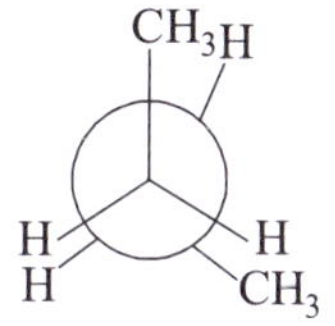

(3) 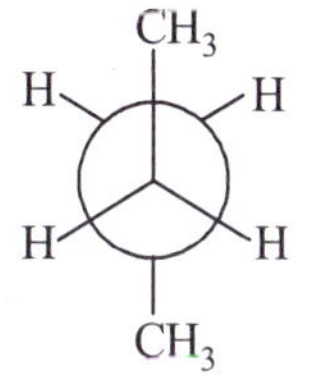

(4) 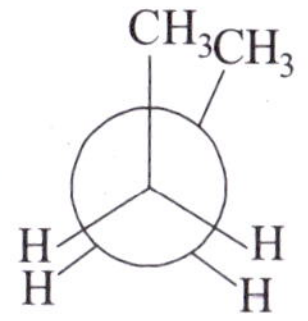

17. Arrange the following conformational isomers of *n*-butane in order of their increasing potential energy: [Aug. 31, 2021 (II)]

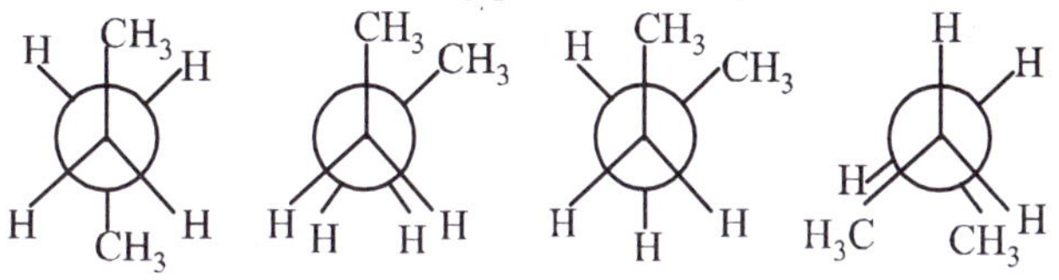

I II III IV

(1) II < III < IV < I (2) I < IV < III < II
(3) II < IV < III < I (4) I < III < IV < II

18. In the following skew conformation of ethane, H' – C – C – H'' dihedral angle is: [April 12, 2019 (II)]

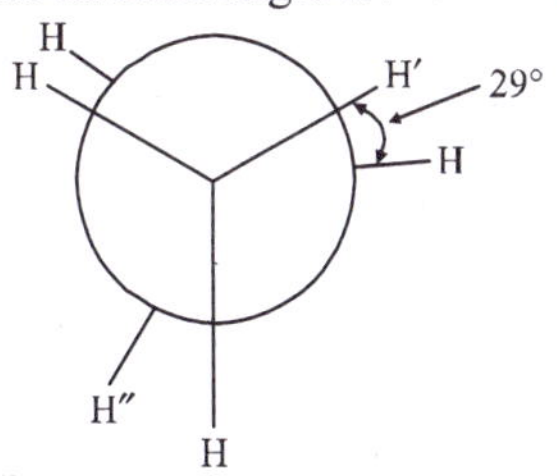

(1) 58° (2) 149° (3) 151° (4) 120°

19. 25 g of an unknown hydrocarbon upon burning produces 88 g of CO_2 and 9 g of H_2O. This unknown hydrocarbon contains: [April 12, 2019 (II)]
(1) 20 g of carbon and 5 g of hydrogen
(2) 22 g of carbon and 3 g of hydrogen
(3) 24 g of carbon and 1 g of hydrogen
(4) 18 g of carbon and 7 g of hydrogen

20. At 300 K and 1 atmospheric pressure, 10 mL of a hydrocarbon required 55 mL of O_2 for complete combustion, and 40 mL of CO_2 is formed. The formula of the hydrocarbon is : [April 10, 2019 (I)]
(1) C_4H_{10} (2) C_4H_6 (3) C_4H_7Cl (4) C_4H_8

21. Which of these factors does not govern the stability of a conformation in acyclic compounds ? [April 10, 2019 (II)]
(1) Steric interactions
(2) Torsional strain
(3) Electrostatic forces of interaction
(4) Angle strain

22. The major product obtained in the photo catalysed bromination of 2-methylbutane is:
[Online April 12, 2014; Same Online May 19, 2012, 2005]
(1) 1-bromo-2-methylbutane
(2) 1-bromo-3-methylbutane
(3) 2-bromo-3-methylbutane
(4) 2-bromo-2-methylbutane

23. Arrange in the correct order of stability (decreasing order) for the following molecules : [Online April 22, 2013]

(I) 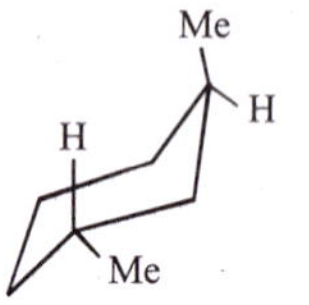

(II) 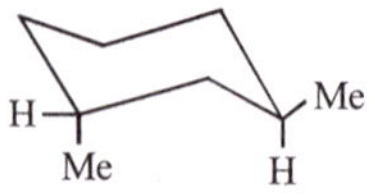

(III) 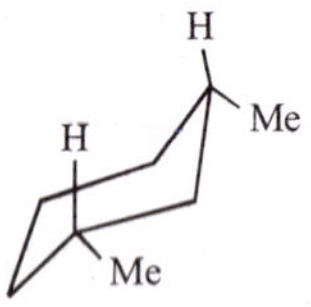

(IV) 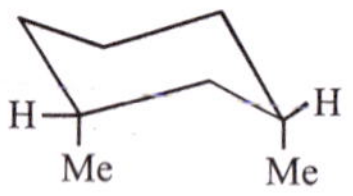

(1) (I) > (II) > (III) > (IV) (2) (IV) > (III) > (II) > (I)
(3) (I) > (II) > (III) > (IV) (4) (III) > (I) > (II) > (IV)

24. Which branched chain isomer of the hydrocarbon with molecular mass 72u gives only one isomer of mono substituted alkyl halide ? [2012; Same 2003]
(1) Tertiary butyl chloride (2) Neopentane
(3) Isohexane (4) Neohexane

25. Which one of the following conformations of cyclohexane is chiral? [2007]
(1) Boat (2) Twist boat
(3) Rigid (4) Chair

26. Increasing order of stability among the three main conformations (i.e., Eclipse, Anti, Gauche) of 2-fluoroethanol is [2006]
(1) Eclipse, Anti, Gauche (2) Anti, Gauche, Eclipse
(3) Eclipse, Gauche, Anti (4) Gauche, Eclipse, Anti

27. Of the five isomeric hexanes, the isomer which can give two monochlorinated compounds is [2005]
(1) 2-methylpentane (2) 2, 2-dimethylbutane
(3) 2, 3-dimethylbutane (4) n-hexane

28. On mixing a certain alkane with chlorine and irradiating it with ultraviolet light, it forms only one monochloroalkane. This alkane could be [2003]
(1) pentane (2) isopentane
(3) neopentane (4) propane

NUMERICAL VALUE QUESTIONS (NVQs)

29. An alkane (Y) requires 8 moles of oxygen for complete combustion and on chlorination with $Cl_2/h\nu$, (Y) gives only one monochlorinated product (Z). The total number of primary carbon atoms in (Y) is______. [April 06, 2026 (II)]

30. RMgI when treated with ice cold water liberated a gas which occupied 1.4 dm^3/g at STP. The gas produced is further reacted with iodine in presence of HIO_3 to give compound (X). Compound (X) in presence of Na and dry ether produced compound (Y). Molar mass of compound (Y) is _______g mol^{-1}. (Nearest integer) [April 05, 2026 (II)]

31. Consider the isomers of hydrocarbon with molecular formula C_5H_{10}. These isomers do not decolourise $KMnO_4$ solution. These isomers are subjected to chlorination with chlorine in presence of light to give monochloro compounds. The total number of monochloro compounds (structural isomers only) formed is _____. [April 02, 2026 (I)]

32. The number of different chain isomers for C_7H_{16} is _____. [April 4, 2024 (I)]

33. Number of alkanes obtained on electrolysis of a mixture of CH_3COONa and C_2H_5COONa is _____. [Jan. 31, 2024 (I)]

34. Number of isomeric products formed by monochlorination of 2-methylbutane in presence of sunlight is _______. [Jan. 31, 2024 (II)]

35. Number of bromo derivatives obtained on treating ethane with excess of Br_2, in diffused sunlight is _____. [April 06, 2023 (I)]

36. Maximum number of isomeric monochloro derivatives which can be obtained from 2,2,5,5- tetramethylhexane by chlorination is __________ [Jan. 24, 2023 (II)]

37. Total number of isomers (including stereoisomers) obtained on monochlorination of methylcyclohexane is ________. [July 26, 2022 (II)]

38. The total number of monobromo derivatives formed by the alkanes with molecular formula C_5H_{12} is (excluding stereoisomers) ______. [July 25, 2022 (II)]

39. The dihedral angle in staggered form of Newman projection of 1,1,1-Trichloroethane is .... degree. (Round off to the nearest integer) [July 27, 2021 (II)]

Topic 2 - Alkenes

MULTIPLE CHOICE QUESTIONS (MCQs)

40. An alkene (X) on ozonolysis followed by reduction gives following products.

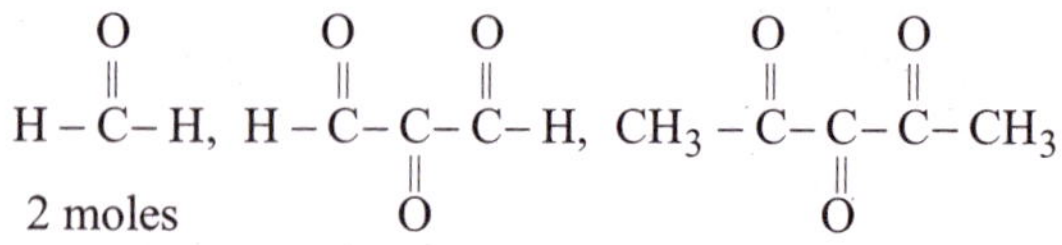

The alkene (X) is : [April 04, 2026 (I)]

(1) 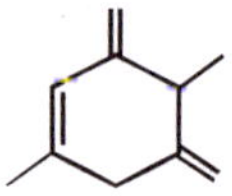
(2) 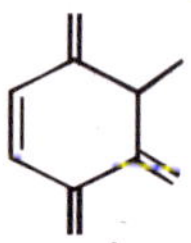
(3) 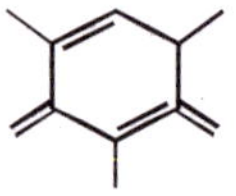
(4) 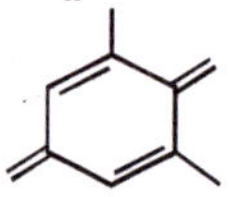

41. $Ph-CH=CH_2 \xrightarrow[HBr]{(PhCOO)_2}$ Product
Consider the above reaction [Jan. 28, 2026 (I)]
A. The reaction proceeds through a more stable radical intermediate.

B. The role of peroxide is to generate $\dot{H}$ (hydrogen radical).
C. During this reaction, benzene is formed as a by-product.
D. 1-Bromo-2-phenylethane is formed as the minor product.
E. The same reaction in absence of peroxide proceeds via carbocation intermediate.

Identify the correct statements. Choose the **correct** answer from the options given below:
(1) A & E Only (2) A, B & D Only
(3) C, D & E Only (4) A, C & E Only

42. Arrange the following alkenes in decreasing order of stability. [Jan. 24, 2026 (I)]

I.  II. 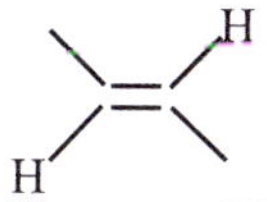

III. 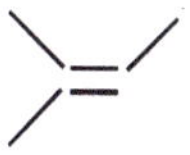 IV. 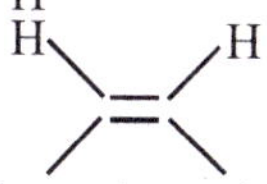

Choose the **correct** answer from the options given below:
(1) III > I > II > IV (2) III > II > I > IV
(3) I > III > II > IV (4) I > III > IV > II

43. 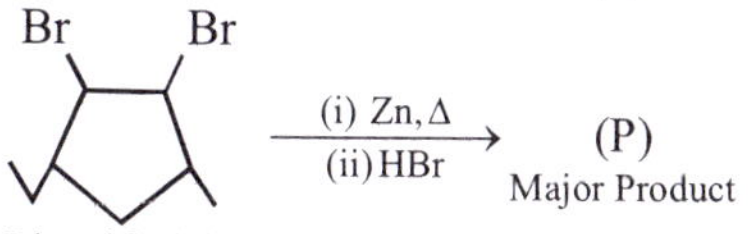

Identify (P) [Jan. 23, 2026 (II)]

(1) 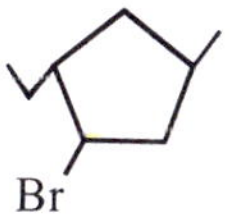

(2) 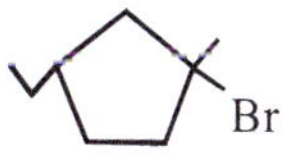

(3) 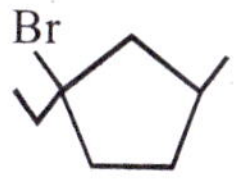

(4) 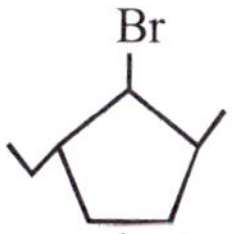

44. The number of optically active products obtained from the complete ozonolysis of the given compound is : [April 7, 2025 (II)]

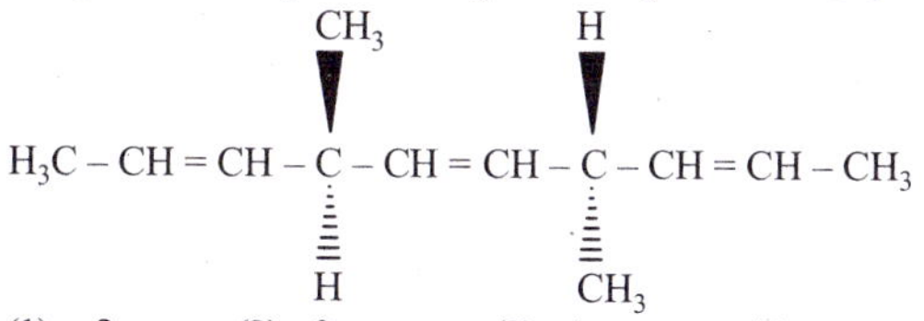

(1) 2 (2) 0 (3) 1 (4) 4

45. The reactions which cannot be applied to prepare an alkene by elimination, are [April 7, 2025 (I)]

A. 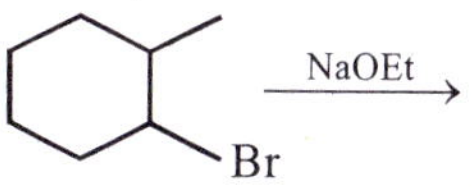

B. $CH_3-CH_2-\underset{Br}{\underset{|}{CH}}-CH_3 \xrightarrow{\text{KOH (aq.)}}$

C. $CH_3-\overset{CH_3}{\overset{|}{\underset{CH_3}{\underset{|}{C}}}}-Br \xrightarrow{\text{NaOMe}}$

D. 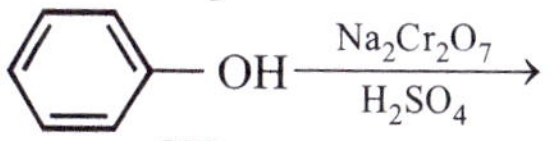

E. $CH_3-\overset{CH_3}{\overset{|}{\underset{CH_3}{\underset{|}{C}}}}-OH \xrightarrow[573\text{ K}]{Cu}$

Choose the **correct** answer from the option given below :
(1) B & E Only (2) B, C & D Only
(3) A, C & D Only (4) B & D Only

46. Given below are two statements : [April 7, 2025 (I)]

Statement I : Ozonolysis followed by treatment with Zn, H_2O of cis-2-butene gives ethanal.

Statement II : The production obtained by ozonolysis followed by treatment with Zn, H_2O of 3, 6-dimethyloct-4-ene has no chiral carbon atom. In the light of the above statements, choose the **correct** answer from the options given below

(1) Both Statement I and Statement II are true
(2) Statement I is false but Statement II are true
(3) Statement I is true but Statement II is false
(4) Both Statement I and Statement II are false

47. Which compound would give 3-methyl-6- oxoheptanal upon ozonolysis ? [April 3, 2025 (I)]

(1) 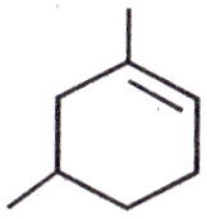 (2) 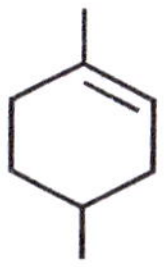

(3) 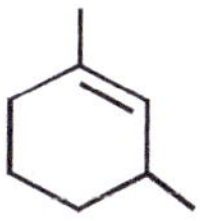 (4) 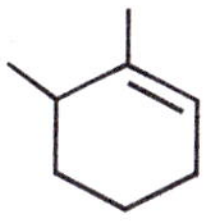

48. A molecule ("P") on treatment with acid undergoes rearrangement and gives ("Q") ("Q") on ozonolysis followed by reflux under alkaline condition gives ("R"). The structure of ("R") is given below : [Jan. 28, 2025 (I)]

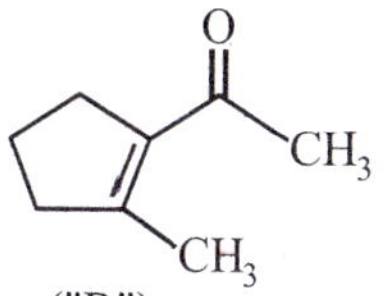

("R")

The structure of ("P") is

(1) 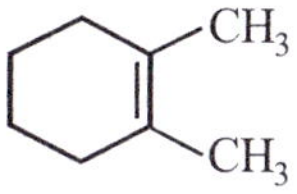

(2) 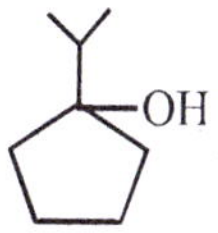

(3) 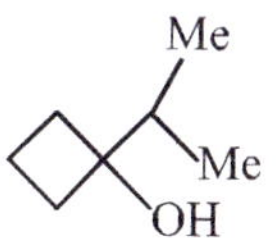

(4) 

49. Following are the four molecules "P", "Q", "R" and "S". Which one among the four molecules will react with H-Br(aq) at the fastest rate? [Jan. 24, 2025 (I)]

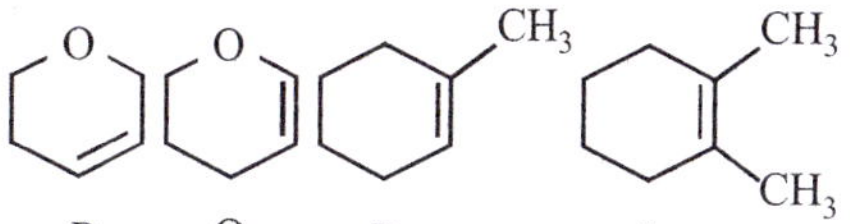

P Q R S

(1) S (2) Q (3) R (4) P

50. Match List-I with List-II [Jan. 23, 2025 (II)]

List-I
(Isomers of $C_{10}H_{14}$)

(A) 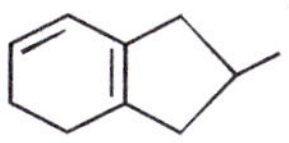 (B) 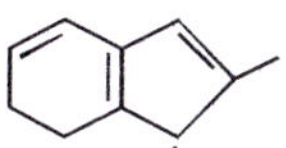

(C) 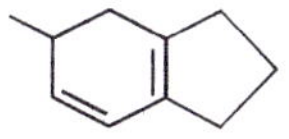 (D) 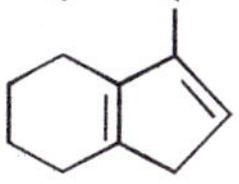

List-II
(Ozonolysis product)

(I) 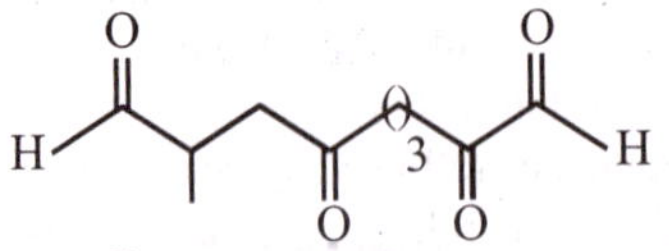

(II) 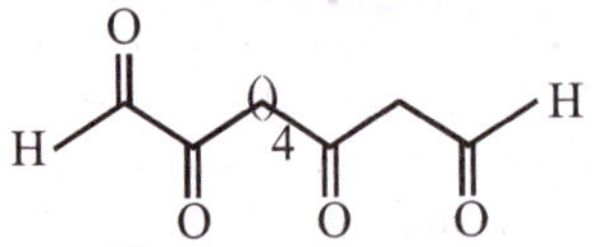

(III) 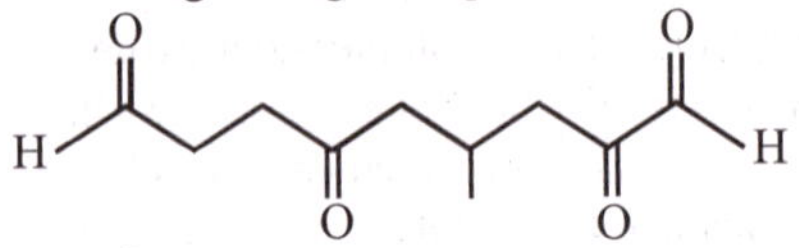

(IV) 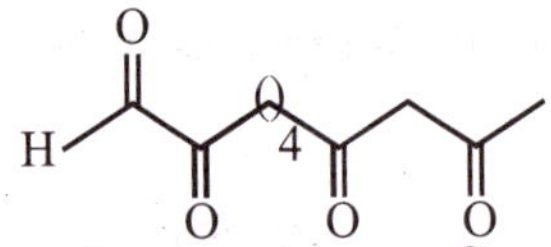

Choose the correct answer from the options given
(1) (A)-(II), (B)-(III), (C)-(I), (D)-(IV)
(2) (A)-(III), (B)-(IV), (C)-(I), (D)-(II)
(3) (A)-(III), (B)-(II), (C)-(I), (D)-(IV)
(4) (A)-(I), (B)-(IV), (C)-(III), (D)-(II)

51. The incorrect statements regarding geometrical isomerism are: [Jan. 22, 2025 (I)]
(A) Propene shows geometrical isomerism.
(B) Trans isomer has identical atoms/group on the opposite sides of the double bond.
(C) Cis-but-2-ene has higher dipole moment than trans-but-2-ene.
(D) 2-methylbut-2-ene shows two geometrical isomers.
(E) Trans-isomer has lower melting point that cis isomer.
Choose the correct answer from the options given below:
(1) (A), (D) and (E) only (2) (C), (D) and (E) only
(3) (B) and (C) only (4) (A) and (E) only

52. Number of σ and π bonds present in ethylene molecule is respectively : [April 5, 2024 (I)]
(1) 3 and 1 (2) 5 and 2 (3) 4 and 1 (4) 5 and 1

53. Given below are two statements : one is labelled as Assertion (A) and the other is labelled as Reason (R). [April 5, 2024 (I)]
Assertion (A): Cis form of alkene is found to be more polar than the trans form
Reason (R): Dipole moment of trans isomer of 2-butene is zero.
In the light of the above statements, choose the correct answer from the options given below :
(1) Both (A) and (R) are true but (R) is NOT the correct explanation of (A)
(2) (A) is true but (R) is false
(3) Both (A) and (R) are true and (R) is the correct explanation of (A)
(4) (A) is false but (R) is true

54. 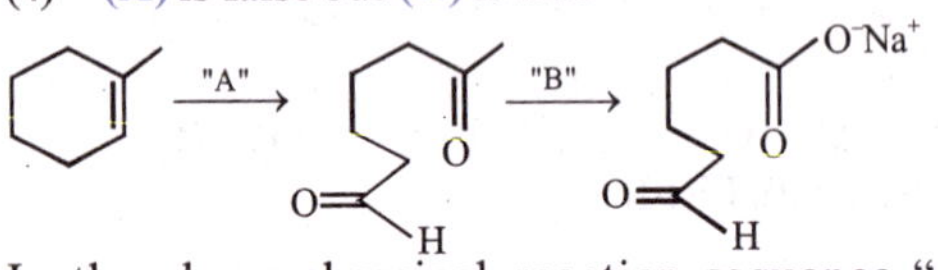

In the above chemical reaction sequence "A" and "B" respectively are : [April 4, 2024 (II)]
(1) O_3, Zn/H_2O and $NaOH(alc.)/I_2$
(2) H_2O, H^+ and $NaOH(alc.)/I_2$
(3) H_2O, H^+ and $KMnO_4$
(4) O_3, Zn/H_2O and $KMnO_4$

55. Major product of the following reaction is – [Jan. 31, 2024 (II)]

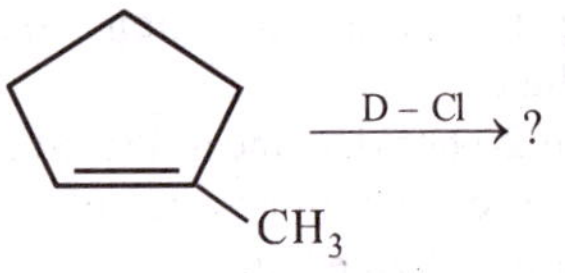

(1) 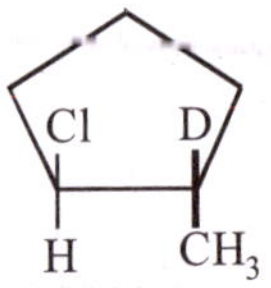
(2) 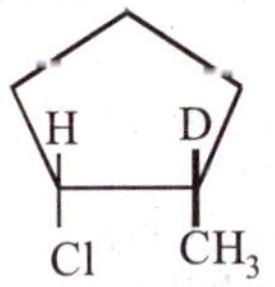
(3) 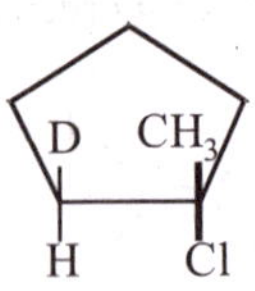
(4) 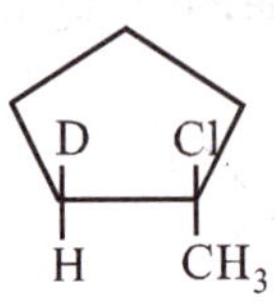

56. Products A and B formed in the following set of reactions are [Jan. 30, 2024 (II)]

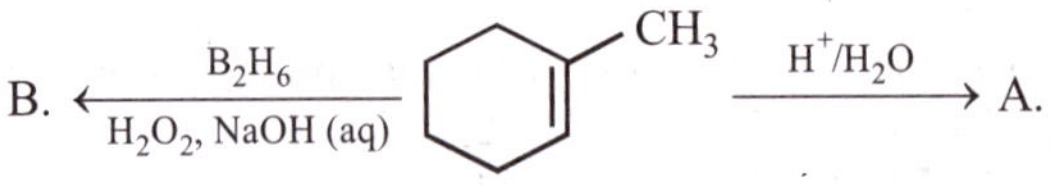

(1) A = 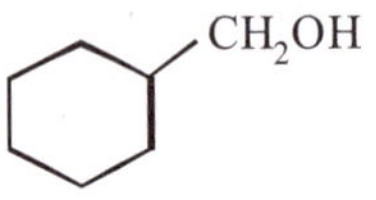 B = 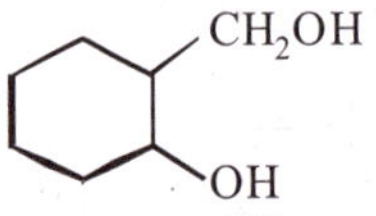
(2) A = 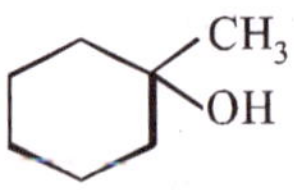 B = 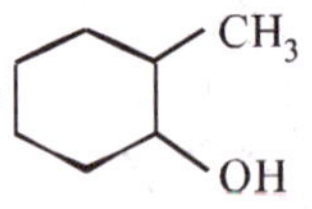
(3) A = 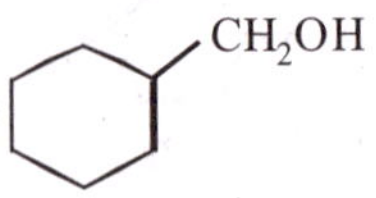 B = 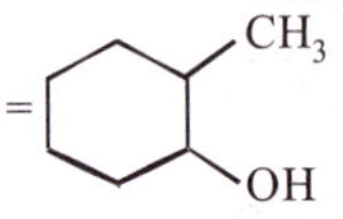
(4) A = 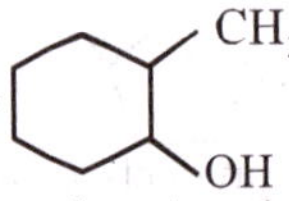 B = 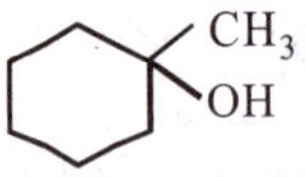

57. Identify product A and product B : [Jan. 29, 2024 (I)]

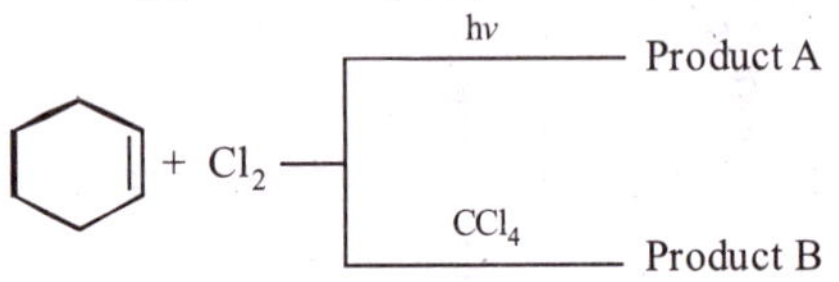

(1) A : 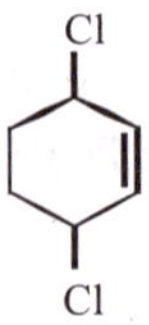 B : 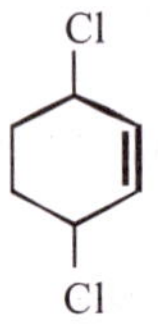
(2) A : 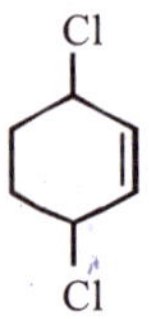 B : 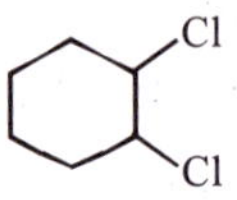
(3) A : 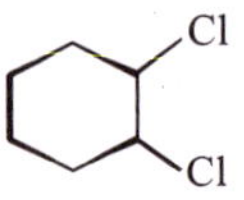 B : 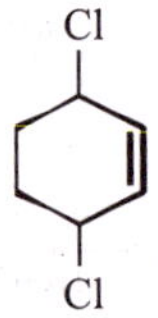
(4) A : 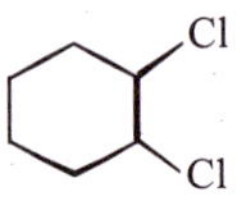 B : 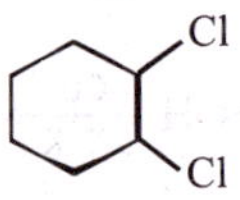

58. Which one of the following will show geometrical isomerism ? [Jan. 29, 2024 (II)]

(1) 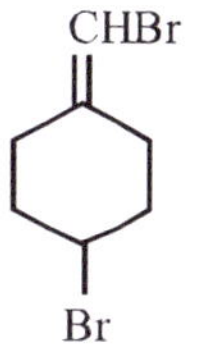 (2) 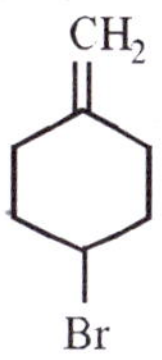

(3) 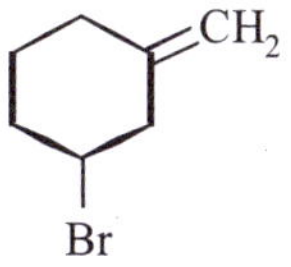 (4) 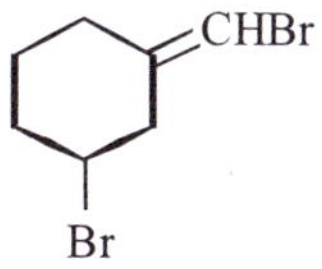

59. The product formed in the following multistep reaction is:

$CH_3-CH-CH_2 \xrightarrow[\text{(iii) PCC} \quad \text{(iv) } CH_3MgBr]{\text{(i) } B_2H_6 \quad \text{(ii) } H_2O_2,\ NaOH}$ [April 15, 2023 (I)]

(1) $CH_3-CH_2-CH(OH)-CH_3$

(2) $CH_3-CH_2-CH_2-CH_2-OH$

(3) $CH_3-CH_2-C(=O)-OCH_3$

(4) $CH_3-C(OH)(CH_3)-CH_3$

60. $2-\text{Hexene} \xrightarrow[\text{(ii) } H_2O]{\text{(i) } O_3} \text{Products}$ [April 12, 2023 (I)]

The two products formed in above reaction are -

(1) Butanoic acid and acetic acid
(2) Butanal and acetic acid
(3) Butanal and acetaldehyde
(4) Butanoic acid and acetaldehyde

61. Choose the correct set of reagents for the following conversion [Jan. 31, 2023 (I)]

trans $(Ph-CH=CH-CH_3) \rightarrow$ cis $(Ph-CH=CH-CH_3)$

(1) Br_2, alc KOH, $NaNH_2$, Na(Liq NH_3)
(2) Br_2, alc KOH, $NaNH_2$, H_2 Lindlar Catalyst
(3) Br_2, aq KOH, $NaNH_2$, H_2 Lindlar Catalyst
(4) Br_2, aq KOH, $NaNH_2$, Na(Liq NH_3)

62. A hydrocarbon 'X' with formula C_6H_8 uses two moles of H_2 on catalytic hydrogenation of its one mole. On ozonolysis, 'X' yields two moles of methane dicarbaldehyde. The hydrocarbon 'X' is : [Jan. 31, 2023 (II)]

(1) hexa-1, 3, 5-triene
(2) 1-methylcyclopenta-1, 4-diene
(3) cyclohexa-1, 3-diene
(4) cyclohexa-1, 4-diene

63. The major products 'A' and 'B', respectively, are

$'A' \xleftarrow[H_2SO_4]{\text{Cold}} H_3C-C(CH_3)=CH_2 \xrightarrow[80°C]{H_2SO_4} 'B'$

[Jan. 30, 2023 (I)]

(1) $H_3C-C(CH_3)(OSO_3H)-CH_3$ & $CH_3-C(CH_3)=CH-C(CH_3)_2-CH_3$

(2) $CH_3-C(CH_3)=CH-C(CH_3)_2-CH_3$ & $CH_3-C(CH_3)(OSO_3H)-CH_3$

(3) $CH_3-CH(CH_3)=CH_2CH_2-CH(CH_3)-CH_3$ & $H_3C-C(CH_3)(OSO_3H)-CH_3$

(4) $H_3C-C(CH_3)(OSO_3H)-CH_3$ & $CH_3-CH(CH_3)-CH_2CH_2-HC(CH_3)-CH_3$

64. The one giving maximum number of isomeric alkenes on dehydrohalogenation reaction is (excluding rearrangement) [Jan. 29, 2023 (II)]

(1) 1-Bromo-2-methylbutane
(2) 2-Bromopropane
(3) 2-Bromopentane
(4) 2-Bromo-3,3-dimethylpentane

65. In the following given reaction 'A' is [Jan. 24, 2023 (I)]

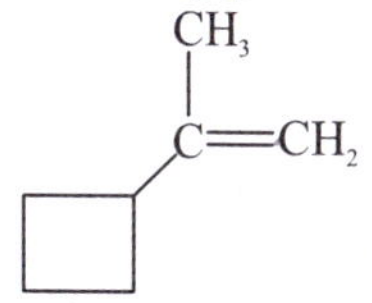 + HBr ⟶ 'A' major product

(1) 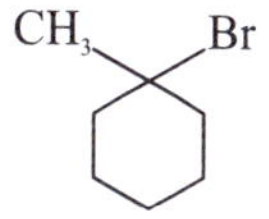 (2) 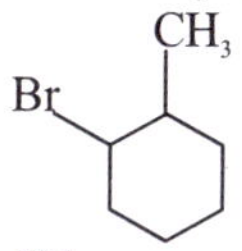

(3) 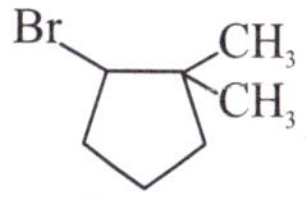 (4) 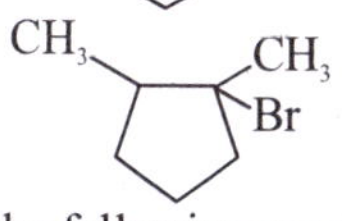

66. Choose the correct option for the following reactions. [July 28, 2022 (I)]

$B \xleftarrow[H_2O_2/OH^{\ominus}]{(BH_3)_2} H_3C-C(CH_3)_2-CH=CH_2 \xrightarrow[NaBH_4]{Hg(OAc)_2,\ H_2O} A$

(1) 'A' and 'B' are both Markovnikov addition products.
(2) 'A' is Markovnikov product and 'B' is anti- Markovnikov product.
(3) 'A' and 'B' are both anti-Markovnikov products.
(4) 'B' is Markovnikov and 'A' is anti-Markovnikov product.

67. Major product 'B' of the following reaction sequence is: [July 27, 2022 (II)]

$CH_3-C(CH_3)=CH-CH_3 \xrightarrow[CH_3OH]{Br_2} A \xrightarrow{HI} B$ (major product)

(1) $CH_3-C(HO)(CH_3)-CH(Br)-CH_3$ (2) $CH_3-C(I)(CH_3)-CH(Br)-CH_3$

(3) $CH_3-C(Br)(CH_3)-CH(OH)-CH_3$ (4) $CH_3-C(Br)(CH_3)-CH(I)-CH_3$

68. A compound 'A' on reaction with 'X' and 'Y' produces the same major product but different byproduct 'a' and 'b'. Oxidation of 'a' gives a substance produced by ants. **[July 25, 2022 (I)]**

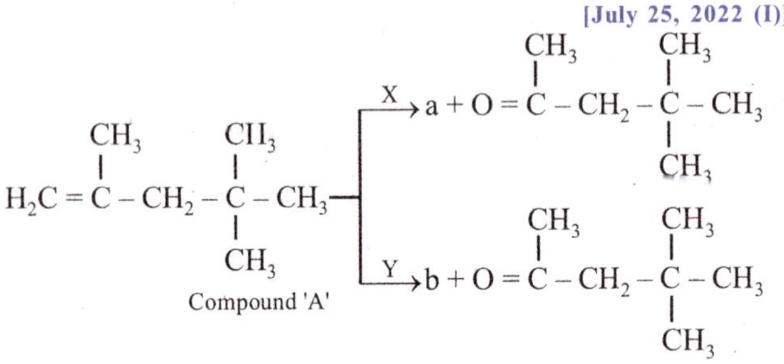

'X' and 'Y' respectively are
(1) $KMnO_4/H^+$ and dil. $KMnO_4$, 273 K
(2) $KMnO_4$(dilute), 273 K and $KMnO_4/H^+$
(3) $KMnO_4/H^+$ and O_3, H_2O/Zn
(4) O_3, H_2O/Zn and $KMnO_4/H^+$

69. Two isomers 'A' and 'B' with molecular formula C_4H_8 give different products on oxidation with $KMnO_4$ in acidic medium. Isomer 'A' on reaction with $KMnO_4/H^+$ results in effervescence of a gas and gives ketone. The compound 'A' is **[June 29, 2022 (I)]**
(1) But-1-ene (2) *cis*-But-2-ene
(3) *trans*-But-2ene (4) 2-methyl propene

70. 'A' and 'B' respectively are:

$$A \xrightarrow[(2) Zn-H_2O]{(1) O_3} \text{Ethane}-1,2-\text{dicarbaldehyde} + \text{Glyoxal/Oxaldehyde}$$

$$B \xrightarrow[(2) Zn-H_2O]{(1) O_3} 5-\text{Oxohexanal}$$ **[June 27, 2022 (I)]**

(1) 1-methylcyclohex-1, 3-diene & cyclopentene
(2) Cyclohex-1, 3-diene & cyclopentene
(3) 1-methylcyclohex-1,4-diene & 1-methylcyclopent-1-ene
(4) Cyclohex-1,3-diene & 1-methylcyclopent-1-ene

71. The products formed in the following reaction.

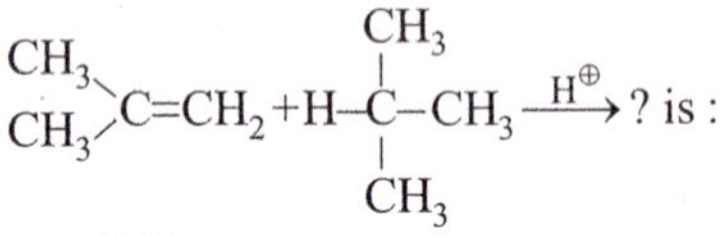

[June 25, 2022 (I)]

(1) 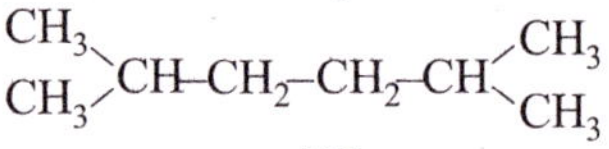
(2) 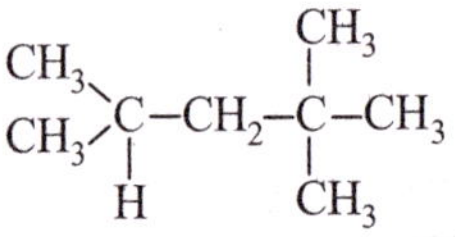
(3) 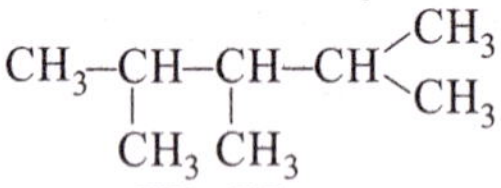
(4) 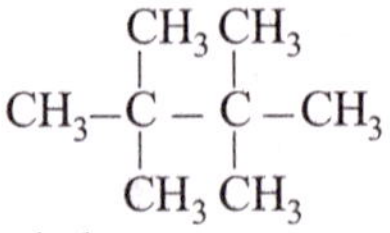

72. Given below are two statements. **[June 24, 2022 (II)]**
Statement I : The presence of weaker π - bonds make alkenes less stable than alkanes.
Statement II : The strength of the double bond is greater than that of carbon-carbon single bond.
In the light of the above statements, choose the *correct* answer from the options given below.
(1) Both Statement I and Statement II are correct.
(2) Both Statement I and Statement II are incorrect.
(3) Statement I is correct but Statement II is incorrect.
(4) Statement I is incorrect but Statement II is correct.

73. The major product formed in the following reaction is :

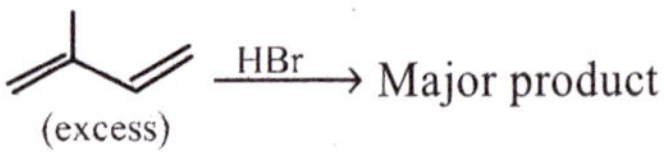

[Aug. 26, 2021 (I)]

(1) 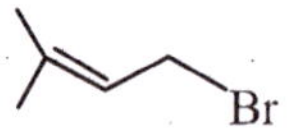
(2) 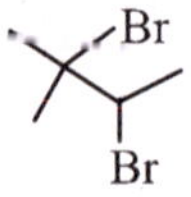
(3) 
(4) 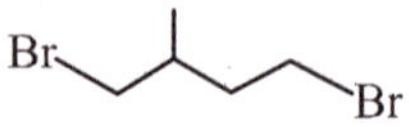

74. An organic compound 'A' C_4H_8 on treatment with $KMnO_4/H^+$ yields compound 'B' C_3H_6O.
Compound 'A' also yield compound 'B' an ozonolysis. Compound 'A' is : **[July 25, 2021 (I)]**
(1) 2-Methylpropene (2) 1-Methylcyclopropane
(3) But-2-ene (4) Cyclobutane

75. Which of the following molecules does not show stereo-isomerism ? **[July 22, 2021 (II)]**
(1) 3,4-Dimethylhex-3-ene (2) 3-Methylhex-1-ene
(3) 3-Ethylhex-3-ene (4) 4-Methylhex-1-ene

76. 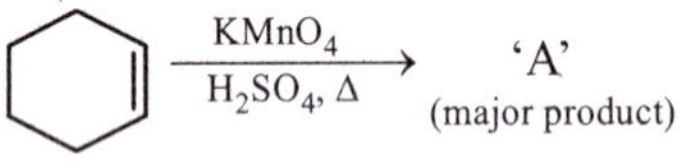

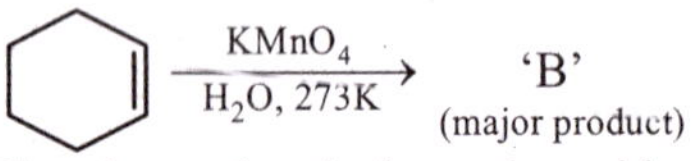

For above chemical reactions, identify the correct statement from the following: **[July 20, 2021 (I)]**
(1) Both compound 'A' and compound 'B' are dicarboxylic acids
(2) Both compound 'A' and compound 'B' are diols
(3) Compound 'A' is diol and compound 'B' is dicarboxylic acid
(4) Compound 'A' is dicarboxylic acid and compound 'B' is diol

77. 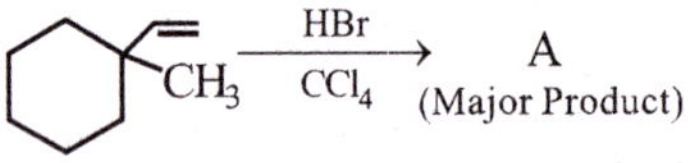 **[March 17, 2021 (I)]**

Product "A" in the above chemical reaction is

(1) 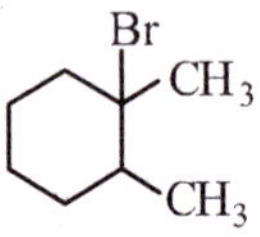
(2) 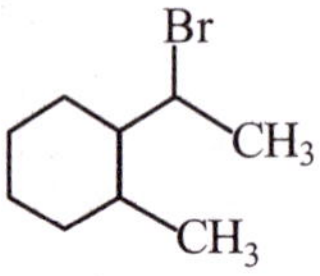
(3) 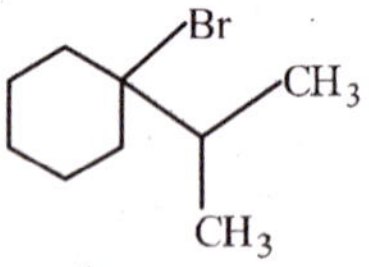
(4) 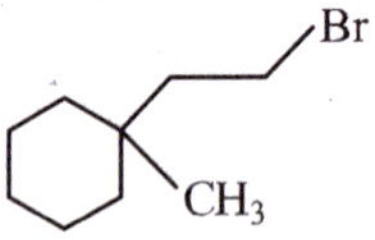

78. Choose the **correct** statement regarding the formation of carbocation A and B given **[March 17, 2021 (II)]**
$CH_3-CH_2-CH=CH_2+HBr$

$\rightarrow CH_3-CH_2-CH_2-\overset{+}{C}H_2+Br^-$ "A"

$\rightarrow CH_3-CH_2-\overset{+}{C}H-CH_3+Br^-$ "B"

(1) Carbocation A is more stable and formed relatively at slow rate
(2) Carbocation A is more stable and formed relatively at faster rate
(3) Carbocation B is more stable and formed relatively at slow rate
(4) Carbocation B is more stable and formed relatively at faster rate

79. 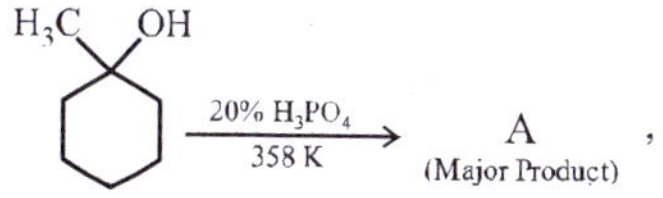

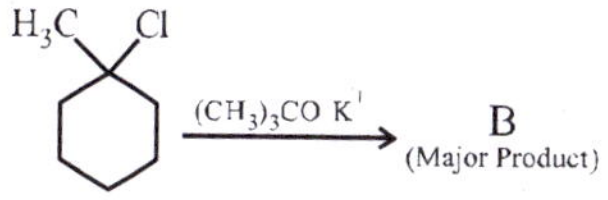

[March 16, 2021 (I)]

The products "A" and "B" formed in above reactions are

(1) 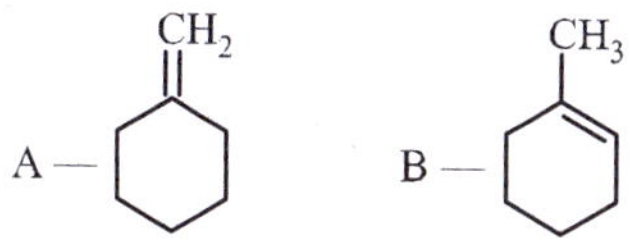

(2) 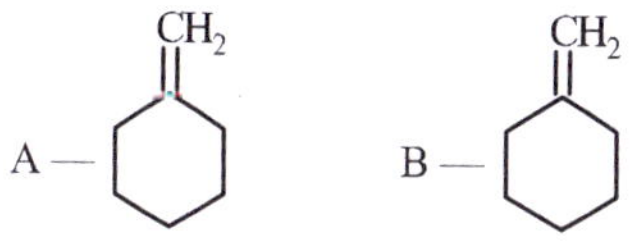

(3) 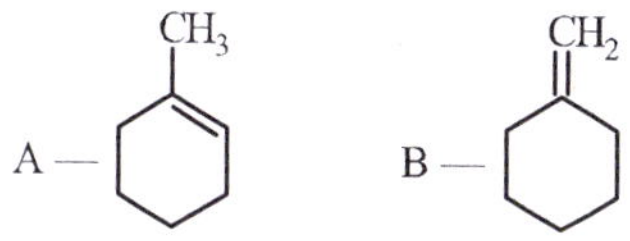

(4) 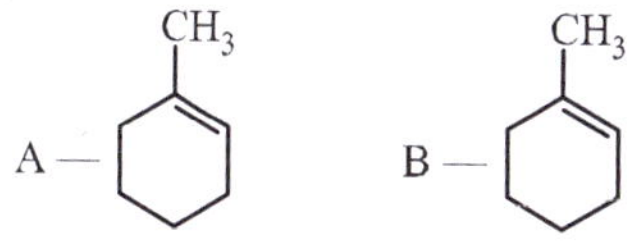

80. 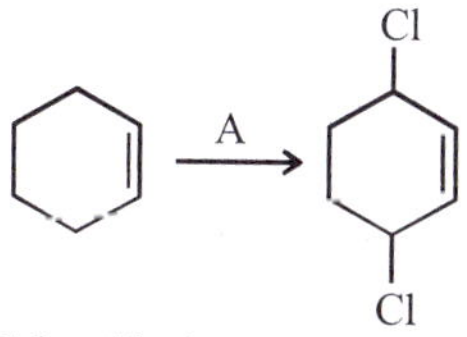

Identify the reagent(s) 'A' and condition(s) for the reaction [March 16, 2021 (II)]

(1) A = HCl ; Anhydrous $AlCl_3$
(2) A = Cl_2 ; UV light
(3) A = Cl_2 ; dark, Anhydrous $AlCl_3$
(4) A = HCl, $ZnCl_2$

81. In $\overset{1}{C}H_2 = \overset{2}{C} = \overset{3}{C}H - \overset{4}{C}H_3$ molecule, the hybridization of carbon 1, 2, 3 and 4 respectively, are : [Feb. 26, 2021 (II)]

(1) sp^2, sp^2, sp^2, sp^3 (2) sp^2, sp, sp^2, sp^3
(3) sp^3, sp, sp^3, sp^3 (4) sp^2, sp^3, sp^2, sp^3

82. The major product of the following reaction is :

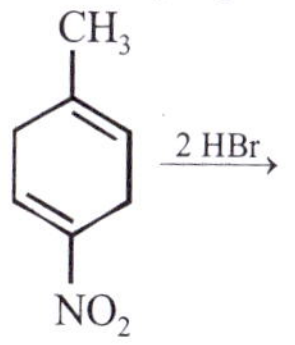

[Sep. 06, 2020 (I)]

(1) 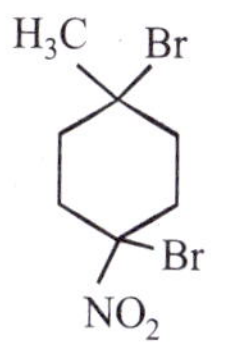 (2) 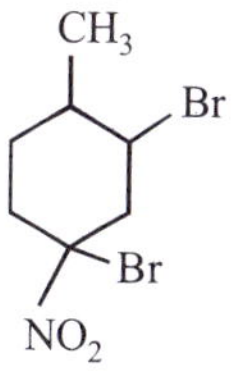

(3) 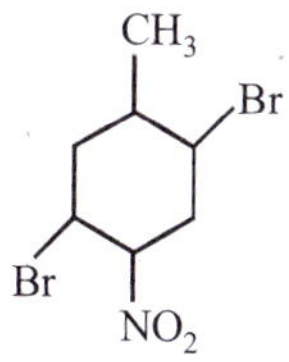 (4) 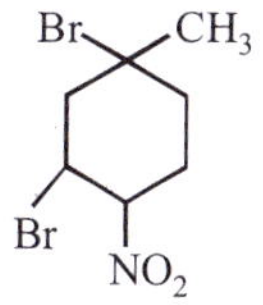

83. Which of the following compounds shows geometrical isomerism? [Sep. 06, 2020 (I)]

(1) 2-methylpent-2-ene (2) 4-methylpent-2-ene
(3) 4-methylpent-1-ene (4) 2-methylpent-1-ene

84. The increasing order of the boiling point of the major products A, B and C of the following reactions will be: [Sep. 06, 2020 (II)]

(A) 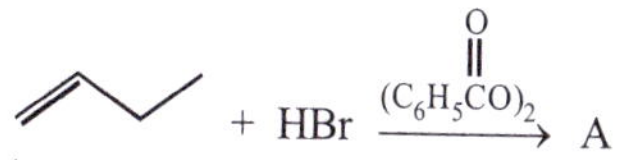

(B) 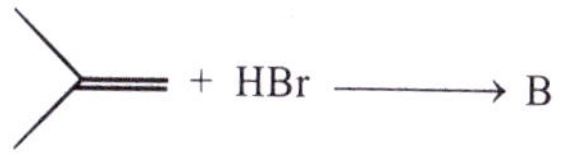

(C) 

(1) B < C < A (2) C < A < B
(3) A < B < C (4) A < C < B

85. The major product formed in the following reaction is :

$CH_3CH = CHCH(CH_3)_2 \xrightarrow{HBr}$ [Sep. 05, 2020 (II)]

(1) $CH_3CH(Br)CH_2CH(CH_3)_2$
(2) $CH_3CH_2CH(Br)CH(CH_3)_2$
(3) $Br(CH_2)_3CH(CH_3)_2$
(4) $CH_3CH_2CH_2C(Br)(CH_3)_2$

86. Among the following compounds, geometrical isomerism is exhibited by : [Sep. 05, 2020 (II)]

(1) 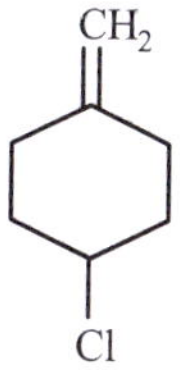 (2) 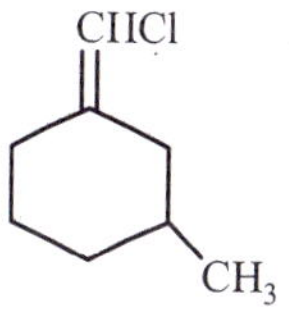

(3) 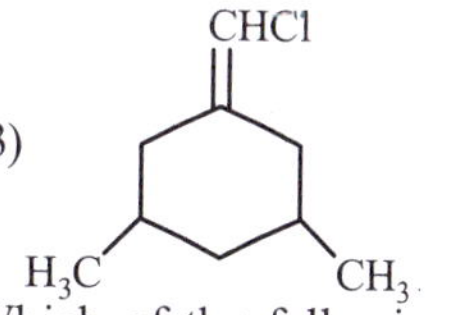 (4) 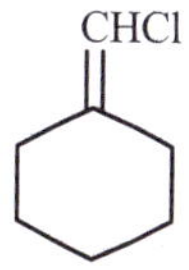

87. Which of the following compounds produces an optically inactive compound on hydrogenation? [Sep. 03, 2020 (I)]

(1) 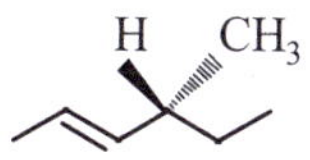 (2) 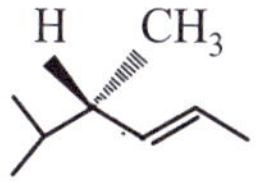

(3) 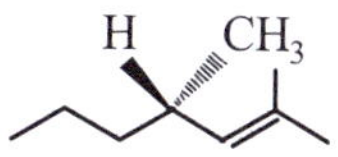 (4) 

88. The major product in the following reaction is :

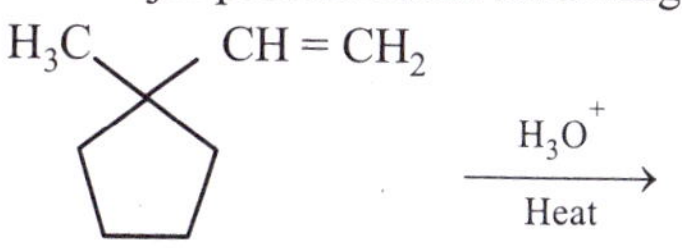

[Sep. 02, 2020 (I)]

(1) 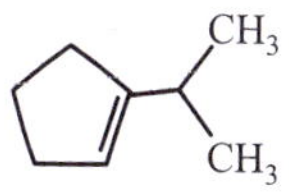 (2) 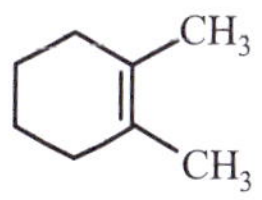

(3) 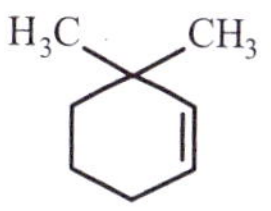 (4) 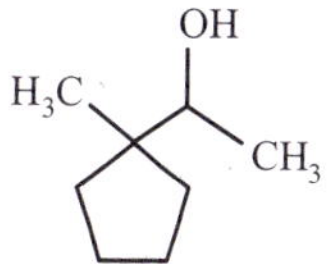

89. The correct order of heat of combustion for following alkadienes is: [Jan. 09, 2020 (I)]

(A) 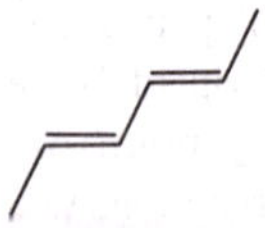

(B) 

(C) 

(1) (A) < (B) < (C) (2) (A) < (C) < (B)
(3) (C) < (B) < (A) (4) (B) < (C) < (A)

90. Which of the following has the shortest C–Cl bond? [Jan. 09, 2020 (II)]

(1) Cl–CH=CH_2 (2) Cl–CH=CH–NO_2
(3) Cl–CH=CH–CH_3 (4) Cl–CH=CH–OCH_3

91. Which of the following reactions will not produce a racemic product? [Jan. 09, 2020 (II)]

(1) $CH_3-\overset{O}{\overset{\|}{C}}CH_2CH_3 \xrightarrow{HCN}$

(2) 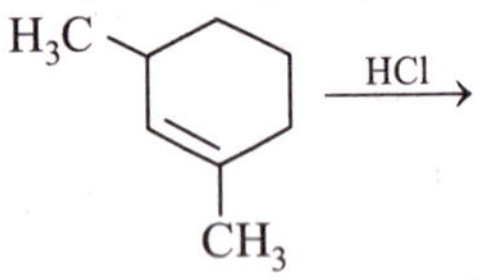

(3) $CH_3CH_2CH=CH_2 \xrightarrow{HBr}$

(4) $CH_3-\underset{H}{\overset{CH_3}{C}}-CH=CH_2 \xrightarrow{HCl}$

92. The major product [B] in the following sequence of reactions is: [Jan. 08, 2020 (II)]

$$CH_3-\underset{CH(CH_3)_2}{C}=CH-CH_2CH_3 \xrightarrow[(ii)\ H_2O_2,\ OH^{\ominus}]{(i)\ B_2H_6} [A] \xrightarrow[\Delta]{dil.\ H_2SO_4} [B]$$

(1) $CH_2=\underset{CH(CH_3)_2}{C}-CH_2CH_2CH_3$

(2) $CH_3-\underset{\underset{H_3C\ \ CH_3}{\overset{\|}{C}}}{C}-CH_2CH_2CH_3$

(3) $CH_3-\underset{CH(CH_3)_2}{C}=CH-CH_2CH_3$

(4) $CH_3-\underset{CH(CH_3)_2}{CH}-CH=CH-CH_3$

93. The major products A and B in the following reactions are: [Jan. 08, 2020 (I)]

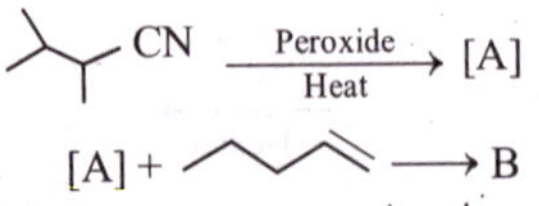

(1) 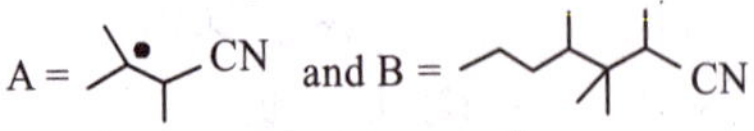

(2) 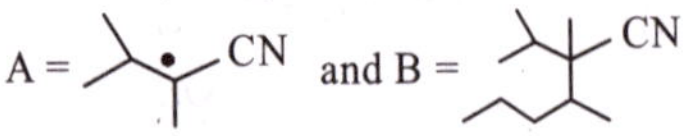

(3) 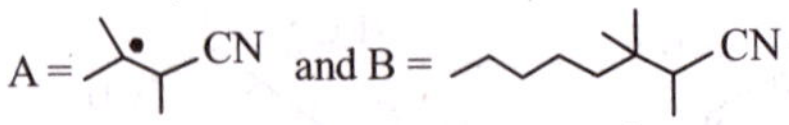

(4) 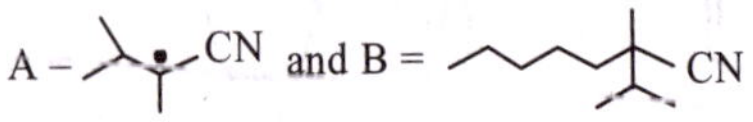

94. The major product of the following addition reaction is

$H_3C-CH=CH_2 \xrightarrow{Cl_2/H_2O}$ [April 12, 2019 (I), 2016]

(1) $H_3C-\underset{Cl}{CH}-\underset{OH}{CH_2}$ (2) $H_3C-\underset{OH}{CH}-\underset{Cl}{CH_2}$

(3) 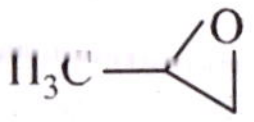 (4) 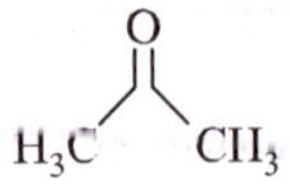

95. But-2-ene on reaction with alkaline $KMnO_4$ at elevated temperature followed by acidification will give: [April 12, 2019 (I)]

(1) $CH_3-\underset{OH}{CH}-\underset{OH}{CH}-CH_3$
(2) one molecule of CH_3CHO and one molecule of CH_3COOH
(3) 2 molecules of CH_3COOH
(4) 2 molecules of CH_3CHO

96. Which one of the following alkenes when treated with HCl yields majorly an anti Markovnikov product? [April 8, 2019 (II)]

(1) $CH_3O-CH=CH_2$ (2) $Cl-CH=CH_2$
(3) $H_2N-CH=CH_2$ (4) $F_3C-CH=CH_2$

97. The major product of the following reaction is: [Jan. 12, 2019 (II)]

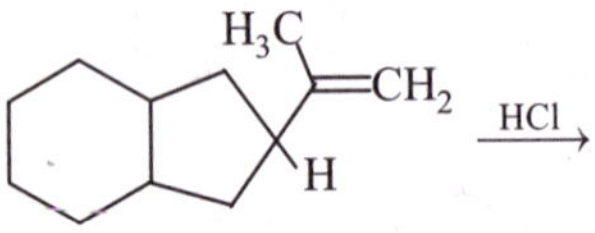

(1) 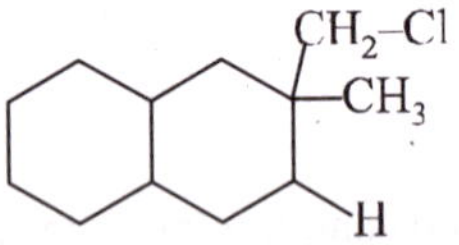

(2) 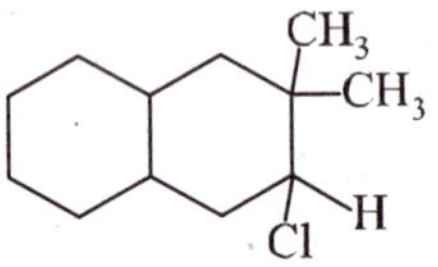

(3) 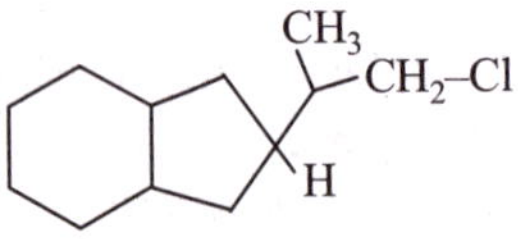

(4) 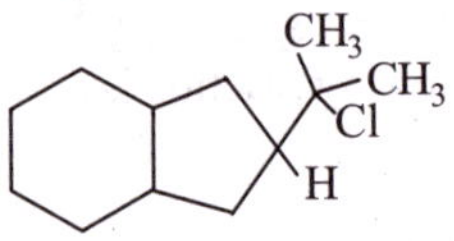

98. The major product of the following reaction is: [Online April 16, 2018]

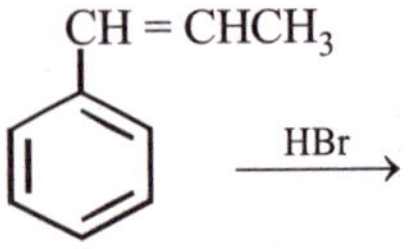

(1) 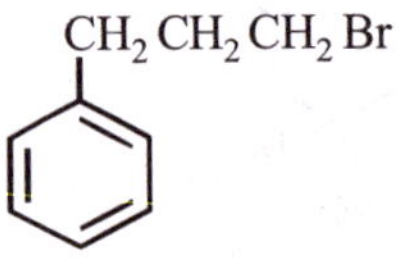 (2) 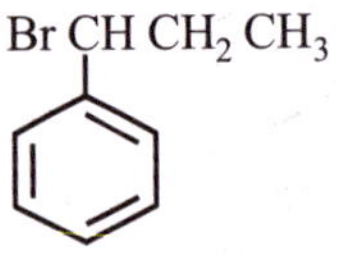

(3) 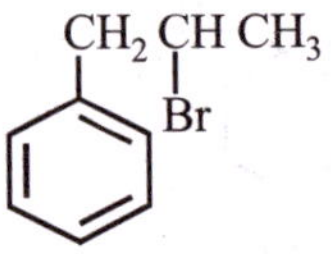 (4) 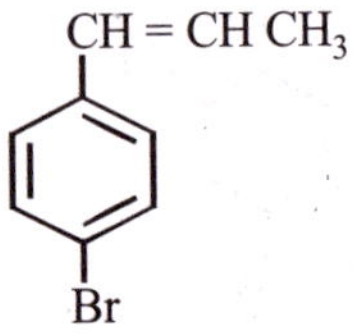

99. The *trans*-alkenes are formed by the reduction of alkynes with: [2018; Same Online April 11, 2014; 2008]
(1) H_2-Pd/C, $BaSO_4$ (2) $NaBH_4$
(3) Na/liq. NH_3 (4) Sn - HCl

100. 3-Methyl pent-2-ene on reaction with HBr in presence of peroxide forms an addition product. The number of possible stereoisomers for the product is : [2017]
(1) Six (2) Zero (3) Two (4) Four

101. Which of the following compounds is most reactive to an aqueous solution of sodium carbonate ? [Online April 10, 2016 (S), Same Online April 9, 2017]
(1) 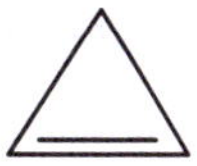 (2) 
(3)  (4) 

102. The major product of the following reaction is : [Online April 9, 2017]

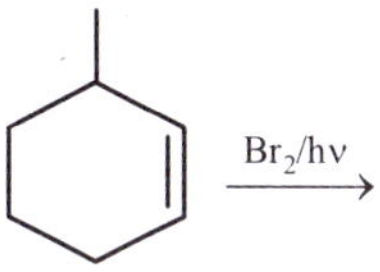

(1) 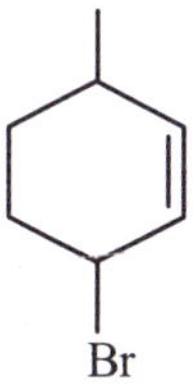

(2) 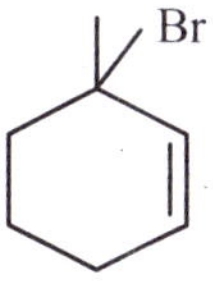

(3) 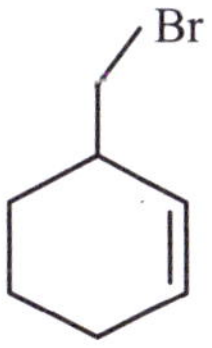

(4) 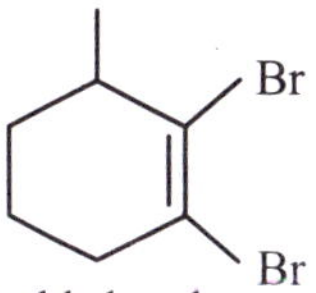

103. In the following structure, the double bonds are marked as I, II, III and IV [Online April 9, 2017]

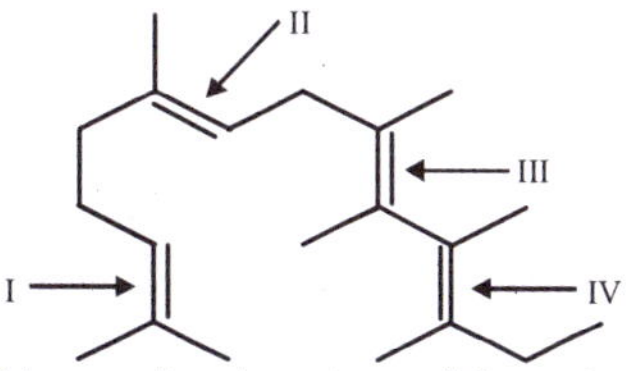

Geometrical isomerism is not possible at site (s) :
(1) III (2) I (3) I and III (4) III and IV

104. The product of the reaction given below is: [2016]

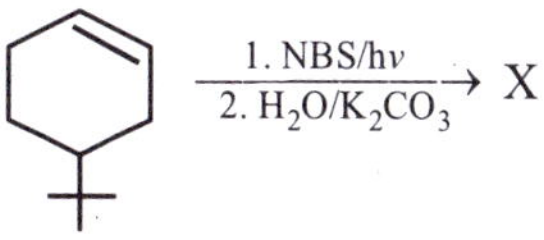

(1) 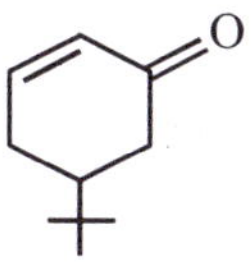

(2) 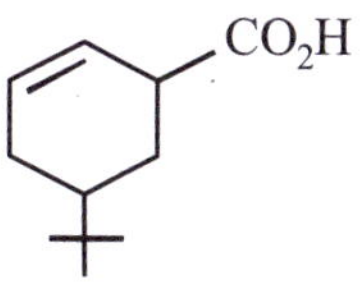

(3) 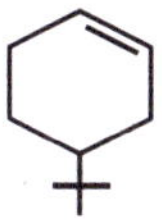
(4) 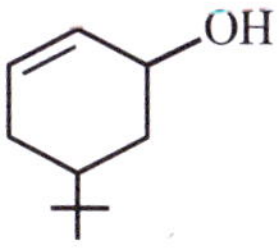

105. The reaction of propene with HOCl ($Cl_2 + H_2O$) proceeds through the intermediate: [2016]
(1) $CH_3-CH(OH)-CH_2^+$
(2) $CH_3-CHCl-CH_2^+$
(3) $CH_3-CH^+-CH_2-OH$
(4) $CH_3-CH^+-CH_2-Cl$

106. Bromination of cyclohexene under conditions given below yields : [Online April 10, 2016]

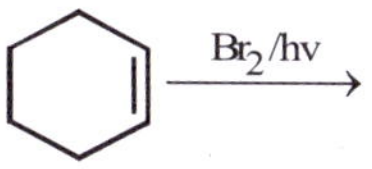

(1) 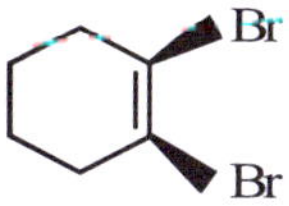

(2) 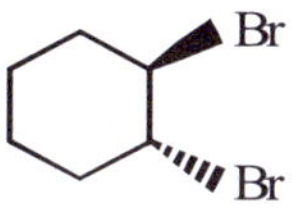

(3) 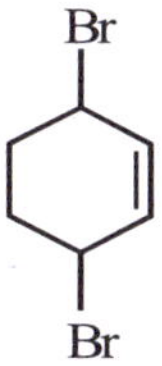

(4) 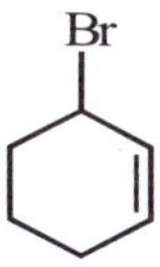

107. What is the major product expected from the following reaction ? [Online April 11, 2015]

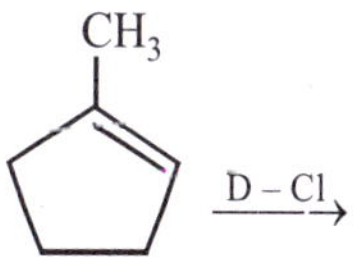

Where D is an isotope of hydrogen

(1) 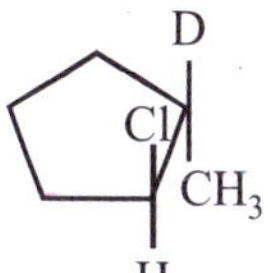

(2) 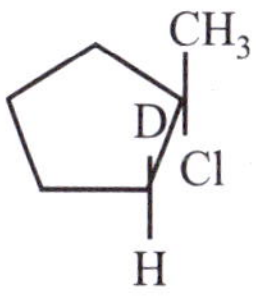

(3) 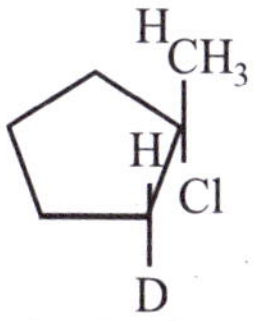

(4) 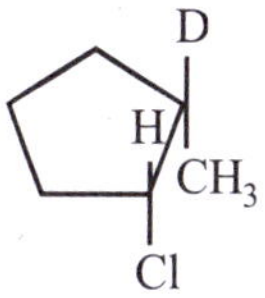

108. Which of the following compounds will exhibit geometrical isomerism? [2015; Same 2002]
(1) 2 - Phenyl -1 - butene
(2) 1, 1 - Diphenyl - 1 - propene
(3) 1 - Phenyl - 2 - butene
(4) 3 - Phenyl -1 - butene

109. The gas liberated by the electrolysis of dipotassium succinate solution is: [Online April 11, 2014]
(1) Ethane (2) Ethyne
(3) Ethene (4) Propene

110. In the hydroboration - oxidation reaction of propene with diborane, H_2O_2 and NaOH, the organic compound formed is: [Online April 9, 2014]
(1) CH_3CH_2OH (2) $CH_3CHOHCH_3$
(3) $CH_3CH_2CH_2OH$ (4) $(CH_3)_3COH$

111. 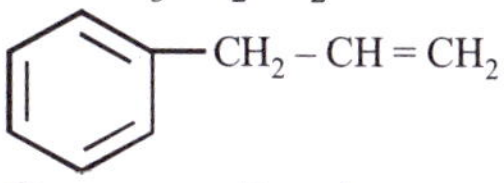

On mercuration-demercuration produces the major product: [Online April 12, 2014]

(1) 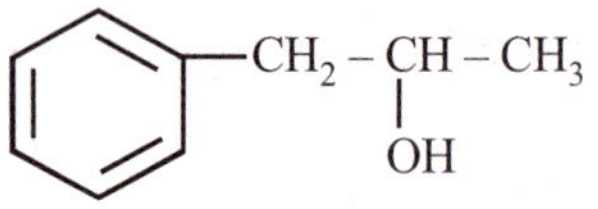

(2) 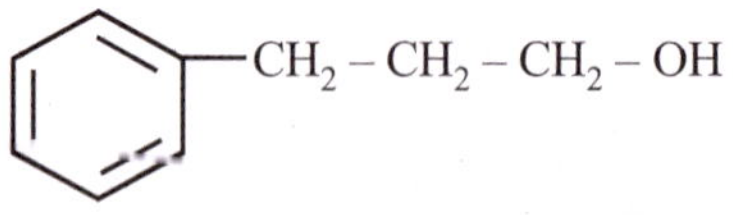

(3) 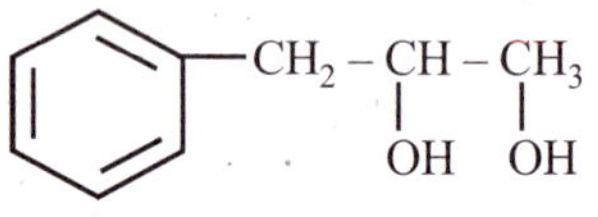

(4) 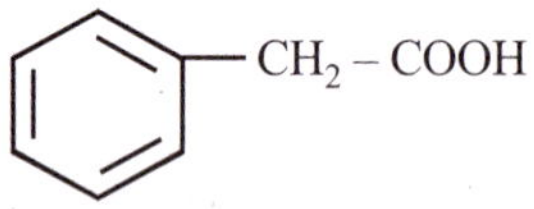

112. In the presence of peroxide, HCl and HI do not give anti-Markownikoff's addition of alkenes because: **[Online April 12, 2014]**
(1) One of the steps is endothermic in HCl and HI
(2) Both HCl and HI are strong acids
(3) HCl is oxidizing and the HI is reducing
(4) All the steps are exothermic in HCl and HI

113. The addition of HI in the presence of peroxide catalyst does not follow anti-Markovnikov's rule because :
(1) HI is a strong reducing agent. **[Online April 9, 2013]**
(2) H-I bond is too strong to be broken homolytically.
(3) I atom combines with H atom to give back HI.
(4) Iodine atom is not reactive enough to add across a double bond.

114. Consider the following sequence of reactions

$$CH_3CH=CH_2 \xrightarrow[700K]{Cl_2} A \xrightarrow[420K, 12atm]{Na_2CO_3} B \xrightarrow[(ii)\ NaOH]{(i)\ HOCl} C$$

Compound 'C' is **[Online May 26, 2012]**
(1) $CH_2OH-CHOH-CH_2OH$
(2) $CH_3CH(OH)COONa$
(3) $HOCH_2-CH=CH_2$
(4) $CH_3CH(OH)COCl$

115. Ozonolysis of an organic compound 'A' produces acetone and propionaldehyde in equimolar mixture. Identify 'A' from the following compounds: **[2011 RS, Same 2010, 2009]**
(1) 1 – Pentene
(2) 2 – Pentene
(3) 2 – Methyl – 2 – pentene
(4) 2 – Methyl – 1 – pentene

116. One mole of a symmetrical alkene on ozonolysis gives two moles of an aldehyde having a molecular mass of 44 u. The alkene is **[2010]**
(1) propene (2) 1-butene
(3) 2-butene (4) ethene

117. In the following sequence of reactions, the alkene affords the compound 'B'

$$CH_3-CH=CH-CH_3 \xrightarrow{O_3} A \xrightarrow[Zn]{H_2O} B.$$

The compound B is **[2008]**
(1) CH_3CH_2CHO (2) CH_3COCH_3
(3) $CH_3CH_2COCH_3$ (4) CH_3CHO

118. Reaction of one molecule of HBr with one molecule of 1, 3-butadiene at 40 °C gives predominantly **[2005]**
(1) 1-bromo-2-butene under kinetically controlled conditions
(2) 3-bromobutene under thermodynamically controlled conditions
(3) 1-bromo-2-butene under thermodynamically controlled conditions
(4) 3-bromobutene under kinetically controlled conditions

119. Butene-1 may be converted to butane by reaction with **[2003]**
(1) Sn – HCl (2) Zn – Hg
(3) Pd/H_2 (4) Zn – HCl

NUMERICAL VALUE QUESTIONS (NVQs)

120. A compound 'X' absorbs 2 moles of hydrogen and 'X' upon oxidation with $KMnO_4|H^+$ gives

$CH_3-\underset{\underset{O}{\|}}{C}-CH_3$, $CH_3-\underset{\underset{O}{\|}}{C}-OH$ and

$CH_3-\underset{\underset{O}{\|}}{C}-CH_2CH_2-\underset{\underset{O}{\|}}{C}-OH$

The total number of σ bonds present in the compound 'X' is ________. **[Jan. 23, 2025 (II)]**

121. The major products from the following reaction sequence are product A and product B. **[April 6, 2024 (I)]**

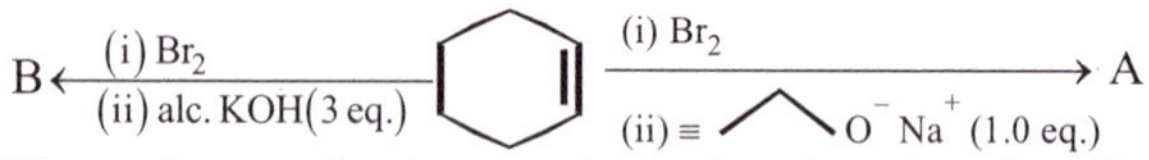

The total sum of π electrons in product A and product B are ______ (nearest integer)

122. Number of geometrical isomers possible for the given structure is / are ___________. **[Jan. 30, 2024 (II)]**

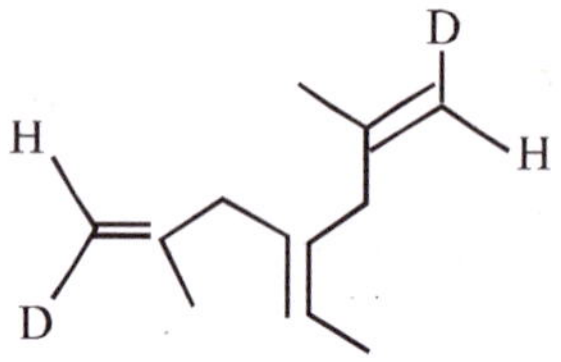

123. 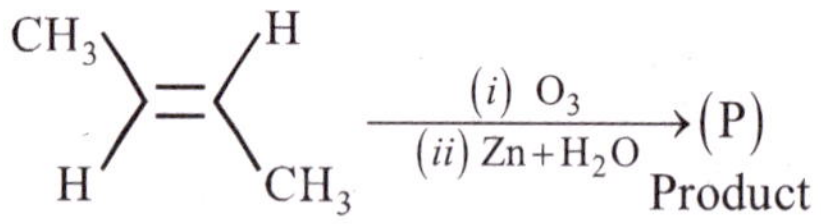

Consider the given reaction. The total number of oxygen atom/s present per molecule of the product (P) is _____. **[Jan. 29, 2024 (I)]**

124. The number of possible isomeric products formed when 3-chloro-1-butene reacts with HCl through carbocation formation is _____ **[April 11, 2023 (II)]**

125. Molar mass of the hydrocarbon (X) which on ozonolysis consumes one mole of O_3 per mole of (X) and gives one mole each of ethanal and propanone is _______ g mol^{-1}
(Molar mass of C : 12 g mol^{-1}, H : 1 g mol^{-1}) **[April 08, 2023 (I)]**

126. 17 mg of a hydrocarbon (M.F. $C_{10}H_{16}$) takes up 8.40 mL of the H_2 gas measured at 0°C and 760 mm of Hg. Ozonolysis of the same hydrocarbon yields **[Jan. 29, 2023 (I)]**

$CH_3-\underset{\underset{O}{\|}}{C}-CH_3$, $H-\underset{\underset{O}{\|}}{C}-H$, $H-\underset{\underset{O}{\|}}{C}-CH_2-CH_2-\underset{\underset{O}{\|}}{C}-\underset{\underset{O}{\|}}{C}-H$

The number of double bond/s present in the hydrocarbon is ________.

127. A hydrocarbon 'X' is found to have molar mass of 80. A 10.0 mg of compound 'X' on hydrogenation consumed 8.40 mL of H_2 gas (measured at STP). Ozonolysis compound 'X' yields only formaldehyde and dialdehyde. The total number of fragments/molecules produced from the ozonolysis of compound 'X' is ________. [June 30, 2022 (I)]

128. The major product 'A' of the following given reaction has _____ sp^2 hybridized carbon atoms. [June 24, 2022 (I)]

$$2,7-\text{Dimethyl}-2,6-\text{octadiene} \xrightarrow{H^+} \underset{\text{Major Product}}{A}$$

129. 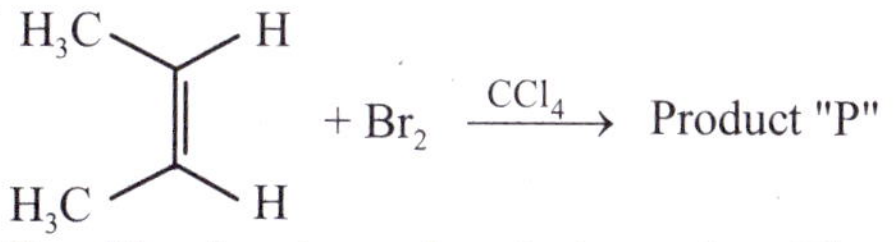

Consider the above chemical reaction. The total number of stereoisomers possible for product 'P' is [July 25, 2021 (II)]

Topic 3 - Alkynes

Multiple Choice Questions (MCQs)

130. Given below are two statements for the following reaction sequence. [Jan. 28, 2026 (I)]

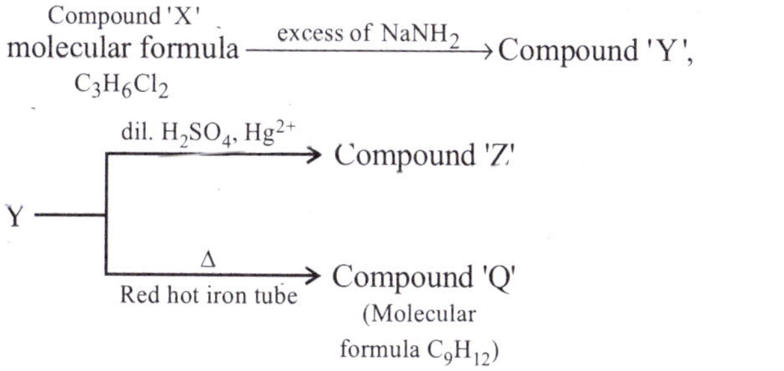

Statement I: Compound 'Z' will give yellow precipitate with NaOI.

Statement II: Compound 'Q' has two different types of 'H' atoms (aromatic : aliphatic) in the ratio 1 : 3.

In the light of the above statements, choose the **correct** answer from the option given below:

(1) Statement I is true but Statement II is false
(2) Both Statement I and Statement II are true
(3) Statement I is false but Statement II is true
(4) Both Statement I and Statement II are false

131. But-2-yne and hydrogen (one mole each) are separately treated with (i) Pd/C and (ii) Na/liq. NH_3 to give the products X and Y respectively. [Jan. 23, 2026 (I)]

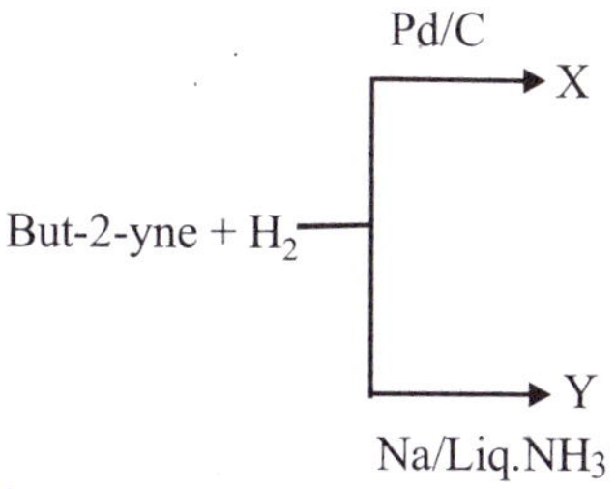

Identify the **incorrect** statements.

A. X and Y are stereoisomers.
B. Dipole moment of X is zero
C. Boiling point of X is higher than Y.
D. X and Y react with $O_3/Zn + H_2O$ to give different products.

Choose the **incorrect** answer from the options given below:

(1) B and C only (2) B and D only
(3) A and B only (4) A and C only

132. Total number of sigma (σ) ______ and pi(π) _____ bonds respectively present in hex-1-en-4-yne are : [Jan. 29, 2025 (II)]

(1) 13 and 3 (2) 11 and 3
(3) 3 and 13 (4) 14 and 3

133. Identify product [A], [B] and [C] in the following reaction sequence : [Jan. 28, 2025 (II)]

$$CH_3-C\equiv CH \xrightarrow[H_2]{Pd/C} [A] \xrightarrow[(ii)\,Zn, H_2O]{(i)\,O_3} [B]+[C]$$

(1) [A] : $CH_3-CH=CH_2$, [B] : CH_3CHO, [C] : HCHO
(2) [A] : $CH_2=CH_2$, [B] : $H_3C-\overset{O}{\overset{\|}{C}}-CH_3$ [C] : HCHO
(3) [A] : $CH_3-CH=CH_2$, [B] : CH_3CHO, [C] : CH_3CH_2OH
(4) [A] : $CH_3CH_2CH_3$, [B] : CH_3CHO, [C] : HCHO

134. The **incorrect** statement regarding ethyne is [April 9, 2024 (II)]

(1) The C–C bonds in ethyne is shorter than that in ethene
(2) Both carbons are sp hybridised
(3) Ethyne is linear
(4) The carbon-carbon bonds in ethyne is weaker than that in ethene

135. In the given reactions identify A and B [Feb. 1, 2024 (II)]

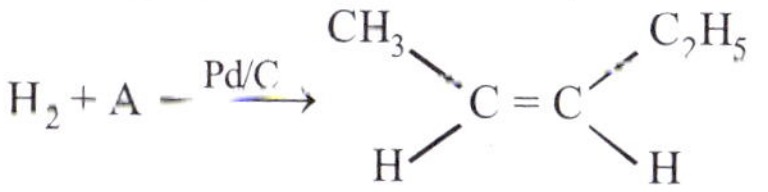

$$CH_3-C\equiv C-CH_3+H_2 \xrightarrow{Na/Liquid\ NH_3} \text{"B"}$$

(1) A : n – Pentane B : Cis – 2 – butene
(2) A : 2 – Pentyne B : Cis – 2 – butene
(3) A : n – Pentane B : trans – 2 – butene
(4) A : 2 – Pentyne B : trans – 2 – butene

136. Compound A formed in the following reaction reacts with B gives the product C. Find out A and B

$$CH_3-C\equiv CH+Na \longrightarrow A \xrightarrow{B} \underset{(C)}{CH_3-C\equiv C-CH_2-\underset{\underset{CH_3}{|}}{CH_2}}+NaBr$$

[Jan. 30, 2024 (I)]

(1) $A=CH_3-C\equiv C\overset{-}{C}\overset{+}{Na}$, $B=CH_3-CH_2-CH_2-Br$
(2) $A=CH_3-CH=CH_2$, $B=CH_3-CH_2-CH_2-Br$
(3) $A=CH_3-CH_2-CH_3$, $B=CH_3-C\equiv CH$
(4) $A=CH_3-C\equiv C\overset{-}{C}\overset{+}{Na}$; $B=CH_3-CH_2-CH_3$

137. But-2-yne is reacted separately with one mole of hydrogen as shown below: [Feb. 01, 2023 (I)]

$$\underline{B} \xleftarrow[liq\ NH_3]{Na} \underset{+H_2}{CH_3-C\equiv C-CH_3} \xrightarrow[\Delta]{Pd/C} \underline{A}$$

(A) A is more soluble than B.
(B) The boiling point & melting point of A are higher and lower than B respectively.
(C) A is more polar than B because dipole moment of A is zero.
(D) Br_2 adds easily to B than A.

Identify the incorrect statements from the options given below:

(1) A and B only (2) B, C & D only

(3) A, C & D only (4) B and C only

138. What will be the major product of following sequence of reactions? [June 27, 2022 (II)]

$n\text{-Bu}-\equiv \xrightarrow[\text{(ii) Lindlar cat, } H_2]{\text{(i) } n\text{-BuLi, } n\ C_5H_{11}Cl}$

(1) 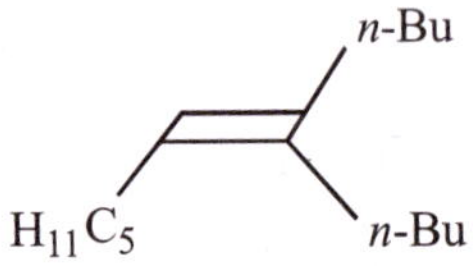

(2) 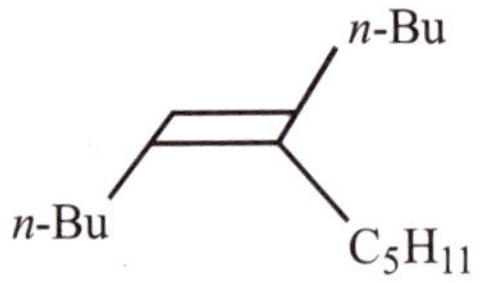

(3) 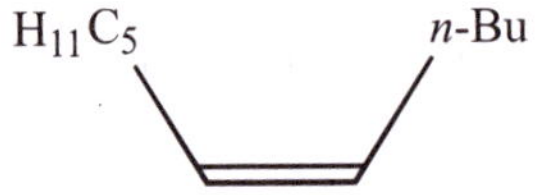

(4) 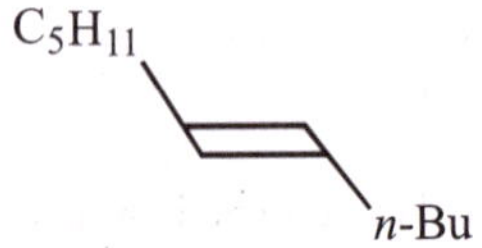

139. An unsaturated hydrocarbon X on ozonolysis gives A. Compound A when warmed with ammonical silver nitrate forms a bright silver mirror along the sides of the test tube. The unsaturated hydrocarbon X is: [March 16, 2021 (II)]

(1) 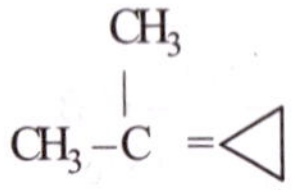 (2) $CH_3-C(CH_3)=C(CH_3)-CH_3$

(3) $HC\equiv C-CH_2-CH_3$ (4) $CH_3-C\equiv C-CH_3$

140. The major product obtained from the following reaction is: [Sep. 06, 2020 (I)]

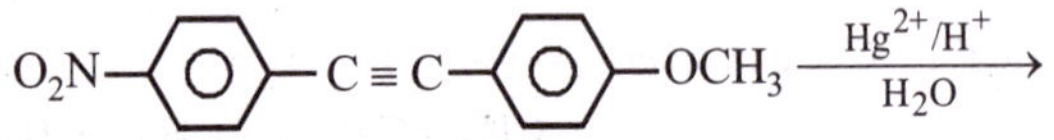

(1) 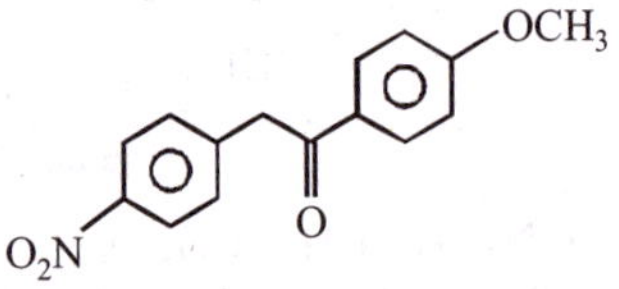

(2) 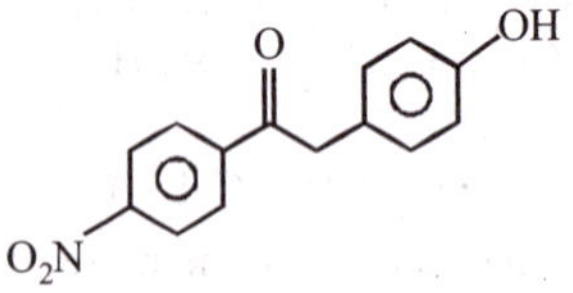

(3) 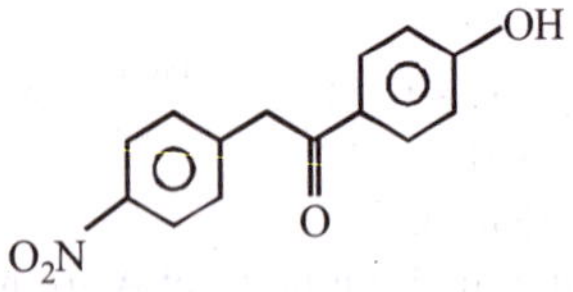

(4) 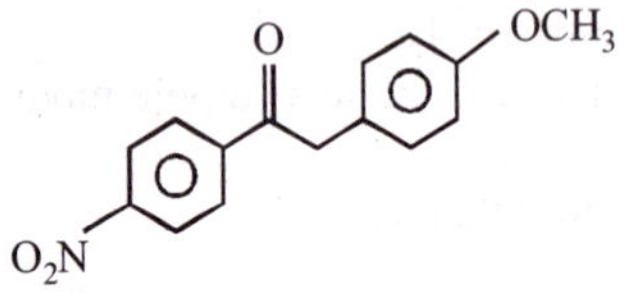

141. The major product (Y) in the following reactions is: [Jan. 09, 2020 (I)]

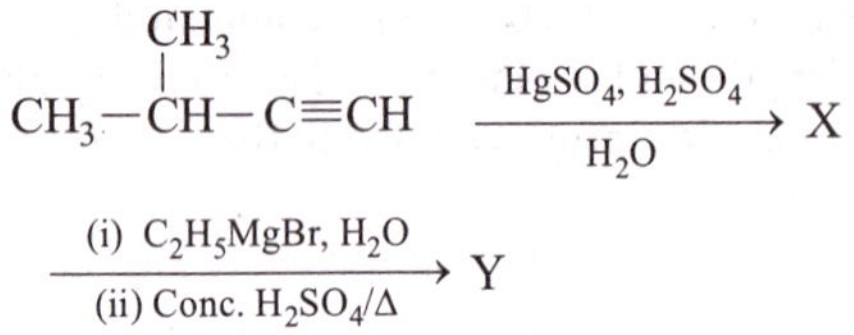

(1) 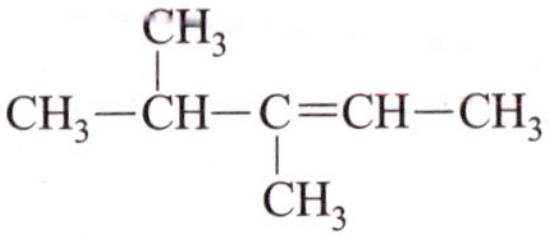

(2) 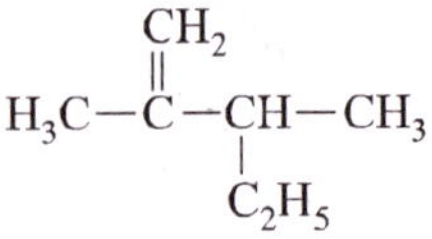

(3) 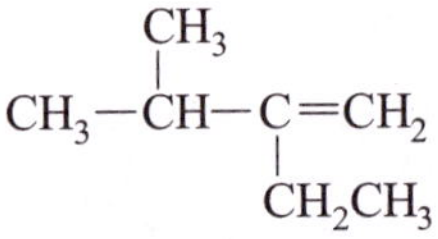

(4) 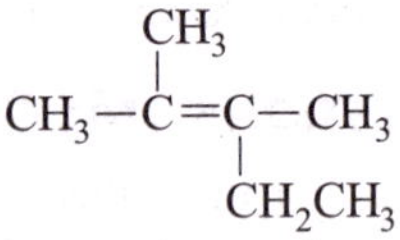

142. The increasing order of the reactivity of the following compounds towards electrophilic aromatic substitution reactions is : [April 10, 2019 (I)]

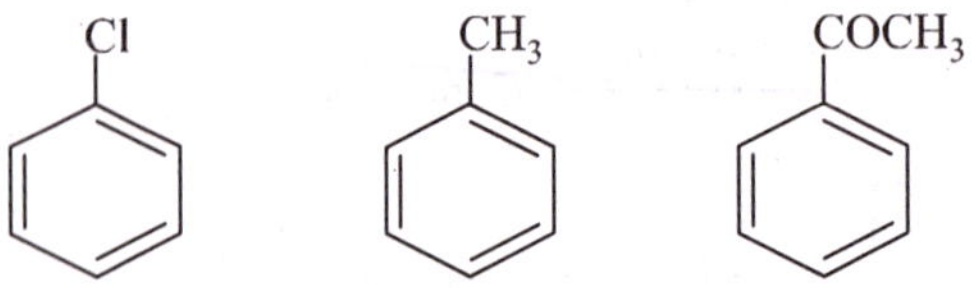

(I) (II) (III)

(1) II < I < III (2) III < II < I

(3) III < I < II (4) I < III < II

143. The major product of the following reaction is:

$CH_3C\equiv CH \xrightarrow[\text{(ii) DI}]{\text{(i) DCl (1 equiv.)}}$ [April 9, 2019 (I); Same 2007]

(1) $CH_3CD(I)CHD(Cl)$ (2) $CH_3CD(Cl)CHD(I)$

(3) $CH_3CD_2CH(Cl)(I)$ (4) $CH_3C(I)(Cl)CHD_2$

144. The correct order for acid strength of compounds $CH\equiv CH$, $CH_3-C\equiv CH$ and $CH_2=CH_2$ is as follows: [Jan. 12, 2019 (I)]

(1) $CH\equiv CH > CH_2=CH_2 > CH_3-C\equiv CH$

(2) $CH_3-C\equiv CH > CH\equiv CH > CH_2=CH_2$

(3) $CH_3-C\equiv CH > CH_2=CH_2 > HC\equiv CH$

(4) $HC\equiv CH > CH_3-C\equiv CH > CH_2=CH_2$

145. When 2-butyne is treated with H_2/Lindlar's catalyst, compound X is produced as the major product and when treated with Na/ liq. NH_3 it produces Y as the major product. Which of the following statements is correct? [Online April 15, 2018 (II)]

(1) Y will have higher dipole moment and higher boiling point than X

(2) Y will have higher dipole moment and lower boiling point than X

(3) X will have lower dipole moment and lower boiling point than Y

(4) X will have higher dipole moment and higher boiling point than Y

146. The reagent needed for the given conversion is

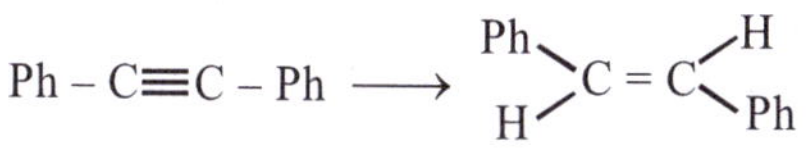

[Online April 11, 2014]

(1) Cat. hydrogenation (2) H_2/ Lindlar Cat.
(3) Li/NH_3 (4) $LiAlH_4$

147. Which one of the following class of compounds is obtained by polymerization of acetylene? [Online April 9, 2014]

(1) Poly-yne (2) Poly-ene
(3) Poly-ester (4) Poly-amine

148. The hydration of propyne results in formation of [Online May 26, 2012]

(1) Acetone (2) Propanol-1
(3) Propene (4) Propanal

149. 2-Hexyne gives *trans*-2-hexene on treatment with: [2012]

(1) Pt/H_2 (2) Li/NH_3
(3) $Pd/BaSO_4$ (4) $LiAlH_4$

150. The hydrocarbon which can react with sodium in liquid ammonia is [2008]

(1) $CH_3CH_2CH_2C\equiv CCH_2CH_2CH_3$
(2) $CH_3CH_2C\equiv CH$
(3) $CH_3CH=CHCH_3$
(4) $CH_3CH_2C\equiv CCH_2CH_3$

151. The treatment of CH_3MgX with $CH_3C\equiv C-H$ produces [2008]

(1) $CH_3-CH=CH_2$ (2) $CH_3C\equiv C-CH_3$
(3) $CH_3-\overset{H}{C}=\overset{H}{C}-CH_3$ (4) CH_4

152. Which of the following reactions will yield 2, 2-dibromopropane? [2007]

(1) $CH_3-CH=CH_2+HBr\rightarrow$
(2) $CH_3-C\equiv CH+2HBr\rightarrow$
(3) $CH_3CH=CHBr+HBr\rightarrow$
(4) $CH\equiv CH+2HBr\rightarrow$

153. Which one of the following has minimum boiling point? [2004]

(1) 1 - Butene (2) 1 - Butyne
(3) *n*- Butane (4) Isobutane

154. What is the product when acetylene reacts with hypochlorous acid? [2002]

(1) CH_3COCl (2) $ClCH_2CHO$
(3) Cl_2CHCHO (4) $ClCH_2COOH$.

155. Which of these will not react with acetylene? [2002]

(1) dil. NaOH (2) ammonical $AgNO_3$
(3) Na in liq. NH_3 (4) HCl.

NUMERICAL VALUE QUESTIONS (NVQs)

156. The sum of sigma (σ) and pi(π) bonds in Hex-1,3-dien-5-yne is ____________. [Jan. 29, 2025 (I)]

157. The compound with molecular formula C_6H_6, which gives only one monobromo derivative and takes up four moles of hydrogen per mole for complete hydrogenation has _______ π electrons. [Jan. 22, 2025 (II)]

158. The major product of the following reaction is P.

$$CH_3C\equiv C-CH_3\xrightarrow[\text{(ii) dil.KmnO}_4\text{ 273K}]{\text{(i) Na / liq. NH}_3}\text{'P'}$$

Number of oxygen atoms present in product 'P' is _______ (nearest integer). [April 6, 2024 (I)]

159. In bromination of propyne, with bromine 1, 1, 2, 2-tetrabromopropane is obtained in 27% yield. The amount of 1, 1, 2, 2-tetrabromopropane obtained from 1 g of bromine in this reaction is ______ $\times 10^{-1}$ g. (Nearest integer) [July 29, 2022 (I)]
(Molar mass : Bromine = 80 g/mol)

160. The number of sigma bonds in [July 25, 2021 (I)]

$$H_3C-\underset{\displaystyle H}{\underset{|}{C}}=CH-C\equiv C-H \text{ is } \_\_\_\_\_\_\_.$$

161. Consider the following chemical reaction.

$$CH\equiv CH\xrightarrow[\text{(2) CO, HCl, AlCl}_3]{\text{(1) Red hot Fe tube, 873K}}\text{Product}$$

The number of sp^2 hybridized carbon atom(s) present in the product is ______ . [Feb. 25, 2021 (I)]

162. In the following sequence of reactions the maximum number of atoms present in molecule 'C' in one plane is _______.

$$A\xrightarrow[\text{Cu tube}]{\text{Red hot}}B\xrightarrow[\text{Anhydrous AlCl}_3]{\text{CH}_3\text{Cl (1 eq.)}}C$$

(A is a lowest molecular weight alkyne) [Jan. 08, 2020 (II)]

Topic 4 - Aromatic Hydrocarbons

MULTIPLE CHOICE QUESTIONS (MCQs)

163. The total number of aromatic compounds/species from the following is [April 08, 2026 (II)]

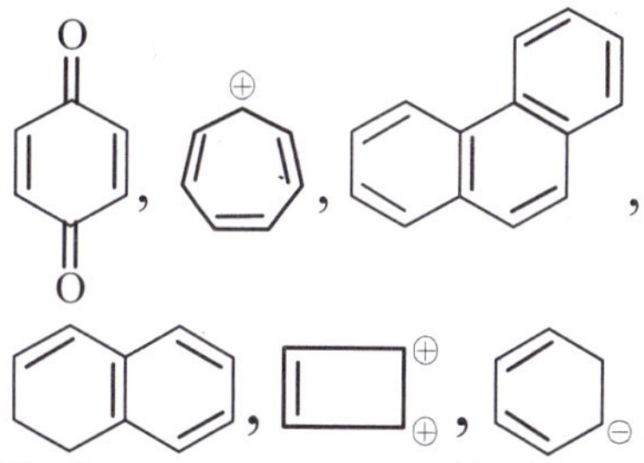

(1) 6 (2) 4
(3) 3 (4) 5

164. The major product of which of the following reaction is not obtained by rearrangement reaction? [April 08, 2026 (II)]

(1) 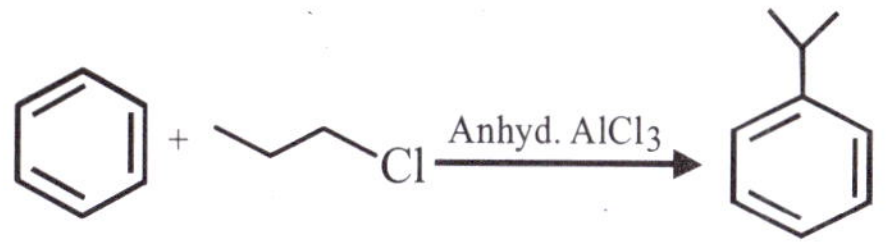

(2) 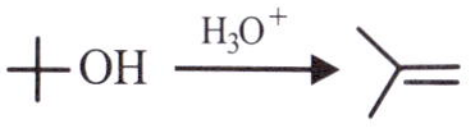

(3) $CH_3(CH_2)_4CH_3\xrightarrow[\text{HCl}]{\text{Anhyd. AlCl}_3}CH_3-\underset{\displaystyle CH_3}{\underset{|}{C}H}(CH_2)_2CH_3$

(4) 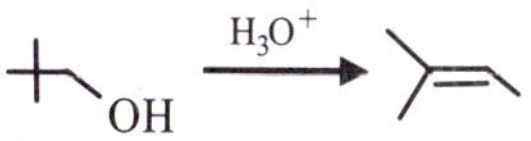

165. For the following Friedel Craft's alkylation reaction, which of the statements are correct ?

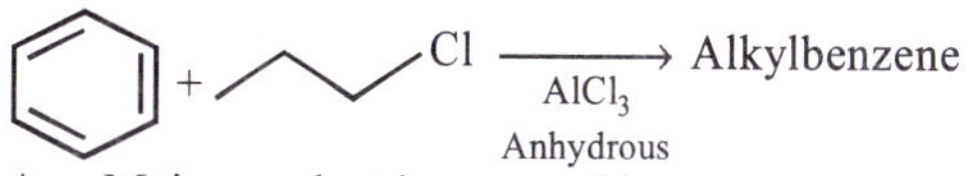

A. Major product is n-propyl benzene.
B. iso-propyl carbocation intermediate is also generated

C. Multiple substitution is inevitable.
D. Introducing electron-donating substituent on benzene will not produce any alkylbenzene.

Choose the **correct** answer from the options given below: [April 05, 2026 (I)]

(1) A and D only (2) B and C only
(3) A and C only (4) B and D only

166. 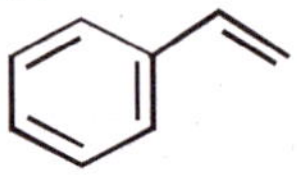
(i) Br_2 / $CHCl_3$
(ii) $NaNH_2$ excess
(iii) CH_3I
(iv) H_2, $Na/NH_3(\ell)$

Compound (X) is subjected to the sequence of reactions as shown above. Molar mass of the major product (Y) formed is ______ g mol^{-1}.
(Given molar mass in g mol^{-1} C : 12, H : 1 , O : 16) [April 04, 2026 (II)]

(1) 90 (2) 118
(3) 160 (4) 125

167. Match the **LIST-I** with **LIST-II** [April 04, 2026 (II)]

	List-I (Reaction)		List-II (Mechanism)
A.	Williamson Synthesis	I.	Electrophilic addition
B.	Friedel Craft Reaction	II.	Free radical substitution
C.	Bromination of vinyl benzene	III.	Nucleophilic substitution
D.	Chlorination of toluene in light	IV.	Electrophilic substitution

Choose the correct answer from the options given below:

(1) A-III, B-I, C-II, D-IV (2) A-III, B-IV, C-II, D-I
(3) A-III, B-IV, C-I, D-II (4) A-I, B-III, C-IV, D-II

168. The compound (X) on [April 02, 2026 (II)]

(i) on heating in the presence of anhydrous $AlCl_3$ and HCl gas gives 2, 4–dimethyl pentane
(ii) aromatization gives toluene and
(iii) cyclisation gives methyl cyclohexane

The **correct** name of compound (X) is :

(1) Hept-2-ene (2) Hept-1,3,5-triene
(3) Heptane (4) Hept-2,4,6-triene

169. Which of the following reaction is NOT correctly represented ? [Jan. 28, 2026 (II)]

(1) 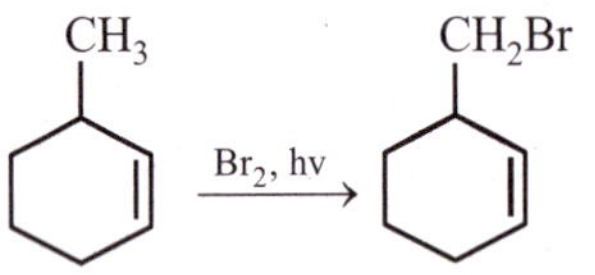

(2) 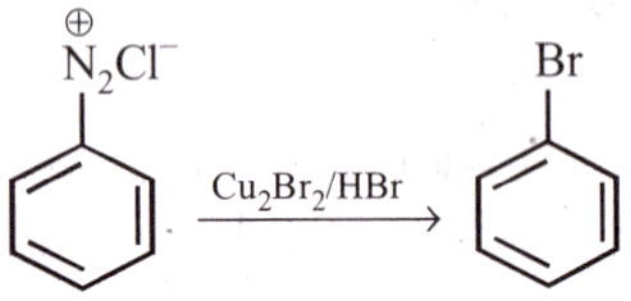

(3) 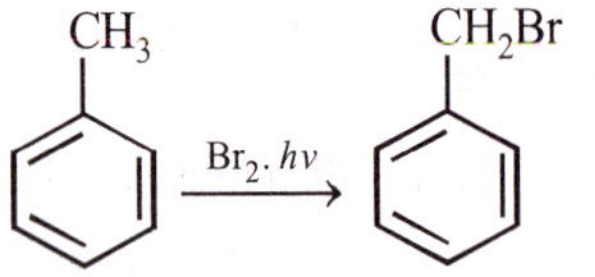

(4) 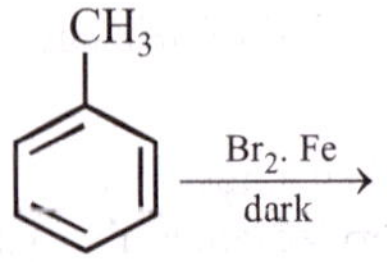 *Ortho* and *para* Bromotoluenes

170. Identify the **correct** statements : [Jan. 28, 2026 (II)]

The presence of $-NO_2$ group in benzene ring

A. activates the ring towards electrophilic substitutions.
B. deactivates the ring towards electrophilic substitutions.
C. activates the ring towards nucleophilic substitutions.
D. deactivates the ring towards nucleophilic substitutions.

(1) B and D Only (2) C and A Only
(3) A and D Only (4) B and C Only

171. Consider the following compounds [Jan. 23, 2026 (I)]

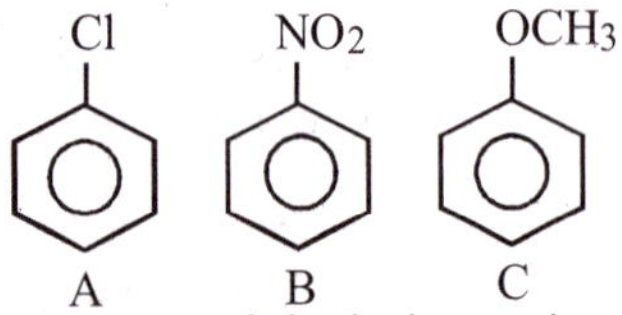

Arrange these compounds in the increasing order of reactivity with nitrating mixture.

(1) C<A<B (2) B<C<A
(3) C<B<A (4) B<A<C

172. Given below are two statements : [Jan. 22, 2026 (I)]

Statement I : Benzene is nitrated to give nitrobenzene, which on further treatment with CH_3COCl / $AlCl_3$ will give

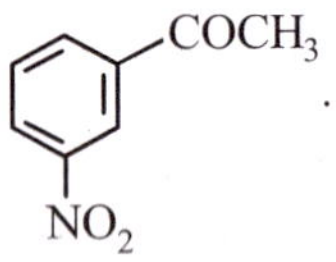

Statement II : NO_2 group is a *m*-directing and deactivating group.

In the light of the above statements, choose the most appropriate answer from the options given below.

(1) Statement I is correct but Statement II is incorrect.
(2) Both Statement I and Statement II are correct.
(3) Statement I is incorrect but Statement II is correct.
(4) Both Statement I and Statement II is are incorrect.

173. Identify A in the following reaction. [Jan. 21, 2026 (I)]

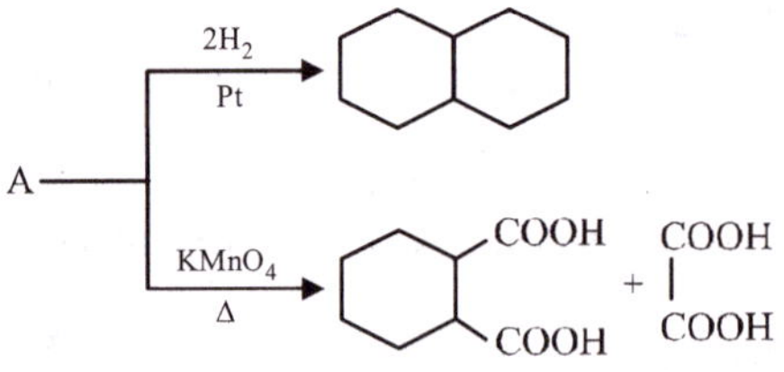

(1) 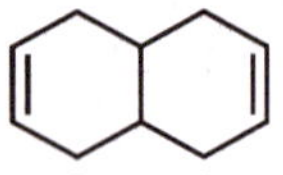 (2) 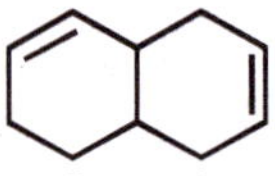

(3) 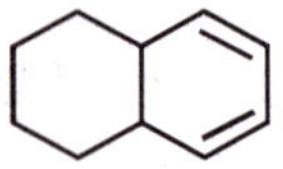 (4) 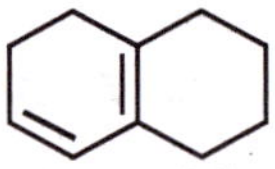

174. Choose the correct set of reagents for the following conversion. [April 8, 2025 (II)]

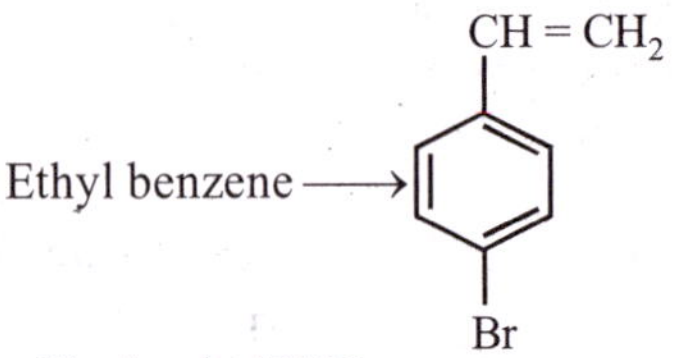

(1) Br_2/Fe ; Cl_2, Δ ; alc. KOH
(2) Cl_2/Fe ; Br_2/anhy.$AlCl_3$; aq. KOH
(3) Br_2/anhy.$AlCl_3$; Cl_2, Δ ; aq. KOH
(4) Cl_2/anhy.$AlCl_3$; Br_2/Fe ; alc. KOH

175. Benzene is treated with oleum to produce compound (*X*) which when further heated with molten sodium hydroxide followed by acidification produces compound (*Y*). The compound *Y* is treated with zinc metal to produce compound (*Z*). Identify the structure of compound (*Z*) from the following option. [April 4, 2025 (I)]

(1) 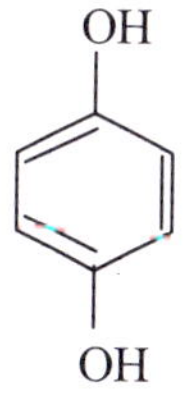 (2) 

(3) 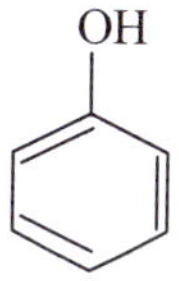 (4) 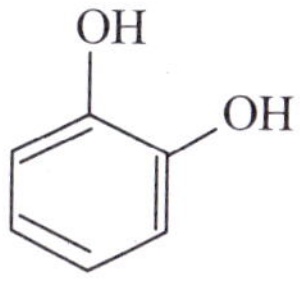

176. Designate whether each of the following compounds is aromatic or not aromatic. [April 2, 2025 (I)]

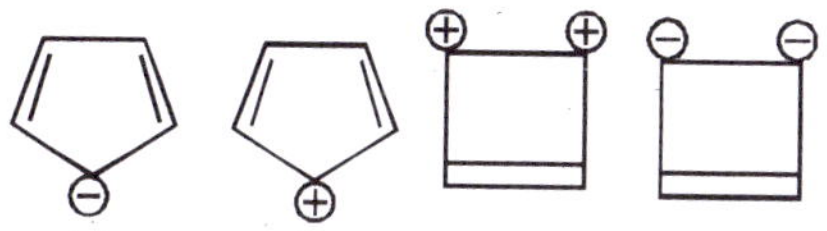

(a) (b) (c) (d)

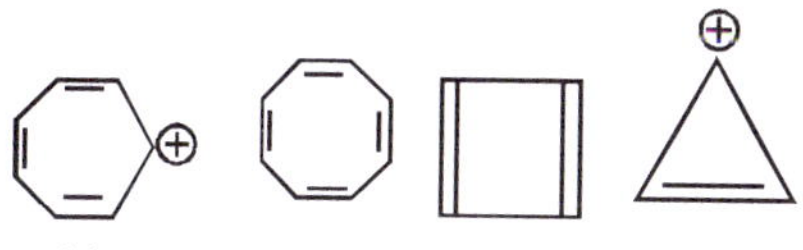

(e) (f) (g) (h)

(1) e, g aromatic and a, b, c, d, f, h not aromatic
(2) b, e, f, g aromatic and a, c, d, h not aromatic
(3) a, b, c, d aromatic and e, f, g, h not aromatic
(4) a, c, d, e, h aromatic and b, f, g not aromatic

177. Given below are two statements :

Statement (I) : On nitration of *m*-xylene with HNO_3, H_2SO_4 followed by oxidation, 4-nitrobenzene-1, 3-dicarboxylic acid is obtained as the major product.

Statement (II) : CH_3 group is o/p-directing while-NO_2 group is m-directing group.

In the light of the above statements, choose the correct answer from the options given below : [Jan. 29, 2025 (II)]

(1) Both Statement I and Statement II are false
(2) Statement I is false but Statement II is true
(3) Both Statement I and Statement II are true
(4) Statement I is true but Statement II is false

178. Identify correct statement/s : [Jan. 24, 2025 (II)]

(A) —OCH_3 and —$NHCOCH_3$ are activating group
(B) —CN and —OH are meta directing group
(C) —CN and —SO_3H are meta directing group
(D) Activating groups act as ortho – and para directing groups
(E) Halides are activating groups

Choose the correct answer from the options given below :

(1) (A), (C) and (D) only (2) (A), (B) and (E) only
(3) (A) only (4) (A) and (C) only

179. The correct stability order of the following species/molecules is: [Jan. 23, 2025 (I)]

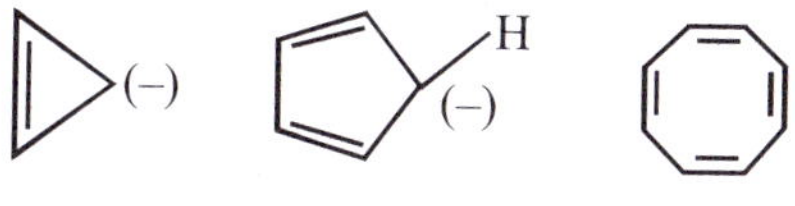

p q r

(1) q > r > p (2) r > q > p
(3) q > p > r (4) p > q > r

180. Which of the following molecule / species is most stable? [Jan. 30, 2024 (I)]

(1)  (2) 

(3)  (4) 

181. 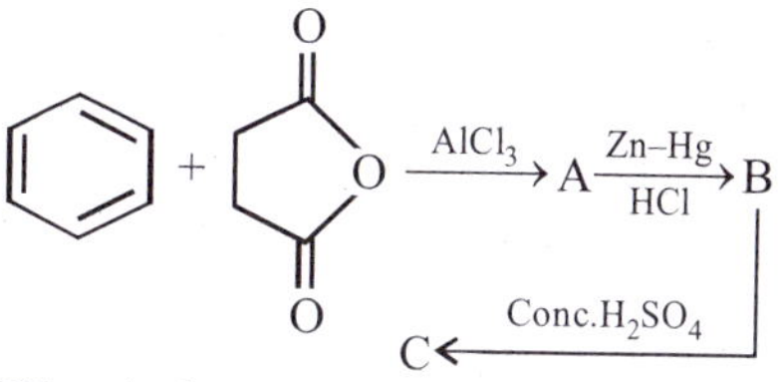

What is the structure of C ? [April 9, 2024 (I)]

(1) 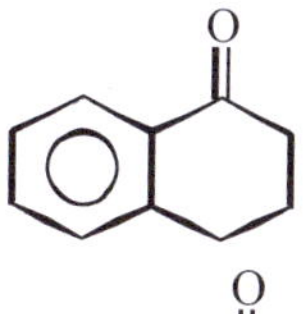

(2) 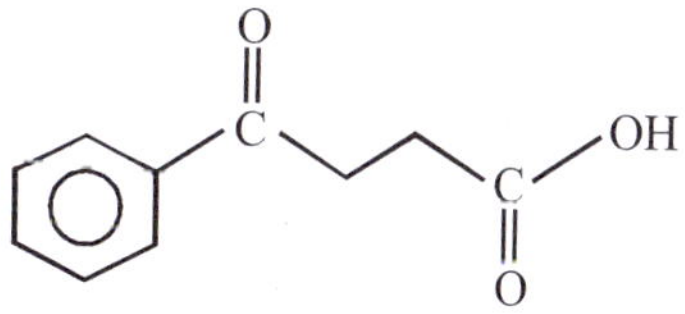

(3) 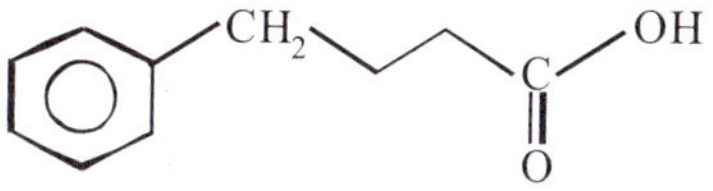

(4) 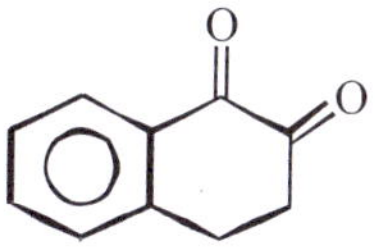

182. Which of the following are aromatic? [April 8, 2024 (I)]

A. 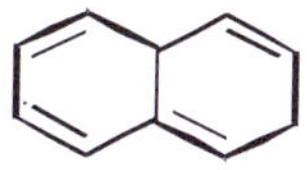 B. 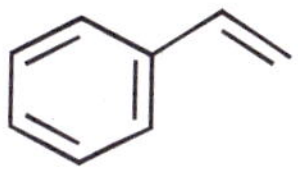

C. 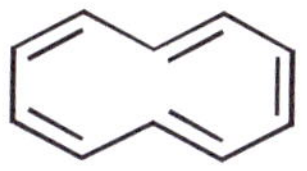 D. 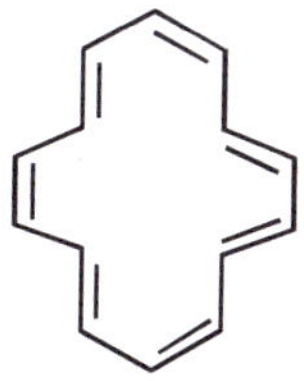

(1) B and D only (2) A and C only
(3) A and B only (4) C and D only

183. Given below are two statements : [April 5, 2024 (I)]

Statement I: Nitration of benzene involves the following step –

$$H-\overset{\overset{H}{|}}{\underset{\cdot\cdot}{O}}{}^{\ominus}-NO_2 \rightleftharpoons H_2O + \overset{\oplus}{N}O_2$$

Statement II: Use of Lewis base promotes the electrophilic substitution of benzene.

In the light of the above statements, choose the **most appropriate** answer from the options given below :

(1) Both **Statement I** and **Statement II** are incorrect
(2) **Statement I** is correct but **Statement II** is incorrect
(3) Both **Statement I** and **Statement II** are correct
(4) **Statement I** is incorrect but **Statement II** is correct

184. Correct order of stability of carbanion is [April 4, 2024 (II)]

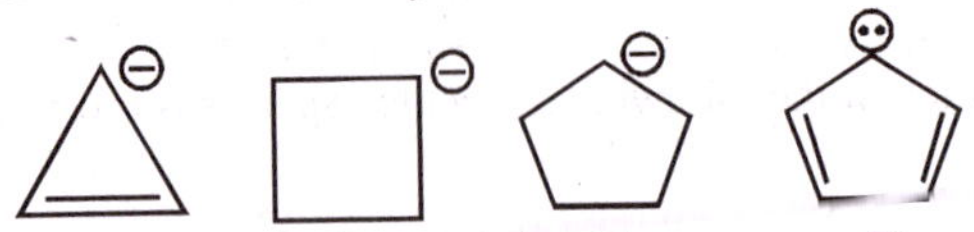

A B C D

(1) C > B > D > A (2) A > B > C > D
(3) D > A > C > B (4) D > C > B > A

185. The set of meta directing functional groups from the following sets is: [Feb. 1, 2024 (II)]
(1) – CN, – NH_2, – NHR, – OCH_3
(2) – CN, – CHO, – $NHCOCH_3$, – COOR
(3) – NO_2, – NH_2, – COOH, – COOR
(4) – NO_2, – CHO, – SO_3H, – COR

186. Which of the following compound will most easily be attacked by an electrophile? [Feb. 1, 2024 (I)]

(1) 
(2) 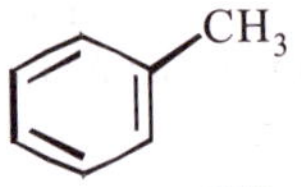

(3) 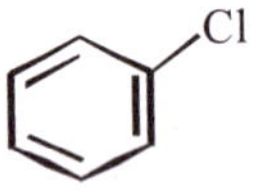

(4) 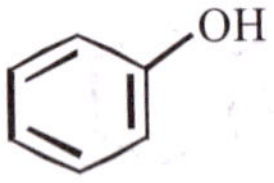

187. Identify major product 'P' formed in the following reaction.

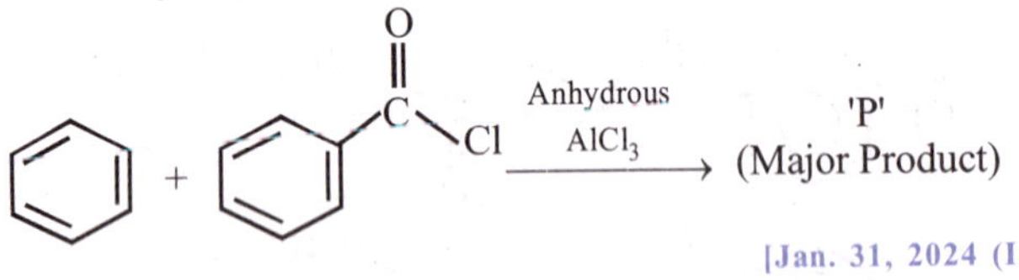

[Jan. 31, 2024 (II)]

(1) 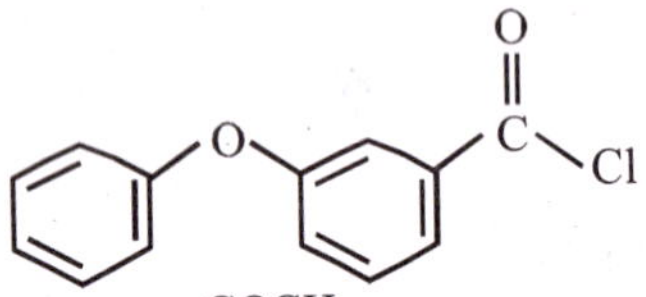

(2) 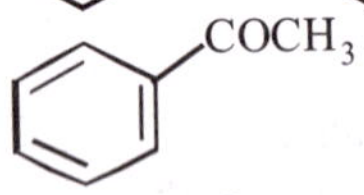

(3) 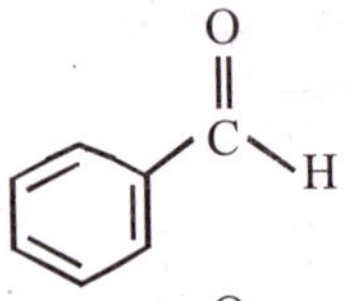

(4) 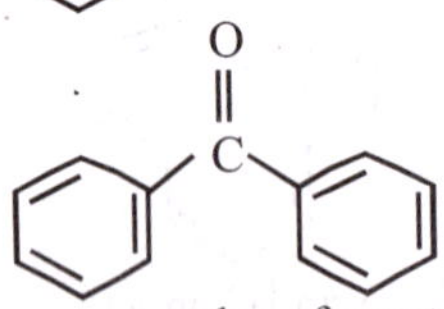

188. The correct order of reactivity in electrophilic substitution reaction of the following compounds is : [Jan. 31, 2024 (II)]

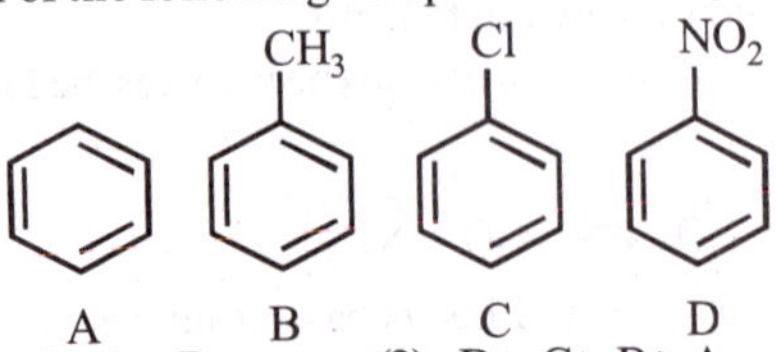

A B C D

(1) B > C > A > D (2) D > C > B > A
(3) A > B > C > D (4) B > A > C > D

189. Given below are two statements: [April 13, 2023 (II)]
Statement I : Tropolone is an aromatic compound and has 8π electrons.
Statement II : π electrons of > C = O group in tropolone is involved in aromaticity.

In the light of the above statements, choose the correct answer from the options given below:
(1) Both Statement I and Statement II are true
(2) Statement I is true but Statement II is false
(3) Statement I is false but Statement II is true
(4) Both Statement I and Statement II are false

190. The major product 'P' formed in the given reaction is

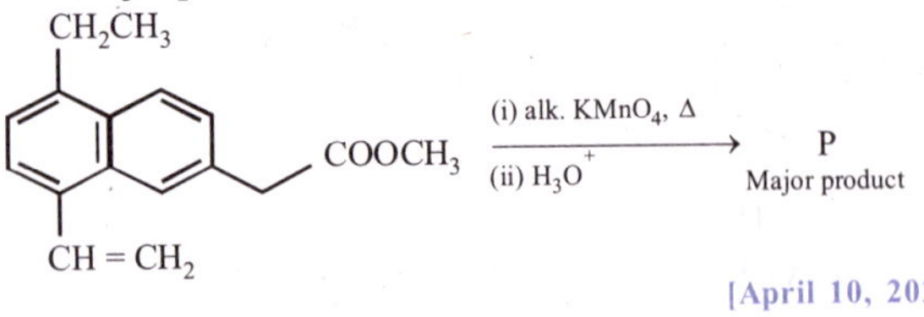

[April 10, 2023 (I)]

(1) 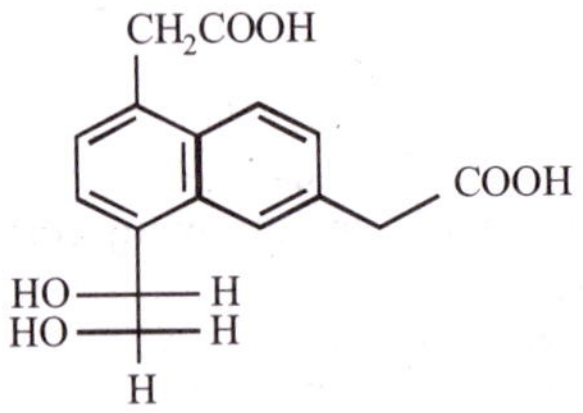

(2) 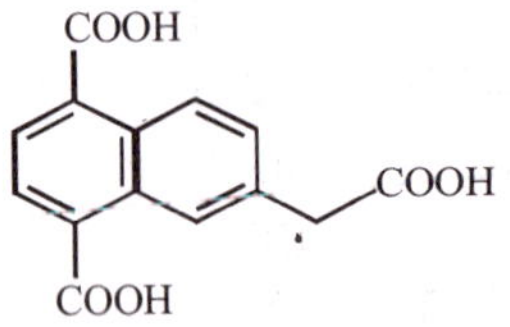

(3) 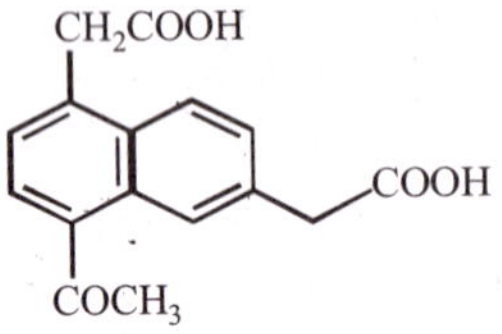

(4) 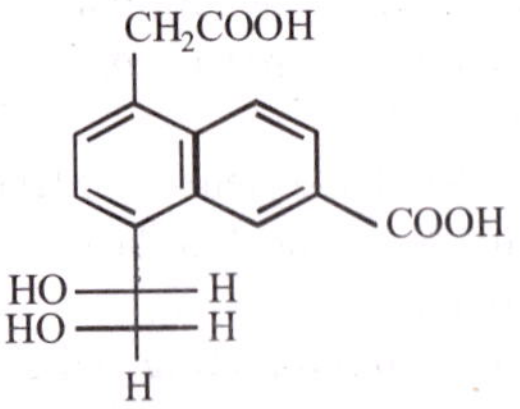

191. The major product 'P' formed in the given reaction is: [April 10, 2023 (II)]

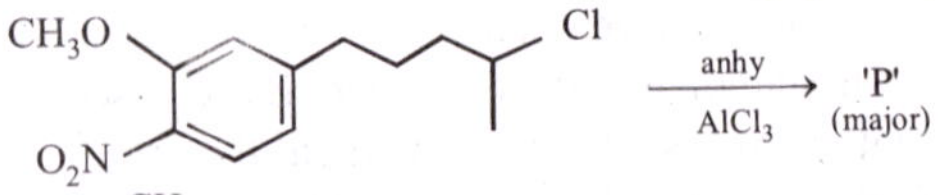

(1) 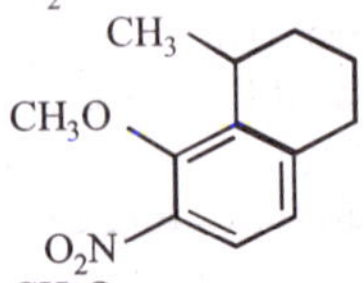

(2) 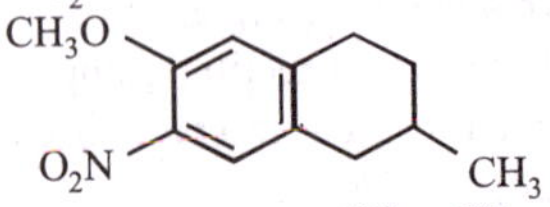

(3) 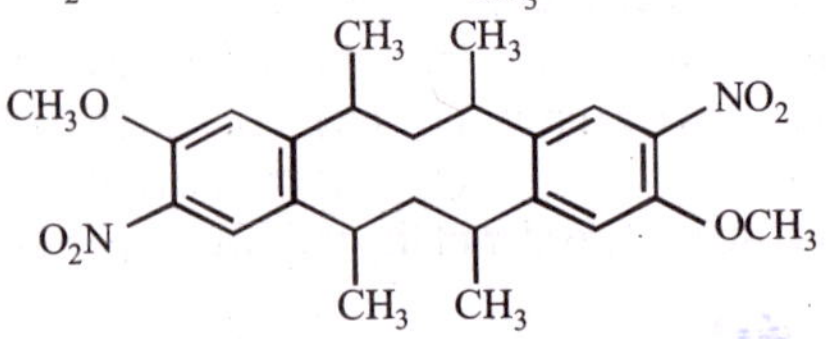

(4) 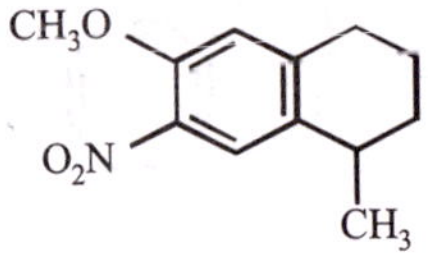

192. 'X' is : 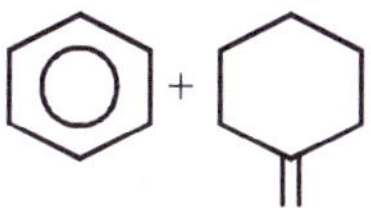 $\xrightarrow[\Delta]{HF}$ X Major product **[Feb. 01, 2023 (II)]**

(1) 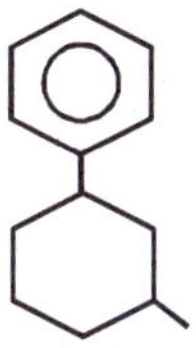 (2) 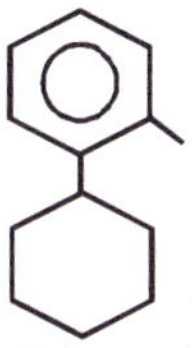 (3) 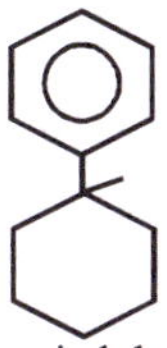 (d) 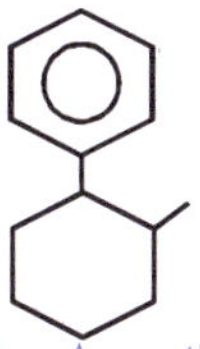

193. Given below are two statements. One is labelled as **Assertion A** and the other is labelled as **Reason R**. **[July 27, 2022 (I)]**

Assertion A : [6] Annulene. [8] Annulene and cis –[10] Annulene, are respectively aromatic, non-aromatic and aromatic.

[6] Annulene  [8] Annulene 

cis-[10] Annulene 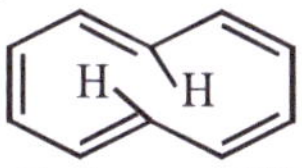

Reason R : Planarity is one of the requirements of aromatic systems.

In the light of the above statements, choose the **most appropriate** answer from the options given below.

(1) Both A and R are correct and R is the correct explanation of A.
(2) Both A and R are correct but R is NOT the correct explanation of A.
(3) A is correct but R is not correct.
(4) A is not correct but R is correct.

194. Which of the following compounds is **not** aromatic? **[July 26, 2022 (I)]**

(1)  (2) 

(3) 

(4) 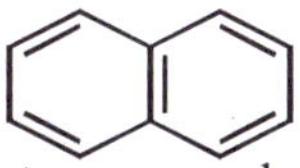

195. Which of the following is not an example of benzenoid compound ? **[July 26, 2022 (II)]**

(1) 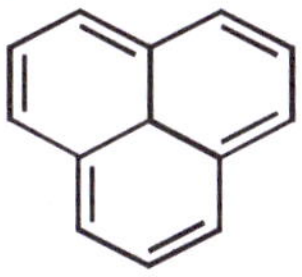 (2) 

(3) 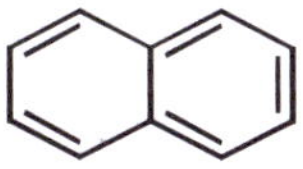 (4) 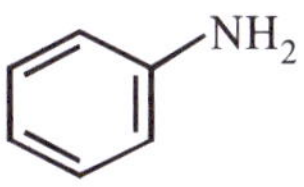

196.  $+ CH_3CH_2CH_2Cl \xrightarrow{\text{Anhydrous, } AlCl_3}$ 'A' Major Product

The stable carbocation formed in the above reaction is : **[June 29, 2022 (II)]**

(1) $CH_3CH_2\overset{\oplus}{C}H_2$ (2) $CH_3\overset{\oplus}{C}H_2$

(3) $CH_3-\overset{\oplus}{C}H-CH_3$ (4) 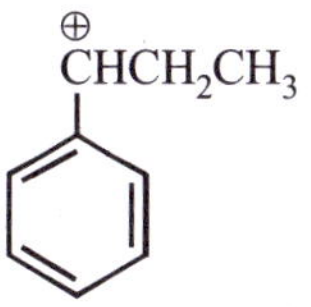

197. Two isomers (A) and (B) with molar mass 184 g/mol and elemental composition C, 52.2%; H, 4.9% and Br 42.9% gave benzoic acid and *p*-bromobenzoic acid, respectively on oxidation with $KMnO_4$. Isomer 'A' is optically active and gives a pale yellow precipitate when warmed with alcoholic $AgNO_3$. Isomer 'A' and 'B' are, respectively : **[June 29, 2022 (II)]**

(1) $H_3C-CHBr-C_6H_5$ and 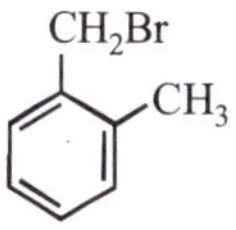

(2) 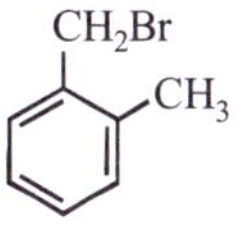

and 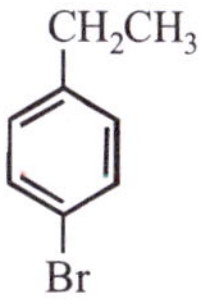

(3) $H_3C-CHBr-C_6H_5$ and 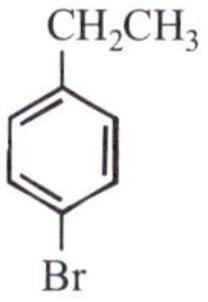

(4) 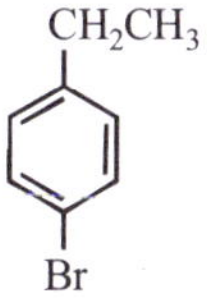

and $H_3C-CHBr-C_6H_5$

198. Which of the following structures are aromatic in nature? **[June 28, 2022 (I)]**

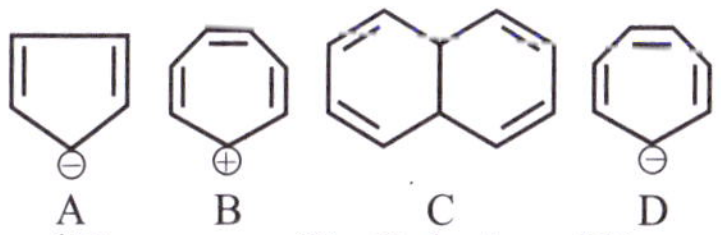

(1) A,B,C and D (2) Only A and B
(3) Only A and C (4) Only B, C and D

199. Which of the following is most stable? **[June 27, 2022 (II)]**

(1) 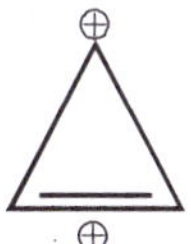 (2) 

(3) 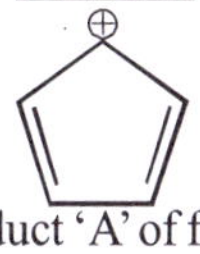 (4) 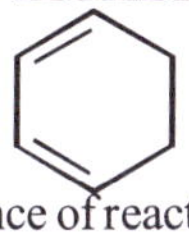

200. Product 'A' of following sequence of reactions is Ethylbenzene $\xrightarrow[\text{(b) } Cl_2, \Delta \text{ (c) alc. KOH}]{\text{(a) } Br_2 \text{. Fe}}$ 'A' (Major product) **[June 27, 2022 (II)]**

(1) 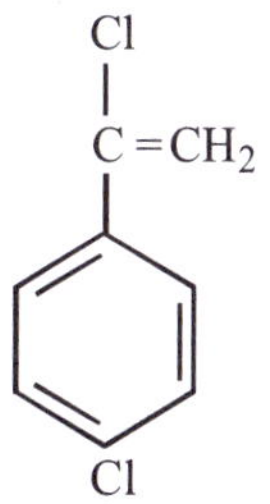

(2) 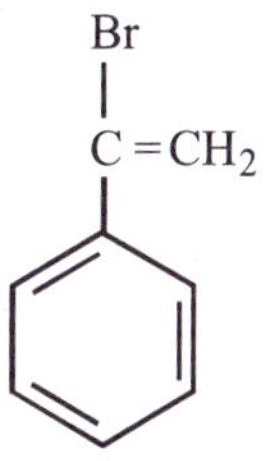

(3) 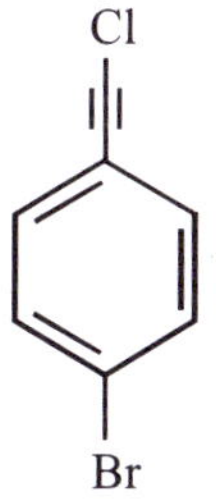

(4) 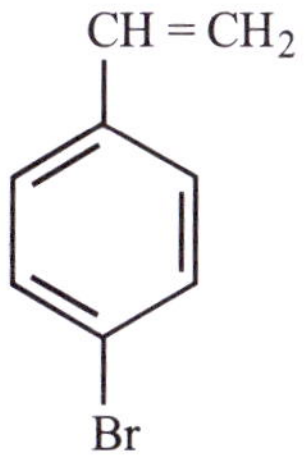

201. Halogenation of which one of the following will yield m-substituted product with respect to methyl group as a major product? **[June 26, 2022 (II)]**

(1) 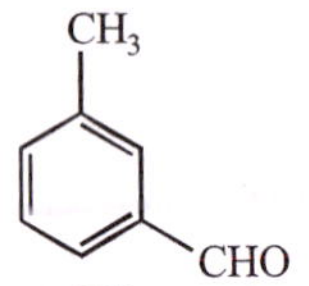
(2) 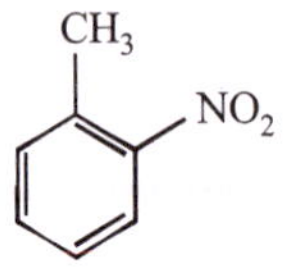
(3) 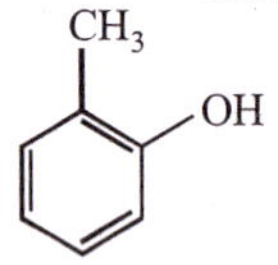
(4) 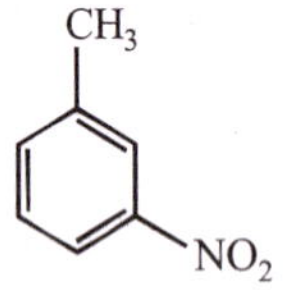

202. Which one of the following compounds is aromatic in nature ? **[Sep. 1, 2021 (II)]**

(1) 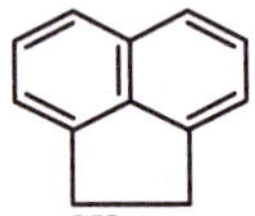
(2) 
(3) 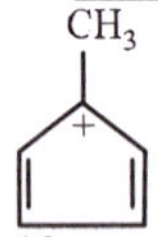
(4) 

203. Identify correct A, B and C in the reaction sequence given below : **[Aug. 31, 2021 (II)]**

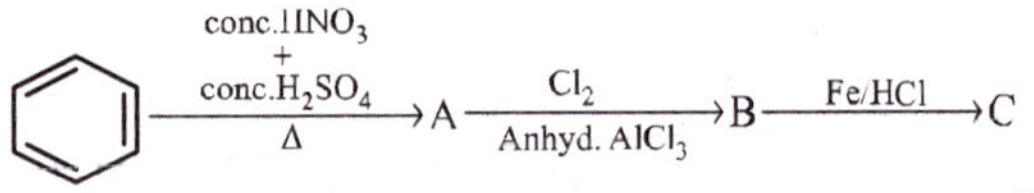

(1) 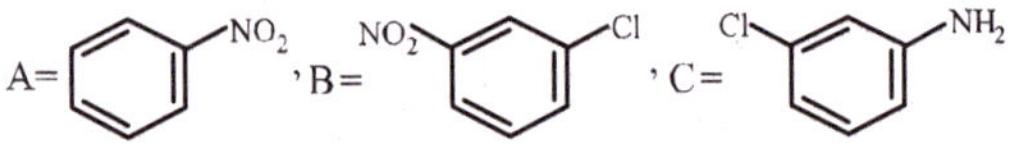
(2) 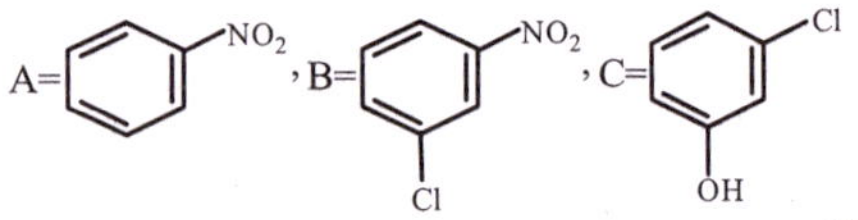
(3) 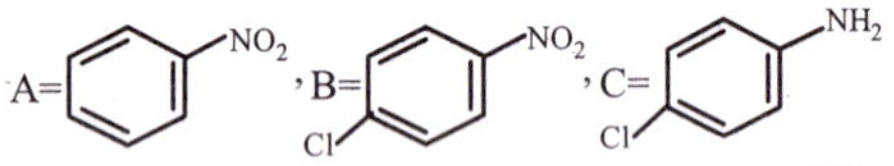
(4) 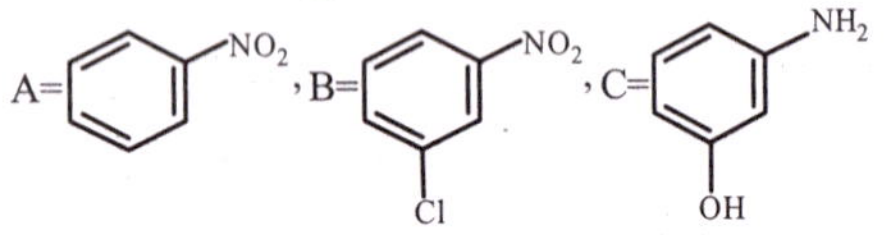

204. Which one of the following compounds is not aromatic? **[Aug. 26, 2021 (II); Same March 16, 2021 (I)]**

(1) 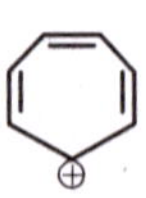
(2) 
(3) 
(4) 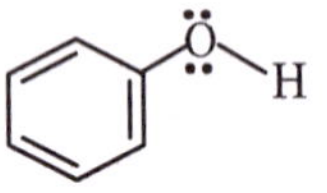

205. Benzene on nitration gives nitrobenzene in presence of HNO_3 and H_2SO_4 mixture, where : **[July 20, 2021 (II)]**
(1) both H_2SO_4 and HNO_3 act as a bases
(2) HNO_3 acts as an acid and H_2SO_4 acts as a base
(3) both H_2SO_4 and HNO_3 act as an acids
(4) HNO_3 acts as a base and H_2SO_4 acts as an acid

206. Among the following, the aromatic compounds are :

(A) 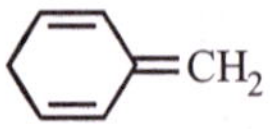
(B) 
(C) 
(D) 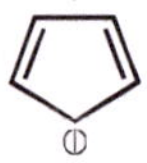

Choose the correct answer from the following options : **[March 16, 2021 (I)]**
(1) (A), (B) and (C) only (2) (B), (C) and (D) only
(3) (B) and (C) only (4) (A) and (B) only

207. For the given reaction :

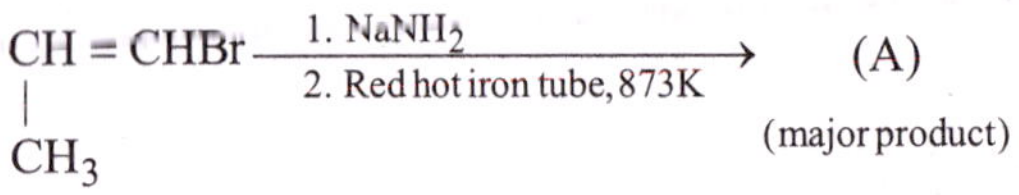

What is 'A'? **[Feb. 26, 2021 (I)]**

(1) 
(2) 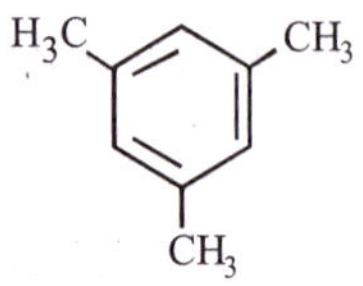
(3) $CH_3CH_2CH_2NH_2$
(4) $CH(CH_3)=CH-NH_2$

208. The correct sequence of reagents used in the preparation of 4-bromo-2-nitroethyl benzene from benzene is: **[Feb. 25, 2021 (II)]**
(1) $CH_3COCl/AlCl_3$, Zn-Hg/HCl, $Br_2/AlBr_3$, HNO_3/H_2SO_4
(2) HNO_3/H_2SO_4, $Br_2/AlCl_3$, $CH_3COCl/AlCl_3$, Zn-Hg/HCl
(3) $CH_3COCl/AlCl_3$, $Br_2/AlBr_3$, HNO_3/H_2SO_4, Zn/HCl
(4) $Br_2/AlBr_3$, $CH_3COCl/AlCl_3$, HNO_3/H_2SO_4, Zn/HCl

209. The number of sp^2 hybrid orbitals in a molecule of benzene is: **[Jan. 09, 2020 (II)]**
(1) 24 (2) 6 (3) 18 (4) 12

210. Consider the following reactions:

(A) 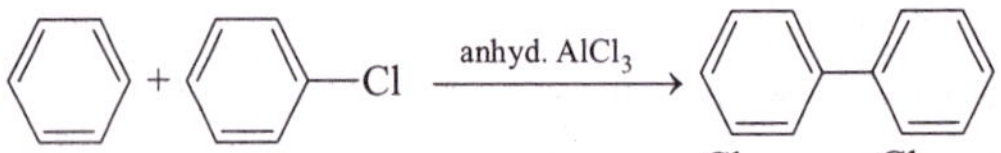
(B) 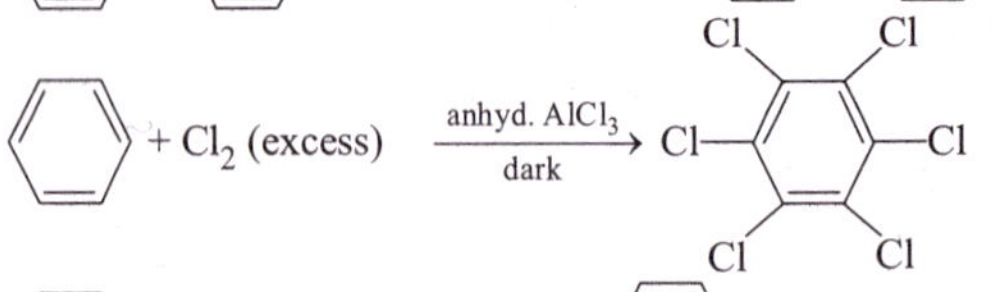
(C) 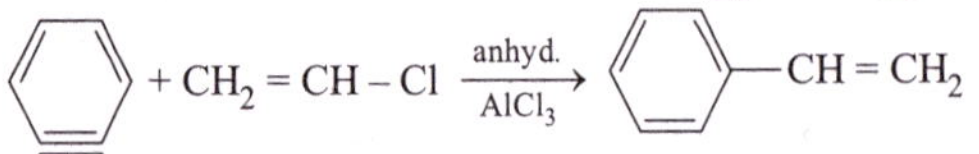
(D) 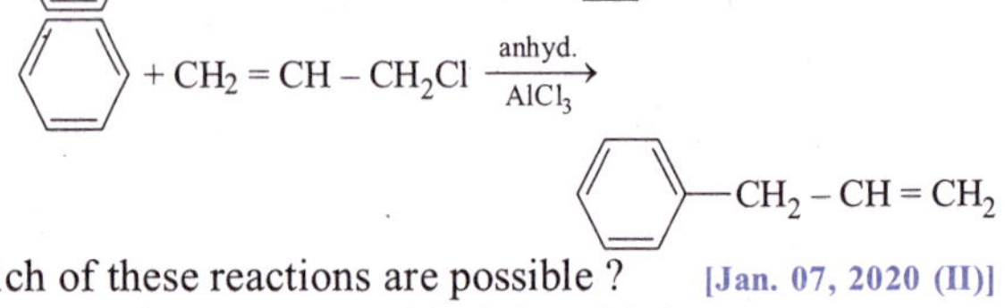

Which of these reactions are possible ? **[Jan. 07, 2020 (II)]**
(1) (A) and (B) (2) (A) and (D)
(3) (B), (C) and (D) (4) (B) and (D)

211. Which of the following compounds is not aromatic? **[Jan. 9, 2019 (II)]**

(1) 
(2) 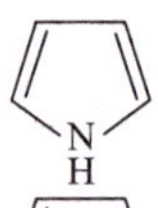
(3) 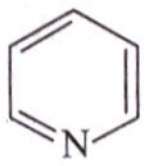
(4) 

212. The increasing order of reactivity of the following compounds towards aromatic electrophilic substitution reaction is: **[April 9, 2019 (I)]**

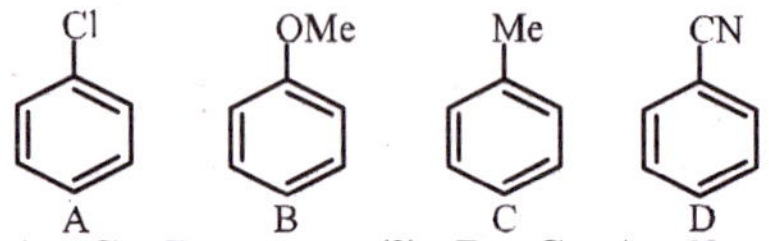

(1) D < A < C < B (2) B < C < A < D
(3) A < B < C < D (4) D < B < A < C

213. Which compound (s) out of the following is/are not aromatic? [Jan. 11, 2019 (I), Jan. 9, 2019 (II)]

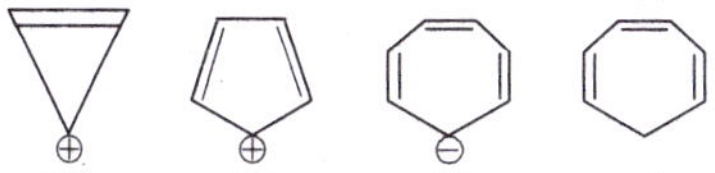

(A) (B) (C) (D)

(1) (B), (C) and (D) (2) (C) and (D)
(3) (B) (4) (A) and (C)

214.

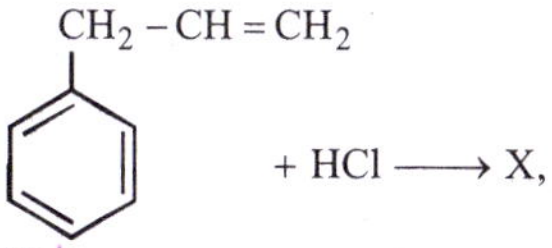

X is : [Online April 16, 2018 Same Online April 9, 2013]

(1) 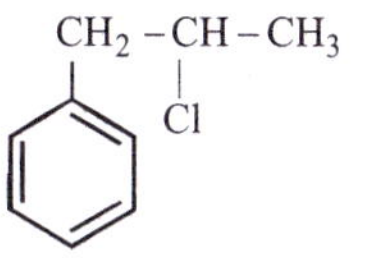 (2) 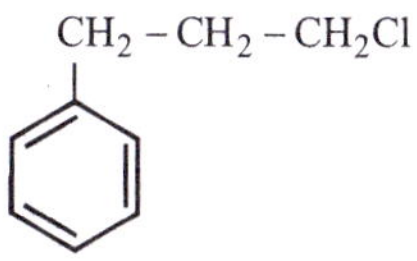

(3) 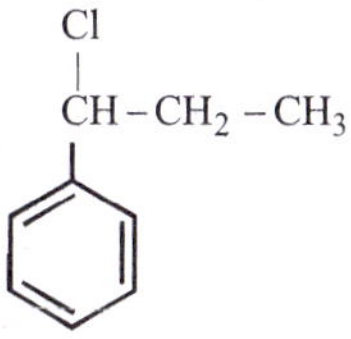 (4) 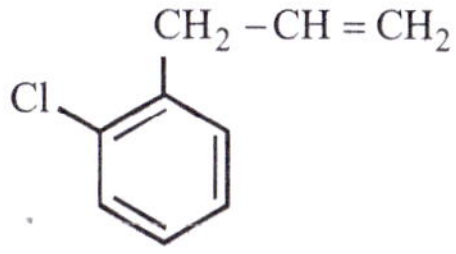

215. Which of the following compounds will show highest dipole moment ? [Online April 9, 2017]

(I) 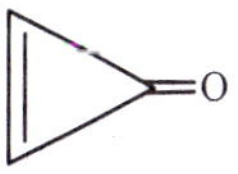 (II) 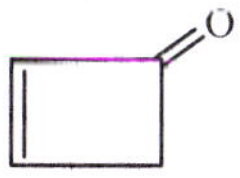

(III) 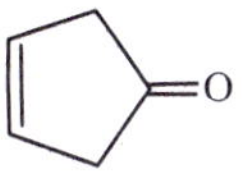 (IV) 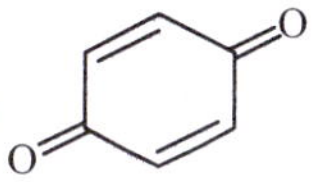

(1) (I) (2) (II) (3) (III) (4) (IV)

216. Which of the following compounds will not undergo Friedel Craft's reaction with benzene ? [Online April 8, 2017]

(1) 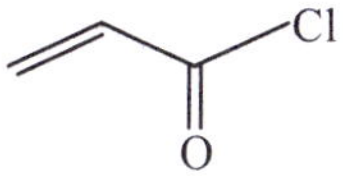 (2) 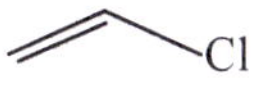

(3)  (4) 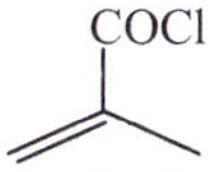

217. Which of the following would not give 2-phenylbutane as the major product in a Friedel-Crafts alkylation reaction ? [Online April 22, 2013]

(1) 1-Butene + HF
(2) 2-Butanol + H_2SO_4
(3) Butanoyl chloride + $AlCl_3$, then Zn, HCl
(4) Butyl chloride + $AlCl_3$

218. Which of the following compounds are antiaromatic

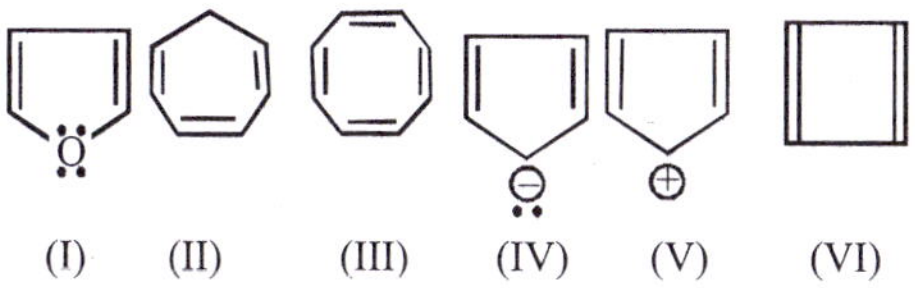

(I) (II) (III) (IV) (V) (VI)

[Online May 26, 2012]

(1) (I) and (V) (2) (II) and (V)
(3) (I) and (IV) (4) (III) and (VI)

219. In the given reaction, [Online May 12, 2012]

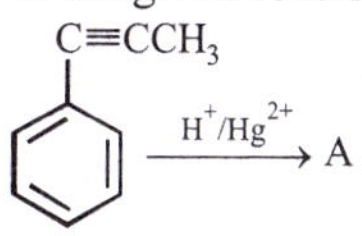

the product 'A' is

(1) 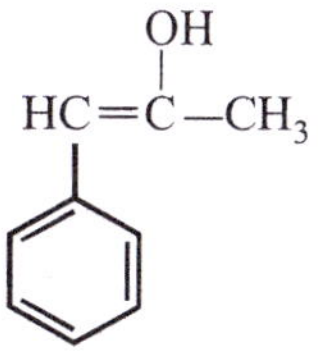 (2) 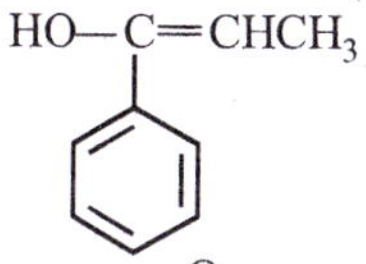

(3) 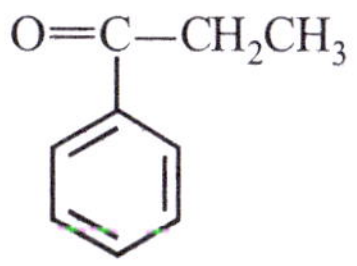 (4) 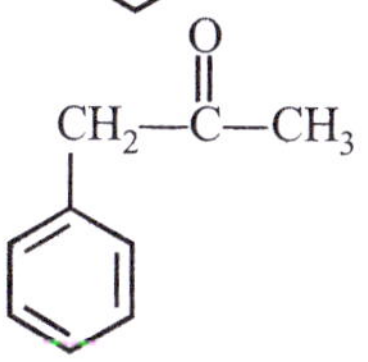

220. The product of the reaction between ethyl benzene and *N*-bromosuccinamide is [Online May 19, 2012]

(1) 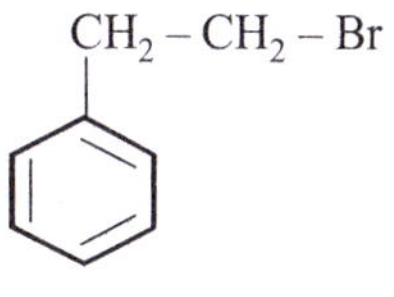 (2) 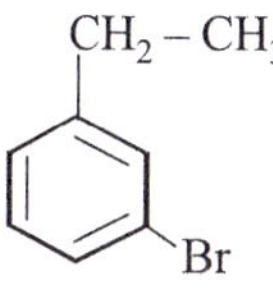

(3) 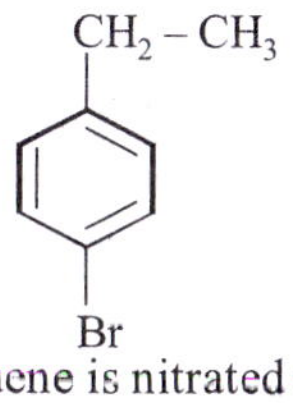 (4) 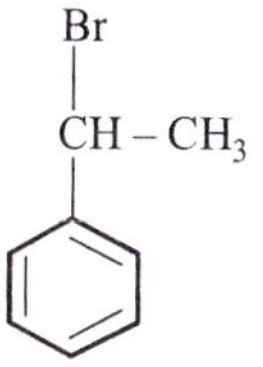

221. Toluene is nitrated and the resulting product is reduced with tin and hydrochloric acid. The product so obtained is diazotised and then heated wth cuprous bromide. The reaction mixture so formed contains [2008]

(1) mixture of *o*- and *p*-bromotoluenes
(2) mixture of *o*- and *p*-dibromobenzenes
(3) mixture of *o*- and *p*-bromoanilines
(4) mixture of *o*- and *m*-bromotoluenes

NUMERICAL VALUE QUESTIONS (NVQs)

222. Consider the following sequence of reactions to give the major product (X)

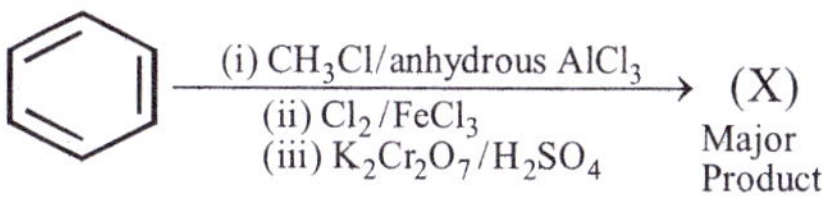

P g of the major product (X) formed is reacted with $NaHCO_3$ soluion to liberate a gas which occupied 11.2 dm^3 at STP. [April 04, 2026 (I)]

P = ______. g.

(Given molar mass in g mol^{-1} H : 1, C : 12, O : 16, Cl : 35.5)

223. Consider the following reaction sequence. [Jan. 21, 2026 (I)]

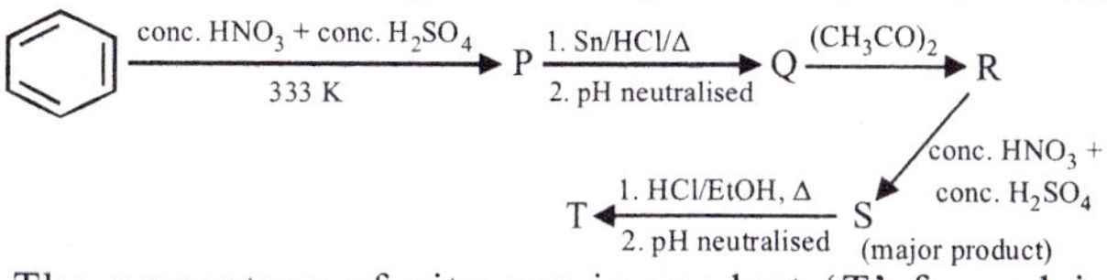

The percentage of nitrogen in product 'T' formed is ______%. (Nearest integer)

(Given molar mass in g mol^{-1} H:1, C:12, N:14, O:16)

224. Isomeric hydrocarbons → negative Baeyer's test (Molecular formula C_9H_{12}).

The total number of isomers from above with four different non-aliphatic substitution sites is ______

[Jan. 29, 2025 (II)]

225. Total number of aromatic compounds among the following compounds is ________. [April 8, 2024 (II)]

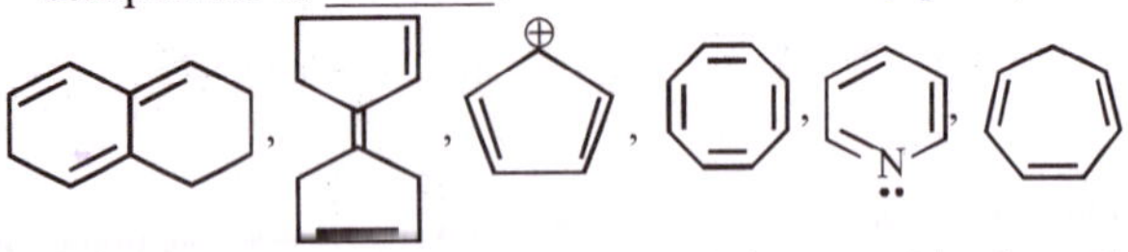

226. Total number of deactivating groups in aromatic eletrophilic susbstitution reaction among the following is ________. [Feb. 1, 2024 (I)]

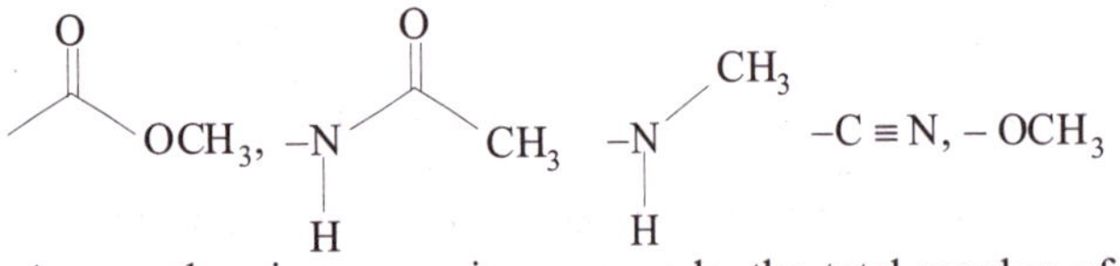

227. Among the given organic compounds, the total number of aromatic compounds is ______. [Jan. 27, 2024 (I)]

(A) 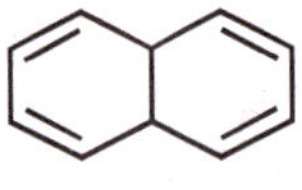 (B) 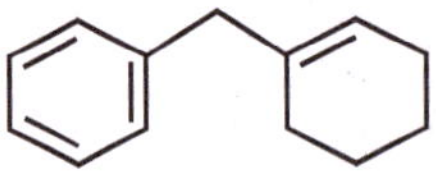

(C) 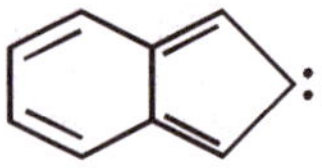 (D) 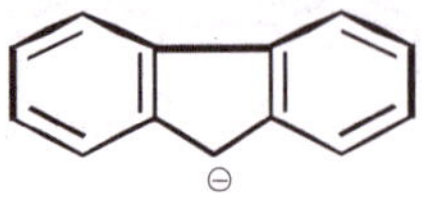

228. In the presence of sunlight, benzene reacts with Cl_2 to give product, X. The number of hydrogens in X is ________. [July 26, 2022 (I)]

229. 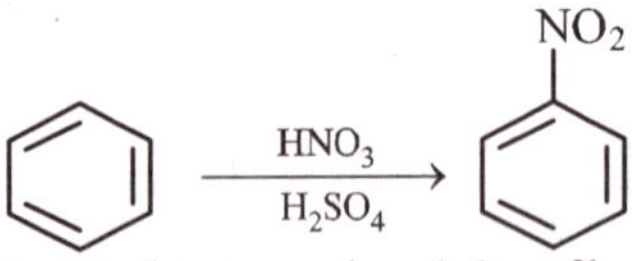

In the above reaction, 3.9 g of benzene on nitration gives 4.92 g of nitrobenzene. The percentage yield of nitrobenzene in the above reaction is __________%. [March 17, 2021 (I)]
(Round off to the nearest integer)
(Given atomic mass : C : 12.0 u, H : 1.0 u, O : 16.0 u, N : 14.0 u)

Part-B (JEE Advanced)

Topic 1- Alkanes

1 MCQs with One Correct Answer

1. The bond energy (in kcal mol–1) of a C–C single bond is approximately [2010]
(a) 1 (b) 10 (c) 100 (d) 1000
2. On monochlorination of 2-methylbutane, the total number of chiral compounds formed is [2004S]
(a) 2 (b) 4 (c) 6 (d) 8
3. In the given conformation, if C_2 is rotated about $C_2 - C_3$ bond anticlockwise by an angle of 120° then the conformation obtained is [2004S]

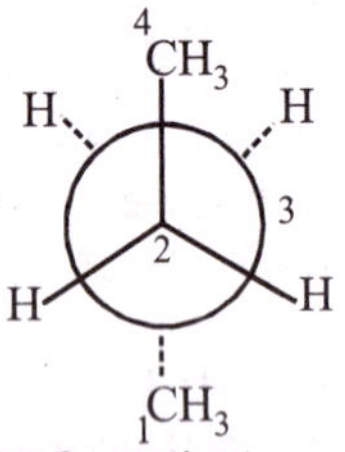

(a) fully eclipsed conformation
(b) partially eclipsed conformation
(c) gauche conformation
(d) staggered conformation
4. Consider the following reaction [2002S]

$$H_3C-\underset{D}{\underset{|}{C}}H-\underset{CH_3}{\underset{|}{CH}}-CH_3+\dot{B}r \longrightarrow 'X' + HBr$$

Identify the structure of the major product 'X'

(a) $H_3C-\underset{D}{\underset{|}{CH}}-\underset{CH_3}{\underset{|}{CH}}-\dot{C}H_2$

(b) $H_3C-\underset{D}{\underset{|}{C}}H-\underset{CH_3}{\underset{|}{\dot{C}}}-CH_3$

(c) $H_3C-\underset{D}{\underset{|}{\dot{C}}}-\underset{CH_3}{\underset{|}{CH}}-CH_3$

(d) $H_3C-\dot{C}H-\underset{CH_3}{\underset{|}{CH}}-CH_3$

5. When cyclohexane is poured on water, it floats, because: [1997 - 1 Mark]
(a) cyclohexane is in 'boat' form
(b) cyclohexane is in 'chair' form
(c) cyclohexane is in 'crown' form
(d) cyclohexane is less dense than water.
6. Which of the following will have least hindered rotation about carbon-carbon bond? [1987 - 1 Mark]
(a) Ethane (b) Ethylene
(c) Acetylene (d) Hexachloroethane
7. The compound with the highest boiling point is [1982 - 1 Mark]
(a) *n*-hexane (b) *n*-pentane
(c) 2, 2-dimethylpropane (d) 2-methylbutane
8. Marsh gas mainly contains [1980]
(a) C_2H_2 (b) CH_4 (c) H_2S (d) CO

2 Integer Value Answer

9. The total number of stereoisomers that can exist for M is [Adv. 2015]

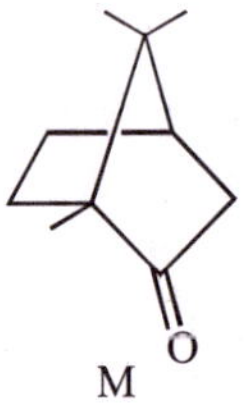

M

10. The total number(s) of stable conformers with non-zero dipole moment for the following compound is (are) [Adv. 2014]

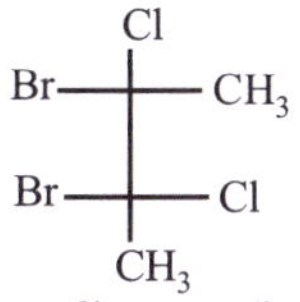

11. The maximum number of isomers (including stereoisomers) that are possible on monochlorination of the following compound is [2011]

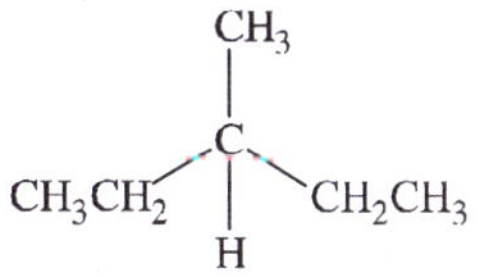

5 True / False

12. Photobromination of 2-methylpropane gives a mixture of 1-bromo-2-methylpropane and 2-bromo-2-methylpropane in the ratio of 9: 1. [1993 - 1 Mark]

6 MCQs with One or More than One Correct Answer

13. Which of the following reactions produce(s) propane as a major product ? [Adv. 2019]

(a) 

(b) 

(c) 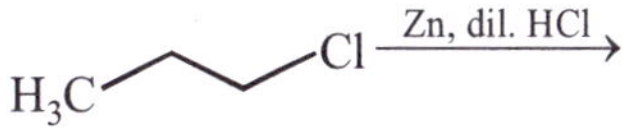

(d) 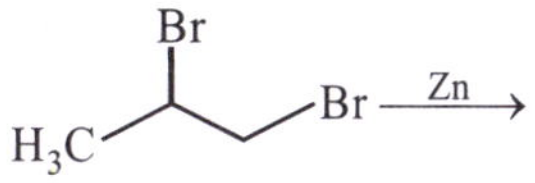

14. Which of the given statement(s) about N, O, P and Q with respect to M is (are) correct ? [2012]

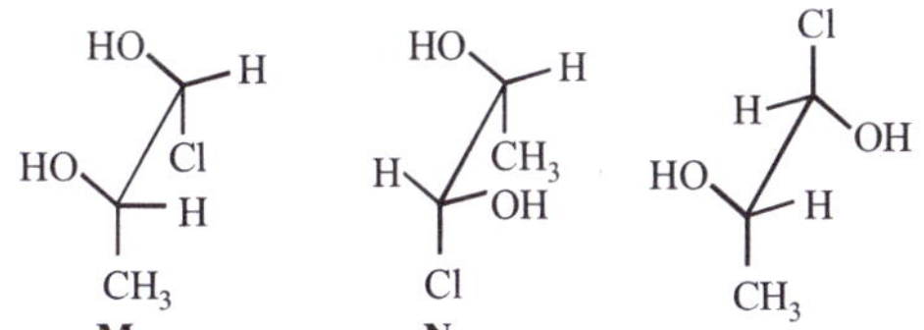

M N O

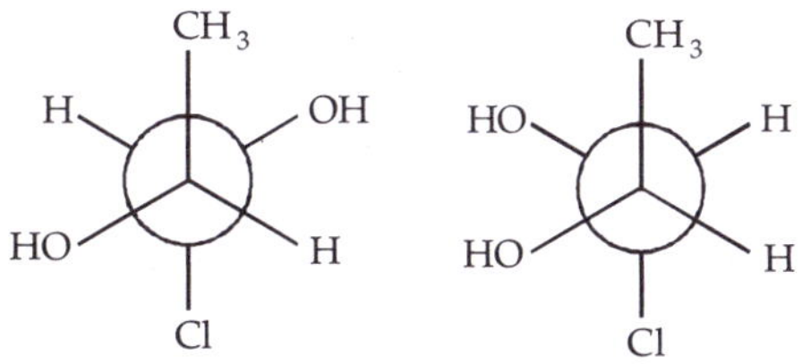

P Q

(a) M and N are non-mirror image stereoisomers
(b) M and O are identical
(c) M and P are enantiomers
(d) M and Q are identical

15. In the Newman projection for 2,2-dimethylbutane

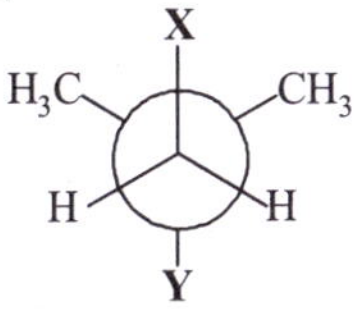

X and Y can respectively be [2010]

(a) H and H (b) H and C_2H_5
(c) C_2H_5 and H (d) CH_3 and CH_3

16. 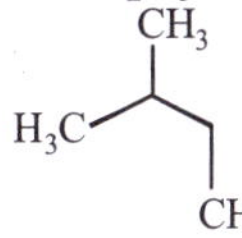 $\xrightarrow{Cl_2,\ h\nu}$ N (isomeric products); $C_5H_{11}Cl$ $\xrightarrow{\text{fractional distillation}}$ M (isomeric products)

Identify N and M [2006 - 5M, –1]

(a) 6, 4 (b) 6, 6 (c) 4, 4 (d) 3, 3

10 Subjective Problems

17. Draw Newman projection of relatively less stable staggered form of *n*-butane. The reason of low stability of this form is van der Waals repulsion, torsional strain, or both. [2004 - 2 Marks]

18. Optically active 2-iodobutane on treatment with NaI in acetone gives a product which does not show optical activity. Explain briefly. [1995 - 2 Marks]

19. *n*-Butane is produced by the monobromination of ethane followed by the Wurtz reaction. Calculate the volume of ethane at NTP required to produce 55 g *n*-butane, if the bromination takes place with 90 per cent yield and the Wurtz reaction with 85 per cent yield. [1989 - 3 Marks]

20. Give reason of the following : [1983 - 1 Mark]
Methane does not react with chlorine in the dark.

Topic 2 - Alkenes

1 MCQs with One Correct Answer

21. The number of optically active products obtained from the complete ozonolysis of the given compound is: [2012]

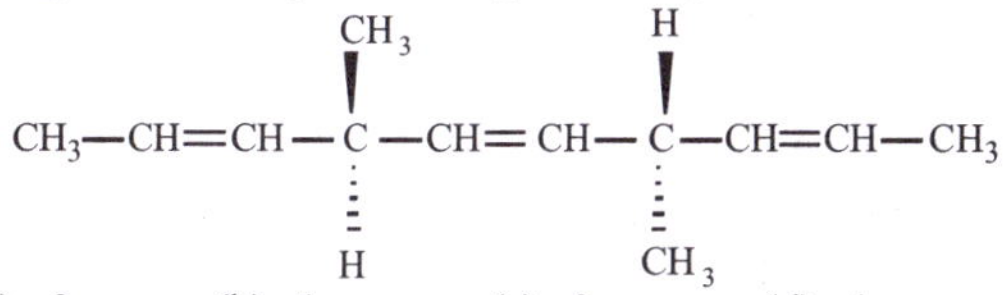

(a) 0 (b) 1 (c) 2 (d) 4

22. In allene (C_3H_4), the type(s) of hybridisation of the carbon atoms is (are) : [2012]

(a) sp and sp^3 (b) sp and sp^2
(c) only sp^3 (d) sp^2 and sp^3

23. Cyclohexene on ozonolysis followed by reaction with zinc dust and water gives compound E. Compound E on further treatment with aqueous KOH yields compound F. Compound F is [2007]

(a) 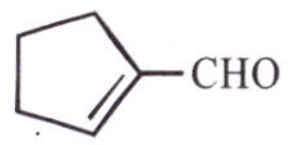 (b) 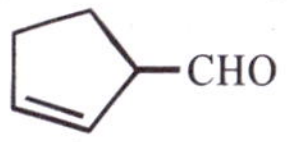

(c) 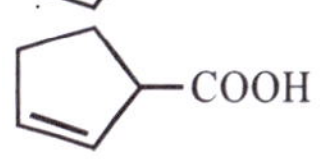 (d) 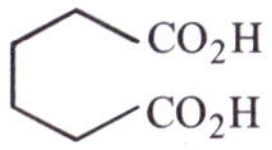

24. Identify the product, *P* in the following reaction :

$CH_3-CH=CH_2+NOCl \longrightarrow P$ [2006 - 3M, –1]

(a) $CH_3-CH(NO)-CH_2(Cl)$ (b) $CH_3-CH(Cl)-CH_2(NO)$

(c) $CH_2(NO)-CH_2-CH_2(Cl)$ (d) $CH_3-CH_2-CH(NO)(Cl)$

25. The nodal plane in the π-bond of ethene is located in
(a) the molecular plane
[2002S]
(b) a plane parallel to the molecular plane
(c) a plane perpendicular to the molecular plane which bisects the carbon - carbon σ -bond at right angle
(d) a plane perpendicular to the molecular plane which contains the carbon - carbon σ -bond.

26. The reaction of propene with HOCl proceeds via the addition of [2001S]
(a) H^+ in the first step
(b) Cl^+ in the first step
(c) OH^- in the first step
(d) Cl^+ and OH^- in a single step

27. In the presence of peroxide, hydrogen chloride and hydrogen iodide do not give *anti*-Markovnikov addition to alkenes because [2001S]
(a) both are highly ionic
(b) one is oxidizing and the other is reducing
(c) one of the steps is endothermic in both the cases
(d) all the steps are exothermic in both the cases

28. Which one of the following will react fastest with H_2 under catalytic hydrogenation condition ? [2000S]

(a) 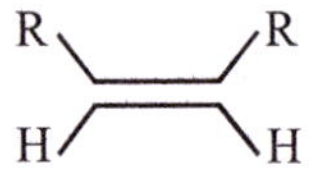 (b) 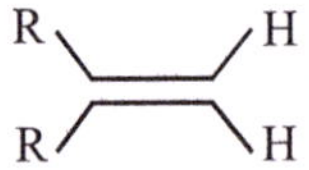

(c) 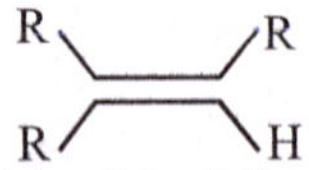 (d) 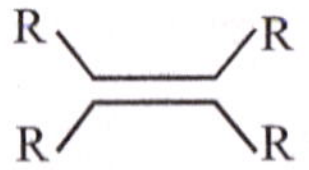

29. Which of the following compounds will exhibit geometrical isomerism ? [2000S]
(a) 1-Phenyl-2-butene (b) 3-Phenyl-1-butene
(c) 2-Phenyl-1-butene (d) 1,1-Diphenyl-1-propene

30. *anti*-Markovnikoff addition of HBr is not observed in :
(a) propene (b) 1-butene [1985 - 1 Mark]
(c) but-2-ene (d) pent-2-ene

31. Baeyer's reagent is : [1984 - 1 Mark]
(a) alkaline permanganate solution
(b) acidified permanganate solution
(c) neutral permanganate solution
(d) aqueous bromine solution

32. Which of the following compounds will exhibit *cis-trans* (geometrical) isomerism? [1983]
(a) 2-butene (b) 2-butyne
(c) 2-butanol (d) butanal

33. Which of the following decolourises alkaline $KMnO_4$ solution [1980]
(a) C_3H_8 (b) C_2H_4 (c) CH_4 (d) CCl_4

2 Integer Value Answer

34. The number of -CH_2- (methylene) groups in the product formed from the following reaction sequence is ________.

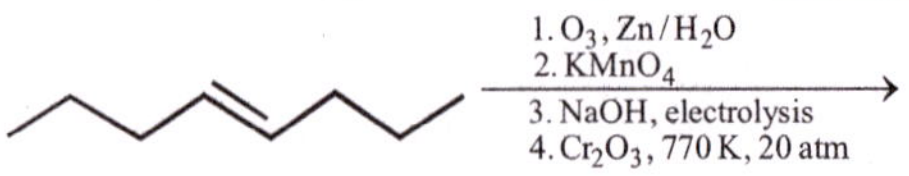

[Adv. 2022]

35. The total number of chiral molecules formed from one molecule of **P** on complete ozonolysis (O_3, Zn/H_2O) is ________. [Adv. 2022]

36. The maximum number of possible isomers (including stereoisomers) which may be formed on *mono*-bromination of 1-methylcyclohex-1-ene using Br_2 and UV light is _____. [Adv. 2021]

37. The total number of cyclic structural as well as stereo isomers possible for a compound with the molecular formula C_5H_{10} is [2009]

4 Fill in the Blanks

38. 1, 3-Butadiene with bromine in molar ratio of 1 : 1 generates predominantly............... . [1997 - 1 Mark]

39. Addition of water to acetylenic compounds is catalyzed by.........and............ [1993 - 1 Mark]

40. Kolbe electrolysis of potassium succinate gives CO_2 and .... [1993 - 1 Mark]

6 MCQs with One or More than One Correct Answer

41. The correct statement(s) for the following addition reactions is (are) [Adv. 2017]

(i) 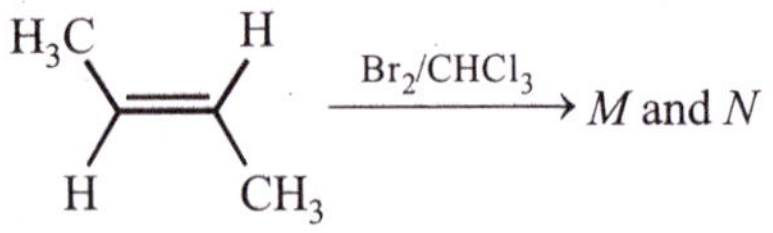

(ii) 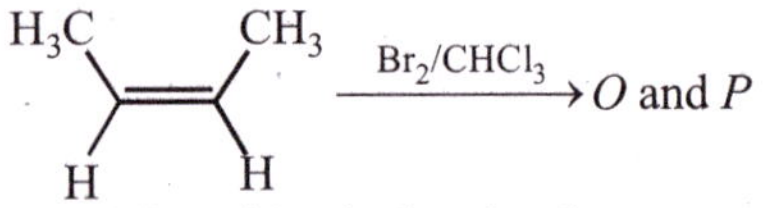

(a) *O* and *P* are identical molecules
(b) (*M* and *O*) and (*N* and *P*) are two pairs of diastereomers
(c) (*M* and *O*) and (*N* and *P*) are two pairs of enantiomers
(d) Bromination proceeds through *trans*-addition in both the reactions

42. Compound(s) that on hydrogenation produce(s) optically inactive compound(s) is (are) [Adv. 2015]

(a) 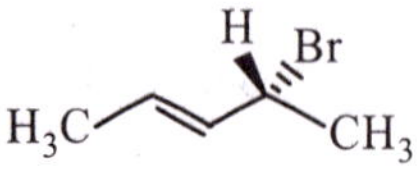 (b) 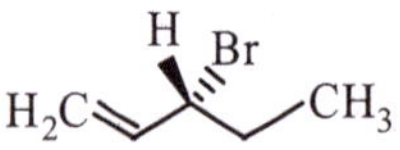

(c) 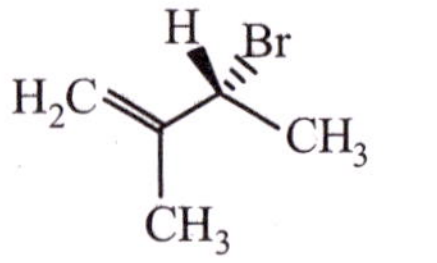 (d) 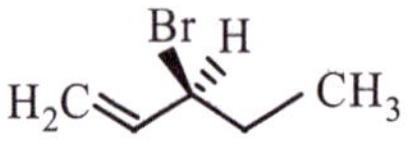

43. In the following reaction, the major product is [Adv. 2015]

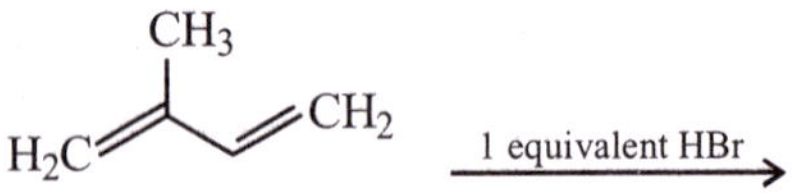

(a) 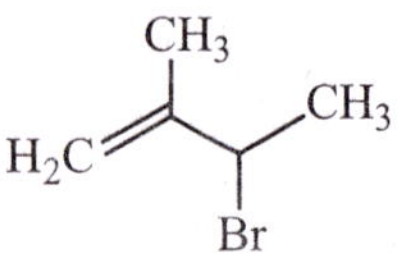 (b) 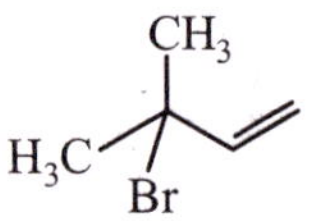

(c) 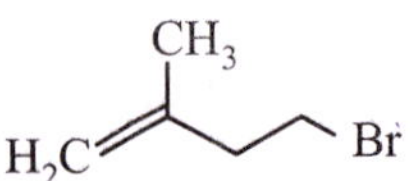 (d) 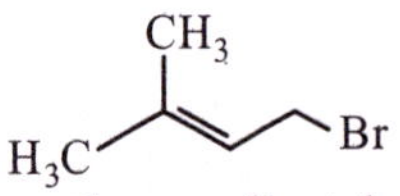

44. Which one of the following has the smallest heat of hydrogenation per mole? [1993 - 1 Mark]
(a) 1-butene (b) *trans*-2-butene
(c) *cis* -2-butene (d) 1,3-butadiene

7 Match the Following

45. **List-I** contains various reaction sequences and **List-II** contains the possible products.

Match each entry in **List-I** with the appropriate entry in **List-II** and choose the correct option. [Adv. 2024]

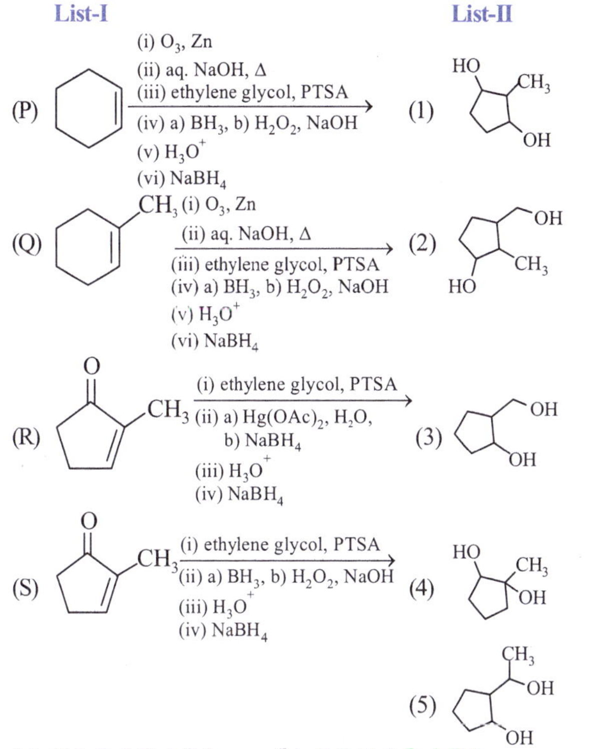

(a) P-3, Q-5, R-4, S-1 (b) P-3, Q-2, R-4, S-1
(c) P-3, Q-5, R-1, S-4 (d) P-5, Q-2, R-4, S-1

9 Assertion and Reason/Statement Type Questions

STATEMENT-1 (Assertion) and **STATEMENT-2 (Reason)**. Each question has 4 choices (a), (b), (c) and (d) out of which ONLY ONE is correct. Mark your answer as

(a) If both Statement-1 and Statement-2 are correct, and Statement -2 is the correct explanation of the Statement -2.
(b) If both Statement-1 and Statement-2 are correct, but Statement -2 is not the correct explanation of the Statement -1.
(c) If Statement -1 is correct but Statement -2 is incorrect.
(d) If Statement -1 is incorrect but Statement -2 is correct.

46. **Assertion :** Addition of bromine to *trans*–2–butene yields *meso*–2,3–dibromobutane.
Reason : Bromine addition to an alkene is an electrophilic addition. [2001S]

47. **Assertion :** 1-Butene on reaction with HBr in the presence of a peroxide produces 1-bromobutane.
Reason : It involves the formation of a primary radical. [2000S]

48. **Assertion :** Addition of Br_2 to 1-butene gives two optical isomers.
Reason : The product contains one asymmetric carbon. [1998 - 2 Marks]

10 Subjective Problems

49. An alkene (A) $C_{16}H_{16}$ on ozonolysis gives only one product (B) C_8H_8O. Compund (B) on reaction with NaOH/I_2 yields sodium benzoate. Compound (B) reacts with KOH/NH_2NH_2 yielding a hydrocarbon (C) C_8H_{10}. Write the structures of compounds (B) and (C). Based on this information, two isomeric structures can be proposed for alkene (A). Write their structures and identify the isomer which on catalytic hydrogenation (H_2/Pd – C) gives a racemic mixture. [2001 - 5 Marks]

50. Complete the following reactions with appropriate structures of products/reagents.

(*i*) $C_6H_5CH{=}CH_2 \xrightarrow{Br_2} [A] \xrightarrow[\text{(ii) } CH_3I]{\text{(i) } NaNH_2 \text{ (3.0 equiv.)}} [B]$ [1998 - 2 + 2 Marks]

(*ii*) 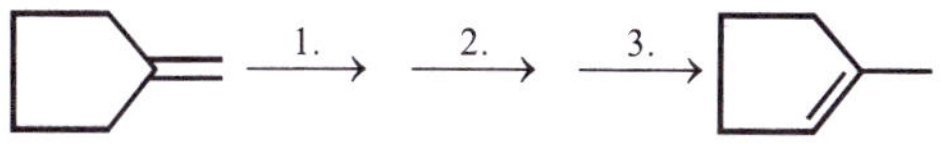

[1999 - 3 Marks]

51. The hydrocarbon *A*, adds one mole of hydrogen in the presence of a platinum catalyst to form *n*-hexane. When *A* is oxidized vigorously with $KMnO_4$, a single carboxylic acid, containing three carbon atoms, is isolated. Give the structure of *A* and explain. [1997 - 2 Marks]

52. Give the structures of the major organic products from 3-ethyl-2-pentene under each of the following reaction conditions. [1996 - 3 Marks]
(a) HBr in the presence of peroxide
(b) Br_2/H_2O
(c) $Hg(OAc)_2/H_2O$; $NaBH_4$

53. An organic compound **E**(C_5H_8) on hydrogenation gives compound **F**(C_5H_{12}). Compound **E** on ozonolysis gives formaldehyde and 2-ketopropanal. Deduce the structure of compound **E**. [1995 - 2 Marks]

54. 1, 4–Pentadiene reacts with excess of HCl in the presence of benzoyl peroxide to give compound *X* which upon reaction with excess of Mg in dry ether forms *Y*. Compound *Y* on treatment with ethyl acetate followed by dilute acid yields *Z*. Identify the structures of compounds *X*, *Y* and *Z*. [1995 - 4 Marks]

55. When gas *A* is passed through dry KOH at low temperature, a deep red coloured compound *B* and a gas *C* are obtained. The gas *A*, on reaction with but-2-ene, followed by treatment with Zn/H_2O yields acetaldehyde. Identify A, B and C. [1994 - 3 Marks]

56. Identify, **D**(C_6H_{12}), an optically active hydrocarbon which on catalytic hydrogenation gives an optically inactive compound, C_6H_{14}. [1993 - 1 Mark]

57. Identify, **B**(C_4H_8) which adds on HBr in the presence and in the absence of peroxide to give the same product, C_4H_9Br. [1993 - 1 Mark]

58. How would you distinguish between cyclohexane and cyclohexene. [1988 - 1 Mark]

59. 2-Methylpropene can be converted into isobutyl bromide by hydrogen bromide, is true under what conditions? [1984 - 1 Mark]

60. Give reasons for the following :
(i) Propene reacts with HBr to give isopropyl bromide but does not give *n*-propyl bromide. [1983 - 1 Mark]
(ii) The central carbon-carbon bond in 1, 3 – butadiene is shorter than that in *n*-butane. [1998 - 2 Marks]

61. State with balanced equations, what happens when propene is bubbled through a hot aqueous solution of potassium permanganate. [1982 - 1 Mark]

62. Write the structural formula of the major product in the following case : [1997 - 1 Mark]

 $+ CHBr_3 + t-BuOK \longrightarrow$ ----

Topic 3 - Alkynes

1 MCQs with One Correct Answer

63. The synthesis of 3-octyne is achieved by adding a bromoalkane into a mixture of sodium amide and an alkyne. The bromoalkane and alkyne respectively are [2010]
(a) $BrCH_2CH_2CH_2CH_2CH_3$ and $CH_3CH_2C\equiv CH$
(b) $BrCH_2CH_2CH_3$ and $CH_3CH_2CH_2C\equiv CH$
(c) $BrCH_2CH_2CH_2CH_2CH_3$ and $CH_3C\equiv CH$
(d) $BrCH_2CH_2CH_2CH_3$ and $CH_3CH_2C\equiv CH$

64. Which of the following is used for the conversion of 2-hexyne into *trans*-2-hexene? [2004S]
(a) $H_2/Pd/BaSO_4$ (b) H_2, PtO_2
(c) $NaBH_4$ (d) $Li\text{-}NH_3/C_2H_5OH$

65. $Ph-C\equiv C-CH_3 \xrightarrow{Hg^{2+}/H^+} A$. *A* is: [2003S]

(a) 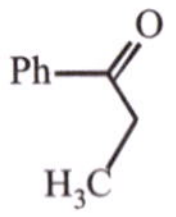 (b) 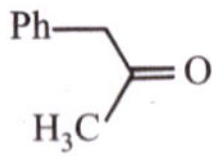

(c) 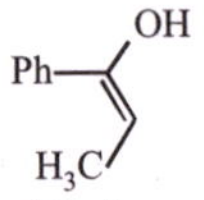 (d) 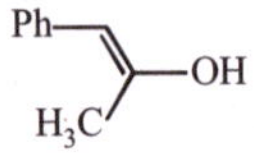

66. Identify the reagent from the following list which can easily distinguish between 1-butyne and 2-butyne [2002S]
(a) bromine, CCl_4 (b) H_2, Lindlar catalyst
(c) dilute H_2SO_4, $HgSO_4$ (d) ammonical Cu_2Cl_2 solution

67. 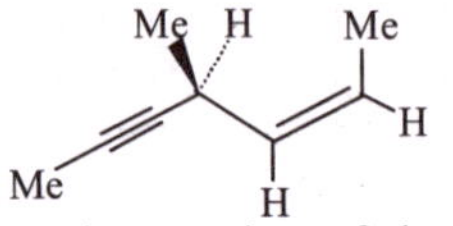

Hydrogenation of the above compound in the presence of poisoned palladium catalyst gives [2001S]
(a) an optically active compound
(b) an optically inactive compound
(c) a racemic mixture
(d) a diastereomeric mixture

68. Propyne and propene can be distinguished by [2000S]
(a) conc. H_2SO_4 (b) Br_2 in CCl_4
(c) dil. $KMnO_4$ (d) $AgNO_3$ in ammonia

69. The product(s) obtained via oxymercuration ($HgSO_4 + H_2SO_4$) of 1- butyne would be [1999 - 2 Marks]
(a) $CH_3-CH_2-\overset{O}{\overset{||}{C}}-CH_3$
(b) $CH_3-CH_2-CH_2-CHO$
(c) $CH_3-CH_2-CHO+HCHO$
(d) $CH_3\,CH_2\,COOH+HCOOH$

70. Acidic hydrogen is present in : [1985 - 1 Mark]
(a) ethyne (b) ethene (c) benzene (d) ethane

71. When propyne is treated with aqueous H_2SO_4 in presence of $HgSO_4$ the major product is [1983 - 1 Mark]
(a) propanal
(b) propyl hydrogensulphate
(c) acetone
(d) propanol

3 Numeric / New Stem Based Questions

72. Treatment of buta-1,3-diyne with $NaNH_2$ (2 equivalents), followed by reaction with excess of trans-$CH_3-CH=CH-CH_2-Br$ gives X as the major product. The maximum number of carbon atoms that are collinear (in a straight line) in X is ______. [Adv. 2026]

4 Fill in the Blanks

73. The starting material for the manufacture of polyvinyl chloride is obtained by reacting HCl with .... [1983 - 1 Mark]

74. Acetylene is treated with excess sodium in liquid ammonia. The product is reacted with excess methyl iodide. The final product is .... [1983 - 1 Mark]

75. .... is most acidic. (Ethane, Ethene, Ethyne) [1981 - 1 Mark]

5 True / False

76. Moist ethylene can be dried by passing it through concentrated sulphuric acid. [1982 - 1 Mark]

6 MCQs with One or More than One Correct Answer

77. Consider the following transformations of a compound P.

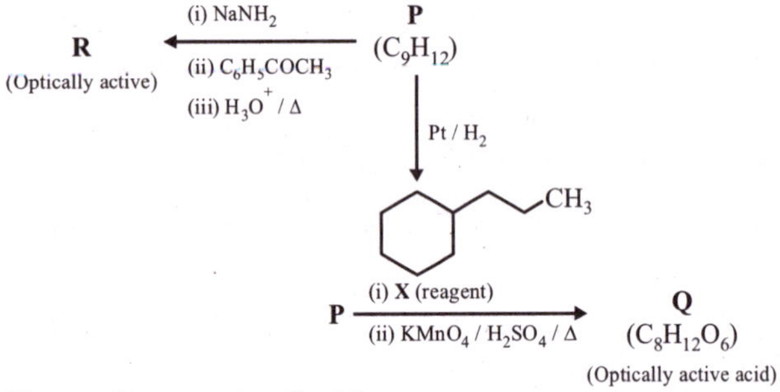

Choose the correct option(s). [Adv. 2020]

(a) P is 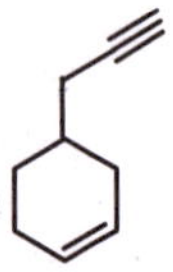 (b) X is Pd-C/quinoline/H_2

(c) P is 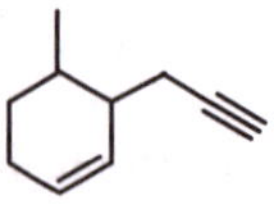 (d) R is 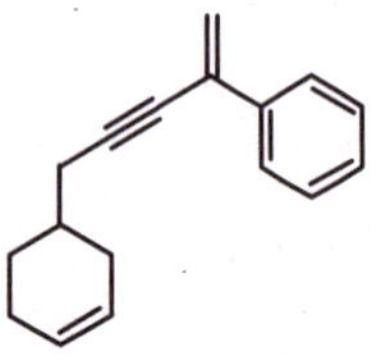

8 Comprehension Based Questions

Passage-I

Schemes 1 and 2 describe sequential transformation of alkynes *M* and *N*. Consider only the major products formed in each step for both the schemes.

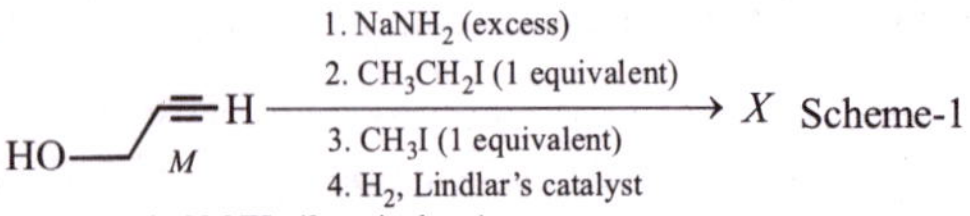

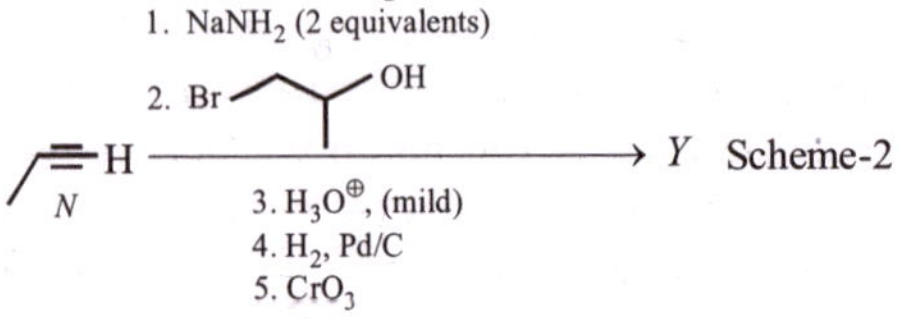

78. The product X is [Adv. 2014]

(a)

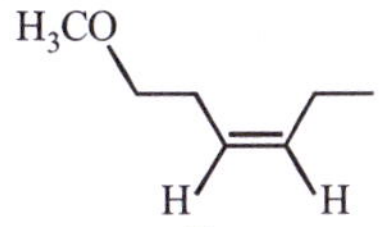

(b)

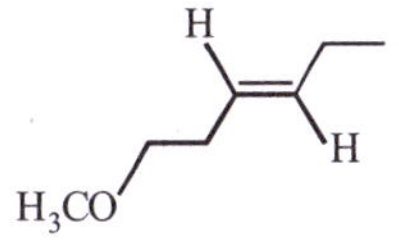

(c)

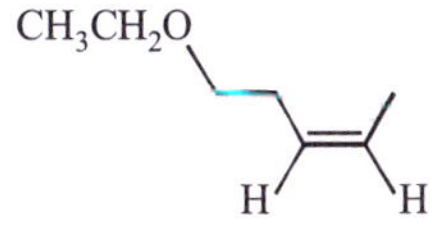

(d)

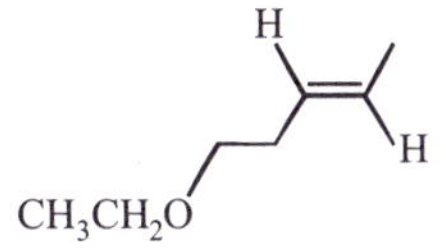

79. The correct statement with respect to product Y is [Adv. 2014]

(a) It gives a positive Tollen's test and is a functional isomer of X

(b) It gives a positive Tollen's test and is a geometrical isomer of X

(c) It gives a positive iodoform test and is a functional isomer of X

(d) It gives a positive iodoform test and is a geometrical isomer of X

Passage-II

An acyclic hydrocarbon P, having molecular formula C_6H_{10}, gave acetone as the only organic product through the following sequence of reactions, in which Q is an intermediate organic compound.

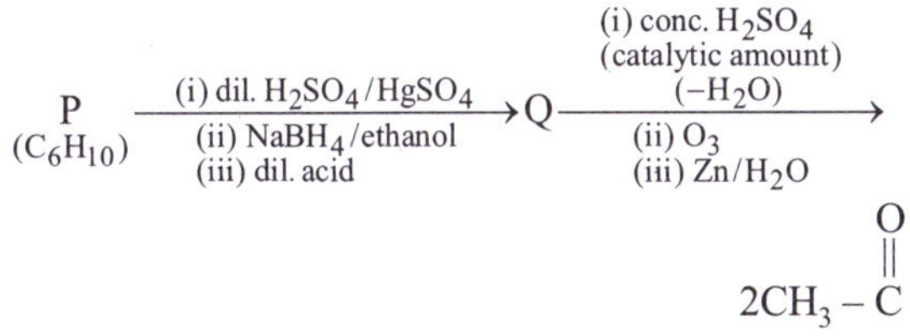

[2011 - I]

80. The structure of compound P is

(a) $CH_3CH_2CH_2CH_2C\equiv CH$

(b) $CH_3CH_2C\equiv CCH_2CH_3$

(c)

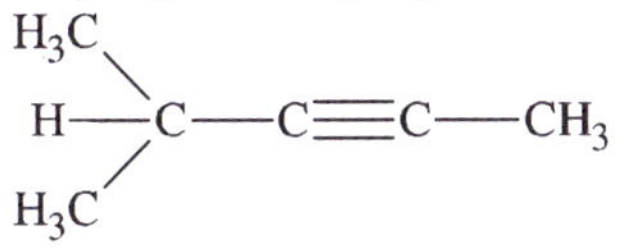

(d)

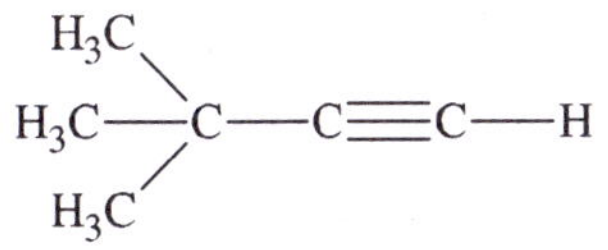

81. The structure of the compound Q is

(a)

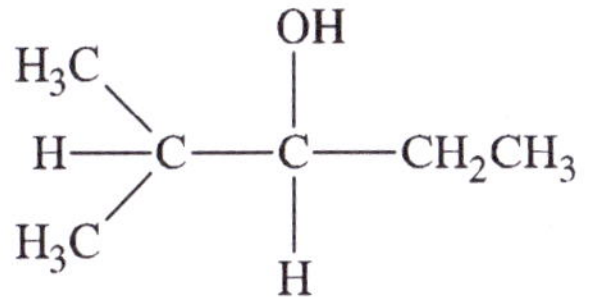

(b)

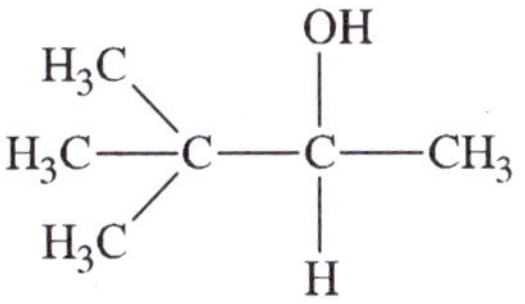

(c)

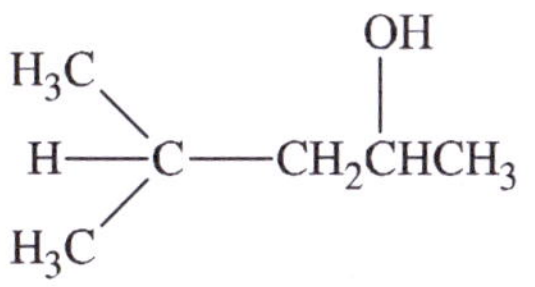

(d)

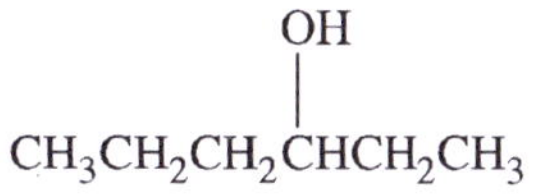

10 Subjective Problems

82. Write down the heterogeneous catalyst involved in the polymerisation of ethylene. [2003 - 2 Marks]

83. Identify X, Y and Z in the following synthetic scheme and write their structures.

$$CH_3CH_2C\equiv C-H \xrightarrow[\text{(ii) } CH_3CH_2Br]{\text{(i) } NaNH_2} X \xrightarrow{H_2/Pd-BaSO_4} Y \xrightarrow{\text{alkaline } KMnO_4} Z$$

Is the compound Z optically active? Justify your answer. [2002 - 5 Marks]

84. A hydrocarbon A, of the formula C_8H_{10}, on ozonolysis gives compound $B(C_4H_6O_2)$ only. The compound B can also be obtained from the alkyl bromide, $C(C_3H_5Br)$ upon treatment with magnesium in dry ether, followed by carbon dioxide and acidification. Identify A, B and C and also give equations for the reactions. [1996 - 3 Marks]

85. Draw the stereochemical structures of the products in the following reactions : [1994 - 4 Marks]

$$R-C\equiv C-R \xrightarrow[\text{Lindlar catalyst}]{H_2}$$

86. How would you distinguish between 2-butyne and 1-butyne. [1985 - 1 Mark]

87. A certain hydrocarbon A was found to contain 85.7 percent carbon and 14.3 per cent hydrogen. This compound consumes 1 molar equivalent of hydrogen to give a saturated hydrocarbon B. 1.00 g of hydrocarbon A just decolourized 38.05 g of a 5 per cent solution (by weight) of Br_2 in CCl_4. Compound A, on oxidation with concentrated $KMnO_4$, gave compound C (molecular formula C_4H_8O) and acetic acid. Compound C could easily be prepared by the action of acidic aqueous mercuric sulphate on 2- butyne. Determine the molecular formula of A and deduce the structure of A, B and C. [1984 - 6 Marks]

88. 'Ethyne and its derivatives will give white precipitate with ammonical silver nitrate solution', is true under what conditions. [1984 - 1 Mark]

89. Outline the reaction sequence for the conversion of ethene to ethyne (the number of steps should not be more than two). [1981 - 1 Mark]

90. Write the structural formula of the major product in each of the following cases :

(*i*) the compound obtained by the hydration of ethyne is treated with dilute alkali [1981 - ½ Mark]

(*ii*)

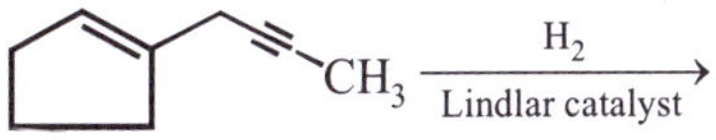

[2000 - 1 Mark]

91. Give one characteristic test which would distinguish. CH_4 from C_2H_2 [1979]

Topic 4 - Aromatic Hydrocarbons

1 MCQs with One Correct Answer

92. Identify the correct order of reactivity in electrophilic substitution reactions of the following compounds **[2002S]**

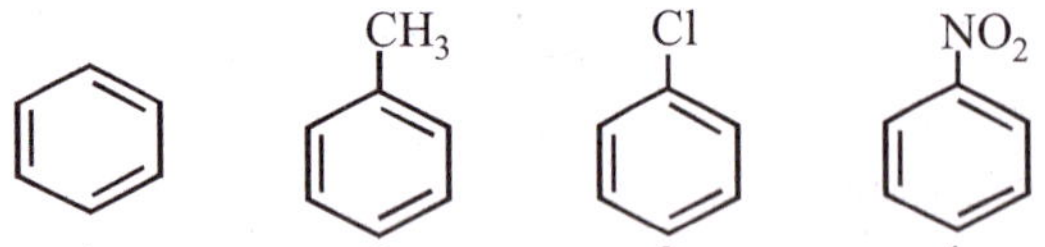

1 2 3 4

(a) $1>2>3>4$ (b) $4>3>2>1$
(c) $2>1>3>4$ (d) $2>3>1>4$

93. Which of the following compounds does not dissolve in conc. H_2SO_4 even on warming? **[1983 - 1 Mark]**

(a) ethylene (b) benzene
(c) hexane (d) aniline

94. Among the following, the compound that can be most readily sulphonated is **[1982]**

(a) benzene (b) nitrobenzene
(c) toluene (d) chlorobenzene

95. The bond order of individual carbon-carbon bonds in benzene is **[1981]**

(a) one (b) two
(c) between one and two (d) one and two, alternately

2 Integer Value Answer

96. Among the following, the number of aromatic compound(s) is **[Adv. 2017]**

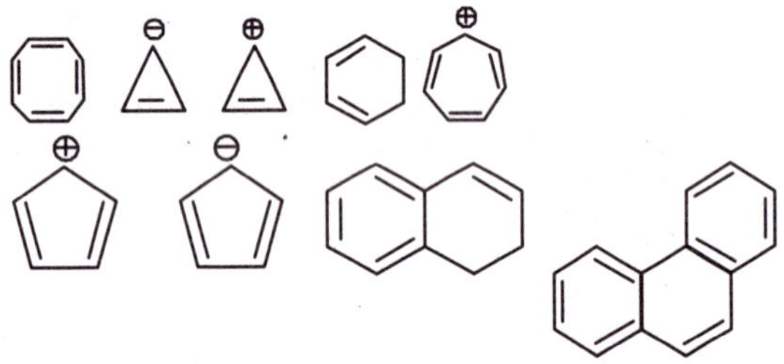

3 Numeric / New Stem Based Questions

Question Stem for Question Nos. 97 and 98 **[Adv. 2026]**

Consider the following reaction sequence in which **J**, **K**, **L** and **M** are the major products.

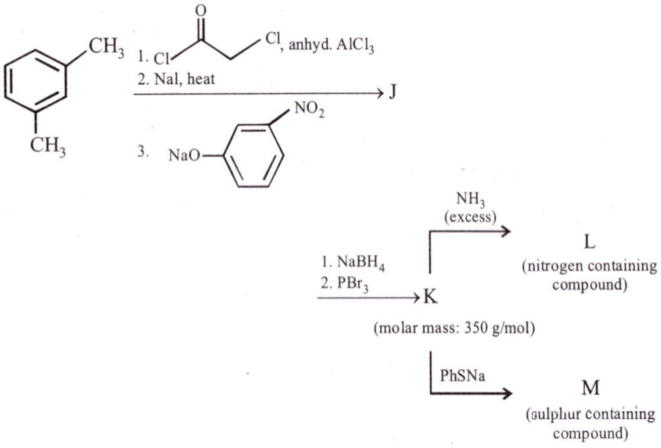

Given:

Atomic mass (in amu): H : 1, C : 12, N : 14, O : 16, S : 32, Br : 80, Ba : 137

97. The volume of 1 M aqueous H_2SO_4 required to completely neutralize the ammonia evolved from 5.72 g of **L** in Kjeldahl's method of nitrogen estimation is ____ mL.

98. In sulphur estimation by Carius method, the amount of $BaSO_4$ formed from 3.79 g of **M** is ____ g.

4 Fill in the Blanks

99. The bond dissociation energy needed to form the benzyl radical from toluene is........than the formation of the methyl radical from methane. **[1994 - 1 Mark]**

100. Kolbe electrolysis of potassium succinate gives CO_2 and .... **[1993 - 1 Mark]**

101. .... ring is most strained. **[1981]**

(Cyclopropane, Cyclobutane, Cyclopentane)

5 True / False

102. An electron donating substituent in benzene orients the incoming electrophilic group to the meta position. **[1987]**

6 MCQs with One or More than One Correct Answer

103. The reaction sequence(s) that would lead to *o*-xylene as the major product is (are) **[Adv. 2021]**

(a) 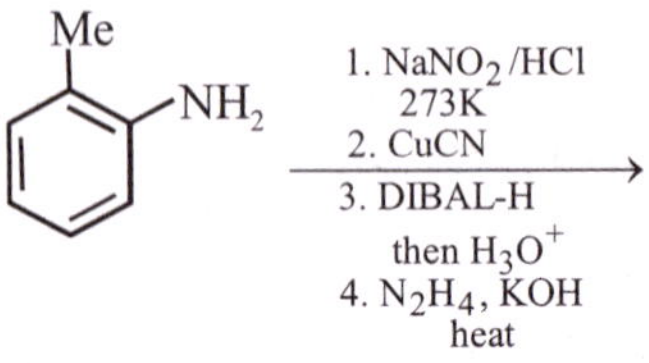

(b) 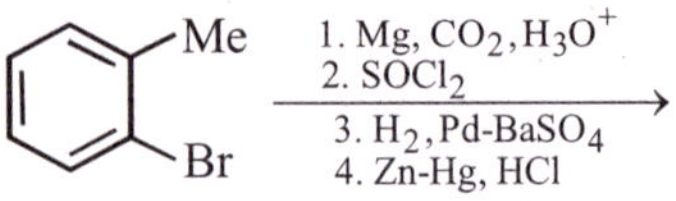

(c) 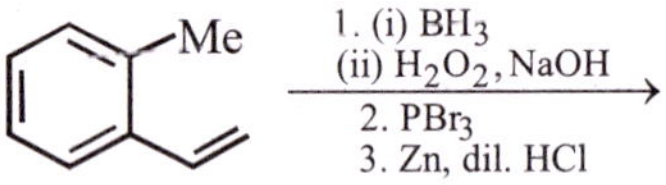

(d) 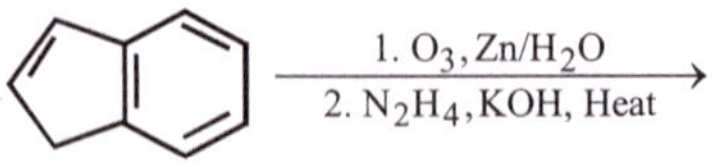

104. Choose the correct option(s) that give(s) an aromatic compound as the major product. **[Adv. 2019]**

(a) 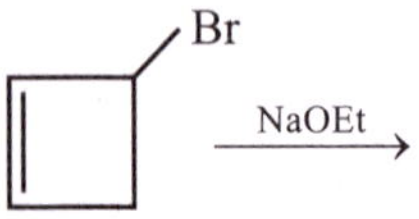

(b) 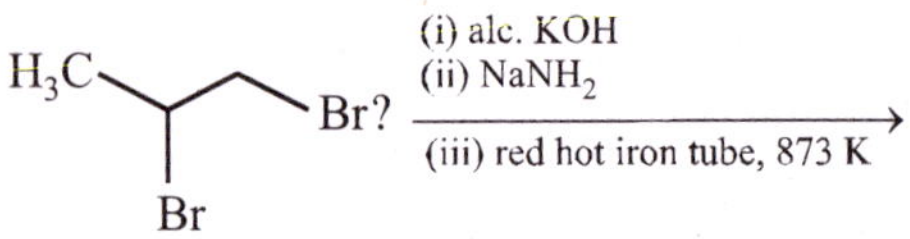

(c) 

(d) 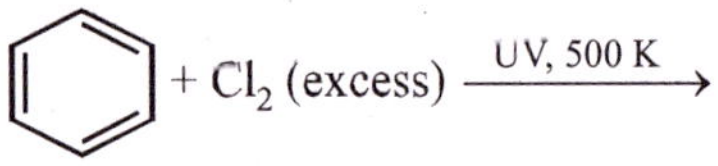

105. Among the following, reaction(s) which gives (give) tert-butyl benzene as the major product is (are) **[Adv. 2016]**

(a) 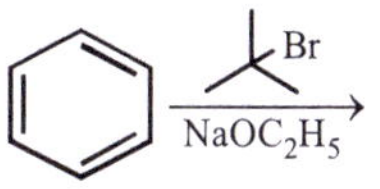
(b) 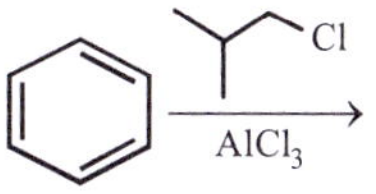

(c) 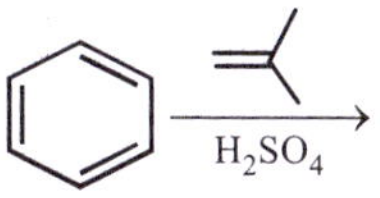
(d) 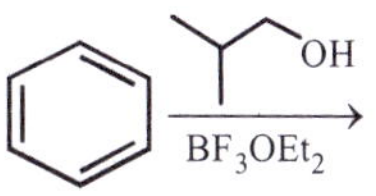

106. The major product U in the following reactions is

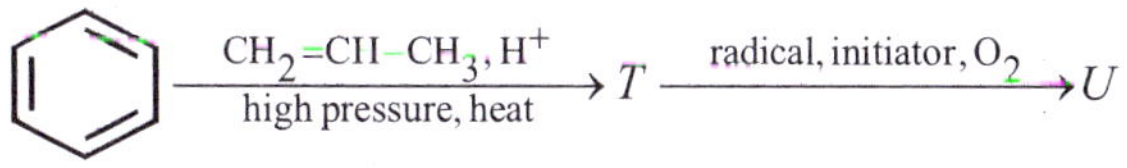

[Adv. 2015]

(a) 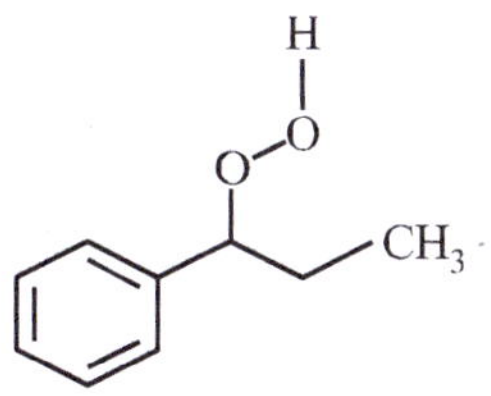

(b) 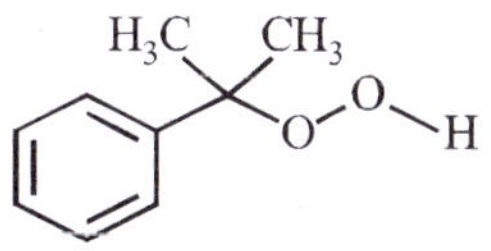

(c) 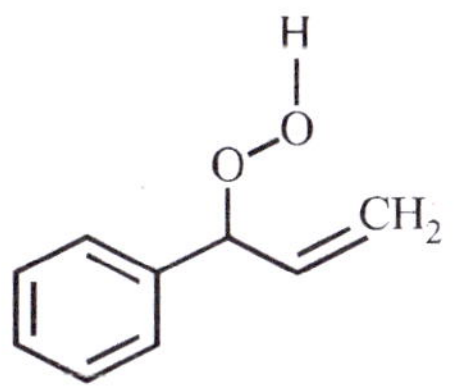

(d) 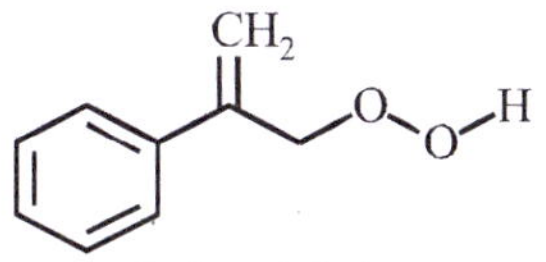

107. Among P, Q, R and S, the aromatic compound(s) is/are **[Adv. 2013-I]**

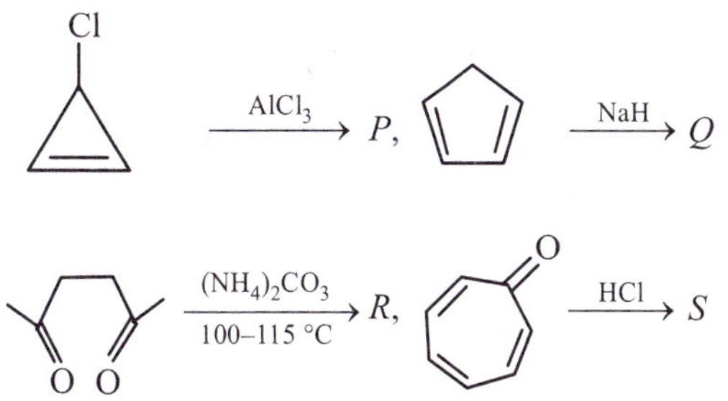

(a) P (b) R (c) Q (d) S

108. Toluene, when treated with Br_2/Fe, gives p-bromotoluene as the major product because CH_3 group **[1999 - 3 Marks]**

(a) is para directing
(b) is meta directing
(c) activates the ring by hyperconjugation
(d) deactivates the ring

109. An aromatic molecule will **[1999]**

(a) have $4n$ π electrons
(b) have $(4n+2)$ π electrons
(c) be planar
(d) be cyclic

10 Subjective Problems

110. Give reasons for the following :

(i) *tert*-Butylbenzene does not give benzoic acid on treatment with acidic $KMnO_4$. **[2000 - 1 Mark]**

(ii) Toluene reacts with bromine in the presence of light to give benzyl bromide while in presence of $FeBr_3$ it gives p-bromotoluene. Give explanation for the above observations. **[1996 - 2 Marks]**

(iii) Although benzene is highly unsaturated, normally it does not undergo addition reaction . **[1983 - 1 Mark]**

111. Write the structural formula of the major product in each of the following cases :

(i) 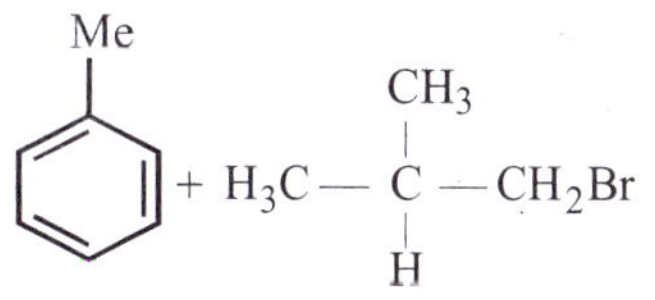

+ Anhyd. $AlCl_3$ $\longrightarrow$ --- **[1997 - 1 Mark]**

(ii) $C_6H_5C_2H_5 \xrightarrow[\text{2. NaCN}]{\text{1.}Br_2\text{, Heat,Light}}$ **[1994 - 1 Mark]**

(iii) $C_6H_6 + (CH_3)_2CHCH_2OH \xrightarrow{H_2SO_4}$ **[1994 - 1 Mark]**

(iv)  $+ (CH_3)_2CHCH_2Cl \xrightarrow{AlCl_3}$ **[1992 - 1 Mark]**

112. Show the steps to carry out the following transformations.

(i) Ethylbenzene → benzene **[1998 - 2 Marks]**

(ii) Ethylbenzene → 2- phenylpropionic acid. **[1998 - 3 Marks]**

113. Give reason of the following :

Normally, benzene gives electrophilic substitution reaction rather than electrophilic addition reaction although it has double bonds. **[1994]**

114. Arrange the following in :

benzene, toluene, methoxybenzene, chlorobenzene in increasing order of reactivity towards sulphonation with fuming sulphuric acid. **[1988]**

115. Write down the reactions involved in the preparation of the following, using the reagents indicated against it in parenthesis.

Ethylbenzene from benzene [C_2H_5OH, PCl_5, anhydrous $AlCl_3$]. **[1984 - 2 Marks]**

Answer Key

Part-A (JEE Main)																			
1	(2)	24	(2)	47	(2)	70	(4)	93	(4)	116	(3)	139	(3)	162	(13)	185	(4)	208	(1)
2	(4)	25	(2)	48	(2 or 4)	71	(2)	94	(2)	117	(4)	140	(1)	163	(3)	186	(4)	209	(3)
3	(3)	26	(1)	49	(2)	72	(1)	95	(3)	118	(3)	141	(4)	164	(2)	187	(4)	210	(4)
4	(4)	27	(3)	50	(2)	73	(1)	96	(4)	119	(3)	142	(3)	165	(2)	188	(4)	211	(1)
5	(2)	28	(3)	51	(1)	74	(1)	97	(4)	120	(27)	143	(4)	166	(2)	189	(2)	212	(1)
6	(4)	29	(4)	52	(4)	75	(3)	98	(2)	121	(8)	144	(4)	167	(3)	190	(2)	213	(1)
7	(4)	30	(30)	53	(3)	76	(4)	99	(3)	122	(4)	145	(4)	168	(3)	191	(4)	214	(3)
8	(2)	31	(14)	54	(1)	77	(3)	100	(4)	123	(1)	146	(3)	169	(1)	192	(3)	215	(1)
9	(3)	32	(9)	55	(3) or (4)	78	(4)	101	(3)	124	(4)	147	(2)	170	(4)	193	(4)	216	(2)
10	(3)	33	(3)	56	(2)	79	(3)	102	(2)	125	(70)	148	(1)	171	(4)	194	(3)	217	(3)
11	(1)	34	(6)	57	(2)	80	(2)	103	(2)	126	(3)	149	(2)	172	(3)	195	(1, 2)	218	(4)
12	(1)	35	(9)	58	(4)	81	(2)	104	(4)	127	(4)	150	(2)	173	(3)	196	(3)	219	(3)
13	(1)	36	(3)	59	(1)	82	(4)	105	(4)	128	(2)	151	(4)	174	(1)	197	(3)	220	(4)
14	(3)	37	(12)	60	(1)	83	(2)	106	(4)	129	(2)	152	(2)	175	(2)	198	(2)	221	(1)
15	(4)	38	(8)	61	(2)	84	(1)	107	(2, 3)	130	(2)	153	(4)	176	(4)	199	(1)	222	(78)
16	(3)	39	(60)	62	(4)	85	(4)	108	(3)	131	(3)	154	(3)	177	(3)	200	(4)	223	(20)
17	(4)	40	(4)	63	(1)	86	(2, 3)	109	(3)	132	(1)	155	(1)	178	(1)	201	(3)	224	(2)
18	(2)	41	(4)	64	(3)	87	(4)	110	(3)	133	(1)	156	(15)	179	(1)	202	(4)	225	(1)
19	(3)	42	(3)	65	(4)	88	(2)	111	(1)	134	(4)	157	(8)	180	(1)	203	(1)	226	(2)
20	(2)	43	(2)	66	(2)	89	(1)	112	(4)	135	(4)	158	(2)	181	(1)	204	(3)	227	(3)
21	(4)	44	(2)	67	(2)	90	(2)	113	(4)	136	(1)	159	(3)	182	(1)	205	(4)	228	(6)
22	(4)	45	(4)	68	(4)	91	(4)	114	(1)	137	(2)	160	(10)	183	(2)	206	(3)	229	(80)
23	(4)	46	(3)	69	(4)	92	(2)	115	(3)	138	(3)	161	(7)	184	(4)	207	(2)		
Part-B (JEE Advanced)																			
1	(c)	10	(3)	23	(a)	32	(a)	42	(b, d)	65	(a)	74	2-butyne	93	(c)	102	False		
2	(b)	11	(8)	24	(b)	33	(b)	43	(d)	66	(d)	75	Ethyne	94	(c)	103	(a, b)		
3	(c)	12	False	25	(a)	34	(0)	44	(b)	67	(b)	76	False	95	(c)	104	(b, c)		
4	(b)	13	(a, c)	26	(b)	35	(2)	45	(a)	68	(d)	77	(b, c)	96	(4)	105	(b, c, d)		
5	(d)	14	(a, b, c)	27	(c)	36	(13)	46	(b)	69	(a)	78	(a)	97	(10.00)	106	(b)		
6	(a)	15	(b, d)	28	(a)	37	(7)	47	(c)	70	(a)	79	(c)	98	(2.33)	107	(a, b, c, d)		
7	(a)	16	(a)	29	(a)	39	H_2SO_4, $HgSO_4$	48	(a)	71	(c)	80	(d)	99	(Less)	108	(a, c)		
8	(b)	21	(a)	30	(c)	40	ethylene	63	(d)	72	(6.00)	81	(b)	100	ethylene	109	(b, c, d)		
9	(2)	22	(b)	31	(a)	41	(b, d)	64	(d)	73	C_2H_2	92	(c)	101	Cyclopropane				

Chapter-14

Environmental Chemistry

Part-A (JEE Main)

Topic 1 - Air Pollution

Multiple Choice Questions (MCQs)

1. Given below are two statements: [July 27, 2022 (II)]
Statement I : The non bio-degradable fly ash and slag from steel industry can be used by cement industry.
Statement II : The fuel obtained from plastic waste is lead free.
In the light of the above statements, choose the **most appropriate** answer from the options given below:
(1) Both Statement I and Statement II are correct.
(2) Both Statement I and Statement II are incorrect.
(3) Statement I is correct but Statement II is incorrect.
(4) Statement I is incorrect but Statement II is correct.

2. $\dot{C}l + CH_4 \rightarrow A + B$
A and B in the above atmospheric reaction step are: [July 26, 2022 (I)]
(1) C_2H_6 and Cl_2 (2) $\dot{C}HCl_2$ and H_2
(3) $\dot{C}H_3$ and HCl (4) C_2H_6 and HCl

3. The photochemical smog does not generally contain: [July 25, 2022 (I)]
(1) NO (2) NO_2 (3) SO_2 (4) HCHO

4. The acid that is believed to be mainly responsible for the damage of Taj Mahal is [June 29, 2022 (I)]
(1) Sulfuric acid (2) Hydrofluoric acid
(3) Phosphoric acid (4) Hydrochloric acid

5. Sulphur dioxide is one of the components of polluted air. SO_2 is also a major contributor to acid rain. The correct and complete reaction to represent acid rain caused by SO_2 is : [June 29, 2022 (II)]
(1) $2SO_2 + O_2 \rightarrow 2SO_3$ (2) $SO_2 + O_3 \rightarrow SO_3 + O_2$
(3) $SO_2 + H_2O_2 \rightarrow H_2SO_4$ (4) $2SO_2 + O_2 + 2H_2O \rightarrow 2H_2SO_4$

6. Correct statement about photo-chemical smog is :
(1) It occurs in humid climate. [June 28, 2022 (II)]
(2) It is a mixture of smoke, fog and SO_2
(3) It is reducing smog.
(4) It results from reaction of unsaturated hydrocarbons.

7. Given below are two statements: [June 27, 2022 (I)]
Statement I: Classical smog occurs in cool humid climate. It is a reducing mixture of smoke, fog and sulphur dioxide
Statement II: Photochemical smog has components, ozone, nitric oxide, acrolein, formaldehyde, PAN etc.
In the light of above statements, choose the **most appropriate** answer from the options given below.
(1) Both Statement I and Statement II are correct.
(2) Both Statement I and Statement II are incorrect.
(3) Statement I is correct but statement II is incorrect.
(4) Statement I is incorrect but Statement II is correct.

8. Polar stratospheric clouds facilitate the formation of: [June 26, 2022 (I)]
(1) $ClONO_2$ (b) HOCl (3) ClO (4) CH_4

9. Some gases are responsible for heating of atmosphere (green house effect). Identify from the following the gaseous species which does not cause it. [June 24, 2022 (II)]
(1) CH_4 (2) O_3 (3) H_2O (4) N_2

10. The gas 'A' is having very low reactivity reaches to stratosphere. It is non–toxic and non–flammable but dissociated by UV—radiations in stratosphere. The intermediates formed initially from the gas 'A' are : [Aug. 27, 2021 (I)]
(1) $\dot{C}lO + \dot{C}F_2Cl$ (2) $\dot{C}lO + \dot{C}H_3$
(3) $\dot{C}H_3 + \dot{C}F_2Cl$ (4) $\dot{C}l + \dot{C}F_2Cl$

11. The Statements that are TRUE : [March 18, 2021 (I)]
(A) Methane leads to both global warming and photochemical smog
(B) Methane is generated from paddy fields
(C) Methane is a stronger global warming gas than CO_2
(D) Methane is a part of reducing smog
Choose the **most appropriate** answer from the options given below
(1) (A), (B), (C) only (2) (A), (B), (D) only
(3) (A) and (B) only (4) (B), (C), (D) only

12. Reducing smog is a mixture of [March 17, 2021 (I)]
(1) Smoke, fog and $CH_2 = CH - CHO$
(2) Smoke, fog and SO_2
(3) Smoke, fog and N_2O_3
(4) Smoke, fog and O_3

13. The type of pollution that gets increased during the day time and in the presence of O_3 is [March 16, 2021 (I)]
(1) Reducing smog (2) Acid rain
(3) Global warming (4) Oxidising smog

14. The gas released during anaerobic degradation of vegetation may lead to: [February 24, 2021 (I)]
(1) Ozone hole (2) Acid rain
(3) Corrosion of metals (4) Global warming and cancer

15. Thermal power plants can lead to : [Sep. 03, 2020 (I)]
(1) Acid rain (2) Blue baby syndrome
(3) Ozone layer depletion (4) Eutrophication

16. The **incorrect** statement(s) among (A) - (D) regarding acid rain is (are) : [Sep. 03, 2020 (II)]

(A) It can corrode water pipes.
(B) It can damage structures made up of stone.
(C) It cannot cause respiratory ailments in animals
(D) It is not harmful for trees
(1) (A), (C) and (D) (2) (C) only
(3) (A), (B) and (D) (4) (C) and (D)

17. The statement that is **not** true about ozone is: [Sep. 02, 2020 (I)]
(1) in the stratosphere, CFCs release chlorine free radicals ($\dot{C}l$) which reacts with O_3 to give chlorine dioxide radicals.
(2) in the atmosphere, it is depleted by CFCs.
(3) in the stratosphere, it forms a protective shield against UV radiation.
(4) it is a toxic gas and its reaction with NO gives NO_2.

18. Among the gases (1)–(5), the gases that cause greenhouse effect are: [Online April 09, 2017, Jan. 08, 2020 (I)]
(A) CO_2 (B) H_2O (C) CFCs
(D) O_2 (E) O_3
(1) (A), (B), (C) and (D) (2) (A), (B), (C) and (E)
(3) (A) and (D) (4) (A), (C), (D) and (E)

19. The correct set of species responsible for the photochemical smog is : [April 12, 2019 (I)]
(1) N_2, NO_2 and hydrocarbons
(2) CO_2, NO_2, SO_2 and hydrocarbons
(3) NO, NO_2, O_3 and hydrocarbons
(4) N_2, O_2, O_3 and hydrocarbons

20. The primary pollutant that leads to photochemical smog is: [April 12,.2019 (II)]
(1) acrolein (2) nitrogen oxides
(3) ozone (4) sulphur dioxide

21. The regions of the atmosphere, where clouds are formed and where we live, respectively, are : [April 10, 2019 (I)]
(1) Troposphere and Stratosphere
(2) Stratosphere and Troposphere
(3) Troposphere and Troposphere
(4) Stratosphere and Stratosphere

22. Air pollution that occurs in sunlight is : [April 10, 2019 (II)]
(1) reducing smog (2) acid rain
(3) oxidizing smog (4) fog

23. The layer of atmosphere between 10 km to 50 km above the sea level is called as: [April 9, 2019 (II)]
(1) troposphere (2) thermosphere
(3) stratosphere (4) mesosphere

24. Excessive release of CO_2 into the atmosphere results in: [April 9, 2019 (I)]
(1) global warming (2) polar vortex
(3) formation of smog (4) depletion of ozone

25. Which is wrong with respect to our responsibility as a human being to protect our environment ? [April 8, 2019 (I)]
(1) Restricting the use of vehicles
(2) Avoiding the use of floodlighted facilities
(3) Setting up compost tin in gardens.
(4) Using plastic bags.

26. **Assertion :** Ozone is destroyed by CFCs in the upper stratosphere.
Reason : Ozone holes increase the amount of UV radiation reaching the earth. [April 8, 2019 (I)]
(1) Assertion and reason are incorrect.
(2) Assertion and reason are both correct, and the reason is the correct explanation for the assertion.
(3) Assertion and reason are correct, but the reason is not the explanation for the assertion.
(4) Assertion is false, but the reason is correct.

27. The molecule that has minimum/no role in the formation of photochemical smog, is: [Jan. 12, 2019 (I); Similar Online April 10, 2015]
(1) N_2 (2) $CH_2=O$ (3) O_3 (4) NO

28. The compound that is NOT a common component of photochemical smog is : [Jan. 12, 2019 (II); Similar Jan. 11, 2019 (I)]
(1) O_3 (2) $H_3C-\underset{\underset{O}{\|}}{C}-OONO_2$
(3) $CH_2=CHCHO$ (4) CF_2Cl_2

29. The upper stratosphere consisting of the ozone layer protects us from the sun's radiation that falls in the wavelength region [Jan. 12, 2019 (II)]
(1) 200 – 315 nm (2) 400 – 550 nm
(3) 0.8 – 1.5 nm (4) 600 – 750 nm

30. Taj Mahal is being slowly disfigured and discoloured. This is primarily due to : [Jan. 11, 2019 (II)]
(1) Global warming (2) Acid rain
(3) Water pollution (4) Soil pollution

31. The higher concentration of which gas in air can cause stiffness of flower buds? [Jan. 11, 2019 (II)]
(1) NO_2 (2) CO_2 (3) SO_2 (4) CO

32. The reaction that is NOT involved in the ozone layer depletion mechanism in the stratosphere is: [Jan. 10, 2019 (II)]
(1) $CF_2Cl_2(g) \xrightarrow{uv} \dot{C}l(g) + \dot{C}F_2Cl(g)$
(2) $\dot{C}lO(g) + O(g) \longrightarrow \dot{C}l(g) + O_2(g)$
(3) $CH_4 + 2O_3 \longrightarrow 3CH_2 = O + 3H_2O$
(4) $HOCl(g) \xrightarrow{hv} \dot{O}H(g) + \dot{C}l(g)$

33. Identify the pollutant gases largely responsible for the discoloured and lustreless nature of marble of the Taj Mahal. [Online April 8, 2017]
(1) O_3 and CO_2 (2) CO_2 and NO_2
(3) SO_2 and NO_2 (4) SO_2 and O_3

34. Which of the following statements about the depletion of ozone layer is correct? [Online April 11, 2014]
(1) The problem of ozone depletion is less serious at poles because NO_2 solidifies and is not available for consuming $ClO^{\bullet}$ radicals.
(2) The problem of ozone depletion is more serious at poles because ice crystals in the clouds over poles act as catalyst for photochemical reactions involving the decomposition of ozone by $Cl^{\bullet}$ and $ClO^{\bullet}$ radicals.
(3) Freons, chlorofluorocarbons, are inert. Chemically, they do not react with ozone in stratosphere.
(4) Oxides of nitrogen also do not react with ozone in stratosphere.

35. Global warming is due to increase of: [Online April 12, 2014]
(1) methane and nitrous oxide in atmosphere
(2) methane and CO_2 in atmosphere
(3) methane and O_3 in atmosphere
(4) methane and CO in atmosphere

36. The gas leaked from a storage tank of the Union Carbide plant in Bhopal gas tragedy was : [2013]
(1) Methyl isocyanate (2) Methylamine
(3) Ammonia (4) Phosgene

37. Green house gases can be arranged in 'Global Warming Potential' sequence as [Online May 7, 2012]
(1) $N_2O > CFC > CH_4 > CO_2$ (2) $CFC > N_2O > CH_4 > CO_2$
(3) $CFC > CO_2 > N_2O > CH_4$ (4) $CO_2 > CFC > N_2O > CH_4$

38. Which one of the following depletes ozone layer? [Online May 26, 2012]
(1) CO (2) NO and freons
(3) SO_2 (4) CO_2

39. Identify the incorrect statement from the following : [2011RS; Similar 2008]
(1) Ozone absorbs the intense ultraviolet radiation of the sun.
(2) Depletion of ozone layer is because of its chemical reactions with chlorofluoro alkanes.
(3) Ozone absorbs infrared radiation.
(4) Oxides of nitrogen in the atmosphere can cause the depletion of ozone layer.

40. The smog is essentially caused by the presence of [2004]
(1) oxides of sulphur and nitrogen
(2) O_2 and N_2
(3) O_2 and O_3
(4) O_3 and N_2

Topic 2 - Water and Soil Pollution

Multiple Choice Questions (MCQs)

41. Given below are two statements : [July 28, 2022 (I)]
Statement I : In polluted water values of both dissolved oxygen and BOD are very low.
Statement II : Eutrophication results in decrease in the amount of dissolved oxygen.
In the light of the above statements, choose the **most appropriate** answer from the options given below :
(1) Both Statement I and Statement II are true.
(2) Both Statement I and Statement II are false.
(3) Statement I is true but Statement II is false.
(4) Statement I is false but Statement II is true.

42. Match List I with List II

List I Pollutant	List II Disease /sickness
(A) Sulphate (>500 ppm)	(I) Methemoglobinemia
(B) Nitrate (>50 ppm)	(II) Brown mottling of teeth
(C) Lead (> 50 ppb)	(III) Laxative effect
(D) Fluoride (>2 ppm)	(IV) Kidney damage

Choose the correct answer from the options given below: [July 27, 2022 (I)]
(1) A–IV, B–I, C–II, D–III (2) A–III, B–I, C–IV, D–II
(3) A–II, B–IV, C–I, D–III (4) A–II, B–IV, C–III, D–I

43. Match List-I with List-II [July 26, 2022 (II)]

List-I (Pollutant)	List-II (Source)
(A) Microorganisms	(I) Strip mining
(B) Plant nutrients	(II) Domestic sewage
(C) Toxic heavy metals	(III) Chemical fertilizer
(D) Sediment	(IV) Chemical factory

(1) A-II, B-III, C-IV, D-I
(2) A-II, B-I, C-IV, D-III
(3) A-I, B-IV, C-II, D-III
(4) A-I, B-IV, C-III, D-II

44. Match List-I with List-II. [July 25, 2022 (II)]

List-I	List-II
(A) Sulphate	(I) Pesticide
(B) Fluoride	(II) Bending of bones
(C) Nicotine	(III) Laxative effect
(D) Sodium arsenite	(IV) Herbicide

Choose the correct answer from the options given below:
(1) A-II, B-III, C-IV, D-I (2) A-IV, B-III, C-II, D-I
(3) A-III, B-II, C-I, D-IV (4) A-III, B-II, C-IV, D-I

45. Which one of the following chemicals has **not** been used for the dry cleaning of clothes? [June 30, 2022 (I)]
(1) Tetrachloroethene (2) Liquid CO_2
(3) Ethanol (4) H_2O_2.

46. The measured BOD values for four different water samples (A-D) are as follows:
A = 3 ppm: B = 18 ppm: C-21 ppm: D=4 ppm. The water samples which can be called as highly polluted with organic wastes, are [June 26, 2022 (II)]
(1) A and B (2) A and D
(3) B and C (4) B and D

47. The eutrophication of water body results in :
(1) loss of Biodiversity [June 25, 2022 (I)]
(2) breakdown of organic matter
(3) increase in biodiversity
(4) decrease in BOD.

48. Given below are two statements: one is labelled as **Assertion A** and the other is labelled as **Reason R**
Assertion A: Polluted water may have a value of BOD of the order of 17 ppm.
Reason R: BOD is a measure of oxygen required to oxidise both the biodegradable and non-biodegradable organic material in water.
In the light of the above statements, choose the most appropriate answer from the options given below. [June 25, 2022 (II)]
(1) Both A and R are correct and R is the correct explanation of A.
(2) Both A and R are correct but R is NOT the correct explanation of A.
(3) A is correct but R is not correct.
(4) A is not correct but R is correct.

49. Which one of the following elemental forms is not present in the enamel of the teeth? [June 24, 2022 (I)]
(1) Ca^{2+} (2) P^{3+} (3) F^- (4) P^{5+}

50. Water sample is called cleanest on the basis of which one of the BOD values given below. [Sep. 1, 2021 (II)]
(1) 11 ppm (2) 15 ppm (3) 3 ppm (4) 21 ppm

51. BOD values (in ppm) for clean water (A) and polluted water (B) are expected respectively : [Aug. 31, 2021 (I); Similar Jan. 12, 2019 (I); Jan. 10 2019 (I); Online April 15, 2018 (II); Online May 12, 2012]
(1) A > 50, B < 27 (2) A > 25, B < 17
(3) A < 5, B > 17 (4) A > 15, B > 47

52. Which one of the following statements is NOT correct ? [July 27, 2021 (I)]
(1) Eutrophication indicates that water body is polluted
(2) The dissolved oxygen concentration below 6 ppm inhibits fish growth
(3) Eutrophication leads to increase in the oxygen level in water
(4) Eutrophication leads to anaerobic conditions

53. Green chemistry in day–to–day life is in the use of:
(1) Chlorine for bleaching of paper [July 20, 2021 (I)]
(2) Large amount of water alone for washing clothes
(3) Tetrachloroethene for laundry
(4) Liquified CO_2 for dry cleaning of clothes

54. The presence of soluble fluoride ion upto 1 ppm concentration in drinking water, is : [Sep. 06, 2020 (I)]
(1) harmful for teeth (2) harmful to skin
(3) harmful to bones (4) safe for teeth

55. The condition that indicates a polluted environment is: [Sep. 05, 2020 (I)]
(1) eutrophication
(2) 0.03% of CO_2 in the atmosphere
(3) BOD value of 5 ppm
(4) pH of rain water to be 5.6

56. Biochemical Oxygen Demand (BOD) is the amount of oxygen required (in ppm): [Jan. 09, 2020 (II)]
(1) for sustaining life in a water body.
(2) by bacteria to break-down organic waste in a certain volume of a water sample.
(3) for the photochemical breakdown of waste present in 1 m^3 volume of a water body.
(4) by anaerobic bacteria to breakdown inorganic waste present in a water body.

57. The maximum prescribed concentration of copper in drinking water is: [April 8, 2019 (II)]
(1) 5 ppm (2) 0.05 ppm (3) 0.5 ppm (4) 3 ppm

58. The concentration of dissolved oxygen (DO) in cold water can go upto : [Jan. 11, 2019 (I)]
(1) 14 ppm (2) 8 ppm (3) 10 ppm (4) 16 ppm

59. A water sample has ppm level concentration of the following metals: Fe = 0.2; Mn = 5.0; Cu = 3.0; Zn = 5.0. The metal that makes the water sample unsuitable for drinking is: [Jan. 9, 2019 (I)]
(1) Cu (2) Mn (3) Fe (4) Zn

60. Which of the following conditions in drinking water causes methemoglobinemia? [Jan. 9, 2019 (II)]
(1) > 50 ppm of lead (2) > 50 ppm of chloride
(3) > 50 ppm of nitrate (4) > 100 ppm of sulphate

61. The pH of rain water, is approximately: [Jan. 9, 2019 (II)]
(1) 5.6 (2) 7.5 (3) 7.0 (4) 6.5

62. The recommended concentration of fluoride ion in drinking water is up to 1ppm as fluoride ion is required to make teeth enamel harder by converting $[3Ca_3(PO_4)_2 \cdot Ca(OH)_2]$ to : [2018]
(1) $[CaF_2]$ (2) $[3(CaF_2) \cdot Ca(OH)_2]$
(3) $[3Ca_3(PO_4)_2 \cdot CaF_2]$ (4) $[3\{(Ca(OH)_2\} \cdot CaF_2]$

63. A water sample has ppm level concentration of following anions
$F^- = 10$; $SO_4^{2-} = 100$; $NO_3^- = 50$
the anion/anions that make/makes the water sample unsuitable for drinking is/are : [2017]
(1) only NO_3^- (2) both SO_4^{2-} and NO_3^-
(3) only F^- (4) only SO_4^{2-}

64. BOD stands for : [Online April 9, 2016]
(1) Biochemical Oxidation Demand
(2) Biological Oxygen Demand
(3) Biochemical Oxygen Demand
(4) Bacterial Oxidation Demand

65. The concentration of fluoride, lead, nitrate and iron in a water sample from an underground lake was found to be 1000 ppb, 40 ppb, 100 ppm and 0.2 ppm, respectively. This water is unsuitable for drinking due to high concentration of : [2016]
(1) Nitrate (2) Iron (3) Fluoride (4) Lead

66. Addition of phosphate fertilisers to water bodies causes: [Online April 11, 2015]
(1) increase in amount of dissolved oxygen in water
(2) deposition of calcium phosphate
(3) increase in fish population
(4) enhanced growth of algae

67. What is DDT among the following ? [2012]
(1) Greenhouse gas
(2) A fertilizer
(3) Biodegradable pollutant
(4) Non-biodegradable pollutant

Answer Key

Part-A (JEE Main)

1	(1)	8	(2)	15	(1)	22	(3)	29	(1)	36	(1)	43	(1)	50	(3)	57	(4)	64	(3)
2	(3)	9	(4)	16	(4)	23	(3)	30	(2)	37	(2)	44	(3)	51	(3)	58	(3)	65	(1)
3	(3)	10	(4)	17	(1)	24	(1)	31	(3)	38	(2)	45	(3)	52	(3)	59	(2)	66	(4)
4	(1)	11	(1)	18	(2)	25	(4)	32	(3)	39	(3)	46	(2)	53	(4)	60	(3)	67	(4)
5	(4)	12	(2)	19	(3)	26	(3)	33	(3)	40	(1)	47	(1)	54	(4)	61	(1)		
6	(4)	13	(4)	20	(2)	27	(1)	34	(2)	41	(4)	48	(3)	55	(1)	62	(3)		
7	(1)	14	(4)	21	(3)	28	(4)	35	(2)	42	(2)	49	(2)	56	(2)	63	(3)		

Chapter-1

HINTS & SOLUTIONS

Some Basic Concepts of Chemistry

Part-A (JEE Main)

1. (3)
No. of atoms = No. of moles × atomicity × N_A
(A) No. of atoms in 1.8 mg water

$$= \frac{1.8\times10^{-3}}{18}\times N_A\times 3 = 3\times10^{-4}N_A$$

(B) No. of atoms in 9.8 mg H_2SO_4

$$= \frac{9.8\times10^{-3}}{98}\times N_A\times 7 = 7\times10^{-4}\times N_A$$

(C) No. of atoms in 1.8 mg carbon

$$= \frac{1.8\times10^{-3}}{12}\times N_A = 1.5\times10^{-4}\times N_A$$

(D) No. of atoms in 5.85 mg NaCl

$$= \frac{5.85\times10^{-3}}{58.5}\times N_A\times 2 = 2\times10^{-4}N_A$$

Correct match: A-III, B-IV, C-II, D-I

2. (4)
Number of atoms = Mole × Atomicity × N_A

(A) N_{atoms} in oxygen $= \frac{2}{32}\times 2\times N_A = 0.125N_A$

(B) N_{atoms} in $SO_2 = \frac{4}{64}\times 3\times N_A = 0.1875N_A$

(C) N_{atoms} in $O_2 = \frac{1400\text{ ml}}{22400}\times 2\times N_A = 0.125N_A$

(D) N_{atoms} in He $= \frac{0.05}{22.4}\times 1\times N_A = 0.002N_A$

(E) N_{atoms} in $H_2 = 0.0625\times 2\times N_A = 0.125N_A$

3. (4) Total number of atoms = No. of moles × Atomicity × N_A
(A) 2 moles of cyclohexane (C_6H_{12})
$= 2\times 18\times N_A = 36\,N_A$
(B) 684 g of Sucrose ($C_{12}H_{22}O_{11}$)

$$= \frac{684}{342}\times 45\times N_A = 90N_A$$

(C) 90.8 L of H_2 at STP

$$= \frac{90.8}{22.7}\times 2\times N_A = 8N_A$$

Correct order = B > A > C.

4. (2) At STP, 1 mole of gas = 22.7 L of volume
So, no. of moles in 1.4187 L of SO_2 will be

$$\text{Moles} = \frac{V\text{ L}}{22.7\text{ L}} \Rightarrow \frac{1.4187\text{ L}}{22.7\text{ L}} = 0.0633$$

No. of molecules = Moles × 6.023×10^{23}
$= 0.0633\times 6.023\times10^{23} = 3.812\times10^{22}$.

5. (2) Number of atoms =

$$\frac{\text{Mass in grams}}{\text{Molar Mas(g / mol)}}\times N_A$$

Therefore for the given mass, element having the least molar mass will have the highest number of atoms.
$M_{Po} = 209$, $M_{Pr} = 141$, $M_{Pb} = 207$, $M_{Pt} = 195$
⇒Pr has the highest number of atoms.

6. (4)
(A)Weight is the force exerted by gravity on an object.
(B)Mass of a substance is the amount of matter present in it.

7. (1) $(\text{moles})_{\text{initial}} = \frac{x\times10^{-3}}{44}$

$$(\text{moles})_{\text{removal}} = \frac{10^{21}}{6.02\times10^{23}}$$

$(\text{moles})_{\text{left}} = (\text{moles})_{\text{initial}} - (\text{moles})_{\text{removal}}$

$$2.8\times10^{-3} = \frac{x\times10^{-3}}{44} - \frac{10^{21}}{6.02\times10^{23}} \Rightarrow x = 196.2\text{ mg}$$

8. (2) A candela is the luminous intensity of a source that emits monochromatic radiation of frequency radiation of frequency 540×10^{12} Hz and has a radiant intensity in that direction of

$\frac{1}{683}$ w / sr. It is unit of Candela.

9. (4) According to dalton's atomic theory matter is made of indivisible atoms. All the atoms of a given element are same and have same mass.

10. (1) In Compound atoms of different elements combine in fixed ratio by mass.

11. (4) Number of moles of

$$O_2 = \frac{2.8375}{22.7} = 0.125$$

⇒ Number of molecules = $0.125\,N_A = 7.525\times10^{22}$

12. (4) 16g CH_4 = 1 mole CH_4 contains $10\times6.02\times10^{23}$ electrons
$= 60.2\times10^{23}$
1g H_2 = 0.5 mole H_2 gas occupy 11.35 litre volume at STP
1 mole of N_2 = 28g
0.5 mole of SO_2 = 32g

13. (3) No. of Fe atoms in 3.3g of hemoglobin

$$= \frac{0.34}{100}\times\frac{3.3}{56}\times 6.022\times10^{23} = 1.206\times10^{20}$$

14. (3) Let total volume = 1000 mL = 1L
total mass of solution = 1460 g

mass of HCl $= \frac{35}{100}\times 1460$; moles of HCl

$$\text{Molarity} = \frac{\text{no. of moles}}{\text{Volume (1 L)}} = \frac{35\times1460}{100\times36.5} = 14$$

$$\text{So molarity} = \frac{14\text{ moles}}{1\text{L}} = 14\text{ M}$$

15. (3)

16. (3) Except (c) all postulates were given by Dalton.

17. (4) 5 mol AB_2 weighs 125 g
∴ $AB_2 = 25$ g/mol ⇒ $M_A + 2M_B = 25$
10 mol A_2B_2 weighs 300 g
∴ $A_2B_2 = 30$ g/mol ⇒ $2M_A + 2M_B = 30$
∴ Molar mass of A(M_A) = 5 g or 5×10^{-3} kg
Molar mass of B (M_B) = 10 g or 10×10^{-3} kg

18. (4) CH_4 has one atom of carbon among 5 atoms (1C + 4H).

$$\therefore\ \text{Mole \% of C} = \frac{1}{5}\times 100 = 20\ \%$$

19. (3) No. of moles of H_2O (n_1) $= \frac{18}{18} = 1$

No. of moles of NaOH (n_2) $= \frac{8}{40} = \frac{1}{5}$

$$\text{Mole fraction of NaOH} = \frac{n_2}{n_2+n_1} = \frac{\frac{1}{5}}{\frac{1}{5}+1} = 0.167$$

$$\text{Molality} = \frac{\text{No. of moles of solute}}{\text{Mass of solvent (kg)}} = \frac{1}{5}\times\frac{1000}{18} =$$

11.11 m

20. (3) As we know,

$$\text{Molarity} = \frac{\text{No. of moles of sugar}}{\text{Volume of solution (in L)}}$$

$$\Rightarrow\quad 0.1 = \frac{\text{No. of moles of sugar}}{2\text{L}}$$

So, no. of moles of sugar = 0.2 mole
∴ Mass of sugar = No. of moles of sugar × Molar mass of sugar
$= 0.2\times 342 = 68.4$ g

21. (4) Given percentage of chlorine in hydrocarbon = 3.55%
i.e., 100 g of chlorohydrocarbon has 3.55 g of chlorine.
1 g of chlorohydrocarbon will have

$\frac{3.55}{100} = 0.0355$ g of chlorine.

Atomic wt. of Cl = 35.5 g/mol
Number of moles of Cl

$$= \frac{0.0355\text{ g}}{35.5\text{ g/mol}} = 0.001\text{ mol}$$

Number of atoms of Cl = 0.001 mol × 6.023×10^{23} mol^{-1} = 6.023×10^{20}

22. (1)

23. (4) The compound undergoing combustion is an alkane. Hence, the combustion reaction can be written as

$$\underset{5L}{C_nH_{2n+2}} + \underset{25L}{\left(\frac{3n+1}{2}\right)O_2} \longrightarrow n\,CO_2 + (n+1)\,H_2O$$

Volumes are measured at constant T and P, hence according to Avogadro's law:
Volume $\propto$ mole

1 L alkane requires = $\frac{3n+1}{2}$ L of O_2

5 L alkane requires = $\left(\frac{3n+1}{2}\right)$5 L of O_2

$\Rightarrow \left(\frac{3n+1}{2}\right)5 = 25$ L (given) $\therefore$ $n = 3$

Hence, alkane is propane (C_3H_8).

24. (2) 2 moles of water softner require 1 mole of Ca^{2+} ion

So, 1 mole of water softner require $\frac{1}{2}$ mole of Ca^{2+} ion

Thus, $\frac{1}{2\times 206} = \frac{1}{412}$ mol/g will be maximum uptake.

25. (2) From the given data,
Mass of solution = 1000 + 120 = 1120 g

$\because d = \frac{\text{mass}}{V}$ or $V = \frac{\text{mass}}{d}$

Volume of solution $V = \frac{1120}{1.12} = 1000$ mL = 1 litre

Now molarity $(M) = \frac{\text{Mass}}{\text{Mol. mass}\times V(\text{L})}$

$= \frac{120}{60\times 1} = 2$ M

26. (2) 1 mole of water contains
= 16 g of oxygen
$\therefore$ 3.6 mole of water of oxygen contains
= 16 × 3.6 = 57.6 g of oxygen

27. (3) The relation between molarity (M) and molality (m) is

$d = M\left(\frac{1}{m} + \frac{M_2}{1000}\right)$, M_2 = Mol. mass of solute

On putting value $1.252 = 3\left(\frac{1}{m} + \frac{58.5}{1000}\right)$

On solving, $m = 2.79$

28. (1)

29. (1) From molarity equation
$M_1V_1 + M_2V_2 = MV$

$M = \frac{M_1V_1 + M_2V_2}{V} = \frac{2\times 10 + 0.5\times 200}{210}$

$M = \frac{120}{210} = 0.57$ M

30. (1)

31. (4)

	O_3	CO	O_2
MW :	48	28	32
given mass :	16	28	16
No. of 'O' atoms :	$\frac{16}{48}\times 3$ = 1	$\frac{28}{28}\times 1$ = 1	$\frac{16}{32}\times 2$ = 1

Ratio of 'O' atoms = 1 : 1 : 1

32. (1) ppm = $\frac{\text{mass of solute (g)}}{\text{mass of solution (g)}}\times 10^6$

$= \frac{0.2}{500}\times 10^6 = 400$ ppm

33. (3) $\underset{78g}{C_6H_6} + HNO_3 \longrightarrow \underset{123g}{C_6H_5NO_2} + H_2O$

Now since, 78 g of benzene on nitration give = 123 g nitrobenzene
Hence 5 g of benzene on nitration give

$= \frac{123}{78}\times 5 = 7.88$ g

The nearest answer is (c), i.e., theoretical yield = 7.88 g

34. (4) 1 Mole of $Mg_3(PO_4)_2$ contains 8 moles of oxygen atoms
$\therefore$ 8 mole of oxygen atoms $\equiv$ 1 mole of $Mg_3(PO_4)_2$
0.25 mole of oxygen atom

$\equiv \frac{1}{8}\times 0.25$ mole of $Mg_3(PO_4)_2$

$= 3.125\times 10^{-2}$ mole of $Mg_3(PO_4)_2$

35. (2)

36. (2) Moles of urea present in 100 mL of sol.

$= \frac{6.02\times 10^{20}}{6.02\times 10^{23}}$ mol

$\therefore M = \frac{6.02\times 10^{20}\times 1000}{6.02\times 10^{23}\times 100} = 0.01\,M$

[$\because M$ = Moles of solute present in 1L of solution]

37. (4)

$Ba(OH)_2$		HCl
N_1V_1	=	N_2V_2
$N_1\times 25$	=	0.1×35
N_1	=	0.14

Since, $Ba(OH)_2$ is diacidic base

Hence, $N_1 = M_1\times 2$ or $M_1 = \frac{N_1}{2}$; $M_1 = 0.07$ M

38. (3) Among all the given options, molarity changes with temperature because the term molarity involves volume which increases on increasing temperature.

39. (1) No. of moles of Fe = $\frac{558.5}{55.85} = 10$ mol;

Therefore, no. of atoms = 10 N_A
No. of moles of C in 60 g of C = 60/12 = 5 mol.
$\therefore$ No of 'C' atoms = 5N_A

40. (372)
Molar mass = 372 g
$\therefore$ 0.1 mole = 372×10^{-1} g

41. (11.00) Thickness of coating = 0.05 cm
Area of coating = 0.05 m² = 0.05 × 10⁴ cm²
Volume = 0.05 × 0.05 × 10⁴ cm³ = 25 cm³
Mass of Ag = 25 cm³ × 7.9 g cm⁻³ = 197.5 g

Mol of Ag = $\frac{197.5\text{ g}}{108\text{ g mol}^{-1}} = 1.828$ mol

Atoms of Ag = $1.828\times 6.022\times 10^{23} = 11.008\times 10^{23}$

42. (2) $\underset{1\,mol}{X} + \underset{1\,mol}{Y} + \underset{0.05\,mol}{3Z} \rightleftharpoons XYZ_3$

Z is act as a limiting reagent

1 mol of $XYZ_3 = \frac{0.05}{3}$ mol of Z.

Mass of $XYZ_3 = \frac{0.05}{3}\times(10 + 20 + 30\times 3) = 2$g

43. (24) Let the weight of Mg in the extract = w g

$\frac{w}{2000}\times 10^6 = 48$

[Assuming 1000 mL ≈ 1000 kg for water]

$x = 96\times 10^{-3}$

$n_{Mg} = \frac{\text{weight}}{\text{molar mass}} = \frac{96\times 10^{-3}}{24} = 0.004$

Number of Mg atom = $0.004\times 6.02\times 10^{23}$
$= 24.08\times 10^{20} \approx 24\times 10^{20}$

44. (46)

	N_2	+ $3H_2$	$\longrightarrow$ $2NH_3$
initial	56L		–
	–10L		+20L
final	46L		20L

From the stoichiometry for 2L production of ammonia 1L of N_2 gas is used.
Final volume of N_2 = 56 – 10 = 46L

45. (2) Moles of A_2B = Moles of AB_3 = 0.15

$\frac{w}{2a+b} = \frac{w}{a+3b}$ $\Rightarrow 2a + b = a + 3b$ $\Rightarrow a = 2b.$

46. (5418)
M.M. of $C_7H_5N_3O_6$ = 84 + 5 + 42 + 96 = 227

Moles of $C_7H_5N_3O_6 = \frac{681}{227} = 3$

Moles of N atoms = 3 × 3 = 9
Number of N atoms = $9\times 6.02\times 10^{23} = 5418\times 10^{21}$.

47. (2) No. of atoms =

$\frac{8}{23}\times 6.02\times 10^{23} = 2.09\times 10^{23} \approx 2\times 10^{23}$

48. (14.00)
Mass percent of HNO_3 = 63%
Thus, 100 g of nitric acid solution contains 63 g of nitric acid by mass.

No. of moles = $\frac{63\text{g}}{63\text{g mol}^{-1}} = 1$

Volume of 100 g of nitric acid solution

$= \frac{\text{Mass}}{\text{Density}} = \frac{100\text{ g}}{1.4\text{ g/mL}} = 71.4$ mL

Molarity = $\frac{\text{No. of moles}}{\text{volume (mL)}}\times 1000$

$= \frac{1}{71.4}\times 1000 = 14$ M

49. (2) Given:-
Fe = 69.9%
O = 30.1%
Let mass of the compound is 100 g
Hence, Fe = 69.9 g and O = 30.1 g

Now, Mol of Fe = $\frac{69.9}{56} = 1.25$

$\because$ Atomic mass of Fe = 56 g mol⁻¹.

Mol of O = $\frac{30.1}{16} = 1.88$

$\because$ Atomic mass of O = 16 g mol⁻¹

Molar Ratio of Fe = $\frac{1.25}{1.25} = 1$

Molar Ratio of O = $\frac{1.88}{1.25} = 1.5$

Simplest Molar Ratio of Fe = 1 × 2 = 2
Simplest Molar Ratio of O = 1.5 × 2 = 3
$\therefore$ Empirical formula is Fe_2O_3

50. (2) In the combustion of organic compound, all carbon in CO_2 and all hydrogen in H_2O comes from organic compound

$$C_xH_yO_z + O_2 \longrightarrow \underset{0.307\,g}{CO_2} + \underset{0.127\,g}{H_2O}$$

Weight of Carbon in $CO_2 = \frac{12}{44} \times 0.307 = 0.0837\,g$

All carbon in CO_2 comes from organic compound.

Weight of Hydrogen in $H_2O = \frac{2}{18} \times 0.127$
$= 0.0141\,g$

(all hydrogen in water comes from organic compound) % of hydrogen in the organic compound

$= \frac{0.0141}{0.21} \times 100 = 6.719\% \sim 6.72\%$

Weight of Oxygen in the organic compound
$= 0.210 - (0.0837 + 0.0141)$
$= 0.1122$

% of Oxygen in the organic compound

$= \frac{0.1122}{0.21} \times 100 = 53.41\%$

51. (1) Moles of carbon $= n_{CO_2} = \frac{1.46}{44} = 0.033$

Mass of carbon $= W_C = 0.033 \times 12$

Moles of Hydrogen $= 2 \times n_{H_2O}$

$= 2 \times \frac{0.567}{18} = 0.063$

Mass of Hydrogen $W_H = 0.063$ g
Mass of Oxygen (O) $= 1 - (W_C + W_H)$
$= 1 - (0.033 \times 12 + 0.063) = 0.541$ g

Moles of Oxygen $= \frac{0.541}{16} = 0.033$

Empirical formula $= CH_2O$
Empirical formula mass = 30

52. (3) C : H : O

$\frac{54.2}{12} : \frac{9.2}{1} : \frac{36.6}{16}$

4.516 : 9.2 : 2.287

$\frac{4.516}{2.287} : \frac{9.2}{2.287} : \frac{2.287}{2.287}$

1.97 : 4.02 : 1

$C_2H_4O \Rightarrow$ Empirical formula
E.F. mass = 24 + 4 + 16 = 44
and molar mass = 132

$n = \frac{132}{44} = 3$

Hence molecular formula $= (C_2H_4O)_3 = C_6H_{12}O_3$

53. (4) Let the molecular formula is $C_xH_yO_z$.
Given, % C = 42.1, % H = 6.4, % O = 51.5

$\%C = \frac{12x}{342} \times 100 = 42.1 \Rightarrow x = 12$

$\%H = \frac{y}{342} \times 100 = 6.4 \Rightarrow y = 22$

$\%O = \frac{16z}{342} \times 100 = 51.5 \Rightarrow z = 11$

Hence, molecular is $C_{12}H_{22}O_{11}$

54. (1) 1.187 g $CaCO_3$ + 1.023 g $MgCO_3$

$$\underset{x\,g}{CaCO_3} \xrightarrow{\Delta} CaO + CO_2$$

Moles of $CaCO_3$ decomposed
= moles of CaO formed

$\therefore \frac{x}{100}$ = moles of CaO formed

Weight of CaO formed $= \frac{x}{100} \times 56$ ...(1)
{CaO = 40 + 16 = 56}

$$\underset{(2.21-x)\,gm}{MgCO_3} \xrightarrow{\Delta} MgO + CO_2$$

Moles of $MgCO_3$ decomposed = moles of MgO formed $\frac{(2.21-x)}{84}$ = moles of MgO formed.

Weight of MgO $= \frac{2.21-x}{84} \times 40$ ...(2)
{MgO = 24 + 16 = 40}
Add equation 1 and 2.

$\therefore \frac{x}{100} \times 56 + \frac{(2.21-x)}{84} \times 40 = 1.152$

$0.56x + \frac{88.4 - 40x}{84} = 1.152$

$47.04x + 88.4 - 40x = 96.768$
$7.04x = 8.368 \Rightarrow x = 1.188$
$CaCO_3 = x = 1.188$ g
$MgCO_3 = 2.21 - 1.188 = 1.022$ g

55. (1) Molecular, weight of metal chloride

$= \frac{0.57}{100} \times 22700 = 129.39$

weight of Cl = 129.39 × 0.55 = 71.1645

$\therefore$ Mole of Cl $= \frac{71.1645}{35.5} \cong 2$, Hence MCl_2

56. (4)

C	74%	$\frac{74}{12} = 6.16$	$\frac{6.16}{1.23} = 5$
N	17.3%	$\frac{17.3}{14} = 1.23$	$\frac{1.23}{1.23} = 1$
H	8.7%	$\frac{8.7}{1} = 8.7$	$\frac{8.7}{1.23} = 7$

Emperical formula $= C_5NH_7$
Emperical weight
$= (12 \times 5) + (14 \times 1) + (1 \times 7) = 81$ amu
Given, molecular weight = 162 g/mol
Multiplying factor

$= \frac{\text{Molecular weight}}{\text{Emperical weight}} = \frac{162}{81} = 2$

Molecular formula
$= 2 \times$ (Emperical formula) $C_{10}N_2H_{14}$

57. (4) Given mass of organic compound
= 120 g
mass of CO_2(g) = 330g; mass of H_2O (l) = 270 g

mass of carbon $= n_{CO_2} \times 12 = \frac{330}{44} \times 12 = 90$ g

% of carbon $= \frac{90}{120} \times 100 = 75\%$

mass of hydrogen $= n_{H_2O} \times 2 = \frac{270}{18} \times 2 = 30$ g

% of hydrogen $= \frac{30}{120} \times 100 = 25\%$

58. (3) Weight of H in 210 g of H_2O

$= \frac{210}{18} \times 2 = 23.333$ g

% of H $= \frac{23.333}{750} \times 100 = 3.111 \simeq 3.$

59. (2) 44 g of CO_2 contains 12 g of C.
So, 2.64 g of CO_2 contains

$\frac{12}{44} \times 2.64 = 0.72$ g C.

1.08 g of H_2O contains $\frac{2}{18} \times 1.08 = 0.12$ g H.

$\therefore$ Mass of oxygen present = 1.80 – (0.72 + 0.12) = 0.96 g

% of O $= \frac{0.96}{1.80} \times 100 = 53.33\%$.

60. (2)

61. (4)

Element	Relative mass	Relative mole	Simplest whole number ratio
C	6	$\frac{6}{12} = 0.5$	1
H	1	$\frac{1}{1} = 1$	2

So, $X = 1$, $Y = 2$
Equation for combustion of C_XH_Y

$$C_XH_Y + \left(X + \frac{Y}{4}\right)O_2 \longrightarrow XCO_2 + \frac{Y}{2}H_2O$$

Oxygen atoms required $= 2\left(X + \frac{Y}{4}\right)$

As mentioned, $Z = \frac{1}{2} \times$ oxygen atom required.

$\left(X + \frac{Y}{4}\right) = Z$

Now putting the values of X and Y

$\Rightarrow \left(1 + \frac{2}{4}\right) = Z \Rightarrow Z = 1.5$

$\therefore$ Molecule $(C_XH_YO_Z)$ can be written as $C_1H_2O_{3/2}$
$\Rightarrow C_2H_4O_3$

62. (3) Percentage (by mass) of elements given in the body of a healthy human adult is :-
Oxygen = 61.4%, Carbon = 22.9%,
Hydrogen = 10.0% and Nitrogen = 2.6%
$\because$ Total weight of person = 75 kg

$\therefore$ Mass due to ^1H is $= 75 \times \frac{10}{100} = 7.5$ kg

If ^1H atoms are replaced by ^2H atoms.
Mass gain by person would be = 7.5 kg

63. (3)

$\frac{V \times 45}{100} + \frac{(800 - V) \times 20}{100} = \frac{800 \times 29.875}{100}$

$\Rightarrow \frac{9V}{20} + 160 - \frac{V}{5} = 239 \Rightarrow V = 316$

64. (4) 375 ml of 20% oxygen corresponds to 375 × 20 = 75ml
The ratio of the volume of oxygen used for combustion to values of hydrocarbon $= \frac{75}{15}$

This ratio $\frac{5}{1}$ can be obtained with propane (C_3H_8)

equation: $C_3H_8 + 5O_2 \rightarrow 3CO_2 + 4H_2O$

65. (2) $X_2O_3 \Rightarrow X^{3+}O^{2-}$

$X_2Cl_3 \Rightarrow X^{3+}Cl^{2-}$

$X_2(SO_4)_3 \Rightarrow X^{3+}SO_4^{2-}$

$XPO_4 \Rightarrow X^{3+}PO_4^{3-}$

Since Cl^{2-} does not exist. So, X_2Cl_3 is incorrect. The correct formula should be XCl_3.

66. (4)

67. (2)

$$\underset{208\,g}{BaCl_2} + \underset{98\,g}{H_2SO_4} \longrightarrow \underset{233\,g}{BaSO_4} + \underset{73\,g}{2HCl}$$

Mass of $BaCl_2$ in solution $= 100 \times \frac{20.8}{100} = 20.8$ g

Mass of H_2SO_4 in solution $= 50 \times \frac{9.8}{100} = 4.9$ g

98 g of H_2SO_4 reacts with 208 g $BaCl_2$

4.9 g H_2SO_4 will react with $\frac{208}{98} \times 4.9$

$= 10.4$ g $BaCl_2$

H_2SO_4 reacts as a limiting reagent because $BaCl_2$ is given in excess.

98 g H_2SO_4 produces = 233 g $BaSO_4$

4.9 g H_2SO_4 will produce $= \frac{233}{98} \times 4.9$

$= 11.65$ g $BaSO_4$

68. (4) $\because$ 18 g H_2O contains = 2 g H

$\therefore 0.72$ g H_2O contains $= \frac{2}{18} \times 0.72$ g $= 0.08$ g H

$\because 44$ g CO_2 contains = 12 g C

$\therefore 3.08$ g CO_2 contains $= \frac{12}{44} \times 3.08 = 0.84$ g C

$\therefore C : H = \frac{0.84}{12} : \frac{0.08}{1}$; $0.07 : 0.08 = 7 : 8$

$\therefore$ Empirical formula $= C_7H_8$

69. (3) Mol. wt = 2 × vapour density

= 2 × 94.8 = 189.6 g

Since 74.75% is chlorine,

therefore, 189.6 g metal chloride contains

$= \frac{74.75}{100} \times 189.6 = 141.72$ g chlorine

Number of atoms of chlorine

$= \frac{141.72}{35.5} = 3.99 \approx 4$

Hence, formula of metal chlorine is MCl_4.

70. (3)

71. (66) $C_6H_{10} \xrightarrow{Br_2} C_6H_{10}Br_2$

C : Br ratio in $C_6H_{10}Br_2 = 6 : 2 = 3 : 1$

Molecular mass of $C_6H_{10}Br_2$

$12 \times 6 + 10 + 160 = 242$

% of Br $= \frac{160}{242} \times 100$

% of Br = 66.11% ≈ 66%

72. (65) Molecular formula of Thyroxine = $C_{15}H_{11}O_4NI_4$

Molecular mass of Thyroxine

$= 15 \times 12 + 11 \times 1 + 16 \times 4 + 14 \times 1 + 127 \times 4$

$= 777$

% of Iodine $= \frac{\text{mass of iodine}}{\text{mass of thyroxine}} \times 100 = \frac{508}{777} \times 100$

= 65.38 % ≈ 65%

73. (12)

Organic compound $\xrightarrow[\Delta]{CuO} CO_2 + H_2O$

$n_{CO_2} = \frac{220 \times 10^{-3}}{44} = 5 \times 10^{-3}$ moles

Mass of carbon $= 5 \times 10^{-3} \times 12$

Percent composition of carbon

$= \frac{5 \times 10^{-3} \times 12}{500 \times 10^{-3}} \times 100 = 12\%$

$\therefore$ Correct answer is 12

74. (3) Mass of carbon $= \frac{80 \times 90}{100} = 72$ gm

Number of C-atoms $= \frac{72}{12} = 6$

Mass of hydrogen $= \frac{80 \times 10}{100} = 8$ gm

Number of H-atoms $= \frac{8}{1} = 8$

So molecular formula C_6H_8

$D.U = \frac{2C + 2 + N - H - X}{2}$

C = Carbon, N = Nitrogen

H = Hydrogen, X = Halogen

$= \frac{2 \times 6 + 2 - 8}{2} = \frac{14 - 8}{2} = \frac{6}{2} = 3$

75. (56) Moles of $CO_2 = \frac{0.22}{44} = \frac{1}{200}$

$\therefore$ Moles of carbon = (Moles of CO_2) × 1 $= \frac{1}{200}$

$\therefore$ wt. of C $= \frac{1}{200} \times 12 = 0.06$

% of C $= \frac{0.06}{W} \times 100 = 24$

(W = Wt. of Organic Compound) ⇒ W = 0.25

Moles of $H_2O = \frac{0.126}{18} = 0.007$

$\therefore$ Moles of H atom = 2 × 0.007 = 0.014

% of Hydrogen $= \frac{0.014 \times 1}{W} \times 100$

$= \frac{0.014 \times 1}{0.25} \times 100 = 5.6 = 56 \times 10^{-1}$

76. (36) Total mass of sugar in mixture of 25% of 200 and 40% of 500 g

Mass of Sugar = (0.25 × 200) + (0.40 × 500)

= 50 + 200 = 250 g

Total mass of solution = 200 + 500 = 700 g

Mass of sugar in solution

$= \frac{250}{700} \times 100 = 35.7\% \approx 36\%$

77. (59) Let the amount of M^{+2} in the sample be x then the amount of M^{+3} in the sample $= 0.83 - x$

$\therefore 2x + 3(0.83 - x) = 2 \Rightarrow x = 0.49$

$\% M^{2+} = \frac{0.49}{0.83} \times 100 = 59\%$

78. (25) Mol. wt of uracil, $C_4N_2H_4O_2 = 112$

$\% N = \frac{28}{112} \times 100 = 25\%$

79. (12)

Element	Percentage	Mole	Mole Ratio
C	85.8	$\frac{85.5}{12} = 7.15$	1
H	14.2	$\frac{14}{1} = 14.2$	2

Empirical formula $(CH_2)_n$

Empirical mass × n = Molecular Mass

14 × n = 84

n = 6

$\therefore$ Molecular formula C_6H_{12}

80. (46) Mole of CO_2 = Moles of C

$= \frac{\text{Weight of } CO_2}{\text{Molar mass of } CO_2} = \frac{0.793}{44}$

Weight of 'C' = moles × molar mass of carbon

$= \frac{0.793}{44} \times 12 = 0.216$ g

1 mole of H_2O = 2 moles of hydrogen.

Moles of $H_2O = \frac{\text{Weight}}{\text{Molar mass}} = \frac{0.442}{18}$

Moles of 'H' $= \frac{0.442}{18} \times 2$

Weight of 'H' $= \frac{0.442}{18} \times 2 \times 1 = 0.049$ g

$\therefore$ Weight of 'O'

= 0.492 – 0.216 – 0.049 = 0.227 g

% of 'O' $= \frac{0.227}{0.492} \times 100 = 46.13\%$

81. (2)

$\% H = \frac{\text{weight of Hydrogen}}{\text{Total weight of substance}} = \frac{7.5}{116} \times 100 = 6.5$

$\% O = \frac{\text{weight of Oxygen}}{\text{Total weight of substance}} = \frac{60}{116} \times 100 = 51.7$

$\% C = \frac{\text{weight of Carbon}}{\text{Total weight of Compound}} = \frac{48.5}{116} \times 100 = 41.8$

Relative mole percentage :

H	$\frac{6.5}{1} = 6.5$	$\frac{6.5}{3.25} = 2$
O	$\frac{51.7}{16} = 3.25$	$\frac{3.25}{3.25} = 1$
C	$\frac{41.8}{12} = 3.5$	$\frac{3.5}{3.25} = 1.07$

Emperical formula = CH_2O

$C_2H_4O_2 = 2(CH_2O)$

$CH_2O = 1(CH_2O)$

82. **(18)** 44g of CO_2 contains 12 g of C

0.2 g of CO_2 contains $\left(\frac{12}{44}\times 0.2\right)$ g of C, which comes from 0.3 g of organic compound

$\therefore$ Percentage of carbon in the given compound

$= \frac{12\times 0.2}{44\times 0.3}\times 100 = 18.18\%$

83. **(5)** Mass ratio of C : H is $4:1 \Rightarrow 12:3$ and C : O is $3:4 \Rightarrow 12:16$

So,

	mass	mole	mole ratio
C	12	1	1
H	3	3	3
O	16	1	1

Empirical formula $\Rightarrow CH_3O$

As compound is saturated acyclic, so molecular formula is $C_2H_6O_2$ (CH_2OH-CH_2OH)

$$\underset{2\text{ mole}}{C_2H_6O_2} + \underset{5\text{ mole}}{\frac{5}{2}O_2(g)} \longrightarrow 2CO_2(g) + 3H_2O(l)$$

$\therefore$ Number of moles of O_2 required to oxidise 2 moles of (X) = 5.

84. **(3)**

$$Zn(s) + H_2SO_4(aq) \rightarrow ZnSO_4(aq) + H_2(g)$$

Number of moles of Zn $= \frac{20}{65} = 0.307$

Number of moles of H_2SO_4

$= \frac{1.3\times 50\times 0.5}{98} = 0.331$

Number of moles of Zn = Number of moles of H_2 (Zn is the limiting reagent)

$\therefore$ Number of moles of $H_2 = \frac{20}{65} = 0.307$

Volume of $H_2 = 0.307\times 22.4 = 6.88$ L

85. **(2)** $Ag_2CO_3(s) \xrightarrow{\Delta} 2Ag(s) + CO_2(g) + \frac{1}{2}O_2(g)$

Molar mass of $Ag_2CO_3 = 2(108) + 12 + 3(16)$
= 276 g/mol

Moles of $Ag_2CO_3 = \frac{2.76}{276} = 0.01$ mol

From the equation,

1 mol $Ag_2CO_3 \rightarrow$ 2 mol Ag

0.01 mol $Ag_2CO_3 \rightarrow$ 0.02 mol Ag

Mass of residue (Ag) = $0.02\times 108 = 2.16$ g

86. **(1)** Mass of carbon present in 0.25 g of CO_2 produced $= \frac{0.25}{44}\times 12$

Given that mass % of carbon = 25%

Mass % of carbon $= \frac{\text{Mass of carbon}}{x}\times 100$

$\Rightarrow x = \frac{\frac{0.25\times 12}{44}\times 100}{25} = \frac{12}{44}$ g

$\Rightarrow x = 0.2727 = 272.7\times 10^{-3}$ g $= 273\times 10^{-3}$ g.

87. **(2)** Firstly, determine the empirical formula of hydrocarbon. This can be done using carbon to hydrogen ratio.

Given ratio is C : H = 12 : 1

Moles of C $= \frac{12}{12} = 1$, Moles of H $= \frac{1}{1} = 1$

Hence, empirical formula is CH and since it contains 2 carbon atoms:

Therefore, molecular formula of hydrocarbon is C_2H_2.

Balanced chemical equation:

$$C_2H_2(g) + \frac{5}{2}O_2(g) \longrightarrow 2CO_2(g) + H_2O(l)$$

Molar mass of $C_2H_2 = 26$ g/mol

Moles of $C_2H_2 = \frac{3.38}{26} = 0.13$ moles

As, 1 mole of C_2H_2 produces 2 moles of CO_2.

Moles of CO_2 produced = $0.13\times 2 = 0.26$ moles

Mass of CO_2 produced = 0.26 mole × 44 g/mol = 11.44 g.

88. **(4)** Balanced chemical equation:

$3Fe(s) + 4H_2O(g) \rightarrow Fe_3O_4(s) + 4H_2(g)$

No. of moles of water $= \frac{18}{18} = 1$

No. of moles of Fe required $= \frac{3}{4}$

Mass of Fe converted $= \frac{3}{4}\times 56 = 42$ g

89. **(1)** Density of HCl solution (d) = 1.13 g/ml

V = 300 ml

Wt. of HCl solution = 300 × 1.13 = 339 g

Wt. of HCl $= 339\times \frac{38.55}{100} = 130.68$ g

$$CaCO_3 + 2HCl \rightarrow CaCl_2 + H_2O + CO_2$$

$\frac{90}{100} = 0.90$ mole $\frac{130.68}{36.5} = 3.58$ mole

Moles of HCl rcmained = 1.78 mole.

Mass of HCl remained $1.78\times 36.5 = 64.97$ g.

90. **(3)** $Cl_2 + 2KOH \rightarrow KCl + KClO + H_2O$

	Cl_2	2KOH	KCl	KClO	H_2O
t = 0	1 mol	4 mol			
t_f	0	2 mol	1 mol	1 mol	

$[OH^-] = \frac{4-2}{2} = 1$ M

$[Cl^-] = \frac{1}{2}$ M = 0.5 M

$[ClO^-] = \frac{1}{2}$ M = 0.5 M

91. **(3)** A + 2B $\longrightarrow AB_2$

	A	2B
	$\frac{36}{60}$	$\frac{56}{80}$
Given Mole	0.6 mole	0.7 mole
at time t left unreacted	0.25 mole –	0.35 mole

(A) Limiting reagent is B.

(B) Molecular wt. of AB_2
= 60 + 2 × 80 = 220 g/mol

Moles of AB_2 formed = 0.35 mol

Weight of AB_2 = 220 × 0.35 = 77 g

(C) Molecular weight of AB_2 = 220 g/mol

(D) 0.25 moles of A is left unreacted.

Weight of A remaining is = 0.25 × 60 = 15 g.

92. **(1)** $2Al(s) + 6HCl(aq) \rightarrow 2Al^{3+}(aq) + 6Cl^-(aq) + 3H_2(g)$

Mole of H_2 produced = 2 × mole of HCl used

$= \frac{2}{3}\times$ mole of Al used

Mole of H_2 produced $= \frac{22.4}{11.2} = 2$ moles

93. **(1)** Reaction involve as below:

$$MnO_2 + 4HCl \rightarrow MnCl_2 + Cl_2 + 2H_2O$$

Mole of $MnO_2 = \frac{8.7}{87} = 0.1$

Since one mole of Cl_2 produced by one mole of MnO_2.

Thus, mole of Cl_2 produced by 0.1 mole MnO_2 weight of $Cl_2 = 0.1\times 71 = 7.1$ g

94. **(3)** $Ca(s) + 2HCl(g) \rightarrow CaCl_2(s) + H_2(g)$

Moles of Ca $= \frac{\text{Mass of Ca}}{\text{Molar mass of Ca}} = \frac{14.0}{40\text{ g/mol}}$
= 0.35 mol

From balanced chemical equation, 1 mole of Ca produces 1 mol of H_2

Moles of Ca = Moles of H_2 produced = 0.35

Volume of H_2(g) evolved

= Moles of H_2 × Molar volume at STP

= 0.35 × 22.4 = 7.84 L

Weight of $CaCl_2$ = Moles of $CaCl_2$ × Molar volume of $CaCl_2$

= 0.35 × 111 = 38.85 gm

So statement (c) is wrong.

95. **(3)**

$$C_xH_y(g) + \left(x+\frac{y}{4}\right)O_2(g) \rightarrow xCO_2(g) + \frac{y}{2}H_2O(l)$$

	$C_xH_y(g)$	$O_2(g)$	$xCO_2(g)$	$H_2O(l)$
t=0	80	264	0	–
$t=t_{final}$	–	$264-80\left(x+\frac{y}{4}\right)80x$	80x	–

$264 - 80\left(x+\frac{y}{4}\right) + 80x = 224$

$264 - \frac{80y}{4} = 224$

$40 = \frac{80y}{4} \Rightarrow y = 2$

$264 - 80\left(x+\frac{y}{4}\right) = 64$

$264 - 80\left(x+\frac{1}{2}\right) = 64$

$264 - 80x - 40 = 64 \Rightarrow x = 2$

96. **(3)** $Mg + 2HCl \rightarrow MgCl_2 + H_2$

Volume of H_2 evolved = 220 mL

Moles of H_2 at STP $= \frac{\text{Volume}}{\text{Molar Volume at STP}}$

$= \frac{220\times 10^{-3}\text{ L}}{22.4\text{ L}}$ = moles of Mg used

$\therefore$ Mass of Mg used $= \frac{220\times 10^{-3}}{22.4}\times 24$ g/mol

$= 235.7\times 10^{-3}$ g

= 235.7 mg ≈ 236 mg

97. **(3)** $CaCO_3 + 2HCl \rightarrow CaCl_2 + CO_2 + H_2O$

Moles of $CaCO_3 = \frac{1000}{100} = 10$

Moles of HCl $= 0.76\times \frac{250}{1000}$

= 0.19 (Limiting reagent)

Moles of $CaCl_2$ formed $= \frac{0.19}{2}$

Mass of $CaCl_2 = \frac{0.19}{2}\times 111 = 10.545$ g

98. (1) The balanced chemical reaction is

$5I^- + IO_3^- + 6H^+ \rightarrow 3I_2 + 3H_2O$

A.200 ml of 0.1 M KI contains 0.02 mol KI

$\frac{\text{Mole of KI}}{5} = \text{mole of } KIO_3$

$\frac{0.02}{5} = 0.004$

B. 0.02 mol of KI reacts with 0.008 mol H_2SO_4.

C.0.5 L of KI produce 0.013 mol I_2.

D.Equivalent weight

$= \frac{\text{molecular mass}}{\text{change in oxidation number}}$

99. (3) % w/w of HNO_3 = 75%

meaning 100 gm of solution contains 75g of HNO_3

and $\left(\frac{\text{mass}}{\text{volume}}\right)_{\text{solution}} = 1.25 = \frac{100\text{ g}}{V}$

V_{ml} of 100 gm solution $= \frac{100}{1.25}$ ml

$\because$ 75 gm of HNO_3 present in $\frac{100}{1.25}$ ml solution

$\therefore$ 30 gm of HNO_3 present in

$\frac{100}{1.25 \times 75} \times 30 = 32$ ml solution

100. (1)

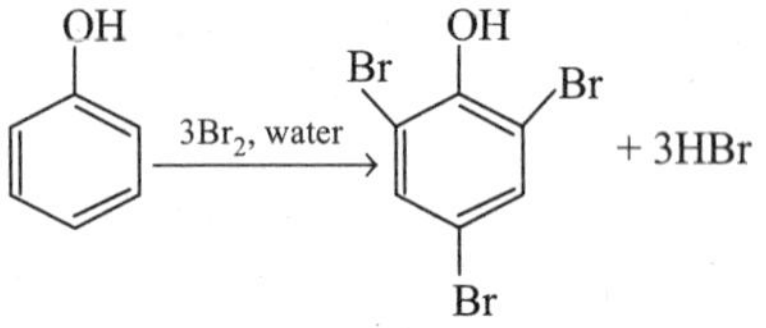

Moles of phenol $= \frac{2}{94} = 0.021$

1 mole of phenol required 3 moles of bromine

$\therefore$ Moles of bromine = 0.021 × 3 = 0.064

$\therefore$ Mass of bromine = 0.064 × 160 = 10.22 g

101. (4) 3M NaCl, $d_{sol} = 1.25$ g/mL

$\text{Molality} = \frac{C_M \times 1000}{1000d - C_M \times M_{Solute}}$

$\frac{3 \times 1000}{(1000 \times 1.25) - (3 \times 58.5)}$

$= \frac{3000}{1250 - 175.5} = 2.79$ molal

102. (1) $CH_3 - C \equiv CH + \underset{(\text{excess})}{Na} \rightarrow$

$CH_3 - C \equiv \overset{-}{C}\overset{+}{N}a + \frac{1}{2}H_2 \uparrow$

$CH_3 - C \equiv CH + NaNH_2 \rightarrow CH_3C \equiv \overset{-}{C}\overset{+}{N}a + NH_3$

4 gm   0.1 mol

$\frac{4}{40} = 0.1$ mol = 2240 mL

Statement I is correct but statement II is incorrect

103. (2) $CH_4 + 2O_2 \rightarrow CO_2 + 2H_2O$

mol of $CO_2 = \frac{11}{44} = \frac{1}{4}$ mol

Since, stoichiometry is same for CH_4 & CO_2

Thus, mol of CH_4 required $= \frac{1}{4} = 0.25$ mol

104. (3) Assuming, : Mass of solvent ≈ Mass of solution

Case I: $0.25 = \frac{W_1}{62} \times \frac{1000}{500}$

Case II: $0.25 = \frac{W_2}{62} \times \frac{1000}{250} \rightarrow \frac{W_1}{W_2} = \frac{2}{1}$

105. (3) $\underset{20g}{N_2(g)} + \underset{5g}{3H_2(g)} \rightleftharpoons 2NH_3(g)$

$\frac{20}{28}$ mol $\frac{5}{2}$ mol

For 3 mol of $H_2 \equiv$ 1 mol of N_2 required

For $\frac{5}{2}$ mol of $H_2 \equiv \frac{5}{6} = 0.83$ mol of N_2 req.

Moles of $N_2 = \frac{20}{28} = 0.71$;

$\therefore$ N_2 is the limiting reagent.

$n(NH_3) = 2 \times n(N_2) = 2 \times \frac{20}{28} = 1.42$ moles.

106. (3) $4HNO_3(l) + 3KCl(s) \rightarrow Cl_2(g) + NOCl(g) + 2H_2O(g) + 3KNO_3(g)$

$n_{KNO_3} = \frac{110}{101} = 1.09$

from stoichiometry

3 Mole KNO_3 requires = 4 mole HNO_3

1.09 mole KNO_3 requires $= \frac{4}{3} \times 1.09$

= 1.45 moles

weight of HNO_3 = 1.45 × 63 = 91.5g

107. (4) $C(s) + O_2(g) \rightarrow CO_2(g) + 400$ kJ

$C(s) + \frac{1}{2}O_2(g) \rightarrow CO(g) + 100$kJ

weight of coal = 0.6 kg = 0.6 × 10^3 g

weight of pure carbon = 0.6 × 10^3 × $\frac{60}{100}$

$n_C = \frac{0.6 \times 10^3 \times 60}{100} \times \frac{1}{12} = \frac{6 \times 60}{12}$

40% of C is converted into CO_2,

Energy released $= \frac{6 \times 60}{12} \times \frac{40}{100} \times 400 = 4800$ kJ

60% of C is converted into CO,

energy released $= \frac{6 \times 60}{12} \times \frac{60}{100} \times 100 = 1800$ kJ

Total energy released = 4800 + 1800 = 6600 kJ

108. (2) In 100 g solution, mass of carbon = 10.8 g

250 g solution, mass of carbon

$= \frac{10.8 \times 250}{100} = 27$ g

Mass of C in one mole of glucose ($C_6H_{12}O_6$) = 12 × 6 = 72 g

72 g of C is present in = 180 g glucose

27 g of C is present in $= \frac{180}{72} \times 27$

= 67.5 g glucose

Mass of solvent (water) = 250 – 67.5 = 182.5 g

Molality of solution $= \frac{(67.5/180)\text{ mol}}{(182.5/1000)\text{ kg}} = 2.05$ m

109. (3) 16 moles of NaOH neutralizes 16 moles of H^+

Source of $H^+ \rightarrow$ 2 moles HCl + 1 mol H_2SO_4.

$\therefore$ 1 mol SO_2Cl_2 is producing 4 moles H^+ ions

$\therefore$ No. of moles of $SO_2Cl_2 = \frac{16}{4} = 4$ moles.

110. (3) Molecular weight of Fe_3O_4 = 236 g/mol

Moles of $Fe_3O_4 = \frac{4.640 \times 10^3}{232} = 20$

Moles of CO

$= \left[\frac{\text{given mass}}{\text{molar mass}}\right] = \frac{2.52 \times 10^3}{28} = 90$

So limiting Reagent

= Fe_3O_4 (20 mole Fe_3O_4 required 80 mole CO)

So moles of Fe formed = 60

Weight of Fe = (moles × molar mass) = 60 × 56 = 3360 g

111. (3) $C_xH_y + \left(x + \frac{y}{4}\right)O_2 \rightarrow xCO_2 + \frac{y}{2}H_2O$

$C_{15}H_{30} + \frac{45}{2}O_2 \rightarrow 15CO_2 + 15H_2O$

Mass of fuel = 0.756 × 1000 g

No. of moles of fuel

$= \frac{0.756 \times 1000}{210}$; M.W. of $C_{15}H_{30} = 210$

Wt. of oxygen $= \frac{0.756 \times 1000}{210} \times \frac{45}{2} \times 32 = 2592$ g

Wt. of $CO_2 = \frac{0.756 \times 1000}{210} \times 15 \times 44 = 2376$ g

112. (3) $NH_2CONH_2 + 2NaOH \longrightarrow Na_2CO_3 + 2NH_3$

1 mol of urea ≡ 2 mol of NH_3

60 g of urea ≡ 2 mol of NH_3

0.6 g of urea $= \frac{2}{60} \times 0.6$ mol = 0.02 mol of NH_3

For neutralisation: mol of NH_3 = mol of HCl

in option (c), no. of moles of HCl

= 0.1 × 0.2 = 0.02

113. (1) (1) 1 g of Fe requires $= \frac{3 \times 32}{4 \times 56} = 0.43$ g of oxygen

(2) 1 g of P requries $= \frac{5 \times 32}{31 \times 4} = 1.3$ g of oxygen

(3) 1 g of C_3H_8 requires $= \frac{5 \times 32}{44} = 3.6$ g of O_2

(4) 1 g of Mg requires $= \frac{32}{2 \times 24}$ g = 0.66 g of O_2

114. (1) According to the stoichiometry of balanced equation, 28 g N_2 reacts with 6 g H_2.

$\underset{\substack{1\text{ mole} \\ 28\text{ g}}}{N_2} + \underset{\substack{3\text{ mole} \\ 6\text{ g}}}{3H_2} \longrightarrow 2NH_3$

$\therefore$ For 56 g of N_2, 12 g of H_2 is required.

115. (1) Given: 0.27 g fatty acid is present in 100 cm^3 of hexane

$\therefore$ 10 mL of aqueous solution contains only 0.027 g acid.

Volume of 0.027 g acid $= \frac{0.027}{0.9}$ mL

$\therefore \pi r^2 h = \frac{0.027}{0.9}$ (given $r = 10$ cm, $\pi = 3$)

$\therefore h = 10^{-4}$ cm $= 10^{-6}$ m

116. (N) **Oxalic acid** **Sodium hydroxide**

$N_1V_1 = N_2V_2$

$(2 \times M_1)V_1 = M_2V_2$

$2 \times 0.5 \times 50 = M_2 \times 25$

$M_2 = 2$

$\text{Molarity} = \frac{\text{No. of moles}}{\text{Vol.(L)}}$

$2 = \frac{\text{No. of moles}}{50/1000} \Rightarrow \text{No. of moles} = \frac{1}{10} = 0.1$

$\text{No. of moles} = \frac{W}{\text{Molar mass}}$

$W = 0.1 \times 40 = 4$ g

No option is correct.

117. (4)

$2NaHCO_3 + H_2C_2O_4 \longrightarrow Na_2C_2O_4 + 2CO_2 + 2H_2O$

Moles of CO_2 evolved $= \frac{0.25}{25 \times 10^3} = 10^{-5}$

$\therefore$ Moles of $NaHCO_3 = 10^{-5}$

$\therefore$ Mass of $NaHCO_3 = 84 \times 10^{-5}$ g $= 0.84 \times 10^{-3}$ g $= 0.84$ mg

$\therefore$ % by weight $= 0.84 \times \frac{100}{10} = 8.4\%$

118. (1) 25 mL of HCl solution requires 30 mL of 0.1 M Na_2CO_3 solution.

$\because N_1V_1 = N_2V_2$

$\therefore 25 \times N_1 = 30 \times 0.2$ (0.1 M Na_2CO_3 = 0.2 N Na_2CO_3)

$N_1 = \frac{6}{25} = 0.24$ N

Now, HCl solution is titrated with NaOH solution.

$\therefore V \times 0.24 = 30 \times 0.2 \Rightarrow V = 25$ mL

119. (3) For the given reaction:

$2\,C_{57}H_{110}O_6(s) + 163\,O_2(g) \rightarrow 114\,CO_2(g) + 110\,H_2O(l)$

Moles of $C_{57}H_{110}O_6 = \frac{445}{890} = 0.5$

Now, moles of water $= \frac{110}{2} \times 0.5 = 27.5$

$\therefore$ Mass of water $= 27.5 \times 18 = 495$ g

120. (4)

121. (2) Given chemical equation:

$\underset{1\,g}{M_2CO_3} + 2\,HCl \rightarrow 2\,MCl + H_2O + \underset{0.01186\,mol}{CO_2}$

moles of $M_2CO_3 \equiv$ moles of CO_2

$\frac{1}{\text{Molar mass of } M_2CO_3} = 0.01186$

$\therefore$ Molar mass of $M_2CO_3 = \frac{1}{0.01186}$

Molar mass = 84.3 g/mol

122. (4)

$\underset{\substack{2\,mol \\ 1\,mol}}{2\,H_3AsO_4} + 5\,H_2S \xrightarrow{\text{Conc. HCl}} \underset{\substack{1\,mol \\ ½\,mol}}{As_2S_5} + 8\,H_2O$

$\because$ The molar mass of H_3AsO_4 is $3 \times 1 + 79 + 4 \times 16 = 142$ g/mol

$\therefore$ Number of moles of $H_3AsO_4 = \frac{35.5}{142} = 0.25$ mol.

$\therefore$ Number of moles of $As_2S_5 = \frac{0.25}{2} = 0.125$ mol.

123. (1) Applying law of equivalence

Equivalent of acid = Equivalent of base

Equivalent of acid = Normality × volume = $0.1 \times V$

Another formula of equivalence

= n factor × number of moles

$\therefore$ Equivalent of base = n factor of OH^- × moles of OH^-

$= 1 \times 0.04 \Rightarrow 0.1 \times V = 1 \times 0.04$

$V = 0.4$ L $= 0.4 \times 1000 = 400$ mL.

124. (4) Weight of hydrated $BaCl_2 = 61$ g

Weight of anhydrous $BaCl_2 = 52$ g

Loss in mass = 9 g

Assuming $BaCl_2 \cdot xH_2O$ as hydrate,

Mass of $H_2O = 9$ g

Moles of $H_2O = \frac{9}{18} = 0.5$ mol

Gram molecular mass of anhydrous $BaCl_2 = 208$ g/mol.

Moles of anhydrous $BaCl_2 = \frac{52}{208} = 0.25$

$\because$ 0.25 moles of $BaCl_2$ combines with = 0.5 moles of H_2O

$\therefore$ 1 mole of $BaCl_2$ combines with = 2 moles of H_2O

125. (1) $A + 2B + 3C \rightleftharpoons AB_2C_3$ ...... (i)

No. of moles of A $= \frac{6.0\text{ g}}{60\text{ g / mol}} = 0.1$ mol

No. of moles of B $= \frac{6.00 \times 10^{23}}{6 \times 10^{23}} = 1$ mol

No. of moles of C = 0.036

Therefore, C is the limiting reagent.

The number of moles of product formed

$= \frac{0.036}{3} = 0.012$ mol

The expression for the molar mass is

$\text{Molar mass} = \frac{\text{mass of product}}{\text{mole of product}}$

$60 + (2 \times x) + (3 \times 80) = \frac{4.8}{0.012}$

$x = 50$ amu

Hence, atomic mass of B is 50 amu.

126. (2)

127. (1) Molality = Moles of solute / Mass of solvent in kg

considering the density of water = 1g/mL

$\text{Molality} = \frac{0.01/60}{0.3} = \frac{0.01}{60 \times 0.3}; = 5.55 \times 10^{-4}$ m

128. (4) Since molarity of solution is 3.60 M. It means 3.6 moles of H_2SO_4 is present in its 1 litre solution.

Mass of 3.6 moles of H_2SO_4 = Moles × Molecular mass = 3.6×98 g = 352.8 g

$\therefore$ 1000 mL solution has 352.8 g of H_2SO_4

Given that 29 g of H_2SO_4 is present in = 100 g of solution

$\therefore$ 352.8 g of H_2SO_4 is present in

$= \frac{100}{29} \times 352.8$ g of solution = 1216 g of solution

$\text{Density} = \frac{\text{Mass}}{\text{Volume}} = \frac{1216}{1000}$

= 1.216 g/mL = 1.22 g/mL

129. (1) $2Al(s) + 6HCl(aq) \rightarrow 2Al^{3+}(aq) + 6Cl^-(aq) + 3H_2(g)$

$\because$ 6 moles of HCl produces = 3 moles of H_2 = 3×22.4 L of H_2 at S.T.P

$\therefore$ 1 mole of HCl produces $= \frac{3 \times 22.4}{6}$ L of H_2 at S.T.P = 11.2 L of H_2 at STP

130. (1)

Apply the formula $d = M\left(\frac{1}{m} + \frac{M_2}{1000}\right)$

$\therefore 1.02 = 2.05\left(\frac{1}{m} + \frac{60}{1000}\right)$

On solving we get, m = 2.288 mol/kg

131. (2) From the molarity equation

$M_1V_1 + M_2V_2 = MV$

Let M be the molarity of final mixture,

$M = \frac{M_1V_1 + M_2V_2}{V}$ where $V = V_1 + V_2$

$M = \frac{480 \times 1.5 + 520 \times 1.2}{480 + 520} = 1.344$ M

132. (1) H_3PO_3 KOH

$N_1V_1 = N_2V_2$

Note: H_3PO_3 is dibasic,

$\therefore$ Normality = 2 × Molarity

$20 \times 0.2 = 0.1 \times V_2$

$\therefore V_2 = 40$ mL

133. (1) $2BCl_3 + 3H_2 \rightarrow 2B + 6HCl$

or $BCl_3 + \frac{3}{2}H_2 \rightarrow B + 3HCl$

Now, since 10.8 g boron requires hydrogen

$= \frac{3}{2} \times 22.4$ L at S.T.P

Hence, 21.6 g boron requires hydrogen

$= \frac{3}{2} \times \frac{22.4}{10.8} \times 21.6 = 67.2$ L at S.T.P.

134. (17) General balanced chemical equation for combustion of an alkane:

$C_nH_{2n+2} + \left(\frac{3n+1}{2}\right)O_2 \rightarrow nCO_2 + (n+1)H_2O$

Given 1 mol of alkane requires 8 moles of O_2

$\frac{3n+1}{2} = 8 \Rightarrow 3n + 1 = 16$

$\Rightarrow 3n = 15 \Rightarrow n = 5$

So, alkane is $C_5H_{12} \Rightarrow 17$

135. (73) $C_xH_yO_z + O_2 \rightarrow CO_2 + H_2O$

t = 0: 0.25 gm

t_{eq} –: 0.18 gm (potash tube), 0.15 gm ($CaCl_2$ tube)

Mass of H_2O = Mass gain in $CaCl_2$ tube = 0.15 g

Mass of CO_2 = Mass gain in potash tube = 0.18 g

Mass of 'C' in 0.18 g of $CO_2 = \frac{0.18}{44} \times 12$

$= 0.049 \simeq 0.05$ gm

Mass of 'H' in 0.15 g of $H_2O = \frac{0.15}{18} \times 2$

$= 0.016 \simeq 0.017$ gm

Mass of 'O' in A = 0.25 – 0.05 – 0.017 = 0.1833 gm

Mass % of 'O' in A $= \frac{0.1833}{0.25} \times 100 = 73.32\%$.

136. (10)

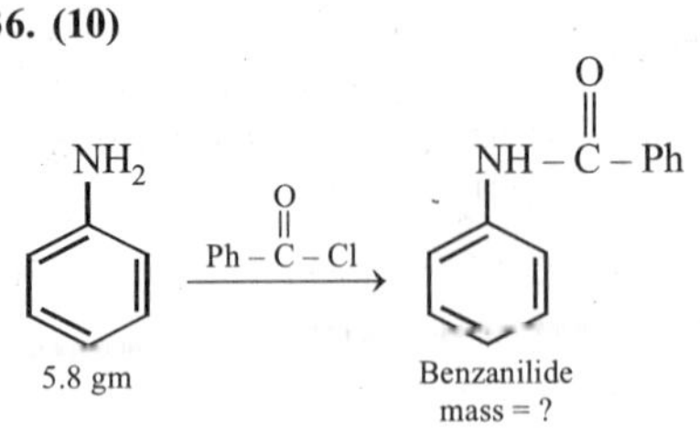

Moles of aniline $= \frac{5.8}{93} = 0.0623$

Moles of benzanilide $= 0.0623 \times \frac{82}{100} = 0.051$

$\therefore$ Mass of benzanilide $= 0.051 \times 197 = 10.06$.

137. (1)

138. (138)

$C_4H_{10}(g) + \frac{13}{2}O_2 \rightarrow 4CO_2(g) + 5H_2O(l)$

Moles of butane (C_4H_{10})

$= \frac{174 \times 1000}{58} = 3 \times 10^3$ moles

moles of $O_2 = \frac{320 \times 1000}{32} = 10 \times 10^3$ moles

1 mol of C_4H_{10} reacts with $\frac{13}{2}$ mol of O_2,

3×10^3 moles of C_4H_{10} react with

$= \frac{13}{2} \times 3 \times 10^3$ mols of O_2

Hence O_2 is the limiting reagent.

$\therefore \frac{13}{2}$ mol of O_2 form 5 mol of H_2O,

10×10^3 mol of O_2 form $= \frac{5 \times 2}{13} \times 10 \times 10^3$ mol of H_2O

$\therefore$ moles of H_2O formed $= \frac{10}{13} \times 10 \times 10^3$

Volume of H_2O formed $= \frac{10}{13} \times 10 \times 18$ L

$= 138.46$ L

$\therefore$ density of water $= 1\text{ g cm}^{-3}$

139. (63) $CaCO_3 \rightarrow CaO + CO_2$

mass of $CaCO_3 = \frac{150 \times 75}{100} = 112.5$ kg

$= 112500$g

$n_{CaCO_3} = \frac{112500\text{ g}}{100\text{ gmol}^{-1}} = 1125$

So moles of CaO = 1125

$\Rightarrow$ mass of CaO $= \frac{1125 \times 56}{1000} = 63$kg

$\therefore$ Correct answer is 63.

140. (80) Number of moles of CO_2 produced

$= \frac{1.46}{44}$

Number of moles of CO_2 produced

= Number of moles of carbon

$\therefore$ mass of carbon produced

$= \frac{1.46}{44} \times 12 = 0.398$g % by mass of carbon

$= \frac{0.398}{0.5} \times 100 = 79.63\% \approx 80\%$.

141. (125) Assuming 100 gm solution contain 70 gm solute.

Volume of 100 gm solution will be $\frac{100}{1.25}$ ml.

Molarity $= \frac{70/70}{100/1.25} \times 1000 = 12.5$ or 125×10^{-1}

142. (420)

Moles of $Fe_3O_4 = \frac{2.32 \times 10^3 \times 10^3}{232}$

$= 10000$ mol $= 10^4$ mol

Moles of CO $= \frac{2.8 \times 10^2 \times 10^3}{28}$

$= 10000$ mol $= 10^4$ mol

$\underset{10^4\text{ mol}}{Fe_3O_4} + \underset{10^4\text{ mol}}{4CO} \rightarrow 3Fe + 4CO_2$

CO is limiting reagent

Moles of Fe $= \frac{3}{4} \times 10^4$

Mass of Fe $= \frac{3}{4} \times \frac{10^4 \times 56}{1000}$ kg $= 420$ kg

143. (61) $\underset{x\text{ g}}{C_6H_5COOH} + NaHCO_3 \rightarrow$

$C_6H_5COO^- Na^+ + H_2O + \underset{11.2L}{CO_2}$

mole of C_6H_5COOH = mole of $CO_2 = \frac{11.2}{22.4} = 0.5$

mass of $C_6H_5COOH = x = 0.5 \times 122 = 61$ g

144. (153)

$4Al + 3O_2 \longrightarrow 2Al_2O_3$

$\frac{81}{27} = 3$ mol $\quad \frac{128}{32} = 4$ mol

Limiting reagent is Al.

$\therefore$ mole of Al_2O_3 formed $= \frac{1}{2} \times 3$ mol

$\therefore$ weight of Al_2O_3 formed $= \frac{3}{2} \times 102 = 153$ gm

145. (100)

Organic compound $\xrightarrow{\text{combustion}} \underset{0.9\text{ gm}}{H_2O}$

$\therefore$ Mole of $H_2O = \frac{0.9}{18} = 0.05$ mole

$\therefore$ Mole of H in $H_2O = 0.05 \times 2 = 0.1$ mol

= mol of H in 0.01 mole Organic compound

$\therefore$ Weight of H atom in 0.01 mole compound

$= 0.1 \times 1 = 0.1$ gm

$\therefore$ Weight of H atom in one mole compound

$= \frac{0.1}{0.01} = 10$ gm $\because$ wt. % of H

$= \frac{\text{wt. of H in one mole compound}}{\text{Molar mass of compound}} \times 100$

$10 = \frac{10}{M} \times 100 \therefore \boxed{M = 100}$

146. (57) $M_FV_F = M_1V_1 + M_2V_2$

$M_F = \frac{M_1V_1 + M_2V_2}{V_1 + V_2}$

$= \frac{2 \times 20 + 0.5 \times 400}{420} = 0.571$M

$= 57.1 \times 10^{-2}$M $= 57$

147. (45) moles of HCl $= 4 \times 10^{-3}$

moles of $Ca(OH)_2$ reacted with HCl $= 2 \times 10^{-3}$

Initial amount of $Ca(OH)_2 = 4 \times 10^{-3}$ mol

(as per the given information)

Moles of $CO_2 = 2 \times 10^{-3}$

Volume $= 2 \times 10^{-3} \times 0.0821 \times 273 = 44.8 \times 10^{-3}$ Lit

$= 44.8\text{ cm}^3$

Nearest integer = 45

148. (24) $\underset{72\text{ mmol}}{3PbCl_2} + \underset{50\text{ mmol}}{2(NH_4)_3PO_4} \longrightarrow$

$Pb_3(PO_4)_2 + 6NH_4Cl$

3 mol $PbCl_2 \equiv 1$ mol $Pb_3(PO_4)_2$

72×10^{-3} mol $PbCl_2 = \frac{1}{3} \times 72 \times 10^{-3}$ mol $Pb_3(PO_4)_2$

$= 24 \times 10^{-3}$ mol

149. (14) $\frac{10}{22400}$ mole of hydrocarbon gives $\frac{40}{22400}$ moles of CO_2

1 mole of hydrocarbon gives $\frac{40}{224} \times \frac{22400}{14} =$

4mole of CO_2

So no. of moles of carbon = 4

$\frac{10}{22400}$ mole of hydrocarbon gives $\frac{50}{22400}$ mole of H_2O

1 mole of hydrocarbon gives $\frac{50}{22400} \times \frac{22400}{10}$

$H_2O = 5$ mole of H_2O

No. of moles of H = 10

So, C : H $\therefore C_4H_{10}$

$4 : 10 \therefore 4 + 10 = 14$

Total no. of atoms = 4 + 10 = 14

150. (50) $\underset{1\text{ mol}}{CH_4(g)} + 2O_2(g) \rightarrow \underset{44\text{ g}}{CO_2(g)} + 2H_2O(l)$

44g CO_2 needs 1 mol CH_4

22g CO_2 needs $= \frac{1}{44} \times 22 = 50 \times 10^{-2}$ mol

$= 0.5$ mol.

151. (4) $M_aV_a = M_bV_b = M_b$

or $M_{NaOH} = \frac{M_aV_a}{V_b}$

$\therefore$ 'n' factor $= 2\frac{50 \times 0.5 \times 2}{25} = 2$ mM

$\Rightarrow n_{NaOH} = M_{NaOH} \times V_{NaOH} = 2 \times 50 = 100$ mmol

$= 100 \times 10^{-3}$ mol $= 10 \times 10^{-2}$ mol

$\Rightarrow$ Mass $= n_{NaOH} \times$ Molar mass of NaOH

$= (10 \times 10^{-2}) \times 40 = 400 \times 10^{-2} = 4$ g

152. (8) $CH_4 + 2O_2 \rightarrow CO_2 + 2H_2O$

Moles of $CO_2 = \frac{22}{44} = 0.5$

So, required moles of $CH_4 = 0.5$

Mass $= 0.5 \times 16 = 8$ gm

153. (7) $M = \frac{n_{NaOH}}{V_{sol}(\text{in L})} \Rightarrow 3 = \frac{(84/40)}{V}$

$\Rightarrow V = 0.7\text{ L} = 7 \times 10^{-1}$ L

154. (135)

$\underset{\text{(Aniline MM = 93)}}{C_6H_5NH_2} + CH_3-\overset{O}{\overset{\|}{C}}-O-\overset{O}{\overset{\|}{C}}-CH_3 \rightarrow$

$\underset{\text{(Acetanilide MM = 135)}}{C_6H_5NH-\overset{O}{\overset{\|}{C}}-CH_3} + CH_3COOH$

$n_{Acetanillide} = n_{Aniline}$

$\frac{m}{135} = \frac{9.3}{93} \Rightarrow m = 13.5\text{ g} = 135 \times 10^{-1}$ gm

155. (100)

$$\underset{1\,g}{M_2CO_3} + \underset{Excess}{2HCl} \rightarrow \underset{0.02\ mole}{2MCl} + H_2O + \underset{0.01\ mole}{CO_2}$$

From principle of atomic conservation of carbon atom,

Mole of $M_2CO_3 \times 1$ = Mole of $CO_2 \times 1$

$$\frac{1\,g}{\text{molar mass of } M_2CO_3} = 0.01 \times 1$$

$\therefore$ Molar mass of M_2CO_3 = 100 g/mole

156. (224) $Mg + 2HCl \rightarrow MgCl_2 + H_2\uparrow$

Number of moles $= \frac{2.4}{24} = 0.1$ mole

1 mole of gas at STP $\Rightarrow$ 22.4 L

$\therefore$ 0.1 mole of gas = 0.1 × 22.4 = 2.24

$= 224 \times 10^{-2}$ L.

157. (11) Percentage of Carbon

$$= \frac{12}{44} \times \frac{\text{mass of } CO_2 \text{ formed}}{\text{mass of compound taken}} \times 100$$

$$60 = \frac{12}{44} \times \frac{\text{mass of } CO_2 \text{ formed}}{0.5} \times 100$$

Mass of CO_2 formed $= \frac{60 \times 44 \times 0.5}{12 \times 100}$ g

$= 1.1$ g $= 11 \times 10^{-1}$ g

158. (1) Hint: First, always check the limiting reagent.

$3BaCl_2 + 2Na_3PO_4 \rightarrow Ba_3(PO_4)_2 + 6NaCl$

5 moles 2 moles

Na_3PO_4 is limiting reagent.

2 mole Na_3PO_4 gives 1 mole of $Ba_3(PO_4)_2$.

159. (44) 44 g of CO_2 contain 12 g of carbon

$\therefore$ 1g of CO_2 has $\frac{12}{44}$ g of carbon

$\Rightarrow$Weight of C in 0.792 g CO_2

$= \frac{12}{44} \times 0.792 = 0.216$

% of C in compound $= \frac{0.216}{0.492} \times 100$

$= 43.90\%. \approx 44$

160. (4) $Zn + 2HCl \rightarrow ZnCl_2 + H_2\uparrow$

Moles of Zn used $= \frac{11.5}{65.4}$ = Moles of H_2 evolved

Volume of $H_2 = \frac{11.5}{65.4} \times 22.7L = 3.99L \approx 4L$

161. (227)

162. (1362) Mole of CO_2

$= 0.2\,M \times (300 \times 10^{-3})L$

$= 0.06$ Mole

Volume of 0.06 mole CO_2 at S.T.P

$= 0.06 \times 22.7 = 1.362\,L = 1362$ mL

163. (150) $\underset{68g}{2H_2O_2} \longrightarrow 2H_2O + \underset{22.72}{O_2}$

$\therefore$ Strength in g/L $= \frac{68}{22.7} \times 50 = 149.7 \approx 150$

164. (200) Let M is the molar mass of the compound (g/mol)

mass of compound = 0.01 × $M_{comp.}$

where $M_{comp.}$ = molar mass of compound

Mass of carbon in the compound

$= 0.01 \times M_{comp.} \times \frac{60}{100}$

moles of carbon $= \frac{0.01\,M_{comp.}}{12} \times \frac{60}{100}$

moles of CO_2 from combustion $= \frac{4.4}{44}$

It should be equal to the moles of carbon in the compound.

$$\Rightarrow \frac{0.01\,M_{comp.}}{12} \times \frac{60}{100} = \frac{4.4}{44}$$

$$M_{comp.} = \frac{4.4}{44} \times \frac{100}{60} \times \frac{12}{0.01} = 200 \text{ g/mol}$$

165. (15) $\underset{1\,mol}{Na} + H_2O \rightarrow \underset{1\,mol}{NaOH} + \frac{1}{2}H_2$

0.69 g of Na forms $= \frac{0.69}{23} = 0.03$ mol of NaOH

For complete neutralization of 0.03 mol of NaOH 0.03 mol = (0.03 × 36.5)g of HCl is required.

Density of HCl (d) = 73 g/L

$\therefore$ Volume of HCl required $= \frac{0.03 \times 36.5}{73}$ L

$= 1.5 \times 10^{-2}$ L = 15 mL

166. (4) A : $Fe_{0.93}O \rightarrow Fe_2O_3$

$$\text{n-factor} = \left(3 - \frac{200}{93}\right) \times 0.93$$

n-factor = 0.79

B : 2x + (0.93 – x) × 3 = 2

x = 0.79

$Fe^{2+} = 0.79$, $Fe^{3+} = 0.21$

D : % $Fe^{2+} = \frac{0.79}{0.93} \times 100 = 85\%$; $Fe^{3+} = 15\%$

167. (3810)

moles of $Cu_2O = \frac{2.86 \times 10^3}{143} = 20$

moles of $Cu_2S = \frac{4.77 \times 10^3}{159} = 30$

Cu_2O is limiting reagent

$2Cu_2O + Cu_2S \longrightarrow 6Cu + SO_2$

Mass of copper = 60 × 63.5 = 3810 g

168. (143) CNG contains methane gas

$CH_4(g) + 2O_2(g) \rightarrow CO_2(g) + 2H_2O\ (l)$

$\frac{100}{16} = 6.25$ $\frac{208}{32} = 6.5$

from the stoichiometry O_2 is limiting reagent

2 moles of O_2 gives = 1 mole CO_2;

6.5 moles of O_2 gives $= \frac{6.5}{2}$ moles of CO_2

Mass of CO_2 = moles × molar mass

$= \frac{6.5}{2} \times 44 = 143$g

169. (225) The balanced chemical reaction is –

$CH_4 + 2O_2 \longrightarrow CO_2 + 2H_2O$

Moles of water produced $(n_{H_2O}) = \frac{81}{18} = 4.5$

from the above reaction;

2 mole of H_2O requires = 1 mole of CH_4

4.5 mole of H_2O requires

$= \frac{4.5}{2} = 2.25 = 225 \times 10^{-2}$ mole of CH_4

170. (1143)

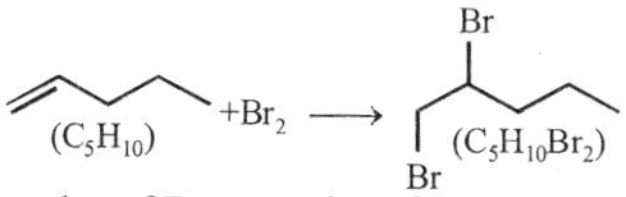

moles of Br_2 = moles of C_5H_{10}

$$\Rightarrow \frac{w}{160} = \frac{5}{70} \Rightarrow w = \frac{5 \times 160}{70} \text{g}$$

$= 11.428$ g $= 1142.8 \times 10^{-2}$ g.

171. (0) Moles of H_2SO_4 in solution A = 50% of original solution = 0.02/2 = 0.01 m mol.

Moles of H_2SO_4 in final solution = 0.01 + 0.01

= 0.02 m mol = 0.00002 × 10^3 m mol

Answer to the nearest integer = 0 (zero).

172. (25) There will be min. one molecule of glycine.

0.30 % glycine ≡ 75 g

1% of protein $\equiv \frac{75}{0.30}$ g

100% of protein $\equiv \frac{75}{0.30} \times 100$ g

= 25000 g = 25 × 10^3 g

173. (3)

$$\underset{\substack{0.1\ M \\ 10\ mL \\ 30\ mL}}{\text{Acid (A)}} + \underset{\substack{(M(OH)_2) \\ 0.05\ M \\ 30}}{\text{Base}} \longrightarrow \text{Salt} + H_2O$$

At equivalence point, equivalent of acid = equivalent of base

$0.1 \times 10 \times n = 30 \times 0.05 \times 2 \rightarrow n = 3.$

174. (3)

175. (525) $3Pb(NO_3)_2 + Cr_2(SO_4)_3 \rightarrow 3PbSO_4 + 2Cr(NO_3)_3$

m moles of $Pb(NO_3)_2$ = 35 × 0.15 = 5.25 m moles

m moles of $Cr_2(SO_4)_3$ = 20 × 0.12 = 2.4 m moles

By stoichiometry, required no. of m moles of $Pb(NO_3)_2$

= 3 × 2.4 = 7.2 m. moles

Hence, $Pb(NO_3)_2$ is limiting reagent, m moles of $PbSO_4$ formed = 5.25 m moles = 525 × 10^{-5} moles

176. (13) $Na_2O + H_2O \rightarrow 2NaOH$

Moles of $Na_2O = \frac{20}{62}$ moles;

M.W. of Na_2O = 62 g/mol

Moles of NaOH formed $= \frac{20}{62} \times 2 = \frac{40}{62}$

$$[NaOH] = \frac{40 \times 1000}{62 \times 500} = 1.29M = 13 \times 10^{-1}M$$

177. (19)

$$C_3H_8(g) + 5O_2(g) \rightarrow 3CO_2(g) + 4H_2O\ (l)$$

$t = 0$: $\frac{100}{44}$ mol, $\frac{1000}{32}$ mol, 0, 0

$t = \infty$: 0, 19.9 mol, 6.81 mol, 9.08 mol

Mole fraction of CO_2 in the final reaction mixture (heterogeneous),

$$x_{CO_2} = \frac{n_{CO_2}}{n_{O_2} + n_{CO_2} + n_{H_2O}} = \frac{6.81}{19.9 + 6.81 + 9.08}$$

$= 0.19 = 19 \times 10^{-2}$

178. (5) Consider 1 L = 1000 mL of solution

Mass of solution = Vol. × density

= 1.2 × 1000 g = 1200 g

Mass of 1L solvent (water) = 1000 g

$(\because \text{density} = 1 \text{ g cm}^{-3})$

$\Rightarrow$Mass of NaOH = (1200 – 1000) g = 200 g

$\Rightarrow$Moles of NaOH $= \dfrac{200\text{ g}}{40\text{ g/mol}} = 5$ mol

$\Rightarrow$ Molality $= \dfrac{5\text{ mol}}{1\text{ kg}} = 5$ m

179. (18) milli-equivalents of Fe^{2+}
= milli-equivalents of $K_2Cr_2O_7$
$\Rightarrow$ millimoles $\times$ *n*-factor of Fe^{2+} = millimoles $\times$ *n*-factor of $K_2Cr_2O_7$.
$x \times 10 \times 1 = 0.02 \times 15 \times 6; x = 0.18 = 18 \times 10^{-2}$ M

180. (464)

$$C_4H_{10} + \frac{13}{2}O_2 \longrightarrow 4\,CO_2 + 5\,H_2O$$

Moles of $H_2O = \dfrac{72}{18} = 4$

Moles of C_4H_{10} used $= \dfrac{1}{5} \times 4$

Weight of C_4H_{10} used $= \dfrac{4}{5} \times 58 = 46.4$ g

46.4 g = 464×10^{-1} g

181. (4) Concentration of glucose in blood = 0.72g/L

$= \dfrac{0.72}{180} = 4 \times 10^{-3}$ mol / L

182. (226) Millimoles of NaOH
= Volume (mL) × Molarity
= 250 mL × 0.5 = 125 m mol
Millimoles of HCl = 500 mL × 1 = 500 m mol
Now reaction is

	NaOH +	HCl →	NaCl +	H_2O
$t = 0$	125	500	0	0
$t = t$	0	375	125	125

So, millimoles of HCl left = 375
$= 375 \times 10^{-3}$ moles No. of HCl molecules
$= 6.022 \times 10^{23} \times 375 \times 10^{-3}$
$= 225.8 \times 10^{21} \approx 226 \times 10^{21}$

183. (1) Let moles of NaOH or Na_2CO_3 be *a*.
Given mixture is equimolar, thus
NaOH : *a* mol
Na_2CO_3 : *a* mol
Now, $W_{NaOH} + W_{Na_2CO_3} = 4$

$\Rightarrow 40 \cdot a + 106 \cdot a = 4 \Rightarrow a = \dfrac{4}{146}$ mol

Therefore, mass of NaOH

$= \dfrac{4}{146} \times 40\text{ g} = 1.095 \approx 1$

184. (16)

$$(HO)CH_2\text{–}CH_2CHO \xrightarrow[64\%]{\Delta} C_3H_4O + H_2O$$

M.W. 74 56
Let *x* g 7.8 g

$\dfrac{x}{74}$ mol $\dfrac{x}{74} \times 0.64$

$\therefore \dfrac{x}{74} \times 0.64 \text{ mol} = \dfrac{7.8}{56}$ moles $\Rightarrow x = 16.10 \approx 16$

185. (3)

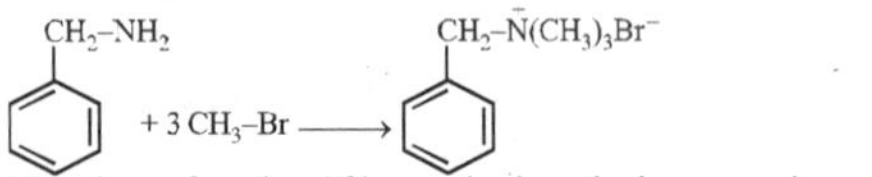

Number of moles of benzyl trimethyl ammonium bromide $= \dfrac{23}{230} = 0.1$

$\therefore$ No. of moles of bromomethane consumed
$= 3 \times 0.1 = 3 \times 10^{-1}$

186. (18) General combustion reaction of ethane is

$$C_xH_y + \left(x + \frac{y}{4}\right)O_2 \longrightarrow xCO_2 + \frac{y}{2}H_2O$$

Then for ethane is,

$$C_2H_6 + \frac{7}{2}O_2 \longrightarrow 2CO_2 + 3H_2O$$

1 mol 3 mol

$\dfrac{3}{30}$ mol = 0.1 0.3 mol

$\therefore$ Number of molecules of water
$= 0.3 \times 6.023 \times 10^{23}$
$= 18.069 \times 10^{22} \approx 18 \times 10^{22}$

187. (1) Volume of Cl_2 at STP = 20 mL
No. of moles of chlorine gas

$= \dfrac{20\text{ mL}}{22400\text{ mL/mol}}$

No. of molecules of Cl_2 gas

$= \dfrac{20}{22400} \times 6.023 \times 10^{23}$

Hence, number of Cl atoms

$= 2 \times \dfrac{20}{22400} \times 6.023 \times 10^{23}$

$= 1.08 \times 10^{21} \approx 1 \times 10^{21}$

188. (24)

$$\underset{n=1}{6Fe^{2+}} + \underset{n=6}{Cr_2O_7^{2-}} + 14H^+ \rightarrow 2Cr^{3+} + 6Fe^{3+} + 7H_2O$$

By law of equivalence,
M_{eq} of Fe^{2+} = M_{eq} of $Cr_2O_7^{2-}$
$1 \times M \times 15 = 0.03 \times 6 \times 20$
$M = 0.24 = 24 \times 10^{-2}$

189. (13) Mass of Na^+ in 50 mL = 70 × 50 mg

Millimoles of $Na^+ = \dfrac{70 \times 50}{23}$

Weight of $NaNO_3 = \dfrac{70 \times 50 \times 85 \times 10^{-3}}{23} = 12.9 \approx 13$ g

190. (2) $M = \dfrac{4.5/90}{250/1000} = 0.2 = 2 \times 10^{-1}$.

191. (8)

$$C_xH_y(g) + \left(x + \frac{y}{4}\right)O_2(g) \rightarrow x\,CO_2(g) + \frac{y}{2}H_2O(l)$$

V $6V$ $4V$

$V_{O_2} = 6 \times V_{C_xH_y}$ (given)

$\Rightarrow V\left(x + \dfrac{y}{4}\right) = 6V \quad \Rightarrow \left(x + \dfrac{y}{4}\right) = 6$ ...(i)

$Vx = 4V$ (given) $\Rightarrow x = 4$ ...(ii)

From relation (i) and (ii): $4 + \dfrac{y}{4} = 6 \Rightarrow y = 8$.

192. (18) Complete combustion of hydrocarbons can be represented by the following reaction.

$$C_xH_y + \left(x + \frac{y}{4}\right)O_2 \longrightarrow x\,CO_2 + \frac{y}{2}H_2O$$

For propane, combustion reaction is:

$$C_3H_8 + \left(3 + \frac{8}{4}\right)O_2 \longrightarrow 3\,CO_2 + \frac{8}{2}H_2O$$

$\therefore C_3H_8 + 5\,O_2 \longrightarrow 3\,CO_2 + 4\,H_2O$

Similarly, for butane, Combustion reaction is

$$C_4H_{10} + \left(4 + \frac{10}{4}\right)O_2 \longrightarrow 4\,CO_2 + \frac{10}{2}H_2O$$

$\therefore C_4H_{10} + \dfrac{13}{2}O_2 \longrightarrow 4\,CO_2 + 5\,H_2O$

$\because$ For 1 mol of C_4H_{10} required $O_2 = \dfrac{13}{2}$ mol

$\therefore$ For 2 mol of C_4H_{10} required O_2

$= \dfrac{13}{2} \times 2 = 13$ mol

Number of moles of O_2 required = 5 + 13 = 18

193. (3400)

Moles of $N_2 = \dfrac{2800}{28} = 100$ and Moles of H_2

$= \dfrac{1000}{2} = 500$

	$N_2(g)$ +	$3H_2(g)$ →	$2NH_3(g)$
Initial mole	100	500	0
Final mole	0	200	200

Mass of NH_3 formed = 200 × 17 = 3400 g

194. (10)

Normality $= \dfrac{\text{No. of equivalents of solute}}{\text{Volume of solution (in L)}}$

$0.1 = \dfrac{1.43}{\dfrac{(106 + 18x)}{2} \times 0.1} \Rightarrow \dfrac{106 + 18x}{2} = 143$

$\Rightarrow 18x = 286 - 106 = 180 \Rightarrow x = 10.$

195. (47)
Let total mole of solution = 1
So, mole of glucose = 0.1
Mole of H_2O = 0.9

% (w/w) of $H_2O = \left[\dfrac{0.9 \times 18}{0.9 \times 18 + 0.1 \times 180}\right] \times 100$
$= 47.368 = 47.37.$

196. (25)
Number of mole of *x*

$= \dfrac{6.022 \times 10^{22}}{6.022 \times 10^{23}} = \dfrac{10}{\text{Molar mass of } x}$

So molar mass of *x* = 100 g

Molarity $= \dfrac{5}{100 \times 2} = 0.025$ M.

197. (10)
Phosphinic acid is hypophosphorous acid (H_3PO_2), a monobasic acid.
$NaOH + H_3PO_2 \longrightarrow NaH_2PO_2 + H_2O$
For neutralisation, $(N_1V_1)_{acid} = (N_2V_2)_{base}$
$0.1 \times 10 = 0.1 \times V_{NaOH}$
$V_{NaOH} = 10$ mL

198. (10.00)

Concentration $= \dfrac{10.3 \times 10^{-3}}{1.03 \times 1000\text{ g}} \times 10^6 = 10$ ppm

199. (4.96) 10 ppm $= \dfrac{\text{Mass of Fe (in g)}}{100 \times 1000\text{ g}} \times 10^6$ ppm

$\Rightarrow$ Mass of Fe = 1 g
Molar mass of $FeSO_4 \cdot 7H_2O$ = 278
56 g of iron present in 1 mole of $FeSO_4 \cdot 7H_2O$

1 g of Fe present in $\dfrac{278}{56}$ g of salt = 4.96 g

200. (2130)
$NaClO_3(s) + Fe(s) \longrightarrow NaCl(s) + FeO(s) + O_2(g)$
Moles of $NaClO_3$ = Moles of O_2

Moles of $O_2 = \dfrac{PV}{RT} = \dfrac{1 \times 492}{0.082 \times 300} = 20$ moles

Mass of $NaClO_3$ = 20 × 106.5 = 2130 g

Part-B (JEE Advanced)

1. (a) Atomic weight in gms = 6.023×10^{23} atoms = 1 Mole atoms

(a) Number of atoms in 24 g of C

$= \frac{24}{12} \times 6.023 \times 10^{23} = 2 \times 6.023 \times 10^{23}$ atom

= 2 mole atoms

(b) Number of atoms in 56 g of Fe

$= \frac{56}{56} \times 6.023 \times 10^{23} = 6.023 \times 10^{23}$ atom

= 1 mole atoms

(c) Number of atoms in 27 g of Al

$= \frac{27}{27} \times 6.023 \times 10^{23} = 6.023 \times 10^{23}$ atom

= 1 mole atoms

(d) Number of atoms in 108 g of Ag

$= \frac{108}{108} \times 6.023 \times 10^{23} = 6.023 \times 10^{23}$ atom

= 1 mole atoms

∴ 24 g of C has maximum number of atoms.

2. (d)

(i) Mass of one electron = 9.108×10^{-31} kg

(ii) 1 mole of electron = 6.023×10^{23} electrons

Weight of 1 mole of electron = Mass of one electron × Avogadro Number $= 9.108 \times 10^{-31} \times 6.023 \times 10^{23}$ kg

∴ No. of moles of electrons in 1 kg

$= \frac{1}{9.108 \times 10^{-31} \times 6.023 \times 10^{23}}$

$= \frac{1}{9.108 \times 6.023} \times 10^{8}$

3. (a)

(a) 18 g of $H_2O = 6.02 \times 10^{23}$ molecules of H_2O

∴ 36 g of $H_2O = 2 \times 6.02 \times 10^{23}$ molecules of H_2O $= 12.04 \times 10^{23}$ molecules of H_2O

(b) 28 g of CO = 6.02×10^{23} molecules of CO

(c) 46 g of $C_2H_5OH = 6.02 \times 10^{23}$ molecules of C_2H_5OH

(d) 108 g of $N_2O_5 = 6.02 \times 10^{23}$ molecules of N_2O_5

∴ 54 g of $N_2O_5 = \frac{1}{2} \times 6.02 \times 10^{23}$ molecules of $N_2O_5 = 3.01 \times 10^{23}$ molecules of N_2O_5

∴ 36 g of water has highest number of molecules.

4. (a) No. of e^- in C = 6 and in O = 8

∴ Total no. of e^- in $CO_2 = 6 + 8 \times 2 = 22$

5. (d) $4Al + 3O_2 \longrightarrow 2Al_2O_3$

At. wt. of Al = 27

Thus 4 × 27 g of Al reacts with oxygen = 3 × 32g

∴ 27 g of Al reacts with oxygen $= \frac{3 \times 32}{4 \times 27} \times 27\text{ g} = 24\text{ g}$

6. (4) $R = N_A \times k$

$= 6.023 \times 10^{23} \times 1.380 \times 10^{-23}$

= 8.312 which has 4 significant figures

7. (55.55) 1 litre water = 1 kg *i.e.* 1000 g water

($\because d = 1000\ \text{kg/m}^3$) $\equiv \frac{1000}{18} = 55.55$ moles of water

So, molarity of water = **55.55M**

8. (4.14 g) 1 Mole = 6.023×10^{23} molecules = Molecular weight in gms.

Weight of 6.023×10^{23} (Avogadro's number) molecules of $CuSO_4.5H_2O$ = Molecular wt. of $CuSO_4.5H_2O$ = 249 g.

∴ Weight of 1×10^{22} molecules of $CuSO_4.5H_2O$

$= \frac{249 \times 1 \times 10^{22}}{6.023 \times 10^{23}} = \mathbf{4.14\ g}$

9. Carbon (C – 12)

10. (6.02×10^{24})

18 mL H_2O = 18 g H_2O (∵ density of water = 1 g/cc) = 1 mole of H_2O.

1 Mole of $H_2O = 10 \times 6.02 \times 10^{23}$ electrons

(∵ Number of electrons present in one molecule of water = 2 + 8 = 10) = 6.02×10^{24} electrons

11. (i) Volume of virus = $\pi r^2 \ell$ (Volume of cylinder)

(ii) Mass of single virus $= \frac{\text{Volume}}{\text{Sp. volume}}$

(iii) Molecular mass of virus = Mass of single virus × 6.02×10^{23}

Volume of virus = $\pi r^2 \ell$

$= \frac{22}{7} \times \frac{150}{2} \times \frac{150}{2} \times 10^{-16} \times 5000 \times 10^{-8}$

$= 0.884 \times 10^{-16}\ \text{cm}^3$

Weight of one virus $= \frac{0.884 \times 10^{-16}}{0.75}$ g

$= 1.178 \times 10^{-16}$ g

∴ Mol. wt. of virus = $1.178 \times 10^{-16} \times 6.02 \times 10^{23}$ $= \mathbf{7.09 \times 10^{7}}$

12. (a)

	$CaCl_2$	≡ $CaCO_3$	≡ $MgCl_2$
M.wt.	111	100	95

From this it is evident, that

111 mg $CaCl_2$ will give $CaCO_3$ = 100 mg

∴ 1 mg $CaCl_2$ will give $CaCO_3 = \frac{100}{111}$ mg = 0.90 mg

95 mg $MgCl_2$ gives $CaCO_3$ = 100 mg

∴ 1 mg $MgCl_2$ gives $CaCO_3 = \frac{100}{95}$ mg = 1.05 mg

∴ Total $CaCO_3$ formed by 1 mg $CaCl_2$ and 1 mg $MgCl_2$ = 0.90 + 1.05 = 1.95 mg

∴ Amount of $CaCO_3$ present per litre of water = 1.95 mg

∴ wt of 1 mL of water = 1g = 10^3 mg

∴ wt of 1000 mL of water = $10^3 \times 10^3 = 10^6$ mg

∴ Total hardness of water in terms of parts of $CaCO_3$ per 10^6 parts of water by weight = **1.95 parts.**

(b) Eq wt of $Ca^{++} = \frac{\text{Mol.wt}}{\text{Charge}} = \frac{40}{2} = 20$

$Ca^{2+} + Na_2CO_3 \longrightarrow CaCO_3 + 2Na^+$

1 milliequivalent of Ca^{2+} = 20 mg

1 milliequivalent of Na_2CO_3 is required to soften 1 litre of hard water.

(c)

$$\underset{2 \times 24 = 48\text{g}}{2Mg} + \underset{32\text{g}}{O_2} \longrightarrow \underset{2(24+16) = 80\text{g}}{2MgO}$$

∵ 32 g of O_2 reacts with = 48 g Mg

∴ 0.5 g of O_2 reacts with $= \frac{48}{32} \times 0.5 = 0.75$ g

Weight of unreacted Mg = 1.00 – 0.75 = 0.25 g

Thus, Mg is left in excess.

Weight of MgO formed $= \frac{80}{48} \times 0.75 = 1.25$ g

$$\underset{(40\text{ g})}{MgO} + H_2SO_4 \longrightarrow MgSO_4 + H_2O$$

According to reaction,

∵ 40 g MgO is dissolved in 1000 mL of 1 N H_2SO_4

∴ 40 g MgO is dissolved in 2000 mL 0.5 N H_2SO_4

∴ 1.25 MgO is dissolved in

$= \frac{2000 \times 1.25}{40}$ mL of 0.5 N H_2SO_4

= 62.5 mL of 0.5N H_2SO_4

13. (i) No. of C atoms in 14g of ^{14}C $= 6.02 \times 10^{23}$

∴ No. of C atom in 7 mg (7/1000g) of ^{14}C

$= \frac{6.02 \times 10^{23} \times 7}{14 \times 1000} = 3.01 \times 10^{20}$

No. of neutrons in 1 carbon atom = 7

∴ Total no. of neutrons in 7 mg of $^{14}C = 3.01 \times 10^{20} \times 7 = 21.07 \times 10^{20}$

Wt of 1 neutron = wt of 1 hydrogen atom

$= \frac{1}{6.02 \times 10^{23}}$ g

∴ Wt of $3.01 \times 10^{20} \times 7$ neutrons

$= \frac{3.0 \times 10^{20} \times 7}{6.02 \times 10^{23}} = \mathbf{3.5 \times 10^{-3} g}$

14. (b) Molecular Formula = n × Empirical formula

Solution : Since the molecular formula is n times the empirical formula, therefore, different compounds having the same empirical formula must have different molecular weights.

15. (c) Let mass of oxygen = 1g, Then mass of nitrogen = 4g

Mol. wt. of N_2 = 28g, Mol. wt. of O_2 = 32g

28 g of N_2 has = 6.02×10^{23} molecules of nitrogen

4 g of N_2 has $= \frac{6.02 \times 10^{23}}{28} \times 4$ molecules of nitrogen $= \frac{6.02 \times 10^{23}}{7}$ molecules of nitrogen

32 g of O_2 has = 6.02×10^{23} molecules of oxygen

∴ 1g of O_2 has $= \frac{6.02 \times 10^{23}}{32} \times 1$

$= \frac{6.02 \times 10^{23}}{32}$ molecules of oxygen

Thus, ratio of molecules of oxygen : nitrogen

$= \frac{6.02 \times 10^{23}/32}{6.02 \times 10^{23}/7} = 7 : 32$

16. (c) No. of nitrogen atoms

$= \frac{\text{Mass in grams}}{\text{Atomic wt.}} = \frac{28}{14} = 2$

No. of oxygen atoms $= \frac{\text{Mass in grams}}{\text{Atomic wt.}} = \frac{80}{16} = 5$

∴ Formula of compound is N_2O_5.

17. (15.05) In pure iron oxide (FeO), iron and oxygen are present in the ratio 1 : 1.

However, here number of Fe^{2+} present = 0.93

or No. of Fe^{2+} ions missing = 0.07

Since each Fe^{2+} ion has 2 positive charge, the total number of charge due to missing (0.07) Fe^{2+} ions = 0.07 × 2 = 0.14

To maintain electrical neutrality, 0.14 positive charge is compensated by the presence of Fe^{3+} ions. Now since, replacement of one Fe^{2+} ion by one Fe^{3+} ion increases one positive charge, 0.14 positive charge must be compensated by the presence of 0.14 Fe^{3+} ions.

In short, 0.93 Fe^{2+} ions have 0.14 Fe^{3+} ions

$100\ Fe^{2+}$ ions have $= \frac{0.14}{0.93} \times 100 = \mathbf{15.05\%}$

Alternate Solution:

Let the sample contains 93 iron ions and 100 oxide ions.

Total negative charge on oxide ions $= 2 \times 100 = 200$

Let no. of ions of iron as $fe^{3+} = x$

$\therefore$ No. of Fe^{2+} ions $= 93 - x$

$3 \times x + (93 - x)2 = 200 \quad 3x - 186 - 2x = 200 \Rightarrow x = 14$

percentage of iron as $fe^{3+} = \frac{14}{93} \times 100 = 15.05$

18. (24) According to problem, three atoms of M combine with 2 atoms of N.

$\therefore$ Formula of compound is M_3N_2 (Where M is the metal)

Equivalent wt of N $= \frac{14}{3}$ ($\because$ valency of N in compound is 3)

$\because$ 28 g N combines with = 72 g metal

$\therefore$ 14/3 g N combines with $= \frac{72}{28} \times \frac{14}{3} = 12$ g

$\therefore$ Eq. wt. of metal = 12

At wt of metal = Eq. wt × valency $= 12 \times 2 = \mathbf{24}$

[Valency of metal = 2]

19. (b, c)

Let M represents the respective molecular mass.

(a) If empirical formula of compound 3 is P_3Q_4,

$$\frac{3M_P}{3M_P + 4M_Q} = \frac{40}{100} \Rightarrow 9M_P = 8M_Q$$

If empirical formula of compound 2 is P_3Q_5,

$$\% \text{ of } P = \frac{3M_P}{3M_P + 5M_Q} \times 100$$

$$= \frac{8/3M_Q}{8/3M_Q + 5M_Q} \times 100 = 34.78$$

Hence, option (a) is incorrect.

(b) If empirical formula of compound 3 is P_3Q_2,

$$\frac{3M_P}{3M_P + 2M_Q} = \frac{40}{100} \Rightarrow 4M_Q = 9M_P$$

$$\Rightarrow \quad M_Q = \frac{9}{4} \times 20 = 45$$

Hence, option (b) is correct.

(c) If empirical formula of compound 2 is PQ,

$$\frac{M_P}{M_P + M_Q} = \frac{44.4}{100} \Rightarrow 5M_P = 4M_Q$$

If empirical formula of compound 1 is P_5Q_4 then weight % of P : Q = 1 : 1

Hence, option (c) is correct.

(d) If empirical formula of compound 1 is P_3Q,

$$\% P = \frac{2M_P}{2M_P + M_Q} \times 100 = 50 \Rightarrow \quad 2M_P = M_Q$$

Hence, atomic weight of P and Q cannot be 70 and 35, respectively.

20. Given $P = 1$ atm $V = 1$ L, $T = 127\,°C$ $= 127 + 273 = 400$ K

$PV = nRT$ (Ideal gas equation)

or $n = \frac{PV}{RT} = \frac{1 \times 1}{0.082 \times 400} = 0.0304$

$$\text{Mol. wt} = \frac{\text{Mass}}{\text{Moles}} = \frac{2.8}{0.0304} = 92.10$$

Element	wt. of element	Relative no. of atoms	Ratio of of atoms	Whole no. of atoms
C	10.5	10.5/12 = 0.875	0.875/0.875 = 1	1 × 7 = 7
H	1.0	1.0/1 = 1	1/0.875 = 1.14	1.14 × 7 = 8

$\therefore$ Empirical formula $= C_7H_8$

Empirical formula wt $= 12 \times 7 + 1 \times 8 = 92$

$$n = \frac{\text{Molecular wt}}{\text{Empirical formula wt}} = \frac{92.10}{92} = 1$$

Molecular formula = n × empirical formula $= 1(C_7H_8) = \mathbf{C_7H_8}$

21. (d) Let the fraction of M^{2+} and M^{3+} be:-

$$M^{+2} = \frac{X}{3}, M^{+3} = \frac{2X}{3}$$

So, total of O.N. of all atoms would be :

$$\left[\frac{X}{3}(+2) + \frac{2X}{3}(+3)\right] + 2(+3) + 4(-2) = 0$$

$$\Rightarrow \frac{2X}{3} + 2X + 6 - 8 = 0 \Rightarrow \frac{8X}{3} = 2$$

$$\Rightarrow X = \frac{6}{8} = \frac{3}{4} = 0.75$$

22. (a) Write the reaction for chemical change during reaction and equate moles of products formed.

$[Co(NH_3)_5SO_4]$ Br has ionisable Br^- ions & $[Co(NH_3)_5\ Br]\ SO_4$ has ionisable SO_4^{--} ion.

Mixture X has 0.02 mol. of $[Co(NH_3)_5SO_4]$ Br and 0.02 mol of $[Co(NH_3)_5Br]\ SO_4$ in 2 L of solution

$\therefore$ Conc. of $[Co(NH_3)_5SO_4]$ Br and $[Co(NH_3)_5Br]SO_4$ = 0.01 mol/L for each of them.

(i) 1 L mixture of X + excess $AgNO_3 \rightarrow Y$

$$\underset{0.01\text{ mol/L soluble}}{[Co(NH_3)_5SO_4]\,Br} + \underset{\text{excess}}{AgNO_3} \longrightarrow$$

$$[Co(NH_3)_5SO_4]NO_3 + \underset{0.01\text{mol}}{\underset{(Y)}{AgBr}}$$

$$\left[Ag^+ + Br^- \rightarrow AgBr\right]$$

$\therefore$ No. of moles of $Y = 0.01$

(ii) Also 1 L mixture of X + excess $BaCl_2$

$$\rightarrow Z\ \underset{0.01\text{ mol/L soluble}}{[Co(NH_3)_5Br]SO_4} + \underset{\text{excess}}{BaCl_2}$$

$$\rightarrow [Co(NH_3)_5Br]Cl_2 + \underset{0.01\text{mol}}{\underset{(Z)}{BaSO_4}}$$

$$\left[Ba^{++} + SO_4^{=} \rightarrow BaSO_4\right]$$

$\therefore$ moles of $Z = 0.01$.

23. (a) Equivalents of $H_2C_2O_4 \cdot 2H_2O$ = Equivalents of NaOH (At equivalence point)

Strength of $H_2C_2O_4 \cdot 2H_2O$ (in g/L) $= \frac{6.3}{250/1000} =$ 25.2 g/L

Normality of $H_2C_2O_4 \cdot 2H_2O = \frac{\text{Strength}}{\text{Eq. wt}} = \frac{25.2}{63} = 0.4$N

$$\left\{\text{Eq. wt. of oxalic acid} = \frac{\text{Mol. wt}}{2} = \frac{126}{2} = 63\right\}$$

Using normality equation :

$N_1V_1 = N_2V_2(H_2C_2O_4 \cdot 2H_2O)$ (NaOH)

$$0.4 \times 10 = 0.1 \times V_2 \text{ or } V_2 = \frac{0.4 \times 10}{0.1} = 40 \text{ mL.}$$

24. (b) (i) Find change in oxidation number of Cr atom.

(ii) Eq. wt. $= \frac{\text{Molecular wt.}}{\text{change in O.N.}}$

In iodometry, $K_2Cr_2O_7$ liberates I_2 from iodides (NaI or KI). Thus, it is titrated with $Na_2S_2O_3$ solution.

$2Na_2S_2O_3 + I_2 \rightarrow 2NaI + Na_2S_4O_6$

O.N. of Cr changes from + 6 (in $K_2Cr_2O_7$) to +3.

i.e. +3 change for each Cr atom

$$Cr_2O_7^{--} + 14H^+ + 6e^- \rightarrow 2Cr^{3+} + 7H_2O$$

Thus, one mole of $K_2Cr_2O_7$ accepts 6 mole of electrons.

$$\therefore \text{ Equivalent weight} = \frac{\text{Molecular weight}}{6}$$

25. (d) (i) H_3PO_3 is dibasic acid as it contains two –OH groups.

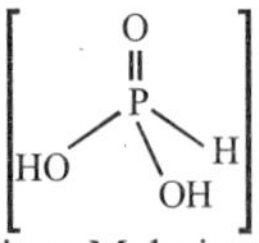

(ii) Normality = Molarity × basicity of acid.

(iii) Basicity of $H_3PO_3 = 2$

$\therefore$ Normality $= 0.3 \times 2 = 0.6$

26. (b) For equivalent weight of $MnSO_4$ to be half of its molecular weight, change in oxidation state must be equal to 2. It is possible only when oxidation state of Mn in product is + 4. Since oxidation state of Mn in $MnSO_4$ is + 2. So, MnO_2 is correct answer.

In MnO_2, O.S. of Mn = +4

$\therefore$ Change in O.S. of Mn $= +4 - (+2) = +2$

27. (d) Molality – Moles of solute/mass of solvent in kg

$\because$ Molality does not involve volume term.

$\therefore$ It is independent of temperature.

28. (a) Molality

$$= \frac{\text{Number of moles of solute}}{\text{Mass of solvent in kg}}$$

A molal solution is one which contains one mole of solute per 1000 g of solvent. $\left\{\because 1m = \frac{1\text{mole}}{1\text{kg}}\right\}$

29. (d)

(i) Write balanced chemical equation for chemical change.

(ii) Find limiting reagent.

(iii) Amount of product formed will be determined by amount of limiting reagent.

The balanced equation is :

$$3BaCl_2 + 2Na_3PO_4 \rightarrow Ba_3(PO_4)_2 + 6NaCl$$

No. of moles :	$BaCl_2$	Na_3PO_4	$Ba_3(PO_4)_2$
	0.5	0.2	
	3mole	2mole	1mole

Limiting reagent is Na_3PO_4 (0.2 mol), $BaCl_2$ is in excess. From the above equation :

2.0 moles of Na_3PO_4 yields $Ba_3(PO_4)_2$ = 1 mole

$\therefore$ 0.2 moles of Na_3PO_4 will yield $Ba_3(PO_4)_2$ $= \frac{1}{2} \times 0.2 = \mathbf{0.1\ mol.}$

30. (a) The change involved is

$$MnO_4^- + e^- \longrightarrow MnO_4^{2-}$$

i.e. it involves only one electron

$$\text{Eq.wt} = \frac{\text{Mol.wt}}{\text{No.of } e^- \text{ involved}} = \frac{M}{1} = M$$

[$\because$ Mol. wt. $= M$]

31. (a) $\underset{2.76}{Ag_2CO_3} \xrightarrow{\Delta} Ag_2O + CO_2$

$\downarrow \Delta$

$\underset{\text{(residue)}}{2Ag} + \frac{1}{2}O_2$

Ag_2O is thermally unstable and decompose on heating liberating oxygen.

Mol. wt. of $Ag_2CO_3 = 108 \times 2 + 12 + 16 \times 3 = 276$ g

$\therefore$ 276 g of Ag_2CO_3 on heating gives residue $= 2 \times 108 = 216$ g of Ag

$\therefore$ 2.76 g of Ag_2CO_3 on heating gives

$= \frac{216}{276} \times 2.76 = 2.16$ g of Ag

32. (222)

$$4(CH_3)_2SiCl_2 + 4H_2O \longrightarrow \underset{(X)}{(CH_3)_8Si_4O_4} + 8HCl$$

4 moles of $(CH_3)_2SiCl_2$ produces, 0.75 mol of X.

$\therefore$ mass of product 'X' $= 0.75 \times 296 = 222$ g

33. (18) $2\overset{+7}{KMnO_4} + 5H_2S + 3H_2SO_4 \rightarrow K_2SO_4 + 2\overset{+2}{MnSO_4} + 5S + 8H_2O$

x = 8 (moles of H_2O produced)

y = 14 – 4 = 10 (number of electrons involved)

x + y = 10 + 8 = 18

34. (1.875)

$$5Fe^{2+} + MnO_4^- + 8H^+ \rightarrow 5Fe^{3+} + Mn^{2+} + 4H_2O$$

For 25 mL

meq of Fe^{2+} = meq of MnO_4^-

$\rightarrow 5Fe^{3+} + Mn^{2+} + 4H_2O = 12.5 \times 0.03 \times 5$

For 250 mL

M moles of $Fe^{2+} = \frac{12.5 \times 0.03 \times 5 \times 250}{25}$

Moles of $Fe^{2+} = \frac{18.75}{1000}$ mol

$= 18.75 \times 10^{-3} = 1.875 \times 10^{-3} \Rightarrow x = 1.875$

Weight of $Fe^{2+} = 1.875 \times 10^{-2} \times 56 = 1.05$ g

% Purity of $Fe^{2+} = \frac{1.05}{5.6} \times 100 = 18.75\%$ y = 18.75

35. (18.75)

36. (2992)

$$\underset{\substack{1584\,g \\ =12\,mole}}{(NH_4)_2SO_4} + Ca(OH)_2 \rightarrow \underset{\substack{gypsum (M=172) \\ 12\,mole}}{CaSO_4.2H_2O} + \underset{24\,mole}{2NH_3}$$

$$\underset{952\,g = 4\,mole}{NiCl_2.6H_2O} + \underset{24\,mole}{6NH_3} \rightarrow \underset{\substack{(M=232) \\ 4\,mole}}{[Ni(NH_3)_6]Cl_2} + 6H_2O$$

Total mass $= 12 \times 172 + 4 \times 232 = 2992$ g

37. (6.47)

$PbS + O_2 \longrightarrow Pb + SO_2$

No. of moles of $O_2 = \frac{10^3}{32}$

Moles of Pb formed $= \frac{10^3}{32}$

$\therefore$ Mass of Pb formed $= \frac{10^3}{32} \times 207 = 6468.75$ g

$= 6.46875$ kg $= 6.47$ kg

38. (8.09)

Use molarity equation to find volume of H_2SO_4 solution.

$$\underset{\substack{63.5+12+48.98 \\ =123.5\,g}}{CuCO_3} + \underset{98\,g}{H_2SO_4} \rightarrow CuSO_4 + H_2O + CO_2\uparrow$$

$\therefore$ For 123.5 gms of Cu(II) carbonate 98 g of H_2SO_4 are required.

For 0.5 gms of Cu(II) carbonate weight of H_2SO_4 reqd.

$= \frac{98 \times 0.5}{123.5}$ g $= 0.39676$ g H_2SO_4

Weight of required $H_2SO_4 = 0.39676$ g

Weight of solute in grams

$= \frac{\text{Mol.wt.} \times \text{Molarity} \times \text{Volume in mL}}{1000}$

$0.39676 = \frac{98 \times 0.5 \times V}{1000}$ or

$V = \frac{0.39676 \times 1000}{90 \times 0.5}$ mL

Volume of H_2SO_4 solution = **8.097 mL**

39. (85) Reaction involved titration is

$$\underset{1\,mole}{KIO_3} + \underset{2\,moles}{2KI} + 6HCl \rightarrow 3ICl + 3KCl + 3H_2O$$

20 mL. of stock KI solution $\equiv$ 30 mL. of $\frac{M}{10}$ KIO_3 solution

Molarity of KI solution $= \frac{30 \times 1 \times 2}{20 \times 10} = \frac{3}{10}$

Millimoles in 50 mL. of KI solution $= 50 \times \frac{3}{10} = 15$

Millimoles of KI left unreacted with $AgNO_3$ solution

$= 2 \times 50 \times \frac{1}{10} = 10$

$\therefore$ Millimoles of KI reacted with $AgNO_3$ – 15 10 = 5

Millimoles of $AgNO_3$ present in $AgNO_3$ solution = 5

Molecular weight of $AgNO_3 = 170$

$\therefore$ Wt. of $AgNO_3$ in the solution $= 5 \times 10^{-3} \times 170 = 0.850$ g

% $AgNO_3$ in the sample $= \frac{0.850}{1} \times 100 =$ **85%**

40. (6.0) Mass of Fe_2O_3 in the sample

$= \frac{55.2}{100} \times 1 = 0.552$ g

Number of moles of $Fe_2O_3 = \frac{0.552}{159.8} = 3.454 \times 10^{-3}$

Number of moles of Fe^{3+} ions $= 2 \times 3.454 \times 10^{-3}$

$= 6.9 \times 10^{-3}$ mol $= 6.90$ mmol

Since, only 1 electron is exchanged in the conversion of Fe^{3+} to Fe^{2+}, the molecular mass is the same as equivalent mass.

$\therefore$ Amount of Fe^{2+} ion in 100 mL. of sol. = 6.90 meq

Volume of oxidant used for 100 mL of Fe^{2+} sol. $= 17 \times 4 = 68$ mL.

Amount of oxidant used $= 68 \times 0.0167$ mmol $= 1.1356$ mmol

Let the number of electrons taken by the oxidant $= n$

$\therefore$ No. of meq. of oxidant used $= 1.1356 \times n$

Thus $1.1356 \times n = 6.90 \Rightarrow n = \frac{6.90}{1.1356} =$ **6**

41. (10.43)

Molality $= \frac{\text{Mass of solute / M. wt. of solute}}{\text{Mass of solvent in kg}}$

Mass of H_2SO_4 in 100 ml of 93% H_2SO_4 solution = 93 g

$\therefore$ Mass of H_2SO_4 in 1000 ml of the H_2SO_4 solution = 930 g

Mass of 1000 ml H_2SO_4 solution $= 1000 \times 1.84 = 1840$ g

Mass of water in 1000 ml of solution = 1840 – 930 g = 910 g = 0.910 kg

Moles of $H_2SO_4 = \frac{\text{Wt. of } H_2SO_4}{\text{Mol. wt. of } H_2SO_4} = \frac{930}{98}$

$\therefore$ Moles of H_2SO_4 in 1 kg of water

$= \frac{930}{98} \times \frac{1}{0.910} = 10.43$ mol

$\therefore$ Molality of solution = **10.43m**

42. (6.5)

No. of equivalents of $KMnO_4$

= No. of equivatents of hydrazine sulphate.

$$\overset{(-2)}{N_2H_4} \longrightarrow \overset{(0)}{N_2}$$

Change in oxidation state for each $N_2H_4 = 2 \times 2 = 4$

Equivalent weight of $N_2H_6SO_4 = \frac{130}{4} = 32.5$

Normality of $KMnO_4 = 5 \times 450$ ($\because$ valence factor = 5)

Number of equivalents of $KMnO_4$

= *n*-factor × no. of moles $= 5 \times \frac{20}{50 \times 1000}$

$= \frac{1}{500}$ and if weight of hydrazin sulphate be *x* g then equivalents of hydrazine sulphate $= \frac{x}{32.5}$

$\therefore \frac{1}{500} = \frac{x}{32.5}$ or $x = \frac{32.5}{500} = 0.065$ g

Hence, wt. of $N_2H_6SO_4$ in 10 mL solution = 0.065 g

$\therefore$ Wt. of $N_2H_6SO_4$ in 1000 mL solution = **6.5 g**

43. (39.6)

Given $2NH_2OH + 4Fe^{3+}$ .....(i)

$\rightarrow N_2O + H_2O + 4Fe^{2+} + 4H^+$

and $MnO_4^- + 5Fe^{2+} + 8H^+$ .....(ii)

$\rightarrow Mn^{2+} + 5Fe^{3+} + 4H_2O$

$\therefore 10NH_2OH + 4MnO_4^- + 12H^+$

$\rightarrow 5N_2O + 21H_2O + 4Mn^{2+}$

[On multiplying (i) by 5 and (ii) by 4 and then adding the resulting equations]

Molecular weight of NH_2OH = 33

Thus 4000 mL of 1M MnO_4^- would react with NH_2OH = 330g

$\therefore$ 12 mL of 0.02 M $KMnO_4$ would react with

$NH_2OH = \frac{330 \times 12 \times 0.02}{400}$ g

$\therefore$ Amount of NH_2OH present in 1000 mL of diluted solution $= \frac{330 \times 12 \times 0.02 \times 1000}{4000 \times 50}$ g

Since, 10 mL of sample of hydroxylamine is diluted to one litre

$\therefore$ Amount of hydroxyl amine in one litre of original solution

$= \frac{330 \times 0.02 \times 12 \times 1000}{4000 \times 50} \times \frac{1000}{10}$ g = **39.6 g**

44. (0.58) The complete oxidation under acidic conditions can be represented as follows:

$$5H_2O_2 + 2MnO_4^- + 6H^+ \rightarrow 5O_2 + 2Mn^{2+} + 8H_2O$$

Since 34 g of H_2O_2 = 2000 mL of 1N . H_2O_2

$\left(\because \text{Eq. wt or } H_2O_2 = \frac{34}{2}\right)$

∴ 34 g of H_2O_2 = 2000 mL of 1N $KMnO_4$

$[\because N_1V_1 = N_2V_2]$

or $\frac{X}{100}$ g of $H_2O_2 = \frac{2000 \times X}{100 \times 34}$ mL of 1N $KMnO_4$

Therefore the unknown normality

$= \frac{2000 \times X}{34 \times 100 \times X} = \frac{10}{17}$ or **0.588 N**

45. (12.15)

(i) Find the volume of CO_2 at NTP
(ii) Find molecular wt. of metal carbonate
(iii) Find the wt. of metal
(iv) Calculate equivalent weight of metal

Given $P_1 = 700$ mm, $P_2 = 760$ mm, $V_1 = 1336$ mL, $V_2 = ?$

$T_1 = 300$ K, $T_2 = 273$ K

$\frac{P_1V_1}{T_1} = \frac{P_2V_2}{T_2}$, or $V_2 = \frac{P_1V_1T_2}{P_2T_1}$

$= \frac{700 \times 1336 \times 273}{760 \times 300}$

$= 1119.78$ mL $= 1.12$ L at NTP

∵ 1.12 L of CO_2 is given by carbonate = 4.215 g

Molecular weight of metal carbonate

$= \frac{4.215}{1.12} \times 22.4 = 84.3$

Metal carbonate is $MCO_3 = M + 12 + 48 = M + 60$

Atomic weight of M = 84.3 – 60 = 24.3

Eq. wt. of metal $= \frac{1}{2} \times$ M. wt. $= \frac{1}{2} \times 24.3 = 12.15$

46. (4.87)

Write the balance chemical equation and use mole concept for limiting reagent.

$AgNO_3 + NaCl \longrightarrow NaNO_3 + AgCl$
170 g   58.5 g   143.5 g

From the given data, we find $AgNO_3$ is limiting reagent as NaCl is in excess.

∵ 170.0 g of $AgNO_3$ precipitates AgCl = 143.5 g

∴ 5.77 g of $AgNO_3$ precipitates AgCl

$= \frac{143.5}{170.0} \times 5.77 =$ **4.87 g**

47. (59.33) $3MnO_2 \longrightarrow Mn_3O_4 + O_2$

3 (54.9 + 32)   (3 × 54.9 + 64)
= 260.7 g   = 228.7 g

Let the amount of pyrolusite ignited = 100.00 g

∴ Wt. of MnO_2 = 80 g (80% of 100 g = 80 g)

Wt. of SiO_2 and other inert substances = 15 g

Wt. of water = 100 – (80 + 15) = 5 g

According to equation,

260.7 g of MnO_2 gives = 228.7 g of Mn_3O_4

∴ 80 g of MnO_2 gives $= \frac{228.7}{260.7} \times 80 = 70.2$ g of Mn_3O_4

During ignition, H_2O present in pyrolusite is removed while silica and other inert substances remain as such.

∴ Total wt. of the residue = 70.2 + 15 = 85.2 g

Calculation of % of Mn in ignited Mn_3O_4

$3Mn = Mn_3O_4$

3 × 54.9 = 164.7 g   3 × 54.9 + 64 = 228.7g

Since, 228.7 g of Mn_3O_4 contains 164.7 g of Mn

70.2 g of Mn_3O_4 contains $= \frac{164.7}{228.7} \times 70.2$

– 50.55 g of Mn

Weight of residue = 85.2 g

Hence, percentage of Mn is the ignited sample

$= \frac{50.55}{85.2} \times 100 =$ **59.33%**

48. (b) Write reaction for titration between Na_2CO_3 and HCl.

Method :

$Na_2CO_3\,(aq) + HCl\,(aq)$
(yellow colour with HPh *i.e.* phenolphthalein)

$\longrightarrow NaHCO_3 + NaCl\,(aq)$
(no colour with HPh *i.e* phenolphthalein)

(Half neutralisation)

$Na_2CO_3(aq) + 2HCl(aq) \rightarrow 2NaCl + H_2O + CO_2$

(Complete neutralisation)

$NaHCO_3 + HCl$
(yellow colour with methyl orange)

$\longrightarrow NaCl + H_2O + CO_2$
(red colour with methyl orange)

From these reaction it is clear that

(i) 2 moles of HCl are required for complete neutralization of Na_2CO_3.

(ii) Titre value using phenolphthalein corresponds only to neutralisation of Na_2CO_3 to $NaHCO_3$, *i.e.* half of value required by Na_2CO_3 solution.

(iii) Titre value using methyl orange corresponds to complete neutralisation of Na_2CO_3

∴ Both S and E are correct but S is not correct explanation of E.

49. Molality $= \frac{\text{Moles of solute}}{\text{Mass of solvent in kg}}$

$= \frac{\text{wt. of solute in gram / M.wt. of solute}}{\text{Mass of solvent in kg}}$

Molality $= \frac{3/30}{250/1000} =$ **0.4m**

50. Write the balanced chemical reaction for change and apply mole concept.

The given reactions are

$MnO_2 \downarrow + Na_2C_2O_4 + 2H_2SO_4$

$\longrightarrow MnSO_4 + CO_2 + Na_2SO_4 + 2H_2O$

∴ Meq. of MnO_2 ≡ Meq of $Na_2C_2O_4$

$= 10 \times 0.2 \times 2 = 4$

∴ mM of MnO_2

$= \frac{4}{2} = 2$ $\left[\begin{array}{l} Mn^{4+} + 2e \rightarrow Mn^{2+} \\ \therefore \text{Valance factor of } MnO_2 = 2 \end{array}\right]$

Now $2KMnO_4 + 3MnSO_4 + 2H_2O$

$\longrightarrow 5MnO_2 \downarrow + K_2SO_4 + 2H_2O$

Since, eq. wt. of MnO_2 is derived from $KMnO_4$ and $MnSO_4$ both, thus it is better to proceed by mole concept

mM of $KMnO_4$ ≡ mM of MnO_2 × (2/5) = 4/5

Also $5H_2O_2 + 2KMnO_4 + 3H_2SO_4$

$\longrightarrow 2MnSO_4 + K_2SO_4 + 8H_2O + 5O_2$

∴ mM of H_2O_2 = mM of $KMnO_4 \times \frac{5}{2} = \frac{4}{5} \times \frac{5}{2} = 2$

$\therefore M_{H_2O_2} \times 20 = 2$ or $M_{H_2O_2} =$ **0.1**

$2KMnO_4 + 5H_2O_2 + 3H_2SO_4$

$\longrightarrow K_2SO_4 + 2MnSO_4 + 8H_2O + 5O_2$

$2KMnO_4 + 3MnSO_4 + 2H_2O$

$\longrightarrow 5MnO_2 + 2H_2SO_4 + K_2SO_4$

$MnO_2 + Na_2C_2O_4 + 2H_2SO_4$

$\longrightarrow MnSO_4 + 2CO_2 + Na_2SO_4 + 2H_2O$

51. Write the reactions taking place, balance them and equate moles of I_2 and $Na_2S_2O_3$.

$KIO_3 + 5KI \rightarrow 3K_2O + 3I_2$ *i.e.*, $2I^{5+} + 10e^- \rightarrow I_2^0$

$2I^- \rightarrow I_2^0 + 2e^-$

Now liberated I_2 reacts with $Na_2S_2O_3$

$I_2 + 2e^- \rightarrow 2I^-$

$2S_2O_3^{2-} \rightarrow S_4O_6^{2-} + 2e^-$

∴ millimole ratio of $I_2 : S_2O_3 = 1 : 2$

Thus, m mole of I_2 liberated

= m mole of $Na_2S_4O_6$ used $\times \frac{1}{2} = 45 \times M \times \frac{1}{2}$

[*M* is molarity of thiosulphate]

Also m mole of $KIO_3 = \frac{0.1}{214} \times 1000$

Now m mole ratio of $KIO_3 : I_2 = 1 : 3$

Thus, $\frac{(0.1/214) \times 1000}{(45M/2)} = \frac{1}{3}$

$\therefore M = \frac{0.1 \times 1000 \times 3 \times 2}{214 \times 45} =$ **0.062**

Alternate Solution:

Let the original sample contains *x* millimol of Fe_3O_4 and *y* millimol of Fe_2O_3

$Fe_3O_4 + I^- \rightarrow 3Fe^{2+} + I_2$ (*n* factor = 2)

$Fe_3O_3 + I^- \rightarrow 2fe^{2+} + I_2$ (*n* = 2)

⇒ M_{eq} of I_2 formed = Me_q ($Fe_3O_4 + Fe_2O_3$)

⇒ M_{eq} of hypo required

⇒ $2 \times 2y = 11 \times 0.5 \times 5 = 27.5$ ... (i)

Now, total millimol of Fe^{2+} formed = $3x + 2y$

In the reaction,

$Fe^{2+} + MnO_4^- + H^+ \rightarrow Fe^{3+} + Mn^{2+}$.(*n* – factor of $Fe^{2+} = 1$)

M_{eq} of $MnO_4^- = M_{eq}$ of Fe^{2+}

$= 3x \times 2y = 12.8 \times 0.25 \times 5 \times 2 = 32$ ... (ii)

Solving equation (i) & (ii),

we get *x* = 4.5 and *y* = 9.25

Mass of $Fe_3O_4 = \frac{4.5}{1000} \times 232 = 1.044\%$

% of Fe_3O_4 $\frac{1.044}{3} \times 100 = 34.80\%$

Mass of $Fe_2O_3 = \frac{9.25}{1000} \times 160 = 1.48\%$

% mass of $Fe_2O_3 = \frac{1.48}{3} \times 100 = 49.33\%$

52. Find the milliequivalents and equate them as per data given in question.

For $Fe_3O_4 \rightarrow 3FeO$

$2e + Fe_3^{(8/3)+} \rightarrow 3Fe^{2+}$

Thus, valence factor for Fe_3O_4 is 2 and for FeO is 2/3.

For, $Fe_2O_3 \rightarrow 2FeO$; $2e + Fe_2^{3+} \rightarrow 2Fe^{2+}$ ...(1)

Thus, valence factor for Fe_2O_3 is 2 and for FeO is 1.

Let Meq.of Fe_3O_4 and Fe_2O_3 be *a* and *b* respectively.

∴ Meq. of Fe_3O_4 + Meq. Fe_2O_3 = Meq. of I_2 liberated = Meq. of hypo used

$$a + b = \frac{11 \times 0.5 \times 100}{20} = 27.5$$

Now, the Fe^{2+} ions are again oxidised to Fe^{3+} by $KMnO_4$.

Note that in the change

$Fe^{2+} \rightarrow Fe^{3+} + e^-$; valence factor of Fe^{2+} is 1.

Thus, Meq. of Fe^{2+} (from Fe_3O_4) + Meq. of Fe^{2+} (from Fe_2O_3) = Meq. of $KMnO_4$ used .... (2)

If valence factor for Fe^{2+} is 2/3 from Eq. (1), then Meq. of Fe^{2+} (from Fe_3O_4) = *a*

If valence factor for Fe^{2+} is 1

then Meq. of Fe^{2+} (from Fe_3O_4) = 3*a*/2 ...(3)

Similarly, from Eq. (2), Meq. of Fe^{2+} from (Fe_2O_3) = *b*.

∴ $3a/2 + b = 0.25 \times 5 \times 12.8 \times 100/50 = 32$

or $3a + 2b = 64$ ....(4)

From Eqs. (3) and (4)

Meq. of $Fe_3O_4 = a = 9$ & Meq. of $Fe_2O_3 = b = 18.5$

$$\therefore W_{Fe_3O_4} = \frac{9 \times 232}{2 \times 1000} = 1.044\,g$$

$$\text{and } W_{Fe_2O_3} = \frac{18.5 \times 160}{2 \times 1000} = 1.48\,g$$

$$\therefore \% \text{ of } Fe_3O_4 = \frac{1.044 \times 100}{3} = \mathbf{34.8}$$

$$\text{and } \% \text{ of } Fe_2O_3 = \frac{1.48 \times 100}{3} = \mathbf{49.33}$$

53. The formula of Glauber's salt is $Na_2SO_4.10H_2O$

Molecular mass of $Na_2SO_4.10H_2O$

$= [2 \times 23 + 32.1 + 4 \times 16] + 10(1.01 \times 2 + 16)$

$= 322.3\ g\ mol^{-1}$

Weight of the Glauber's salt taken = 80.575 g

Out of 80.575 g of salt, weight of anhydrous

$$Na_2SO_4 = \frac{142.1}{322.3} \times 80.575 = 35.525\ g$$

Number of moles of Na_2SO_4 per dm^3 of the

$$\text{solution} = \frac{35.525}{142.1} = 0.25$$

Molarity of the solution = **0.25 M**

Density of solution = 1077.2 kgm^{-3}

$$= \frac{1077.2 \times 10^3}{10^6}\ gm\ cm^{-3} = 1.0772\ g\ cm^{-3}$$

Total weight of sol = $V \times d = 1\ dm^3 \times d$

$= 1000\ cm^3 \times 1.0772\ gcm^{-3} = 1077.2\ g$

Weight of water = 1077.2 – 35.525 = 1041.67 g

$$\text{Molality of sol.} = \frac{0.25}{1041.67\ g} \times 1000\ g = 0.2399 =$$

0.24 m

$$\text{Number of moles of water in the solution} = \frac{1041.67}{18}$$

= 57.87

Mole fraction of Na_2SO_4

$$= \frac{\text{No. of moles of } Na_2SO_4}{\text{Total number of moles}} = \frac{0.25}{0.25 + 57.87}$$

$= 0.0043 = \mathbf{4.3 \times 10^{-3}}$

54. Calculation of number of moles in 45 mL. of 0.025 M $Pb(NO_3)_2$

$$\text{Moles of } Pb(NO_3)_2 = 0.25 \times \frac{45}{1000} = 0.01125$$

∴ Initial moles of Pb^{2+} = 0.01125

Moles of NO_3^- = 0.01125 × 2 = 0.02250

[1 mole $Pb(NO_3)_2$ ≡ 2 moles of NO_3]

Calculation of number of moles in 25 mL. of 0.1 M chromic sulphate

Moles of chromic sulphate $Cr_2(SO_4)_3$

$$= 0.1 \times \frac{25}{1000} = 0.0025 \text{ moles}$$

Moles of SO_4^{2-} = 0.0025 × 3 = 0.0075 [1 Mole of chromic sulphate ≡ 3 moles of SO_4^{2-}]

Moles of $PbSO_4$ formed = **0.0075** [SO_4^{2-} is totally consumed]

Moles of Pb^{2+} left = 0.01125 – 0.0075 = 0.00375

Moles of NO_3^- left = 0.02250 [NO_3^- remain unreacted]

Moles of chromium ions = 0.0025 × 2 = 0.005

Total volume of the solution = 45 + 25 = 70 mL.

∴ Molar concentration of the species left

$$(i)\quad Pb^{2+} = \frac{0.00375}{70} \times 1000 = \mathbf{0.05357\ M}$$

$$(ii)\quad NO_3^- = \frac{0.0225}{70} \times 1000 = \mathbf{0.3214\ M}$$

$$(iii)\quad Cr^{3+} = \frac{0.005}{70} \times 1000 = \mathbf{0.0714\ M}$$

55. 1.5 g of sample require = 150 mL. of $\frac{M}{10}$ HCl

$$\therefore 2\text{ g of sample require} = \frac{150 \times 2}{1.5}\text{ mL. of } \frac{M}{10}\text{ HCl}$$

= 200 mL. of $\frac{M}{10}$ HCl

On heating, the sample, only $NaHCO_3$ undergoes decomposition as given below.

$2NaHCO_3 \rightarrow Na_2CO_3 + H_2O + CO_2\uparrow$

2 moles 1 mole 1 mole

2 equ.

Neutralisation of the sample with HCl takes place as given below.

$NaHCO_3 + HCl \rightarrow NaCl + H_2O + CO_2$

1 eq. 1 eq.

$Na_2CO_3 + 2HCl \rightarrow 2NaCl + H_2O + CO_2$

1 mole 2 mole

2 eq. 2 eq.

Hence, 2 g sample ≡ 200 ml. of M/10 HCl

= 200 ml. of N/10 HCl = 20 meq = 0.020 eq

Number of moles of CO_2 formed, *i.e.*

$$n = \frac{PV}{RT} = \frac{750}{760} \times \frac{123.9}{1000} \times \frac{1}{0.082 \times 298} = 0.005$$

Moles of $NaHCO_3$ in the sample (2 g) = 2× 0.005 = 0.01

Equivalent of $NaHCO_3$ = 0.01

Wt. of $NaHCO_3$ = 0.01 × 84 = 0.84 g

$$\% \text{ of } NaHCO_3 = \frac{0.84 \times 100}{2} = \mathbf{42\%}$$

Equivalent of Na_2CO_3 = 0.02 – 0.01 = 0.01

Wt. of Na_2CO_3 = 0.01 × 53 = 0.53 g

$$\therefore \% \text{ of } Na_2CO_3 = \frac{0.53 \times 100}{2} = \mathbf{26.5\%}$$

∴ % of Na_2SO_4 in the mixture = 100 – (42 + 26.5) = **31.5%**

56. In the given problem, a solution containing Cu^{2+} and $C_2O_4^{2-}$ is titrated first with $KMnO_4$ and then with $Na_2S_2O_3$ in presence of KI. In titration with $KMnO_4$, it is the $C_2O_4^{2-}$ ions that react with the MnO_4^- ions. The concerned balanced equation may be written as given below.

$$2MnO_4^- + 5C_2O_4^{2-} + 16H^+ \rightarrow 2Mn^{2+} + 10CO_2 + 8H_2O$$

Thus, according to the above reaction

2 mmol MnO_4^- ≡ 5 mmol $C_2O_4^{2-}$

However,

No. of mmol of MnO_4^- used in titration = Vol. in ml × M = 22.6 × 0.02 = 0.452 mmol MnO_4^-

Since 2 mmol MnO_4^- ≡ 5 mmol $C_2O_4^{2-}$

$$0.452\text{ mmol } MnO_4^- \equiv \frac{5}{2} \times 0.452 = 1.130\text{ mmol } C_2O_4^{2-}$$

Titration with $Na_2S_2O_3$ in the presence of KI.

Here Cu^{2+} react and the reactions involved during titration are :

$2Cu^{2+} + 2I^- \rightarrow 2Cu^+ + I_2$, $2S_2O_3^{2-} + I_2 \rightarrow 2I^- + S_4O_6^{2-}$

Thus $2Cu^{2+} \equiv I_2 \equiv 2S_2O_3^{2-}$

No. of m mol of $S_2O_3^{2-}$ used in titration

= 0.05 × 11.3 = 0.565 mmol $S_2O_3^{2-}$

Now since 2 mmol $S_2O_3^{2-}$ ≡ 2 mmol Cu^{2+}

0.565 mmol $S_2O_3^{2-}$ ≡ 0.565 mmol Cu^{2+}

∴ Molar ratio of Cu^{2+} to $C_2O_4^{2-}$

$$= \frac{0.565\text{ mmol}}{1.130\text{ mmol}} = \mathbf{1:2}$$

Balanced equations in two cases

Case I. $Mn^{+7} + 5e^- \rightarrow Mn^{+2}$

$C_2^{+3} \rightarrow 2C^{+4} + 2e^-$

Case II. $2Cu^{+2} + 2e^- \rightarrow Cu_2^+$

$2I^- \rightarrow I_2 + 2e^-$

and $I_2 + 2e^- \rightarrow 2I^-$

$2S_2^{+2} \rightarrow S_4^{+3/2} + 2e^-$

Alternate Solution

With $KMnO_4$, oxalate ion is oxidised only as'

$5C_2O_4^{2-} + 2MnO_4^- + 16H^+ \rightarrow 2Mn^{2+} + 10\,CO_2 + 8\,H_2O$

Let, x Millimol of $C_2O_4^{2-}$ ion is present, then

M_{eq} of C_2O_4 = M_{eq} of MnO_4^-

⇒ $2x = 0.02 \times 5 \times 22.6 \Rightarrow x = 1.13$

At the later stage, with I, Cu^{2+} is reduced as;

$2\,Cu^{2+}\ 4I^- \rightarrow 2\,CuI + I_2$

and $I_2 + 2S_2O_3^{2-} \rightarrow 2I^- + S_4O_6^{2-}$

Let, there be *x* millimol of Cu^{2+}

⇒ M_{eq} of Cu^{2+} = M_{eq} of I_2 = M_{eq} of hypo

⇒ $x = 11.3 \times 0.05 = 0.565$

$\Rightarrow$ Moles of Cu^{2+}: moles of $C_2O_4^{2-}$
$\Rightarrow 0.565 : 1.13 \Rightarrow 1:2$

57. No. of equivalents of $KMnO_4$
= No. of equivatents of reducing agents.

Case I. Reaction of NaOH with $H_2C_2O_4$ and $NaHC_2O_4$.

(*i*) $H_2C_2O_4 + 2NaOH \rightarrow Na_2C_2O_4 + 2H_2O$
(*ii*) $NaHC_2O_4 + NaOH \rightarrow Na_2C_2O_4 + H_2O$

Number of milliequivalents of NaOH $= N \times V = 3.0 \times 0.1 = 0.3$
$\therefore$ Combined normality of the mixture titrated with

$$NaOH = \frac{0.3}{10} = 0.03$$

Case II. Reaction of $C_2O_4^-$ ion and $KMnO_4$

(*iii*) $5C_2O_4^- + MnO_4^- + 16H^+ \rightarrow 2Mn^{2+} + 10CO_2 + 8H_2O$

$KMnO_4$ will react in same manner with both $NaHC_2O_4$ and $H_2C_2O_4$ as it can be seen from the above reaction.

Number of milliequivalents of $KMnO_4 = 4.0 \times 0.1 = 0.4$
$\therefore$ Combined normality of the mixture titrated with

$$KMnO_4 = \frac{0.4}{10} = 0.04$$

The difference (0.04 N – 0.03 N = 0.01 N) is due to $NaHC_2O_4$

The total normality of $NaHC_2O_4$ will be = 0.01 + 0.01 = 0.02 N

From equation (*ii*) in case I.
Eq. wt. of $NaHC_2O_4 = 112$
Amount of $NaHC_2O_4$ in one litre of solution formed
$= 0.01 \times 112 =$ **1.12 g** and amount of $H_2C_2O_4$
$= 2.02 -$ Wt. of $NaHC_2O_4 = 2.02 - 1.12 =$ **0.90 g**

58. Let the amount of $NaNO_3$ in the mixture = *x* g
$\therefore$ The amount of $Pb(NO_3)_2$ in the mixture = $(5 - x)$ g
Heating effect of sodium nitrate and lead nitrate

$$\underset{2(23+14+48)=170\text{ g}}{2NaNO_3} \xrightarrow{\Delta} 2NaNO_2 + \underset{2\times16=32\text{ g}}{O_2}$$

$$\underset{2(207+28+96)=662\text{ g}}{2Pb(NO_3)_2} \xrightarrow{\Delta} 2PbO_2 + \underbrace{\underset{4(14+32)=184\text{ g}}{4NO_2} + \underset{2\times16=32\text{ g}}{O_2}}_{216\text{ g}}$$

Now since, 170 g of $NaNO_3$ gives = 32 g of O_2
$\therefore$ *x* g of $NaNO_3$ gives

$$= \frac{32}{170} \times x \text{ g of } O_2$$

Similarly, 662 g of $Pb(NO_3)_2$ gives
= 216 g of gases ($NO_2 + O_2$)
$(5 - x)$ g of $Pb(NO_3)_2$ gives

$$= \frac{216}{662} \times (5 - x) \text{ g of gases}$$

Actual loss, on heating, is 28% of

$$5 \text{ g of mixture} = \frac{5 \times 28}{100} = 1.4 \text{ g}$$

$$\therefore \frac{32x}{170} + \frac{216}{662} \times (5 - x) = 1.4$$

$\Rightarrow x = 1.676$ g
Wt. of $NaNO_3$ = **1.676 g** and
Wt. of $Pb(NO_3)_2 = 5 - 1.676$ g = **3.324 g**

59. No. of equivalents of $KMnO_4$ in neutral medium = No. of equivalents of reducing agent. Assuming that $KMnO_4$ shows the following changes during its oxidising nature.

Acidic medium $Mn^{7+} + n_1e^- \rightarrow Mn^{a+}$ $\therefore n_1 = 7 - a$
Neutral medium $Mn^{7+} + n_2e^- \rightarrow Mn^{b+}$ $\therefore n_2 = 7 - b$
Alkaline medium $Mn^{7+} + n_3e^- \rightarrow Mn^{c+}$ $\therefore n_3 = 7 - c$

Let *V* ml. of reducing agent be used for $KMnO_4$ in different medium.
$\therefore$ Meq. of reducing agent
= Meq. of $KMnO_4$ in acid medium
Meq. of $KMnO_4$ in neutral medium
= Meq. of $KMnO_4$ in alkaline medium
$\Rightarrow 1 \times n_1 \times 20 = 1 \times n_2 \times 33.4 = 1 \times n_3 \times 100$
$\Rightarrow n_1 = 1.667 n_2 = 5 n_3$
Since n_1, n_2 and n_3 are integers and n_1 is not greater than 7
$\therefore \quad n_3 = 1$
Hence, $n_1 = 5$ and $n_2 = 3$
$\therefore$ Different oxidation states of Mn in
Acidic medium $Mn^{7+} + 5e^- \rightarrow Mn^{a+}$ or $a = +2$
Neutral medium $Mn^{7+} + 3e^- \rightarrow Mn^{b+}$ or $b = +4$
Alkaline medium $Mn^{7+} + 1e^- \rightarrow Mn^{c+}$ or $c = +6$

Further, same volume of reducing agent is treated with $K_2Cr_2O_7$, and therefore
Meq. of reducing agent = Meq. of $K_2Cr_2O_7$
$1 \times 5 \times 20 = 1 \times 6 \times V$ [$\because Cr^{+6} + 6e^- \rightarrow 2Cr^{+3}$]
$V =$ **16.66 mL** $\quad \therefore$ 1M = 6 × 1N

60. (i) Weight of sugar syrup = 214.2 g
Weight of sugar in the syrup = 34.2 g
$\therefore$ Weight of water in the syrup = 214.2 – 34.2 = 180.0 g
Mol. wt. of sugar, $C_{12}H_{22}O_{11} = 342$

$$\therefore \text{Molal concentration} = \frac{34.2 \times 1000}{342 \times 180} = \mathbf{0.56}$$

(ii) Mol. wt. of water, $H_2O = 18$

$$\therefore \text{Mole fraction of sugar} = \frac{34.2/342}{180/18 + 34.2/342}$$

$$= \frac{0.1}{10 + 0.1} = \frac{0.1}{10.1} = \mathbf{0.0099}$$

61. (i) From the given half-cell reaction,

$$\text{Here, Eq. wt. of } NaBrO_3 = \frac{\text{Mol. wt.}}{6} = \frac{151}{6} = 25.17$$

[$\because$ number of electron involved = 6]

Now we know that
Meq. = Normality × Vol. in mL = 85.5 × 0.672 = 57.456

$$\text{Also, Meq.} = \frac{W_{NaBrO_3}}{\text{Eq. wt.}_{NaBrO_3}} \times 1000$$

$$\Rightarrow \frac{W_{NaBrO_3}}{25.17} \times 1000 = 57.456 \text{ g}$$

$$\therefore \quad W_{NaBrO_3} = \mathbf{1.446\ g}$$

$$\text{Molarity of } NaBrO_3 = \frac{\text{Normality}}{\text{Valence factor}}$$

$$= \frac{0.672}{6} = \mathbf{0.112\ M}$$

(ii) From the given half-cell reaction,

$$\text{Eq. wt. of } NaBrO_3 = \frac{\text{Mol. wt.}}{5} = \frac{151}{5} = 30.2$$

[Number of electron involved per $BrO_3^- = \frac{10}{2} = 5$]

Thus, the amount of $NaBrO_3$ required for preparing 1000 mL of 1 N $NaBrO_3$ = 30.2 g
$\therefore$ The amount of $NaBrO_3$ required for preparing 85.5 mL of 0.672 N $NaBrO_3$

$$= \frac{30.2 \times 0.672 \times 85.5}{1000} = \mathbf{1.7532\ g}$$

$$\text{Also, Molarity} = \frac{0.672}{5} = \mathbf{0.1344\ M}$$

62. (i) Find normality of acid mixture and $Na_2CO_3 \cdot 10H_2O$. Equate them to find volume of H_2SO_4.

(ii) Meq. of $H_2SO_4 = V \times N = \frac{V \times N}{1000}$ eq.

(iii) Equivalent of SO_4^{2-} = equivalents of H_2SO_4 × Eq. wt. of SO_4^{--}
$N \times V$(mL.) = meq.

Acid mixture contains 5 mL of 8N, HNO_3, 4.8 mL of 5N, HCl and say, '*V*' mL of 17 M ≡ 34 N, H_2SO_4.
[1M H_2SO_4 = 2N H_2SO_4]

$$N \text{ of the acid mixture} = \frac{\text{meq. (total) of acid}}{\text{mL. of solution}}$$

$$= \frac{5 \times 8 + 4.8 \times 5 + V \times 34}{2000}$$

[Total volume = 2 L = 2000 mL]

$$\text{or, } N_{mixture} = \frac{64 + 34V}{2000}$$

$$\because \text{Eq. of wt. of } Na_2CO_3.10H_2O = \frac{\text{Mol. wt.}}{2}$$

$$= \frac{106 + 180}{2} = 143$$

$$N \text{ of } Na_2CO_3 = \frac{\text{Meq. of } Na_2CO_3}{\text{Volume of solution (mL)}}$$

$$= \frac{\frac{1}{143}}{\frac{100}{1000}} = \frac{1}{143} \times \frac{1000}{100} = 0.069\text{N}$$

$N_1V_1 = N_2V_2$ or $\underset{\text{(acid)}}{30 \times N_{mixture}} = \underset{\text{(sod. carbonate)}}{42.9 \times 0.069}$

$$\therefore N_{mixture} = \frac{42.9 \times 0.069}{30} = 0.0986\,N$$

$$\text{Hence } \frac{64 + 34V}{2000} = 0.0986$$

$64 + 34V = 0.0986 \times 2000$, $64 + 34V = 197.2$
$34V = 197.2 - 64.0 = 133.2$

$$\therefore \text{or } V = \frac{133.2}{34} = 3.9 \text{ mL.}$$

Hence, meq. of $H_2SO_4 = V \times N$ of H_2SO_4
$= 3.9 \times 34 = 132.6$ meq.

$= 0.1326$ eq. of $H_2SO_4 = 0.1326$ eq. of SO_4^{2-}

$= 0.1326 \times 48$ g of SO_4^{2-}

$\left(\because \text{Eq. wt. of } SO_4^{2-} = \frac{32 + 64}{2} = 48\right)$

= **6.3648 g** of SO_4^{2-} are in 3.9 mL of 17M H_2SO_4

63. Equivalents of *A* oxidised = Equivalents of *A* reduced.

Since in acidic medium, A^{n+} is oxidised to AO_3^-, the change in oxidation state from

(+5) to (+*n*) = 5 – *n* [$\because$ O.S. of A in $AO_3^- = +5$]

$\therefore$ Total number of electrons that have been given

out during oxidation of 2.68×10^{-3} moles of $A^{n+} = 2.68 \times 10^{-3} \times (5-n)$

Thus the number of electrons added to reduce 1.61×10^{-3} moles of MnO_4^- to Mn^{2+}, *i.e.* (+7) to (+2) $= 1.61 \times 10^{-3} \times 5$

[Number of electrons involved $= +7-(+2) = 5$]

$\therefore \quad 1.61 \times 10^{-3} \times 5 = 2.68 \times 10^{-3} \times (5-n)$

$5-n = \frac{1.61 \times 5}{2.68}$ or $n = 5 - \frac{8.05}{2.68} \approx \mathbf{2}$

64. (i) Mole fraction $= \frac{\text{Moles of substance}}{\text{Total moles}}$

(ii) 1 mole of $Na_2S_2O_3$ gives 2 moles of Na^+ and 1 mole of $S_2O_3^{2-}$

Molecular wt. of sodium thiosulphate solution $(Na_2S_2O_3) = 23 \times 2 + 32 \times 2 + 16 \times 3 = 158$

(i) The percentage by weight of $Na_2S_2O_3$

$= \frac{\text{wt of } Na_2S_2O_3}{\text{wt of solution}} \times 100 = \frac{3 \times 158 \times 100}{1000 \times 1.25}$

$= \mathbf{37.92}$ [Wt. of $Na_2S_2O_3$ = Molarity × Mol wt]

(ii) Mass of 1 litre solution $= 1.25 \times 1000$ g $= 1250$ g [$\because$ density = 1.25g/l]

Mole fraction of $Na_2S_2O_3$

$= \frac{\text{Number of moles of } Na_2S_2O_3}{\text{Total number of moles}}$

Moles of water $= \frac{1250 - 158 \times 3}{18} = 43.1$

Mole fraction of $Na_2S_2O_3 = \frac{3}{3+43.1} = 0.065$

(iii) 1 mole of sodium thiosulphate $(Na_2S_2O_3)$ yields 2 moles of Na^+ and 1 mole of $S_2O_3^{2-}$.

Molality of $Na_2S_2O_3 = \frac{3 \times 1000}{776} = 3.87$

Molality of $Na^+ = 3.87 \times 2 = \mathbf{7.74m}$

65. Weight of MCO_3 and BaO = 4.08 g (given)

Weight of residue = 3.64 g (given)

$\therefore$ Weight of CO_2 evolved on heating

$= (4.08 - 3.64)\text{ g} = 0.44\text{ g} = \frac{0.44}{44} = 0.01$ mole

Number of moles of $MCO_3 \equiv 0.01$ mole

$$\left[\because MCO_3 \xrightarrow{\text{heat}} MO + CO_2\right]$$

Volume of 1N HCl in which residue is dissolved = 100 mL

Volume of 1N HCl used for dissolution $= (100 - 2.5 \times 16)$ mL = 60 ml

$= \frac{60}{1000} = 0.06$ equivalents

The chemical equation for dissolution can be written as

$$\underbrace{BaO + MO}_{\text{Residue}} + 4HCl \longrightarrow BaCl_2 + MCl_2 + 2H_2O$$

[Number of moles of BaO and MO = 1 + 1 = 2]

Number of moles of BaO + Number of moles of MO $= \frac{0.06}{2} = 0.03$

Number of moles of BaO $= (0.03 - 0.01)$ $= 0.02$ moles

Molecular weight of BaO = 138 + 16 = 154

$\therefore$ Weight of BaO $= (0.02 \times 154)$ g = 3.08 g

Weight of $MCO_3 = (4.08 - 3.08) = 1.0$ g

Since weight of 0.01 mole of $MCO_3 = 1.0$ g

$\therefore$ Mol. wt. of $MCO_3 = \frac{1}{0.01} = 100$

Hence, atomic weight of unknown M = (100 – 60) = **40**

The atomic weight of metal is **40** so the metal M is **Ca**.

66. Following reactions take place-

$$3MnSO_4 \cdot 4H_2O \xrightarrow{\text{heat}}$$

$$\underset{\text{(residue)}}{Mn_3O_4} + 4H_2O\uparrow + 3SO_2\uparrow + O_2\uparrow$$

$$Mn_3O_4 + 2FeSO_4 + 4H_2SO_4 \longrightarrow Fe_2(SO_4)_3 + 3MnSO_4 + 4H_2O$$

Milliequivalents of $FeSO_4$ in 30 mL of 0.1N $FeSO_4$ $= 30 \times 0.1 = 3$ m. eq.

According to problem step (iv)

25 mL of $KMnO_4$ reacts with = 3 m eq of $FeSO_4$

Thus in step (iii) of the problem,

50 mL of $KMnO_4$ reacts with $= \frac{3}{25} \times 50$ m.eq. of $FeSO_4$ = 6 meq of $FeSO_4$

Milli eq. of 100 mL of 0.1N $FeSO_4 = 100 \times 0.1$ = 10 m eq.

$FeSO_4$ which reacted with $Mn_3O_4 = (10-6) = 4$ m eq.

Milli eq of $FeSO_4$ = Milli eq. of Mn_3O_4

($\because$ Milli eq of oxidising agent and reducing agent are equal)

$\because Mn_3O_4 \equiv 3MnSO_4.4H_2O$

$\therefore$ 1 Meq of Mn_3O_4 = 3 Meq of $MnSO_4 . 4H_2O$

$\therefore$ 4 Meq of Mn_3O_4 = 12 Meq of $MnSO_4 . 4H_2O$

Eq. wt of $MnSO_4.4H_2O = \frac{\text{Mol wt.}}{2} = \frac{223}{2} = 111.5$

Wt of $MnSO_4.4H_2O$ in sample $= 12 \times 111.5$ = 1338 mg = **1.338g.**

67. Weight of AgCl formed = 2.567 g

Amount of AgCl formed due to MCl = 1.341 g

($\because$ NaCl does not decompose on heating to 300 °C)

$\therefore$ Weight of AgCl formed due to NaCl = 2.567 – 1.341 = 1.226 g

$$\begin{Bmatrix} NaCl + AgNO_3 \rightarrow AgCl + NaNO_3 \\ MCl + AgNO_3 \rightarrow AgCl + MNO_3 \end{Bmatrix}$$

$\because$ 143.5 g of AgCl is obtained from NaCl = 58.5 g

$\therefore$ 1.226 g of AgCl is obtained from NaCl

$= \frac{58.5}{143.5} \times 1.226 = 0.4997$ g

$\therefore$ Wt of MCl in 1 g of mixture = 1.000 – 0.4997 = 0.5003 g

$\because$ 1.341 g of AgCl is obtained from MCl = 0.5003 g

$\therefore$ 143.5g of AgCl is obtained from MCl

$= \frac{0.5003}{1.341} \times 143.5 = 53.53$ g

$\therefore$ Molecular weight of MCl = **53.53**

68. Volume of oxygen taken = 30 mL,

Volume of unused oxygen = 15 mL

Volume of O_2 used = Volume of O_2 added – Volume of O_2 left = 30 – 15 = 15 mL

Volume of CO_2 produced

= Volume of gaseous mixture after explosion – Volume of unused oxygen

or Volume of CO_2 produced = 25 – 15 = 10 mL

Volume of hydrocarbon = 5 mL

General equation for combustion of a hydrocarbon is as follows -

$$\underset{\substack{\text{(Hydrocarbon)} \\ 5\text{ mL}}}{C_xH_y} + \underset{5\left(x+\frac{y}{4}\right)\text{mL}}{\left(x+\frac{y}{4}\right)O_2} \longrightarrow \underset{5x}{xCO_2} + \frac{y}{2}H_2O$$

$\therefore$ Volume of CO_2 produced = 5*x*, Since Volume of CO_2 = 10 mL

$\therefore$ $5x = 10 \Rightarrow x = 2$, Volume of O_2 used = 15 mL

$\therefore 5\left(x+\frac{y}{4}\right) = 15 \Rightarrow x + \frac{y}{4} = 3$

$\Rightarrow 2 + \frac{y}{4} = 3 \quad (\because x = 2) \Rightarrow 8 + y = 12 \therefore y = 4$

Hence, Molecular formula of hydrocarbon is $\mathbf{C_2H_4}$.

69.

Given conditions	N.T.P conditions
$P_1 = 0.92$ atm.	$P_2 = 1$ atm.
$V_1 = 1.20$ litres	$V_2 = ?$
$T_1 = 0 + 273 = 273$ K	$T_2 = 273$ K

Applying ideal gas equation, $\frac{P_1V_1}{T_1} = \frac{P_2V_2}{T_2}$

$\frac{0.92 \times 1.20}{273} = \frac{1 \times V_2}{273}$, $V_2 = \frac{0.92 \times 1.20 \times 273}{273 \times 1}$ litres = 1.104 litres = 1104 mL

The relevant chemical equations are

(i) $\underset{\substack{2 \times 27 \\ = 54\text{ g}}}{2Al} + 6HCl \longrightarrow 2AlCl_3 + \underset{\substack{3 \times 22400 \\ = 67200 \text{ mL at NTP}}}{3H_2}$

(ii) $\underset{24\text{ g}}{Mg} + 2HCl \longrightarrow MgCl_2 + \underset{22400\text{ mL at NTP}}{H_2}$

Wt. of alloy = 1 g

Let the wt. of aluminium in alloy = *x* g

$\therefore$ Wt. of magnesium in alloy = (1 – *x*) g

According to equation (i)

54 g of Al = 67200 mL of H_2 at N.T.P

$\therefore$ *x* g of Al $= \frac{67200}{54} \times x = 1244.4x$ mL of H_2 at N.T.P

Similarly, from equation (ii)

24 g of Mg = 22400 mL of H_2 at N.T.P

$(1-x)$ g of Mg $= \frac{22400}{24} \times (1-x) = 933.3(1-x)$ mL of H_2

Hence, total vol. of H_2 collected at N.T.P $= 1244.4x + 933.3(1-x)$ mL

But total vol. of H_2 as calculated above = 1104 mL

$\therefore \quad 1244.4x + 933.3(1-x) = 1104$ mL

$\Rightarrow \quad x = 0.5487$

Hence, 1 g of alloy contains Al = 0.5487 g

$\therefore$ Percentage of Al in alloy $= \frac{0.5487 \times 100}{1}$

= 54.87%

% of Mg in alloy = 100 – 54.87 = **45.13%**

Chapter-2

Structure of Atom

Part-A (JEE Main)

1. **(2)**

(i) Wavelength of A (λ) = 400 nm.

(ii) Wavelength of B (λ) = $\frac{c}{v} = \frac{3\times10^8}{10^{16}}$

$= 3 \times 10^{-8} = 30 \times 10^{-9} = 30$ nm.

(iii) Wavelength of C (λ) = $\frac{1}{\bar{v}} = \frac{1}{10^4} = 10^{-4}$ cm

$= 10^{-6}$ m = 1000 nm

Here $\lambda_c > \lambda_A > \lambda_B$

Energy (E) $\propto \frac{1}{\lambda}$

So $E_B > E_A > E_C$

2. **(2)** $KE_{max} = E - \phi$ where E = energy

ϕ = work function

$(KE_{max})_1 = 6 - \phi_1$ ...(1)

$(KE_{max})_2 = 6 - \phi_2$ ...(2)

divide eq. (1) by eq. (2)

$$\frac{(KE_{max})_1}{(KE_{max})_2} = \frac{2.642}{1} = \frac{6-\phi_1}{6-\phi_2}$$

$\phi_1 : \phi_2 = 1 : 2$ (given)

$$\therefore \frac{2.642}{1} = \frac{6-\phi_1}{6-2\phi_1} \Rightarrow \phi_2 = 2\phi_1$$

On solving the equation, we get

$\phi_1 = 2.3$ eV and $\phi_2 = 4.6$ eV.

3. **(2)**

(A) $_{12}Mg^{24}$ represents 12 protons and 12 neutrons.

So statement A is incorrect

(B) $\lambda = \frac{c}{v} = \frac{3\times10^8}{4.5\times10^{15}} = 6.67 \times 10^{-8}$ m.

So statement B is correct

(C) $\frac{E_1}{E_2} = \frac{hc/\lambda_1}{hc/\lambda_2} = \frac{\lambda_2}{\lambda_1}$

$\frac{E_1}{E_2} = \frac{300}{900} = \frac{1}{3}$

So statement C is incorrect

(D) No. of photons = $\frac{\text{Energy}}{hc/\lambda}$

$= \frac{1\times2000\times10^{-12}}{6.634\times10^{-34}\times3\times10^8} = 1.004 \times 10^{16}$.

So D is correct statement.

4. **(4)** $\lambda = \frac{c}{v} = \frac{3\times10^8}{5.16\times10^{14}} \Rightarrow \lambda = 581.39$ nm

A. λ_{Photon} is near & below yellow light therefore it can show photoelectric effect.

B. If intensity of light decreases photocurrent decreases.

C. Red light will not produce photoelectric effect.

D. $v_{Blue} > v_{yellow}$ so photoelectric current will be produced.

E. White light contain all frequencies so it will show photo electric current.

$\therefore$ Correct statements are A, B, D & E.

5. **(1)** There is a characteristic minimum frequency, or threshold frequency, for each metal below which the photoelectric effect is not seen. The ejected electrons leave with a specific amunt of kinetic energy at a frequency $v > v_0$ with an increase in light frequency of these electron kinetic energies also rise.

6. **(4)** Both assertion and reason are correct but reason is not the correct explanation of assertion.

7. **(3)** Energy of one mole of photons

$= \frac{hc}{\lambda} \times N_A$

$= \frac{6.63\times10^{-34}\times3\times10^8}{300\times10^{-9}} \times 6.02\times10^{23}$

$= 399.13 \times 10^3$ Joule/mole = 399 kJ/mole

8. **(3)** $W = hv = 6.6 \times 10^{-34} \times 1.3 \times 10^{15}$

$= 8.58 \times 10^{-19}$ J

9. **(4)** Remember the formula

$p^2 = 2\,m\,K.E.$ $\quad K.E. = \frac{hc}{\lambda}$

In photoelectric effect, $\frac{hc}{\lambda} = w + KE$

Given that KE of ejected photoelectron is very high in comparison to work function *w*.

$\frac{hc}{\lambda} = KE$

$\frac{hc}{\lambda} = \frac{1}{2}mV^2\left(\frac{m}{m}\right) \Rightarrow \frac{hc}{\lambda} = \frac{1}{2}\cdot\frac{m^2V^2}{m}$

$\frac{hc}{\lambda} = \frac{p^2}{2m}$

New wavelength $\frac{hc}{\lambda_1} = \frac{(1.5p)^2}{2m} \Rightarrow \lambda_1 = \frac{4}{9}\lambda$

10. **(3)** $E = hv = \frac{hc}{\lambda}$

$E = \frac{6.626\times10^{-34}\times3\times10^8}{4000\times10^{-10}} = 4.97\times10^{-19}$ J

$= \frac{4.97\times10^{-19}}{1.6\times10^{-19}} = 3.1$ eV

$KE = \frac{1}{2}mv^2 = \frac{1}{2}\times9\times10^{-31}\,kg\times(6\times10^5\,ms^{-1})^2$

$= 1.62 \times 10^{-19}$ J = 1 eV

According to photoelectric effect,

$K.E. = hv - hv_0$

$hv_0 = hv - K.E.$

Work function $(W_0) = E - K.E. = 3.1 - 1 = 2.1$ eV

11. **(2, 3)** After ejection of e^-, further increase in frequency also increases K.E. Thus, graph (2) is incorrect not all v release e^-, thus (3) is incorrect.

$K.E. = hv - hv_0$

where, v = Frequency of incident radiation

v_0 = Threshold frequency

KE is independent of intensity but it depends on frequency of light. Intensity is directly proportional to the no. of electrons emitted.

12. **(3)** $\lambda = 250$ nm

$E = \frac{hc}{\lambda}$

$= \frac{6.626\times10^{-34}\times3\times10^8}{250\times10^{-9}} / 1.60\times10^{-19} = 4.96$ eV

KE = stopping potential = 0.5 eV

$E = W_0 + K.E.; 4.96 = W_0 + 0.5$

$W_0 = 4.46 \approx 4.5$ eV

13. **(3)** $E = hv = 6.63 \times 10^{-34} \times 2.47 \times 10^{15}$

$= 1.640 \times 10^{-18}$ J

14. **(4)** Energy = $N_A hv$

$495.5 \times 10^3 = 6.023 \times 10^{23} \times 6.6 \times 10^{-34} \times v$

$v = \frac{495.5\times10^3}{6.023\times10^{23}\times6.6\times10^{-34}} = 12.4\times10^{14}\,s^{-1}$

$= 1.24 \times 10^{15}\ s^{-1}$.

15. **(3)** The kinetic energy of the ejected electron is given by the equation

$hv = hv_0 + \frac{1}{2}mv^2 \quad \because v = \frac{c}{\lambda}$

or $\frac{hc}{\lambda} = \frac{hc}{\lambda_0} + \frac{1}{2}mv^2$

$\frac{1}{2}mv^2 = \frac{hc}{\lambda} - \frac{hc}{\lambda_0} = hc\left(\frac{\lambda_0-\lambda}{\lambda\lambda_0}\right)$

$\therefore v^2 = \frac{2hc}{m}\left(\frac{\lambda_0-\lambda}{\lambda\lambda_0}\right)$ or $v = \sqrt{\frac{2hc}{m}\left(\frac{\lambda_0-\lambda}{\lambda\lambda_0}\right)}$

16. **(4)** Energy required to break single Cl – Cl bond

$= \frac{242\times10^3}{6.023\times10^{23}} = \frac{hc}{\lambda} = \frac{6.626\times10^{-34}\times3\times10^8}{\lambda}$

$\therefore \lambda = \frac{6.626\times10^{-34}\times3\times10^8\times6.023\times10^{23}}{242\times10^3}$

$= 0.4947 \times 10^{-6}$ m = 494.7 nm

17. **(3)** A – II, B – IV, C – III, D – I.

18. **(1)** In hydrogen spectrum, maximum lines of Balmer series lies in visible region.

19. **(1)** The lines falling in the visible region comprise Balmer series. Hence, the third line from red ($n_2 = 3$) would be $n_1 = 2, n_2 = 5$, i.e. $5 \rightarrow 2$.

20. **(3)** Energy of photon is given by:

$E_{photon} = \frac{hc}{\lambda}$

$\therefore \quad E \propto \frac{1}{\lambda}$ (hc = constant)

$\Rightarrow \frac{E_1}{E_2} = \frac{\lambda_2}{\lambda_1} \Rightarrow \frac{E_1}{E_2} = \frac{6000}{2000}$

$\Rightarrow \frac{E_1}{E_2} = 3$

21. (5) Photoelectric equation :

$E_{photon} = W + K.E.$

Energy of a photon : $E = \frac{hc}{\lambda}$

Convert work from eV to Joules

$W = 2.3 \text{ eV} \times (1.6 \times 10^{-19} \text{ J/eV})$

$K.E. = 2.8 \times 10^{-20} \text{ J} = 0.28 \times 10^{-19} \text{ J}$

Total energy of incident photon

$E = W + K.E. = 3.96 \times 10^{-19} \text{ J}$

$$\lambda = x = \frac{hc}{E} = \frac{(6.6\times10^{-34}\text{ Js})\times(3.0\times10^{8}\text{ m/s})}{3.96\times10^{-19}\text{ J}}$$

$x = 5 \times 10^{-7} = 500 \times 10^{-9} \text{ m} = 5 \times 10^{2} \text{ nm}.$

22. (494) $E = h\nu = \frac{hc}{\lambda}$ For 1 atom

For 1 mol

$$E = \frac{N_A hc}{\lambda}$$

$$= \frac{6.022\times10^{23}\times6.62\times10^{-34}\times3\times10^{8}}{242\times10^{-9}} = 494 \text{ kJ}$$

23. (3) $E = h\nu$

$$= \frac{6.62\times10^{-34}\times3\times10^{8}}{400\times10^{-9}} \times 6.24\times10^{18} \text{ eV}$$

$= 3.1$ eV.

$\Rightarrow$ Li, Na, K will show photoelectric effect under given condition.

24. (0) A blackbody can emit and absorb all the wavelengths

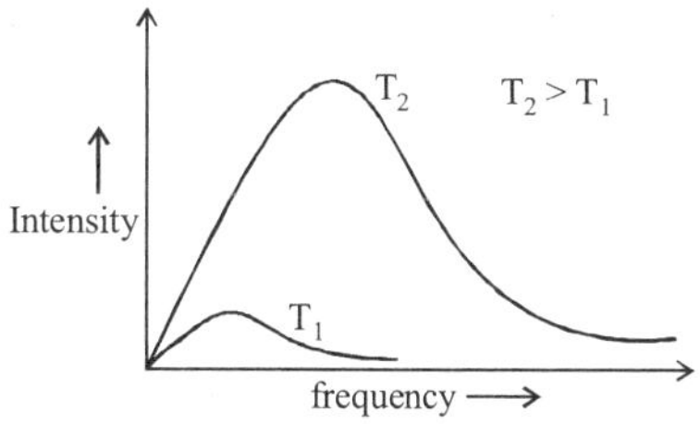

$\Rightarrow$ (A), (B), (C), (D) are correct.

25. (1) Statement (B) is incorrect as the emission spectra of atoms in the gas phase does not show a continuous spread of wavelength.

26. (798) For one photon $E = h\nu$

For one mole of photons,

$E = 6.023 \times 10^{23} \times 6.626 \times 10^{-34} \times 2 \times 10^{12}$

$= 798.16 \text{ J} \approx 798 \text{ J}$

27. (2) The spectrum of Black body radiation is explained using quantization of energy. With increase in temperature, peak of spectrum shifts to shorter wavelength or higher frequency. For above graph $\rightarrow T_1 > T_2 > T_3 > T_4$.

A black body does not have particles undergoing SHM.

$$\frac{T_1}{\nu_1} = \frac{T_2}{\nu_2} = \frac{T_3}{\nu_3} = \text{constant}$$

28. (2) $\lambda = \frac{h}{\sqrt{2m(K.E.)}}$; $K.E. = \frac{h^2}{2m\lambda^2}$

$$K.E. = \frac{h^2}{2m\lambda^2} = \frac{43.9\times10^{-68}}{2\times9.1\times10^{-31}\times10.89\times10^{-20}}$$

$K.E. = 2.215 \times 10^{-18}$ J

$E_{abs} = E_{req} + K.E.$

Now, $E_{req} = -13.6 \text{ eV} = -13.6 \times 1.602 \times 10^{-19}$ J

$$\frac{E_{abs}}{E_{req}} = 1 + \frac{K.E.}{E_{req}} = 1 + \frac{2.215\times10^{-18}}{13.6\times1.602\times10^{-19}} = 2.0166$$

29. (10) Maximum number of spectral lines

$$= \frac{\Delta n(\Delta n + 1)}{2} = \frac{4\times5}{2} = 10$$

30. (300) The minimum energy required to remove an electron from the metal surface is called work function.

$$E = h\nu = \frac{hc}{\lambda}$$

$$\phi = 6.63\times10^{-19}\text{ J} = \frac{hc}{\lambda} = \frac{6.63\times10^{-34}\times3\times10^{8}}{\lambda}$$

$\Rightarrow \lambda = 3 \times 10^{-7} \text{ m} = 300 \text{ nm}$

31. (2) Total energy per sec. = 50 J

$$\because E = nh\nu = \frac{nhc}{\lambda}$$

$$50 = \frac{n\times6.63\times10^{-34}\times3\times10^{8}}{795\times10^{-9}};\ n \simeq 2\times10^{20}$$

32. (50) Power of infrared range = 1 mW = 10^{-3}W or 10^{-3} Js^{-1}

$\therefore$ In 0.1 second the energy emitted = 10^{-4}J

$$\because E = nh\nu = nh\frac{c}{\lambda} \Rightarrow \text{then; } 10^{-4} = n\times\frac{hc}{\lambda}$$

$$\Rightarrow 10^{-4} = \frac{n\times6.63\times10^{-34}\times3\times10^{8}}{1000\times10^{-9}}$$

$$\Rightarrow n = 5.02\times10^{14} = 50.2\times10^{13}$$

33. (5) Einstein equation, $E = \phi + K.E._{max}$

$$h\nu = h\nu_0 + \frac{1}{2}mv^2 \Rightarrow \frac{hc}{\lambda} = h\nu_0 + \frac{1}{2}mv^2$$

$$\Rightarrow \frac{6.63\times10^{-34}\times3\times10^{8}}{500\times10^{-9}}$$

$$= 6.63\times10^{-34}\times4.3\times10^{14} + \frac{1}{2}mv^2$$

$$\Rightarrow \frac{6.63\times30\times10^{-20}}{5} = 6.63\times4.3\times10^{-20} + \frac{1}{2}mv^2$$

$$\Rightarrow 11.271\times10^{-20} = \frac{1}{2}\times9\times10^{-31}\times v^2$$

$v = 5 \times 10^{5}$ ms^{-1}.

34. (2) Total energy provided by source per second $= \frac{1000}{10} = 100$ J

Energy required to eject electron

$$= \frac{hc}{\lambda} = \frac{6.626\times10^{-34}}{400\times10^{-9}}\times3\times10^{8}\text{ J}$$

Number of electrons ejected

$$= \frac{100}{\frac{6.626\times10^{-34}\times3\times10^{8}}{400\times10^{-9}}}$$

$$= \frac{400\times10^{-7}\times10^{26}}{6.626\times3} = 2.01\times10^{20}$$

35. (181) $E = \frac{hc}{\lambda}$

$$= \frac{6.63\times10^{-34}\times3\times10^{8}}{663\times10^{-9}}\text{ J} = 3\times10^{-19}\text{ J/atom}$$

Ionisation energy per mol

$= 3 \times 10^{-19} \times 10^{-3} \times 6.02 \times 10^{23}$ kJ mol^{-1}

$= 180.6$ kJ $mol^{-1} \approx 181$ kJ mol^{-1}

36. (222)

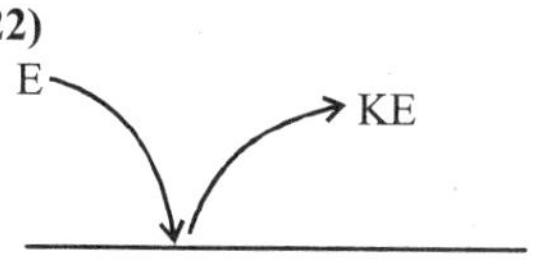

Metal (Work function = E_0)

$E = E_0 + (KE)_{max}$.

$$\frac{hc}{\lambda} = 4.41\times10^{-19} + KE$$

$$\frac{6.63\times10^{-34}\times3\times10^{8}}{300\times10^{-9}} = 4.41\times10^{-19} + KE$$

So, $(KE)_{max} = 6.63 \times 10^{-19} - 4.41 \times 10^{-19}$

$= 2.22 \times 10^{-19} \text{ J} = 222 \times 10^{-21}$ J

37. (4) $r_{species} = a_0 \frac{n^2}{Z}$

$$\Rightarrow 70.53 = 52.9\left(\frac{n^2}{Z}\right)$$

$$\Rightarrow \frac{n^2}{Z} = \frac{70.53}{52.9} = 1.33 = \frac{4}{3}$$

From above equation, n = 2 and $Z = 3$

Therefore : Li^{+2}, 2

38. (1) Using the Rydberg formula

$$\frac{1}{\lambda} = RZ^2\left(\frac{1}{n_1^2} - \frac{1}{n_2^2}\right)$$

For hydrogen; Z = 1

$$\frac{1}{x} = R(1)^2\left(\frac{1}{1^2} - \frac{1}{\infty^2}\right) = R \quad ...(1)$$

For Helium, Z = 2

$$\frac{1}{\lambda} = R(2)^2\left(\frac{1}{2^2} - \frac{1}{3^2}\right) = 4R\frac{5}{36} \quad ...(2)$$

Dividing (1) by (2)

$$\frac{\lambda}{x} = \frac{36}{20R}\frac{R}{1} = \frac{9}{5}$$

$$\therefore \lambda = \frac{9}{5}x$$

39. (2) $r_n = \frac{n^2}{z} \times a_0$

[a_0 = Bohr's radius = 0.529Å]

(A) H (1st orbit) $r_n = \frac{1\times1}{1}a_0 \Rightarrow r_n = a_0$

(B) He^{+} (1st orbit) $r_n = \frac{1\times1}{2}a_0 \Rightarrow r_n = 0.5a_0$

(C) He^{+} (2nd orbit) $r_n = \frac{2\times2}{2}a_0 \Rightarrow r_n = 2a_0$

(D) Li^{2+} (1st orbit)

$$r_n = \frac{1\times1}{3}a_0 \Rightarrow r_n = 0.33a_0$$

(E) Be^{3+} (2nd orbit) $r_n = \frac{2\times2}{4}a_0 \Rightarrow r_n = a_0$

H (1st orbit) and Be^{+3} (2nd orbit) have identical radii.

40. (2) Ist line of Balmer series ($n = 3$ to $n = 2$)

$$\bar{\nu}_1 = R_H(z)^2\left[\frac{1}{2^2} - \frac{1}{3^2}\right]$$

Ist line of Brackett series ($n = 5$ to $n = 4$)

$$\bar{v}_2 = R_H(z)^2\left[\frac{1}{4^2} - \frac{1}{5^2}\right]$$

$$\therefore \frac{\bar{v}_1}{\bar{v}_2} = \frac{500}{81} = 5:0.81.$$

41. (3) $Li^{2+}(g) \rightarrow Li^{3+}(g) + e^-$

Given IE of H = 2.18×10^{-18} J atom^{-1}

$E_{Li^{2+}} = E_H \times Z^2$

$= 2.18 \times 10^{-18} \times (3)^2 = 1.962 \times 10^{-17}$ atom^{-1}

42. (1) $\frac{1}{\lambda} = R_H Z^2\left(\frac{1}{Z^2} - \frac{1}{n^2}\right)$

For first line ($3 \rightarrow 2$)

Thus $\frac{1}{\lambda} = R \times 1\left[\frac{1}{4} - \frac{1}{9}\right] = \frac{5R}{36}$

For second line ($4 \rightarrow 2$)

$\frac{1}{\lambda} = R \times 1\left[\frac{1}{4} - \frac{1}{16}\right] = \frac{3R}{16}$

For third line ($5 \rightarrow 2$)

$\frac{1}{\lambda} = R \times 1\left[\frac{1}{4} - \frac{1}{25}\right] = \frac{21R}{100}$

43. (4) $n_1 \rightarrow$ lower energy level

$n_2 \rightarrow$ higher energy level

$n_1 + n_2 = 4$, $n_2 - n_1 = 2$

Hence $n_1 = 1$, $n_2 = 3$

Rydberg's formula :

$$\frac{1}{\lambda} = R_H Z^2\left[\frac{1}{n_1^2} - \frac{1}{n_2^2}\right]$$

$$\frac{1}{\lambda} = R_H(3)^2\left[\frac{1}{1^2} - \frac{1}{3^2}\right]$$

$$\frac{1}{\lambda} = 8R_H$$

$$\lambda = \frac{1}{8R_H}$$

$$\lambda = \frac{1}{8 \times 1.1 \times 10^5}$$

$$\lambda = \frac{1000}{8.8} \times 10^{-8} \text{ cm}$$

$\lambda = 113.63 \times 10^{-8}$ cm

$\lambda = 1.14 \times 10^{-6}$ cm

44. (4) According to Bohr's theory for hydrogen & hydrogen like species, the energy of stationary state is given by:-

$$E_a = -2.18 \times 10^{-18}\frac{Z^2}{n^2} \text{ J / atom.}$$

Where, Z = tomic number,

For 3rd orbit of Li^{+2} ion, n = orbit

$$= -2.18 \times 10^{-18} \times \frac{3^2}{3^2} = -2.18 \times 10^{-18} \text{ J}$$

45. (4) Transition of first Balmer line

$n_1 = 2$; $n_2 = 3$

$$\Delta E = x = 13.6(1)^2\left[\frac{1}{2^2} - \frac{1}{3^2}\right] \quad \text{...(i)}$$

Transition of second Balmer line

$n_1 = 2$; $n_2 = 4$

$$\Delta E = 13.6(1)^2\left[\frac{1}{2^2} - \frac{1}{4^2}\right] \quad \text{...(ii)}$$

Divide equation (ii) by equation (i)

$$\frac{\Delta E}{x} = \frac{\frac{1}{4} - \frac{1}{16}}{\frac{1}{4} - \frac{1}{9}}$$

$$\Rightarrow \frac{\Delta E}{x} = \frac{\frac{3}{16}}{\frac{5}{36}} \Rightarrow \frac{\Delta E}{x} = \frac{27}{20}$$

$\therefore \Delta E = 1.35x$.

46. (2) $E_n = -R_H\left(\frac{Z^2}{n^2}\right)$ J

$$\Delta E = 2.18 \times 10^{-18} \times 1^2\left[\frac{1}{1^2} - \frac{1}{2^2}\right] \text{ Joule}$$

$= 1.635 \times 10^{-18}$ Joule

$= 1.635 \times 10^{-18} \times 6.02 \times 10^{23}$ Joule/mole

$= 9.835 \times 10^5$ Joule/mole

47. (4) $\Delta E = 13.6Z^2\left(\frac{1}{n_1^2} - \frac{1}{n_2^2}\right)$

Series	n_1	n_2
(A) Paschen (1st line)	3	4
(B) Balmer (2nd line)	2	4
(C) Paschen (3rd line)	3	6
(D) Bracket (4th line)	4	8

So correct ascending order of energy of above lines is:

D < A < C < B

48. (1) $\frac{1}{\lambda} = RZ^2\left(\frac{1}{n_1^2} - \frac{1}{n_2^2}\right)$

For 1st line of lyman series in H-atom

$$\frac{1}{\lambda} = R(1)^2\left(\frac{1}{1^2} - \frac{1}{2^2}\right) \Rightarrow \frac{1}{\lambda} = \frac{3R}{4}$$

for 2nd line of Balmer series of He^+

$$\frac{1}{\lambda'} = R(2)^2\left(\frac{1}{2^2} - \frac{1}{4^2}\right) \Rightarrow \frac{1}{\lambda'} = \frac{3R}{4}$$

As the value of λ and λ' is equal so frequency of these lines will be also equal.

49. (4) The statement given in option (d) is not in agreement with quantum mechanical model of an atom.

50. (3) $r \propto n^2 \Rightarrow \frac{r'}{r} = \frac{n'^2}{n^2}$

Thus, $\frac{r_3}{r_1} = \frac{9}{1} \Rightarrow \frac{r_8}{r_4} = \frac{64}{16} = 4$

$$\frac{r_6}{r_4} = \left(\frac{6}{4}\right)^2 = \frac{9}{4} \Rightarrow \frac{r_4}{r_2} = \left(\frac{4}{2}\right)^2 = 4$$

51. (2) The value of energy becomes more and more negative with increase of Z. They do not follow a linear relationship. Hence (2) is the correct option.

52. (1) $\lambda = 900$ nm

$= 900 \times 10^{-7}$ cm $= 9 \times 10^{-5}$ cm

$R_H = 10^5$ cm^{-1}; For H-atom (Z = 1)

According to Rydberg eq.

$$\frac{1}{\lambda} = R_H Z^2 \times \left(\frac{1}{n_1^2} - \frac{1}{n_2^2}\right)$$

$$\Rightarrow \frac{1}{\lambda \times R_H} = \frac{1}{n_1^2} - \frac{1}{n_2^2}$$

$$\Rightarrow \frac{1}{9 \times 10^{-5} \text{ cm} \times 10^5 \text{ cm}^{-1}} = \left(\frac{1}{n_1^2} - \frac{1}{n_2^2}\right)$$

$$\Rightarrow \frac{1}{n_1^2} - \frac{1}{n_2^2} = \frac{1}{9}$$

It is possible when $n_1 = 3$, $n_2 = \infty$

Possible series : $\infty \rightarrow 3$

53. (4) $r = a_0\frac{n^2}{Z} = a_0 \cdot \frac{(2)^2}{2} = 2a_0$.

54. (1) According to Bohr's model the angular momentum is quantised and equal to $\frac{nh}{2\pi}$. Heisenberg uncertainty principle explains orbital concept, which is based on probability of finding electron.

55. (4) $E_n = \frac{-2.18 \times 10^{-18} Z^2}{n^2}$

i.e. $E_n \propto \frac{1}{n^2} \Rightarrow$ Energy in the 3rd Bohr orbit is $\frac{1}{9}$th of energy in first Bohr orbit.

56. (1) $\frac{1}{\lambda_H} = R(1)^2\left[\frac{1}{n_1^2} - \frac{1}{n_2^2}\right]$ $(n_1 < n_2)$

$$\frac{1}{\lambda_{He^+}} = R(2)^2\left[\frac{1}{2^2} - \frac{1}{4^2}\right]$$

Given $\lambda_H = \lambda_{He^+}$

$$R(1)^2\left[\frac{1}{n_1^2} - \frac{1}{n_2^2}\right] = R(4)\left[\frac{1}{2^2} - \frac{1}{4^2}\right]$$

$$\Rightarrow \frac{1}{n_1^2} - \frac{1}{n_2^2} = \frac{1}{1^2} - \frac{1}{2^2}$$

On comparing $n_1 = 1$ and $n_2 = 2$ or $n_i = 2$, and $n_f = 1$

57. (2) For H : $\frac{1}{\lambda} = R_H \times 1^2\left(\frac{1}{1^2} - \frac{1}{\infty^2}\right)$ ...(i)

$$\frac{1}{\lambda_{He^+}} = R_H \times 2^2 \times \left(\frac{1}{2^2} - \frac{1}{3^2}\right) \quad \text{...(ii)}$$

From equation (i) & (ii) $\frac{\lambda_{He^+}}{\lambda} = \frac{9}{5}$

$$\lambda_{He^+} = \lambda \times \frac{9}{5} \Rightarrow \lambda_{He^+} = \frac{9\lambda}{5}$$

58. (3) $r_n = \frac{52.9n^2}{Z}$ pm

Li^{2+} : $r_2 = x = k \times \frac{2^2}{3} = \frac{4k}{3}$

Be^{3+} : $r_3 = y = k \times \frac{3^2}{4} = \frac{9k}{4}$

$$\frac{y}{x} = \frac{9}{4} \times \frac{3}{4} = \frac{27}{16} \Rightarrow y = \frac{27}{16}x$$

59. (2) $r = 0.529 \times \frac{n^2}{Z}$ Å

$r_3 = 0.529 \times \frac{3^2}{1}$; $r_4 = 0.529 \times \frac{4^2}{1}$

$$\frac{r_4}{r_3} = \frac{4^2}{3^2} = \frac{16}{9} \Rightarrow r_4 = \frac{16r_3}{9}$$

60. **(3)** According to Bohr's model : $v_n \propto \frac{Z}{n}$

Z = Atomic no. (corresponds to +ve charge)
n = Principal quantum no.
This relation suggest that when the Z increases, the velocity will also increase, so statement I is wrong and as 'n' decreases, velocity will also increase. So statement II is correct.

61. **(4)**

62. **(3)** $r = 0.529\frac{n^2}{Z}$ Å

For Li^{2+},

$$(r_{Li^{2+}})_{n=4} - (r_{Li^{2+}})_{n=3} = \frac{0.529}{3}[4^2 - 3^2] = \Delta R_1$$

For He^+,

$$(r_{He^+})_{n=4} - (r_{He^+})_{n=3} = \frac{0.529}{2}[4^2 - 3^2] = \Delta R_2$$

$$\frac{\Delta R_1}{\Delta R_2} = \frac{2}{3}$$

63. **(3)** Shortest wavelength

→ Max. energy (∞ → 1)

For Lyman series of H atom,

$$\frac{1}{\lambda_1} = R_H(1)^2\left[\frac{1}{1} - 0\right] \Rightarrow \frac{1}{\lambda_1} = R_H \Rightarrow R_H = \frac{1}{\lambda_1}$$

For Balmer series of He^+,

$$\frac{1}{\lambda} = R_H(2)^2\left[\frac{1}{2^2} - \frac{1}{3^2}\right] \Rightarrow \frac{1}{\lambda} = R_H(4)\left(\frac{9-4}{36}\right)$$

$$\Rightarrow \frac{1}{\lambda} = \frac{5R_H}{9} \Rightarrow \lambda = \frac{9}{5R_H} = \frac{9\lambda_1}{5}$$

64. **(3)** $r = \frac{a_0 n^2}{Z}$

For Li^{2+}, $r = \frac{a_0(2)^2}{3} = \frac{4a_0}{3}$

65. **(2)** In the Balmer series of H–atom, the transition takes place from the higher orbital to n = 2. Therefore, the longest wavelength corresponds to $n_1 = 2$ and $n_2 = 3$. As wavelength decreases, energy increases or we can say that n_2 increases. The energy difference between two consecutive lines decreases or the lines slowly converge. Hence, statement I, II, III are the correct statements among the given options.

66. **(3)** $r = \frac{a_0 n^2}{Z}$

For Li^{2+}, $r = \frac{a_0(2)^2}{3} = \frac{4a_0}{3}$

67. **(1)** For determined shortest wavelength, $n_2 = \infty$

For Lyman series, $\bar{v}_L = \frac{1}{\lambda_L} = R\left[\frac{1}{(1)^2} - \frac{1}{\infty^2}\right]$

For Paschen series, $\bar{v}_P = \frac{1}{\lambda_P} = R\left[\frac{1}{(3)^2} - \frac{1}{\infty^2}\right]$

$$\frac{\bar{v}_L}{\bar{v}_P} = \frac{\lambda_P}{\lambda_L} = 9$$

68. **(2)**

69. **(4)** The total energy of the electron is minimum at a distance of a_0 from the nucleus for 1s orbital.

70. **(3)** According to Bohr's model energy in n^{th} state

$$= -13.6 \times \frac{Z^2}{n^2} \text{ eV}$$

For second excited state, of He^+, $n = 3$

$$\therefore E_3(He^+) = -13.6 \times \frac{2^2}{3^2} \text{ eV} = -6.04 \text{ eV}$$

71. **(3)** For emission spectrum,

$$\bar{v} = R_H\left(\frac{1}{n_f^2} - \frac{1}{n_i^2}\right) \Rightarrow \bar{v} = R_H\left(\frac{1}{n^2} - \frac{1}{8^2}\right)$$

$$\Rightarrow \bar{v} = \frac{R_H}{n^2} - \frac{R_H}{64}$$

Comparing to $y = mx + c$, we get

$x = \frac{1}{n^2}$ and $m = +R_H$ (slope)

72. **(2)** When temperature is increased, black body emits high energy radiation from higher wavelength to lower wavelength.

73. **(2)** $r = 0.529 \times \frac{n^2}{Z}$ Å

r=211.6pm=2.11 Å $\Rightarrow 0.529 \times \frac{n^2}{1} = 2.11$ Å [Z=1]

$n = 2$ (Balmer series)

74. **(4)** For Lyman series (shortest wavelength)

$n_1 = 1, n_2 = \infty$

$$\frac{1}{\lambda} = RZ^2\left(\frac{1}{n_1^2} - \frac{1}{n_2^2}\right)$$

$$\Rightarrow \frac{1}{A} = 1^2 R\left(\frac{1}{1} - \frac{1}{\infty}\right) \Rightarrow \frac{1}{A} = R$$

Longest wavelength = 1st line

$n_1 = 3, n_2 = 4$

$$\frac{1}{\lambda} = RZ^2\left(\frac{1}{3^2} - \frac{1}{4^2}\right) \Rightarrow \frac{1}{\lambda} = \frac{R7}{36}$$

$$R = \frac{1}{A}$$

$$\frac{1}{\lambda} = \frac{\frac{1}{A} \times 7}{36} \Rightarrow \frac{1}{\lambda} = \frac{7}{36A} \Rightarrow \lambda = \frac{36A}{7}$$

75. **(4)** Radius of nth Bohr orbit in H-atom $= 0.53\, n^2$Å

Radius of II Bohr orbit = 0.53 × $(2)^2$ = 2.12 Å

76. **(1)** Total energy $= -\frac{13.6}{n^2}Z^2$eV ; where n = 2, 3, 4 ...

Putting $n = 2$

$$E_T = -\frac{13.6}{4} = -3.4 \text{ eV}$$

77. **(3)** For Li^{2+} ion,

$$E = -13.6 \times \frac{Z^2}{n^2} \text{ eV} = -13.6 \times \frac{(3)^2}{(2)^2}$$

$$= \frac{-13.6 \times 9}{4} = -30.6 \text{ eV}$$

78. **(1)** $\Delta E = -2.0 \times 10^{-18} \times \left(\frac{1}{2^2} - \frac{1}{1^2}\right)$

$$= -2.0 \times 10^{-18} \times \frac{-3}{4} = 1.5 \times 10^{-18} \text{ J}$$

$$\Delta E = \frac{hc}{\lambda}$$

$$\lambda = \frac{hc}{\Delta E} = \frac{6.625 \times 10^{-34} \text{ Js} \times 3 \times 10^8 \text{ ms}^{-1}}{1.5 \times 10^{-18} \text{ J}}$$

$= 1.325 \times 10^{-7}$ m

79. **(4)** Total energy of a revolving electron is the sum of its kinetic and potential energy.
Total energy = K.E. + P. E.

$$= \frac{Ke^2}{2r} + \left(-\frac{Ke^2}{r}\right) = -\frac{Ke^2}{2r}$$

80. **(1)** $\Delta E = 2.178 \times 10^{-18}\left(\frac{1}{1^2} - \frac{1}{2^2}\right) = \frac{hc}{\lambda}$

$$\Rightarrow 2.178 \times 10^{-18} \times \frac{3}{4} = \frac{hc}{\lambda}$$

$$= \frac{6.62 \times 10^{-34} \times 3 \times 10^8}{\lambda}$$

$$\lambda = \frac{6.62 \times 10^{-34} \times 3 \times 10^8 \times 4}{2.178 \times 10^{-18} \times 3}$$

$= 1.214 \times 10^{-7}$m

81. **(1)** $\bar{v} = RZ^2\left(\frac{1}{2^2} - \frac{1}{3^2}\right) = R\left(\frac{1}{4} - \frac{1}{9}\right)$

$$= \frac{5R}{36}$$

82. **(1)** $\nu = R\left(\frac{1}{n_1^2} - \frac{1}{n_2^2}\right)$Hz

$$\nu = 3.29 \times 10^{15}\left(\frac{1}{2^2} - \frac{1}{\infty^2}\right) = 8.22 \times 10^{14} \text{s}^{-1}$$

83. **(1)**

84. **(2)** I. E $= \frac{Z^2}{n^2} \times 13.6$eV ...(i)

or $\frac{I_1}{I_2} = \frac{Z_1^2}{n_1^2} \times \frac{n_2^2}{Z_2^2}$ ...(ii)

Given $I_1 = -19.6 \times 10^{-18}$ J/atom, $Z_1 = 2$, $n_1 = 1$, $Z_2 = 3$ and $n_2 = 1$

Substituting these values in equation (ii).

$$-\frac{19.6 \times 10^{-18}}{I_2} = \frac{4}{1} \times \frac{1}{9}$$

or $I_2 = -19.6 \times 10^{-18} \times \frac{9}{4}$

$= -4.41 \times 10^{-17}$ J/atom

85. **(4)** The energy required to excite an electron in an atom of hydrogen from n = 1 to n = 2 is ΔE (difference in energy E_2 and E_1).
Values of E_2 and E_1 are,

$$E_2 = \frac{-1.312 \times 10^6 \times (1)^2}{(2)^2} = -3.28 \times 10^5 \text{ J mol}^{-1}$$

$E_1 = -1.312 \times 10^6$ J mol^{-1}

$\therefore \Delta E = E_2 - E_1 = [-3.28 \times 10^5] - [-1.312 \times 10^6]$ J mol^{-1}

$= (-3.28 \times 10^5 + 1.312 \times 10^6)$ J mol^{-1}

$= 9.84 \times 10^5$ J mol^{-1}

86. **(2)** Angular momentum of an electron in n^{th} orbital is given by, $mvr = \frac{nh}{2\pi}$

For $n = 5$, we have

Angular momentum of electron $= \frac{5h}{2\pi} = \frac{2.5h}{\pi}$

87. (3) $\frac{1}{\lambda}=R\left(\frac{1}{n_1^2}-\frac{1}{n_2^2}\right)$

$\frac{1}{\lambda}=1.097\times10^7\left(\frac{1}{1}-\frac{1}{\infty}\right)=1.097\times10^7$

$\lambda=91.15\times10^{-9}\,m\approx91\,nm$

88. (1) 2nd excited state will be the 3rd energy level

$E_n=\frac{13.6}{n^2}eV$

or $E_3=\frac{13.6}{9}eV=1.51eV.$

89. (54) For lowest energy spectral line in Lyman series of hydrogen,

$Z=1, n_1=1, n_2=2$

$\Delta E(L_1)=13.6\times1^2\left(\frac{1}{1^2}-\frac{1}{2^2}\right)$

$=13.6\times\frac{3}{4}=10.2\ eV$

For lowest energy spectral line in Balmer series of hydrogen, $Z=1, n_1=2, n_2=3$.

$\Delta E(B_1)=13.6\times(1)^2\left(\frac{1}{(2)^2}-\frac{1}{(3)^2}\right)$

$=13.6\frac{5}{4\times9}=1.889$

$\frac{\Delta E(L_1)}{\Delta E(B_1)}=\frac{10.2}{1.889}=5.3999=54\times10^{-1}.$

90. (54)

E_1 of H-atom $=-13.6$ eV

E_2 of $Be^{3+}=\frac{-13.6\times Z^2}{n^2}$

$=\frac{-13.6\times(4)^2}{(2)^2}=-54.4\ eV$

$|E_2|$ of $Be^{3+}=54$ eV

91. (5) Given : $\frac{\lambda}{4}=1.5$ pm

$\Rightarrow\lambda=1.5\times4$ pm

$=6\times10^{-12}$ m

$\lambda v=C$

$6\times10^{-12}\times v=3\times10^8$

$\Rightarrow v=5\times10^{19}$ Hz

92. (1724)

Wave number $=\frac{1}{\lambda}=\frac{1}{5800\times10^{-8}cm}=17241\ cm^{-1}$

$1724.1\times10\ cm^{-1}\Rightarrow x=1724.$

93. (34) KE $=-$Total energy

$=-(-3.4)=3.4=34\times10^{-1}$

$x=34.$

94. (22) From Bohr's theory we have

$V_n=2.18\times10^6\times\frac{Z}{n}\ ms^{-1}$

$=21.8\times10^5\times\frac{1}{1}\approx22\times10^5$ (nearest)

95. (8) On considering wave the nature of electron.

$2\pi r_n=n\lambda_d$

$2\pi a_0\frac{n^2}{Z}=n\lambda_d\left\{r_n=a_0\frac{n^2}{Z}\right\}$

$2\pi a_0\frac{4^2}{1}=4\lambda_d$

$\lambda_d=8\pi a_0$

96. (10) 5th excited state $\Rightarrow n_1=6$

1st excited state $\Rightarrow n_2=2,$

$\Delta n=n_1-n_2=6-2=4$

$\Rightarrow$ Maximum number of spectral lines

$=\frac{\Delta n(\Delta n+1)}{2}=\frac{4(4+1)}{2}=10$

97. (1) $\Delta E=R_HZ^2\left(\frac{1}{n_1^2}-\frac{1}{n_2^2}\right)\ (n_1<n_2)$

$1.47\times10^{-17}=2.18\times10^{-18}\times9\left[\frac{1}{n^2}-\frac{1}{(n+1)^2}\right]$

$\frac{1.47}{1.96}=\frac{3}{4}=\frac{1}{n^2}-\frac{1}{(n+1)^2}$

So, n = 1

98. (270) $r\propto\frac{n^2}{Z}\ \therefore\ r_{He^+}=r_H\times\frac{n^2}{Z}$

$r_{He^+}=0.6\times\frac{(3)^2}{2}=2.7\ Å=270\ pm$

99. (492) For first line of Paschen series

$\frac{1}{(\lambda_1)_p}=R_HZ^2\left(\frac{1}{9}-\frac{1}{16}\right)$

For second line of Paschen series

$\frac{1}{(\lambda_2)_p}=R_HZ^2\left(\frac{1}{9}-\frac{1}{25}\right)$

$\frac{(\lambda_2)_p}{(\lambda_1)_p}=\frac{\frac{7}{16\times9}}{\frac{16}{25\times9}}=\frac{25\times7}{16\times16}$

$(\lambda_2)_P=\frac{25\times7}{16\times16}\times720$

$(\lambda_2)_P=492$ nm

100. (N) We can not calculate I.E. of lithium atom using Bohr's model because it is not a single e^- species.

101. (3155) $mvr=\frac{nh}{2\pi}$

$K.E.=\frac{1}{2}mv^2\ \Rightarrow\ K.E.=\frac{n^2h^2}{8\pi^2mr^2}$

For $n=2, r=4a_0$

$K.E.=\frac{4h^2}{8\pi^2m(4a_0)^2}=\left(\frac{4}{8\pi^2\times16}\right)\frac{h^2}{ma_0^2}$

Comparing with given value,

$\left(\frac{4}{8\pi^2\times16}\right)=\frac{1}{x}\Rightarrow x=315.507$

$\Rightarrow 10x=3155$ (nearest integer)

102. (2)

(A) Heisenberg uncertainty principle is applicable to all sub-atomic particles, including electrons.

(B) $2p_x$ has less value of n thus has lesser size than $3p_x$.

(C) 2*s* orbital of Li is lower in energy than 2*s* orbital of H due to nuclear charge and electron shielding.

(D) Cr is exceptional case. Half filled 3*d* is more stable in Cr.

Correct statements are A, B and D.

103. (2) $\Delta x.\Delta p\geq\frac{h}{4\pi}\Rightarrow\Delta p^2\geq\frac{h}{4\pi}$

$\Rightarrow m\Delta V\geq\sqrt{\frac{h}{4\pi}}\Rightarrow\Delta V\geq\sqrt{\frac{h}{4\pi}}\times\frac{1}{m}$

$\Rightarrow\Delta V\geq\sqrt{\frac{h}{4\pi}}\times\frac{1}{m}$

$\Rightarrow\Delta V\geq\sqrt{\frac{h}{\pi}}\times\frac{1}{2m}$

104. (2) $2\pi r_n=n\lambda$, $r_n=n^2a_0=(2)^2a_0=4a_0$

$2\pi(4a_0)=n\lambda\ =\lambda=\frac{8\pi a_0}{n}$

105. (1) de-Broglie equation:

$\lambda=\frac{h}{p}\ \Rightarrow\ \lambda\propto\frac{1}{p}$

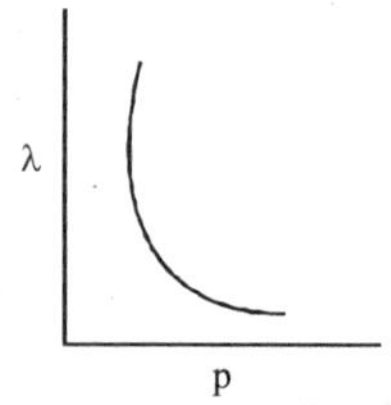

106. (3) $(r_3)_H=\frac{a_0n^2}{Z}=a_0\times3^2=9a_0$

$2\pi r=n\lambda\Rightarrow2\pi\times9a_0=3\lambda\ \Rightarrow\ \lambda=6\pi a_0$

107. (4) (1), (2) and (3) are according to quantum theory but (4) is statement of kinetic theory of gases.

108. (4) $2\pi r=n\lambda$

$r=\frac{n^2a_0}{Z}$

$2\pi\times\frac{4^2}{1}a_0=4\lambda\Rightarrow\lambda=2\pi\times\frac{4}{1}a_0\Rightarrow\lambda=8\pi a_0$

109. (4) Given $\lambda=1.5\pi a_0$

$n\lambda=2\pi r$ ...(i)

Radii of stationary states (r) is expressed as:

$r=a_0\frac{n^2}{Z}$ ...(ii)

From eqn (i) and (ii)

$n\lambda=\frac{2\pi a_0n^2}{Z}$; $\lambda=\frac{2\pi a_0n}{Z}$

$1.5\pi a_0=2\pi a_0\frac{n}{Z}\ \Rightarrow\ \frac{n}{Z}=\frac{1.5}{2}=0.75$

110. (4) According to de-Broglie wavelength equation,

$\lambda=\frac{h}{mV}\ \Rightarrow\ \lambda\propto\frac{1}{V}$

According to photoelectric effect,

$h\nu-h\nu_0=\frac{1}{2}mV^2;\ \nu-\nu_0=\frac{1}{2}\times\frac{mV^2}{h}$

$\nu-\nu_0\propto V^2\ \Rightarrow\ V\propto(\nu-\nu_0)^{1/2}$

$\therefore\ \lambda\propto\frac{1}{(\nu-\nu_0)^{1/2}}$

111. (2) First Bohr orbit of H atom has radius r = 0.529 Å

Also, the angular momentum is quantised.

$mvr = \frac{h}{2\pi} \Rightarrow 2\pi r = \frac{h}{mv} = \lambda$

$\therefore \lambda = 2\pi \times 0.529$ Å

112. (2) As electron of charge 'e' is passed through '*V*' volt, kinetic energy of electron will be eV.

Wavelength of electron wave (λ) $= \frac{h}{\sqrt{2m \cdot KE}}$

$\lambda = \frac{h}{\sqrt{2meV}} \quad \therefore \frac{h}{\lambda} = \sqrt{2meV}$

113. (4) Kinetic energy of any particle $= \frac{3}{2}kT$

Also K.E. $= \frac{1}{2}mv^2$

$\frac{1}{2}mv^2 = \frac{3}{2}kT \Rightarrow v^2 = \frac{3kT}{m} \Rightarrow v = \sqrt{\frac{3kT}{m}}$

de-Broglie wavelength $= \lambda = \frac{h}{mv} = \frac{h}{m\sqrt{\frac{3kT}{m}}}$

$\lambda = \frac{h}{\sqrt{3kTm}}; \quad \lambda \propto \frac{1}{\sqrt{m}}$

Mass of electron < mass of neutron

λ (electron) > λ (neutron)

114. (1) de-Broglie wavelength (λ) $= \frac{h}{mv}$

$= \frac{6.63 \times 10^{-34} \text{ Js}}{6.63 \times 10^{-3} \text{ kg} \times 100 \text{ m/s}} = 10^{-33}$ m

115. (2) $\lambda = \frac{h}{mv}$

$h = 6.6 \times 10^{-34}$ Js

$m = 1000$ kg

$v = 36$ km/hr $= \frac{36 \times 10^3}{60 \times 60}$ m/sec $= 10$ m/sec

$\therefore \lambda = \frac{6.6 \times 10^{-34}}{10^3 \times 10} = 6.6 \times 10^{-38}$ m

116. (2) de-Broglie wavelength is given by :

$\lambda = \frac{h}{mV}$ ... (i)

K.E. $= \frac{1}{2}mV^2 \Rightarrow V^2 = \frac{2K.E}{m} \Rightarrow V = \sqrt{\frac{2K.E}{m}}$

Substituting this in equation (i),

$\lambda = \frac{h}{m}\sqrt{\frac{m}{2K.E}} \Rightarrow \lambda = h\sqrt{\frac{1}{2m(K.E)}}$

i.e., $\boxed{\lambda \propto \frac{1}{\sqrt{K.E}}} \rightarrow$ important relation

$\therefore$ When *KE* become 4 times, wavelength become 1/2.

117. (3) $r_n = a_0 n^2$

$r = a_0 \times (3)^2 = 9a_0$

$mvr = \frac{nh}{2\pi}; \quad mv = \frac{nh}{2\pi r} = \frac{3h}{2\pi \times 9a_0} = \frac{h}{6\pi a_0}$

$\lambda = \frac{h}{mv} = \frac{n}{h} \times 6\pi a_0 = 6\pi a_0$

118. (1) $\lambda = \frac{h}{mv} = \frac{6.63 \times 10^{-34}}{1.67 \times 10^{-27} \times 1 \times 10^3}$

$= 3.97 \times 10^{-10}$ m $= 0.397$ nm.

119. (2) According to Heisenberg uncertainty principle,

$\Delta x \cdot m\Delta v = \frac{h}{4\pi}; \quad \Delta x = \frac{h}{4\pi m \Delta v}$

Here $\Delta v = \frac{600 \times 0.005}{100} = 0.03$ m/s

So, $\Delta x = \frac{6.6 \times 10^{-34}}{4 \times 3.14 \times 9.1 \times 10^{-31} \times 0.03} = 1.92 \times 10^{-3}$ m

120. (1) Given $m = 9.1 \times 10^{-31}$ kg

$h = 6.6 \times 10^{-34}$ Js

$\Delta v = \frac{300 \times 0.001}{100} = 0.003 \text{ ms}^{-1}$

From Heisenberg's uncertainity principle

$\Delta x - \frac{6.62 \times 10^{-34}}{4 \times 3.14 \times 0.003 \times 9.1 \times 10^{-31}}$

$= 1.92 \times 10^{-2}$ m

121. (4) $\lambda = \frac{h}{mv} = \frac{6.63 \times 10^{-34}}{60 \times 10^{-3} \times 10}$

$= 1.105 \times 10^{-33} \approx 10^{-33}$ m

122. (3) $\Delta x.\Delta p = \frac{h}{4\pi}$; or $\Delta x.m\,\Delta v = \frac{h}{4\pi}$

$\therefore \Delta v = \frac{6.62 \times 10^{-34}}{4 \times 3.14 \times 0.025 \times 10^{-5}}$

$= 2.1 \times 10^{-28} \text{ ms}^{-1}$

123. (2) $\lambda = \frac{h}{\sqrt{2m\,(eV)}}$

$\frac{\lambda_{m_1}}{\lambda_{m_2}} = \frac{1/\sqrt{2 \times 1\,(eV)}}{\sqrt{1/2 \times 4\,(eV)}} = 2$

$\lambda_{m_1} = 2\lambda_{m_2}$

$\therefore \quad x = 2$

124. (58) $m\Delta V.\Delta x = \frac{h}{4\pi} \Rightarrow \Delta V = \frac{h}{m\Delta x\, 4\pi}$

$= \frac{6.626 \times 10^{-34}}{9.1 \times 10^{-31} \times 10^{-15} \times 4 \times 3.14}$

$= 57.97 \times 10^9$ m/sec.

125. (661) $\lambda = \frac{h \times V}{mv \times V}; \quad \lambda = \frac{hV}{mV^2}$

$\Rightarrow \frac{mV^2}{h} = \frac{V}{\lambda} = v$

$K.E = R_H\left(\frac{1}{n^2}\right) \quad \therefore n = 1$

$\frac{1}{2}mV^2 = R_H$

$mV^2 = 2R_H$

$mV^2 = 2R_H = 2 \times 2.18 \times 10^{-18}$ J

$= 4.36 \times 10^{-18}$ J

$h = 6.6 \times 10^{-34}$ J.s (given)

$\Rightarrow v = \frac{4.36 \times 10^{-18}}{6.6 \times 10^{-34}} = 660.60 \times 10^{13}$ Hz

$\approx 661 \times 10^{13}$ Hz

126. (7)

$\lambda_d = \frac{h}{mv} = \frac{h}{\sqrt{2mKE}} = \frac{6.6 \times 10^{-34}}{\sqrt{2 \times 9 \times 10^{-31} \times 4.5 \times 10^{-29}}}$

$= \frac{6.6 \times 10^{-34}}{\sqrt{9^2 \times 10^{-60}}}$

$= \frac{6.6 \times 10^{-34}}{9 \times 10^{-30}} = \frac{6.6}{9} \times 10^{-4}$

$= 7.3 \times 10^{-5}$ m $\sim 7 \times 10^{-5}$ m.

127. (2) de Broglie equation

$p = \frac{h}{\lambda}$ or

$\lambda = \frac{h}{mv} = \frac{6 \times 10^{-34}}{9 \times 10^{-31} \times 10^3} [\because v_e = 1000 \text{ m/s}]$

$= \frac{2}{3} \times 10^{-6}$ m $= 0.6667$ µm $= 666.7$ nm

The characteristic of electrons emitted do not depend on the nature of the material of electrodes.

The cathode ray starts from cathode and move towards anode.

The nature of the emitted electrons is independent on the nature of the gas present in cathode ray tube.

128. (548) $\Delta x \times \Delta p_x \geq \frac{h}{4\pi}$

$\Rightarrow 2a_0 \times m\Delta v_x = \frac{h}{4\pi}$

(for minimum uncertainty)

$\Rightarrow \Delta v_x = \frac{h}{4\pi} \times \frac{1}{2a_0} \times \frac{1}{m}$

$= \frac{6.63 \times 10^{-34}}{4 \times 3.14 \times 2 \times 52.9 \times 10^{-12} \times 9.1 \times 10^{-31}}$

$= 548273 \text{ ms}^{-1} = 548.273 \text{ km s}^{-1}$

129. (1758) $\lambda_e = \lambda_n \Rightarrow \frac{h}{p_e} = \frac{h}{p_n} \Rightarrow p_e = p_n$

$\Rightarrow m_e v_e = m_p v_n = (9.1 \times 10^{-31})v_e = (1.6 \times 10^{-27})\,v_n$

$\Rightarrow v_e = \frac{1.6 \times 10^{-27}}{9.1 \times 10^{-31}} \times v_n \Rightarrow 1758.2 \times v_n$

130. (22) $\Delta v = 2.4 \times 10^{-26} \text{ ms}^{-1}$; $\Delta x = 10^{-7}$ m

$\therefore \Delta p \cdot \Delta x = \frac{h}{4\pi} \quad \therefore m\Delta v \cdot \Delta x = \frac{h}{4\pi}$

$\Rightarrow m \times 2.4 \times 10^{-26} \times 10^{-7} = \frac{6.626 \times 10^{-34}}{4 \times \pi}$

$\Rightarrow m = 0.02198$ kg $\Rightarrow m = 21.98$ g ≈ 22g.

131. (58) $\Delta x \cdot \Delta p = \frac{h}{4\pi}$

$\Delta x \cdot m\Delta v = \frac{h}{4\pi}$

$\Delta v = 5 \times 10^6 \times \frac{0.02}{100}$

$\Delta v = 10^3$ m/s

$\Delta x = \frac{h}{4\pi \times m \cdot \Delta v} = \frac{6.63 \times 10^{-34}}{4 \times 3.14 \times 9.1 \times 10^{-31} \times 10^3}$

$= 5.8 \times 10^{-8}$ m $= 58 \times 10^{-9}$ m

132. (6) de-Broglie wave-length of electron:

$\lambda_e = \frac{h}{\sqrt{2m(KE)}} \quad \left\{\begin{array}{l} \because e^- \text{ is accelerated} \\ \text{from rest} \\ \Rightarrow KE = q \times V \end{array}\right.$

$$\lambda = \frac{h}{\sqrt{2mqV}}$$

$$= \frac{6.63\times10^{-34}}{\sqrt{2\times1.6\times10^{-19}\times9.1\times10^{-31}\times40\times10^{3}}}$$

$= 0.614 \times 10^{-11}\,m = 6.14 \times 10^{-12}\,m$

133. (9) Incident energy of photon

= Work function of metal + K.E of photoelectron

$h\nu = h\nu_0 + KE$ ...(i)

Incident energy of photon

$$= \frac{hc}{\lambda} = \frac{6.63\times10^{-34}\times3\times10^{8}}{248\times10^{-9}\times1.6\times10^{-19}}\,eV = 5\,eV$$

Now using equation (i),

$5 = 3 + KE$

$KE = 2.0\,eV$

Now, $\lambda = \frac{h}{\sqrt{2m\ KE}}$

$$= \frac{6.63\times10^{-34}}{\sqrt{2\times9.1\times10^{-31}\times2\times1.6\times10^{-19}}}$$

$= 8.68\times10^{-10}\,m \approx 9\,Å$

134. (1) According to Heisenberg uncertainty principle,

$$\Delta x\cdot\Delta p \geq \frac{h}{4\pi}$$

$$\Delta x = \frac{h}{4\pi m\Delta v} \qquad (\because \Delta p = m\Delta v)$$

$$\Delta v = 90\times\frac{5}{100} = 4.5\ m/s$$

$$\Delta x = \frac{6.63\times10^{-34}\,Js}{4\times3.14\times10\times10^{-3}\,kg\times4.5\,m/s}\left[1J = \frac{kg.m^2}{s^2}\right]$$

$= 1.173 \times 10^{-33}\,m = 1 \times 10^{-33}\,m$

135. (2) $\lambda = \frac{h}{\sqrt{2mqV}}$

$$\frac{\lambda_{Li}}{\lambda_P} = \sqrt{\frac{m_P(e)V}{m_{Li}(3e)(V)}}; \quad m_{Li} = 8.3m_P$$

$$\frac{\lambda_{Li}}{\lambda_P} = \sqrt{\frac{1}{8.3\times3}} = \frac{1}{5} = 0.2 = 2\times10^{-1}$$

136. (4) For a given orbital

Number of Radial nodes is calculated by $n - \ell - 1$

Number of Angular nodes or nodal planes is calculated by ℓ

For 2s $\Rightarrow n = 2, \ell = 0$

1 radial node + zero nodal plane

For 3s $\Rightarrow n = 3, \ell = 0$

2 radial node + zero nodal plane

For 3p $\Rightarrow n = 3, \ell = 1$

1 radial node + 1 nodal plane

for 4d $\Rightarrow n = 4, \ell = 2$

1 radial node + 2 nodal plane

137. (2) Statement (A) and (C) are correct statements.

Energies of the orbitals in the same subshell decreases with increase in atomic number.

Orbital	No. of radial node $(n - l - 1)$
5f	$5 - 3 - 1 = 1$
6s	$6 - 0 - 1 = 5$
4d	$4 - 2 - 1 = 5$
5p	$5 - 1 - 1 = 3$
5d	$5 - 2 - 1 = 2$

138. (4) Order of energy is decided by $(n + l)$ rule. More the value of $(n + l)$, more will be the energy of that particular orbital.

If $(n + l)$ value will be same then more the value of n, more will be the energy.

(A) $n = 3, l = 2, m = +1; (n + l) = 5$

(B) $n = 4, l = 0, m = 0; (n + l) = 4$

(C) $n = 6, l = 1, m = 0; (n + l) = 7$

(D) $n = 5, l = 1, m = +1; (n + l) = 6$

(E) $n = 2, l = 1, m = +1; (n + l) = 3$

E < B < A < D < C

139. (4) Cr : $1s^2 2s^2 2p^6 3s^2 3p^6 4s^1 3d^5$

$\Rightarrow$ 19th e^- belongs to 4s subshell

$n = 4; l = 0; m = 0; s = +\frac{1}{2}$ or $-\frac{1}{2}$

140. (1) At spherical node

$\psi_r = 0$

141. (4)

(A) n = 5

$l = 0$ $m_l = 0$

$l = 1$ $m_l = -1, 0, 1 \Rightarrow$ 2 electrons

$l = 2$ $m_l = -2, -1, 0, 1, 2 \Rightarrow$ 2 electrons

$l = 3$ $m_l = -3, -2, -1, 0, 1, 2, 3 \Rightarrow$ 2 electrons

$l = 4$ $m_l = -4, -3, -2, -1, 0, 1, 2, 3, 4 \Rightarrow$ 2 electrons

Total number of electrons = 8

(B) $n = 3, l = 2, m_l = -1,$

$m_s = +\frac{1}{2} \Rightarrow$ only 1 electron is possible

142. (2) Element with atomic number 9 is Fluorine

F(9) = $1s^2$ $2s^2$ $2p^5$

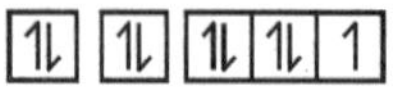

(A)5 electrons can be up-spin $\left[m_s = +\frac{1}{2}\right]$ and 4 electrons can be down spin $\left[m_s = +\frac{1}{2}\right]$

(B)Unpaired electron can be in anyone of p_x, p_y or p_z orbital

(C)Last electron is in 2p subshell with $n = 2, l = 1$

(D)Angular node for s-orbital = 0 while for each p-orbital = 1

$\therefore$ Sum of all angular node = 3

143. (2) Extra stability of half filled is due to :

(i) Symmetrical distribution of electrons

(ii) Large exchange energy

(iii) Smaller coulombic repulsion

(iv) Smaller shielding of electrons by one another.

144. (3) Cr : $1s^2 2s^2 2p^6 3s^2 3p^6 3d^5 4s^1$

Electrons having $\ell = 1 \Rightarrow 12$

Electrons having $\ell = 2 \Rightarrow 5$

145. (4)

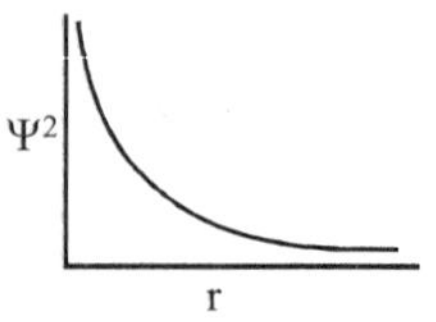

1. Probability density $(\Psi)^2$ is maximum at nucleus.
2. Electron can exist upto infinity from nucleus.
3. Statement is true.
4. Energy of electron is maximum at infinite distance from nucleus. Total energy of an electron is given by $\frac{-13.6z^2}{n^2}$. As 'n' increases, energy also increases.

146. (4) Orbital angular momentum

$= \sqrt{\ell(\ell+1)}\frac{h}{2\pi}$

$\therefore$ For 2s orbital : $\ell = 0$

Orbital angular momentum = 0

$\therefore$ For 2p orbital : $\ell = 1$

Orbital angular momentum $= \sqrt{\ell(\ell+1)}\frac{h}{2\pi}$

$= \sqrt{2}\frac{h}{2\pi}$

147. (4) Subsidiary quantum number or azimuthal quantum number ℓ defines shape of orbital.

The given surface diagram is for $2p_x$ orbital.

148. (3) In single electron system an electron's energy only depends on 'n' (principal quantum number)

So energy of 2s = 2p and energy of 3d < 4s

149. (4) Orbital

A : $n = 1, l = 0, m_l = 0$ 1s

B : $n = 2, l = 0, m_l = 0$ 2s

C : $n = 2, l = 1, m_l = 1$ 2p

D : $n = 3, l = 2, m_l = 1$ 3d

E : $n = 3, l = 2, m_l = 0$ 3d

When there are no electric and magnetic fields, all the 3d orbitals remain degenerate.

150. (4) In hydrogen atom the orbitals in a shell are degenerate meaning the energy depends only on 'n'

$\therefore E_{3p_x} = E_{3d_{x^2-y^2}} = E_{3d_{z^2}}$

151. (3) For a shell, total number of orbitals = n^2

Magnetic quantum number have values $(-l$ to $+l)$ including 0.

152. (4) No spectral line will be observed for a $2p_x \rightarrow 2p_y$ transition because $2p_x$ and $2p_y$ orbitals are degenerate orbitals.

153. (4) Energy level can be determined by comparing $(n + l)$ values

(A) $n = 4, l = 1 \Rightarrow (n + l) = 5$

(B) $n = 4, l = 2 \Rightarrow (n + l) = 6$

(C) $n = 3, l = 1 \Rightarrow (n + l) = 4$

(D) $n = 3, l = 2 \Rightarrow (n + l) = 5$

(E) $n = 4, l = 0 \Rightarrow (n + l) = 4$

For same value of $(n + l)$, orbital having higher value of n, will have more energy. Thus,

(B) > (A) > (D) > (E) > (C)

154. (3) For 3p orbital

$l = 1, n = 3$

$\therefore$ Radial node $= (n - l - 1) = (3 - 1 - 1) = 1$

155. (2) $n = 4, l = 0, m = 0, s = +½$

$_{19}K = 1s^2\ 2s^2\ 2p^6\ 3s^2\ 3p^6\ 4s^1$

Outermost orbital is 4s, containing one electron

$\therefore$ For s, $l = 0, m = 0$

single electron will have spin as + ½

$\therefore s = +½$

156. (1) Statement-I is true but statement-II is false. For single electron species the energy depends upon principal quantum number only.

157. (4) Rubidium (Z = 37)
Electronic configuration
$= [Zn]\, 4p^6\, 5s^1 n = 5$
For s, $l = 0$

$m_l = 2l + 1 = 1$ value i.e. $0 \Rightarrow S = +\frac{1}{2}$
The correct set of four quantum number for the given atom is 5, 0, 0, $+\frac{1}{2}$.

158. (2) 5f orbital not buried as 4f orbitals so e^- present in 5f orbital experience less nuclear attraction than e^- present in 4f orbital. Hence electrons of 5f orbital can take part in bonding to a far greater extent.

159. (2) (A)n = 3; l = 0; n + l = 3
(B)n = 4; l = 0; n + l = 4
(C)n = 3; l = 1; n + l = 4
(D)n = 3; l = 2; n + l = 5
As per Hund's rule energy is given by (n + l) value. If value of (n + l) remains same then energy is given by n only.
Therefore the order of energy is : D > B > C > A

160. (2) The number of electrons in the orbitals of sub-shell of $n = 4$ are 4s (2), 4p (6), 4d (10), 4f (14), total 32 electrons.

161. (4) At node $\Psi_{2s} = 0$

$\therefore\ 2 - \frac{r_0}{a_0} = 0 \Rightarrow r_0 = 2a_0$

162. (2) Lower the value of (n + l) for an orbital, lower is its energy. If two orbitals have same value of (n + l) then lower value of n will have the lower energy
order is D < A < B < C

163. (4) Atomic orbital is characterised by n, l, m.

164. (1)

165. (1)
Higher (n + l) value, higher the energy; if same (n + l) value, then higher n value, higher the energy.
Thus : D > B > C > A

166. (3) n = 3, l = 3 is not possible as l = 0, 1 ... (n – 1)

167. (1) for 4p
$n = 4, l = 1$
Total number of nodes $= n - 1 = 4 - 1 = 3$
Total radial nodes $= n - l - 1 = 4 - 1 - 1 = 2$

168. (4) Total number of nodes (angular + radial) $= n - 1$

169. (2) $\psi^2(r) \Rightarrow$ always positive
no. of radial nodes $= n - \ell - 1$
For 2s, n = 2, $\ell = 0$
no. of radial nodes = 2 – 0 – 1 = 1

170. (3)

171. (1) Radial node $= n - l - 1 = 4 - 2 - 1 = 1$
Angular node (l) = 2.

172. (2) Both e^- belong to $3d$-orbitals, which are degenerate.

173. (3)
$3s$ orbital has no angular nodes and two radial nodes.

174. (3) Number of radial nodes $= n - l - 1$
Number of angular nodes $= l$
For $5d$; $n = 5, l = 2$
$5d$ orbital has two radial nodes and two angular nodes.

175. (1)

	6s	4f	5d	6p
n + l	6+0	4+3	5+2	6+1
	⇓	⇓	⇓	⇓
	6	7	7	7

Thus, order of orbitals filled are
6s < 4f < 5d < 6p.

176. (3) Probability density (ψ^2) can be zero for $3p$ orbital other than infinite distance. It has one radial node.
Thus, statement (3) is correct.

177. (3)

178. (2) For $n = 4$ possible values of $l = 0$, 1, 2, 3; only $l = 2$ and $l = 3$ can have $m = -2$. So possible subshells are 2 (4d and 4f).

179. (2) The possible number of orbitals in a shell in term of 'n' is n^2. All orbitals can have m_s = +1/2
$\therefore n = 5\,;\, n^2 = 25$

180. (1) As the value of Z (atomic number) increases, energy of orbitals decreases (becomes more –ve value).
$\therefore$ Order of energy of 2s orbital is H > Li > Na > K.

181. (1) Probability of finding an electron will have maximum value at both 'a' and 'c'. There is zero probability of finding an electron at 'b'.

182. (2) The given probability density curve is for 2s orbital due to the presence of only one radial node. 1s and 2p orbital do not have any radial node and 3s orbital has two radial nodes. Hence, option (2) is correct.

183. (4) Iso-electronic species differ in size due to different effective nuclear charge.

184. (3)

				$n + \ell$
(I)	$n = 4$	$\ell = 2$	$4d$	6
(II)	$n = 3$	$\ell = 2$	$3d$	5
(III)	$n = 4$	$\ell = 1$	$4p$	5
(IV)	$n = 3$	$\ell = 1$	$3p$	4

The energy of an atomic orbital increases with increasing $n + \ell$. For identical values of $n + \ell$, energy increases with increasing value of n. Therefore the correct order of energy is:
$3p < 3d < 4p < 4d$
IV II III I

185. (1)

186. (2) Number of orbitals in a shell $= n^2 = (5)^2 = 25$.

187. (2) According to Aufbau principle, the sequence of filling electrons in sixth period is
$6s - 4f - 5d - 6p$
i.e., $(ns) \to (n-2)f \to (n-1)d \to np$

188. (1) The electronic configuration of Rubidium (Rb = 37) is

$1s^2 2s^2 2p^6 3s^2 3p^6 3d^{10} 4s^2 4p^6 5s^1$
The last electron enters in 5s orbital.

Hence, $n = 5, l = 0, m = 0, s = \pm\frac{1}{2}$

189. (3) $n = 3, l = 2$ means $3d$ orbital

+2	+1	0	–1	–2

In an atom, only one orbital can have the value $m_l = +2$.

190. (2) (i) $n = 5$ means $l = 0, 1, 2, 3, 4$
Since $m = +1$,
hence total no. of electrons will be
= 0(from s) + 2(from p) + 2(from d) + 2(from f) + 2(from g) = 0 + 2 + 2 + 2 + 2 = 8
(ii) $n = 2, l = 1, m_l = -1, m_s = -1/2$ represent 2p orbital with one electron.

191. (3) Among isoelectronic species ionic radii increases as the negatives charge increases. Order of ionic radii $Ca^{2+} < K^+ < Cl^- < S^{2-}$
The number of electrons remains the same but nuclear charge increases with increase in the atomic number causing decrease in size.

192. (2) (i) $4p$ (ii) $4s$
(iii) $3d$ (iv) $3p$
According to Bohr Bury's $(n + l)$ rule, increasing order of energy will be (iv) < (ii) < (iii) < (i).
Note: If the two orbitals have same value of $(n + l)$ then the orbital with lower value of n will be filled first.

193. (2) Species having same number of electrons are isoelectronic. On calculating the number of electrons in each species given here, we get.
CN^- (6 + 7 + 1 = 14); N_2 (7 + 7 = 14)
O_2^{2-} (8 + 8 + 2 = 18) ; C_2^{2-} (6 + 6 + 2 = 14)
O_2^- (8 + 8 + 1 = 17) ; NO^+ (7 + 8 – 1 = 14)
CO (6 + 8 = 14) ; NO (7 + 8 = 15)
From the above calculation we find that all the species listed in choice (2) have 14 electrons each so it is the correct answer.

194. (3) (1) $n = 3, l = 0$ means $3s$-orbital and $n + l = 3$
(2) $n = 3, l = 1$ means $3p$-orbital $n + l = 4$
(3) $n = 3, l = 2$ means $3d$-orbital $n + l = 5$
(4) $n = 4, l = 0$ means $4s$-orbital $n + l = 4$
Increasing order of energy among these orbitals is $3s < 3p < 4s < 3d$
$\therefore$ $3d$ has highest energy.

195. (3) (1) $N^{3-} = 7 + 3 = 10e^-$, $O^{2-} \longrightarrow 8 + 2 = 10e^-$
$F^- = 9 + 1 = 10e^-$, $S^{2-} \longrightarrow 16 + 2 = 18e^-$
(not isoelectronic)
(2) $Li^+ = 3 - 1 = 2e^-$, $Na^+ = 11 - 1 = 10e^-$,
$Mg^{2+} = 12 - 2 = 10e^-$
$Ca^{2+} = 20 - 2 = 18e^-$ (not isoelectronic)
(3) $K^+ = 19 - 1 = 18e^-$, $Cl^- = 17 + 1 = 18e^-$,
$Ca^{2+} = 20 - 2 = 18e^-$, $Sc^{3+} = 21 - 3 = 18e^-$
(isoelectronic)
(4) $Ba^{2+} = 56 - 2 = 54e^-$, $Sr^{2+} = 38 - 2 = 36e^-$
$K^+ = 19 - 1 = 18e^-$, $Ca^{2+} = 20 - 2 = 18e^-$
(not isoelectronic)

196. (2)

197. (2) (A) 4p (B) 4s (C) 3d (D) 3p
According to $(n + l)$ rule, increasing order of energy $(D) < (B) < (C) < (A)$.

198. (1) The energy of an orbital is given by $(n + l)$ rule. $(n + l)$ value for option (E) and (D) is (3 + 2) = 5 hence they will have same energy, since their n values are also same.

199. (1) The possible quantum numbers for 4f electron are

$n = 4, l = 3, m = -3, -2 -1, 0, 1, 2, 3$ and $s = \pm\frac{1}{2}$
Of various possiblities, only option (a) is possible.

200. (3) $_{19}K^+$, $_{20}Ca^{2+}$, $_{21}Sc^{3+}$, $_{17}Cl^-$ each contains 18 electrons.

201. (2) Electronic configuration of Cr atom (Z =24)
$= 1s^2 2s^2 2p^6 3s^2 3p^6 3d^5 4s^1$
when $l = 1$, p-subshell,
Numbers of electrons = 12
when $l = 2$, d-subshell,
Numbers of electrons = 5

202. (3) $Fe^{2+}(26 - 2 = 24) = 1s^2 2s^2 2p^6 3s^2 3p^6 4s^0 3d^6$
Hence, no. of d electrons retained is 6.
[Two 4s electron are removed.]

203. (1) For s-electron, $l = 0$
∴ Orbital angular momentum

$$= \sqrt{0(0+1)}\frac{h}{2\pi} = 0$$

204. (1) N^{3-}, F^- and Na^+ contain 10 electrons each.

205. (6) $n = 4; |m_l| = 1$ and $m_s = -\frac{1}{2}$

4s	4p	4d	4f
0	–1 0 +1	–2 –1 0 +1 +2	–3 –2 –1 0 +1 +2 +3

Total orbitals with $m_l = +1$ or $-1 = 6$ orbitals

One orbital have one $-\frac{1}{2}e^-$ spin

Total e^- with $n = 4; |m_l| = 1; m_s = -\frac{1}{2} = 6$

206. (4) For $n = 4, \ell = 3, 2, 1, 0$

	4s	4p	4d
$m_\ell \Rightarrow$	0	–1 0 +1	–2 –1 0 +1 +2
	ℓ	ℓ	ℓ

4f
–3 –2 –1 0 +1 +2 +3
ℓ

So answer is 4.

207. (16) n = 4 can have, 4s4p4d4f
Total electrons261014

Total electrons with $s = +\frac{1}{2}$ 1357

Thus, total electrons with $n = 4$ and $s = +\frac{1}{2} = 16$

208. (0) Orbital angular momentum =

$$\sqrt{l(l+1)}\frac{h}{2\pi}$$

Value of l for s orbital = 0
⇒orbital angular momentum = 0.

209. (3) A, D and E statements are correct.
For 2s orbital, the probability density first decreases and then increases.
At any distance from nucleus the probability density of finding electron is never zero and it always have some finite value.
Number of angular nodes = ℓ
Probability density of p-orbital is zero at the nucleus.

210. (3) Radial node = $n - \ell - 1$

Orbital	$n - \ell - 1$
7s	7 – 0 – 1 = 6
7p	7 – 1 – 1 = 5
6s	6 – 0 – 1 = 5
8p	8 – 1 – 1 = 6
8d	8 – 2 – 1 = 5

So, Answer is 3.

211. (5) p_x, p_y, p_z, d_{z^2}, and $d_{x^2-y^2}$ are axial orbitals.

212. (2)

213. (0) $Zn^+ \rightarrow 1s^2 2s^2 2p^6 3s^2 3p^6 3d^{10} 4s^1$.
Outermost electron is in 4s subshell, so $m = 0$

214. (0) $Ga^+ : 1s^2 2s^2 2p^6 3s^2 3p^6 3d^{10} 4s^2$
The azimuthal quantum number for the valence electrons (4s-subshell) of Ga^+ ion is zero(0).

215. (0) Given : $n = 4$ and $m_l = -3$
Hence, l value must be 3.
∴Number of radial nodes
$= n - l - 1 = 4 - 3 - 1 = 0$

216. (3) Possible values of $l = 0, 1, 2, 3, 4$
If $l = 0, m = 0$
$l = 1, m = \{-1, 0, +1\}$
$l = 2, m = \{-2, -1, 0, +1, +2\}$
$l = 3, m = \{-3, -2, -1, 0, +1, +2, +3\}$
$l = 4, m = \{-4, -3, -2, -1, 0, +1, +2, +3, +4\}$
5d, 5f and 5g-sub shell contains one-one orbital having ($m = +2$).

Part-B (JEE Advanced)

1. (b) Average atomic mass of Fe

$$= \frac{(54\times 5)+(56\times 90)+(57\times 5)}{100} = 55.95$$

2. (d) Rutherford's experiment was actually α -particle scattering experiment. α -Particle is doubly positively charged helium ion, i.e., He - nucleus.

3. (c) **X-rays** can ionise gases and cannot get deflected by electric and magnetic fields, wavelength of these rays is 150 to 0.1Å. Thus, the wavelength of X-rays is shorter than that of U.V. rays.

4. (c) Difference in the energy of the energy levels involved in the transition.

5. (a) Isotones have same number of neutrons. All atoms in triad (a) have same number of neutrons (i.e., $A - Z = 8$).

6. (d) $E = \frac{hc}{\lambda}$; $\lambda_1 = 2000$ Å; $\lambda_2 = 4000$ Å;

so $\frac{E_1}{E_2} = \frac{\lambda_2}{\lambda_1} = \frac{4000}{2000} = 2$

7. (b) Electrons in an atom occupy the extra nuclear region.

8. (b) The following is the increasing order of wavelength or decreasing order of energy of electromagnetic radiations :

cosmic rays
γ-rays
X-rays
UV-rays
Visible
Infra-red radiation
Micro waves
Radio waves

(↓ Increasing λ, Decreasing ν)

Among given choices, radiowaves have maximum wavelength.

9. (b) The radius of nucleus is of the order of 1.5×10^{-13} to 6.5×10^{-13} cm or 1.5 to 6.5 Fermi (1 Fermi = 10^{-13} cm)

10. (d)

$\frac{e}{m}$ for neutron $= \frac{0}{1} = 0$; α-particle $= \frac{2}{4} = 0.5$;

proton $= \frac{1}{1} = 1$; electron $= \frac{1}{1/1837} = 1837$

11. (a) According to Rutherford's experiment. "The central part consisting of whole of the positive charge and most of the mass, called nucleus, is extremely small in size compared to the size of the atom."

12. (c) Rutherford's scattering experiment led to the discovery of nucleus.

13. (d) No. of neutrons
= Mass number – Atomic number = 70 – 30 = 40.

14. (22.8) For maximum energy, $n_1 = 1$ and $n_2 = \infty$

$$\frac{1}{\lambda} = R_H Z^2\left(\frac{1}{n_1^2} - \frac{1}{n_2^2}\right)$$

Since R_H is a constant and transition remains the same

$$\frac{1}{\lambda} \propto Z^2 ; \frac{\lambda_{He}}{\lambda_H} = \frac{Z_H^2}{Z_{He}^2} = \frac{1}{4}$$

Hence, $\lambda_{He} = \frac{1}{4}\times 91.2 = \mathbf{22.8\ nm}$

15. photons

16. isobars

17. neutrons

18. 1.66 × 10^{-27} kg
Mass of hydrogen atom

$$= \frac{\text{Atomic mass of hydrogen}}{\text{Avogadro number}} = \frac{1.008}{6.02\times 10^{23}}$$

$= 0.166 \times 10^{-23}$ g $= 1.66 \times 10^{-27}$ kg

19. True : β-particles are deflected more than α-particles because they have very-very large e/m value as compared to α-particles due to the fact that electrons are much lighter than He^{2+} species.

20. False : Gamma rays are electromagnetic radiations of wavelengths 10^{-9} cm to 10^{-10} cm.

21 **(b, d)** In tritium (the isotope of hydrogen) nucleus there is one proton and 2 neutrons. $\therefore n + p = 3$. In deuterium nucleus there is one proton and one neutron $\therefore n + p = 2$.

22. **(a, c)** α-particles pass through because most part of the atom is empty.

23. **(a, c)** Because they have isotopes with different masses. The average atomic mass is the weighted mean of their presence in nature; e.g., Cl^{35} and Cl^{37} are present in ratio 3 : 1 in nature.

$$\text{So } A = \frac{35 \times 3 + 37 \times 1}{4} = 35.5$$

24. **(b, d)** ${}^{77}_{33}As$ and ${}^{78}_{34}Se$ have same number of neutrons ($= A - Z$) as ${}^{76}_{32}Ge$.

25. Let the % of isotope with At. wt. 10.01 $= x$

$\therefore$ % of isotope with At. wt. 11.01 $= (100 - x)$

$$\text{At. wt. of boron} = \frac{x \times 10.01 + (100 - x) \times 11.01}{100}$$

$$\Rightarrow 10.81 = \frac{x \times 10.01 + (100 - x) \times 11.01}{100} \quad \therefore x = 20$$

Hence, % of isotope with At. wt. 10.01 = 20%

$\therefore$ % of isotope with At. wt. 11.01 = 100 – 20 = **80%**.

26. **(b)** K.E. of electron in n^{th} Bohr's orbit is given by :

$$\text{K.E.} = 13.6 \frac{Z^2}{n^2} \text{ eV/atom}$$

$$n = 1 \text{ (H-atom)} \rightarrow \text{K.E.} \propto \frac{1^2}{1^2} = 1$$

$$n = 1 \text{ (He}^+ \text{ ion)} \rightarrow \text{K.E.} \propto \frac{2^2}{1^2} = 4$$

$$n = 2 \text{ (He}^+ \text{ ion)} \rightarrow \text{K.E.} \propto \frac{2^2}{2^2} = 1$$

$$n = 2 \text{ (Li}^{2+} \text{ ion)} \rightarrow \text{K.E.} \propto \frac{3^2}{2^2} = \frac{9}{4}$$

Thus, K.E. is highest for first orbit of He^+.

27. **(c)** As per Bohr's postulate,

$$mvr = \frac{nh}{2\pi} \quad \text{So, } v = \frac{nh}{2\pi mr}$$

$$KE = \frac{1}{2} mv^2 \quad \text{So, KE} = \frac{1}{2} m \left(\frac{nh}{2\pi mr} \right)^2$$

$$\text{Since, } r = \frac{a_0 \times n^2}{z}$$

So, for 2nd Bohr orbit

$$r = \frac{a_0 \times 2^2}{1} = 4a_0$$

$$KE = \frac{1}{2} m \left(\frac{2^2 h^2}{4\pi^2 m^2 \times (4a_0)^2} \right) = \frac{h^2}{32\pi^2 m a_0^2}$$

28. **(d)** $r_n = 0.529 \dfrac{n^2}{Z}$ Å

For hydrogen, $n = 1$ and $Z = 1$; $\therefore r_H = 0.529$

For Be^{3+}, $n = 2$ and $Z = 4$;

$$\therefore r_{Be^{3+}} = \frac{0.529 \times 2^2}{4} = 0.529$$

29. **(b)** Bohr model can explain spectrum of atoms/ions containing one electron only.

30. **(d)** Energy is emitted when electron falls from higher energy level to lower energy level and energy is absorbed when electron moves from lower level to higher level.

1*s* is the lowest energy level of electron in an atom.

$\therefore$ An electron in 1*s* level of hydrogen can absorb energy but cannot emit energy.

31. **(30)** For single electron system,

$$r_n = 52.9 \times \frac{n^2}{Z} \text{ pm}$$

$$105.8 = \frac{52.9 \times n_1^2}{2} \therefore n_1^2 = 4 \Rightarrow n_1 = 2$$

$$26.45 = \frac{52.9 \times n_2^2}{2} \therefore n_2 = 1$$

So, transition is from 2 to 1.

$$\text{Now } \frac{hc}{\lambda} = R_H Z^2 \left(\frac{1}{n_1^2} - \frac{1}{n_2^2} \right)$$

$\lambda = 30 \times 10^{-9}$ m = 30 nm

32. **(4)** Energy associated with incident photon $= \dfrac{hc}{\lambda}$

$$E = \frac{6.6 \times 10^{-34} \times 3 \times 10^8}{300 \times 10^{-9}} \text{ J}$$

$$= \frac{6.6 \times 10^{-34} \times 3 \times 10^8}{300 \times 10^{-9} \times 1.6 \times 10^{-19}} \text{ eV} = 4.16 \text{ eV}$$

Photoelectric effect can take place only when

$E_{photon} > \phi$

Thus, number of metals showing photoelectric effect will be 4 (*i.e.* Li, Na, K and Mg).

33. **(3.00)** $\dfrac{1}{\lambda} = R_H Z^2 \left[\dfrac{1}{n_1^2} - \dfrac{1}{n_2^2} \right]$

$$\text{For } X^{a+} = \frac{1}{\lambda} = R_H (Z_x)^2 \left\{ \frac{1}{1^2} - \frac{1}{2^2} \right\} \quad ...(1)$$

$$\text{For } Y^{b+} = R_H (Z_y)^2 \left\{ \frac{1}{2^2} - \frac{1}{4^2} \right\} \quad ..(2)$$

Equation (1) ÷ Equation (2)

$$9 = \frac{Z_x^2}{Z_y^2} \times \frac{3/4}{3/16}, \quad \frac{Z_x^2}{Z_y^2} = \frac{9}{4}, \quad \frac{Z_x}{Z_y} = \frac{3}{2}$$

For minimum values $= Z_x = 3, Z_y = 2$

X^{a+} and Y^{b+} are hydrogen like species.

So 'a' should be 2 and 'b' should be 1.

Lowest possible value of (a + b) = 2 + 1 = 3.

34. **(27419)** The shortest wavelength transition in the Balmer series corresponds to the transition $n = 2 \rightarrow n = \infty$. Hence, $n_1 = 2, n_2 = \infty$

$$\bar{v} = R_H \left(\frac{1}{n_1^2} - \frac{1}{n_2^2} \right) = (109677 \text{cm}^{-1}) \left(\frac{1}{2^2} - \frac{1}{\infty^2} \right)$$

= **27419.25 cm^{-1}**

35. **(2055)** E_n of H $= \dfrac{-21.76 \times 10^{-19}}{n^2}$ J

$$\therefore E_n \text{ of He}^+ = \frac{-21.76 \times 10^{-19}}{n^2} \times Z^2 \text{ J}$$

$$\therefore E_3 \text{ of He}^+ = \frac{-21.76 \times 10^{-19} \times 4}{9} \text{ J}$$

Hence, energy equivalent to E_3 must be supplied to remove the electron from 3rd orbit of He^+. Wavelength corresponding to this energy can be determined by applying the relation.

$$E = \frac{hc}{\lambda} \quad \text{or } \lambda = \frac{hc}{E}$$

$$= \frac{6.625 \times 10^{-34} \times 3 \times 10^8 \times 9}{21.76 \times 10^{-19} \times 4}$$

$= 2055 \times 10^{-10}$ m = **2055 Å**

36. **(1220)** (i) Energy of nth orbit $= E_n - \dfrac{E_1}{n^2}$

(ii) Difference in energy $= E_1 - E_2 = h\nu = \dfrac{hc}{\lambda}$

$$\text{or} \quad \lambda = \frac{hc}{E_1 - E_2}$$

Given $E_1 = 2.17 \times 10^{-11}$

$$\therefore \text{Energy of second orbit} = E_2 = \frac{2.17 \times 10^{-11}}{2^2}$$

$= 0.5425 \times 10^{-11}$ erg

$\Delta E = E_1 - E_2 = 2.17 \times 10^{-11} - 0.5425 \times 10^{-11}$

$= 1.6275 \times 10^{-11}$ erg

$$\lambda = \frac{6.62 \times 10^{-27} \times 3 \times 10^{10}}{1.6275 \times 10^{-11}} = 12.20 \times 10^{-6} \text{ cm}$$

= **1220 Å**

37. **(660)**

$$\Delta E = E_3 - E_2 = h\nu = \frac{hc}{\lambda} \text{ or } \lambda = \frac{hc}{E_3 - E_2}$$

Given $E_2 = -5.42 \times 10^{-12}$ erg,

$E_3 = -2.41 \times 10^{-12}$ erg

$$\therefore \quad \lambda = \frac{6.626 \times 10^{-27} \times 3 \times 10^{10}}{-2.41 \times 10^{-12} - (-5.42 \times 10^{-12})}$$

($h = 6.2 \times 10^{-27}$ erg)

$$= \frac{19.878 \times 10^{-17}}{3.01 \times 10^{-12}} = 6.604 \times 10^{-5} \text{ cm} = \textbf{660 nm}$$

38. (a, b, d)

(a) The energy of subshells in multi electron system is compared on the basis of $(n + l)$

$(n+l)$ value for E_{2s} (Li) $= 2+0=2$
$(n+l)$ value for E_{2p} (Li) $= 2+1=3$
E_{2s} (Li) $<$ E_{2p} (Li)
(b) For hydrogen like atom energy of subshell depends on the value of n.
E_{2s} (H) $=$ E_{2p} (H)
(d) For Li, Z = 3 more effective nuclear charge pulling its 2s electron more closer and they have more negative lower energy.

39. (a, d) The energy of an electron on Bohr orbits of hydrogen atoms is given by the expression

$$E_n = -\frac{\text{Constant}}{n^2}$$

Where n takes only integral values. For the first Bohr orbit, $n = 1$ and it is given that $E_1 = -13.6$ eV

Hence $E_n = -\frac{13.6\text{eV}}{n^2}$ of the given values of energy, only -3.4 eV and -1.5 eV can be obtained by substituting $n = 2$ and 3 respectively in the above expression.

40. (d) $r \propto \frac{n^2}{Z}$ or $r = 0.529 \times \frac{n^2}{Z}$; (I), (T)

$|L| \propto n$ or $mvr = \frac{nh}{2\pi}$; (II), (S)

41. Determination of number of moles of hydrogen gas,

$$n = \frac{PV}{RT} = \frac{1\times 1}{0.082\times 298} = 0.0409$$

The concerned reaction is $H_2 \longrightarrow 2H$; $\Delta H = 436$ kJ mol^{-1}
Energy required to bring 0.0409 moles of hydrogen gas to atomic state $= 436 \times 0.0409 =$ **17.83 kJ**
Calculation of total number of hydrogen atoms in 0.0409 mole of H_2 gas.
1 mole of H_2 gas has 6.02×10^{23} molecules

0.0409 mole of H_2 gas $= \frac{6.02\times 10^{23}}{1} \times 0.0409$ molecules

Since, 1 molecule of H_2 gas has 2 hydrogen atoms
$6.02 \times 10^{23} \times 0.0409$ molecules of H_2 gas
$= 2 \times 6.02 \times 10^{23} \times 0.0409 = 4.92 \times 10^{22}$ atoms of hydrogen
Since, energy required to excite an electron from the ground state to the next excited state is given by

$$E = 13.6\left(\frac{1}{n_1^2} - \frac{1}{n_2^2}\right)\text{eV}$$

$$= 13.6\times\left(\frac{1}{1} - \frac{1}{4}\right) = 13.6\times\frac{3}{4} = 10.2\text{eV}$$

$= 1.632 \times 10^{-21}$ kJ
Therefore, energy required to excite 4.92×10^{22} electrons
$= 1.632 \times 10^{-21} \times 4.92 \times 10^{22}$ kJ $= 8.03 \times 10 = 80.3$ kJ
Therefore, total energy required
$= 17.83 + 80.3 =$ **98.17 kJ**

42. Work done while bringing an electron infinitely slowly from infinity to proton of radius a_0 is given as follows

$$W = -\frac{e^2}{4\pi\varepsilon_0 . a_0}$$

This work done is equal to the total energy of an electron in its ground state in the hydrogen atom. At this stage, the electron is not moving and do not possess any K.E., so this total energy is equal to the potential energy.

$$\text{T.E.} = \text{P. E} + \text{K. E.} = \text{P. E.} = -\frac{e^2}{4\pi\varepsilon_0 . a_0} \quad ...(1)$$

In order the electron to be captured by proton to form a ground state hydrogen atom it should also attain $\text{K.E.} = \frac{e^2}{8\pi\varepsilon_0 a_0}$

(It is given that magnitude of K.E. is half the magnitude of P.E. Note that P.E. is –ve and K.E is +ve)

$$\therefore \text{ T.E} = \text{P. E.} + \text{K. E.} = -\frac{e^2}{4\pi\varepsilon_0 a_0} + \frac{e^2}{8\pi\varepsilon_0 a_0}$$

or $\quad \text{T.E.} = -\frac{e^2}{8\pi\varepsilon_0 a_0}$

$\text{P.E.} = 2\times\text{T.E.} = 2\times\frac{-e^2}{8\pi\epsilon_0 a_0}$ or

$$\text{P.E.} = \frac{-e^2}{4\pi\epsilon_0 a_0}$$

43. Bond energy of $I_2 = 240$ kJ mol^{-1}
$= 240 \times 10^3$ J mol^{-1}

$$= \frac{240\times 10^3}{6.023\times 10^{23}} \text{ J molecule}^{-1}$$

$= 3.984 \times 10^{-19}$ J molecule^{-1}

Energy absorbed $= \frac{hc}{\lambda}$

$$= \frac{6.626\times 10^{-34}\text{ Js}\times 3\times 10^8\text{ ms}^{-1}}{4500\times 10^{-10}\text{ m}}$$

$= 4.417 \times 10^{-19}$ J
Kinetic energy = Absorbed energy – Bond energy
$\therefore$ Kinetic energy $= 4.417 \times 10^{-19} - 3.984 \times 10^{-19}$ J
$= 4.33 \times 10^{-20}$ J
$\therefore$ Kinetic energy of each atom of iodine

$$= \frac{4.33\times 10^{-20}}{2} = \mathbf{2.165\times 10^{-20}\ J}$$

44. Number of waves $= \frac{n(n-1)}{2}$

where n = Principal quantum number or number of orbit

$$\text{Number of waves} = \frac{3(3-1)}{2} = \frac{3\times 2}{2} = 3$$

45. For He^+ ion, we have

$$\frac{1}{\lambda} = Z^2 R_H\left[\frac{1}{n_1^2} - \frac{1}{n_2^2}\right]$$

$$= (2)^2 R_H\left[\frac{1}{(2)^2} - \frac{1}{(4)^2}\right] = R_H\frac{3}{4} \quad ...(i)$$

Now for hydrogen atom $\frac{1}{\lambda} = R_H\left[\frac{1}{n_1^2} - \frac{1}{n_2^2}\right] \quad ...(ii)$

Equating equations (i) and (ii), we get

$$\frac{1}{n_1^2} - \frac{1}{n_2^2} = \frac{3}{4}$$

Obviously, $n_1 = 1$ and $n_2 = 2$
Hence, the transition $\boldsymbol{n = 2}$ to $\boldsymbol{n = 1}$ in hydrogen atom will have the same wavelength as the transition, $n = 4$ to $n = 2$ in He^+ species.

46. $\Delta E = RhcZ^2\left(\frac{1}{n_1^2} - \frac{1}{n_2^2}\right)$

Here, $R = 1.0967 \times 10^7$ m^{-1}
$h = 6.626 \times 10^{-34}$ J sec, $c = 3 \times 10^8$ m/sec
$n_1 = 1, n_2 = 2$ and for H-atom, $Z = 1$
$E_2 - E_1 = 1.0967 \times 10^7 \times$

$$6.626\times 10^{-34}\times 3\times 10^8\left(\frac{1}{1} - \frac{1}{4}\right)$$

$$\Delta E = 1.0967\times 6.626\times 3\times\frac{3}{4}\times 10^{-19}\text{ J}$$

$= 16.3512 \times 10^{-19}$ J

$$= \frac{16.3512\times 10^{-19}}{1.6\times 10^{-19}}\text{ eV} = \mathbf{10.22\ eV}$$

$$\Delta E = \frac{hc}{\lambda} = RhcZ^2\left(\frac{1}{n_1^2} - \frac{1}{n_2^2}\right)$$

$$\frac{1}{\lambda} = RZ^2\left(\frac{1}{1} - \frac{1}{4}\right) = RZ^2\times\frac{3}{4}$$

Given, $\lambda = 3 \times 10^{-8}$ m

$$\therefore \frac{1}{3\times 10^{-8}} = 1.0967\times Z^2\times\frac{3}{4}\times 10^7$$

$$\therefore Z^2 = \frac{10^8\times 4}{3\times 3\times 1.0967\times 10^7} = \frac{40}{9\times 1.0967} \approx 4$$

$\therefore Z = \mathbf{2}$
So, it corresponds to He^+ which has 1 electron like hydrogen.

47. To calculate the energy required to remove electron from atom, $n = \infty$ is to be taken.
Energy of an electron in the nth orbit of hydrogen is given by

$$E = -21.7\times 10^{-12}\times\frac{1}{n^2}\text{ ergs}$$

$\therefore\ \Delta E = -21.7 \times 10^{-12}\left(\frac{1}{2^2} - \frac{1}{\infty^2}\right)$

$= -21.7 \times 10^{-12}\left(\frac{1}{4} - 0\right) = -21.7 \times 10^{-12} \times \frac{1}{4}$

$= \mathbf{-5.42 \times 10^{-12}}$ **ergs**

Now we know that $\Delta E = h\nu$

$\therefore\quad \Delta E = \frac{hc}{\lambda}\ \left(\because \nu = \frac{c}{\lambda}\right)$ or $\lambda = \frac{hc}{\Delta E}$

Substituting the values,

$\lambda = \frac{6.627 \times 10^{-27} \times 3 \times 10^{10}}{5.42 \times 10^{-12}} = \mathbf{3.67 \times 10^{-5}}$ **cm**

48. (a) According to de-Broglie's equation

$\lambda = \frac{h}{p} = \frac{h}{mv}$

Given, $h = 6.6 \times 10^{-34}$ Js, $m = 200 \times 10^{-3}$ kg

$v = \frac{5}{60 \times 60}$ m/s $\lambda = \frac{6.6 \times 10^{-34}}{200 \times 10^{-3} \times 5/(60 \times 60)}$

$= 2.38 \times 10^{-10}$ m

49. (d) As packet of energy equal to $h\nu$; as wave having frequency ν.

50. (30) $p = \frac{h}{\lambda} \Rightarrow \frac{6.6 \times 10^{-34}\,\text{kgm}^2/\text{s}^2}{330 \times 10^{-9}\,\text{m}}$

$= \frac{4 \times 10^{-3}\,\text{kg mol}^{-1}}{6 \times 10^{23}\,\text{mol}^{-1}} \times v\ (p = m \times v)$

$v = 0.3$ m/s $= 30$ cm/s

51. (5) Since,

$\lambda = \frac{h}{mV} = \frac{h}{\sqrt{2M\,K.E}}$ (since K.E. $\propto T$)

$\Rightarrow\ \lambda \propto \frac{1}{\sqrt{MT}}$

For two gases,

$\frac{\lambda_{He}}{\lambda_{Ne}} = \sqrt{\frac{M_{Ne}T_{Ne}}{M_{He}T_{He}}} = \sqrt{\frac{20}{4} \times \frac{1000}{200}} = 5$

52. orbitals

53. Heisenberg, de-Broglie

54. (a, b, c)

(a) Uncertainty principle rules out existence of definite paths or trajectories of electron and other similar particles.

So, option (a) is correct.

(b) Shell or orbit more near to nucleus has less energy than far away.

So, option (b) is also correct.

(c) $E = -13.6\frac{Z^2}{n^2}$ eV/atom

So, $n = 1$ has most negative energy.

So, option (c) is also correct.

(d) $V_e = V_0 \times \frac{Z}{n}$

When n increase velocity decreases.

So, option (d) is incorrect.

55. For hydrogen atom, $Z = 1$, $n = 1$

$v = 2.18 \times 10^6 \times \frac{Z}{n}\text{ms}^{-1} = 2.18 \times 10^6\,\text{ms}^{-1}$

de Broglie wavelength,

$\lambda = \frac{h}{mv} = \frac{6.626 \times 10^{-34}}{9.1 \times 10^{-31} \times 2.18 \times 10^6}$

$= \mathbf{3.34 \times 10^{-10}}$ **m** $= 3.3$ Å

For $2p$, $l = 1$

$\therefore\quad$ Orbital angular momentum $= \sqrt{l(l+1)}\frac{h}{2\pi} =$

$\sqrt{2}\,\frac{\boldsymbol{h}}{\boldsymbol{2\pi}}$

56. $\lambda = \frac{h}{mu} = \frac{6.627 \times 10^{-34}}{0.1 \times 100}$

or $\lambda = \mathbf{6.627 \times 10^{-35}}$ **m** $=$ $\mathbf{6.627 \times 10^{-25}}$ **Å**

57. ψ^2_{2s} = probability of finding electron within $2s$ sphere

$\psi^2_{2s} = 0$ (at node)

($\because$ probability of finding an electron is zero at node)

$\therefore\ 0 = \frac{1}{32\pi}\left(\frac{1}{a_0}\right)^3\left(2 - \frac{r_0}{a_0}\right)^2 . e^{-\frac{2r_0}{a_0}}$

(Squaring the given value of ψ_{2s})

or $\left[2 - \frac{r_0}{a_0}\right] = 0;\ \therefore\ 2 = \frac{r_0}{a_0};\ \mathbf{2a_0} = r_0$

58. (a) Number of radial nodes $= (n - l - 1)$

For $3s$: $n = 3, l = 0$ (Number of radial node $= 2$)

For $2p$: $n = 2, l = 1$ (Number of radial node $= 0$)

59. (c) Only two e^- can exist in the same orbital.

60. (d) The quantum numbers $+1/2$ and $-1/2$ for the electron spin can represent any one among clockwise or anticlockwise spin direction. But if one value represents clockwise spin then the other value will represent anticlockwise spin.

61. (b) $3d^5 4s^1$ system is more stable than $3d^4 4s^2$, hence former is the ground state configuration.

62. (a) p_x orbital being dumbell shaped, have number of nodal planes = 1, in yz plane. The electron density is only along x-axis (xy and zx planes), thus, in yz-plane, there will be zero electron density.

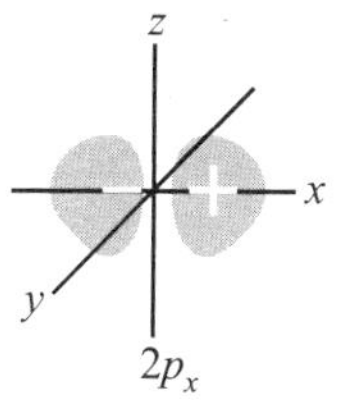

$2p_x$

63. (a) The two guiding rules to arrange the various orbitals in the increasing energy are:

(i) Energy of an orbital increases with increase in the value of $n + l$.

(ii) Of orbitals having the same value of $n + l$, the orbital with lower value of n has lower energy.

Thus, for the given orbitals, we have

(i) $n + l = 4 + 1 = 5$ (ii) $n + l = 4 + 0 = 4$

(iii) $n + l = 3 + 2 = 5$ (iv) $n + l = 3 + 1 = 4$

Hence, the order of increasing energy is

(iv) < (ii) < (iii) < (i)

64. (a) The expression for orbital angular momentum is Angular momentum

$= \sqrt{l(l+1)}\left(\frac{h}{2\pi}\right)$

For d orbital, $l = 2$.

Hence, $L = \sqrt{2(2+1)}\left(\frac{h}{2\pi}\right) = \sqrt{6}\left(\frac{h}{2\pi}\right)$

65. (b) Orbital angular momentum $(mvr) = \frac{h}{2\pi}\sqrt{l(l+1)}$

For $2s$ orbital, l (azimuthal quantum number) $= 0$

$\therefore$ Orbital angular momentum $= 0$.

66. (c) Total nodes $= n - l$

No. of radial nodes $= n - l - 1$

No. of angular nodes $= l$

For $3p$ sub-shell, $n = 3$, $l = 1$

$\therefore$ No. of radial nodes $= n - l - 1 = 3 - 1 - 1 = 1$

$\therefore$ No. of angular nodes $= l = 1$

67. (c) Electronic configuration of chlorine is [Ne] $3s^2, 3p^5$

$\therefore$ Unpaired electron is found in $3p$ sub-shell.

$\therefore\ n = 3, l = 1, m - 1$

68. (a) Exactly half filled orbitals are more stable than nearly half filled orbitals.

Cr (At. no. 24) has configuration [Ar] $3d^5, 4s^1$.

69. (c) Configuration ns^2, np^5 means it requires only one electron to attain nearest noble gas configuration. So, it will be most electronegative element among given choices.

70. (b) According to Aufbau principle, the orbital of lower energy ($2s$) should be fully filled before the filling of orbital of higher energy starts.

71. (c) If $l = 2$, $m \neq -3$, m will vary from -2 to $+2$. *i.e.* possible values of m are $-2, -1, 0, +1$ and $+2$.

72. (a) Rb has the configuration : $1s^2\ 2s^2p^6\ 3s^2p^6d^{10}\ 4s^2p^6\ 5s^1$; so $n = 5, l = 0, m = 0$ and $s = +½$ is correct set of quantum numbers for valence shell electron of Rb.

73. (d) One p-orbital can accommodate up to two electrons with opposite spin while p-subshell can accommodate upto six electrons.

74. (a) The principal quantum number (n) is related to the size of the orbital ($n = 1, 2, 3$.....)

75. (3) In one electron system, all orbitals of a shell are degenerate.

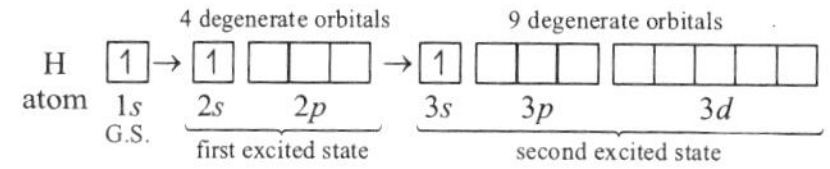

H^- ion: [↿⇂] (1s, G.S.) → [↿] [↿] [ ][ ][ ] (1s 2s 2p, first excited state) → [↿] [ ] [↿][ ][ ] (1s 2s 2p, second excited state)

In case of many electron system, different orbitals of a shell are non-degenerate. Hence, in the second excited state, only three *p*-orbitals (2*p*) are degenerate.

76. (6) $|m_l| = 1$ means m_l can be +1 and –1.

For $n = 4$, the total number of possible orbitals are :

4s	4p	4d	4f
0	–1, 0, +1	–2, –1, 0, +1, +2	–3, –2, –1, 0, +1, +2, +3

Thus, total number of orbitals having $|m_l| = 1$ is 6.

The number of electrons with $s = -1/2$ is 6.

77. (9)

Maximum number of orbitals when $n = 3$ is $n^2 = 3^2 = 9$

$\therefore$ Number of electrons with $m_s = -\frac{1}{2}$ will be 9.

78. (10) For $n = 3$ and $l = 2$ (*i.e.*, 3*d* orbital), the values of *m* varies from –2 to +2, *i.e.* –2, –1, 0, +1, +2 and for each '*m*' there are 2 values of '*s*', *i.e.* +½ and –½.

$\therefore$ Maximum no. of electrons in all the five *d*-orbitals is **10**.

79. $4s^1, 3d^5$;

The electronic configuration of Cr is : $1s^2, 2s^2, 2p^6, 3s^2, 3p^6, 4s^1, 3d^5$.

$\therefore$ Outermost electronic configuration is $3d^5, 4s^1$.

80. orientation in space

81. antiparallel; or opposite

82. False : The orbital $3d_{x^2-y^2}$ lie along *X* and *Y* axis where electron density is maximum.

83. True : In case of hydrogen (single electron system), energy of electron depends only on the principal quantum number. Thus, 4*s* is in higher energy level than 3*d*.

84. False : The outer electronic configuration of the ground state chromium atom is $3d^5 4s^1$, as half filled orbitals are more stable than nearly half filled orbitals.

85. (a, c) Given, azimuthal quantum no. (*l*) = 2 (*d*-subshell)

Magnetic quantum no.(*m*) = 0 (zero), which is for d_{z^2} orbital.

$$E = -13.6\frac{z^2}{n^2} = -13.6 \times \frac{2^2}{n^2} = -3.4$$

$$13.6 \times \frac{2^2}{n^2} = 3.4$$

$n^2 = 4^2 \Rightarrow n = 4$

Radial node $= n - l - 1 = 4 - 2 - 1 = 1$

Angular node $= l = 2$

Wave function corresponds to $\psi_{4,2,0}$. It represents $4d_{z^2}$-orbital which has only one radial node and two angular nodes. It experiences nuclear charge of 2e units.

86. (a, d) According to Hund's rule pairing of electrons starts only when each of the orbital in a sub shell has one electron each of parallel spin.

$\therefore$ (a) and (d) are correct ground state electronic configurations of nitrogen atom in ground state.

87. (a, b, c)

(a) $_{24}Cr = 1s^2, 2s^2 2p^6, 3s^2 3p^6 3d^5, 4s^1 = [Ar] 3d^5, 4s^1$

(b) For magnetic quantum number (*m*), negative values are possible.

For *s* - subshell, *l* = 0, hence *m* = 0

for *p* - subshell, *l* = 1, hence *m* = –1, 0, +1

(c) $_{47}Ag = 1s^2, 2s^2\ 2p^6, 3s^2 3p^6 3d^{10}, 4s^2\ 4p^6\ 4d^{10}, 5s^1$

Hence, 23 electrons have a spin of one type and 24 of the opposite type.

(d) Oxidation state of N in HN_3 is –1/3.

88. (c) No. of radial nodes = *n*–*l*–1, For 2*s* orbital, *n*=2, *l*=0

$\therefore$ No. of radial nodes = 2 – 0 – 1 = 1

The plotted graph is correct for 2*s*-orbital, as wave function changes its sign at node.

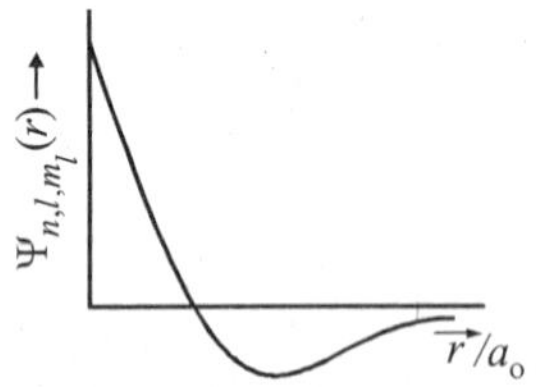

89. (a) E.C. of H : $1s^1$; for 1*s* orbital

$$\psi_{n,l,m} \propto \left(\frac{Z}{a_0}\right)^{3/2} e^{-(Zr/a_0)}$$

For *s*-orbital, θ and ϕ cannot be a part of wave function expression. Hence, this is correct.

For 1*s* orbital of hydrogen like species: $E \propto -\frac{1}{n^2}$

Then, $E_4 - E_2 = \left(\frac{1}{2}\right)^2 - \left(\frac{1}{4}\right)^2 = \frac{3}{16}$

$E_6 - E_2 = \left(\frac{1}{2}\right)^2 - \left(\frac{1}{6}\right)^2 = \frac{8}{36}$

$\therefore (E_4 - E_2) = \frac{27}{32} \times (E_6 - E_2)$

90. (c) In the wave function (ψ) expression for 1*s*-orbital of He^+, there should be no angular part (θ).

91. (A) - (q); (B) - (p, q, r, s); (C) - (p, q, r); (D) - (p, q, r)

(A) Orbital angular momentum $L = \sqrt{l(l+1)}\frac{h}{2\pi}$,

i.e., *L* depends on azimuthal quantum number only.

(B) To describe a hydrogen like one-electron wave function, three quantum numbers *n*, *l* and *m* are required. Further, to obey Pauli principle, fourth quantum number *s* is also required.

(C) To define size, shape and orientation of atomic orbitals, *n*, *l* and *m* are required respectively.

(D) Probability density (ψ^2) of an electron can be determined from the value of *n*, *l* and *m*.

92. (A) - (r); (B) - (q); (C) - (p); (D) - (s)

(A) $\frac{V_n}{K_n} = \frac{-Kze^2/r}{Kze^2/2r} = -2;$

where $K = \frac{1}{4\pi\varepsilon_0}$

(B) $r_n = \frac{n^2 h^2 \varepsilon_0}{\pi m Z e^2}, E_n = \frac{me^4}{8\varepsilon_0^2 h^2 n^2}$

$\Rightarrow E_n \propto \frac{1}{n^2} \propto \frac{1}{r_n} \Rightarrow r_n \propto \frac{1}{E_n}$

$\Rightarrow r_n \propto (E_n)^{-1} \Rightarrow x = -1$

(C) Angular momentum of electron in lowest (1*s*) orbital

$= \sqrt{l(l+1)}\frac{h}{2\pi} = \sqrt{0(0+1)}\frac{h}{2\pi} = 0;$

(D) $r_n = \frac{a_0 n^2}{Z} \Rightarrow \frac{1}{r_n} \propto Z^1 \Rightarrow y = 1$

For 93-95: The spherically symmetric state S_1 of Li^{2+} with one radial node is 2*s*. Upon absorbing light, the ion gets excited to state S_2, which also has one radial node. The energy of electron in S_2 is same as that of H-atom in its ground state.

$\therefore E_n = \frac{Z^2}{n^2} E_1$ where E_1 is the energy of H-atom in the ground state $= \frac{(3)^2 E_1}{n^2}$ for Li^{2+}

$E_n = E_1 \Rightarrow n = 3$

$\therefore$ State S_2 of Li^{2+} having one radial node is 3*p*.

Orbital angular momentum quantum number of 3*p* is 1.

Energy of state $S_1 = \frac{(3)^2}{(2)^2} E_1 = 2.25 E_1$

93. (b)

94. (c)

95. (b)

96. Ground state electronic configuration of Si

3*s*: [↑↓] $3p_x\ 3p_y\ 3p_z$: [↑][↑][ ]

In a subshell, single e^- occupied orbitals must have parallel spins.

Chapter-3

Classification of Elements and Periodicity in Properties

Part-A (JEE Main)

1. **(1)** $n = 45, e = 36, p = 35$
mass number $= n + p = 80$
X = Br; Period 4
Br_2 exist as liquid.

2. **(2)**
(A) Pnictogen ⇒ Mc (Moscovium); Atomic No. = 115
(B) Chalcogen ⇒ Lv (Livermorium); Atomic No. = 116
(C) Halogen ⇒ Ts (Tennessine); Atomic No. = 117
(D) Noble gas ⇒ Og (Oganesson); Atomic No. = 118

3. **(4)** According to Law of octaves the elements were arranged in the increasing order of their atomic weight. Lothar Meyer plotted the physical properties such as atomic volume, melting point and boiling point against their respective atomic weight.

4. **(1)** Properties of elements are periodic function of their atomic number. Elements having similar outer electronic configuration are arranged in same group. Number of elements in a period is not equal to number of atomic orbitals available in energy level that is being filled.
Hence, A, C & E are incorrect.

5. **(3)**

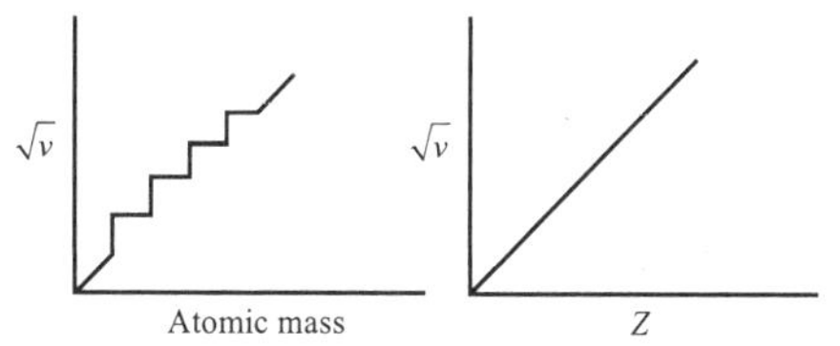

According to Henry Moseley's experiment, $\sqrt{v}$ vs z is a straight line.

6. **(1)** Palladium (Pd) ⇒ 5th period
Iridium (Ir), Osmium (Os), Platinum (Pt)
⇒ 6th Period

7. **(2)** (A) $_7$N: $[He]2s^2 2p^3$.
(B) $_{16}$S: $[Ne]2s^2 3p^4$
(C) $_{35}$Br: $[Ar]3d^{10}4s^2 4p^5$
(D) $_{36}$Kr: $[Ar]3d^{10}4s^2 4p^6$

8. **(2)** $\sqrt{v} \propto Z$

9. **(3)** According to Henry Moseley
$\sqrt{v} \propto Z$ So, $n = \frac{1}{2}$

10. **(4)** A - IV, B - II, C - I, D - III

11. **(2)**
$33 \Rightarrow [Ar]3d^{10}4s^2 4p^3$ Metalloid
$53 \Rightarrow [Kr]4d^{10}5s^2 5p^5$ Non-metal
$83 \Rightarrow [Xe]4f^{14}5d^{10}6s^2 6p^3$ Metal

12. **(4)**

13. **(2)**

14. **(11)** Unununium refers to the element with symbol and atomic number 111. Its IUPAC name is roentgenium.
It belongs to group number 11, so n = 11.

15. **(1)**
(A) $2s^2 \Rightarrow$ Be (B) $2s^2 2p^1 \Rightarrow$ B
(C) $2s^2 2p^3 \Rightarrow$ N (D) $2s^2 2p^6 \Rightarrow$ Ne
IE order: B < Be < N < Ne
Ionisation energy decreases across the period due to increase in effective nuclear charge.
A-II, B-III, C-IV, D-I

16. **(3)** Ionisation enthalpy increases as we move left to right thus extreme left element (group I) have lowest ionisation enthalpy.
On moving left to right negative electron gain enthalpy increase thus extreme right element (group 17) have highest negative electron gain enthalpy (ΔH_{eg}).

17. **(1) Statement I :**
[Al_2O_3, Cr_2O_3] is Amphoteric that means it act as acidic or basic both. [CO, N_2O] is Neutral oxide [Na_2O, V_2O_3] is Basic. [Cl_2O_7, Mn_2O_7] is highly covalent and Acidic oxide
Statement II :
Most basic oxide : Na_2O - it is ionic oxide of alkali metal. Most acidic oxide : Cl_2O_7 - it is anhydride of strong acid $HClO_4$.

18. **(3)** Since successive ionization energy increases thus IE_2 (Mg) > IE_1 (Mg).

19. **(3) Basic oxides:** Typically metal oxides of group 1 and 2. Eg.: Na_2O, K_2O
Acidic oxides: Typically non-metal oxides with high oxidation states. Eg.: Cl_2O_7
Amphoteric oxides: Certain metal/metalloid oxides that react with both acids and bases. Eg.: Al_2O_3, As_2O_3
Neutral oxides: Non-metal oxides that react with neither. Eg.: N_2O, NO, CO.
Option (3) contains all types of oxides.

20. **(4)**
(i) Na < Mg < Cl < Ar → 1st ionization energy, However, the statement gives the reverse order.
(ii) Among B, Ca, Al, Fe. Ca has maximum 3rd ionization energy among given due to inert gas configuration.

21. **(3)** Least metallic character : F with 7 valence electrons.
Most metallic character : P with 5 valence electrons.

22. **(3)** In a period moving from left to right atomic size decreases.

23. **(3)** To compare second ionization potential, configuration of of mono-cationis observed:

C^+	N^+	O^+	F^+
[He] $2s^2 2p^3$	[He] $2s^2 2p^2$	$[He]2s^2 2p^3$ Half-filled stable	$[He]2s^2 2p^4$

2nd IE order O > F > N > C.

24. **(4)** Metallic character decreases on moving from left to right, but it increases on moving down the group.
Anionic radius is greater than atomic radius but cationic radius is always less than atomic radius for any element.

25. **(1)**

$$\underset{1s^2, 2s^2p^6, 3s^1}{Na} \xrightarrow{IE_1} \underset{\substack{1s^2, 2s^2p^6 \\ \text{(stable configuration)}}}{Na^+} \xrightarrow{IE_2}$$

So IE_2 of Na will be high

$$\underset{1s^2, 2s^2p^6, 3s^2}{Mg} \xrightarrow{IE_1} \underset{1s^2, 2s^2p^6, 3s^1}{Mg^+} \xrightarrow{IE_2} \underset{\substack{1s^2, 2s^2p^6 \\ \text{(Stable)}}}{Mg^{2+}}$$

So IE_2 of Mg is not high
So statement I is true
O^{2-} (1.40 Å) is larger than F^- (1.33 – 1.36Å)
So statement II is false.

26. **(1)** In general on moving from left to right in a period ionization energy increases as Z_{eff} increases.
Al < Si < S < P < Cl
Ionisation energy of phosphorus is more because of half filled stable configuration.

27. **(2)** The order of first ionization enthalpy values is
C < O < N < F
$\xrightarrow{2p^2 2p^4 2p^3 2p^5}$
The order of magnitude of electron gain enthalpy values is
S > Se > Te > Po > O
$|\Delta_{eg}H|$ 200 195 190 174 141 (in kJ/mol)
So, both statement-I and statement-II are correct.

28. **(2)**

Electronegativity order : N > P > As > Sb
(N ↑ Most electronegative; Sb ↑ Least electronegative)

X = $NX_2O_3 = N_2O_3$ (Acidic)
Y = $SbY_2O_3 = Sb_2O_3$ (Amphoteric)
Statement-I is true.

$$BCl_3 + 3H_2O \longrightarrow B(OH)_3 + 3HCl$$

In boron due to non-availability of d orbitals, boron is unable to expand its octet. Therefore the maximum covalence of boron cannot exceed 4, hence $[B(OH)_4]^-$ and $[B(H_2O)_6]^{3+}$ are not formed.
So, statement-II is false.

29. **(3)** EH_3 reacts with K_2HgI_4 (Nessler's reagent). This is a character test for Ammonia.
Nitrogen forms binary cation NH_4^+.

Element E is N, the species is NH_4^+, among B, C, N and O, N has highest first ionization energy. Nitrogen has a higher IE than oxygen because its 2p subshell is exactly half filled making it exceptionally stable.

30. (2) For Al and Mg

$$Al < Mg \text{ and } Al^{3+} < Mg^{2+}$$

Correct order of size is $Mg > Al > Mg^{2+} > Al^{3+}$

Atomic size depends mainly upon $Z_{effective}$ and shell number.

Generally on moving down the group electron affinity decreases and on moving across the period electron affinity increases.

In the periodic table Cl has maximum electron affinity. Halogens have higher electron affinity than chalcogens.

$$Cl > Br > S > O$$

31. (1)

⇒ Metallic character generally decreases as you move from left to right across a period or increases as you move down a group. The correct order of metallic character is

$$K > Mg > Al > B$$

Thus statement C is incorrect.

32. (3) First ionisation energy of an element generally decreases down the group and increases from left to right along a period. Therefore, element having atomic number 87 has the lowest first ionisation energy.

33. (2) In a period the atomic radius decreases from left to right Al comes after Mg in period so Mg should have smaller radius.

34. (1) (A)The process of adding an electron to a neutral gaseous atom is not always exothermic it may be exothermic or endothermic.

(D) In CCl_4 ⟶

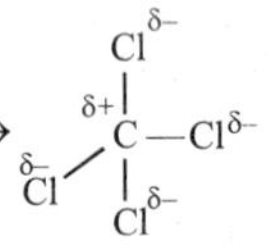

Due to partialy positive charge z_{eff}↑, EN↑

So, EN of C ⇒ $CCl_4 > CH_4$

(E)Cs is most electropositive among the elements of group I.

35. (1) TeO_2 exhibits oxidising properties while TeH_2 act as reducing agent and acidic in nature.

36. (4)

(A)$1s^2 2s^2 2p^3$ ⇒ N(B) $1s^2 2s^2 2p^4$ ⇒ O

(C)$1s^2 2s^2 2p^5$ ⇒ F(D) $1s^2 2s^2 2p^2$ ⇒ C

Order of electronegativity: $F > O > N > C$

$\therefore C > B > A > D$

37. (1) B is smallest so it forms B_2O_3

38. (4) For isoelectronic species, the cations with the greater positive charge will have a smaller radius because of the greater attraction of the electrons to the nucleus. Anion with greater negative charge will have the larger radius. Therefore statement I is correct.

Electron gain enthalpy of fluorine is less negative that of chlorine due to interelectronic repulsion.

Hence the order will be $Cl > F > Br > I$. Therefore statement II is correct.

39. (3) Atomic radii is increases moving down a group in the perodic table.

Correct order of atomic radii : $Be < Mg > Al > Si$

40. (2) The IE_4 is suddenly very high therefore element must have 3 valence e^-(s) and it belongs to group 13.

41. (1) Statement-I: False but Statement-II is true.

On moving left to right in periodic table non-metallic character increases and we know that non-metal oxides are acidic in nature.

Non metallic character ↑ Acidic strength of oxide↑

Element on extreme left are metallic and form basic oxides.

42. (3) Ionic radii – $Al^{3+} < Mg^{2+} < Na^+ < F^-$

Ionisation energy – $B < C < O < N$

Metallic character – $B < Al < Mg < K$

Electronegativity – $Si < P < S < Cl$

43. (1)

	Li	Be	B	C	N	O	F
(E.N.) =	1	1.5	2	2.5	3	3.5	4.0

On pauling scale

	Na	Mg	Al	Si	P	S	Cl
(E.N.) =	0.9	1.2	1.5	1.8	2.1	2.5	3.0

Correct order $Mg < Al < B < N$

44. (3) Elements with highest electronegativity ⇒ F, O

Elements with largest atomic size ⇒ Fr, Ra

Elements which shows properties of both metal and non-metals i.e. metalloids ⇒ Ge, As

Elements with highest negative electron gainenthalpy ⇒ Cl, S.

45. (2) Non metallic nature ∝ Electronegativity. Across the period ionisation energy increase due to increase in effective nuclear charge.

High electronegativity difference results in ionic bond formation.

Oxides of metals are generally basic and that of non-metals are acidic in nature.

46. (4) For process (A), (B), (C) electron affinity value are negative.

Be = –18, N = –20.3,

$O \rightarrow O^{2-} = [141 + (-844)] = -703$

Na = +53, Al = +44

47. (2) (A)-(III), (B)-(I), (C)-(IV), (D)-(II)

48. (1)

(A) Covalent radius of group 14 elements increases from C to Pb.

(B) Electronegativity decreases from C to Si and thereafter it remains more or less constant.

49. (4) All species are isoelectronic because each has 10 e^-. Nuclear charge depends on number of protons so all have different nuclear charge. More the number of protons high is the nuclear charge and small is the ionic radius. So order of ionic radius is :

(Ionic radius) $O^{-2} > F^- > Na^+ > Mg^{+2}$

50. (1) Greater the effective nuclear charge smaller will be its size, so statement I is correct but statement II is incorrect as anions are larger than parent atom.

51. (4) $B > Tl > Ga > Al > In$

52. (1)

(i) Na < Li < Cl < F

	Na	Li	Cl	F
$I.E._1$ in kJ/mol	496	520	1256	1681

(ii) Na < Li < F < Cl

	Na	Li	F	Cl
$\Delta_{eg}H$ in kJ/mol	–53	–60	–328	–349

53. (2) Correct order of I^{st} IE

$Li < B < Be < C < O < N < F < Ne$

$E < C < A < B < D$

54. (4) Isoelectronic species have equal number of electrons but they have different number of protons in the nucleus, Nuclear charge depends on the number of protons, which affect the size of isoelectronic species, Higher the nuclear charge smaller is the size.

55. (3) Statement I is false but Statement II is true.

p-Block element contains metal, non-metal and metalloid but *d*-block element contain only metals.

56. (1)

Element		$\Delta_{eg}H$ (kJ/mol)
F	–	333
S	–	200
Br	–	325
Ar	+	96

57. (2) A, B and D only

Down the group the atomic size and ionic radii increases and along the rows the atomic radii and ionic radii decreases.

∴ Order of atomic radii : $B' < A' < D' < C'$

∴ Order of ionic radii : $B'^+ < A'^+ < D'^+ < C'^+$

∴ Order of metallic character : $B' < A' < D' < C'$

58. (2) As to Bi only a small increase in covalent radius is observed, due to the presence of completely filled *d* and *f* orbitals in heavier members.

59. (3) In a period from left to right, the reactivity first decreases and then increases from Group 1 alkali metals to Group 17 halogens and then decreases to group 18.

Thus, Statement I is false.

Group 1 metals form basic oxide like Na_2O and group 17 non-metals form acidic oxide like Cl_2O_7.

Thus, Statement II is true.

60. (1) Across a period, increasing nuclear charge outweights the shielding. Consequently, the outermost electrons are held more and more tightly and the first ionization enthalpy increases across a period.

61. (2) The electronegativity of group 14 elements from Si to Pb is approximately same. Therefore statement I is false and statement II is true.

62. (4) $_{13}Al = 1s^2 2s^2 2p^6 3s^2 3p^1$, $_7N = 1s^2 2s^2 2p^3$

$_{14}Si = 1s^2 2s^2 2p^6 3s^2 3p^2$, $_6C = 1s^2 2s^2 2p^2$

Removal of one electron from the valence shell is most difficult in nitrogen as it disturbs the stable half-filled p-subshell configuration and therefore its first ionization energy is highest.

63. (3) s-block elements are highly reactive and found in combined state due to very low ionization energies. Thus, I is correct but II is false.

64. (3) K^+ (18), Cl^- (18), Ca^{2+}(18), Sc^{3+} (18)

65. (2) NO is a neutral oxide.

Al_2O_3 is a amphoteric oxide.

Cl has more negative electron gain enthalpy than F.

Electronegativity depends on the effective nuclear charge and distance of outer most electron from the nucleus. Bonded atom can affect the bond polarity only.

66. (1) According to Fajan's rule Sb^{+5} has more polarising power than Sb^{+3}. Therefore $SbCl_5$ is more covalent than $SbCl_3$.

Stability of higher oxides of halogen is primarily due to

Higher oxidation state, more electronegative halogen and resonance stabilization.

67. (3) First I.E.

F > N > O > C > Be > B > Li

Generally along a period ionisation energy increases. In case of N and Be ionisation energy increases due to stable electronic configuration.

68. (4) Assertion and Reason are correct and Reason is correct explanation. Hence, correct order is :

∵ Successive I.E. always increases.

69. (3) On moving from top to bottom metallic character increases while on moving from left to right metallic decreases. Hence, correct order is :

K > Ca > Be.

70. (4) Br > C > At > P

71. (2) Cl has the most negative ΔH_{eg} among all the elements and Ne has the most positive ΔH_{eg}.

72. (3) As electropositive character of element increases, basic character of its oxide increases.

$$\underbrace{B_2O_3}_{\text{acidic}} < \underbrace{Al_2O_3 < Ga_2O_3}_{\text{amphoteric}} < \underbrace{In_2O_3 < Tl_2O_3}_{\text{basic}}$$

73. (4) In isoelectronic species, size $\propto \dfrac{1}{Z_{eff}}$.

Hence correct increasing order of ionic radii is : $Ca^{2+} < K^+ < Cl^- < S^{2-}$

74. (2) Both (A) and (R) are true but (R) is not the correct explanation of (A).

75. (1) Metallic character increases down the group and decreases along the period.

76. (3) I.E. order: Na < Al < Mg < Si

∴ 496 < IE (Al) < 737 < 786

Only option (3), matches the condition.

77. (3) Rb and Cs have nearly same electron gain enthalpy $\Delta H_{eg} = -46$ kJ mol^{-1}.

Ar and Kr have same electron gain enthalpy, $\Delta H_{eg} = +96$ kJ mol^{-1}.

78. (1) In the periodic table; on moving down the group, the size of atom increases and tendency to remove an electron increases. Hence, metallic character increases and on moving left to right in a period, metallic character decreases.

Therefore; the decreasing order for metallic character is Na > Mg > Be > Si > P.

79. (4) $_{31}Ga$ is in group 13 with e^- configuration [Ar] $3d^{10}4s^2 4p^1$ whereas $_{30}Zn$ has e^- configuration as [Ar] $3d^{10}4s^2$. Removal of e^- from $4p$ orbital is easier than that from $4s$ orbital. Hence, first I.E. of Ga is lower than that of Zn.

80. (2) All elemen.ts have $n = 3$. Along the period I.E. increases but half-filled orbital stability of P results increase in I.E. Thus, P > S.

81. (1)

Li ↘ Be Diagonal relationship

Na   Mg

$Li^+ \rightarrow$ Maximum hydration enthalpy in group 1 due to small size. So 'B' is Mg.

82. (4) For chlorine, the atomic radius and covalent radius are taken as same. Hence, statement I is incorrect.

In case of anionic species, extra electrons are added in the parent atom, so effective nuclear charge decrease. Hence, anionic radius always greater than their parent atomic radius.

83. (4) On moving from left to right along the period size of atom decrease, hence ionization energy increases.

Nitrogen have half-filled p-orbital and beryllium have completely filled s-orbital so their ionization energy are greater than oxygen and boron, respectively. So the order is :

$$\underset{(2p^3)}{N} > \underset{(2p^4)}{O} > \underset{(2s^2)}{Be} > \underset{(2p^1)}{B}$$

84. (3) Down the group, the size of atom increases, so electron gain enthalpy decreases.

In case of oxygen, due to its small size, which leads to electron-electron repulsion, results in less electron gain enthalpy than sulphur.

85. (1) Ionisation energy ⇒ N > O.

$N \rightarrow 1s^2 2s^2 2p^3$

| ↑ | ↑ | ↑ | } half filled subshell

$O \rightarrow 1s^2 2s^2 2p^4$

| ↑↓ | ↑ | ↑ | } more than half filled

In oxygen atom, two of the four 2p electrons must occupy the same 2p orbital resulting in an increased electron – electron repulsion.

86. (3) Non-metals, metals generally form acidic oxide, basic oxides, respectively.

SO_3, $SiO_2 \rightarrow$ Acidic oxide

CaO → Basic oxide

$Al_2O_3 \rightarrow$ Amphoteric oxide

87. (2) Halogens generally form acidic oxides. Group 1 elements generally form basic oxides.

Cl_2O_7 – Acidic oxide; Na_2O – Basic oxide

Al_2O_3 – Amphoteric oxide; N_2O – Neutral oxide

88. (4) For isoelectronic species, ionic radii decrease with increase in nuclear charge. Ionic radius of O^{2-} is more than that of Mg^{2+}. Both O^{2-} and Mg^{2+} are isoelectronic with 10 electrons.

89. (1) For the isoelectronic species the ionic radius increase with increase of negative charged and ionic radius decrease with increase of positive charged.

$N^{3-} > O^{2-} > F^- > Na^+ > Mg^{2+}$ (Radii)

90. (4)

Element	Be	Mg	Ca	Sr
M.P(K)	1560	924	1124	1062

91. (4) Metalloids have intermediate properties between metals and non-metals.

Sc, Pb, Bi are metals Te is a metalloid.

92. (4) Isoelectronic species have same no. of electrons, e.g., Al^{+3}, O^{2-}, Mg^{+2} all have 10 electrons.

93. (2) The magnitude of e^- gain enthalpy decreases down the group but its value for Cl is more than that of F.

Element	F	Cl	Te	Po
$\Delta_{eg}H$(kJ/mol)	–328	–349	–190	–174

94. (2) Hg, Ga, Cs are liquid at near room temperature but Ag in solid.

95. (3) Al^{3+}, Mg^{2+} and Na^+ are isoelectronic ionic species. As positive charge increases, ionic size decreases.

The order of size of Na^+ and K^+ is $Na^+ < K^+$.

∴ Order of ionic radii is : $Al^{3+} < Mg^{2+} < Na^+ < K^+$.

96. (2) $B(OH)_3$ and H_3PO_3 are acidic.

$Al(OH)_3$ and $Be(OH)_2$ are amphoteric.

NaOH and $Ca(OH)_2$ are basic.

97. (2) Li ↘ Be ↘ B ↘

Na   Mg   Al   Si

B–Si, Be–Al and Li–Mg exhibit diagonal relationship,

whereas Li–Na belongs to same group.

98. (3) Ionization energy generally increases moving from left to right across the period.

So, the order is Na < Mg < ..... Cl < Ar.

99. (3) CaO–Basic, SiO_2–Acidic, B_2O_3–Acidic, N_2O–Neutral, BaO–Basic

100. (4) Correct order of electron gain enthalpy is S > Se > Te > O

Electron gain enthalpy of oxygen is even less than Te because of inter-electronic repulsion and effective nuclear charge in the oxygen atom.

101. (2) Mg and P exhibit abnormal behaviour due to extra stability of half-filled and filled electronic configuration.

First ionisation enthalpy order is :

Al < Mg < Si < S < P

102. (2)

Group Number						
3	4	5	6	7	...	18
$_{21}Sc$				$_{25}Mn$		$_{36}Kr$
$_{39}Y$			$_{42}Mo$			$_{54}Xe$
$_{57}La$	$_{72}Hf$					

103. (3) All are isoelectronic species, so more is the Z_{eff}, less will be the ionic size.

∴ Correct order of ionic radii is :

$Al^{3+} < Mg^{2+} < Na^+ < F^- < O^{2-} < N^{3-}$

Among isoelectronic species ionic radii increases as the negatives charge increases.

The number of electrons remains the same but nuclear charge increases with increase in the atomic number causing decrease in size.

104. (2)

$Ar(g) + e^- \longrightarrow Ar^-(g)$ (endothermic)

$H(g) + e^- \longrightarrow H^-(g)$ (exothermic)

$Na(g) \longrightarrow Na^+(g) + e^-$ (endothermic)

$O^-(g) + e^- \longrightarrow O^{2-}(g)$ (endothermic)

105. (2)

	O^{2-}	F^-	Na^+	Mg^{2+}
	Z	89	11	12
No. of e^-	10	10	10	10

In isoelectronic species greater is $Z_{effe.}$ smaller is radius so order is $O^{2-} > F^- > Na^+ > Mg^{2+}$.

106. (1) Charge/ radius ratio of Be and Al is same because of diagonal relationship. Remaining statements are correct.

107. (3) As difference in 3rd and 4th ionisation energies is high, so atom contains 3 valence electrons.

108. (4) On moving left to right along a period in the periodic table, atomic radius decreases while electronegativity, electron gain enthalpy and ionisation enthalpy increases.

109. (2) On moving left to right in a period, the acidic character of oxides increases.

3rd period element oxides.

Na_2O MgO (Basic) Al_2O_3 (Amphoteric) SiO_2 P_2O_5 Cl_2O_7 (Acidic)

So, X has minimum atomic number while Z has maximum atomic number.

Thus, the correct order of the atomic number is :

X < Y < Z

110. (3) $_5B : 1s^2 2s^2 2p^1$; $_4Be : 1s^2 2s^2$

First ionisation enthalpy of B is lower than Be because Be has a stable electronic configuration. It required more energy to remove the first electron from 2s (in Be) than 2p (in B) because 2s e^- has more penetration power than 2p. Therefore options (I), (II) and (III) are correct. Atomic radius of B is less than Be.

111. (3) Generally, non-metal oxides are acidic in nature and metal oxides are basic in nature, Al_2O_3 is amphoteric.

112. (4) A large difference between first and second ionisation enthalpies (4560–496 = 4064 kJ mol^{-1}) confirms the metal to be an alkali metal and thus is monovalent and form hydroxide of the type M(OH).

$MOH + HCl \rightarrow MCl + H_2O$

1 mol   1 mol

$2\,MOH + H_2SO_4 \rightarrow M_2SO_4 + H_2O$

1 mol   1/2 mol

113. (3) Generally, electronegativity decreases down the group as the size increases. This can also be formulated as:

$$\text{Electronegativity} \propto \frac{1}{\text{size}}$$

114. (1) All given elements belong to period III and generally their ionisation energy will increase along the period but Mg will show higher ionisation potential compared to Al due to its stable configuration. Thus, correct order of ionisation energy will be : Na < Al < Mg < Si.

Ionisation energy (kJ/mol) of the given metals are Na : 496; Al : 577; Mg : 737; Si : 786

115. (3) Correct increasing order of atomic radii is : F < O < C < Cl < Br

116. (3) Fluorine has lower electron gain enthalpy value than chlorine because of its small size. The small size of the fluorine atoms results in strong interelectronic repulsion in the 2p orbital.

117. (2) Generally, electron affinity decreases on moving down a group. Chlorine has more electron affinity than F because of very small size of fluorine. Therefore chlorine, sulphur and Li has higher electron affinity among given groups.

118. (4) Nuclear charge : B > Be

Be = $1s^2 2s^2$ (more stable)

B = $1s^2 2s^2 2p^1$

∴ Ionisation energy of Be is greater than B due to ns^2 outer electronic configuration.

119. (1) Atomic numbers of N, O, F and Na are 7, 8, 9 and 11 respectively. Therefore, total number of electrons in each of N^{3-}, O^{2-}, F^-, and Na^+ are 10 and hence, they are isoelectronic.

120. (4) Alkali metals have high difference in first and second ionisation energy as they achieve stable noble gas configuration after first ionisation.

121. (1) On going down the group, size increases while going from left to right in a period, size decreases. So order is:

C < S < Al < Cs.

122. (2) Correct order of electronegativity values of the elements is

Si > Al; S > P; Se > Te; Ge > Ga.

123. (4) The second electron gain enthalpy of oxygen is positive as energy has to be added for the addition of another electron.

124. (3) All species are isoelectronic ($10e^-$).

In isoelectronic series, when negative charge increases the radius of ion increases.

∴ $Mg^{2+} < Na^+ < F^- < O^{2-}$

125. (4)

126. (N) All the given statements are true.

127. (1) Isoelectronic species have same no. of electrons.

Ions ⇒	O^{2-}	F^-	Na^+	Mg^{2+}
	8 + 2	9 + 1	11 – 1	12 – 2
No. of e^- ⇒	10,	10,	10,	10
Therefore	O^{2-},	F^-,	Na^+,	Mg^{2+}

are isoelectronic.

128. (3) The smaller the atomic size, larger is the value of IP. Further, the atoms having half filled or fully filled orbitals, are comparatively more stable. Hence, more energy is required to remove the electron from such atoms.

129. (3) Due to large difference of 2nd and 3rd ionisation enthalpy, both A and B will belong to group-2. Due to low values of IE for B than that of A, B will be below A.

130. (2) Alkali metals have the lowest ionization energy in each period, on the other hand, Sc is a d - block element.

Transition metals have smaller atomic radii and higher nuclear charge, leading to high ionisation energy.

131. (2) Due to extra stable half-filled p-orbital electronic configurations of Group 15 elements, larger amount of energy is required to remove electrons compared to Group 16 elements.

132. (1) For isoelectronic species, size of anion increases as negative charge increases. Thus the correct order is (1).

133. (1) For isoelectronic species, ionic radii decreases as the negative charge on ion decreases. Further on moving down in a group, ionic radii increases. Hence, the correct order is : $O^{2-} < N^{3-} < S^{2-} < P^{3-}$

Trick: Ionic radii ∝ Negative charge on ion.

134. (3) On moving down in a group atomic radii increases.

135. (2) In O_2^{2-}, there are two atoms, therefore, it will be largest in size.

136. (3) On moving down a group, size increases, hence ionisation enthalpy decreases. Hence Se < S and Ba < Ca. Further, Ar being an inert gas has maximum IE.

137. (4) On moving down in a group, atomic radii increases due to successive addition of extra, shell. Hence, order is :

O < S < Se

Further, As is in group 15 having one less electron in its p orbital compared to group 16, hence As has higher atomic radii than Se.

i.e., O < S < Se < As

138. (2) The second ionization potential means removal of electron from unit charged cation.

$C^+ = 1s^2 2s^2 2p^1$; $N^+ = 1s^2 2s^2 2p^2$

$O^+ = 1s^2 2s^2 2p^3$; $F^+ = 1s^2 2s^2 2p^4$

Therefore, the correct order is :

$O > F > N > C$

139. (1) Due to stable $2s^2\, 2p^3$ configuration (half filled p-orbital). Nitrogen atom has highest energy.

140. (3) Among isoelectronic species ionic radii increases as the negatives charge increases.

Order of ionic radii $Ca^{2+} < K^+ < Cl^- < S^{2-}$

The number of electrons remains the same but nuclear charge increases with increase in the atomic number causing decrease in size.

141. (4)

Number of moles of chlorine $= \frac{1}{35.5}$ mol

Given, $1\ eV = 23.06\ kcal\ mol^{-1}$

$3.7\ eV = 3.7 \times 23.06\ kcal\ mol^{-1}$

i.e., 1 mole of chlorine releases energy

$= 3.7 \times 23.06$ kcal

$\therefore\ \frac{1}{35.5}$ mole of chlorine will release energy

$= \frac{1}{35.5} \times 3.7 \times 23.06\ kcal = 2.4\ kcal$

142. (2) Correct order of basic strength for group is hydride should be $NH_3 > PH_3 > AsH_3 > SbH_3$, it is because as the size of central atom increases electron density on the central atom decreases.

143. (2)

144. (4) All the given species contains 10 e^- each i.e. isoelectronic. For isoelectronic species anion having high negative charge is largest in size and the cation having high positive charge is smallest.

145. (2) The correct order of ionisation enthalpies is

$F > P > S > B$

146. (2) On moving down in group I from Li to Cs, ionisation enthalpy decreases, hence the reactivity increases. The halogens are most reactive elements due to their low bond dissociation energy, high electron affinity and high enthalpy of hydration of halide ion. Hence, their reactivity decreases with increase in atomic number.

147. (3) In a period the value of ionisation potential increases from left to right with breaks where the atoms have somewhat stable configuration. In the given list N has half-filled stable orbitals. Hence N has highest ionisation energy. Thus the correct order is

$B < C < O < N$

and not as given in option (3)

148. (1) SnO_2 is amphoteric as it reacts with both acid and base.

$SnO_2 + 2H_2SO_4 \longrightarrow Sn(SO_4)_2 + 2H_2O$

$SnO_2 + 2KOH \longrightarrow K_2SnO_3 + H_2O$

149. (4) As the size decreases, the basic nature of oxides changes to acidic nature, i.e., acidic nature increases.

$\underset{\text{Acidic}}{SO_2 > P_2O_3} > \underset{\text{Weak acidic}}{SiO_2} > \underset{\text{Amphoteric}}{Al_2O_3}$

SO_2 and P_2O_3 are acidic as their corresponding acids are H_2SO_3 and H_3PO_3.

150. (1) O^- ion exerts a force of repulsion on the incoming electron. Hence energy is required to overcome it.

151. (1)

152. (4) Na_2O (basic), SO_2 and B_2O_3 (acidic) and ZnO is amphoteric

153. (6) A_2B

Lowest oxidation state of B is –2 means it can accept 2 electrons maximum to complete its octet, so it should have six valence electrons.

154. (5) Isoelectronic species O^{2-}, F^-, Mg^{2+}, Na^+, Al^{3+} (all have $10e^-$)

155. (3)

Oxides	Nature
Cl_2O_7	Acidic
CO	Neutral
PbO_2	Amphoteric
N_2O	Neutral
NO	Neutral
Al_2O_3	Amphoteric
SiO_2	Acidic
N_2O_5	Acidic
SnO_2	Amphoteric

156. (5) $Na_2O + H_2O \rightarrow 2NaOH$

$Cl_2O_7 + H_2O \rightarrow 2HClO_4 \Rightarrow 1 + 4 = 5.$

157. (4) Acidic oxides are N_2O_3, NO_2, Cl_2O_7, SO_2

Basic oxides are CaO, Na_2O

Neutral oxides are CO, NO, N_2O

Part-B (JEE Advanced)

1. (d) The electronic configuration of the given ions are as follows.

${}_{12}Mg^{2+} = 1s^2$, (No unpaired electron)

${}_{22}Ti^{3+} = 1s^2, 2s^2 2p^6, 3s^2 3p^6 3d^1$ (One unpaired electron)

${}_{23}V^{3+} = 1s^2, 2s^2 2p^6, 3s^2 3p^6 3d^2$ (Two unpaired electrons)

${}_{26}Fe^{2+} = 1s^2, 2s^2 2p^6, 3s^2 3p^6 3d^6$ (Four unpaired electrons)

2. (d) The electrons are not filled in d-subshell monotonically with increase in atomic number, among transition elements. It breaks at chromium and copper.

3. (a, c, d) Periodic table does not help to predict the stable valency states of the elements.

4. (a) Non-metallic oxides are acidic and metallic oxides are basic. Thus, the acidic order is $CaO < CuO < H_2O < CO_2$.

5. (b) Effective nuclear charge (*i.e.* Z/e ratio) decreases from F^- to N^{3-}. Hence the radii follows the order:

$F^- < O^{2-} < N^{3-}$. Z/e for $F^- = 9/10 = 0.9$, for $O^{2-} = 8/10 = .8$, for $N^{3-} = 7/10 = 0.7$

6. (a) Non-metallic oxides are acidic and acidic character decreases with increase in metallic character.

7. (a)

(i) Hydrogen bonding increases the boiling point.

(ii) Hydrogen bonds are formed in compounds having F or O or N with hydrogen. S, Se, Te cannot undergo hydrogen bond formation because of their larger size and lower electronegativity values.

8. (b)

(i) Ion having half filled or full filled orbital have extra stability.

(ii) Larger the size of cation more will be its stability Pb^{2+} ($5d^{10}\, 6s^2$), has the most stable +2 oxidation state because here the d-orbital is completely filled and is more stable than Fe^{2+} ($3d^6$). Again Ag^+ ($4d^{10}$) is more stable as here again the d-orbital is completely filled and Ag^{2+} is not easily obtained. Pb^{2+} is more stable compared to Sn^{2+} ($4d^{10}\, 5s^2$) because of its large size.

9. (b) Ionisation energy increases with increasing atomic number in a period, while it decreases on moving down a group. IE of element with electronic configuration (d) is lowest because of its biggest size. Among the remaining three elements of the same period (3rd), IE of element with electronic configuration (b) is the highest due to greater stability of the exactly half-filled 3 p-subshell.

10. (b) Nitrogen, being smallest in size, can give up its lone pair of electrons most easily.

11. (d) For isoelectronic ions, ionic size $\propto \frac{1}{\text{atomic number}}$

Species	No. of e^-	At. No.
N^{-3}	10	7
O^{-2}	10	8
F^-	10	9
Na^+	10	11

$\therefore$ Na^+ is smallest in size.

12. (a) First ionisation potential increases from left to right in a period.

IE_1 of Mg is higher than that of Na because of increased nuclear charge and also than that of Al because in Mg, a 3 *s*-electron has to be removed while in Al, it is the 3 *p*-electron. The IE_1 of Si is, however, higher than those of Mg and Al because of its increased nuclear charge. Thus, the overall order is : Na < Mg > Al < Si.

13. **(c)** Electronegativity increases on moving from left to right in a period and decreases on moving from top to bottom in a group.

Si and P are placed in the 3rd period while C and N are placed in the 2nd period. Elements in 2nd period have higher electronegativities than those in the 3rd period. Since, N has smaller size and higher nuclear charge than C, its electronegativity is higher than that of C. Similarly, the electronegativity of P is higher than that of Si. Thus, the overall order is : Si, P, C, N.

14. **(a)** Ionisation potential of nitrogen is more than that of oxygen. This is because nitrogen has more stable **half-filled** *p*-orbitals. (N = $1s^2, 2s^2, 2p^3$, O = $1s^2, 2s^2, 2p^4$)

15. **(a)** (i) Noble gases do not have covalent radii. They have only van der Waals radii.

(ii) Covalent radii is always smaller than corresponding van der Waals radii.

16. **(c)** Amongst B, C, N and O; N has the highest first ionization energy, because of its half filled 2*p* orbital which is more stable.

17. **(c)**

C – $1s^2\,2s^2\,p^2$	C^+ – $1s^2\,2s^2\,p^1$
N – $1s^2\,2s^2\,p^3$	N^+ – $1s^2\,2s^2\,p^2$
O – $1s^2\,2s^2\,p^4$	O^+ – $1s^2\,2s^2\,p^3$
F – $1s^2\,2s^2\,p^5$	F^+ – $1s^2\,2s^2\,p^4$

As for IE_2, the electron in all the cases is to be removed from 2*p* orbital, so it must follow the order : C < N < O < F (*i.e.* increase across a period)

But in case of O^+, the 2*p* orbital is half-filled and is more stable as compared to others. So, the orderbecomes : O > F > N > C

18. **(9)** By observing the values of ionization enthalpy for atomic number (*n* + 2), it is observed that $I_2 >> I_1$. Which shows that number of valence shell electrons is 1 for atomic number (*n* + 2). Therefore element with atomic number (*n* + 2) should be an alkali metal.

For atomic number (*n* + 3), $I_3 >> I_2$, which shows that it will be an alkaline earth metal. All the observations suggests that atomic number (*n* + 1) should be a noble gas and atomic number (*n*) should belong to halogen family. Since *n* < 10; hence *n* = 9.

19. **(2)** Fluorine generally shows 0 and –1 oxidation states while sodium shows 0 and +1 oxidation state.

20. **Electronegativity**

21. **Electron affinity**

22. **False :** On moving down the group 13 (III) A the basic nature of hydroxides increases. The basic nature increases as the element becomes more electropositive or acquires more metallic character when we move down a group.

23. **False :** Halogens have high electron affinities which decrease as we move down the group. However, fluorine has lower value of E.A. than chlorine which is due to its small size and more repulsion between the electron added and electrons already present. Hence, the order is :

Cl > F > Br.

24. **False :** *Ionisation energy decreases on moving down in group IA from* Li *to* Cs, the reducing property should **increase** in the same order, *i.e.*, from Li to Cs which is found to be so, except an *anomaly in lithium which is found to be the strongest reducing agent*; because of its **higher oxidation potential** (E^o).

25. **(a, b)** C = $1s^2\,2s^2\,2p^2$, $C^+ = 1s^2\,2s^2\,2p^1$

B = $1s^2\,2s^2\,2p^1$, $B^+ = 1s^2\,2s^2$

(a) 2nd IE order: B > C

(b) For isoelectronic species smaller the charge more will be the radius.

Ionic Radius : $Na^+ > Mg^{2+} > Al^{3+}$

(c) Density of K = 0.86 g/cm^3;

Density of Na = 0.97g/cm^3

This is because 'd' orbital present in potassium which increases the volume of potassium, so density will decreases.

(d) Bond dissociation energy

H–H	>	F – F
435.8 kJ/mol		158.8 kJ/mol

This is because in F_2 each fluorine atom have three non bonding electrons so there is strong lone pair repulsion which make the bond weak.

26. **(a, d)**

NO ⇒ Neutral

B_2O_3 ⇒ Acidic

CrO ⇒ Basic

All other oxides are amphoteric.

27. **(d)** Higher the (+) charge, smaller will be radii.

28. **(a, b)** For dissolution, Hydration energy > Lattice energy.

$BaSO_4$ is sparingly soluble in water because its hydration energy is lesser than the lattice energy and thus, ions are not separated from each other. On the contrary in Na_2SO_4, the hydration energy is more than its lattice energy. Thus, ions are separated from each other and pass in solution state.

29. **(c)** The first ionisation energy of Be ($1s^22s^2$) is greater than that of B($1s^22s^2p^1$) because removal of electron from 2*s* orbital is tough compare to that from 2*p* orbital. This is because 2*s*-orbital is closer to nucleus than the 2*p*-orbital. Reason is incorrect as 2*p*-orbital is higher in energy.

30. (*i*) $Al^{3+} < Mg^{2+} < Li^+ < K^+$

Al^{3+} and Mg^{2+} are isoelectronic species, so in these, size decreases with increase in atomic number because increase in atomic number increases Z_{eff}

Due to anomalous behaviour and diagonal relationship, the ionic radius of Li^+ is slightly bigger than Mg^{2+}.

(*ii*) Increasing order of basic character :

$NiO < MgO < SrO < K_2O < Cs_2O$

The basic character of oxides increases when we move down the group. So, $K_2O < Cs_2O$ and MgO < SrO.

Further, higher the group number lesser is the basic character. Hence, NiO is the least basic.

(*iii*) Increasing order of ionic size :

$Mg^{2+} < Na^+ < F^- < O^{2-} < N^{3-}$

All the above ions are isoelectronic having 10 electrons each.

In such a case, the greater the nuclear charge, the greater is the attraction for electrons and smaller is the ionic radius. Hence, N^{3-} has the highest and Mg^{2+} has the least ionic size.

(*iv*) $Ca^{2+} < Ar < Cl^- < S^{2-}$; All of these are isoelectronic. In such cases, the greater the nuclear charge, the greater is the attraction for electrons and smaller is ionic size.

$$\text{ionic radius} \propto \frac{1}{\text{effective nuclear charge}}$$

(*v*) The first ionization potential of the 3rd period elements follows the order :

Na < Al < Mg < Si

Ionisation energy **increases across a period** but not regularly. Mg ($1s^2, 2s^2p^6, 3s^2$) is more stable because the electron is to be removed from 3*s* which is difficult as compared to Al ($1s^2, 2s^2p^6, 3s^2p^1$) where electron is to be removed from 3*p*.

(*vi*) $\overset{+1}{Na_2}O_2 < \overset{+2}{Mg}O < \overset{+2}{Zn}O < \overset{+5}{P_2}O_5$

Among oxides, the acidic strength increases with oxidation state. So Na_2O_2 is least acidic and P_2O_5 is most acidic. Further Na_2O_2 and MgO are basic, ZnO is amphoteric and P_2O_5 is acidic.

(*vii*) $O^{2-} > F^- > Na^+ > Mg^{2+}$

All the above ions are isoelectronic having 10 electron each.

In such case, the greater the nuclear charge, the greater is the attraction for electrons and smaller is the ionic radius. Hence, O^{2-} has the highest and Mg^{2+} has the least ionic size.

31. C ($1s^22s^2p^2$) has high nuclear charge than B ($1s^22s^2p^1$). Thus, carbon has higher IE value than Boron.

Further, for second ionization energy (IE_2) in C^+ ($1s^22s^2p^1$) the electron is to be removed from 2*p* which is easy as compared to B^+ ($1s^22s^2$), where it has to be removed from 2*s*.

Chemical Bonding and Molecular Structure

Part-A (JEE Main)

1. (1) 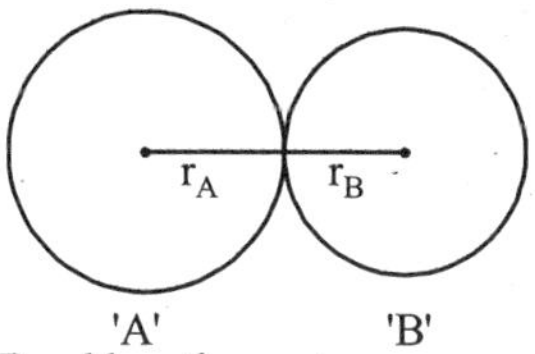

Covalent Bond length = $r_A + r_B$
Total length of AB molecule = $2(r_A + r_B)$

2. (2) $C-H(A) = 107$ pm
$C\equiv N(D) = 116$ pm
$C-O(B) = 143$ pm
$C=O(C) = 121$ pm
So the correct order of covalent bond length is $A < D < C < B$.

3. (3) KCl is an ionic compound while others (PH_3, O_2, B_2H_6, and H_2SO_4) are covalent compounds.

4. (N)
(1) $NO^+ = 7 + 8 - 1 = 14\ e^-$ and $O_2 = 16\ e^-$
i.e., not isoelectronic.
(2) Boron forms only covalent compounds. This is due to its extremely high ionisation energy.
(3) Compounds of Tl^+ are much more stable than those of Tl^{3+}.
(4) $LiAlH_4$ is a versatile reducing agent in organic synthesis.

5. (1) The value of lattice energy depends on the charge present on the ions and the inter-ionic distance between them.

6. (2) For A

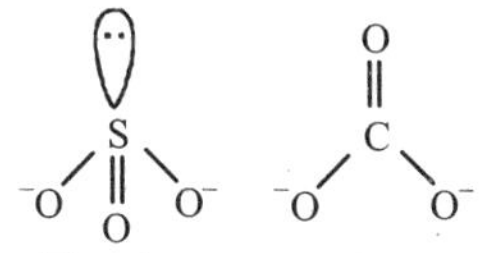

Trigonal Pyramidal — Trigonal Planar

Hence, do not have similar Lewis dot structures.
For E
MnO_4^{2-} have one unpaired electron whereas in CrO_4^{2-} it does not have any unpaired electron.
Hence, do not have similar Lewis dot structures.

7. (3) NF_3 ;

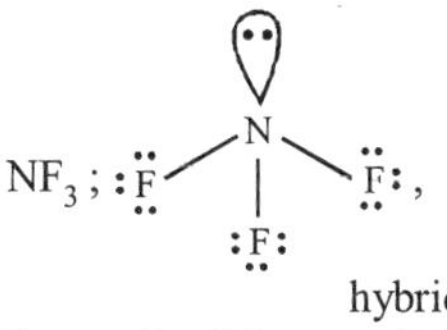

hybridization : sp^3
Number of lone pair of electrons in $NF_3 = 10$
Bond angle in $NF_3 \approx 102°$

8. (3) Consider the structure of HNO_3

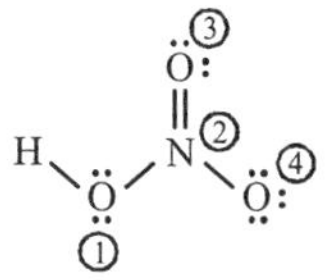

Formal charge = valence electrons – non bonding electrons $-\left(\frac{\text{bonding electrons}}{2}\right)$

① $O: 6-4-\left(\frac{4}{2}\right) = 0$ ② $N: 5-0-\left(\frac{8}{2}\right) = 1$

③ $O: 6-4-\frac{4}{2} = 0$ ④ $O: 6-6-\frac{2}{2} = -1$

9. (3) 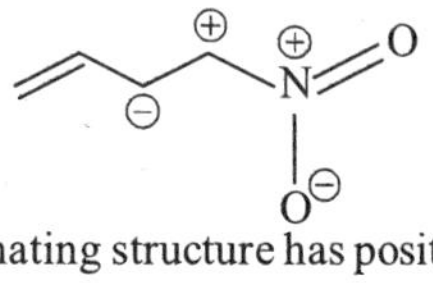

This resonating structure has positive charge on adjacent–N atoms, so it is least stable.

10. (1) Due to resonance bond length is identical in ozone. Therefore statement I is true and statement II is false.

11. (1)
$NO = 7e^-$, $NO_2 = 7e^-$ } incomplete octet with odd electron
$BCl_3 = 6e^-$, $AlCl_3 = 6e^-$ } incomplete octet
$H_2SO_4 = 12e^-$, $PCl_5 = 10e^-$ } molecule with expanded octet
$CCl_4 = 8e^-$, $CO_2 = 8e^-$ } molecules obeying octet rule.

12. (4) The difference in energy between the actual structure and the lowest energy resonance structure of the given compound is resonance energy.

13. (3) The real structure of any chemical species is the hybrid of weighted average of all possible resonating structures.

14. (N) Order should be : $C < A < B < D$

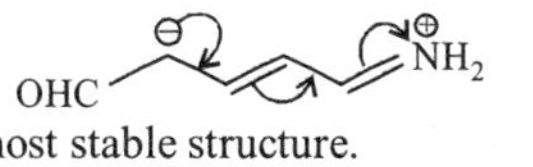

D is the most stable structure.

15. (2) NO $\rightarrow$ Odd electron molecule
$H_2SO_4 \rightarrow$ Expanded octet of S

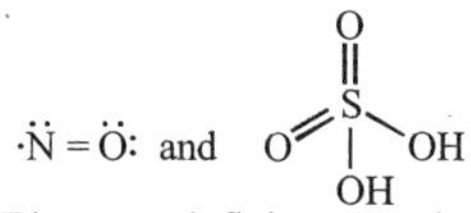

16. (3) Electron deficient species have less than 8 electrons (or two electrons for H) in their valence (incomplete octet). B_2H_6, BCl_3 have incomplete octet.

17. (4)

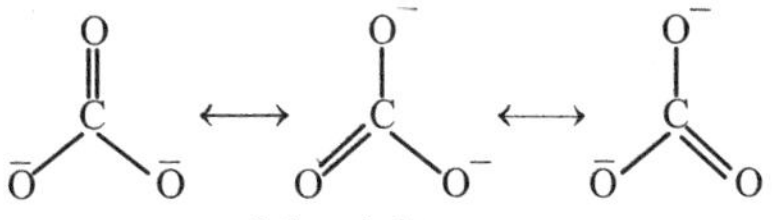

3 Canonical structures

18. (98) $\Delta H_{R.E} = \Delta H_{F(exp)} - \Delta H_{F(Theo)}$
$\Delta H_{F(exp)}$ for $X_2Y_{(g)} = 80$ kJ/mole (given)

$X_{2(g)} + \frac{1}{2}Y_{2(g)} \rightarrow X_2Y_{(g)}$

$$\Delta H_{F(Theo)} = \left(BE_{X\equiv X} + \frac{1}{2}BE_{Y=Y}\right) - (BE_{X=X} + BE_{X=Y})$$

$$= \left(940 + \frac{1}{2}\times 500\right) - (410 + 602) = 178 \text{ kJ/mole}$$

$\Delta H_{R.E} = 80 - 178$
$= -98$ kJ/mol
$|\Delta H_{R.E}| = 98$ kJ/mol

19. (12) $\ddot{O} = \ddot{N} - \ddot{O}:^-$, 12 electron

20. (6) According to octet rule central atom should have 8 valence electrons.

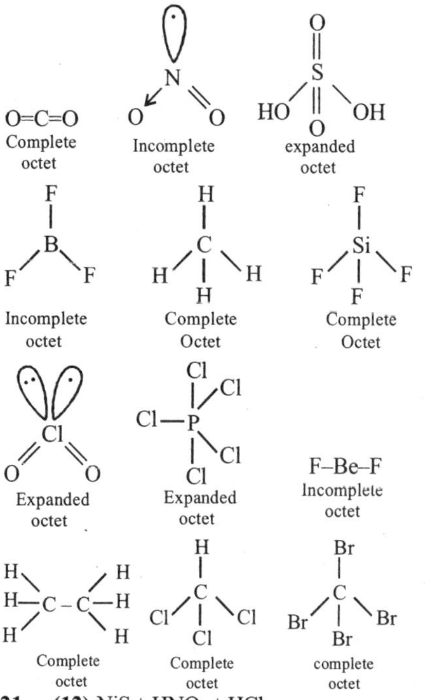

21. (12) $NiS + HNO_3 + HCl \longrightarrow NiCl_2 + S + NO + H_2O$
(A)
$NiCl_2 + NH_4OH +$ Dimethylglyoxime $\longrightarrow$ $Ni(dmg)_2 + NH_4Cl_4 + H_2O$
(A) (B)

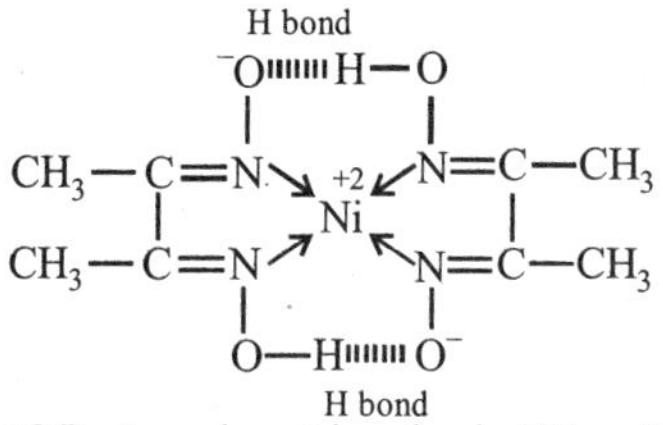

Total 12 Protons do not involve in H-bonding.

22. (8) Structure of NO_2^- is : 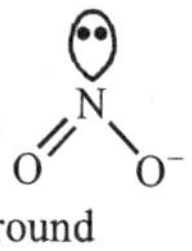

Number of valence electrons around N atom = 8

23. (3) ICl_4^-, BrF_3 and NO_2^+ do not have odd number of electron.

24. (6) Total no of lone pairs on oxygen atoms = 6

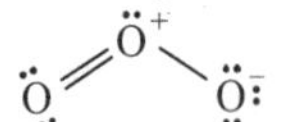

25. (4) $MgCl_2$, $AlCl_3$, $CuCl_2$ are water soluble at room temperature.
AgCl, Hg_2Cl_2 are sparingly soluble in water
$PbCl_2$ is soluble in hot water.
So X = $CuCl_2$, Y = AgCl, Z = $PbCl_2$

26. (2)
(A) Dipole moment : $NF_3 < NH_3$
(B) BeH_2 is 'sp' hybridized, linear molecule with zero dipole moment.
(C) $O_2^{-2} \Rightarrow$ bond order = 1 (18 electrons)
$F_2 \Rightarrow$ bond order = 1 (18 electrons)
(D) Formal charge on central oxygen atom in O_3
$= 6 - \frac{6}{2} - 2 = +1$
(E) In NO_2; nitrogen does not follow octet rule as it has an odd electron on it.

27. (3)

Molecule	Dipole moment
H_2S	0.95
H_2O	1.85
NF_3	0.23 (minimum)
NH_3	1.47
$CHCl_3$	1.04

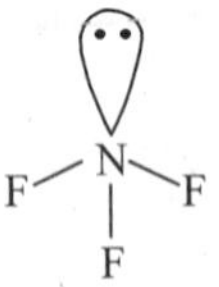

Number of lone pair of electrons on central atom of NF_3 = 1.

28. (1) Statement I:
Bond energy order is $Cl_2 > Br_2 > F_2 > I_2$
Bond Dissociation enthalpy
$Cl_2 - 242.6$, $Br_2 - 192.8$, $F_2 - 158.8$, $I_2 - 151.1$
Bond energy increases with increase in bond order.
Statement II:
Correct order of covalent character
According to the Fajan's rule, higher the charge on cation, greater is the covalent character.

$\overset{2+}{PbCl_2} < \overset{4+}{PbCl_4}$

$\overset{6+}{UF_6} > \overset{4+}{UF_4}$

$\overset{4+}{SnCl_4} > \overset{2+}{SnCl_2}$

29. (3) Statement I :
$CH_3 - CH = CH - CHO$ has conjugation between the double bond and the carbonyl group, which leads to resonance stabilization and contributes to greater charge separation across the molecule. So the dipole moment is greater than Butanal.
Statement II :

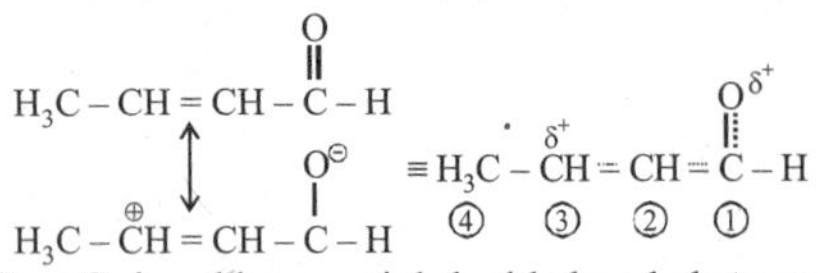

$C_1 - C_2$ bond has partial double bond character that means lesser bond length than $C_1 - C_2$ bond of other compound.
Statement-II is false.

30. (1) Difference between ionization energy & electron gain enthalpy increases, ionic character increases.
∴ The correct order of ionic character is:
EB > EA > EC > ED

31. (1) Increasing order of Dipole moment
$NF_3 <$ HBr $< H_2S < CHCl_3$
μ = 0.24D 0.76D 0.95 1.04D

32. (3) Order of bond angle is $ClF_3 < PF_3 < BF_3$

33. (1)

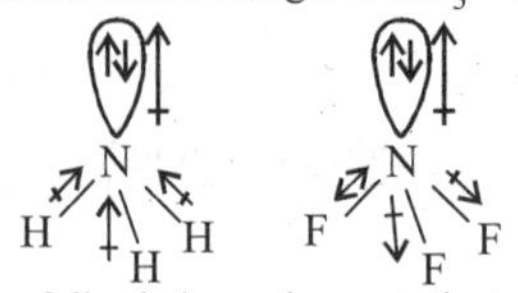

directions of dipole in such a way that net dipole moment of NH_3 is higher than NF_3 as μ of lone-pair is in same direction as the resultant dipole moment of N – H bonds.
So, both (A) & (R) are true, & (R) is the correct explanation of (A).

34. (3) CH_4 and PF_5, $\mu_{net} = 0$ (non-polar)

μ_{NH_3} > μ_{NF_3}
Vector addition of bond moment & lone pair moment; Vector subtraction of bond moment & lone pair moment

35. (3) Ionic character of a molecule increases with increase in the electronegativity difference between the atoms. The correct increasing order of ionic character is:
$N_2 < SO_2 < ClF_3 < K_2O <$ LiF

36. (2) The ionic character decrease as we move from left to right in the periodic table and it increases as we move top to bottom in a group.
∴ Ionic character follows the order as
KCl > $BaCl_2 > CoCl_2 >$ AgCl
∴ least ionic is AgCl.

37. (2)

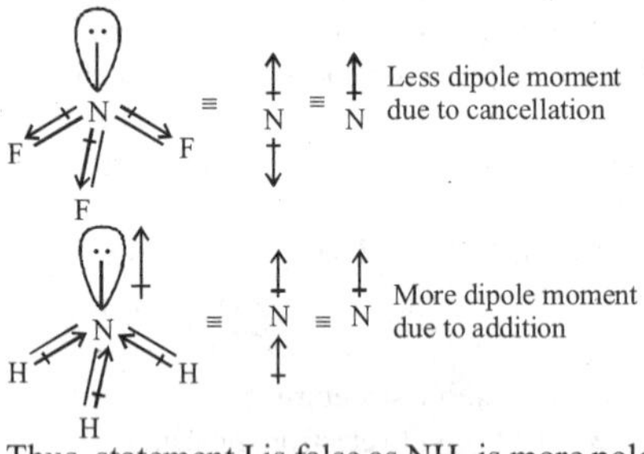

Thus, statement I is false as NH_3 is more polar. Statement II is also false as dipoles due to lone pair and bond pairs are aligned in the same direction in NH_3 but in opposite direction in NF_3.

38. (4)

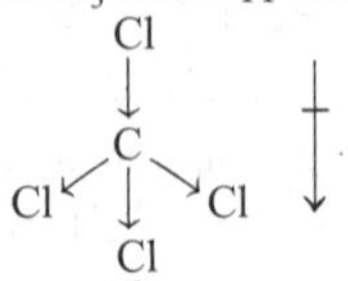

$\mu \neq 0$
$CHCl_3$ is polar molecule and rest all molecules are non-polar due to cancellation of bond dipoles due to symmetry.

39. (1) Among the given compounds, the following compound has the highest dipole moment because both the +ve and –ve ends acquire aromaticity.

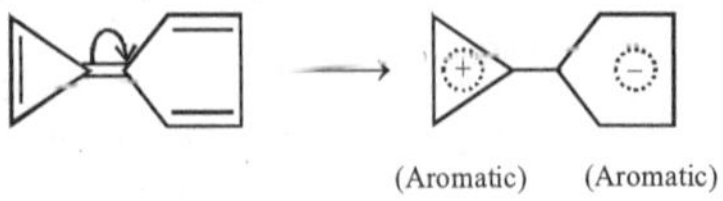

40. (1) SO_2 and NH_3 are polar molecules. They are constituent particles of polar molecular solids. AlN and MgO are ionic compounds.

41. (3)

42. (2) Statement I:

$\overset{\delta+}{H} \longrightarrow \overset{\delta-}{F}$ (crossed arrow)

Tail is on the positive atom and arrow head on the negative atom.
Statement II: The crossed arrow symbolises the direction of the shift of electron density in the molecule.

43. (3)

44. (2) According to Fajan's rule,
Covalent character ∝ size of Anion
As the cation is same in all the options. Therefore covalent character will depend upon the size of anion. Covalent character :
A < B < C < D; Size of anion : $F^- < Cl^- < Br^- < I^-$

45. (3) According to Fajan's rule, small size of cation and large size of anion will have high covalent character. So the order is:
LiCl > NaCl > KCl > CsCl

46. (4)
$\mu_{CCl_4} = \mu_{CH_4} = 0$ due to symmetrical structure
but $\mu_{CHCl_3} \neq 0$. So dipole moment order is :
$CHCl_3 > CH_4 = CCl_4$

47. (3)

48. (2) Dipole moment (μ) = $q \times d$
$\Rightarrow$1 D $\approx 10^{-18}$ esu cm
0.38×10^{-18} esu cm = $q \times (1.617 \times 10^{-8}$cm)
$q = 2.35 \times 10^{-11}$esu

So, fractional charge $= \frac{\text{Partial charge}}{\text{Total charge}} = \frac{q}{Q}$

$= \frac{2.35 \times 10^{-11}\text{esu}}{4.802 \times 10^{-10}\text{esu}} = 0.049 \approx 0.05$

49. (2) According to Fajans rules, small and highly charged cation has greatest covalent character while large cation with small charge has greatest ionic character.

50. (4) Given $e = 1.60 \times 10^{-19}$ C
$d = 9.17 \times 10^{-11}$ m
From $\mu = e \times d$
$\mu = 1.60 \times 10^{-19} \times 9.17 \times 10^{-11} = 14.672 \times 10^{-30}$

% ionic character = $\frac{\text{Observed dipole moment}}{\text{Dipole moment for 100\% ionic bond}}$

$= \frac{6.104 \times 10^{-30}}{14.672 \times 10^{-30}} \times 100 = 41.5\%$

51. (2) The geometry of IF_5 is square pyramidal with an unsymmetric charge distribution, therefore this molecule is polar.

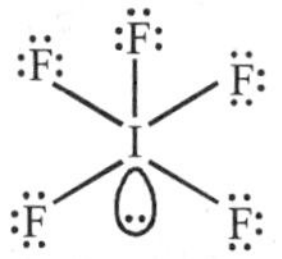

52. (1) In 1, 2-dichlorobenzene, the two dipoles are at 60° (i.e., unsymmetric). Thus, it possesses dipole moment.

53. (4) CCl_4 is a non-polar molecule and it has symmetrical tetrahedral structure. Although each of the C-Cl bond is polar but the resultant of all four dipole moments is zero.

54. (2)

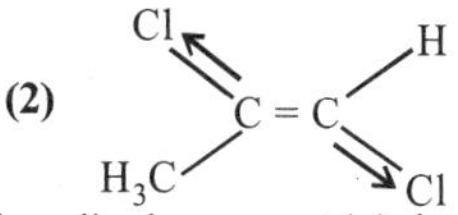

Very low dipole moment (μ) due to $C-CH_3$ bond.

55. (3) In nitrogen molecule, both the nitrogen atoms have same electronegativity. So it has zero polarity and hence, less tendency to break away and forms ions.

56. (4) Smaller the size and higher the charge, more will be the polarising power of the cation. Since the order of the size of cation is $K^+ > Ca^{2+} > Mg^{2+} > Be^{2+}$, so the correct order of polarising power is : $K^+ < Ca^{2+} < Mg^{2+} < Be^{2+}$

57. (2) Both NO_2 and O_3 have angular shape and hence, will have net dipole moment.

58. (6)

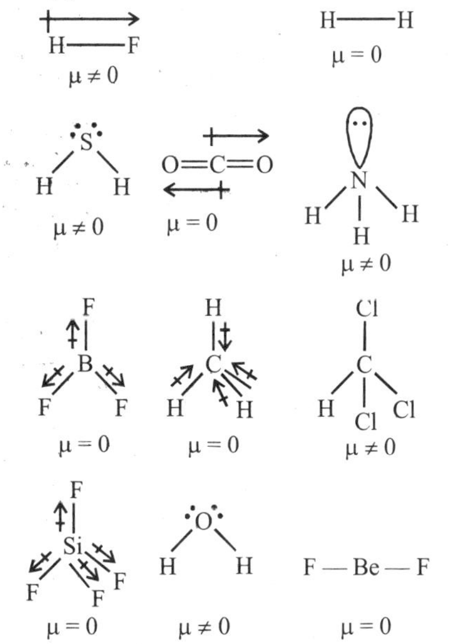

59. (5) Polar molecule have non-zero dipole moment.

Polar molecule NF_3, H_2O, H_2S, HBr, HCl

Non-polar molecules have zero dipole moment.

Non Polar molecule : $BeCl_2$, BCl_3, XeF_4, CCl_4, CO_2, H_2

(μ = 0)

So answer is 5.

60. (2.5) Given : μ = 1.2 D

$1\ D = 10^{-18}$ esu cm $= 10^{-10}$ esu Å

d = 1Å

$\mu = q \times d \Rightarrow 1.2\ D = q \times 1Å$

1.2×10^{-10} esu Å $= q \times 1Å$

$\therefore q = 1.2 \times 10^{-10}$ esu

Charge on electron $= 4.8 \times 10^{-10}$ esu

$$\text{Fractional charge} = \frac{\text{partial charge (q)}}{\text{Total charge }(q_o)}$$

$$= \frac{1.2\times10^{-10}\text{esu}}{4.802\times10^{-10}\text{esu}} = 0.25 = 2.5\times10^{-1}$$

61. (3) Non-polar molecules (Due to symmetry):

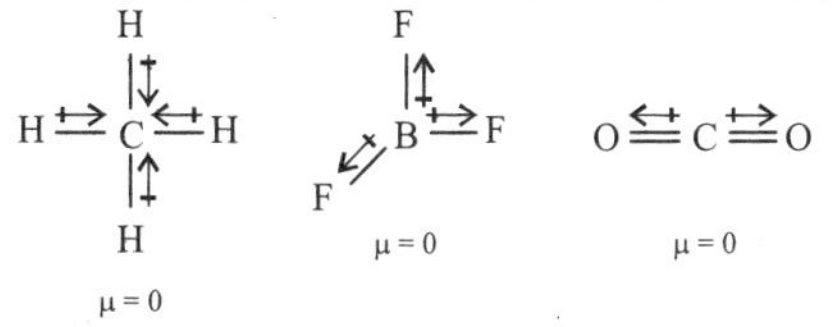

62. (4) The non-polar molecules are CO_2, H_2, CH_4 and BF_3

63. (3) (A), (B) and (C) are factors which affect the percent covalent character of the ionic bond according to Fajan's rule

64. (3) Symmetrical molecules have zero dipole moment like BF_3, BeF_2, SiF_4, CCl_4 and PF_5. Unsymmetrical molecules have net dipole moment like – NF_3, $CHCl_3$ and H_2S

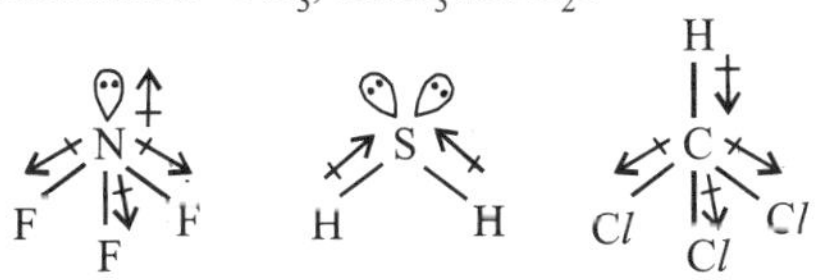

65. (3) BeF_2, BF_3 and $CCl_4 \Rightarrow \mu_{net} = 0$

H_2O, NH_3 and HCl $\Rightarrow \mu_{net} \neq 0$

66. (1) $2BrF_3 \rightleftharpoons BrF_2^+ + BrF_4^-$

$BrF_2^+ \Rightarrow sp^3$

[Bond pair = 2, lone pair = 2] ⇒ bent

$BrF_4^- \Rightarrow sp^3d^2$

[Bond pair = 4, lone pair = 2] ⇒ sqaure planar

67. (3)

Species	Bond angle
Cl_2O	111°
H_2O	104.5°
OF_2	103°

Thus order is, $OF_2 < H_2O < Cl_2O$

Statement I is true.

SiF_4 is covalent, thus statement II is false.

68. (4) SO_2, SO_3, SF_6 and SF_4 do not follow octet rule, having 10 e^-, 10 e^-, 12 e^- and 10 e^- respectively.

NH_3, BrF_5, SF_4 have one lone pair on central atom. ClF_3, XeF_4 and H_2O have two lone pair on central atom.

69. (4) $2Al(s) + 2NaOH(aq) + 6H_2O(l) \longrightarrow 2Na^+[Al(OH)_4]^-(aq) + 3H_2O(g)$

So, statement I is false.

$ClO_3^{\ominus}$ is sp^3 and pyramidal.

$ICl_4^{\ominus}$ is sp^3d^2 and square planar.

$IBr_2^{\ominus}$ is sp^3d and linear.

So, Statement II is true.

70. (2) Two molecules are isostructural if they possess exact same molecular geometry (shape).

Species **Structure/Shape**

SF_4 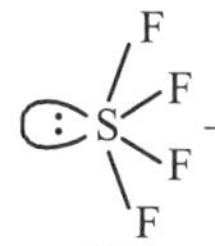 → see-saw

$IF_4^{\oplus}$ 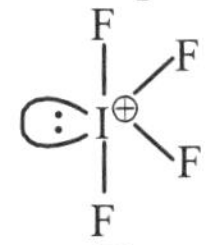 → see-saw

$BrF_4^{\ominus}$ 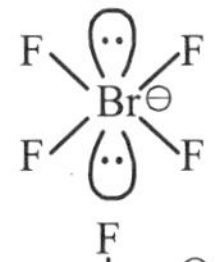 → square planar

XeO_2F_2 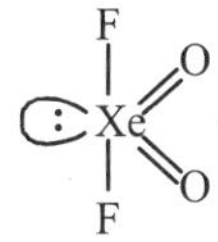 → see-saw

CH_4 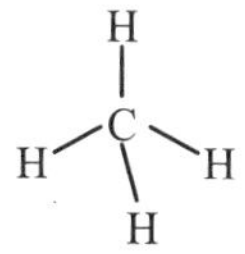 → tetrahedral

XeF_4  → square planar

Both IF_4^+ and XeO_2F_2 have a see-saw structure identical to SF_4.

71. (2)

(A) XeO_3 & NH_3 = (II) 3 bond pair + 1 lone pair = Pyramidal

(B) XeF_2 & I_3^- = (III) 2 bond pair + 3 lone pair = Linear

(C) XeO_2F_2 & SF_4 = (IV) 4 bond pair + 1 lone pair = See saw

(D) $XeOF_4$ & BrF_5 = (I) 5 bond pair + 1 lone pair = Square pyramidal

A–(II), B–(III), C–(IV), D–(I)

72. (1)

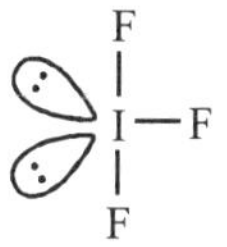

Hybridisation sp^3d (T-shape)

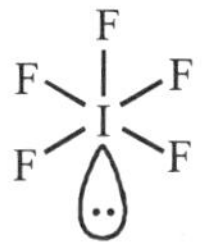

Hybridisation sp^3d^2 (Squar pyramidal)

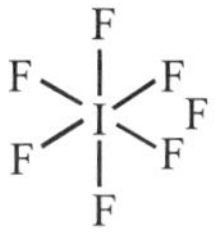

Hybridisation sp^3d^3 (Pentagonal bipyramidal)

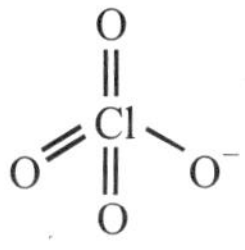

Hybridisation sp^3 (Tetrahedral)

73. (2) $XF_3 = BF_3$; sp^2 (Lewis acid)

$YF_3 = NF_3$; sp^3 (Lewis base)

74. (3) Statement-I is false.

SF_4 (See-saw)

XeF_4 (square planar),

$[PtCl_4]^{2-}$ (square planar),

$[NiCl_4]^{2-}$ (Tetrahedral),

$[Ni(CN)_4]^{2-}$ (square planar),

SeF_4 (See-saw)

NH_4^+ (Tetrahedral).

Only two species have tetrahedral geometry.

Statement I is false.

NO_2 (seven electrons on N)

BeH_2 (four electrons on Be)

BF_3 (six electrons on B)

$AlCl_3$ (six electrons on Al)

All the four species have incomplete octet around central atom.

Statement-II is true.

75. (1) $SO_2 \Rightarrow 2\sigma$ bond + 1 lone pair $\Rightarrow sp^2$

$NO_2 \Rightarrow 2\sigma$ bond + 1 lone pair $\Rightarrow sp^2$

$NO_3^- \Rightarrow 2\sigma$ bond $\Rightarrow sp$

76. (2)

(A) 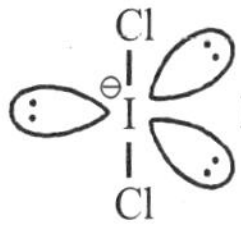 B.P : L.P = 2 : 3

(B) 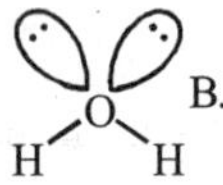 B.P : L.P = 2 : 2

(C) 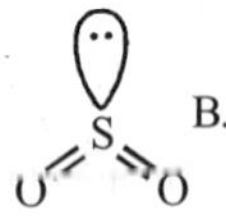 B.P : L.P = 4 : 1

(D) 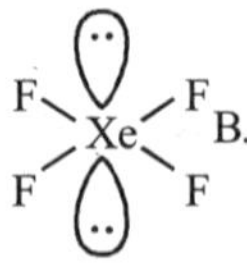 B.P : L.P = 4 : 2

77. (2) Statement II is incorrect since lone pairs cannot occupy axial positions. So structure I is most stable.

78. (1) PF_5 : 5 bond pair + 0 $\ell p \rightarrow sp^3d$

SF_6 : 6 bond pair + 0 $\ell p \rightarrow sp^3d^2$

$Ni(CO)_4$: Ni (0)

In presence of ligand field :-

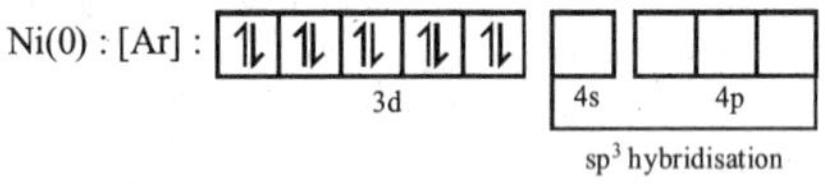

$[PtCl_4]^{2-}$: Pt^{2+}

In presence of ligand field :-

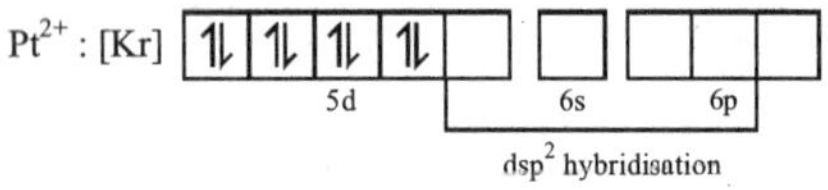

79. (3) 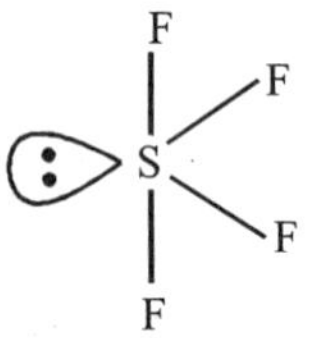

$(S-F)_{axial} > (S-F)_{equitorial}$

SF_4 has 4 bond pairs of electrons and 1 lone pair of electrons.

Hybridisation of SF_4 is sp^3d

80. (1) 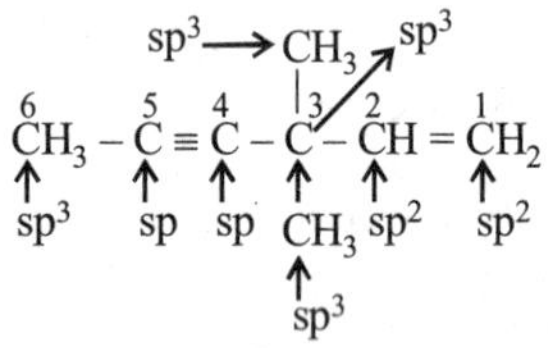

sp^3 C = 4, sp^2 C = 2, sp C = 2

81. (1) Given A is rarest, monoatomic, non-radioactive p-block element and form AX_4Y type of molecule.

∴ It is concluded that it is Xe.

It is given that electronegativity of A is less than X & Y

It is given that electronegativity of X & Y is highest and second highest respectively among all element.

∴ X & Y are flourine & oxygen respectively

∴ Compound is $XeOF_4$ with square pyramidal shape.

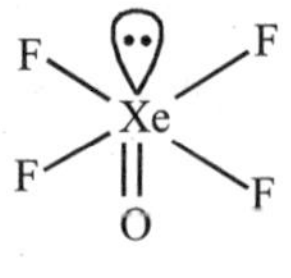

82. (4)

Molecule	Hybridisation	Dipole Moment	Lone pair on the central atom
SO_2	sp^2	Non-zero	1
NF_3	sp^3	Non-zero	1
NH_3	sp^3	Non-zero	1
XeF_2	sp^3d	zero	3
ClF_3	sp^3d	Non-zero	2
SF_4	sp^3d	Non-zero	1

83. (1)

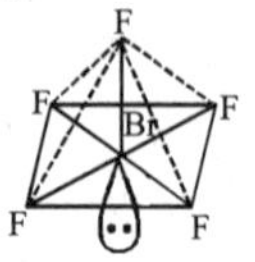

Square pyramidal

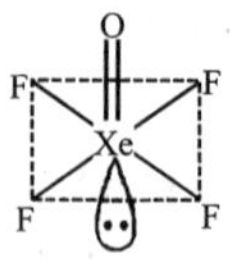

Square pyramidal

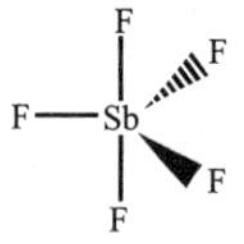

Trigonal Bipyramidal

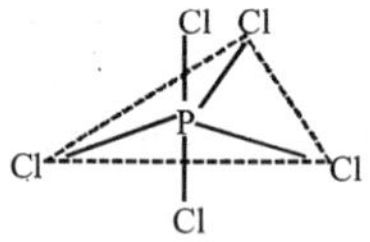

Trigonal Bipyramidal

BrF_5 : Square pyramidal due to one lone pair on Br

$XeOF_4$: Square pyramidal due to one lone pair on Xe

SbF_5 : Trigonal bipyramidal

PCl_5 : Trigonal bipyramidal

84. (2) 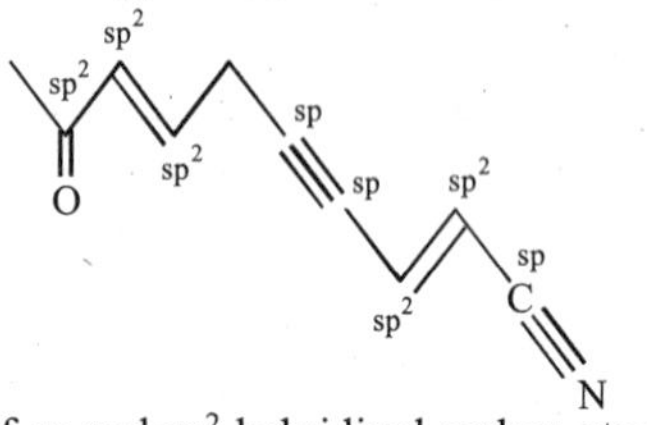

Number of sp and sp^2 hybridised carbon atom are 3 and 5.

85. (1)

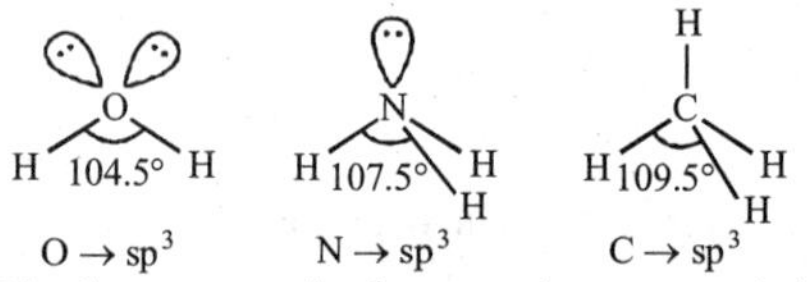

Dipole moment is due to, electronegativity difference and symmetry.

So order is $H_2O > NH_3 > CH_4$

H_2O & NH_3 are Lewis Bases.

NH_3 act as Lowry-Bronsted base.

Hence, A, B & C are correct.

86. (1) $BF_3 \rightarrow sp^2(3bp)$

$NO_2^- \rightarrow sp^2.(2bp + 1lp)$

$H_2O \rightarrow sp^3(2bp + 2lp)$

$NO_2 \rightarrow sp^2(2bp + 1\ e^-$ in hybrid orbital)

$NH_2^- \rightarrow sp^3(2bp + 2lp)$

87. (4) Oxidation state of an element in a particular compound is defined by the charge acquired by its atom on the basis of **electronegativity** consideration from other atoms in molecule. $d\pi$-$d\pi$ and $p\pi$-$d\pi$ are more prevalent in 3rd period or later.

88. (2) Antibonding molecular orbitals are formed by destructive interference of wave functions.

(ABMO) $\sigma^* = \psi_A - \psi_B$

89. (3) A-(III), B-(I), C-(IV), D-(II)

90. (3) I $\begin{cases} PF_5 - sp^3d \\ Br\ F_5 - sp^3d^2 \end{cases}$

II $\begin{cases} SF_6 \quad sp^3d^2 \\ [Co(NH_3)_6]^{3+} \quad d^2sp^3 \end{cases}$

Hybridisation of P in PF_5 is sp^3d but the hybridisation of Br in BrF_5 is sp^3d^2. So, statement-1 is false.

Hybridisation of S in SF_6 is sp^3d^2 but the hybridisation of Co^{3+} in $[Co(NH_3)_6]^{3+}$ is d^2sp^3 and NH_3 is a strong field ligand forcing the unpaired electrons to pair up. So, statement-II false.

91. (2) A-(III), B-(IV), C-(II), D-(I)

92. (2) A-(III), B-(I), C-(II), D-(IV)

93. (4) A-III, B-IV, C-I, D-II

94. (3) A–(IV), B–(I), C–(II), D–(III)

95. (3)

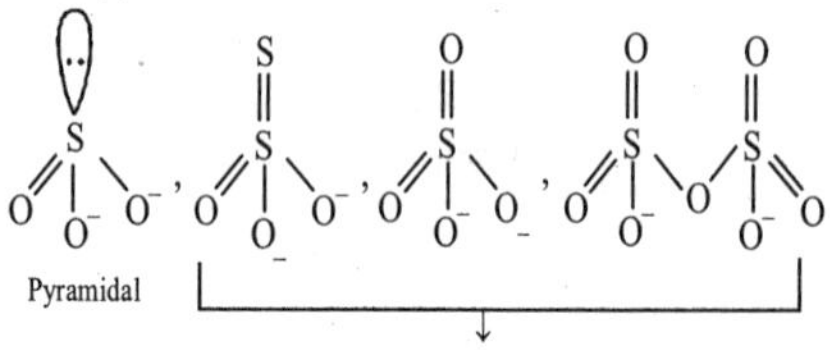

Tetrahedral with respect to central atom

96. (1)

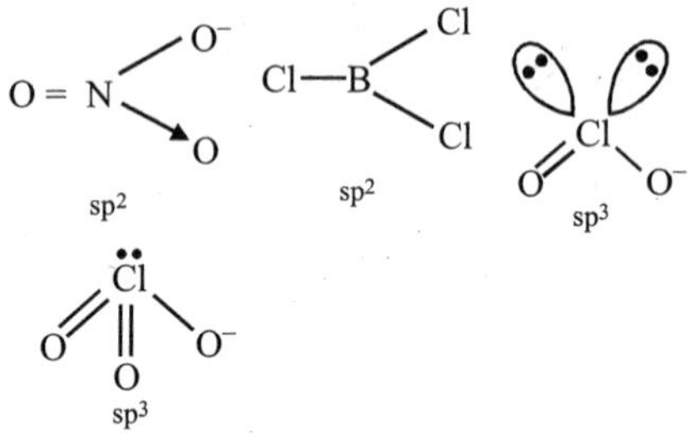

97. (3)

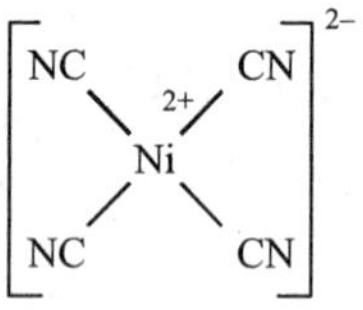

Square Planar

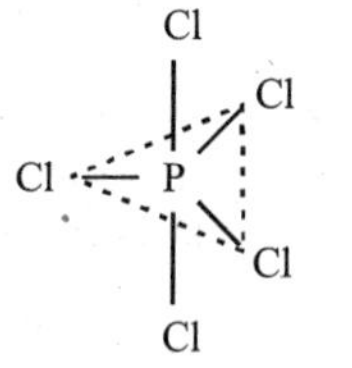

Trigonal bipyramidal

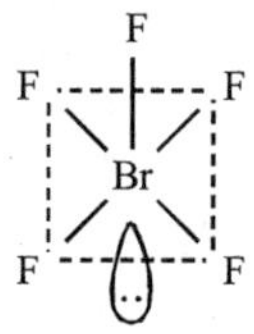

Square Pyramidal

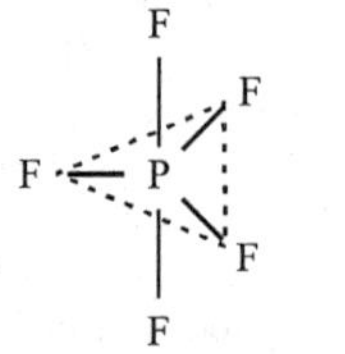

Trigonal bipyramidal

98. (4) (A) – (IV), (B) – (III), (C) – (I), (D) – (II)

99. (2) 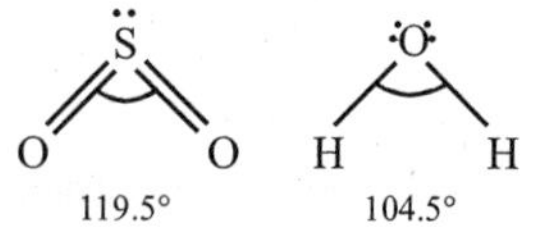

Both are bent in shape. (V-shaped)

Bond angle of SO_2 (sp^2) is greater than that of H_2O (sp^3) due to higher repulsion between bond pair and lone pair electrons in H_2O.

100. (1) A-(III), B-(II), C-(I), D-(IV)

101. (4)

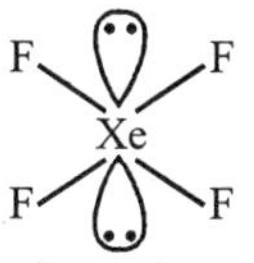

Square planar

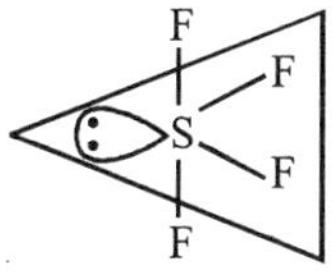

See-Saw

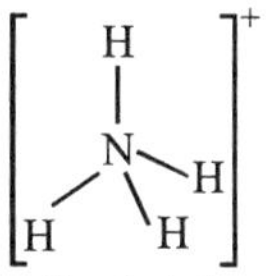

Tetrahedral

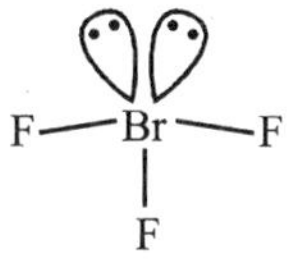

Bent T-Shaped

102. (2) A - IV, B - III, C - II, D - I

103. (4)

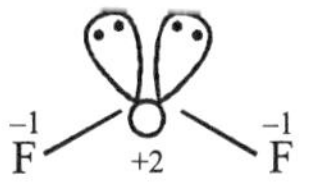

Molecule is bent 'V' shaped and oxidation state of oxygen is (+2).

104. (2)

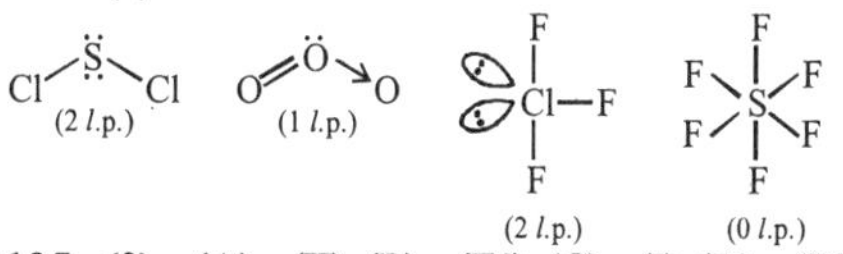

105. (3) (A) – (II), (B) – (IV), (C) – (I), (D) – (III)

106. (2) Isoelectronic species have same number of electrons valence

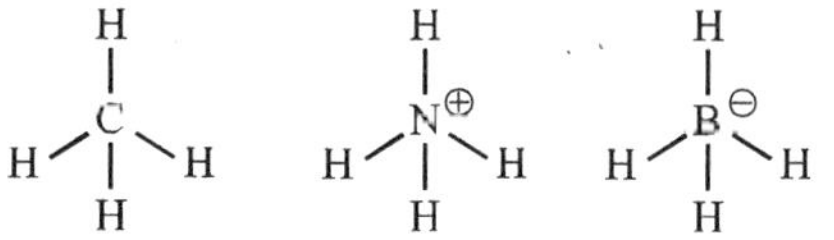

All have steric number 4.

All are tetrahedral and each have 10 electrons.

107. (3) As Fluorine is more electronegative it occupies the axial position and lone pairs occupy equitorial position

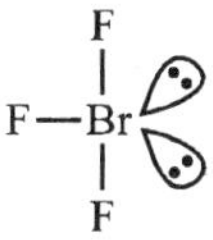

Steric no. = 5 (sp^3 d), lone pair = 2

Bent T shape.

108. (1) For sulphur steric number = 4 + 1 = 5

Hybridization = sp^3d, shape – see-saw

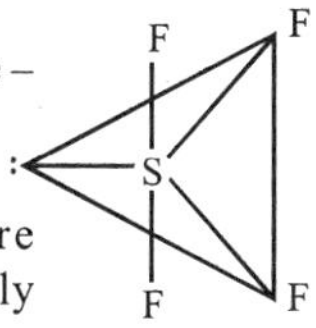

Fluorine being more electronegative preferentially occupy the axial position.

Structure is :

109. (3) The sp^3d hybridization in PCl_5 is a combination of $[sp^2 + pd]$ mixing.

sp^2 – hybrid orbitals form equitorial bonds and pd hybrid orbitals forms axial bonds. Hence, axial bonds are longer and weaker than equitorial bonds due to absence of s-character.

110. (2) (A) – (III), (B) – (IV), (C) – (I), (D) – (II)

111. (2) (A) → (III), (B) → (IV), (C) → (I), (D) → (II)

112. (4) NO_2^- 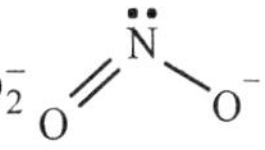 $2\sigma + 1$ lp $\therefore sp^2$

$NO_2^+ \Rightarrow O = \overset{\oplus}{N} = O \; 2\sigma + 0$ lp $\therefore$ sp

$NH_4^+ \Rightarrow$ 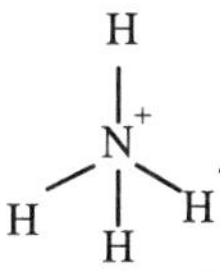 $4\sigma + 0 \; \therefore sp^3$

113. (3) For 3 bond pairs and 2 lone pairs of e^-, the geometry will be trigonal bi-pyramidal and shape is T-shaped.

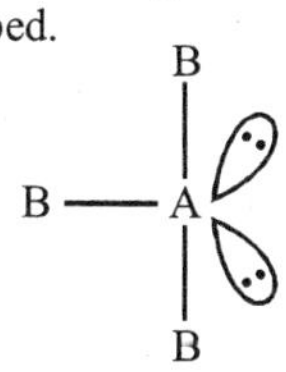

(AB_3)

114. (3)

Species	Hybridisation (shape)	Bond length
XeF_4	sp^3d^2 (Sq. planar)	All bond lengths equal
BF_4^-	sp^3 (Tetrahedral)	All bond lengths equal
SF_4	sp^3d (See-saw)	Axial B.L. > Equatorial B.L.
SiF_4	sp^3 (Tetrahedral)	All bond lengths equal

115. (4) H_2O - 104.5° (sp^3 with 2 lone pair at O)

NH_3 - 107° (sp^3 with 1 lone pair at N)

CH_4 - 109.5° (sp^3)

H_2S - 92° (sp^3 with 2 lone pair at S)

Lone pair-bond pair repulsion in H_2S will increase because 'S' has lower electronegativity than 'O'. So there will be lesser electron density on 'S' and thus H–S–H bond angle will be smaller than H_2O.

116. (1) In case of tetrahedral, rectangular planar and square planar geometries, there are only bond pair of electrons by four bonds. Square pyramidal geometry is possible with four bond pairs and one lone pair of e^-. Hence, it will be polar due to dipole moment of these e^-.

117. (1)

(i) XeF_5^- Steric No. = Bond pair + Lone Pair = (5 + 2) = 7

So, hybridisation is = sp^3d^3

and structure is pentagonal planar.

(ii) XeO_3F_2 Steric No. = 5

So, hybridisation is = sp^3d

and structure is trigonal bipyramidal.

118. (2) SF_4

Bond pair = 4

Lone pair = 1

Steric number = 5,

So, hybridisation is sp^3d.

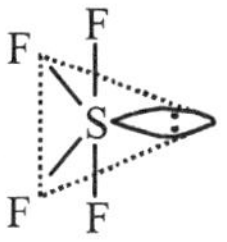

Geometry is trigonal bipyramidal but shape is "See-Saw".

119. (4) ICl_5 is sp^3d^2 hybridised

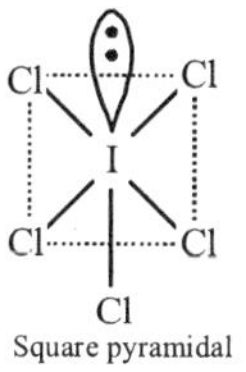

Square pyramidal

(5 bp, 1 lp)

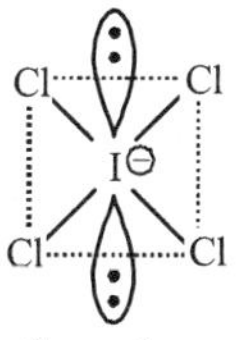

Square planar

ICl_4^- is sp^3d^2 hybridised (4 bp, 2 lp)

120. (1)

Species	Hybridisation
ICl_2^-	sp^3d
ICl_4^-	sp^3d^2
BrF_2^-	sp^3d
IF_6^-	sp^3d^3

121. (1) Structure of XeO_3F_2

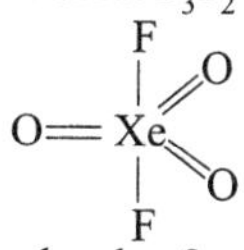

So, bond pairs = 5, π bonds = 3 and lone pairs = 0

122. (1) 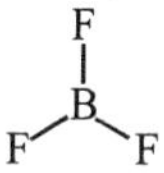 B is sp^2, bond angle = 120°

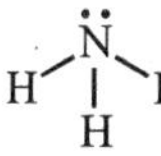 N is sp^3 with 1 lp, bond angle = 107°

 P is sp^3 with 1 lp, when central atom size↑, bond angle ↓, $\therefore NH_3 > PF_3$

I_3^- is sp^3d (linear), bond angle = 180°

$\therefore$ Dereasing order of bond angle is

$$I_3^- > BF_3 > NH_3 > PF_3$$

123. (2) $BF_3 \longrightarrow BF_4^-$

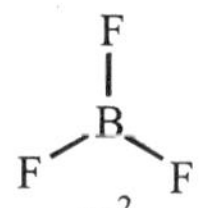

sp^2

Trigonal planar

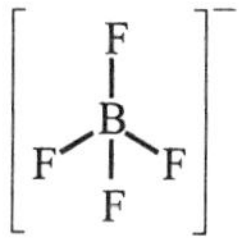

sp^3

Tetrahedral

124. (1) NF_3 has trigonal pyramidal shape. N atom has one lone pair and three bond pairs of electrons. The electron pair geometry is tetrahedral and molecular geometry is trigonal pyramidal. The bond angles are lower than tetrahedral bond angles due to lone pair - bond pair repulsions. N atom is sp^3 hybridised.

125. (4) $XeOF_2$ $XeOF_4$

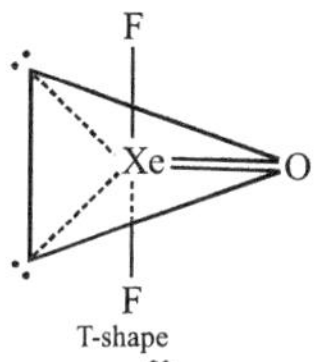

T-shape

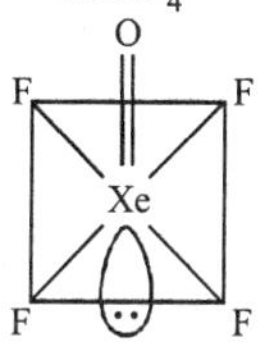

Square pyramidal

126. (3)

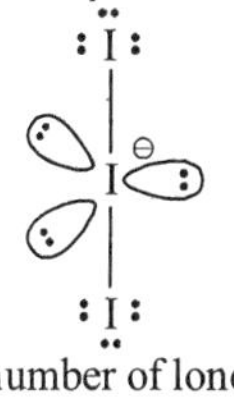

$\therefore$ Total number of lone pair of electrons is 9.

127. (2)

	Group	Hybridisation	Structure
(1)	BF_3	sp^2	Triangular Planar (T.P.)
	NF_3	sp^3	Tetrahedral (Td)
	CO_3^{2-}	sp^2	T.P
(2)	CO_3^{2-}	sp^2	T.P
	NO_3^-	sp^2 T.P	(All have same hybridisation)
	SO_3	sp^2 T.P	
(3)	NH_3	sp^3 Td	
	SO_3	sp^2	T.P
	CO_3^{2-}	sp^2	T.P
(4)	NCl_3	sp^3	Td
	BCl_3	sp^2	T.P
	SO_3	sp^2	T.P

128. (4)

PF_5 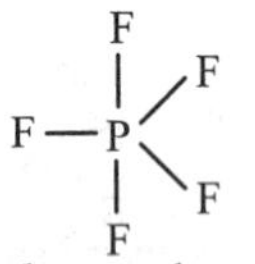 sp^3d

129. (2) More the number of lone pairs on central atom, the greater is the contraction caused in the angle between bond pairs. In CH_4 there is no lone pair of electrons, hence bond angle is greatest.

130. (4) $ClF_3 \longrightarrow$ Hybridisations sp^3d

$XeOF_2 \longrightarrow$ Hybridisation sp^3d

$XeF_3^+ \longrightarrow$ Hybridisation sp^3d

All molecules have sp^3d hybridisation and 2 lone pairs. Hence all have identical shape (T-shape).

131. (3) NO_2^+ sp hybridisation

NO_2^- sp^2 hybridisation

NO_3^- sp^2 hybridisation

The Lewis structure of NO_2 shows a bent molecular geometry with trigonal planar electron pair geometry, hence the hybridization will be sp^2.

132. (1)

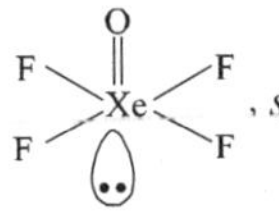 , sp^3d^2, no. of lone pair = 1

Square Pyramidal

133. (2) The structure of CaC_2 is

$Ca^{2+}[:C\equiv C:]^{2-}$,

i.e, one σ and two π bonds.

134. (4) $H \overset{\sigma}{—} C \overset{\pi,\,\sigma}{\underset{\pi}{\equiv}} N$

Therefore, HCN has 2 π and 2 σ bonds.

135. (4) All have tetrahedral structure.

136. (1) The structure of IF_6^- is distorted octahedral.

This is due to presence of a "weak" lone pair.

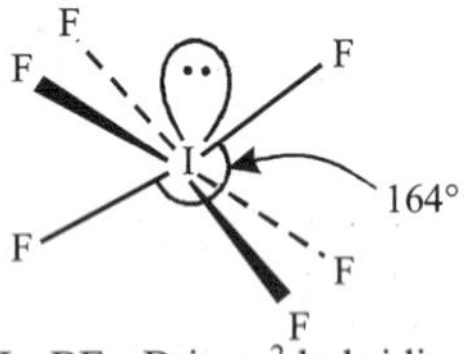

137. (2) In BF_3, B is sp^2 hybridized with one empty p_z orbital. The empty p_z orbital of BF_3 can be filled by lone pair of molecules such as NH_3. When this occurs a tetrahedral molecule or ion is formed which is sp^3 hybridized.

138. (1) XeF_4 has square planar structure, while NH_4^+, BF_4^- and CCl_4 have tetrahedral structure.

139. (1)

Species	Hybridisation	Shape
NI_3	sp^3	Trigonal pyramidal
I_3^-	sp^3d	Linear
SO_3^{2-}	sp^3	Trigonal pyramidal
NO_3^-	sp^2	Trigonal planar

Hence, NI_3 and SO_3^{2-} have same shape.

140. (3) PF_5 is trigonal bipyramidal

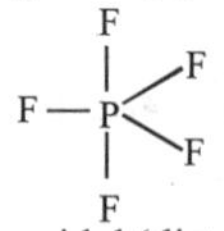

BrF_5 is square pyramidal (distorted)

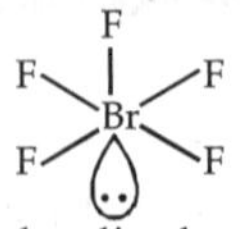

141. (2) The delocalised $p\pi - p\pi$ bonding between filled p-orbital of F and vacant p-orbital of B leads to shortening of B – F bond length which results in higher bond dissociation energy of the B – F bond.

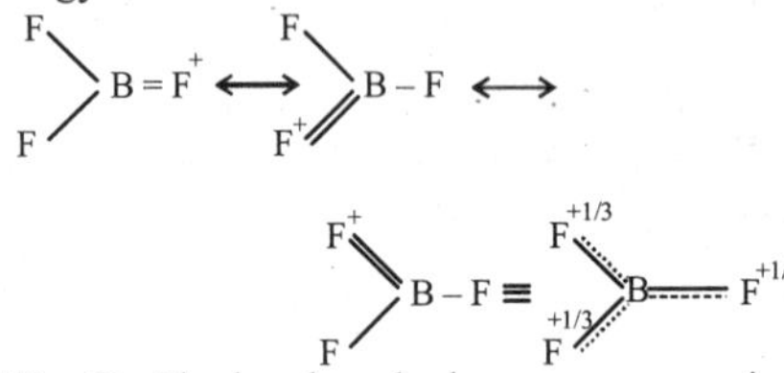

142. (2) The bond angle decreases on moving down the group due to decrease in bond pair-bond pair repulsion.

NH_3	PH_3	AsH_3	SbH_3	BiH_3
107°	94°	92°	91°	90°

143. (2)

144. (4)

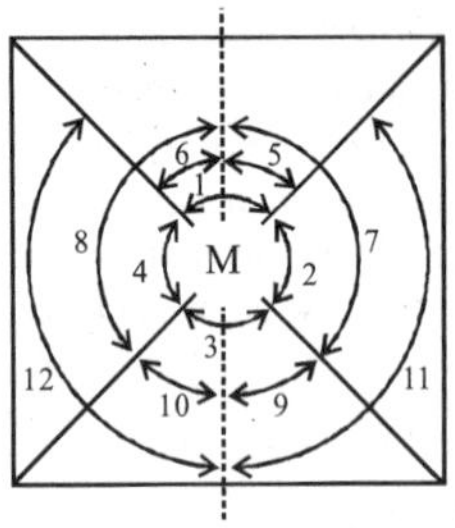

sp^3d^2 Hybridisation

Number of 90° angle between bonds = 12

145. (1) XeF_4 (sp^3d^2, square planar),

$[Ni(CN)_4]^{2-}$ (dsp^2, square planar),

BF_4^- (sp^3, tetrahedral), SF_4 (sp^3d, see saw shaped)

146. (1) Both XeF_2 and CO_2 have a linear structure.

F — Xe — F   O = C = O

147. (1) Hybridisation = No. of σ bonds + lone pair of e^- with central atom.

(1) For AlH_3,
Hybridisation of Al atom = 3 + 0 = 3 = sp^2
For AlH_4^-,
Hybridisation of Al atom = 4 + 0 = 4 = sp^3

(2) For H_2O,
Hybridisation of O atom = 2 + 2 = 4 = sp^3
For H_3O^+, Hybridisation of O atom = 3 + 1 = 4 = sp^3

(3) For NH_3, Hybridisation of N atom = 3 + 1 = 4 = sp^3
For NH_4^+, Hybridisation of N atom = 4 + 0 = 4 = sp^3

Thus, hybridisation changes only in option (1).

148. (1) In NH_3 and BF_4^-, the hybridisation is sp^3 and the bond angle is almost 109° 28'.

149. (4)

	Species	Hybridisation
(1)	BrF_5	sp^3d^2
(2)	$XeF_5^{\ominus}$	sp^3d^3
(3)	$BF_4^{\ominus}$	sp^3
(4)	$ICl_4^{\ominus}$	sp^3d^2
(5)	XeF_4	sp^3d^2
(6)	SF_4	sp^3d
(7)	$NH_4^{\oplus}$	sp^3
(8)	ClF_3	sp^3d
(9)	XeF_2	sp^3d
(10)	$ICl_2^{\ominus}$	sp^3d

Number of species having sp^3d hybridisation are 4.

150. (6) 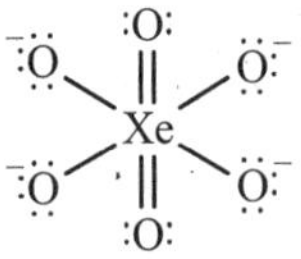

For central metal Xe,

σ Bond pairs of e^- around Xe = 6

Lone pairs of e^- around Xe = 0.

151. (6) Linear species are

Cl – Be – Cl, O = C = O, $N^- = N^+ = N^-$

(sp) (sp) (sp)

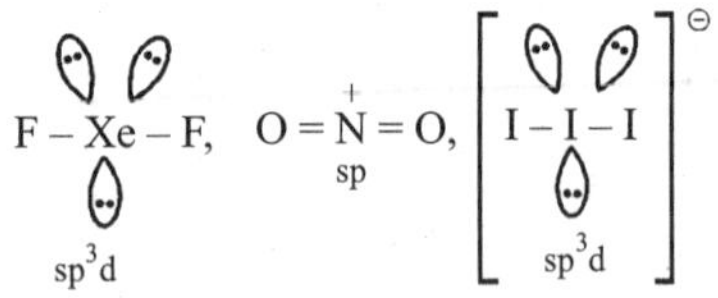

152. (6) SO_2, C_2H_4, BCl_3, HCHO, C_6H_6, BF_3 are sp^2 hybridised central atom

153. (3)

PF_5: 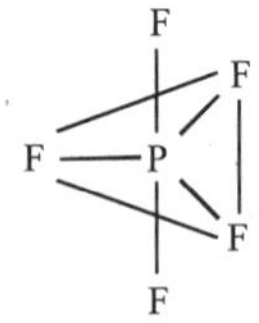 Trigonal bipyramidal

BrF_5: 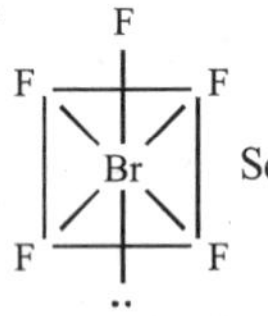 Square Pyramidal

PCl_5: 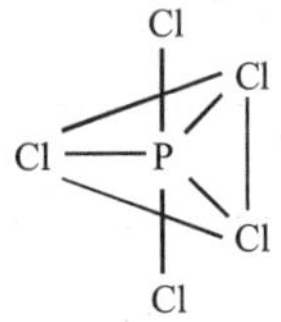 Trigonal bipyramidal

$[PtCl_4]^{2-}$: 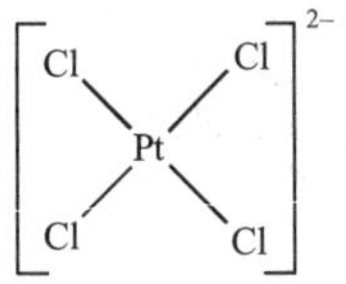 Square Planar

BF_3: 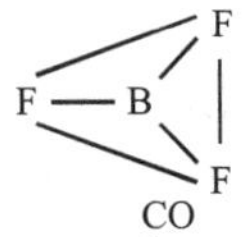 Trigonal Planar

$Fe(CO)_5$: 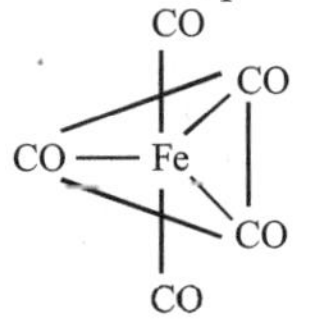 Trigonal bipyramidal

154. (3) $NH_3 \rightarrow sp^3$ $SO_2 \rightarrow sp^2$
$SiO_2 \rightarrow sp$ $BeCl_2 \rightarrow sp$
$CO_2 \rightarrow sp$ $H_2O \rightarrow sp^3$
$CH_4 \rightarrow sp^3$ $BF_3 \rightarrow sp^2$

155. (8) According to molecular orbital theory, number of molecular orbitals formed is equal to number of atomic orbitals combined, so when a diatomic molecule forms by combining of 2s and 2p orbitals 8-molecular orbitals are formed.
Two molecular orbital $\sigma 2s$ and $\sigma^* 2s$
Six molecular orbitals $\sigma 2p_z$, $\sigma^* 2p_z$
$\pi 2p_x$, $\pi 2p_y$ and $\pi^* 2p_x$, $\pi^* 2p_y$

156. (4)

Molecule No. of	H_2O	O_3	SF_4	ClF_3	NH_3	BrF_5	XeF_4
lone pair on central atom	2	1	1	2	1	1	2

157. (3)

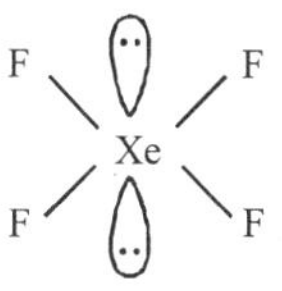

[2 lone pair]

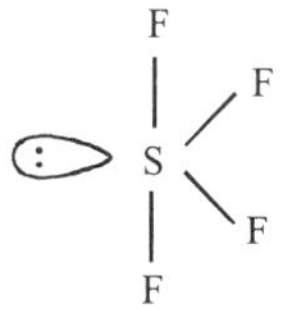

[1 lone pair]

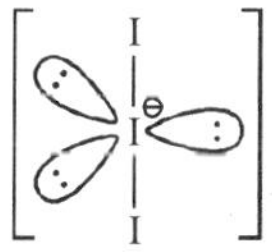

[3 lone pair]

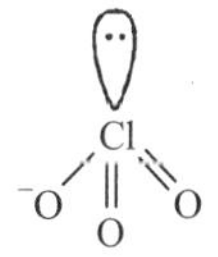

[1 lone pair]

158. (4) H_2O, CO, N_2, XeF_4, has two lone pair of electrons.

159. (3) $N_3^- =$ linear (4 b.p., 0 l.p.)
$NO_2^- =$ bent (3 b.p., 1 l.p.)
$I_3^- =$ linear (2 b.p., 3 l.p.)
$O_3 =$ bent (3 b.p., 1 l.p.)
$SO_2 =$ bent (4 b.p., 1 l.p.)

160. (4) XeF_4, BrF_4^-, $[Cu(NH_3)_4]^{+2}$, $[PtCl_4]^{2-}$ has square planar shape.
$[FeCl_4]^{2-}$, SiF_4 BrF_4^- has tetrahedral shape.
$SF_4 \rightarrow$ Trigonal bipyramidal geometry, See – Saw Shape.

161. (3) PF_5 : sp^3d (0 lone pair)
BrF_4^- : sp^3d^2 (2 lone pair)
IF_5 : sp^3d^2 (1 lone pair)
BrF_5 : sp^3d^2 (1 lone pair)
$XeOF_4$: sp^3d^2 (1 lone pair)
ICl_4^- : sp^3d^2 (2 lone pair)

162. (5)

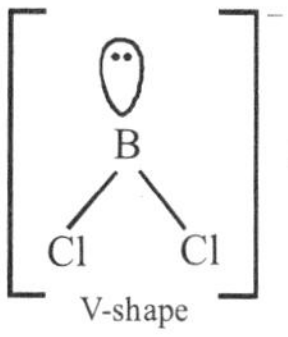

V-shape

;Cl – Be – Cl; Linear

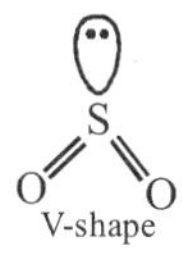

V-shape

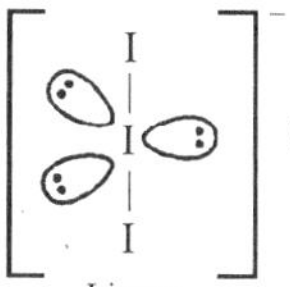

Linear

;

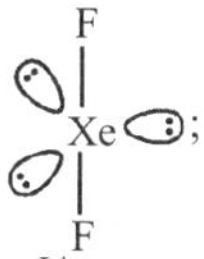

Linear

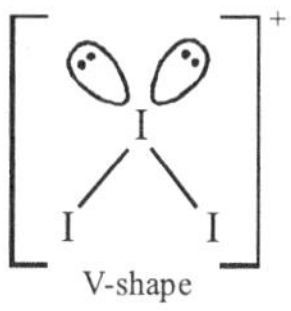

V-shape

O = C = O Linear   O = C = C = C = O Linear

163. (6)

SO_3	—	sp^2, Planar
BF_3	—	sp^2, Planar
NO_3^-	—	sp^2, Planar
SF_4	—	sp^3d, Non-planar
H_2O_2		sp^3, Non planar
PCl_3	—	sp^3, Non-planar
$[Al(OH)_4]^-$	—	sp^3, Non-planar
XeF_4	—	sp^3d^2, Planar
XeO_3	—	sp^3, Non-planar
PH_4^+	—	sp^3, Non-planar

164. (4)

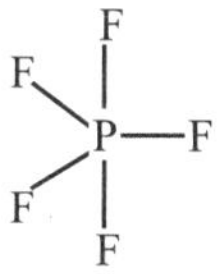

sp^3d

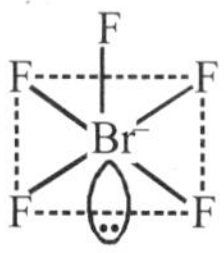

sp^3d^2

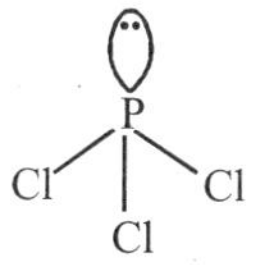

sp^3

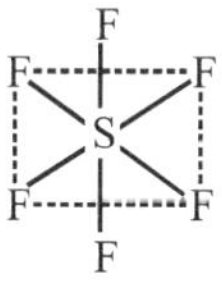

sp^3d^2

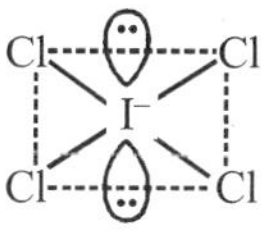

sp^3d^2

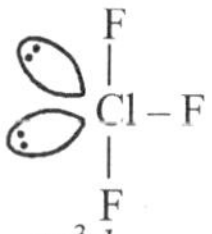

sp^3d

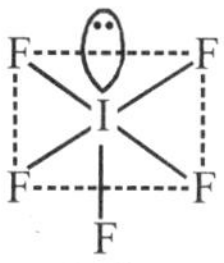

sp^3d^2

165. (1) S.N = 5[5 b.p + 0 *l*. p]
$PF_5 \Rightarrow sp^3d$ hybridisation

166. (2)

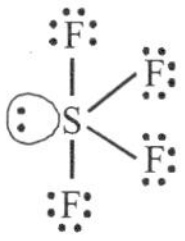

l.p. = 1

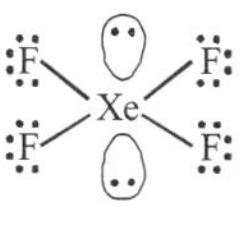

l.p. = 2

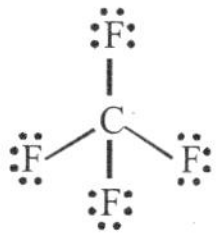

l.p. = 0

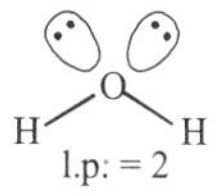

l.p. = 2

167. (3)

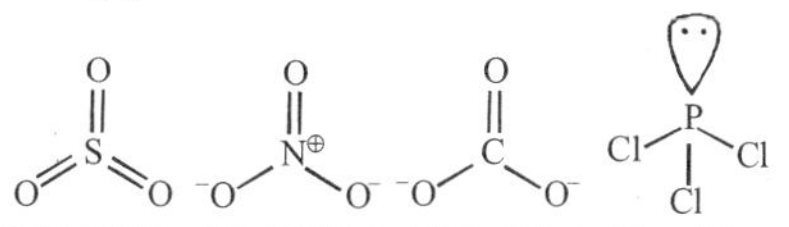

Trigonal planar Trigonal planar Trigonal planar Pyramidal

Hence, non-pyramidal species are SO_3, NO_3^- and CO_3^{2-}.

168. (2) T-shaped molecule means central atom has 3 sigma bond and 2 lone pairs of electron.

169. (3) I_3^- : It is sp^3d hydridised.

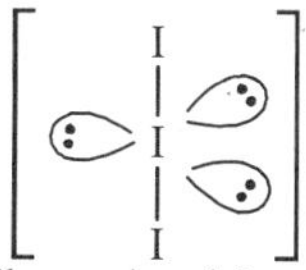

The number of lone pairs of electron on the central atom is 3.

170. (2) Short Trick:

$$\text{Lone pairs} = \frac{\text{Valence electron of central atom} + \text{charge of ion} - \text{Total bonded electron}}{2}$$

So, only ClF_3 and XeF_4 have two lone pairs.

171. (3)

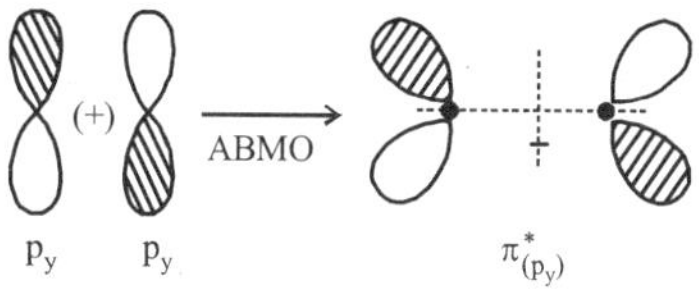

Destructive Interference

172. (3)

Species	Bond order	Unpaired electrons
$O_2^{\oplus}$	2.5	1
O_2	2	2
O_2^-	1.5	1
O_2^{2-}	1	0

Since, Bond length $\propto \frac{1}{\text{Bond order}}$
Thus, statement I is true.
The sequence of unpaired electrons is

$$O_2 > O_2^+ - O_2^- > O_2^{2-}$$

Thus, statement II is false.

173. (1) Among the four, BF_4^-, SiF_4 and XeF_4 have identical bond lengths.
SF_4 (has see saw structure) has unequal bond length.

Molecule	O_2^+	O_2^-	O_2^{2-}	F_2
B.O.	2.5	1.5	1	1

Thus O^+_2 has highest bond order.

174. (3) $O_2^+(15\,e) \rightarrow$ Paramagnetic

$$\text{Bond order} = \frac{\text{Bondng e} - \text{Antibonding e}}{2}$$

$$= \frac{(10-5)}{2} = 2.5$$

$N_2^-(15e) \rightarrow$ Paramagnetic

$$\text{Bond order} = \frac{(10-5)}{2} = 2.5$$

N_2^+ (13e) $\rightarrow$ Paramagnetic

$$\text{Bond order} = \frac{(9-4)}{2} = 2.5$$

O_2^- (17e) $\rightarrow$ Paramagnetic

$$\text{Bond order} = \frac{(10-7)}{2} - 1.5$$

N_2^{2-} (16e) $\rightarrow$ Paramagnetic

$$\text{Bond order} = \frac{(10-6)}{2} = 2$$

175. (4)

		No. of unpaired electrons
(A)	O_2	2
(B)	N_2	0
(C)	F_2	0
(D)	S_2	2
(E)	Cl_2	0

If species contain unpaired electron than it is paramagnetic.
So A & D are paramagnetic. While B, C or E are diamagnetic.

176. (3)

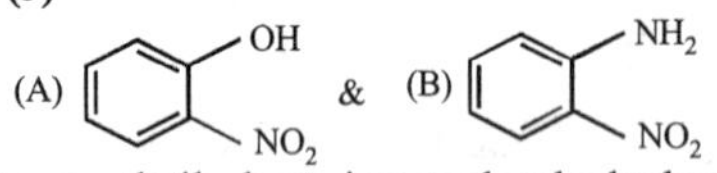

are steam volatile due to intramolecular hydrogen bonding. Compound (C) and (D) shows intermolecular hydrogen bonding.

177. (N)

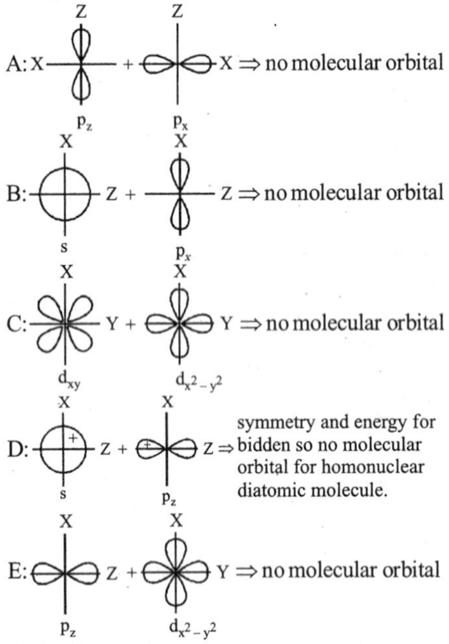

178. (2)
(B) Intramolecular H bonding is present in o-nitrophenol.

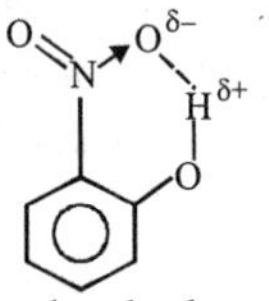

(C) Intermolecular hydrogen bonding is present in HF.

179. (1)

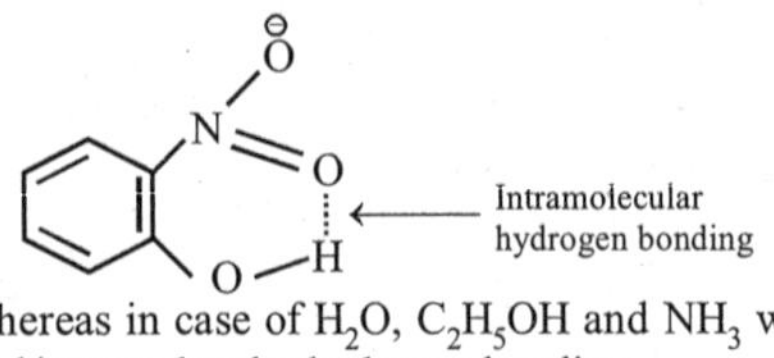

Whereas in case of H_2O, C_2H_5OH and NH_3 we find intermolecular hydrogen bonding.

180. (4) Statement I is false but statement II is true.
A π bonding MO has larger electron density above and below the inter nuclear axis.

181. (2) (B)Combining atomic orbitals should have maximum overlap.
(D)They have same symmetry about the molecular axis.

182. (4)
(A) NF_3 has trigonal pyramidal shape.
(B) Bond order $\Rightarrow N_2 > O_2$
Bond length $\Rightarrow N_2 < O_2$
(C) Isoelectronic species have same bond order due to same total no. of electrons.
(D) Dipole moment $H_2O > H_2S$ due to Electronegativity difference.

183. (2) NO is paramagnetic with BO = 2.5, NO^+ is diamagnetic with BO = 3

184. (1) Acetylide ion $\rightarrow C_2^{2-}$ $(\bar{C} \equiv \bar{C})$
Bond order = 3 & Diamagnetic
NO^+, $14e^- \rightarrow$ Bond order = 3 & Diamagnetic

185. (1) For O_2^{2-}: 18 electrons ⇒ Bond order : 1
CO : 14 electrons ⇒ Bond order : 3
NO^+ : 14 electrons ⇒ Bond order : 3
Short trick : for N_2 bond order is 3, no. of electrons are 14
∴ for 18 electrons bond order
$= [3-(18-14) \times 0.5] = 3-2 = 1$

186. (1)
$N_2 \Rightarrow$ 0 unpaired e^- in HOMO
$N_2^+ \Rightarrow$ 1 unpaired e^- in HOMO
$O_2 \Rightarrow$ 2 unpaired e^- in HOMO
$O_2^+ \Rightarrow$ 1 unpaired e^- in HOMO

187. (1) H-bonding stabilises the structure.

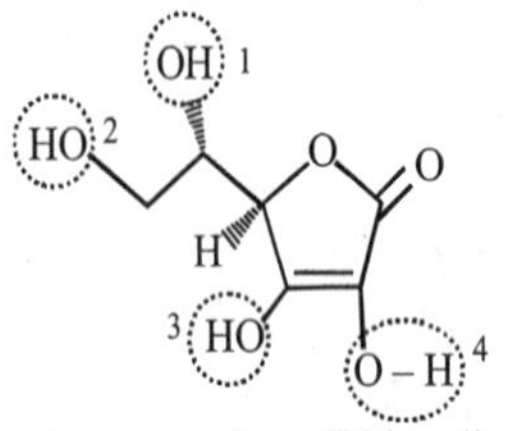

There is maximum number of H-bonding between 1, 2, 3 and 4 –OH groups

188. (2) Ice > Liquid water > Impure water
Due to impurity extent of H-Bonding decreases.

189. (1) Zero overlapping is occur due to improper orientation of orbitals and it is an out of phase overlaping.

190. (3) (A)-(III), (B)-(I), (C)-(IV), (D)-(II)

191. (3) The strength of H–bonding depends upon two factors:
(1) Electronegativity difference between the hydrogen atom and electronegative atom.
(2) Number of H-bonding interaction.
So the order of H-bonding is $CH_4 < HCN < NH_3$.

192. (1)

Bond order: $O_2^{2-} < O_2^- < O_2 < O_2^+$

193. (3) Bond strength ∝ Bond order
Removal of electron from antibonding MO increases B.O.
NO and O_2 have valence e^- in π* orbital.

M.O. of NO^+: $\sigma_{1s}^2\, \sigma_{1s}^{*2}\, \sigma_{2s}^2 \sigma_{2s}^{*2}\, \sigma_{2p_z}^2\, \pi_{2p_x}^2\, \pi_{2p_y}^2$

194. (2) $O_2^{2-} < N_2^{2-} < C_2^{2-}$

195. (1) According to the electronic configuration of O_2^- ion is

$$\sigma 1s^2\, \sigma^* 1s^2 \sigma 2s^2 \sigma^* 2s^2 \sigma 2p_z^2 \left(\pi 2p_x^2 = \pi 2p_y^2\right) \left(\pi^* 2p_x^2 = \pi^* 2p_y^1\right)$$

$$\text{Bond order} = \frac{10-7}{2} = 1.5$$

It also exhibit paramagnetic character due to the presence of unpaired electron.

196. (4) (A) - (III); (B) - (IV); (C) - (I); (D) - (II)

197. (2) Be_2 does not exist due to zero bond order

198. (2) When two H-atoms come closer then initially due to attraction P.E. is –ve, which decreases more as atoms come closer and after reaching to a minimum value as repulsion starts dominating, P.E., increases.

199. (1)

$$2\,PCl_5\,(g) \longrightarrow \underset{\text{Tetrahedral}}{[PCl_4]^+}\ \underset{\text{Octahedral}}{[PCl_6]^-}(s)$$

200. (1)
(1) $[Ni(CN)_4]^{2-} = dsp^2$
(2) $BrF_5 = sp^3d^2$
(3) $XeF_4 = sp^3d^2$
(4) $[CrF_6]^{3-} = d^2sp^3$

201. (4) Molecular orbital configuration for NO is

$$\sigma 1s^2\, \sigma^* 1s^2\, \sigma 2s^2\, \sigma^* 2s^2\, \sigma 2p_z^2\, (\pi 2p_x^2 = \pi 2p_y^2) \left(\pi^* 2p_x^1 = \pi 2p_y^{*0}\right)$$

Species	Bond order
NO^+	3
NO^{2+}	2.5
NO^-	2
NO	2.5

Bond strength is directly proportional to the bond order, so NO^- has minimum bond strength.

202. (1) $\mu = \sqrt{n(n+2)}$ B.M.
$1.73 = \sqrt{n(n+2)}$
$n = 1$

$O_2^+ - \sigma 1s^2\, \sigma^* 1s^2\, \sigma 2s^2\, \sigma^* 2s^2\, \sigma 2p_z^2\, (\pi 2p_x^2 = \pi 2p_y^2)\, (\pi^* 2p_x^1 = \pi^* 2p_y^0)$

$O_2^- - \sigma 1s^2\, \sigma^* 1s^2\, \sigma 2s^2\, \sigma^* 2s^2\, \sigma 2p_z^2\, (\pi 2p_x^2 = \pi 2p_y^2)\, (\pi^* 2p_x^2 = \pi^* 2p_y^1)$

203. (2) Total number of electrons in $CN^- = 6 + 7 + 1 = 14$
∴ Molecular orbital distribution

$$\sigma 1s^2\, \sigma^* 1s^2\, \sigma 2s^2\, \sigma^* 2s^2 \begin{bmatrix} \pi 2p_x^2 \\ \pi 2p_y^2 \end{bmatrix} \sigma 2p_z^2$$

$$\therefore \text{Bond order} = \frac{10-4}{2} = 3$$

CN^- is diamagnetic because all electrons are paired.

204. (3) Electronic configuration of O_2 is
$\sigma 1s^2 \sigma^* 1s^2 \sigma 2s^2 \sigma^* 2s^2 \sigma 2p_z^2$
$(\pi 2p_x^2 = \pi 2p_y^2)(\pi^* 2p_x^1 = \pi^* 2p_y^1)$
When an electron is added in O_2 to form O_2^-, the incoming electron goes to $\pi^* 2p_x$ or $\pi^* 2p_y$ orbital.

205. (2) The molecules with no unpaired electrons are diamagnetic.

Molecule	No. of unpaired electrons
NO	1
CO	Zero
O_2	2
B_2	2

Since CO has no unpaired electron. Hence CO is diamagnetic.

206. (1) Configuration of C_2^-

$\sigma 1s^2 \sigma^* 1s^2\, \sigma 2s^2 \sigma^* 2s^2\, (\pi 2p_x^2 = \pi 2p_y^2)$

Configuration of C_2^-

$\sigma 1s^2 \sigma^* 1s^2\, \sigma 2s^2 \sigma^* 2s^2\, (\pi 2p_x^2 = \pi 2p_y^2)\, \sigma 2p_z^1$

Bond order

$= \dfrac{\text{No. of bonding } e^- - \text{No. of antibonding } e^-}{2}$

C_2 has *s-p* mixing and the HOMO is $\pi 2p_x = \pi 2p_y$ and LUMO is $\sigma 2p_z$. So, the extra electron will occupy bonding molecular orbital and this will lead to increase in bond order.

So, C_2^- has more bond order than C_2.

207. (4) Bond length $\propto \dfrac{1}{\text{Bond order}}$

and diamagnetic species has no unpaired electron in their molecular orbitals.

No. of unpaired electrons	Bond order	Magnetic character
C_2^{2-}	03	diamagnetic
N_2^{2-}	22	paramagnetic
O_2^{2-}	01	diamagnetic
O_2	22	paramagnetic

$\therefore$ C_2^{2-} has maximum bond order, least bond length and is diamagnetic.

208. (4) $N_2^+ = 13e^- = \sigma 1s^2 \sigma^* 1s^2 \sigma 2s^2 \sigma^* 2s^2$

$\left(\pi 2p_x^2 = \pi 2p_y^2\right) \sigma 2p_z^1$

$\text{B.O.} = \dfrac{\text{Bonding electrons} - \text{Antibonding electrons}}{2}$

$\text{B.O.} = \dfrac{9-4}{2} = 2.5 = 2\pi \text{ bond} + 0.5\, \sigma \text{ bond}$

209. (2) Electronic configuratios of Li_2^+ and Li_2^-:

$Li_2^+ : \sigma 1s^2 \sigma^* 1s^2 \sigma 2s^1$

$Li_2^- : \sigma 1s^2 \sigma^* 1s^2 \sigma 2s^2 \sigma^* 2s^1$

Now,

Bond order of $Li_2^+ = \dfrac{1}{2}(3-2) = \dfrac{1}{2}$

Bond order of $Li_2^- = \dfrac{1}{2}(4-3) = \dfrac{1}{2}$

Li_2^+ have less electron in antibonding ortibal than Li_2^-. So Li_2^+ is stable and Li_2^- is unstable.

210. (1) In the resonance structure of hydrogen azide, it can be seen that number of N - N bonds for bond (I) ≤ 2.

$H-\ddot{N} \overset{(I)}{=} \overset{+}{N} \overset{(II)}{=} \ddot{\bar{N}} \longleftrightarrow H-\ddot{\bar{N}} \overset{(I)}{-} \overset{+}{N} \overset{(II)}{\equiv} \ddot{N}$

Hence for bond (I), bond order will be < 2 whereas for bond (II), its bond order will be > 2.

211. (4) An antibonding π orbital best describes the given diagram of a molecular orbital. Two orbitals laterally overlap to form π bond. Out of phase combination of these two *p*-orbitals give π^* MO.

212. (4) Total electrons in

$N_2^+ = (7\times 2) - 1 = 13$

$N_2^+ \to KK\sigma(2s)^2 \sigma^*(2s)^2 \pi(2p_x)^2 = \pi(2p_y)^2 \sigma(2p_z)^1$

Number of electron in $\sigma 2p_z$ is 1

213. (4)

Species	No. of e⁻s	Elec. conf.	Bond order
He_2^+	$(4-1=3)$	$\sigma 1s^2 \sigma^* 1s^1$	$\dfrac{2-1}{2} = 0.5$
H_2^-	$(2+1=3)$	$\sigma 1s^2 \sigma^* 1s^1$	$\dfrac{2-1}{2} = 0.5$
H_2^{2-}	$(2+2=4)$	$\sigma 1s^2 \sigma^* 1s^2$	$\dfrac{2-2}{2} = 0$
He_2^{2+}	$(4-2=2)$	$\sigma 1s^2$	$\dfrac{2-0}{2} = 1$

Molecule having zero bond order will not be a viable molecule.

214. (4) $B_2 \Rightarrow KK \cdot \sigma(2s)^2 \sigma^*(2s)^2 \cdot \pi(2p_x)^1$

$= \pi(2p_y)^1$ paramagnetic

215. (2) $CO(14) \to$

$KK\sigma 2s^2 \pi 2p_x^2 = \pi 2p_y^2 \sigma 2p_z^2 \sigma^* 2s^2$

No unpaired electron, hence diamagnetic.

216. (2) Assertion is correct but reason is in correct. Bonding MO shows constructive interference of the combining electron waves.

217. (3) The bond order of N_2, O_2, and O_2^- are 3, 2 and 1.5 respectively.

Since higher bond order implies higher bond dissociation energy, hence the correct order will be $N_2 > O_2 > O_2^-$

218. (2) The molecular orbital configuration of the molecules given is

Total no. of electrons in NO = 7(N) + 8(O) = 15

Hence E.C. of NO = $KK\sigma(2s)^2 \sigma^*(2s)^2 \sigma 2p_z^2$

$\pi(2p_x)^2 \pi(2p_y)^2 \pi^*(2p_x)^1$

Due to presence of one unpaired electron NO is paramagnetic. Except NO all are diamagnetic due to absence of unpaired electrons.

219. (2)

$O_2^-(17) \Rightarrow KK(\sigma 2s)^2 (\sigma 2s^2)(\sigma 2p_z)^2$

$(\pi 2p_x)^2 = (\pi 2p_y)^2 (\pi^* 2p_x)^2 = (\pi^* 2p_y)^1$

One unpaired electron - paramagnetic.

220. (1) Nitric oxide is paramagnetic in the gaseous state because of the presence of one unpaired electron in its outermost MO.

The electronic configuration of NO is

$\sigma KK\sigma 2s^2\, \sigma^* 2s^2 \sigma 2p_z^2\, \pi 2p_x^2 = \pi 2p_y^2\, \pi^* 2p_x^1 = \pi^* 2p_y^0$

221. (1) For NO

Total no. of electrons = 15

B.O = 2.5

Magnetic behaviour = paramagnetic

For NO^+

Total no. of electrons = 14

B.O = 3

Magnetic behaviour = diamagnetic

222. (1, 2) The molecular orbital structures of C_2 and N_2 are

$N_2 = KK\sigma 2s^2 \sigma^* 2s^2 \pi 2p_x^2 = \pi 2p_y^2 \sigma 2p_z^2$

$C_2 = KK\sigma 2s^2 \sigma^* 2s^2 \pi 2p_x^2 = \pi 2p_y^2$

Both N_2 and C_2 have paired electrons, hence they are diamagnetic.

223. (1)

(1) $ONCl = 8+7+17 = 32e^-$, $ONO^- = 8+7+8+1 = 24e^-$ } not isoelectronic

224. (3) $H_2^{2+} = \sigma 1s^0 \sigma^* 1s^0$

Bond order for $H_2^{2+} = \dfrac{1}{2}(0-0) = 0$

$He_2 = \sigma 1s^2 \sigma^* 1s^2$

Bond order for $He_2 = \dfrac{1}{2}(2-2) = 0$

So both H_2^{2+} and He_2 do not exist.

225. (2) $Li_2 = \sigma 1s^2 \sigma^* 1s^2 \sigma 2s^2$; $\therefore$ Bond order $= \dfrac{1}{2}(4-2) = 1$

$Li_2^+ = \sigma 1s^2 \sigma^* 1s^2 \sigma 2s^1$; B.O. $= \dfrac{1}{2}(3-2) = 0.5$

$Li_2^- = \sigma 1s^2 \sigma^* 1s^2 \sigma 2s^2 \sigma^* 2s^1$;

B.O. $= \dfrac{1}{2}(4-3) = 0.5$

The bond order of Li_2^+ and Li_2^- is same but Li_2^+ is more stable than Li_2^- because Li_2^+ has 2 electrons in antibonding orbitals whereas Li_2^- has 3 electrons in antibonding orbitals.

226. (4) The bond length follows the order :

$O_2^+ < O_2 < O_2^- < O_2^{2-}$

According to this, the possible values are :

1.12 Å, 1.21 Å, 1.30 Å, 1.49 Å

227. (N) None of the given option is correct. The molecular orbital configuration of the given molecules is

$H_2 = \sigma 1s^2 \sigma 1s^0$ (no anti-bonding electron)

$Li_2 = \sigma 1s^2 \sigma^* 1s^2 \sigma 2s^2$ (two anti-bonding electrons)

$B_2 = \sigma 1s^2 \sigma^* 1s^2 \sigma 2s^2 \sigma^* 2s^2 \left\{\pi 2p_y^1 = \pi 2p_z^1\right\}$

(4 anti-bonding electrons)

Though the bond order of all the species are same (B.O = 1) but stability is different. This is due to difference of no. of anti-bonding electrons. Higher the no. of anti-bonding electrons, lower is the stability. Hence, the correct order is $H_2 > Li_2 > B_2$

228. (2)

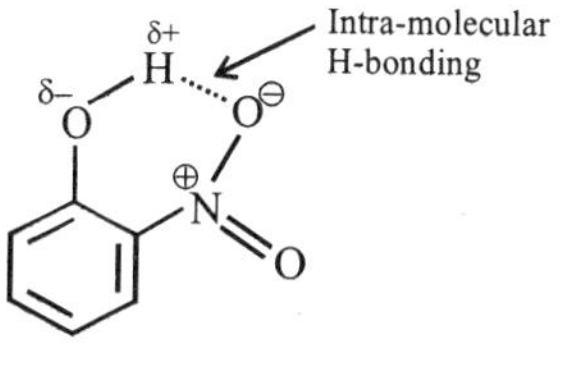

229. (2)

230. (4) Since, bond order $\propto \dfrac{1}{\text{Bond length}}$

$\therefore$ Bond length is shortest in O_2^{2+}.

231. (1) For any species to have same bond order we can expect them to have same number of electrons. Calculating the number of electrons in various species.

$O_2^-(8+8+1=17)$; $CN^-(6+7+1=14)$

$NO^+(7+8-1=14)$; $CN^+(6+7-1=12)$

We find CN^- and NO^+ both have 14 electrons, so they have same bond order.

232. (2) Diamagnetic species have no unpaired electrons whereas paramagnetic species has one or more unpaired electrons.
For electronic configuration of O_2^+, O_2 and O_2^{2-}
O_2 and O_2^+ have 2 and 1 unpaired electron respectively, while O_2^{2-} has no unpaired electron

$NO \rightarrow KK\sigma 2s^2\sigma^* 2s^2\sigma 2p_z^2\ \pi 2p_x^2 = \pi 2p_y^2$

$\pi^* 2p_x^1 = \pi^* 2p_y^0$

Thus No has one unpaired electron

233. (3) Greater the difference between electronegativity of bonded atoms, stronger will be bond. Since F is most electronegative, hence F – H ...... F is the strongest bond.

234. (3) The distribution of electrons in MOs is as follows :
N_2^+(electrons 13)

$\Rightarrow KK\sigma 2s^2\sigma^* 2s^2\sigma 2p_z^2\pi 2p_x^2 = \pi 2p_y^1$

$\Rightarrow O_2$(electron 16) $\Rightarrow KK\ \sigma 2s^2\ \sigma^* 2s^2\ \sigma 2p_x^2$
$\pi 2p_x^2 = \pi 2p_y^2\ \pi^* 2p_x^1 = \pi^* 2p_y^1$
$\Rightarrow O_2^{2-}$ (electron 18) $\Rightarrow KK\ \sigma 2s^2\ \sigma^* 2s^2\ \sigma 2p_x^2$
$\pi 2p_x^2 = \pi 2p_y^2\ \pi^* 2p_x^2 = \pi^* 2p_y^2$
B_2 (electrons 10) $\Rightarrow KK\ \sigma 2s^2\ \sigma^* 2s^2\ \sigma 2p_x^2$
$KK\ \sigma 2s^2\ \sigma^* 2s^2\ \pi 2p_x^1 = \pi 2p_y^1$
Only O_2^{2-} does not contain any unpaired electron.

235. (3) A diamagnetic substance contains no unpaired electron. H_2 is diamagnetic as it contains all paired electrons

$H_2 = \sigma 1s^2$, $H_2^+ = \sigma 1s^1$, $H_2^- = \sigma 1s^2\sigma 1s$
(diamagnetic) (paramagnetic) (paramagnetic)

$He_2^+ = \sigma 1s^2\ \sigma^* 1s$
(paramagnetic)

236. (2) Bond order of NO^+ (3) is higher than that of NO (2.5). Thus bond length of NO^+ will be shorter.

237. (3) In ether, there is no H-bonding while alcohols have intermolecular H-bonding.

238. (2) The correct order of their increasing bond strength is $O_2^{2-} < O_2^- < O_2 < O_2^+$

239. (6) Molecules having unpaired electrons are paramagnetic which are O_2, O_2^+, O_2^-, NO, NO_2, $K_2[NiCl_4]$.

240. (6) $O_2(16e^-) : (\sigma_{1s})^2\ (\sigma^*_{1s})^2 (\sigma_{2s})^2\ (\sigma^*_{2s})^2$

$(\sigma_{2p_z})^2 \left[(\pi_{2p_x})^2 = (\pi_{2p_y})^2\right], \left[(\pi^*_{2p_x})^1 = (\pi^*_{2p_y})^1\right]$

Number of e^{-1} present in (π^*) of $O_2 = 2$Number of e^{-1} present in (π^*) of $O_2^+ = 1$Number of e^{-1} present in (π^*) of $O_2^- = 3$

So total e^- in $(\pi^*) = 2 + 1 + 3 = 6$

241. (4) One unpaired e^- is present in :

$C_2^-; O_2^-; H_2^+; He_2^+$

N_2, O_2^{2-} and CN^- have zero while O_2 has two unpaired e^{-1}.
These calculations can be done using the molecular electronic configurations of these species.
for other species:
$\sigma 1s < \sigma^* 1s < \sigma 2s < \sigma^* 2s < \pi 2p_x = \pi 2p_y$
$< \sigma 2pz_1 < \pi 2p_x = \pi^* 2p_y < \sigma^* 2p_z$

For O_2, O_2^- and O_2^- :
$\sigma 1s < \sigma^* 1s < \sigma 2s < \sigma^* 2s < \sigma 2p_z < \pi 2p_x$
$= \pi 2p_y < \pi^* 2p_x = \pi^* 2p_y < \sigma^* 2p_z$

242. (2) $C_2\ (12e^-) : \sigma 1s^2, \sigma^* 1s^2, \sigma 2s^2, \sigma^* 2s^2$

$\left[\pi ?p_x^2 = \pi ?p_y^2\right]$

$B.O. = \frac{8-4}{2} = 2$

$O_2(16e^-) : \sigma 1s^2, \sigma^* 1s^2, \sigma 2s^2, \sigma^* 2s^2, \sigma 2pz^2$

$\left[\pi 2p_x^2 = \pi 2p_y^2\right]\left[\pi^* 2p_x^1 = \pi^* 2p_y^1\right]$

$B.O. = \frac{10-6}{2} = 2,$

$Be_2\ (8e^-) : \sigma 1s^2, \sigma^* 1s^2, \sigma 2s^2, \sigma^* 2s^2$

$B.O. = \frac{4-4}{2} = 0$, molecule not exist

$Li_2\ (6e^-) : \sigma 1s^2, \sigma^* 1s^2, \sigma 2s^2$

$B.O. = \frac{4-2}{2} = 1$, One bond formed

$Ne_2\ (20e^-) : \sigma 1s^2, \sigma^* 1s^2, \sigma 2s^2, \sigma^* 2s^2, \sigma 2pz^2$

$\left[\pi 2p_x^2 = \pi 2p_y^2\right]\left[\pi^* 2p_x^2 = \pi^* 2p_y^2\right]\sigma^* 2p_z^2$

$B.O. = \frac{10-10}{2} = 0$, molecule not exist

$N_2\ (14e^-) : \sigma 1s^2, \sigma^* 1s^2, \sigma 2s^2, \sigma^* 2s^2$

$\left[\pi 2p_x^2 = \pi 2p_y^2\right]\sigma 2p_z^2$

$B.O. = \frac{10-4}{2} = \frac{6}{2} = 3$

$He_2\ (4e^-) : \sigma 1s^2, \sigma^* 1s^2$

$B.O. = \frac{2-2}{2} = 0$, Molecule not exist.

243. (5)

$CH_3OH, H_2O,$ 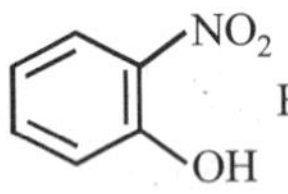 HF, NH_3

can show hydrogen bonding.

244. (2) According to M.O.T.
$O_2 \rightarrow$ no. of unpaired electrons = 2
$O_2^- \rightarrow$ no. of unpaired electron = 1
$NO \rightarrow$ no. of unpaired electron = 1
$CN^- \rightarrow$ no. of unpaired electron = 0
$O_2^{2-} \rightarrow$ no. of unpaired electron = 0

245. (1)

Magnetic	Behaviour	Bond order
$H_2(2)$	Diamagnetic	1
He_2^+ (3)	Paramagnetic	0.5
O_2^- (17)	Paramagnetic	1.5
N_2^{2-} (16)	Paramagnetic	2
O_2^{2-} (18)	Diamagnetic	1
F_2(18)	Diamagnetic	1
Ne_2^+ (19)	Paramagnetic	0.5
B_2(10)	Paramagnetic	1

Only B_2 has bond order 1 and paramagnetic in the given options.

246 (4) $2s \longrightarrow \sigma 2s$, $\boxed{\sigma^* 2s}$
$2p \longrightarrow \sigma 2p_z, \pi 2p_x, \pi 2p_y,$
$\boxed{\sigma^* 2p_z, \pi^* 2p_z, \pi^* 2p_y}$

247. (22)

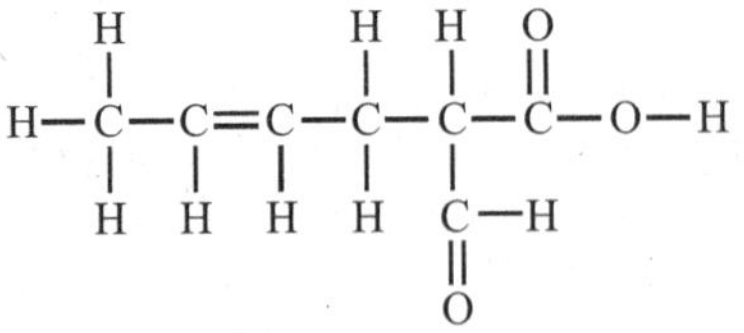

σ-bonds = 19, π-bonds = 3, total = 19 + 3 = 22

248. (6) $CO \Rightarrow \overset{-}{C} \equiv \overset{+}{O} : BO = 3$

$NO^+ \Rightarrow N \equiv O^+ : BO = 3$

249. (4) The molecular orbital configuration of B_2 is :
$(\sigma 1s)^2 (\sigma^* 1s)^2 (\sigma 2s)^2 (\sigma^* 2s)^2 (\pi 2p_x)^1 (\pi 2p_y)^1$
The molecular orbital configuration of C_2^- is :
$(\sigma 1s)^2 (\sigma^* 1s)^2 (\sigma 2s)^2 (\sigma^* 2s)^2 (\pi 2p_x)^2 (\pi 2p_y)^2 (\sigma 2p_z)^1$
The molecular orbital configuration O_2^+ is:
$(\sigma 1s)^2 (\sigma^* 1s)^2 (\sigma 2s)^2 (\sigma^* 2s)^2 (\sigma 2p_z)^2$
$(\pi 2p_x)^2 (\pi 2p_y)^2 (\pi^* 2p_x)^1$
The molecular orbital configuration of He_2^+ is :
$(\sigma 1s)^2 (\sigma^* 1s)^1$
Hence, B_2, C_2^-, O_2^+ and He_2^+ are paramagnetic species.

250. (3) CN^-, NO^+ and O_2^{2+} all have bond order 3.

251. (2)

252. (0) Molecular orbital configuration of O_2^{2-} is :

$\sigma 1s^2\ \sigma^* 1s^2 \sigma 2s^2 \sigma^* 2s^2 \sigma 2p_z^2 \pi \left(2p_x^2 = \pi 2p_y^2\right)$

$\left(\pi^* 2p_x^2 = \pi^* 2p_y^2\right)$

Total unpaired electron in O_2^{2-} is zero.

253. (0) Bond order of CO = 3
Due to isolectronic species, bond order of $NO^+ = 3$

Difference $= 0 = \frac{x}{2} \Rightarrow x = 0$

254. (10) M.O. Configuration of O_2^{2-} $((18\bar{e})$

$\sigma 1s^2\ \sigma^* 1s^2\ \sigma 2s^2\ \sigma^* 2s^2\ \sigma 2p_z^2 \pi 2p_x^2 = \pi 2p_y^2$

$\pi^* 2p_x^2 = \pi^* 2p_y^2$

Total B.M.O electrons = 10

255. (15) Bond order

$= \frac{\text{No. of } e^- \text{s in BMO} - \text{No. of } e^- \text{s in ABMO}}{2} = 2.5$

$\Rightarrow$No. of e^-s in BMO – No. of e^-s in AMBO = 5
As AX is diatomic molecule (neutral), the only possible case is NO.
Total number of electrons = 15
Note : Total number of electrons equal to 13 will also have the 2.5 bond order. But in this case neutral diatomic molecule will not be possible.

Part-B (JEE Advanced)

1. **(c)**

(i) P_2O_5 will be more covalent than other metallic oxides.

(ii) Oxidation state of Mn is + 7 in Mn_2O_7, oxidation state of Cr in CrO_3 is + 6 and oxidation state of Mn is + 2 in MnO.

$\therefore$ MnO is most ionic.

P_2O_5, being a non-metallic oxide will definitely be more covalent than the other metallic oxides. Further, we know that higher the polarising power of the cation (higher for higher oxidation state of the similar size cations) more will be the covalent character. Here Mn is in +7 O.S in Mn_2O_7, Cr in +6 in CrO_3 and Mn in +2 in MnO. So, MnO is the most ionic and Mn_2O_7 is the most covalent.

2. **(d)** No. of e^- in $CH_3^+ = 6 + 3 - 1 = 8$

No. of e^- in $H_3O^+ = 3 + 8 - 1 = 10$

No. of e^- in $NH_3 = 7 + 3 = 10$

No. of e^- in $CH_3^- = 6 + 3 + 1 = 10$

$\therefore$ H_3O^+, NH_3 and CH_3^- are isoelectronic.

3. **(d)** In covalent bonds, between two identical non-metal, atoms share the pair of electrons equally between them, *e.g.* : F_2, O_2, N_2.

4. **(c)** Ionic bond or electrovalent between Cu^{2+} and SO_4^{2-}, covalent and coordinate in

SO_4^{2-}; 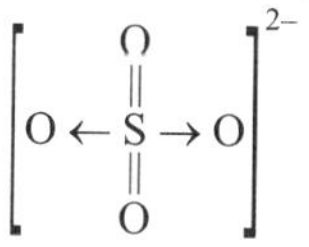 ion.

5. **(b)** $\because$ It forms hydrogen bonds with water.

6. **(a)** H_2 H–H

7. **(a)** $X^+ Y^-$

$\because$ Electropositive elements forms cation and electronegative elements forms anion.

8. **(c)** In KCN, ionic bond is present between K^+ and CN^- and covalent bonds are present between carbon and nitrogen C≡N.

9. **(4)** B_2H_6 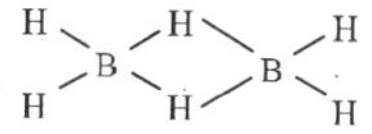

N_2O $\overset{-}{N}=\overset{+}{N}=O \longleftrightarrow N\equiv\overset{+}{N}-\overset{-}{O} \longleftrightarrow \overset{-}{N}-\overset{+}{N}\equiv\overset{+}{O}$

N_2O_4 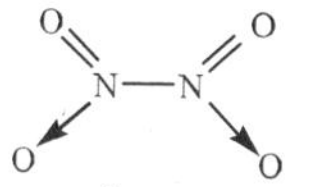

$H_2S_2O_3$ 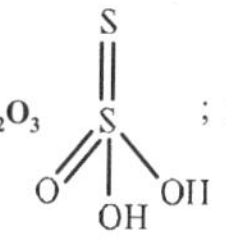 ; $H_2S_2O_8$ 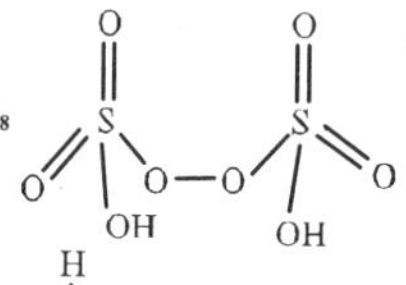

$B_3N_3H_6$ 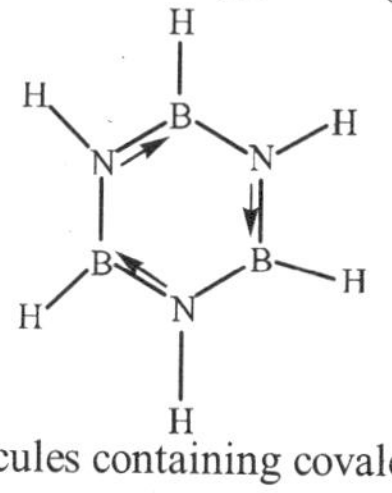

Total No. of molecules containing covalent bond between two atoms of the same kind are 4.

10. Three centred two electron bonds or **banana bond;**

The formation of three centred two electron bond is due to one empty sp^3 orbital of one of the B atom, 1*s* orbital of the bridge hydrogen atom and one of the sp^3 (filled) orbital of the other B-atom. This forms a delocalized orbital covering the three nuclei giving the shape of a banana. Thus, also known as banana bonds.

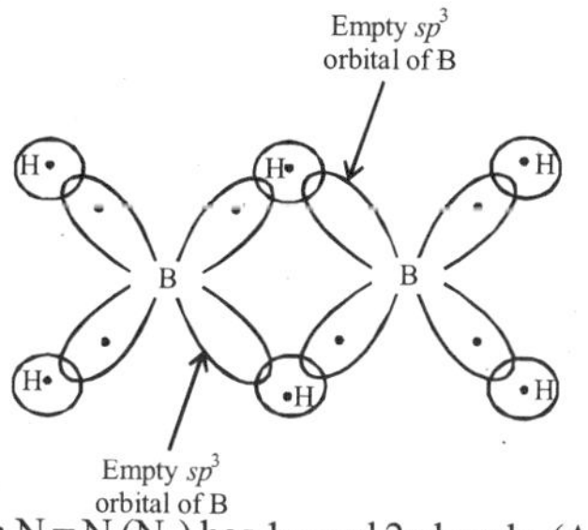

11. **2;** N≡N (N_2) has 1σ and 2π bonds. (A triple bond consists of 1σ and 2π bonds)

12. **True :** Sigma bond is formed by the overlapping of two *s*-orbitals or one *s* and one *p* or the two *p* orbitals of the two different atoms. Thus, linear overlap of two *p*-orbitals results in formation of a σ– bond.

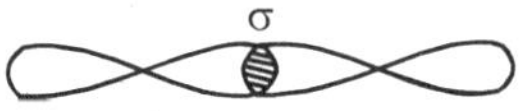

Axial p-p overlapping

13. **(c)** LiCl is a covalent compound due to the large size of the anion (Cl^-) and small size of Li^+ cation. This generates large amount of polarisation in bond.

14. **(b)** In N_2, similar atoms are linked to each other and thus, there is no polarity.

15. **(a)** Isoelectronic species have same number of electrons.

Electrons in CO = 6 + 8 = 14

Electrons in CN^- = 6 + 7 + 1 = 14

16. **(c)** N≡N :N⋮⋮N:

17. **(d)** $\because$ after forming the bonds, C has only 6 e^- in its valence shell.

18. **(6)** Polar molecules will be attracted/ deflected near charged comb.

Polar molecules : HF, H_2O, NH_3, H_2O_2, $CHCl_3$, C_6H_5Cl (6-polar molecules)

Nonpolar molecules : O_2, CCl_4, C_6H_6

19. **(a, d)**

(a) CO_2, C_2H_4 and HCl follow octet rule.

(b) O_3 and HCl and follow octet rule.

(c) None of them follow octet rule.

(d) CO_2, O_3 and C_2H_4 follow octet rule.

Correct answer is (a) and (d).

20. **(a)** Both assertion and reason are correct. The reason explains the assertion as the central O-atom cannot have more than 8 electrons (octet).

21. O_3; 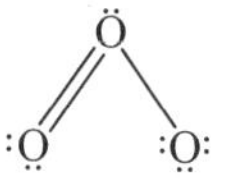

$COCl_2$; 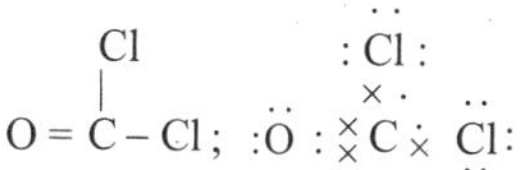

22. **Lewis dot structure Neutral molecule**

O_2^{2-}; F_2; 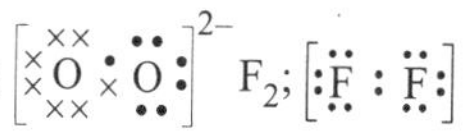

CO_3^{2-}; SO_3; CN^-; N_2; NCS^-; CO_2; 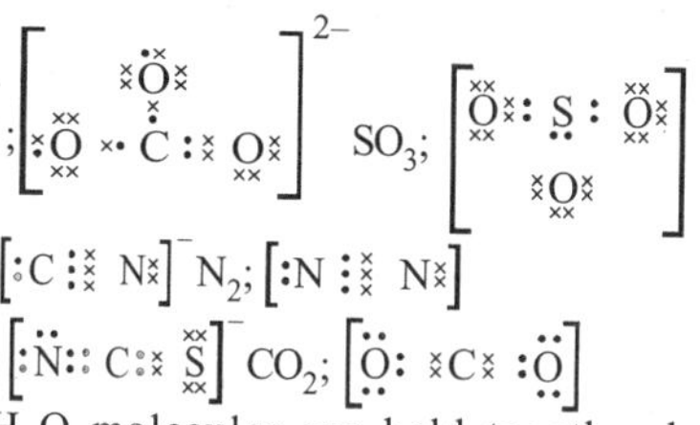

23. H_2O molecules are held together by hydrogen bonding which is stronger force of attraction but H_2S molecules are held together by van der Waals forces of attraction, which are weaker forces. As a result, water molecules come closer and exist in liquid state.

24. **(a)**

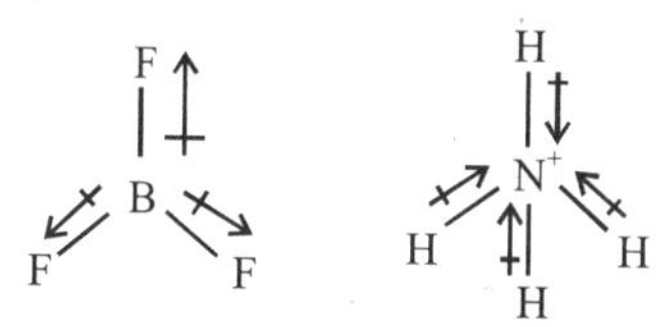

$\mu_{net} = 0$ $\mu_{net} = 0$

$\mu_{net} = 0.8 \times 10^{-30}$ Cm = 0.23 *D* $\mu_{net} = 4.90 \times 10^{-30}$ Cm = 1.47 *D*

μ_{net} order $BF_3 = NH_4^+ < NF_3 < NH_3$

25. **(a)** Hybridisation of S in H_2S

$= \frac{1}{2}(6 + 2 + 0 - 0) = 4$

$\therefore$ S has sp^3 hybridisation and 2 lone pair of electrons in H_2S.

$\therefore$ It has angular geometry and so it has non-zero value of dipole moment.

26. **(d)** Critical temperature of water is higher than O_2 because H_2O molecule has dipole moment which is due to its V-shape.

27. **(c)**

(a) 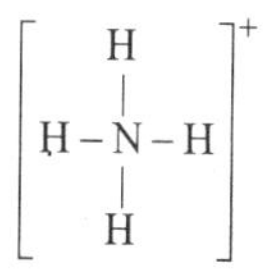 $[Cl]^-$ – It has ionic and non-polar covalent bond

(b) H – C ≡ N - It has ionic and polar covalent bonds.

(c) 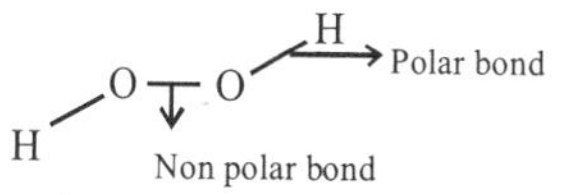

It has polar and non polar both type of covalent bonds.

(d) 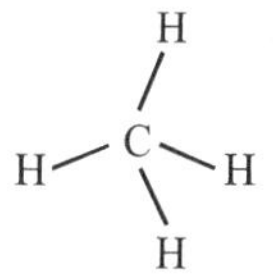

It has non polar covalent bonds only.

28. **(b)** The zero dipole moment of BF_3 is due to its symmetrical (triangular planar) structure. The three fluorine atoms lie at the corners of an equilateral triangle with boron at the centre.

The vectorial addition of the dipole moments of the three bonds gives a net sum of zero because the resultant of any two dipole moments is equal and opposite to the third.

29. (b) In regular tetrahedral structure, dipole moment of one bond is cancelled by opposite dipole moment of the other bonds.

30. (b) ∴ Carbon tetrachloride having regular geometry and identical atoms attached to bonds has zero dipole moment.

31. (c) In trigonal planar geometry (for sp^2 hybridisation), the vector sum of two bond moments is equal and opposite to the dipole moment of third bond.

32. (80.09) Dipole moment, $\mu = e \times d$ coulombs metre

For KCl $d = 2.6 \times 10^{-10}$ m

For complete separation of unit charge (electronic charge)

$(e) = 1.602 \times 10^{-19}$ C

Hence $\mu = 1.602 \times 10^{-19} \times 2.6 \times 10^{-10} = 4.1652 \times 10^{-29}$ Cm

$\mu_{KCl} = 3.336 \times 10^{-29}$ Coulomb meter (given)

∴ % Ionic character of KCl

$$= \frac{3.336 \times 10^{-29}}{4.1652 \times 10^{-29}} \times 100$$

$= \mathbf{80.09\%}$

33. False : The C – F distance is less than the C – Cl, although the former involves more charge separation. However, here bond distance has more dominating effect causing dipole moment of CH_3Cl to be more than that of CH_3F.

34. False : The presence of polar bonds in a polyatomic molecule does not always lead to a definite dipole moment. This is because the dipole moment is a vector quantity and when the bond moment of one bond is cancelled by the equal but opposite bond moment due to other bond(s), the molecule has zero dipole moment, *e.g.* CO_2, CH_4, CCl_4 *etc.*

35. False : Symmetrical molecules with polar bonds have zero dipole moment.

36. (b, c) Dipole moment (μ) value of BF_3, SF_6, $BeCl_2$, CO_2, BCl_3 is zero.

37. (b, c) Alkanes (a) and (d) don't have dipole moment because of symmetry in them.

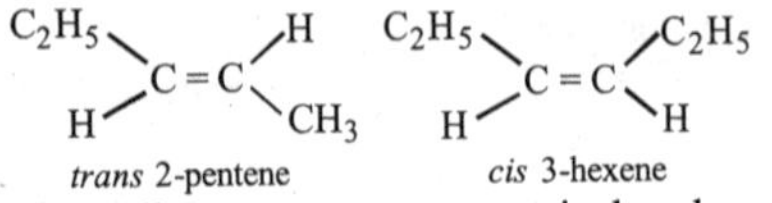

trans 2-pentene   *cis* 3-hexene

These alkenes are not symmetrical and so, they have dipole moment.

38. (d) OSF_2: $H = \frac{6+2}{2} = 4$.

It has 1 lone pair.

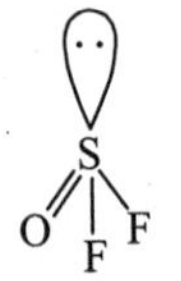

(Shape is trigonal pyramidal)

The shapes of SO_3, BrF_3 and SiO_3^{2-} are triangular planar, see-saw and triangular planar respectively.

39. (d)

ClO_3^- (1 lone pair)

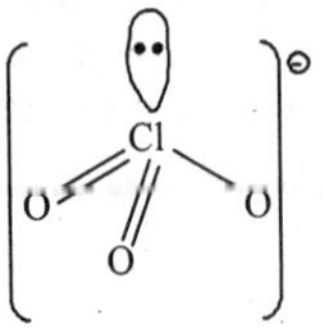

Pyramidal

XeF_4 (2 lone pair)

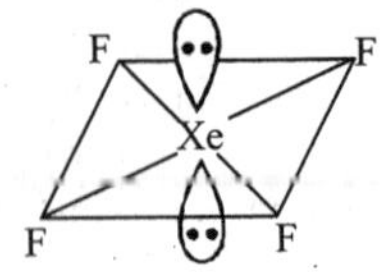

Square planar

SF_4 (1 lone pair)

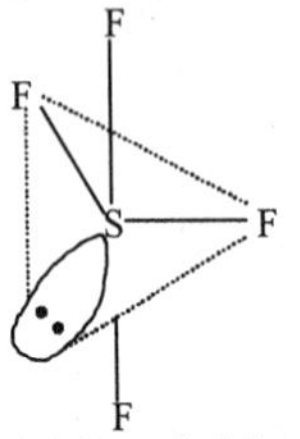

Irregular tetrahedral

I_3^- (3 lone pair)

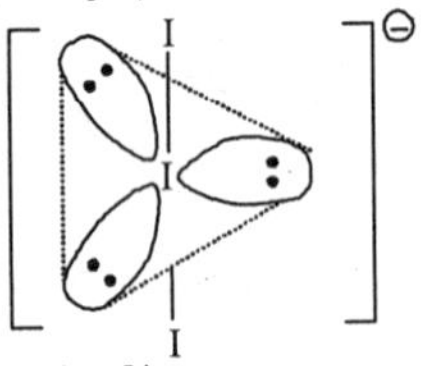

Linear

40. (a) Isoelectronic species have same number of electrons and isostructural species have same type of hybridisation at central atom.

NO_3^-; No. of $e^- = 7 + 8 \times 3 + 1 = 32$, hybridisation of N in NO_3^- is sp^3

CO_3^{2-}; No. of $e^- = 6 + 8 \times 3 + 2 = 32$, hybridisation of C in CO_3^{2-} is sp^3

ClO_3^-; No. of $e^- = 17 + 8 \times 3 + 1 = 42$, hybridisation of Cl in ClO_3^- is sp^3

SO_3; No. of $e^- = 16 + 8 \times 3 = 40$, hybridisation of S in SO_3 is sp^2

∴ NO_3^- and CO_3^{2-} are isostructural and isoelectronic.

41. (a) $H_3N \rightarrow BF_3$ where both N, B are attaining tetrahedral geometry.

42. (b) $H = \frac{1}{2}[V + M - C + A]$

Hybridisation of N in $NH_3 = \frac{1}{2}[5 + 3 - 0 + 0] = 4$

∴ sp^3

Hybridisation of Pt in $[PtCl_4]^{2-} = \frac{1}{2}[2 + 4 - 0 + 2] = 4$

∴ dsp^2

Hybridisation of P in $PCl_5 = \frac{1}{2}[5 + 5 - 0 + 0] = 5$

∴ sp^3d

Hybridisation of B in $BCl_3 = \frac{1}{2}[3 + 3 - 0 + 0] = 3$

∴ sp^2

43. (b) For NO_2^+ : $H = \frac{1}{2}(5 + 0 + 0 - 1) = 2$;

∴ sp hybridisation

For NO_3^- : $H = \frac{1}{2}(5 + 0 + 1 - 0) = 3$;

∴ sp^2 hybridisation

For NH_4^+ : $H = \frac{1}{2}(5 + 4 + 0 - 1) = 4$;

∴ sp^3 hybridisation

44. (d)

$$H = \frac{1}{2}\left[\begin{matrix}\text{No. of electrons} \\ \text{in valence} \\ \text{shell}(V)\end{matrix} + \begin{matrix}\text{No. of mono-} \\ \text{valent atoms} \\ (M)\end{matrix} + \begin{matrix}\text{charge} \\ \text{on} \\ \text{anion (A)}\end{matrix} - \begin{matrix}\text{charge} \\ \text{on} \\ \text{cation (C)}\end{matrix}\right]$$

For SF_4 : $H = \frac{1}{2}(6 + 4 + 0 - 0) = 5$, ∴ S is sp^3d hybridised in SF_4. Thus SF_4 has 5 hybrid orbitals of which only four are used by F, leaving one lone pair of electrons on sulphur.

For CF_4 : $H = \frac{1}{2}[4 + 4 + 0 - 0] = 4$ ∴ sp^3 hybridisaion

Since, all the four orbitals of carbon are involved in bond formation, no lone pair is present on C having four valence electrons

For XeF_4 : $H = \frac{1}{2}(8 + 4 + 0 - 0) = 6$, ∴ sp^3d^2 hybridization of the six hybrid orbitals, four form bond with F, leaving behind two lone pairs of electrons on Xe.

45. (b) $H = \frac{1}{2}(3 + 3 + 0 - 0) = 3$

∴ Boron, in BF_3, is sp^2 hybridised leading to trigonal planar shape.

46. (b) $O = \ddot{S} = O$

$H = 2 + 1 = 3 \Rightarrow sp^2$ hybridisation

47. (c) Structure of a molecule can be ascertained by knowing the number of hybrid bonds in the molecule. Thus

In NF_3: $H = \frac{1}{2}(5 + 3 - 0 + 0) = 4$

Thus, N in NF_3 is sp^3 hybridized as 4 orbitals are involved in bonding.

In NO_3^- : $H = \frac{1}{2}(5 + 0 - 0 + 1) = 3$

Thus, N in NO_3^- is sp^2 hybridized as 3 orbitals are involved in bonding

In BF_3: $H = \frac{1}{2}(3 + 3 - 0 + 0) = 3$

Thus, B in BF_3 is sp^2 hybridized and 3 orbitals are involved in bonding

In H_3O^+: $H = \frac{1}{2}(6 + 3 - 1 + 0) = 4$

Thus, O in H_3O^+ is sp^3 hybridized as 4 orbitals are involved in bonding.

Thus, isostructural pairs are [NF_3, H_3O^+] and [NO_3^-, BF_3].

48. (a)

For ClO_2^-, we have, $H = \frac{1}{2}(7 + 0 - 0 + 1)$

$\Rightarrow H = 4$ or sp^3 hybridisation as 4 orbitals are involved.

49. (c) 4σ bonds – sp^3 hybridisation

3σ and 1π bonds – sp^2 hybridisation

2σ and 2π bonds – sp hybridisation

[For hybridization only σ-bonds are considered]

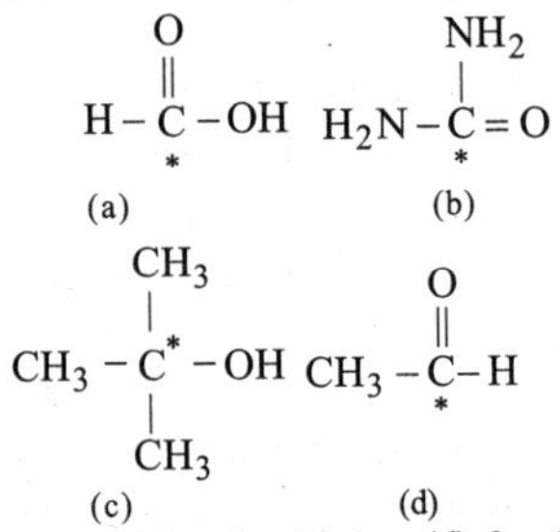

(a) (b) (c) (d)

(a) 3σ, 1π (b) 3σ, 1π (c) 4σ (d) 3σ, 1π

∴ $(CH_3)_3COH$ has 4σ bonds and thus it has sp^3 hybridisation.

50. (a) In PCl_3, P is sp^3 hybridised and has one lone pair of electrons, hence it is pyramidal in shape.

51. (c) From amongst given species PH_3, NH_3 and SbH_3 are all sp^3 hybridised. Their central atom has both bond pair as well as lone pair of electrons. The lone pair occupy the fourth orbital. CH_3^+ has only three pairs of electrons so it is sp^2 hybridised.

52. (a) Compound having *sp* hybridisation will have linear shape.
∴ CO_2 or (O=C=O) which has C in *sp* hybrid state has linear shape.

53. (c) $H = \frac{1}{2}(V + M - C + A)$

For SO_2, $H = \frac{1}{2}(6 + 0 + 0 - 0) = 3$

∴ sp^2 hybridisation.

54. (b) *sp* type of hybridization involves the intermixing of one *s* and one *p* (say p_x) orbitals to give two equivalent hybrid orbitals, known as *sp* hybrid orbitals.
The two *sp* hybrid orbitals are directed in a straight line with an angle of 180° **(collinear orbitals)**. The other two *p* orbitals (say p_y and p_z) remain pure.

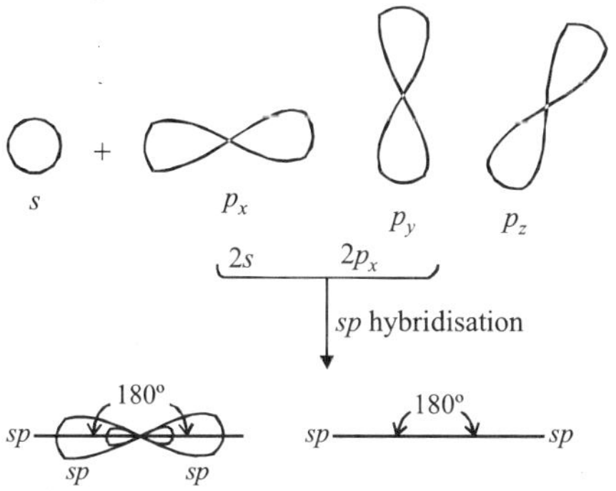

55. (5)

I_3^+ :

sp^3 ;

SiO_4^{4-} :
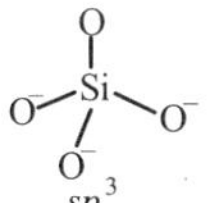
sp^3

SO_2Cl_2 :
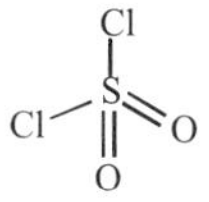
sp^3 ;

XeF_2 :

sp^3d

SF_4 :

sp^3d ;

ClF_3 :
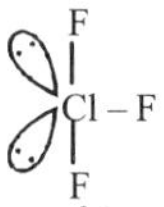
sp^3d

$[Ni(CO)_4]$:
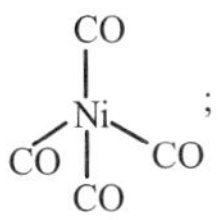
sp^3 ;

XeO_2F_2 :
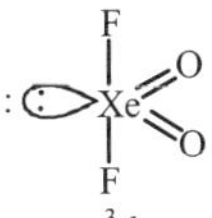
sp^3d

$[PtCl_4]^{2-}$:
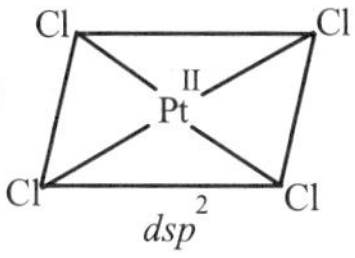
dsp^2

XeF_4 :
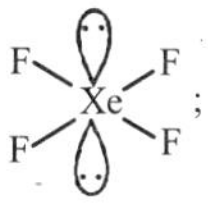
sp^3d ;

$SOCl_2$:
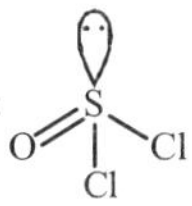
sp^3

56. (6)

Species	No. of valence e⁻ with central atom	No. of σ-bonds	No. of lone pairs
$[TeBr_6]^{2-}$	6	6	1
$[BrF_2]^+$	6	2	2
SNF_3	5	4	0
$[XeF_3]^-$	8+1	3	3

Sum of number of lone pairs = 1 + 2 + 0 + 3 = 6

57. (4) Cl – Be – Cl N ≡ N – N̈:

Hybridization *sp*Hybridisation *sp*
Structure linearStructure linear

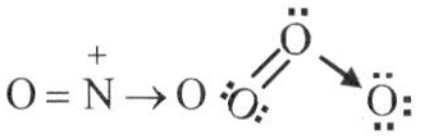

Hybridisation *sp*Hybridisation sp^2
Structure LinearStructure Trigonal planar

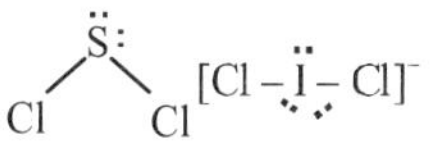

Hybridisation sp^3Hybridisation sp^3d
Structure AngularStructure linear

$[I-I-I]^-$ F–Xe–F

Hybridisation sp^3dHybridisation sp^3d
Structure LinearStructure Linear

$\overset{-}{N} = \overset{+}{N} = O$

Hybridisation *sp*
Structure Linear

Only $BeCl_2$, N_3^-, N_2O and NO_2 are linear with *sp*-hybridisation.

58. (4)

XeF_4 :
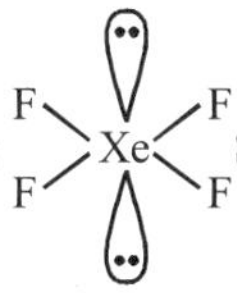
Square planar (sp^3d^2)

SF_4 :
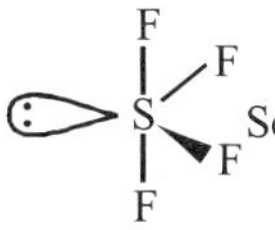
See-saw (sp^3d)

SiF_4 :
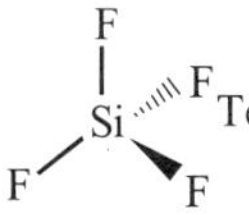
Tetrahedral (sp^3)

BF_4^- :
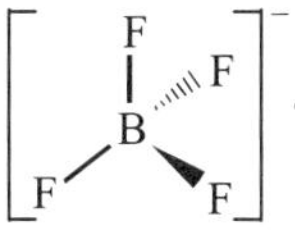
Tetrahedral (sp^3)

BrF_4^- :
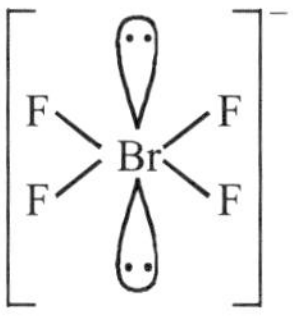
Square planar (sp^3d^2)

$[Cu(NH_3)_4]^{2+}$:
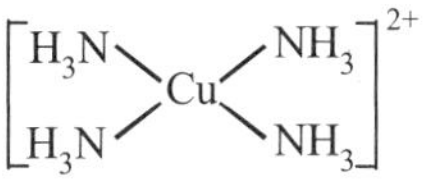
Square planar (dsp^2)

$[FeCl_4]^{2-}$:
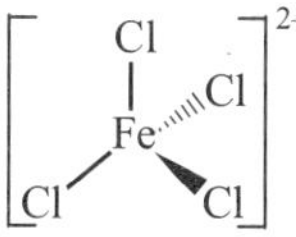
Tetrahedral (sp^3)

$[CoCl_4]^{2-}$:
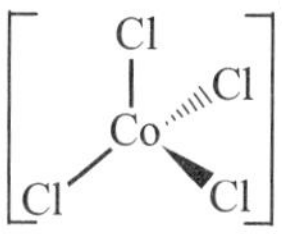
Tetrahedral (sp^3)

$[PtCl_4]^{2-}$:
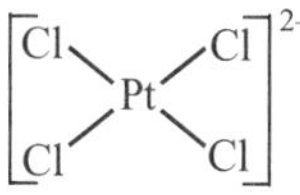
Square planar (dsp^2)

59. (4) According to VSEPR theory, number of electron pairs around central atom (Br) are 6. Five are bond pairs and one is lone pair.
Its geometry is octahedral but due to lone pair – bond pair repulsion, the four fluorine atoms at equatorial positions are forced towards the axial fluorine atom, thus reducing F–Br–F angle from 90° to 84.8°.
The $F_{eq} - Br - F_{eq}$ angle will remain 90°.

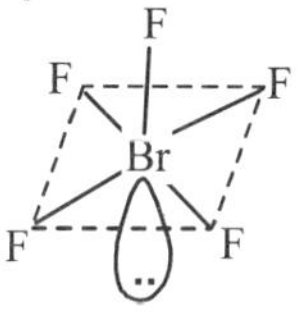

60. Planar; $\overset{+}{C}H_3$ is a carbocation and such a species has a planar shape due to sp^2 hybridisation.

61. CO_2; Bond angle in CH_4 is 109°28', in H_2O it is 104.5° and in CO_2 it is 180°. So, it is maximum in case of CO_2.

62. sp^3; Hybridisation (*H*) = $\frac{1}{2}$ [No. of valence electron in central atom + No. of monovalent atoms – Charge on cation + Charge on anion]
For N in NH_4^+, hybridisation (*H*)

$= \frac{1}{2}(5 + 4 - 1 + 0) = 4$

∴ sp^3 hybridisation.

63. False : sp^2 hybrid orbitals do not have equal *s* and *p* character. They have 33.3% *s*-character and 66.7% *p*-character.

64. False : Only two orbitals are used since C in benzene is in sp^2 hybridised state.

65. True : $SnCl_2$ has 2 bond pairs and one lone pair of electrons. It is sp^2 hybridised and is trigonal planar in shape.

66. (b, c) CompoundNumber of lone pairs on central atom
$BrF_5 \rightarrow 1$
$ClF_3 \rightarrow 2$
$XeF_4 \rightarrow 2$
$SF_4 \rightarrow 1$

67. (b, c, d) $[O=N=O]^+$; $[N≡C–O]^-$; S=C=S
It can be seen from the structure shown above that CS_2 being *sp* hybridized has a linear shape and other two molecules are isoelectronic to CS_2, so they are also linear. $SnCl_2$ and SO_2 are sp^2 hybridised and are not linear.

68. **(a, c)** CO_2, $HgCl_2$ and C_2H_2 have linear structure (*sp* hybridization), while $SnCl_2$ is trigonal planar (sp^2 hybridisation). NO_2 has angular structure (*V*-shape).

69. **(a)**

Species	Structure	Shape	Hybridisation
$SOCl_2$	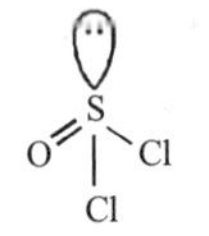	(Pyramidal)	sp^3
$XeOF_4$	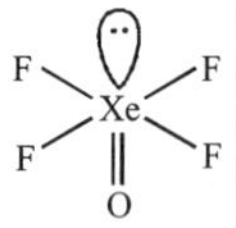	(Square pyramidal)	sp^3d^2
ClF_3	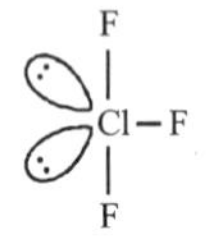	(T-shape)	sp^3d
ClF_5	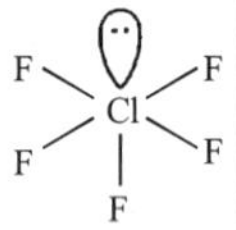	(Square pyramidal)	sp^3d^2
$XeF_5^{\oplus}$	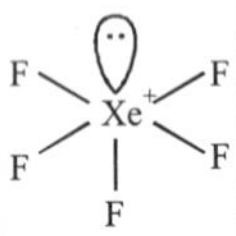	(Square pyramidal)	sp^3d^2
SO_3^{2-}	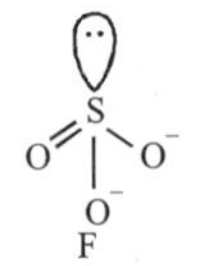	(Pyramidal)	sp^3
$XeF_3^{\oplus}$	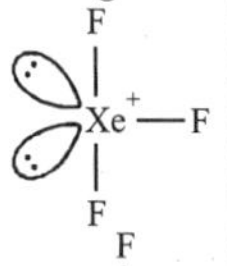	(T-shape)	sp^3d
SF_4	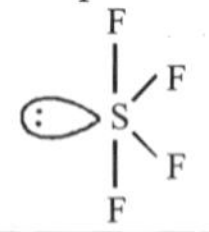	(See-saw)	sp^3d

$\therefore$ See – Saw = 1
T-shape = 2
Trigonal planar = 0
Square pyramidal = 3

70. **(b)**

XeF_2: 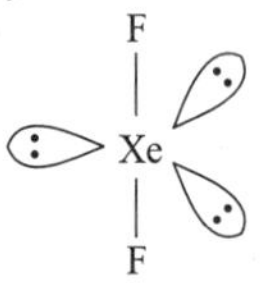 Hybridisation – sp^3d
Geometry – Trigonal bipyramidal
L.P. = 3

XeF_4: 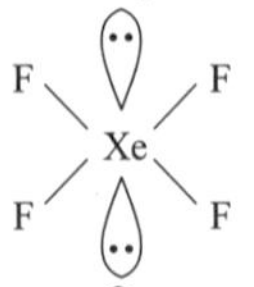 Hybridisation – sp^3d^2
Geometry – Octahedral
L.P. = 2

XeO_3: 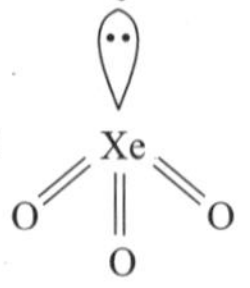 Hybridisation – sp^3
Geometry – Tetrahedral
L.P. = 1

XeO_3F_2: 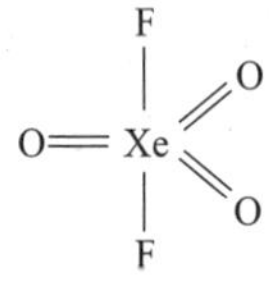 Hybridisation – sp^3d
Geometry – Trigonal bipyramidal
L.P. = 0

Hence, correct match is:
$P \to 5; Q \to 3; R \to 2; S \to 4$

71. First determine the total number of electron pairs around the central atom.

$$XeF_4 = \frac{N}{2} = \frac{8+4}{2} = 6$$

Thus, in XeF_4, Xe is sp^3d^2 hybridised. The structure of the molecule is octahedral and shape is square planer with two lone pair of electrons.

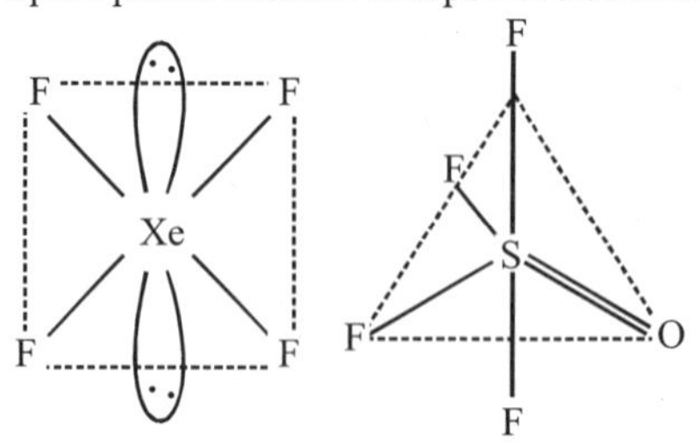

For OSF_4: $\frac{N}{2} = \frac{6+4}{2} = 5$

Thus, the central atom (S) is sp^3d hybridised leading to trigonal bipyramidal structure with no lone pair of electrons.

72. 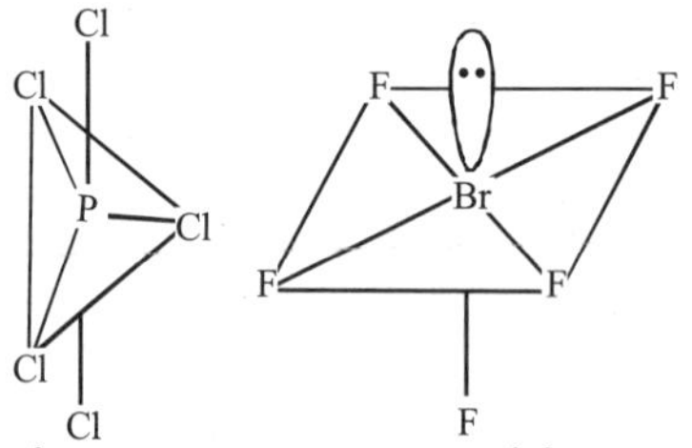

PCl_5: sp^3d Trigonal bipyramidal BrF_5 : sp^3d^2 Square pyramidal

To decrease lp - bp repulsions, lone pair takes the axial position in BrF_5.

73. In H_2S, no. of hybrid orbitals

$$= \frac{1}{2}(6+2-0+0) = 4$$

Hence, here sulphur is sp^3 hybridised, so

$${}_{16}S = 1s^2, 2s^2 2p^6, \underbrace{3s^2 3p_x^2 3p_y^1 3p_z^1}_{sp^3 \text{ hybridisation}}$$

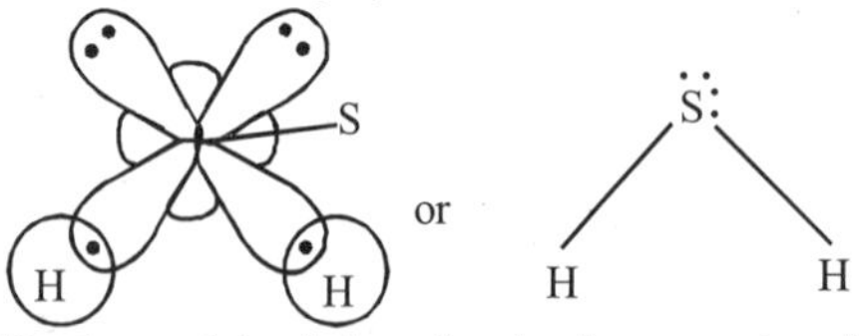

Due to repulsion betwen lp - lp; the geometry of H_2S is distorted from tetrahedral to V-shape.

In PCl_3, no. of hybrid orbitals

$$= \frac{1}{2}[5+3-0+0] = 4$$

Hence, here P shows sp^3 - hybridisation

$${}_{15}P = 1s^2, 2s^2 2p^6, \underbrace{3s^2 3p_x^1 3p_y^1 3p_z^1}_{sp^3 \text{ hybridisation}}$$

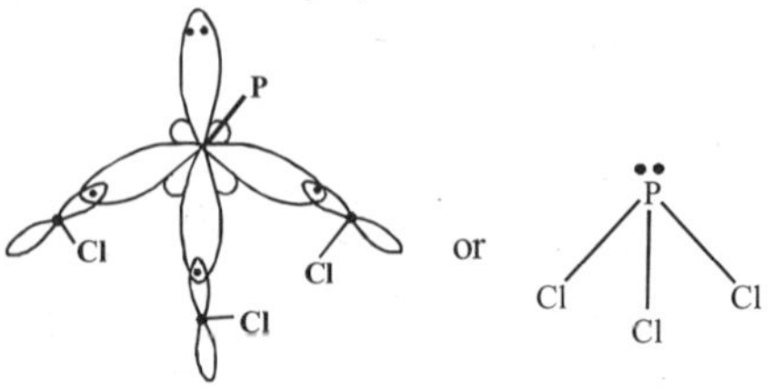

Thus, due to repulsion between lp - bp, geometry is distorted from tetrahedral to pyramidal.

74. The structure of OF_2 is similar to H_2O and involves sp^3 hybridization on O atom. The bond angle in F – O – F is not exactly 109°28′, but distorted (103°) due to presence of lone pair of electrons on O as well as F leading to V shape or tetrahedral positions with two positions occupied by lone pair of electrons of the molecule.

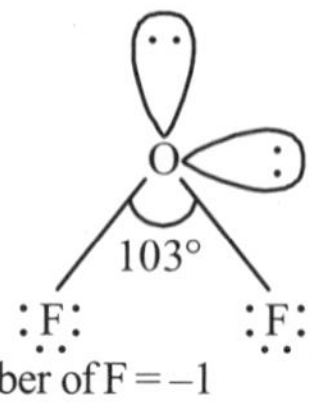

Oxidation number of F = –1
$\therefore$ Oxidation number of O = +2

75. **(c)** $F_2(18e^-)$; $\sigma 1s^2\, \sigma^* 1s^2\, \sigma 2s^2\, \sigma^* 2s^2$
$\sigma 2p_z^2\, \pi 2p_x^2\, \pi 2p_y^2\, \pi^* 2p_x^2\, \pi^* 2p_y^2\, \sigma^* 2p_z$

76. **(c)** $C_2 = \sigma 1s^2\, \sigma^* 1s^2\, \sigma 2s^2\, \sigma^* 2s^2\, \sigma 2p_z^2\, \pi 2p_x^1\, \pi 2p_y^1$

When there is no mixing of 2*s* and 2*p* atomic orbitals, the energy of $\sigma 2p_z$ molecular orbital will be low.

Thus, only C_2 will be paramagnetic.

77. **(a)** If Hund's rule is violated then pairing of e^- in the same molecular orbital (of same energy) can take place. Complying with Hund's rule: M.O. configuration: $\sigma 1s^2, \sigma^* 1s^2, \sigma 2s^2, \sigma^* 2s^2, \pi 2p_x^1, \pi 2p_y^1$ Molecular orbital configuration of $B_2(10)$ violating Hund's rule : $\sigma 1s^2, \sigma^* 1s^2, \sigma 2s^2, \sigma^* 2s^2,$

$\pi 2p_y^2$

Bond order of $B_2 = \frac{6-4}{2} = 1$, B_2 will be diamagnetic.

78. **(a)** Molecular electronic configuration of

$CO: \sigma 1s^2, \sigma^* 1s^2, \sigma 2s^2, \sigma^* 2s^2, \pi 2p_y^2$
$= \pi 2p_z^2, \sigma 2p_x^2$

Therefore, bond order $= \frac{N_b - N_a}{2} = \frac{10-4}{2} = 3$

NO^- : $\sigma 1s^2, \sigma^* 1s^2, \sigma 2s^2, \sigma^* 2s^2, \sigma 2p_z^2,$
$\pi 2p_y^2 = \pi 2p_x^2$, $\{\pi^* 2p_y^1 = \pi^* 2p_x^1$

Bond order $= \frac{10-6}{2} = 2$

$\therefore NO^-$ has different bond order from that in CO.

79. **(d)**

(a) In Na_2O_2, we have O_2^{2-} ion. Number of valence eIctrons of the two oxygen in O_2^{2-} ion = $8 \times 2 + 2 = 18$ which are present as follows :

$\sigma 1s^2, \sigma^* 1s^2, \sigma 2s^2, \sigma^* 2s^2, \sigma 2p_z^2,$

$\pi 2p_y^2 = \pi 2p_x^2,$

$\pi^* 2p_y^2 = \pi^* 2p_x^2$

$\therefore$ Number of unpaired electrons = 0, hence, O_2^{2-} is diamagnetic.

(b) No. of valence electrons of all atoms in $O_3 = 6 \times 3 = 18$.
Thus, it also does not have any unpaired electron, hence, it is diamagnetic.
(c) No. of valence electrons of all atom in $N_2O = 2 \times 5 + 6 = 16$. Hence, here also all electrons are paired. So, it is diamagnetic.
(d) In KO_2, we have O_2^- No. of valence electrons of all atoms in $O_2^- = 2 \times 6 + 1 = 13$,
It has one unpaired electron, hence it is paramagnetic.

80. (b)

$$O_2 : \sigma 1s^2, \sigma^* 1s^2, \sigma 2s^2, \sigma^* 2s^2, \sigma 2p_x^2, \begin{cases}\pi 2p_y^2 \\ \pi 2p_z^2\end{cases}, \begin{cases}\pi^* 2p_y^1 \\ \pi^* 2p_z^1\end{cases}$$

$$\text{Bond order} = \frac{10-6}{2} = 2$$

(two unpaired electrons in antibonding molecular orbital)

$$O_2^+ : \sigma 1s^2, \sigma^* 1s^2, \sigma 2s^2, \sigma^* 2s^2, \sigma 2p_x^2, \begin{cases}\pi 2p_y^2, \\ \pi 2p_z^2,\end{cases} \begin{cases}\pi^* 2p_y^1 \\ \pi^* 2p_z^0\end{cases}$$

$$\text{Bond order} = \frac{10-5}{2} = 2.5$$

(One unpaired electron in antibonding molecular orbital)
Hence, O_2 as well as O_2^+ is paramagnetic, and bond order of O_2^+ is greater than that of O_2.

81. (c) O_2^- is the only species having unpaired electron.

82. (b) Write configuration of all species. Half filled and full filled orbitals are more stable as compared to nearly half filled and nearly full filled orbitals.
$Li^- = 1s^2, 2s^2$; $Be^- = 1s^2, 2s^2, 2p^1$
$B^- = 1s^2, 2s^2, 2p^2$; $C^- = 1s^2, 2s^2, 2p^3$
$\therefore$ Be^- will be least stable. It has lowest I.E.

83. (a) Number of electrons in each species are

$CN^- = 6 + 7 + 1 = 14$, $CO = 6 + 8 = 14$

$NO^+ = 7 + 8 - 1 = 14$
Each of the species has 14 electrons which are distributed in MOs as below

$$\sigma 1s^2, \sigma^* 1s^2, \sigma 2s^2, \sigma^* 2s^2, \quad \pi 2p_y^2 = \pi 2p_z^2, \sigma 2p_x^2$$

$$\text{Bond order} = \frac{10-4}{2} = 3$$

84. (d)

(i) $\text{Bond length} \propto \dfrac{1}{\text{Bond order}}$

(ii) Bond order is calculated by either the help of molecular orbital theory or by resonance.
(i) Bond order of CO as calculated by molecular orbital theory = 3
(ii) Bond order of CO_2 (by resonance method)

$$= \frac{\text{No. of bonds in all possible sides}}{\text{No. of resonating structure}} = \frac{4}{2} = 2$$

(iii) Bond order in CO_3^{2-} (by resonance method)

$$= \frac{4}{3} = 1.33$$

$\therefore$ Order of bond length of C – O is $CO < CO_2 < CO_3^{2}$

85. (b) Calcium carbide is an ionic compound (Ca^{2+} C^{2-}) which produces acetylene on reacting with water. Thus, the structure of C^{2-} is $[C \equiv C]^{2-}$. It has one σ and two π bonds.

86. (d) O_2 = Oxygen (Z = 8) has following molecular orbital configuration of O_2.
O_2 (16e⁻) = $\sigma 1s^2$, $\sigma^* 1s^2$, $\sigma 2s^2$, $\sigma^* 2s^2$, $\sigma 2p_x^2$, $\pi 2p_y^2 = \pi 2p_z^2$, $\pi^* 2p_y^1 = \pi^* 2p_z^1$ i.e., 2 unpaired and 14 paired electrons.

87. (b) H_2O molecule can form four hydrogen bonds per molecule, two via lone pairs and two via hydrogen atoms.

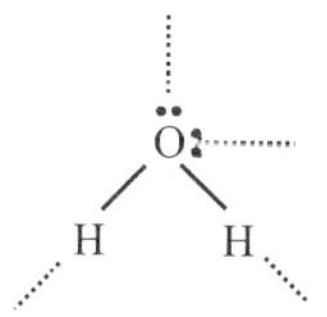

88. (a) O_2^- (17e⁻) – $KK \sigma 2s^2 \sigma^* 2s^2 \sigma 2p_x^2$, $\pi 2p_y^2 = \pi 2p_z^2$, $\pi^* 2p_y^2 = \pi^* 2p_z^1$
Thus, O_2^- has one unpaired electron; hence it is paramagnetic. Other species have no unpaired electron. All of them have 14 electrons.

89. (a) Compounds having F or O or N attached to H form hydrogen bond.

CH_3CH_2OH	$C_2H_5-O-C_2H_5$	CH_3CH_2Cl
ethanol	diethyl ether	ethyl chloride

$$CH_3-\underset{|}{\overset{CH_3}{\overset{|}{N}}}-CH_3$$
Trimethyl amine

90. (a) Write the electronic configuration of each species according to molecular orbital theory.

NO(15e⁻)– $\sigma 1s^2, \sigma^* 1s^2, \sigma 2s^2, \sigma^* 2s^2, \sigma 2p_x^2$,
$\pi 2p_y^2 = \pi 2p_z^2, \pi^* 2p_y^1 = \pi^* 2p_z^0$
1 unpaired electron.

CO (14e⁻)– $\sigma 1s^2, \sigma^* 1s^2, \sigma 2s^2, \sigma^* 2s^2$,
$\pi 2p_y^2 = \pi 2p_z^2, \sigma 2p_x^2$
no unpaired electron

CN⁻ (14e⁻) – $\sigma 1s^2, \sigma^* 1s^2, \sigma 2s^2, \sigma^* 2s^2$,
$\pi 2p_y^2 = \pi 2p_z^2, \sigma 2p_x^2$

O_2 (16e⁻) – $\sigma 1s^2$, $\sigma^* 1s^2$, $\sigma 2s^2$, $\sigma^* 2s^2$, $\sigma 2p_x^2$, $\pi 2p_y^2 = \pi 2p_z^2$, $\pi^* 2p_y^1 = \pi^* 2p_z^1$;
Two unpaired electrons.

91. (c) $\therefore$ F – H ········· F bond is strongest due to largest difference in electronegativity of atoms and smallest size of F atom.

92. (d) Hydrogen bonding is formed in those compounds in which F or O or N atoms are attached to hydrogen atom.
$\because$ HCl does not have F or N or O
$\therefore$ It does not form hydrogen bond.

93. (6) $2AgNO_3(s) \xrightarrow{\Delta} 2Ag(s) + 2NO_2(g) + O_2(g)$
NO_2 has one unpaired e⁻ and O_2 has two unpaired e⁻.
Molecular orbital configuration of O_2:

$$\sigma 1s^2 \sigma^* 1s^2 \sigma 2s^2 \sigma^* 2s^2$$

$$\sigma 2p_z^2 \, \pi 2p_x^2 \, \pi 2p_y^2 \, \pi 2^* p_x^1 \, \pi^* 2p_y^1$$

Total number of e^- present in antibonding molecular orbitals = 6.

94. (6) ($H_2, Li_2, Be_2, C_2, N_2, F_2$)
H_2:$\sigma 1s^2$ (Diamagnetic)
He_2^+:$\sigma 1s^2, \sigma^* 1s^1$ (Paramagnetic)
Li_2:$\sigma 1s^2, \sigma^* 1s^2, \sigma 2s^2$ (Diamagnetic)
Be_2:$\sigma 1s^2, \sigma^* 1s^2, \sigma 2s^2, \sigma^* 2s^2$ (Diamagnetic)
B_2:$\sigma 1s^2, \sigma^* 1s^2, \sigma 2s^2, \sigma^* 2s^2$,
$\pi 2p_x^1 = \pi 2p_y^1$ (Paramagnetic)
C_2:$\sigma 1s^2, \sigma^* 1s^2, \sigma 2s^2, \sigma^* 2s^2$,
$\pi 2p_x^2 = \pi 2p_y^2$ (Diamagnetic)
N_2:$\sigma 1s^2, \sigma^* 1s^2, \sigma 2s^2, \sigma^* 2s^2$,
$\pi 2p_x^2 = \pi 2p_y^2, \sigma 2p_z^2$ (Diamagnetic)
O_2^-:$\sigma 1s^2, \sigma^* 1s^2, \sigma 2s^2, \sigma^* 2s^2, \sigma 2p_z^2, \pi 2p_x^2$
$= \pi 2p_y^2, \pi^* 2p_x^2 = \pi^* 2p_y^1$ (Paramagnetic)
F_2:$\sigma 1s^2, \sigma^* 1s^2, \sigma 2s^2, \sigma^* 2s^2, \sigma 2p_z^2, \pi 2p_x^2$
$= \pi 2p_y^2, \pi^* 2p_x^2 = \pi^* 2p_y^2$ (Diamagnetic)

95. Increases, decreases;
$\therefore$ Bond order in N_2 = 3 and Bond order in $N_2^+ = 2.5$
Thus, conversion of N_2 to N_2^+ decreases bond order (from 3 to 2.5) and hence, increases the N – N bond distance.
Bond order in O_2 = 2 and Bond order in $O_2^+ = 2.5$
Thus, conversion of O_2 to O_2^+ increases bond order (from 2 to 2.5) hence, decrease in O – O bond distance.

96. $\mathbf{H-\underset{\underset{O}{\|}}{C}-OH}$ and $\mathbf{CH_3-\overset{\overset{O}{\|}}{C}-OH}$ form strongest hydrogen bonds because of largest difference in electronegativities of bonded atoms.

97. (b, c) For Ne_2 molecule the molecular orbital configuration is :-

$$\sigma 1s^2\, \sigma^* 1s^2\, \sigma 2s^2\, \sigma^* 2s^2\, \sigma 2pz^2\, \pi 2p_x^2 = \pi 2p_y^2\, \pi^* 2p_x^2 = \pi^* 2p_y^2\, \sigma^* 2p_z^2$$

- As bond order $= \frac{1}{2}(N_b - N_a)$

  ($\because N_b$ = No. of electrons in bonding molecular orbital & N_a = No. of eletrons in anti bonding molecular orbital)

- So, Bond order $= \frac{1}{2}(10 - 10) = 0$
- For F_2 molecule the molecular orbital configuration is:-

  $$\sigma 1s^2\, \sigma^* 1s^2\, \sigma 2s^2\, \sigma^* 2s^2\, \sigma 2p_z^2\, \pi 2p_x^2 = \pi 2p_y^2\, \pi^* 2p_x^2 = \pi^* 2p_y^2$$

  Hence, HOMO (Highest Occupied Molecular Orbital) is

  $$\pi^* 2p_x^2 = \pi^* 2p_y^2$$

98. (a, d) Overlap of two $2p_z$ orbitals.

(a)

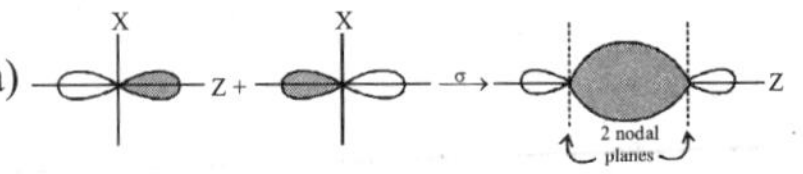

(b)

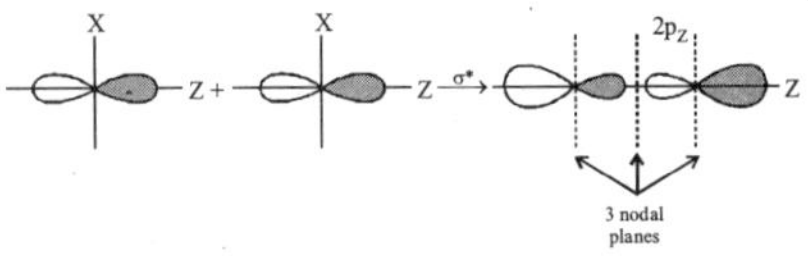

(c)

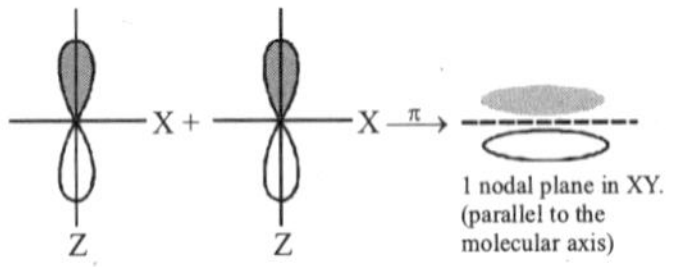

(d)

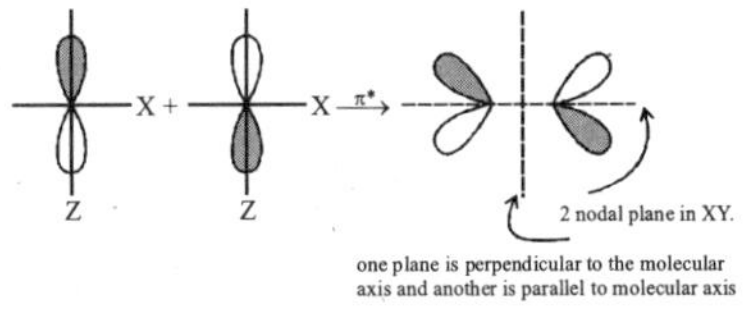

99. (a, c)

(a) The molecular orbital energy configuration of C_2^{2-} is

$$\sigma_{1s}^2, \sigma_{1s}^{*2}, \sigma_{2s}^2, \sigma_{2s}^{*2}, \pi_{2p_x}^2 = \pi_{2p_y}^2, \sigma_{2p_z}^2$$

In the MO of C_2^{2-}, there is no unpaired electron hence, it is diamagnetic.

(b) Bond order of O_2^{2+} is 3 and O_2 is 2 therefore bond length of O_2 is greater than O_2^{2+}.

(c) The molecular orbital energy configuration of N_2^+ is

$$\sigma_{1s}^2, \sigma_{1s}^{*2}, \sigma_{2s}^2, \sigma_{2s}^{*2}, \pi_{2p_x}^2 = \pi_{2p_y}^2, \sigma_{2p_z}^1$$

Bond order of $N_2^+ = \frac{1}{2}(9-4) = 2.5$

The molecular orbital energy configuration of N_2^- is

$$\sigma_{1s}^2, \sigma_{1s}^{*2}, \sigma_{2s}^2, \sigma_{2s}^{*2}, \pi_{2p_x}^2 = \pi_{2p_y}^2, \sigma_{2p_z}^2, \pi_{2p_x}^{*1}$$

Bond order of $N_2^- = \frac{1}{2}(10-5) = 2.5$

(d) He_2^+ has less energy in comparison to two isolated He atoms because some energy is released during the formation of He_2^+ from 2 He atoms.

100. (c) When two same phase overlap, it forms bonding molecular orbital otherwise antibonding molecular orbital.

Also axial overlap produces σ-bond and sideways overlap produces π-bond. For example :

σ-antibonding orbital

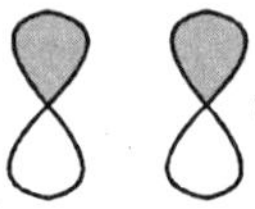

π -bonding orbital

P. 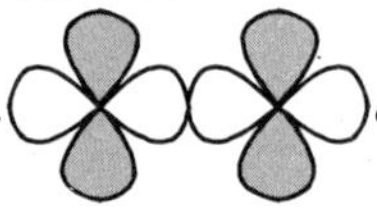 $d-d$ (σ bonding)

Q. 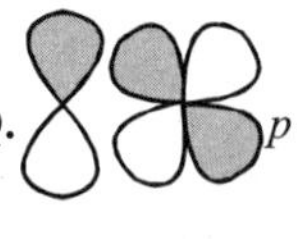 $p-d$ (π bonding)

R. 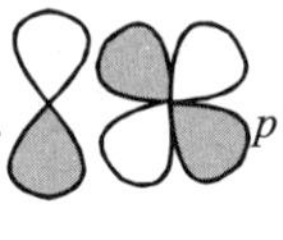 $p-d$ (π antibonding)

S.  $d-d$ (σ antibonding)

101. (A)–p, r, t; (B)– s, t; (C)–p, q; (D)–p, q, s

(A) $B_2 : \sigma 1s^2, \sigma^* 1s^2, \sigma 2s^2, \sigma^* 2s^2, \pi 2p_x^1, \pi 2p_y^1$

Addition of e^- will increase the B.O., hence it will undergo reduction.

(B) $N_2 : \sigma 1s^2, \sigma^* 1s^2, \sigma 2s^2, \sigma^* 2s^2, \pi 2p_x^2, \pi 2p_y^2, \pi 2p_z^2, \pi^* 2p_x^\circ, \pi^* 2p_y^\circ$

Addition or removal of e^- will decrease the B.O., hence it will not undergo oxidation or reduction.

(D) $O_2 : \sigma 1s^2, \sigma^* 1s^2, \sigma 2s^2, \sigma^* 2s^2, \pi 2p_z^2, \pi 2p_x^2, \pi 2p_y^2, \pi^* 2p_x^1, \pi^* 2p_y^1$

Removal of e^- will increase the B.O., hence it will undergo oxidation easily.

(C) Similarly, removal of e^- will increase the B.O. for O_2^- also. Hence it will undergo oxidation easily.

102. (i) Increasing order of bond dissociation energy.

$F_2 < Cl_2 < O_2 < N_2$

Fluorine-fluorine bond energy is less than the Cl–Cl because of larger repulsion between the non-bonded electrons of the two smaller fluorine atoms. Oxygen having two lone pairs of electrons on each atom exert less repulsion than that of chlorine or fluorine each having three lone pairs of electrons. Nitrogen having only one lone pair of electrons exert minimum repulsion, hence it is the most stable.

(ii) H-bonding is an electrostatic attractive force between covalently bonded hydrogen atom of one molecule and an electronegative atom (F, O, N). Further, higher the electronegativity and smaller the size of the atom, the stronger is the hydrogen bond.

Although Cl has the same electronegativity as nitrogen, it does not form effective hydrogen bonds. This is because of its larger size than that of N with the result its electrostatic attractions are weak. Similarly, sulphur forms a very weak hydrogen bond due to its low electronegativity, although oxygen present in the same group forms a strong hydrogen bond.

Hence, the order is $S < Cl < N < O < F$

(iii) In KO_2, O_2 is present as O_2^-, while in $\overset{+1}{O_2}(\overset{+3}{As}\overset{-4}{F_4})$, O_2 is present as O_2^+. Write down the MO configuration of O_2, O_2^- and O_2^+.

$O_2 : \sigma 1s^2, \sigma^* 1s^2, \sigma 2s^2, \sigma^* 2s^2, \sigma 2p_x^2,$
$\pi 2p_y^2 = \pi 2p_z^2,$
$\pi^* 2p_y^1 = \pi^* 2p_z^1.$

Thus, the bond order = 2

O_2^- : Same as above except $\pi^* 2p_y^2$, $\pi^* 2p_z^1$ in place of $\pi^* 2p_y^1, \pi^* 2p_z^1$.

Thus, the bond order in $O_2^- = 1.5$

O_2^+ : Same as in O_2 except $\pi^* 2p_y^1 = \pi^* 2p_z^0$ in place of $\pi^* 2p_y^1, \pi^* 2p_z^1$.

$\therefore$ Bond order in $O_2^+ = 2.5$

$\therefore$ Bond order in the three species is $O_2^+ > O_2 > O_2^-$

or $O_2[AsF_4] > O_2 > KO_2$

Bond length order : $KO_2 > O_2 > O_2[AsF_4]$

103. Benzene has 12σ and 3π bonds.

Chapter-5

States of Matter

Part-A (JEE Main)

1. **(2)** According to ideal gas equation

$PV = nRT \Rightarrow PV = \frac{w}{M}RT$

$\Rightarrow P = \frac{RT}{M} \times d$

$\Rightarrow P \propto d$ and $P \propto T$

2. **(4)** Mass of liquid = 135 – 40 = 95 g

Volume of liquid $= \frac{\text{mass}}{\text{density}} = \frac{95}{0.95}$ mL

= 100 mL = 0.1 L

Mass of ideal gas = 40.5 – 40g = 0.5g

$PV = nRT$

$0.82 \times 0.1 = \left(\frac{0.5}{M}\right) \times 0.082 \times 250$

$\Rightarrow M = 125$

3. **(3)** For London dispersion forces,

$E \propto \rightarrow r^{-6}$; So, $x = -6$

4. **(1)** $P_{gas} = \frac{n_{gas}RT}{V}$

As n, T and V are constant, so,

$P_{H_2} = P_{O_2} = P_{He} = 2$ atm

So, $P_{Total} = P_{H_2} + P_{O_2} + P_{He} = 6$ atm

5. **(2)** As temperature increases, density of gas will decrease.

Trick: Either of graph I and II is not correct.

6. **(4)**

7. **(1)** Ethyl acetate is polar molecule, so dipole-dipole interaction and London dispersion will be present in its liquid state.

8. **(4)** Among given intermolecular forces, ionic interactions are strongest forces. Thus, correct order is ion-ion > ion-dipole > dipole-dipole

9. **(4)** Ideal gas equation: $PV = nRT$

After putting the values, we get

$200 \times 10 = (0.5 + x) \times R \times 1000$

(total no. of moles = 0.5 + x)

$\frac{2000}{1000} = 0.5R + xR$

$\Rightarrow \quad 2 = \left(\frac{1}{2} + x\right)R \Rightarrow \frac{2}{R} = \frac{1}{2} + x$

$\Rightarrow \quad x = \frac{4 - R}{2R}$

10. **(2)** At 27 °C or 300 K, number of moles of an ideal gas = n_1

At T_2 K, number of moles of the ideal gas = n_2

Number of moles escaped $= \frac{2n_1}{5}$

$n_2 = n_1 - \frac{2n_1}{5} = \frac{3n_1}{5}$

$PV = nRT$ (Ideal gas equation)

At constant volume and pressure,

$n \propto \frac{1}{T}$; $n_1T_1 = n_2T_2$

$T_2 = \frac{n_1}{n_2}T_1 \Rightarrow T_2 = \frac{n_1}{\frac{3n_1}{5}} \times T_1 = \frac{5}{3} \times 300 = 500$ K

11. **(3)**

$\because$ At constant temperature and pressure, $d \propto M$

$\therefore$ $d_1/d_2 = M_1/M_2$

(Here d_1 and M_1 are density and molecular mass of ammonia whereas d_2 and M_2 are density and molecular mass of hydrogen chloride)

$d_1/d_2 = 17/36.5 : d_1/d_2 = 0.46$

12. **(2)** The real gases show deviation from ideality at low temperature, high pressure and low volume.

13. **(3)** Density (ρ) $= \frac{PM}{RT}$ (1 bar = 0.987 atm)

$\rho_{N_2} = \frac{4 \times 0.987 \text{ atm} \times 28 \text{ g/mol}}{R \times 300 \text{ K}}$

Let the molar mass of gas be x

$\rho_{gas} = \frac{2 \times 0.987 \text{ atm} \times x}{R \times 300 \text{ K}}$

Given $\rho_{gas} = \rho_{N_2} \times 2$

$\frac{2 \times 0.987 \text{ atm} \times x}{R \times 300 \text{ K}} = \frac{4 \times 0.987 \text{ atm} \times 28 \text{ g/mol}}{R \times 300} \times 2$

$\therefore x = 112$ g/mol.

14. **(1)** In the given case, initial moles and final moles are equal $(n_T)_i = (n_T)_f$

$\frac{P_iV}{RT_1} + \frac{P_iV}{RT_1} = \frac{P_fV}{RT_1} + \frac{P_fV}{RT_2}$

$2\frac{P_i}{T_1} = \frac{P_f}{T_1} + \frac{P_f}{T_2}$; $\quad P_f = 2P_i\left(\frac{T_2}{T_1 + T_2}\right)$

15. **(1)** According to Boyle's law

$\frac{V_1}{V_2} = \frac{P_2}{P_1}$; $\frac{750}{V_2} = \frac{360}{840}$

$V_2 = 1750$ mL = 1.750 L

16. **(2)** According to Graham's law of diffusion:

$\frac{r_1}{r_2} = \sqrt{\frac{d_2}{d_1}}$ or $\frac{r_1}{r_2} = \sqrt{\frac{m_2}{m_1}}$ $\left[\because d = \frac{\text{Mol.wt}}{2}\right]$

Since rate of diffusion

$= \frac{\text{Vol. of gas diffused } (V)}{\text{Time taken for diffusion} (t)}$

$\therefore \quad \frac{r_1}{r_2} = \frac{V_1/t_1}{V_2/t_2}$ or $\frac{r_1}{r_2} = \frac{V_1/t_1}{V_2/t_2} \Rightarrow \sqrt{\frac{m_2}{m_1}}$

$= \frac{20/60}{V_2/30} \Rightarrow \sqrt{\frac{32}{64}} = \frac{20}{60} \times \frac{30}{V_2} \Rightarrow V_2 = 14.1$

17. **(2)** According to Boyle's law, PV = constant

$\therefore$ $\log P + \log V$ = constant

$\log P = -\log V$ + constant

Hence, the plot of $\log P$ vs $\log V$ is straight line with negative slope.

18. **(1)**

19. **(2)** Hydrogen bond is a type of strong electrostatic dipole-dipole interaction and dependent on the inverse cube of distance between the molecules.

20. **(4)** From the ideal gas equation :

$PV = nRT$

The volume occupied by H_2O molecule in vapour phase is $(1 - 10^{-4})$ dm$^3 \approx 1$ L

Given, $P = 3170$ Pa $\approx 3170 \times 10^{-5}$ atm

$R = 8.314$ JK^{-1} mol^{-1}

$\Rightarrow \quad \frac{8.314 \text{ JK}^{-1}\text{mol}^{-1} \times 1.02 \text{ atm}}{101.3 \text{ J}}$

$= 0.0821$ L atm K^{-1}mol^{-1}

Thus, $T = 300$ K

$n = \frac{PV}{RT} = \frac{3170 \times 10^{-5} \text{atm} \times 1}{0.0821 \text{ Latm K}^{-1}\text{mol}^{-1} \times 300 \text{ K}}$

$= 1.28 \times 10^{-3}$ mol

21. **(3)** Value of gas constant

$(R) = 0.082$ L atm K^{-1} mol^{-1}

$= 8.314 \times 10^7$ ergs K^{-1}mol^{-1}

$= 8.314$ JK^{-1}mol^{-1} = 1.987 cal K^{-1}mol^{-1}

[For conversion, these terms can be used 1.0 L atm = 101.3J; 1 J = 10^7 ergs; 1 J = 0.239 cal; or 1 cal = 4.184 J]

22. **(80)** p_{Total} = 25 bar; p_{Ne} = 20 bar

$p_{Ne} = \chi_{Ne} \times p_{Total} \Rightarrow 20 = \chi_{Ne} \times 25$

$\Rightarrow \quad \chi_{Ne} = \frac{20}{25} = \frac{4}{5} \Rightarrow \frac{n_{Ne}}{n_{Total}} = \frac{4}{5}$

$\Rightarrow \quad \frac{200/20}{n_{Total}} = \frac{4}{5} \Rightarrow n_{Total} = 10 \times \frac{5}{4} = \frac{50}{4}$

$\Rightarrow \quad n_{Total} = \frac{50}{4} \quad \Rightarrow \frac{x}{32} + \frac{200}{20} = \frac{50}{4}$

$\Rightarrow \quad x = 80$ g

23. **(8)** PV = n_{mix}RT

$n_{mix} = \frac{6 \times 12.5}{0.083 \times 300} \approx 3$

Let mole of He = x

Mole of H_2 = 3 – x

4x + 2(3 – x) = 10

x = 2 mol

Mass of He = 8g

24. **(22)** At 27°C, moist gas pressure,

$(p_g^o)_1$ + v.p. of water = 4

or, $(p_g^o)_1 = (4 - 0.4)$ atm. = 3.6 atm.
After doubling the volume at constant T,
$(p_g^o)_2 = 1.8$ atm
∴ The new pressure of the moist gas
= (1.8 + 0.4) atm
$= 2.2 = 22 \times 10^{-1}$ atm

25. (1655)

Moles of $C_3H_8 = \frac{\text{weight}}{\text{Molar mass}} = \frac{11}{44}$
= 0.25 moles
Assuming ideal gas equation
PV = nRT
$\Rightarrow 2 \times 10^6 \times 2 \times 10^{-3} = 0.25 \times 8.3 \times T$
$\Rightarrow$ T = 1927.710 K = 1927.710 − 273.15
= 1654.56°C

26. (152)
Assuming ideal behaviour PV = nRT

$$PV = \frac{m}{M}RT \Rightarrow M = \frac{m}{V}\frac{RT}{P} = \frac{dRT}{P}$$

760 mm Hg ⇒ 1 atm

$$P = \frac{100}{760}\text{atm}, T = 257 + 273 = 530\text{ K},$$

d = 0.46 gm/L

$$\text{So } M = \frac{0.46 \times 0.082 \times 530}{100} \times 760$$

= 151.93 g ≈ 152g

27. (4) Using the ideal gas equation PV = nRT

$$1.5 \times 416 = \frac{100}{M} \times 0.083 \times 300$$

M = 3.99 ≈ 4

28. (32) $\frac{P_1}{T_1} = \frac{P_2}{T_2}$

$$\frac{30}{300} = \frac{P_2}{318} \Rightarrow P_2 = \frac{30}{300} \times 318 = 32$$

29. (45) Given: Ideal gas A and H_2 gas at same pressure and volume.
From ideal gas equation,
$n_1T_1 = n_2T_2$

$$\frac{2}{\text{molar mass} \times 2 \text{ of A}} \times 300 = \frac{0.2}{2} \times 200$$

molar mass × 2 of A = 45 g/mole

30. (2) Initial mass of gas = 29 − 14.8 = 14.2 kg
Final mass of the gas = 23 − 14.8 = 8.2 kg
Ideal gas equation is
$pV = nRT$

$$3.47 \times V = \left(\frac{14.2 \times 10^3}{M}\right) \times R \times T \quad ...(i)$$

$$p \times V = \left(\frac{8.2 \times 10^3}{M}\right) \times R \times T \quad ...(ii)$$

Dividing equation (i) by equation (ii):

$$\frac{3.47}{p} = \frac{14.2}{8.2}; p = 2.0 \text{ atm}$$

31. (26) $PV = nRT$

$$PV = \frac{\text{Mass}}{\text{Molar mass}} \times RT$$

$$\text{Mass} = \frac{PV}{RT} \times \text{molar mass}$$

$$= \frac{1 \times 4 \times 10^6}{0.083 \times 300} \times 16 = 25.7 \times 10^5 \text{ g} \simeq 26 \times 10^5 \text{ g}$$

32. (927) $\frac{p_1}{T_1} = \frac{p_2}{T_2}$

Given, $p_1 = 300$ kPa = 300×10^3 Pa
$T_1 = 27°C = 27 + 273 = 300$ K
$p_2 = 1.2 \times 10^6$ Pa
$T_2 = ?$

$$\frac{300 \times 10^3}{300} = \frac{1.2 \times 10^6}{T_2}$$

$$T_2 = \frac{1.2 \times 10^6}{10^3} = 1200\text{ K} = 1200 - 273 = 927\text{ °C}$$

33. (150) Number of moles of $CH_4 = \frac{6.4}{16} = 0.4$

Number of moles of $CO_2 = \frac{8.8}{44} = 0.2$

Total number of moles of the mixture = 0.6

Now, $P = \frac{nRT}{V}$

$$= \frac{0.6 \text{ mol} \times 8.314 \text{ J mol}^{-1}\text{ K}^{-1} \times 300\text{ K}}{10\text{ L}}$$

$$= 149.65 \text{ J/L} = \frac{149.65 \text{ N m}}{10^{-3}\text{ m}^3}$$

$= 149.65 \times 10^3$ Pa [∵ 1 Pascal = $\frac{N}{m^2}$; 1 J = N.m]

$$= \frac{149.65 \times 10^3}{1000}\text{kPa} \simeq 150 \text{ kPa}$$

34. (70) Using : $\frac{P}{T}$ = Constant $\frac{P_1}{T_1} = \frac{P_2}{T_2}$

$$\Rightarrow T_2 = \frac{40 \times 300}{35} = 342.86 \text{ K}$$

$$\therefore T_2 = 69.86 \text{ °C} \approx 70 \text{ °C}.$$

35. (5) Moles of $C_2H_2(g) = \frac{4.75}{26}$ mol.

T = 50 + 273 = 323 K

$$P = \frac{740}{760}\text{ atm}; R = 0.0826\frac{\text{L atm}}{\text{mol K}}$$

$PV = nRT$

$$\Rightarrow V = \frac{nRT}{P} = \frac{4.75}{26} \times \frac{0.0826 \times 323 \times 760}{740}$$

$$\Rightarrow V = \frac{96314.078}{19240} = 5.0059 \text{ L} \simeq 5\text{L}.$$

36. (750) At constant temperature and number of moles,

$P_1V_1 = P_2V_2$

$$P_1 = 48 \times 10^{-3}\text{ bar}; V_1 = \frac{4}{3}\pi(3)^3$$

$$V_2 = \frac{4}{3}\pi(12)^3$$

$$P_2 = \frac{P_1V_1}{V_2} = \frac{48 \times 10^{-3} \times (3)^3}{(12)^3}$$

$$= \frac{48 \times 10^{-3}}{64} = 7.5 \times 10^{-4} = 750 \times 10^{-6}\text{ bar}$$

37. (4) $V_{rms} > V_{average} > V_{mp}$

$$\sqrt{\frac{3RT}{M}} > \sqrt{\frac{8RT}{\pi M}} > \sqrt{\frac{2RT}{M}}$$

38. (1) $V_{mp} = \sqrt{\frac{2RT}{M}} \therefore V_{mp} \propto \left(\frac{T}{M}\right)^{\frac{1}{2}}$

From curve,
$(V_{mp})_I < (V_{mp})_{II} < (V_{mp})_{III}$

$$(V_{mp})_{N_2} \propto \sqrt{\frac{300}{28}};$$

$$(V_{mp})_{O_2} \propto \sqrt{\frac{400}{32}}; (V_{mp})_{H_2} \propto \sqrt{\frac{300}{2}}$$

$\therefore (V_{mp})_{N_2} < (V_{mp})_{O_2} < (V_{mp})_{H_2}$
(under given condition)

39. (1) $rms = \sqrt{\frac{3RT}{M}}$

rms for $N_2, u = \sqrt{\frac{3RT}{28}}$

After dissociation, $u' = \sqrt{\frac{3R \times 2T}{14}}$

$$\frac{u}{u'} = \sqrt{\frac{1}{4}} = \frac{1}{2}$$

$u' = 2u$

40. (3) At high pressure, real gas particles can be easily compressed.

41. (2) Number of moles of $O_2 = \frac{m}{32}$

Number of moles of $N_2 = \frac{4m}{28} = \frac{m}{7}$

∴ Ratio of moles or molecules = $\frac{m}{32} : \frac{m}{7} = 7 : 32$

42. (4) $V_{rms} = \sqrt{\frac{3RT}{M}}$

$V_{rms}(O_2) = V_{rms}(He)$

$$\sqrt{\frac{3RT_{O_2}}{M_{O_2}}} = \sqrt{\frac{3RT_{He}}{M_{He}}}$$

or $\frac{T_{O_2}}{M_{O_2}} = \frac{T_{He}}{M_{He}}$

$$\therefore T_{O_2} = \frac{300 \times 32}{4} = 2400 \text{ K}$$

43. (3) Most probable speed (C*) = $\sqrt{\frac{2RT}{M}}$

Average Speed $(\bar{C}) = \sqrt{\frac{8RT}{\pi M}}$

Root mean square velocity (C) = $\sqrt{\frac{3RT}{M}}$

$$C^* : \bar{C} : C = \sqrt{\frac{2RT}{M}} : \sqrt{\frac{8RT}{\pi M}} : \sqrt{\frac{3RT}{M}}$$

$$= 1 : \sqrt{\frac{4}{\pi}} : \sqrt{\frac{3}{2}} = 1 : 1.128 : 1.225$$

44. (4) r.m.s. velocity $V_{rms} = \sqrt{\frac{3RT}{M}}$

i.e., $\frac{V_1}{V_2} = \sqrt{\frac{T_1}{T_2}}$

$$\frac{5 \times 10^4}{10 \times 10^4} = \frac{1}{2} = \sqrt{\frac{T_1}{T_2}}; \therefore T_2 = 4T_1$$

45. (4) Molecules move very fast in all directions in a straight line by colliding with each other but with different velocity.

46. **(1)** $r \propto \frac{P}{\sqrt{M}}$

$$\frac{r_A}{r_B} = \frac{P_A}{P_B}\sqrt{\frac{M_B}{M_A}}$$

47. **(3)**

48. **(3)** At higher temperature more molecules have higher velocities and less molecules have lower velocities. It is clear that with the increase in temperature the most probable velocity increases but the fraction of such molecules decreases.

49. **(1)**

$$\frac{\text{K.E of neon at } 40\,°\text{C}}{\text{K.E of neon at } 20\,°\text{C}} = \frac{\frac{3}{2}K\times 313}{\frac{3}{2}K\times 293} = \frac{313}{293}$$

50. **(2)** According to kinetic theory of gases, between two successive collisions, molecules travel in a straight line path but show haphazard motion due to collisions.

51. **(1)** Critical temperature indicates the ease of liquefaction of gas more the critical temp. of gas more intermolecular attractive forces and greater is the ease of liquefaction hence greater is the adsorption.

52. **(1)** If values of 'b' for two gases are same but values of 'a' are different, then the gas having a larger value of 'a' will occupy lesser volume. Since, it will have larger force of attraction, therefore, lesser distance between its moleucles. If values of 'a' for two gases are same but values of 'b', are different then the smaller value of 'b' will occupy lesser volume, therefore, will be more compressible.

53. **(1)** Critical temperature $= \frac{8a}{27Rb}$

Ar $\Rightarrow \frac{a}{b} = 0.4$; Ne $\Rightarrow \frac{a}{b} = 0.12$

Kr $\Rightarrow \frac{a}{b} = 5.1$; Xe $\Rightarrow \frac{a}{b} = 0.82$

Value of $\frac{a}{b}$ is highest for Kr. Therefore, Kr has highest value of critical temperature.

54. **(1)** $Z = 1 + \frac{Pb}{RT}$

[Where Z (compressibility factor) $= PV/RT$]

Slope of Z vs P curve (straight line) $= \frac{b}{RT}$

$\therefore$ Higher the value of b, more steep will be the curve. Constant 'b' value depends on size of atoms or molecules.

55. **(2)** Compressibility factor is given by,

$$Z = \frac{PV}{RT};\ \therefore\ \frac{Z_A}{Z_B} = \frac{P_A V_A}{P_B V_B}$$

Given $Z_A = 3Z_B$; $V_A = 2V_B$

$$\frac{3Z_B}{Z_B} = \frac{P_A \times 2V_B}{P_B \times V_B} \Rightarrow 3 = \frac{P_A}{P_B}\times 2$$

$2P_A = 3P_B$

56. **(1)**

57. **(2)** Compressibility factor $(Z) = \frac{PV}{RT}$

van der Waals equation,(For one mole of real gas)

$$(p + \frac{a}{V^2})(V - b) = RT$$

At low pressure, volume is very large and hence, correction term b can be neglected in comparison to very large volume of V, i.e., $V - b \approx V$

$$\left(p + \frac{a}{V^2}\right)V = RT;\ pV + \frac{a}{V} = RT$$

$$pV = RT - \frac{a}{V};\ \frac{PV}{RT} = 1 - \frac{a}{VRT}$$

Hence, $Z = 1 - \frac{a}{VRT}$

58. **(1)** At low pressure and high temperature, the effect of $\frac{a}{V^2}$ and b is negligible, hence $pV = nRT$.

59. **(1)** In van der Waals equation, 'b' is for volume correction.

60. **(25)** Real gas equation = P(V – nb) = nRT

or, PV – Pnb = nRT

or, PV = nRT + Pnb

or, $\frac{PV}{nRT} = 1 + \frac{Pb}{RT}$ or, $Z = 1 + \frac{Pb}{RT}$

When Z = 2, $\frac{Pb}{RT} = 1$.

or, $b = \frac{RT}{P} = \frac{0.083\times 298}{99}\,\text{L mol}^{-1}$

$= 0.25\,\text{L mol}^{-1} = 25\times 10^{-2}\,\text{L mol}^{-1}$

61. **(1)** $P(V_m\ \ b) = RT$

$$\Rightarrow PV_m - Pb = RT \Rightarrow \frac{PV_m}{RT} = 1 + \frac{Pb}{RT}$$

$$\Rightarrow Z = 1 + \frac{Pb}{RT} \Rightarrow \left(\frac{\partial Z}{\partial P}\right)_T = \frac{b\times 1}{RT}\ \therefore x = 1$$

Part-B (JEE Advanced)

1. **(b)** Use Grahams' law of diffusion

$$\frac{r_{He}}{r_{CH_4}} = \sqrt{\frac{M_{CH_4}}{M_{He}}} = \sqrt{\frac{16}{4}} = 2$$

2. **(c)** For positive deviation: $PV = nRT + nPb$

$$\Rightarrow \frac{PV}{nRT} = 1 + \frac{Pb}{RT}$$

Thus, the factor nPb is responsible for increasing the PV value, above ideal value. b is actually the effective volume of molecule. So, it is the finite size of molecules that leads to the origin of b and hence, positive deviation at high pressure.

3. **(c)** Find the volume by either $V = RT/P$ ($PV = RT$) or $P_1V_1 = P_2V_2$ and and match it with the values given in graph to find correct answer. Volume of 1 mole of an ideal gas at 273 K and 1 atm is 22.4 L and that at 373 K and 1 atm pressure is calculated as ;

$$V = \frac{RT}{P} = \frac{0.082\times 373}{1} = 30.58\text{ L} \simeq 30.6\text{ L}$$

4. **(c)** Mass of 1 L of vapour = volume × density = 1000 × 0.0006 = 0.6 g

V of liquid water $= \frac{\text{mass}}{\text{density}} = \frac{0.6}{1} = 0.6\text{ cm}^3$

5. **(c)** For an ideal-gas behaviour, the molecules of a gas should be far apart. The factors favouring this condition are high temperature and low pressure.

6. **(b)** Under identical conditions,

$$\frac{r_1}{r_2} = \sqrt{\frac{M_2}{M_1}}$$

As rate of diffusion is also inversely proportional to time, we will have,

$$\frac{t_2}{t_1} = \sqrt{\frac{M_2}{M_1}} \qquad [M_1 = 2 \text{ and } t_1 = 5]$$

(a) Thus, For He, $t_2 = \sqrt{\frac{4}{2}}(5\text{s}) = 5\sqrt{2}s \neq 10\text{s}$;

(b) For O_2, $t_2 = \sqrt{\frac{32}{2}}(5\text{s}) = 20\text{s}$ (as given)

(c) For CO, $t_2 = \sqrt{\frac{28}{2}}(5\text{s}) \neq 25\text{s}$;

(d) For CO_2, $t_2 = \sqrt{\frac{44}{2}}(5\text{s}) \neq 55\text{s}$

7. **(b)**

	$N_2O_4(g)$ $\rightleftharpoons$	$2NO_2(g)$
At start	100/92 mol = 1.08 mol	0
At equilibrium	80/92 mol = 0.86 mol	20/46 mol =0.43 mol

According to ideal gas equation, at two conditions

At 300 K; $P_0V = n_0RT_0$

$1\times V = 1.08\times R\times 300$ ...(i)

At 600 K; $P_1V = n_1RT_1$

$P_1\times V = (0.86 + 0.43)\times R\times 600$ ...(ii)

Divide (ii) by (i), $\frac{P_1}{1} = \frac{1.29\times 600}{1.08\times 300}$;

$P_1 = \frac{1.29\times 2}{1.08} = 2.38\text{ atm.} \simeq 2.4\text{ atm.}$

8. **(a)** Due to increase in the temperature, the kinetic energy of the gas molecules increases resulting in an increase in average molecular speed. The molecules are bombarded to the walls of the container with a greater velocity resulting in an increase in pressure.

9. **(b)** $d = \frac{PM}{RT}$

It means density of gas is directly proportional to pressure and inversely proportional to temperature. Density of neon will be maximum at highest pressure and lowest temperature.

10. **(a)** $\frac{r_{CH_4}}{r_x} = 2 = \sqrt{\frac{M_x}{M_{CH_4}}} = \sqrt{\frac{M_x}{16}}$,

or $M_x = 64$

11. **(b)** Rate of diffusion $\propto \sqrt{\frac{1}{\text{Molecular mass}}}$

$\because$ Molecular mass of HCl > molecular mass of NH_3

$\therefore$ HCl diffuses at slower rate and white ammonium chloride is first formed near HCl bottle.

12. **(d)** $\frac{r_1}{r_2} = \sqrt{\frac{d_2}{d_1}} = \sqrt{\frac{M_2}{M_1}}$

13. (b) Pressure exerted by hydrogen will be proportional to its mole fraction.

$$\text{Mole fraction of } H_2 = \frac{\frac{w}{2}}{\frac{w}{16}+\frac{w}{2}} = \frac{8}{9}$$

14. (b) The temperature at which a real gas behaves like an ideal gas is called Boyle's temperature or Boyle's point.

15. (2.22) $P_A = 5$ bar, $V_A = 1\ m^3$, $T_A = 400$ K; $P_B = 1$ bar, $V_B = 3\ m^3$, $T_B = 300$ K

$$\Rightarrow n_A = \frac{P_AV_A}{RT_A} \text{ and } n_B = \frac{P_BV_B}{RT_B}\quad n_A = \frac{5}{400R}$$

and $n_B = \dfrac{3}{300R}$

Let the new volume of compartment A and B respectively be $(1+x)$ and $(3-x)$.

$\dfrac{V_A}{n_A} = \dfrac{V_B}{n_B}$ (at cons. T and P)

$$\Rightarrow \frac{(1+x)}{(5/400R)} = \frac{(3-x)}{(3/300R)} \Rightarrow x = \frac{11}{9}$$

$$V_A = 1+x = 1+\frac{11}{9} = \left(\frac{20}{9}\right) = 2.22$$

16. (4) Diffusion coefficient $\propto \lambda\mu$

Since $\lambda \propto \dfrac{T}{P}$ and $\mu \propto \sqrt{T}$

$\therefore$ Diffusion coefficient $\propto \dfrac{T\sqrt{T}}{P}$

$$\text{Thus } \frac{D_i}{D_f} = \frac{\frac{T\sqrt{T}}{P}}{\frac{4T\sqrt{4T}}{2P}} = \frac{1}{(4\times 2)/2} = \frac{1}{4}$$

or $\dfrac{D_f}{D_i} = \dfrac{4}{1}$

17. (7) $P_{He} = 1 - 0.68 = 0.32$ atm, $n = 0.1$

$$V = \frac{nRT}{P} = \frac{0.1\times 0.0821\times 273}{0.32} = 7\text{ L}$$

18. (9.80)

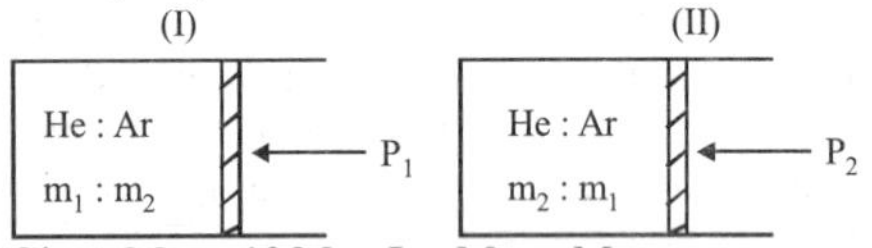

Given $M_{Ar} = 10\,M_{He}$, Let $M_{He} = M$

$P_1 = 5P_2$

$V_1 = V_2 = V$ (Volume of the gas mixture in both the cylinders are equal)

$T_1 = T_2 = T$

Applying ideal gas equation : $PV = nRT$

$$\text{vessel (I)} \Rightarrow P_1V = \left[\frac{m_1}{M}+\frac{m_2}{10M}\right]RT$$

$$\text{vessel (II)} \Rightarrow P_2V = \left[\frac{m_2}{M}+\frac{m_1}{10M}\right]RT$$

$$\text{Dividing (II)/(I)} \Rightarrow \frac{P_2}{P_1} = \frac{\frac{m_2}{M}+\frac{m_1}{10M}}{\frac{m_1}{M}+\frac{m_2}{10M}} = \frac{1}{5}$$

$$\Rightarrow 5m_2 + \frac{m_1}{2} = m_1 + \frac{m_2}{10}$$

$$\Rightarrow 5m_2 - \frac{m_2}{10} = m_1 - \frac{m_1}{2}$$

$$\Rightarrow \frac{49m_2}{10} = \frac{m_1}{2} \Rightarrow \frac{m_1}{m_2} = 9.8$$

19. (–5242.42) Potential energy of H-atom is taken as zero when electron and nucleus are at infinite distance.

$\therefore$ P.E. of a H-atom with electron in its ground state $= 2\times(-13.6\text{ eV}) = -27.2$ eV (from Bohr's Model)

At internuclear distance 'd_0' electron-electron repulsion and nuclues-nucleus repulsion are absent.

P.E. of two H-atom $= -2\times 27.2$ eV $= -54.4$ eV

$$= \frac{54.4\times 1.6\times 10^{-19}\times 6.023\times 10^{23}}{1000}\text{ kJ/mol}$$

$= -5242.42$ kJ/mol

20. (4.53)

	PCl_5	$\rightleftharpoons$	PCl_3	+	Cl_2
Initial moles	1		0		0
Moles at eq.	1–0.4		0.4		0.4

$\therefore$ Total moles at equilibrium $= 1 - 0.4 + 0.4 + 0.4 = 1.4$

$$\text{Also } \frac{\text{Normal mol. wt. of } PCl_5}{\text{Exp. mol. wt. of } PCl_5} = 1+\alpha = 1.4$$

$$\text{or } \frac{208.5}{\text{Exp. mol. wt. of } PCl_5} = 1.4$$

$$\therefore \text{Exp. mol. wt. of } PCl_5 \text{ or m. wt. of mixture} = \frac{208.5}{1.4}$$

Now using, $PV = \dfrac{w}{m}RT$ for mixture

$$d = \frac{w}{V} = \frac{Pm}{RT} = \frac{1\times 208.5}{1.4\times 0.082\times 400} = \mathbf{4.53\ g/litre}$$

21. (0.4) $N_2O_5(g) \rightleftharpoons 2NO_2(g) + \frac{1}{2}O_2(g)$

	N_2O_5	NO_2	O_2
Initial pressure	600	0	0
Final pressure	600-*P*	2*P*	*P*/2

$P \propto$ moles when V and T are constant.

(where moles equivalent to pressure P are decomposed)

Total pressure $= 600 - P + 2P + P/2$

$= 960$ mm of Hg $\therefore P = 240$ mm Hg

Thus, moles of N_2O_5 decomposed $= \dfrac{240}{600} = \mathbf{0.4}$

22. (2.197) Since the pressures of gases are different, and the temperature is constant, the rate at which molecules of the two gases diffuse is directly proportional to the pressure. This rate of diffusion is also directly proportional to the distance travelled by the gas. Hence, $l \propto r \propto \dfrac{P}{\sqrt{M}}$

r_1 (of HCl gas) at pressure $P = 60 = \dfrac{kP}{\sqrt{36.5}}$ .....(i)

and r_2 (of NH_3) at 1 atm. pressure $= 40 = \dfrac{k\times 1}{\sqrt{17}}$ .....(ii)

From (i) and (ii)

$$\frac{r_1}{r_2} = \frac{60}{40} = \frac{kP}{\sqrt{36.5}}\times\frac{\sqrt{17}}{k\times 1}$$

$$P = \frac{60}{40}\times\frac{\sqrt{36.5}}{\sqrt{17}} = \mathbf{2.197\ atm}$$

23. (85.2) Let NH_3 diffuse through $= x$ cm

HCl diffuses through $= y$ cm

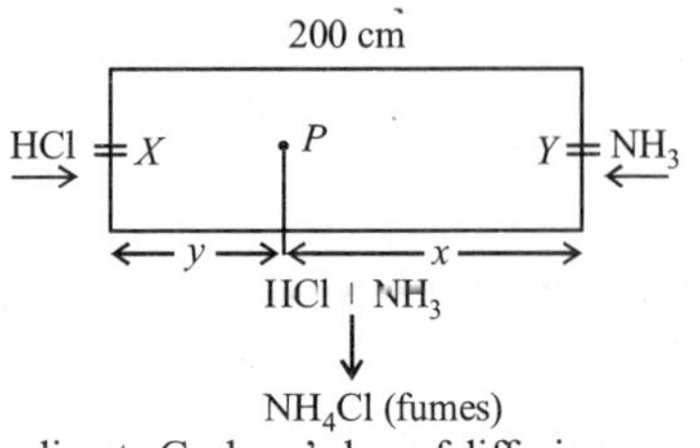

According to Graham's law of diffusion

$$\frac{x}{y} = \sqrt{\frac{\text{Mol. wt HCl}}{\text{Mol. wt of } NH_3}} = \sqrt{\frac{36.5}{17}} = \sqrt{2.14} = 1.465$$

$x = 1.465\,y$ ...(1)

$x + y = 200$ cm ...(2)

From these equations; $y = 85.2$ cm

Distance between P and X $= y =$ **85.2 cm.**

24. (3.42) $PV = \dfrac{m}{M}RT$

$$\Rightarrow P = \frac{m}{V}\times\frac{RT}{M} = d\frac{RT}{M}\quad \therefore d = \frac{MP}{RT}$$

Substituting the value, we get

$$d = \frac{17\times 5}{0.082\times 303} = \mathbf{3.42\ g/litre}$$

25. (0.25) $\left(\because 5.6\,l \equiv \frac{1}{4}\text{ mole of an ideal gas}\right)$

Using ideal gas equation

$$PV = nRT = \frac{1}{4}RT = 0.25\,RT$$

26. Inversely, time; $\left[\dfrac{r_1}{r_2} = \sqrt{\dfrac{d_2}{d_1}} = \dfrac{t_2}{t_1} = \sqrt{\dfrac{M_2}{M_1}}\right]$.

27. R. $[\because C_p - C_v = R]$

28. False : An ideal gas cannot be liquefied as there exists no intermolecular attraction between molecules.

29. (c, d)

(a) Separation energy of point charge and dipole becomes zero faster due to $\dfrac{1}{r^3}$ dependence.

(b) The average potential energy of two rotating polar molecules that are separated by a distance r has $\dfrac{1}{r^6}$ dependence. It is $\dfrac{1}{r^3}$ for stationary dipoles.

(c) The dipole-induced dipole average interaction energy depends on the dipole moment and polarizability.

(d) Nonpolar molecule also have force of attraction that is London force.

30. (b) For gas A, $a = 0$, $Z = 1 + \dfrac{Pb}{RT}$ implies Z varies linearly with pressure.

For gas B, $b = 0$, $Z = 1 - \dfrac{a}{VRT}$. Hence, Z does not vary linearly with pressure.

Given the intersection data for gas C, it is possible to find the values of 'a' and 'b'. All van der Waal gases, like gas C, give positive slope at high pressures.

31. (c) According to Graham's law of diffusion for two gases undergoing diffusion at different pressures through same hole

$$\frac{r_A}{r_B} = \sqrt{\frac{M_B}{M_A}} \times \frac{P_A}{P_B}$$

$$\left(r \propto P \times \sqrt{\frac{1}{M}} \quad \text{At constant temperature} \right)$$

32. (d) Pressure exerted by H_2 is proportional to its mole fraction.

$$\text{Mole fraction of } H_2 = \frac{\frac{W}{2}}{\frac{W}{2}+\frac{W}{30}} = \frac{30}{32} = \frac{15}{16}$$

33. (a, b) At constant temp., when gas expands, the K.E. of the molecules remains the same, but the pressure decreases. $\left(\because P \propto \frac{1}{V} \right)$

34. (b) No work is required to tear apart the molecules due to the absence of attractive forces in an ideal gas.

35. (c) According to Graham's law of diffusion, if all conditions are identical,

$r = \frac{1}{\sqrt{M}}$ As in this question, all conditions are identical for X and Y, then

$$\frac{r_X}{r_Y} = \sqrt{\frac{M_X}{M_Y}} \Rightarrow \frac{d}{24-d} = \sqrt{\frac{40}{10}} = 2$$

$\Rightarrow d = 48 - 2d \Rightarrow 3d = 48 \Rightarrow d = 16$ cm

36. (d) The general formula of mean free path (λ) is $\lambda = \frac{RT}{\sqrt{2}\pi d^2 N_A p}$

(d = diameter of molecule, p = pressure inside the vessel)

Since d and p are same for both gases, ideally their λ are same. Hence, it must be the higher drift speed of X due to which it is facing more collisions per second with the inert gas in comparison to gas Y. Hence, X faces more resistance from inert gas than Y and hence, covers lesser distance than that predicted by Graham's law.

37. (a) $d = 0.36 \text{ kg m}^{-3} = 0.36$ g/L

(i) From Graham's Law of diffusion

$$\frac{r_v}{r_{O_2}} = \sqrt{\frac{M_{O_2}}{M_v}}; \; 1.33 = \sqrt{\frac{32}{M_v}}$$

$$\therefore M_v = \frac{32}{(1.33)^2} = \mathbf{18.09};$$

(ii) Thus, $0.36\text{g} = \frac{0.36}{18.09}$ mol

$\frac{0.36}{18.09}$ mol occupies 1 L volume, so 1 mol occupies $\frac{18.09}{0.36}$ L = 50.25 L

(iii) Assuming ideal behaviour, the volume of the vapour can be calculated by

$$\frac{V_1}{T_1} = \frac{V_2}{T_2} \Rightarrow V_2 = 22.4 \times \frac{500}{273} = \mathbf{41.025\ L}$$

Compressibility factor (Z)

$$= \frac{(PV)_{\text{obs}}}{(PV)_{\text{ideal}}} = \frac{1 \times 50.25}{1 \times 41.025} = \mathbf{1.224}$$

(iv) Z is greater than unity, hence it is the short range repulsive force that would dominate.

($\because$ actual density is less than given density)

(b) $E = \frac{3}{2}KT = \frac{3}{2} \times \frac{8.31}{6.02 \times 10^{23}} \times 1000$

$= \mathbf{2.07 \times 10^{-20}}$ **J per molecule**

($\because K$, Boltzmann constant $= R/N$)

38. We know that

$$\frac{r_1}{r_2} = \sqrt{\frac{M_2}{M_1}} \times \frac{P_1}{P_2} \text{ or } \frac{n_1}{t_1} \times \frac{t_2}{n_2} = \sqrt{\frac{M_2}{M_1}} \times \frac{P_1}{P_2}$$

$$\text{or } \frac{1}{38} \times \frac{57}{1} = \sqrt{\frac{M}{28}} \times \frac{0.8}{1.6} \quad \therefore M = 252$$

$\left[\begin{array}{l} \text{Xe} + (\text{F})x = 252 \\ 131 + 19x = 252 \;; \quad \therefore x = 6. \end{array} \right.$

Thus, compound of xenon with fluorine is $\mathbf{XeF_6}$.

39. (I) Given $P = 1$ atm, $w = 12$ g; $T = (t + 273)$K; $V = V$ litre

(II) If $T = t + 10 + 273 = t + 283$ K ; $V = V$ litre,

$$P = 1 + \frac{10}{100} = 1.1 \text{ atm}$$

Using gas equation, $PV = \frac{w}{m}RT$

Case I. $1 \times V = \frac{12}{m}R(t + 273)$ ... (1)

Case II. $1.1 \times V = \frac{12}{m}R(t + 283)$ ...(2)

From (1) and (2), $t = -173°\text{C}$ or $t = \mathbf{100\ K}$

Also from (1), on substituting t and m (120), $V = \mathbf{0.82}$ **litre**

40. Weight of liquid = 148 – 50 = 98 g

Volume of liquid $= \frac{98}{0.98} = 100$ mL = volume of vessel

It means, vessel of 100 mL contains ideal gas at 760 mm Hg at 300 K.

Weight of gas = 50.5 – 50 = 0.5g

using, $PV = nRT = \frac{w}{m}RT$

$$\frac{760}{760} \times \frac{100}{1000} = \frac{0.5}{m} \times 0.082 \times 300 \quad \left[n = \frac{0.5}{m} \right]$$

$\therefore$ Molecular weight of gas (m) = **123**

41. Let the volume of ethane in mixture = x litre

$\therefore$ Volume of ethene = $(40 - x)$ litre

Combustion reactions of ethane and ethene are :

(i) $C_2H_6(g) + 3\frac{1}{2}O_2(g) \longrightarrow 2CO_2(g) + 3H_2O(l)$

or $2C_2H_6(g) + 7O_2(g) \longrightarrow 4CO_2(g) + 6H_2O(l)$

(ii) $C_2H_4(g) + 3O_2(g) \rightarrow 2CO_2(g) + 2H_2O(l)$

Volume of O_2 required for complete combustion of ethane $= \frac{7x}{2}$ [For x litres]

Volume of O_2 required for complete combustion of ethene $= (40 - x) \times 3$ [For $(40 - x)$ L]

$\therefore$ Total volume of O_2 required $= \frac{7x}{2} + (40 - x)3$ L

Calculation of number of moles (n)

$P = 1$ atm, $V = \frac{7x}{2} + (40 - x)3$ L;

$R = 0.082 \text{ L atm K}^{-1}\text{mol}^{-1}$; $T = 400$ K

Since

$$n = \frac{PV}{RT} = \frac{1 \times \left[\frac{7x}{2} + (40 - x)3 \right]}{0.082 \times 400} = \frac{7x + (40 - x)6}{2 \times 0.082 \times 400}$$

Mass of n moles of O_2

$$= \left[\frac{7x + (40 - x)6}{2 \times 0.082 \times 400} \right] \times 32 = 130$$

$$\text{or } 130 = \left[\frac{7x + 240 - 6x}{65.6} \right] \times 32$$

$\Rightarrow 8528 = 32x + 240 \times 32 \Rightarrow 32x = 848$

$$\Rightarrow \text{or } x = \frac{848}{32} = 26.5$$

Hence, mole fraction (%) of ethane

$$= \frac{26.5}{40} \times 100 = \mathbf{66.25\%}$$

Mole fraction (%) of ethene = **33.75%**

42. Mixture — Krypton

$r_{\text{mix}} = 1.16$ — $r_{\text{Kr}} = 1$

$M_{\text{mix}} = ?$ — $M_{\text{Kr}} = 84$

We know that

$$\frac{r_{\text{mix}}}{r_{\text{Kr}}} = \sqrt{\frac{M_{\text{Kr}}}{M_{\text{mix}}}} \text{ or } \frac{1.16}{1} = \sqrt{\frac{84}{M_{\text{mix}}}} \text{ or}$$

$$(1.16)^2 = \frac{84}{M_{\text{mix}}} \Rightarrow M_{\text{mix}} = \frac{84}{(1.16)^2} = 62.426$$

Determination of the composition of the equilibrium mixture:

Let the fraction of Cl_2 molecules dissociated at equlibrium = x

	Cl_2	$\rightleftharpoons$ 2Cl	Total
Initially	1	0	1
At equilibrium	$1 - x$	$2x$	$1 - x + 2x$ $= 1 + x$

$$\because \frac{\text{Normal molecular mass}}{\text{Experimental molecular mass}} = 1 + x$$

$$\therefore \frac{71}{64.426} = 1 + x \;\therefore x = 0.137 = \mathbf{13.7\%}.$$

43. (i) He does not react with oxygen.

(ii) KOH absorbs only CO_2.

When the mixture of CO, CH_4 and He gases (20 mL) are exploded by an electric discharge with excess of O_2, He gas remains as such and the other reactions involved are :

$$\underset{a}{CO(g)} + \underset{a/2}{\frac{1}{2}O_2(g)} \rightarrow \underset{a}{CO_2(g)} \quad ...(i)$$

$$\underset{b}{CH_4(g)} + \underset{2b}{2O_2(g)} \rightarrow \underset{b}{CO_2(g)} + 2H_2O(l) \quad ...(ii)$$

Let the volumes of CO and CH_4 to be 'a' mL and 'b' mL in the mixture then,

Volume of He gas = $[20 - (a + b)]$ mL

For the initial contraction of 13 mL,

Volume of left hand side in the above reactions – 13 = Volume of right hand side.

$$\therefore \quad [20 - (a + b)] + \left(a + \frac{1}{2}a \right) + (b + 2b) - 13$$

$= [20 - (a + b)] + a + b$ [neglect the volume of H_2O (l)]

(Since for gases, volume α no. of moles)

$$\therefore \frac{1}{2}a + 2b = 13 \text{ or } a + 4b = 26 \quad ...(iii)$$

The CO_2 produced above in reactions (i) & (ii), $(a + b)$ mL, reacts with KOH sol for a further contraction of 14 mL.

$$\underset{(a+b)}{CO_2(g)} + 2KOH(l) \rightarrow K_2CO_3(l) + H_2O(l)$$

$\therefore a + b = 14$ ...(iv)

Solving (iii) & (iv) we get, $a = 10$ mL and $b = 4$ mL

$\therefore CH_4 = \frac{4}{20}\times 100 = \mathbf{20\%}$, $CO = \frac{10}{20}\times 100 = \mathbf{50\%}$

and He = 100 – (20 + 50) = **30%**

44. Partial pressure = Mole fraction × Total pressure

$\therefore \quad p_{He} = x_{He}\times P = \frac{4}{5}\times 20 = 16$ bar

$\left[\text{mole fraction of He} = \frac{4}{5}\right]$

$\therefore \quad p_{CH_4} = 20 - 16 = 4$ bar

Now applying the formula

$$\frac{r_{He}}{r_{CH_4}} = \frac{P^0_{He}}{P^0_{CH_4}}\sqrt{\frac{M_{CH_4}}{M_{He}}} = \frac{16}{4}\sqrt{\frac{16}{4}} = \frac{16}{4}\sqrt{4}$$

$\therefore \; r_{He} : r_{CH_3} = 8 : 1$

$\therefore$ Composition of the mixture (He : CH_4) effusing out = **8 : 1**

45. *Calculation of volume of gas :*

Weight of cylinder with gas = 29.0 kg

Weight of empty cylinder = 14.8 kg

$\therefore$ Weight of gas in the cylinder = 14.2 kg

Pressure in cylinder = 2.5 atm

$\therefore$ No. of moles (n) in 14.2 kg (14.2×10^3 g) of butane

$$n = \frac{\text{Wt. of butane}}{\text{Mol. wt. of butane}} = \frac{14.2\times 10^3}{58} = 244.83 \text{ mol}$$

Applying gas equation,

$$V = \frac{nRT}{P} = \frac{244.83\times 0.0821\times 300}{2.5} = 2412 \text{ litres}$$

[27 °C = 273 + 27 = 300 K]

Calculation of pressure in cylinder after use.

Weight of cylinder after use = 23.2 kg

Weight of empty cylinder = 14.8 kg

$\therefore$ Wt. of unused gas = 8.4 kg $= \frac{8.4\times 10^3}{58}$ moles of butane

$$\text{Thus } P = \frac{nRT}{V} = \frac{8.4\times 10^3\times 0.0821\times 300}{58\times 2412}$$

= **1.478 atm** [V = 2412 L]

Calculation of volume of used gas at 2.5 atm and 27°C.

Weight of used gas = 14.2 – 8.4 = 5.8 kg

Pressure under normal usage conditions = 1 atm

$$V = \frac{nRT}{P} = \frac{5.8\times 10^3}{58}\times\frac{0.0821\times 300}{1}$$

$\left[\because n = \frac{5.8}{58}\right]$ = 2463 litres = **2.463 m³**

46. Using gas equation; $PV = nRT$

Total no. of moles of gases in the mixture (n)

$$= \frac{PV}{RT} = \frac{6\times 3}{0.0821\times 300} = 0.7308 \text{ mol.}$$

Thus, no. of moles of unknown gas = 0.7308 – 0.7 = 0.0308 mol.

Now we know that

$$\frac{r_1}{r_2} = \frac{\text{moles of hydrogen gas}}{\text{moles of unknown gas}} = \frac{0.7}{0.0308}$$

Also we know that $\frac{r_1}{r_2} = \sqrt{\frac{M_2}{M_1}}$

$$\therefore M_2 = \left(\frac{r_1}{r_2}\right)^2 M_1 \text{ or } M_2 = \left(\frac{0.7}{0.0308}\right)^2\times 2$$

= **1033**

47. $2NO + O_2 \rightarrow 2NO_2 \rightarrow N_2O_4$

Calculating the number of moles of NO and O_2 by applying the formula, $n = \frac{PV}{RT}$

Moles of NO in the larger flask $= \frac{1.053\times 0.250}{0.082\times 300}$

= 0.0107 [250 mL = 0.250 L]

Moles of O_2 in the smaller flask

$= \frac{0.789\times 0.100}{0.082\times 300} = 0.0032$ [100 mL = 0.100 L]

The reaction takes place as follows.

	2NO	+	O_2	$\longrightarrow$	N_2O_4
Mols before reaction	0.0107		0.0032		0
Mole after reaction	(0.0107 – 2 × .0032)		0		0.0032

Moles of NO left = 0.0107 – 0.0064 = **0.0043**

Oxygen will be completely changed into NO_2 which in turn is completely converted into N_2O_4 which solidifies at 262 K. Hence at 220 K, the dimer is in the solid state and only NO present in excess will remain in the gaseous state occupying volume equal to 350 mL.

Hence, pressure (P) of NO gas left

$$= \frac{nRT}{V} = \frac{0.0043\times 0.082\times 220}{0.350} = \mathbf{0.221 \text{ atm}}$$

[Total volume = 0.250 + 0.100 = 0.350 L]

48. Applying the general gas equation

$$PV = nRT = \frac{m}{M}RT$$

Here, Mol. wt. of acetylene *i.e.*, C_2H_2 (M) = 26,

$P = \frac{740}{760}$ atm, T = 50 °C = 50 + 273 = 323 K

$$\therefore V = \frac{mRT}{MP} \text{ or } V = \frac{5\times 0.082\times 323\times 760}{26\times 740} = \mathbf{5.23 L}$$

49. Volume of balloon

$$= \frac{4}{3}\pi r^3 = \frac{4}{3}\times\frac{22}{7}\times\left(\frac{21}{2}\right)^3 = 4.85 \text{ L}$$

Let no. of balloons to be filled = n

$\therefore$ Total volume occupied by n balloons = $4.851\times n$

Volume of H_2 present in cylinder = 2.82 L (given)

$\therefore$ Total volume of H_2 at NTP = $(4.851n + 2.82)$L

P_1 = 1 atm P_2 = 20 atm

$V_1 = 4.85\times n + 2.82$ L V_2 = 2.82 L

T_1 = 273 K T_2 = 300 K

$$\frac{P_1V_1}{T_1} = \frac{P_2V_2}{T_2} \text{ or } \frac{1\times(4.85\times n + 2.82)}{273} = \frac{20\times 2.82}{300} \quad \therefore n = \frac{48.504}{4.851}\simeq 10$$

In equilibrium with balloon, cylinder will hold 2.82 L of gas (it will not empty completely).

50. First we should calculate the number of moles of the gas under the given conditions by the relation $PV = nRT$

Here $P = 7.6\times 10^{-10}$ mm Hg

$$= \frac{7.6\times 10^{-10}}{760} \text{ atm.} = 1\times 10^{-12} \text{ atm.}$$

V = 1 litre, T = 273 + 0 = 273K, R = 0.082 litre atm./K/mol

Putting the values in equation

$$n = \frac{PV}{RT} = \frac{1\times 10^{-12}\times 1}{0.082\times 273} \text{ moles}$$

Now since 1 mole = 6.023×10^{23} molecules

$$\frac{10^{-12}}{0.082\times 273} \text{ moles}$$

$$= \frac{6.023\times 10^{23}\times 10^{-12}}{0.082\times 273} \text{ molecules}$$

= **2.7×10^{10} molecules**

51. From ideal gas equation,

$$PV = nRT \Rightarrow PV = \left(\frac{m}{M}\right)RT \text{ or } M = m\frac{RT}{PV}$$

Let the molecular wt. of A and B be M_A and M_B respectively.

Then $M_A = 2\frac{RT}{1\times V}$; $M_B = \frac{3\times RT}{0.5\times V}$

$$\therefore \frac{M_A}{M_B} = \frac{2RT}{V}\times\frac{0.5V}{3RT} = \frac{2\times 0.5}{3} = \frac{1}{3}$$

Therefore, the ratio $M_A : M_B$ = **1 : 3**

52. Following reaction takes places in tube

$C + CO_2 \longrightarrow 2CO$

Volume of mixture of CO and CO_2 = 1L

Let volume of CO_2 in mixture = x

$\therefore$ Volume of CO in mixture = $2x$

$\therefore$ Original volume of CO in mixture = $1 - x$

Total volume of gases after reaction = $(1 - x) + 2x$ = $1 + x$

$1 + x = 1.6$ ($\because$ It is given total volume after reaction = 1.6 L)

$\therefore x$ = 0.6 L $\therefore$ Volume of CO_2 = 0.6 L

Volume of CO = 0.4 L

CO_2 : CO = **3 : 2**

53. Given, moles $= \frac{\text{mass}}{\text{mol.wt}}$

mass of gas = 3.7g, mass of hydrogen = 0.184g

T_1 = 298 K, T_2 = 17°C = 273 + 17 = 290 K

Moles of $H_2 = n_1 = \frac{\text{Mass}}{\text{M. wt.}} = \frac{0.184}{2} = 0.092$

Moles of gas $= n_2 = \frac{\text{Mass}}{\text{M. wt.}} = \frac{3.7}{M}$

For hydrogen $P_1V_1 = n_1RT_1$ ....(i)

For gas $P_1V_1 = n_2RT_2$ ....(ii)

($\because$ Pressure and volume of gas are same)

$\therefore$ From equation (i) and equation (ii)

$$\frac{P_1V_1}{P_1V_1} = \frac{n_1RT_1}{n_2RT_2} \text{ or } 1 = \frac{0.092\times 298}{n_2\times 290}$$

$$\text{or } n_2 = \frac{0.092\times 298}{290} \text{ or } \frac{3.7}{M} = \frac{0.092\times 298}{290}$$

$$\text{or } \frac{3.7}{M} = 0.0945 \therefore M = \frac{3.7}{0.0945} = \mathbf{39.15}$$

54. **(d)** Given,

W_X = 10 g

P_X = 2 atm

W_Y = 80 g

$P_Y = P_{total} - P_X$

$\Rightarrow$ 6 – 2 = 4 atm

$$\text{As } v_{rms} = \sqrt{\frac{3RT}{M}}, \frac{(v_{rms})_X}{(v_{rms})_Y} = \sqrt{\frac{M_Y}{M_X}}$$

As we know, from ideal gas equation $PV = nRT$

If volume and temperature remains same we have.

$$P_XV = \frac{W_X}{M_X}RT \; ; \; P_YV = \frac{W_Y}{M_Y}RT$$

$M_X \propto \frac{W_X}{P_X}$; $M_Y \propto \frac{W_Y}{P_Y}$

$\frac{(v_{rms})_X}{(v_{rms})_Y} = \sqrt{\frac{W_Y}{P_Y} \cdot \frac{P_X}{W_X}}$

$= \sqrt{\frac{80}{4} \times \frac{2}{10}} = \sqrt{4} = \frac{2}{1} = 2:1$

55. (b) Graph shows symmetrical distribution of speed therefore, the most probable and the average speed should be same in square of speeds. Higher speed will give more contribution, therefore the root mean square speed must be greater than the average speed.

56. (c) Average KE $= E = \frac{1}{2} M u_{rms}^2$

$\therefore \quad u_{rms}^2 = \frac{2E}{M}$ or $u_{rms} = \sqrt{\frac{2E}{M}}$

57. (d) $U_{RMS} = \sqrt{\frac{3RT}{M}}$ Using ideal gas equation,

$PV = nRT = \frac{w}{M} RT$; $\frac{RT}{M} = \frac{PV}{w} = \frac{P}{d}$ where d is the density of the gas

$\therefore \quad U_{RMS} = \sqrt{\frac{3P}{d}}$ at constant pressure,

$U_{RMS} \propto \frac{1}{\sqrt{d}}$

58. (c) $U_{rms} = \sqrt{\frac{3RT}{M}}$

$\Rightarrow \sqrt{\frac{3RT_{H_2}}{2}} = \sqrt{7}\sqrt{\frac{3RT_{N_2}}{28}}$;

$\therefore \quad T_{N_2} = 2T_{H_2}$ or $T_{N_2} > T_{H_2}$

59. (c) The expression of root mean square speed is $U_{rms} = \sqrt{\frac{3RT}{M}}$

Hence,

$$\frac{U_{rms}(H_2)}{U_{rms}(O_2)} = \left[\frac{3R(50K)/(2\,g\,mol^{-1})}{3R(800K)/(32\,g\,mol^{-1})}\right]^{1/2} = 1$$

60. (a) The mean free path, $\lambda = \frac{1}{\sqrt{2}\pi a^2 N}$

or $\lambda \alpha \frac{1}{a^2}$, where a = molecular diameter

$\therefore$ Smaller the molecular diameter, longer the mean free path. Hence, H_2 is the answer.

61. (d) Pressure exerted by the gas,

$P = \frac{1}{3}\frac{mnu^2}{V}$ ...(1)

Here, u = root mean square velocity

m = mass of a molecule, n – No. of molecules of the gas. Hence, (a) & (b) are clearly wrong.

Again $u^2 = \frac{3RT}{M}$ [explained from (1)]

Here, M = Molecular wt. of the gas;

Hence, (c) is wrong

Further, Average K.E. $= \frac{3}{2} KT$; Hence, (d) is true.

62. (a)

$U_{av} = \sqrt{\frac{8RT}{\pi M}}$; $\left\{\begin{array}{l} T_1 = 27 + 273 = 300\text{ K} \\ T_2 = 927 + 273 = 1200\text{ K}\end{array}\right\}$

$\frac{U_{av_1}}{U_{av_2}} = \sqrt{\frac{T_1}{T_2}}$ or $\frac{0.3}{U_{av_2}} = \sqrt{\frac{300}{1200}}$

or $\frac{0.3}{U_{av_2}} = \sqrt{\frac{1}{4}}$ or U_{av_2} **= 0.6 m/sec.**

63. (b) Average kinetic energy depends only on temperature and does not depend upon the nature of the gas. ($\because$ K.E. = 3/2 KT)

64. (a) $U_{rms} : U_{av} = \sqrt{\frac{3RT}{M}} : \sqrt{\frac{8RT}{\pi M}}$

or $\sqrt{3} : \sqrt{\frac{8}{\pi}} = 1.086 : 1$

65. (4) v_{rms} of $X = \sqrt{\frac{3RT_x}{M_x}}$;

v_{mp} of $Y = \sqrt{\frac{2RT_y}{M_y}}$

Given $v_{rms} = v_{mp} \Rightarrow \sqrt{\frac{3RT_x}{M_x}} = \sqrt{\frac{2RT_y}{M_y}}$

$\Rightarrow M_y = \frac{2RT_y M_x}{3RT_x} = \frac{2 \times 60 \times 40}{3 \times 400} = 4$

66. (434) $C_{rms} = \sqrt{\frac{3RT}{M}}$, $C_{av} = \sqrt{\frac{8RT}{\pi M}}$

$\frac{C_{rms}}{C_{av}} = \sqrt{\frac{3RT}{M}} \times \sqrt{\frac{\pi M}{8RT}} = \sqrt{\frac{3\pi}{8}} = 1.085$

$C_{rms} = 1.085 \times C_{av} = 1.085 \times 400 =$ **434 ms^{-1}**

67. (1 : 16)

$\frac{K.E._{O_2}}{K.E._{H_2}} = \frac{(3/2)n_{O_2}RT}{(3/2)n_{H_2}RT} = \frac{n_{O_2}}{n_{H_2}} = \frac{8/32}{8/2} = \frac{1}{16}$

68. (900) Energy of one mole of an ideal monoatomic gas $= \frac{3}{2}RT$

$R = 8.314\text{ J K}^{-1}\text{ mole}^{-1} = 1.99\text{ cal K}^{-1}\text{mole}^{-1}$

$T = 27°C = 273 + 27 = 300$ K

$\Rightarrow \quad E = \frac{3}{2} \times 1.99 \times 300 = 900$ cal

69. False : K.E. $= \frac{3}{2} KT$, and cannot be zero at 0 °C or 273 K.

70. False : The pressure exerted by the gas is the result of collisions of the molecules on the walls of the container.

71. (a, b, c)

According to kinetic theory of gases, all gases at a given temperature have same average kinetic energy.

$E_{av} = \frac{3}{2}RT$; $E_{av} \propto T$ (absolute temp)

Root mean square velocity is directly proportional to square root of absolute temperature and inversely proportional to square root of molecular weight of the gas.

$U_{rms} \propto \sqrt{T}$ (absolute temp) : $U_{rms} \propto \frac{1}{\sqrt{M}}$

72. (a, b, c, d)

According to kinetic theory of gases :

(a) Collision between the molecules as well as with the walls of the container is perfectly elastic in nature.

(b) Momentum is defined as $m \cdot \bar{\mu}$. Hence, heavier molecules transfer more momentum to the walls of the container.

(c) According to Maxwell-Boltzmann distribution of molecular speed, very few molecules have either very high or very low speeds.

(d) A gas molecule moves in a straight line unless it collides with another molecule or walls of the container with constant velocities.

73. (a) At constant volume.

$P \propto T$ (from $PV = nRT$)

$U_{rms} = \sqrt{\frac{3RT}{M}} \quad \therefore U_{rms} \propto \sqrt{T}$ ...... (i)

Collision frequency is directly proportional to U_{rms}.

Thus collision frequency $\propto \sqrt{T}$

Hence, on increasing the collosion, frequency, pressure increases.

74. Given $V = 1\text{ L} = 10^{-3}\text{m}^3$, $P = 7.57 \times 10^{-3}\text{ Nm}^{-2}$, $R = 8.314$ J, $n = 2 \times 10^{21}/6.023 \times 10^{23}$ moles

$PV = nRT$ or $T = \frac{PV}{nR}$

$= \frac{7.57 \times 10^{-3} \times 10^{-3} \times 6.023 \times 10^{23}}{2 \times 10^{21} \times 8.31} = 274.13$ K

$U_{rms} = \sqrt{\frac{3RT}{M}} = \sqrt{\frac{3 \times 8.314 \times 274.13}{28 \times 10^{-3}}}$ m/s

$= 494.15$ m/s

$\frac{U_{mp}}{U_{rms}} = 0.82$ (given)

$\therefore U_{mp} = 0.82 \times U_{rms} = 0.82 \times 494.15 = 405.2$ m/sec.

75. Average velocity $= \sqrt{\frac{8RT}{\pi M}}$

and Most probable velocity $= \sqrt{\frac{2RT}{M}}$

Given -For CO_2

Average velocity at T_1 = Most probable velocity at T_2

$= 9 \times 10^4$ cm/sec $= \frac{9 \times 10^4}{100}$ m/sec. $= 9 \times 10^2$ m/sec.

$\therefore 9 \times 10^2 = \sqrt{\frac{8 \times 8.314 \times T_1}{3.14 \times 44 \times 10^{-3}}}$ ...(A)

[Average velocity at T_1K]

and $9 \times 10^2 = \sqrt{\frac{2 \times 8.314 \times T_2}{44 \times 10^{-3}}}$ ...(B)

[Most probable velocity at T_2K]

On solving, T_1 = **1682.5 K**, T_2 = **2143.4 K**

76. $U_{rms} = \sqrt{\frac{3RT}{M}}$

Given $T = 20\,°C = 20 + 273 = 293$ K

$R = 8.314 \times 10^7$ erg per degree per mol

M (of O_3) = 48

$\therefore \quad U_{rms} = \sqrt{\frac{3 \times 8.314 \times 10^7 \times 293}{48}}$

$= 3.9 \times 10^4$ cm sec^{-1}

77. Total kinetic energy $= n\,(3/2\ RT)$
where, n = Number of moles of the gas, R = Gas constant, T = Absolute temperature
Molecular weight of methane,
$CH_4 = 12 + 4 \times 1 = 16$
$\therefore$ Number of moles of methane in 8.0 g of methane
$= \dfrac{8.0}{16.0} = 0.5$
$R = 8.314$ joules/K/mole, $T = 27 + 273 = 300$ K
$\therefore$ Total kinetic energy of the molecules in 8.0 g of methane at 27 °C $= n \times 3/2\ RT = 0.5 \times 3/2 \times 8.314 \times 300 = 1870.65$ joules.
$\therefore$ Average kinetic energy $= \dfrac{1870.65}{6.023 \times 10^{23} \times 0.5}$
$= \mathbf{6.21 \times 10^{-21}}$ **joules/molecule**

78. **(d)**
- A solution of CH_3OH and water shows positive deviation, from Raoult's law, it means by adding CH_3OH, intermolecular force of attraction decreases and hence, surface tension decreases.
- By adding KCl in water, intermolecular force of attraction bit increases, so surface tension increases by small value.
- By adding surfactant like $CH_3(CH_2)_{11}OSO_3^-\ Na^+$, surface tension decreases rapidly and after forming micelle it slightly increases.

79. **(b)**

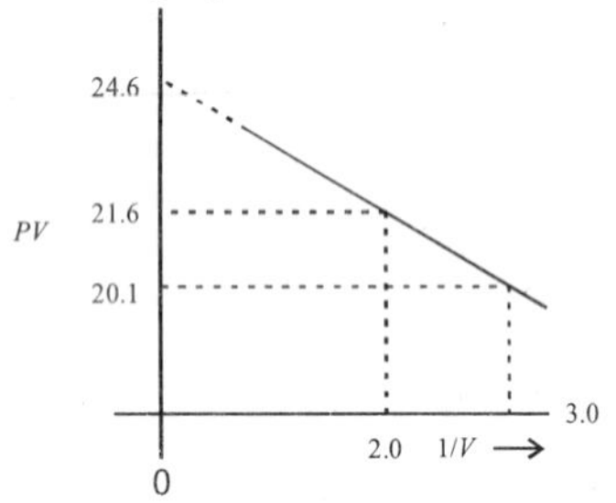

$\left(P + \dfrac{a}{V^2}\right)(V) = RT$
$PV + a/V = RT;\ PV = RT - a(1/V)$
$y = RT - a(x)$
So, slope $= a = \dfrac{21.6 - 20.1}{3 - 2} = \dfrac{1.5}{1} = 1.5$

80. **(b)** Correction factor for attractive force for n moles of real gas is given by the term mentioned in (b).

81. **(b)** Upon increase of temperature, the internal energy of water or any system increases resulting in decrease in intermolecular force and hence, decrease in surface tension. Surface tension decreases with increase in mobility due to increase in temperature.

82. **(b)** $(PV)_{Observed}/(PV)_{Ideal} < 1$
$\Rightarrow V_{obs} < V_{ideal},\ V_{obs} < 22.4$ litre.

83. **(b)** The compressibility factor of a gas is defined as $Z = \dfrac{pV_m}{RT}$
For an ideal gas, $pV_m = RT$. Hence $Z = 1$

84. **(c)** 'a' is directly related to forces of attraction. Hence, greater the value of 'a', more easily the gas is liquified.

85. **(c)** $\left(P + \dfrac{a}{V^2}\right)(V - b) = RT$; Here $\left(P + \dfrac{a}{V^2}\right)$ represents the intermolecular forces.

86. **(100)**
$Z = 0.5,\ V_m = 0.4$ L/mol, $T = 800$ K, $P = x$ atm.
$Z = \dfrac{PV_m}{RT} \Rightarrow \dfrac{x(0.4)}{0.08 \times 800} = 0.5 \Rightarrow x = 80$
For ideal gas, $PV_m = RT$
$\Rightarrow V_m = \dfrac{RT}{P} = \dfrac{0.08 \times 800}{80} = 0.8\ \text{L mol}^{-1} = y$
Now, $\dfrac{x}{y} = \dfrac{80}{0.8} = 100.$

87. **(-7.11)** Van der Waals equation
$\left(P + \dfrac{a}{V_m^2}\right)(V_m - b) = RT$
$(P V_m^2 + a)(V_m - b) = V_m^2 RT$
$P V_m^3 + aVm - Pb V_m^2 - ab = V_m^2 RT$
$P V_m^3 - [Pb + RT]\ V_m^2 + aV_m - ab = 0$
so ratio of coefficient of V_m^2 & V_m is
$\dfrac{-[Pb + RT]}{a} = \dfrac{-[300 \times 0.06 + 0.082 \times 300]}{6}$
$= \dfrac{-[18 + 24.6]}{6} = -7.1$

88. **(0.99)** van der Waals' equation for one mole of a gas is
$\left[P + \dfrac{a}{V^2}\right](V - b) = RT$ ...(1)
Given that volume occupied by CO_2 molecules, 'b' = 0
Hence, (1) becomes $\left[P + \dfrac{a}{V^2}\right]V = RT$
or $P = \dfrac{RT}{V} - \dfrac{a}{V^2}$
Using $R = 0.082$, $T = 273$K, $V = 22.4$ L for 1 mole of an ideal gas at 1 atm pressure.
$\therefore P = \dfrac{0.082 \times 273}{22.4} - \dfrac{3.592}{(22.4)^2} = \mathbf{0.9922\ atm.}$

89. **False :** The constant 'a' reflects the intermolecular attraction between gaseous molecules. The constant 'b' reflects the actual volume of one mole of gaseous molecules.

90. **(c)** $P(V - b) = RT \Rightarrow PV - Pb = RT$
$\Rightarrow \dfrac{PV}{RT} = \dfrac{Pb}{RT} + 1 \Rightarrow Z = 1 + \dfrac{Pb}{RT}$
Hence $Z > 1$ at all pressures.
This means , repulsive tendencies will be dominant when interatomic distances are small.
This means, interatomic potential is never negative but becomes positive at small interatomic distances.

91. **(a, c)** Van der Waals equation is
$\left[P + \dfrac{n^2 a}{V^2}\right](V - nb) = nRT$ (For n moles of a gas) a, b are van der Waals constants.
At low pressure, when the gas occoupies large volume, the intermolecular distance, between gaseous molecules is quite large and in such case, there is no significant role played by intermolecular forces and thus the gas behaves like an ideal gas. Hence, (a) is correct.
Under high pressure, the intermolecular distance decreases and the intermolecular forces play a significant role and the gas shows a devation from ideal behaviour.
Thus, (b) is *not* correct.
a, b i.e. the van der Waals coefficients defined on the nature of gas and are independent of temperature, so (c) is correct.
The pressure $\left(P + \dfrac{n^2 a}{V^2}\right)$ is not lower than P so, (d) is not correct.

92. **(A) : (p) and (s)** Because 200 atm pressure is very large. For H_2 gas, at very high pressure $Z > 1$.
(B) : (r) Since $P \sim 0$, it means very low pressure, so ideal behaviour is observed.
(C) : (p) and (q) Since P is 1 atm, Z for CO_2 would be less than 1 and attractive forces will be dominant.
(D) : (r) In real gas with very high molar volume, molecules will be very far apart from each other due to which van der Waals forces as well as actual volume occupied by molecules will be negligible.

93. **(c)** The value of 'a' indicates the magnitude of attractive forces between gas molecules.
Value of 'a' $\propto$ size of molecule.
$\therefore$ inert gas will have minimum value of 'a' followed by H_2O, C_6H_6 and $C_6H_5CH_3$

94. **(a)** 'a' indicates the magnitude of the attractive forces among the gas molecules, which increases in NH_3 due to H-bonding.

95. The van der Waals equation (for one mole) of a real gas is
$\left(P + \dfrac{a}{V_m^2}\right)(V_m - b) = RT$;
$PV_m - Pb + \dfrac{a}{V_m} - \dfrac{ab}{V_m^2} = RT$
$PV_m = RT + Pb - \dfrac{a}{V_m} + \dfrac{ab}{V_m^2}$ ...(i)
To calculate the intercept $P \to 0$, hence $V_m \to \infty$ due to which the last two terms on the right side of the equation (i) can be neglected.
$\therefore PV_m = RT + Pb$, comparing this with $y = mx + c$
intercept = $\boldsymbol{RT}$

96. Compressibility factor, $Z = \dfrac{PV}{RT}$
$0.5 = \dfrac{100 \times V}{0.082 \times 273}$ $\quad\therefore V = 0.1119$ L
Further when volume of a gas molecule is negligible, van der Waal's equation becomes
$\left(P + \dfrac{a}{V^2}\right)(V - 0) = RT$
or $PV = RT - \dfrac{a}{V}$ or $a = RTV - PV^2$
Substituting the values,
$a = (0.082 \times 0.1119 \times 273) - (100 \times 0.1119 \times 0.1119)$
$= \mathbf{1.253\ atm\ L^2\ mol^{-2}}$

97. van der Waals equation for n moles of gas is, $\left[P + \dfrac{n^2 a}{V^2}\right][V - nb] = nRT$
Given $V = 4$ litre; $P = 11.0$ atm, $T = 300$ K;
$b = 0.05$ litre mole^{-1}, $n = 2$
Thus, $\left[11 + \dfrac{2^2 a}{4^2}\right][4 - 2 \times 0.05] = 2 \times 0.082 \times 300$
$\therefore a = \mathbf{6.46\ atm\ litre^2\ mol^{-2}}$

Chapter-6

Thermodynamics

Part-A (JEE Main)

1. **(3)** A-II, B-I, C-III, D-IV

2. **(3)** The magnitude of work depends on how close the process is to reversible.
In case of multistage expansion or compression it is more closer to reversible so, more work in expansion and less work in compression.
In case of single stage expansion or compression it is more closer to irreversible so, less work in expansion and more work in compression.

3. **(2)** Helium (g) is monoatomic gas so

$$C_p = \frac{5}{2}R = 20.8 \text{ J mol}^{-1}\text{K}^{-1}$$

Copper is solid so $C_p = 3R = 24.9$ J mol^{-1} K^{-1}
Bromine is liquid and diatomic so it has higher heat capacity due to more degree of freedom (Translational + rotational + vibrational interaction)
The correct order :
Bromine (l) > Copper (s) > Helium (g)

4. **(4)** From first law of thermodynamics

$\Delta U = q + w$

Constant volume $w = 0$

$\Delta U = q$

Also, $C_p = C_v + R$
C_p = Heat capacity at constant pressure
C_v = Heat capacity at constant volume
R = Universal gas constant.

5. **(4)** For a cyclic process,

$\Delta U_{x\to y} = -\Delta U_{y\to x}$

(Internal energy is a state function)
For forward process $X \to Y$
$w_1 = -18$ J
$q_1 = +10$ J
$\Delta U = q_1 + w_1$

$\Delta U_{X\to Y} = 10 + (-18) = -8$ J

For reverse process:

$\Delta U_{X\to Y} = 8$ J

$\Delta U_{Y\to X} = -(-8\text{ J}) = 8$ J

$q_2 = -6$ J

$\Delta U = q_2 + w_2$

$8 = -6 + w_2$

$w_2 = 14$ J.

6. **(1)** For isothermal reversible process, $\Delta U = \Delta H = 0$

$$W_{iso} = -P_1V_1 \ln\frac{P_1}{P_2}$$

$$W_{iso} = -0.5\times10^6\times20\times10^{-3}\, \ell n\,\frac{0.5}{0.2}$$

$W_{iso} = -0.5\times10^6\times20\times10^{-3}\times2.303\times(0.6989-0.3010)$
$W \simeq -9.1$kJ
$q = -W = 9.1$kJ

7. **(4)** A-II, B-I, C-III, D-IV

8. **(2)** $\underset{5°C}{H_2O(l)} \to \underset{100°C}{H_2O(l)} \rightleftharpoons \underset{100°C}{H_2O(g)}$

Due to expansion
w = –ve
As heat is given to system, so q = +ve and internal energy of gas will be more than internal energy of liquid.
So, ΔU = +ve

9. **(2)** In (A) work done (W) in reversible process,

$$W = -nRT\ln\frac{V_2}{V_1}$$

$$= \frac{-2\times8.314\times300}{1000}\times\ln\left(\frac{20}{2}\right)\text{kJ} = -11.5\text{ kJ}$$

In (B) workdone in irreversible process
$W = -P_{ext}[V_2 - V_1] = -3[3-1] = -6$ kJ
In (C) change in internal energy

$\Delta U = nC_V\Delta T$

$$= 1\times\frac{3}{2}\times\frac{8.314\times320}{1000}\text{kJ} = 3.99$$

In (D) change in enthalpy

$\Delta H = nC_P\Delta T$

$$= 1\times\frac{5}{2}\times\frac{8.314\times337}{1000}\text{kJ} = 7\text{ kJ}$$

10. **(4)** The area under the PV-graph is the maximum work done. (According to NTA answer key, do not consider the signs)
So, graph '2' denotes isomeric process, hence the work done is zero.
Graph '1' & '3' shows both expansion as well as contraction, on the other hand, graph '4' shows only expansion. Hence, maximum work is shown by graph '4'.

11. **(4)** $(10\text{L}, 300\text{K}) \xrightarrow{n=1} (20\text{L}, 300\text{K})$

$$-q = w = -nRT\ln\frac{V_2}{V_1} = -8.3\times300\times\ln\left(\frac{20}{10}\right)$$

$= -1.718$ kJ $\Rightarrow q = 1.718$ kJ
$w = -1.718$ kJ
$\Delta U = 0$ ($\because \Delta T = 0$; Isothermal process)

12. **(4)** Both solutions are having same comp.osition. Which is 1 mole of 'x' in 1'ℓ' water, so all the intensive properties will remain same, but as total amount is greater in solution '1' compared to solution '2'. Therefore extensive properties will be different hence Gibbs free energy will be different.

13. **(2)**
$|W_{\text{reversible expansion}}| > |W_{\text{irreversible expansion}}|$
$|W_{\text{reversible expansion}}| = |W_{\text{reversible Compression}}|$
$|W_{\text{irreversible expansion}}| < |W_{\text{reversible Compression}}|$

14. **(4)** It is free expansion of gas as $\Delta T = 0$ for the gas indicating vaccum in B
$\Rightarrow P_{ext} = 0$, w = 0, q = 0 and $\Delta U = 0$

15. **(1)** **NTA Ans. (4)** $q_p = n\times C_p\times\Delta T$

$$\Rightarrow 500 = 0.5\times\frac{5}{2}\times8.3(T_f - 298)$$

$$\left[C_p \text{ for ideal manoatomic gas} = \frac{5}{2}R\right]$$

$\Rightarrow T_f \simeq 346.2$ K

$$\frac{\Delta H}{\Delta U} = \frac{C_p\Delta T}{C_v\Delta T} = \frac{C_p}{C_v} = \left(\frac{5}{3}\right)$$

$$\left[C_v \text{ for ideal manoatomic gas} = \frac{3}{2}R\right]$$

$$\Rightarrow \Delta U = \frac{C_v}{C_p}\times\Delta H = \frac{3}{5}\times500 = 300\text{ J}$$

[$q = \Delta H$ at constant p]

16. **(4)** As internal energy 'U' is a state function, its net change over a complete cycle must be zero in a cyclic process
$\therefore \Delta U$ case (I) = ΔU case (II) = ΔU case (III)

17. **(1)** Thermally insulated $\Rightarrow q = 0$
from 1st law
$\Delta U = q + w$; $\Delta U = w$
$w > 0$, $\Delta U > 0$

18. **(4)** $\Delta U = q + w$
For a adiabatic process, $\boxed{q = 0}$
$\Delta U = w = -p_{ex}\Delta v$
For free expansion of ideal gas $p_{ex} = 0$
$\boxed{\Delta U = w = 0}$
As $\Delta U = C\,\Delta T = 0$, $\boxed{\Delta T = 0}$

19. **(1)** Adiabatic boundary does not allow heat exchange thus heat generated in container can't escape out thereby increasing the temperature.
In case of Diathermic container, heat flow can occur to maintain the constant temperature.

20. **(1)** H = U + PV (By definition)
$\Delta H = \Delta U + \Delta(PV)$ at constant pressure
$\Delta H = \Delta U + P\Delta V$

21. **(4)** In expansion against vacuum,
$P_{ext} = 0$; $w = -P_{ext}\Delta V = 0$

22. **(4)**
(A) For ideal gas, U and H are function of temperature.
$H = U + PV$ but for ideal gas, $PV = nRT$.
Hence, $H = U + nRT$ or H depends only on temperature.
(B) Compressibility factor for an ideal gas is 1.
(C) $C_P - C_V = R$
(D) $\Delta U = C_V dT$ for all processes.

23. **(3)** $w = -P\Delta V = -(1\text{ bar})\times(9\text{ L})$
$= -(10^5\text{ Pa})\times(9\times10^{-3})\text{ m}^3 = -9\times10^2$ N.m
[$\because$ 1 Pa = 1N/m^2]
$= -900$ J $= -0.9$ kJ

24. (1) We know that heat and work are not state functions but $q + w = \Delta U$ is a state function. $H - TS = G$ is also a state function.

25. (3) $w = +10$ kJ
$q = -2$ kJ
$\Delta U = q + w = -2 + 10 = 8$ kJ

26. (2) From first law of thermodynamics, $\Delta U = q + w$
For adiabatic process, $q = 0 \Rightarrow \Delta U = w$
For isothermal process, $\Delta U = 0 \Rightarrow q = -w$
For cyclic process, $\Delta U = 0 \Rightarrow q = -w$
For isochoric process, $w = 0 \Rightarrow \Delta U = q$

27. (3) $\Delta U = n C_v \Delta T = 5 \times 28 \times 100 = 14$ kJ
$\Delta(PV) = nR(T_2 - T_1) = 5 \times 8 \times 100 = 4$ kJ

28. (1) For ideal gas, C_p is independent of pressure.
For polyatomic ideal gas, C_v increases slowly with increase in T.
Thus, C_p will not change with pressure.

29. (1) Isothermal expansion, $PV_m = K$ (graph - C)

$P = \frac{K}{V_m}$ (graph - A)

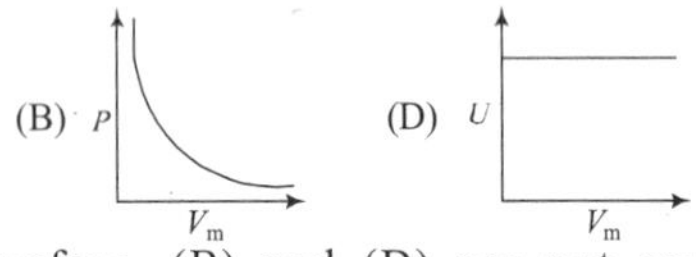

Therefore, (B) and (D) are not correct representation.

30. (3) We know that,
$w = -P_{ext}(V_f - V_i)$
$w = -4\ Nm^{-2}(1 - 5)m^3$
$w = 16\ Nm \Rightarrow 16$ J
For isothermal compression, $\Delta U = q + w$
$\Rightarrow q = -w = -16$ J ($\because \Delta U = 0$ for isothermal process)
$q = nC_P\Delta T$

So, $16 = \frac{1 \times 24 J \times \Delta T}{mol\ K}$

$\therefore$ Change in temperature, $\Delta T = \frac{2}{3}$ K

31. (3) For reversible isothermal expansion,

$w = -nRT \ln\frac{V}{V_1} \Rightarrow |w| = nRT \ln\frac{V}{V_1}$

$|w| = nRT(\ln V - \ln V_1)$
$|w| = nRT \ln V - nRT \ln V_1$
Comparing with $y = mx + c$, $m = nRT$ and $c = -nRT \ln V_1$
So, slope of curve 2 is more than curve 1 and intercept of curve 2 is more negative than curve 1.

32. (2) $\Delta U_{AB} = q_{AB} + W_{AB}$
$= 2 + (-5) = -3$ kJ/mol
$\Delta U_{BC} = -5$ kJ/mol
For cyclic process, $\Delta U = 0$
$\Delta U_{AB} + \Delta U_{BC} + \Delta U_{CA} = 0$
$\Delta U_{CA} = -\Delta U_{AB} - \Delta U_{BC}$
$\Delta U_{CA} = -(-3) - (-5) = 8$ kJ/mol
$\Delta U_{CA} = q_{CA} + W_{CA}$
$8 = q_{CA} + 3 \Rightarrow q_{CA} = +5$ kJ/mol
Heat absorbed has positive sign.

33. (3) From 1st law of thermodynamics:
$\Delta U = q + w$
For adiabatic process : $q = 0$
$\therefore \Delta U = w$

34. (4) $A \xrightarrow{q=+5,\ w=-8J} B$ (and back from B to A)

$\Delta U_{AB} = q + w = +5 + (-8) = -3$ J
$q = -3,\ \Delta U_{BA} = +3$
$\Delta U_{BA} = q + w \Rightarrow 3 = -3 + w \Rightarrow w = +6$
(work done on the system).

35. (1) $2H_2O_2(l) \rightleftharpoons 2H_2O(l) + O_2(g)$

$w = -P_{ext}(\Delta V) = -n_{O_2}RT$
$\because$ 100 mole H_2O_2 on decomposition give 50 mole O_2.
$\therefore w = -(50)(8.3)(300) = -124500$ J $= -124.5$ kJ.

36. (1) Process is isothermal reversible expansion, hence
$\Delta U = 0$, therefore $q = -w$.
Since $q = +208$ J, $w = -208$ J

37. (3) For isothermal reversible expansion.

$w = -nRT \ln\frac{V_2}{V_1}$

38. (4) Given $\Delta H = 41$ kJ mol^{-1}
$= 41000$ J mol^{-1}
T = 100 °C = 273 + 100 = 373 K

$H_2O(l) \xrightarrow{\Delta} H_2O(g)$
$\Delta n_g = 1 - 0 = 1$
$\Delta U = \Delta H - \Delta nRT = 41000 - (1 \times 8.314 \times 373)$
$= 37898.88$ J mol$^{-1} \simeq 37.9$ kJmol^{-1}

39. (3) According to first law of thermodynamics
$\Delta U = q + w$
An isolated system is adiabatic. This means $\Delta q = 0$. The first law in this case yields
$\Delta U = w$..... (i)
For expansion, W is negative and hence ΔU is negative. This means T_f is less than T_i in both the cases.
For the same expansion of volume, the work done in reversible process is greater than that in ireversible one because the system has to work against friction etc. Thus,
$\Rightarrow T_{f\,irreversible} > T_{f\,reversible}$

40. (3) $w = -P\Delta V = -10^5$
$[(1 \times 10^{-2}) - (1 \times 10^{-3})]$
$= -900$ J

41. (3) For a cyclic process, the net change in the internal energy is zero because the change in internal energy does not depend on the path.

42. (4) According to first law of thermodynamics, energy can neither be created nor be destroyed although it can be converted from one form to another.
Hence, the given data does not violate 1st law of thermodynamics but it violates 2nd law of thermodynamics.

43. (2) The heat required to raise the temperature of body by 1K is called thermal capacity or heat capacity.

44. (304) The gas is monoatomic as $\bar{C}_V = \frac{3}{2}R$ and the reversible work done is calculated from the area under the curve.
$|W| = P \times \Delta V$
$= 3 \times (2000 - 1000)$
$= 3000$ mL . atm $= 3$ L. atm $= 3 \times 101.3$ J
$= 303.9 \approx 304$ J

45. (55) Work done for reversible isothermal expansion is given as:

$w = -nRT \ln\left(\frac{V_2}{V_1}\right)$

$= -1 \times 0.08206 \times 291.15 \ln\left(\frac{100}{10}\right) \approx 55$ L atm

46. (38) $H_2O(\ell) \rightleftharpoons H_2O(g)$ $\Delta H^o_{vap} = 40.79$ kJ/mole

$\Delta H^o_{vap} = \Delta U^o_{vap} + \Delta n_g RT$

$40.79 = \Delta U^o_{vap} + \frac{1 \times 8.3 \times 373.15}{1000}$

$\Delta U^o_{vap} = 40.70 - 3.097 = 37.60 \approx 38$

47. (274) $\Delta U = q + w$ where $(q = 0)$
$nC_v\Delta T = -P_{ext}(V_2 - V_1)$
$V_2 = 2V_1$ (given)
$\frac{nRT_2}{P_2} = \frac{2nRT_1}{P_1}$

$P_1 = 5, T_1 = 298, P_2 = \frac{5T_2}{2 \times 298}$

$n\frac{5}{2}R(T_2 - T_1) = -1\left(\frac{nRT_2}{P_2} - \frac{nRT_1}{P_1}\right)$

Put $T_1 = 298$ and $P_2 = \frac{5T_2}{2 \times 298}$

We get $T_2 = 274.16$ K
$T_2 \approx 274$ K

48. (200)
For isothermal irreversible process, $\Delta U = 0$
$\Rightarrow q = -w$
$-w = -[-P_{ext}(V_2 - V_1)]$
$q = 5(20 - 60) = -200$ atm-L.

49. (28721)
In Isothermal and reversible expansion the work done is negative as the work is done by the system.

$\therefore W = -2.303\ nRT \log\left(\frac{V_2}{V_1}\right)$

$= -2.303 \times 5 \times 8.314 \times 300 \log\left(\frac{100}{10}\right)$

$= -28720.7 \approx -28721$ J.

50. (200) Work done is given by area enclosed by triangle ABC.

Area of $\Delta ABC = \frac{1}{2} \times AC \times AB$

$= \frac{1}{2}(30\ kPa - 10kPa) \times (30dm^3 - 10dm^3)$

$= \frac{1}{2}(20\ kPa)(20dm^3) = 200\ kPa\ dm^3$

$= 200\left(\frac{1}{100}bar\right)(L) = 2$ L bar

$= 200$ J {1 L bar = 100 J}

51. (1200)
Using, first law of thermodynamics,
$\Delta U = Q + W$,
$\Delta U = 0$: Process is isothermal
$Q = -W$
$W = -P_{ext}\Delta V$ (Irreversible)
$= -80 \times 10^3 (45 - 30) \times 10^{-3} = -1200$ J

52. (4) Extensive ⇒ Mole, Volume, Gibbs free energy.
Intensive ⇒ Molar mass, Molar heat capacity, Molarity, E° cell.

53. (4) A → Endothermic (Atomisation)
B → Endothermic (Atomisation)
C → Endothermic (Vapourisation)
D → Exothermic (Combustion)
E → Endothermic (Dissolution)

54. (1200) Power of heater = 60 W
Total energy emitted = 60 × 100 = 6000 J
Heat capacity × temp difference = 6000

$$\text{Heat capacity} = \frac{6000}{5} = 1200 JK^{-1}$$

55. (0) For an, ideal gas internal energy is a function of temperature and for isothermal process temperature is constant
∴ The change in internal energy is zero.

56. (150) $q = 0, \Delta U = w \Rightarrow \Delta U = nCv\Delta T$
$1 \times 20 \times [T_2 - 300] = -3000$
$T_2 - 300 = -150 \Rightarrow T_2 = 150 K.$

57. (620) 1 → 2 ⇒ Isobaric process
2 → 3 ⇒ Isochoric process
3 → 1 ⇒ Isothermal process

$$W = W_{1\to2} + W_{2\to3} + W_{3\to1}$$
$$-\left(-P(V_2 - V_1) + 0 + \left[-P_1V_1 \ln\left(\frac{V_2}{V_1}\right)\right]\right)$$
$$= \left[-1\times(40-20) + 0 + \left[-1\times 20 \ln\left(\frac{20}{40}\right)\right]\right]$$
$= -20 + 20 \ln 2 = -20 + 20 \times 2.3 \times 0.3$
$= -6.2$ Lbar (1L bar = 100 J)
$\Rightarrow |W| = 6.2$ L bar = 620 J

58. (3) State Variable is a variable which is independent of the path of change of state. Internal energy, volume and enthalpy are state variable.

59. (2) $C_{m,p} - C_{m,v} = R$
$\Rightarrow 20.785 - C_{m,v} = 8.314$
$\Rightarrow C_{m,v} = 12.471 \ J K^{-1} mol^{-1}$
$$\Delta U = n C_{m,v} \Delta T \Rightarrow n = \frac{5000}{12.471\times(500-300)}$$
$\Rightarrow n = 2$

60. (1718) $w = -2.303\, nRT \log \frac{V_2}{V_1}$
$$P_1V_1 = P_2V_2 \Rightarrow \frac{V_2}{V_1} = \frac{P_1}{P_2} = \frac{6}{3} = 2$$
$w = -2.303 \times 1 \times 8.3 \times 300 \times \log(2) = -1718.1$ J

61. (195)
$$\text{Moles of } N_2O = \frac{\text{mass}}{\text{molar mass}} = \frac{2.2}{44} = \frac{1}{20}$$
$$\Delta H = nC_p\Delta T = \frac{1}{20}\times 100\,(270-310)$$
$$= \frac{1}{20}\times 100\,(-40) = -200 \text{ J}$$
$\Delta U = q_p + w[q_p \Rightarrow$ heat given at constant pressure]
(work done) $w = -P_{ext}$
$$\Delta V = -1\frac{(167.75-217.1)}{1000}\times 101.3 J$$
$w = +5J, \quad q_p = \Delta H$
$\Delta U = -200 + 5 = -195J$

62. (0) For free expansion : $w = -P_{ext}dV$
$P_{ext} = 0; w = 0$; Isothermal expansion, $\Delta U = 0$
$\Delta U = q + w = 0$
$q = -w$ (for free expansion)
Thus, $\Delta U = q - q = 0.$

63. (8630) $n = 5$ mol; $T = 300$ K; $V_1 = 10$ L; $V_2 = 20$ L
Work done in isothermal process is :
$$w_{rev} = -nRT \ln\frac{V_2}{V_1} = -5\times 8.3\times 300 \ln\frac{20}{10}$$
$= -8630.38$ J

64. (84) Assume the system attains a final temperature of T (such that $300 < T < 60$)
$$\Rightarrow \begin{pmatrix}\text{Heat lost by}\\ N_2 \text{ of container}\\ \text{I}\end{pmatrix} = \begin{pmatrix}\text{Heat gained by}\\ N_2 \text{ of container}\\ \text{II}\end{pmatrix}$$
$\Rightarrow n_I C_m(300 - T) = n_{II} C_m(T - 60)$
$$\Rightarrow \left(\frac{2.8}{28}\right)(300-T) = \frac{0.2}{28}(T-60)$$
$\Rightarrow T = 284$ K (final temperature)
$$\Rightarrow (n_I + n_{II})_{final} = \left(\frac{3.0}{28}\right)$$
$$\Rightarrow \frac{P}{RT}(V_I + V_{II}) = \frac{3.0 \text{ g}}{28 \text{ g/mol}}$$
$$P = \left(\frac{3}{28}\text{mol}\right)\times 8.31\frac{\text{J}}{\text{mol K}}\times\frac{284 \text{ K}}{3\times10^{-3}}\times 10^{-5}\frac{\text{bar}}{\text{Pa}}$$
$\Rightarrow 84.28 \times 10^{-2}$ bar.

65. (38) Given equation is
$H_2O(l) \to H_2O(g) : \Delta H = 41 \text{ kJ mol}^{-1}$
⇒ From the relation : $\Delta H = \Delta U + \Delta n_g RT$
$$\Rightarrow 41\frac{\text{kJ}}{\text{mol}} = \Delta U + (1)\times\frac{8.3}{1000}\times 373$$
$\Delta U = 41 - 3.0959 = 38$ kJ/mol

66. (31) $H_2O_{(g)} \rightleftharpoons H_2O_{(g)}$
$\Delta H = \Delta U + \Delta n_g RT; \Delta H - \Delta U = \Delta n_g RT$
∴ $\Delta n_g RT = 1$ mol × 8.31 J/mol-k × 373 K
$= 3099.63 \text{ J} \cong 31 \times 10^2$ J

67. (50) $q = +150$ J
$w = -200$ J
$\Delta U = q + w \Rightarrow \Delta U = 150 + (-200) = -50.$

68. (2218) $Fe + 2HCl \longrightarrow FeCl_2 + H_2$
$$\text{No. of moles of Fe} = \frac{50}{55.85}\text{ moles}$$
$$\text{No. of moles of } H_2 \text{ produced} = \frac{50}{55.85}\text{ moles}$$
Work done $= -P_{ext}\cdot\Delta V = -\Delta n_g RT$
$$= -\frac{50}{55.85}\times 8.314\times 298 \approx -2218 \text{ J}$$
word done by gas = −2218J

69. (48.00)
Work done is given by the area under the curve
$$|w| = \frac{1}{2}\times(a+b)\times h$$
$$\therefore |w| = \frac{1}{2}(6+10)\frac{N}{M^2}\times 6\,m^3 = 48 \text{ J}$$
= 48 J = 48 Nm

70. (6.25) $\Delta U = nC_v\Delta T$
$5000 = 4 \times C_v(500 - 300)$
$C_v = 6.25 \ JK^{-1} mol^{-1z}$

71. (4) $2Al(s) + 6HCl(aq) \to$
$Al_2Cl_6(aq) + 3H_2(g); \Delta_rH_1 = -1200$ kJ/mol
To eliminate $3H_2(g)$
$[H_2(g) + Cl_2(g) \to 2HCl(g)] \times 3$
$\Delta_rH_2 = -164 \times 3$ kJ/mol
$[HCl(g) + aq \to HCl(aq) \times 6$
$\Delta_rH_3 = -83 \times 6$ kJ/mol
$Al_2Cl_6(aq) \to Al_2Cl_6(s) + aq \ \Delta_rH_4 = +663$ kJ/mol
$2Al(s) + 3Cl_2 \to Al_2Cl_6(s) \ \Delta_fH$
$\Delta_fH = \Delta_rH_1 + \Delta_rH_2 + \Delta_rH_3 + \Delta_rH_4$
$= -1200 - 164 \times 3 - 83 \times 6 + 663 = -1527$ kJ/mol

72. (2) $CH_4(g) \to C(g) + 4H(g); \Delta_rH = x$ kJ/mol
$C_2H_6(g) \to 2C(g) + 6H(g); \ \Delta_rH = y$ kJ/mol
$1000x = 4 \times E_{C-H}$
$1000y = 1 \times E_{C-C} + 6 + E_{C-H}$
Rearranging :
$$E_{C-C} = \left[y - \frac{3x}{2}\right]\times 1000 = \frac{hc}{\lambda}, N_A$$
⇒ ('λ') wavelength of photon
$$= \frac{hc\,N_A}{[4y - 6x]\times 250}$$

73. (1) Formation of methane :
$\Delta H = -X \text{ kJ mol}^{-1}$
$C(s) + 2H_2(g) \longrightarrow CH_4(g)$
$\Delta_fH = \Delta H_{sub}$ carbon + 2× bond enthalpy of H – H – 4 (BE of C – H)
$-X = Y + 2Z - 4$ bond enthalpy of C – H
$-X - Y - 2Z = -4$ bond enthalpy of C – H
$$\therefore \quad \text{Bond enthalpy of C – H} = \frac{X+Y+2Z}{4}$$

74. (4) For standard state pressure should be 1 bar and temperature is specified only.
$$\Rightarrow \left(\Delta H_f^o\right)_{O_2(g)} = 0$$
The enthalpy of formation of O_2 is zero even at high temperature.

75. (3)
$$KCl(s) + H_2O(l) \xrightarrow{\Delta H_{sol.}} K^+(aq) + Cl^{-1}(aq)$$

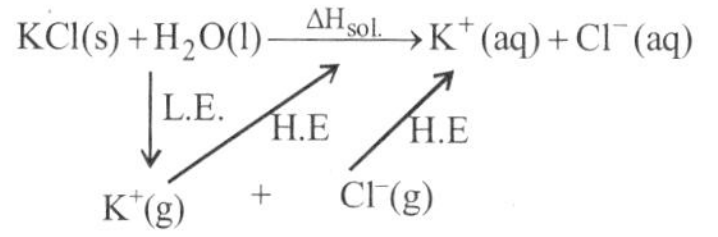

$\Delta H_{solution} = (-\Delta H_{L.E.}) + \Delta H_{Hydration}$
$= -(-z) + (-x - y) = z - (x + y)$

76. (2)

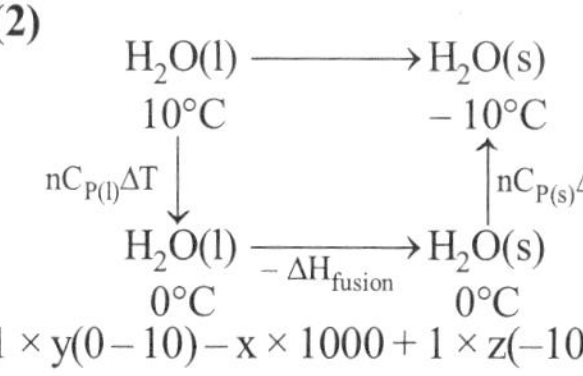

$\Delta H = 1 \times y(0 - 10) - x \times 1000 + 1 \times z(-10° - 0°)$
$\Delta H = -10(100x + y + z)$ Joule.

77. (2) ΔH is negative so it means dissolution of gas HCl(g) is exothermic, hence (I) is incorrect.
$HCl(g) + 10H_2O(l) \to HCl.10H_2O$...(1)
$$\Delta H_1 = -69.01\frac{\text{kJ}}{\text{mol}}$$
$HCl(g) + 40H_2O(l) \to HCl.40H_2O$ ...(2)
$$\Delta H_2 = -72.79\frac{\text{kJ}}{\text{mol}}$$
Hence heat of solution depends the upon amount of solvent present.
$HCl.10H_2O + 30H_2O(l) \to HCl.40H_2O$

So Heat of dilution = –72.79 – (–69.01)

$= -3.78\ \frac{\text{kJ}}{\text{mol}}$

Hence option (3) is incorrect.

For heat of formation reactant should be in elemental form hence option (4) is incorrect

78. (4) Subtracting eq. (iii) from (ii) we get eq (i)

79. (2)

$S(g) + \frac{3}{2}O_2(g) \rightarrow SO_3(g) \quad \Delta H_1 = -2x$

$SO_2(g) + \frac{1}{2}O_2(g) \rightarrow SO_3(g) \quad \Delta H_2 = -y$

$\Delta H_f(SO_2) = (\Delta H_1) - \Delta H_2 = (-2x) - (-y)$

= y – 2x. and is shorter than a single bond but longer than a double bond as bond order is 1.5

80. (2) Combustion of glucose takes place as:

$C_6H_{12}O_6(s) + 6O_2(g) \rightarrow 6CO_2(g) + 6H_2O(\ell)$

Moles of glucose in 900g of glucose

$= \frac{900}{180} = 5$ mol

1 mole glucose requires 6 moles of O_2

5 moles of glucose require 30 moles of O_2

Mass of O_2 required = 30 × 32 = 960 g.

81. (1)

$C(\text{graphite}) + \frac{1}{2}O_2(g) \rightarrow CO(g)$ .....(i)

$C(\text{graphite}) + O_2(g) \rightarrow CO_2(g);$ (ii) $\Delta H_1^\circ = -y$ kJ / mol

$CO_2(g) \rightarrow CO(g) + \frac{1}{2}O_2(g);$ (iii) $\Delta H_2^\circ = \frac{x}{2}$ kJ / mol

eq. (i) = eq. (ii) + eq. (iii)

$\therefore \Delta H^\circ = \frac{x}{2} - y = \frac{x-2y}{2}$

82. (3) $\Delta_r H = \Sigma\Delta_c H(\text{Reactant}) - \Sigma\Delta_c H(\text{Product})$ = 3 × (– 1300) – (– 3268) = – 632 kJ mol^{-1}

83. (3) (i) $C_2H_6(g) + \frac{7}{2}O_2(g) \longrightarrow 2CO_2(g) + 3H_2O(l)$

ΔH = – 1560 KJ

(ii) $C(s) + O_2(g) \rightarrow CO_2(g)$ ΔH = – 394 KJ

(iii) $H_2(g) + \frac{1}{2}O_2 \rightarrow H_2O(l)$ ΔH = –286 KJ.

Heat of combustion

$2C(s) + 2H_2(g) \rightarrow C_2H_6(g)$ ΔH = ΔHf($2H_6$)

$= \Sigma\Delta_f H_{(\text{products})} - \Sigma\Delta_f H_{(\text{reactants})}$

$\Delta_c H(C_2H_6, g) = 2\Delta_c H(C, \text{graphite}) + 3\Delta_c H(H_2, g) - \Delta_f H(O_2, g) - \Delta_f H(C_2H_6, g)$

$\Rightarrow -1560 = 2(-394) + 3(-286) - 0 - \Delta_f H(C_2H_6, g)$

$\Rightarrow \Delta_f H(C_2H_6, g) = -86$ KJ mol^{-1}

84. (3) $\Delta_{\text{sol.}} H^\circ = \Delta_{\text{lattice}} H^\circ + \Delta_{\text{Hyd.}} H^\circ$

$4 = 788 + \Delta_{\text{Hyd.}} H^\circ$

$\Delta_{\text{Hyd.}} H^\circ = -784$ kJ mol^{-1}

85. (4) A-B bond has highest intermolecular potential energy among the given molecules. Hence, it is strongest bond and has maximum bond enthalpy.

86. (3) $\Delta H_{\text{atomisation}} = \Delta H_{\text{vap}}$ + Bond energy

Hence, $x > y$.

87. (3) $I_2(s) \longrightarrow I_2(g)$

Heat of reaction depend upon temperature, i.e., it varies with temperature, as given by Kirchoff's equation,

$$\Delta H_{T_2} = \Delta H_{T_1} + \int_{T_1}^{T_2} \Delta C_p dT$$

where $\Delta C_p = C_p$ of product – C_p of reactant

$\therefore \Delta C_p = 0.031 - 0.055 = -0.024$ cal/g

Now, $\Delta H_{T_2} - \Delta H_{T_1} = \Delta C_p(T_2 - T_1)$

$\Delta H_{(250)} - \Delta H_{(200)} = -0.024\ (523 - 473)$

$\Delta H_{(250)} = 24 - 50 \times 0.024 = 22.8$ cal/g

88. (1)

$C_7H_{16}(l) + 11\,O_2(g) \xrightarrow{\Delta} 7\,CO_2(g) + 8\,H_2O(l)$

$\Delta H - \Delta U = \Delta n_g RT; \because \Delta n_g = 7 - 11 = -4$

$\therefore \Delta H - \Delta U = -4RT$

89. (1) Given: $n = 3$; $T_1 = 300$; $T_2 = 1000$

$C_p = 23 + 0.01T$

The relation between ΔH and C_p is

$$\Delta H = \int_{T_1}^{T_2} nC_p dT \quad ...(i)$$

After putting all variable values in eq. (i), we get

$$\Delta H = n\int_{300}^{1000}(23 + 0.01T)dT$$

$$= 3\left[23T + \frac{0.01T^2}{2}\right]_{300}^{1000}$$

$= 3[23(1000 - 300) + \frac{0.01}{2}(1000^2 - 300^2)]$

$= 61950$ J = 61.95 kJ ≈ 62 kJ

90. (1) Equation (i) can be obtained by adding equations (ii) and equation (iii).

$C(\text{graphite}) + \frac{1}{2}O_2(g) \rightarrow CO(g);$

$\Delta H^\circ_1 = y$ kJ mol^{-1} ...(ii)

$CO(g) + \frac{1}{2}O_2(g) \rightarrow CO_2(g); \Delta H^\circ_2$

$= z$ kJ mol^{-1} ...(iii)

$C(\text{graphite}) + O_2(g) \rightarrow CO_2(g); \Delta H^\circ_3$

$= x$ kJ mol^{-1} ...(i)

$\because \Delta H^\circ_3 = \Delta H^\circ_1 + \Delta H^\circ_2 \Rightarrow x = y + z$

91. (4)

$C_6H_6(l) + \frac{15}{2}O_2(g) \rightarrow 6CO_2(g) + 3H_2O(l)$

$\Delta n_g = 6 - \frac{15}{2} = -\frac{3}{2} \Rightarrow \Delta H = \Delta U + \Delta n_g RT$

$= -3263.9 + \left(-\frac{3}{2}\right) \times 8.314 \times 10^{-3} \times 298$

$= -3263.9 + (-3.71) = -3267.6$ kJ mol^{-1}

92. (2) $\Delta H = \Delta U + \Delta n_g RT$

$2\,HI(g) \rightarrow H_2(g) + I_2(g)$; $\Delta n_g = (1 + 1) - 2 = 0$

$\therefore \Delta H = \Delta U$

93. (3) $C(\text{graphite}) + O_2(g) \longrightarrow CO_2(g);$

$\Delta_r H^\circ_1 = -393.5$ kJ/mol^{-1} ...(i)

$H_2(g) + \frac{1}{2}O_2(g) \longrightarrow H_2O(l);$

$\Delta_r H^\circ_2 = -285.8$ kJ/mol^{-1} ...(ii)

$CO_2(g) + 2\,H_2O(l) \longrightarrow CH_4(g) + 2\,O_2(g);$

$\Delta_r H^\circ_3 = +890.3$ kJ/mol^{-1} ...(iii)

$C(\text{graphite}) + 2H_2(g) \longrightarrow CH_4(g);$

$\Delta_r H = ?$ ...(iv)

[Eq. (i) + Eq. (iii)] + [2 × Eq. (ii)] = Eq. (iv)

$\therefore [\Delta H_1 + \Delta H_3] + [2 \times \Delta H_2] = \Delta H$

[(–393.5) + (890.3)] + [2(–285.8)] = –74.8 kJ/mol

94. (4) A; $\Delta H = \Delta U + \Delta n_g RT$

$\Delta H = -3RT$ and $\Delta n_g = n_p - n_r = 0 - 1 = -1$

$\Delta H = \Delta U + \Delta n_g RT$

$\Rightarrow -3RT = \Delta U - RT \Rightarrow -2RT = \Delta U$

$|\Delta H| > |\Delta U|$

95. (3) In order to calculate the enthalpy change for H_2O at 5 °C to ice at –5 °C, we need to calculate the enthalpy change of all the transformations involved in the process.

(1) Enthalpy change of 1 mol, H_2O (l), at 5 °C → 1 mol, H_2O (l), 0 °C

(2) Enthalpy change of 1 mol, H_2O (l), at 0 °C → 1 mol, H_2O (s) (ice), 0 °C

(3) Enthalpy change of 1 mol, ice (s), at 0 °C → 1 mol, ice (s), –5 °C

Total $\Delta H = C_p\,[H_2O\,(l)]\,\Delta T + \Delta H$ freezing $+ C_p\,[H_2O\,(s)]\,\Delta T$

= (75.3 J mol^{-1} K^{-1}) (0–5)K + (–6 × 10^3 J mol^{-1}) + (36.8 J mol^{-1} K^{-1}) (–5–0) K

ΔH = – 6.56 kJ mol^{-1} (exothermic process)

96. (2) Given $C(s) + O_2(g) \rightarrow CO_2(g);\ \Delta H_1$ = –393.5 kJ mol^{-1} ...(i)

$CO(g) + \frac{1}{2}O_2(g) \rightarrow CO_2(g);\ \Delta H_2$

= –283.5 kJ mol^{-1} ...(ii)

∴ Heat of formation of CO = $\Delta H_1 - \Delta H_2$

= –393.5 – (–283.5) = –110 kJ

97. (2) In CH_4, 4 × $BE_{(C-H)}$ = 360 kJ/mol

∴ $BE_{(C-H)}$ = 90 kJ/mol

In C_2H_6, $BE_{(C-C)}$ + 6 × $BE_{(C-H)}$ = 620 kJ/mol

∴ $BE_{(C-C)}$ = 620 – (6 × 90) = 80 kJ/mol

$\therefore BE_{(C-C)} = \frac{80 \times 10^3}{6.023 \times 10^{23}}$ J/mol

Now, $E = \frac{hc}{\lambda}$

$\therefore \lambda = \frac{6.626 \times 10^{-34} \times 3 \times 10^8 \times 6.023 \times 10^{23}}{80 \times 10^3}$

λ = 1.49 × 10^{-6} m (∵ 1 nm = 10^{-9} m)

∴ λ = 1.49 × 10^3 nm

98. (3) Given $\frac{1}{2}N_2 + \frac{3}{2}H_2 \rightleftharpoons NH_3;$

$\Delta_f H = -46.0$ kJ / mol

$H + H \rightleftharpoons H_2; \Delta_f H = -436$ kJ / mol

$N + N \rightleftharpoons N_2; \Delta_f H = -712$ kJ / mol

$\Delta_f H(NH_3) = \frac{1}{2}\Delta H_{N-N} + \frac{3}{2}\Delta H_{H-H} - 3\Delta H_{N-H}$

$-46 = \frac{1}{2}(712) + \frac{3}{2}(436) - 3\Delta H_{N-H}$

$\Delta H_{N-H} = 352$ kJ/mol

99. (3) To calculate average enthalpy of C – H bond in methane, following informations are needed:

(i) Dissociation energy of H_2

$\frac{1}{2}H_2(g) \longrightarrow H(g); \Delta H = x$ (suppose)

(ii) Sublimation energy of C(graphite) to C(g)

$C(\text{graphite}) \longrightarrow C(g); \Delta H = y$ (Suppose)

Given

$C(\text{graphite}) + 2\,H_2(g) \longrightarrow CH_4(g);$

$\Delta H = -74.9$ kJ mol^{-1}

100. (3) Given, $H_2(g) + \frac{1}{2}O_2(g) \longrightarrow H_2O(l)$;

$\Delta H^\circ = -285.9$ kJ mol^{-1} ... (i)

$H_2(g) + \frac{1}{2}O_2(g) \longrightarrow H_2O(g)$;

$\Delta H^\circ = -241.8$ kJ mol^{-1} ... (ii)

We have to calculate

$H_2O(l) \longrightarrow H_2O(g)$; $\Delta H^\circ = ?$

On substracting eqn. (ii) from eqn. (i) we get,

$H_2O(l) \rightarrow H_2O(g); \Delta H^\circ = -241.8 - (-285.9) = 44.1$ kJmol^{-1}

101. (3) Applying Hess's Law,

$\Delta_f H^\circ = \Delta_{sub} H + \frac{1}{2}\Delta_{diss} H + \text{I.E.} + \text{E.A} + \Delta_{lattice} H$

$-617 = 161 + 520 + 77 + \text{E.A.} + (-1047)$

$\text{E.A.} = -617 + 289 = -328$ kJ mol^{-1}

102. (3) For an alkaline earth metal M :

$MO(s) + H_2O(l) \rightarrow M(OH)_2(s)$

$\Delta_r H = \Delta_f H[M(OH)_2] - \Delta_f H[MO] - \Delta_f H[H_2O]$

$\Delta_f H[H_2O] = -285$ kJ/mol

$\Delta_f H[MgO] = -602$ kJ/mol, $\Delta_f H[CaO] = -635$ kJ/mol

$\Delta_f H[SrO] = -592$ kJ/mol, $\Delta_f H[BaO] = -548$ kJ/mol

$\Delta_f H[Mg(OH)_2] = -925$ kJ/mol, $\Delta_f H[Ca(OH)_2] = -985$ kJ/mol

$\Delta_f H[Sr(OH)_2] = -959$ kJ/mol, $\Delta_f H[Ba(OH)_2] = -945$ kJ/mol

After calculation,

$\Delta_r H[Mg(OH)_2] = -38$ kJ/mol, $\Delta_r H[Ca(OH)_2] = -65$ kJ/mol

$\Delta_r H[Sr(OH)_2] = -82$ kJ/mol, $\Delta_r H[Ba(OH)_2] = -111$ kJ/mol

The order of liberated heat per gram is:

$Ba(OH)_2 > Sr(OH)_2 > Ca(OH)_2 > Mg(OH)_2$

103. (1) $N_2(g) + \frac{1}{2}O_2 \rightarrow N_2O(g)$

$N \equiv N(g) + \frac{1}{2}(O = O) \rightarrow \overset{-}{N} = \overset{+}{N} = O(g)$

$\Delta_f H^\circ$ = [Energy required for breaking of bonds] − [Energy released for forming of bonds]

$= (\Delta H_{N\equiv N} + \frac{1}{2}\Delta H_{O=O}) - (\Delta H_{N=N} + \Delta H_{N=O})$

$= (946 + \frac{1}{2} \times 498) - (418 + 607) = 170$ kJ mol^{-1}

Resonance energy = observed $\Delta_f H^\circ$ − calculated $\Delta_f H^\circ$

$82 - 170 = -88$ kJ mol^{-1}

104. (4) $\Delta H = \Delta U + \Delta n_g RT$

For the reaction $\Delta n_g = 12 - 15 = -3$

$\Delta H - \Delta U = -3 \times 8.314 \times 300 = -7482.6$ J mol^{-1}

105. (4) $NH_4OH(l) + HCl(l) \longrightarrow NH_4Cl(s) + H_2O(l)$

$\Delta H_1 = -51.46$ kJ/mol ... (i)

$NaOH(l) + HCl(l) \longrightarrow NaCl(s) + H_2O(l)$

$\Delta H_2 = -55.90$ kJ/mol ... (ii)

On subtracting eq. (ii) from eq. (i)

$NH_4OH(l) \longrightarrow NH_4^+(aq) + OH^-(aq)$

Enthalpy of ionisation

$= -51.46 - (-55.90) = 4.44$ kJ/mol

106. (3)

$4NO_2(g) + O_2(g) \rightarrow 2N_2O_5(g)$, $\Delta_r H = -111$ kJ

$\Delta_r H' \downarrow$ (to $2N_2O_5(s)$); $2N_2O_5(s) \nearrow 2 \times 54$ kJ mol^{-1} (to $2N_2O_5(g)$)

$\Delta_r H' = \Delta_r H - 54 = -111 - 108 = -219$ kJ

107. (2) $C_2H_5OH(l) + 3O_2(g) \longrightarrow 2CO_2(g) + 3H_2O(l)$

$\Delta n_g = 2 - 3 = -1 \Rightarrow \Delta U = \Delta H - \Delta n_g RT$

$= -1366.5 + 0.8314 \times 3 = -1364$ kJ

108. (2) $N_2 + 3H_2 \longrightarrow 2NH_3$

$\Delta H = 2 \times (-46.0)$ kJ mol^{-1}

Let x be the bond enthalpy of N – H bond then

[**Note :** Enthalpy of formation or bond formation enthalpy is given which is negative but the given reaction involves bond breaking hence values should be taken as positive.]

$\Delta H = \Sigma$ Bond energies of reactants − Σ Bond energies of products

$2 \times (-46) = 712 + 3 \times (436) - 6x \Rightarrow x = +352$ kJ/mol

109. (1) $H_2O(l) \longrightarrow H^+(aq) + OH^-(aq)$;

$\Delta H = 57.32$ kJ ... (i)

$H_2(g) + \frac{1}{2}O_2(g) \longrightarrow H_2O(l)$;

$\Delta H = -286.20$ kJ ... (ii)

By adding equation (i) & (ii) we get,

$H_2(g) + \frac{1}{2}(O_2) \longrightarrow H^+(aq) + OH^-(aq)$

$\Delta H = 57.32 - 286.2 = -228.88$ kJ

Here $\Delta_f H^\circ$ of $H^+(aq.) = 0$

$\therefore$ $\Delta_f H^\circ$ of $OH^- = -228.88$ kJ

110. (2) The energy involved in the conversion of $\frac{1}{2}Cl_2(g)$ to $Cl^-(aq)$ is given by

$\Delta H = \frac{1}{2}\Delta_{diss}H^\circ_{Cl_2} + \Delta_{eg}H^\circ_{Cl} + \Delta_{hyd}H^\circ_{Cl}$

Substituting various values from given data, we get

$\Delta H = \left(\frac{1}{2} \times 240\right) + (-349) + (-381)$

$= (120 - 349 - 381) = -610$ kJmol^{-1}

111. (1) $I_2(s) + Cl_2(g) \longrightarrow 2ICl(g)$

$\Delta H = [\Delta H^2_{I(s) \rightarrow I(g)} + \Delta H_{I-I} + \Delta H_{Cl-Cl}] - 2[\Delta H_{I-Cl}]$

$= 62.76 + 151.0 + 242.3 - 2 \times 211.3 = 33.46$ kJ

$\Delta H^\circ_f(ICl) = \frac{33.46}{2} = 16.73$ kJ / mol

112. (4) For the reaction,

$C(s) + \frac{1}{2}O_2(g) \longrightarrow CO(g)$

$\Delta H = \Delta U + \Delta n_g RT$ or $\Delta H - \Delta U = \Delta n_g RT$

$\Delta H - \Delta U = \frac{1}{2} \times 8.314 \times 298 = 1238.78$ J mol^{-1}

113. (2) $\Delta H = \Delta U + \Delta n_g RT$ for

$N_2 + 3H_2 \longrightarrow 2NH_3$

$\Delta n_g = 2 - 4 = -2 \quad \therefore \Delta H = \Delta U - 2RT$

or $\Delta U = \Delta H + 2RT \quad \therefore \Delta U > \Delta H$

114. (N) $\frac{1}{2}X_2 + \frac{1}{2}Y_2 \longrightarrow XY$, $\Delta H = (-200)$ J

Let x be the bond dissociation energy of X_2. Then

$\Delta H_f = -200 = \frac{1}{2}\Delta H_{X-X} + \frac{1}{2}\Delta H_{Y-Y} - \Delta H_{X-Y}$

$-400 = x + 0.5x - 2x$

$\Rightarrow x = \frac{400}{0.5} = 800$ kJ mol^{-1}

115. (3) $CH_2 = CH_2(g) + H_2(g) \longrightarrow CH_3 - CH_3(g)$

Enthalpy change = Bond energy of reactants − Bond energy of products.

$\Delta H = 1(C=C) + 4(C-H) + 1(H-H) - 1(C-C) - 6(C-H)$

$= 1(C=C) + 1(H-H) - 1(C-C) - 2(C-H)$

$= 615 + 435 - 347 - 2 \times 414$

$= 1050 - 1175 = -125$ kJ.

116. (2) Enthalpy change for a reaction does not depend upon the nature of intermediate reaction steps.

117. (1126) Change in enthalpy for a reaction can be calculated as:

$\Delta_r H^\circ = \sum \Delta H_f^\circ(\text{Products}) - \sum \Delta H_f^\circ(\text{Reactants})$

$\Delta_r H^\circ = 2 \times \Delta H_f^\circ(SO_2, g) + 2 \times \Delta H_f^\circ(H_2O, l) - 2 \times \Delta H_f^\circ(H_2S, g)$,

$\Delta_r H^\circ = 2 \times (-297) + 2 \times (-286) - 2(-20.1)$

$= -594 - 572 + 40.2$

$\Delta_r H^\circ = -1125.8$ kJ/mol.

$|\Delta_r H^\circ| = 1125.8$ kJ/mol.

118. (4) First ionization enthalpy of Lithium is:-

$Li(g) \rightarrow Li^+(g) + 1e^-$ $IE_1 = 520$ kJ mol^{-1}

Second ionization enthalpy of Lithium is:

$Li^+(g) \rightarrow Li^{2+}(g) + 1e^-$ $IE_2 = 7297$ kJ mol^{-1}

Total Energy per mole:-

Total IE = $IE_1 + IE_2 = 520$ kJ mol^{-1} + 7297 kJ mol^{-1} = 7817 kJ mol^{-1}.

Moles of Lithium $= \frac{3.5 \times 10^{-3} \text{ g}}{7 \text{ g mol}^{-1}} = 0.5 \times 10^{-3}$ mol

$= 5 \times 10^{-4}$ mol

Total Energy Required:

Energy = Moles × Total Ionization Energy

$\therefore$ Energy = $(5 \times 10^{-4}$ mol$) \times (7817$ kJ mol$^{-1})$

$\therefore$ Energy = 3.90 kJ ≈ 4 kJ

119. (3)

$C_2H_5OH(l) + 3O_2(g) \rightarrow 2CO_2(g) + 3H_2O(l)$;

$\Delta U^0_{combustion} = \frac{-q}{\text{moles}} = \frac{-99.472}{3.365/46}$

$\Rightarrow -1359.8$ kJ/mole

As, $\Delta H^0_r = \Delta H^0_C$

and, $\Delta H^0_C = \Delta U^0 + \Delta n_g RT$

$= -1359.8 + \frac{(-1) \times 8.314 \times 298.15}{1000}$

$= -1362.27$ kJ/mol

$\Delta H^0_C = 2 \times \Delta H^0_f CO_2 + 3 \times \Delta H^0_f H_2O - \Delta H^0_{f C_2H_5OH}$

$-1362.27 = [2 \times (-393.5) + 3 \times (-285.8)]$

$-1362.27 = -787 - 857.4 - \Delta H^0_{f[C_2H_5OH(l)]}$

$\Delta H^0_{f[C_2H_5OH(l)]} = -2.82 \times 10^2$ kJ / mol

120. (1031)

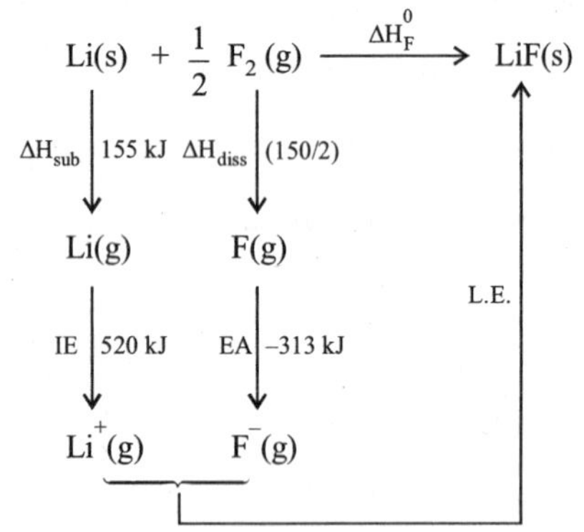

$\Delta H_F^0 = \Delta H_{sub} + IE + \Delta H_{diss} - EA + U$

$-594 = 155 + 520 + \frac{150}{2} - 313 + (L.E.)$

L.E. = –1031 kJ/mol

121. (25)

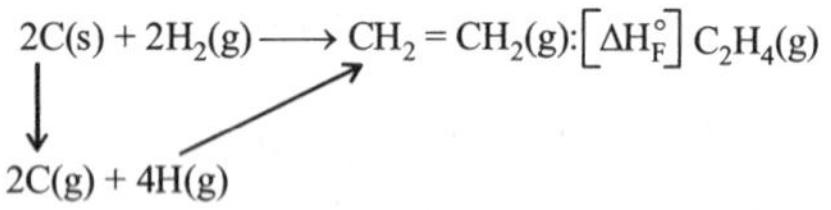

$[\Delta H_F^\circ]_{C_2H_4(g)} = 2 \times [\Delta H_{sub}^\circ]_{C(s)} + 2 \times \Delta H_{H-H}^\circ$

$-1 \times \Delta H_{C=C}^\circ - 4 \times \Delta H_{C-H}^\circ$

$\Rightarrow [\Delta H_F^\circ]_{C_2H_4(g)}$
$= (2 \times 710) + (2 \times 436) - 611 - 4 \times 414$

$\Rightarrow [\Delta H_F^\circ]_{C_2H_4(g)} = 25$ kJ/mol

122. (48) $\Delta H_f[CO_2(g)] = -393.5$ kJ / mol

$\Delta H_f[H_2O(l)] = -286.0$ kJ / mol

$\Delta H_c[C_6H_6] = -3267.0$ kJ / mol

$\Delta H_f(C_6H_6) = (?)$

$C_6H_6 + \frac{15}{2}O_2(g) \longrightarrow 6CO_2(g) + 3H_2O(l)$

$\Delta H_R = \Delta H_C = \Sigma\Delta H_f(P) - \Sigma\Delta H_f(R)$

$-3267 = [6 \times (-393.5) + 3(-286)] -$
$[\Delta H_f(C_6H_6) + \Delta H_f(O_2)]$

$\Rightarrow \Delta H_f(C_6H_6) = 48$ kJ/mol

123. (466)

$\frac{1}{2}H_2(g) \to H(g);\ \Delta_f H(H(g)) = 220$ kJ / mol

$\frac{1}{2}O_2(g) \to O(g);\ \Delta_f H(O(g)) = 250$ kJ / mol

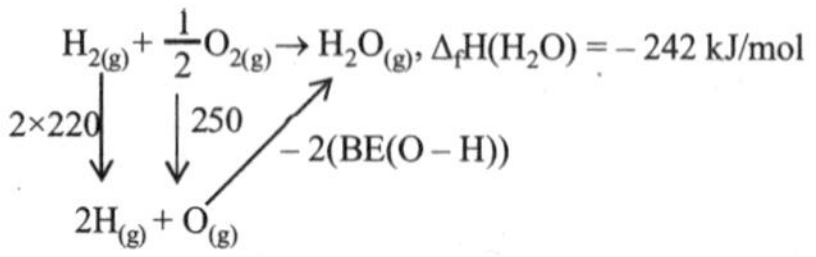

$\Delta_f(H_2O) = -242 = 440 + 250 - 2(B.E.(O-H))$

$\Rightarrow$ 2.B.E. (O – H) = 932 kJ/mol

$\Rightarrow$ B.E. (O – H) = 644 kJ/mol

124. (200)

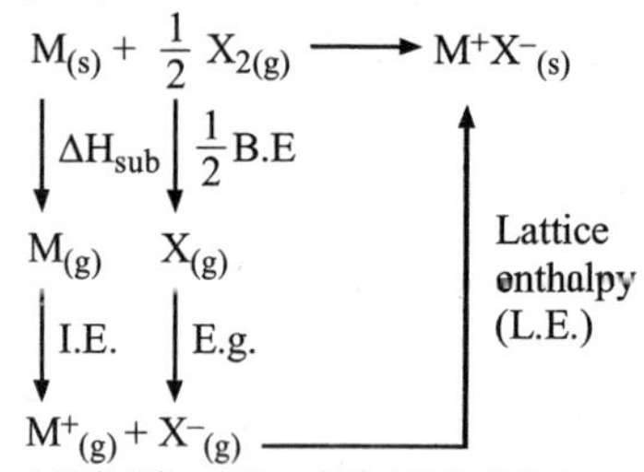

$\therefore$ $\Delta H_f(MX) = \Delta H_{sub}(M) + I.E.(M) +$
$\frac{1}{2}$[B.E.(X – X)] + EG(X) + L.E. (MX)

$-400 = (100) + (500) + \frac{1}{2}$ (B.E.)
+ (–300) + (–800) (energy released)

$\therefore$ B.E. = 200 kJ mol^{-1}

125. (95) $2C_{(graphite)} + 3H_2(g) \to C_2H_6(g)\ \Delta H_f = ?$

$C_2H_6(g) + \frac{7}{2}O_2(g) \to 2CO_2(g) + 3H_2O(l)$

$\Delta H_1^\circ = -1550$

$C_{(graphite)} + O_2(g) \to CO_2(g)\ \Delta H_2^\circ = -393.5$

$H_2(g) + \frac{1}{2}O_2(g) \to H_2O(l)\ \Delta H_3^\circ = -286$

$\Delta H_f^\circ = 2\Delta H_2^\circ + 3\Delta H_3^\circ - \Delta H_1^\circ = -95$ kJ/mole.

126. (150) Combustion taking place as :

$C_6H_5COOH(s) + \frac{15}{2}O_2(g) \to 7CO_2(g) + 3H_2O(\ell)$

$\Delta n_g = 7 - \frac{15}{2} = -\frac{1}{2}$

$\Delta H = \Delta U + \Delta n_g RT$

$= -321.30 - \frac{1}{2}R \times 300$

= (– 321.30 – 150 R) kJ

On comparing with (–321.30 – xR) = x = 150.

127. (6535)

$C_6H_6(l) + \frac{15}{2}O_2(g) \to 6CO_2(g) + 3H_2O(l)$

$\Delta_f H = H_P - H_R$

$\Delta H_c = 6\Delta H_f(CO_2) + 3\Delta H_f(H_2O) - \Delta H_f(C_6H_6)$

$= 6 \times (-393.5) + 3(-286) - 48.5$

= –3267.5 kJ/mol

For 2 mol of benzene = – 6535 kJ/mol

x = 6535

128. (125) $C_2H_4(g) + H_2(g) \to C_2H_6(g)$

$\Delta_r H$ = [BE(C = C) + 4BE (C – H) + BE (H – H)]
– [BE(C – C) + 6BE (C – H)]

$\Delta_r H$ = BE(C = C) + BE(H – H) – BE(C – C)
– 2BE(C – H)

$= 615 + 435 - 347 - 2 \times 414$

= – 125 kJ

129. (492) Let's take:-

$2Fe(s) + \frac{3}{2}O_2(g) \longrightarrow Fe_2O_3(s)$ ... (1)

$C(s) + \frac{1}{2}O_2(g) \longrightarrow CO(g)$ ... (2)

$3C(s) + Fe_2O_3(s) \longrightarrow 2Fe(s) + 3CO(g)$... (3)

Equation (3) can be obtained by subtracting (1) from (3) × (2).

Thus, $\Delta H_3^\circ = (3 \times (-110)) - (-822)$

$= +492$ kJ mol^{-1}

130. (56) $\Delta H^0_{vap.} = 30.5$ kJ mol^{-1}, M_{CCl_4}
$= 154.82$ g mol^{-1}

$\Rightarrow$ Heat required for 154 g CCl_4 = 30.5 kJ

$\Rightarrow$ Heat required for 284 g $CCl_4 = \frac{30.5 \times 284}{154}$

= 56.24 kJ.

131. (800) $A_2 + B_2$
$\to 2AB;\ \Delta H_f^\circ = -200$ kJ mol^{-1}

$\Rightarrow \Delta H_f^\circ(AB) = -200$ kJ mol^{-1}

$\therefore \Delta H_R^\circ$ for reaction $A_2 + B_2 \to$
2AB is –400 kJ mol^{-1}

Given: Bond Enthalpy of A_2, B_2 and AB is 1 : 0.5 : 1

Assuming bond enthalpy of A_2 be x kJ mol^{-1}

$\therefore$ Bond enthalpy B_2 = 0.5 x kJ mol^{-1}

$\therefore$ Bond enthalpy AB = (x) kJ mol^{-1}

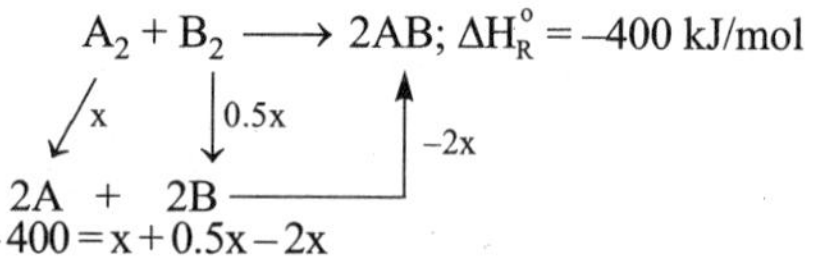

$-400 = x + 0.5x - 2x$

$-400 = -0.5x\ \ \therefore x = 800$ kJ/mol

132. (4)

$Fe_2O_3 + 2Al \longrightarrow Al_2O_3 + 2Fe$

Molar mass: 160g 27g

$(\Delta H_f^\circ)_{reaction} = [(\Delta H_f^\circ)_{Al_2O_3} + 2(\Delta H_f^\circ)_{Fe}] -$
$[(\Delta H_f^\circ)_{Fe_2O_3} + 2(\Delta H_f^\circ)_{Al}]$

$= [-1700 + 0] - [-840 + 0] = -860$ kJ/mol

Total mass of mixture = Fe_2O_3 + Al (1 : 2 molar ratio)

$= (160) + (2 \times 27) = 214$ g/mol

Heat evolved per gram $= \frac{860}{214} = 4$ kJ / g

133. (278)

$(\Delta H_f)_{C_2H_5OH_{(l)}} = \sum(\Delta H_{comb})_{reactant}$
$- \sum(\Delta H_{comb})_{product}$

$= 2 \times (-393.5) + 3(-241.8) - (-1234.7)$

$= -277.7$ kJ/mol ≈ 278 kJ/mol

134. (499) (i) $H_2O(g) \to H_2(g) + \frac{1}{2}O_2(g)$

$\Delta H_1^\circ = 242$ kJ / mol ...(1)

Multiply by 2

(ii) $H_2(g) + O_2(g) \to 2OH,\ \Delta H_2^\circ = 78$ kJ mol^{-1}

(iii) $H_2(g) \to 2H,\ \Delta H_3^\circ = 436$ kJ mol^{-1}

$2H_2O \to 2H + 2OH, \Delta H_4^\circ = 998$ KJmol^{-1}

Divide by 2

$\Rightarrow H_2O \to H + OH$

$\Delta H^\circ = \frac{998}{2} = 499$ kJ mol^{-1}

135. (610)

$\frac{1}{2}Cl_2(g) \to Cl(g) \to Cl^-(g) \to Cl^-(aq)$

$\Delta H_{reaction} = \frac{1}{2} \times 240 + (-350) + (-380) = -610$

136. (173) $\Delta_r H = \Sigma H_p - \Sigma H_R$

$CCl_4(g) + 2H_2O(g) \longrightarrow CO_2(g) + 4HCl(g)$

$= [(-394 + 4 \times (-92)] - [(-105) + (2 \times -242)]$

$\doteq -173$ kJ/mol

137. (925) Let, volume of C_2H_4 is x litre

$C_2H_4 + 3O_2 \rightarrow 2CO_2 + 2H_2O$

Initial x   –2x

Final –

$CH_4 + 2O_2 \rightarrow CO_2 + 2H_2O$

Initial (16.8 – x)   –

Final –   (16.8 – x)

Total volume of $CO_2 = 2x + 16.8 - x$

$\Rightarrow 28 - 16.8 + x$

$x = 11.2$ L

Volume of $C_2H_4 = 11.2$ L, Volume of CH_4

$= (16.8 - 11.2)L = 5.6$ L

$nCH_4 = \frac{5.6}{22.4} = 0.25$ mol

$nC_2H_4 = \frac{11.2}{22.4} = 0.5$ mol

$\therefore$ Heat evolved $= 0.25 \times (-900) + 0.5(-1400)$

$= -925$ kJ

138. (360)

$C_6H_{12}O_6(s) + 6O_2 \rightarrow 6CO_2(g) + 6H_2O(l)$

50% energy used to convert $H_2O(l)$ into $H_2O(g)$

$= \frac{1800}{2} = 900$kJ

The energy required to evaporate 1 mole of water $= 45$ kJ

$\Rightarrow 900 = n_{H_2O} \times 45 \Rightarrow n_{H_2O} = \frac{900}{45} = 20$ moles

$W_{H_2O} = 20 \times 18 = 360$ g

139. (54)

	HNO_3	NaOH
	600 mL × 0.2 M	400 mL × 0.1 M
	= 120 m mol	= 40 m mol

$HNO_3 + NaOH \rightarrow NaNO_3 + H_2O$

	HNO_3	NaOH	$NaNO_3$
Start	120	40	
End	80	0	40 m mol

$\Delta_r H = 40 \times 10^{-3}$ mol $\times 57 \times 10^3 \frac{J}{mol} = 2280$ J

$mS\Delta T = 2280 \Rightarrow 1000\text{ g} \times 4.2 \times \Delta T = 2280$

[Total volume of the solution

$= 600 + 400 = 1000$ mL]

Considering the density of water = 1g/mL.

$\Delta T = \frac{2280}{4.2} \times 10^{-3} = \frac{22800}{42} \times 10^{-3}$

$= 542.86 \times 10^{-3}$

$= 54.286 \times 10^{-2}$ K $= 54.286 \times 10^{-2}$ °C

140. (57) $\Delta H = \Delta U + \Delta n_g RT$

$H_2F_2(g) \longrightarrow H_2(g) + F_2(g)$

$\Delta n_g = n_P(g) - n_R(g) = 2 - 1 = 1$

$\Delta H = -59.6 + 1 \times 8.314 \times 300 \times 10^{-3}$

$= -57.10$ kJ mol^{-1}.

141. (117) Molar mass of $NH_3 = 17$ g/mol

moles of NH_3 $\frac{85}{17} = 5$ moles

For 1 mole enthalpy change = 23.4 kJ

5 moles enthalpy change = 23.4 × 5 = 117 kJ

142. (–103.7)

$C_3H_8 + 5O_2 \rightarrow 3CO_2 + 4H_2O$ ...(1)

$\Delta H_1 = -2220.0$ kJ/mol

$C + O_2 \rightarrow CO_2$ ...(2)

$\Delta H_2 = -393.5$ kJ/mol

$H_2 + \frac{1}{2}O_2 \rightarrow H_2O$ ...(3)

$\Delta H_3 = -285.8$ kJ/mol

Formation of C_3H_8 (1 mole):

$3C + 4H_2 \rightarrow C_3H_8$ ...(4)

Eq. (4) obtained by

3 × Eq (2) + 4 × Eq. (3) – Eq. (1)

$= 3 \times (-393.5) + 4 \times (-285.8) - (-2220.0)$

$= -1180.5 - 1143.2 + 2220.0 = -103.7$ kJ/mol

143. (2) $\Delta H_{neutaralization} = -57.3$ kJ/mol

In case of acetic acid

$\Delta H = \Delta H_{ionisation} + \Delta H_{neutralization}$

$-55.3 = \Delta H_{ionisation} - 57.3$

$\Delta H_{ioni} = 2$ kJ/mol

144. (600) $Mg(s) + \frac{1}{2}O_2(g) \longrightarrow MgO(s)$

$\Delta H = \Delta U + \Delta n_g RT$

$\Delta n_g = \Sigma n_P - \Sigma n_R = 0 - \frac{1}{2} = -\frac{1}{2}$

$-601.70 \times 10^3 = \Delta U - \frac{1}{2} \times 8.3 \times 300$

-601.70 kJ $= \Delta U - 1.245$ kJ

$\Delta U = -600.455$ kJ ≈ -600 kJ

145. (38) $H_2O(l) \rightarrow H_2O(g)$

$\Delta n_g = \sum n_p - \sum n_r = 1 - 0 = 1$

$\Delta H = \Delta U + \Delta n_g RT$

$\Delta U = \Delta H - \Delta n_g RT$

$= 41.1 - \frac{1 \times 8.31 \times 373}{1000}$ kJ/mol $= 38$ kJ/mol

146. (82)

HCl	+	NaOH	→	NaCl	+	H_2O
0.04		0.03		–		–
0.01		–		0.03		0.03

Heat released (q) = 0.03 × 57.1 × 1000 = 1713 J

$q = mc\Delta T$

1713 J = 500 g × 4.18 × ΔT

v = 500 mL

$\Delta T = \frac{1713}{500 \times 4.18} = 0.8196 \approx 82 \times 10^{-2}$ K

147. (3) Standard electrode potential represents the overall change:

$M(s) \rightarrow M(g)$ sublimation enthalpy

$M(g) \rightarrow M^+(g) + e^-$ Ionization enthalpy

$M^+(g) + H_2O \rightarrow M^+(aq)$ hydration enthalpy

148. (718)

$\Delta_f H^\circ_{KCl} = \Delta_{sub}H^\circ_{(K)} + \Delta_{ionization}H^\circ_{(K)} + \frac{1}{2}\Delta_{bond}H^\circ_{(Cl_2)} + \Delta_{electron\ gain}H^\circ_{(Cl)} + \Delta_{lattice}H^\circ_{(KCl)}$

$\Rightarrow -436.7 = 89.2 + 419.0 + \frac{1}{2}(243.0) + \{-348.6\} + \Delta_{lattice}H^\circ_{(KCl)}$

$\Rightarrow \Delta_{lattice}H^\circ_{(KCl)} = -717.8$ kJ mol^{-1}

$\simeq 718$ kJ mol^{-1}

149. (82) Moles of H^+,

$(n_{H^+}) = \frac{400 \times 0.2}{1000} \times 2 = 0.16$

$(n_{OH^-}) = \frac{600 \times 0.1}{1000} = 0.06$

(Limiting reagent)

Total volume of solution = (400 + 600) mL

= 1000 mL or 1000 cm^3

Enthalpy of neutralization, i.e.,

heat given out by 1 mole = 57.1 kJ

$\therefore$ Heat given out by 0.06 mole $= 57.1 \times 10^3 \times 0.06$ J

Now, if ΔT is increase in temperature

then, $0.06 \times 57.1 \times 10^3$

$= \left(1000\text{ cm}^3 \times 1.0\frac{g}{cm^3}\right) \times 4.18 \times \Delta T$

$\therefore \Delta T = 0.8196$ K $= 81.96 \times 10^{-2}$ K $\approx 82 \times 10^{-2}$ K

150. (101) $\Delta H_{sub} = \Delta H_{fus.} + \Delta H_{vap.} = 2.8 + 98.2$

$= 101$ kJ/mol

151. (21) Heat of combustion per mol (for 12 g) $= -2.48 \times 10^2$ kJ

Heat generated for 1 g of graphite

$= \frac{-2.48 \times 10^2}{12} = -20.66$ kJ ≈ 21 kJ

152. (128)

$\Delta_r H = \Sigma B.E_{Reactant} - \Sigma B.E_{Products}$

$\Delta_r H = [(B.E)_{C-C} + 6 \times (B.E)_{C-H}] - [(B.E)_{C=C} + 4 \times (B.E)_{C-H} + (B.E)_{H-H}]$

$= [347 + 6 \times 414] - [611 + 4 \times 414 + 436]$

$= 128$ kJ mol^{-1}

153. (309) $SF_6(g) \rightarrow S(g) + 6F(g)$

$\Delta H^\circ = \Delta H^\circ_f(S) + 6\Delta H^\circ_f(F) - \Delta H^\circ_f(SF_6)$

$= 275 + 6 \times 80 - (-1100) = 1855$ kJ mol^{-1}

Also, $\Delta H^\circ = 6\Delta H_{S-F}$

$\therefore \Delta H_{S-F} = \frac{1855}{6} = 309.17 \simeq 309$ kJ mol^{-1}

154. (5576)

$$Na(g) + Br(g) \longrightarrow NaBr(s)$$

$IE_1 \downarrow$ (for Na), $\Delta H_{eg_1} \downarrow$ (for Br) to $\boxed{\overset{+}{Na}(g) + \overset{-}{Br}(g)}$, then LE $\uparrow$ to NaBr (s)

$\Delta H_{formation} = IE_1 + \Delta H_{eg_1} + LE$

$= 495.8 + (-325) + (-728.4)$

$= -557.6$ kJ/mol $= -5576 \times 10^{-1}$ kJ/mol.

155. (855)

$3HC \equiv CH(g) \rightarrow C_6H_6(l);\ \Delta G^0 = -RT\ln K$

$\Rightarrow \Delta G^0 = \Sigma(\Delta G^0_f)_P - \Sigma(\Delta G^0_f)_R$

$\Rightarrow -RT\ln K = 1 \times (-1.24 \times 10^5) - (-3 \times 2.04 \times 10^5)$

$\Rightarrow -2.303 \times R \times T\log K = 4.88 \times 10^5$

$\Rightarrow \log K = -\frac{4.88 \times 10^5}{2.303 \times 8.314 \times 298}$

$= -85.52 = -855 \times 10^{-1}$

$\Rightarrow$ So; magnitude of 10gk $= 855 \times 10^{-1}$

156. (189494)

$\Delta H = \Delta U + \Delta n_g RT$; $n = \frac{90}{18} = 5$ mol

$H_2O(l) \rightleftharpoons H_2O(g)$; $\Delta n = 1$

$41000 = \Delta U + 1 \times 8.314 \times 373$

$\Rightarrow \Delta U = 37898.875$ J

For 5 moles, $\Delta U = 37898.87 \times 5 = 189494$ J

157. (– 326400)

$C_2H_5OH(l) + 3O_2(g) \longrightarrow 2CO_2(g) + 3H_2O(l)$

$\Delta H = \Delta U + \Delta n_g RT$ 1

$\Rightarrow -327\times10^3 = \Delta U + (-1)\times 2\times 300$

$\Rightarrow \Delta U = -327\times10^3 + 600$

$\therefore \Delta U = -326400$ cal

158. (–192.5) $C(s) + O_2(g) \rightarrow CO_2(g)$;
$\Delta_C H^0[C(\text{graphite})] = -286.0$ kJ/mol ...(i)

$H_2(g) + \frac{1}{2}O_2(g) \rightarrow H_2O(g)$;
$\Delta_C H^0[H_2(g)] = -393.5$ kJ/mol ...(ii)

$C_2H_6(g) + \frac{7}{2}O_2(g) \rightarrow 2\,CO_2(g) + 3H_2O(g)$;
$\Delta_C H^0[C_2H_6(g)] = -1560$ kJ/mol ...(iii)

The reaction of formation of ethane is
$2C(s) + 3H_2(g) \rightarrow C_2H_6(g)$

$\therefore \Delta_f H^0$ of $C_2H_6(g) = 2\times\Delta_C{}^0H[C_{\text{graphite}}] + 3\times\Delta_C H^0[H_2(g)] - \Delta_C H^0[C_2H_6(g)]$

$= 2\times(-286.0) + 3(-393.5) - (-1560)$

$= -192.5$ kJ/mol

159. (2)

$$\underset{268K}{\text{Ice}} \xrightarrow{\Delta S_1} \underset{273K}{\text{Ice}} \xrightleftharpoons{\Delta S_2} \underset{273K}{\text{Water}} \xrightarrow{\Delta S_3} \underset{373K}{\text{Water}}$$

$$\xrightleftharpoons{\Delta S_4} \underset{373K}{\text{Water vapour}} \xrightarrow{\Delta S_5} \underset{383K}{\text{Water vapour}}$$

$\Delta S_{\text{overall}} = \Delta S_1 + \Delta S_2 + \Delta S_3 + \Delta S_4 + \Delta S_5$

$\Delta S_1 = \int_{268}^{273} \frac{C_{p,m}dT}{T}$

$\Delta S_2 = \frac{\Delta H_{m\,\text{fusion}}}{273}$; $T_f = 273$ K

$\Delta S_3 = \int_{273}^{373} \frac{C_{p,m}dT}{T}$

$\Delta S_4 = \frac{\Delta H_{m\,\text{vaporisation}}}{373}$; $T_b = 373$ K

$\Delta S_5 = \int_{373}^{383} \frac{C_{p,m}dT}{T}$

160. (2) (A) $dG = VdP - SdT$
Constant pressure

$dG = -SdT \Rightarrow \left(\frac{\partial G}{\partial T}\right)_P = -S$

(B) $dH = (q)_P = nC_pdT \Rightarrow \left(\frac{\partial H}{\partial T}\right)_P = C_P$ (for 1 mole)

(C) $dG = VdP - SdT$
At constant temperature $\Rightarrow dG = VdP$

$\left(\frac{\partial G}{\partial P}\right)_T = V$

(D) $dU = nC_vdT = (q)_v$

$\left(\frac{\partial U}{\partial T}\right)_V = C_V$ (for 1 mole)

161. (1) A system at higher temperature has greater entropy (randomness). S and ΔS are related with T as:

$S_T = \int_0^T \frac{nC\cdot dT}{T}$ and $\Delta S = \int \frac{dq}{T}$

Thus, both S and ΔS are function of temperature.

162. (1) Final temperature

$= \frac{T_1 + T_2}{2}$, let $T_2 > T_1$

(blocks have same mass)

$\therefore dS = \frac{dq}{T} = \frac{C_p dT}{T}$ $\therefore \Delta S = C_p \ln\left(\frac{T_f}{T_i}\right)$

$\therefore \Delta S_{\text{Total}} = C_p \ln\left(\frac{T_1+T_2}{2T_1}\right) + C_p \ln\left(\frac{T_1+T_2}{2T_2}\right)$

$= C_p \ln\left[\frac{(T_1+T_2)^2}{4T_1T_2}\right]$

163. (4) In the process of synthesis of ammonia from N_2 and H_2, number of moles decreases which implies that the change in entropy will be negative.

$N_2(g) + 3H_2(g) \rightleftharpoons 2NH_3(g)$

164. (4)

$$\underset{273K}{H_2O(s)} \xrightarrow{\Delta S_1} \underset{273K}{H_2O(l)} \xrightarrow{\Delta S_2} \underset{373K}{H_2O(l)} \xrightarrow{\Delta S_3} \underset{373K}{H_2O(g)} \xrightarrow{\Delta S_4} \underset{383K}{H_2O(g)}$$

$\Delta S_1 = \frac{\Delta H_{\text{fus}}}{273} = \frac{334}{273} = 1.22$; $\Delta S_2 = m\,C \ln\left(\frac{T_2}{T_1}\right)$

$= 4.2\times\ln\left(\frac{373}{273}\right) = 4.2\times2.303\times(\log 373 - \log 273)$

$= 4.2\times2.303\times(2.572 - 2.436) = 1.31$

$\Delta S_3 = \frac{\Delta H_{\text{vap}}}{373} = \frac{2491}{373} = 6.68$

$\Delta S_4 = m\,C\ln\left(\frac{T_2}{T_1}\right) = 2\times\ln\left(\frac{383}{373}\right)$

$= 2\times2.303\times(\log 383 - \log 373)$

$= 2\times2.303\times(2.583 - 2.572) = 0.05$

$\therefore$ Total entropy change

$\Delta S = 1.28 + 6.68 + 1.31 + 0.05 = 9.26$ kJ kg^{-1} K^{-1}

165. (2)
(1) When diamond is converted into graphite (it is heated to 1500 °C) entropy is increased, $\Delta S > 0$
(2) When pressure increases then molecules of gas will come closer and intermolecular distance decreases, so entropy will also decrease, $\Delta S < 0$
(3) When we increase the temperature of a gas then randomness is increased as the kinetic energy is gained by molecules. So, $\Delta S > 0$
(4) H_2 molecule is converted into atoms, the no. of particles increases. Thus entropy will increase, $\Delta S > 0$

166. (2) Given, $C_p = 10$ cals at 1000 K

$T_1 = 1000$ K. $T_2 = 100$ K, $m = 32$ g, $\therefore n = \frac{32}{32} = 1$

$\Delta S = ?$

At constant pressure

$\Delta S = nC_p \ln\frac{T_2}{T_1} = 2.303\times C_p \log\frac{T_2}{T_1}$

$= 2.303\times10\log\frac{100}{1000} = -23.03$ cal deg^{-1}

167. (2)

$\Delta S^\circ = S^\circ_{CO_2} + 2\times S^\circ_{H_2O} - \left(S^\circ_{CH_4} + 2\times S^\circ_{O_2}\right)$

$= (213.6 + 2\times69.9) - (186.2 + 2\times205.2)$

$= -242.8$ J K^{-1} mol^{-1}.

168. (4) $\Delta S = \frac{\Delta H}{T}$; $T = \frac{\Delta H}{\Delta S} = \frac{0.3344}{0.836} = 0.4$K

169. (2) For isothermal process ($\Delta T = 0$) for one mole

$\Delta S = R\ln\frac{P_1}{P_2} = 8.314\ln 2$

$= 8.314\times0.693 = 5.76$

170. (3)

$\frac{1}{2}X_2 + \frac{3}{2}Y_2 \longrightarrow XY_3$; $\Delta H^\circ = -30$ kJ

Calculating ΔS° for the above reaction, we get

$\Delta S^\circ = 50 - \left[\frac{1}{2}\times60 + \frac{3}{2}\times40\right]$

$= 50 - (30 + 60) = -40$ JK^{-1}

At equilibrium, $T\Delta S^\circ = \Delta H^\circ$ [$\because \Delta G^\circ = 0$]

$\therefore T\times(-40$ JK$^{-1})$

$= -30\times1000$ J [$\because$ 1 kJ = 1000J]

or $T = \frac{-30\times1000}{-40} = 750$ K

171. (400) At equilibrium $\Delta G_{PT} = 0$ and $\Delta G = \Delta H - T\Delta S$

$\Rightarrow \Delta H_{vap} = T\Delta S_{vap} \Rightarrow 30\times1000 = T\times75 \Rightarrow T = 400$K

172. (2000) Applying Gibbs Helmholtz equation
$\Delta G = \Delta H - T\Delta S$
For reaction to be spontaneous $\Delta G = -$ ve
For limiting case $\Delta G = 0$
$\Delta H = T\Delta S$

$T = \frac{\Delta H}{\Delta S} = \frac{400\text{kJ/mol}}{0.2\text{kJ/mol} - \text{K}} = 2000$ K

above 2000 K reaction become spontaneous.

173. (1070) $\Delta S = \frac{\Delta H}{T_{mp}}$

$28.4 = \frac{30.4\times1000}{T_{mp}} \Rightarrow T_{mp} = 1070.422$ K.

174. (3) $\Delta S_{\text{System}} = nR\ln\left(\frac{V_2}{V_1}\right)$

$= 1\times8.314\ln\left(\frac{3}{2}\right)$

$\Delta S_{\text{System}} = 3.37$
$\Delta S_{\text{Surr.}} = 3.37$

175. (2) $\Delta G = \Delta H - T\Delta S$
A : ΔG (J mol^{-1}) $= -25\times10^3 + 80\times300$
$\Rightarrow \Delta G = -$ve
B : ΔG (J mol^{-1}) $= -22\times10^3 - 40\times300$
$\Rightarrow \Delta G = -$ve
C : ΔG (J mol^{-1}) $= 25\times10^3 + 300\times50$
$\Rightarrow \Delta G = +$ve
D : ΔG (J mol^{-1}) $= 22\times10^3 - 20\times300$
$\Rightarrow \Delta G = +$ve
$\Rightarrow$ Processes C and D are non-spontaneous.

176. (336) Given chemical reaction:

$A \xrightarrow[T=300K]{} B$ $\Delta G = -49.4$ kJ / mol

$\Delta H = 51.4$ kJ/mol; $\Delta S = ?$

$\Rightarrow$ From the relation $\Delta G = \Delta H - T\Delta S$

$\Rightarrow \Delta S = \frac{\Delta H - \Delta G}{T}$

$= \frac{[51.4 - (-49.4)]\times1000}{300}\ \frac{\text{J}}{\text{mol K}}$

$\Rightarrow \Delta S = 336$ JK^{-1} mol^{-1}.

177. (200) $A + B \rightleftharpoons C + D$
For a reaction to be spontaneous, $\Delta_r G^\circ < 0$
$\Delta_r G^\circ = \Delta_r H^\circ - T\Delta_r S^\circ$
$\Rightarrow \Delta_r H^\circ - T\Delta_r S^\circ < 0$; $80\times1000 - T\times2T < 0$
$\Rightarrow T^2 > 40000$; $T > 200$ $\therefore T_{min} = 200$ K

178. (2) $\Delta_r G^\ominus = \Delta_r H^\ominus - T\cdot\Delta_r S^\ominus$ ...(i)

$\Delta_r H^\ominus = [(2\times 42)-(80+8)] = -4$ kJ/mol

$\Delta_r S^\ominus = (400-250-140) = +10$ J/K.mol

Putting the value of $\Delta H^\ominus$, T and $\Delta S^\ominus$ in eq. (i)

$\Delta_r G^\ominus = -4000 - 600\times(10)$

$= -10{,}000$ J/mol $= -10$ kJ/mol.

179. (3) $\overset{+4}{PbO_2} + Pb \rightarrow 2\overset{+2}{PbO}$

Pb^{2+} is more stable hence reaction will be spontaneous. So $\Delta G^\circ_r(1)$ is negative, $\Delta_r G^\circ(1) < 0$.

$\overset{+4}{SnO_2} + Sn \rightarrow 2\overset{+2}{SnO}$

Sn^{+2} is less stable, so reaction will be non-spontaneous hence $\Delta G^\circ_r(2)$ is positive, $\Delta_r G^\circ(2) > 0$.

180. (3) For reaction to be spontaneous according to IInd law: $\Delta G < 0$

$\Rightarrow \Delta H - T\Delta S < 0$ (ΔH or ΔS are + ve)

$\Rightarrow T > \left(\dfrac{\Delta H}{\Delta S}\right) = T_e$

($\Delta G = -$ ve at high temperature) $\Rightarrow T > T_e$

181. (1) $N_2(g) + 3H_2(g) \rightleftharpoons 2NH_3(g)$

$\Delta S_r = -$ve, $\Delta H_r = -$ve

$\dfrac{-\Delta H_r^\circ}{T} \Rightarrow$ first decreases then constant

$\dfrac{-\Delta G_R^\circ}{T} \Rightarrow$ first decreases then constant

182. (1) Reaction is spontaneous at relatively high temperature and non-spontaneous at low temperature $\Delta G = \Delta H - T\Delta S$.

It is only possible when ΔH and ΔS both are positive as $\Delta H < T\Delta S$

183. (3) $\because \Delta G = \Delta H \quad T\Delta S$

For spontaneity of reaction : $\Delta G = -$ve

ΔG is +ve for (B) and – ve for (C).

184. (3) $\Delta G = 0$ for reversible reaction

$\Delta G = -$ve for spontaneous reaction

$\Delta G = +$ve for non-spontaneous reaction

Therefore option (3) is not correct.

185. (2) (A) First law is given by $\Delta U = Q + W$

If we apply reversible work at constant P.

$\Rightarrow \Delta U = Q - P\Delta V$

(D) $\Delta H = \Delta U + \Delta nRT$

186. (1) $\Delta G = \Delta H - T\Delta S$

$\because \Delta S_g > \Delta S_l > \Delta S_s$

$\therefore$ On melting the entropy increases and ΔG becomes more negative and hence it becomes easier to reduce metal.

187. (2) (A) – (II), (B) – (III), (C) – (IV), (D) – (I)

188. (3) A reaction is spontaneous if ΔG_{sys} is negative.

$\Delta G_{sys} = \Delta H_{sys} - T\Delta S_{sys}$

A reaction will be spontaneous at all temperatures if ΔH_{sys} is negative and ΔS_{sys} is positive.

189. (4) At 315 K; $\Delta G^\circ = 120 - \dfrac{3}{8}T$

$\Delta G^\circ = 120 - \dfrac{3}{8}\times 315 =$ positive

Since ΔG° is positive then $K_{eq} < 1$.

So $\dfrac{[Y]}{[X]} < 1$.

$\therefore [X] > [Y]$

190. (3) $MgO(s) + C(s) \rightarrow Mg(s) + CO(g)$

For a reaction to be spontaneous, $\Delta G < 0$

$\Delta H^\circ - T\Delta S^\circ < 0 \Rightarrow T > \dfrac{\Delta H^\circ}{\Delta S^\circ}$

$T > \dfrac{491.1\times 1000}{198}$; $T > 2480.3$ K

191. (2) $\Delta G^\circ = A - BT$; A and B are non-zero constants

$\therefore \Delta G^\circ = \Delta H^\circ - T\Delta S^\circ = A - BT$

$\therefore$ Reaction will be endothermic if $\Delta H^\circ > 0$.

Hence, $A > 0$

192. (3) $\Delta H = 200$ J mol^{-1}

$\Delta S = 40$ JK^{-1} mol^{-1}

For spontaneous reaction,

$\Delta G < 0$

$\Delta H - T\Delta S < 0$; $\Delta H < T\Delta S$

$\dfrac{\Delta H}{\Delta S} < T$; $\dfrac{200}{40} < T$

$5 < T$

So, minimum temperature is 5 K

193. (1) From thermodynamics relation.

$\Delta G^\circ = -RT\ln K$

$\Delta H^\circ - T\Delta S^\circ = RT\ln K$

$-\dfrac{\Delta H^\circ}{RT} + \dfrac{T\Delta S^\circ}{RT} = \ln K$ or $\ln K = -\dfrac{\Delta H^\circ}{RT} + \dfrac{\Delta S^\circ}{R}$

For exothermic reaction, $\Delta H^\circ = -$ve

slope $= -\dfrac{\Delta H^\circ}{R} = +$ve

So from graph, lines should be A & B.

194. (3)

	A_2	$\rightleftharpoons$	2A
initial	1		0
final	(1 – 0.2)		(2 × 0.2)

The equilibrium constant,

$K = \dfrac{[A]^2}{[A_2]}$; $K = \dfrac{[0.4]^2}{[0.8]} = 0.2$

$\Delta G^\circ = -RT\ln K = -8.314$ JK^{-1} mol^{-1} $\times 320$ K $\times \ln 0.2$

$= 4281$ J/mol

195. (4) $S(\ell) \longrightarrow S(g)$; $K_p = P_{vap}$

$\Delta G^o_{rxn} = \Delta_f G^\circ(\text{vapour}) - \Delta_f G^\circ(\text{liquid})$

$\Delta G^o_{rxn} = 103 - 100.7 = 2.3$ kcal/mol

$= 2300$ cal/mol

$\Delta G^o_{rxn} = -RT\ln K_p$

2300 cal/mol $= -2$ cal/mol/K × 500 K × $\ln K_p$

$\ln K_p = -2.3$

$\log_{10} K_p = -1$

$K_p =$ Antilog $(-1) = 0.1$ atm

$\therefore$ Vapour pressure of liquid 'S' at 500 K is approximately equal to 0.1 atm.

196. (2) (i) $2Fe_2O_3(s) \rightarrow 4Fe(s) + 3O_2(g)$;

$\Delta_r G^\circ = +1487.0$ kJ mol^{-1}

(ii) $2CO(g) + O_2(g) \rightarrow 2CO_2(g)$;

$\Delta_r G^\circ = -514.4$ kJ mol^{-1}

Multiply reaction (ii) with 3, we get

(iii) $6CO(g) + 3O_2(g) \rightarrow 6CO_2(g)$;

$\Delta_r G^\circ = 3\times(-514.4) = -1543.2$ kJmol^{-1}

When we add reaction (i) and reaction (iii), we get reaction (iv) :

(iv) $2Fe_2O_3(s) + 6CO(g) \rightarrow 4Fe(s) + 6CO_2(g)$

Free energy change, $\Delta_r G^\circ$ for the reaction will be $1487.0 - 1543.2 = -56.2$ kJ mol^{-1}

197. (2) $\Delta H = nC_p\Delta T = 0$

means ΔH is constant if $\Delta T = 0$

$\Delta S = nR\ln\left(\dfrac{V_f}{V_i}\right) \geq 0$ if $V_f > V_i$

198. (4) $\Delta G = \Delta H - T\Delta S$

For a spontaneous reaction $\Delta G = -$ve which is possible when both ΔH and ΔS are positive at high temperature and hence, the reaction becomes spontaneous.

199. (3) $\Delta G^\circ = \Delta H^\circ - T\Delta S^\circ$;

$-RT\ln K = \Delta H^\circ - T\Delta S^\circ$

$\ln K = -\dfrac{\Delta H^\circ - T\Delta S^\circ}{RT}$

200. (2) At equilibrium $\Delta G = 0$

Hence, $\Delta G = \Delta H - T_e\Delta S = 0$

$\therefore \Delta H = T_e\Delta S$ or $T_e = \dfrac{\Delta H}{\Delta S}$

For a spontaneous reaction,

ΔG must be negative which is possible only if $\Delta H < T\Delta S$

or $T > \dfrac{\Delta H}{\Delta S}$; $T_e < T$

201. (1) $\Delta G^\circ = \Delta H^\circ - T\Delta S^\circ$

For a spontaneous reaction $\Delta G^\circ < 0$

or $\Delta H^\circ - T\Delta S^\circ < 0 \Rightarrow T > \dfrac{\Delta H^\circ}{\Delta S^\circ}$

$\Rightarrow T > \dfrac{179.3\times 10^3}{160.2} > 1117.9\text{K} \approx 1118\text{K}$

202. (2)

203. (1) For spontaneous reaction, $dS > 0$ and $dG < 0$.

204. (3) $\Delta G^\circ = -RT\ln K_c$ or $-\Delta G^\circ = RT\ln K_c$

205. (2) For an endothermic reaction, $\Delta H > 0$

Therefore, options (a) and (c) are not possible.

At freezing point (T_f), reaction is non-spontaneous.

$\therefore \Delta G > 0 \Rightarrow \Delta H - T_f.\Delta S > 0 \Rightarrow \Delta S < \dfrac{\Delta H}{T_f}$

At boiling point (T_b), reaction is spontaneous.

$\therefore \Delta G < 0 \Rightarrow \Delta H - T_b.\Delta S < 0$

$\Rightarrow \Delta S > \dfrac{\Delta H}{T_b} \Rightarrow \Delta S > 0$

206. (1) $C + O_2 \Rightarrow CO_2$; $\Delta H_1 = -393$ J. ..(i)

$2Zn + O_2 \Rightarrow 2ZnO$; $\Delta H_2 = -412$ J ...(ii)

On the basis of given values oxidation of zinc is more feasible than carbon therefore will carbon will oxidise Zn.

207. (34) The relationship between Gibbs free energy, enthalpy and entropy is:

$\Delta G^\circ = \Delta H^\circ - T\Delta S^\circ$ ...(1)

The relationship between Gibbs free energy and equilibrium constant is

$\Delta G^\circ = -RT\ln K_{eq}$ ...(2)

Equating (1) and (2) we get:

$-RT\ln K_{eq} = \Delta H^\circ - T\Delta S^\circ$

$\Delta S^\circ = \dfrac{\Delta H^\circ + RT\ln K_{eq}}{T}$

$\therefore \quad \Delta S^\circ = \frac{\Delta H^\circ}{T} + R \ln K_{eq}$

$R \ln K_{eq} = R \times 2.3 \log(1.8 \times 10^{-7})$
$= 8.3 \times 2.3 [\log(18 \times 10^{-8})]$
$= 19.09 [\log(2 \times 3^2) - 8]$
$= 19.09 [0.3 + 2(0.48) - 8]$
$\therefore R \ln K_{eq} = -128.66\ JK^{-1}\ mol^{-1}$

For, $\frac{\Delta H^\circ}{T} = \frac{28400}{300} = \frac{284}{3} = 94.67\ JK^{-1}mol^{-1}$

$\therefore \Delta S^\circ = 94.67 - 128.6 = -33.9\ JK^{-1}\ mol^{-1}$
$\sim -34\ JK^{-1}\ mol^{-1}$

$\therefore$ The magnitude is $34\ JK^{-1}\ mol^{-1}$

208. (230)

$$\log K_{P2} - \log K_{P1} = \frac{\Delta H^o}{2.303R}\left(\frac{1}{T_1} - \frac{1}{T_2}\right)$$

$$\Rightarrow 3.5 - 2.5 = \frac{\Delta H^o}{2.303R}(0.06 - 0.05)$$

$$\Rightarrow \frac{\Delta H^o}{R} = \frac{2.303}{0.01}$$

$$\Rightarrow \frac{\Delta H^o}{R} = 230.3 \approx 230$$

209. (727) Transition temparature is the temperature at which two phases or states of a substance are in equilibrium.

$\Delta G^\circ = 0$ (eq. at standard pressure)
$105 - 35 \log T = 0$
$\log T = 3$
$T = 1000\ K$
Convert temperature from kelvin to celsius.
$T = 727^\circ C$.

210. (2850) $\Delta H^\circ_{rxn} = 55\ kJ/mol, T = 298\ K$

$\Delta S^\circ_{rxn} = 175\ J/k\ mol$

$\Delta G^\circ_{rxn} = \Delta H^\circ_{rxn} - T\Delta S^\circ_{rxn}$

$\Rightarrow \Delta G^\circ_{rxn} = 55000\ J/mol - 298$
$K \times 175\ J/K\ 1\ mol$

$\Rightarrow \Delta G^\circ_{rxn} = 55000 - 52150$

$\Rightarrow \Delta G^\circ_{rxn} = 2850\ J/mol$

211. (2) For, Spontaneous process $\Delta G < 0$
For, Equilibrium $\Delta G = 0$
For, Non-spontaneous process $\Delta G > 0$
Statement B and C are correct.

212. (300) $\Delta G = \Delta H - T\Delta S = 0$ at equilibrium
$\Rightarrow -165 \times 10^3 - T \times (-550) = 0 \Rightarrow T = 300K$.

213. (710) $N_2O_4 \rightleftharpoons 2NO_2$

	N_2O_4	$2NO_2$
$t = 0$	1	mol
$t = t$	(1-0.5) mol = 0.5 mol	0.5 × 2 mol = 1 mol

Total pressure = 1 atm.

$$K_P = \frac{\left(\frac{1}{1.5}\times 1\right)^2}{\left(\frac{0.5}{1.5}\times 1\right)} = \frac{1}{0.75} = \frac{100}{75} = 1.33$$

$\Delta G^0 = -RT\ln K_p = -8.31 \times 300 \times \ln(1.33)$
$= -8.31 \times 300 \times \log_{10}1.33 \times \log_e 10$
$= -8.31 \times 300 \times 0.1239 \times 2.3 = -710.43\ J\ mol^{-1}$.

214. (5) $\Delta G = \Delta H - T\Delta S$
$\Delta G = -57.8 - 298 \times (-176 \times 10^{-3}) = -5\ kJ\ mol^{-1}$

215. (964) $T_{min} = \left(\frac{\Delta^o H}{\Delta^o S}\right)$

$\Delta^o H_{rxn} = \left[\Delta^o_f H(Fe) + \Delta^o_f H(CO)\right] - \left[\Delta^o_f H(FeO) + \Delta^o_f H(C(graphite))\right]$
$= [0 - 110.5] - [-266.3 + 0] = 155.8\ kJ/mol$
$\Delta^o S_{rxn} = [\Delta^o S(Fe) + \Delta^o S(CO)] - [\Delta^o S(FeO) + \Delta^o S(C(graphite))]$
$= [27.28 + 197.6] - [57.49 + 5.74]$
$= 161.65\ J/mol\text{-}K$

$$T_{min} = \frac{155.8 \times 10^3\ J/mol}{161.65 J/mol - K} = 963.8\ K$$

$\approx 964\ K$ (nearest integer)

216. (– 13538)

From $\Delta H^\circ = \Delta U^\circ + \Delta n_g RT$

$\Delta H^\circ = -20 \times 1000 - 1 \times 8.314\ J/mol.$
$K \times 298K = -22477.57\ J$

$\Delta G^\circ = \Delta H^\circ - T\Delta S^\circ = -22477.57 -$
$(298 \times (-30)) = -13538\ J$

217. (–2.70)
$\Delta U = 2.1\ kcal = 2.1 \times 10^3\ cal$
$\Delta n_g = 2$
$\Delta H = \Delta U + \Delta n_g RT$
$= 2.1 \times 10^3 + 2 \times 2 \times 300 = 2100 + 1200 = 3300\ cal$
$\Delta G = \Delta H - T\Delta S$
$= 3300 - 300 \times 20 = 3300 - 6000$
$= -2700\ cals = -2.7\ kcal$

Part-B (JEE Advanced)

1. (b) $n = 0.5\ mol, P_0 = 2\ bar, T = 600\ K$

$(2\ bar, 600\ K) \xrightarrow[P_{ext} = P\ bar\ (2 < P < 8)]{①} (P\ bar, 600\ K) \xrightarrow[P_{ext} = 8\ bar]{②} (8\ bar, 600\ K)$

For ideal gas, $PV = nRT$

$$V_0 = \frac{nRT}{2}, V_1 = \frac{nRT}{P}, V_2 = \frac{nRT}{8}$$

$W_1 = -P_{ext}(V_1 - V_0)$

$$W_1 = -P\left(\frac{nRT}{P} - \frac{nRT}{2}\right) = -nRT\left(1 - \frac{P}{2}\right)$$

$$W_2 = -P_{ext}(V_2 - V_1) = -8\left(\frac{nRT}{8} - \frac{nRT}{P}\right)$$

$$= -nRT\left(1 - \frac{8}{P}\right)$$

$$W_{min} = W_1 + W_2 = -nRT\left(1 - \frac{P}{2} + 1 - \frac{8}{P}\right)$$

$$= -nRT\left(2 - \frac{P}{2} - \frac{8}{P}\right)$$

for minimum value of W

$P = 4\ bar$

$$So, -\frac{1}{2} \times R \times 600\left(2 - \frac{4}{2} - \frac{8}{4}\right)$$

$$\frac{1}{2} \times 600R[-2 + 2 + 2]$$

$W_{min} = 600R$

2. (c) From 1st law of thermodynamics
$q_{sys} = \Delta U - w = 0 - [-P_{ext}.\Delta V]$
$= 3.0\ atm \times (2.0\ L - 1.0\ L) = 3.0\ L\text{-}atm$

$$\therefore \Delta S_{surr} = \frac{(q_{rev})_{surr}}{T} = -\frac{q_{sys}}{T}$$

$$= -\frac{3.0 \times 101.3\ J}{300\ K} = -1.013\ J/K$$

3. (b) In general, the molar heat capacity for any process is given by

$C = C_v + \frac{R}{1 - \gamma}$, when PV^γ = constant

Here $\frac{P}{V} = 1$, *i.e.* PV^{-1} = constant $\Rightarrow \gamma = -1$

For monoatomic gas, $C_v = \frac{3}{2}R$

$$\therefore C = \frac{3}{2}R + \frac{R}{1 - (-1)} = \frac{3}{2}R + \frac{R}{2} = \frac{4R}{2} = 2R.$$

4. (d) At constant external pressure, for an irreversible adiabatic process,
$C_v(T_2 - T_1) + P_e(V_2 - V_1) = 0$

$\frac{3}{2} \times R(T_2 - T_1) = -1(2 - 1)$

$T_2 = T_1 - \frac{2}{3R}$

$\therefore T_2 = T - \frac{2}{3 \times 0.082}$

5. (c) $\Delta H = nC_p\Delta T$ solution; since $\Delta T = 0$ so, $\Delta H = 0$

6. (c)
$\Delta H = \Delta U + P_2V_2 - P_1V_1$ Given, $\Delta U = 30.0\ L\ atm$
$P_1 = 2.0\ atm, V_1 = 3.0\ L, T_1 = 95\ K$
$P_2 = 4.0\ atm, V_2 = 5.0\ L, T_2 = 245\ K$
$\Delta H = \Delta U + P_2V_2 - P_1V_1$
$= 30 + (4 \times 5) - (2 \times 3) = 30 + 20 - 6 = 44\ L\ atm.$

7. (a) Work is not a state function because it depends upon the path followed.

8. (c) In a reversible process, the driving and the opposite forces are nearly equal, hence the system and the surroundings always remain in equilibrium with each other.

9. (8120) $X \rightarrow Y$ is an isothermal process an ideal gas:
$\Delta H = 0, \Delta U = 0$
$Y \rightarrow Z$ is an isochoric process

$\Rightarrow \quad \Delta V = 0$

$\therefore \quad W = 0$

$\Delta U = nC_{v.m}(T_2 - T_1)$
$= 5 \times 12(415 - 335) = 4800$ J

$\Delta H = \Delta U + \Delta(PV)$
$= \Delta U + nR\Delta T = 4800 + 5 \times 8.3 \times (415 - 335)$
$= 8120$ J

10. (7) For A → B (Reversible adiabatic)

$T_1 V_1^{\gamma-1} = T_2 V_2^{\gamma-1}$

$\Rightarrow 600(V_1)^{2/3} = 60(V_2)^{2/3} \quad \left(\because \gamma = \frac{5}{3}\right)$

$\Rightarrow 10 = \left(\frac{V_2}{10}\right)^{2/3}$

$V_{total} = E_{AB} + q_{BC} = 0 + q_{BC} = q_{BC}$

$q_{BC} = RT_2 \ln 10$

$\Rightarrow q_{BC} = 60 R \ln 10 = 60 R \ln \frac{V_3}{V_2}$

[∵ B → C is reversible isothermal]

$\Rightarrow 60 R \ln 10 = 60 R \ln \left(\frac{V_3}{10^{5/2}}\right)$

$\Rightarrow \log 10 = \log V_3 - \frac{5}{2}$

$\Rightarrow \log V_3 = \frac{7}{2} \Rightarrow 2\log V_3 = 7$

11. (8) At T_1 K: $A(g) \rightleftharpoons P(g)$

	t = 0	6
At eq, T_1K	$6 - x$	$x = 4$ (from plot)
At eq, T_2K	$6 - y$	$y = 2$ (from plot)

$\Rightarrow$ At T_1 K: $K_{P_1} = \frac{4}{2} = 2$

$\Rightarrow$ At T_2 K: $K_{P_2} = \frac{2}{4} = \frac{1}{2}$

Now, $\Delta G_2^\circ = -RT_2 \ln K_{P_2} = -RT_2 \ln \frac{1}{2} = RT_2 \ln 2$

$\Delta G_1^\circ = -RT_1 \ln K_{P_1} = RT_1 \ln 2$

Given : $\Delta G_2^\circ - \Delta G_1^\circ = RT_2 \ln 2 + RT_1 \ln 2$
$= RT_2 \ln 2 + 2RT_2 \ln 2$
$= 3RT_2 \ln 2 = RT_2 \ln x \; (T_1 = 2T_2)$
$\Rightarrow x = 2^3 = 8$

12. (10)

Process (I) ⇒ (Adiabatic reversible)

$\frac{\Delta U}{R} = 450 - 2250$

$\Delta U = -1800R$

$W_I = \Delta U = -1800 R$

Process (II) ⇒ (Reversible isothermal process)

$T_1 = 900$ K

Calculation of T_2 after reversible adiabatic process

$\Delta U = nC_v dT$

$\Rightarrow \quad -1800 R = 1 \times \frac{5}{2} R(T_2 - 900)$

$T_2 = 180$ K

$W_{II} = -nRT_2 \ln \frac{V_3}{V_2} = W_I$

$\Rightarrow \quad -1 \times R \times 180 \ln \frac{V_3}{V_2} = -1800R$

$\Rightarrow \quad \ln \frac{V_3}{V_2} = 10$

13. (2) $w_d = \left(-4 \times \frac{3}{2}\right) + (-1 \times 1) + \left(-\frac{1}{2} \times \frac{5}{2}\right)$

$= -\left(6 + 1 + \frac{5}{4}\right)$

$w_d = -\frac{33}{4}$ L atm

$w_s = -2.303\, RT \log \frac{5.5}{1/2} = -2.303\, PV \log 11$

$w_s = -4.606 \times 1.04 = -4.8$ L atm

$\frac{w_d}{w_s} = \frac{-\frac{33}{4}}{-4.8} = 1.72 \simeq 2.0$

14. (9) Energy released by combustion of 3.5 g gas
$= 2.5 \times (298.45 - 298)$ kJ

Energy released by 1 mole of gas

$= \frac{2.5 \times 0.45}{3.5/28} = 9 \text{ kJmol}^{-1}$

15. (– 29.88) $H_2O(l) \rightarrow H_2(g) + \frac{1}{2}O_2(g)$

$H_2O(l)$	$H_2(g)$	$\frac{1}{2}O_2(g)$
⇓	⇓	⇓
144g	–	–
⇓	⇓	⇓
$\frac{144}{18}$	8 mol	$\frac{8}{2} = 4$ mol

$W_{expansion} = -P\Delta V$ (Here – ve sign indicates that work is done by the system)

As per ideal gas equation : -

PV = nRT

$P\Delta V = \Delta n_g RT$ where, Δn_g is calculated as the difference between the total number of moles of gaseous products & total number of moles of gaseous reactants

$\therefore W_{expansion} = \Delta n_g RT \Rightarrow \Delta n_g = 12$

$\Rightarrow W_{expansion} = \frac{-12 \times 8.3 \times 300}{1000}$ kJ

$\therefore W_{expansion} = -29.88$ kJ.

16. (0.31)

$\alpha \rightarrow \beta$ and $S^0_{\beta(600)} - S^0_{\alpha(600)} = 6 - 5 = 1$

(from graph)

$S^0_{\alpha(600)} - S^0_{\alpha(300)} + C_{P(\alpha)} \ln \frac{600}{300}$

$S^0_{\beta(600)} = S^0_{\beta(300)} + C_{P(\beta)} \ln \frac{600}{300}$

$S^0_{\beta(600)} - S^0_{\alpha(600)} = S^0_{\beta(300)} - S^0_{\alpha(300)} + C_{P(\beta)} - C_{P(\alpha)} \ln 2$

$6 - 5 = S^0_{\beta(300)} - S^0_{\alpha(300)} + [1 \times \ln 2]$

$1 = S^0_{\beta(300)} - S^0_{\alpha(300)} + 0.69$

So $S^0_{\beta(300)} - S^0_{\alpha(300)} = 0.31 \text{ J mol}^{-1}\text{ K}^{-1}$

17. (300) As the phase transition temperature is 600 K

So at 600 K ; $\Delta G^\circ_{rxn} = 0$

So $\Delta H^\circ_{reaction\,(600)} = T\,\Delta S^\circ_{reaction\,(600)}$

$\Delta H^\circ_{(600)} = 600 \times 1 = 600$ Joule/mole

So, $\Delta H_{600} - \Delta H_{300} = \Delta C_P (T_2 - T_1)$
$= 1 \times 300 = 300 \text{ J mol}^{-1}$

$\Rightarrow \Delta H_{300} = \Delta H_{600} - 300 = 600 - 300 = 300 \text{ J mol}^{-1}$

18. (166.28) Slope $= \frac{dy}{dx} = \frac{d(\ln K)}{d(10^4/T)}$

$\Rightarrow \quad \frac{-\Delta H^\phi}{10^4 R} = \frac{-7-(-3)}{12-10}$

$\Rightarrow \quad \Delta H^\phi = 2 \times 10^4 \times R = 2 \times 10^4 \times 8.314$ J/mol
$= 166.28$ kJ/mol

19. (141.34) Form the plot when, $\frac{10^4}{T} = 10$

[$T = 1000$ K]

$\ln\left(\frac{p_z}{1}\right) = -3$

$\Delta G^\circ = -RT \ln K \Rightarrow \Delta H^\phi - T\Delta S^\phi = -RT \ln \frac{P_z}{p^\phi}$

Substituting in following equation :

$\ln\left(\frac{p_z}{1}\right) = -\frac{\Delta H^\ominus}{RT} + \frac{\Delta S^\ominus}{R}$

We get, $-3 = -\frac{2 \times 10^4 \times R}{R \times 1000} + \frac{\Delta S^\ominus}{R}$

$\Rightarrow \quad \Delta S^\ominus = 17R \Rightarrow \Delta S^\ominus = 17 \times 8.314$ J/K-mol

$\Rightarrow \quad \Delta S^\ominus = 141.34$ J/K-mol

20. (935.00) $SnO_2(s) + C(s) \longrightarrow Sn(s) + CO_2(g)$

$\Delta_r H^\circ = [-394] - [-581] = 187$ kJ/mole
$= 187 \times 10^3$ J/mol

$\Delta_r S^\circ = [52 + 210] - [56 + 6] = 200 \text{ JK}^{-1}\text{ mol}^{-1}$

$T = \frac{\Delta_r H^\circ}{\Delta_r S^\circ} = \frac{187 \times 10^3}{200} = 935$ K

21. (557)

$\Delta H = \Delta U + \Delta(PV) = \Delta U + V\Delta P \quad (\because \Delta V = 0)$

or

$\Delta U = \Delta H - V\Delta P = -560 - [1(40 - 70) \times 0.1]$

$= -560 + 3 = -557 \text{ kJ mol}^{-1}$

So, the magnitude is **557** kJ mol⁻¹.

22. (115.87) $\quad C_p - C_v = R$

$\Rightarrow \quad C_p = 12.48 + 8.31 = 20.794 \text{ J mol}^{-1}\text{ K}^{-1}$

$\gamma = \frac{C_p}{C_v} = \frac{20.794}{12.48} = 1.67$

For a reversible adiabatic process,
$T_1V_1^{\gamma-1} = T_2V_2^{\gamma-1}$

$$\Rightarrow T_2 = T_1\left(\frac{V_1}{V_2}\right)^{\gamma-1} \Rightarrow T_2 = 300\left(\frac{1.25}{2.50}\right)^{1.67-1}$$

$\Rightarrow T_2 = 188.55$ K
Number of moles of gas,

$$n = \frac{PV_1}{RT_1} \Rightarrow n = \frac{1\times 1.25}{0.0821\times 300} = 0.05$$

Enthalpy change at constant pressure,
$\Delta H = n{\cdot}C_p{\cdot}\Delta T$
$= 0.05 \times 20.794 \times (300 - 188.55)$
$= 115.87$ Joule.

23. (319.1)
100 g of glucose = 1560 kJ

$$\text{Energy utilised in body} = \frac{50}{100}\times 1560 = 780 \text{ kJ}$$

Energy to be given out = 1560 – 780 = 780 kJ
Enthalpy of evaporation of water = 44 kJ/mole = 44 kJ/18 g of water [1 mole H_2O = 18 g water]
Hence, amount water to be perspired to avoid

$$\text{storage of energy} = \frac{18}{44}\times 780 = \mathbf{319.1\ g}$$

24. extensive (because its value depends on quantity of substance)
25. isolated
26. True;

Heat capacity	Monoatomic	Diatomic
C_v	$3R/2$	$5R/2$
C_p	$5R/2$	$7R/2$

Thus, the heat capacity of diatomic gas is higher than that of a monoatomic gas.

27. True; It only tells that if the heat gained by one end would be exactly equal to heat lost by the other. It does not predict the direction.

28. (a, b, d) From state I to II (Reversible isothermal expansion):
$T \to$ constant, $\Delta V \to$ +ve, $\Delta S \to$ +ve, $\Delta H \to 0$
$\Delta P \to$ –ve
From state II to III (Reversible adiabatic expansion) :
$q \to 0$, $\Delta V \to$ +ve, $\Delta S \to$ constant
$\Delta H \to$ –ve, $\Delta P \to$ –ve, $\Delta T \to$ –ve
∴ Plots (a), (b), (d) are correct while (c) is wrong as from state II to III, H is decreasing.

29. (a, b, c)
P-V work done is applicable for reversible isobaric as well as isothermal and adiabatic process.

$$w = -\int P_{ext}\cdot dV$$

For van der Waals equation,

$$P_{ext} = P = \left(\frac{RT}{v-b} - \frac{a}{v^2}\right)$$

$$w = -\int dv\left(\frac{RT}{v-b} - \frac{a}{v^2}\right) \quad \text{..... (i)}$$

Equation (i) is not applicable to irreversible process. Therefore work done is calculated assuming pressure is constant throughout the process.

30. (b, c)
A– C ⇒ isochoric process
A– B ⇒ isothermal process
B– C ⇒ isobaric process

(a) $q_{AC} = \Delta U_{AC} = nC_{V,m}(T_2 - T_1) = \Delta U_{BC}$

$$W_{AB} = -nRT_1 \ln\left(\frac{V_2}{V_1}\right) \text{ (pressure is not constant)}$$

(b) $W_{BC} = -P_2(V_1 - V_2) = P_2(V_2 - V_1)$
$q_{BC} = \Delta H_{BC} = nC_{P,m}(T_2 - T_1) = \Delta H_{AC}$
(c) $\Delta H_{CA} = nC_{P,m}(T_1 - T_2)$
(d) $\Delta U_{CA} = nC_{V,m}(T_1 - T_2)$
$\Delta H_{CA} < \Delta U_{CA}$ since both are negative ($T_1 < T_2$) and $C_{p,m} > C_{V,m}$

31. (a, b, c)
(a)

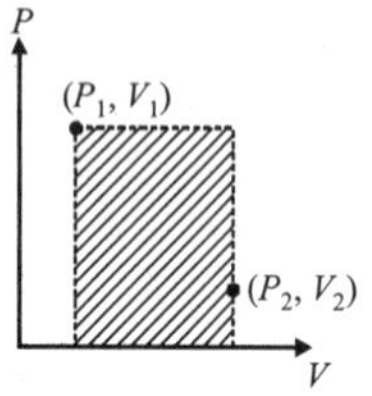

During irreversible compression, maximum work is done on the gas (corresponding to shaded area) when $P_1 = P_2$
(d) When $T_1 = T_2 \Rightarrow \Delta U = nC_V\Delta T = 0$
In reversible adiabatic expansion, $T_2 < T_1$.
∴ ΔT = –ve and also ΔU = –ve
(b) In free expansion, $P_{ext} = 0$, ∴ $W = 0$
From I[st] law of thermodynamics,
$\Delta U = q + W$
∴ $\Delta U = q$
If expansion is carried out isothermally, $\Delta U = 0$
Hence $q = 0$
∴ It is adiabatic process.
If carried out adiabatically ($q = 0$), ∴ $\Delta U = 0$
∴ It is an isothermal process.
(c) During adiabatic expansion, the final temperature is less than the initial temperature. Therefore, the final volume in adiabatic expansion will also be less than the final volume in isothermal expansion. This can be graphically shown as:

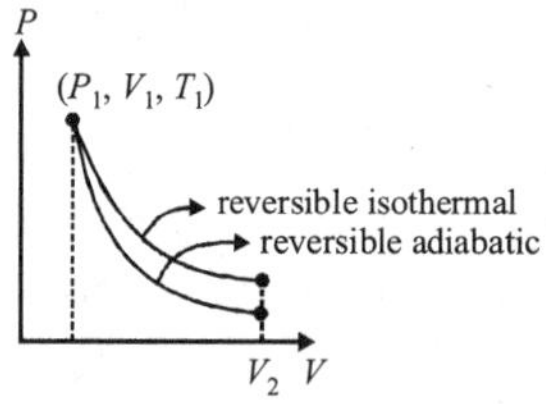

The magnitude of work done by the gas is equal to the area under the curve. As seen from the figure, the area under curve in reversible isothermal is more. Hence, the magnitude of work done is lesser in adiabatic reversible expansion as compared to the corresponding work in isothermal expansion.

32. (a, b, c)
Since the vessel is thermally insulated, $q = 0$
Further since, $P_{ext} = 0$, so $w = 0$, hence $\Delta U = 0$
Therefore, $\Delta T = 0$, $T_2 = T_1$, and $P_2V_2 = P_1V_1$
However, the process is adiabatic irreversible, so we can't apply $P_2V_2^{\gamma} = P_1V_1^{\gamma}$.

33. (a, c, d)
$T_1 = T_2$ because process is isothermal.
Work done in adiabatic process is less than in isothermal process because area covered by isothermal curve is more than the area covered by the adiabatic curve.

In adiabatic process expansion occurs by using internal energy, hence, it decreases while in isothermal process temperature remains constant, that's why no change in internal energy.

34. (a, c)
$\Delta S_{X\to Z} = \Delta S_{X\to Y} + \Delta S_{Y\to Z}$ [Entropy is a state function, hence additive]
$w_{X\to Y\to Z} = w_{X\to Y}$ [Work done in $Y \to Z$ is zero because it is an isochoric process].

35. (a, b) Mass independent properties (molar conductivity and electromotive force) are intensive properties. Resistance and heat capacity are mass dependent, hence extensive properties.

36. (a, d) Internal energy and molar enthalpy are state functions. Work (reversible or irreversible) is a path function.

37. (b, d) Properties independent of mass are intensive properties. Hence, (b) and (d) which are independent of mass are the obvious choices.

38. A- (r, t); B -(p, q, s); C -(p, q, s); D- (p, q, s, t)
(A) → r, t
$H_2O(l) \rightleftharpoons H_2O(s)$
It is at equilibrium at 273 K and 1 atm.
So, ΔS_{sys} is negative.
As it is equilibrium process, so $\Delta G = 0$.
(B) → p, q, s
Expansion of 1 mole of an ideal gas in vacuum under isolated condition
Hence, $w = 0$ and $q_p = C_p dT$ (∵ $dT = 0$)
⇒ $q = 0$
$\Delta U = C_v dT$ (∵ $dT = 0$) $\Delta U = 0$
(C) → p, q, s
Mixing of two ideal gases at constant temperature
Hence, $\Delta T = 0$
∴ $q = 0$; $\Delta U = 0$
also $w = 0$ ($\Delta U = q + w$)
(D) → p, q, s, t
Reversible heating and cooling of gas follows same path; also initial and final position is same.

Hence, $\left.\begin{matrix} q=0 \\ w=0 \end{matrix}\right\}$ Path same $\left.\begin{matrix} \Delta U=0 \\ \Delta G=0 \end{matrix}\right\}$ State function

39. (a) Stability of free radical $\propto \dfrac{1}{\text{Bond energy}}$

(P) 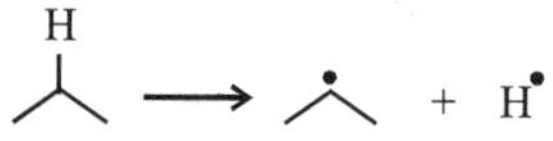

2° free radical

(Q) $Ph - CH_2 - H \longrightarrow Ph - \dot{C}H_2 + H^{\bullet}$
Most stable due to resonance

(R) $CH_2 = CH - H \longrightarrow CH_2 = \dot{C}H + H^{\bullet}$
Less stable

(S) $CH \equiv C - H \longrightarrow CH \equiv \dot{C} + H^{\bullet}$
More % s-Character decreases stability of free radical

Q require least BDE and S required maximum BDE.
So, order of BDE is Q < P < R < S.

40. (d)

$$CH_4 \longrightarrow {}^{\bullet}CH_3 + H^{\bullet} \quad \Delta H = 105 \text{ kcal/mol}$$

$$Cl_2 \longrightarrow Cl^{\bullet} + Cl^{\bullet} \quad \Delta H = 58 \text{ kcal/mol}$$

$$Cl^{\bullet} + {}^{\bullet}CH_3 \longrightarrow CH_3-Cl \quad \Delta H = -85 \text{ kcal/mol}$$

$$Cl^{\bullet} + H^{\bullet} \longrightarrow HCl \quad \Delta H = -103 \text{ kcal/mol}$$

$$CH_4 + Cl_2 \longrightarrow CH_3-Cl + HCl \quad \Delta H = -25 \text{ kcal/mol}$$

Initiation step is endothermic, hence option (a) is wrong.
Propagation step involving ${}^{\bullet}CH_3$ formation is endothermic, hence option (b) is wrong.
Propagation step involving CH_3Cl formation is exothermic, hence option (c) is wrong.
So, overall raction is exothermic with $\Delta H^{o} = -25$kcal/mol, hence option (d) is correct.

41. (c) $K \rightarrow L \Rightarrow V$ increasing at constant P
Hence, T increases (Heating).
$L \rightarrow M \Rightarrow P$ decreasing at constant V
Hence, T decreases (Cooling),
$M \rightarrow N \Rightarrow V$ decreasing at constant P
Hence, T decreases (Cooling),
$N \rightarrow K \Rightarrow P$ increasing at constant V
Hence, T increases (Heating).

42. (b) L to M and N to K, both are having constant volume, therefore, these processes are isochoric.

43. (a) Statement 1 is true because it is not possible to convert whole of heat to work. For such a conversion, we need an efficiency of 100% but so far, we have not been able to get such a machine (carnot engine).
Statement 2 is true because it is not possible to convert the whole of heat absorbed from a reservoir into work. Some of the heat is always given to the sink.
Also, statement 2 is correct explanation for statement 1.
Thus, the correct choice is option (a).

44. (b) Assertion : For isothermal expansion, $\Delta T = 0 \Rightarrow \Delta U = 0$
For an ideal gas, work done against vacuum is zero, *i.e.* $W = 0$
Now, $\Delta Q = \Delta U + W \Rightarrow \Delta Q = 0$
Thus, assertion is correct.
Reason : By kinetic theory of ideal gases, the volume occupied by the molecules of an ideal gas is zero.
Thus, reason is correct, but it is not the correct explanation of the assertion.

45. For adiabatic process, $W = P(V_2 - V_1)$
Here $P_1 = 1$ bar, $P_2 = 100$ bar, $V_1 = 100$ mL, $V_2 = 99$ mL; For adiabatic process, $q = 0 \therefore \Delta U = W$
$\Delta U = q + W = q - P(V_2 - V_1)$
$= 0 - 100\,(99 - 100) =$ **100 bar mL**
$\Delta H = \Delta U + \Delta(PV) = \Delta U + (P_2V_2 - P_1V_1)$
$= 100 + [(100 \times 99) - (1 \times 100)]$
$= 100 + (9900 - 100) =$ **9900 bar mL**

46. Helium molecule is monoatomic so it has just three degrees of freedom corresponding to the three translational motion at all temperature and hence C_v value is always 3/2 R. Hydrogen molecule is diatomic which are not rigidly held, so they vibrate about a well defined average separation. For hydrogen molecule, we have rotational and vibrational motion both besides translational motion. Contribution from vibrational motion is not appreciable at low temperature but increases from 0 to R on raising temperature.

47. (i)

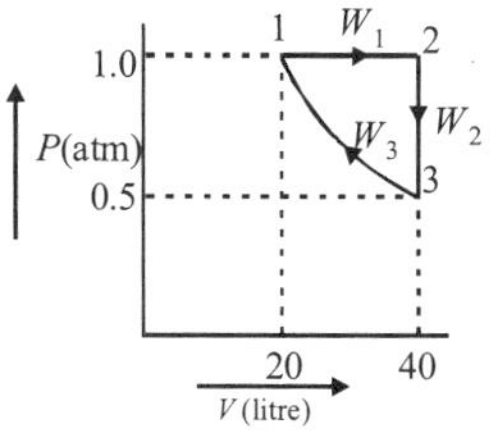

(ii) $PV = nRT$

$$T = \frac{PV}{nR} = \frac{0.5 \times 40}{2 \times 0.082} = 121.95 \text{ K}$$

Total work $(W) = W_1 + W_2 + W_3$

$$= -P\Delta V + 0 + 2.303nRT \log \frac{V_2}{V_1}$$

$$= -1 \times 20 + 2.303 \times 2 \times 0.082 \times 121.95 \log 2$$

$= -20 + 13.87 =$ **– 6.13 L atm**
Since the system has returned to its initial state i.e. the process is cyclic, so $\Delta U = 0$
$\Delta U = q + W = 0,$
so $q = -W = -(-6.13)$ L.atm $=$ **620.7 J**
In a cyclic process, heat absorbed is completely converted into work.
(iii) Entropy is a state function and since the system has returned to its initial state, so $\boldsymbol{\Delta S = 0}$. Similarly $\boldsymbol{\Delta H = 0}$ and $\boldsymbol{\Delta U = 0}$ for the same reason.

48. If heat is absorbed at constant pressure, then $q_p = \Delta E - (-P\Delta V)$ or $q_p = E_2 - E_1 - [-P(V_2 - V_1)]$
or $q_p = (E_2 + PV_2) - (E_1 + PV_1) = H_2 - H_1 = \Delta H$

49. (c) The standard enthalpy of the combustion of glucose can be calculated by the eqn.
$C_6H_{12}O_6(s) + 6O_2(g) \rightarrow 6CO_2(g) + 6H_2O(l)$
$\Delta H_C = 6 \times \Delta H_f(CO_2) + 6 \times \Delta H_f(H_2O) - \Delta H_f(C_6H_{12}O_6)$
$\Delta H^{o} = 6(-400) + 6(-300) - (-1300)$
$= -2900$ kJ/mol
For one gram of glucose, enthalpy of combustion

$$\Delta H^{o} = -\frac{2900}{180} = -16.11 \text{ kJ/g}$$

50. (b) The species in its elemental form has zero standard molar enthalpy of formation at 298 K. At 298K, Cl_2 is gas while Br_2 is liquid.

51. (b) ΔH_f° is the enthalpy change when 1 mole of the substance is formed from its elements in their standard states.
In (a) carbon is present in diamond however, standard state of carbon is graphite. Again, in (d) CO (g) is involved so it can't be the right option. Further in (c) 2 moles of NH_3 are generated. Hence, the correct option is (b).

52. (b)

$$CO_2(g) + H_2(g) \longrightarrow CO(g) + H_2O(g), \Delta H = ?$$

$$\Delta H = \sum \Delta H_f \text{ (Product)} - \sum \Delta H_f \text{ (reactant)}$$

Given, $\Delta H_f CO_2(g) = -393.5$ kJ/mol
$\Delta H_f CO(g) = -110.5$ kJ/mol
$\Delta H_f H_2O(g) = -241.8$ kJ/mol

$$\therefore \Delta H = [\Delta H_f CO(g) + \Delta H_f H_2O(g)] - [\Delta H_f CO_2(g) + \Delta H_f H_2(g)]$$

$$= [-110.5 + (-241.8)] - [-393.5 + 0] \quad [\because \Delta H_f(H_2)(g) = 0]$$

$= 41.2$ kJ mol^{-1}

53. (d) $\Delta H = \Delta E + \Delta nRT$ For $\Delta H \neq \Delta E, \Delta n \neq 0$
Where Δn = no. of moles of gaseous products – no. of moles of gaseous reactants
(a) $\Delta n = 2 - 2 = 0$
(b) $\Delta n = 0$ ($\because$ they are either in solid or liquid state)
(c) $\Delta n = 1 - 1 = 0$ ($\because$ C is in solid state)
(d) $\Delta n = 2 - 4 = -2$
$\therefore$ (d) is correct answer

54. (a) Heat capacity at constant volume $(q_v) = \Delta E$
Heat capacity of constant pressure $(q_p) = \Delta H$
$\Delta H = \Delta E + \Delta nRT$ or $\Delta H - \Delta E = \Delta nRT$
Δn = no. of moles of gaseous products – no. of moles of gaseous reactants
$= 12 - 15 = -3$
$\Delta H - \Delta E = -3 \times 8.314 \times 298$ J $= -7.43$ kJ.

55. (90.39) $2Hg(g) + O_2(g) \rightarrow 2HgO(s)$
$\Delta_r H^{\circ} = 2\Delta_f H^{\circ}$ (HgO, s) $- 2\Delta_f H^{\circ}$(Hg, g) $- \Delta_f H^{\circ}(O_2, g)$
$= 2\Delta_f H^{\circ}$ (HgO, s) $- 2\Delta_f H^{\circ}$ (Hg, g) ... (i)
$[\because \Delta_f H^{\circ}(O_2, g) = 0]$
Now, $\Delta_r H^{\circ}$ is the heat evolved by bomb calorimeter due to the occurrence of the reaction at constant volume.
$\therefore -(Q_v)_r = \Delta_r U$
$\therefore \Delta_r H = \Delta_r U + \Delta n_g RT = -C\Delta T + \Delta n_g RT$
[where C – the heat capacity of calorimeter = 20 kJ/K at 298 K]
$= -[20(312.8 - 298)] - 3\,RT$
$= -296 - 3 \times 8.3 \times 10^{-3} \times 298$ kJ $= -303.42$ kJ
Hence, from eq. (i)
$-303.42 = 2\Delta_f H^{\circ}$ (HgO, s) $- 2 \times 61.32$
or, $2\Delta_f H^{\circ}$ (HgO, s) $= -180.78$ kJ
$\therefore$ Standard Molar Enthalpy of formation of HgO

$$= \frac{-180.78}{2} = -90.39 \text{ kJ} \Rightarrow |X| = 90.39$$

56. (–2035)
The chemical reaction for combustion of diborane is
$B_2H_6(g) + 3O_2(g) \longrightarrow B_2O_3(s) + 3H_2O(g), \Delta H = ?$
For this the enthalpy change can be calculated in the following way.

$$\Delta_c H = [\Delta_f H_{B_2O_3(s)} + 3\Delta_f H_{H_2O(g)}]; - \Delta_f H_{B_2H_6(g)} \quad (\because \Delta_f H_{O_2} = 2)$$

$\Delta_f H_{H_2O(g)}$ can be obtained by adding $\Delta_r H_{H_2O(l)}$ and $\Delta_r H_{H_2O(g)}$, *i.e.* $-286 + 44 = -242$ kJ mol^{-1}

$\Delta H = [-1273 + 3 \times (-242)] - 36 \text{ kJ mol}^{-1}$
$= -1273 - 726 - 36$
$= \mathbf{-2035 \text{ kJ mol}^{-1}}$

57. (309.16)
Given $S(s) + 3F_2(g) \rightarrow SF_6(g)$; $\Delta H = -1100$ kJ .....(i)
$S(s) \rightarrow S(g)$; $\Delta H = 275$ kJ .....(ii)
$1/2\, F_2(g) \rightarrow F(g)$; $\Delta H = 80$ kJ .....(iii)
To get $SF_6(g) \rightarrow S(g) + 6F(g)$ we can proceed as (ii) + 6 × (iii) - (i)
$\therefore SF_6(g) \rightarrow S(g) + 6F(g)$; $\Delta H = 1855$ kJ
Thus, average bond energy for S-F bond
$= \frac{1855}{6} = \mathbf{309.16 kJ}$

58. (–2091.32)

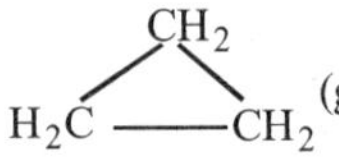

(g) $\rightarrow CH_3CH = CH_2(g)$;
$\Delta H = -33.0$ kJ ...(i)
$C(s) + O_2(g) \rightarrow CO_2(g)$; $\Delta H = -393.5$ kJ ...(ii)
$H_2(g) + 1/2O_2(g) \rightarrow H_2O(l)$; $\Delta H = -285.8$ kJ ...(iii)
$3C(s) + 3H_2(g) \rightarrow CH_3–CH = CH_2(g)$;
$\Delta H = 20.42$ kJ ...(iv)
The required reaction is

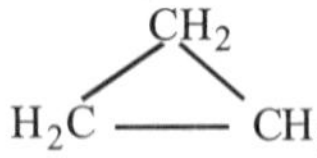

$+ \frac{9}{2}O_2 \longrightarrow 3CO_2 + 3H_2O;\ \Delta H = ?$

To calculate the value of ΔH follow the following steps.
(i) + 3 × (ii) + 3 × (iii)] – (iv):

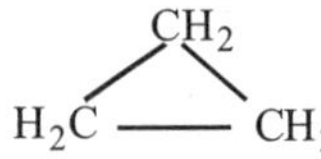

$+ (9/2)O_2 \rightarrow 3CO_2 + 3H_2O$;
$\Delta H = \mathbf{-2091.32 \text{ kJ}}$

59. (–266)
The required thermochemical equation is

$C(g) + 4H(g) + O(g) \longrightarrow CH_3OH(l);\ \Delta H_f = ?$

$$\Delta H_f = \left[\Delta H_{C(s) \rightarrow C(g)} + 2\Delta H_{H-H} + \frac{1}{2}\Delta H_{O=O}\right]$$
$$-\left[3\Delta H_{C-H} + \Delta H_{C-O} + \Delta H_{O-H} + \Delta H_{vap.CH_3OH}\right]$$
$= [715 + 2 \times 436 + 249] - [3 \times 415 + 356 + 463 + 38]$
$= \mathbf{-266 \text{ kJ mol}^{-1}}$

60. (–152) Standard enthalpy of hydrogenation of cyclohexene (-119kJ mol^{-1}) means the enthalpy of hydrogenation of one C = C double bond. Now benzene has three C = C double bonds, the enthalpy of the reaction would be = 3 × (–119) = -357 kJ mol^{-1}

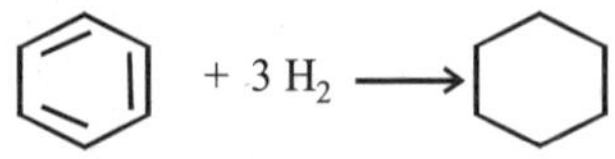

Actual enthalpy of the reaction can be evaluated as follows.

$\Delta H_{(Reaction)} = \Delta H_f^\circ(\text{Product}) - \Delta H_f^\circ(\text{Reactants})$
$= -156 - (49 + 0) = -205 \text{ kJ mol}^{-1}$

$\therefore$ Resonance energy = $\Delta H_{Exp} - \Delta H_{cal}$
$= -357 - (-205) = \mathbf{-152 kJ\ mol^{-1}}$

61. (–72) $nCH_2 = CH_2 \rightarrow (CH_2 - CH_2)_n$
During the polymerisation of ethylene, one mole of ethylene breaks *i.e.* one C = C double bond breaks and the two CH_2 – groups are linked with C – C single bonds thus, forming three single bonds (two single bonds are formed when each CH_2 – group of ethylene links with one CH_2 – group of another ethylene molecule). But in the whole unit of polymer, number of single C–C bonds formed/mole of ethylene is 2.

$\text{+}CH_2 - CH_2\text{+}\text{+}CH_2 - CH_2\text{+}\text{+}CH_2 - CH_2\text{+}\text{+}CH_2 - CH_2\text{+}$

e.g. Number of single bonds formed by 4 moles of ethylene = 8
Energy released = Energy due to formation of 2 C–C single bonds
= 2 × 331 = 662 kJ/mol of ethylene
Energy absorbed = Energy due to dissociation of 1 C=C double bond
= 590 kJ/mol of ethylene
$\therefore$ Enthalpy of polymerisation/mol of ethylene or
$\Delta H_{polymerisation} = 590 - 662$ kJ/mol
$= \mathbf{-72 \text{ kJ/mole}}$

62. (–55.7) From the given data, we can write:
(*i*) $H_2 + \frac{1}{2}O_2 \rightarrow H_2O; \Delta H_1 = -285.8$ kJ/mol
(*ii*) $CH_4 + 2O_2 \rightarrow CO_2 + 2H_2O$;
$\Delta H_2 = -890$ kJ mol
(*iii*) $C_2H_6 + \frac{7}{2}O_2 \rightarrow 2CO_2 + 3H_2O$;
$\Delta H_3 = -1560$ kJ/mol
(*iv*) $C(s) + O_2 \rightarrow CO_2; \Delta H_4 = -393.5$ kJ/mol
(*v*) $3C(s) + 4H_2 \rightarrow C_3H_8(g); \Delta H_5 = -103.8$ kJ/mol
The required reaction is $C_3H_8(g) + H_2(g) \rightarrow C_2H_6(g) + CH_4(g), \Delta H = ?$
It can be obtained by the following calculations.
3 × (*iv*)–(*v*) + 5(*i*) – (*iii*) – (*ii*)
In other words, $\Delta H = 3\Delta H_4 - \Delta H_5 + 5\Delta H_1 - \Delta H_2 - \Delta H_3$
$\therefore \Delta H = 3(-393.5) - (-103.8) + 5(-285.8) + 890 + 1560$
$= -2609.5 + 2553.8 = \mathbf{-55.7 \text{ kJ/mol}}$

63. (50.90) Combustion of C_2H_4 and CH_4 takes place as follows:
$C_2H_4 + 3O_2 \rightarrow 2CO_2 + 2H_2O$
1 vol. 2 vol.
$CH_4 + 2O_2 \rightarrow CO_2 + 2H_2O$
1 vol. 1 vol.
Let the vol. of CH_4 in mixture = x L
$\therefore$ Vol. of C_2H_4 in the mixture = $(3.67 - x)$ L
Vol. of CO_2 produced by x L of CH_4 = x L and Vol. of CO_2 produced by $(3.67 - x)$ L of C_2H_4 = 2 $(3.67 - x)$ L
$\therefore$ Total vol. of CO_2 produced = $x + 2(3.67 - x)$
or $6.11 = x + 2(3.67 - x)$ or $x = 1.23$ L
$\therefore$ Vol. of CH_4 in the mixture = 1.23 L
and Vol. of C_2H_4 in the mixture = 3.67 – 1.23 = 2.44 L
Vol. of CH_4 per litre of the mixture
$= \frac{1.23}{3.67} = 0.335$ L
Vol. of C_2H_4 per litre of the mixture
$= \frac{2.44}{3.67} = 0.665$ L
Now, we know that volume of 1 mol. of any gas at
25 °C (298 K) $= \frac{22.4 \times 298}{273} = 24.45$ L
[$\because$ Volume at NTP = 22.4 L]
Heat evolved due to combustion of 0.335 L of CH_4
$= -\frac{0.335 \times 891}{24.45} = -12.20$ kJ
[given, heat evolved by combustion of 1 L = 891 kJ]
Similarly, heat evolved due to combustion of 0.665
L of $C_2H_4 = -\frac{0.665 \times 1423}{24.45} = -38.70$ kJ
$\therefore$ Total heat evolved = 12.20 + 38.70 = **50.90 kJ**

64. (–121) The required reaction is
$C_6H_{10}(g) + H_2(g) \rightarrow C_6H_{12}(g), \Delta H_1 = ?$ ...(1)
Cyclohexene Cyclohexane
The given facts can be written as :
$H_2(g) + ½O_2(g) \rightarrow H_2O(l)$,
$\Delta H_2 = 241$ kJ/mol ...(2)
$C_6H_{10}(g) + \frac{17}{2}O_2(g) \rightarrow 6CO_2(g) + 5H_2O$;
$\Delta H_3 = -3800$ kJ/mol ...(3)
$C_6H_{12}(g) + 9O_2(g) \rightarrow 6CO_2(g) + 6H_2O$,
$\Delta H_4 = -3920$ kJ/mol ...(4)
The required reaction (1) can be obtained by adding equations (2) and (3) and subtracing (4) from the sum of (2) and (3).
$C_6H_{10}(g) + H_2(g) \rightarrow C_6H_{12}(g)$
$\Delta H_1 = (\Delta H_2 + \Delta H_3) - \Delta H_4$
$= [-241 + (-3800)] - (-3920)$
$= (-241 - 3800) - (-3920)$
$= -4041 + 3920 = \mathbf{-121 \text{ kJ/mole}}$

65. (3.94)
$Fe_2O_3 + 2Al \rightarrow 2Fe + Al_2O_3$
2 × 56 + 48 = 160 2 × 27 = 54
Heat of reaction = 399 – 199 = 200 kcal [Al and Fe are in their standard states]
Total weight of reactants = 160 + 54 = 214 g
$\therefore$ Fuel value/gram $= \frac{200}{214} = 0.9346$ kcal/g
Volume of Al $= \frac{54}{2.7} = 20$ cc
Volume of $Fe_2O_3 = \frac{160}{5.2} = 30.77$ cc
Total volume = 20 + 30.77 = 50.77 cc
$\therefore$ Fuel value per cc $= \frac{200}{50.77} = \mathbf{3.94 \text{ kcal/cc}}$

66. (–372.0) The required chemical reaction.
$2C_2H_6 + 7O_2 \longrightarrow 4CO_2 + 6H_2O; \Delta H = x$
Note that since 2 moles of ethane are reacting, the ΔH of the reaction will be ½ x.
The thermochemical equations for the given data are written as below.

(i) $C(s)+O_2(g) \longrightarrow CO_2(g); \Delta H=-94.1$ kcal
(ii) $H_2(g)+½O_2(g) \longrightarrow H_2O(g);$
$\Delta H=-68.3$ kcal
(iii) $2C(s)+3H_2(g) \longrightarrow C_2H_6(g);$
$\Delta H=-21.1$ kcal
We know that $\Delta H = H_{Products} - H_{Reactants}$

$\Delta H = 4\Delta H_{CO_2} + 6\Delta H_{H_2O} - (2\Delta H_{C_2H_6} + 7\Delta H_{O_2})$

$\Delta H=4\times(-94.1)+6\times(-68.3)-(2\times(-21.1)+0)$
$=-376.4-409.8+42.2=-744.0$ kcal/2 mole of ethane
= –372.0 kcal/mole of ethane

67. (–22)

Bond	H – H	Cl – Cl	H – Cl
ΔH disso.	104 kcal	58 kcal	103 kcal

Formation of hydrogen chloride can be represented as
$H-H+Cl-Cl \rightarrow 2H-Cl$
Thus, the reaction involves
Cleavage of one H – H bond, ΔH = 104 kcal
Cleavage of one Cl – Cl bond, ΔH = 58 kcal
Formation of two H – Cl bonds,
$\Delta H = 2\times(-103)$ kcal
$\therefore \Delta H$ of the reaction $= (104+58)-2(103)$
$= 162-206 = -44$ kcal
Now, since the enthalpy of formation of a compound is the change in heat content accompanied in the formation of one mole of the compound, the enthalpy of formation of HCl gas

$= -\frac{44}{2} =$ **– 22 kcal**

68. (41.104) The given data can be written as follows

(i) $H_2(g)+\frac{1}{2}O_2(g)\rightarrow H_2O(l);$
$\Delta H=-68.3$ kcal

(ii) $C_2H_2(g)+\frac{5}{2}O_2(g)\rightarrow H_2O(l)+2CO_2(g);$
$\Delta H=-310.6$ kcal
(iii) $C_2H_4(g)+3O_2(g)\rightarrow 2H_2O(l)+2CO_2(g);$
$\Delta H=-337.2$ kcal
The required thermochemical equation is
$C_2H_2(g)+H_2(g)\rightarrow C_2H_4(g)$
The required equation can be obtained by subtracting equation (*iii*) from the sum of equations (*i*) and (*ii*), thus ΔH of the required equation can be calculated as below.
$\Delta H=[-68.3+(-310.6)]-(-337.2)$
$=[-68.3-310.6]+337.2=-378.9+337.2$
$=-41.7$ kcal
ΔE, the heat of reaction for the hydrogenation of acetylene at constant volume is given by :
$\Delta E = \Delta H - \Delta nRT$
Here Δn = Moles of the gaseous products – Moles of the gaseous reactants
$=1-(1+1)=-1$
Substituting the values of ΔH, Δn, R and T in
$\Delta E = \Delta H - \Delta nRT = -41.7-(-1\times 2\times 10^{-3}\times 298)$

$\left[\because R = 2\text{cal / degree / mole} = 2\times 10^{-3}\text{ kcal / deg/ mole}\right]$

$=-41.7+0.596=$ **41.104 kcal**

69. (54.20) The required equation is :
$2C(s)+H_2(g)\rightarrow C_2H_2; \quad \Delta H=?$
Write the thermochemical equations for the given data

(i) $C_2H_2(g)+\frac{5}{2}O_2(g)\rightarrow 2CO_2(g)+H_2O(l);$
$\Delta H=-310.62$ kcal
(ii) $C(s)+O_2(g)\rightarrow CO_2(g)$; $\Delta H=-94.05$ kcal

(iii) $H_2(g)+\frac{1}{2}O_2(g)\rightarrow H_2O(l);$
$\Delta H=-68.32$ kcal
For getting the above required reaction, we will have to
(a) Bring C_2H_2 in the product that can be done by reversing the equation (i) to give equation (iv).
(b) Multiply equation (ii) by 2 to get 2C atoms in the reactants and thus, equation (v) is obtained.
(c) Keep equation (iii) as such.
(d) Add equations (iv), (v) and (iii).

(iv) $2CO_2+H_2O\rightarrow C_2H_2+\frac{5}{2}O_2;$
$\Delta H=310.62$ kcal
(v) $2C+2O_2\rightarrow 2CO_2; \Delta H=-188.10$ kcal

(iii) $H_2+\frac{1}{2}O_2\rightarrow H_2O;$
$\Delta H=-68.32$ kcal
On adding,
$2C+H_2\rightarrow C_2H_2; \quad \Delta H=54.20$ kcal
Hence, the standard heat of formation of $C_2H_2(g)$ = **54.20 kcal**

Alternate Solution:
(54.20) The standard state formation reaction of $C_2H_2(g)$ is

$2\,C(g)+H_2(g)\rightarrow C_2H_2(g),\ \Delta H^o_f=?$

$\Delta H^o_r\ \sum \Delta H^o_{comb}$ (reactant – $\sum \Delta H^o_{comb}$ (product)
$=-2\times 94.05-68.32-(-310.62)$

$\Delta H^o_f\,(C_2H_2)=54.2$ K cal

70. (101.19)
The required reaction in terms of dissociation energy is
$OH(g)\rightarrow O(g)+H(g); \quad \Delta H=?$
This equation can be achieved by (a) reversing the equation (*i*), (b) dividing equation (*ii*) and (*iii*) each by 2, and (c) adding the three resulting equations.

$OH(g)\rightarrow \frac{1}{2}H_2(g)+\frac{1}{2}O_2(g);\ \Delta H=+10.06$ kcal
[Reversing eq (i)]

$\frac{1}{2}H_2(g)\rightarrow H(g) \qquad \Delta H=-52.09$ kcal

$\left[\frac{1}{2}\text{Eq (ii)}\right]\ \frac{1}{2}O_2(g)\rightarrow O(g);\ \Delta H=-59.16$ kcal

$\left[\frac{1}{2}\text{Eq (iii)}\right]$

$OH(g)\rightarrow O(g)+H(g); \Delta H=-101.19$ kcal (adding)
Thus, one mole of OH(g) needs 101.19 kcal of energy to break into oxygen and hydrogen gaseous atoms. Hence, the bond energy of O–H bond is **101.19 kcal**.

71. zero; In a closed vessel, $\Delta V=0$
72. endothermic
73. (a, c) Enthalpy of formation is the enthalpy change for formation of 1 mole of substance from its elements present in the most stable natural form.
74. (b, c, d) All combustion reactions are exothermic in nature.
(b) Decomposition reactions are endothermic in nature.
(c) $C_2H_6(g)\longrightarrow C_2H_4(g)+H_2(g)$
More stable compound is converting into less stable compound. Thus, reaction is endothermic.
(d) Graphite $\longrightarrow$ Diamond
More stable allotrope is converting into less stable allotrope. Thus, reaction is endothermic.
75. (c)
(P) In physisorption energy is released and randomness decreases, $x(g) \rightarrow x(s)$
hence $\Delta S<0$, $\Delta H<0$
(Q) Diamond $\rightarrow$ Graphite
$\Rightarrow \Delta S>0,\ \Delta H<0$
Graphite has weaker van der waal forces of attraction between layers and hence, greater disorder and graphite is more stable in normal condition so energy is released.
(R) In denaturation of protein energy is absorbed and entropy increases hence,
$\Delta H>0$ [Heat absorption]
$\Delta S>0$ [relative bonds between polymer protein chain break]
(S) $CH_3CH=CH_2(g)\rightarrow \Delta(g)$
$\Delta H>0$ [Cyclopropane is unstable due to angular strain]
$\Delta S>0$ [$\because$ open chain has more freedom of rotation]
76. For ionisation $\Delta H_{Hydration} > \Delta H_{ionisation}$
Total hydration energy of Al^{3+} and $3Cl^-$ ions of $AlCl_3$ ($\Delta H_{hydration}$)
= (Hydration energy of Al^{3+} + 3 × Hydration energy of Cl^-)
$=[-4665+3(-381)]$kJ mole^{-1} $=-5808$ kJ mole^{-1}
This amount of energy is more than that required for the ionisation of Al into Al^{3+} (Ionisation energy of Al to Al^{3+}= 5137 kJ mol^{-1}). Due to this reason, $AlCl_3$ becomes ionic in aqueous solution. In aqueous solution it exists in ionic form as below
$AlCl_3+6H_2O \rightarrow [Al(H_2O)_6]^{3+}+3Cl^-$
$AlCl_3+aq.\rightarrow AlCl_3(aq.); \ \Delta H=?$
ΔH = Energy released during hydration – Energy used during ionisation
$=-4665-3\times 381+5137=-671$ kJ/mol
Thus, formation of ions will take place.
77. Combustion of CH_4 and C_4H_{10} takes place as follows
$CH_4+2O_2\rightarrow CO_2+2H_2O, \Delta H=-809$ kJ mol^{-1}
$C_4H_{10}+13/2O_2\rightarrow 4CO_2+5H_2O,$
$\Delta H=-2878$ kJ mol^{-1}
In order to get the same calorific output due to C_4H_{10},

the rate of supply of butane $=\frac{x\times 809}{2878}$
= 0.281 *x* L/hr

Rate of supply of oxygen $= 0.28\,x \times \frac{13}{2} \times 3$

$= \mathbf{5.481\,x\ L/hr}$

78. For C_3H_8: $3C + 4H_2 \rightarrow C_3H_8;\ \Delta H_1 = ?$

For C_2H_6: $2C + 3H_2 \rightarrow C_2H_6;\ \Delta H_2 = ?$

$\therefore\ \Delta H_1 = -[2(C–C) + 8(C–H)] + [3C_{s\rightarrow g} + 4(H–H)]$ ...(1)

$\therefore\ \Delta H_2 = -[1(C–C) + 6(C–H)] + [2C_{s\rightarrow g} + 3(H–H)]$ ...(2)

Let bond energy of C–C be x kcal and bond energy of C–H be y kcal

$\therefore$ By eq. (1) $\Delta H_1 = -(2x + 8y) + [3 \times 172 + 4 \times 104]$ ...(3)

$\Delta H_2 = -(x + 6y) + [2 \times 172 + 3 \times 104]$ ...(4)

Also given $C + O_2 \rightarrow CO_2$;

$\Delta H = -94.0$ k cal ...(5)

$H_2 + ½O_2 \rightarrow H_2O;\ \Delta H = -68.0$ kcal ...(6)

$C_2H_6 + (7/2)O_2 \rightarrow 2CO_2 + 3H_2O$;

$\Delta H = -372$ k cal ...(7)

$C_3H_8 + 5O_2 \rightarrow 3CO_2 + 4H_2O$;

$\Delta H = -530$ k cal ...(8)

By inspection method : $2 \times (5) + 3 \times (6) - (7)$ gives

$2C + 3H_2 \rightarrow C_2H_6;\ \Delta H_2 = -20$ k cal ...(9)

and $3 \times (5) + 4 \times (6) - (8)$ gives

$3C + 4H_2 \rightarrow C_3H_8;\ \Delta H_1 = -20$ k cal ...(10)

$\therefore$ By eq. (3), (4), (9) and (10)

$x + 6y = 676$

$2x + 8y = 956$

$\therefore\ x = 82$ k cal and $y = 99$ k cal

Bond energy of C–C bond = **82 k cal**

and Bond energy of C–H bond = **99 k cal**

79. (b) Given conditions are boiling conditions for water due to which system is in equilibrium.

$H_2O(l) \rightleftharpoons H_2O(g)$

$\Delta S_{total} = 0$

$\Delta S_{system} + \Delta S_{surroundings} = 0$

$\Delta S_{system} = -\Delta S_{surroundings}$

For process, $\Delta S_{system} > 0$

$\Delta S_{surroundings} < 0$

80. (b) $\Delta S = \frac{\Delta Q_{rev.}}{T}$; $75 = \frac{30 \times 10^3}{T}$

$\therefore\ T = 400$ K

81. (a)

C(graphite) $\rightarrow$ C(diamond) (Isothermally)

$\Delta_r G° = \Delta G°(\text{diamond}) - \Delta G°(\text{graphite})$

$= 2.9 - 0 = 2.9$ kJ mol^{-1}

Gibbs free energy is the maximum useful work, then

$-\Delta G° = w_{max} = \Delta PV$

$-2.9 \times 10^3 = -\Delta P \times 2 \times 10^{-6}$

$\Delta P = \frac{2.9 \times 10^3}{2 \times 10^{-6}} = 1.45 \times 10^9$ Pa $= 1.45 \times 10^9 \times 10^{-5}$ bar

$= 1.45 \times 10^4$ bar = 14500 bar

$P = \Delta P + P_0 = 14500 + 1 = 14501$ bar

82. (a) Since, liquid is passing into gaseous phase so entropy will increase and at 373 K during the phase transformation, it remains at equilibrium. So, $\Delta G = 0$.

83. (b) $A \rightleftharpoons B$

$\Delta G° = \Delta H° - T\Delta S°$, $\Delta G° = -2.303\ RT\log_{10} K$

$-2.303\ RT\log_{10} K = \Delta H° - T\Delta S°$

$\Rightarrow\ 2.303\ RT\log_{10} K = T\Delta S° - \Delta H°$

$\log_{10} K = \frac{T\Delta S° - \Delta H°}{2.303RT}$

$= \frac{298 \times 10 + 54.07 \times 1000}{2.303 \times 8.314 \times 298} = 10$

84. (– 14.6)

(i) $2Cu(s) + \frac{1}{2}O_2(g) \longrightarrow Cu_2O(s) : \Delta G°$

$= -78$ kJ/mol

(ii) $H_2(g) + \frac{1}{2}O_2(g) \longrightarrow H_2O(g),\ \Delta G°$

$= -178$ kJ/mol

(i) – (ii) then

$2Cu(s) + H_2O(g) \longrightarrow Cu_2O(s) + H_2(g)$

$\Delta G° = -78 + 178 = 100$ kJ/mol $= 10^5$ J/mol

Now for the above reaction

$$\Delta G = \Delta G° + RT\ln\left(\frac{P_{H_2}}{P_{H_2O}}\right)$$

To prevent the above reaction: $\Delta G \geq 0$

$$\Delta G° + RT\ln\left(\frac{P_{H_2}}{P_{H_2O}}\right) \geq 0$$

$$\Rightarrow\ 10^5 + 8 \times 1250\ln\left(\frac{P_{H_2}}{P_{H_2O}}\right) \geq 0$$

$\ln P_{H_2} \geq -10 + \ln P_{H_2O}$

Now, $P_{H_2O} = X_{H_2O} \times P_{total} = 0.01 \times 1 = 10^{-2}$

$\Rightarrow\ \ln P_{H_2} \geq -10 - 2\ln 10$

$\Rightarrow \ln P_{H_2} \geq -14.6$ (Given $\ln 10 = 2.3$)

$\therefore$ Minimum $\ln P_{H_2} = -14.6$

85. (b, d) $\Delta S_{surr} = \frac{\Delta H}{T_{surr}}$

For endothermic reaction, if T_{surr} increases, unfavourable change in entropy of the surroundings decreases.

For exothermic reaction, if T_{surr} increases, favourable change in entropy of the surroundings decreases.

86. A – p, r, s ; B – r, s ; C – t ; D – p, q, t

(A) $CO_2(s) \rightarrow CO_2(g)$

It is phase transition. The process is endothermic (sublimation). Gas is produced, so entropy increases.

(B) On heating $CaCO_3$ decomposes. So, process is endothermic. The entropy increases as gaseous product is formed.

(C) $2H\bullet \rightarrow H_2(g)$

Entropy decreases as number of gaseous particles decreases.

(D) The transition between different allotropes is considered as phase transtion.

White and red P are allotropes.

Due to polymeric nature of red P, its entropy is less than that of white P.

Red P is more stable than white.

So ΔH is –ve.

87. (i) $N_2O_4(g) \rightleftharpoons 2NO_2(g)$

Initially $p_{N_2O_4} = p_{NO_2} = 10$

Reaction quotient $= \frac{(p_{NO_2})^2}{p_{N_2O_4}} = \frac{100}{10} = 10$

$\Delta G° = 2\Delta G°_{f(NO_2)} - \Delta G°_{f(N_2O_4)} = 100 - 100 = 0$

$\Delta G = \Delta G° - 2.303\ RT\log K_p$

$= 0 - 2.303 \times 298\ \log 10 = \mathbf{5.70\ KJ/mol.}$

(ii) The negative value of ΔG indicates that the reaction is spontaneous and will lie in the right direction, (forward).

88. $\Delta G° = -2.303RT\log\frac{[\text{Product}]}{[\text{Reactant}]}$

Calculation of ΔG values :

Thus for the equilibrium $B \rightleftharpoons A$

$\Delta G_1° = (-2.303 \times 8.314 \times 448)\log\frac{1.3}{95.2}$

or $\Delta G_1° = \mathbf{15.992\ kJ\ mol^{-1}}$

Similarly for the equilibrium $B \rightleftharpoons C$

$\Delta G_2° = (-2.303 \times 8.314 \times 448)\log\frac{3.5}{95.2}$

or $\Delta G_2° = \mathbf{12.312\,kJ\ mol^{-1}}$

Similarly for equilibrium, $A \rightleftharpoons C$

$\Delta G_3° = -8.314 \times 448 \times 2.303 \times \log_{10}\frac{3.5}{1.3}$

$= -3.688$ kJ mole^{-1}

Hence, we have that

$B \rightleftharpoons A,\ \Delta G_1° = +15.992$ kJ mole^{-1}

$B \rightleftharpoons C,\ \Delta G_2° = +12.312$ kJ mole^{-1}

$A \rightleftharpoons C,\ \Delta G_3° = -3.688$ kJ mole^{-1}

Thus, the correct order of stability, $\boldsymbol{B > C > A}$

89. For following reaction

$CO(g) + \frac{1}{2}O_2(g) \longrightarrow CO_2(g)$

$\Delta G°$ can be calculated as follows :

$\Delta G° = \Delta G_p° - \Delta G°_R$

$= \left[\Delta G°CO_2 - \left(\Delta G°CO + \frac{1}{2}\Delta G°O_2\right)\right]$

$= -394.4 - (-137.2 + \frac{1}{2} \times 0) = -257.2$ kJ mol^{-1}

Since,

$\Delta G° = \Delta H° - T\Delta S°$ or $-257.2 = \Delta H° - 300(0.094)$

$\therefore\ \Delta H° = -285.4$ kJ/mol

$\Delta H°$ is negative, so the reaction is exothermic and since $\Delta G°$ is negative so the reaction is spontaneous.

Chapter-7

Equilibrium

Part-A (JEE Main)

1. (3) $x_2 + y_2 \rightleftharpoons 2xy,$

$$K' = \frac{1}{K_1} = \frac{1}{2.5\times10^5} = \frac{1}{25\times10^4}$$

$$\frac{1}{2}x_2 + \frac{1}{2}y_2 \rightleftharpoons xy, K'' = \left(\frac{1}{25\times10^4}\right)^{1/2}$$

$$= \frac{1}{5\times10^2} \quad ..(i)$$

$xy + \frac{1}{2}z_2 \rightleftharpoons xyz, K_2 = 5\times10^{-3}$ ...(ii)

On adding equation (i) & (ii),

$$\frac{1}{2}x_2 + \frac{1}{2}y_2 + \frac{1}{2}z_2 \rightleftharpoons xyz,$$

$$K_3 = \frac{1}{5\times10^2}\times5\times10^{-3} = 1\times10^{-5}$$

2. (2) $A_xB_y \rightleftharpoons xA^{y+} + yB^{x-}$

$C(1-\alpha) \quad xC\alpha \quad yC\alpha$

Dissociation constant of weak electrolyte

$$K = \frac{(xC\alpha)^x (yC\alpha)^y}{C(1-\alpha)}$$

As $\alpha << 1 \Rightarrow (1-\alpha) = 1$

$K = x^x y^y C^{x+y-1}\alpha^{x+y}$

Degree of dissociation (α) given by

$$\alpha = \left(\frac{K}{C^{x+y-1}x^x y^y}\right)^{\frac{1}{x+y}}.$$

3. (1) $A \rightleftharpoons B$

	A	B
t = 0	1 mol	–
t at eq.	1 – x	x

$$K_C = \frac{x/10}{\frac{1-x}{10}} = 4;\ \frac{x}{1-x} = 4;\ x = 4-4x$$

$$\therefore x = \frac{4}{5} = 0.8$$

Using ideal gas equation, PV = nRT

$$\text{Now, } P_{He} = \frac{1\times0.082\times400}{10} = 3.28 \text{ atm}$$

$$\text{and } P_B = \frac{0.8\times0.082\times400}{10} = 2.624 \text{ atm}$$

4. (2) $A(g) \rightleftharpoons B(g) + C(g)$

	A	B	C
t = 0	a	0	0
t_{eq}	a – x	x	x

Partial pressure $p_A = \frac{a-x}{a+x}\times p$

$$p_B = \frac{x}{a+x}\times p$$

$$p_C = \frac{x}{a+x}\times p$$

$$K_p = \frac{p_B\cdot p_C}{p_A}$$

$$\Rightarrow K_p = \frac{\left(\frac{x}{a+x}\right)\times p\times\left(\frac{x}{a+x}\right)\times p}{\left(\frac{a-x}{a+x}\right)\times p}$$

$$\Rightarrow K_p = \frac{\left(\frac{x}{a+x}\times p\right)^2}{\left(\frac{a-x}{a+x}\right)\times p} = \frac{x^2p}{a^2-x^2}.$$

5. (4) $A_2(g) + B_2(g) \rightleftharpoons C(g)$; $K_{eq} = 2.7\times10^{-5}$

$$\frac{1}{3}A_2(g) + \frac{1}{3}B_2(g) \rightleftharpoons \frac{1}{3}C(g);\ K'_{eq}$$

$$K'_{eq} = (K_{eq})^{\frac{1}{3}}$$

$$K'_{eq} = (2.7\times10^{-5})^{\frac{1}{3}} = 3\times10^{-2}.$$

6. (4) $A(g) \rightleftharpoons B(g)$

(At equilibrium) 0.5 M 0.375 M

$$K_{eq} = \frac{[B]_{eq}}{[A]_{eq}} = \frac{0.375}{0.5} = 0.75$$

Now 0.1 mole of A is added so reaction will move in forward direction.

	A (g)	⇌	B (g)
At *t*	0.5 + 0.1		0.375
(At equilibrium)	0.6 – *x*		0.375 M

$$K_{eq} = 0.75 = \frac{0.375+x}{0.6-x}$$

$0.45 - 0.75x = 0.375 + x$

$1.75x = 0.075$

$$x = \frac{0.075}{1.75} = \frac{3}{70} = 0.043$$

Moles of A = 0.6 – 0.043 = 0.557

Moles of B = 0.375 + 0.043 = 0.418

7. (1) $P_2(g) + Q_2(g) \rightleftharpoons 2PQ(g)$

$t = t_{eq}$ 2 mol 2 mol 2 mol

$$K_{eq} = \frac{2^2}{2\times2} = 1$$

Now 1 mole of each P_2 and Q_2 is added.

So, reaction will move in forward direction.

$P_2(g) + Q_2(g) \rightleftharpoons 2PQ(g)$

$t = t'_{eq}$ 3 – x 3 – x 2 + 2x

$$K_C = 1 = \frac{(2+2x)^2}{(3-x)(3-x)}$$

$$\Rightarrow \frac{2+2x}{3-x} = 1 \Rightarrow 2+2x = 3-x$$

$$\Rightarrow x = \frac{1}{3}$$

At new equilibrium :

$$\text{Moles of } P_2 = 3-x = 3-\frac{1}{3} = \frac{8}{3} = 2.67$$

$$\text{Moles of } Q_2 = 3-x = \frac{8}{3} = 2.67$$

$$\text{Moles of } PQ = 3-x = \frac{8}{3} = 2.67$$

8. (4) For reaction (i) : $K_P > K_c$

So, $\Delta n_g > 0$

$y - x > 0.$

$85.87 = 2.586\,(0.0821\times400)^{y-x}$

On solving, $y - x \simeq 1$

For reaction (ii) : $K_P < K_c$

$y - x < 0.$

$0.862 = 28.62(0.0821\times400)^{y-x}$

On solving, $y - x \simeq -1$

$\therefore$ Among options (4) fits this.

9. (3) At melting point when ice melts, supplied heat is utilised to overcome intermolecular attraction within the molecules so temperature remain constant, and kinetic energy also does not change.

10. (4) On addition of inert gas at constant P and T, reaction moves in the direction of greater number of moles so it will shift in forward direction, so $[PCl_5]$ decrease and $[PCl_3]$ and $[Cl_2]$ will increase.

11. (4) A catalyst cannot alter the equilibrium constant (K_c) of the reaction. Therefore statement I is correct. A homogenous catalyst cannot alter the equilibrium composition of a system. Hence statement II is incorrect.

12. (2) According to Le-Chatelier's principle, increasing pressure favours the reaction to move in backward direction. Which results in reduction in the dissociation of A hence α decreases.

13. (3) $AB_2(g) \rightleftharpoons AB(g) + \frac{1}{2}B_2(g)$

	AB_2	AB	B_2
Initial	1	0	0
at equ.	(1–x)	x	$\frac{x}{2}$

$$\text{Total} = 1 - x + x + \frac{x}{2} \text{ moles} = \left(1+\frac{x}{2}\right) \text{moles}$$

$$K_p = \frac{\left(\frac{xp}{1+\frac{x}{2}}\right)\times\left(\frac{\frac{x}{2}\cdot p}{1+\frac{x}{2}}\right)^{1/2}}{\frac{(1-x)p}{\left(1+\frac{x}{2}\right)}}$$

(p = total pressure at equilibrium)

($p_i = x_i \times p_{total}$)

$$\Rightarrow x << 1 \Rightarrow 1+\frac{x}{2} \approx 1 \text{ and } 1-x \approx 1$$

$$\therefore K_p = \frac{(xp)\cdot\left(\frac{xp}{2}\right)^{1/2}}{p}$$

$$\Rightarrow K_p^2=\frac{x^2\cdot p^2\cdot\frac{xp}{2}}{p^2}\Rightarrow K_p^2=x^2\cdot\frac{xp}{2}$$

$$\Rightarrow x^3=\frac{2K_p^2}{p}\Rightarrow x=\sqrt[3]{\frac{2K_p^2}{p}}$$

14. (4)

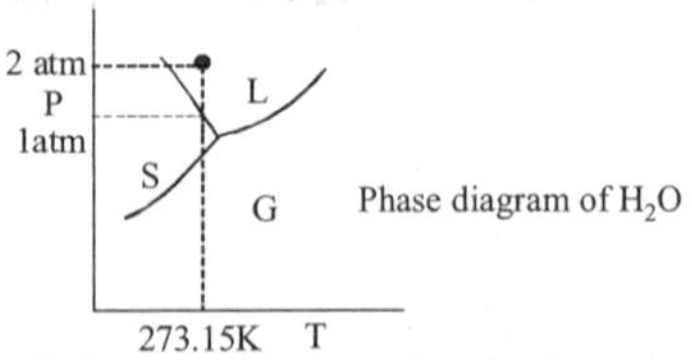

If pressure is increased by two time leads to the complete conversion of ice into water (liquid) form. Ice, being less dense than liquid water, melts under higher pressure.

15. (2) $H_2+I_2 \rightleftharpoons 2HI$

Concentration of H_2 and I_2 decreases untill equilibrium condition and concentration of HI increases till equilibrium condition and after equilibrium concentration of all the reactant and products remain constant.

16. (4)

$$N_2O_5(g)\longrightarrow 2NO_2(g)+\frac{1}{2}O_2(g)$$

t=0	P_0	–	–
t=eq.	P_0-x	$2x$	$\frac{x}{2}$

$$x=\frac{P_0}{2};\ P_{total}=\left(P_0-\frac{P_0}{2}\right)+P_0+\frac{P_0}{4}=\frac{7}{4}P_0$$

17. (2) Let's take 1 mole of X_2Y.

$$X_2Y(g)\rightleftharpoons X_2(g)+\frac{1}{2}Y_2(g)$$

Initial	1.0	0.0	0.0
Final	$1-x$ mole	x mole	$\frac{x}{2}$ mole

$$\therefore\ P_{X_2Y}=\frac{1-x}{1+\frac{x}{2}}\times P\quad (\text{Using } P_i=x_i\times P_{total})$$

$$P_{X_2}=\frac{x}{1+\frac{x}{2}}\times P\Rightarrow P_{Y_2}=\frac{x/2}{1+\frac{x}{2}}\times P$$

$$\therefore\ K_p=\left(\frac{x}{1+\frac{x}{2}}P\right)\left(\frac{x}{2\left(1+\frac{x}{2}\right)}P\right)^{\frac{1}{2}}\Bigg/\left(\frac{1-x}{1+\frac{x}{2}}\right)\times P$$

$$\therefore\ K_P=\left(\frac{x}{1-x}\right)\left(\frac{x}{2\left(1+\frac{x}{2}\right)}\right)^{\frac{1}{2}}\times P^{\frac{1}{2}}$$

$\because$ x is very very small

$$\therefore\ K_P=\frac{x^{3/2}}{(2)^{\frac{1}{2}}}\times P^{\frac{1}{2}}\ \therefore\ x^{\frac{3}{2}}=\frac{K_P\times 2^{\frac{1}{2}}}{P^{\frac{1}{2}}}$$

$$\therefore\ x^3=\frac{K_p^2\times 2}{P}\Rightarrow x=\left(\frac{K_p^2\times 2}{P}\right)^{\frac{1}{3}}$$

18. (2) $CO_2(g)+C(s)\rightleftharpoons 2CO(g)$

0.5 –	0.5 – x	2x

$P_{total}=0.5+x=0.8=x=0.3$

$$K_p=\frac{(0.6)^2}{0.2}-1.8$$

19. (3) On adding given reactions we get desired reaction.

$X\rightleftharpoons Y;\ K_1=1$ ...(1)

$Y\rightleftharpoons Z;\ K_2=2$ ...(2)

$Z\rightleftharpoons W;\ K_3=4$ ...(3)

$X\rightleftharpoons W\ K_1.K_2.K_3$

$K=1\times 2\times 4$

$K=8$

Remember: When reaction (1, 2, 3) added, the rate constant will be multiply $k_1\times k_2\times k_3$

20. (3) $K_P=K_C(RT)^{\Delta n_g}$

$\Delta n_g=0\ \therefore K_P=K_c$

$$K_p=\frac{[HI]^2}{[H_2][I_2]}=1=10\times 10^{-1}$$

$\Rightarrow\ 1=x\times 10^{-1}\Rightarrow x=10$

21. (4) $CO(g)+\frac{1}{2}O_2(g)\rightleftharpoons CO_2(g)$

K_P and K_C are related as

$$K_P=K_C(RT)^{\Delta n_g}$$

$$\Delta n_g=n_P-n_R=1-1\frac{1}{2}=-\frac{1}{2}$$

$$\Delta n_g \text{ for the given reaction}=-\frac{1}{2}\ \therefore\ \frac{K_P}{K_C}=\frac{1}{\sqrt{RT}}$$

22. (4) $SO_3(g)\rightleftharpoons SO_2(g)+\frac{1}{2}O_2(g)$

On reversing the reaction and multiplying by 2 we get desired reaction. Thus,

$$K'_c=\left(\frac{1}{K_c}\right)^2=\left(\frac{1}{4.9\times 10^{-2}}\right)^2,\ K'_c=416.49$$

23. (4) $Fe^{3+}(aq)+SCN^-(aq)\rightleftharpoons[FeSCN]^{2+}(aq)$

$$K_c=\frac{[FeSCN^{2+}]}{[Fe^{3+}][SCN^-]}$$

24. (2) $\therefore K_P=\dfrac{\alpha^{3/2}P^{1/2}}{(2+\alpha)^{1/2}(1-\alpha)}$

$$A\rightleftharpoons B+\frac{C}{2}$$

t = 0	1	–	–
t = t	$(1-\alpha)$	α	$\frac{\alpha}{2}$

$$P_t=1-\alpha+\alpha+\frac{\alpha}{2}=\left(1+\frac{\alpha}{2}\right)$$

$$P_A=\frac{1-\alpha}{\left(1+\frac{\alpha}{2}\right)}\cdot P\qquad P_B=\frac{\alpha}{\left(1+\frac{\alpha}{2}\right)}\cdot P$$

$$P_C=\frac{\frac{\alpha}{2}}{\left(1+\frac{\alpha}{2}\right)}\cdot P$$

$$K_P=\frac{P_B\cdot(P_C)^{1/2}}{P_A}$$

$$=\frac{\frac{\alpha}{\left(1+\frac{\alpha}{2}\right)}\cdot P\times\left(\frac{\frac{\alpha}{2}\cdot P}{\left(1+\frac{\alpha}{2}\right)}\right)^{1/2}}{\frac{1-\alpha}{\left(1+\frac{\alpha}{2}\right)}\cdot P}$$

$$=\frac{\left(\frac{2\alpha}{2+\alpha}\right)\cdot P\times\left[\left(\frac{\alpha}{2+\alpha}\right)P\right]^{1/2}}{\frac{2(1-\alpha)\cdot P}{2+\alpha}}$$

$$=\frac{P^{1/2}\cdot(\alpha)(\alpha)^{1/2}}{(2+\alpha)^{1/2}(1-\alpha)}\Rightarrow \boxed{K_P=\frac{P^{1/2}\cdot(\alpha)^{3/2}}{(2+\alpha)^{1/2}(1-\alpha)}}$$

25. (1) $A_2B_3(aq.)\rightleftharpoons 2A^{3+}(aq)+3B^{2-}(aq)$

$c(1-\alpha)$	$2c\alpha$	$3c\alpha$

$$K_{eq}=\frac{[A^{3+}]^2[B^{2-}]^3}{[A_2B_3]}=\frac{4c^2\alpha^2\times 27c^3\alpha^3}{c(1-\alpha)},\alpha<<1$$

$$\Rightarrow K_{eq}=\frac{108c^5\alpha^5}{c},\alpha=\left(\frac{K_{eq}}{108c^4}\right)^{\frac{1}{5}}$$

26. (3) $2A(g)\rightleftharpoons 2B(g)+C(g)\ K_1$ ...(i)

$\frac{3}{2}A(g)\rightleftharpoons\frac{3}{2}B(g)+\frac{3}{4}C(g)\ K_2$ ...(ii)

eq. (ii) is $\frac{3}{4}$ times of eq. (i), hence, $K_2=(K_1)^{\frac{3}{4}}$

27. (1) $PCl_5=5$ mole; $Ar=4$ mole

$$P_{Total}=\frac{9\times 0.082\times 610}{100}=4.5\text{ atm}$$

$P_y=X_yP_T$ — P_T: Total pressure; X_y: mole fraction of any species y

$$X_{PCl_5}=\frac{5}{9};\quad X_{Ar}=\frac{4}{9}$$

$$P_{PCl_5}=\frac{5\times 4.5}{9}=2.5\text{ atm};$$

$$P_{Ar}=\frac{4\times 4.5}{9}=2\text{ atm}$$

$PCl_5\rightleftharpoons PCl_3+Cl_2$

2.5 – P	P	P

At eqm., $P_{total}=2.5-P+P+P+P_{Ar}=6$ atm

$P=1.5$ atm

$$K_p=\frac{1.5\times 1.5}{1}=2.25$$

28. (2) $A(g)\rightleftharpoons B(g)+\frac{1}{2}C(g)$

Initial :	P_i	0	0
At eq. :	$P_i(1-\alpha)$	$P_i.\alpha$	$P_i\frac{\alpha}{2}$

Equilibrium pressure $(p)=p_i\left(1-\alpha+\alpha+\frac{\alpha}{2}\right)$

$$p=P_i\times\left(1+\frac{\alpha}{2}\right)\Rightarrow\therefore p_A=\left(\frac{1-\alpha}{1+\frac{\alpha}{2}}\right)p$$

$$p_B=\left(\frac{\alpha}{1+\frac{\alpha}{2}}\right)p;\ p_C=\left(\frac{\frac{\alpha}{2}}{1+\frac{\alpha}{2}}\right)p$$

$$\therefore K=\frac{p_C^{\frac{1}{2}}\times p_B}{p_A}\Rightarrow K=\frac{\alpha^{\frac{3}{2}}p^{\frac{1}{2}}}{(2+\alpha)^{\frac{1}{2}}(1-\alpha)}$$

29. (3)

$$K_p=K_c(RT)^{\Delta n_g}=K_c(RT)^{1-3/2}-K_c(RT)^{-1/2}$$

$$\Rightarrow K_c=K_p(RT)^{1/2}$$

30. (4) $N_2(g)+3H_2(g) \rightleftharpoons 2NH_3(g)$; K_c

$2NH_3(g) \rightleftharpoons N_2(g)+3H_2(g)$; $\frac{1}{K_c}$

$NH_3(g) \rightleftharpoons \frac{1}{2}N_2(g)+\frac{3}{2}H_2(g)$; $\frac{1}{K_c^{1/2}}$

$\frac{1}{K_c^{1/2}} = \frac{1}{(64)^{1/2}} = \frac{1}{8}$

31. (2) At equilibrium, rate of forward reaction = Rate of backward reaction.

32. (4) $A \rightleftharpoons B+C$; $K_{eq}^{(1)}$ ...(i)

$B+C \rightleftharpoons P$; $K_{eq}^{(2)}$ ...(ii)

On adding equations (i) and (ii), we get

$A \rightleftharpoons P$

K_{eq} (overall) $= K_{eq}^{(1)} \cdot K_{eq}^{(2)}$

33. (4) Equilibrium constant

$K_c = \frac{[B]}{[A]} = \frac{11}{6} \approx 2$

34. (1) We know that, $K_p = K_c.(RT)^{\Delta n_g}$

$\therefore$ If $\Delta n_g \neq 0$ then $K_p \neq K_c$

Now, $2C(s)+O_2(g) \rightleftharpoons 2CO(g)$

$\Delta n_g = 1$ So, $K_p = K_c(RT)^1$

Hence, $K_p \neq K_c$

35. (3) Given, $S+O_2 \rightleftharpoons SO_2$ ... (i) ; $K_1 = 10^{52}$

$2S+3O_2 \rightleftharpoons 2SO_3$ ... (ii) ; $K_2 = 10^{129}$

$2SO_2+O_2 \rightleftharpoons 2SO_3$ ... (iii) ; K = ?

To get equation (iii) follow (ii) – 2 (i),

$2S+3O_2 \rightarrow 2SO_3 K_2 = 10^{129}$

$-(2S+2O_2 \rightarrow 2SO_2$ $K_1' = 10^{104})$

$O_2 \rightarrow 2SO_3 - 2SO_2$ $K = 10^{129-104}$

or $2SO_2+O_2 \rightarrow 2SO_3$ $K = 10^{25}$

36. (2) $A(s) \longrightarrow B(g) + C(g)$
$\quad\quad P_1 \quad P_1+P_2$

$K_{P_1} = P_B \times P_C$

$P_1(P_1+P_2) = x$ ...(i)

$D(s) \longrightarrow C(g) + E(g)$
$\quad\quad P_1+P_2 \quad P_2$

$K_{P_2} = P_C \times P_E$

$(P_1+P_2)P_2 = y$ ...(ii)

Adding (i) and (ii)

$\therefore \; P_1(P_1+P_2)+P_2(P_1+P_2) = x+y$

$\Rightarrow (P_1+P_2)(P_1+P_2) = (P_1+P_2)^2 = x+y$

$\Rightarrow P_1+P_2 = \sqrt{x+y}$

$\therefore$ Total pressure $(P_T) = P_C+P_B+P_E$

$(P_1+P_2)+P_1+P_2 = 2(P_1+P_2)$

$P_T = 2(\sqrt{x+y})$

37. (1)

	A	+ 2B	⇌ 2C	+ D
$t=0$	2	3	0	0
t_{eq}	$2-x$	$3-2x$	$2x$	x

Given, $3-2x = 2-x; x = 1$

$\therefore [C]=2, [D]=1, [A]=1, [B]=1$

$\therefore K_c = \left\{\frac{2^2 \times 1}{1^2 \times 1}\right\} = 4$

38. (1)

$2NH_3(g) \rightleftharpoons N_2(g)+3H_2(g)$, $K = \frac{1}{K_P}$
$\quad p_{NH_3} \quad\quad p_{N_2} \quad\quad p_{H_2}$

$\therefore K = \frac{1}{K_P} = \frac{P_{N_2}(P_{H_2})^3}{(P_{NH_3})^2}$ ...(i)

$\Rightarrow P_{total}(P) = P_{N_2}+P_{H_2}+P_{NH_3}$

$\approx P_{N_2}+P_{H_2}$ ($\because P_{NH_3} << P_{total}$, remaining NH_3 at equilibrium is negligible)

Now, Partial pressure of N_2 = mole fraction of $N_2 \times P = \frac{1}{4}P$;

Partial pressure of $H_2 = \frac{3}{4}P$

From eq (i), $\frac{1}{K_P} = \frac{\left(\frac{1}{4}P\right)\left(\frac{3}{4}P\right)^3}{(P_{NH_3})^2}$

$\frac{(P_{NH_3})^2}{K_P} = \frac{P}{4}\cdot\frac{P^3}{4^3}\cdot 3^3$; $\frac{(P_{NH_3})^2}{K_P} = \frac{P^4}{4^4}\cdot 3^3$

$(P_{NH_3})^2 = K_P\cdot\frac{P^4}{4^4}\cdot 3^3$; $P_{NH_3} = \left[K_P\cdot\frac{P^4}{4^4}\cdot 3^3\right]^{1/2}$

$P_{NH_3} = \frac{3^{3/2}\cdot P^2\cdot K_P^{1/2}}{16}$

39. (4) Concerned reaction:

$NH_4SH \longrightarrow NH_3(g)+H_2S(g)$

Initial moles $= \frac{5.1}{51} = 0.1$ mol

Moles at equillibrium

$NH_4SH \longrightarrow NH_3(g)+H_2S(g)$
$0.1(1-0.3) \quad 0.1\times0.3 \quad 0.1\times0.3$

$\therefore K_c = [NH_3][H_2S] = \left(\frac{0.03}{3}\right)^2 = 10^{-4}$

$K_p = K_c(RT)^{\Delta n}g$

$= 10^{-4}\times(0.082\times600)^2 = 0.242$ atm^2

40. (2) $K_p = K_c\,(RT)^{\Delta n_g}$

Δn_g = No. of gaseous moles of products – No. of gaseous moles of reactants

$\frac{K_p}{K_c} = (RT)^{\Delta n_g}$

$N_2(g)+O_2(g) \rightleftharpoons 2NO(g)$, $\Delta n_g = 0$

$\frac{K_p}{K_c} = (24.62\ dm^3\ atm\ mol^{-1})^0 = 1$

$N_2O_4(g) \rightleftharpoons 2NO_2(g)$, $\Delta n_g = 1$

$\frac{K_p}{K_c} = 24.62\ dm^3\ atm\ mol^{-1}$

$N_2(g)+3H_2(g) \rightleftharpoons 2NH_3(g)$, $\Delta n_g = -2$

$\frac{K_p}{K_c} = (24.62\ dm^3\ atm\ mol^{-1})^{-2}$

$= 1.65\times10^{-3}\ dm^{-6}\ atm^{-2}\ mol^2$

41. (3) Given: $A_2+B_2 \xrightleftharpoons{K_1} 2AB$

$\Rightarrow 2AB \rightleftharpoons A_2+B_2; K = \frac{1}{K_1}$

$6AB \xrightleftharpoons{K_1} 3A_2+3B_2; K_2 = \left(\frac{1}{K_1}\right)^3$

The relation between K_1 and K_2 is $K_2 = K_1^{-3}$

42. (1) Initially 2 moles of CO are present.

At equilibrium, 1 mole of CO is present

Hence, 2 – 1 = 1 moles of CO has reacted.

1 mole of CO will react with 1 mole of Cl_2 to form 1 mole of $COCl_2$.

3 – 1 = 2 moles of Cl_2 remains at equilibrium

The equilibrium constant

$K_c = \frac{[COCl_2]}{[CO][Cl_2]} = \frac{\frac{1\ mol}{5L}}{\frac{1\ mol}{5L}\times\frac{2\ mol}{5L}} = 2.5$

43. (1) $XY(s) \rightleftharpoons X(g)+Y(g)$ (Assume)
$\quad\quad\quad p \quad\quad p$

At eqm.,

Total pressure $= 2p = 10$ bar

$\therefore p = 5$; Now $K_p = (p_X)(p_Y) = p^2 = 25$.

44. (3)

	$N_2O_4(g)$	⇌ $2NO_2(g)$
t = 0	1 mol	0
t = eqm.	1 – 0.2 = 0.8 mol	0.2 × 2 = 0.4 mol
MW :	92	46

From ideal gas equation, PV = nRT

P = 1 atm, T = 300K, $n = n_T = 0.8+0.4 = 1.2$

$\therefore V = \frac{nRT}{P} = 1.2\times0.082\times300\ L = 29.52\ L$

Total mass inside the vessel

$= 0.8\times(MW)_{N_2O_4}+0.4\,(MW)_{NO_2}$

$= (0.8\times92)+(0.4\times46) = 92$ g

$\therefore$ Density of the equilibrium mixture,

$d = \frac{m}{v} = \frac{92}{29.52} = 3.11$ g / L

45. (2) $SO_2(g)+\frac{1}{2}O_2(g) \rightleftharpoons SO_3(g)$

$K_p = K_c(RT)^x$

where $x = \Delta n_g$ = number of moles of gaseous product – number of moles of gaseous reactants

$= 1-\left(1+\frac{1}{2}\right) = 1-\frac{3}{2} = -\frac{1}{2}$

46. (4)

	2HI	⇌ H_2	+ I_2
At t = 0	c	0	0
At eqm.	c–cα	$\frac{c\alpha}{2}$	$\frac{c\alpha}{2}$

Now, $K_c = \frac{\left(\frac{c\alpha}{2}\right)\left(\frac{c\alpha}{2}\right)}{(c-c\alpha)^2}$

$K_c = \frac{\alpha^2}{4(1-\alpha)^2}$; $K_c = \frac{(0.5)^2}{4(1-0.5)^2}$

$K_c = 0.25$

47. (2)

$NH_2COONH_4(s) \rightleftharpoons 2NH_3(g)+CO_2(g)$

$K_p = \frac{(p_{NH_3})^2\times(p_{CO_2})}{p_{NH_2COONH_4}}$

$= (p_{NH_3})^2\times(p_{CO_2})$

As evident by the reaction, NH_3 and CO_2 are formed in molar ratio of 2 : 1. Thus if P is the total pressure of the system at equilibrium, then

$p_{NH_3} = \frac{2}{3}\times P \quad p_{CO_2} = \frac{1}{3}\times P$

$$K_p = \left(\frac{2P}{3}\right)^2 \times \frac{P}{3} = \frac{4P^3}{27}$$

Given, $K_p = 2.9 \times 10^{-5}$

$$\therefore\ 2.9\times10^{-5} = \frac{4P^3}{27}$$

$$P = \left(\frac{2.9\times10^{-5}\times27}{4}\right)^{1/3} = 5.82\times10^{-2}\text{atm}$$

48. (1) $K_p = K_c(RT)^{\Delta n_g}$

For the reaction

$$CO(g) + \frac{1}{2}O_2(g) \rightleftharpoons CO_2(g)$$

$$\Delta n_g = 1 - \left(1+\frac{1}{2}\right) = -\frac{1}{2}$$

$$\therefore\ K_p = \frac{K_c}{\sqrt{RT}};\quad \frac{K_p}{K_c} = \frac{1}{\sqrt{RT}}$$

49. (4) To calculate the value of K_4 in the given equation we should apply :

eqn. (ii) + eqn.(iii) × 3 – eqn. (i)

Hence, $K_4 = \frac{K_2K_3^3}{K_1}$

50. (4) For the reaction

$N_2 + O_2 \longrightarrow 2NO$ $\quad Kc = 4\times10^{-4}$

Hence for the reaction

$$NO \longrightarrow \frac{1}{2}N_2 + \frac{1}{2}O_2$$

$$K'_c = \frac{1}{\sqrt{K_c}} = \frac{1}{\sqrt{4\times10^{-4}}} = 50$$

51. (2) (I) $\quad N_2 + 2O_2 \overset{K_1}{\rightleftharpoons} 2NO_2$

$$K_1 = \frac{[NO_2]^2}{[N_2][O_2]^2} \quad ...(i)$$

(II) $2NO_2 \overset{K_2}{\rightleftharpoons} N_2 + 2O_2$

$$K_2 = \frac{[N_2][O_2]^2}{[NO_2]^2} \quad ...(ii)$$

(III) $NO_2 \overset{K_3}{\rightleftharpoons} \frac{1}{2}N_2 + O_2$

$$K_3 = \frac{[N_2]^{1/2}[O_2]}{[NO_2]}$$

$$\therefore\ (K_3)^2 = \frac{[N_2][O_2]^2}{[NO_2]^2} \quad ...(iii)$$

∴ From equation (i), (ii) and (iii)

$$K_1 = \frac{1}{K_2} = \frac{1}{(K_3)^2}$$

52. (3) Reaction (iii) can be obtained by adding reactions (i) and (ii) therefore $K_3 = K_1 . K_2$

Hence (3) is the correct answer.

53. (3)

	$2AB_3(g)$ $\rightleftharpoons$	$A_2(g)$ +	$3B_2(g)$
at $t = 0$	8	0	0
at eqm.	(8 – 2 × 2) = 4	2 / 2	3 × 2 / 6
conc.	4/V	2/V	6/V

$$\text{now } K_c = \frac{[A_2][B_2]^3}{[AB_3]^2} = \frac{2/1\times[6/1]^3}{[4/1]^2} = 27$$

54. (4)

	$N_2O_4(g) \rightleftharpoons$	$2NO_2(g)$
Initial moles	1	0
Moles at eqm.	(1 – α)	2α

(α = degree of dissociation)

Total no. of moles at eqm.

$= (1-\alpha) + 2\alpha = (1+\alpha)$

$$p_{N_2O_4} = \frac{(1-\alpha)}{(1+\alpha)}\times P$$

$$p_{NO_2} = \frac{2\alpha}{(1+\alpha)}\times P$$

$$K_p = \frac{(p_{NO_2})^2}{p_{N_2O_4}} = \frac{\left(\frac{2\alpha}{(1+\alpha)}\times P\right)^2}{\left(\frac{1-\alpha}{1+\alpha}\right)\times P}$$

$$K_p = \frac{\frac{4\alpha^2 P}{(1+\alpha)^2(1-\alpha)}}{1+\alpha}$$

$$\Rightarrow\ K_p = \frac{4\alpha^2 P}{(1+\alpha)(1-\alpha)} = \frac{4\alpha^2 P}{1-\alpha^2}$$

Given, $K_p = 2$, $P = 0.5$ atm

$$\therefore\ 2 = \frac{4\alpha^2\times0.5}{1-\alpha^2}$$

$\alpha = 0.707 \approx 0.71 \Rightarrow$ 71 % dissociation

55. (3)

	O_2 +	$2SO_2$ $\rightleftharpoons$	$2SO_3$
At t = 0	1 mol	2 mol	0
At eqm.	(1 – α)	2(1 – α)	2α

Given at equilibrium,

$2\alpha = 1.6$ ∴ $\alpha = 0.8$

$$K_c = \frac{(2\alpha/v)^2}{\frac{(1-\alpha)}{v}\times\left(\frac{2-2\alpha}{v}\right)^2} = \frac{(2\alpha)^2}{(1-\alpha)(2-2\alpha)^2}$$

[Since V = 1 l]

$$\therefore K_c = \frac{(0.8)^2}{(1-0.8)\,(1-0.8)^2} = \frac{0.64}{0.002}$$

$K_c = 80$

56. (1) $X \rightleftharpoons 2Y$

1 $\quad$ 0

$1-x \quad 2x \rightarrow$ Total pressure = 1 + x

$$\text{for } K_{p_1} = \frac{(2x)^2}{1-x}\frac{p_1}{1+x} = \frac{4(x^2)}{(1-x)}\frac{p_1}{1+x}$$

Hence, $Z \rightleftharpoons P + Q$

$1-x \quad x \quad x$

$$\text{for } K_{p_2} = \frac{(x)^2}{(1-x)}\left(\frac{p_2}{1+x}\right)$$

$$\frac{K_{p_1}}{K_{p_2}} = \frac{4\times p_1}{p_2} \Rightarrow \frac{1}{9} = \frac{4\times p_1}{p_2} \Rightarrow \frac{p_1}{p_2} = \frac{1}{36}$$

57. (3) $SO_3(g) \rightleftharpoons SO_2(g) + \frac{1}{2}O_2(g)$

$$K_c = \frac{[SO_2][O_2]^{1/2}}{[SO_3]} = 4.9\times10^{-2};$$

On taking the square of the above reaction

$$\frac{[SO_2]^2[O_2]}{[SO_3]^2} = 24.01\times10^{-4}$$

Now for $2SO_2(g) + O_2(g) \rightleftharpoons 2SO_3$

$$K'_c = \frac{[SO_3]^2}{[SO_2]^2[O_2]} = \frac{1}{24.01\times10^{-4}} = 416$$

58. (3) $PCl_5(g) \rightleftharpoons PCl_3(g) + Cl_2(g)$

$1-x \qquad x \qquad x$

Total moles after dissociation

$1 - x + x + x = 1 + x$

p_{PCl_3} = Mole fraction of PCl_3 × Total pressure

$$= \left(\frac{x}{1+x}\right)P$$

59. (4) For the reaction:-

$2NO_2(g) \rightleftharpoons 2NO(g) + O_2(g)$; $\Delta n_g = 1$

Hence it is clear that, $K_p > K_c$

60. (1)

	$NH_4HS(s) \rightleftharpoons$	$NH_3(s)$ +	$H_2S(g)$
At t = 0		0.5 atm	0 atm
At eqm.		0.5 + x atm	x atm.

Then $0.5 + x + x = 2x + 0.5$

$= 0.84$ (given) $x = 0.17$ atm.

$p_{NH_3} = 0.5 + 0.17 = 0.67$ atm ;

$p_{H_2S} = 0.17$ atm

$K_p = p_{NH_3}\times p_{H_2S} = 0.67\times0.17\ \text{atm}^2$

$= 0.11\ \text{atm}^2$

61. (3) $K_p = K_c(RT)^{\Delta n}$;

Here $\Delta n = 1 - 2 = -1$

$$\therefore \frac{K_p}{K_c} = \frac{1}{RT}$$

62. (4) For $P_4(s) + 5O_2(g) \rightleftharpoons P_4O_{10}(s)$

$$K_c = \frac{1}{(O_2)^5}.$$

[Since solids have unit concentration]

63. (2) $K_c = \frac{[NO_2]^2}{[N_2O_4]} = \frac{[1.2\times10^{-2}]^2}{[4.8\times10^{-2}]}$

$= 3\times10^{-3}$ mol/L

64. (3) $K_p = K_c(RT)^{\Delta n}$;

$$\Delta n = 1 - \left(1+\frac{1}{2}\right) = 1 - \frac{3}{2} = -\frac{1}{2}.$$

$$\therefore\ \frac{K_p}{K_c} = (RT)^{-1/2}$$

65. (4) $CaCO_3(s) \rightleftharpoons CaO(s) + CO_2(g)$,

$K_{P_1} = 0.08$ atm

$K_{P_1} = P_{CO_2} = 0.08$ atm

$C(s) + CO_2(g) \rightleftharpoons 2CO(g)$, $K_{P_2} = 2$

$$K_{P_2} = \frac{P_{CO}^2}{P_{CO_2}} = 2 \Rightarrow \frac{P_{CO}^2}{0.08} = 2$$

$\Rightarrow P_{CO}^2 = 16\times10^{-2}$ atm.

$\Rightarrow P_{CO} = 4\times10^{-1}$ atm.

66. (20) $2A(g) \longrightarrow 4B(g) + C(g)$

	p°		
t = 30 min	$p^\circ - 2x$	4x	x
t = ∞	0	$2p^\circ$	$p^\circ/2$

$p_\infty = p_B + p_C$

Given that $p_\infty = 600$ mm Hg

$p_\infty = 2p^\circ + p^\circ/2 = 600$

$5p^\circ/2 = 600$

$p^\circ = 240$ mm Hg

At t = 30 min, $p_{total} = 300 = p^\circ - 2x + 4x + x$

$300 = 240 + 3x$

3x = 60

x = 20 mm Hg

Therefore, pressure of C(g) = 20 mm Hg at 30 minutes.

67. (15) $X_2(g) + Y_2(g) \rightleftharpoons 2Z(g)$

$$K_c = \frac{(9)^2}{3 \times 3} = 9$$

Now 10 moles of Z are added then reaction will move in backward direction.

$$\underset{3+X}{X_2(g)} + \underset{3+X}{Y_2(g)} \rightleftharpoons \underset{19-2X}{2Z(g)}$$

$$K_c = \frac{(19-2X)^2}{(3+X)(3+X)} = 9$$

$\Rightarrow \dfrac{19-2X}{3+X} = 3$

$\Rightarrow 19 - 2X = 9 + 3X$

$\Rightarrow 10 = 5X$

$\Rightarrow X = 2$

$\therefore$ At equilibrium moles of Z = 19 – 2 × 2 = 15 moles

68. (125) $NH_3(g) \rightleftharpoons \frac{1}{2}N_2(g) + \frac{3}{2}H_2(g)$

Initial moles t = 0	1 mole	–	–
Mole of at t = t_{eq} equilibrium	1 – α	α/2	3α/2

Total moles at equilibrium

$$= (1-\alpha) + \frac{\alpha}{2} + \frac{3\alpha}{2} = (1+\alpha)$$

$$k_P = \frac{\left(\frac{\alpha}{2}\right)^{1/2}\left(\frac{3\alpha}{2}\right)}{(1-\alpha)}\left[\frac{P_T}{1+\alpha}\right]^1$$

[$\because P_T$ = Total pressure = $\sqrt{3}$ atm]

$$9 = \frac{\left(\frac{\alpha}{2}\right)^{1/2}\left(\frac{3\alpha}{2}\right)}{(1-\alpha)} \times \frac{(3)^{1/2}}{1+\alpha}$$

$$9 = \frac{9\left(\frac{\alpha}{2}\right)^2}{1-\alpha^2}$$

$$1-\alpha^2 = \frac{\alpha^2}{4}$$

$$\frac{5\alpha^2}{4} = 1$$

$\alpha^2 = 0.8$

$\alpha = (0.8)^{1/2}$

$$\alpha = \left[\frac{1}{0.8}\right]^{-1/2}$$

$\alpha = [125 \times 10^{-2}]^{-1/2}$

x = 125.

69. (5) $H_2O(g) \rightleftharpoons H_2(g) + \frac{1}{2}O_2(g)$

t = 0	1 mole	–	–
t = t_{eq}	1–α	α	$\frac{\alpha}{2}$

$$n_T = 1 + \frac{\alpha}{2} = 1 \; (\alpha << 1)$$

$$K_p = \frac{P_{H_2}P_{O_2}^{1/2}}{P_{H_2O}} = \frac{(\alpha \cdot P)\left(\frac{\alpha}{2}P\right)^{\frac{1}{2}}}{(1-\alpha)P} \text{ (given P = 1)}$$

$$\therefore \; 8 \times 10^{-3} = \frac{\alpha^{3/2}}{\sqrt{2}}$$

$\Rightarrow \alpha^{3/2} = 8\sqrt{2} \times 10^{-3} \Rightarrow \alpha^3 = 128 \times 10^{-6}$

$\alpha = 5.03 \times 10^{-2} \approx = 5 \times 10^{-2}$

70. (74) $CO(g) + 2H_2(g) \rightleftharpoons CH_3OH(g)$

t = 0	0.1 mol	a mol	–
t_{eq}	0.1 – x	a – 2x	x = 0.04
	= 0.06	= a – 0.08	

V = 2L, T = 500 K, P_{total} = 5 bar

$$P_{total} = n_{total} \times \frac{RT}{V}$$

$$\Rightarrow 5 = (0.06 + a - 0.08 + 0.04) \times \frac{0.08 \times 500}{2}$$

$\Rightarrow$ a = 0.23 mol

$$n_{Total} = 0.25 = \frac{1}{4} \text{ mol}$$

$$Kp = \frac{X_{CH_3OH}}{X_{CO} \times X_{H_2}^2} \times \frac{1}{(P_T)^2} = \frac{0.04}{0.06 \times (0.15)^2} \times \left[\frac{1/4}{5}\right]^2$$

$$= \frac{4}{6 \times (0.15)^2 \times 16} \times \frac{1}{25} = 0.074 = 74 \times 10^{-3}$$

71. (962) Initial pressure of N_2O_5 (P) = $\frac{nRT}{V}$

$$= \frac{\frac{37.8}{108} \times 0.082 \times 500}{1} = 14.35 \text{ bar}$$

$2N_2O_5 \rightleftharpoons 2N_2O_4 + O_2$

t = 0	14.35		
t = eq	14.35–2P	2P	P

P_{total} at equilibrium = 14.35 + P = 18.65

$\Rightarrow$ P = 4.3

$\Rightarrow P_{N_2O_5} = 5.75$ bar ; $P_{N_2O_4} = 8.6$ bar

$P_{O_2} = 4.3$ bar

$$k_P = \frac{(8.6)^2 \times (4.3)}{(5.75)^2} = 9.619 = x \times 10^{-2}$$

x = 961.9 ≈ 962

72. (2) $N_2O_4 (g) \rightleftharpoons 2NO_2 (g)$

K_p = 0.492 atm

T = 300 K

K_c = ?

$K_P = K_c (RT)^{\Delta n_g}$

$\Delta n_g = n_P - n_r = 2 - 1 = 1$

$\therefore K_P = K_c . RT$

$$\Rightarrow K_c = \frac{K_P}{RT} = \frac{0.492}{0.082 \times 300}$$

$= 0.02 = 2 \times 10^{-2}$

73. (417) $N_2 (g) + 3H_2 (g) \rightleftharpoons 2NH_3$

$$\Rightarrow K_c = \frac{[NH_3]^2}{[N_2][H_2]^3} = \frac{(1.5 \times 10^{-2})^2}{(2 \times 10^{-2})(3 \times 10^{-2})^3}$$

$$= \frac{2.25 \times 10^{-4}}{(2 \times 10^{-2})(27 \times 10^{-6})} = \frac{2.25 \times 10^{-4}}{54 \times 10^{-8}}$$

= 416.67

74. (44) $CO(g) + H_2O(g) \rightleftharpoons CO_2(g) + H_2(g)$

t = 0;	1 mol	1 mol	0	0
at eqm.;	1–x	1–x	x	x

at equilibrium 40% by mass water reacts with CO

x = 0.4, 1 – x = 0.6

$$K_C = \frac{[CO_2][H_2]}{[CO][H_2O]} = \frac{0.4 \times 0.4}{0.6 \times 0.6} = 0.44$$

$K_C \times 10^2 = 44$

75. (1) $H_2(g) + I_2(g) \rightleftharpoons 2HI(g)$

t = 0 ;	4.5	4.5	–
t_{eq} ;	3	3	3

$$K_c = \frac{[HI]^2}{[H_2][I_2]} = \frac{(3)^2}{3 \times 3} = \frac{9}{9} = 1$$

76. (3) (C) at equilibrium the value of parameters become constant at a given temperature and not equal. Therefore statement (C) is incorrect.

77. (48.6) $PCl_3 + Cl_2 \rightleftharpoons PCl_5$

	0.2M	(0.1+0.2)M	0.4M
Eq^m.	0.2–x	0.3–x	0.4+x

$$\frac{(0.4+x)}{(0.2-x)(0.3-x)} = 20$$

$\Rightarrow x \simeq 0.086$

$[PCl_5]_{eq} = 0.4 + 0.086 = 0.486M = 48.6 \times 10^{-2}$ M

78. (12) $X \rightleftharpoons Y(g) + Z(g)$; $K_{P_1} = 3$

At t = 0,	a	0	0
At equilibrium,	a – αa	αa	αa

$\therefore \; K_{P_1} = \dfrac{p_Z \, p_Y}{p_X}$; Total pressure = P_1

$$= \frac{\left(\frac{\alpha}{1+\alpha} \times P_1\right)^2}{\frac{1-\alpha}{1+\alpha}P_1} = 3, \text{ or,}$$

$$\frac{\alpha^2 P_1}{(1+\alpha)(1-\alpha)} = 3; \; \therefore \; \frac{\alpha^2 P_1}{1-\alpha^2} = 3$$

$A \rightleftharpoons 2B \quad K_{P_2} = 1$

At t = 0,	a	0
At equilibrium,	a – αa	2αa

$$\therefore \; K_{P_2} = \frac{\left(\frac{2\alpha}{1+\alpha} \times P_2\right)^2}{\frac{1-\alpha}{1+\alpha} \times P_2} \text{ or } \frac{4\alpha^2 P_2}{1-\alpha^2} = 1$$

$$\frac{K_{P_1}}{K_{P_2}} = \frac{P_1}{4P_2} \text{ or}$$

$$\frac{P_1}{4P_2} = 3/1 \;\; \therefore \; P_1 : P_2 = 12 : 1$$

79. (1) $SO_2(g) + \frac{1}{2}O_2(g) \rightleftharpoons SO_3(g)$

$K_P = 2 \times 10^{12}$ at 300 K

$$\Delta n_g = 1 - \left(1 + \frac{1}{2}\right) = 1 - \frac{3}{2} = -\frac{1}{2}$$

$K_P = K_c \times (RT)^{\Delta n_g}$

$\Rightarrow 2 \times 10^{12} = K_c \times (0.082 \times 300)^{-1/2}$

$K_c = 9.92 \times 10^{12} = 0.992 \times 10^{13} \approx 1 \times 10^{13}$

80. (2) $H_2O(g) \rightleftharpoons H_2(g) + \frac{1}{2}O_2(g)$

At equilibrium : $P[1-\alpha]$ $\quad P\alpha \quad \frac{P\alpha}{2}$

$$K_p = \frac{(P_{H_2})(P_{O_2})^{1/2}}{P_{H_2O}}$$

$$\frac{(P\alpha)\left(\frac{P\alpha}{2}\right)^{1/2}}{P[1-\alpha]} = 2 \times 10^{-3}$$

since α is negligible w.r.t 1 so P = 1 and $1-\alpha \approx 1$

$$\frac{\alpha\sqrt{\alpha}}{\sqrt{2}} = 2 \times 10^{-3}$$

$\alpha^{3/2} = 2^{3/2} \times 10^{-3}$

$\alpha = 2^{3/2 \times 2/3} \times 10^{-3 \times 2/3}$

$\alpha = 2 \times 10^{-2} \Rightarrow \%\alpha = 2\%$

81. (4) $N_2(g) + 3H_2(g) \rightleftharpoons 2NH_3(g),$

$K_1 = 4 \times 10^5$ ...(i)

$N_2(g) + O_2(g) \rightleftharpoons 2NO(g), K_2 = 1.6 \times 10^{12}$ ..(ii)

$H_2(g) + \frac{1}{2}O_2(g) \rightleftharpoons H_2O(g),$

$K_3 = 1.0 \times 10^{-13}$ ...(iii)

By applying (ii) + 3 × (iii) – (i), we get

$$2NH_3(g) + \frac{5}{2}O_2(g) \rightleftharpoons 2NO(g) + 3H_2O(g)$$

$$k_{eq} = \frac{k_2 \times k_3^3}{k_1} = \frac{1.6 \times 10^{12} \times (10^{-13})^3}{4 \times 10^5}$$

$$= \frac{1.6}{4} \times 10^{-32} = 4 \times 10^{-33}$$

82. (2) $2NO + O_2 \rightleftharpoons 2NO_2$

	2NO	O_2	$2NO_2$
Initial:	2	1	–
Change:	$2-2\alpha$	$1-\alpha$	2α
Equilibrium:	1.2	0.6	0.8

$$K_p = \frac{(x_{NO_2}P)^2}{(x_{NO}P)^2 (x_{O_2}P)}$$

$$K_p = \frac{\left(\frac{0.8}{2.6}\right)^2}{\left(\frac{1.2}{2.6}\right)^2\left(\frac{0.6}{2.6}\right)} = 1.925$$

83. (2) $Cu(s) + 2Ag^+(aq) \rightleftharpoons$

$Cu^{2+}(aq) + 2Ag(s);\ 2 \times 10^{15} = k$

$\frac{1}{2}Cu^{2+}(aq) + Ag^+(s) \rightleftharpoons \frac{1}{2}Cu(s) + Ag^+(aq);\ K$

$$K = \frac{1}{\sqrt{2 \times 10^{15}}} = 2.23 \times 10^{-8}$$

84. (29) moles of water $= \frac{0.90}{18}$

By ideal gas equation PV = nRT

$$V = \frac{nRT}{P} = \frac{0.90 \times 0.082 \times 300 \times 760}{18 \times 32}$$

$= 29.21 \approx 29$ L

85. (125) $2NOCl(g) \rightleftharpoons 2NO(g) + Cl_2(g)$

	$2NOCl(g)$	$2NO(g)$	$Cl_2(g)$
$t=0$	2M	–	–
$t=t_{eq}$	$(2-x)$M	x M	$\frac{x}{2}$ M

$\because x = 0.4M$

$\therefore [NOCl]_{eq} = 2.0 - 0.4 = 1.6$ M; $[Cl_2]_{eq} = 0.2$ M

$$\Rightarrow K_c = \frac{[NO]^2[Cl_2]}{[NOCl]^2} = \frac{[0.4]^2[0.2]}{[1.6]^2}$$

$= 125 \times 10^{-4}$.

86. (1107) Given : 2 mole of N_2 gas was present as inert gas. Equilibrium pressure = 2.46 atm

	$PCl_5(g)$	$PCl_3(g)$	$Cl_2(g)$
$t=0$	5	0	0
t = Eqm.	$5-x$	x	x

From ideal gas equation, $PV = nRT$

$2.46 \times 200 = (5 - x + x + x + 2) \times 0.082 \times 600$

$\Rightarrow x = 3$

$$K_P = \frac{\left(\frac{3}{10} \times 2.46\right)\left(\frac{3}{10} \times 2.46\right)}{\left(\frac{2}{10} \times 2.46\right)}$$

$= 1.107 = 1107 \times 10^{-3}$

87. (4) Let moles of NH_3 added = a

$\therefore$ concentration of $NH_3 = \frac{a}{2}$ mol/L

	$Ag^+(aq.)$	$+ 2NH_3(aq.)$	$\rightleftharpoons Ag(NH_3)_2^+(aq.)$
$t=0$	0.8	$\left(\frac{a}{2}\right)$	0
$t=\infty$	5×10^{-8}	$\left(\frac{a}{2} - 1.6\right)$	0.8

$$K_f = \frac{0.8}{(5 \times 10^{-8})\left(\frac{a}{2} - 1.6\right)^2} = 10^8$$

$\Rightarrow \frac{a}{2} - 1.6 = 0.4 \Rightarrow a = 4$

88. (6) Moles of NH_4HS initially taken

$$= \frac{5.1\text{ g}}{51\text{ g/mol}} = 0.1\text{ mol}$$

Volume of vessel = 2 L

	$NH_4HS(s)$	$NH_3(g)$	$H_2S(g)$
Initial	0.1 mol	0	0
At eqm.	$0.1(1-0.2)$	0.1×0.2	0.1×0.2

$\Rightarrow$ Partial pressure of each component gaseous

$$P = \frac{nRT}{V} = \frac{0.1 \times 0.2 \times 0.082 \times 300}{2} = 0.246\text{ atm}$$

$\Rightarrow K_P = P_{NH_3} \times P_{H_2S} = (0.246)^2 = 0.060516$

$= 6.05 \times 10^{-2}$ or 6×10^{-2}

89. (182)

$A + B \rightleftharpoons C + D : K_{eq} = 100$

	A	B	C	D
t=0	1	1	1	1
t=$t_{eq.}$	(1–x)	(1–x)	(1+x)	(1+x)

$$K_{eq} = 100 = \frac{(1+x)(1+x)}{(1-x)(1-x)} \Rightarrow 100 = \left(\frac{1+x}{1-x}\right)^2$$

$$10 = \left(\frac{1+x}{1-x}\right) \Rightarrow x = \frac{9}{11}$$

$$[D] = (1+x) = 1 + \frac{9}{11} = \frac{20}{11}$$

$= 1.8181 = 181.81 \times 10^{-2} \simeq 182 \times 10^{-2}$.

90. (0)

$$[PtCl_4]^{2-} + H_2O \underset{k_b}{\overset{k_f}{\rightleftharpoons}} [Pt(H_2O)Cl_3]^- + Cl^-$$

$$\frac{-d[PtCl_4]^{-2}}{dt} = k_f[PtCl_4^{2+}] = k_b$$

$[Pt(H_2O)Cl_3] - [Cl^-]$

$$\Rightarrow K_{eq} = \frac{k_f}{k_b} = \frac{4.8 \times 10^{-5}}{2.4 \times 10^{-3}} = 0.02 = 2 \times 10^{-2}$$

91. (1400)

$PCl_5(g) \rightleftharpoons PCl_3(g) + Cl_2(g);\ K_c = 1.844$

	PCl_5	PCl_3	Cl_2
t = 0	3 moles	0	0
t = ∞	$(3-x)$	x	x

$$\Rightarrow \frac{[PCl_3][Cl_2]}{[PCl_5]} = \frac{x^2}{3-x} = 1.844$$

$\Rightarrow x^2 + 1.844x - 5.532 = 0$

$$x = \frac{-b \pm \sqrt{D}}{2a} = \frac{-1.844 + \sqrt{(1.844)^2 + 4 \times 5.532}}{2}$$

$\cong 1.604$

$\Rightarrow$ Moles of $PCl_5 = 3 - 1.604$

$\cong 1.396 \simeq 1.4 \approx 1400 \times 10^{-3}$

92. (16) $K_p = (P_{O_2})^{1/2}$

$4 = (P_{O_2})^{1/2} \Rightarrow P_{O_2} = 16$

93. (875)

	$2SO_2(g)$	$+ O_2(g)$	$\rightarrow 2SO_3(g)$
Initial :	250 m bar	750 m bar	0
Final :	–250 m bar	–125 m bar	250 m bar
After completion	0	625 m bar	250 m bar

$\therefore$ Final total pressure = 625 + 250 = 875 m bar

94. (25) $A + B \rightleftharpoons 2C$

A	B	2C
1	1	1
1 – x	1 – x	1 + 2x

$$K = \frac{[C]_{eq}^2}{[A]_{eq}[B]_{eq}} = \frac{(1+2x)^2}{(1-x)(1-x)}$$

$$100 = \left(\frac{1+2x}{1-x}\right)^2$$

$$\left(\frac{1+2x}{1-x}\right) = 10$$

$$x = \frac{3}{4}\ [C]eq. = 1 + 2x = \frac{15}{6} = 25 \times 10^{-1}M$$

95. (2) $K_P = K_C(RT)^{\Delta n_g}$

$N_2O_4(g) \rightleftharpoons 2NO_2(g)$

$$\Delta n_g = 1\ K_C = \frac{K_P}{(RT)} = \frac{47.9}{(0.083 \times 288)} \approx 2$$

96. (172) $2SO_2(g) + O_2(g) = 2SO_3(g)$

$$K_P = \frac{(p_{SO_{3(g)}})^2}{(p_{SO_{2(g)}})^2 \times p_{O_{2(g)}}} = \frac{(43 \times 10^3)^2}{(45 \times 10^3)^2 \times 530}$$

$$= \frac{(43)^2}{(45)^2 \times 530 \times 10^{-3}} = 172 \times 10^{-2}$$

97. (354) $N_2O_4(g) \rightleftharpoons 2NO_2(g)$

$\Delta n_g = 2 - 1 = 1$

Relation between K_P and K_C is

$K_P = K_C(RT)^{\Delta n_g}$

$\Rightarrow 600.1 = 20.4(0.0831 \times T)^1$

$\Rightarrow T = \dfrac{600.1}{20.4 \times 0.0831} = 354$ K

98. (72) $AB_2(g) \rightleftharpoons A(g) + 2B(g)$

t = 0	1	–	–
t = eqm.	1 – α	α	2α

∴ No. of moles at equilibrium

$= 1 + 2\alpha = \dfrac{PV}{RT} = \dfrac{1.9 \times 25}{0.08206 \times 300} \approx 1.93$

∴ $\alpha = 0.465$

∴ $P_{AB_2} = \dfrac{1-\alpha}{1+2\alpha} \times P_T$

$= \dfrac{1-0.465}{1+2\times0.465} \times 1.9 \approx 0.53$ atm

$P_A = \dfrac{\alpha}{1+2\alpha} \times P_T = \dfrac{0.465}{1.93} \times 1.9 \approx 0.46$ atm

$P_B = \dfrac{2\alpha}{1+2\alpha} \times P_T = \dfrac{2\times0.465}{1.93} \times 1.9 \approx 0.91$ atm

$K_P = \dfrac{P_A \cdot (P_B)^2}{P_{AB_2}}$

$= \dfrac{0.46 \times (0.91)^2}{0.53} \approx 0.72 \approx 72 \times 10^{-2}$

99. (5) $Cl_2(g) \rightleftharpoons 2Cl(g)$

Let mol of both of Cl_2 and Cl be x.

$P_{Cl} = \dfrac{x}{2x} \times 1 = \dfrac{1}{2}$ and $P_{Cl_2} = \dfrac{x}{2x} \times 1 = \dfrac{1}{2}$

$\therefore K_p = \dfrac{\left(\frac{1}{2}\right)^2}{\frac{1}{2}} = \dfrac{1}{2} = 0.5 = 5 \times 10^{-1}$.

100. (16)

	X +	Y	⇌ 2Z
t = 0	1 mol	1.5 mol	0.5 mol
$t = t_{eq}$	1 – a	1.5 – a	0.5 + 2a

$0.5 + 2a = 1;\ a = 0.25$

X +	Y	⇌ 2Z
0.75	1.25	1

$K_{eq} = \dfrac{[Z]^2}{[X][Y]} = \dfrac{1}{0.75 \times [1.25]} = \dfrac{x}{15}$

$\Rightarrow x = \dfrac{15}{(0.75 \times 1.25)} = 16$

101. (1)

(A) $\Delta_r G° = G°_B - G°_A = +ve$

(B) $\Delta_r G° = +ve$, N_2O_4 will partially dissociate into NO_2.

(C) For reverse reaction.
It is partially completed as there is equilibrium at E.

(D) For 1 mole N_2O_4; $\Delta_r G° = -0.84$ kJ mol^{-1}

(E) For 2 mole NO_2; $\Delta G° = -5.4 - 0.84 = -6.25$ kJ mol^{-1}.

102. (N)

(A) Concentration of only products increases.

(B) Equilibrium shifts in forward direction to decrease pressure.

(C) K_c remains constant at constant T.

(D) K_c is constant because of constant T and concentrations of only product change.

103. (1) For reaction :

$Fe_2O_3(s) + 3CO(g) \rightleftharpoons Fe(\ell) + 3CO_2(g)$

Addition or removal of $Fe_2O_3(s)$ and $Fe(\ell)$ will not affect K_C and hence equilibrium

$$K_C = \frac{[CO_2]^3}{[CO]^3}$$

104. (1, 4) If the volume is kept constant and an inert gas such as He is added which does not take part in the reaction the equilibrium remain undisturbed. If P is constant, equilibrium will shift towards the forward direction.

105. (3) $\Delta G° = -RT \ln K$,

$T_1 = 25\ °C,\ K_1 = 10$ $\Delta G°$ at

$T_1 = -8.314 \times 298 \times 2.303 \times \log 10$

$= -5.71$ kJ/mol

$\Delta G°$ at $T_2 = -8.314 \times 373 \times 2.303 \times \log(100)$

$= -14.29$ kJ/mol

$\ln\left[\dfrac{K_2}{K_1}\right] = \dfrac{\Delta H}{R}\left[\dfrac{1}{T_1} - \dfrac{1}{T_2}\right]$

$\ln 10 = \dfrac{\Delta H}{8.314}\left[\dfrac{1}{298} - \dfrac{1}{373}\right]$

$\Delta H° = 28.4$ kJ/mol

106. (3)

(i) As reaction is endothermic (ΔHz = +ve) so on decrease in temperature equilibrium will shift towards reactant side.

(ii) On increase in pressure by adding inert gas (N_2) at same temperature, no shifting will take place. The equilibrium changes only if the added gas is a reactant or product involved in the reaction.

107. (4) $\Delta G° = -RT \ln K$

∴ If $K > 1$ then $\Delta G° < 0$

If $K < 1$ then $\Delta G° > 0$

If $K = 1$ then $\Delta G° = 0$

108. (1) Equilibrium constant has no relation with catalyst. Catalyst only affects the rate of the reaction.

Catalyst, V_2O_5 in the given reaction, is used to speed up the reaction.

109. (3) $\Delta G = \Delta G° + RT \ln Q$

At equilibrium; $\Delta G = 0$ and $Q = K_{eq}$

$\Rightarrow \Delta G° = -2.303$ RT $\log K_w$

$= -2.303 \times 8.314 \times 298 \times \log 10^{-14}$

$= 79.9$ kJ/mol ≈ 80 kJ/mol

110. (2) As volume ↑ pressure ↓, reaction proceeds in that direction where the number of moles of gases increases.

111. (1) Reaction at equilibrium

$$N_2O_4 \rightleftharpoons 2NO_2$$

According to Le chatelier's principle-

(1) addition of an inert gas at constant pressure will increase volume or decrease moles per unit volume. Thus, equilibrium shifts towards more number of molecules.

112. (4)

Perturbation	Shifts reaction towards
Removal of CO	Left
Removal of CO_2	Right
Addition of CO_2	Left
Addition of Fe_2O_3	No change

(This is a solid compound. Its concentration has no effect on the equilibrium.)

113. (3)

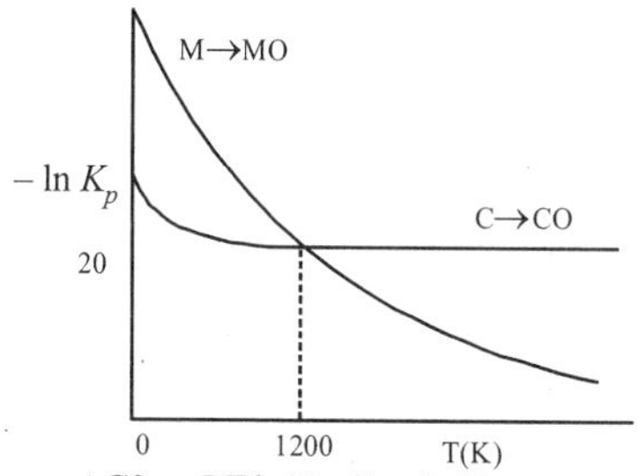

$\Delta G° = -RT \ln K_p$, So $-\ln K_p$ represents $\Delta G°$ here and thus the feasibility or spontaneity of a reaction.

At T < 1200 K, oxidation of carbon is more favourable than oxidation of M.

Hence, the following reactions will take place:

$C(s) + \frac{1}{2}O_2(g) \rightarrow CO(g)$

$MO(s) \rightarrow M(s) + \frac{1}{2}O_2(g)$

Adding these two reactions:

$C(s) + MO(s) \rightarrow M(s) + CO(g)$

114. (3) Given, $\Delta H° = -29.8$ kJ mol^{-1}

$\Delta S° = -0.100$ kJ K^{-1}

From the equation

$\Delta G° = \Delta H° - T\Delta S° = -29.8 - (298 \times (-0.100))$

$= -29.8 + 29.8 = 0$

Now, $\Delta G° = -2.303\ RT \log K_{eq}$

$0 = -2.303$ RT $\log K_{eq}$ ∴ $K_{eq} = 1$.

115. (4) $\Delta G° - 2494.2$J

$$2A \rightleftharpoons B + C.$$

$[A] = \dfrac{1}{2}, [B] = 2, [C] = \dfrac{1}{2}$

$Q = \dfrac{[B][C]}{[A]^2} = \dfrac{2 \times 1/2}{\left(\frac{1}{2}\right)^2} = 4$

$\Delta G° = -2.303$ RT $\log K_c$.

$2494.2\ J = -2.303 \times (8.314\ J/K/mol) \times (300K) \log K_c$

$\Rightarrow \log K_c = -0.4341 \Rightarrow K_c = 0.37;$

∴ $Q > K_c$.

116. (2)

$\Delta G_r^{\circ} = 2\Delta G°_f(NO_2) - 2\Delta G°_f(NO) - \Delta G°_f(O_2)$

$2\Delta G°_f(NO_2) = \Delta G_r^{\circ} + 2\Delta G°_f(NO) + \Delta G°_f(O_2)$

At equilibrium, $\Delta G° = -RT \ln K_p$

$\Delta G°_f(O_2) = 0$ and $\Delta G°_f(NO) = 86.6$ kJ/mol

$\therefore \Delta G°_f(NO_2) = \dfrac{1}{2}[2 \times 86600 - R(298) \ln(1.6 \times 10^{12})]$

117. (4) So, equilibrium will shift in forward direction.

118. (4) On adding inert gas at constant volume, the total pressure of the system is increased, but the concentration of each reactant and product remains the same. Hence, there is no effect on the state of equilibrium.

119. (1) The reaction given is an exothermic reaction thus accordingly to Le-Chatalier's principle lowering of temperature, addition of F_2 and/or Cl_2 favour the forward direction or the production of ClF_3.

120. (2)

121. (1)

122. (8) $X_2Y_4(g) \rightleftharpoons 2XY_2(g)$

Initial: 1, 0

Equilibrium: $(1-\alpha)$, 2α

Total moles at equilibrium = $(1+\alpha)$

Total moles at equilibrium = 1 + 0.75 = 1.75

Total pressure at equilibrium = 1 atm

$$P_{X_2Y_4} = \frac{1-\alpha}{1+\alpha}\times P_{total} = \frac{1-0.75}{1+0.75}\times 1 = \frac{0.25}{1.75} = \frac{1}{7}$$

$$P_{XY_2} = \frac{2\alpha}{1+\alpha}\times P_{total} = \frac{2\times 0.75}{1+0.75} = \frac{1.5}{1.75} = \frac{6}{7}$$

$$K_P = \frac{(P_{XY_2})^2}{P_{X_2Y_2}} = \frac{(6/7)^2}{(1/7)} = \frac{36/49}{1/7} = \frac{36}{7}$$

$\Delta G^0 = -RT \ln K_P$

$= -2.303\times 8.3\times 600 \log\frac{36}{7}$

$= -2.303 \times 8.3 \times 600 (2\log 2 + 2\log 3 - \log 7)$

$= -2.303 \times 8.3 \times 600 (2\times 0.3 + 2\times 0.48 - 0.84)$

$= -2.303 \times 8.3 \times 600 \times 0.72$

$= -19.11 \times 432 = -8255.5\ J\ mol^{-1}$

$\Delta G^0 \approx -8.25\ kJ\ mol^{-1}$

Magnitude of $\Delta G^0 = 8\ kJ\ mol^{-1}$

123. (2474)

$2N_2O_5(g) \rightleftharpoons 2N_2O_4(g) + O_2(g)$

	$2N_2O_5$	$2N_2O_4$	O_2
$t=0$	1	–	–
$t_{eqm.}$	$1-2x$	$2x$	x
	$1-0.5$	0.5	0.25
	0.5	0.5	0.25

Total moles = 1.25

As, $p_i = \frac{n_i}{n_{total}}\times P$

At eqm. partial pressure for individual gases are :

$\frac{0.5}{1.25}\times 2 \quad \frac{0.5}{1.25}\times 2 \quad \frac{0.25}{1.25}\times 2$

$$K_P = \frac{P^2_{N_2O_4}\cdot P_{O_2}}{P^2_{N_2O_5}} = 0.4$$

$\Delta G^0 = -RT \ln K_P$

$= -2.303 \times 8.314 \times 323 \log (4/10)$

$= -2.303 \times 8.314 \times 323 - 0.4 = 2473.8\ J/mol.$

124. (33) $\Delta G = -RT \ln K_P$

$-1.664 \times 10^3 = -8.3 \times 300 \ln K_P$

$\ln K_P = 0.693$

$K_P = 2,\ P_{total} = 1$

$A_2(g) \rightleftharpoons 2A(g)$

Initial mol: 1, 0

Equilibrium mol: $1-\alpha$, 2α

Total mol = $1 - \alpha + 2\alpha = 1 + \alpha$

$$K_P = \frac{[A]^2 P_{total}}{A_2} \text{(using } p_i = x_i \times P_{total})$$

$$2 = \frac{4\alpha^2\times 1}{1-\alpha^2}$$

$$\alpha = \frac{1}{\sqrt{3}}$$

$$\alpha = \left(\frac{100}{3}\times 10^{-2}\right) = (33.33\times 10^{-2})^{1/2}$$

125. (700) $\frac{1}{2}X_2 + \frac{5}{2}Y_2 \rightleftharpoons XY_5$

$$\Delta S^0_{Rxn} = (\Delta S^0_{xy_5}) - \left(\frac{1}{2}\Delta S^0_{X_2} + \frac{5}{2}\Delta S^0_{Y_2}\right)$$

$$\Delta S^0_{Rxn} = 110 - \left[\left(\frac{1}{2}\times 70\right) + \left(\frac{5}{2}\times 50\right)\right]$$

$\Delta S^0_{Rxn} = 110 - 160 = -50\ JK^{-1}\ mol^{-1}$

$\Delta G^0 = 0$ at equilibrium; $\Delta G^0 = \Delta H^0 - T\Delta S^0$

$0 = -35000 - T(-50);\ T = 700\ K$

126. (57) Given: T = 300 K, K = 10

$\Delta G = -RT \ln K = -2.303\ RT \log K$

$= -2.303 \times 8.314 \times 300 \log 10$

$= -5744\ J/mol = -5.744\ kJ/mol$

$= -57 \times 10^{-1}\ kJ/mol$

127 . (163)

$\frac{3}{2}O_2(g) \rightleftharpoons O_3(g) \cdot K_P = 2.47\times 10^{-29}$

$\Delta_r G^{\ominus} = -RT \ln Kp$

$= -8.314 \times 10^{-3} \times 298 \times \ln(2.47\times 10^{-29})$

$= -8.314 \times 10^{-3} \times 298 \times (-65.87)$

$= 163.19\ kJ$

128. (37) $\Delta G° = \Delta H° - T\Delta S°$

$= 77.2 \times 10^3 - 400 \times 122 = 28400\ J$

$\Delta G° = -2.303\ RT \log K$

$\Rightarrow 28400 = -2.303 \times 8.314 \times 400 \log K$

$\Rightarrow \log K = -3.708 = -37.08 \times 10^{-1}$

129. (10) $\Delta G° = \Delta H° - T\Delta S°$

$\Rightarrow \Delta G° = [(-54070) - (10\times 298)]$

Also, $\Delta G° = (-2.303\ RT \log K)$

$\Rightarrow [(-54070) - (10\times 298)]$

$= (-2.303 \times 8.134 \times 298 \log K)$

$\Rightarrow \log K = 10$

130. (3) The yield of SO_3 at equilibrium will be due to :

(B) Increasing pressure

(C) Adding more SO_2

(D) Adding more O_2

131. (6) $\Delta G° = -RT \ln K_{eq}$

$$\text{and } K_{eq} = \frac{K_f}{K_b} = \frac{10^3}{10^2} = 10$$

$\therefore \Delta G = -RT \ln 10 = -(8.3 \times 300 \times 2.3)$

$= -5.7\ kJ\ mole^{-1} \approx 6\ kJ\ mol^{-1}$.

132. (2735) $HI \rightleftharpoons \frac{1}{2}H_2 + \frac{1}{2}I_2$

t_i: 1, –, –

t_{eq}: $1-0.4$, $\frac{0.4}{2}$, $\frac{0.4}{2}$

(Total no. of moles = 0.6 + 0.2 + 0.2 = 1.0)

$$K_c = \frac{(0.2)^{\frac{1}{2}}(0.2)^{\frac{1}{2}}}{1-0.4} = \frac{0.2}{0.6} = \frac{1}{3}$$

[Pressure at eqm. = 1]

$\Delta G = \Delta G° + RT \ln K = 0$ [$\therefore$ at Equilibrium $\Delta G = 0$]

$\Delta G° = -RT \ln$

$$K = -8.31\times 300\times 2.3\times \log\left(\frac{1}{3}\right) = 2735\ J/mol$$

133. (747) $2O_3(g) \rightleftharpoons 3O_2(g)$

$t = 0$: 2, 0

$t = $ eq.: $2 - 1 = 1$, 3/2

Total moles at equilibrium = 1 + 3/2 = 5/2

Total pressure = 1 atm. (given)

Partial pressure = mole fracrtion × total pressure

$$p_{O_3} = \frac{1}{5/2}\times 1 = \frac{2}{5};\ p_{O_2} = \frac{3/2}{5/2}\times 1 = \frac{3}{5}$$

$$K_p = \frac{(P_{O_2})^3}{(P_{O_3})^2} = \frac{(3/5)^3}{(2/5)^2} = 1.35$$

Free energy change,

$\Delta G = -RT \ln K_p = -8.3\times 300\times \ln 1.35 = -747\ J\ mol^{-1}$

134. (20)

$\Delta G° = -2.303\ RT \log K = -RT \ln K_{eq}$

$$\ln K_{eq} = \frac{\Delta G°}{RT} = \frac{9.478\times 10^3}{8.314\times 495}$$

$\ln K_{eq} = 2.303$

$\because \ln K_{eq} = \ln 10 \therefore K_{eq} = 10$

Now,

	A(g)	$\rightleftharpoons$ B(g)
$t=0$	22	0
$t=t$	$22-x$	x

$$K_{eq} = \frac{[B]}{[A]} \Rightarrow 10 = \frac{x}{22-x} \Rightarrow x = 20$$

135. (166) $2A(g) \rightleftharpoons A_2(g)$

For homogeneous gas phase reaction, $K_{eq} = K_p$

$\Delta G° = -RT \ln K_p$

$\Rightarrow 25.2\times 10^3 = -8.3\times 400\times 2.3 \log K_p$

$$\log K_p = \frac{-25.2\times 10^3}{8.3\times 400\times 2.3} \Rightarrow \log K_p = -3.3$$

K_p = antilog (–3.3)

$K_p = 10^{-3.3};\ K_p = 10^{-3}.\ 10^{-0.3}$

$K_p = 0.501 \times 10^{-3} = 5 \times 10^{-4}\ \because K_p = K_c(RT)^{-1}$

$\therefore K_c = 5\times 10^{-4}\times 8.3\times 400 = 1.66 = 166\times 10^{-2}$

136. (1380) $\Delta G° = RT \ln K_p$

$= -R(300)\ln(10)^2 = -R(300\times 2\times 2.3)$

$\Delta G° = -1380R.$

137. (2) pH = 1 $\Rightarrow [H^+] = 10^{-1}$

If equal volume of water is added concentration will become half

$$[H^+]_{sol} = \frac{10^{-1}}{2} = 0.05$$

pH = –log 0.05 = 1.3

138. (4) $HA \rightleftharpoons H^{\oplus} + A^{\ominus}$

	HA	$H^{\oplus}$	$A^{\ominus}$
t = 0	a		
t = t	a(1–x)	ax	ax

$$K_a = (ax)\frac{(x)}{1-x};[H^+] = ax$$

$$-\log(K_a) = -\log(ax) - \log\left(\frac{x}{1-x}\right)$$

$$pK_a = pH - \log\left(\frac{x}{1-x}\right)$$

$$pH - pK_a = \log\left(\frac{x}{1-x}\right)$$

139. (1) With increase in temperature, K_w of water increases

So, degree of dissociation of water increases.

$\therefore$ pH as well as pOH of water decreases.

At 80°C, pH $\approx$ 6.93.

140. (2) Strong acid have weak conjugate base.

Acidic strength :

H–Cl > CH_3COOH > H_2O > R–OH

Conjugate base strength :

$Cl^- < CH_3COO^- < \bar{O}H < RO^-$

141. (4)

(A) pH of 10^{-8} M HCl is in acidic range (6.98).

(B) Conjugate base of $H_2PO_4^-$ is HPO_4^{2-}.

(C) K_w increases with increasing temperature, as the temperature increases, the dissociation of water increases.

(D) At half neutralisation point, half of the acid is present in the form of salt.

$$pH = pK_a + \log\frac{1}{1} = pK_a$$

142. (2) $\Delta[H^+] = 1000$

$\Delta pH = -\log \Delta[H^+] = -\log 10^3$

$= -3$; (–ve) sign indicates decrease in pH.

143. (2) $[H^+] = \dfrac{(n_{H^+})_{HCl} + (n_{H^+})_{H_2SO_4}}{\text{Total volume}}$

$= \dfrac{(M_{HCl} \times n_{factor,HCl} \times V_{HCl}) + M_{H_2SO_4} \times n_{f_{H_2SO_4}} \times V_{H_2SO_4}}{600}$

$[H^+] = \dfrac{(0.01 \times 1 \times 200) + (0.01 \times 2 \times 400)}{600}$

$= \dfrac{2+8}{600} = \dfrac{10}{600} = \dfrac{1}{60}$

$pH = -\log\left[\dfrac{1}{60}\right] = 1.78$

144. (3) Phenolphathalein have colour change range between pH 8.3 and 10.5.

145. (3) Phenolphthalein dissociate in basic medium

$$\underset{\text{(colourless)}}{HPh(aq)} \rightleftharpoons \underset{\text{(Pink)}}{H^+ + Ph^-}$$

146. (4) Assertion : The amphoteric nature of water can be explained by Bronsted – Lowry concept.

Reason : $\underset{\text{Acid}}{H_2S} + \underset{\text{Base}}{H_2O} \rightleftharpoons H_3O^+ + HS^-$

$\underset{\text{Acid}}{H_2O} + \underset{\text{Base}}{NH_3} \rightleftharpoons NH_4^+ + \bar{O}H$

147. (2) The chemical species having capability to donate electron pair is called Lewis base. PCl_5 cannot function as a Lewis base as the central atom P does not have lone pair of electrons.

148. (2) The solution with lowest pH will have the highest value of pOH.

149. (3) At equivalence point, pH is 7 and pH increases with addition of NaOH. So correct graph is (3).

150. (2) Temperature plays a significant role on pH measurements. As the temperature rises, molecular vibrations increase which results in greater ability of water to ionise and form more hydrogen ions. As a result, the pH will drop. So assertion is incorrect.

The dissociation of water molecules into ions is bond breaking and is therefore an endothermic process. So reason is also incorrect.

151. (2)

(A) $H_2SO_4 + NaOH \rightarrow NaHSO_4 + H_2O$

	H_2SO_4	NaOH	$NaHSO_4$	H_2O
Initial mol	0.04	0.04	0	0
mol at eqm.	0	0	0.04	0.04

$NaHSO_4 \rightarrow Na^+ + H^+ + SO_4^{2-}$

Total volume of the solution = (400+400) mL = 800 mL = 0.8 L

$[H^+] = \dfrac{0.04}{0.80} = 0.05M \Rightarrow pH = -\log[H^+]$

$\Rightarrow pH = 1.3$

(B) Ionic product of water increases with increase in temperature because ionisation of water is endothermic.

(C) $HA \rightleftharpoons H^+ + A^-$

	HA	H^+	A^-
Initial	C	0	0
At eqm.	C (1 – a)	Ca	Ca

Given pH = 5 $\Rightarrow -\log[H^+] = 5$

$\therefore [H^+] = 10^{-5}$

As we know,

$\Rightarrow K_a = \dfrac{C\alpha^2}{1-\alpha} \Rightarrow 10^{-5} = \dfrac{C\alpha^2}{1-\alpha} = \dfrac{C.\alpha}{(1-\alpha)}$

$10^{-5} = 10^{-5}\dfrac{\alpha}{1-\alpha} \Rightarrow \alpha = \dfrac{1}{2}$, *i.e.*, 50%

152. (2)

153. (3) Due to the resonance stabilisation of the conjugate base, $CH(CN)_3$ is the strongest acid amongst the given compounds.

154. (2) 75 mL $\frac{M}{5}$ HCl + 25 mL $\frac{M}{5}$ NaOH

25 mL $\frac{M}{5}$ NaOH will neutralise 25 mL $\frac{M}{5}$ HCl

75–25 = 50 mL $\frac{M}{5}$ HCl will remain.

Total volume will be 75 + 25 = 100 mL

When 50 mL $\frac{M}{5}$ HCl is diluted to 100 mL, molarity will reduce to half, i.e., $\frac{M}{10}$.

$[H^+] = [HCl] = \dfrac{1}{10}$

$pH = -\log_{10}[H^+] = -\log_{10}\dfrac{1}{10} = 1$

155. (4)

Due to incomplete octet of B, it can act as a Lewis acid.

156. (4) $NH_3 + HCl \longrightarrow NH_4Cl$

Moles of HCl = $0.2 \times 25 \times 10^{-3} = 0.005$

Moles of $NH_3 = 0.2 \times 50 \times 10^{-3} = 0.01$

Excess $NH_3 = 0.01 - 0.005 = 0.005$ moles

1 mole ammonia = 1 mole NH_4Cl

0.005 $NH_3 \equiv$ 0.005 NH_4Cl

Total volume = $V_{HCl} + V_{NH_3}$

= 25 + 50 = 75 mL

$[NH_3] = [NH_4Cl] = \dfrac{0.005 \text{ mole}}{75 \times 10^{-3} \text{ L}} = 0.066$ M

$pOH = pK_b + \log\dfrac{[NH_4Cl]}{[NH_3]}$

$pOH = 4.75 + \log\dfrac{[0.066]}{[0.066]} \Rightarrow pOH = 4.75$

$pH = 14 - pOH \Rightarrow pH = 9.25$

157. (2) $\underset{\text{Hydrazoic acid}}{N_3H} \rightleftharpoons N_3^- + H^+$

158. (4) $\because$ pH = 1 ; $[H^+] = 10^{-1} = 0.1$ M

pH = 2 ; $[H^+] = 10^{-2} = 0.01$ M $\therefore$ $M_1 = 0.1, V_1 = 1$

$M_2 = 0.01, V_2 = ?$

From $M_1V_1 = M_2V_2$

$\Rightarrow 0.1 \times 1 = 0.01 \times V_2 \Rightarrow V_2 = 10$ L

$\therefore$ Volume of water added = 10 – 1 = 9 L

159. (4) Given $[OH^-] = 5 \times 10^{-2}$

$\therefore pOH = -\log 5 \times 10^{-2} = -\log 5 + 2\log 10 = 1.30$

pH = 14 – pOH = 14 – 1.30 = 12.70

160. (1) Metal halide on hydrolysis with water form corresponding hydroxides.

The basic strength of hydroxide increases as we move down in a group.

Hence, $Be(OH)_2$ will have lowest pH.

161. (4) Given at 330 K, $K_w = 10^{-13.6}$

$pK_w = pH + pOH \Rightarrow 13.6 = pH + pOH$

$pOH = -\log 10^{-4} \Rightarrow pOH = 4$

$\therefore pH = 13.6 - 4 = 9.6$

162. (1) The species with the greatest proton affinity will be the strongest base, and its conjugate acid will be the weakest acid.

Since NH_3 is a very weak acid, NH_2^- must be a very strong base. Therefore, the correct order of proton affinity is

$I^- < F^- < HS^- < NH_2^-$

163. (3) The correct order of acidic strength of the given species is

$$\underset{\text{(iv)}}{HSO_3F} > \underset{\text{(ii)}}{H_3O^+} > \underset{\text{(iii)}}{HSO_4^-} > \underset{\text{(i)}}{HCO_3^-}$$

or (i) < (iii) < (ii) < (iv)

164. (1) $pH = -\log[H^+] = \log\dfrac{1}{[H^+]}$

$5.4 = \log\dfrac{1}{[H^+]}$

On Solving, $[H^+] = 3.98 \times 10^{-6}$

165. (1) Conjugate acid-base pair differ by only one proton.

$$OH^- \longrightarrow H^+ + O^{2-}$$

166. (4) $\underset{\text{Acid}}{H_2PO_4^-} \xrightarrow{-H^+} \underset{\text{Conjugate base}}{HPO_4^{2-}}$

Conjugate acid-base differs by H^+.

167. (2)

168. (4) The rain water after thunderstorm contains dissolved acid and therefore, the pH is less than rain water without thunderstorm.

169. (1) $(HSO_4)^-$ can accept and donate a proton

$(HSO_4)^- + H^+ \rightarrow H_2SO_4$ (acting as base)

$(HSO_4)^- - H^+ \rightarrow SO_4^{2-}$. (acting as acid)

170. (3) pH = 10

$\Rightarrow$ pOH = 14 – pH = 14 – 10 = 4

$\Rightarrow [OH^-] = 10^{-4}$

Number of moles of $OH^- = 10^{-4}$

Number. of moles of $Mg(OH)_2$

$= \dfrac{10^{-4}}{2} = 5 \times 10^{-5}$

$\Rightarrow$ mass of $Mg(OH)_2 = n_{Mg(OH)_2} \times M_{mg(OH)_2}$

$= 5 \times 10^{-5} \times 58 \times 10^3$ mg = 2.9 mg $\approx$ 3 mg.

171. (9000) $(M_1 \times V_1) = (M_2 \times V_2)$
$10^{-1} \times 1 = 10^{-2} \times V_2$
$V_2 = 10L$
Water added = 10 – 1 = 9 Litre = 9000 mL

172. (186) Total milimoles of H^+
$= (600 \times 0.01) + (400 \times 0.01 \times 2) = 14$
$[H^+] = \frac{14}{1000} = 14 \times 10^{-3}$
$pH = -\log[H^+] \Rightarrow pH = 3 - \log 14$
$= 1.86 = 186 \times 10^{-2}$

173. (5) Considering complete dissociation of $Ca(OH)_2$
$$\underset{1\,mol}{Ca(OH)_2} \rightarrow Ca^{2+} + \underset{2\,mol}{2OH^-}$$
pH = 12, ∴ pOH = 14 – 12 = 2
(Considering T = 298 K)
∴ $[OH^-] = 10^{-2}$ (M)
$[OH^-] = 10^{-2}$ (M)
$= 10^{-2}$ mol OH^- in 1000 mL
∴ In 100 mL aqueous solution has $\frac{10^{-2} \times 100}{1000}$
$= 10^{-3}$ mol of OH^-
2 mol of OH^- is produced by 1 mol $Ca(OH)_2$
10^{-3} mol OH^- is produced by $\frac{1}{2} \times 10^{-3}$ mol
$= 0.5 \times 10^{-3}$ mol = 0.5 m mol
$= 5 \times 10^{-1}$ m mol. of $Ca(OH)_2$

174. (9079) 0.02 mol HCl neutralize 0.02 mol of NH_3.
Rest $n_{NH_3} = 0.1 - 0.02 = 0.08$
$n_{NH_4Cl} = n_{NH_4^+} = 0.1 + 0.02 = 0.12$
$$pOH = pK_b + \log\frac{[NH_4^+]}{[NH_3]} = 4.745 + \log\frac{0.12}{0.08}$$
$= 4.745 + \log\frac{3}{2} = 4.745 + 0.477 - 0.301$
$pOH = 4.921; pH = 14 - pH = 9.079 = 9079 \times 10^{-3}$

175. (12) millimole of NaOH = 0.24 × 25
∴ millimole of acid = 0.24 × 25
⇒ mass of acid = 0.24 × 25 × 24.2 mg
for pure acid,
$V = \frac{w}{d}$; (d = 1.21 kg/L = 1.21 g/mL)
$\therefore V = \frac{0.24 \times 25 \times 24.2}{1.21} \times 10^{-3}$
$= 120 \times 10^{-3}$ mL $= 12 \times 10^{-2}$ mL

176. (11) 0.001 M NaOH
$[OH^-] = 10^{-3}$
$pOH = -\log[OH^-] \Rightarrow pOH = 3$
$pH + pOH = 14 \Rightarrow pH = 11$

177. (6021)
$$HCl(aq.) + NaOH(aq.) \rightarrow NaCl(aq.) + H_2O(\ell)$$

	HCl	NaOH	NaCl	H_2O
Initial	50 m mol	30 m mol	–	–
Final	20 m mol		30	30

$[HCl] = \frac{20}{80} = \frac{1}{4} M = 2.5 \times 10^{-1} M$
$pH = -\log 2.5 \times 10^{-1} = 1 - 0.33979 = 0.6021$
⇒ $pH = 6021 \times 10^{-4}$.

178. (37) $CO_2 + H_2O \longrightarrow H_2CO_3$
30 bar .... 1 mol/L
3 bar .... 0.1 mol/L

$$H_2CO_3 \rightleftharpoons H^+ + HCO_3^-$$

	H_2CO_3	H^+	HCO_3^-
t = 0	0.1	0	0
at Eqm.	0.1(1–α)	0.1α	0.1α

$4.0 \times 10^{-7} = \frac{0.1\alpha^2}{1-\alpha}$
As (1 – α) ~ 1
$\Rightarrow \alpha^2 = 4 \times 10^{-6} \Rightarrow \alpha = 2 \times 10^{-3}$
$[H^+] = C\alpha = 2 \times 10^{-4}$ M
$pH = 4 - \log 2 = 3.7 \simeq 37 \times 10^{-1}$

179. (10.60) $M_{H_2SO_4} = \frac{9.8}{98 \times 100} = 10^{-3}$ M
$M_{NaOH} = \frac{4}{40 \times 100} = 10^{-3}$ M
After neutralisation $[OH^-]$ can be calculated as
$$[OH^-] = \frac{(40 \times 10^{-3}) - (2 \times 10^{-3} \times 10)}{50}$$
$\Rightarrow [OH^-] = \frac{2}{5} \times 10^{-3}$
$\Rightarrow pOH = -\log[OH^-] \Rightarrow pOH = 3.397$
pH = 14 – pOH = 14 – 3.397 = 10.603

180. (5.23)
$3g\ CH_3COOH = \frac{3}{60} = 0.05$ mol = 50 m mol
No. of millimoles = Molarity × Volume in mL
250 mL of 0.1 M HCl = 250 × 0.1 = 25 m mol
500 mL solution = 50 m mol CH_3COOH
20 mL solution $= \frac{50}{500} \times 20 = 2$ m mol CH_3COOH
500 mL solution contains = 25 m mol HCl
20 mL solution contains $= \frac{25}{500} \times 20 = 1$ m mol HCl
$\frac{1}{2}$ mL of 5M NaOH $= \frac{1}{2} \times 5 = 2.5$ m mol NaOH
$HCl + NaOH \longrightarrow NaCl + H_2O$
1 2.5 1
Remaining NaOH = 2.5–1 = 1.5 m mol
$CH_3COOH + NaOH(\text{remaining}) \rightarrow CH_3COONa + Water$

CH_3COOH	NaOH	CH_3COONa	Water
2	1.5	0	0
0.5	0	1.5	–

$pH = pK_a + \log\frac{1.5}{0.5} = 4.75 + \log 3$
= 4.75 + 0.48 = 5.23

181. (3)
(A) $H^+(aq) + OH^-(aq) \rightarrow H_2O(\ell)$
t = 0 2.5 mmol 4 mmol
t = ∞ 0 1.5 mmol
Final solution, $(OH^-) = \frac{1.5\text{ mmol}}{35\text{ ml}} = \frac{1.5}{35}$
$pOH = -\log\frac{1.5}{35}$
$pH = 14 + \log\left(\frac{1.5}{35}\right)$ (Basic solution)
(B) $H^+(aq) + OH^-(aq) \longrightarrow H_2O(\ell)$
t = 0 0.2 m mol 0.2 m mol
Final pH = 7 (complete neutralization)
(C) $H^+(aq) + OH^-(aq) \longrightarrow H_2O(\ell)$
t = 0 2 mmol 1 mmol
t = ∞ 1 mmol 0
Final pH < 7 (acidic solution)
So, order of pH values :
(C) < (B) < (A)

182. (3) $HCl \longrightarrow H^+ + Cl^-$
0.1 M – –
– 0.1M 0.1M
Initial concentration of H^+ = 0.1 M
Let x be the amount of H_2A that dissociates.
At equilibrium,
$$\underset{0.1-x}{H_2A} \rightleftharpoons \underset{x+0.1}{H^+} + \underset{x}{HA^-}$$
(Due to common ion effect, x is very less)
(**Note:** $K_{a2} = 1 \times 10^{-13}$ which is very small constant i.e. it consumes negligible amount of HA^-, hence ignored)
$K_{a_1} = \frac{[H^+][HA^-]}{[H_2A]}$
Given $K_{a_1} = 8.1 \times 10^{-8}$
$K_{a_1} = \frac{0.1 \times [HA^-]}{0.1} \Rightarrow [HA^-] = K_{a_1} = 8.1 \times 10^{-8}$.

183. (1)

184. (3) Concentration of salt, NH_4Cl, produced by neutralization of NH_4OH and HCl.
NH_4Cl, no. of moles = 2m mol = 2×10^{-3} mol
$[NH_4Cl] = \frac{2\text{m mol}}{(20+40)\text{ml}} = \frac{1}{30}M$
$pH = 7 - \frac{1}{2}pK_b - \frac{1}{2}\log C$
$= 7 - \frac{1}{2} \times (+5) - \frac{1}{2}\log\left(\frac{1}{30}\right)$
$= 7 - \frac{5}{2} + \frac{1.48}{2} = 5.24$

185. (4)
$H_2C_2O_4 \rightleftharpoons H^+ + HC_2O_4^-\quad K_{a_1}$ ...(i)
$H_2C_2O_4 \rightleftharpoons H^+ + C_2O_4^{2-}\quad K_{a_2}$ ...(ii)
$H_2C_2O_4 \rightleftharpoons 2H^+ + C_2O_4^{2-}\quad K_{a_3}$ ...(iii)
addition of eq. (i) with eq. (ii) gives eq (iii),
So $K_{a_3} = K_{a_1} \times K_{a_2}$

186. (4) Sodium acetate is a salt of strong base and weak acid.
$\therefore pH = 7 + \frac{1}{2}pK_a + \frac{1}{2}\log c$ where
$pK_a = -\log K_a$
$= 7 + \frac{5}{2} - \frac{1}{2} = 9.0$

187. (3) Higher the value of K_a, higher will be the acidic nature. Further, since CN^-, F^- and NO_2^- are conjugate base of the acids HCN, HF and HNO_2 respectively, hence the correct order of base strength will be :
$F^- < NO_2^- < CN^-$
(∵ stronger the acid, weaker will be its conjugate base)

188. (4)
Conc. of CH_3COOH = 5 g in 500 mL = 10 g/L
$[CH_3COOH] = \frac{10}{60}M = \frac{1}{6}M$
[MW of CH_3COOH = 60]

Concentration of CH_3COONa is computed as under.
conc. = 7.5 g in 500 mL = 15 g / L

$[CH_3COONa] = \frac{15}{82} M$ [MW of $CH_3COONa = 82$]

$pK_a = -\log K_a = -\log(1.75 \times 10^{-5}) = 4.76$

$pH = pK_a + \log\frac{[salt]}{[acid]} = 4.76 + \log\frac{15/82}{1/6} = 4.8$

189. (3) $[H^+] = C\alpha, \alpha = \frac{[H^+]}{C}$

or $\alpha = \frac{10^{-3}}{0.1} = 10^{-2}$
[Since pH = 3; $[H^+] = 10^{-3}$]

$K_a = C\alpha^2 = 0.1 \times 10^{-2} \times 10^{-2} = 10^{-5}$

190. (2) $CN^-(aq) + H_2O(l) \rightleftharpoons HCN(aq) + OH^-(aq)\ K_b = 1.6 \times 10^{-5}$

$\underset{base}{OH^-(aq)} + \underset{acid}{H_2O(l)} \rightleftharpoons \underset{conj.acid}{H_2O(l)} + \underset{conj.base}{OH^-(aq)}$

$K_b = 1$

$\underset{base}{H_2O(l)} + \underset{acid}{H_2O(l)} \rightleftharpoons \underset{conj.\ acid}{H_3O^+(aq)} + \underset{conj.\ base}{OH^-(aq)}$

$K_b = 1.0 \times 10^{-14}$
Hence, the relative base strength:
$OH^- > CN^- > H_2O$

191. (4) pH = 5 means
$[H^+] = 10^{-5}$

	HA	⇌	H^+	+	A^-
t = 0	c		0		0
t_{eq}	c (1 – α)		cα		cα

$K_a = \frac{[H^+][A^-]}{[HA]} = \frac{(c\alpha)^2}{c(1-\alpha)} = \frac{[H^+]^2}{c - [H^+]}$

But, $[H^+] << c$

$\therefore K_a = [H^+]^2 = (10^{-5})^2 = 10^{-10}$

192. (2) The salt (AB) given is a salt of weak acid and weak base. Hence the pH can be calculated by the following formula

$\therefore pH = 7 + \frac{1}{2}pK_a - \frac{1}{2}pK_b = 7 + \frac{1}{2}(3.2) - \frac{1}{2}(3.4) = 6.9$

193. (1)

(i) $\underset{acid_1}{H_3PO_4} + \underset{base_2}{H_2O} \longrightarrow \underset{acid_2}{H_3O^+} + \underset{base_1}{H_2PO_4^-}$

(ii) $\underset{acid_1}{H_2PO_4^-} + \underset{base_2}{H_2O} \longrightarrow \underset{base_1}{HPO_4^{2-}} + \underset{acid_2}{H_3O^+}$

(iii) $\underset{base_1}{H_2PO_4^-} + \underset{acid_2}{OH^-} \longrightarrow \underset{acid_1}{H_3PO_4} + \underset{base_2}{O^{2-}}$

Hence only in (ii) reaction, $H_2PO_4^-$ is acting as an acid.

194. (3)

195. (3) In aqueous solution, BA(salt) hydrolyses to give

$BA + H_2O \rightleftharpoons \underset{base}{BOH} + \underset{acid}{HA}$

Now pH is given by

$pH = \frac{1}{2}pK_w + \frac{1}{2}pK_a - \frac{1}{2}pK_b$

Substituting given values, we get

$pH = \frac{1}{2}(14 + 4.80 - 4.78) = 7.01$

196. (4) $H_2A \rightleftharpoons H^+ + HA^-$

$\therefore K_1 = 1.0 \times 10^{-5} = \frac{[H^+][HA^-]}{[H_2A]}$

$HA^- \rightleftharpoons H^+ + A^-$

$\therefore K_2 = 5.0 \times 10^{-10} = \frac{[H^+][A^-]}{[HA^-]}$

For the reaction, $H_2A \rightleftharpoons 2H^+ + A^{2-}$

$K = \frac{[H^+]^2[A^{2-}]}{[H_2A]} = K_1 \times K_2$

$= (1.0 \times 10^{-5}) \times (5 \times 10^{-10}) = 5 \times 10^{-15}$

197. (3) $pH = \frac{1}{2}[pK_a - \log c]$

$pH = \frac{1}{2}\left[4 - \log 10^{-2}\right]$

pH = 3

198. (100) $H_2X \rightleftharpoons H^+ + HX^-$,
$0.1 - x \quad x + y \quad x - y$

$2.5 \times 10^{-8} = \frac{(x+y)(x-y)}{0.1\ x}$

$HX^- \rightleftharpoons H^+ + X^{2-}$,
$x - y \quad x + y \quad y$

$1 \times 10^{-13} = \frac{(x+y)(y)}{x-y}$

For very weak acids

Approximate: $K_{a_1} >> K_{a_2} \Rightarrow$ So $x >> y$.

$x + y \approx x, \quad x - y \approx x$

$10^{-13} = \frac{x \cdot y}{x} \Rightarrow y = 10^{-13}$

$[X^{2-}] = 10^{-13} \Rightarrow [X^{2-}] = 100 \times 10^{-15}$

199. (70) $A_2 + B_2 \xrightarrow{500K} 2AB \ \log K = 2.2$

$\Delta H° = (2 \times 32) - (6 + x) = (58 - x)$ kJ

$\Delta S° = (2 \times 222) - (146 + 280) = 18$ Joule

$\Delta G° = -RT \ln K$

$\Delta G° = -\frac{8.314 \times 500 \times 2.2 \times 2.303}{1000}$

$\Delta G° = -21.06$

$\Delta H° - T\Delta S° = -21.06$

$58 - x - 500\left(\frac{18}{1000}\right) = -21.06$

x = 70.06 kJ/mol.

200. (25) After Dilution
pH = 6 ⇒ $[H^+] = 10^{-6}$

$[H^+] = \sqrt{k_a \cdot C}$

$10^{-6} = \sqrt{4 \times 10^{-10} \times C} \Rightarrow \frac{10^{-12}}{4 \times 10^{-10}} = C$

$0.25 \times 10^{-2} = C \Rightarrow 25 \times 10^{-4} = C$

∴ Value of x is 25.

201. (7) Given pH = 9 ⇒ pOH = 14 – 9 = 5;

$[OH]^\ominus = 10^{-5} M$

$C_2H_5NH_2(aq) + H_2O \rightleftharpoons C_2H_2NH_3^+ + OH^\ominus$

Initial C = 10^{-3} M
Final C (1 – α) Cα Cα

⇒ C = 10^{-3} = 10^{-5} = 10^{-5} as 1 – α ≈ 1

Now, $K_b = \frac{[C_2H_5NH_3^+][OH]^\ominus}{[C_2H_5NH_2]}$

$\Rightarrow K_b = \frac{10^{-5} \times 10^{-5}}{10^{-3}} = 10^{-7}$

202. (7) $K_a(CH_3COOH) = 1.8 \times 10^{-5}$
$pK_a = -\log(1.8 \times 10^{-5}) = 4.7$
$K_b(NH_4OH) = 1.8 \times 10^{-5}$, pK_b
$= -\log(1.8 \times 10^{-5}) = 4.7$

$pH = 7 + \frac{1}{2}[pK_a - pK_b] = 7 + \frac{1}{2}[4.7 - 4.7]$

pH = 7

203. (27) K_a of Butyric acid = 2×10^{-5}
$pK_a = -\log K_a = -\log(2 \times 10^{-5}) = 4.7$
The pH of butyric acid is calculated as:

$pH = \frac{1}{2}pK_a - \frac{1}{2}\log c = \frac{1}{2}(4.7) - \frac{1}{2}\log(0.2)$

= 2.35 + 0.35 = 2.7
pH = 27 × 10^{-1}.

204. (2) $HCl \longrightarrow \underset{0.1\ M}{H^+} + \underset{0.1\ M}{Cl^-}$

$\underset{0.01(1-\alpha)}{HA} \rightleftharpoons \underset{0.1 + 0.01\ \alpha}{H^+} + \underset{0.01\alpha}{A^-} \quad K_a = 2.0 \times 10^{-6}$

$K_a = \frac{[H^+][A^-]}{[HA]}$

$2 \times 10^{-6} = \frac{0.1 \times 0.01\alpha}{0.01(1-\alpha)} \simeq \frac{0.1 \times 0.01\alpha}{0.01}$ [considering, $0.1 + 0.1\alpha \approx 0.1$ since α is very small]

α = 2.0 × 10^{-5}

205. (1) k_{a_1} of $H_2SO_3 >> k_{a_2}$ of H_2SO_3

∴ The contribution of H^+ from 2nd dissociation of H_2SO_3 can be neglected.

	H_2SO_3	⇌	H^+	+	HSO_3^-
t=0	c		0		0
t=t	c(1–α)		cα		cα

$\Rightarrow \frac{c\alpha^2}{1-\alpha} = 1.7 \times 10^{-2} \Rightarrow \frac{0.588\alpha^2}{1-\alpha} = 1.7 \times 10^{-2}$

$\Rightarrow 58.8\alpha^2 + 1.7\alpha - 1.7 = 0$

$\alpha = \frac{-1.7 + \sqrt{1.7^2 + 4 \times 1.7 \times 58.8}}{2 \times 58.8} = 0.156$

$[H^+] = c\alpha = 0.156 \times 0.588 = 0.092$

$pH = -\log[H^+] = 1.036 \approx 1$

206. (7.0) Since, $(NH_4)_3PO_4$ is salt of weak acid (H_3PO_4) and weak base (NH_4OH).

$pH = 7 + \frac{1}{2}(pK_a - pK_b)$

$= 7 + \frac{1}{2}(5.23 - 4.75) = 7.24 \approx 7.0$

207. (1)

$M_3A_2(s) \rightleftharpoons 3M^{+2}(aq) + 2A^{-3}(aq)$

	$M_3A_2(s)$	$3M^{+2}(aq)$	$2A^{-3}(aq)$
Molarity	$\frac{x}{y}$	–	–
	–	$3\frac{x}{y}$	$2\frac{x}{y}$

$K_{sp} = [M^{+2}]^3 [A^{-3}]^2$
$K_{sp} = [3x/y]^3 [2x/y]^2$

$S = \frac{x}{y}$ mol L^{-1}

$K_{sp} = (3S)^3 \times (2S)^2 = 108 S^5 = 108\left(\frac{x}{y}\right)^5$

$[A^{-3}] = \frac{2x}{y}$

$\frac{K_{sp}}{[A^{-3}]} = \frac{108 \times x^5 \times y}{y^5 \times 2x} = \frac{54x^4}{y^4} \Rightarrow \frac{[A^{-3}]}{K_{sp}} = \frac{y^4}{54x^4}$

208. (2) Neutralization occurs at equivalence point

so, at equivalence point

Moles of CH_3COOH = Moles of NaOH

$M_1V_{1\,(acid)} = M_2V_{2\,(base)}$

$M_1 \times 20 = 0.1 \times 28.4$

$M_1 = \frac{0.1 \times 28.4}{20} = 0.142$ M

	CH_3COOH +	NaOH →	CH_3COONa +	H_2O
Initial conc.	0.142 M	0.1 M	–	–
Millimoles	0.142 × 20 = 2.84	0.1 × 14.2 = 1.42	–	–
After Neutralisation millimoles left	1.42	0	1.42	–

$pH = pK_a + \log\left(\frac{[Salt]}{[Acid]}\right)$

$= 4.75 + \log\left(\frac{1.42}{1.42}\right) = 4.75 + \log(1)$

$\therefore$ pH = 4.75.

209. (2) For 0.2 M, 20 ml HX solution.

For weak acid with $\alpha <<< 1 : [H^+] = \sqrt{k_a \cdot c}$

pH of weak acid $= \frac{1}{2}[pk_a - \log c]$

$= \frac{1}{2}[4 - \log 5 - \log 0.2]$

$= \frac{1}{2}[4 - \log 5 + \log 5] = 2$

Initially,

$$HX + NaOH \longrightarrow NaX + H_2O$$

HX	NaOH
20 ml, 0.2 M	10 ml, 0.2 M
4m mole	2m mole
2m mole	2m mole

Using the Henderson-Hasselbach equation :

$pH = pK_a + \log\frac{salt}{acid}$

$pH = 3.3 + \log\frac{2}{2} = 3.3$

210. (4) For Ag_2CrO_4

$$Ag_2CrO_4 \rightleftharpoons 2Ag^+ + CrO_4^{2-}$$
$$\qquad\qquad 2S_1 \quad S_1$$

$(2S_1)^2(S_1) = 32x$ (given)

$4S_1^3 = 32x \Rightarrow S_1 = 2x^{1/3}$

For AgBr

$$AgBr \rightleftharpoons Ag^+ + Br^-$$
$$\qquad (S_2) \quad (S_2)$$

$(S_2)(S_2) = 4y$ (given)

$S_2^2 = 4y \Rightarrow S_2 = 2y^{1/2}$

Ratio $\left(\frac{S_1}{S_2}\right) = \frac{\sqrt[3]{x}}{\sqrt{y}}$

211. (3)

	HCl	+ B	→ BH⁺ + Cl⁻
	0.02M	0.02M	
	x ml	y ml	
T_f	0	0.02y−0.02x	0.02x

$pOH = pK_b + \log\left[\frac{Salt}{Base}\right] \Rightarrow 5 = 5.699 + \log\left[\frac{Salt}{Base}\right]$

$\frac{x}{y-x} = \frac{1}{5} \Rightarrow 6x = y \Rightarrow 7x = 100$

$x = \frac{100}{7}$ ml = 14.3 ml

and $y = \frac{600}{7}$ ml = 85.7 ml

212. (2) $pOH = pK_b + \log\frac{Salt}{Base}$

$4.75 = 4.75 + \log\frac{Salt}{Base}$

Milimole of [Salt] = milimole of [Base]

$$NH_4OH + HCl \longrightarrow NH_4Cl + H_2O$$

NH_4OH	HCl	NH_4Cl
0.2M, 0.5L 100 mmole	0.1M, 0.5L 50 mmole	
50 mmole	–	50 mmole

Millimoles of NH_4OH = millimoles of NH_4Cl.

213. (3) When equal volumes are mixed molarity reduces to half.

For precipitation $Q_{SP} = [A^{+2}] [Y^-]^2 > K_{SP}$

(1) $Q_{SP} = (1.8 \times 10^{-3})\left(\frac{5}{2} \times 10^{-4}\right)^2 < K_{SP}$

(2) $Q_{SP} = (10^{-4})(0.4 \times 10^{-3})^2 < K_{SP}$

(3) $Q_{SP} = (10^{-2})(10^{-2})^2 > K_{SP}$

(4) $Q_{SP} = \left(\frac{1.5}{2} \times 10^{-4}\right)\left(\frac{1.5}{2} \times 10^{-3}\right)^2 < K_{SP}$

214. (2) K_{sp} values determine solubility. Based on the K_{sp} values and salt analysis cation identification, the order of K_{sp} value is:

HgS < PbS < AgBr < $Ca(OH)_2$

Lower K_{sp} means lower solubility.

215. (1)

$$Cr(OH)_3(s) \rightleftharpoons Cr^{+3}(aq) + 3OH^-(aq)$$

At eq: $\qquad s \qquad 3s$

$K_{sp} = (s).(3s)^3 = 27s^4$

$27s^4 = 1.6 \times 10^{-30}$; $s = \left(\frac{1.6}{27} \times 10^{-30}\right)^{1/4}$

216. (1) Condition for precipitation $Q_{ip} > K_{sp}$

For $[A(OH)_2]$

$[A^{2+}][OH^-]^2 > 9 \times 10^{-10}$

$[A^{+2}] = 1$ M

$\Rightarrow [OH^-] > 3 \times 10^{-5}$ M

For $[B(OH)_3]$

$[B^{3+}][OH^-]^3 > 27 \times 10^{-18}$

$[B^{3+}] = 1$M

$\Rightarrow [OH^-] > 3 \times 10^{-6}$ M

So, $B(OH)_3$ will precipitate before $A(OH)_2$

217. (1)

$$Zr_3(PO_4)_4(s) \rightleftharpoons 3Zr^{+4}(aq) + 4PO_4^{-3}(aq)$$

– 3s 4s $K_{sp} = (3s)^3(4s)^4 = 27s^3 \times 256\, s^4$

$= 6912\, s^7 \Rightarrow s = \left(\frac{K_{sp}}{6912}\right)^{\frac{1}{7}}$

218. (2) $AB_2(s) \rightleftharpoons A^{+2}(aq) + 2B^-(aq)$

$K_{sp} = [A^{+2}][B^-]^2$

$= 1.2 \times 10^{-4} \times (2.4 \times 10^{-4})^2$

$= 6.91 \times 10^{-12}\, M^3$

219. (1)

220. (4) Due to common ion effect reaction goes backward and precipitation occurs.

221. (3) $Ca_3(PO_4)_2 \rightleftharpoons 3Ca^{2+} + 2PO_4^{3-}$

$S = \left(\frac{W \times 1000}{M \times 100}\right) = \frac{W \times 10}{M}$

$\therefore K_{sp} = [Ca^{2+}][PO_4^{3-}]$

$= (3S)^3 (2S)^2 = 108S^5$

$= 108 \times \left(\frac{W \times 10}{M}\right)^5 = 108 \times 10^5\left(\frac{W}{M}\right)$

$= 1.08 \times 10^7\left(\frac{W}{M}\right)$

$K_{sp} = 10^7\left(\frac{W}{M}\right)^5$

222. (3) pH + pOH = 14

pOH = 14 − 8.26

$pOH = pK_b + \log\frac{[NH_4^+]}{[NH_3]}$

$\Rightarrow 5.74 = 4.74 + \log\frac{[NH_4^+]}{0.2} \Rightarrow [NH_4^+] = 2M$

Molar mass of NH_4Cl = 53.5 g mol^{-1}

Amount of NH_4Cl = 2 × 53.5 = 107 g.

223. (4) In deionized water no common ion effect will take place so solubility of AgCl will be maximum in it.

224. (2) According to Henderson-Hasselbalch equation

$pH = pK_a + \log\frac{[Salt]}{[Acid]}$

$4 = 5 - \log 1.3 + \log\frac{[CH_3CH_2COO^-]}{[CH_3CH_2COOH]}$

$\log\frac{[CH_3CH_2COO^-]}{[CH_3CH_2COOH]} = \log 1.3 - 1 = \log\frac{1.3}{10}$

$\therefore \log A - \log(B) = \log\left(\frac{A}{B}\right)$

$\frac{[CH_3CH_2COO^-]}{[CH_3CH_2COOH]} = 0.13$

225. (1) $Bi_2S_3 \rightleftharpoons 2Bi^{3+} + 3S^{2-}$
$\qquad\qquad 2s \qquad 3s$

$K_{sp} = (2s)^2(3s)^3 = 108\,(s)^5$

$(s)^5 = \frac{1.08 \times 10^{-73}}{108} \Rightarrow s = 10^{-15}$

226. (2) Let s be the solubility of $Ca(OH)_2$ in water

$Ca(OH)_2 \rightleftharpoons Ca^{2+} + 2OH^-$
$\qquad\qquad s \qquad 2s$

$K_{sp} = [Ca^{2+}][OH^-]^2 = s \times (2s)^2$

$\Rightarrow 5.5\times10^{-6} = 4s^3$

$\Rightarrow s^3 = \frac{5.5}{4}\times10^{-6}$

$s = \left[\frac{5.5}{4}\right]^{\frac{1}{3}}\times10^{-2} = 1.11\times10^{-2}$

227. (1) $AgCN \rightleftharpoons Ag^+ + CN^-$

Solubility of AgCN is x. pH = 3 ⇒ $[H^+] = 10^{-3}$

$HCN \rightleftharpoons H^+ + CN^-$

$K_a = \frac{[H^+][CN^-]}{[HCN]}$

$\Rightarrow \frac{[10^{-3}][CN^-]}{[HCN]} = 6.2\times10^{-10}$

$\Rightarrow \frac{[CN^-]}{[HCN]} = 6.2\times10^{-7}$

Each CN– hydrolyses to give one HCN as:

$CN^- + H_2O \rightleftharpoons HCN + OH^-$

HCN is a weak acid, thus, [HCN] >> [CN⁻]

$x = [Ag^+] = [CN^-] + [HCN]$

$\Rightarrow x = [Ag^+] \approx [HCN]$

$\Rightarrow [CN^-] = x \times 6.2 \times 10^{-7}$

$K_{sp}(AgCN) = [Ag^+][CN^-]$

$\Rightarrow 2.2 \times 10^{-16} = (x)(x \times 6.2 \times 10^{-7})$

$\Rightarrow x^2 = 3.55 \times 10^{-10} \Rightarrow x = 1.88 \times 10^{-5}$

228. (4) $HCl + CH_3COONa \rightarrow CH_3COOH + NaCl$

	HCl	CH_3COONa	CH_3COOH	NaCl
Millimoles (initial)	10	20	0	0
Millimoles (Final)	0	10	10	10

Buffer solution contains CH_3COONa (10 millimole) and CH_3COOH (10 millimole) which is a acidic buffer.

229. (4) CuS gets precipitated very quickly due to low K_{sp} value.

230. (3) $PbCl_2(s) \rightleftharpoons Pb^{2+}(aq) + 2Cl^-(aq)$

Given; $K_{sp} = 1.6 \times 10^{-5}$

$[Pb^{2+}] = \frac{300\times0.134}{400} = 0.1005$; $[Cl^-] = \frac{100\times0.4}{400} = 0.1$

$Q = [Pb^{2+}][Cl^-]^2 = 0.1005 \times (0.1)^2 = 1.005 \times 10^{-3}$

Therefore, $Q > K_{sp}$

231. (2) $Cr(OH)_3 \rightarrow \underset{s}{Cr^{3+}} + \underset{3s}{3OH^-}$

$K_{sp} = s\cdot(3s)^3$

$\Rightarrow 6 \times 10^{-31} = 27\cdot s^4$; $s = \left(\frac{6}{27}\times10^{-31}\right)^{1/4}$

$[OH^-] = 3s = 3 \times \left(\frac{6}{27}\times10^{-31}\right)^{1/4}$

$= (18 \times 10^{-31})^{1/4}$ M

232. (2) From the given curve,

if [X] = 1 mM then [Y] = 2 mM ∴ Salt is XY_2

$K_{sp} = [X][Y]^2 = (10^{-3})(2 \times 10^{-3})^2 = 4 \times 10^{-9} M^3$

233. (4) $Cd(OH)_2 \rightleftharpoons \underset{s}{Cd^{2+}} + \underset{2s}{2OH^-}$

At equilibrium, $K_{sp} = s(2s)^2 = 4s^3$

$\Rightarrow K_{sp} = 4 \times (1.84 \times 10^{-5})^3$

Solubility in buffer solution having pH = 12

$[OH^-] = 10^{-2}$

$Cd(OH)_2 \rightleftharpoons \underset{s'}{Cd^{2+}} + \underset{2s+10^{-2}\approx10^{-2}}{2OH^-}$

$\therefore K_{sp} = 4 \times (1.84 \times 10^{-5})^3 = s'(10^{-2})^2$

$\Rightarrow s' = \frac{24.9\times10^{-15}}{10^{-4}} = 2.49\times10^{-10}$ M

234. (3) Let the solubility of $Al(OH)_3$ in 0.2M NaOH solution be s.

Then,

$Al(OH)_3 \rightleftharpoons \underset{s}{Al^{3+}} + \underset{3s}{3OH^-}$

and $\underset{0.2M}{NaOH} \rightleftharpoons \underset{0.2M}{Na^+} + \underset{0.2M}{OH^-}$

$[Al^{3+}] = s$ and $[OH^-] = 3s + 0.2 » 0.2$

$K_{sp} = 2.4 \times 10^{-24} = [Al^{3+}][OH]^3$

$2.4 \times 10^{-24} = s(0.2)^3$

$s = \frac{2.4\times10^{-24}}{8\times10^{-3}} = 3\times10^{-22}$ mol / L

235. (4) $pH = 7 - \frac{1}{2}pK_b - \frac{1}{2}\log C$

$pH = 7 - \frac{5}{2} - \frac{1}{2}(\log 2\times10^{-2}) = 5.35$

236. (2) $Zr_3(PO_4)_4 \rightleftharpoons \underset{3S}{3Zr^{4+}} + \underset{4S}{4PO_4^{3-}}$

$K_{sp} = [Zr^{4+}]^3[PO_4^{3-}]^4 = (3S)^3(4S)^4$

$K_{sp} = 6912\ S^7 \Rightarrow S = \left(\frac{K_{sp}}{6912}\right)^{1/7}$

237. (2) m. mol of H_2SO_4

= 20 × 0.1 = 2

m. mol of NH_4OH = 30 × 0.2 = 6

	H_2SO_4	$+2NH_4OH$	$\rightarrow (NH_4)_2SO_4 + 2H_2O$
Initial	2 m mol	6 m mol	0
Final	(0) = 0 m mol	(6 – 2 × 2) = 2 m mol	2 m mol

$[\overset{+}{N}H_4]$ (from $(NH_4)_2SO_4$) = 2 × 2 = 4 m mol

Total volume = 30 + 20 = 50 mL

$pOH = pK_b + \log\left[\frac{Salt}{Base}\right] = 4.7 + \log\frac{4/50}{2/50}$

= 4.7 + log 2 = 5

pH = 14 – pOH = 14 – 5 = 9

238. (3) As $AgNO_3$ dissociates completely, therefore in 0.1 M $AgNO_3$ solution, $[Ag^+]$ = 0.1 M

$AgNO_3 \longrightarrow \underset{0.1}{Ag^+} + \underset{0.1}{NO_3^-}$

$Ag_2CO_3 \rightleftharpoons \underset{0.1+2s}{2Ag^+} + \underset{s}{CO_3^{2-}}$

$K_{sp} = [Ag^+]^2[CO_3^{2-}]$

$8 \times 10^{-12} = (0.1 + 2s)^2 \times s$

$0.01\,s = 8 \times 10^{-12}$; $(0.1 + 2s \times 0.1)$

$s = 8 \times 10^{-10}$ M

239. (1)

$\underset{14.08\text{ m mol}}{Na_2SO_4} + Ca(OH)_2 \rightarrow CaSO_4 + \underset{100\text{ m mol}}{2NaOH}$

m mol of $Na_2SO_4 = \frac{2\times1000}{142} = 14.08$ m mol

m mol of $CaSO_4$ formed = 14.08 m mol

Mass of $CaSO_4$ formed = 14.08 × 10^{-3} × 136 = 1.9 g

m mol of NaOH ≈ 28 m mol = 0.028 mol

$Ca(OH)_2 \rightleftharpoons \underset{s}{Ca^{2+}} + \underset{2s+0.028}{2OH^-}$

Value of 's' will be negligible in comparison to 0.028.

$[OH^-] = \frac{0.028\text{ mol}}{0.1\text{ L}} = 0.28\text{ mol L}^{-1}$

240. (2)

241. (2)

$H_2S \rightleftharpoons 2H^+ + S^{2-}$, $K_{a_1}.K_{a_2} = K_{eq}$

H_2S is a very weak acid and HCl is a strong acid. Assuming that all $[H^+]$ will be from dissociation of HCl, $[H^+]$ = 0.2M

$\therefore \frac{[H^+]^2[S^{2-}]}{[H_2S]} = 1\times10^{-7}\times1.2\times10^{-13}$

$\frac{[0.2]^2[S^{2-}]}{[0.1]} = 1.2\times10^{-20}$; $[S^{2-}] = 3 \times 10^{-20}$

242. (3) For final concentration of SO_4^{2-} in $BaSO_4$ solution,

$M_1V_1 = M_2V_2$

$1 \times 50 = M_2 \times 500$; $M_2 = \frac{1}{10}$

For just precipitation, Ionic product = K_{sp}

$[Ba^{2+}][SO_4^{2-}] = K_{sp}(BaSO_4)$

$[Ba^{2+}] \times \frac{1}{10} = 10^{-10}$

$[Ba^{2+}] = 10^{-9}$ M in 500 mL solution

For $[Ba^{2+}]$ in original solution,

$V_1 = 500 - 50 = 450$ mL

$\Rightarrow M_1 \times 450 = 10^{-9} \times 500$

[where M_1 = Molarity of original solution]

$M_1 = \frac{500}{450}\times10^{-9} = 1.11\times10^{-9}$ M

243. (4) $(K_{sp})_{PbCl_2} = 3.2\times10^{-8} = 32\times10^{-9}$

$PbCl_2 \rightleftharpoons \underset{s}{Pb^{2+}} + \underset{2s}{2Cl^-}$; $K_{sp} = [Pb^{2+}][Cl^-]^2$

$K_{sp} = 4s^3 = 32 \times 10^{-9} \Rightarrow s = 2 \times 10^{-3}$ M

$\frac{w}{M.W.}\times\frac{1}{V} = 2\times10^{-3}$

$\Rightarrow \frac{0.1}{278}\times\frac{1}{V} = 2\times10^{-3}$

$\Rightarrow V = \frac{0.1\times1000}{278\times2} = 0.18$ L

244. (3) $HA \rightleftharpoons H^+ + A^-$

(Unionized, weak acid and common ion effect)

$HA + NaOH \longrightarrow NaA + H_2O$

$NaA \longrightarrow Na^+ + A^-$ (ionized)

$K_a = \frac{[H^+][A^-]}{[HA]}$

Given, pH = 6, $[H^+] = 1 \times 10^{-6}$

From Henderson Hasselbalch equation

$[H^+] = \frac{K_a[Acid]}{[Salt]}$

$\Rightarrow \frac{[Salt]}{[Acid]} = \frac{K_a}{[H^+]} = \frac{10^{-5}}{10^{-6}} = 10:1$

245. (4) Solutions which resist the change in the value of pH when small amount of acid or base is added to them are known as buffers.

246. (3)
$2HNO_3(aq) + [Ag(NH_3)_2]^+(aq) + Cl^-(aq) \longrightarrow AgCl(s)\downarrow + 2NH_4^+(aq) + 2NO_3^-(aq)$

247. (3) $\left[Zr_3(PO_4)_4\right] \rightleftharpoons \underset{3s}{3Zr^{4+}} + \underset{4s}{4PO_4^{3-}}$

$K_{sp} = (3s)^3(4s)^4 = 6912\, s^7$. $\therefore\ s = \left(\frac{K_{sp}}{6912}\right)^{1/7}$

248. (1) Conc. of $Na_2CO_3 = 1.0 \times 10^{-4}$ M
$\therefore\ [CO_3^{2-}] = 1.0 \times 10^{-4}$ M
i.e., $s = 1.0 \times 10^{-4}$ M
At equilibrium, $[Ba^{2+}][CO_3^{2-}] = K_{sp}$ of $BaCO_3$

$$[Ba^{2+}] = \frac{K_{sp}}{[CO_3^{2-}]} = \frac{5.1\times10^{-9}}{1.0\times10^{-4}} = 5.1 \times 10^{-5}\text{ M}$$

249. (2) The correct order of solubility is:
$BaSO_4 > Hg_2Cl_2 > CrCl_3 > Cr_2(SO_4)_3$

250. (1) When ionic product, i.e., the product of the concentration of ions in the solution exceeds the value of solubility product, formation of precitpiate occurs.

$CaF_2 \rightleftharpoons Ca^{2+} + 2F^-$
Ionic product $= [Ca^{2+}][F^-]^2$
$= (1\times10^{-2})(1\times10^{-6}) = 1\times10^{-8}$
In this case,
Ionic product $(1\times 10^{-8}) >$ solubility product (1.7×10^{-10})

251. (4) $PbI_2 \rightleftharpoons \underset{s}{Pb^{2+}} + \underset{2s}{2I^-}$
$K_{sp} = s \times (2s)^2 = 4s^3$
$= 4\times\left(\frac{0.7}{461.2}\right)^3 = 14.0\times10^{-9}$

252. (1) Let solubility of AgCl = s mol/L
$AgCl \rightleftharpoons Ag^+ + Cl^-$
i.e., $K_{sp(AgCl)} = s \times s$
$KCl \longrightarrow K^+ + Cl^-$ 0.1
$[Cl^-]$ from KCl = 0.1 M
Total $[Cl^-]$ in solution = s + 0.1
$K_{sp}(AgCl) = [Ag^+][Cl^-] = s\,(s + 0.1)$
$1.0 \times 10^{-10} = s(s + 0.1)$
$1.0 \times 10^{-10} = s^2 + 0.1s$
$1.0 \times 10^{-10} = 0.1s$ (as $s^2 << 1$)
$s = 1.0 \times 10^{-9}$ mol/L

253. (2)
$Cr(OH)_3(s) \rightleftharpoons \underset{s}{Cr^{3+}(aq.)} + \underset{3s}{3OH^-(aq.)}$
$27s^4 = K_{sp}$
$s = \left(\frac{K_{sp}}{27}\right)^{1/4} = \left(\frac{1.6\times10^{-30}}{27}\right)^{1/4}$

254. (2) $AgBr \rightleftharpoons Ag^+ + Br^-$;
$K_{sp} = [Ag^+][Br^-]$
For precipitation to occur : Ionic product > Solubility product

$$[Br^-] = \frac{K_{sp}}{[Ag^+]} = \frac{5\times10^{-13}}{0.05} = 10^{-11}$$

i.e., precipitation just starts when 10^{-11} moles of KBr is added to 1L $AgNO_3$ solution.
$\therefore$ Number of moles of Br^- needed from KBr $= 10^{-11}$
$\therefore$ Mass of KBr $= 10^{-11} \times 120 = 1.2 \times 10^{-9}$ g

255. (2) $Mg(OH)_2 \rightleftharpoons Mg^{2+} + 2OH^-$;
$K_{sp} = [Mg^{2+}][OH^-]^2$
$1.0 \times 10^{-11} = 10^{-3} \times [OH^-]^2$
$[OH^-] = \sqrt{\frac{10^{-11}}{10^{-3}}} = 10^{-4}$ $\therefore$ pOH = 4
$\therefore$ pH + pOH = 14 $\therefore$ pH = 10

256. (1) $\underset{1\times10^{-4}M}{Na_2CO_3} \longrightarrow \underset{1\times10^{-4}M}{2Na^+} + \underset{1\times10^{-4}M}{CO_3^{2-}}$
$K_{sp(BaCO_3)} = [Ba^{2+}][CO_3^{2-}]$
$[Ba^{2+}] = \frac{5.1\times10^{-9}}{1\times10^{-4}} = 5.1\times10^{-5}$ M

257. (3)
$AgIO_3 \rightleftharpoons \underset{s}{Ag^+} + \underset{s}{IO_3^-}$;
$K_{sp} = [Ag^+][IO_3^-] = s\times s = s^2$
Given $K_{sp} = 1 \times 10^{-8}$
$\therefore\ s = \sqrt{K_{sp}} = \sqrt{1\times10^{-8}}$
$= 1.0 \times 10^{-4}$ mol/L $= 1.0 \times 10^{-4} \times 283$ g/L
($\because$ Molecular mass of $AgIO_3 = 283$)
$= \frac{1.0\times10^{-4}\times283\times100}{1000}$ g/100 mL
$= 2.83 \times 10^{-3}$ g / 100 mL

258. (4) For acidic buffer,
$pH = pK_a + \log\left[\frac{salt}{acid}\right] \Rightarrow pH = 4.5 + \log\frac{[salt]}{[acid]}$
As HA is 50% ionized so [salt] = [acid]
$\therefore$ pH = 4.5 $\because$ pH + pOH = 14
pOH = 14 – pH = 14 – 4.5 = 9.5

259. (3) $MX_2 \rightleftharpoons \underset{s}{M^{2+}} + \underset{2s}{2X^-}$
Where s is the solubility of MX_2
then $K_{sp} = 4s^3$
$4 \times 10^{-12} = 4s^3$
or $s = 1 \times 10^{-4}$

260. (4) $MX_4 \rightleftharpoons \underset{S}{M^{4+}} + \underset{4S}{4X^-}$
$K_{sp} = [s][4s]^4 = 256\, s^5 \Rightarrow \therefore s = \left(\frac{K_{sp}}{256}\right)^{1/5}$

261. (4) $AB_2 \rightleftharpoons A^{2+} + 2B^-$
$[A] = 1.0 \times 10^{-5}$, $[B] = [2.0 \times 10^{-5}]$
$K_{sp} = [B]^2[A] = [2\times10^{-5}]^2[1.0\times10^{-5}] = 4\times10^{-15}$

262. (1)

263. (1) $Mg(OH)_2 \rightarrow \underset{x}{[Mg^{2+}]} + \underset{2x}{2[OH^-]}$
$K_{sp} = [Mg][OH]^2 = [x][2x]^2 = x.4x^2 = 4x^3$.

264. (756) Total volume after mixing = 255 ml
$[NH_4OH] = \frac{5\times0.1}{255}$, $[NH_4Cl] = \frac{250\times0.1}{255}$
$pOH = pK_b + \log\frac{[NH_4Cl]}{[NH_4OH]}$
$pOH = 4.74 + \log\left(\frac{250\times0.1}{5\times0.1}\right) = 6.44$
pH = 14 – pOH = 14 – 6.44 = 7.56 = 756×10^{-2}

265. (4) $H_2S \rightarrow 2H^+ + S^{2-}$
$K_{a1} \times K_{a2} = 1.0 \times 10^{-21}$
Concentration of $[S^{2-}]$ to start precipitation of
$YS \Rightarrow \frac{K_{sp}(YS)}{[Y^{2+}]} = \frac{4\times10^{-14}}{0.01} = 4\times10^{-14}$

For H_2S: $K_{a_1} \times K_{a_2} = \frac{[H^+]^2\cdot[S^{2-}]^1}{H_2S}$
$1.0\times10^{-21} = \frac{[H^+]^2\times[4\times10^{-14}]}{0.1}$
$[H^+]^2 = \frac{1}{4}\times10^{-8}$
$[H^+] = \frac{1}{2}\times10^{-4}$
$pH = -\log\left[\frac{1}{2}\times10^{-4}\right] = 4.3 \approx 4$

266. (48) $pOH = pK_b + \log\frac{[NH_4^+]}{[NH_3]}$
$pOH = 4.745 + \log\frac{0.1}{0.1} \Rightarrow pOH = 4.745$
$\Rightarrow$ pH = 14 – pOH = 9.255.
on adding 0.05 mole PCl

$NH_3 +$	$H^\oplus$	$\rightarrow$	$NH_4^\oplus$
0.1	0.05		0.1
0.05	0		0.15

$pOH' = = pK_b + \log\frac{0.15}{0.05}$
$= 4.745 + 0.477 = 5.222$
pH' = 14 – pOH' = 14 – 5.222 = 8.778
ΔpH = pH – pH' = 9.225 – 8.778 = 0.477
$= 47.7 \times 10^{-2} \approx 48 \times 10^{-2}$

267. (9) $K_{sp} = 10^{-11}$
$[Mg^{2+}][OH^-]^2 = 10^{-11}$
$0.10 \times [OH^-]^2 = 10^{-11} \Rightarrow [OH^-] = 10^{-5}$
$[H^+][OH^-] = 10^{-14}$
$[H^+] = \frac{10^{-14}}{10^{-5}} = 10^{-9} \Rightarrow pH = -\log 10^{-9} = 9$
$\Rightarrow$ pH = 9

268. (100) Using Henderson-Hasselbalch equation:
$pH = pKa + \log\frac{[Salt]}{[Acid]}$
$\Rightarrow 4.5 = 4.2 + \log\frac{[V_{salt}]}{[V_{acid}]}$
$\Rightarrow \log\left(\frac{V_{salt}}{V_{acid}}\right) = 0.3 \Rightarrow \frac{V_{salt}}{V_{acid}} \cong 2$
Since $V_{salt} + V_{acid} = 300$ mL, $V_{acid} = 100$ mL.

269. (458)

	CH_3COOH +	NaOH →	CH_3COONa +	H_2O
Initially,	5 mmol	2mmol	0	0
after reaction,	3 mmol	0	2 mmole	2 mmole

$pH = pK_a + \log_{10}\frac{[salt]}{[acid]}$
$pH = 4.76 + \log_{10}\frac{2}{3} = 4.58 = 458 \times 10^{-2}$

270. (5) $\left[Ba^{2+}\right] = \frac{25\times0.05}{50} = 0.025$ M
$\left[F^-\right] = \frac{25\times0.02}{50} = 0.01$ M
$\left[Ba^{+2}\right]\left[F^-\right]^2 = 25\times10^{-7}$
$K_{sp} = 5 \times 10^{-7}$ (given)
$\text{Ratio} = \frac{\left[Ba^{+2}\right]\left[F^-\right]^2}{K_{sp}} = 5$

271. (233) $K_2SO_4 \longrightarrow 2K^+ + SO_4^{2-}$
0.1 M 0.2 M 0.1M

$BaSO_4 \rightleftharpoons Ba^{+2} + SO_4^{2-}$
a – S S S + 0.1 ≈ 0.1

$K_{SP} = S \times 10^{-1}$

$\Rightarrow 1 \times 10^{-10} = S \times 10^{-1} \Rightarrow S = 10^{-9} \times 233 \text{ g L}^{-1}$

So, Answer : 233

272. (10) $AgCl(s) \rightarrow Ag^+(aq.) + Cl^-(aq.)$

$K_{sp} = S^2 = \left(\frac{1.43}{143.4} \times 10^{-3}\right)^2 = 10^{-10}; -\log K_{sp} = 10$

273. (10) $pH = pK_a + \log \frac{0.2}{0.02}$

$5 = pK_a + 1 \Rightarrow pK_a = 4$

$K_a = 10^{-4} = 10 \times 10^{-5}$ $\therefore x = 10$

274. (85)

Concentration of calcium lactate = 0.005 M

Concentration of lactate ion = (2 × 0.005) M.

Calcium lactate is a salt of weak acid + strong base

∴ Salt hydrolysis will take place.

$pH = 7 + \frac{1}{2}(pK_a + \log C)$

$= 7 + \frac{1}{2}(5 + \log(2 \times 0.005))$

$= 7 + \frac{1}{2}[5 - 2\log 10] = 7 + \frac{1}{2} \times 3 = 8.5 = 85 \times 10^{-1}$

275. (282) $K_{sp} = S^2$

$S = \sqrt{K_{sp}} = \sqrt{8 \times 10^{-28}} = 2\sqrt{2} \times 10^{-14}$

$= 2.82 \times 10^{-14}$

$= 282 \times 10^{-16}$

276. (0) Solubility of $CaF_2 = 2.34 \times 10^{-3} \times 10$ g/L

$s = \frac{2.34}{78} \times 10^{-2} \text{ mol/L}$

$s = 3 \times 10^{-4}$ mol/L

$CaF_2 \rightleftharpoons Ca^{2+} + 2F^-$
s 2s

$K_{sp} = s(2s)^2 = 4(3 \times 10^{-4})^3$

$= 0.0108 \times 10^{-8} (\text{mol/L})^3$

277. (476)

Moles of $CH_3COOH = 50 \times 0.1 = 5$ m mole

moles of $NaOH = 25 \times 0.1 = 2.5$ m mole

$NaOH + CH_3COOH \longrightarrow CH_3COONa + H_2O$
2.5 m mole 5 m mole
0 2.5 m mole 2.5 m mole

Total volume of this buffer solution = 75 mL

$pH = pK_a + \log\left(\frac{2.5/75}{2.5/75}\right)$

$\Rightarrow pH = 4.76 = 476 \times 10^{-2}$

278. (2)

$Zn(OH)_2(s) \rightleftharpoons Zn^{2+}(aq) + 2OH^-(aq)$ s

$(0.1 + 2s) \simeq 0.1$

$K_{sp} = s(0.1)^2$

$2 \times 10^{-20} = s \times 10^{-2} \Rightarrow s = 2 \times 10^{-18}$

Therefore, x = 2

279. (108) $A_3B_2 \rightleftharpoons 3A^{2+}(aq) + 2B^{3-}(aq)$
3s 2s

$K_{sp} = (3s)^3(2s)^2 = 108s^5$

$K_{sp} = 108\left(\frac{x}{M}\right)^5 \left(\because \frac{x}{M} = s\right)$

(Given) $K_{sp} = a\left(\frac{x}{M}\right)^5$

On comparing, a = 108

280. (3) $NH_3 + H_2O \rightleftharpoons NH_4^+ + OH^-$

$[NH_3] = 2 \times 0.0210 = 0.042$ m mol

$NH_4Cl \longrightarrow NH_4^+ + Cl^-$

$[NH_4Cl] = 5 \times 0.0504 = 0.252$ m mol

By Henderson – Hasselbalch equation,

$pOH = pK_b + \log\frac{[Salt]}{[Base]}$

$= -\log(1.8 \times 10^{-5}) + \log\left(\frac{0.252}{0.042}\right) = 4.74 + 0.77$

pOH = 5.51

$[OH^-] = 10^{-5.51} = 3.09 \times 10^{-6} \simeq 3 \times 10^{-6}$

281. (10) Henderson – Hasselbalch equation,

$pH = pK_a + \log\frac{[Salt]}{[Acid]}$

$\Rightarrow 5.74 = 4.74 + \log\frac{[CH_3COONa]}{1}$

$5.74 - 4.74 = \log[CH_3COONa] - \log 1$

$1 = \log[CH_3COONa] - 0$

$[CH_3COONa] = 10^1 \approx 10$ M

282. (64) In pure water, $s = 8 \times 10^{-4}$

$K_{sp} = s^2 = (8 \times 10^{-4})^2 = 64 \times 10^{-8}$

In 0.01 M H_2SO_4,

$H_2SO_4 \longrightarrow 2H^+ + SO_4^{2-}$
0.02 0.01

$CdSO_4 \longrightarrow Cd^{2+} + SO_4^{2-}$
s s

Total conc. of $SO_4^{2-} = 0.01 + s$

$K_{sp} = s(s + 0.01)$

$K_{sp} = s \cdot (0.01)$ (neglecting s^2)

$s = \frac{64 \times 10^{-8}}{0.01} = 64 \times 10^{-6}$

283. (50) For A_2X, $A_2X \longrightarrow 2A^+ + X^{2-}$
$2s_1$ s_1

$K_{sp} = 4s_1^3 \Rightarrow s_1 = \sqrt[3]{\frac{K_{sp}}{4}} = 10^{-4}$

For MX, $MX \longrightarrow M^{2+} + X^{2-}$
s_2 s_2

$K_{sp} = s_2^2 \Rightarrow s_2 = \sqrt{K_{sp}} = 2 \times 10^{-6}$

$\frac{s_1}{s_2} = \frac{s(A_2X)}{s(MX)} = \frac{10^{-4}}{2 \times 10^{-6}} = 50.$

284. (141) Given : $[K_{sp}]_{PbI_2} = 8 \times 10^{-9}$

(I) $Pb(NO_3)_2 \rightarrow Pb^{2+}(aq) + 2NO_3^-(aq)$

At t = 0 0.1 M – –

At t = t – 0.1 M 0.2 M

(II) $PbI_2(s) \rightleftharpoons Pb^{+2}(aq) + 2I^-(aq)$
s 2s

$[Pb^{2+}] = s + 0.1 \simeq 0.1$

Now: $K_{sp} = 8 \times 10^{-9} = [Pb^{+2}][I^-]^2$

$\Rightarrow 8 \times 10^{-9} = 0.1 \times (2s)^2$

$\Rightarrow 8 \times 10^{-8} = 4s^2 \Rightarrow s = \sqrt{2} \times 10^{-4}$

$\Rightarrow s = 141 \times 10^{-6} M \Rightarrow x = 141.$

285. (2.0) $AB_2 \rightleftharpoons A^{2+}(aq) + 2B^-(aq)$
s 2s

$K_{sp} = 4s^3 = 3.2 \times 10^{-11}$

$\Rightarrow s^3 = 8 \times 10^{-12} \Rightarrow s = 2 \times 10^{-4}$

Part-B (JEE Advanced)

1. (c) $R \rightleftharpoons P$

$t = 0$ $[R]_0$ 0

$t = t$ $[R]_0 - x$ x

$K_{eq} = \frac{[P]}{[R]} = \frac{x}{[R]_0 - x} = \frac{K_f}{K_b} = \frac{1}{4}$

$\therefore 4x = [R]_0 - x$

or, $[R]_0 = 5x$ or, $x = \frac{[R]_0}{5} = [P]$

$\therefore [R] = [R]_0 - \frac{[R]_0}{5} = \frac{4[R]_0}{5}$

For, $\frac{[R]}{[R]_0} = \frac{4[R]_0}{5} \times \frac{1}{R_0} = 0.8$

$\therefore \frac{[P]}{[R]_0} = \frac{[R]_0}{5} \times \frac{1}{R_0} = 0.2$

Given condition satisfy option '*c*'.

2. (d) At constant temperature, K_p remains constant.

With change of pressure, *x* will change in such a way that K_p remains a constant.

3. (d) At initial stage of reaction, concentration of each product will increase and hence, *Q* will increase.

4. (d) $K_p = K_c \cdot (RT)^{\Delta n}$

$\Delta n = 2 - 4 = -2;\ K_c = \frac{K_p}{(RT)^{\Delta n}}$

$= \frac{1.44 \times 10^{-5}}{(0.082 \times 773)^{-2}}$, (*R* in L.atm.K^{-1} mole^{-1}).

5. (d) As all the reactants and products are present in aqueous form in (d), so it is a reversible reaction. In others, either solid or gas is generated, which is insoluble or volatile and hence, makes the reaction unidirectional.

6. (a) Statement (a) is correct and the rest statements are wrong. K_p depends only on temperature, hence at constant temp. K_p will not change.

7. (d) Liquid $\xrightleftharpoons{\text{b.p.}}$ Vapour at equilibrium

(a) Inter-molecular forces are more in liquid phase than that in gas phase.

(b) Due to difference in inter-molecular forces, the potential energy is different.

(c) Due to difference in potential energy, the total energy of the molecules are different.

(d) Kinetic energy depends on temperature only. Hence, both the molecules of liquid and solid have same kinetic energy.

8. **(d)** Only temperature affects the equilibrium constant.

9. **(8.93)**

$$Fe^{2+}(aq) + S^{2-}(aq) \rightleftharpoons FeS(s)\ (K_c = 1.6 \times 10^{17})$$

$t = 0$	0.03M	0.1 M
At equilibrium	$(0.03 - x)$	$(0.1 - x)$

Since,

$K_c >> 10^3$; $0.03 - x = 0$

$\therefore x = 0.03$ and $0.1 - x = 0.07$

$$K_c = \frac{1}{(0.07) \times [Fe^{2+}]} = 1.6 \times 10^{17}$$ (conc. of solid is taken as 1)

$$[Fe^{2+}] = \frac{1}{0.07 \times 1.6} \times 10^{-17} = \frac{250}{28} \times 10^{-17}$$

$= 8.93 \times 10^{-17}$ (i)

comparing (i) with given value in question we get, $Y = 8.93$.

10. **(1.886)**

	$A_2(g)$ +	$B_2(g)$ ⇌	$2AB(g)$
At start	1	2	0
At equ.	$1 - x$	$2 - x$	$2x$

$$\therefore [A_2] = \frac{1-x}{3},\ [B_2] = \frac{2-x}{3},\ [AB] = \frac{2x}{3}$$

$$\therefore K = \frac{(2x/3)^2}{[(1-x)/3][(2-x)/3]} = 50$$

On solving we get, $23x^2 - 75x + 50 = 0; x = 2.317$ or 0.943

The value 2.317 is inadmissable because initial concentration of reactants is 2 moles and so $x = 0.943$

$\therefore$ Moles of AB formed $= 2 \times 0.943 =$ **1.886**

11. **No change;** (K_p of an equilibrium reaction is independent of the pressure of the system.)

12. $\boldsymbol{K_p = K_c (RT)^{\Delta n}}$**;**

Here Δn = No. of moles of gaseous products – no. of moles of gaseous reactants

R = gas constant, and T = absolute temperature.

13. **False :** K for $A_2 + B_2 \rightleftharpoons 2AB$ is $\dfrac{[AB]^2}{[A_2][B_2]}$

K' for $AB \rightleftharpoons \frac{1}{2} A_2 + \frac{1}{2} B_2$ is $\dfrac{[A_2]^{1/2}[B_2]^{1/2}}{[AB]}$

or $(K')^2 = \dfrac{[A_2][B_2]}{[AB]^2} = \dfrac{1}{K}$ $\therefore$ $K' = \sqrt{\dfrac{1}{K}}$

14. **(b)** Initially, on increasing temperature, rate of reaction will increase, so % yield will also increase with time. But at equilibrium, % yield at high temperature (T_2) would be less than at T_1 as reaction is exothermic so the graph is represented by option (b).

15. **(a, b, d)**

(a) $\Delta H = C_{P(rxn)} \Delta T$

Hence, enthalpy depends on temperature.

(b) $CaCO_3(s) \rightleftharpoons CaO(s) + CO_2(g)$

$K_p = P_{CO_2}$

(c) K_{eq} depends only on temperature and not on Pressure.

(d) Enthalpy of reaction is independent of the catalyst. Catalyst changes activation energy.

16. **(b)**

	$X_2(g)$ ⇌	$2X(g)$
Initial mole :	1	0
moles at equilibrium :	$\left(1 - \frac{\beta_{eqbm}}{2}\right)$	β_{eqbm}
Partial pressure :	$\dfrac{\left(1 - \frac{\beta_{eqbm}}{2}\right)}{\left(1 + \frac{\beta_{eqbm}}{2}\right)} \times P$	$\dfrac{\beta_{eqbm}}{\left(1 + \frac{\beta_{eqbm}}{2}\right)} P$

$$\therefore K_p = \frac{(P_x)^2}{P_{x_2}} = \frac{\beta_{eqbm}^2 P}{\left(1 - \frac{\beta_{eqbm}^2}{4}\right)}$$

$$\Rightarrow K_p = \frac{4\beta_{eqbm}^2 P}{\left(4 - \beta_{eqbm}^2\right)}$$

Since, $P = 2$ bar So, $\boxed{K_p = \dfrac{8\beta_{eqbm}^2}{\left(4 - \beta_{eqbm}^2\right)}}$

17. **(c)**

(a) Correct statement.

As on decrease in pressure, reactions moves in direction where no. of gaseous molecules increase.

(b) Correct statement

At the start of reaction, $Q_P < K_P$ so dissociation of X_2 take place spontaneously. $\Delta G0° > 0$ but at start $\Delta G << 0$.

(c) Incorrect statement as

$$K_P = \frac{8\beta_{eq}^2}{4 - \beta_{eq}^2} = \frac{8 \times (0.7)^2}{4 - (0.7)^2} > 1$$

at equilibrium, $\Delta G° = -RT \ln K_P$

If $K_P > 1$, then $\Delta G°$ is negative.

But it is given that, $\Delta G°$ is positive.

(d) $K_P < 1$ and $K_P = K_C (RT)^{\Delta n}$; $\Delta n = 1$

$$\Rightarrow K_C = \frac{K_P}{RT} \Rightarrow K_C < 1;\ RT > 1$$

18.

	$NH_4HS(s)$ ⇌	$NH_3(g)$ +	$H_2S(g)$
Initial moles	$\frac{3.06}{51}$	0	0
Moles at eq.	$\frac{3.06}{51} \times \frac{70}{100}$	$\frac{3.06}{51} \times \frac{30}{100}$	$\frac{3.06}{51} \times \frac{30}{100}$

Given $V = 2$ litre, $T = 300$ K, $\Delta n = 2 - 0 = 2$

$\therefore K_c = [NH_3][H_2S]$

$$= \frac{3.06 \times 30}{51 \times 100 \times 2} \times \frac{3.06 \times 30}{51 \times 100 \times 2}$$

$=$ **8.1 × 10**$^{-5}$ **mol**^{2} **litre**$^{-2}$

Also $K_p = K_c (RT)^{\Delta n} = 8.1 \times 10^{-5} (0.082 \times 300)^2$

$=$ **4.90 × 10**$^{-2}$ **atm**^{2}

Addition of more NH_4HS on this equilibrium will cause no effect because concentration of NH_4HS is not involved in formula of K_p or K_c.

19.

	$2AB_2(g)$ ⇌	$2AB(g)$ +	$B_2(g)$
Initial mole	1	0	0
Moles at equb	$1 - x$	x	$\frac{x}{2}$

$$\text{Total moles at equb.} = 1 - x + x + \frac{x}{2} = 1 + \frac{x}{2}$$

$$= \frac{2 + x}{2}$$

$$p_{AB_2} = \frac{(1-x)}{(2+x)/2} P = \frac{2(1-x)}{(2+x)} P$$

$$p_{AB} = \frac{x}{(2+x)/2} P = \frac{2x}{(2+x)} P$$

$$p_{B_2} = \frac{x/2}{(2+x)/2} P = \frac{x}{(2+x)} P$$

$$\therefore K_p = \frac{(p_{AB})^2 (p_{B_2})}{(p_{AB_2})^2}$$

$$= \frac{\left[\frac{2x}{(2+x)} P\right]^2 \left[\frac{x}{(2+x)} P\right]}{\left[\frac{2(1-x)}{(2+x)} P\right]^2} = \frac{x^3 P}{(2+x)(1-x)^2}$$

$$K_p \approx \frac{x^3}{2} P \quad \text{or} \quad x = \left[\frac{2K_p}{P}\right]^{1/3}$$

20. (i)

	$CO(g)$ +	$2H_2(g)$ ⇌	$CH_3OH(g)$
Moles at start	0.15	a	0
Moles at equb.	$(0.15 - x)$	$(a - 2x)$	0.08
or	$(0.15 - 0.08)$	$(a - 0.16)$	0.08

$\therefore$ Total moles at equb. $= 0.15 - 0.08 + a - 0.16 + 0.08 = a - 0.01$

Total moles at equilibrium can also be calculated from the following relation,

$$n = \frac{PV}{RT} = \frac{8.5 \times 2.5}{0.0821 \times 750} = 0.345$$

$\therefore 0.345 = a - 0.01$ or $a = 0.355$

Thus, Moles of CO at equilibrium $= 0.15 - 0.08 = 0.07$

Moles of H_2 at equilibrium $= 0.355 - 0.16 = 0.195$

Moles of CH_3OH at equilibrium $= 0.08$

Substituting the values in the relation,

$$K_c = \frac{[CH_3OH]}{[H_2]^2 [CO]}$$

$$= \frac{0.08/2.5}{(0.195/2.5)^2 \times (0.07/2.5)}$$

$=$ **187.85 mole**$^{-2}$ **litre**^{2} [$\because V = 2.5$ L]

Calculation of K_p

$K_p = K_c (RT)^{\Delta n} = 187.85 \times (0.0821 \times 750)^{-2} =$ **0.05 atm**$^{-2}$ [$\because \Delta n = -2$]

(ii) *Calculation of final pressure when there is no reaction*

Moles of CO $= 0.15$; Moles of $H_2 = 0.355$

$\therefore$ Total moles $= 0.15 + 0.355 = 0.505$

$PV = nRT$

$P \times 2.5 = 0.505 \times 0.0821 \times 750$

$\Rightarrow P =$ **12.438 atm.**

21. Let the total number of moles of all gases at equilibrium point $= n$

$P = 4.92$ atm. $V = 5$L

$R = 0.0821$ atm. L mol^{-1} K^{-1} $T = 273 + 327 = 600$K

By applying the formula $PV = nRT$

$$n = \frac{PV}{RT} = \frac{4.92 \times 5}{0.0821 \times 600} = 0.5 \text{ moles}$$

(i) *Calculation of the number of moles of the individual gases at equilibrium point.*

No. of moles of CH_3OH formed = 0.1 (Given)

$\therefore$ No. of moles of CO (also) = 0.1

[$\because$ moles of CO = moles of CH_3OH formed]

Hence, No. of moles of $H_2 = 0.5 - (0.1 + 0.1) = 0.3$

$\therefore$ Molar concentration of various species will be

$[CH_3OH] = [CO] = \frac{0.1}{5} = 0.02; [H_2] = \frac{0.3}{5} = 0.06$

$$\therefore K_c = \frac{[CH_3OH]}{[CO][H_2]^2} = \frac{0.02}{0.02 \times (0.06)^2}$$

$= \mathbf{277.78\ mol^{-2}\ L^2}$

(ii) *Calculation of K_p*. We know that $K_p = K_c \times (RT)^{\Delta n} = 277.78 \times (0.0821 \times 600)^{-2}$ $(\Delta n = 1 - 3 = -2)$

$$= \frac{277.78}{2426.54} = \mathbf{0.1144\ atm^{-2}}$$

22. Since the reaction is carried out at constant volume, change in partial pressure of a species will be directly proportional to the change in its amount. Hence, we can write

	$2SO_2(g)$ +	$O_2(g)$ ⇌	$2SO_3(g)$
Initial pressure	0	2 atm	1 atm
Equb. pressure	$2x$	2 atm + x	1 atm − $2x$

Where $2x$ is the change in partial pressure of SO_3 at equilibrium.

Substituting the expression of partial pressure in the expression. For K_p, we get

$$K_p = \frac{(p_{SO_3})^2}{(p_{SO_2})^2 (p_{O_2})} \text{ or } 900\ atm^{-1}$$

$$= \frac{(1\ atm - 2x)^2}{(2x)^2\ (2\ atm + x)}$$

Assuming x is very small as compared to 1

$$900\ atm^{-1} = \frac{1\ atm^2}{(4x^2)atm^2\ (2\ atm)}$$

On usual calculations, $x = 0.0118$ atm

Thus, $p_{SO_2} = 2x = 2 \times 0.0118\ atm = \mathbf{0.0236\ atm}$

$p_{O_2} = 2\ atm + x = 2 + 0.0118 = \mathbf{2.0118\ atm}$

$p_{SO_3} = 1\ atm - 2x = 1 - 0.0236 = \mathbf{0.9764\ atm}$

23. (i)

	N_2O_4 ⇌	$2NO_2$
Before dissociation	1	0
After dissociation	$1 - \alpha$	2α

$\therefore$ Total moles $= 1 - \alpha + 2\alpha = 1 + \alpha$

$$\therefore K_p = \frac{(p_{NO_2})^2}{p_{N_2O_4}} = \frac{\left(\frac{2\alpha}{1+\alpha} \times P\right)^2}{\left(\frac{1-\alpha}{1+\alpha} \times P\right)}$$

where P is total pressure

$$K_p = \frac{\left(\frac{2 \times 0.25 \times 1}{1 + 0.25}\right)^2}{\left(\frac{1 - 0.25}{1 + 0.25} \times 1\right)} = \mathbf{0.266\ atm}\ [\because \alpha = 0.25]$$

(ii) $$K_p = \frac{4\alpha^2 P^2 \times (1+\alpha)}{(1+\alpha)^2 (1-\alpha) \times P}$$

$$= \frac{4\alpha^2 P}{(1+\alpha)(1-\alpha)}$$

$$0.266 = \frac{4\alpha^2 \times 0.1}{1 - \alpha^2} \Rightarrow \alpha = 0.63$$

$\therefore$ Percentage dissociation = **63 %**

24. Initial concentration of each gas = 1 mole

Let the No. of moles of NO_2 reacted at equilibrium $= x$

Then, $SO_2(g) + NO_2(g) \rightleftharpoons SO_3(g) + NO_2(g)$

At equilibrium $(1 - x)$ $(1 - x)$ $(1 + x)$ $(1 + x)$

Now we know that, $\frac{[SO_3][NO]}{[SO_2][NO_2]} = K_c$

or $\frac{\{(1+x)/V\}\{(1+x)/V\}}{\{(1-x)/V\}\{(1-x)/V\}} = 16$ or $\frac{(1+x)^2}{(1-x)^2} = 16$

$(\because V = 1L)$

or $\frac{1+x}{1-x} = 4$ or $1 + x = 4 - 4x$ or $5x = 3$,

$x = \frac{3}{5} = 0.6$

$\therefore$ Thus, the concentration of NO at equilibrium $= 1 + x = 1 + 0.6 = \mathbf{1.6\ moles}$

Concentration of NO_2 at equilibrium $= 1 - x = 1 - 0.6 = \mathbf{0.4\ moles}$

25. Let x be the degree of dissociation of $PCl_5(g)$, then

	$PCl_5(g)$ ⇌	$PCl_3(g)$ +	$Cl_2(g)$
Initial	3	0	1
At equilibrium	$3(1 - x)$	$3x$	$1 + 3x$

$\therefore$ Total number of moles at equilibrium $= 3(1 - x) + 3x + 1 + 3x = 3(1 + x) + 1$

Using the gas equation : $PV = nRT$

$$\therefore n = \frac{PV}{RT}$$

Here, $P = 2.05$ atm., $V = 100$ litres, $R = 0.082$ atm/deg.,

$T = 273 + 227 = 500$ K

$$\therefore n = \frac{2.05 \times 100}{0.082 \times 500} = 5 \quad \therefore 3(1 + x) + 1 = 5$$

Hence perecentage dissociation of $PCl_5 = 0.333 \times 100 = \mathbf{33.3\%}$

Calculation of K_p for the reaction :

$$K_p = \frac{[PCl_3][Cl_2]}{[PCl_5]}$$

$$= \frac{\left[\frac{3xP}{3(1+x)+1}\right]\left[\frac{(1+3x)P}{3(1+x)+1}\right]}{\left[\frac{3(1-x)}{3(1+x)+1}P\right]}$$

$$= \frac{3x(3x+1)}{4+3x} \times \frac{P}{3(1-x)}$$

$$= \frac{(3x^2 + x) \times P}{(4+3x)(1-x)} = \frac{x(3x+1) \times P}{(4+3x)(1-x)}$$

Substituting, $x = 1/3$ and $P = 2.05$ atm., we get

$$K_p = \frac{\frac{1}{3}\left(3 \times \frac{1}{3} + 1\right) \times 2.05}{\left(4 + 3 \times \frac{1}{3}\right)\left(1 - \frac{1}{3}\right)} = \frac{4.1}{10} = \mathbf{0.41}$$

26.

	$N_2(g)$ +	$3H_2(g)$ ⇌	$2NH_3(g)$
Initial moles	1	3	0
Eq. moles	$1 - 0.0025$	$3 - 0.0075$	2×0.0025
Eq. conc.	$\frac{1 - 0.0025}{4}$	$\frac{3 - 0.0075}{4}$	$\frac{2 \times 0.0025}{4}$

Now we know that $K_c = \frac{[NH_3]^2}{[N_2][H_2]^3}$

Since 0.0025 and 0.0075 are very small, 1–0.0025 and 3–0.0075 may be taken as 1 and 3 respectively. Substitute the various values

$$K_c = \frac{\left(\frac{2 \times 0.0025}{4}\right)^2}{\left[\frac{1}{4}\right]\left[\frac{3}{4}\right]^3} = \frac{0.0025 \times 0.0025}{4} \times \frac{4 \times 4 \times 4 \times 4}{3 \times 3 \times 3} = \mathbf{1.48 \times 10^{-5}\ litre^2\ mol^{-2}}$$

For the equilibrium,

$$\frac{1}{2}N_2(g) + \frac{3}{2}H_2(g) \rightleftharpoons NH_3(g)$$

$$K_c' = \frac{[NH_3]}{[N_2]^{1/2}[H_2]^{3/2}} = \sqrt{K_c}$$

$$= \sqrt{(1.48 \times 10^{-5}\ litre^2\ mol^{-2})}$$

$= \mathbf{3.82 \times 10^{-3}\ litre\ mol^{-1}}$

27. (a) In a reversible reaction, catalyst speeds up both the forward and backward reactions to the same extent, so (c) is wrong. At equilibrium,

$\Delta G = G_{products} - G_{reactants} = 0$

$\Rightarrow 2G_{NH_3} - (G_{N_2} + 3G_{H_2}) = 0$

or $2G_{NH_3} = G_{N_2} + 3G_{H_2}$

On adding N_2, reaction will move to forward direction because of Le-Chatelier's principle.

28. (d) At constant temperature K_p or K_c remains constant.

For the equilibria : $N_2O_4(g) \rightleftharpoons 2NO_2(g)$

$\Delta n = 1$

Since, temperature is constant so K_c or K_p will remain constant. Further, since volume is halved, the pressure will be doubled so α will decrease so as to maintain the constancy of K_c or K_p. From Le-Chatelier's principle reaction will move backwards.

29. (a) The given reaction will be exothermic in nature due to the formation of three X - Y bonds from the gaseous atoms. The reaction is also accompanied with the decrease in the gaseous species (*i.e.* Δn is negative). Hence, the reaction will be affected by both temperature and pressure. The use of catalyst does not affect the equilibrium concentrations of the species in the chemical reaction.

30. (b)

(i) According is Le-Chateliers principle, exothermic, reactions are favoured at low temperature.

(ii) According to Le-Chateliers principle, the reaction in which $n < 0$, are favoured at high pressure.

Given

$2SO_2(g)+O_2(g) \rightleftharpoons 2SO_3(g)+\text{Heat}$

$\because$ It is exothermic reaction

$\therefore$ Yield of SO_3 is maximum at low temperature

$n = 2-3 = -1$ or $n < 0$

$\therefore$ Yield of SO_3 is maximum at high pressure

31. **(0.25)** $A \rightleftharpoons B$

$$K_{eq} = \frac{P_B}{P_A}$$

$$K_{1000k} = \frac{10}{1} = 10\ ;\ K_{2000k} = \frac{100}{1} = 100$$

Now,

$$\frac{\Delta G^\circ_{1000k}}{\Delta G^\circ_{2000k}} = \frac{(-RT\ln k_{eq})_{1000k}}{(-RT\ln k_{eq})_{2000k}} = \frac{1000\times\ln 10}{2000\times \ln 100} = 0.25$$

32. **True :** Lower the pressure, lower will be boiling point. More liquid will vapourise and temperature decreases.

33. **(d)** $CO_{(g)}+H_2O_{(g)} \rightleftharpoons CO_{2(g)}+H_{2(g)}$

$$K_C = \frac{[CO_2][H_2]}{[CO][H_2O]}$$

A catalyst simply helps in attaining the equilibrium earlier.

Addition of inert gas has no effect on a reaction because in it, $\Delta n = 0$.

This equilibrium is not based upon volume because in it, $\Delta n = 0$.

On increasing the amount of CO, K_C should decrease but it is constant at constant temperature, so for maintaining the constant value of K_C, the amount of CO_2 increases.

34. **(b, c, d)**

(b) Introducing an inert gas at constant pressure will increase the volume. As a result, concentration of reactants and products will decrease. Thus, the reaction will move in forward direction to increase the concentration.

(c) Increasing the volume of container, will move the reaction in forward direction.

(d) Introducing PCl_5 at constant volume will favour the forward direction to decrease the concentration of PCl_5.

35. **(a)** At constant volume, concentrations do not change.

36. **(c, d)** $2NaNO_3(s) \rightleftharpoons 2NaNO_2(s)+O_2(g)$

According to Le-Chatelier principle, an increase in pressure always favours the reaction, where volume or moles decrease (*i.e.* reverse direction). As heat is added, *i.e.* reaction is endothermic and is supported in forward direction with increase in temperature. $NaNO_3$ and $NaNO_2$ both are solid. Thus, they will not effect the position of equilibrium.

37. **(a, b, c, d)** The reaction is exothermic, hence increasing temperature will favour backward reaction (*i.e.* conc. of C_2H_4 increases). Removing H_2 and adding C_2H_6 favours backward reaction. By decreasing the pressure, reaction will move towards preparing more moles, *i.e.*, backward reaction.

38. **(d)** We know that for every chemical reaction at equilibrium, change in Gibbs free energy ($\Delta G = 0$) is zero. However, charge in standard Gibbs free energy (ΔG°) may or may not be zero. Thus, statement 1 is False.

For a spontaneous reaction, at constant temperature and pressure, the reaction proceeds in the direction in which ΔG is < 0, *i.e.*, in the direction of decreasing Gibbs energy (*G*). So, statement 2 is True.

39. **(d)** Statement-1 is clearly wrong in context to Le-Chateliers principle, which states that "increase in temperature shifts the equilibrium in the forward direction of those reactions which proceed with absorption of heat (**endothermic reactions**), and in the backward direction of those reactions which proceed with the evolution of heat (**exothermic reactions**)."

Statement -2 is clearly true again according to Le-chatelier principle.

40. **(b)** Let the solubility of MX at pH 7 = s

$$\underset{}{MX(s)} \rightleftharpoons \underset{s}{M^+} + \underset{s}{X^-}$$

$K_{sp} = s^2$ ....(1)

Let the solubility of MX at pH = 2 = s_1

$$MX(s) \rightleftharpoons \underset{s_1}{M^+} + \underset{s_1}{X^-}$$

$$\underset{s_1}{X^-} + \underset{10^{-2}}{H^+} \rightarrow \underset{10^{-2}\quad s_1}{HX}$$

$$\frac{1}{K_a} = \frac{[HX]}{[X^-];[H^+]} = \frac{s_1}{[X^-]\times 10^{-2}}$$

$$\frac{s_1^2}{10^{-2}} = \frac{K_{sp}}{K_a} \quad ....(2)$$

Divide equation (2) by equation (1)

$$\frac{s_1^2}{s^2\times 10^{-2}} = \frac{K_{sp}}{K_a}\times\frac{1}{K_{sp}};$$

$$\frac{10^{-3}}{(10^{-4})^2\times 10^{-2}} = \frac{1}{K_a}$$

$K_a = 10^{-4} \Rightarrow pK_a = 4$

41. **(a)** For oxyacids containing similar central atom, the acid strength increases with the increase in the number of oxygen atom attached to the central atom and not attached to any other atom. Higher the oxidation number of the central atom, higher is the acidity of the species. Thus, acidity follows the order

Oxi. No. of Cl $\underset{+1}{HClO} < \underset{+3}{HClO_2} < \underset{+5}{HClO_3} < \underset{+7}{HClO_4}$

42. **(b)** The characteristics of the given solutions are:

NaCl – neutral solution

NH_4Cl – slightly acidic due to the following reaction

$$NH_4^+ + H_2O \rightleftharpoons NH_4OH + H^+$$

NaCN – slightly alkaline due to the following reaction

$$CN^- + H_2O \rightleftharpoons HCN + OH^-$$

HCl – highly acidic

The pH of the solution will follow the order highly acidic < slightly acidic < neutral < slightly alkaline *i.e.*

$$HCl < NH_4Cl < NaCl < NaCN$$

43. **(a)** Among oxyacids of the same type formed by different elements, acidic nature increases with increasing electronegativity. In general, the strength of oxyacids decreases as we go down the family in the periodic table.

HOCl (I) > HOBr (II) > HOI (III)

44. **(c)** Salts of weak base and strong acid get hydrolysed in aqueous solution forming an acidic solution.

$$\underset{\text{(weak)}}{AlCl_3 + 3H_2O} \longrightarrow \underset{\text{(strong)}}{Al(OH)_3} + \underset{\text{(acidic)}}{3HCl}$$

45. **(d)**

(a) It is not correct answer because 100 ml M/10 HCl will completely neutralise 100 ml M/10 NaOH and the solution will be neutral.

(b) After neutralisation resultant solution will be acidic due to presence of excess of HCl.

(c) After netralisation resultant solution will be basic due to presence of excess of NaOH.

(d) M. eq. of HCl = 75 N/5 = 15 Meq

$$\text{M.eq. of NaOH} = 25\times\frac{1}{5} = 5\text{ Meq}$$

$$\therefore \text{M. eq. of HCl left} = 10 \therefore [HCl] = \frac{10}{100} = M/10$$

$$\therefore pH = -\log[H^+] = -\log\left[\frac{1}{10}\right] = 1$$

46. **(d)** Since HCl is stronger than CH_3COOH hence acts as acid. On the other hand Cl^- is a stronger base than $CH_3COOH_2^+$ and is the conjugate base of HCl.

$$\underset{acid_1}{HCl} + \underset{base_2}{CH_3COOH} \rightleftharpoons \underset{base_1}{Cl^-} + \underset{acid_2}{CH_3COOH_2^+}$$

47. **(a)** Due to hydrolysis of $FeCl_3$, backward reaction will not take place.

$$FeCl_3 + 6H_2O \longrightarrow [Fe(H_2O)_6]^{3+}(aq) + 3Cl^-(aq)$$

$$[Fe(H_2O)_6]^{3+}(aq) + H_2O(l) \longrightarrow [Fe(H_2O)_5(OH)]^{2+}(aq) + H_3O^+(aq)$$

48. **(a)**

(i) Higher the electronegatively of central atom, higher will be the acidic strength.

(ii) In case of same atom, higher the value of oxidation state of the metal, higher will be its acidic strength.

The electronegativity of Cl > S.

Oxidation no. of Cl in $ClO_3(OH) = +7$

Oxidation no. of Cl in $ClO_2(OH) = +5$

Oxidation no. of S in $SO(OH)_2 = +4$

Oxidation no. of S in $SO_2(OH)_2 = +6$

$\therefore$ $ClO_3(OH)$ is the strongest acid.

49. **(d)** In acidic medium, weak acids are unionized due to common ion effect and they are completely ionised in alkaline medium.

Aspirin (or acetyl salicylic acid) is unionised in stomach (where pH is 2-3) and is completely ionised in small intestine (when pH is 8).

50. **(d)** (a) is a neutral solution due to both cationic and anionic hydrolysis ($K_a = K_b = 1.8\times 10^{-5}$); (b) is acidic solution due to cationic hydrolysis; (c) is acidic solution due to cationic hydrolysis; (d) is basic solution due to anionic hydrolysis.

51. (c) CaO, $CaCO_3$ and $Ca(OH)_2$ dissolve in CH_3COOH due to formation of $(CH_3COO)_2Ca$. But CaC_2O_4 does not dissolve as CH_3COO^- is a stronger conjugate base than $C_2O_4^{2-}$.

52. (c) Electron acceptors or elements having incomplete octet are Lewis acids.
$BeCl_2$ exist in polymeric forms and has no electron deficiency, here not a lweis acid.

53. (a) Base + $H^+ \rightarrow$ (conjugate acid)

NH_2^- (base) + $H^+ \rightarrow NH_3$ (conjugate acid)

54. (d) The pH of the solution at the equivalence point will be greater than 7 due to salt hydrolysis. So an indicator giving colour in basic medium will be suitable.
Phenolphthalein is a good indicator if the base is strong because strong base immediately changes the pH at end point.

55. (b) For pure water, $[H_3O^+] = [OH^-]$
$\Rightarrow K_w = 10^{-6} \times 10^{-6} = 10^{-12}$

56. (a)
(i) Lower the oxidation state of central atom, weaker will be oxy acid.
(ii) Weaker the acid, stronger will be its conjugate base.

Oxidation state of Cl in HClO is + 1, in $HClO_2$ is + 3, in $HClO_3$ is + 5, and in $HClO_4$ is + 7
$\therefore$ HClO is the weakest acid and so its conjugate base ClO^- is the strongest Bronsted base.

57. (d)
(i) pH of acid cannot be more than 7.
(ii) While calculating pH in such case, consider contribution of $[H^+]$ from water also.
Molar conc. of HCl = 10^{-8}. (given)

$\therefore$ pH = 8. But this cannot be possible as pH of an acidic solution can not be more than 7. So, we have to consider $[H^+]$ coming from H_2O.
Total $[H^+] = [H^+]_{HCl} + [H^+]_{H_2O}$

Ionisation of H_2O : $H_2O \rightleftharpoons H^+ + OH^-$
$K_w = 10^{-14} = [H^+][OH^-]$
Let x be the conc. of $[H^+]$ from H_2O
or $[H^+] = x = [OH^-]$ [$\because [H^+] = [OH]^-$ in water]
$\therefore 10^{-14} = (x + 10^{-8})(x)$ or $x = 9.5 \times 10^{-8}$ M

[For quadratic equation $x = \dfrac{-b \pm \sqrt{4ac}}{2a}$]

$\therefore$ Total $[H^+] = 10^{-8} + 9.5 \times 10^{-8}$
$= 10.5 \times 10^{-8}$ or pH
$= -\log(10.5 \times 10^{-8}) = \mathbf{6.98}$
It is between 6 and 7.
Trick : pH of acidic solution cannot be more than 7. Thus, only option (d) is possible.

58. (b) In molten state, the cations and anions become free and flow of current is due to migration of these ions in opposite directions in the electric field.

59. (3) $B + HA \longrightarrow BH^+ + A^-$
Volume of HA used till equivalence point = 6 mL
At half of equivalence point, solution will be basic buffer with B and BH^+

$$\therefore \quad pOH = pK_b + \log\frac{[BH^+]}{[B]}$$

At half equivalence point $[BH^+] = [B]$
$\therefore \quad pOH = pK_b = 14 - 11 = 3$

60. (3) KCN, K_2CO_3 and LiCN are the salts of weak acid and strong base. So, their aqueous solutions turns red litmus paper blue.

61. (6) Diprotic acids are H_2SO_4, H_3PO_3, H_2CO_3, $H_2S_2O_7$, H_2CrO_4 and H_2SO_3.

62. (8) pH of sodium salt of weak acid

$$= \frac{1}{2}(pK_w + pK_a + \log C) = \frac{1}{2}(14 + 4 - 2) = 8$$

63. (9) $HA + NaOH \longrightarrow NaA + H_2O$
At the end point, the solution contains only Na*A* whose concentration is 0.1/2 = 0.05 M
Since, the salt Na*A* is formed by strong alkali (NaOH) and weak acid H*A* (indicated by its low K_a value), its pH can be evaluated by the following relation.

$$pH = \frac{1}{2}(pK_w + pK_a + \log C)$$

$$= \frac{1}{2}(14 + 5.3010 + (-1.3010)) = 9$$

64. (2) Given $K_a = 1 \times 10^{-5}$ $\therefore pK_a = 5$
The two conditions when colour indicator will be visible are derived by

$$pH = pK_a + \log\frac{[In^-]}{[HIn]}$$

(i) pH = 5 + log 10 = 6
(ii) pH = 5 + log 0.1 = 4
Thus, minimum change in pH = **2**

65. (10.02) H_2CO_3 is a weak dibasic acid. Since NaOH is a strong acid H_2CO_3 reacts completely as a monobasic acid.

$H_2CO_3 \underset{pK_{a_1} = 6.37}{\rightleftharpoons} H^+ + HCO_3^-$; $HCO_3^- \underset{pK_{a_2} = 10.32}{\rightleftharpoons} H^+ + CO_3^{2-}$

	H_2CO_3 + NaOH ⟶	$NaHCO_3$ + H_2O
Initial:	0.01 mol	0.01 mol
After mixing :	– –	0.01 mol

Final components in mixture :
$NaHCO_3$ = 0.01 mol + 0.01 mol = 0.02 mol
Na_2CO_3 = 0.01 mol
Now,

$$\underset{\text{Acid}}{HCO_3^-} + H_2O \rightleftharpoons H_3O^+ + \underset{\text{conjugated base}}{CO_3^{2-}}$$

Applying Henderson – Hasselbalch equation,

$$pH = pK_{a_2} + \log\frac{[CO_3^{2-}]}{[HCO_3^-]}$$

$$= 10.32 + \log\frac{0.01/0.1}{0.02/0.1}$$

$$= 10.32 + \log\frac{1}{2} = 10.32 - 0.3 = 10.02$$

66. (4.86) Amount of SO_2 in atmosphere $= \dfrac{10}{10^6}$
$= 10 \times 10^{-6}$
Molar concentration of SO_2 present in water
= Amount of SO_2 × Solubility of SO_2 in water
$= 10 \times 10^{-6} \times 1.3653$ mole $L^{-1} = 1.3653 \times 10^{-5}$ M
Writing the concerned chemical equation

	H_2SO_3 ⟶	H^+ +	HSO_3^-
Initial conc.	1.3653×10^{-5}	0	0
Molar conc. at equb.	$1.3653 \times 10^{-5} - x$	x	x

Therefore $K_a = \dfrac{x^2}{(1.3653 \times 10^{-5} - x)}$

$\Rightarrow 10^{-1.92} = \dfrac{x^2}{(1.3653 \times 10^{-5} - x)}$

(pK_a = 1.92, $\therefore K_a = 10^{-1.92}$)

$\Rightarrow 1.2 \times 10^{-2} = \dfrac{x^2}{(1.3653 \times 10^{-5} - x)}$

$x^2 = 1.2 \times 10^{-2}(1.3653 \times 10^{-5} - x)$
On solving, $x = 1.364 \times 10^{-5}$
Therefore, pH = – log (1.364×10^{-5}) = **4.865**

67. (11.30)
HCl + → NaCl + H_2O
Meg. before reaction $200 \times 10^{2-}$ $300 \times 10^{2-}$
Meg. after reaction 0 $100 \times 10^{2-}$ $200 \times 10^{2-}$ $200 \times 10^{2-}$
pH of HCl = 2, pH of NaOH = 12
$\therefore$ [HCl] = 10^{-2} M, $\therefore$ [NaOH] = 10^{-2} M

$\therefore [OH^-] = \dfrac{100 \times 10^{-2}}{500} = 2 \times 10^{-3}$ or p[OH] = – log (2×10^{-1})
$\therefore$ pOH = 2.6989; $\therefore$ pH = 14 – 2.6989 = **11.3010**

68. (0.983) $H_2S \rightleftharpoons H^+ + HS^-$

$\therefore K_1 = \dfrac{[H^+][HS^-]}{[H_2S]}$ Further, $HS^- \rightleftharpoons H^+ + S^{2-}$

$\therefore K_2 = \dfrac{[H^+][S^{2-}]}{[HS^-]}$

Dissociation constant of H_2S, $K = K_1 \times K_2$
i.e. $K = 1 \times 10^{-7} \times 1.3 \times 10^{-13} = 1.3 \times 10^{-20}$
Now we know that
$K_{sp} = [M^{2+}][S^{2-}]$ $\Rightarrow 6 \times 10^{-21} = 0.05 \times [S^{2-}]$

$[S^{2-}] = \dfrac{6 \times 10^{-21}}{0.05} = 1.2 \times 10^{-19}$

Substituting the various values in the following relation

$$K = \frac{[H^+]^2[S^{2-}]}{[H_2S]}$$

$$1.3 \times 10^{-20} = \frac{[H^+]^2[1.2 \times 10^{-19}]}{0.1}$$

$\because [H_2S] = 0.1$ M

$$[H^+]^2 = \frac{1.3 \times 10^{-20} \times 0.1}{1.2 \times 10^{-19}}$$

$$[H^+] = \sqrt{\frac{1.3 \times 10^{-20} \times 0.1}{1.2 \times 10^{-19}}} = 1.04 \times 10^{-1}$$

pH = –log $[H^+]$; pH = –log (1.04×10^{-1})
= 1.0 – log 1.04 = 1.0 – 0.017 = **0.983**

69. I_2. $\because$ electron acceptors are Lewis acids.

70. amphoteric; because amphoteric substances show properties of both acids and basic.

$\underset{\text{(amphoteric)}}{Al_2O_3} + \underset{\text{(acid)}}{6HF} \longrightarrow 2AlF_3 + 3H_2O$ $[Al^{3+}3F^-]$

$\underset{\text{(amphoteric)}}{Al_2O_3} + \underset{\text{(base)}}{2NaOH} + 3H_2O \longrightarrow 2NaAl(OH)_4$

$[Na^+[Al(OH)_4]^-]$

71. SO_4^{2-} ; Conjugate base = Acid – H^+

$\therefore$ Conjugate base of HSO_4^- is SO_4^{2-}

72. False : $AlCl_3$ is a Lewis acid (although they do not have a proton, aprotic) because it accepts electrons (octet being incomplete).

73. (b, c) (a) pH of 1×10^{-8} M is below 7 because it is an acid.

(b) $H_2PO_4^- + H_2O \rightleftharpoons HPO_4^{2-} + H_3O^+$

(c) $H_2O + H_2O \rightleftharpoons OH^- + H_3O^+$

K (Auto protolysis constant of water *i.e.* ionic product of water) increases with temperature.

(d) For half neutralisation of a weak acid by a strong base, $pH = pK_a + \log\dfrac{[Salt]}{[Acid]}$

$[Salt] = [Acid]$, $\therefore$ $pH = pK_a$

74. (a) Let the heat capacity of insulated beaker be C.

Mass of aqueous content in expt.

$1 = (100 + 100) \times 1 = 200$ g

$\Rightarrow \pm$ Total heat capacity $= (C + 200 \times 4.2)$ J/K

Moles of acid, base neutralised in expt.

$1 = 0.1 \times 1 = 0.1$

$\Rightarrow$ Heat released in expt. $1 = 0.1 \times 57 = 5.7$ kJ $= 5.7 \times 1000$ J

$\Rightarrow 5.7 \times 1000 = (C + 200 \times 4.2) \times \Delta T.$

$5.7 \times 1000 = (C + 200 \times 4.2) \times 5.7$

$\Rightarrow (C + 200 \times 4.2) = 1000$

In second experiment, $n_{CH_3COOH} = 0.2$, $n_{NaOH} = 0.1$

Total mass of aqueous content = 200 g

$\Rightarrow$ Total heat capacity $= (C + 200 \times 4.2) = 1000$

$\Rightarrow$ Heat released $= 1000 \times 5.6 = 5600$ J.

Overall, only 0.1 mol of CH_3COOH undergo neutralization.

$$\Rightarrow \Delta H_{neutralization} \text{ of } CH_3COOH = \frac{-5600}{0.1}$$

$= -56000$ J/mol $= -56$ KJ/mol.

$\Rightarrow \Delta H_{dissociation}$ of $CH_3COOH = 57 - 56 = 1$ kJ/mol

75. (b) Final solution contain 0.1 mole of CH_3COOH and CH_3COONa each.

Hence, it is a buffer solution.

$$pH = pK_a + \log\frac{[CH_3COO^-]}{[CH_3COOH]}$$

$$= 5 - \log 2 + \log\frac{0.1}{0.1} = 4.7$$

76. (c) Among oxyacids, the acidic character increases with increase in oxidation state of the central atom.

O.S. of N in $HNO_3 = +5$

O.S. of N in $HNO_2 = +3$

Thus, HNO_3 is stronger acid than HNO_2. Hence, assertion is correct.

Structure of HNO_2 : H—O—N=O;

Structure of HNO_3 : 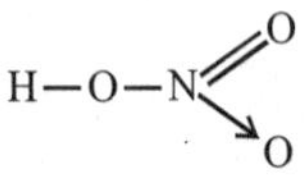

The statement-1 is true but the statement-2 is wrong as can be clearly seen from the above structures.

77. $BaO > B_2O_3 > CO_2 > SO_3 > Cl_2O_7$

Oxide basicity decreases in a period but increases in a group.

78. (*i*) The volume being doubled by mixing the two solutions, the molarity of each component will be halved *i.e.* $[CH_3COOH] = 0.1$ M, $[HCl] = 0.1$ M.

HCl being a strong acid will remain completely ionised and hence H^+ ion concentration furnished by it will be 0.1 M. This would exert common ion effect on the dissociation of acetic acid, (a weak acid).

	$CH_3COOH \rightleftharpoons$	CH_3COO^- +	H^+
At start	C	0	0
At equilibrium	$C(1-\alpha)$	$C\alpha$	$C\alpha + 0.1$

$$K_a = \frac{C\alpha(C\alpha + 0.1)}{C(1-\alpha)} = \frac{C\alpha^2 + 0.1\alpha}{(1-\alpha)}$$

Since α is very very small, $C\alpha^2$ can be neglected and $(1-\alpha)$ can be taken as unity

$\therefore K_a = 0.1\ \alpha$ or

$$\alpha = \frac{K_a}{0.1} = \frac{1.75 \times 10^{-5}}{0.1} = \mathbf{1.75 \times 10^{-4}}$$

$[H^+]_{Total} = 0.1 + C\alpha$, $C\alpha$ is negligible as compared to 0.1.

$\therefore [H^+]_{Total} = 0.1$ $\qquad \therefore$ pH = **1**

(*ii*) $6g\ NaOH = \dfrac{6}{40} = 0.15 mol$

0.1 mole of NaOH will be consumed by 0.1 mole of HCl. Thus, 0.05 mole of NaOH will react with acetic acid according to the equation.

	CH_3COOH +	$NaOH \rightarrow$	CH_3COONa +	H_2O
Initial moles	0.1 mol	0.05 mol	0	0
At equilibrium	0.05 mol	0 mol	0.05 mol	0.05 mol

Thus, solution of acetic acid and sodium acetate will become acidic buffer. So, pH of the buffer will be

$$pH = pK_a + \log\frac{[salt]}{[acid]}$$

$$= -\log\left(1.75 \times 10^{-5}\right) + \log 1 = \mathbf{4.75}$$

79.

$NH_4^+(aq) + H_2O(l) \rightleftharpoons NH_4OH(aq) + H^+(aq)$; K_a
(water)

$= 5.6 \times 10^{-10}$

$NH_4^+(aq) + \overline{O}H(aq) \underset{K_r}{\overset{K_f}{\rightleftharpoons}} NH_3(aq) + H_2O(l)$; K_f
(acid) (base) (conj. base) (conj. acid)

$= 3.4 \times 10^{10}$

$NH_3(aq) + H_2O(l) \underset{K_r}{\overset{K_f}{\rightleftharpoons}} NH_4^+(aq) + \overline{O}H(aq)$; $K_r = ?$
(base) (water)

For above reaction, the dissociation constant base, $K_b(NH_3) = K_r/K_f$

$$\Rightarrow \frac{K_w}{K_a(NH_4^+)} = \frac{K_r}{K_f} \Rightarrow$$

$$\frac{1 \times 10^{-14}}{5.6 \times 10^{-10}} = \frac{K_r}{3.4 \times 10^{10}} \Rightarrow K_r = 6.07 \times 10^5.$$

80. $H_2O < CH_3-\ddot{O}H < {}^-OH < {}^-OCH_3$

Weaker the base, stronger is its conjugate acids

$H_3O^+ > CH_3\overset{+}{O}H_2 > H_2O > CH_3OH$ (Decreasing acidic order of the conjugate bases.)

NOTE: Acidic order in aqueous medium: $CH_3OH > H_2O > C_2H_5OH$.

81. Among oxyacids of the same element, acidic nature increases with its oxidation number, *e.g.*,

	HOCl <	HOClO <	$HOClO_2$ <	$HOClO_3$
O.N. of Cl	+1	+3	+5	+7

82. $N_2 < O_2 < F_2 < Cl_2$

i.e., $N \equiv N < O = O < F - F < Cl - Cl$

As the number of bonds increases the bond length decreases.

So $N_2 < O_2 <$ halogens. Among F_2 and Cl_2, bond length of Cl_2 will be higher because of higher atomic radii.

83. (i) From the dissociation of weak acid *HA*,

$HA \rightleftharpoons H^+ + A^-$

It α is the degree of ionization of the acid *HA*, then $[H^+] = 0.1\ \alpha$ $\quad$ [$\because$ the acid is decimolar]

$[A^-] = 0.1\ \alpha$; $[HA] = 0.1(1-\alpha)$

$$\text{Therefore, } K_a = \frac{[H^+][A^-]}{[HA]} = \frac{0.1\alpha \times 0.1\alpha}{0.1(1-\alpha)}$$

$$= \frac{0.1\alpha \times 0.1\alpha}{0.1} \quad \text{(since acid is weak, } 1 - \alpha = 1)$$

$K_a = 0.1\ \alpha^2$ or $4.9 \times 10^{-8} = 0.1\ \alpha^2$

$$\text{or } \alpha^2 = \frac{4.9 \times 10^{-8}}{0.1} \quad \text{or } \alpha = 7 \times 10^{-4}$$

$\therefore$ Percentage ionization $= 100 \times 7 \times 10^{-4}$ $= \mathbf{7 \times 10^{-2}\%}$

(*ii*) Calculation of pH

$[H^+] = 0.1\alpha = 0.1 \times 7 \times 10^{-4}$ mole/litre $= 7 \times 10^{-5}$ mole/litre

Now since $pH = -\log[H^+] = -\log[7 \times 10^{-5}]$ $= 5 - \log 7 = 5 - 0.8451 = \mathbf{4.1549}$

(*iii*) Concentration of OH^- in decimolar solution

$[H^+] = 7 \times 10^{-5}$ mole per litre

Now, $K_w = [H^+][OH^-]$ or 1.0×10^{-14} $= 7 \times 10^{-5} \times [OH^-]$

$$\therefore [OH^-] = \frac{1 \times 10^{-14}}{7 \times 10^{-5}}$$

$= \mathbf{1.43 \times 10^{-10}}$ **mole per litre**

84. Phenolphthalein indicates half neutralization.

$Na_2CO_3 + H^+ \longrightarrow NaHCO_3 + Na^+$ ...(i)

Methyl orange indicates complete neutralisation

$NaHCO_3 + H^+ \longrightarrow Na^+ + H_2O + CO_2$ ...(ii)

$\therefore$ Volume of 0.1M H_2SO_4 required for complete neutralisation $= 2 \times 2.5 = 5.0$ mL

0.1 M $H_2SO_4 \equiv 0.2$ N H_2SO_4

[For H_2SO_4 molarity = 2 × normality]

($\because$ Mol. wt. of $H_2SO_4 = 98$, and eq. wt. of $H_2SO_4 = 49$)

$\therefore$ 0.2 M $H_2SO_4 \equiv 0.4$ NH_2SO_4

N_1 = normality of Na_2CO_3,

V_1 = volume of Na_2CO_3 = 10 mL,

N_2 = normality of H_2SO_4 = 0.2,

V_2 = volume of H_2SO_4 = 5.0 mL

$\therefore N_1V_1 - N_2V_2 \Rightarrow N_1 \times 10 = 0.2 \times 5 \therefore N_1$

$= \frac{0.2 \times 5}{10} = 0.1\text{ N}$

$\therefore$ Eq. wt. of $Na_2CO_3 = \frac{1}{2} \times$ molecular weight

$= \frac{106}{2} = 53$

Strength of $Na_2CO_3 = 53 \times 0.1 =$ **5.3 g/L**
[$\because$ strength = normality × Eq. wt]

For neutralization with methyl orange, volume of 0.2 M H_2SO_4 used = 2.5 mL = 2.5 mL of 0.4 N H_2SO_4
= 5 mL of 0.2 N H_2SO_4 [$\because N_1V_1 = N_2V_2$]

From 5 mL of 0.2 N H_2SO_4, 2.5 mL is used for neutralising $NaHCO_3$ formed during first half neutralization Na_2CO_3.

$\therefore$ Volume of 0.2N H_2SO_4 used for neutralisation of $NaHCO_3$ present in original solution = 5.0 – 2.5 = 2.5 mL

$\therefore N_1V_1 = N_2V_2$

where N_1 = Normality of $NaHCO_3$,
N_2 = Normality of H_2SO_4 = 0.2,
V_1 = Volume of $NaHCO_3$ = 10 mL, V_2 = Volume of H_2SO_4 = 2.5 mL

$N_1V_1 = N_2V_2 \Rightarrow N_1 \times 10 = 0.2 \times 2.5$

$N_1 = \frac{0.2 \times 2.5}{10} = 0.05\text{ N}$

Eq. wt. of $NaHCO_3 = 84$

$\therefore$ Strength of $NaHCO_3 = 84 \times 0.05 =$ **4.2 g/L**

85. (d) Let the weak manoacidic base be BOH, then the reaction that occurs during titration is $BOH + HCl \rightarrow BCl + H_2O$

Equilibrium: $\underset{C(1-h)}{B^+} + H_2O \rightleftharpoons \underset{C.h}{BOH} + \underset{C.h}{H^+}$

Using the normality equation, $\underset{(acid)}{N_1V_1} = \underset{(base)}{N_2V_2}$

Substituting various given values, we get

$\frac{2}{15} \times V_1 = 2.5 \times \frac{2}{5}$ or $V_1 = 2.5 \times \frac{2}{5} \times \frac{15}{2}$

$= 2.5 \times 3 = 7.5\text{ mL}$

Then the concentration of BCl in resulting solution is given by

$[BCl] = \frac{\frac{2}{15} \times 2.5}{10} = \frac{2}{10}$ or 0.1 M

Since $K_h = \frac{K_w}{K_b}$ $\therefore K_h = \frac{1 \times 10^{-14}}{1 \times 10^{-12}} = 10^{-2}$

Thus $K_h = \frac{0.1h^2}{(1-h)}$ or $10^{-2} = \frac{0.1h^2}{(1-h)}$

or $10^{-2} - 10^{-2}h = 0.1h^2$ or $0.1h^2 + 10^{-2}h - 10^{-2} = 0$

(Solving this quadratic equation for h, we get)

$h = \frac{-10^{-2} \pm \sqrt{(10^{-2})^2 + 4 \times 10^{-1} \times 10^{-2}}}{2 \times 0.1}$

$\left[x = \frac{-b \pm \sqrt{b^2 - 4ac}}{2a}\right]$

$= \frac{-10^{-2} \pm \sqrt{10^{-4} + 4 \times 10^{-3}}}{2 \times 0.1}$

$= \frac{-0.01 \pm \sqrt{.0001 + 0.004}}{0.2}$

$= \frac{-0.01 \pm \sqrt{0.0041}}{0.2}$

$= \frac{-0.01 \pm 0.64}{0.2} = \frac{0.54}{0.2}$

[Neglecting the negative term]

–0.27

$\therefore [H^+] = C.h = 0.1 \times 0.27 = 2.7 \times 10^{-2}\text{ M}$

86. (b) For basic buffer pH is more than 7.

	CH_3NH_2	+ HCl	$\longrightarrow CH_3NH_3^+Cl^-$
Initial moles	0.1	0.08	0
Moles after mixing	0.02	0	0.08

As it is a basic buffer solution.

$pOH = pK_b + \log\frac{0.08}{0.02} = -\log 5 \times 10^{-4} + \log 4$

$= 3.30 + 0.602 = 3.902$

$pH = 14 - 3.902 = 10.09$; $[H^+] = 7.99 \times 10^{-11}$
$\approx 8 \times 10^{-11}\text{ M}$

87. (c) For isotonic solutions, osmotic pressure is same.

$\underset{(0.004-x)}{Na_2SO_4} \rightleftharpoons \underset{2x}{2Na^+} + \underset{x}{SO_4^{2-}}$

Since both solutions are isotonic, therefore, 0.004 + 2x = 0.01 ; $x = 3 \times 10^{-3}$

$\therefore$ % Dissociation $= \frac{3 \times 10^{-3}}{0.004} \times 100 = 75\%$

88. (b) $h = \sqrt{\frac{K_w}{K_a \times c}} = \sqrt{\frac{10^{-14}}{10^{-5} \times 0.1}} = 10^{-4}$;

Hence, % hydrolysis = $10^{-4} \times 100 = 0.01$

89. (b)

$\underset{(1-\alpha)\quad c}{H_2O} \rightleftharpoons \underset{\alpha c}{H^+} + \underset{\alpha c}{OH^-}$

$\alpha = 1.9 \times 10^{-7}$; Density of water $= \frac{1.0\text{ gm}}{\text{cm}^3}$

$\therefore c = \frac{1}{18} \times 1000 = 55.56$ moles/l

$\therefore [H^+] = 55.56 \times 1.9 \times 10^{-9} = 1.055 \times 10^{-7}$

[$\because 1.9 \times 10^{-7}\% = 1.9 \times 10^{-9}$]

$\therefore K_w = [H^+][OH^-] = (1.055 \times 10^{-7})^2 = 1.0 \times 10^{-14}$

90. (c) The equilibrium constant for the nuetralization of a weak acid with a strong base is given by $K = K_a/K_w = \frac{1.0 \times 10^{-4}}{1.0 \times 10^{-14}}$

$= 1.0 \times 10^{10}$

91. (2.23) Because concentration of H^+ ion from weak acid (HX) is less so, we need to consider the H^+ ion from the self ionization of water also.

$\underset{(10^{-3}-x)}{HX(aq)} \rightleftharpoons \underset{(x+y)}{H^+(aq)} + \underset{x}{X^-(aq)}$

$H_2O(l) \rightleftharpoons \underset{(x+y)}{H^+(aq)} + \underset{x}{\overline{O}H(aq)}$

Now, K_a for HX is : -

$K_a = \frac{(x+y)(x)}{(10^{-3}-x)} = 4 \times 10^{-11}$

Approximation $= (10^{-3} - x) \approx 10^{-3}$

Hence, $K_a = \frac{(x+y)(x)}{10^{-3}} = 4 \times 10^{-11}$ ... (1)

$K_w = (x+y)(y) = 10^{-14}$ ... (2)

Adding (1) & (2) we get

$\rightarrow (x+y)^2 = 5 \times 10^{-14}$

$\rightarrow (x+y) = [H^+] = \sqrt{5} \times 10^{-7}$ M

$= 2.23 \times 10^{-7}$ M

92. (11.5) $pK_b = 4.70$, $\cdot$ $K_b = 10^{-4.7}$

Now we know that, $[OH^-] = \sqrt{K_b \times c}$

$\therefore [OH^-] = \sqrt{10^{-4.7} \times 0.5} = 3.158 \times 10^{-3}\text{ M}$

Now we know that, pOH = –log[OH^-]

pOH = –log $3.158 \times 10^{-3} = 2.5$ or,

pH = 14 – 2.5 = **11.5**

93. (6.5) For ammonium formate which is a salt of weak acid with weak base, we know that

$pH = \frac{1}{2}[pK_w + pK_a - pK_b] = \frac{1}{2}[14 + 3.8 - 4.8] =$

6.5

94. (a) As ester hydrolysis is first order with respect $[H^+]$.

$R_{HA} = K[H^+]_{HA}$ [ester]

$R_{HX} = K[H^+]_{HX}$ [ester]

$\therefore \frac{R_{HA}}{R_{HX}} = \frac{[H^+]_{HA}}{[H^+]_{HX}}$;

$\frac{1}{100} = [H^+]_{HA} = 0.01$

$\underset{\substack{1-0.01 \\ \approx 1}}{HA} \rightleftharpoons \underset{0.01}{H^+} + \underset{0.01}{A^-}$

$K_a = \frac{0.01 \times 0.01}{1} = 10^{-4}$

95. (d) P – 1; Q – 5; R – 4; S – 1

(P) $NaOH + CH_3COOH \longrightarrow CH_3COONa + H_2O$
m mole 10×0.1 $\quad$ 20×0.1
= 1 m.mol $\quad$ = 2 m.mol

$\therefore$ Solution contains 1 m. mol CH_3COOH & 1 m.mol CH_3COONa in 30 mL solution.

It is a Buffer solution. Hence, pH does not change with dilution.

(Q) $NaOH + CH_3COOH \longrightarrow CH_3COONa + H_2O$
m mole 20×0.1 $\quad$ 20×0.1
= 2 m.mol $\quad$ = 2 m.mol

$\therefore$ Solution contains 2 m. mol of CH_3COONa in 40 mL solution (salt of weak acid and strong base)

For salts of weak acid and strong base : $[H^+]_{initial} = \sqrt{\frac{K_wK_a}{C}}$

On dilution upto 80 mL, new conc. will be $= \frac{C}{2}$

$\therefore [H^+]_{new} = \sqrt{\frac{K_wK_a}{C/2}} = [H^+]_{initial} \times \sqrt{2}$

(R)HCl	+	NH_3	$\longrightarrow$	NH_4Cl
m mole		20×0.1		20×0.1
= 2m.mol				= 2 m.mol

∴ Solution contains 2 m. mol of NH_4Cl in 40 mL solution (salt of strong acid and weak base)
For salts of strong acid and weak base,

$$[H^+]_{\text{initial}} = \sqrt{\frac{K_w C}{K_b}}$$

On dilution upto 80 mL, new conc. will be $= \frac{C}{2}$.

$$\therefore [H^+]_{\text{new}} = \sqrt{\frac{K_w C}{K_b 2}} = \frac{[H^+]_{\text{initial}}}{\sqrt{2}}$$

(S) $Ni(OH)_2(s) \longrightarrow Ni^{2+} + 2OH^-$
∵ it is sparingly soluble salt
∴ On dilution $[OH^-]$ conc. in saturated solution of $Ni(OH)_2$ remains constant
∴ $[H^+]_{\text{new}} = [H^+]_{\text{initial}}$

96. **Case I.** Write the concerned chemical reaction

	BOH	+	HCl	→	BCl	+	H_2O
Moles before reaction	x		$0.1 \times 5 = 0.5$		0		0
Moles after 0.5 reaction	$(x - 0.5)$		0		0.5		
∴ Molar concentration	$\frac{x-0.5}{V}$		$\frac{0.5}{V}$		$\frac{0.5}{V}$		

Since, the solution represents a basic buffer, following Hendersen equation can be applied.

$$pOH = -\log K_b + \log \frac{[\text{Salt}]}{[\text{Base}]}$$

$$14 - 10.04 = -\log K_b + \log \frac{0.5}{(x-0.5)} \quad ...(i)$$

Case II.

	BOH	+	HCl →	BCl	+	H_2O
Moles at start	x		$0.1 \times 20 = 2$	0		0
Moles after adding 20 ml. HCl	$(x - 2)$		0	2		2 of 0.1N
∴ Molar concentration	$1\ \frac{x-2}{V_1}$		0	$\frac{2}{V_1}$		$\frac{2}{V_1}$

Again the solution is acting as basic buffer

$$\therefore pOH = -\log K_b + \log \frac{[\text{Salt}]}{[\text{Base}]}$$

$$14 - 9.14 = -\log K_b + \log \frac{2}{(x-2)} \quad ...(ii)$$

Subtracting (i) by (ii):

$$0.9 = \log\left\{\left(\frac{2}{x-2}\right) \times \left(\frac{x-0.5}{0.5}\right)\right\}$$

$$\Rightarrow \log 8 = \log\left\{\frac{4(x-0.5)}{(x-2)}\right\} \Rightarrow x = 3.5$$

Substituting this value in eq. (ii) :

$$4.86 = -\log K_b + \log\left(\frac{2}{3.5-2}\right)$$

$\Rightarrow \log K_b = -4.735 \Rightarrow \log K_b = 1.84 \times 10^{-5}$.

97. **Case I.**

	CH_3COOH	⇌	CH_3COO^-	+	H^+
At start	1		0		0
At equib.	$1 - \alpha$		α		α

$$[H^+] = c\alpha = c\sqrt{\frac{K_a}{c}} = \sqrt{K_a c}$$

$$\therefore [H^+] = \sqrt{1.8 \times 10^{-5} \times 1} = 4.24 \times 10^{-3}\ \text{M}$$

Thus pH $= -\log [H^+] = -\log 4.24 \times 10^{-3} = \mathbf{2.3724}$

Case II. pH after dilution $= 2 \times$ original pH
$= 2 \times 2.3724 = 4.7448$
Let conc. after dilution $= c_1$
and degree of dissociation $= \alpha_1$
Since pH $= -\log [H^+]$
$4.7448 = -\log [H^+]$
$[H^+] = 1.8 \times 10^{-5} = c_1\alpha_1 \therefore c_1\alpha_1 = 1.8 \times 10^{-5}$
Dissociation constant

$$\text{Since } K_a = \frac{[CH_3COO^-][H^+]}{[CH_3COOH]}$$

$$= \frac{(c_1\alpha_1)(c_1\alpha_1)}{c_1[1-\alpha_1]} = \frac{c_1\alpha^2}{(1-\alpha_1)}$$

$$1.8 \times 10^{-5} = \frac{1.8 \times 10^{-5} \times \alpha_1}{1-\alpha_1} \quad \therefore \alpha_1 = 0.5$$

Substituting the value of α_1 in the following relation
$c_1\alpha_1 = [H^+]; c_1 \times 0.5 = 1.8 \times 10^{-5}$

$$c_1 = \frac{1.8 \times 10^{-5}}{0.5} = 3.6 \times 10^{-5}\ \text{M}$$

The number of moles of CH_3COOH before and after dilution will be same
∴ Mole of CH_3COOH before dilution = Mole of CH_3COOH after dilute
(∵ Mole $= M \times V_{\text{in litre}}$)
$1 \times 1 = 3.6 \times 10^{-5} \times V \Rightarrow V = \mathbf{2.78 \times 10^4}$ **litres**

98. (i) Amount of HCl added = 0.20 mole
$[H^+] = 0.2$ g litre^{-1}
Added H^+ ions will combine with the acetate ions forming acetic acid with the result, concentration of acetate ions will decrease while that of acetic acid will increase.

	CH_3COO^-	+ HCl ⟶	CH_3COOH	+ Cl^-
before reaction	1	0.2	1	0
after reaction	0.8	0	1.2	0.2

∴ Concentration of acetate ions after adding 0.20 mole of HCl.
$[CH_3COO^-] = 1.0 - 0.2 = 0.8$ mole
Similarly, concentration of acetic acid,
$[CH_3COOH] = 1.0 + 0.2 = 1.2$ mole

$$\text{Now, pH} = -\log K_a + \log \frac{[\text{Salt}]}{[\text{Acid}]}$$

$$= -\log 1.8 \times 10^{-5} + \log \frac{0.8}{1.2}$$

pH $= 4.7447 + 0.3010 - 0.4771 = \mathbf{4.5686}$

(ii) Amount of HCl added = 0.20 mole
Out of 0.2 mole of $[H^+]$ added, 0.1 mole will combine with 0.1 mole of CH_3COO^- forming 0.1 mole of CH_3COOH.

	CH_3COO^-	+ H^+ →	CH_3COOH	+ Cl^-
before reaction	0.1	0.2	0.1	0
after reaction	0	0.1	0.2	0.1

∴ Total concentration of acetic acid $[CH_3COOH] = 0.1 + 0.1 = 0.2$ mole
In presence of $[H^+]$, CH_3COOH will not ionize. Therefore, pH of the solution will be due to the presence of H^+ of HCl, *i.e.* $0.2 - 0.1 = 0.1$ mole HCl
pH $= -\log [H^+] = -\log [0.1] = \mathbf{1}$

99. For acidic buffer

$$pH = -\log K_a + \log \frac{[\text{Salt}]}{[\text{Acid}]}$$

Calculation of concentration of HCOOH.
Here, $c = 0.2$ M; $[H^+] = 6.4 \times 10^{-3}$

$$[H^+] = c\alpha \text{ or } \alpha = \frac{[H^+]}{c} \Rightarrow \alpha = \frac{6.4 \times 10^{-3}}{0.2}$$

$= 3.2 \times 10^{-2}$
Since the degree of dissociation is very low (3.2×10^{-2}), it can be neglected and hence [HCOOH] can be taken as 0.2 M.
Calculation of concentration of $HCOO^-$, $[HCOO^-]$
It can be obtained in the following manner :

	HCOONa	⇌	$HCOO^-$	+	Na^+
At start	1		0		0
At equb.	0.25		0.75		0.75

∴ $[HCOO^-] = 0.75$

$$\text{For acidic buffer pH} = -\log K_a + \log \frac{[\text{Salt}]}{[\text{acid}]}$$

$$= -\log 2.4 \times 10^{-4} + \log \frac{0.75}{0.20} = \mathbf{4.19}$$

100. (i) Find the moles of each species after reaction.

$$\text{(ii)} \quad pH = -\log K_a + \log\left[\frac{\text{salt}}{\text{Acid}}\right]$$

Given, NaOH 0.2 M, 20 mL; CH_3COOH 0.2 M, 50 mL
$K_a = 1.8 \times 10^{-5}$
V of 0.2M NaOH required to make $pH = 4.74 = ?$
From the chemical reaction

$$\underset{50\text{ mL}}{CH_3COOH} + \underset{20\text{ mL}}{NaOH} \longrightarrow \underbrace{CH_3COONa + H_2O}_{70\text{ mL}}$$

It is evident that 70 mL of the product will contain
(i) 30 mL of 0.2 M unused CH_3COOH
[unused $CH_3COOH = 50 - 20 = 30$ mL]
(ii) 20 mL of CH_3COONa.
∴ No. of moles of CH_3COOH in solution

$$= \frac{0.2}{1000} \times 30 = 0.006 \text{ mole}$$

Similarly, No. of moles of CH_3COONa solution

$$= \frac{0.2}{1000} \times 20 = 0.004 \text{ moles}$$

$$pH = -\log K_a + \log \frac{[Salt]}{[Acid]}$$

Substituting the values of the various values

$$pH = -\log 1.8 \times 10^{-5} + \log \frac{0.004}{0.006}$$

$= 4.7447 - 0.1761 = \mathbf{4.5686}$

Calculation of the additional volume of 0.2 M NaOH required to make pH of solution 4.74.

$$pH = -\log K_a + \log \frac{[Salt]}{[Acid]}$$

$$\therefore \log \frac{[Salt]}{[Acid]} = 0.0047 \text{ or } \frac{[Salt]}{[Acid]} = \frac{1}{1.011}$$

Let x mL. be the volume of additional 0.2 M NaOH added to make the pH of the solution 4.74. This will further neutralise x mL of 0.2 M CH_3COOH and produce x mL of 0.2 M sodium acetate. The resulting solution $(70 + x)$ mL will now contain

(i) $(30 - x)$ mL of 0.2 M acetic acid.

(ii) $(20 + x)$ mL of 0.2 M sodium acetate.

Number of moles of acetic acid in $(70 + x)$ mL.

$$\text{solution} = \frac{0.2}{1000} \times (30 - x) = 2 \times 10^{-4}(30 - x)$$

Number of moles of CH_3COONa in $(70 + x)$ mL.

$$\text{solution} = \frac{0.2}{1000} \times (20 + x) = 2 \times 10^{-4}(20 + x)$$

$$\text{Therefore, } \frac{[Salt]}{[Acid]} = \frac{2 \times 10^{-4}(20 + x)}{2 \times 10^{-4}(30 - x)}$$

$$= \frac{20 + x}{30 - x} \text{ or } 1.001x + x = 30 - 20.22;$$

$2.011x = 9.78$ or $x = 4.86$

Therefore, the additional volume of 0.2 M NaOH required to make the pH of the solution 4.74 is **4.86 mL.**

Alternate Solution:

M mol of NaOH = 20 × 0.2 = 4

M Mol of acetic acid = 50×0.2 = 10

After neutralisation, buffer solution is formed which contain.

$$\text{No. of moles of } CH_3COOH = \frac{0.2}{1000} \times 30$$

= 0.006 mols = 6 millimol

$$\text{No. of moles of } CH_3COONa = \frac{0.2}{1000} \times 20$$

= 0.004 moles = 4 millimol

$$pH = pKa + \log \frac{(CH_3COONa)}{CH_3COOH}$$

$$pH = \log(1.8 \times 10^{-5}) + \log \frac{4}{6}$$

pH = 4.56

Now, let x M Mol of NaOH is further added, so that pH of resulting buffer solution is 4.74

Now, Buffer solution contains $(4+x)$ M Mol of CH_3COONa and $(6-x)$ M Mol of CH_3COOH.

$$4.74 = -\log(1.8 \times 10^{-5}) + \log \frac{(4+x)}{6-x}$$

$x = 4.86$

V = 4.86 ~ 5 mL of NaOH

101. Suppose the number of moles of sodium propionate = x

$$\text{Then } pH = pK_a + \log \frac{[Salt]}{[Acid]}$$

$$4.75 = -\log(1.34 \times 10^{-5}) + \log\left(\frac{x}{0.02}\right)$$

$x = 0.7536 \times 0.02 = \mathbf{1.5072 \times 10^{-2}\ mol}$

$$\underset{0.01 \text{ mole}}{HCl} \rightleftharpoons \underset{0.01 \text{ mole}}{H^+} + Cl^-$$

When 0.01 mole of HCl is added, there is (0.01 + 0.02) M of propionic acid and (0.015 – 0.010) M of propionate. Therefore,

$$pH = -\log(1.34 \times 10^{-5}) + \log \frac{0.005}{0.03} = \mathbf{4.09}$$

The pH of a 0.010 molar HCl solution $= -\log 10^{-2} = \mathbf{2}$

102. (d) $MX \longrightarrow \underset{s}{M^+} + \underset{s}{X^-}$ (Where s is the solubility)

Then $K_{sp} = s^2$ or $s = \sqrt{K_{sp}}$

Similarly for $MX_2 \longrightarrow \underset{s}{M^{2+}} + \underset{2s}{2X^-}$

$K_{sp} = s \times (2s)^2 = 4s^3$ or $s = \left[\frac{K_{sp}}{4}\right]^{\frac{1}{3}}$

and for $M_3X \longrightarrow \underset{3s}{3M^+} + \underset{s}{X^{-3}}$

$K_{sp} = (3s)^3 \times s = 27s^4$ or $s = \left[\frac{K_{sp}}{27}\right]^{\frac{1}{4}}$

From the given values of K_{sp} for MX, MX_2 and M_3X, we can find the solubilities of those salts at temperature, T.

Solubility of $MX = \sqrt{4 \times 10^{-8}} = 2 \times 10^{-4}$

Solubility of $MX_2 = \left[\frac{3.2 \times 10^{-14}}{4}\right]^{\frac{1}{3}}$

or $\left[\frac{32}{4} \times 10^{-15}\right]^{\frac{1}{3}}$

$= \left[8 \times 10^{-15}\right]^{\frac{1}{3}}$ or 2×10^{-5}

Solubility of $M_3X = \left[\frac{2.7 \times 10^{-15}}{27}\right]^{\frac{1}{4}}$

$= \left[10^{-16}\right]^{\frac{1}{4}}$ or 10^{-4}

Thus, the solubilities are in the order $MX > M_3X > MX_2$

i.e. the correct answer is (d).

103. (a) $A_pB_q(s) \rightleftharpoons \underset{pS}{pA^{+q}} + \underset{qS}{qB^{-p}}$

$L_S = (pS)^p.(qS)^q = p^p.q^q.S^{(p+q)}$

104. (d) In case of alkaline earth hydroxides solubility increases on moving down the group. $Be(OH)_2$ has lowest solubility and hence, lowest solubility product.

105. (a) For a precipitation to occur

Solubility product < Ionic product

Given $K_{sp} = 1.8 \times 10^{-10}$

Calculating ionic products in each

$$\text{Ionic product} = [Ag^+][Cl^-] = \frac{10^{-4}}{2} \times \frac{10^{-4}}{2}$$

$= 2.5 \times 10^{-9}$ which is greater than K_{sp} (1.8×10^{-10}).

106. (a) For a basic buffer,

$$pH = 14 - pK_b - \log \frac{[salt]}{[base]}$$

$$pH = 14 - pK_b - \log \frac{[salt]}{[base]} = 14 - (-\log 10^{-10}) - \log 1 \Rightarrow pH = 4$$

107. (b) For precipitation to occur, ionic product > solubility products

Given, $K_{sp}CaF_2 = 1.7 \times 10^{-10}$

$$CaF_2 \rightleftharpoons Ca^{2+} + 2F^-$$

Ionic product of $CaF_2 = [Ca^{2+}][F^-]^2$

Calculate I.P. in each case

(a) I.P. of $CaF_2 = (10^{-4}) \times (10^{-4})^2 = 10^{-12}$

(b) I.P. of $CaF_2 = (10^{-2}) \times (10^{-3})^2 = 10^{-8}$

(c) I.P. of $CaF_2 = (10^{-5}) \times (10^{-3})^2 = 10^{-11}$

(d) I.P. of $CaF_2 = (10^{-3}) \times (10^{-5})^2 = 10^{-13}$

∵ I.P > solubility in choice (b) only.

∴ ppt of CaF_2 is obtained in case of choice (b) only.

108. (a) Acidic buffer is mixture of weak acid and its salt with common anion.

(a) $CH_3COOH + CH_3COONH_4$ is acidic buffer.

(b) $NH_4Cl + NH_4OH$ is basic buffer.

(c) $H_2SO_4 + Na_2SO_4$ is not buffer because both the compounds are strong electrolytes.

(d) $NaCl + NaOH$ is not buffer solution because both compounds are strong electrolytes.

109. (6)

$$H_2SO_4(aq) \longrightarrow \underset{1M}{H^+(aq)} + \underset{1M}{HSO_4^-(aq)}$$

(as K_{a_1} is very large)

$$Na_2SO_4(aq) \longrightarrow 2Na^+(aq) + \underset{0.018\,M}{SO_4^{2-}(aq)}$$

(as Na_2SO_4 is a salt)

Now, due to common ion effect, dissociation of HSO_4^- will be suppressed.

$$\underset{1M}{HSO_4^-(aq)} \rightleftharpoons \underset{1M}{H^+(aq)} + \underset{0.018\,M}{SO_4^{2-}(aq)} \quad \text{(at start)}$$

$$Q_C = \frac{[H^+][SO_4^{2-}]}{[HSO_4^-]}$$

$$\Rightarrow Q_C = \frac{1\times0.018}{1} = 0.018$$

$K_C = K_{a_2} = 1.2\times10^{-2} = 0.012$

Thus, $Q_C > K_C$ and the reaction will move in backward direction.

$$\underset{(1+x)}{HSO_4^-(aq)} \rightleftharpoons \underset{(1-x)}{H^+(aq)} + \underset{(0.018-x)}{SO_4^{2-}(aq)}$$

$$K_{a_2} = \frac{(1-x)(0.018-x)}{(1+x)} = 0.012$$

Assuming x to be very small,

$1-x \approx 1$ and $1+x \approx 1$.

$\Rightarrow \quad 0.018 - x = 0.012 \;\Rightarrow\; x = 0.006$

$[SO_4^{2-}] = 0.018 - x = 0.012$ mol/L

$$PbSO_4(s) \rightleftharpoons \underset{\substack{0\\S}}{Pb^{2+}(aq)} + \underset{\substack{0.012\\S+0.012}}{SO_4^{2-}(aq)}$$

$K_{sp} = S(S+0.012)$

Since, S is very small, $S + 0.012 \approx 0.012$

$\Rightarrow \quad 1.6\times10^{-8} = S\times0.012$

$\Rightarrow \quad S = 1.33\times10^{-6} = X\times10^{-Y} \Rightarrow Y = 6.$

110. (7) Let the solubility of AgCl is x mol litre^{-1} and that of CuCl is y mol litre^{-1}

$$AgCl \rightleftharpoons \underset{x}{Ag^+} + \underset{x}{Cl^-}$$

$$CuCl \rightleftharpoons \underset{y}{Cu^+} + \underset{y}{Cl^-}$$

$\therefore \quad K_{sp}$ of AgCl $= [Ag^+][Cl^-]$

$1.6\times10^{-10} = x(x+y)$ ...(i)

Similarly, K_{sp} of CuCl $= [Cu^+][Cl^-]$

$1.6\times10^{-6} = y(x+y)$ ...(ii)

On solving, (i) and (ii)

$[Ag^+] = 1.6\times10^{-7} \therefore x = 7$

111. (3.95)

Number of mmol $Ba^{2+} = 2.00$ mmol

Number of mmol $IO_3^- = 10.00$ mmol

$$\underset{\substack{2\\0}}{Ba(NO_3)_2} + \underset{\substack{10\\6}}{2NaIO_3} \rightleftharpoons \underset{\substack{0\\2}}{Ba(IO_3)_2} + \underset{\substack{0\\4}}{2NaNO_3}$$

Number of excess $NaIO_3 = 6$ mmol

$$\underset{6\text{ mmol}}{NaIO_3} \longrightarrow Na + \underset{6\text{ mmol}}{IO_3^-}$$

$$\therefore [IO_3^-] = \frac{6\text{ mmol}}{300\text{ ml}} = 0.02\text{ M}$$

$$Ba(IO_3)_2 \rightleftharpoons \underset{s}{Ba^{2+}} + \underset{2s}{2IO_3^-}$$

Total solubility of $[IO_3^-] = 0.02 + 2s$

$Ba(IO_3^-) << 0.02$

$[IO_3^-] = 0.02$

$\therefore K_{sp} = [Ba^{2+}][IO_3^-]^2$

$1.58\times10^{-9} = s\times(0.02)^2$

$$\therefore s = \frac{1.58\times10^{-9}}{4\times10^{-4}} = 0.395\times10^{-5}$$

$s = 3.95\times10^{-6}$.

112. (0.20) $ZnS(s) \rightleftharpoons Zn^{2+}(aq) + S^{2-}(aq)$

$[S^{2-}]_{max}$ to prevent precipitation is given by

$$[S^{2-}]_{max} = \frac{1.25\times10^{-22}}{0.05} = 2.5\times10^{-21}\text{ M}$$

$H_2S \rightleftharpoons 2H^+ + S^{-2}$

$$K_1 \cdot K_2 = \frac{[H^+]^2[S^{2-}]}{[H_2S]}$$

$$K_{Net} = 10^{-21} = \frac{[H^+]^2\times2.5\times10^{-21}}{0.1}$$

$$[H^+]^2 = \frac{1}{25} \Rightarrow [H^+] = \frac{1}{5}\text{ M} = 0.2\text{ M}$$

113. (4.47) $S = \sqrt{K_{sp}\left(\frac{[H^+]}{K_a} + 1\right)}$

$$= \sqrt{20\times10^{-10}\left(\frac{10^{-3}}{10^{-8}} + 1\right)} \simeq \sqrt{2\times10^{-5}}$$

$= 4.47\times10^{-3}$ M

Alternate solution :

Let the solubility of salt AB be x.

$$AB \rightleftharpoons \underset{x}{A^+} + \underset{x-y}{B^-}$$

$K_{sp} = 2\times10^{-10} = x(x-y)$..... (i)

Association constant of weak acid HA,

$K_a' = 1/K_a = 10^8$

$$\underset{10^{-3}}{H^+} + \underset{x-y}{B^-} \rightleftharpoons \underset{y}{HB}$$

Let concentration of HB at equilibrium be y. It is given that pH of solution is 3 which means $[H^+] = 10^{-3}$

$$10^8 = \frac{y}{10^{-3}\times(x-y)} \quad \text{..... (ii)}$$

$$\Rightarrow y = \frac{x}{1+10^{-5}}$$

Putting value of y in equation (i) :

$$x\left(x - \frac{x}{1+10^{-5}}\right) = 2\times10^{-10}$$

$\Rightarrow x^2 = 2\times10^{-5} + 2\times10^{-10} \Rightarrow x^2 \approx 2\times10^{-5}$

$\Rightarrow \quad x = \sqrt{20\times10^{-6}} \Rightarrow x = 4.47\times10^{-3}$

Hence, $Y = 4.47$

114. (78.36) Volume of blood = 10 mL (given)

$[H_2CO_3]$ in blood = 2 M(given)

$[NaHCO_3]$ to be added = 5 M(given)

Let volume of $NaHCO_3$ added in 10 mL blood = VmL

$$\therefore [H_2CO_3] \text{ in blood mixture} = \frac{2\times10}{(V+10)}$$

$$[NaHCO_3] \text{ in blood mixture} = \frac{5\times V}{(V+10)}$$

$$\therefore \text{ pH} = pK_a + \log\frac{[Salt]}{[Acid]}$$

$$\text{or } 7.4 = -\log 7.8\times10^{-7} + \log\frac{5V/(V+10)}{20/(V+10)}$$

$\therefore$ **$V = 78.36$ mL**

115. True : Overal disolution of NaOH is exothermic but near saturation point, the change in dissolution enthalpy is positive. Hence, from the definition of solubility, it will increase with increase in temperature.

116. (b) $Ag_2CrO_4 \rightleftharpoons 2Ag^+ + CrO_4^{2-}$

$K_{sp} = 1.1\times10^{-12} = [Ag^+]^2[CrO_4^{-2}]$

$1.1\times10^{-12} = [0.1]^2[s]$; $s = 1.1\times10^{-10}$

117. (c, d) Any solution of a weak acid and its salt with strong base acts as an acidic buffer solution.

If volume of HNO_3 solution added is less as compared to that of CH_3COONa solution, it results in the formation of an acidic buffer solution.

CH_3COONa	+ HNO_3 ⟶	CH_3COOH	+ $NaNO_3$
Excess reagent	limiting		
MV	MV'	–	–
$M(V-V')$	0	MV'	MV' ($V' < V$)

118. (a, c) A buffer solution is prepared by mixing a weak acid/base with salt of its conjugate base/acid.

119. $Pb(OH)_2 \rightleftharpoons Pb^{2+} + 2OH^-$

(given) 6.7×10^{-6}M $\quad 6.7\times10^{-6}$M $\quad 2\times6.7\times10^{-6}$M

$\therefore K_{sp} = [Pb^{2+}][OH^-]^2 = (6.7\times10^{-6})(2\times6.7\times10^{-6})^2 = 1.203\times10^{-15}$

The buffer solution pH = 8 (given)

$\therefore$ pOH = 6 or $[OH^-] = 10^{-6}$

Thus, in this buffer we have, $[Pb^{2+}][OH^-]^2 = 1.203\times10^{-15}$

or $[Pb^{2+}]\times[10^{-6}]^2 = 1.203\times10^{-15}$

$\therefore [Pb^{2+}] =$ **1.203×10^{-3} mol litre^{-1}**

120. For AgCl; $AgCl \rightleftharpoons Ag^+ + Cl^-$

$\therefore K_{sp} = [Ag^+][Cl^-]$ .....(i)

Again it is given that

$[Ag(NH_3)_2]^+ \rightleftharpoons Ag^+ + 2NH_3$;

$K_c = 6.2\times10^{-8}$

or $Ag^+ + 2NH_3 \rightleftharpoons [Ag(NH_3)_2]^+$;

$$K_f = \frac{1}{6.2\times10^{-8}} = \frac{10^8}{6.2} \therefore K_f = \frac{[Ag(NH_3)_2]^+}{[Ag^+][NH_3]^2} \text{ or}$$

$$[Ag^+] = \frac{[Ag(NH_3)_2]^+}{K_f[(NH_3)_2]}$$

Since the formation constant of the complex is very high, most of the $[Ag^+]$ which dissolves must be converted into complex and each Ag^+ dissolved also requires dissolution of Cl^-.

$\therefore [Cl^-] = [Ag(NH_3)_2]^+$ and let it be c M

Equation (i) becomes

$$K_{sp} = \frac{[Ag(NH_3)_2]^+}{K_f[NH_3]^2}\times c \Rightarrow K_{sp} = \frac{c}{K_f[1]^2}\times c$$

$$\Rightarrow c^2 = K_{sp}\times K_f[1]^2 = 1.8\times10^{-10}\times\frac{10^8}{6.2}\times(1)^2$$

$\Rightarrow c^2 = \dfrac{1.8\times10^{-2}}{6.2} = 0.2903\times10^{-2}$

or $c = 0.538\times10^{-1} =$ **0.0538 M**

121. The concerned chemical reaction is

$$2AgCl + Na_2CO_3 \longrightarrow Ag_2CO_3 + 2NaCl$$

Calculation of $[Ag^+]$ left in the solution :

$K_{sp}(Ag_2CO_3) = [Ag^+]^2[CO_3^{2-}]$

$$[Ag^+] = \sqrt{\frac{8.2\times10^{-12}}{1.5}} = 2.34\times10^{-6}\,M$$

Concentration of Cl^- left = 0.0026 g/L

$$= \frac{0.0026}{35.5}\,mol/L = 7.33\times10^{-5}\,M$$

$\therefore K_{sp}(AgCl) = [Ag^+][Cl^-]$

$= (2.34\times10^{-6})(7.33\times10^{-5}) =$ **1.71×10^{-10}**

122. Consider common ion effect

Conc. of Ag^+ ions = Conc. of $AgNO_3 = 0.03$ M

Most of these Ag^+ ions will be present in the form of $[Ag(CN)_2]^-$.

0.03 M $AgNO_3$ requires 2×0.03 M

= 0.06 M CN^- to form $[Ag(CN)_2]^-$

$\therefore$ Conc. of free CN^- at equilibrium will be $0.1 - 0.06 = 0.04$ M

$$[Ag(CN)_2] \rightleftharpoons Ag^+ + 2CN^-$$

$$\therefore K = \frac{[Ag^+][CN^-]^2}{[Ag(CN)_2]};\ 4.0\times10^{-9}$$

$$= \frac{[Ag^+][0.04]^2}{0.03}\ [Ag^+] = \frac{4.0\times10^{-19}\times0.03}{(0.04)^2}$$

= **7.5×10^{-18} M**

123. Let the solubility of $Ca(OH)_2$ in pure water = S moles/litre

$$\underset{}{Ca(OH)_2} \rightleftharpoons \underset{S}{Ca^{2+}} + \underset{2S}{2OH^-}$$

Then $K_{sp} = [Ca^{2+}][OH^-]^2$

$4.42\times10^{-5} = S\times(2S)^2;\ 4.42\times10^{-5} = 4S^3$

$S = 2.224\times10^{-2} = 0.0223$ moles litre^{-1}

$\therefore$ No. of moles of Ca^{2+} ions in 500 mL. of solution

$$= \lambda = \frac{0.0223\times500}{1000} = 0.01115$$

Now when 500 mL of saturated solution is mixed with 500 mL of 0.4M NaOH, the resultant volume is 1000 mL. The molarity of OH^- ions in the resultant solution would therefore be 0.2 M.

$$\therefore [Ca^{2+}] = \frac{K_{sp}}{[OH^-]^2} = \frac{4.42\times10^{-5}}{(0.2)^2}$$

= 0.001105 M

Thus, No. of moles of Ca^{2+} or $Ca(OH)_2$ precipitated = 0.01115 − 0.001105 = 0.010045

Mass of $Ca(OH)_2$ precipitated

$= 0.010045\times74 = 0.7433$ g = **743.3 mg**

[mole wt. of $Ca(OH)_2 = 74$]

124. $Ag_2CO_3 + K_2C_2O_4 \rightarrow Ag_2C_2O_4 + K_2CO_3$

	Ag_2CO_3	$K_2C_2O_4$	$Ag_2C_2O_4$	K_2CO_3
Moles at start	Excess	0.1520	0	0
Moles after reaction		0.1520 − 0.0358 = 0.1162	0.0358	0.0358

Molar concentration of $K_2C_2O_4$ or $C_2O_4^{2-}$ left unreacted

$$= \frac{0.1162}{0.5} = 0.2324\ \text{moles}\ L^{-1}[\because 500\,mL = 0.5\,L]$$

$[K_2CO_3] = [CO_3^{2-}]$ at equilibrium

$$= \frac{0.0358}{0.5} = 0.07156\ \text{moles}\ L^{-1}$$

Given that K_{sp} for $Ag_2C_2O_4 = 1.29\times10^{-11}$ mol^3 L^{-3} at 25°C

So, $[Ag^+]^2[C_2O_4^{2-}] = 1.29\times10^{-11}$

or $[Ag^+]^2\times0.2324 = 1.29\times10^{-11}$

$$\text{Hence } [Ag^+]^2 = \frac{1.29}{0.2324}\times10^{-11}$$

Then K_{sp} for Ag_2CO_3

$$= [Ag^+]^2[CO_3^{2-}] = \frac{1.29\times10^{-11}}{0.2324}\times0.0716$$

= **3.974×10^{-12} mol^3 L^{-3}**

125. p(OH) for basic buffer

$$= pK_b + \log\left(\frac{\text{Salt}}{\text{Base}}\right)$$

We know that

$$pOH = pK_b + \log\frac{[\text{Salt}]}{[\text{Base}]} \quad \text{or} \quad -\log 1.8\times10^{-5} + \log\frac{0.25}{0.05}$$

$pOH = 5 - \log 1.8 + \log 5 = 5.6989 - 0.2552$

$-\log[OH^-] = 5.4437;\ \log[OH^-] = -5.4437$

$[OH^-] = 3.5999\times10^{-6}$ [Taking antilog]

K_{sp} for $Mg(OH)_2 = [Mg^{2+}][OH^-]^2$

$6\times10^{-10} = [Mg^{2+}][3.5999\times10^{-6}]^2$

$$[Mg^{2+}] = \frac{6\times10^{-10}}{12.9598\times10^{-12}} = 0.4629\times10^2$$

= **46.29 mole ion/L**

K_{sp} for $Al(OH)_3 = [Al^{3+}][OH^-]^3$

$6\times10^{-32} = [Al^{3+}](3.5999\times10^{-6})^3$

$$[Al^{3+}] = \frac{6\times10^{-32}}{(3.5999\times10^{-6})^3}$$

= **1.286×10^{-15} mol ion/L**

126. $pH = pK_a + \log\left(\dfrac{\text{Salt}}{\text{Acid}}\right)$

If x moles of HCl are added then they will combine with NaCN to form x moles of very weak acid HCN.

	NaCN	+ HCl	→ NaCl	+ HCN
At equilibrium :	$(0.01-x)$	x	x	x

For an acidic buffer,

$$pH = -\log K_a + \log\frac{[\text{Salt}]}{[\text{Acid}]}$$

$$\therefore 8.5 = -\log 4.1\times10^{-10} + \log\left(\frac{0.01-x}{x}\right)$$

$x = 8.85\times10^{-3}$ M = **8.85×10^{-3} moles of HCl**

127. Solubility of $Mg(OH)_2$ in water

$$S = 9.57\times10^{-3}\ \text{g/litre} = \frac{9.57\times10^{-3}}{58}$$

$= 1.65\times10^{-4}$ mole/litre

[$\because$ Molar mass for $Mg(OH)_2 = 58$]

$$Mg(OH)_2 \rightleftharpoons Mg^{2+} + 2OH^-$$

$K_{sp} = (S)(2S)^2 = 4S^3 = 4(1.65\times10^{-4})^3$

$= 1.8\times10^{-11}$ approx.

Calculation of solubility of $Mg(OH)_2$, say, x, in $Mg(NO_3)_2$ or $[Mg^{2+}] = x + 0.02$; $[OH^-] = x$

$K_{sp} = [Mg^{2+}][2OH^-]^2$ or $1.8\times10^{-11} = (x + 0.02)(2x)^2$

Neglecting x in comparison to 0.02 (common ion effect)

$$\Rightarrow 4x^2 = \frac{1.8\times10^{-11}}{0.02} = 9\times10^{-10}$$

or $2x = 3\times10^{-5}$

$x = 1.5\times10^{-5}$ moles/litres

$= 1.5\times58\times10^{-5} =$ **8.7×10^{-4} g/litre.**

128. For precipitation to occur ionic product > K_{sp}.

Mixture solution contains 0.1 M Ag^+ and 0.1 M Hg_2^{2+}.

K_{sp} of $Hg_2I_2 = 2.5\times10^{-26}$ is much smaller than K_{sp} of AgI which is 8.5×10^{-17}.

$[I^-]$ concentration needed to precipitate Hg_2I_2 is calculated as :

$$Hg_2I_2 \rightleftharpoons Hg_2^{2+} + 2I^-$$

$$[I^-] = \sqrt{\frac{K_{sp}}{[Hg_2^{2+}]}} = \sqrt{\frac{2.5\times10^{-26}}{0.1}}$$

= **5.0×10^{-13} M**

Similarly, $[I^-]$ concentration needed to precipitate AgI is : $AgI \rightleftharpoons Ag^+ + I^-$

$$[I^-] = \frac{K_{sp}}{[Ag^+]} = \frac{8.5\times10^{-17}}{0.1} = 8.5\times10^{-16}\,M$$

Since $[I^-]$ concentration needed to ppt. AgI is smaller than that needed to ppt. Hg_2I_2, AgI is completely precipitated first. AgI starts precipitation with $[I^-] = 8.5\times10^{-16}$ M. However, Hg_2I_2 starts precipitating with AgI only when molar concentration of I^- reaches **5.0×10^{-13} M.**

$[Ag^+]$ left when Hg_2I_2 begins to ppt. is given by

$$\frac{K_{sp}\ \text{of AgI}}{[I^-]_{Hg_2I_2}} = \frac{8.5\times10^{-17}}{5.0\times10^{-13}} = 1.7\times10^{-4}\,M$$

$$\text{Thus \% } [Ag^+] \text{ left unprecipitated} = \frac{1.7\times10^{-4}}{0.1}\times100 = 0.17\%$$

Hence % Ag^+ precipitated = **99.83%**

Chapter-8

Redox Reaction

Part-A (JEE Main)

1. (1) Reducing power is inversely proportional to its reduction potential so order will be

$$Al > Cr > Fe^{2+} > Co^{2+}$$

2. (3) A^{2+} has the tendency to lose electrons so it is reducing in nature while B^{4+} has the tendency to gain electrons so it is oxidising in nature.

3. (1) Oxidising agent is a species which it self undergo reduction. Out of the given option in N^{3-} ion 'N' is present in its lowest possible oxidation state hence it cannot be reduced further because of which it cannot act as an oxidising agent.

4. (1) In H_3PO_4 oxidation state of P is +5, which cannot be oxidised further to a higher oxidation state.

Hence, it cannot act as reducing agent.

5. (4) $\overset{0}{N_2} + \overset{0}{O_2} \longrightarrow 2\,\overset{+2\,-2}{NO}$ (Redox reaction)

6. (1) In the reaction

$$\overset{+4}{XeF_4} + \overset{+1}{O_2F_2} \longrightarrow \overset{+6}{XeF_6} + \overset{0}{O_2}$$

(Xe: Oxidation; O: Reduction)

7. (2) Copper when exposed to moist air having CO_2, it gets superficially coated with a green layer of basic carbonate $CuCO_3 \cdot Cu(OH)_2$.

8. (3) $2\,\overset{-1}{HI} + H_2\overset{+6}{S}O_4 \longrightarrow \overset{0}{I_2} + \overset{+4}{S}O_2 + 2\,H_2O$

9. (8) $PbS + 4O_3 \rightarrow PbSO_4 + 4O_2$

$x = 4, y = 4 \therefore x + y = 4 + 4 = 8$

($O_3 \rightarrow O_2 + O$ so we need $4O_3$ for $PbSO_4$)

10. (2) Potassium shows +1 state in all its oxides, superoxides and peroxides.

11. (3) $[Cr(H_2O)_6]Cl_3$

$\Rightarrow x + 6 \times 0 + (-1) \times 3 = 0; x = +3$

$[Cr(C_6H_6)_2]$

$x + 2 \times 0 = 0\,; x = 0$

$K_2[Cr(CN)_2(O)_2(O_2)(NH_3)]$

[Here $(O)_2$ is oxo form and (O_2) is peroxo form]

$2 \times 1 + x + 2 \times (-1) + 2 \times (-2) + (-2) + 0 = 0$

$x = +6$

12. (4)

$$\overset{-2}{PbS} + 4H_2\overset{-1}{O_2} \longrightarrow Pb\overset{+6}{S}O_4 + 4H_2\overset{-2}{O}$$

(O: Reduction; S: Oxidation)

Notice that the oxidation state of oxygen goes from –1 on the H_2O_2 to –2 on the H_2O means H_2O_2 is being reduced. On the other hand the oxidation state of sulfur is going from –2 on the PbS to +6 on the $PbSO_4$. i.e Sulfur is being oxidised.

13. (4) CrO_2Cl_2

Let O. No. of Cr = x

$\therefore\ x + 2(-2) + 2(-1) = 0$

$x - 4 - 2 = 0; x = +6$

14. (3) In SO_3^{2-}

$x + 3(-2) = -2; x = +4$

In $S_2O_4^{2-}$

$2x + 4(-2) = -2$

$2x = 6;\ \ x = +3$

In $S_2O_6^{2-}$

$2x + 6(-2) = -2$

$2x = 10;\ \ x = +5$

Hence the correct order is :

$S_2O_4^{2-} < SO_3^{2-} < S_2O_6^{2-}$

15. (3)

$$H_2\overset{+4}{S}O_3(aq) + Sn^{4+}(aq) + H_2O(l) \longrightarrow Sn^{2+}(aq) + H\overset{+6}{S}O_4^-(aq) + 3H^+$$

Hence H_2SO_3 is the reducing agent because it undergoes oxidation.

16. (1) If an electronegative element is in its lowest possible oxidation state in a compound or in free state. It can function as a powerful reducing agent, e.g., I^-.

17. (3)

Compounds	O.S. of Sulphur
SO_3	+6
H_2SO_3	+4
$SOCl_2$	+4
SF_4	+4
$BaSO_4$	+6
$H_2S_2O_7$	+6

18. (6) CrO_2Cl_2

$[x - 4 - 2] = 0$

Oxidation State = +6

19. (6) $(NH_4)_3PO_4 \cdot 12MoO_3$

Let X = oxidation state of Mo in MoO_3

$X + [(-2) \times 3] = 0$

$X = +6$

20. (5) $\overset{+7}{IO_4^-} + H_2O_2 \rightarrow \overset{+5}{IO_3^-} + O_2 + H_2O$

21. (6) The oxidation states of iron in these compounds will be -

In A, $x + 5(-1) + (-1) = -4 \Rightarrow x = +2$

In B, $y + 4(-2) = -4 \Rightarrow y = +4$

In C, $z = 0$

The sum of oxidation states will be =4+2+0=6.

22. (19)

	Compound	Oxidation state of transition element
(i)	$K_2Cr_2O_7$	$x = +6$
(ii)	$KMnO_4$	$y = +7$
(iii)	K_2FeO_4	$z = +6$

So, $(x + y + z) = 6 + 7 + 6 = 19$.

23. (1)

24. (2)

Chlorine is in its maximum oxidation state, so disproportionation not possible in ClO_4^-

25. (3) In the reaction of $S_2O_3^{2-}$ with I_2, average oxidation state of sulphur changes + 2 to + 2.5 In the reaction of $S_2O_3^{2-}$ with Br_2, average oxidation state of sulphur changes from + 2 to + 6. Both I_2 and Br_2 are oxidation (oxidising agent)

26. (2) F_2 can not disproportionate because fluorine does not show positive oxidation state however Cl_2, Br_2 and I_2 undergoes disproportionation reaction.

27. (4) A→(IV), B→(I), C→(II), D→(III)

28. (4)

$$Cr_2O_7^{2-} + 14H^+ + 6e^- \longrightarrow 2Cr^{3+} + 7H_2O$$

X = 14, Y = 6, Z = 7, A = Cr^{3+}

29. (1) In a disproportionation reaction an element in one oxidation state is simultaneously oxidised and reduced.

$Cu^+ \longrightarrow Cu^{2+} + Cu$

$$3\overset{+6}{MnO_4^{2-}} + 4H^+ \rightarrow 2\overset{+7}{MnO_4^-} + \overset{+4}{MnO_2} + 2H_2O$$

$$2\overset{+7}{KMnO_4} \longrightarrow K_2\overset{+6}{MnO_4} + \overset{+4}{MnO_2} + O_2$$

(Not a disproportionation Reaction)

30. (1) Statement I is correct but statement II is incorrect.

In ClO_4^- molecule. The oxidation state of Cl is +7 hence it's in its maximum oxidation state and cannot undergo disproportionation reaction.

31. (2) $Cl_2 + 2OH^- \longrightarrow Cl^- + ClO^- + H_2O$

$\therefore\ a = 1, b = 2, c = 1, d = 1$

32. (4) Number of atoms of iodine on reactant side = number of atoms of Iodine on product side

$2 + x = 6 \times 2 \therefore \Rightarrow x = 10$

$2IO_3^- + 10I^- + 12H^+ \rightarrow 6I_2 + 6H_2O$

33. (1) Zn displaced Au Reduction and Oxidation both are taking place.

34. (3) $2H_2\overset{-1}{O_2} \longrightarrow 2H_2\overset{-2}{O} + \overset{0}{O_2}$

(Disproportionation reaction)

- $2\overset{+4}{NO_2} + H_2O \longrightarrow H\overset{+5}{NO_3} + H\overset{+3}{NO_2}$

  (Disproportionation reaction)

- $\overset{+7}{MnO_4^-} + 4H^+ + 3e^- \longrightarrow \overset{+4}{MnO_2} + 2H_2O$

  (Reduction reaction)

- $\overset{+6}{MnO_4^{2-}} + 4H^+ \longrightarrow 2\overset{+7}{MnO_4^-} + \overset{+4}{MnO_2} + 2H_2O$

  (Disproportionation reaction)

35. (1) In disproportion reaction an element in one oxidation state is simultaneously oxidised and reduced.

$$3\,\overset{+6}{MnO_4^{2-}} + 4H^+ \rightarrow 2\,\overset{+7}{MnO_4^-} + \overset{+4}{MnO_2} + 2H_2O$$

(Reduction; Oxidation)

36. (1) In BrO_4^-, Br is in highest oxidation state (+7). So it cannot oxidise further and hence, it cannot show disproportionation reaction.

37. (4) $\overset{(+1)}{CuBr} \longrightarrow \overset{(0)}{Cu} + \overset{(+2)}{CuBr_2}$

It is an example of disproportionation reaction, as Cu undergoes both oxidation and reduction.

38. (1)

39. (1) Reaction involved:

$\overset{+3}{C_2}O_4^{2-} \longrightarrow 2\overset{+4}{C}O_2 + 2e^-$

∴ The number of electrons involved in producing one mole of CO_2 is 1.

40. (3) The reaction given is

$Cr_2O_7^{2-} + Fe^{2+} + C_2O_4^{2-} \longrightarrow Cr^{3+} + Fe^{3+} + CO_2$

$\overset{+6}{C}rO_7^{2-} \longrightarrow 2Cr^{3+}$

On balancing

$14H^+ + Cr_2O_7^{2-} + 6e^- \longrightarrow 2Cr^{3+} + 7H_2O$ ......(i)

$4Fe^{2+} \longrightarrow 4Fe^{3+} + 4e^-$ ......(ii)

$C_2O_4^{2-} \longrightarrow 2CO_2 + 2e^-$ ....(iii)

On adding all three equations, we get

$Cr_2O_7^{2-} + Fe^{2+} + C_2O_4^{2-} + 14H^+ \longrightarrow 2Cr^{3+} + Fe^{3+} + 2CO_2 + 7H_2O$

The reduction of $Cr_2O_7^{2-}$ to $2Cr^{3+}$ requires 6 electrons because on balancing the equation, we will see 2 moles of chromium ion goes from +6 to +3.

41. (1)

42. (3) $3Na_2HAsO_3 + NaBrO_3 + 6HCl \rightarrow 3H_3AsO_4 + NaBr + 6NaCl$

Hence, X = 3, Y = 1 and Z = 6

43. (13) $NaCl + K_2Cr_2O_7 + H_2SO_4 \rightarrow \underset{(A)}{CrO_2Cl_2} + KHSO_4 + NaHSO_4 + H_2O$

$\underset{(A)}{CrO_2Cl_2} + NaOH \rightarrow \underset{(B)}{Na_2CrO_4} + NaCl + H_2O$

$\underset{(B)}{Na_2CrO_4} + H_2SO_4 + H_2O_2 \longrightarrow \underset{(C)}{CrO_5} + Na_2SO_4 + H_2O$

X = 2, Y = 5 and Z = 6

X + Y + Z

2 + 5 + 6 = 13.

44. (8) $2MnO_4^- + 5C_2O_4^{2-} + 16H^+ \longrightarrow 2Mn^{+2} + 10CO_2 + 8H_2O$

From the above balanced equation.

2 mole of MnO_4^- requires 16 H^+ ion to form CO_2.

∴ 1 mole of

$MnO_4^- = \frac{16 \times 1}{2} H^+$ ion to form CO_2

= 8 moles of H^+ ions required

45. (8) $2MnO_4^- + 6I^- + 4H_2O \longrightarrow 3I_2 + 2MnO_2 + 8OH^-$

∴ z = 8

46. (6) Species that have elements that are in intermediate oxidation states undergo disproportionation. Such species are:

$\overset{-1}{H_2O_2}, \overset{+5}{ClO_3^-}, \overset{0}{P_4}, \overset{0}{Cl_2}, \overset{+1}{Cu^+}, \overset{+4}{NO_2}$

47. (23) $Cr_2O_7^{2-} + 14H^+ + 6Fe^{2+} \rightarrow 2Cr^{3+} + 6Fe^{3+} + 7H_2O$

X = 14, Y = 2, Z = 7

Hence (X + Y + Z) = 14 + 2 + 7 = 23

48. (16) Writing the half reaction,

Oxidation half reaction: $MnO_4^- \longrightarrow Mn^{2+}$

Balancing oxygen: $MnO_4^- \longrightarrow Mn^{2+} + 4H_2O$

Balancing hydrogen:

$MnO_4^- + 8H^+ \longrightarrow Mn^{2+} + 4H_2O$

Balancing charge:

$MnO_4^- + 8H^+ + 5e^- \longrightarrow Mn^{2+} + 4H_2O$

Reducing half reaction:

$C_2O_4^{2-} \longrightarrow CO_2$

Balancing carbon and charge:

$C_2O_4^{2-} \longrightarrow 2CO_2 + 2e^-$

Balancing charge in both oxidation and reduction reaction,

$MnO_4^- + 8H^+ + 5e^- \longrightarrow Mn^{2+} + 4H_2O \times 2$

$C_2O_4^{2-} \longrightarrow 2CO_2 + 2e^- \quad \times 5$

$2MnO_4^- + 5C_2O_4^{2-} + 16H^+ \longrightarrow 2Mn^{2+} + 10CO_2 + 8H_2O$

So, we get

$b = 5; c = 16; x = 2; y = 10; z = 8.$

49. (12) $16e^- + S_8 \longrightarrow 8S^{2-}$

$12H_2O + \overset{0}{S_8} \longrightarrow 4\overset{+2}{S_2}O_3^{2-} + 24H^+ + 16e^-$

$2S_8 + 12H_2O \longrightarrow 8S^{2-} + 4S_2O_3^{2-} + 24H^+$

For balancing in basic medium add OH^- equal to H^+.

$2S_8 + 12H_2O + 24OH^- \longrightarrow 8S^{2-} + 4S_2O_3^{2-} + 24H_2O$

$2S_8 + 24OH^- \longrightarrow 8S^{2-} + 4S_2O_3^{2-} + 12H_2O$

$S_8 + 12OH^- \longrightarrow 4S^{2-} + 2S_2O_3^{2-} + 6H_2O$

∴ $a = 12$.

50. (19) $2Fe^{2+} + H_2\overset{-1}{O_2} \longrightarrow 2Fe^{3+} + 2\overset{-2}{O}H^-$

$2\overset{+7}{Mn}O_4^- + 6H^+ + 5H_2O_2 \rightarrow 2Mn^{2+} + 8H_2O + 5O_2$

∵ $x = 2, y = 2, x' = 2, y' = 8, z' = 5$

∴ $x + y + x' + y' + z' = 19$

Part-B (JEE Advanced)

1. I^- (∵ I_2 is weakest oxidising agent)

2. True : Cu^+ is the intermediate oxidation state between Cu^{++} and Cu. If the reduction potential from the intermediate oxidation state to the lower one is more positive than from the higher to the intermediate, then the intermediate state will undergo disproportionation.

$$Cu^{++} \xrightarrow{+0.15V} Cu^+ \xrightarrow{+0.52V} Cu$$

3. False : Copper metal does not reduces Fe^{2+} in an acidic medium, because the $E°$ value for Cu is more.

4. (A - p, s); (B - r); (C - p, q); (D - p).

A → p, s; The reaction is redox reaction because the O.N. of O in O_2^- is –0.5 and that in O_2 is zero. In O_2^{2-} is –1.0. It involves reduction oxidation reaction. Since, here a part of molecule is oxidised and a part is reduced, so it is disproportionation.

B → r; The structure of $Cr_2O_7^{2-}$ is given below

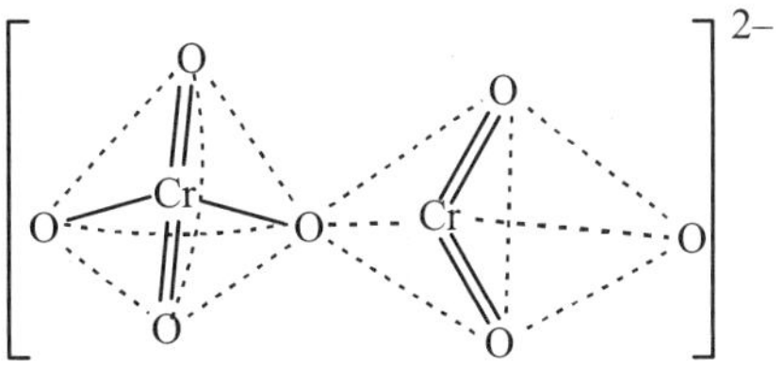

[In any solution dichromate ions and chromate ions exist in equilibrium. In alkali solution, dichromate ions are converted into chromate ions and on acidification chromate ions are converted back into dichromate ion.]

C → p, q; The reaction is

$2MnO_4^- + 6H^+ + 5NO_2^- \longrightarrow 2Mn^{2+} + 3H_2O + 5NO_3^-$

In involves change in O.N of Mn from + 7 (in MnO_4^-) to + 2 (in Mn^{2+}). So, Mn is reduced and NO_2^- is oxidised to NO_3^-; it is a redox reaction. The structure of NO_3^- (one of the products) is **trigonal planar**.

$3Fe^{2+} + NO_3^- + 2H_2SO_4 \longrightarrow 3Fe^{3+} + NO + 2SO_4^{2-} + 2H_2O$

D → p, It is a **redox reaction.**

5. (d) The following reaction occurs:

$6Fe^{2+} + Cr_2O_7^{2-} + 14H^+ \longrightarrow 6Fe^{3+} + 2Cr^{3+} + 7H_2O$

From the above equation, we find that Mohr's salt ($FeSO_4 \cdot (NH_4)_2SO_4 \cdot 6H_2O$) and dichromate reacts in 6 : 1 molar ratio.

6. (b)

The highest O.S. of an element is equal to the number of its valence electrons

(a) $[Fe(CN)_6]^{3-}$, O.N. of Fe = + 3, $[Co(CN)_6]^{3-}$, O.N. of Co = + 3

(b) CrO_2Cl_2, O.N. of Cr = +6, (Highest O.S. of Cr) $[MnO_4]^-$ O.N of Mn = + 7 (Highest O.S. of Mn)

(c) TiO_3, O.N. of Ti = + 6, MnO_2 O.N. of Mn = +4

(d) $[Co(CN)_6]^{3-}$, O.N. of Co = + 3, MnO_3, O.N. of Mn = + 6

7. (d)

Oxidation state of Mn in MnO_4^- = + 7

Oxidation state of Cr in $Cr(CN)_6^{3-}$ = + 3

Oxidation state of Ni in NiF_6^{2-} = + 4

Oxidation state of Cr in CrO_2Cl_2 = + 6

8. (a)

O.N. of S in $S_8 = 0$; O.N. of S in $S_2F_2 = +1$; O.N. of S in $H_2S = -2$;

9. (b) $BaO_2 + H_2SO_4 \rightarrow BaSO_4 + H_2O_2$

Oxygen is the most electronegative element in the reaction and has the oxidation states of –1 (in H_2O_2) and –2 (in $BaSO_4$). In H_2O_2, peroxo ion is present.

10. (c) $2 + 2(2 + x - 4) = 0$ [$\because$ $Ba(H_2PO_2)_2$ is neutral molecule]

or $2x - 2 = 0 \Rightarrow x = +1$

11. (b) Sum of oxidation state of all atoms in neutral compound is zero. Let the oxidation state of iron in the complex ion

$[Fe(H_2O)_5(NO)]^{2+}SO_4^{2-}$ be x; then

$x + 5 \times 0 + 0 = +2. \quad \therefore x = +2$

12. (c) The sum of oxidation states of all atoms in compound is zero. Calculation of O.S. of C in CH_2O.

$x + 2 + (-2) = 0 \Rightarrow x = 0$

13. (c)

$N_2H_4 \rightarrow Y + 10e^-$, Calculation of O.S. of N in N_2H_4:

$2x + 4 = 0 \Rightarrow x = -2$

The two nitrogen atoms will balance the charge of 10 *e*. Hence, oxidation state of N will increase by +5, *i.e.* from –2 to +3.

14. (6) Br_3O_8

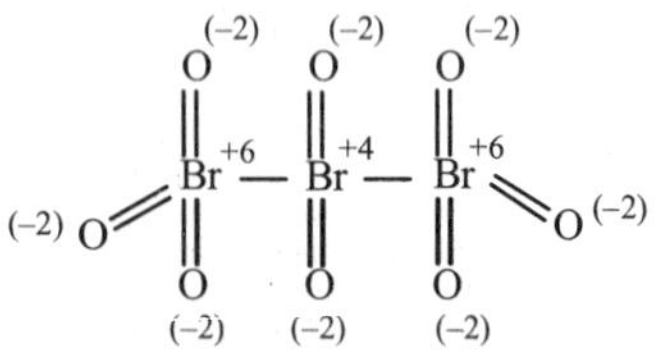

Number of atoms with zero oxidation state = 0

F_2O

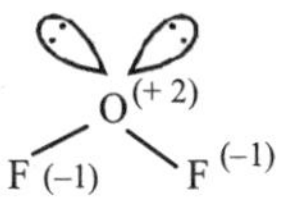

Number of atom with zero oxidation state = 0

$H_2S_4O_6$

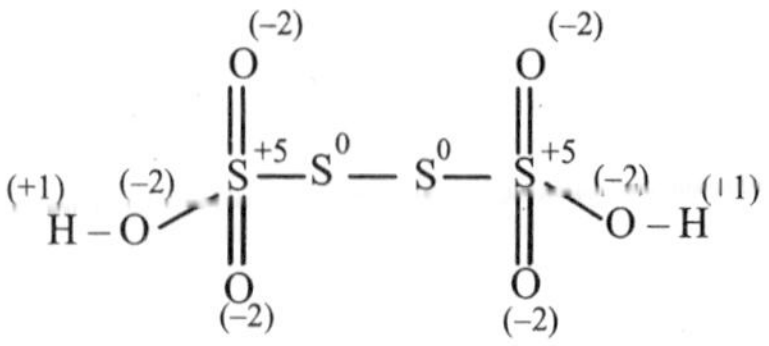

Number of atoms with zero oxidation state = 2

$H_2S_5O_6$

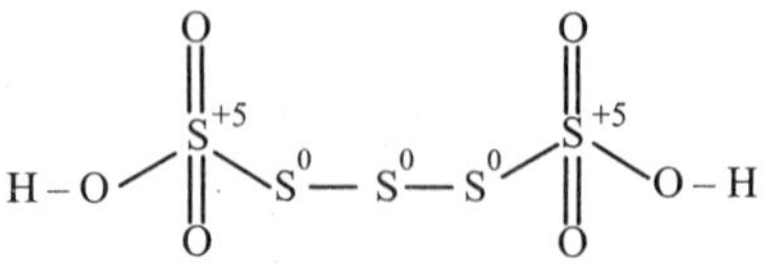

Number of atoms where zero oxidation state = 3

C_3O_2

Number of atoms with zero oxidation state = 1

Sum = 2 + 3 + 1 = 6

15. (5)

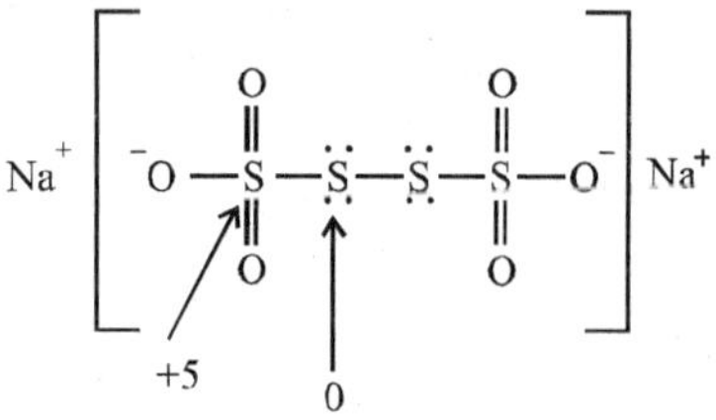

Difference in oxidation number = 5 – 0 = 5

16. Sum of oxidation states of all atoms (elements) in a neutral compound is zero.

As $YBa_2Cu_3O_7$ is neutral.

$(+3) + 2(+2) + 3(x) + 7(-2) = 0$

or $3 + 4 + 3x - 14 = 0$; $x = +\frac{7}{3}$

17. $HI < I_2 < ICl < HIO_4$; O.N. of I in $I_2 = 0$, HI = –1, ICl = +1, HIO_4 = +7.

18. (c)

(i) $3\overset{+1}{Cl}O^-(aq) \rightarrow \overset{+5}{Cl}O_3^- + 2\overset{-1}{Cl}^-(aq)$

It is disproportionation reaction because Cl is both oxidised (+ 1 to + 5) and reduced (+ 1 to – 1) during reaction.

19. (a)

Balance the reaction by ion electron method.

Oxidation reaction : $C_2O_4^{-2} \rightarrow 2CO_2 + 2e^-] \times 5$

Reduction reaction :

$MnO_4^- + 8H^+ + 5e^- \rightarrow Mn^{2+} + 4H_2O] \times 2$

Net reaction :

$$2MnO_4^- + 16H^+ + 5C_2O_4^{2-} \longrightarrow 2Mn^{2+} + 10CO_2 + 8H_2O$$

20. (a, b, d) Balancing the chemical equation by half-reaction method.

$I^- + ClO_3^- + H_2SO_4 \longrightarrow Cl^- + HSO_4^- + I_2$

$2I^- \longrightarrow I_2 + 2e^- \times 3$

$ClO_3^- + 6H^+ + 6e^- \longrightarrow Cl^- + 3H_2O$

$ClO_3^- + 6I^- + 6H^+ \longrightarrow 3I_2 + Cl^- + 3H_2O$

Adding $6HSO_4^-$ to both sides.

$$ClO_3^- + 6I^- + 6H_2SO_4 \longrightarrow 3I_2 + Cl^- + 6HSO_4^- + 3H_2O$$

21. Balance the atoms as well as charges by ion electron/oxidation number method.

(i) $6Ag^+ + AsH_3 + 3H_2O \longrightarrow 6Ag + H_3AsO_3 + 6H^+$

(ii) $ClO_3^- + 6I^- + 6H_2SO_4 \longrightarrow Cl^- + 6HSO_4^- + 3I_2 + 3H_2O$

(iii) $4S + 6OH^- \rightarrow 2S^{2-} + S_2O_3^{2-} + 3H_2O$

(iv) $2Mn^{2+} + 5PbO_2 + 4H^+ \longrightarrow 2MnO_4^- + 2H_2O + 5Pb^{2+}$

(v) $Cl_2 + 2OH^- \longrightarrow Cl^- + ClO^- + H_2O$

(vi) $2Ce^{3+} + S_2O_8^{2-} \longrightarrow 2SO_4^{2-} + 2Ce^{4+}$

(vii) $2HNO_3 + 6HCl \longrightarrow 2NO + 3Cl_2 + 4H_2O$

(viii) $Cr_2O_7^{2-} + 3C_2H_4O + 8H^+ \longrightarrow 3C_2H_4O_2 + 2Cr^{3+} + 4H_2O$

(ix) $4Zn + NO_3^- + 10H^+ \longrightarrow 4Zn^{2+} + NH_4^+ + 3H_2O$

22. Balance the reactions by ion electron method.

(i) $Cu_2O + 2H^+ \longrightarrow 2Cu^{2+} + H_2O + 2e^-] \times 3$ ......(1)

$NO_3^- + 4H^+ + 3e^- \rightarrow NO + 2H_2O] \times 2$ .....(2)

$3Cu_2O + 14H^+ + 2NO_3^- \longrightarrow 6Cu^{2+} + 2NO + 7H_2O$

(ii) $K_4[Fe(CN)_6] + 6H_2SO_4 + 6H_2O \longrightarrow 2K_2SO_4 + FeSO_4 + 3(NH_4)_2SO_4 + 6CO$

(iii) $C_2H_5OH + 4I_2 + 8OH^- \longrightarrow CHI_3 + HCO_3^- + 5I^- + 6H_2O$

Chapter-9

Hydrogen

Part-A (JEE Main)

1. **(3)** Electrolysis of warm aqueous $Ba(OH)_2$ solution using nickel electrode yields high purity (>99.95%) dihydrogen.

2. **(1)**

3. **(2)** Around 55% of hydrogen consumed in ammonia production.

4. **(1)** The bond dissociation energy of D_2 is greater than H_2 and therefore, D_2 reacts slower than H_2.

5. **(3)** Radioactive isotope of hydrogen is tritium, ${}_1H^3$

No. of neutrons $(A - Z) = 3 - 1 = 2$

No. of electrons = 1

6. **(3)** 7. **(4)** 8. **(4)**

$$CO(g) + H_2O(g) \xrightarrow[\text{Catalyst}]{673K} CO_2(g) + H_2(g)$$

This reaction is called water gas shift reaction.

9. **(2)** $Zn + 2NaOH \rightarrow Na_2ZnO_2 + H_2$

$Zn + 2HCl \rightarrow ZnCl_2 + H_2$

NaOH and HCl reacts with a certain amount of zinc to produce equal number of moles of H_2.

10. **(1)**

	${}_1^1H$	${}_1^2H(D)$	${}_1^3H(T)$
Number of neutron	0	1	2
	(x)	(y)	(z)

Total number of neutrons in three isotopes of hydrogen $= 0 + 1 + 2 = 3$

11. **(3)**

12. **(1)**

(i) Saline hydrides with water produces H_2 gas.

(ii) $3LiAlH_4 + 4BF_3 \rightarrow 2B_2H_6 + 3LiF + 3AlF_3$

(iii) PH_3 is electron rich whereas CH_4 is electron precise hydride.

(iv) HF and CH_4 are molecular hydrides as they are covalent molecules.

13. **(4)**

14. **(3)** NaH is an ionic hydride which is also known as saline hydride.

15. **(1)** The reaction between I_2 and H_2 requires catalyst, whereas all other halogens react with H_2 without a catalyst.

16. **(4)** H_2 is a highly inflammable gas.

17. **(4)** Very pure hydrogen can be prepared by the action of water on sodium hydride.

$NaH + H_2O \longrightarrow NaOH + H_2\uparrow$

18. **(1)**

19. **(2)** Clark's Method Reaction

$Ca(HCO_3)_2 + Ca(OH)_2 \rightarrow 2CaCO_3 + 2H_2O$

$Mg(HCO_3)_2 + 2Ca(OH)_2 \rightarrow 2CaCO_3 + Mg(OH)_2 + 2H_2O$

20. **(3)**

21. **(1)**

$$Mg(HCO_3)_2 \xrightarrow{\text{Boil}} Mg(OH)_2\downarrow + 2CO_2\uparrow$$

$$Ca(HCO_3)_2 \xrightarrow{\text{Boil}} CaCO_3\downarrow + H_2O + CO_2\uparrow$$

The ppt formed can be removed by filtration

22. **(1)**

23. **(4)** Solubility of gas (O_2) in liquid decreases with the increase in temperature.

24. **(1)**

25. **(1)** Clark's method is used to remove temporary hardness, using lime water (or) $Ca(OH)_2$ from water.

26. **(1)** In distilled water, there are only neutral water molecules therefore, it does not conduct electricity.

27. **(2)** Synthetic resin method is more efficient than zeolite process as it can exchange both cations as well as anions.

28. **(1)**

29. **(4)** The gaseous mixture of CO and H_2 is called water gas or syn gas (synthesis gas).

30. **(3)** In $CuSO_4{\cdot}5H_2O$, four H_2O molecules are directly coordinated to the central metal ion while one H_2O molecule is hydrogen bonded.

31. **(4)** Moles of $Ca(HCO_3)_2 = \frac{0.81}{162} = 0.005$

Moles of $Mg(HCO_3)_2 = \frac{0.73}{146} = 0.005$

Hardness in terms of $CaCO_3$ in ppm

$$= \frac{(0.005 + 0.005) \times 100}{100} \times 10^6 = 10^4 \text{ ppm}$$

32. **(3)**

33. **(1)** There is extensive intermolecular hydrogen bonding in the condensed phase instead of intramolecular H-bonding.

34. **(2)** Heavy water acts as moderator.

35. **(2)**

36. **(2)** Heavy water is D_2O hence

number of electrons = 2 + 8 = 10

number of protons = 10

Mass number of $D_2O = 4 + 16 = 20$,

Hence number of neutrons

= Mass number – number of protons

= 20 – 10 = 10

37. **(100.00)**

10^{-3} molar $MgSO_4 \equiv 10^{-3}$ moles of $MgSO_4$ present in 1 L solutions.

10^{-3} M $MgSO_4 \equiv 10^{-3}$ M $CaCO_3$

10^{-3} M $CaCO_3 = 10^{-3} \times 100$g $CaCO_3$ in 1L water

$$\text{ppm}_{(\text{in term of }CaCO_3)} = \frac{10^{-3} \times 100}{1000} \times 10^6$$

$$\text{ppm}_{(\text{in term of }CaCO_3)} = 100 \text{ ppm}$$

38. **(3) Statement I:** It is correct.

Statement II: Hydrogen bonding in H_2O_2 is more than that in D_2O. Therefore, due to more intermolecular forces, the density of H_2O_2 is more than that of D_2O.

39. **(1)**

(1) Urea acts as stablising agent. It prevents the decomposition of H_2O_2.

40. **(4)** $6H^+ + 2MnO_4^- + 5H_2O_2 \rightarrow 2Mn^{2+} + 8H_2O + 5O_2$

41. **(1)** In acidic medium MnO_4^- forms Mn^{2+} by reacting with H_2O_2.

42. **(3)**

43. **(2)** Possible reactions in basic medium are

$$Mn^{2+} + H_2O_2 \longrightarrow Mn^{4+} + 2OH^-$$

$$I_2 + H_2O_2 + 2OH^- \longrightarrow 2I^- + 2H_2O + O_2$$

$PbS \rightarrow PbSO_4$ is possible in acidic medium,

$$PbS + 4H_2O_2 \longrightarrow PbSO_4 + 4H_2O$$

44. **(2)**

(A) H_2O_2 is an important chemical used in pollution control treatment of domestic and industrial effluents.

(B) It has both oxidising and reducing properties.

(C) It has open-book like structure and both hydroxyl groups lie in different planes.

(D) It is miscible in water due to hydrogen bonding.

45. **(1)** H_2O_2 has open book like structure, which is non-planar. It is a colourless viscous liquid but in large quantity appears blue in colour.

46. **(1)** For H_2O_2

$$\text{Molarity} = \frac{\text{Volume strength}}{11.2} = \frac{5.6}{11.2} = 0.5 \text{ M}$$

No. of moles in 1L = 0.5

Mass of $H_2O_2 = 0.5 \times 34 = 17$ g

In 100 g (=100 mL), mass of $H_2O_2 = 1.7$ g

47. **(1)**

48. **(2)** 11.2 V strength of H_2O_2 means, 11.2 L of O_2 is liberated at STP.

$$H_2O_2 \longrightarrow H_2O + \frac{1}{2}O_2$$

11.2 L of O_2 at STP = 0.5 mol

$\therefore$ No. of moles of H_2O_2 = 1 mol

i.e., 1 L of given H_2O_2 solution has 1 mole of H_2O_2 (i.e., 34 g)

$$\text{Strength} = \frac{34}{1000} \times 100 = 3.4\%$$

49. **(4)**

50. **(3)** $[Fe(CN)_6]^{4-} + \frac{1}{2}H_2O_2 + H^+ \longrightarrow [Fe(CN)_6]^{3-} + H_2O$

$$[Fe(CN)_6]^{3-} + \frac{1}{2}H_2O_2 + OH^- \longrightarrow [Fe(CN)_6]^{4-} + H_2O + \frac{1}{2}O_2$$

51. **(3)** H_2O_2 has oxidizing and reducing properties both.

52. **(1)** H_2O_2 acts as a reducing agent only in presence of strong oxidising agents (i.e., MnO_4^-) in acidic as well as alkaline medium.

$$2KMnO_4 + 3H_2SO_4 + 5H_2O_2 \longrightarrow K_2SO_4 + 2MnSO_4 + 8H_2O + 5O_2$$

53. **(100)**

Molarity of H_2O_2 solution

$$= \left\{\frac{\text{Volume strength}}{11.2}\right\}$$

Volume strength $= 8.9 \times 11.2 = 99.68\ V \simeq 100\ V$

Part-B (JEE Advanced)

1. **(a)** The alkali metals are highly electropositive elements, hence, the hydrides, hydrogen exist as hydride ion H^-. This is the reason why hydrogen liberate at anode.

2. **(c)** $2Ca^{2+} + Na_2[Na_4(PO_3)_6] \longrightarrow 4Na^+ + \underset{\text{water soluble}}{Na_2[Ca_2(PO_3)_6]}$

3. **(c)** Heavy water is D_2O, deuterium oxide.

4. **(b)** Temporary hardness of water is due to presence of bicarbonates of Ca and Mg and it is removed by adding $Ca(OH)_2$ to hard water and precipitating these soluble bicarbonates in the form of insoluble salts.

$$Ca(HCO_3)_2 + Ca(OH)_2 \longrightarrow CaCO_3\downarrow + 2H_2O$$

$$Mg(HCO_3)_2 + 2Ca(OH)_2 \longrightarrow 2CaCO_3\downarrow + Mg(OH)_2\downarrow + 2H_2O$$

5. **(b, c, d)** Temporary hardness is due to bicarbonates of calcium and magnesium. Temporary hardness can be removed by Clark's process, which involves the addition of slaked lime, $Ca(OH)_2$. Washing soda (Na_2CO_3) removes both the temporary and permanent hardness by converting soluble calcium and magnesium compounds into insoluble carbonates.

$$Ca(HCO_3) + Ca(OH)_2 \longrightarrow 2CaCO_3\downarrow + 2H_2O$$

$$Ca(HCO_3) + Na_2CO_3 \longrightarrow CaCO_3\downarrow + 2NaHCO_3$$

NaOCl can remove the hardness as :

$$2OCl^- + 2H_2O \rightleftharpoons 2HOCl + 2OH^-$$

$$Ca(HCO_3)_2 + 2OH^- \longrightarrow CaCO_3\downarrow + CO_3^{2-} + 2H_2O$$

6. **(a)** $CO_2 + H_2O \rightleftharpoons H_2CO_3 \rightleftharpoons H^+ + HCO_3^- \rightleftharpoons H^+ + CO_3^{2-}$

7. **(b, d)** $Na_2Al_2Si_2O_8\cdot xH_2O + Ca^{2+} \rightarrow CaAl_2Si_2O_8\cdot xH_2O + 2Na^+$

$Na_2Al_2Si_2O_8\cdot xH_2O + Mg^{2+} \rightarrow MgAl_2Si_2O_8\cdot xH_2O + 2Na^+$

8. **(a)** $KIO_4 + H_2O_2 \rightarrow KIO_3 + H_2O + O_2$

Thus, H_2O_2 is acting as a reducing agent

$2NH_2OH + H_2O_2 \rightarrow N_2 + 4H_2O$

Here H_2O_2 is acting as an oxidising agent

9. **(b)** Volume strength = Normality $\times$ 5.6

$= 1.5 \times 5.6 = 8.4\ L$

10. **(1.344)** By law of equivalence:

Meq. of H_2O_2 = Meq. of $Na_2S_2O_3$ = Meq. of I_2 = Meq. of KI

$$\frac{w}{17} \times 1000 = 20 \times 0.3$$

$\therefore$ $w = 0.102$ g (equating Meq. in 25 mL solution)

$$H_2O_2 \rightarrow H_2O + \frac{1}{2}O_2$$

$$\therefore \text{ Volume of } O_2 = \frac{11200 \times 0.102}{34} = 33.6\ \text{mL}$$

$$\therefore \text{ Volume strength} = \frac{33.6}{25} = \mathbf{1.344}$$

11. **(4.48)**

$$\underset{\substack{34\ g\\ 5\ cm^3\ \text{or mL}}}{H_2O_2} + H_2SO_4 + \underset{\substack{\text{Acidified}\\ \text{KI sol}}}{2KI} \longrightarrow K_2SO_4 + \underset{\substack{254\ g\\ 0.508\ g}}{I_2} + 2H_2O$$

i.e. 254 g of I_2 is released by 34 g H_2O_2

$\therefore$ 0.508 g of I_2 will be released by

$$= \frac{34}{254} \times 0.508 = 0.068\ g$$

5 mL of H_2O_2 sol. contains 0.068 g of H_2O_2.

$$\therefore \text{ 1 mL of } H_2O_2 \text{ sol contains } \frac{0.068}{5}\ g\ H_2O_2$$

The strength of H_2O_2 is generally calculated in terms of **volume strength**. According to which, 10 volume of H_2O_2 means that 1 mL of H_2O_2 sol gives 10 mL of O_2 at STP.

$$\underset{2\times 34\ g}{2H_2O_2} \longrightarrow 2H_2O + \underset{\substack{32\ g\ \text{or}\\ 22{,}400\ \text{mL at STP}}}{O_2}$$

i.e., 68 g of H_2O_2 gives 22,400 mL of O_2 at STP

or 1 mL of H_2O_2 sol

or $\frac{0.068}{5}$ g of H_2O_2

gives $\frac{22{,}400}{68} \times \frac{0.068}{5} = 4.48$ mL

or 1 mL of H_2O_2 sol gives 4.48 mL of O_2 *i.e.* strength of H_2O_2 sol is **4.48 volumes.**

12. Example of oxidising character of H_2O_2 in alkaline medium

$$2Cr(OH)_3 + 4NaOH + 3H_2O_2 \rightarrow 2Na_2CrO_4 + 8H_2O$$

Here, Cr^{3+} (Cr is a first row transition metal) is oxidised to Cr^{6+}.

Example of reducing character of H_2O_2 in alkaline medium:

$$2K_3[Fe(CN)_6] + 2KOH + H_2O_2 \rightarrow 2K_4[Fe(CN)_6] + 2H_2O + O_2$$

Here, Fe^{3+} (Fe is a first row transition metal) is reduced to Fe^{2+}.

13.

$$2KMnO_4 + 3H_2SO_4 \rightarrow K_2SO_4 + 2MnSO_4 + 3H_2O + 5[O]$$

$$[H_2O_2 + [O] \rightarrow H_2O + O_2] \times 5$$

$$2KMnO_4 + 5H_2O_2 + 3H_2SO_4 \rightarrow K_2SO_4 + 2MnSO_4 + 8H_2O + 5O_2$$

14. (*i*) The oxidation state of oxygen in H_2O_2 (*i.e.* –1) can be changed to 0 or –2 *i.e.* oxygen in H_2O_2 exists in an intermediate oxidation state with respect to O_2 and O^{2-}. Hence, it acts both as an oxidising and reducing agent.

(*ii*) H_2O_2 is a better oxidising agent than H_2O because oxidation number of oxygen in H_2O_2 is –1 and that in water it is –2. So, H_2O_2 easily reduces to –2 oxidation number.

15. The mixture of N_2H_4 and H_2O_2 (in presence of Cu (II) catalyst) is used as a rocket propellant because the reaction is highly exothermic and large volume of gases are evolved, which can propel a rocket.

$$\underset{\text{Hydrazine}}{N_2H_4} + \underset{\text{Hydrogen peroxide}}{2H_2O_2} \xrightarrow{Cu(II)} N_2(g)\uparrow + 4H_2O(g)\uparrow$$

16. Ferricyanide is oxidised to ferrocyanide on treatment with alkali:

$$2K_3[Fe(CN)_6] + H_2O_2 + 2KOH \rightarrow 2K_4[Fe(CN)_6] + 2H_2O + O_2$$

Chapter-10

The *s*-Block Elements

Part-A (JEE Main)

1. **(3)** $4LiNO_3 \rightarrow 2Li_2O + 4NO_2 + O_2$
$2NaNO_3 \rightarrow 2NaNO_2 + O_2$

2. **(1)** Low solubility of LiF in water is due to high lattice enthalpy.

3. **(3)**

4. **(1)** Density of the alkali metals increases down the group.
∴ The correct order: $Li < K < Na < Rb < Cs$

5. **(2)** Superoxide is paramagnetic due 1 unpaired electron.

6. **(1)**

7. **(1)**
(A) Both LiCl and $MgCl_2$ are soluble in ethanol.
(B) Li and Mg do not form superoxide.
(C) LiF has highest lattice energy among alkali metal fluorides.
(D) Li_2O is least soluble in water than other alkali metal oxides due to high lattice energy.

8. **(4)**

	Colour	λ/nm
Li	Crimson red	670.8
Na	Yellow	589.2
Rb	Red violet	780.0
Cs	Blue	455.5

Wavelength increases from voilet to red colour.

9. **(1)** Lithium has smallest size among the alkali metal. So charge to radius ratio of Li is very high, due to which the lithium salts are hydrated in nature.
Li due to smallest size in its group has highest polarizing power which is related to covalent character.

10. **(3)**
(1) Standard enthalpy of formation for alkali metal bromides becomes more negative on descending down the group.
(2) In case of CsI, lattice energy is less, but Cs^+ is having less hydration enthalpy due to which it is less soluble in water.
(3) For alkali metal fluorides, the solubility in water increases from lithium to caesium. LiF is least soluble in water.
(4) Standard enthalpy of formation for LiF is most negative among alkali metal fluorides.

11. **(1)**

12. **(2)** Metallic sodium does not react with but-2-yne due to absence of acidic hydrogen in it.

13. **(4)** Sodium metal on dissolution in liquid ammonia gives a deep blue solution due to the ammoniated electrons.

14. **(3)** The order of conductivity of ions in H_2O is
$Cs^+ > Rb^+ > K^+ > Na^+$

15. **(2)** On heating in excess air, Li form oxide, sodium form peroxide while K, Rb and Cs form superoxide.

16. **(4)** Li and Mg do not form solid bicarbonate, but react with N_2 to give nitrides.

17. **(2)** $Rb \xrightarrow{O_2\,(excess)} \underset{X}{RbO_2}$
Superoxides are stronger oxidising agents than peroxides. Thus they give both H_2O_2 and O_2 with water.
$$2RbO_2 + 2H_2O \longrightarrow 2RbOH + H_2O_2 + O_2$$

18. **(3)**

19. **(2)** Amongst the given alkali metals, only lithium can react with N_2 in air to form lithium nitride.

20. **(3)**

Compounds	Nature
KO_2	Superoxide
BaO_2	Peroxide
SiO_2	Oxide
CsO_2	Superoxide

21. **(4)** Covalent character increases, when the cation has small size and high charge density.
Among all these, Mg^{2+} size is smallest, so $MgCl_2$ tends to be more covalent in nature.

22. **(2)**

23. **(4)** The lattice enthalpy of alkali metals follow the order :
$$Li > Na > K > Rb$$
Hence, the correct order of solubility is :
$$LiF < NaF < KF < RbF$$

24. **(1)** NaCl is expected to have maximum melting point among the alkali metal chlorides.

25. **(3)** Reactivity with water increases on moving down the Group from Li to Cs as Li < Na < K < Rb < Cs. This is due to increase in electropositive character. Hence, Rb reacts most vigorously with water.

26. **(4)**

27. **(1)** The alkali metals dissolve in liquid ammonia without evolution of hydrogen. The metal loses electrons and combine with ammonia molecule.
$M \rightarrow M^+ \text{(in liquid ammonia)} + e^- \text{(ammoniated)}$
$M + (x+y)NH_3 \rightarrow [M(NH_3)_x]^+ + \underset{\text{Solvated electron}}{e^-(NH_3)_y}$
It is ammoniated electron which is responsible for blue colour.

28. **(1)** Among the given alkali metals, only cesium (Cs) is used as electrode in the photoelectric cell due to its lowest ionisation energy.

29. **(2)**

30. **(1)** Baking soda $\rightarrow NaHCO_3$
(Sodium Hydrogen Carbonate)
Washing soda $\rightarrow Na_2CO_3.10H_2O$
(Sodium Carbonate decahydrate)
Caustic soda $\rightarrow$ NaOH (Sodium Hydroxide)

31. **(1)** Sodium hydroxide is generally prepared commercially by electrolysis of sodium chloride in Castner Kellner cell.
at cathode : $Na + e^- \xrightarrow{Hg} Na - \text{amalgam}$
Anode : $Cl^- \longrightarrow \frac{1}{2}Cl_2 + e^-$
The Na-amalgam is treated with water to give sodium hydroxide and hydrogen gas :
$2Na\,(amalgam) + H_2O \rightarrow 2NaOH + H_2 + 2Hg$

32. **(1)** Baking soda or $NaHCO_3$ is used in the fire extinguishers.

33. **(3)** Lithium nitrate decomposes into its oxide on heating.
$$4LiNO_3 \xrightarrow{\Delta} 2Li_2O + 4NO_2 + O_2$$

34. **(1)** $\underset{(X)}{Na_2CO_3\cdot 10H_2O} \longrightarrow \underset{(Y)}{Na_2CO_3\cdot H_2O} + 9H_2O$
$\underset{(Y)}{Na_2CO_3\cdot H_2O} \longrightarrow \underset{(Z)}{Na_2CO_3} + H_2O$
X = Washing soda; Z = Soda ash

35. **(2)** $SiCl_4 + LiAlH_4 \rightarrow LiCl + AlCl_3 + SiH_4$

36. **(4)** Na_2O_2 is the peroxide of sodium not super oxide. The formula of sodium superoxide is NaO_2.

37. **(1)**

38. **(1)**

39. **(4)** **Statement I:** Due to vacant orbitals in Be and Al in their chlorides, they act as Lewis acid.
Statement II: Hydroxide of Al dissolve in excess of NaOH to form aluminate ions.
$Al(OH)_3 + NaOH \rightarrow NaAlO_2 + 2H_2O$
Similarly,
$Be(OH)_2 + 2NaOH \rightarrow Na_2BeO_2 + 2H_2O$

40. **(3)** In case of group 2, down the group both atomic mass and atomic size increase. Upto calcium increase in atomic size is more dominating and after calcium increase in atomic mass dominates so the order of density is
$$Sr > Be > Mg > Ca$$

41. **(4)** Be has very high ionization enthalpy and the energy of flame is not sufficient to excite the electrons to the higher level.

42. **(4)** Highest ionic mobility corresponds to lowest extent of hydration and biggest size of gaseous ion.
Hence, Sr^{2+} has the highest ionic mobility in its aqueous solution.

43. **(1)** $Ba \Rightarrow [Xe]6s^2$
Ca ⇒ Calcium oxalate is insoluble (sparingly soluble) in water
Li ⇒ LiCl is soluble in organic solvents like pyridineNa ⇒ NaOH is a very strong monoacidic base

44. **(3)** $BeSO_4$ and $MgSO_4$ are readily soluble in water due to greater hydration enthalpies of Be^{2+} and Mg^{2+} ions.

45. **(4)** Solubility of $BeSO_4$ is highest among the given metal sulphates
$BeSO_4 > MgSO_4 > CaSO_4 > SrSO_4 > BaSO_4$

Solubility order for hydroxide is
$Be(OH)_2 < Mg(OH)_2 < Ca(OH)_2 < Sr(OH)_2 < Ba(OH)_2$
Thus $BeSO_4$ is soluble and $Be(OH)_2$ is insoluble.
BeO does not form rock salt like structure.
In the solid state, it adopts the hexagonal wurtzite structure form while in the vapour phase, it is present as discrete diatomic covalent molecules, due to its covalent nature.

46. (2) LiCl is soluble in pyridine due to its covalent nature.

47. (3) $\underset{(A)}{3Mg + N_2} \longrightarrow \underset{(B)}{Mg_3N_2} \xrightarrow{H_2O} 3Mg(OH)_2 + 2NH_3$

$4NH_3 + CuSO_4 + 2H_2O \longrightarrow \underset{\text{(Dark blue coloured)}}{[Cu(NH_3)_4(OH_2)_2]SO_4}$

48. (2) Thermal stability of alkaline earth metal carbonates increases down the group.
$MgCO_3 < CaCO_3 < SrCO_3 < BaCO_3$

49. (1)

50. (2) Mg burns in air and produces a mixture of nitride and oxide.

51. (3) $BeCl_2$ in vapour phase exists as dimer (below 1200 K temperature) whereas, in solid state $BeCl_2$ has chain structure.

52. (3) Be^{2+} has maximum polarising power among given cations. Therefore, BeX_2 would be most covalent alkaline earth metal halides among the given halides.

53. (3) Be is transparent to X-rays, so it is used in making X-ray tube windows.

54. (4) The chances of formation of hydrate decreases with the decrease in the charge density down the group. This is why, $Ba(NO_3)_2$ does not crystallise with water molecules.

55. (1)

56. (4) $BeSO_4$ has its hydration enthalpy greater than its lattice enthalpy. This is why it is soluble in water. Other than this $MgSO_4$ also soluble in water due to greater hydration energy. Compounds given in other option are poorly soluble in water.

57. (3) The amount of heating required depends on the extent of polarisation. More polarization required less energy. As the positive ions get larger down the group, their ionic character increases. Therefore, the hydroxides become more thermally stable down the group.

58. (1)

$Mg_3N_2 + 6H_2O \rightleftharpoons 3Mg(OH)_2 + 2NH_3$

59. (1) As we move down the group, the lattice energy of carbonates remains approximately the same. However, the hydration energy of the metal cations decrease from Be^{2+} to Ba^{2+}, hence the solubility of carbonates of the alkaline earth metals decreases down the group.

60. (2)

61. (4) Sulphates of alkaline earth metals are sparingly soluble or almost not soluble in water whereas $BeSO_4$ is soluble in water due to high degree of solvation. $Be(OH)_2$ is insoluble in water but soluble in NaOH.

$BeO + 2NaOH \longrightarrow Na_2BeO_2 + H_2O$

62. (2) BeO and $Be(OH)_2$ are amphoteric in nature while BaO and $Sr(OH)_2$ are basic in nature.

63. (2) Gypsum ($CaSO_4 \cdot 2H_2O$) is used to enhance setting time in portland cement.

64. (2) The balanced chemical reaction is
$2BeCl_2 + LiAlH_4 \longrightarrow 2BeH_2 + LiCl + AlCl_3$

65. (4) The common method for the preparation of hydrogen peroxide is reaction with BaO_2 and H_2SO_4.

$BaO_2 + H_2SO_4 \rightarrow BaSO_4 + H_2O_2$

66. (3)

67. (N)

(A) For temporary hardness,
$Mg(HCO_3)_2 \xrightarrow{\Delta} Mg(OH)_2\downarrow + 2CO_2\uparrow$
Hence, (A) is false.

(R) The solubilities (s) of both $Mg(OH)_2$ and $MgCO_3$ are very low. The values of solubility products are :
$K_{sp}(Mg(OH)_2) = 4s^3$ and $K_{sp}(MgCO_3) = s_1^2$
Because $s << 1$, thus $K_{sp}(Mg(OH)_2) < K_{sp}(MgCO_3)$
Hence, (R) is also false.

68. (2) Calgon is sodium hexametaphosphate, a polymeric compound also called as Graham's salt.
Silicon is the 2nd most abundant element which is absent in calgon.
$Ca^{2+} + Na_4P_6O_{18}^{2-} \longrightarrow [Na_2CaP_6O_{18}]^{2-} + 2Na^+$
Hence, it does not remove Ca^{2+} ion by precipitation.

69. (2)

(I) $Ca(OH)_2$ is used in white wash.

(II) NaCl is used in preparation of washing soda (Na_2CO_3).

(III) $CaSO_4 \cdot \frac{1}{2}H_2O$ is used for making casts of statues.

(IV) $CaCO_3$ is used as an antacid.

70. (3) Gypsum on heating to 393 K forms plaster of Paris.

$CaSO_4 \cdot 2H_2O \xrightarrow{393\,K} CaSO_4 \cdot 0.5H_2O + 1.5H_2O$

71. (4)

(i) $Na_2CO_3 \cdot 10H_2O \rightarrow$ Solvay process

(ii) $Mg(HCO_3)_2 \rightarrow$ Temporary hardness

(iii) $NaOH \rightarrow$ Castner-Kellner process

(iv) $Ca_3Al_2O_6 \rightarrow$ Portland cement

72. (1) Quick lime is commercial name of CaO.

73. (3)

$CaCO_3 \xrightarrow{\Delta} \underset{\text{Basic oxide}}{CaO} + \underset{\text{Acidic oxide}}{CO_2\uparrow}$

74. (1) Gypsum is $CaSO_4 \cdot 2H_2O$.

75. (1) Setting of cement is exothermic process which develops interlocking crystals of hydrated silicates.

76. (2) The reaction is

$BeO + HF + NH_3 \rightarrow \underset{[A]}{(NH_4)_2[BeF_4]} \xrightarrow{\Delta} 2NH_4F + \underset{[B]}{BeF_2}$

Oxidation state of Be in A is (+2)

Part-B (JEE Advanced)

1. (b) When light falls on sodium surface, valence e^- gets excited and then returns back. The emission of photons creates a metallic lustre on the surface.

2. (a) Acidic and basic salts cannot exist together.
Since $NaHCO_3$ is an acid salt of H_2CO_3, it reacts with NaOH to form Na_2CO_3 and H_2O.
$NaHCO_3 + NaOH \rightarrow Na_2CO_3 + H_2O$

3. (b) $Na_2O_2 + H_2SO_4\,(20\%\text{ ice cold}) \rightarrow Na_2SO_4 + H_2O_2$

4. (d) The free ammoniated electrons make the solution of Na in liquid NH_3 a very powerful reducing agent.
The ammonical solution of an alkali metal is rather favoured as a reducing agent than its aqueous solution because in aqueous solution the alkali metal being highly electropositive evolves hydrogen from water (thus H_2O acts as an oxidisng agent) while its solution in ammonia is quite stable, provided no catalyst (transition metal) is present.

5. Hydrogen, anode.

$NaH \xrightarrow{\text{electricity}} \underset{\text{cathode}}{Na^+} + \underset{\text{anode}}{2H^+}$

$Na^+ \xrightarrow{+e^-} Na$; $\quad 2H^+ \xrightarrow{-2e^-} H_2$

6. of solvated electrons.

7. False : Sodium when burnt in excess of oxygen gives sodium peroxide (Na_2O_2) and not sodium oxide.
$4Na + O_2\,(\text{limited}) \rightarrow 2Na_2O$;
$2Na + O_2\,(\text{excess}) \rightarrow Na_2O_2$

8. True : The metallic bonding decreases with increase in atomic size and thus, the tendency to show metallic bonding among alkali metals decreases from Li to Cs. Close packing of atoms in crystal lattice decreases from Li to Cs, resulting in an increase in softness.

9. (a, b) $4Na + O_2\,(\text{limited}) \xrightarrow{\Delta} 2Na_2O$

$2Na + O_2\,(\text{excess}) \xrightarrow{\Delta} Na_2O_2$

10. (a, b) Blue colour is due to the presence of solvated (ammoniated) electrons, while electrical conductance is due to the presence of ions.

$M \longrightarrow M^+ + e^-$
$M^+ + xNH_3 \longrightarrow [M(NH_3)_x]^+$
$e^- + yNH_3 \longrightarrow [e(NH_3)_y]^-$

on adding

$M + (x+y)NH_3 \longrightarrow \underset{\text{Ammoniated cation}}{[M(NH_3)_x]^+} + \underset{\text{Ammoniated electron}}{[e(NH_3)_y]^-}$

Note: Sodium in liquid ammonia forms $NaNH_2$ only in presence of a catalyst like Pt black, iron oxide, etc.

$[e(NH_3)_y]^- \xrightarrow{Fe_2O_3} NH_2^- + H_2 + (y-1)NH_3$.

11. (b) Statement-1 is correct. Statement-2 is also correct but not the correct explanation becuase blue colour of the solution is due to the solvated electrons.

12. LiF has more ionic character while LiI has more covalent character. The latter is due to the greater polarizability of larger iodide ion than that of fluoride ion.

13. (*iv*) $2K_3[Fe(CN)_6] + H_2O_2 + 2KOH \rightarrow 2K_4[Fe(CN)_6] + 2H_2O + O_2$

(*iii*) $NaCl + NH_4OH + CO_2 \rightarrow NH_4Cl + NaHCO_3$

14. Potassium carbonate cannot be manufactured by Solvay process, since, unlike sodium hydrogen carbonate, potassium hydrogen carbonate is rather too soluble in water to be precipitated like $NaHCO_3$.

15. (a)

$MgCl_2 + 2NaHCO_3 \longrightarrow \underset{\text{(soluble)}}{MgHCO_3} + 2NaCl$

$MgHCO_3 \xrightarrow{\text{heat}} MgCO_3 \downarrow + H_2O + CO_2$

16. (b) In going from top to bottom in a group, the first ionization potential decreases, thus $Be > Mg > Ca$

17. (b) The increasing thermal stability is

$\underset{(IV)}{BeCO_3} < \underset{(II)}{MgCO_3} < \underset{(III)}{CaCO_3} < \underset{(I)}{K_2CO_3}$

Increasing size of cation decreases its polarization ability towards carbonate, making the compound more stable.

18. (d) Glauber's salt is $Na_2SO_4 \cdot 10\,H_2O$.

19. (a) Ca is obtained by electrolysis of molten mixture of $CaCl_2$ mixed with CaF_2.

20. (c) Peroxide can produce hydrogen peroxide

$BaO_2 \cdot 8H_2O + H_2SO_4 \longrightarrow BaSO_4 + H_2O_2 + 8H_2O$

21. (b) $CaO + H_2O \xrightarrow{\text{Hissing sound}} Ca(OH)_2 + \text{Heat}$

$CaO + CO_2 \longrightarrow CaCO_3$

22. higher effective nuclear charge.

23. Anhydrous HCl

$\because$ All the water of crystallisation cannot be removed by heating hydrated $MgCl_2$. HCl checks the hydrolysis of $MgCl_2$ by its own water of crystalization.

24. False : Although 4 molecules of water of crystallisation are removed by heating, the remaining two react with $MgCl_2$ as per the equation given below :

$MgCl_2 + 2H_2O \rightarrow MgO + 2HCl + H_2O$

In order to avoid this to happen, $MgCl_2.2H_2O$ is dehydrated in presence of HCl gas, which checks, (being in excess) the hydrolysis of $MgCl_2$ by its own water of crystallisation.

25. (a)

$Mg^{2+} + NH_3 + HPO_4^{2-} \longrightarrow Mg(NH_4)PO_4$

26. (a) H_2O_2 (hydrogen peroxide); SrO_2 and BaO_2 (Barium peroxide) contain peroxide ions $\left(O_2^{2-}\right)$.

27. $BeCl_2$ is hydrolysed due to high polarising power and presence of vacant *p*-orbitals in Be-atom.

$(Be = 1s^2, 2s^2 2p_x^0\ 2p_y^0\ 2p_z^0)$

28. When chlorine is passed over dry slaked line, pleaching powder is provided as shown in the reaction,

$Ca(OH)_2 + Cl_2 \rightarrow CaOCl_2 + H_2O$

29. $SrSO_4 > CaSO_4 > MgSO_4 > BeSO_4$ (Based upon size of cation or ionic character)

30. Smaller the size of cation, higher will be hydration tendency because **hydration energy of cation is inversely proportional to size of cation**. The size of alkaline earth metal ions are smaller than the size of alkali metal ions. So, in crystalline form, the salts of alkaline earth metals have more water molecules than those of alkali metals.

31. $\underset{'A'}{3M} + N_2 \rightarrow \underset{'B'}{M_3N_2}$

$\underset{'B'}{M_3N_2} + 6H_2O \rightarrow \underset{'C'}{3M(OH)_2} + \underset{'D'}{2NH_3}$

$\underset{'C'}{M(OH)_2} + CO_2 \rightarrow \underset{'D'}{MCO_3} + H_2O$

M may be either Ca or Ba.

It is not magnesium because $Mg(OH)_2$ has very low solubility in water.

If we consider Ba as *M* then *A* is **Ba**, *B* is $\mathbf{Ba_3N_2}$, *C* is $\mathbf{Ba(OH)_2}$, *D* is $\mathbf{BaCO_3}$.

32. N^{3-} being smaller in size and high charge present on it make it more susceptible to hydrolysis:

$N^{3-} + 3H_2O \rightarrow NH_3 + 3OH^-$

Cl^- being a weak conjugate base does not undergo hydrolysis.

Mg^{2+} get hydrolysed as :

$Mg^{2+} + 2H_2O \rightleftharpoons Mg(OH)_2 + 2H^+$

33. $Ca_5(PO_4)_3F + 5H_2SO_4 + 10H_2O \xrightarrow{\Delta} 3H_3PO_4 + 5CaSO_4 \cdot 2H_2O + HF$

34. In sea water, Mg exists as $MgCl_2$.

On treating sea water with slaked lime, $Mg(OH)_2$ is obtained.

$\underset{\text{in sea water}}{MgCl_2} + \underset{\text{slaked lime}}{Ca(OH)_2} \rightarrow Mg(OH)_2 \downarrow + CaCl_2$

On reacting $Mg(OH)_2$ with HCl, $MgCl_2$ is obtained.

$Mg(OH)_2 + 2HCl \rightarrow MgCl_2 + H_2O$

From $MgCl_2$, Mg is obtained by reduction of $MgCl_2$ with CaC_2.

$MgCl_2 + CaC_2 \rightarrow Mg + CaCl_2 + 2C$

35. $CaCO_3 + CO_2 + H_2O \rightarrow \underset{\text{Calcium bicarbonate}}{Ca(HCO_3)_2}$

Suspension of lime stone is $CaCO_3$.

Chapter-11

The *p*-Block Elements (Group-13 and 14)

Part-A (JEE Main)

1. (1) Aluminium is more electropositive than Thallium because its standard electrode potential

$\left(E^{\circ}_{Al^{3+}/Al} = -1.66V\right)$ is more negative than that of Thallium $\left(E^{\circ}_{Tl^{3+}/Tl} = 1.26V\right)$.

A more negative electrode potential indicates a greater tendency to lose electrons.

Boron has a very small atomic size, resulting in extremely high ionization enthalpies for its first three electrons ($IE_1 + IE_2 + IE_3 \sim 6887$ kJ/mol). This high energy requirement prevents the formation of B^{3+} ions, leading to form covalent compounds.

2. (2) Order of second ionization energy for boron family elements is:

B (2427 kJ/mol) > **Ga** (1979 kJ/mol) > **Al** (1816 kJ/mol)

Order of first ionization energy for carbon family elements is:

Si (786 kJ/mol) > **Ge** (761 kJ/mol) > **Pb** (715 kJ/mol) > **Sn** (708 kJ/mol)

3. (3)

(A) $B^{+3} < Al^{+3} < Ga^{+3} < In^{+3} < Tl^{+3}$: Ionic size

(B) $B > Tl > In > Ga > Al$: EN

(C) $B > Tl > Ga > Al > In$: IE_1

(D Trichlorides and triiodides of group 13th elements are covalent in nature due to Fajan's rule.

4. (2)

A. Tl^{+} is more stable than Tl^{3+}. So Tl^{3+} is a powerful oxidising agent and easily get reduced to Tl^{+}

B. Al^{3+} is highly stable. $E^{o}_{Al^{3+}}Al = -1.66$ V.

So it is difficult to reduce Al^{3+}.

C. Tl^{3+} is unstable.

D. Tl^{+} is more stable than Tl^{3+} due to inert pair effect.

E. Al^{3+} and Tl^{+} are highly stable.

5. (4) IE order, $B > Tl > Ga > Al > In$

6. (1) order of atomic radius is $B < Ga < Al < In \leq Tl$

For Ionic radius incorrect pair is $Al^{3+} < Ga^{3+}$

(X) is Ga. Atomic number of Ga is 31

7. (1)

Atomic Radius :	$Tl > In > Al > Ga > B$
Electronegativity :	$B > Tl > In > Ga > Al$
Density :	$Tl > In > Ga > Al > B$
First Ionisation energy:	$B > Tl > Ga > Al > In$

8. (2) The metallic radius order of Al & Ga is $B < Ga < Al < In < Tl$

This is due to poor shielding by *d*-subshell electrons in Ga.

The ionic radius order of Al^{3+} & Ga^{3+} is :

$B^{3+} < Al^{3+} < Ga^{3+} < In^{3+} < Tl^{3+}$ (No effect of poor sheilding as *d*-electrons are removed)

9. (3)

10. (1) The relative stability of +1 oxidation state progressively increases down the group due to inert pair effect.

∴ Stability of + 1 oxidation state of group 13 is : $Al^{+1} < Ga^{+1} < In^{+1} < Tl^{+1}$

11. (4) Gallium is the only metal with a low melting point (29.7°C) and high boiling point (1500 – 2000°C). All these properties make Ga ideal for high temperature thermometers.

12. (4)

13. (4) The most abundant isotope of boron is ^{11}B

Number of neutrons in $^{11}B = 6$; $x = 6$

$B + O_2 \xrightarrow{\Delta} B_2O_3$

Oxidation state of B in $B_2O_3 = +3$. So, $y = 3$

Hence $x + y = 9$

14. (1) (A) Group-13 size: $B < Ga < Al < In \simeq Tl$

(B) In Group-13, Al has exceptional lower EN

EN : $B > Tl > In > Ga > Al$

(C) $2Al + 6HCl \rightarrow 2AlCl_3 + 3H_2$

$Al + HNO_3 \rightarrow$ No reaction

(Due to protective oxide layer)

(D) +1 oxidation state is observed mainly for heavier Group-13 element like T1.

B and Al do not show +1 oxidation state

(E) $[Al(H_2O)_6]^{3+} \rightarrow$ Hybridisation : sp^3d^2.

15. (4) Both statement I and statement II are true.

16. (1) Aluminium chloride in acidified aqueous solution from octahedral complex with water molecules.

$AlCl_3 \xrightarrow{\text{acidified water}} [Al(H_2O)_6]^{3+}(aq) + 3Cl^-(aq)$

Al^{3+} utilises its vacant 3s, 3p and 3d orbitals to form sp^3d^2 hybridized octahedral complex.

17. (2) Solid Boron has very strong crystalline lattice so its melting point unusually high in group 13 element due to polymeric nature of the structure.

18. (1)

Number of covalent bond formed by Boron is 4

Oxidation number of fluorine is –1,

Oxidation number of $B + (4 \times (-1)) = -1$,

Thus, Oxidation number of $B = +3$

19. (2) Boron is non-metallic in nature. It is extremely hard and black coloured solid. It exists in many allotropic forms. Due to very strong crystalline lattice, boron has unusually high melting point and boiling point.

20. (2) Li, and Cs reacts vigorously with water. Bromine changes in vapour state in boiling water (BP = 58°C). Ga do not reacts with water at 100°C.

21. (4) Down the group extent of back bonding, reduces leading to more Lewis acidic strength.

$BF_3 > BCl_3 > BBr_3 > BI_3$ (extent of back bonding)

(2p – 2p) (2p – 3p) (2p – 4p) (2p – 5p)

∴ $BF_3 < BCl_3 < BBr_3 < BI_3$ (lewis acidic character)

22. (2) The first ionization energies are as follows:

B : 801 kJ/mol; Al : 577 kJ/mol; Ga : 579 kJ/mol;

Ga(31) : $[Ar]3d^{10}4s^24p^1$

23. (2) Expansion of octet not possible for 'B' as it has outermost shell n = 2.

24. (2) Ga has very low m.pt. and its periodic position is close to metalliods.

25. (3) SnH_4 is non planar molecular hydride. It has tetraheadral shape

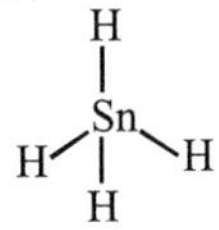

26. (1)

27. (1) Electronic configuration of Tl (81) = Xe $4f^{14}\,5d^{10}\,6s^2\,6p^1$

Due to inert pair effect, Tl is more stable in +1 oxidation state.

Hence, TlI_3 and CsI_3 are isomorphous.

28. (3) $[SiCl_6]^{2-}$ does not exist due to steric crowding of surrounding atoms and less interaction between some pair of chloride is an and Si^{++}.

29. (4) PbO_2 is strong oxidizing agent because Pb^{+4} is not stable and can be easily reduced to Pb^{+2}. PbO_2 is used in lead storage batteries. It is also amphoteric in nature.

30. (1) Moving down the group, stability of lower oxidation state increases, so the order will be $Al < Ga < In < Tl$.

31. (1) SiH_4 : Electron precise hydride

B_2H_6, GaH_3 and AlH_3 are electron deficient

32. (2) Be and Al show diagonal relationship due to which these two elements have similar electronegativity.

33. (1) Due to the inert pair effect, thallium exists in more than one oxidation state. Also, for thallium + 1 oxidation state is more stable than +3 oxidation state.

34. (3) The maximum valency of beryllium is +2, while that of aluminium is +3.

35. (3) EN of 'Tl' is 1.8 that is same as 'Ge'

'E' = Tl

Tl^{+3} acts as oxidizing agent.

Stability order : $Tl^{+3} < Tl^{+1}$ (due to inert pair effect)

The standard electrode potential value

$E°(Tl^{3+}\,Tl) = +1.26$ V.

36. (1) C^{13} is not radioactive. C^{14} is radioactive.

37. (1) $[SiO_2, CO_2]$, $[GeO, GeO_2]$ are acidic in nature.

$[SnO, SnO_2]$, $[PbO, PbO_2]$ are amphoteric in nature.

So there are 2 such pairs where both oxides are amphoteric in nature.

BF_3 is an electron deficient molecule because the boron atom has only six valence electrons (an incomplete octet). Lewis acids act as (electron pair acceptor) and has sp^2 hybridization with trigonal planar geometry and it forms an adduct with NH_3 by accepting the lone pair from nitrogen atom.

38. (1) Statement I: Generally left to right I.E. increases in a period.
Statement II : Melting point of group 14 elements is higher than group 13 elements.

39. (2) Among Al, In, Tl, Ge, Sn, Pb, the metal having highest IE_1 is of Ge 761 and lowest IE_1 is of In. 558
Most stable oxidation state of Ge is +4 and In is +3

40. (1) Order of I.E. (in kJ/mol)

C >	Si >	Ge >	Sn <	Pb
1086	786	761	708	715

41. (4) Order of M.P. of group 14 : C > Si > Ge > Pb > Sn
Sn has lowest melting point due to distorted structure and weaker bonding.

42. (4) Due to inert pair effect more stable ions do not lose electrons but accept electron i.e. Pb^{4+}, Tl^{+3}

43. (2) Germanium, silicon and copper oxide are semi conductor while graphite is a conductor.

44. (4) Statement I is true but statement II is false.

45. (4) GeO - Acidic
GeO_2 - Acidic
SiO_2 - Acidic   CO - Neutral
SnO_2 - Amphoteric PbO_2 - Amphoteric

46. (1) The order of catenation is C >> Si > Ge ≈ Sn.

Bond	C – C	Si – Si	Ge – Ge	Sn – Sn
Bond Enthalpy (kJ/mol)	348	297	260	240

47. (2) The catenation property among 14th group elements is based on bond enthalpy value of bond between the same element. The decreasing order of bond enthalpy values is
Bond enthalpy

C – C >	Si – Si >	Ge – Ge ≈	Sn – Sn
348	297	260	240

in (kJ/mol)
∴ Decreasing order of catenation is
C > Si > Ge ≈ Sn

48. (1) Buckminster fullerene is a type of fullerene with the formula C_{60}. It has a cage like fused ring structure that resembles a soccer ball, made of 20 hexagons and 12 pentagons. Each carbon atom has three bonds.

49. (4) Catenation power of the elements decreases as we move down in the group. Therefore, Pb does not show catenation property.

50. (4) Carbon-carbon bond length is maximum in diamond because diamond has all single bonds while graphite, C_{70} and C_{60} have single and double bonds.

51. (4) Fullerene (C_{60}) contains 20 hexagons (six membered) rings and 12 pentagons (five membered rings):

52. (2) Due to the small size of carbon atom, effective lateral overlapping between 2*p* and 2*p* occurs.

53. (2, 4) BCl_3 and $AlCl_3$, both have vacant *p*-orbital and incomplete octet, thus they behave as Lewis acids.
$SiCl_4$ can accept lone pair of electron in *d*-orbital of silicon, hence it can act as Lewis acid.
Although the most suitable answer is (4). However, both options (2) and (4) can be considered as correct answers.

54. (2)

$$\%\text{ of } p\text{- in graphite } (sp^2) = \frac{2}{3}\times 100 = 67\%$$

$$\%\text{ of } p\text{- in diamond } (sp^3) = \frac{3}{4}\times 100 = 75\%$$

55. (2) In diamond, each C-atom is covalently bonded to four other C-atoms to give a tetrahedral unit, so it shows sp^3 hybridisation. Therefore, each C-atom forms four sigma bonds with neighbouring C-atoms. In diamond, each C-atom utilizes its four unpaired electrons in bond formation. These bonding electrons are localized. Due to this reason diamond is a bad conductor of electricity.
In graphite, each C-atom is covalently bonded to three C-atoms to give trigonal geometry. Each C-atom in graphite is sp^2-hybridized. Three out of four valence electrons of each C-atom are used in bond formation while the fourth electron is free to move in the structure of graphite. Due to this reason, graphite is a good conductor of electricity.

56. (1) The linking of identical atoms with each other to form long chains is called catenation. However, this property decreases from carbon to lead. Decrease of this property is associated with M-M bond energy, which decreases from carbon to lead.

57. (4) The correct order is:
$SiX_2 < GeX_2 < SnX_2 < PbX_2$

58. (1) Metal atom in the lower oxidation state forms ionic bond whereas in the higher oxidation state forms covalent bond. Higher oxidation state means small size and high polarizing power and hence, greater the covalent character. So, MCl_2 is more ionic than MCl_4.

59. (3) Graphite has a two dimensional sheet like structure. These various sheets are held together by van der Waals force of attraction.

Part-B (JEE Advanced)

1. (b) Atomic radii increases on moving down a group. However, due to poor shielding effect of *d* orbital, atomic radius of Ga is smaller than Al (**anomaly**). Thus, the correct order is Ga < Al < In < Tl.

2. (b) *cis*-1,2-diol forms chelated complex ion with the product, $[B(OH)_4]^-$ causing the reaction to proceed in forward direction.

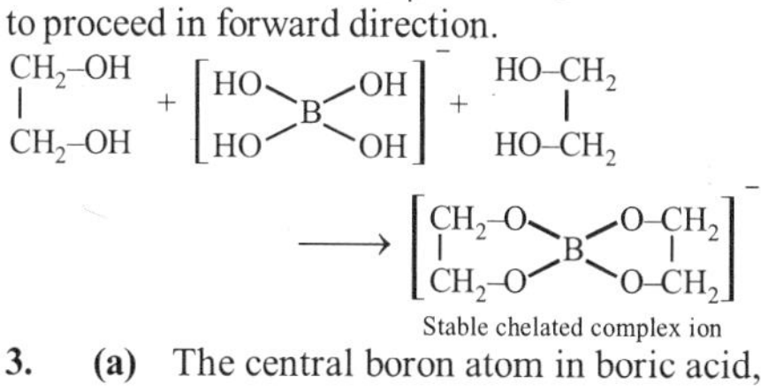

Stable chelated complex ion

3. (a) The central boron atom in boric acid, H_3BO_3 is electron-deficient.
Boric acid is a Lewis acid with one *p*–orbital vacant. There is no d-orbital of suitable energy in boron atom. So, it can accommodate only one additional electron pair in its outermost shell.

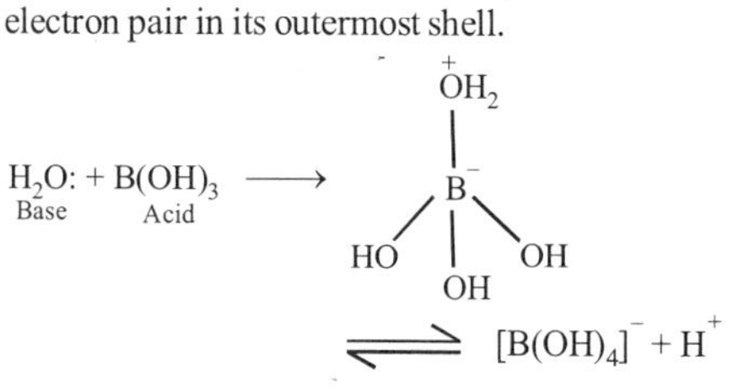

4. (c) $AlCl_3$ exists as a dimer (Al_2Cl_6). It is a strong Lewis acid as it has an incomplete octet and has a tendency to gain electrons. $AlCl_3$ undergoes hydrolysis easily and forms an acidic solution. $AlCl_3 + 3H_2O \rightarrow Al(OH)_3 + 3HCl$
Option (c) is true that $AlCl_3$ sublimes at 100°C under vacuum. $AlCl_3$ is a **Lewis acid**.

5. (6) $3B_2H_6 + 18CH_3OH \rightarrow 6B(OCH_3)_3 + 18H_2$

6. (6) Coordination number of Al is 6. It exists in *ccp* lattice with 6 coordinate layer structure.

7. sodium hydroxide;
$Al + 2NaOH + 2H_2O \rightarrow 2NaAlO_2 + 3H_2$

8. False :

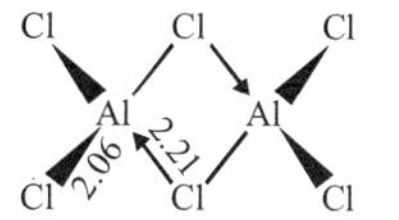

Structure of Al_2Cl_6

Bond distance between aluminium-chlorine bond forming bridge is greater (2.21 Å) than the distance between aluminium-chlorine bond present in the end (2.06 Å).

9. (a, b, c)
(a) $\underset{\text{amorphous}}{2B} + 2NH_3 \longrightarrow 2BN + 3H_2$
(b) $3B_2H_6 + 6NH_3 \longrightarrow 2B_3N_3H_6 + 12H_2 \xrightarrow{\Delta} (BN)_x$
(c) $B_2O_3 + 2NH_3 \xrightarrow{\Delta} 2BN + 3H_2O$
(d) $HBF_4 + NH_3 \longrightarrow [NH_4]^+[BF_4]^-$

10. (a, b, d)
(a) Structure of $Al_2(CH_3)_6$

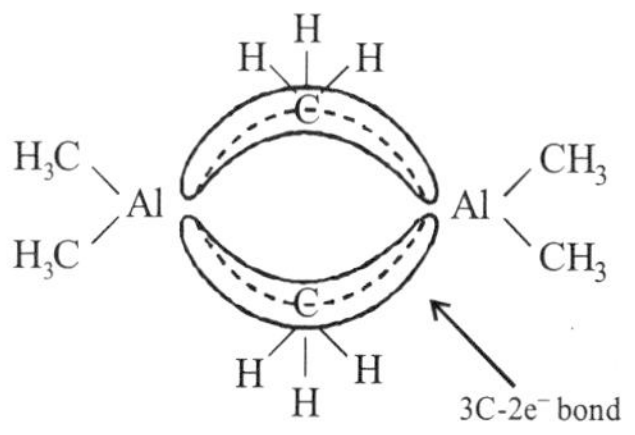

(b) Structure of B_2H_6

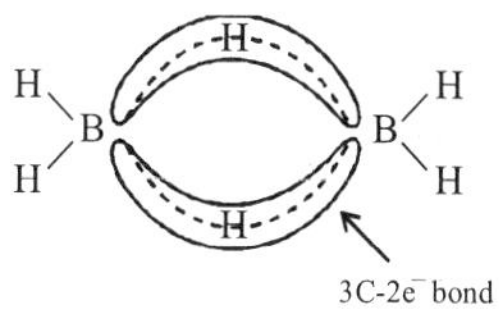

(c) Structure of Al_2Cl_6

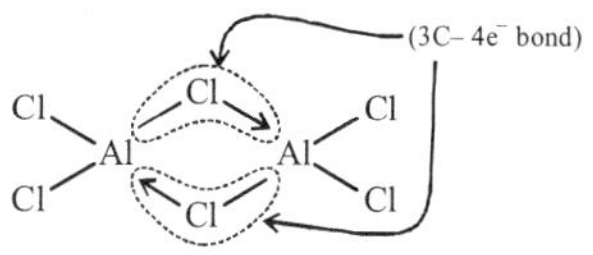

(d) Towards hard Lewis bases such as water, the Lewis acidity would decrease down the group as the hardness of acceptor element decreases, i.e., $BCl_3 > AlCl_3$.

11. (a, c, d) Structure of borax

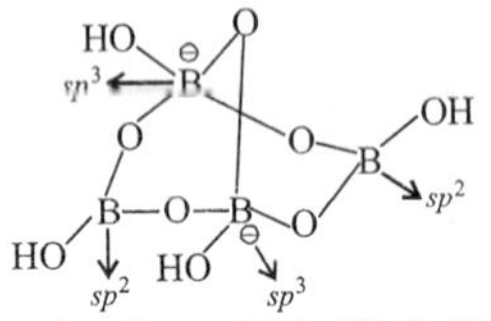

Correct formula of borax is $Na_2[B_4O_5(OH)_4]\cdot 8H_2O$

(a) Borax has tetranuclear $[B_4O_5(OH)_4]^{2-}$ unit
(b) Only two 'B' atom lie in same plane
(c) two Boron are sp^2 and two are sp^3 hybridised.
(d) one terminal hydroxide per boron atom.

12. (b, d) H_3BO_3 does not undergo self ionization. However, it acts as a weak acid in water (hence it is a weak electrolyte in water).

$$H_3BO_3 + H_2O \rightarrow B(OH)_4^- + H^+$$

Addition of *cis*-diols (*e.g.*, ethylene glycol) to aqueous solution of orthoboric acid leads to complex formation, thus acidity of aqueous solution of orthoboric acid is increased.

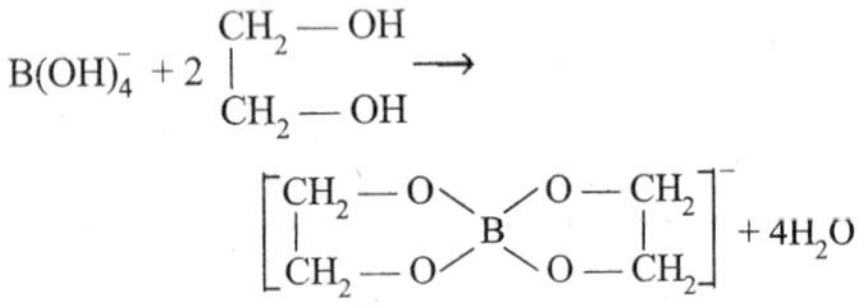

It arranges to planar sheets due to H-bonding.

13. (A)-(q), (B)-(s), (C)-(p), (D)-(r)

(A) Hydrolysis of $BiCl_3$

$$Bi^{3+}(aq) + Cl^-(aq) + H_2O(l) \longrightarrow BiO^+Cl^-(aq) + H_2(g)$$

(B) $Al + 3NaOH \longrightarrow Al(OH)_3\downarrow$ (white ppt)

$$Al(OH)_3 + NaOH(excess) \rightleftharpoons NaAlO_2 + 2H_2O$$

Upon dilution by water, we will again get the white ppt of $Al(OH)_3$.

(C) Pyrosilicates are formed from 2 moles of orthosilicates as :

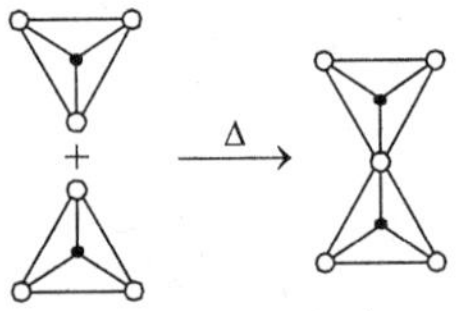

$$2\left([SiO_4]^{4-}\right) \xrightarrow{\Delta} \left([Si_2O_7]^{6-}\right)$$

(D) $Na_2B_4O_7 + 2HCl + 5H_2O \rightarrow 2NaCl + 4B(OH)_3$

14. (a) Both statements are true. Boron forms only covalent compounds (bonds) because small sized B ion polarizes the corresponding anion largely.

15. (c) Orthoboric acid (H_3BO_3) is soluble in water and behaves as weak monobasic acid. It does not donate protons like most acids, but rather it accepts OH^- ions. It is, therefore, Lewis acid, and is better written as $B(OH)_3$.

$B(OH)_3 + 2H_2O \rightleftharpoons H_3O^+ + [B(OH)_4]^-$; $pK_a = 9.25$

16. (c) Assertion is correct because $Al(OH)_3$ can react with acid and bases.

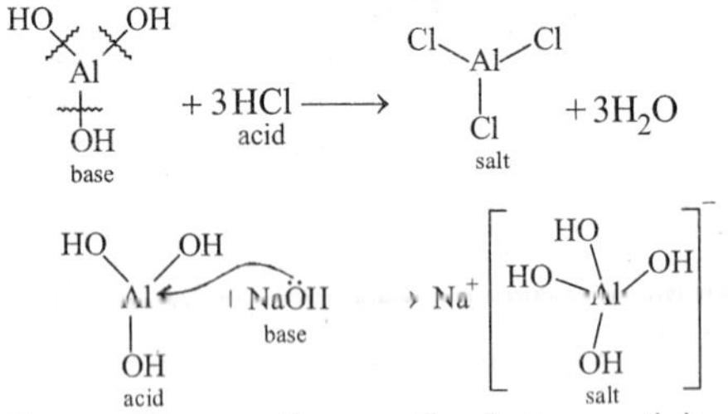

Reason is incorrect because the electronegativity difference between Al–O and O–H are different.

17. HF is weakly dissociated, while KF is highly dissociated giving a high concentration of F^- which leads to the formation of soluble AlF_6^{3-}.

$AlF_3 + 3\,KF \rightarrow K_3[AlF_6]$

Since BF_3 is more acidic than AlF_3, it pulls out F^- from AlF_6^{3-} reprecipitating AlF_3.

$K_3[AlF_6] + 3BF_3 \rightarrow 3KBF_4 + AlF_3\downarrow$

18. When hot concentrated HCl is added to borax ($Na_2B_4O_7.10H_2O$) the sparingly soluble H_3BO_3 is formed which on subsequent heating gives B_2O_3 which is reduced to boron on heating with Mg, Na or K.

$$Na_2B_4O_7(\text{anhydrous}) + 2HCl(\text{hot, conc.}) \longrightarrow 2NaCl + H_2B_4O_7$$

$$H_2B_4O_7 + 5H_2O \longrightarrow 4H_3BO_3\downarrow$$

$$2H_3BO_3 \xrightarrow{\text{strong heating}} B_2O_3 + 3H_2O$$

$$B_2O_3 + 6K \longrightarrow 2B + 3K_2O \text{ or}$$

$$B_2O_3 + 6Na \longrightarrow 2B + 3Na_2O \text{ or}$$

$$B_2O_3 + 3Mg \longrightarrow 2B + 3MgO$$

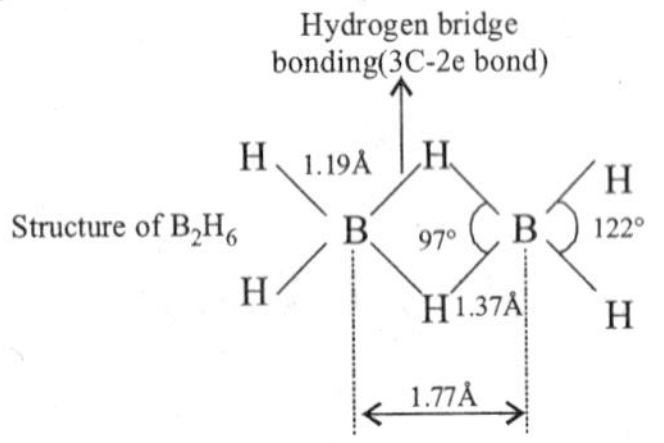

$$B_2H_6 + HCl \longrightarrow B_2H_5Cl + H_2$$

[Normally this reaction takes place in the presence of Lewis acid ($AlCl_3$)]

19. Since B_2O_3 is formed by reaction of (*Y*) with air, (*Y*) therefore should be B_2H_6 in which % of hydrogen is 21.72. The compound (*X*) on reduction with $LiAlH_4$ gives B_2H_6. Thus it is boron trihalide. The reactions are shown as:

$$\underset{(X)}{4BX_3} + 3LiAlH_4 \longrightarrow \underset{(Y)}{2B_2H_6} + 3LiX + 3AlX_3$$

(X = Cl or Br)

$$\underset{(Y)}{B_2H_6} + 3O_2 \longrightarrow B_2O_3 + 3H_2O + \text{heat}$$

Structure of B_2H_6 is as follows:

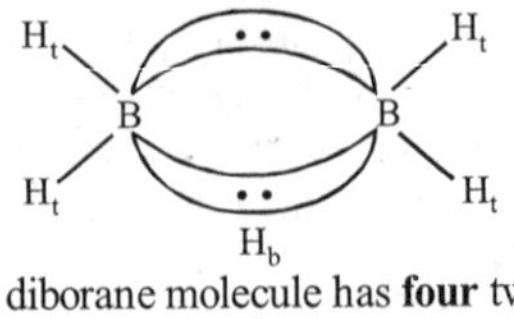

Thus, the diborane molecule has **four** two-centre-two-electron bonds ($2c-2e$ bonds) also called usual bonds and **two** three-centre-two-electron bonds ($3c-2e$) also called **banana bonds**. Hydrogen attached to usual and banana bonds are called H_t (terminal H) and H_b (bridged H) respectively.

20. $Al + NaOH \xrightarrow{aq.} NaAlO_2 \xrightarrow{2H_2O} Al(OH)_3 \xrightarrow{heat} Al_2O_3\,(\text{Alumina})$

21. (*i*) $Al_2S_3 + 6H_2O \rightarrow 2Al(OH)_3\downarrow + \underset{(\text{foul odour})}{3H_2S\uparrow}$

Foul odour, on damping of Al_2S_3 is due to formation of H_2S gas, which smells like rotten eggs.

(*ii*) $2Al + 2NaOH \rightarrow \underset{\text{Sodium meta-aluminate}}{2NaAlO_2} + H_2$

22. In excess of NaOH, the hydroxide of Al becomes soluble due to the formation of meta-aluminate.

23. (*i*) $HCO_3^- + Al^{3+} \rightarrow Al(OH)_3 + CO_3^{2-}$

(*ii*) $AlBr_3 + K_2Cr_2O_7 + H_3PO_4 \rightarrow K_3PO_4 + AlPO_4 + H_2O + Br_2 + Cr^{3+}$

24. (b) Two dimensional sheet structures are formed when three oxygen atoms of each $[SiO_4]^{4-}$ tetrahedral are shared.

25. (c) It appears at the first sight that Me_2SiCl_2 on hydrolysis will produce $Me_2Si(OH)_2$ which ultimately upon loss of water, will form $Me_2Si{=}O$. But silicon atom, because of its very large size in comparison to oxygen, is unable to form π-bond. Thus, the product of hydrolysis is polymeric in nature.

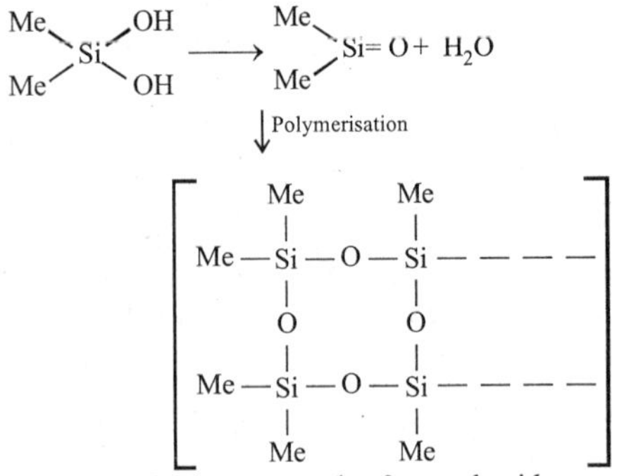

26. (a) CO is an example of neutral oxide.

27. (d) Due to inert pair effect, Pb^{4+} is not a stable oxidation state. Both Pb^{4+} and I^- are large in size. It is difficult for the atoms to come close together to form the compound.

28. (b) Silicon is used in solar cells.

29. (b) Graphite shows moderate electrical conductivity due to the presence of unpaired or free fourth valence electron on each carbon atom.

30. (c) Earlier the graphite present in pencil was thought to be lead, as it resembles to graphite in appearance.

31. (3) Total cationic charge = Total anionic charge

$2n + 6 + 24 = 36 \Rightarrow n = 3$

32. inert-pair effect ; When ns^2 electrons of outermost shell do not participate in bonding it is called inert pair and the effect is called inert pair effect.

33. glass

34. Fullerene

35. Trialkylchlorosilanol; The hydrolysis of R_3SiCl, yields $R_3Si(OH)$ which condenses to give $R_3\,Si-O-SiR_3$

$R_3\,Si\,O-\boxed{H + H-O}-Si\,R_3 \rightarrow R_3{-}Si{-}O{-}Si{-}R_3$

36. Silicones

37. True : The property of catenation in carbon is due to the fact that in carbon atom, the number of valence electrons (4) is equal to the number of valence orbitals (one $2s$ + three $2p$). Hence, carbon in the tetravalent state is fully saturated, *i.e.,* it has neither any vacant orbital nor any lone pair of electrons on its atom due to which the C – C bond is extremely stable.
The reason for greater tendency of carbon for catenation than that in silicon may further be explained by the fact that the C – C bond energy is approximately of the same magnitude as the energies of the bond between C and other elements. On the other hand, the Si Si bond is weaker than the bonds between silicon and other elements.

38. True : In diamond, each carbon atom is in sp^3 hybridised state and is linked to four other neighbouring carbon atoms held at the corners of a regular tetrahedron by covalent bonds. Owing to very strong covalent bonds by which the atoms are held together, **diamond is the hardest substance known**. Graphite has a two dimensional sheet like structure and carbon in sp^2 hybridised state is attached to three other carbon atoms by three σ bonds forming a hexagonal planar structure. Due to wide separation and weak interlayer bonds, the two adjacent layers can easily slide over each other; hence **graphite is soft**.

39. True : Graphite is better lubricant on the moon than on the earth because of lack of gravitation pull on the moon, where friction is already less than earth.

40. False : CCl_4 is most stable as compared to other tetrachlorides of the group because of its high thermal stability.

41. False : CCl_4 gives phosgene with superheated steam

$CCl_4 + H_2O \rightarrow COCl_2 + 2HCl$

42. False : PbO_2 is a dioxide and it *does not* give hydrogen peroxide when it reacts with a dilute acid.

$PbO_2 + 4HCl \rightarrow PbCl_2 + Cl_2 + 2H_2O$

43. (a, b) (a) $SnCl_2 . 2H_2O$ is a reducing agent since Sn^{2+} tends to convert into Sn^{4+}.

(b) $SnO_2 + 2KOH + 2H_2O \longrightarrow K_2[Sn(OH)_6]$

(c) In presence of conc. HCl, $PbCl_2$ exists as $H_2[PbCl_4]$

$PbCl_2 + 2HCl \longrightarrow \underset{\text{Chloroplumbous acid}}{H_2[PbCl_4]}$

(d) $\underset{\substack{(2PbO\ +\ PbO_2)\\ \text{(mixture of oxides)}}}{Pb_3O_4} + 4HNO_3 \longrightarrow PbO_2 + 2Pb(NO_3)_2 + 2H_2O$

It is not a redox reaction.

44. (c, d) $SnCl_2 + Cl^- \rightarrow \underset{(X)}{SnCl_3^-}$

$SnCl_2 + Me_3N \rightarrow \underset{(Y)}{SnCl_2[NMe_3]}$

There is a coordinate bond between NMe_3 and $SnCl_2$ due to sharing of lone pairs of NMe_3 with $SnCl_2$.

Structure of *X*, *Y*, *Z* are respectively :

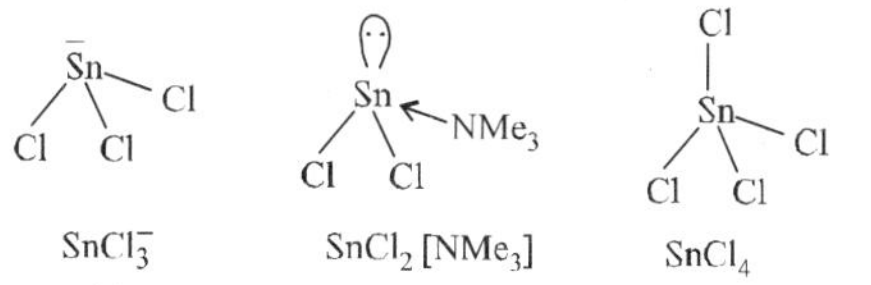

$SnCl_3^-$ (*X*) $SnCl_2[NMe_3]$ (*Y*) $SnCl_4$ (*Z*)

$SnCl_2 + 2CuCl_2 \rightarrow \underset{(Z)}{SnCl_4} + 2CuCl$

45. (b) $(CH_3)_2SiCl_2$ form linear polymer on hydrolysis and $(CH_3)_3SiCl$ is a chain terminator.

46. (b, d)

(a) Diamond is harder than graphite.

(b) Graphite is good conductor of electricity as each carbon is attached to three C-atoms leaving one valency free, which is responsible for electrical conduction, while in diamond, all the four valencies of carbon are satisfied, hence insulator.

(c) Diamond is better thermal conductor than graphite. Whereas electrical conduction is due to availability of free electrons; thermal conduction is due to transfer of thermal vibrations from atom to atom. A compact and precisely aligned crystal like diamond thus facilitates fast movement of heat.

(d) In graphite, C – C bond acquires double bond character, hence higher bond order than in diamond.

47. (A)-(t), $Pb(N_3)_2$ is an explosive

(B)-(s), Al_2O_3 is used to prepare artificial gem.

(C)-(v), Extraction of copper involves self-reduction process.

(D)-(q), Fe_3O_4 is a magnetic material.

48. A-(c)-(ii); B-(e)-(iii); C-(a)-(iv); D-(b)-(v); E-(d)-(i).

49. (c) In group 14 elements, the lower (and not higher) oxidation states are more stable for heavier members of the group due to inert pair effect.
Thus, Pb^{4+} is less stable as compared to Sn^{4+} (lead is heavier than Tin). Therefore, Pb^{4+} acts as a strong oxidising agent than Sn^{4+}. Hence, statement 1 is false and statement 2 is true.

50. (c) $SiCl_4$ undergoes hydrolysis due to the presence of empty *d* orbitals in the valence shell of Si, while C has no vacant *d* orbitals to accommodate electron pairs donated by water molecules during hydrolysis.
$SiCl_4$ is a covalent compound because the difference in electronegativity of Si and C is not high.

51. (i) $SiCl_4$ + 2Mg (or Zn) $\longrightarrow$ Si + $2MgCl_2$ (or $ZnCl_2$)

(ii) $SiCl_4 + 2CH_3MgCl \longrightarrow (CH_3)_2SiCl_2 + 2MgCl_2$

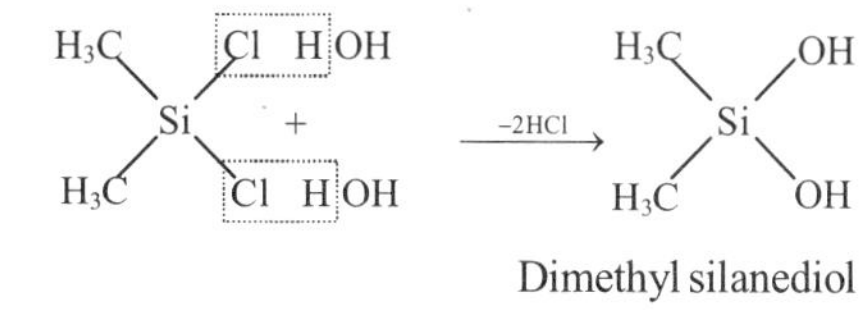

Dimethyl silanediol

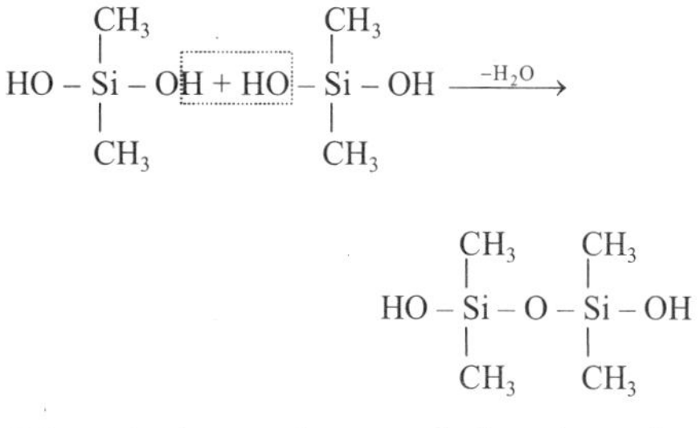

Polymerisation continues on both ends to give linear silicone.

(*iii*) $SiCl_4 + 2Mg \longrightarrow Si + 2MgCl_2$

$Si + Na_2CO_3 \longrightarrow Na_2SiO_3 + C$

OR

$SiCl_4 + 4H_2O \longrightarrow Si(OH)_2 + 4HCl$

$Si(OH)_2 \xrightarrow{\text{heat}} SiO_2 + 2H_2O$

$SiO_2 + 2NaOH \longrightarrow Na_2SiO_3 + H_2O$

52. In cyclic $Si_3O_9^{6-}$, three tetrahedra of SiO_4^{2-} are joined together sharing two oxygen atoms per tetrahedron.

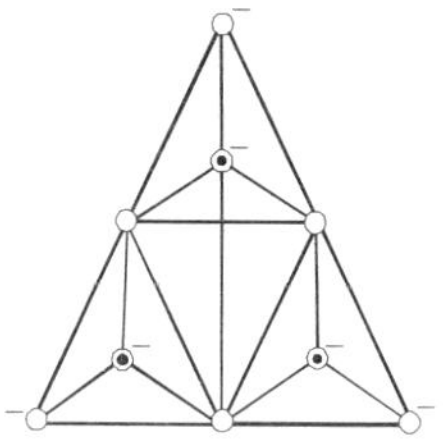

Structure of $Si_3O_9^{6-}$

Dark circle represents Si and open circle represents oxygen atom/ion

53. $Sn + 2KOH + 4H_2O \rightarrow K_2Sn(OH)_6 + 2H_2$

54. (i) $SnCl_4 + 2C_2H_5Cl + Na \rightarrow C_4H_{10} + Na_2[SnCl_6]$

(ii) $3SiCl_4 + \underset{\text{(molten)}}{4Al} \xrightarrow{\Delta} 4AlCl_3 \uparrow + 3Si$

55. (i) In $(SiH_3)_3N$, lone pair of electrons on nitrogen is involved in $p\pi - d\pi$ back bonding, while in $(CH_3)_3N$ no such $p\pi - d\pi$ back bonding is possible because of absence of *d* orbitals in carbon, so $(CH_3)_3N$ is more basic than $(SiH_3)_3N$.

(ii) In $MgCl_2$, Mg is *sp* hybridised while in $SnCl_2$, Sn is sp^2 hybridised (hence the molecule is angular).

(iii) In graphite, out of four valence electrons, only three form covalent bonds (sp^2 hybridisation) with three other carbon atoms. This forms hexagonal rings as sheets of one atom thickness. These sheets are held together by weak attractive forces. One electron of each carbon atom is free and this enables these thin sheets to slide over one another. For this reason, graphite is a soft material with lubricating properties.

(iv) Solid CO_2 is technically known as **dry ice** because it sumblimes without leaving any stain on surface.

(v) Carbon exists in various allotropic forms like diamond, graphite, coal, *etc.* Diamond consists of a three-dimensional structure of sp^3 hybridised carbon atoms bonded through very strong covalent bonds. It makes it hard and useful as an abrasive.
Graphite, on the other hand, is made up of a two dimensional sheet like structure made of sp^2 hybridised carbon atoms. These layers of carbon atoms are held together by relatively weak van der Waals forces and can, therefore, slip over one another imparting lubricating properties to graphite.

56.

(i) $Sn + 4HNO_3 \longrightarrow \underset{\text{Metastanic acid}}{H_2SnO_3} + 4NO_2 + H_2O$

(ii) $2Al + 2NaOH + 2H_2O \longrightarrow \underset{\text{Sod. aluminate}}{2NaAlO_2} + 3H_2$

Chapter-12

Organic Chemistry-Some Basic Principles and Techniques

Part-A (JEE Main)

1. (3) Given :-
Molar mass = 72 g mol^{-1}
General formula for alkanes : C_nH_{2n+2}
Molar mass = $12n + (2n + 2) = 72$
$\Rightarrow \quad 14n + 2 = 72$
$\Rightarrow \quad 14n = 70 \Rightarrow n = 5$
So, the formula is C_5H_{12} and isomer of pentane is

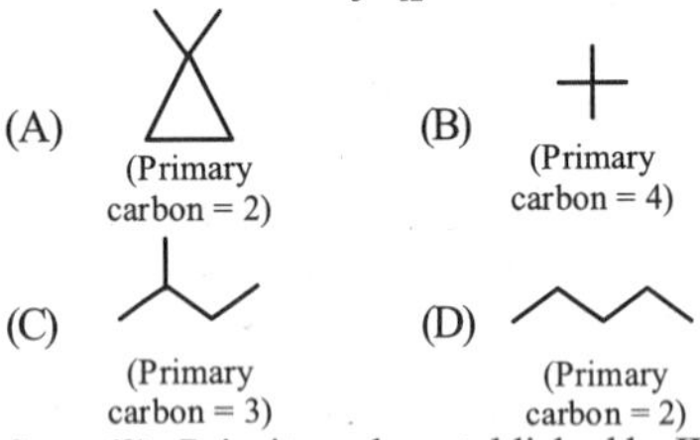

2. (3) Priority order established by IUPAC :
–COOH > –SO_3H > –COOR > –COX > –CHO > >C = O > –OH > –NH_2 > –C ≡ C– > >C = C<

3. (1)
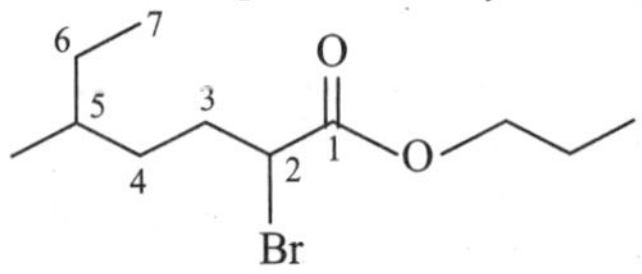

n-Propyl-2-Bromo-5-Methylheptanoate

4. (4)
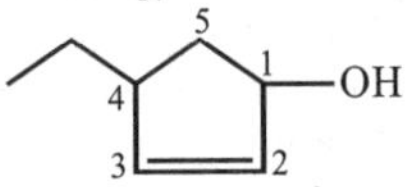

4-Ethylcyclopent-2-en-1-ol

5. (3)
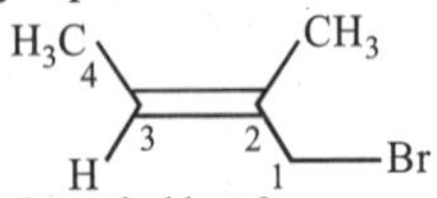

1-Bromo-2-methyl but-2-ene

6. (4)

$$\underset{7}{HC} \equiv \underset{6}{C} - \underset{5}{CH_2} - \underset{4}{\overset{OH}{\overset{|}{C}}H} - \underset{3}{CH_2} - \underset{2}{CH} = \underset{1}{CH_2}$$

Hept-1-en-6-yn-4-ol

7. (1)
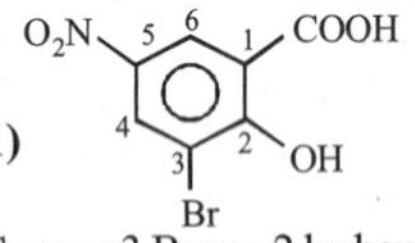

IUPAC name : 3-Bromo-2-hydroxy-5-nitrobenzoic acid.

8. (3) (A)-(II), (B)-(III), (C)-(IV), (D)-(I)

9. (3)

Alkane	2°H
(1) 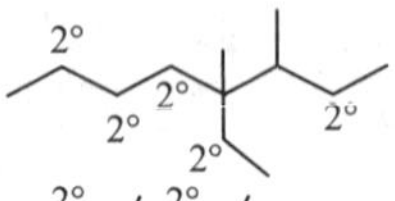	10
(2) 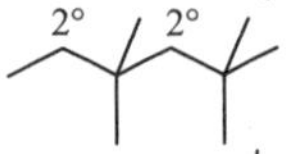	4
(3) 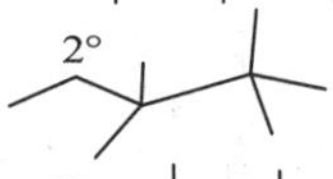	2
(4) 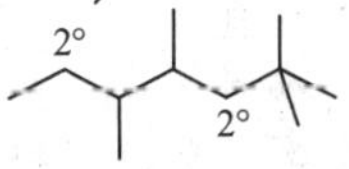	4

10. (4)
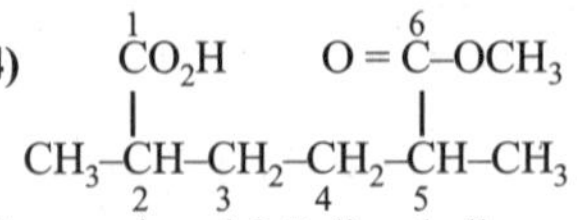

6-Methoxycarbonyl-2,5-dimethylhexanoic acid.

11. (3)
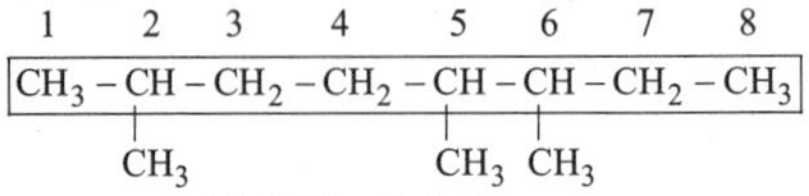

2,5,6-Trimethyloctane

12. (2) Statement-I is incorrect
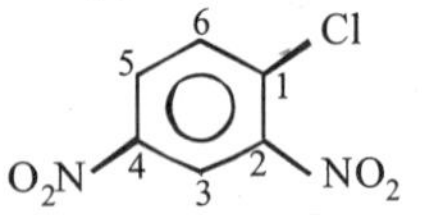

Correct IUPAC name is 1-chloro-2, 4-dinitrobenzene.

13. (2) Group present in sulphonic acids is
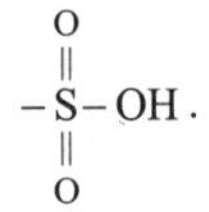

14. (3)
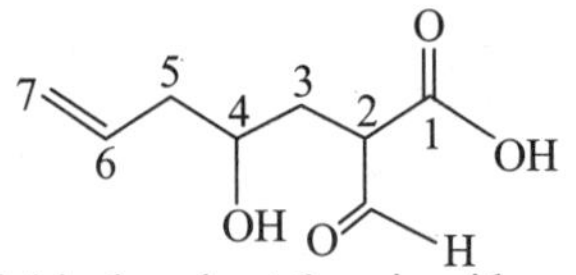

2-formyl-4-hydroxyhept-6-enoic acid.

15. (3)
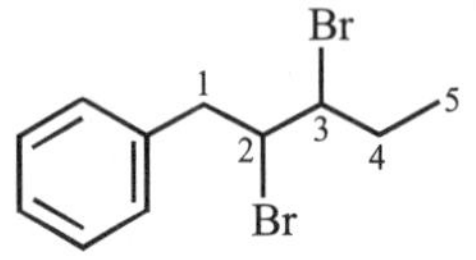

2, 3-Dibromo-1-phenylpentane.

16. (1) 7-Hydroxyheptan-2-one is correct IUPAC name.

17. (3)
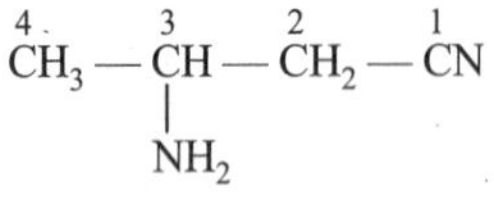

3 – Aminobutane nitrile
Priority order = –CN > –NH_2.

18. (4)
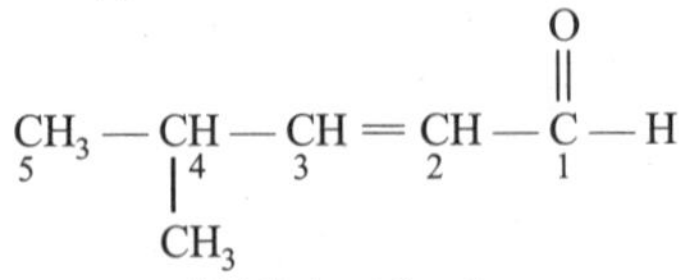

4-Methylpent-2-enal

19. (2) The given compound has a six-membered ring with a double bond and a hydroxyl group.
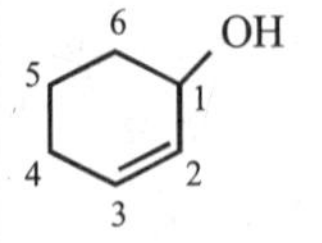

IUPAC Name = cyclohex-2-en-1-ol.

20. (4) CH(OH) $(CN)_2$ is
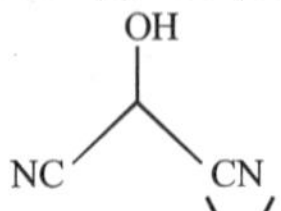

21. (2)
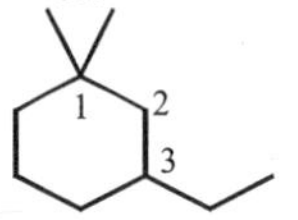

More branched carbon gets lower number

3-Ethyl 1, 1 -dimethylcyclohexane

22. (4) For Aromaticity, it must fallow Huckel's rule as per which it must have $(4n + 2)\pi$ electrons and all C atoms sp^2.

 is Alicyclic (cyclic + Aliphatic)

23. (2)
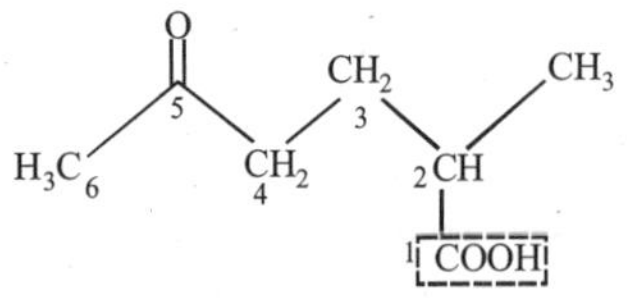

2–Methyl–5–oxohexanoic acid

24. (3) (A) – (III), (B) – (IV), (C) – (I), (D) – (II)

25. (2)
—SO_3H > —COCl > —CONH$_2$ > —CN

26. (3)
- Carbon skeleton with maximum number of carbon atoms is chosen (7).
- Carbonyl group gets priority over nitro group.

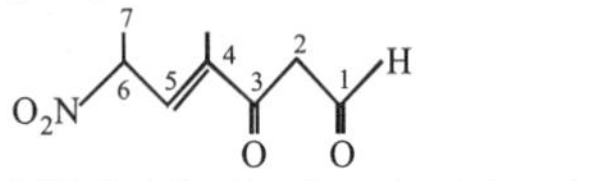

4-Methyl-6-nitro-3-oxohept-4-enal

27. (4)
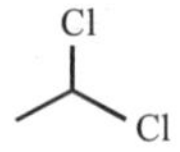

1, 1-Dichloroethane is ethylidene chloride.

28. (3)
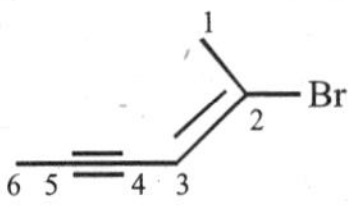

(2*E*)-2-Bromohex-2-en-4-yne

29. (4) Mesityl oxide (common name) is the aldol condensation product of acetone. Its structure and IUPAC name are
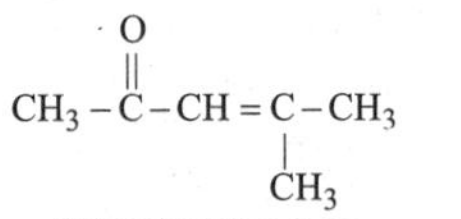

4-Methylpent-3-en-2-one

30. (3)
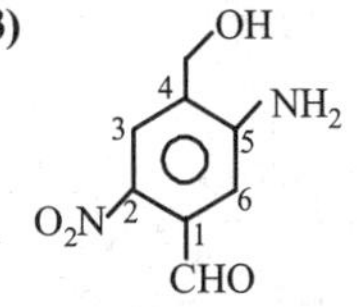

5-Amino-4-hydroxymethyl-2-nitrobenzaldehyde

31. (2)

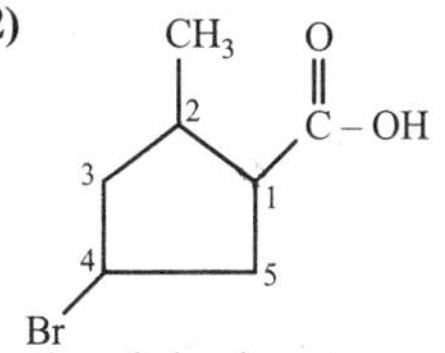

4-Bromo-2-methylcyclopentane carboxylic acid

32. (3)

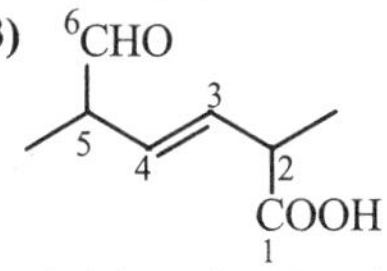

2,5-dimethyl-6-oxo-hex-3-enoic acid

33. (4) $\overset{1}{H_2C}=\overset{2}{CH}-\overset{3}{CH}(CH_3)-\overset{4}{CH}(CH_2CH_2CH_3)-\overset{5}{CH}(CH_3)-\overset{6}{C}\equiv\overset{7}{CH}$

3,5-dimethyl-4-propylhept-1-en-6-yne

34. (2)

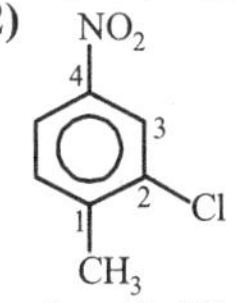

2-Chloro-1-methyl-4-nitrobenzene

35. (3)

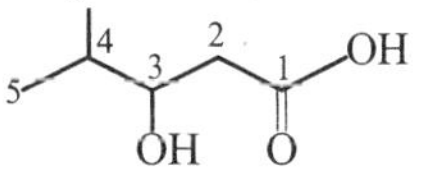

3-Hydroxy-4-methylpentanoic acid

36. (4) IUPAC name: 4-Bromo-3-methylpent-2-ene

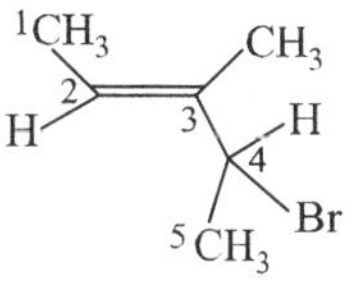

37. (4)

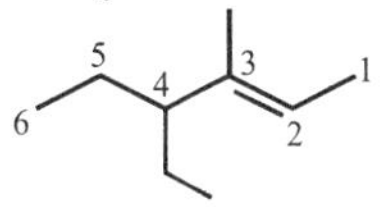

4-Ethyl-3-methylhex-2-ene

38. (2)

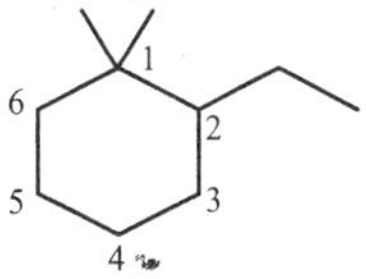

2-Ethyl-1,1-dimethylcyclohexane

39. (2) $\overset{5}{H_3C}-\overset{4}{C}(CH_3)_2-\overset{3}{CH_2}-\overset{2}{CH}=\overset{1}{CH_2}$

4, 4 - Dimethylpentene

40. (2) $\overset{sp^2}{H_2C}=\overset{sp}{C}=\overset{sp^2}{CH_2}$

41. (1) In 1, 2-dichlorobenzene, the two dipoles are at 60° (i.e., unsymmetric). Thus, it possesses dipole moment.

42. (1) $\overset{1}{CH_3}-\overset{2}{C}H=\overset{3}{C}H-\overset{4}{C}\equiv\overset{5}{C}-\overset{6}{C}H_2\overset{7}{C}H_3$

(*E*) – 2 – hepten – 4 – yne

43. (4) 1, 2-Epoxy propane is the correct IUPAC name of given compound.

44. (2)

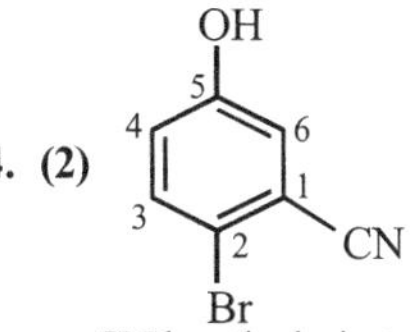

Since – CN has the hgiest priority, hence theparent compound is a nitrile.

∴ IUPAC name of the compound is 2-bromo-5-hydroxybenzonitrile.

45. (1) $\overset{1}{H_3C}-\overset{2}{C}(CH_3)_2-\overset{3}{CH_3}$

Neopentane
or 2, 2- Dimethylpropane

46. (1) The correct order of priority for the given functional group is:

$-COOH > -SO_3H > -\overset{O}{\overset{\|}{C}}-NH_2 > -\overset{O}{\overset{\|}{C}}-H$

47. (1) $\overset{7}{CH_3}-\overset{6}{CH_2}-\overset{5}{CH_2}-\overset{4}{C}(CH_3)_2-\overset{3}{CH}(CH_2CH_3)-\overset{2}{CH_2}-\overset{1}{CH_3}$

3-Ethyl-4,4-dimethylheptane

48. (1)

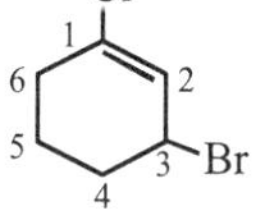

3-Bromo-1-chlorocyclohexene

49. (4)

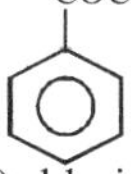

is known as benzne carbonyl (benzoyl) chloride.

50. (1)

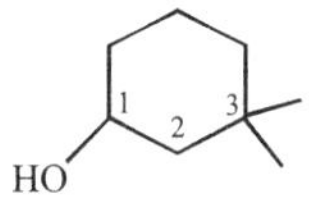

3, 3-Dimethyl-1-cyclohexanol

51. (1) $\overset{sp^3}{H_3C}-\underset{sp^2}{\overset{O}{\overset{\|}{C}}}-\overset{sp^3}{CH_3}$; $\overset{sp^3}{CH_3}-\underset{sp^2}{\overset{O}{\overset{\|}{C}}}-OH$;

Acetone; Acetic acid

$\overset{sp^3}{CH_3}-\overset{sp}{C}\equiv N$; $\overset{sp^3}{CH_3}-\underset{sp^2}{\overset{O}{\overset{\|}{C}}}-NH_2$

Acetonitrile; Acetamide

52. (1) $C_nH_{2n}O_2$ is general formula for carboxylic acids.

53. (3) $\overset{1}{CH_3}-\overset{2}{\overset{O}{\overset{\|}{C}}}-\overset{3}{CH}(CH_3)-\overset{4}{CH_3}$

3-Methyl-2-butanone

54. (3) The correct name is 3-methylbutan-2-ol.

55. (2) Only 2 compounds from the given options contain sulphur as heteroatom i.e. Thiophene and Cysteine.

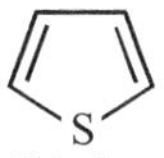

Thiophene

$HS-CH_2-CH(NH_2)-COOH$

Cysteine

While tyrosine, furan has oxygen and pyridine, pyrole has N as heteroatom.

56. (3) Statement-I is true and Statement-II is false.

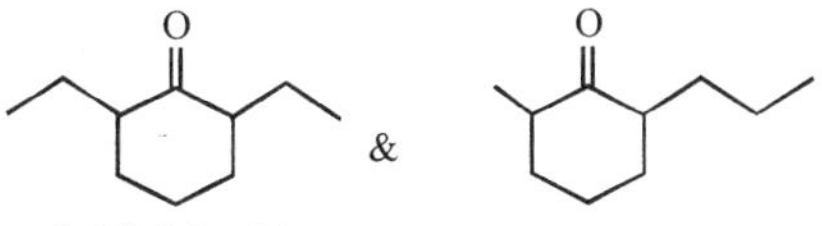

2, 6-diethyl cyclohexanone & 6-methyl-2-n-propylcyclo hexanone

are metamers because the alkyl chains are different on either side of the functional group in the molecule.

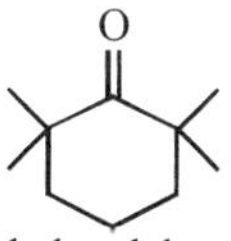

2, 2, 6, 6- tetra methyl cyclohexanone do not have α – H due to which it doesn't exhibit keto-enol tautomerism.

57. (1)

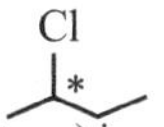

sec-Butyl chloride (2 Chlorobutane) is optically active

Molecular formula ⇒ C_4H_9Cl

Molar mass = 48 + 9 + 35.5 = 92.5

$\%\text{ of C} = \frac{48}{92.5}\times 100 = 51.89\%$

58. (1) A-III, B-I, C-IV, D-II

59. (3)

(A)

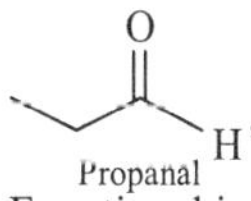

, and

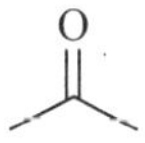

are

Propanal; Propanone

Functional isomers because they have different functional groups.

(B) $CH_3-CH_2-O-CH_2-CH_3$ and
Ethoxyethane
$CH_3-CH_2-CH_2-O-CH_3$ are metamers.
Methoxypropane

(C) $CH_3-CH=CH-CH_3$ does not exhibit optical isomerism because it lacks a chiral center.

(D) $CH_3-CH=CH-CH_3$ and
(But–2–ene)

$CH_3-CH_2-CH=CH_2$
(But–1–ene)

differ only in position of double bond, so they are positional isomers.

(E)

and

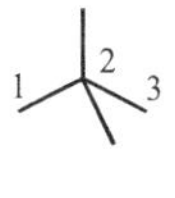

are chain isomers.

Pentane 2, 2-Dimethylpropane

So only (A), (B) and (E) are correct.

60. (3)

A. Metamers

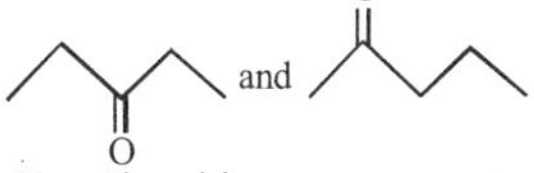

B. Functional isomers

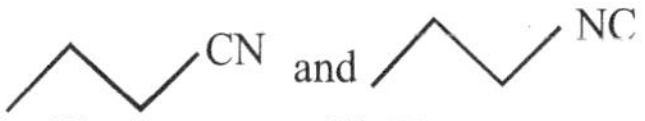

61. (2) **Statement–I** is True

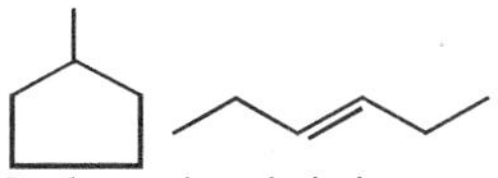

Both are ring chain isomers

Statement–II → True

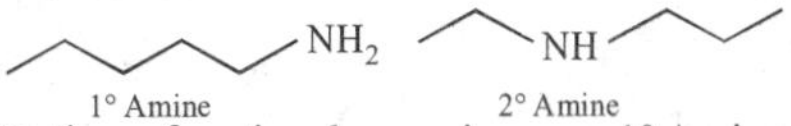

Both are functional group isomers. 1° Amine and 2° Amine are different functional groups.

62. (4) $CH_3 - CH(OH) - CH = CH - CH_3$

It has 4 stereoisomers $\begin{bmatrix} \text{R cis} & \text{R trans} \\ \text{S cis} & \text{S trans} \end{bmatrix}$

63. (1)

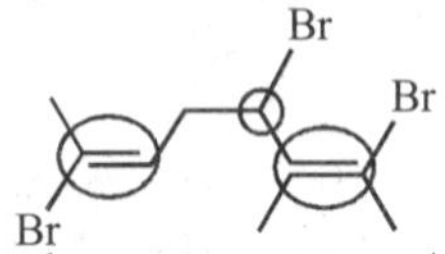

There are three stereo – centers (n = 3)
So, No. of stereoisomer = $2^n = 2^3 = 8$

64. (4) Metamer are the isomers having same molecular formula and same functional group but different alkyl or arly groups on either side of functional group.

65. (2) cis-2-butene and trans-2-butene are stereoisomers.
trans-2-butene is more stable than cis-2-butene.
cis-2-butene has more dipole moment than trans-2-butene.
cis-2-butene and trans-2-butene are not interconvertible at room temperature due to rotational barrier.

66. (4)

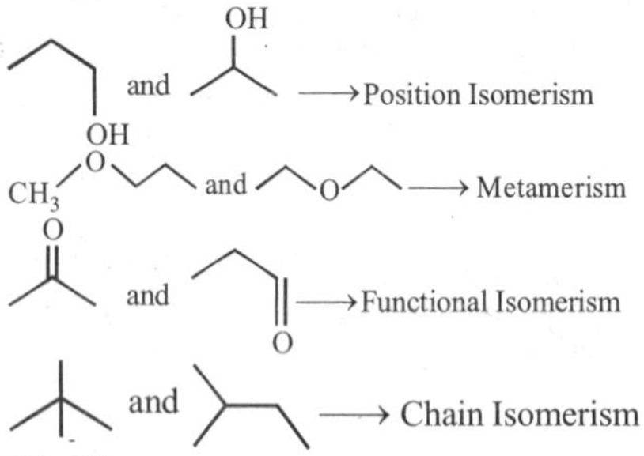

67. (4)
A.Propanamine N–Methylethanamine

$H_3C{-}CH_2{-}CH_2{-}NH_2$

$C_2H_5{-}NH{-}CH_3$ ⇒ Functional isomers

B.Hexan–2–one Hexan–3–one

$CH_3{-}CO{-}(CH_2)_3\,CH_3$

$CH_3{-}CH_2{-}CO{-}(CH_2)_2\,CH_3$

⇒ Metamers

C.Ethanamide Hydroxyethanimine

$H_3C{-}C(=O){-}NH_2$ $CH_3{-}C(OH){=}NH$ ⇒ Tautomers

D. *o*–Nitrophenol *p*–nitrophenol
⇒ Positional isomers

68. (4) For a compound with formula
⇒$C_aH_bN_cO_dX_e$

Degree of unsaturation = $\frac{1}{2}(2 + 2a - b + c - e)$

For C_3H_6O, $a = 3$; $b = 6$

Degree of unsaturation = $\frac{1}{2}(2 + 2 \times 3 - 6) = 1$

So, given compound contains one double bond. Hence, possible isomer can be written as follows :

$CH_3 - CH_2 - \overset{O}{\overset{\|}{C}} - H$ and $CH_3 - \overset{O}{\overset{\|}{C}} - CH_3$

They are functional group isomers.

69. (4)

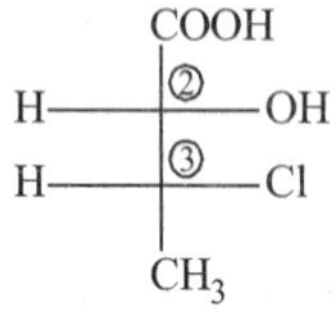

At (2),

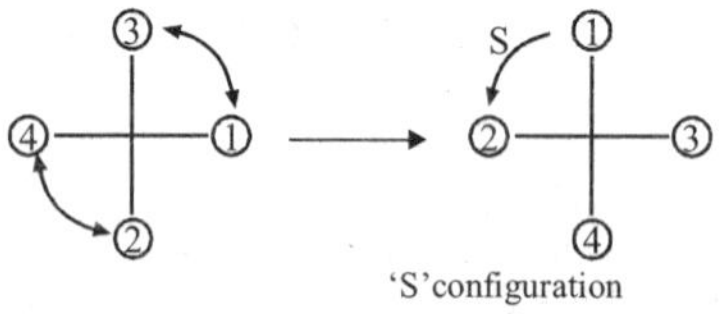

At (3),

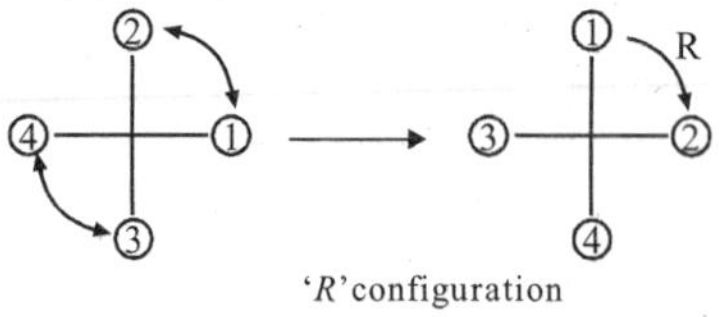

70. (4)

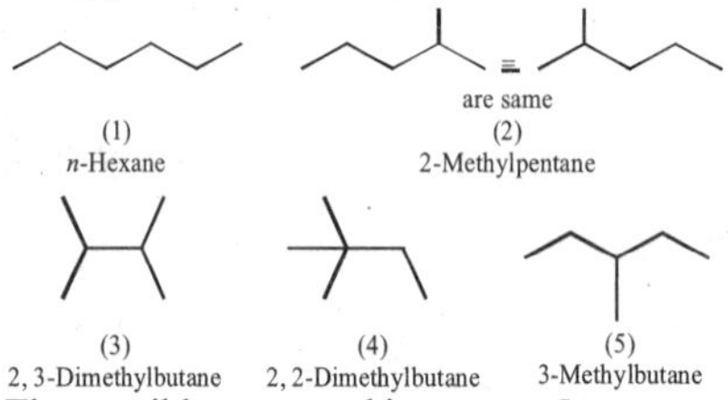

The possible structural isomers = 5

71. (1) Pentan-2-one and pentan-3-one are positional isomers. (2), (3), (4) contain different functional groups, *viz* aldehyde and ketone. These exhibit functional group isomerism.

72. (4)

73. (3) Optically active compounds contain an asymmetric (chiral) carbon atom (a carbon atom attached to four different atoms or groups). Therefore, all acids except maleic acid exhibit optical isomerism.

74. (4) Maleic acid and fumaric acids are geometrical isomers.

Maleic acid	Fumaric acid
H – C – COOH ‖ H – C – COOH	H – C – COOH | HOOC – C – H

75. (2) Five cyclic structures are possible for C_4H_6. These are as follows.

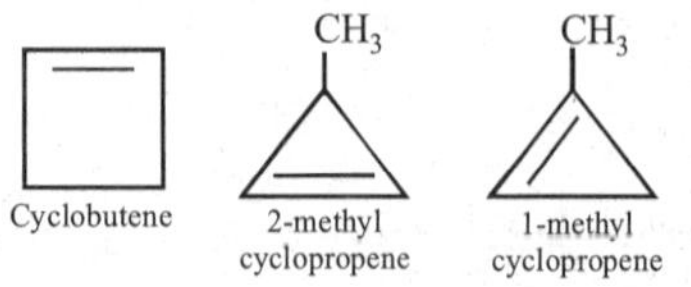

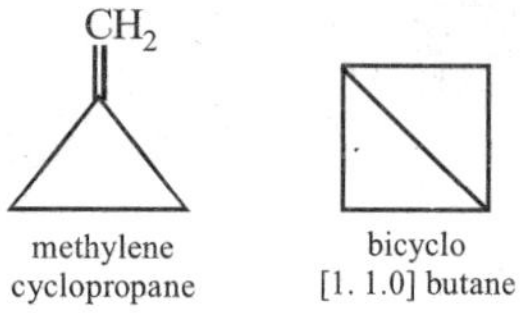

76. (3) $H_2C = HC - \overset{*}{C}H(CH_3) - CH_2 - CH_3$

3- Methyl-1-pentene

77. (2)

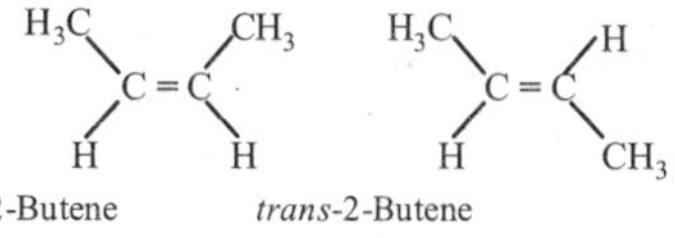

78. (2) $CH_3 - CH = CH - \overset{*}{C}H(OH)CH_3$

It exhibits both geometrical as well as optical isomerism.
cis - R *cis* - S
trans - R *trans* - S

79. (2) The organic compounds which have chiral carbon atom (a carbon atom attached to four different groups or atoms) and do not have plane of symmetry rotate plane polarised light.

$HO - \overset{*}{C}(CHO)(CH_2OH) - H$ (* is asymmetric carbon)

80. (3) $CH_3 - CH(Cl) - CH(Cl) - CH_3$, 2, 3-Dichlorobutane

exhibits optical isomerism due to the presence of two asymmetric carbon atoms.

81. (2) Compounds containing two similar chiral C-atoms have plane of symmetry and can exist in *meso* form too.

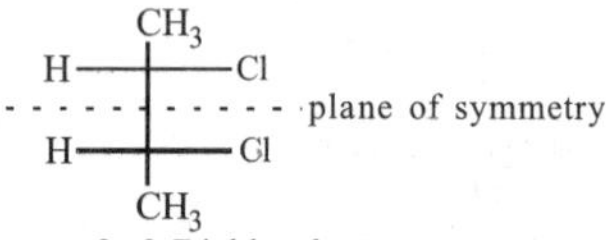

meso-2, 3-Dichlorobutane

82. (3) 1-chloropentane is not chiral while others are chiral in nature

(a) $ClCH_2\overset{*}{C}H(CH_3)CH_2CH_2CH_3$ (1 2 3 4 5)

(b) $H_3C\overset{*}{C}H(Cl)CH_2CH_2CH_3$ (1 2 3 4 5)

(c) $ClCH_2\,CH_2\,CH_2\,CH_2\,CH_2$ (1 2 3 4 5)

(d) $CH_3CH(CH_3)\overset{*}{C}H(Cl)CH_2CH_3$ (1 2 3 4 5)

83. (1) only I and II are chiral compounds

84. (3) Both differ in the arrangement of groups in space, therefore grouped under sterio-isomerism.

85. (4) A mixture of equal amount of two enantiomers is called a racemic mixture.

86. (4) Stereoisomerism, isomers differ in the arrangement of groups in space. The two structures show sterioisomerism. Structure I shows geometrical isomerism as it contains two different atoms or groups H and CH_3 attached to each carbon containing double bond.

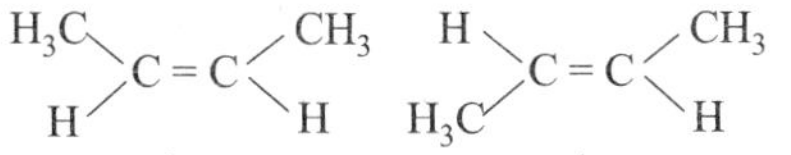

cis-butene *trans*-butene

Structure II shows optical isomerism as it contains a chiral carbon (attached to four different groups) atom.

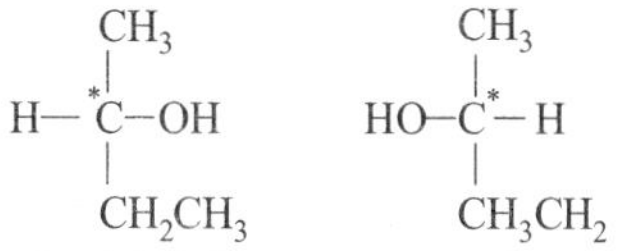

Butyl alcohol (two enantiomers)

87. (8) Molecular formula = C_9H_{12}

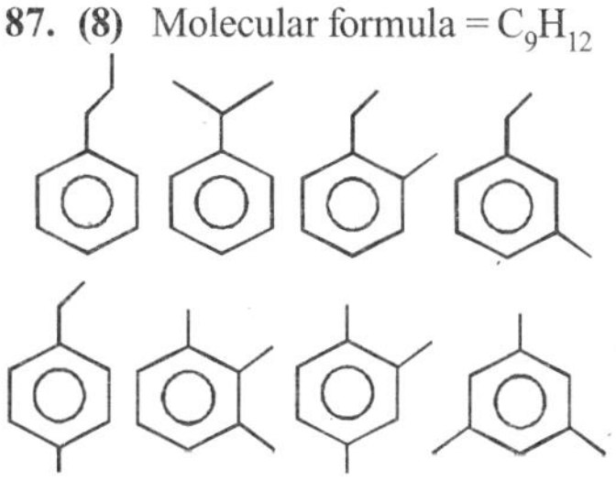

88. (4)

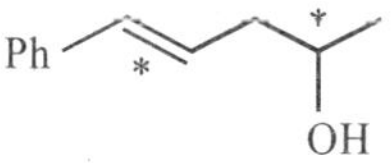

n(stereogenic center) = 2, 2^2 = 4 stereoisomers are possible.
(two geometrical and two optical)

89. (6) Dimethylcyclopentane has following isomers :

1, 1-Dimethylcyclopentane no stereoisomer (i)

1, 2-Dimethylcyclopentane will show stereoisomerism (ii)

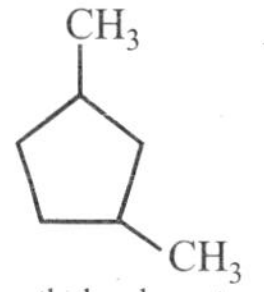

1, 3-Dimethylcyclopentane will show stereoisomerism (iii)

(ii)

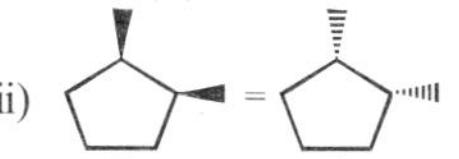

cis *cis*
cis –1, 2 – Dimethylcyclopentane (Identical)

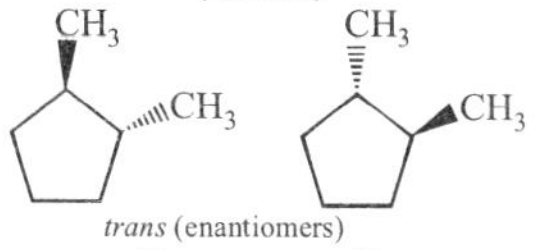

trans (enantiomers)

(iii)

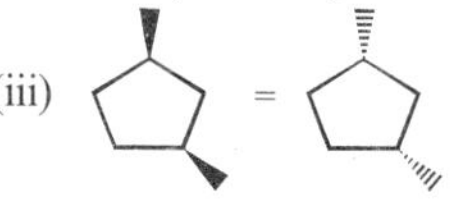

cis–1, 3 – Dimethylcyclopentane (Identical)

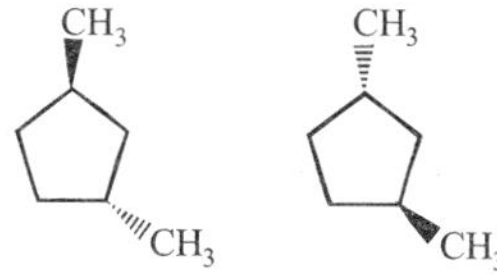

trans (enantiomers)

So total number of stereoisomers are 3[from (ii)] + 3[from (iii)] = 6

90. (6) C_5H_{10}

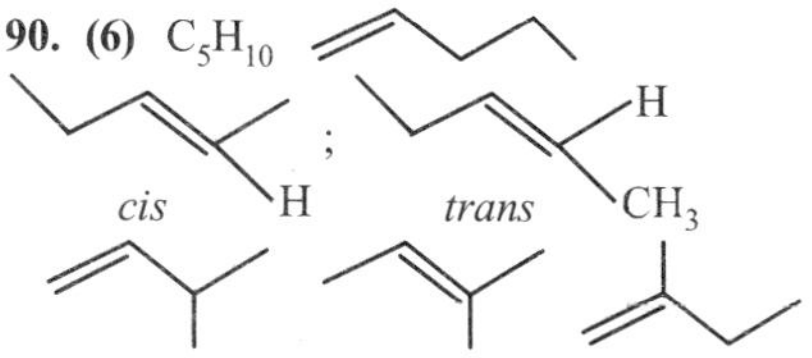

Total 6 isomers are possible.

91. (3) Statement I: True
Fractional distillation is used to differentiate liquids whose difference in boiling points is not much.
Statement II: False
Reason: As vapours rise, they become richer in the lower-
boiling component. (more volatile component)

92. (1) Sugar is soluble in alcohol not NaCl
So, they are separated by solvent extraction (or differential extraction).
Statement I is true.
Rose essence oil is steam volatile so it is separated by steam distillation.
Thus statement II also true.

93. (4)

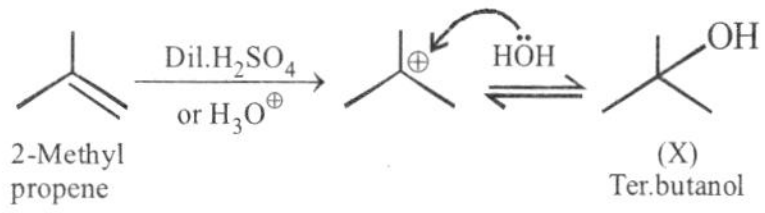

The product (X) formed is tertiary butanol which is more polar than the reactant 2-Methyl propene. As more polar compounds have lower R_f value. So, R_f of tertiary butanol is less than 2-Methyl propene.

94. (1) More the number of α hydrogen more will be the stability of alkene through hyper-conjugation.

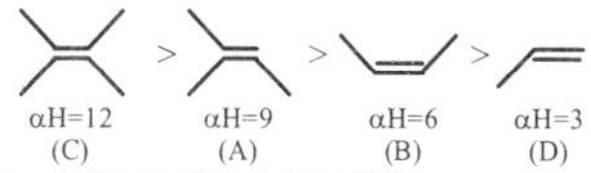

95. (2) A-II, B-IV, C-I, D-III

96. (3)

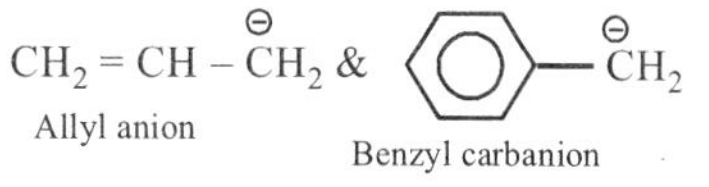

Allyl and Benzyl carbanions are strongly resonance stabilized due to delocalisation of negative charge over π-system.

97. (3) Electron withdrawing power determines how strongly e^- density is pulled from the compound affecting its acidity or basicity.
Increasing order:

$$-I < -COOH < -CN < -NO_2.$$
(d) (b) (a) (c)

98. (1) Electron donating groups (+R) stabilize positive charges, while electron withdrawing groups (–R) stabilize negative charges.

99. (4) Predominant site towards attack of electrophile is 'u' because of +M nature of nitrogen, making it electron rich.

100. (4) Glycerol has a very high boiling point and it tends to decompose at or near its boiling point. Separating it from water via simple distillation at atmospheric pressure is difficult. Instead, it is separated by distillation under reduced pressure.

101. (3)

$$\% \text{ of S} = \frac{32}{233} \times \frac{\text{wt. of BaSO}_4}{\text{wt. of compound}} \times 100$$

$$\% S = \frac{32}{233} \times \frac{0.4813}{0.314} \times 100$$

$$= 21.052\%$$

102. (1)

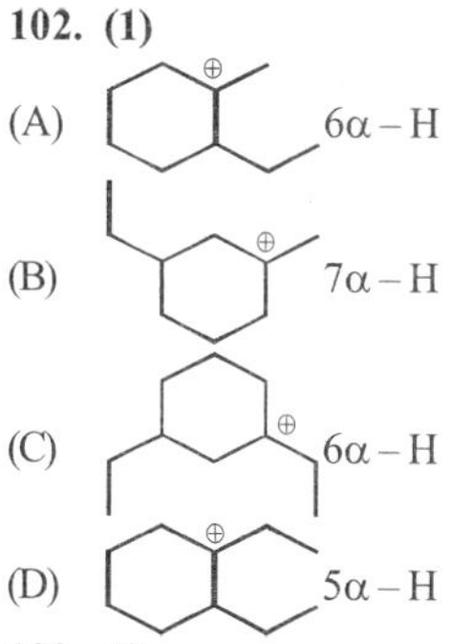

103. (4)

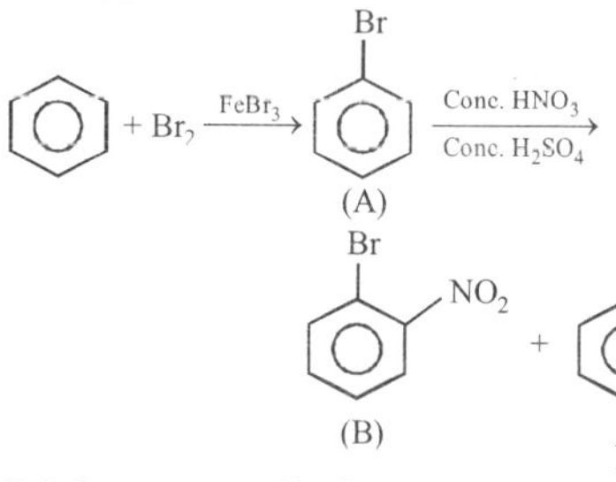

B & C are separated by fractional distillation method. In this technique vapours of liquid mixture are passed through a fractionating coloumn before condensation. Vapour of liquid with higher boiling point will condense before the other liquid with lower b.pt.

104. (3) $CH_2 = CH^-$, $CH_3 - CH_2^-$, $CH \equiv C^-$
Stability ∝ % s character
Order of stability
$CH \equiv C^- > CH_2 = CH^- > CH_3 - CH_2^-$

105. (4) Statement B & D are not true.
+E effect is shown with an E^+ and –E effect with a Nu^-

$$\text{Heat of hydrogenation} \propto \frac{1}{\text{Stability}}$$

$$\propto \frac{1}{\text{R groups on alkene}}$$

106. (3) Electron withdrawing groups (–CHO, –Br) increase stability of carbanion and electron donating groups (–CH_3, –OCH_3) decrease stability of carbanion.

107. (1) Molar mass of AgCl = 108 + 35.5 = 143.5 g mol^{-1}
143.5 g of AgCl contain 35.5 g of chlorine.

∴ 0.5253 g of AgCl contains $\frac{35.5 \times 0.5253}{143.5}$ g of chlorine

$$\therefore \text{Percentage of chlorine} = \frac{35.5 \times 0.5253 \times 100}{143.5 \times 0.2425}$$

$$= 53.58\%$$

108. (1) $(CH_3)_3C^+$ has nine C – H bonds. It shows nine hyperconjugation interactions. In $\overset{+}{C}H_3$, vacant p-orbital is perpendicular to the plane in which C – H bonds lie, hence cannot overlap with it. Thus $\overset{+}{C}H_3$ lacks hyperconjugative stability. Hence statement I is true but statement II false.

109. (4) Statement I: False
Sublimation is used to separate sublimable compounds from non-sublimable impurities.
Statement II: False
Increasing the external pressure raises the boiling point.

110. (3) 1 M of 15 mL H_2SO_4 = 1 M of 30 mL NH_3
1000 mL of 1 M ammonia contains 14g of nitrogen
30 mL of 1 M ammonia contains

$$= \frac{30}{1000} \times 14 = 0.42$$

$$\% \text{ weight of 'N'} = \frac{0.42}{1} \times 100 = 42\%.$$

111. (1)

$$\% \text{ of P} = \frac{\text{Mole of } Mg_2P_2O_7 \times 2 \times 31}{W_{(\text{unknown compound})}} \times 100$$

$$= \frac{\left(\frac{1.79}{222} \times 2 \times 31\right)}{1} \times 100 = 49.99\% \approx 50\%.$$

112. (3)

$$\text{Percentage of S} = \frac{n_{BaSO_4} \times 32}{W_{(\text{unknown comp.})}} \times 100$$

$$= \frac{1.2 \times 32}{233} \times \frac{100}{0.75} = 21.97\%$$

113. (4)
(A) Carbocation → sp^2 hybridised carbon with empty p-orbital
(B) Carbon free radical → sp^2/sp^3 hybridised carbon with one unpaired electron.
(C) Nuecleophile → species that can supply a pair of electron. i.e electron rich species.
(D) Electrophile → species that can receive a pair of electron i.e electron deficient species.

114. (1)
(A) Aniline – H_2O : Steam Distillation (Aniline is steam volatile)
(B) Glycerol from spent-lye in soap industry: Distillation under reduced pressure
(C) Different fraction of crude oil in petroleum industry : Fractional distillation
(D) $CHCl_3$ – Aniline : Simple distillation

115. (2)

(A) 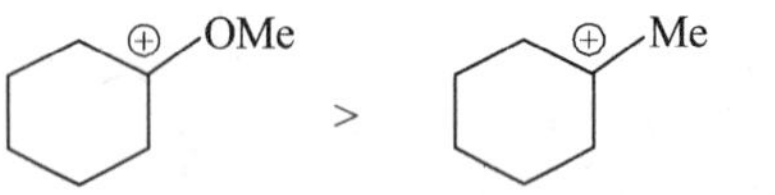
Stable due to resonance effect (+M)

(B) 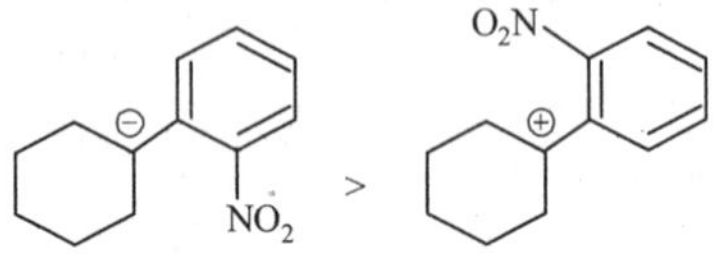
Stable due to –I and –M effect of $-NO_2$, which stabilized the carbanion

(C) 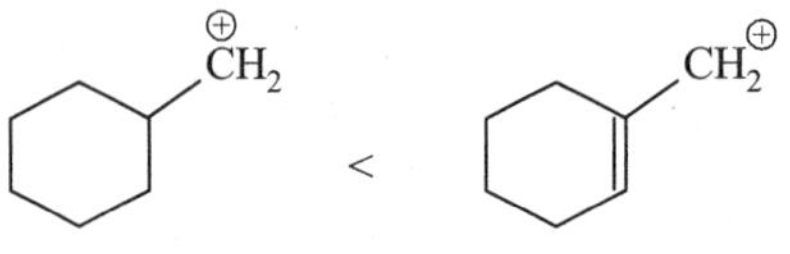
Stable due to resonance effect

(D) 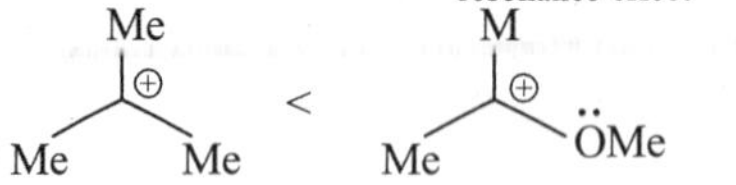
Stable due to +M effect of $-OCH_3$

116. (3) Hyperconjugation is permanent effect because external reagent is not required, so Statement-I is false. Statement-II is true because greater is the number of alkyl groups greater is the number of α-H, so more hyperconjugation which results in more stability of carbocations.

117. (1) Pressure of N_2 gas evolved = Total pressure – Aqueous tension

$$= 715 - 15 = 700 \text{ mm Hg} = \frac{700}{760} \text{ atm.}$$

$$\therefore \text{Moles of } N_2 \text{ evolved} = \frac{PV}{RT}$$

$$= \frac{700 \times 60 \times 10^{-3}}{760 \times 0.0821 \times 300} = 0.0022 \text{ mol}$$

∴ Mass of N_2 evolved = 0.0022 × 28 = 0.062 g
∴ Mass % of nitrogen in compound

$$= \frac{\text{wt. of nitrogen}}{\text{wt. of compound}} \times 100$$

$$= \frac{0.062}{0.4} \times 100 = 15.5\% \cong 15.71\%.$$

118. (4) $P_{N2} = P_{\text{moist gas}}$ – aqueous tension
= (715 – 15) mm Hg
= 700 mm Hg
V_{N2} = 60 mL = 0.060 L
T = 300 K

$$n_{N_2} = \frac{PV}{RT} = \frac{700 \times 0.06}{760 \times 0.0821 \times 300} = 0.022 \text{ mol}$$

Mass of N_2 = 0.0224 × 28 = 0.0628

$$\%N_2 = \frac{0.0628}{0.5} \times 100 = 12.58\%$$

119. (4) Boiling point of aniline is 547 K and B.P. of $CHCl_3$ is 334 K. So, they are separated by simple distillation.

120. (4) 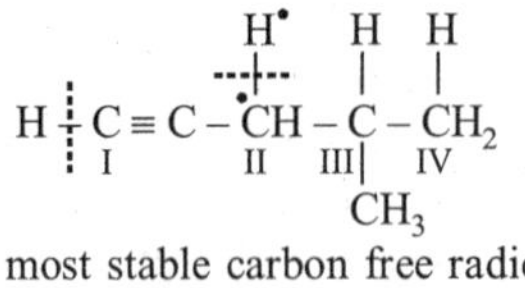

II gives most stable carbon free radical due to resonance stabilization.
I gives least stable carbon radical due to no stability factor.

121. (2) Statement-I is true. In partition chromatography, stationary phase is thin liquid film present in the inert support.
Statement-II is false. Because stationary phase in paper chromatography is water.

122. (4) A reagent that brings an electron pair to the reactive site is called a nucleophile.
As total six nucleophiles are there from the given options.

NH_3, PhSH, $(H_3C)_2S$, $CH_2=CH_2$, $\overset{\ominus}{O}H$, $>NCH_3$

123. (1) The purification process that involves a solid changing into vapour and then back into solid is called sublimation.

124. (4)

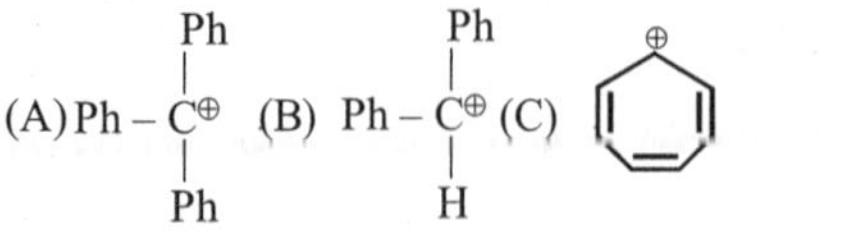

(D) $H_3C - CH_2 - \overset{\oplus}{C}H - CH_3$
Structure C is aromatic hence it is most stable.
Structure A has more resonance structures.
Structure B has less resonance structures compared to A.
Structure D have only hyperconjugation.
Preference order will be: Aromaticity > Resonance > Hyper conjugation
D < B < A < C

125. (4) In Dumas method nitrogen present in organic compound is converted into N_2 gas whose volumetric analysis gives the percentage of nitrogen atom in the organic compound.

126. (2) Carbocation intermediate is stabilised by +I, +M & hyperconjugation effect. Since in option (b) carbocation is in conjugation with stronger +M group $-OCH_3$ hence it will be most stable.

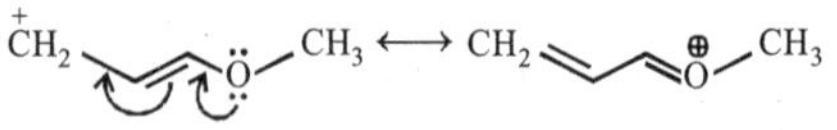

127. (1) Due to +M effect of –OMe at para position.

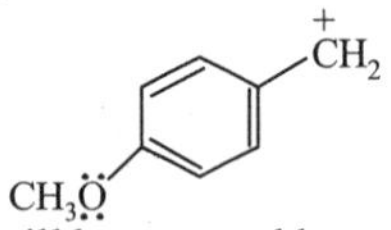

will be most stable among these

128. (3) Organic compounds are purified based on their nature and impurities present in them.

129. (2) (1) Neutral structures are more stable than charged ones. Therefore (I) is more stable than (II) and (III).
(2) +ve charge on less electronegative atom is more stable i.e., $C^{\oplus}$ is more stable than $O^{\oplus}$
∴ Order is (I) > (II) > (III)

130. (1) Carbocation 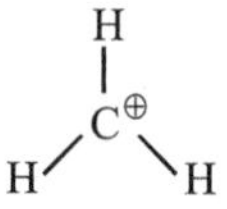
Trigonal planar

131. (1) Nitrogen present in pyridine can not be estimated by Kjeldahl method as the nitrogen present in pyridine can not be easily converted into ammonium sulphate.

132. (2) Aniline can be purified by steam distillation because aniline is steam volatile and immisible with water. Therefore option (b) is incorrect.

133. (2) Methoxy group strongly activates benzene ring towards electrophilic aromatic substitution due to +R effect. Methyl group weakly activates benzene ring due to its +I and +H effects. CF_3 group strongly deactivates the benzene ring due to –I and–H effects.
order as
Resonance > Hyper conjugation > Induction effect

Therefore, the correct order of the given compounds towards electrophilic aromatic substitution is

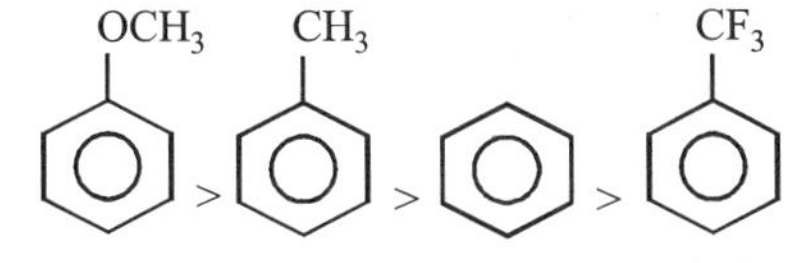

(III) (I) (II) (IV)

134. (3) Organic compounds run slower than solvent in thin layer chromatography. R_f is not an integral value. It is the ratio of distance travelled by the organic compound to that of solvent. R_f value of a polar compound is smaller than that of non-polar compound as the polar compound is retained more by the adsorbent than non-polar compound.

135. (1)

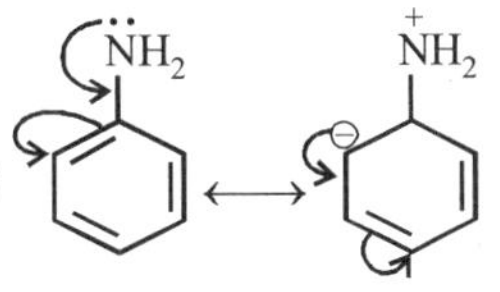

+ R effect (– $\ddot{N}H_2$ is a electron donating group)

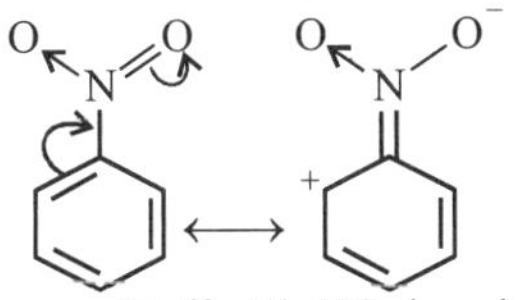

– R effect (– NO_2 is a electron withdrawing group)

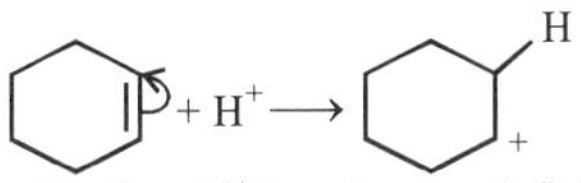

+ E effect (H^+ is a electron deficient species)

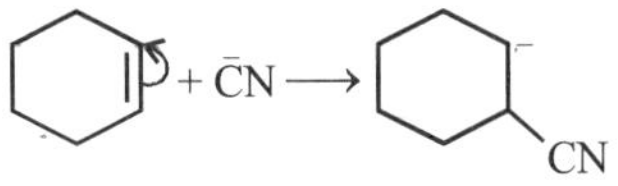

– E effect ($\bar{C}N$ is a electron rich species)

136. (4) Hydrogen ion (H^+) shows positive electromeric effect.

137. (3) The most common polar and acidic support used is adsorption chromatography is silicagel. The surface silanol groups adsorb polar compound and work particularly well for basic substances. Alumina is the example of polar and basic adsorbent that is used in adsorption chromatography.

138. (2) $CuSO_4$ acts as a catalytic agent in Kjeldahl's method for estimation of nitrogen.

139. (3) Only B

Ionic reactions with organic compounds proceed through heterolytic bond cleavage.

$$A — B \xrightarrow{\text{Heterolysis}} A^{\oplus} + B^{\ominus}$$

140. (3) –COOH

Negative R effect is shown by:

–COOH, –CHO, >C=O, –CN, $–NO_2$

Positive R effect is shown by:

–X, –OH, –OR, –OCOR, $–NH_2$, –NHR, $–NR_2$, –NHCOR

141. (3) Carbocation has sextet of electrons. It has incomplete octet so it is a electrophile.

142. (1) Principle used in chromotography is adsorption.

143. (2) A-IV, B-III, C-II, D-I

144. (4) Since the fragrance of flowers is due to some steam volatile organic compounds and they are insoluble in water at room temperature, therefore the suitable method for the extraction will be steam distillation.

145. (4) Column Chromatography → based on absorption.

Sublimation → based on volatile nature of components.

Distillation → based on relative boiling points

Differential extraction → based on relative solubilities.

146. (3)

$$(CH_3)_3\overset{+}{C} > (CH_3)_2\overset{+}{C}H > CH_3-\overset{+}{C}H_2 > \overset{+}{C}H_3$$

tertiary secondary primary

This is due to dispersion of positive charge on more atoms available.

147. (1) The resonance effect is defined as 'the polarity produced in the molecule by the interaction of two π-bonds or between a π-bond and a lone pair of electrons present on an adjacent atom'.

148. (4) (C) < (D) < (B) < (A)

SP^3 SP^3 SP^2 SP

Due to the relative electronegativities of the carbon atoms of the corresponding hydrogens and their hybridization states, we get the order of acidic strengths as:

149. (3) Column and thin layer chromatography (TLC) are both based on the principle of differential absorption. Paper chromatography is based on the principle of differential partitioning of the components of a mixture.

150. (4)

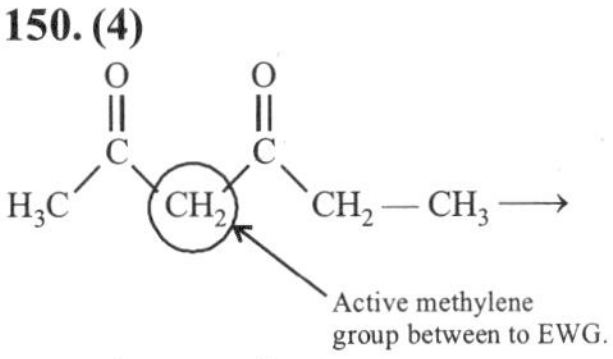

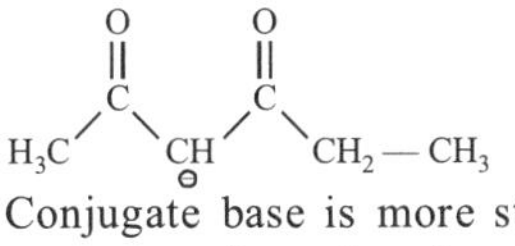

Conjugate base is more stable due to more resonance of negative charge, that results in its dispersal.

151. (1) I > II > III, since neutral resonating structures are more stable than charged resonating structures II > III, since stability of structure (II) with –ve charge on more electronegative atom (o) is higher.

152. (4) Steam distillation is used for those liquids which are insoluble in water, containing non-volatile impurities and are steam volatile.

153. (2) Higher the electron density on benzene ring, higher is reactivity towards electrophilic substitution reaction

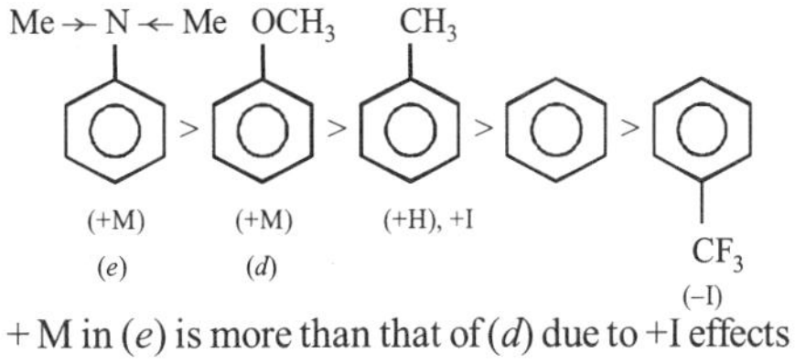

+M in (*e*) is more than that of (*d*) due to +I effects of Me groups.

154. (2) In paper chromatography, a special quality of paper known as chromatography paper is used. Paper contains water trapped in it, which acts as the stationary phase.

155. (1)

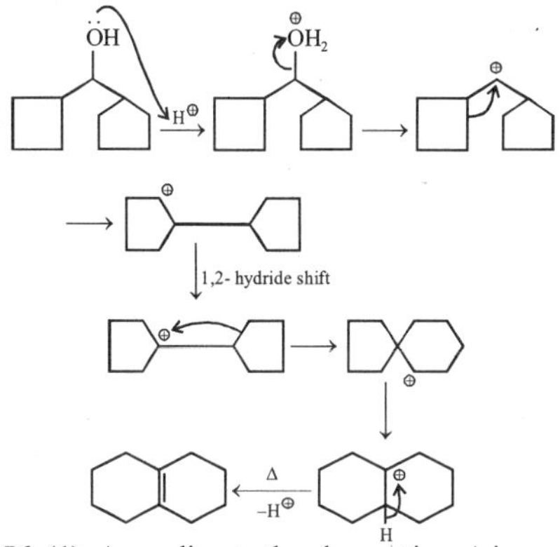

156. (1) According to the observation, A is more stable and interacts with the mobile phase more than C, and C is more drawn to the mobile phase than B. Hence, the correct order of elution in the silica gel column chromatography is – B < C < A or order of elution A, C, then B.

157. (3) Benzene becomes more reactive towards EAS when any substituent raises the electron density.

∴ Correct order

c < a < b < d

158. (1)

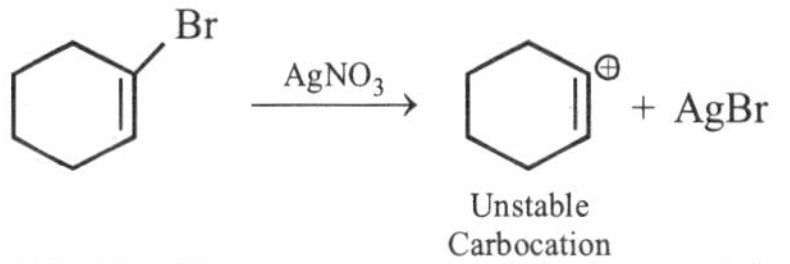

159. (4) If any component eluted second then it means that its R_f value is low and its adsorption is stronger

$$R_f = \frac{\text{distance covered by substance from base line}}{\text{total distance covered by solvent from base line}}$$

160. (4) $X \xrightarrow{Na_2O_2} Y \xrightarrow{BaCl_2} \underset{[BaSO_4]}{Z}$

Methionine: $C_5H_{11}NO_2S$ (Sulphur containing organic compound)

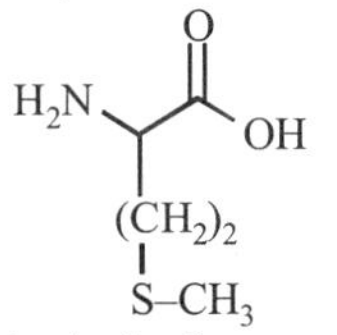

The given test checks for the quantity of sulphur.

$$\underset{(X)}{C_5H_{11}NO_2S} \xrightarrow{Na_2O_2} \underset{(Y)}{H_2SO_4} \xrightarrow{BaCl_2} \underset{(Z)}{BaSO_4}$$

161. (2) → Hydride affinity means low stability. Thus, higher the stability of a carbocation lower will be its hydride affinity. Stability order of cations is : C < A < B < D

'A' is allylic carbocation, 'B' is resonance stabilized but –Ph is elec. withdrawing. C is stablized through hyper conjugation, D is stabilized by +I effect. Thus the decreasing order of hydride affinity is

C > A > B > D

162. (4) Na + C + N → NaCN

$$6NaCN + FeSO_4 \rightarrow Na_4[Fe(CN)_6] + Na_2SO_4$$

$Na_4[Fe(CN)_6] + Fe^{3+} \rightarrow Fe_4[Fe(CN)_6]_3$
(Prussian blue)

Sulphur detection by Sodium nitroprusside
$Na_2[Fe(CN)_5NO] + Na_2S \rightarrow Na_4[Fe(CN)_5NOS]$
[Purple]

Phosphorus detection by ammonium molybdate
$Na_3PO_4 + 3HNO_3 \rightarrow H_3PO_4 + 3NaNO_3$
$H_3PO_4 + 12(NH_4)_2MoO_4 + 21HNO_3 \rightarrow$
$(NH_4)_3PO_4 \cdot 12MoO_3 + 21NH_4NO_3 + 12H_2O$
(canary yellow)

Halogens give specific coloured ppt with $AgNO_3(aq)$
$NaCl + AgNO_3(aq) \rightarrow AgCl + NaNO_3$
(White)
$NaBr + AgNO_3(aq) \rightarrow AgBr + NaNO_3$
(Pale yellow)
$NaI + AgNO_3(aq) \rightarrow AgI + NaNO_3$
(Yellow)

163. (1) Compound 'a' is more polar than 'b' and 'c', compound 'c' is least polar. Compound 'c' comes out first.

164. (3) Density of CH_2Cl_2 is greater than H_2O. Therefore these can be separated by differential solvent extraction.
Due to H-bonding in p-nitrophenol it can be separated from other component by column chromatography.
Due to different boiling point of kerosene and naphthalene, these can be separated by fractional distillation.
NaCl (ionic compound) and $C_6H_{12}O_6$ can be separated by crystallisation.

165. (2) The nitrogen containing organic compound, when heated with CuO in an atmosphere of CO_2, yields free N_2 in addition to CO_2 and H_2O.
Traces of nitrogen oxides formed, if any, are reduced to nitrogen by passing the gaseous mixture over heated copper gauze.

166. (2) Acidity of an acid depends upon the stability of its conjugate base

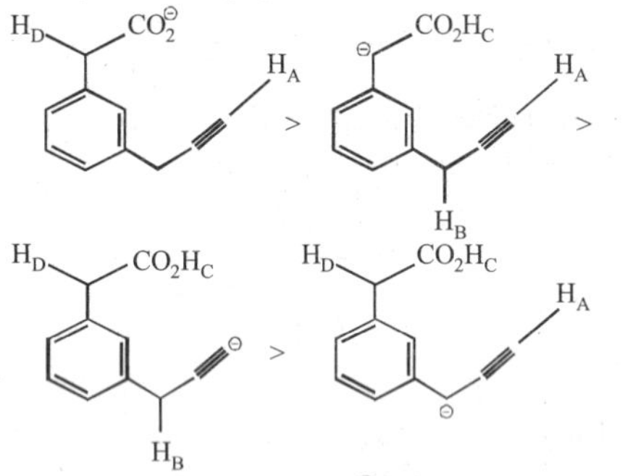

167. (1)

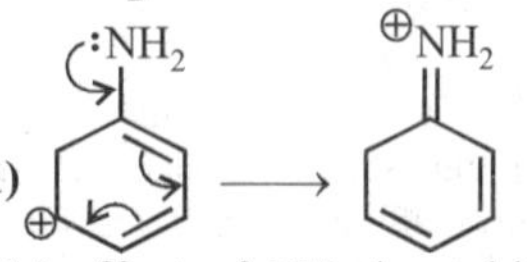

The +M effect of NH_2 is stabilizing the carbocation.

168. (2) A-(III), B-(IV), C-(I), D-(II).

169. (1) Thin layer chromoatography (TLC) is type of adsorption chromatography, which is used for the separation of substance of a mixture using a thin layer of an adsorbent coated on glass plate.
A thin layer of an adsorbent (silica gel) or (alumina) is spread over a glass plate of suitable size. Hence Assertion (A) is correct and Reason (R) is correct explanation of (A)

170. (4)

171. (4) Preparative thin layer chromatography is the best method of separation on a macro scale (0.05 – 1g) by using a thick layer of adsorbent (0.5 – 5) mm.

172. (4) *d* is most stable due to maximum extended conjugation.

(1) 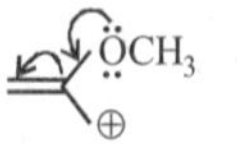

Positive charge not dispersed

(2) 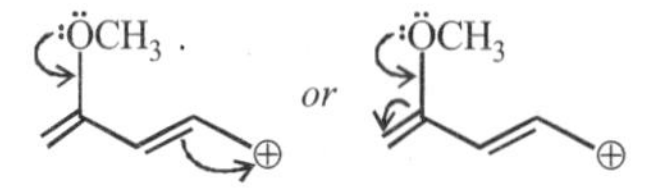

(3) 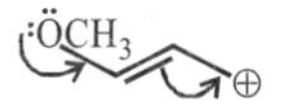

(4) 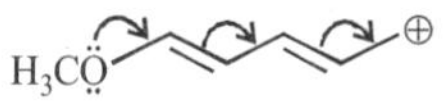

Conjugation is more extended than in (2) and (3)

173. (1) 188 g AgBr contains = 80 g Br

$$0.36 \text{ g AgBr contains } = \frac{80}{188} \times 0.36 = 0.1532 \text{ g}$$

$$\% \text{ Br} = \frac{0.1532}{0.45} \times 100 = 34.04$$

174. (3) Ninhydrin, iodine and UV rays are used for identification of spot.

175. (1) Two immiscible liquids with different densities can be separated by using this funnel.

176. (2)

177. (1)

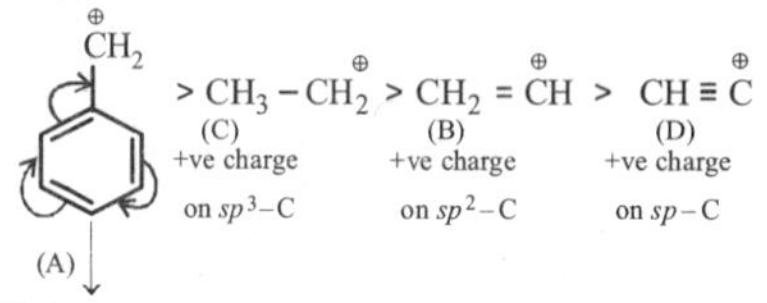

Stable due to resonance

178. (4) Vacuum distillation or reduced pressure distillation method is used to purify compounds that can decompose before reaching their boiling point at atmospheric pressure.

179. (4)

(1) $CH_3 - CH_2 - \ddot{O} - CH = CH_2 \rightleftharpoons CH_3 - CH_2 - \overset{\oplus}{O} = CH - \overset{\ominus}{C}H_2$

(2) 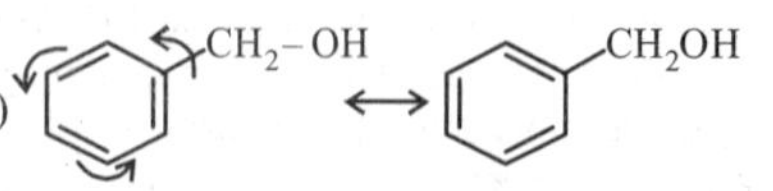

(3) $CH_3 - CH_2 - CH_2 - \overset{O}{\overset{\|}{C}} - \ddot{N}H_2 \rightleftharpoons CH_3 - CH_2 - CH_2 - \overset{\ominus O}{\overset{|}{C}} = \overset{\oplus}{N}H_2$

(4) $CH_3 - CH_2 - CH = CHCH_2 - NH_2$
There is no conjugation of double bond with lone pair of electrons.

180. (3) (A) and (B) are in resonance.

(A) 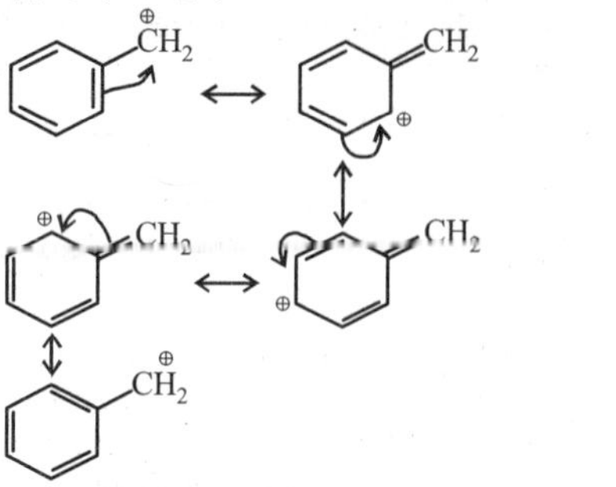

(B) 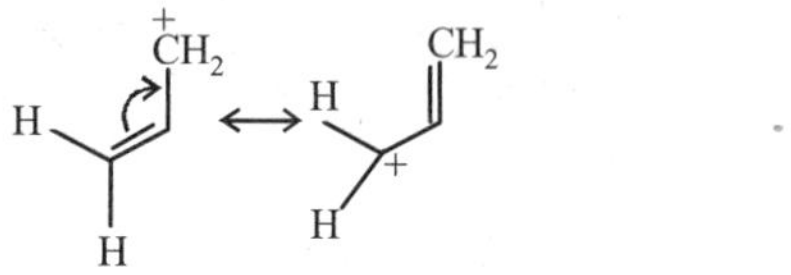

The positive charge in (C) and (D) is not in conjugation with π electrons.

181. (2) 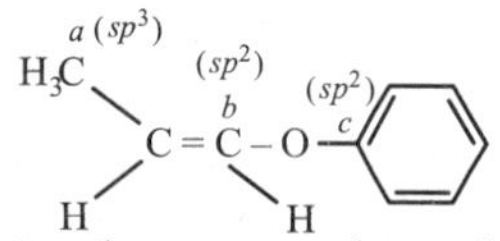

182. (4) In chromatography technique, the purification of compound is independent of physical state of the pure compound.

183. (1) Carius method is used for the estimation of halogens in organic compounds.

184. (1) (1) 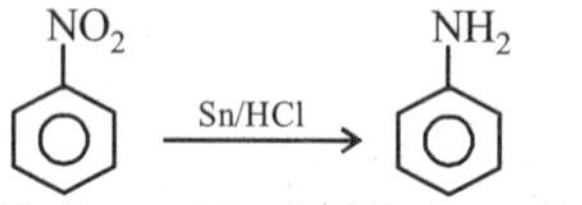

(Product will give positive Kjeldhal test due to presence of $-NH_2$ group)

(2) 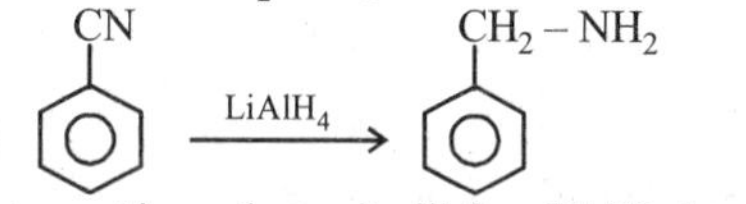

(N-present in product so it will show Kjeldhal test)

(3) 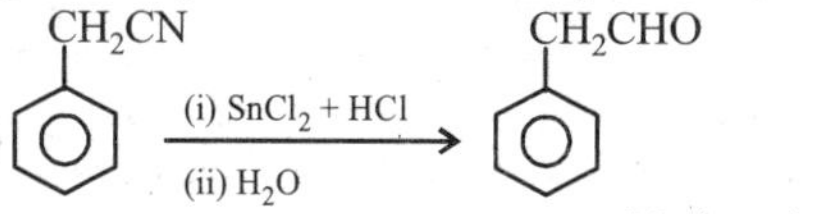

(N absent)

(4) 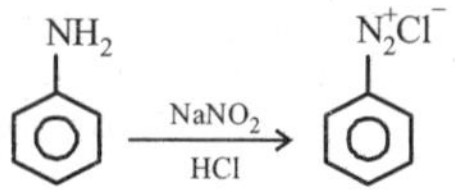

($-N_2^+$ does not show Kjeldhal test)

Kjeldahl's method is not applicable to (i) compounds which do not have N-atom (ii) compounds containing nitrogen in the ring (iii) compounds containing nitrogen directly linked to oxygen atom (*e.g.*, NO_2) (iv) compounds having $-N = N-$ gp.

185. (4) Acidic strength ∝ –I, –M effect. Due to strong – I, and –M effect of three – $COOCH_3$ groups, it has most acidic hydrogen.

186. (4) Glycerol can be separated from spent-lye (alkaline glycerol containing liquid) in soap industry by using reduced pressure distillation technique because glycerol decomposes at its boiling point.

187. (3) Since acidic order of their conjugate acid to basicity is opposite to the acidic order of their conjugate acids.

$H-C\equiv N > HC\equiv CH > H_2C=CH-CH_3 > CH_4 > CH_3-\underset{CH_3}{\overset{CH_3}{C}}-H$

Increasing basicity order of the given species will be

$\overset{\ominus}{C}\equiv N$ (v) $< HC\equiv\overset{\ominus}{C}$ (ii) $< H_2C=CH-\overset{\ominus}{C}H_2$ (iii) $< \overset{\ominus}{C}H_3$ (iv) $< H_3C-\underset{CH_3}{\overset{CH_3}{C^{\ominus}}}$ (i)

188. (1) Liquid having lower boiling point comes out first in fractional distillation. Simple distillation can't be used as boiling point difference is very small.

3-Methylpentane will show greater boiling point (63°C) comparative to isohexane due to symmetrical structure. Therefore, isohexane will distill out first.

189. (2) Electron withdrawing group like (NO_2) increases stability of alkoxide ion by dispersal of negative charge.

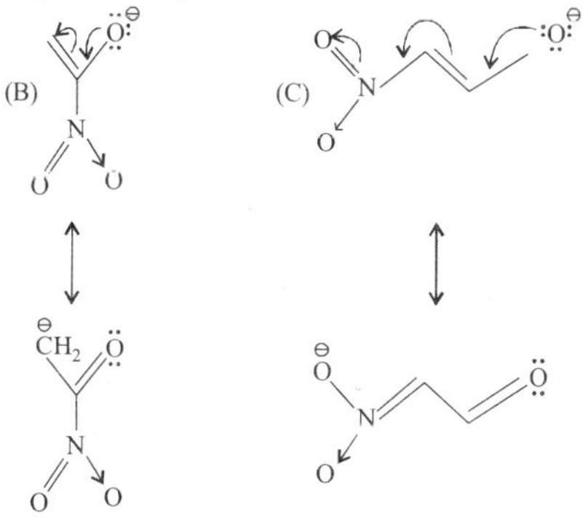

conjugation and –I effect of $-NO_2$ | extended conjugation and –M, –I effect of $-NO_2$

Hence, (C) is more stable than (B) which is more stable than (A),(A) is least stable because negative charge is dispersed only due to – I effect.

190. (4) In all four nucleophiles, oxygen is the nucleophilic atom. Higher the basicity of a base, stronger will be its nucleophilic character. Further, a stronger base has a weaker conjugate acid. The acidic order of the conjugate acid of the given bases is:

$H_3O^+ > CH_3SO_3H > CH_3COOH > H_2O$

Thus, the increasing order of the basic nature or nucleophilicity of the given species is:

$\underset{(ii)}{H_2O} < \underset{(iii)}{CH_3SO_3^-} < \underset{(i)}{CH_3COO^-} < \underset{(iv)}{OH^-}$

191. (2) Mole of bromine $= \dfrac{0.08}{80} = 10^{-3}$ mole

Molar mass of compound is given by the following equation,

$$\frac{0.172}{M} = 10^{-3} \Rightarrow M = \frac{0.172}{10^{-3}} = 172\ g$$

$\because$ Molar mass of C_6H_6NBr

$= (6\times 12) + (1\times 6) + (1\times 14) + (1\times 80) = 172$ g.

172 g corresponds to the molar mass of (b).

Thus option (2) is the correct structure of compound.

192. (3) Kjeldahl's method can not be used for nitrogen determination of compounds having nitro group or azo group or nitrogen present in rings as the nitrogen of these compounds can not be converted to $(NH_4)_2SO_4$ under the condition of this method.

193. (3) Compounds which are able to form strong H-bond with the stationary phase (silica gel) will come last with the mobile phase.

H-bonding order :

Aniline > Benzanilide > Acetophenone

B > A > C

$\therefore$ Sequence of obtained compounds is (C, A, B)

194. (4) In column chromatograph; a solid adsorbent is packed on a column and a solution containing number of solute particles is allowed to flow down the column. The solute molecules get adsorbed on the surface of adsorbent and move through column at different rates based on differential adsorption of the substances on the solid phase.

195. (3) In chromatography, R_f represents retardation factor.

$$R_f = \frac{\text{Distance moved by the substance from baseline}}{\text{Distance moved by the solvent from baseline}}$$

$\therefore$ Higher R_f value means lower adsorption.

196. (2) Enolic form of acetylacetone (b) is quite stable due to H-bonding which leads to stable 6-membered ring.

197. (2) (A) Water and sugar do not react chemically, and sugar dissolves completely in water on heating. On cooling sugar solution, sugar recrystallises back **(Q)**.

(B) Aniline is insoluble in water, but it is steam volatile. Hence, it can be separated from water by steam distillation **(R)**.

(C)Toluene is insoluble as well as non-volatile in steam. Hence, it can be separated by differential extraction method **(S)**.

198. (3)

$$[C_xH_yN_z] \xrightarrow[\text{method}]{\text{Dumas}} 6CO_2 + 4H_2O + N_2$$

Mol of $CO_2 = 6$, so mol of C is = 6

Mol of $H_2O = 4$, so mol of H is = 8

Mol of $N_2 = 1$, so mol of N is = 2

$\therefore$ Formula is $C_6H_8N_2$.

199. (1) Due to higher density of dichloromethane than water, DCM would be in the lower layer and water will form the upper layer in the separating funnel.

200. (2) Benzaldehyde is an adsorbate, alumina is an adsorbent (stationary phase) and acetonitrile (a solvent) is in mobile phase.

201. (3)

(A) Charcoal treatment removes coloured impurity through adsorption.

(B) Steam distillation separates the mixture of *o*-nitrophenol and *p*-nitrophenol. The *o*-nitrophenol is steam volatile (due to intramolecular hydrogen bonding), and the *para* isomer is not volatile.

(C) Fractional distillation separates crude naphtha. Naphtha is a flammable liquid hydrocarbon mixture.

(D) Distillation under reduced pressure separates mixture of glycerol and sugar. Vacuum distillation lowers the boiling point and prevents decomposition.

202. (2) The R_f (retardation factor) value is the ratio of the Solute's distance travelled to the Solvent's distance travelled.

Polar molecules will attract more toward adsorbent and cover smaller distance from the baseline.

A has maximum R_f so it will less adsorbed on the adsorbent surface and travel faster and will be eluded first.

C has minimum R_f value. So it is more strongly adsobred and less soluble in mobile phase.

203. (4) Negatively charged ions of same atom are more nucleophilic than their neutral counterpart.

Nucleophilicity $\propto$ electro density on donor atom

$\propto$ size of donor atom (in gas)

$\propto \dfrac{1}{\text{electronegativity of atom}}$ (for period)

204. (N) Cabocation (B) is most stable as it is stabilised by resonance. Carbocation (A) is stabilised by hyperconjugation due to 4α hydrogen. Carbocation (C) is least stable due –I effector oxygen atom.

So order will be B > A > C

205. (2) Kjeldahl's method is not applicable for compounds containing nitrogen in nitro, azo groups and nitrogen in ring, as N of these compounds does not change to ammonium sulphate under these conditions.

206. (3)

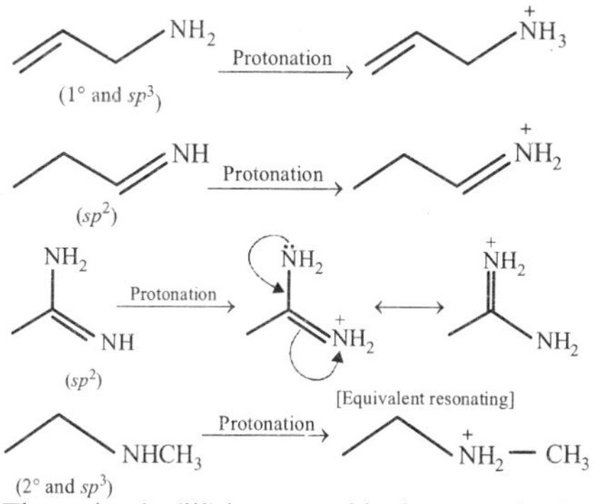

The cation in (iii) is very stable due to equivalent resonating structures, making the parent (iii) most basic. In (ii), N is sp^2 hybridised makeup it electronegative which decreases its electron decreasing character. Consequently (i) is least basic. Among (i) and (iv), (i) being 1° is more basic than the 2° amine (iv).

Hence, correct order of basicity will be : (ii) < (i) < (iv) < (iii).

207. (3) Among the substituents attached to the given compounds, fluorine has maximum electronegativity, so it will pull electron pair towards itself. In option (2), the two F groups are attached opposite to each other, thus net dipole moment will cancel out and reduce its polarity. In option (4), the F groups are attached in slightly opposite direction. Thus, this also decreases its polarity. But in option (3), the compound has the two F groups along same direction. Thus, net dipole moment will increase in this direction and therefore, it will exhibit maximum polarity.

208. (2) The lone pair of electrons on O_2 is involved in resonance with C = C. Hence O_2 will not be protonated.
The lone pair of electrons on O_5 is not involved in resonance with C = C. Hence, O_5 will be protonated. Chloride ion will then attack least substituted C atom (C_6)

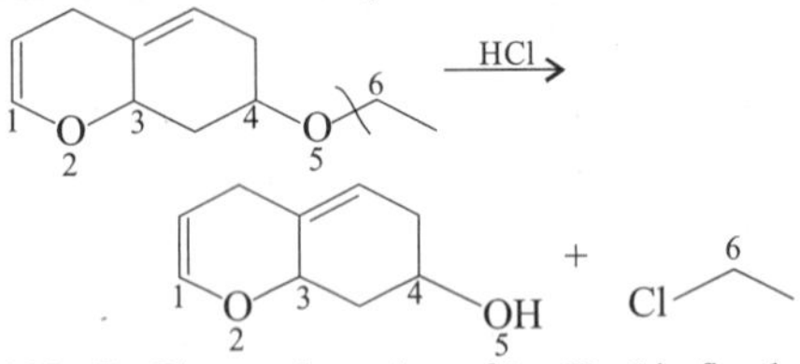

209. (2) Since, adsorption of I > II, I is firmly attached to column (stationary phase). Hence, it moves slowly and will cover little distance. II is loosely attached to column (stationary phase). Hence, it moves faster and will cover large distance.

210. (2) Partition chromatography is the process of separation whereby the components of the mixture get distributed into two phases that may be liquid-liquid, liquid-gas but not solid-gas. Thus, it is not possible that the stationory phase is a finally divided solid adsorbent in partition chromatography.

211. (1)
(I) C_2H_5OHHydrogen bondingHydrogen bonding > dipole-dipole
(II) C_2H_5ClDipole-dipole
C_2H_5Cl is more polar than $C_2H_5OCH_3$ so,
dipole-dipole C_2H_5Cl > dipole-dipole $C_2H_5OCH_3$
(III) $C_2H_5CH_3$Weak van der Waals forces
(IV) $C_2H_5OCH_3$ Dipole-dipole
Thus, boiling point order is:
(III) < (IV) < (II) < (I).

212. (2) When the given mixture is shaken with 1 M HCl, amine gets protonated and becomes cation ($R_2\overset{\oplus}{N}H_2$), which does not dissolve in organic solvent but usually dissolve in H_2O due to its charge. So, shaking with aqueous HCl will pull amines into the aqueous phase and leave all other compounds in organic layer.

213. (4) 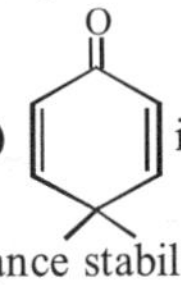 is nonaromatic and hence, least reasonance stabilized, whereas other three are aromatic.

214. (2) Since S_N1 reactions involve the formation of carbocation as intermediate in the rate determining step, more the stability of carbocation higher will be the reactivity of alkyl halides towards S_N1 route.
Since stability of carbocations follows order.

$$\underset{1^\circ \text{ carbocation}}{CH_3-CH_2-\overset{\oplus}{C}H_2} < \underset{2^\circ \text{ carbocation}}{CH_3-\overset{\oplus}{C}H-CH_2-CH_3}$$

$$< p-H_3CO-C_6H_4-\overset{\oplus}{C}H_2$$

Max. stable due to +M effect of $-OCH_3$ group.

215. (2) Percentage of nitrogen $= \dfrac{1.4 \times NV}{W}$

$\because$ NV = milliequivalent of H_2SO_4 reacted with NH_3

Number of milliequivalent of H_2SO_4 taken

$= 60 \times \dfrac{1}{10} \times 2 = 12$

Number of milliequivalent of NaOH' used

$= 20 \times \dfrac{1}{10} = 2$

So, the no. of milliequivalent of H_2SO_4 reacted with $NH_3 = 12 - 2 = 10$

Percentage of nitrogen $= \dfrac{1.4 \times 10}{1.4} = 10\%$

216. (4) In both the molecules the bond moments are not cancelling with each other and hence, the molecules have a resultant dipole.

217. (4) $Ph_3\dot{C}$ is more stable than $(CH_3)_3\dot{C}$ because resonance stabilisation effect in $Ph_3\dot{C}$ is more pronounced as compared to hyperconjugation stabilisation effect in $(CH_3)_3\dot{C}$.
Overall stability order among free radical is :
Triphenylmethyl > benzyl > allyl > tertiary alkyl > secondary > primary > methyl > vinyl

218.(3) Carbocations are planar, hence can be attacked on either side to form racemic mixture.

$$\underset{(-)}{\underset{|}{\underset{Ph}{}}}Cl-\underset{\underset{Ph\,(-)}{|}}{CH}-CH_3 \xrightarrow[\text{Toluene}]{SbCl_5}$$

$$\underset{\text{(carbocation)}}{Ph-\overset{\oplus}{C}H-CH_3} + SbCl_6^- \longrightarrow$$

$$\underset{(d+\ell)\text{ mixture}}{Ph-\underset{\underset{Cl}{|}}{CH}-CH_3} + SbCl_5$$

219. (4) Higher stability of allyl and benzyl carbocations is due to dispersal of positive charge by resonance.

$$CH_2=CH-\overset{\oplus}{C}H_2 \longleftrightarrow \overset{\oplus}{C}H_2-CH=CH_2$$

Resonating structures of allyl carbocation

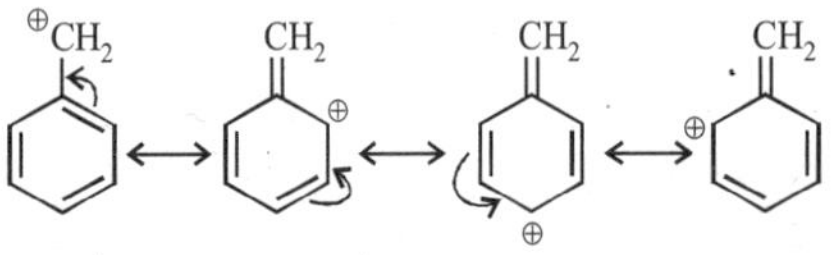

Resonating structures of benzyl carbocation

Whereas in alkyl carbocations, dispersal of positive charge on different carbon atoms is due to inductive effect. Hence, the correct order of stability will be:

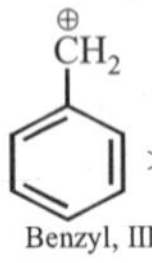 $> CH_2=CH-\overset{\oplus}{C}H_2 > CH_3-CH_2-\overset{\oplus}{C}H_2$

Benzyl, III   Allyl, I   Propyl, II

220. (1) 3° Carbocations are most stable.

221. (1) – Cl and – CH_3 groups are *o* and *p* directing. They are electron releasing due to + M and hyperconjugation effects. Further, since such groups increase electron density in the nucleus, they facilitate further electrophilic substitution and hence, known as activating group. The activating effect of these groups is in order of – CH_3 > –X but chlorine exceptionally deactive the ring due to strong – I effect. Hence, it is difficult to carry out substitution in chlorobenzene than in benzene.
Further, – NO_2 is a deactivating group, hence deactivates the benzene nucleus, i.e., hinder the further substitution. Thus, nitrobenzene undergoes electrophilic substitution with a great difficulty.
Hence, the correct order will be.

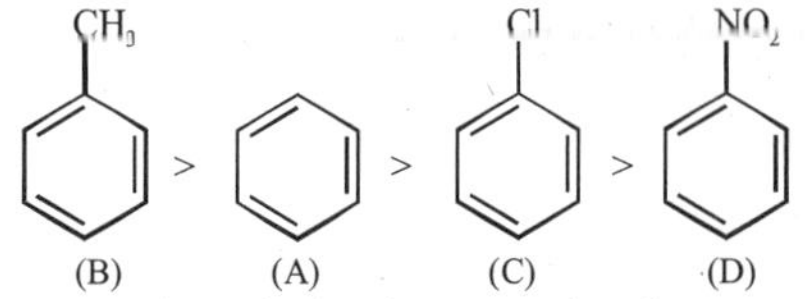

(B)   (A)   (C)   (D)

222. (2) Since the leaving group breaks away as a base, it is easier to displace weaker bases as compared to stronger bases. Thus less basic the substituent, the more easily it is displaced.
The basic strength of the given groups is in order:
$I^- < Br^- < Cl^-$
Thus, the order of halogen leaving groups is:
$I^- > Br^- > Cl^-$

223. (4) $-OCH_3$ and $-CH_3$ groups are activating group while $-CF_3$ is a deactivating group. Thus order is

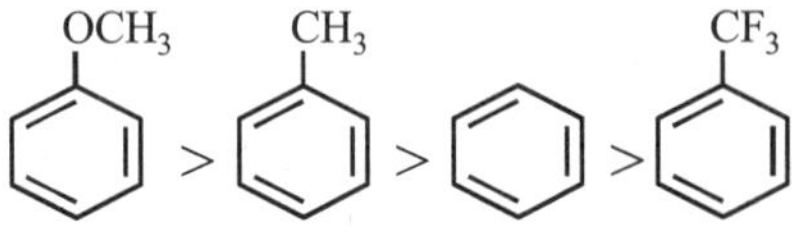

(i)   (ii)   (iii)   (iv)

224. (1) Ethers, due to absence of delocalized pair of electrons do not show resonance.

225. (4) The corresponding conjugate acids are:

$$\underset{\text{Most acidic}}{CH_3COOH} > CH \equiv CH > NH_3 > \underset{\text{Least acidic}}{RH}$$

$\therefore$ the correct order of basicity is:

$$RCOO^- < CH \equiv C^- < \bar{N}H_2 < R^-$$

226. (2)

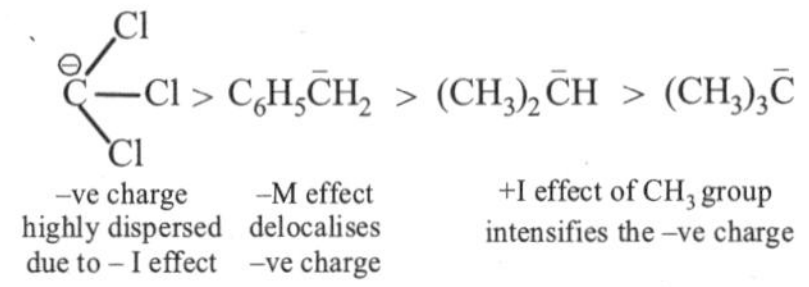

227. (2) In option (2) the complex formed is with benzene whereas in other cases it is formed with nitrobenzene with $-NO_2$ group in different positions (*o*-, *m*-, *p*-).
Due to strong –I effect of $-NO_2$ group, it will increase the positive charge inside the ring. The complex formed with nitrobenzene in any position of $-NO_2$ group is less stable than the complex formed with benzene, so the most stable complex has lowest energy.

228. (1) Nitro group is electron withdrawing group, so it deactivates the ring towards electrophilic substitution.

229. (2) The order of stability of free radicals is:

$$(C_6H_5)_3\dot{C} > (C_6H_5)_2\dot{C}H > (CH_3)_3\dot{C} > (CH_3)_2\dot{C}H$$

The stabilisation of first two is due to resonance and last two is due to hyperconjugation.

230. (3)

$$\underset{e^-\text{s are delocalised}}{CH_3COO^-} < C_6H_5O^- < OH^- < \underset{\text{Max. } e^- \text{ density on O}}{\bar{O}CH_3}$$

231. (4) Compounds bearing a quaternary nitrogen atom, when heated in presence of base like OH^- undergo E2 reaction with the formation of less substituted alkene as a major product with the loss β-hydrogen atom (**Hoffmann elimeniation**, or **Hoffmann rule**).

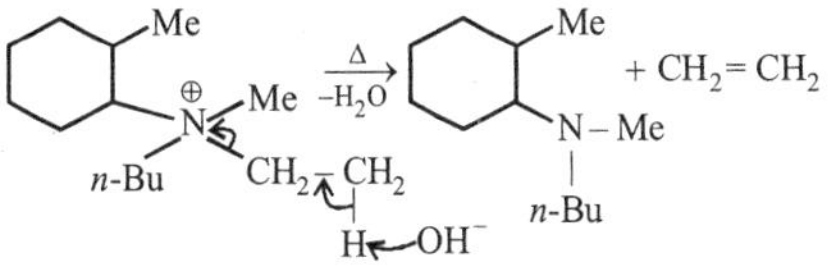

Note: Less sterically hindered β-hydrogen is removed.

232. (4) Free radicals are electrically neutral, unstable and very reactive on account of the presence of odd electrons.

233. (2) Decreasing acid order of the four acids is (D) > (A) > (C) > (B)
Hence decreasing order of nucleophilicity is (B) > (C) > (A) > (D)

234. (4) Poor the nucleophile, better is the leaving ability of the group. Among the given species, Cl^- is the most poor neucleophile, hence it is the best leaving group among the given options.

235. (4) In carboxylic acids, presence of electron withdrawing substituent, e.g., $-NO_2$ disperses the negative charge of the anion and stabilises it and hence, increases the acidity of the parent acid. Further, *o*-isomer will have higher acidity than corresponding *m*-and *p*-isomers due to ortho and high inductive and mesomeric effect of $-NO_2$ group. Hence, the correct order is:

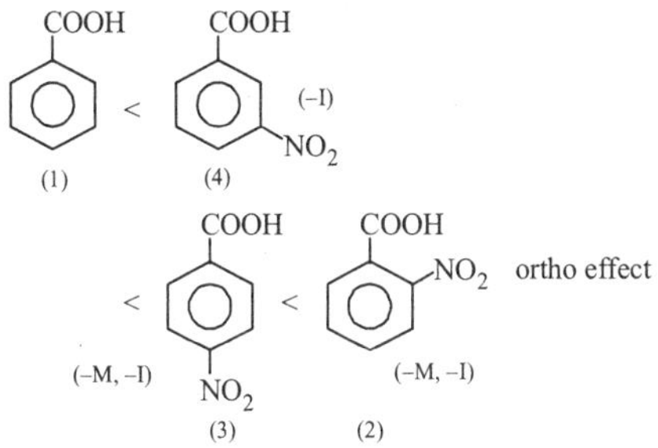

236. (4) Lone pair of electrons present on the nitrogen of benzyl amine is not involved in resonance.

237. (2) $HCOO^-$ exists in following resonating structures

$$H-\overset{\overset{O}{||}}{C}-O^- \longleftrightarrow H-\overset{\overset{O^\ominus}{|}}{C}=O$$

Hence in it both the carbon oxygen bonds are found equal.

238. (2) In molecules (a), (c) and (d), the carbon atom has a multiple bond, only (b) has sp^3 hybridisation.

239. (2) $-CH_3$ group has +I effect, as number of $-CH_3$ group increases, the inductive effect increases.
Therefore, the correct order is:

$$CH_3-CH_2- < (CH_3)_2CH- < (CH_3)_3C-$$

240. (2) $CH_3-CH=CH-\overset{\overset{O}{||}}{CH}$ (III)
↓
Non polar R.S.
More No. of covalent bonds

$CH_3-\overset{\oplus}{CH}-CH=\overset{\overset{O^\ominus}{|}}{CH}$ (II)
↓
Having – ve charge on more electronegative atom

$CH_3-\overset{\ominus}{CH}-CH=\overset{\overset{O^\oplus}{|}}{CH}$ (I)
↓
Having – ve charge on less electronegative atom

∴ Stability order III > II > I

241. (42)

$$\text{Moles of } BaSO_4 \text{ formed} = \frac{0.6}{231} = 0.0026$$

Moles of $BaSO_4$ = Moles of S

$$\text{Mass of sulphur } (W_s) = \frac{0.6}{231} \times 32$$

$$\text{\% of suphur (S)} = \frac{0.6 \times 32}{231 \times 0.2} \times 100$$

$$= \frac{1920}{46.2} = 41.558 \approx 42$$

242. (435)

$$\text{\% of S} = \frac{\text{molar mass of S}}{\text{molar mass of } BaSO_4} \times \frac{m}{w} \times 100$$

m = mass of $BaSO_4$ ppt
w = mass of organic compound
$w = 2 \times 10^{-3} \times 76 = 0.152$ gm

$$\text{\% of S} = \frac{32}{233} \times \frac{0.4813}{0.152} \times 100 = 43.487$$

$$\text{In form of } 10^{-1} = \frac{43.487}{10^{-1}} \times 10^{-1}$$

$= 434.87 \times 10^{-1} \approx 435. \times 10^{-1}$

243. (5) Given mass of compound = 2.0 g
Mass of AgBr formed = 3.36 g
Carbon % = 26.7%
Moles of Br :

$$n_{Br} = \frac{3.36}{108 + 80} = 0.01787 \text{ moles}$$

$m_{Br} = 0.01787 \times 80 = 1.43$ g
Moles of Carbon :

$$m_C = 2.0 \times \frac{26.7}{100} = 0.534 \text{ g}$$

$$\text{moles} = \frac{0.534}{12} = 0.0445 \text{ moles.}$$

Moles of Hydrogen :
$m_H = 2.0 - (1.43 + 0.534) = 0.036$ g

$$n_H = \frac{0.036}{1} = 0.036 \text{ moles.}$$

For empirical formula,
C : H : Br = 0.0445 : 0.036 : 0.01787
Making it in simplest ratios,
C : H : Br = 5 : 4 : 2
∴ Empirical formula = $C_5H_4Br_2$.

244. (4) $C_xH_yBr_z \rightarrow CO_2$
1g 1.32 g

$$\text{\% C} = \frac{1.32 \times 12}{44 \times 1} \times 100 = 36\%$$

$C_xH_yBr_z \rightarrow AgBr$
0.53g 0.75g

$$\text{\% Br} = \frac{0.75 \times 80}{188 \times 0.53} \times 100 = 60.2\%$$

% H = 100 – (36 + 60.2) ⇒ %H ≃ 4

245. (15) $P_{N_2} = (P - P_{aq})$

$$= (715 - 15)\text{ mm} = \frac{700}{760} \text{ atm}$$

$$V_{N_2} = 70 \text{ mL} = \frac{70}{1000} \text{ L}$$

$$n_{N_2} = \frac{PV}{RT} = \frac{\left(\frac{700}{760}\right) \times \left(\frac{70}{1000}\right)}{0.0821 \times 300}$$

$$W_{N_2} = \frac{700}{760} \times \frac{\frac{70}{1000}}{0.0821 \times 300} \times 28$$

$$\text{\%N} = \frac{W_{N_2}}{0.5} \times 100$$

$$= \frac{\frac{700}{760} \times \frac{70/100}{0.0821 \times 300} \times 28}{0.5} \times 100 = 14.64\% \approx 15$$

246. (43) In chloroform layer test for halogens, appearance of violet colour in $CHCl_3$ layer indicates iodine.
Mass of AgI (m) = 0.12
Mass of organic compound (W) = 0.15

$$\text{\% of I} = \frac{\text{Atomic weight of I}}{\text{Molecular weight of AgI}} \times \frac{m}{W} \times 100$$

$$= \frac{127}{235} \times \frac{0.12}{0.15} \times 100 \text{ \% of I} = 43.23\% \approx 43\%$$

247. (18) Volume of nitrogen gas liberated (V)

$$= 50 \text{ mL} = \frac{50}{1000} \text{ L}$$

Pressure (P) = 715 mm of Hg
Temperature (T) = 300 K
Aqueous tension = 15 mm of Hg

P_{N_2} = Pressure (P) – aqueous tension

$$= 715 - 15 = 700 \text{ mm of Hg} = \frac{700}{760} \text{ atm}$$

$$n_{N_2} = \frac{P_{N_2} \cdot V}{RT} = \frac{700}{760} \times \frac{50}{1000} \times \frac{1}{0.082 \times 300}$$

$n_N = 2 \times n_{N_2}$
Mass of N = $2 \times n_N \times 14$

$$\text{\%N} = \frac{\text{mass of N}}{\text{mass of organic compound}} \times 100$$

$$\text{\%N} = \frac{700}{760} \times \frac{50}{1000} \times \frac{2 \times 14}{0.0821 \times 300} \times \frac{1000}{292} \times 100$$

%N = 18%

248. (20) Partial pressure of N_2 = P_{total} – Aqueous tension = (900 – 15) = 885 mm Hg

$$\text{Mole of } N_2 = \frac{\left(\frac{885}{760} \times 0.15\right)}{(0.0821 \times 300)} = 0.0071 \text{ moles}$$

mass of nitrogen = n × molar mass of N_2 = 0.0071 × 28 % of nitrogen in organic compound

$= \frac{(0.0071 \times 28)}{1} \times 100 = 19.88\,\% \approx 20\,\%$

249. (109) moles of compound $= \frac{0.42}{86} = 0.00488$

∴ moles of $N_2 = 0.00488$

Volume of N_2 at STP = 0.00488 × 22.4 = 0.1093 L = 109.3 mL ≈ 109 mL

250. (275) Original Compound ⟶ $BaSO_4$

0.20 g 0.40 g

$= \frac{0.40}{233}$ mol ($BaSO_4$) $= \frac{0.40}{233}$ mol (Sulphur)

$= \frac{0.40}{233} \times 32$ g (Sulphur)

$\%S = \frac{\text{mass of S}}{\text{mass of organic compound}} \times 100$

$= \frac{\frac{0.40 \times 32}{233} \times 100}{0.20} = 27.5\%$ or $275 \times 10^{-1}\%$

251. (1655)

	C:	Cl:	H:	O
% mass	14.5	64.46	1.8	19.24
Molar ratio	$\frac{14.5}{12}$ = 1.2	$\frac{64.46}{35.5}$ = 1.8	$\frac{1.8}{1}$ = 1.8	$\frac{19.24}{16}$ = 1.2
Minimum integral ratio	2	3	3	2

Empiricial formula = $C_2H_3Cl_3O_2$

Mass = 165.5

Mass = 1655×10^{-1}

252. (255)

$\% \text{ Bromine} = \frac{\text{Molar Mass of Bromine}}{\text{Molar Mass of Silver Bromide}} \times \frac{\text{Weight of AgBr}}{\text{Weight of sample}} \times 100$

$= \frac{80}{188} \times \frac{0.15}{0.25} \times 100 = \frac{4800}{188} = 25.53 = 255 \times 10^{-1}$

253. (40) Millimoles of $BaSO_4 = \frac{466}{233} = 2$ m mol

= m moles of S

$\% \text{ S} = \frac{m_1 \times 32}{m \times 233} \times 100 = \frac{n \times 32}{m} \times 100$

$\Rightarrow \frac{\frac{466}{233} \times 32}{160} \times 100 = 40\%$

254. (20) millimoles of AgCl $= \frac{143.5}{143.5} = 1$

Millimoles of Cl = 1

Mass of Cl = 35.5×10^{-3} g

% by mass of Cl $= \frac{35.5}{180} \times 100 = 19.72\%$

Nearest integer = 20%

255. (4)

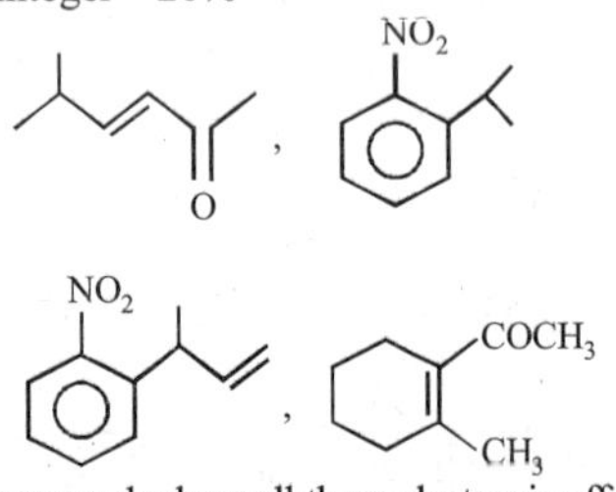

These compounds show all these electronic effects.

256. (15)

$R_f = \frac{\text{Distance moved by substance from base line}}{\text{Distance moved by solvent from base line}}$

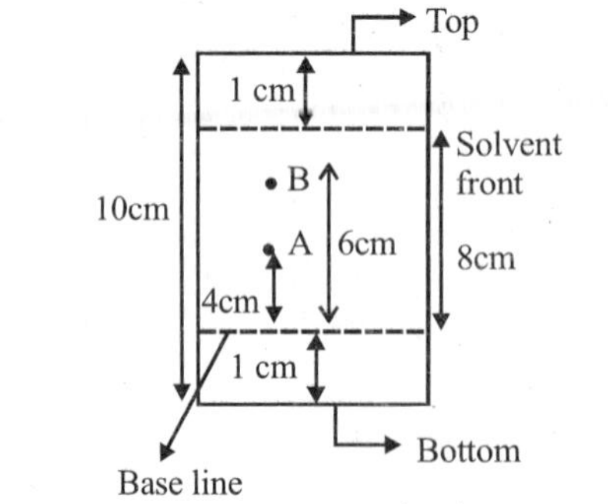

$(R_f)_A = \frac{4}{8}, (R_f)_B = \frac{6}{8} \Rightarrow \frac{(R_f)_B}{(R_f)_A} = \frac{6}{8} \times \frac{8}{4}$

$\Rightarrow (R_f)_B = 1.5\,(R_f)_A$

x = 15

257. (5) 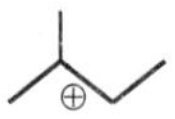 has 8 α − H

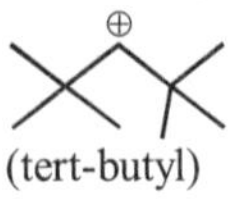 no α − H $\overset{+}{C}H_3$ no α − H

(tert-butyl)

, no hyperconjugable H

$\overset{\oplus}{C}H_2 - OCH_3$ no α − H

 9 α −H $\overset{\oplus}{N} - CH_2$ no α − H

5 carbocations are not stabilised by hyperconjugation.

greater the α − H greater will be its hyperconjugating structure.

258. (50)

$R_f = \frac{\text{Distance travelled by the substance from base line (x)}}{\text{Distance travelled by the solvent from base line (y)}}$

$\frac{(R_f)_A}{(R_f)_C} = \frac{\frac{5}{12.5}}{\frac{10}{12.5}} = 0.5 \Rightarrow = 50 \times 10^{-2} \Rightarrow x = 50$

259. (56)

m = 1 gm, M = 2 M, V = 10 ml

$\% \text{ of N} = \frac{14 \times M \times 2V}{1000} \times \frac{100}{m}$

$= \frac{14 \times 2 \times 2 \times 10}{1000} \times \frac{100}{1}$

= 56%

260. (7)

$R_f = \frac{\text{Distance moved by organic compound}}{\text{Distance moved by solvent}}$

$= \frac{3.5 \text{ cm}}{5 \text{ cm}} = 0.7 \Rightarrow 7 \times 10^{-1}$

261. (4) $-NO_2$, $-C \equiv N$, $-COOH$, $-COR$ are meta directing.

262. (40) mole of AgBr $= \frac{0.376}{188}$

mole of Br^- = mole of AgBr $= \frac{0.376}{188}$

mass of $Br^- = \frac{0.376}{188} \times 80$

$\% \text{ of } Br^- = \frac{0.376 \times 80}{188 \times 0.4} \times 100 = 40\%$

263. (25) More R_f, less its polarity

$R_f = \frac{\text{Distance travelled by compound 'X'}}{\text{Distance travelled by solvent 'Y'}}$

$= \frac{2}{8} = 0.25 = 25 \times 10^{-2}$

264. (5)

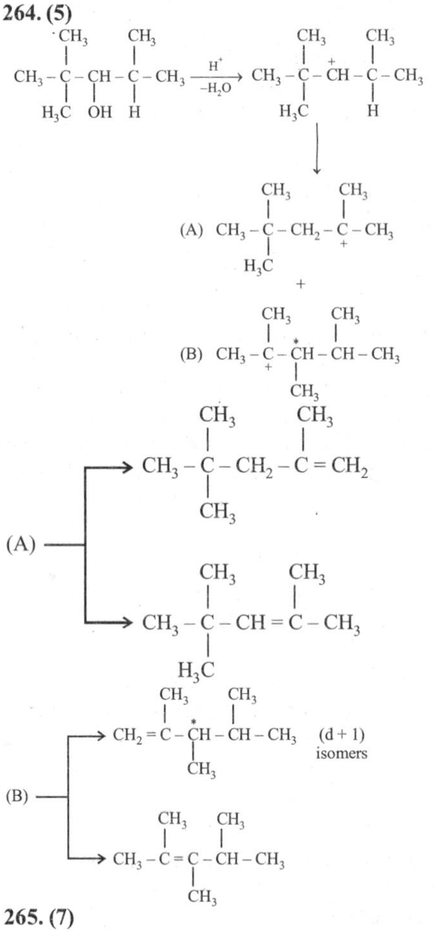

265. (7)

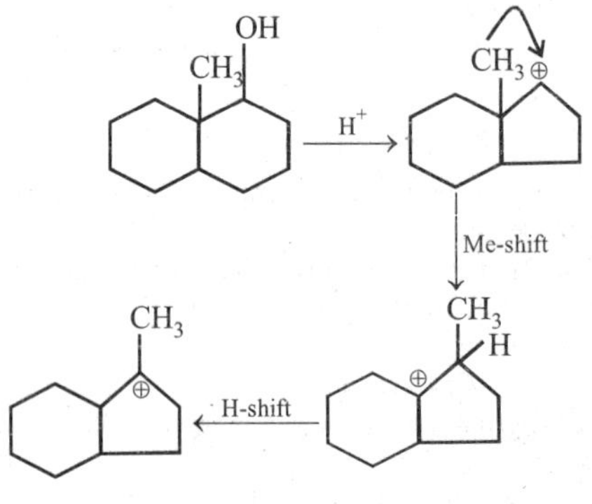

including this structure there are 7 Hyperconjugation structure (Number of

Hyperconjugation structure = no. α – hydrogens +1)

266. (6)

$$R_f = \frac{\text{Distance moved by the substance from base line}}{\text{Distance moved by the solvent from base line}}$$

$$= \frac{3.0\text{ cm}}{5.0\text{ cm}} = 0.6 \text{ or } 6\times10^{-1}$$

267. (42) 233 g of $BaSO_4$ contains 32g of S

$\therefore 1.4439\text{ g } BaSO_4 = \left(\frac{32}{233}\times 1.4439\right)$ g of S

$= \frac{32\times1.4439}{233}$ g of S Obtained from 0.471 g of organic compound.

$$\therefore \%\text{ of 'S'} = \frac{32\times1.4439\times100}{233\times0.471} = 42.10.$$

268. (2) Benzylic and tertiary carbocations are stable.

269. (22) V = 22.78 mL, T = 280 K

$P_{total} = 759$ mmHg

$P_{N_2} = 759 - 14.2 = 744.8$ mmHg

$$n_{N_2} = \frac{744.8\times22.78}{760\times1000\times0.082\times280} = 0.00097$$

$$W_{Nitrogen} = 0.0097\times28 = 0.02716\ (n_{N_2}\times M_{N_2})$$

$$\%N = \frac{\text{Wt. of N}}{\text{Wt. of sample}} = \frac{0.02716}{0.125}\times1000 = 21.728$$

270. (1) Possible structures of acyclic neutral compound is

$$\underset{sp^3}{CH_3} - \underset{sp^2}{CH} = \underset{sp^2}{CH} - \underset{sp}{C} \equiv N \text{ or } \underset{sp^3}{CH_3} - \overset{\overset{CN}{|}}{CH} = CH_2$$

is one.

$\therefore$ No of sp^3 hybridized carbon in either of the compounds.

271. (2)

$$R_f = \frac{\text{distance travelled by the component}}{\text{distance travelled by the solvent}}$$

$$\frac{R_{f_A}}{R_{f_B}} = \frac{\frac{2.08}{3.25}}{\frac{1.05}{3.25}} = \frac{2.08}{1.05} \approx 2$$

272. (30) Overall yield of complete reaction

$$= \frac{60}{100}\times\frac{50}{100} = 0.3 \Rightarrow \%\text{ yield} = 30\%$$

273. (56) The formula used in Kjeldahl's method

$$\%\text{ of 'N'} = \frac{1.4\,(N_1V_1)}{W}$$

N_1 = Normality of acid = 2 × 2 (N)
V_1 = Volume of acid used = 2.5 mL
W = Mass of organic compound = 0.25 g

$$\%\text{ of 'N'} = \frac{1.4\times2.5\times2\times2}{0.25} = 56$$

274. (64) Basicity of $H_2SO_4 = 2$
M_{eq} of H_2SO_4 used by $NH_3 = 12.5\times1\times2 = 25$
% of N in the compound

$$= \frac{25\times10^{-3}\times14\times100}{0.55} = 63.6 \approx 64\%$$

275. (40) Weight of organic compound = 0.25 g
Molar mass of AgCl = 143.5g/mol
143.5 g AgCl contains 35.5g of Cl
0.40 g of AgCl contains

$$= \frac{35.5\times0.4}{143.5} = 0.099\text{ g of Cl}$$

⇒Percentage of Cl in organic compound

$$= \frac{0.099}{0.25}\times100 = 39.6 \approx 40\%$$

276. (3)

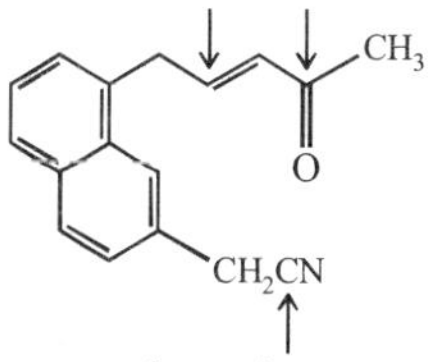

277. (7) $C_2H_7N + \left(2x+\frac{y}{2}\right)CuO \longrightarrow$

$xCO_2 + \frac{y}{2}H_2O + \frac{z}{2}N_2 + \left(2x+\frac{y}{2}\right)Cu$ ...(i)

$C_xH_yN_z + \left(2x+\frac{y}{2}\right)CuO \longrightarrow$

$xCO_2 + \frac{y}{2}H_2O + \frac{z}{2}N_2 + \left(2x+\frac{y}{2}\right)Cu$ ...(ii)

Comparing (i) & (ii), we get : $x = 2, y = 7, z = 1$

278. (1125) Moles of N in N, N -dimethylaminopentane

$$= \left(\frac{57.5}{115}\right) = 0.5\text{ mol}$$

$$C_7H_{17}N + \frac{45}{2}CuO \rightarrow 7CO_2 + \frac{17}{2}H_2O + \frac{1}{2}N_2 + \frac{45}{2}Cu$$

For 0.5 moles of $C_7H_{17}N$ number of moles of CuO required

$$= \frac{1}{2}\left(14+\frac{17}{2}\right) = \frac{45}{4} = 1125\times10^{-2}$$

279. (42) Molecular mass of $BaSO_4$ = 233 g
∵ 233 g $BaSO_4$ contains → 32 g sulphur
∴ 1.44 g $BaSO_4$ contains

$\frac{32}{233}\times1.44$ g sulphur

Given : 0.471 g of organic compound

$$\%\text{ of S} = \frac{32\times1.44}{233\times0.471}\times100 = 41.98\% \approx 42\%$$

280. (19) $\underset{5.0\text{ g}}{\text{Organic compound}} + Cl_2 \longrightarrow A$

$$\underset{0.5\text{ g}}{\text{Compound A}} \xrightarrow{AgNO_3} \underset{0.3849\text{ g}}{AgCl}$$

$$\text{Moles of AgCl} = \text{Moles of Cl} = \frac{0.3849}{143.3}$$

$$\text{Mass of chlorine} = \frac{0.3849}{143.3}\times35.5 = 0.095\text{ g}$$

$$\%\text{ of chlorine in A} = \frac{0.095}{0.5}\times100 = 19\%$$

281. (4) Retardation factor (R_f)

$$= \frac{\text{Distance moved by the substance from base line}}{\text{Distance moved by the solvent from base line}}$$

$$= \frac{2}{5} = 4\times10^{-1}$$

282. (19) Number of moles of

$$N_2 = \frac{758-14}{760}\times\frac{30\times10^{-3}}{0.0821\times287}$$

$$= 1.246\times10^{-3}\text{ mol}$$

Mass of $N_2 = 1.246\times10^{-3}\times28$ g% of N_2

$$= \frac{1.246\times10^{-3}\times28}{0.1840}\times100 = 18.96\% \approx 19\%$$

283. (50)
Mass of organic compound = 1.6 g
Mass of AgBr = 1.88 g

$$\text{Moles of Br} = \text{Moles of AgBr} = \frac{1.88}{188} = 0.01$$

Mass of Br = 0.01 × 80 = 0.80 g

$$\%\text{ of Br} = \frac{0.80\times100}{1.60} = 50\%$$

Part-B (JEE Advanced)

1. **(b)** – CN has highest priority. The nearest locant is at 2nd position.
2. **(b)** Carboxylic acids are named as *oyl chlorides*.
3. **(d)** $CH_3 - \overset{\overset{CH_3}{|}}{CH} - CH_2 - CH_2CH_3$ (Isopropyl group boxed)

   2-Methylpentane
4. **(b)** $\overset{1}{C}H_2 = \overset{2}{C}H - \overset{\overset{CH_3}{3|}}{C}H - \overset{4}{C}H_3$ (3-Methyl-1-butene)
5. **(c)** $\overset{4}{C}H_3 - \overset{\overset{CH_3}{3|}}{\underset{\underset{CH_3}{|}}{C}} - \overset{2}{C}H = \overset{1}{C}H_2$

   3, 3-dimethylbut-1-ene
6. **Butane-1, 4-dioic acid.**
7. **Hyperconjugation.**
8. **vicinal, adjacent or stable, different.**
9. **(a, b)** IUPAC name

   (a) 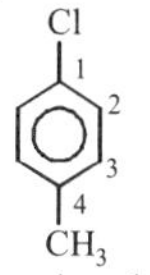
   1-Chloro-4-methylbenzene

   (b) 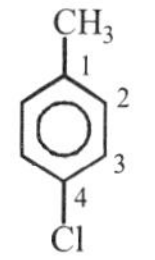
   4-Chlorotoluene
10. (*i*) IUPAC name is

    $H_3C - \underset{\underset{H_3C}{|}}{N} - \overset{\overset{CH_3}{|}}{\underset{\underset{\underset{4}{C}H_2 - \underset{5}{C}H_3}{|}}{\overset{3}{C}}} - \overset{2}{C}H_2 - \overset{1}{C}H_3$

    3–(*N*, *N*-dimethylamino)–3–methylpentane.

    (*ii*) 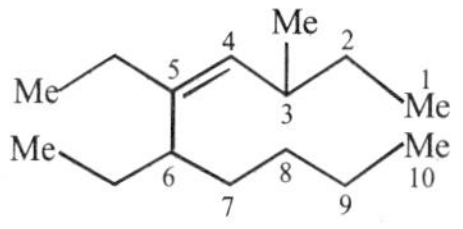
    5, 6-diethyl-3-methyl-4-decene

    (*iii*) $\overset{5}{C}H_3\overset{4}{C}H_2\overset{3}{C}H = \overset{2}{C}H\overset{1}{C}OOH$
    Pent-2-en-1-oic acid

11. $^5CH_3-{}^4\underset{\displaystyle Cl}{\underset{|}{C}}H-{}^3CH={}^2CH-{}^1CH_3$

12. (b)

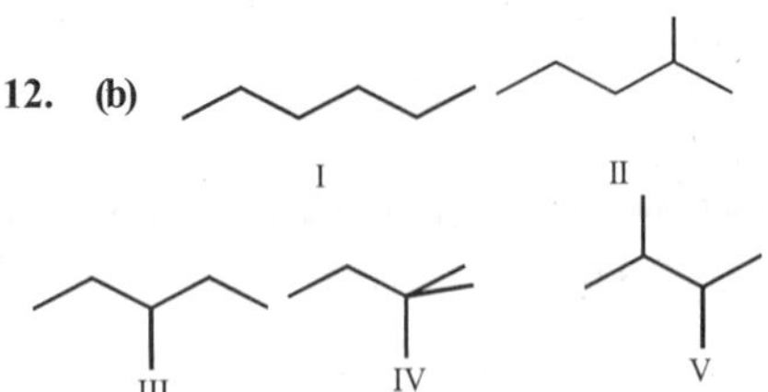

13. (d) Number of isomers (six) can be derived by keeping the position of any one halogen (say Br) fixed and changing the position of the other halogen one by one.

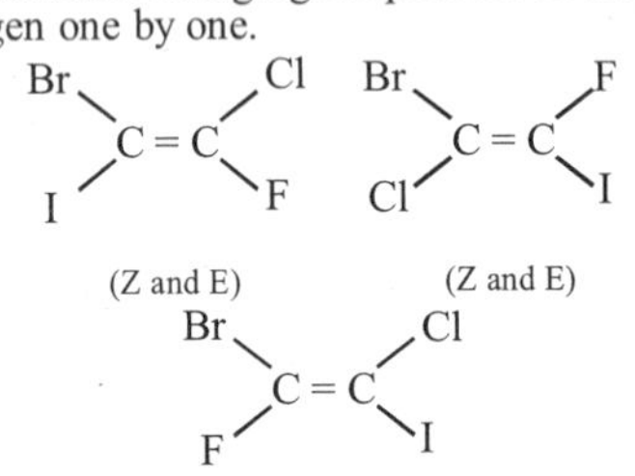

14. (b) A compound which consists of at least one asymmetric carbon atom is capable of showing the phenomenon of optical isomerism. The structure cannot show geometrical isomerism as one of the carbons along the double bond has identical group (methyl). Tautomerism is not possible because of the absence of –CO group.

15. (a) Stereoisomers which are mirror image of each other are *enantiomers* and the one which are not mirror images are *diasteromers*. **Conformation** of the molecule is the spatial arrangement of the atoms of a given molecular structure that are obtained merely by rotation about a sigma bond in the molecule.

16. (d) $\underset{\text{dimethyl ether}}{CH_3-O-CH_3}$ is an isomer of $\underset{\text{ethanol}}{CH_3CH_2OH}$

17. (d) The first three are isomers of diethyl ether, $C_2H_5OC_2H_5$ $(C_4H_{10}O)$.

$\underset{n\text{-Propylmethylether}}{n\text{-}C_3H_7OCH_3}$, $\underset{\text{2-Methylpropan-2-ol}}{CH_3-C(CH_3)(OH)-CH_3}$,

$\underset{\text{Butan-1-ol}}{n\text{-}C_4H_9OH}$, $\underset{\text{Butanone } (C_4H_8O)}{CH_3\overset{O}{\overset{\|}{C}}CH_2CH_3}$

18. (b, c)

(a)

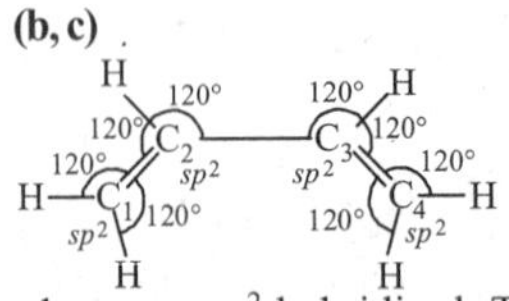

All carbons are sp^2 hybridised. Thus, they will form trigonal planar structure. But the conformer about C2–C3 bond can be in another plane. Hence, the molecule is non-planar in some of its conformations.

(b)

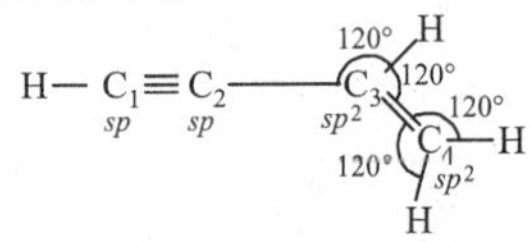

All carbons are either in *sp* or sp^2 hybridisation. Due to the presence of C2–C3 single bond, we can rotate the molecule about this bond. If we rotate the groups attached to C3 carbon, the molecule still remains planar as the C1–C2 part of the molecule is linear. The alkyne group has infinite rotational symmetry along its axis. After rotation the molecular plane will be different but the molecule will always remain in a plane.

(c)

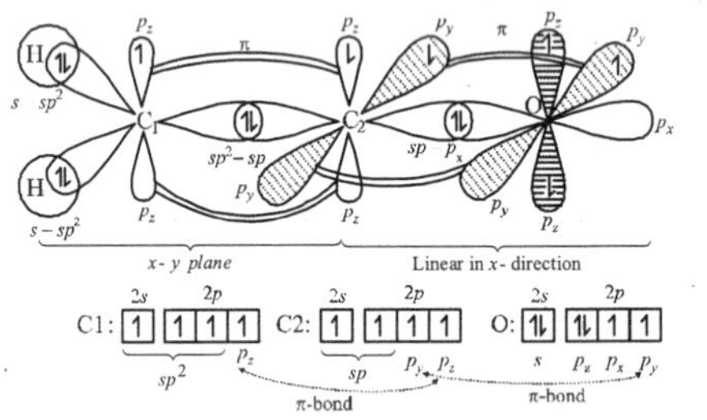

C1 – carbon is sp^2 hybridised. Thus, it forms 3σ-bonds in *x-y* plane. It also forms a π-bond which must be perpendicular to this plane. Therefore, it will use the remaining p_z-orbital to form π-bond with C2-carbon.

C2 – carbon is *sp*-bybridised. Thus,it forms two linear *sp*-orbitals to form two σ-bonds, *i.e.* one with carbon and the other with oxygen. The remaining p_y orbital will form a π-bond with oxygen. Oxygen atom will use its $p_x$orbital to form σ-bond with carbon, *i.e.* $(sp-p_x)$ σ-bond. Its p_y-orbital is used in making π-bond and the p_z-orbital is already full. Hence, the C2 – O part of the molecule is linear in *x*-direction.

Hence, we can say that the whole moleucle is planar.

(d) The given compound is allene, which is non-planar as explained in page no.

19. (b, c, d) E and F ; and also E and G differ in position of atom (H), so these are tautomers (not resonating structures). Geometrical isomers are also diastereomers.

20. (a, c, d)

For a carbonyl compound to show tautomerism, it must have at least one H at the α – carbon atom.

(a)

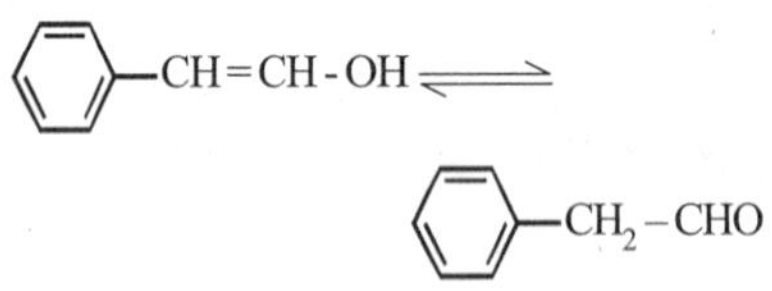

(enol form) (keto form)

(b)

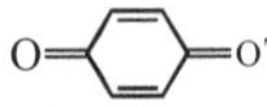

Tautomerism is not possible

(c)

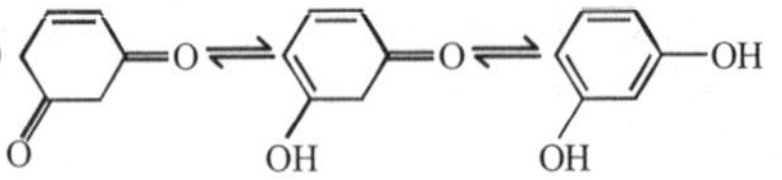

(d)

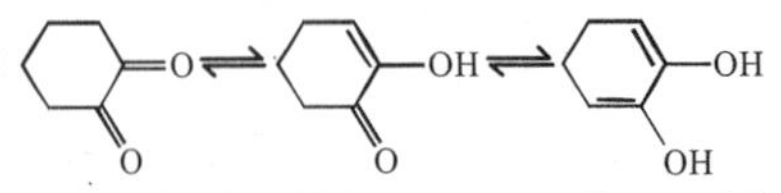

21. (a, d) Option (a) In *n*-butane, Cl can add at either the first or second carbon giving two isomers.

Option (b) : $CH_3-\underset{CH_3}{CH}-CH_2-\underset{CH_3}{CH}-CH_3$ will give three isomers with Cl group at either of the CH_3 groups, second C-atom and third C-atom.

Option (c) Benzene forms only one single derivative.

Option (d) : $CH_3-\underset{CH_3}{CH}-CH_3$ will again give two isomers with Cl at either one of the CH_3 groups or at the central C-atom.

22.

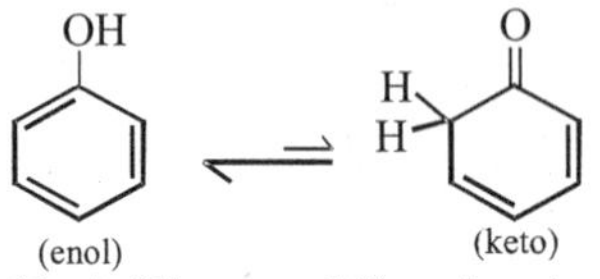

(enol) (keto)

23. (b) Acidity ∝ stability of conjugate Base

In case of compound (B) the conjugate base generated is 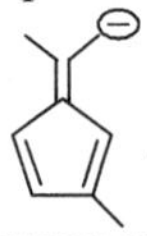 . This conjugate base generated undergoes resonance in which one of the resonating structures generated is 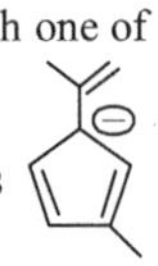 which is an aromatic compound as per Huckel's rule.

24. (d) Heterolytic fission occurs when the two atoms differ considerably in their electronegativities.

O – H bond undergoes cleavage most readily because O and H differ markedly in their electronegativity and further oxygen being highly electronegative can accommodate the negative charge more effectively developed after the cleavage.

25. (a) The bond angle in sp^3, sp^2 and *sp* hybridisad carbon atoms is respectively 109° 28', 120° and 180°.

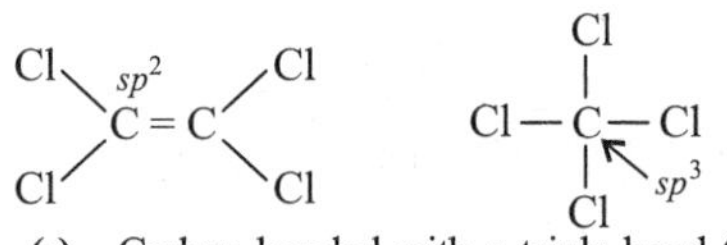

26. (c) Carbon bonded with a triple bond (*i.e.* C_1) is *sp* hybridised. Carbon bonded with a double bond (C_2) is sp^2 hybridised.

27. (c)

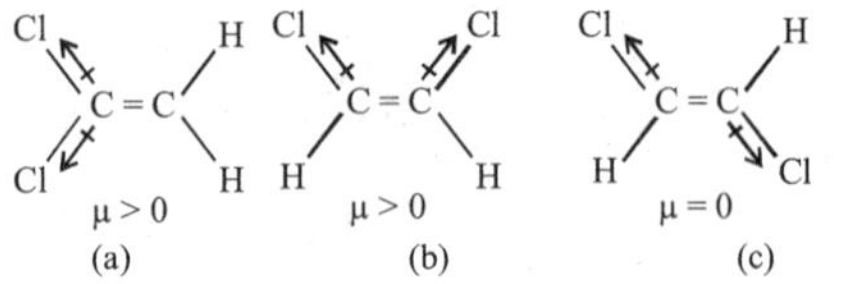

[**Note :** dipole moment is a vector quantity].

28. (d) $\overset{4}{C}H_3-\overset{3}{C}H=\overset{2}{C}=\overset{1}{C}H_2$

Hybridisation in $C_1=sp^2, C_2=sp, C_3=sp^2, C_4=sp^3$.

29. (a) The bond length decreases in the order.

$$\underset{\text{alkane}}{sp^3-sp^3} > \underset{\text{alkene}}{sp^2-sp^2} > \underset{\text{alkyne}}{sp-sp}$$

On the basis of the size of the hybrid orbitals, *sp* orbital should form the shortest and sp^3 orbital the longest bond with other atom.

30. (6)

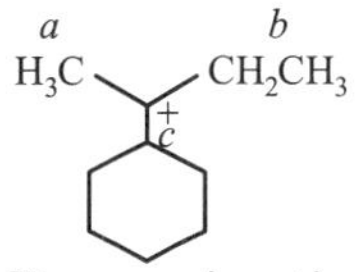

a = 3 Hyperconjugative H's; b = 2 Hyperconjugative H's
c = 1 Hyperconjugative H

31. (280) Nylon – 6,6 is a condensation polymer of Hexamethylene diamine & adipic acid.
So, the reaction is :-

$$nH_2N-(CH_2)_6-NH_2 + nHOOC-(CH_2)_4-COOH$$
Hexamethylenediamine   Adipic acid

$$\xrightarrow{H_2O} (-\underset{|}{N}H-(CH_2)_6-NH-\overset{O}{\overset{\|}{C}}-(CH_2)_4-\overset{O}{\overset{\|}{C}}-)_n$$
Nylon – 6,6

Here, the monomer 'X' is Hexamethylenediamine which gives the positive carbylamine test
Now as per Dumas Method:-

$$C_xH_yN_z + (2x + \frac{y}{2})CuO \rightarrow$$
$$xCO_2 + \frac{y}{2}H_2O + \frac{z}{2}N_2 + (2x + \frac{y}{2})Cu$$

For the monomer 'X' which is Hexamethylenediamine or $C_6H_{16}N_2$ then

$$\frac{z}{2}N_2 = \frac{2}{2}N_2 = 1 \text{ mole}$$

Given 10 mol in this eaction so, amount of N_2 gas evolved is $\Rightarrow 10 \times 28 = 280$ g

32. ***sp*; $Ag-C\equiv C-Ag$**

33. propadiene; $\overset{sp^2}{H_2C} = \overset{sp}{C} = \overset{sp^2}{CH_2}$

34. ***tert*-Butyl carbonium ion** is most stable due to hyperconjugation and +I effect of the three methyl groups.

35. (a, b, c)

(I) Ph–CH(Ph)–Ph, $pk_a = 33.3$
(II) 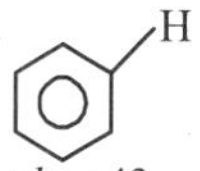 $pk_a = 43$
(III) $H-CH_3$, $pk_a = 50$
(IV) 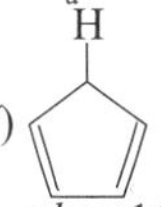 $pk_a = 16$
(V) $CH\equiv CH$, $pk_a = 25$

(a) $Ph-\underset{|}{CH}(Ph)-Ph \rightleftharpoons Ph-\overset{-}{C}(Ph)-Ph + H^+$
I   (Resonance stabilised anion of I)

(b) 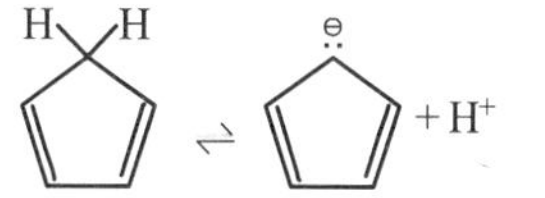
IV   Aromatic

(c) $-NO_2$ is –I group (electron withdrawing group), which increases acid strength of III.
(d) Acid strength order : IV > V > I > II > III

36. (a)

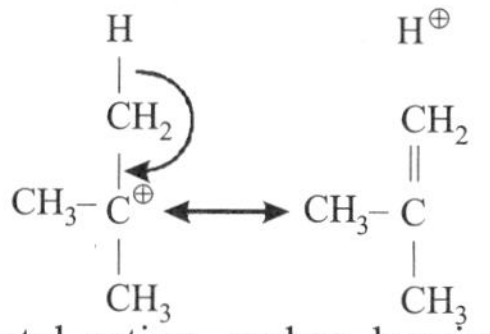

In *tert*- butyl cation, carbon bearing positive charge has one vacant *p*-orbital hence it is σ–p (empty) conjugation or hyperconjugation.

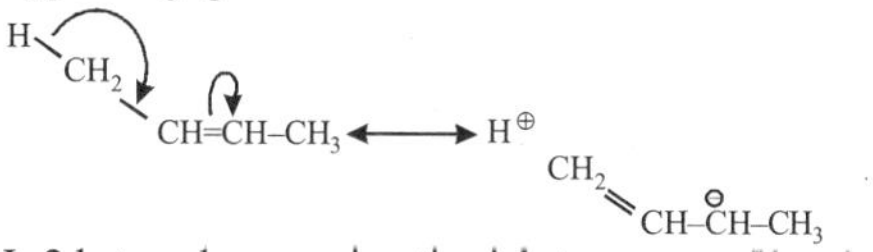

In 2-butene, hyperconjugation is between σ→π* bond.

37. (b, c) (b) and (c) being antiaromatic, are unstable at room temperature.

38. (d) Order of acidic strength
$CH_3OH > CH\equiv CH > C_6H_6 > C_2H_6$; CH_3OH is most acidic because O is more electronegative than C and capable of accommodating negative charge in CH_3O^-. Although alcohols are neutral towards the litmus paper.

39. (a) Conjugate base of strong acid is weak while conjugate base of a weak acid is strong. Acidic strength of acids, of the given conjugate bases $HOH > CH\equiv CH > NH_3 > CH_3-CH_3$. Hence, the order of strength of bases, $CH_3\overline{C}H_2 > \overline{N}H_2 > CH\equiv C^- > OH^-$

40. (b,d) 1, 4-Dichlorobenzene (*p* dichlorobenzene) and *trans*-1, 2-dichloroethene have zero dipole moment because of their symmetrical structures.

41. (a, b, c)
Resonating structures differ in bonding pattern.

42. (i)

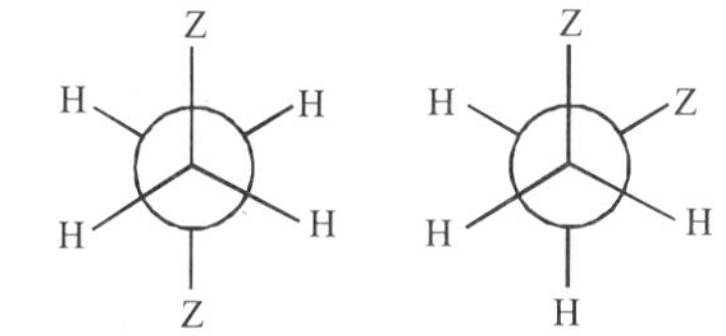

Anti conformer   Gauche conformer

Given, mole fraction of anti conformer = 0.82
∴ mole fraction of gauche conformer = 0.18

$$\mu_{ob.} = \mu_{anti} \times \chi_{anti} + \mu_{gauche} \times \chi_{gauche}$$
$$1 = \mu_{(anti)} \times 0.82 + \mu_{(gauche)} \times 0.18$$
$$1 = 0 \times 0.82 + \mu_{(gauche)} \times 0.18 \quad [\because \mu_{(anti)} = 0]$$
$$\therefore\ 1 = \mu_{(gauche)} \times 0.18$$
$$\mu_{(gauche)} = \frac{1}{0.18} = \mathbf{5.55\ D}$$

(ii)

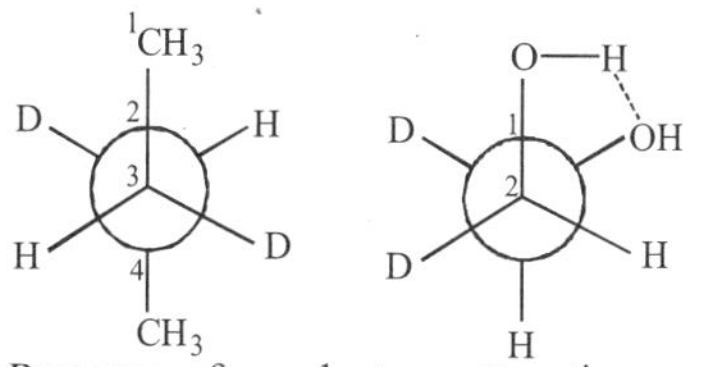

43. Presence of an electron-attracting group increases acidity of the compound. Thus, acidity order:

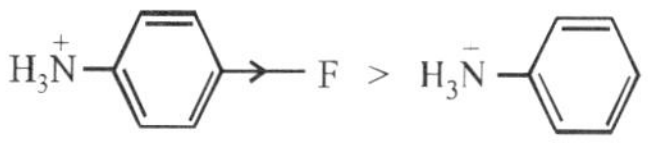

44.

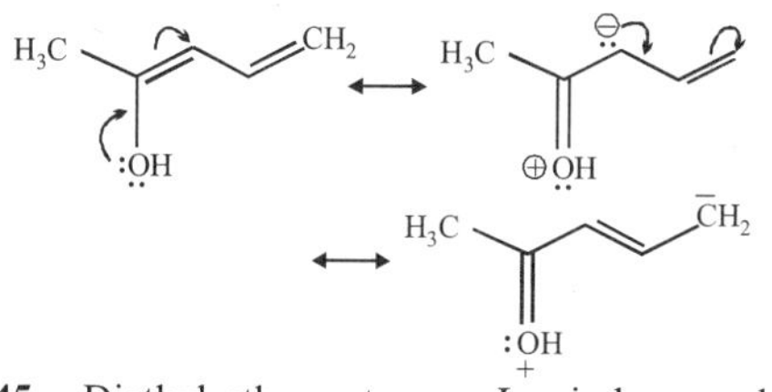

45. Diethyl ether acts as a Lewis base and anhydrous $AlCl_3$ as a Lewis acid.
Anyhydrous $AlCl_3$ is more soluble in diethyl ether because the oxygen atom of ether donates its pair of electrons to the vacant orbital of electron deficient aluminium of $AlCl_3$ through the formation of coordinate bond. In case of hydrated $AlCl_3$, aluminium is not electron deficient as oxygen atom of water molecule has already donated its pair of electrons to meet the electron deficiency of aluminium.

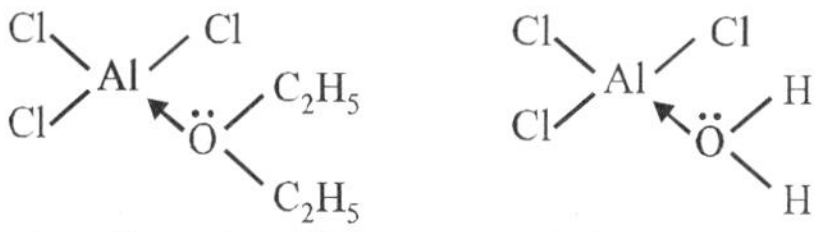

46. (i) $CH\equiv C^-$, C^- is *sp* hybridised and more electronegative than the $\overline{C}H$ of $CH_2=\overline{C}H$ which is sp^2 hybridised. Thus, the former can better accommodate electron pair and hence, less basic.
(ii) In biphenyl, one of the phenyl groups acts as electron donor and the other electron acceptor due to mesomeric effect. This makes it more reactive than benzene.

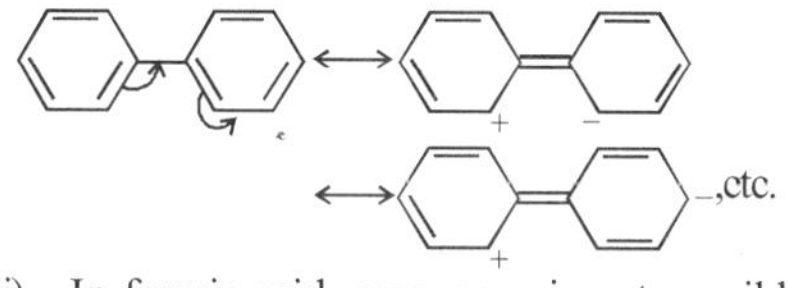

(iii) In formic acid, resonance is not possible, with the result, there are two types of C – O bonds. In sodium formate, resonance is possible, so both of the C – O bonds have same bond length.

Formic acid   Sodium formate (Resonating hybrid)

47.
Isobutane < *n*-Butane < *n*-Butyl chloride < *n* - Butanol
Less van der Waals forces   More   dipole-dipole attraction   H-bonding

Straight chain alkane isomer has higher boiling point than the isomeric branched chain isomer because the former isomer has larger surface area which leads to large van der Waals attractive forces.

48. (i)

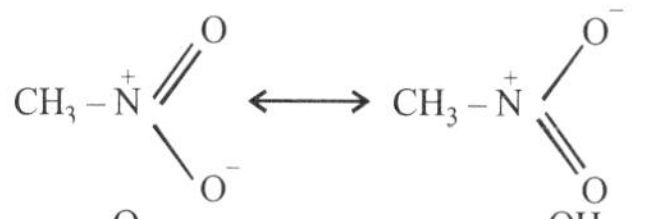

(ii)

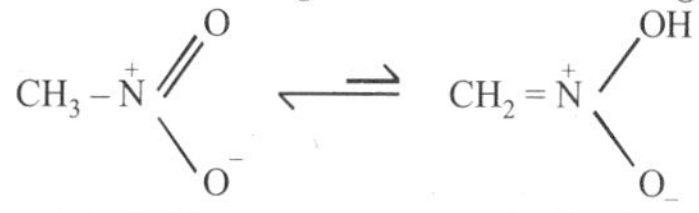

(*nitro* form)   (*aci*-form)

Chapter-13

Hydrocarbons

Part-A (JEE Main)

1. (2)

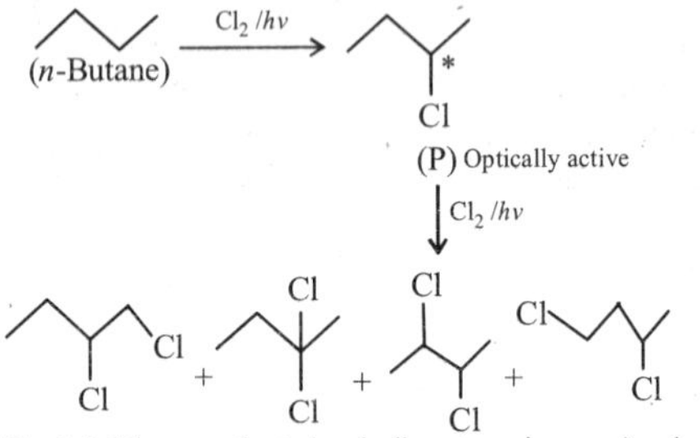

Total dichloro products (excluding stereoisomers) = 4

2. (4) Methane can not be prepared by Kolbe's electrolysis process because it gives alkanes containing even number of carbon atoms.

So. statement-I is false

$$2CH_3-\overset{O}{\overset{\|}{C}}-ONa \xrightarrow{\text{Kolbe electrolysis}} \underset{\text{Ethane}}{CH_3-CH_3}+2CO_2+2NaOH+H_2$$

Methane can not be prepared by Wurtz reaction because it also gives alkanes containing even number of carbon atoms.

$$2R-X+2Na \xrightarrow[\text{ether}]{\text{dry}} R-R+2NaX$$

3. (3) Reactants of the given reactions are :

(A) $CH_4+O_2 \xrightarrow[\Delta]{Mo_2O_3} HCHO$

(B) $2CH_3CH_3+3O_2 \xrightarrow[\Delta]{(CH_3COO)_2Mn} CH_3COOH$

(C)

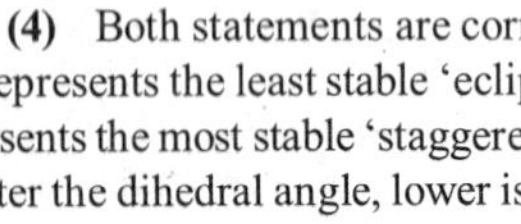

(D) $2CH_4+O_2 \xrightarrow[523K,\ 100\ atm]{Cu} CH_3-OH$

(E)

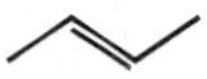

$\xrightarrow{KMnO_4/H^+} 2CH_3COOH$

Only (C) and (D) give alcohol.

∴ Option (3) is correct.

4. (4) Both statements are correct.

(X) represents the least stable 'eclipsed' and (Y) represents the most stable 'staggered' conformer. Greater the dihedral angle, lower is the torsional strain.

5. (2)

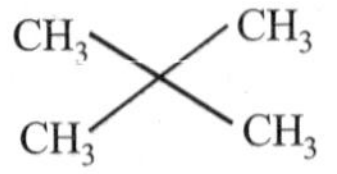

All the hydrogen in neopentane are equivalent so it will form only one monosubstituted derivative.

Melting point :

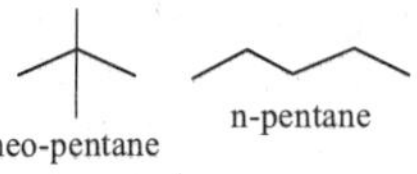

6. (4) $CH_3-\overset{*}{C}H(Cl)-CH_2(Cl) \xrightarrow[hv]{Cl_2}$ "x"

$CH_3-\overset{*}{C}H(Cl)-CH(Cl)-Cl + CH_3-C(Cl)(Cl)-CH_2-Cl + Cl-CH_2-CH(Cl)-CH_2-Cl$

7. (4)

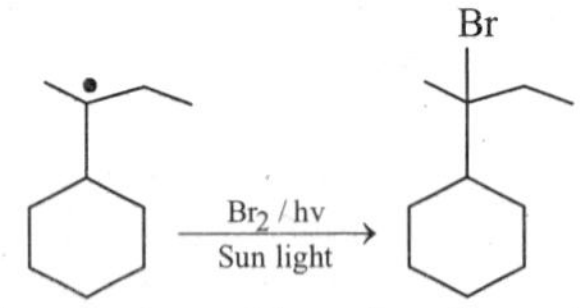

Formation of more stable tertiary free radical.

8. (2) Structure of given compound is

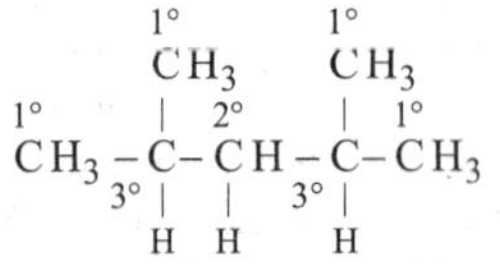

Only one 2° carbon is present in this compound.

9. (3) Eclipsed conformation is the least stable conformation of ethane.

The electron clouds of the C–H bond come closer to each other resulting in increase in electron cloud repulsion, more energy and less stability.

10. (3) n-Alkanes on heating in presence of anhydrous $AlCl_3$ and hydrogen chloride gas isomerise to branched chain alkanes. The major product is *ter*-alkane (most stable).

11. (1) $C_xH_y+\left(x+\frac{y}{4}\right)O_2 \to xCO_2+\frac{y}{2}H_2O$

$\frac{y}{2}=4 \Rightarrow y=8;\ x+\frac{8}{4}=11 \Rightarrow x=9$

∴ Hydrocarbon will be = C_9H_8

12. (1)

$$C_xH_y+\left(x+\frac{y}{4}\right)O_2 \to xCO_2+\frac{y}{2}H_2O$$

$x+\frac{y}{4}=9.5$ ...(i) & $\frac{y}{2}=3$ ...(ii)

By solving equation (i) and (ii) we get.

$x=8, y=6$

∴ The molecular formula of A is C_8H_6.

13. (1) The staggered conformation has lowest van der waal and torsional strain. Hence it must be most stable.

14. (3)

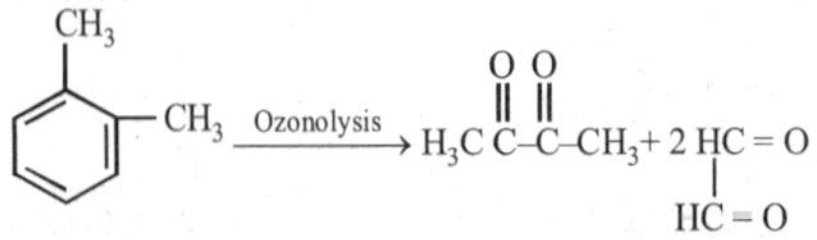

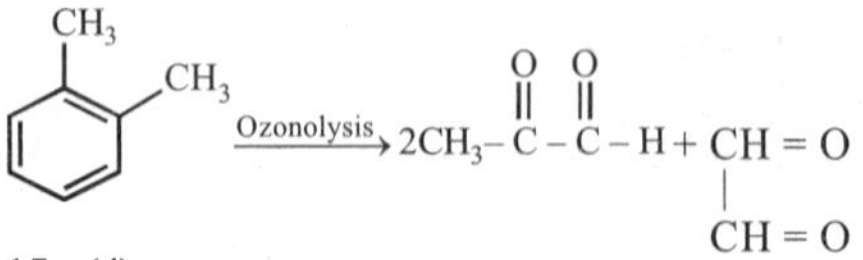

15. (4)

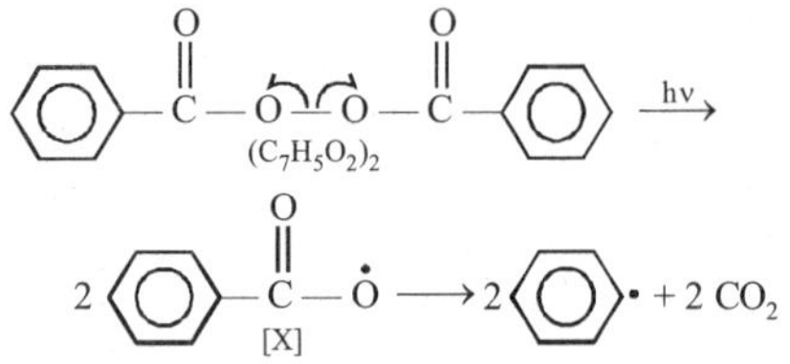

16. (3) Conformations (1) and (3) are staggered, while (2) and (4) are eclipsed. In the anti-conformation (3), the dihedral angle between two methyl groups is largest (180°).

17. (4) In II and IV, both $-CH_3$ groups are at 60°. In IV, there is more repulsion between $-CH_3$ and $-H$ than that in II between –H and –H. Potential energy $\propto \frac{1}{\text{Stability}}$

Stability order : I > III > IV > II

So, Potential energy : II > IV > III > I

18. (2) A dihedral angle is the angle between two C–H bonds projected on a plane orthogonal to the C–C bond. In the given conformation, the dihedral angle H′–C–C–H″ is the sum of $H'-C-C-H_a + H_a-C-C-H''$ i.e. 29 + 120 = 149°

19. (3) Let the hydrocarbon be C_xH_y.

C_xH_y+	$\left(x+\frac{y}{4}\right)O_2 \longrightarrow$	xCO_2+	$\frac{y}{2}H_2O$
25g	(excess)	88g	9g
		2 mol	$\frac{1}{2}$mol

$x=2 \quad y=1$

∵ $x=2$ and $y=1$, the hydrocarbon will be C_2H.

Hence, 2 mol carbon contains 24 g and 1 mol hydrogen contains 1g.

20. (2) Let the hydrocarbon be C_xH_y.

	C_xH_y+	$\left(x+\frac{y}{4}\right)O_2 \longrightarrow$	xCO_2+	$\frac{y}{2}H_2O$
Before combustion:	10 mL	55 mL	0	
After combustion:	0	$55-10\left(x+\frac{y}{4}\right)$	$10x$	

Volume of CO_2, $10x=40$; $x=4$

$55-10\left(x+\frac{y}{4}\right)=0$; $y=6$

∴ Hydrocarbon is C_4H_6

21. (4) In acyclic or open chain compounds, angle strain is absent.

22. (4) The order of substitution in different alkanes is : 3° > 2° > 1°

Thus, the bromination of 2-methylbutane mainly gives 2-bromo - 2 - methylbutane.

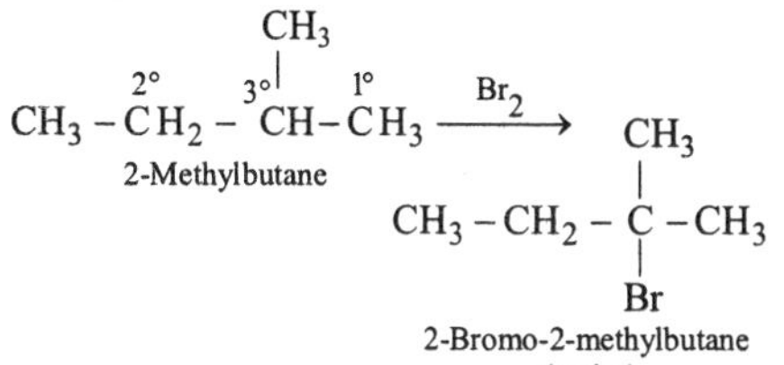

2-Bromo-2-methylbutane (major)

23. (4)

24. (2) In neopentane (molecular mass = 72), all hydrogen atoms are equivalent.

25. (2) Chiral conformation will not have plane of symmetry. Since twist boat does not have plane of symmetry, it is chiral.

26. (1) Due to hydrogen bonding between H and F, gauche conformation is most stable, hence the correct order is Eclipse, Anti, Gauche.

27. (3) 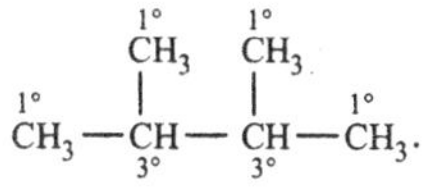

Since it contains only two types of H-atoms, hence it will give only two monochlorinated compounds.

28. (3) In neopentane all H atoms are equivalent (1°).

29. (4)

$$C_nH_{2n+2} + \left[\frac{3n+1}{2}\right]O_2 \rightarrow nCO_2 + (n+1)\,H_2O$$

8 moles of oxygen is required for complete combustion of alkane having 'n' carbon atom

$$\frac{3n+1}{2} = 8 \Rightarrow n = 5$$

∴ Alkane is C_5H_{12} which gives only one monochlorination product. Thus structure will be

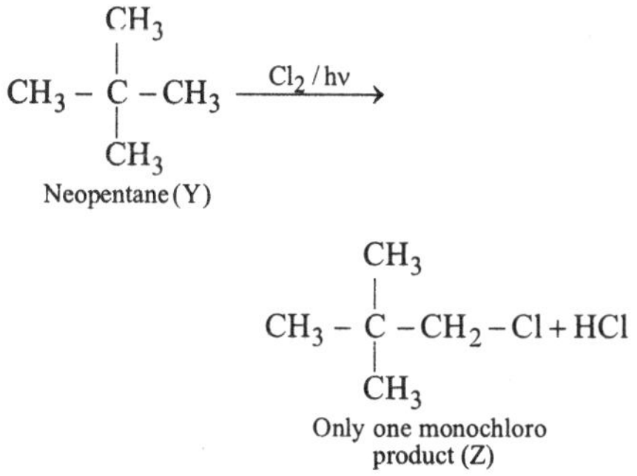

Total number of primary carbon atoms = 4.

30. (30)

$$RMgI \xrightarrow[H_2O]{\text{Ice cold}} R-H + Mg\begin{matrix} I \\ OH \end{matrix}$$

(R–H: Gas)

1.4 L gas weighs 1 g at STP

$$22.4\ \text{L/mol gas weigh} = \frac{1}{1.4\ \text{L g}^{-1}} \times 22.4\ \text{L mol}^{-1}$$

= 16 g/mol.

Hence liberated gas is = CH_4 (Methane)

$$CH_4 + I_2 \xrightarrow{HIO_3} \underset{\boxed{X}}{CH_3-I} + HI$$

$$\underset{\boxed{X}}{2CH_3I} \xrightarrow[\text{(Wurtz reaction)}]{\text{Na/Dry ether}} \underset{\boxed{Y}}{CH_3-CH_3}$$

Molar mass of compound (Y) is = 30 g/mol.

31. (14)

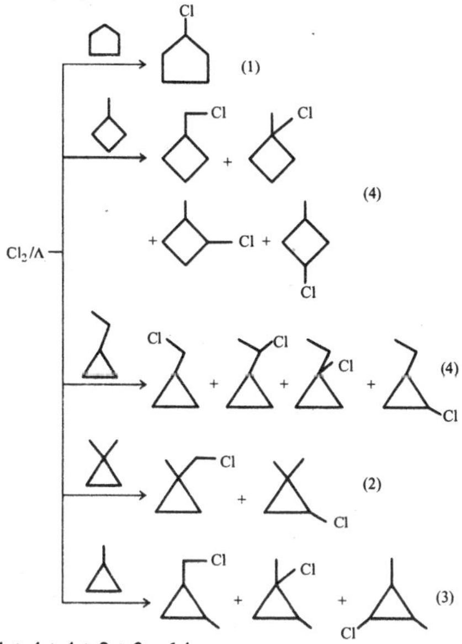

1 + 4 + 4 + 2 + 3 = 14

32. (9) Different chain iosmers of C_7H_{16} are :

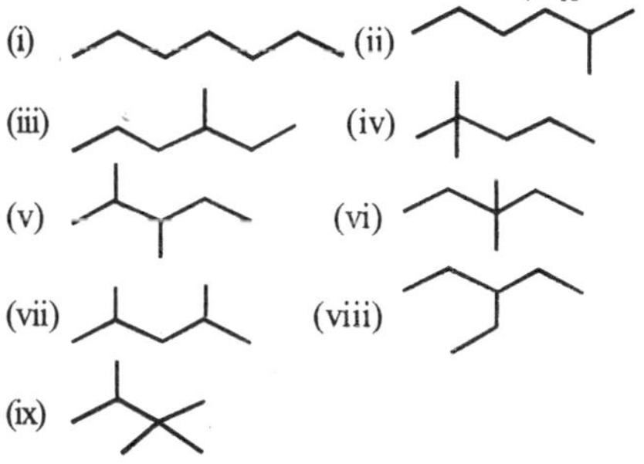

33. (3) $CH_3COONa \xrightarrow{\text{Electrolysis}} \dot{C}H_3$

$C_2H_5COONa \xrightarrow{\text{Electrolysis}} \dot{C}_2H_5$

$2\dot{C}_2H_5 \longrightarrow CH_3-CH_2-CH_2-CH_3$ (1)

$2\dot{C}H_3 \longrightarrow CH_3-CH_3$ (2)

$\dot{C}H_3 + \dot{C}_2H_5 \longrightarrow CH_3-CH_2-CH_3$ (3)

34. (6)

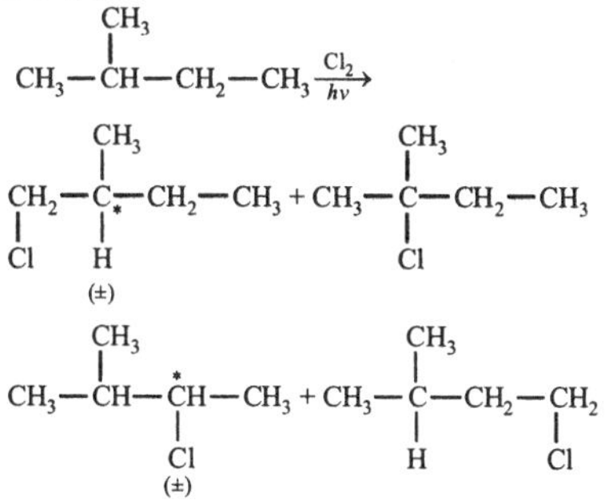

35. (9) $CH_3-CH_3 + Br_2\ (\text{Excess}) \xrightarrow{hv}$

Monobromo, Dibromo, Tribromo

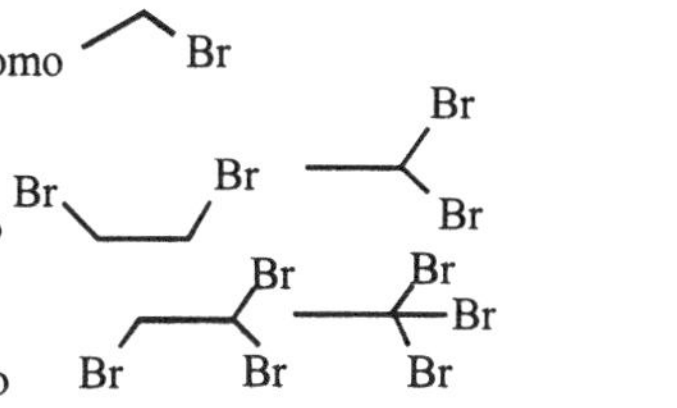

Tretrabromo, Pentabromo, Hexabromo

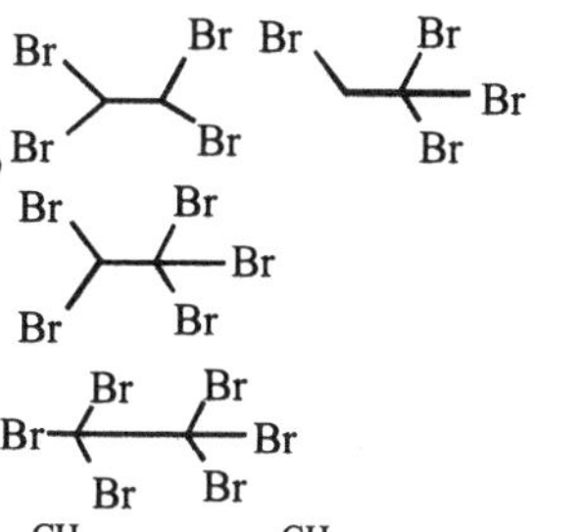

36. (3)

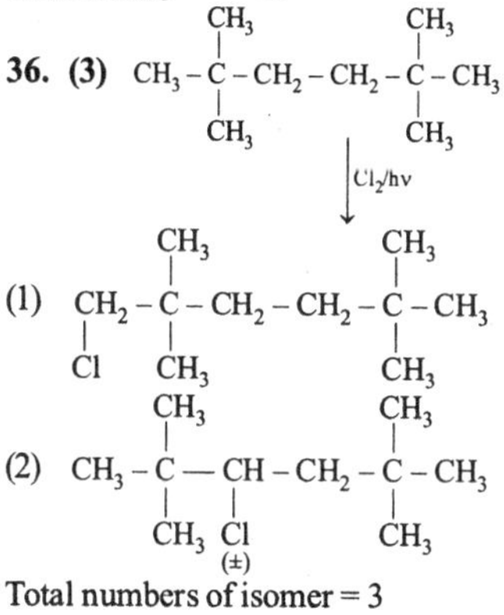

Total numbers of isomer = 3

37. (12)

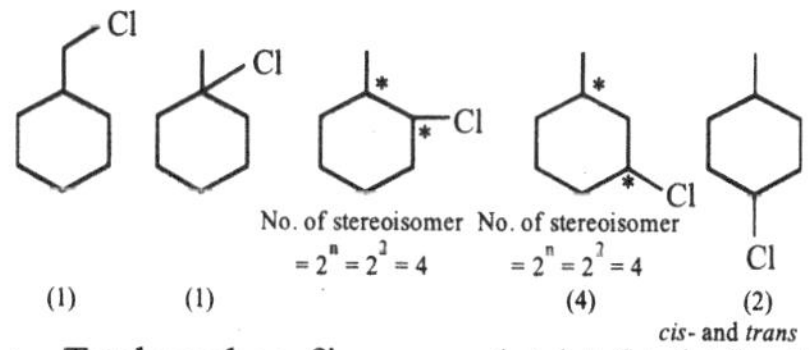

⇒ Total number of isomers = 1 + 1 + 4 + 4 + 2 = 12

38. (8) Number of monohalogen derivatives are equal to the number of chemically different hydrogen atoms present in the compound. Possible number of structures for C_5H_{12} are-

(i) $\overset{a}{C}H_3-\overset{b}{C}H_2-\overset{c}{C}H_2-\overset{b}{C}H_2-\overset{a}{C}H_3$

3-types of hydrogen ∴ 3-*Products*

(ii) $\overset{a}{C}H_3-\overset{b}{C}H(\overset{a}{C}H_3)-\overset{c}{C}H_2-\overset{d}{C}H_3$

4-types of hydrogen ∴ 4-*Products*

(iii) $CH_3-C(CH_3)_2-CH_3$ (all a)

1-type of hydrogen ∴ 1-*Product*

∴ Total number of monobromo derivatives in the hydrocarbon having molecular formula C_5H_{12} = 3 + 4 + 1 = 8

39. (60) Staggered form is produced when the rearrangement of atoms or group takes place by an angle of 60°. 1, 1, 1 trichloroethane (CCl_3-CH_3) in eclipsed form on rotation by 60° gives staggered form.

40. (4)

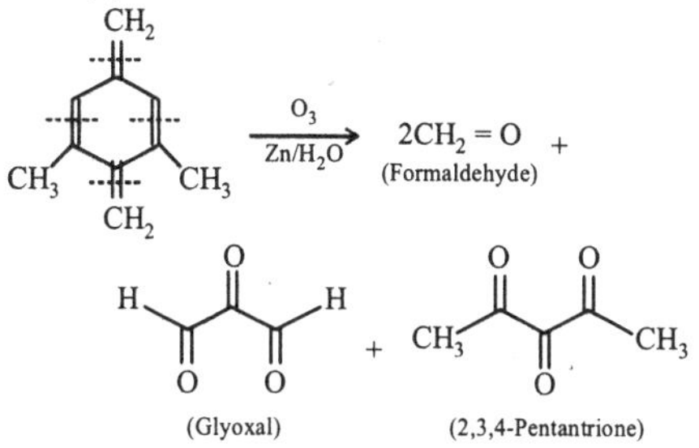

41. (4) $Ph-CH=CH_2 \xrightarrow[HBr]{(PhCOO)_2} Ph-CH_2-CH_2-Br$

- Anti Markovnikov addition
- Reaction follows free radical addition in presence of peroxide
- In absence of peroxide, it will follow carbocation mechanism
- Benzene is formed as by-product

$Ph-\overset{O}{\overset{||}{C}}-O-O-\overset{O}{\overset{||}{C}}-Ph \rightarrow 2Ph-\overset{O}{\overset{||}{C}}-\dot{O}$

$Ph-\overset{O}{\overset{||}{C}}-\dot{O} \longrightarrow \dot{P}h + CO_2$

$\dot{P}h - HBr \longrightarrow \underset{Benzene}{Ph-H} + \dot{B}r$

$Ph-CH=CH_2 + \dot{B}r \longrightarrow Ph-\dot{C}H-CH_2-Br$ (2°, Free radical (more stable)) $\xrightarrow{HBr} Ph-CH_2-CH_2-Br$ (Main product)

42. (3)

(i) Stability order can be explained on the basis of hyperconjugation; more the number of α–H in an alkene, more will be hyperconjugative structures and hence higher will be the stability of that alkene.

(ii) In case, when the alkene shows geometrical isomerism, *trans*-isomer will be more stable than the *cis*-isomer (steric factor).

Thus stability order is

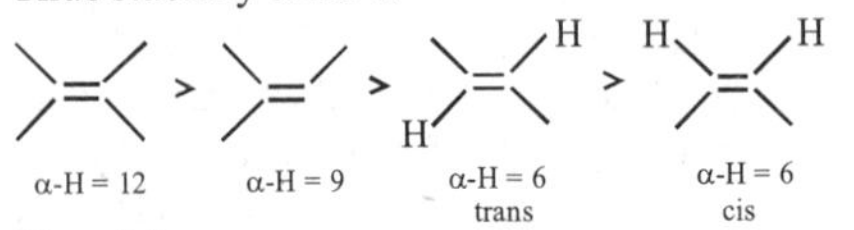

43. (2)

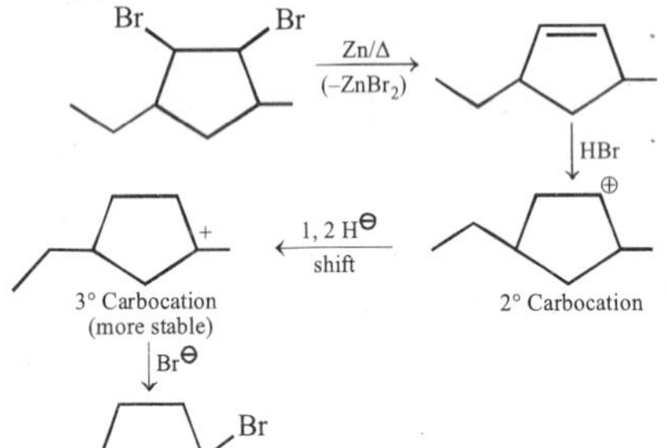

44. (2)

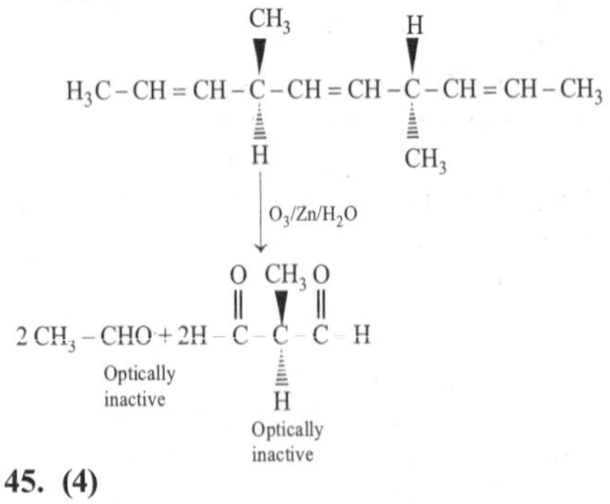

45. (4)

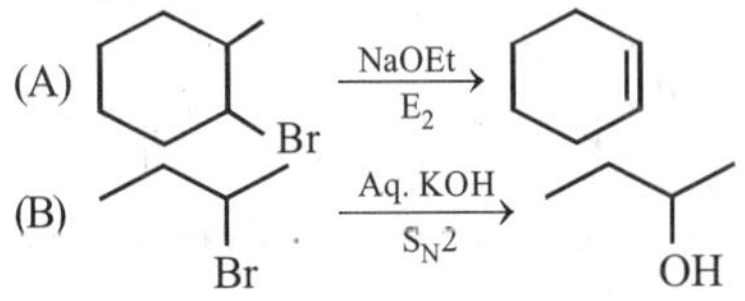

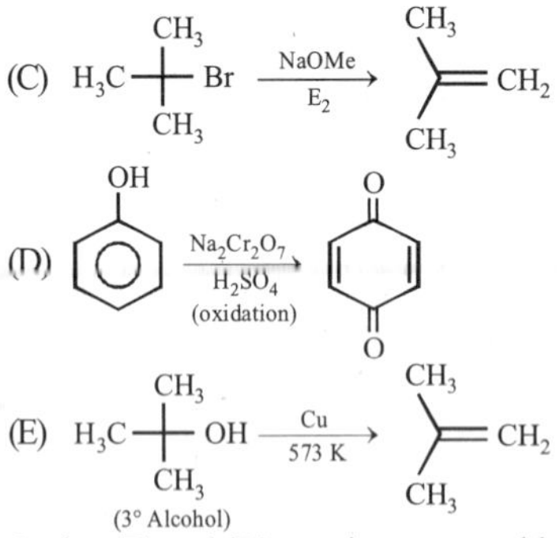

Option (B) and (D) reaction are not able to form alkene as a product.

46. (3)

Statement-I :

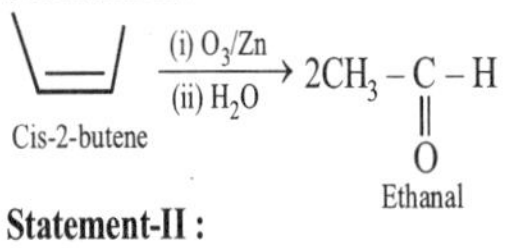

Statement-II :

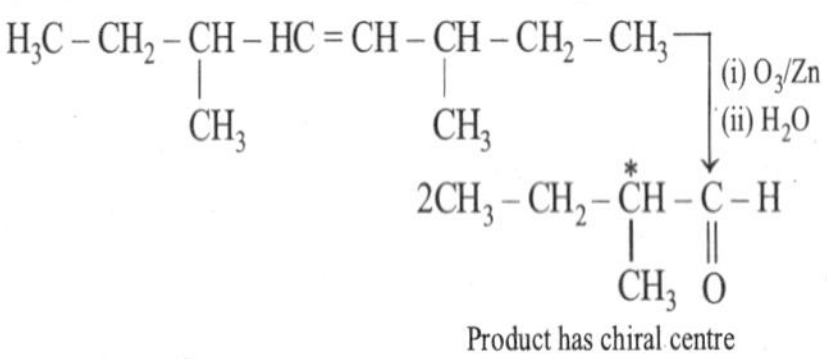

47. (2)

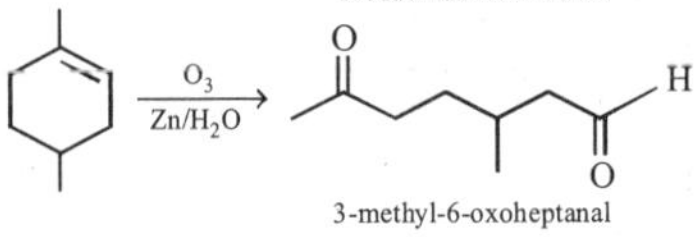

48. (2 or 4)

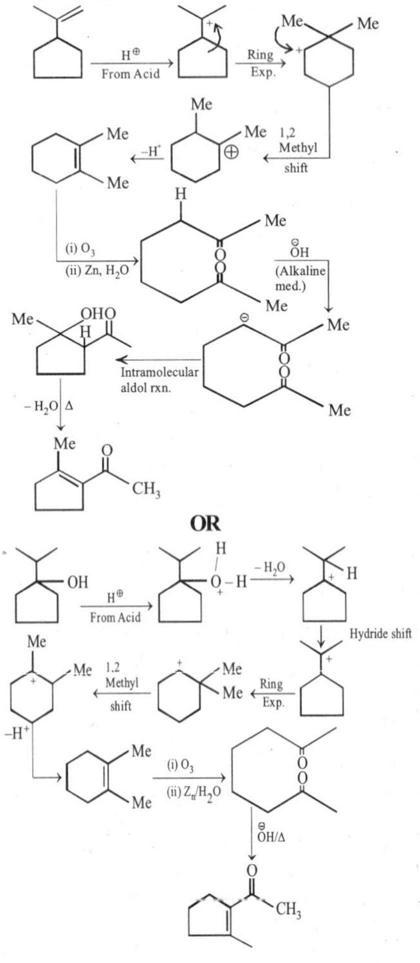

Note: In question about molecule "p" is not clear whether it is an alcohol or alkene and according to question rearrangement product is asked hence answer is either (b) or (d). Alkene also undergoes rearrangement in presence of acid and so option (b) also fulfills all conditions.

49. (2) Addition of H-Br(aq) to alkene follows electrophilic addition mechanism. In the rate determining step a carbocation intermediate is formed. Among P, Q, R & S compound Q will form most stable carbocation intermediate since it is resonance stabilized and so Q reacts fastest.

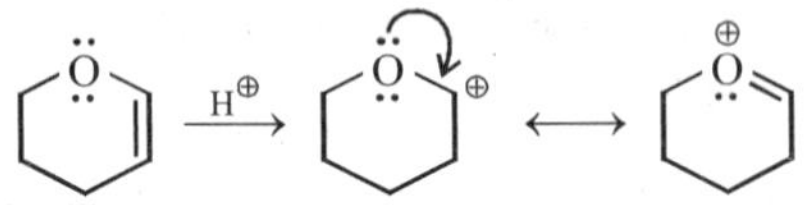

50. (2)

51. (1) (A) $CH_3-CH=CH_2$. Geometric isomer is not possible.

(B) Trans isomer has identical atoms/groups on the opposite side of double bond.

(C) 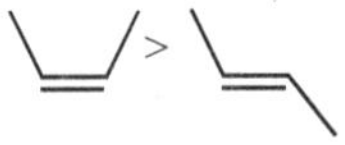 (dipole moment)

(D) $H_3C-\underset{CH_3}{\underset{|}{C}}=CH-CH_3$ (does not show GI)

2 – methylbut – 2 – ene

(E)  (Melting point due to symmetry)

52. (4) Ethylene has 5σ bonds and 1π bond.

53. (3) In *cis*-alkenes, the two bulky groups are on the same side, hence faces more repulsion leading to its unstability than the *trans*-alkenes. Dipole moment is a vector quantity and for a molecule net dipole moment is the vector sum of all bond dipoles hence dipole moment of *Cis* form is greater than *Trans* form.

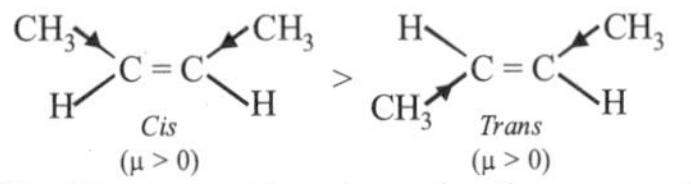

54. (1) First reaction is reductive ozonolysis and second is iodoform reaction.

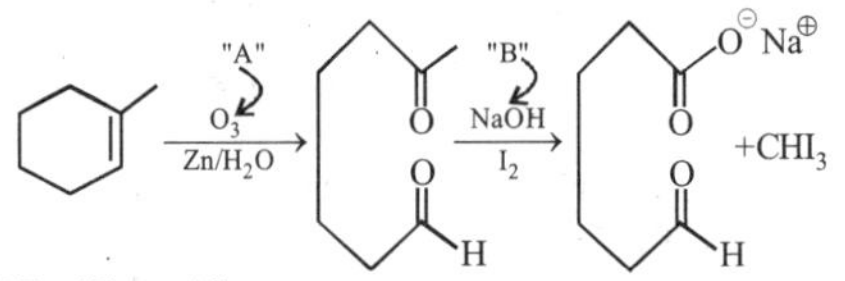

55. (3) or (4)

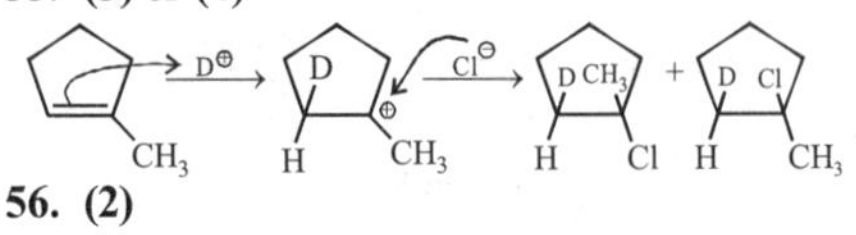

56. (2)

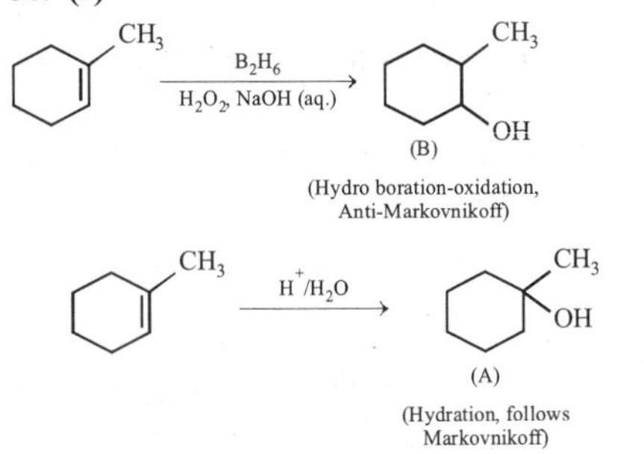

57. (2)

58. (4)

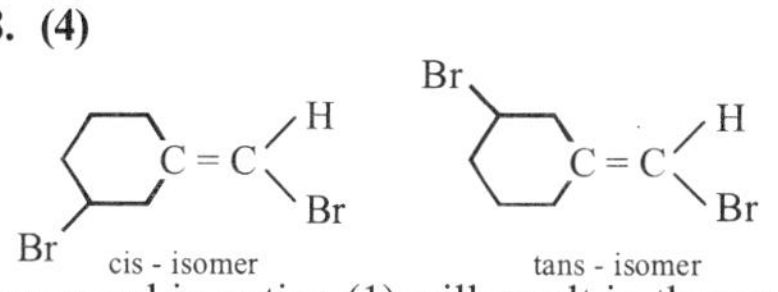

Compound in option (1) will result in the same compound when rotation along the C = C double bond is performed (formation of two such isomers, not an actual rotation along the double bond). Same will be for compound in (4).

For (3), we cannot decide the cis- and trans-isomers as these are two H-atoms on one doubly-bonded carbon atom.

59. (1)

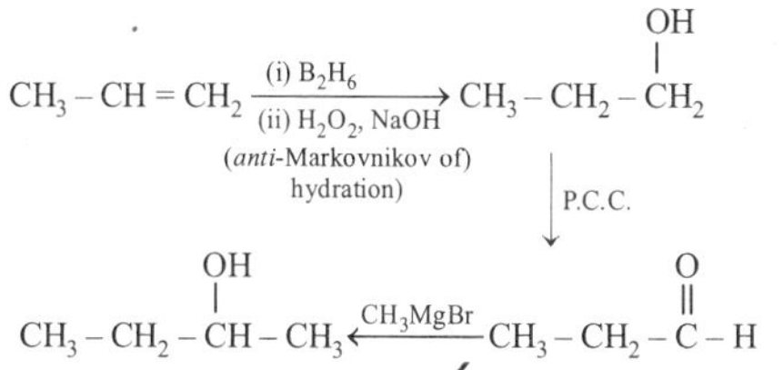

60. (1) 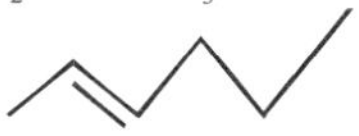

$\xrightarrow[\text{(ii) } H_2O]{\text{(i) } O_3} CH_3COOH + CH_3CH_2CH_2COOH$
Acetic acid   Butanoic acid

It is oxidative ozonolysis.

61. (2)

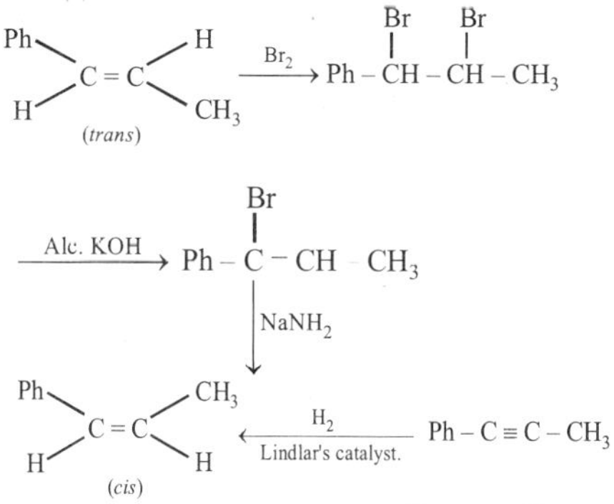

62. (4)

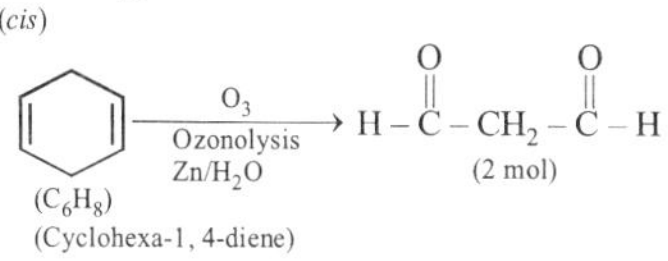

63. (1)

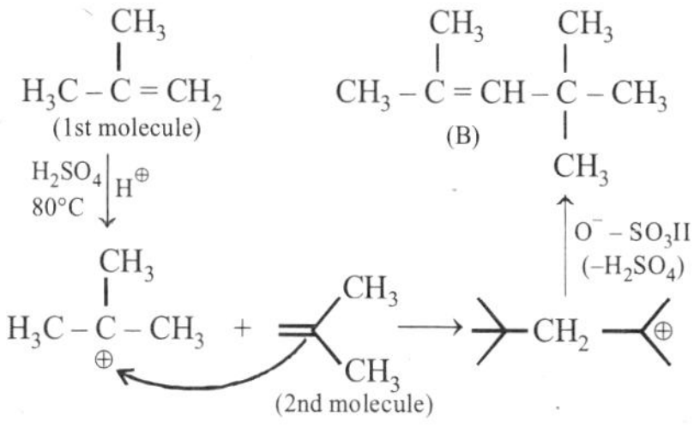

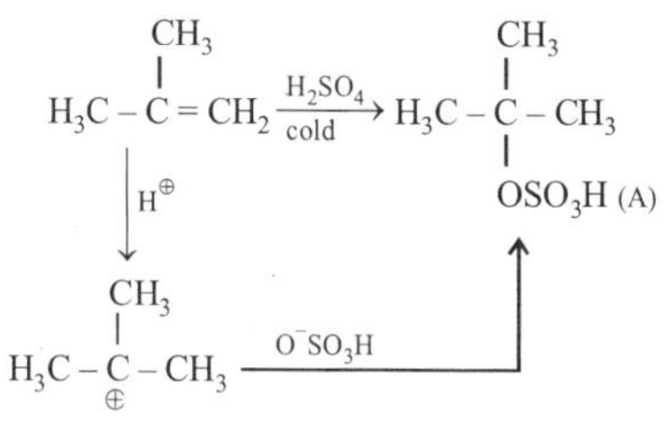

64. (3)

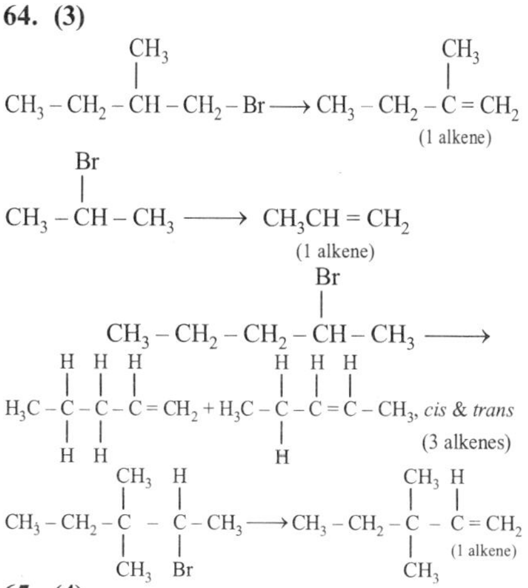

65. (4)

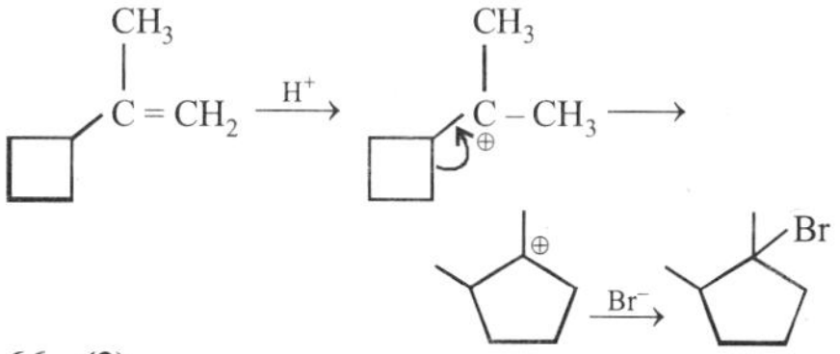

66. (2)

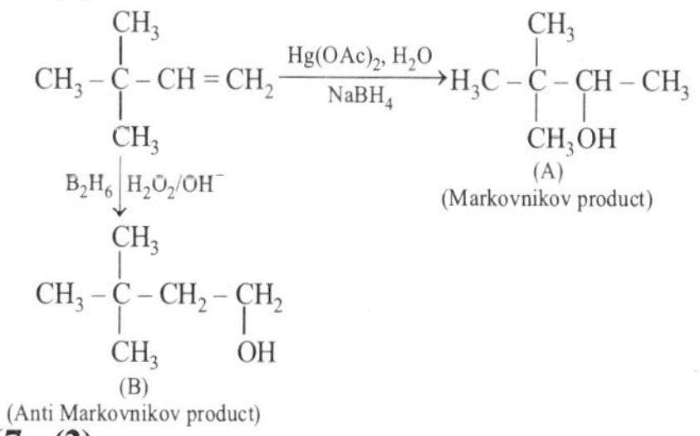

67. (2)

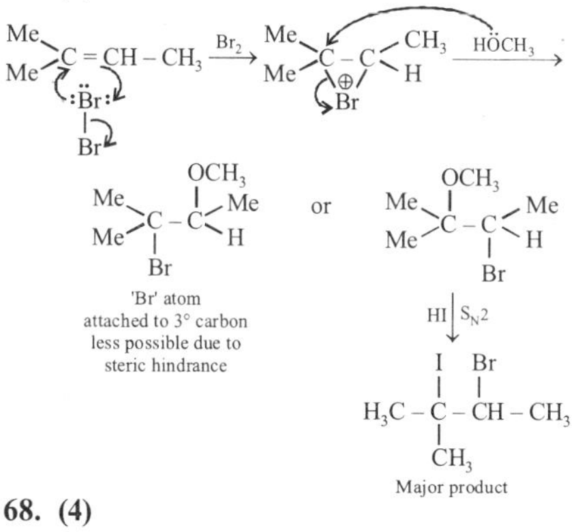

68. (4)

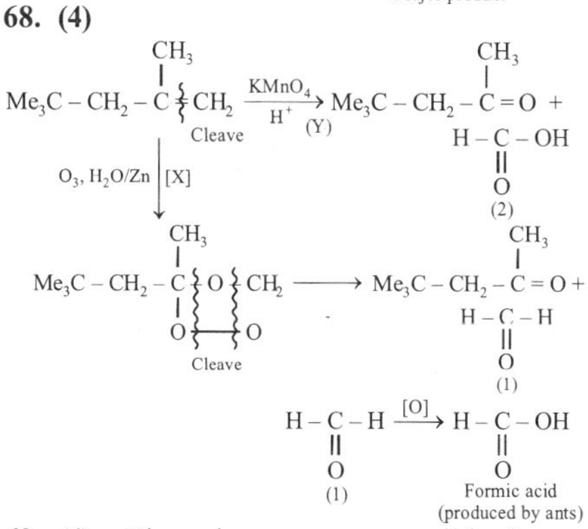

69. (4) Three isomers are possible for an oxidisable molecule having molecular formula C_4H_8.

$CH_3CH=CHCH_3$ (But-2-ene) $CH_3CH_2CH=CH_2$ (But-1-ene) $CH_3-\underset{}{C}(CH_3)=CH_2$ (2-Methylpropene)

Oxidation with $KMnO_4$ to give a ketone along with evolution of effervescences is possibly only with 2- methylpropene

$\underset{(A)}{CH_3-C(CH_3)=CH_2} \xrightarrow{KMnO_4/H^+} CH_3-C(CH_3)=O + CO_2\uparrow$

Thus isomer A should be 2 – methylpropene.

70. (4)

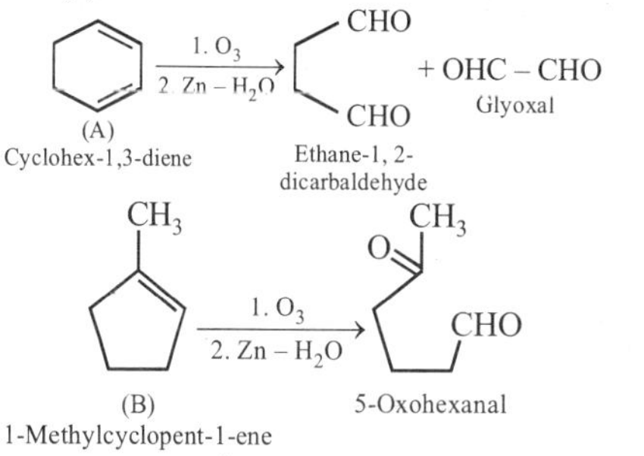

71. (2)

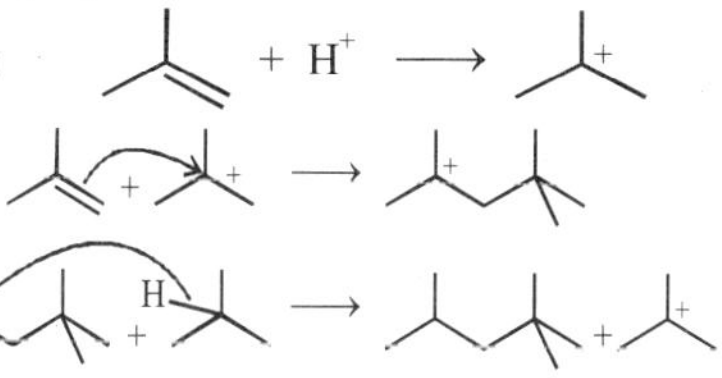

72. (1) Both the statements are correct.

73. (1)

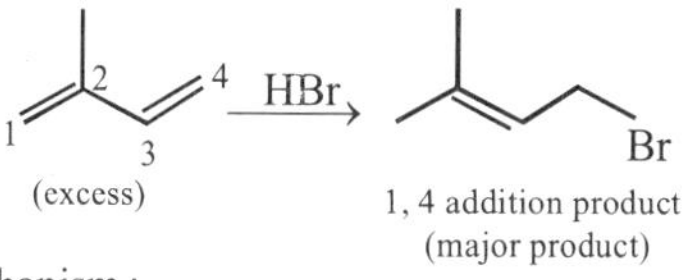

Mechanism :

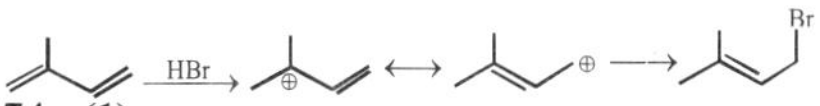

74. (1)

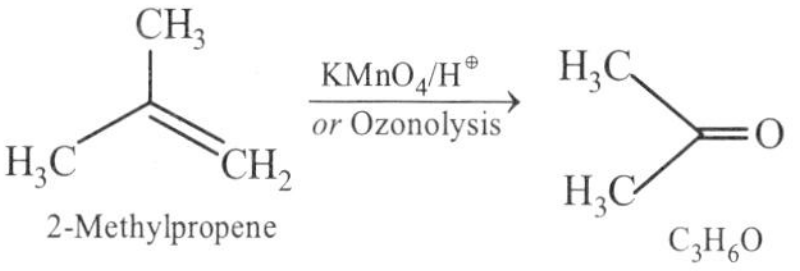

75. (3)

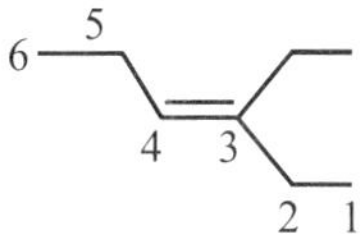

76. (4)

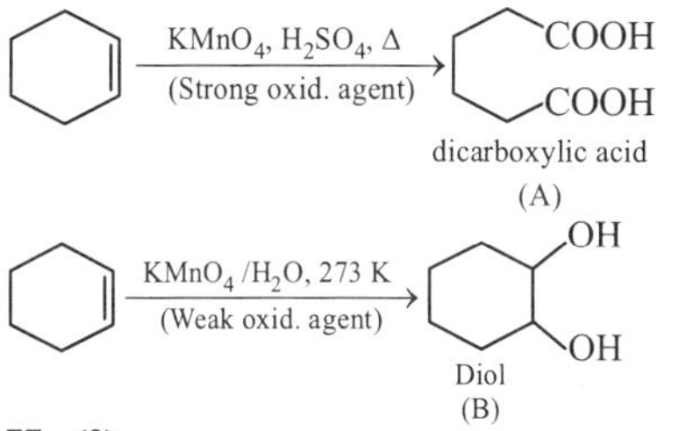

77. (3)

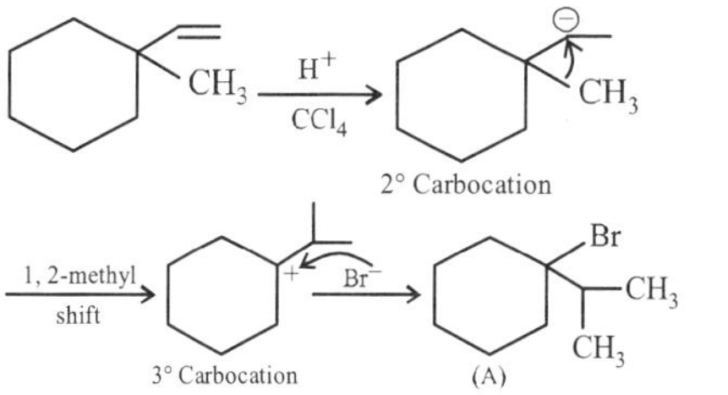

78. (4) Carbocation B is more stable as it is secondary carbocation having more number of α-hydrogens and having greater +I effect.

∴ Carbocation B formed at a faster rate than carbocation A.

79. (3)

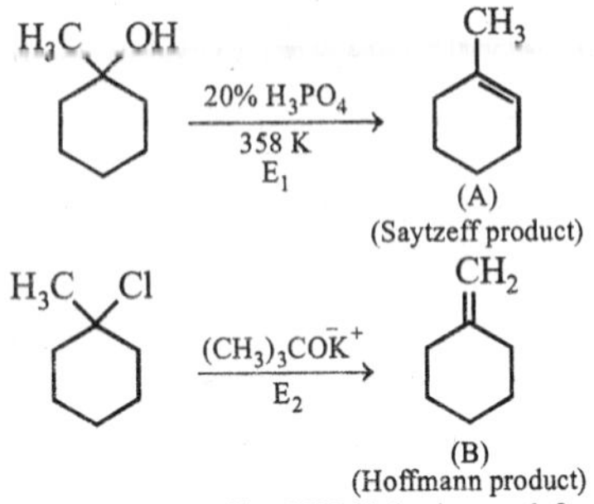

80. (2) The reagent, Cl_2; UV light is used for free radical allylic substitution.

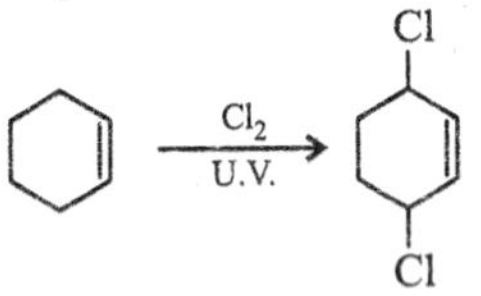

81. (2) $\underset{sp^2}{CH_2} = \underset{sp}{C} = \underset{sp^2}{CH} - \underset{sp^3}{CH_3}$

82. (4)

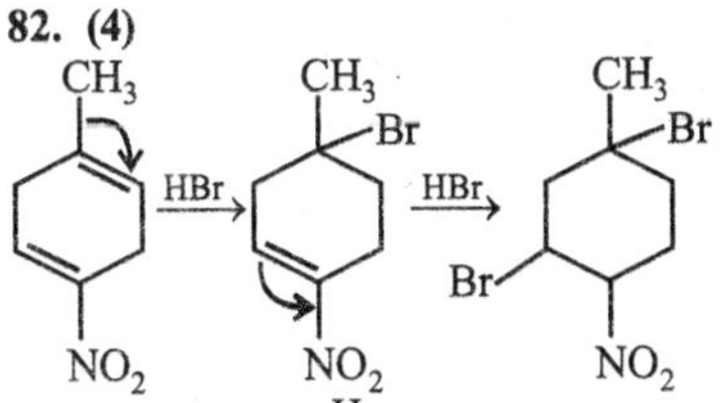

83. (2)

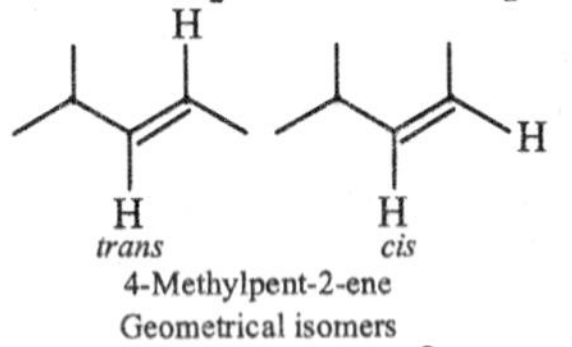

84. (1)

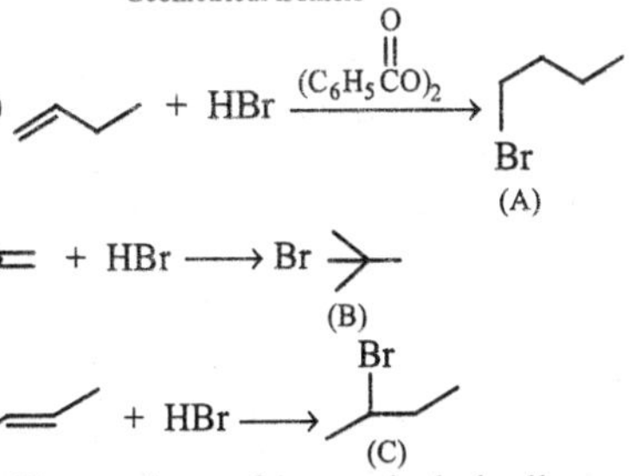

The boiling points of isomeric haloalkanes decrease with increase in branching.

So order of B.P. is : A > C > B.

85. (4) $CH_3CH = CHCH(CH_3)_2 \xrightarrow{H^+} CH_3CH_2-\overset{+}{C}H-\underset{CH_3}{\underset{|}{C}HCH_3} \xrightarrow{1,2\text{-}\bar{H}\text{ shift}}$

2° carbocation

$CH_3CH_2-CH_2-\underset{CH_3}{\underset{|}{\overset{+}{C}CH_3}} \xrightarrow{Br^-}$

3° carbocation

$CH_3CH_2-CH_2-\underset{CH_3}{\underset{|}{\overset{Br}{\overset{|}{C}CH_3}}}$

86. (2, 3)

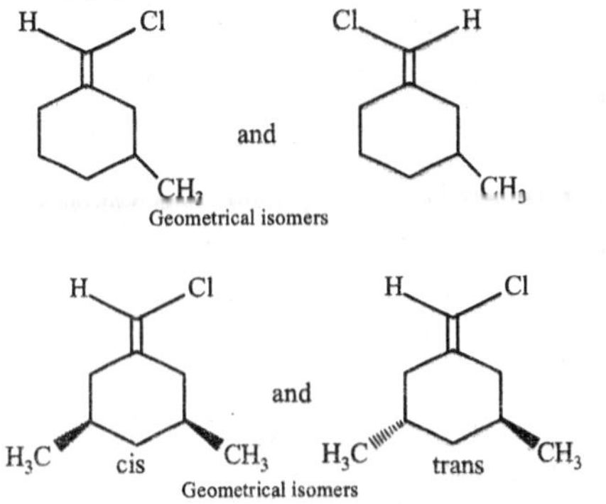

87. (4)

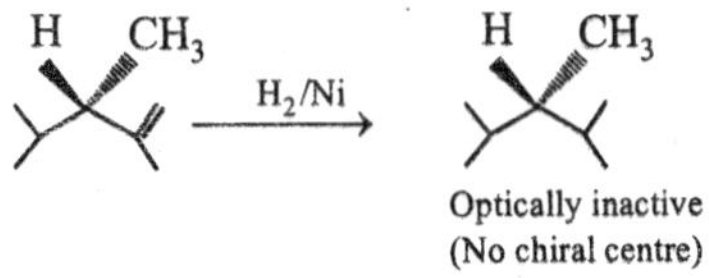

88. (2)

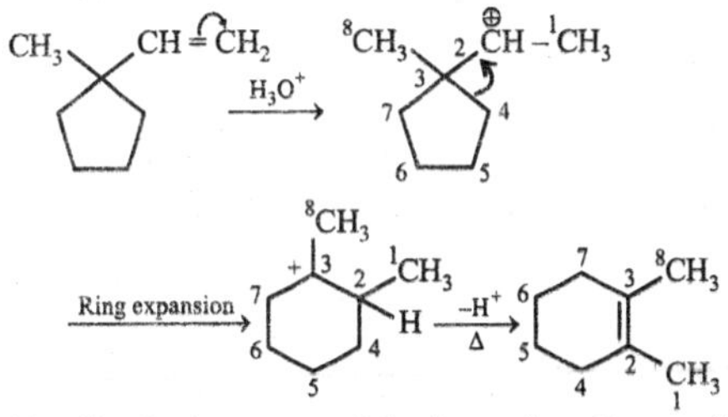

89. (1) In isomers of hydrocarbon heat of combustion is inversely proportional to the stability. Between two geometrical isomers *trans* isomer is more stable than *cis*.

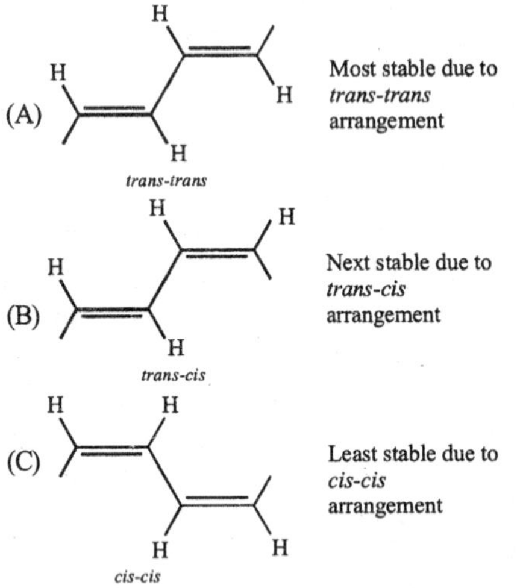

90. (2) In Cl–CH=CH–NO_2 double bond character in carbon-chlorine bond is maximum due to resonance and so the bond length is shortest.

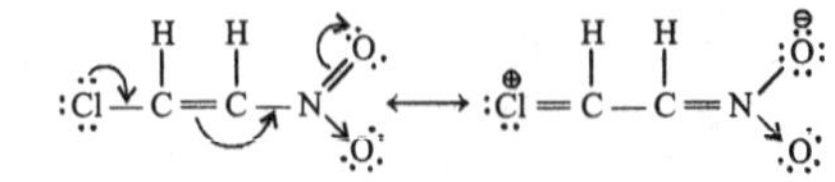

91. (4)

$(CH_3)_2CH-CH=CH_2 \xrightarrow{HCl} (CH_3)_2CH-\overset{+}{C}H-CH_3$ (Cl)

$\xrightarrow{\text{Hydride shift}} CH_3-\underset{CH_3}{\underset{|}{\overset{+}{C}}}-CH_2-CH_3 \rightarrow CH_3-\underset{CH_3}{\underset{|}{\overset{Cl}{\overset{|}{C}}}}-CH_2-CH_3$

Carbocation (planar) (±)

There is no chiral carbon in the product of this reaction.

92. (2)

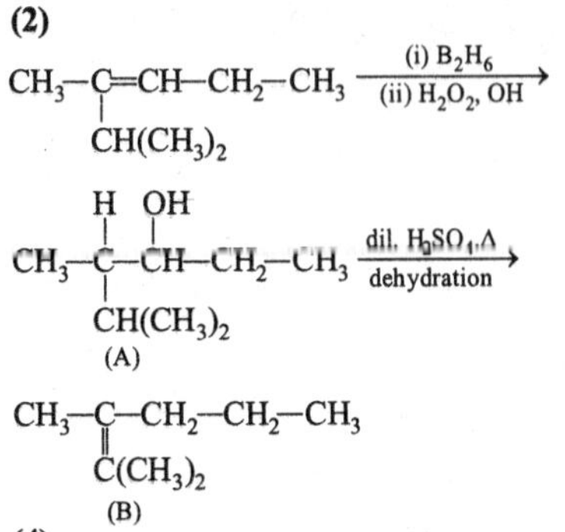

93. (4)

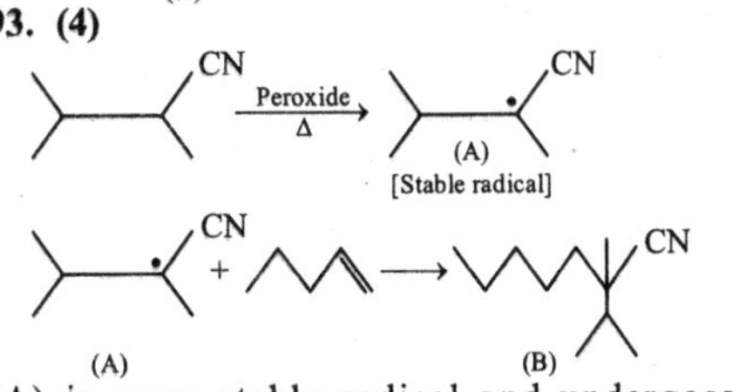

(A) is more stable radical and undergoes Markovnikov addition to form (B).

94. (2)

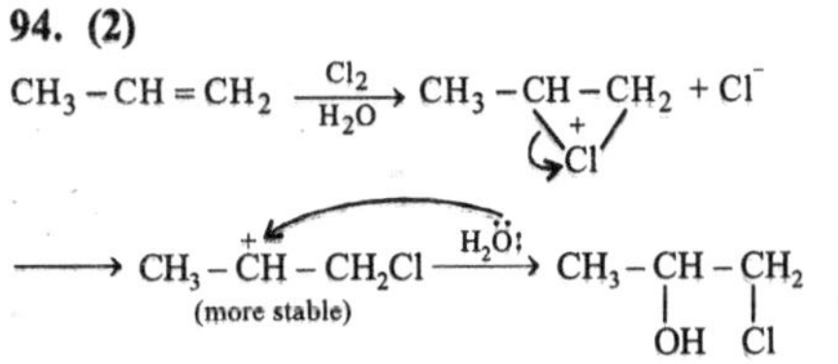

95. (3)

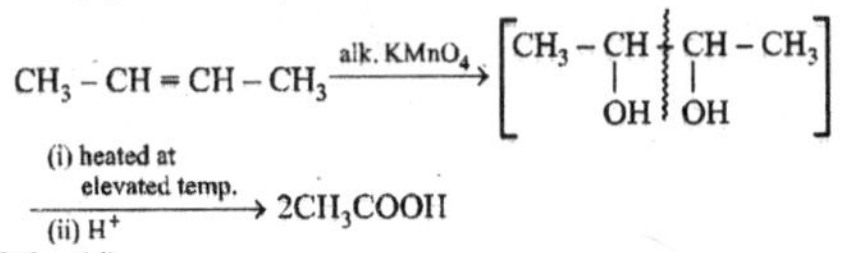

96. (4)

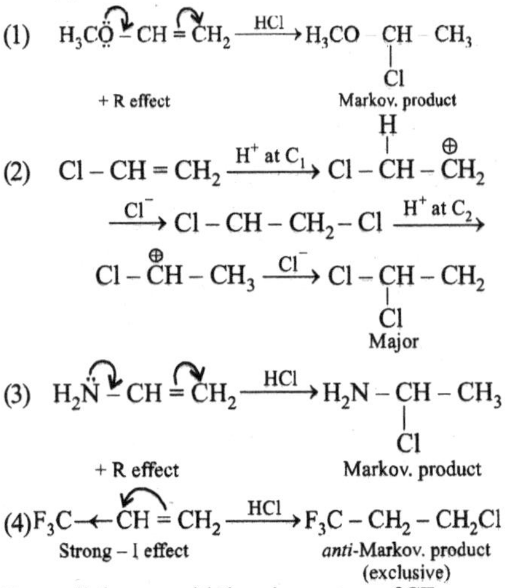

Due to lighter e^- withdrawing nature of CF_3 group it show antimarkornikoff product.

97. (4)

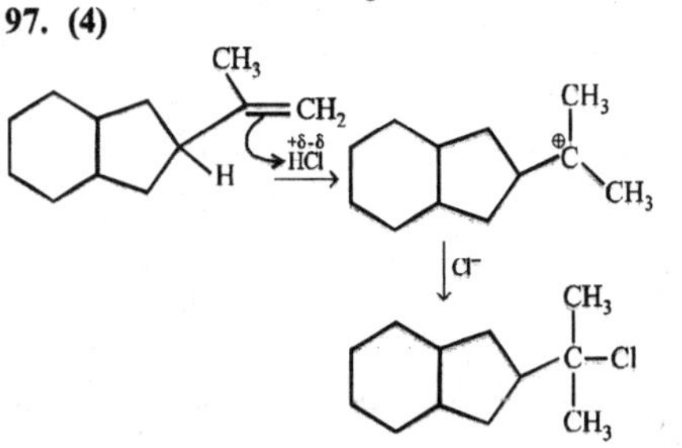

98. (2)

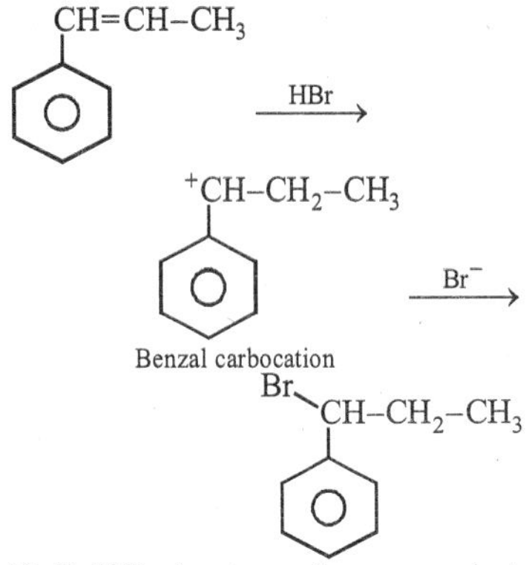

99. (3) Na/liq NH_3 gives trans alkene as product.

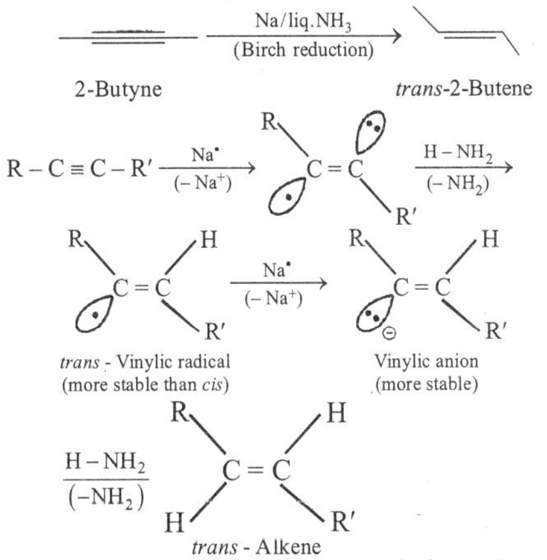

100. (4) Addition of HBr on 3-methylpent-2-ene in presence of peroxide takes place according to *anti*-Markownikov's rule.

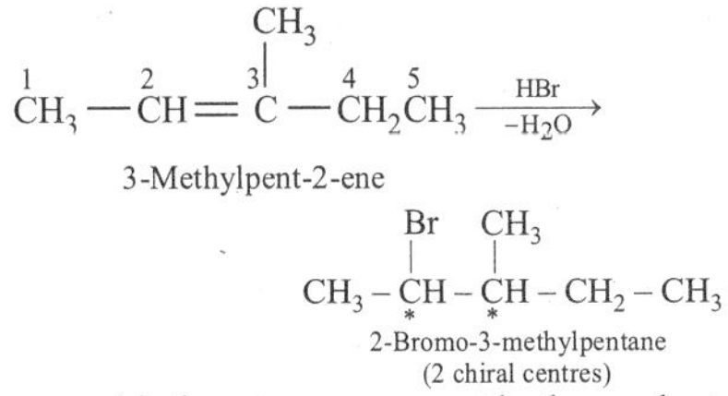

Since two chiral centres are present in the product, four stereoisomers ($2^n = 2^2 = 4$) are possible.

101. (3)

(1)

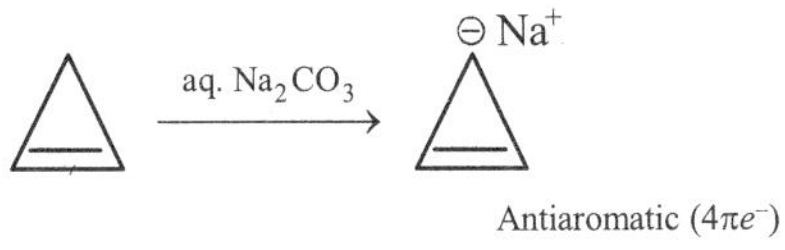

Unstable.

(2)

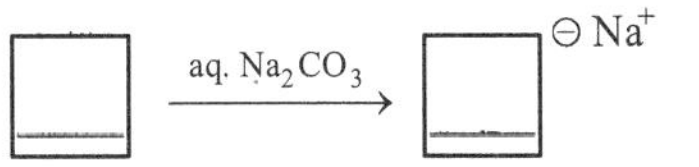

Antiaromatic ($4\pi e^-$) **Unstable.**

(3)

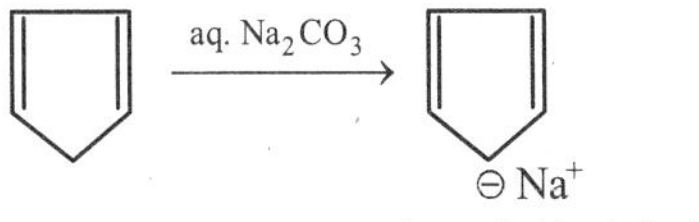

Aromatic ($6\pi e^-$) **Stable.**

(4)

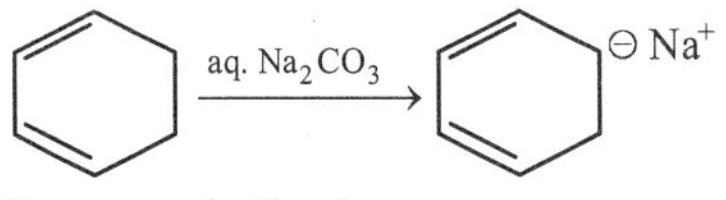

Homoaromatic ($6\pi e^-$)

Delocalization of electrons in (4) is not possible, so less stable than (3).

102. (2) $Br_2 \xrightarrow{h\nu} 2\dot{B}r$

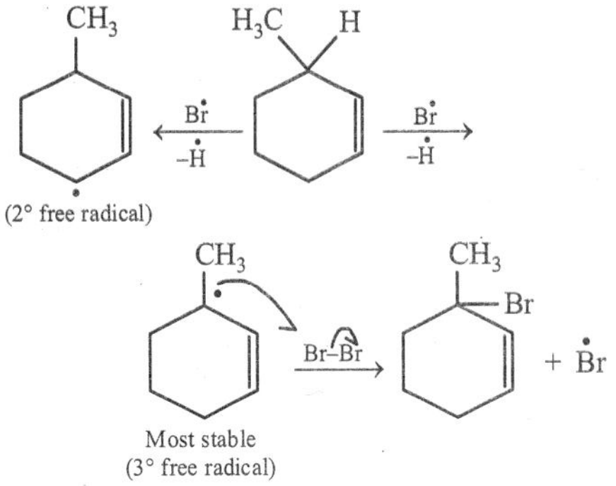

103. (2)

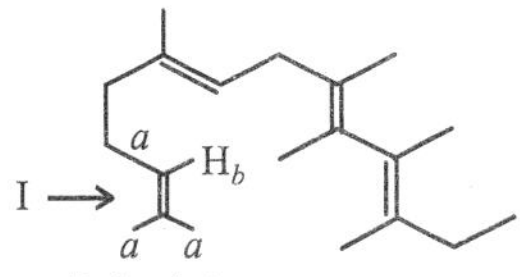

Both substituents are same (G.I. not possible)

104. (4) N – bromosuccinimide results into bromination at allylic and benzylic positions.

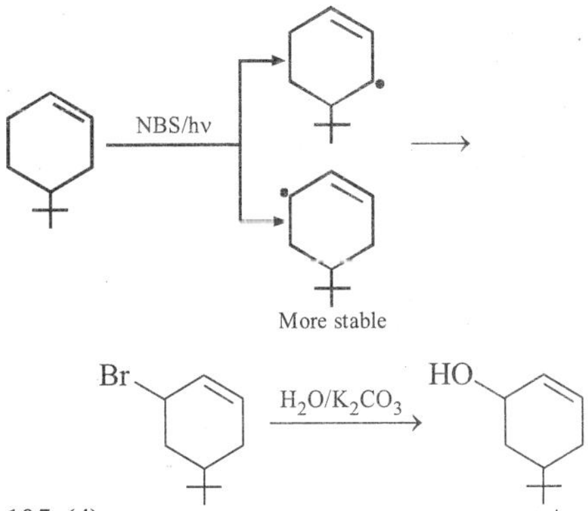

105. (4)

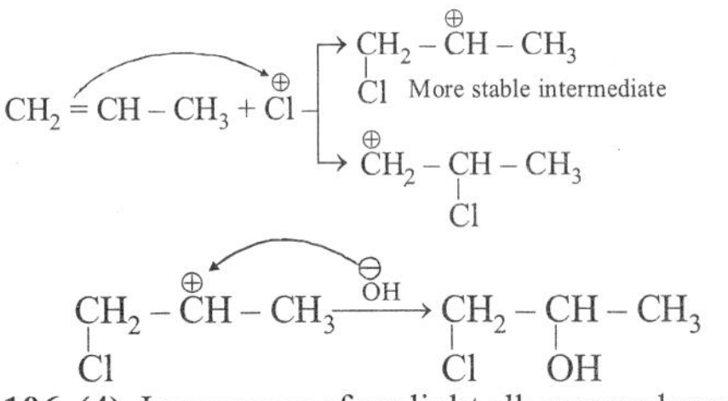

106. (4) In presence of sunlight alkenes undergo free radical substitution.

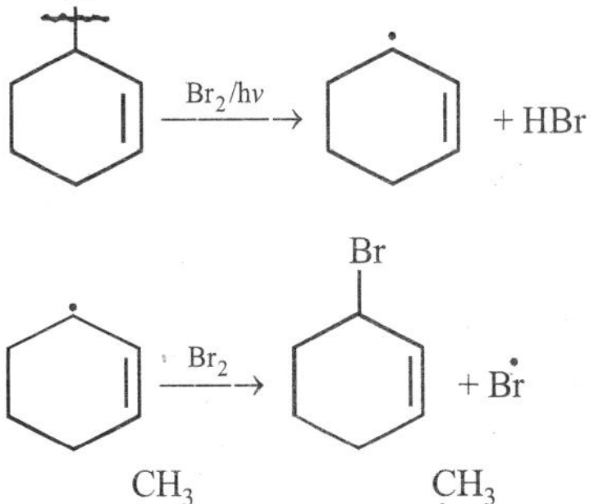

107. (2, 3)

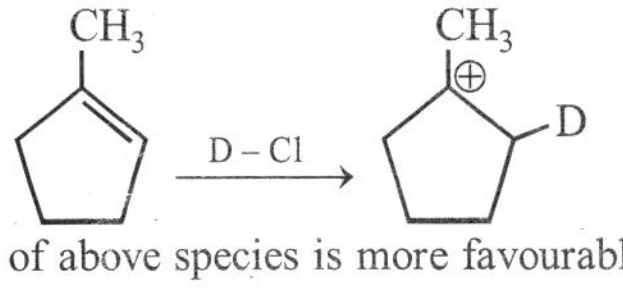

Formation of above species is more favourable. The stability order of carbocations is

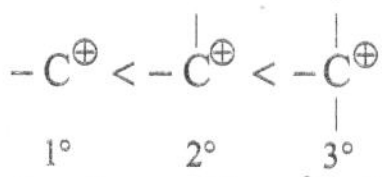

Now, carbocation formed is sp^2 hybridised that is triangular planar as shown below

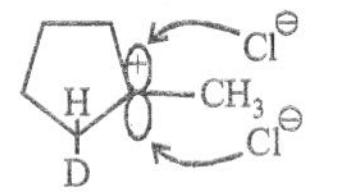

$Cl^{\ominus}$ ion can attack either from above or below the plane of the molecule, so there is probability of resultant product as shown below i.e., both product will form.

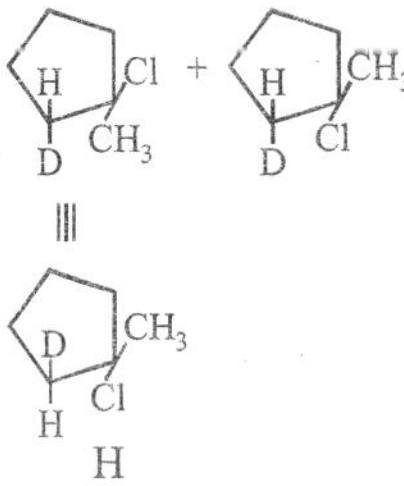

108. (3) $H_3C-\overset{H}{\overset{|}{C}}=CH-CH_2Ph$

109. (3) 110. (3) 111. (1) 112. (4)

113. (4) HI does not exhibit peroxide effect. HI bond although dissociates easily into iodine radicals, they being bigger in size are not much reactive but recombine together to form iodine molecule.

114. (1)

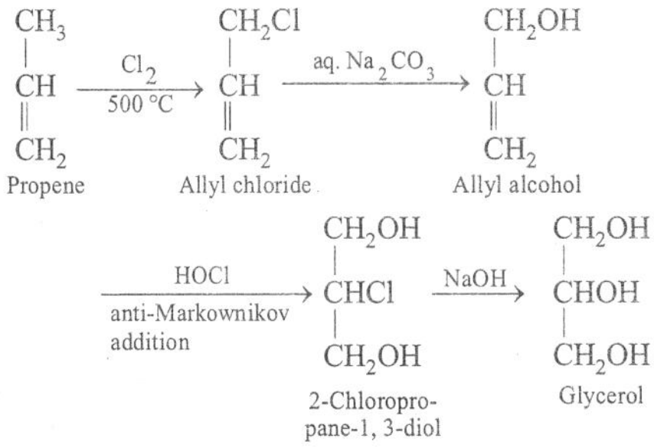

115. (3) From the products formed it is clear that the compound has 5 carbon atoms with a double bond.

116. (3) $CH_3CH=CHCH_3 \xrightarrow{O_3}$
2-Butene

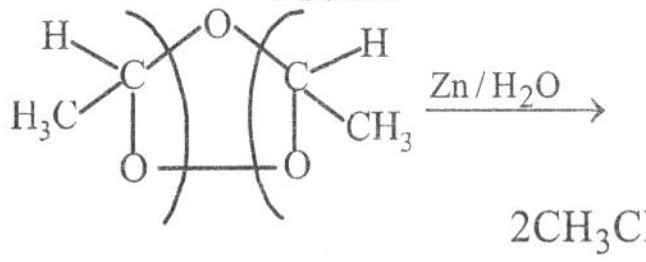

$2CH_3CHO + H_2O_2$

117. (4)

118. (3)

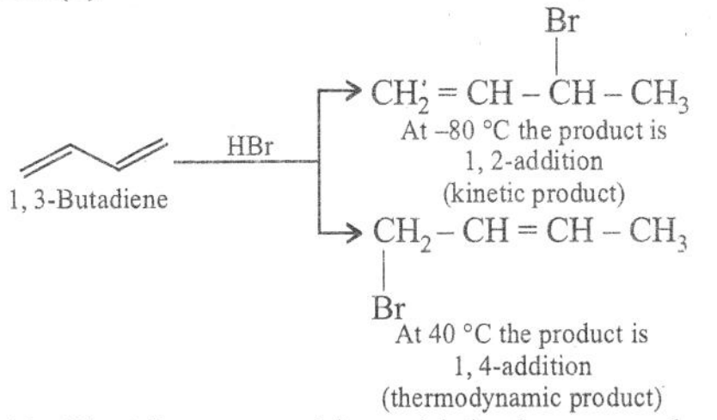

119. (3) Alkenes combine with hydrogen under pressure and in presence of a catalyst (Ni, Pt or Pd) and form alkanes.

Butene - 1 $\xrightarrow{H_2/Pd}$ Butane.

120. (27)

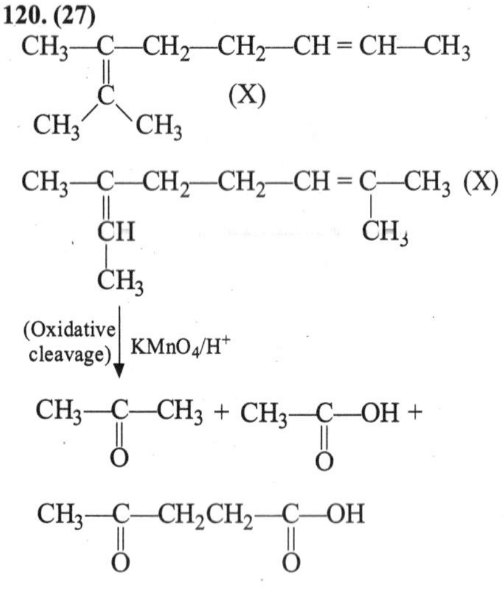

Thus, there are 27 σ-bonds in either of the 'X' molecules giving these products upon oxidative cleavage by $KMnO_4$.

121. (8)

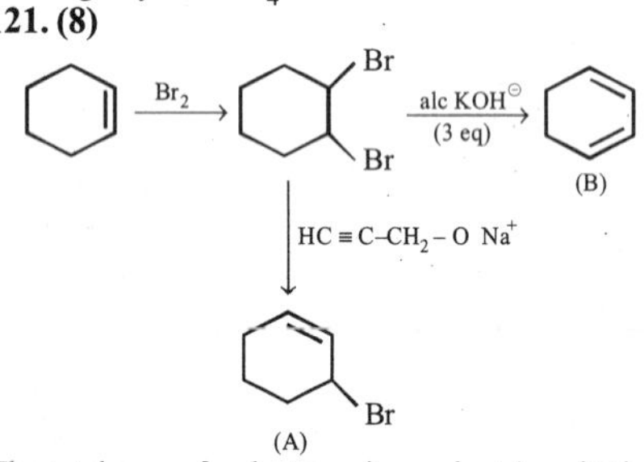

The total sum of π electrons in product A and B is 8.

122. (4) We can use the E/Z nomenclature for determining the geometrical isomers of this compound.

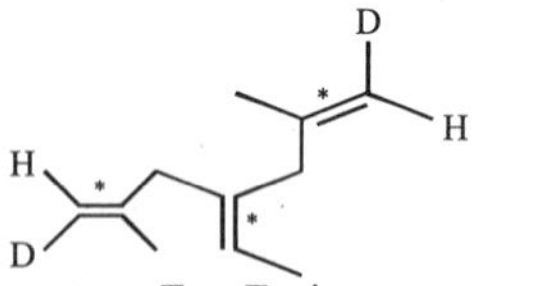

First stereocentre = E or Z – isomer.
Second stereocentre = E or Z – isomer.
Third stereocentre = We don't count it as the compound has the same two substituents and we have done it already once for one double bond.
Thus, total geometrical isomers = 4

123. (1)

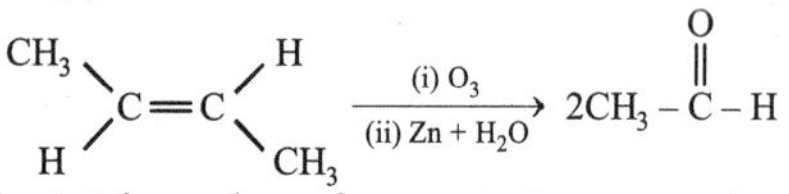

The total number of oxygen atom present per molecule of the product is 1.

124. (4)

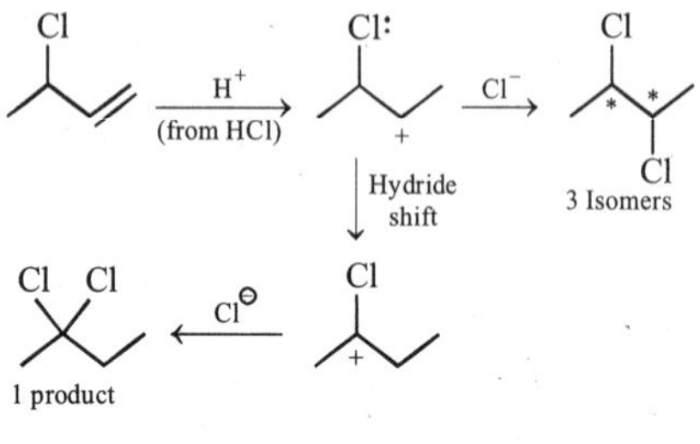

Total Possible Isomeric product = 1 + 3 = 4

125. (70) Hydrocarbon (X) $\xrightarrow{\text{Ozonolysis}}$

$$CH_3-CHO + O=C(CH_3)_2$$

Thus the hydrocarbon (X) is $CH_3CH=C(CH_3)_2$. Its molar mass of the hydrocarbon (X) is 70 g/mol.

126. (3) Moles of hydrocarbon $= \frac{17 \times 10^{-3}}{136}$

$= 1.25 \times 10^{-4}$

Moles of H_2 gas $= \frac{8.4 \times 10^{-3}}{22.4} = 0.375 \times 10^{-3}$

$= 3.75 \times 10^{-4}$

Hydrogen molecule used for 1 molecule of hydrocarbon

$= \frac{3.75 \times 10^{-4}}{1.25 \times 10^{-4}} = 3$

Thus, hydrocarbon must contain 3 double bonds. On the basis of the ozonolysis products of the hydrocarbon, it has following structure.

$$CH_3-C(CH_3)=CH-CH_2-CH_2-C(=CH_2)-CH=CH_2$$

127. (4) Moles of X $= \frac{10\text{ mg}}{80} = 0.125$ m mol.

Moles consumed of $H_2 = \frac{8.4}{22.4} = 0.375$ m mol.

$\frac{n_{H_2}}{n_X} = \frac{0.375}{0.125} = 3$

So, the compound X have 3 double bonds. Ozonolysis of the compound yields only formaldehyde and dialdehyde, so the compound should be

$$H_2C=CH-CH=CH-C(H)=CH_2$$

Molecular mass = (12 × 6) + 1 × 8 = 72 + 8 = 80 amu

Ozonolysis form:

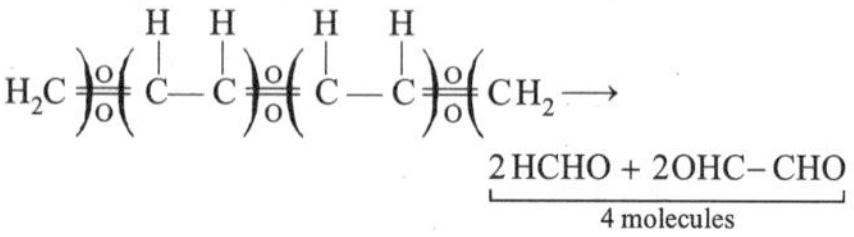

128. (2)

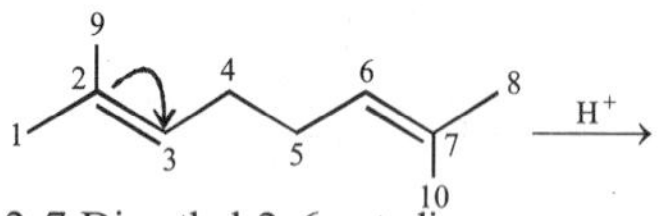

2, 7-Dimethyl-2, 6-octadiene

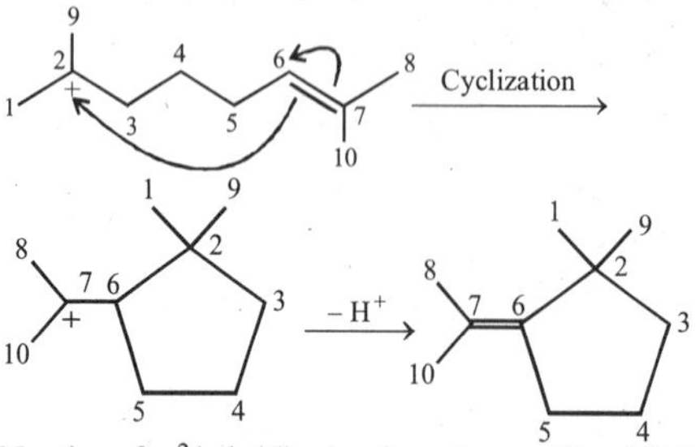

Number of sp^2 hybridised carbon atoms = (6 and 7) 2

129. (2)

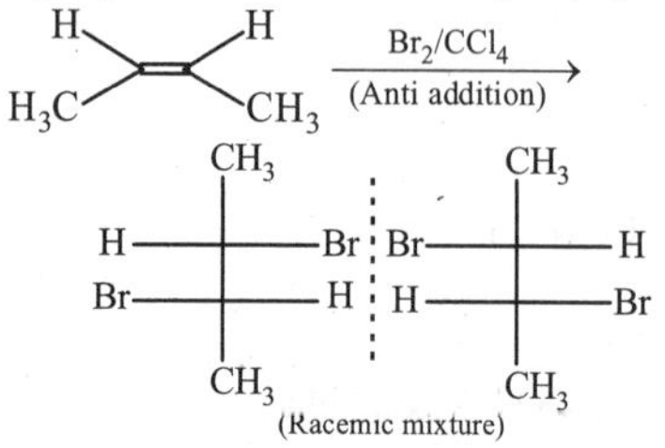

130. (2)

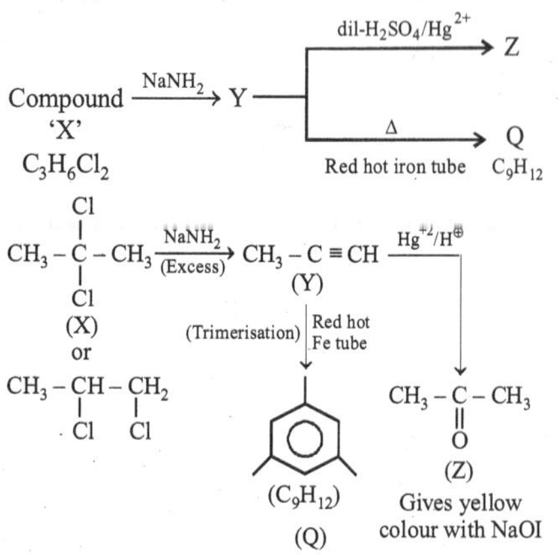

No. of aromatic H atoms in Q = 3
No. of aliphatic H atoms in Q = 9
∴ Ratio is 1 : 3.

131. (3)

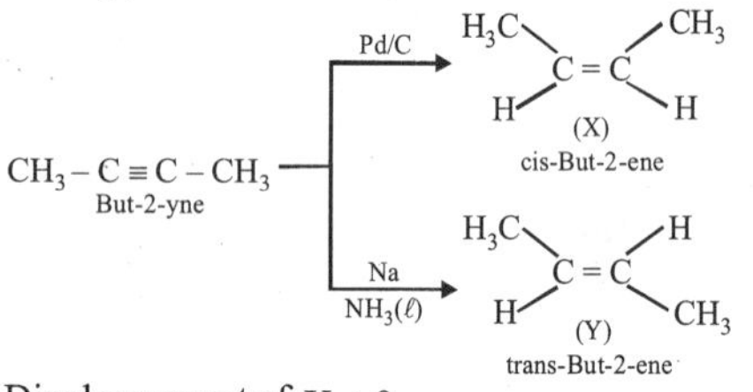

Dipole moment of X ≠ 0

$$\underset{\text{(Cis/Trans)}}{\text{But}-2-\text{ene}} \xrightarrow[\text{(ii) Zn/}H_2O]{\text{(i) }O_3} \underset{\text{(Acetaldehyde)}}{2CH_3-CHO}$$

132. (1)

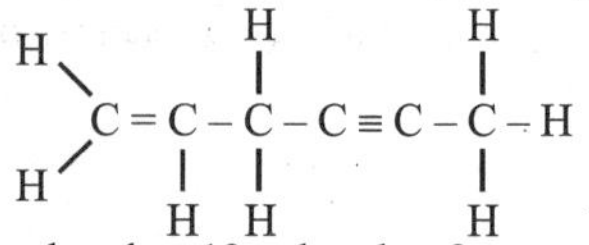

σ bonds = 13, π bonds = 3

133. (1)

$$CH_3-C\equiv CH \xrightarrow[H_2]{Pd/C} \underset{[A]}{CH_3-CH=CH_2}$$

$$\xrightarrow[\text{(ii)Zn},H_2O]{\text{(i)}O_3} \underset{[B]}{CH_3-CH=O} + \underset{[C]}{HCHO}$$

134. (4) The carbon-carbon bonds in ethyne is stronger than that in ethene.
(H–C=C–H) Ethyne is linear and carbon atoms are sp-hybridised.
Number of oxygen atoms present in product 'P' is 2.

135. (4) A : 2 – Pentyne B : trans-2-butene

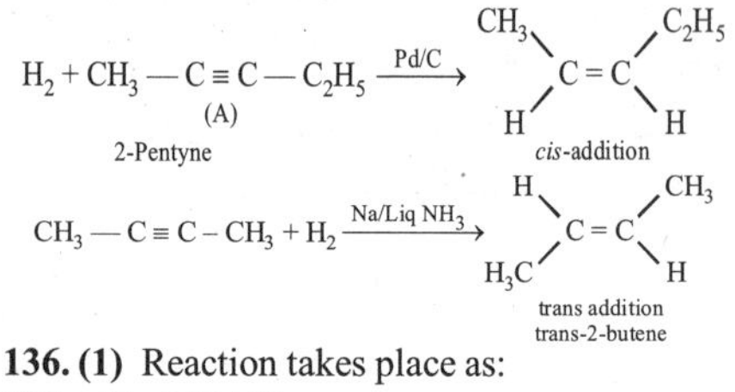

136. (1) Reaction takes place as:

$CH_3-C\equiv CH + Na \longrightarrow$

$$\underset{(A)}{CH_3-C\equiv C^-\ Na^+} + H_2\uparrow$$

$$CH_3-C\equiv C^-Na^+ + \underset{(B)}{CH_3CH_2CH_2-Br} \longrightarrow$$

$$\underset{(C)}{CH_3-C\equiv C-CH_2-CH_2-CH_3} + NaBr$$

137. (2) Since *cis*-butene (A) is more polar than *trans*-butene (B) it has higher boiling point.

However, *trans*-butene has better packing than *cis*-butene. So, *trans*-butene (B) has higher melting point than the *cis*-isomer (A).

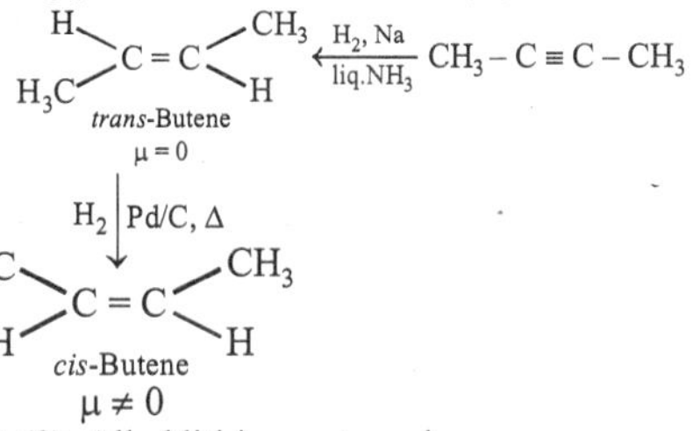

138. (3) Alkyl lithium act as a base.

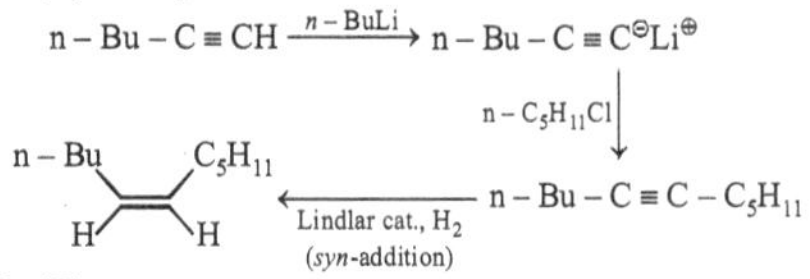

139. (3)

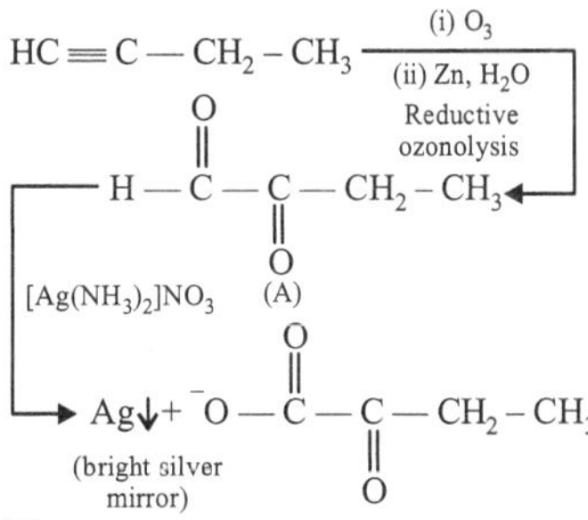

140. (1)

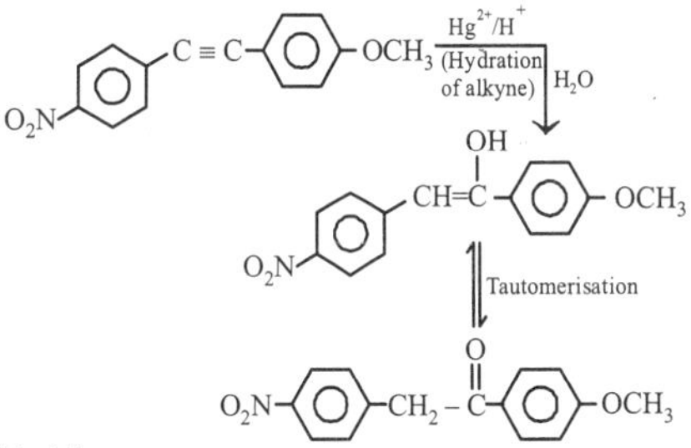

141. (4)

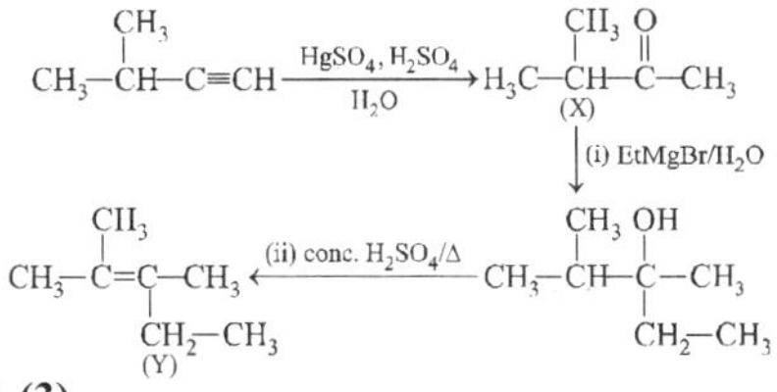

142. (3)

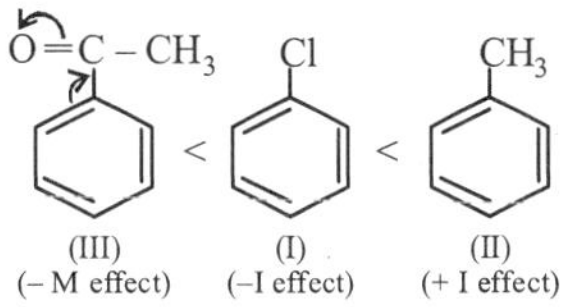

143. (4)

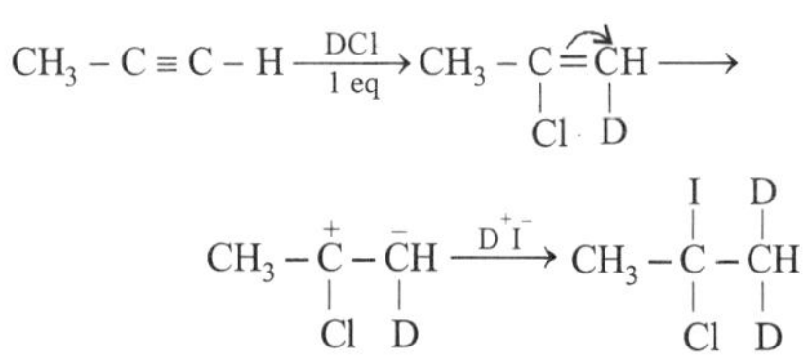

Due to combined +I effect of $-CH_3$ and +M effect of –Cl, positive charge is more stable at the second carbon.

144. (4)

$HC \equiv CH > H_3C \rightarrow C \equiv CH > CH_2 = CH_2$ (sp ... sp^2)

145. (4) When but-2-yne is treated with H_2/Lindlar's catalyst, compound X (*cis*-but-2-ene) is produced as the major product; and when treated with Na/liq NH_3 it produces Y (*trans* -but-2-ene) as the major product. *Cis*-isomer(X) will have higher dipole moment and higher boiling point than *trans* (Y).

146. (3) With Lithium in liquid ammonia, trans-alkene is almost an exclusive product.

147. (2) $3\,CH \equiv CH \longrightarrow C_6H_6$ (poly-ene)
Acetylene Benzene

148. (1) $CH_3 - C \equiv CH + H_2O \xrightarrow[HgSO_4]{H_2SO_4}$
Propyne

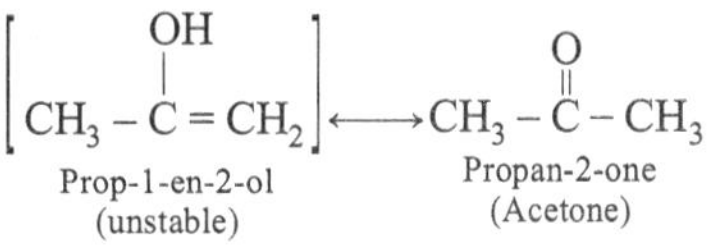

149. (2)

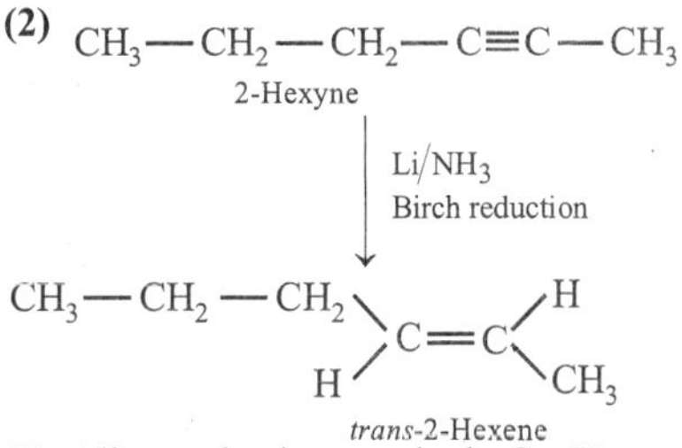

150. (2) Alkynes having terminal $-C \equiv H$ react with Na in liquid ammonia to yield H_2 gas

$$CH_3CH_2C \equiv CH \xrightarrow[\text{liquid } NH_3]{\text{Na in}} CH_3CH_2C \equiv C^-Na^+ + \frac{1}{2}H_2(g)$$

151. (4) $CH_3MgX + CH_3 - C \equiv C - H \longrightarrow CH_3 - C \equiv CMgX + CH_4(g)$

152. (2) $CH_3 - C \equiv CH + HBr \longrightarrow$

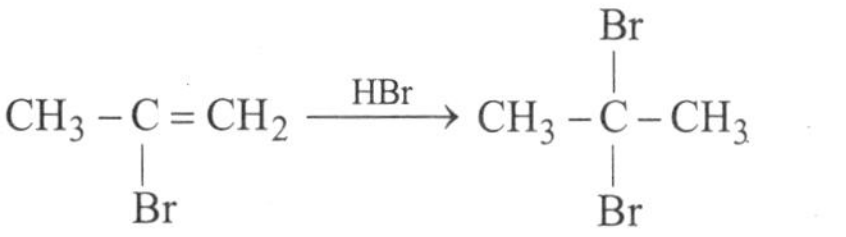

2,2-Dibromopropane

153. (4) Among isomeric alkanes, the straight chain isomer has higher boiling point due to large surface area than the branched chain isomer.

154. (3)

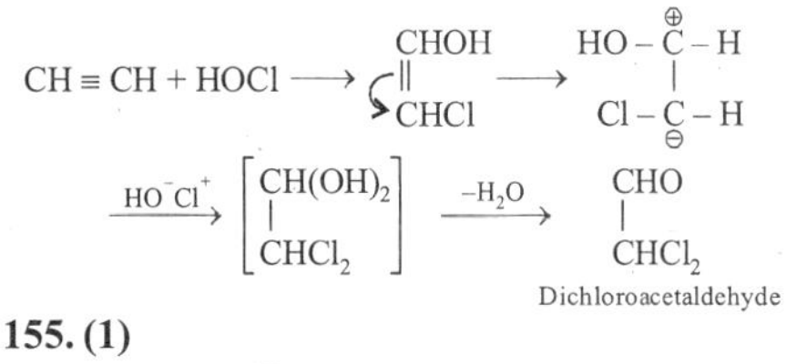

155. (1)

156. (15)

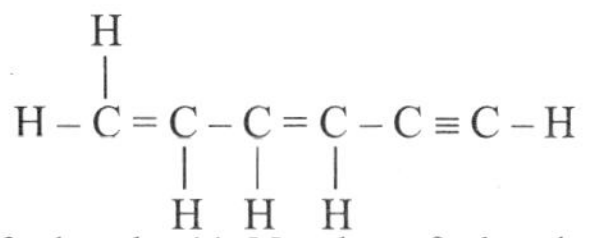

Number of σ bond = 11, Number of π bond = 4

$\therefore\ \sigma + \pi = 11 + 4 = 15$

157. (8)

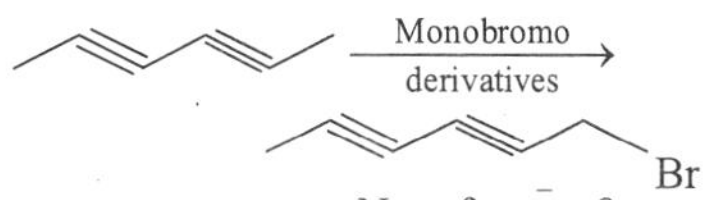

No. of π e^- = 8

158. (2)

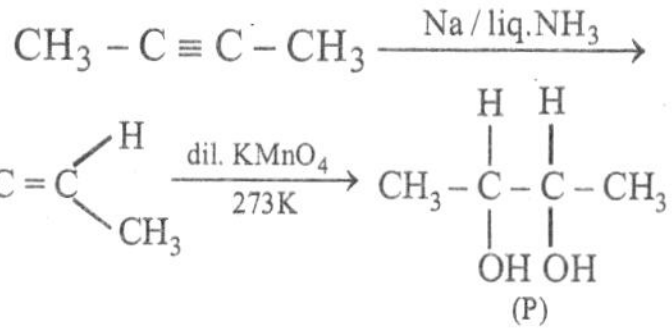

159. (3)

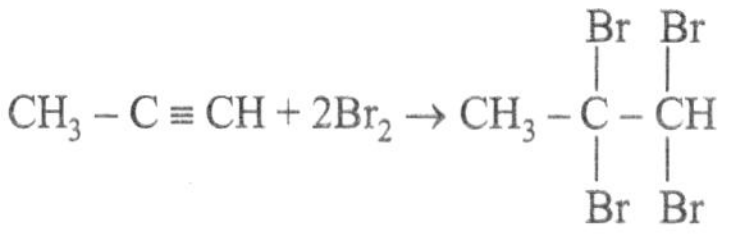

(molar mass = 360)

2×160 g Br_2 produces = 360×0.27 g Compound

$$1 \text{ g } Br_2 \text{ produces} = \frac{360 \times 0.27}{2 \times 160} \text{ g}$$

$= 0.30375 = 3.0375 \times 10^{-1}$

160. (10)

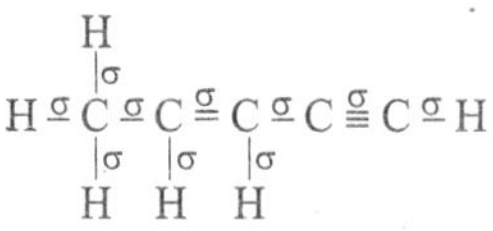

Numbers of σ bonds = 10

161. (7)

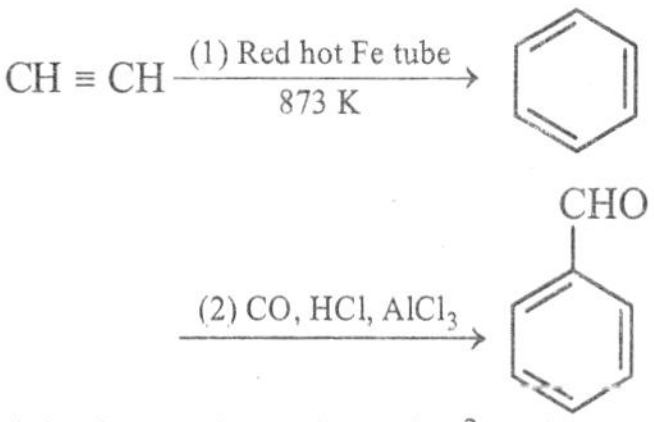

In benzaldehyde, total number of sp^2 carbons are 7.

162. (13)

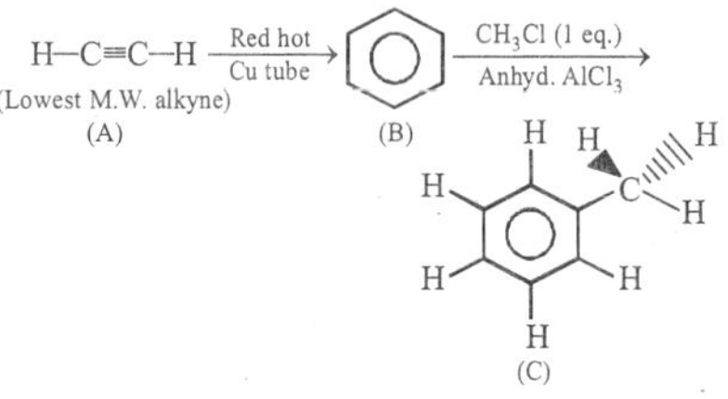

Number of atoms present in molecule (C) in one plane = 13

163. (3) For a compound to be aromatic it should be planar and follow Huckel's rule of (4n + 2)π electrons

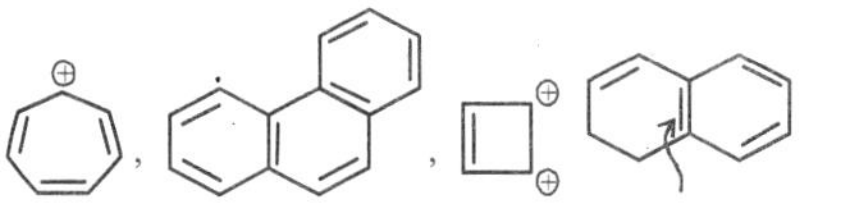

Double bond not on periphery i.e., non Huckle π bond. So does not participate in aromaticity.

164. (2) This is an example of elimination reaction.

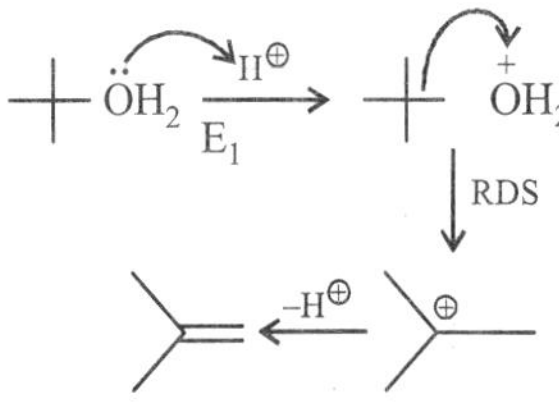

Product obtained without rearrangement.

165. (2)

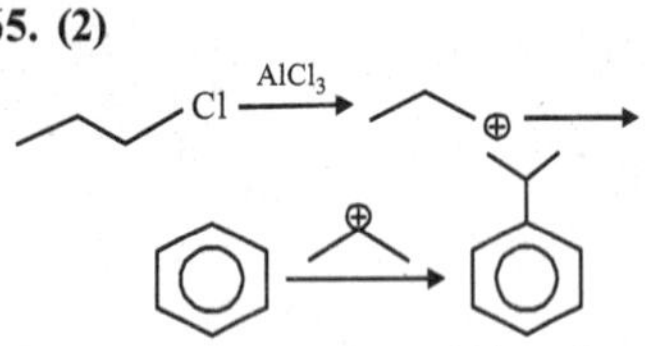

Major product

166. (2)

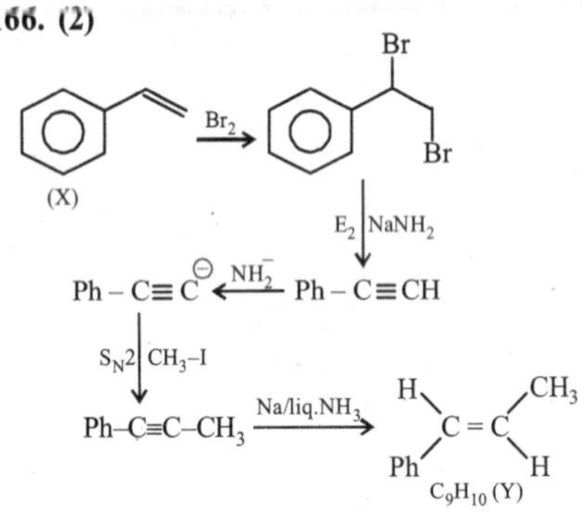

Molar mass of Y = 118 g mol^{-1}.

167. (3) Williamson synthesis:

$$R-X+R-O-Na \longrightarrow R-O-R+NaX$$

Here alkoxide ion $(R-O^{\ominus})$ acts as nucleophile and attacks on alkyl halide (R – X).

Friedel Craft reaction:

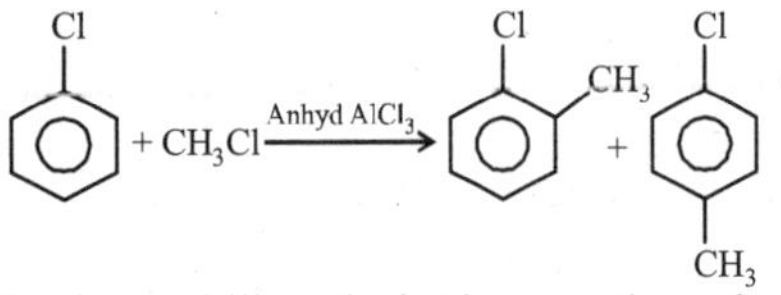

It is electrophilic substitution reaction where CH_3^+/R^+ acts as an electrophile and substitutes the hydrogen of electron rich phenyl ring.

Bromination of vinyl benzene:

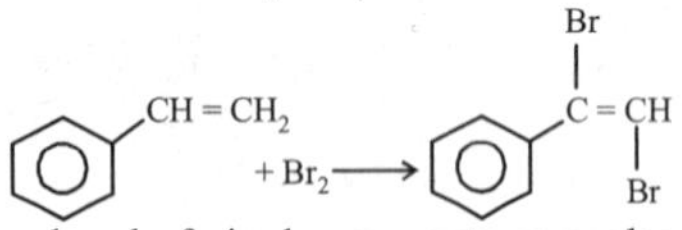

Here π-bond of vinyl group acts as nucleophile while Br^+ acts as electrophile.

Chlorination of toluene in light:

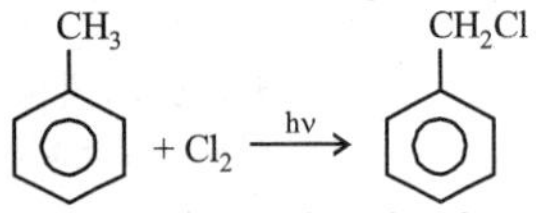

This reaction proceeds via free radical substitution.

168. (3)

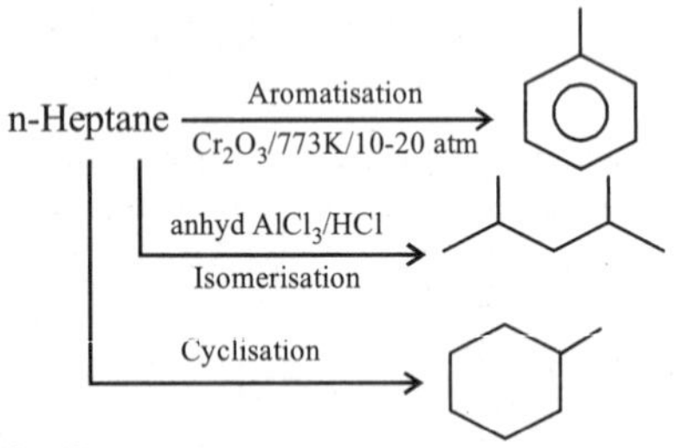

169. (1)

(1)

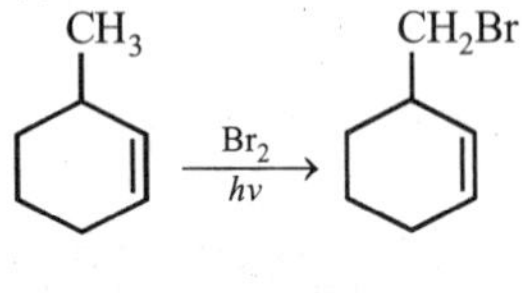

(2)

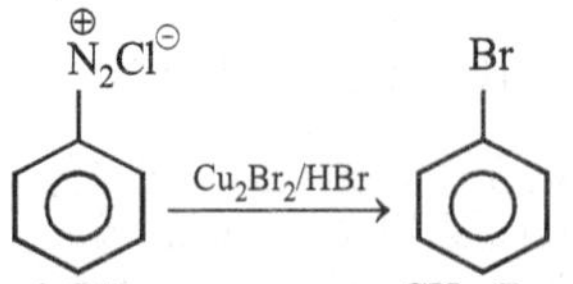

(3)

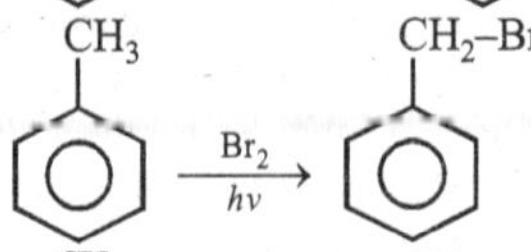

(4)

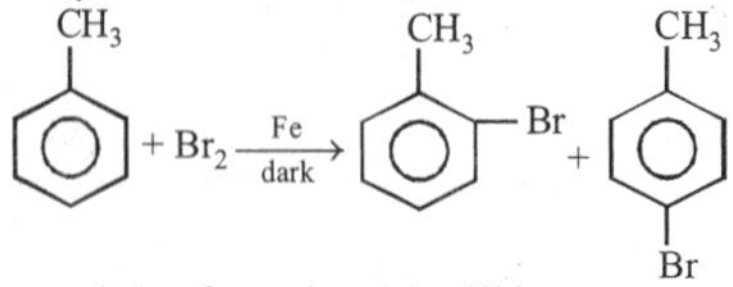

Major product of reaction (a) will be

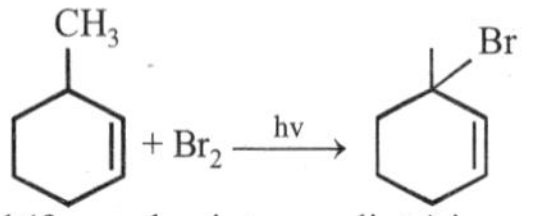

3° Radical (formed as intermediate) is more stable.

170. (4) Presence of NO_2 group in benzene ring deactivates ring towards electrophilic substitution reaction due to – M effect and activates ring towards nucleophilic substitution.

171. (4) In Ph–OCH_3, –OCH_3 is a electron donating group (+M).
In Ph–NO_2, –NO_2 is a strong electron withdrawing group (–M).
In Ph–Cl, –Cl is an electron withdrawing group.

172. (3) Nitrobenzene does not give Fridel-Craft acylation since it is highly deactivating group towards electrophilic substitution.

173. (3)

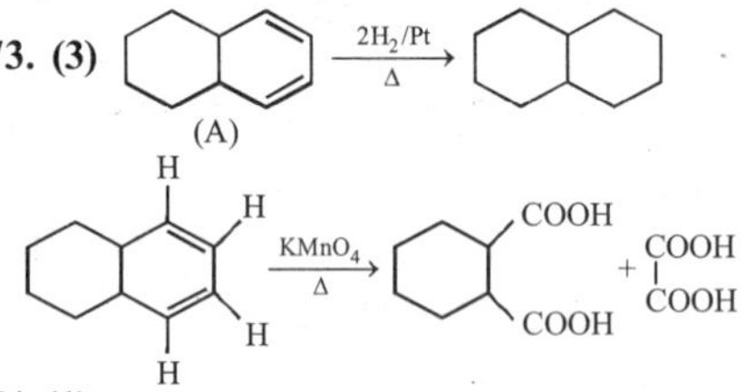

174. (1)

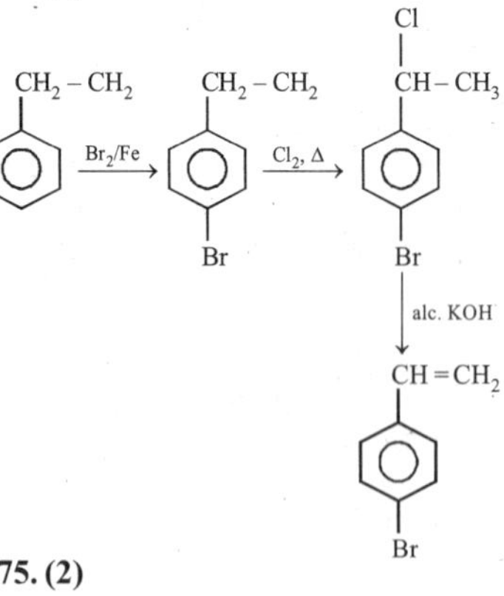

175. (2)

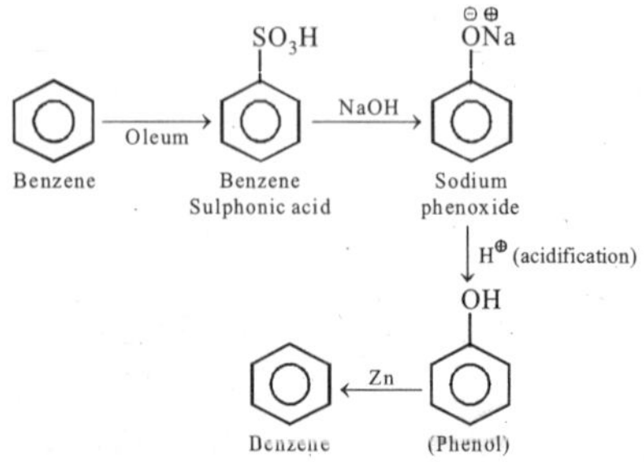

176. (4) Aromatic compounds :

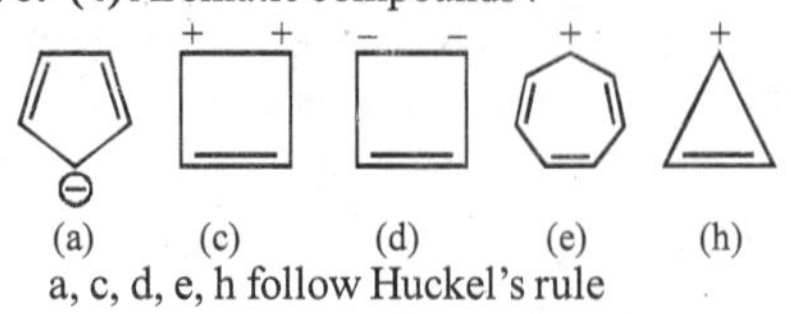

a, c, d, e, h follow Huckel's rule

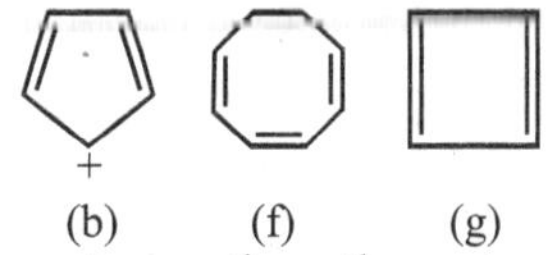

b, f, g, are not aromatic, as these compounds do not follow Huckel's rule

177. (3) –CH_3 group increase electron density at o/p position while –NO_2 group decrease electron density at ortho and para position.

178. (1) (B) – CN is meta directing but –OH is ortho/para directing.
(E) Halides are deactivating groups.

179. (1) q is aromatic r is non-aromatic (non-planar structure) p is antiaromatic so, order of stability is :

Aromatic > non-aromatic > Antiaromatic
(q) (r) (p)

180. (1) Species follow Huckel's rule are aromatic in nature and aromatic systems are stable.

 follow (4n + 2)π rule, it has 2π electrons in conjugation. So it is must stable.

181. (1)

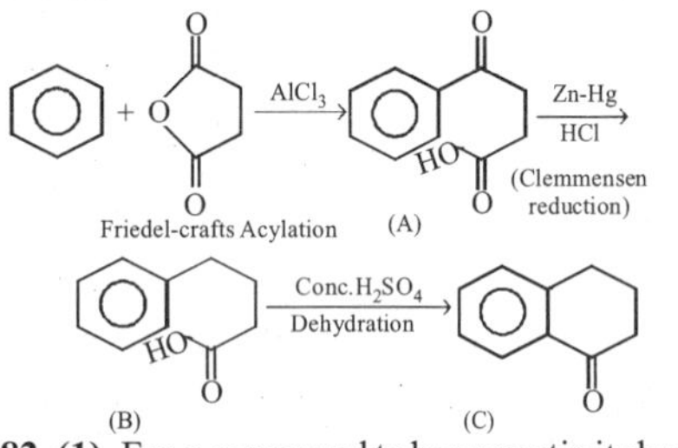

182. (1) For a compound to be aromatic it should be planar and follow Huckel's rule of (4n +2)π electrons.

A. 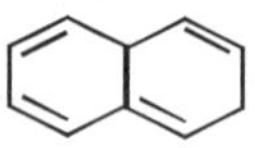 → Non aromatic

B. 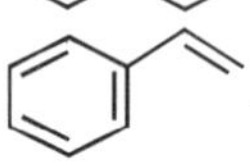 → Aromatic

C. 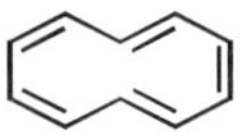 → Non aromatic

D.  → Aromatic

183. (2) In nitration of benzene concentrated H_2SO_4 and HNO_3 is used as reagent which generates electrophile $\overset{\oplus}{N}O_2$ in following steps:

$$H_2SO_4 + HNO_3 \rightleftharpoons HSO_4^- + H-\overset{\oplus}{O}(H)-NO_2$$

$$\rightleftharpoons HSO_4^- + H_2O + NO_2^{\oplus}$$

Lewis acids can promote the formation of electrophiles not Lewis base

184. (4) As we know compound (4) is aromatic and the compound (1) is anti-aromatic. Hence compound (4) is most stable and compound (1) is least stable among these in compound (2) and (3) carbon atom having charge is sp^3 hybridised. On the basis of angle strain theory compound (3) is more stable than compound (2).

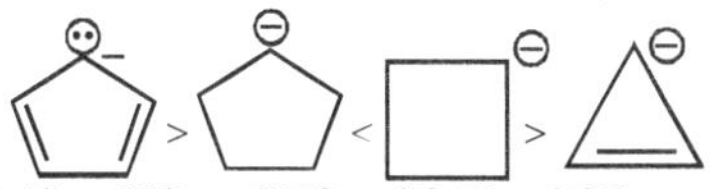

185. (4) $-NO_2, -CHO, -SO_3H, -COR$.

Ortho and para directing groups are:

$-NH_2$, $-NHR$, $-NHCOCH_3$, $-OCH_3$, $-CH_3$, $-C_2H_5$ etc.

meta directing groups are:

$-NO_2, -CN, -CHO, -COR, -COOH, -COOR, -SO_3H$ etc.

186. (4) Order of increasing reactivity towards electrophilic substitution reaction is:

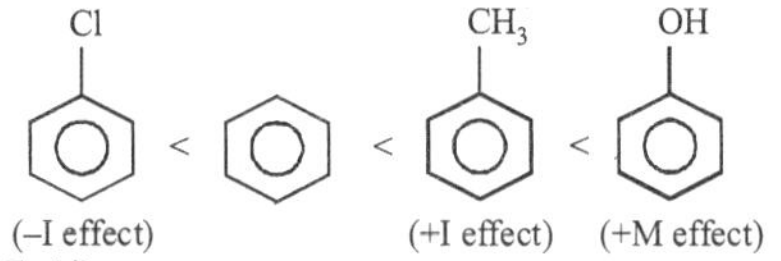

187. (4)

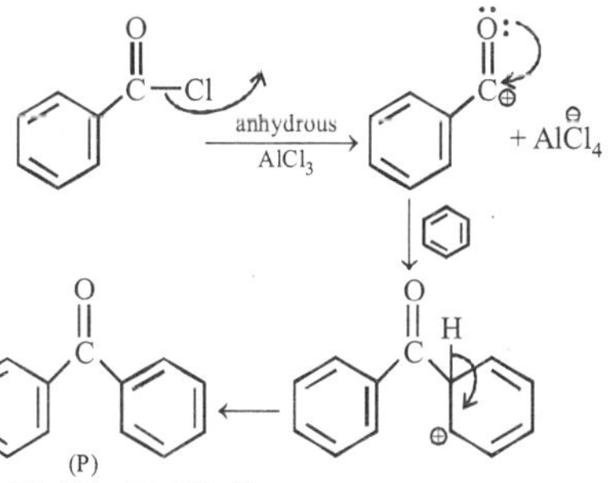

188. (4) B > A > C > D

- CH_3 will show +I and +M effect.
- Cl will show –I and +M effect but inductive effect will be more influential.
- NO_2 will show –I and –M effect.

189. (2) Tropolone is an aromatic compound and has 8π electrons ($6\pi e^-$ are endocyclic and $2\pi e^-$ are exocyclic) but 2π electrons of $>C=O$ group in tropolone are not involved in aromaticity.

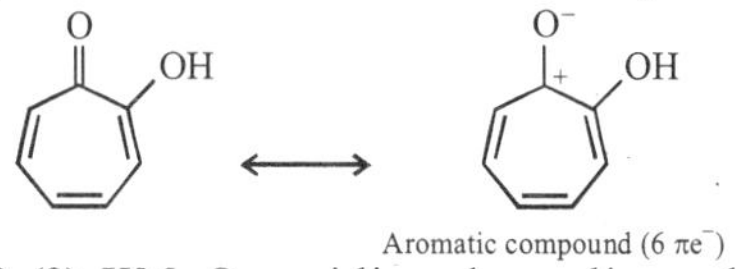

190. (2) $KMnO_4$ oxidises benzylic carbon containing atleast one α–hydrogen atom to –COOH.

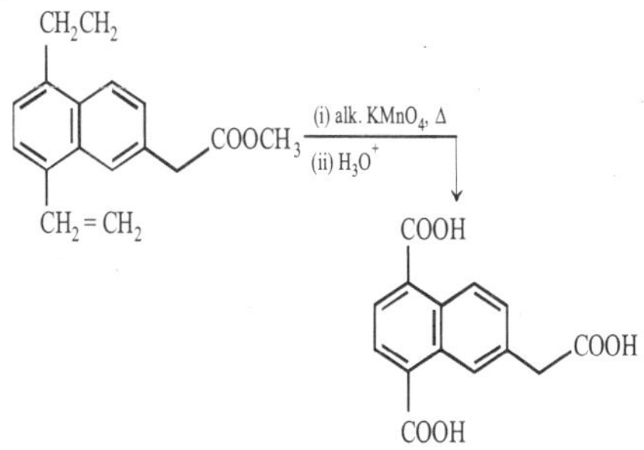

191. (4) $AlCl_3$ gives rise to a carbcation here. Carbocations in the side chains of a ring undergo cyclization or ring formation.

192. (3)

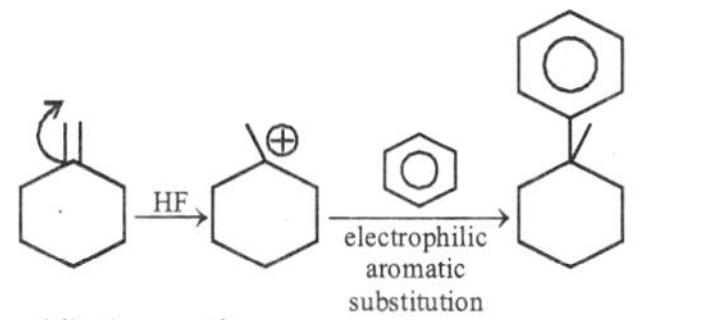

193. (4) Assertion:

Annulene	Huckle's rule	Planarity	Aromaticity
[6]	4n + 2 = 6, n = 1	Planar	Aromatic
[8]	4n + 2 ≠ 8, n = 1	Non-planar	Non-aromatic
cis – [10]	4n + 2 = 10, n = 2	Non-Planar	Non-aromatic

[8] annulene is non-aromatic because it does not obey Huckel's rule and it is non-planar.

Reason: It is correct. Although it has $(4n + 2)\pi$ electrons, i.e. 10, but it is not planar.

194. (3) Due to the steric repulsion between the two hydrogen atoms this molecule can not achieve planarity, hence compound is not aromatic.

195. (1, 2) Benzenoid compound contain only benzene rings. In option (1) one central carbon is sp^3 hybridised so no benzene ring is present.

196. (3)

$$CH_3-CH_2CH_2Cl \xrightarrow{AlCl_3} CH_3-CH_2-\overset{+}{C}H_2$$

$$\downarrow \text{Hydride shift}$$

$$CH_3-\overset{+}{C}H-CH_3 \text{ (more stable)}$$

197. (3)

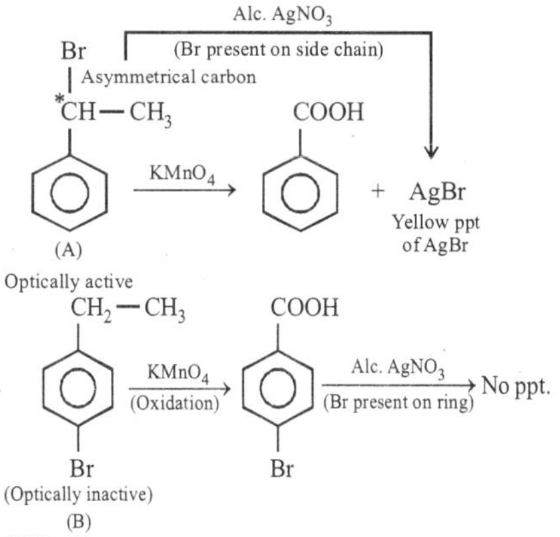

198. (2)

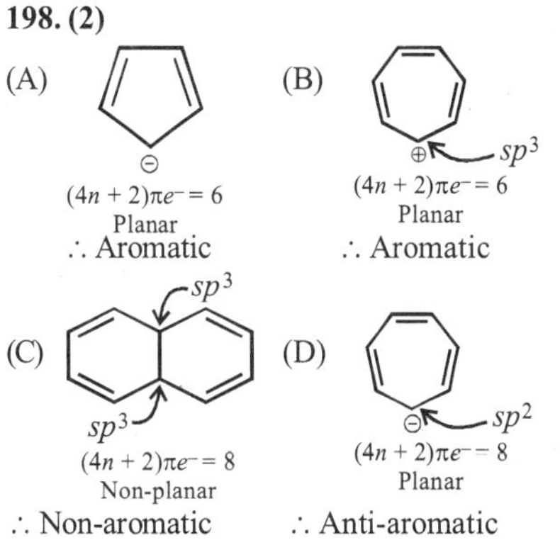

199. (1) Planar molecules having $(4n + 2)\pi e^-$ are aromatic in nature.

(1)  is aromatic ion.

(2)  is antiaromatic ion.

(3)  is antiaromatic ion.

(4)  is non-aromatic.

Hence (1) being aromatic, is most stable.

200. (4)

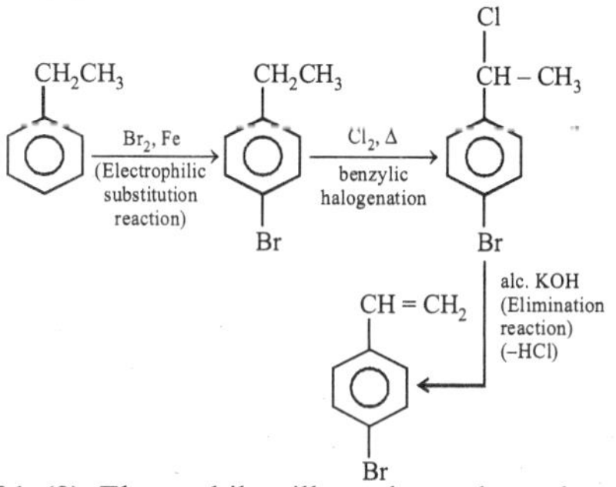

201. (3) Electrophile will attack at ortho and para positions with respect to better electron releasing group (ERG)

–OH donates through +M and $-CH_3$ donates through +H

$$\text{ERG}: -OH > -CH_3$$

Para position with respect to OH group and it will be meta position with respect to $-CH_3$ group.

202. (4) In option a, saturated five membered ring therefore does not exhibit aromaticity as a whole

(1) 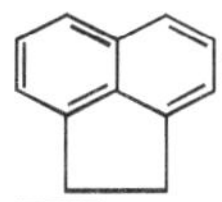 not aromatic

(2)  ⇒ $4\pi e^-$ in ring conjugation
⇒ Anti-aromatic

(3) 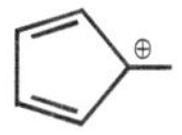 CH_3 ⇒ $4\pi e^-$ in ring conjugation
⇒ Anti-aromatic

(4) 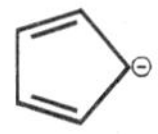 ⇒ $6\pi e^-$ in ring conjugation
⇒ Aromatic

203. (1)

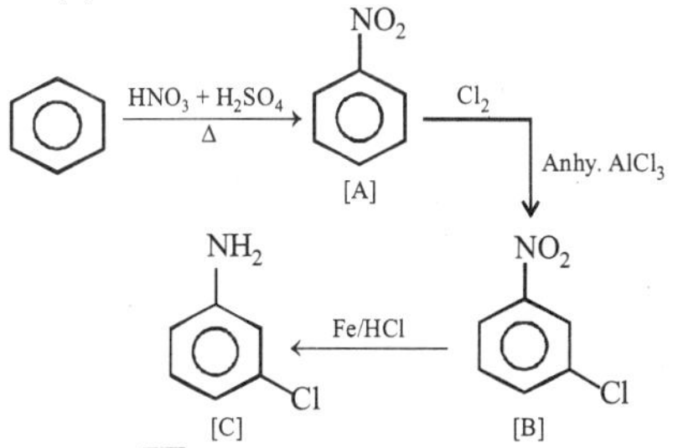

204. (3)  : Non-aromatic

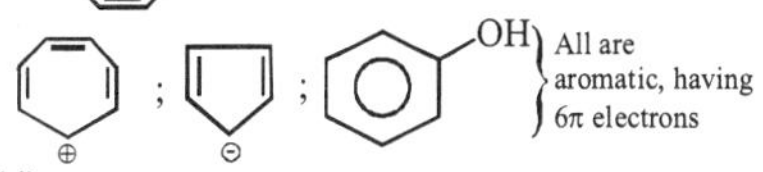

205. (4)

206. (3) To be aromatic, a molecule must meet the following conditions :

(i) It must be a ring.

(ii) It must be planar.

(iii) It must have in each atom of the ring a *p*-orbital that's orthogonal to the plane of the ring. In other words, no atom in the ring can be sp^3 hybridized.

(iv) It must have a Huckel number of π electrons, following the $4n + 2$ rule. If the molecules meets the first three conditions, but only has $4n$ π electrons, the molecule is anti-aromatic.

If the molecule fails any or all of the first three conditions, then the molecule is non-aromatic.

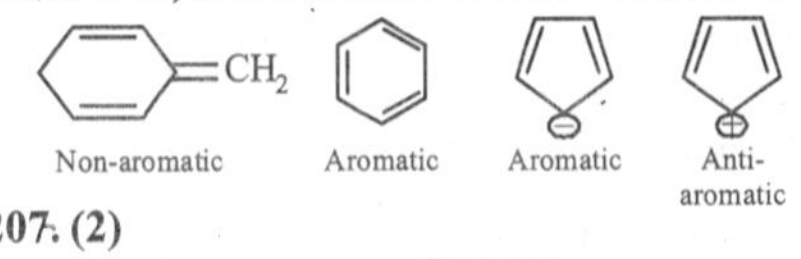

207. (2)

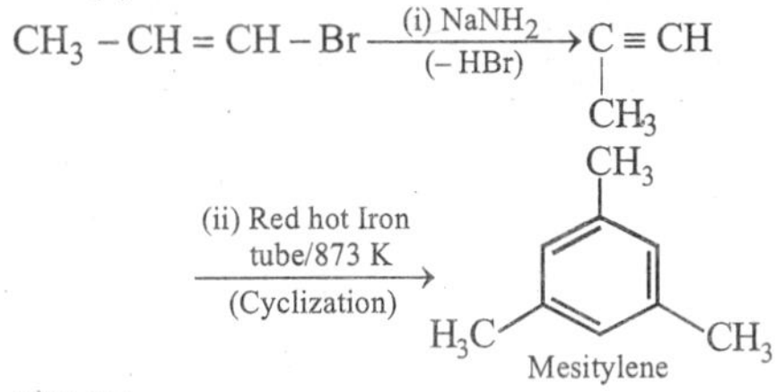

208. (1)

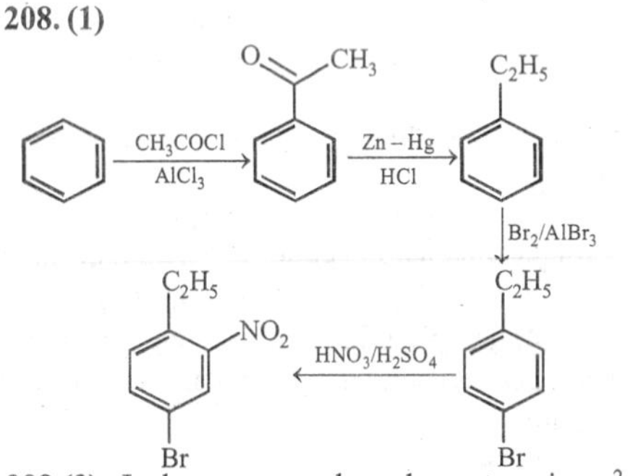

209.(3) In benzene, each carbon atom is sp^2 hybridised. Therefore, total (3 × 6) 18 sp^2 hybrid orbitals are present in benzene.

210. (4) Vinyl halides and aryl halides are unreactive towards Friedel Craft's reaction. Therefore, reactions (A) and (C) are not possible.

211. (1) Compounds (2), (3) and (4) are containing $6\pi e^-$ in complete conjugation and are aromatic.

Compound (1) is anti-aromatic as it has $4\pi e^-$ in complete conjugation and doesn't obey Hückel rule.

212. (1)

213. (1)  is aromatic; $(4\times 0+2)=2\,\pi e^-$ in conjugation

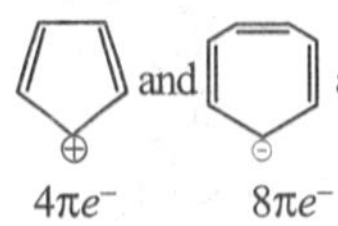 and are anti-aromatic due to $4\,n\pi e^-$.

 is non-aromatic due to the presence of sp^3 carbon.

214. (3)

215. (1) 

A dipolar resonance structure has aromatic character in the ring and would be expected to make a major contribution to the overall structure. Conjugation is possible only in option –(I).

216. (2) Hence, formation of carbocation is not possible in the case of vinyl halide.

217. (3) In (3) Friedal-Craft's alkylation reaction will give propyl phenyl ketone which further on Clemmensen's reduction will give butylbenzene.

$$C_6H_6 + CH_3CH_2CH_2COCl \xrightarrow{AlCl_3} C_6H_5COCH_2CH_2CH_3$$

Butanoyl chloride

$$\xrightarrow{Zn\text{-}Hg/HCl} C_6H_5CH_2CH_2CH_2CH_3$$

Butylbenzene

218. (4) For a compound to be aromatic it must have $(4n + 2)\,\pi e^-$ where n is an integer. (III) have 8 and (VI) have 4 e^-. Hence are antiaromatic

219. (3)

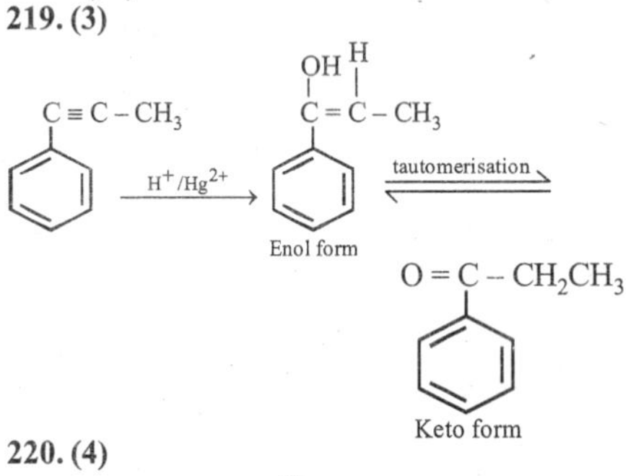

220. (4)

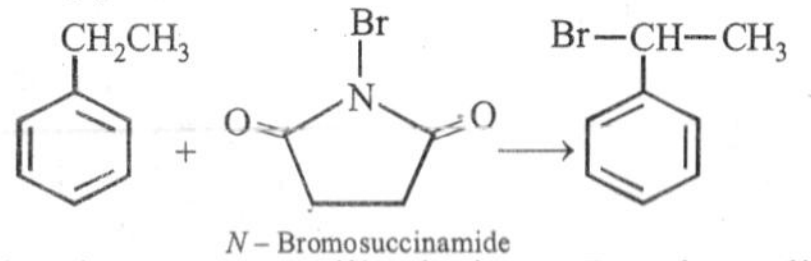

The given reagent will substitute –Br at benzylic position. (Allylic bromination).

221. (1)

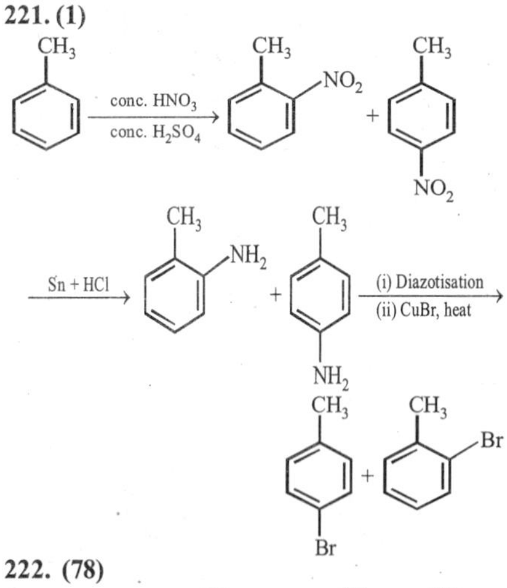

222. (78)

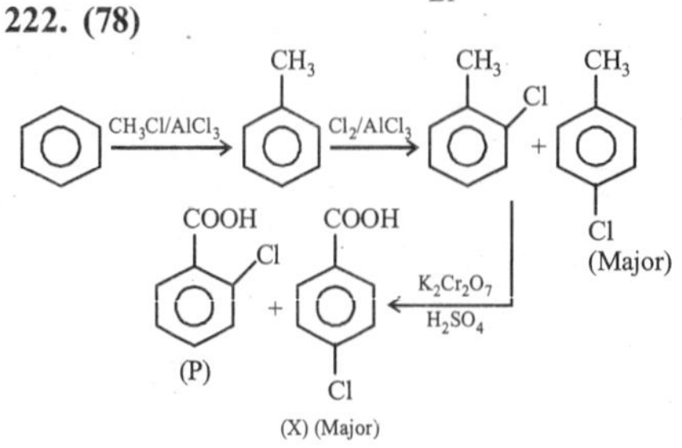

$$\text{Moles of P} = \frac{11.2}{22.4} = 0.5 \text{ mole}$$

$$M_{C_7H_5O_2Cl} = 156.5 \text{ gm}$$

$$W_{P(C_7H_5O_2Cl)} = \frac{156.5}{2} = 78.25 \approx 78 \text{ grams.}$$

223. (20)

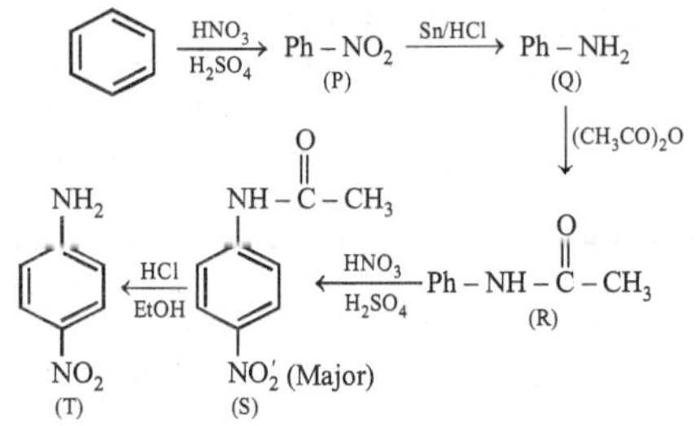

$$\text{Mol. wt of T} = 6\times 12 + (6\times 1) + (2\times 14) + (2\times 16) = 138$$

$$\%N = \frac{28}{138}\times 100 = 20.29\%.$$

224. (2) $\text{D.O.V.} = \dfrac{2C+2-H}{2} = 6$

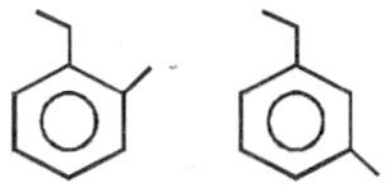

Above two isomers of C_9H_{12} have four different sites for aromatic electrophilic substitution reaction.

225. (1) 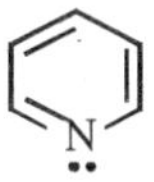

226. (2)

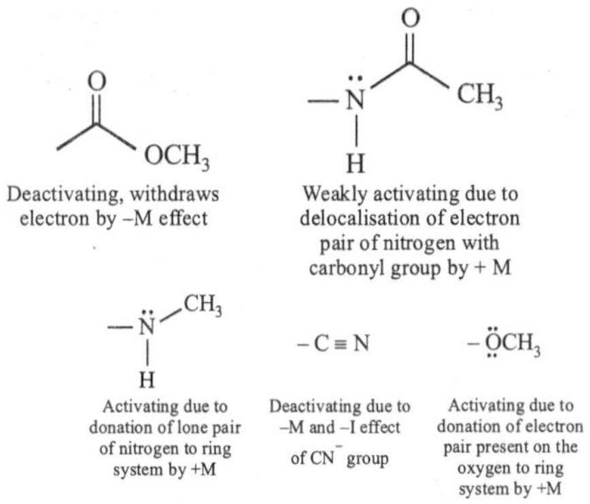

227. (3) B, C and D are aromatic.

B has one aromatic benzene ring, C has two rings fused and conjugated with all atoms sp^2-hybridised, D has three rings in conjugation.

All three have $(4n + 2)\pi$ electrons.

228. (6) 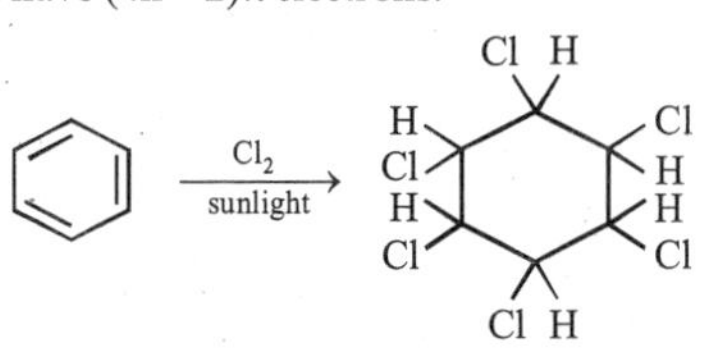

229. (80)

$$\text{Moles of benzene} = \frac{3.9}{78} = 0.05$$

$$\text{Moles of nitrobenzene} = \frac{4.92}{123} = 0.04$$

Percentage yield of nitrobenzene

$$= \frac{0.04}{0.05}\times 100 = 80\%$$

Part-B (JEE Advanced)

1. **(c)** C – C bond energy = 348 kJ/mol = $\frac{348}{4.2}$ kcal/mol
= 82.85 kcal/mol ≈ 100 kcal/mol.

2. **(b)**
(i) Chlorination at C-2 and C-4 produces no chiral compounds
(ii) Chlorination at C-3 produces a chiral carbon marked with star (*d* and *l* form).
(iii) Chlorination at C-1 also produces a chiral carbon marked with star (*d* and *l* form).

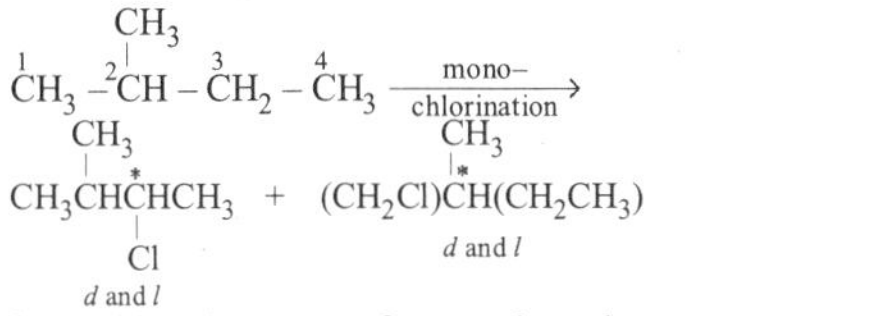

3. **(c)** Any conformation between two extreme positions i.e. eclipsed and staggered is known as gauche or skew form.

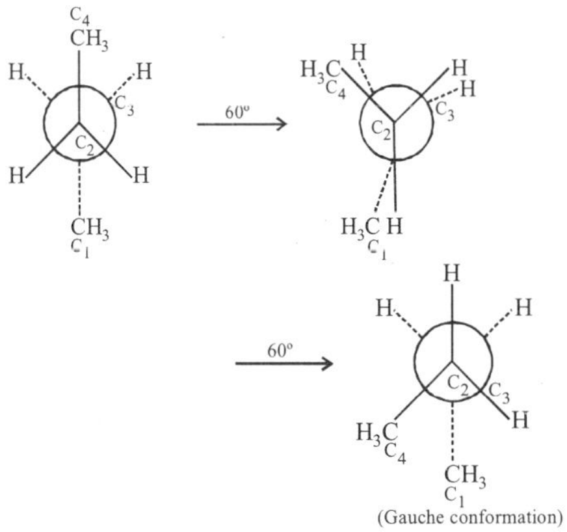

4. **(b)** $Br^{\bullet}$ is less reactive and more selective and so the most stable free radical (3°) will be the major product.

5. **(d)**

6. **(a)** Ethylene has restricted rotation [due to C = C], acetylene has no rotation [due to C ≡ C], hexachloroethane has more rotation than ethylene but less than ethane because of greater size of the substituent (chlorine) than in ethane (substituent is hydrogen).

7. **(a)** In a homologous series, higher the number of C-atoms, higher is the b.p.

8. **(b)**

9. **(2)** The molecule cannot show geometrical isomerism, so only its mirror image will be the other stereoisomer.

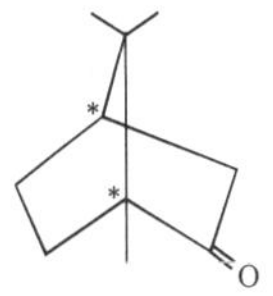

10. **(3)**

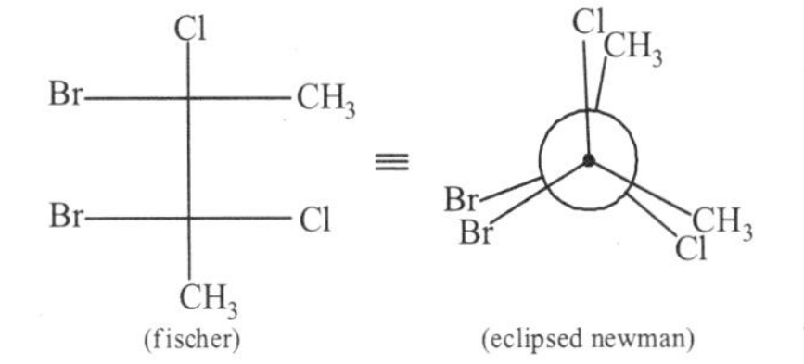

A fischer projection always represents an eclipsed conformation which is least stable. Now, converting this to stable staggered conformation by keeping one carbon atom fixed and other by rotating, we get

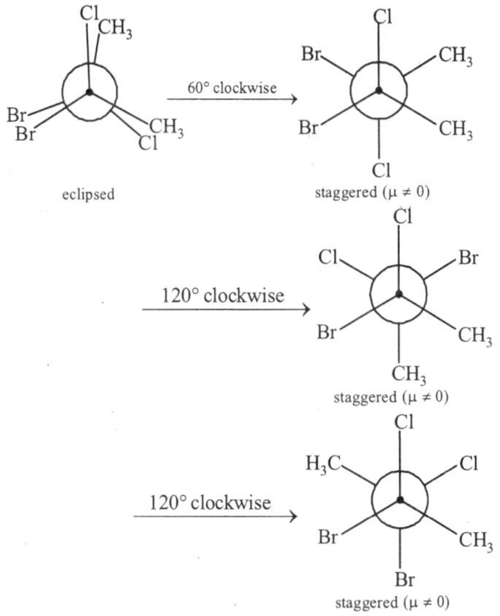

Hence, only 3 different staggered conformers are possible.

11. **(8)**

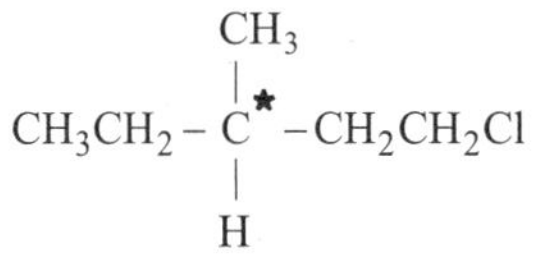

Enantiomeric pair = 2

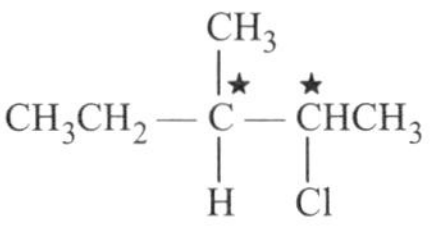

Two enantiomeric pairs = 4

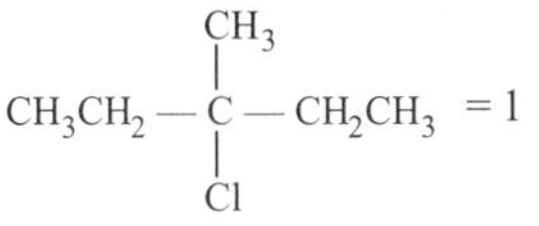

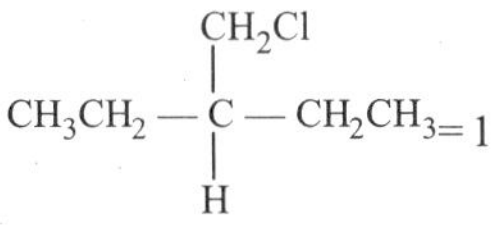

Total = 2 + 4 + 1 + 1 = 8

12. **False :** Bromine is less reactive, hence it is more selective and thus 3° hydrogen will be removed more easily than the 1° hydrogen leading to 2-bromo-2-methylpropane as the main product.

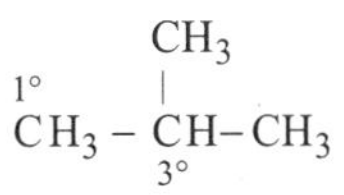

13. **(a, c)**
(a)

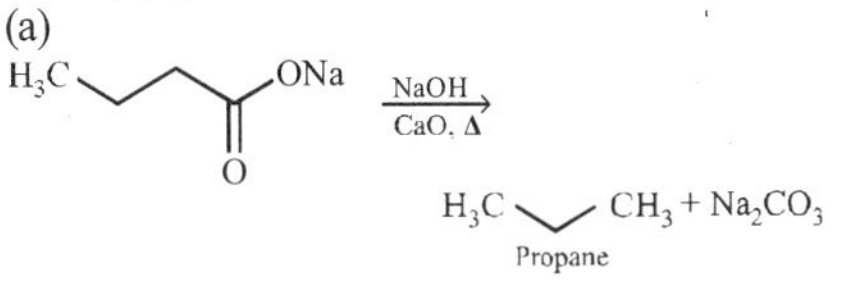

This process of elimination of carbon dioxide from a carboxylic acid is known as decarboxylation.
(b)

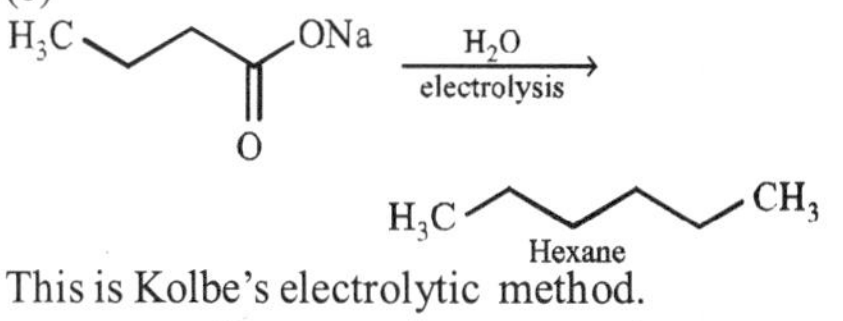

This is Kolbe's electrolytic method.
(c)

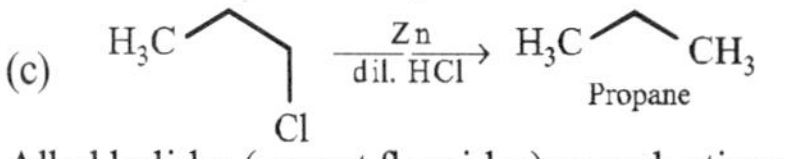

Alkyl halides (except fluorides) on reduction with zinc and dilute hydrochloric acid give alkanes.
(d)

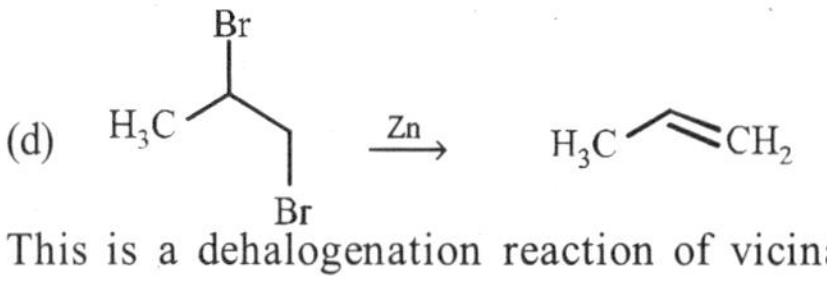

This is a dehalogenation reaction of vicinal dihalide. In this reaction, an alkene is produced on treatment of dihalide with Zn metal by loosing a molecule of ZnX_2.

14. **(a, b, c)** The given conformation are of the following compound

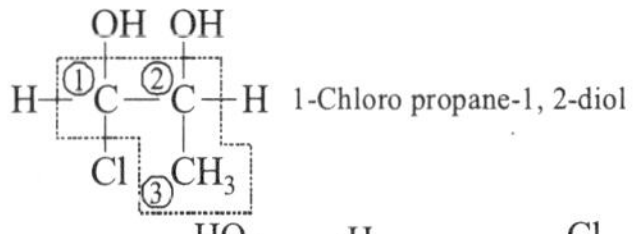

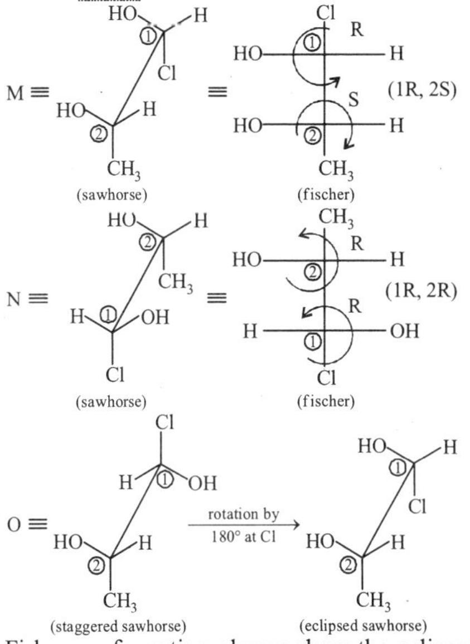

Fisher conformation always show the eclipsed form of molecule. Thus, we have to convert the given staggered conformation of 'O' to an eclipsed conformation.

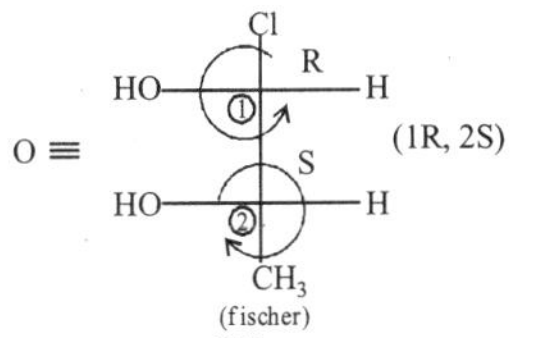

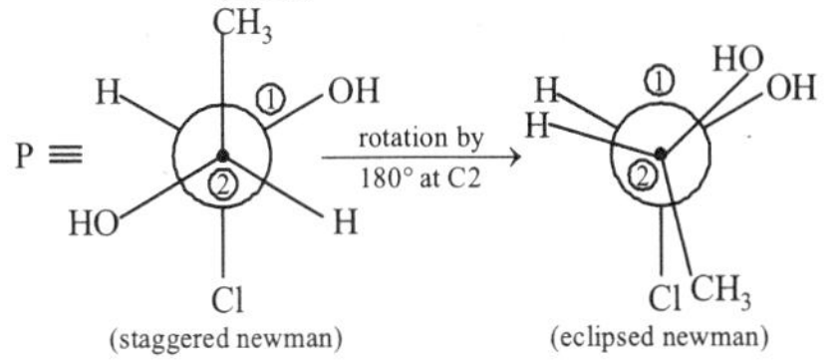

P ≡

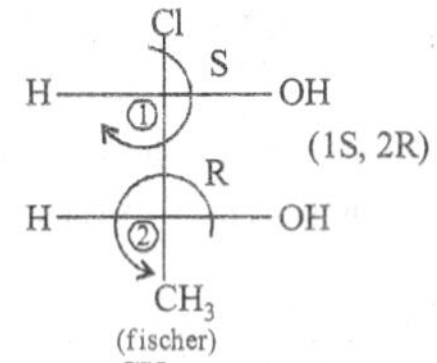

Q ≡

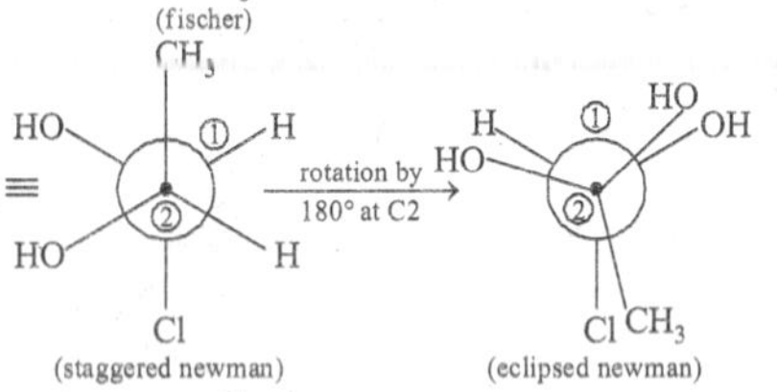

Q ≡

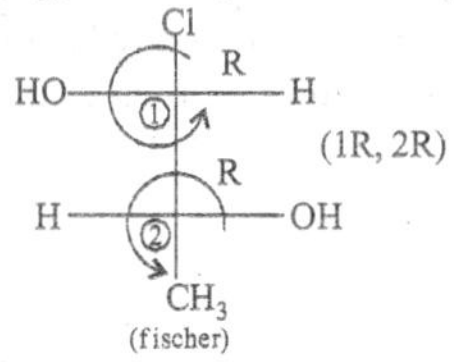

Based on the above structures, we can say that

(a) M and N are diastereomers or non-mirror image stereoisomers.

(b) M and O are identical.

(c) M and P are enantiomers.

(d) M and Q are diastereomers and therefore not identical.

15. (b, d) Structural formula of 2, 2-dimethylbutane is

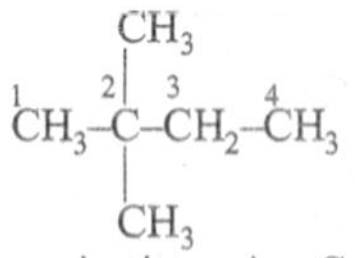

(I) Newman projection using C_1–C_2 bond

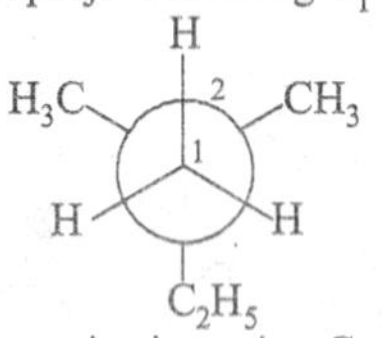

(II) Newman projection using C_3–C_2 bond

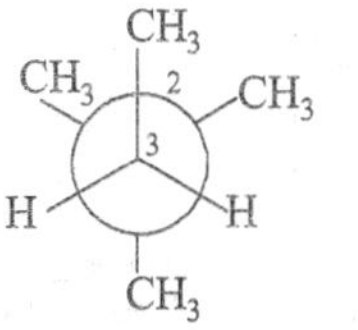

16. (a)

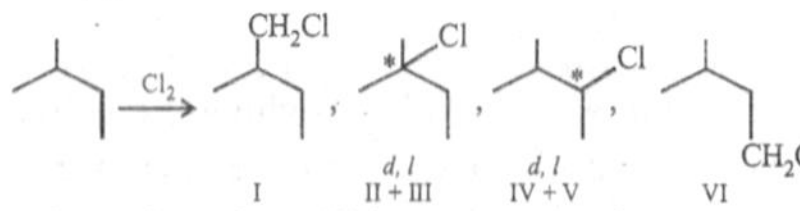

So, the value of *N* will be $1+2+2+1=6$.

Since, enantiomers have nearly same physical properties, II and III as well as IV and V can't be separated. Hence, the number of isomers (*M*) will be

$1+1+1+1=4$.

17.

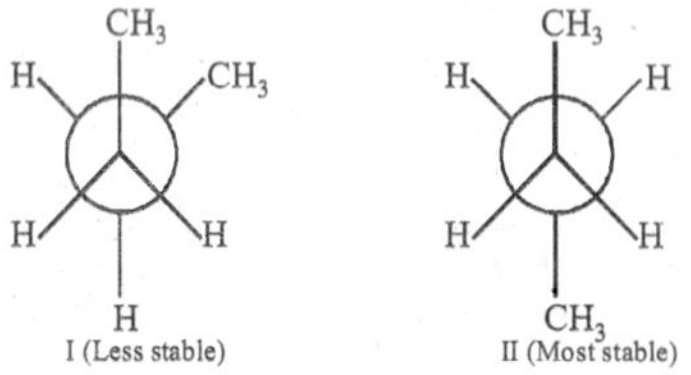

I (Less stable) II (Most stable)

Staggered conformations of *n*-butane

Newman projection formulae II is the most stable because the bulky groups (CH_3) are at maximum possible distance from each other. Structure I is relatively less stable because the two CH_3 groups are close to each other leading to van der Waal's repulsion between the two methyl groups.

18. In this reaction, the leaving group and attacking nucleophile both are $I^{\ominus}$.

Due to strong nucleophile ($I^{\ominus}$) and polar-aprotic solvent (CH_3–CO–CH_3), the reaction will proceed via S_N2 mechanism.

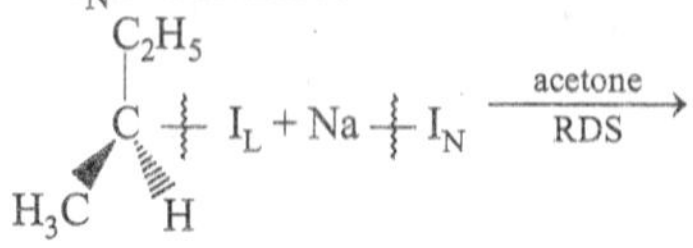

d-2-iodobutane

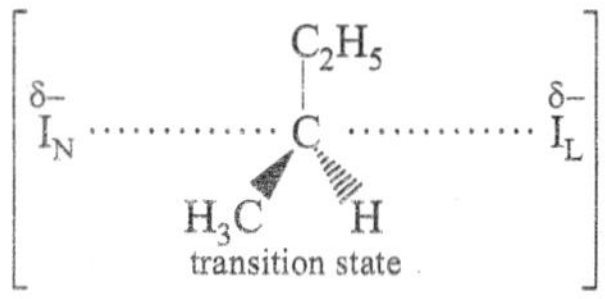

transition state

$+ Na^{\oplus} \xrightarrow{RDS}$

$I_N^{\ominus}$ is attacking nucleophile and $I_L^{\ominus}$ is leaving group. Thus, attacking nucleophile can attack only from the near side.

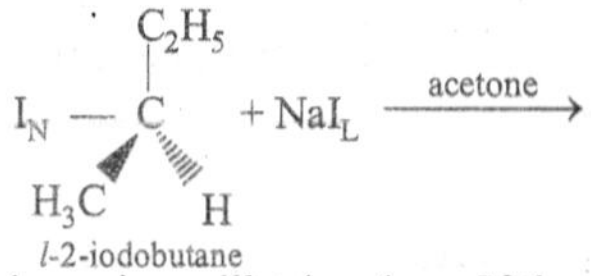

l-2-iodobutane

Hence, inversion will take place. If the starting product was *d*-2-iodobutane then the above compound will be *l*-2-iodobutate.

Now, the same process will happen again and again and the final product will be a equimolar mixture of *d*- and *l*-forms *i.e.* a recemic mixture. Therefore, the final product will not show optical activity.

19. $2C_2H_6 \xrightarrow{\text{monobromination}} 2C_2H_5Br$ (yield 90%) (given)

$2C_2H_5Br \xrightarrow[\text{reaction}]{\text{Wurtz}} C_4H_{10} + 2HBr$ (yield 85%) (given)

Moles of *n*-butane to be produced

$= \dfrac{55\text{ g}}{58\text{ g mol}^{-1}} = 0.948\text{ mol}$

($\because$ molecular mass of $C_4H_{10} = 58$)

Amount of C_2H_5Br (100%) required to obtain 0.948 mol. of $C_4H_{10} = 2 \times 0.948$ mol.

Hence, the amount of C_2H_5Br (85%) required

$= \dfrac{2 \times 0.948 \times 100}{85}$ mol. ...(1)

[$\because$ yield is 85% only]

Further, 1 mole of C_2H_6 gives one mole of C_2H_5Br, hence number of moles of C_2H_6 reqd. for C_2H_5Br in (1)

$= \dfrac{2 \times 0.948 \times 100 \times 100}{85 \times 90}$ mol. $= 2.48$ mol

[$\because$ yield is 90%]

$\therefore$ Required volume of ethane at NTP

$= 22400 \times 2.48 = 55552\text{ mL} =$ **55.55 litres**

20. Chlorination of methane is a free radical substitution reaction.

In dark, chlorine is unable to be converted into free radicals, hence the reaction does not occur.

21. (a)

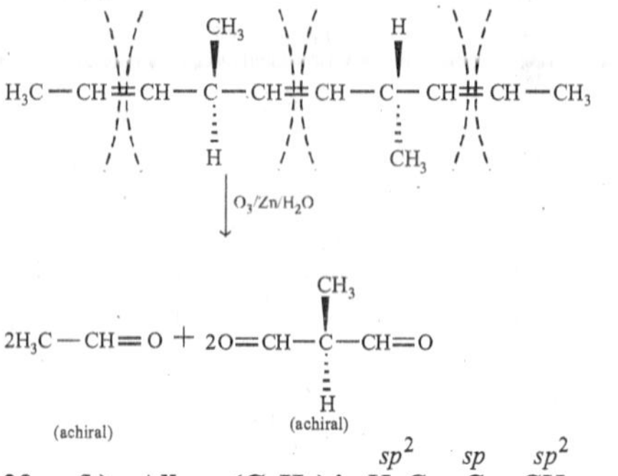

22. (b) Allene (C_3H_4) is $\overset{sp^2}{H_2C} = \overset{sp}{C} = \overset{sp^2}{CH_2}$

23. (a)

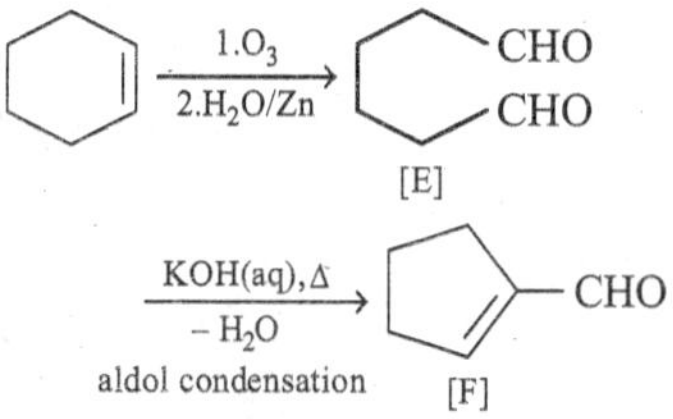

24. (b) Nitrosyl chloride adds on olefins according to Markovnikov's rule, where NO^+ constitutes the positive part of the addendum.

$CH_3CH = CH_2 + NOCl \longrightarrow CH_3\underset{Cl}{CH}\underset{NO}{CH_2}$

25. (a) The π bond is formed by the sideways overlapping of two *p*-orbitals of the two carbon atoms.

The molecular plane does not have any π electron density as the *p*-orbitals are perpendicular to the plane containing the ethene molecule. The nodal plane in the π-bond of ethene is located in the molecular plane.

26. (b) Alkenes undergo electrophilic addition reaction.

With HOCl.

$HOCl \longrightarrow OH^- + Cl^+$

$CH_3 \rightarrow CH = CH_2 \xrightarrow[1^{st}\text{ step}]{Cl^+} CH_3 - \overset{+}{C}H - CH_2Cl$

More stable

or $CH_3 \rightarrow CH = CH_2 \xrightarrow[1^{st}\text{ step}]{OH^-} CH_3 - \underset{OH}{CH} - \overset{+}{C}H_2$

Less stable

So, it is the Cl^+ that attacks in the first step

$CH_3 - \overset{+}{C}H - CH_2Cl \xrightarrow[2^{nd}\text{ step}]{OH^-} CH_3\overset{OH}{C}H - CH_2Cl$

27. (c) Peroxide effect is effective only in case of HBr and not in case of HCl and HI.

Step: I (a) $R - O - O - R \xrightarrow{\Delta} 2RO^{\bullet}$;

Step: I (b) $RO\cdot + H-X \longrightarrow RO-H + X\cdot$

Step: II $R'CH=CH_2 + X^{\cdot} \longrightarrow$

$$R\dot{C}H-CH_2X + R'CH(X)-\dot{C}H_2$$
(more stable) (less stable)

$$R'-\dot{C}H-CH_2X + HX \longrightarrow R'-CH_2-CH_2X + X\cdot$$

For HCl, step–I (b) is endothermic while step–II is exothermic but for HI, step–I(b) is exothermic while step–II is endothermic. Since one of the propagating step is endothermic, the reaction does not occur.

28. (a) The relative rates of hydrogenation decreases with increase of steric hindrance.

$$\underset{(b)}{R_2C=CH_2} > \underset{(a)}{RCH=CHR} > \underset{(c)}{R_2C=CHR} > \underset{(d)}{R_2C=CR_2}$$

Among the four olefins, (a) and (b) are less stable (less no. of (α – H). Further in (a), the bulky alkyl groups are on same side of the double bond, hence (a) is more reactive than (b).

29. (a) $PhCH_2CH=CHCH_3$ will exhibit geometrical isomerism because in others one of the doubly bonded carbon atom has two similar groups.

30. (c) *anti*-Markovnikoff's addition of HBr is observed only with unsymmetrical alkenes *i.e.*, options a, b, and d.

$\underset{(a)}{CH_3CH=CH_2}$ $\underset{(b)}{CH_2=CHCH_2CH_3}$

$\underset{(c)}{CH_3CH=CHCH_3}$ $\underset{(d)}{CH_3CH=CHCH_2CH_3}$

31. (a) Cold alkaline $KMnO_4$

32. (a) *cis-trans* – Isomerism is due to restricted rotation either due to carbon-carbon double bond or due to cyclic structure.

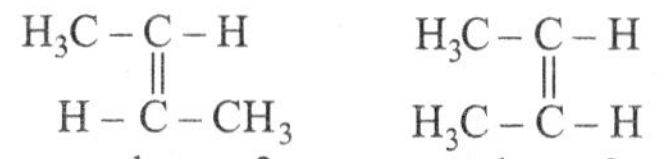

trans-butene-2 *cis*-butene-2

33. (b) Unsaturated hydrocarbons decolourise alk. $KMnO_4$ solution; C_2H_4 ($H_2C=CH_2$) is an alkene.

34. (0)

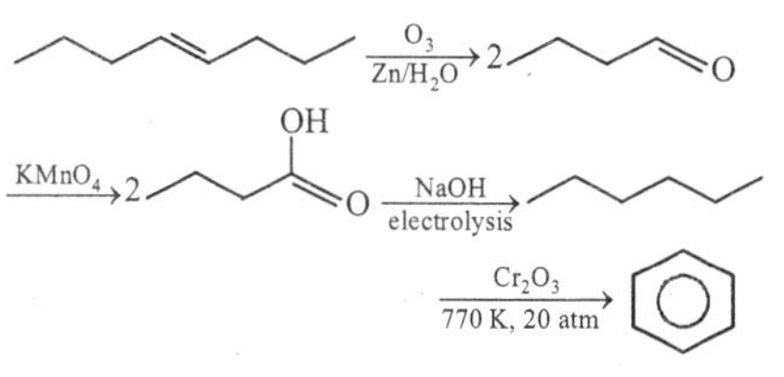

Number of $-CH_2-$ groups in the product = 0.

35. (2)

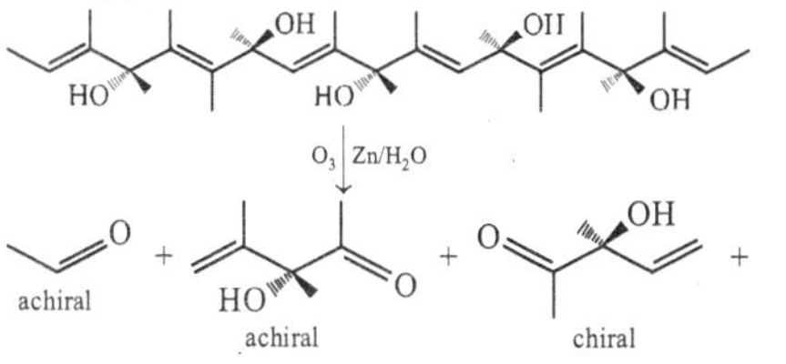

36. (13)

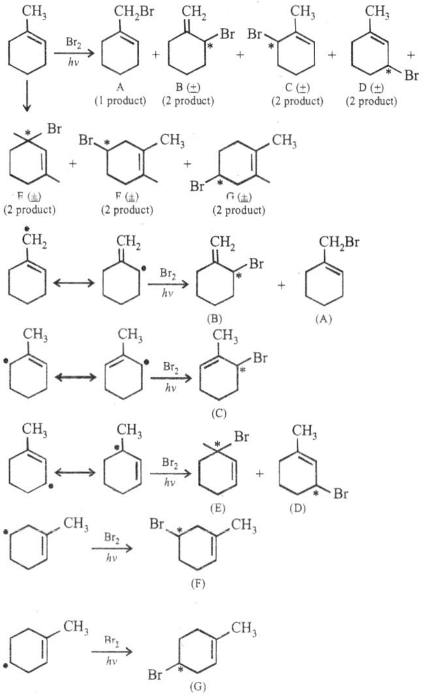

So, total 13 possible isomers are formed.

37. (7) The seven possible cyclic structural and stereoisomers are

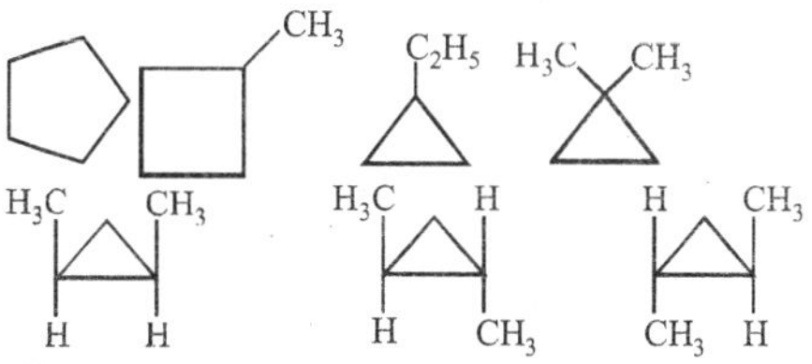

38. 3, 4-dibromo-1-butene (at low temperature) or **1, 4- dibromo-2-butene** (at high temperature). These products are formed by 1, 2-addition and 1-4 addition respectively.

39. H_2SO_4, $HgSO_4$

$$\begin{array}{l} CH_2COOK \\ | \\ CH_2COOK \end{array} \longrightarrow \begin{array}{l} CH_2COO^- \\ | \\ CH_2COO^- \end{array} + 2K^+$$

$$\downarrow$$

$$CH_2=CH_2 + 2CO_2$$

40. ethylene

41. (b, d)

(b) *M* and *N* are identical. Further, *O* and *P* are enantiomers. Hence, *M* and *O* and *N* and *P* are two sets of diastereomers.

(d) Bromination proceeds through *trans*- or *anti*-addition on alkenes, *i.e.*, in both the reactions. Thus upon addition of bromine, *trans*-alkenes give *meso*-product, while *cis*-alkenes give enantomeric pair.

(i)

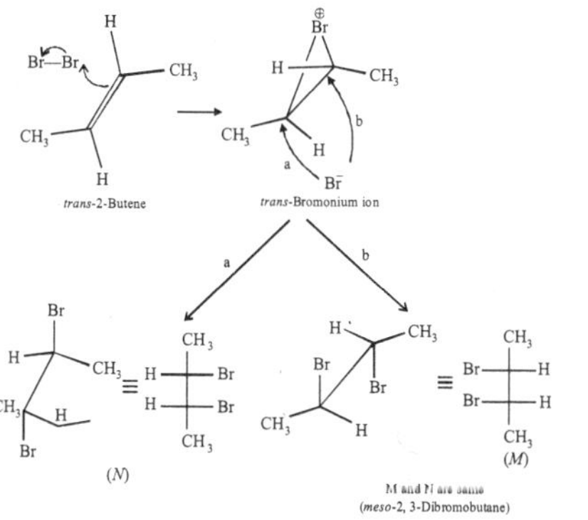

(ii)

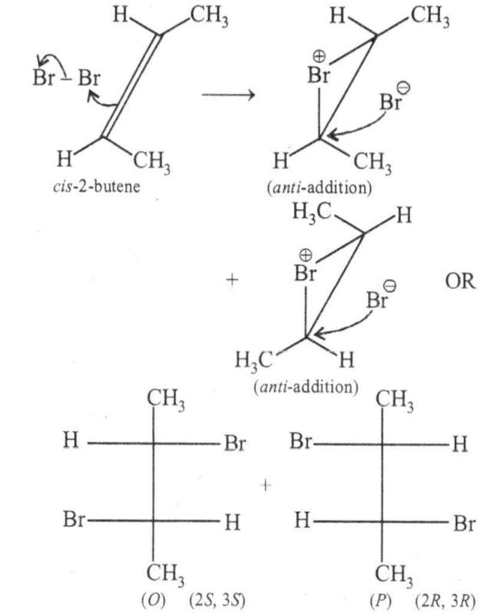

Pair of enantiomers

42. (b, d)

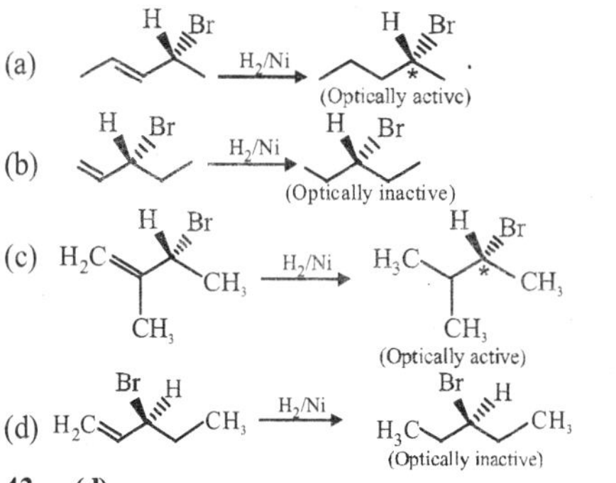

43. (d)

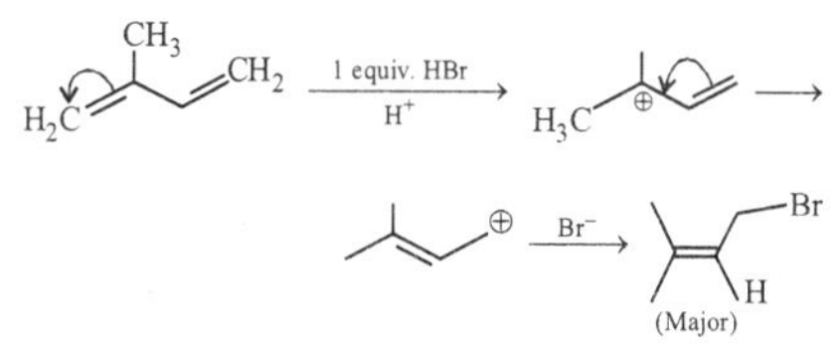

more substituted alkene is more stable hence it is the major product.

44. (b) Heat of hydrogenation is related to stability of molecules; higher the stability, lower is the heat of hydrogenation.

Butadiene, $CH_2=CHCH=CH_2$ has two double bonds so its heat of hydrogenation will be more than the other three.

Alkenes follow the following order of stability

$$R_2C=CR_2 > R_2C=CHR > RCH=CHR$$
tetrasubstituted (most stable due to hyper conjugative sturctures)

$$> RCH=CH_2 > CH_2=CH_2$$
Unsubstituted (least stable due to no hyperconjugation)

Thus here, stability order of the given monoalkenes is

$$CH_3CH=CHCH_3 > CH_3CH_2CH=CH_2$$
Butene-2 (*trans*- and *cis*-) Butene-1

The *trans*-2-butene is more stable than the *cis*- because in the *cis*-isomer the two bulky groups are crowded together with the result, it has more van der Waal's strain than the *trans*-isomer.

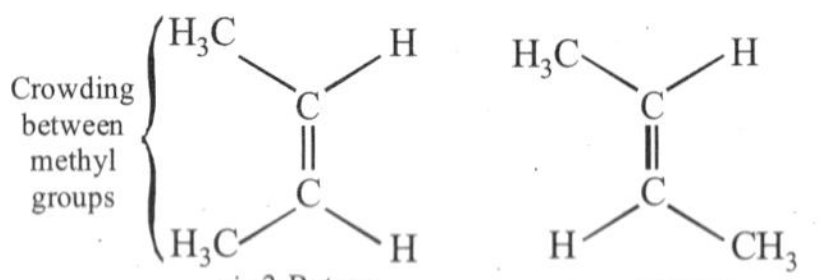

cis-2-Butene   *trans*-2-Butene (Less van der Waal's strain hence more stable)

45. (a)

(P)
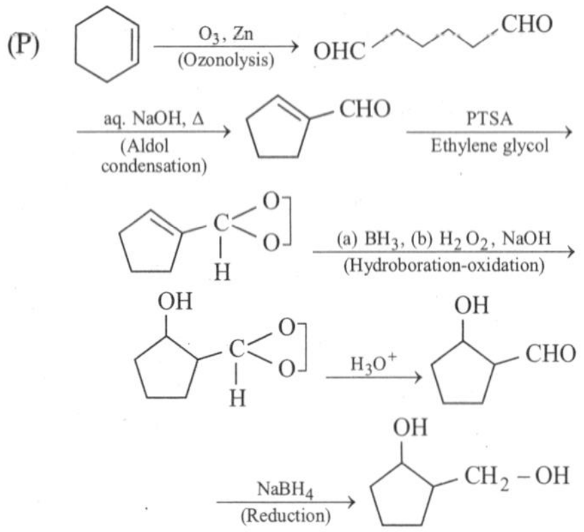

(Q)
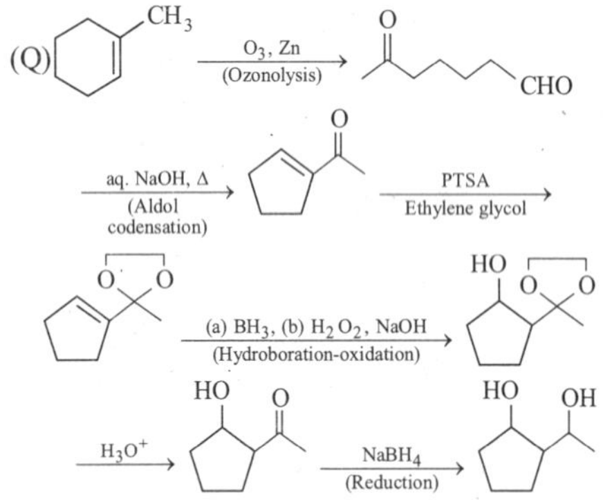

(R)
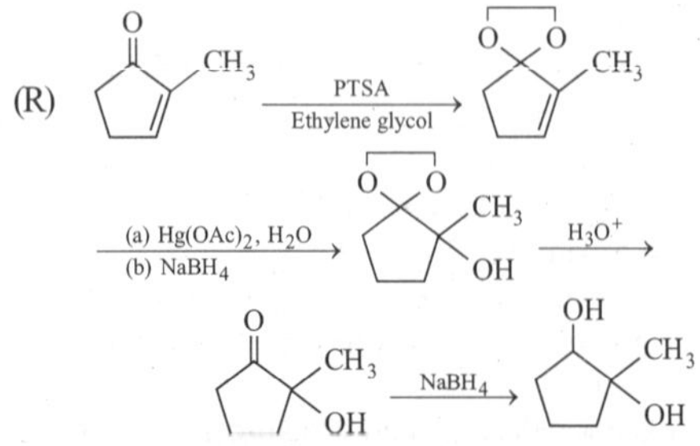

(S)
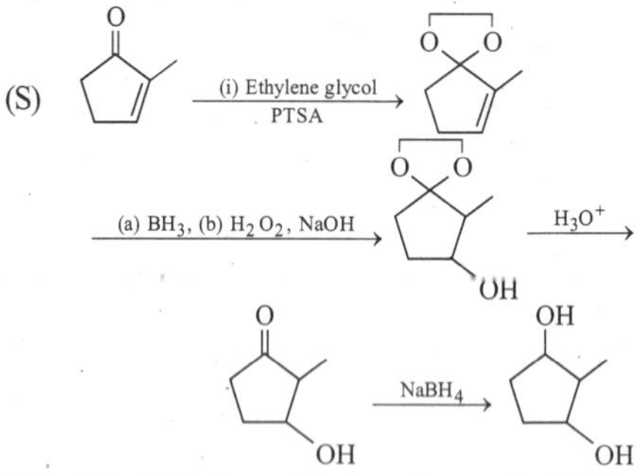

46. (b) With *trans*-2-butene, the product of Br_2 addition is optically inactive.
Even though, both assertion and reason are correct the correct reason for the formation of *meso*–2,3–dibromobutane from *trans*–2–butene is *anti* addition of Br_2.

47. (c) In presence of peroxide, addition of HBr on alkenes takes place via free radicals. Here, assertion is correct but reasoning is incorrect. Here, two free radical are formed, 2° *free radical, being more stable, governs the product leading to* 1-*bromobutane.*

$$CH_3CH_2CHBr\dot{C}H_2 \xleftarrow{Br^{\bullet}} CH_3CH_2CH=CH_2$$
1° free radical (less stable)

$$CH_3CH_2CH=CH_2 \xrightarrow{Br^{\bullet}} CH_3CH_2\dot{C}HCH_2Br$$
2° free radical (more stable)

48. (a) $CH_3CH_2-CH=CH_2+Br_2 \longrightarrow CH_3CH_2\overset{*}{C}HCH_2Br$ (with Br on the starred carbon)

49.
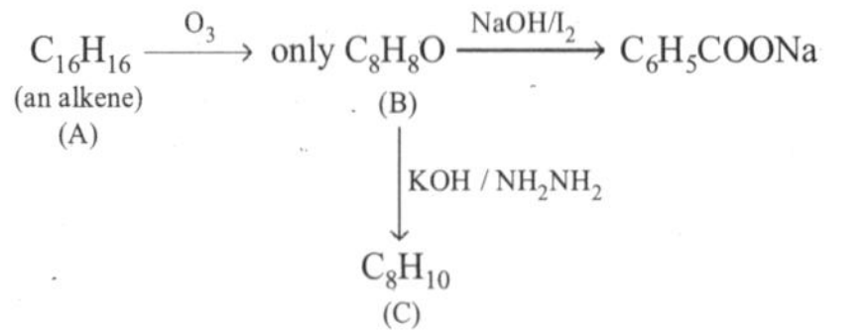

(i) Conversion of B (C_8H_8O) to sodium benzoate involves iodoform reaction, hence (B) must contain – $COCH_3$ group leading to $C_6H_5COCH_3$ (C_8H_8O) as its molecular formula.
(ii) Since, the given alkene gives only one product (B) or C_6H_5–$COCH_3$ on ozonolysis, so the given alkene must be a symmetrical alkene containing a double bond in centre. Thus, the alkene (A) must have following structure:

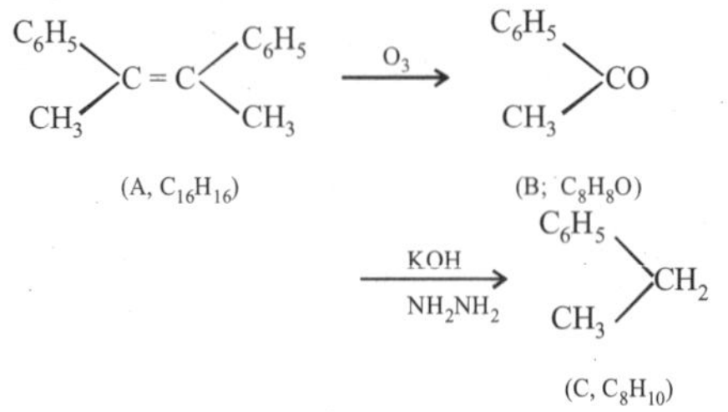

Isomeric structures of A

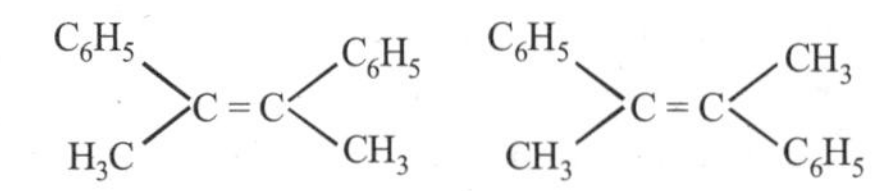
cis   *trans*

Since catalytic hydrogenation of alkenes takes place in (*syn*-) manner; hence recemic mixture will be formed by the *trans*-isomer.

50. (i)
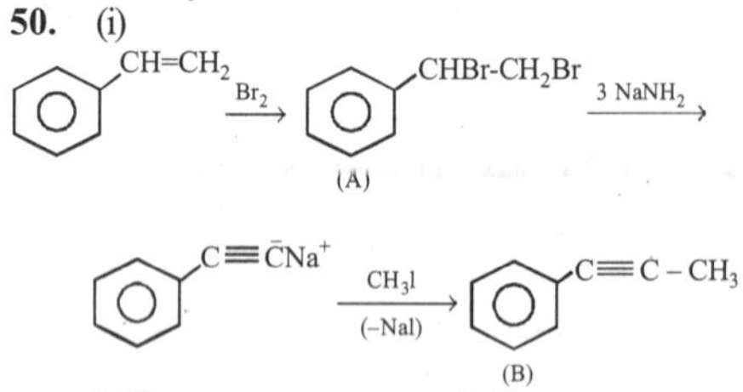

(ii)
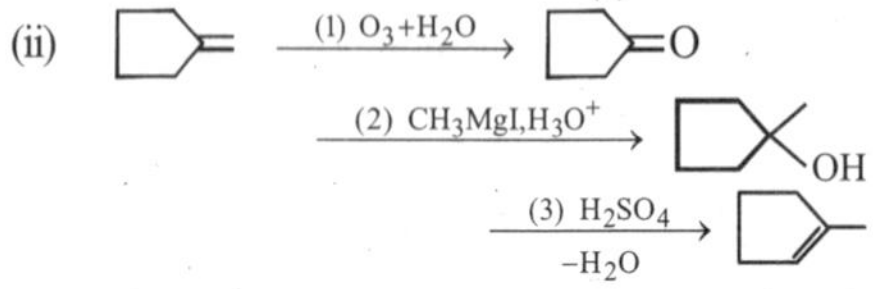

51. (i) It should be an alkene as it adds one mole of H_2 to form *n*-hexane.
(ii) The C_6 alkene should be symmetrical because on oxidation it gives a single carboxylic acid having three carbon atoms.

$$\underset{n\text{-Hexane}}{CH_3(CH_2)_4CH_3} \xleftarrow{H_2} \underset{(A)}{CH_3CH_2CH=CHCH_2CH_3} \xrightarrow[KMnO_4]{(O)} 2CH_3CH_2COOH$$

52. (a) $CH_3CH_2-\underset{CH_2CH_3}{\underset{|}{C}}=CH-CH_3$ (unsymmetrical)

$$\xrightarrow[\text{(Peroxide effect)}]{\text{HBr/Peroxide}} CH_3CH_2-\underset{CH_2CH_3}{\underset{|}{CH}}-\overset{Br}{\overset{|}{CH}}-CH_3$$

(b) $CH_3CH_2-\underset{CH_2CH_3}{\underset{|}{C}}=CH-CH_3$

$$\xrightarrow{Br_2/H_2O} CH_3CH_2-\underset{CH_2CH_3}{\underset{|}{\overset{OH}{\overset{|}{C}}}}-\overset{Br}{\overset{|}{CH}}-CH_3$$

(c) $CH_3CH_2-\underset{CH_2CH_3}{\underset{|}{C}}=CH-CH_3$

$$\xrightarrow[NaBH_4]{Hg(OAc)_2/H_2O} CH_3CH_2-\underset{CH_2CH_3}{\underset{|}{\overset{OH}{\overset{|}{C}}}}-\overset{H}{\overset{|}{CH}}-CH_3$$

53. Summary of the given reactions

$$\underset{(F)}{C_5H_{12}} \xleftarrow{H_2} \underset{(E)}{C_5H_8} \xrightarrow{\text{ozonolysis}} HCHO + \underset{\text{2–Ketopropanal}}{CH_3-\overset{O}{\overset{||}{C}}-CHO}$$

Since, hydrogenation of (E) to (F) takes up two molecules of hydrogen, it indicates the presence of two double bonds in (E) which is further supported by its ozonolysis to form two products having three carbonyl groups. Further, structure of ozonolysis product leads to following structure to compound (E).

$$CH_3-\underset{(F)}{\overset{CH_3}{\overset{|}{C}H}}-CH_2-CH_3 \xleftarrow{2H_2} CH_3-\underset{(E)}{\overset{CH_3}{\overset{||}{C}}}-CH=CH_2$$

$$\downarrow O_3$$

$$\underset{\text{2-Ketopropanal}}{CH_3-\overset{O}{\overset{||}{C}}-CHO} + \underset{\text{Formaldehyde}}{2CH_2O}$$

54. (i) 1, 4-Pentadiene reacts with HCl in presence of benzoyl peroxide in Markownikoff's way.

Peroxide effect applies to HBr only.

(ii) Grignard reagent reacts with ethyl acetate to form ketones, or *ter*-alcohol if Grignard reagent is taken in excess.

Thus, the given reactions can be written as below.

$$H_2C=CH-CH_2-CH=CH_2$$

$$\xrightarrow[(C_6H_5CO)_2O]{\text{excess HCl}} \underset{(X)}{H_3C-\underset{Cl}{\underset{|}{C}H}-CH_2-\underset{Cl}{\underset{|}{C}H}-CH_3}$$

$$\xrightarrow{\text{Mg in dry ether}} \underset{(Y)}{CH_3-\underset{MgCl}{\underset{|}{C}H}-CH_2-\underset{MgCl}{\underset{|}{C}H}-CH_3}$$

$$\xrightarrow{CH_3COOC_2H_5} \underset{(Z)}{CH_3-\underset{COCH_3}{\underset{|}{C}H}-CH_2-\underset{COCH_3}{\underset{|}{C}H}-CH_3}$$

55. $A \xrightarrow[\text{low temp.}]{\text{dry KOH}} \underset{\text{deep red}}{B} + C\uparrow$

$A \xrightarrow[\text{but-2-ene}]{Zn/H_2O} CH_3CHO$

The reaction of gas (A) with but-2-ene followed by treatment with Zn/H_2O gives CH_3CHO. This shows that the gas (A) is ozone (O_3).

$CH_3CH=CHCH_3 + O_3 \longrightarrow$

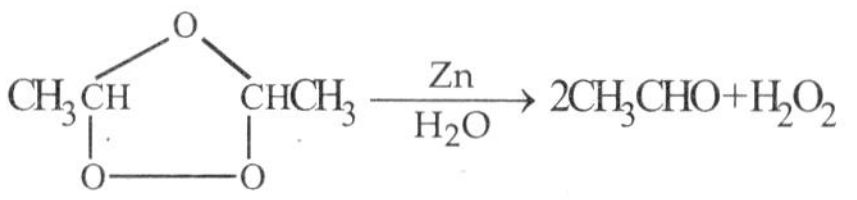

Reaction of ozone with KOH.

$$\underset{(A)}{4KOH + 4O_3} \longrightarrow \underset{\text{Pot. ozonide, (B) (Red colour)}}{4KO_3} + 2H_2O + \underset{(C)}{O_2\uparrow}$$

56. An optically active hydrocarbon D (C_6H_{12}) should have an asymmetric C-atom and C_6H_{14} will have **no** asymmetric C-atom, hence D would be **3-methylpentene-1**.

$$\underset{(D)\ \text{Optically active}(C_6H_{12})}{CH_3-CH_2-\overset{*}{C}H(CH_3)-CH=CH_2}$$

$$\xrightarrow{H_2} \underset{\text{Optically inactive }(C_6H_{14})}{CH_3CH_2-CH(CH_3)-CH_2CH_3}$$

57. B should be a symmetric alkene (**butene-2**) $CH_3CH=CHCH_3$ gives same product $CH_3-CH(Br)-CH_2-CH_3$ in presence /absence of peroxide.

58. Cyclohexene gives positive response to *bromine water test* and *Baeyer's test* while cyclohexane does not respond to these reagents.

59. Under normal conditions, *ter*-butyl bromide is formed, isobutyl bromide is formed in presence of peroxide.

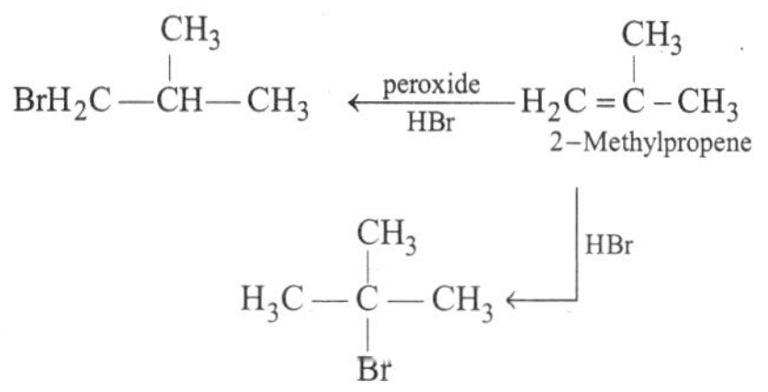

60. (i) Addition of unsymmetrical addendum (HBr in present case) to unsymmetrical olefin ($CH_3CH = CH_2$, in present case) takes place according to Markownikoff rule.

$$CH_3CH=CH_2 + HBr \rightarrow CH_3.CHBr.CH_3$$

Propene *iso*-Propyl bromide

(ii) 1, 3 - Butadiene is a conjugated diene and is a reasonance hybrid:

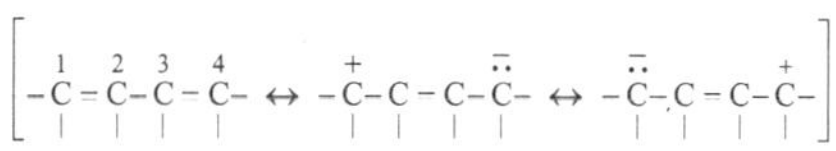

Thus resonance induces some double bond character in the central C-C bond leading to the shortening of this bond. **Alternatively**, all the four C atoms of 1, 3–butadiene are sp^2 hybridised and thus their C – C bond length will be lower than that of n-butane in which all the four C atoms are sp^3 hybridised.

61.

$$\underset{\text{Propene}}{CH_3-CH=CH_2} + H_2O + [O] \xrightarrow{KMnO_4} \underset{\text{Propene glycol}}{CH_3-\overset{OH}{\overset{|}{C}H}-\overset{OH}{\overset{|}{C}H_2}}$$

[Colour of $KMnO_4$ is discharged]

62.

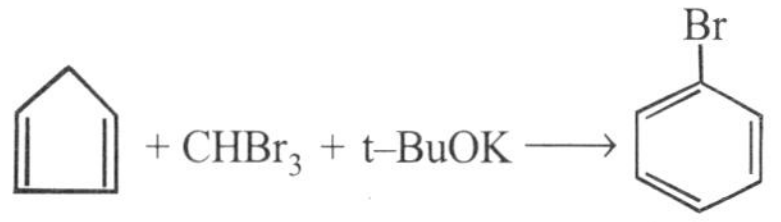

63. (d) Only (d) can form 3-octyne

$$CH_3CH_2C\equiv CH \xrightarrow[-NH_3]{NaNH_2}$$

$$CH_3CH_2C\equiv C^-Na^+ \xrightarrow[(S_N2)]{CH_3CH_2CH_2CH_2Br}$$

$$CH_3CH_2C\equiv CCH_2CH_2CH_2CH_3 + NaBr$$

64. (d) $H_2/Pd/BaSO_4$ reduces an alkyne to *cis*-alkene, H_2/Pt reduces it to alkane, $NaBH_4$ does not reduce an alkyne. Reduction of an alkyne by active metal in liq. NH_3 gives *trans*-alkene.

65. (a) Reaction proceeds through carbocation intermediate:

$$Ph-C\equiv C-CH_3 + H^+ \longrightarrow$$

$$Ph-\overset{+}{C}=CH-CH_3 \xrightarrow{H_2O}$$

Resonance stabilised

$$\underset{\text{Unstable enol}}{Ph-\overset{OH}{\overset{|}{C}}=CH-CH_3} \xrightleftharpoons[\text{(Tautomerisation)}]{} Ph-\overset{O}{\overset{||}{C}}-CH_2-CH_3$$

66. (d) 1-Butyne has terminal hydrogen which is acidic whereas in 2-butyne there is no terminal hydrogen. Thus, 2-butyne will not react with ammonical Cu_2Cl_2, while 1-butyne, will give red ppt. with ammonical cuprous chloride

67. (b) Addition on triple bond takes place by the *syn*-addition of hydrogen due to Lindlar catalyst.

Since the configuration of the double bond already present is *cis*, the compound formed will not remain chiral and hence, optically inactive.

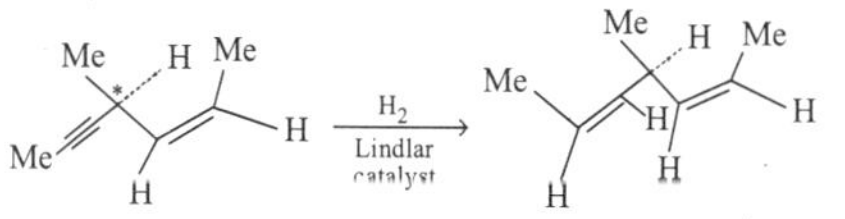

meso-compound

68. (d) In propyne ($CH_3C\equiv CH$), the terminal hydrogen is acidic and reacts with ammonical $AgNO_3$.

69. (a) Hydration of alkynes via mercuration takes place in accordance with Markovnikov's rule.

$$CH_3CH_2C\equiv CH \xrightarrow[HgSO_4/H_2SO_4]{+2H_2O} \left[CH_3CH_2-\underset{OH}{\underset{|}{\overset{OH}{\overset{|}{C}}}}-CH_3\right] \xrightarrow{-H_2O}$$

$$CH_3CH_2-\underset{O}{\underset{||}{C}}-CH_3$$

70. (a) Acidic hydrogen is present in alkynes, having H on the triply bonded C-atom. They can be easily removed by means of a strong base.

71. (c) $CH_3C\equiv CH + H_2O \xrightarrow[HgSO_4]{H_2SO_4}$

$$[CH_3CH(OH)=CH_2] \rightarrow \underset{\text{acetone}}{CH_3COCH_3}$$

72. (6.00)

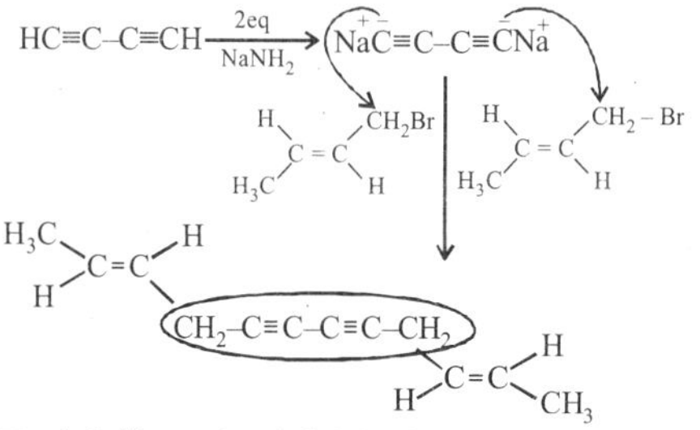

Total Collinear (straight) Carbons = 6

73. C_2H_2

$$CH\equiv CH + HCl \longrightarrow CH_2=CHCl \longrightarrow PVC$$

74. 2-butyne

$$CH\equiv CH + Na \xrightarrow[\text{Liq } NH_3]{} CH\equiv CNa$$

$$\xrightarrow{NaNH_2} NaC\equiv CNa$$

$$NaC\equiv CNa + 2CH_3I \rightarrow \underset{\text{2-Butyne}}{CH_3C\equiv C\,CH_3}$$

75. Ethyne, because of the high s character of the carbon of the – C ≡ H bond in ethyne (sp hybridisation).

76. False : Ethylene reacts with sulphuric acid to form ethyl hydrogen sulphate. It can be dried by passing it through phosphorus pentoxide.

77. (b, c)

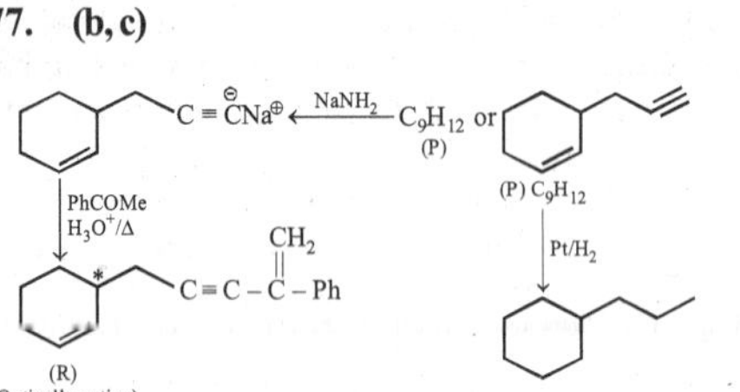

(R)
(Optically active)

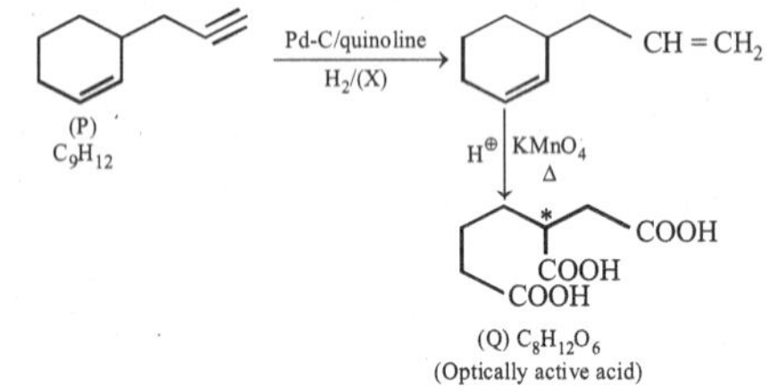

78. (a)

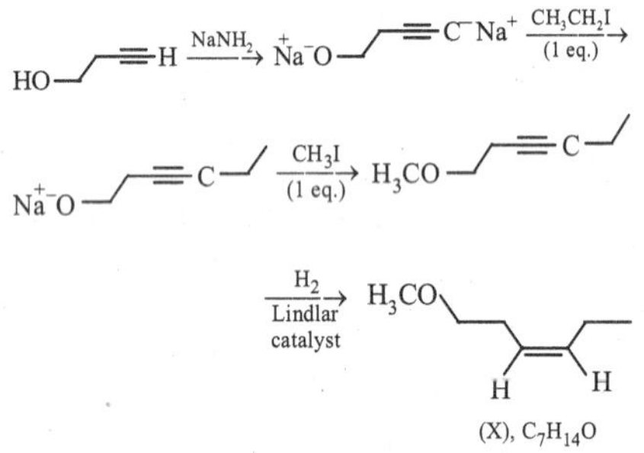

Note : Lindlar's catalsyt will form the *cis*-isomer.

79. (c)

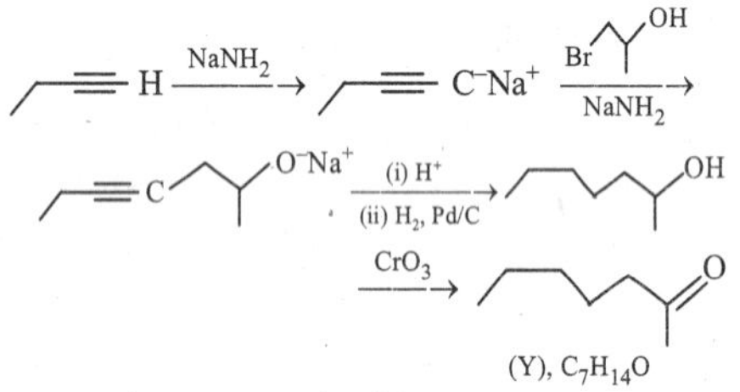

Note: The compounds with $-\overset{\overset{\displaystyle O}{||}}{C}-CH_3$ group, gives positive iodoform test.

For 80 & 81

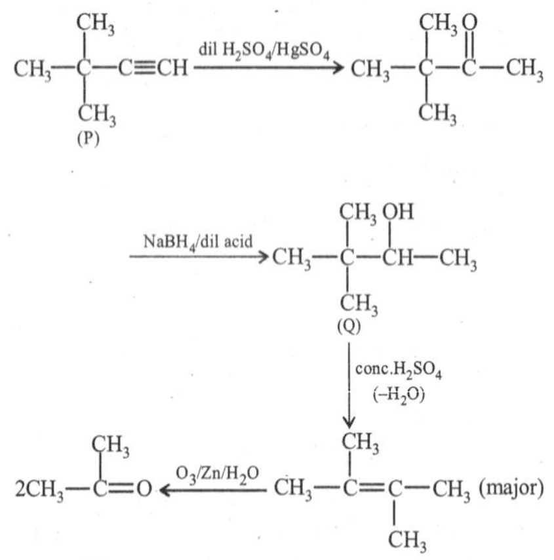

80. (d) 81. (b)

82. Ziegler – Natta catalyst ($R_3Al + TiCl_4$)

83. Meso forms are optically inactive.

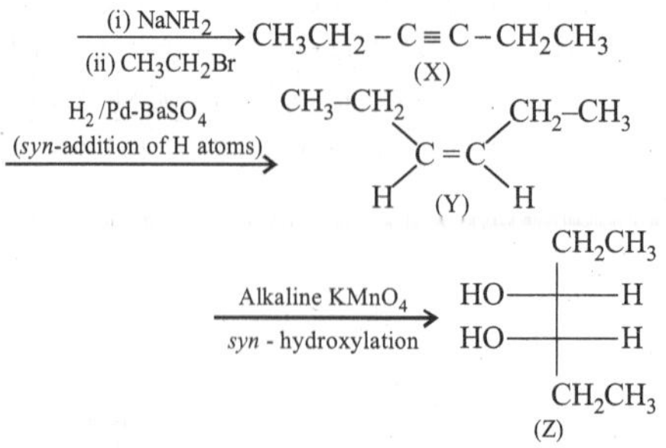

(Z) is in meso form having plane of symmetry. The upper half molecule is mirror image of the lower half molecule. The molecule is, therefore, optically inactive due to internal compensation.

84. Summary of the given facts

$$\underset{(A)}{C_8H_{10}} \xrightarrow[\text{(ii)hydrolysis}]{\text{(i)}O_3} \underset{(B)}{C_4H_6O_2} \xleftarrow[\text{(ii)}CO_2,\ \text{(iii)}H^+]{\text{(i) Mg in dry ether}} \underset{(C)}{C_3H_5Br}$$

Since compound (B) is obtained from compound (C) C_3H_5Br through reaction with Mg and CO_2, it seems that compound (B) is a carboxylic acid formed via the formation of Grignard reagent. Hence, compound (C) should be an alkyl halide having three carbon atoms. The alkyl halide (C) is unsaturated (indicated by number of hydrogen atoms) which is present in the form of ring and thus (C) should be bromocyclopropane.

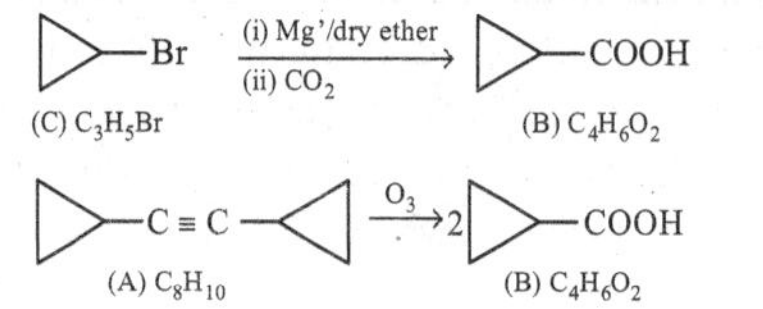

85. $R-C\equiv C-R \xrightarrow[\text{Lindlar catalyst}]{H_2}$

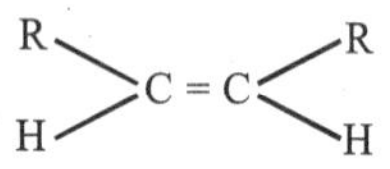

(*cis*-alkene)

(i) **Lindlar's catalyst** is Pd supported over $CaCO_3$ which is partially poisoned by $(CH_3COO)_2Pb$. It can restrict the hydrogenation of alkyne to alkene stage. It yields a *cis*-alkene.

86. (i) By amm. $AgNO_3$ or by acidic-H tests : Terminal alkynes give white precipitate with amm. $AgNO_3$ or red ppt. with amm. Cu_2Cl_2 (H atom attached on *sp* hybridized carbon is acidic).

$$2CH_3CH_2C\equiv CH + Ag_2O \rightarrow 2CH_3CH_2C\equiv CAg + H_2O$$

1-Butyne

$$CH_3-C\equiv C-CH_3 + Ag_2O \rightarrow \text{No reaction}$$

2-Butyne

Only terminal alkynes respond to these reactions.

87. Calculation of molecular formula of *A*.

Element	Percentage of atoms	Relative no. ratio	Simplest whole
C	85.7	$\frac{85.7}{12} = 74.14$	$\frac{7.14}{7.14} = 1$
H	14.3	$\frac{14.3}{1} = 14.3$	$\frac{14.3}{7.14} = 2$

$\therefore$ Empirical formula of $A = CH_2$

Determination of molecular weight of A

1 g of *A* consumes = 38.05 g of 5% Br_2 (in CCl_4)

$$= \frac{38.05 \times 5}{100} \text{ g of } 100\%\ Br_2 = 1.90 \text{ g of } 100\%\ Br_2.$$

Now since, 1.90 g of Br_2 is consumed by 1 g of compound *A*.

$\therefore$ 160 g (1 mole) of Br_2 will be consumed by

$$= \frac{1}{1.90} \times 160 = 84.2 \text{ g of A} = 84.0 \text{ (app.)g of A}$$

$\therefore$ Molecular weight of *A* = **84**

$$\text{Hence, } n = \frac{84}{12+2} = 6$$

$\therefore$ Molecular formula of $A = (CH_2)_6 = \mathbf{C_6H_{12}}$

Since, the hydrocarbon *A* consumes 1 molar equivlaent of hydrogen, it must contain one double bond. Oxidation of compound *A* with $KMnO_4$ to form compound C (C_4H_8O) and acetic acid indicates = $CH.CH_3$ fragment in *A*, *i.e.*

$$\underset{A(C_6H_{12})}{C_4H_8 = CHCH_3} \xrightarrow{KMnO_4} \underset{C}{C_4H_8O} + \underset{\text{Acetic acid}}{CH_3COOH}$$

Now the fragment C_4H_8 of *A* on oxidation forms the compound '*C*' (C_4H_8O) which may be easily obtained from butyne-2 and acidic aq. $HgSO_4$, the compound '*C*' must be ethylmethyl ketone.

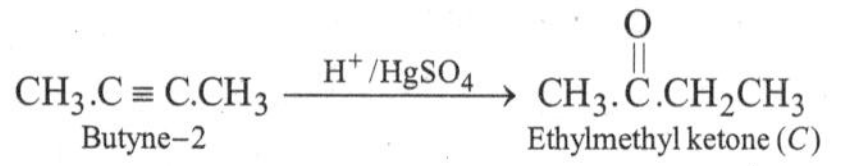

The formation of ketone '*C*' (uthyl methyl ketone) and acetic acid from C_4H_8 fragment of '*A*' can be explained by the following structure of *A*.

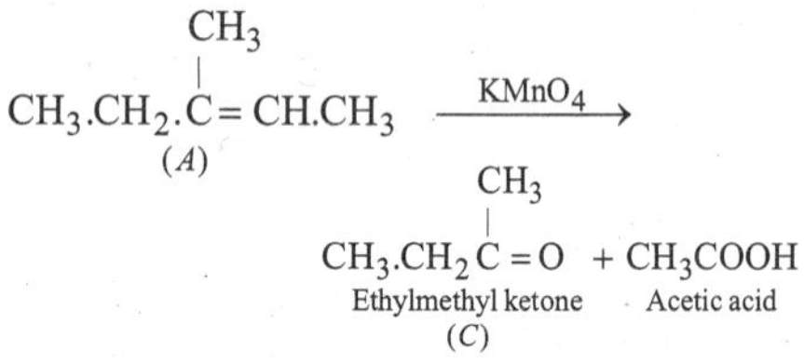

Hence, formation of '*B*' can be represented as below.

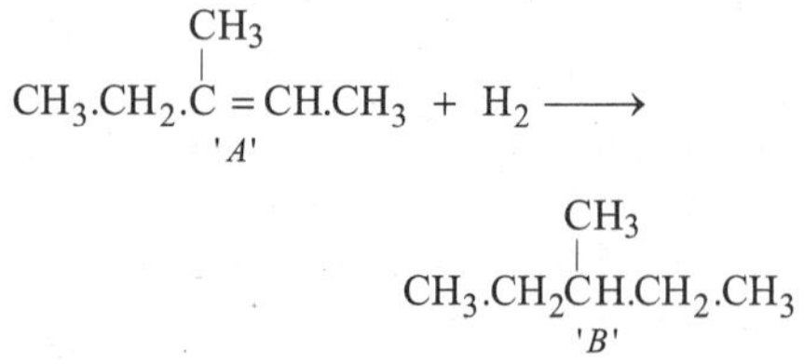

88. Ethyne ($HC\equiv CH$) and only those derivatives which have at least one acetylenic hydrogen atom ($\equiv C-H$) *i.e.* terminal alkynes, will give white precipitate with ammonical silver nitrate solution.

89.

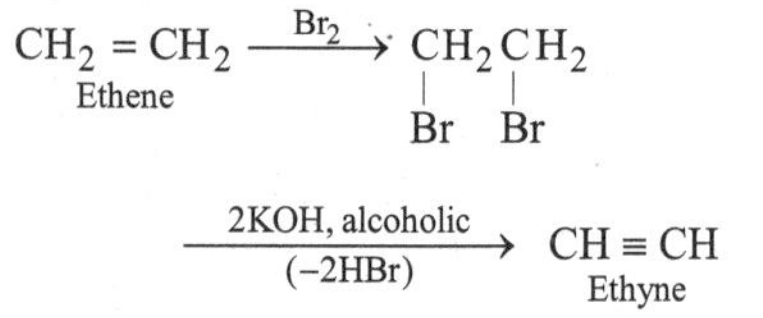

90. (*i*) $CH \equiv CH \xrightarrow{\text{hydration}}$
Ethyne

$CH_3CHO \xrightarrow[\text{Aldol cond.}]{OH^-} CH_3\underset{}{\overset{OH}{|}}{CH}CH_2CHO$
3-Hydroxybutanal

$\xrightarrow{-H_2O} CH_3CH=CHCHO$

(*ii*) 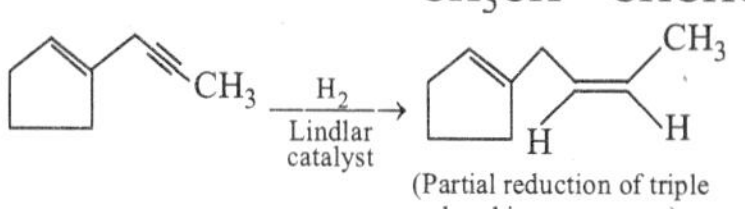

(Partial reduction of triple bond in *syn*-manner)

91. Bromine water test : C_2H_2 decolourises bromine water while CH_4 does not decolourise bromine water.

92. (c)

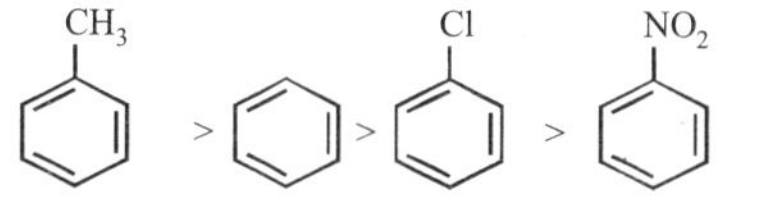

(Activated by + I, and hyperconjugation) Deactivated due to –I   Deactivated due to –I, –M, –E

93. (c) $CH_2=CH_2+H_2SO_4 \rightarrow CH_3CH_2OSO_3H$

$C_6H_6+H_2SO_4 \rightarrow C_6H_5SO_3H+H_2O$

$C_6H_{14}+H_2SO_4 \rightarrow$ No reaction

$C_6H_5NH_2+H_2SO_4 \rightarrow C_6H_5\overset{+}{N}H_3HSO_4^-$

Only hexane does not dissolve in conc. H_2SO_4 even on warming.

94. (c) $-NO_2$, –Cl and –OH are electron-attracting or withdrawing groups due to –M, –E and/or –I effects whereas $-CH_3$ shows +I effect (electron releasing).

Because of the + I effect of the CH_3 group, toluene has the highest electron density in the *o*- and *p*-positions and hence, can be most readily sulphonated.

95. (c) The phenomenon of resonance gives identical bonding and hence identical bond lengths.

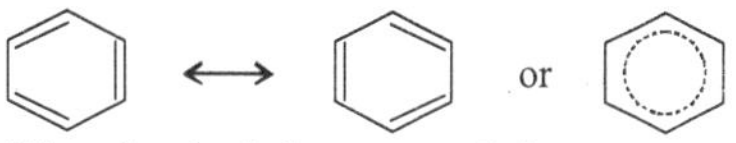

C – C bond order in benzene = 1.5

96. (4)

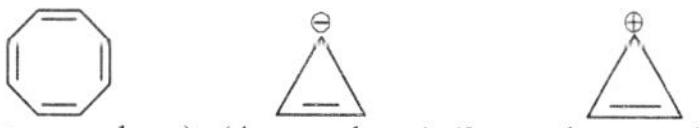

($8\pi e^-$s, non-planar) ($4\pi e^-$s, planar) ($2\pi e^-$s, planar, conjugated)
Non-aromatic   Anti-aromatic   Aromatic

($4\pi e^-$s non-planar) ($6\pi e^-$s, planar) ($4\pi e^-$s, conjugated,
Non-aromati   Anti-aromatic   Aromatic
cconjugated,planar)

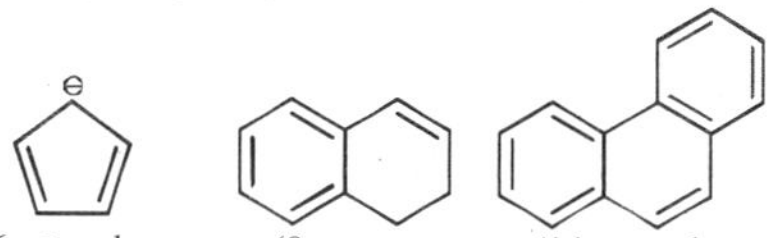

($6\pi e^-$s, planar, conjugated) ($8\pi e^-$s, non-planar, conjugated) ($14\pi e^-$s, planar,
Aromaticnon-conjugated) Aromatic
non-Aromatic

97. (10.00)

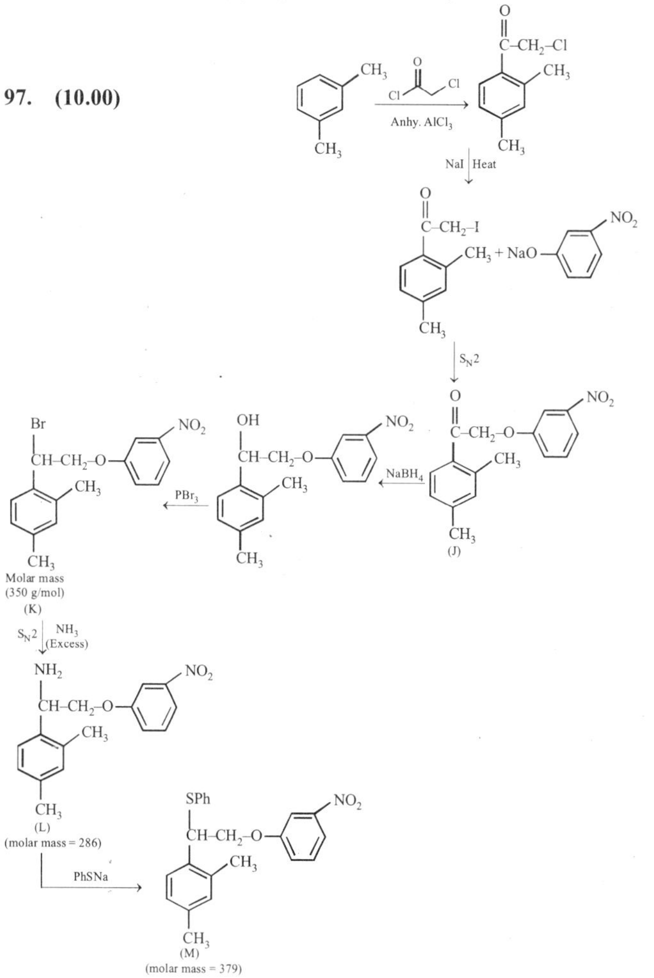

286 gm 'L' requires $=\frac{1}{2}$ mole H_2SO_4

$\therefore$ 572 gm 'L' requires = 1 mole H_2SO_4

Hence 5.72 gm of 'L' = 0.02 mole of compound 'L'

0.02 mole 'L' requires = 0.01 mole H_2SO_4

$$\frac{\text{Molarity} \times V(\text{ml})}{1000}=0.01 \Rightarrow V_{(\text{mL})}=10$$

98. (2.33) Molecular formula of M = $C_{22}H_{21}NO_3S$

Molar mass of M = 379 gm/mol

Given mass of M = 3.79 gm

Hence Moles of M $=\frac{3.79}{379}=0.01$

Molecule of M contains 1 sulphur atom so 0.01 mole of $BaSO_4$ is formed

Mass of $BaSO_4$ formed = 0.01 × 233 = 2.33 g

99. less; Stability of free radical

$$\propto \frac{1}{\text{Bond dissociation energy}}$$

Benzyl ($C_6H_5\dot{C}H_2$) free radical is more stable than methyl ($\dot{C}H_3$) free radical because of hyperconjugation (no bond resonance).

100. ethylene

$$\begin{array}{l} CH_2COOK \\ | \\ CH_2COOK \end{array} \longrightarrow \begin{array}{l} CH_2COO^- \\ | \\ CH_2COO^- \end{array} + 2K^+$$
$$\downarrow$$
$$CH_2=CH_2+2CO_2$$

101. cyclopropane, because it has maximum deviation, from the normal bond angle of 109°28' present in alkanes. In cyclopropane bond angle is 60°, therefore deviation *d*,

$$d=\frac{1}{2}(109°28'-60°).$$

102. False : An electron-donating group increases the electron density in *o*- and *p*-positions due to +M, and/or +I effects and hence, orients the new electrophile to *o*- and *p*-positions.

103. (a, b)

(a)

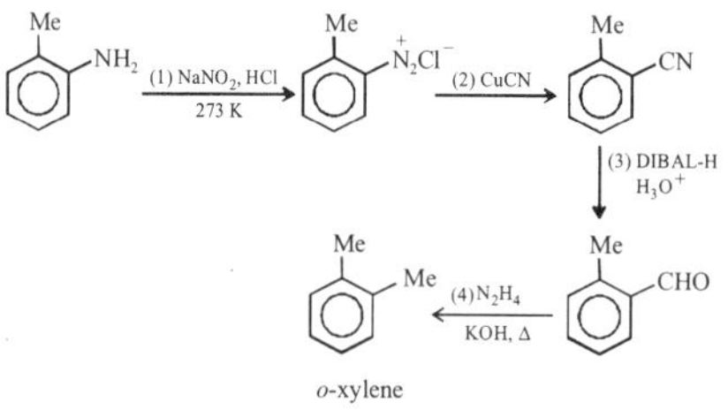

(b)

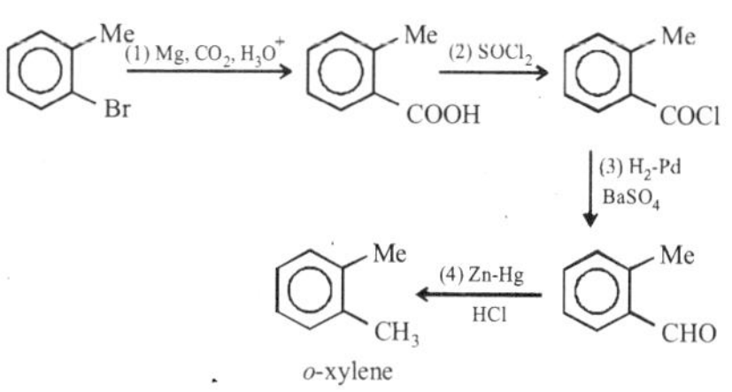

(c)
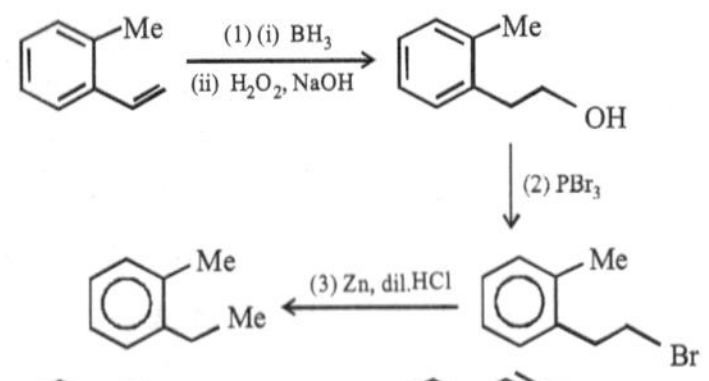

(d)
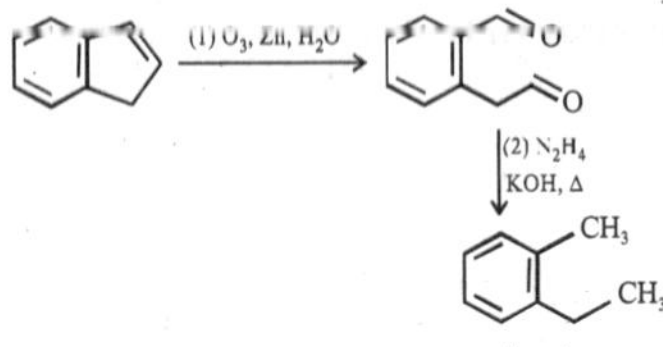

o-xylene is not found

104. (b, c)

(a)
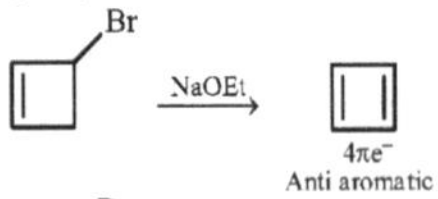

(b)
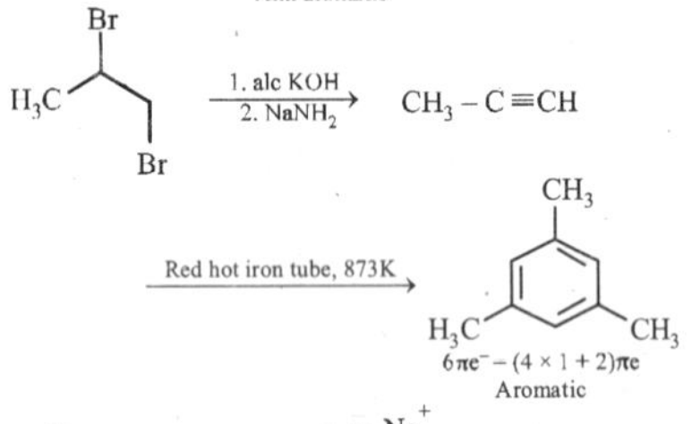

(c)
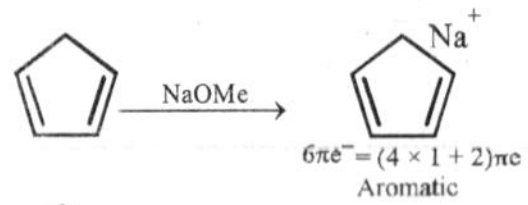

(d)
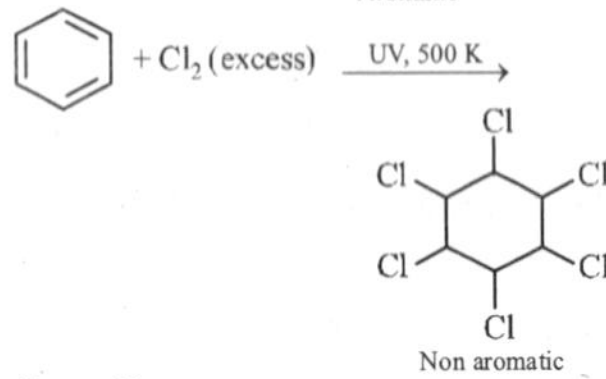

105. (b, c, d)

(a)
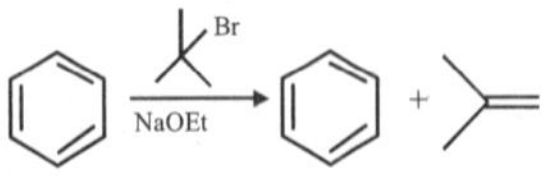

(b)
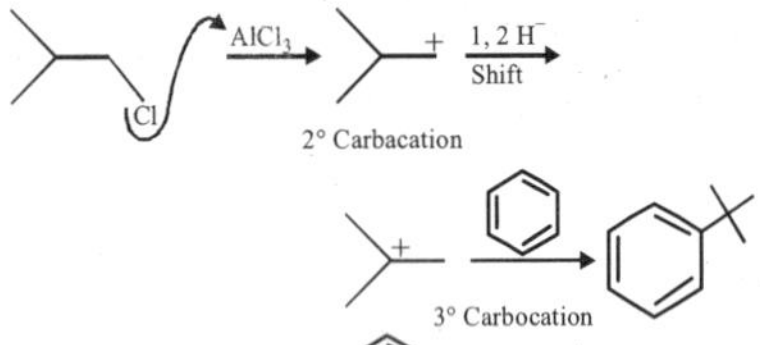

(c)
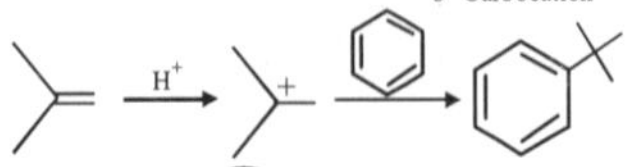

(d)
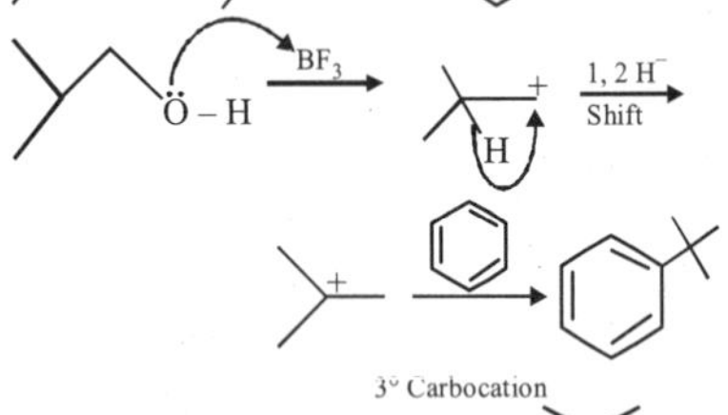

106. (b)
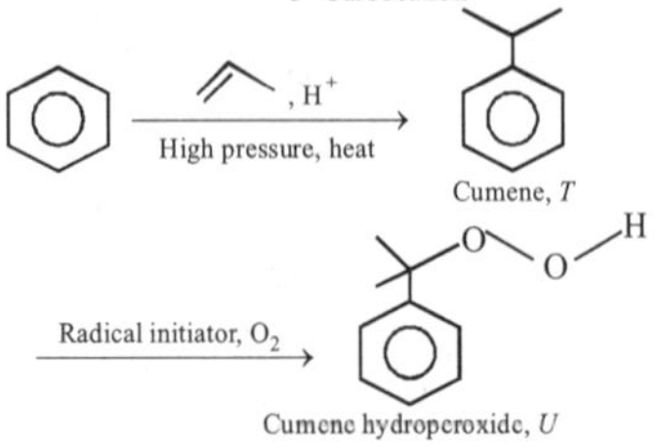

107. (a, b, c, d)
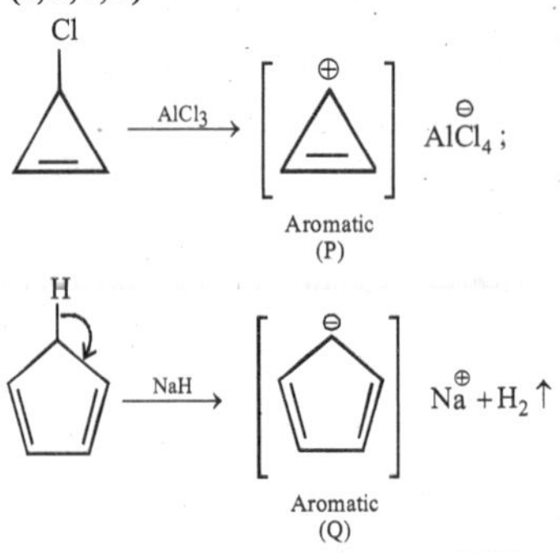

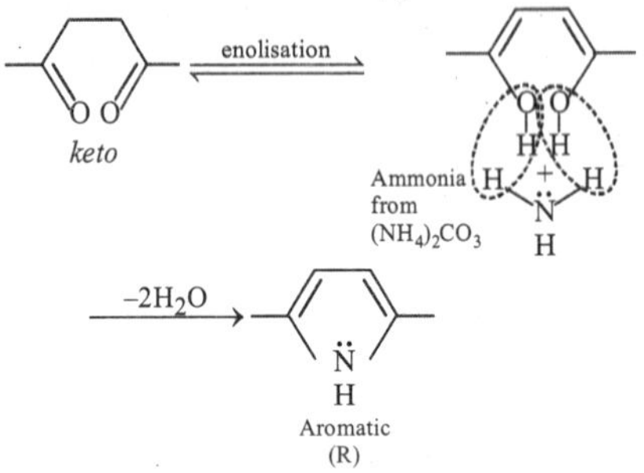

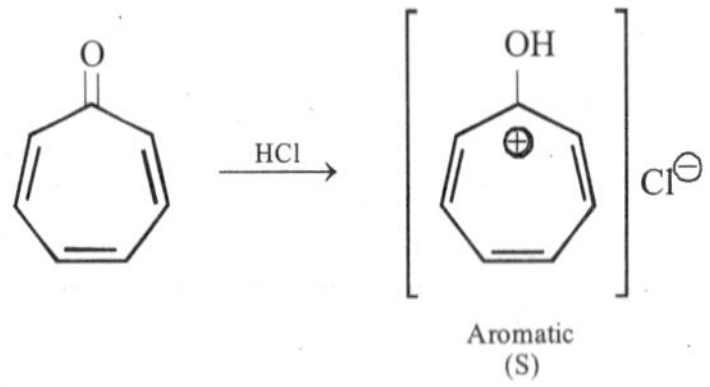

108. (a, c)
Hyperconjugation in toluene also activates the benzene ring for electrophilic substitution.

Methyl group is activating group and activates *o*- and *p*-positions for electrophilic substitution.

109. (b, c, d) An aromatic species will have : $(4n + 2)\ \pi$ electrons (by Huckel's Rule), planar structure (to exhibit resonance) and cyclic structure

110. (i) *tert*-Butylbenzene does not give benzoic acid on treatment with acidic $KMnO_4$ because it does not contain benzylic hydrogen or α- hydrogen atom on the key carbon atom.

(ii) In presence of light, toluene undergoes *side chain bromination* through a free radical mechanism.

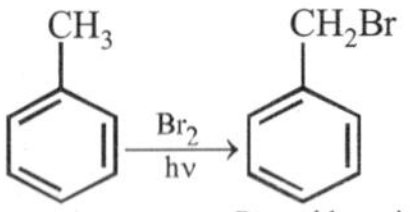

Benzyl bromide

In presence of $FeBr_3$, toluene undergoes *electrophilic substitution* in the benzene ring.

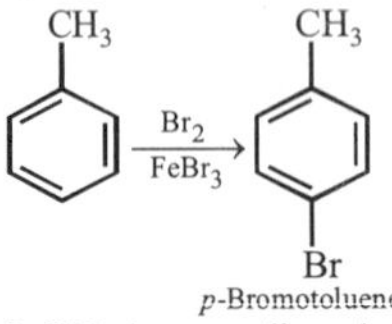

p-Bromotoluene

[$-CH_3$ is *o*-, *p*-directing]

111. (i)
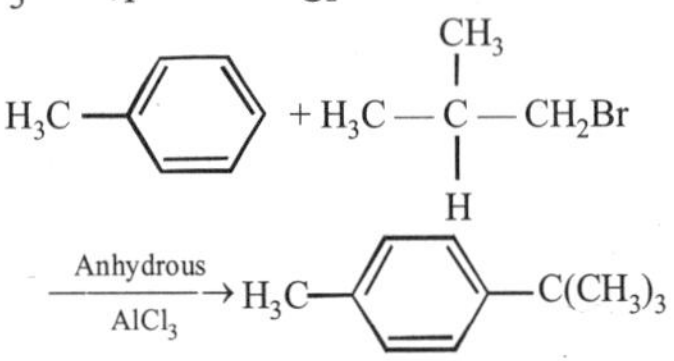

(ii) $C_6H_5C_2H_5 \xrightarrow[\text{(HVZ reaction)}]{Br_2,\ \Delta,\ \text{light}} C_6H_5-\underset{Br}{CH}-CH_3 \xrightarrow{NaCN}$

$C_6H_5-\underset{CN}{CH}-CH_3$

2–Phenylpropanenitrile

(iii) $C_6H_6 + (CH_3)_2CHCH_2OH \xrightarrow{H_2SO_4}$

$C_6H_5(CH_3)_3$

tert–Butylbenzene

Explanation :

$(CH_3)_2CHCH_2OH \xrightarrow{H^+} (CH_3)_2CH\overset{+}{C}H_2$

1° Carbocation

$\xrightarrow[\text{to}]{\text{rearranges}} (CH_3)_3C^+ \xrightarrow{C_6H_6} C_6H_5C(CH_3)_3$

3° Carbocation

(iv)
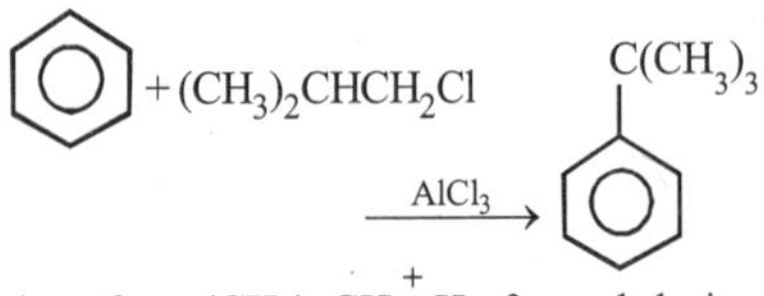

1° carbocation, $(CH_3)_2CH\overset{+}{C}H_2$ formed during reaction rearranges to the more stable, 3° carbocation, $(CH_3)_3\overset{+}{C}$ and hence the above product is formed.]

112. (i)
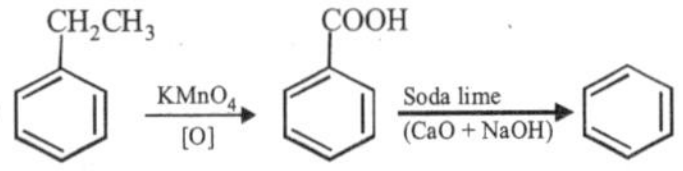

(ii)
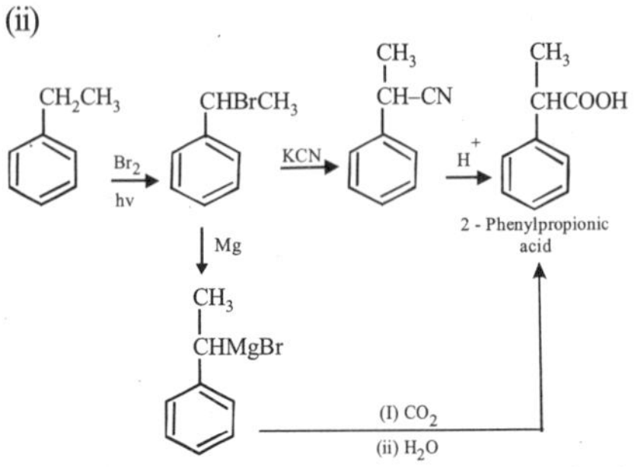

113. Benzene gives electrophilic substitution reactions rather than electrophilic addition reactions because it will have a stable benzene ring in the product, whereas electrophilic addition on benzene destroys the benzenoid ring.

114. $-OCH_3$ and $-CH_3$ groups are activating group while $-Cl$ is a deactivating group for electrophilic substitution.

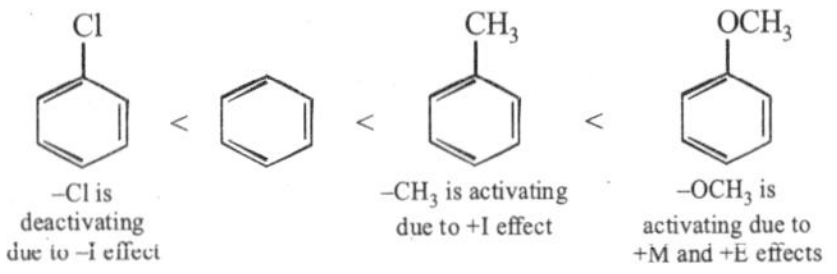

115. $[C_2H_5OH + PCl_5 \longrightarrow C_2H_5Cl]$

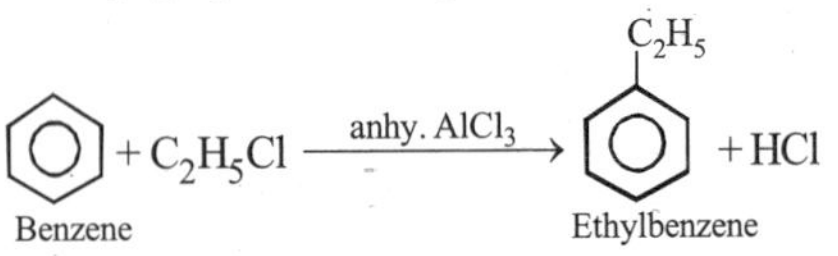

Chapter-14

Environmental Chemistry

Part-A (JEE Main)

1. (1)

(I) Fly ash and slag from steel industry are utilised by cement industry. 15% Fly ash and 15% steel smag is optimum mix in road construction.

(II) Fuel obtained from plastic waste has high octane rating. It contains no lead and it is known as green fuel.

Both statement (I) & (II) are correct.

2. (3) $\dot{C}l + CH_4 \longrightarrow \underset{\text{(A)}}{\dot{C}H_3} + \underset{\text{(B)}}{HCl}$

Methyl radical

3. (3) Photochemical smog contains nitrogen dioxide (NO_2), Ozone (O_3), PAN (peroxyacetylnitrate), and compounds containing –CHO group.

4. (1) $CaCO_3(s) + H_2SO_4(aq) \rightarrow CaSO_4(s) + H_2O(l) + CO_2(g)$

5. (4) SO_2 is atmospherically oxidized in presence of water to give sulphuric acid

$SO_2 + O_2 + 2H_2O \rightarrow 2H_2SO_4$ (Acid rain)

6. (4) Photochemical smog is oxidising in nature and results from the action of sunlight on unsaturated hydrocarbons and nitrogen oxide.

7. (1)

8. (2) $ClONO_2(g) + H_2O(g) \rightarrow HOCl(g) + HNO_3(g)$

9. (4) CH_4, O_3 and H_2O causes global warming in Tropospheric level.

N_2 does not cause global warming.

10. (4) The gas 'A' is likely to be freon (CF_2Cl_2) which is non-reactive, non-flammable and non-toxic. In stratosphere, CFCs get broken down by powerful UV radiation releasing $Cl^{\bullet}$.

$$CF_2Cl_2(g) \xrightarrow{\text{U.V.}} Cl^{\bullet}(g) + {}^{\bullet}CF_2Cl(g)$$

11. (1)

(i) CH_4 causes global warming and also a constituent of photochemical smog.

(ii) CH_4 is generated from paddy fields.

(iii) Due to high heat capacity, CH_4 is a stronger global warming causing gas than CO_2.

(iv) Methane is not a part of reducing smog.

12. (2) Reducing or classical smog is a mixture of smoke, fog and SO_2.

13. (4) Ozone is strong oxidising agent and can react with the unburnt hydrocarbons in the polluted air to produce chemicals.

14. (4) The gas, CH_4 evolved due to anaerobic degradation of vegetation which causes global warming and cancer.

15. (1)

16. (4) Acid rain is harmful for trees and animal life. It can also cause respiratory ailments in animals.

17. (1)

18. (2)

19. (3) Photochemical smog contains oxides of nitrogen, ozone and hydrocarbons. It is also called oxidation smog.

20. (2) In photochemical smog, NO_2 and hydrocarbons are primary pollutants whereas ozone and acrolein are secondary pollutants.

21. (3)

22. (3)

23. (3) Stratosphere lies between 10-50 km above sea level.

24. (1) Global warming is caused by the emission of green house gases. 72% of the totally emitted green house gases is CO_2. Therefore, excessive release of CO_2 is the main cause of global warming.

25. (4)

26. (3) CFC's are responsible for depletion of ozone layer.

Due to depletion of ozone layer more UV radiation filters into troposphere.

27. (1) NO, O_3 and HCHO are responsible for the formation of photochemical smog while N_2 has no role in photochemical smog.

28. (4) Acrolein, PAN (produced by the oxidation of unburnt hydrocarbons present in the polluted air) and ozone are the main components of photochemical smog. While CF_2Cl_2 is not a common component of photochemical smog.

29. (1)

30. (2)

31. (3) High concentration of SO_2 leads to stiffness of flower buds.

32. (3) The reaction, $CH_4 + 2O_3 \rightarrow 3CH_2=O + 3H_2O$, is not involved in the ozone layer depletion mechanism.

33. (3) SO_2 and NO_2 are responsible for the discoloured and lustureless nature of marble.

34. (2)

35. (2) The main gas responsible for the global warming is carbon dioxide. Other contributors include methane released from landfills and agriculture.

36. (1)

37. (2)

38. (2) NO and freons are responsible for ozone layer depletion.

39. (3) The ozone layer, existing between 20 to 35 km above the earth's surface, shield the earth from the harmful U. V. radiations from the sun.

Depletion of ozone is caused by oxides of nitrogen.

40. (1) Photochemical smog is caused by oxides of sulphur and nitrogen.

41. (4)

42. (2) A – III, B – I, C – IV, D – II

43. (1) PollutantSource

A.MicroorganismsDomestic sewage

B.Plant nutrientsChemical fertilizer

C.Toxic heavy metalsChemical factory

D.SedimentStrip mining

44. (3) Sulphate have laxative effect. Nictoine works as natural pesticides. Sodium Arsinite acts as herbicide Fluoride causes bending of bones.

45. (3) Dry-cleaning means the cleaning of clothes without using water. The solvent should be such that it evaporates easily.

Ethanol is not used for dry cleaning of clothes.

46. (2) BoD values indicates the amount of organic matter present in water.

Clean water should have B.O.D. < 5 ppm

Highly polluted water have B.O.D. > 17 ppm

47. (1) Nutrient enriched water bodies support a dense plant population which can deprive oxygen from animal life. This results in biodiversity.

48. (3) (A) : Clean water have BOD less than 5 ppm while highly polluted water has BOD greater or equal to 17 ppm.

(R) : BOD is measure of oxygen required to oxidise only bio-degradable organic matter.

49. (2) Calcium (Ca^{2+}) and phosphate (PO_4^{3-}) are the major components of teeth enamel.

50. (3)

Water	BOD
Clean water	< 5 ppm
Polluted water	≥ 17 pmm

51. (3) BOD values of clean water (A) is less than 5 ppm and BOD values of polluted water (B) is greater than 17ppm.

52. (3) Eutrophication leads to decrease in oxygen level of water. Therefore 3rd statement is incorrect.

53. (4)

54. **(4)** Fluoride (F^-) ion conc. upto 1 ppm makes the enamel of teeth much harder by converting hydroxyapatite to fluorapatite.

$$\underset{\text{hydroxyapatite}}{[3(Ca_3(PO_4)_2)\cdot Ca(OH)_2]} \xrightarrow{F^-} \underset{\text{Fluorapatite (harder)}}{[3(Ca_3(PO_4)_2)\cdot CaF_2]}$$

But above 2 ppm cause brown mottling of teeth.

55. **(1)** The process in which nutrient enriched water bodies support a dense plant population which kill animal life by depriving it of oxygen results in subsequent loss of biodiversity is known as Eutrophication. So, this condition indicates a polluted environment.

56. **(2)** By definition.

57. **(4)** Maximum prescribed concentration of Cu in drinking water is 3 ppm.

58. **(3)** 10 ppm

59. **(2)** The water sample containing Mn = 5 ppm is unsuitable for drinking as the prescribed level for Mn in drinking water is 0.05 ppm.

60. **(3)** Methemoglobinemia is caused by drinking water which is contaminated with nitrations.

61. **(1)** pH of rain water is around 5.6.

62. **(3)**

$$\underset{\text{Hydroxyapatite}}{[3Ca_3(PO_4)_2\cdot Ca(OH)_2]} + \underset{\text{(drinking water upto 1ppm)}}{2F^-} \longrightarrow \underset{\text{Fluorapatite (Harder teeth enamel)}}{[3Ca_3(PO_4)_2\cdot CaF_2] + 2OH^-}$$

63. **(3)** Concentration of F^- in drinking water above 2 ppm cause brown mottling of teeth.

64. **(3)** BOD stands for Biochemical Oxygen Demand.

65. **(1)** The maximum limit of nitrate in drinking water is 50 ppm. Excess nitrate in drinking water can cause disease such as methemoglobinemia ('blue baby' syndrome).

66. **(4)** Phosphate fertilisers can enhance growth of algae.

67. **(4)** DDT is a non-biodegradable pollutant.

Chapter-1

The Solid State

Part-A (JEE Main)

Topic 1 - Classification of Solids, Crystal Structure and Properties of Solids

MULTIPLE CHOICE QUESTIONS (MCQs)

1. A hard substance melts at high temperature and is an insulator in both solid and in molten state. This solid is most likely to be a/an: [March 18, 2021 (II)]
(1) Metallic solid (2) Ionic solid
(3) Molecular solid (4) Covalent Solid

2. 'X' melts at low temperature and is a bad conductor of electricity in both liquid and solid state. X is: [Jan. 09, 2020 (I)]
(1) Zinc sulphide (2) Mercury
(3) Silicon carbide (4) Carbon tetrachloride

3. The parameters of the unit cell of a substance are $a = 2.5, b = 3.0, c = 4.0, \alpha = 90°, \beta = 120°, \gamma = 90°$.
The crystal system of the substance is : [July 27, 2021 (I); Similar Online April 12, 2014]
(1) Hexagonal (2) Orthorhombic
(3) Monoclinic (4) Triclinic

4. Which primitive unit cell has unequal edge lengths ($a \neq b \neq c$) and all axial angles different from 90°? [Jan. 10, 2019 (I)]
(1) Triclinic (2) Hexagonal
(3) Monoclinic (4) Tetragonal

5. The one that is extensively used as a piezoelectric material is: [Jan. 9, 2019 (I)]
(1) tridymite (2) amorphous silica
(3) quartz (4) mica

6. All of the following share the same crystal structure except. [Online April 15, 2018 (II)]
(1) RbCl (2) NaCl (3) CsCl (4) LiCl

7. Which of the following exists as covalent crystals in the solid state ? [2013]
(1) Iodine (2) Silicon
(3) Sulphur (4) Phosphorus

8. The radius of a calcium ion is 94 pm and of the oxide ion is 146 pm. The possible crystal structure of calcium oxide will be [Online May 12, 2012]
(1) tetrahedral (2) trigonal
(3) octahedral (4) pyramidal

9. Among the following which is the best description of water in the solid phase? [Online May 7, 2012]
(1) Covalent solid (2) Molecular solid
(3) Ionic solid (4) Network solid

Topic 2 - Close Packed Structure, Packing Efficiency and Imperfection in Solids

MULTIPLE CHOICE QUESTIONS (MCQs)

10. An element X has a body centred cubic (bcc) structure with a cell edge of 200 pm. The density of the element is 5 g cm^{-3}. The number of atoms present in 300 g of the element X is ______ . [June 30, 2022 (I)]
Given : Avogadro Constant, $N_A = 6.0 \times 10^{23}$ mol^{-1}.
(1) $5N_A$ (2) $6N_A$ (3) $15N_A$ (4) $25N_A$

11. The incorrect statement about the imperfections in solids is: [June 28, 2022 (I)]
(1) Schottky defect decreases the density of the substance.
(2) Interstitial defect increases the density of the substance.
(3) Frenkel defect does not alter the density of the substance.
(4) Vacancy defect increases the density of the substance.

12. Match items of List–I with those of List–II :

List–I (Property)	List–II (Example)
(A) Diamagnetism	(i) MnO
(B) Ferrimagnetism	(ii) O_2
(C) Paramagnetism	(iii) NaCl
(D) Antiferromagnetism	(iv) Fe_3O_4

Choose the most appropriate answer from the options given below : [Aug. 27, 2021 (I)]
(1) (A)–(ii), (B)–(i), (C)–(iii), (D)–(iv)
(2) (A)–(i), (B)–(iii), (C)–(iv), (D)–(ii)
(3) (A)–(iii), (B)–(iv), (C)–(ii), (D)–(i)
(4) (A)–(iv), (B)–(ii), (C)–(i), (D)–(iii)

13. Given below are two statements.
Statement I : Frenkel defects are vacancy as well as interstitial defects.
Statement II : Frenkel defect leads to colour in ionic solids due to presence of F-centres.
Choose the **most appropriate** answer for the statements from the options given below : [Aug. 26, 2021 (I)]
(1) Statement I is false but Statement II is true
(2) Both Statement I and Statement II are true
(3) Statement I is true but Statement II is false
(4) Both Statement I and Statement II are false

14. Which of the following compounds is likely to show both Frenkel and Schottky defects in its crystalline form? [Jan. 08, 2020 (II)]
(1) AgBr (2) CsCl (3) KBr (4) ZnS

15. A crystal is made up of metal ions 'M_1' and 'M_2' and oxide ions. Oxide ions form a *ccp* lattice structure. The cation 'M_1' occpies 50% of octahedral voids and the cation 'M_2' occupies 12.5% of tetrahedral voids of oxide lattice. The oxidation numbers of 'M_1' and 'M_2' are, respectively : **[Sep. 06, 2020 (II)]**
(1) +2, +4 (2) +1, +3 (3) +3, +1 (4) +4, +2

16. An element crystallises in a face-centred cubic (fcc) unit cell with cell edge a. The distance between the centres of two nearest octahedral voids in the crystal lattice is : **[Sep. 05, 2020 (II)]**
(1) $\frac{a}{\sqrt{2}}$ (2) a (3) $\sqrt{2}a$ (4) $\frac{a}{2}$

17. A diatomic molecule X_2 has a body-centred cubic (bcc) structure with a cell edge of 300 pm. The density of the molecule is 6.17 g cm^{-3}. The number of molecules present in 200 g of X_2 is : **[Sep. 05, 2020 (I)]**
(Avogadro constant $(N_A) = 6 \times 10^{23}\ mol^{-1}$)
(1) $40\ N_A$ (2) $8\ N_A$ (3) $4\ N_A$ (4) $2\ N_A$

18. The ratio of number of atoms present in a simple cubic, body centered cubic and face centered cubic structure are, respectively : **[April 12, 2019 (II)]**
(1) 8 : 1 : 6 (2) 1 : 2 : 4 (3) 4 : 2 : 1 (4) 4 : 2 : 3

19. An element has a face-centred cubic (fcc) structure with a cell edge of a. The distance between the centres of two nearest tetrahedral voids in the lattice is : **[April 12, 2019 (I)]**
(1) $\sqrt{2}a$ (2) a (3) $\frac{a}{2}$ (4) $\frac{3}{2}a$

20. Element 'B' forms ccp structure and 'A' occupies half of the octahedral voids, while oxygen atoms occupy all the tetrahedral voids. The structure of bimetallic oxide is : **[April 8, 2019 (I)]**
(1) A_2BO_4 (2) AB_2O_4 (3) A_2B_2O (4) A_4B_2O

21. Consider the bcc unit cells of the solids 1 and 2 with the position of atoms are shown below. The radius of atom B is twice that of atom A. The unit cell edge length is 50% more in solid 2 than in 1. What is the approximate packing efficiency in solid 2? **[April 8, 2019 (II)]**

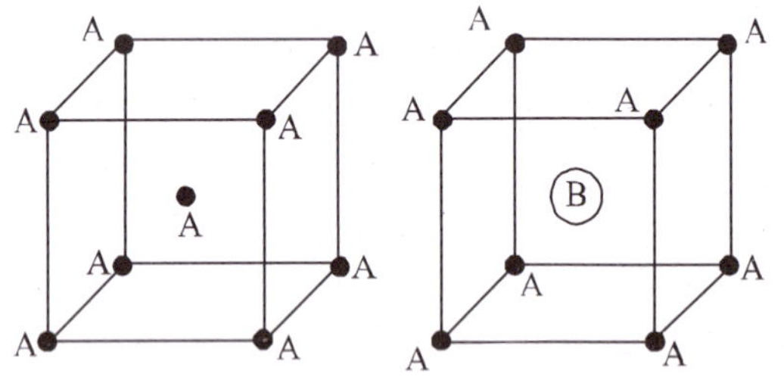

Solid 1 Solid 2

(1) 45% (2) 75% (3) 90% (4) 65%

22. The radius of the largest sphere which fits properly at the centre of the edge of a body centred cubic unit cell is : (Edge length is represented by 'a') **[Jan. 11, 2019 (II)]**
(1) $0.027a$ (2) $0.047a$
(3) $0.134a$ (4) $0.067a$

23. A compound of formula A_2B_3 has the hcp lattice. Which atom forms the hcp lattice and what fraction of tetrahedral voids is occupied by the other atoms: **[Jan. 10, 2019 (II)]**
(1) hcp lattice – A, $\frac{2}{3}$ Tetrahedral voids – B
(2) hcp lattice – A, $\frac{1}{3}$ Tetrahedral voids – B
(3) hcp lattice – B, $\frac{2}{3}$ Tetrahedral voids – A
(4) hcp lattice – B, $\frac{1}{3}$ Tetrahedral voids – A

24. At 100 °C, copper (Cu) has fcc unit cell structure with cell edge length of x Å. What is the approximate density of Cu (in g cm^{-3}) at this temperature?
[Atomic Mass of Cu = 63.55 u] **[Jan. 9, 2019 (II)]**
(1) $\frac{205}{x^3}$ (2) $\frac{105}{x^3}$ (3) $\frac{211}{x^3}$ (4) $\frac{422}{x^3}$

25. Which type of 'defect' has the presence of cations in the interstitial sites? **[2018]**
(1) Schottky defect (2) Vacancy defect
(3) Frenkel defect (4) Metal deficiency defect

26. Which of the following arrangements shows the schematic alignment of magnetic moments of antiferromagnetic substance? **[Online April 15, 2018 (I)]**
(1) ↑↓↓↓↓↑
(2) ↑↑↑↑↑↑
(3) ↑↑↓↑↑↓
(4) ↑↓↑↓↑↓

27. A metal crystallises in a face centred cubic structure. If the edge length of its unit cell is 'a', the closest approach between two atoms in metallic crystal will be: **[2017]**
(1) $2a$ (2) $2\sqrt{2}a$ (3) $\sqrt{2}a$ (4) $\frac{a}{\sqrt{2}}$

28. Sodium metal crystallizes in a body centred cubic lattice with a unit cell edge of 4.29Å. The radius of sodium atom is approximately: **[2015], [2012 (S)]**
(1) 5.72 Å (2) 0.93 Å (3) 1.86 Å (4) 3.22 Å

29. CsCl crystallises in body centered cubic lattice. If 'a' is its edge length then which of the following expressions is correct? **[2014]**
(1) $r_{Cs^+} + r_{Cl^-} = 3a$ (2) $r_{Cs^+} + r_{Cl^-} = \frac{3a}{2}$
(3) $r_{Cs^+} + r_{Cl^-} = \frac{\sqrt{3}}{2}a$ (4) $r_{Cs^+} + r_{Cl^-} = \sqrt{3}a$

30. Copper crystallises in fcc with a unit length of 361pm. What is the radius of copper atom?
[Online April 25, 2013; Similar Online April 9, 2013]
(1) 157 pm (2) 128 pm (3) 108 pm (4) 181 pm

31. In a face centered cubic lattice, atoms A are at the corner points and atoms B at the face centered points. If atom B is missing from one of the face centered points, the formula of the ionic compound is:
[Online April 9, 2014; Similar Online April 23, 2013]
(1) AB_2 (2) A_5B_2 (3) A_2B_3 (4) A_2B_5

32. Which one of the following statements about packing in solids is **incorrect** ? **[Online April 22, 2013]**
(1) Coordination number in bcc mode of packing is 8.

(2) Coordination number in hcp mode of packing is 12.
(3) Void space in hcp mode of packing is 32%.
(4) Void space is ccp mode of packing is 26%.

33. Among the following, the incorrect statement is **[Online May 26, 2012]**
(1) Density of crystals remains unaffected due to Frenkel defect.
(2) In bcc unit cell, the void space is 32%.
(3) Density of crystals decreases due to Schottky defect.
(4) Electrical conductivity of semiconductors and metals increases with increase in temperature.

34. A solid has a 'bcc' structure. If the distance of nearest approach between two atoms is 1.73Å, the edge length of the cell is **[Online May 7, 2012]**
(1) 314.20 pm (2) 1.41 pm
(3) 200 pm (4) 216 pm

35. Ammonium chloride crystallizes in a body centred cubic lattice with edge length of unit cell of 390 pm. If the size of chloride ion is 180 pm, the size of ammonium ion would be **[Online May 19, 2012]**
(1) 174 pm (2) 158 pm (3) 142 pm (4) 126 pm

36. The edge length of a face centered cubic cell of an ionic substance is 508 pm. If the radius of the cation is 110 pm, the radius of the anion is **[2010]**
(1) 288 pm (2) 398 pm (3) 618 pm (4) 144 pm

37. Percentages of free space in cubic close packed structure and in body centered packed structure are respectively **[2010]**
(1) 30% and 26% (2) 26% and 32%
(3) 32% and 48% (4) 48% and 26%

38. In a compound, atoms of element Y form *ccp* lattice and those of element X occupy 2/3rd of tetrahedral voids. The formula of the compound will be **[2008]**
(1) X_4Y_3 (2) X_2Y_3 (3) X_2Y (4) X_3Y_4

39. Total volume of atoms present in a face-centred cubic unit cell of a metal is (r is atomic radius) **[2006]**
(1) $\frac{12}{3}\pi r^3$ (2) $\frac{16}{3}\pi r^3$ (3) $\frac{20}{3}\pi r^3$ (4) $\frac{24}{3}\pi r^3$

40. What type of crystal defect is indicated in the diagram below? **[2004]**

$Na^+Cl^-Na^+Cl^-Na^+Cl^-$
Cl^- □ Cl^-Na^+ □ Na^+
Na^+Cl^- □ $Cl^-Na^+Cl^-$
$Cl^-Na^+Cl^-Na^+$ □ Na^+

(1) Interstitial defect
(2) Schottky defect
(3) Frenkel defect
(4) Frenkel and Schottky defects

41. How many unit cells are present in a cube-shaped ideal crystal of NaCl of mass 1.00 g? **[2003]**
[Atomic masses : Na = 23, Cl = 35.5]
(1) 5.14×10^{21} unit cells (2) 1.28×10^{21} unit cells
(3) 1.71×10^{21} unit cells (4) 2.57×10^{21} unit cells

42. Na and Mg crystallize in bcc and fcc type crystals respectively, then the number of atoms of Na and Mg present in the unit cell of their respective crystal is **[2002]**
(1) 4 and 2 (2) 9 and 14 (3) 14 and 9 (4) 2 and 4.

NUMERICAL VALUE QUESTIONS (NVQs)

43. Metal deficiency defect is shown by $Fe_{0.93}O$. In the crystal, some Fe^{2+} cations are missing and loss of positive charge is compensated by the presence of Fe^{3+} ions. The percentage of Fe^{2+} ions in the $Fe_{0.93}O$ crystals is ________ . (Nearest integer) **[June 27, 2022 (I)]**

44. Ionic radii of cation A^+ and anion B^- are 102 and 181 pm respectively. These ions are allowed to crystallize into an ionic solid. This crystal has cubic close packing for B^-. A^+ is present in all octahedral voids. The edge length of the unit cell of the crystal AB is _____ pm. (Nearest integer) **[July 29, 2022 (I)]**

45. Metal M crystallizes into a fcc lattice with the edge length of 4.0×10^{-8} cm. The atomic mass of the metal is _______ g/mol. (Nearest integer) **[July 28, 2022 (II)]**
(Use : $N_A = 6.02 \times 10^{23}$ mol^{-1}, density of metal, M = 9.03 g cm^{-3})

46. An element M crystallises in a body centred cubic unit cell with a cell edge of 300 pm. The density of the element is 6.0 g cm^{-3}. The number of atoms present in 180 g of the element is ______ $\times 10^{23}$. (Nearest integer) **[July 28, 2022 (I)]**

47. In a solid AB. A atoms are in ccp arrangement and B atoms occupy all the octahedral sites. If two atoms from the opposite faces are removed, then the resultant stoichiometry of the compound is A_xB_y. The value of x is_______. [nearest integer] **[June 26, 2022 (II)]**

48. The distance between Na^+ and Cl^- ions in solid NaCl of density 43.1 g cm^{-3} is __________ $\times 10^{-10}$ m. (Nearest Integer)
(Given : $N_A = 6.02 \times 10^{23}$ mol^{-1}) **[June 25, 2022 (I)]**

49. Atoms of element X form hcp lattice and those of element Y occupy $\frac{2}{3}$ of its tetrahedral voids. The percentage of element X in the lattice is_________ (Nearest integer) **[June 24, 2022 (I)]**

50. The empirical formula for a compound with a cubic close packed arrangement of anions and with cations occupying all the octahedral sites in A_xB. The value of x is _________ (Integer answer) **[Aug. 31, 2021 (II)]**

51. A copper complex crystallising in a ccp lattice with a cell edge of 0.4518 nm has been revealed by employing X-ray diffraction studies. The density of a copper complex is found to be 7.62 g cm^{-3}.
The molar mass of copper complex is ______ g mol^{-1}. (Nearest integer)
[Given : $N_A = 6.022 \times 10^{23}$ mol^{-1}]
[July 22, 2021 (II); Similar Jan. 11, 2019 (I)]

52. Diamond has a three dimensional structure of C atoms formed by covalent bonds. The structure of diamond has face centred cubic lattice where 50% of the tetrahedral voids are also occupied by carbon atoms. The number of carbon atoms present per unit cell of diamond is _______. **[July 20, 2021 (II)]**

53. KBr is doped with 10^{-5} mole percent of $SrBr_2$. The number of cationic vacancies in 1 g of KBr crystal is_____ $\times 10^{14}$. (Round off to the Nearest Integer).
[Atomic Mass : K : 39.1 u, Br : 79.9 u, $N_A = 6.023 \times 10^{23}$] **[March 17, 2021 (II)]**

54. A certain element crystallises in a bcc lattice of unit cell edge length 27 Å. If the same element under the same conditions crystallises in the fcc lattice, the edge length of the unit cell in Å will be ____.
(Round off to the Nearest Integer).

[Assume each lattice point has a single atom]

[Assume $\sqrt{3} = 1.73, \sqrt{2} = 1.41$] **[March 16, 2021 (I)]**

55. Ga (atomic mass 70 u) crystallizes in a hexagonal close packed structure. The total number of voids in 0.581 g of Ga is ________ $\times 10^{21}$. (Round off to the Nearest Integer).
[Given : $N_A = 6.023 \times 10^{23}$] **[March 16, 2021 (II)]**

56. The number of octahedral voids per lattice site in a lattice is ______. (Rounded off to the nearest integer) **[Feb. 26, 2021 (II)]**

57. The unit cell of copper corresponds to a face centered cube of edge length 3.596 Å with one copper atom at each lattice point. The calculated density of copper in kg/m^3 is _____.
[Molar mass of Cu : 63.54 g; Avogadro Number = 6.022×10^{23}] **[Feb. 25, 2021 (II)]**

58. The coordination number of an atom in a body-centered cubic structure is ______ . **[Feb. 24, 2021 (I)]**
[Assume that the lattice is made up of atoms.]

59. An element with molar mass 2.7×10^{-2} kg mol^{-1} forms a cubic unit cell with edge length 405 pm. If its density is 2.7×10^3 kg m^{-3}, the radius of the element is approximately ________ $\times 10^{-12}$ m (to the nearest integer). **[Sep. 03, 2020 (I)]**

Part-B (JEE Advanced)

Topic 1 - Classification of Solids, Crystal Structure and Properties of Solids

7 Match the Following

1. Match the crystal system/unit cells mentioned in Column I with their characteristic features mentioned in Column II. Indicate your answer by darkening the appropriate bubbles of the 4 × 4 matrix given in the ORS. **[2007]**

Column-I	Column-II
(A) simple cubic and face-centered cubic parameters	(p) have these parameters, $a = b = c$ and $\alpha = \beta = \gamma$
(B) cubic and rhombohedral	(q) are two crystal systems
(C) cubic and tetragonal	(r) have only two crystallo-graphic angles of 90°
(D) hexagonal and monoclinic	(s) belong to same crystal system

Topic 2 - Close Packed Structure, Packing Efficiency and Imperfection in Solids

1 MCQs with One Correct Answer

2. Atom X occupies the fcc lattice sites as well as alternate tetrahedral voids of the same lattice. The packing efficiency (in %) of the resultant solid is closest to **[Adv. 2022 (II)]**
(a) 25 (b) 35 (c) 55 (d) 75

3. The arrangement of X^- ions around A^+ ion in solid AX is given in the figure (not drawn to scale). If the radius of X^- is 250 pm, the radius of A^+ is **[Adv. 2013]**

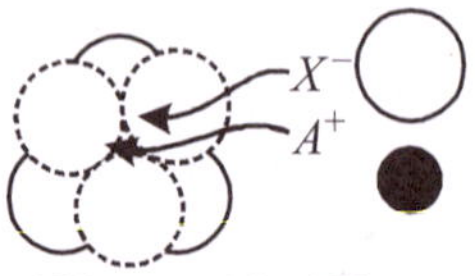

(a) 104 pm (b) 125 pm (c) 183 pm (d) 57 pm

4. A compound M_pX_q has cubic close packing (*ccp*) arrangement of X. Its unit cell structure is shown below. The empirical formula of the compound is **[2012-I]**

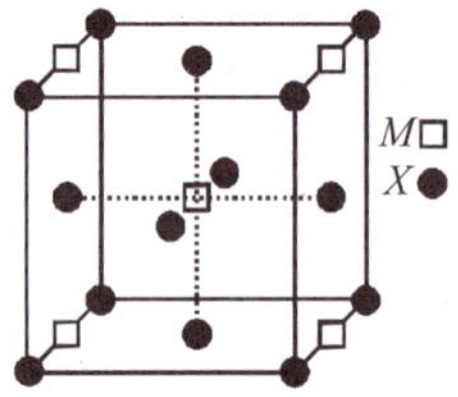

(a) MX (b) MX_2 (c) M_2X (d) M_5X_{14}

5. The packing efficiency of the two-dimensional square unit cell shown below is : **[2010]**

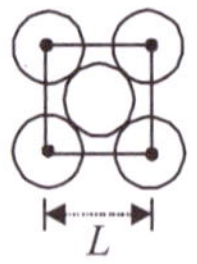

(a) 39.27% (b) 68.02% (c) 74.05% (d) 78.54%

6. A substance A_xB_y crystallizes in a face centred cubic (FCC) lattice in which atoms 'A' occupy each corner of the cube and atoms 'B' occupy the centres of each face of the cube. Identify the correct composition of the substance A_xB_y **[2002S]**
(a) AB_3
(b) A_4B_3
(c) A_3B
(d) Compostion cannot be specified

7. In a solid 'AB' having the NaCl structure, 'A' atoms occupy the corners of the cubic unit cell. If all the face-centered atoms along one of the axes are removed, then the resultant stoichiometry of the solid is **[2001S]**
(a) AB_2 (b) A_2B (c) A_4B_3 (d) A_3B_4

8. The coordination number of a metal crystallizing in a hexagonal close-packed structure is **[1999 - 2 Marks]**
(a) 12 (b) 4 (c) 8 (d) 6

9. CsBr has *bcc* structure with edge length 4.3. The shortest inter ionic distance between Cs$^+$ and Br$^-$ is : **[1995S]**
(a) 3.72 (b) 1.86 (c) 7.44 (d) 4.3

2 Integer Value Answer

10. Consider an ionic solid MX with NaCl structure. Construct a new structure (Z) whose unit cell is constructed from the unit cell of MX following the sequential instructions given below. Neglect the charge balance. **[Adv. 2018]**

(i) Remove all the anions (X) except the central one
(ii) Replace all the face centered cations (M) by anions (X)
(iii) Remove all the corner cations (M)
(iv) Replace the central anion (X) with cation (M)

The value of $\left(\frac{\text{number of anions}}{\text{number of cations}}\right)$ in Z is____.

11. A crystalline solid of a pure substance has a face-centred cubic structure with a cell edge of 400 pm. If the density of the substance in the crystal is 8 g cm^{-3}, then the number of atoms present in 256 g of the crystal is $N \times 10^{24}$. The value of N is [Adv. 2017]

12. The number of hexagonal faces that are present in a truncated octahedron is [2011]

3 Numeric / New Stem Based Questions

13. The density (in g cm^{-3}) of the metal which forms a cubic close packed (ccp) lattice with an axial distance (edge length) equal to 400 pm is _____.

Use: Atomic mass of metal = 105.6 amu and Avogadro's constant $= 6 \times 10^{23}$ mol^{-1} [Adv. 2025]

14. A metal crystallises into two cubic phases, face centered cubic (FCC) and body centred cubic (BCC), whose unit cell lengths are 3.5 and 3.0 Å, respectively. Calculate the ratio of densities of FCC and BCC. [1999 - 3 Marks]

15. Chromium metal crystallizes with a body centred cubic lattice. The length of the unit cell edge is found to be 287 pm. Calculate the atomic radius. What would be the density of chromium in g/cm^3? [1997 - 3 Marks]

16. Sodium metal crystallizes in body centred cubic lattice with the cell edge, $a = 4.29$Å. What is the radius of sodium atom? [1994 - 2 Marks]

6 MCQs with One or More than One Correct Answer

17. Atoms of metals x, y, and z form face-centred cubic (fcc) unit cell of edge length L_x, body-centred cubic (bcc) unit cell of edge length L_y, and simple cubic unit cell of edge length L_z, respectively. [Adv. 2023]

If $r_z = \frac{\sqrt{3}}{2} r_y$; $r_y = \frac{8}{\sqrt{3}} r_x$; $M_z = \frac{3}{2} M_y$ and $M_z = 3M_x$, then the correct statement(s) is (are)

[Given: M_x, M_y, and M_z are molar masses of metals x, y, and z, respectively. r_x, r_y, and r_z are atomic radii of metals x, y, and z, respectively.]

(a) Packing efficiency of unit cell of x > Packing efficiency of unit cell of y > Packing efficiency of unit cell of z
(b) $L_y > L_z$
(c) $L_x > L_y$
(d) Density of x > Density of y

18. The cubic unit cell structure of a compound containing cation M and anion X is shown below. When compared to the anion, the cation has smaller ionic radius. Choose the correct statement(s). [Adv. 2020]

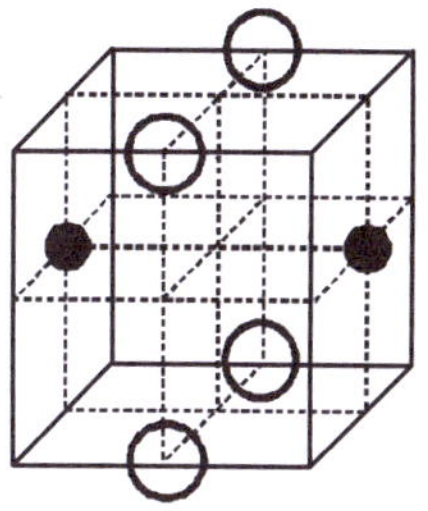

(a) The empirical formula of the compound is MX.
(b) The cation M and anion X have different coordination geometries.
(c) The ratio of M-X bond length to the cubic unit cell edge length is 0.866.
(d) The ratio of the ionic radii of cation M to anion X is 0.414.

19. The CORRECT statement(s) for cubic close packed (ccp) three dimensional structure is (are) [Adv. 2016]
(a) The number of the nearest neighbours of an atom present in the topmost layer is 12
(b) The efficiency of atom packing is 74%
(c) The number of octahedral and tetrahedral voids per atom are 1 and 2, respectively
(d) The unit cell edge length is $2\sqrt{2}$ times the radius of the atom

20. If the unit cell of a mineral has cubic close packed (ccp) array of oxygen atoms with m fraction of octahedral holes occupied by aluminium ions and n fraction of tetrahedral holes occupied by magnesium ions, m and n, respectively, are [Adv. 2015]

(a) $\frac{1}{2}, \frac{1}{8}$ (b) $1, \frac{1}{4}$ (c) $\frac{1}{2}, \frac{1}{2}$ (d) $\frac{1}{4}, \frac{1}{8}$

21. The correct statement(s) regarding defects in solids is (are) [2009S]
(a) Frenkel defect is usually favoured by a very small difference in the sizes of cation and anion
(b) Frenkel defect is a dislocation defect
(c) Trapping of an electron in the lattice leads to the formation of F-centre
(d) Schottky defects have no effect on the physical properties of solids

22. Which of the following statement(s) is (are) correct? [1998 - 2 Marks]
(a) The coordination number of each type of ion in CsCl crystal is 8.
(b) A metal that crystallizes in bcc structure has a coordination number of 12.
(c) A unit cell of an ionic crystal shares some of its ions with other unit cells.
(d) The length of the unit cell in NaCl is 552 pm. ($r_{Na^+} = 95$ pm; $r_{Cl^-} = 181$ pm).

8 Comprehension Based Questions

Passage

In hexagonal systems of crystals, a frequently encountered arrangement of atoms is described as a hexagonal prism. Here, the top and bottom of the cell are regular hexagons and three atoms are sandwiched in between them. A space-filling model of this structure, called hexagonal close-packed (HCP), is constituted of a sphere on a flat surface surrounded in the same plane by six identical spheres as closely as possible. Three spheres are then placed over the first layer so that they touch each other and represent the second layer. Each one of these three spheres touches three spheres of the bottom layer. Finally, the second layer is covered with third layer that is identical to the bottom layer in relative position. Assume radius of every sphere to be 'r'.

23. The number of atoms in the HCP unit cell is [2008]
(a) 4 (b) 6 (c) 12 (d) 17

24. The volume of this *hcp* unit cell is – [2008]
(a) $24\sqrt{2}r^3$ (b) $16\sqrt{2}r^3$ (c) $12\sqrt{2}r^3$ (d) $\frac{64}{3\sqrt{3}}r^3$

25. The empty space in this *hcp* unit cell is [2008]
(a) 74% (b) 47.6% (c) 32% (d) 26%

9 Assertion and Reason/Statement Type Questions

Each question contains **STATEMENT-1 (Assertion)** and **STATEMENT-2 (Reason)**. Each question has 4 choices (a), (b), (c) and (d) out of which ONLY ONE is correct. Mark your answer as

(a) If both Statement -1 and Statement -2 are correct, and Statement -2 is the correct explanation of the Statement -2.
(b) If both Statement -1 and Statement -2 are correct, but Statement -2 is not the correct explanation of the Statement -1.
(c) If Statement -1 is correct but Statement -2 is incorrect.
(d) If Statement -1 is incorrect but Statement -2 is correct.

26. **Assertion :** In any ionic solid [*MX*] with Schottky defects, the number of positive and negative ions are same.
Reason : Equal number of cation and anion vacancies are present. [2001S]

10 Subjective Problems

27. The edge length of unit cell of a metal having molecular weight 75 g/mol is 5Å which crystallizes in cubic lattice. If the density is 2 g/cc then find the radius of metal atom ($N_A = 6 \times 10^{23}$). Give the answer in pm. [2006 - 6M]

28. In face centred cubic (*fcc*) crystal lattice, edge length is 400 pm. Find the diameter of greatest sphere which can be fit into the interstitial void without distortion of lattice. [2005 - 2 Marks]

29. A compound *AB* has rock salt type structure. The formula weight of *AB* is 6.023 *Y* amu, and the closest *A* – *B* distance is $Y^{1/3}$ nm, where *Y* is an arbitrary number. [2004 - 2 Marks]
(a) Find the density of lattice
(b) If the density of lattice is found to be 20 kg m^{-3}, then predict the type of defect.

30. You are given marbles of diameter 10 mm. They are to be placed such that their centres are lying in a square bound by four lines each of length 40 mm. What will be the arrangements of marbles in a plane so that maximum number of marbles can be placed inside the area? Sketch the diagram and derive expression for the number of molecules per unit area. [2003 - 2 Marks]

31. The figures given below show the location of atoms in three crystallographic planes in an FCC lattice. Draw the unit cell for the corresponding structure and identify these planes in your diagram. [2000 - 3 Marks]

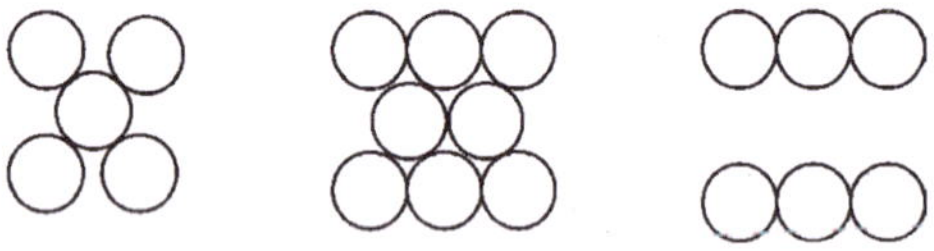

32. A metallic element crystallises into a lattice containing a sequence of layers of ABABAB..... Any packing of spheres leaves out voids in the lattice. What percentage by volume of this lattice is empty space? [1996 - 3 Marks]

33. The density of mercury is 13.6 g/mL. Calculate approximately the diameter of an atom of mercury assuming that each atom is occupying a cube of edge length equal to the diameter of the mercury atom. [1983 - 3 Marks]

Answer Key

Part-A (JEE Main)

1	(4)	7	(2)	13	(3)	19	(3)	25	(3)	31	(4)	37	(2)	43	(85)	49	(43)	55	(15)
2	(4)	8	(3)	14	(1)	20	(2)	26	(4)	32	(3)	38	(1)	44	(566)	50	(1)	56	(1)
3	(3)	9	(2)	15	(1)	21	(3)	27	(4)	33	(4)	39	(2)	45	(87)	51	(106)	57	(9076)
4	(1)	10	(4)	16	(1)	22	(4)	28	(3)	34	(3)	40	(2)	46	(22)	52	(8)	58	(8)
5	(3)	11	(4)	17	(4)	23	(4)	29	(3)	35	(2)	41	(4)	47	(3)	53	(5)	59	(143)
6	(3)	12	(3)	18	(2)	24	(4)	30	(2)	36	(4)	42	(4)	48	(1)	54	(33)		

Part-B (JEE Advanced)

1	A-p, s; B-p,q; C-q; D-q, r.	4	(b)	7	(d)	10	(3)	13	(11)	16	(1.86)	19	(b, c, d)	22	(a, c, d)	25	(d)	
2	(b)	5	(d)	8	(a)	11	(2)	14	(1.259)	17	(a, b, d)	20	(a)	23	(b)	26	(a)	
3	(a)	6	(a)	9	(a)	12	(8)	15	(124.27)	18	(a, c)	21	(b, c)	24	(a)			

Chapter-2

Solutions

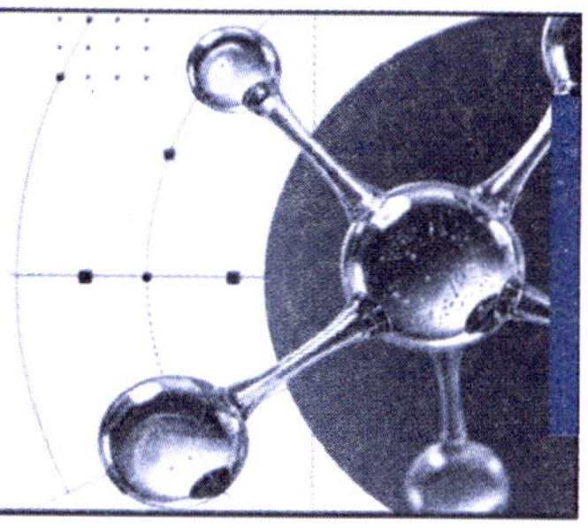

Part-A (JEE Main)

Topic 1 - Concentration of Solutions and Solubility

MULTIPLE CHOICE QUESTIONS (MCQs)

1. Given below are two statements : **[April 08, 2026 (II)]**

Given : Molar mass of C, H, O, Cl are 12, 1, 16 and 35.5 g mol^{-1}, respectively.

Statement I: In 30% (w/w) solution of methanol in CCl_4 (at T K), the mole fraction of CCl_4 is equal to 0.33.

Statement II: Mixture of methanol and CCl_4 shows positive deviation deviation from Raoult's law.

In the light of the above statements, choose the **correct** answer from the option given below :

(1) Both Statement I and Statement II are true
(2) Both Statement I and Statement II are false
(3) Statement I is true but Statement II are false
(4) Statement I is false but Statement II are true

2. What is the mole fraction of water in 10% by weight (w/w) of aqueous urea solution ?

[Given: Molar mass of H, O, C and N are 1, 16, 12 and 14 g mol^{-1} respectively] **[April 05, 2026 (I)]**

(1) 0.825 (2) 0.032
(3) 0.867 (4) 0.967

3. At 298 K, the mole percentage of N_2(g) in air is 80%. Water is in equilibrium with air at a pressure of 10 atm. What is the mole fraction of N_2(g) in water at 298 K ? (K_H for N_2 is 6.5×10^7 mm Hg) **[Jan. 24, 2026 (II)]**

(1) 1.23×10^{-7} (2) 1.17×10^{-4}
(3) 9.35×10^5 (4) 9.35×10^{-5}

4. At T(K), 100 g of 98% H_2SO_4 (w/w) aqueous solution is mixed with 100 g of 49% H_2SO_4 (w/w) aqueous solution. What is the mole fraction of H_2SO_4 in the resultant solution?

(**Given :** Atomic mass H = 1 u ; S = 32 u; O = 16 u)

(Assume that temperature after mixing remains constant) **[Jan. 22, 2026 (II)]**

(1) 0.9 (2) 0.1 (3) 0.337 (4) 0.663

5. Given below are two statements:

Statement I : The Henry's law constant K_H is constant with respect to variations in solution's concentration over the range for which the solutions is ideally dilute.

Statement II : K_H does not differ for the same solute in different solvents. **[Jan. 22, 2026 (I)]**

In the light of the above statements, choose the **correct** answer from the options.

(1) Statement I is false but Statement II is true.
(2) Statement I is true but Statement II is false.
(3) Both Statement I and Statement II are true.
(4) Both Statement I and Statement II are false.

6. Consider a solution of CO_2 (g) dissolved in water in a closed container. **[Jan. 22, 2026 (I)]**

Which one of the following plots correctly represents variation of log (partial pressure of CO_2 in vapour phase above water) [y-axis] with log (mole fraction of CO_2 in water) [x-axis] at 25°C ?

(1) 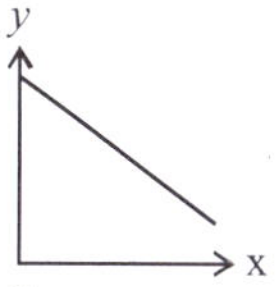

(2) 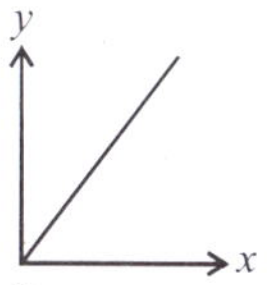

(3) 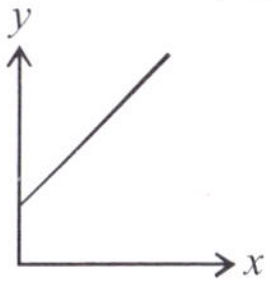

(4) 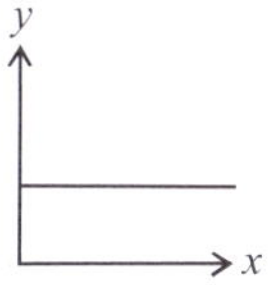

7. At the sea level, the dry air mass percentage composition is given as nitrogen gas : 70.0, oxygen gas : 27.0 and argon gas : 3.0. If total pressure is 1.15 atm, then calculate the ratio of followings respectively : **[April 7, 2025 (I)]**

(i) partial pressure of nitrogen gas to partial pressure of oxygen gas
(ii) partial pressure of oxygen gas to partial pressure of argon gas

(Given : Molar mass of N, O and Ar are 14, 16, and 40 g mol–1 respectively)

(1) 4.26, 19.3 (2) 2.59, 11.85
(3) 5.46, 17.8 (4) 2.96, 11.2

8. 'x' g of NaCl is added to water in a beaker with a lid. The temperature of the system is raised from 1°C to 25°C. Which out of the following plots, is best suited for the change in the molarity (M) of the solution with respect to temperature ? [Consider the solubility of NaCl remains unchanged over the temperature range] **[April 2, 2025 (II)]**

(1) 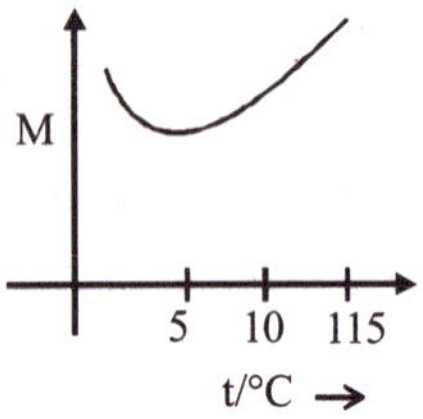

(2) 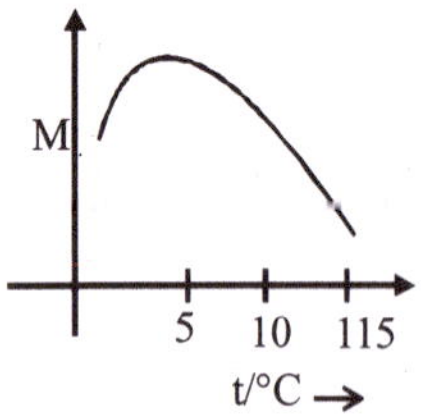

(3) 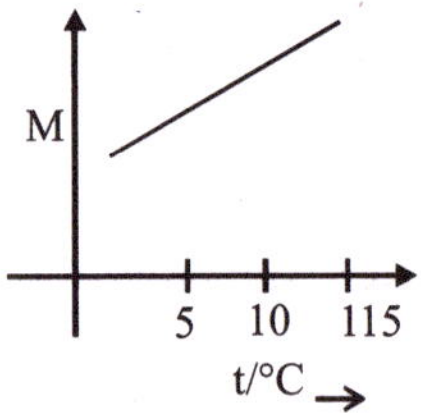

(4) 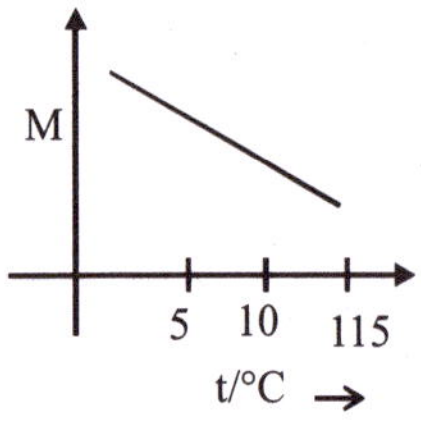

9. Which of the following graph correctly represents the plots of K_H at 1 bar gases in water versus temperature ? **[April 2, 2025 (I)]**

(1) 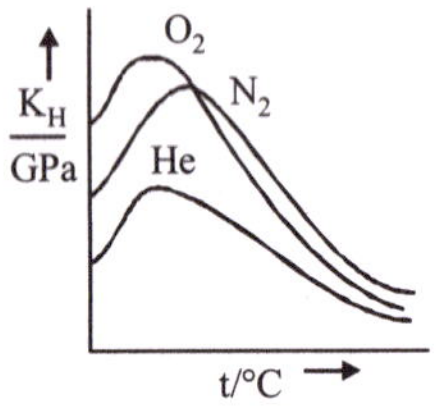

(2) 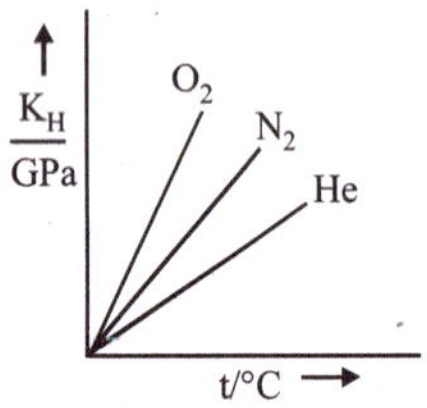

(3) 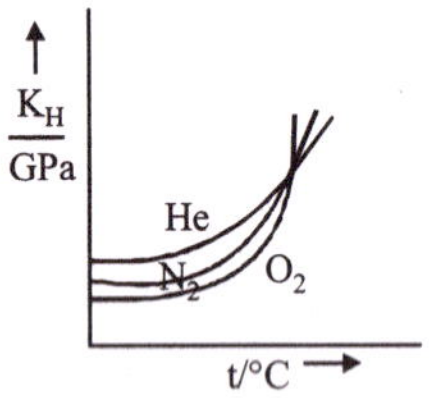

(4) 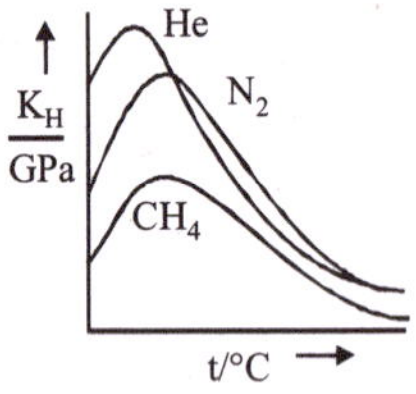

10. The density of 'x' M solution ('x' molar) of NaOH is 1.12 g mL^{-1}. while in molality, the concentration of the solution is 3 m (3 molal). Then x is **[April 6, 2024 (I)]**

(Given : Molar mass of NaOH is 40 g/mol)

(1) 3.5 (2) 3.0 (3) 3.8 (4) 2.8

11. Molality (m) of 3 M aqueous solution of NaCl is:

(Given : Density of solution = 1.25 g mL^{-1}, Molar mass in g mol^{-1} (Na-23, Cl-35.5) **[April 6, 2024 (II)]**

(1) 2.90m (2) 2.79m (3) 1.90m (4) 3.85m

12. The Molarity (M) of an aqueous solution containing 5.85 g of NaCl in 500 mL water is : **[April 4, 2024 (I)]**

(Given : Molar Mass Na : 23 and Cl : 35.5 $gmol^{-1}$)

(1) 20 (2) 0.2 (3) 2 (4) 4

13. If a substance 'A' dissolves in solution of a mixture of 'B' and 'C' with their respective number of moles as n_A, n_B and n_C, Mole fraction of C in the solution is : **[Jan. 30, 2024 (II)]**

(1) $\frac{n_C}{n_A \times n_B \times n_C}$ (2) $\frac{n_C}{n_A + n_B + n_C}$

(3) $\frac{n_C}{n_A - n_B - n_C}$ (4) $\frac{n_B}{n_A + n_B}$

14. The quantity which changes with temperature is: **[Jan. 27, 2024 (II)]**

(1) Molarity (2) Mass percentage

(3) Molality (4) Mole fraction

15. A solution is prepared by adding 2g of "X" of 1 mole of water. Mass percent of "X" in the solution is : **[April 11, 2023 (II)]**

(1) 20% (2) 5% (3) 2% (4) 10%

16. Given below are two statements: one is labelled as **Assertion A** and the other is labelled as **Reason R.**

Assertion A: 3.1500g of hydrated oxalic acid dissolved in water to make 250.0 mL solution will result in 0.1 M oxalic acid solution.

Reason R: Molar mass of hydrated oxalic acid is 126 g mol^{-1}. **[April 10, 2023 (II)]**

In the light of the above statements, choose the correct answer from the options given below:

(1) Both A and R are true but R is NOT the correct explanation of A.

(2) A is false but R is true.

(3) A is true but R is false.

(4) Both A and R are true and R is the correct explanation of A.

17. Given below are two statements : one is labelled as **Assertion (A)** and the other is labelled as **Reason (R)**

Assertion (A) : At 10°C, the density of a 5M solution of KCl [atomic masses of K and Cl are 39 & 35.5 g mol^{-1}] is x g mL^{-1}. The solution is cooled to –21°C. The molality of the solution will remain unchanged.

Reason (R) : The molality of a solution does not change with temperature as mass remains unaffected with temperature.

In the light of the above statements, choose the correct answer from the options given below: **[June 27, 2022 (I)]**

(1) Both (A) and (R) are true and (R) is the correct explanation of (A).

(2) Both (A) and (R) are true but (R) is not the correct explanation of (A).

(3) (A) is true but (R) is false.

(4) (A) is false but (R) is true.

18. The mole fraction of a solvent in aqueous solution of a solute is 0.8. The molality (in mol kg^{-1}) of the aqueous solution is : **[April 12, 2019 (I)]**

(1) 13.88×10^{-2} (2) 13.88×10^{-1}

(3) 13.88 (4) 13.88×10^{-3}

19. What would be the molality of 20% (mass/mass) aqueous solution of KI? (molar mass of KI = 166 g mol^{-1}) **[April 9, 2019 (II)]**

(1) 1.08 (2) 1.35 (3) 1.48 (4) 1.51

20. A solution of sodium sulfate contains 92 g of Na^+ ions per kilogram of water. The molality of Na^+ ions in that solution in mol kg^{-1} is: **[Jan. 9, 2019 (I)]**

(1) 12 (2) 4 (3) 8 (4) 16

21. The solubility of N_2 in water at 300 K and 500 torr partial pressure is 0.01 g L^{-1}. The solubility (in g L^{-1}) at 750 torr partial pressure is : **[Online April 9, 2016]**

(1) 0.0075 (2) 0.005 (3) 0.02 (4) 0.015

22. The molarity of a solution obtained by mixing 750 mL of 0.5(M) HCl with 250 mL of 2(M) HCl will be : **[2013]**

(1) 0.875 M (2) 1.00 M (3) 1.75 M (4) 0.975 M

23. The density of a solution prepared by dissolving 120 g of urea (mol. mass = 60 u) in 1000 g of water is 1.15 g/mL. The molarity of this solution is : [2012]
(1) 0.50 M (2) 1.78 M (3) 1.02 M (4) 2.05 M

NUMERICAL VALUE QUESTIONS (NVQs)

24. Sea water, which can be considered as a 6 molar (6 M) solution of NaCl, has a density of 2 g mL^{-1}. The concentration of dissolved oxygen (O_2) in sea water is 5.8 ppm. Then the concentration of dissolved oxygen (O_2) in sea water, is $x \times 10^{-4}$ m.
x = _______. (Nearest integer)
Given: Molar mass of NaCl is 58.5 g mol^{-1}
Molar mass of O_2 is 32 g mol^{-1}. [April 4, 2025 (II)]

25. Fortification of food with iron is done using $FeSO_4 \cdot 7H_2O$.
The mass in grams of the $FeSO_4 \cdot 7H_2O$ required to achieve 12 ppm of iron in 150 kg of wheat is _____(Nearest integer)
[Given : Molar mass of Fe, S and O respectively are 56, 32 and 16 g mol^{-1}] [April 4, 2025 (I)]

26. Molarity (M) of an aqueous solution containing x g of anhyd. $CuSO_4$ in 500 mL solution at 32 °C is 2×10^{-1} M. Its molality will be _______ $\times 10^{-3}$ m. (nearest integer).
[Given density of the solution = 1.25 g/mL.] [April 9, 2024 (I)]

27. Molality of an aqueous solution of urea is 4.44 m. Mole fraction of urea in solution is $x \times 10^{-3}$.
Value of x is _______. (integer answer) [April 8, 2024 (II)]

28. A solution is prepared by adding 1 mole ethyl alcohol in 9 mole water. The mass percent of solute in the solution is _______ (Integer Answer) [April 8, 2024 (II)]
(Given : Molar mass in g mol^{-1} Ethyl alcohol : 46, water : 18)

29. The molarity of 1 L orthophosphoric acid (H_3PO_4) having 70% purity by weight (specific gravity 1.54 g cm^{-3}) is _______M.
(Molar mass of H_3PO_4 = 98 g mol^{-1}) [Jan. 31, 2024 (II)]

30. The mass of sodium acetate (CH_3COONa) required to prepare 250 mL of 0.35 M aqueous solution is __________ g. (Molar mass of CH_3COONa is 82.02 g mol^{-1}) [Jan. 30, 2024 (I)]

31. A solution of H_2SO_4 is 31.4% H_2SO_4 by mass and has a density of 1.25 g/mL.
The molarity of the H_2SO_4 solution is _____ M (nearest integer)
(Given molar mass of H_2SO_4 = 98g mol^{-1}] [Jan. 29, 2024 (I)]

32. Molality of 0.8 M H_2SO_4 solution (density 1.06 g cm^{-3}) is ____ $\times 10^{-3}$ m. [Jan. 29, 2024 (II)]

33. The density of 3 M solution of NaCl is 1.0 g mL^{-1}. Molality of the solution is _________ $\times 10^{-2}$m. (Nearest integer).
Given: Molar mass of Na and Cl is 23 and 35.5 g mol^{-1} respectively. [Feb. 01, 2023 (I)]

34. The molality of a 10% (v/v) solution of di-bromine solution in CCl_4 (carbon tetrachloride) is 'x'.
x = ______ $\times 10^{-2}$ M. (Nearest integer)
[Given: Molar mass of Br_2 = 160 g mol^{-1}
atomic mass of C = 12 g mol^{-1}
atomic mass of Cl = 35.5 g mol^{-1}
density of dibromine = 3.2 g cm^{-3}
density of CCl_4 = 1.6 g cm^{-3}] [Feb. 01, 2023 (II)]

35. Some amount of dichloromethane (CH_2Cl_2) is added to 671.141 mL of chloroform ($CHCl_3$) to prepare 2.6×10^{-3} M solution of CH_2Cl_2 (DCM).
The concentration of DCM is ______ ppm (by mass).
Given: Atomic mass : C = 12; H : 1; Cl = 35.5 density of $CHCl_3$ 1.49 g cm^{-3}. [Jan. 30, 2023 (I)]

36. 5 g of NaOH was dissolved in deionized water to prepare a 450 mL stock solution. What volume (in *mL*) of this solution would be required to prepare 500 mL of 0.1 M solution ?
Given : Molar Mass of Na, O and H is 23, 16 and 1g mol^{-1} respectively. [Jan. 24, 2023 (I)]

37. The number of units, which are used to express concentration of solutions from the following is _________. [Jan. 24, 2023 (II)]
Mass percent, Mole, Mole fraction, Molarity, ppm, Molality

38. When 800 mL of 0.5 M nitric acid is heated in a beaker, its volume is reduced to half and 11.5 g of nitric acid is evaporated. The molarity of the remaining nitric acid solution is $x \times 10^{-2}$ M. (Nearest Integer) [July 26, 2022 (I)]
(Molar mass of nitric acid is 63 g mol^{-1})

39. If 80 g of copper sulphate $CuSO_4 \cdot 5H_2O$ is dissolved in deionised water to make 5 L of solution. The concentration of the copper sulphate solution is $x \times 10^{-3}$ mol L^{-1}. The value of x is __________. [Sep. 1, 2021 (II)]
[Atomic masses Cu: 63.54u, S: 32u, O: 16u, H: 1u]

40. The molarity of the solution prepared by dissolving 6.3 g of oxalic acid ($H_2C_2O_4.2H_2O$) in 250 mL of water in mol L^{-1} is $x \times 10^{-2}$. The value of x is ________. (Nearest integer)
[Atomic mass : H : 1.0, C : 12.0, O : 16.0] [Aug. 31, 2021 (I)]

41. An aqueous KCl solution of density 1.20 g mL^{-1} has a molality of 3.30 mol kg^{-1}. The molarity of the solution in mol L^{-1} is ________ (Nearest integer)
[Molar mass of KCl = 74.5]
[Aug. 26, 2021(I); Same March 16, 2021 (I)]
[Atomic masses : K : 39.0 u; O : 16.0 u; H : 1.0 u]

42. 100 mL of Na_3PO_4 solution contains 3.45 g of sodium. The molarity of the solution is _____$\times 10^{-2}$ mol L^{-1}. (Nearest integer)
[Atomic Masses - Na : 23.0 u, O : 16.0 u, P : 31.0 u]
[Aug. 26, 2021 (II)]

43. The mole fraction of a solute in a 100 molal aqueous solution is ____ $\times 10^{-2}$.
(Round off to the Nearest Integer).
[Given : Atomic masses : H : 1.0 u, O : 16.0 u]
[March 17, 2021 (I)

44. A 6.50 molal solution of KOH (aq.) has a density of 1.89 g cm^{-3}. The molarity of the solution is ____ mol dm^{-3}. (Round off to the Nearest Integer). [March 16, 2021 (I)]

Topic 2 - Vapour Pressure, Laws of Solutions and Ideal, Non-ideal Solutions

MULTIPLE CHOICE QUESTIONS (MCQs)

45. When 0.25 moles of a non-volatile, non-ionizable solute was dissolved in 1 mole of a solvent the vapor pressure of solution was x% of vapor pressure of pure solvent. What is x% ? **[April 06, 2026 (I)]**

(1) 50% (2) 60% (3) 70% (4) 80%

46. At T(K), 2 moles of liquid A and 3 moles of liquid B are mixed. The vapour pressure of ideal solution formed is 320 mm Hg. At this stage, one mole of A and one mole of B are added to the solution. The vapour pressure is now measured as 328.6 mm Hg. The vapour pressure (in mm Hg) of A and B are respectively: **[Jan. 28, 2026 (I)]**

(1) 300, 200 (2) 600, 400
(3) 400, 300 (4) 500, 200

47. Two liquids A and B form an ideal solution at temperature T K. At T K, the vapour pressures of pure A and B are 55 and 15 kNm^{-2} respectively. What is the mole fraction of A in solution of A and B in equilibrium with a vapour in which the mole fraction of A is 0.8? **[Jan. 24, 2026 (II)]**

(1) 0.5217 (2) 0.480 (3) 0.663 (4) 0.340

48. 'W' g of non-volatile electrolyte solid solute of molar mass 'M' g mol^{-1} when dissolved in 100 mL water, decreases vapour pressure of water from 640 mm Hg to 600 mm Hg. If aqueous solution of the electrolyte boils at 375 K and K_b for water is 0.52 K kg mol^{-1}, then the mole fraction of the electrolyte solute (x_2) in the solution can be expressed as

(Given : density of water = 1 g/mL and boiling point of water = 373 K) **[Jan. 24, 2026 (I)]**

(1) $\frac{1.3}{8} \times \frac{W}{M}$ (2) $\frac{16}{2.6} \times \frac{W}{M}$

(3) $\frac{2.6}{16} \times \frac{M}{W}$ (4) $\frac{1.3}{8} \times \frac{M}{W}$

49. Which one of the following graphs accurately represents the plot of partial pressure of CS_2 vs its mole fraction in a mixture of acetone and CS_2 at constant temperature ? **[Jan. 23, 2026 (I)]**

(1)
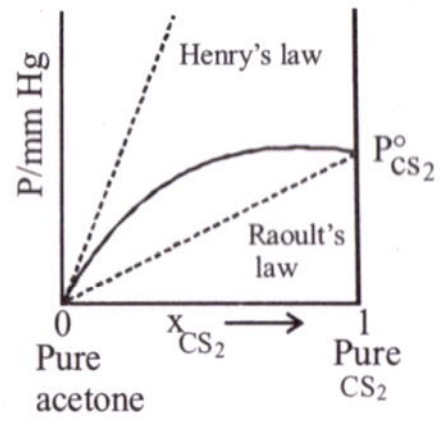

(2)
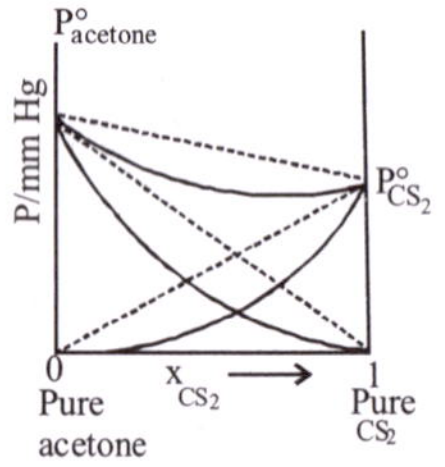

(3)
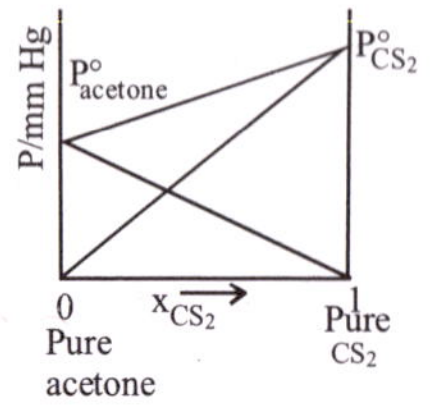

(4)
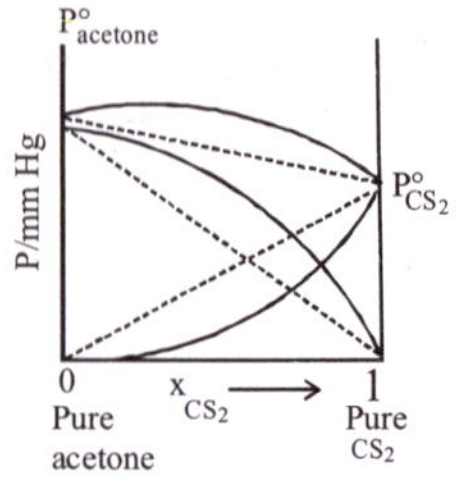

50. Which of the following binary mixture does not show the behaviour of minimum boiling azeotropes ? **[April 8, 2025 (II)]**

(1) $H_2O + CH_3COC_2H_5$ (2) $C_6H_5OH + C_6H_5NH_2$
(3) $CS_2 + CH_3COCH_3$ (4) $CH_3OH + CHCl_3$

51. Match **List-I** with **List-II** **[April 7, 2025 (II)]**

List-I	List-II
(A) Solution of chloroform and acetone	(I) Minimum boiling azeotrope
(B) Solution of ethanol and water	(II) Dimerizes
(C) Solution of benzene and toluene	(III) Maximum boiling azeotrope
(D) Solution of acetic acid in benzene	(IV) $\Delta V_{mix} = 0$

Choose the **correct** answer from the options given below:

(1) (A)-(III), (B)-(I), (C)-(IV), (D)-(II)
(2) (A)-(II), (B)-(IV), (C)-(I), (D)-(III)
(3) (A)-(III), (B)-(IV), (C)-(I), (D)-(II)
(4) (A)-(II), (B)-(I), (C)-(IV), (D)-(III)

52. Liquid A and B form an ideal solution. The vapour pressure of pure liquids A and B are 350 and 750 mm Hg respectively at the same temperature. If x_A and x_B are the mole fraction of A and B in solution while y_A and y_B are the mole fraction of A and B in vapour phase then : **[April 7, 2025 (II)]**

(1) $\frac{x_A}{x_B} < \frac{y_A}{y_B}$ (2) $\frac{x_A}{x_B} = \frac{y_A}{y_B}$

(3) $\frac{x_A}{x_B} > \frac{y_A}{y_B}$ (4) $(x_A - y_A) < (x_B - y_B)$

53. A solution is made by mixing one mole of volatile liquid A with 3 moles of volatile liquid B. The vapour pressure of pure A is 200 mm Hg and that of the solution is 500 mm Hg. The vapour pressure of pure B and the least volatile component of the solution, respectively, are : **[April 2, 2025 (I)]**

(1) 1400 mm Hg, A (2) 1400 mm Hg, B
(3) 600 mm Hg, B (4) 600 mm Hg, A

54. Consider a binary solution of two volatile liquid components 1 and 2. x_1 and y_1 are the mole fractions of component 1 in liquid and vapour phase, respectively. The slope and intercept of the linear plot of $\frac{1}{x_1}$ vs $\frac{1}{y_1}$ are given respectively as: **[Jan. 23, 2025 (II)]**

(1) $\frac{P_1^0}{P_2^0}, \frac{P_2^0 - P_1^0}{P_2^0}$ (2) $\frac{P_2^0}{P_1^0}, \frac{P_1^0 - P_2^0}{P_2^0}$

(3) $\frac{P_1^0}{P_2^0}, \frac{P_1^0 - P_2^0}{P_2^0}$ (4) $\frac{P_2^0}{P_1^0}, \frac{P_2^0 - P_1^0}{P_2^0}$

55. Identify the mixture that shows positive deviations from Raoult's Law **[Jan. 31, 2024 (I)]**

(1) $CHCl_3 + (CH_3)_2CO$ (2) $CHCl_3 + C_6H_6$
(3) $(CH_3)_2CO + C_6H_5NH_2$ (4) $(CH_3)_2CO + CS_2$

56. A solution of two miscible liquids showing negative deviation from Raoult's law will have : **[Jan. 27, 2024 (I)]**

(1) increased vapour pressure, increased boiling point
(2) increased vapour pressure, decreased boiling point
(3) decreased vapour pressure, decreased boiling point
(4) decreased vapour pressure, increased boiling point

57. For a solution of the gases A, B, C and D in water at 298 K, the values of Henry's law constant (K_H) are 30.40, 2.34, 1.56×10^{-5} and 0.513 k bar respectively. In the given graph, the lines marked as 'p' and 's' correspond respectively to: **[June 30, 2022 (I)]**

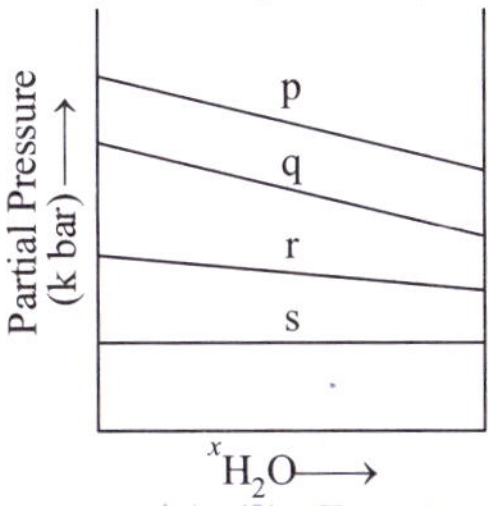

(1) A and C (2) B and A (3) D and A (4) C and D

58. Henry's constant (in kbar) for four gases α, β, γ and δ in water at 298 K is given below : **[Sep. 03, 2020 (I)]**

	α	β	γ	δ
K_H	50	2	2×10^{-5}	0.5

(density of water = 10^3 kg m^{-3} at 298 K)

This table implies that :

(1) α has the highest solubility in water at a given pressure
(2) solubility of γ at 308 K is lower than at 298 K
(3) The pressure of a 55.5 molal solution of γ is 1 bar
(4) The pressure of a 55.5 molal solution of δ is 250 bar

59. An open beaker of water in equilibrium with water vapour is in a sealed container. When a few grams of glucose are added to the beaker of water, the rate at which water molecules :

(1) leaves the vapour increases **[Sep. 02, 2020 (I)]**
(2) leaves the solution increases
(3) leaves the solution decreases
(4) leaves the vapour decreases

60. A graph of vapour pressure and temperature for three different liquids *X*, *Y*, and *Z* is shown below:

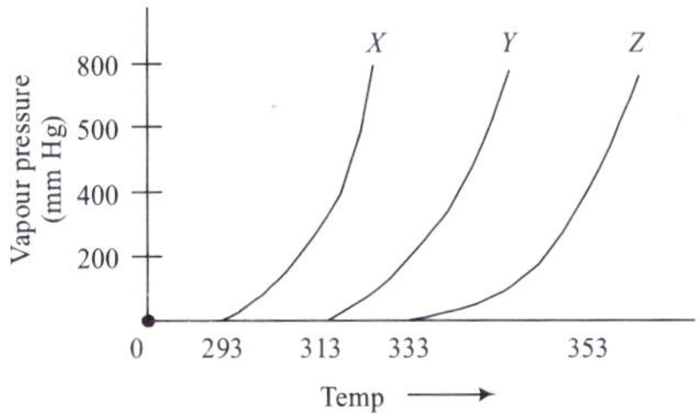

The following inferences are made: **[Jan. 08, 2020 (I)]**

(A) *X* has higher intermolecular interactions compared to Y.
(B) *X* has lower intermolecular interactions compared to Y.
(C) *Z* has lower intermolecular interactions compared to Y.

The correct inference(s) is/are:

(1) (A) and (C) (2) (A)
(3) (B) (4) (C)

61. At 35 °C, the vapour pressure of CS_2 is 512 mm Hg and that of acetone is 344 mm Hg. A solution of CS_2 in acetone has a total vapour pressure of 600 mm Hg. The false statement amongst the following is: **[Jan. 07, 2020 (I)]**

(1) Raoult's law is not obeyed by this system
(2) a mixture of 100 mL CS_2 and 100 mL acetone has a volume < 200 mL
(3) CS_2 and acetone are less attracted to each other than to themselves
(4) heat must be absorbed in order to produce the solution at 35 °C

62. Two open beakers one containing a solvent and the other containing a mixture of that solvent with a non volatile solute are together sealed in a container. Over time: **[Jan. 07, 2020 (II)]**

(1) the volume of the solution increases and the volume of the solvent decreases
(2) the volume of the solution decreases and the volume of the solvent increases
(3) the volume of the solution and the solvent does not change
(4) the volume of the solution does not change and the volume of the solvent decreases

63. Liquid 'M' and liquid 'N' form an ideal solution. The vapour pressures of pure liquids 'M' and 'N' are 450 and 700 mmHg, respectively, at the same temperature. Then correct statement is:
(x_M = Mole fraction of 'M' in solution; **[April 9, 2019 (I)]**
x_N = Mole fraction of 'N' in solution;
y_M = Mole fraction of 'M' in vapour phase;
y_N = Mole fraction of 'N' in vapour phase)

(1) $\frac{x_M}{x_N} = \frac{y_M}{y_N}$ (2) $(x_M - y_M) < (x_N - y_N)$

(3) $\frac{x_M}{x_N} < \frac{y_M}{y_N}$ (4) $\frac{x_M}{x_N} > \frac{y_M}{y_N}$

64. The vapour pressures of pure liquids A and B are 400 and 600 mmHg, respectively at 298 K. On mixing the two liquids, the sum of their initial volumes is equal to the volume of the final mixture. The mole fraction of liquid B is 0.5 in the mixture. The vapour pressure of the final solution, the mole fractions of components A and B in vapour phase, respectively are: **[April 8, 2019 (I)]**

(1) 450 mmHg, 0.4, 0.6 (2) 500 mmHg, 0.5, 0.5
(3) 450 mmHg, 0.5, 0.5 (4) 500 mmHg, 0.4, 06

65. For the solution of the gases w, x, y and z in water at 298 K, the Henry's law constants (K_H) are 0.5, 2, 35 and 40 kbar, respectively. The correct plot for the given data is : **[April 8, 2019 (II)]**

(1) 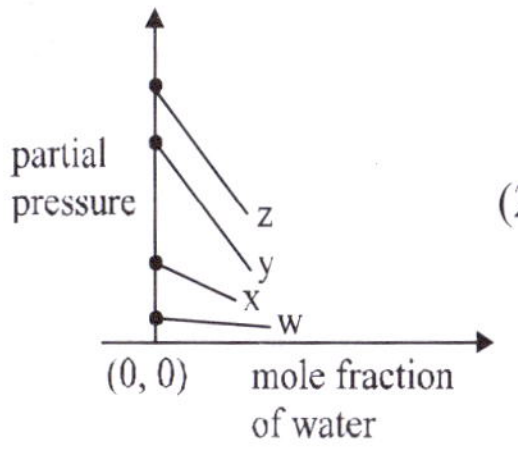

(2) 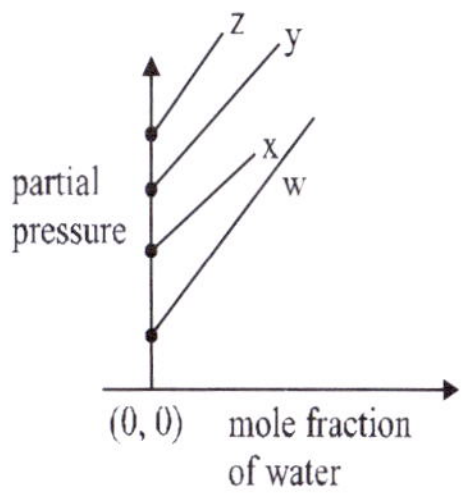

(3) 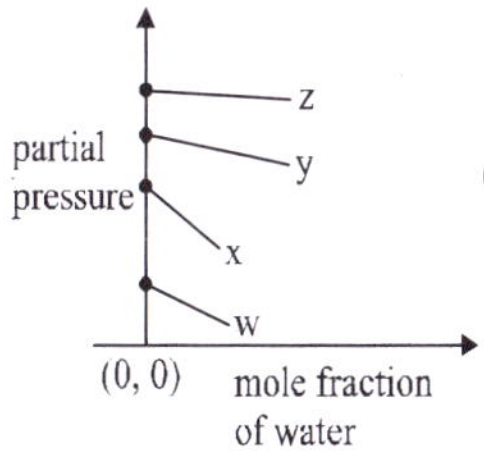

(4) 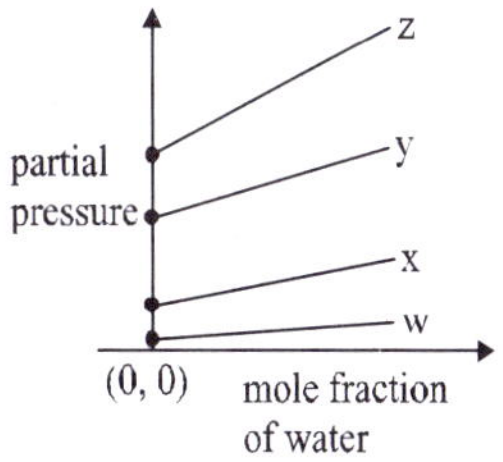

66. Liquids A and B form an ideal solution in the entire composition range. At 350 K, the vapor pressures of pure A and pure B are 7×10^3 Pa and 12×10^3 Pa, respectively. The composition of the vapour is in equilibrium with a solution containing 40 mole percent of A at this temperature is: **[Jan. 10, 2019 (I)]**

(1) $x_A = 0.37; x_B = 0.63$ (2) $x_A = 0.28; x_B = 0.72$
(3) $x_A = 0.4; x_B = 0.6$ (4) $x_A = 0.76; x_B = 0.24$

67. Which one of the following statements regarding Henry's law is not correct? [Jan. 9, 2019 (I)]
(1) Higher the value of K_H at a given pressure, higher is the solubility of the gas in liquids.
(2) Different gases have different K_H (Henry's law constant) values at the same temperature.
(3) The partial pressure of the gas in vapour phase is proportional to the mole fraction of the gas in the solution.
(4) The value of K_H increases with increase of temperature and K_H is function of the nature of the gas.

68. A solution is prepared by mixing 8.5 g of CH_2Cl_2 and 11.95 g of $CHCl_3$. If vapour pressure of CH_2Cl_2 and $CHCl_3$ at 298 K are 415 and 200 mmHg respectively, the mole fraction of $CHCl_3$ in vapour form is : [Online April 9, 2017]
(Molar mass of Cl = 35.5 g mol^{-1})
(1) 0.162 (2) 0.675 (3) 0.325 (4) 0.486

69. The solubility of N_2 in water at 300 K and 500 torr partial pressure is 0.01 g L^{-1}. The solubility (in g L^{-1}) at 750 torr partial pressure is : [Online April 9, 2016]
(1) 0.0075 (2) 0.005 (3) 0.02 (4) 0.015

70. A solution at 20 °C is composed of 1.5 mol of benzene and 3.5 mol of toluene. If the vapour pressure of pure benzene and pure toluene at this temperature are 74.7 torr and 22.3 torr, respectively, then the total vapour pressure of the solution and the benzene mole fraction in equilibrium with it will be, respectively : [Online April 10, 2015]
(1) 35.8 torr and 0.280 (2) 38.0 torr and 0.589
(3) 30.5 torr and 0.389 (4) 30.5 torr and 0.480

71. Choose the correct statement with respect to the vapour pressure of a liquid among the following: [Online April 19, 2014]
(1) Increases linearly with increasing temperature
(2) Increases non-linearly with increasing temperature
(3) Decreases linearly with increasing temperature
(4) Decreases non-linearly with increasing temperature

72. For an ideal solution of two components A and B, which of the following is true? [Online April 19]
(1) $\Delta H_{mixing} < 0$ (zero)
(2) $\Delta H_{mixing} > 0$ (zero)
(3) A – B interaction is stronger than A – A and B – B interactions
(4) A– A, B – B and A – B interactions are identical.

73. Vapour pressure of pure benzene is 119 torr and that of toluene is 37.0 torr at the same temperature. Mole fraction of toluene in vapour phase which is in equilibrium with a solution of benzene and toluene having a mole fraction of toluene 0.50, will be : [Online April 23, 2013]
(1) 0.137 (2) 0.237 (3) 0.435 (4) 0.205

74. Liquids A and B form an ideal solution. At 30 °C, the total vapour pressure of a solution containing 1 mol of A and 2 mol of B is 250 mm Hg. The total vapour pressure becomes 300 mm Hg when 1 more mol of A is added to the first solution. The vapour pressures of pure A and B at the same temperature are [Online May 19, 2012]
(1) 150, 450 mm Hg (2) 125, 150 mm Hg
(3) 450, 150 mm Hg (4) 250, 300 mm Hg

75. On mixing, heptane and octane form an ideal solution. At 373 K, the vapour pressures of the two liquid components (heptane and octane) are 105 kPa and 45 kPa respectively. Vapour pressure of the solution obtained by mixing 25.0 g of heptane and 35 g of octane will be (molar mass of heptane = 100 g mol^{-1} and of octane = 114 g mol^{-1}) [2010]
(1) 72.0 kPa (2) 36.1 kPa
(3) 96.2 kPa (4) 144.5 kPa

76. A binary liquid solution is prepared by mixing *n*-heptane and ethanol. Which one of the following statements is correct regarding the behaviour of the solution? [2009]
(1) The solution is non-ideal, showing – ve deviation from Raoult's Law.
(2) The solution is non-ideal, showing + ve deviation from Raoult's Law.
(3) n-Heptane shows + ve deviation while ethanol shows – ve deviation from Raoult's Law.
(4) The solution formed is an ideal solution.

77. Two liquids X and Y form an ideal solution. At 300 K, vapour pressure of the solution containing 1 mol of X and 3 mol of Y is 550 mmHg. At the same temperature, if 1 mol of Y is further added to this solution, vapour pressure of the solution increases by 10 mmHg. Vapour pressure (in mmHg) of X and Y in their pure states will be, respectively:
(1) 300 and 400 (2) 400 and 600 [2009]
(3) 500 and 600 (4) 200 and 300

78. At 80 °C, the vapour pressure of pure liquid 'A' is 520 mm Hg and that of pure liquid 'B' is 1000 mm Hg. If a mixture solution of 'A' and 'B' boils at 80 °C and at 1 atm pressure, the amount of 'A' in the mixture is (1 atm = 760 mm Hg) [2008]
(1) 52 mol percent (2) 34 mol percent
(3) 48 mol percent (4) 50 mol percent

79. A mixture of ethyl alcohol and propyl alcohol has a vapour pressure of 290 mm at 300 K. The vapour pressure of propyl alcohol is 200 mm. If the mole fraction of ethyl alcohol is 0.6, its vapour pressure (in mm) at the same temperature will be
(1) 360 (2) 350 (3) 300 (4) 700 [2007]

80. Equal masses of methane and oxygen are mixed in an empty container at 25 °C. The fraction of the total pressure exerted by oxygen is [2007]
(1) 1/2 (2) 2/3 (3) $\frac{1}{3} \times \frac{273}{298}$ (4) 1/3

81. Benzene and toluene form nearly ideal solution. At 20°C, the vapour pressure of benzene is 75 torr and that of toluene is 22 torr. The partial vapour pressure of benzene at 20°C for a solution containing 78 g of benzene and 46 g of toluene in torr is [2005]
(1) 53.5 (2) 37.5 (3) 25 (4) 50

82. Equimolar solutions in the same solvent have [2005]
(1) Different boiling and different freezing points
(2) Same boiling and same freezing points
(3) Same freezing point but different boiling points
(4) Same boiling point but different freezing points

83. For which of the following parameters, the structural isomers C_2H_5OH and CH_3OCH_3 would be expected to have the same values? (Assume ideal behaviour).
(1) Boiling points [2004]
(2) Vapour pressure at the same temperature
(3) Heat of vaporization
(4) Gaseous densities at the same temperature and pressure

84. Which of the following liquid pairs shows a positive deviation from Raoult's law? [2004]
(1) Water - nitric acid (2) Benzene - methanol
(3) Water - hydrochloric acid (4) Acetone - chloroform

85. If liquids A and B form an ideal solution [2003]
(1) the entropy of mixing is zero
(2) the free energy of mixing is zero
(3) the free energy as well as the entropy of mixing is zero
(4) the enthalpy of mixing is zero

86. In mixture A and B components show -ve deviation as [2002]
(1) $\Delta V_{mix} > 0$
(2) $\Delta H_{mix} < 0$
(3) $A-B$ interaction is weaker than $A-A$ and $B-B$ interaction
(4) $A-B$ interaction is stronger than $A-A$ and $B-B$ interaction.

NUMERICAL VALUE QUESTIONS (NVQs)

87. A non-volatile, non-electrolyte solid solute when dissolved in 40 g of a solvent, the vapour pressure of the solvent decreased from 760 mm Hg to 750 mm Hg. If the same solution boils at 320 K, then the number of moles of the solvent present in the solution is _____. (Nearest integer) [April 04, 2026 (I)]
[Given : boiling point of the pure solvent = 319.5 K, K_b of the solvent = 0.3 K kg mol^{-1}]

88. Two liquids A and B form an ideal solution. At 320 K, the vapour pressure of the solution, containing 3 mol of A and 1 mol of B is 500 mm Hg. At the same temperature, if 1 mol of A is further added to this solution, vapour pressure of the solution increases by 20 mm Hg. Vapour pressure (in mm Hg) of B in the pure state is ______. (Nearest integer) [Jan. 23, 2026 (II)]

89. The vapour pressure of 30% (w/v) aqueous solution of glucose is ________ mm Hg at 25°C. [April 15, 2023 (I)]
[Given : The density of 30% (w/v), aqueous solution of glucose is 1.2 g cm^{-3} and vapour pressure of pure water is 24 mm Hg.]
(Molar mass of glucose is 180 g mol^{-1}.)

90. The vapour pressure vs. temperature curve for a solution solvent system is shown below. [April 08, 2023 (I)]

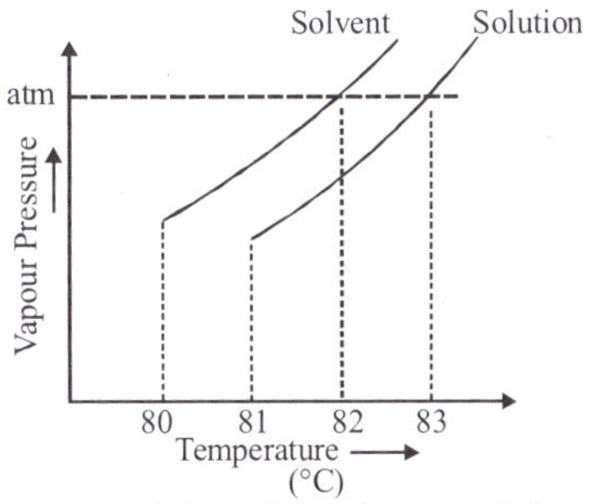

The boiling point of the solvent is _____ °C

91. The total pressure of a mixture of non-reacting gases X (0.6 g) and Y (0.45 g) in a vessel is 740 mm of Hg. The partial pressure of the gas X is ________ mm of Hg. (Nearest Integer)
(Given : molar mass X = 20 and Y = 45 g mol^{-1}) [Jan. 31, 2023 (I)]

92. The Total pressure observed by mixing two liquid A and B is 350 mm Hg when their mole fractions are 0.7 and 0.3 respectively. [Jan. 24, 2023 (II)]
The Total pressure becomes 410 mm Hg if the mole fractions are changed to 0.2 and 0.8 respectively for A and B. The vapour pressure of pure A is ______ mm Hg. (Nearest integer)
Consider the liquids and solutions behave ideally.

93. If O_2 gas is bubbled through water at 303 K, the number of millimoles of O_2 gas that dissolve in 1 litre of water is______. (Nearest integer)
(Given : Henry's Law constant for O_2 at 303 K is 46.82 k bar and partial pressure of O_2 = 0.920 bar) [July 29, 2022 (I)]
(Assume solubility of O_2 in water is too small, nearly negligible)

94. A gaseous mixture of two substances A and B, under a total pressure of 0.8 atm is in equilibrium with an ideal liquid solution. The mole fraction of substance A is 0.5 in the vapour phase and 0.2 in the liquid phase. The vapour pressure of pure liquid A is _______ atm. (Nearest integer) [July 28, 2022 (II)]

95. A mixture of hydrogen and oxygen contains 40% hydrogen by mass when the pressure is 2.2 bar. The partial pressure of hydrogen is ________ bar. (Nearest Integer) [July 26, 2022 (I)]

96. The vapour pressures of two volatile liquids A and B at 25°C are 50 Torr and 100 Torr, respectively. If the liquid mixture contains 0.3 mole fraction of A, then the mole fraction of liquid B in the vapour phase is $\frac{x}{17}$. The value of x is ________. [June 28, 2022 (I)]

97. A company dissolves 'X' amount of CO_2 at 298 K in 1 litre of water to prepare soda water X = ____ × 10^{-3}g. (nearest integer)
(Given: partial pressure of CO_2 at 298 K = 0.835 bar.
Henry's law constant for CO_2 at 298 K = 1.67 kbar. Atomic mass of H, C and O is 1, 12 and 16 g mol^{-1}, respectively) [June 24, 2022 (II)]

98. CO_2 gas is bubbled through water during a soft drink manufacturing process at 298 K. If CO_2 exerts a partial pressure of 0.835 bar then x m mol of CO_2 would dissolve in 0.9 L of water. The value of x is ______. (Nearest integer)
(Henry's law constant for CO_2 at 298 K is 1.67 × 10^3 bar) [July 25, 2021 (I)]

99. At 20 °C, the vapour pressure of benzene is 70 torr and that of methyl benzene is 20 torr. The mole fraction of benzene in the vapour phase at 20 °C above an equimolar mixture of benzene and methyl benzene is _____ × 10^{-2}.
(Nearest integer) [July 20, 2021 (I)]

100. The vapour pressures of A and B at 25 °C are 90 mm Hg and 15 mm Hg respectively. If A and B are mixed such that the mole fraction of A in the mixture is 0.6, then the mole fraction of B in the vapour phase is $x \times 10^{-1}$. The value of x is _____. (Nearest integer) [July 20, 2021 (II)]

101. The oxygen dissolved in water exerts a partial pressure of 20 kPa in the vapour above water. The molar solubility of oxygen in water is _________ × 10^{-5} mol dm^{-3}.
(Round off to the Nearest Integer).
[Given : Henry's law constant = K_H = 8.0 × 10^4 kPa for O_2.
Density of water with dissolved oxygen = 1.0 kg dm^{-3}] [March 17, 2021 (I)]

102. At 363 K, the vapour pressure of A is 21 kPa and that of B is 18 kPa. One mole of A and 2 moles of B are mixed. Assuming that this solution is ideal, the vapour pressure of the mixture is __________ kPa.
(Round off to the Nearest Integer). [March 16, 2021 (II)]

103. At 300 K, the vapour pressure of a solution containing 1 mole of n-hexane and 3 moles of n-heptane is 550 mm of Hg. At the same temperature, if one more mole of n-heptane is added to this solution, the vapour pressure of the solution increases by 10 mm of Hg. What is the vapour pressure in mm Hg of n-heptane in its pure state ________? [Sep. 04, 2020 (I)]

Topic 3 - Colligative Properties and Abnormal Molecular Masses

MULTIPLE CHOICE QUESTIONS (MCQs)

104. Which of the following are **not correct** ? [April 08, 2026 (II)]

A. For water, magnitude of K_b is more than the magnitude of K_f

B. The elevation in boiling point of water when a non- volative solute is added to it is larger in magnitude than its depression in freezing point.

C. Osmotic pressure measurement is preferred over any other colligative property to determine molar mass of proteins and polymers.

D. The dimerised form of benzoic acid in benzene is

$$C_6H_5-\overset{\overset{O}{\|}}{C}-OH \ldots\ldots O=\overset{\overset{OH}{|}}{C}-C_6H_5$$

Choose the **correct** answer from the options given below:

(1) A and B only (2) A and D only
(3) A, B and D only (4) A, C and D only

105. Given below are two statements: [April 06, 2026 (II)]

Chamber 1		Chamber 2
18 g glucose in 100 mL aqueous solution	Semipermeable membrane	30 g glucose in 250 mL aqueous solution

Statement I: H_2O molecules move from the chamber 1 to chamber 2.

Statement II: The osmotic pressure of a solution prepared by dissolving 50 mg of potassium sulphate (molar mass = 174 g/mol) in 2 L of water (at 27°C) is 0.0107 bar. (Given: R = 0.083dm^3 bar K^{-1} mol^{-1} and assume complete dissociation of electrolyte)

In the light of the above statements, choose the **correct** answer from the options given below :

(1) Both Statement-I and Statement-II are true
(2) Both Statement-I and Statement-II are false
(3) Statement-I is true but Statement-II is false
(4) Statement-I is false Statement-II is true

106. At 27° C, 0.1 M, 1 L $K_4[Fe(CN)_6]$ aqueous solution and 0.1 M, 1 L $FeCl_3$ aqueous solution are placed in a container separated by a semi permeable membrane AB. Assume complete dissociation of both the solutes. Which of the following statement is **correct** ? [April 04, 2026 (II)]

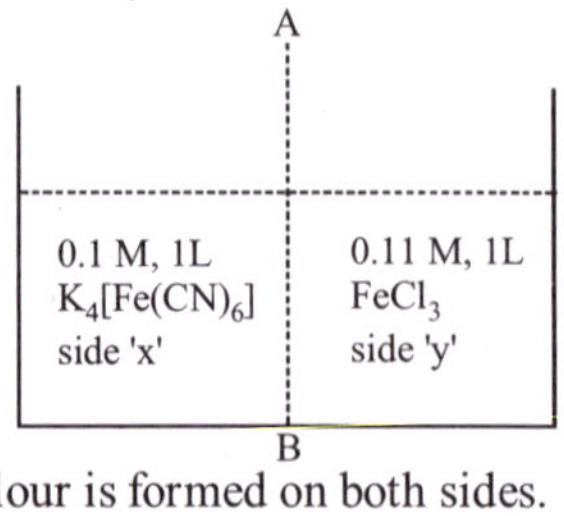

(1) Blue colour is formed on both sides.
(2) Ionic solutes in aqueous solution can pass through semi-permeable membrane.
(3) Solution on side 'y' is hypotonic.
(4) To cause the reverse flow of solvent during osmosis, external pressure (any value) should be applied to side 'x'.

107. Solution A is prepared by dissolving 1 g of a protein (molar mass = 50000 g mol^{-1}) in 0.5 L of water at 300 K. Its osmotic pressure is x bar. Solution B is made by dissolving 2 g of same protein in 1 L of water at 300 K. Osmotic pressure of solution B is y bar. Entire solution of A is mixed with entire solution of B at same temperature. The osmotic pressure of resultant solution is z bar. x, y and z respectively are :
(R = 0.083 L bar mol^{-1} K^{-1}) [April 02, 2026 (II)]

(1) 9.96×10^{-4}; 9.96×10^{-4}; 9.96×10^{-4}
(2) 9.96×10^{-4}; 9.96×10^{-4}; 19.92×10^{-4}
(3) 4.98×10^{-4}; 4.98×10^{-4}; 9.96×10^{-4}
(4) 4.98×10^{-4}; 4.98×10^{-4}; 4.98×10^{-4}

108. 19.5 g of fluoro acetic acid (molar mass = 78 g mol^{-1}) is dissolved in 500g of water at 298 K. The depression in the freezing point of water was 1°C. What is K_a of fluoro acetic acid ? (For water, K_f = 1.86 K kg mol^{-1}). Assume molarity and molality to have same values. [April 02, 2026 (I)]

(1) 10^{-6} (2) 4×10^{-4} (3) 3×10^{-5} (4) 3×10^{-3}

109. Consider the following aqueous solutions.

I. 2.2 g Glucose in 125 mL of solution.
II. 1.9 g Calcium chloride in 250 mL of solution.
III. 9.0 g Urea in 500 mL of solution.
IV. 20.5 g Aluminium sulphate in 750 mL of solution.

The **correct** increasing order of boiling point of these solutions will be [Given: Molar mass in g mol^{-1}: H = 1, C = 12, N = 14, O = 16, Cl = 35.5, Ca = 40, Al = 27 and S = 32] [Jan. 28, 2026 (II)]

(1) I < II < III < IV (2) III < I < II < IV
(3) II < III < I < IV (4) II < III < IV < I

110. A solution is prepared by dissolving 0.3 g of a non-volatile non-electrolyte solute 'A' of molar mass 60 g mol^{-1} and 0.9 g of a non-volatile non-electrolyte solute 'B' of molar mass 180 g mol^{-1} in 100 mL H_2O at 27°C. Osmotic pressure of the solution will be [Jan. 24, 2026 (I)]
[Given : R = 0.082 L atm K^{-1} mol^{-1}]

(1) 1.23 atm (2) 2.46 atm (3) 0.82 atm (4) 1.47 atm

111. Elements P and Q form two types of non-volatile, non-ionizable compounds PQ and PQ_2. When 1 g of PQ is dissolved in 50 g of solvent 'A'. ΔT_b was 1.176 K while when 1 g of PQ_2 is dissolved in 50 g of solvent 'A', ΔT_b was 0.689 K. (K_b of 'A' = 5 K kg mol^{-1}). The molar masses of elements P and Q (in g mol^{-1}) respectively, are: [Jan. 21, 2026 (I)]

(1) 70, 110 (2) 65, 145 (3) 60, 25 (4) 25, 60

112. $HA(aq) \rightleftharpoons H^+(aq) + A^-(aq)$

The freezing point depression of a 0.1 m aqueous solution of a monobasic weak acid HA is 0.20°C.
The dissociation constant for the acid is
Given : [April 8, 2025 (II)]
$K_f(H_2O)$ = 1.8 K kg mol^{-1}, molality ≡ molarity

(1) 1.38×10^{-3} (2) 1.1×10^{-2}
(3) 1.90×10^{-3} (4) 1.89×10^{-1}

113. Given below are two statements : [April 4, 2025 (II)]

Statement I : Molal depression constant K_f is given by $\frac{M_1 R T_f}{\Delta S_{fus}}$, where symbols have their usual meaning.

Statement II : K_f for benzene is less than the K_f for water.

In the light of the above statements, choose the **most appropriate answer** from the options given below :

(1) Statement I is incorrect but Statement II is correct
(2) Both Statement I and Statement II are incorrect
(3) Both Statement I and Statement II are correct
(4) Statement I is correct but Statement II is incorrect

114. *XY* is the membrane / partition between two chambers 1 and 2 containing sugar solutions of concentration c_1 and c_2 ($c_1 > c_2$) mol L^{-1}. For the reverse osmosis to take place identify the correct condition
(Here p_1 and p_2 are pressures applied on chamber 1 and 2) [April 4, 2025 (I)]

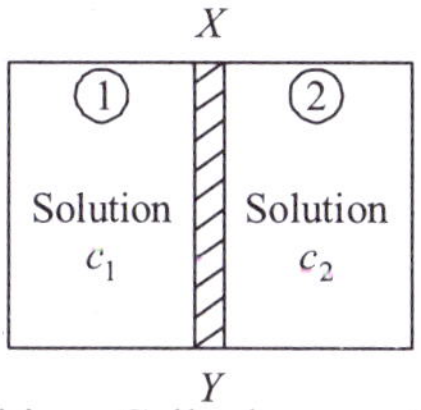

(A) Membrane/Partition ; Cellophane, $p_1 > \pi$
(B) Membrane/Partition ; Porous. $p_2 > \pi$
(C) Membrane/Partition ; Parchment paper, $p_1 > \pi$
(D) Membrane/Partition : Cellophane, $p_2 > \pi$
Choose the **correct** answer from the option given below :
(1) B and D only (2) A and D only
(3) A and C only (4) C only

115. 2 moles each of ethylene glycol and glucose are dissolved in 500 g of water. The boiling point of the resulting solution is : [April 3, 2025 (I)]
(Given : Ebullioscopic constant of water = 0.52 K kg mol^{-1})
(1) 379.2 K (2) 377.3 K (3) 375.3 K (4) 277.3 K

116. Given below are two statements : [Jan. 29, 2025 (II)]
Statement I : NaCl is added to the ice at 0°C, present in the ice cream box to prevent the melting of ice cream.
Statement II : On addition of NaCl to ice at 0°C, there is a depression in freezing point.
In the light of the above statements, choose the correct answer from the options given below :
(1) Statement I is false but Statement II is true
(2) Both Statement I and Statement II are true
(3) Both Statement I and Statement II are false
(4) Statement I is true but Statement II is false

117. 1.24 g of AX_2 (molar mass 124 g mol^{-1}) is dissolved in 1 kg of water to form a solution with boiling point of 100.0156°C, while 25.4 g of AY_2 (molar mass 250 g mol^{-1}) in 2 kg of water constitutes a solution with a boiling point of 100.0260°C. [Jan. 29, 2025 (I)]
$K_b(H_2O) = 0.52$ K kg mol^{-1}
Which of the following is correct ?
(1) AX_2 and AY_2 (both) are completely unionised.
(2) AX_2 and AY_2 (both) are fully ionised.
(3) AX_2 is completely unionised while AY_2 is fully ionised.
(4) AX_2 is fully ionised while AY_2 is completely unionised.

118. Assume a living cell with 0.9% $\left(\frac{\omega}{\omega}\right)$ of glucose solution (aqueous). This cell is immersed in another solution having equal mole fraction of glucose and water.
(Consider the data upto first decimal place only)
The cell will : [Jan. 28, 2025 (II)]
(1) shrink since soluton is 0.5 % $\left(\frac{\omega}{\omega}\right)$
(2) shrink since solution is 0.45% $\left(\frac{\omega}{\omega}\right)$ as a result of association of glucose molecules (due to hydrogen bonding)
(3) swell up since solution is 1% $\left(\frac{\omega}{\omega}\right)$
(4) Show no change in volume since solution is 0.9% $\left(\frac{\omega}{\omega}\right)$

119. What is the freezing point depression constant of a solvent, 50 g of which contain 1 g non volatile solute (molar mass 256 g mol^{-1}) and the decrease in freezing point is 0.40 K? [Jan. 28, 2025 (I)]
(1) 5.12 K kg mol^{-1} (2) 4.43 K kg mol^{-1}
(3) 1.86 K kg mol^{-1} (4) 3.72 K kg mol^{-1}

120. Consider the given plots of vapour pressure (VP) vs temperature (T/K) Which amongst the following options is correct graphical representation showing ΔT_f, depression in the freezing point of solvent in a solution ? [Jan. 24, 2025 (I)]

(1)
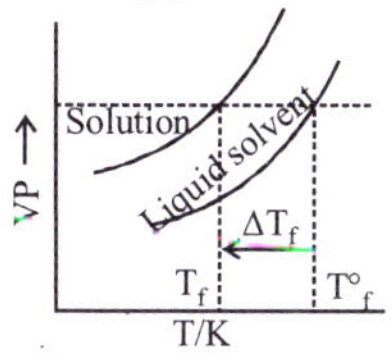

(2)
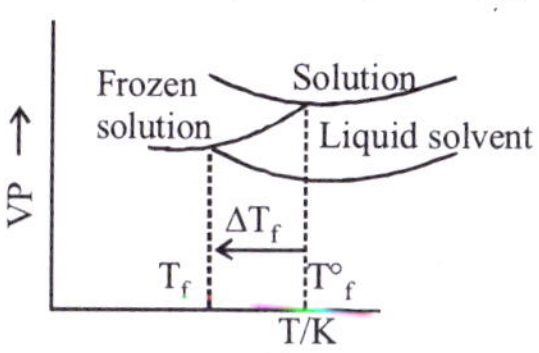

(3)
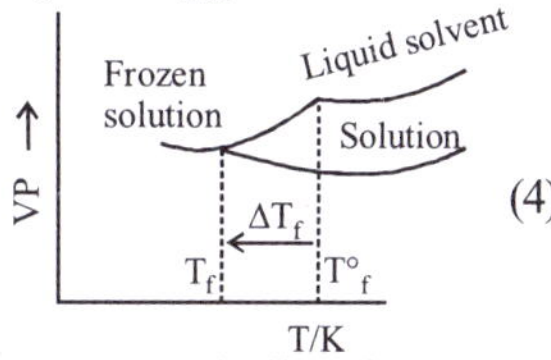

(4)
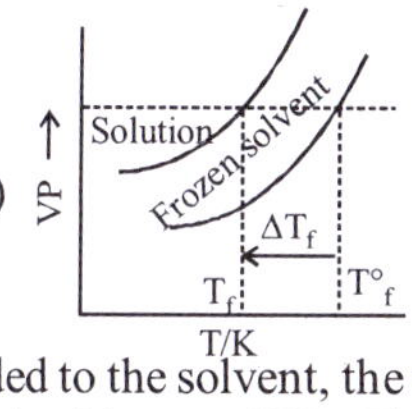

121. When a non-volatile solute is added to the solvent, the vapour pressure of the solvent decreases by 10 mm of Hg. The mole fraction of the solute in the solution is 0.2. What would be the mole fraction of the solvent if decrease in vapour pressure is 20 mm of Hg? [Jan. 23, 2025 (II)]
(1) 0.6 (2) 0.4 (3) 0.2 (4) 0.8

122. $CrCl_3 \cdot xNH_3$ can exist as a complex. 0.1 molal aqueous solution of this complex shows a depression in freezing point of 0.558°C. Assuming 100% ionisation of this complex and coordination number of Cr is 6, the complex will be
(Given $K_f = 1.86$ K kg mol^{-1}) [Jan. 23, 2025 (I)]
(1) $[Cr(NH_3)_6]Cl_3$ (2) $[Cr(NH_3)_4Cl_2]Cl$
(3) $[Cr(NH_3)_5Cl]Cl_2$ (4) $[Cr(NH_3)_3Cl_3]$

123. Arrange the following solutions in order of their increasing boiling points. [Jan. 22, 2025 (I)]
(i) 10^{-4} M NaCl (ii) 10^{-4} M Urea
(iii) 10^{-3} M NaCl (iv) 10^{-2} M NaCl
(1) (ii) < (i) < (iii) < (iv) (2) (ii) < (i) < (iii) < (iv)
(3) (i) < (ii) < (iii) < (iv) (4) (iv) < (iii) < (i) < (ii)

124. 0.05M $CuSO_4$ when treated with 0.01M $K_2Cr_2O_7$ gives green colour solution of $Cu_2Cr_2O_7$. The two solutions are separated as shown below:
[SPM : Semi Permeable Membrane]

$K_2Cr_2O_7$	$CuSO_4$

Side X SPM Side Y

Due to osmosis : [April 9, 2024 (I)]
(1) Green colour formation observed on side Y.
(2) Green colour formation observed on side X.
(3) Molarity of $K_2Cr_2O_7$ solution is lowered.
(4) Molarity of $CuSO_4$ solution is lowered.

125. We have three aqueous solution of NaCl labelled as 'A' 'B' and 'C' with concentration 0.1 M, 0.01 M and 0.001 M, respectively. The value of van't Hoff factor (i) for these solution will be in the order [Feb. 1, 2024 (I)]
(1) $i_A < i_B < i_C$ (2) $i_A < i_C < i_B$
(3) $i_A = i_B = i_C$ (4) $i_A > i_B > i_C$

126. What happens to freezing point of benzene when small quantity of napthalene is added to benzene? [Jan. 30, 2024 (I)]
(1) Increases
(2) Remains unchanged
(3) First decreases and then increases
(4) Decreases

127. The solution from the following with highest depression in freezing point / lowest freezing point is [Jan. 30, 2024 (II)]
(1) 180 g of acetic acid dissolved in water
(2) 180 g of acetic acid dissolved in benzene
(3) 180 g of benzoic acid dissolved in benzene
(4) 180 g of glucose dissolved in water

128. What weight of glucose must be dissolved in 100 g of water to lower the vapour pressure by 0.20 mm Hg?
(Assume dilute solution is being formed)
Given : Vapour pressure of pure water is 54.2 mm Hg at room temperature. Molar mass of glucose is 180 g mol^{-1}. [April 11, 2023 (II)]
(1) 4.69 g (2) 3.59 g (3) 2.59 g (4) 3.69 g

129. Evaluate the following statements for their correctness. [Jan. 31, 2023 (II)]
(A) The elevation in boiling point temperature of water will be same for 0.1 M NaCl and 0.1 M urea.
(B) Azeotropic mixtures boil without change in their composition
(C) Osmosis always takes place from hypertonic to hypotonic solution
(D) The density of 32% H_2SO_4 solution having molarity 4.09 M is approximately 1.26g mL^{-1}
(E) A negatively charged sol is obtained when KI solution is added to silver nitrate solution.
Choose the correct answer from the options given below:
(1) B, D, and E only (2) A, B, and D only
(3) A and C only (4) B and D only

130. Match List I with List II. [Jan. 29, 2023 (II)]

List I		List II	
A.	van't Hoff factor, i	I.	Cryoscopic constant
B.	k_f	II.	Isotonic solutions
C.	Solutions with same osmotic pressure	III.	Normal molar mass / Abnormal molar mass
D.	Azeotropes	IV.	Solutions with same composition of vapour above it

Choose the correct answer from the options given below:
(1) A-III, B-I, C-II, D-IV (2) A-III, B-II, C-I, D-IV
(3) A-III, B-I, C-IV, D-II (4) A-I, B-III, C-II, D-IV

131. In the depression of freezing point experiment [Jan. 24, 2023 (I)]
A. Vapour pressure of the solution is less than that of pure solvent
B. Vapour pressure of the solution is more than that of pure solvent
C. Only solute molecules solidify at the freezing point
D. Only solvent molecules solidify at the freezing point
(1) A and D only (2) B and C only
(3) A and C only (4) A only

132. Boiling point of a 2% aqueous solution of a non-volatile solute A is equal to the boiling point of 8% aqueous solution of a non-volatile solute B. The relation between molecular weights of A and B is. [July 27, 2022 (I)]
(1) $M_A = 4M_B$ (2) $M_B = 4M_A$
(3) $M_A = 8M_B$ (4) $M_B = 8M_A$

133. The depression in freezing point observed for a formic acid solution of concentration 0.5 mL L^{-1} is 0.0405°C. Density of formic acid is 1.05 g mL^{-1}. The Van't Hoff factor of the formic acid solution is nearly:
(Given for water k_f = 1.86 K kg mol^{-1}) [July 25, 2022 (I)]
(1) 0.8 (2) 1.1 (3) 1.9 (4) 2.4

134. Two solutions A and B are prepared by dissolving 1g of non-volatile solutes X and Y, respectively in 1 kg of water. The ratio of depression in freezing points for A and B is found to be 1 : 4. The ratio of molar masses of X and Y is: [July 25, 2022 (II)]
(1) 1 : 4 (2) 1 : 0.25 (3) 1 : 0.20 (4) 1 : 5

135. Solute A associates in water. When 0.7 g of solute A is dissolved in 42.0 g of water, it depresses the freezing point by 0.2°C. The percentage association of solute A in water, is
[Given: Molar mass of A = 93 g mol^{-1}. Molal depression constant of water is 1.86 K kg mol^{-1}] [June 25, 2022 (II)]
(1) 50% (2) 60% (3) 70% (4) 80%

136. A set of solutions is prepared using 180 g of water as a solvent and 10 g of different non-volatile solutes A, B and C. The relative lowering of vapour pressure in the presence of these solutes are in the order [Given, molar mass of A = 100 g mol^{-1}; B = 200 g mol^{-1}; C = 10,000 g mol^{-1}] [Sep. 06, 2020 (II)]
(1) B > C > A (2) C > B > A
(3) A > B > C (4) A > C > B

137. A solution is prepared by dissolving 0.6 g of urea (molar mass = 60 g mol^{-1}) and 1.8 g of glucose (molar mass = 180 g mol^{-1}) in 100 mL, of water at 27 °C. The osmotic pressure of the solution is : [April 12, 2019 (II)]
(R = 0.08206 L atm K^{-1} mol^{-1})
(1) 8.2 atm (2) 2.46 atm (3) 4.92 atm (4) 1.64 atm

138. At room temperature, a dilute solution of urea is prepared by dissolving 0.60 g of urea in 360 g of water. If the vapour pressure of pure water at this temperature is 35 mmHg, lowering of vapour pressure will be : [April 10, 2019 (I)]
(molar mass of urea = 60 g mol^{-1})
(1) 0.027 mmHg (2) 0.028 mmHg
(3) 0.017 mmHg (4) 0.031 mmHg

139. 1 g of a non-volatile non-electrolyte solute is dissolved in 100 g of two different solvents A and B whose ebullioscopic constants are in the ratio of 1 : 5. The ratio of the elevation in their boiling points, $\frac{\Delta T_b(A)}{\Delta T_b(B)}$, is : [April 10, 2019 (II)]
(1) 5 : 1 (2) 10 : 1 (3) 1 : 5 (4) 1 : 0.2

140. The osmotic pressure of a dilute solution of an ionic compound XY in water is four times that of a solution of 0.01 M $BaCl_2$ in water. Assuming complete dissociation of the given ionic compounds in water, the concentration of XY (in mol L^{-1}) in solution is: [April 9, 2019 (I)]
(1) 4×10^{-2} (2) 6×10^{-2} (3) 4×10^{-4} (4) 16×10^{-4}

141. Molal depression constant for a solvent is 4.0 K kg mol^{-1}. The depression in the freezing point of the solvent for 0.03 mol kg^{-1} solution K_2SO_4 is: [April 9, 2019 (II)]
(Assume complete dissociation of the electrolyte)
(1) 0.18 K (2) 0.24 K (3) 0.12 K (4) 0.36 K

142. Freezing point of a 4% aqueous solution of X is equal to freezing point of 12% aqueous salution of Y. If molecular weight of X is A, then molecular weight of Y is: [Jan. 12, 2019 (I)]
(1) 3A (2) 2A (3) A (4) 4A

143. Molecules of benzoic acid (C_6H_5COOH) dimerise in benzene. 'w' g of the acid dissolved in 30 g of benzene shows a depression in freezing point equal to 2 K. If the percentage association of the acid to form dimer in the solution is 80, then w is :
(Given that K_f = 5 K kg mol^{-1}, Molar mass of benzoic acid = 122 g mol^{-1}) [Jan. 12, 2019 (II)]
(1) 2.4 g (2) 1.0 g (3) 1.5 g (4) 1.8 g

144. The freezing point of a diluted milk sample is found to be $-0.2\ °C$, while it should have been $-0.5\ °C$ for pure milk. How much water has been added to pure milk to make the diluted sample? [Jan. 11, 2019 (I)]
(1) 1 cup of water to 2 cups of pure milk
(2) 3 cups of water to 2 cups of pure milk
(3) 1 cup of water to 3 cups of pure milk
(4) 2 cups of water to 3 cups of pure milk

145. K_2HgI_4 is 40% ionised in aqueous solution. The value of its van't Hoff factor (i) is : [Jan. 11, 2019 (II)]
(1) 1.6 (2) 1.8 (3) 2.0 (4) 2.2

146. Elevation in the boiling point for 1 molal solution of glucose is 2 K. The depression in the freezing point for 2 molal solution of glucose in the same solvent is 2 K. The relation between K_b and K_f is: [Jan. 10, 2019 (II)]
(1) $K_b = 1.5\,K_f$ (2) $K_b = K_f$
(3) $K_b = 0.5\,K_f$ (4) $K_b = 2\,K_f$

147. A solution containing 62 g ethylene glycol in 250 g water is cooled to $-10\ °C$. If K_f for water is $1.86\ K\ kg\ mol^{-1}$, the amount of water (in g) separated as ice is: [Jan. 9, 2019 (II)]
(1) 48 (2) 32 (3) 64 (4) 16

148. For 1 molal aqueous solution of the following compounds, which one will show the highest freezing point? [2018]
(1) $[Co(H_2O)_6]Cl_3$ (2) $[Co(H_2O)_5Cl]Cl_2.H_2O$
(3) $[Co(H_2O)_4Cl_2]Cl.2H_2O$ (4) $[Co(H_2O)_3Cl_3].3H_2O$

149. The mass of non-volatile, non electrolyte solute (molar mass = $50\ g\ mol^{-1}$) needed to be dissolved in 114 g octane to reduce its vapour pressure to 75%, is ________. [Online April 16, 2018]
(1) 37.5 g (2) 75 g (3) 150 g (4) 50 g

150. Two 5 molal solutions are prepared by dissolving a non-electrolyte, non-volatile solute separately in the solvents X and Y. The molecular weights of the solvents are M_X and M_Y, respectively where $M_X = \frac{3}{4} M_Y$. The relative lowering of vapour pressure of the solution in X is "m" times that of the solution in Y. Given that the number of moles of solute is very small in comparison to that of solvent, the value of "m" is: [Online April 15, 2018 (II)]
(1) $\frac{3}{4}$ (2) $\frac{1}{2}$ (3) $\frac{1}{4}$ (4) $\frac{4}{3}$

151. 5 g of Na_2SO_4 was dissolved in x g of H_2O. The change in freezing point was found to be 3.82°C. If Na_2SO_4 is 81.5% ionised, the value of x [Online April 8, 2017]
(K_f for water = $1.86°C\ kg\ mol^{-1}$) is approximately :
(molar mass of S = $32\ g\ mol^{-1}$ and that of Na = $23\ g\ mol^{-1}$)
(1) 15 g (2) 25 g (3) 45 g (4) 65 g

152. The freezing point of benzene decreases by 0.45 °C when 0.2g of acetic acid is added to 20 g of benzene. If acetic acid associates to form a dimer in benzene, percentage association of acetic acid in benzene will be : [2017]
(K_f for benzene = $5.12\ K\ kg\ mol^{-1}$)
(1) 64.6% (2) 80.4% (3) 74.6% (4) 94.6%

153. An aqueous solution of a salt MX_2 at certain temperature has a van't Hoff factor of 2. The degree of dissociation for this solution of the salt is : [Online April 10, 2016]
(1) 0.50 (2) 0.33 (3) 0.67 (4) 0.80

154. 18 g glucose ($C_6H_{12}O_6$) is added to 178.2 g water. The vapour pressure of water (in torr) for this aqueous solution is: [2016, 2006]
(1) 752.4 (2) 759.0 (3) 7.6 (4) 76.0

155. The vapour pressure of acetone at 20 °C is 185 torr. When 1.2 g of a non-volatile substance was dissolved in 100 g of acetone at 20 °C, its vapour pressure was 183 torr. The molar mass (g mol^{-1}) of the substance is : [2015]
(1) 128 (2) 488 (3) 32 (4) 64

156. Determination of the molar mass of acetic acid in benzene using freezing point depression is affected by : [Online April 11, 2015]
(1) partial ionization (2) dissociation
(3) complex formation (4) association

157. Consider separate solutions of 0.500 M C_2H_5OH(aq), 0.100 M $Mg_3(PO_4)_2$ (aq), 0.250 M KBr (aq) and 0.125 M Na_3PO_4 (aq) at 25 °C. Which statement is **true** about these solutions, assuming all salts to be strong electrolytes?
(1) They all have the same osmotic pressure. [2014]
(2) 0.100 M $Mg_3(PO_4)_2$(aq) has the highest osmotic pressure.
(3) 0.125 M Na_3PO_4(aq) has the highest osmotic pressure.
(4) 0.500 M C_2H_5OH(aq) has the highest osmotic pressure.

158. The observed osmotic pressure for a 0.10 M solution of $Fe(NH_4)_2(SO_4)_2$ at 25 °C is 10.8 atm. The expected and experimental (observed) values of van't Hoff factor (i) will be respectively: [Online April 19, 2014]
($R = 0.082\ L\ atm\ K^{-1}\ mol^{-1}$)
(1) 5 and 4.42 (2) 4 and 4.00
(3) 5 and 3.42 (4) 3 and 5.42

159. 12 g of a nonvolatile solute dissolved in 108 g of water produces the relative lowering of vapour pressure of 0.1. The molecular mass of the solute is : [Online April 9, 2013]
(1) 80 (2) 60 (3) 20 (4) 40

160. A molecule M associates in a given solvent according to the equation M $\rightleftharpoons$ $(M)_n$. For a certain concentration of M, the van't Hoff factor was found to be 0.9 and the fraction of associated molecules was 0.2. The value of n is: [Online April 22, 2013]
(1) 3 (2) 5 (3) 2 (4) 4

161. How many grams of methyl alcohol should be added to 10 litre tank of water to prevent its freezing at 268 K ? (K_f for water is $1.86\ K\ kg\ mol^{-1}$) [Online April 25, 2013]
(1) 880.07 g (2) 899.04 g (3) 886.02 g (4) 868.06 g

162. K_f for water is $1.86\ K\ kg\ mol^{-1}$. If your automobile radiator holds 1.0 kg of water, how many grams of ethylene glycol ($C_2H_6O_2$) must you add to get the freezing point of the solution lowered to $-2.8\ °C$?
(1) 72 g (2) 93 g (3) 39 g (4) 27 g [2012]

163. A solution containing 0.85 g of $ZnCl_2$ in 125.0 g of water freezes at $-0.23\ °C$. The apparent degree of dissociation of the salt is (K_f for water = $1.86\ K\ kg\ mol^{-1}$, atomic mass: Zn = 65.3 and Cl = 35.5) [Online May 12, 2012]
(1) 1.36% (2) 73.5% (3) 7.35% (4) 2.47%

164. The freezing point of a 1.00 m aqueous solution of HF is found to be $-1.91°C$. The freezing point constant of water, K_f is 1.86 K kg mol^{-1}. The percentage dissociation of HF at this concentration is [Online May 26, 2012]
(1) 30% (2) 10% (3) 5.2% (4) 2.7%

165. A 5% solution of cane sugar (molar mass 342) is isotonic with 1% of a solution of an unknown solute. The molar mass of unknown solute in g/mol is : [2011 RS]
(1) 171.2 (2) 68.4 (3) 34.2 (4) 136.2

166. If sodium sulphate is considered to be completely dissociated into cations and anions in aqueous solution, the change in freezing point of water (ΔT_f), when 0.01 mol of sodium sulphate is dissolved in 1 kg of water, is ($K_f = 1.86\ K\ kg\ mol^{-1}$).
(1) 0.372 K (2) 0.0558 K [2010]
(3) 0.0744 K (4) 0.0186 K

167. The vapour pressure of water at 20 °C is 17.5 mm Hg. If 18 g of glucose ($C_6H_{12}O_6$) is added to 178.2 g of water at 20 °C, the vapour pressure of the resulting solution will be [2008]
(1) 17.325 mm Hg (2) 15.750 mm Hg
(3) 16.500 mm Hg (4) 17.675 mm Hg

168. A 5.25% solution of a substance is isotonic with a 1.5% solution of urea (molar mass = 60 g mol^{-1}) in the same solvent. If the densities of both the solutions are assumed to be equal to 1.0 g cm^{-3}, molar mass of the substance will be [2007]
(1) 210.0 g mol^{-1} (2) 90.0 g mol^{-1}
(3) 115.0 g mol^{-1} (4) 105.0 g mol^{-1}.

169. Which one of the following aqueous solutions will exihibit highest boiling point ? [2004]
(1) 0.015 M urea (2) 0.01 M KNO_3
(3) 0.01 M Na_2SO_4 (4) 0.015 M glucose

170. In a 0.2 molal aqueous solution of a weak acid HX, the degree of ionization is 0.3. Taking K_f for water as 1.85, the freezing point of the solution will be nearest to [2003]
(1) –0.360 °C (2) –0.260 °C (3) +0.480 °C (4) – 0.480 °C

171. Which one of the following statements is **false?** [2004]
(1) The correct order of osmotic pressure for 0.01 M aqueous solution of each compound is
$BaCl_2 > KCl > CH_3COOH >$ sucrose
(2) The osmotic pressure (π) of a solution is given by the equation π = MRT, where M is the molarity of the solution
(3) Raoult's law states that the vapour pressure of a component over a solution is proportional to its mole fraction
(4) Two sucrose solutions of same molality prepared in different solvents will have same freezing point depression

172. A pressure cooker reduces cooking time for food because [2003]
(1) boiling point of water involved in cooking is increased
(2) the higher pressure inside the cooker crushes the food material
(3) cooking involves chemical changes helped by a rise in temperature
(4) heat is more evenly distributed in the cooking space

173. Freezing point of an aqueous solution is (–0.186) °C. Elevation of boiling point of the same solution is K_b = 0.512 °C, K_f = 1.86°C, find the increase in boiling point. [2002]
(1) 0.186 °C (2) 0.0512 °C
(3) 0.092 °C (4) 0.2372 °C.

Numerical Value Questions (NVQs)

174. 20 g hemoglobin in a 1 L aqueous solution (A) at 300 K is separated from pure water by semi permeable membrane. At equilibrium the height of solution in a tube dipped in a solution (A) is found to be 80.0 mm higher than the tube dipped in water.
The molar mass of hemoglobin is _____ kg mol^{-1}. (Nearest integer) [April 05, 2026 (II)]
(Given: g = 10 ms^{-2}, R = 8.3 kPa dm^3 K^{-1} mol^{-1}, density of solution = 1000 kg m^{-3})

175. A substance 'X' (1.5 g) dissolved in 150 g of a solvent 'Y' (molar mass = 300 g mol^{-1}) led to an elevation of the boiling point by 0.5 K. The relative lowering in the vapour pressure of the solvent 'Y' is _________ × 10^{-2}. (Nearest integer)
[Given : K_b of the solvent = 5.0 K kg mol^{-1}]
Assume the solution to be dilute and no association or dissociation of X takes place in solution. [Jan. 21, 2026 (II)]

176. The osmotic pressure of a living cell is 12 atm at 300K. The strength of sodium chloride solution that is isotonic with the living cell at this temperature is _______ g L^{-1}. (Nearest integer)
Given : R = 0.08 L atm K^{-1} mol^{-1}
Assume complete dissociation of NaCl
(Given : Molar mass of Na and Cl are 23 and 35.5 g mol^{-1} respectively.) [Jan. 21, 2026 (II)]

177. When 1 g each of compounds AB and AB_2 are dissolved in 15 g of water separately, they increased the boiling point of water by 2.7 K and 1.5 K respectively. The atomic mass of A (in amu) is ______ × 10^{-1}(Nearest integer)
(Given : Molal boiling point elevation constant is 0.5 K kg mol^{-1}) [April 2, 2025 (II)]

178. The percentage dissociation of a salt (MX_3) solution at given temperature (van't Hoff factor i = 2) is.................. % (Nearest integer) [April 7, 2025 (I)]

179. If A_2B is 30% ionised in an aqueous solution, then the value of van't Hoff factor (i) is _____ × 10^{-1}. [Jan. 29, 2025 (I)]

180. The observed and normal masses of compound MX_2 are 65.6 and 164 respectively. The percent degree of ionisation of MX_2 is _________%. (Nearest integer) [Jan. 24, 2025 (II)]

181. The vapour pressure of pure benzene and methyl benzene at 27°C is given as 80 Torr and 24 Torr, respectively. The mole fraction of methyl benzene in vapour phase, in equilibrium with an equimolar mixture of those two liquids (ideal solution) at the same temperature is____ × 10^{-2} (nearest integer) [April 9, 2024 (II)]

182. A solution containing 10g of an electrolyte AB_2 in 100g of water boils at 100.52°C. The degree of ionization of the electrolyte (α) is _____ × 10^{-1}.
(nearest integer) [April 8, 2024 (I)]
[Given : Molar mass of AB_2 = 200g mol^{-1}. K_b (molal boiling point elevation const. of water) = 0.52 K kg mol^{-1}, boiling point of water = 100°C; AB_2 ionises as $AB_2 \rightarrow A^{2+} + 2B^-$]

183. Consider the dissociation of the weak acid HX as given below
$HX(aq) \rightleftharpoons H^+(aq) + X^-(aq)$, $K_a = 1.2 \times 10^{-5}$ [April 6, 2024 (I)]
[K_a : dissociation constant]
The osmotic pressure of 0.03 M aqueous solution of HX at 300 K is _____ × 10^{-2} bar (nearest integer).
[Given : R = 0.083 L bar Mol| K^{-1}]

184. When 'x' × 10^{-2} mL methanol (molar mass = 32 g; density = 0.792 g/cm^3) is added to 100 mL water (density = 1 g/cm^3), the following diagram is obtained. [April 6, 2024 (II)]

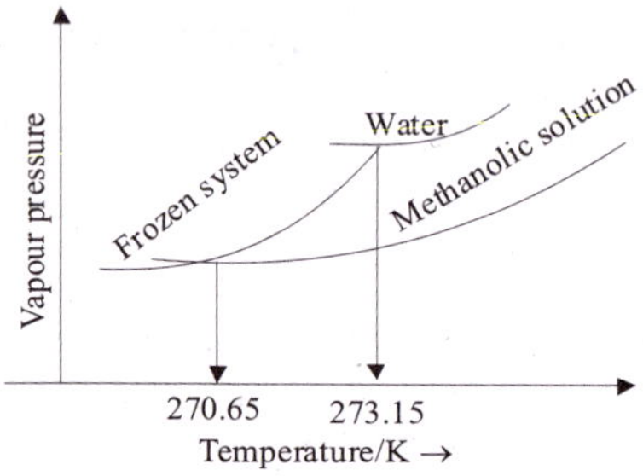

x =.................(nearest integer)
[Given: Molal freezing point depression constant of water at 273.15 K is 1.86 K kg mol^{-1}]

185. An artificial cell is made by encapsulating 0.2 M glucose solution within a semipermeable membrane. The osmotic pressure developed when the artificial cell is placed within a 0.05 M solution of NaCl at 300 K is ______ $\times 10^{-1}$ bar. (Nearest Integer) [April 5, 2024 (I)]
[Given : $R = 0.083$ L bar mol^{-1} K^{-1}]
Assume complete dissociation of NaCl

186. Considering acetic acid dissociates in water, its dissociation constant is 6.25×10^{-5}. If 5 mL of acetic acid is dissolved in 1 litre water, the solution will freeze at $-x \times 10^{-2}$ °C, provided pure water freezes at 0 °C.
x = ________. (Nearest integer) [April 5, 2024 (II)]
Given : $(K_f)_{water} = 1.86$ K kg mol^{-1}.
density of acetic acid is 1.2 g mL^{-1}
molar mass of water = 18 g mol^{-1}.
molar mass of acetic acid = 60 g mol^{-1}.
density of water = 1 g cm^{-3}
Acetic acid dissociates as
$CH_3COOH \rightleftharpoons CH_3COO^{\ominus} + H^{\oplus}$

187. 2.5 g of a non-volatile, non-electrolyte is dissolved in 100 g of water at 25°C. The solution showed a boiling point elevation by 2°C. Assuming the solute concentration in negligible with respect to the solvent concentration, the vapour pressure of the resulting aqueous solution is _______ mm of Hg (nearest integer)
[Given : Molal boiling point elevation constant of water (K_b) = 0.52 K. kg mol^{-1},
1 atm pressure = 760 mm of Hg, molar mass of water = 18 g mol^{-1}] [April 4, 2024 (I)]

188. 2.7 Kg of each of water and acetic acid are mixed, The freezing point of the solution will be –x °C. Consider the acetic acid does not dimerise in water, nor dissociates in water x = ______(nearest integer)
[Given : Molar mass of water = 18 g mol^{-1}, acetic acid = 60 g mol^{-1}] [April 4, 2024 (II)]
$K_{f(H_2O)}$: 1.86 K kg mol^{-1}
$K_{f(\text{acetic acid})}$: 3.90 K kg mol^{-1}
freezing point : H_2O = 273 K, acetic acid = 290 K]

189. Mass of ethylene glycol (antifreeze) to be added to 18.6 kg of water to protect the freezing point at – 24°C is _____ kg (Molar mass in g mol^{-1} for ethylene glycol 62 K_f of water = 1.86 K kg mol^{-1}). [Feb. 1, 2024 (II)]

190. The osmotic pressure of a dilute solution is 7×10^5 Pa at 273K. Osmotic pressure of the same solution at 283K is _____ $\times 10^4$ Nm^{-2}. [Jan. 29, 2024 (I)]

191. Solution of 12 g of non - electrolyte (A) prepared by dissolving it in 1000 mL of water exerts the same osmotic pressure as that of 0.05 M glucose solution at the same temperature. The empirical formula of A is CH_2O. The molecular mass of A is ______ g. (Nearest integer) [April 13, 2023 (I)]

192. Sea water contains 29.25% NaCl and 19% $MgCl_2$ by weight of solution. The normal boiling point of the sea water is _______ °C (Nearest integer) [April 13, 2023 (II)]
Assume 100% ionization for both NaCl and $MgCl_2$
Given : $K_b(H_2O) = 0.52$ K kg mol^{-1}
Molar mass of NaCl and $MgCl_2$ is 58.5 and 95 g mol^{-1} respectively.

193. 80 mole percent of $MgCl_2$ is dissociated in aqueous solution. The vapour pressure of 1.0 molal aqueous solution of $MgCl_2$ at 38°C is _______ mm Hg. (Nearest integer) [April 12, 2023 (I)]
Given : Vapour pressure of water at 38°C is 50 mm Hg

194. 0.004 M K_2SO_4 solution is isotonic with 0.01 M glucose solution. Percentage dissociation of K_2SO_4 is _______ (Nearest integer) [April 11, 2023 (I)]

195. If the degree of dissociation of aqueous solution of weak monobasic acid is determined to be 0.3, then the observed freezing point will be _____ % higher than the expected/theoretical freezing point. (Nearest integer) [April 10, 2023 (I)]

196. An aqueous solution of volume 300 cm^3 contains 0.63 g of protein. The osmotic pressure of the solution at 300 K is 1.29 mbar. The molar mass of the protein is ______ g mol^{-1}
Given : R = 0.083 L bar $K^{-1}mol^{-1}$ [April 10, 2023 (II)]

197. If the boiling points of two solvents X and Y (having same molecular weights) are in the ratio 2 : 1 and their enthalpy of vaporizations are in the ratio 1 : 2, then the boiling point elevation constant of X is m times the boiling point elevation constant of Y. The value of m is _____ (Nearest integer) [April 08, 2023 (II)]

198. Mass of Urea (NH_2CONH_2) required to be dissolved in 1000 g of water to reduce the vapour pressure of water by 25% is .... g. (Nearest integer)
Given: Molar mass of N, C, O and H are 14, 12, 16 and 1 g mol^{-1} respectively. [April 06, 2023 (I)]

199. Consider the following pairs of solution which will be isotonic at the same temperature. The number of pairs of solutions is/are .... [April 06, 2023 (II)]
A. 1 M aq. NaCl and 2 M aq. Urea
B. 1 M aq. $CaCl_2$ and 1.5 M aq. KCl
C. 1.5 M aq. $AlCl_3$ and 2 M aq. Na_2SO_4
D. 2.5 M aq. KCl and 1 M aq. $Al_2(SO_4)_3$

200. 25 mL of an aqueous solution of KCl was found to require 20 mL of 1 M $AgNO_3$ solution when titrated using K_2CrO_4 as an indicator. What is the depression in freezing point of KCl solution of the given concentration? ______ (Nearest integer).
(Given: K_f = 2.0 K kg mol^{-1}) [Feb. 01, 2023 (I)]
Assume 1) 100% ionization and
2) density of the aqueous solution as 1 g ML^{-1}

201. 20% of acetic acid is dissociated when its 5 g is added to 500 mL of water. The depression in freezing point of such water is ______ $\times 10^{-3}$°C. [Feb. 01, 2023 (II)]
Atomic mass of C, H and O are 12, 1 and 16 a.m.u. respectively.
[Given : Molal depression constant and density of water are 1.86 K kg mol^{-1} and 1 g cm^{-3} respectively.

202. At 27°C, a solution containing 2.5 g of solute in 250.0 mL of solution exerts an osmotic pressure of 400 Pa. The molar mass of the solute is ____ g mol^{-1} (Nearest integer)
(Given : R = 0.083 L bar K^{-1} mol^{-1}) [Jan. 31, 2023 (I)]

203. A solution containing 2 g of a non-volatile solute in 20 g of water boils at 373.52 K. The molecular mass of the solute is _______ g mol^{-1}. (Nearest integer) [Jan. 30, 2023 (I)]
Given, water boils at 373 K, K_b for water = 0.52 kg mol^{-1}.

204. Lead storage battery contains 38% by weight solution of H_2SO_4. The van't Hoff factor is 2.67 at this concentration. The temperature in Kelvin at which the solution in the battery will freeze is ____ (Nearest integer).
Given $K_f = 1.8$ K kg mol^{-1}. **[Jan. 30, 2023 (II)]**

205. Solid Lead nitrate is dissolved in 1 litre of water. The solution was found to boil at 100.15°C. When 0.2 mol of NaCl is added to the resulting solution, it was observed that the solution froze at –0.8° C. The solubility product of $PbCl_2$ formed is ______ × 10^{-6} at 298 K. (Nearest integer) **[Jan. 29, 2023 (I)]**
Given : $K_b = 0.5$ K kg mol^{-1} and $K_f = 1.8$ kg mol^{-1}.
Assume molality to be equal to molarity in all cases.

206. The osmotic pressure of solutions of PVC in cyclohexanone at 300 K are plotted on the graph.
The molar mass of PVC is __________ g mol^{-1}
(Nearest integer) **[Jan. 25, 2023 (I)]**

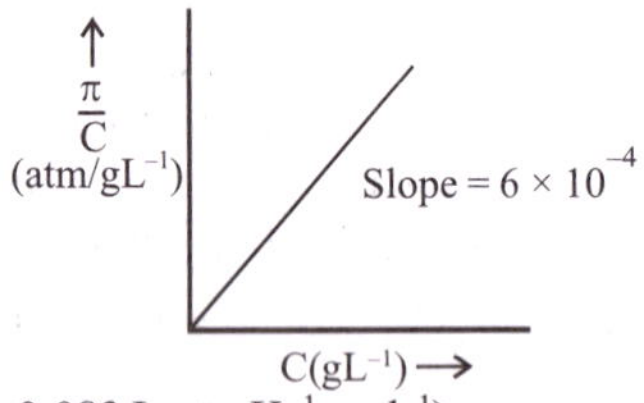

(Given : R = 0.083 L atm K^{-1} mol^{-1})

207. The number of pairs of the solution having the same value of the osmotic pressure from the following is ________.
(Assume 100% ionization) **[Jan. 25, 2023 (II)]**
A. 0.500 M C_2H_5OH (aq) and 0.25 M KBr (aq)
B. 0.100 M $K_4[Fe(CN)_6]$ (aq) and 0.100 M $FeSO_4(NH_4)_2SO_4$ (aq)
C. 0.05 M $K_4[Fe(CN)_6]$ (aq) and 0.25 M NaCl (aq)
D. 0.15 M NaCl (aq) and 0.1 M $BaCl_2$ (aq)
E. 0.02 M KCl. $MgCl_2$. $6H_2O$ (aq) and 0.05 M KCl (aq)

208. 1.80 g of solute A was dissolved in 62.5 cm^3 of ethanol and freezing point of the solution was found to be 155.1 K. The molar mass of solute A is __ g mol^{-1}.
[Given: Freezing point of ethanol is 156.0 K.
Density of ethanol is 0.80 g cm^{-3}.
Freezing point depression constant of ethanol is 2.00 K kg mol^{-1}] **[July 29, 2022 (II)]**

209. 150 g of acetic acid was contaminated with 10.2 g ascorbic acid ($C_6H_8O_6$) to lower down its freezing point by (x × 10^{-1})°C. The value of x is ________. (Nearest integer)
[Given $K_f = 3.9$ K kg mol^{-1}; Molar mass of ascorbic acid = 176 g mol^{-1}] **[July 28, 2022 (I)]**

210. When a certain amount of solid A is dissolved in 100 g of water at 25°C to make a dilute solution, the vapour pressure of the solution is reduced to one-half of that of pure water. The vapour pressure of pure water is 23.76 mmHg. The number of moles of solute A added is________.
(Nearest integer) **[July 27, 2022 (II)]**

211. The elevation in boiling point for 1 molal solution of non-volatile solute A is 3K. The depression in freezing point for 2 molal solution of A in the same solvent is 6 K. The ratio of K_b and K_f i.e., K_b/K_f is 1 : X. The value of X is [nearest integer] **[July 26, 2022 (II)]**

212. 1.2 mL of acetic acid is dissolved in water to make 2.0 L of solution. The depression in freezing point observed for this strength of acid is 0.0198 °C. The percentage of dissociation of the acid is ________. (Nearest integer)
[Given : Density of acetic acid is 1.02 g mL^{-1}
Molar mass of acetic acid is 60 g mol^{-1}
$K_f(H_2O)$ = 1.85 K kg mol^{-1}] **[June 29, 2022 (I)]**

213. Elevation in boiling point for 1.5 molal solution of glucose in water is 4K. The depression in freezing point for 4.5 molal solution of glucose in water is 4K. The ratio of molal elevation constant to molal depression constant (K_b/K_f) is____. **[June 29, 2022 (II)]**

214. 2.5 g of protein containing only glycine ($C_2H_5NO_2$) is dissolved in water to make 500 mL of solution. The osmotic pressure of this solution at 300 K is found to be 5.03 × 10^{-3} bar. The total number of glycine units present in the protein is ____
(Given : R = 0.083 L bar K^{-1} mol^{-1}) **[June 28, 2022 (II)]**

215. 2g of a non-volatile non-electrolyte solute is dissolved in 200 g of two different solvents A and B whose ebullioscopic constants are in the ratio of 1 : 8. The elevation in boiling points of A and B are in the ratio $\frac{x}{y}$ (x : y). The value of y is__________ (Nearest integer) **[June 27, 2022 (I)]**

216. A solution containing 2.5 × 10^{-3} kg of a solute dissolved in 75 × 10^{-3} kg of water boils at 373.535 K. The molar mass of the solute is ______ g mol^{-1}.
[nearest integer] (Given: K_b (H_2O) = 0.52 K Kg mol^{-1}, boiling point of water = 373.15 K) **[June 27, 2022 (II)]**

217. A 0.5 percent solution of potassium chloride was found to freeze at –0.24°C. The percentage dissociation of potassium chloride is ________. (Nearest integer)
(Molal depression constant for water is 1.80 K kg mol^{-1} and molar mass of KCl is 74.6 g mol^{-1}) **[June 26, 2022 (I)]**

218. The osmotic pressure exerted by a solution prepared by dissolving 2.0 g of protein of molar mass 60 kg mol^{-1} in 200 mL of water at 27°C is _______ Pa. [integer value]
(Use R = 0.083 L bar mol^{-1} K^{-1}) **[June 26, 2022 (II)]**

219. The osmotic pressure of blood is 7.47 bar at 300 K. To inject glucose to a patient intravenously, it has to be isotonic with blood. The concentration of glucose solution in gL^{-1} is _____(Molar mass of glucose = 180 g mol^{-1}
R = 0.083 L bar K^{-1} mol^{-1}) (Nearest integer) **[June 24, 2022 (I)]**

220. 1.22 g of an organic acid is separately dissolved in 100 g of benzene (K_b = 2.6 K kg mol^{-1}) and 100 g of acetone (K_b = 1.7 K kg mol^{-1}). The acid is known to dimerize in benzene but remain as a monomer in acetone. The boiling point of the solution in acetone increases by 0.17°C.
The increase in boiling point of solution in benzene in °C is x × 10^{-2}. The value of x is _______.(Nearest integer)
[Atomic mass : C = 12.0, H = 1.0, O = 16.0] **[Aug. 31, 2021 (II)]**

221. 1 kg of 0.75 molal aqueous solution of sucrose can be cooled up to –4 °C before freezing. The amount of ice (in g) that will be separated out is ________. (Nearest integer)
[Given : $K_f(H_2O)$ = 1.86 K kg mol^{-1}] **[Aug. 27, 2021 (I)]**

222. 40 g of glucose (Molar mass = 180) is mixed with 200 mL of water. The freezing point of solution is_____ K. (Nearest integer) **[Aug. 27, 2021 (II)]**
[Given : K_f = 1.86 K kg mol^{-1} ; Density of water = 1.00 g cm^{-3}; Freezing point of water = 273.15 K]

223. 83 g of ethylene glycol dissolved in 625 g of water. The freezing point of the solution is _______K.
(Nearest integer)
[Use : Molal Freezing point depression constant of water = 1.86 K kg mol^{-1}]
Freezing Point of water = 273 K
Atomic masses : C : 12.0 u, O : 16.0 u, H : 1.0 u] **[Aug. 26, 2021 (II)]**

224. Of the following four aqueous solutions, total number of those solutions whose freezing point is lower than that of 0.10 M C_2H_5OH is ________ (Integer answer) [Aug. 26, 2021 (I)]

(i) 0.10 M $Ba_3(PO_4)_2$ (ii) 0.10 M Na_2SO_4
(iii) 0.10 M KCl (iv) 0.10 M Li_3PO_4

225. 1.46 g of a biopolymer dissolved in a 100 mL water at 300 K exerted an osmotic pressure of 2.42×10^{-3} bar. The molar mass of the biopolymer is ____ $\times 10^4$ g mol^{-1}. (Round off to the nearest integer)

[Use : $R = 0.083$ L bar mol^{-1} K^{-1}] [July 27, 2021 (I)]

226. In a solvent 50% of an acid HA dimerizes and the rest dissociates. The van't Hoff factor of the acid is _____ $\times 10^{-2}$.

(Round off to the nearest integer) [July 27, 2021 (II)]

227. When 3.00 g of a substance 'X' is dissolved in 100 g of CCl_4, it raises the boiling point by 0.60 K. The molar mass of the substance 'X' is ________ g mol^{-1}.

(Nearest integer).

[Given K_b for CCl_4 is 5.0 K kg mol^{-1}] [July 25, 2021 (II)]

228. 2 molal solution of a weak acid HA has a freezing point of –3.885 °C. The degree of dissociation of this acid is _____ $\times 10^{-3}$.

(Rounded off to the nearest integer).

[Given : Molal depression constant of water $= 1.85$ K kg mol^{-1}, Freezing point of pure water = 0 °C]

[March 18, 2021 (I)]

229. A solute A dimerizes in water. The boiling point of a 2 molal solution of A is 100.52 °C. The percentage association of A is . (Round off to the nearest integer).

[Use : K_b for water = 0.52 K kg mol^{-1}, Boiling point of water = 100 °C] [March 18, 2021 (II)]

230. A 1 molal $K_4Fe(CN)_6$ solution has a degree of dissociation of 0.4. Its boiling point is equal to that of another solution which contains 18.1 weight percent of a non-electrolytic solute A. The molar mass of A is ________ u. (Round off to the nearest integer).

[Density of water = 1.0 g cm^{-3}] [March 17, 2021 (II)]

231. AB_2 is 10% dissociated in water to A^{2+} and B^-. The boiling point of a 10.0 molal aqueous solution of AB_2 is _____ °C. (Round off to the Nearest Integer)

[Given : Molal elevation constant of water $K_b = 0.5$ K kg mol^{-1} boiling point of pure water = 100°C] [March 16, 2021 (I)]

232. When 12.2 g of benzoic acid is dissolved in 100 g of water, the freezing point of solution was found to be –0.93 °C ($K_f(H_2O) = 1.86$ K kg mol^{-1}). The number (n) of benzoic acid molecules associated (assuming 100% association) is ________. [Feb. 26, 2021 (II)]

233. 224 mL of $SO_2(g)$ at 298 K and 1 atm is passed through 100 mL of 0.1 M NaOH solution. The non-volatile solute produced is dissolved in 36 g of water. The lowering of vapour pressure of solution (assuming the solution is dilute) $\left(P^{\circ}_{(H_2O)} = 24 \text{ mm of Hg}\right)$ is $x \times 10^{-2}$ mm of Hg, the value of x is ______. (Integer answer) [Feb. 26, 2021 (I)]

234. If a compound AB dissociates to the extent of 75% in an aqueous solution, the molality of the solution which shows a 2.5 K rise in the boiling point of the solution is ____ molal. (Rounded-off to the nearest integer)

[$K_b = 0.52$ K kg mol^{-1}] [Feb. 25, 2021 (II)]

235. 1 molal aqueous solution of an electrolyte A_2B_3 is 60% ionised. The boiling point of the solution at 1 atm is _______ K. (Rounded-off to the nearest integer)

[Given K_b for $(H_2O) = 0.52$ K kg mol^{-1}] [Feb. 25, 2021 (I)]

236. When 9.45 g of $ClCH_2COOH$ is added to 500 mL of water, its freezing point drops by 0.5 °C. The dissociation constant of $ClCH_2COOH$ is $x \times 10^{-3}$. The value of x is ________. (Rounded off to the nearest integer)

[$K_{f(H2O)} = 1.86$ K kg mol^{-1}] [Feb. 24, 2021 (I)]

237. C_6H_6 freezes at 5.5 °C. The temperature at which a solution 10 g of C_4H_{10} in 200 g of C_6H_6 freeze is ________ °C. (The molal freezing point depression constant of C_6H_6 is 5.12 °C/m.) [Feb. 24, 2021 (II)]

238. The elevation of boiling point of 0.10 m aqueous $CrCl_3.xNH_3$ solution is two times that of 0.05 m aqueous $CaCl_2$ solution. The value of x is ______.

[Assume 100% ionisation of the complex and $CaCl_2$, coordination number of Cr as 6, and that all NH_3 molecules are present inside the coordination sphere] [Sep. 06, 2020 (I)]

239. The osmotic pressure of a solution of NaCl is 0.10 atm and that of a glucose solution is 0.20 atm. The osmotic pressure of a solution formed by mixing 1 L of the sodium chloride solution with 2 L of the glucose solution is $x \times 10^{-3}$ atm. x is __________. (nearest integer) [Sep. 04, 2020 (II)]

240. If 250 cm^3 of an aqueous solution containing 0.73 g of a protein A is isotonic with one litre of another aqueous solution containing 1.65 g of a protein B, at 298 K, the ratio of the molecular masses of A and B is __________ $\times 10^{-2}$ (to the nearest integer). [Sep. 03, 2020 (II)]

241. How much amount of NaCl should be added to 600 g of water ($\rho = 1.00$ g/mL) to decrease the freezing point of water to –0.2 °C? _______. (The freezing point depression constant for water = 2 K kg mol^{-1}) [Jan. 09, 2020 (I)]

242. A cylinder containing an ideal gas (0.1 mol of 1.0 dm^3) is in thermal equilibrium with a large volume of 0.5 molal aqueous solution of ethylene glycol at its freezing point. If the stoppers S_1 and S_2 (as shown in the figure) are suddenly withdrawn, the volume of the gas in litres after equilibrium is achieved will be _______.

(Given, K_f (water) = 2.0 K kg mol^{-1}, $R = 0.08$ dm^3 atm K^{-1} mol^{-1}) [Jan. 09, 2020 (II)]

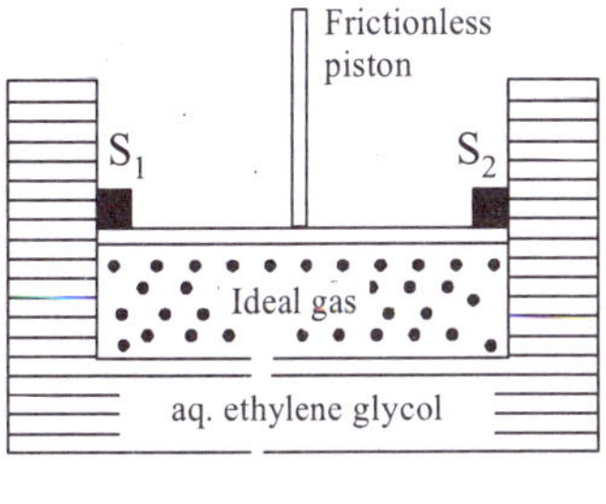

Part-B (JEE Advanced)

Topic 1 - Concentration of Solutions and Solubility

1 MCQs with One Correct Answer

1. Dissolving 120 g of urea (mol. wt. 60) in 1000 g of water gave a solution of density 1.15 g/mL. The molarity of the solution is [2011]
(a) 1.78 M (b) 2.00 M (c) 2.05 M (d) 2.22 M

2. The Henry's law constant for the solubility of N_2 gas in water at 298 K is 1.0×10^5 atm. The mole fraction of N_2 in air is 0.8. The number of moles of N_2 from air dissolved in 10 moles of water at 298 K and 5 atm pressure is [2009]
(a) 4.0×10^{-4} (b) 4.0×10^{-5} (c) 5.0×10^{-4} (d) 4.0×10^{-6}

2 Integer Value Answer

3. Liquids A and B form ideal solution over the entire range of composition. At temperature T, equimolar binary solution of liquids A and B has vapour pressure 45 Torr. At the same temperature, a new solution of A and B having mole fractions x_A and x_B, respectively, has vapour pressure of 22.5 Torr. The value of x_A/x_B in the new solution is________.
(given that the vapour pressure of pure liquid A is 20 Torr at temperature T) [Adv. 2018]

4. The mole fraction of a solute in a solution is 0.1. At 298 K, molarity of this solution is the same as its molality. Density of this solution at 298 K is 2.0 g cm^{-3}. The ratio of the molecular weights of the solute and solvent, $\left(\frac{MW_{solute}}{MW_{solvent}}\right)$, is [Adv. 2016]

5. A compound H_2X with molar weight of 80g is dissolved in a solvent having density of 0.4 g mL^{-1}. Assuming no change in volume upon dissolution, the molality of a 3.2 molar solution is [Adv. 2014]

6. 29.2% (w/w) HCl stock solution has a density of 1.25 g mL^{-1}. The molecular weight of HCl is 36.5 g mol^{-1}. The volume (mL) of stock solution required to prepare a 200 mL solution of 0.4 M HCl is : [2012]

3 Numeric / New Stem Based Questions

7. The mole fraction of urea in an aqueous urea solution containing 900 g of water is 0.05. If the density of the solution is 1.2 g cm^{-3}, the molarity of urea solution is ___ (Given data: Molar masses of urea and water are 60 g mol^{-1} and 18 g mol^{-1}, respectively) [Adv. 2019]

8. 0.5 g of fuming H_2SO_4 (Oleum) is diluted with water. This solution is completely neutralized by 26.7 mL of 0.4 N NaOH. Find the percentage of free SO_3 in the sample of oleum. [1980]

10 Subjective Problems

9. The vapour pressure of a dilute aqueous solution of glucose ($C_6H_{12}O_6$) is 750 mm of mercury at 373 K. Calculate (i) molality and (ii) mole fraction of the solution. [1989 - 3 Marks]

10. A bottle of commercial sulphuric acid (density 1.787 g/mL) is labelled as 86 percent by weight. What is the molarity of the acid? What volume of the acid has to be used to make 1 litre of 0.2 M H_2SO_4? [1979]

11. What is the molarity and molality of a 13% solution (by weight) of sulphuric acid with a density of 1.02 g/mL? To what volume should 100 mL of this acid be diluted in order to prepare a 1.5 N solution? [1978]

Topic 2 - Vapour Pressure, Laws of Solutions and Ideal, Non-ideal Solutions

1 MCQs with One Correct Answer

12. For a dilute solution, Raoult's law states that : [1985 - 1 Mark]
(a) the lowering of vapour pressure is equal to the mole fraction of solute.
(b) the relative lowering of vapour pressure is equal to the mole fraction of solute.
(c) the relative lowering of vapour pressure is proportional to the amount of solute in solution.
(d) the vapour pressure of the solution is equal to the mole fraction of solvent.

13. An azeotropic solution of two liquids has boiling point lower than either of them when it [1981 - 1 Mark]
(a) shows negative deviation from Raoult's law
(b) shows no deviation from Raoult's law
(c) shows positive deviation from Raoult's law
(d) is saturated

3 Numeric / New Stem Based Questions

14. In a solvent S, a compound B is partially dissociated into C and D as given below: [Adv. 2026]

$$B \rightleftharpoons 2C + 2D$$

B, C and D are non-volatile in nature. The molar mass of B is 10 times the molar mass of S. The standard boiling point and the standard enthalpy of vaporization of S are 400 K and 10R J mol^{-1}, respectively (R is the gas constant in J K^{-1} mol^{-1}). A solution of B in S with an initial concentration of B as 0.25% (mass/mass) has a boiling point of 408 K at 1 bar pressure. In this solution, the mole percent of B that has been dissociated is ______.

15. Liquids A and B form ideal solution for all compositions of A and B at 25°C. Two such solutions with 0.25 and 0.50 mole fractions of A have the total vapor pressures of 0.3 and 0.4 bar, respectively. What is the vapor pressure of pure liquid B in bar? [Adv. 2020]

16. The molar volume of liquid benzene (density = 0.877 g mL^{-1}) increases by a factor of 2750 as it vaporises at 20°C and that of liquid toluene (density = 0.867 g mL^{-1}) increases by a factor of 7720 at 20°C. A solution of benzene and toluene at 20°C has a vapour pressure of 46.0 Torr. Find the mole fraction of benzene in the vapour above the solution. [1996 - 3 Marks]

17. The vapour pressure of pure benzene is 639.7 mm of mercury and the vapour of a solution of a solute in benzene at the same temperature is 631.9 mm of mercury. Calculate the molality of the solution. [1981 - 3 Marks]

6 MCQs with One or More than One Correct Answer

18. For a solution formed by mixing liquids L and M, the vapour pressure of L plotted against the mole fraction of M in solution is shown in the following figure. Here x_L and x_M represent mole fractions of L and M, respectively, in the solution. The correct statement(s) applicable to this system is (are) [Adv. 2017]

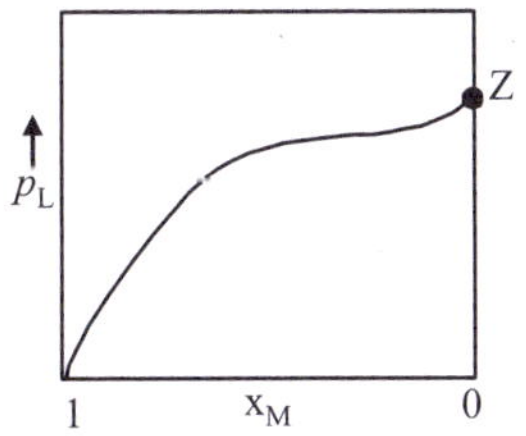

(a) The point Z represents vapour pressure of pure liquid M and Raoult's law is obeyed from $x_L = 0$ to $x_L = 1$
(b) The point Z represents vapour pressure of pure liquid L and Raoult's law is obeyed when $x_L \to 1$
(c) The point Z represents vapour pressure of pure liquid M and Raoult's law is obeyed when $x_L \to 0$
(d) Attractive intermolecular interactions between L-L in pure liquid L and M-M in pure liquid M are stronger than those between L-M when mixed in solution.

19. Mixture(s) showing positive deviation from Raoult's law at 35 °C is (are) [Adv. 2016]
(a) carbon tetrachloride + methanol
(b) carbon disulphide + acetone
(c) benzene + toluene
(d) phenol + aniline

20. Benzene and naphthalene form an ideal solution at room temperature. For this process, the true statement(s) is(are) [Adv. 2013]
(a) ΔG is positive (b) ΔS_{system} is positive
(c) $\Delta S_{surroundings} = 0$ (d) $\Delta H = 0$

10 Subjective Problems

21. The vapour pressure of ethanol and methanol are 44.5 mm and 88.7 Hg respectively. An ideal solution is formed at the same temperature by mixing 60 g of ethanol with 40 g of methanol. Calculate the total vapour pressure of the solution and the mole fraction of methanol in the vapour. [1986 - 4 Marks]

22. 'Two volatile and miscible liquids can be separated by fractional distillation into pure component', is true under what conditions? [1984 - 1 Mark]

23. An organic compound $C_xH_{2y}O_y$ was burnt with twice the amount of oxygen needed for complete combustion to CO_2 and H_2O. The hot gases when cooled to 0°C and 1 atm. pressure, measured 2.24 liters. The water collected during cooling weighed 0.9 g. The vapour pressure of pure water at 20°C is 17.5 mm Hg and is lowered by 0.104 mm when 50 g of the organic compound is dissolved in 1000 g of water. Give the molecular formula of the organic compound. [1983 - 5 Marks]

Topic 3 - Colligative Properties and Abnormal Molecular Masses

1 MCQs with One Correct Answer

24. Pure water freezes at 273 K and 1 bar. The addition of 34.5 g of ethanol to 500 g of water changes the freezing point of the solution. Use the freezing point depression constant of water as 2 K kg mol^{-1}. The figures shown below represent plots of vapour pressure (V.P.) versus temperature (T). [molecular weight of ethanol is 46 g mol^{-1}] Among the following, the option representing change in the freezing point is [Adv. 2017]

(a)
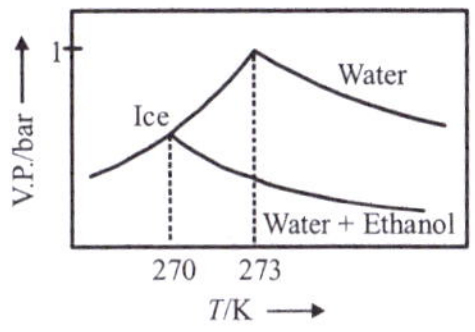

(b)
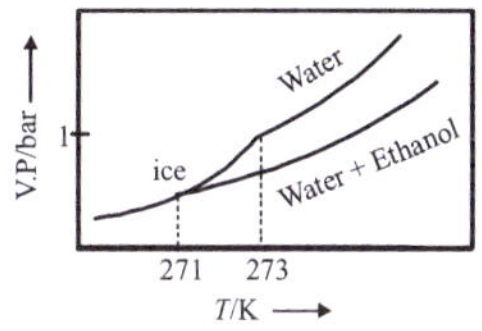

(c)
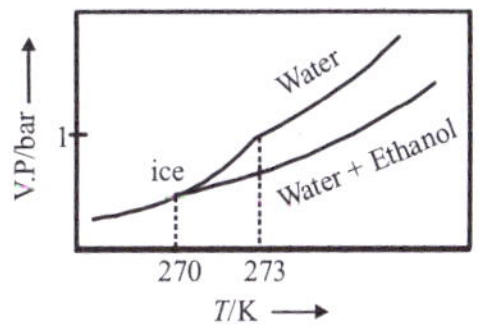

(d)
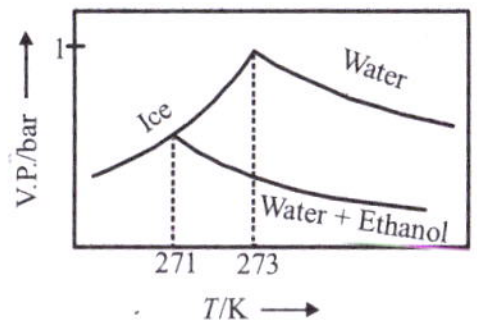

25. For a dilute solution containing 2.5 g of a non-volatile non-electrolyte solute in 100 g of water, the elevation in boiling point at 1 atm pressure is 2°C. Assuming concentration of solute is much lower than the concentration of solvent, the vapour pressure (mm of Hg) of the solution is (take $K_b = 0.76$ K kg mol^{-1}
(a) 724 (b) 740 (c) 736 (d) 718 [2012]

26. The freezing point (in °C) of a solution containing 0.1 g of $K_3[Fe(CN)_6]$ (Mol. wt. 329) in 100 g of water ($K_f = 1.86$ K kg mol^{-1}) is [2011]
(a) -2.3×10^{-2} (b) -5.7×10^{-2}
(c) -5.7×10^{-3} (d) -1.2×10^{-2}

27. When 20 g of naphthoic acid ($C_{11}H_8O_2$) is dissolved in 50 g of benzene ($K_f = 1.72$ K kg mol^{-1}), a freezing point depression of 2K is observed. The van't Hoff factor (i) is [2007]
(a) 0.5 (b) 1 (c) 2 (d) 3

28. The elevation in boiling point of a solution of 13.44 g of $CuCl_2$ in 1 kg of water using the following information will be (Molecular weight of $CuCl_2$ = 134.4 and $K_b = 0.52$ K molal^{-1}) [2005S]
(a) 0.16 (b) 0.05 (c) 0.1 (d) 0.2

29. During depression of freezing point in a solution, the following are in equilibrium [2003S]
(a) liquid solvent, solid solvent (b) liquid solvent, solid solute
(c) liquid solute, solid solute (d) liquid solute, solid solvent

30. The molecular weight of benzoic acid in benzene as determined by depression in freezing point method corresponds to: [1996 - 1 Mark]
(a) ionization of benzoic acid.
(b) dimerization of benzoic acid.
(c) trimerization of benzoic acid.
(d) solvation of benzoic acid.

31. 0.2 molal acid HX is 20% ionised in solution. $K_f = 1.86$ K molality^{-1}. The freezing point of the solution is : [1995S]
(a) 0.45 (b) -0.90 (c) -0.31 (d) -0.53

32. The freezing point of equimolal aqueous solutions will be highest for : [1990 - 1 Mark]
(a) $C_6H_5NH_3Cl$ (b) $Ca(NO_3)_2$
(c) $La(NO_3)_3$ (d) $C_6H_{12}O_6$ (glucose)

33. Which of the following 0.1 M aqueous solutions will have the lowest freezing point? [1989 - 1 Mark]
(a) Potassium sulphate (b) Sodium chloride
(c) Urea (d) Glucose

34. When mercuric iodide is added to the aqueous solution of potassium iodide then [1987 - 1 Mark]
(a) freezing point is raised.
(b) freezing point is lowered.
(c) freezing point does not change.
(d) boiling point does not change.

2 Integer Value Answer

35. Vessel-1 contains w_2 g of a non-volatile solute X dissolved in w_1 g of water. Vessel-2 contains w_2 g of another non-volatile solute Y dissolved in w_1 g of water. Both the vessels are at the same temperature and pressure. The molar mass of X is 80% of that of Y. The van't Hoff factor for X is 1.2 times of that of Y for their respective concentrations.
The elevation of boiling point for solution in Vessel-1 is _____ % of the solution in Vessel-2. [Adv. 2024]

36. 50 mL of 0.2 molal urea solution (density = 1.012 g mL^{-1} at 300 K) is mixed with 250 mL of a solution containing 0.06 g of urea. Both the solutions were prepared in the same solvent. The osmotic pressure (in Torr) of the resulting solution at 300 K is _______. [Adv. 2023]
[Use: Molar mass of urea = 60 g mol^{-1}; gas constant, R = 62 L Torr K^{-1} mol^{-1} ; Assume, $\Delta_{mix}H = 0$, $\Delta_{mix}V = 0$]

37. An aqueous solution is prepared by dissolving 0.1 mol of an ionic salt in 1.8 kg of water at 35 °C. The salt remains 90% dissociated in the solution. The vapour pressure of the solution is 59.724 mm of Hg. Vapor pressure of water at 35 °C is 60.000 mm of Hg. The number of ions present per formula unit of the ionic salt is ________. [Adv. 2022]

38. If the freezing point of a 0.01 molal aqueous solution of a cobalt(III) chloride-ammonia complex (which behaves as a strong electrolyte) is –0.0558°C, the number of chloride(s) in the coordination sphere of the complex is
[K_f of water = 1.86 K kg mol^{-1}] [Adv. 2015]

39. MX_2 dissociates into M^{2+} and X^- ions in an aqueous solution, with a degree of dissociation (α) of 0.5. The ratio of the observed depression of freezing point of the aqueous solution to the value of the depression of freezing point in the absence of ionic dissociation is [Adv. 2014]

3 Numeric / New Stem Based Questions

Question Stem for Question Nos. 40 and 41 [Adv. 2026]

Two volatile liquids A and B form an ideal solution. Consider a 5 molal solution of B in A inside a closed container having a total vapour pressure of 100 mm Hg at 300 K. The vapour pressure of pure A at 300K is 105 mm Hg. Assume that A and B behave as ideal gases in the vapour phase.

Given: The gas constant R = 0.08 L atm K^{-1} mol^{-1}

Molar mass of A is 50 g mol^{-1}

Molar mass of B is 57 g mol^{-1}

Density of liquid B at 300 K is 0.5 g/mL

1 atm = 760 mm Hg

40. At 300 K, the ratio of the molar volume of pure B in vapour phase to its molar volume in liquid phase is ____.

41. The mole fraction of B in vapour phase which is in equilibrium with this solution is _____.

42. At 300 K, an ideal dilute solution of a macromolecule exerts osmotic pressure that is expressed in terms of the height (h) of the solution (density = 1.00 g cm^{-3}) where h is equal to 2.00 cm. If the concentration of the dilute solution of the macromolecule is 2.00 g dm^{-3}, the molar mass of the macromolecule is calculated to be X × 10^4 g mol^{-1}. The value of **X** is ______.
Use: Universal gas constant (R) = 8.3 J K^{-1} mol^{-1} and acceleration due to gravity (g) = 10 m s^{-2} [Adv. 2025]

Question Stem for Question No. 43 and 44

The boiling point of water in a 0.1 molal silver nitrate solution (solution A) is x °C. To this solution A, an equal volume of 0.1 molal aqueous barium chloride solution is added to make a new solution B. The difference in the boiling points of water in the two solutions A and B is y × 10^{-2} °C.
(Assume : Densities of the solutions A and B are the same as that of water and the soluble salts dissociate completely.)
Use: Molal elevation constant (Ebullioscopic Constant), K_b = 0.5 K kg mol^{-1}; Boiling point of pure water as 100°C.)

43. The value of x is ________. [Adv. 2021]

44. The value of |y| is ________. [Adv. 2021]

45. On dissolving 0.5 g of a non-volatile non-ionic solute to 39 g of benzene, its vapour pressure decreases from 650 mm Hg to 640 mm Hg. The depression of freezing point of Benzene (in K) upon addition of the solute is _____
(Given data : Molar mass and the molal freezing point depression constant of benzene are 78 g mol^{-1} and 5.12 K kg mol^{-1}, respectively) [Adv. 2019]

46. The plot given below shows P — T curves (where P is the pressure and T is the temperature) for two solvents **X** and **Y** and isomolal solutions of NaCl in these solvents. NaCl completely dissociates in both the solvents. [Adv. 2018]

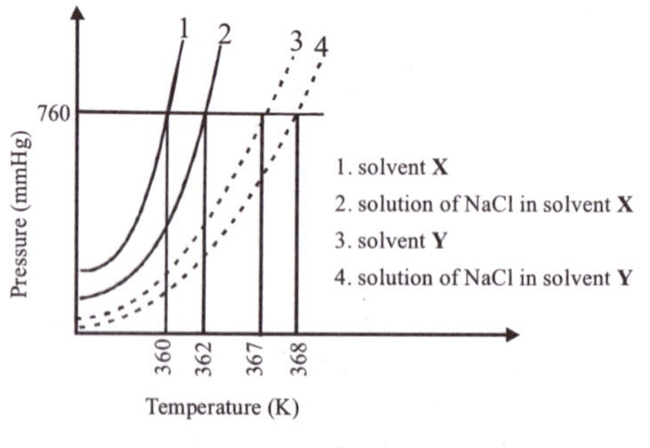

On addition of equal number of moles of a non-volatile solute S in equal amount (in kg) of these solvents, the elevation of boiling point of solvent X is three times that of solvent Y. Solute S is known to undergo dimerization in these solvents. If the degree of dimerization is 0.7 in solvent Y, the degree of dimerization in solvent X is______.

47. 75.2 g of C_6H_5OH (phenol) is dissolved in a solvent of K_f = 14. If the depression in freezing point is 7 K then find the % of phenol that dimerises. [2006 - 6 Marks]

48. To 500 cm^3 of water, 3.0 × 10^{-3} kg of acetic acid is added. If 23% of acetic acid is dissociated, what will be the depression in freezing point? K_f and density of water are 1.86 K kg^{-1} mol^{-1} and 0.997 g cm^{-3}, respectively. [2000 - 3 Marks]

49. A solution of a non-volatile solute in water freezes at – 0.30°C. The vapour pressure of pure water at 298 K is 23.51 mm Hg and K_f for water is 1.86 K kg mol^{-1}. Calculate the vapour pressure of this solution at 298 K. [1998 - 4 Marks]

50. Addition of 0.643 g of a compound to 50 mL. of benzene (density : 0.879 g/mL.) lowers the freezing point from 5.51°C to 5.03°C. If K_f for benzene is 5.12 K kg mol^{-1}, calculate the molecular weight of the compound. [1992 - 2 Marks]

51. The degree of dissociation of calcium nitrate in a dilute aqueous solution, containing 7.0 g. of the salt per 100 g of water at 100°C is 70%. If the vapour pressure of water at 100°C is 760 mm, calculate the vapour pressure of the solution. [1991 - 4 Marks]

52. The vapour pressure of pure benzene at a certain temperature is 640 mm Hg. A non-volatile non-electrolyte solid weighing 2.175 g is added to 39.0 g of benzene. The vapour pressure of the solution is 600 mm Hg. What is the molecular weight of the solid substance? [1990 - 3 Marks]

4 Fill in the Blanks

53. Given that ΔT_f is the depression in freezing point of the solvent in a solution of a non-volatile solute of molality, m, the quantity $\lim_{m\to 0}(\Delta T_f / m)$ is equal to .... [1994 - 1 Mark]

6 MCQs with One or More than One Correct Answer

54. In the depression of freezing point experiment, it is found that the [1999 - 3 Marks]
(a) vapour pressure of the solution is less than that of pure solvent
(b) vapour pressure of the solution is more than that of pure solvent
(c) only solute molecules solidify at the freezing point
(d) only solvent molecules solidify at the freezing

9 Assertion and Reason/Statement Type Questions

Properties such as boiling point, freezing point and vapour pressure of a pure solvent change when solute molecules are added to get homogeneous solution. These are called colligative properties. Application of colligative properties are very useful in day-to-day life. One of its example is the use of ethylene glycol and water mixture as anti-freezing liquid in the radiator of automobiles.

A solution M is prepared by mixing ethanol and water. The mole fraction of ethanol in the mixture is 0.9

Given : Freezing point depression constant of water (K_f^{water}) = 1.86 K kg mol^{-1}

Freezing point depression constant of ethanol ($K_f^{ethanol}$) = 2.0 K kg mol^{-1}

Boiling point elevation constant of water (K_b^{water}) = 0.52 K kg mol^{-1}

Boiling point elevation constant of ethanol ($K_b^{ethanol}$) = 1.2 K kg mol^{-1}

Standard freezing point of water = 273 K

Standard freezing point of ethanol = 155.7 K

Standard boiling point of water = 373 K

Standard boiling point of ethanol = 351.5 K

Vapour pressure of pure water = 32.8 mm Hg

Vapour pressure of pure ethanol = 40 mm Hg

Molecular weight of water = 18 g mol^{-1}

Molecular weight of ethanol = 46 g mol^{-1}

In answering the following questions, consider the solution to be ideal dilute solutions and solutes to be non-volatile and non-dissociative.

55. The freezing point of the solution M is [2008 - 3 Marks]
(a) 268.7 K (b) 268.5 K (c) 234.2 K (d) 150.9 K

56. The vapour pressure of the solution M is [2008 - 3 Marks]
(a) 39.3 mm Hg (b) 36.0 mm Hg
(c) 29.5 mm Hg (d) 28.8 mm Hg

57. Water is added to the solution M such that the mole fraction of water in the solution becomes 0.9. The boiling point of this solution is [2008 - 3 Marks]
(a) 380.4 K (b) 376.2 K
(c) 375.5 K (d) 354.7 K

10 Subjective Problems

58. 1.22g of benzoic acid is dissolved in 100 g of acetone and 100 g of benzene separately. Boiling point of the solution in acetone increases by 0.17° C, while that in the benzene increases by 0.13° C; K_b for acetone and benzene is 1.7 K kg mol^{-1} and 2.6 K kg mol^{-1} respectively. Find molecular weight of benzoic acid in two cases and justify your answer. [2004 - 2 Marks]

59. Match the following if the molecular weights of X, Y and Z are same. [2003 - 2 Marks]

	Boiling Point	K_b
X	100	0.63
Y	27	0.53
Z	253	0.98

60. Nitrobenzene is formed as the major product along with a minor product in the reaction of benzene with a hot mixture of nitric acid and sulphuric acid. The minor product consists of carbon : 42.86%, hydrogen : 2.40%, nitrogen : 16.67%, and oxygen: 38.07% (i) Calculate the empirical formula of the minor product. (ii) When 5.5 g of the minor product is dissolved in 45 g of benzene, the boiling point of the solution is 1.84 °C higher than that of pure benzene. Calculate the molar mass of the minor product and determine its molecular and structural formula. (Molal boiling point elevation constant of benzene is 2.53 K kg mol^{-1}.) [1999 - 10 Marks]

61. What weight of the non-volatile solute, urea ($NH_2-CO-NH_2$) needs to be dissolved in 100g of water, in order to decrease the vapour pressure of water by 25%? What will be the molality of the solution? [1993 - 3 Marks]

Answer Key

Part-A (JEE Main)

1	(1)	**26**	(164)	**51**	(1)	**76**	(2)	**101**	(1389)	**126**	(4)	**151**	(3)	**176**	(15)	**201**	(372)	**226**	(125)
2	(4)	**27**	(74)	**52**	(3)	**77**	(2)	**102**	(19)	**127**	(1)	**152**	(4)	**177**	(25)	**202**	(62250)	**227**	(250)
3	(4)	**28**	(22)	**53**	(4)	**78**	(4)	**103**	(600)	**128**	(4)	**153**	(1)	**178**	(33)	**203**	(100)	**228**	(50)
4	(3)	**29**	(11)	**54**	(1)	**79**	(2)	**104**	(3)	**129**	(4)	**154**	(1)	**179**	(16)	**204**	(243)	**229**	(100)
5	(2)	**30**	(7)	**55**	(4)	**80**	(4)	**105**	(4)	**130**	(1)	**155**	(4)	**180**	(75)	**205**	(13)	**230**	(85)
6	(3)	**31**	(4)	**56**	(4)	**81**	(4)	**106**	(3)	**131**	(1)	**156**	(4)	**181**	(23)	**206**	(41,500)	**231**	(106)
7	(4)	**32**	(815)	**57**	(1)	**82**	(2)	**107**	(1)	**132**	(2)	**157**	(1)	**182**	(5)	**207**	(4)	**232**	(2)
8	(2)	**33**	(3.64)	**58**	(4)	**83**	(4)	**108**	(4)	**133**	(3)	**158**	(1)	**183**	(76)	**208**	(80)	**233**	(24)
9	(4)	**34**	(139)	**59**	(1, 3)	**84**	(2)	**109**	(1)	**134**	(2)	**159**	(3)	**184**	(543)	**209**	(15)	**234**	(3)
10	(2)	**35**	(148)	**60**	(3)	**85**	(4)	**110**	(2)	**135**	(4)	**160**	(3)	**185**	(25)	**210**	(6)	**235**	(375)
11	(2)	**36**	(180)	**61**	(2)	**86**	(4)	**111**	(4)	**136**	(3)	**161**	(3)	**186**	(19)	**211**	(1)	**236**	(35)
12	(2)	**37**	(5)	**62**	(1)	**87**	(5)	**112**	(1)	**137**	(3)	**162**	(2)	**187**	(707)	**212**	(5)	**237**	(1)
13	(2)	**38**	(54)	**63**	(4)	**88**	(200)	**113**	(4)	**138**	(3)	**163**	(2)	**188**	(31)	**213**	(3)	**238**	(5)
14	(1)	**39**	(64)	**64**	(4)	**89**	(23)	**114**	(3)	**139**	(3)	**164**	(4)	**189**	(15)	**214**	(330)	**239**	(167)
15	(4)	**40**	(20)	**65**	(1)	**90**	(82)	**115**	(2)	**140**	(2)	**165**	(2)	**190**	(72.56)	**215**	(8)	**240**	(177)
16	(4)	**41**	(3)	**66**	(2)	**91**	(555)	**116**	(2)	**141**	(4)	**166**	(2)	**191**	(240)	**216**	(45)	**241**	(1.75)
17	(1)	**42**	(50)	**67**	(1)	**92**	(314)	**117**	(4)	**142**	(1)	**167**	(1)	**192**	(116)	**217**	(98)	**242**	(2.18)
18	(3)	**43**	(64)	**68**	(3)	**93**	(1)	**118**	(N)	**143**	(1)	**168**	(1)	**193**	(48)	**218**	(415)		
19	(4)	**44**	(9)	**69**	(4)	**94**	(2)	**119**	(1)	**144**	(2)	**169**	(3)	**194**	(75)	**219**	(54)		
20	(2)	**45**	(4)	**70**	(2)	**95**	(2)	**120**	(3)	**145**	(2)	**170**	(4)	**195**	(30)	**220**	(13)		
21	(4)	**46**	(4)	**71**	(2)	**96**	(14)	**121**	(1)	**146**	(4)	**171**	(4)	**196**	(40535)	**221**	(518)		
22	(1)	**47**	(1)	**72**	(4)	**97**	(1222)	**122**	(3)	**147**	(3)	**172**	(1)	**197**	(8)	**222**	(271)		
23	(4)	**48**	(1)	**73**	(2)	**98**	(25)	**123**	(1)	**148**	(4)	**173**	(2)	**198**	(1111)	**223**	(269)		
24	(2)	**49**	(1)	**74**	(3)	**99**	(77)	**124**	(4)	**149**	(3)	**174**	(62)	**199**	(4)	**224**	(4)		
25	(9)	**50**	(2)	**75**	(1)	**100**	(1)	**125**	(1)	**150**	(1)	**175**	(3)	**200**	(3.4)	**225**	(15)		

Part-B (JEE Advanced)

1	(c)	**7**	(2.98)	**16**	(0.735)	**25**	(a)	**31**	(a)	**37**	(5)	**43**	(100.1)	**49**	(23.44)	**55**	(d)		
2	(a)	**8**	(3.84)	**17**	(0.156)	**26**	(a)	**32**	(d)	**38**	(1)	**44**	(2.5)	**50**	(156.056)	**56**	(b)		
3	(19)	**12**	(b)	**18**	(b, d)	**27**	(a)	**33**	(a)	**39**	(2)	**45**	(1.03)	**51**	(746.3)	**57**	(b)		
4	(9)	**13**	(c)	**19**	(a, b)	**28**	(a)	**34**	(a)	**40**	(2000.00)	**46**	(0.05)	**52**	(65.25)				
5	(8)	**14**	(33.16)	**20**	(b, d)	**29**	(a)	**35**	(150)	**41**	(0.16)	**47**	(75)	**53**	(K_f)				
6	(8)	**15**	(0.02)	**24**	(c)	**30**	(b)	**36**	(682)	**42**	(2.49)	**48**	(0.228)	**54**	(a, d)				

Chapter-3

Electrochemistry

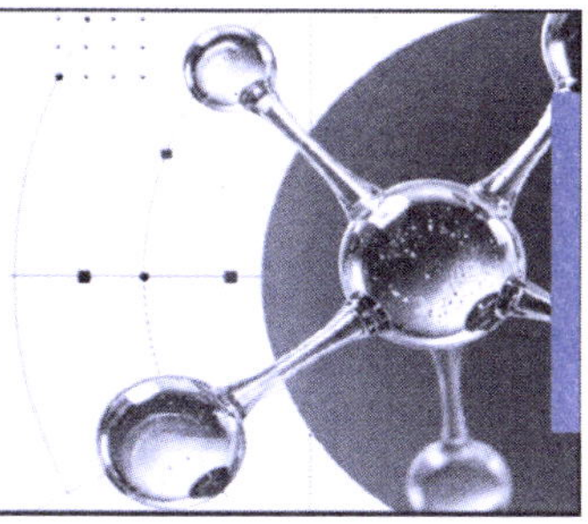

Part-A (JEE Main)

Topic 1 - Cells and Electrode Potential, Nernst Equation

Multiple Choice Questions (MCQs)

1. Given at 298 K : $E^{\ominus}_{Fe^{2+}/Fe} = X$ Volt

$E^{\ominus}_{Fe^{3+}/Fe} = Y$ Volt

The $E^{\ominus}_{Fe^{3+}/Fe^{2+}}$ in Volt at 298 K is given by : [April 08, 2026 (II)]

(1) $2X-3Y$ (2) $3Y-2X$

(3) $3Y+2X$ (4) $Y+X$

2. For a general redox reaction [April 06, 2026 (II)]

Anode: $Red_1 \rightarrow Ox_1^{n_1^+} + n_1e^-$

Cathode: $Ox_2 + n_2e^- \rightarrow Red_2^{n_2^-}$

Which of the following statement is **incorrect** ?

(1) The overall reaction can be written as

$n_2 Red_1 + n_1 Ox_2 \rightleftharpoons n_2 Ox_1^{n1+} + n_1 Red_2^{n2+}$

(2) The electrons do not appear in the overall reaction because electrons produced at the anode are consumed at the cathode.

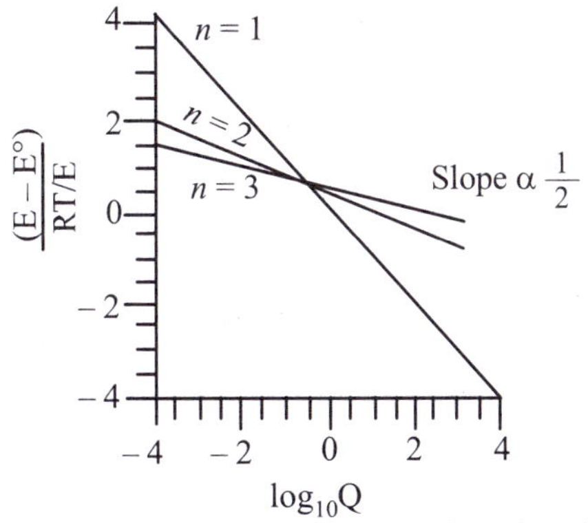

(3) Here n is the number of electrons transferred in redox reaction.

(4) If the reaction is carried out reversibly, the electrical work done is equal to the ratio of charge and potential difference through which charge is moved.

3. One half cell in a voltaic cell is constructed by dipping silver rod in $AgNO_3$ solution of unknown concentration, other half cell is Zn rod dipped in 1 molar solution of $ZnSO_4$. [April 05, 2026 (II)]

A voltage of 1.60 V is measured at 298 K for this cell. What is the concentration of Ag^+ ions used in terms of log x(x = [Ag^+]) ?

$E^{\ominus}_{Zn^{2+}/Zn} = -0.76V, E^{\ominus}_{Ag^+/Ag} = +0.80V$

$\frac{2.303RT}{F} = 0.059V$

(1) $\frac{2}{3.9}$ (2) $\frac{4}{5.9}$ (3) $\frac{2.9}{2}$ (4) $\frac{5.9}{4}$

4. An electrochemical cell is constructed using half cells in the direction of spontaneous change $Fe(OH)_2(s) + 2e^- \rightarrow Fe(s) + 2OH^-(aq)$ $E^0 = -0.88V$ and $AgBr(s) + e^- Ag(s) + Br^-(aq)$ $E^0 = +0.07V$

Which of the following option is correct ? [April 02, 2026 (I)]

(1) Overall reaction $Fe(s) + 2OH^{-1}(aq) + 2AgBr(s) \rightleftharpoons Fe(OH)_2(s) + 2Ag(s) + 2Br^-(aq)$

(2) $E^0_{cell} = -0.95V$

(3) Fe is reduced in the electrochemical cell

(4) E^0_{cell} is an extensive property

5. 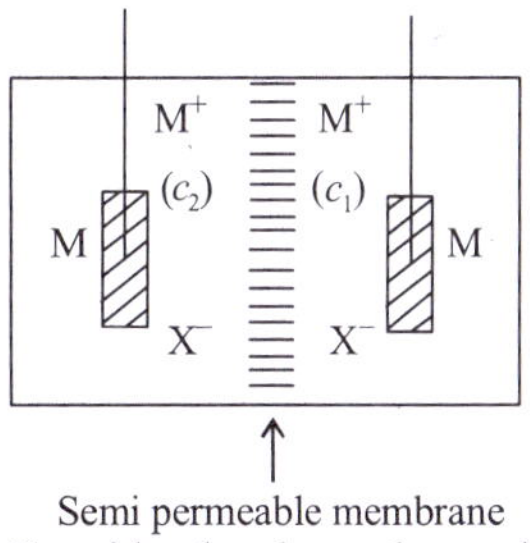

Consider the above electrochemical cell where a metal electrode (M) is undergoing redox reaction by forming M^+ ($M \rightarrow M^+ + e^-$). The cation M^+ is present in two different concentrations c_1 and c_2 as shown above. Which of the following statement is correct for generating a positive cell potential? [Jan. 23, 2026 (II)]

(1) If c_1 is present at anode, then $c_1 = c_2$

(2) If c_1 is present at cathode, then $c_1 < c_2$

(3) If c_1 is present at cathode, then $c_1 > c_2$

(4) If c_1 is present at anode, then $c_1 > c_2$

6. In the given electrochemical cell, $Ag(s)|AgCl(s)|$ $FeCl_2(aq)$, $FeCl_3(aq)|Pt(s)$ at 298 K, the cell potential (E_{cell}) will increase when: [Jan. 23, 2026 (I)]

(A) Concentration of Fe^{2+} is increased

(B) Concentration of Fe^{3+} is decreased

(C) Concentration of Fe^{2+} is decreased

(D) Concentration of Fe^{3+} is increased

(E) Concentration of Cl^- is increased

Choose the **correct** answer from the options given below:

(1) A and B only (2) A and E only

(3) B only (4) C, D and E only

7. For a closed circuit Daniell cell, which of the following plots is the accurate one at a given temperature? **[Jan. 21, 2026 (II)]**

(1) 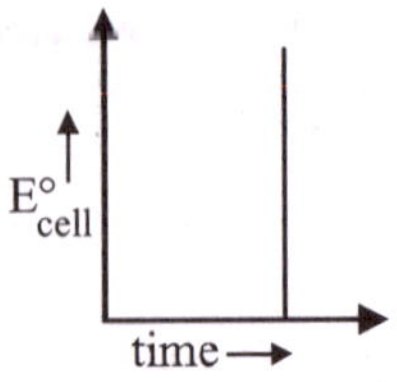

(2) 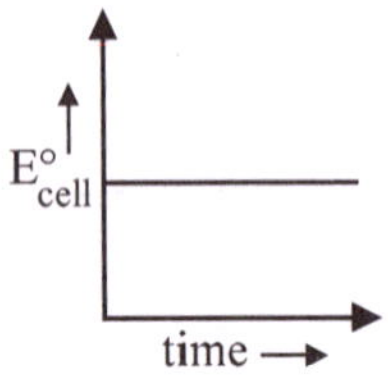

(3) 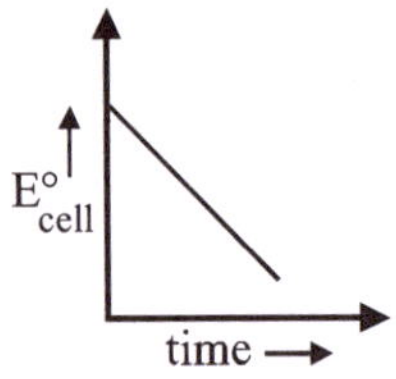

(4) 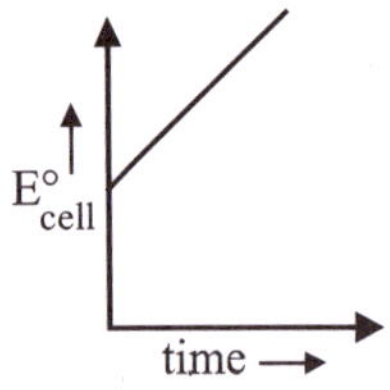

8. For a $Mg \mid Mg^{2+}(aq) \parallel Ag^+(aq) \mid Ag$ the correct Nernst Equation is : **[Jan. 29, 2025 (I)]**

(1) $E_{cell} = E^o_{cell} - \frac{RT}{2F}\ln\frac{[Ag^+]}{[Mg^{2+}]}$

(2) $E_{cell} = E^o_{cell} + \frac{RT}{2F}\ln\frac{[Ag^+]^2}{[Mg^{2+}]}$

(3) $E_{cell} = E^o_{cell} - \frac{RT}{2F}\ln\frac{[Mg^{2+}]}{[Ag^+]}$

(4) $E_{cell} = E^o_{cell} - \frac{RT}{2F}\ln\frac{[Ag^+]^2}{[Mg^{2+}]}$

9. The standard reduction potential values of some of the p-block ions are given below. Predict the one with the strongest oxidising capacity. **[Jan. 29, 2025 (I)]**

(1) $E^\ominus_{Sn^{4+}/Sn^{2+}} = +1.15V$ (2) $E^\ominus_{Tl^{3+}/Tl} = +1.26V$

(3) $E^\ominus_{Al^{3+}/Al} = -1.66V$ (4) $E^\ominus_{Pb^{4+}/Pb^{2+}} = +1.67V$

10. Based on the data given below: **[Jan. 24, 2025 (II)]**

$E^0_{Cr_2O_7^{2-}/Cr^{3+}} = 1.33\text{ V}$ $E^0_{Cl_2/Cl^-} = 1.36\text{ V}$

$E^0_{MnO_4^-/Mn^{2+}} = 1.51\text{ V}$ $E^0_{Cr^{3+}/Cr} = -0.74\text{ V}$

the strongest reducing agent is :

(1) Mn^{2+} (2) Cr

(3) MnO_4^- (4) Cl^-

11. For the given cell **[Jan. 24, 2025 (I)]**

$Fe^{2+}(aq) + Ag^+(aq) \rightarrow Fe^{3+}(aq) + Ag(s)$

The standard cell potential of the above reaction is Given :

$Ag^+ + e^- \rightarrow Ag$ $E^0 = xV$

$Fe^{2+} + 2e^- \rightarrow Fe$ $E^0 = yV$

$Fe^{3+} + 3e^- \rightarrow Fe$ $E^0 = zV$

(1) $x + y - z$ (2) $x + 2y - 3z$

(3) $y - 2x$ (4) $x + 2y$

12. Standard electrode potentials for a few half cells are mentioned below: **[Jan. 23, 2025 (II)]**

$E^o_{Cu^{2+}/Cu} = 0.34V,\ E^o_{Zn^{2+}/Zn} = -0.76V$

$E^o_{Ag^+/Ag} = 0.80V,\ E^o_{Mg^{2+}/Mg} = -2.37V$

Which one of the following cells gives the most negative value of ΔG°?

(1) $Zn|Zn^{2+}(1M)||Ag^+(1M)|Ag$

(2) $Zn|Zn^{2+}(1M)||Mg^{2+}(1M)|Mg$

(3) $Ag|Ag^+(1M)||Mg^{2+}(1M)|Mg$

(4) $Cu|Cu^{2+}(1M)||Ag^+(1M)|Ag$

13. $FeO_4^{2-} \xrightarrow{+2.0V} Fe^{3+} \xrightarrow{0.8V} Fe^{2+} \xrightarrow{-0.5V} Fe^0$

In the above diagram, the standard electrode potentials are given in volts (over the arrow).

The value of $E^\ominus_{FeO_4^{2-}/Fe^{2+}}$ is **[Jan. 23, 2025 (I)]**

(1) 1.7V (2) 1.2V (3) 2.1V (4) 1.4V

14. The emf of cell $Tl\left|\begin{matrix}Tl^+\\(0.001M)\end{matrix}\right|\left|\begin{matrix}Cu^{2+}\\(0.01M)\end{matrix}\right|Cu$ is 0.83 V at 298 K. It could be increased by : **[April 8, 2024 (II)]**

(1) increasing concentration of Tl^+ ions

(2) increasing concentration of both Tl^+ and Cu^{2+} ions

(3) decreasing concentration of both Tl^+ and Cu^{2+} ions

(4) increasing concentration of Cu^{2+} ions

15. The reaction ; **[April 8, 2024 (II)]**

$\frac{1}{2}H_{2(g)} + AgCl_{(s)} \rightarrow H^+_{(aq)} + Cl^-_{(aq)} + Ag_{(s)}$

Occurs in which of the following galvanic cell :

(1) $Pt|H_{2(g)}|HCl_{(soln.)}|AgCl_{(s)}|Ag$

(2) $Pt|H_{2(g)}|HCl_{(soln.)}|AgNO_{3(aq)}|Ag$

(3) $Pt|H_{2(g)}|KCl_{(soln.)}|AgCl_{(s)}|Ag$

(4) $Ag|AgCl_{(s)}|KCl_{(soln.)}|AgNO_{3(aq.)}|Ag$

16. For the electro chemical cell

$M|M^{2+}||X|X^{2-}$ **[April 5, 2024 (II)]**

If $E^0_{(M^{2+}/M)} = 0.46V$ and $E^0_{(X/X^{2-})} = 0.34V.$

Which of the following is **correct** ?

(1) $E_{cell} = -0.80\text{ V}$

(2) $M + X \rightarrow M^{2+} + X^{2-}$ is a spontaneous reaction

(3) $M^{2+} + X^{2-} \rightarrow M + X$ is a spontaneous reaction

(4) $E_{cell} = 0.80\text{ V}$

17. What pressure (bar) of H_2 would be required to make emf of hydrogen electrode zero in pure water at 25°C ? **[April 4, 2024 (I)]**

(1) 10^{-14} (2) 10^{-7} (3) 1 (4) 0.5

18. One of the commonly used electrode is calomel electrode. Under which of the following categories calomel electrode comes ? **[April 4, 2024 (I)]**

(1) Metal – Insoluble Salt – Anion electrodes

(2) Oxidation – Reduction electrodes

(3) Gas – Ion electrodes
(4) Metal ion – Metal electrodes

19. The reaction [April 08, 2023 (I)]

$$\frac{1}{2}H_2(g) + AgCl(s) \rightleftharpoons H^+(aq) + Cl^-(aq) + Ag(s)$$

occurs in which of the given galvanic cell.
(1) $Pt|H_2(g)|KCl(sol^n)|AgCl(s)|Ag$
(2) $Pt|H_2(g)|HCl(sol^n)|AgCl(s)|Ag$
(3) $Ag|AgCl(s)|KCl(sol^n)|AgCl(s)|Ag$
(4) $Pt|H_2(g)|HCl(sol^n)|AgNO_3(sol^n)|Ag$

20. The standard electrode potential (M^{3+}/M^{2+}) for V, Cr, Mn & Co are –0.26 V, –0.41 V, + 1.57 V and +1.97 V, respectively. The metal ions which can liberate H_2 from a dilute acid are [Jan. 29, 2023 (I)]
(1) V^{2+} and Mn^{2+} (2) Cr^{2+} and CO^{2+}
(3) V^{2+} and Cr^{2+} (4) Mn^{2+} and Co^{2+}

21. In which of the following half cells, electrochemical reaction is pH dependent? [June 30, 2022 (I)]
(1) $Pt | Fe^{3+}, Fe^{2+}$ (2) $MnO_4^- | Mn^{2+}$
(3) $Ag | AgCl | Cl^-$ (4) $\frac{1}{2}F_2 | F^-$

22. The $\left(\frac{\partial E}{\partial T}\right)_P$ of different types of half cells are as follows:

A	B	C	D
1×10^{-4}	2×10^{-4}	0.1×10^{-4}	0.2×10^{-4}

(Where E is the electromotive force)
Which of the above half cells would be preferred to be used as reference electrode ? [June 26, 2022 (I)]
(1) A (2) B (3) C (4) D

23. The correct order of reduction potentials of the following pairs is
A. Cl_2/Cl^- B. I_2/I^- C. Ag^+/Ag D. Na^+/Na E. Li^+/Li
Choose the correct answer from the options given below. [June 25, 2022 (II)]
(1) A > C > B > D > E (2) A > B > C > D > E
(3) A > C > B > E > D (4) A > B > C > E > D

24.
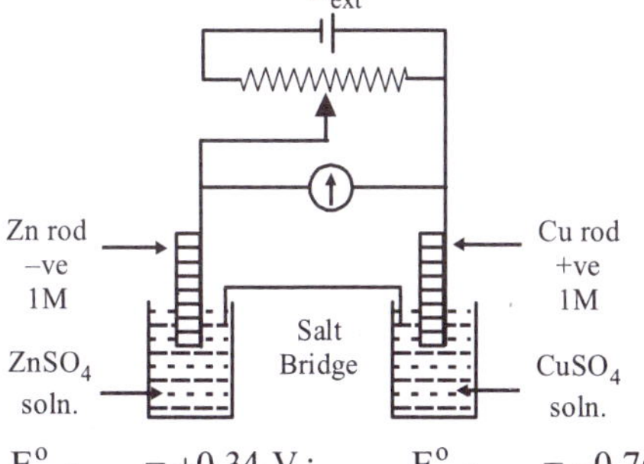

$E^o_{Cu^{2+}|Cu} = +0.34$ V; $E^o_{Zn^{2+}|Zn} = -0.76$ V

Identify the **incorrect** statement from the options below for the above cell : [Sep. 04, 2020 (I)]
(1) If $E_{ext} > 1.1$ V, e^- flows from Cu to Zn
(2) If $E_{ext} > 1.1$ V, Zn dissolves at Zn electrode and Cu deposits at Cu electrode
(3) If $E_{ext} < 1.1$ V, Zn dissolves at anode and Cu deposits at cathode
(4) If $E_{ext} = 1.1$ V, no flow of e^- or current occurs

25. For the given cell; [Sep. 06, 2020 (II)]
$Cu(s)|Cu^{2+}(C_1M)||Cu^{2+}(C_2M)|Cu(s)$
change in Gibbs energy (ΔG) is negative, if :
(1) $C_1 = C_2$ (2) $C_2 = \frac{C_1}{\sqrt{2}}$
(3) $C_1 = 2C_2$ (4) $C_2 = \sqrt{2}C_1$

26. Given that the standard potentials (E^0) of Cu^{2+}/Cu and Cu^+/Cu are 0.34 V and 0.522 V respectively, the E^0 of Cu^{2+}/Cu^+ is: [Jan. 07, 2020 (I)]
(1) +0.182 V (2) +0.158 V
(3) –0.182 V (4) –0.158 V

27. The standard Gibbs energy for the given cell reaction in kJ mol^{-1} at 298 K is:
$Zn(s) + Cu^{2+}(aq) \rightarrow Zn^{2+}(aq) + Cu(s)$,
$E^\circ = 2$ V at 298 K
(Faraday's constant, F = 96000 C mol^{-1}) [April 9, 2019 (I)]
(1) –384 (2) 384 (3) 192 (4) –192

28. Given : [April 12, 2019 (I)]
$Co^{3+} + e^- \rightarrow Co^{2+}$; $E^o = +1.81$ V
$Pb^{4+} + 2e^- \rightarrow Pb^{2+}$; $E^o = +1.67$ V
$Ce^{4+} + e^- \rightarrow Ce^{3+}$; $E^o = +1.61$ V
$Bi^{3+} + 3e^- \rightarrow Bi$; $E^o = +0.20$ V
Oxidizing power of the species will increase in the order:
(1) $Ce^{4+} < Pb^{4+} < Bi^{3+} < Co^{3+}$ (2) $Bi^{3+} < Ce^{4+} < Pb^{4+} < Co^{3+}$
(3) $Co^{3+} < Ce^{4+} < Bi^{3+} < Pb^{4+}$ (4) $Co^{3+} < Pb^{4+} < Ce^{4+} < Bi^{3+}$

29. Calculate the standard cell potential (in V) of the cell in which following reaction takes place :
$Fe^{2+}(aq) + Ag^+(aq) \rightarrow Fe^{3+}(aq) + Ag(s)$
Given that, [April 8, 2019 (II)]
$E^0_{Ag^+/Ag} = x$ V; $E^0_{Fe^{2+}/Fe} = y$ V; $E^0_{Fe^{3+}/Fe} = z$ V
(1) $x - z$ (2) $x - y$
(3) $x + 2y - 3z$ (4) $x + y - z$

30. Given that $E^\circ_{O_2/H_2O} = +1.23$ V;
$E^\circ_{S_2O_8^{2-}/SO_4^{2-}} = 2.05$ V; $E^\circ_{Br_2/Br^-} = +1.09$ V
$E^\circ_{Au^{3+}/Au} = +1.4$ V
The strongest oxidising agent is : [April 8, 2019 (I)]
(1) Au^{3+} (2) O_2 (3) $S_2O_8^{2-}$ (4) Br_2

31. The standard electrode potential E^0 and its temperature coefficient $\left(\frac{dE^0}{dT}\right)$ for a cell are 2 V and -5×10^{-4} VK^{-1} at 300 K respectively. The cell reaction is:
$Zn(s) + Cu^{2+}(aq) \rightarrow Zn^{2+}(aq) + Cu$
The standard reaction enthalpy ($\Delta_r H^0$) at 300 K in kJ mol^{-1} is,
[Use R = 8 $JK^{-1}mol^{-1}$ and F = 96,000 C mol^{-1}] [Jan. 12, 2019 (I)]
(1) –412.8 (2) –384.0 (3) 192.0 (4) 206.4

32. Given the equilibrium constant K_C of the reaction :
$Cu(s) + 2Ag^+(aq) \rightarrow Cu^{2+}(aq) + 2Ag(s)$ is 10×10^{15}. Calculate the E^0_{cell} of this reaction at 298 K.
$\left[2.303\frac{RT}{F} \text{ at } 298 \text{ K} = 0.059 \text{ V}\right]$ [Jan. 11, 2019 (II)]
(1) 0.04736 mV (2) 0.4736 mV
(3) 0.4736 V (4) 0.04736 V

33. For the cell $Zn(s) | Zn^{2+}(aq) || M^{x+}(aq) | M(s)$, different half cells and their standard electrode potentials are given below:

$M^{x+}(aq)/M(s)$	$Au^{3+}(aq)/Au(s)$	$Ag^+(aq)/Ag(s)$	$Fe^{3+}(aq)/Fe^{2+}(aq)$	$Fe^{2+}(aq)/Fe(s)$
$E^o_{M^{x+}/M}/(V)$	1.40	0.80	0.77	–0.44

If $E^o_{Zn^{2+}/Zn} = -0.76$ V, which cathode will give a maximum value of E°_{cell} per electron transferred? [Jan. 11, 2019 (I)]

(1) Ag^+/Ag (2) Fe^{3+}/Fe^{2+}
(3) Au^{3+}/Au (4) Fe^{2+}/Fe

34. In the cell, $Pt(s) | H_2(g, 1 bar)/HCl(aq)||AgCl(s)/Ag(s)|Pt(s)$, the cell potential is 0.92 V when a 10^{-6} molal HCl solution is used. The standard electrode potential of $(AgCl/Ag, Cl^-)$ electrode is: [Jan. 10, 2019 (II)]

$$\left\{\text{Given}: \frac{2.303RT}{F} = 0.06 \text{ V at } 298 \text{ K}\right\}$$

(1) 0.94 V (2) 0.76 V (3) 0.40 V (4) 0.20 V

35. Consider the following reduction processes:

$Zn^{2+} + 2e^- \rightarrow Zn(s); E^\circ = -0.76$ V
$Ca^{2+} + 2e^- \rightarrow Ca(s); E^\circ = -2.87$ V
$Mg^{2+} + 2e^- \rightarrow Mg(s); E^\circ = -2.36$ V
$Ni^{2+} + 2e^- \rightarrow Ni(s); E^\circ = -0.25$ V

The reducing power of the metals increases in the order: [Jan. 10, 2019 (I)]

(1) $Ca < Zn < Mg < Ni$ (2) $Ni < Zn < Mg < Ca$
(3) $Zn < Mg < Ni < Ca$ (4) $Ca < Mg < Zn < Ni$

36. If the standard electrode potential for a cell is 2 V at 300 K, the equilibrium constant (K) for the reaction

$$Zn(s) + Cu^{2+}(aq) \rightleftharpoons Zn^{2+}(aq) + Cu(s)$$

at 300 K is approximately
($R = 8\ JK^{-1}\ mol^{-1}$, $F = 96000\ C\ mol^{-1}$) [Jan. 9, 2019 (II)]

(1) e^{-80} (2) e^{-160} (3) e^{320} (4) e^{160}

37. To find the standard potential of M^{3+}/M electrode, the following cell is constituted: $Pt/M|M^{3+}(0.001\ mol\ L^{-1})||Ag^+(0.01\ mol\ L^{-1})|Ag$
The emf of the cell is found to be 0.421 volt at 298 K. The standard potential of half reaction $M^{3+} + 3e^- \rightarrow M$ at 298 K will be: (Given $E^\circ_{Ag^+/Ag}$ at 298 K = 0.80 Volt) [Online April 9, 2017]

(1) 0.38 Volt (2) 0.32 Volt
(3) 1.28 Volt (4) 0.66 Volt

38. Given,

$E^o_{Cl_2/Cl^-} = 1.36$ V, $E^o_{Cr^{3+}/Cr} = -0.74$ V,

$E^o_{Cr_2O_7^{2-}/Cr^{3+}} = 1.33$ V, $E^o_{MnO_4^-/Mn^{2+}} = 1.51$ V.

Among the following, the strongest reducing agent is [2017, Same 23, April, 2013]

(1) Cr (2) Mn^{2+} (3) Cr^{3+} (4) Cl^-

39. Consider the following standard electrode potentials (E° in volts) in aqueous solution : [Online April 8, 2017]

Element	M^{3+}/M	M^+/M
Al	–1.66	+0.55
Tl	+1.26	–0.34

Based on these data, which of the following statements is correct?

(1) Tl^+ is more stable than Al^{3+}
(2) Al^+ is more stable than Al^{3+}
(3) Tl^+ is more stable than Al^+
(4) Tl^{3+} is more stable than Al^{3+}

40. What is the standard reduction potential (E°) for $Fe^{3+} \rightarrow Fe$? Given that : [Online April 8, 2017]

$Fe^{2+} + 2e^- \rightarrow Fe;\ E^o_{Fe^{2+}/Fe} = -0.47$V

$Fe^{3+} + e^- \rightarrow Fe^{2+};\ E^o_{Fe^{3+}/Fe^{2+}} = +0.77$V

(1) –0.057 V (2) +0.057 V
(3) +0.30 V (4) –0.30 V

41. What will occur if a block of copper metal is dropped into a beaker containing a solution of 1M $ZnSO_4$? [Online April 9, 2016]

(1) The copper metal will dissolve with evolution of oxygen gas
(2) The copper metal will dissolve with evolution of hydrogen gas
(3) No reaction will occur
(4) The copper metal will dissolve and zinc metal will be deposited.

42. At 298 K, the standard reduction potentials are 1.51 V for $MnO_4^- | Mn^{2+}$, 1.36 V for $Cl_2|Cl^-$, 1.07 V for $Br_2|Br^-$, and 0.54 V for $I_2|I^-$. At pH = 3, permanganate is expected to oxidize: $\left(\frac{RT}{F} = 0.059 \text{ V}\right)$ [Online April 11, 2015]

(1) Cl^-, Br^- and I^- (2) Br^- and I^-
(3) Cl^- and Br^- (4) I^- only

43. A variable, opposite external potential (E_{ext}) is applied to the cell $Zn|Zn^{2+}$ (1 M) || Cu^{2+} (1 M) | Cu, of potential 1.1 V. When $E_{ext} < 1.1$ V and $E_{ext} > 1.1$ V, respectively electrons flow from : [Online April 10, 2015]

(1) anode to cathode in both cases
(2) cathode to anode and anode to cathode
(3) anode to cathode and cathode to anode
(4) cathode to anode in both cases

44. Given

$Fe^{3+}(aq) + e^- \rightarrow Fe^{2+}(aq); E^\circ = +0.77$ V
$Al^{3+}(aq) + 3e^- \rightarrow Al(s); E^\circ = -1.66$ V
$Br_2(aq) + 2e^- \rightarrow 2Br^-; E^\circ = +1.09$ V

Considering the electrode potentials, which of the following represents the correct order of reducing power? [Online April 11, 2014]

(1) $Fe^{2+} < Al < Br^-$ (2) $Br^- < Fe^{2+} < Al$
(3) $Al < Br^- < Fe^{2+}$ (4) $Al < Fe^{2+} < Br^-$

45. The standard electrode potentials $\left(E^\circ_{M^+/M}\right)$ of four metals A, B, C and D are –1.2 V, 0.6 V, 0.85 V and –0.76 V, respectively. The sequence of deposition of metals on applying potential is: [Online April 9, 2014]

(1) A, C, B, D (2) B, D, C, A
(3) C, B, D, A (4) D, A, B, C

46. Given below are the half-cell reactions:

$$Mn^{2+} + 2e^- \rightarrow Mn;\ E^\circ = -1.18\text{V}$$

$$2\left(Mn^{3+} + e^- \rightarrow Mn^{2+}\right);\ E^\circ = +1.51\text{V}$$

The E° for $3Mn^{2+} \rightarrow Mn + 2Mn^{3+}$ will be: [2014]

(1) –2.69 V; the reaction will not occur
(2) –2.69 V; the reaction will occur
(3) –0.33 V; the reaction will not occur
(4) –0.33 V; the reaction will occur

47. Electrode potentials (E°) are given below :
$Cu^+/Cu = +0.52$ V,
$Fe^{3+}/Fe^{2+} = +0.77$ V,
$\frac{1}{2}I_2(s)/I^- = +0.54$ V,
$Ag^+/Ag = +0.88$V.
Based on the above potentials, strongest oxidizing agent will be : [Online April 9, 2013]
(1) Cu^+ (2) Fe^{3+} (3) Ag^+ (4) I_2

48. Given : $E^\circ_{Cr^{3+}/Cr} = -0.74$ V; $E^\circ_{MnO_4^-/Mn^{2+}} = 1.51$ V
$E^\circ_{Cr_2O_7^{2-}/Cr^{3+}} = 1.33$ V; $E^\circ_{Cl/Cl^-} = 1.36$ V
Based on the data given above, strongest oxidising agent will be : [2013]
(1) Cl (2) Cr^{3+} (3) Mn^{2+} (4) MnO_4^-

49. The standard reduction potentials for Zn^{2+}/Zn, Ni^{2+}/Ni and Fe^{2+}/Fe are –0.76, –0.23 and –0.44 V respectively.
The reaction $X + Y^{2+} \longrightarrow X^{2+} + Y$ will be spontaneous when: [2012]
(1) X = Ni, Y = Fe (2) X = Ni, Y = Zn
(3) X = Fe, Y = Zn (4) X = Zn, Y = Ni

50. Given
$E^\circ_{Cu^{2+}/Cu} = 0.34$V, $E^\circ_{Cu^{2+}/Cu^+} = 0.15$V
Standard electrode potential for the half cell Cu^+/Cu is [Online May 26, 2012]
(1) 0.38 V (2) 0.53 V (3) 0.19 V (4) 0.49 V

51. The Gibbs energy for the decomposition of Al_2O_3 at 500 °C is as follows :
$\frac{2}{3}Al_2O_3 \rightarrow \frac{4}{3}Al + O_2, \Delta_r G = +966\ kJ\ mol^{-1}$
The potential difference needed for electrolytic reduction of Al_2O_3 at 500 °C is at least [2010]
(1) 4.5 V (2) 3.0 V (3) 2.5 V (4) 5.0 V

52. Given: [2009]
$E^\circ_{Fe^{3+}/Fe} = -0.036$V, $E^\circ_{Fe^{2+}/Fe} = -0.439$V
The value of standard electrode potential for the change,
$Fe^{3+}(aq) + e^- \longrightarrow Fe^{2+}(aq)$ will be:
(1) 0.385 V (2) 0.770 V
(3) –0.270 V (4) –0.072 V

53. Given $E^\circ_{Cr^{3+}/Cr} = -0.72$ V,
$E^\circ_{Fe^{2+}/Fe} = -0.42$ V. The potential for the cell
$Cr|Cr^{3+}(0.1M)||Fe^{2+}(0.01\ M)|Fe$ is [2003 Same 2008]
(1) –0.26 V (2) 0.336 V (3) –0.339 V (4) 0.26 V

54. The cell, $Zn\,|\,Zn^{2+}(1\ M)\,||\,Cu^{2+}(1\ M)\,|\,Cu$ ($E^\circ_{cell} = 1.10$ V) was allowed to be completely discharged at 298 K. The relative concentration of Zn^{2+} to Cu^{2+} $\left(\frac{[Zn^{2+}]}{[Cu^{2+}]}\right)$ is [2007]
(1) 9.65×10^4 (2) antilog (24.08)
(3) 37.3 (4) $10^{37.3}$.

55. For a spontaneous reaction, the ΔG, equilibrium constant (K) and E°_{cell} will be respectively [2005]
(1) –ve, >1, –ve (2) –ve, <1, –ve
(3) +ve, >1, –ve (4) –ve, >1, +ve

56. The standard e.m.f. of a cell involving one electron change is found to be 0.591 V at 25 °C. The equilibrium constant of the reaction is ($F = 96{,}500\ C\ mol^{-1}$; $R = 8.314\ JK^{-1}\ mol^{-1}$) [2004]
(1) 1.0×10^{10} (2) 1.0×10^5
(3) 1.0×10^1 (4) 1.0×10^{30}

57. In a cell that utilises the reaction,
$Zn(s) + 2H^+(aq) \rightarrow Zn^{2+}(aq) + H_2(g)$, addition of H_2SO_4 to cathode compartment, will [2004]
(1) increase the E and shift equilibrium to the right
(2) lower the E and shift equilibrium to the right
(3) lower the E and shift equlibrium to the left
(4) increase the E and shift equilibrium to the left

58. Consider the following E° values
$E^\circ_{Fe^{3+}/Fe^{2+}} = +0.77$V; $E^\circ_{Sn^{2+}/Sn} = -0.14$V
Under standard conditions the potential for the reaction
$Sn(s) + 2Fe^{3+}(aq) \rightarrow 2Fe^{2+}(aq) + Sn^{2+}(aq)$ is [2004]
(1) 0.91 V (2) 1.40 V (3) 1.68 V (4) 0.63 V

59. The $E^\circ_{M^{3+}/M^{2+}}$ values for Cr, Mn, Fe and Co are – 0.41, + 1.57, + 0.77 and + 1.97V respectively. For which one of these metals the change in oxidation state from +2 to +3 is easiest?
(1) Fe (2) Mn (3) Cr (4) Co [2004]

60. Standard reduction electrode potentials of three metals A, B & C are respectively + 0.5 V, – 3.0 V & –1.2 V. The reducing powers of these metals are [2003]
(1) A > B > C (2) C > B > A
(3) A > C > B (4) B > C > A

61. For a cell reaction involving a two-electron change, the standard e.m.f. of the cell is found to be 0.295 V at 25 °C. The equilibrium constant of the reaction at 25°C will be [2003]
(1) 29.5×10^{-2} (2) 10
(3) 1×10^{10} (4) 1×10^{-10}

62. For the redox reaction : [2003]
$Zn(s) + Cu^{2+}(0.1M) \rightarrow Zn^{2+}(1M) + Cu(s)$
taking place in a cell, E°_{cell} is 1.10 volt. E_{cell} for the cell will be $\left(2.303\frac{RT}{F} = 0.0591\right)$
(1) 1.80 volt (2) 1.07 volt
(3) 0.82 volt (4) 2.14 volt

63. What will be the emf for the given cell ?
$Pt\,|\,H_2(P_1)\,|\,H^+(aq)\,||\,H_2(P_2)\,|\,Pt$ [2002]
(1) $\frac{RT}{F}\log_e\frac{P_1}{P_2}$ (2) $\frac{RT}{2F}\log_e\frac{P_2}{P_1}$
(3) $\frac{RT}{F}\log_e\frac{P_2}{P_1}$ (4) None of these

64. Which of the following reaction is possible at anode? [2002]
(1) $2Cr^{3+} + 7H_2O \rightarrow Cr_2O_7^{2-} + 14H^+$
(2) $F_2 \rightarrow 2F^-$
(3) $(1/2)O_2 + 2H^+ \rightarrow H_2O$
(4) none of these.

NUMERICAL VALUE QUESTIONS (NVQs)

65. An electrochemical cell. consist of the following two redox couples, **[April 04, 2026 (II)]**

$M^{x+}(aq)/M(s)\ [E^{\ominus}_{red} = +0.15V]$ and

$Fe^{3+}(aq)/Fe(s)\ [E^{\ominus}_{red} = -0.036V]$

The cell EMF (E_{cell}) is recorded to be 0.2057 V.

If the reaction quotient of the electrochemical reaction is found to be 10^{-2}, then the value of x is _______ (Nearest integer)

[Given : M is a p-block metal and $\frac{2.303RT}{F} = 0.059V$]

66. A volume of x mL of 5 M $NaHCO_3$ solution was mixed with 10 mL of 2 M H_2CO_3 solution to make an electrolytic buffer. If the same buffer was used in the following electrochemical cell to record a cell potential of 235.3 mV, then the value of x = _________ mL (nearest integer).

$Sn(s)\,|\,Sn(OH)_6^{2-}\ (0.5\ M)\,|\,HSnO_2^-\ (0.05\ M)\,|\,OH^-\,|\,Bi_2O_3(s)\,|\,Bi(s)$

Consider upto one place of decimal for intermediate calculations **[Jan. 28, 2026 (II)]**

Given : $E^{\circ}_{HSnO_2^-|Sn(OH)_6^{2-}} = -0.9V$

$E^{\circ}_{Bi_2O_3|Bi} = -0.44V$

$pKa_{(H_2CO_3)} = 6.11$

$\frac{2.303RT}{F} = 0.059V$

Anti log (1.29) = 19.5

67. Consider the following redox reaction taking place in acidic medium **[Jan. 28, 2026 (I)]**

$BH_4^-(aq) + ClO_3^-(aq) \rightarrow H_2BO_3^-(aq) + Cl^-(aq)$

If the Nernst equation for the above balanced reaction is

$$E_{cell} = E^{\circ}_{cell} - \frac{RT}{nF}\ell n\, Q,$$

Then the value of n is _________. (Nearest integer)

68. Consider the following electrochemical cell : $Pt\,|\,O_2(g)\,(1bar)\,|\,HCl(aq)\,||\,M^{2+}(aq, 1.0\ M)\,|\,M(s)$ The pH above which, oxygen gas would start to evolve at anode is _______ (nearest integer).

[Given $E^0_{M^{2+}/M} = 0.994V$, $E^0_{O_2/H_2O} = 1.23V$ standard reduction potential and $\frac{RT}{F}(2.303) = 0.059$ V at the given condition] **[Jan. 22, 2026 (II)]**

69. Consider the following electrochemical cell at 298K

$$Pt\,|\,HSnO_2^-(aq)\,|\,Sn(OH)_6^{2-}(aq)\,|\,Bi_2O_3(s)\,|\,Bi(s)$$

If the reaction quotient at a given time is 10^6, then the cell EMF (E_{cell}) is ________ $\times 10^{-1}$ V (Nearest integer).

Given the standard half-cell reduction potential as

$E^0_{Bi_2O_3/Bi,\,OH^-} = -0.44V$

$E^0_{Sn(OH)_6^{2-}/HSnO_2^-,\,OH} = -0.90V$ **[Jan. 22, 2026 (I)]**

70. MX is a sparingly soluble salt that follows the given solubility equilibrium at 298 K

$MX(s) \rightleftharpoons M^+(aq)\,X^-(aq);\ K_{sp} = 10^{-10}$

If the standard reduction potential for

$M^+(aq) \xrightarrow{+e^-} M(s)$ is $(E^{\ominus}_{M^+/M}) = 0.79V$, then the value of the standard reduction potential for the metal/metal insoluble salt electrode $E^{\ominus}_{X^-/MX(s)/M}$ is _________ mV. (nearest integer)

[Given : $\frac{2.303RT}{F} = 0.059V$] **[Jan. 21, 2026 (II)]**

71. Consider the following half cell reaction

$Cr_2O_7^{2-}(aq) + 6e^- + 14H^+(aq) \rightarrow 2Cr^{3+}(aq) + 7H_2O(l)$

The reaction was conducted with the ratio of $\frac{[Cr^{3+}]^2}{[Cr_2O_7^{2-}]} = 10^{-6}$.

The pH value at which the EMF of the half cell will become zero is _________ .

(nearest integer value)

[Given : standard half cell reduction potential

$E^0_{Cr_2O_7^{2-},H^+/Cr^{3+}} = 1.33V, \frac{2.303RT}{F} = 0.059V$] **[April 8, 2025 (II)]**

72. 1 Faraday electricity was passed through Cu^{2+} (1.5 M, 1 L)/Cu and 0.1 Faraday was passed through Ag^+ (0.2 M, 1 L)/Ag electrolytic cells. After this the two cells were connected as shown below to make an electrochemical cell. The emf of the cell thus formed at 298 K is- **[April 7, 2025 (I)]**

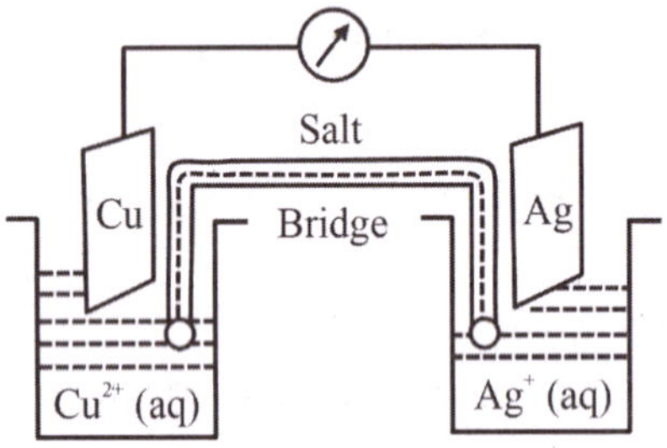

Given : $E^0_{Cu^{2+}/Cu} = 0.34V$

$E^0_{Ag^+/Ag} = 0.8V$

$\frac{2.303RT}{F} = 0.06V$

73. Consider the following electrochemical cell at standard condition.

$Au(s)|QH_2,Q|NH_4X(0.01M)||Ag^+(1M)|Ag(s)$

$E_{cell} = +0.4V$

The couple QH_2/Q represents quinhydrone electrode, the half cell reaction is given below **[April 2, 2025 (I)]**

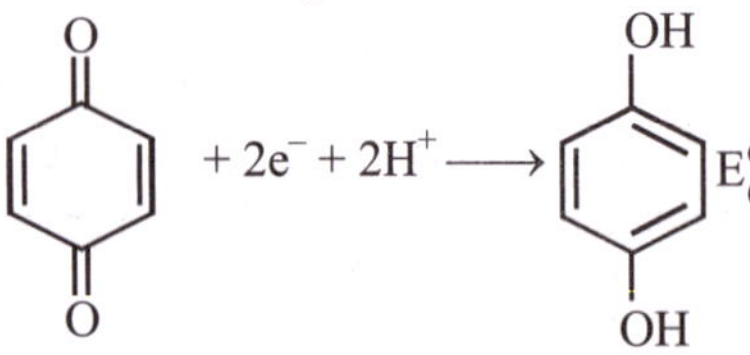

$E^0_{Q/QH_2} = +0.7V$

(Q) (QH$_2$)

[Given : $E^{\circ}_{Ag^+/Ag} = +0.8V$ and $\frac{2.303RT}{F} = 0.06V$]

The pK_b value of the ammonium halide salt (NH_4X) used here is ______. (nearest integer)

74. The standard reduction potentials at 298 K for the following half cells are given below : [April 9, 2024 (I)]

$Cr_2O_7^{2-} + 14H^+ + 6e^- \rightarrow 2Cr^{3+} + 7H_2O, E° = 1.33V$

$Fe^{3+}(aq) + 3e^- \rightarrow Fe \quad E° = -0.04V$

$Ni^{2+}(aq) + 2e^- \rightarrow Ni \quad E° = -0.25V$

$Ag^+(aq) + e^- \rightarrow Ag \quad E° = 0.80V$

$Au^{3+}(aq) + 3e^- \rightarrow Au \quad E° = 1.40V$

Consider the given electrochemical reactions,

The number of metal(s) which will be oxidized be $Cr_2O_7^{2-}$, in aqueous solution is __________.

75. Consider the following redox reaction:

$$MnO_4^- + H^+ + H_2C_2O_4 \rightleftharpoons Mn^{2+} + H_2O + CO_2$$

The standard reduction potentials are given as below ($E°_{red}$):

$E°_{MnO_4^-/Mn^{2+}} = +1.51V$

$E°_{CO_2/H_2C_2O_4} = -0.49\ V$

If the equilibrium constant of the above reaction is given as $K_{eq} = 10^x$, then the value of x = _________ (nearest integer) [Feb. 1, 2024 (II)]

76. The potential for the given half cell at 298 K is (–) ______ × 10^{-2} V. [Feb. 1, 2024 (I)]

$$2H^+(aq) + 2e^- \longrightarrow H_2(g)$$

$$[H^+] = 1M, P_{H_2} = 2\ atm$$

(Given : 2.303RT/F = 0.06 V, log2 = 0.3)

77. The hydrogen electrode is dipped in a solution of pH = 3 at 25°C. The potential of the electrode will be –______ × 10^{-2} V.

$\left(\frac{2.303RT}{F} = 0.059V\right)$ [Jan. 27, 2024 (II)]

78. At 298 K, the standard reduction potential for Cu^{2+}/Cu electrode is 0.34 V.

Given : $K_{sp}\ Cu(OH)_2 = 1 \times 10^{-20}$

Take $\frac{2.303RT}{F} = 0.059V$

The reduction potential at pH = 14 for the above couple is (–)x × 10^{-2} V. The value of x is ______. [April 13, 2023 (II)]

79. The number of correct statements from the following is : [April 11, 2023 (II)]

A. E_{cell} is an intensive parameter.

B. A negative $E^{\ominus}$ means that the redox couple is a stronger reducing agent than the H^+/H_2 couple.

C. The amount of electricity required for oxidation or reduction depends on the stoichiometry of the electrode reaction.

D. The amount of chemical reaction which occurs at any electrode during electrolysis by a current is proportional to the quantity of electricity passed through the electrolyte.

80. In an electrochemical reaction of lead, at standard temperature, if $E^0_{(Pb^{2+}/Pb)}$ = m Volt and $E^0_{(Pb^{4+}/Pb)}$ = n Volt, then the value of $E^0_{(Pb^{2+}/Pb^{4+})}$ is given by m – xn. The value of x is _______.

(Nearest integer) [April 11, 2023 (I)]

81. $FeO_4^{2-} \xrightarrow{+2.2V} Fe^{3+} \xrightarrow{+0.70V} Fe^{2+} \xrightarrow{-0.45V} Fe^0$

$E^{\ominus}_{FeO_4^{2-}/Fe^{2+}}$ is x × 10^{-3} V. The value of x is ______ [April 10, 2023 (I)]

82. The number of incorrect statements from the following is _____ [April 08, 2023 (II)]

(A) The electrical work that a reaction can perform at constant pressure and temperature is equal to the reaction Gibbs energy.

(B) E^0_{cell} is dependent on the pressure

(C) $\frac{dE^0_{cell}}{dT} = \frac{\Delta_r S^0}{nF}$

(D) A cell is operating reversibly if the cell potential is exactly balanced by an opposing source of potential difference.

83. The standard reduction potential at 298 K for the following half cells are given below:- [April 06, 2023 (II)]

$NO_3^- + 4H^+ + 3e^- \rightarrow NO(g) + 2H_2O \quad E^0 = 0.97V$

$V^{2+}(aq) + 2e^- \rightarrow V \quad E^0 = -1.19V$

$Fe^{3+}(aq) + 3e^- \rightarrow Fe \quad E^0 = -0.04V$

$Ag^+(aq) + e^- \rightarrow Ag(s) \quad E^0 = 0.80V$

$Au^{3+}(aq) + 3e^- \rightarrow Au(s) \quad E^0 = 1.40V$

The number of metal(s) which will be oxidized by NO_3^- is aqueous solution is ________.

84. At what pH, given half cell MnO_4^- (0.1 M) | Mn^{2+} (0.001 M) will have electrode potential of 1.282 V? _____ (Nearest Integer) [Feb. 01, 2023 (I)]

Given $E^0_{MnO_4^-|Mn^{2+}} = 1.54V, \frac{2.303RT}{F} = 0.059V$

85. The logarithm of equilibrium constant for the reaction $Pd^{2+} + 4Cl^- \rightleftharpoons PdCl_4^{2-}$ is _______ (Nearest integer)

Given : $\frac{2.303RT}{F} = 0.06V$

$Pd^{2+}(aq) + 2e^- \rightleftharpoons Pd(s) \quad E° = 0.83V$

$PdCl_4^{2-}(aq) + 2e^- \rightleftharpoons Pd(s) + 4Cl^-(aq)$

$E° = 0.65\ V$ [Jan. 31, 2023 (I)]

86. Consider the cell [Jan. 30, 2023 (I)]

$$Pt(s)|H_2(g, 1\ atm)|H^+(aq, 1M)||Fe^{3+}(aq), Fe^{2+}(aq)|Pt(s)$$

When the potential of the cell is 0.712 V at 298 K, the ratio $[Fe^{2+}]/[Fe^{3+}]$ is _______.

(Nearest integer)

Given: $Fe^{3+} + e^- = Fe^{2+}, E°_{Fe^{3+}, Fe^{2+}pt} = 0.771$

$\frac{2.303\ RT}{F} = 0.06V$

87. The electrode potential of the following half cell at 298 K $X|X^{2+}(0.001\ M)||Y^{2+}(0.01\ M)|Y$ is _____ × 10^{-2} V (Nearest integer).

Given : $E^0_{X^{2+}|X} = -2.36V$; $E^0_{Y^{2+}|Y} = +0.36V$

$\frac{2.303RT}{F} = 0.06V$ [Jan. 30, 2023 (II)]

The equilibrium constant for the reaction $Zn(s) + Sn^{2+}(aq) \rightleftharpoons Zn^{2+}(aq) + Sn(s)$ is 1×10^{20} at 298 K. The magnitude of standard electrode potential of Sn/Sn^{2+} if $E^\circ_{Zn^{2+}/Zn} = -0.76$ V is ______ $\times 10^{-2}$ V. (Nearest integer)

Given: $\frac{2.303\ RT}{F} = 0.059V$

$Pt(s)\ H_2(g)\ (1bar)\ |H^+(aq)\ (1M)\ ||M^{3+}(aq), M^+(aq)|\ Pt(s)$

The E_{cell} for the given cell is 0.1115 V at 298 K

when $\frac{[M^+(aq)]}{[M^{3+}(aq)]} = 10^a$ The value of a is ______

Given: $E^\circ_{M^{3+}/M^+} = 0.2$ V $\quad \frac{2.303\ RT}{F} = 0.059$ V

Consider the cell

$Pt(s)\ |H_2(s)(1atm)\ |H^+(aq, [H^+] = 1)||\ Fe^{3+}(aq), Fe^{2+}(q)|\ Pt(s)$

Given: $E^\circ_{Fe^{3+}/Fe^{2+}} = 0.771$ V and $E^\circ_{H^+/\frac{1}{2}H_2} = 0$ V, T = 298 K

If the potential of the cell is 0.712 V the ratio of concentration of Fe^{2+} to Fe^{3+} is ______ (Nearest integer)

At 298 K, a 1 litre solution containing 10 mmol of $Cr_2O_7^{2-}$ and 100 mmol of Cr^{3+} shows a pH of 3.0.

Given: $Cr_2O_7^{2-} \rightarrow Cr^{3+}$; $E^0 = 1.330$ V and $\frac{2.303RT}{F} = 0.059V$

The potential for the half cell reaction is $x \times 10^{-3}$ *V*. The value of x is ______.

For a cell, $Cu(s)\ |Cu^{2+}(0.001M)\ ||\ Ag^+(0.01M)|\ Ag(s)$ the cell potential is found to be 0.43 V at 298 K. The magnitude of standard electrode potential for Cu^{2+}/Cu is ___ $\times 10^{-2}$ V.

$\left[\text{Given}: E^\ominus_{Ag^+/Ag} = 0.80 \text{ V and } \frac{2.303RT}{F} = 0.06 \text{ V}\right]$

The cell potential for $Zn|Zn^{2+}(aq)||\ Sn^{x+}|\ Sn$ is 0.801 V at 298 K. The reaction quotient for the above reaction is 10^{-2}. The number of electrons involved in the given electrochemical cell reaction is ______.

(Given: $E^\circ_{Zn^{2+}|Zn} = -0.763V$, $E^\circ_{Sn^{x+}|Sn} = +0.008$ V and $\frac{2.303RT}{F} = 0.06$ V)

The cell potential for the given cell at 298 K

$Pt\ |\ H_2(g, 1\ bar)\ |\ H^+(aq)\ ||\ Cu^{2+}(aq)\ |\ Cu(s)$ is 0.31V. The pH of the acidic solution is found to be 3, whereas the concentration of Cu^{2+} is 10^{-x} M. The value of x is ______.

(Given $E^\circ_{Cu^{2+}/Cu} = 0.34V$ and $\frac{2.303RT}{F} = 0.06V$)

For the given reactions

$Sn^{2+} + 2e^- \rightarrow Sn$

$Sn^{4+} + 4e^- \rightarrow Sn$

The electrode potentials are; $E^\circ_{Sn^{2+}/Sn} = -0.140$ V and $E^\circ_{Sn^{4+}/Sn} = 0.010$ V. The magnitude of standard electrode potential for Sn^{4+}/Sn^{2+} i.e. $E^\circ_{Sn^{4+}/Sn^{2+}}$ is ______ $\times 10^{-2}$ V. (Nearest integer)

For the reaction taking place in the cell:

$Pt(s)\ |\ H_2(g)\ |\ H^+(aq)\ ||\ Ag^+(aq)\ |\ Ag(s)$

$E^\circ_{Cell} = +0.5332$ V.

The value of $\Delta_f G^0$ is ______ kJ mol^{-1}.

(in nearest integer)

$Cu(s) + Sn^{2+}(0.001M) \rightarrow Cu^{2+}(0.01M) + Sn(s)$

The Gibbs free energy change for the above reaction at 298 K is $x \times 10^{-1}$ kJ mol^{-1};

The value of x is______. [nearest integer]

[Given $E^\circ_{Cu^{2+}/Cu} = 0.34V$; $E^\circ_{Sn^{2+}/Sn} = -0.14V$; $F = 96500C\ mol^{-1}$]

In a cell, the following reactions take place

$Fe^{2+} \rightarrow Fe^{3+} + e^-$ $E^\circ_{Fe^{3+}/Fe^{2+}} = 0.77$ V

$2I^- \rightarrow I_2 + 2e^-$ $E^\circ_{I_2/I^-} = 0.54$ V

The standard electrode potential for the spontaneous reaction in the cell is $x \times 10^{-2}$ V 298 K. The value of x is ______. (Nearest Integer)

The cell potential for the following cell

$Pt\ |H_2(g)\ |H^+(aq)\ ||Cu^{2+}(0.01M)\ |Cu(s)$

is 0.576 V at 298 K. The pH of the solution is ___.

(Nearest integer) (Given: $E^\circ Cu^{2+}/Cu = 0.34V$ and) $\frac{2.303RT}{F} = 0.06V$)

Consider the following cell reaction:

$$Cd(s) + Hg_2SO_4(s) + \frac{9}{5}H_2O(l) \rightleftharpoons CdSO_4.\frac{9}{5}H_2O(s) + 2Hg(l)$$

The value of E^0_{cell} is 4.315 V at 25°C. If $\Delta H^\circ = -825.2$ kJ mol^{-1}, the standard entropy change ΔS° in J K^{-1} is ______. (Nearest integer)

[Given: Faraday constant = 96487 C mol^{-1}]

For the galvanic cell,

$Zn(s) + Cu^{2+}(0.02\ M) \rightarrow Zn^{2+}(0.04\ M) + Cu(s)$,

E_{cell} = ______ $\times 10^{-2}$ V. (Nearest integer)

[Use: $E^0_{Cu/Cu^{2+}} = -0.34$ V, $E^0_{Zn/Zn^{2+}} = +0.76$ V, $\frac{2.303RT}{F} = 0.059$ V]

For the cell

$Cu(s)\ |\ Cu^{2+}(aq)\ (0.1M)\ ||\ Ag^+(aq)\ (0.01M)\ |\ Ag(s)$

the cell potential $E_1 = 0.3095$ V

For the cell

$Cu(s)\ |\ Cu^{2+}(aq)\ (0.01\ M)\ ||\ Ag^+(aq)\ (0.001\ M)\ |\ Ag(s)$

the cell potential = ______ $\times 10^{-2}$ V. (Round off the nearest integer).

[Use: $\frac{2.303\ RT}{F} = 0.059$]

Consider the cell at 25 °C

$Zn\ |\ Zn^{2+}(aq), (1\ M)\ ||\ Fe^{3+}(aq), Fe^{2+}(aq)\ |\ Pt(s)$

The fraction of total iron present as Fe^{3+} ion at the cell potential of 1.500 V is x $\times 10^{-2}$. The value of x is ______. (nearest integer)

$\left(\text{Given}: E^0_{Fe^{3+}/Fe^{2+}} = 0.77 \text{ V}, E^0_{Zn^{2+}/Zn} = -0.76 \text{ V}\right)$

Assume a cell with the following reaction

$Cu(s) + 2Ag^+(1 \times 10^{-3} M) \rightarrow Cu^{2+}(0.250 M) + 2Ag(s)$

$E^{\theta}_{cell} = 2.97$ V

E_{cell} for the above reaction is _______ V. (Nearest integer)

[Given : log 2.5 = 0.3979, T = 298 K]

Emf of the following cell at 298 K in V is $x \times 10^{-2}$.

$Zn|Zn^{2+}(0.1 M)||Ag^+(0.01 M)|Ag$

The value of x is ________. (Rounded off to the nearest integer)

[Given : $E^{\theta}_{Zn^{2+}/Zn} = -0.76V$;

$E^{\theta}_{Ag^+/Ag} = +0.80V$; $\frac{2.303RT}{F} = 0.059$]

Copper reduces NO_3^- into NO and NO_2 depending upon the concentration of HNO_3 in solution. (Assuming fixed $[Cu^{2+}]$ and $P_{NO} = P_{NO_2}$), the HNO_3 concentration at which the thermodynamic tendency for reduction of NO_3^- into NO and NO_2 by copper is same is 10^x M. The value of 2x is ______. (Rounded-off to the nearest integer)

[Given: $E^{\circ}_{Cu^{2+}/Cu} = 0.34$ V, $E^{\circ}_{NO_3^-/NO} = 0.96$ V,

$E^{\circ}_{NO_3^-/NO_2} = 0.79$ V and at 298 K, $\frac{RT}{F}(2 \cdot 303) = 0.059$]

The magnitude of the change in oxidising power of the MnO_4^- / Mn^{2+} couple is $x \times 10^{-4}$ V, if the H^+ concentration is decreased from 1 M to 10^{-4} M at 25 °C. (Assume concentration of MnO_4^- and Mn^{2+} to be same on change in H^+ concentration). The value of x is ________. (Rounded off to the nearest integer)

$$\left[\text{Given}: \frac{2.303RT}{F} = 0.059\right]$$

An oxidation-reduction reaction in which 3 electrons are transferred has a ΔG^0 of 17.37 kJ mol^{-1} at 25°C. The value of E^0_{cell} (in V) is ______ $\times 10^{-2}$. (1 F = 96,500 C mol^{-1})

The photoelectric current from Na (work function, w_0 = 2.3 eV) is stopped by the output voltage of the cell $Pt(s)|H_2(g, 1 bar)|HCl(aq. pH = 1)|AgCl(s)|Ag(s)$.

The pH of aq. HCl required to stop the photoelectric current from K(w_0 = 2.25 eV), all other conditions remaining the same, is __________ $\times 10^{-2}$ (to the nearest integer).

Given,

$2.303\frac{RT}{F} = 0.06$ V; $E^0_{AgCl|Ag|Cl^-} = 0.22$ V

For the disproportionation reaction,

$2Cu^+(aq) \rightleftharpoons Cu(s) + Cu^{2+}(aq)$ at 298 K, ln K (where K is the equilibrium constant) is _______ $\times 10^{-1}$.

Given

$(E^0_{Cu^{2+}/Cu^+} = 0.16$ V ; $E^0_{Cu^+/Cu} = 0.52$ V ; $\frac{RT}{F} = 0.025)$

The Gibbs energy change (in J) for the given reaction at $[Cu^{2+}] = [Sn^{2+}] = 1$ M and 298 K is :

$Cu(s) + Sn^{2+}(aq) \rightarrow Cu^{2+}(aq) + Sn(s)$;

$(E^0_{Sn^{2+}|Sn} = -0.16$ V, $E^0_{Cu^{2+}|Cu} = 0.34$ V,

Take F = 96500 C mol^{-1})

For an electrochemical cell,

$Sn(s)|Sn^{2+}(aq, 1M)||Pb^{2+}(aq, 1M)|Pb(s)$

the ratio $\frac{[Sn^{2+}]}{[Pb^{2+}]}$ when this cell attains equilibrium is ________.

Given $E^{\circ}Sn^{2+}/Sn = -0.14V$, $E^{\circ}Pb^{2+}/Pb = 0.13V$,

$\frac{2.303RT}{F} = 0.06$

What would be the electrode potential for the given half cell reaction at pH=5? __________.

$2H_2O \rightarrow O_2 + 4H^+ + 4e^-$; $E^0_{red} = 1.23V$

(R = 8.314 J mol^{-1} K^{-1}; Temp = 298 K; oxygen under std. atm. pressure of 1 bar)

MULTIPLE CHOICE QUESTIONS (MCQs)

Consider the following data :

	$\wedge^0_m$
$BaCl_2$	X_1
H_2SO_4	X_2
HCl	X_3

$BaSO_4$ is sparingly soluble in water. If the conductivity of the saturated $BaSO_4$ solution is x S cm^{-1} then the solubility product of $BaSO_4$ can be given as.

(Here $\wedge_m = \wedge^0_m$)

(1) $\frac{10^6 x^2}{\alpha^2(x_1 + x_2 - 2x_3)^2}$ (2) $\frac{x^2}{(x_1 + x_2 - 2x_3)^2}$

(3) $\frac{\alpha^2(x_1 + x_2 - 2x_3)^2}{10^6 x^2}$ (4) $\frac{x^2}{(x_1 + x_2 + 2x_3)^2}$

Given below are two statements :

Mohr's salt is composed of only three types of ions-ferrous, ammonium and sulphate.

If the molar conductance at infinite dilution of ferrous, ammonium and sulphate ions are x_1, x_2 and x_3 S cm^2 mol^{-1}, respectively then the molar conductance for Mohr's salt solution at infinite dilution would be given by $x_1 + x_2 + 2x_3$.

In the light of the given statements, choose the answer from the options given below :

(1) Both statements I and Statement II are false

(2) Statement I is false but Statement II is true

(3) Statement I is true but Statement II are false

(4) Both statements I and Statement II are true

116. Correct order of limiting molar conductivity for cations in water at 298 K is : **[April 3, 2025 (I)]**

(1) $H^+ > Na^+ > K^+ > Ca^{2+} > Mg^{2+}$ (2) $H^+ > Ca^{2+} > Mg^{2+} > K^+ > Na^+$
(3) $Mg^{2+} > H^+ > Ca^{2+} > K^+ > Na^+$ (4) $H^+ > Na^+ > Ca^{2+} > Mg^{2+} > K^+$

117. The molar conductivity of a weak electrolyte when plotted against the square root of its concentration, which of the following is expected to be observed? **[Jan. 29, 2025 (I)]**

(1) A small decrease in molar conductivity is observed at infinite dilution.
(2) A small increase in molar conductivity is observed at infinite dilution.
(3) Molar conductivity increases sharply with increase in concentration.
(4) Molar conductivity decreases sharply with increase

118. The molar conductivity for electrolytes A and B are plotted against $C^{1/2}$ as shown below. Electrolytes A and B respectively are : **[April 9, 2024 (I)]**

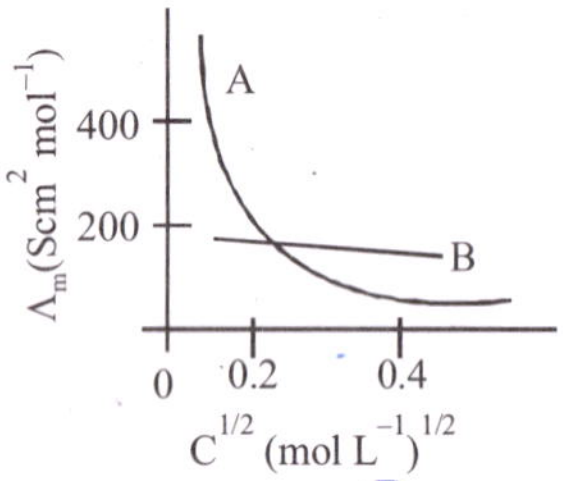

	A	B
(1)	Weak electrolyte	weak electrolyte
(2)	Strong electrolyte	strong electrolyte
(3)	Weak electrolyte	strong electrolyte
(4)	Strong electrolyte	weak electrolyte

119. Which out of the following is a correct equation to show change in molar conductivity with respect to concentration for a weak electrolyte, if the symbols carry their usual meaning:

(1) $\Lambda^2_m C - K_a \Lambda^{o2}_m + K_a \Lambda_m \Lambda^\circ_m = 0$ **[April 9, 2024 (II)]**

(2) $\Lambda_m - \Lambda^\circ_m + AC^{\frac{1}{2}} = 0$

(3) $\Lambda_m - \Lambda^\circ_m - AC^{\frac{1}{2}} = 0$

(4) $\Lambda^2_m C + K_a \Lambda^{o2}_m - K_a \Lambda_m \Lambda^\circ_m = 0$

120. A conductivity cell with two electrodes (dark side) are half filled with infinitely dilute aqueous solution of a weak electrolyte. If volume is doubled by adding more water at constant temperature, the molar conductivity of the cell will - **[April 6, 2024 (I)]**

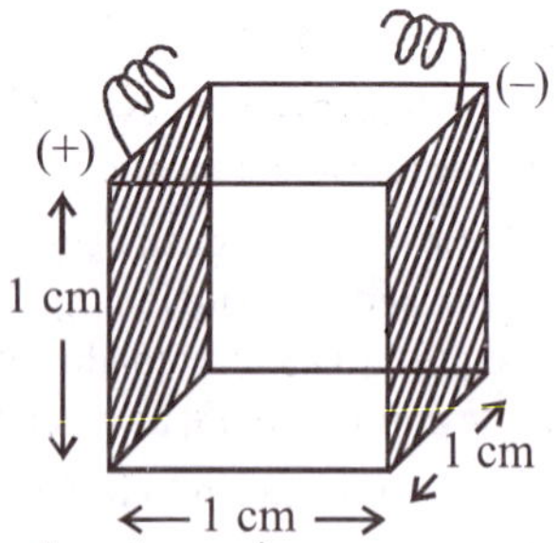

(1) increase sharply
(2) remain same or can not be measured accurately
(3) decrease sharply
(4) depend upon type of electrolyte

121. Molar ionic conductivities of divalent cation and anion are 57 S $cm^2 mol^{-1}$ and 73 S $cm^2 mol^{-1}$ respectively. The molar conductivity of solution of an electrolyte with the above cation and anion will be : **[April 5, 2024 (I)]**

(1) 65 S $cm^2 mol^{-1}$ (2) 130 S $cm^2 mol^{-1}$
(3) 187 S $cm^2 mol^{-1}$ (4) 260 S $cm^2 mol^{-1}$

122. For a strong electrolyte, a plot of molar conductivity against $(concentration)^{1/2}$ is a straight line, with a negative slope, the correct unit for the slope is **[April 4, 2024 (II)]**

(1) $S\,cm^2\,mol^{-3/2}\,L^{1/2}$ (2) $S\,cm^2\,mol^{-1}\,L^{1/2}$
(3) $S\,cm^2\,mol^{-3/2}\,L$ (4) $S\,cm^2\,mol^{-3/2}\,L^{-1/2}$

123. Identify the factor from the following that **does not** affect electrolytic conductance of a solution. **[Jan. 31, 2024 (I)]**

(1) The nature of solvent used.
(2) Concentration of the electrolyte.
(3) The nature of the electrode used.
(4) The nature of the electrolyte added.

124. Given below are two statements:

Statement I: For KI, molar conductivity increases steeply with dilution.

Statement II: For carbonic acid, molar conductivity increases slowly with dilution.

In the light of the above statements, choose the correct answer from the options given below: **[July 27, 2022 (II)]**

(1) Both Statement I and Statement II are true.
(2) Both Statement I and Statement II are false.
(3) Statement I is true but Statement II is false.
(4) Statement I is false but Statement II is true.

125. The molar conductivity of a conductivity cell filled with 10 moles of 20 mL NaCl solution is Λ_{m_1} and that of 20 moles another identical cell having 80 mL NaCl solution is Λ_{m_2}. The conductivities exhibited by these two cells are same. The relationship between Λ_{m_2} and Λ_{m_1} is **[July 25, 2022 (II)]**

(1) $\Lambda_{m_2} = 2\Lambda_{m_1}$ (2) $\Lambda_{m_2} = \Lambda_{m_1}/2$
(3) $\Lambda_{m_2} = \Lambda_{m_1}$ (4) $\Lambda_{m_2} = 4\Lambda_{m_1}$

126. Given below are two statements : **[Aug. 26, 2021 (I)]**

Statement I : The limiting molar conductivity of KCl (strong electrolyte) is higher compared to that of CH_3COOH (weak electrolyte).

Statement II : Molar conductivity decreases with decrease in concentration of electrolyte.

In the light of the above statements, choose the **most appropriate** answer from the options given below :

(1) Statement I is true but Statement II is false.
(2) Statement I is false but Statement II is true.
(3) Both Statement I and Statement II are true.
(4) Both Statement I and Statement II are false.

127. The variation of molar conductivity with concentration of an electrolyte (X) in aqueous solution is shown in the given figure.

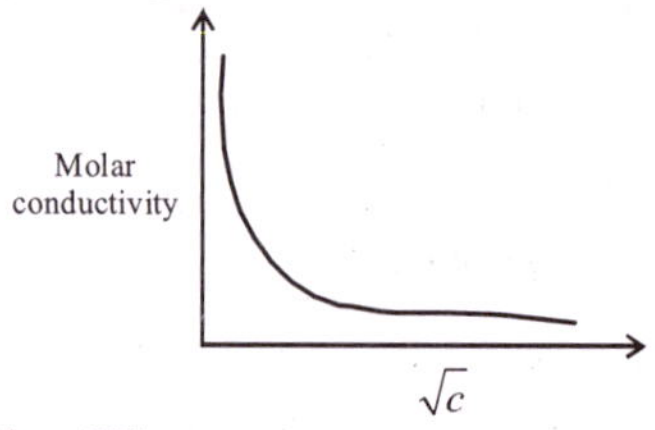

The electrolyte X is : **[Sep. 05, 2020 (II)]**

(1) HCl (2) NaCl (3) KNO_3 (4) CH_3COOH

128. Let C_{NaCl} and C_{BaSO_4} be the conductances (in S) measured for saturated aqueous solutions of NaCl and $BaSO_4$, respectively, at a temperature T.
Which of the following is **false**? [Sep. 03, 2020 (I)]
(1) Ionic mobilities of ions from both salts increase with T
(2) $C_{BaSO_4}(T_2) > C_{BaSO_4}(T_1)$ for $T_2 > T_1$
(3) $C_{NaCl}(T_2) > C_{NaCl}(T_1)$ for $T_2 > T_1$
(4) $C_{NaCl} >> C_{BaSO_4}$ at a given T

129. The equation that is incorrect is: [Jan. 07, 2020 (II)]
(1) $(\Lambda_m^0)_{NaBr} - (\Lambda_m^0)_{NaCl} = (\Lambda_m^0)_{KBr} - (\Lambda_m^0)_{KCl}$
(2) $(\Lambda_m^0)_{KCl} - (\Lambda_m^0)_{NaCl} = (\Lambda_m^0)_{KBr} - (\Lambda_m^0)_{NaBr}$
(3) $(\Lambda_m^0)_{H_2O} = (\Lambda_m^0)_{HCl} + (\Lambda_m^0)_{NaOH} - (\Lambda_m^0)_{NaCl}$
(4) $(\Lambda_m^0)_{NaBr} - (\Lambda_m^0)_{NaI} = (\Lambda_m^0)_{KBr} - (\Lambda_m^0)_{NaBr}$

130. The decreasing order of electrical conductivity of the following aqueous solutions is :
0.1 M Formic acid (A), 0.1 M Acetic acid (B),
0.1 M Benzoic acid (C). [April 12, 2019 (II)]
(1) A > C > B (2) C > B > A
(3) A > B > C (4) C > A > B

131. Consider the statements S1 and S2 :
S1 : Conductivity always increases with decrease in the concentration of electrolyte.
S2 : Molar conductivity always increases with decrease in the concentration of electrolyte.
The correct option among the following is : [April 10, 2019 (I)]
(1) Both S1 and S2 are wrong
(2) S1 is wrong and S2 is correct
(3) Both S1 and S2 are correct
(4) S1 is correct and S2 is wrong

132. Which one of the following graphs between molar conductivity (Λ_m) versus $\sqrt{C}$ is correct? [April 10, 2019 (II)]

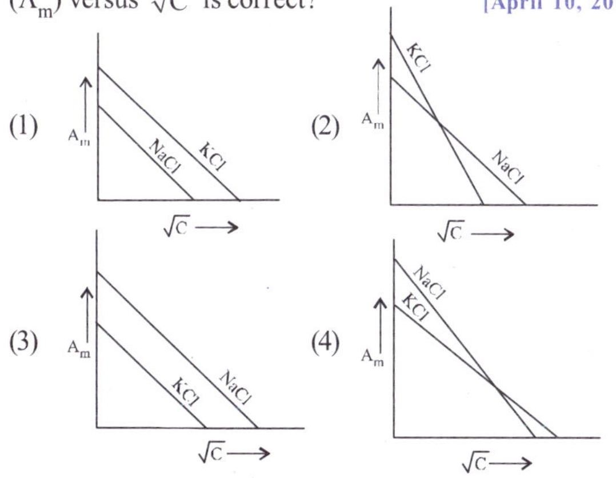

133. Λ_m° for NaCl, HCl and NaA are 126.4, 425.9 and 100.5 S cm^{2}mol^{-1}, respectively. If the conductivity of 0.001 M HA is 5×10^{-5} S cm^{-1}, degree of dissociation of HA is : [Jan. 12, 2019 (II)]
(1) 0.50 (2) 0.25 (3) 0.125 (4) 0.75

134. The equivalent conductance of NaCl at concentration C and at infinite dilution are λ_C and λ_∞, respectively. The correct relationship between λ_C and λ_∞ is given as:
(Where the constant B is positive) [2014]
(1) $\lambda_C = \lambda_\infty + (B)C$ (2) $\lambda_C = \lambda_\infty - (B)C$
(3) $\lambda_C = \lambda_\infty - (B)\sqrt{C}$ (4) $\lambda_C = \lambda_\infty + (B)\sqrt{C}$

135. Resistance of 0.2 M solution of an electrolyte is 50 Ω. The specific conductance of the solution is 1.4 S m^{-1}. The resistance of 0.5 M solution of the same electrolyte is 280 Ω. The molar conductivity of 0.5 M solution of the electrolyte in S m^2 mol^{-1} is: [2006, Same 2011(S)]
(1) 5×10^{-4} (2) 5×10^{-3} (3) 5×10^3 (4) 5×10^2

136. The equivalent conductances of two strong electrolytes at infinite dilution in H_2O (where ions move freely through a solution) at 25 °C are given below :
$\Lambda^\circ_{CH_3COONa} = 91.0$ S cm^2 / equiv.
$\Lambda^\circ_{HCl} = 426.2$ S cm^2 / equiv.
What additional information/ quantity one needs to calculate Λ° of an aqueous solution of acetic acid? [2007]
(1) Λ° of chloroacetic acid ($ClCH_2COOH$)
(2) Λ° of NaCl
(3) Λ° of CH_3COOK
(4) the limiting equivalent conductance of $H^+ (\lambda^\circ_{H^+})$.

137. The molar conductivities Λ^o_{NaOAc} and Λ^o_{HCl} at infinite dilution in water at 25°C are 91.0 and 426.2 S cm^2/mol respectively. To calculate $\Lambda^o_{CH3COOH}$, the additional value required is [2006]
(1) Λ^o_{NaOH} (2) Λ^o_{NaCl} (3) $\Lambda^o_{H_2O}$ (4) Λ^o_{KCl}

138. Resistance of a conductivity cell filled with a solution of an electrolyte of concentration 0.1 M is 100 Ω. The conductivity of this solution is 1.29 S m^{-1}. Resistance of the same cell when filled with 0.2 M of the same solution is 520 Ω. The molar conductivity of 0.2 M solution of electrolyte will be [2006]
(1) 1.24×10^{-4} S m^2 mol^{-1} (2) 12.4×10^{-4} S m^2 mol^{-1}
(3) 124×10^{-4} S m^2 mol^{-1} (4) 1240×10^{-4} S m^2 mol^{-1}

139. The highest electrical conductivity of the following aqueous solutions is of [2005]
(1) 0.1 M difluoroacetic acid (2) 0.1 M fluoroacetic acid
(3) 0.1 M chloroacetic acid (4) 0.1 M acetic acid

140. The limiting molar conductivities Λ° for NaCl, KBr and KCl are 126, 152 and 150 S cm^2 mol^{-1} respectively. The Λ° for NaBr is
(1) 278 S cm^2 mol^{-1} (2) 176 S cm^2 mol^{-1} [2004]
(3) 128 S cm^2 mol^{-1} (4) 302 S cm^2 mol^{-1}

141.

Electrolyte:	KCl	KNO_3	HCl	NaOAc	NaCl
Λ^∞(S cm^2 mol^{-1}) :	149.9	145	426.2	91	126.5

Calculate Λ^∞_{HOAc} using appropriate molar conductances of the electrolytes listed above at infinite dilution in H_2O at 25 °C [2004]
(1) 217.5 (2) 390.7 (3) 552.7 (4) 517.2

142. Conductivity (unit Siemen's S) is directly proportional to area of the vessel and the concentration of the solution in it and is inversely proportional to the length of the vessel then the unit of the constant of proportionality is [2002]
(1) Sm mol^{-1} (2) Sm^2 mol^{-1}
(3) S^{-2}m^2 mol (4) S^{2}m^2 mol^{-2}.

NUMERICAL VALUE QUESTIONS (NVQs)

143. At 298 K, the molar conductivity of x% (w/w) MX solution (aqueous) is 123.5 S cm^2 mol^{-1}. The conductance of same solution is 1.9×10^{-3} S. The value of x is ______ $\times 10^{-2}$. [April 05, 2026 (II)]
(Given : cell constant = 1.3 cm^{-1}; molar mass of MX is 75 g mol^{-1}, density of aqueous solution of MX at 298 K is 1.0 g mL^{-1})

For strong electrolyte Λ_m increases slowly with dilution and can be represented by the equation $\Lambda_m = \Lambda^\circ_m - Ac^{1/2}$.
Molar conductivity values of the solutions of strong electrolyte AB at 18°C are given below :

c [mol L^{-1}]	0.04	0.09	0.16	0.25
Λ_m [S cm^2 mol^{-1}]	96.1	95.7	95.3	94.9

The value of constant A based on the above data [in S cm^2 mol^{-1}/(mol/L)$^{1/2}$] unit is ________.

Molar conductivity of a weak acid HQ of concentration 0.18 M was found to be 1/30 of the molar conductivity of another weak acid HZ with concentration of 0.0 of M. If $\lambda^0_{Q^-}$ happened to be equal with $\lambda^0_{Z^-}$, then the difference of the pK_a values of the two weak acids (pK_a(HQ) – pK_a(HZ)) is ______ (Nearest integer).
[Given : degree of dissociation (α) <<1 for both weak acids, λ° : limiting molar conductivity of ions]

The pH and conductance of a weak acid (HX) was found to be 5 and 4×10^{-5} S, respectively. The conductance was measured under standard condition using a cell where the electrode plates having a surface area of 1 cm^2 were at a distance of 15 cm apart. The value of the limiting molar conductivity is ________ S m^2 mol^{-1}. (nearest integer)
(Degree of dissociation of the weak acid (a) << 1)

The molar conductance of an infinitely dilute solution of ammonium chloride was found to be 185 S cm^2 mol^{-1} and the ionic conductance of hydroxyl and chloride ions are 170 and 70 S cm^2 mol^{-1}, respectively. If molar conductance of 0.02 M solution of ammonium hydroxide is 85.5 S cm^2 mol^{-1}, its degree of dissociation is given by $x \times 10^{-1}$. The value of x is _____. (Nearest integer)

0.2 % (w/v) solution of NaOH is measured to have resistivity 870.0 mΩ m. The molar conductivity of the solution will be ______ $\times 10^2$ mS dm^2 mol^{-1}. (Nearest integer)

Given below is the plot of the molar conductivity vs $\sqrt{\text{concentration}}$ for KCl in aqueous solution.

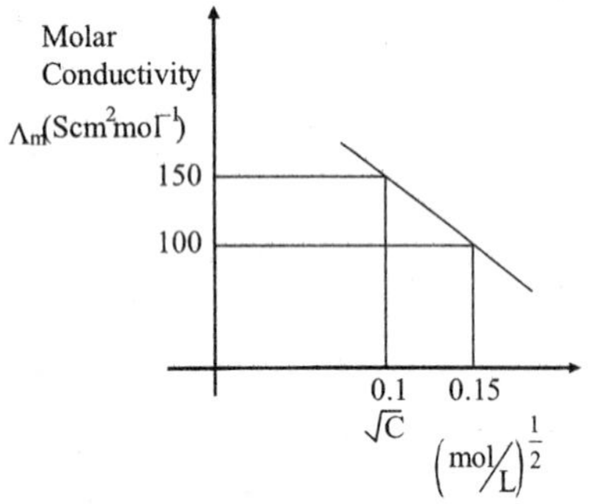

If, for the higher concentration of KCl solution, the resistance of the conductivity cell is 100Ω, then the resistance of the same cell with the dilute solution is 'x' Ω.
The value of x is _________ (Nearest integer)

The values of conductivity of some materials at 298.15 K in Sm^{-1} are 2.1×10^3, 1.0×10^{-16}, 1.2×10, 3.91, 1.5×10^{-2}, 1×10^{-7}, 1.0×10^3. The number of conductors among the materials is ______.

The number of correct statements from the following is _______

(A) Conductivity always decreases with decrease in concentration for both strong and weak electrolytes.
(B) The number of ions per unit volume that carry current in a solution increases on dilution.
(C) Molar conductivity increases with decrease in concentration.
(D) The variation in molar conductivity is different for strong and weak electrolytes.
(E) For weak electrolytes, the change in molar conductivity with dilution is due to decrease in degree of dissociation.

The specific conductance of 0.0025 M acetic acid is 5×10^{-5} S cm^{-1} at a certain temperature. The dissociation constant of acetic acid is ______ $\times 10^{-7}$. (Nearest integer)
Consider limiting molar conductivity of CH_3COOH as 400 S cm^2 mol^{-1}

1×10^{-5} M $AgNO_3$ is added to 1 L of saturated solution of AgBr. The conductivity of this solution at 298 K is ____ $\times 10^{-8}$ S m^{-1}.
[Given : K_{SP}(AgBr) = 4.9×10^{-13} at 298 K
$\lambda^0_{Ag^+} = 6 \times 10^{-3}$ S m^2 mol^{-1}
$\lambda^0_{Br^-} = 8 \times 10^{-3}$ S m^2 mol^{-1}
$\lambda^0_{NO_3^-} = 7 \times 10^{-3}$ S m^2 mol^{-1}]

The resistivity of a 0.8 M solution of an electrolyte is 5×10^{-3} Ωcm. Its molar conductivity is ________ $\times 10^4$ Ω^{-1} cm^2 mol^{-1}. (Nearest integer)

Following figure shows dependence of molar conductance of two electrolytes on concentration. Λ^0_m is the limiting molar conductivity.

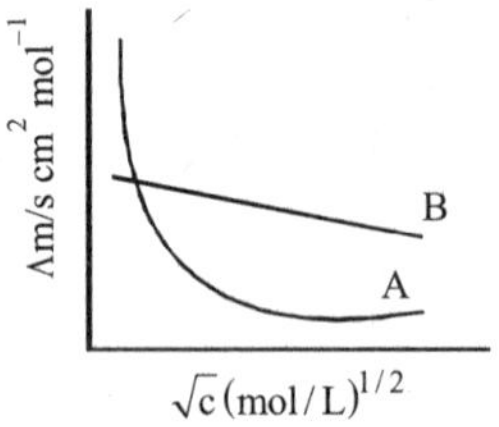

The number of incorrect statement(s) from the following is ________.

(A) Λ^0_m for electrolyte A is obtained by extrapolation
(B) For electrolyte B, Λ_m vs $\sqrt{c}$ graph is a straight line with intercept equal to Λ^0_m
(C) At infinite dilution, the value of degree of dissociation approach zero for electrolyte B.
(D) Λ^0_m for any electrolyte A or B can be calculated using λ° for individual ions.

Resistance of a conductivity cell (cell constant 129 m^{-1}) filled with 74.5 ppm solution of KCl is 100 Ω (labelled as solution 1). When the same cell is filled with KCl solution of 149 ppm, the resistance is 50 Ω (labelled as solution 2). The ratio of molar conductivity of solution 1 and solution 2 is i.e. $\frac{\Lambda_1}{\Lambda_2} = x \times 10^{-3}$.
The value of x is _____. (Nearest integer)
Given, molar mass of KCl is 74.5 g mol^{-1}

The solubility product of a sparingly soluble salt A_2X_3 is 1.1×10^{-23}. If specific conductance of the solution is 3×10^{-5} S m^{-1}, the limiting molar conductivity of the solution is $x \times 10^{-3}$ S m^2 mol^{-1}.
The value of x is __________.

The limiting molar conductivities of NaI, $NaNO_3$ and $AgNO_3$ are 12.7, 12.0 and 13.3 mS m^2 mol^{-1}, respectively (all at 25°C). The limiting molar conductivity of AgI at this temperature is _____ mS m^2 mol^{-1}.

The resistance of conductivity cell containing 0.01 M KCl solution at 298 K is 1750 Ω. If the conductivity of 0.01 M KCl solution at 298 K is 0.152×10^{-3} S cm^{-1}, then the cell constant of the conductivity cell is_______ $\times 10^{-3}$ cm^{-1}.

The conductivity of a weak acid HA of concentration 0.001 mol L^{-1} is 2.0×10^{-5} S cm^{-1}. If $\mathring{\Lambda}_m$ (HA) = 190 S cm^2 mol^{-1}, the ionization constant (K_a) of HA is equal to ____ $\times 10^{-6}$. (Round off to the Nearest Integer)

The resistance of a conductivity cell with cell constant 1.14 cm^{-1}, containing 0.001 M KCl at 298 K is 1500 Ω. The molar conductivity of 0.001 M KCl solution at 298 K in S cm^2 mol^{-1} is_______. (Integer answer)

The molar conductivities at infinite dilution of barium chloride, sulphuric acid and hydrochloric acid are 280, 860 and 426 S cm^2 mol^{-1} respectively. The molar conductivity at infinite dilution of barium sulphate is ______ S cm^2 mol^{-1}. (Round off to the nearest integer).

A KCl solution of conductivity 0.14 Sm^{-1} shows a resistance of 4.19 Ω in a conductivity cell. If the same cell is filled with an HCl solution, the resistance drops to 1.03 Ω. The conductivity of the HCl solution is ______ $\times 10^{-2}$ S m^{-1}. (Round off to the nearest integer).

A 5.0 m mol dm^{-3} aqueous solution of KCl has a conductance of 0.55 mS when measured in a cell of cell constant 1.3 cm^{-1}. The molar conductivity of this solution is ______ m Sm^2 mol^{-1}. (Round off to the nearest integer).

MULTIPLE CHOICE QUESTIONS (MCQs)

Given below are two statements :

1 M aqueous solution of each of $Cu(NO_3)_2$, $AgNO_3$, $Hg_2(NO_3)_2$; $Mg(NO_3)_2$ are electrolysed using inert electrodes,

$$E^{\theta}_{Ag^+/Ag} = 0.80\text{ V},\ E^{\theta}_{Hg_2^{2+}/Hg} = 0.79\text{ V},$$

$$E^{\theta}_{Cu^{2+}/Cu} = 0.24\text{ V and } E^{\theta}_{Mg^{2+}/Mg} = -2.37\text{ V}$$

With increasing voltage, the sequence of deposition of metals on the cathode will be Ag, Hg and Cu

Magnesium will not be deposited at cathode instead oxygen gas will be evolved at the cathode.

In the light of the above statement, choose the from the options given below

(1) Both statement I and statement II are incorrect
(2) Statement I is correct but statement II is incorrect
(3) Both statement I and statement II are correct
(4) Statement I is incorrect but statement II is correct

O_2 gas will be evolved as a product of electrolysis of :

(A) an aqueous solution of $AgNO_3$ using silver electrodes.
(B) an aqueous solution of $AgNO_3$ using platinum electrodes.
(C) a dilute solution of H_2SO_4 using platinum electrodes.
(D) a high concentration solution of H_2SO_4 using platinum electrodes.

Choose the correct answer from the options given below :

(1) (B) and (C) only (2) (A) and (D) only
(3) (B) and (D) only (4) (A) and (C) only

Which of the following electrolyte can be used to obtain $H_2S_2O_8$ by the process of electrolysis?

(1) Dilute solution of sodium sulphate
(2) Dilute solution of sulphuric acid
(3) Concentrated solution of sulphuric acid
(4) Acidified dilute solution of sodium sulphate.

A solution of aluminium chloride is electrolysed for 30 minutes using a current of 2A. The amount of the aluminium deposited at the cathode is ___ .

[Given : molar mass of aluminium and chlorine are 27 g mol^{-1} and 35.5 g mol^{-1} respectively, Faraday constant = 96500 C mol^{-1}].

(1) 1.660 g (2) 1.007 g (3) 0.336 g (4) 0.441 g

How can an electrochemical cell be converted into an electrolytic cell ?

(1) Applying an external opposite potential greater than E^0_{cell}
(2) Reversing the flow of ions in salt bridge.
(3) Applying an external opposite potential lower than E^0_{cell}.
(4) Exchanging the electrodes at anode and cathode.

The quantity of silver deposited when one coulomb charge is passed through $AgNO_3$ solution:

(1) 0.1 g atom of silver
(2) 1 chemical equivalent of silver
(3) 1 g of silver
(4) 1 electrochemical equivalent of silver

The standard electrode potential of M^+/M in aqueous solution does not depend on

(1) Ionisation of a solid metal atom
(2) Sublimation of a solid metal
(3) Ionisation of a gaseous metal atom
(4) Hydration of a gaseous metal ion

The product obtained from the electrolytic oxidation of acidified sulphate solutions, is :

(1) HSO_4^- (2) HO_3SOOSO_3H
(3) HO_2SOSO_2H (4) HO_3SOSO_3H

250 mL of a waste solution obtained from the workshop of a goldsmith contains 0.1 M $AgNO_3$ and 0.1 M AuCl. The solution was electrolyzed at 2 V by passing a current of 1 A for 15 minutes. The metal/metals deposited will be :

$$\left(E^0_{Ag^+/Ag} = 0.80\text{ V},\ E^0_{Au^+/Au} = 1.69\text{ V}\right)$$

(1) only gold
(2) silver and gold in proportion to their atomic weights
(3) only silver
(4) silver and gold in equal mass proportion

A solution of Ni $(NO_3)_2$ is electrolysed between platinum electrodes using 0.1 Faraday electricity. How many mole of Ni will be deposited at the cathode?

(1) 0.05 (2) 0.20 (3) 0.15 (4) 0.10

The anodic half-cell of lead-acid battery is recharged using electricity of 0.05 Faraday. The amount of $PbSO_4$ electrolyzed in g during the process is :

(Molar mass of $PbSO_4$ = 303 g mol^{-1})

(1) 22.8 (2) 15.2 (3) 7.6 (4) 11.4

How long (approximate) should water be electrolysed by passing through 100 amperes current so that the oxygen released can completely burn 27.66 g of diborane? (Atomic weight of B = 10.8 u)

(1) 6.4 hours (2) 0.8 hours
(3) 3.2 hours (4) 1.6 hours

177. When 9.65 ampere current was passed for 1.0 hour into nitrobenzene in acidic medium, the amount of *p*-aminophenol produced is _______. [Online April 16, 2018]
(1) 109.0 g (2) 98.1 g (3) 9.81 g (4) 10.9 g

178. When an electric current is passes through acidified water, 112 mL of hydrogen gas at N.T.P. was collected at the cathode in 965 seconds. The current passed, in ampere, is : [Online April 15, 2018 (I)]
(1) 2.0 (2) 0.1 (3) 0.5 (4) 1.0

179. Two Faraday of electricity is passed through a solution of $CuSO_4$. The mass of copper deposited at the cathode is (at. mass of Cu = 63.5 amu) [2015]
(1) 2g (2) 127 g (3) 0 g (4) 63.5 g

180. A current of 10.0 A flows for 2.00 h through an electrolytic cell containing a molten salt of metal X. This results in the decomposition of 0.250 mol of metal X at the cathode. The oxidation state of X in the molten salt is: (F = 96,500 C) [Online April 9, 2014]
(1) 1+ (2) 2+ (3) 3+ (4) 4+

181. How many electrons would be required to deposit 6.35 g of copper at the cathode during the electrolysis of an aqueous solution of copper sulphate? (Atomic mass of copper = 63.5 u, N_A = Avogadro's constant): [Online April 12, 2014]
(1) $\frac{N_A}{20}$ (2) $\frac{N_A}{10}$ (3) $\frac{N_A}{5}$ (4) $\frac{N_A}{2}$

182. A solution of copper sulphate ($CuSO_4$) is electrolysed for 10 minutes with a current of 1.5 amperes. The mass of copper deposited at the cathode (at. mass of Cu = 63u) is: [Online April 25, 2013]
(1) 0.3892 g (2) 0.2938 g
(3) 0.2398 g (4) 0.3928 g

183. A battery is constructed of Cr and $Na_2Cr_2O_7$. The unbalanced chemical equation when such a battery discharges is following:

$$Na_2Cr_2O_7 + Cr + H^+ \rightarrow Cr^{3+} + H_2O + Na^+$$

If one Faraday of electricity is passed through the battery during the charging, the number of moles of Cr^{3+} removed from the solution is [Online May 7, 2012]
(1) $\frac{4}{3}$ (2) $\frac{1}{3}$ (3) $\frac{3}{3}$ (4) $\frac{2}{3}$

184. The standard potentials of Ag^+/Ag, $Hg_2^{2+}/2Hg$, Cu^{2+}/Cu and Mg^{2+}/Mg electrodes are 0.80, 0.79, 0.34 and –2.37 V, respectively. An aqueous solution which contains one mole per litre of the salts of each of the four metals is electrolyzed. With increasing voltage, the correct sequence of deposition of the metals at the cathode is [Online May 19, 2012]
(1) Ag, Hg, Cu, Mg (2) Cu, Hg, Ag only
(3) Ag, Hg, Cu only (4) Mg, Cu, Hg, Ag

185. Aluminium oxide may be electrolysed at 1000 °C to furnish aluminium metal (At. Mass = 27 amu; 1 Faraday = 96,500 Coulombs). The cathode reaction is: – $Al^{3+} + 3e^- \rightarrow Al°$
To prepare 5.12 kg of aluminium metal by this method we require [2005]
(1) 5.49×10^1 C of electricity
(2) 5.49×10^4 C of electricity
(3) 1.83×10^7 C of electricity
(4) 5.49×10^7 C of electricity

186. When during electrolysis of a solution of $AgNO_3$ 9650 coulombs of charge pass through the electroplating bath, the mass of silver deposited on the cathode will be [2003]
(1) 10.8 g (2) 21.6 g (3) 108 g (4) 1.08 g

187. When the sample of copper with zinc impurity is to be purified by electrolysis, the appropriate electrodes are [2002]

	Cathode	Anode
(1)	pure zinc	pure copper
(2)	impure sample	pure copper
(3)	impure zinc	impure sample
(4)	pure copper	impure sample.

NUMERICAL VALUE QUESTIONS (NVQs)

188. Electricity is passed through an acidic solution of Cu^{2+} till all the cu^{2+} was exhausted, leading to the deposition of 300 mg of Cu metal. However, a current of 600 mA was continued to pass through the same solution for another 28 minutes by keeping the total volume of the solution fixed at 200 mL. The total volume of oxygen evolved at STP during the entire process is ______ mL. (Nearest integer) [Jan. 24, 2026 (I)]
[Given : $Cu^{2+}(aq) + 2e^- \rightarrow Cu(s); E^0_{red} = +0.34V$
$O_2(g) + 4H^+ + 4e^- \rightarrow 2H_2O; E^0_{red} = +1.23V$
Molar mass of Cu = 63.54 g mol^{-1}
Molar mass of O_2 = 32 g mol^{-1}
Faraday Constant = 96500 C mol^{-1}
Molar volume at STP = 22.4L]

189. Electrolysis of 600 mL aqueous solution of NaCl for 5 min changes the pH of the solution to 12.
The current in Amperes used for the given electrolysis is ____. (Nearest integer). [Jan. 28, 2025 (II)]

190. The amount of electricity in Coulomb required for the oxidation of 1 mol of H_2O to O_2 is ________ $\times 10^5$C. [Feb. 1, 2024 (II)]

191. One Faraday of electricity liberates $x \times 10^{-1}$ gram atom of copper from copper sulphate, x is _____. [Jan. 31, 2024 (I)]

192. A constant current was passed through a solution of $AuCl_4^-$ ion between gold electrodes. After a period of 10.0 minutes, the increase in mass of cathode was 1.314g. The total charge passed through the solution ____ $\times 10^{-2}$ F.
(Given atomic mass of Au = 197) [Jan. 29, 2024 (II)]

193. The mass of zinc produced by the electrolysis of zinc sulphate solution with a steady current of 0.015 A for 15 minutes is _____ $\times 10^{-4}$ g. (Atomic mass of zinc = 65.4 amu) [Jan. 29, 2024 (I)]

194. The mass of silver (Molar mass of Ag : 108 $gmol^{-1}$, displaced by a quantity of electricity which displaces 5600 mL of O_2 at S.T.P. will be ______ g. [Jan. 27, 2024 (I)]

195. A metal surface of 100 cm^2 area has to be coated with nickel layer of thickness 0.001 mm. A current of 2A was passed through a solution of $Ni(NO_3)_2$ for 'x' seconds to coat the desired layer. The value of x is ______ (Nearest integer)
(ρ_{Ni} (density of Nickel) is 10 gmL^{-1}, Molar mass of Nickel is 60 $gmol^{-1}$ F = 96500 C mol^{-1}) [April 13, 2023 (I)]

196. The amount of charge in F (Faraday) required to obtain one mole of iron from Fe_3O_4 is ________. (Nearest Integer) [July 26, 2022 (I)]

197. A dilute solution of sulphuric acid is electrolysed using a current of 0.10 A for 2 hours to produce hydrogen and oxygen gas. The total volume of gases produced at STP is ________ cm^3. (Nearest integer) [Given : Faraday constant F = 96500 C mol^{-1} at STP, molar volume of an ideal gas is 22.7 L mol^{-1}] [June 29, 2022 (I)]

198. A solution of $Fe_2(SO_4)_3$ is electrolyzed for 'x' min with a current of 1.5 A to deposit 0.3482 g of Fe. The value of x is ________. [nearest integer] [June 25, 2022 (II)]
Given : 1 F = 96500 C mol^{-1}
Atomic mass of Fe = 56 g mol^{-1}

199. Consider the following reaction
$MnO_4^- + 8H^+ + 5e^- \rightarrow Mn^{2+} + 4H_2O, E° = 1.51$ V.
The quantity of electricity required in Faraday to reduce five moles of MnO_4^- is ________ . [Feb. 26, 2021 (I)]

200. Potassium chlorate is prepared by the electrolysis of KCl in basic solution $6\,OH^- + Cl^- \rightarrow ClO_3^- + 3\,H_2O + 6\,e^-$
If only 60% of the current is utilized in the reaction, the time (rounded to the nearest hour) required to produce 10 g of $KClO_3$ using a current of 2 A is ______.
(Given : F= 96,500 C mol^{-1}; molar mass of $KClO_3$ = 122 g mol^{-1})
[Sep. 06, 2020 (I) Same July 20, 2021 (II)]

201. An acidic solution of dichromate is electrolyzed for 8 minutes using 2A current. As per the following equation
$$Cr_2O_7^{2-} + 14H^+ + 6e^- \rightarrow 2Cr^{3+} + 7H_2O$$
The amount of Cr^{3+} obtained was 0.104 g. The efficiency of the process (in %) is (Take : F = 96000 C, At. mass of chromium = 52) ____________. [Sep. 03, 2020 (II)]

202. 108 g of silver (molar mass 108 g mol^{-1}) is deposited at cathode from $AgNO_3$(aq) solution by a certain quantity of electricity. The volume (in L) of oxygen gas produced at 273 K and 1 bar pressure from water by the same quantity of electricity is ________. [Jan. 09, 2020 (I)]

Topic 4 - Commercial Cells and Corrosion

MULTIPLE CHOICE QUESTIONS (MCQs)

203. On charging the lead storage battery, the oxidation state of lead changes from x_1 to y_1 at the anode and from x_2 to y_2 at the cathode. The values of x_1, y_1, x_2, y_2 are respectively: [April 4, 2025 (I)]
(1) +4, +2, 0, +2 (2) +2, 0, +2, +4
(3) 0, +2, +4, +2 (4) +2, 0, 0, +4

204. The standard cell potential ($E^{\ominus}_{cell}$) of a fuel cell based on the oxidation of methanol in air that has been used to power television relay station is measured as 1.21 V. The standard half cell reduction potential for O_2 ($E^\circ_{O_2/H_2O}$) is 1.229 V.
Choose the correct statement: [April 3, 2025 (II)]
(1) The standard half cell reduction potential for the reduction of CO_2 ($E^\circ_{CO_2/CH_3OH}$) is 19 mV
(2) Oxygen is formed at the anode.
(3) Reactants are fed at one go to each electrode.
(4) Reduction of methanol takes place at the cathode.

205. Match List-I with List-II : [Jan. 29, 2025 (II)]

List-I (Applications)	List-II (Batteries/Cell)
(A) Transistors	(I) Anode - Zn/Hg ; Cathode - HgO + C
(B) Hearing aids	(II) Hydrogen fuel cell
(C) Invertors	(III) Anode – Zn; Cathode - Carbon
(D) Apollo space ship	(IV) Anode – Pb ; Cathode – Pb | PbO_2

Choose the correct answer from the options given below :
(1) (A)-(III), (B)-(I), (C)-(IV), (D)-(II)
(2) (A)-(III), (B)-(II), (C)-(IV), (D)-(I)
(3) (A)-(IV), (B)-(III), (C)-(II), (D)-(I)
(4) (A)-(II), (B)-(III), (C)-(IV), (D)-(I)

206. Given below are two statements : [Jan. 22, 2025 (II)]
Statement I : Corrosion is an electrochemical phenomenon in which pure metal acts as an anode and impure metal as a cathode.
Statement II : The rate of corrosion is more in alkaline medium than in acidic medium.
In the light of the above statements, choose the correct answer from the options given below :
(1) Both Statement I and Statement II are false
(2) Statement I is false but Statement II is true
(3) Both Statement I and Statement II are true
(4) Statement I is true but Statement II is false

207. Match List-I with List-II. [April 9, 2024 (II)]

List - I (Cell)	List - II (Use/Property/Reaction)
(A) Leclanche cell	(I) Converts energy of combustion into electrical energy
(B) Ni-Cd cell	(II) Does not involve any ion in solution and is used in hearing aids
(C) Fuel cell	(III) Rechargeable
(D) Mercury cell	(IV) Reaction at anode $Zn \rightarrow Zn^{2+} + 2e^-$

Choose the correct answer from the options given below:
(1) A-(I), B-(II), C-(III), D-(IV)
(2) A-(III), B-(I), C-(IV), D-(II)
(3) A-(IV), B-(III), C-(I), D-(II)
(4) A-(II), B-(III), C-(IV), D-(I)

208. The reaction at cathode in the cells commonly used in clocks involves. [April 5, 2024 (I)]
(1) reduction of Mn from +4 to +3
(2) oxidation of Mn from +3 to +4
(3) reduction of Mn from +7 to +2
(4) oxidation of Mn from + 2 to +7

209. Fuel cell, using hydrogen and oxygen as fuels, [April 4, 2024 (II)]
(A) has been used in spaceship
(B) has as efficiency of 40% to produce electricity
(C) uses aluminium as catalysts
(D) is eco-friendly
(E) is actually a type of Galvanic cell only
(1) A, B, C only (2) A, B, D only
(3) A, B, D, E only (4) A, D, E only

210. The metals that are employed in the battery industries are [Jan. 31, 2024 (I)]
A. Fe B. Mn C. Ni D. Cr
E. Cd
Choose the correct answer from the options given below:
(1) A, B, C and D only (2) B, C and E only
(3) B, D and E only (4) A, B, C, D and E

211. Which of the following statements is not correct about rusting of iron? [Jan. 27, 2024 (II)]
(1) Coating of iron surface by tin prevents rusting, even if the tin coating is peeling off.
(2) When pH lies above 9 or 10, rusting of iron does not take place.
(3) Dissolved acidic oxides SO_2, NO_2 in water act as catalyst in the process of rusting.
(4) Rusting of iron is envisaged as setting up of electrochemical cell on the surface of iron object.

For lead storage battery pick the correct statements

A. During charging of battery, $PbSO_4$ on anode is converted into PbO_2
B. During charging of battery, $PbSO_4$ on cathode is converted into PbO_2
C. Lead storage battery, consists of grid of lead packed with PbO_2 as anode
D. Lead storage battery has ~ 38% solution of sulphuric acid as an electrolyte

Choose the correct answer from the options given below:

(1) B, D only (2) B, C, D only
(3) A, B, D only (4) B, C only

Match List-I with List-II.

List-I	List-II
(A) $Cd(s) + 2Ni(OH)_3(s) \rightarrow CdO(s) + 2Ni(OH)_2(s) + H_2O(l)$	(I) Primary battery
(B) $Zn(Hg) + HgO(s) \rightarrow ZnO(s) + Hg(l)$	(II) Discharging of secondary battery
(C) $2PbSO_4(s) + 2H_2O(l) \rightarrow Pb(s) + PbO_2(s) + 2H_2SO_4(aq)$	(III) Fuel cell
(D) $2H_2(g) + O_2(g) \rightarrow 2H_2O(l)$	(IV) Charging of secondary battery

Choose the correct answer from the options given below :

(1) (A) – (I), (B) – (II), (C) – (III), (D) – (IV)
(2) (A) – (IV), (B) – (I), (C) – (II), (D) – (III)
(3) (A) – (II), (B) – (I), (C) – (IV), (D) – (III)
(4) (A) – (II), (B) – (I), (C) – (III), (D) – (IV)

In a fuel cell methanol is used as fuel and oxygen gas is used as an oxidizer. The reaction is

$$CH_3OH(l) + 3/2O_2(g) \longrightarrow CO_2(g) + 2H_2O(l)$$

At 298 K, standard Gibb's energies of formation for $CH_3OH(l)$, $H_2O(l)$ and and CO_2 (g) are –166.2, –237.2 and –394.4 kJ mol^{-1} respectively. If standard enthalpy of combustion of methanol is – 726 kJ mol^{-1}, efficiency of the fuel cell will be:

(1) 87% (2) 90% (3) 97% (4) 80%

In a hydrogen-oxygen fuel cell, combustion of hydrogen occurs to

(1) produce high purity water
(2) create potential difference between two electrodes
(3) generate heat
(4) remove adsorbed oxygen from elctrode surfaces

Numerical Value Questions (NVQs)

Consider the following two half-cell reactions along with the standard reduction potential given :

$$CO_2 + 6H^+ + 6e^- \rightarrow CH_3OH + H_2O \quad E^o_{red} = 0.02\ V$$

$$\frac{1}{2}O_2 + 2H^+ + 2e^- \rightarrow H_2O \quad E^o_{red} = 1.23\ V$$

The fuel cell was set up using the above two reactions such that the cell operates under the standard condition of 1 bar pressure and 298 K temperature. The fuel cell works with 80% efficiency. If the work derived from the cell using 1 mol of CH_3OH is used to compress an ideal gas isothermally against a constant pressure of 1kPa, then the change in the volume of the gas, ΔV = ________ m^3. (Nearest integer)

Given : F = 96500 C mol^{-1}

Part-B (JEE Advanced)

MCQs with One Correct Answer

For the following cell,

$Zn(s) | ZnSO_4(aq) || CuSO_4(aq) | Cu(s)$

when the concentration of Zn^{2+} is 10 times the concentration of Cu^{2+}, the expression for ΔG (in J mol^{-1}) is [F is Faraday constant; R is gas constant; T is temperature; E^0 (cell) = 1.1 V]

(a) $1.1F$ (b) $2.303RT - 2.2F$
(c) $2.303RT + 1.1F$ (d) $-2.2F$

For the following electrochemical cell at 298 K,

$Pt(s) | H_2(g, 1\ bar) | H^+(aq, 1\ M) || M^{4+}(aq), M^{2+}(aq) | Pt(s)$

$$E_{cell} = 0.092\ V \text{ when } \frac{[M^{2+}(aq)]}{[M^{4+}(aq)]} = 10^x.$$

Given : $E^o_{M^{4+}/M^{2+}} = 0.151$ V; $2.303\dfrac{RT}{F} = 0.059$ V

The value of x is

(a) –2 (b) –1 (c) 1 (d) 2

Consider the following cell reaction:

$2Fe(s) + O_2(g) + 4H^+(aq) \rightarrow 2Fe^{2+}(aq) + 2H_2O(l)$; $E° = 1.67V$

At $[Fe^{2+}] = 10^{-3}$ M, $P(O_2) = 0.1$ atm and pH = 3, the cell potential at 25°C is

(a) 1.47V (b) 1.77V (c) 1.87V (d) 1.57V

The rusting of iron takes place as follows

$2H^+ + 2e^- + ½O_2 \longrightarrow H_2O(l)$; $E° = +1.23$ V

$Fe^{2+} + 2e^- \longrightarrow Fe(s)$; $E° = -0.44$ V

Calculate $\Delta G°$ for the net process

(a) –322 kJ mol^{-1} (b) –161 kJ mol^{-1}
(c) –152 kJ mol^{-1} (d) –76 kJ mol^{-1}

The emf of the cell

$Zn | Zn^{2+}(0.01\ M) || Fe^{2+}(0.001\ M) | Fe$

at 298 K is 0.2905 then the value of equilibrium constant for the cell reaction is

(a) $e^{\frac{0.32}{0.0295}}$ (b) $10^{\frac{0.32}{0.0295}}$ (c) $10^{\frac{0.26}{0.0295}}$ (d) $10^{\frac{0.32}{0.0591}}$

In the electrolytic cell, flow of electrons is from

(a) Cathode to anode in solution
(b) Cathode to anode through external supply
(c) Cathode to anode through internal supply
(d) Anode to cathode through internal supply

Standard electrode potential data are useful for understanding the suitability of an oxidant in a redox titration. Some half cell reactions and their standard potentials are given below :

$MnO_4^-(aq) + 8H^+(aq) + 5e^- \rightarrow Mn^{2+}(aq) + 4H_2O(l)$ $E° = 1.51$ V

$Cr_2O_7^{2-}(aq) + 14H^+(aq) + 6e^- \rightarrow 2Cr^{3+}(aq) + 7H_2O(l)$ $E° = 1.38$ V

$Fe^{3+}(aq) + e^- \rightarrow Fe^{2+}(aq)$ $E° = 0.77$ V

$Cl_2(g) + 2e^- \rightarrow 2Cl^-(aq)$ $E° = 1.40$ V

Identify the only incorrect statement regarding the quantitative estimation of aqueous $Fe(NO_3)_2$

(a) MnO_4^- can be used in aqueous HCl
(b) $Cr_2O_7^{2-}$ can be used in aqueous HCl
(c) MnO_4^- can be used in aqueous H_2SO_4
(d) $Cr_2O_7^{2-}$ can be used in aqueous H_2SO_4

Saturated solution of KNO_3 is used to make 'salt–bridge' because

(a) velocity of K^+ is greater than that of NO_3^-
(b) velocity of NO_3^- is greater than that of K^+
(c) velocities of both K^+ and NO_3^- are nearly the same
(d) KNO_3 is highly soluble in water

For the electrochemical cell, $M | M^+ \| X^- | X, E°(M^+/M) = 0.44$V and $E°(X/X^-) = 0.33$V.

From this data one can deduce that

(a) $M + X \rightarrow M^+ + X^-$ is the spontaneous reaction
(b) $M^+ + X^- \rightarrow M + X$ is the spontaneous reaction
(c) $E_{cell} = 0.77$V
(d) $E_{cell} = -0.77$ V

A gas X at 1 atm is bubbled through a solution containing a mixture of 1 M Y and 1 M Z at 25°C. If the reduction potential of $Z > Y > X$, then,

(a) Y will oxidize X and not Z
(b) Y will oxidize Z and not X
(c) Y will oxidize both X and Z
(d) Y will reduce both X and Z

The standard reduction potentials of Cu^{2+} | Cu and Cu^{2+}| Cu^+ are 0.337 V and 0.153 respectively. The standard electrode potential of Cu^+ |Cu half cell is

(a) 0.184V (b) 0.827V (c) 0.521V (d) 0.490V

A dilute aqueous solution of Na_2SO_4 is electrolyzed using platinum electrodes. The products at the anode and cathode are:

(a) O_2, H_2 (b) $S_2O_8^{2-}$, Na (c) O_2,Na (d) $S_2O_8^{2-}, H_2$

The standard oxidation potentials, $E°$, for the half reactions are as

$Zn = Zn^{2+} + 2e^-; E° = +0.76$ V

$Fe = Fe^{2+} + 2e^-; E° = +0.41$ V

The EMF for the cell reaction :

$Fe^{2+} + Zn \rightarrow Zn^{2+} + Fe$

(a) −0.35 V (b) +0.35 V (c) +1.17 V (d) −1.17 V

A solution of sodium sulphate in water is electrolysed using inert electrodes. The products at the cathode and anode are respectively

(a) H_2, O_2 (b) O_2, H_2 (c) O_2, Na (d) O_2, SO_2

The reaction :

$\frac{1}{2} H_2(g) + AgCl(s) \rightarrow H^+(aq) + Cl^-(aq) + Ag(s)$

occurs in the galvanic cell

(a) Ag | AgCl(s) | KCl (soln) | $AgNO_3$ (soln) | Ag
(b) Pt | H_2(g) | HCl (soln) | $AgNO_3$ (soln) | Ag
(c) Pt | H_2(g) | HCl (soln) | AgCl(s) | Ag
(d) Pt | H_2(g) | KCl (soln) | AgCl(s) | Ag

A solution containing one mole per litre of each $Cu(NO_3)_2$; $AgNO_3$; $Hg_2(NO_3)_2$; is being electrolysed by using inert electrodes. The values of standard electrode potentials in volts (reduction potentials) are :

$Ag/Ag^+ = +0.80$, $2Hg/Hg_2^{++} = +0.79$

$Cu/Cu^{++} = +0.34$, $Mg/Mg^{++} = -2.37$

With increasing voltage, the sequence of deposition of metals on the cathode will be :

(a) Ag, Hg, Cu, Mg (b) Mg, Cu, Hg, Ag
(c) Ag, Hg, Cu (d) Cu, Hg, Ag

The standard reduction potentials at 298 K for the following half reactions are given against each

$Zn^{2+}(aq) + 2e \rightleftharpoons Zn(s)$ −0.762

$Cr^{3+}(aq) + 2e \rightleftharpoons Cr(s)$ −0.740

$2H^+(aq) + 2e \rightleftharpoons H_2(g)$ 0.000

$Fe^{3+}(aq) + 2e \rightleftharpoons Fe^{2+}(aq)$ 0.770

which is the strongest reducing agent?

(a) Zn(s) (b) Cr(s) (c) H_2(g) (d) Fe^{2+}(aq)

Integer Value Answer

For the electrochemical cell,

$Mg(s) | Mg^{2+}$ (aq, 1 M) $\|$ Cu^{2+} (aq, 1 M) | Cu(s)

the standard emf of the cell is 2.70 V at 300 K. When the concentration of Mg^{2+} is changed to x M, the cell potential changes to 2.67 V at 300 K. The value of x is________.

(given, $\frac{F}{R} = 11500 \text{ K V}^{-1}$, where F is the Faraday constant and R is the gas constant, ln 10=2.30)

All the energy released from the reaction $X \rightarrow Y, \Delta_r G° = -193$ kJ mol^{-1} is used for oxidizing M^+ as $M^+ \rightarrow M^{3+} + 2e^-$, $E° = -0.25$ V

Under standard conditions, the number of moles of M^+ oxidized when one mole of X is converted to Y is

[$F = 96500$ C mol^{-1}]

Numeric / New Stem Based Questions

The reduction potential (E^0, in V) of $MnO_4^-(aq)/Mn(s)$ is ______.

[Given:

$[E^0_{(MnO_4^-(aq)/MnO_{2(s)})} = 1.68$ V; $E^0_{(MnO_2(s)/Mn^{2+}(aq))} = 1.21$ V;

$E^0_{(Mn^{2+}(aq)/Mn(s))} = -1.03$ V]

Consider an electrochemical cell:

$A(s) | A^{n+}$ (aq, 2 M) |B^{2n+} (aq, 1 M) | B(s).

The value of $\Delta H°$ for the cell reaction is twice that of $\Delta G°$ at 300 K. If the emf of the cell is zero, the $\Delta S°$ (in J K^{-1} mol^{-1}) of the cell reaction per mole of B formed at 300 K is______.

(Given: ln(2) = 0.7, R (universal gas constant) = 8.3 J K^{-1} mol^{-1}. H, S and G are enthalpy, entropy and Gibbs energy, respectively.)

Two students use same stock solution of $ZnSO_4$ and a solution of $CuSO_4$. The emf of one cell is 0.03 V higher than the other. The conc. of $CuSO_4$ in the cell with higher emf value is 0.5 M. Find out the conc. of $CuSO_4$ in the other cell (2.203 RT/F = 0.06).

The standard reduction potential for Cu^{2+}| Cu is +0.34 V. Calculate the reduction potential at pH = 14 for the above couple. K_{sp} of $Cu(OH)_2$ is 1.0×10^{-19}

24. An excess of liquid mercury is added to an acidified solution of 1.0×10^{-3} M Fe^{3+}. It is found that 5% of Fe^{3+} remains at equilibrium at 25°C. Calculate $E^\circ_{Hg_2^{2+}|Hg}$, assuming that the only reaction that occurs is

$2Hg + 2Fe^{3+} \longrightarrow Hg_2^{2+} + 2Fe^{2+}$.

(Given $E^\circ_{Fe^{3+}|Fe^{2+}} = 0.77$ V.) **[1995 - 4 Marks]**

4 Fill in the Blanks

25. The more .... the standard reduction potential, the .... is its ability to displace hydrogen from acids. **[1986 - 1 Mark]**

5 True / False

26. The dependence of electrode potential for the electrode M^{n+}/M with concentration under STP conditions is given by the expression : $E = E^\circ + \frac{0.0591}{n}\log_{10}[M^{n+}]$ **[1993 - 1 Mark]**

6 MCQs with One or More than One Correct Answer

27. The correct option(s) about entropy (S) is(are)
[R = gas constant, F = Faraday constant, T = Temperature] **[Adv. 2022]**
(a) For the reaction, $M(s) + 2H^+(aq) \rightarrow H_2(g) + M^{2+}(aq)$, if $\frac{dE_{cell}}{dT} = \frac{R}{F}$, then the entropy change of the reaction is R (assume that entropy and internal energy changes are temperature independent).
(b) The cell reaction, $Pt(s) | H_2(g, 1bar) | H^+(aq, 0.01M) || H^+(aq, 0.1M) | H_2(g, 1bar) | Pt(s)$, is an entropy driven process.
(c) For racemization of an optically active compound, $\Delta S > 0$.
(d) $\Delta S > 0$, for $[Ni(H_2O)_6]^{2+} + 3\,en \rightarrow [Ni(en)_3]^{2+} + 6H_2O$ (where en = ethylenediamine).

28. Some standard electrode potentials at 298 K are given below: **[Adv. 2021]**

Pb^{2+}/Pb	-0.13 V
Ni^{2+}/Ni	-0.24 V
Cd^{2+}/Cd	-0.40 V
Fe^{2+}/Fe	-0.44 V

To a solution containing 0.001 M of X^{2+} and 0.1 M of Y^{2+}, the metal rods X and Y are inserted (at 298 K) and connected by a conducting wire. This resulted in dissolution of X. The correct combination(s) of X and Y, respectively, is (are)
(Given: Gas constant, $R = 8.314\ J\ K^{-1}\ mol^{-1}$, Faraday constant, $F = 96500\ C\ mol^{-1}$)
(a) Cd and Ni (b) Cd and Fe
(c) Ni and Pb (d) Ni and Fe

29. In a galvanic cell, the salt bridge **[Adv. 2014]**
(a) Does not participate chemically in the cell reaction
(b) Stops the diffusion of ions from one electrode to another
(c) Is necessary for the occurrence of the cell reaction
(d) Ensures mixing of the two electrolytic solutions

30. For the reduction of NO_3^- ion in an aqueous solution, E° is + 0.96V. Values of E° for some metal ions are given below

$V^{2+}(aq) + 2e^- \rightarrow V$	$E^\circ = -1.19$ V
$Fe^{3+}(aq) + 3e^- \rightarrow Fe$	$E^\circ = -0.04$ V
$Au^{3+}(aq) + 3e^- \rightarrow Au$	$E^\circ = +1.40$ V
$Hg^{2+}(aq) + 2e^- \rightarrow Hg$	$E^\circ = +0.86$ V

The pair(s) of metals that is (are) oxidized by NO_3^- in aqueous solution is (are) **[2009]**
(a) V and Hg (b) Hg and Fe
(c) Fe and Au (d) Fe and V

31. The standard reduction potential values of three metallic cations, X, Y and Z are 0.52, –3.03 and –1.18 V respectively. The order of reducing power of the corresponding metals is
(a) $Y > Z > X$ (b) $X > Y > Z$ **[1998 - 2 Marks]**
(c) $Z > Y > X$ (d) $Z > X > Y$

7 Match the Following

32. The standard reduction potential data at 25°C is given below : **[Adv. 2013]**
$E^\circ(Fe^{3+}, Fe^{2+}) = +0.77$ V; $E^\circ(Fe^{2+}, Fe) = -0.44$ V; $E^\circ(Cu^{2+}, Cu) = +0.34$ V; $E^\circ(Cu^+, Cu) = +0.52$ V
$E^\circ[O_2(g) + 4H^+ + 4e^- \rightarrow 2H_2O] = +1.23$ V; $E^\circ[O_2(g) + 2H_2O + 4e^- \rightarrow 4OH^-] = +0.40$ V
$E^\circ(Cr^{3+}, Cr) = -0.74$ V; $E^\circ(Cr^{2+}, Cr) = -0.91$ V
Match E° of the redox pair in List-I with the values given in List-II and select the correct answer using the code given below the lists:

	List-I		List-II
P.	$E^\circ(Fe^{3+}, Fe)$	1.	-0.18 V
Q.	$E^\circ(4H_2O \rightleftharpoons 4H^+ + 4OH^-)$	2.	-0.4 V
R.	$E^\circ(Cu^{2+} + Cu \rightarrow 2Cu^+)$	3.	-0.04 V
S.	$E^\circ(Cr^{3+}, Cr^{2+})$	4.	-0.83 V

Codes :

	P	Q	R	S
(a)	4	1	2	3
(b)	2	3	4	1
(c)	1	2	3	4
(d)	3	4	1	2

8 Comprehension Based Questions

Passage-I

The electrochemical cell shown below is a concentration cell.
$M | M^{2+}$ (saturated solution of a sparingly soluble salt, MX_2) $|| M^{2+}$ (0.001 mol dm^{-3}) $| M$.
The emf of the cell depends on the difference in concentrations of M^{2+} ions at the two electrodes. The emf of the cell at 298 K is 0.059 V. **[2012]**

33. The value of ΔG (kJ mol^{-1}) for the given cell is (take 1F = 96500 C mol^{-1})
(a) –5.7 (b) 5.7 (c) 11.4 (d) –11.4

Passage-II

The concentration of potassium ions inside a biological cell is atleast twenty times higher than the outside. The resulting potential difference across the cell is important in several processes such as transmission of nerve impulses and maintaining the ion balance. A simple model for such a concentration cell involving a metal M is
$M(s) | M^+(aq; 0.05\ molar) || M^+(aq; 1\ molar) | M(s)$
For the above electrolytic cell the magnitude of the cell potential $|E_{cell}| = 70$ mV. **[2010]**

34. For the above cell
(a) $E_{cell} < 0; \Delta G > 0$ (b) $E_{cell} > 0; \Delta G < 0$
(c) $E_{cell} < 0; \Delta G^\circ > 0$ (d) $E_{cell} > 0; \Delta G^\circ < 0$

35. If the 0.05 molar solution of M^+ is replaced by a 0.0025 molar M^+ solution, then the magnitude of the cell potential would be
(a) 35mV (b) 70mV (c) 140mV (d) 700mV

Passage: III

Redox reactions play a pivotal role in chemistry and biology. The values of standard redox potential (E°) of two half-cell reactions decide which way the reaction is expected to proceed. A simple example is a Daniell cell in which zinc goes into solution and copper gets deposited. Given below are a set of half-cell reactions (acidic medium) along with their $E°$ (V with respect to normal hydrogen electrode) values. Using this data, obtain the correct explanations to questions given. [2007]

$I_2 + 2e^- \rightarrow 2I^-$ $E° = 0.54$
$Cl_2 + 2e^- \rightarrow 2Cl^-$ $E° = 1.36$
$Mn^{3+} + e^- \rightarrow Mn^{2+}$ $E° = 1.50$
$Fe^{3+} + e^- \rightarrow Fe^{2+}$ $E° = 0.77$
$O_2 + 4H^+ + 4e^- \rightarrow 2H_2O$ $E° = 1.23$

36. Among the following, identify the correct statement.
(a) Chloride ion is oxidised by O_2
(b) Fe^{2+} is oxidised by iodine
(c) Iodide ion is oxidised by chlorine
(d) Mn^{2+} is oxidised by chlorine

37. While Fe^{3+} is stable, Mn^{3+} is not stable in acid solution because
(a) O_2 oxideses Mn^{2+} to Mn^{3+}
(b) O_2 oxideses both Mn^{2+} to Mn^{3+} and Fe^{2+} to Fe^{3+}
(c) Fe^{3+} oxideses H_2O to O_2
(d) Mn^{3+} oxideses H_2O to O_2

38. Sodium fusion extract, obtained from aniline, on treatment with iron (II) sulphate and H_2SO_4 in presence of air gives a Prussian blue precipitate. The blue colour is due to the formation of
(a) $Fe_4[Fe(CN)_6]_3$ (b) $Fe_3[Fe(CN)_6]_2$
(c) $Fe_4[Fe(CN)_6]_2$ (d) $Fe_3[Fe(CN)_6]_3$

Passage-IV

Tollen's test is given by aldehydes.

$Ag^+ + e^- \longrightarrow Ag;\ E°_{red} = +0.800\ V$

$C_6H_{12}O_6 + H_2O \longrightarrow C_6H_{12}O_7 + 2H^+ + 2e^-;\ E°_{ox} = -0.05V$

Gluconic acid

$[Ag(NH_3)_2]^+ + e^- \longrightarrow Ag + 2NH_3;\ E°_{red} = 0.373V$

Given $\frac{2.303RT}{F} = 0.0591\ \&\ \left(\frac{F}{RT}\right) = 38.92V^{-1}$

39. Calculate ($\ln K$) for [2006 - 5M, –2]

$$C_6H_{12}O_6 + 2Ag^+ + H_2O \longrightarrow C_6H_{12}O_7 + 2H^+ + 2Ag$$

(a) 55.6 (b) 29.6 (c) 66 (d) 58.38

40. On adding NH_3, pH of the solution increases to 11 then, identify the effect on potential of half-cell [2006 - 5M, –2]
(a) E_{ox} increased from $E°_{ox}$ by 0.65 V
(b) E_{ox} decreased from $E°_{ox}$ by 0.65 V
(c) E_{red} increased from $E°_{red}$ by 0.65 V
(d) E_{red} decreased from $E°_{red}$ by 0.65 V

41. NH_3 is used in this reaction rather than any other base. Select the correct statement out of the following [2006 - 5M, –2]
(a) $[Ag(NH_3)_2]^+$ is a weaker oxidizing agent than Ag^+
(b) to dissolve the insoluble silver oxide formed under the reaction conditions
(c) Ag precipitates gluconic acid as its silver salt
(d) NH_3 changes the standard reduction potential of $[Ag(NH_3)_2]^+$

10 Subjective Problems

42. (a) For the reaction

$$Ag^+(aq) + Cl^-(aq) \rightleftharpoons AgCl(s)$$

Given :

Species	ΔG_f° (kJ/mol)
$Ag^+(aq)$	+77
$Cl^-(aq)$	– 129
AgCl (s)	– 109

Write the cell representation of above reaction and calculate $E°_{cell}$ at 298 K. Also find the solubility product of AgCl.

(b) If 6.539×10^{-2} g of metallic zinc is added to 100 mL saturated solution of AgCl. Find the value of $\log_{10}\frac{[Zn^{2+}]}{[Ag^+]^2}$.

How many moles of Ag will be precipitated in the above reaction. Given that [2005 - 6 Marks]

$Ag^+ + e^- \longrightarrow Ag;\ E° = 0.80\ V;$
$Zn^{2+} + 2e^- \longrightarrow Zn;\ E° = -0.76\ V$
(It was given that Atomic mass of Zn = 65.39)

43. Find the equilibrium constant for the reaction,

$In^{2+} + Cu^{2+} \longrightarrow In^{3+} + Cu^+$ at 298 K

given : $E°_{Cu^{2+}/Cu^+} = 0.15\ V;\ E°_{In^{2+}/In^+} = -0.40\ V,$

$E°_{In^{3+}/In^+} = -0.42\ V$ [2004 - 4 Marks]

44. The standard potential of the following cell is 0.23 V at 15°C and 0.21 V at 35°C. [2001 - 10 Marks]

Pt | H_2(g) | HCl(aq) | AgCl(s) | Ag(s)

(i) Write the cell reaction.
(ii) Calculate $\Delta H°$ and $\Delta S°$ for the cell reaction by assuming that these quantities remain unchanged in the range 15°C to 35°C.
(iii) Calculate the solubility of AgCl in water at 25°C.
Given : The standard reduction potential of the $Ag^+(aq)/Ag(s)$ couple is 0.80 V at 25°C.

45. The following electrochemical cell has been set up.
Pt(1) | Fe^{3+}, Fe^{2+} (a = 1) | Ce^{4+}, Ce^{3+} (a = 1) | Pt (2)
$E°(Fe^{3+}, Fe^{2+}) = 0.77\ V : E°(Ce^{4+}/Ce^{3+}) = 1.61\ V$
If an ammeter is connected between the two platinum electrodes, predict the direction of flow of current. Will the current increase or decrease with time? [2000 - 2 Marks]

46. A cell, Ag | Ag^+ || Cu^{2+} | Cu, initiallly contains 1 M Ag^+ and 1 M Cu^{2+} ions. Calculate the change in the cell potential after the passage of 9.65 A of current for 1 h. [1999 - 6 Marks]

47. Find the solubility product of a saturated solution of Ag_2CrO_4 in water at 298 K if the emf of the cell Ag | Ag^+ (satd. Ag_2CrO_4 soln.) || Ag^+(0.1M) | Ag is 0.164 V at 298 K. [1998 - 6 Marks]

48. Calculate the equilibrium constant for the reaction, $2Fe^{3+} + 3I^- \rightleftharpoons 2Fe^{2+} + I_3^-$. The standard reduction potentials in acidic conditions are 0.77 V and 0.54 V respectively for Fe^{3+} | Fe^{2+} and I_3^- | I^- couples. [1998 - 3 Marks]

49. Calculate the equilibrium constant for the reaction

$Fe^{2+} + Ce^{4+} \rightleftharpoons Fe^{3+} + Ce^{3+}$ [1997 - 2 Marks]

(given $E°_{Ce^{4+}/Ce^{3+}} = 1.44V;\ E°_{Fe^{3+}/Fe^{2+}} = 0.68V;$)

50. Although aluminium is above hydrogen in the electrochemical series, it is stable in air and water. Explain. [1994 - 1 Mark]

51. The Edison storage cells is represented as
Fe(s) | FeO(s) | KOH(aq) | Ni_2O_3(s) | Ni(s)
The half-cell reactions are :

$Ni_2O_3(s) + H_2O(l) + 2e^- \rightleftharpoons 2NiO_{(s)} + 2OH^-;\ E° = +0.40V$

$FeO(s) + H_2O(l) + 2e^- \rightleftharpoons Fe(s) + 2OH^-;\ E° = -0.87V$

(i) What is the cell reaction ?

(*ii*) What is the cell e.m.f ? How does it depend on the concentration of KOH?

(*iii*) What is the maximum amount of electrical energy that can be obtained from one mole of Ni_2O_3?

The standard reduction potential of the Ag^+ /Ag electrode at 298 K is 0.799 V. Given that for AgI, $K_{sp} = 8.7 \times 10^{-17}$, evaluate the potential of the Ag^+/Ag electrode in a saturated solution of AgI. Also calculate the standard reduction potential of the I^-/AgI/Ag electrode.

The standard reduction potential for the half-cell

$NO_3^-(aq) + 2H^+(aq) + e \rightarrow NO_2(g) + H_2O$ is 0.78 V.

(i) Calculate the reduction potential in 8 M H^+

(ii) What will be the reduction potential of the half-cell in a neutral solution? Assume all the other species to be at unit concentration.

An aqueous solution of NaCl on electrolysis gives $H_2(g)$, $Cl_2(g)$ and NaOH according to the reaction :

$2Cl^-(aq) + 2H_2O = 2OH^-(aq) + H_2(g) + Cl_2(g)$.

A direct current of 25 amperes with a current efficiency of 62% is passed through 20 litres of NaCl solution (20% by weight). Write down the reactions taking place at the anode and the cathode. How long will it take to produce 1 kg of Cl_2? What will be the molarity of the solution with respect to hydroxide ion? (Assume no loss due to evaporation.)

For the galvanic cell.

Ag | AgCl(s), KCl(0.2 M) || KBr (0.001M), AgBr(s) | Ag

Calculate the EMF generated and assign correct polarity to each electrode for a spontaneous process after taking into account the cell reaction at 25°C.

[$K_{sp}(AgCl) = 2.8 \times 10^{-10}$; $K_{sp}(AgBr) = 3.3 \times 10^{-13}$]

Zinc granules are added in excess to a 500 mL. of 1.0 M nickel nitrate solution at 25°C until the equilibrium is reached. If the standard reduction potential of Zn^{2+} | Zn and Ni^{2+} | Ni are –0.75 V and –0.24 V respectively, find out the concentration of Ni^{2+} in solution at equilibrium.

The standard reduction potential of Cu^{++}/Cu and Ag^+/Ag electrodes are 0.337 and 0.799 volt respectively. Construct a galvanic cell using these electrodes so that its standard e.m.f. is positive. For what concentration of Ag^+ will the e.m.f. of the cell, at 25°C, be zero if the concentration of Cu^{++} is 0.01 M?

The standard reduction potential at 25°C of the reaction, $2H_2O + 2e^- \rightleftharpoons H_2 + 2OH^-$ is –0.8277V. Calculate the equilibrium constant for the reaction $2H_2O \rightleftharpoons H_3O^+ + OH^-$ at 25°C.

A cell contains two hydrogen electrodes. The negative electrode is in contact with a solution of 10^{-6} M hydrogen ions. The EMF of the cell is 0.118 V at 25°C. Calculate the concentration of hydrogen ions at the positive electrode.

The EMF of a cell corresponding to the reaction :

$Zn(s) + 2H^+(aq) \rightarrow Zn^{2+} + (0.1\ M) + H_2\ (g)\ (1\ atm.)$

is 0.28 volt at 25°C.

Write the half-cell reactions and calculate the pH of the solution at the hydrogen electrode.

$E^{o}_{Zn^{2+}/Zn} = -0.76$ volt; $E^{o}_{H^+/H_2} = 0$

Consider the cell

$Zn \mid Zn^{2+}(aq)\ (1.0\ M) \parallel Cu^{2+}(aq)\ (1.0\ M) \mid Cu$.

The standard reduction potentials are :

+0.350 volts for $2e^- + Cu^{2+}(aq) \rightarrow Cu$ and –0.763 volts for $2e^- + Zn^{2+}(aq) \rightarrow Zn$

(i) Write down the cell reaction.

(ii) Calculate the emf of the cell.

(iii) Is the cell reaction spontaneous or not?

(a) 19 g of molten $SnCl_2$ is electrolysed for some time. Inert electrodes are used. 0.119 g of Sn is deposited at the cathode. No substance is lost during the electrolysis. Find the ratio of the weights of $SnCl_2$: $SnCl_4$ after electrolysis.

(b) A hot solution of NaCl in water is electrolysed. Iron electrodes are used. Diaphragm cell is not used. Give equations for all the chemical reactions that take place during electrolysis.

(c) Find the charge in coulombs of 1 gram ion of N^{3-}.

MCQs with One Correct Answer

At 300 K, the molar conductivities of the aqueous solutions of three salts at two different concentrations are given below:

$NaNO_3$	0.01	111
	0.04	101
NaCl	0.01	117
	0.04	107
$AgNO_3$	0.01	125
	0.04	116

The conductivity of a saturated aqueous solution of AgCl is 1.40×10^{-6} S cm^{-1} at 300 K. If the solubility of AgCl in water at 300 K is X mol L^{-1}, then $\log_{10}(X^{-1})$ is

(Assume that AgCl dissolved in water ionizes completely and that the molar conductivity of saturated AgCl solution is equal to its limiting molar conductivity.)

(a) 3 (b) 4 (c) 5 (d) 6

Plotting $1/\Lambda_m$ against $c\Lambda_m$ for aqueous solutions of a monobasic weak acid (HX) resulted in a straight line with y–axis intercept of P and slope of S. The ratio P/S is

[Λ_m = molar conductivity, Λ_m° = limiting molar conductivity

c = molar concentration, K_a = dissociation constant of HX]

(a) $K_a\Lambda_m^{\circ}$ (b) $K_a\Lambda_m^{\circ}/2$ (c) $2K_a\Lambda_m^{\circ}$ (d) $1/(K_a\Lambda_m^{\circ})$

Molar conductivity of aqueous solution of sodium stearate, which behaves as a strong electrolyte is recorded at varying concentration (C) of sodium stearate. Which one of the following plots provides the correct representation of micelle formation in the solution?

(Critical micelle concentration (CMC) is marked with an arrow in the figures)

(a) 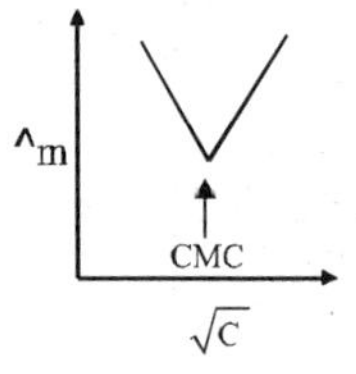

(b) 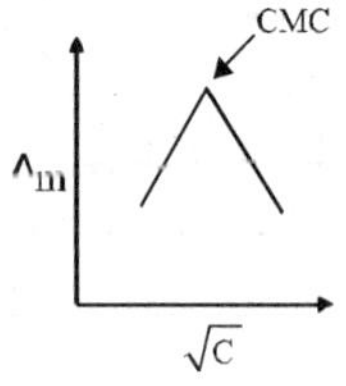

(c) 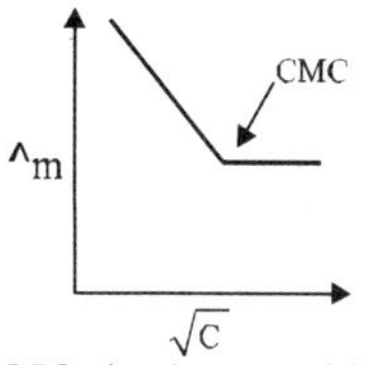 (d) 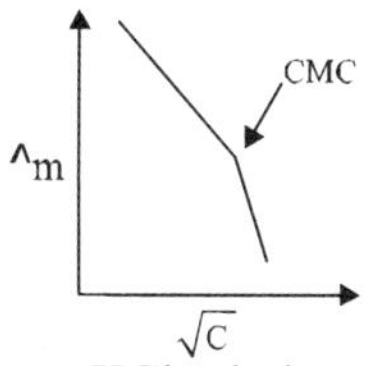

AgNO$_3$(aq.) was added to an aqueous KCl solution gradually and the conductivity of the solution was measured. The plot of conductance (Λ) versus the volume of AgNO$_3$ is

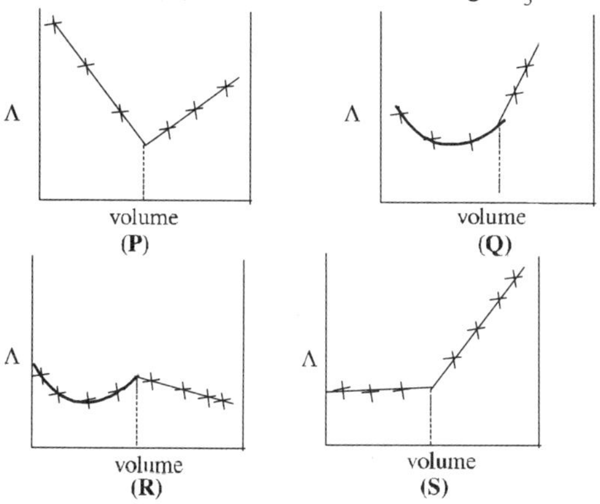

(a) (P) (b) (Q) (c) (R) (d) (S)

The correct order of equivalent conductance at infinite dilution of LiCl, NaCl and KCl is

(a) LiCl > NaCl > KCl (b) KCl > NaCl > LiCl
(c) NaCl > KCl > LiCl (d) LiCl > KCl > NaCl

Integer Value Answer

Consider the strong electrolytes Z_mX_n, U_mY_p and V_mX_n. Limiting molar conductivity (Λ^0) of U_mY_p and V_mX_n are 250 and 440 S cm^2 mol^{-1}, respectively. The value of $(m + n + p)$ is _______.

Given:

Ion	Z^{n+}	U^{p+}	V^{n+}	X^{m-}	Y^{m-}
$\lambda^0\left(\text{S cm}^2\ \text{mol}^{-1}\right)$	50.0	25.0	100.0	80.0	100.0

λ^0 is the limiting molar conductivity of ions.

The plot of molar conductivity (Λ) of Z_mX_n vs $c^{1/2}$ is given below.

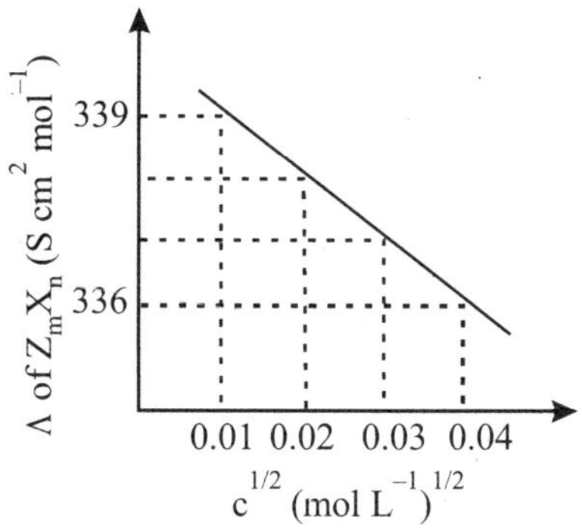

The conductance of a 0.0015 M aqueous solution of a weak monobasic acid was determined by using a conductivity cell consisting of platinized Pt electrodes. The distance between the electrodes is 120 cm with an area of cross section of 1 cm^2. The conductance of this solution was found to be 5×10^{-7} S. The pH of the solution is 4. The value of limiting molar conductivity (Λ_m°) of this weak monobasic acid in aqueous solution is $Z \times 10^2$ S cm^2 mol^{-1}. The value of Z is

The molar conductivity of a solution of a weak acid HX (0.01 M) is 10 times smaller than the molar conductivity of a solution of a weak acid HY (0.10 M). If $\lambda^0_{X^-} \approx \lambda^0_{Y^-}$ the difference in their pK_a values, $pK_a(HX) - pK_a(HY)$, is (consider degree of ionization of both acids to be <<1)

Numeric / New Stem Based Questions

At 298 K, the limiting molar conductivity of a weak monobasic acid is 4×10^2 S cm^2 mol^{-1}. At 298 K, for an aqueous solution of the acid the degree of dissociation of α and the molar conductivity is $\times 10^2$ S cm^2 mol^{-1}. At 298 K, upon 20 times dilution with water, the molar conductivity of the solution becomes 3 $\times 10^2$ S cm^2 mol^{-1}.

The value of α is ____.

The value of is ____.

Fill in the Blanks

The electrical conductivity of a solution of acetic acid will be .... if a solution of sodium hydroxide is added.

MCQs with One or More than One Correct Answer

An aqueous solution of hydrazine (N_2H_4) is electrochemically oxidized by O_2, thereby releasing chemical energy in the form of electrical energy. One of the products generated from the electrochemical reaction is N_2(g).

Choose the correct statement(s) about the above process:

(a) OH^- ions react with N_2H_4 at the anode to form N_2(g) and water, releasing 4 electrons to the anode.
(b) At the cathode, N_2H_4 breaks to N_2(g) and nascent hydrogen released at the electrode reacts with oxygen to form water.
(c) At the cathode, molecular oxygen gets converted to OH^-.
(d) Oxides of nitrogen are major by-products of the electrochemical process.

Match the Following

In a conductometric titration, small volume of titrant of higher concentration is added stepwise to a larger volume of titrate of much lower concentration, and the conductance is measured after each addition.

The limiting ionic conductivity (Λ_0) values (in mS m^2 mol^{-1}) for different ions in aqueous solutions are given below:

Ag^+	K^+	Na^+	H^+	NO_3^-	Cl^-	SO_4^{2-}	OH^-	CH_3COO^-
6.2	7.4	5.0	35.0	7.2	7.6	16.0	19.9	4.1

For different combinations of titrates and titrants given in , the graphs of 'conductance' versus 'volume of titrant' are given in .

Match each entry in with the appropriate entry in and choose the correct option.

(P) Titrate: KCl
Titrant: AgNO$_3$

(1) 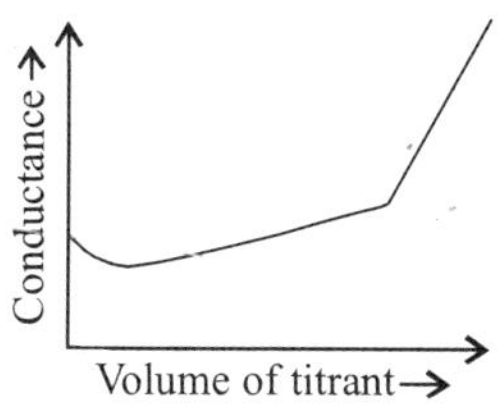

(Q) Titrate: $AgNO_3$
Titrant: KCl

(2) 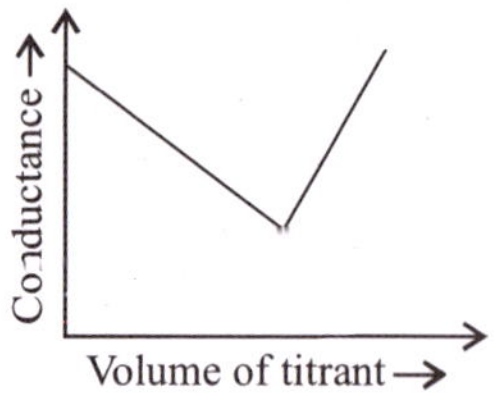

(R) Titrate: NaOH
Titrant: HCl

(3) 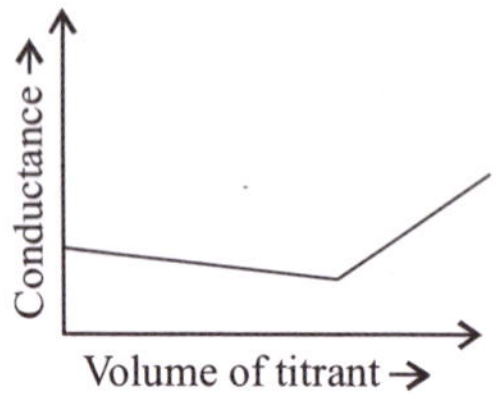

(S) Titrate: NaOH
Titrant: CH_3COOH

(4) 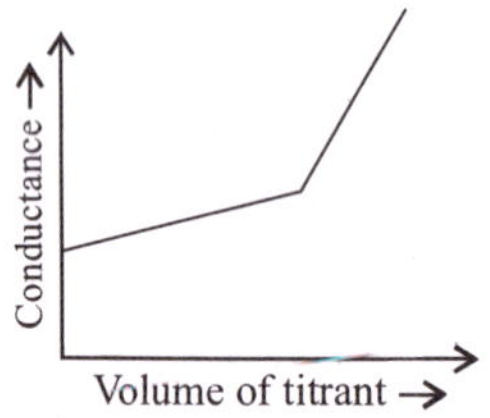

(5) 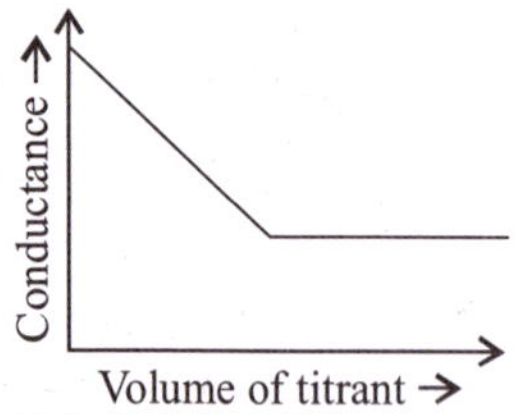

(a) P-4, Q-3, R-2, S-5 (b) P-2, Q-4; R-3; S-1
(c) P-3, Q-4, R-2, S-5 (d) P-4; Q-3; R-2; S-1

76. An aqueous solution of X is added slowly to an aqueous solution of Y as shown in List I. The variation in conductivity of these reactions is given in List II. Match list I with List II and select the correct answer using the code given below the lists : **[Adv. 2013]**

	List-I		List-II
P.	$(C_2H_5)_3N + CH_3COOH$ (X, Y)	1.	Conductivity decreases and then increases
Q.	$KI\,(0.1M) + AgNO_3(0.01M)$ (X, Y)	2.	Conductivity decreases and then does not change much
R.	$CH_3COOH + KOH$ (X, Y)	3.	Conductivity increases and then does not change much
S.	$NaOH + HI$ (X, Y)	4.	Conductivity does not change much and then increases

Codes :

	P	Q	R	S
(a)	3	4	2	1
(b)	4	3	2	1
(c)	2	3	4	1
(d)	1	4	3	2

77. Match the following, choosing one item from column X and one from column Y. **[1982 - 2 Marks]**

	X		Y
(i)	neutrons	(p)	Kohlrausch
(ii)	molecular speed	(q)	van der Waals
(iii)	intermolecular forces	(r)	Maxwell
(iv)	conductance of ions	(s)	Chadwick

10 Subjective Problems

78. We have taken a saturated solution of AgBr. K_{sp} of AgBr is 12×10^{-14}. If 10^{-7} mole of $AgNO_3$ are added to 1 litre of this solution, find conductivity (specific conductance) of this solution in terms of 10^{-7} S m^{-1} units. Given, Molar conductance of Ag^+, Br^- and NO_3^- are 6×10^{-3} Sm^2mol^{-1}, 8×10^{-3} Sm^2mol^{-1} and 7×10^{-3} Sm^2mol^{-1}. **[2006 - 6M]**

Topic 3 - Electrolytic Cells and Electrolysis

1 MCQs with One Correct Answer

79. Electrolysis of dilute aqueous NaCl solution was carried out by passing 10 milli ampere current. The time required to liberate 0.01 mol of H_2 gas at the cathode is (1 Faraday = 96500 C mol^{-1}) **[2008S]**

(a) 9.65×10^4 sec (b) 19.3×10^4 sec
(c) 28.95×10^4 sec (d) 38.6×10^4 sec

80. The electric charge for electrode deposition of one gram equivalent of a substance is : **[1984 - 1 Mark]**

(a) one ampere per second.
(b) 96,500 coloumbs per second.
(c) one ampere for one hour.
(d) charge on one mole of electrons.

81. Faraday's laws of electrolysis are related to the **[1983 - 1 Mark]**

(a) atomic number of the reactants.
(b) atomic number of the anion.
(c) equivalent weight of the electrolyte.
(d) speed of the cation.

3 Numeric / New Stem Based Questions

82. An electrochemical cell is fueled by the combustion of butane at 1 bar and 298 K. Its cell potential is $\frac{X}{F} \times 10^3$ volts, where F is the Faraday constant. The value of X is _____.
Use: Standard Gibbs energies of formation at 298 K are: $\Delta_f G^\circ_{CO_2} = -394$ kJ mol^{-1}; $\Delta_f G^\circ_{water} = -237$ kJ mol^{-1}; $\Delta_f G^\circ_{butane} = -18$ kJ mol^{-1} **[Adv. 2025]**

83. In an electrochemical cell, dichromate ions in aqueous acidic medium are reduced to Cr^{3+}. The current (in amperes) that flows through the cell for 48.25 minutes to produce 1 mole of Cr^{3+} is _____.
Use: 1 Faraday = 96500 C mol^{-1} **[Adv. 2025]**

84. Consider a 70% efficient hydrogen-oxygen fuel cell working under standard conditions at 1 bar and 298 K. Its cell reaction is **[Adv. 2020]**

$$H_2(g) + \frac{1}{2}O_2(g) \rightarrow H_2O(l).$$

The work derived from the cell on the consumption of 1.0×10^{-3} mol of $H_2(g)$ is used to compress 1.00 mol of a monoatomic ideal gas in a thermally insulated container. What is the change in the temperature (in K) of the ideal gas?

The standard reduction potentials for the two half-cells are given below.

$$O_2(g)+4H^+(aq)+4e^- \rightarrow 2H_2O(l),\ E^0=1.23\,V,$$

$$2H^+(aq)+2e^- \rightarrow H_2(g),\ E^0=0.00\,V$$

Use $F = 96500\ C\ mol^{-1}$, $R = 8.314\ J\ mol^{-1}\ K^{-1}$.

85. A current of 1.70 A is passed through 300.0 mL of 0.160 M solution of a $ZnSO_4$ for 230 sec. with a current efficiency of 90%. Find out the molarity of Zn^{2+} after the deposition of Zn. Assume the volume of the solution to remain constant during the electrolysis. **[1991 - 4 Marks]**

86. A 100 watt, 110 volt incandescent lamp is connected in series with an electrolyte cell containing cadmium sulphate solution. What weight of cadmium will be deposited by the current flowing for 10 hours? **[1987 - 5 Marks]**

87. How long a current of 3 ampere has to be passed through a solution of silver nitrate to coat a metal surface of 80 cm^2 with a 0.005 mm thick layer? Density of silver is 10.5 g/cm^3. **[1985 - 3 Marks]**

88. The density of copper is 8.94 g/mL. Find out the number of coulombs needed to plate an area 10 cm × 10 cm to a thickness 10^{-2} cm using $CuSO_4$ solution as electrolyte. **[1979]**

8 Comprehension Based Questions

Chemical reactions involve interaction of atoms and molecules. A large number of atoms/molecules (approximately 6.023×10^{23}) are present in a few grams of any chemical compound varying with their atomic/molecular masses. To handle such large numbers conveniently, the mole concept was introduced. This concept has implications in diverse areas such as analytical chemistry, biochemistry, electrochemistry and radiochemistry. The following example illustrates a typical case, involving chemical/electrochemical reaction, which requires a clear understanding of the mole concept.

A 4.0 molar aqueous solution of NaCl is prepared and 500 mL of this solution is electrolysed. This leads to the evolution of chlorine gas at one of the electrodes (atomic mass : Na = 23, Hg = 200; 1 Faraday = 96500 coulombs).

89. The total number of moles of chlorine gas evolved is **[2007]**
(a) 0.5 (b) 1.0 (c) 2.0 (d) 3.0

90. If the cathode is a Hg electrode, the maximum weight (g) of amalgam formed from this solution is **[2007]**
(a) 200 (b) 225 (c) 400 (d) 446

91. The total charge (coulombs) required for complete electrolysis is **[2007]**
(a) 24125 (b) 48250 (c) 96500 (d) 193000

10 Subjective Problems

92. Copper sulphate solution (250 mL) was electrolysed using a platinum anode and a copper cathode. A constant current of 2 mA was passed for 16 minutes. It was found that after electrolysis, the absorbance of the solution was reduced to 50% of its original value. Calculate the concentration of copper sulphate in the solution to begin with. **[2000 - 3 Marks]**

93. How many grams of silver could be plated out on a serving tray by electrolysis of a solution containing silver in +1 oxidation state for a period of 8.0 hours at a current of 8.46 amperes? What is the area of the tray if the thickness of the silver plating is 0.00254 cm? Density of silver is 10.5 g/cm^3. **[1997 - 3 Marks]**

94. Chromium metal can be plated out from an acidic solution containing CrO_3 according to the following equation.

$$CrO_3(aq)+6H^+(aq)+6e^- \rightarrow Cr(s)+3H_2O$$

Calculate (i) how many grams of chromium will be plated out by 24,000 coulombs and (ii) how long will it take to plate out 1.5 g of chromium by using 12.5 amp current. **[1993 - 2 Marks]**

95. Calculate the quantity of electricity that would be required to reduce 12.3 g of nitrobenzene to aniline, if the current efficiency for the process is 50 per cent. If the potential drop across the cell is 3.0 volts, how much energy will be consumed? **[1990 - 3 Marks]**

96. An acidic solution of Cu^{2+} salt containing 0.4 g of Cu^{2+} is electrolysed until all the copper is deposited. The electrolysis is continued for seven more minutes with the volume of solution kept at 100 mL. and the current at 1.2 amp. Calculate the volume of gases evolved at NTP during the entire electrolysis. **[1989 - 5 Marks]**

97. In a fuel cell, hydrogen and oxygen react to produce electricity. In the process, hydrogen gas is oxidised at the anode and oxygen at the cathode. If 67.2 litre of H_2 at STP react in 15 minutes, what is the average current produced? If the entire current is used for electro deposition of copper from copper (II) solution, how many grams of copper will be deposited?

Anode reaction : $H_2 + 2OH^- \rightarrow 2H_2O + 2e^-$ **[1988 - 4 Marks]**

Cathode reaction : $\frac{1}{2}O_2 + H_2O + 2e^- \rightarrow 2OH^-$.

98. During the discharge of a lead storage battery, the density of sulphuric acid fell from 1.294 to 1.139 g/mL. Sulphuric acid of density 1.294 g/mL is 39% by weight and that of 1.139 g/mL is 20% H_2SO_4 by weight. The battery holds 3.5 litres of the acid and the volume remained practically constant during the discharge.

Calculate the number of ampere-hours for which the battery must have been used. The charging and discharging reactions are : **[1986 - 5 Marks]**

Anode :

$$Pb + SO_4^{2-} = PbSO_4 + 2e^-\ \text{(discharging)}$$

Cathode :

$$PbO_2 + 4H^+ + SO_4^{2-} + 2e^- = PbSO_4 + 2H_2O\ \text{(discharging)}$$

Note : Both the reactions take place at the anode and cathode respectively during discharge. Both reaction get reverse during charging.

99. In an electrolysis experiment, current, was passed for 5 hours through two cells connected in series. The first cell contains a solution of gold and the second contains copper sulphate solution. 9.85 g of gold was deposited in the first cell. If the oxidation number of gold is +3, find the amount of copper deposited on the cathode of the second cell. Also calculate the magnitude of the current in amperes. (1 faraday = 96,500 coulombs) **[1983 - 3 Marks]**

Answer Key

Part-A (JEE Main)

1	(2)	**23**	(1)	**45**	(3)	**67**	(24)	**89**	(3)	**111**	(96500)	**133**	(3)	**155**	(2)	**177**	(3)	**199**	(25)
2	(4)	**24**	(2)	**46**	(1)	**68**	(4)	**90**	(10)	**112**	(2.15)	**134**	(3)	**156**	(1000)	**178**	(4)	**200**	(11)
3	(2)	**25**	(4)	**47**	(3)	**69**	(4)	**91**	(917)	**113**	(1.52)	**135**	(1)	**157**	(3)	**179**	(4)	**201**	(60)
4	(1)	**26**	(2)	**48**	(4)	**70**	(200)	**92**	(34)	**114**	(1)	**136**	(2)	**158**	(14)	**180**	(3)	**202**	(5.68)
5	(3)	**27**	(1)	**49**	(4)	**71**	(10)	**93**	(4)	**115**	(3)	**137**	(2)	**159**	(266)	**181**	(3)	**203**	(2)
6	(4)	**28**	(2)	**50**	(2)	**72**	(400)	**94**	(7)	**116**	(2)	**138**	(2)	**160**	(12)	**182**	(2)	**204**	(1)
7	(2)	**29**	(3)	**51**	(3)	**73**	(6)	**95**	(16)	**117**	(4)	**139**	(1)	**161**	(760)	**183**	(3)	**205**	(1)
8	(2)	**30**	(3)	**52**	(2)	**74**	(3)	**96**	(51)	**118**	(3)	**140**	(3)	**162**	(288)	**184**	(3)	**206**	(4)
9	(4)	**31**	(1)	**53**	(4)	**75**	(338.9)	**97**	(983)	**119**	(1)	**141**	(2)	**163**	(57)	**185**	(4)	**207**	(3)
10	(2)	**32**	(3)	**54**	(4)	**76**	(1)	**98**	(23)	**120**	(2)	**142**	(2)	**164**	(14)	**186**	(1)	**208**	(1)
11	(2)	**33**	(1)	**55**	(4)	**77**	(18)	**99**	(5)	**121**	(2)	**143**	(15)	**165**	(2)	**187**	(4)	**209**	(4)
12	(1)	**34**	(4)	**56**	(1)	**78**	(25)	**100**	(25)	**122**	(1)	**144**	(4)	**166**	(1)	**188**	(111)	**210**	(2)
13	(1)	**35**	(2)	**57**	(1)	**79**	(4)	**101**	(109)	**123**	(3)	**145**	(2)	**167**	(3)	**189**	(2)	**211**	(1)
14	(4)	**36**	(4)	**58**	(1)	**80**	(2)	**102**	(28)	**124**	(2)	**146**	(6)	**168**	(3)	**190**	(1.93)	**212**	(1)
15	(1)	**37**	(2)	**59**	(3)	**81**	(1825)	**103**	(24)	**125**	(1)	**147**	(3)	**169**	(1)	**191**	(5)	**213**	(3)
16	(3)	**38**	(1)	**60**	(4)	**82**	(1)	**104**	(3)	**126**	(4)	**148**	(23)	**170**	(4)	**192**	(2)	**214**	(3)
17	(1)	**39**	(3)	**61**	(3)	**83**	(3)	**105**	(147)	**127**	(4)	**149**	(150)	**171**	(1)	**193**	(45.75)	**215**	(2)
18	(1)	**40**	(1)	**62**	(2)	**84**	(3)	**106**	(4)	**128**	(N)	**150**	(4)	**172**	(2)	**194**	(108)	**216**	(560)
19	(2)	**41**	(3)	**63**	(4)	**85**	(6)	**107**	(3776)	**129**	(4)	**151**	(3)	**173**	(1)	**195**	(161)		
20	(3)	**42**	(2)	**64**	(1)	**86**	(10)	**108**	(– 6)	**130**	(1)	**152**	(66)	**174**	(1)	**196**	(3)		
21	(2)	**43**	(3)	**65**	(2)	**87**	(275)	**109**	(142)	**131**	(2)	**153**	(13.4)	**175**	(3)	**197**	(127)		
22	(3)	**44**	(2)	**66**	(78)	**88**	(17)	**110**	(144)	**132**	(1)	**154**	(25)	**176**	(3)	**198**	(20)		

Part-B (JEE Advanced)

1	(b)	**9**	(b)	**17**	(a)	**26**	(False)	**34**	(b)	**63**	(c)	**71**	(0.22)	**80**	(d)	**88**	(27171.96)		
2	(d)	**10**	(a)	**18**	(10)	**27**	(b, c, d)	**35**	(c)	**64**	(a)	**72**	(0.86)	**81**	(c)	**89**	(b)		
3	(d)	**11**	(c)	**19**	(4)	**28**	(a, b, c)	**36**	(c)	**65**	(d)	**74**	(a, c)	**82**	(105.5)	**90**	(d)		
4	(a)	**12**	(a)	**20**	(0.77)	**29**	(a)	**37**	(d)	**66**	(d)	**75**	(c)	**83**	(100)	**91**	(d)		
5	(b)	**13**	(b)	**21**	(– 11.62)	**30**	(a, b, d)	**38**	(a)	**67**	(b)	**76**	(a)	**84**	(13.32)				
6	(c)	**14**	(a)	**22**	(0.05)	**31**	(a)	**39**	(d)	**68**	(7)	**77**	(i)-(s); (ii)-(r); (iii)-(q); (iv)-(p)	**85**	(0.154)				
7	(a)	**15**	(c)	**23**	(–0.22)	**32**	(d)	**40**	(a)	**69**	(6)			**86**	(19.06)				
8	(c)	**16**	(c)	**24**	(0.792)	**33**	(d)	**41**	(b)	**70**	(3)	**79**	(b)	**87**	(125.09)				

Chapter-4

Chemical Kinetics and Nuclear Chemistry

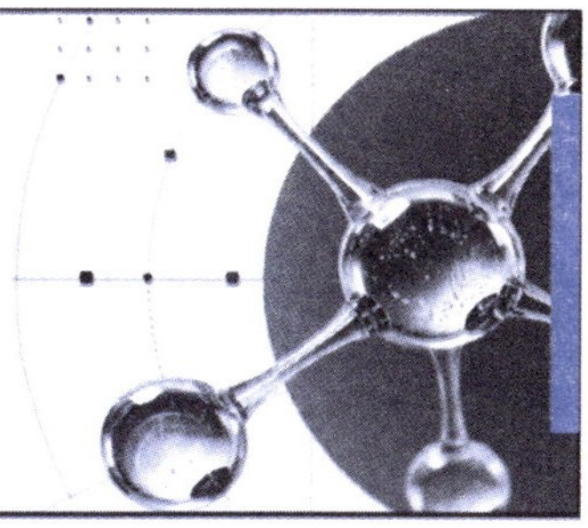

Part-A (JEE Main)

Topic 1- Rate of Reaction, Rate Laws and Rate Constant

MULTIPLE CHOICE QUESTIONS (MCQs)

1. Observe the following reactions at T (K).
I. A $\rightarrow$ products. [Jan. 23, 2026 (II)]
II. $5Br^-(aq) + BrO_3^-(aq) + 6H^+(aq) \rightarrow 3Br_2(aq) + 3H_2O(l)$
Both the reactions are started at 10.00 am. The rates of these reactions at 10.10 am are same. The value of $-\frac{\Delta[Br^-]}{\Delta t}$ at 10.10 am is 2×10^{-4} mol L^{-1} Min^{-1}. The concentration of A at 10.10 am is 10^{-2} mol L^{-1}. What is the first order rate constant (in min^{-1}) of reaction I?
(1) 2×10^{-3} (2) 10^{-3} (3) 10^{-2} (4) 4×10^{-3}

2. A $\rightarrow$ products (First order reaction). [Jan. 22, 2026 (I)]
Three sets of experiment were performed for a reaction under similar experimental conditions.
Run 1 $\Rightarrow$ 100 mL of 10 M solution of reactant A
Run 2 $\Rightarrow$ 200 mL of 10 M solution of reactant A
Run 3 $\Rightarrow$ 100 mL of 10 M solution of reactant A + 100 mL of H_2O added.
The correct variation of rate of reaction is
(1) Run 1 = Run 2 = Run 3 (2) Run 3 < Run 1 = Run 2
(3) Run 3 < Run 1 < Run 2 (4) Run 1 < Run 2 < Run 3

3. Decomposition of A is a first order reaction at T(K) and is given by $A(g) \rightarrow B(g) + C(g)$. In a closed 1 L vessel, 1 bar A(g) is allowed to decompose at T(K). After 100 minutes, the total pressure was 1.5 bar. What is the rate constant (in min^{-1}) of the reaction? ($\log 2 = 0.3$) [Jan. 21, 2026 (II)]
(1) 6.9×10^{-1} (2) 6.9×10^{-3}
(3) 6.9×10^{-2} (4) 6.9×10^{-4}

4. In a first order decomposition reaction, the time taken for the decomposition of reactant to one fourth and one eighth of its initial concentration are t_1 and t_2 (s), respectively.
The ratio $\frac{t_1}{t_2}$ will : [April 8, 2025 (II)]
(1) $\frac{4}{3}$ (2) $\frac{3}{2}$ (3) $\frac{3}{4}$ (4) $\frac{2}{3}$

5. $A(g) \rightarrow B(g) + C(g)$ is a first order reaction. [April 7, 2025 (II)]

Time	T	∞
P_{system}	P_t	P_∞

The reaction was started with reactant *A* only. Which of the following expression is correct for rate constant *k* ?
(1) $k = \frac{1}{t}\ln\frac{2(P_\infty - P_t)}{P_t}$ (2) $k = \frac{1}{t}\ln\frac{P_\infty}{P_t}$
(3) $k = \frac{1}{t}\ln\frac{P_\infty}{2(P_\infty - P_t)}$ (4) $k = \frac{1}{t}\ln\frac{P_\infty}{(P_\infty - P_t)}$

6. Reaction $A(g) \rightarrow 2B(g) + C(g)$ is a first order reaction. It was started with pure A [April 7, 2025 (I)]

t/min	Pressure of system at time t/mm Hg
10	160
∞	240

Which of the following option is incorrect ?
(1) Initial pressure of A is 80 mm Hg
(2) The reaction never goes to completion
(3) Rate constant of the reaction is 1.693 min^{-1}
(4) Partial pressure of A after 10 minute is 40 mm Hg

7. Rate law for a reaction between *A* and *B* is given by
$R = k[A]^n[B]^m$ [April 4, 2025 (I)]
If concentration of *A* is doubled and concentration of *B* is halved from their initial value, the ratio of new rate of reaction to the initial rate of reaction $\left(\frac{r_2}{r_1}\right)$ is
(1) $2^{(n-m)}$ (2) $(n-m)$
(3) $(m+n)$ (4) $\frac{1}{2^{m+n}}$

8. For a reaction $A \xrightarrow{K_1} B \xrightarrow{K_2} C$
If the rate of formation of B is set to be zero then the concentration of B is given by : [April 8, 2024 (II)]
(1) $K_1K_2[A]$ (2) $(K_1 - K_2)[A]$
(3) $(K_1 + K_2)[A]$ (4) $(K_1/K_2)[A]$

9. For the reaction $2A+3B+\frac{3}{2}C \rightarrow 3P$, which statement is correct? [Sep. 03, 2020 (II)]
(1) $\frac{dn_A}{dt}=\frac{3}{2}\frac{dn_B}{dt}=\frac{3}{4}\frac{dn_C}{dt}$ (2) $\frac{dn_A}{dt}=\frac{dn_B}{dt}=\frac{dn_C}{dt}$
(3) $\frac{dn_A}{dt}=\frac{2}{3}\frac{dn_B}{dt}=\frac{4}{3}\frac{dn_C}{dt}$ (4) $\frac{dn_A}{dt}=\frac{2}{3}\frac{dn_B}{dt}=\frac{3}{4}\frac{dn_C}{dt}$

10. For the reaction
$2H_2(g) + 2NO(g) \rightarrow N_2(g) + 2H_2O(g)$ the observed rate expression is, rate = $k_f[NO]^2[H_2]$. The rate expression for the reverse reaction is: [Jan. 07, 2020 (II)]
(1) $k_b[N_2][H_2O]^2$ (2) $k_b[N_2][H_2O]^2/[NO]$
(3) $k_b[N_2][H_2O]$ (4) $k_b[N_2][H_2O]^2/[H_2]$

11. In the following reaction : $xA \rightarrow yB$
$$\log_{10}\left[-\frac{d[A]}{dt}\right]=\log_{10}\left[\frac{d[B]}{dt}\right]+0.3010$$
'A' and 'B' respectively can be: [April 12, 2019 (I)]
(1) *n*-Butane and Iso-butane (2) C_2H_2 and C_6H_6
(3) C_2H_4 and C_4H_8 (4) N_2O_4 and NO_2

12. NO_2 required for a reaction is produced by the decomposition of N_2O_5 in CCl_4 as per the equation,
$2N_2O_5(g) \rightarrow 4NO_2(g)+O_2(g)$.
The initial concentration of N_2O_5 is 3.00 mol L^{-1} and it is 2.75 mol L^{-1} after 30 minutes. The rate of formation of NO_2 is : [April 12, 2019 (II)]
(1) 4.167×10^{-3} mol L^{-1} min^{-1}
(2) 1.667×10^{-2} mol L^{-1} min^{-1}
(3) 8.333×10^{-3} mol L^{-1} min^{-1}
(4) 2.083×10^{-3} mol L^{-1} min^{-1}

13. For a reaction scheme $A \xrightarrow{k_1} B \xrightarrow{k_2} C$, if the rate of formation of B is set to be zero then the concentration of B is given by : [April 8, 2019 (II)]
(1) $(k_1-k_2)[A]$ (2) $k_1k_2[A]$
(3) $(k_1+k_2)[A]$ (4) $\left(\frac{k_1}{k_2}\right)[A]$

14. For the reaction $2A+B \rightarrow C$, the values of initial rate at different reactant concentrations are given in the table below.
The rate law for the reaction is :

[A](mol L^{-1})	[B](mol L^{-1})	Initial Rate(mol $L^{-1}s^{-1}$)
0.05	0.05	0.045
0.10	0.05	0.090
0.20	0.10	0.72

[April 8, 2019 (I)]
(1) Rate = $k[A][B]^2$ (2) Rate = $k[A]^2[B]^2$
(3) Rate = $k[A][B]$ (4) Rate = $k[A]^2[B]$

15. For an elementary chemical reaction, $A_2 \underset{k_{-1}}{\overset{k_1}{\rightleftharpoons}} 2A$, the expression for $\frac{d[A]}{dt}$ is: [Jan. 10, 2019 (II)]
(1) $k_1[A_2]-k_{-1}[A]^2$ (2) $2k_1[A_2]-k_{-1}[A]^2$
(3) $k_1[A_2]+k_{-1}[A]^2$ (4) $2k_1[A_2]-2k_{-1}[A]^2$

16. The reaction of ozone with oxygen atoms in the presence of chlorine atoms can occur by a two step process shown below : [Online April 9, 2016]
$O_3(g)+Cl^*(g) \rightarrow O_2(g)+ClO^*(g)$ ...(i)
$[K_i = 5.2\times10^9$ L mol^{-1} $s^{-1}]$
$ClO^*(g)+O^*(g) \rightarrow O_2(g)+Cl^*(g)$ ...(ii)
$[K_{ii} = 2.6\times10^{10}$ L mol^{-1} $s^{-1}]$
The closest rate constant for the overall reaction $O_3(g)+O^*(g) \rightarrow 2O_2(g)$ is .
(1) 1.4×10^{20} L mol^{-1} s^{-1} (2) 3.1×10^{10} L mol^{-1} s^{-1}
(3) 5.2×10^9 L mol^{-1} s^{-1} (4) 2.6×10^{10} L mol^{-1} s^{-1}

17. The rate law for the reaction below is given by the expression k [A][B]
$A+B \rightarrow$ Product
If the concentration of B is increased from 0.1 to 0.3 mole, keeping the value of A at 0.1 mole, the rate constant will be: [Online April 10, 2016]
(1) 3*k* (2) 9*k* (3) *k*/3 (4) *k*

18. $A+2B \rightarrow C$, the rate equation for this reaction is given as Rate = k[A][B]. [Online April 11, 2015]
If the concentration of A is kept the same but that of B is doubled what will happen to the rate itself?
(1) halved (2) the same
(3) doubled (4) quadrupled

19. For the non - stoichiometric reaction 2A + B → C + D, the following kinetic data were obtained in three separate experiments, all at 298 K.

Initial Concentration [A]	Initial Concentration [B]	Initial rate of formation of C (mol L^{-1} s^{-1})
0.1 M	0.1 M	1.2×10^{-3}
0.1 M	0.2 M	1.2×10^{-3}
0.2 M	0.1 M	2.4×10^{-3}

The rate law for the formation of C is: [2014]
(1) $\frac{dC}{dt}=k[A][B]$ (2) $\frac{dC}{dt}=k[A]^2[B]$
(3) $\frac{dC}{dt}=k[A][B]^2$ (4) $\frac{dC}{dt}=k[A]$

20. In the reaction of formation of sulphur trioxide by contact process, $2SO_2+O_2 \rightleftharpoons 2SO_3$, the rate of reaction was measured as: $\frac{d[O_2]}{dt}=-2.5\times10^{-4}$ mol $L^{-1}s^{-1}$. The rate of reaction in terms of $[SO_2]$ in mol L^{-1} s^{-1} will be: [Online April 11, 2014]
(1) -1.25×10^{-4} (2) -2.50×10^{-4}
(3) -3.75×10^{-4} (4) -5.00×10^{-4}

21. For the reaction, $2N_2O_5 \rightarrow 4NO_2+O_2$, the rate equation can be expressed in two ways $-\frac{d[N_2O_5]}{dt}=k[N_2O_5]$ and $+\frac{d[NO_2]}{dt}=k'[N_2O_5]$
k and *k'* are related as: [Online April 11, 2014]
(1) $k=k'$ (2) $2k=k'$ (3) $k=2k'$ (4) $k=4k'$

22. For the reaction, [Online April 19, 2014]
$3A+2B \rightarrow C+D$,
the differential rate law can be written as:
(1) $\frac{1}{3}\frac{d[A]}{dt}=\frac{d[C]}{dt}=k[A]^n[B]^m$
(2) $-\frac{d[A]}{dt}=\frac{d[C]}{dt}=k[A]^n[B]^m$
(3) $+\frac{1}{3}\frac{d[A]}{dt}=-\frac{d[C]}{dt}=k[A]^n[B]^m$
(4) $-\frac{1}{3}\frac{d[A]}{dt}=\frac{d[C]}{dt}=k[A]^n[B]^m$

23. The instantaneous rate of disappearance of MnO_4^- ion in the following reaction is $4.56 \times 10^{-3}\ Ms^{-1}$
$2MnO_4^- + 10I^- + 16H^+ \rightarrow 2Mn^{2+} + 5I_2 + 8H_2O$
The rate of appearance of I_2 is : [Online April 9, 2013]
(1) $4.56 \times 10^{-4}\ Ms^{-1}$ (2) $1.14 \times 10^{-2}\ Ms^{-1}$
(3) $1.14 \times 10^{-3}\ Ms^{-1}$ (4) $5.7 \times 10^{-3}\ Ms^{-1}$

24. Reaction rate between two substances A and B is expressed as following:
rate $= k[A]^n[B]^m$ [Online May 7, 2012, Offline 2003]
If the concentration of A is doubled and concentration of B is made half of initial concentration, the ratio of the new rate to the earlier rate will be:
(1) $m+n$ (2) $n-m$ (3) $\frac{1}{2^{(m+n)}}$ (4) $2^{(n-m)}$

25. Consider the reaction :
$Cl_2(aq) + H_2S(aq) \rightarrow S(s) + 2H^+(aq) + 2Cl^-(aq)$
The rate equation for this reaction is,
rate $= k[Cl_2][H_2S]$
Which of these mechanisms is/are consistent with this rate equation? [2010]
A. $Cl_2 + H_2S \rightarrow H^+ + Cl^- + Cl^+ + HS^-$ (slow)
$Cl^+ + HS^- \rightarrow H^+ + Cl^- + S$ (fast)
B. $H_2S \rightleftharpoons H^+ + HS^-$ (fast equilibrium)
$Cl_2 + HS^- \rightarrow 2Cl^- + H^+ + S$ (Slow)
(1) B only (2) Both A and B
(3) Neither A nor B (4) A only

26. For a reaction $\frac{1}{2}A \rightarrow 2B$, rate of disappearance of 'A' is related to the rate of appearance of 'B' by the expression [2008]
(1) $-\frac{d[A]}{dt} = \frac{1}{2}\frac{d[B]}{dt}$ (2) $-\frac{d[A]}{dt} = \frac{1}{4}\frac{d[B]}{dt}$
(3) $-\frac{d[A]}{dt} = \frac{d[B]}{dt}$ (4) $-\frac{d[A]}{dt} = 4\frac{d[B]}{dt}$

27. The differential rate law for the reaction [2002]
$H_2 + I_2 \rightarrow 2HI$ is
(1) $-\frac{d[H_2]}{dt} = -\frac{d[I_2]}{dt} = -\frac{d[HI]}{dt}$
(2) $\frac{d[H_2]}{dt} = \frac{d[I_2]}{dt} = \frac{1}{2}\frac{d[HI]}{dt}$
(3) $\frac{1}{2}\frac{d[H_2]}{dt} = \frac{1}{2}\frac{d[I_2]}{dt} = -\frac{d[HI]}{dt}$
(4) $-2\frac{d[H_2]}{dt} = -2\frac{d[I_2]}{dt} = \frac{d[HI]}{dt}$

28. Units of rate constant of first and zero order reactions in terms of molarity M unit are respectively [2002]
(1) sec^{-1}, $Msec^{-1}$ (2) sec^{-1}, M
(3) $Msec^{-1}$, sec^{-1} (4) M, sec^{-1}.

Numerical Value Questions (NVQs)

29. For the reaction $A \rightarrow B$ the following graph was obtained. The time required (in seconds) for the concentration of A to reduce to 2.5 g L^{-1} (if the initial concentration of A was 50 g L^{-1}) is ________.
(Nearest integer) [April 2, 2025 (II)]
Given: log 2 = 0.3010

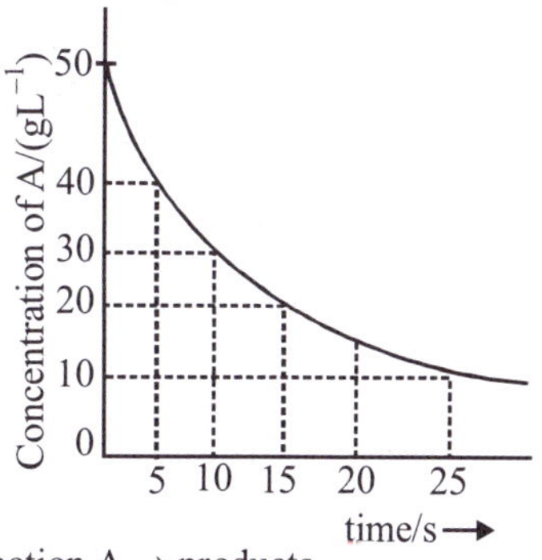

30. For the reaction $A \rightarrow$ products.

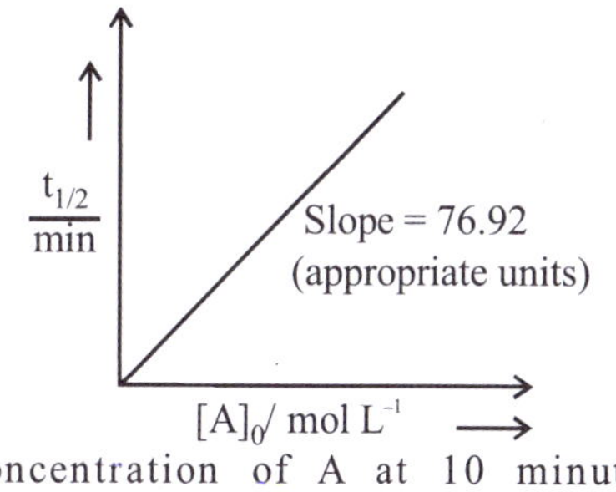

The concentration of A at 10 minutes is ____ $\times 10^{-3}$ mol L^{-1} (nearest integer).
The reaction was started with 2.5 mol L^{-1} of A. [April 2, 2025 (I)]

31. Given below are two statements :
Statement I : The rate law for the reaction $A + B \rightarrow C$ is rate (r) $= k[A]^2[B]$. When the concentration of both A and B is doubled, the reaction rate is increased "x" times.
Statement II :

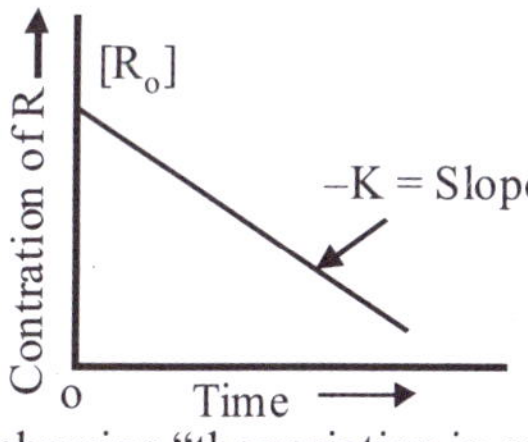

The figure is showing "the variation in concentration against time plot" for a "y" order reaction.
The value of x + y is _________. [April 9, 2024 (I)]

32. Consider the following first order gas phase reaction at constant temperature $A(g) \rightarrow 2B(g) + C(g)$
If the total pressure of the gases is found to be 200 torr after 23 sec. and 300 torr upon the complete decomposition of A after a very long time, then the rate constant of the given reaction is____ $\times 10^{-2}\ s^{-1}$ (nearest integer)
[Given : $\log_{10}(2) = 0.301$] [April 9, 2024 (II)]

33. NO_2 required for a reaction is produced by decomposition of N_2O_5 in CCl_4 as by equation [Jan. 30, 2024 (II)]
$2N_2O_5(g) \rightarrow 4NO_2(g) + O_2(g)$
The initial concetration, of N_2O_5 is 3 mol L^{-1} and it is 2.75 mol L^{-1} after 30 minutes.
The rate of formation of NO_2 is $x \times 10^{-3}$ mol L^{-1} min^{-1}, value of x is ____________. (nearest integer)

34. $A(g) \rightarrow 2B(g) + C(g)$ is a first order reaction. The initial pressure of the system was found to be 800 mm Hg which increased to 1600 mm Hg after 10 min. The total pressure of the system after 30 min will be ______ mm Hg. (Nearest integer)
[April 13, 2023 (II)]

35. $KClO_3 + 6FeSO_4 + 3H_2SO_4 \rightarrow KCl + 3Fe_2(SO_4)_3 + 3H_2O$
The above reaction was studied at 300 K by monitoring the concentration of $FeSO_4$ in which initial concentration was 10 M and after half an hour became 8.8 M. The rate of production of $Fe_2(SO_4)_3$ is ______ $\times 10^{-6}$ mol L^{-1} s^{-1}.
(Nearest integer) [April 11, 2023 (I)]

36. A first order reaction has the rate constant, $k = 4.6 \times 10^{-3}$ s^{-1}. The number of correct statement/s from the following is/are ______. [Jan. 25, 2023 (II)]
Given : log 3 = 0.48
A. Reaction completes in 1000 s.
B. The reaction has a half-life of 500 s.
C. The time required for 10% completion is 25 times the time required for 90% completion.
D. The degree of dissociation is equal to $(1 - e^{-kt})$.
E. The rate and the rate constant have the same unit.

37. The reaction between X and Y is first order with respect to X and zero order with respect to Y. [July 29, 2022 (I)]

Experiment	[X] mol L^{-1}	[Y] mol L^{-1}	Initial rate mol L^{-1} min^{-1}
I.	0.1	0.1	2×10^{-3}
II.	L	0.2	4×10^{-3}
III.	0.4	0.4	$M \times 10^{-3}$
IV.	0.1	0.2	2×10^{-3}

Examine the data of table and calculate ratio of numerical values of M and L. (Nearest integer)

38. For a given chemical reaction $\gamma_1 A + \gamma_2 B \rightarrow \gamma_3 C + \gamma_4 D$
Concentration of C changes from 10 m mol dm^{-3} to 20 m mol dm^{-3} in 10 seconds. Rate of appearance of D is 1.5 times the rate of disappearance of B which is twice the rate of disappearance A. The rate of appearance of D has been experimentally determined to be 9 m mol dm^{-3} s^{-1}. Therefore the rate of reaction is ________ m mol dm^{-3} s^{-1}. (Nearest Integer) [June 25, 2022 (I)]

39. The reaction that occurs in a breath analyser, a device used to determine the alcohol level in a person's blood stream is [Aug. 27, 2021 (I)]
$$2K_2Cr_2O_7 + 8H_2SO_4 + 3C_2H_6O \rightarrow 2Cr_2(SO_4)_3 + 3C_2H_4O_2 + 2K_2SO_4 + 11H_2O$$
If the rate of appearance of $Cr_2(SO_4)_3$ is 2.67 mol min^{-1} at a particular time, the rate of disappearance of C_2H_6O at the same time is ________ mol min^{-1}. (Nearest integer)

40. For a chemical reaction $A \rightarrow B$, it was found the concentration of B increased by 0.2 mol L^{-1} in 30 min. The average rate of the reaction is ________ $\times 10^{-1}$ mol L^{-1} h^{-1}. (in nearest integer) [July 25, 2021 (II)]

41. The inactivation rate of a viral preparation is proportional to the amount of virus. In the first minute after preparation, 10% of the virus is inactivated. The rate constant for viral inactivation is __________ $\times 10^{-3}$ min^{-1}. (Nearest integer)
[Use : ln 10 = 2.303 ; $\log_{10} 3 = 0.477$;
property of logarithm : $\log x^y = y \log x$] [July 20, 2021 (I)]

42. The reaction $2A + B_2 \rightarrow 2AB$ is an elementary reaction. For a certain quantity of reactants. If the volume of the reaction vessel is reduced by a factor of 3, the rate of the reaction increases by a factor of _______. (Round off to the nearest integer). [March 17, 2021 (II)]

Topic 2 - Order of Reaction and Half Life Period

MULTIPLE CHOICE QUESTIONS (MCQs)

43. First order gas phase reaction
$A \rightarrow B + C$
p_i = initial pressure of gas A, p_t = total pressure of the reaction mixture at time t.
Expression of rate constant (k) is [April 05, 2026 (I)]
(1) $\frac{1}{t}\ln\frac{p_i}{2p_i - p_t}$ (2) $\frac{1}{t}\ln\frac{2p_i}{p_i - p_t}$
(3) $\frac{1}{t}\ln\frac{p_i}{3p_i - 2p_t}$ (4) $\frac{1}{t}\ln\frac{3p_i}{4p_i - p_t}$

44. Consider the given graph showing variation of reactant concentration with time.
Three different reactions were started with identical initial concentration of reactants. Which of the following statement is correct ? [April 04, 2026 (II)]

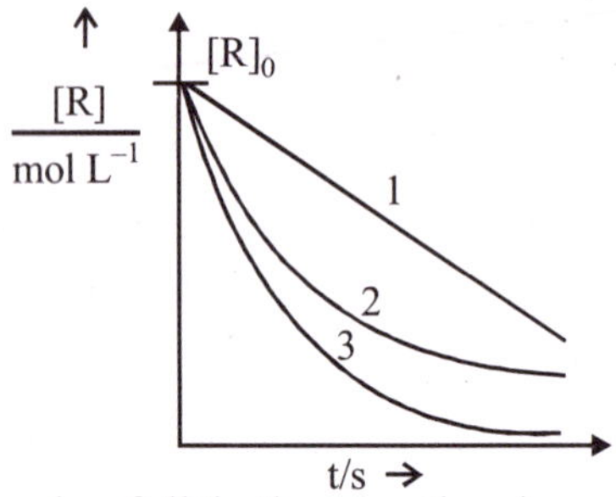

(1) The order of all the three reactions is same.
(2) The rate constant of reaction 3 is larger than the rate constant of reaction 2 if the order of reaction is same for both.
(3) The SI unit of rate constant of reaction 1 is s^{-1}.
(4) Thermal decomposition of HI on gold surface is an example of reaction 2.

45. Consider the first order reaction $R \rightarrow P$.
The fraction of molecules decomposed in the given first order reaction can be expressed as [April 04, 2026 (I)]
(1) $1 - e^{k_1 t}$ (2) $1 + e^{k_1 t}$
(3) $1 + e^{-k_1 t}$ (4) $1 - e^{-k_1 t}$

46. Consider the reaction $aX \rightarrow bY$, for which the rate constant at 30°C is 1×10^{-3} mol^{-1} Ls^{-1}. Which of the following statements are true ? [April 02, 2026 (II)]
A. When concentration of 'X' is increased to four times, the rate of reaction becomes 16 times.
B. The reaction is a second order reaction.
C. The half-life period is independent of the concentration of X.
D. Decomposition of N_2O_5 is an example of the above reaction.
E. 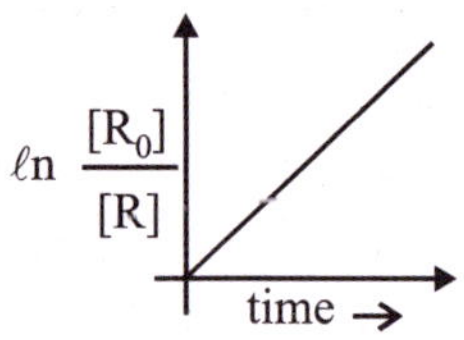

is valid for the above reaction.

Choose the correct answer from the options given below:
(1) A and B only (2) A, B and C only
(3) A, B, D and E only (4) C and D only

47. $t_{100\%}$ is the time required for the 100% completion of the reaction while $t_{1/2}$ is the time required for 50% of the reaction to be completed. Which of the following option correctly represents the relation between $t_{100\%}$ and $t_{1/2}$ for zero and first order reactions respectively? **[April 02, 2026 (I)]**
(1) $t_{100\%} = (t_{1/2})^2$ and $t_{100\%} = (t_{1/2})^{-\infty}$
(2) $t_{100\%} = 2t_{1/2}$ and $t_{100\%} = (t_{1/2})^{\infty}$
(3) $t_{100\%} = 2t_{1/2}$ and $t_{100\%} = (2t_{1/2})^2$
(4) $t_{100\%} = (t_{1/2})^{\infty}$ and $t_{100\%} = 2t_{1/2}$

48. An organic compound undergoes first order decomposition.

The time taken for decomposition to $\left(\frac{1}{8}\right)^{th}$ and $\left(\frac{1}{10}\right)^{th}$ of its initial concentration are $t_{1/8}$ and $t_{1/10}$ respectively. **[Jan. 28, 2026 (I)]**

What is the value of $\frac{t_{1/8}}{t_{1/10}} \times 10$?

($\log 2 = 0.3$)
(1) 9 (2) 0.9 (3) 3 (4) 30

49.
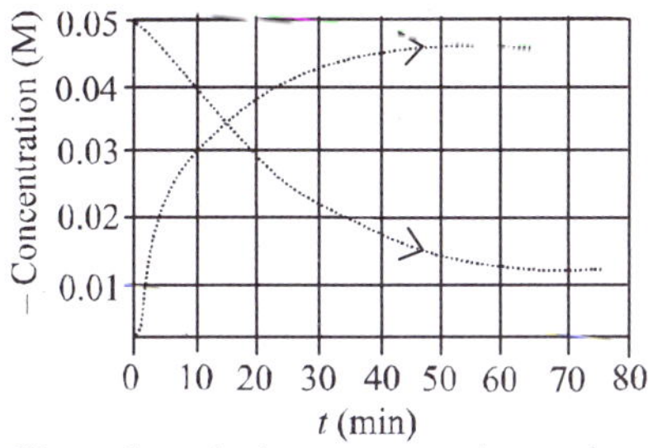

Given above is the concentration *vs* time plot for a dissociation reaction : $A \rightarrow nB$.
Based on the data of the initial phase of the reaction (initial 10 min), the value of n is .... **[Jan. 23, 2026 (II)]**
(1) 4 (2) 3 (3) 2 (4) 5

50. A person's wound was exposed to some bacteria and then bacteria growth started to happen at the same place. The wound was later treated with some antibacterial medicine and the rate of bacterial decay (r) was found to be proportional with the square of the existing number of bacteria at any instance. Which of the following set of graphs correctly represents the 'before' and 'after' situation of the application of the medicine?
[Given : N = No. of bacteria, t = time, bacterial growth follows I[st] order kinetics.] **[April 7, 2025 (I)]**

(1)
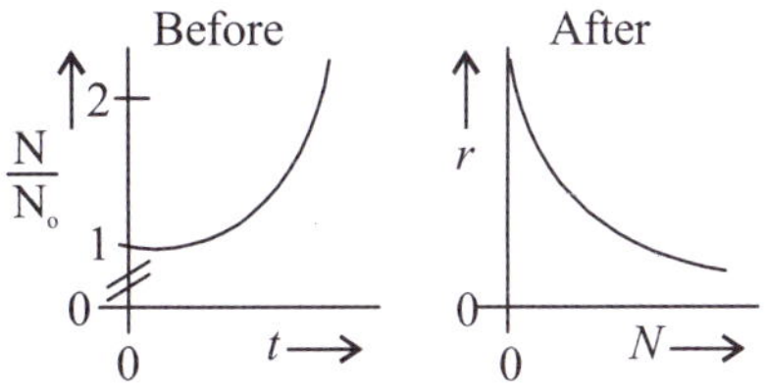

(2)
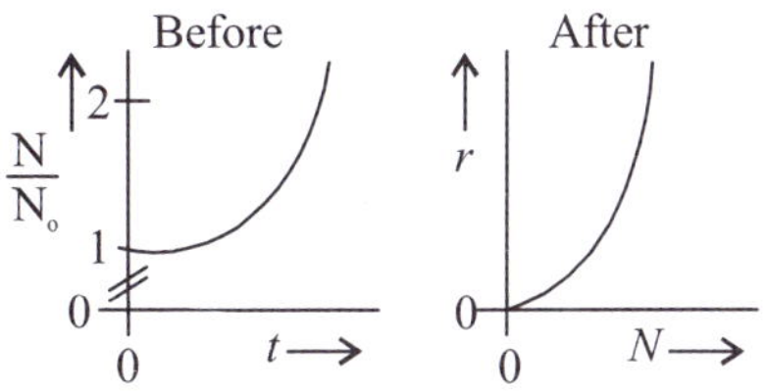

(3)
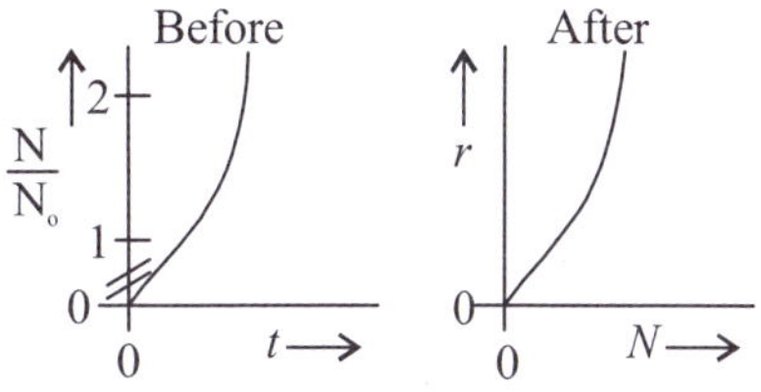

(4)
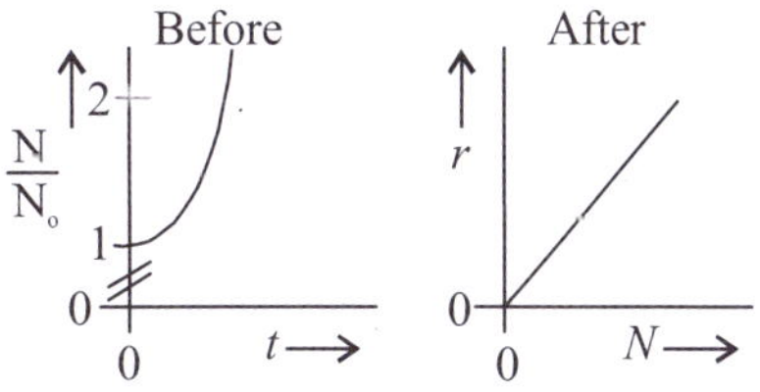

51. Half life of zero order reaction $A \rightarrow$ product is 1 hour, when initial concentration of reaction is 2.0 mol L^{-1}. The time required to decrease concentration of A from 0.50 to 0.25 mol L^{-1} is:
(1) 0.5 hour (2) 4 hour **[April 4, 2025 (II)]**
(3) 15 min (4) 60 min

52. In a reaction $A + B \rightarrow C$, initial concentrations of A and B are related as $[A]_0 = 8[B]_0$. The half lives of A and B are 10 min and 40 min. respectively. If they start to disappear at the same time, both following first order kinetics, after how much time will the concentration of both the reactants be same? **[April 3, 2025 (I)]**
(1) 60 min (2) 80 min (3) 20 min (4) 40 min

53. Drug X becomes ineffective after 50% decomposition. The original concentration of drug in a bottle was 16 mg/mL which becomes 4 mg/mL in 12 months. The expiry time of the drug in months is _____.
Assume that the decomposition of the drug follows first order kinetics. **[Jan. 29, 2025 (II)]**
(1) 12 (2) 2 (3) 3 (4) 6

54. The reaction $A_2 + B_2 \rightarrow 2AB$ follows the mechanism

$$A_2 \underset{k_{-1}}{\overset{k_1}{\rightleftharpoons}} A + A \text{ (fast)}$$

$$A + B_2 \xrightarrow{k_2} AB + B \text{ (slow)}$$

$$A + B \rightarrow AB \text{ (fast)}$$ **[Jan. 29, 2025 (I)]**

The overall order of the reaction is :
(1) 1.5 (2) 3 (3) 2.5 (4) 2

55. For bacterial growth in a cell culture, growth law is very similar to the law of radioactive decay. Which of the following graphs is most suitable to represent bacterial colony growth? **[Jan. 28, 2025 (II)]**

(1)
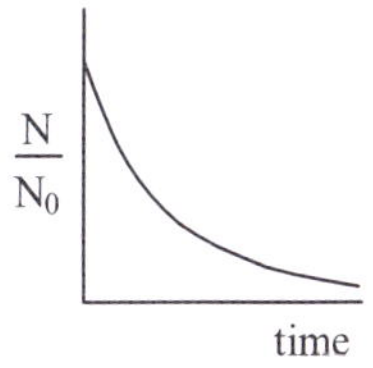

(2)
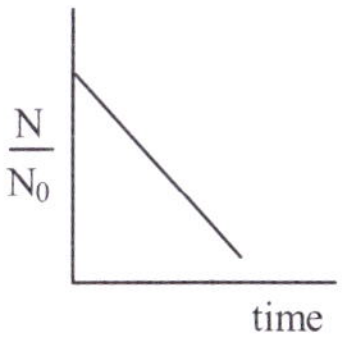

(3)
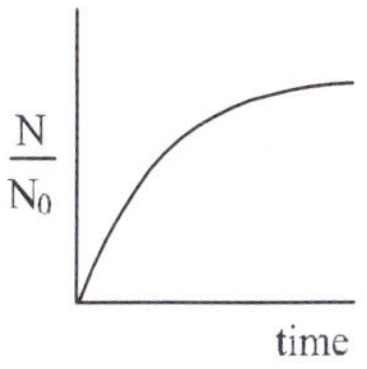

(4)
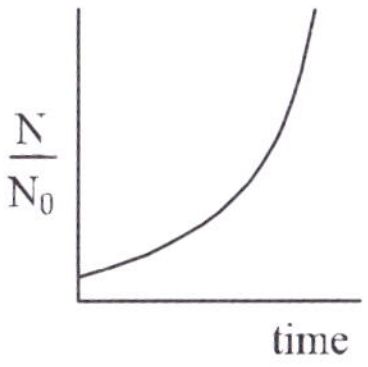

56. consider the elementary reaction **[Jan. 28, 2025 (II)]**
$A(g)+B(g) \rightarrow C(g)+D(g)$

If the volume of reaction mixture is suddenly reduced to $\frac{1}{3}$ of its initial volume, the reaction rate will become 'x' times of the original reaction rate. The value of x is :

(1) $\frac{1}{9}$ (2) 9 (3) $\frac{1}{3}$ (4) 3

57.

$[A]_0$ / mol L^{-1}	$t_{1/2}$ / min
0.100	200
0.025	100

For a given reaction $R \rightarrow P$, $t_{1/2}$ is related to $[A]_0$ as given in table : **[Jan. 28, 2025 (I)]**
Given : log 2 = 0.30
Which of the following is true ?

(A) The order of the reaction is $\frac{1}{2}$.
(B) If $[A]_0$ is 1M, then $t_{1/2}$ is $200\sqrt{10}$ min
(C) The order of the reaction changes to 1 if the concentration of reactant changes from 0.100 M to 0.500 M.
(D) $t_{1/2}$ is 800 min for $[A]_0 = 1.6$ M

Choose the correct answer from the options given below :
(1) A and C only (2) A and B only
(3) A, B and D only (4) C and D only

58. Given below are two statements : **[Jan. 24, 2025 (II)]**

Statement(I) :

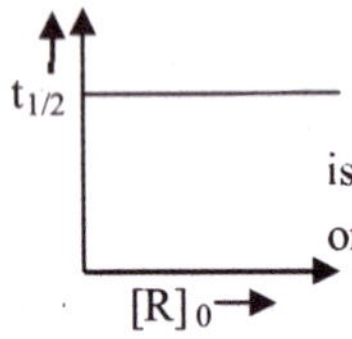

is valid for first order reaction.

Statement(II) :

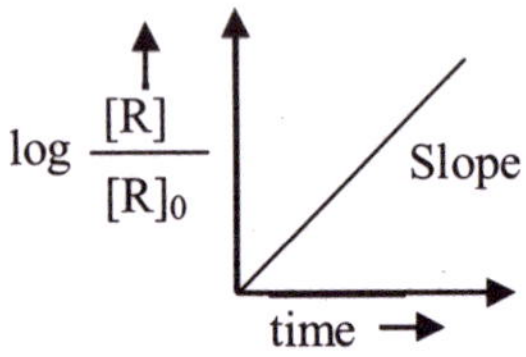

Slope = $\frac{k}{2.303}$ is valid for first order reaction

In the light of the above statements, choose the correct answer from the options given below :
(1) Both Statement I and Statement II are false
(2) Statement I is false but Statement II is true
(3) Both Statement I and Statement II are true
(4) Statement I is true but Statement II is false

59. Which of the following graphs most appropriately represents a zero order reaction ? **[Jan. 23, 2025 (II)]**

(1)
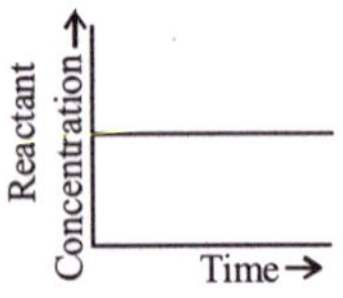

(2)
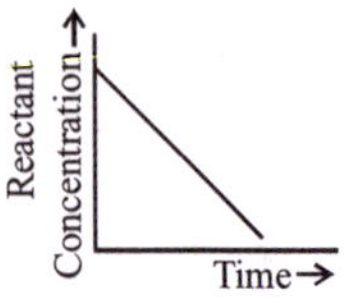

(3)
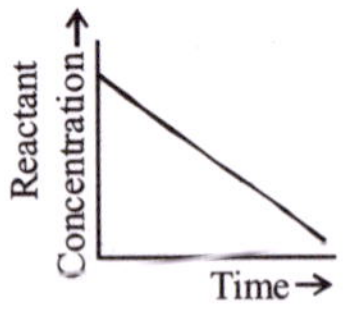

(4)
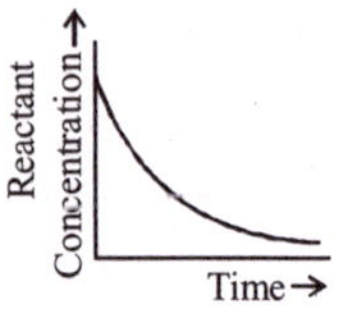

60. Which of the following statement is not true for radioactive decay ? **[Jan. 22, 2025 (I)]**
(1) Amount of radioactive substance remained after three half lives is $\frac{1}{8}$th of original amount.
(2) Decay constant does not depend upon temperature.
(3) Decay constant increases with increase in temperature.
(4) Half life is ln 2 times of $\frac{1}{\text{rate constant}}$.

61. Integrated rate law equation for a first order gas phase reaction is given by (where P_i is initial pressure and P_t is total pressure at time t) **[Jan. 31, 2024 (I)]**

(1) $k = \frac{2.303}{t} \times \log \frac{(2P_i - P_t)}{P_i}$

(2) $k = \frac{2.303}{t} \times \frac{P_i}{(2P_i - P_t)}$

(3) $k = \frac{2.303}{t} \times \log \frac{2P_i}{(2P_i - P_t)}$

(4) $k = \frac{2.303}{t} \times \log \frac{P_i}{(2P_i - P_t)}$

62. For a chemical reaction A + B → Product, the order is 1 with respect to A and B. **[April 11, 2023 (II)]**

Rate mol $L^{-1}s^{-1}$	**[A]** mol L^{-1}	**[B]** mol L^{-1}
0.10	20	0.5
0.40	x	0.5
0.80	40	y

What is the value of x and y?
(1) 80 and 2 (2) 40 and 4
(3) 160 and 4 (4) 80 and 4

63. A student has studied the decomposition of a gas AB_3 at 25°C. He obtained the following data. **[Jan. 24, 2023 (II)]**

p(mm Hg)	50	100	200	400
Relative $t_{1/2}$(s)	4	2	1	0.5

The order of the reaction is
(1) 0.5 (2) 2 (3) 1 (4) 0 (zero)

64. At 30°C, the half life for the decomposition of AB_2 is 200 s and is independent of the initial concentration of AB_2. The time required for 80% of the AB_2 to decompose is (Given: log 2 = 0.30; log 3 = 0.48) **[July 26, 2022 (II)]**
(1) 200 s (2) 323 s (3) 467 s (4) 532 s

65. For a first order reaction, the time required for completion of 90% reaction is 'x' times the half life of the reaction. The value of 'x' is **[June 24, 2022 (II)]**
(Given: ln 10 = 2.303 and log 2 = 0.3010)
(1) 1.12 (2) 2.43 (3) 3.32 (4) 33.31

66. For a reaction of order n, the unit of the rate constant is : **[July 27, 2021 (I)]**
(1) $mol^{1-n} L^{1-n} s$ (2) $mol^{1-n} L^{2n} s^{-1}$
(3) $mol^{1-n} L^{n-1} s^{-1}$ (4) $mol^{1-n} L^{1-n} s^{-1}$

67. Consider the following reactions
$A \rightarrow P1$; $B \rightarrow P2$; $C \rightarrow P3$; $D \rightarrow P4$,

The order of the above reactions are (i), (ii), (iii), and (iv), respectively. The following graph is obtained when log [rate] vs. log[conc.] are plotted : [Sep. 06, 2020 (I)]

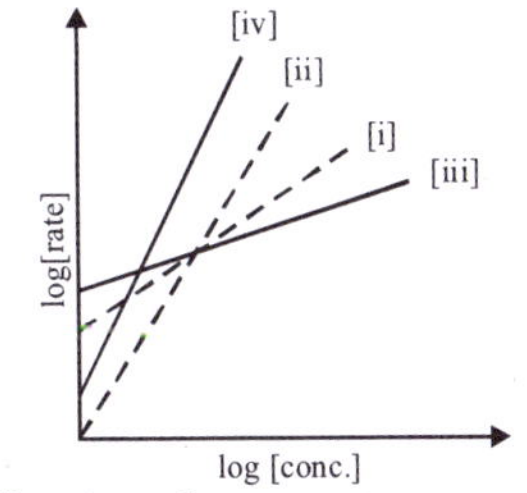

Among the following, the correct sequence for the order of the reactions is:

(1) (iv) > (i) > (ii) > (iii) (2) (i) > (ii) > (iii) > (iv)
(3) (iii) > (i) > (ii) > (iv) (4) (iv) > (ii) > (i) > (iii)

68. A flask contains a mixture of compounds A and B. Both compounds decompose by first-order kinetics. The half-lives for A and B are 300 s and 180 s, respectively. If the concentrations of A and B are equal initially, the time required for the concentration of A to be four times that of B (in s) is: (Use ln 2 = 0.693) [Sep. 05, 2020 (I)]

(1) 180 (2) 900 (3) 300 (4) 120

69. It is true that : [Sep. 03, 2020 (I)]

(1) A second order reaction is always a multistep reaction
(2) A zero order reaction is a multistep reaction
(3) A first order reaction is always a single step reaction
(4) A zero order reaction is a single step reaction

70. The results given in the below table were obtained during kinetic studies of the following reaction :

$2A + B \rightarrow C + D$ [Sep. 02, 2020 (II)]

Experiment	[A]/ $molL^{-1}$	[B]/ $molL^{-1}$	Initial rate/ $molL^{-1} min^{-1}$
I	0.1	0.1	6.00×10^{-3}
II	0.1	0.2	2.40×10^{-2}
III	0.2	0.1	1.20×10^{-2}
IV	X	0.2	7.20×10^{-2}
V	0.3	Y	2.88×10^{-1}

X and Y in the given table are respectively :

(1) 0.4, 0.4 (2) 0.4, 0.3 (3) 0.3, 0.4 (4) 0.3, 0.3

71. A bacterial infection in an internal wound grows as N′(t) = N_0exp(t), where the time t is in hours. A dose of antibiotic, taken orally, needs 1 hour to reach the wound. Once it reaches there, the bacterial population goes down as $\frac{dN}{dt} = -5N^2$. What will be the plot of $\frac{N_0}{N}$ vs. t after 1 hour? [April 10, 2019 (I)]

(1) 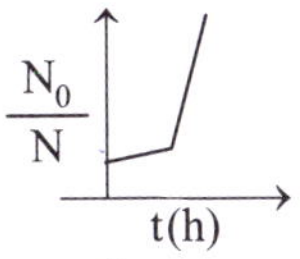

(2) 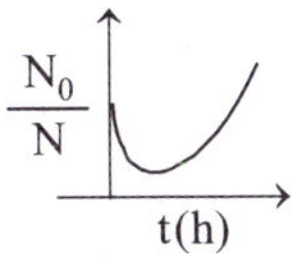

(3) 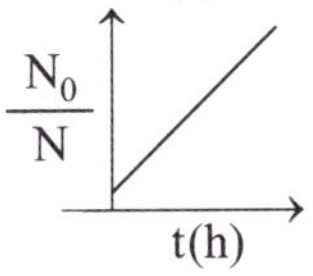

(4) 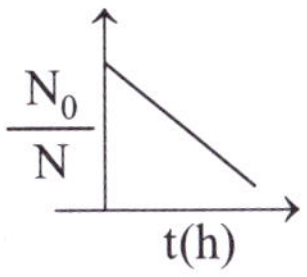

72. The given plots represents the variation of the concentration of a reactant R with time for two different reactions (i) and (ii). The respective orders of the reactions are: [April 9, 2019 (I)]

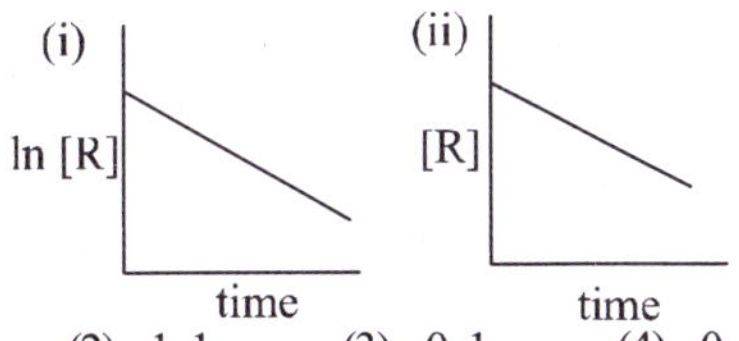

(1) 1, 0 (2) 1, 1 (3) 0, 1 (4) 0, 2

73. Decomposition of X exhibits a rate constant of 0.05 μg/year. How many years are required for the decomposition of 5 μg of X into 2.5 μg? [Jan. 12, 2019 (I)]

(1) 50 (2) 25 (3) 20 (4) 40

74. The reaction $2X \rightarrow B$ is a zeroth order reaction. If the initial concentration of X is 0.2 M, the half-life is 6 h. When the initial concentration of X is 0.5 M, the time required to reach its final concentration of 0.2 M will be : [Jan. 11, 2019 (II)]

(1) 9.0 h (2) 12.0 h (3) 18.0 h (4) 7.2 h

75. The following results were obtained during kinetic studies of the reaction; [Jan. 9, 2019 (I)]

$2A + B \rightarrow$ Products

Experiment	[A] (in mol L^{-1})	[B] (in mol L^{-1})	Initial Rate of reaction (in mol L^{-1} min^{-1})
I	0.10	0.20	6.93×10^{-3}
II	0.10	0.25	6.93×10^{-3}
III	0.20	0.30	1.386×10^{-2}

The time (in minutes) required to consume half of A is:

(1) 5 (2) 10 (3) 1 (4) 100

76. For the reaction, $2A + B \rightarrow$ products, when the concentrations of A and B both were doubled, the rate of the reaction increased from 0.3 mol L^{-1} s^{-1} to 2.4 mol $L^{-1}s^{-1}$. When the concentration of A alone is doubled, the rate increased from 0.3 mol $L^{-1}s^{-1}$ to 0.6 mol $L^{-1}s^{-1}$. Which one of the following statements is correct?

(1) Total order of the reaction is 4 [Jan. 9, 2019 (II)]
(2) Order of the reaction with respect to B is 2
(3) Order of the reaction with respect to B is 1
(4) Order of the reaction with respect to A is 2

77. At 518 °C, the rate of decomposition of a sample of gaseous acetaldehyde, initially at a pressure of 363 Torr, was 1.00 Torr s^{-1} when 5% had reacted and 0.5 Torr s^{-1} when 33% had reacted. The order of the reaction is : [2018]

(1) 2 (2) 3 (3) 1 (4) 0

78. If 50% of a reaction occurs in 100 seconds and 75% of the reaction occurs in 200 seconds, the order of this reaction is: [Online April 16, 2018]

(1) 2 (2) 3 (3) Zero (4) 1

79. N_2O_5 decomposes to NO_2 and O_2 and follows first order kinetics. After 50 minutes, the pressure inside the vessel increases from 50 mm Hg to 87.5 mm Hg. The pressure of the gaseous mixture after 100 minutes at constant temperature will be ________. [Online April 15, 2018 (I)]

(1) 136.25 mm Hg (2) 106.25 mm Hg
(3) 175.0 mm Hg (4) 116.25 mm Hg

80. For a first order reaction, $A \rightarrow P$, $t_{1/2}$ (half-life) is 10 days. The time required for $\frac{1}{4}^{th}$ conversion of A (in days) is: ($\ln 2 = 0.693$, $\ln 3 = 1.1$). [Online April 15, 2018 (II)]

(1) 3.2 (2) 2.5 (3) 4.1 (4) 5

81. Decomposition of H_2O_2 follows a first order reaction. In fifty minutes the concentration of H_2O_2 decreases from 0.5 to 0.125 M in one such decomposition. When the concentration of H_2O_2 reaches 0.05 M, the rate of formation of O_2 will be: **[2016]**
(1) $2.66\ L\ min^{-1}$ at STP (2) $1.34 \times 10^{-2}\ mol\ min^{-1}$
(3) $6.96 \times 10^{-2}\ mol\ min^{-1}$ (4) $6.93 \times 10^{-4}\ mol\ min^{-1}$

82. The reaction $2N_2O_5(g) \rightarrow 4NO_2(g) + O_2(g)$ follows first order kinetics. The pressure of a vessel containing only N_2O_5 was found to increase from 50 mm Hg to 87.5 mm Hg in 30 min. The pressure exerted by the gases after 60 min. will be (assume temperature remains constant) : **[Online April 10, 2015]**
(1) 106.25 mm Hg (2) 150 mm Hg
(3) 125 mm Hg (4) 116.25 mm Hg

83. Higher order (>3) reactions are rare due to : **[2015]**
(1) shifting of equilibrium towards reactants due to elastic collisions
(2) loss of active species on collision
(3) low probability of simultaneous collision of all the reacting species
(4) increase in entropy and activation energy as more molecules are involved

84. The half-life period of a first order reaction is 15 minutes. The amount of substance left after one hour will be: **[Online April 9, 2014]**
(1) $\frac{1}{4}$ of the original amount
(2) $\frac{1}{8}$ of the original amount
(3) $\frac{1}{16}$ of the original amount
(4) $\frac{1}{32}$ of the original amount

85. The rate constant of a zero order reaction is $2.0 \times 10^{-2}\ mol\ L^{-1}\ s^{-1}$. If the concentration of the reactant after 25 seconds is 0.5 M. What is the initial concentration ? **[Online April 23, 2013]**
(1) 0.5 M (2) 1.25 M (3) 12.5 M (4) 1.0 M

86. A radioactive isotope having a half - life period of 3 days was received after 12 days. If 3 g of the isotope is left in the container, what would be the initial mass of the isotope ? **[Online April 25, 2013]**
(1) 12 g (2) 36 g (3) 48 g (4) 24 g

87. For a first order reaction $(A) \rightarrow$ products, the concentration of A changes from 0.1 M to 0.025 M in 40 minutes.
The rate of reaction when the concentration of A is 0.01 M is : **[2012]**
(1) 1.73×10^{-5} M/min (2) 3.47×10^{-4} M/min
(3) 3.47×10^{-5} M/min (4) 1.73×10^{-4} M/min

88. In a chemical reaction A is converted into B. The rates of reaction, starting with initial concentrations of A as 2×10^{-3} M and 1×10^{-3} M, are equal to $2.40 \times 10^{-4}\ Ms^{-1}$ and $0.60 \times 10^{-4}\ Ms^{-1}$ respectively. The order of reaction with respect to reactant A will be **[Online May 12, 2012]**
(1) 0 (2) 1.5 (3) 1 (4) 2

89. For a reaction $A \rightarrow$ Products, a plot of $\log t_{1/2}$ versus $\log a_0$ is shown in the figure. If the initial concentration of A is represented by a_0, the order of the reaction is **[Online May 19, 2012]**

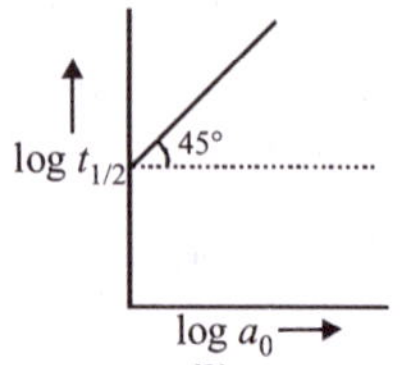

(1) one (2) zero (3) two (4) three

90. The time for half life period of a certain reaction $A \longrightarrow$ Products is 1 hour. When the initial concentration of the reactant 'A' is 2.0 mol L^{-1}, how much time does it take for its concentration to come from 0.50 to 0.25 mol L^{-1} if it is a zero order reaction ? **[2010]**
(1) 4 h (2) 0.5 h (3) 0.25 h (4) 1 h

91. The half life period of a first order chemical reaction is 6.93 minutes. The time required for the completion of 99% of the chemical reaction will be ($\log 2 = 0.301$) **[2009]**
(1) 23.03 minutes (2) 46.06 minutes
(3) 460.6 minutes (4) 230.03 minutes

92. Consider the reaction, $2A + B \rightarrow$ products. When concentration of B alone was doubled, the half-life did not change. When the concentration of A alone was doubled, the rate increased by two times. The unit of rate constant for this reaction is **[2007]**
(1) s^{-1} (2) $L\ mol^{-1}\ s^{-1}$ (3) no unit (4) $mol\ L^{-1}\ s^{-1}$.

93. A radioactive element gets spilled over the floor of a room. Its half-life period is 30 days. If the initial velocity is ten times the permissible value, after how many days will it be safe to enter the room? **[2007]**
(1) 100 days (2) 1000 days
(3) 300 days (4) 10 days.

94. The following mechanism has been proposed for the reaction of NO with Br_2 to form NOBr :

$$NO(g) + Br_2(g) \rightleftharpoons NOBr_2(g)$$
$$NOBr_2(g) + NO(g) \longrightarrow 2NOBr(g)$$

If the second step is the rate determining step, the order of the reaction with respect to NO(g) is **[2006]**
(1) 3 (2) 2 (3) 1 (4) 0

95. A reaction was found to be second order with respect to the concentration of carbon monoxide. If the concentration of carbon monoxide is doubled, with everything else kept the same, the rate of reaction will **[2006]**
(1) increase by a factor of 4 (2) double
(3) remain unchanged (4) triple

96. A reaction involving two different reactants can never be **[2005]**
(1) bimolecular reaction (2) second order reaction
(3) first order reaction (4) unimolecular reaction

97. $t_{1/4}$ can be taken as the time taken for the concentration of a reactant to drop to 3/4 of its initial value. If the rate constant for a first order reaction is k, the $t_{1/4}$ can be written as **[2005]**
(1) $0.75/k$ (2) $0.69/k$ (3) $0.29/k$ (4) $0.10/k$

98. The rate equation for the reaction $2A + B \rightarrow C$ is found to be: rate $= k[A][B]$. The correct statement in relation to this reaction is that the **[2004]**
(1) rate of formation of C is twice the rate of disappearance of A
(2) $t_{1/2}$ is a constant
(3) unit of k must be s^{-1}
(4) value of k is independent of the initial concentrations of A and B

99. For the reaction system : **[2003]**

$$2NO(g) + O_2(g) \rightarrow 2NO_2(g)$$

volume is suddenly reduced to half of its value by increasing the pressure on it. If the reaction is of first order with respect to O_2 and second order with respect to NO, the rate of reaction will
(1) diminish to one-eighth of its initial value
(2) increase to eight times of its initial value
(3) increase to four times of its initial value
(4) diminish to one-fourth of its initial value

100. The half-life of a radioactive isotope is three hours. If the initial mass of the isotope were 256 g, the mass of it remaining undecayed after 18 hours would be **[2003]**
(1) 8.0 g (2) 12.0 g (3) 16.0 g (4) 4.0 g

101. The integrated rate equation is **[2002]**
$Rt = \log C_0 - \log C_t$.
The straight line graph is obtained by plotting
(1) time vs $\log C_t$ (2) $\frac{1}{\text{time}}$ vs C_t
(3) time vs C_t (4) $\frac{1}{\text{time}}$ vs $\frac{1}{C_t}$

102. In a first order reaction, the concentration of the reactant, decreases from 0.8 M to 0.4 M in 15 minutes. The time taken for the concentration to change from 0.1 M to 0.025 M is **[2004]**
(1) 7.5 minutes (2) 15 minutes
(3) 30 minutes (4) 60 minutes

103. The half-life of a radioisotope is four hours. If the initial mass of the isotope was 200 g, the mass remaining after 24 hours undecayed is **[2004]**
(1) 3.125 g (2) 2.084 g
(3) 1.042 g (4) 4.167 g

104. For the reaction $A + 2B \rightarrow C$, rate is given by $R = [A][B]^2$ then the order of the reaction is **[2002]**
(1) 3 (2) 6 (3) 5 (4) 7

105. If half-life of a substance is 5 yrs, then the total amount of substance left after 15 years, when initial amount is 64 grams is **[2002]**
(1) 16 grams (2) 2 grams
(3) 32 grams (4) 8 grams

NUMERICAL VALUE QUESTIONS (NVQs)

106. Sucrose hydrolyses in acidic medium into glucose and fructose by first order rate law with $t_{1/2} = 3$ hour.
The percentage of sucrose remaining after 6 hours is ____. (Nearest integer) **[April 06, 2026 (I)]**
(Given : log 2 = 0.3010 and log 3 = 0.4771)

107. For a reaction A → P at T K, the half life ($t_{1/2}$) is plotted as a function of initial concentration $[A]_0$ of A as given below:

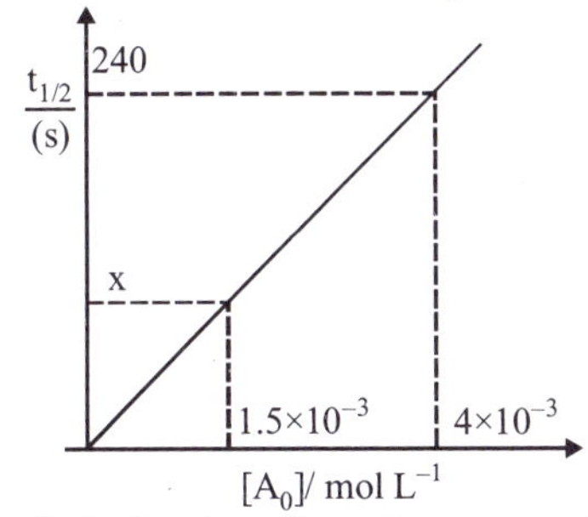

The value of x in the given figure is ______ s (Nearest integer) **[April 05, 2026 (II)]**

108. If the half life of a first order reaction is 6.93 minutes then the time required for completion of 99% of the reaction will be_______minutes.
(Given : log 2 = 0.3010) **[April 05, 2026 (I)]**

109. For a first order reaction A → B

t/min	[A]/M
0	0.6500
x	0.0650
20	0.00065

x = _______ min. (Nearest integer) **[April 04, 2026 (II)]**

110. The half-life of ^{65}Zn is 245 days. After x days, 75% of original activity remained. The value of x in days is _______. (Nearest integer)
(Given : log 3 = 0.4771 and log 2 = 0.3010) **[Jan. 24, 2026 (II)]**

111. For the thermal decomposition of reaction AB(g), the following is constructed.

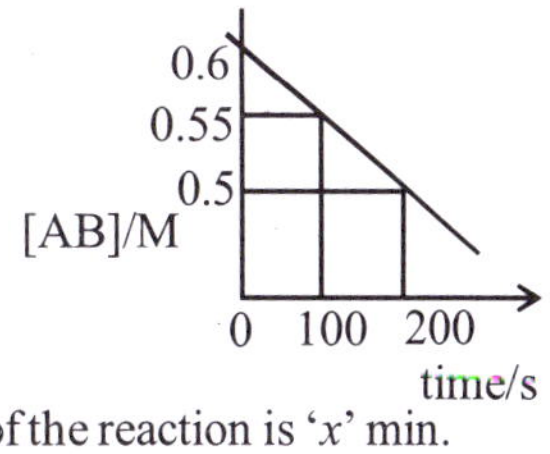

The half life of the reaction is '*x*' min.
x = ________ min. (Nearest integer) **[Jan. 23, 2026 (I)]**

112. For the thermal decomposition of $N_2O_5(g)$ at constant volume, the following table can be formed, for the reaction mentioned below : **[Jan. 23, 2025 (I)]**
$2N_2O_5(g) \rightarrow 2N_2O_4(g) + O_2(g)$

S.No.	Time / s	Total pressure / (atm)
1.	0	0.6
2.	100	'x'

x = ________ $\times 10^{-3}$ atm [nearest integer]
Given : Rate constant for the reaction is $4.606 \times 10^{-2}\ s^{-1}$.

113. Consider the following reaction
A + B → C **[April 8, 2024 (I)]**
The time taken for A to become 1/4th of its initial concentration is twice the time taken to become 1/2 of the same. Also, when the change of concentration of B is plotted against time, the resulting graph gives a straight line with a negative slope and a positive intercept on the concentration axis.
The overall order of the reaction is ____.

114. Time required for 99.9% completion of a first order reaction is _____ times the time required for completion of 90% reaction.(nearest integer). **[April 6, 2024 (I)]**

115. Consider the two different first order reactions given below
A + B → C (Reaction 1) **[April 6, 2024 (II)]**
P → Q (Reaction 2)
The ratio of the half life of Reaction 1 : Reaction 2 is 5 : 2. If t_1 and t_2 represent the time taken to complete $2/3^{rd}$ and $4/5^{th}$ of Reaction 1 and Reaction 2, respectively, then the value of the ratio $t_1 : t_2$ is _______ $\times 10^{-1}$ (nearest integer).
[Given: $\log_{10}(3) = 0.477$ and $\log_{10}(5) = 0.699$]

116. During Kinetic study of reaction 2A + B → C + D, the following results were obtained : **[April 5, 2024 (I)]**

EXP	A[M]	B[M]	initial rate of formation of D
I	0.1	0.1	6.0×10^{-3}
II	0.3	0.2	7.2×10^{-2}
III	0.3	0.4	2.88×10^{-1}
IV	0.4	0.1	2.40×10^{-2}

Based on above data, overall order of the reaction is ....

117. Consider the following single step reaction in gas phase at constant temperature.
$2A(g) + B(g) \rightarrow C(g)$

The initial rate of the reaction is recorded as r_1 when the reaction starts with 1.5 atm pressure of A and 0.7 atm pressure of B. After some time, the rate r_2 is recorded when the pressure of C becomes 0.5 atm. The ratio $r_1 : r_2$ is ________ $\times 10^{-1}$. (Nearest integer) [April 5, 2024 (II)]

118. Consider the following transformation involving first order elementary reaction in each step at constant temperature as shown below. [April 4, 2024 (I)]

$$A + B \underset{\text{Step 3}}{\overset{\text{Step 1}}{\rightleftharpoons}} C \xrightarrow{\text{Step 2}} P$$

Some details of the above reaction are listed below.

Step energy	Rate constant (sec^{-1})	Activation $(kJ\ mol^{-1})$
1.	k_1	300
2.	k_2	200
3.	k_3	E_{a3}

If the overall rate constant of the above transformation (k) is given as $k = \frac{k_1 k_2}{k_3}$ and the overall activation energy (E_a) is 400 kJ mol^{-1}, then the value of E_{a3} is ______ kJ mol^{-1} (nearest integer)

119. Consider the following reaction, the rate expression of which is given below [April 4, 2024 (II)]
$A + B \rightarrow C$
$rate = k[A]^{1/2}[B]^{1/2}$
The reaction is initiated by taking 1M concentration A and B each. If the rate constant (k) is $4.6 \times 10^{-2}\ s^{-1}$, then the time taken for A to become 0.1 M is_____sec. (nearest integer)

120. The ratio of $\frac{^{14}C}{^{12}C}$ in a piece of wood is $\frac{1}{8}$ part that of atmosphere. If half of ^{14}C is 5730 years, the age of wood sample is _______ years. [Feb. 1, 2024 (I)]

121. The following data were obtained during the first order thermal decomposition of a gas A at constant volume:
$A(g) \rightarrow 2B(g) + C(g)$

S. NO	Time/S	Total pressure/(atm)
1.	0	0.1
2.	115	0.28

The rate constant of the reaction is ______ $\times 10^{-2} s^{-1}$ (nearest integer) [Feb. 1, 2024 (II)]

122. $r = k[A]$ for a reaction, 50% of A is decomposed in 120 minutes. The time taken for 90% decomposition of A is _____ minutes. [Jan. 31, 2024 (II)]

123. The rate of First order reaction is 0.04 mol $L^{-1} s^{-1}$ at 10 minutes and 0.03 mol $L^{-1} s^{-1}$ at 20 minutes after initiation. Half life on the reaction is _______ minutes. (Given log2 = 0.3010, log3 = 0.4771) [Jan. 30, 2024 (I)]

124. The half-life of radioisotope bromine- 82 is 36 hours. The fraction which remains after one day is _____ $\times 10^{-2}$. (Given antilog 0.2006 = 1.587) [Jan. 29, 2024 (II)]

125. Consider the following data for the given reaction
$2HI(g) \rightarrow H_2(g) + I_2(g)$ [Jan. 27, 2024 (I)]

	1	2	3
HI (mol L^{-1})	0.005	0.01	0.02
Rate (mol $L^{-1}s^{-1}$)	7.5×10^{-4}	3.0×10^{-3}	1.2×10^{-2}

The order of the reaction is _________.

126. Time required for completion of 99.9% of a First order reaction is ________ times of half life $(t_{1/2})$ of the reaction. [Jan. 27, 2024 (II)]

127. $t_{87.5}$ is the time required for the reaction to undergo 87.5% completion and t_{50} is the time required for the reaction to undergo 50% completion. The relation between $t_{87.5}$ and t_{50} for a first order reaction is $t_{87.5} = x \times t_{50}$
The value of x is _______ (Nearest integer) [April 13, 2023 (I)]

128. The reaction $2NO + Br_2 \rightarrow 2NOBr$
takes places through the mechanism given below :
$NO + Br_2 \Leftrightarrow NOBr_2$ (fast); $NOBr_2 + NO \rightarrow 2NOBr$ (slow)
The overall order of the reaction is ______. [April 12, 2023 (I)]

129. A molecule undergoes two independent first order reactions whose respective half lives are 12 min and 3 min. If both the reactions are occurring then the time taken for the 50% consumption of the reactant is _______ min. (Nearest integer) [April 10, 2023 (I)]

130. The rate constant for a first order reaction is 20 min^{-1}. The time required for the initial concentration of the reactant to reduce to its $\frac{1}{32}$ level is ______ $\times 10^{-2}$ min. (Nearest integer) [Jan. 31, 2023 (II)]
(Given : ln10 = 2.303 log 2 = 0.3010)

131. If compound A reacts with B following first order kinetics with rate constant $2.011 \times 10^{-3}\ s^{-1}$. The time taken by A (in seconds) to reduce from 7 g to 2 g will be ________. (Nearest Integer) [Jan. 30, 2023 (I)]
[log5 = 0.698, log 7 = 0.845, log 2 = 0.301]

132. An organic compound undergoes first order decomposition. If the time taken for the 60% decomposition is 540 s, then the time required for 90% decomposition will be is _______ s. (Nearest integer).
Given : ln 10 = 2.3; log 2 = 0.3 [Jan. 30, 2023 (II)]

133. The number of incorrect statement/s from the following is _________ [April 10, 2023 (II)]
A. The successive half lives of zero order reactions decreases with time.
B. A substance appearing as reactant in the chemical equation may not affect the rate of reaction.
C. Order and molecularity of a chemical reaction can be a fractional number.
D. The rate constant units of zero and second order reaction are mol $L^{-1} s^{-1}$ and $mol^{-1} Ls^{-1}$ respectively.

134. A and B are two substances undergoing radioactive decay in a container. The half life of A is 15 min and that of B is 5 min. If the initial concentration of B is 4 times that of A and they both start decaying at the same time, how much time will it take for the concentration of both of them to be same? ______ min. [Feb. 01, 2023 (I)]

135. $A \rightarrow B$
The above reaction is of zero order. Half life of this reaction is 50 min. The time taken for the concentration of A to reduce to one-fourth of its initial value is ____ min. (Nearest integer) [Feb. 01, 2023 (II)]

136. For certain chemical reaction $X \rightarrow Y$, the rate of formation of product is plotted against the time as shown in the figure. The number of Correct statement/s from the following is _______ [Jan. 29, 2023 (I)]

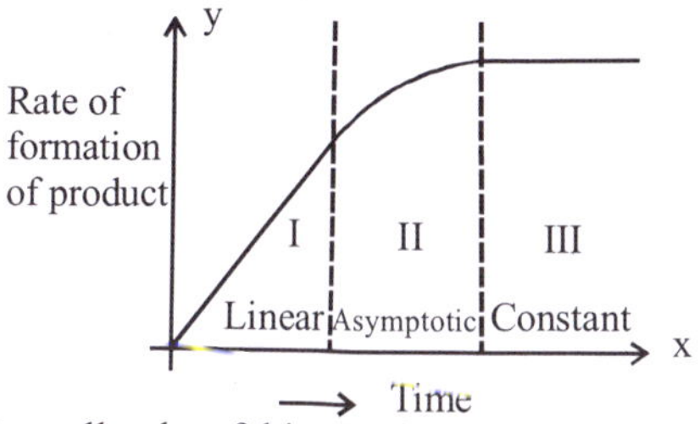

(A) Over all order of this reaction is one
(B) Order of this reaction can't be determined
(C) In region-I and III, the reaction is of first and zero order respectively
(D) In region-II, the reaction is of first order
(E) In region-II, the order of reaction is in the range of 0.1 to 0.9.

137. For conversion of compound $A \rightarrow B$, the rate constant of the reaction was found to be 4.6×10^{-5} L mol^{-1} s^{-1}. The order of the reaction is ______ . **[Jan. 29, 2023 (II)]**

138. For the first order reaction $A \rightarrow B$ the half life is 30 min. The time taken for 75% completion of the reaction is_____min. (Nearest integer)
Given : $\log 2 = 0.3010$
$\log 3 = 0.4771$
$\log 5 = 0.6989$ **[Jan. 25, 2023 (I)]**

139. A flask is filled with equal moles of A and B. The half lives of A and B are 100 s and 50 s respectively and are independent of the initial concentration. The time required for the concentration of A to be four times that of B is _________s. **[June 26, 2022 (I)]**
(Given : $\ln 2 = 0.693$)

140. Assuming 1 μg of trace radioactive element X with a half life of 30 years is absorbed by a growing tree. The amount of X remaining in the tree after 100 years is___$\times 10^{-1}$ μg.
[Given : $\ln 10 = 2.303$; $\log 2 = 0.30$] **[July 29, 2022 (II)]**

141. For the given first order reaction
$A \rightarrow B$ the half life of the reaction is 0.3010 min. The ratio of the initial concentration of reactant to the concentration of reactant at time 2.0 min will be equal to ____________. (Nearest integer) **[July 28, 2022 (I)]**

142. $2NO + 2H_2 \rightarrow N_2 + 2H_2O$
The above reaction has been studied at 800°C. The related data are given in the table below **[July 27, 2022 (I)]**

Reaction serial number	Initial pressure of H_2 / kPa	Initial Pressure of NO/kPa	Initial rate $\left(\frac{-dp}{dt}\right)$ / (kPa/s)
1	65.6	40.0	0.135
2	65.6	20.1	0.033
3	38.6	65.6	0.214
4	19.2	65.6	0.106

The order of the reaction with respect to NO is____

143. [A] → [B]
Reactant Product
If formation of compound [B] follows the first order of kinetics and after 70 minutes the concentration of [A] was found to be half of its initial concentration. Then the rate constant of the reaction is $x \times 10^{-6} s^{-1}$. The value of x is_____. (Nearest integer) **[July 27, 2022 (II)]**

144. For a reaction $A \rightarrow 2B + C$ the half lives are 100 s and 50 s when the concentration of reactant A is 0.5 and 1.0 mol L^{-1} respectively. The order of the reaction is ______. (Nearest Integer) **[July 26, 2022 (I)]**

145. The half life for the decomposition of gaseous compound A is 240 s when the gaseous pressure was 500 Torr initially. When the pressure was 250 Torr, the half life was found to be 4.0 min. The order of the reaction is _________. (Nearest integer) **[July 25, 2022 (I)]**

146. For the decomposition of azomethane,
$CH_3N_2CH_3(g) \rightarrow CH_3CH_3(g) + N_2(g)$ a first order reaction, the variation in partial pressure with time at 600 K is given as

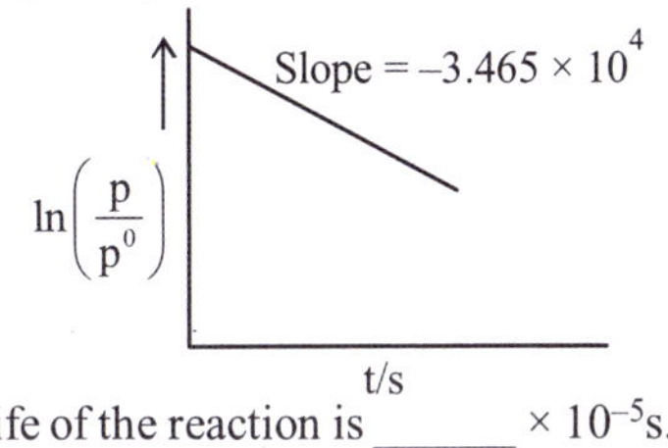

The half life of the reaction is ______ $\times 10^{-5}$s. [Nearest integer] **[July 25, 2022 (II)]**

147. For a first order reaction $A \rightarrow B$, the rate constant, $k = 5.5 \times 10^{-14} s^{-1}$. The time required for 67% completion of reaction is $x \times 10^{-1}$ times the half life of reaction. The value of x is ________(Nearest integer)
(Given : $\log 3 = 0.4771$) **[June 28, 2022 (I)]**

148. A radioactive element has a half life of 200 days. The percentage of original activity remaining after 83 days is ______. (Nearest integer)
(Given : antilog 0.125 = 1.333, antilog 0.693 = 4.93) **[June 28, 2022 (II)]**

149. At 345 K, the half life for the decomposition of a sample of a gaseous compound initially at 55.5 kPa was 340 s. When the pressure was 27.8 kPa, the half life was found to be 170 s. The order of the reaction is__________.
[integer answer] **[June 25, 2022 (II)]**

150. The following data was obtained for chemical reaction given below at 975 K.
$$2NO(g) + 2H_2(g) \rightarrow N_2(g) + 2H_2O(g)$$

	[NO] mol L^{-1}	$[H_2]$ mol L^{-1}	Rate mol $L^{-1}s^{-1}$
(A)	8×10^{-5}	8×10^{-5}	7×10^{-9}
(B)	24×10^{-5}	8×10^{-5}	2.1×10^{-8}
(C)	24×10^{-5}	32×10^{-5}	8.4×10^{-8}

The order of the reaction with respect to NO is ______.
[Integer answer] **[Aug. 26, 2021 (I)]**

151. For a first order reaction, the ratio of the time for 75% completion of a reaction to the time for 50% completion is ________.
(Integer answer) **[Aug. 31, 2021 (I)]**

152. For the first order reaction $A \rightarrow 2B$, 1 mole of reactant A gives 0.2 moles of B after 100 minutes. The half life of the reaction is ...... min. (Round off to the nearest integer).
[Use : $\ln 2 = 0.69$, $\ln 10 = 2.3$
Properties of logarithms : $\ln x^y = y \ln x$;
$\ln\left(\frac{x}{y}\right) = \ln x - \ln y$]

153. $N_2O_5(g) \rightarrow 2NO_2(g) + \frac{1}{2}O_2(g)$
In the above first order reaction, the initial concentration of N_2O_5 is 2.40×10^{-2} mol L^{-1} at 318 K. The concentration of N_2O_5 after 1 hour was 1.60×10^{-2} mol L^{-1}. The rate constant of the reaction at 318 K is _______ $\times 10^{-3}$ min^{-1}. (Nearest integer) **[July 22, 2021 (II)]**

154. $PCl_5(g) \rightarrow PCl_3(g) + Cl_2(g)$
In the above first order reaction, the concentration of PCl_5 reduces from initial concentration 50 mol L^{-1} to 10 mol L^{-1} in 120 minutes at 300 K. The rate constant for the reaction at 300 K is $x \times 10^{-2}$ min^{-1}. The value of x is ________.
[Given log 5 = 0.6989] **[July 20, 2021 (II)]**

155. $2NO(g) + Cl_2(g) \rightleftharpoons 2NOCl(s)$
This reaction was studied at –10 °C and the following data was obtained **[March 18, 2021 (I)]**

run	$[NO]_0$	$[Cl_2]_0$	r_0
1	0.10	0.10	0.18
2	0.10	0.20	0.35
3	0.20	0.20	1.40

$[NO]_0$ and $[Cl_2]_0$ are the initial concentrations and r_0 is the initial reaction rate.
The overall order of the reaction is ________.
(Rounded off to the nearest integer).

156. A reaction has a half life of 1 min. The time required for 99.9% completion of the reaction is ________ min.
(Round off to the nearest integer).
[Use : ln 2 = 0.69; ln 10 = 2.3] **[March 18, 2021 (II)]**

157. For a certain first order reaction, 32% of the reactant is left after 570 s. The rate constant of this reaction is ________ $\times 10^{-3}$ s^{-1}.
(Round off to the nearest integer). **[March 17, 2021 (I)]**
[Given $\log_{10}2 = 0.301$, ln10 = 2.303]

158. A and B decompose via first order kinetics with half lives 54.0 min and 18.0 min respectively. Starting from an equimolar non reactive mixture of A and B, the time taken for the concentration of A to become 16 times that of B is ________ min.
(Round off to the nearest integer). **[March 16, 2021 (II)]**

159. Sucrose hydrolyses in acid solution into glucose and fructose following first order rate law with a half-life of 3.33 h at 25 °C. After 9 h, the fraction of sucrose remaining is *f*. The value of

$\log_{10}\left(\frac{1}{f}\right)$ is ________ $\times 10^{-2}$.

(Rounded off to the nearest integer)
[Assume : ln 10 = 2.303, ln 2 = 0.693] **[Feb. 24, 2021 (II)]**

160. Gaseous cyclobutene isomerizes to butadiene in a first order process which has a '*k*' value of 3.3×10^{-4} s^{-1} at 153 °C. The time in minutes it takes for the isomerization to proceed 40% to completion at this temperature is ________. (Rounded off to the nearest integer) **[Feb. 24, 2021 (I)]**
(Round off to the nearest integer) **[July 27, 2021 (II)]**

161. If 75% of a first order reaction was completed in 90 minutes, 60% of the same reaction would be completed in approximately (in minutes) ________.
(Take : log 2 = 0.30; log 2.5 = 0.40) **[Sep. 04, 2020 (I)]**

162. During the nuclear explosion, one of the products is ^{90}Sr with half life of 6.93 years. If 1 μg of ^{90}Sr was absorbed in the bones of a newly born baby in place of Ca, how much time, in years, is required to reduce it by 90% if it is not lost metabolically.
[Jan. 07, 2020 (I)]

Topic 3 - Theories of Rate of Reaction

MULTIPLE CHOICE QUESTIONS (MCQs)

163. Given below are two statements : **[April 08, 2026 (II)]**
R = 8.314 J K^{-1} mol^{-1} and 1 cal = 4.2 J

Statement I: When Ea = 12.6 kcal/mol, the room temperature rate constant is doubled by a 10°C increase in temperature (298 K to 308 K)

Statement II: For a first order reaction $A \rightarrow B$.

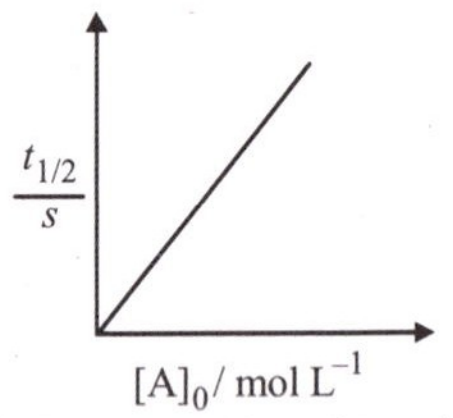

Here $[A]_0$ is the initial concentration of A and $t_{1/2}$ is half-life of reaction.
In the light of the above statements, choose the **correct** answer from the option given below :
(1) Both Statement I and Statement II are true
(2) Both Statement I and Statement II are false
(3) Statement I is true but Statement II are false
(4) Statement I is false but Statement II are true

164. The plot of $\log_{10}K$ vs $\frac{1}{T}$ gives a straight line. The intercept and slope respectively are (where K is equilibrium constant).
[Jan. 28, 2026 (II)]

(1) $\frac{2.303R}{\Delta H^\circ}, \frac{2.303R}{\Delta S^\circ}$ (2) $\frac{\Delta S^\circ}{2.303R}, -\frac{\Delta H^\circ}{2.303R}$
(3) $-\frac{\Delta S^\circ R}{2.303}, -\frac{\Delta H^\circ R}{2.303}$ (4) $-\frac{\Delta H^\circ}{2.303R}, -\frac{\Delta S^\circ}{2.303R}$

165. At 27°C in presence of a catalyst, activation energy of a reaction is lowered by 10 kJ mol^{-1}. The logarithm ratio of $\frac{k(\text{catalysed})}{k(\text{uncatalysed})}$ is ...
(Consider that the frequency factor for both the reactions is same) **[Jan. 24, 2026 (I)]**
(1) 17.41 (2) 1.741 (3) 3.482 (4) 0.1741

166. $A \rightarrow D$ is an endothermic reaction occurring in three steps (elementary). **[Jan. 24, 2026 (I)]**
(i) $A \rightarrow B$ ΔH_i = +ve (ii) $B \rightarrow C$ ΔH_{ii} = –ve
(iii) $C \rightarrow D$ ΔH_{iii} = –ve
Which of the following graphs between potential energy (y-axis) vs reaction coordinate (x-axis) correctly represents the reaction profile of $A \rightarrow D$?

(1)
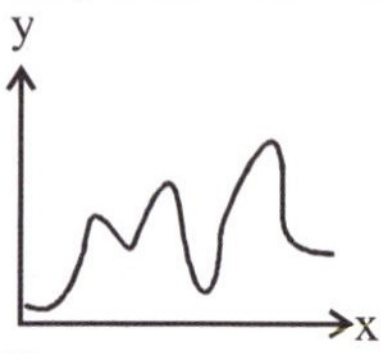

(2)

(3)
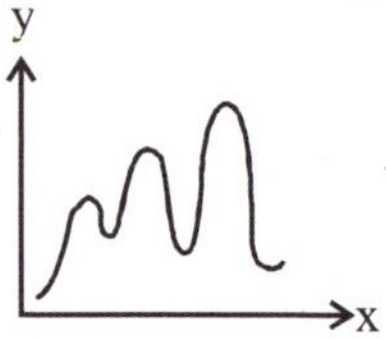

(4)
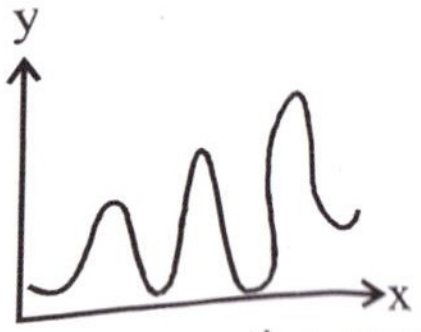

167. Correct statements regarding Arrhenius equation among the following are : **[Jan. 22, 2026 (II)]**
(A) Factor $e^{-Ea/RT}$ corresponds to fraction of molecules having kinetic energy less than Ea.
(B) At a given temperature, lower the Ea, faster is the reaction.

(C) Increase in temperature by about 10°C doubles the rate of reaction.

(D) Plot of log k vs $\frac{1}{T}$ gives a straight line with slope $= -\frac{E_a}{R}$.

Choose the **correct** answer from the options given below:

(1) B and D only (2) A and B only
(3) A and C only (4) B and C only

168. Consider the following plots of log of rate constant k ($\log k$) vs $\frac{1}{T}$ for three different reactions. The correct order of activation energies of these reactions is **[April 4, 2025 (II)]**

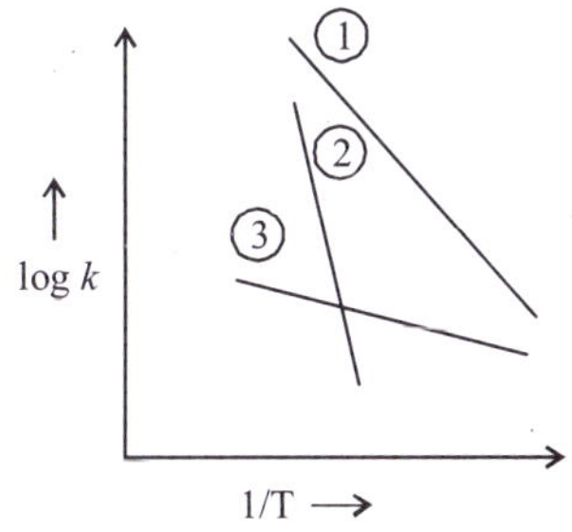

(1) $Ea_2 > Ea_1 > Ea_3$ (2) $Ea_1 > Ea_3 > Ea_2$
(3) $Ea_1 > Ea_2 > Ea_3$ (4) $Ea_3 > Ea_2 > Ea_1$

169. For $A_2 + B_2 \rightleftharpoons 2AB$

E_a for forward and backward reaction are 180 and 200 kJ mol^{-1} respectively **[April 4, 2025 (I)]**

If catalyst lowers E_a for both reaction by 100 kJ mol^{-1}.

Which of the following statement is correct?

(1) Catalyst does not alter the Gibbs energy change of a reaction.
(2) Catalyst can cause non-spontaneous reactions to occur.
(3) The enthalpy change for the reaction is +20 kJ mol^{-1}.
(4) The enthalpy change for the catalysed reaction is different from that of uncatalysed reaction.

170. Consider the following statements related to temperature dependence of rate constants. Identify the correct statements, **[April 3, 2025 (II)]**

A. The Arrhenius equation holds true only for an elementary homogenous reaction.
B. The unit of A is same as that of k in Arrhenius equation.
C. At a given temperature, a low activation energy means a fast reaction.
D. A and E_a as used in Arrhenius equation depend on temperature.
E. When $E_a >> RT$, A and E_a become interdependent.

Choose the **correct** answer from the options given below:

(1) A, C and D Only (2) B, D and E Only
(3) B and C Only (4) A and B Only

171. Reactant A converts to product D through the given mechanism (with the net evolution of heat) :

$A \rightarrow B$ slow ; $\Delta H = +ve$
$B \rightarrow C$ fast ; $\Delta H = -ve$
$C \rightarrow D$ fast ; $\Delta H = -ve$

Which of the following represents the above reaction mechanism ? **[April 2, 2025 (II)]**

(1)
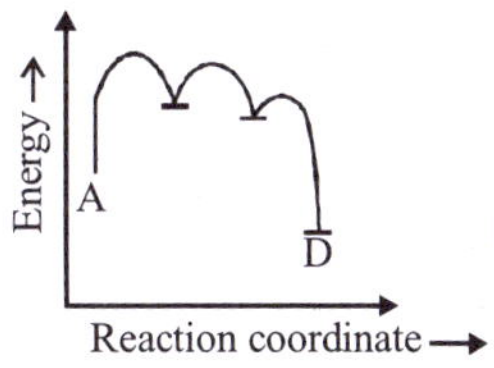

(2)
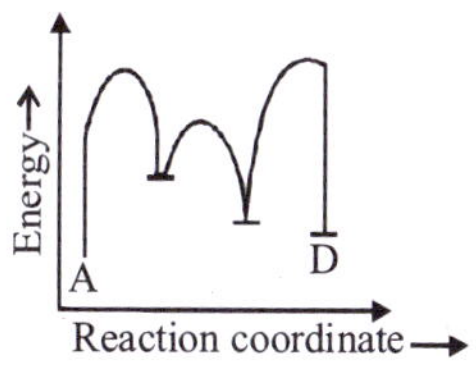

(3)
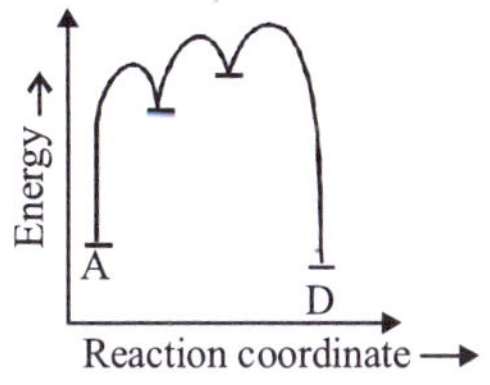

(4)
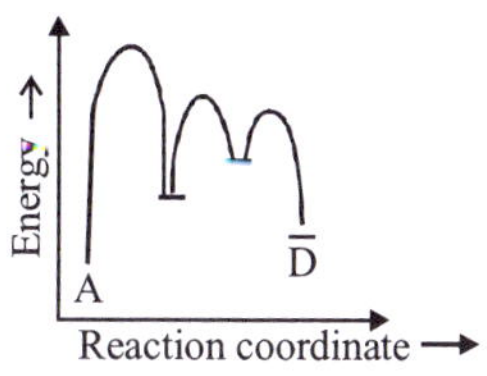

172. Consider the given figure and choose the correct option: **[Jan. 22, 2025 (II)]**

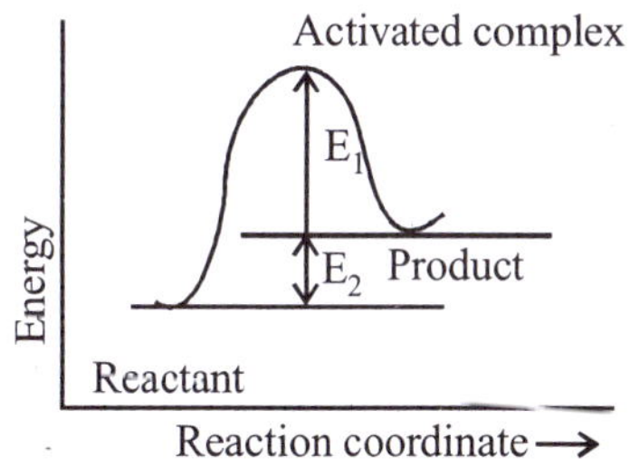

(1) Activation energy of backward reaction is E_1 and product is more stable than reactant.
(2) Activation energy of forward reaction is $E_1 + E_2$ and product is more stable than reactant.
(3) Activation energy of forward reaction is $E_1 + E_2$ and product is less stable than reactant.
(4) Activation energy of both forward and backward reaction is $E_1 + E_2$ and reactant is more stable than product.

173. The correct reaction profile diagram for a positive catalyst reaction. **[April 08, 2023 (II)]**

(1)
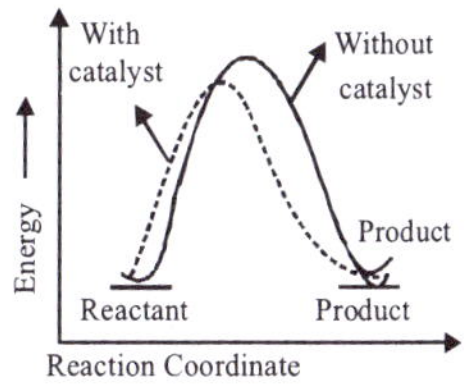

(2)
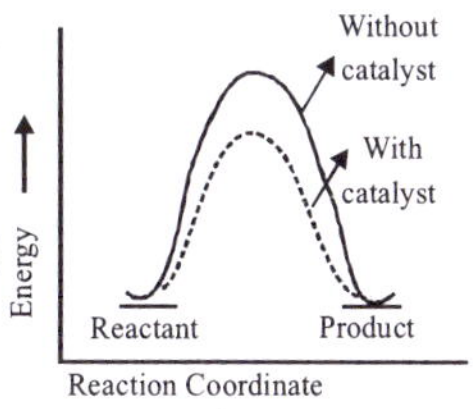

(3)
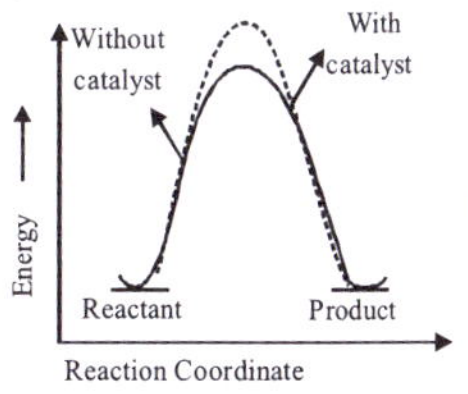

(4)
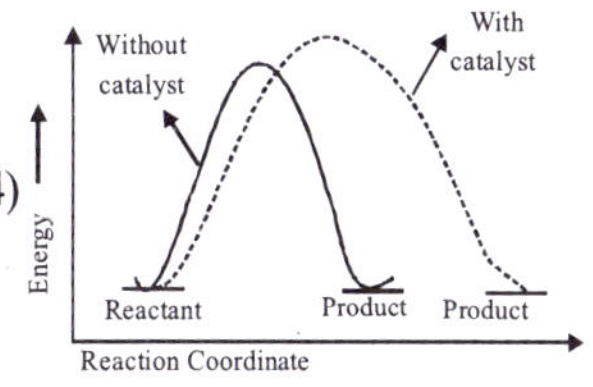

174. Consider the following reaction that goes from A to B in three steps as shown below:- **[April 06, 2023 (II)]**

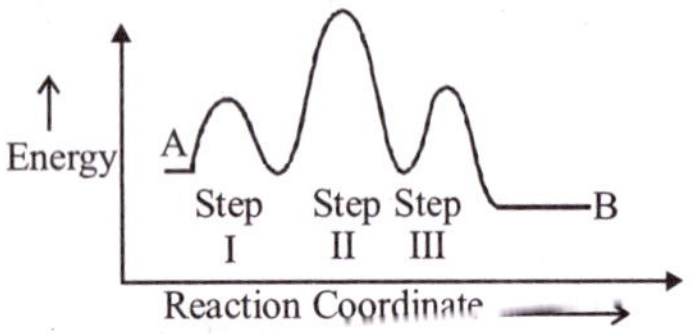

Choose the correct

Number of intermediates	Number of Activated complexs	Rate determining step
(1) 3	2	II
(2) 2	3	II
(3) 2	3	I
(4) 2	3	III

175. The rate constant (k) of a reaction is measured at different temperatures (T), and the data are plotted in the given figure. The activation energy of the reaction in kJ mol^{-1} is : (R is gas constant) **[Sep. 05, 2020 (II)]**

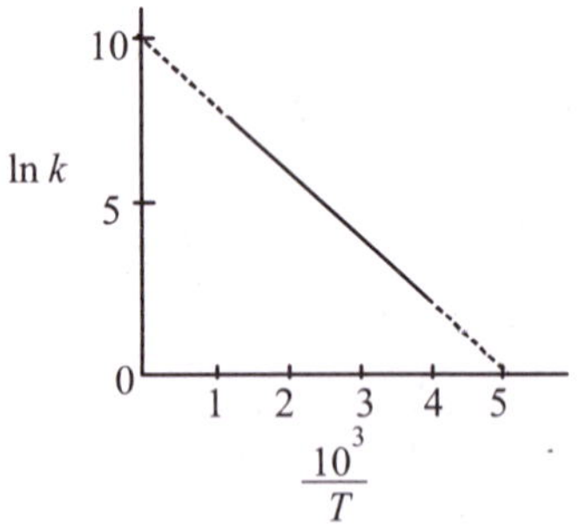

(1) $2/R$ (2) $1/R$ (3) R (4) $2R$

176. For following reactions:

$A \xrightarrow{700\,K} \text{Product}$ $\quad A \xrightarrow[\text{catalyst}]{500\,K} \text{Product}$

it was found that the E_a is decrease by 30 kJ/mol in the presence of catalyst. If the rate remains unchanged, the activation energy for catalysed reaction is (Assume pre exponential factor is same): **[Jan. 09, 2020 (I)]**

(1) 75 kJ/mol (2) 105 kJ/mol
(3) 135 kJ/mol (4) 198 kJ/mol

177. The rate of a certain biochemical reaction at physiological temperature (T) occurs 10^6 times faster with enzyme than without. The change in the activation energy upon adding enzyme is: **[Jan. 08, 2020 (I)]**

(1) –6(2.303)RT (2) –6RT
(3) +6(2.303)RT (4) +6RT

178. Consider the following plots of rate constant versus $\frac{1}{T}$ for four different reactions. Which of the following orders is correct for the activation energies of these reactions? **[Jan. 08, 2020 (II)]**

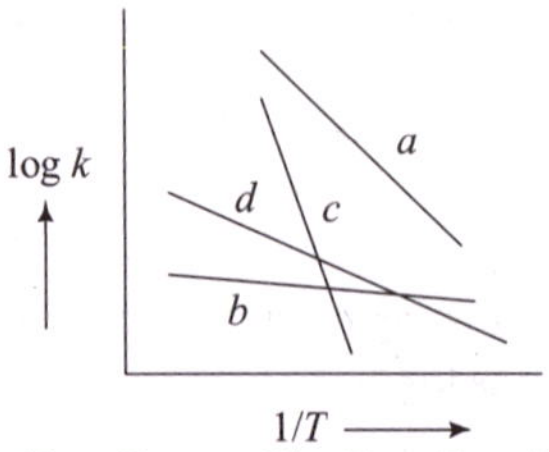

(1) $E_b > E_a > E_d > E_c$ (2) $E_a > E_c > E_d > E_b$
(3) $E_c > E_a > E_d > E_b$ (4) $E_b > E_d > E_c > E_a$

179. For the reaction of H_2 with I_2, the rate constant is 2.5×10^{-4} dm^3 mol^{-1} s^{-1} at 327 °C and 1.0 dm^3 mol^{-1} s^{-1} at 527 °C. The activation energy for the reaction, in kJ mol^{-1} is :
($R = 8.314\,J\,K^{-1}\,mol^{-1}$) **[April 10, 2019 (II)]**

(1) 166 (2) 150 (3) 72 (4) 59

180. Consider the given plot of enthalpy of the following reaction between A and B, $A + B \rightarrow C + D$.
Identify the incorrect statement. **[April 9, 2019 (II)]**

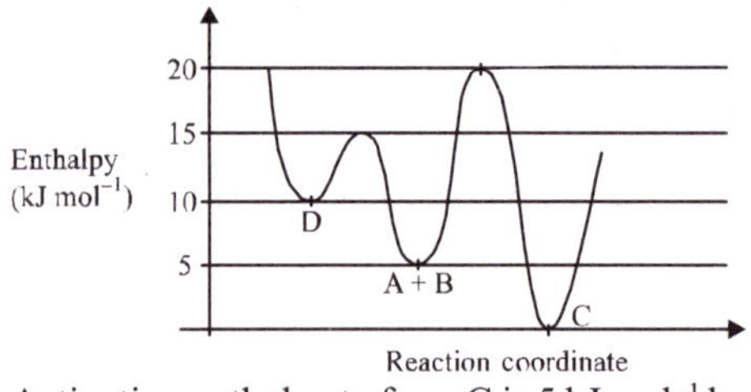

(1) Activation enthalpy to form C is 5 kJ mol^{-1} less than that to form D.
(2) C is the thermodynamically stable product.
(3) D is kinetically stable product.
(4) Formation of A and B from C has highest enthalpy of activation.

181. For a reaction, consider the plot of ln k versus 1/T given in the figure. If the rate constant of this reaction at 400 K is 10^{-5} s^{-1}, then the rate constant at 500 K is : **[Jan. 12, 2019 (II)]**

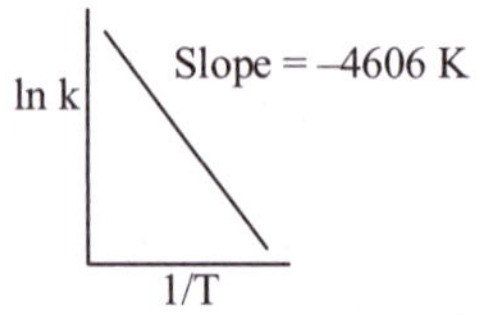

(1) 10^{-6} s^{-1} (2) 2×10^{-4} s^{-1}
(3) 10^{-4} s^{-1} (4) 4×10^{-4} s^{-2}

182. If a reaction follows the Arrhenius equation, the plot ln k vs $\frac{1}{(RT)}$ gives straight line with a gradient (–y) unit. The energy required to activate the reactant is: **[Jan. 11, 2019 (I)]**

(1) y/R unit (2) y unit
(3) yR unit (4) – y unit

183. Consider the given plots for a reaction obeying Arrhenius equation (0 °C < T < 300 °C) : (k and E_a are rate constant and activation energy, respectively) **[Jan. 10, 2019 (I)]**

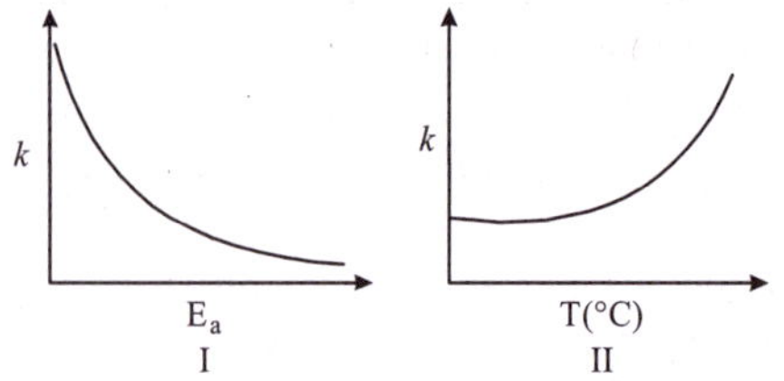

Choose the correct option:

(1) I is right but II is wrong (2) Both I and II are correct
(3) I is wrong but II is right (4) Both I and II are wrong

184. Two reactions R_1 and R_2 have identical pre-exponential factors. Activation energy of R_1 exceeds that of R_2 by 10 kJ mol^{-1}. If k_1 and k_2 are rate constants for reactions R_1 and R_2 respectively at 300 K, then $\ln(k_2/k_1)$ is equal to :
($R = 8.314\,J\,mol^{-1}K^{-1}$) **[2017]**

(1) 8 (2) 12 (3) 6 (4) 4

185. The rate of a reaction A doubles on increasing the temperature from 300 to 310 K. By how much, the temperature of reaction B should be increased from 300 K so that rate doubles if activation energy of the reaction B is twice to that of reaction A. **[Online April 8, 2017]**

(1) 9.84 K (2) 4.92 K (3) 2.45 K (4) 19.67 K

186. The rate of a reaction quadruples when the temperature changes from 300 to 310 K. The activation energy of this reaction is : (Assume activation energy and pre-exponential factor are independent of temperature; $\ln 2 = 0.693$; $R = 8.314\ J\ mol^{-1}\ K^{-1}$) **[Online April 9, 2017]**

(1) $107.2\ kJ\ mol^{-1}$ (2) $53.6\ kJ\ mol^{-1}$
(3) $26.8\ kJ\ mol^{-1}$ (4) $214.4\ kJ\ mol^{-1}$

187. For the equilibrium, $A(g) \rightleftharpoons B(g)$, ΔH is -40 kJ/mol. If the ratio of the activation energies of the forward (E_f) and reverse (E_b) reactions is $\frac{2}{3}$ then : **[Online April 11, 2015]**

(1) $E_f = 80$ kJ/mol; $E_b = 120$ kJ/mol
(2) $E_f = 60$ kJ/mol; $E_b = 100$ kJ/mol
(3) $E_f = 30$ kJ/mol; $E_b = 70$ kJ/mol
(4) $E_f = 70$ kJ/mol; $E_b = 30$ kJ/mol

188. The rate coefficient (k) for a particular reactions is $1.3 \times 10^{-4}\ M^{-1}\ s^{-1}$ at 100 °C, and $1.3 \times 10^{-3}\ M^{-1}\ s^{-1}$ at 150 °C. What is the energy of activation (E_a) (in kJ) for this reaction? (R = molar gas constant = $8.314\ JK^{-1}\ mol^{-1}$) **[Online April 12, 2014]**

(1) 16 (2) 60 (3) 99 (4) 132

189. The reaction $X \rightarrow Y$ is an exothermic reaction. Activation energy of the reaction for X into Y is $150\ kJ\ mol^{-1}$. Enthalpy of reaction is $135\ kJ\ mol^{-1}$. The activation energy for the reverse reaction, $Y \rightarrow X$ will be : **[Online April 22, 2013]**

(1) $280\ kJ\ mol^{-1}$ (2) $285\ kJ\ mol^{-1}$
(3) $270\ kJ\ mol^{-1}$ (4) $15\ kJ\ mol^{-1}$

190. The rate of a reaction doubles when its temperature changes from 300 K to 310 K. Activation energy of the reaction will be : ($R = 8.314\ JK^{-1}\ mol^{-1}$ and $\log 2 = 0.301$) **[2013]**

(1) $53.6\ kJ\ mol^{-1}$ (2) $48.6\ kJ\ mol^{-1}$
(3) $58.5\ kJ\ mol^{-1}$ (4) $60.5\ kJ\ mol^{-1}$

191. The activation energy for a reaction which doubles the rate when the temperature is raised from 298 K to 308 K is **[Online May 26, 2012]**

(1) $59.2\ kJ\ mol^{-1}$ (2) $39.2\ kJ\ mol^{-1}$
(3) $52.9\ kJ\ mol^{-1}$ (4) $29.5\ kJ\ mol^{-1}$

192. A reactant (A) froms two products : **[2011RS]**

$A \xrightarrow{k_1} B$, Activation energy E_{a1}
$A \xrightarrow{k_2} C$, Activation energy E_{a2}

If $E_{a_2} = 2E_{a_1}$, then k_1 and k_2 are related as :

(1) $k_2 = k_1 e^{E_{a_1}/RT}$ (2) $k_2 = k_1 e^{E_{a_2}/RT}$
(3) $k_1 = Ak_2 e^{E_{a_1}/RT}$ (4) $k_1 = 2k_2 e^{E_{a_2}/RT}$

193. Rate of a reaction can be expressed by Arrhenius equation as : **[2006]**

$$k = A\, e^{-E_a/RT}$$

In this equation, E_a represents

(1) the total energy of the reacting molecules at a temperature, T
(2) the fraction of molecules with energy greater than the activation energy of the reaction
(3) the energy above which all the colliding molecules will react
(4) the energy below which all the colliding molecules will react

194. A schematic plot of $\ln K_{eq}$ versus inverse of temperature for a reaction is shown below **[2005]**

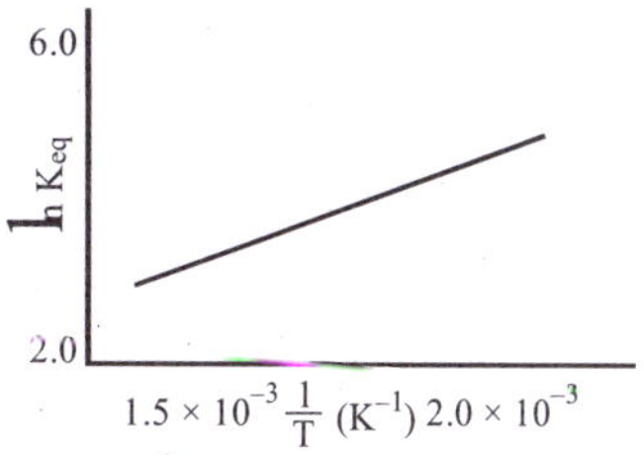

The reaction must be

(1) highly spontaneous at ordinary temperature
(2) one with negligible enthalpy change
(3) endothermic
(4) exothermic

195. Consider an endothermic reaction $X \rightarrow Y$ with the activation energies E_b and E_f for the backward and forward reactions, respectively. In general **[2005]**

(1) there is no definite relation between E_b and E_f
(2) $E_b = E_f$
(3) $E_b > E_f$
(4) $E_b < E_f$

196. In respect of the equation $k = Ae^{-E_a/RT}$ in chemical kinetics, which one of the following statements is correct ? **[2003]**

(1) A is adsorption factor
(2) E_a is energy of activation
(3) R is Rydberg's constant
(4) k is equilibrium constant

NUMERICAL VALUE QUESTIONS (NVQs)

197. Decomposition of a hydrocarbon follows the equation $k = (5.5 \times 10^{11}\ s^{-1})\ e^{\frac{-28000K}{T}}$. The activation energy of reaction is______ $kJ\ mol^{-1}$. (Nearest Integer)
Given : $R = 8.3\ JK^{-1}\ mol^{-1}$ **[April 06, 2026 (II)]**

198. For reaction $A \rightarrow P$, rate constant $k = 1.5 \times 10^3\ s^{-1}$ at 27°C. If activation energy for the above reaction is $60\ kJ\ mol^{-1}$, then the temperature (in °C) at which rate constant, $k = 4.5 \times 10^3\ s^{-1}$ is ______. (Nearest integer)
Given : $\log 2 = 0.30$, $\log 3 = 0.48$, $R = 8.3\ J\ K^{-1}\ mol^{-1}$, $\ln 10 = 2.3$ **[April 02, 2026 (I)]**

199. $A \rightarrow B$ (first reaction)
$C \rightarrow D$ (second reaction)
Consider the above two first-order reactions. The rate constant for first reaction at 500 K is double of the same at 300 K. At 500 K, 50% of the reaction becomes complete in 2 hour. The activation energy of the second reaction is half of that of first reaction. If the rate constant at 500 K of the second reaction becomes double of the rate constant of first reaction at the same temperature; then rate constant for the second reaction at 300 K is ________ $\times 10^{-1}$ $hour^{-1}$ (nearest integer). **[Jan. 28, 2026 (II)]**

200. Consider $A \xrightarrow{K_1} B$ and $C \xrightarrow{K_2} D$ are two reactions. If the rate constant (k_1) of the $A \rightarrow B$ reaction can be expressed by the following equation $\log_{10} k$

$= 14.34 - \frac{1.5 \times 10^4}{T/K}$ and activation energy of $C \rightarrow D$ reaction (Ea_2) is $\frac{1}{5}$ th of the $A \rightarrow B$ reaction (Ea_1), then the value of (Ea_2) is ______ kJ mol^{-1}. (Nearest Integer) [Jan. 22, 2026 (II)]

201. The temperature at which the rate constants of the given below two gaseous reactions become equal is ____________ K. (Nearest integer).

$$X \longrightarrow Y \quad k_1 = 10^6 e^{\frac{-30000}{T}}$$

$$P \longrightarrow Q \quad k_2 = 10^4 e^{\frac{-24000}{T}}$$

Given : ln 10 = 2.303 [Jan. 22, 2026 (I)]

202. Pre-exponential factors of two different reactions of same order are identical. Let activation energy of first reaction exceeds the activation energy of second reaction by 20 kJ mol^{-1}. If k_1 and k_2 are the rate constants of first and second reaction respectively at 300 K, then $ln \frac{k_2}{k_1}$ will be ...... (nearest integer)

$[R = 8.3\ J K^{-1} mol^{-1}]$ [Jan. 21, 2026 (I)]

203. Consider a complex reaction taking place in three steps with rate constants k_1, k_2 and k_3 respectively. The overall rate constant k is given by the expression $k = \sqrt{\frac{k_1 k_3}{k_2}}$. If the activation energies of the three steps are 60, 30 and 10 kJ mol^{-1} respectively, then the overall energy of activation in kJ mol^{-1} is ________. (Nearest integer) [Jan. 24, 2025 (II)]

204. $A \rightarrow B$

The molecule A changes into its isomeric form B by following a first order kinetics at a temperature of 1000 K. If the energy barrier with respect to reactant energy for such isomeric transformation is 191.48 kJ mol^{-1} and the frequency factor is 10^{20}, the time required for 50%, molecules of A to become B is ______ picoseconds (nearest integer). [Jan. 22, 2025 (I)]

$[R = 8.314\ J K^{-1} mol^{-1}]$

205. For a reaction taking place in three steps at same temperature, overall rate constant $K = \frac{K_1 K_2}{K_3}$. If E_{a_1}, E_{a_2} and E_{a_3} are 40, 50 and 60 kJ/mol respectively, the overall E_a is ______ kJ/mol. [Jan. 29, 2024 (I)]

206. The number of given statement/s which is/are correct is____ [April 08, 2023 (I)]

(A) The stronger the temperature dependence of the rate constant, the higher is the activation energy.

(B) If a reaction has zero activation energy, its rate is independent of temperature.

(C) The stronger the temperature dependence of the rate constant, the smaller is the activation energy.

(D) If there is no correlation between the temperature and the rate constant then it means that the reaction has negative activation energy.

207. For the adsorption of hydrogen on platinum, the activation energy is 30 kJ mol^{-1} and for the adsorption of hydrogen on nickel, the activation energy is 41.4 kJ mol^{-1}. The logarithm of the ratio of the rates of chemisorption on equal areas of the metals at 300 K is .... (Nearest integer)

Given: ln 10 = 2.3 R = 8.3 $JK^{-1} mol^{-1}$ [April 06, 2023 (I)]

208. $A \rightarrow B$ [Jan. 31, 2023 (I)]

The rate constants of the above reaction at 200 K and 300K are 0.03 min^{-1} and 0.05 min^{-1} respectively. The activation energy for the reaction is ________ J (Nearest integer)

(Given : In 10 = 2.3

R = 8.3 J $K^{-1} mol^{-1}$

log5 = 0.70

log3 = 0.48

log2 = 0.30

209. The number of correct statement/s from the following is ________. [Jan. 24, 2023 (I)]

A. Larger the activation energy, smaller is the value of the rate constant.

B. The higher is the activation energy, higher is the value of the temperature coefficient.

C. At lower temperatures, increase in temperature causes more change in the value of k than at higher temperature.

D. A plot of ln k vs $\frac{1}{T}$ is a straight line with slope equal to $-\frac{Ea}{R}$

210. For a reaction, given below is the graph of ln k vs $\frac{1}{T}$. The activation energy for the reaction is equal to ______ cal mol^{-1}. (Nearest integer) [July 28, 2022 (II)]

(Given : R = 2 cal $K^{-1} mol^{-1}$)

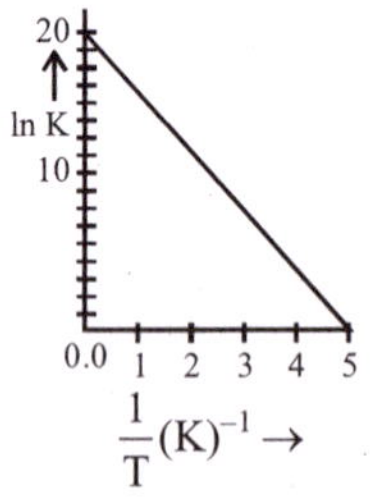

211. For the reaction $P \rightarrow B$, the values of frequency factor A and activation energy E_A are $4 \times 10^{13}\ s^{-1}$ and 8.3 kJ mol^{-1} respectively. If the reaction is of first order, the temperature at which the rate constant is $2 \times 10^{-6}\ s^{-1}$ is ______ $\times 10^{-1}$ K. [June 30, 2022 (I)]

(Given : ln 10 = 2.3, $R = 8.3\ J K^{-1} mol^{-1}$, log 2 = 0.30)

212. The activation energy of one of the reactions in a biochemical process is 532611 J mol^{-1}. When the temperature falls from 310 K to 300 K, the change in rate constant observed is $k_{300} = x \times 10^{-3} k_{310}$. The value of x is ________. [Given: ln10 = 2.3 , R = 8.3 J $K^{-1} mol^{-1}$] [June 29, 2022 (I)]

213. The equation $k = (6.5 \times 10^{12}\ s^{-1})\ e^{-26000 K/T}$ is followed for the decomposition of compound A. The activation energy for the reaction is ____ kJ mol^{-1}. [nearest integer]

(Given: $R = 8.314\ J K^{-1} mol^{-1}$) [June 29, 2022 (II)]

214. The rate constant for a first order reaction is given by the following equation:

$$\ln k = 33.24 - \frac{2.0 \times 10^4 \text{ K}}{T}$$

The Activation energy for the reaction is given by ______ $kJ\ mol^{-1}$. (In nearest integer)

(Given: $R = 8.3\ J\ K^{-1}\ mol^{-1}$) [June 27, 2022 (I)]

215. It has been found that for a chemical reaction with rise in temperature by 9K the rate constant gets doubled. Assuming a reaction to be occurring at 300 K, the value of activation energy is found to be ______ $kJ\ mol^{-1}$. [nearest integer] [June 27, 2022 (II)]

(Given $\ln 10 = 2.3$, $R = 8.3\ JK^{-1}mol^{-1}$, $\log 2 = 0.30$)

216. Catalyst A reduces the activation energy for a reaction by 10 kJ mol^{-1} at 300 K. The ratio of rate constants, $\frac{k_t,\ \text{catalysed}}{k_t,\ \text{Uncatalysed}}$ is e^x. The value of x is _____. [nearest integer]

[Assume that the pre-exponential factor is same in both the cases. Given $R = 8.31\ J\ K^{-1}\ mol^{-1}$] [June 26, 2022 (II)]

217. The rate constants for decomposition of acetaldehyde have been measured over the temperature range 700 – 1000 K. The data has been analysed by plotting $\ell n\ k$ vs $\frac{10^3}{T}$ graph. The value of activation energy for the reaction is___ $kJ\ mol^{-1}$. (Nearest integer) (Given : $R = 8.31\ J\ K^{-1}\ mol^{-1}$)

[June 24, 2022 (I)]

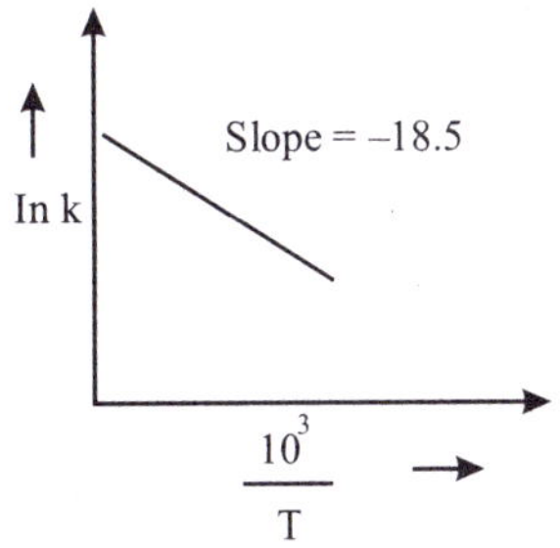

218. According to the following figure, the magnitude of the enthalpy change of the reaction

$A + B \rightarrow M + N$ in $kJ\ mol^{-1}$

is equal to ________. (Integer answer) [Aug. 31, 2021 (I)]

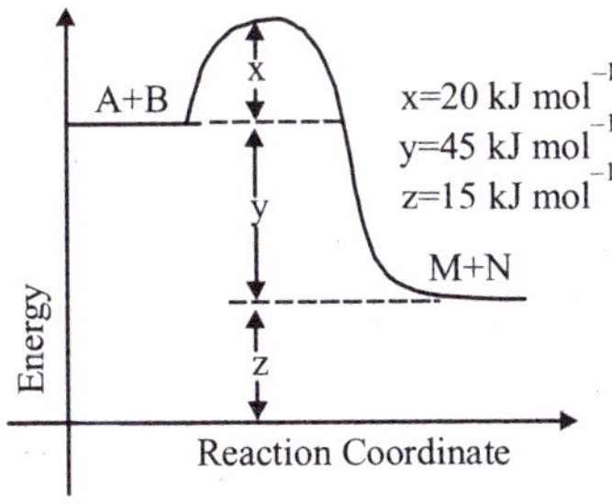

219. For the reaction $A \rightarrow B$, the rate constant k(in s^{-1}) is given by

$$\log_{10} k = 20.35 - \frac{(2.47 \times 10^3)}{T}$$

The energy of activation in $kJ\ mol^{-1}$ is _______.

(Nearest integer) [Given : $R = 8.314\ J\ K^{-1}\ mol^{-1}$]

[Aug. 31, 2021 (II)]

220. The first order rate constant for the decomposition of $CaCO_3$ at 700 K is $6.36 \times 10^{-3} s^{-1}$ and activation energy is 209 $kJ\ mol^{-1}$. Its rate constant (in s^{-1}) at 600 K is $x \times 10^{-6}$. The value of x is _______. (Nearest integer)

[Given $R = 8.31\ J\ K^{-1}\ mol^{-1}$; $\log 6.36 \times 10^{-3} = -2.19$, $10^{-4.79} = 1.62 \times 10^{-5}$] [Aug. 27, 2021 (II)]

221. The decomposition of formic acid on gold surface follows first order kinetics. If the rate constant at 300 K is $1.0 \times 10^{-3}\ s^{-1}$ and the activation energy $E_a = 11.488\ kJ\ mol^{-1}$, the rate constant at 200 K is _____ $\times 10^{-5}\ s^{-1}$. (Round off to the Nearest Integer) [March 16, 2021 (I)]

222. An exothermic reaction $X \rightarrow Y$ has an activation energy 30 $kJ\ mol^{-1}$. If energy change ΔE during the reaction is –20 kJ, then the activation energy for the reverse reaction in kJ is ________. [Feb. 26, 2021 (I)]

223. If the activation energy of a reaction is 80.9 $kJ\ mol^{-1}$, the fraction of molecules at 700 K, having enough energy to react to form products is e^{-x}. The value of x is ________.

[Use $R = 8.31\ J\ K^{-1}\ mol^{-1}$] [Feb. 26, 2021 (II)]

224. For the reaction, $aA + bB \rightarrow cC + dD$, the plot of log k vs $\frac{1}{T}$ is given below:

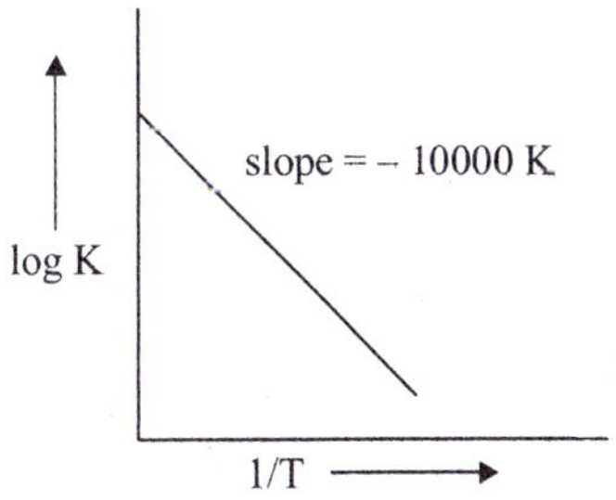

The temperature at which the rate constant of the reaction is $10^{-4}s^{-1}$ is ________ K.

(Rounded-off to the nearest integer)

[Given: The rate constant of the reaction is $10^{-5}s^{-1}$ at 500 K.]

[Feb. 25, 2021 (I)]

225. The rate constant of a reaction increases by five times on increase in temperature from 27 °C to 52 °C. The value of activation energy in $kJ\ mol^{-1}$ is _______. (Rounded-off to the nearest integer)

[R = 8.314 J K–1 mol–1] [Feb. 25, 2021 (II)]

226. The rate of a reaction decreased by 3.555 times when the temperature was changed from 40°C to 30°C. The activation energy (in $kJ\ mol^{-1}$) of the reaction is _______. Take;

$R = 8.314\ J\ mol^{-1}\ K^{-1}$ $\ln 3.555 = 1.268$ [Sep. 06, 2020 (II)]

227. The number of molecules with energy greater than the threshold energy for a reaction increases five fold by a rise of temperature from 27°C to 42°C. Its energy of activation in J/ mol is __________. (Take $\ln 5 = 1.6094$; $R = 8.314\ J\ mol^{-1}K^{-1}$)

[Sep. 04, 2020 (II)]

228. A sample of milk splits after 60 min. at 300 K and after 40 min. at 400 K when the population of *lactobacillus acidophilus* in it doubles. The activation energy (in kJ/mol) for this process is closest to _____.

(Given, $R = 8.3\ J\ mol^{-1}\ K^{-1}$, $\ln\left(\frac{2}{3}\right) = -0.4$, $e^{-3} = 0.04$)

[Jan. 09, 2020 (II)]

Part-B (JEE Advanced)

Topic 1- Rate of Reaction, Rate Laws and Rate Constant

1 MCQs with One Correct Answer

1. For the elementary reaction $M \rightarrow N$, the rate of disappearance of M increases by a factor of 8 upon doubling the concentration of M. The order of the reaction with respect to M is [Adv. 2014]
(a) 4 (b) 3 (c) 2 (d) 1

2. The reaction, $A \rightarrow$ Product, follows first order kinetics. In 40 minutes the concentration of A changes from 0.1 to 0.025 M. The rate of reaction, when concentration of A is 0.01 M is [2004S]
(a) 1.73×10^{-4} M min^{-1} (b) 3.47×10^{-5} M min^{-1}
(c) 3.47×10^{-4} M min^{-1} (d) 1.73×10^{-5} M min^{-1}

3. In a first order reaction the concentration of reactant decreases from 800 mol/dm^3 to 50 mol/dm^3 in 2×10^4 sec. The rate constant of reaction in sec^{-1} is: [2003S]
(a) 2×10^4 (b) 3.45×10^{-5}
(c) 1.386×10^{-4} (d) 2×10^{-4}

4. Consider the chemical reaction, $N_2(g) + 3H_2(g) \rightarrow 2NH_3(g)$. The rate of this reaction can be expressed in terms of time derivative of concentration of $N_2(g)$, $H_2(g)$ or $NH_3(g)$. Identify the correct relationship amongst the rate expressions. [2002S]
(a) Rate $= -d[N_2]/dt = -1/3 d[H_2]/dt = 1/2 d[NH_3]/dt$
(b) Rate $= -d[N_2]/dt = -3d[H_2]/dt = 2d[NH_3]/dt$
(c) Rate $= d[N_2]/dt = 1/3 d[H_2]/dt = 1/2 d[NH_3]/dt$
(d) Rate $= -d[N_2]/dt = -d[H_2]/dt = d[NH_3]/dt$

5. If 'I' is the intensity of absorbed light and 'C' is the concentration of AB for the photochemical process, $AB + h\nu \longrightarrow AB^*$, the rate of formation of AB^* is directly proportional to [2001S]
(a) C (b) I (c) I^2 (d) C.I

6. The rate constant for the reaction, $2N_2O_5 \rightarrow 4NO_2 + O_2$, is 3.0×10^{-5} sec^{-1}. If the rate is 2.40×10^{-5} mol litre^{-1} sec^{-1}, then the concentration of N_2O_5 (in mol litre^{-1}) is [2000S]
(a) 1.4 (b) 1.2 (c) 0.04 (d) 0.8

7. The specific rate constant of a first order reaction depends on the [1983 - 1 Mark]
(a) concentration of the reactant
(b) concentration of the product
(c) time
(d) temperature

8. The rate constant of a reaction depends on [1981 - 1 Mark]
(a) temperature
(b) initial concentration of the reactants
(c) time of reaction
(d) extent of reaction

2 Integer Value Answer

9. In dilute aqueous H_2SO_4, the complex diaquodioxalatoferrate(II) is oxidized by MnO_4^-. For this reaction, the ratio of the rate of change of $[H^+]$ to the rate of change of $[MnO_4^-]$ is [Adv. 2015]

4 Fill in the Blanks

10. For the reaction $N_2(g) + 3H_2(g) \longrightarrow 2NH_3(g)$, under certain conditions of temperature and partial pressure of the reactants, the rate of formation of NH_3 is 0.001 kg h^{-1}. The rate of conversion of H_2 under the same condition is .... kg h^{-1}. [1994 - 1 Mark]

11. The rate of chemical change is directly proportional to .... [1985 - 1 Mark]

6 MCQs with One or More than One Correct Answer

12. For the following reaction $2X + Y \xrightarrow{k} P$
the rate of reaction is $\frac{d[P]}{dt} = k[X]$. Two moles of X are mixed with one mole of Y to make 1.0 L of solution. At 50 s, 0.5 mole of Y is left in the reaction mixture. The correct statement(s) about the reaction is(are)
(Use: $\ln 2 = 0.693$) [Adv. 2021]
(a) The rate constant, k, of the reaction is 13.86×10^{-4} s^{-1}.
(b) Half-life of X is 50s.
(c) At 50 s, $-\frac{d[X]}{dt} = 13.86 \times 10^{-3}$ mol L^{-1}s^{-1}.
(d) At 100 s, $-\frac{d[Y]}{dt} = 3.46 \times 10^{-3}$ mol L^{-1}s^{-1}.

13. The rate law for the reaction : [1988 - 1 Mark]
$$RCl + NaOH(aq.) \rightarrow ROH + NaCl$$
is given by, Rate $= k_1$ [RCl]. The rate of the reaction will be
(a) doubled on doubling the concentration of sodium hydroxide.
(b) halved on reducing the concentration of alkyl halide to one half.
(c) increased on increasing the temperature of the reaction.
(d) unaffected by increasing the temperature of the reaction.

10 Subjective Problems

14. For the given reactions, $A + B \rightarrow$ Products, following data were obtained. [2004 - 2 Marks]

	$[A_0]$	$[B_0]$	R_0 (mol L^{-1} s^{-1})
1.	0.1	0.2	0.05
2.	0.2	0.2	0.10
3.	0.1	0.1	0.05

(a) Write the rate law expression
(b) Find the rate constant

15. The rate constant for an isomerisation reaction, $A \rightarrow B$ is 4.5×10^{-3} min^{-1}. If the initial concentration of A is 1 M, calculate the rate of the reaction after 1 h. [1999 - 4 Marks]

16. The progress of the reaction, $A \rightleftharpoons nB$, with time, is presented in figure given below. Determine

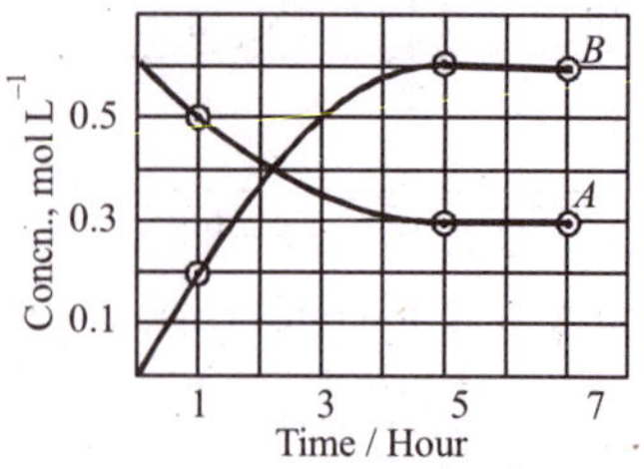

(i) the value of n
(ii) the equilibrium constant, K and
(iii) the initial rate of conversion of A. [1994 - 3 Marks]

Topic 2 - Order of Reaction and Half Life Period

1 MCQs with One Correct Answer

17. In the reaction,

$P+Q \longrightarrow R+S$

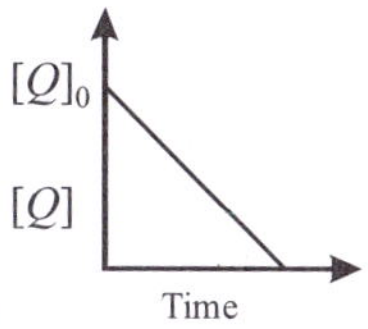

The time taken for 75% reaction of P is twice the time taken for 50% reaction of P. The concentration of Q varies with reaction time as shown in the figure. The overall order of the reaction is [Adv. 2013]

(a) 2 (b) 3 (c) 0 (d) 1

18. Under the same reaction conditions, initial concentration of 1.386 mol dm^{-3} of a substance becomes half in 40 seconds and 20 seconds through first order and zero order kinetics, respectively. Ratio (k_1/k_0) of the rate constant for first order (k_1) and zero order (k_0) of the reaction is – [2008]

(a) 0.5 $mol^{-1}\ dm^3$ (b) 1.0 mol dm^{-3}

(c) 1.5 mol dm^{-3} (d) 2.0 $mol^{-1}\ dm^3$

19. Consider a reaction $aG + bH \rightarrow$ Products. When concentration of both the reactants G and H is doubled, the rate increases by eight times. However, when concentration of G is doubled keeping the concentration of H fixed, the rate is doubled. The overall order of the reaction is [2007]

(a) 0 (b) 1 (c) 2 (d) 3

20. Which one of the following statement for order of reaction is not correct? [2005S]

(a) Order can be determined experimentally

(b) Order of reaction is equal to sum of the powers of concentration terms in differential rate law.

(c) It is not affected with the stoichiometric coefficient of the reactants

(d) Order cannot be fractional.

2 Integer Value Answer

21. A sample initially contains only U-238 isotope of uranium. With time, some of the U-238 radioactively decays into Pb-206 while the rest of it remains undisintegrated. [Adv. 2024]
When the age of the sample is $P \times 10^8$ years, the ratio of mass of Pb-206 to that of U-238 in the sample is found to be 7. The value of P is _____.
[**Given:** Half-life of U-238 is 4.5×10^9 years; $\log_e 2 = 0.693$]

22. Consider the following reaction,

$2H_2(g) + 2NO(g) \rightarrow N_2(g) + 2H_2O(g)$

which follows the mechanism given below: [Adv. 2024]

$2NO(g) \underset{k_{-1}}{\overset{k_1}{\rightleftharpoons}} N_2O_2(g)$ (fast equilibrium)

$N_2O_2(g) + H_2(g) \xrightarrow{k_2} N_2O(g) + H_2O(g)$ (slow reaction)

$N_2O(g) + H_2(g) \xrightarrow{k_3} N_2(g) + H_2O(g)$ (fast reaction)

The order of the reaction is _______.

23. An organic compound undergoes first-order decomposition. The time taken for its decomposition to 1/8 and 1/10 of its initial concentration are $t_{1/8}$ and $t_{1/10}$ respectively. What is the value of $\left[\dfrac{t_{1/8}}{t_{1/10}}\right] \times 10$?

$(\log_{10} 2 = 0.3)$ [2012]

24. The concentration of R in the reaction $R \rightarrow P$ was measured as a function of time and the following data is obtained:

$[R]$ (molar)	1.0	0.75	0.40	0.10
t(min.)	0.0	0.05	0.12	0.18

The order of reaction is [2010]

3 Numeric / New Stem Based Questions

25. Consider a reaction A + R → Product. The rate of this reaction is measured to be k[A][R]. At the start of the reaction, the concentration of R, $[R]_0$, is 10-times the concentration of A, $[A]_0$. The reaction can be considered to be a pseudo first order reaction with assumption that $k[R] = k^1$ is constant. Due to this assumption, the relative error (in %) in the rate when this reaction is 40 % complete, is _______. [Adv. 2025]
[k and k' represent corresponding rate constants]

26. Consider the kinetic data given in the following table for the reaction $A + B + C \rightarrow$ product. [Adv. 2019]

Experiment No.	$[A]$ (mol dm^{-3})	$[B]$ (mol dm^{-3})	$[C]$ (mol dm^{-3})	Rate of reaction (mol dm^{-3})
1	0.2	0.1	0.1	6.0×10^{-5}
2	0.2	0.2	0.1	6.0×10^{-5}
3	0.2	0.1	0.2	1.2×10^{-4}
4	0.3	0.1	0.1	9.0×10^{-5}

The rate of the reaction for $[A] = 0.15$ mol dm^{-3}, $[B] = 0.25$ mol dm^{-3} and $[C] = 0.15$ mol dm^{-3} is found to be $Y \times 10^{-5}$ mol $dm^{-3}\ s^{-1}$. The value of Y is _____

27. The decomposition reaction

$2N_2O_5(g) \xrightarrow{\Delta} 2N_2O_4(g) + O_2(g)$ is started in a closed cylinder under isothermal isochoric condition at an initial pressure of 1 atm. After $Y \times 10^3$s, the pressure inside the cylinder is found to be 1.45 atm. If the rate constant of the reaction is $5 \times 10^{-4}\ s^{-1}$, assuming ideal gas behaviour, the value of Y is _____. [Adv. 2019]

28. The rate of a first-order reaction is 0.04 mol $litre^{-1}\ s^{-1}$ at 10 minutes and 0.03 mol $litre^{-1}\ s^{-1}$ at 20 minutes after initiation. Find the half-life of the reaction. [2001 - 5 Marks]

29. The gas phase decomposition of dimethyl ether follows first order kinetics.

$CH_3-O-CH_3(g) \rightarrow CH_4(g) + H_2(g) + CO(g)$

The reaction is carried out in a constant volume container at 500°C and has a half life of 14.5 minutes. Initially, only dimethyl ether is present at a pressure of 0.40 atmosphere. What is the total pressure of the system after 12 minutes? Assume ideal gas behaviour. [1993 - 4 Marks]

4 Fill in the Blanks

30. The hydrolysis of ethyl acetate in .... medium is a .... order reaction. [1986 - 1 Mark]

5 True / False

31. For a first order reaction, the rate of the reaction doubles as the concentration of the reactant (s) doubles. [1986 - 1 Mark]

6 MCQs with One or More than One Correct Answer

32. For a first-order reaction R → P at a given temperature, k is the rate constant. For this reaction, at the given temperature, the concentrations of R and P at a time t are [R] and [P], respectively. The correct graphical representation(s) for this reaction is(are) [Adv. 2026]

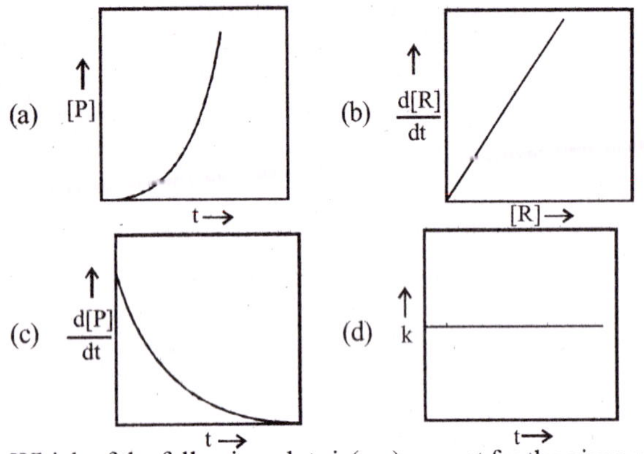

33. Which of the following plots is(are) correct for the given reaction?
($[P]_0$ is the initial concentration of **P**) **[Adv. 2020]**

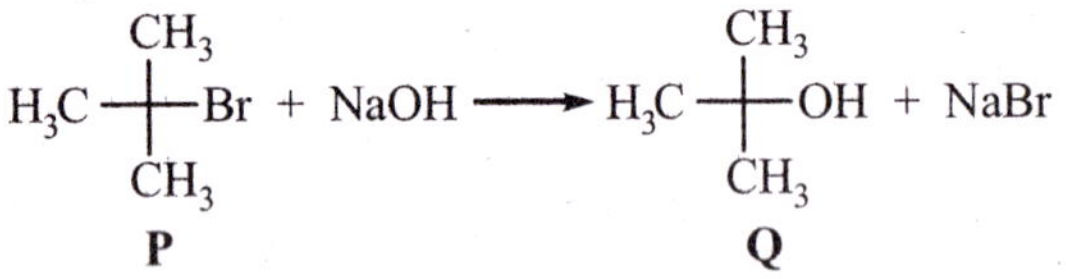

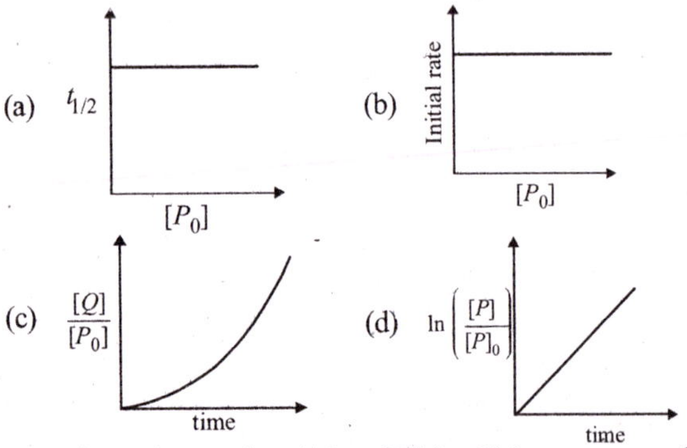

34. For a first order reaction $A(g) \rightarrow 2B(g) + C(g)$ at constant volume and 300 K, the total pressure at the beginning ($t = 0$) and at time t are P_0 and P_t, respectively. Initially, only A is present with concentration $[A]_0$, and $t_{1/3}$ is the time required for the partial pressure of A to reach $1/3^{rd}$ of its initial value. The correct option(s) is (are) (Assume that all these gases behave as ideal gases) **[Adv. 2018]**

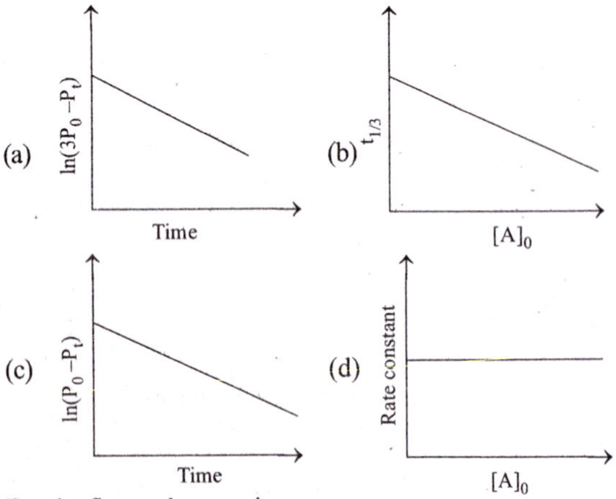

35. For the first order reaction **[2011]**

$$2N_2O_5(g) \longrightarrow 4NO_2(g) + O_2(g)$$

(a) the concentration of the reactant decreases exponentially with time
(b) the half-life of the reaction decreases with increasing temperature
(c) the half-life of the reaction depends on the initial concentration of the reactant
(d) the reaction proceeds to 99.6% completion in eight half life duration

36. The following statement(s) is (are) correct: **[1999 - 3 Marks]**
(a) A plot of log K_p versus $1/T$ is linear
(b) A plot of log $[X]$ versus time is linear for a first order reaction, $X \rightarrow P$
(c) A plot of P versus $1/T$ is linear at constant volume
(d) A plot of P versus $1/V$ is linear at constant temperature

37. For a first order reaction, **[1998 - 2 Marks]**
(a) the degree of dissociation is equal to $(1 - e^{-kt})$
(b) a plot of reciprocal concentration of the reactant vs. time gives a straight line.
(c) the time taken for the completion of 75% reaction is thrice the $t_{1/2}$ of the reaction
(d) the pre-exponential factor in the Arrhenius equation has the dimension of time, T^{-1}.

7 Match the Following

38. Match the rate expressions in LIST-I for the decomposition of X with the corresponding profiles provided in LIST-II. Xs and k are constants having appropriate units. **[Adv. 2022]**

LIST-I

(I) rate $= \dfrac{k[X]}{X_s + [X]}$ under all possible initial concentrations of X

(II) rate $= \dfrac{k[X]}{X_s + [X]}$ where initial concentrations of X are much less than X_s.

(III) rate $= \dfrac{k[X]}{X_s + [X]}$ where initial concentrations of X are much higher than X_s.

(IV) rate $= \dfrac{k[X]^2}{X_s + [X]}$ where initial concentration of X are much higher than X_s.

LIST-II

(P) (Q) (R) (S) (T)

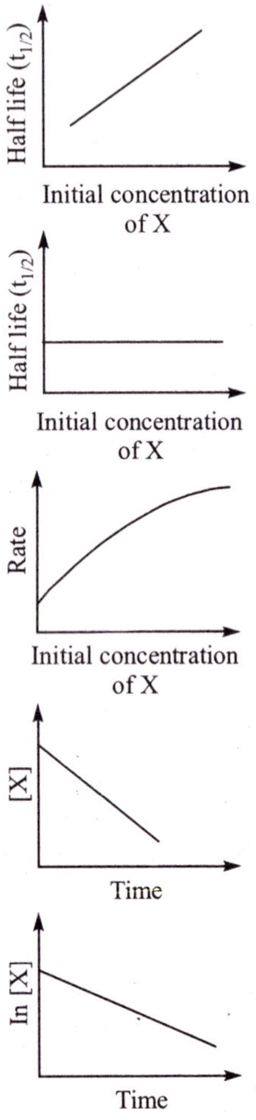

(a) I → P; II → Q; III → S; IV → T
(b) I → R; II → S; III → S; IV → T
(c) I → P; II → Q; III → Q; IV → R
(d) I → R; II → S; III → Q; IV → R

10 Subjective Problems

39. At constant temperature and volume, X decomposes as [2005 - 4 Marks]

$2X(g) \longrightarrow 3Y(g) + 2Z(g)$; P_X is the partial pressure of X.

Observation No.	Time (in minute)	P_X (in mm of Hg)
1	0	800
2	100	400
3	200	200

(i) What is the order of reaction with respect to X?
(ii) Find the rate constant.
(iii) Find the time for 75% completion of the reaction.
(iv) Find the total pressure when pressure of X is 700 mm of Hg.

40. The vapour pressure of the two miscible liquids (A) and (B) are 300 and 500 mm of Hg respectively. In a flask, 10 moles of (A) is mixed with 12 moles of (B). However, as soon as (B) is added, (A) starts polymerizing into a completely insoluble solid. The polymerization follows first-order kinetics. After 100 minutes, 0.525 mole of a solute is dissolved which arrests the polymerization completely. The final vapour pressure of the solution is 400 mm of Hg. Estimate the rate of constant of the polymerization reaction. Assume negligible volume change on mixing and polymerization and ideal behaviour for the final solution. [2001 - 10 Marks]

41. The decomposition of N_2O_5 according to the equation :
$2N_2O_5(g) \rightarrow 4NO_2(g) + O_2(g)$ [1991 - 6 Marks]
is a first order reaction. After 30 min. from the start of the decomposition in a closed vessel, the total pressure developed is found to be 284.5 mm of Hg and on complete decomposition, the total pressure is 584.5 mm of Hg. Calculate the rate constant of the reaction.

42. A first order reaction has $k = 1.5 \times 10^{-6}$ per second at 200°C. If the reaction is allowed to run for 10 hours, what percentage of the initial concentration would have changed in the product? What is the half life of this reaction? [1987 - 5 Marks]

43. While studying the decomposition of gaseous N_2O_5, it is observed that a plot of logarithm of its partial pressure versus time is linear. What kinetic parameters can be obtained from this observation? [1985 - 2 Marks]

44. Rate of a reaction $A + B \rightarrow$ products, is given below as a function of different initial concentrations of A and B: [1982 - 4 Marks]

$[A]$ (mol/l)	$[B]$ (mol/l)	Initial rate (mol/l/min)
0.01	0.01	0.005
0.02	0.01	0.010
0.01	0.02	0.005

Determine the order of the reaction with respect to A and with respect to B. What is the half-life of A in the reaction?

Topic 3 - Theories of Rate of Reaction

1 MCQs with One Correct Answer

45. Plots showing the variation of the rate constant (k) with temperature (T) are given below. The plot that follows Arrhenius equation is [2010]

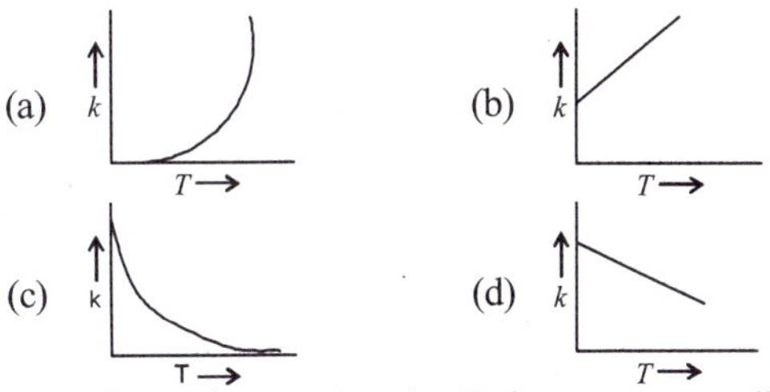

46. For a first order reaction $A \rightarrow P$, the temperature (T) dependent rate constant (k) was found to follow the equation $\log k = -(2000)\frac{1}{T} + 6.0$. The pre-exponential factor A and the activation energy E_a, respectively, are [2009]
(a) $1.0 \times 10^6\ s^{-1}$ and 9.2 kJ mol^{-1}
(b) $6.0\ s^{-1}$ and 16.6 kJ mol^{-1}
(c) $1.0 \times 10^6\ s^{-1}$ and 16.6 kJ mol^{-1}
(d) $1.0 \times 10^6\ s^{-1}$ and 38.3 kJ mol^{-1}

47. The rate constant, the activation energy and the Arrhenius parameter of a chemical reaction at 25°C are $3.0 \times 10^{-4} s^{-1}$, 104.4 kJ mol^{-1} and $6.0 \times 10^{14}\ s^{-1}$ respectively. The value of the rate constant as $T \rightarrow \infty$ is, [1996 - 1 Mark]
(a) $2.0 \times 10^{18}\ s^{-1}$ (b) $6.0 \times 10^{14}\ s^{-1}$
(c) infinity (d) $3.6 \times 10^{30}\ s^{-1}$

48. A catalyst is a substance which [1983 - 1 Mark]
(a) increases the equilibrium concentration of the product
(b) changes the equilibrium constant of the reaction
(c) shortens the time to reach equilibrium
(d) supplies energy to the reaction

2 Integer Value Answer

49. The plot of log k_f versus 1/T for a reversible reaction A (g) $\rightleftharpoons$ P (g) is shown. [Adv. 2023]

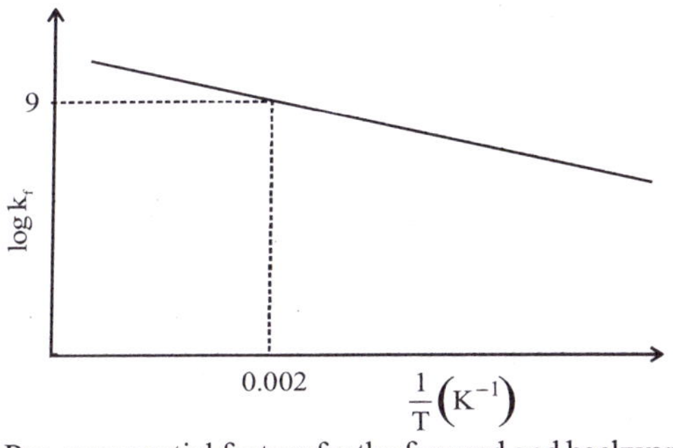

Pre–exponential factors for the forward and backward reactions are $10^{15}\ s^{-1}$ and $10^{11}\ s^{-1}$, respectively. If the value of log K for the reaction at 500 K is 6, the value of $|\log k_b|$ at 250 K is ______.
[K = equilibrium constant of the reaction
k_f = rate constant of forward reaction
k_b = rate constant of backward reaction]

3 Numeric / New Stem Based Questions

50. Consider the following reversible reaction,
$A(g) + B(g) \rightarrow AB(g)$.
The activation energy of the backward reaction exceeds that of the forward reaction by $2RT$ (in J mol^{-1}). If the pre-exponential factor of the forward reaction is 4 times that of the reverse reaction, the absolute value of ΔG° (in J mol^{-1}) for the reaction at 300 K is______.
(Given; ln(2) = 0.7, RT = 2500 J mol^{-1} at 300 K and G is the Gibbs energy) [Adv. 2018]

51. A hydrogenation reaction is carried out at 500 K. If same reaction is carried out in the presence of a catalyst at the same rate, the temperature required is 400 K. Calculate the activation energy of the reaction if the catalyst lowers the activation barrier by 20 kJ mol^{-1}. **[2000 - 3 Marks]**

52. At 380°C, the half-life period for the first order decomposition of H_2O_2 is 360 min. The energy of activation of the reaction is 200 kJ mol^{-1}. Calculate the time required for 75% decomposition at 450°C. **[1995 - 4 Marks]**

53. In the Arrhenius equation for a certain reaction, the value of A and E_a (activation energy) are 4×10^{13} sec^{-1} and 98.6 kJ mol^{-1} respectively. If the reaction is of first order, at what temperature will its half-life period be ten minutes? **[1990 - 3 Marks]**

4 Fill in the Blanks

54. In the Arrhenius equation, $k = A \exp(-E_a/RT)$, A may be termed as the rate constant at .... **[1997 - 1 Mark]**

5 True / False

55. The rate of an exothermic reaction increases with increasing temperature. **[1990 - 1 Mark]**

56. Catalyst does not affect the energy of activation in a chemical reaction. **[1989 - 1 Mark]**

57. Catalyst makes a reaction more exothermic. **[1987 - 1 Mark]**

6 MCQs with One or More than One Correct Answer

58. For a reaction, $A \rightleftharpoons P$, the plots of [A] and [P] with time at temperatures T_1 and T_2 are given below. **[Adv. 2018]**

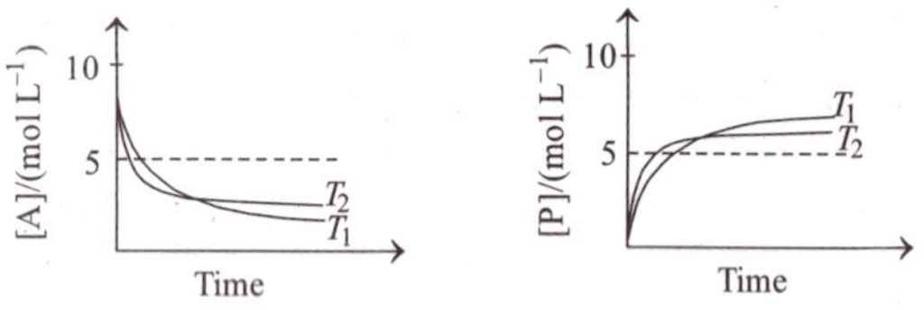

If $T_2 > T_1$, the correct statement(s) is (are)

(Assume $\Delta H°$ and $\Delta S°$ are independent of temperature and ratio of ln$K$ at T_1 to ln$K$ at T_2 is greater than $\frac{T_2}{T_1}$. Here H, S, G and K are enthalpy, entropy, Gibbs energy and equilibrium constant, respectively.)

(a) $\Delta H° < 0, \Delta S° < 0$
(b) $\Delta G° < 0, \Delta H° > 0$
(c) $\Delta G° < 0, \Delta S° < 0$
(d) $\Delta G° < 0, \Delta S° > 0$

59. In a bimolecular reaction, the steric factor P was experimentally determined to be 4.5. The correct option(s) among the following is(are) **[Adv. 2017]**

(a) The activation energy of the reaction is unaffected by the value of the steric factor
(b) Experimentally determined value of frequency factor is higher than that predicted by Arrhenius equation
(c) Since P = 4.5, the reaction will not proceed unless an effective catalyst is used
(d) The value of frequency factor predicted by Arrhenius equation is higher than that determined experimentally

60. According to the Arrhenius equation, **[Adv. 2016]**

(a) a high activation energy usually implies a fast reaction.
(b) rate constant increases with increase in temperature. This is due to a greater number of collisions whose energy exceeds the activation energy.
(c) higher the magnitude of activation energy, stronger is the temperature dependence of the rate constant.
(d) the pre–exponential factor is a measure of the rate at which collisions occur, irrespective of their energy.

61. A catalyst : **[1984 - 1 Mark]**

(a) increases the average kinetic energy of reacting molecules
(b) decreases the activation energy
(c) alters the reaction mechanism
(d) increases the frequency of collisions of reacting species

9 Assertion and Reason/Statement Type Questions

Each question contains **STATEMENT-1 (Assertion)** and **STATEMENT-2 (Reason)**. Each question has 4 choices (a), (b), (c) and (d) out of which ONLY ONE is correct. Mark your answer as

(a) If both Statement -1 and Statement -2 are correct, and Statement -2 is the correct explanation of the Statement -2.
(b) If both Statement -1 and Statement -2 are correct, but Statement -2 is not the correct explanation of the Statement -1.
(c) If Statement -1 is correct but Statement -2 is incorrect.
(d) If Statement -1 is incorrect but Statement -2 is correct.

62. **Statement-1 :** For each ten degree rise of temperature the specific rate constant is nearly doubled.

Statement-2 : Energy-wise distribution of molecules in a gas is an experimental function of temperature. **[1989 - 2 Marks]**

10 Subjective Problems

63. The rate constant of a reaction is 1.5×10^7 s^{-1} at 50°C and 4.5×10^7 s^{-1} at 100°C. Evaluate the Arrhenius parameters A and E_a. **[1998 - 5 Marks]**

64. The rate constant for the first order decomposition of a certain reaction is described by the equation

$$\log(k) = 14.34 - \frac{1.25 \times 10^4 K}{T}$$ **[1997 - 5 Marks]**

(i) What is the energy of activation for this reaction?
(ii) At what temperature will its half-life period be 256 minutes?

65. From the following data for the reaction between A and B. **[1994 - 5 Marks]**

	[A], mol lit^{-1}	[B], mol lit^{-1}	Initial rate mole lit^{-1} s^{-1} at	
			300 K	320 K
I	2.5×10^{-4}	3.0×10^{-5}	5.0×10^{-4}	2.0×10^{-3}
II	5.0×10^{-4}	6.0×10^{-5}	4.0×10^{-3}	–
III	1.0×10^{-3}	6.0×10^{-5}	1.6×10^{-2}	–

Calculate

(i) the order of the reaction with respect to A and with respect to B,
(ii) the rate constant at 300K
(iii) the energy of activation, and
(iv) the pre-exponential factor

66. A first order reaction $A \rightarrow B$, requires activation energy of 70kJ mol^{-1}. When a 20% solution of A was kept at 25°C for 20 minutes, 25% decomposition took place. What will be the percent decomposition in the same time in a 30% solution maintained at 40°C? Assume that activation energy remains constant in this range of temperature. **[1993 - 4 Marks]**

67. Two reactions (i) $A \rightarrow$ products, (ii) $B \rightarrow$ products, follows first order kinetics. The rate of the reaction : (*i*) is doubled when the temperature is raised from 300K to 310K. The half life for this reaction at 310K is 30 minutes. At the same temperature B decomposes twice as fast as A. If the energy of activation for the reaction, (*ii*) is half that of reaction (i), calculate the rate constant of the reaction (ii) at 300K. **[1992 - 3 Marks]**

68. A first order reaction is 50% complete in 30 minutes at 27°C and in 10 minutes at 47°C. Calculate the reaction rate constant at 27°C and the energy of activation of the reaction in kJ/mole. [1988 - 3 Marks]

Topic 4 – Nuclear Chemistry

1 MCQs with One Correct Answer

69. Bombardment of aluminium by α-particle leads to its artificial disintegration in two ways, (i) and (ii) as shown. Products X, Y and Z respectively are, [2011]

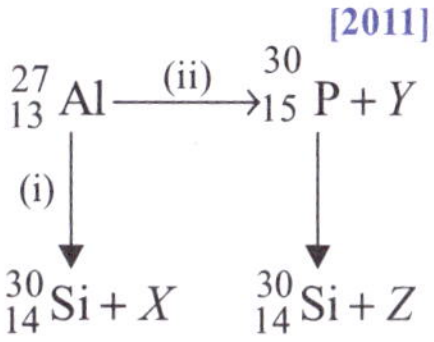

(a) proton, neutron, positron
(b) neutron, positron, proton
(d) proton, positron, neutron
(d) positron, proton, neutron

70. A positron is emitted from $^{23}_{11}Na$. The ratio of the atomic mass and atomic number of the resulting nuclide is [2007]
(a) 22/10 (b) 22/11 (c) 23/10 (d) 23/12

71. ^{23}Na is the more stable isotope of Na. Find out the process by which $^{24}_{11}Na$ can undergo radioactive decay [2003S]
(a) β^- emission (b) α emission
(c) β^+ emission (d) K electron capture

72. The number of neutrons accompanying the formation of $^{139}_{54}Xe$ and $^{94}_{38}Sr$ from the absorption of a slow neutron by $^{235}_{92}U$, followed by nuclear fission is, [1999 - 2 Marks]
(a) 0 (b) 2 (c) 1 (d) 3

73. $^{27}_{13}Al$ is a stable isotope, $^{29}_{13}Al$ is expected to disintegrate by
(a) α-emission (b) β-emission [1996 - 1 Mark]
(c) positron emission (d) proton emission

74. The half-life period of a radioactive element is 140 days. After 560 days, one gram of the element will reduced to : [1986 - 1 Mark]
(a) $\frac{1}{2}$ g (b) $\frac{1}{4}$ g (c) $\frac{1}{8}$ g (d) $\frac{1}{16}$ g

75. The radiations from a naturally occurring radioactive substance, as seen after deflection by a magnetic field in one direction, are :
(a) definitely alpha rays [1984 - 1 Mark]
(b) definitely beta rays
(c) both alpha and beta rays
(d) either alpha or beta rays

76. If uranium (mass number 238 and atomic number 92) emits an α-particle, the product has mass no. and atomic no. [1981 - 1 Mark]
(a) 236 and 92 (b) 234 and 90
(c) 238 and 90 (d) 236 and 90

2 Integer Value Answer

77. A closed vessel with rigid walls contains 1 mol of $^{238}_{92}U$ and 1 mol of air at 298 K. Considering complete decay of $^{238}_{92}U$ to $^{206}_{82}Pb$, the ratio of the final pressure to the initial pressure of the system at 298 K is [Adv. 2015]

78. The periodic table consists of 18 groups. An isotope of copper, on bombardment with protons, undergoes a nuclear reaction yielding element X as shown below. To which group, element X belongs in the periodic table? [2012]

$$^{63}_{29}Cu + {}^{1}_{1}H \rightarrow 6\,{}^{1}_{0}n + {}^{4}_{2}\alpha + 2\,{}^{1}_{1}H + X$$

79. The number of neutrons emitted when $^{235}_{92}U$ undergoes controlled nuclear fission to $^{142}_{54}Xe$ and $^{90}_{38}Sr$ is [2010]

80. The total number of α and β particles emitted in the nuclear reaction $^{238}_{92}U \rightarrow {}^{214}_{82}Pb$ is [2009]

3 Numeric / New Stem Based Questions

81. $^{238}_{92}U$ is known to undergo radioactive decay to form $^{206}_{82}Pb$ by emitting alpha and beta particles. A rock initially contained 68×10^{-6} g of $^{238}_{92}U$. If the number of alpha particles that it would emit during its radioactive decay of $^{238}_{92}U$ to $^{206}_{82}Pb$ in three half-lives is $Z \times 10^{18}$, then what is the value of Z? [Adv. 2020]

82. One of the hazards of nuclear explosion is the generation of ^{90}Sr and its subsequent incorporation in bones. This nuclide has a half-life of 28.1 years. Suppose one microgram was absorbed by a new-born child, how much ^{90}Sr will remain in his bones after 20 years? [1995 - 2 Marks]

4 Fill in the Blanks

83. A radioactive nucleus decays emitting one alpha and two beta particles; the daughter nucleus is .... of the parent. [1989 - 1 Mark]

84. The number of neutrons in the parent nucleus which gives $^{14}_{7}N$ on beta emission is .... [1985 - 1 Mark]

85. An element $^{A}_{Z}M$ undergoes an α-emission followed by two successive β-emissions. The element formed is .... [1982 - 1 Mark]

5 True / False

86. In β-emission from a nucleus, the atomic number of the daughter element decreases by one. [1990 - 1 Mark]

6 MCQs with One or More than One Correct Answer

87. In the decay sequence, [Adv. 2019]

$$^{238}_{92}U \xrightarrow{-x_1} {}^{234}_{90}Th \xrightarrow{-x_2} {}^{234}_{91}Pa \xrightarrow{-x_3} {}^{234}Z \xrightarrow{-x_4} {}^{230}_{90}Th$$

x_1, x_2, x_3 and x_4 are particles/radiation emitted by the respective isotopes. The correct option(s) is (are)
(a) Z is an isotope of uranium
(b) x_1 will deflect towards negatively charged plate
(c) x_3 is γ-ray
(d) x_2 is β-ray

88. A plot of the number of neutrons (N) against the number of protons (P) of stable nuclei exhibits upward deviation from linearity for atomic number, $Z > 20$. For an unstable nucleus having N/P ratio less than 1, the possible mode(s) of decay is (are) [Adv. 2016]
(a) β^-–decay (β emission)
(b) orbital or K–electron capture
(c) neutron emission
(d) β^+–decay (positron emission)

89. In the nuclear transmutation [Adv. 2013]

$${}^{9}_{4}Be + X \longrightarrow {}^{8}_{4}Be + Y$$

(X, Y) is (are)

(a) (γ, n) (b) (p, D) (c) (n, D) (d) (γ, p)

90. Decrease in atomic number is observed during [1998 - 2 Marks]

(a) alpha emission (b) beta emission
(c) positron emission (d) electron capture.

91. Nuclear reactions accompanied with emission of neutron(s) are : [1988 - 1 Mark]

(a) ${}^{27}_{13}Al + {}^{4}_{2}H \rightarrow {}^{30}_{15}P$ (b) ${}^{12}_{6}C + {}^{1}_{1}H \rightarrow {}^{13}_{7}N$

(c) ${}^{30}_{15}P \rightarrow {}^{30}_{14}Si + {}^{0}_{1}e$ (d) ${}^{241}_{96}Am + {}^{4}_{2}He \rightarrow {}^{244}_{97}Bk + {}^{0}_{1}e$

8 Comprehension Based Questions

Several short-lived radioactive species have been used to determine the age of wood or animal fossils. One of the most interesting substances is ${}_6C^{14}$ (half-life 5760 years) which is used in determining the age of carbon-bearing materials (*e.g.* wood, animal fossils, *etc.*). Carbon-14 is produced by the bombardment of nitrogen atoms present in the upper atmosphere with neutrons (from cosmic rays).

$${}_7N^{14} + {}_0n^1 \longrightarrow {}_6C^{14} + {}_1H^1$$

Thus carbon-14 is oxidised to CO_2 and eventually ingested by plants and animals. The death of plants or animals put an end to the intake of C^{14} from the atmosphere. After this the amount of C^{14} in the dead tissues starts decreasing due to its disintegration as per the following reaction :

$${}_6C^{14} \longrightarrow {}_7N^{14} + {}_{-1}\beta^0$$

The C^{14} isotope enters the biosphere when carbon dioxide is taken up in plant photosynthesis. Plants are eaten by animals, which exhale C^{14} as CO_2. Eventually, C^{14} participates in many aspects of the carbon cycle. The C^{14} lost by radioactive decay is constantly replenished by the production of new isotopes in the atmosphere. In this decay-replenishment process, a dynamic equilibrium is established whereby the ratio of C^{14} to C^{12} remains constant in living matter. But when an individual plant or an animal dies, the C^{14} isotope in it is no longer replenished, so the ratio decreases as C^{14} decays. So, the number of C^{14} nuclei after time t (after the death of living matter) would be less than in a living matter. The decay constant can be calculated using the following formula,

$$t_{1/2} = \frac{0.693}{\lambda}$$

The intensity of the cosmic rays have remain the same for 30,000 years. But since some years the changes in this are observed due to excessive burning of fossil fuel and nuclear tests.

92. Why do we use the carbon dating to calculate the age of the fossil? [2006 - 5M, –2]

(a) Rate of exchange of carbon between atmosphere and living is slower than decay of C^{14}
(b) It is not appropriate to use C^{14} dating to determine age
(c) Rate of exchange of C^{14} between atmosphere and living organism is so fast that an equilibrium is set up between the intake of C^{14} by organism and its exponential decay
(d) none of the above

93. What should be the age of the fossil for meaningful determination of its age? [2006 - 5M, –2]

(a) 6 years
(b) 6000 years
(c) 60,000 years
(d) can be used to calculate any age

94. A nuclear explosion has taken place leading to increase in concentration of C^{14} in nearby areas. C^{14} concentration is C_1 in nearby areas and C_2 in areas far away. If the age of the fossil is determined to be T_1 and T_2 at the respective places then [2006 - 5M, –2]

(a) The age of the fossil will increase at the place where explosion has taken place and $T_1 - T_2 = \frac{1}{\lambda}\ln\frac{C_1}{C_2}$

(b) The age of the fossil will decrease at the place where explosion has taken place and $T_1 - T_2 = \frac{1}{\lambda}\ln\frac{C_1}{C_2}$

(c) The age of fossil will be determined to be same

(d) $\frac{T_1}{T_2} = \frac{C_1}{C_2}$

9 Assertion and Reason/Statement Type Questions

Each question contains **STATEMENT-1 (Assertion)** and **STATEMENT-2 (Reason)**. Each question has 4 choices (a), (b), (c) and (d) out of which ONLY ONE is correct. Mark your answer as

(a) If both Statement -1 and Statement -2 are correct, and Statement -2 is the correct explanation of the Statement -2.
(b) If both Statement -1 and Statement -2 are correct, but Statement -2 is not the correct explanation of the Statement -1.
(c) If Statement -1 is correct but Statement -2 is incorrect.
(d) If Statement -1 is incorrect but Statement -2 is correct.

95. **Statement-1 :** The plot of atomic number (y-axis) versus number of neutrons (x-axis) for stable nuclei shows a curvature towards x-axis from the line of 45° slope as the atomic number is increased.

Statement-2 : Proton-proton electrostatic repulsions begin to overcome attractive forces involving protons and neutrons in heavier nuclides. [2008]

96. **Assertion :** Nuclide ${}^{30}Al_{13}$ is less stable than ${}^{40}Ca_{20}$ [1998-2Marks]

Reason : Nuclides having odd number of protons and neutrons are generally unstable.

10 Subjective Problems

97. Complete and balance the following reactions.

(*i*) ${}_{92}Th^{234} \longrightarrow .... + 7\ {}_2He^4 + 6\ {}_{-1}\beta^0$ [2004 - 1 Mark]

(*ii*) ${}_{92}U^{235} + {}_0n^1 \longrightarrow ....+ {}_{52}Te^{137} + {}_{40}Zr^{92}$ [2005 - 1 Mark]

(*iii*) ${}_{34}Se^{86} \longrightarrow 2\ {}_{-1}e^0 + ...$ [2005 - 1 Mark]

98. ${}^{64}Cu$ (half-life = 12.8 h) decays by β^- emission (38%), β^+ emission (19%) and electron capture (43%). Write the decay products and calculate partial half-lives for each of the decay processes. [2002 - 5 Marks]

99. $^{238}_{92}U$ is radioactive and it emits α and β particles to form $^{206}_{82}Pb$. Calculate the number of α and β particles emitted in this conversion. An ore of $^{238}_{92}U$ is found to contain $^{238}_{92}U$ and $^{206}_{82}Pb$ in the weight ratio of 1:0.1. The half-life period of $^{238}_{92}U$ is 4.5×10^9 years. Calculate the age of the ore. [2000 - 5 Marks]

100. Write a balanced equation for the reaction of ^{14}N with α-particle. [1997 - 1 Mark]

101. ^{227}Ac has a half-life of 21.8 years with respect to radioactive decay. The decay follows two parallel paths, one leading to ^{227}Th and the other to ^{223}Fr. The percentage yields of these two daughter nuclides are 1.2 and 98.8 respectively. What are the decay constants (λ) for each of the separate paths? [1996 - 2 Marks]

102. The nucleidic ratio, ^3_1H to ^1_1H in a sample of water is $8.0 \times 10^{-18} : 1$. Tritium undergoes decay with a half life period of 12.3 years. How many tritium atoms would 10.0 g of such a sample contain 40 years after the original sample is collected? [1992 - 4 Marks]

103. An experiment requires minimum beta activity product at the rate of 346 beta particles per minute. The half life period of $^{99}_{42}Mo$, which is a beta emitter is 66.6 hours. Find the minimum amount of $^{99}_{42}Mo$ required to carry out the experiment in 6.909 hours. [1989 - 5 Marks]

104. $^{234}_{90}Th$ disintegrates to give $^{206}_{82}Pb$ as the final product. How many alpha and beta particles are emitted during this process? [1986 - 2 Marks]

105. Radioactive decay is a first order process. Radioactive carbon in wood sample decays with a half life of 5770 years. What is the rate constant (in years^{-1}) for the decay? What fraction would remain after 11540 years? [1984 - 3 Marks]

Answer Key

Part-A (JEE Main)

1	(4)	24	(4)	47	(2)	70	(3)	93	(1)	116	(3)	139	(200)	162	(23.03)	185	(2)	208	(2520)
2	(3)	25	(4)	48	(1)	71	(3)	94	(2)	117	(315)	140	(1)	163	(3)	186	(1)	209	(3)
3	(2)	26	(2)	49	(2)	72	(1)	95	(1)	118	(100)	141	(100)	164	(2)	187	(1)	210	(8)
4	(4)	27	(4)	50	(2)	73	(1)	96	(4)	119	(50)	142	(2)	165	(2)	188	(2)	211	(225)
5	(3)	28	(1)	51	(3)	74	(3)	97	(3)	120	(17190)	143	(165)	166	(3)	189	(2)	212	(1)
6	(3)	29	(47)	52	(4)	75	(2)	98	(4)	121	(2)	144	(2)	167	(4)	190	(1)	213	(216)
7	(1)	30	(2435)	53	(4)	76	(2)	99	(2)	122	(399)	145	(1)	168	(1)	191	(3)	214	(166)
8	(4)	31	(8)	54	(1)	77	(1)	100	(4)	123	(24)	146	(2)	169	(1)	192	(3)	215	(59)
9	(3)	32	(3)	55	(4)	78	(4)	101	(1)	124	(63)	147	(16)	170	(3)	193	(3)	216	(4)
10	(4)	33	(17)	56	(2)	79	(2)	102	(3)	125	(2)	148	(75)	171	(1)	194	(4)	217	(154)
11	(3)	34	(2200)	57	(3)	80	(3)	103	(1)	126	(10)	149	(0)	172	(3)	195	(4)	218	(45)
12	(2)	35	(333)	58	(4)	81	(4)	104	(1)	127	(3)	150	(1)	173	(2)	196	(2)	219	(47)
13	(4)	36	(7)	59	(2)	82	(1)	105	(4)	128	(3)	151	(2)	174	(2)	197	(232)	220	(16)
14	(1)	37	(40)	60	(3)	83	(3)	106	(25)	129	(2)	152	(654)	175	(4)	198	(41)	221	(10)
15	(4)	38	(1)	61	(4)	84	(3)	107	(90)	130	(17)	153	(7)	176	(1)	199	(5)	222	(50)
16	(1)	39	(4)	62	(1)	85	(4)	108	(46)	131	(623)	154	(1)	177	(1)	200	(57)	223	(14)
17	(4)	40	(4)	63	(2)	86	(3)	109	(7)	132	(1350)	155	(3)	178	(3)	201	(1303)	224	(526)
18	(3)	41	(106)	64	(3)	87	(2)	110	(102)	133	(1)	156	(10)	179	(1)	202	(8)	225	(52)
19	(4)	42	(27)	65	(3)	88	(4)	111	(10)	134	(15)	157	(2)	180	(1)	203	(20)	226	(100)
20	(4)	43	(1)	66	(3)	89	(2)	112	(897)	135	(75)	158	(108)	181	(3)	204	(69)	227	(84297.48)
21	(2)	44	(2)	67	(4)	90	(3)	113	(1)	136	(2)	159	(81)	182	(2)	205	(30)	228	(3.98)
22	(4)	45	(4)	68	(2)	91	(2)	114	(3)	137	(2)	160	(26)	183	(2)	206	(2)		
23	(2)	46	(1)	69	(2)	92	(2)	115	(17)	138	(60)	161	(60)	184	(4)	207	(2)		

Part-B (JEE Advanced)

1	(b)	9	(8)	22	(3)	31	True	45	(a)	53	(311.35)	62	(a)	76	(b)	87	(a, b, d)	95	(a)
2	(c)	12	(b,c, d)	23	(9)	32	(c, d)	46	(d)	55	True	69	(a)	77	(9)	88	(b, d)	96	(c)
3	(c)	13	(b, c)	24	(0)	33	(a)	47	(b)	56	False	70	(c)	78	(8)	89	(a, b)		
4	(a)	17	(d)	25	(4.17)	34	(a, d)	48	(c)	57	False	71	(a)	79	(3)	90	(a,c,d)		
5	(b)	18	(a)	26	(6.75)	35	(a, b, d)	49	(5)	58	(a, c)	72	(d)	80	(8)	91	(a, d)		
6	(d)	19	(d)	27	(2.30)	36	(a, b, d)	50	(– 8500)	59	(a, b)	73	(b)	81	(1.2)	92	(c)		
7	(d)	20	(d)	28	(24.14)	37	(a, d)	51	(100)	60	(b, c, d)	74	(d)	82	(0.061)	93	(b)		
8	(a)	21	(143)	29	(0.749)	38	(a)	52	(20.34)	61	(b, c)	75	(d)	86	False	94	(a)		

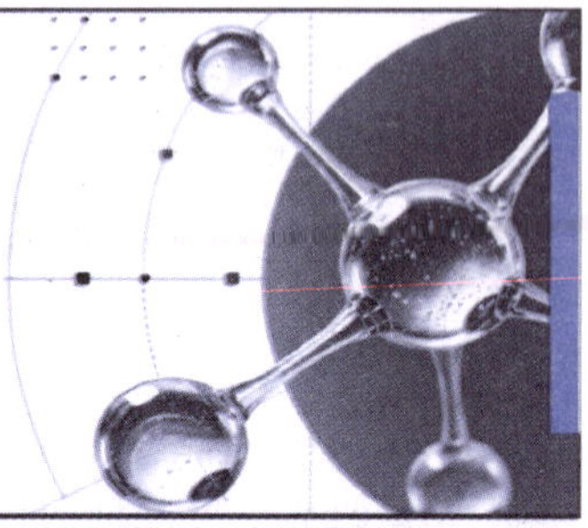

Chapter-5

Surface Chemistry

Part-A (JEE Main)

Topic 1 - Adsorption

MULTIPLE CHOICE QUESTIONS (MCQs)

1. Given below are two statements. One is labelled as **Assertion A** and the other is labelled as **Reason R**.

Assertion A : Activated charcoal adsorbs SO_2 more efficiently than CH_4.

Reason R : Gases with lower critical temperatures are readily adsorbed by activated charcoal.

In the light of the above statements, choose the correct answer from the options given below. [July 27, 2022 (I)]

(1) Both A and R are correct and R is the correct explanation of A.

(2) Both A and R are correct but R is NOT the correct explanation of A.

(3) A is correct but R is not correct.

(4) A is not correct but R is correct.

2. Select the graph that correctly describes the adsorption isotherms at two temperatures T_1 and T_2 ($T_1 > T_2$) for a gas:

(x – mass of the gas adsorbed ; m – mass of adsorbent; P – pressure)] [Aug. 31, 2021 (I); Similar Sep. 02, 2020 (I); Sep. 05, 2020 (II)]

(1) 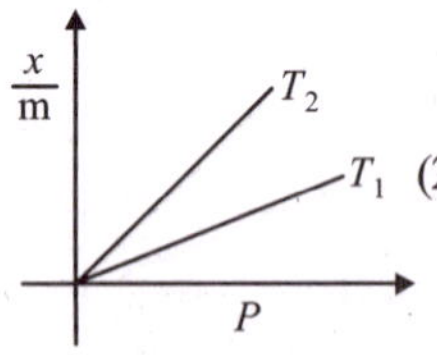

(2) 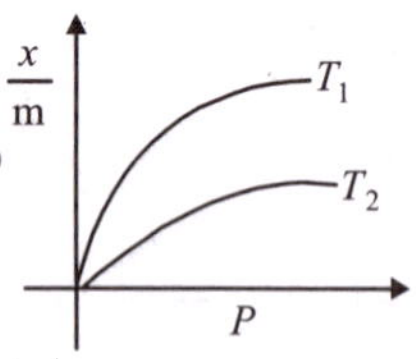

(3) 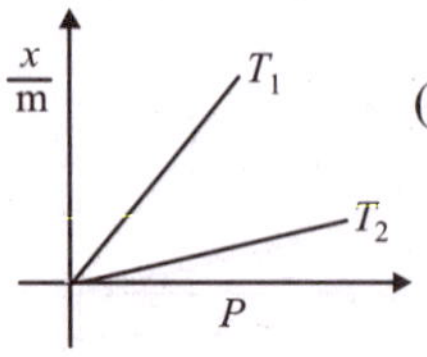

(4) 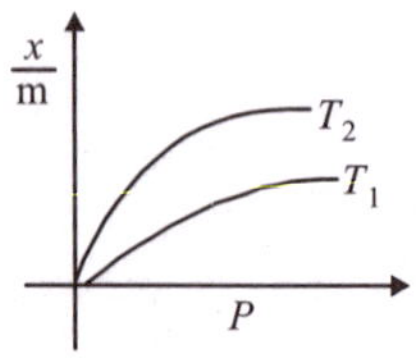

3. Which one of the following is correct for the adsorption of a gas at a given temperature on a solid surface? [Aug. 26, 2021 (I)]

(1) $\Delta H > 0, \Delta S > 0$ (2) $\Delta H > 0, \Delta S < 0$

(3) $\Delta H < 0, \Delta S < 0$ (4) $\Delta H < 0, \Delta S > 0$

4. In Freundlich adsorption isotherm at moderate pressure, the extent of adsorption $\left(\frac{x}{m}\right)$ is directly proportional to P^{χ}. The value of χ is: [Feb. 25, 2021 (I)]

(1) zero (2) 1 (3) ∞ (4) $\frac{1}{n}$

5. In Freundlich adsorption isotherm, slope of AB line is: [February 24, 2021 (I)]

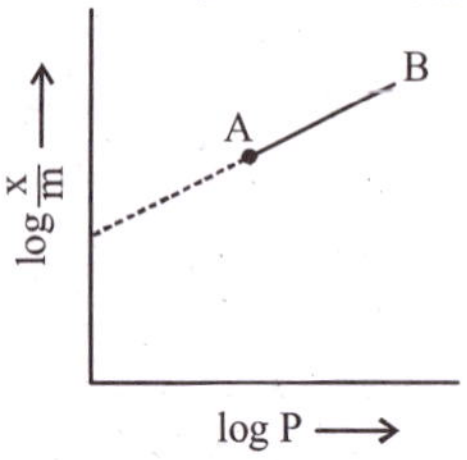

(1) log n with ($n > 1$)

(2) n with (n, 0.1 to 0.5)

(3) $\log\frac{1}{n}$ with ($n < 1$)

(4) $\frac{1}{n}$ with $\left(\frac{1}{n} = 0 \text{ to } 1\right)$

6. Amongst the following statements regarding adsorption, those that are valid are : [Sep. 02, 2020 (II)]

(A) ΔH becomes less negative as adsorption proceeds.

(B) On a given adsorbent, ammonia is adsorbed more than nitrogen gas.

(C) On adsorption, the residual force acting along the surface of the adsorbent increases.

(D) With increase in temperature, the equilibrium concentration of adsorbate increases.

(1) (D) and (A) (2) (B) and (C)

(3) (A) and (B) (4) (C) and (D)

7. A mixture of gases O_2, H_2 and CO are taken in a closed vessel containing charcoal. The graph that represents the correct behaviour of pressure with time is: [Jan. 09, 2020 (II)]

(1) 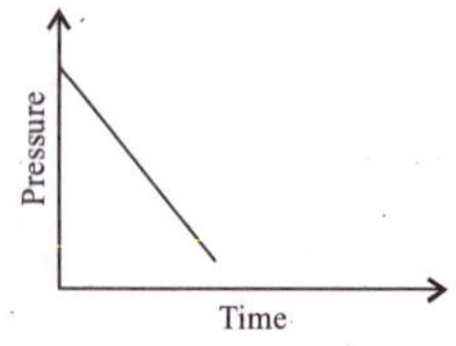

(2) 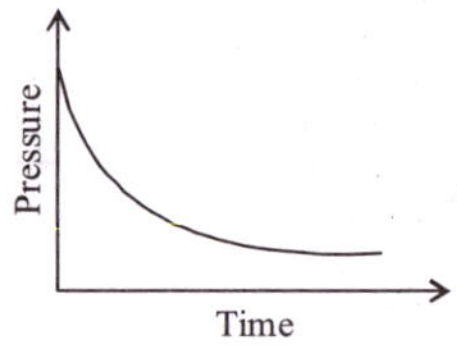

(3) 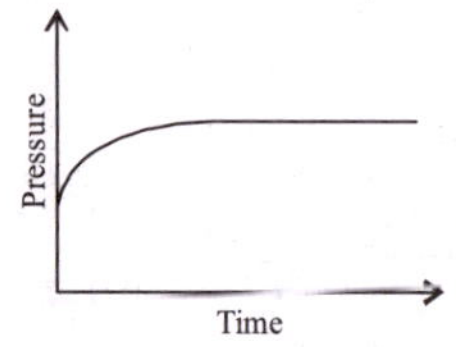

(4) 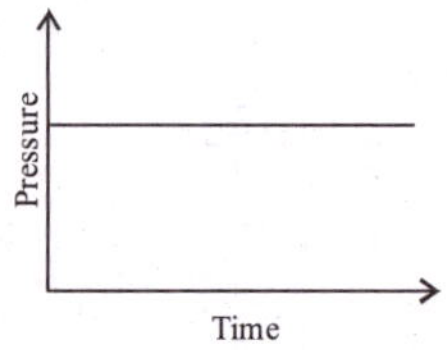

8. A gas undergoes physical adsorption on a surface and follows the given Freundlich adsorption isotherm equation $\frac{x}{m}=kp^{0.5}$. Adsorption of the gas increases with : [April 10, 2019 (I)]
(1) Decrease in p and increase in T
(2) Decrease in p and decrease in T
(3) Increase in p and decrease in T
(4) Increase in p and increase in T

9. 10 mL of 1 mM surfactant solution forms a monolayer covering 0.24 cm^2 on a polar substrate. If the polar head is approximated as a cube, what is its edge length? [April 9, 2019 (II)]
(1) 1.0 pm (2) 2.0 pm (3) 0.1 nm (4) 2.0 nm

10. Adsorption of a gas follows Freundlich adsorption isotherm. x is the mass of the gas adsorbed on mass m of the adsorbent. The plot of $\frac{x}{m}$ versus $\log p$ is shown in the given graph. $\frac{x}{m}$ is proportional to : [Jan. 9, 2019 (I) (S)]

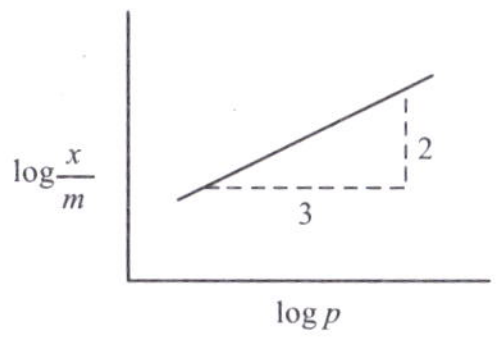

(1) $p^{2/3}$
(2) $p^{3/2}$
(3) p^3
(4) p^2

11. Given gas critical temperature/K for
H_2 CH_4 CO_2 SO_2 is
33 190 304 630
On the basis of data given above, predict which of the following gases shows least adsorption on a definite amount of charcoal? [Jan. 12, 2019 (I)]
(1) SO_2 (2) CH_4 (3) CO_2 (4) H_2

12. Which one of the following is not a property of physical adsorption? [Online April 16, 2018; Similar 2003]
(1) Higher the pressure, more the adsorption
(2) Greater the surface area, more the adsorption
(3) Lower the temperature, more the adsorption
(4) Unilayer adsorption occurs

13. If x gram of gas is adsorbed by m gram of adsorbent at pressure P, the plot of $\log \frac{x}{m}$ versus $\log P$ is linear. The slope of the plot is: (n and k are constants and $n > 1$)
[Online April 15, 2018 (II); Similar 2016; May 7, 2012]
(1) $\log k$ (2) $\frac{1}{n}$ (3) $2k$ (4) n

14. Adsorption of a gas on a surface follows Freundlich adsorption isotherm. Plot of $\log \frac{x}{m}$ versus $\log p$ gives a straight line with slope equal to 0.5, then : [Online April 9, 2017]
($\frac{x}{m}$ is the mass of the gas adsorbed per gram of adsorbent)
(1) Adsorption is independent of pressure.
(2) Adsorption is proportional to the pressure.
(3) Adsorption is proportional to the square root of pressure.
(4) Adsorption is proportional to the square of pressure.

15. A particular adsorption process has the following characteristics : (i) It arises due to van der Waals forces and (ii) it is reversile. Identify the correct statement that describes the above adsorption process : [Online April 9, 2016]
(1) Adsorption is monolayer.
(2) Adsorption increases with increase in temperature.
(3) Enthalpy of adsorption is greater than 100 kJ mol^{-1}
(4) Energy of activation is low.

16. The following statements relate to the adsorption of gases on a solid surface. Identify the incorrect statement among them : [Online April 10, 2015]
(1) Enthalpy of adsorption is negative
(2) Energy appears as heat
(3) On adsorption, the residual forces on the surface are increased
(4) Entropy of adsorption is negative

17. 3 g of activated charcoal was added to 50 mL of acetic acid solution (0.06 N) in a flask. After an hour it was filtered and the strength of the filtrate was found to be 0.042 N. The amount of acetic acid adsorbed (per gram of charcoal) is: [2015]
(1) 42 mg (2) 54 mg (3) 18 mg (4) 36 mg

18. According to Freundlich adsorption isotherm which of the following is correct? [2012]
(1) $\frac{x}{m} \propto p^0$ (2) $\frac{x}{m} \propto p^1$
(3) $\frac{x}{m} \propto p^{1/n}$ (4) All the above are correct for different ranges of pressure

19. The correct statement for both the processes of physisorption and chemisorption is [Online May 12, 2012]
(1) both are endothermic
(2) chemisorption is endothermic but physisorption is exothermic
(3) both are exothermic
(4) physisorption is endothermic but chemisorption is exothermic.

20. Which of the following statements is incorrect regarding physisorptions? [2009]
(1) More easily liquefiable gases are adsorbed readily.
(2) Under high pressure, it results into multi-molecular layer on adsorbent surface.
(3) Enthalpy of adsorption ($\Delta H_{adsorption}$) is low and positive.
(4) It occurs because of van der Waals forces.

21. In Langmuir's model of adsorption of a gas on a solid surface [2006]
(1) the mass of gas striking a given area of surface is proportional to the pressure of the gas
(2) the mass of gas striking a given area of surface is independent of the pressure of the gas
(3) the rate of dissociation of adsorbed molecules from the surface does not depend on the surface covered
(4) the adsorption at a single site on the surface may involve multiple molecules at the same time

22. The formation of gas at the surface of tungsten due to adsorption is the reaction of order [2002]
(1) 0 (2) 1 (3) 2 (4) insufficient data.

NUMERICAL VALUE QUESTIONS (NVQs)

23. Among the following the number of curves not in accordance with Freundlich adsorption isotherm is _______. [July 28, 2022 (II)]

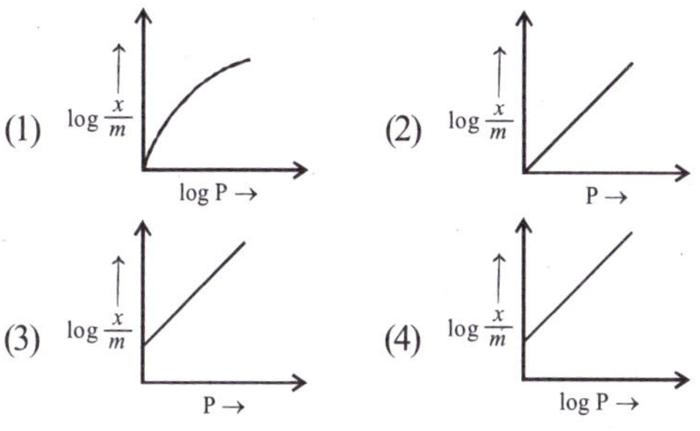

24. 100 mL of 0.3 M acetic acid is shaken with 0.8 g of wood charcoal. The final concentration of acetic acid in the solution after adsorption is 0.125 M. The mass of acetic acid adsorbed per gram of carbon is ________ $\times 10^{-4}$g.
(Given : Molar mass of acetic acid = 60 g mol^{-1})
[June 30, 2022 (I)]

25. 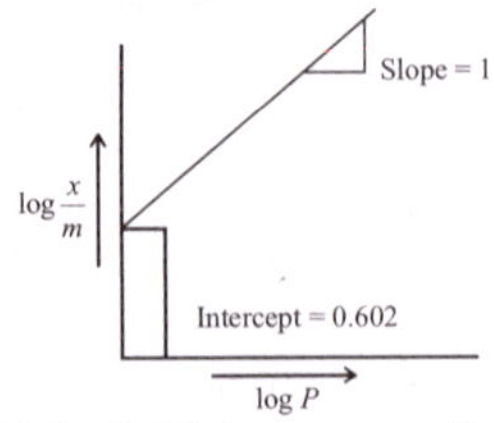

If the initial pressure of a gas is 0.03 atm, the mass of the gas adsorbed per gram of the adsorbent is _____ $\times 10^{-2}$ g.
[June 27, 2022 (II)]

26. 2.0 g of H_2 gas is adsorbed on 2.5 g of platinum powder at 300 K and 1 bar pressure. The volume of the gas adsorbed per gram of the adsorbent is ________ mL.
(Given : R = 0.083 L bar K^{-1} mol^{-1}) **[June 26, 2022 (I)]**

27. When 200 mL of 0.2 M acetic acid is shaken with 0.6 g of wood charcoal, the final concentration of acetic after adsorption is 0.1 M. The mass of acetic acid adsorbed per garm of carbon is__ g.
[June 24, 2022 (II)]

28. CH_4 is adsorbed on 1 g charcoal at 0 °C following the Freundlich adsorption isotherm. 10.0 mL of CH_4 is adsorbed at 100 mm of Hg, whereas 15.0 mL is adsorbed at 200 mm of Hg. The volume of CH_4 adsorbed at 300 mm of Hg is 10^x mL. The value of x is ______ $\times 10^{-2}$. (Nearest integer)
[Use $\log_{10}2 = 0.3010$, $\log_{10}3 = 0.4771$] **[Aug. 31, 2021 (II)]**

29. CO_2 gas adsorbs on charcoal following Freundlich adsorption isotherm. For a given amount of charcoal, the mass of CO_2 adsorbed becomes 64 times when the pressure of CO_2 is doubled. The value of n in the Freundlich isotherm equation is ________ $\times 10^{-2}$. (Round off to the Nearest Integer) **[July 27, 2021 (I)]**

30. 3.12 g of oxygen is adsorbed on 1.2 g of platinum metal. The volume of oxygen adsorbed per gram of the adsorbent at 1 atm and 300 K in L is ________ .
[$R = 0.0821$ L atm K^{-1} mol^{-1}] **[Feb. 26, 2021 (I)]**

31. For Freundlich adsorption isotherm, a plot of log (x/m) (y-axis) and log p (x - axis) gives a straight line. The intercept and slope for the line is 0.4771 and 2, respectively. The mass of gas, adsorbed per gram of adsorbent if the initial pressure is 0.04 atm, is ______ $\times 10^{-4}$g.
(log 3 = 0.4771) **[Sep. 06, 2020 (II)]**

Topic 2 - Catalysis and Theories of Catalysis

MULTIPLE CHOICE QUESTIONS (MCQs)

32. Match List-I with List-II. **[July 28, 2022 (I)]**

List-I (Reaction)	List-II (Catalyst)
(A) $4NH_3(g)\ 5O_2(g) \rightarrow 4NO(g) + 6H_2O(g)$	(I) NO(g)
(B) $N_2(g) + 3H_2(g) \rightarrow 2NH_3(g)$	(II) $H_2SO_4(l)$
(C) $C_{12}H_{22}O_{11}(aq) + H_2O(l) \rightarrow C_6H_{12}O_6$ (Glucose) + $C_6H_{12}O_6$ (Fructose)	(III) Pt(s)
(D) $2SO_2(g) + O_2(g) \rightarrow 2SO_3(g)$	(IV) Fe(s)

Choose the correct answer from the options given below :
(1) (A) – (II), (B) – (III), (C) – (I), (D) – (IV)
(2) (A) – (III), (B) – (II), (C) – (I), (D) – (IV)
(3) (A) – (III), (B) – (IV), (C) (II), (D) – (I)
(4) (A) – (III), (B) – (II), (C) – (IV), (D) – (I)

33. Match List-I with List - II

List-I	List-II
(A) $N_2(g) + 3H_2(g) \rightarrow 2NH_3(g)$	(I) Cu
(B) $CO(g) + 3H_2(g) \rightarrow CH_4(g) + H_2O(g)$	(II) $Cu/ZnO - Cr_2O_3$
(C) $CO(g) + H_2(g) \rightarrow HCHO(g)$	(III) $Fe_xO_y + K_2O + Al_2O_3$
(D) $CO(g) + 2H_2(g) \rightarrow CH_3OH(g)$	(IV) Ni

[July 25, 2022 (I)]
Choose the **correct** answer from the options given below:
(1) (A) - (II), (B) - (IV), (C) - (I), (D) - (III)
(2) (A) - (II), (B) - (I), (C) - (IV), (D) - (III)
(3) (A) - (III), (B) - (IV), (C) - (I), (D) - (II)
(4) (A) - (III), (B) - (I), (C) - (IV), (D) - (II)

34. Match List - I with List - II **[July 26, 2022 (I)]**

List - I (Processes/Reactions)	List - II (Catalyst)
(A) $2SO_2(g) + O_2(g) \rightarrow 2SO_3(g)$	(I) Fe(s)
(B) $4NH_3(g) + 5O_2(g) \rightarrow 4NO(g) + 6H_2O(g)$	(II) Pt(s) – Rh(s)
(C) $N_2(g) + 3H_2(g) \rightarrow 2NH_3(g)$	(III) V_2O_5
(D) Vegetable oil(l) + $H_2 \rightarrow$ Vegetable ghee(s)	(IV) Ni(s)

Choose the **correct** answer from the options given below:
(1) (A) – (III), (B) – (I), (C) – (II), (D) – (IV)
(2) (A) – (III), (B) – (II), (C) – (I), (D) – (IV)
(3) (A) – (IV), (B) – (III), (C) – (I), (D) – (II)
(4) (A) – (IV), (B) – (II), (C) – (III), (D) – (I)

35. Match List-I with List-II.

List - I (Enzyme)	List - II (Conversion of)
(A) Invertase	(I) Starch into maltose
(B) Zymase	(II) Maltose into glucose
(C) Diastase	(III) Glucose into ethanol
(D) Maltase	(IV) Cane sugar into glucose

Choose the **most appropriate** answer from the options given below : **[June 26, 2022 (II)]**
(1) A-III, B-IV, C-II, D-I (2) A-III, B-II. C-I, D-IV
(3) A-IV, B-III, C-I, D-II (4) A-IV, B-II, C-III, D-I

36. Match List - I with List - II. **[June 25, 2022 (II)]**

List - I	List - II
(A) Zymase	(I) Stomach
(B) Diastase	(II) Yeast
(C) Urease	(III) Malt
(D) Pepsin	(IV) Soyabean

Choose the correct answer from the options given below:
(1) A-II, B-III, C-I, D-IV (2) A-II, B-III, C-IV, D-I
(3) A-III, B-II, C-IV, D-I (4) A-III, B-II, C-I, D-IV

37. Which one of the following statements is not true about enzymes ? **[July 20, 2021 (II)]**
(1) Enzymes are non-specific for a reaction and substrate.
(2) Almost all enzymes are proteins.

(3) Enzymes work as catalysts by lowering the activation energy of a biochemical reaction.
(4) The action of enzymes is temperature and pH specific

38. Which of the following is Lindlar catalyst? [March 16, 2021 (I)]
(1) Cold dilute solution of $KMnO_4$
(2) Partially deactivated palladised charcoal
(3) Zinc chloride and HCl
(4) Sodium and Liquid NH_3

39. Which of the following is not an example of heterogeneous catalytic reaction? [Jan. 10, 2019 (I)]
(1) Ostwald's process (2) Combustion of coal
(3) Hydrogenatoin of vegetable oils (4) Haber's process

40. Match the catalysts to the correct processes : [2015]

Catalyst	Process
(A) $TiCl_4$	(i) Wacker process
(B) $PdCl_2$	(ii) Ziegler - Natta polymerization
(C) $CuCl_2$	(iii) Contact process
(D) V_2O_5	(iv) Deacon's process

(1) (A) - (ii), (B) - (iii), (C) - (iv), (D) - (i)
(2) (A) - (iii), (B) - (i), (C) - (ii), (D) - (iv)
(3) (A) - (iii), (B) - (ii), (C) - (iv), (D) - (i)
(4) (A) - (ii), (B) - (i), (C) - (iv), (D) - (iii)

41. Identify the correct statement regarding enzymes [2004]
(1) Enzymes are specific biological catalysts that cannot be poisoned
(2) Enzymes are normally heterogeneous catalysts that are very specific in their action
(3) Enzymes are specific biological catalysts that can normally function at very high temperatures (T~1000K)
(4) Enzymes are specific biological catalysts that possess well-defined active sites

Topic 3 - Colloids and Emulsions

Multiple Choice Questions (MCQs)

42. 100 mL of 5% (w/v) solution of NaCl in water was prepared in 250 mL beaker. Albumin from the egg was poured into NaCl solution and stirred well. This resulted in a/ an : [July 29, 2022 (I)]
(1) Lyophilic sol (2) Lyophobic sol
(3) Emulsion (4) Precipitate

43. Given below are two statements: one is labelled as Assertion (A) and the other is labelled as Reason (R). [July 27, 2022 (II)]
Assertion (A) : Dissolved substances can be removed from a colloidal solution by diffusion through a parchment paper.
Reason (R) : Particles in a true solution cannot pass through parchment paper but the collodial particles can pass through the parchment paper.
In the light of the above statements, choose the correct answer from the options given below:
(1) Both (A) and (R) are correct and (R) is the correct explanation of (A).
(2) Both (A) and (R) are correct but (R) is NOT the correct explanation of (A).
(3) (A) is correct but (R) is not correct.
(4) (A) is not correct but (R) is correct.

44. Given below are two statements : one is labelled as Assertion A and the other is labelled as Reason R.
Assertion A : Finest gold is red in colour, as the size of the particles increases, it appears purple then blue and finally gold.
Assertion R : The colour of the colloidal solution depends on the wavelength of light scattered by the dispersed particles.
In the light of the above statements, choose the **most appropriate** answer from the options given below : [July 26, 2022 (II)]
(1) Both A and R are true and R is the correct explanation of A.
(2) Both A and R are true but R is NOT the correct explanation of A.
(3) A is true but R is false.
(4) A is false but R is true.

45. For micelle formation, which of the following statements are correct? [July 25, 2022 (II)]
(A) Micelle formation is an exothermic process.
(B) Micelle formation is an endothermic process.
(C) The entropy change is positive.
(D) The entropy change is negative.
(1) A and D only (2) A and C only
(3) B and C only (4) B and D only

46. Which of the following is a **correct** statement ?
(1) Brownian motion destabilises sols. [June 29, 2022 (I)]
(2) Any amount of dispersed phase can be added to emulsion without destabilising it.
(3) Mixing two oppositely charged sols in equal amount neutralises charges and stabilises colloids.
(4) Presence of equal and similar charges on colloidal particles provides stability to the colloidal solution.

47. A 42.12% (w/v) solution of NaCl causes precipitation of a certain sol in 10 hours. The coagulating value of NaCl for the sol is [Given : Molar mass : Na = 23.0 g mol^{-1}; Cl = 35.5 g mol^{-1}] [June 29, 2022 (II)]
(1) 36 mmol L^{-1} (2) 36 mol L^{-1}
(3) 1440 mol L^{-1} (4) 1440 mmol L^{-1}

48. The Zeta potential is related to which property of colloids"
(1) Colour [June 28, 2022 (I)]
(2) Tyndall effect
(3) Charge on the surface of colloidal particles
(4) Brownian movement

49. Match List-I with List-II.

List-I	List-II
(A) Negatively charged sol	(I) $Fe_2O_3 \cdot xH_2O$
(B) Macromolecular colloid	(II) CdS sol
(C) Positively charged sol	(III) Starch
(D) Cheese	(IV) a gel

Choose the correct answer from the options given below :
(1) (A) – (II), (B) – (III), (C) – (IV), (D) – (I) [June 28, 2022 (II)]
(2) (A) – (II), (B) – (I), (C) – (III), (D) – (IV)
(3) (A) – (II), (B) – (III), (C) – (I), (D) – (IV)
(4) (A) – (I), (B) – (III), (C) – (II), (D) – (IV)

50. Match List-I with List-II [June 27, 2022 (I)]

List-I	List-II
(A) Lyophilic colloid	(I) Liquid-liquid colloid
(B) Emulsion	(II) protective colloid
(C) Positively charged	(III) $FeCl_3$ + NaOH
(D) Negatively charged	(IV) $FeCl_3$ + hot water colloid

Choose the correct answer from the options given below:

(1) (A) – (II), (B) – (I), (C) – (IV), (D) – (III)
(2) (A) – (III), (B) – (I), (C) – (IV), (D) – (II)
(3) (A) – (II), (B) – (I), (C) – (III), (D) – (IV)
(4) (A) – (III), (B) – (II), (C) – (I), (D) – (IV)

51. Incorrect statement for Tyndall effect is : **[June 25, 2022 (I)]**

(1) The refractive indices of the dispersed phase and the dispersion medium differ greatly in magnitude.
(2) The diameter of the dispersed particles is much smaller than the wavelength of the light used.
(3) During projection of movies in the cinemas hall, Tyndall effect is noticed.
(4) It is used to distinguish a true solution from a colloidal solution.

52. Using very little soap while washing clothes, does not serve the purpose of cleaning of clothes because **[June 25, 2022 (I)]**

(1) soap particles remain floating in water as ions
(2) the hydrophobic part of soap is not able to take away grease
(3) the micelles are not formed due to concentration of soap, below its CMC value
(4) colloidal structure of soap in water is completely disturbed.

53. Given below are two statements :

Statement I : Emulsions of oil in water are unstable and sometimes they separate into two layers on standing.

Statement II : For stabilisation of an emulsion, excess of electrolyte is added.

In the light of the above statements, choose the **most appropriate** answer from the options given below : **[June 24, 2022 (I)]**

(1) Both Statement I and Statement II are correct.
(2) Both Statement I and Statement II are incorrect.
(3) Statement I is correct but Statement II is incorrect.
(4) Statement I is incorrect but Statement II is correct.

54. Lyophilic sols are more stable than lyophobic sols because:

(1) there is a strong electrostatic repulsion between the negatively charged colloidal particles. **[Aug. 27, 2021 (II)]**
(2) the colloidal particles have positive charge.
(3) the colloidal particles have no charge.
(4) the colloidal particles are solvated.

55. The sol given below with negatively charged colloidal particles is : **[Aug. 26, 2021 (II)]**

(1) $FeCl_3$ added to hot water
(2) KI added to $AgNO_3$ solution
(3) $AgNO_3$ added to KI solution
(4) $Al_2O_3.xH_2O$ in water

56. When silver nitrate solution is added to potassium iodide solution then the sol produced is : **[July 22, 2021 (II)]**

(1) AgI/I^- (2) AgI/Ag^+
(3) KI/NO_3^- (4) $AgNO_3/NO_3^-$

57. The charges on the colloidal CdS sol. and TiO_2 sol. are, respectively **[March 18, 2021 (II)]**

(1) positive and positive (2) positive and negative
(3) negative and positive (4) negative and positive

58. A colloidal system consisting of a gas dispersed in a solid is called a/an **[March 17, 2021 (I)]**

(1) aerosol (2) foam (3) solid sol (4) gel

59. For the coagulation of a negative sol, the species below, that has the highest flocculating power is **[March 17, 2021 (II)]**

(1) Ba^{2+} (2) PO_4^{3-} (3) SO_4^{2-} (4) Na^+

60. The **INCORRECT** statements below regarding colloidal solutions is **[March 16, 2021 (II)]**

(1) The flocculating power of Al^{3+} is more than that of Na^+.
(2) A colloidal solution shows Brownian motion of colloidal particles.
(3) An ordinary filter paper can stop the flow of colloidal particles.
(4) A colloidal solution shows colligative properties.

61. The nature of charge on resulting colloidal particles when $FeCl_3$ is added to excess of hot water is: **[Feb. 26, 2021 (II)]**

(1) Sometimes positive and sometimes negative
(2) Negative
(3) Neutral
(4) Positive

62. Most suitable salt which can be used for efficient clotting of blood will be : **[February 24, 2021 (II)]**

(1) $NaHCO_3$ (2) $FeSO_4$
(3) $Mg(HCO_3)_2$ (4) $FeCl_3$

63. Kraft temperature is the temperature: **[Sep. 06, 2020 (I)]**

(1) below which the aqueous solution of detergents starts freezing.
(2) below which the formation of micelles takes place.
(3) above which the aqueous solution of detergents starts boiling.
(4) above which the formation of micelles takes place.

64. Identify the correct molecular picture showing what happens at the critical micellar concentration (CMC) of an aqueous solution of a surfactant (● polar head, ∿ non-polar tail, • water). **[Sep. 05, 2020 (I)]**

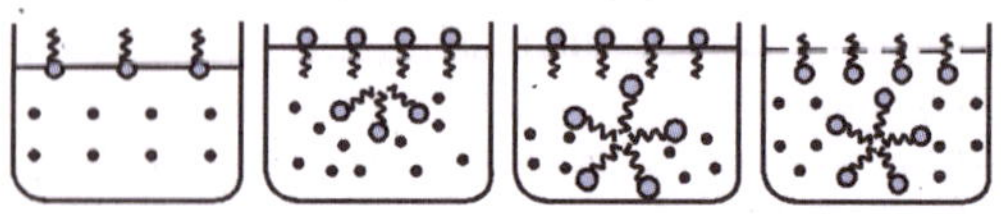

(A) (B) (C) (D)

(1) (D) (2) (B) (3) (A) (4) (C)

65. A sample of red ink (a colloidal suspension) is prepared by mixing eosin dye, egg white, HCHO and water. The component which ensures stabilitiy of the ink sample is: **[Sep. 04, 2020 (II)]**

(1) Egg white (2) Water
(3) HCHO (4) Eosin dye

66. Tyndall effect is observed when : **[Sep. 03, 2020 (I); Similar 2017]**

(1) The diameter of dispersed particles is much larger than the wavelength of light used.
(2) The diameter of dispersed particles is much smaller than the wavelength of light used.
(3) The refractive index of dispersed phase is greater than that of the dispersion medium.
(4) The diameter of dispersed particles is similar to the wavelength of light used.

67. An ionic micelle is formed on the addition of : **[Sep. 03, 2020 (II)]**

(1) liquid diethyl ether to aqueous NaCl solution
(2) excess water to liquid

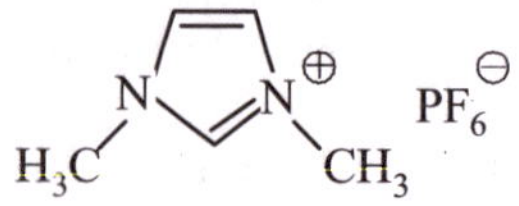

excess water to liquid

(3)

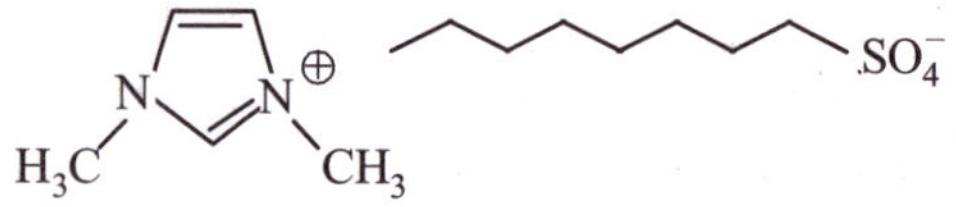

(4) sodium stearate to pure toluene

68. Which of the following is used for the preparation of colloids ? [Sep. 02, 2020 (I)]
(1) Ostwald Process (2) Van Arkel Method
(3) Bredig's Arc Method (4) Mond Process

69. The size of a raw mango shrinks to a much smaller size when kept in a concentrated salt solution. Which one of the following processes can explain this ? [Sep. 02, 2020 (II)]
(1) Osmosis (2) Dialysis
(3) Diffusion (4) Reverse osmosis

70. As per Hardy-Schulze formulation, the flocculation values of the following for ferric hydroxide sol are in the order: [Jan. 08, 2020 (I)]
(1) $K_3[Fe(CN)_6] < K_2CrO_4 < KBr = KNO_3 = AlCl_3$
(2) $K_3[Fe(CN)_6] < K_2CrO_4 < AlCl_3 < KBr < KNO_3$
(3) $AlCl_3 > K_3[Fe(CN)_6] > K_2CrO_4 > KBr = KNO_3$
(4) $K_3[Fe(CN)_6] > AlCl_3 > K_2CrO_4 > KBr > KNO_3$

71. Peptization is a : [April 12, 2019 (I)]
(1) process of bringing colloidal molecule into solution
(2) process of converting precipitate into colloidal solution
(3) process of converting a colloidal solution into precipitate
(4) process of converting soluble particles to form colloidal solution

72. Among the following, the INCORRECT statement about colloids is: [April 12, 2019 (II)]
(1) They can scatter light.
(2) They are larger than small molecules and have high molar mass.
(3) The osmotic pressure of a colloidal solution is of higher order than the true solution at the same concentration.
(4) The range of diameters of colloidal particles is between 1 and 1000 nm.

73. The correct option among the following is : [April 10, 2019 (II)]
(1) Colloidal medicines are more effective because they have small surface area.
(2) Addition of alum to water makes it unfit for drinking
(3) Colloidal particles in lyophobic sols can be precipitated by electrophoresis.
(4) Brownian motion in colloidal solution is faster if the viscosity of the solution is very high.

74. The aerosol is a kind of colloid in which: [April 9, 2019 (I)]
(1) solid is dispersed in gas
(2) gas is dispersed in solid
(3) gas is dispersed in liquid
(4) liquid is dispersed in water

75. Among the following, the false statement is : [Jan. 12, 2019 (II)]
(1) It is possible to cause artificial rain by throwing electrified sand carrying charge opposite to the one on clouds from an aeroplane.
(2) Tyndall effect can be used to distinguish between a colloidal solution and a true solution.
(3) Lyophilic sol can be coagulated by adding an electrolyte.
(4) Latex is a colloidal solution of rubber particles which arc positively charged.

76. An example of solid sol is : [Jan. 11, 2019 (I)]
(1) Paint (2) Gem stones
(3) Butter (4) Hair cream

77. Among the colloids cheese (C), milk (M) and smoke (S), the correct combination of the dispersed phase and dispersion medium, respectively is : [Jan. 11, 2019 (II)]
(1) C : liquid in solid; M : liquid in solid; S : solid in gas
(2) C : liquid in solid; M : liquid in liquid; S : solid in gas
(3) C : solid in liquid; M : liquid in liquid; S : gas in solid
(4) C : solid in liquid; M : solid in liquid; S : solid in gas

78. Haemoglobin and gold sol are examples of: [Jan. 10, 2019 (II)]
(1) positively and negatively charged sols, respectively
(2) positively charged sols
(3) negatively charged sols
(4) negatively and positively charged sols, respectively

79. For coagulation of arsenious sulphide sol, which one of the following salt solution will be most effective? [Jan. 9, 2019 (II); Similar 2013]
(1) $BaCl_2$ (2) $AlCl_3$ (3) NaCl (4) Na_3PO_4

80. Which of the following statements about colloids is false? [Online April 15, 2018 (I)]
(1) When silver nitrate solution is added to potassium iodide solution, a negatively charged colloidal solution is formed
(2) Freezing point of colloidal solution is lower than true solution at same concentration of a solute
(3) Colloidal particles can pass through ordinary filter paper
(4) When excess of electrolyte is added to colloidal solution, colloidal particle will be precipitated

81. Among the following, correct statement is : [Online April 8, 2017]
(1) Brownian movement is more pronounced for smaller particles than for bigger–particles.
(2) Sols of metal sulphides are lyophilic.
(3) Hardy Schulze law states that bigger the size of the ons, the greater is its coagulating power.
(4) One would expect charcoal to adsorb chlorine more than hydrogen sulphide.

82. The most appropriate method of making egg–albumin sol is : [Online April 9, 2016]
(1) Break an egg carefully and transfer the transparent part of the content to 100 mL of 5% w/V saline solution and stir well.
(2) Keep the egg in boiling water for 10 minutes. After removing the shell, transfer the yellow part of the content to 100 mL of 5% w/V saline solution and homogenize with a mechanical shaker.
(3) Keep the egg in boiling water for 10 minutes. After removing the shell, transfer the white part of the content to 100 mL of 5% w/V saline solution and homogenize with a mechanical shaker
(4) Break an egg carefully and transfer only the yellow part of the content to 100 mL of 5% w/V saline solution and stir well.

83. Gold numbers of some colloids are : Gelatin : 0.005 – 0.01, Gum arabic : 0.15 – 0.25; Oleate : 0.04 – 1.0; Starch : 15 – 25. Which among these is a better protective colloid ? [2008 (S), Online April 10, 2016]
(1) Gelatin (2) Starch (3) Oleate (4) Gum arabic

84. Under ambient conditions, which among the following surfactants will form micelles in aqueous solution at lowest molar concentration? [Online April 11, 2015]
(1) $CH_3-(CH_2)_8-COO^-Na^+$ (2) $CH_3(CH_2)_{11}\overset{\oplus}{N}(CH_3)_3Br^-$
(3) $CH_3-(CH_2)_{13}-OSO_3^-Na^+$ (4) $CH_3(CH_2)_{15}\overset{\oplus}{N}(CH_3)_3Br^-$

85. The migration of dispersion medium under the influence of an electric potential is called : **[Online April 9, 2013]**
(1) Cataphoresis (2) Electroosmosis
(3) Electrophoresis (4) Sedimentation

86. Flocculation value of $BaCl_2$ is much less than that of KCl for sol A and flocculation value of Na_2SO_4 is much less than that of NaBr for sol B. The correct statement among the following is : **[Online April 22, 2013]**
(1) Both the sols A and B are negatively charged.
(2) Sol A is positively charged and Sol B is negatively charged.
(3) Both the sols A and B are positively charged.
(4) Sol A is negatively charged and sol B is positively charged.

87. Smoke is an example of : **[Online April 23, 2013]**
(1) Solid dispersed in solid
(2) Gas dispersed in liquid
(3) Solid dispersed in gas
(4) Gas dispersed in solid

88. On addition of 1 mL of 10% NaCl solution to 10 mL gold sol in the presence of 0.025 g of starch, the coagulation is just prevented. Starch has gold number **[Online May 19, 2012]**
(1) 2.5 (2) 25 (3) 0.25 (4) 0.025

89. Fog is a colloidal solution of **[Online May 19, 2012]**
(1) solid particles dispersed in gas
(2) solid particles dispersed in a liquid
(3) liquid particles dispersed in gas
(4) gaseous particles dispersed in a liquid

90. The volume of a colloidal particle, V_C as compared to the volume of a solute particle in a true solution V_S, could be **[2005]**
(1) $\frac{V_C}{V_S} \simeq 10^3$ (2) $\frac{V_C}{V_S} \simeq 10^{-3}$
(3) $\frac{V_C}{V_S} \simeq 10^{23}$ (4) $\frac{V_C}{V_S} \simeq 1$

91. The disperse phase in colloidal iron (III) hydroxide and colloidal gold is positively and negatively charged, respectively. Which of the following statements is NOT correct ? **[2005]**
(1) Coagulation in both sols can be brought about by electrophoresis
(2) Mixing the sols has no effect
(3) Sodium sulphate solution causes coagulation in both sols
(4) Magnesium chloride solution coagulates, the gold sol more readily than the iron (III) hydroxide sol

NUMERICAL VALUE QUESTIONS (NVQs)

92. The flocculation value of HCl for arsenic sulphide sol. is 30 m mol L^{-1}. If H_2SO_4 is used for the flocculation of arsenic sulphide, the amount, in grams, of H_2SO_4 in 250 mL required for the above purpose is ________. **[Jan. 07, 2020 (II)]**
(molecular mass of H_2SO_4 = 98 g/mol)

Part-B (JEE Advanced)

Topic 1 - Adsorption

1 MCQs with One Correct Answer

1. Methylene blue, from its aqueous solution, is adsorbed on activated charcoal at 25°C. For this process, the correct statement is **[Adv. 2013]**
(a) The adsorption requires activation at 25°C
(b) The adsorption is accompanied by a decrease in enthalpy
(c) The adsorption increases with increase of temperature
(d) The adsorption is irreversible

2. Adsorption of gases on solid surface is generally exothermic because **[2004S]**
(a) enthalpy is positive (b) entropy decreases
(c) entropy increases (d) free energy increases

3. Rate of physisorption increases with **[2003S]**
(a) decrease in temperature (b) increase in temperature
(c) decrease in pressure (d) decrease in surface area

2 Integer Value Answer

4. To form a complete monolayer of acetic acid on 1g of charcoal, 100 mL of 0.5 M acetic acid was used. Some of the acetic acid remained unadsorbed. To neutralize the unadsorbed acetic acid, 40 mL of 1 M NaOH solution was required. If each molecule of acetic acid occupies $P \times 10^{-23}$ m^2 surface area on charcoal, the value of P is _____. **[Adv. 2024]**

[**Use given data:** Surface area of charcoal = 1.5×10^2 m^2g^{-1}; Avogadro's number (N_A) = 6.0×10^{23} mol^{-1}]

5. 20% of surface sites are occupied by N_2 molecules. The density of surface site is 6.023×10^{14} cm^{-2} and total surface area is 1000 cm^2. The catalyst is heated to 300 K while N_2 is completely desorbed into a pressure of 0.001 atm and volume of 2.46 cm^3. Find the number of active sites occupied by each N_2 molecule. **[2005 - 4 Marks]**

6. Adsorption of phenol from its aqueous solution on to fly ash obeys Freundlich isotherm. At a given temperature, from 10 mg g^{-1} and 16 mg g^{-1} aqueous phenol solutions, the concentrations of adsorbed phenol are measured to be 4 mg g^{-1} and 10 mg g^{-1}, respectively. At this temperature, the concentration (in mg g^{-1}) of adsorbed phenol from 20 mg g^{-1} aqueous solution of phenol will be _____.
Use: $\log_{10} 2 = 0.3$ **[Adv. 2025]**

3 Numeric / New Stem Based Questions

7. At a given temperature, 0.45 g of acetic acid in 50 mL of water is shaken with 1.0 g of charcoal and the pH of the resulting solution is 3.0. Assume, the adsorption of acetic acid from the aqueous solution by charcoal follows Freundlich isotherm,

$$\frac{x}{m} = kc^{1/n}$$

If the plot of $\log_{10}(x/m)$ against $\log_{10}C$ gives a straight line with slope 1, the value of k in L mol^{-1} is ______.
Given : The molar mass of acetic acid is 60 g mol^{-1}.
The acid dissociation constant of acetic acid is 1.0×10^{-5} at the given temperature.
x is the mass (in grams) of acetic acid adsorbed.

m is the mass (in grams) of charcoal.
C is the equilibrium concentration of acetic acid in the solution after the adsorption is complete.
k and n are constants for acetic acid–charcoal system at the given temperature. **[Adv. 2026]**

6 MCQs with One or More than One Correct Answer

8. The correct option(s) related to adsorption processes is(are)
(a) Chemisorption results in a unimolecular layer.
(b) The enthalpy change during physisorption is in the range of 100 to 140 kJ mol^{-1}.
(c) Chemisorption is an endothermic process.
(d) Lowering the temperature favors physisorption processes. **[Adv. 2022]**

9. When O_2 is adsorbed on a metallic surface, electron transfer occurs from the metal to O_2. The true statement(s) regarding this adsorption is(are) **[Adv. 2015]**
(a) O_2 is physisorbed
(b) Heat is released
(c) Occupancy of π^*2p of O_2 is increased
(d) Bond length of O_2 is increased

10. The given graphs/data I, II, III and IV represent general trends observed for different physisorption and chemisorption processes under mild conditions of temperature and presure. Which of the following choice(s) about I, II, III and IV is (are) correct? **[2012]**

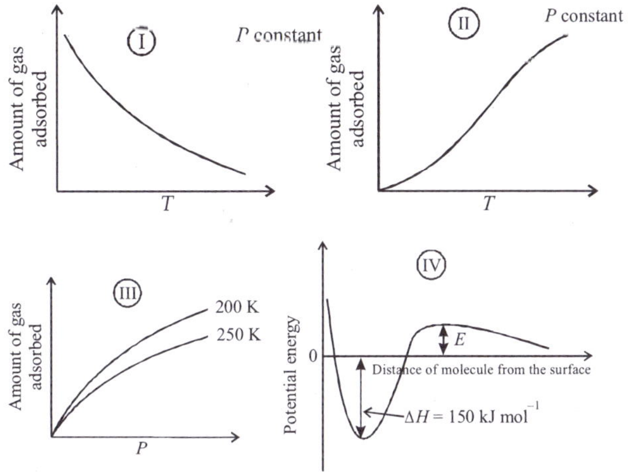

(a) I is physisorption and II is chemisorption
(b) I is physisorption and III is chemisorption
(c) IV is chemisorption and II is chemisorption
(d) IV is chemisorption and III is chemisorption

11. The correct statement(s) pertaining to the adsorption of a gas on a solid surface is (are) **[2011]**
(a) Adsorption is always exothermic
(b) Physisorption may transform into chemisorption at high temperature
(c) Physisorption increases with increasing temperature but chemisorption decreases with increasing temperature
(d) Chemisorption is more exothermic than physiosorption, however it is very slow due to higher energy of activation

10 Subjective Problems

12. 1 g of charcoal adsorbs 100 mL 0.5 M CH_3COOH to form a monolayer, and thereby the molarity of CH_3COOH reduces to 0.49. Calculate the surface area of the charcoal adsorbed by each molecule of acetic acid. Surface area of charcoal = 3.01×10^2 m^2/g. **[2003 - 2 Marks]**

Topic 2 - Catalysis and Theories of Catalysis

1 MCQs with One Correct Answer

13. Identify the correct statement regarding enzymes **[2004]**
(a) Enzymes are specific biological catalysts that cannot be poisoned
(b) Enzymes are normally heterogeneous catalysts that are very specific in their action
(c) Enzymes are specific biological catalysts that can normally function at very high temperatures (T~1000K)
(d) Enzymes are specific biological catalysts that possess well-defined active sites

Topic 3 - Colloids and Emulsions

1 MCQs with One Correct Answer

14. Consider the following statements related to colloids. **[Adv. 2023]**
(I) Lyophobic colloids are not formed by simple mixing of dispersed phase and dispersion medium
(II) For emulsions, both the dispersed phase and the dispersion medium are liquid.
(III) Micelles are produced by dissolving a surfactant in any solvent at any temperature.
(IV) Tyndall effect can be observed from a colloidal solution with dispersed phase having the same refractive index as that of the dispersion medium.
The option with the correct set of statements is
(a) (I) and (II) (b) (II) and (III)
(c) (III) and (IV) (d) (II) and (IV)

15. Among the electrolytes Na_2SO_4, $CaCl_2$, $Al_2(SO_4)_3$ and NH_4Cl, the most effective coagulating agent for Sb_2S_3 sol is **[2009S]**
(a) Na_2SO_4 (b) $CaCl_2$ (c) $Al_2(SO_4)_3$ (d) NH_4Cl

16. Among the following, the surfactant that will form micelles in aqueous solution at the lowest molar concentration at ambient condition is : **[2008S]**
(a) $CH_3(CH_2)_{15}N^+(CH_3)_3Br^-$ (b) $CH_3(CH_2)_{11}OSO_3^- Na^+$
(c) $CH_3(CH_2)_6COO^-Na^+$ (d) $CH_3(CH_2)_{11}N^+(CH_3)_3Br^-$

17. Lyophilic sols are **[2005S]**
(a) Irreversible sols
(b) They are prepared from inorganic compound
(c) Coagulated by adding electrolytes
(d) Self-stabilizing

6 MCQs with One or More than One Correct Answer

18. The correct statement(s) related to colloids is(are) **[Adv. 2021]**
(a) The process of precipitating colloidal sol by an electrolyte is called peptization.

(b) Colloidal solution freezes at higher temperature than the true solution at the same concentration.
(c) Surfactants form micelle above critical micelle concentration (CMC). CMC depends on temperature.
(d) Micelles are macromolecular colloids.

19. The correct statement(s) about surface properties is (are) [Adv. 2017]
(a) Adsorption is accompanied by decrease in enthalpy and decrease in entropy of the system
(b) The critical temperatures of ethane and nitrogen are 563 K and 126 K, respectively. The adsorption of ethane will be more than that of nitrogen on same amount of activated charcoal at a given temperature
(c) Cloud is an emulsion type of colloid in which liquid is dispersed phase and gas is dispersion medium
(d) Brownian motion of colloidal particles does not depend on the size of the particles but depends on viscosity of the solution.

20. Choose the correct reason(s) for the stability of the lyophobic colloidal particles. [2012]
(a) Preferential adsorption of ions on their surface from the solution.
(b) Preferential adsorption of solvent on their surface from the solution.
(c) Attraction between different particles having opposite charges on their surface.
(d) Potential difference between the fixed layer and the diffused layer of opposite charges around the colloidal particles.

9 Assertion and Reason/Statement Type Questions

Each question contains STATEMENT-1 (Assertion) and STATEMENT-2 (Reason). Each question has 4 choices (a), (b), (c) and (d) out of which ONLY ONE is correct. Mark your answer as
(a) If both Statement -1 and Statement -2 are correct, and Statement -2 is the correct explanation of the Statement -2.
(b) If both Statement -1 and Statement -2 are correct, but Statement -2 is not the correct explanation of the Statement -1.
(c) If Statement -1 is correct but Statement -2 is incorrect.
(d) If Statement -1 is incorrect but Statement -2 is correct.

21. **Assertion :** Micelles are formed by surfactant molecules above the critical micellar concentration (CMC).
Reason : The conductivity of a solution having surfactant molecules decreases sharply at the CMC. [2007]

Answer Key

Part-A (JEE Main)

1	(3)	11	(4)	21	(1)	31	(48)	41	(4)	51	(2)	61	(4)	71	(2)	81	(1)	91	(2)
2	(4)	12	(4)	22	(1)	32	(3)	42	(1)	52	(3)	62	(4)	72	(3)	82	(1)	92	(0.37)
3	(3)	13	(2)	23	(3)	33	(3)	43	(3)	53	(3)	63	(4)	73	(3)	83	(1)		
4	(4)	14	(3)	24	(13125)	34	(2)	44	(1)	54	(4)	64	(1)	74	(1)	84	(4)		
5	(4)	15	(4)	25	(12)	35	(3)	45	(3)	55	(3)	65	(1)	75	(4)	85	(2)		
6	(3)	16	(3)	26	(9960)	36	(2)	46	(4)	56	(1)	66	(4)	76	(2)	86	(4)		
7	(2)	17	(3)	27	(2)	37	(1)	47	(N)	57	(3)	67	(3)	77	(2)	87	(3)		
8	(3)	18	(4)	28	(128)	38	(2)	48	(3)	58	(3)	68	(3)	78	(1)	88	(2)		
9	(2)	19	(3)	29	(17)	39	(2)	49	(3)	59	(1)	69	(1)	79	(2)	89	(3)		
10	(1)	20	(3)	30	(2)	40	(4)	50	(1)	60	(3)	70	(1)	80	(2)	90	(1)		

Part-B (JEE Advanced)

1	(b)	3	(a)	5	(2)	7	(1.50)	9	(b, c, d)	11	(a, b, d)	14	(a)	16	(a)	18	(b, c)	20	(a, d)
2	(b)	4	(2500)	6	(15.62)	8	(a, d)	10	(a, c)	13	(d)	15	(c)	17	(d)	19	(a, b)	21	(b)

Chapter-6

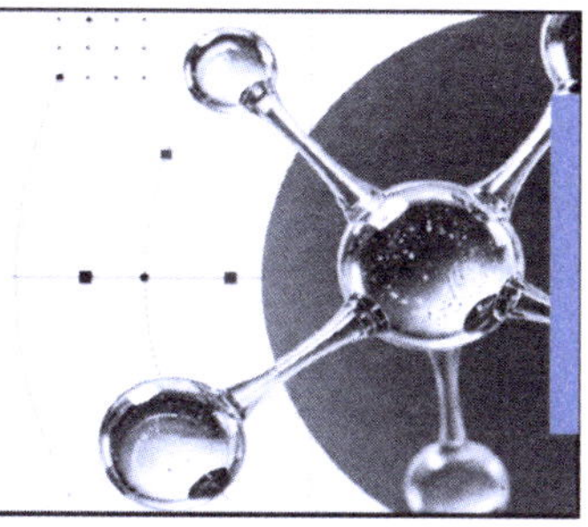

General Principles and Process of Isolation of Elements

Part-A (JEE Main)

Topic 1 - Occurrence of Metals and Metallurgical Processes

MULTIPLE CHOICE QUESTIONS (MCQs)

1. $\Delta G°$ vs T plot for the formation of MgO, involving reaction $2Mg + O_2 \rightarrow 2MgO$, will look like: [June 30, 2022 (I)]

(1) 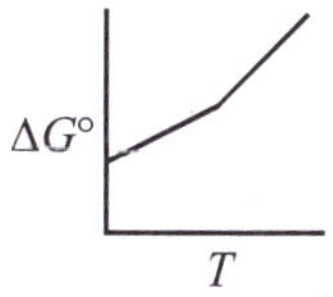

(2) 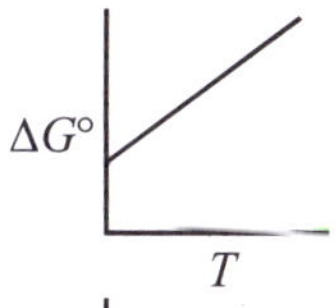

(3) 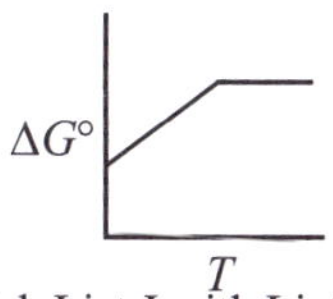

(4) 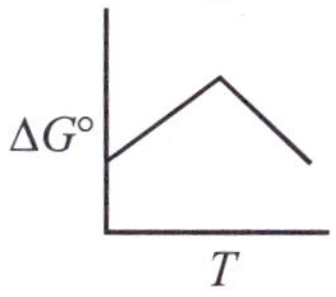

2. Match List-I with List-II. [June 29, 2022 (II)]

List-I (One)	List-II (Composition)
(A) Siderite	(I) $FeCO_3$
(B) Malachite	(II) $CuCO_3 \times Cu(OH)_2$
(C) Sphalerite	(III) ZnS
(D) Calamine	(IV) $ZnCO_3$

Choose the correct answer from the options given below:
(1) A-I, B-II, C-III, D-IV (2) A-III, B-IV, C-II, D-I
(3) A-IV, B-III, C-I, D-II (4) A-I, B-II, C-IV, D-III

3. In isolation of which one of the following metals from their ores, the use of cyanide salt is not commonly involved? [June 29, 2022 (I)]
(1) Zinc (2) Gold (3) Silver (4) Copper

4. In the metallurgical extraction of copper, following reaction is used: $FeO + SiO_2 \rightarrow FeSiO_3$ [June 28, 2022 (II)]
FeO and $FeSiO_3$ respectively are
(1) gangue and flux (2) flux and slag
(3) slag and flux (4) gangue and slag

5. Given are two statements one is labelled as Assertion A and other is labelled as Reason R. [June 28, 2022 (I)]
Assertion A : Magnesium can reduce Al_2O_3 at a temperature below 1350°C, while above 1350°C aluminium can reduce MgO.
Reason R : The melting and boiling points of magnesium are lower than those of aluminium.
In light of the above statements. choose most appropriate answer from the options given below:
(1) Both A and R are correct and R is correct explanation of A.
(2) Both A and R are correct. but R is NOT the correct explanation of A.
(3) A is correct but R is not correct.
(4) A is not correct but R is correct.

6. Statement I : Leaching of gold with cyanide ion in absence of air/O_2 leads to cyano complex of Au (III).
Statement II : Zinc is oxidized during the displacement reaction carried out for gold extraction.
In the light of the above statements, choose the correct answer from the options given below. [June 27, 2022 (II)]
(1) Both Statement I and Statement II are correct.
(2) Both Statement I and Statement II are incorrect.
(3) Statement I is correct but Statement II is incorrect.
(4) Statement I is incorrect but Statement II is correct.

7. Match List-I with List-II.

List-I	List-II
(A) Concentration of gold ore	(I) Aniline
(B) Leaching of alumina	(II) NaOH
(C) Froth stabiliser	(III) SO_2
(D) Blister copper	(IV) NaCN

[June 27, 2022 (I)]
Choose the correct answer from the options given below.
(1) (A) – (IV), (B) – (III), (C) – (II), (D) – (I)
(2) (A) – (IV), (B) – (II), (C) – (I), (D) – (III)
(3) (A) – (III), (B) – (II), (C) – (I), (D) – (IV)
(4) (A) – (II), (B) – (IV), (C) – (III), (D) – (I)

8. Given below are two statements : [June 26, 2022 (I)]
Statement I : According to the Ellingham diagram, any metal oxide with higher $\Delta G°$ is more stable than the one with lower $\Delta G°$.
Statement II : The metal involved in the formation of oxide placed lower in the Ellingham diagram can reduce the oxide of a metal placed higher in the diagram.
In the light of the above statements, choose the most appropriate answer from the options given below :
(1) Both Statement I and Statement II are correct.
(2) Both Statement I and Statement II are incorrect.
(3) Statement I is correct but Statement II is incorrect.
(4) Statement I is incorrect but Statement II is correct.

9. The role of depressants in Froth Flotation method is to [June 26, 2022 (II)]
(1) selectively prevent one component of the ore from coming to the froth.
(2) reduce the consumption of oil for froth formation.
(3) stabilize the froth.
(4) enhance non-wettability of the mineral particles.

10. Leaching of gold with dilute aqueous solution of NaCN in presence of oxygen gives complex [A], which on reaction with zinc forms the elemental gold and another complex [B]. [A] and [B], respectively are :- [June 25, 2022 (I)]
(1) $[Au(CN)_4]^-$ and $[Zn(CN)_2(OH)_2]^{2-}$
(2) $[Au(CN)_2]^-$ and $[Zn(OH)_4]^{2-}$
(3) $[Au(CN)_2]^-$ and $[Zn(CN)_4]^{2-}$
(4) $[Au(CN)_4]^{2-}$ and $[Zn(CN)_6]^{4-}$

11. Match List-I with List-II [June 24, 2022 (I)]

List-I	List-II
(A) Sphalerite	(I) $FeCO_3$
(B) Calamine	(II) PbS
(C) Galena	(III) $ZnCO_3$
(D) Siderite	(IV) ZnS

Choose the most appropriate answer from the options given below:
(1) (A) - (IV), (B) - (III), (C) - (II), (D) - (I)
(2) (A) - (IV), (B) - (I), (C) - (II), (D) - (III)
(3) (A) - (II), (B) - (III), (C) - (I), (D) - (IV)
(4) (A) - (III), (B) - (IV), (C) - (II), (D) - (I)

12. In metallurgy the term "gangue" is used for: [July 29, 2022 (I)]
(1) Contamination of undesired earthy materials.
(2) Contamination of metals, other than desired metal
(3) Minerals which are naturally occuring in pure form
(4) Magnetic impurities in an ore.

13. In liquation process used for tin (Sn), the metal : [July 29, 2022 (II)]
(1) is reacted with acid
(2) is dissolved in water
(3) is brought to molten form which is made to flow on a slope
(4) is fused with NaOH

14. Which of the reaction is suitable for concentrating ore by leaching process ? [July 28, 2022 (I)]
(1) $2Cu_2S + 3O_2 \rightarrow 2Cu_2O + 2SO_2$
(2) $Fe_3O_4 + CO \rightarrow 3FeO + CO_2$
(3) $Al_2O_3 + 2NaOH + 3H_2O \rightarrow 2Na[Al(OH)_4]$
(4) $Al_2O_3 + 6Mg \rightarrow 6MgO + 4Al$

15. The metal that is not extracted from its sulphide ore is : [July 26, 2022 (II)]
(1) Aluminium (2) Iron
(3) Lead (4) Zinc

16. The compound(s) that is (are) removed as slag during the extraction of copper is: [July 25, 2022 (I)]
(A) CaO (B) FeO (C) Al_2O_3 (D) ZnO
(E) NiO
Choose the **correct** answer from the options given below:
(1) (C), (D) only (2) (A), (B), (E) only
(3) (A), (B) only (4) (B) only

17. Calamine and Malachite, respectively, are the ores of : [Sep. 1, 2021 (II); Similar April 11, 2015; April 19, 2014]
(1) Nickel and Aluminium (2) Zinc and Copper
(3) Copper and Iron (4) Aluminium and Zinc

18. Given below are two statements. [Aug. 26, 2021 (I)]
Statement I : The choice of reducing agents for metals extraction can be made by using Ellingham diagram, a plot of ΔG vs temperature.
Statement II : The value of ΔS increases from left to right in Ellingham diagram.
In the light of the above statements, choose the **most appropriate** answer from the options given below:
(1) Both Statement I and Statement II are true
(2) Statement I is false but Statement II is true
(3) Both Statement I and Statement II are false
(4) Statement I is true but Statement II is false

19. Match List-I with List-II : [Aug. 27, 2021 (II)]

List-I (Name of ore/mineral)	List-II (Chemical formula)
(A) Calamine	(i) ZnS
(B) Malachite	(ii) $FeCO_3$
(C) Siderite	(iii) $ZnCO_3$
(D) Sphalerite	(iv) $CuCO_3 \cdot Cu(OH)_2$

Choose the **most appropriate** answer from the options given below :
(1) (A)-(iii), (B)-(iv), (C)-(ii), (D)-(i)
(2) (A)-(iii), (B)-(iv), (C)-(i), (D)-(ii)
(3) (A)-(iv), (B)-(iii), (C)-(i), (D)-(ii)
(4) (A)-(iii), (B)-(ii), (C)-(iv), (D)-(i)

20. The statement that is INCORRECT about Ellingham diagram is [July 27, 2021 (I)]
(1) provides idea about the reaction rate.
(2) provides idea about free energy change.
(3) provides idea about changes in the phases during the reaction.
(4) provides idea about reduction of metal oxide.

21. In the leaching of alumina from bauxite, the ore expected to leach out in the process by reacting with NaOH is : [July 25, 2021 (I)]
(1) TiO_2 (2) Fe_2O_3 (3) ZnO (4) SiO_2

22. Match List I with List II : (Both having metallurgical terms) [July 25, 2021 (II)]

List-I	List-II
(A) Concentration of Ag ore	(i) Reverberatory furnace
(B) Blast furnace	(ii) Pig iron
(C) Blister copper	(iii) Leaching with dilute NaCN solution
(D) Froth floatation method	(iv) Sulfide ores

Choose the correct answer from the options given below:
(1) (A)–(iii), (B)–(ii), (C)–(i), (D)–(iv)
(2) (A)–(iii), (B)–(iv), (C)–(i), (D)–(ii)
(3) (A)–(iv), (B)–(i), (C)–(iii), (D)–(ii)
(4) (A)–(iv), (B)–(iii), (C)–(ii), (D)–(i)

23. Sulphide ion is soft base and its ores are common for metals. [July 22, 2021 (II)]
(A) Pb (B) Al (C) Ag (D) Mg
Choose the **correct** answer from the options given below:
(1) (A) and (C) only (2) (A) and (D) only
(3) (A) and (B) only (4) (C) and (D) only

24. Consider two chemical reactions (A) and (B) that take place during metallurgical process : [July 20, 2021 (II)]
(A) $ZnCO_3(s) \xrightarrow{\Delta} ZnO(s) + CO_2(g)$
(B) $2ZnS(s) + 3O_2(g) \xrightarrow{\Delta} 2ZnO(s) + 2SO_2(g)$
The **correct** option of names given to them respectively is:
(1) (A) is calcination and (B) is roasting
(2) Both (A) and (B) are producing same product so both are roasting
(3) Both (A) and (B) are producing same product so both are calcination
(4) (A) is roasting and (B) is calcination

25. The chemical that is added to reduce the melting point of the reaction mixture during the extraction of aluminium is [March 18, 2021 (I); Similar May 12, 2012]
(1) Cryolite (2) Calamine
(3) Kaolite (4) Bauxite

26. Ellingham diagram is a graphical representation of:
(1) ΔH vs T (2) ΔG vs P [Feb. 25, 2021 (I)]
(3) ΔG vs T (4) $(\Delta G - T\Delta S)$ vs T

27. Which of the following ore is concentrated using group 1 cyanide salt? [February 24, 2021 (I)]
(1) Sphalerite (2) Calamine
(3) Siderite (4) Malachite

28. The INCORRECT statement is : [Sep. 06, 2020 (I)]
(1) bronze is an alloy of copper and tin.
(2) cast iron is used to manufacture wrought iron.
(3) german silver is an alloy of zinc, copper and nickel.
(4) brass is an alloy of copper and nickel.

29. For a reaction, $4\,M(s) + nO_2(g) \rightarrow 2M_2O_{2n}(s)$, the free energy change is plotted as a function of temperature. The temperature below which the oxide is stable could be inferred from the plot as the point at which : [Sep. 06, 2020 (II)]
(1) the slope change from negative to positive
(2) the free energy change shows a change from negative to positive value
(3) the slope changes from positive to negative
(4) the slope changes from positive to zero

30. An Ellingham diagram provides information about: [Sep. 05, 2020 (I)]
(1) the conditions of pH and potential under which a species is thermodynamically stable.
(2) the temperature dependence of the standard Gibbs energies of formation of some metal oxides.
(3) the pressure dependence of the standard electrode potentials of reduction reactions involved in the extraction of metals.
(4) the kinetics of the reduction process.

31. The processes of calcination and roasting in metallurgical industries, respectively, can lead to : [Sep. 04, 2020 (II)]
(1) Global warming and photochemical smog
(2) Global warming and acid rain
(3) Photochemical smog and ozone layer depletion
(4) Photochemical smog and global warming

32. Among statements (1) - (4), the **correct** ones are : [Sep. 04, 2020 (I); Similar April 8, 2019 (II)]
(A) Lime stone is decomposed to CaO during the extraction of iron from its oxides.
(B) In the extraction of silver, silver is extracted as an anionic complex.
(C) Nickel is purified by Mond's process.
(D) Zr and Ti are purified by van Arkel method.
(1) (A), (B), (C) and (D) (2) (A), (C) and (D) only
(3) (B), (C) and (D) only (4) (C) and (D) only

33. According to the following diagram, A reduces BO_2 when the temperature is: [Jan. 09, 2020 (I)]

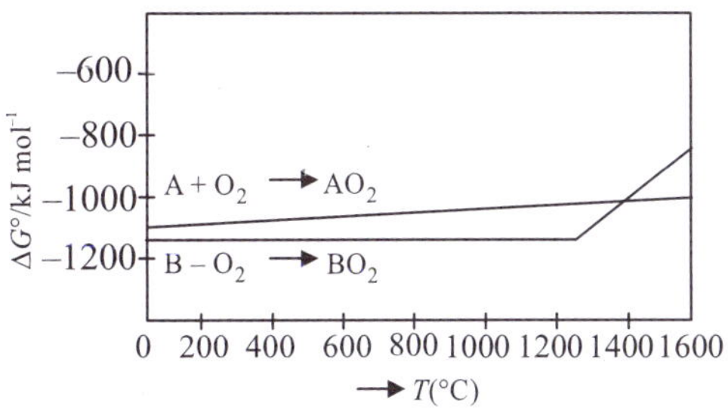

(1) $< 1400\,°C$ (2) $> 1400\,°C$
(3) $> 1200\,°C$ but $< 1400\,°C$ (4) $< 1200\,°C$

34. Among the reactions (A) - (D), the reaction(s) that does/do not occur in the blast furnace during the extraction of iron is/are:
(A) $CaO + SiO_2 \rightarrow CaSiO_3$ (B) $3Fe_2O_3 + CO \rightarrow 2Fe_3O_4 + CO_2$
(C) $FeO + SiO_2 \rightarrow FeSiO_3$ (D) $FeO \rightarrow Fe + \frac{1}{2}O_2$
[Jan. 08, 2020 (II)]
(1) (A) (2) (A) and (D)
(3) (C) and (D) (4) (D)

35. The idea of froth floatation method came from a person X and this method is related to the process Y of ores, X and Y, respectively, are : [April 12, 2019 (I)]
(1) fisher woman and concentration
(2) washer woman and concentration
(3) fisher man and reduction
(4) washer man and reduction

36. The correct statement is : [April 12, 2019 (II)]
(1) leaching of bauxite using concentrated NaOH solution gives sodium aluminate and sodium silicate.
(2) the Hall-Heroult process is used for the production of aluminium and iron.
(3) pig iron is obtained from cast iron.
(4) the blistered appearance of copper during the metallurgical process is due to the evolution of CO_2.

37. The correct statement is : [April 10, 2019 (II)]
(1) aniline is a froth stabilizer.
(2) zincite is a carbonate ore.
(3) sodium cyanide cannot be used in the metallurgy of silver.
(4) zone refining process is used for the refining of titanium.

38. The one that is not a carbonate ore is: [April 9, 2019 (II)]
(1) malachite (2) calamine
(3) siderite (4) bauxite

39. The ore that contains the metal in the form of fluoride is:
(1) cryolite (2) malachite [April 9, 2019 (I)]
(3) magnetite (4) sphalerite

40. **Assertion**: For the extraction of iron, haematite ore is used.
Reason: Haematite is a carbonate ore of iron.
(1) Only the reason is correct [April 9, 2019 (II)]
(2) Both the assertion and reason are correct, but the reason is not the correct explanation for the assertion.
(3) Both the assertion and reason are correct and the reason is the correct explanation for the assertion.
(4) Only the assertion is correct.

41. With respect to an ore, Ellingham diagram helps to predict the feasibility of its [April 8, 2019 (I)]
(1) Electrolysis (2) Zone refining
(3) Vapour phase refining (4) Thermal reduction

42. In the Hall-Heroult process, aluminium is formed at the cathode. The cathode is made out of: [Jan. 12, 2019 (I)]
(1) Pure aluminium (2) Carbon
(3) Copper (4) Platinum

43. The pair that does NOT require calcination is : [Jan. 12, 2019 (II)]
(1) ZnO and MgO (2) ZnO and $Fe_2O_3 \cdot xH_2O$
(3) $ZnCO_3$ and CaO (4) Fe_2O_3 and $CaCO_3 \cdot MgCO_3$

44. The reaction that does NOT define calcination is:
(1) $Fe_2O_3 \cdot xH_2O \xrightarrow{\Delta} Fe_2O_3 + xH_2O$ [Jan. 11, 2019 (II)]
(2) $2Cu_2S + 3O_2 \xrightarrow{\Delta} 2Cu_2O + 2SO_2$
(3) $ZnCO_3 \xrightarrow{\Delta} ZnO + CO_2$
(4) $CaCO_3 \cdot MgCO_3 \xrightarrow{\Delta} CaO + MgO + 2CO_2$

45. Match the ores (column-A) with the metals (column-B) : [Jan. 11, 2019 (I)]

(Column-A) Ores	(Column-B) Metals
(I) Siderite	(A) Zinc
(II) Kaolinite	(B) Copper
(III) Malachite	(C) Iron
(IV) Calamine	(D) Aluminium

(1) (I) - (A); (II) - (B); (III) - (C); (IV) - (D)
(2) (I) - (C); (II) - (D); (III) - (B); (IV) - (A)
(3) (I) - (C); (II) - (D); (III) - (A); (IV) - (B)
(4) (I) - (B); (II) - (C); (III) - (D); (IV) - (A)

46. Hall-Heroult's process is given by: [Jan. 10, 2019 (I)]
(1) $Cu^{2+}(aq) + H_2(g) \rightarrow Cu(s) + 2H^+(aq)$
(2) $Cr_2O_3 + 2Al \rightarrow Al_2O_3 + 2Cr$
(3) $2Al_2O_3 + 3C \rightarrow 4Al + 3CO_2$
(4) $ZnO + C \xrightarrow{\text{Coke, 1673 K}} Zn + CO$

47. The electrolytes usually used in the electroplating of gold and silver, respectively, are: [Jan. 10, 2019 (II)]
(1) $[Au(CN)_2]^-$ and $[Ag(CN)_2]^-$ (2) $[Au(CN)_2]^-$ and $[AgCl_2]^-$
(3) $[Au(OH)_4]^-$ and $[Ag(OH)_2]^-$ (4) $[Au(NH_3)_2]^+$ and $[Ag(CN)_2]^-$

48. The correct statement regarding the given Ellingham diagram is: [Jan. 9, 2019 (II)]

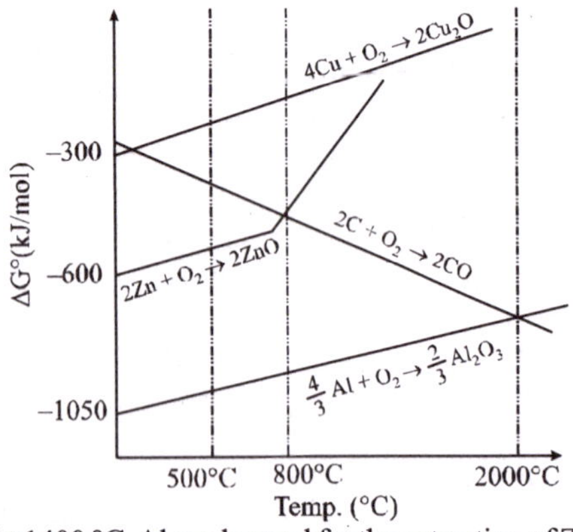

(1) At 1400 °C, Al can be used for the extraction of Zn from ZnO
(2) At 500 °C, coke can be used for the extraction of Zn from ZnO
(3) Coke cannot be used for the extraction of Cu from Cu_2O.
(4) At 800 °C, Cu can be used for the extraction of Zn from ZnO

49. The ore that contains both iron and copper is: [Jan. 9, 2019 (I)]
(1) copper pyrites (2) malachite
(3) dolomite (4) azurite

50. In the extraction of copper from its sulphide ore, metal is finally obtained by the oxidation of cuprous sulphide with : [Online April 16, 2018 (II)]
(1) SO_2 (2) Fe_2O_3 (3) Cu_2O (4) CO

51. In the leaching method, bauxite ore is digested with a concentrated solution of NaOH that produces 'X'. When CO_2 gas is passed through the aqueous solution of 'X', a hydrated compound 'Y' is precipitated. 'X' and 'Y' respectively are. [Online April 15, 2018 (II)]
(1) $Na[Al(OH)_4]$ and $Al_2(CO_3)_3 \cdot xH_2O$
(2) $Al(OH)_3$ and $Al_2O_3 \cdot xH_2O$
(3) $NaAlO_2$ and $Al_2(CO_3)_3 \cdot xH_2O$
(4) $Na[Al(OH)_4]$ and $Al_2O_3 \cdot xH_2O$

52. Which one of the following ores is best concentrated by froth floatation method? [2016; Similar 2004]
(1) Galena (2) Malachite (3) Magnetite (4) Siderite

53. Extraction of copper by smelting uses silica as an additive to remove : [Online April 10, 2016]
(1) Cu_2O (2) FeS (3) FeO (4) Cu_2S

54. In the context of the Hall - Heroult process for the extraction of Al, which of the following statements is false ? [2015]
(1) Al^{3+} is reduced at the cathode to form Al
(2) Na_3AlF_6 serves as the electrolyte
(3) CO and CO_2 are produced in this process
(4) Al_2O_3 is mixed with CaF_2 which lowers the melting point of the mixture and brings conductivity

55. In the isolation of metals, calcination process usually results in : [Online April 10, 2015]
(1) metal hydroxide (2) metal sulphide
(3) metal oxide (4) metal carbonate

56. The form of iron obtained from blast furnace is: [Online April 9, 2014]
(1) Steel (2) Cast Iron
(3) Pig Iron (4) Wrought Iron

57. Calcination is the process in which: [Online April 9, 2013]
(1) ore is heated above its melting point to expel H_2O or CO_2 or SO_2
(2) ore is heated below its melting point to expel volatile impurities
(3) ore is heated above its melting point to remove S, As and Sb as SO_2, As_2O_3 and Sb_2O_3 respectively
(4) ore is heated below its melting point to expel H_2O or CO_2

58. In Goldschmidt aluminothermic process which of the following reducing agents is used: [Online April 22, 2013]
(1) calcium (2) coke (3) Al-powder (4) sodium

59. Which of the oxide groups among the following cannot be reduced by carbon? [Online May 7, 2012]
(1) Cu_2O, SnO_2 (2) CaO, K_2O
(3) PbO, Fe_3O_4 (4) Fe_2O_3, ZnO

60. Extraction of zinc from zinc blende is achieved by
(1) roasting followed by self-reduction [Online May 19, 2012]
(2) electrolytic reduction
(3) roasting followed by reduction with carbon
(4) roasting followed by reduction with another metal

61. The substance used as froth stabilisers in froth-floatation process is [Online May 26, 2012]
(1) Potassium ethyl xanthate (2) Aniline
(3) Sodium cyanide (4) Copper sulphate

62. Which of the following factors is of no significance for roasting sulphide ores to oxides and not subjecting the sulphide ores to carbon reduction directly? [2008]
(1) Metal sulphides are thermodynamically more stable than CS_2
(2) CO_2 is thermodynamically more stable than CS_2
(3) Metal sulphides are less stable than the corresponding oxides
(4) CO_2 is more volatile than CS_2

63. Aluminium is extracted by the electrolysis of [2002]
(1) bauxite
(2) alumina
(3) alumina mixed with molten cryolite
(4) molten cryolite.

64. The metal extracted by leaching with a cyanide is [2002]
(1) Mg (2) Ag (3) Cu (4) Na

NUMERICAL VALUE QUESTIONS (NVQs)

65. Among the following ores Bauxite, Siderite, Cuprite, Calamine, Haematite, Kaolinite, Malachite, Magnetite, Sphalerite,

Limonite, Cryolite, the number of principal ores if (of) iron is_______. [July 27, 2022 (II)]

66. (a) Baryte, (b) Galena, (c) Zinc blende and (d) Copper pyrites. How many of these minerals are sulphide based? [June 24, 2022 (II)]

Topic 2 - Purification and Uses of Metals

MULTIPLE CHOICE QUESTIONS (MCQs)

67. Which of the following methods are not used to refine any metal? [July 27, 2022 (I)]
(A) Liquation (B) Calcination
(C) Electrolysis (D) Leaching
(E) Distillation
Choose the correct answer from the options given below:
(1) B and D only (2) A, B, D and E only
(3) B, D and E only (4) A, C and E only

68. Refining using liquation method is the most suitable for metals with: [July 26, 2022 (I)]
(1) Low melting point
(2) High boiling point
(3) High electrical conductivity
(4) Less tendency to be soluble in melts than impurities

69. Given below are two statements.
Statement I: Pig iron is obtained by heating cast iron with scrap iron.
Statement II: Pig iron has a relatively lower carbon content than that of cast iron.
In the light of the above statements, choose the correct answer from the options given below. [July 25, 2022 (II)]
(1) Both statement I and statement II are correct.
(2) Both statement I and statement II are not correct.
(3) Statement I is correct but statement II is not correct.
(4) Statement I is not correct but statement II is correct.

70. Given below are two statements.
Statement I: During electrolytic refining, blister copper deposits precious metals
Statement II: In the process of obtaining pure copper by electrolysis method, copper blister is used to make the anode.
In the light of the above statements, choose the correct answer from the options given below. [June 25, 2022 (II)]
(1) Both Statement I and Statement II are true.
(2) Both Statement I and Statement II are false.
(3) Statement I is true but Statement II is false.
(4) Statement I is false but Statement II is true.

71. Which of the following chemical reactions represents Hall-Heroult Process? [June 24, 2022 (II)]
(1) $Cr_2O_3 + 2Al \rightarrow Al_2O_3 + 2Cr$
(2) $2Al_2O_3 + 3C \rightarrow 4Al + 3CO_2$
(3) $FeO + CO \rightarrow Fe + CO_2$
(4) $2[Au(CN)_2]^-_{(aq)} + Zn(s) \rightarrow 2Au(s) + [Zn(CN_4)]^{2-}$

72. Given below are two statements, one is labelled as Assertion (A) and the other is labelled as Reason (R).
Assertion (A) : Aluminium is extracted from bauxite by the electrolysis of molten mixture of Al_2O_3 with cryolite.
Reason (R) : The oxidation state of Al in cryolite is +3.
In the light of the above statements, choose the most appropriate answer from the options given below :
(1) (A) is true but (R) is false [Aug. 31, 2021 (I)]
(2) (A) is false but (R) is true.
(3) Both (A) and (R) are correct and (R) is the correct explanation of (A)
(4) Both (A) and (R) are correct but (R) is not the correct explanation of (A)

73. Which refining process is generally used in the purification of low melting metals? [Aug. 27, 2021 (I); Similar Jan. 07, 2020 (II)]
(1) Chromatographic method
(2) Liquation
(3) Electrolysis
(4) Zone refining

74. Match List-I with List-II. [Feb. 26, 2021 (II)]

List - I	List - II
(A) Sodium Carbonate	(I) Deacon
(B) Titanium	(II) Castner-Kellner
(C) Chlorine	(III) van-Arkel
(D) Sodium hydroxide	(IV) Solvay

Choose the correct answer from the options given below:
(1) (A) → (I), (B) → (III), (C) → (IV), (D) → (II)
(2) (A) → (III), (B) → (II), (C) → (I), (D) → (IV)
(3) (A) → (IV), (B) → (I), (C) → (II), (D) → (III)
(4) (A) → (IV), (B) → (III), (C) → (I), (D) → (II)

75. The element that can be refined by distillation is : [Sep. 06, 2020 (II)]
(1) nickel (2) zinc
(3) tin (4) gallium

76. Boron and silicon of very high purity can be obtained through : [Sep. 05, 2020 (II)]
(1) liquation (2) zone refining
(3) vapour phase refining (4) electrolytic refining

77. Cast iron is used for the manufacture of :
(1) wrought iron and pig iron [Sep. 02, 2020 (II)]
(2) pig iron, scrap iron and steel
(3) wrought iron, pig iron and steel
(4) wrought iron and steel

78. The purest form of commercial iron is: [Jan. 07, 2020 (I)]
(1) pig iron (2) wrought iron
(3) cast iron (4) scrap iron and pig iron

79. Match the refining methods (Column-I) with metals (Column-II). [April 10, 2019 (I)]

Column-I (Refining methods)	Column-II (Metals)
(A) Liquation	(i) Zr
(B) Zone Refining	(ii) Ni
(C) Mond Process	(iii) Sn
(D) van Arkel Method	(iv) Ga

(1) (A) – (iii); (B) – (i); (C) – (ii); (D) – (iv)
(2) (A) – (ii); (B) – (iii); (C) – (iv); (D) – (i)
(3) (A) – (iii); (B) – (iv); (C) – (ii); (D) – (i)
(4) (A) – (ii); (B) – (iv); (C) – (i); (D) – (iii)

80. The metal that cannot be obtained by electrolysis of an aqueous solution of its salts is: [2014]
(1) Ag (2) Ca (3) Cu (4) Cr

81. Which method of purification is represented by the following equation ? [2012]

$$Ti(s) + 2I_2(g) \xrightarrow{523\ K} TiI_4(s)$$

$$TiI_4(g) \xrightarrow{1700\ K} Ti(s) + 2I_2(g)$$

(1) Zone refining (2) Cupellation
(3) Polling (4) van Arkel

NUMERICAL VALUE QUESTIONS (NVQs)

82. In the electrolytic refining of blister copper, the total number of main impurities, from the following, removed as anode mud is ________ [Aug. 31, 2021 (II); Similar 2005]

Pb, Sb, Se, Te, Ru, Ag, Au and Pt

Part-B (JEE Advanced)

Topic 1 - Occurrence of Metals and Metallurgical Processes

MULTIPLE CHOICE QUESTIONS (MCQs)

1. Calamine, malachite, magnetite and cryolite, respectively, are [Adv. 2019]
(a) $ZnCO_3$, $CuCO_3.Cu(OH)_2$, Fe_3O_4, Na_3AlF_6
(b) $ZnSO_4$, $Cu(OH)_2$, Fe_3O_4, Na_3AlF_6
(c) $ZnSO_4$, $CuCO_3$, Fe_2O_3, AlF_3
(d) $ZnCO_3$, $CuCO_3$, Fe_2O_3, Na_3AlF_6

2. In the cyanide extraction process of silver from argentite ore, the oxidising and reducing agents used are [2012]
(a) O_2 and CO respectively
(b) O_2 and Zn dust respectively
(c) HNO_3 and Zn dust respectively
(d) HNO_3 and CO respectively

3. Oxidation states of the metal in the minerals haematite and magnetite, respectively, are [2011]
(a) II, III in haematite and III in magnetite
(b) II, III in haematite and II in magnetite
(c) II in haematite and II, III in magnetite
(d) III in haematite and II, III in magnetite

4. Extraction of zinc from zinc blende is achieved by [2007]
(a) electrolytic reduction
(b) roasting followed by reduction with carbon
(c) roasting followed by reduction with another metal
(d) roasting followed by self-reduction

5. Which ore contains both iron and copper? [2005S]
(a) Cuprite (b) Chalcocite
(c) Chalcopyrite (d) Malachite

6. In the process of extraction of gold, [2003S]

$$\text{Roasted gold ore} + CN^- + H_2O \xrightarrow{O_2} [X] + OH^-$$

$$[X] + Zn \longrightarrow [Y] + Au$$

Identify the complexes [X] and [Y]
(a) $X = [Au(CN)_2]^-$, $Y = [Zn(CN)_4]^{2-}$
(b) $X = [Au(CN)_4]^{3-}$, $Y = [Zn(CN)_4]^{2-}$
(c) $X = [Au(CN)_2]^-$, $Y = [Zn(CN)_6]^{4-}$
(d) $X = [Au(CN)_4]^-$, $Y = [Zn(CN)_4]^{2-}$

7. Which of the following process is used in the extractive metallurgy of magnesium? [2002S]
(a) fused salt electrolysis
(b) self reduction
(c) aqueous solution electrolysis
(d) thermite reduction

8. Electrolytic reduction of alumina to aluminium by Hall-Heroult process is carried out [2000S]
(a) in the presence of NaCl
(b) in the presence of fluorite
(c) in the presence of cryolite which forms a melt with lower melting temperature
(d) in the presence of cryolite which forms a melt with higher melting temperature

9. The chemical processes in the production of steel from haematite ore involve [2000S]
(a) reduction
(b) oxidation
(c) reduction followed by oxidation
(d) oxidation followed by reduction

10. The chemical composition of 'slag' formed during the smelting process in the extraction of copper is [2001S]
(a) $Cu_2O + FeS$ (b) $FeSiO_3$
(c) $CuFeS_2$ (d) $Cu_2S + FeO$

11. In the commercial electrochemical process for aluminium extraction the electrolyte used is [1999 - 2 Marks]
(a) $Al(OH)_3$ in NaOH solution
(b) an aqueous solution of $Al_2(SO_4)_3$
(c) a molten mixture of Al_2O_3 and Na_3AlF_6
(d) a molten mixture of AlO(OH) and $Al(OH)_3$

12. In the alumino-thermite process, aluminium acts as [1983 - 1 Mark]
(a) an oxidizing agent (b) a flux
(c) a reducing agent (d) a solder

13. In the metallurgy of iron, when limestone is added to the blast furnace, the calcium ion ends up in [1982 - 1 Mark]
(a) slag (b) gangue
(c) metallic calcium (d) calcium carbonate

14. Copper can be extracted from [1978]
(a) Kupfernical (b) Dolomite
(c) Malachite (d) Galena

4 Fill in the Blanks

15. In extractive metallurgy of zinc, partial fusion of ZnO with coke is called .... and reduction of the ore to the molten metal is called ....
(smelting, calcining, roasting, sintering) [1988 - 1 Mark]

16. In the basic Bessemer process for the manufacture of steel, the lining of the converter is made of .... The slag formed consists of...... [1980]

17. In the thermite process .... is used as reducing agent. [1980]

18. Casseterite is ore of........... [1980]

6 MCQs with One or More than One Correct Answer

19. The treatment of galena with HNO_3 produces a gas that is [Adv. 2022]
(a) paramagnetic (b) bent in geometry
(c) an acidic oxide (d) colorless

20. The correct option(s) related to the extraction of iron from its ore in the blast furnace operating in the temperature range 900 – 1500 K is(are) [Adv. 2022]
(a) Limestone is used to remove silicate impurity.
(b) Pig iron obtained from blast furnace contains about 4% carbon.
(c) Coke (C) converts CO_2 to CO.
(d) Exhaust gases consist of NO_2 and CO.

21. The correct statement(s) related to the metal extraction processes is(are) [Adv. 2021]
(a) A mixture of PbS and PbO undergoes self-reduction to produce Pb and SO_2.
(b) In the extraction process of copper from copper pyrites, silica is added to produce copper silicate.
(c) Partial oxidation of sulphide ore of copper by roasting, followed by self-reduction produces blister copper.
(d) In cyanide process, zinc powder is utilized to precipitate gold from $Na[Au(CN)_2]$.

22. Which among the following statement(s) is(are) true for the extraction of aluminium from bauxite? [Adv. 2020]
(a) Hydrated Al_2O_3 precipitates, when CO_2 is bubbled through a solution of sodium aluminate.
(b) Addition of Na_3AlF_6 lowers the melting point of alumina.
(c) CO_2 is evolved at the anode during electrolysis.
(d) The cathode is a steel vessel with a lining of carbon.

23. The cyanide process of gold extraction involves leaching out gold from its ore with CN^- in the presence of Q in water to form R. Subsequently, R is treated with T to obtain Au and Z. Choose the correct option(s) [Adv. 2019]
(a) Q is O_2 (b) T is Zn
(c) Z is $[Zn(CN)_4]^{2-}$ (d) R is $[Au(CN)_4]^-$

24. Extraction of copper from copper pyrite ($CuFeS_2$) involves [Adv. 2016]
(a) crushing followed by concentration of the ore by froth-flotation
(b) removal of iron as slag
(c) self-reduction step to produce 'blister copper' following evolution of SO_2
(d) refining of 'blister copper' by carbon reduction

25. Upon heating with Cu_2S, the reagent(s) that give copper metal is/are [Adv. 2014]
(a) $CuFeS_2$ (b) CuO (c) Cu_2O (d) $CuSO_4$

26. The carbon-based reduction method is **NOT** used for the extraction of [Adv. 2013]
(a) Tin from SnO_2
(b) Iron from Fe_2O_3
(c) Aluminium from Al_2O_3
(d) Magnesium from $MgCO_3.CaCO_3$

27. Extraction of metal from the ore **cassiterite** involves [2011]
(a) carbon reduction of an oxide ore
(b) self-reduction of a sulphide ore
(c) removal of copper impurity
(d) removal of iron impurity

28. The major role of fluorspar (CaF_2), which is added in small quantities in the electrolytic reduction of alumina dissolved in fused cryolite (Na_3AlF_6), is [1993 - 1 Mark]
(a) as a catalyst
(b) to make the fused mixture very conducting
(c) to lower the temperature of the melt
(d) to decrease the rate of oxidation of carbon at the anode.

29. Of the following, the metals that cannot be obtained by electrolysis of the aqueous solution of their salts are : [1990 - 1 Mark]
(a) Ag (b) Mg (c) Cu (d) Al
(e) Cr

30. In the electrolysis of alumina, cryolite is added to : [1986 - 1 Mark]
(a) lower the melting point of alumina
(b) increase the electrical conductivity
(c) minimise the anode effect
(d) remove impurities from alumina

7 Match the Following

31. Match the anionic species given in **Column I** that are present in the ore(s) given in **Column II**. [Adv. 2015]

Column-I	Column-II
(A) Carbonate	(p) Siderite
(B) Sulphide	(q) Malachite
(C) Hydroxide	(r) Bauxite
(D) Oxide	(s) Calamine
	(t) Argentite

32. Match the conversions in **Column I** with the type(s) of reaction(s) given in **Column II**. [2008 - 6M]

Column I	Column II
(A) $PbS \rightarrow PbO$	(p) roasting
(B) $CaCO_3 \rightarrow CaO$	(q) calcination
(C) $ZnS \rightarrow Zn$	(r) carbon reduction
(D) $Cu_2S \rightarrow Cu$	(s) self reduction

33. Match the extraction processes listed in **Column I** with metals listed in **Column II** : [2006 - 6M]

Column I	Column II
(A) Self reduction	(p) Lead
(B) Carbon reduction	(q) Silver
(C) Complex formation and displacement by metal	(r) Copper
(D) Decomposition of iodide	(s) Boron

34. Match the following choosing one item from column **X** and the appropriate item from column **Y**: [Multiple concept, 1986 - ½×8 = 4 Marks]

X	Y
(*i*) Lewis acid	(a) K electron capture
(*ii*) Philosopher's wool	(b) Zinc ore
(*iii*) Electrophile	(c) HCHO
(*iv*) Preservative	(d) NH_4^+
(*v*) Electron emission	(e) Small proton to neutron ratio
(*vi*) Bronsted acid	(f) SO_3
(*vii*) Black jack	(g) BF_3
(*viii*) X-ray emission	(h) ZnO

35. Match the following, choosing one item from column X and the appropriate item from column Y. [1983 - 2 Marks]

X		Y	
(i)	Al	(a)	Calamine
(ii)	Cu	(b)	Cryolite
(iii)	Mg	(c)	Malachite
(iv)	Zn	(d)	Carnallite

8 Comprehension Based Questions

Copper is the most noble of the first row transition metals and occurs in small deposits in several countries. Ores of copper include chalcanthite ($CuSO_4.5H_2O$), atacamite ($Cu_2Cl(OH)_3$), cuprite (Cu_2O), copper glance (Cu_2S) and malachite ($Cu_2(OH)_2CO_3$). However, 80% of the world copper production comes from the ore chalcopyrite ($CuFeS_2$). The extraction of copper from chalcopyrite involves partial roasting, removal of iron and self-reduction. [2010]

36. Partial roasting of chalcopyrite produces
(a) Cu_2S and FeO (b) Cu_2O and FeO
(c) CuS and Fe_2O_3 (d) Cu_2O and Fe_2O_3

37. Iron is removed from chalcopyrite as
(a) FeO (b) FeS (c) Fe_2O_3 (d) $FeSiO_3$

38. In self-reduction, the reducing species is
(a) S (b) O^{2-} (c) S^{2-} (d) SO_2

10 Subjective Problems

39. Some reactions of two ores, A_1 and A_2 of the metal M are given below. [2004 - 4 Marks]

$$[A_1] \xrightarrow{\text{calcination}} \underset{\text{Black}}{[C]}\downarrow + CO_2 + H_2O$$

$$[A_1] \xrightarrow{\text{KI/HCl}} [D]\downarrow + I_2$$

$$[A_2] \xrightarrow{\text{roasting}} [G]\uparrow + M;$$

$$[G] + K_2Cr_2O_7 \xrightarrow{H^+} \text{Green solution}$$

Identify A_1, A_2, M, C, D, and G, and explain using the required chemical reactions.

40. Write down reactions involved in the extraction of Pb. What is the oxidation number of lead in litharge? [2003 - 2 Marks]

41. When the ore haematite is burnt in air with coke around 2000°C along with lime, the process not only produces steel but also produces a silicate slag that is useful in making building materials such as cement. Discuss the same and show through balanced chemical equations. [1998 - 4 Marks]

42. State with balanced equations what happens when :
(i) Write balanced equations for the extraction of copper from copper pyrites by self-reduction. [1990 - 2 Marks]
(ii) Write balanced equations for the extraction of silver from silver glance by cyanide process. [1988 - 1 Mark]

43. Give the equations for the recovery of lead from Galena by air reduction. [1987 - 1 Mark]

44. What is the actual reducing agent of haematite in blast furnace? [1987 - 1 Mark]

45. Give reasons for the following :
(i) Why is chalcocite roasted and not calcinated during recovery of copper? [1987 - 1 Mark]
(ii) Metals can be recovered from their ores by chemical methods. [1984 - 1 Mark]

46. Give balanced equations for extraction of silver from its sulphide ore [1982 - 2 Marks]

47. Give balanced equations for the extraction of aluminium from bauxite by electrolysis. [1982 - 2 Marks]

48. Write the matching pairs: [1980]

Bleaching agent	Aluminium
Smelling salt	Carbon
Cryolite	Tin
Bell metal	Ammonium carbonate
Fluorspar	Ammonium phosphate
Fertilizer	Calcium
Anthracite	Chlorine

Examples :
Bleaching agent – Chlorine
Smelling salt – Ammonium carbonate

49. (a) Write the chemical equations involved in the extraction of lead from galena by self reduction process.
(b) Match the following extraction processes with the appropriate metals listed below : [1979]

(i)	Silver	(A)	Fused salt electrolysis
(ii)	Calcium	(B)	Carbon reduction
(iii)	Zinc	(C)	Carbon monoxide reduction
(iv)	Iron	(D)	Amalgamation
(v)	Copper	(E)	Self reduction

Topic 2 - Purification and Uses of Metals

6 MCQs with One or More than One Correct Answer

50. The correct statement(s) related to processes involved in the extraction of metals is(are) [Adv. 2023]
(a) Roasting of malachite produces cuprite.
(b) Calcination of calamine produces zincite.
(c) Copper pyrites is heated with silica in a reverberatory furnace to remove iron.
(d) Impure silver is treated with aqueous KCN in the presence of oxygen followed by reduction with zinc metal.

51. The electrochemical extraction of aluminum from bauxite ore involves [Adv. 2022]
(a) the reaction of Al_2O_3 with coke (C) at a temperature > 2500 °C.
(b) the neutralization of aluminate solution by passing CO_2 gas to precipitate hydrated alumina ($Al_2O_3 \cdot 3H_2O$).
(c) the dissolution of Al_2O_3 in hot aqueous NaOH.
(d) the electrolysis of Al_2O_3 mixed with Na_3AlF_6 to give Al and CO_2.

52. Copper is purified by electrolytic refining of blister copper. The correct statement(s) about this process is(are) [Adv. 2015]
(a) Impure Cu strip is used as cathode
(b) Acidified aqueous $CuSO_4$ is used as electrolyte
(c) Pure Cu deposits at cathode
(d) Impurities settle as anode-mud

10 Subjective Problems

53. Magnesium oxide is used for the lining of steel making furnace. [1987 - 1 Mark]

54. High purity metals can be obtained by zone refining method. [1984 - 1 Mark]

Answer Key

Part-A (JEE Main)																	
1	(1)	**10**	(3)	**19**	(1)	**28**	(4)	**37**	(1)	**46**	(3)	**55**	(3)	**64**	(2)	**73**	(2)	**82**	(6)
2	(1)	**11**	(1)	**20**	(1)	**29**	(2)	**38**	(4)	**47**	(1)	**56**	(3)	**65**	(4)	**74**	(4)		
3	(4)	**12**	(1)	**21**	(4)	**30**	(2)	**39**	(1)	**48**	(1)	**57**	(4)	**66**	(3)	**75**	(2)		
4	(4)	**13**	(3)	**22**	(1)	**31**	(2)	**40**	(4)	**49**	(1)	**58**	(3)	**67**	(1)	**76**	(2)		
5	(2)	**14**	(3)	**23**	(1)	**32**	(1)	**41**	(4)	**50**	(3)	**59**	(2)	**68**	(1)	**77**	(4)		
6	(4)	**15**	(1)	**24**	(1)	**33**	(2)	**42**	(2)	**51**	(4)	**60**	(3)	**69**	(2)	**78**	(2)		
7	(2)	**16**	(4)	**25**	(1)	**34**	(3)	**43**	(1)	**52**	(1)	**61**	(2)	**70**	(1)	**79**	(3)		
8	(4)	**17**	(2)	**26**	(3)	**35**	(2)	**44**	(2)	**53**	(3)	**62**	(3)	**71**	(2)	**80**	(2)		
9	(1)	**18**	(4)	**27**	(1)	**36**	(1)	**45**	(2)	**54**	(2)	**63**	(3)	**72**	(4)	**81**	(4)		
Part-B (JEE Advanced)																			
1	(a)	**6**	(a)	**11**	(c)	**20**	(a, b, c)	**25**	(b, c, d)	**30**	(a, b)	**36**	(b)	**52**	(b, c, d)				
2	(b)	**7**	(a)	**12**	(c)	**21**	(a, c, d)	**26**	(c, d)	**31**	A-(p, q, s), B-(t), C-(q, r), D-(r)	**37**	(d)						
3	(d)	**8**	(c)	**13**	(a)	**22**	(a, b, c, d)	**27**	(a, c, d)	**32**	(A) - p; (B) - q; (C) - p, r; (D) - p, s	**38**	(c)						
4	(b)	**9**	(c)	**14**	(c)	**23**	(a, b, c)	**28**	(b, c)	**33**	(A)- p, r; (B) - p, r; (C) - q; (D) - s	**50**	(b, c, d)						
5	(c)	**10**	(b)	**19**	(a, d)	**24**	(a, b, c)	**29**	(b, d)	**35**	(c)	**51**	(b, c, d)						

Chapter-7

The *p*-Block Elements (Group 15, 16, 17 & 18)

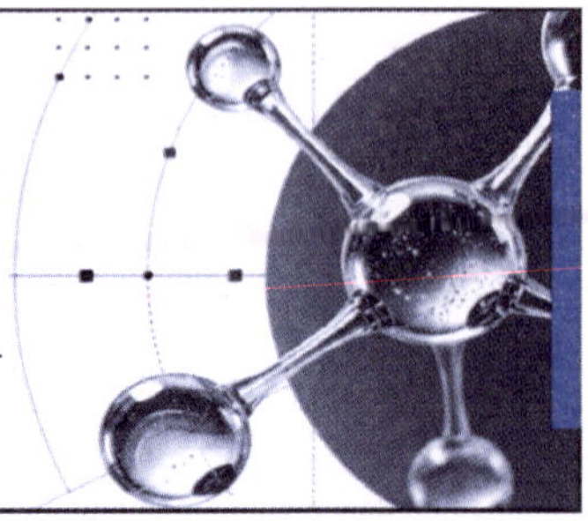

Part-A (JEE Main)

Topic 1- Nitrogen Family

Multiple Choice Questions (MCQs)

1. Find the **correct** statements related to group 15 hydrides. [April 08, 2026 (II)]

   A. Reducing nature increases from NH_3 to BiH_3

   B. Tendency to donate lone pair of electron decreases from NH_3 to BiH_3

   C. The stability of hydrides decreases from NH_3 to BiH_3

   D. HEH bond angle decreases from NH_3 to SbH_3 (E = Elements of group 15)

   Choose the **correct** answer from the options given below:

   (1) A and B only (2) B and C only
   (3) A, B, C and D (4) A, C and D only

2. Correct statements from the following are: [April 05, 2026 (I)]

   A. Nitrogen in oxidation states from +1 to +4 disproportionate in acid medium.

   B. Nitrogen has the ability of form $d\pi$-$p\pi$ multiple bonds with itself and other elements with small size and high electronegativity.

   C. N-N single bond is stronger than P-P single bond.

   D. Nitrogen has highest density in its group due to small size.

   E. The maximum covalency of nitrogen is four since it has only four valence orbitals for bonding.

   Choose the correct answer the options given below:

   (1) B, C, and D only (2) C, D, and E only
   (3) A, C and E only (4) A and E only

3. Regarding the hydrides of group 15 elements EH_3 (E = N, P, As, Sb), select the correct statement from the following:

   A. The stability of hydrides decreases down the group.

   B. The basicity of hydrides decreases down the group.

   C. The reducing character increases down the group.

   D. The boiling point increases down the group.

   Choose the **correct** answer from the options given below: [Jan. 28, 2026 (I)]

   (1) A, B & C only (2) A & D only
   (3) A, B, C & D (4) B & C only

4. Elements X and Y belong to Group 15. The difference between the electronegativity values of 'X' and phosphorus is higher than that of the difference between phosphorus and 'Y'. 'X' and 'Y' are respectively [Jan. 23, 2026 (II)]

   (1) N and As (2) As and Bi
   (3) Bi and N (4) As and Sb

5. Given below are two statements :

   Statement I : Nitrogen forms oxides with +1 to +5 oxidation states due to the formation of $p\pi$ - $p\pi$ bond with oxygen .

   Statement II : Nitrogen does not form halides with +5 oxidation state due to the absence of *d*-orbital in it.

   In the light of given statements, choose the **correct** answer from the options given below. [April 4, 2025 (I)]

   (1) Statement I is true but Statement II is false
   (2) Both Statement I and Statement II are false
   (3) Statement I is false but Statement II is true
   (4) Both Statement I and Statement II are true

6. Given below are two statements [April 3, 2025 (I)]

   Statement I : The N–N single bond is weaker and longer than that of P–P single bond

   Statement II : Compounds of group 15 elements in +3 oxidation states readily undergo disproportionation reactions.

   In the light of above statements, choose the **correct** answer from the options given below.

   (1) Statement I is true but statement II is false
   (2) Both statement I and statement II are false
   (3) Statement I is false but statement II is true
   (4) Both statement I and statement II are true

7. First ionisation enthalpy values of first four group 15 elements are given below. Choose the correct value for the element that is a main component of apatite family : [Jan. 29, 2025 (II)]

   (1) 1012 kJ mol^{-1} (2) 1402 kJ mol^{-1}
   (3) 834 kJ mol^{-1} (4) 947 kJ mol^{-1}

8. The incorrect statements among the following is [Jan. 23, 2025 (I)]

   (1) PH_3 shows lower proton affinity than NH_3.
   (2) PF_3 exists but NF_5 does not.
   (3) NO_2 can dimerise easily.
   (4) SO_2 can act as an oxidizing agent, but not as a reducing agent.

9. The maximum covalency of a non-metallic group 15 element 'E' with weakest E–E bond is : [Jan. 22, 2025 (II)]

   (1) 5 (2) 3 (3) 6 (4) 4

10. Identify the **incorrect** statements about group 15 elements: [April 8, 2024 (II)]

    (A) Dinitrogen is a diatomic gas which acts like an inert gas at room temperature.

    (B) The common oxidation states of these elements are –3, +3 and +5.

    (C) Nitrogen has unique ability to form $p\pi - p\pi$ multiple bonds.

(D) The stability of +5 oxidation states increases down the group.
(E) Nitrogen shows a maximum covalency of 6.
Choose the **correct** answer from the options given below.
(1) (A), (B), (D) only (2) (A), (C), (E) only
(3) (B), (D), (E) only (4) (D) and (E) only

11. Given below are two statements: One is labelled as **Assertion (A)** and the other is labelled as **Reason (R).**
Assertion (A) : PH_3 has lower boiling point than NH_3.
Reason (R) : In liquid state NH_3 molecules are associated through vander Waal's forces, but PH_3 molecules are associatd through hydrogen bonding.
In the light of the above statements, choose the most appropriate answer from the options given below: [Feb. 1, 2024 (I)]
(1) Both (A) and (R) are correct but (R) is not the correct explanation of (A)
(2) (A) is not correct but (R) is correct
(3) Both (A) and (R) are correct and (R) is the correct explanation of (A)
(4) (A) is correct but (R) is not correct.

12. The strongest reducing agent among the following is: [Feb. 1, 2024 (II)]
(1) SbH_3 (2) NH_3 (3) BiH_3 (4) PH_3

13. Choose the correct statements about the hydrides of group 15 elements. [Jan. 30, 2024 (II)]
A. The stability of the hydrides decreases in the order $NH_3 > PH_3 > AsH_3 > SbH_3 > BiH_3$.
B. The reducing ability of the hydride increases in the order $NH_3 < PH_3 < AsH_3 < SbH_3 < BiH_3$.
C. Among the hydrides, NH_3 is strong reducing agent while BiH_3 is mild reducing agent.
D. The basicity of the hydrides increases in the order $NH_3 < PH_3 < AsH_3 < SbH_3 < BiH_3$.
Choose the most appropriate from the options given below:
(1) B and C only (2) C and D only
(3) A and B only (4) A and D only

14. Given below are two statements : [April 11, 2023 (I)]
Statement I : Methane and steam passed over a heated Ni catalyst produces hydrogen gas.
Statement II : Sodium nitrite reacts with NH_4Cl to give H_2O, N_2 and NaCl.
In the light of the above statements, choose the most appropriate answer from the options given below :
(1) Both the Statements I and II are correct
(2) Both the Statements I and II are incorrect
(3) Statement I is incorrect but Statement II is correct
(4) Statement I is correct but Statement II is incorrect

15. Dinitrogen is a robust compound, but reacts at high altitude to form oxides. The oxide of nitrogen that can damage plant leaves and retard photosynthesis is : [July 29, 2022 (II)]
(1) NO (2) NO_3^- (3) NO_2 (4) NO_2^-

16. Dinitrogen and dioxygen, the main constituents of air do not react with each other in atmosphere to form oxides of nitrogen because [July 28, 2022 (II)]
(1) N_2 is unreactive in the condition of atmosphere.
(2) Oxides of nitrogen are unstable.
(3) Reaction between them can occur in the presence of a catalyst.
(4) The reaction is endothermic and require very high temperature.

17. $A \xrightarrow{573\ K} \text{Red phosphorus} \xrightarrow[\text{under pressure}]{\text{heat; 803 K}} B$
Red phosphorus is obtained by heating "A" at 573 K, and can be converted to "B" by heating at 803 K under pressure.
A and B, respectively, are [June 30, 2022 (I)]
(1) β-black phosphorus and white phosphorus.
(2) white phosphorus and β-black phosphorus.
(3) α-black phosphorus and white phosphorus.
(4) white phosphorus and α-black phosphorus.

18. Nitrogen gas is obtained by thermal decomposition of [June 28, 2022 (I)]
(1) $Ba(NO_3)_2$ (2) $Ba(N_3)_2$
(3) $NaNO_2$ (4) $NaNO_3$

19. Given below are two statements : [June 28, 2022 (I)]
Statement I : The pentavalent oxide of group-15 element E_2O_5 is less acidic than trivalent oxide. E_2O_3 of the same element.
Statement II : The acidic character of trivalent oxide of group 15 elements E_2O_3 decreases down the group.
In light of the above statements choose **most appropriate** answer from the options given below:
(1) Both Statement I and Statement II are true.
(2) Both Statement I and Statement II are false.
(3) Statement I true but Statement II is false.
(4) Statement I is false but Statement II is true.

20. The gas produced by treating an aqueous solution of ammonium chloride with sodium nitrite is [June 27, 2022 (II)]
(1) NH_3 (2) N_2 (3) N_2O (4) Cl_2

21. PCl_5 is well known but NCl_5 is not. Because. [June 24, 2022 (II)]
(1) nitrogen is less reactive than phosphorous.
(2) nitrogen doesn't have d-orbitals in its valence shell.
(3) catenation tendency is weaker in nitrogen than phosphorous.
(4) size of phosphorous is larger than nitrogen.

22. The most stable trihalide of nitrogen is: [June 24, 2022 (I)]
(1) NF_3 (2) NCl_3 (3) NBr_3 (4) NI_3

23. Which one of the following is formed (mainly) when red phosphorus is heated in a sealed tube at 803 K? [Aug. 27, 2021 (II)]
(1) White phosphorus (2) Yellow phosphorus
(3) β-Black phosphorus (4) α-Black phosphorus

24. Which one of the following group-15 hydride is the strongest reducing agent? [July 22, 2021 (II)]
(1) AsH_3 (2) BiH_3 (3) PH_3 (4) SbH_3

25. A group 15 element, which is a metal and forms a hydride with strongest reducing power among group 15 hydrides. The element is [March 16, 2021 (I)]
(1) As (2) P (3) Bi (4) Sb

26. The correct statement with respect to dinitrogen is: [Sep. 06, 2020 (I)]
(1) N_2 is paramagnetic in nature.
(2) it can combine with dioxygen at 25 °C.
(3) liquid dinitrogen is not used in cryosurgery.
(4) it can be used as an inert diluent for reactive chemicals.

27. Reaction of ammonia with excess Cl_2 gives : [Sep. 05, 2020 (II)]
(1) NH_4Cl and N_2 (2) NH_4Cl and HCl
(3) NCl_3 and NH_4Cl (4) NCl_3 and HCl

28. The number of pentagons in C_{60} and trigons (triangles) in white phosphorous, respectively, are: [April 10, 2019 (II)]
(1) 20 and 3 (2) 12 and 4
(3) 12 and 3 (4) 20 and 4

29. The compound that **does not** produce nitrogen gas by the thermal decomposition is : [2018]
(1) $Ba(N_3)_2$ (2) $(NH_4)_2Cr_2O_7$
(3) NH_4NO_2 (4) $(NH_4)_2SO_4$

30. For per gram of reactant, the maximum quantity of N_2 gas is produced in which of the following thermal decomposition reactions? (Given: Atomic wt. : Cr = 52 u, Ba = 137 u). [Online April 15, 2018 (II)]
(1) $Ba(N_3)_2(s) \rightarrow Ba(s) + 3N_2(g)$
(2) $(NH_4)_2Cr_2O_7(s) \rightarrow N_2(g) + 4H_2O(g) + Cr_2O_3(s)$
(3) $2NH_3(g) \rightarrow N_2(g) + 3H_2(g)$
(4) $2NH_4NO_3(s) \rightarrow 2N_2(g) + 4H_2O(g) + O_2(g)$

31. A metal 'M' reacts with nitrogen gas to afford 'M_3N'. 'M_3N' on heating at high temperature gives back 'M' and on reaction with water produces a gas 'B'. Gas 'B' reacts with aqueous solution of $CuSO_4$ to form a deep blue compound. 'M' and 'B' respectively are : [Online April 8, 2017]
(1) Li and NH_3 (2) Ba and N_2
(3) Na and NH_3 (4) Al and N_2

32. The molecule having smallest bond angle is : [2012]
(1) NCl_3 (2) $AsCl_3$
(3) $SbCl_3$ (4) PCl_3

33. Regular use of the following fertilizers increases the acidity of soil? [2007]
(1) Ammonium sulphate (2) Potassium nitrate
(3) Urea (4) Superphosphate of lime.

34. Among the following mixtures, dipole-dipole as the major interaction, is present in [2006]
(1) KCl and water
(2) benzene and carbon tetrachloride
(3) benzene and ethanol
(4) acetonitrile and acetone

35. Which one of the following substances has the highest proton affinity? [2003]
(1) H_2S (2) NH_3 (3) PH_3 (4) H_2O

36. In case of nitrogen, NCl_3 is possible but not NCl_5 while in case of phosphorus, PCl_3 as well as PCl_5 are possible. It is due to
(1) availability of vacant *d* orbitals in P but not in N [2002]
(2) lower electronegativity of P than N
(3) lower tendency of H-bond formation in P than N
(4) occurrence of P in solid while N in gaseous state at room temperature.

NUMERICAL VALUE QUESTIONS (NVQs)

37. A group 15 element forms $d\pi - d\pi$ bond with transition metals. It also forms hydride, which is a strongest base among the hydrides of other group members that form $d\pi - d\pi$ bond. The atomic number of the element is _____. [Jan. 28, 2025 (II)]

Topic 2 - Oxygen Family

MULTIPLE CHOICE QUESTIONS (MCQs)

38. **Statement I :** The correct order of electronegativity of fluorine, oxygen nitrogen is F > O > N.
Statement II : The oxidation state of oxygen in OF_2 is +2 and in Na_2O is –2.
In the light of the above statements, choose the **correct** answer from the options given below [April 05, 2026 (I)]
(1) Both Statement I and Statement II are true
(2) Both Statement I and Statement II are false
(3) Statement I is true but Statement II is false
(4) Statement I is false but Statement II is true

39. Given below are two statements : [April 04, 2026 (I)]
Statement I : The covalency of oxygen is generally two but it can exceed upto four. The oxidation state of oxygen in SO_2 is –2 and in OF_2 it is +2.
Statement II : The anomalous behaviour of oxygen when compared to the other elements of group 16 is due to its small size and high electronegativity.
In the light of the above statements, choose the **correct** answer from the options given below :
(1) Both Statement I and Statement II are true
(2) Both Statement I and Statement II are false
(3) Statement I is true but Statement II is false
(4) Statement I is false but Statement II is true

40. Given below are two statements : [April 8, 2025 (II)]
Statement I : H_2Se is more acidic than H_2Te.
Statement II : H_2Se has higher bond enthalpy for dissociation than H_2Te.
In the light of the above statements, choose the **correct** answer from the options given below.
(1) Both statement I and Statement II are false.
(2) Both statement I and Statement II are true.
(3) Statement I is true but Statement II is false.
(4) Statement I is false but Statement II is true.

41. The large difference between the melting and boiling points of oxygen and sulphur may be explained on the basis of [Jan. 24, 2025 (I)]
(1) Atomic size (2) Atomicity
(3) Electronegativity (4) Electron gain enthalpy

42. Given below are two statements: one is labelled as **Assertion (A)** and the other is labelled as **Reason (R).** [April 9, 2024 (I)]
Assertion (A) : Both rhombic and monoclinic sulphur exist as S_8 while oxygen exists as O_2.
Reason (R) : Oxygen forms $p\pi$-$p\pi$ multiple bonds with itself and other elements having small size and high electronegativity like C, N, which is not possible for sulphur.
In the light of the above statements, choose the **most appropriate** answer from the options given below :
(1) Both **(A)** and **(R)** are correct and **(R)** is the correct explanation of **(A).**
(2) Both **(A)** and **(R)** are correct but **(R)** is not the correct explanation of **(A).**
(3) **(A)** is correct but **(R)** is not correct.
(4) **(A)** is not correct but **(R)** is correct.

43. Given below are two statements: One is labelled as Assertion A and the other is labelled as Reason R:

Assertion A : H_2Te is more acidic than H_2S.

Reason R : Bond dissociation enthalpy of H_2Te is lower than H_2S. [Jan. 30, 2024 (II)]

In the light of the above statements, choose the most appropriate from the options given below:

(1) Both A and R are true but R is NOT the correct explanation of A.
(2) Both A and R are true and R is the correct explanation of A.
(3) A is false but R is true.
(4) A is true but R is false.

44. Anomalous behavior of oxygen is due to its [Jan. 29, 2024 (II)]

(1) small size and high electronegativity
(2) small size and low electronegativity
(3) large size and high electronegativity
(4) large size and low electronegativity

45. Choose the correct statements from the following [Jan. 31, 2024 (II)]

A. All group 16 elements form oxides of general formula EO_2 and EO_3 where E = S, Se, Te and Po. Both the types of oxides are acidic in nature.
B. TeO_2 is an oxidising agent while SO_2 is reducing in nature.
C. The reducing property decreases from H_2S to H_2Te down the group.
D. The ozone molecule contains five lone pairs of electrons.

Choose the correct answer from the options given below:

(1) A and D only (2) B and C only
(3) C and D only (4) A and B only

46. Given below are two statements: [Jan. 27, 2024 (II)]

Statement I : Oxygen being the first member of group 16 exhibits only –2 oxidation state.

Statement II : Down the group 16 stability of +4 oxidation state decreases and +6 oxidation state increases.

In the light of the above statements, choose the most appropriate answer from the options given below:

(1) Statement I is correct but Statement II is incorrect
(2) Both Statement I and Statement II are correct
(3) Both Statement I and Statement II are incorrect
(4) Statement I is incorrect but Statement II is correct

47. Bond dissociation energy of E–H bond of the "H_2E" hydrides of group 16 elements (given below), follows order. [Jan. 30, 2023 (II)]

(A) O (B) S (C) Se (D) Te

(1) A > B > C > D (2) A > B > D > C
(3) B > A > C > D (4) D > C > B > A

48. The correct order of melting points of hydrides of group 16 elements is : [June 26, 2022 (I)]

(1) $H_2S < H_2Se < H_2Te < H_2O$
(2) $H_2O < H_2S < H_2Se < H_2Te$
(3) $H_2S < H_2Te < H_2Se < H_2O$
(4) $H_2Se < H_2S < H_2Te < H_2O$

49. Reaction of an inorganic sulphite X with dilute H_2SO_4 generates compound Y. Reaction of Y with NaOH gives X. Further, the reaction of X with Y and water affords compound Z. Y and Z, respectively, are: [Sep. 06, 2020 (II)]

(1) SO_2 and Na_2SO_3 (2) SO_3 and $NaHSO_3$
(3) SO_2 and $NaHSO_3$ (4) S and Na_2SO_3

50. If the boiling point of H_2O is 373 K, the boiling point of H_2S will be : [Sep. 03, 2020 (I)]

(1) less than 300 K
(2) equal to 373 K
(3) more than 373 K
(4) greater than 300 K but less than 373 K

51. Identify the incorrect statement: [Online April 10, 2016]

(1) The S–S–S bond angles in the S_8 and S_6 rings are the same.
(2) Rhombic and monoclinic sulphur have S_8 molecules.
(3) S_2 is paramagnetic like oxygen
(4) S_8 ring has a crown shape.

Numerical Value Questions (NVQs)

52. Number of oxygen atoms present in chemical formula of fuming sulphuric acid is _______. [April 9, 2024 (II)]

53. Among the following allotropic forms of sulphur, the number of allotropic forms, which will show paramagnetism is ________. [Feb. 24, 2021 (II)]

(A) α-sulphur (B) β-sulphur (C) S_2-form

Topic 3 - Halogen Family

Multiple Choice Questions (MCQs)

54. Given below are two statements : [April 02, 2026 (I)]

Statement I : Oxidising power of halogens decreases in the order $F_2 > Cl_2 > Br_2 > I_2$, which is the basis of "Layer test".

Statement II : "Layer test" to identify Br_2 and I_2 in aqueous solution involves the oxidation of bromide or iodide into Br_2 or I_2 respectively with Cl_2, which is a type of displacement redox reaction.

In the light of the above statements, choose the correct answer from the options given below :

(1) Both Statement I and Statement II are true.
(2) Both Statement I and Statement II are false.
(3) Statement I is true but Statement II is false.
(4) Statement I is false but Statement II is true.

55. Given below are two statements : [Jan. 28, 2026 (II)]

Statement I : The increasing order of boiling point of hydrogen halides is HCl < HBr < HI < HF.

Statement II : The increasing order of melting point of hydrogen halides is HCl < HBr < HF < HI.

In the light of the above statements, choose the correct answer from the options given below :

(1) Both Statement I and Statement II are true.
(2) Statement I is true but Statement II is false.
(3) Both Statement I and Statement II are false.
(4) Statement I is false but Statement II is true.

56. Given below are two statements : [Jan. 22, 2026 (I)]

Statement I : The halogen that makes longest bond with hydrogen in HX, has the smallest covalent radius in its group.

Statement II : A group 15 element's hydride EH_3 has the lowest boiling point among corresponding hydrides of other group 15 elements. The maximum covalency of that element E is 4.

In the light of the above statements, choose the correct answer from the options given below.

(1) Both Statement I and Statement II are true.
(2) Statement I is false but Statement II is true.
(3) Both Statement I and Statement II are false.
(4) Statement I is true but Statement II is false.

57. The property/properties that show irregularity in first four elements of group-17 is/are : [April 2, 2025 (I)]
(A) Covalent radius (B) Electron affinity
(C) Ionic radius (D) First ionization energy
Choose the **correct** answer from the options given below:
(1) B and D only (2) A and C only
(3) B only (4) A, B, C and D

58. The F^- ions make the enamel on teeth much harder by converting hydroxyapatite (the enamel on the surface of teeth) into much harder fluoroapatite having the formula. [April 9, 2024 (I)]
(1) $[3(Ca_3(PO_4)_2).CaF_2]$ (2) $[3(Ca_2(PO_4)_2).Ca(OH)_2]$
(3) $[3(Ca_2(PO_4)_3).CaF_2]$ (4) $[3(Ca_3(PO_4)_2).Ca(OH)_2]$

59. Number of elements from the following that CANNOT form compounds with valencies which match with their respective group valencies is ______.
B, C, N, S, O, F, P, Al, Si [April 4, 2024 (I)]
(1) 7 (2) 5 (3) 6 (4) 3

60. Identify the incorrect pair from the following : [Jan. 29, 2024 (I)]
(1) Carnallite - $KCl . MgCl_2 . 6H_2O$
(2) Cryolite - Na_3AlF_6
(3) Fluoroapatite - 3 $Ca_3(PO_4)_2.CaF_2$
(4) Fluorspar - BF_3

61. Given below are two statements [Jan. 29, 2024 (II)]
Statement I : Fluorine has most negative electron gain enthalpy in its group.
Statement II : Oxygen has least negative electron gain enthalpy in its group.
In the light of the above statements, choose the **most appropriate** from the options given below:
(1) Statement I is false but Statement II is true
(2) Both Statement I and Statement II are true
(3) Both Statement I and Statement II are false
(4) Statement I is true but Statement II is false

62. Element not showing variable oxidation state is :
(1) Bromine (2) Iodine [Jan. 27, 2024 (I)]
(3) Chlorine (4) Fluorine

63. The correct group of halide ions which can be oxidised by oxygen in acidic medium is [April 13, 2023 (II)]
(1) Br^- only (2) Cl^-, Br^- and I^- only
(3) Br^- and I^- only (4) I^- only

64. For electron gain enthalpies of the elements denoted as $\Delta_{eg}H$, the incorrect option is : [Feb. 01, 2023 (II)]
(1) $\Delta_{eg}H(Cl) < \Delta_{eg}H(F)$ (2) $\Delta_{eg}H(Se) < \Delta_{eg}H(S)$
(3) $\Delta_{eg}H(Te) < \Delta_{eg}H(Po)$ (4) $\Delta_{eg}H(I) < \Delta_{eg}H(At)$

65. The bond dissociation energy is highest for [Jan. 29, 2023 (I)]
(1) Cl_2 (2) I_2 (3) Br_2 (4) F_2

66. Given below are two statements: one is labelled as **Assertion A** and the other is labelled as **Reason R**.
Assertion A : Flourine forms one oxoacid.
Reason R : Flourine has smallest size amongst all halogens and is highly electronegative.
In the light of the above statements, choose the **most appropriate** answer from the options given below. [June 27, 2022 (II)]
(1) Both A and R are correct and R is the correct explanation of A.
(2) Both A and R are correct but R is NOT the correct explanation of A.
(3) A is correct but R is not correct.
(4) A is not correct but R is correct.

67. The correct order of bond dissociation enthalpy of halogens is: [Feb. 25, 2021 (II)]
(1) $Cl_2 > Br_2 > F_2 > I_2$ (2) $F_2 > Cl_2 > Br_2 > I_2$
(3) $Cl_2 > F_2 > Br_2 > I_2$ (4) $I_2 > Br_2 > Cl_2 > F_2$

68. Arrange the following bonds according to their average bond energies in descending order: C – Cl, C – Br, C – F, C – I
(1) C – F > C – Cl > C – Br > C – I [Jan. 08, 2020 (II)]
(2) C – Br > C – I > C – Cl > C – F
(3) C – I > C – Br > C – Cl > C – F
(4) C – Cl > C – Br > C – I > C – F

69. HF has highest boiling point among hydrogen halides, because it has: [April 9, 2019 (II)]
(1) strongest van der Waals' interactions
(2) lowest ionic character
(3) strongest hydrogen bonding
(4) lowest dissociation enthalpy

70. The non–metal that does not exhibit positive oxidation state is : [Online April 9, 2016]
(1) Chlorine (2) Iodine
(3) Fluorine (4) Oxygen

71. Aqueous solution of which salt will not contain ions with the electronic configuration $1s^2 2s^2 2p^6 3s^2 3p^6$?
(1) NaF (2) KBr [Online April 10, 2016]
(3) NaCl (4) CaI_2

72. Which among the following is the most reactive ? [2015]
(1) I_2 (2) ICl (3) Cl_2 (4) Br_2

73. Shapes of certain interhalogen compounds are stated below. Which one of them is not correctly stated?
(1) IF_7 : pentagonal bipyramid [Online April 11, 2014]
(2) BrF_5 : trigonal bipyramid
(3) BrF_3 : planar T-shaped
(4) ICl_3 : planar dimeric

74. Electron gain enthalpy with negative sign of fluorine is less than that of chlorine due to : [Online April 9, 2013]
(1) High ionization enthalpy of fluorine
(2) Smaller size of chlorine atom
(3) Smaller size of fluorine atom
(4) Bigger size of 2*p* orbital of fluorine

75. The correct order of the thermal stability of hydrogen halides (H–X) is [2005]
(1) HI > HCl < HF > HBr (2) HCl < HF > HBr < HI
(3) HF > HCl > HBr > HI (4) HI < HBr > HCl < HF

76. Which among the following factors is the most important in making fluorine, the strongest oxidizing halogen? [2004]
(1) Hydration enthalpy (2) Ionization enthalpy
(3) Electron affinity (4) Bond dissociation energy

NUMERICAL VALUE QUESTIONS (NVQs)

77. The sum of lone pairs present on the central atom of the interhalogen IF_5 and IF_7 is ______ [April 10, 2023 (I)]

78. The number of interhalogens from the following having square pyramidal structure is : [July 28, 2022 (I)]
ClF_3, IF_7, BrF_5, BrF_3, I_2Cl_6, IF_5, ClF, ClF_5

79. Amongst the following the number of oxide(s) which are paramagnetic in nature is [July 27, 2022 (I)]

$Na_2O, KO_2, NO_2, N_2O, ClO_2, NO, SO_2, Cl_2O$

Topic 4 - Noble Gases

Multiple Choice Questions (MCQs)

80. Which statements are **NOT TRUE** about XeO_2F_2? [Jan. 23, 2026 (II)]

A. It has a see-saw shape

B. Xe has 5 electron pairs in its valence shell in XeO_2F_2.

C. The O–Xe–O bond angle is close to 180°.

D. The F–Xe–F bond angle is close to 180°.

E. Xe has 16 valence electrons in XeO_2F_2.

Choose the **correct** answer from the options given below.

(1) B, C and E only (2) B and D only
(3) A and D only (4) B, D and E only

81. Given below are two statements [Jan. 31, 2024 (I)]

Statement I : Noble gases have very high boiling points.

Statement II : Noble gases are monoatomic gases. They are held together by strong dispersion forces. Because of this they are liquefied at very low temperature. Hence, they have very high boiling points.

In the light of the above statements, choose the **correct answer** from the options given below:

(1) Statement I is false but Statement II is true
(2) Both Statement I and Statement II are true
(3) Statement I is true but Statement II is false
(4) Both Statement I and Statement II are false

82. Inert gases have positive electron gain enthalpy. Its correct order is [Jan. 25, 2023 (I)]

(1) $Xe < Kr < Ne < He$ (2) $He < Ne < Kr < Xe$
(3) $He < Xe < Kr < Ne$ (4) $He < Kr < Xe < Ne$

83. Match List-I with List-II: [July 25, 2022 (II)]

List-I (Molecule)	List-II (hybridization; shape)
(A) XeO_3	(I) sp^3d; linear
(B) XeF_2	(II) sp^3; pyramidal
(C) $XeOF_4$	(III) sp^3d^3;distorted octahedral
(D) XeF_6	(IV) sp^3d^2; sqaure pyramidal

Choose the correct answer from the options given below:

(1) A-II, B-I, C-IV, D-III (2) A-II, B-IV, C-III, D-I
(3) A-IV, B-II, C-III, D-I (4) A-IV, B-II, C-I, D-III

84. Match List–I with List–II : [Aug. 27, 2021 (I)]

List–I (Species)	List–II (Number of lone pairs of electrons on the central atom)
(A) XeF_2	(i) 0
(B) XeO_2F_2	(ii) 1
(C) XeO_3F_2	(iii) 2
(D) XeF_4	(iv) 3

Choose the **most appropriate** answer from the options given below :

(1) (A)–(iv), (B)–(i), (C)–(ii), (D)–(iii)
(2) (A)–(iii), (B)–(iv), (C)–(ii), (D)–(i)
(3) (A)–(iii), (B)–(ii), (C)–(iv), (D)–(i)
(4) (A)–(iv), (B)–(ii), (C)–(i), (D)–(iii)

85. The noble gas that does NOT occur in the atmosphere is : [April 10, 2019 (II)]

(1) He (2) Kr (3) Ne (4) Ra

86. In XeO_3F_2, the number of bond pair(s), π-bond(s) and lone pair(s) on Xe atom respectively are: [Online April 15, 2018 (II)]

(1) 5, 3, 0 (2) 5, 2, 0 (3) 4, 2, 2 (4) 4, 4, 0

87. Which intermolecular force is most responsible in allowing xenon gas to liquefy? [Online April 9, 2016]

(1) Instantaneous dipole–induced dipole
(2) Ion–dipole
(3) Ionic
(4) Dipole – dipole

88. The geometry of $XeOF_4$ by VSEPR theory is : [Online April 10, 2015]

(1) pentagonal planar (2) octahedral
(3) square pyramidal (4) trigonal bipyramidal

89. Which one has the highest boiling point ? [2015]

(1) Kr (2) Xe (3) He (4) Ne

90. Which has trigonal bipyramidal shape ? [Online April 23, 2013]

(1) $XeOF_4$ (2) XeO_3 (3) XeO_3F_2 (4) $XeOF_2$

91. Trigonal bipyramidal geometry is shown by : [Online April 9, 2013]

(1) XeO_2F_2 (2) XeO_3F_2
(3) $FXeOSO_2F$ (4) $[XeF_8]^{2-}$

92. The compound of xenon with zero dipole moment is [Online May 19, 2012]

(1) XeO_3 (2) XeF_4 (3) $XeOF_4$ (4) XeO_2

93. Which of the following has maximum number of lone pairs associated with Xe ? [2011RS]

(1) XeF_4 (2) XeF_6 (3) XeF_2 (4) XeO_3

94. In XeF_2, XeF_4, XeF_6 the number of lone pairs on Xe are respectively [2002]

(1) 2, 3, 1 (2) 1, 2, 3 (3) 4, 1, 2 (4) 3, 2, 1.

95. Which one of the following statement regarding helium is **incorrect** ? [2004]

(1) It is used to produce and sustain powerful superconducting magnets
(2) It is used as a cryogenic agent for carrying out experiments at low temperatures
(3) It is used to fill gas balloons instead of hydrogen because it is lighter and non-inflammable
(4) It is used in gas-cooled nuclear reactors

Numerical Value Questions (NVQs)

96. The number of species from the following carrying a single lone pair on central atom Xenon is ____:

$XeF_5^+, XeO_3, XeO_2F_2, XeF_5^-, XeO_3F_2, XeOF_4, XeF_4$ [April 08, 2023 (II)]

97. The sum of number of lone pairs of electrons present on the central atoms of [July 25, 2022 (II)]

XeO_3, $XeOF_4$ and XeF_6 is ______.

Part-B (JEE Advanced)

Topic 1- Nitrogen Family

1 MCQs with One Correct Answer

1. The correct order of ONO bond angle in the given species is [Adv. 2026]
(a) $NO_2^+ < NO_2 < NO_3^- < NO_2^-$
(b) $NO_2^- < NO_3^- < NO_2 < NO_2^+$
(c) $NO_3^- < NO_2^- < NO_2 < NO_2^+$
(d) $NO_2^- < NO_3^- < NO_2^+ < NO_2$

2. The heating of NH_4NO_2 at 60 – 70 °C and NH_4NO_3 at 200 – 250°C is associated with the formation of nitrogen containing compounds X and Y, respectively. X and Y, respectively, are [Adv. 2025]
(a) N_2 and N_2O (b) NH_3 and NO_2
(c) NO and N_2O (d) N_2 and NH_3

3. The species formed on fluorination of phosphorus pentachloride in a polar organic solvent are [Adv. 2024]
(a) $[PF_4]^+[PF_6]^-$ and $[PCl_4]^+[PF_6]^-$
(b) $[PCl_4]^+[PCl_4F_2]^-$ and $[PCl_4]^+[PF_6]^-$
(c) PF_3 and PCl_3
(d) PF_5 and PCl_3

4. At room temperature, disproportionation of an aqueous solution of *in situ* generated nitrous acid (HNO_2) gives the species [Adv. 2024]
(a) H_3O^+, NO_3^- and NO (b) H_3O^+, NO_3^- and NO_2
(c) H_3O^+, NO^- and NO_2 (d) H_3O^+, NO_3^- and N_2O

5. The order of the oxidation state of the phosphorus atom in H_3PO_2, H_3PO_4, H_3PO_3 and $H_4P_2O_6$ is [Adv. 2017]
(a) $H_3PO_3 > H_3PO_2 > H_3PO_4 > H_4P_2O_6$
(b) $H_3PO_4 > H_3PO_2 > H_3PO_3 > H_4P_2O_6$
(c) $H_3PO_4 > H_4P_2O_6 > H_3PO_3 > H_3PO_2$
(d) $H_3PO_2 > H_3PO_3 > H_4P_2O_6 > H_3PO_4$

6. The product formed in the reaction of $SOCl_2$ with white phosphorous is [Adv. 2014]
(a) PCl_3 (b) SO_2Cl_2 (c) SCl_2 (d) $POCl_3$

7. Concentrated nitric acid, upon long standing, turns yellow brown due to the formation of [Adv. 2013]
(a) NO (b) NO_2 (c) N_2O (d) N_2O_4

8. The reaction of white phosphorus with aqueous NaOH gives phosphine along with another phosphorus containing compound. The reaction type; the oxidation states of phosphorus in phosphine and the other product are respectively [2012]
(a) redox reaction; – 3 and – 5
(b) redox reaction; + 3 and + 5
(c) disproportionation reaction; – 3 and + 5
(d) disproportionation reaction; – 3 and + 3

9. Which ordering of compounds is according to the decreasing order of the oxidation state of nitrogen? [2012]
(a) HNO_3, NO, NH_4Cl, N_2
(b) HNO_3, NO, N_2, NH_4Cl
(c) HNO_3, NH_4Cl, NO, N_2
(d) NO, HNO_3, NH_4Cl, N_2

10. Extra pure N_2 can be obtained by heating [2011]
(a) NH_3 with CuO (b) NH_4NO_3
(c) $(NH_4)_2Cr_2O_7$ (d) $Ba(N_3)_2$

11. The reaction of P_4 with X leads selectively to P_4O_6. The X is [2009]
(a) Dry O_2
(b) A mixture of O_2 and N_2
(c) Moist O_2
(d) O_2 in the presence of aqueous NaOH

12. The percentage of π-character in the orbitals forming P – P bonds in P_4 is [2007]
(a) 25 (b) 33 (c) 50 (d) 75

13. When PbO_2 reacts with conc. HNO_3, the gas evolved is
(a) NO_2 (b) O_2 (c) N_2 (d) N_2O [2005S]

14. Blue liquid which is obtained on reacting equimolar amounts of two gases at –30°C is? [2005S]
(a) N_2O (b) N_2O_3 (c) N_2O_4 (d) N_2O_5

15. Which is the most thermodynamically stable allotropic form of phosphorus? [2005S]
(a) red (b) white (c) black (d) yellow

16. For H_3PO_3 and H_3PO_4, the correct choice is: [2003S]
(a) H_3PO_3 is dibasic and reducing
(b) H_3PO_3 is dibasic and non-reducing
(c) H_3PO_4 is tribasic and reducing
(d) H_3PO_3 is tribasic and non-reducing

17. Ammonia can be dried by [2000S]
(a) conc. H_2SO_4 (b) P_4O_{10}
(c) CaO (d) anhydrous $CaCl_2$

18. The number of P – O – P bonds in cyclic metaphosphoric acid is [2000S]
(a) zero (b) two (c) three (d) four

19. In compounds of type ECl_3, where E = B, P, As or Bi, the angles Cl – E– Cl for different E are in the order [1999 - 2 Marks]
(a) B > P = As = Bi (b) B > P > As > Bi
(c) B < P = As = Bi (d) B < P < As < Bi

20. On heating ammonium dichromate, the gas evolved is
(a) oxygen (b) ammonia [1999 - 2 Marks]
(c) nitrous oxide (d) nitrogen

21. One mole of calcium phosphide on reaction with excess water gives [1999 - 2 Marks]
(a) one mole of phosphine
(b) two moles of phosphoric acid
(c) two moles of phosphine
(d) one mole of phosphorus pentoxide

22. Amongst the trihalides of nitrogen which one is least basic?
(a) NF_3 (b) NCl_3 [1987 - 1 Mark]
(c) NBr_3 (d) NI_3

23. Which of the following oxides of nitrogen is a coloured gas? [1987 - 1 Mark]
(a) N_2O (b) NO (c) N_2O_5 (d) NO_2

24. The bonds present in N_2O_5 are : [1986 - 1 Mark]
(a) only ionic (b) covalent and coordinate
(c) only covalent (d) covalent and ionic

25. Nitrogen dioxide cannot be obtained by heating : [1985 - 1 Mark]
(a) KNO_3 (b) $Pb(NO_3)_2$ (c) $Cu(NO_3)_2$ (d) $AgNO_3$

26. Ammonia can be dried by [1980]
(a) Conc. H_2SO_4 (b) P_2O_5
(c) Anhydrous $CuSO_4$ (d) none

27. The reddish brown coloured gas formed when nitric oxide is oxidised by air is [1979]
(a) N_2O_5 (b) N_2O_4 (c) NO_2 (d) N_2O_3

2 Integer Value Answer

28. The total number of lone pairs of electrons in N_2O_3 is [Adv. 2015]

29. Among the following, the number of compounds than can react with PCl_5 to give $POCl_3$ is [2011]
O_2, CO_2, SO_2, H_2O, H_2SO_4, P_4O_{10}

3 Numeric / New Stem Based Questions

30. Dissolving 1.24 g of white phosphorous in boiling NaOH solution in an inert atmosphere gives a gas Q. The amount of $CuSO_4$ (in g) required to completely consume the gas Q is ________. [Adv. 2022]
[Given: Atomic mass of H = 1, O = 16, Na = 23, P = 31, S = 32, Cu = 63]

4 Fill in the Blanks

31. In P_4O_{10}, the number of oxygen atoms bonded to each phosphorus atom is .... [1992 - 1 Mark]
32. The basicity of phosphorous acid (H_3PO_3) is .... [1990 - 1 Mark]
33. .... phosphorus is reactive because of its highly strained tetrahedral structure. [1987 - 1 Mark]
34. The lowest possible oxidation state of nitrogen is [1980]

5 True / False

35. Nitric oxide, though an odd electron molecule, is diamagnetic in liquid state. [1991 - 1 Mark]
36. The H-N-H bond angle in NH_3 is greater than the H-As-H bond angle is AsH_3. [1984 - 1 Mark]
37. Red phosphorus is less volatile than white phosphorus because the former has a tetrahedral structure. [1982 - 1 Mark]

6 MCQs with One or More than One Correct Answer

38. The compound(s) with P-H bond(s) is(are) [Adv. 2025]
(a) H_3PO_4 (b) H_3PO_3
(c) $H_4P_2O_7$ (d) H_3PO_2
39. The correct statement(s) related to oxoacids of phosphorous is(are) [Adv. 2021]
(a) Upon heating, H_3PO_3 undergoes disproportionation reaction to produce H_3PO_4 and PH_3.
(b) While H_3PO_3 can act as reducing agent, H_3PO_4 cannot.
(c) H_3PO_3 is a monobasic acid.
(d) The H atom of P–H bond in H_3PO_3 is not ionizable in water.
40. Based on the compounds of group 15 elements, the correct statement(s) is (are) [Adv. 2018]
(a) Bi_2O_5 is more basic than N_2O_5
(b) NF_3 is more covalent than BiF_3
(c) PH_3 boils at lower temperature than NH_3
(d) The N–N single bond is stronger than the P–P single bond
41. The compound(s) which generate(s) N_2 gas upon thermal decomposition below 300 °C is (are) [Adv. 2018]
(a) NH_4NO_3 (b) $(NH_4)_2Cr_2O_7$
(c) $Ba(N_3)_2$ (d) Mg_3N_2
42. The nitrogen containing compound produced in the reaction of HNO_3 with P_4O_{10} [Adv. 2016]
(a) can also be prepared by reaction of P_4 and HNO_3
(b) is diamagnetic
(c) contains one N-N bond
(d) reacts with Na metal producing a brown gas
43. The nitrogen oxide(s) that contain(s) N-N bond(s) is(are)
(a) N_2O (b) N_2O_3 (c) N_2O_4 (d) N_2O_5 [2009]
44. A solution of colourless salt H on boiling with excess NaOH produces a non-flammable gas. The gas evolution ceases after sometime. Upon addition of Zn dust to the same solution, the gas evolution restarts. The colourless salt (s) H is (are) [2008]
(a) NH_4NO_3 (b) NH_4NO_2 (c) NH_4Cl (d) $(NH_4)_2SO_4$
45. Ammonia, on reaction with hypochlorite anion, can form [1999 - 3 Marks]
(a) NO (b) NH_4Cl (c) N_2H_4 (d) HNO_2
46. White phosphorus (P_4) has [1998 - 2 Marks]
(a) six P-P single bonds
(b) four P-P single bonds
(c) four lone pairs of electrons
(d) PPP angle of 60°
47. Sodium nitrate decomposes above 800° C to give [1998 - 2 Marks]
(a) N_2 (b) O_2
(c) NO_2 (d) Na_2O
48. Nitrogen(I) oxide is produced by: [1989 - 1 Mark]
(a) thermal decomposition of ammonium nitrate
(b) disproportionation of N_2O_4
(c) thermal decomposition of ammonium nitrite
(d) interaction of hydroxylamine and nitrous acid.

7 Match the Following

49. Match the reactions (in the given stoichiometry of the reactants) in List-I with one of their products given in List-II and choose the correct option. [Adv. 2023]

List-I	List-II
(P) $P_2O_3 + 3H_2O \rightarrow$	(1) $P(O)(OCH_3)Cl_2$
(Q) $P_4 + 3NaOH + 3H_2O \rightarrow$	(2) H_3PO_3
(R) $PCl_5 + CH_3COOH \rightarrow$	(3) PH_3
(S) $H_3PO_2 + 2H_2O + 4AgNO_3 \rightarrow$	(4) $POCl_3$
	(5) H_3PO_4

(a) $P \rightarrow 2; Q \rightarrow 3; R \rightarrow 1; S \rightarrow 5$
(b) $P \rightarrow 3; Q \rightarrow 5; R \rightarrow 4; S \rightarrow 2$
(c) $P \rightarrow 5; Q \rightarrow 2; R \rightarrow 1; S \rightarrow 3$
(d) $P \rightarrow 2; Q \rightarrow 3; R \rightarrow 4; S \rightarrow 5$
50. Match the following, choosing one item from column X and the appropriate item from column Y. [1983 - 2 Marks]

X	Y
(i) Haber	(a) Activation energy
(ii) Graham	(b) Diffusion of gases
(iii) Arrhenius	(c) Octet rule
(iv) Lewis	(d) Ammonia synthesis

8 Comprehension Based Questions

There are some deposits of nitrates and phosphates in earth's crust. Nitrates are more soluble in water. Nitrates are difficult to reduce under the laboratory conditions but microbes do it easily. Ammonia forms large number of complexes with transition metal ions. Hybridization easily explains the ease of sigma donation capability of NH_3 and PH_3. Phosphine is a flammable gas and is prepared from white phosphorous. [2008]
51. Among the following, the correct statement is
(a) Phosphates have no biological significance in humans
(b) Between nitrates and phosphates, phosphates are less abundant in earth's crust
(c) Between nitrates and phosphates, nitrates are less abundant in earth's crust
(d) Oxidation of nitrates is possible in soil
52. Among the following, the correct statement is
(a) Between NH_3 and PH_3, NH_3 is a better electron donor because the lone pair of electrons occupies spherical *s*-orbital and is less directional
(b) Between NH_3 and PH_3, PH_3 is a better electron donor because the lone pair of electrons occupies sp^3 orbital and is more directional
(c) Between NH_3 and PH_3, NH_3 is a better electron donor because the lone pair of electrons occupies sp^3 orbital and is more directional
(d) Between NH_3 and PH_3, PH_3 is a better electron donor because the lone pair of electrons occupies spherical *s*-orbital and is less directional

53. White phosphorus on reaction with NaOH gives PH_3 as one of the products. This is a
(a) dimerization reaction
(b) disproportionation reaction
(c) condensation reaction
(d) precipitation reaction

9 Assertion and Reason/Statement Type Questions

Each question contains **STATEMENT-1 (Assertion)** and **STATEMENT-2 (Reason)**. Each question has 4 choices (a), (b), (c) and (d) out of which ONLY ONE is correct. Mark your answer as
(a) If both Statement -1 and Statement -2 are correct, and Statement -2 is the correct explanation of the Statement -2.
(b) If both Statement -1 and Statement -2 are correct, but Statement -2 is not the correct explanation of the Statement -1.
(c) If Statement -1 is correct but Statement -2 is incorrect.
(d) If Statement -1 is incorrect but Statement -2 is correct.

54. **Statement-1 :** Although PF_5, PCl_5 and PBr_5 are known, the pentahalides of nitrogen have not been observed
Statement-2 : Phosphorus has lower electronegativity than nitrogen. **[1994 - 2 Marks]**

10 Subjective Problems

55. Write the structures of $(CH_3)_3N$ and $(Me_3Si)_3N$. Are they isostructural? Justify your answer. **[2005 - 2 Marks]**

56. How many grams of CaO are required to neutralize 852 g of P_4O_{10}? Draw structure of P_4O_{10} molecule. **[2005 - 2 Marks]**

57. Give reason(s) why elemental nitrogen exists as a diatomic molecule whereas elemental phosphorus as a tetraatomic molecule. **[2000 - 2 Marks]**

58. The Haber process can be represented by the following scheme;

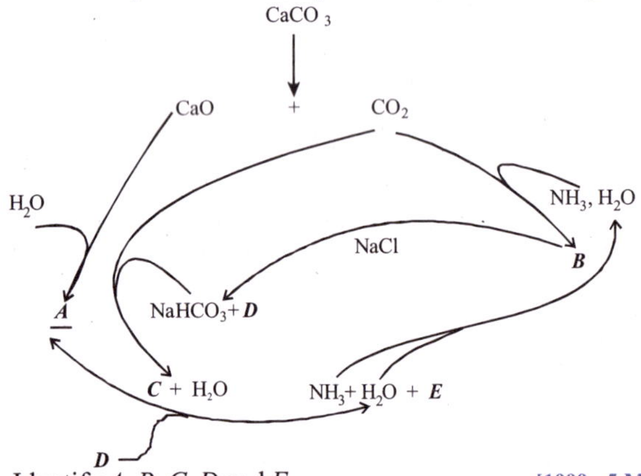

Identify *A*, *B*, *C*, *D* and *E*. **[1999 - 5 Marks]**

59. Reaction of phosphoric acid with $Ca_5(PO_4)_3F$ yields a fertilizer "triple superphosphate". Represent the same through balanced chemical equation. **[1998 - 2 Marks]**

60. A soluble compound of a poisonous element *M*, when heated with Zn/H_2SO_4, gives a colourless and extremely poisonous gaseous compound *N*, which on passing through a heated tube gives a silvery mirror of element *M*. Identify *M* and *N*. **[1997 - 2 Marks]**

61. Draw the structure of P_4O_{10} and identify the number of single and double P—O bonds. **[1996 - 3 Marks]**

62. Complete and balance the following chemical reactions :
(*i*) Red phosphorus is reacted with iodine in presence of water.
$P + I_2 + H_2O \rightarrow .... + ...$ **[1992 - 1 Mark]**
(*ii*) Anhydrous potassium nitrate is heated with excess of metallic potassium. **[1992 - 1 Mark]**
$KNO_3(s) + K(s) \rightarrow .... + ...$

63. Arrange the following in : **[1991 - 1 Mark]**
Increasing order of extent of hydrolysis :
$CCl_4, MgCl_2, AlCl_3, PCl_5, SiCl_4$

64. Write down the balanced equations for the reactions when:
(*i*) a mixture of potassium chlorate, oxalic acid and sulphuric acid is heated; **[1985 - 1 Mark]**
(*ii*) ammonium sulphate is heated with a mixture of nitric oxide and nitrogen dioxide. **[1985 - 1 Mark]**

65. Write down the resonance structures of nitrous oxide. **[1985 - 2 Marks]**

OR

Write the two resonance structures of N_2O that satisfy the octet rule. **[1990 - 1 Mark]**

66. State with balanced equations what happens when :
(*i*) $P_4O_{10} + PCl_5 \rightarrow$ **[1998 - 1 Mark]**
(*ii*) Phosphorus is treated with concentrated nitric acid. **[1997 - 1 Mark]**

OR

Manufacture of phosphoric acid from phosphorus. **[1997 - 1 Mark]**

(*iii*) Elemental phosphorus reacts with conc. HNO_3 to give phosphoric acid. **[1991 - 1 Mark]**
(*iv*) Nitrogen is obtained in the reaction of aqueous ammonia with potassium permanganate. **[1991 - 1 Mark]**
(*v*) Sodium nitrite is produced by absorbing the oxides of nitrogen in aqueous solution of washing soda. **[1991 - 1 Mark]**
(*vi*) Aqueous solution of sodium nitrate is heated with zinc dust and caustic soda solution. **[1990 - 1 Mark]**
(*vii*) Write balanced equations for the preparation of ammonium sulphate from gypsum, ammonia and carbon dioxide. **[1990 - 1 Mark]**
(*viii*) Write balanced equations for the preparation of phosphine from CaO and white phosphorus. **[1990 - 2 Marks]**
(*ix*) Hypophosphorous acid is heated. **[1989 - 1 Mark]**
(*x*) Phosphorus reacts with nitric acid to give equimolar ratio of nitric oxide and nitrogen dioxide. **[1988 - 1 Mark]**
(*xi*) Dilute nitric acid is slowly reacted with metallic tin. **[1987 - 1 Mark]**
(*xii*) White phosphorous (P_4) is boiled with a strong solution of sodium hydroxide in an inert atmosphere. **[1982/87 - 1 Mark]**

67. Give reasons for the following :
(*i*) The experimentally determined N – F bond length in NF_3 is greater than the sum of the single covalent bond radii of N and F. **[1995 - 2 Marks]**
(*ii*) Ammonium chloride is acidic in liquid ammonia solvent. **[1991 - 1 Mark]**
(*iii*) Phosphine has lower boiling point than ammonia. **[1989 - 1 Mark]**
(*iv*) H_3PO_3 is a dibasic acid. **[1989 - 1 Mark]**
(*v*) Orthophosphorus acid is not tribasic acid. **[1987 - 1 Mark]**
(*vi*) A bottle of liquor ammonia should be cooled before opening the stopper. **[1983 - 1 Mark]**
(*vii*) Orthophosphoric acid, H_3PO_4, is tribasic, but phosphorous acid, H_3PO_3, is dibasic. **[1982 - 1 Mark]**

68. Give structural formula for the following :
(*i*) Phosphorous acid, H_3PO_3 **[1981 - 1 Mark]**
(*ii*) Pyrophosphoric acid, $H_4P_2O_7$ **[1981 - 1 Mark]**

Topic 2 - Oxygen Family

1 MCQs with One Correct Answer

69. In the following reaction sequence in aqueous solution, the species X, Y and Z, respectively, are [Adv. 2016]

$$S_2O_3^{2-} \xrightarrow{Ag^+} \underset{\text{clear solution}}{X} \xrightarrow{Ag^+} \underset{\text{white precipitate}}{Y} \xrightarrow{\text{with time}} \underset{\text{black precipitate}}{Z}$$

(a) $[Ag(S_2O_3)_2]^{3-}$, $Ag_2S_2O_3$, Ag_2S
(b) $[Ag(S_2O_3)_3]^{5-}$, Ag_2SO_3, Ag_2S
(c) $[Ag(SO_3)_2]^{3-}$, $Ag_2S_2O_3$, Ag
(d) $[Ag(SO_3)_3]^{3-}$, Ag_2SO_4, Ag

70. Aqueous solution of $Na_2S_2O_3$ on reaction with Cl_2 gives – [2008]
(a) $Na_2S_4O_6$ (b) $NaHSO_4$
(c) NaCl (d) NaOH

71. Which of the following is not oxidized by O_3? [2005S]
(a) KI (b) $FeSO_4$ (c) $KMnO_4$ (d) K_2MnO_4

72. The acid having O – O bond is [2004S]
(a) $H_2S_2O_3$ (b) $H_2S_2O_6$ (c) $H_2S_2O_8$ (d) $H_2S_4O_6$

73. The number of S – S bonds in sulphur trioxide trimer (S_3O_9) is [2001S]
(a) three (b) two (c) one (d) zero

74. Sodium thiosulphate is prepared by [1996 - 1 Mark]
(a) reducing Na_2SO_4 solution with H_2S
(b) boiling Na_2SO_3 solution with S in alkaline medium
(c) neutralising $H_2S_2O_3$ solution with NaOH
(d) boiling Na_2SO_3 solution with S in acidic medium

75. Hydrolysis of one mole of peroxodisulphuric acid produces
(a) two moles of sulphuric acid [1996 - 1 Mark]
(b) two moles of peroxomonosulphuric acid
(c) one mole of sulphuric acid and one mole of peroxomonosulphuric acid
(d) one mole of sulphuric acid, one mole of peroxomonosulphuric acid and one mole of hydrogen peroxide.

76. H_2SO_4 cannot be used to prepare HBr from NaBr as it : [1995S]
(a) reacts slowly with NaBr (b) oxidises HBr
(c) reduces HBr (d) disproportionates HBr

77. The compound which gives off oxygen on moderate heating is : [1986 - 1 Mark]
(a) cupric oxide (b) mercuric oxide
(c) zinc oxide (d) aluminium oxide

78. A gas that cannot be collected over water is : [1985 - 1 Mark]
(a) N_2 (b) O_2 (c) SO_2 (d) PH_3

79. Which of the following is coloured [1980]
(a) NO (b) N_2O (c) SO_3 (d) None

2 Integer Value Answer

80. The total number of compounds having at least one bridging oxo group among the molecules given below is______.
N_2O_3, N_2O_5, P_4O_6, P_4O_7, $H_4P_2O_5$, $H_5P_3O_{10}$, $H_2S_2O_3$, $H_2S_2O_5$ [Adv. 2018]

3 Numeric / New Stem Based Questions

81. The amount of water produced (in g) in the oxidation of 1 mole of rhombic sulphur by conc.HNO_3 to a compound with the highest oxidation state of sulphur is ____ (Given data : Molar mass of water = 18g mol^{-1}) [Adv. 2019]

4 Fill in the Blanks

82. The lead chamber process involves oxidation of SO_2 by atomic oxygen under the influence of .... as catalyst. [1992 - 1 Mark]

6 MCQs with One or More than One Correct Answer

83. The compound(s) having peroxide linkage is(are) [Adv. 2024]
(a) $H_2S_2O_7$ (b) $H_2S_2O_8$
(c) $H_2S_2O_5$ (d) H_2SO_5

84. The correct statement(s) about O_3 is(are)
(a) O—O bond lengths are equal [Adv. 2013-II]
(b) Thermal decomposition of O_3 is endothermic
(c) O_3 is diamagnetic in nature
(d) O_3 has a bent structure

85. Which of the following halides react(s) with $AgNO_3$(aq) to give a precipitate that dissolves in $Na_2S_2O_3$(aq)?
(a) HCl (b) HF [2012]
(c) HBr (d) HI

8 Comprehension Based Questions

Passage-1

Upon heating $KClO_3$ in the presence of catalytic amount of MnO_2, a gas W is formed. Excess amount of W reacts with white phosphorus to give X. The reaction of X with pure HNO_3 gives Y and Z.

86. W and X are, respectively [Adv. 2017]
(a) O_3 and P_4O_6 (b) O_2 and P_4O_6
(c) O_2 and P_4O_{10} (d) O_3 and P_4O_{10}

87. Y and Z are, respectively
(a) N_2O_3 and H_3PO_4 (b) N_2O_5 and HPO_3
(c) N_2O_4 and HPO_3 (d) N_2O_4 and H_3PO_3

Passage-2

The reactions of Cl_2 gas with cold-dilute and hot-concentrated NaOH in water give sodium salts of two (different) oxoacids of chlorine, P and Q, respectively. The Cl_2 gas reacts with SO_2 gas, in presence of charcoal, to give a product R. R reacts with white phosphorus to give a compound S. On hydrolysis, S gives an oxoacid of phosphorus, T. [Adv. 2013]

88. P and Q, respectively, are the sodium salts of
(a) Hypochlorus and chloric acids
(b) Hypochlorus and chlorus acids
(c) Chloric and perchloric acids
(d) Chloric and hypochlorus acids

89. R, S and T respectively, are
(a) SO_2Cl_2, PCl_5 and H_3PO_4 (b) SO_2Cl_2, PCl_3 and H_3PO_3
(c) $SOCl_2$, PCl_3 and H_3PO_2 (d) $SOCl_2$, PCl_5 and H_3PO_4

10 Subjective Problems

90. $(B) \xleftarrow{NaBr + MnO_2} (A) \xrightarrow{Conc.HNO_3} (C)$

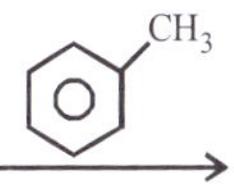

$\longrightarrow$ (D) (explosive product)

Identify the missing compounds. Give the equation from A to B and A to C. [2005 - 4 Marks]

91. Identify the following: [2003 - 4 Marks]

$$Na_2CO_3 \xrightarrow{SO_2} A \xrightarrow{Na_2CO_3} B \xrightarrow[\Delta]{\text{Elemental S}} C \xrightarrow{I_2} D$$

Also mention the oxidation state of S in all the compounds.

92. In the contact process for industrial manufacture of sulphuric acid, some amount of sulphuric acid is used as a starting material. Explain briefly. What is the catalyst used in the oxidation of SO_2? [1999 - 4 Marks]

93. In the following equation, [1999 - 6 Marks]
$A + 2B + H_2O \rightarrow C + 2D$
($A = HNO_2$, $B = H_2SO_3$, $C = NH_2OH$). Identify D. Draw the structures of A, B, C and D.

94. Thionyl chloride can be synthesized by chlorinating SO_2 using PCl_5. Thionyl chloride is used to prepare anhydrous ferric chloride starting from its hexahydrated salt. Alternatively, the anhydrous ferric chloride can also be prepared from its hexahydrated salt by treating with 2, 2 – dimethoxypropane. Discuss all this using balanced chemical equations. [1998 - 6 Marks]

95. $PbS \xrightarrow[\text{air}]{\text{heat in}} A + PbS \xrightarrow{B} Pb + SO_2$; Identify *A* and *B*. [1991 - 2 Marks]

96. Write the two resonance structures of ozone which satisfy the octet rule. [1991 - 1 Mark]

97. Mention the products formed in the following :
Sulphur dioxide gas, water vapour and air are passed over heated sodium chloride. [1986 - 1 Mark]

98. Arrange the following in :
CO_2, N_2O_5, SiO_2, SO_3 in the order of increasing acidic character. [1988 - 1 Mark]

99. What happens when :
(*i*) hydrogen sulphide is bubled through an aqueous solution of sulphur dioxide. [1985 - 1 Mark]
(*ii*) tin is treated with concentrated nitric acid. [1985 - 1 Mark]
(*iii*) Pb_3O_4 is treated with nitric acid. [1985 - 1 Mark]

100. Show with equations, how the following compound is prepared (equations need not be balanced) : sodium thiosulphate from sodium sulphite. [1982 - 1 Mark]

101. State with balanced equations what happens when :
Sulphur is precipitated in the reaction of hydrogen sulphide with sodium bisulphite solution. [1991 - 1 Mark]

102. Give reasons for the following :
(*i*) Sulphur dioxide is a more powerful reducing agent in an alkaline medium than in acidic medium. [1992 - 1 Mark]
(*ii*) Valency of oxygen is generally two whereas sulphur shows valency of two, four and six. [1988 - 1 Mark]
(*iii*) Sulphur melts to a clear mobile liquid at 119°C, but on further heating above 160°C, it becomes viscous. [1981 - 1 Mark]

Topic 3 - Halogen Family

1 MCQs with One Correct Answer

103. The complete hydrolysis of ICl, ClF_3 and BrF_5, respectively, gives [Adv. 2025]
(a) IO^-, ClO_2^- and BrO_3^- (b) IO_3^-, ClO_2^- and BrO_3^-
(c) IO^-, ClO^- and BrO_2^- (d) IO_3^-, ClO_4^- and BrO_2^-

104. The reaction of $HClO_3$ with HCl gives a paramagnetic gas, which upon reaction with O_3 produces [Adv. 2022]
(a) Cl_2O (b) ClO_2 (c) Cl_2O_6 (d) Cl_2O_7

105. A colorless aqueous solution contains nitrates of two metals, X and Y. When it was added to an aqueous solution of NaCl, a white precipitate was formed. This precipitate was found to be partly soluble in hot water to give a residue P and a solution Q. The residue P was soluble in aq. NH_3 and also in excess sodium thiosulfate. The hot solution Q gave a yellow precipitate with KI. The metals X and Y, respectively, are [Adv. 2020]
(a) Ag and Pb (b) Ag and Cd
(c) Cd and Pb (d) Cd and Zn

106. The products obtained when chlorine gas reacts with cold and dilute aqueous NaOH are : [2017]
(a) ClO^- and ClO_3^- (b) ClO_2^- and ClO_3^-
(c) Cl^- and ClO^- (d) Cl^- and ClO_2^-

107. Which one of the following species is not a pseudohalide? [1997 - 1 Mark]
(a) CNO^- (b) $RCOO^-$ (c) OCN^- (d) NNN^-

108. KF combines with HF to form KHF_2. The compound contains the species. [1996 - 1 Mark]
(a) K^+, F^- and H^+ (b) K^+, F^- and HF
(c) K^+ and $[HF_2]^-$ (d) $[KHF]^+$ and F^-

109. Which of the following statements is correct for $CsBr_3$?
(a) It is a covalent compound. [1996 - 1 Mark]
(b) It contains Cs^{3+} and Br^- ions.
(c) It contains Cs^+ and Br_3^- ions
(d) It contains Cs^+, and Br^- and lattice Br_2 molecule

110. Bromine can be liberated from potassium bromide solution by the action of [1987 - 1 Mark]
(a) Iodine solution (b) Chlorine water
(c) Sodium chloride (d) Potassium iodide

111. Chlorine acts as a bleaching agent only in presence of [1983 - 1 Mark]
(a) dry air (b) moisture
(c) sunlight (d) pure oxygen

112. HBr and HI reduce sulphuric acid, HCl can reduce $KMnO_4$ and HF can reduce [1981 - 1 Mark]
(a) H_2SO_4 (b) $KMnO_4$
(c) $K_2Cr_2O_7$ (d) none of the above

113. A solution of KBr is treated with each of the following. Which one would liberate bromine [1980]
(a) Cl_2 (b) HI (c) I_2 (d) SO_2

114. Which of the following is most stable to heat [1980]
(a) HCl (b) HOCl (c) HBr (d) HI

2 Integer Value Answer

115. Ozonolysis of ClO_2 produces an oxide of chlorine. The average oxidation state of chlorine in this oxide is ___. [Adv. 2021]

116. Reaction of Br_2 with Na_2CO_3 in aqueous solution gives sodium bromide and sodium bromate with evolution of CO_2 gas. The number of sodium bromide molecules involved in the balanced chemical equation is [2011]

4 Fill in the Blanks

117. Solubility of iodine in water is greatly increased by the addition of iodide ions because of the formation of .... [1994 - 1 Mark]

118. .... acid gives hypo .... ion. [1988 - 1 Mark]
(hydrobromic, hypobromous, perbromic, bromide, bromite, perbromate)

119. The increase in the solubility of iodine in an aqueous solution of potassium iodide is due to the formation of .... [1982 - 1 Mark]

120. .... is a weak acid. (HF, HCl, HI) [1981 - 1 Mark]

121. Iodine reacts with hot NaOH solution. The products are NaI and .... [1980]

5 True / False

122. HBr is a stronger acid than HI because of hydrogen bonding. [1993 - 1 Mark]

123. In aqueous solution, chlorine is a stronger oxidizing agent than fluorine [1984 - 1 Mark]

124. Dil. HCl oxidizes metallic Fe to Fe^{2+}. [1983 - 1 Mark]

6 MCQs with One or More than One Correct Answer

125. Correct statement(s) about the compounds X, Y and Z is(are)
MnO_2 + Conc. HCl $\longrightarrow$ $MnCl_2$ + X + H_2O
(greenish yellow gas)

$$NH_3 + \underset{(excess)}{X} \longrightarrow Y + HCl$$

$$X + \underset{(excess)}{F_2} \xrightarrow{573\ K} Z$$ [Adv. 2026]

(a) X is used for sterilizing drinking water.
(b) Y has a planar structure.
(c) Z is used in the enrichment of ^{235}U.
(d) Y is a stronger Lewis base than ammonia.

126. With respect to hypochlorite, chlorate and perchlorate ions, choose the correct statement(s). [Adv. 2020]
(a) The hypochlorite ion is the strongest conjugate base.
(b) The molecular shape of only chlorate ion is influenced by the lone pair of electrons of Cl.
(c) The hypochlorite and chlorate ions disproportionate to give rise to identical set of ions.
(d) The hypochlorite ion oxidizes the sulfite ion.

127. The colour of the X_2 molecules of group 17 elements changes gradually from yellow to violet down the group. This is due to [Adv. 2017]
(a) The physical state of X_2 at room temperature changes from gas to solid down the group
(b) Decrease in ionization energy down the group
(c) Decrease in π^* - σ^* gap down the group
(d) Decrease in HOMO-LUMO gap down the group

128. The correct statement(s) about the oxoacids, $HClO_4$ and HClO, is (are) [Adv. 2017]
(a) The central atom in both $HClO_4$ and HClO is sp^3 hybridized
(b) $HClO_4$ is more acidic than HClO because of the resonance stabilization of its anion
(c) $HClO_4$ is formed in the reaction between Cl_2 and H_2O
(d) The conjugate base of $HClO_4$ is weaker base than H_2O

129. The compounds used as refrigerant are [1990 - 1 Mark]
(a) NH_3 (b) CCl_4 (c) CF_4 (d) CF_2Cl_2
(e) CH_2F_2

8 Comprehension Based Questions

Bleaching powder and bleach solution are produced on a large scale and used in several household products. The effectiveness of bleach solution is often measured by iodometry. [2012 - II]

130. Bleaching powder contains a salt of an oxoacid as one of its components. The anhydride of that oxoacid is
(a) Cl_2O (b) Cl_2O_7 (c) ClO_2 (d) Cl_2O_6

131. 25 mL of household solution was mixed with 30 mL of 0.50 M KI and 10 mL of 4N acetic acid. In the titration of the liberated iodine, 48 mL of 0.25 N $Na_2S_2O_3$ was used to reach the end point. The molarity of the household bleach solution is
(a) 0.48 M (b) 0.96 M (c) 0.24 M (d) 0.024 M

9 Assertion and Reason/Statement Type Questions

Each question contains STATEMENT-1 (Assertion) and STATEMENT-2 (Reason). Each question has 4 choices (a), (b), (c) and (d) out of which ONLY ONE is correct. Mark your answer as
(a) If both Statement -1 and Statement -2 are correct, and Statement -2 is the correct explanation of the Statement -2.
(b) If both Statement -1 and Statement -2 are correct, but Statement -2 is not the correct explanation of the Statement -1.
(c) If Statement -1 is correct but Statement -2 is incorrect.
(d) If Statement -1 is incorrect but Statement -2 is correct.

132. Statement-1: F atom has less electron affinity than Cl atom.
Statement-2: Additional electrons are repelled more effectively by $3p$ electrons in Cl atom than by $2p$ electrons in F atom. [1998 - 2 Marks]

10 Subjective Problems

133. Give an example of oxidation of one halide by another halogen. Explain the feasibility of the reaction. [2000 - 2 Marks]

134. Complete the following chemical equations :
(a) $KI + Cl_2 \rightarrow$ (b) $KClO_3 + I_2 \rightarrow$
Justify the formation of the products in the above reactions. [1996 - 2 Marks]

135. Gradual addition of KI solution to $Bi(NO_3)_3$ solution initially produces a dark brown precipitate which dissolves in excess of KI to give a clear yellow solution. Write chemical equations for the above reactions. [1996 - 2 Marks]

136. Complete and balance the following chemical reactions :
(iii) $NH_3 + NaOCl \rightarrow$ .... + ... [1993 - 1 Mark]

137. Mention the products formed in the following :
(*i*) Chlorine gas is bubbled through a solution of ferrous bromide. [1986 - 1 Mark]
(*ii*) Iodine is added to a solution of stannous chloride. [1986 - 1 Mark]

138. Arrange the following in :
(*i*) HOCl, $HOClO_2$, $HOClO_3$, HOClO in increasing order of thermal stability. [1988 - 1 Mark]
(*ii*) increasing bond strength [1986 - 1 Mark]
HCl, HBr, HF, HI

139. State with balanced equations what happens when :
(*i*) Sodium iodate is added to a solution of sodium bisulphite. [1990 - 1 Marks]
(*ii*) Sodium chlorate reacts with sulphur dioxide in dilute sulphuric acid medium. [1989 - 1 Mark]
(*iii*) Sodium bromate reacts with fluorine in presence of alkali. [1989 - 1 Mark]
(*iv*) Iodate ion reacts with bisulphite ion to liberate iodine. [1988 - 1 Mark]
(*v*) Sodium iodate is treated with sodium bisulphite solution. [1982 - 1 Mark]

140. Give reasons for the following :
(*i*) Bond dissociation energy of F_2 is less than that of Cl_2. [1992 - 1 Mark]
(*ii*) Fluorine cannot be prepared from fluorides by chemical oxidation. [1985 - 1 Mark]
(*iii*) Anhydrous HCl is a bad conductor of electricity but aqueous HCl is a good conductor; [1985 - 1 Mark]
(*iv*) In the preparation of hydrogen iodide from alkali iodides, phosphoric acid is preferred to sulphuric acid [1982 - 1 Mark]

141. Write balanced equation involved in the preparation of
(i) Anhydrous aluminium chloride from alumina.
(ii) Bleaching powder from slaked lime.
(iii) Tin metal from cassiterite
(iv) Chlorine from sodium chloride. [1979]

142. Account for the following. Limit your answer to two sentences
(*i*) Hydrogen bromide cannot be prepared by action of concentrated sulphuric acid or sodium bromide.
(*ii*) When a blue litmus paper is dipped into a solution of hypochlorous acid, it first turns red and then later gets decolourised. [1979]

Topic 4 - Noble Gases

1 MCQs with One Correct Answer

143. Under ambient conditions, the total number of gases released as products in the final step of the reaction scheme shown below is **[Adv. 2014]**

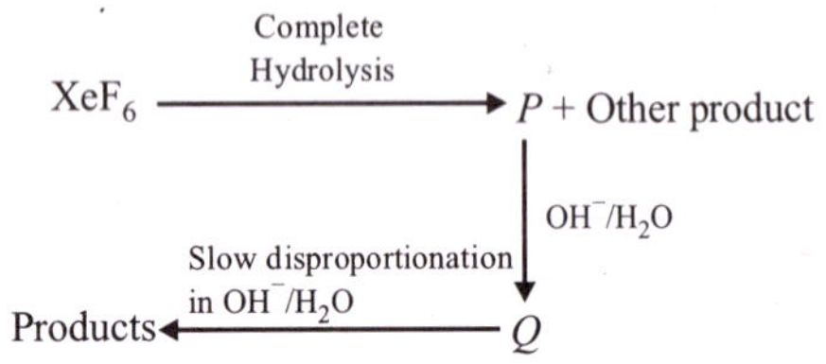

(a) 0 (b) 1 (c) 2 (d) 3

144. The shape of XeO_2F_2 molecule is **[2012]**

(a) trigonal bipyramidal (b) square planar
(c) tetrahedral (d) see-saw

145. Total number of lone pair of electrons with central atom in $XeOF_4$ is **[2004S]**

(a) 0 (b) 1 (c) 2 (d) 3

3 Numeric / New Stem Based Questions

146. The reaction of Xe and O_2F_2 gives a Xe compound **P**. The number of moles of HF produced by the complete hydrolysis of 1 mol of **P** is _______. **[Adv. 2023]**

147. At 143 K, the reaction of XeF_4 with O_2F_2, produces a xenon compound *Y*. The total number of lone pair(s) of electron present on the whole molecule of *Y* is ______ **[Adv. 2019]**

6 MCQs with One or More than One Correct Answer

148. Correct statement(s) about the compounds **P**, **Q** and **R** is(are)

$$\underset{(1:5 \text{ ratio})}{Xe(g) + F_2(g)} \xrightarrow{873\text{ K, 7 bar}} \mathbf{P}$$

$$\mathbf{P} + O_2F_2 \xrightarrow{143\text{ K}} \mathbf{Q} + O_2$$

$$\mathbf{Q} + H_2O \xrightarrow{\text{complete hydrolysis}} \mathbf{R} + HF$$ **[Adv. 2026]**

(a) **P** has two lone pairs of electrons on the central atom.
(b) **Q** has a perfect octahedral geometry.
(c) **Q** can act as a fluorinating agent.
(d) The molecular structure of **R** is trigonal pyramidal.

149. Reaction of PtF_6 with oxygen (O_2) gas results in the formation of an ionic compound, X^+Y^-. Correct statement(s) is (are) **[Adv. 2026]**

(a) The bond order of X^+ is 1.5.
(b) Valence d-orbitals of the metal ion in X^+Y^- has 5 electrons.
(c) PtF_6 acts as an oxidant in this reaction.
(d) PtF_6 acts as a fluorinating agent in this reaction.

7 Match the Following

150. The unbalanced chemical reactions given in **List-I** show missing reagent or condition (?) which are provided in **List-II**. Match **List-I** with **List-II** and select the correct answer using the code given below the lists **[Adv. 2013-II]**

	List-I		List-II
P.	$PbO_2 + H_2SO_4 \xrightarrow{?} PbSO_4 + O_2 +$ other product	1.	NO
Q.	$Na_2S_2O_3 + H_2O \xrightarrow{?} NaHSO_4 +$ other product	2.	I_2
R.	$N_2H_4 \xrightarrow{?} N_2 +$ other product	3.	Warm
S.	$XeF_2 \xrightarrow{?} Xe +$ other product	4.	Cl_2

Codes :

	P	Q	R	S
(a)	4	2	3	1
(b)	3	2	1	4
(c)	1	4	2	3
(d)	3	4	2	1

151. All the compounds listed in **Column I** react with water. Match the result of the respective reactions with the appropriate options listed in **Column II**. **[2010]**

Column-I	Column-II
(A) $(CH_3)_2SiCl_2$	(p) Hydrogen halide formation
(B) XeF_4	(q) Redox reaction
(C) Cl_2	(r) Reacts with glass
(D) VCl_5	(s) Polymerization
	(t) O_2 formation

8 Comprehension Based Questions

The noble gases have closed-shell electronic configuration and are monoatomic gases under normal conditions. The low boiling points of the lighter noble gases are due to weak dispersion forces between the atoms and the absence of other interatomic interactions.

The direct reaction of xenon with fluorine leads to a series of compounds with oxidation numbers +2, +4 and +6. XeF_4 reacts violently with water to given XeO_3. The compounds of xenon exhibit rich stereochemistry and their geometries can be deduced considering the total number of electron pairs in the valence shell.

152. Argon is used in arc welding because of its **[2007]**

(a) low reactivity with metal
(b) ability to lower the melting point of metal
(c) flammability
(d) high calorific value

153. The structure of XeO_3 is **[2007]**

(a) linear (b) planar
(c) pyramidal (d) T-shaped

154. XeF_4 and XeF_6 are expected to be **[2007]**

(a) oxidizing (b) reducing
(c) unreactive (d) strongly basic

10 Subjective Problems

155. Write balanced equations for the reactions of the following compounds with water : **[2002 - 5 Marks]**

(i) Al_4C_3 (ii) CaNCN
(iii) BF_3 (iv) NCl_3
(v) XeF_4

156. Draw the molecular structures of XeF_2, XeF_4 and XeO_2F_2 indicating the location of lone pair(s) of electrons. **[2000 - 3 Marks]**

Answer Key

Part-A (JEE Main)																			
1	(3)	**11**	(4)	**21**	(2)	**31**	(1)	**41**	(2)	**51**	(1)	**61**	(1)	**71**	(1)	**81**	(4)	**91**	(2)
2	(4)	**12**	(3)	**22**	(1)	**32**	(3)	**42**	(3)	**52**	(7)	**62**	(4)	**72**	(2)	**82**	(3)	**92**	(2)
3	(1)	**13**	(3)	**23**	(4)	**33**	(1)	**43**	(2)	**53**	(1)	**63**	(4)	**73**	(2)	**83**	(1)	**93**	(3)
4	(1)	**14**	(1)	**24**	(2)	**34**	(4)	**44**	(1)	**54**	(1)	**64**	(2)	**74**	(3)	**84**	(4)	**94**	(4)
5	(4)	**15**	(3)	**25**	(3)	**35**	(2)	**45**	(4)	**55**	(1)	**65**	(1)	**75**	(3)	**85**	(4)	**95**	(3)
6	(2)	**16**	(4)	**26**	(4)	**36**	(1)	**46**	(3)	**56**	(3)	**66**	(1)	**76**	(1)	**86**	(1)	**96**	(4)
7	(1)	**17**	(4)	**27**	(4)	**37**	(15)	**47**	(1)	**57**	(3)	**67**	(1)	**77**	(1)	**87**	(1)	**97**	(3)
8	(4)	**18**	(2)	**28**	(2)	**38**	(1)	**48**	(1)	**58**	(1)	**68**	(1)	**78**	(3)	**88**	(3)		
9	(4)	**19**	(4)	**29**	(4)	**39**	(1)	**49**	(3)	**59**	(4)	**69**	(3)	**79**	(4)	**89**	(2)		
10	(4)	**20**	(2)	**30**	(3)	**40**	(4)	**50**	(1)	**60**	(4)	**70**	(3)	**80**	(1)	**90**	(2)		
Part-B (JEE Advanced)																			
1	(b)	**12**	(d)	**23**	(d)	**38**	(b, d)	**49**	(d)	**75**	(c)	**87**	(b)	**111**	(b)	**127**	(c, d)	**148**	(a, c, d)
2	(a)	**13**	(b)	**24**	(b)	**39**	(a, b, d)	**51**	(c)	**76**	(b)	**88**	(a)	**112**	(d)	**128**	(a, b, d)	**149**	(b, c)
3	(b)	**14**	(b)	**25**	(a)	**40**	(a, b, c)	**52**	(c)	**77**	(b)	**89**	(a)	**113**	(a)	**129**	(a, d)	**150**	(d)
4	(a)	**15**	(c)	**26**	(d)	**41**	(b, c)	**53**	(b)	**78**	(c)	**103**	(a)	**114**	(a)	**130**	(a)	**152**	(a)
5	(c)	**16**	(a)	**27**	(c)	**42**	(b, d)	**54**	(b)	**79**	(d)	**104**	(c)	**115**	(6)	**131**	(c)	**153**	(c)
6	(a)	**17**	(c)	**28**	(8)	**43**	(a, b, c)	**69**	(a)	**80**	(6)	**105**	(a)	**116**	(5)	**132**	(c)	**154**	(a)
7	(b)	**18**	(c)	**29**	(4)	**44**	(a,b)	**70**	(b)	**81**	(288)	**106**	(c)	**122**	False	**143**	(c)		
8	(c)	**19**	(b)	**30**	(2.38)	**45**	(c)	**71**	(c)	**83**	(b, d)	**107**	(b)	**123**	False	**144**	(d)		
9	(b)	**20**	(d)	**35**	True	**46**	(a, c, d)	**72**	(c)	**84**	(a, c, d)	**108**	(c)	**124**	True	**145**	(b)		
10	(d)	**21**	(c)	**36**	True	**47**	(a, b, d)	**73**	(d)	**85**	(a, c, d)	**109**	(c)	**125**	(a, c)	**146**	(4)		
11	(b)	**22**	(a)	**37**	False	**48**	(a, d)	**74**	(b)	**86**	(c)	**110**	(b)	**126**	(a, b, d)	**147**	(19)		

Chapter-8

The *d*-and *f*-Block Elements

Part-A (JEE Main)

Topic 1- Electronic Configuration and Properties of Transition Elements

MULTIPLE CHOICE QUESTIONS (MCQs)

1. Given below are two statements for catalytic properties of transition metals : [April 08, 2026 (II)]

Statement I: First row transition metals which act as catalyst utilise their 3d electrons only for formation of bonds between reactant molecules and atoms on the surface of catalyst.

Statement II: There is increase in the concentration of reactants on the surface of catalyst which strengthens the bonds in reacting molecules.

In the light of the above statements, choose the correct answer from the option given below :

(1) Both Statement I and Statement II are correct
(2) Both Statement I and Statement II are incorrect
(3) Statement I is correct but Statement II are incorrect
(4) Statement I is incorrect but Statement II are correct

2. The correct order of first $(\Delta_i H_1)$ and second $(\Delta_i H_2)$ ionisation enthalpy values of Cr and Mn are :

A. $\Delta_i H_1 : Cr > Mn$ B. $\Delta_i H_2 : Cr > Mn$
C. $\Delta_i H_1 : Mn > C$ D. $\Delta_i H_2 : Mn > Cr$

Choose the correct answer from the options given below :

(1) A and B only (2) B and C only [April 06, 2026 (II)]
(3) A and D only (4) C and D only

3. The correct statements among the following are : [April 06, 2026 (I)]

A. Basic vanadium oxide is used in the manufacture of H_2SO_4.
B. The spin-only magnetic moment value of the transition metal halide employed in Ziegler-Natta polymerization is 2.84 BM.
C. The p-block metal compound employed in Ziegler-Natta polymerization has the metal in +3 oxidation state.
D. The number of electrons present in the outer most '*d*' orbital of metal halide employed in Wacker process is 8.

Choose the correct answer from the options given below :

(1) A and B only (2) A, C and D only
(3) C and D only (4) B, C and D only

4. Which of the following is NOT a physical or chemical characteristics of interstitial compounds? [April 05, 2026 (I)]

(1) They have high melting points, higher than those of pure metals.
(2) They are very soft ionic in nature.
(3) They retain metallic conductivity.
(4) They are chemically inert and usually non-stoichiometric.

5. Consider |x| is the difference in oxidation states of Mn in highest manganese fluoride and highest manganese oxide. The ions with |x| number of unpaired electrons from the following are: [April 04, 2026 (II)]

A. Sc^{3+} B. Zn^{2+} C. V^{2+} D. Fe^{2+}
E. Co^{2+}

Choose the correct answer from the options given below:

(1) A and B Only (2) C, D and E Only
(3) C and E Only (4) B and E Only

6. Given below are two statements :

Statement-I: The first ionization enthalpy of Cr is lower than that of Mn. [Jan. 22, 2026 (II)]

Statement-II: The second and third ionization enthalpies of Cr are higher than those of Mn.

In the light of the above statements, choose the correct answer from the options given below :

(1) Both Statement-I and Statement-II are false.
(2) Statement-I is true but Statement-II is false.
(3) Both Statement-I and Statement-II are true.
(4) Statement-I is false but Statement-II is true.

7. The first transition series metal 'M' has the highest enthalpy of atomisation in its series. One of its aquated ion (M^{n+}) exists in green colour. The nature of the oxide formed by the above M^{n-} ion is : [April 7, 2025 (I)]

(1) neutral (2) acidic (3) basic (4) amphoteric

8. The number of valence electrons present in the metal among Cr, Co, Fe and Ni which has the lowest enthalpy of atomisation is [April 7, 2025 (I)]

(1) 8 (2) 9 (3) 6 (4) 10

9. The incorrect relationship in the following pairs in relation to ionisation enthalpies is : [April 4, 2025 (II)]

(1) $Mn^{+} < Cr^{+}$ (2) $Mn^{+} < Mn^{2+}$
(3) $Fe^{2+} < Fe^{3+}$ (4) $Mn^{2+} < Fe^{2+}$

10. Pair of transition metal ions having the same number of unpaired electrons is : [April 4, 2025 (I)]

(1) V^{2+}, Co^{2+} (2) Ti^{2+}, Co^{2+}
(3) Fe^{3+}, Cr^{2+} (4) Ti^{3+}, Mn^{2+}

11. Given below are two statements : [April 3, 2025 (II)]

Statement I : CrO_3 is a stronger oxidizing agent than MoO_3

Statement II : Cr(VI) is more stable than Mo(VI)

In the light of the above statements, choose the correct answer from the options given below

(1) Statement I is false but Statement II is true
(2) Statement I is true but Statement II is false
(3) Both Statement I and Statement II are true
(4) Both Statement I and Statement II are false

12. The correct option with order of melting points of the pairs (Mn, Fe), (Tc, Ru) and (Re, Os) is : [Jan. 29, 2025 (I)]
(1) Fe < Mn, Ru < Tc and Re < Os
(2) Mn < Fe, Tc < Ru and Re < Os
(3) Mn < Fe, Tc < Ru and Os < Re
(4) Fe < Mn, Ru < Tc and Os < Re

13. Consider '*n*' is the number of lone pair of electrons present in the equatorial position of the most stable structure of ClF_3. The ions from the following with '*n*' number of unpaired electrons are : [Jan. 28, 2025 (I)]
(A) V^{3+} (B) Ti^{3+} (C) Cu^{2+} (D) Ni^{2+}
(E) Ti^{2+}
Choose the correct answer from the options given below :
(1) A and C only (2) A, D and E only
(3) B and C only (4) B and D only

14. Match List-I with List-II. [Jan. 24, 2025 (II)]

List-I (Transition metal ion)	List-II (Spin only magnetic moment (B.M.))
(A) Ti^{3+}	(I) 3.87
(B) V^{2+}	(II) 0.00
(C) Ni^{2+}	(III) 1.73
(D) Sc^{3+}	(IV) 2.84

Choose the correct answer from the options given below :
(1) (A)-(III), (B)-(I), (C)-(II), (D)-(IV)
(2) (A)-(III), (B)-(I), (C)-(IV), (D)-(II)
(3) (A)-(IV), (B)-(II), (C)-(III), (D)-(I)
(4) (A)-(II), (B)-(IV), (C)-(I), (D)-(III)

15. Match List-I with List-II. [Jan. 23, 2025 (II)]

List-I	List-II
(A) Bronze	(I) Cu, Ni
(B) Brass	(II) Fe, Cr, Ni, C
(C) UK silver coin	(III) Cu, Zn
(D) Stainless Steel	(IV) Cu, Sn

Choose the correct answer from the options given below:
(1) (A)-(IV), (B)-(II), (C)-(III), (D)-(I)
(2) (A)-(IV), (B)-(III), (C)-(I), (D)-(II)
(3) (A)-(III), (B)-(I), (C)-(IV), (D)-(II)
(4) (A)-(III), (B)-(IV), (C)-(II), (D)-(I)

16. The correct set of ions (aqueous solution) with same colour from the following is : [Jan. 23, 2025 (I)]
(1) V^{2+}, Cr^{3+}, Mn^{3+} (2) Zn^{2+}, V^{3+}, Fe^{3+}
(3) Ti^{4+}, V^{4+}, Mn^{2+} (4) $Sc^{3+}, Ti^{3+}, Cr^{2+}$

17. The electronic configuration of Cu(II) is $3d^9$ whereas that of Cu(I) is $3d^{10}$. Which of the following is correct ?
(1) Cu(II) is less stable [April 9, 2024 (I)]
(2) Stability of Cu(I) and Cu(II) depends on nature of copper salts
(3) Cu(II) is more stable
(4) Cu(I) and Cu(II) are equally stable

18. Given below are two statements : [April 9, 2024 (II)]
Statement I : The higher oxidation states are more stable down the group among transition elements unlike p-block elements.
Statement II : Copper can not liberate hydrogen from weak acids.
In the light of the above statements, choose the correct answer from the options given below :
(1) Both Statement I and Statement II are false
(2) Statement I is false but Statement II is true
(3) Both Statement I and Statement II are true
(4) Statement I is true but Statement II is false

19. Iron (III) catalyses the reaction between iodide and persulphate ions, in which [April 8, 2024 (I)]
A. Fe^{3+} oxidises the iodide ion
B. Fe^{3+} oxidises the persulphate ion
C. Fe^{2+} reduces the iodide ion
D. Fe^{2+} reduces the persulphate ion
Choose the most appropriate answer from the options given below:
(1) B and C only (2) B only
(3) A only (4) A and D only

20. Arrange the following elements in the increasing order of number of unpaired electrons in it. [April 6, 2024 (II)]
(A) Sc (B) Cr (C) V (D) Ti
(E) Mn
Choose the correct answer from the options given below:
(1) (C)<(E)<(B)<(A)<(D) (2) (B) < (C) < (D) < (E) < (A)
(3) (A)<(D)<(C)<(B)<(E) (4) (A) < (D) < (C) < (E) < (B)

21. The metal that shows highest and maximum number of oxidation state is: [April 5, 2024 (I)]
(1) Fe (2) Mn (3) Ti (4) Co

22. The number of ions from the following that have the ability to liberate hydrogen from a dilute acid is _________.
Ti^{2+}, Cr^{2+} and V^{2+} [April 5, 2024 (II)]
(1) 0 (2) 2 (3) 3 (4) 1

23. The element which shows only one oxidation state other than its elemental form is : _______. [April 4, 2024 (I)]
(1) Cobalt (2) Scandium
(3) Titanium (4) Nickel

24. A first row transition metal in its +2 oxidation state has a spin-only magnetic moment value of 3.86 BM. The atomic number of the metal is [April 4, 2024 (II)]
(1) 25 (2) 26 (3) 22 (4) 23

25. The transition metal having highest 3rd ionisation enthalpy is: [Feb. 1, 2024 (II)]
(1) Mn (2) Fe (3) Cr (4) V

26. Given below are two statements : one is labelled as Assertion (A) and the other is labelled as Reason (R).
Assertion (A) : In aqueous solutions Cr^{2+} is reducing while Mn^{3+} is oxidising in nature.
Reason (R) : Extra stability to half filled electronic configuration is observed than incompletely filled electronic configuration.
In the light of the above statements, choose the most appropriate answer from the options given below: [Feb. 1, 2024 (II)]
(1) (A) is true but (R) is false
(2) Both (A) and (R) are true and (R) is the correct explanation of (A)
(3) Both (A) and (R) are true but (R) is not the correct explanation of (A)
(4) (A) is false but (R) is true

27. Match List - I with List - II [Jan. 30, 2024 (I)]

List - I Species	List - II Electronic distribution
(A) Cr^{+2}	(I) $3d^8$
(B) Mn^+	(II) $3d^3 4s^1$
(C) Ni^{+2}	(III) $3d^4$
(D) V^+	(IV) $3d^5 4s^1$

Choose the correct answer from the options given below:
(1) (A) - (I), (B) - (II), (C) - (III), (D) - (IV)
(2) (A) - (III), (B) - (IV), (C) - (I), (D) - (II)
(3) (A) - (IV), (B) - (III), (C) - (I), (D) - (II)
(4) (A) - (II), (B) - (I), (C) - (IV), (D) - (III)

28. Match List I with List II [Jan. 29, 2024 (I)]

LIST I (Substances)		LIST II (Element Present)	
A.	Ziegler catalyst	I.	Rhodium
B.	Blood Pigment	II.	Cobalt
C.	Wilkinson catalyst	III.	Iron
D.	Vitamin B_{12}	IV.	Titanium

Choose the correct answer from the options given below:
(1) A-II, B-IV, C-I, D-III (2) A-IV, B-III, C-I, D-II
(3) A-II, B-III, C-IV, D-I (4) A-III, B-II, C-IV, D-I

29. Which of the following statements are correct about Zn, Cd and Hg ? [Jan. 29, 2024 (II)]
A. They exhibit high enthalpy of atomization as the d-subshell is full.
B. Zn and Cd do not show variable oxidation state while Hg shows +I and +II.
C. Compounds of Zn, Cd and Hg are paramagnetic in nature.
D. Zn, Cd and Hg are called soft metals.
Choose the **most appropriate** from the options given below:
(1) B, C only (2) A, D only
(3) C, D only (4) B, D only

30. Which of the following electronic configuration would be associated with the highest magnetic moment ? [Jan. 27, 2024 (I)]
(1) $[Ar]\,3d^7$ (2) $[Ar]\,3d^8$ (3) $[Ar]\,3d^3$ (4) $[Ar]\,3d^6$

31. Choose the correct option having all the elements with d^{10} electronic configuration from the following: [Jan. 27, 2024 (II)]
(1) $^{27}Co, ^{28}Ni, ^{26}Fe, ^{24}Cr$ (2) $^{29}Cu, ^{30}Zn, ^{48}Cd, ^{47}Ag$
(3) $^{46}Pd, ^{28}Ni, ^{26}Fe, ^{24}Cr$ (4) $^{28}Ni, ^{24}Cr, ^{26}Fe, ^{29}Cu$

32. Identity the incorrect pair from the following: [Jan. 27, 2024 (II)]
(1) Photography - AgBr
(2) Polythene preparation – $TiCl_4, Al(CH_3)_3$
(3) Haber process - Iron
(4) Wacker process – Pt Cl_2

33. Which of the following statements are correct? [April 10, 2023 (I)]
(A) The M^{3+}/M^{2+} reduction potential for iron is greater than manganese
(B) The higher oxidation states of first row d-block elements get stabilized by oxide ion.
(C) Aqueous solution of Cr^{2+} can liberate hydrogen from dilute acid.
(D) Magnetic moment of V^{2+} is observed between 4.4 – 5.2 BM
Choose the correct answer from the options given below:
(1) (B), (C) only (2) (C), (D) only
(3) (A), (B), (D) only (4) (A), (B) only

34. In chromyl chloride, the number of d-electrons present on chromium is same as in (Given at no. of Ti : 22, V : 23, Cr : 24, Mn : 25, Fe : 26) [April 08, 2023 (I)]
(1) Ti(III) (2) Fe(III) (3) V(IV) (4) Mn(VII)

35. Given below are two statements: one is labelled as **Assertion (A)** and the other is labelled as **Reason (R).**
Assertion (A): Cu^{2+} in water is more stable than Cu^+.
Reason (R): Enthalpy of hydration for Cu^{2+} is much less than that of Cu^+.
In the light of the above statements, choose the **correct** answer from the options given below [Feb. 01, 2023 (II)]
(1) Both (A) and (R) are correct but (R) is not the correct explanation of (A)
(2) (A) is not correct but (R) is correct
(3) Both (A) and (R) are correct and (R) is the correct explanation of (A)
(4) (A) is correct but (R) is not correct

36. The set of correct statements is: [Jan. 29, 2023 (II)]
(i) Manganese exhibits +7 oxidation state in its oxide.
(ii) Ruthenium and Osmium exhibit +8 oxidation in their oxides.
(iii) Sc shows +4 oxidation state which is oxidizing in nature.
(iv) Cr shows oxidising nature in +6 oxidation state.
(1) (ii) and (iii) (2) (i), (ii) and (iv)
(3) (i) and (iii) (4) (ii), (iii) and (iv)

37. In following pairs, the one in which both transition metal ions are colourless is : [July 29, 2022 (I)]
(1) Sc^{3+}, Zn^{2+} (2) Ti^{4+}, Cu^{2+}
(3) V^{2+}, Ti^{2+} (4) Zn^{2+}, Mn^{2+}

38. Which of the following 3d-metal ion will give the lowest enthalpy of hydration ($\Delta_{hyd}H$) when dissolved in water ? [July 29, 2022 (II)]
(1) Cr^{2+} (2) Mn^{2+} (3) Fe^{2+} (4) Co^{2+}

39. Which of the following has least tendency to liberate H_2 from mineral acids ? [July 28, 2022 (I)]
(1) Cu (2) Mn (3) Ni (4) Zn

40. The electronic configuration of Pt (atomic number 78) is: [June 29, 2022 (I)]
(1) $[Xe]\,4f^{14}\,5d^9\,6s^1$ (2) $[Kr]\,4f^{14}\,5d^{10}$
(3) $[Xe]\,4f^{14}\,5d^{10}$ (4) $[Xe]\,4f^{14}\,5d^8\,6s^2$

41. The reaction of zinc with excess of aqueous alkali, evolves hydrogen gas and gives : [July 29, 2022 (I)]
(1) $Zn(OH)_2$ (2) ZnO
(3) $[Zn(OH)_4]^{2-}$ (4) $[ZnO_2]^{2-}$

42. In $3d$ series, the metal having the highest M^{2+}/M standard electrode potential is [June 27, 2022 (II)]
(1) Cr (2) Fe (3) Cu (4) Zn

43. Among the following, which is the strongest oxidizing agent ? [June 25, 2022 (I)]
(1) Mn^{3+} (2) Fe^{3+} (3) Ti^{3+} (4) Cr^{3+}

44. Identify the element for which electronic configuration in +3 oxidation state is $[Ar]3d^5$: [Sep. 1, 2021 (II)]
(1) Ru (2) Mn
(3) Co (4) Fe

45. The set having ions which are coloured and paramagnetic both is - [July 22, 2021 (II)]
(1) Cu^{2+}, Cr^{3+}, Sc^+ (2) $Cu^{2+}, Zn^{2+}, Mn^{4+}$
(3) Sc^{3+}, V^{5+}, Ti^{4+} (4) $Ni^{2+}, Mn^{7+}, Hg^{2+}$

46. The common positive oxidation states for an element with atomic number 24, are [March 17, 2021 (II)]
(1) +1 and +3 to +6 (2) +1 to +6
(3) +2 to +6 (4) +1 and +3

47. What is the spin-only magnetic moment value (BM) of a divalent metal ion with atomic number 25, in its aqueous solution? [March 17, 2021 (I)]
(1) 5.92 (2) 5.26 (3) zero (4) 5.0

48. In which of the following pairs, the outer most electronic configuration will be the same? [Feb. 25, 2021 (I)]
(1) Ni^{2+} and Cu^+ (2) Fe^{2+} and Co^+
(3) Cr^+ and Mn^{2+} (4) V^{2+} and Cr^+

49. The major components of German Silver are: [Feb. 25, 2021 (II)]
(1) Cu, Zn and Ni (2) Zn, Ni and Ag
(3) Ge, Cu and Ag (4) Cu, Zn and Ag

50. The third ionization enthalpy is minimum for: [Jan. 08, 2020 (I)]
(1) Co (2) Fe (3) Ni (4) Mn

51. For the following Assertion and Reason, the correct option is:
Assertion: For hydrogenation reactions, the catalytic activity increases from Group 5 to Group 11 metals with maximum activity shown by Group 7-9 elements.
Reason: The reactants are most strongly adsorbed on group 7-9 elements. [Jan. 08, 2020 (II)]
(1) The assertion is true, but the reason is false.
(2) Both assertion and reason are false.
(3) Both assertion and reason are true and the reason is the correct explanation for the assertion.
(4) Both assertion and reason are true but the reason is not the correct explanation for the assertion.

52. The atomic radius of Ag is closest to: [Jan. 07, 2020 (I)]
(1) Au (2) Ni (3) Cu (4) Hg

53. The pair that has similar atomic radii is : [April 12, 2019 (II)]
(1) Mn and Re (2) Ti and Hf
(3) Sc and Ni (4) Mo and W

54. Consider the hydrated ions of Ti^{2+}, V^{2+}, Ti^{3+}, and Sc^{3+}. The correct order of their spin-only magnetic moments is: [April 10, 2019 (I)]
(1) $V^{2+} < Ti^{2+} < Ti^{3+} < Sc^{3+}$ (2) $Sc^{3+} < Ti^{3+} < Ti^{2+} < V^{2+}$
(3) $Ti^{3+} < Ti^{2+} < Sc^{3+} < V^{2+}$ (4) $Sc^{3+} < Ti^{3+} < V^{2+} < Ti^{2+}$

55. The correct order of the first ionization enthalpies is : [April 10, 2019 (II)]
(1) Ti < Mn < Zn < Ni (2) Ti < Mn < Ni < Zn
(3) Mn < Ti < Zn < Ni (4) Zn < Ni < Mn < Ti

56. Match the catalysts **(Column-I)** with products **(Column-II)**. [April 9, 2019 (I)]

Column-I (Catalyst)	Column-II (Product)
(A) V_2O_5	(i) Polyethylene
(B) $TiCl_4/Al(Me)_3$	(ii) ethanol
(C) $PdCl_2$	(iii) H_2SO_4
(D) Iron Oxide	(iv) NH_3

(1) (A)-(iii); (B)-(iv); (C)-(i); (D)-(ii)
(2) (A)-(ii); (B)-(iii); (C)-(i); (D)-(iv)
(3) (A)-(iii); (B)-(i); (C)-(ii); (D)-(iv)
(4) (A)-(iv); (B)-(iii); (C)-(ii); (D)-(i)

57. The statement that is INCORRECT about the interstitial compounds is : [April 8, 2019 (II)]
(1) they are chemically reactive.
(2) they are very hard.
(3) they have metallic conductivity.
(4) they have high melting points.

58. The element that usually does **NOT** show variable oxidation states is: [Jan. 11, 2019 (I)]
(1) Cu (2) Ti (3) Sc (4) V

59. The highest value of the calculated spin only magnetic moment (in BM) among all the transition metal complexes is : [Jan. 9, 2019 (I)]
(1) 5.92 (2) 6.93 (3) 3.87 (4) 4.90

60. The transition element that has lowest enthalpy of atomisation is: [Jan. 9, 2019 (II)]
(1) Fe (2) Cu (3) V (4) Zn

61. Which of the following ions does **not** liberate hydrogen gas on reaction with dilute acids ? [Online April 9, 2017]
(1) Ti^{2+} (2) V^{2+} (3) Cr^{2+} (4) Mn^{2+}

62. Which one of the following species is stable in aqueous solution ? [Online April 9, 2016]
(1) Cr^{2+} (2) MnO_4^{2-} (3) MnO_4^{3-} (4) Cu^{+}

63. The reaction of zinc with dilute and concentrated nitric acid, respectively, produces: [2016]
(1) NO and N_2O (2) NO_2 and N_2O
(3) N_2O and NO_2 (4) NO_2 and NO

64. Chloro compound of Vanadium has only spin magnetic moment of 1.73 BM. This Vanadium chloride has the formula: [Online April 9, 2014]
(1) VCl_2 (2) VCl_4 (3) VCl_3 (4) VCl_5

65. Which one of the following exhibits the large number of oxidation states? [Online April 12, 2014]
(1) Ti (22) (2) V (23) (3) Cr (24) (4) Mn (25)

66. Identify incorrect statement :
(1) Cu_2O is colourless. [Online April 23, 2013]
(2) Copper (I) compounds are colourless except when colour results from charge transfer.
(3) Copper (I) compounds are diamagnetic.
(4) Cu_2S is black.

67. Which of the following arrangements does not represent the correct order of the property stated against it ? [2013]
(1) $V^{2+} < Cr^{2+} < Mn^{2+} < Fe^{2+}$: paramagnetic behaviour
(2) $Ni^{2+} < Co^{2+} < Fe^{2+} < Mn^{2+}$: ionic size
(3) $Co^{3+} < Fe^{3+} < Cr^{3+} < Sc^{3+}$: stability in aqueous solution
(4) Sc < Ti < Cr < Mn : number of oxidation states

68. Four successive members of the first row transition elements are listed below with atomic numbers. Which one of them is expected to have the highest $E^\circ_{M^{3+}/M^{2+}}$ value ? [2013]
(1) Cr (Z = 24) (2) Mn (Z = 25)
(3) Fe (Z = 26) (4) Co (Z = 27)

69. The element with which of the following outer electron configuration may exhibit the largest number of oxidation states in its compounds : [Online April 9, 2013]
(1) $3d^5 4s^2$ (2) $3d^8 4s^2$ (3) $3d^7 4s^2$ (4) $3d^6 4s^2$

70. Iron exhibits +2 and + 3 oxidation states. Which of the following statements about iron is incorrect ? [2012]
(1) Ferrous oxide is more basic in nature than the ferric oxide.
(2) Ferrous compounds are relatively more ionic than the corresponding ferric compounds.
(3) Ferrous compounds are less volatile than the corresponding ferric compounds.
(4) Ferrous compounds are more easily hydrolysed than the corresponding ferric compounds.

71. Which pair of elements with the given atomic numbers is expected to have similar properties? [Online May 19, 2012]
(1) 40, 72 (2) 20, 36 (3) 10, 28 (4) 11, 12

72. The correct order of $E^\circ_{M^{2+}/M}$ values with negative sign for the four successive elements Cr, Mn, Fe and Co is [2010]
(1) Mn > Cr > Fe > Co (2) Cr < Fe > Mn > Co
(3) Fe > Mn > Cr > Co (4) Cr > Mn > Fe > Co

73. Of the following outer electronic configurations of atoms, the highest oxidation state is achieved by which one of them ? [2004]
(1) $(n-1)d^3\, ns^2$ (2) $(n-1)d^5\, ns^1$
(3) $(n-1)d^8\, ns^2$ (4) $(n-1)d^5\, ns^2$

74. Which of the following ions has the maximum magnetic moment? [2002]
(1) Mn^{+2} (2) Fe^{+2} (3) Ti^{+2} (4) Cr^{+2}.

NUMERICAL VALUE QUESTIONS (NVQs)

75. Number of paramagnetic ions among the following d- and f-block metal ions is ______. [April 02, 2026 (II)]
Mn^{2+}, Cu^{2+}, Zn^{2+}, Yb^{2+}, Sc^{3+}, La^{3+}, Gd^{3+}, Lu^{3+}, Ti^{4+}, Ce^{4+}
(Atomic number of Mn = 25, Cu = 29, Zn = 30, Yb = 70, Sc = 21, La = 57, Gd = 64, Lu = 71, Ti = 22, Ce = 58)

76. Among the following oxides of 3d elements, the number of mixed oxides are __________. [Jan. 22, 2026 (II)]
Ti_2O_3, V_2O_4, Cr_2O_3, Mn_3O_4, Fe_3O_4, Fe_2O_3, Co_3O_4

77. Among, Sc, Mn, Co and Cu, identify the element with highest enthalpy of atomisation. The spin only magnetic moment value of that element in its +2 oxidation state is ______ BM (in nearest integer). [April 3, 2025 (II)]

78. The spin-only magnetic moment value of M^{n+-} ion formed among Ni, Zn, Mn and Cu that has the least enthalpy of atomisation is ______. (in nearest integer)
Here n is equal to the number of diamagnetic complexes among $K_2[NiCl_4]$, $[Zn(H_2O)_6]Cl_2$, $K_3[Mn(CN)_6]$ and $[Cu(PPh_3)_3I]$ [April 2, 2025 (II)]

79. Niobium (Nb) and ruthenium (Ru) have "x" and "y" number of electrons in their respective 4d orbitals. The value of x + y is ____. [Jan. 22, 2025 (II)]

80. A transition metal 'M' among Sc, Ti, V, Cr, Mn and Fe has the highest second ionisation enthalpy. The spin only magnetic moment value of M^+ ion is ______ BM (Near integer)
(Given atomic number Sc : 21, Ti : 22, V : 23, Cr : 24, Mn : 25, Fe : 26) [April 9, 2024 (II)]

81. The 'spin only' magnetic moment value of MO_4^{2-} is _____ BM. (Where M is a metal having least metallic radii. among Sc, Ti, V, Cr, Mn and Zn). [April 8, 2024 (I)]
(Given atomic number : Sc = 21, Ti = 22, V = 23, Cr = 24, Mn = 25 and Zn = 30)

82. Among VO_2^+, MnO_4^- and $Cr_2O_7^{2-}$, the spin-only magnetic moment value of the species with least oxidising ability is .... BM (Nearest integer). [April 6, 2024 (II)]
(Given atomic member V = 23, Mn = 25, Cr = 24)

83. The spin only magnetic moment value of the ion among Ti^{2+}, V^{2+}, Co^{3+} and Cr^{2+}, that acts as strong oxidising agent in aqueous solution is .... BM (Near integer).
(Given atomic numbers : Ti : 22, V : 23, Cr : 24, Co : 27) [April 5, 2024 (I)]

84. In a borax bead test under hot condition, a metal salt (one from the given) is heated at point B of the flame, resulted in green colour salt bead. The spin-only magnetic moment value of the salt is .... BM (Nearest integer)
[Given atomic number of Cu = 29, Ni = 28, Mn = 25, Fe = 26] [April 5, 2024 (I)]

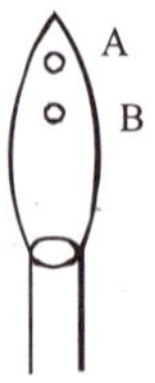

85. A first row transition metal with highest enthalpy of atomisation, upon reaction with oxygen at high temperature forms oxides of formula M_2O_n (where n = 3,4,5). The 'spin-only' magnetic moment value of the amphoteric oxide from the above oxides is ___ BM (near integer) [April 4, 2024 (II)]
(Given atomic number : Sc : 21, Ti : 22, V : 23, Cr : 24, Mn : 25, Fe : 26, Co : 27, Ni : 28, Cu : 29, Zn : 30)

86. The spin-only magnetic moment value of the compound with strongest oxidizing ability among MnF_4, MnF_3 and MnF_2 is ______ B.M. [nearest integer] [July 26, 2022 (II)]

87. Among Co^{3+}, Ti^{2+}, V^{2+} and Cr^{2+} ions, one if used as a reagent cannot liberate H_2 from dilute mineral acid solution, its spin-only magnetic moment in gaseous state is _______ B.M. (Nearest integer) [July 25, 2022 (I)]

88. The spin-only magnetic moment value of M^{3+} ion (in gaseous state) from the pairs Cr^{3+}/Cr^{2+}, Mn^{3+}/Mn^{2}, Fe^{3+}/Fe^{2+} and Co^{3+}/Co^{2+} that has negative standard electrode potential is B.M. [Nearest integer] [July 25, 2022 (II)]

89. The number of statement(s) correct from the following for copper (at no. 29) is/are ______.
(A) Cu (II) complexes are always paramagnetic
(B) Cu (I) complexes are generally colourless
(C) Cu (I) is easily oxidized
(D) In Fehling solution, the active reagent has Cu (I) [June 27, 2022 (I)]

90. Number of electrons that Vanadium (Z = 23) has in *p*-orbitals is equal to ______ [July 22, 2021 (II)]

91. In the ground state of atomic Fe(Z = 26), the spin-only magnetic moment is ________ $\times 10^{-1}$ BM.
(Round off to the nearest integer).
[Given : $\sqrt{3} = 1.73$, $\sqrt{2} = 1.41$] [March 17, 2021 (II)]

Topic 2 - Compounds of Transition Elements

MULTIPLE CHOICE QUESTIONS (MCQs)

92. Correct statements from the following are [April 04, 2026 (I)]
A. Potassium dichromate is an oxidising agent and it oxidises $FeSO_4$ to $Fe_2(SO_4)_3$ in acidic medium.
B. Sodium dichromate can be used as primary standard in volumetric estimation.
C. CrO_4^{2-} and $Cr_2O_7^{2-}$ are interconvertible in aqueous solution by varying the pH of the solution.
D. Cr – O – Cr bond angle in $Cr_2O_7^{2-}$ is 126°.
Choose the **correct** answer from the options given below :
(1) A, B and C only (2) A, C and D only
(3) A and C only (4) B and D only

93. Which of the following sets includes all the species that will change the orange colour of $K_2Cr_2O_7$ in acidic medium ? [April 02, 2026 (I)]
(1) Fe^{2+}, Sn^{2+}, I^-, S^{2-} (2) S^{2-}, Fe^{3+}, I^-, $C_2O_4^{2-}$
(3) Fe^{2+}, NO_2^-, SO_2, Sn^{4+} (4) Fe^{3+}, SO_4^{2-}, S^{2-}, Sn^{4+}

94. Consider the following statements about manganate and permanganate ions. Identify the **correct** statements : [Jan. 28, 2026 (II)]
A. The geometry of both manganate and permanganate ions is tetrahedral.
B. The oxidation states of Mn in manganate and permanganate are +7 and +6, respectively.
C. Oxidation of Mn (II) salt by peroxodisulphate gives manganate ion as the final product.
D. Manganate ion is paramagnetic and permanganate ion is diamagnetic.
E. Acidified permanganate ion reduces oxalate, nitrite and iodide ions.

Choose the **correct** answer from the options given below:
(1) A, C and D Only (2) A, B and C Only
(3) A, D and E Only (4) A and D Only

95. "X" is an oxoanion of the lightest element of group 7 (in the periodic table). The metal is in +6 oxidation state in "X". The colour of the potassium salt of X is **[Jan. 24, 2026 (II)]**
(1) green (2) purple (3) yellow (4) orange

96. The oxidation state of chromium in the final product formed in the reaction between KI and acidified $K_2Cr_2O_7$ solution is: **[Jan. 23, 2026 (II)]**
(1) +4 (2) +3 (3) +2 (4) +6

97. On heating a mixture of common salt and $K_2Cr_2O_7$ in equal amount along with concentrated H_2SO_4 in a test tube, a gas is evolved. Formula of the gas evolved and oxidation state of the central metal atom in the gas respectively are: **[Jan. 21, 2026 (II)]**
(1) CrO_2Cl_2 and +5 (2) CrO_2Cl_2 and +6
(3) $Cr_2O_2Cl_2$ and +6 (4) $Cr_2O_2Cl_2$ and +3

98. Given below are some of the statements about Mn and Mn_2O_7. Identify the correct statements **[Jan. 21, 2026 (II)]**
A. Mn forms the oxide Mn_2O_7 in which Mn is in its highest oxidation state.
B. Oxygen stabilizes the Mn in higher oxidation states by forming multiple bonds with Mn
C. Mn_2O_7 is an ionic oxide.
D. The structure of Mn_2O_7 consists of one bridged oxygen.
Choose the **correct** answer from the options given below:
(1) A, B, C and D (2) A, B and D Only
(3) A, C and D Only (4) A, B and C Only

99. MnO_4^{2-}, in acidic medium, disproportionates to: **[Jan. 21, 2026 (I)]**
(1) Mn_2O_7 and MnO_2 (2) MnO_4^- and MnO
(3) MnO_4^- and MnO_2 (4) Mn_2O_7 and MnO

100. Consider the following reactions.
$PbCl_2 + K_2CrO_4 \rightarrow A + 2KCl$
(Hot solution)
$A + NaOH \rightleftharpoons B + Na_2CrO_4$
$PbSO_4 + 4CH_3COONH_4 \rightarrow (NH_4)_2SO_4 + X$
In the above reactions, A, B and X are respectively. **[Jan. 21, 2026 (I)]**
(1) $Na_2[Pb(OH)_2]$, $PbCrO_4$ and $(NH_4)_2[Pb(CH_3COO)_4]$
(2) $PbCrO_4$, $Na_2[Pb(OH)_4]$ and $[Pb(NH_3)_4]SO_4$
(3) $Na_2[Pb(OH)_2]$, $PbCrO_4$ and $[Pb(NH_3)_4]SO_4$
(4) $PbCrO_4$, $Na_2[Pb(OH)_4]$ and $(NH_4)_2[Pb(CH_3COO)_4]$

101. The amphoteric oxide among V_2O_3, V_2O_4 and V_2O_5 upon reaction with alkali leads to formation of an oxide anion. The oxidation state of V in the oxide anion is : **[Jan. 28, 2025 (II)]**
(1) +3 (2) +7 (3) +5 (4) +4

102. Which of the following oxidation reactions are carried out by both $K_2Cr_2O_7$ and $KMnO_4$ in acidic medium ? **[Jan. 28, 2025 (I)]**
(A) $I^- \rightarrow I_2$ (B) $S^{2-} \rightarrow S$
(C) $Fe^{2+} \rightarrow Fe^{3+}$ (D) $I^- \rightarrow IO_3^-$
(E) $S_2O_3^{2-} \rightarrow SO_4^{2-}$
Choose the **correct** answer from the options given below :
(1) B, C and D only (2) A, D and E only
(3) A, B and C only (4) C, D and E only

103. Preparation of potassium permanganate from MnO_2 involves two step process in which the 1st step is a reaction with KOH and KNO_3 to produce **[Jan. 24, 2025 (I)]**
(1) $K_4[Mn(OH)_6]$ (2) K_3MnO_4
(3) $KMnO_4$ (4) K_2MnO_4

104. Consider the following reactions **[Jan. 23, 2025 (II)]**
$$K_2Cr_2O_7 \xrightarrow[-H_2O]{KOH} [A] \xrightarrow[-H_2O]{H_2SO_4} [B] + K_2SO_4$$
The products [A] and [B], respectively are:
(1) $K_2Cr(OH)_6$ and Cr_2O_3 (2) K_2CrO_4 and Cr_2O_3
(3) K_2CrO_4 and $K_2Cr_2O_7$ (4) K_2CrO_4 and CrO

105. The equilibrium $Cr_2O_7^{2-} \rightleftharpoons 2CrO_4^{2-}$ is shifted to the right in: **[April 8, 2024 (II)]**
(1) an acidic medium (2) a basic medium
(3) a weakly acidic medium (4) a neutral medium

106. Given below are two statements: **[April 8, 2024 (II)]**
Statement (I): Fusion of MnO_2 with KOH and an oxidising agent gives dark green K_2MnO_4.
Statement (II): Manganate ion on electrolytic oxidation in alkaline medium gives permanganate ion.
In the light of the above statements, choose the **correct** answer from the options given below:
(1) Statement I is false but Statement II is true
(2) Both Statement I and Statement II is false
(3) Statement I is true but Statement II is false
(4) Both Statement I and Statement II is true

107. Which of the following compounds show colour due to d - d transition? **[Feb. 1, 2024 (II)]**
(1) $K_2Cr_2O_7$ (2) $CuSO_4.5H_2O$
(3) $KMnO_4$ (4) K_2CrO_4

108. Identify correct statements from below: **[Jan. 31, 2024 (I)]**
A. The chromate ion is square planar.
B. Dichromates are generally prepared from chromates.
C. The green manganate ion is diamagnetic.
D. Dark green coloured K_2MnO_4 disproportionates in a neutral or acidic medium to give permanganate.
E. With increasing oxidation number of transition metal, ionic character of the oxides decreases.
Choose the correct answer from the options given below:
(1) B, C, D only (2) A, D, E only
(3) A, B, C only (4) B, D, E only

109. Choose the correct statements from the following **[Jan. 31, 2024 (II)]**
A. Mn_2O_7 is an oil at room temperature
B. V_2O_4 reacts with acid to give VO_2^{2+}
C. CrO is a basic oxide
D. V_2O_5 does not react with acid
Choose the correct answer from the options given below:
(1) A, B and D only (2) A and C only
(3) A, B and C only (4) B and C only

110. The orange colour of $K_2Cr_2O_7$ and purple colour of $KMnO_4$ is due to **[Jan. 30, 2024 (II)]**
(1) Charge transfer transition in both.
(2) $d \rightarrow d$ transitions in $KMnO_4$ and charge transfer transitions in $K_2Cr_2O_7$.
(3) $d \rightarrow d$ transitions in $K_2Cr_2O_7$ and charge transfer transitions in $KMnO_4$.
(4) $d \rightarrow d$ transitions in both

111. Alkaline oxidative fusion of MnO_2 gives "A" which on electrolytic oxidation in alkaline solution produces B. A and B respectively are **[Jan. 30, 2024 (II)]**
(1) Mn_2O_7 and MnO_4^- (2) MnO_4^{2-} and MnO_4^-
(3) Mn_2O_3 and MnO_4^{2-} (4) MnO_4^{2-} and Mn_2O_7

112. In chromyl chloride test for confirmation of Cl^- ion, a yellow solution is obtained. Acidification of the solution and addition of amyl alcohol and 10% H_2O_2 turns organic layer blue indicating formation of chromium pentoxide. The oxidation state of chromium in that is **[Jan. 29, 2024 (I)]**
(1) +5 (2) +3 (3) +10 (4) +6

113. In alkaline medium, MnO_4^- oxidises I^- to **[Jan. 29, 2024 (I)]**
(1) IO^- (2) IO_4^- (3) I_2 (4) IO_3^-

114. $KMnO_4$ decomposes on heating at 513K to form O_2 along with **[Jan. 29, 2024 (I)]**
(1) Mn & KO_2 (2) K_2MnO_4 & MnO_2
(3) K_2MnO_4 & Mn (4) MnO_2 & K_2O_2

115. Yellow compound of lead chromate gets dissolved on treatment with hot NaOH solution. The product of lead formed is a : **[Jan. 27, 2024 (I)]**
(1) Tetraanionic complex with coordination number six
(2) Neutral complex with coordination number four
(3) Dianionic complex with coordination number six
(4) Dianionic complex with coordination number four

116. Given below are two statements **[April 10, 2023 (I)]**
Statement I : Aqueous solution of $K_2Cr_2O_7$ is preferred as a primary standard in volumetric analysis over $Na_2Cr_2O_7$ aqueous solution
Statement II : $K_2Cr_2O_7$ has a higher solubility in water than $Na_2Cr_2O_7$
In the light of the above statements, choose the correct answer from the options given below:
(1) Both Statement I and Statement II are true
(2) Both Statement I and Statement II are false
(3) Statement I is true but Statement II is false
(4) Statement I is false but Statement II is true

117. Which halogen is known to cause the reaction given below:
$2Cu^{2+} + 4X^- \rightarrow Cu_2X_2(s) + X_2$ **[April 08, 2023 (I)]**
(1) Only Iodine (2) Only Bromine
(3) All halogens (4) Only Chlorine

118. During the reaction of permanganate with thiosulphate, the change in oxidation of manganese occurs by value of 3. Identify which of the below medium will favour the reaction.
(1) aqueous acidic **[April 06, 2023 (II)]**
(2) aqueous neutral
(3) both aqueous acidic and neutral
(4) both aqueous acidic and faintly alkaline.

119. Highest oxidation state of Mn is exhibited in Mn_2O_7. The correct statements about Mn_2O_7 are. **[Feb. 01, 2023 (I)]**
(A) Mn is tetrahedrally surrounded by oxygen atoms.
(B) Mn is octahedrally surrounded by oxygen atoms.
(C) Contains Mn-O-Mn bridge.
(D) Contains Mn-Mn bond.
Choose the correct answer from the options given below:
(1) A and D only (2) A and C only
(3) B and D only (4) B and C only

120. The correct order of basicity of oxides of vanadium is **[Jan. 31, 2023 (I)]**
(1) $V_2O_3 > V_2O_4 > V_2O_5$ (2) $V_2O_3 > V_2O_5 > V_2O_4$
(3) $V_2O_5 > V_2O_4 > V_2O_3$ (4) $V_2O_4 > V_2O_3 > V_2O_5$

121. When Cu^{2+} ion is treated with KI, a white precipitate, X appears in solution. The solution is titrated with sodium thiosulphate, the compound Y is formed. X and Y respectively are **[Jan. 31, 2023 (I)]**
(1) X= Cu_2I_2 Y= $Na_2S_4O_5$ (2) X= Cu_2I_2 Y= $Na_2S_4O_6$
(3) X= CuI_2 Y= $Na_2S_4O_3$ (4) X= CuI_2 Y= $Na_2S_4O_6$

122. $KMnO_4$ oxidises I^- in acidic and neutral/faintly alkaline solution, respectively to **[Jan. 30, 2023 (II)]**
(1) I_2 & IO_3^- (2) IO_3^- & I_2 (3) IO_3^- & IO_3^- (4) I_2 & I_2

123. Potassium dichromate acts as a strong oxidizing agent in acidic solution. During this process, the oxidation state changes from **[Jan. 25, 2023 (II)]**
(1) +3 to +1 (2) +6 to +3 (3) +2 to +1 (4) +6 to +2

124. $K_2Cr_2O_7$ paper acidified with dilute H_2SO_4 turns green when exposed to **[Jan. 24, 2023 (II)]**
(1) Carbon dioxide (2) Sulphur trioxide
(3) Hydrogen sulphide (4) Sulphur dioxide

125. In neutral or faintly alkaline medium, $KMnO_4$ being a powerful oxidant can oxidize, thiosulphate almost quantitatively, to sulphate. In this reaction overall change in oxidation state of manganese will be : **[July 29, 2022 (I)]**
(1) 5 (2) 1 (3) 0 (4) 3

126. Dihydrogen reacts with CuO to give **[June 28, 2022 (I)]**
(1) CuH_2 (2) Cu (3) Cu_2O (4) $Cu(OH)_2$

127. In neutral or alkaline solution, MnO_4^- oxidises thiosulphate to: **[July 27, 2022 (II)]**
(1) $S_2O_7^{2-}$ (2) $S_2O_8^{2-}$ (3) SO_3^{2-} (4) SO_4^{2-}

128. Given below are two statements:
Statement I : Iron (III) catalyst, acidified $K_2Cr_2O_7$ and neutral $KMnO_4$ have the ability to oxidise I^- to I_2 independently.
Statement II: Manganate ion is paramagnetic in nature and involves $p\pi - p\pi$ bonding.
In the light of the above statements, choose the **correct** answer from the options given below: **[July 27, 2022 (I)]**
(1) Both statement I and statement II are ture
(2) Both statement I and statement II are false
(3) Statement I is true but statement II is false
(4) Statement I is false but statement II is true

129. The total number of Mn = O bonds in Mn_2O_7 is ____ **[July 27, 2022 (I)]**
(1) 4 (2) 5 (3) 6 (4) 3

130. Potassium permanganate on heating at 513 K gives a product which is :
(1) paramagnetic and colourless **[Aug. 27, 2021 (II)]**
(2) diamagnetic and green
(3) diamagnetic and colourless
(4) paramagnetic and green

131. Given below are two statements :
Statement I : Potassium permanganate on heating at 573 K forms potassium manganate.
Statement II : Both potassium permanganate and potassium manganate are tetrahedral and paramagnetic in nature.
In the light of the above statements, choose the **most appropriate** answer from the options given below : **[March 17, 2021 (I)]**
(1) Statement I is false but statement II is true
(2) Both statement I and statement II are false
(3) Both statement I and statement II are true
(4) Statement I is true but statement II is false

132. The incorrect statement(s) among (1) - (3) is (are) :
(A) W(VI) is more stable than Cr(VI). **[Sep. 04, 2020 (II)]**
(B) In the presence of HCl, permanganate titrations provide satisfactory results.
(C) some lanthanoid oxides can be used as phosphorus.
(1) (B) and (C) only (2) (A) and (B) only
(3) (B) only (4) (A) only

133. The **incorrect** statement is : [Sep. 03, 2020 (II)]
(1) Manganate and permanganate ions are tetrahedral
(2) In manganate and permanganate ions, the π-bonding takes place by overlap of p-orbitals of oxygen and d-orbitals of manganese
(3) Manganate and permanganate ions are paramagnetic
(4) Manganate ion is green in colour and permanganate ion is purple in colour

134. Thermal decomposition of a Mn compound (X) at 513 K results in compound Y, MnO_2 and a gaseous product. MnO_2 reacts with NaCl and concentrated H_2SO_4 to give a pungent gas Z. X, Y, and Z, respectively, are : [April 12, 2019 (II)]
(1) $KMnO_4$, K_2MnO_4 and Cl_2 (2) K_2MnO_4, $KMnO_4$ and SO_2
(3) K_3MnO_4, K_2MnO_4 and Cl_2 (4) K_2MnO_4, $KMnO_4$ and Cl_2

135. $A \xrightarrow{4\,KOH,\,O_2} \underset{(Green)}{2B} + 2H_2O$

$B \xrightarrow{4\,HCl} \underset{(Purple)}{2C} + MnO_2 + 2H_2O$

$2C \xrightarrow{H_2O,\,KI} 2A + KOH + D$

In the above sequence of reactions, A and D, respectively, are: [Jan. 11, 2019 (II)]
(1) KI and $KMnO_4$ (2) MnO_2 and KIO_3
(3) KIO_3 and MnO_2 (4) KI and K_2MnO_4

136. When XO_2 is fused with an alkali metal hydroxide in presence of an oxidizing agent such as KNO_3, a dark green product is formed which disproportionates in acidic solution to afford a dark purple solution. X is: [Online April 16, 2018]
(1) Mn (2) Cr (3) V (4) Ti

137. Which of the following compounds is metallic and ferromagnetic? [2016]
(1) VO_2 (2) MnO_2 (3) TiO_2 (4) CrO_2

138. Which of the following statements is false ? [Online April 11, 2015]
(1) $Na_2Cr_2O_7$ is less soluble than $K_2Cr_2O_7$
(2) $Na_2Cr_2O_7$ is primary standard in volumetry
(3) CrO_4^{2-} is tetrahedral in shape
(4) $Cr_2O_7^{2-}$ has a Cr–O–Cr bond

139. The equation which is balanced and represents the correct product(s) is: [2014]
(1) $Li_2O + 2KCl \longrightarrow 2LiCl + K_2O$
(2) $\left[CoCl(NH_3)_5\right]^{+} + 5H^{+} \longrightarrow Co^{2+} + 5NH_4^{+} + Cl^{-}$
(3) $\left[Mg(H_2O)_6\right]^{2+} + (EDTA)^{4-} \xrightarrow{\text{excess NaOH}} \left[Mg(EDTA)\right]^{2+} + 6H_2O$
(4) $CuSO_4 + 4KCN \longrightarrow K_2\left[Cu(CN)_4\right] + K_2SO_4$

140. Which series of reactions correctly represents chemical reactions related to iron and its compound? [2014]
(1) $Fe \xrightarrow{dil.\,H_2SO_4} FeSO_4 \xrightarrow{H_2SO_4,\,O_2} Fe_2(SO_4)_3 \xrightarrow{heat} Fe$
(2) $Fe \xrightarrow{O_2,\,heat} FeO \xrightarrow{dil.\,H_2SO_4} FeSO_4 \xrightarrow{heat} Fe$
(3) $Fe \xrightarrow{Cl_2,\,heat} FeCl_3 \xrightarrow{heat,\,air} FeCl_2 \xrightarrow{Zn} Fe$
(4) $Fe \xrightarrow{O_2,\,heat} Fe_3O_4 \xrightarrow{CO,\,600^\circ C} FeO \xrightarrow{CO,\,700^\circ C} Fe$

141. Which of the following is not formed when H_2S reacts with acidic $K_2Cr_2O_7$ solution? [Online April 9, 2014]
(1) $CrSO_4$ (2) $Cr_2(SO_4)_3$ (3) K_2SO_4 (4) S

142. When a small amount of $KMnO_4$ is added to concentrated H_2SO_4, a green oily compound is obtained which is highly explosive in nature. Compound may be: [Online April 23, 2013]
(1) $MnSO_4$ (2) Mn_2O_7 (3) MnO_2 (4) Mn_2O_3

143. The oxidation state of chromium in the final product formed by the reaction between KI and acidified potassium dichromate solution is: [2005]
(1) +3 (2) +2 (3) +6 (4) +4

144. Heating mixture of Cu_2O and Cu_2S will give [2005]
(1) Cu_2SO_3 (2) $CuO + CuS$
(3) $Cu + SO_3$ (4) $Cu + SO_2$

145. Ammonia forms the complex ion $[Cu(NH_3)_4]^{2+}$ with copper ions in alkaline solutions but not in acidic solutions. What is the reason for it ? [2003]
(1) In acidic solutions, protons coordinate with ammonia molecules forming NH_4^+ ions and thus NH_3 molecules are not available
(2) In alkaline solutions insoluble $Cu(OH)_2$ is precipitated which is soluble in excess of any alkali
(3) Copper hydroxide is an amphoteric substance
(4) In acidic solutions hydration protects copper ions

146. A red solid is insoluble in water. However it becomes soluble if some KI is added to water. Heating the red solid in a test tube results in liberation of some violet coloured fumes and droplets of a metal appear on the cooler parts of the test tube. The red solid is [2003]
(1) HgI_2 (2) HgO
(3) Pb_3O_4 (4) $(NH_4)_2Cr_2O_7$

147. Which one of the following nitrates will leave behind a metal on strong heating ? [2003]
(1) Copper nitrate (2) Manganese nitrate
(3) Silver nitrate (4) Ferric nitrate

148. What would happen when a solution of potassium chromate is treated with an excess of dilute nitric acid? [2003]
(1) $Cr_2O_7^{2-}$ and H_2O are formed
(2) CrO_4^{2-} is reduced to +3 state of Cr
(3) CrO_4^{2-} is oxidized to +7 state of Cr
(4) Cr^{3+} and $Cr_2O_7^{2-}$ are formed

149. When $KMnO_4$ acts as an oxidising agent and ultimately forms $[MnO_4]^{-2}$, MnO_2, Mn_2O_3, Mn^{+2} then the number of electrons transferred in each case respectively is [2002]
(1) 4, 3, 1, 5 (2) 1, 5, 3, 7 (3) 1, 3, 4, 5 (4) 3, 5, 7, 1.

Numerical Value Questions (NVQs)

150. X and Y are the number of electrons involved, respectively during the oxidation of I^- to I_2 and S^{2-} to S by acidified $K_2Cr_2O_7$. The value of X + Y is ______. [Jan. 24, 2026 (I)]

151. $KMnO_4$ acts as an oxidising agent in acidic medium. 'X' is the difference between the oxidation states of Mn in reactant and product. 'Y' is the number of 'd' electrons present in the brown red precipitate formed at the end of the acetate ion test with neutral ferric chloride. The value of $X + Y$ is ______. [April 4, 2025 (I)]

152. Consider the following reactions
$A + NaCl + H_2SO_4 \rightarrow CrO_2Cl_2 + \text{Side Products}$
Little amount
$CrO_2Cl_{2(Vapour)} + NaOH \rightarrow B + NaCl + H_2O$
$B + H^+ \rightarrow C + H_2O$
The number of terminal 'O' present in the compound 'C' is ______. [April 3, 2025 (I)]

153. The molar mass of the water insoluble product formed from the fusion of chromite ore ($FeCr_2O_4$) with Na_2CO_3 in presence of O_2 is ____ g mol^{-1}. [Jan. 29, 2025 (I)]

154. The spin only magnetic moment (μ) value (B.M.) of the compound with strongest oxidising power among Mn_2O_3, TiO and VO is ______ B.M. (Nearest integer). [Jan. 28, 2025 (II)]

155. The difference in the 'spin-only' magnetic moment values of $KMnO_4$ and the manganese product formed during titration of $KMnO_4$ against oxalic acid in acidic medium is _____ BM. (nearest integer) [April 6, 2024 (I)]

156. Among CrO, Cr_2O_3 and CrO_3, the sum of spin-only magnetic moment values of basic and amphoteric oxides is ______ 10^{-2} BM (nearest integer).
(Given atomic number of Cr is 24) [April 6, 2024 (I)]

157. The fusion of chromite ore with sodium carbonate in the presence of air leads to the formation of products A and B along with the evolution of CO_2. The sum of spin-only magnetic moment values of A and B is ______ B.M. (Nearest integer)
(Given atomic number : C : 6, Na : 11, O : 8, Fe : 26, Cr : 24) [April 5, 2024 (II)]

158. Consider the following reaction
$MnO_2 + KOH + O_2 \rightarrow A + H_2O$.
Product 'A' in neutral or acidic medium disproportionate to give products 'B' and 'C' along with water. The sum of spin-only magnetic moment values of B and C is ________ BM. (nearest integer) [April 4, 2024 (I)]
(Given atomic number of Mn is 25)

159. In the reaction of potassium dichromate, potassium chloride and sulfuric acid (conc.), the oxidation state of the chromium in the product is (+)________. [Jan. 31, 2024 (II)]

160. The total change in the oxidation state of manganese involved in the reaction of $KMnO_4$ and potassium iodide in the acidic medium is ______ [April 15, 2023 (I)]

161. In alkaline medium, the reduction of permanganate anion involves a gain of _______ electrons. [April 10, 2023 (II)]

162. The sum of oxidation state of the metals in $Fe(CO)_5$, VO^{2+} and WO_3 is ______: [April 08, 2023 (II)]

163. The number of electrons involved in the reduction of permanganate to manganese dioxide in acidic medium is ________. [Jan. 30, 2023 (I)]

164. The disproportionation of MnO_4^{2-} in acidic medium resulted in the formation of two manganese compounds A and B. If the oxidation state of Mn in B is smaller than that of A, then the spin-only magnetic moment (μ) value of B in BM is ___________. (Nearest integer) [July 28, 2022 (I)]

165. In the titration of $KMnO_4$ and oxalic acid in acidic medium, the change in oxidation number of carbon at the end point is_____ [July 27, 2022 (I)]

166. The oxidation state of manganese in the product obtained in a reaction of potassium permanganate and hydrogen peroxide in basic medium is______. [July 27, 2022 (II)]

167. The number of terminal oxygen atoms present in the product *B* obtained from the following reaction is ________.
$FeCr_2O_4 + Na_2CO_3 + O_2 \rightarrow A + Fe_2O_3 + CO_2$
$A + H^+ \rightarrow B + H_2O + Na^+$ [June 29, 2022 (I)]

168. An acidified manganate solution undergoes disproportionation reaction. The spin-only magnetic moment value of the product having manganese in higher oxidation state is _________ B.M. (Nearest integer) [June 29, 2022 (I)]

169. Acidified potassium permanganate solution oxidises oxalic acid. The spin-only magnetic moment of the manganese product formed from the above reaction is ______ B.M. (Nearest Integer) [June 27, 2022 (I)]

170. The difference in oxidation state of chromium in chromate and dichromate salts is ________ [June 24, 2022 (I)]

171. Manganese (VI) has ability to disproportionate in acidic solution. The difference in oxidation states of two ions it forms in acidic solution is______ [June 24, 2022 (II)]

172. The spin–only magnetic moment value of the most basic oxide of vanadium among V_2O_3, V_2O_4 and V_2O_5 is ________ B.M. (Nearest Integer) [June 26, 2022 (I)]

173. Dichromate ion is treated with base, the oxidation number of Cr in the product formed is ____. [Feb. 26, 2021 (I)]

174. In mildly alkaline medium, thiosulphate ion is oxidized by MnO_4^- to "A". The oxidation state of sulphur in "A" is ________. [Feb. 26, 2021 (II)]

175. In basic medium CrO_4^{2-} oxidises $S_2O_3^{2-}$ to form SO_4^{2-} and itself changes into $Cr(OH)_4^-$. The volume of 0.154 M CrO_4^{2-} required to react with 40 mL of 0.25 M $S_2O_3^{2-}$ is _______ mL. (Rounded-off to the nearest integer) [Feb. 25, 2021 (I)]

176. The sum of the total number of bonds between chromium and oxygen atoms in chromate and dichromate ions is ________. [Jan. 09, 2020 (II)]

177. Consider the following reactions:
$NaCl + K_2Cr_2O_7 + H_2SO_4 (Conc.) \rightarrow (A) + \text{Side products}$
$(1) + NaOH \rightarrow (B) + \text{Side products}$
$(2) + H_2SO_4 (dilute) + H_2O_2 \rightarrow (C) + \text{Side products}$
The sum of the total number of atoms in one molecule each of (A), (B) and (C) is _______. [Jan. 07, 2020 (II)]

Topic 3 - Lanthanoids and Actinoids

MULTIPLE CHOICE QUESTIONS (MCQs)

178. Given below are two statements : [April 08, 2026 (II)]
Statement I: The number of pairs among [Ti^{4+}, V^{2+}], [V^{2+}, Mn^{2+}], [Mn^{2+}, Fe^{3+}] and [V^{2+}, Cr^{2+}] in which both ions are coloured is 3.
Statement II: The number pairs among [La^{3+}, Yb^{2+}], [Lu^{3+}, Ce^{4+}] and [Ac^{3+}, Lr^{3+}] ions in which both are diamagnetic is 3.
In the light of the above statements, choose the **correct** answer from the option given below:
(1) Both Statement I and Statement II are correct
(2) Both Statement I and Statement II are incorrect
(3) Statement I is correct but Statement II are incorrect
(4) Statement I is correct but Statement II are incorrect

179. Pairs of elements with the same number of electrons in their respective 4f orbital are
[Atomic number, Eu-63, Gd-64, Dy-66, Ho-67, Tm-69, Yb-70, Lu-71, Hf-72] **[April 04, 2026 (II)]**
A. (Eu and Gd) B. (Dy and Ho)
C. (Yb and Hf) D. (Lu and Tm)
Choose the correct answer from the options given below:
(1) B and C Only (2) A and B Only
(3) A and D Only (4) A and C Only

180. The **correct** statements among the following are,
A. Mo(VI) and W(VI) are less stable than Cr(VI).
B. Ce^{4+} and Tb^{4+} are oxidant while Eu^{2+} and Yb^{2+} are reductant.
C. Cm and Am have seven unpaired electrons.
D. Actinoid contraction is greater from element to element than lanthanoid contraction.
Choose the **correct** answer from the options given below: **[April 04, 2026 (I)]**
(1) A and B Only (2) C and D Only
(3) B and D Only (4) A and C Only

181. Given below are two statements : **[Jan. 28, 2026 (I)]**
Statement I : The number of pairs, from the following, in which both the ions are coloured in aqueous solution is 3.
[Sc^{3+}, Ti^{3+}], [Mn^{2+}, Cr^{2+}], [Cu^{2+}, Zn^{2+}] and [Ni^{2+}, Ti^{4+}]
Statement II : Th^{4+} is the strongest reducing agent among Th^{4+}, Ce^{4+}, Gd^{3+} and Eu^{2+}.
In the light of the above statements, choose the **correct** answer from the options given below
(1) Statement I is true but Statement II is false
(2) Statement I is false but Statement II is true
(3) Both Statement I and Statement II are false
(4) Both Statement I and Statement II are true

182. Which of the following ions is the strongest oxidizing agent ? **[Jan. 24, 2025 (I)]**
(Atomic Number of Ce = 58, Eu = 63, Tb = 65, Lu = 71]
(1) Lu^{3+} (2) Eu^{2+} (3) Tb^{4+} (4) Ce^{3+}

183. Lanthanoid ions with $4f^7$ configuration are: **[Jan. 22, 2025 (I)]**
(A) Eu^{2+} (B) Gd^{3+} (C) Eu^{3+} (D) Tb^{3+}
(E) Sm^{2+}
Choose the correct answer from the options given below:
(1) (A) and (B) only (2) (A) and (D) only
(3) (B) and (E) only (4) (B) and (C) only

184. The electronic configuration of Einsteinium is :
(Given atomic number of Einsteinium = 99) **[April 9, 2024 (II)]**
(1) [Rn] $5f^{12}6d^0 7s^2$ (2) [Rn] $5f^{11}6d^0 7s^2$
(3) [Rn] $5f^{13}6d^0 7s^2$ (4) [Rn] $5f^{10}6d^0 7s^2$

185. The number of element from the following that do not belong to lanthanoids is : **[April 6, 2024 (I)]**
Eu, Cm, Er, Tb, Yb and Lu
(1) 3 (2) 4 (3) 1 (4) 5

186. Diamagnetic Lanthanoid ions are: **[Jan. 30, 2024 (I)]**
(1) Nd^{3+} & Eu^{3+} (2) La^{3+} & Ce^{4+}
(3) Nd^{3+} & Ce^{4+} (4) Lu^{3+} & Eu^{3+}

187. Which of the following acts as a strong reducing agent? (Atomic number : Ce = 58, Eu = 63, Gd = 64, Lu = 71) **[Jan. 29, 2024 (II)]**
(1) Lu^{3+} (2) Ce^{4+} (3) Gd^{3+} (4) Eu^{2+}

188. The electronic configuration for Neodymium is :
[Atomic Number for Neodymium 60] **[Jan. 27, 2024 (I)]**
(1) [Xe] $4f^4 6s^2$ (2) [Xe] $5f^4 7s^2$
(3) [Xe] $4f^6 6s^2$ (4) [Xe] $4f^1 5d^1 6s^2$

189. Given below are two statements: **[Jan. 27, 2024 (II)]**
Statement (I) : In the Lanthanoids, the formation of Ce^{+4} is favoured by its noble gas configuration.
Statement (II) : Ce^{+4} is a strong oxidant reverting to the common +3 state.
In the light of the above statements, choose the most appropriate answer from the options given below:
(1) Statement I is false but Statement II is true
(2) Both Statement I and Statement II are true
(3) Statement I is true but Statement II is false
(4) Both Statement I and Statement II are false

190. Strong reducing and oxidising agents among the following, respectively, are **[April 06, 2023 (I)]**
(1) Ce^{4+} and Eu^{2+} (2) Ce^{4+} and Tb^{4+}
(3) Ce^{3+} and Ce^{4+} (4) Eu^{2+} and Ce^{4+}

191. The pair of lanthanides in which both elements have high third - ionization energy is: **[April 13, 2023 (I)]**
(1) Eu, Gd (2) Eu, Yb (3) Lu, Yb (4) Dy, Gd

192. Nd^{2+} = ________ **[Jan. 31, 2023 (I)]**
(1) $4f^2 6s^2$ (2) $4f^4$ (3) $4f^3$ (4) $4f^4 6s^2$

193. Which of the following elements have half-filled f-orbitals in their ground state ? **[Jan. 31, 2023 (II)]**
(Given : atomic number
Sm = 62; Eu = 63; Tb = 65; Gd = 64, Pm = 61)
A. Sm B. Eu C. Tb D. Gd
E. Pm
Choose the correct answer from the options given below:
(1) B and D only (2) A and E only
(3) A and B only (4) C and D only

194. Which one amongst the following are good oxidizing agents?
A. Sm^{2+} B. Ce^{2+} C. Ce^{4+} D. Tb^{4+}
Choose the most appropriate answer from the options given below: **[Jan. 24, 2023 (II)]**
(1) C only (2) D only
(3) A and B only (4) C and D only

195. Which of the following pair is not isoelectronic species?
(At. no. Sm, 62; Er, 68; Yb, 70; Lu, 71; Eu, 63; Tb, 65; Tm, 69)
(1) Sm^{2+} and Er^{3+} (2) Yb^{2+} and Lu^{3+} **[July 28, 2022 (II)]**
(3) Eu^{2+} and Tb^{4+} (4) Tb^{2+} and Tm^{4+}

196. Which one of the lanthanoids given below is the most stable in divalent form? **[June 28, 2022 (I)]**
(1) Ce (Atomic Number 58) (2) Sm (Atomic Number 62)
(3) Eu (Atomic Number 63) (4) Yb (Atomic Number 70)

197. The *f*-orbitals are half and completely filled, respectively in lanthanide ions
(Given: Atomic no. Eu, 63; Sm, 62; Tm, 69; Tb, 65; Yb, 70; Dy, 66]
(1) Eu^{2+} and Tm^{2+} (2) Sm^{2+} and Tm^{3+} **[June 27, 2022 (II)]**
(3) Tb^{4+} and Yb^{2+} (4) Dy^{3+} and Yb^{3+}

198. The most common oxidation state of Lanthanoid elements is +3. Which of the following is likely to deviate easily from +3 oxidation state? **[June 26, 2022 (II)]**
(1) Ce (At. No. 58) (2) La (At. No. 57)
(3) Lu (At. No. 71) (4) Gd (At. No. 64)

199. Cerium (IV) has a noble gas configuration. Which of the following is correct statement about it? **[June 25, 2022 (I)]**
(1) It will not prefer to undergo redox reactions.
(2) It will prefer to gain electron and act as an oxidizing agent
(3) It will prefer to give away an electron and behave as reducing agent
(4) It acts as both, oxidizing and reducing agent.

200. Which one of the following lanthanides exhibits +2 oxidation state with diamagnetic nature?
(Given Z for Nd = 60, Yb = 70, La = 57, Ce = 58) **[Aug. 31, 2021 (I)]**
(1) Nd (2) Yb (3) La (4) Ce

201. The Eu^{2+} ion is a strong reducing agent in spite of its ground state electronic configuration **[Aug. 31, 2021 (II)]**
(outermost) : [Atomic number of Eu = 63]
(1) $4f^7 6s^2$ (2) $4f^6$ (3) $4f^7$ (4) $4f^6 6s^2$

202. Given below are two statements : **[March 16, 2021 (I)]**
Statement I : The E° value for Ce^{4+}/Ce^{3+} is + 1.74 V.
Statement II : Ce is more stable in Ce^{4+} state than Ce^{3+} state.
In the light of the above statements, choose the **most appropriate** answer from the options given below.
(1) Both statement I and statement II are correct
(2) Statement I is correct but statement II is incorrect
(3) Both statement I and statement II are incorrect
(4) Statement I is incorrect but statement II is correct

203. Arrange the following metal complex/compounds in the increasing order of spin only magnetic moment. Presume all the three, high spin system. **[March 16, 2021 (II)]**
(Atomic numbers Ce = 58, Gd = 64 and Eu = 63.)
(A) $(NH_4)_2[Ce(NO_3)_6]$
(B) $Gd(NO_3)_3$ and
(C) $Eu(NO_3)_3$
(1) (A) < (C) < (B) (2) (A) < (B) < (C)
(3) (C) < (A) < (B) (4) (B) < (A) < (C)

204. The lanthanoid that does NOT show + 4 oxidation state is: **[Sep. 06, 2020 (I)]**
(1) Dy (2) Ce (3) Eu (4) Tb

205. Mischmetal is an alloy consisting mainly of: **[Sep. 06, 2020 (II)]**
(1) lanthanoid metals
(2) actinoid and transition metals
(3) lanthanoid and actinoid metals
(4) actinoid metals

206. The correct electronic configuration and spin-only magnetic moment (BM) of Gd^{3+}(Z = 64), respectively, are: **[Sep. 05, 2020 (I)]**
(1) [Xe] $4f^7$ and 8.9 (2) [Xe] $4f^7$ and 7.9
(3) [Xe] $5f^7$ and 8.9 (4) [Xe] $5f^7$ and 7.9

207. The electronic configurations of bivalent europium and trivalent cerium are:
(atomic number : Xe = 54, Ce = 58, Eu = 63) **[Jan. 09, 2020 (I)]**
(1) [Xe] $4f^2$ and [Xe] $4f^7$
(2) [Xe] $4f^7$ and [Xe] $4f^1$
(3) [Xe] $4f^7 6s^2$ and [Xe] $4f^2 6s^2$
(4) [Xe] $4f^4$ and [Xe] $4f^9$

208. The highest possible oxidation states of uranium and plutonium, respectively, are : **[April 10, 2019 (II)]**
(1) 6 and 7 (2) 6 and 4 (3) 7 and 6 (4) 4 and 6

209. The maximum number of possible oxidation states of actinoides are shown by: **[April 9, 2019 (II)]**
(1) Nobelium (No) and lawrencium (Lr)
(2) Actinium (Ac) and thorium (Th)
(3) Berkelium (Bk) and californium (Cf)
(4) Neptunium (Np) and plutonium (Pu)

210. The lanthanide ion that would show colour is: **[April 8, 2019 (I)]**
(1) Gd^{3+} (2) Sm^{3+} (3) La^{3+} (4) Lu^{3+}

211. The correct order of atomic radii is : **[Jan. 12, 2019 (II)]**
(1) N > Ce > Eu > Ho (2) Ho > N > Eu > Ce
(3) Ce > Eu > Ho > N (4) Eu > Ce > Ho > N

212. The effect of lanthanoid contraction in the lanthanoid series of elements by and large means: **[Jan. 10, 2019 (I)]**
(1) increase in both atomic and ionic radii
(2) decrease in atomic radii and increase in ionic radii
(3) decrease in both atomic and ionic radii
(4) increase in atomic radii and decrease in ionic radii

213. Which is not the correct statement?
(At. nos. Ce = 58, Lu = 71, La = 57, Yb = 70)
(1) Colour of Yb^{3+} ion is pink. **[Online May 7, 2012]**
(2) La^{3+} is diamagnetic.
(3) Ce^{4+} has f^0 configuration.
(4) Lu^{3+} had f^{14} configuration.

214. The number of unpaired electrons in Gadolinium [Z = 64] is **[Online May 26, 2012]**
(1) 3 (2) 8 (3) 6 (4) 2

215. Magnetic moment of Gd^{3+} ion (Z = 64) is **[Online May 12, 2012; Same May 26, 2012]**
(1) 3.62 BM (2) 9.72 BM (3) 7.9 BM (4) 10.60 BM

216. Which of the following forms stable + 4 oxidation state? **[Online May 26, 2012]**
(1) La (Z = 57) (2) Eu (Z = 63) (3) Ce (Z = 58) (4) Gd (Z = 64)

217. Knowing that the chemistry of lanthanoids (Ln) is dominated by its + 3 oxidation state, which of the following statements is incorrect? **[2009]**
(1) The ionic size of Ln (III) decreases in general with increasing atomic number
(2) Ln (III) compounds are generally colourless.
(3) Ln (III) hydroxides are mainly basic in character.
(4) Because of the large size of the Ln (III) ions, the bonding in its compounds is predominantly ionic in character.

218. Larger number of oxidation states are exhibited by the actinoids than those by the lanthanoids, the main reason being **[2008]**
(1) $4f$ orbitals more diffused than the $5f$ orbitals
(2) lesser energy difference between $5f$ and $6d$ than between $4f$ and $5d$ orbitals
(3) more energy difference between $5f$ and $6d$ than between $4f$ and $5d$ orbitals
(4) more reactive nature of the actionids than the lanthanoids

219. Identify the incorrect statement among the following: **[2007]**
(1) $4f$ and $5f$ orbitals are equally shielded.
(2) *d*-Block elements show irregular and erratic chemical properties among themselves.
(3) La and Lu have partially filled *d*-orbitals and no other partially filled orbitals.
(4) The chemistry of various lanthanoids is very similar.

220. The actinoids exhibit more number of oxidation states in general than the lanthanoids. This is because **[2007]**
(1) the $5f$ orbitals extend further from the nucleus than the $4f$ orbitals
(2) the $5f$ orbitals are more buried than the $4f$ orbitals
(3) there is a similarity between $4f$ and $5f$ orbitals in their angular part of the wave function
(4) the actinoids are more reactive than the lanthanoids.

221. Lanthanoid contraction is caused due to **[2006; Same 2005]**
(1) the same effective nuclear charge from Ce to Lu
(2) the imperfect shielding on outer electrons by $4f$ electrons from the nuclear charge
(3) the appreciable shielding on outer electrons by $4f$ electrons from the nuclear charge
(4) the appreciable shielding on outer electrons by $5d$ electrons from the nuclear charge

222. Which of the following factors may be regarded as the main cause of lanthanide contraction? [2005]
(1) Greater shielding of 5d electrons by 4f electrons
(2) Poorer shielding of 5d electrons by 4f electrons
(3) Effective shielding of one of the 4f electrons by another in the subshell
(4) Poor shielding of one of the 4f electrons by another in the subshell

223. The lanthanide contraction is responsible for the fact that [2005]
(1) Zr and Zn have the same oxidation state
(2) Zr and Hf have about the same radius
(3) Zr and Nb have similar oxidation state
(4) Zr and Y have about the same radius

224. The radius of La^{3+} (Atomic number of La = 57) is 1.06Å. Which one of the following given values will be closest to the radius of Lu^{3+} (Atomic number of Lu = 71) ? [2003]
(1) 1.40 Å (2) 1.06 Å (3) 0.85 Å (4) 1.60 Å

225. A reduction in atomic size with increase in atomic number is a characteristic of elements of [2003]
(1) d-block (2) f-block
(3) radioactive series (4) high atomic masses

226. Most common oxidation states of Ce (cerium) are [2002]
(1) +2, +3 (2) +2, +4 (3) +3, +4 (4) +3, +5.

227. Arrange Ce^{+3}, La^{+3}, Pm^{+3} and Yb^{+3} in increasing order of their ionic radii. [2002]
(1) $Yb^{+3} < Pm^{+3} < Ce^{+3} < La^{+3}$ (2) $Ce^{+3} < Yb^{+3} < Pm^{+3} < La^{+3}$
(3) $Yb^{+3} < Pm^{+3} < La^{+3} < Ce^{+3}$ (4) $Pm^{+3} < La^{+3} < Ce^{+3} < Yb^{+3}$.

228. Cerium (Z = 58) is an important member of the lanthanoids. Which of the following statements about cerium is **incorrect**? [2004; Same 2002]
(1) The +4 oxidation state of cerium is not known in solutions.
(2) The +3 oxidation state of cerium is more stable than the +4 oxidation state
(3) The common oxidation states of cerium are +3 and +4
(4) Cerium (IV) acts as an oxidizing agent

Numerical Value Questions (NVQs)

229. Number of colourless lanthanoid ions among the following is ________. [April 9, 2024 (I)]
Eu^{3+}, Lu^{3+}, Nd^{3+}, La^{3+}, Sm^{3+}

230. Total number of ions from the following with noble gas configuration is ________. [Jan. 27, 2024 (II)]
Sr^{2+} (Z = 38), Cs^+ (Z = 55), La^{2+} (Z = 57) Pb^{2+} (Z = 82), Yb^{2+} (Z = 70) and Fe^{2+} (Z = 26)

231. The number of f electrons in the ground state electronic configuration of Np (Z = 93) is ________. (Nearest integer) [Aug. 27, 2021 (I)]

232. The number of 4f electrons in the ground state electronic configuration of Gd^{2+} is ________.
[Atomic number of Gd = 64]
[Aug. 26, 2021 (I); Same Sep. 05, 2020 (I)]

233. Number of electrons present in 4f orbital of Ho^{3+} ion is ________. (Given Atomic No. of Ho = 67) [July 25, 2021 (II)]

Part-B (JEE Advanced)

Topic 1- Electronic Configuration and Properties of Transition Elements

1 MCQs with One Correct Answer

1. In the scheme given below, X and Y, respectively, are

$$\text{Metal halide} \xrightarrow{\text{aq. NaOH}} \text{White precipitate (P)} + \text{Filtrate (Q)}$$

$$P \xrightarrow[\text{heat}]{\text{aq. } H_2SO_4,\ PbO_2\text{ (excess)}} X \text{ (a coloured species in solution)}$$

$$Q \xrightarrow[\text{warm}]{MnO(OH)_2,\ \text{conc. } H_2SO_4} Y \text{ (gives blue–coloration with KI–starch paper)}$$

(a) CrO_4^{2-} and Br_2 (b) MnO_4^{2-} and Cl_2 [Adv. 2023]
(c) MnO_4^- and Cl_2 (d) $MnSO_4$ and HOCl

2. Which of the following combination will produce H_2 gas? [Adv. 2017]
(a) Fe metal and conc. HNO_3
(b) Cu metal and conc. HNO_3
(c) Zn metal and NaOH(aq)
(d) Au metal and NaCN(aq) in the presence of air

3. The colour of light absorbed by an aqueous solution of $CuSO_4$ is: [2012]
(a) orange-red (b) blue-green
(c) yellow (d) violet

4. Native silver metal forms a water soluble complex with a dilute aqueous solution of NaCN in the presence of [2008]
(a) nitrogen (b) oxygen
(c) carbon dioxide (d) argon

5. $CuSO_4$ decolourises on addition of KCN, the product formed is [2006 - 3M, –1]
(a) Cu^{2+} get reduced to form $[Cu(CN)_4]^{3-}$
(b) $[Cu(CN)_4]^{2-}$
(c) CuCN
(d) $Cu(CN)_2$

6. Anhydrous ferric chloride is prepared by [2002S]
(a) heating hydrated ferric chloride at a high temperature in a stream of air
(b) heating metallic iron in a stream of dry chlorine gas
(c) reaction of metallic iron with hydrochloric acid
(d) reaction of metallic iron with nitric acid

7. An aqueous solution of $FeSO_4$, $Al_2(SO_4)_3$ and chrome alum is heated with excess of Na_2O_2 and filtered. The materials obtained are : [1996 - 1 Mark]
(a) a colourless filtrate and a green residue
(b) a yellow filtrate and a green residue
(c) a yellow filtrate and a brown residue
(d) a green filtrate and a brown residue

8. Which compound does not dissolve in hot, dilute HNO_3? [1996 - 1 Mark]
(a) HgS (b) PbS (c) CuS (d) CdS

9. Which pair gives Cl_2 at room temperature? [1995S]
(a) HCl(conc) + $KMnO_4$ (b) NaCl + $H_2SO_{4(conc)}$
(c) NaCl + MnO_2 (d) NaCl + $HNO_{3(conc)}$

10. Which one is solder ? [1995S]
(a) Cu & Pb (b) Zn & Cu (c) Pb & Sn (d) Fe & Zn

11. Zinc-copper couple that can be used as a reducing agent is obtained by : [1984 - 1 Mark]
(a) mixing zinc dust and copper gauze
(b) zinc coated with copper
(c) copper coated with zinc
(d) zinc and copper wires welded together

12. Iron is rendered passive by treatment with concentrated [1982 - 1 Mark]
(a) H_2SO_4 (b) H_3PO_4 (c) HCl (d) HNO_3

13. Sodium thiosulphate is used in photography because of its [1981 - 1 Mark]
(a) reducing behaviour
(b) oxidising behaviour
(c) complex forming behaviour
(d) reaction with light

14. How many unpaired electrons are present in Ni^{2+}? [1981 - 1 Mark]
(a) 0 (b) 2 (c) 4 (d) 8

15. Which of the following dissolve in hot conc. NaOH solution [1980]
(a) Fe (b) Zn (c) Cu (d) Ag

16. One of the constituent of German silver is [1980]
(a) Ag (b) Cu (c) Mg (d) Al

17. Which of the following is the weakest base [1980]
(a) NaOH (b) $Ca(OH)_2$ (c) KOH (d) $Zn(OH)_2$

18. When same amount of zinc is treated separately with excess of sulphuric acid and excess of sodium hydroxide, the ratio of volume of hydrogen evolved is [1979]
(a) 1 : 1 (b) 1 : 2 (c) 2 : 1 (d) 9 : 4

2 Integer Value Answer

19. Consider the following list of reagents: [Adv. 2014]
Acidified $K_2Cr_2O_7$, alkaline $KMnO_4$, $CuSO_4$, H_2O_2, Cl_2, O_3, $FeCl_3$, HNO_3 and $Na_2S_2O_3$.
The total number of reagents that can oxidise aqueous iodide to iodine is

20. The oxidation number of Mn in the product of alkaline oxidative fusion of MnO_2 is [2009 - 4 Marks]

4 Fill in the Blanks

21. Silver jewellery items tarnish slowly in the air due to their reaction with .... [1997 - 1 Mark]

22. The salts .... and .... are isostructural.
($FeSO_4.7H_2O$, $CuSO_4.5H_2O$, $MnSO_4.4H_2O$, $ZnSO_4.7H_2O$) [1988-1 Mark]

23. Silver chloride is sparingly soluble in water because its lattice energy is greater than .... energy. [1987 - 1 Mark]

24. Galvanization of iron denotes coating with .... [1983 - 1 Mark]

25. Mn^{2+} can be oxidised to MnO_4^- by ....
(SnO_2, PbO_2, BaO_2) [1981 - 1 Mark]

26. AgCN dissolves in excess KCN solution to give the complex compound .... [1980]

5 True / False

27. Dipositive zinc exhibits paramagnetism due to loss of two electrons from 3*d*-orbital of neutral atom. [1987 - 1 Mark]

28. Silver chloride is more soluble in very concentrated sodium chloride solution than in pure water. [1984 - 1 Mark]

29. Silver fluoride is fairly soluble in water. [1982 - 1 Mark]

6 MCQs with One or More than One Correct Answer

30. Consider the following reactions (unbalanced) Zn + hot conc. $H_2SO_4 \rightarrow G + R + X$ [Adv. 2019]
Zn + conc. NaOH $\rightarrow T + Q$
$G + H_2S + NH_4OH \rightarrow Z$ (a precipitate) $+ X + Y$
Choose the correct option(s)
(a) The oxidation state of Zn in *T* is + 1
(b) Bond order of *Q* is 1 in its ground state
(c) *Z* is dirty white in colour
(d) *R* is a V-shaped molecule

31. Fe^{3+} is reduced to Fe^{2+} by using [Adv. 2015]
(a) H_2O_2 in presence of NaOH
(b) Na_2O_2 in water
(c) H_2O_2 in presence of H_2SO_4
(d) Na_2O_2 in presence of H_2SO_4

32. The correct statement(s) about Cr^{2+} and Mn^{3+} is (are)
[Atomic numbers of Cr = 24 and Mn = 25] [Adv. 2015]
(a) Cr^{2+} is a reducing agent
(b) Mn^{3+} is an oxidizing agent
(c) Both Cr^{2+} and Mn^{3+} exhibit d^4 electronic configuration
(d) When Cr^{2+} is used as a reducing agent, the chromium ion attains d^5 electronic configuration

33. The pair(s) of reagents that yield paramagnetic species is/are [Adv. 2014]
(a) Na and excess of NH_3
(b) K and excess of O_2
(c) Cu and dilute HNO_3
(d) O_2 and 2-ethylanthraquinol

34. The equilibrium [2011]
$2Cu^{I} \rightleftharpoons Cu^{0} + Cu^{II}$
in aqueous medium at 25°C shifts towards the left in the presence of
(a) NO_3^- (b) Cl^- (c) SCN^- (d) CN^-

35. Reduction of the metal centre in aqueous permanganate ion involves [2011]
(a) 3 electrons in neutral medium
(b) 5 electrons in neutral medium
(c) 3 electrons in alkaline medium
(d) 5 electrons in acidic medium

36. Addition of high proportions of manganese makes steel useful in making rails of railroads, because manganese [1998 - 2 Marks]
(a) gives hardness to steel
(b) helps the formation of oxides of iron
(c) can remove oxygen and sulphur
(d) can show highest oxidation state of +7.

37. Which of the following alloys contain(s) Cu and Zn? [1993 - 1 Mark]
(a) Bronze (b) Brass
(c) Gun metal (d) Type metal

38. The aqueous solutions of the following salts will be coloured in the case of [1990 - 1 Mark]
(a) $Zn(NO_3)_2$ (b) $LiNO_3$
(c) $Co(NO_3)_2$ (d) $CrCl_3$
(e) Potash alum

7 Match the Following

39. Match each of the reactions given in **Column-I** with the corresponding product(s) given in **Column-II**. [2009]

Column-I		Column-II	
(A)	Cu + dil HNO_3	(p)	NO
(B)	Cu + conc HNO_3	(q)	NO_2
(C)	Zn + dil HNO_3	(r)	N_2O
(D)	Zn + conc HNO_3	(s)	$Cu(NO_3)_2$
		(t)	$Zn(NO_3)_2$

40. Match the following, choosing one item from column **X** and one from column **Y**. [Multiple Concepts, 1982 - 3 Marks]

X		Y	
(*i*)	Hg_2Cl_2	(a)	cassiterite
(*ii*)	$(NaPO_3)_n$	(b)	lunar caustic
(*iii*)	NO_3^-	(c)	producer gas
(*iv*)	SnO_2	(d)	water softener

(v) $KCl.MgCl_2.6H_2O$ (e) brown ring test
(vi) $AgNO_3$ (f) carnallite
(vii) $CO + N_2$ (g) calomel

8 Comprehension Based Questions

When a metal rod **M** is dipped into an aqueous colourless concentrated solution of compound **N**, the solution turns light blue. Addition of aqueous NaCl to the blue solution gives a white precipitate **O**. Addition of aqueous NH_3 dissolves **O** and gives an intense blue solution. [2011]

41. The metal rod **M** is
(a) Fe (b) Cu (c) Ni (d) CO
42. The compound **N** is
(a) $AgNO_3$ (b) $Zn(NO_3)_2$
(c) $Al(NO_3)_3$ (d) $Pb(NO_3)_2$
43. The final solution contains
(a) $[Pb(NH_3)_4]^{2+}$ and $[CoCl_4]^{2-}$
(b) $[Al(NH_3)_4]^{3+}$ and $[Cu(NH_3)_4]^{2+}$
(c) $[Ag(NH_3)_2]^{+}$ and $[Cu(NH_3)_4]^{2+}$
(d) $[Ag(NH_3)_2]^{+}$ and $[Ni(NH_3)_6]^{2+}$

9 Assertion and Reason/Statement Type Questions

Each question contains **STATEMENT-1 (Assertion)** and **STATEMENT-2 (Reason)**. Each question has 4 choices (a), (b), (c) and (d) out of which ONLY ONE is correct. Mark your answer as
(a) If both Statement -1 and Statement -2 are correct, and Statement -2 is the correct explanation of the Statement -2.
(b) If both Statement -1 and Statement -2 are correct, but Statement -2 is not the correct explanation of the Statement -1.
(c) If Statement -1 is correct but Statement -2 is incorrect.
(d) If Statement -1 is incorrect but Statement -2 is correct.

44. **Statement-1 :** Zn^{2+} is diamagnetic.
Statement-2 : Two electrons are lost from 4s orbital to form Zn^{2+}. [1998 - 2 Marks]
45. **Statement-1 :** To a solution of potassium chromate if a strong acid is added it changes its colour from yellow to orange.
Statement-2 : The colour change is due to the oxidation of potassium chromate. [1988 - 2 Marks]

10 Subjective Problems

46. Write the chemical reaction involved in developing of a black and white photographic film. An aqueous $Na_2S_2O_3$ solution is acidified to give a milky white turbitity. Identify the product and write the balanced half chemical reaction for it. [2005 - 4 Marks]
47. (i) Write the chemical reactions involved in the extraction of metallic silver from argentite.
(ii) Write the balanced chemical equation for developing photographic films. [2000 - 4 Marks]
48. Write the chemical reaction associated with the 'brown ring test'. [2000 - 2 Marks]
49. Work out the following using chemical equations [1998 - 2 Marks]
In moist air copper corrodes to produce a green layer on the surface.
50. Compare qualitatively the first and second ionisation potentials of copper and zinc. Explain the observation. [1996 - 2 Marks]
51. The acidic, aqueous solution of ferrous ion forms a brown complex in the presence of NO_3^-, by the following two steps. Complete and balance the equations : [1993 - 2 Marks]

$$[Fe(H_2O)_6]^{2+} + NO_3^- + H^+ \rightarrow ......+[Fe(H_2O_6)]^{3+} + H_2O$$

$$[Fe(H_2O)_6]^{2+} + ... \rightarrow ......+ H_2O$$

52. Mention the products formed when zinc oxide is treated with excess of sodium hydroxide solution. [1986 - 1 Mark]
53. State the conditions under which the following preparation is carried out. [1983 - 1 Mark]
Potassium permanganate from manganese hydroxide.
Give the necessary equations which need not be balanced.
54. Give reasons for the following :
(i) CrO_3 is an acid anhydride. [1999 - 2 Marks]
(ii) The species $[CuCl_4]^{2-}$ exists while $[CuI_4]^{2-}$ does not. [1992 - 1 Mark]
(iii) The colour of mercurous chloride, Hg_2Cl_2, changes from white to black when treated with ammonia. [1988 - 1 Mark]
(iv) Zinc and not copper is used for the recovery of metallic silver from complex $[Ag(CN)_2]^-$. Explain. [1987 - 1 Mark]
(v) Most transition metal compounds are coloured. [1986 - 1 Mark]
(vi) Silver bromide is used in photography. [1983 - 1 Mark]
55. Complete the following equation (no balancing is needed):
$SO_2 + MnO_4^- + ... \longrightarrow SO_4^{2-} + Mn^{2+} + ...$ [1981 - 1 Mark]
56. A solution of $FeCl_3$ in water gives a brown precipitate on standing. [1980]
57. State with balanced equations, what happens when
(i) Silver is treated with hot concentrated sulphuric acid.
(ii) Ammonium dichromate is heated.
(iii) Hydrogen sulphide is passed through a solution of potassium permanganate acidified with dilute sulphuric acid. [1979]

Topic 2 - Compounds of Transition Elements

1 MCQs with One Correct Answer

58. One of the products formed from the reaction of permanganate ion with iodide ion in neutral aqueous medium is [Adv. 2025]
(a) I_2 (b) IO_3^- (c) IO_4^- (d) IO_2^-
59. Among the following, the coloured compound is [2008]
(a) CuCl (b) $K_3[Cu(CN)_4]$
(c) CuF_2 (d) $[Cu(CH_3CN)_4]BF_4$
60. Which pair of compounds is expected to show similar colour in aqueous medium? [2005S]
(a) $FeCl_2$ and $CuCl_2$ (b) $VOCl_2$ and $CuCl_2$
(c) $VOCl_2$ and $FeCl_2$ (d) $FeCl_2$ and $MnCl_2$
61. $(NH_4)_2Cr_2O_7$ on heating liberates a gas. The same gas will be obtained by [2004S]
(a) heating NH_4NO_2 (b) heating NH_4NO_3
(c) treating H_2O_2 with $NaNO_2$ (d) treating Mg_3N_2 with H_2O
62. The product of oxidation of I^- with MnO_4^- in alkaline medium is
(a) IO_3^- (b) I_2 (c) IO^- (d) IO_4^- [2004S]
63. When MnO_2 is fused with KOH, a coloured compound is formed, the product and its colour is: [2003S]
(a) K_2MnO_4, purple green (b) $KMnO_4$, purple
(c) Mn_2O_3, brown (d) Mn_3O_4, black
64. In the dichromate anion, [1999 - 2 Marks]
(a) 4 Cr – O bonds are equivalent
(b) 6 Cr – O bonds are equivalent
(c) all Cr – O bonds are equivalent
(d) all Cr – O bonds are nonequivalent

65. Which of the following compounds is expected to be coloured? [1997 - 1 Mark]
(a) Ag_2SO_4 (b) CuF_2
(c) MgF_2 (d) CuCl.

66. The number of moles of $KMnO_4$ that will be needed to react with one mole of sulphite ion in acidic solution is [1997 - 1 Mark]
(a) $\frac{2}{5}$ (b) $\frac{3}{5}$ (c) $\frac{4}{5}$ (d) 1

67. Ammonium dichromate is used in some fireworks. The green coloured powder blown in the air is [1997 - 1 Mark]
(a) CrO_3 (b) Cr_2O_3 (c) Cr (d) $CrO(O_2)$

68. Which compound is formed when excess of KCN is added to aqueous solution of copper sulphate? [1996 - 1 Mark]
(a) $Cu(CN)_2$ (b) $K_2[Cu(CN)_4]$
(c) $K[Cu(CN)_2]$ (d) $K_3[Cu(CN)_4]$

2 Integer Value Answer

69. In neutral or faintly alkaline solution, 8 moles of permanganate anion quantitatively oxidize thiosulphate anions to produce *X* moles of a sulphur containing product. The magnitude of *X* is [Adv. 2016]

6 MCQs with One or More than One Correct Answer

70. Fusion of MnO_2 with KOH in presence of O_2 produces a salt *W*. Alkaline solution of *W* upon electrolytic oxidation yields another salt *X*. The manganese containing ions present in W and *X*, respectively are *Y* and *Z*. Correct statement(s) is (are) [Adv. 2019]
(a) In both *Y* and *Z*, p-bonding occurs between *p*-orbitals of oxygen and *d*-orbitals of manganese
(b) In aqueous acidic solution, *Y* undergoes dispro-portionation reaction to give *Z* and MnO_2
(c) Both *Y* and *Z* are coloured and have tetrahedral shape
(d) *Y* is diamagnetic in nature while *Z* is paramagnetic

71. For the given aqueous reactions, which of the statement (s) is (are) true?

$$\text{excess KI} + K_3[Fe(CN)_6] \xrightarrow{\text{dilute } H_2SO_4} \text{brownish-yellow solution}$$

$\downarrow ZnSO_4$

white precipitate + brownish - yellow filtrate

$\downarrow Na_2S_2O_3$ (filtrate)

colourless solution

(a) The first reaction is a redox reaction. [2012]
(b) White precipitate is $Zn_3[Fe(CN)_6]_2$.
(c) Addition of filtrate to starch solution gives blue colour.
(d) White precipitate is soluble in NaOH solution.

72. Potassium manganate (K_2MnO_4) is formed when [1988 - 1 Mark]
(a) chlorine is passed into aqueous $KMnO_4$ solution
(b) manganese dioxide is fused with potassium hydroxide in air
(c) formaldehyde reacts with potassium permanganate in presence of a strong alkali
(d) potassium permanganate reacts with conc. sulphuric acid

8 Comprehension Based Questions

Paragraph - II

When potassium iodide is added to an aqueous solution of potassium ferricyanide, a reversible reaction is observed in which a complex P is formed. In a strong acidic medium, the equilibrium shifts completely towards P. Addition of zinc chloride to P in a slightly acidic medium results in a sparingly soluble complex Q.

73. The number of moles of potassium iodide required to produce two moles of P is _____. [Adv. 2024]

74. The number of zinc ions present in the molecular formula of Q is _____. [Adv. 2024]

10 Subjective Problems

75.
$$\underset{\text{(Colourless liquid)}}{MCl_4} \xrightarrow{Zn} \underset{(A)}{\text{Purple colour compound;}}$$

M = Transition metal

$$MCl_4 \xrightarrow{\text{moist air}} \underset{\text{white fumes}}{(B)}$$

Identify (*A*), (*B*) and MCl_4. Also explain colour difference between MCl_4 and (*A*). [2005 - 4 Marks]

76. What happens when :
(*i*) aqueous ammonia is added dropwise to a solution of copper sulphate till it is in excess. [1985 - 1 Mark]
(*ii*) $CrCl_3$ solution is treated with sodium hydroxide and then with hydrogen peroxide. [1985 - 1 Mark]

77. State with balanced equations what happens when :
(*i*) Write balanced equations for the reaction of zinc with dilute nitric acid. [1997 - 1 Mark]
(*ii*) Write a balanced equation for the reaction of argentite with KCN and name the products in solution. [1996 - 1 Mark]
(*iii*) $[MnO_4]^{2-} + H^+ \longrightarrow \ldots\ldots\ldots + [MnO_4]^- + H_2O$ [1994 - 1 Mark]
(*iv*) $SO_2(aq) + Cr_2O_7^{2-} + 2H^+ \longrightarrow \ldots\ldots + \ldots\ldots + \ldots\ldots$ [1994 - 1 Mark]
(*v*) $(NH_4)_2S_2O_8 + H_2O + MnSO_4 \rightarrow \ldots. + \ldots\ldots + \ldots\ldots$ [1993 - 1 Mark]
(*vi*) $AgBr + Na_2S_2O_3 \rightarrow \ldots. + \ldots\ldots$ [1993 - 1 Mark]
(*vii*) Potassium dichromate and concentrated hydrochloric acid are heated together. [1992 - 1 Mark]
(*viii*) Na_2CO_3 is added to a solution of copper sulphate. [1992 - 1 Marks]
$CuSO_4 + Na_2CO_3 + H_2O \rightarrow \ldots\ldots\ldots. + Na_2SO_4 + \ldots\ldots\ldots.$
(*ix*) Copper reacts with HNO_3 to give NO and NO_2 in molar ratio of 2 : 1. [1992 - 1 Marks]
$Cu + HNO_3 \rightarrow \ldots\ldots\ldots. + NO + NO_2 + \ldots\ldots\ldots.$.
(*x*) Potassium permanganate is added to a hot solution of manganous sulphate. [1990 - 1 Mark]
(*xi*) Iron reacts with cold dilute nitric acid. [1990 - 1 Mark]
(*xii*) A mixture of potassium dichromate and sodium chloride is heated with concentrated H_2SO_4. [1990 - 1 Mark]
(*xiii*) Write balanced equations for the extraction of copper from copper pyrites by self-reduction. [1990 - 2 Marks]
(*xiv*) Cobalt(II) solution reacts with KNO_2 in acetic acid medium. [1989 - 1 Mark]
(*xv*) Silver chloride is treated with aqueous sodium cyanide and the product thus formed is allowed to react with zinc in alkaline medium. [1989 - 1 Mark]
(*xvi*) Write balanced equations for the extraction of silver from silver glance by cyanide process. [1988 - 1 Mark]
(*xvii*) Gold is dissolved in *aqua regia*. [1987 - 1 Mark]
(*xviii*) Potassium permanganate is reacted with warm solution of oxalic acid in the presence of sulphuric acid. [1987 - 1 Mark]
(*xix*) potassium ferrocyanide is heated with concentrated sulphuric acid; [1985 - 1 Mark]
(*xx*) potassium permanganate interacts with manganese dioxide in presence of potassium hydroxide; [1985 - 1 Mark]
(*xxi*) aqueous solution of potassium chromate and acid are mixed. [1984 - 2 Marks]
(*xxii*) aqueous solution of potassium manganate and acid are mixed. [1984 - 2 Marks]

(*xxiii*) aqueous solution of ferric sulphate and potassium iodide are mixed. **[1984 - 2 Marks]**

(*xxiv*) sulphur dioxide gas is bubbled through an aqueous solution of copper sulphate in presence of potassium thiocyanate. **[1982 - 1 Mark]**

78. A white amorphous powder (A) on heating yields a colourless, non-combustible gas (B) and a solid (C). The latter compound assumes a yellow colour on heating and changes to white on cooling. 'C' dissolves in dilute acid and the resulting solution gives a white precipitate on adding $K_4[Fe(CN)_6]$ solution.
'A' dissolves in dilute HCl with the evolution of gas, which is identical in all respects with 'B'. The gas 'B' turns lime water milky, but the milkiness disappears with the continuous passage of gas. The solution of 'A', as obtained above, gives a white precipitate (D) on the addition of excess of NH_4OH and passing H_2S. Another portion of the solution gives initially a white precipitate (E) on the addition of sodium hydroxide solution, which dissolves on futher addition of the base. Identify the compounds A, B, D, and E **[1979]**

79. A certain inorganic compound (A) on heating loses its water of crystallisation. On further heating, a blackish brown powder (B) and two oxides of sulphur (C and D) are obtained. The powder (B) on boiling with hydrochloric acid gives a yellow solution (E). When H_2S is passed in (E) a white turbidity (F) and an apple green solution (G) are obtained. The solution (E) on treatment with thiocyanate ions gives a blood red coloured compound (H). Identify compounds from (A) to (H). **[1978]**

Topic 3 - Lanthanoids and Actinoids

6 MCQs with One or More than One Correct Answer

80. The pair(s) of diamagnetic ions is(are) **[Adv. 2025]**

(a) La^{3+}, Ce^{4+} (b) Yb^{2+}, Lu^{3+}

(c) La^{2+}, Ce^{3+} (d) Yb^{3+}, Lu^{2+}

Answer Key

Part-A (JEE Main)

1	(2)	**25**	(1)	**49**	(1)	**73**	(4)	**97**	(2)	**121**	(2)	**145**	(1)	**169**	(6)	**193**	(1)	**217**	(2)
2	(2)	**26**	(2)	**50**	(2)	**74**	(1)	**98**	(2)	**122**	(1)	**146**	(1)	**170**	(0)	**194**	(4)	**218**	(2)
3	(3)	**27**	(2)	**51**	(1)	**75**	(3)	**99**	(3)	**123**	(2)	**147**	(3)	**171**	(3)	**195**	(1,4)	**219**	(1)
4	(2)	**28**	(2)	**52**	(1)	**76**	(3)	**100**	(4)	**124**	(4)	**148**	(1)	**172**	(3)	**196**	(3)	**220**	(1)
5	(3)	**29**	(4)	**53**	(4)	**77**	(4)	**101**	(3)	**125**	(4)	**149**	(3)	**173**	(6)	**197**	(3)	**221**	(2)
6	(2)	**30**	(4)	**54**	(2)	**78**	(0)	**102**	(3)	**126**	(2)	**150**	(12)	**174**	(6)	**198**	(1)	**222**	(4)
7	(3)	**31**	(2)	**55**	(2)	**79**	(11)	**103**	(4)	**127**	(4)	**151**	(10)	**175**	(173)	**199**	(2)	**223**	(2)
8	(3)	**32**	(4)	**56**	(3)	**80**	(6)	**104**	(3)	**128**	(2)	**152**	(6)	**176**	(18)	**200**	(2)	**224**	(3)
9	(4)	**33**	(1)	**57**	(1)	**81**	(0)	**105**	(2)	**129**	(3)	**153**	(160)	**177**	(18.00)	**201**	(3)	**225**	(2)
10	(1)	**34**	(4)	**58**	(3)	**82**	(0)	**106**	(4)	**130**	(4)	**154**	(5)	**178**	(1)	**202**	(2)	**226**	(3)
11	(2)	**35**	(1)	**59**	(1)	**83**	(5)	**107**	(2)	**131**	(4)	**155**	(6)	**179**	(4)	**203**	(1)	**227**	(1)
12	(3)	**36**	(2)	**60**	(4)	**84**	(6)	**108**	(4)	**132**	(3)	**156**	(877)	**180**	(3)	**204**	(3)	**228**	(1)
13	(2)	**37**	(1)	**61**	(4)	**85**	(0)	**109**	(2)	**133**	(3)	**157**	(6)	**181**	(3)	**205**	(1)	**229**	(2)
14	(2)	**38**	(2)	**62**	(2)	**86**	(5)	**110**	(1)	**134**	(1)	**158**	(4)	**182**	(3)	**206**	(2)	**230**	(2)
15	(2)	**39**	(1)	**63**	(3)	**87**	(5)	**111**	(2)	**135**	(2)	**159**	(6)	**183**	(1)	**207**	(2)	**231**	(18)
16	(1)	**40**	(1)	**64**	(2)	**88**	(4)	**112**	(4)	**136**	(1)	**160**	(5)	**184**	(2)	**208**	(1)	**232**	(7)
17	(3)	**41**	(3)	**65**	(4)	**89**	(3)	**113**	(4)	**137**	(4)	**161**	(3)	**185**	(3)	**209**	(4)	**233**	(10)
18	(3)	**42**	(3)	**66**	(1)	**90**	(12)	**114**	(2)	**138**	(1,2)	**162**	(10)	**186**	(2)	**210**	(2)		
19	(4)	**43**	(1)	**67**	(1)	**91**	(49)	**115**	(4)	**139**	(2)	**163**	(3)	**187**	(4)	**211**	(4)		
20	(4)	**44**	(4)	**68**	(4)	**92**	(2)	**116**	(3)	**140**	(4)	**164**	(4)	**188**	(1)	**212**	(3)		
21	(2)	**45**	(1)	**69**	(1)	**93**	(1)	**117**	(1)	**141**	(1)	**165**	(1)	**189**	(2)	**213**	(1)		
22	(3)	**46**	(3)	**70**	(4)	**94**	(4)	**118**	(2)	**142**	(2)	**166**	(4)	**190**	(4)	**214**	(2)		
23	(2)	**47**	(1)	**71**	(1)	**95**	(1)	**119**	(2)	**143**	(1)	**167**	(6)	**191**	(2)	**215**	(3)		
24	(4)	**48**	(3)	**72**	(1)	**96**	(2)	**120**	(1)	**144**	(4)	**168**	(0)	**192**	(2)	**216**	(3)		

Part-B (JEE Advanced)

1	(c)	**7**	(c)	**13**	(c)	**19**	(7)	**32**	(a, b, c)	**38**	(c, d)	**45**	(c)	**63**	(a)	**69**	(6)	**80**	(a, b)
2	(c)	**8**	(a)	**14**	(b)	**20**	(6)	**33**	(a, b, c)	**39**	(A) –p, s ; (B) q, s ; (C) –r, t ; (D) –q, t	**58**	(b)	**64**	(b)	**70**	(a, b, c)		
3	(a)	**9**	(a)	**15**	(b)	**28**	(True)	**34**	(b, c, d)	**41**	(b)	**59**	(c)	**65**	(b)	**71**	(a, c, d)		
4	(b)	**10**	(c)	**16**	(b)	**29**	(True)	**35**	(a, d)	**42**	(a)	**60**	(b)	**66**	(a)	**72**	(b, c)		
5	(a)	**11**	(b)	**17**	(d)	**30**	(b, c, d)	**36**	(a, c)	**43**	(c)	**61**	(a)	**67**	(b)	**73**	(2)		
6	(b)	**12**	(d)	**18**	(a)	**31**	(a, b)	**37**	(b, c)	**44**	(b)	**62**	(a)	**68**	(d)	**74**	(3 or 2)		

Chapter-9

Co-ordination Compounds

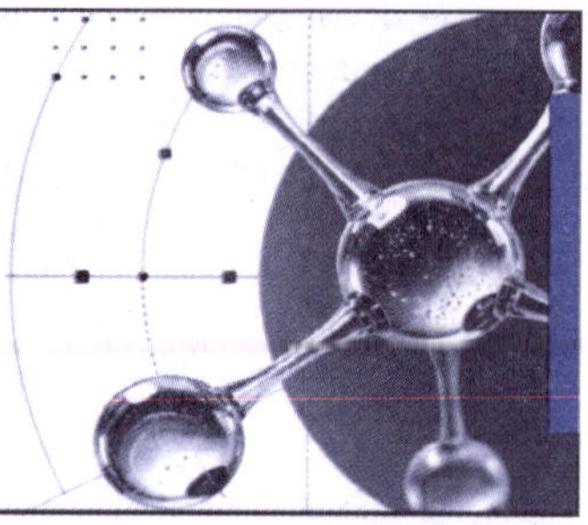

Part-A (JEE Main)

Topic 1 – Werner's Theory, Important terms, Nomenclature and Isomerism of Coordination Compounds

MULTIPLE CHOICE QUESTIONS (MCQs)

1. Which of the following complexes will show coordination isomerism? **[April 02, 2026 (II)]**

(A) $[Ag(NH_3)_2][Ag(CN)_2]$ (B) $[Co(NH_3)_6][Cr(CN)_6]$
(C) $[Co(NH_3)_6][Co(CN)_6]$ (D) $[Fe(NH_3)_6][Co(CN)_6]$
(E) $[Co(NH_3)_6][Fe(CN)_6]$

Choose the **correct** answer from the options given below :

(1) B, C and D Only (2) B, D and E Only
(3) A, C and D Only (4) C, D and E Only

2. Consider a mixture 'X' which is made by dissolving 0.4 mol of $[Co(NH_3)_5SO_4]$ Br and 0.4 mol of $[Co(NH_3)_5Br]SO_4$ in water to make 4 L of solution. When 2L of mixture 'X' is allowed to react with excess of $AgNO_3$, it forms precipitate 'Y'. The rest 2L of mixture 'X' reacts with excess $BaCl_2$ to form precipitate 'Z'. Which of the following statement is CORRECT. **[Jan. 24, 2026 (I)]**

(1) 0.2 mol of 'Z' is formed (2) 'Y' is $BaSO_4$ and 'Z' is Ag Br
(3) 0.4 mol of 'Z' is formed (4) 0.1 mol of 'Y' is formed

3. A first row transition metal (M) does not liberate H_2 gas from dilute HCl. 1 mol of aqueous solution of MSO_4 is treated with excess of aqueous KCN and then $H_2S(g)$ is passed through the solution. The amount of MS (metal sulphide) formed from the above reaction is _______ mol. **[Jan. 22, 2026 (I)]**

(1) 2 (2) 1 (3) 3 (4) 0

4. Given below are two statements : **[April 8, 2025 (II)]**

Statement I : A homoleptic octahedral complex, formed using monodentate ligands, will not show stereoisomerism.

Statement II : cis– and trans– platin are heteroleptic complexes of Pd.

In the light of the above statements, choose the **correct** answer from the options given below.

(1) Both statement I and Statement II are false.
(2) Statement I is false but Statement II is true.
(3) Both statement I and Statement II are true.
(4) Statement I is true but Statement II is false.

5. Match **List-I** with **List-II** **[April 7, 2025 (II)]**

List-I (Complex)		List-II Primary valency	Secondary valency)
(A) $[Co(en)_2Cl_2]Cl$	(I)	3	6
(B) $[Pt(NH_3)_2Cl(NO_2)]$	(II)	3	4
(C) $Hg[Co(SCN)_4]$	(III)	2	6
(D) $[Mg(EDTA)]^{2-}$	(IV)	2	4

Choose the **correct** answer from the options given below:

(1) (A)-(III), (B)-(I), (C)-(II), (D)-(IV)
(2) (A)-(I), (B)-(IV), (C)-(II), (D)-(III)
(3) (A)-(I), (B)-(III), (C)-(II), (D)-(IV)
(4) (A)-(II), (B)-(III), (C)-(IV), (D)-(I)

6. 'X' is the number of acidic oxides among VO_2, V_2O_3, CrO_3, V_2O_5 and Mn_2O_7. The primary valency of cobalt in $[Co(H_2NCH_2CH_2NH_2)_3]_2(SO_4)_3$ is Y. The value of $X + Y$ is: **[April 7, 2025 (II)]**

(1) 5 (2) 4 (3) 2 (4) 3

7. An octahedral complex having molecular composition $Co.5NH_3.Cl.SO_4$ has two isomers A and B. The solution of A gives a white precipitate with $AgNO_3$ solution and the solution of B gives white precipitate with $BaCl_2$ solution. The type of isomerism exhibited by the complex is, **[April 7, 2025 (I)]**

(1) Co-ordinate isomerism (2) Linkage isomerism
(3) Ionisation isomerism (4) Geometrical isomerism

8. Number of stereoisomers possible for the complexes, $[CrCl_3(py)_3]$ and $[CrCl_2(ox)_2]^{3-}$ are respectively (py = pyridine, ox = oxalate) **[April 4, 2025 (I)]**

(1) 3 & 3 (2) 2 & 2 (3) 2 & 3 (4) 1 & 2

9. Identify the homoleptic complexes with odd number of d electrons in the central metal. **[Jan. 29, 2025 (II)]**

(A) $[FeO_4]^{2-}$ (B) $[Fe(CN)_6]^{3-}$
(C) $[Fe(CN)_5NO]^{2-}$ (D) $[CoCl_4]^{2-}$
(E) $[Co(H_2O)_3F_3]$

Choose the correct answer from the options given below :

(1) (B) and (D) only (2) (C) and (E) only
(3) (A), (B) and (D) only (4) (A), (C) and (E) only

10. One mole of the octahedral complex compound $Co(NH_3)_5Cl_3$ gives 3 moles of ions on dissolution in water. One mole of the same complex reacts with excess of $AgNO_3$ solution to yield two moles of $AgCl_{(s)}$. The structure of the complex is: **[Jan. 24, 2025 (I)]**

(1) $[Co(NH_3)_5Cl]Cl_2$ (2) $[Co(NH_3)_4Cl].Cl_2.NH_3$
(3) $[Co(NH_3)_4Cl_2]Cl.NH_3$ (4) $[Co(NH_3)_3Cl_3].2NH_3$

11. The complex that shows Facial – Meridional isomerism is [Jan. 23, 2025 (I)]
(1) $[Co(NH_3)_3Cl_3]$ (2) $[Co(NH_3)_4Cl_2]^+$
(3) $[Co(en)_3]^{3+}$ (4) $[Co(en)_2Cl_2]^+$

12. Identify the homoleptic complex(es) that is/are low spin.
(A) $[Fe(CN)_5NO]^{2-}$ (B) $[CoF_6]^{3-}$
(C) $[Fe(CN)_6]^{4-}$ (D) $[Co(NH_3)_6]^{3+}$
(E) $[Cr(H_2O)_6]^{2+}$ [Jan. 22, 2025 (II)]
Choose the correct answer from the options given below:
(1) (B) and (E) only (2) (A) and (C) only
(3) (C) and (D) only (4) (C) only

13. An octahedral complex with the formula $CoCl_3nNH_3$ upon reaction with excess of $AgNO_3$ solution given 2 moles of AgCl. Consider the oxidation state of Co in the complex is 'x'. The value of "x + n" is _____. [April 8, 2024 (I)]
(1) 3 (2) 6 (3) 8 (4) 5

14. The correct IUPAC name of $[PtBr_2(PMe_3)_2]$ is: [April 6, 2024 (II)]
(1) bis(trimethylphosphine)dibromoplatinum(II)
(2) bis[bromo(trimethylphosphine)]platinum(II)
(3) dibromobis(trimethylphosphine)platinum(II)
(4) dibromodi(trimethylphosphine)platinum(II)

15. The metal atom present in the complex MABXL (where A, B, X and L are unidentate ligands and M is metal) involves sp^3 hybridization. The number of geometrical isomers exhibited by the complex is: [April 5, 2024 (II)]
(1) 4 (2) 0 (3) 2 (4) 3

16. Which of the following complex is homoleptic? [Feb. 1, 2024 (I)]
(1) $[Ni(CN)_4]^{2-}$ (2) $[Ni(NH_3)_2Cl_2]$
(3) $[Fe(NH_3)_4Cl_2]^+$ (4) $[Co(NH_3)_4Cl_2]^+$

17. The correct IUPAC name of K_2MnO_4 is [Jan. 29, 2024 (II)]
(1) Potassium tetraoxidomanganese (VI)
(2) Potassium tetraoxidomanganate (VI)
(3) Dipotassium tetraoxidomanganate (VII)
(4) Potassium tetraoxopermanganate (VI)

18. The total number of stereoisomers for the complex $[Cr(ox)_2ClBr]^{3-}$ (where ox = oxalate) is: [April 13, 2023 (II)]
(1) 2 (2) 3 (3) 1 (4) 4

19. The set which does not have ambidentate ligand(s) is
(1) $C_2O_4^{2-}$, ethylene diammine, H_2O [April 11, 2023 (I)]
(2) $EDTA^{4-}$, NCS^-, $C_2O_4^{2-}$
(3) NO_2^-, $C_2O_4^{2-}$, $EDTA^{4-}$
(4) $C_2O_4^{2-}$, NO_2^-, NCS^-

20. Which of the following complex has a possibility to exist as meridional isomer? [April 11, 2023 (I)]
(1) $[Co(NH_3)_3(NO_2)_3]$ (2) $[Co(en)_3]$
(3) $[Co(en)_2Cl_2]$ (4) $[Pt(NH_3)_2Cl_2]$

21. The IUPAC name of $K_3[Co(C_2O_4)_3]$ is:- [April 06, 2023 (II)]
(1) Potassium trioxalatocobaltate (III)
(2) Potassium tris(oxalato)cobalt (III)
(3) Potassium tris(oxalato) cobaltate (III)
(4) Potassium trioxalatocobalt (III)

22. The complex cation which has two isomers is : [Feb. 01, 2023 (II)]
(1) $[Co(NH_3)_5Cl]^{2+}$ (2) $[Co(NH_3)_5Cl]^+$
(3) $[Co(NH_3)_5NO_2]^{2+}$ (4) $[Co(H_2O)_6]^{3+}$

23. Which of the following are the example of double salt?
(A) $FeSO_4.(NH_4)_2SO_4.6H_2O$ [Feb. 01, 2023 (I)]
(B) $CuSO_4.4NH_3.H_2O$
(C) $K_2SO_4.Al_2(SO_4)_3.24H_2O$
(D) $Fe(CN)_2.4KCN$
Choose the correct answer
(1) A, B and D only (2) A and B only
(3) B and D only (4) A and C only

24. 1 L, 0.02 M solution of $[Co(NH_3)_5SO_4]Br$ is mixed with 1L, 0.02 M solution of $[Co(NH_3)_5Br]SO_4$. The resulting solution is divided into two equal parts (X) and treated with excess $AgNO_3$ solution and $BaCl_2$ solution respectively as shown below:
1 L Solution (X) + $AgNO_3$ solution (excess) → Y
1 L Solution (X) + $BaCl_2$ solution (excess) → Z
The number of moles of Y and Z respectively are
(1) 0.02, 0.02 (2) 0.01, 0.01 [Jan. 30, 2023 (II)]
(3) 0.02, 0.01 (4) 0.01, 0.02

25. Chiral complex from the following is : [Jan. 29, 2023 (I)]
Here en = ethylene diamine
(1) cis – $[PtCl_2(en)_2]^{2+}$ (2) trans – $[PtCl_2(en)_2]^{2+}$
(3) cis – $[PtCl_2(NH_3)_2]$ (4) trans – $[Co(NH_3)_4Cl_2]^+$

26. The primary and secondary valencies of cobalt respectively in $[Co(NH_3)_5Cl]Cl_2$ are : [Jan. 24, 2023 (I)]
(1) 3 and 5 (2) 2 and 6 (3) 2 and 8 (4) 3 and 6

27. Correct formula of the compound which gives a white precipitate with $BaCl_2$ solution, but not with $AgNO_3$ solution, is [June 30, 2022 (I)]
(1) $[Co(NH_3)_5Br]SO_4$ (2) $[Co(NH_3)_5SO_4]Br$
(3) $[Pt(NH_3)_4Cl_2]Br_2$ (4) $[Pt(NH_3)_4Br_2]Cl_2$

28. Given below are two statements. [June 29, 2022 (II)]
Statement I : In $CuSO_4.5H_2O$, Cu–O bonds are present.
Statement II : In $CuSO_4.5H_2O$, ligands coordinating with Cu(II) ion are O-and S-based ligands.
In the light of the above statements, choose the correct answer from the options given below:
(1) Both Statement I and Statement II are correct.
(2) Both Statement I and Statement II are incorrect.
(3) Statement I is correct but Statement II is incorrect.
(4) Statement I is incorrect but Statement II is correct.

29. White precipitate of AgCl dissolves in aqueous ammonia solution due to formation of : [June 25, 2022 (I)]
(1) $[Ag(NH_3)_4]Cl_2$ (2) $[Ag(Cl)_2(NH_3)_2]$
(3) $[Ag(NH_3)_2]Cl$ (4) $[Ag(NH_3)Cl]Cl$

30. The denticity of an organic ligand, biuret is : [Aug. 31, 2021 (I)]
(1) 2 (2) 4 (3) 3 (4) 6

31. The number of geometrical isomers found in the metal complexes $[PtCl_2(NH_3)_2]$, $[Ni(CO)_4]$, $[Ru(H_2O)_3Cl_3]$ and $[CoCl_2(NH_3)_4]^+$ respectively, are : [July 27, 2021 (I)]
(1) 1, 1, 1, 1 (2) 2, 1, 2, 2
(3) 2, 0, 2, 2 (4) 2, 1, 2, 1

32. Which one of the following metal complexes is most stable? [July 25, 2021 (II)]
(1) $[Co(en)(NH_3)_4]Cl_2$ (2) $[Co(en)_3]Cl_2$
(3) $[Co(en)_2(NH_3)_2]Cl_2$ (4) $[Co(NH_3)_6]Cl_2$

33. Consider the complex ions, *trans*-$[Co(en)_2Cl_2]^+$ (A) and *cis*-$[Co(en)_2Cl_2]^+$ (B). The correct statement regarding them is : [Sep. 05, 2020 (II)]
(1) both (A) and (B) cannot be optically active.
(2) (A) can be optically active, but (B) cannot be optically active.
(3) both (A) and (B) can be optically active.
(4) (A) cannot be optically active, but (B) can be optically active.

34. The number of isomers possible for $[Pt(en)(NO_2)_2]$ is : **[Sep. 04, 2020 (I)]**

(1) 2 (2) 4 (3) 1 (4) 3

35. The complex that can show optical activity is : **[Sep. 03, 2020 (I)]**

(1) *trans*-$[Cr(Cl_2)(ox)_2]^{3-}$
(2) *trans*-$[Fe(NH_3)_2(CN)_4]^-$
(3) *cis*-$[Fe(NH_3)_2(CN)_4]^-$
(4) *cis*-$[CrCl_2(ox)_2]^{3-}$ (ox = oxalate)

36. Complex A has a composition of $H_{12}O_6Cl_3Cr$. If the complex on treatment with conc. H_2SO_4 loses 13.5% of its original mass, the correct molecular formula of A is :
[Given : atomic mass of Cr = 52 amu and Cl = 35 amu] **[Sep. 03, 2020 (II)]**

(1) $[Cr(H_2O)_6]Cl_3$
(2) $[Cr(H_2O)_3Cl_3]\cdot 3H_2O$
(3) $[Cr(H_2O)_5Cl]Cl_2\cdot H_2O$
(4) $[Cr(H_2O)_4Cl_2]Cl\cdot 2H_2O$

37. The one that is not expected to show isomerism is :

(1) $[Ni(NH_3)_4(H_2O)_2]^{2+}$ (2) $[Ni(en)_3]^{2+}$ **[Sep. 02, 2020 (II)]**
(3) $[Ni(NH_3)_2Cl_2]$ (4) $[Pt(NH_3)_2Cl_2]$

38. Complex X of composition $Cr(H_2O)_6Cl_n$ has a spin only magnetic moment of 3.83 BM. It reacts with $AgNO_3$ and shows geometrical isomerism. The IUPAC nomenclature of X is: **[Jan. 09, 2020 (I)]**

(1) Hexaaqua chromium (III) chloride
(2) Tetraaquadichlorido chromium (IV) chloride dihydrate
(3) Dichloridotetraaqua chromium (IV) chloride dihydrate
(4) Tetraaquadichlorido chromium (III) chloride dihydrate

39. The isomer(s) of $[Co(NH_3)_4Cl_2]$ that has/have a Cl–Co–Cl angle of 90°, is/are: **[Jan. 09, 2020 (II)]**

(1) meridional and *trans* (2) *cis* and *trans*
(3) *trans* only (4) *cis* only

40. The complex that can show *fac*- and *mer*- isomers is: **[Jan. 08, 2020 (I)]**

(1) $[Co(NH_3)_4Cl_2]^+$ (2) $[Pt(NH_3)_2Cl_2]$
(3) $[CoCl_2(en)_2]$ (4) $[Co(NH_3)_3(NO_2)_3]$

41. Among (1) - (4), the complexes that can display geometrical isomerism are: **[Jan. 08, 2020 (II)]**

(A) $[Pt(NH_3)_3Cl]^+$ (B) $[Pt(NH_3)Cl_5]^-$
(C) $[Pt(NH_3)_2Cl(NO_2)]$ (D) $[Pt(NH_3)_4ClBr]^{2+}$
(1) (B) and (C) (2) (D) and (A)
(3) (C) and (D) (4) (A) and (B)

42. The IUPAC name of the complex $[Pt(NH_3)_2Cl(NH_2CH_3)]Cl$ is: **[Jan. 07, 2020 (I)]**

(1) Diamminechlorido (methanamine) platinum (II)chloride
(2) Diammine (methanamine) chlorido platinum (II)chloride
(3) Diamminechlorido (aminomethane) platinum (II)chloride
(4) Bisammine (methanamine) chlorido platinum (II)chloride

43. The number of possible optical isomers for the complexes MA_2B_2 with sp^3 and dsp^2 hybridized metal atom, respectively, is:
Note: A and B are unidentate neutral and unidentate monoanionic ligands, respectively. **[Jan. 07, 2020 (II)]**

(1) 0 and 2 (2) 2 and 2 (3) 0 and 0 (4) 0 and 1

44. The coordination numbers of Co and Al in $[Co(Cl)(en)_2]Cl$ and $K_3[Al(C_2O_4)_3]$, respectively, are :
(en = ethane-1, 2-diamine) **[April 12, 2019 (II)]**

(1) 5 and 3 (2) 3 and 3 (3) 6 and 6 (4) 5 and 6

45. The species that can have a trans-isomer is :
(en = ethane – 1, 2–diamine, Ox = oxalate) **[April 10, 2019 (I)]**

(1) $[Zn(en)Cl_2]$ (2) $[Pt(en)Cl_2]$
(3) $[Cr(en)_2(Ox)]^+$ (4) $[Pt(en)_2Cl_2]^{2+}$

46. The maximum possible denticities of a ligand given below towards a common transition and inner-transition metal ion, respectively, are: **[April 9, 2019 (II)]**

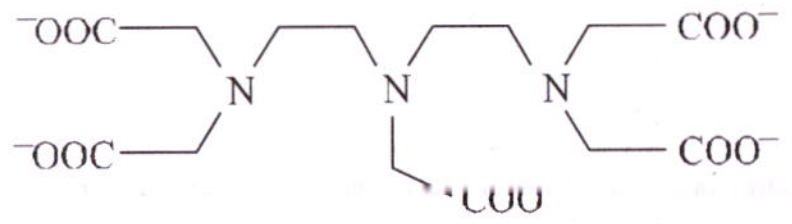

(1) 8 and 6 (2) 6 and 8 (3) 6 and 6 (4) 8 and 8

47. The one that will show optical activity is:
(en = ethane 1, 2-diamine) **[April 9, 2019 (I)]**

(1) 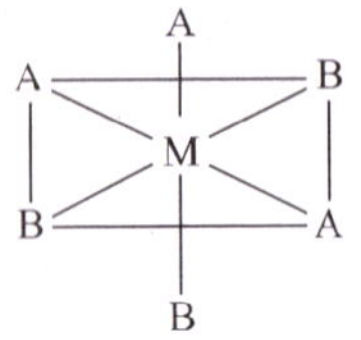

(2) 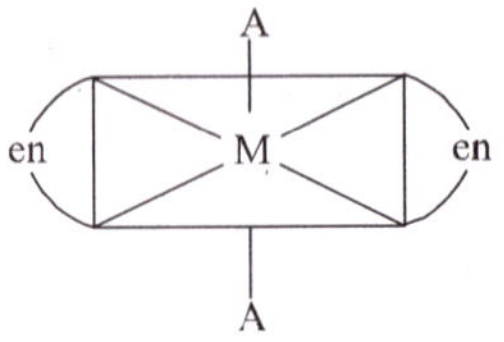

(3) 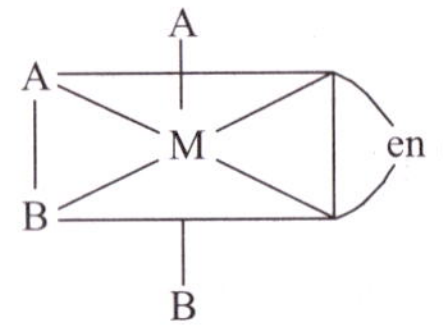

(4) 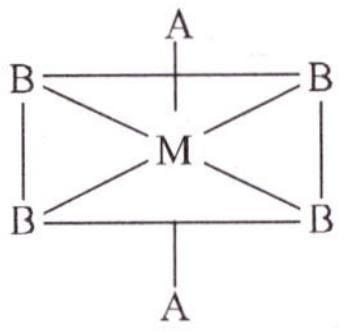

48. The following ligand is: **[April 8, 2019 (I)]**

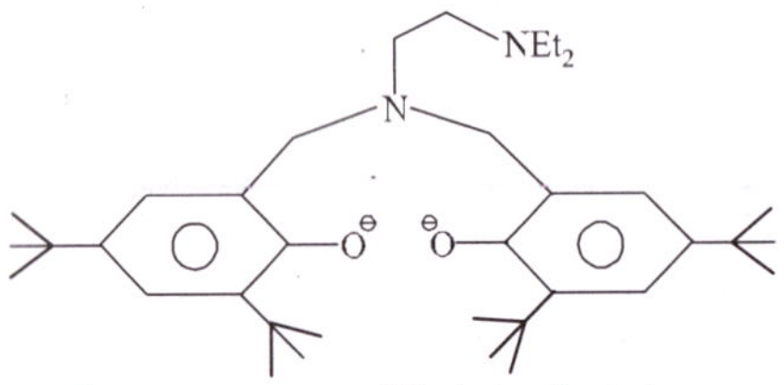

(1) hexadentate (2) tetradentate
(3) bidentate (4) tridentate

49. The coordination number of Th in $K_4[Th(C_2O_4)_4(H_2O)_2]$ is: ($C_2O_4^{2-}$ = oxalato) **[Jan. 11, 2019 (II)]**

(1) 14 (2) 6 (3) 8 (4) 10

50. The total number of isomers for a square planar complex $[M(F)(Cl)(SCN)(NO_2)]$ is: **[Jan. 10, 2019 (I)]**

(1) 16 (2) 8 (3) 4 (4) 12

51. A reaction of cobalt(III) chloride and ethylenediamine in a 1 : 2 mole ratio generates two isomeric products A (violet coloured) and B (green coloured). A can show optical activity, but, B is optically inactive. What type of isomers does A and B represent? **[Jan. 10, 2019 (II)]**

(1) Geometrical isomers (2) Coordination isomers
(3) Linkage isomers (4) Ionisation isomers

52. Consider the following reaction and statements: **[2018]**

$[Co(NH_3)_4Br_2]^+ + Br^- \rightarrow [Co(NH_3)_3Br_3] + NH_3$

(I) Two isomers are produced if the reactant complex ion is a *cis*-isomer.
(II) Two isomers are produced if the reactant complex ion is a *trans* -isomer.
(III) Only one isomer is produced if the reactant complex ion is a *trans* -isomer.
(IV) Only one isomer is produced if the reactant complex ion is a *cis*-isomer.

The correct statements are:

(1) (I) and (II) (2) (I) and (III)
(3) (III) and (IV) (4) (II) and (IV)

53. Which of the following complexes will show geometrical isomerism? [Online April 16, 2018]
(1) Potassium tris(oxalato) chromate (III)
(2) Pentaaquachlorochromium (III) chloride
(3) Aquachlorobis (ethylenediamine) cobalt (II) chloride
(4) Potassium amminetrichloroplatinate (II)

54. The total number of possible isomers for square-planar $[Pt(Cl)(NO_2)(NO_3)(SCN)]^{2-}$ is: [Online April 15, 2018 (II)]
(1) 16 (2) 12 (3) 8 (4) 24

55. On treatment of 100 mL of 0.1 M solution of $CoCl_3 \cdot 6H_2O$ with excess $AgNO_3$; 1.2×10^{22} ions are precipitated. The complex is : [2017]
(1) $[Co(H_2O)_4Cl_2]Cl \cdot 2H_2O$ (2) $[Co(H_2O)_3Cl_3] \cdot 3H_2O$
(3) $[Co(H_2O)_6]Cl_3$ (4) $[Co(H_2O)_5Cl]Cl_2 \cdot H_2O$

56. The pair of compounds having metals in their highest oxidation state is : [Online April 8, 2017]
(1) MnO_2 and CrO_2Cl_2
(2) $[NiCl_4]^{2-}$ and $[CoCl_4]^{2-}$
(3) $[Fe(CN)_6]^{3-}$ and $[Cu(CN)_4]^{2-}$
(4) $[FeCl_4]^-$ and Co_2O_3

57. Which one of the following complexes will consume more equivalents of aqueous solution of $AgNO_3$? [Online April 9, 2016]
(1) $Na_2[CrCl_5(H_2O)]$ (2) $Na_3[CrCl_6]$
(3) $[Cr(H_2O)_5Cl]Cl_2$ (4) $[Cr(H_2O)_6]Cl_3$

58. Which of the following is an example of homoleptic complex? [Online April 10, 2016]
(1) $[Co(NH_3)_6]Cl_3$ (2) $[Pt(NH_3)_2Cl_2]$
(3) $[Co(NH_3)_4Cl_2]$ (4) $[Co(NH_3)_5Cl]Cl_2$

59. The number of geometric isomers that can exist for square planar complex $[Pt (Cl) (py) (NH_3) (NH_2OH)]^+$ is (py = pyridine) : [2015]
(1) 4 (2) 6 (3) 2 (4) 3

60. The correct statement on the isomerism associated with the following complex ions, [Online April 10, 2015]
(A) $[Ni(H_2O)_5NH_3]^{2+}$,
(B) $[Ni(H_2O)_4(NH_3)_2]^{2+}$ and
(C) $[Ni(H_2O)_3(NH_3)_3]^{2+}$ is :
(1) (A) and (B) show only geometrical isomerism
(2) (A) and (B) show geometrical and optical isomerism
(3) (B) and (C) show geometrical and optical isomerism
(4) (B) and (C) show only geometrical isomerism

61. Which of the following name formula combinations is not correct? [Online April 11, 2014]

	Formula	Name
(1)	$K_2[Pt(CN)_4]$	Potassium tetracyanoplatinate (II)
(2)	$[Mn(CN)_5]^{2-}$	Pentacyanomangnate (II) ion
(3)	$K[Cr(NH_3)_2Cl_4]$	Potassium diammine tetrachlorochromate (III)
(4)	$[Co(NH_3)_4(H_2O)I]SO_4$	Tetraammineaquaiodido cobalt (III) sulphate.

62. An octahedral complex with molecular composition $M \cdot 5NH_3 \cdot Cl \cdot SO_4$ has two isomers, A and B. The solution of A gives a white precipitate with $AgNO_3$ solution and the solution of B gives white precipitate with $BaCl_2$ solution. The type of isomerism exhibited by the complex is: [Online April 19, 2014]
(1) Linkage isomerism (2) Ionisation isomerism
(3) Coordinate isomerism (4) Geometrical isomerism

63. Which of the following complex species is not expected to exhibit optical isomerism ? [2013]
(1) $[Co(en)_3]^{3+}$ (2) $[Co(en)_2Cl_2]^+$
(3) $[Co(NH_3)_3Cl_3]$ (4) $[Co(en)(NH_3)_2Cl_2]^+$

64. Type of isomerism which exists between $[Pd(C_6H_5)_2(SCN)_2]$ and $[Pd(C_6H_5)_2(NCS)_2]$ is : [Online April 9, 2013]
(1) Linkage isomerism (2) Coordination isomerism
(3) Ionisation isomerism (4) Solvate isomerism

65. Which among the following will be named as dibromido bis (ethylenediamine) chromium (III) bromide? [2012]
(1) $[Cr(en)_3]Br_3$ (2) $[Cr(en)_2Br_2]Br$
(3) $[Cr(en)Br_4]^-$ (4) $[Cr(en)Br_2]Br$

66. Which one of the following complex ions has geometrical isomers ? [2011]
(1) $[Ni(NH_3)_5Br]^+$ (2) $[Co(NH_3)_2(en)_2]^{3+}$
(3) $[Cr(NH_3)_4(en)_2]^{3+}$ (4) $[Co(en)_3]^{3+}$
(en-ethylenediamine)

67. A solution containing 2.675 g of $CoCl_3 \cdot 6NH_3$ (molar mass = 267.5 g mol^{-1}) is passed through a cation exchanger. The chloride ions obtained in solution were treated with excess of $AgNO_3$ to give 4.78 g of AgCl (molar mass = 143.5 g mol^{-1}). The formula of the complex is (At. mass of Ag = 108 u) [2010]
(1) $[Co(NH_3)_6]Cl_3$ (2) $[CoCl_2(NH_3)_4]Cl$
(3) $[CoCl_3(NH_3)_3]$ (4) $[CoCl(NH_3)_5]Cl_2$

68. Which one of the following has an optical isomer? [2010]
(1) $[Zn(en)(NH_3)_2]^{2+}$ (2) $[Co(en)_3]^{3+}$
(3) $[Co(H_2O)_4(en)]^{3+}$ (4) $[Zn(en)_2]^{2+}$
(en = ethylenediamine)

69. Which of the following shows optical isomerism [2009]
(1) $[Co(en)(NH_3)_2]^{2+}$ (2) $[Co(H_2O)_4(en)]^{3+}$
(3) $[Co(en)_2(NH_3)_2]^{3+}$ (4) $[Co(NH_3)_3Cl]^+$

70. Which of the following pairs represent linkage isomers? [2009]
(1) $[Pd(PPh_3)_2(NCS)_2]$ and $[Pd(PPh_3)_2(SCN)_2]$
(2) $[Co(NH_3)_5NO_3]SO_4$ and $[Co(NH_3)_5SO_4]NO_3$
(3) $[PtCl_2(NH_3)_4]Br_2$ and $[PtBr_2(NH_3)_4]Cl_2$
(4) $[Cu(NH_3)_4][PtCl_4]$ and $[Pt(NH_3)_4][CuCl_4]$

71. The coordination number and the oxidation state of the element 'E' in the complex $[E(en)_2(C_2O_4)]NO_2$ (where (en) is ethylene diamine) are, respectively, [2008]
(1) 6 and 2 (2) 4 and 2 (3) 4 and 3 (4) 6 and 3

72. The IUPAC name for the complex $[Co(NO_2)(NH_3)_5]Cl_2$ is : [2006]
(1) pentaamminenitrito-N-cobalt(II) chloride
(2) pentaamminenitrito-N-cobalt(III) chloride
(3) nitrito-N-pentaamminecobalt(III) chloride
(4) nitrito-N-pentaamminecobalt(II) chloride

73. The oxidation state Cr in $[Cr(NH_3)_4Cl_2]^+$ is [2005]
(1) 0 (2) +1 (3) +2 (4) +3

74. The IUPAC name of the coordination compound $K_3[Fe(CN)_6]$ is [2005]
(1) Tripotassium hexacyanoiron (II)
(2) Potassium hexacyanoiron (II)
(3) Potassium hexacyanoferrate (III)
(4) Potassium hexacyanoferrate (II)

75. The coordination number of a central metal atom in a complex is determined by **[2004]**

(1) the number of ligands around a metal ion bonded by sigma and pi-bonds both

(2) the number of ligands around a metal ion bonded by pi-bonds

(3) the number of ligands around a metal ion bonded by sigma bonds

(4) the number of only anionic ligands bonded to the metal ion.

76. Which one of the following has largest number of isomers? **[2004]**

(1) $[Ir(PR_3)_2H(CO)]^{2+}$ (2) $[Co(NH_3)_5Cl]^{2+}$

(3) $[Ru(NH_3)_4Cl_2]^+$ (4) $[Co(en)_2Cl_2]^+$

(R = alkyl group, en = ethylenediamine)

77. Among the properties (a) reducing (b) oxidising (c) complexing, the set of properties shown by CN^- ion towards metal species is **[2004]**

(1) c, a (2) b, c (3) a, b (4) a, b, c

78. One mole of the complex compound $Co(NH_3)_5Cl_3$, gives 3 moles of ions on dissolution in water. One mole of the same complex reacts with two moles of $AgNO_3$ solution to yield two moles of AgCl (s). The structure of the complex is **[2003]**

(1) $[Co(NH_3)_3Cl_3]\cdot 2NH_3$ (2) $[Co(NH_3)_4Cl_2]Cl\cdot NH_3$

(3) $[Co(NH_3)_4Cl]Cl_2\cdot NH_3$ (4) $[Co(NH_3)_5Cl]Cl_2$

79. In the coordination compound, $K_4[Ni(CN)_4]$, the oxidation state of nickel is **[2003]**

(1) 0 (2) +1 (3) +2 (4) –1

80. The type of isomerism present in nitropentammine chromium (III) chloride is **[2002]**

(1) optical (2) linkage

(3) ionization (4) polymerisation.

NUMERICAL VALUE QUESTIONS (NVQs)

81. An excess of $AgNO_3$ is added to 100 mL of a 0.05 M solution of tetraaquadichloridochromium (III) chloride. The number of moles of AgCl precipitated will be________ $\times 10^{-3}$. (Nearest integer) **[April 06, 2026 (II)]**

82. 5.33 g of $CrCl_3\cdot 6H_2O$, which is a 1 : 3 electrolyte, is dissolved in water and is passed through a cation exchanger. The chloride ions in the eluted solution, on treatment with $AgNO_3$ results in 8.61 g of AgCl. The ratio of moles of complex reacted and moles of AgCl formed is _____ $\times 10^{-2}$. (Nearest integer)
[Molar mass in g mol^{-1} Cr : 52, Ag : 108, Cl : 35.5, H : 1, O : 16] **[April 02, 2026 (I)]**

83. X is the number of geometrical isomers exhibited by $[Pt(NH_3)(H_2O)BrCl]$.
Y is the number of optically inactive isomer(s) exhibited by $[CrCl_2(ox)_2]^{3-}$
Z is the number of geometrical isomers exhibited by $[Co(NH_3)_3(NO_2)_3]$
The value of X + Y + Z is ________. **[Jan. 28, 2026 (I)]**

84. A chromium complex with a formula $CrCl_3.6H_2O$ has a spin only magnetic moment value of 3.87 BM and its solution conductivity corresponds to 1 : 2 electrolyte. 2.75 g of the complex solution was initially passed through a cation exchanger. The solution obtained after the process was reacted with excess of $AgNO_3$. The amount of AgCl formed in the above process is _______ g. (Nearest integer) [Given : Molar mass in g mol^{-1} Cr : 52; Cl : 35.5, Ag : 108, O : 16, H : 1] **[Jan. 24, 2026 (II)]**

85. A metal complex with a formula $MCl_4\cdot 3NH_3$ is involved in sp^3d^2 hybridisation. It upon reaction with excess of $AgNO_3$ solution gives '*x*' moles of AgCl. Consider '*x*' is equal to the number of lone pairs of electron present in central atom of BrF_5. Then the number of geometrical isomers exhibited by the complex is________. **[April 4, 2025 (II)]**

86. The number of optical isomers exhibited by the iron complex (A) obtained from the following reaction is ________.
$FeCl_3 + KOH + H_2C_2O_4 \rightarrow A$ **[April 3, 2025 (I)]**

87. The complex of Ni^{2+} ion and dimethyl glyoxime contains _______ number of Hydrogen (H) atoms. **[Jan. 22, 2025 (II)]**

88. Number of ambidentate ligands among the following is ________. **[April 9, 2024 (I)]**

$NO_2^-, SCN^-, C_2O_4^{2-}, NH_3, CN^-, SO_4^{2-}, H_2O$.

89. Number of complexes which show optical isomerism among the following is ________. **[Jan. 30, 2024 (II)]**
cis – $[Cr(ox)_2Cl_2]^{3-}$, $[Co(en)_3]^{3+}$, cis – $[Pt(en)_2Cl_2]^{2+}$. cis – $[Co(en)_2Cl_2]^+$, trans – $[Pt(en)_2Cl_2]^{2+}$, trans – $[Cr(ox)_2Cl_2]^3$

90. The volume (in mL) of 0.1 M $AgNO_3$ required for complete precipitation of chloride ions present in 20 mL of 0.01 M solution of $[Cr(H_2O)_5Cl]Cl_2$ as silver chloride is _______ **[April 15, 2023 (I)]**

91. Number of ambidentate ligands in a representative metal complex $[M(en)(SCN)_4]$ is [en = ethylenediamine] **[April 06, 2023 (I)]**

92. The denticity of the ligand present in the Fehling's reagent is _______. **[Jan. 29, 2023 (II)]**

93. Total number of moles of AgCl precipitated on addition of excess of $AgNO_3$ to one mole each of the following complexes $[Co(NH_3)_4Cl_2]Cl$, $[Ni(H_2O)_6]Cl_2$, $[Pt(NH_3)_2Cl_2]$ and $[Pd(NH_3)_4]Cl_2$ is **[Jan. 25, 2023 (II)]**

94. Sum of oxidation state (magnitude) and coordination number of cobalt in $Na[Co(bpy)Cl_4]$ is____.

(Given bpy = 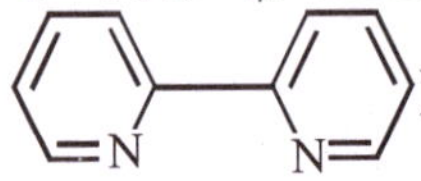) **[July 29, 2022 (II)]**

95. Total number of relatively more stable isomer(s) possible for octahedral complex $[Cu(en)_2(SCN)_2]$ will be __________. **[July 28, 2022 (I)]**

96. The conductivity of a solution of complex with formula $CoCl_3(NH_3)_4$ corresponds to 1 : 1 electrolyte, then the primary valency of central metal ion is______ **[July 27, 2022 (I)]**

97. In the following brown complex, the oxidation state of iron is + ________. **[June 30, 2022 (I)]**

$$[Fe(H_2O)_6]^{2+} + NO \rightarrow \underset{\text{Brown complex}}{[Fe(H_2O)_5(NO)]^{2+}} + H_2O$$

98. For the reaction given below: $CoCl_3.xNH_3 + AgNO_3(aq) \rightarrow$
If two equivalents of AgCl precipitate out, then the value of *x* will be _____. **[June 29, 2022 (II)]**

99. Number of complexes which will exist in cis trans is/are
(A) $CoCl_3\cdot 4NH_3$ (B) $CoCl_3\cdot 5NH_3$
(C) $CoCl_3\cdot 6NH_3$ and (D) $CoCl(NO_3)_2\cdot 5NH_3$ **[June 28, 2022 (II)]**

100. The sum of oxidation states of two silver ions in $[Ag(NH_3)_2][Ag(CN)_2]$ complex is _______. **[Sep. 1, 2021 (II)]**

101. The number of hydrogen bonded water molecule(s) associated with stoichiometry $CuSO_4\cdot 5H_2O$ is _______. **[Aug. 31, 2021 (I)]**

102. 1 mol of an octahedral metal complex with formula $MCl_3 \cdot 2L$ on reaction with excess of $AgNO_3$ gives 1 mol of AgCl. The denticity of Ligand L is ________. (Integer answer) [Aug. 27, 2021 (I)]

103. The number of optical isomers possible for $[Cr(C_2O_4)_3]^{3-}$ is ______. [Aug. 27, 2021 (II)]

104. The ratio of number of water molecules in Mohr's salt and potash alum is ________ $\times 10^{-1}$. (Integer answer) [Aug. 26, 2021 (I)]

105. Indicate the complex/complex ion which did not show any geometrical isomerism : [Aug. 26, 2021 (II)]
(A) $[CoCl_2(en)_2]$ (B) $[Co(CN)_5(NC)]^{3-}$
(C) $[Co(NH_3)_3(NO_2)_3]$ (D) $[Co(NH_3)_4Cl_2]^+$

106. The number of geometrical isomers possible in triamminetrinitrocobalt (III) is X and in trioxalatochromate (III) is Y. Then the value of X + Y is ______. [July 27, 2021 (I)]

107. 3 moles of metal complex with formula $Co(en)_2Cl_3$ gives 3 moles of silver chloride on treatment with excess of silver nitrate. The secondary valency of Co in the complex is _____.
(Round off to the nearest integer) [July 27, 2021 (II)]

108. Three moles of AgCl get precipitated when one mole of an octahedral co-ordination compound with empirical formula $CrCl_3 \cdot 3NH_3 \cdot 3H_2O$ reacts with excess of silver nitrate. The number of chloride ions satisfying the secondary valency of the metal ion is________. [July 25, 2021 (I)]

109. The total number of unpaired electrons present in the complex $K_3[Cr(oxalate)_3]$ is__________. [March 18, 2021 (I)]

110. On complete reaction of $FeCl_3$ with oxalic acid in aqueous solution containing KOH, resulted in the formation of product A. The secondary valency of Fe in the product A is ________. (Round off to the Nearest integer). [March 17, 2021 (II)]

111. The equivalents of ethylene diamine required to replace the neutral ligands from the coordination sphere of the trans-complex of $CoCl_2 \cdot 4NH_3$ is ____
(Round off to the nearest integer). [March 16, 2021 (I)]

112. The number of stereoisomers possible for $[Co(ox)_2(Br)(NH_3)]^{2-}$ is __________. [ox = oxalate] [Feb. 26, 2021 (II)]

113. The total number of coordination sites in ethylenediaminetetraacetate ($EDTA^{4-}$) is ______. [Sep. 05, 2020 (I)]

Topic 2 - Bonding in Coordination Compounds, (Valence Bond Theory and Crystal Field Theory)

MULTIPLE CHOICE QUESTIONS (MCQs)

114. Which of the following sequence of hybridisation, geometry and magnetic nature are correct for the given coordination compounds ? [April 06, 2026 (II)]
A. $[NiCl4]^{2-}$ – sp^3, tetrahedral, paramagnetic
B. $[Ni(NH_3)_6]^{2+}$ – sp^3d^2, octahedral, paramagnetic
C. $[Ni(CO)_4]$ – sp^3, tetrahedral, paramagnetic
D. $[Ni(CN)_4]^{2-}$ – dsp^2, square planar diamagnetic
Choose the correct answer from the options given below :
(1) A, B, C and D (2) B, C and D only
(3) A, C and D only (4) A, B and D only

115. Which of the following are true about the energy of the given d-orbitals of a tetrahedral complex ? [April 06, 2026 (I)]
A. $d_{xy} = d_{xz} > d_{x^2-y^2}$ B. $d_{xy} = d_{yz} > d_{z^2}$
C. $d_{x^2-y^2} > d_{z^2} > d_{xz}$ D. $d_{x^2-y^2} = d_{z^2} > d_{xz}$
Choose the correct answer from the options given below:
(1) A, B and D only (2) A and B only
(3) B and D only (4) B, C and D only

116. Match the List-I with List-II [April 06, 2026 (I)]

List-I Electronic configuration of tetrahedral metal ion		List-II Crystal Field Stabilization Energy (Δt)	
A.	d^2	I.	−0.6
B.	d^4	II.	−0.8
C.	d^6	III.	−1.2
D.	d^8	IV.	−0.4

Choose the correct answer from the options given below:
(1) A-III, B-IV, C-II, D-I (2) A-III, B-I, C-IV, D-II
(3) A-III, B-IV, C-I, D-II (4) A-II, B-I, C-IV, D-III

117. Identify the correct statements from the following:
(A) $[Fe(C_2O_4)_3]^{3-}$ is the most stable complex among $[Fe(OH)_6]^{3-}$, $[Fe(C_2O_4)_3]^{3-}$ and $[Fe(SCN)_6]^{3-}$
(B) The stability of $[Cu(NH_3)_4]^{2+}$ is greater than that of $[Cu(en)_2]^{2+}$
(C) The hybridization of Fe in $K_4[Fe(CN)_6]$ is d^2sp^3
(D) $[Fe(NO_2)_3Cl_3]^3$ exhibits linkage isomerism
(E) NO_2^- and SCN^- ligands are NOT ambidentate ligands
Choose the correct answer from the options given below: [April 05, 2026 (II)]
(1) A, B, C, D and E (2) B, C and D only
(3) A, C and D only (4) A, C and E only

118. Given below are two statements: [April 05, 2026 (II)]
Statement-I: Presence of large number of unpaired electrons in transition metal atoms results in higher enthalpies of their atomisation.
Statement-II: $d_{xy} = d_{xz} = d_{yz} < d_{x^2-y^2} = d_{x^2}$ and $d_{x^2-y^2} = d_{z^2} < d_{xy} = d_{xz} = d_{yz}$ are the d-orbital splittings in $[Fe(H_2O)_6]^{3+}$ and $[Ni(Cl)_4]^{2-}$ complex ions respectively.
In the light of the above statements, choose the correct answer from the options given below :
(1) Both Statement I and Statement II are true
(2) Both Statement I and Statement II are false
(3) Statement I is true but Statement II is false
(4) Statement I is false but Statement II is true

119. Given below are two statements: [April 05, 2026 (I)]
Statement I: Each electron in e_g orbitals destabilizes the orbitals by $+0.6\ \Delta_0$ and each electron in the t_{2g} orbitals stabilizes the orbitals by $-0.4\ \Delta_0$ in an octahedral field on the basis of crystal field theory.
Statement II: All the d-orbitals of the transition meals have the same energy in their free atomic state but when a complex is formed the ligands destroy the degeneracy of these orbitals on the basis of crystal field theory.
In the light of the above statements, choose the correct answer from the options given below.
(1) Both Statement I and Statement II are correct
(2) Both Statement I and Statement II are incorrect
(3) Statement I is correct but Statement II is incorrect
(4) Statement I is incorrect but Statement II is correct

120. Consider the metal complexes $[Ni(en)_3]^{2+}$ (A), $[NiCl_4]^{2-}$(B) and $[Ni(NH_3)_6]^{2+}$(C). Choose the **correct** option by considering the number of unpaired electron present in (A), (B) and (C) respectively and the order of frequency of absorption. [April 04, 2026 (II)]

(1) 2, 2, 2 and (A) > (C) > (B)
(2) 0, 2, 0 and (A) > (C) > (B)
(3) 2, 2, 0 and (B) > (C) > (A)
(4) 2, 2, 2 and (C) > (A) > (B)

121. Match the **List-I** with **List-II.** [April 04, 2026 (I)]

List - I Complex ion		List - II Calculated spin only magnetic moment (BM)	
A.	$[Cr(H_2O)_6]^{2+}$	I.	3.87
B.	$[Co(H_2O)_6]^{2+}$	II.	5.92
C.	$[Cu(H_2O)_6]^{2+}$	III.	4.90
D.	$[Mn(H_2O)_6]^{2+}$	IV.	1.73

Choose the **correct** answer from the options given below:

(1) A-I, B-III, C-IV, D-II (2) A-II, B-I, C-III, D-IV
(3) A-IV, B-II, C-I, D-III (4) A-III, B-I, C-IV, D-II

122. Given below are two statements : [April 02, 2026 (II)]

Statement-I : Among Zn, Mn, Sc and Cu, the energy required to remove the third valence electron is highest for Zn and lowest for Sc.

Statement-II : The correct order of the following complexes in terms of CFSE is

$[Co(H_2O)_6]^{2+} < [Co(H_2O)_6]^{3+} < [Co(en)_3]^{3+}$.

In the light of the above statements, choose the **correct** answer from the options given below :

(1) Both Statement I and Statement II are true.
(2) Both Statement I and Statement II are false.
(3) Statement I is true but Statement II is false.
(4) Statement I is false but Statement II is true.

123. Match the **List-I** with **List-II.** [April 02, 2026 (I)]

List - I Chromium (III) Complexes (en = ethylene diamine)		List - II Δ_0(cm–1)	
A.	$[Cr(CN)_6]^{3-}$	I.	15,060
B.	$[CrF_6]^{3-}$	II.	17,400
C.	$[Cr(H_2O)_6)]^{3+}$	III.	22,300
D.	$[Cr(en)_3]^{3+}$	IV.	26,600

Choose the **correct** answer from the options given below:

(1) A-I, B-II, C-III, D-IV (2) A-II, B-III, C-IV, D-I
(3) A-III, B-IV, C-I, D-II (4) A-IV, B-I, C-II, D-III

124. The correct increasing order of spin-only magnetic moment values of the complex ions

$[MnBr_4]^{2-}$ (A), $[Cu(H_2O)_6]^{2+}$ (B), $[Ni(CN_4)]^{2-}$ (C) and $[Ni(H_2O)_6]^{2+}$ (D) is: [Jan. 28, 2026 (II)]

(1) A = B < C < D (2) A = B < D < C
(3) C = D < B < A (4) C < B < D < A

125. The correct statement among the following is : [Jan. 28, 2026 (I)]

(1) $[Ni(CN)_4]^{2-}$ and $[NiCl_4]^{2-}$ are diamagnetic and $Ni(CO)_4$ is paramagnetic.
(2) $Ni(CO)_4$ and $[NiCl_4]^{2-}$ are diamagnetic and $[Ni(CN)_4]^{2-}$ is paramagnetic.
(3) $Ni(CO)_4$ and $[Ni(CN)_4]^{2-}$ are diamagnetic and $[NiCl_4]^{2-}$ is paramagnetic.
(4) $Ni(CO)_4$ is diamagnetic and $[NiCl_4]^{2-}$ and $[Ni(CN)_4]^{2-}$ are paramagnetic.

126. The wavelength of light absorbed for the following complexes are in the order. [Jan. 24, 2026 (II)]

$\left[Co(NH_3)_6\right]^{3+}$ (I); $\left[Co(H_2O)_6\right]^{3+}$ (II); $\left[Co(CN)_6\right]^{3-}$ (III)

$\left[Co(NH_3)_5(H_2O)\right]^{3+}$ (IV); $\left[CoF_6\right]^{3-}$ (V)

(1) III < I < II < IV < V (2) III < I < IV < V < II
(3) III < IV < I < II < V (4) III < I < IV < II < V

127. Given below are two statements. [Jan. 24, 2026 (I)]

Statements-I: The number of paramagnetic species among $[CoF_6]^{3-}$, $[TiF_6]^{3-}$, V_2O_5 and $[Fe(CN)_6]^{3-}$ is 3.

Statements-II: $K_4[Fe(CN)_6] < K_3[Fe(CN)_6] < [Fe(H_2O)_6]SO_4.H_2O < [Fe(H_2O)_6]Cl_3$ is the correct order in terms of number of unpaired electron(s) in the complexes.

In the light of the above statements, choose the correct answer from the options given below.

(1) Both statement-I and statement-II are true.
(2) Both statement-I and statement-II are false.
(3) Statement-I is true but statement-II is false.
(4) Statement-I is false but statement-II is true.

128. Given below are two statements: [Jan. 24, 2026 (I)]

Statement-I: Hybridisation, shape and spin only magnetic moment of $K_3[Co(CO_3)_3]$ is sp^3d^2, octahedral and 4.9 BM respectively.

Statement-II: Geometry, hybridisation and spin only magnetic moment values (BM) of the ions $[Ni(CN)_4]^{2-}$, $[MnBr_4]^{2-}$ and $[CoF_6]^{3-}$ respectively are square planar, tetrahedral, octahedral: dsp^2, sp^3, sp^3d^2 and 0, 5.9, 4.9.

In the light of the above statements, choose the correct answer from the options given below.

(1) Both statement-I and statement-II are false.
(2) Statement I is false but statement-II is true.
(3) Both statement-I and statement-II are true.
(4) Statement-I is true but statement-II is false.

129. Identify the **CORRECT** set of details from the following: [Jan. 23, 2026 (II)]

A. $[Co(NH_3)_6]^{3+}$: Inner orbital complex ; d^2sp^3 hybridized
B. $[MnCl_6]^{3-}$: Outer orbital complex ; sp^3d^2 hybridized
C. $[CoF_6]^{3-}$: Outer orbital complex ; d^2sp^3 hybridized
D. $[FeF_6]^{3-}$: Outer orbital complex ; sp^3d^2 hybridized
E. $[Ni(CN)_4]^{2-}$: Inner orbital complex ; sp^3 hybridized

Choose the correct answer from the options given below:

(1) C & D only (2) A, B & D only
(3) A, C & E only (4) A, B, C, D & E

130. The statements that are **incorrect** about the nickel (II) complex of dimethylglyoxime are: [Jan. 23, 2026 (I)]

A. It is red in colour
B. It has a high solubility in water at pH = 9
C. The Ni ion has two unpaired d-electrons
D. The N – Ni – N bond angle is almost close to 90°
E. The complex contains four five-membered metallacycles (metal containing rings)

Choose the correct answer from the options given below:

(1) C and E only (2) A, D and B only
(3) B, C and E only (4) C and D only

131. Given below are two statements: [Jan. 23, 2026 (I)]

Statement I: $[CoBr_4]^{2-}$ ion will absorb light of lower energy than $[CoCl_4]^{2-}$ ion.

Statement II: In $[CoI_4]^{2-}$ ion, the energy separation between the two set of d-orbitals is more than $[CoCl_4]^{2-}$ ion.

In the light of the above statements, choose the correct answer from the options given below:
(1) Both Statement I and Statement II are false.
(2) Statement I is true but Statement II is false.
(3) Statement I is false but Statement II is true .
(4) Both Statement I and Statement II are true.

132. $[Ni(PPh_3)_2Cl_2]$ is a paramagnetic complex. Identify the **INCORRECT** statements about this complex. **[Jan. 22, 2026 (II)]**
A. The complex exhibits geometrical isomerism.
B. The complex is white in colour.
C. The calculated spin-only magnetic moment of the complex is 2.84 BM.
D. The calculated CFSE (Crystal Field Stabilization Energy) of Ni in this complex is $-0.8\Delta_0$.
E. The geometrical arrangement of ligands in this complex is similar to that in $Ni(CO)_4$.
Choose the **correct** answer from the options given below:
(1) A and B only (2) A, B and D only
(3) C and D only (4) C, D and E only

133. Consider the transition metal ions Mn^{3+}, Cr^{3+}, Fe^{3+} and Co^{3+} and all form low spin octahedral complexes. The correct decreasing order of unpaired electrons in their respective *d*-orbitals of the complexes is **[Jan. 22, 2026 (I)]**
(1) $Cr^{3+} > Fe^{3+} > Co^{3+} > Mn^{3+}$
(2) $Mn^{3+} > Fe^{3+} > Co^{3+} > Cr^{3+}$
(3) $Fe^{3+} > Co^{3+} > Mn^{3+} > Cr^{3+}$
(4) $Cr^{3+} > Mn^{3+} > Fe^{3+} > Co^{3+}$

134. Given below are two statements: **[Jan. 21, 2026 (II)]**
Statement I: Crystal Field Stabilization Energy (CFSE) of $[Cr(H_2O)_6]^{2+}$ is greater than that of $[Mn(H_2O)_6]^{2+}$.
Statement II: Potassium ferricyanide has a greater spin-only magnetic moment than sodium Ferrocyanide.
In the light of the above statements, choose the **correct** answer from the options given below:
(1) Both Statement I and Statement II are true.
(2) Both Statement I and Statement II are false.
(3) Statement I is true but Statement II is false.
(4) Statement I is false but Statement II is true.

135. Given below are two statements: **[Jan. 21, 2026 (I)]**
Statement-I: Among $[Cu(NH_3)^{2+}$, $[Ni(en)_3]^{2+}$, $[Ni(NH_3)_6]^{2+}$ and $[Mn(H_2O)_6]^{2+}$, $[Mn(H_2O)_6]^{2+}$, has the maximum number of unpaired electrons.
Statement-II: The number of pairs among $\{[NiCl_4]^{2-}, [Ni(CO)_4]\}$, $\{[NiCl_4]^{2-}, [Ni(CN)_4]^{2-}\}$ and $\{[Ni(CO)_4], [Ni(CN)_4]^{2-}\}$ that contain only diamagnetic species is two.
In the light of the above statements, choose the **correct** answer from the options given below:
(1) Statement-I is false but Statement-II is true.
(2) Both Statement-I and Statement-II are true.
(3) Both Statement-I and Statement-II are false.
(4) Statement-I is true but Statement-II is false.

136. Match the **LIST-I** with **LIST-II** **[April 8, 2025 (II)]**

LIST-I (Complex/Species)	**LIST-II (Shape & magnetic moment)**
A. $[Ni(CO)_4]$	I. Tetrahedral, 2.8 BM
B. $[Ni(CN)_4]^{2-}$	II. Square planar, 0 BM
C. $[NiCl_4]^{2-}$	III. Tetrahedral, 0 BM
D. $[MnBr_4]^{2-}$	IV. Tetrahedral, 5.9 BM

Choose the **correct** answer from the options given below:
(1) A-III, B-IV, C-II, D-I (2) A-I, B-II, C-III, D-IV
(3) A-III, B-II, C-I, D-IV (4) A-IV, B-I, C-III, D-II

137. The number of species from the following that are involved in sp^3d^2 hybridization is
$[Co(NH_3)_6]^{3+}$, SF_6, $[CrF_6]^{3-}$, $[CoF_6]^{3-}$, $[Mn(CN)_6]^{3-}$ and $[MnCl_6]^{3-}$ **[April 8, 2025 (II)]**
(1) 5 (2) 6 (3) 4 (4) 3

138. The correct decreasing order of spin only magnetic moment values (BM) of Cu^+, Cu^{2+}, Cr^{2+} and Cr^{3+} ions is : **[April 8, 2025 (II)]**
(1) $Cu^+ > Cu^{2+} > Cr^{3+} > Cr^{2+}$ (2) $Cu^{2+} > Cu^+ > Cr^{2+} > Cr^{3+}$
(3) $Cr^{2+} > Cr^{3+} > Cu^{2+} > Cu^+$ (4) $Cr^{3+} > Cr^{2+} > Cu^+ > Cu^{2+}$

139. The number of unpaired electrons responsible for the paramagnetic nature of the following complex species are respectively :
$[Fe(CN)_6]^{3-}$, $[FeF_6]^{3-}$, $[CoF_6]^{3-}$, $[Mn(CN)_6]^{3-}$ **[April 7, 2025 (II)]**
(1) 1, 5, 4, 2 (2) 1, 5, 5, 2
(3) 1, 1, 4, 2 (4) 1, 4, 4, 2

140. '*X*' is the number of electrons in t_{2g} orbitals of the most stable complex ion among $[Fe(NH_3)_6]^{3+}$, $[FeC_6]^{3-}$, $[Fe(C_2O_4)_3]^{3-}$ and $[Fe(H_2O)_6]^{3+}$. The nature of oxide of vanadium of the type V_2O_X is: **[April 4, 2025 (II)]**
(1) Acidic (2) Neutral
(3) Basic (4) Amphoteric

141. The correct order of $[FeF_6]^{3-}$, $[CoF_6]^{3-}$, $[Ni(CO)_4]$ and $[Ni(CN)_4]^{2-}$ complex species based on the number of unpaired electrons present is : **[April 4, 2025 (II)]**
(1) $[FeF_6]^{3-} > [CoF_6]^{3-} > [Ni(CN)_4]^{2-} > [Ni(CO)_4]$
(2) $[Ni(CN)_4]^{2-} > [FeF_6]^{3-} > [CoF_6]^{3-} > [Ni(CO)_4]$
(3) $[CoF_6]^{3-} > [FeF_6]^{3-} > [Ni(CO)_4] > [Ni(CN)_4]^{2-}$
(4) $[FeF_6]^{3-} > [CoF_6]^{3-} > [Ni(CN)_4]^{2-} = [Ni(CO)_4]$

142. Which one of the following complexes will have $\Delta_0 = 0$ and $\mu = 5.96$ B.M.? **[April 4, 2025 (I)]**
(1) $[Fe(CN)_6]^{4-}$ (2) $[CO(NH_3)_6]^{3+}$
(3) $[FeF_6]^{4-}$ (4) $[Mn(SCN)_6]^{4-}$

143. Identify the diamagnetic octahedral complex ions from below: **[April 3, 2025 (II)]**
A. $[Mn(CN)_6]^{3-}$ B. $[Co(NH_3)_6]^{3+}$
C. $[Fe(CN)_6]^{4-}$ D. $[Co(H_2O)_3F_3]$
Choose the **correct** answer from the options given below:
(1) B and D Only (2) A and D Only
(3) A and C Only (4) B and C Only

144. The correct order of the complexes $[Co(NH_3)_5(H_2O)]^{3+}$ (A), $[Co(NH_3)_6]^{3+}$ (B), $[Co(CN)_6]^{3-}$ (C) and $[CoCl(NH_3)_5]^{2+}$ (D) in terms wavelength of light absorbed is : **[April 3, 2025 (I)]**
(1) D > A > B > C (2) C > B > D > A
(3) D > C > B > A (4) C > B > A > D

145. The metal ions that have the calculated spin only magnetic moment value of 4.9 B.M. are **[April 3, 2025 (I)]**
A. Cr^{2+} B. Fe^{2+} C. Fe^{3+} D. Co^{2+}
E. Mn^{3+}
Choose the **correct** answer from the options given below
(1) A, C and E only (2) A, D and E only
(3) B and E only (4) A, B and E only

146. The type of hybridization and the magnetic property of $[MnCl_6]^{3-}$ are : **[April 2, 2025 (II)]**
(1) d^2sp^3, paramagnetic with four unpaired electrons
(2) sp^3d^2, paramagnetic with four unpaired electrons
(3) d^2sp^3, paramagnetic with two unpaired electrons
(4) sp^3d^2, paramagnetic with two unpaired electrons

147. The d-orbital electronic configuration of the complex among $[Co(en)_3]^{3+}$, $[CoF_6]^{3-}$, $[Mn(H_2O)_6]^{2+}$ and $[Zn(H_2O)_6]^{2+}$ that has the highest CFSE is : **[April 2, 2025 (II)]**

(1) $t_{2g}^6e_g^0$ (2) $t_{2g}^6e_g^4$ (3) $t_{2g}^3e_g^2$ (4) $t_{2g}^4e_g^2$

148. Given below are two statements :

Statement (I) : In octahedral complexes, when $\Delta_o < P$ high spin complexes are formed. When $\Delta_o > P$ low spin complexes are formed.

Statement (II) : In tetrahedral complexes because of $\Delta_t < P$, low spin complexes are rarely formed.

In the light of the above statements, choose the **most appropriate answer** from the options given below : **[April 2, 2025 (I)]**

(1) **Statement I** is correct but **Statement II** is incorrect.
(2) Both **Statement I** and **Statement II** are incorrect
(3) **Statement I** is incorrect but **Statement II** is correct
(4) Both **Statement I** and **Statement II** are correct

149. The calculated spin-only magnetic moments of $K_3[Fe(OH)_6]$ and $K_4[Fe(OH)_6]$ respectively are : **[Jan. 29, 2025 (II)]**

(1) 4.90 and 4.90 B.M. (2) 5.92 and 4.90 B.M.
(3) 3.87 and 4.90 B.M. (4) 4.90 and 5.92 B.M.

150. Match List-I with List-II. **[Jan. 29, 2025 (I)]**

List-I (Complex)	**List-II (Hybridisation & Magnetic characters)**
(A) $[MnBr_4]^{2-}$	(I) d^2sp^3 & diamagnetic
(B) $[FeF_6]^{3-}$	(II) sp^3d^2 & paramagnetic
(C) $[Co(C_2O_4)_3]^{3-}$	(III) sp^3 & diamagnetic
(D) $[Ni(CO)_4]$	(IV) sp^3 & paramagnetic

Choose the correct answer from the options given below :

(1) (A)-(III), (B)-(II), (C)-(I), (D)-(IV)
(2) (A)-(III), (B)-(I), (C)-(II), (D)-(IV)
(3) (A)-(IV), (B)-(I), (C)-(II), (D)-(III)
(4) (A)-(IV), (B)-(II), (C)-(I), (D)-(III)

151. The correct increasing order of stability of the complexes based on Δ_o value is : **[Jan. 29, 2025 (I)]**

(I) $[Mn(CN)_6]^{3-}$ (II) $[Co(CN)_6]^{4-}$
(III) $[Fe(CN)_6]^{4-}$ (IV) $[Fe(CN)_6]^{3-}$

(1) II < III < I < IV (2) IV < III < II < I
(3) I < II < IV < III (4) III < II < IV < I

152. Match List-I with List-II. **[Jan. 28, 2025 (II)]**

List-I (Complex)	**List-II (Hybridisation of central metal ion)**
(A) $[CoF_6]^{3-}$	(I) d^2sp^3
(B) $[NiCl_4]^{2-}$	(II) sp^3
(C) $[Co(NH_3)_6]^{3+}$	(III) sp^3d^2
(D) $[Ni(CN)_4]^{2-}$	(IV) dsp^2

Choose the correct answer from the options given below :

(1) (A)-(I), (B)-(IV), (C)-(III), (D)-(II)
(2) (A)-(III), (B)-(II), (C)-(I), (D)-(IV)
(3) (A)-(I), (B)-(II), (C)-(III), (D)-(IV)
(4) (A)-(III), (B)-(IV), (C)-(I), (D)-(II)

153. The conditions and consequence that favours the t_{2g}^3, e_g^1 configuration in a metal complex are : **[Jan. 24, 2025 (II)]**

(1) weak field ligand, high spin complex
(2) strong field ligand, high spin complex
(3) strong field ligand, low spin complex
(4) weak field ligand, low spin complex

154. When Ethane-1,2-diamine is added progressively to an aqueous solution of Nickel (II) chloride, the sequence of colour change observed will be : **[Jan. 24, 2025 (II)]**

(1) Pale Blue → Blue → Green → Violet
(2) Pale Blue → Blue → Violet → Green
(3) Green → Pale Blue → Blue → Violet
(4) Violet → Blue → Pale Blue → Green

155. Identify the coordination complexes in which the central metal ion has d^4 configuration. **[Jan. 23, 2025 (II)]**

(A) $[FeO_4]^{2-}$
(B) $[Mn(CN)_6]^{3-}$
(C) $[Fe(CN)_6]^{3-}$
(D) $Cr_2(O-\overset{\overset{O}{\|}}{C}-Me)_4(H_2O)_2$
(E) $[NiF_6]^{2-}$

Choose the correct answer from the options given below:

(1) (C) and (E) only (2) (B), (C) and (D) only
(3) (B) and (D) only (4) (A), (B) and (E) only

156. The d-electronic configuration of an octahedral Co(II) complex having magnetic moment of 3.95 BM is: **[Jan. 23, 2025 (I)]**

(1) $t_{2g}^6\ e_g^1$ (2) $t_{2g}^3\ e_g^0$ (3) $t_{2g}^5\ e_g^2$ (4) $e^4t_2^3$

157. The correct order of the following complexes in terms of their crystal field stabilization energies is: **[Jan. 22, 2025 (II)]**

(1) $[Co(NH_3)_4]^{2+} < [Co(NH_3)_6]^{2+} < [Co(en)_3]^{3+} < [Co(NH_3)_6]^{3+}$
(2) $[Co(NH_3)_4]^{2+} < [Co(NH_3)_6]^{2+} < [Co(NH_3)_6]^{3+} < [Co(en)_3]^{3+}$
(3) $[Co(NH_3)_6]^{2+} < [Co(NH_3)_6]^{3+} < [Co(NH_3)_4]^{2+} < [Co(en)_3]^{3+}$
(4) $[Co(en)_3]^{3+} < [Co(NH_3)_6]^{3+} < [Co(NH_3)_6]^{2+} < [Co(NH_3)_4]^{2+}$

158. From the magnetic behaviour of $[NiCl_4]^{2-}$ (paramagnetic) and $[Ni(CO)_4]$ (diamagnetic), choose the correct geometry and oxidation state: **[Jan. 22, 2025 (I)]**

(1) $[NiCl_4]^{2-}$: Ni^{II}, square planar
$[Ni(CO)_4]$: Ni(0), square planar
(2) $[NiCl_4]^{2-}$: Ni^{II}, tetrahedral
$[Ni(CO)_4]$: Ni(0), tetrahedral
(3) $[NiCl_4]^{2-}$: Ni^{II}, tetrahedral
$[Ni(CO)_4]$: Ni^{II}, square planar
(4) $[NiCl_4]^{2-}$: Ni(0), tetrahedral
$[Ni(CO)_4]$: Ni(0), square planar

159. In which of the following complexes the CFSE, Δ_0 will be equal to zero? **[Jan. 22, 2025 (I)]**

(1) $[Fe(NH_3)_6]Br_2$ (2) $[Fe(en)_3]Cl_3$
(3) $K_4[Fe(CN)_6]$ (4) $K_3[Fe(SCN)_6]$

160. Given below are two statements : one is labelled as **Assertion (A)** and the other is labelled as **Reason (R).**

Assertion (A) : The total number of geometrical isomers shown by $[Co(en)_2Cl_2]^+$ complex ion is three

Reason (R) : $[Co(en)_2Cl_2]^+$ complex ion has an octahedral geometry. **[April 9, 2024 (I)]**

In the light of the above statements, choose the **most appropriate** answer from the options given below :

(1) Both **(A)** and **(R)** are correct and **(R)** is the correct explanation of **(A).**
(2) **(A)** is correct but **(R)** is not correct
(3) **(A)** is not correct but **(R)** is correct
(4) Both **(A)** and **(R)** are correct but **(R)** is not the correct explanation of **(A)**

161. Match List-I with List-II. [April 9, 2024 (II)]

List-I	List-II
(A) $K_2[Ni(CN)_4]$	(I) sp^3
(B) $[Ni(CO)_4]$	(II) sp^3d^2
(C) $[Co(NH_3)_6]Cl_3$	(III) dsp^2
(D) $Na_3[CoF_6]$	(IV) d^2sp^3

Choose the correct answer from the options given below:
(1) A-(III), B-(I), C-(II), D-(IV)
(2) A-(III), B-(II), C-(IV), D-(I)
(3) A-(I), B-(III), C-(II), D-(IV)
(4) A-(III), B-(I), C-(IV), D-(II)

162. The coordination environment of Ca^{2+} ion in its complex with $EDTA^{4-}$ is : [April 9, 2024 (II)]
(1) tetrahedral (2) octahedral
(3) square planar (4) trigonal prismatic

163. Given below are two statements: [April 8, 2024 (I)]
Statement I : $N(CH_3)_3$ and $P(CH_3)_3$ can act as ligands to form transition metal complexes.
Statement II: As N and P are from same group, the nature of bonding of $N(CH_3)_3$ and $P(CH_3)_3$ is always same with transition metals.
In the light of the above statements, choose the **most appropriate** answer from the options given below:
(1) Statement I is incorrect but Statement II is correct
(2) Both Statement I and Statement II are correct
(3) Statement I is correct but Statement II is incorrect
(4) Both Statement I and Statement II are incorrect

164. Number of complexes with even number of electrons in t_{2g} orbitals is - [April 8, 2024 (I)]
$[Fe(H_2O)_6]^{2+}$, $[Co(H_2O)_6]^{2+}$, $[Co(H_2O)_6]^{3+}$, $[Cu(H_2O)_6]^{2+}$, $[Cr(H_2O)_6]^{2+}$
(1) 1 (2) 3 (3) 2 (4) 5

165. Match List-I with List-II [April 8, 2024 (II)]

List-I (Complex ion)	List-II (Spin only magnetic moment in B.M.)
(A) $[Cr(NH_3)_6]^{3+}$	(I) 4.90
(B) $[NiCl_4]^{2-}$	(II) 3.87
(C) $[CoF_6]^{3-}$	(III) 0.0
(D) $[Ni(CN)_4]^{2-}$	(IV) 2.83

Choose the **correct** answer from the option given below:
(1) (A)-(I), (B)-(IV), (C)-(II), (D)-(III)
(2) (A)-(IV), (B)-(III), (C)-(I), (D)-(II)
(3) (A)-(II), (B)-(IV), (C)-(I), (D)-(III)
(4) (A)-(II), (B)-(III), (C)-(I), (D)-(IV)

166. Consider the following complexes. [April 6, 2024 (I)]
$[CoCl(NH_3)_5]^{2+}$ (A), $[Co(CN)_6]^{3-}$ (B),
$[Co(NH_3)_5(H_2O)]^{3+}$ (C), $[Cu(H_2O)_4]^{2+}$ (D)
The correct order of A, B, C and D in terms of wavenumber of light absorbed is :
(1) C < D < A < B (2) D < A < C < B
(3) A < C < B < D (4) B < C < A < D

167. Match List-I with List-II [April 6, 2024 (II)]

List-I Tetrahedral Complex	List-II Electronic configuration
(A) $TiCl_4$	(I) e^2, t_2^0
(B) $[FeO_4]^{2-}$	(II) e^4, t_2^3
(C) $[FeCl_4]^-$	(III) e^0, t_2^0
(D) $[CoCl_4]^{2-}$	(IV) e^2, t_2^3

Choose the **correct** answer from the option given below:
(1) (A)-(I), (B)-(III), (C)-(IV), (D)-(II)
(2) (A)-(IV), (B)-(III), (C)-(I), (D)-(II)
(3) (A)-(III), (B)-(IV), (C)-(II), (D)-(I)
(4) (A)-(III), (B)-(I), (C)-(IV), (D)-(II)

168. Which one of the following complexes will exhibit the least paramagnetic behaviour ? [April 5, 2024 (I)]
[Atomic number, Cr = 24, Mn = 25, Fe = 26, Co = 27]
(1) $[Co(H_2O)_6]^{2+}$ (2) $[Fe(H_2O)_6]^{2+}$
(3) $[Mn(H_2O)_6]^{2+}$ (4) $[Cr(H_2O)_6]^{2+}$

169. The correct order of ligands arranged in increasing field strength. [April 5, 2024 (I)]
(1) $Cl^- < {}^-OH < Br^- < CN^-$
(2) $F^- < Br^- < I^- < NH_3$
(3) $Br^- < F^- < H_2O < NH_3$
(4) $H_2O < {}^-OH < CN^- < NH_3$

170. The number of complexes from the following with no electrons in the t_2 orbital is ______.
$TiCl_4$, $[MnO_4]^-$, $[FeO_4]^{2-}$, $[FeCl_4]^-$, $[CoCl_4]^{2-}$ [April 5, 2024 (II)]
(1) 3 (2) 1 (3) 4 (4) 2

171. The correct sequence of ligands in the order of decreasing field strength is : [April 4, 2024 (I)]
(1) $CO > H_2O > F^- > S^{2-}$
(2) ${}^-OH > F^- > NH_3 > CN^-$
(3) $NCS^- > EDTA^{4-} > CN^- > CO$
(4) $S^{2-} > {}^-OH > EDTA^{4-} > CO$

172. Number of complexes from the following with even number of unpaired "d" electrons is_____.
$[V(H_2O)_6]^{3+}$, $[Cr(H_2O)_6]^{2+}$, $[Fe(H_2O)_6]^{3+}$, $[Ni(H_2O)_6]^{3+}$, $[Cu(H_2O)_6]^{2+}$
[Given atomic numbers : V = 23, Cr = 24, Fe = 26, Ni = 28, Cu = 29] [April 4, 2024 (I)]
(1) 2 (2) 4 (3) 5 (4) 1

173. If an iron (III) complex with the formula $[Fe(NH_3)_x(CN)_y]^-$ has no electron in its e_g orbital, then the value of x + y is [April 4, 2024 (II)]
(1) 5 (2) 6 (3) 3 (4) 4

174. Given below are two statements: [Feb. 1, 2024 (I)]
Statement (I) : A solution of $[Ni(H_2O)_6]^{2+}$ is green in colour.
Statement (II) : A solution of $[Ni(CN)_4]^{2-}$ is colourless. In the light of the above statements, choose the most appropriate answer from the options given below:
(1) Both Statement I and Statement II are incorrect
(2) Both Statement I and Statement II are correct
(3) Statement I is incorrect but Statement II is correct
(4) Statement I is correct but Statement II is incorrect.

175. $[Co(NH_3)_6]^{3+}$ and $[CoF_6]^{3-}$ are respectively known as: [Feb. 1, 2024 (II)]
(1) Inner orbital Complex, Spin paired Complex
(2) Spin paired Complex, Spin free Complex
(3) Spin free Complex, Spin paired Complex
(4) Outer orbital Complex, Inner orbital Complex

176. The correct statements from following are: [Jan. 31, 2024 (I)]
A. The strength of anionic ligands can be explained by crystal field theory.
B. Valence bond theory does not give a quantitative interpretation of kinetic stability of coordination compounds.

C. The hybridization involved in formation of $[Ni(CN)_4]^{2-}$ complex is dsp^2.
D. The number of possible isomer(s) of cis-$[PtCl_2(en)_2]^{2+}$ is one

Choose the correct answer from the options given below:
(1) B, C only (2) A, C only
(3) A, D only (4) B, D only

177. Match List I with List II [Jan. 31, 2024 (II)]

LIST - I (Complex ion)		LIST -II (Electronic Configuration)
A. $[Cr(H_2O)_6]^{3+}$	I.	$t_{2g}^2e_g^0$
B. $[Fe(H_2O)_6]^{3+}$	II.	$t_{2g}^3e_g^0$
C. $[Ni(H_2O)_6]^{2+}$	III.	$t_{2g}^3e_g^2$
D. $[V(H_2O)_6]^{3+}$	IV.	$t_{2g}^6e_g^2$

Choose the correct answer from the options given below:
(1) A-III, B-II, C-IV, D-I (2) A-IV, B-I, C-II, D-III
(3) A-IV, B-III, C-I, D-II (4) A-II, B-III, C-IV, D-I

178. Select the option with correct property - [Jan. 31, 2024 (II)]
(1) $[Ni(CO)_4]$ and $[NiCl_4]^{2-}$ both diamagnetic
(2) $[Ni(CO)_4]$ and $[NiCl_4]^{2-}$ both paramagnetic
(3) $[NiCl_4]^{2-}$ diamagnetic, $[Ni(CO)_4]$ paramagnetic
(4) $[Ni(CO)_4]$ diamagnetic, $[NiCl_4]^{2-}$ paramagnetic

179. Consider the following complex ions [Jan. 27, 2024 (I)]
$P = [FeF_6]^{3-}$
$Q = [V(H_2O)_6]^{2+}$
$R = [Fe(H_2O)_6]^{2+}$
The correct order of the complex ions, according to their spin only magnetic moment values (in B.M.) is :
(1) $R < Q < P$ (2) $R < P < Q$
(3) $Q < R < P$ (4) $Q < P < R$

180. The complex with highest magnitude of crystal field splitting energy (Δ_0) is [April 15, 2023 (I)]
(1) $[Cr(OH_2)_6]^{3+}$ (2) $[Ti(OH_2)_6]^{3+}$
(3) $[Fe(OH_2)_6]^{3+}$ (4) $[Mn(OH_2)_6]^{3+}$

181. Which of the following complexes will exhibit maximum attraction to an applied magnetic field? [April 13, 2023 (II)]
(1) $[Zn(H_2O)_6]^{2+}$ (2) $[Co(H_2O)_6]^{2+}$
(3) $[Co(en)_3]^{3+}$ (4) $[Ni(H_2O)_6]^{2+}$

182. Match List I with List II [April 12, 2023 (I)]

	List I Complex		List II CFSE (Δ_0)
A.	$[Cu(NH_3)_6]^{2+}$	I.	–0.6
B.	$[Ti(N_2O)_6]^{3+}$	II.	–2.0
C.	$[Fe(CN)_6]^{3-}$	III.	–1.2
D.	$[NiF_6]^{4-}$	IV.	–0.4

Choose the correct answer from the options given below:
(1) A-I, B-IV, C-II, D-III (2) A-II, B-III, C-I, D-IV
(3) A-I, B-II, C-IV, D-III (4) A-III, B-IV, C-I, D-II

183. The complex that dissolves in water is [April 11, 2023 (I)]
(1) $Fe_4[Fe(CN)_6]_3$ (2) $[Fe_3(OH)_2(OAc)_6]Cl$
(3) $K_3[Co(NO_2)_6]$ (4) $(NH_4)_3[As(Mo_3O_{10})_4]$

184. Given below are two statements, one is labelled as **Assertion A** and the other is labelled as **Reason R**.
Assertion A : $[CoCl(NH_3)_5]^{2+}$ absorbs at lower wavelength of light with respect to $[Co(NH_3)_5(H_2O)]^{3+}$
Reason R: It is because the wavelength of the light absorbed depends on the oxidation state of the metal ion.
In the light of the above statements, choose the correct answer from the options given below : [April 11, 2023 (II)]
(1) A is false but R is true.
(2) A is true but R is false.
(3) Both A and R are true and R is the correct explanation of A.
(4) Both A and R are true and R is NOT the correct explanation of A.

185. The magnetic moment is measured in Bohr Magneton (BM). [April 11, 2023 (II)]
Spin only magnetic moment of Fe in $[Fe(H_2O)_6]^{3+}$ and $[Fe(CN)_6]^{3-}$ complexes respectively is:
(1) 6.92 B.M. in both
(2) 4.89 B.M. and 6.92 B.M.
(3) 3.87 B.M. and 1.732 B.M.
(4) 5.92 B.M. and 1.732 B.M.

186. Match List I with List II. [April 11, 2023 (II)]

	List I Complex		List II Colour
A.	$Mg(NH_4)PO_4$	I.	Brown
B.	$K_3[Co(NO_2)_6]$	II.	White
C.	$MnO(OH)_2$	III.	Yellow
D.	$Fe_4[Fe(CN)_6]_3$	IV.	Blue

Choose the correct answer from the options given below:
(1) A-II, B-III, C-I, D-IV (2) A-III, B-IV, C-II, D-I
(3) A-II, B-IV, C-I, D-III (4) A-II, B-III, C-IV, D-I

187. If Ni^{2+} is replaced by Pt^{2+} in the complex $[NiCl_2Br_2]^{2-}$, which of the following properties are expected to get changed? [April 11, 2023 (II)]
A. Geometry B. Geometrical isomerism
C. Optical isomerism D. Magnetic properties
(1) A, B and C (2) A, B and D
(3) A and D (4) B and C

188. The octahedral diamagnetic low spin complex among the following is [April 10, 2023 (I)]
(1) $[NiCl_4]^{2-}$ (2) $[CoCl_6]^{3-}$
(3) $[CoF_6]^{3-}$ (4) $[Co(NH_3)_6]^{3+}$

189. The correct order of the number of unpaired electrons in the given complexes is [April 10, 2023 (II)]
A. $[Fe(CN)_6]^{3-}$ B. $[FeF_6]^{3-}$
C. $[CoF_6]^{3-}$ D. $[Cr(oxalate)_3]^{3-}$
E. $[Ni(CO)_4]$
Choose the correct answer from the options given below:
(1) $A < E < D < C < B$ (2) $E < A < D < C < B$
(3) $E < A < B < D < C$ (4) $A < E < C < B < D$

190. Match List I with List II [April 10, 2023 (II)]

List I (Complex)		List II (Crystal Field splitting energy (Δ_0))
A. $[Ti(H_2O)_6]^{2+}$	I.	–1.2
B. $[V(H_2O)_6]^{2+}$	II.	–0.6
C. $[Mn(H_2O)_6]^{3+}$	III.	0
D. $[Fe(H_2O)_6]^{3+}$	IV.	–0.8

Choose the correct answer from the options given below:
(1) A-II, B-IV, C-I, D-III (2) A-IV, B-I, C-II, D-III
(3) A-IV, B-I, C-III, D-II (4) A-II, B-IV, C-III, D-I

191. Which of the following complex is octahedral, diamagnetic and the most stable? **[April 08, 2023 (I)]**
(1) $Na_3[CoCl_6]$ (2) $[Ni(NH_3)_6]Cl_2$
(3) $K_3[Co(CN)_6]$ (4) $[Co(H_2O)_6]Cl_2$

192. The correct order of spin only magnetic moments for the following complex ions is **[April 08, 2023 (I)]**
(1) $[Fe(CN)_6]^{3-} < [CoF_6]^{3-} < [MnBr_4]^{2-} < [Mn(CN)_6]^{3-}$
(2) $[Fe(CN)_6]^{3-} < [Mn(CN)_6]^{3-} < [CoF_6]^{3-} < [MnBr_4]^{2-}$
(3) $[MnBr_4]^{2-} < [CoF_6]^{3-} < [Fe(CN)_6]^{3-} < [Mn(CN)_6]^{3-}$
(4) $[CoF_6]^{3-} < [MnBr_4]^{2-} < [Fe(CN)_6]^{3-} < [Mn(CN)_6]^{3-}$

193. Match List I with List II: **[April 08, 2023 (II)]**

	List I (Coordination complex)		List II (Number of unpaired electrons)
A.	$[Cr(CN)_6]^{3-}$	I.	0
B.	$[Fe(H_2O)_6]^{2+}$	II.	3
C.	$[Co(NH_3)_6]^{3+}$	III.	2
D.	$[Ni(NH_3)_6]^{2+}$	IV.	4

Choose the correct answer from the options given below:-
(1) A-II, B-IV, C-I, D-III (2) A-IV, B-III, C-II, D-I
(3) A-III, B-IV, C-I, D-II (4) A-II, B-I, C-IV, D-III

194. Given below are two statements, one is labelled as **Assertion A** and the other is labelled as **Reason R.**
Assertion A: The spin only magnetic moment value for $[Fe(CN)_6]^{3-}$ is 1.74 BM, whereas for $[Fe(H_2O)_6]^{3+}$ is 5.92 BM.
Reason R : In both complexes, Fe is present in +3 oxidation state.
In the light of the above statements, choose the correct answer from the options given below: **[April 06, 2023 (I)]**
(1) Both A and R are true but R is NOT the correct explanation of A
(2) A is false but R is true
(3) A is true but R is false
(4) Both A and R are true and R is the correct explanation of A

195. Which of the following complex will show largest splitting of d-orbitals? **[Feb. 01, 2023 (I)]**
(1) $[Fe(CN)_6]^{3-}$ (2) $[Fe(C_2O_4)_3]^{3-}$
(3) $[FeF_6]^{3-}$ (4) $[Fe(NH_3)_6]^{3+}$

196. Cobalt chloride when dissolved in water forms pink colored complex X which has octahedral geometry. This solution on treating with conc. HCl forms deep blue complex, Y which has a Z geometry. X, Y and Z, respectively, are **[Jan. 31, 2023 (I)]**
(1) X= $[Co(H_2O)_6]^{2+}$, Y= $[CoCl_4]^{2-}$, Z = Tetrahedral
(2) X= $[Co(H_2O_6)]^{2+}$, Y= $[CoCl_6]^{3-}$, Z = Octahedral
(3) X= $[Co(H_2O)_6]^{3+}$, Y= $[CoCl_6]^{3-}$, Z = Octahedral
(4) X= $[Co(H_2O)_4Cl_2]^{+}$, Y= $[CoCl_4]^{2-}$, Z = Tetrahedral

197. Which of the following is correct order of ligand field strength? **[Jan. 30, 2023 (I)]**
(1) $CO < en < NH_3 < C_2O_4^{2-} < S^{2-}$
(2) $S^{2-} < C_2O_4^{2-} < NH_3 < en < CO$
(3) $NH_3 < en < CO < S^{2-} < C_2O_4^{2-}$
(4) $S^{2-} < NH_3 < en < CO < C_2O_4^{2-}$

198. Match List I with List II: **[Jan. 30, 2023 (II)]**

	List I (Complexes)		List II (Hybridisation)
(A)	$[Ni(CO)_4]$	I	sp^3
(B)	$[Cu(NH_3)_4]^{2+}$	II	dsp^2
(C)	$[Fe(NH_3)_6]^{2+}$	III	sp^3d^2
(D)	$[Fe(H_2O)_6]^{2+}$	IV	d^2sp^3

(1) A – II, B – I, C – III, D – IV
(2) A – I, B – II, C – III, D – IV
(3) A – II, B – I, C – IV, D – III
(4) A – I, B – II, C – IV, D – III

199. The Cl – Co – Cl bond angle values in a fac-$[Co(NH_3)_3Cl_3]$ complex is/are: **[Jan. 30, 2023 (II)]**
(1) 90° & 180° (2) 90°
(3) 180° (4) 90° & 120°

200. Correct order of spin only magnetic moment of the following complex ions is:
(Given At. No. Fe: 26, Co:27) **[Jan. 29, 2023 (II)]**
(1) $[FeF_6]^{3-} > [CoF_6]^{3-} > [Co(C_2O_4)_3]^{3-}$
(2) $[Co(C_2O_4)_3]^{3-} > [CoF_6]^{3-} > [FeF_6]^{3-}$
(3) $[FeF_6]^{3-} > [Co(C_2O_4)_3]^{3-} > [CoF_6]^{3-}$
(4) $[CoF_6]^{3-} > [FeF_6]^{3-} > [Co(C_2O_4)_3]^{3-}$

201. Match List I with List II **[Jan. 25, 2023 (II)]**

	List I (Coordination entity)		List II (Wavelength of light absorbed in nm)
A.	$[CoCl(NH_3)_5]^{2+}$	I.	310
B.	$[Co(NH_3)_6]^{3+}$	II.	475
C.	$[Co(CN)_6]^{3-}$	III.	535
D.	$[Cu(H_2O)_4]^{2+}$	IV.	600

Choose the correct answer from the options given below:-
(1) A-IV, B-I, C-III, D-II (2) A-III, B-II, C-I, D-IV
(3) A-III, B-I, C-II, D-IV (4) A- II, B-III, C-IV, D-I

202. The magnetic moment of a transition metal compound has been calculated to be 3.87 B.M. The metal ion is **[Jan. 24, 2023 (I)]**
(1) Cr^{2+} (2) Mn^{2+} (3) V^{2+} (4) Ti^{2+}

203. Which of the following cannot be explained by crystal field theory? **[Jan. 24, 2023 (II)]**
(1) The order of spectrochemical series
(2) Magnetic properties of transition metal complexes
(3) Colour of metal complexes
(4) Stability of metal complexes

204. The hybridization and magnetic behaviour of cobalt ion in $[Co(NH_3)_6]^{3+}$ complex, respectively is **[Jan. 24, 2023 (II)]**
(1) sp^3d^2 and diamagnetic (2) d^2sp^3 and paramagnetic
(3) d^2sp^3 and diamagnetic (4) sp^3d^2 and paramagnetic

205. Octahedral complexes of copper (II) undergo structural distortion (John-Teller). Which one of the given copper (II) complexes will show the maximum structural distortion ?
(en–ethylenediamine; $H_2N – CH_2 – CH_2 – NH_2$) **[July 29, 2022 (II)]**
(1) $[Cu(H_2O)_6]SO_4$ (2) $[Cu(en)(H_2O)_4]SO_4$
(3) *cis*-$[Cu(en)_2Cl_2]$ (4) *trans*-$[Cu(en)_2Cl_2]$

206. Match List-I with List-II

	List-I (Complex)		List-II (Hybridization)
(A)	$Ni(CO)_4$	I.	sp^3
(B)	$[Ni(CN)_4]^{2-}$	II.	sp^3d^2
(C)	$[Co(CN)_6]^{3-}$	III.	d^2sp^3
(D)	$[CoF_6]^{3-}$	IV.	dsp^2

Choose the correct answer from the options given below: **[July 28, 2022 (II)]**
(1) (A) - IV, (B) - I, (C) - III, (D) - II
(2) (A) - I, (B) - IV, (C) - III, (D) - II
(3) (A) - I, (B) - IV, (C) - II, (D) - III
(4) (A) - IV, (B) - (I), (C) - II, (D) - III

207. Low oxidation state of metals in their complexes are common when ligands: **[July 27, 2022 (II)]**
(1) have good π-accepting character
(2) have good σ-donor character
(3) are having good π-donating ability
(4) are having poor σ-donating ability

208. The metal complex that is diamagnetic is (Atomic number : Fe, 26; Cu, 29) **[July 26, 2022 (II)]**
(1) $K_3[Cu(CN)_4]$ (2) $K_2[Cu(CN)_4]$
(3) $K_3[Fe(CN)_4]$ (4) $K_4[FeCl_6]$

209. The correct order of energy of absorption for the following metal complexes is **[July 25, 2-022 (II)]**
A: $[Ni(en)_3]^{2+}$, B: $[Ni(NH_3)_6]^{2+}$, C: $[Ni(H_2O)_6]^{2+}$
(1) C<B<A (2) B<C<A
(3) C<A<B (4) A<C<B

210. Given below are two statements : **[June 28, 2022 (I)]**
Statement I : $[Ni(CN)_4]^{2-}$ is square planar and diamagnetic complex with dsp^2 hybridization for Ni but $[Ni(CO)_4]$ is tetrahedral paramagnetic and with sp^3-hybridication for Ni.
Statement II : $[NiCl_4]^{2-}$ and $[Ni(CO)_4]$ both have same *d*-electron configuration have same geometry and are paramagnetic.
In light the above statements choose the correct answer form the options given below:
(1) Both Statement I and Statement II are true.
(2) Both Statement I and Statement II are false.
(3) Statement I is correct but Statement II is false.
(4) Statement I is incorrect but Statement II is true.

211. Which of the following will have maximum stabilization due to crystal field? **[June 27, 2022 (I)]**
(1) $[Ti(H_2O)_6]^{3+}$ (2) $[Co(H_2O)_6]^{2+}$
(3) $[Co(CN)_6]^{3-}$ (4) $[Cu(NH_3)_4]^{2+}$

212. Arrange the following coordination compounds in the increasing order of magnetic moments. (Atomic numbers: Mn = 25; Fe = 26) **[June 27, 2022 (II)]**
(A) $[FeF_6]^{3-}$ (B) $[Fe(CN)_6]^{3-}$
(C) $[MnCl_6]^{3-}$ (high spin) (D) $[Mn(CN)_6]^{3-}$
(1) A<B<D<C (2) B<D<C<A
(3) A<C<D<B (4) B<D<A<C

213. The metal ion (in gaseous state) with lowest spin-only magnetic moment value is **[June 25, 2022 (II)]**
(1) V^{2+} (2) Ni^{2+} (3) Cr^{2+} (4) Fe^{2+}

214. Transition metal complex with highest value of crystal field splitting (Δ_0) will be **[June 24, 2022 (II)]**
(1) $[Cr(H_2O)_6]^{3+}$ (2) $[Mo(H_2O)_6]^{3+}$
(3) $[Fe(H_2O)_6]^{3+}$ (4) $[Os(H_2O)_6]^{3+}$

215. The Crystal Field Stabilization Energy (CFSE) and magnetic moment (spin-only) of an octahedral aqua complex of a metal ion (M^{Z+}) are $-0.8\ \Delta_0$ and 3.87 BM, respectively. Identify (M^{Z+}): **[Sep. 1, 2021 (II)]**
(1) V^{3+} (2) Cr^{3+} (3) Mn^{4+} (4) Co^{2+}

216. Spin only magnetic moment in BM of $[Fe(CO)_4(C_2O_4)]^+$ is: **[Aug. 31, 2021 (II)]**
(1) 5.92 (2) 0 (3) 1 (4) 1.73

217. Which one of the following complexes is violet in colour? **[Aug. 26, 2021 (I)]**
(1) $[Fe(CN)_6]^{4-}$ (2) $[Fe(SCN)_6]^{4-}$
(3) $Fe_4[Fe(CN)_6]_3 \cdot H_2O$ (4) $[Fe(CN)_5NOS]^{4-}$

218. Arrange the following Cobalt complexes in the order of increasing Crystal Field Stabilization Energy (CFSE) value.
Complexes :

$$\underset{A}{[CoF_6]^{3-}}, \underset{B}{\left[Co(H_2O)_6\right]^{2+}}, \underset{C}{\left[Co(NH_3)_6\right]^{3+}} \text{ and } \underset{D}{\left[Co(en)_3\right]^{3+}}$$

Choose the correct option : **[Aug. 26, 2021 (II)]**
(1) A<B<C<D (2) B<A<C<D
(3) B<C<D<A (4) C<D<B<A

219. The type of hybridisation and magnetic property of the complex $[MnCl_6]^{3-}$, respectively, are : **[July 27, 2021 (I)]**
(1) sp^3d^2 and diamagnetic (2) d^2sp^3 and diamagnetic
(3) d^2sp^3 and paramagnetic (4) sp^3d^2 and paramagnetic

220. Which one of the following species responds to an external magnetic field? **[July 25, 2021 (I)]**
(1) $[Fe(H_2O)_6]^{3+}$ (2) $[Ni(CN)_4]^{2-}$
(3) $[Co(CN)_6]^{3-}$ (4) $[Ni(CO)_4]$

221. The spin only magnetic moments (in BM) for free Ti^{3+}, V^{2+} and Sc^{3+} ions respectively are
(At. No. Sc : 21, Ti : 22, V : 23) **[July 25, 2021 (II)]**
(1) 3.87, 1.73, 0 (2) 1.73, 3.87, 0
(3) 1.73, 0, 3.87 (4) 0, 3.87, 1.73

222. According to the valence bond theory, the hybridization of central metal atom is dsp^2 for which one of the following compounds? **[July 20, 2021 (I)]**
(1) $NiCl_2 \cdot 6H_2O$ (2) $K_2[Ni(CN)_4]$
(3) $[Ni(CO)_4]$ (4) $Na_2[NiCl_4]$

223. The correct order of intensity of colors of the compounds is : **[July 20, 2021 (I)]**
(1) $[Ni(CN)_4]^{2-} > [NiCl_4]^{2-} > [Ni(H_2O)_6]^{2+}$
(2) $[Ni(H_2O)_6]^{2+} > [NiCl_4]^{2-} > [Ni(CN)_4]^{2-}$
(3) $[NiCl_4]^{2-} > [Ni(H_2O)_6]^{2+} > [Ni(CN)_4]^{2-}$
(4) $[NiCl_4]^{2-} > [Ni(CN)_4]^{2-} > [Ni(H_2O)_6]^{2+}$

224. Spin only magnetic moment of an octahedral complex of Fe^{2+} in the presence of a strong field ligand in BM is : **[July 20, 2021 (II)]**
(1) 4.89 (2) 2.82 (3) 0 (4) 3.46

225. Which one of the following species doesn't have a magnetic moment of 1.73 B.M., (spin only value) ? **[July 20, 2021 (II)]**
(1) O_2^+ (2) CuI
(3) $[Cu(NH_3)_4]Cl_2$ (4) O_2^-

226. The hybridization and magnetic nature of $[Mn(CN)_6]^{4-}$ and $[Fe(CN)_6]^{3-}$, respectively are **[Feb. 25, 2021 (I)]**
(1) d^2sp^3 and paramagnetic
(2) d^2sp^3 and diamagnetic
(3) sp^3d^2 and paramagnetic
(4) sp^3d^2 and diamagnetic

227. In which of the following order, the given complex ions are arranged correctly with respect to their decreasing spin only magnetic moment? **[Feb. 25, 2021 (II)]**
(i) $[FeF_6]^{3-}$ (ii) $[Co(NH_3)_6]^{3+}$
(iii) $[NiCl_4]^{2-}$ (iv) $[Cu(NH_3)_4]^{2+}$
(1) (iii) > (iv) > (ii) > (i) (2) (ii) > (iii) > (i) > (iv)
(3) (i) > (iii) > (iv) > (ii) (4) (ii) > (i) > (iii) > (iv)

228. The calculated magnetic moments (spin only value) for species $[FeCl_4]^{2-}$, $[Co(C_2O_4)_3]^{3-}$ and MnO_4^{2-} respectively are : **[Feb. 24, 2021 (II)]**
(1) 5.82, 0 and 0 BM (2) 4.90, 0 and 1.73 BM
(3) 5.92, 4.90 and 0 BM (4) 4.90, 0 and 2.83 BM

229. The species that has a spin-only magnetic moment of 5.9 B.M., is : (T_d = tetrahedral) [Sep. 06, 2020 (I)]
(1) $[Ni(CN)_4]^{2-}$ (square planar)
(2) $[NiCl_4]^{2-}$ (T_d)
(3) $[Ni(CO)_4]$ (T_d)
(4) $[MnBr_4]^{2-}$ (T_d)

230. For a d^4 metal ion in an octahedral field, the correct electronic configuration is : [Sep. 06, 2020 (II)]
(1) $t_{2g}^3 e_g^1$ when $\Delta_0 < P$ (2) $t_{2g}^3 e_g^1$ when $\Delta_0 > P$
(3) $t_{2g}^4 e_g^0$ when $\Delta_0 < P$ (4) $e_g^2 t_{2g}^2$ when $\Delta_0 < P$

231. The values of the crystal field stabilization energies for a high spin d^6 metal ion in octahedral and tetrahedral fields, respectively, are : [Sep. 05, 2020 (I)]
(1) $-0.4\,\Delta_o$ and $-0.6\,\Delta_t$ (2) $-2.4\,\Delta_o$ and $-0.6\,\Delta_t$
(3) $-1.6\,\Delta_o$ and $-0.4\,\Delta_t$ (4) $-0.4\,\Delta_o$ and $-0.27\,\Delta_t$

232. The pair in which both the species have the same magnetic moment (spin only) is: [Sep. 04, 2020 (I); Same 2016]
(1) $[Cr(H_2O)_6]^{2+}$ and $[Fe(H_2O)_6]^{2+}$
(2) $[Co(OH)_4]^{2-}$ and $[Fe(NH_3)_6]^{2+}$
(3) $[Mn(H_2O)_6]^{2+}$ and $[Cr(H_2O)_6]^{2+}$
(4) $[Cr(H_2O)_6]^{2+}$ and $[CoCl_4]^{2-}$

233. The Crystal Field Stabilization Energy (CFSE) of $[CoF_3(H_2O)_3]$ $(\Delta_0 < P)$ is : [Sep. 04, 2020 (II)]
(1) $-0.8\Delta_0 + 2P$ (2) $-0.4\Delta_0$
(3) $-0.8\Delta_0$ (4) $-0.4\Delta_0 + P$

234. The one that can exhibit highest paramagnetic behaviour among the following is : [Sep. 04, 2020 (II)]
gly = glycinato; bpy = 2, 2'-bipyridine
(1) $[Pd(gly)_2]$ (2) $[Fe(en)(bpy)(NH_3)_2]^{2+}$
(3) $[Co(ox)_2(OH)_2]^-$ $(\Delta_0 > P)$ (4) $[Ti(NH_3)_6]^{3+}$

235. The electronic spectrum of $[Ti(H_2O)_6]^{3+}$ shows a single broad peak with a maximum at 20,300 cm^{-1}. The crystal field stabilization energy (CFSE) of the complex ion, in kJ mol^{-1}, is : (1 kJ mol^{-1} = 83.7 cm^{-1}) [Sep. 03, 2020 (I)]
(1) 145.5 (2) 242.5 (3) 83.7 (4) 97

236. The d-electron configuration of $[Ru(en)_3]Cl_2$ and $[Fe(H_2O)_6]Cl_2$, respectively are : [Sep. 03, 2020 (II)]
(1) $t_{2g}^6\, e_g^0$ and $t_{2g}^6\, e_g^0$ (2) $t_{2g}^4\, e_g^2$ and $t_{2g}^6\, e_g^0$
(3) $t_{2g}^6\, e_g^0$ and $t_{2g}^4\, e_g^2$ (4) $t_{2g}^4\, e_g^2$ and $t_{2g}^4\, e_g^2$

237. Consider that a d^6 metal ion (M^{2+}) forms a complex with aqua ligands, and the spin only magnetic moment of the complex is 4.90 BM. The geometry and the crystal field stabilization energy of the complex is : [Sep. 02, 2020 (I)]
(1) octahedral and $-2.4\Delta_0 + 2P$
(2) tetrahedral and $-0.6\,\Delta_t$
(3) octahedral and $-1.6\,\Delta_0$
(4) tetrahedral and $-1.6\Delta_t + 1P$

238. For octahedral Mn(II) and tetrahedral Ni(II) complexes, consider the following statements : [Sep. 02, 2020 (I)]
(I) both the complexes can be high spin.
(II) Ni(II) complex can very rarely below spin.
(III) with strong field ligands, Mn(II) complexes can be low spin.
(IV) aqueous solution of Mn(II) ions is yellow in color.

The correct statements are :
(1) (I) and (II) only (2) (I), (III) and (IV) only
(3) (I), (II) and (III) only (4) (II), (III) and (IV) only

239. Simplified absorption spectra of three complexes ((i), (ii) and (iii)) of M^{n+} ion are provided below; their λ_{max} values are marked as A, B and C respectively. The correct match between the complexes and their λ_{max} values is : [Sep. 02, 2020 (II)]

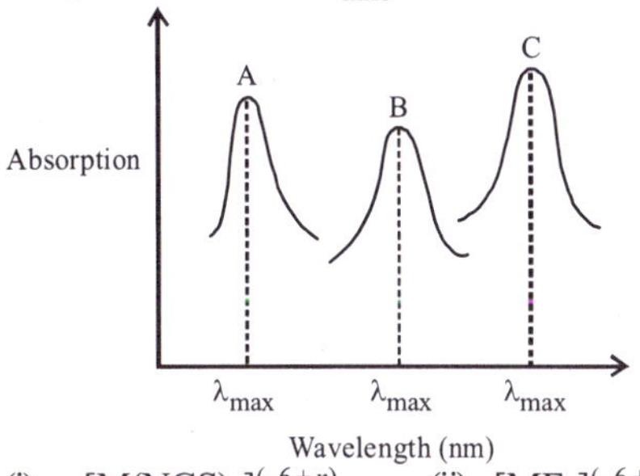

(i) $[M(NCS)_6]^{(-6+n)}$ (ii) $[MF_6]^{(-6+n)}$
(iii) $[M(NH_3)_6]^{n+}$
(1) A-(iii), B-(i), C-(ii) (2) A-(ii), B-(i), C-(iii)
(3) A-(ii), B-(iii), C-(i) (4) A-(i), B-(ii), C-(iii)

240. $[Pd(F)(Cl)(Br)(I)]^{2-}$ has n number of geometrical isomers. Then, the spin-only magnetic moment and crystal field stabilisation energy [CFSE] of $[Fe(CN)_6]^{n-6}$, respectively, are:
[Note : Ignore the pairing energy] [Jan. 09, 2020 (I)]
(1) 2.84 BM and $-1.6\,D_0$ (2) 5.92 BM and 0
(3) 1.73 BM and $-2.0\,D_0$ (4) 0 BM and $-2.4\,D_0$

241. The correct order of the spin-only magnetic moments of the following complexes is: [Jan. 09, 2020 (II)]
(A) $[Cr(H_2O)_6]Br_2$ (B) $Na_4[Fe(CN)_6]$
(C) $Na_3[Fe(C_2O_4)_3]$ $(\Delta_0 > P)$ (D) $(Et_4N)_2[CoCl_4]$
(1) (C) > (A) > (D) > (B) (2) (C) > (A) > (B) > (D)
(3) (A) > (D) > (C) > (B) (4) (B) ≈ (A) > (D) > (C)

242. The correct order of the calculated spin-only magnetic moments of complexes (A) to (D) is: [Jan. 08, 2020 (II)]
(A) $Ni(CO)_4$ (B) $[Ni(H_2O)_6]Cl_2$
(C) $Na_2[Ni(CN)_4]$ (D) $PdCl_2(PPh_3)_2$
(1) (A) ≈ (C) < (B) ≈ (D) (2) (C) < (D) < (B) < (A)
(3) (C) ≈ (D) < (B) < (A) (4) (A) ≈ (C) ≈ (D) < (B)

243. Among the statements (A)-(D), the incorrect ones are:
(A) Octahedral Co(III) complexes with strong field ligands have very high magnetic moments. [Jan. 07, 2020 (II)]
(B) When $\Delta_0 < P$, the d-electron configuration of Co(III) in an octahedral complex is $t_{2g}^4\, e_g^2$.
(C) Wavelength of light absorbed by $[Co(en)_3]^{3+}$ is lower than that of $[CoF_6]^{3-}$.
(D) If the Δ_0 for an octahedral complex of Co(III) is 18,000 cm^{-1}, the Δ_t for its tetrahedral complex with the same ligand will be 16,000 cm^{-1}.
(1) (A) and (D) only (2) (C) and (D) only
(3) (A) and (B) only (4) (B) and (C) only

244. The complex ion that will lose its crystal field stabilization energy upon oxidation of its metal to +3 state is : [April 12, 2019 (I)]

(Phen = 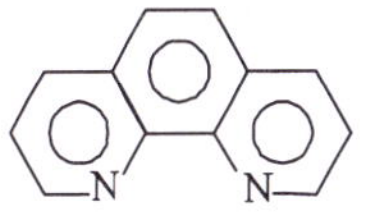

and ignore pairing energy)

(1) $[Co(phen)_3]^{2+}$ (2) $[Ni(phen)_3]^{2+}$
(3) $[Zn(phen)_3]^{2+}$ (4) $[Fe(phen)_3]^{2+}$

245. Complete removal of both the axial ligands (along the z-axis) from an octahedral complex leads to which of the following splitting patterns ? (relative orbital energies not on scale) **[April 12, 2019 (I)]**

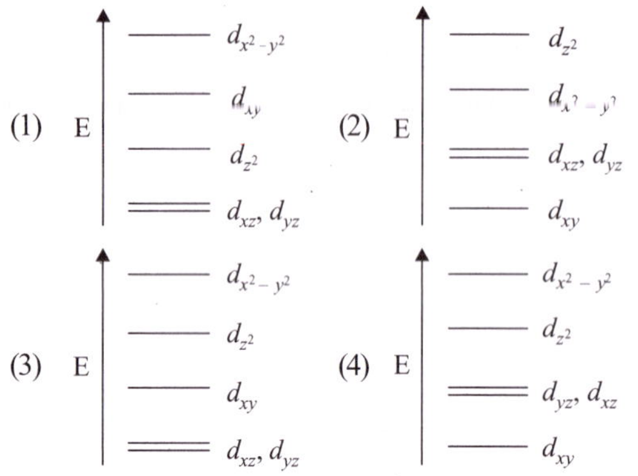

246. Three complexes, $[CoCl(NH_3)_5]^{2+}$(I), $[Co(NH_3)_5H_2O]^{3+}$(II) and $[Co(NH_3)_6]^{3+}$ (III) absorb light in the visible region. The correct order of the wavelength of light absorbed by them is : **[April 10, 2019 (I)]**
(1) (III) > (I) > (II) (2) (III) > (II) > (I)
(3) (II) > (I) > (III) (4) (I) > (II) > (III)

247. The crystal field stabilization energy (CFSE) of $[Fe(H_2O)_6]Cl_2$ and $K_2[NiCl_4]$, respectively are : **[April 10, 2019 (II)]**
(1) $-0.6\,\Delta_o$ and $-0.8\,\Delta_\tau$ (2) $-0.4\,\Delta_o$ and $-0.8\,\Delta_\tau$
(3) $-2.4\,\Delta_o$ and $-1.2\,\Delta_\tau$ (4) $-0.4\,\Delta_o$ and $-1.2\,\Delta_\tau$

248. The INCORRECT statement is : **[April 10, 2019 (II)]**
(1) the gemstone, ruby, has Cr^{3+} ions occupying the octahedral sites of beryl.
(2) the spin-only magnetic moment of $[Ni(NH_3)_4(H_2O)_2]^{2+}$ is 2.83 BM.
(3) the color of $[CoCl(NH_3)_5]^{2+}$ is violet as it absorbs the yellow light.
(4) the spin-only magnetic moments of $[Fe(H_2O)_6]^{2+}$ and $[Cr(H_2O)_6]^{2+}$ are nearly similar.

249. The degenerate orbitals of $[Cr(H_2O)_6]^{3+}$ are: **[April 9, 2019 (I)]**
(1) d_{xz} and d_{yz} (2) d_{yz} and d_{z^2}
(3) d_{z^2} and d_{xz} (4) $d_{x^2-y^2}$ and d_{xy}

250. The correct statements among I to III are: **[April 9, 2019 (II)]**
(I) Valence bond theory cannot explain the color exhibited by transition metal complexes.
(II) Valence bond theory can predict quantitatively the magnetic properties of transition metal complexes.
(III) Valence bond theory cannot distinguish ligands as week and strong field ones.
(1) (II) and (III) only (2) (I), (II) and (III)
(3) (I) and (III) only (4) (I) and (II) only

251. The calculated spin-only magnetic moments (BM) of the anionic and cationic species of $[Fe(H_2O)_6]_2[Fe(CN)_6]$, respectively, are:
(1) 0 and 4.9 (2) 2.84 and 5.92 **[April 8, 2019 (II)]**
(3) 4.9 and 0 (4) 0 and 5.92

252. The metal d-orbitals that are directly facing the ligands in $K_3[Co(CN)_6]$ are : **[Jan. 12, 2019 (I)]**
(1) d_{xy} and $d_{x^2-y^2}$ (2) $d_{x^2-y^2}$ and d_{z^2}
(3) d_{xz}, d_{yz} and d_{z^2} (4) d_{xy}, d_{xz} and d_{yz}

253. The pair of metal ions that can give a spin only magnetic moment of 3.9 BM for the complex $[M(H_2O)_6]Cl_2$, is: **[Jan. 12, 2019 (I)]**
(1) V^{2+} and Co^{2+} (2) V^{2+} and Fe^{2+}
(3) Co^{2+} and Fe^{2+} (4) Cr^{2+} and Mn^{2+}

254. The magnetic moment of an octahedral homoleptic Mn(II) complex is 5.9 BM. The suitable ligand for this complex is :
(1) Ethylenediamine (2) CN^- **[Jan. 12, 2019 (II)]**
(3) NCS^- (4) CO

255. Match the metals **(column I)** with the coordination compound(s)/enzyme(s) **(column II)** : **[Jan. 11, 2019 (I)]**

Column I (Metals)	**Column II (Coordination compound(s)/enzyme(s))**
(A) Co	(i) Wilkinson catalyst
(B) Zn	(ii) Chlorophyll
(C) Rh	(iii) Vitamin B_{12}
(D) Mg	(iv) Carbonic anhydrase

(1) (A)-(iii); (B)-(iv); (C)-(i); (D)-(ii)
(2) (A)-(i); (B)-(ii); (C)-(iii); (D)-(iv)
(3) (A)-(ii); (B)-(i); (C)-(iv); (D)-(iii)
(4) (A)-(iv); (B)-(iii); (C)-(i); (D)-(ii)

256. The difference in the number of unpaired electrons of a metal ion in its high-spin and low-spin octahedral complexes is two. The metal ion is: **[Jan. 10, 2019 (II)]**
(1) Ni^{2+} (2) Fe^{2+} (3) Co^{2+} (4) Mn^{2+}

257. Two complexes $[Cr(H_2O)_6]Cl_3$ (A) and $[Cr(NH_3)_6]Cl_3$ (B) are violet and yellow coloured, respectively. The incorrect statement regarding them is: **[Jan. 9, 2019 (I)]**
(1) Δ_0 values of (A) and (B) are calculated from the energies of violet and yellow light, respectively.
(2) both are paramagnetic with three unpaired electrons.
(3) both absorb energies corresponding to their complementary colors.
(4) Δ_0 value for (A) is less than that of (B).

258. The complex that has highest crystal field splitting energy (Δ) is: **[Jan. 9, 2019 (II)]**
(1) $[Co(NH_3)_5(H_2O)]Cl_3$ (2) $K_2[CoCl_4]$
(3) $[Co(NH_3)_5Cl]Cl_2$ (4) $K_3[Co(CN)_6]$

259. Homoleptic octahedral complexes of a metal ion 'M^{3+}' with three monodentate ligands L_1, L_2 and L_3 absorb wavelengths in the region of green, blue and red respectively. The increasing order of the ligand strength is: **[Jan. 9, 2019 (II)]**
(1) $L_3 < L_1 < L_2$ (2) $L_3 < L_2 < L_1$
(3) $L_1 < L_2 < L_3$ (4) $L_2 < L_1 < L_3$

260. In Wilkinson's catalyst, the hybridisation of central metal ion and its shape are respectively: **[Online April 16, 2018]**
(1) sp^3d, trigonal bipyramidal
(2) d^2sp^3, octahedral
(3) dsp^2, square planar
(4) sp^3, tetrahedral

261. The correct combination is : **[Online April 15, 2018 (I)]**
(1) $[NiCl_4]^{2-}$-square-planar; $[Ni(CN)_4]^{2-}$-paramagnetic
(2) $[Ni(CN)_4]^{2-}$-tetrahedral; $[Ni(CO)_4]$-paramagnetic
(3) $[NiCl_4]^{2-}$-paramagnetic; $[Ni(CO)_4]$-tetrahedral
(4) $[NiCl_4]^{2-}$-dimagnetic; $[Ni(CO)_4]$-square-planar

262. The correct order of spin-only magnetic moments among the following is: (Atomic number: Mn = 25, Co = 27, Ni = 28, Zn = 30) **[Online April 15, 2018 (II)]**
(1) $[ZnCl_4]^{2-} > [NiCl_4]^{2-} > [CoCl_4]^{2-} > [MnCl_4]^{2-}$
(2) $[CoCl_4]^{2-} > [MnCl_4]^{2-} > [NiCl_4]^{2-} > [ZnCl_4]^{2-}$
(3) $[NiCl_4]^{2-} > [CoCl_4]^{2-} > [MnCl_4]^{2-} > [ZnCl_4]^{2-}$
(4) $[MnCl_4]^{2-} > [CoCl_4]^{2-} > [NiCl_4]^{2-} > [ZnCl_4]^{2-}$

263. The transition metal ions responsible for colour in ruby and emerald are, respectively : [Online April 10, 2016]
(1) Co^{3+} and Cr^{3+} (2) Co^{3+} and Co^{3+}
(3) Cr^{3+} and Cr^{3+} (4) Cr^{3+} and Co^{3+}

264. Identify the correct trend given below : [Online April 9, 2016]
(Atomic No. = Ti : 22, Cr : 24 and Mo : 42)
(1) Δ_o of $[Cr(H_2O)_6]^{2+} > [Mo(H_2O)_6]^{2+}$ and Δ_o of $[Ti(H_2O)_6]^{3+} > [Ti(H_2O)_6]^{2+}$
(2) Δ_o of $[Cr(H_2O)_6]^{2+} > [Mo(H_2O)_6]^{2+}$ and Δ_o of $[Ti(H_2O)_6]^{3+} < [Ti(H_2O)_6]^{2+}$
(3) Δ_o of $[Cr(H_2O)_6]^{2+} < [Mo(H_2O)_6]^{2+}$ and Δ_o of $[Ti(H_2O)_6]^{3+} > [Ti(H_2O)_6]^{2+}$
(4) Δ_o of $[Cr(H_2O)_6]^{2+} < [Mo(H_2O)_6]^{2+}$ and Δ_o of $[Ti(H_2O)_6]^{3+} < [Ti(H_2O)_6]^{2+}$

265. The pair having the same magnetic moment is:
[At. No.: Cr = 24, Mn = 25, Fe = 26, Co = 27] [2016]
(1) $[Mn(H_2O)_6]^{2+}$ and $[Cr(H_2O)_6]^{2+}$
(2) $[CoCl_4]^{2-}$ and $[Fe(H_2O)_6]^{2+}$
(3) $[Cr(H_2O)_6]^{2+}$ and $[CoCl_4]^{2-}$
(4) $[Cr(H_2O)_6]^{2+}$ and $[Fe(H_2O)_6]^{2+}$

266. Which of the following compounds is not colored yellow? [2015]
(1) $(NH_4)_3[As(Mo_3O_{10})_4]$ (2) $BaCrO_4$
(3) $Zn_2[Fe(CN)_6]$ (4) $K_3[Co(NO_2)_6]$

267. The color of $KMnO_4$ is due to : [2015]
(1) L → M charge transfer transition
(2) σ - σ* transition
(3) M → L charge transfer transition
(4) d – d transition

268. Which of the following complex ions has electrons that are symmetrically filled in both t_{2g} and e_g orbitals ? [Online April 11, 2015]
(1) $[FeF_6]^{3-}$ (2) $[Mn(CN)_6]^{4-}$
(3) $[CoF_6]^{3-}$ (4) $[Co(NH_3)_6]^{2+}$

269. When concentrated HCl is added to an aqueous solution of $CoCl_2$, its colour changes from reddish pink to deep blue. Which complex ion gives blue colour in this reaction? [Online April 11, 2015]
(1) $[CoCl_4]^{2-}$ (2) $[CoCl_6]^{3-}$
(3) $[CoCl_6]^{4-}$ (4) $[Co(H_2O)_6]^{2+}$

270. An octahedral complex of Co^{3+} is diamagnetic. The hybridisation involved in the formation of the complex is: [Online April 9, 2014]
(1) sp^3d^2 (2) dsp^2 (3) d^2sp^3 (4) sp^3d

271. The correct statement about the magnetic properties of $[Fe(CN)_6]^{3-}$ and $[FeF_6]^{3-}$ is (Z = 26) : [Online April 9, 2014]
(1) both are paramagnetic.
(2) both are diamagnetic.
(3) $[Fe(CN)_6]^{3-}$ is diamagnetic, $[FeF_6]^{3-}$ is paramagnetic.
(4) $[Fe(CN)_6]^{3-}$ is paramagnetic, $[FeF_6]^{3-}$ is diamagnetic.

272. Consider the coordination compound, $[Co(NH_3)_6]Cl_3$. In the formation of this complex, the species which acts as the Lewis acid is: [Online April 11, 2014]
(1) $[Co(NH_3)_6]^{3+}$ (2) Cl^-
(3) Co^{3+} (4) NH_3

273. Among the following species, the one which causes the highest CFSE, Δ_o as a ligand is: [Online April 12, 2014]
(1) CN^- (2) NH_3 (3) F^- (4) CO

274. Which one of the following complexes will most likely absorb visible light? [Online April 12, 2014]
(At nos. Sc = 21, Ti = 22, V = 23, Zn = 30)
(1) $[Sc(H_2O)_6]^{3+}$ (2) $[Ti(NH_3)_6]^{4+}$
(3) $[V(NH_3)_6]^{3+}$ (4) $[Zn(NH_3)_6]^{2+}$

275. Nickel (Z = 28) combines with a uninegative monodentate ligand to form a diamagnetic complex $[NiL_4]^{2-}$. The hybridisation involved and the number of unpaired electrons present in the complex are respectively: [Online April 19, 2014]
(1) sp^3, two (2) dsp^2, zero
(3) dsp^2, one (4) sp^3, zero

276. The octahedral complex of a metal ion M^{3+} with four monodentate ligands L_1, L_2, L_3 and L_4 absorb wavelengths in the region of red, green, yellow and blue, respectively. The increasing order of ligand strength of the four ligands is: [2014]
(1) $L_4 < L_3 < L_2 < L_1$ (2) $L_1 < L_3 < L_2 < L_4$
(3) $L_3 < L_2 < L_4 < L_1$ (4) $L_1 < L_2 < L_4 < L_3$

277. Which of the following is diamagnetic? [Online April 22, 2013]
(1) $[Fe(CN)_6]^{3-}$ (2) $[Co(ox)]^{3-}$
(3) $[FeF_6]^{3-}$ (4) $[CoF_6]^{3-}$

278. The magnetic moment of the complex anion $[Cr(NO)(NH_3)(CN)_4]^{2-}$ is : [Online April 23, 2013]
(1) 5.91 BM (2) 3.87 BM
(3) 1.73 BM (4) 2.82 BM

279. The structure of which of the following chloro species can be explained on the basis of dsp^2 hybridisation ? [Online April 25, 2013]
(1) $PdCl_4^{2-}$ (2) $FeCl_4^{2-}$
(3) $CoCl_4^{2-}$ (4) $NiCl_4^{2-}$

280. Square-planar geometry is shown by [Online May 19, 2012]
(1) $[PtCl_2(NH_3)_2]$ (2) $[NiCl_4]^{2-}$
(3) MnO_4^- (4) CrO_4^{2-}

281. The *d*-electron configurations of Cr^{2+}, Mn^{2+}, Fe^{2+} and Co^{2+} are d^4, d^5, d^6 and d^7 respectively. Which one of the following will exhibit the lowest paramagnetic behaviour? [Online May 7, 2012]
(Atomic no. Cr = 24, Mn = 25, Fe = 26, Co = 27).
(1) $[Co(H_2O)_6]^{2+}$ (2) $[Cr(H_2O)_6]^{2+}$
(3) $[Mn(H_2O)_6]^{2+}$ (4) $[Fe(H_2O)_6]^{2+}$

282. The correct order of ligands in the spectrochemical series is [Online May 19, 2012]
(1) $Cl^- > en > CN^- > NCS^-$ (2) $CN^- > en > NCS^- > Cl^-$
(3) $NCS^- > CN^- > Cl^- > en$ (4) $en > CN^- > Cl^- > NCS^-$

283. Which of the following paramagnetic ions would exhibit a magnetic moment (spin only) of the order of 5 BM?
(At. Nos. Mn = 25, Cr = 24, V = 23, Ti = 22) [Online May 19, 2012]
(1) Mn^{2+} (2) Ti^{2+} (3) V^{2+} (4) Cr^{2+}

284. In which of the following complexes of the Co (at. no. 27), will the magnitude of Δ_0 be the highest? [2008]
(1) $[Co(CN)_6]^{3-}$ (2) $[Co(C_2O_4)_3]^{3-}$
(3) $[Co(H_2O)_6]^{3+}$ (4) $[Co(NH_3)_6]^{3+}$

285. Which of the following has a square planar geometry? [2007]
(1) $[PtCl_4]^{2-}$ (2) $[CoCl_4]^{2-}$
(3) $[FeCl_4]^{2-}$ (4) $[NiCl_4]^{2-}$
(At. nos.: Fe = 26, Co = 27, Ni = 28, Pt = 78)

286. Nickel (Z = 28) combines with a uninegative monodentate ligand X^- to form a paramagnetic complex $[NiX_4]^{2-}$. The number of unpaired electron(s) in the nickel and geometry of this complex ion are, respectively : [2006]
(1) one, square planar (2) two, square planar
(3) one, tetrahedral (4) two, tetrahedral

287. The "spin-only" magnetic moment [in units of Bohr magneton, (μ_B)] of Ni^{2+} in aqueous solution would be (At. No. Ni = 28) [2006]

(1) 6 (2) 1.73 (3) 2.82 (4) 4.90

288. The value of the 'spin only' magnetic moment for one of the following configurations is 2.84 BM. The correct one is [2005]

(1) d^5 (in strong ligand field)
(2) d^3 (in weak as well as in strong fields)
(3) d^4 (in weak ligand fields)
(4) d^4 (in strong ligand fields)

289. Which one of the following cyano complexes would exhibit the lowest value of paramagnetic behaviour ? [2005]

(1) $[Co(CN)_6]^{3-}$ (2) $[Fe(CN)_6]^{3-}$
(3) $[Mn(CN)_6]^{3-}$ (4) $[Cr(CN)_6]^{3-}$

(At. No. : Cr = 24, Mn = 25, Fe = 26, Co = 27)

290. The correct order of magnetic moments (spin only values in B.M.) among the following is [2004]

(1) $[Fe(CN)_6]^{4-} > [MnCl_4]^{2-} > [CoCl_4]^{2-}$
(2) $[MnCl_4]^{2-} > [Fe(CN)_6]^{4-} > [CoCl_4]^{2-}$
(3) $[MnCl_4]^{2-} > [CoCl_4]^{2-} > [Fe(CN)_6]^{4-}$
(4) $[Fe(CN)_6]^{4-} > [CoCl_4]^{2-} > [MnCl_4]^{2-}$

(Atomic nos. : Mn = 25, Fe = 26, Co = 27)

291. Which one of the following complexes is an outer orbital complex ? [2004]

(1) $[Co(NH_3)_6]^{3+}$ (2) $[Mn(CN)_6]^{4-}$
(3) $[Fe(CN)_6]^{4-}$ (4) $[Ni(NH_3)_6]^{2+}$

(Atomic no. : Mn = 25; Fe = 26; Co = 27, Ni = 28)

292. A square planar complex is formed by hybridisation of which atomic orbitals? [2002]

(1) s, p_x, p_y, d_{yz} (2) $s, p_x, p_y, d_{x^2-y^2}$
(3) s, p_x, p_y, d_{z^2} (4) s, p_y, p_z, d_{xy}

293. The most stable ion is [2002]

(1) $[Fe(OH)_5]^{3-}$ (2) $[Fe(Cl)_6]^{3-}$
(3) $[Fe(CN)_6]^{3-}$ (4) $[Fe(H_2O)_6]^{3+}$.

294. When concentrated HCl is added to an aqueous solution of $CoCl_2$, its colour changes from reddish pink to deep blue. Which complex ion gives blue colour in this reaction? [Online April 11, 2015]

(1) $[CoCl_4]^{2-}$ (2) $[CoCl_6]^{3-}$
(3) $[CoCl_6]^{4-}$ (4) $[Co(H_2O)_6]^{2+}$

NUMERICAL VALUE QUESTIONS (NVQs)

295. Number of paramagnetic complexes among the following is _______.

$[MnBr_4]^{2-}$, $[NiCl_4]^{2-}$, $[Ni(CN)_4]^{2-}$, $[Ni(CO)_4]$, $[CoF_6]^{3-}$, $[Fe(CN)_6]^{4-}$, $[Mn(CN)_6]^{3-}$, $[Ti(CN)_6]^{3-}$, $[Cu(H_2O)_6]^{2+}$, $[Co(C_2O_4)_3]^{3-}$ [April 08, 2026 (II)]

296. The total number of unpaired electrons present in the d^3, d^4 (low spin) d^5 (high spin), d^6 (high spin) and d^7 (low spin) octahedral complex systems is _______. [April 05, 2026 (II)]

297. The number of isoelectronic species among Sc^{3+}, Cr^{2+}, Mn^{3+}, Co^{3+} and Fe^{3+} is 'n'. If 'n' moles of AgCl is formed during the reaction of complex with formula $CoCl_3(en)_2NH_3$ with excess of $AgNO_3$ solution, then the number of electrons present in the t_{2g} orbital of the complex is _______. [Jan. 28, 2026 (II)]

298. Total number of unpaired electrons present in the central metal atoms/ions of $[Ni(CO)_4]$, $[NiCl_4]^{2-}$, $[PtCl_2(NH_3)_2]$, $[Ni(CN_4)]^{2-}$ and $[Pt(CN_4)]^{2-}$ is _______. [Jan. 23, 2026 (II)]

299. The crystal field splitting energy of $[Co(oxalate)_3]^{3-}$ complex is 'n' times that of the $[Cr(oxalate)_3]^{3-}$ complex. Here 'n' is _______. [Assume $\Delta_0 >> P$] [Jan. 23, 2026 (I)]

300. Identify the metal ions among Co^{2+}, Ni^{2+}, Fe^{2+}, V^{3+} and Ti^{2+} having a spin-only magnetic moment value more than 3.0 BM. The sum of unpaired electrons present in the high spin octahedral complexes formed by those metal ions is ______. [Jan. 21, 2026 (II)]

301. The number of paramagnetic metal complex species among $[Co(NH_3)_6]^{3+}$, $[Co(C_2O_4)_3]^{3-}$, $[MnCl_6]^{3-}$, $[Mn(CN)_6]^{3-}$, $[CoF_6]^{3-}$, $[Fe(CN)_6]^{3-}$ and $[FeF_6]^{3-}$ with same number of unpaired electrons is ____. [April 7, 2025 (II)]

302. The number of paramagnetic complex among $[FeF_6]^{3-}$, $[Fe(CN)_6]^{3-}$, $[Mn(CN)_6]^{3-}$, $[Co(C_2O_4)_3]^{3-}$, $[MnCl_6]^{3-}$ and $[CoF_6]^{3-}$, which involved d^2sp^3 hybridization is .... [April 7, 2025 (I)]

303. A transition metal (M) among Mn, Cr, Co and Fe has the highest standard electrode potential (M^{3+}/M^{2+}). It forms a metal complex of the type $[M(CN)_6]^{4-}$. The number of electrons present in the e_g orbital of the complex is ______. [April 2, 2025 (I)]

304. Consider the following low-spin complexes $K_3[Co(NO_2)_6]$, $K_4[Fe(CN)_6]$, $K_3[Fe(CN)_6]$, $Cu_2[Fe(CN)_6]$ and $Zn_2[Fe(CN)_6]$. [Jan. 29, 2025 (II)]
The sum of the spin-only magnetic moment values of complexes having yellow colour is ________.
B.M. (answer is nearest integer)

305. Total number of unpaired electrons in the complex ion $[Co(NH_3)_6]^{3+}$ and $[NiCl_4]^{2-}$ is [April 8, 2024 (II)]

306. The number of unpaired d-electrons in $[Co(H_2O)_6]^{3+}$ is _______ [April 4, 2024 (II)]

307. The 'Spin only' Magnetic moment for $[Ni(NH_3)_6]^{2+}$ is ____ × 10^{-1} BM. (given = Atomic number of Ni : 28) [Jan. 31, 2024 (I)]

308. The Spin only magnetic moment value of square planar complex $[Pt(NH_3)_2Cl(NH_2CH_3)]Cl$ is ________ B.M. (Nearest integer) (Given atomic number for Pt = 78) [Jan. 27, 2024 (II)]

309. The homoleptic and octahedral complex of Co^{2+} and H_2O has ______ unpaired electron(s) in the t_{2g} set of orbitals. [April 15, 2023 (I)]

310. The ratio of spin-only magnetic moment values $\mu_{eff}[Cr(CN)_6]^{3-}$ / $\mu_{eff}[Cr(H_2O)_6]^{3+}$ is ________. [April 11, 2023 (I)]

311. In potassium ferrocyanide, there are ____ pairs of electrons in the t_{2g} set of orbitals [April 10, 2023 (I)]

312. For a metal ion, the calculated magnetic moment is 4.90 BM. This metal ion has __________ number of unpaired electons. [April 10, 2023 (II)]

313. The observed magnetic moment of the complex $[Mn(\underline{N}CS)_6]^{x-}$ is 6.06 BM. The numerical value of x is ______: [April 08, 2023 (II)]

314. The spin only magnetic moment of $[Mn(H_2O)_6]^{2+}$ complexes is ______ B.M. (Nearest integer) (Given: Atomic no. of Mn is 25) [Feb. 01, 2023 (II)]

315. If the CFSE of $[Ti(H_2O)_6]^{3+}$ is –96.0 kJ/mol , this complex will absorb maximum at wavelength ____ nm. (nearest integer) Assume Planck's constant (h) = 6.4×10^{-34} Js Speed of light (c) = 3.0×10^8 m / s and Avogadro's constant (N_A) = 6×10^{23}/mol. [Jan. 31, 2023 (II)]

316. How many of the following metal ions have similar value of spin only magnetic moment in gaseous state? ___________
(Given: Atomic number : V, 23 ; Cr, 24 ; Fe, 26 ; Ni, 28)
V^{3+}. Cr^{3+}, Fe^{2+}, Ni^{3+} [Jan. 25, 2023 (I)]

317. The number of paramagnetic species from the following is_____. [Jan. 25, 2023 (I)]
$[Ni(CN)_4]^{2-}$, $[Ni(CO)_4]$, $[NiCl_4]^{2-}$
$[Fe(CN)_6]^{4-}$, $[Cu(NH_3)_4]^{2+}$
$[Fe(CN)_6]^{3-}$ and $[Fe(H_2O)_6]^{2+}$

318. The *d*-electronic configuration of $[CoCl_4]^{2-}$ in tetrahedral crystal field is $e^m t_2^n$. Sum of '*m*' and 'number of unpaired electrons is _______. [Jan. 24, 2023 (I)]

319. $[Fe(CN)_6]^{3-}$ should be an inner orbital complex. Ignoring the pairing energy, the value of crystal field stabilization energy for this complex is (–) _________ Δ_0. (Nearest integer) [July 29, 2022 (I)]

320. The spin only magnetic moment of the complex present in Fehling's reagent is______ B.M. (Nearest integer). [July 27, 2022 (II)]

321. The difference between spin only magnetic moment values of $[Co(H_2O)_6]Cl_2$ and $[Cr(H_2O)_6]Cl_3$ is __________. [July 26, 2022 (I)]

322. Consider the following metal complexes:
$[Co(NH_3)_6]^{3+}$, $[CoCl(NH_3)_5]^{2+}$
$[Co(CN)_6]^{3-}$, $[Co(NH_3)_5(H_2O)]^{3+}$
The spin-only magnetic moment value of the complex that absorbs light with shortest wavelength is _______ B.M. (Nearest integer) [July 25, 2022 (I)]

323. Spin only magnetic moment (μ_s) of $K_3[Fe(CN)_6]$ is _________ B.M. (Nearest integer). [June 30, 2022 (I)]

324. Spin only magnetic moment of $[MnBr_6]^{4-}$ is____ B.M. (round off to the closest integer) [June 29, 2022 (II)]

325. $[Fe(CN)_6]^{4-}$ $[Fe(CN)_6]^{3-}$ $[Ti(CN)_6]^{3-}$
$[Ni(CN)_4]^{2-}$ $[Co(CN)_6]^{3-}$
Among the given complexes, number of paramagnetic complexes is _____. [June 28, 2022 (II)]

326. The spin – only magnetic moment value of an octahedral complex among $CoCl_3.4NH_3$, $NiCl_2.6H_2O$ and $PtCl_4.2HCl$, which upon reaction with excess of $AgNO_3$ gives 2 moles of AgCl is ________ B.M. (Nearest Integer) [June 26, 2022 (I)]

327. Reaction of $[Co(H_2O)_6]^{2+}$ with excess ammonia and in the presence of oxygen results into a diamagnetic product. Number of electrons present in t_{2g} orbitals of the product is _______. [June 26, 2022 (II)]

328. If $[Cu(H_2O)_4]^{2+}$ absorbs a light of wavelength 600 nm for d–d transition, then the value of octahedral crystal field splitting energy for $[Cu(H_2O)_6]^{2+}$ will be ____________ $\times 10^{-21}$ J. (Nearest Integer)
(Given : $h = 6.63 \times 10^{-34}$ Js and $c = 3.08 \times 10^8$ ms^{-1}) [June 25, 2022 (I)]

329. Amongst $FeCl_3$. $3H_2O$, $K_3[Fe(CN)_6]$ and $[Co(NH_3)_6]$ Cl_3, the spin-only magnetic moment value of the inner-orbital complex that absorbs light at shortest wavelength is _________ B.M. [nearest integer] [June 25, 2022 (II)]

330. The spin-only magnetic moment value of B_2^+ species is ________$\times 10^{-2}$ BM. (Nearest integer)
[Given : $\sqrt{3} = 1.73$] [Sep. 1, 2021 (II)]

331. The total number of unpaired electrons present in $[Co(NH_3)_6]Cl_2$ and $[Co(NH_3)_6]Cl_3$ is [July 22, 2021 (II)]

332. The spin-only magnetic moment value for the complex $[Co(CN)_6]^{4-}$ is ________ BM. [July 20, 2021 (I)]
[At. no. of Co = 27]

333. An aqueous solution of $NiCl_2$ was heated with excess sodium cyanide in presence of strong oxidizing agent to form $[Ni(CN)_6]^{2-}$. The total change in number of unpaired electrons on metal centre is ____. [July 20, 2021 (II)]

334. $[Ti(H_2O)_6]^{3+}$ absorbs light of wavelength 498 nm during a d – d transition. The octahedral splitting energy for the above complex is _______ $\times 10^{-19}$J.
(Round off to the nearest integer).
$h = 6.626 \times 10^{-34}$ Js; $c = 3 \times 10^8$ ms^{-1} [March 16, 2021 (II)]

335. The spin only magnetic moment of a divalent ion in aqueous solution (atomic number 29) is [Feb. 25, 2021 (II)]

336. Considering that $\Delta_0 > P$, the magnetic moment (in BM) of $[Ru(H_2O)_6]^{2+}$ would be ___________. [Sep. 05, 2020 (II)]

337. The volume (in mL) of 0.125 M $AgNO_3$ required to quantitatively precipitate chloride ions in 0.3 g of $[Co(NH_3)_6]Cl_3$ is _______.
$M_{[Co(NH_3)_6]Cl_3} = 267.46$ g/mol [Jan. 08, 2020 (I)]
$M_{AgNO_3} = 169.87$ g/mol.

338. Complexes (ML_5) of metals Ni and Fe have ideal square pyramidal and trigonal bipyramidal geometries, respectively. The sum of the 90°, 120° and 180° L-M-L angles in the two complexes is ________. [Jan. 08, 2020 (II)]

Topic 3 - Importance of Coordination Compounds

MULTIPLE CHOICE QUESTIONS (MCQs)

339. The correct statements about metal carbonyls are
A. The metal-carbon bonds in metals carbonyls possess both σ and π-character.
B. Due to synergic bonding interactions between metals and CO ligand the metal-carbon bond becomes weak.
C. The metal-carbon σ bond is formed by the donation of lone pair of electrons on the carbonyl carbon into a vacant orbital of metal.
D. The metal-carbon π bond is formed by the donation of electrons from filled d-orbital of metal into vacant π* orbital of CO.
Choose the correct answer from the options given below: [April 05, 2026 (I)]
(1) A and B only (2) A, C and D only
(3) B and C only (4) A and D only

340. Choose the correct statements from the following [Jan. 30, 2024 (I)]
(A) Ethane – 1, 2 - diamine is a cheating ligand.
(B) Metallic aluminium is produced by electrolysis of aluminium oxide is presence of cryolite.
(C) Cyanide ion is used as ligand for leaching of silver.
(D) Phosphine act as a ligand in Wilkinson catalyst.
(E) The stability constants of Ca^{2+} and Mg^{2+} are similar with EDTA complexes.
Choose the correct answer from the options given below:
(1) (B), (C), (E) only (2) (C), (D), (E) only
(3) (A), (B), (C) only (4) (A), (D), (E) only

341. The coordination geometry around the manganese in decacarbonyldimanganese (0) is : **[Jan. 30, 2024 (II)]**
(1) Octahedral (2) Trigonal bipyramidal
(3) Square pyramidal (4) Square planar

342. In which one of the following metal carbonyls, CO forms a bridge between metal atoms ? **[Jan. 29, 2024 (I)]**
(1) $[Os_3(CO)_{12}]$ (2) $[Ru_3(CO)_{12}]$
(3) $[Co_2(CO)_8]$ (4) $[Mn_2(CO)_{10}]$

343. The mismatched combinations are **[April 13, 2023 (I)]**
A. Chlorophyll - Co
B. Water hardness - EDTA
C. Photography - $[Ag(CN)_2]^-$
D. Wilkinson catalyst - $[(Ph_3P)_3 RhCl]$
E. Chelating ligand - D - Penicillamine
Choose the correct answer from the options given below:
(1) A and C Only (2) A and E Only
(3) D and E Only (4) A, C and E Only

344. To inhibit the growth of tumours, identify the compounds used from the following: **[Jan. 30, 2023 (I)]**
(A) EDTA
(B) Coordination Compounds of Pt
(C) D – Penicillamine
(D) Cis – Platin
Choose the correct answer from the option given below:
(1) B and D Only (2) C and D Only
(3) A and B Only (4) A and C Only

345. The theory that can completely/properly explain the nature of bonding in $[Ni(CO)_4]$ is: **[Jan. 07, 2020 (I)]**
(1) Werner's theory (2) Molecular orbital theory
(3) Crystal field theory (4) Valence bond theory

346. The compound used in the treatment of lead poisoning is : **[April 12, 2019 (II)]**
(1) D-penicillamine (2) desferrioxime B
(3) *Cis*-platin (4) EDTA

347. The compound that inhibits the growth of tumors is : **[April 8, 2019 (II)]**
(1) *trans*-$[Pt(Cl)_2(NH_3)_2]$ (2) *cis*-$[Pd(Cl)_2(NH_3)_2]$
(3) *cis*-$[Pt(Cl)_2(NH_3)_2]$ (4) *trans*-$[Pd(Cl)_2(NH_3)_2]$

348. The number of bridging CO ligand(s) and Co-Co bond(s) in $Co_2(CO)_8$, respectively are: **[Jan. 11, 2019 (II)]**
(1) 2 and 1 (2) 2 and 0 (3) 0 and 2 (4) 4 and 0

349. Wilkinson catalyst is: **[Jan. 10, 2019 (I)]**
(1) $[(Ph_3P)_3IrCl]$ (2) $[(Et_3P)_3RhCl]$
(3) $[(Ph_3P)_3RhCl]$ (4) $[(Et_3P)_3IrCl]$

350. $[Co_2(CO)_8]$ displays : **[Online April 9, 2017]**
(1) one Co – Co bond, six terminal CO and two bridging CO
(2) one Co – Co bond, four terminal CO and four bridging CO
(3) no Co – Co bond, six terminal CO and two bridging CO
(4) no Co – Co bond, four terminal CO and four bridging CO

351. Which molecule/ion among the following cannot act as a ligand in complex compounds ? **[Online April 10, 2015]**
(1) CH_4 (2) CO (3) CN^- (4) Br^-

352. In $Fe(CO)_5$, the Fe – C bond possesses **[2006]**
(1) ionic character (2) σ-character only
(3) π-character (4) both σ and π characters

353. How many EDTA (ethylenediaminetetraacetic acid) molecules are required to make an octahedral complex with a Ca^{2+} ion? **[2006]**
(1) One (2) Two (3) Six (4) Three

354. Coordination compounds have great importance in biological systems. In this context which of the following statements is **incorrect** ? **[2004]**
(1) Cyanocobalamin is B_{12} and contains cobalt
(2) Haemoglobin is the red pigment of blood and contains iron
(3) Chlorophylls are green pigments in plants and contain calcium
(4) Carboxypeptidase - A is an enzyme and contains zinc.

355. $CH_3 – Mg – Br$ is an organo metallic compound due to **[2002]**
(1) Mg – Br bond (2) C – Mg bond
(3) C – Br bond (4) C – H bond

Part-B (JEE Advanced)

Topic 1 - Werner's Theory, Important terms, Nomenclature and Isomerism of Coordination Compounds

1 MCQs with One Correct Answer

1. Among the following options, select the option in which each complex in **Set-I** shows geometrical isomerism and the two complexes in **Set-II** are ionization isomers of each other. [en = $H_2NCH_2CH_2NH_2$] **[Adv. 2024]**
(a) **Set-I:** $[Ni(CO)_4]$ and $[PdCl_2(PPh_3)_2]$
Set-II: $[Co(NH_3)_5Cl]SO_4$ and $[Co(NH_3)_5(SO_4)]Cl$
(b) **Set-I:** $[Co(en)(NH_3)_2Cl_2]$ and $[PdCl_2(PPh_3)_2]$
Set-II: $[Co(NH_3)_6][Cr(CN)_6]$ and $[Cr(NH_3)_6][Co(CN)_6]$
(c) **Set-I:** $[Co(NH_3)_3(NO_2)_3]$ and $[Co(en)_2Cl_2]$
Set-II: $[Co(NH_3)_5Cl]SO_4$ and $[Co(NH_3)_5(SO_4)]Cl$
(d) **Set-I:** $[Cr(NH_3)_5Cl]Cl_2$ and $[Co(en)(NH_3)_2Cl_2]$
Set-II: $[Cr(H_2O)_6]Cl_3$ and $[Cr(H_2O)_5Cl]Cl_2 . H_2O$

2. As per IUPAC nomenclature, the name of the complex $[Co(H_2O)_4(NH_3)_2]Cl_3$ is : **[2012]**
(a) Tetraaquadiaminecobalt (III) chloride
(b) Tetraaquadiamminecobalt (III) chloride
(c) Diaminetetraaquacoblat (II) chloride
(d) Diamminetetraaquacobalt (III) chloride

3. The correct structure of ethylenediaminetetraacetic acid (EDTA) is **[2010]**
(a) 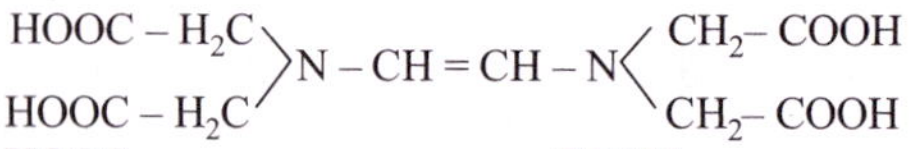
(b) 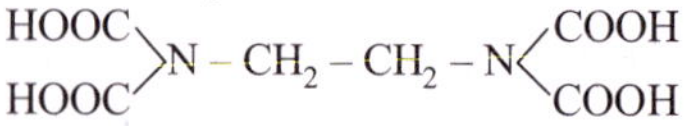
(c) 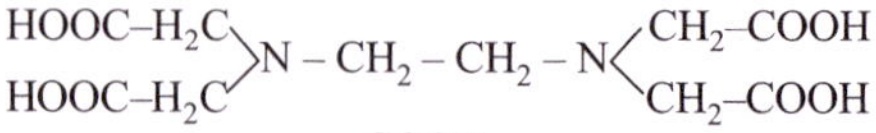
(d) 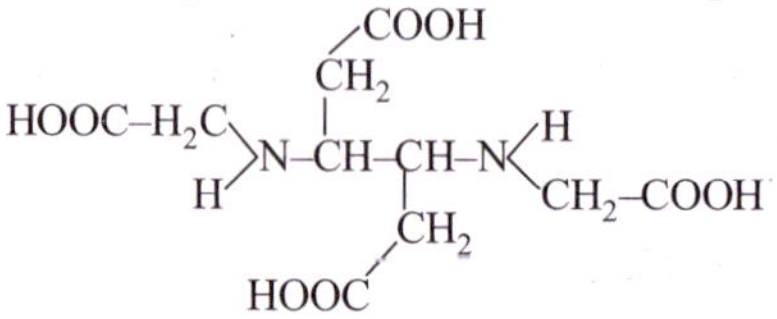

4. The ionisation isomer of $[Cr(H_2O)_4Cl(NO_2)]Cl$ is [2010]
(a) $[Cr(H_2O)_4(O_2N)]Cl_2$ (b) $[Cr(H_2O)_4Cl_2](NO_2)$
(c) $[Cr(H_2O)_4Cl(ONO)]Cl$ (d) $[Cr(H_2O)_4Cl_2(NO_2)].H_2O$

5. The IUPAC name of $[Ni(NH_3)_4][NiCl_4]$ is [2008]
(a) Tetrachloronickel (II) - tetraamminenickel (II)
(b) Tetraamminenickel (II) - tetrachloronickel (II)
(c) Tetraamminenickel (II) - tetrachloronickelate (II)
(d) Tetrachloronickel (II) - tetrachloronickelate (0) Ans. (C)

6. Which kind of isomerism is exhibited by octahedral $Co(NH_3)_4Br_2Cl$? [2005S]
(a) Geometrical and Ionization
(b) Geometrical and Optical
(c) Optical and Ionization
(d) Geometrical only

7. The complex ion which has no '*d*' electron in the central metal atom is [2001S]
(a) $[MnO_4]^-$ (b) $[Co(NH_3)_6]^{3+}$
(c) $[Fe(CN)_6]^{3-}$ (d) $[Cr(H_2O)_6]^{3+}$

8. Which of the following is an organometallic compound? [1997 - 1 Mark]
(a) Lithium methoxide (b) Lithium acetate
(c) Lithium dimethylamide (d) Methyl lithium

2 Integer Value Answer

9. Among $V(CO)_6$, $Cr(CO)_5$, $Cu(CO)_3$, $Mn(CO)_5$, $Fe(CO)_5$, $[Co(CO)_3]^{3-}$, $[Cr(CO)_4]^{4-}$, and $Ir(CO)_3$, the total number of species isoelectronic with $Ni(CO)_4$ is ______.
[Given, atomic number: V = 23, Cr = 24, Mn = 25, Fe = 26, Co = 27, Ni = 28, Cu = 29, Ir = 77] [Adv. 2024]

10. The total number of possible isomers for $[Pt(NH_3)_4Cl_2]Br_2$ is _____. [Adv. 2021]

11. Total number of *cis* N – Mn – Cl bond angles (that is Mn – N and Mn – Cl bonds in *cis* positions) present in a molecule of *cis* –$[Mn(en)_2Cl_2]$ complex is ___
(en = $NH_2CH_2CH_2NH_2$) [Adv. 2019]

12. The number of geometric isomers possible for the complex $[CoL_2Cl_2]^-$ ($L = H_2NCH_2CH_2O^-$) is [Adv. 2016]

13. Among the complex ions, $[Co(NH_2–CH_2–CH_2–NH_2)_2Cl_2]^+$, $[CrCl_2(C_2O_4)_2]^{3-}$, $[Fe(H_2O)_4(OH)_2]^+$, $[Fe(NH_3)_2(CN)_4]^-$, $[Co(NH_2–CH_2–CH_2–NH_2)_2(NH_3)Cl]^{2+}$ and $[Co(NH_3)_4(H_2O)Cl]^{2+}$, the number of complex ion(s) that show(s) *cis-trans* isomerism is [Adv. 2015]

14. In the complex acetylbromidodicarbonylbis (triethylphosphine) iron (II), the number of Fe–C bond(s) is [Adv. 2015]

15. $EDTA^{4-}$ is ethylenediaminetetraacetate ion. The total number of N—Co—O bond angles in $[Co(EDTA)]^{1-}$complex ion is [Adv. 2013]

16. The volume (in mL) of 0.1 M $AgNO_3$ required for complete precipitation of chloride ions present in 30 mL of 0.01 M solution of $[Cr(H_2O)_5Cl]Cl_2$, as silver chloride is close to [2011]

17. Total number of geometrical isomers for the complex $[RhCl(CO)(PPh_3)(NH_3)]$ is [2010]

18. The number of water molecule(s) directly bonded to the metal centre in $CuSO_4.5H_2O$ is [2009 - 4 Marks]

3 Numeric / New Stem Based Questions

19. The total number of all possible isomers for the square planar complex with formula $K[M(NCS)(NO_2)(gly)]$ is _____.
(M = metal ion and gly = $NH_2CH_2COO^-$) [Adv. 2026]

20. Consider that the coordinating atoms of the ligands in cis-$[Co(NH_3)_4Cl_2]Cl$ and mer-$[Co(NH_3)_3Cl_3]$ octahedral complexes are at the vertices of an octahedron. The sum of total number of the triangular faces in both the complexes having one N atom and two Cl atoms at their corners is ____. [Adv. 2026]

4 Fill in the Blanks

21. The IUPAC name of $[Co(NH_3)_6]Cl_3$ is .... [1994 - 1 Mark]

6 MCQs with One or More than One Correct Answer

22. The complex(es), which can exhibit the type of isomerism shown by $[Pt(NH_3)_2Br_2]$, is(are) [en = $H_2NCH_2CH_2NH_2$] [Adv. 2023]
(a) $[Pt(en)(SCN)_2]$ (b) $[Zn(NH_3)_2Cl_2]$
(c) $[Pt(NH_3)_2Cl_4]$ (d) $[Cr(en)_2(H_2O)(SO_4)]^+$

23. The pair(s) of coordination complexes/ions exhibiting the same kind of isomerism is(are) [Adv. 2013]
(a) $[Cr(NH_3)_5Cl]Cl_2$ and $[Cr(NH_3)_4Cl_2]Cl$
(b) $[Co(NH_3)_4Cl_2]^+$ and $[Pt(NH_3)_2(H_2O)Cl]^+$
(c) $[CoBr_2Cl_2]^{2-}$ and $[PtBr_2Cl_2]^{2-}$
(d) $[Pt(NH_3)_3(NO_3)]Cl$ and $[Pt(NH_3)_3Cl]Br$

24. The compound(s) that exhibit(s) geometrical isomerism is (are) [2009]
(a) $[Pt(en)Cl_2]$ (b) $[Pt(en)_2]Cl_2$
(c) $[Pt(en)_2Cl_2]Cl_2$ (d) $[Pt(NH_3)_2Cl_2]$

9 Assertion and Reason/Statement Type Questions

Each question contains **STATEMENT-1 (Assertion)** and **STATEMENT-2 (Reason)**. Each question has 4 choices (a), (b), (c) and (d) out of which ONLY ONE is correct. Mark your answer as
(a) If both Statement -1 and Statement -2 are correct, and Statement -2 is the correct explanation of the Statement -2.
(b) If both Statement -1 and Statement -2 are correct, but Statement -2 is not the correct explanation of the Statement -1.
(c) If Statement -1 is correct but Statement -2 is incorrect.
(d) If Statement -1 is incorrect but Statement -2 is correct.

25. **Statement-1 :** The geometrical isomers of the complex $[M(NH_3)_4Cl_2]$ are optically inactive.
Statement-2 : Both geometrical isomers of the complex $[M(NH_3)_4Cl_2]$ possess axis of symmetry. [2008 - 2 Marks]

10 Subjective Problems

26. *A*, *B*, and *C* are three complexes of chromium (III) with the empirical formula $H_{12}O_6Cl_3Cr$. All the three complexes have water and chloride ion as ligands. Complex *A* does not react with concentrated H_2SO_4, whereas complexes *B* and *C* lose 6.75% and 13.5% of their original mass, respectively, on treatment with concentrated H_2SO_4. Identify *A*, *B* and *C*. [1999 - 6 Marks]

27. Write the formulae of the following complexes :
(*i*) Pentaamminechlorocobalt(III) [1997 - 1 Mark]
(*ii*) Lithium tetrahydroaluminate(III). [1997 - 1 Mark]

28. Write down the IUPAC names of the following compounds:
(*i*) $[Cr(NH_3)_5CO_3]Cl$ [1996 - 1 Mark]
(*ii*) $K_3[Cr(CN)_6]$ [1995 - 1 Mark]
(*iii*) $[Co(NH_3)_5ONO]Cl_2$ [1995 - 1 Mark]

Topic 2 - Bonding in Coordination Compounds, (Valence Bond Theory and Crystal Field Theory)

1 MCQs with One Correct Answer

29. The correct order of the wavelength maxima of the absorption band in the ultraviolet-visible region for the given complexes is [Adv. 2025]
(a) $[Co(CN)_6]^{3-} < [Co(NH_3)_6]^{3+} < [Co(NH_3)_5(H_2O)]^{3+} < [Co(NH_3)_5(Cl)]^{2+}$
(b) $[Co(NH_3)_5(Cl)]^{2+} < [Co(NH_3)_5(H_2O)]^{3+} < [Co(NH_3)_6]^{3+} < [Co(CN)_6]^{3-}$
(c) $[Co(CN)_6]^{3-} < [Co(NH_3)_5(Cl)]^{2+} < [Co(NH_3)_5(H_2O)]^{3+} < [Co(NH_3)_6]^{3+}$
(d) $[Co(NH_3)_6]^{3+} < [Co(CN)_6]^{3-} < [Co(NH_3)_5(Cl)]^{2+} < [Co(NH_3)_5(H_2O)]^{3+}$

30. The reaction of $Pb(NO_3)_2$ and NaCl in water produces a precipitate that dissolves upon the addition of HCl of appropriate concentration. The dissolution of the precipitate is due to the formation of [Adv. 2022]
(a) $PbCl_2$ (b) $PbCl_4$
(c) $[PbCl_4]^{2-}$ (d) $[PbCl_6]^{2-}$

31. The calculated spin only magnetic moments of $[Cr(NH_3)_6]^{3+}$ and $[CuF_6]^{3-}$ in BM, respectively, are (Atomic numbers of Cr and Cu are 24 and 29, respectively) [Adv. 2021]
(a) 3.87 and 2.84 (b) 4.90 and 1.73
(c) 3.87 and 1.73 (d) 4.90 and 2.84

32. The geometries of the ammonia complexes of Ni^{2+}, Pt^{2+} and Zn^{2+} respectively, are [Adv. 2016]
(a) octahedral, square planar and tetrahedral
(b) square planar, octahedral and tetrahedral
(c) tetrahedral, square planar and octahedral
(d) octahedral, tetrahedral and square planar

33. Consider the following complex ions, *P*, *Q* and *R*.
$P = [FeF_6]^{3-}$, $Q = [V(H_2O)_6]^{2+}$ and $R = [Fe(H_2O)_6]^{2+}$
The correct order of the complex ions, according to their spin-only magnetic moment values (in B.M.) is [Adv. 2013]
(a) $R < Q < P$ (b) $Q < R < P$
(c) $R < P < Q$ (d) $Q < P < R$

34. $NiCl_2\{P(C_2H_5)_2(C_6H_5)\}_2$ exhibits temperature dependent magnetic behaviour (paramagnetic/diamagnetic). The coordination geometries of Ni^{2+} in the paramagnetic and diamagnetic states are respectively [2012]
(a) tetrahedral and tetrahedral
(b) square planar and square planar
(c) tetrahedral and square planar
(d) square planar and tetrahedral

35. Among the following complexes (**K–P**)
$K_3[Fe(CN)_6]$ (**K**), $[Co(NH_3)_6]Cl_3$ (**L**), $Na_3[Co(oxalate)_3]$ (**M**), the $[Ni(H_2O)_6]Cl_2$ (**N**), $K_2[Pt(CN)_4]$ (**O**) and $[Zn(H_2O)_6](NO_3)_2$ (**P**) the diamagnetic complexes are [2011]
(a) K, L, M, N (b) K, M, O, P
(c) L, M, O, P (d) L, M, N, O

36. Geometrical shapes of the complexes formed by the reaction of Ni^{2+} with Cl^-, CN^- and H_2O, respectively, are [2011]
(a) octahedral, tetrahedral and square planar
(b) tetrahedral, square planar and octahedral
(c) square planar, tetrahedral and octahedral
(d) octahedral, square planar and octahedral

37. The complex showing a spin-only magnetic moment of 2.82 B.M. is: [2010]
(a) $[Ni(CO)_4]$ (b) $[NiCl_4]^{2-}$
(c) $[Ni(PPh_3)_4]$ (d) $[Ni(CN)_4]^{2-}$

38. The spin only magnetic moment value (in Bohr magneton units) of $Cr(CO)_6$ is [2009]
(a) 0 (b) 2.84 (c) 4.90 (d) 5.92

39. Both $[Ni(CO)_4]$ and $[Ni(CN)_4]^{2-}$ are diamagnetic. The hybridisations of nickel in these complexes, respectively, are [2008]
(a) sp^3, sp^3 (b) sp^3, dsp^2
(c) dsp^2, sp^3 (d) dsp^2, sp^2

40. Among the following metal carbonyls, the C–O bond order is lowest in [2007]
(a) $[Mn(CO)_6]^+$ (b) $[Fe(CO)_5]$
(c) $[Cr(CO)_6]$ (d) $[V(CO)_6]^-$

41. $Ag^+ + NH_3 \rightleftharpoons [Ag(NH_3)]^+; k_1 = 6.8 \times 10^{-3}$
$[Ag(NH_3)]^+ + NH_3 \rightleftharpoons [Ag(NH_3)_2]^+; k_2 = 1.6 \times 10^{-3}$
then the formation constant of $[Ag(NH_3)_2]^+$ is [2006 - 3M, –1]
(a) 6.8×10^{-6} (b) 1.08×10^{-5}
(c) 1.08×10^{-6} (d) 6.8×10^{-5}

42. The spin magnetic moment of cobalt in the compound $Hg[Co(SCN)_4]$ is [2004S]
(a) $\sqrt{3}$ (b) $\sqrt{8}$ (c) $\sqrt{15}$ (d) $\sqrt{24}$

43. The species having tetrahedral shape is [2004S]
(a) $[PdCl_4]^{2-}$ (b) $[Ni(CN)_4]^{2-}$
(c) $[Pd(CN)_4]^{2-}$ (d) $[NiCl_4]^{2-}$

44. The geometry of $Ni(CO)_4$ and $Ni(PPh_3)_2Cl_2$ are [1999 - 2 Marks]
(a) both square planar
(b) tetrahedral and square planar, respectively
(c) both tetrahedral
(d) square planar and tetrahedral, respectively

45. In nitroprusside ion the iron and NO exist as Fe^{II} and NO^+ rather than Fe^{III} and NO. These forms can be differentiated by [1998 - 2 Marks]
(a) estimating the concentration of iron
(b) measuring the concentration of CN^-
(c) measuring the solid state magnetic moment
(d) thermally decomposing the compound.

46. Among the following ions which one has the highest paramagnetism? [1993 - 1 Mark]
(a) $[Cr(H_2O)_6]^{3+}$ (b) $[Fe(H_2O)_6]^{2+}$
(c) $[Cu(H_2O)_6]^{2+}$ (d) $[Zn(H_2O)_6]^{2+}$

47. Amongst $[Ni(CO)_4]$, $[Ni(CN)_4]^{2-}$ and $[NiCl_4]^{2-}$ [1991 - 1 Mark]
(a) $[Ni(CO)_4]$ and $[NiCl_4]^{2-}$ are diamagnetic and $[Ni(CN)_4]^{2-}$ is paramagnetic
(b) $[NiCl_4^{2-}]$ and $[Ni(CN)_4]^{2-}$ are diamagnetic and $[Ni(CO)_4]$ is paramagnetic
(c) $[Ni(CO)_4]$ and $[Ni(CN)_4]^{2-}$ are diamagnetic and $[NiCl_4]^{2-}$ is paramagnetic
(d) $[Ni(CO)_4]$ is diamagnetic and $[NiCl_4]^{2-}$ and $[Ni(CN)_4]^{2-}$ are paramagnetic

48. Amongst the following, the lowest degree of paramagnetism per mole of the compound at 298 K will be shown by [1988 - 1 Mark]
(a) $MnSO_4.4H_2O$ (b) $CuSO_4.5H_2O$
(c) $FeSO_4.6H_2O$ (d) $NiSO_4.6H_2O$

2 Integer Value Answer

49. Among $[Co(CN)_4]^{4-}$, $[Co(CO)_3(NO)]$, XeF_4, $[PCl_4]^+$, $[PdCl_4]^{2-}$, $[ICl_4]^-$, $[Cu(CN)_4]^{3-}$ and P_4 the total number of species with tetrahedral geometry is ______. **[Adv. 2024]**

50. Among the following complexes, the total number of diamagnetic species is __________. **[Adv. 2024]**
$[Mn(NH_3)_6]^{3+}$, $[MnCl_6]^{3-}$, $[FeF_6]^{3-}$, $[CoF_6]^{3-}$, $[Fe(NH_3)_6]^{3+}$, and $[Co(en)_3]^{3+}$
[Given, atomic number: Mn = 25, Fe = 26, Co = 27; en = $H_2NCH_2CH_2NH_2$]

51. Among the species given below, the total number of diamagnetic species is_______.
H atom, NO_2 monomer, O_2^- (superoxide), dimeric sulphur in vapour phase, Mn_3O_4, $(NH_4)_2[FeCl_4]$, $(NH_4)_2[NiCl_4]$, K_2MnO_4, K_2CrO_4 **[Adv. 2018]**

52. For the octahedral complexes of Fe^{3+} in SCN^- (thiocyanato-S) and in CN^- ligand environments, the difference between the spin-only magnetic moments in Bohr magnetons (when approximated to the nearest integer) is
[Atomic number of Fe = 26] **[Adv. 2015]**

3 Numeric / New Stem Based Questions

53. The sum of the spin only magnetic moment values (in B.M.) of $[Mn(Br)_6]^{3-}$ and $[Mn(CN)_6]^{3-}$ is_____. **[Adv. 2025]**

4 Fill in the Blanks

54. The type of magnetism exhibited by $[Mn(H_2O)_6]^{2+}$ ion is .... **[1994 - 1 Mark]**

5 True / False

55. Both potassium ferrocyanide and potassium ferricyanide are diamagnetic. **[1989 - 1 Mark]**

6 MCQs with One or More than One Correct Answer

56. The pair(s) of complexes wherein both exhibit tetrahedral geometry is(are)
(Note: py = pyridine
Given: Atomic numbers of Fe, Co, Ni and Cu are 26, 27, 28 and 29, respectively) **[Adv. 2021]**
(a) $[FeCl_4]^-$ and $[Fe(CO)_4]^{2-}$
(b) $[Co(CO)_4]^-$ and $[CoCl_4]^{2-}$
(c) $[Ni(CO)_4]$ and $[Ni(CN)_4]^{2-}$
(d) $[Cu(py)_4]^+$ and $[Cu(CN)_4]^{3-}$

57. Choose the correct statement(s) among the following: **[Adv. 2020]**
(a) $[FeCl_4]^-$ has tetrahedral geometry.
(b) $[Co(en)(NH_3)_2Cl_2]^+$ has 2 geometrical isomers.
(c) $[FeCl_4]^-$ has higher spin-only magnetic moment than $[Co(en)(NH_3)_2Cl_2]^+$.
(d) The cobalt ion in $[Co(en)(NH_3)_2Cl_2]^+$ has sp^3d^2 hybridization.

58. The correct option(s) regarding the complex $[Co(en)(NH_3)_3(H_2O)]^{3+}$ (en = $H_2NCH_2CH_2NH_2$) is (are) **[Adv. 2018]**
(a) It has two geometrical isomers
(b) It will have three geometrical isomers if bidentate 'en' is replaced by two cyanide ligands
(c) It is paramagnetic
(d) It absorbs light at longer wavelength as compared to $[Co(en)(NH_3)_4]^{3+}$

59. The correct statement(s) regarding the binary transition metal carbonyl compounds is (are) (Atomic numbers: Fe = 26, Ni = 28) **[Adv. 2018]**
(a) Total number of valence shell electrons at metal centre in $Fe(CO)_5$ or $Ni(CO)_4$ is 16
(b) These are predominantly low spin in nature
(c) Metal–carbon bond strengthens when the oxidation state of the metal is lowered
(d) The carbonyl C–O bond weakens when the oxidation state of the metal is increased

60. Addition of excess aqueous ammonia to a pink coloured aqueous solution of $MCl_2.6H_2O(X)$ and NH_4Cl gives an octahedral complex *Y* in the presence of air. In aqueous solution, complex *Y* behaves as 1:3 electrolyte. The reaction of *X* with excess HCl at room temperature results in the formation of a blue coloured complex *Z*. The calculated spin only magnetic moment of *X* and *Z* is 3.87 B.M., whereas it is zero for complex *Y*. Among the following options, which statement(s) is (are) correct? **[Adv. 2017]**
(a) Addition of silver nitrate to *Y* gives only two equivalents of silver chloride
(b) The hybridization of the central metal ion in *Y* is d^2sp^3
(c) *Z* is a tetrahedral complex
(d) When *X* and *Z* are in equilibrium at 0°C, the colour of the solution is pink

61. If the bond length of CO bond in carbon monoxide is 1.128Å, then what is the value of CO bond length in $Fe(CO)_5$? **[2006 - 5M, –1]**
(a) 1.15Å (b) 1.128Å (c) 1.13Å (d) 1.118Å

7 Match the Following

62. Match the electronic configurations in List-I with appropriate metal complex ions in List-II and choose the correct option.
[Atomic number: Fe = 26, Mn = 25, Co = 27] **[Adv. 2023]**

List-I	List-II
(P) $t_{2g}^6e_g^0$	(1) $[Fe(H_2O)_6]^{2+}$
(Q) $t_{2g}^3e_g^2$	(2) $[Mn(H_2O)_6]^{2+}$
(R) $e^2t_2^3$	(3) $[Co(NH_3)_6]^{3+}$
(S) $t_{2g}^4e_g^2$	(4) $[FeCl_4]^-$
	(5) $[CoCl_4]^{2-}$

(a) P → 1; Q → 4; R → 2; S → 3
(b) P → 1; Q → 2; R → 4; S → 5
(c) P → 3; Q → 2; R → 5; S → 1
(d) P → 3; Q → 2; R → 4; S → 1

63. List-I contains metal species and List-II contains their properties. **[Adv. 2022]**

List-I	List-II
(I) $[Cr(CN)_6]^{4-}$	(P) t_{2g} orbitals contain 4 electrons
(II) $[RuCl_6]^{2-}$	(Q) μ(spin-only) = 4.9 BM
(III) $[Cr(H_2O)_6]^{2+}$	(R) low spin complex ion
(IV) $[Fe(H_2O)_6]^{2+}$	(S) metal ion in 4+ oxidation state
	(T) d^4 species

[Given: Atomic number of Cr = 24, Ru = 44, Fe = 26]
Match each metal species in LIST-I with their properties in LIST-II, and choose the correct option
(a) I → R, T; II → P, S; III → Q, T; IV → P, Q
(b) I → R, S; II → P, T; III → P, Q; IV → Q, T
(c) I → P, R; II → R, S; III → R, T; IV → P, T
(d) I → Q, T; II → S, T; III → P, T; IV → Q, R

64. Match each set of hybrid orbitals from List-I with complex(es) given in List-II **[Adv. 2018]**

List-I	List-II
(A) dsp^2	(p) $[FeF_6]^{4-}$
(B) sp^3	(q) $[Ti(H_2O)_3Cl_3]$
(C) sp^3d^2	(r) $[Cr(NH_3)_6]^{3+}$
(D) d^2sp^3	(s) $[FeCl_4]^{2-}$
	(t) $[Ni(CO)_4]$
	(w) $[Ni(CN)_4]^{2-}$

The correct option is

(a) A-t; B-s, w ; C-q, r; D-p

(b) A-t, w; B-s; C- q; D-p, q

(c) A-w; B-s, t ; C-p; D-q, r

(d) A-s, w; B-t, w ; C-p, q; D-r

65. Match each coordination compound in List-I with an appropriate pair of characteristics from List- II and select the correct answer using the code given below the lists.
{en = $H_2NCH_2CH_2NH_2$; atomic numbers : Ti = 22; Cr = 24; Co = 27; Pt = 78} **[Adv. 2014]**

List-I	List-II
(A) $[Cr(NH_3)_4Cl_2]Cl$	(p) Paramagnetic and exhibits ionisation isomerism
(B) $[Ti(H_2O)_5Cl](NO_3)_2$	(q) Diamagnetic and exhibits *cis-trans* isomerism
(C) $[Pt(en)(NH_3)Cl]NO_3$	(r) Paramagnetic and exhibits *cis-trans* isomerism
(D) $[Co(NH_3)_4(NO_3)_2]NO_3$	(s) Diamagnetic and exhibits ionisation isomerism

Code:

	A	B	C	D
(a)	(s)	(q)	(r)	(p)
(b)	(r)	(p)	(s)	(q)
(c)	(q)	(p)	(r)	(s)
(d)	(p)	(r)	(s)	(q)

66. Match the complexes in **Column-I** with their properties listed in **Column-II.** **[2007]**

Column-I	Column-II
(A) $[Co(NH_3)_4(H_2O)_2]Cl_2$	(p) geometrical isomers
(B) $[Pt(NH_3)_2Cl_2]$	(q) paramagnetic
(C) $[Co(H_2O)_5Cl]Cl$	(r) diamagnetic
(D) $[Ni(H_2O)_6]Cl_2$	(s) metal ion with +2 oxidation state

8 Comprehension Based Questions

The coordination number of Ni^{2+} is 4.

$NiCl_2 + KCN(excess) \longrightarrow A$ (cyano complex)

$NiCl_2 + Conc.\ HCl(excess) \longrightarrow B$ (chloro complex)

67. The IUPAC name of *A* and *B* are **[2006 - 5M, –2]**

(a) Potassium tetracyanonickelate (II), potassium tetrachloronickelate (II)

(b) Tetracyanopotassiumnickelate (II), tetrachloropotassiumnickelate (II)

(c) Tetracyanonickel (II), tetrachloronickel (II)

(d) Potassium tetracyanonickel (II), potassium tetrachloronickel (II)

68. Predict the magnetic nature of *A* and *B* **[2006 - 5M, –2]**

(a) Both are diamagnetic

(b) *A* is diamagnetic and *B* is paramagnetic with one unpaired electron

(c) *A* is diamagnetic and *B* is paramagnetic with two unpaired electrons

(d) Both are paramagnetic

69. The hybridization of *A* and *B* are **[2006 - 5M, –2]**

(a) dsp^2, sp^3 (b) sp^3, sp^3

(c) dsp^2, dsp^2 (d) sp^3d^2, d^2sp^3

9 Assertion and Reason/Statement Type Questions

Each question contains **STATEMENT-1 (Assertion)** and **STATEMENT-2 (Reason)**. Each question has 4 choices (a), (b), (c) and (d) out of which ONLY ONE is correct. Mark your answer as

(a) If both Statement -1 and Statement -2 are correct, and Statement -2 is the correct explanation of the Statement -2.

(b) If both Statement -1 and Statement -2 are correct, but Statement -2 is not the correct explanation of the Statement -1.

(c) If Statement -1 is correct but Statement -2 is incorrect.

(d) If Statement -1 is incorrect but Statement -2 is correct.

70. **Statement-1 :** $[Fe(H_2O)_5NO]SO_4$ is paramagnetic.
Statement-2 : The Fe in $[Fe(H_2O)_5NO]SO_4$ has three unpaired electrons. **[2008]**

10 Subjective Problems

71. $Fe^{3+} \xrightarrow[\text{excess}]{SCN^-} \underset{\text{Blood red colouration}}{(A)} \xrightarrow[\text{excess}]{F^-} \underset{\text{colourless}}{(B)}$

What are (A) and (B)? Give IUPAC name of (A). Find the spin only magnetic moment of (B). **[2005 - 4 Marks]**

72. Nickel chloride, when treated with dimethylgyloxime in presence of ammonium hydroxide, a bright red precipitate is obtained. Answer the following. **[2004 - 4 Marks]**

(a) Draw the structure of the complex showing H-bonds

(b) Give oxidation state of nickel and its hybridisation

(c) Predict the magnetic behaviour of the complex

73. Write the IUPAC nomenclature of the given complex along with its hybridisation and structure.

$K_2[Cr(NO)(NH_3)(CN)_4]$, $\mu = 1.73$ BM **[2003 - 4 Marks]**

74. Deduce the structure of $[NiCl_4]^{2-}$ and $[Ni(CN)_4]^{2-}$ considering the hybridization of the metal ion. Calculate the magnetic moment (spin only) of the species. **[2002 - 5 Marks]**

75. A metal complex having composition $Cr(NH_3)_4Cl_2Br$ has been isolated in two forms (*A*) and (*B*). The form (*A*) reacts with $AgNO_3$ to give a white precipitate readily soluble in dilute aqueous ammonia, whereas (*B*) gives a pale yellow precipitate soluble in concentrated ammonia. Write the formula of (*A*) and (*B*) and state the hybridization of chromium in each. Calculate their magnetic moments (spin–only value). **[2001 - 5 Marks]**

76. Draw the structures of $[Co(NH_3)_6]^{3+}$, $[Ni(CN)_4]^{2-}$ and $[Ni(CO)_4]$. Write the hybridisation of atomic orbitals of the transition metal in each case. **[2000 - 4 Marks]**

77. A compound of vanadium has a magnetic moment of 1.73 BM. Work out the electronic configuration of the vanadium ion in the compound. **[1997 - 2 Marks]**

78. Identify the complexes which are expected to be coloured. Explain **[1994 - 2 Marks]**

(i) $[Ti(NO_3)_4]$ (ii) $[Cu(NCCH_3)_4]^+BF_4^-$

(iii) $[Cr(NH_3)_6]^{3+}3Cl^-$ (iv) $K_3[VF_6]$

Answer Key

Part-A (JEE Main)

1	(2)	37	(4)	73	(3)	109	(3)	145	(4)	181	(2)	217	(4)	253	(1)	289	(2)	325	(2)
2	(1)	38	(4)	74	(3)	110	(6)	146	(2)	182	(1)	218	(2)	254	(3)	290	(3)	326	(3)
3	(4)	39	(4)	75	(4)	111	(2)	147	(1)	183	(2)	219	(4)	255	(1)	291	(4)	327	(6)
4	(4)	40	(3)	76	(1)	112	(3)	148	(4)	184	(1)	220	(1)	256	(3)	292	(3)	328	(766)
5	(2)	41	(1)	77	(4)	113	(6)	149	(2)	185	(4)	221	(2)	257	(1)	293	(3)	329	(2)
6	(1)	42	(3)	78	(1)	114	(4)	150	(4)	186	(1)	222	(2)	258	(4)	294	(1)	330	(173)
7	(3)	43	(4)	79	(2)	115	(1)	151	(3)	187	(2)	223	(3)	259	(1)	295	(6)	331	(1)
8	(3)	44	(4)	80	(1)	116	(3)	152	(2)	188	(4)	224	(3)	260	(3)	296	(15)	332	(2)
9	(1)	45	(2)	81	(5)	117	(3)	153	(1)	189	(2)	225	(2)	261	(3)	297	(6)	333	(2)
10	(1)	46	(3)	82	(33)	118	(1)	154	(3)	190	(2)	226	(1)	262	(4)	298	(2)	334	(4)
11	(3)	47	(2)	83	(1)	119	(1)	155	(3)	191	(3)	227	(3)	263	(3)	299	(2)	335	(2)
12	(3)	48	(4)	84	(3)	120	(1)	156	(3)	192	(2)	228	(2)	264	(3)	300	(7)	336	(0)
13	(3)	49	(4)	85	(2)	121	(4)	157	(2)	193	(1)	229	(4)	265	(4)	301	(2)	337	(26.92)
14	(2)	50	(1)	86	(2)	122	(1)	158	(2)	194	(1)	230	(1)	266	(3)	302	(2)	338	(20.0)
15	(1)	51	(2)	87	(14)	123	(4)	159	(4)	195	(1)	231	(1)	267	(1)	303	(1)	339	(2)
16	(2)	52	(3)	88	(3)	124	(4)	160	(3)	196	(1)	232	(1)	268	(1)	304	(0)	340	(3)
17	(2)	53	(2)	89	(4)	125	(3)	161	(4)	197	(2)	233	(2)	269	(1)	305	(2)	341	(1)
18	(1)	54	(4)	90	(4)	126	(4)	162	(2)	198	(4)	234	(3)	270	(3)	306	(3)	342	(3)
19	(1)	55	(1)	91	(4)	127	(1)	163	(3)	199	(2)	235	(4)	271	(1)	307	(28)	343	(1)
20	(1)	56	(4)	92	(4)	128	(3)	164	(2)	200	(1)	236	(3)	272	(3)	308	(0)	344	(1)
21	(3)	57	(1)	93	(5)	129	(2)	165	(3)	201	(2)	237	(2)	273	(4)	309	(1)	345	(2)
22	(4)	58	(4)	94	(9)	130	(3)	166	(2)	202	(3)	238	(3)	274	(3)	310	(1)	346	(4)
23	(2)	59	(4)	95	(3)	131	(2)	167	(4)	203	(1)	239	(1)	275	(2)	311	(3)	347	(3)
24	(1)	60	(2)	96	(3)	132	(2)	168	(1)	204	(4)	240	(3)	276	(2)	312	(4)	348	(1)
25	(4)	61	(2)	97	(1)	133	(4)	169	(3)	205	(1)	241	(3)	277	(2)	313	(4)	349	(3)
26	(1)	62	(3)	98	(5)	134	(1)	170	(1)	206	(2)	242	(4)	278	(4)	314	(6)	350	(1)
27	(3)	63	(1)	99	(1)	135	(4)	171	(1)	207	(1)	243	(1)	279	(1)	315	(480)	351	(2)
28	(3)	64	(2)	100	(2)	136	(3)	172	(1)	208	(1)	244	(4)	280	(1)	316	(2)	352	(4)
29	(1)	65	(2)	101	(1)	137	(4)	173	(2)	209	(1)	245	(4)	281	(1)	317	(4)	353	(1)
30	(3)	66	(1)	102	(2)	138	(3)	174	(2)	210	(2)	246	(4)	282	(2)	318	(7)	354	(3)
31	(2)	67	(2)	103	(2)	139	(1)	175	(2)	211	(3)	247	(2)	283	(4)	319	(2)	355	(2)
32	(4)	68	(3)	104	(5)	140	(4)	176	(1)	212	(2)	248	(1)	284	(1)	320	(2)		
33	(4)	69	(1)	105	(2)	141	(4)	177	(4)	213	(2)	249	(1)	285	(1)	321	(0)		
34	(4)	70	(4)	106	(2)	142	(4)	178	(4)	214	(4)	250	(3)	286	(4)	322	(0)		
35	(4)	71	(2)	107	(6)	143	(4)	179	(3)	215	(4)	251	(1)	287	(3)	323	(2)		
36	(3)	72	(4)	108	(0)	144	(1)	180	(1)	216	(4)	252	(2)	288	(4)	324	(6)		

Part-B (JEE Advanced)

1	(c)	8	(d)	15	(8)	23	(b, d)	33	(b)	40	(d)	47	(c)	55	False	62	(d)	69	(a)
2	(d)	9	(3)	16	(6)	24	(c, d)	34	(c)	41	(b)	48	(b)	56	(a, b, d)	63	(a)	70	(a)
3	(c)	10	(6)	17	(3)	25	(b)	35	(c)	42	(c)	49	(5)	57	(a, c)	64	(c)	72	(a)
4	(b)	11	(6)	18	(4)	29	(a)	36	(b)	43	(d)	50	(1)	58	(a, b, d)	65	(b)		
5	(c)	12	(5)	19	(6.00)	30	(c)	37	(b)	44	(c)	51	(1)	59	(b, c)	66	(A):(p), (q) and (s)		
6	(a)	13	(6)	20	(8.00)	31	(a)	38	(a)	45	(c)	52	(4)	60	(b, c, d)	67	(a)		
7	(a)	14	(3)	22	(c, d)	32	(a)	39	(b)	46	(b)	53	(7.73)	61	(a, c)	68	(c)		

Chapter-10

Haloalkanes and Haloarenes

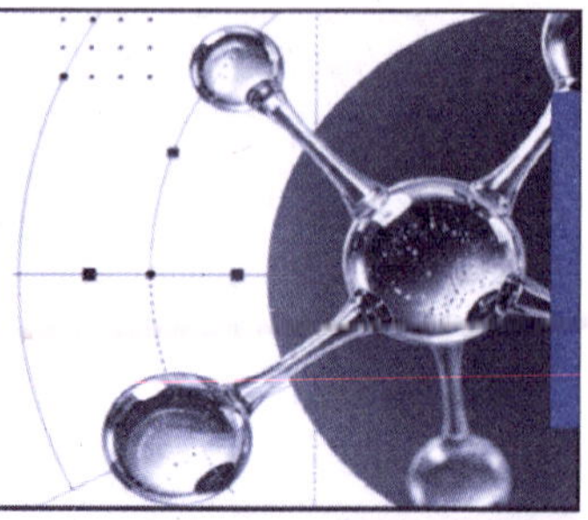

Part-A (JEE Main)

Topic 1- Preparation and Properties of Haloalkanes

MULTIPLE CHOICE QUESTIONS (MCQs)

1. An optically active alkyl bromide C_4H_9Br, reacts with ethanolic KOH to form major compound [A] which reacts with bromine to give compound [B]. Compound [B] reacts with ethanolic KOH and sodamide to give compound [C]. One molecule of water adds to compound [C] on warming with mercuric sulphate and dilute sulphuric acid at 333 K to form compound [D]. The functional group in compound D will be confirmed by: **[April 06, 2026 (II)]**

(1) Haloform test (2) Lucas test
(3) Silver mirror test (4) Benedict test

2. The following structures are **[April 04, 2026 (II)]**

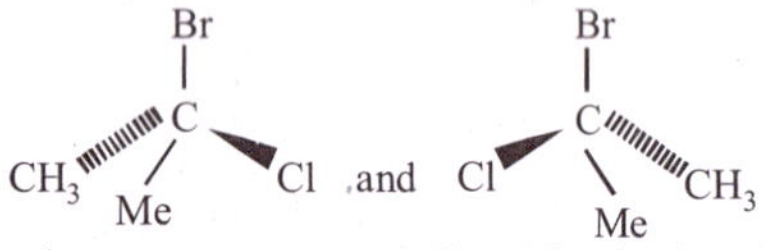

(1) enantiomer (2) Identical molecules
(3) Diastereomers (4) meso compound

3. Correct statements regarding alkyl halides (R–X) among the following are : **[April 02, 2026 (II)]**

(A) Alcohol being less polar solvent as compared to water, alcoholic KOH favours elimination reaction with R–X.
(B) Order of reactivity towards S_N1 mechanism is $C_6H_5–CH_2–Cl > C_6H_5–CHCl–C_6H_5$.
(C) Non substituted aryl halides exhibit properties similar to alkyl halides.
(D) Vinyl chloride is an example of haloalkene and allyl chloride is an example of haloalkyne.
(E) R–Cl can be prepared by reacting R–OH with $SOCl_2$ but Ar–Cl cannot be prepared by reacting Ar–OH with $SOCl_2$.

Choose the **correct** answer from the options given below:

(1) A, B and C Only (2) B and D Only
(3) A and E Only (4) D and E Only

4. Given below are two statements : **[April 02, 2026 (I)]**

Statement (I) : Benzyl chloride reacts faster in S_N1 mechanism than ethyl chloride.

Statement (II) : Ethyl carbocation intermediate is less stabilized by hyperconjugation than benzyl carbocation by resonance.

In the light of the above statements, choose the **correct** answer from the options given below :

(1) Both Statement I and Statement II are true.
(2) Both Statement I and Statement II are false.
(3) Statement I is true but Statement II is false.
(4) Statement I is false but Statement II is true.

5. Given below are two statements : **[Jan. 24, 2026 (I)]**

Statement I: 'C–Cl' bond is stronger in $CH_2=CH-Cl$ than CH_3-CH_2-Cl

Statement II: The given optically active molecule, 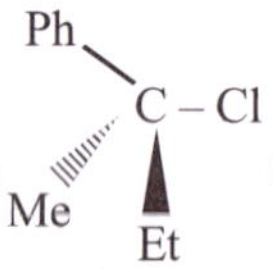 on hydrolysis gives a solution that can rotate the plane polarized light.

In the light of the above statements, choose the correct answer from the options given below

(1) Statement I is false but Statement II is true.
(2) Both Statement I and Statement II are true.
(3) Both Statement I and Statement II are false.
(4) Statement I is true but Statement II is false.

6. Match the List-I with List-II. **[Jan. 24, 2026 (I)]**

List-I (Chloro derivative)		List-II (Example)	
A.	Vinyl Chloride	I.	$CH_2=CH-CH_2Cl$
B.	Benzyl chloride	II.	$CH_3-CH(Cl)CH_3$
C.	Alkyl chloride	III.	$CH_2=CHCl$
D.	Allyl chloride	IV.	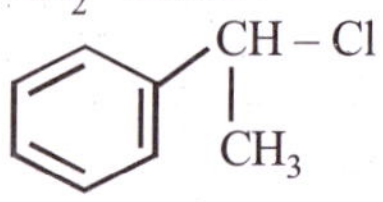

Choose the **correct** answer from the options given below:

(1) A-IV, B-I, C-III, D-II (2) A-III, B-IV, C-I, D-II
(3) A-III, B-IV, C-II, D-I (4) A-I, B-II, C-IV, D-III

7. Consider the following reaction : **[Jan. 22, 2026 (II)]**

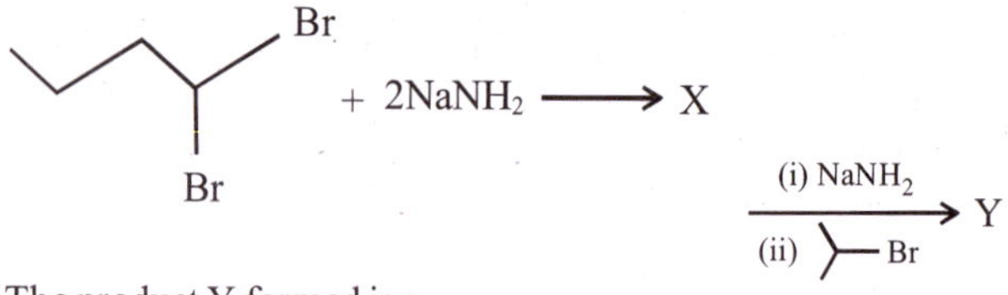

The product Y formed is :

(1) 2-methylhex-2-yne (2) 5-methylhex-2-yne
(3) 2-methylhex-3-yne (4) Isopropylbut-1-yne

8. The correct order of the rate of reaction of the following reactants with nucleophile by S_N1 mechanism is :
(**Given :** Structure I and II are rigid) [Jan. 22, 2026 (I)]

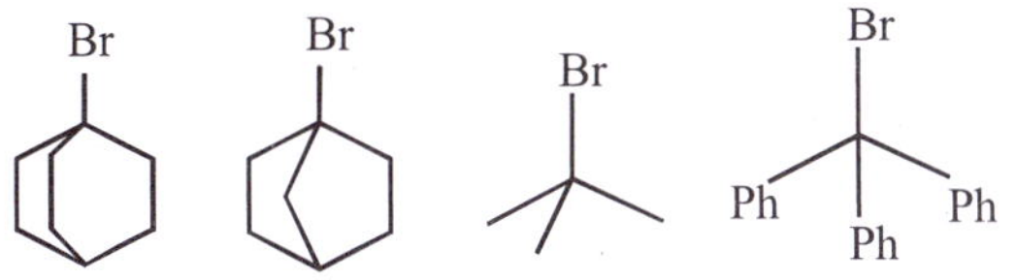

(I) (II) (III) (IV)

(1) IV < III < I < I (2) III < I < II < IV
(3) II < I < III < IV (4) I < II < III < IV

9. The correct order of reactivity of CH_3Br in methanol with the following nucleophiles is [Jan. 22, 2026 (I)]
F^-, I^-, $C_2H_5O^-$ and $C_6H_5O^-$
(1) $I^- > C_6H_5O^- > F^- > C_2H_5O^-$
(2) $I^- > C_2H_5O^- > C_6H_5O^- > F^-$
(3) $I^- > C_2H_5O^- > F^- > C_6H_5O^-$
(4) $I^- > F^- > C_6H_5O^- > C_2H_5O^-$

10. The correct order of the rate of the reaction for the following reaction with respect to nucleophiles is:
$CH_3Br + Nu^{\ominus} \longrightarrow CH_3Nu + Br^{\ominus}$ [Jan. 21, 2026 (II)]
(1) $PhO^- > {}^-OH > CH_3COO^- > ClO_4^-$
(2) $ClO_4^- > CH_3COO^- > {}^-OH > PhO^-$
(3) $CH_3COO^- > PhO^- > {}^-OH > ClO_4^-$
(4) ${}^-OH > PhO^- > CH_3COO^- > ClO_4^-$

11. Given below are two statements : [April 7, 2025 (II)]
Statement (I) :  is more polar than 

Statement (II) : Boiling point of 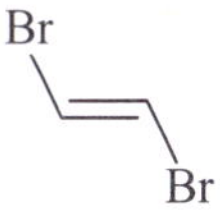 is lower than 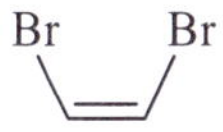 but it is more polar than 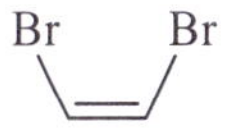

In the light of the above statements, choose the **most appropriate answer** from the options given below :
(1) Statement I is correct but statement II is incorrect
(2) Statement I is incorrect but statement II is correct
(3) Both statement I and statement II are incorrect
(4) Both statement I and statement II are correct

12. Consider the following molecule (*X*). [April 4, 2025 (II)]

The structure of *X* is 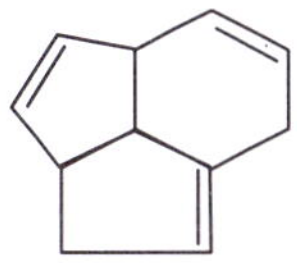

(*X*)

The major product formed when the given molecule(X) treated with HBr (1 eq) is :

(1) 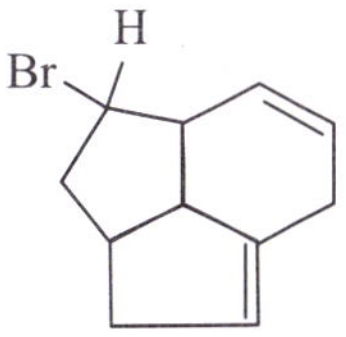 (2) 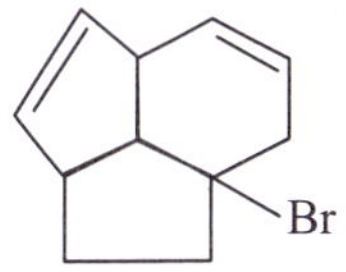

(3) 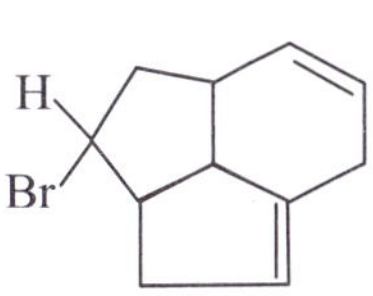 (4) 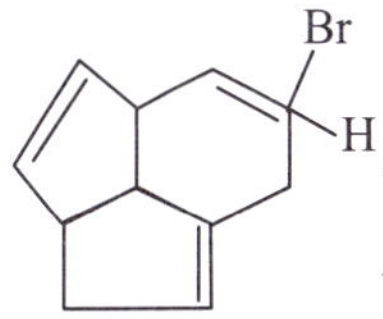

13. Given below are two statements : [April 4, 2025 (II)]
Statement (I) : Alcohols are formed when alkyl chlorides are treated with aqueous potassium hydroxide by elimination reaction.
Statement (II) : In alcoholic potassium hydroxide, alkyl chlorides form alkenes by abstracting the hydrogen from the β-carbon.
In the light of the above statements, choose the **most appropriate answer** from the options given below :
(1) Both Statement I and Statement II are incorrect
(2) Statement I is incorrect but Statement II is correct
(3) Statement I is correct but Statement II is incorrect
(4) Both Statement I and Statement II are correct

14. Predict the major product of the following reaction sequence :- [April 4, 2025 (I)]

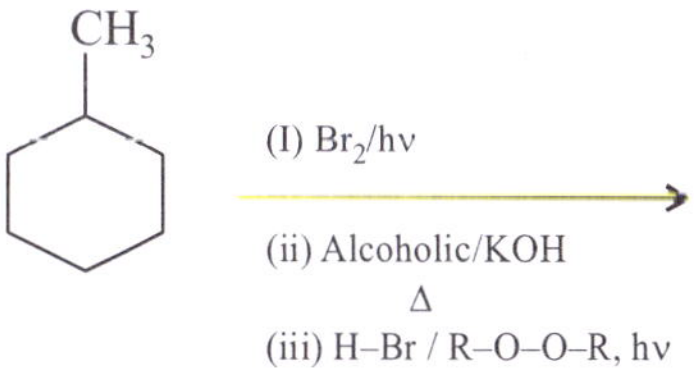

(1) 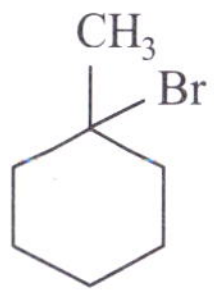 (2) 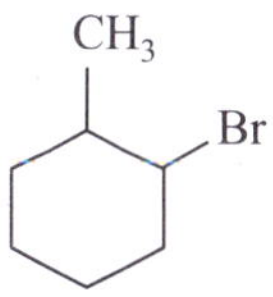

(3) 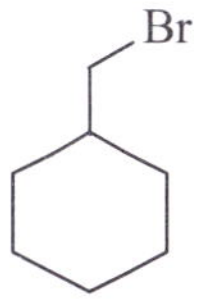 (4) 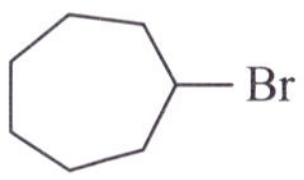

15. An optically active alkyl halide C_4H_9Br [A] reacts with hot KOH dissolved in ethanol and forms alkene [B] as major product which reacts with bromine to give dibromide [C]. The compound [C] is converted into a gas [D] upon reacting with alcoholic $NaNH_2$. During hydration 18 gram of water is added to 1 mole of gas [D] on warming with mercuric sulphate and dilute acid at 333 K to form compound [E]. The IUPAC name of compound [E] is : [April 2, 2025 (I)]
(1) But-2-yne (2) Butan-2-ol
(3) Butan-2-one (4) Butan-1-al

16. Which among the following halides will generate the most stable carbocation in Nucleophillic substitution reaction ? [Jan. 29, 2025 (II)]

(1)  (2) 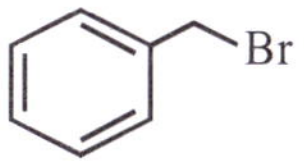

(3) 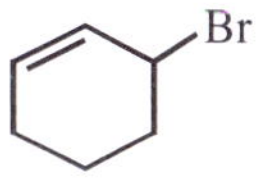 (4) 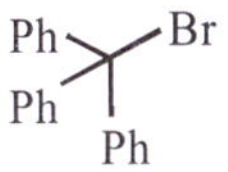

17. The major product of the following reaction is : [Jan. 28, 2025 (II)]

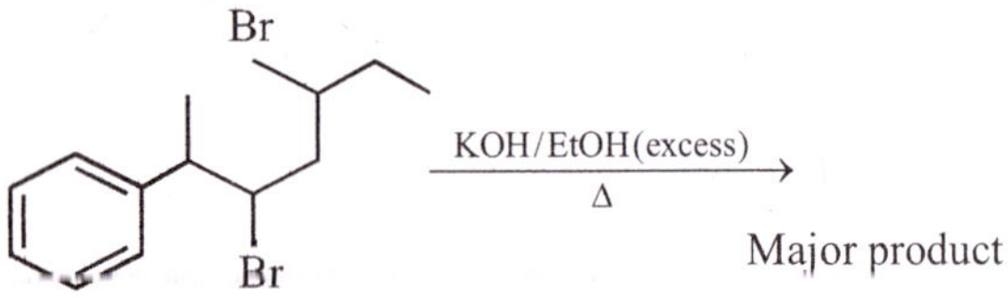

(1) 6-Phenylhepta-2,4-diene
(2) 2-Phenylhepta-2,5-diene
(3) 6-Phenylhepta-3,5-diene
(4) 2-Phenylhepta-2,4-diene

18. The product B formed in the following reaction sequence is : [Jan. 28, 2025 (II)]

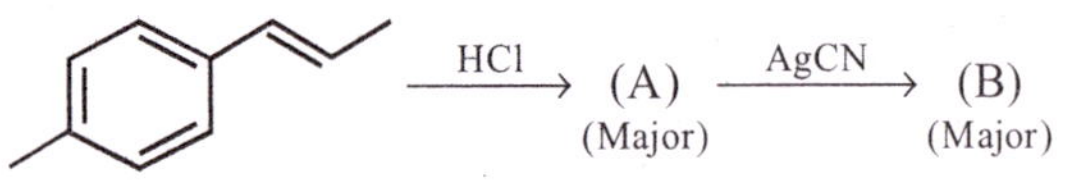

(1) 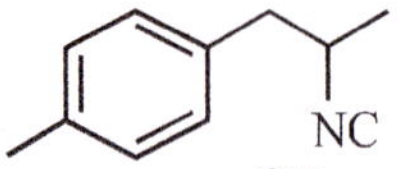

(2) 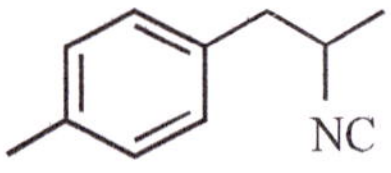

(3) 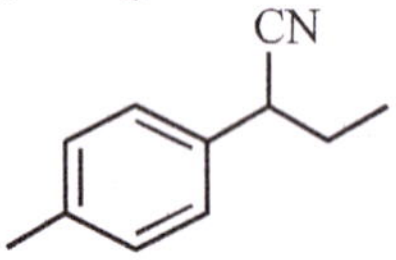

(4) 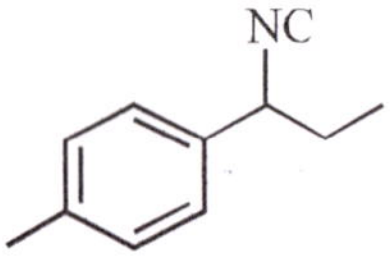

19. The products A and B in the following reactions, respectively are [Jan. 28, 2025 (I)]

$$A \xleftarrow{Ag-NO_2} CH_3-CH_2-CH_2-Br \xrightarrow{AgCN} B$$

(1) $CH_3-CH_2-CH_2-ONO$, $CH_3-CH_2-CH_2-NC$
(2) $CH_3-CH_2-CH_2-ONO$, $CH_3-CH_2-CH_2-CN$
(3) $CH_3-CH_2-CH_2-NO_2$, $CH_3-CH_2-CH_2-CN$
(4) $CH_3-CH_2-CH_2-NO_2$, $CH_3-CH_2-CH_2-NC$

20. Given below are two statements : [Jan. 28, 2025 (I)]

Statement I : $Et_2N-CH_2-CH_2-Cl$ will undergo alkaline hydrolysis at a faster rate than $Et_2CH-CH_2-CH_2-Cl$

Statement II : $Et_2N-CH_2-CH_2-Cl$ intramolecular substitution takes place first by involving lone pair of electrons on nitrogen.

In the light of the above statements, choose the most appropriate answer from the options given below :

(1) Both Statement I and Statement II are incorrect
(2) Statement I is incorrect but statement II is correct
(3) Both Statement I and Statement II are correct
(4) Statement I is correct but Statement II is incorrect

21. The structure of the major product formed in the following reaction is : [Jan. 24, 2025 (II)]

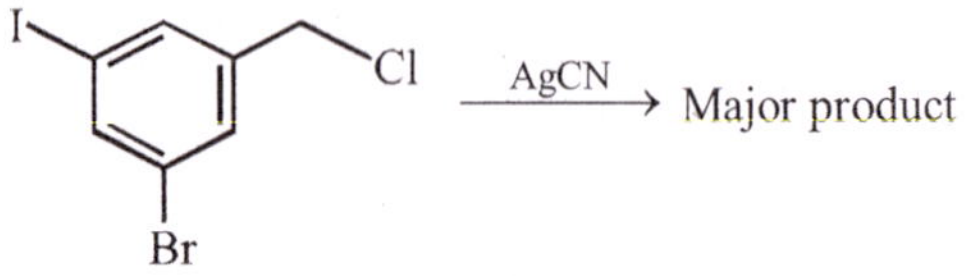

(1) 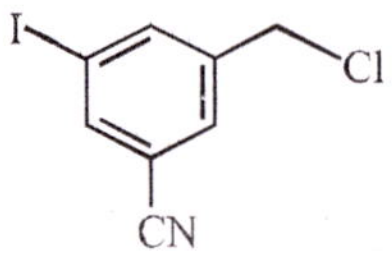

(2) 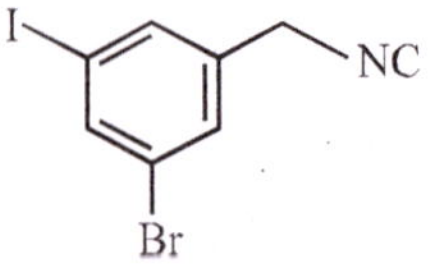

(3) 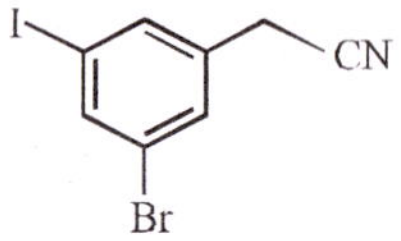

(4) 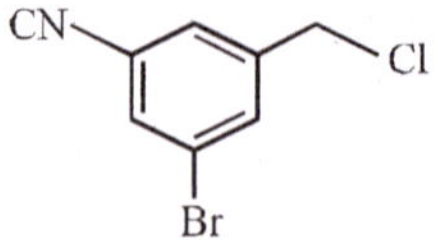

22. Given below are two statements:

Statement-I: The conversion proceeds well in the less polar medium. [Jan. 24, 2025 (I)]

$$CH_3-CH_2-CH_2-CH_2-Cl \xrightarrow{HO^-} CH_3-CH_2-CH_2-CH_2-OH + Cl^-$$

Statement-II: The conversion proceeds well in the more polar medium.

$$CH_3-CH_2-CH_2-CH_2-Cl \xrightarrow{R_3\ddot{N}} CH_3-CH_2-CH_2-CH_2-\overset{\oplus}{N}R_3\ Cl^{\ominus}$$

In the light of the above statements, choose the correct answer from the options given below.

(1) Both statement I and statement II are true
(2) Both statement I and statement II are false
(3) Statement I is false but statement II is true
(4) Statement I is true but statement II is false

23. The ascending order of relative rate of solvolysis of following compounds is [Jan. 23, 2025 (II)]

(A) 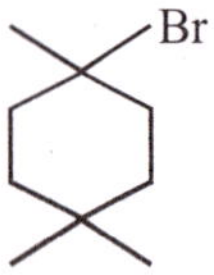

(B) 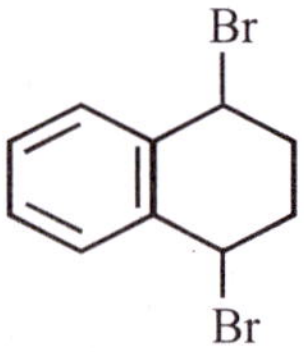

(C) 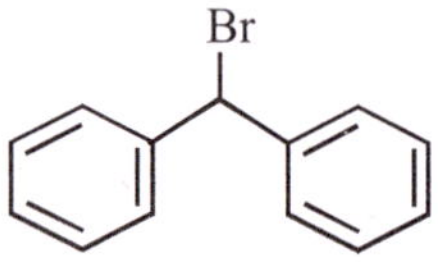

(D) 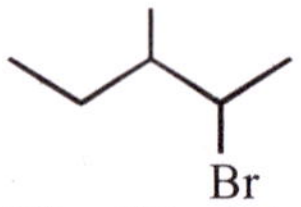

(1) (D) < (A) < (B) < (C)
(2) (C) < (B) < (A) < (D)
(3) (D) < (B) < (A) < (C)
(4) (C) < (D) < (B) < (A)

24. Match the List-I with List-II [Jan. 23, 2025 (I)]

	List-I Name reaction		List-II Product obtainable
A.	Swarts reaction	I.	Ethyl benzene
B.	Sandmeyer's reaction	II.	Ethyl iodide
C.	Wurtz Fittig reaction	III.	Cyanobenzene
D.	Finkelstein reaction	IV.	Ethyl fluoride

Choose the correct answer from the options given below:

(1) A-II, B-III, C-I, D-IV
(2) A-IV, B-I, C-III, D-II
(3) A-IV, B-III, C-I, D-II
(4) A-II, B-I, C-III, D-IV

25. $RBr \xrightarrow[(ii)\ H_2O]{(i)\ Mg,\ dry\ ether}$ 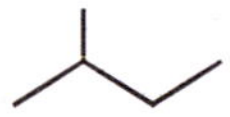
2-Methylbutane

The maximum number of RBr producing 2-methylbutane by above sequence of reactions is ______. (Consider the structural isomers only) [Jan. 22, 2025 (II)]

(1) 4 (2) 5 (3) 3 (4) 1

26. Given below are two statements : **[Jan. 22, 2025 (I)]**

Statement I : CH_3-O-CH_2-Cl will undergo S_N1 reaction though it is a primary halide.

Statement II : $CH_3-\underset{CH_3}{\overset{CH_3}{\underset{|}{\overset{|}{C}}}}-CH_2-Cl$ will not undergo S_N2 reaction very easily though it is a primary halide.

In the light of the above statements, choose the most appropriate answer from the options given below:

(1) Statement I is incorrect but Statement II is correct.
(2) Both Statement I and Statement II are incorrect
(3) Statement I is correct but Statement II is incorrect
(4) Both Statement I and Statement II are correct.

27. Given below are two statements : one is labelled as **Assertion (A)** and the other is labelled as **Reason (R) :** **[April 9, 2024 (I)]**

Assertion (A) : S_N2 reaction of $C_6H_5CH_2Br$ occurs more readily than the S_N2 reaction of CH_3CH_2Br.

Reason (R) : The partially bonded unhybridized p-orbital that develops in the trigonal bipyramidal transition state is stabilized by conjugation with the phenyl ring.

In the light of the above statements, choose the **most appropriate** answer from the options given below :

(1) **(A)** is not correct but **(R)** is correct
(2) Both **(A)** and **(R)** are correct but **(R)** is not the correct explanation of **(A)**
(3) Both **(A)** and **(R)** are correct and **(R)** is the correct explanation of **(A)**
(4) **(A)** is correct but **(R)** is not correct

28. 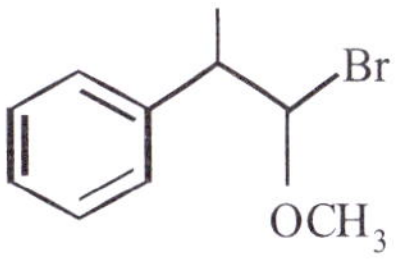

$\xrightarrow[\Delta]{KCN(alc)}$ Major Product 'P'

In the above reaction product 'P' is **[April 9, 2024 (II)]**

(1) 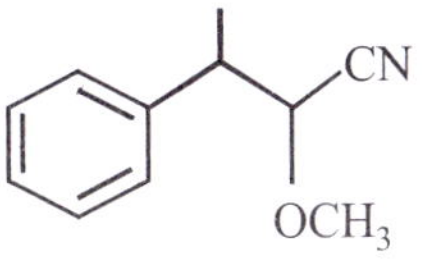

(2) 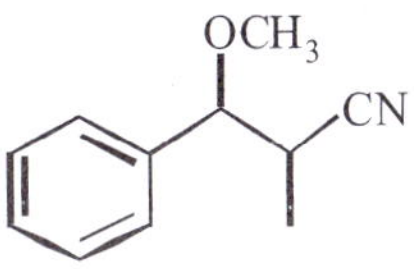

(3) 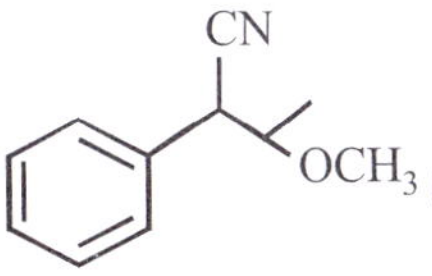

(4) 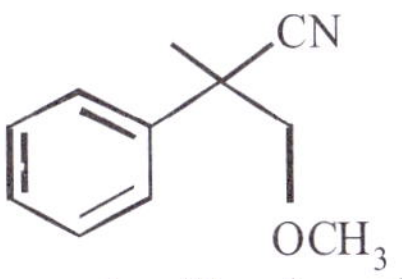

29. Which among the following compounds will undergo fastest S_N2 reaction. **[April 8, 2024 (I)]**

(1) 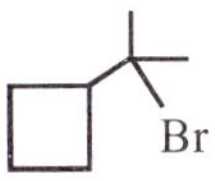

(2) 

(3) 

(4) 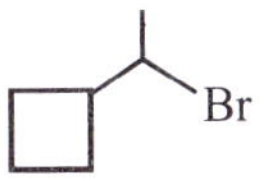

30. Given below are two statements : **[April 8, 2024 (II)]**

Statement (I) : S_N2 reactions are 'stereospecific', indicating that they result in the formation only one stereo-isomers as the product.

Statement (II) : S_N1 reactions generally result in formation of product as racemic mixtures. In the light of the above statements, choose the **correct** answer from the options given below :

(1) **Statement I** is true but **Statement II** is false
(2) **Statement I** is false but **Statement II** is true
(3) Both **Statement I** and **Statement II** is true
(4) Both **Statement I** and **Statement II** is false

31. 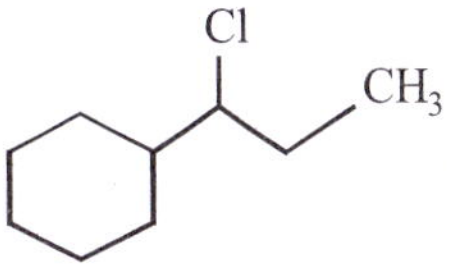

$+NaOH \xrightarrow{H_2O}$ Major Product "A"

Consider the above chemical reaction. Product "A" is: **[April 6, 2024 (II)]**

(1) 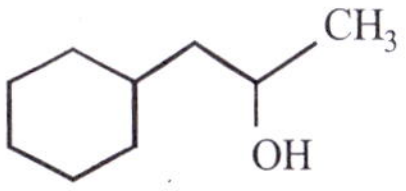

(2) 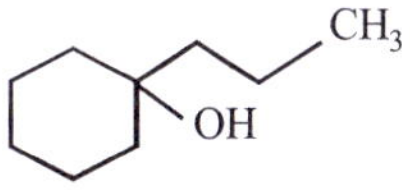

(3) 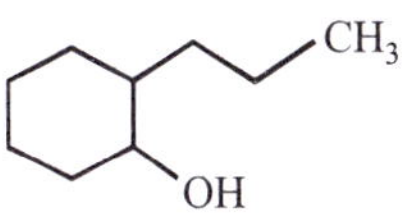

(4) 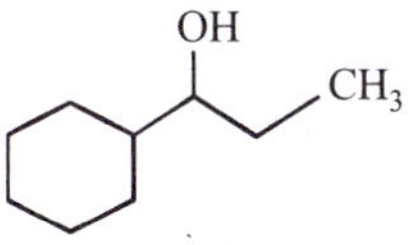

32. Identify the major product in the following reaction.

$\xrightarrow[C_2H_5OH]{\bar{O}H}$ Major Product

[April 5, 2024 (II)]

(1) 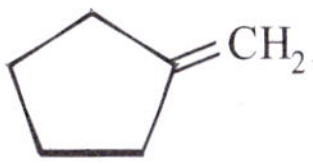

(2) 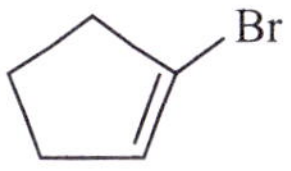

(3) 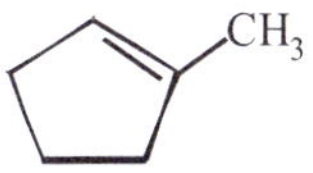

(4) 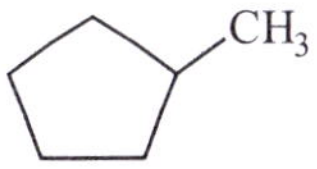

33. Identify the correct set of reagents or reaction conditions 'X' and 'Y' in the following set of transformation.

$$CH_3-CH_2-CH_2-Br \xrightarrow{'X'} \text{Product} \xrightarrow{'Y'} CH_3-\underset{\underset{Br}{|}}{CH}-CH_3$$

[April 4, 2024 (I)]

(1) X = conc.alc. NaOH, 80°C, Y = $Br_2/CHCl_3$
(2) X = dil.aq. NaOH, 20°C, Y = HBr/acetic acid
(3) X = conc.alc. NaOH, 80°C, Y = HBr/acetic acid
(4) X = dil.aq. NaOH, 20°C, Y = $Br_2/CHCl_3$

34. Find out the major product formed from the following reaction. [Me: $-CH_3$] **[April 4, 2024 (II)]**

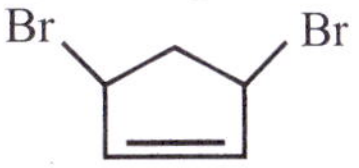

$\xrightarrow{Me_2NH(2equiv)}$

(1) 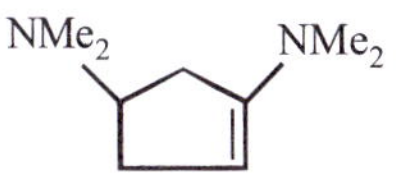

(2) 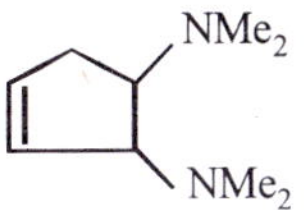

(3) 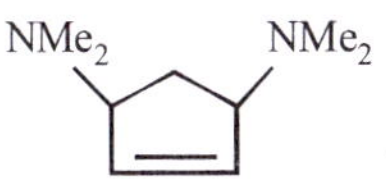

(4) 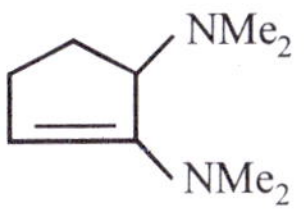

35. $CH_3-CH_2-CH_2-Br + NaOH \xrightarrow{C_2H_5OH}$ Product 'A'

Product A $\xrightarrow[H^+]{H_2O}$ Product "B"

Product A $\xrightarrow[H_2O/H_2O_2/\ OH]{Diborane}$ Product "C"

Consider the above reactions, identify product B and product C. [April 4, 2024 (II)]

(1) B = C = 2-Propanol
(2) B = 2-Propanol C = 1-Propanol
(3) B = 1-Propanol C = 2-Propanol
(4) B = C = 1-Propanol

36. 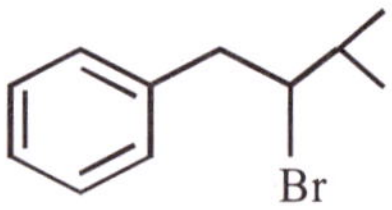

$\xrightarrow[\Delta]{KOH(alc.)}$ major product "P"

Product P is [April 4, 2024 (II)]

(1) 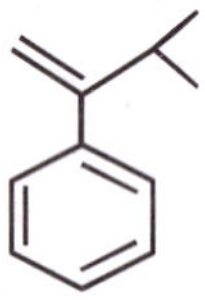
(2) 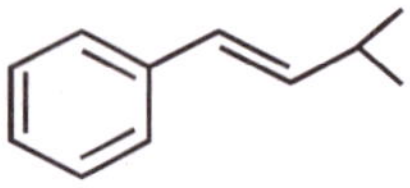
(3) 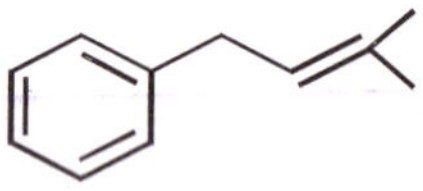
(4) 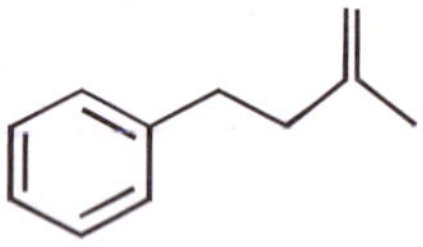

37. Given below are two statements : one is labelled as **Assertion (A)** and the other is labelled as **Reason (R).**

Assertion (A) : Haloalkanes react with KCN to form alkyl cyanides as a main product while with AgCN form isocyanide as the main product.

Reason (R) : KCN and AgCN both are highly ionic compounds.

In the light of the above statements, choose the most appropriate answer from the options given below: [Feb. 1, 2024 (I)]

(1) (A) is correct but (R) is not correct.
(2) Both (A) and (R) are correct but (R) is not the correct explanation of (A)
(3) (A) is not correct but (R) is correct.
(4) Both (A) and (R) are correct and (R) is the correct explanation of (A).

38. The product (C) in the below mentioned reaction is :

$CH_3-CH_2-CH_2-Br \xrightarrow[\Delta]{KOH(alc)} A \xrightarrow{HBr}$

$B \xrightarrow[KOH(aq)]{\Delta} C$ [Jan. 31, 2024 (I)]

(1) Propene (2) Propan-2-ol
(3) Propyne (4) Propan-1-ol

39. Example of vinylic halide is : [Jan. 30, 2024 (I)]

(1) 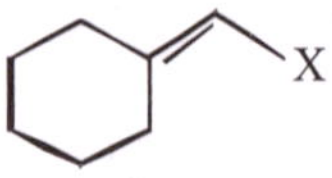

(2) 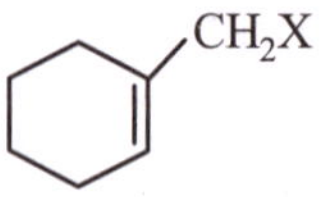

(3) 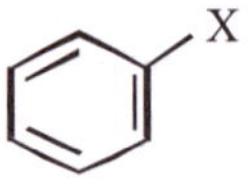

(4) 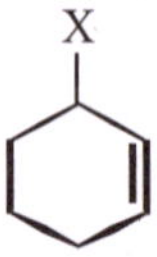

40. Given below are two statements : One is labelled as Assertion **(A)** and the other is labelled as Reason **(R).** [Jan. 30, 2024 (I)]

Assertion (A) : $CH_2=CH-CH_2-Cl$ is an example of allyl halide.

Reason (R) : Allyl halides are the compounds in which the halogen atom is attached to sp^2 hybridised carbon atom.

In the light of the above statements, choose the most **appropriate** answer from the option given below:

(1) **(A)** is true but **(R)** is false
(2) Both **(A)** and **(R)** are true but **(R)** is not the correct explanation of **(A)**
(3) **(A)** is false but **(R)** is true
(4) Both **(A)** and **(R)** are true and **(R)** is the correct explanation of **(A).**

41. Given below are two statements: [Jan. 30, 2024 (II)]

Statement I: High concentration of strong nucleophilic reagent with secondary alkyl halides which do not have bulky substituents will follow S_N2 mechanism.

Statement II: A secondary alkyl halide when treated with a large excess of ethanol follows S_N1 mechanism.

In the light of the above statements, choose the most appropriate from the options given below:

(1) Statement I is true but Statement II is false.
(2) Statement I is false but Statement II is true.
(3) Both Statement I and Statement II are false.
(4) Both Statement I and Statement II are true.

42. Alkyl halide is converted into alkyl isocyanide by reaction with [Jan. 29, 2024 (II)]

(1) NaCN (2) AgCN
(3) NH_4CN (4) KCN

43. The correct statement regarding nucleophilic substitution reaction in a chiral alkyl halide is : [Jan. 27, 2024 (I)]

(1) Retention occurs in S_N1 reaction and inversion occurs in S_N2 reaction.
(2) Racemisation occurs in S_N1 reaction and retention occurs in S_N2 reaction.
(3) Racemisation occurs in both S_N1 and S_N2 reactions.
(4) Racemisation occurs in S_N1 reaction and inversion occurs in S_N2 reaction.

44. Which among the following halide/s will not show S_N1 reaction: [Jan. 27, 2024 (II)]

(A) $H_2C=CH-CH_2Cl$ (B) $CH_3-CH=CH-Cl$

(C) 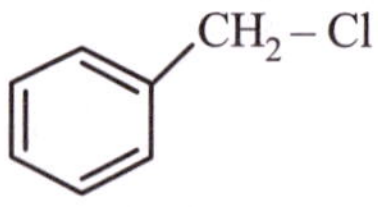

(D) 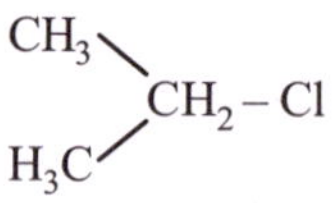

Choose the most appropriate answer from the options given below:

(1) (A), (B) and (D) only (2) (A) and (B) only
(3) (B) and (C) only (4) (B) only

45. 2-Methyl propyl bromide reacts with $C_2H_5O^-$ and gives 'A' whereas on reaction with C_2H_5OH it gives 'B'. The mechanism followed in these reactions and the products 'A' and 'B' respectively are: [April 13, 2023 (I)]

(1) S_N2, A = iso-butyl ethyl ether; S_N1, B = tert-butyl ethyl ether

(2) S_N1, A = tert-butyl ethyl ether; S_N1, B = 2-butyl ethyl ether
(3) S_N1, A = tert-butyl ethyl ether; S_N2, B = iso-butyl ethyl ether
(4) S_N2, A = 2-butyl ethyl ether; S_N2, B = iso-butyl ethyl ether

46. Match List I with List II **[April 13, 2023 (II)]**

1 - Bromopropane is reacted with reagents in List I to give product in List II

	List I - Reagent		List II - Product
A.	KOH (alc)	I.	Nitrile
B.	KCN (alc)	II.	Ester
C.	$AgNO_2$	III.	Alkene
D.	$H_3CCOOAg$	IV.	Nitroalkane

(1) A-IV, B-III, C-II, D-I (2) A-III, B-I, C-IV, D-II
(3) A-I, B-II, C-III, D-IV (4) A-I, B-III, C-IV, D-II

47. 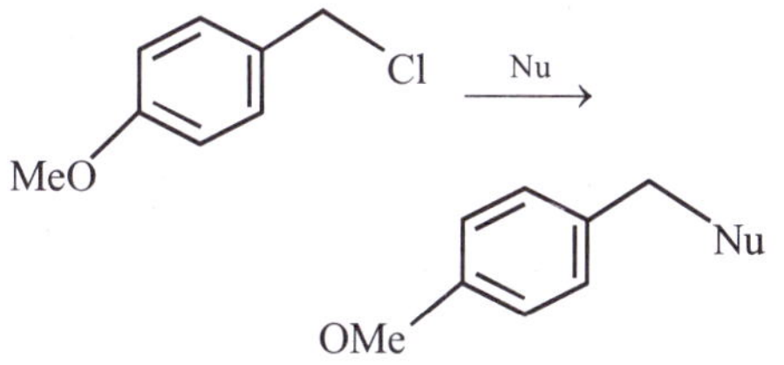 ... (I)

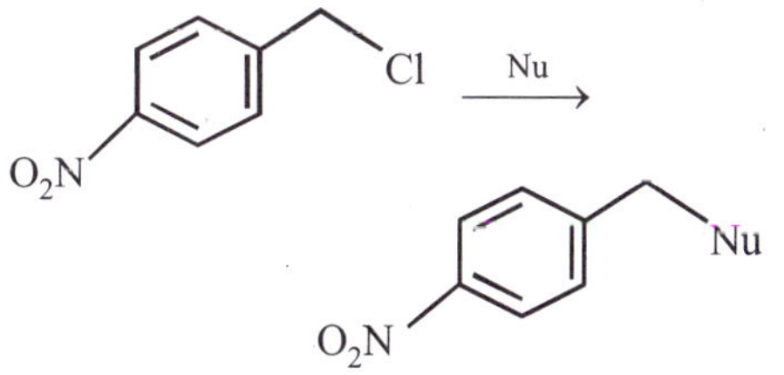 ... (II)

Where Nu = Nucleophile

Find out the correct statement from the options given below for the above 2 reactions. **[April 11, 2023 (I)]**

(1) Reaction (I) is of 2nd order and reaction (II) is of 1st order
(2) Reaction (I) and (II) both are of 2nd order
(3) Reaction (I) is of 1st order and reaction (II) is of 2nd order
(4) Reaction (I) and (II) both are of 1st order

48. Given below are two statements, one is labelled as **Assertion A** and the other is labelled as **Reason R**.

Assertion A : A solution of the product obtained by heating a mole of glycine with a mole of chlorine in presence of red phosphorous generates chiral carbon atom.

Reason R: A molecule with 2 chiral carbons is always optically active.

In the light of the above statements, choose the correct answer from the options given below: **[April 11, 2023 (II)]**

(1) A is false but R is true.
(2) A is true but R is false.
(3) Both A and R are true and R is the correct explanation of A.
(4) Both A and R are true and R is NOT the correct explanation of A.

49. Choose the halogen which is most reactive towards S_N1 reaction in the given compounds (A, B, C & D) **[April 08, 2023 (I)]**

A. 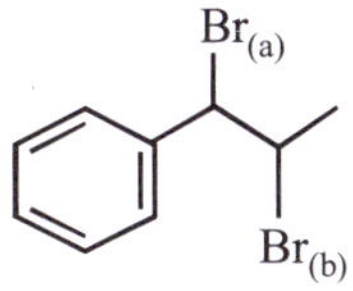 B. 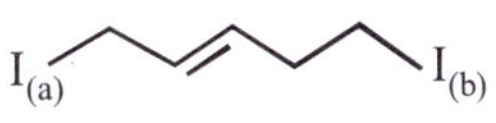

C. 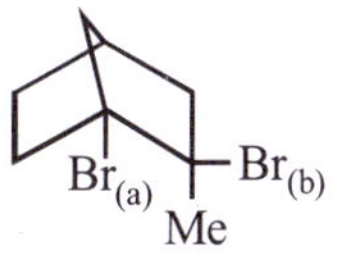 D. 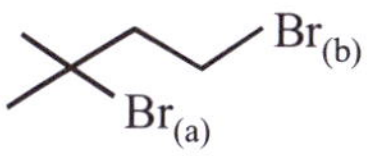

(1) A-$Br_{(b)}$; B-$I_{(b)}$; C-$Br_{(b)}$; D-$Br_{(b)}$
(2) A-$Br_{(a)}$; B-$I_{(a)}$; C-$Br_{(b)}$; D-$Br_{(a)}$
(3) A-$Br_{(b)}$; B-$I_{(a)}$; C-$Br_{(a)}$; D-$Br_{(a)}$
(4) A-$Br_{(b)}$; B-$I_{(a)}$; C-$Br_{(a)}$; D-$Br_{(a)}$

50. For the reaction: **[April 06, 2023 (I)]**

$$RCH_2Br + I^- \xrightarrow{\text{Acetone}} \underset{\text{major}}{RCH_2I} + Br^-$$

The correct statement is:

(1) The transition state formed in the above reaction is less polar than the localised anion.
(2) The reaction can occur in acetic acid also.
(3) The solvent used in the reaction solvates the ions formed in rate determining step.
(4) Br^- can act as competing nucleophile.

51. The graph which represents the following reaction is:

$$(C_6H_5)_3C-Cl \xrightarrow[\text{Pyridine}]{OH^-} (C_6H_5)_3C-OH$$

[Feb. 01, 2023 (II)]

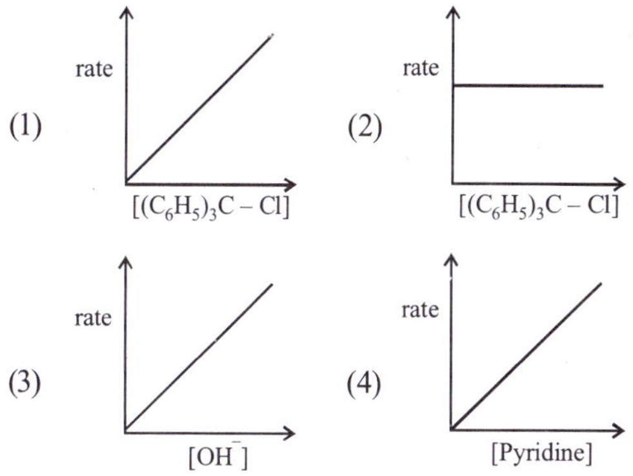

52. Identify the correct order for the given property for following compounds **[Jan. 29, 2023 (I)]**

(A) Boiling Point: 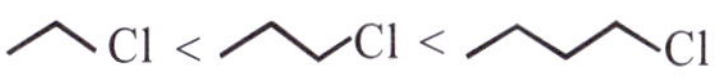
(B) Density 
(C) Boiling Point: 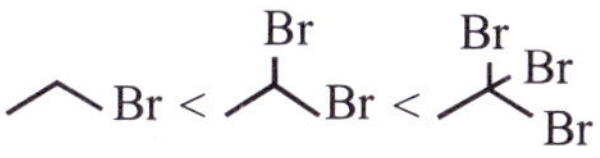
(D) Density: 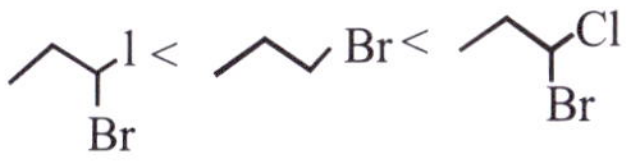
(E) Boiling Point: 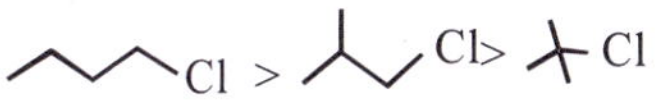

Choose the correct answer from the option given below :-

(1) (B), (C) and (D) only (2) (A), (C) and (E) only
(3) (A), (C) and (D) only (4) (A), (B) and (E) only

53. The isomeric deuterated bromide with molecular formula C_4H_8DBr having two chiral carbon atoms is **[Jan. 25, 2023 (II)]**

(1) 2–Bromo–1–deuterobutane
(2) 2–Bromo–2–deuterobutane

(3) 2–Bromo–3–deuterobutane

(4) 2–Bromo–1–deutero–2–methylpropane

54. **Assertion A** : Hydrolysis of an alkyl chloride is a slow reaction but in the presence of NaI, the rate of the hydrolysis increases. **[Jan. 24, 2023 (I)]**

Reason R : I^- is a good nucleophile as well as a good leaving group.

In the light of the above statements, choose the correct answer from the options given below.

(1) A is false but R is true

(2) A is true but R is false

(3) Both A and R are true and R is the correct explanation of A

(4) Both A and R are true but R is NOT the correct explanation of A

55. (Major Product) (A′) $\xleftarrow[C_2H_5OH - H_2O]{AgCN}$  $\xrightarrow[C_2H_5OH - H_2O]{NaCN}$ (Major Product) (B′)

Considering the above reactions, the compound 'A' and compound 'B' respectively are : **[July 29, 2022 (I)]**

(1) 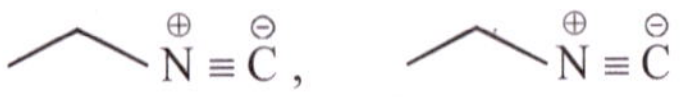

(2) 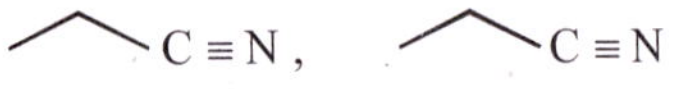

(3) 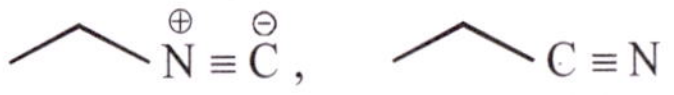

(4) 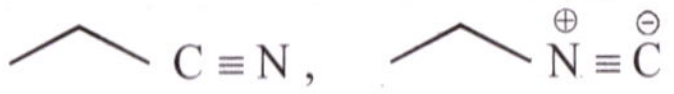

56. Given below are two statements. **[July 29, 2022 (II)]**

Statement I : The compound 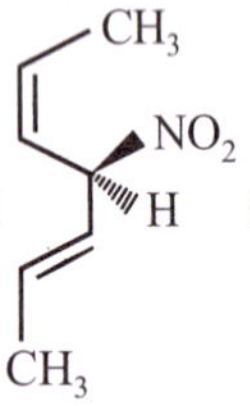

is optically active.

Statement II : 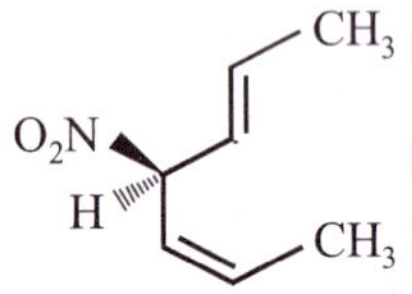

is mirror image of above compound A.

In the light of the above statement, choose the **most appropriate** answer from the options given below.

(1) Both statement I and statement II are correct.

(2) Both statement I and statement II are incorrect.

(3) Statement I is correct but statement II is incorrect.

(4) Statement I is incorrect but statement II is correct.

57. Compound 'A' undergoes following sequence of reactions to give compound 'B'. The correct structure and chirality of compound 'B' is:

[where Et is $-C_2H_5$] **[July 29, 2022 (II)]**

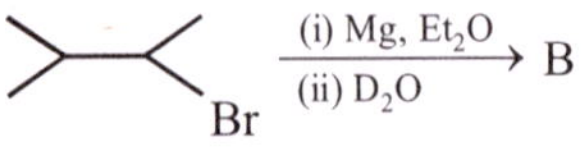

Compound 'A'

(1) 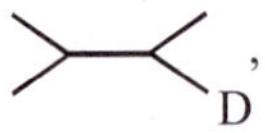

, Achiral

(2) 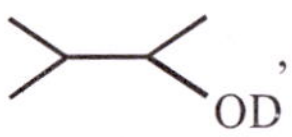

, Chiral

(3) 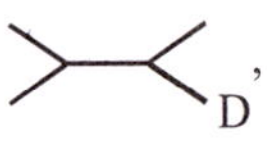

, Chiral

(4) 

, Achiral

58. Major product of the following reaction is

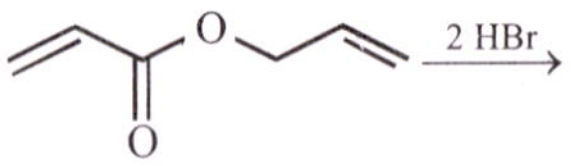

[July 25, 2022 (II)]

(1) 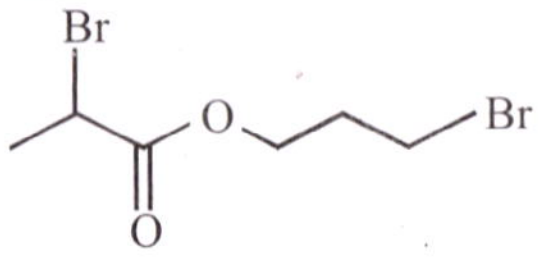

(2) 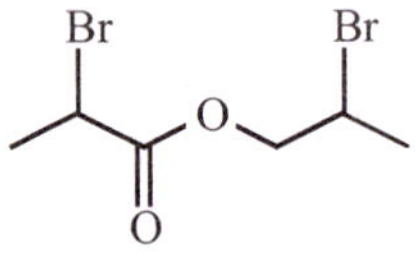

(3) 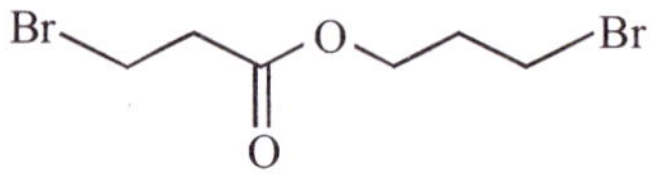

(4) 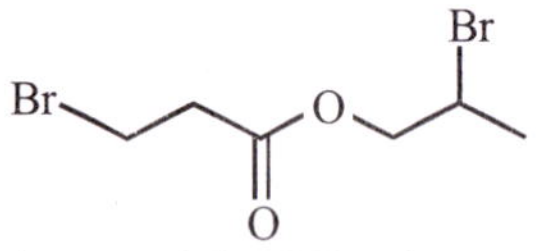

59. Which one of the following compounds is inactive towards S_N1 reaction? **[June 28, 2022 (I)]**

(1) 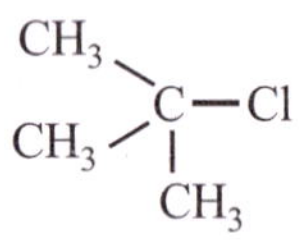

(2) 

(3) 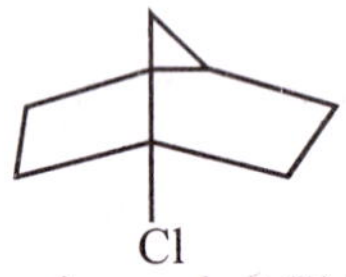

(4) 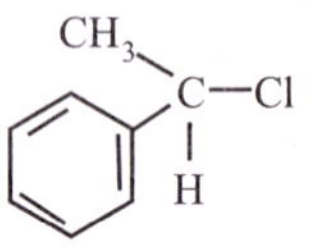

60. The major product (P) in the reaction **[June 28, 2022 (I)]**

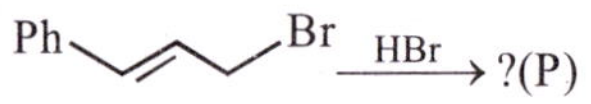

[Ph is $-C_6H_5$] is

(1) 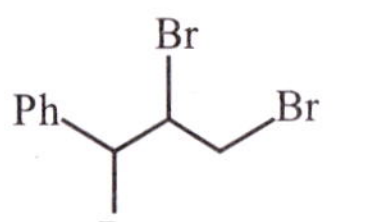

(2) 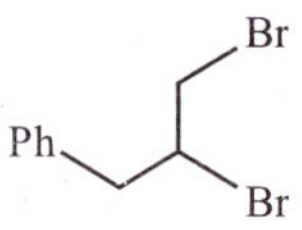

(3) 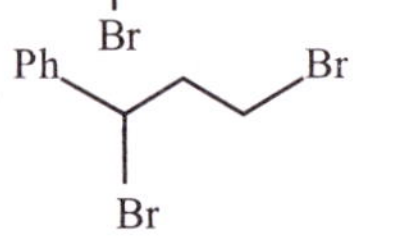

(4) 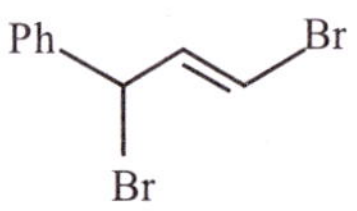

61. The major product of the following reaction is:

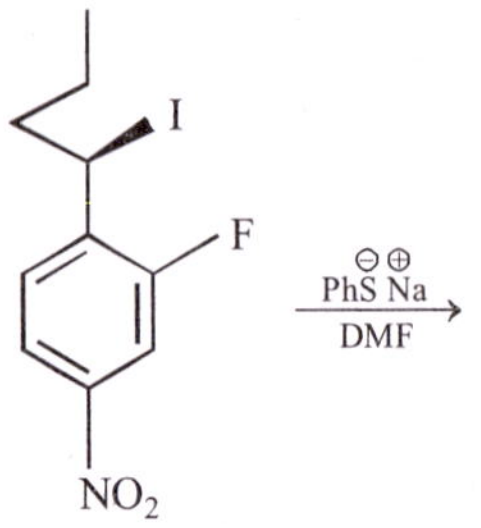

[June 27, 2022 (I)]

(1) 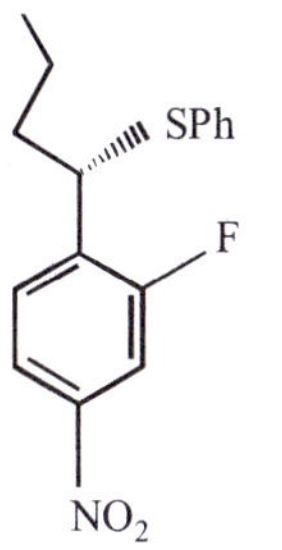
(2) 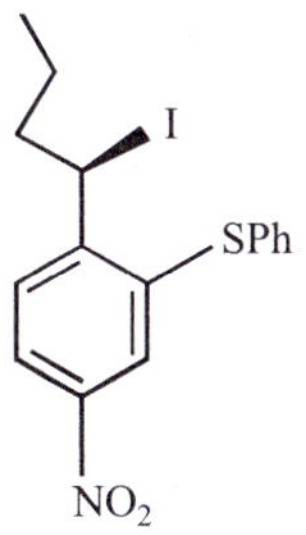

(3) 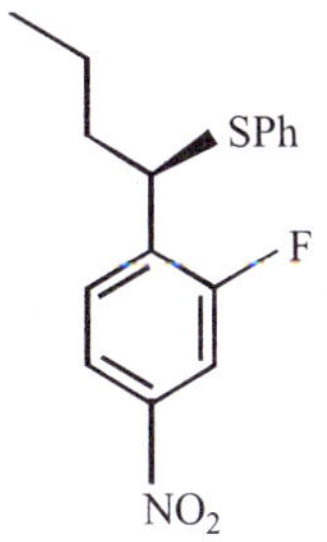
(4) 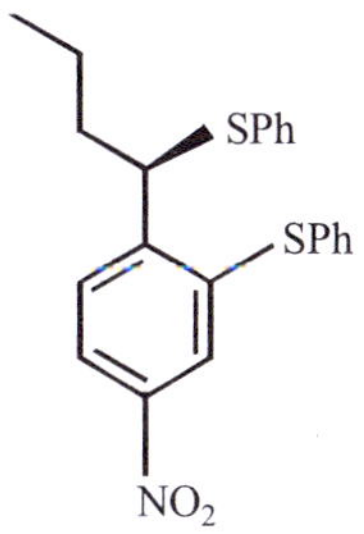

62. The major product in the reaction

$$CH_3-\underset{CH_3}{\overset{CH_3}{C}}-Cl+\overset{+}{K}\,\overset{-}{O}-\underset{CH_3}{\overset{CH_3}{C}}-CH_3 \longrightarrow ?\text{ is}$$

[June 25, 2022 (I)]

(1) t-Butyl ethyl ether (2) 2, 2-Dimethyl butane
(3) 2-Methyl pent-1-ene (4) 2-Methyl prop-1-ene

63. Amongst the following, the major product of the given chemical reaction is [June 25, 2022 (II)]

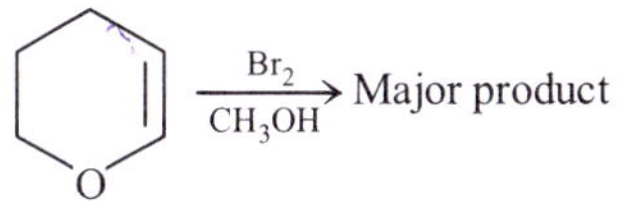

(1) 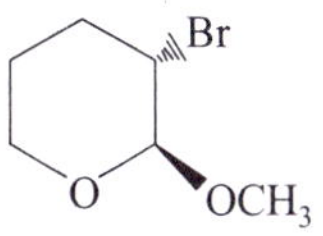
(2) 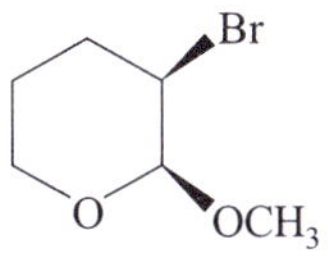
(3) 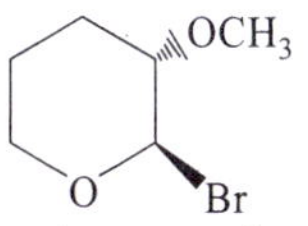
(4) 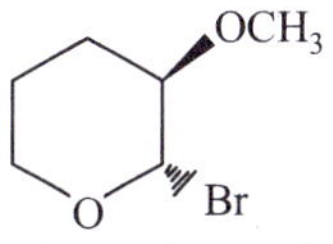

64. In the given reactions sequence, the major product 'C' is :

$$C_8H_{10}\xrightarrow[H_2SO_4]{HNO_3}A\xrightarrow[\Delta]{Br_2}B\xrightarrow[KOH]{\text{alcoholic}}C$$

[June 24, 2022 (I)]

(1) 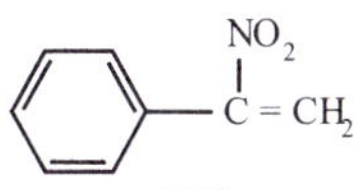
(2) 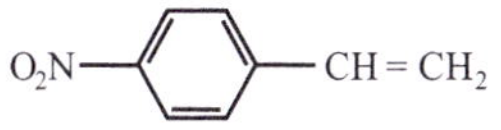
(3) 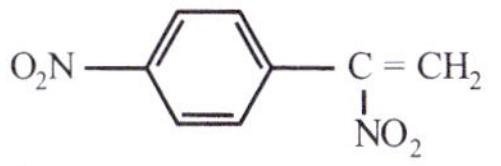
(4) 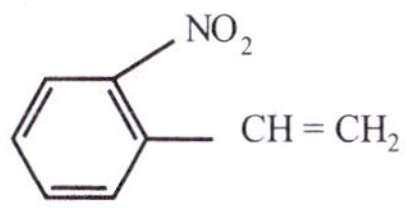

65. The major product of the following reaction is : [Aug. 31, 2021 (II)]

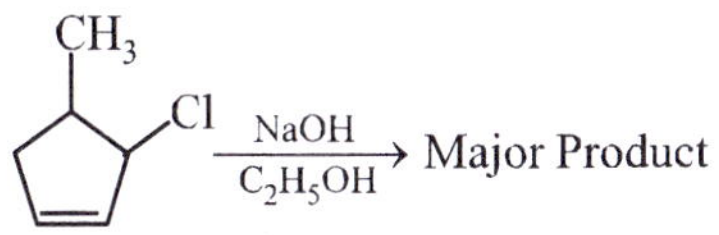

(1) 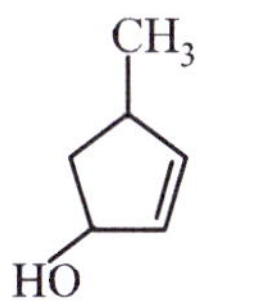
(2) 
(3) 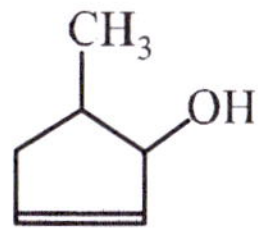
(4) 

66. Ammonolysis of alkyl halides followed by the treatment with NaOH solution can be used to prepare primary, secondary and tertiary amines. The purpose of NaOH in the reaction is [March 16, 2021 (II)]

(1) To remove basic impurities
(2) To activate NH_3 used in the reaction
(3) To remove acidic impurities
(4) To increase the reactivity of alkyl halide

67. The product formed in the first step of the reaction of

$$CH_3-CH_2-\overset{Br}{\overset{|}{C}H}-CH_2-\underset{Br}{\underset{|}{C}H}-CH_3 \text{ with excess Mg/Et}_2\text{O(Et}$$

$=C_2H_5$) is : [Feb. 24, 2021 (I)]

(1) $CH_3-CH_2-CH-CH_2-CH-CH_3$ / $CH_3-CH-CH_2-CH-CH_2-CH_3$ (bonded through the vertical lines)

(2) CH_2 $CH_2-CH-CH_2-CH-CH_2$ / $CH_3-CH_2-CH-CH_2-CH-CH_3$ (bonded through the vertical lines)

(3) 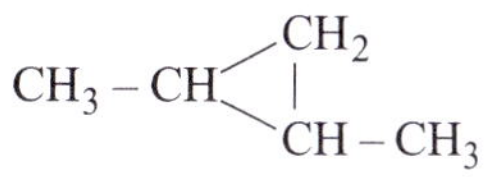

(4) $CH_3CH_2-\overset{MgBr}{\overset{|}{C}H}-CH_2-\underset{MgBr}{\underset{|}{C}H}-CH_3$

68. For the following reactions

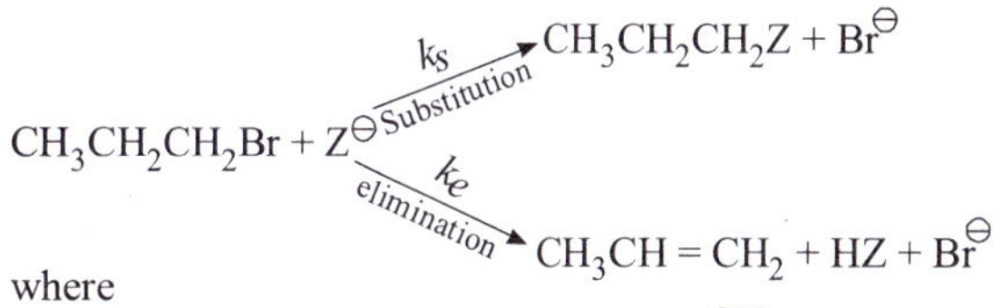

where

$$\overset{\ominus}{Z}=CH_3CH_2\overset{\ominus}{O}\text{ (A) or } H_3C-\underset{CH_3}{\overset{CH_3}{C}}-\overset{\ominus}{O}\text{ (B)},$$

k_s and k_e, are, respectively, the rate constants for substitution and elimination, and $\mu=\frac{k_s}{k_e}$, the correct option is ________.

[Jan. 07, 2020 (II)]

(1) $\mu_B>\mu_A$ and $k_e(A)>k_e(B)$
(2) $\mu_A>\mu_B$ and $k_e(B)>k_e(A)$
(3) $\mu_B>\mu_A$ and $k_e(B)>k_e(A)$
(4) $\mu_A>\mu_B$ and $k_e(A)>k_e(B)$

69. The decreasing order of reactivity of the following organic molecules towards $AgNO_3$ solution is : [Sep. 04, 2020 (I)]

(A)

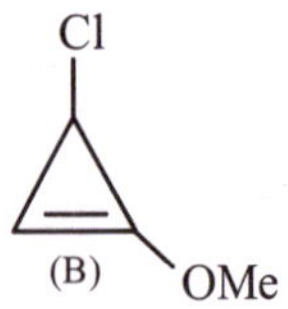
(B)

CH_3CHCH_3 (with Cl on the central C)
(C)

$CH_3CHCH_2NO_2$ (with Cl on the second C)
(D)

(1) (C) > (D) > (A) > (B) (2) (A) > (B) > (D) > (C)
(3) (A) > (B) > (C) > (D) (4) (B) > (A) > (C) > (D)

70. Which of the following compounds will form the precipitate with aq. $AgNO_3$ solution most readily? [Sep. 04, 2020 (II)]

(1) 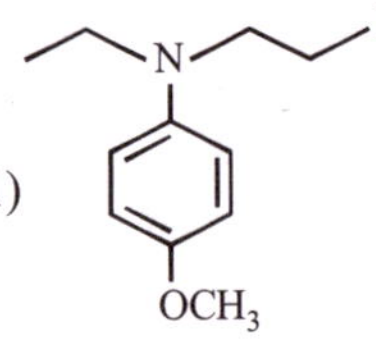
(2) 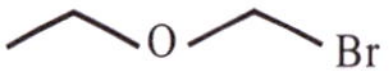
(3) 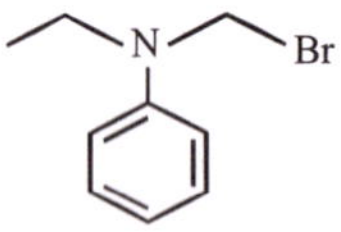
(4) 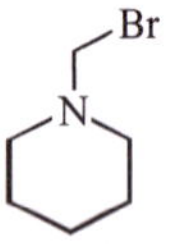

71. Among the following compounds, which one has the shortest C – Cl bond? [Sep. 04, 2020 (II)]

(1) 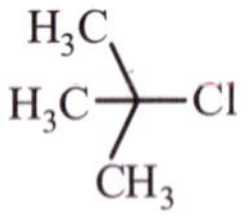
(2) 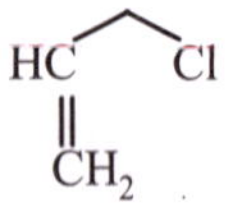
(3) H_3C-Cl
(4) 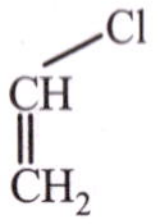

72. The mechanism of S_N1 reaction is given as :

$$R-X \longrightarrow \underset{\text{Ion pair}}{\overset{\oplus}{R}\,\overset{\ominus}{X}} \longrightarrow \underset{\text{Solvent Separated ion pair}}{\overset{\oplus}{R}\,\|\,\overset{\ominus}{X}} \xrightarrow{Y^{\ominus}} R-Y+X^{\ominus}$$

A student writes general characteristics based on the given mechanism as : [Sep. 03, 2020 (I)]

(A) The reaction is favoured by weak nucleophiles.
(B) $\overset{\oplus}{R}$ would be easily formed if the substituents are bulky.
(C) The reaction is accompanied by racemization.
(D) The reaction is favoured by non-polar solvents.

Which observations are correct?

(1) (A) and (B) (2) (A) and (C)
(3) (A), (B) and (C) (4) (B) and (D)

73. The major product in the following reaction is :

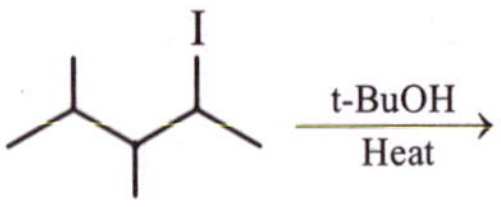

[Sep. 03, 2020 (II)]

(1) 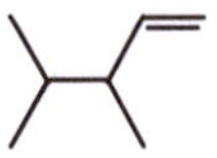
(2) 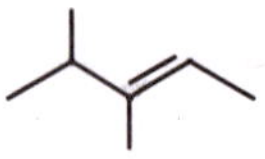
(3) 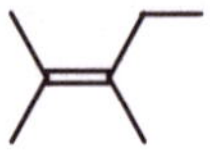
(4) 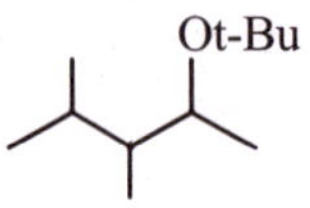

74. Which of the following compounds will show retention in configuration on nucleophilic substitution by OH^- ion ? [Sep. 02, 2020 (I)]

(1) $CH_3-\underset{C_6H_{13}}{\overset{Br}{C}}-H$ (2) $CH_3-\underset{C_6H_5}{CH}-Br$
(3) $CH_3-\underset{CH_3}{CH}-Br$ (4) $CH_3-\underset{C_2H_5}{CH}-CH_2Br$

75. The major product obtained from E2-elimination of 3-bromo-2-fluoropentane is : [Sep. 02, 2020 (II)]

(1) $CH_3CH_2-\overset{Br}{CH}-CH=CH_2$
(2) $CH_3CH_2CH=\underset{CH_3}{C}-F$
(3) $CH_3-CH=CH-\overset{F}{CH}-CH_3$
(4) $CH_3-CH_2-\overset{Br}{C}=CH-CH_3$

76. Consider the reaction sequence given below: [Sep. 02, 2020 (II)]

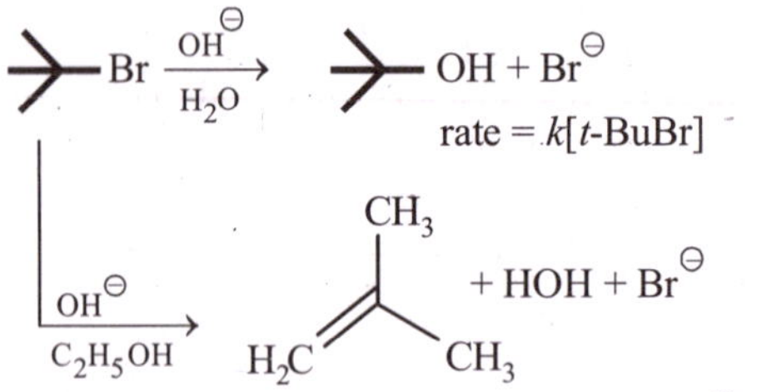

....(1) rate = k[t-BuBr]

....(2) rate = k[t-BuBr][$OH^{\ominus}$]

Which of the following statements is true ?

(1) Changing the base from $OH^{\ominus}$ to $^{\ominus}OR$ will have no effect on reaction (2).
(2) Changing the concentration of base will have no effect on reaction (1).
(3) Doubling the concentration of base will double the rate of both the reactions.
(4) Changing the concentration of base will have no effect on reaction (2).

77. The decreasing order of reactivity towards dehydrohalogenation (E_1) reaction of the following compounds is: [Jan. 08, 2020 (I)]

(A) 
(B) 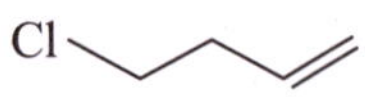
(C) 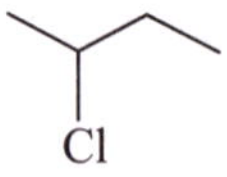
(D) 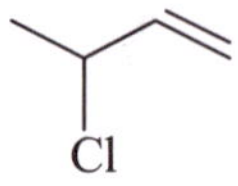

(1) D>B>C>A (2) B>D>A>C
(3) B>D>C>A (4) B>A>D>C

78. Consider the following reactions:

(A) $(CH_3)_3CCH(OH)CH_3 \xrightarrow{\text{conc.}H_2SO_4}$
(B) $(CH_3)_2CHCH(Br)CH_3 \xrightarrow{\text{alc.KOH}}$
(C) $(CH_3)_2CHCH(Br)CH_3 \xrightarrow{(CH_3)_3O^-K^{\oplus}}$
(D) $(CH_3)_2-\underset{OH}{C}-CH_2-CHO \xrightarrow{\Delta}$

Which of these reaction(s) will not produce Saytzeff product? [Jan. 07, 2020 (I)]

(1) (A), (C) and (D) (2) (D) only
(3) (C) only (4) (B) and (D)

79. In the following reaction sequence, structures of A and B, respectively will be: [Jan. 07, 2020 (II)]

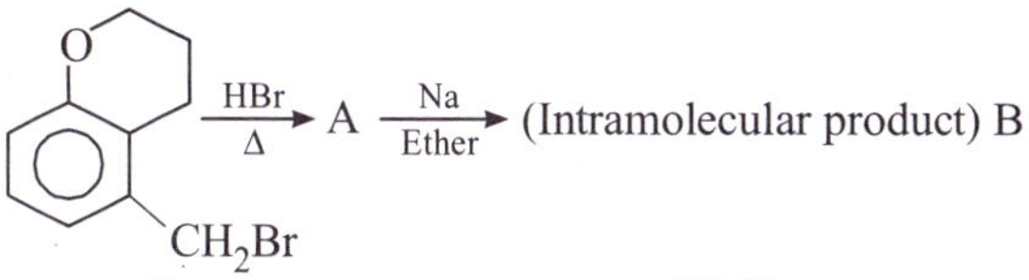

(1) 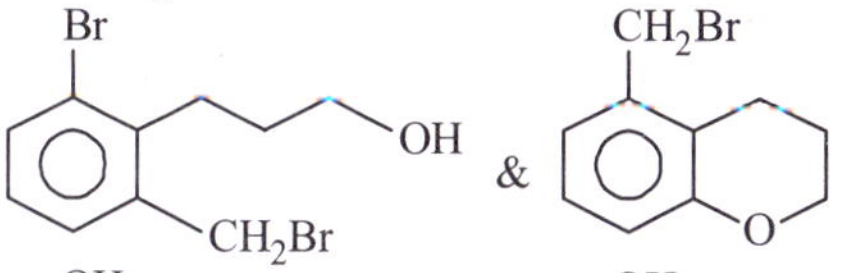

(2) 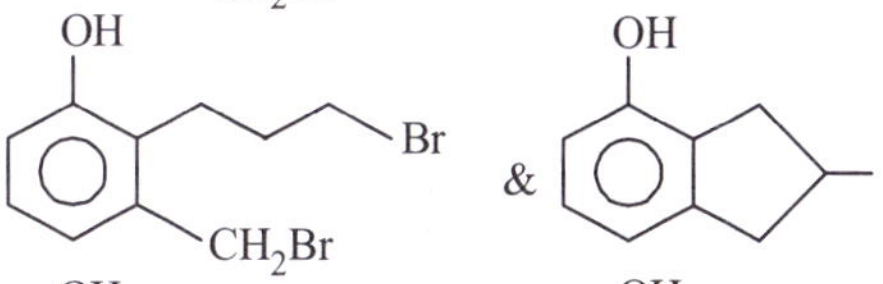

(3) 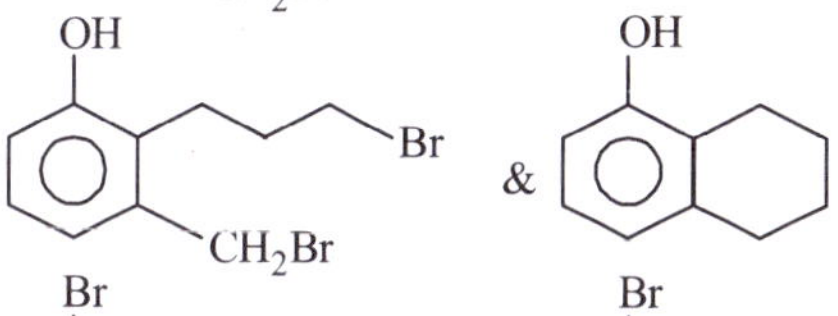

(4) 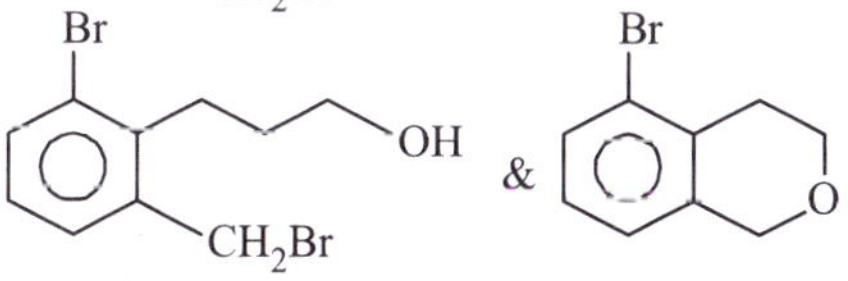

80. An 'Assertion' and a 'Reason' are given below. Choose the correct answer from the following options :

Assertion (A) : Vinyl halides do not undergo nucleophilic substitution easily.

Reason (R) : Even though the intermediate carbocation is stabilized by loosely held **p**-electrons, the cleavage is difficult because of strong bonding. [April 12, 2019 (II)]

(1) Both (A) and (R) are wrong statements.
(2) Both (A) and (R) are correct statements and (R) is the correct explanation of (A).
(3) Both (A) and (R) are correct statements but (R) is not the correct explanation of (A).
(4) (A) is a correct statement but (R) is a wrong statement.

81. Increasing rate of S_N1 reaction in the following compounds is : [April 10, 2019 (I)]

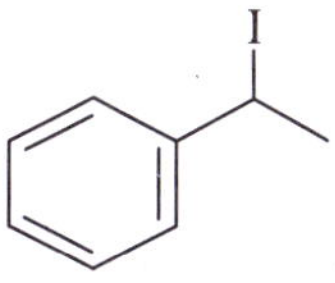

(A)

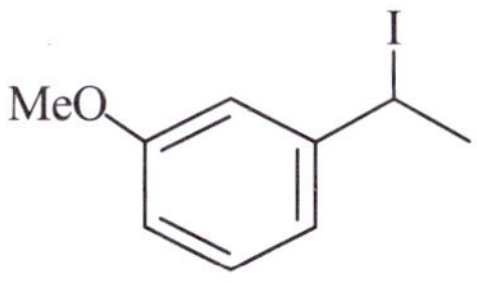

(B)

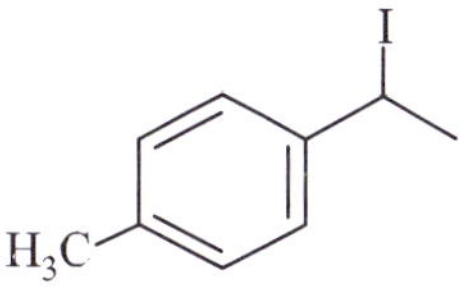

(C)

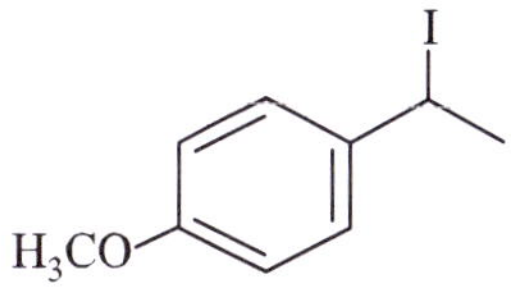

(D)

(1) (A) < (B) < (C) < (D) (2) (B) < (A) < (C) < (D)
(3) (B) < (A) < (D) < (B) (4) (A) < (B) < (D) < (C)

82. The major product of the following reaction is [April 10, 2019 (I)]

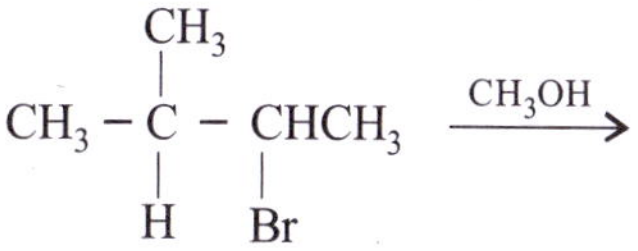

(1) 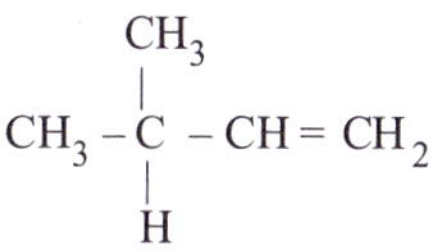
(2) 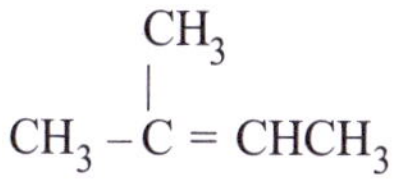
(3) 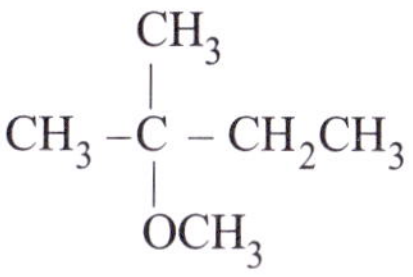
(4) 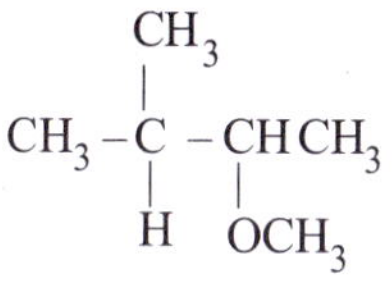

83. The major product 'Y' in the following reaction is : [April 10, 2019 (II)]

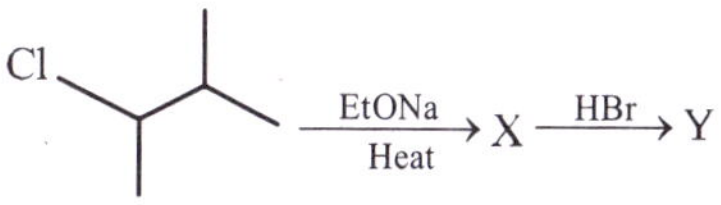

(1) 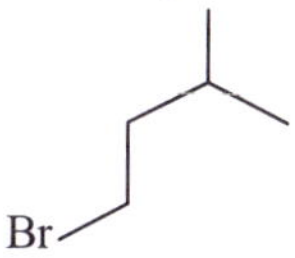
(2) 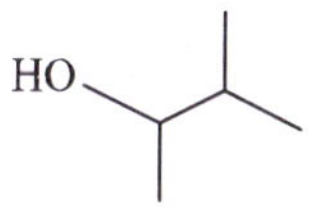
(3) 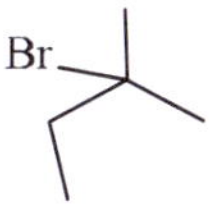
(4) 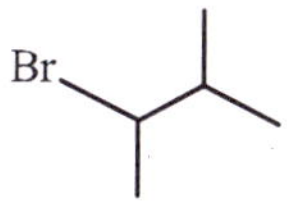

84. Increasing order of reactivity of the following compounds for S_N1 substitution is: [April 9, 2019 (II)]

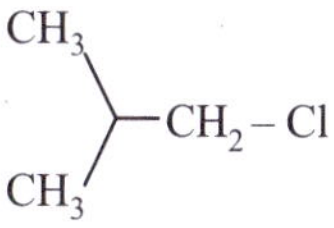

(A)

(B)

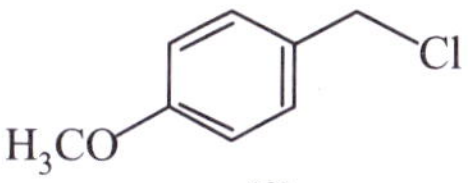

(C)

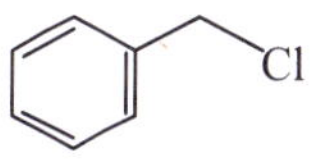

(D)

(1) (B) < (C) < (D) < (A) (2) (B) < (C) < (A) < (D)
(3) (B) < (A) < (D) < (C) (4) (A) < (B) < (D) < (C)

85. Which of the following potential energy (PE) diagrams represents the S_N1 reaction? [April 9, 2019 (II)]

(1) 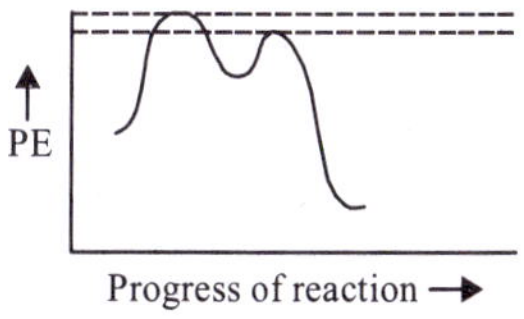

(2) 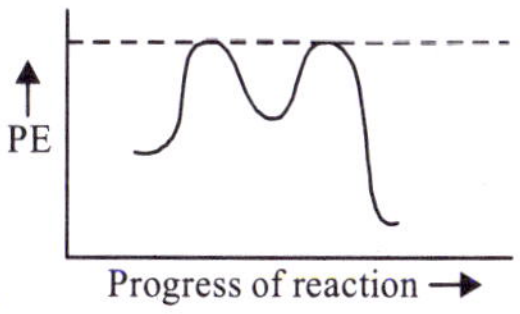

(3) 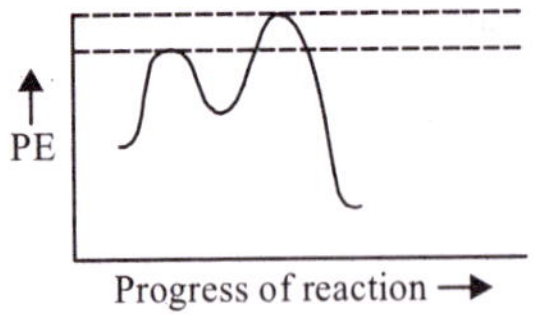

(4) 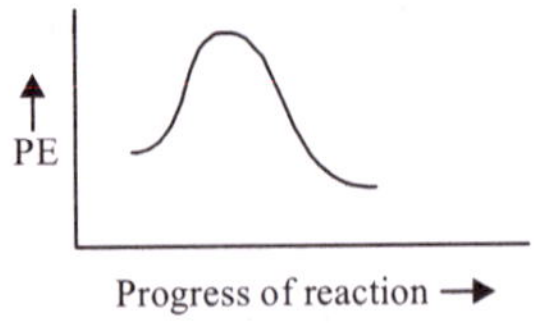

86. The major product of the following reaction is :

$CH_3CH_2\underset{Br}{CH}-\underset{Br}{CH_2} \xrightarrow[\text{(ii) } NaNH_2 \text{ in liq. } NH_3]{\text{(i) alc. KOH}}$ [Jan. 12, 2019 (II)]

(1) $CH_3CH=C=CH_2$
(2) $CH_3CH_2\underset{NH_2}{CH}-\underset{NH_2}{CH_2}$
(3) $CH_3CH=CHCH_2NH_2$
(4) $CH_3CH_2C\equiv CH$

87. The major product of the following reaction is: [Jan. 10, 2019 (I)]

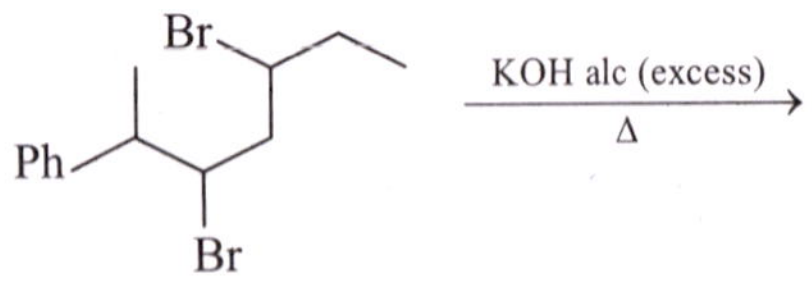

(1) 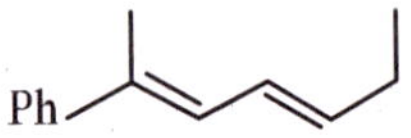 (2) 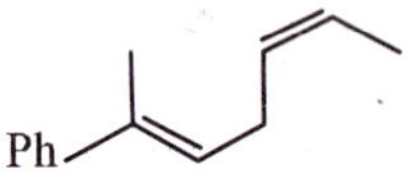

(3) 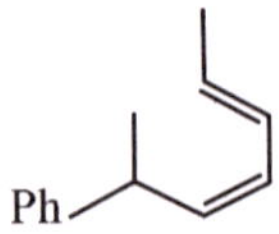 (4) 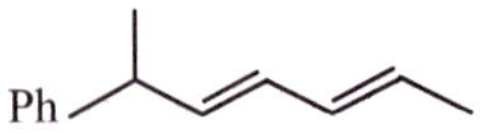

88. The major product of the following reaction is: [Jan. 10, 2019 (II)]

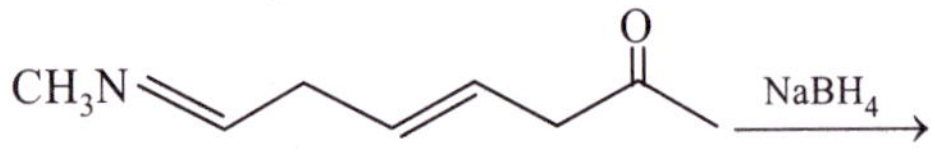

(1) 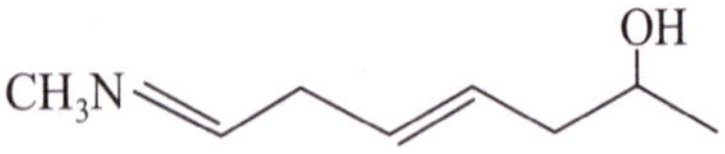

(2) 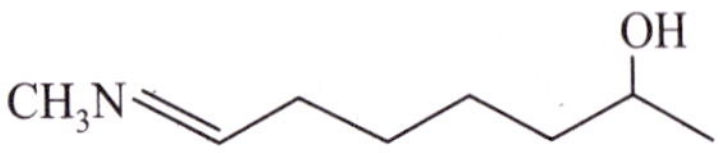

(3) 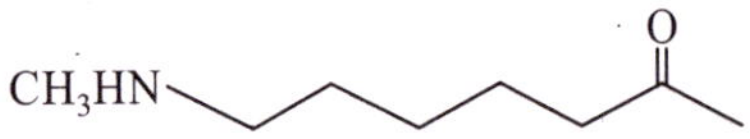

(4) 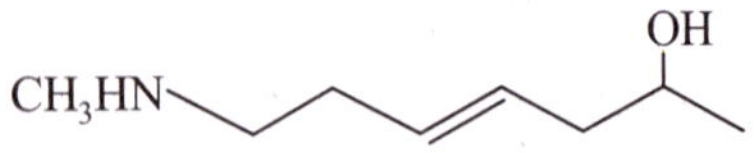

89. Which hydrogen in compound (E) is easily replaceable during bromination reaction in presence of light?

$\underset{\delta}{CH_3}-\underset{\gamma}{CH_2}-\underset{\beta}{CH}=\underset{\alpha}{CH_2}$ [Jan. 10, 2019 (I)]

(1) α - hydrogen (2) γ - hydrogen
(3) δ - hydrogen (4) β - hydrogen

90. The major product of the following reaction is: [Jan. 9, 2019 (I)]

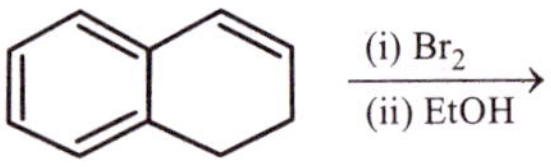

(1) 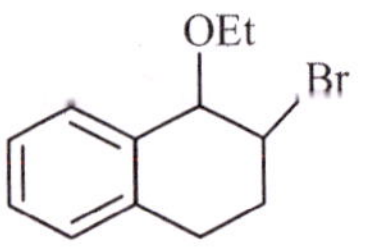 (2) 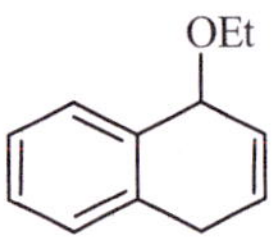

(3) 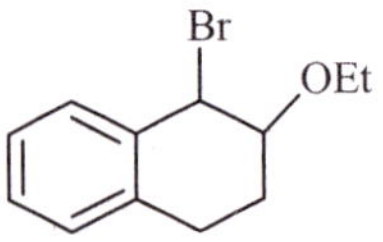 (4) 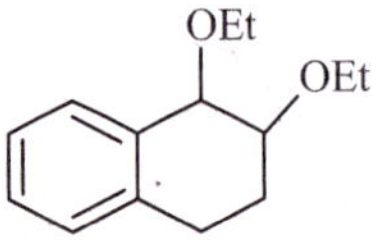

91. Which of the following will most readily give the dehydrohalogenation product? [Online April 15, 2018 (I)]

(1) 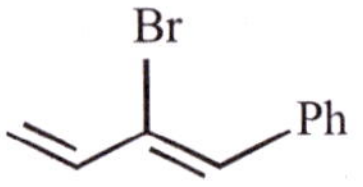 (2) 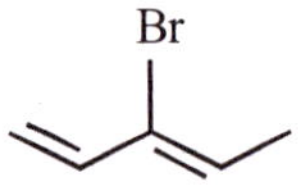

(3) 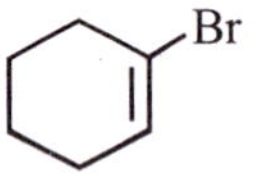 (4) 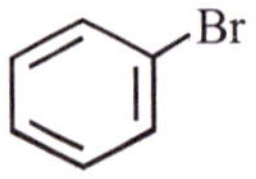

92. The major product of the following reaction is : [2018]

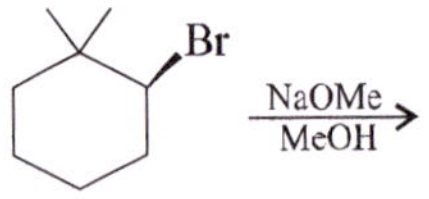

(1) 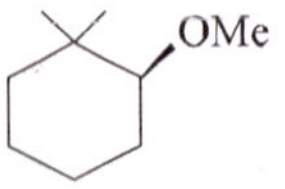 (2) 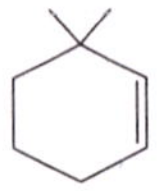

(3) 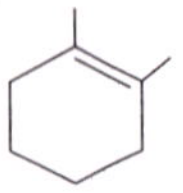 (4) 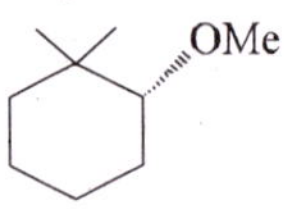

93. The major product of the following reaction is: [Online April 16, 2018]

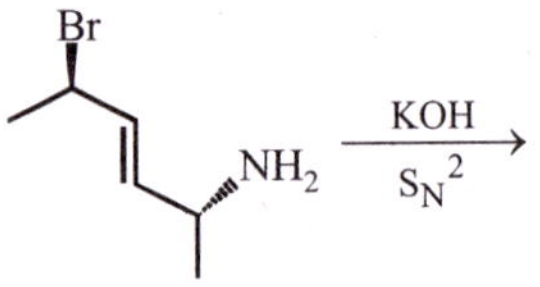

(1) 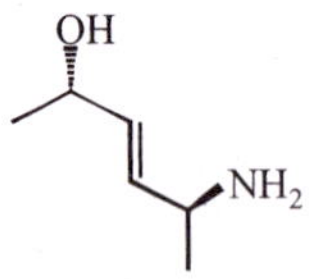 (2) 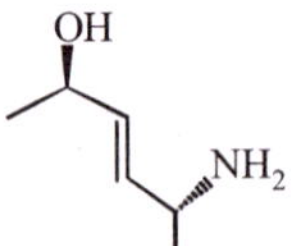

(3) 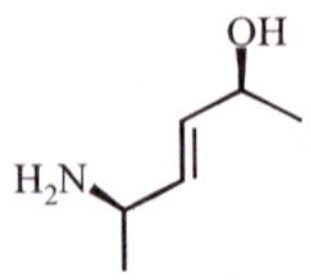 (4) 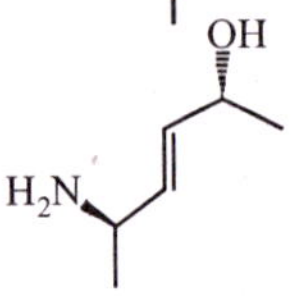

94. The major product of the following reaction is : [Online April 8, 2017]

$CH_3\underset{Br}{CH}CH_2\underset{Br}{CH}CH_2CH_3 \xrightarrow[\text{heat}]{KOH, CH_3OH}$

(1) $CH_2=CHCH_2CH=CHCH_3$
(2) $CH_2=CHCH=CHCH_2CH_3$
(3) $CH_3CH=C=CHCH_2CH_3$
(4) $CH_3CH=CH-CH=CHCH_3$

95. The increasing order of the reactivity of the following halides for the S_N1 reaction is [2017]

$CH_3CHCH_2CH_3$ (with Cl on C-2) (I) $\quad$ $CH_3CH_2CH_2Cl$ (II)

p-H_3CO–C_6H_4–CH_2Cl (III)

(1) (III) < (II) < (I) $\quad$ (2) (II) < (I) < (III)

(3) (I) < (III) < (II) $\quad$ (4) (II) < (III) < (I)

96. The major product of the following reaction is : [Online April 8, 2017]

$$C_6H_5CH_2-\underset{Br}{\overset{CH_3}{C}}-CH_2-CH_3 \xrightarrow[C_2H_5OH]{C_2H_5ONa}$$

(1) $C_6H_5CH_2-\underset{OC_2H_5}{\overset{CH_3}{C}}-CH_2-CH_3$

(2) $C_6H_5CH=\underset{CH_3}{C}-CH_2-CH_3$

(3) $C_6H_5CH_2-\underset{CH_3}{C}=CHCH_3$

(4) $C_6H_5CH_2-\underset{CH_2CH_3}{C}=CH_2$

97. In the following reaction sequence : [Online April 9, 2017]

$$\underset{(C_3H_6Cl_2)}{I} \xrightarrow{KOH(aq)} II \xrightarrow[(ii)\ H_2O/H^+]{(i)\ CH_3MgBr} III \xrightarrow{Anhy.ZnCl_2+Conc.HCl} \text{gives turbidity immediately}$$

The compound I is :

(1) $\underset{Cl}{CH_2}-\underset{Cl}{CH}-CH_3$

(2) $\underset{Cl}{CH_2}-CH_2-\underset{Cl}{CH_3}$

(3) $CH-\underset{Cl}{CH}-CH_2-CH_3$

(4) $CH_3-\underset{Cl}{\overset{Cl}{C}}-CH_3$

98. Which one of the following reagents is not suitable for the elimination reaction ? [Online April 10, 2016]

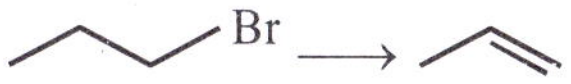

(1) NaI $\quad$ (2) NaOEt/ EtOH

(3) $NaOH/H_2O$ $\quad$ (4) $NaOH/H_2O-EtOH$

99. The synthesis of alkyl fluorides is best accomplished by : [2015]

(1) Finkelstein reaction

(2) Swarts reaction

(3) Free radical fluorination

(4) Sandmeyer's reaction

100. A compound A with molecular formula $C_{10}H_{13}Cl$ gives a white precipitate on adding silver nitrate solution. A on reacting with alcoholic KOH gives compound B as the main product. B on ozonolysis gives C and D. C gives Cannizzaro reaction but not aldol condensation. D gives aldol condensation but not Cannizzaro reaction. A is : [Online April 10, 2015]

(1) $C_6H_5-CH_2-CH_2-CH_2-CH_2-Cl$

(2) $C_6H_5-CH_2-CH_2-\underset{Cl}{CH}-CH_3$

(3) $C_6H_5-CH_2-\underset{Cl}{C}(CH_3)_2$

(4) 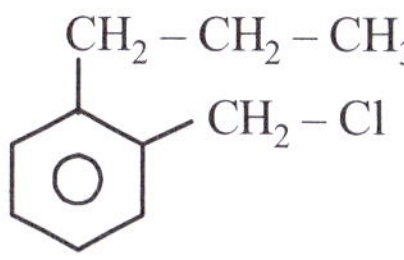

101. For the compounds

CH_3Cl, CH_3Br, CH_3I and CH_3F,

the correct order of increasing C-halogen bond length is: [Online April 9, 2014]

(1) $CH_3F < CH_3Cl < CH_3Br < CH_3I$

(2) $CH_3F < CH_3Br < CH_3Cl < CH_3I$

(3) $CH_3F < CH_3I < CH_3Br < CH_3Cl$

(4) $CH_3Cl < CH_3Br < CH_3F < CH_3I$

102. In S_N2 reactions, the correct order of reactivity for the following compounds: [2014, 2007 (s)]

CH_3Cl, CH_3CH_2Cl, $(CH_3)_2CHCl$ and $(CH_3)_3CCl$ is:

(1) $CH_3Cl > (CH_3)_2CHCl > CH_3CH_2Cl > (CH_3)_3CCl$

(2) $CH_3Cl > CH_3CH_2Cl > (CH_3)_2CHCl > (CH_3)_3CCl$

(3) $CH_3CH_2Cl > CH_3Cl > (CH_3)_2CHCl > (CH_3)_3CCl$

(4) $(CH_3)_2CHCl > CH_3CH_2Cl > CH_3Cl > (CH_3)_3CCl$

103. The order of reactivity of the given haloalkanes towards nucleophile is : [Online April 23, 2013]

(1) RI > RBr > RCl $\quad$ (2) RCl > RBr > RI

(3) RBr > RCl > RI $\quad$ (4) RBr > RI > RCl

104. Which of the following statements is wrong? [Online May 12, 2012]

(1) Ethyl chloride on reduction with Zn-Cu couple and alcohol gives ethane.

(2) The reaction of methyl magnesium bromide with acetone gives butanol-2.

(3) Alkyl halides follow the following reactivity sequence on reaction with alkenes.

R – I > R – Br > R – Cl > R – I

(4) $C_2H_4Cl_2$ may exist in two isomeric forms

105. How many chiral compounds are possible on monochlorination of 2- methylbutane ? [2012]

(1) 8 $\quad$ (2) 2 $\quad$ (3) 4 $\quad$ (4) 6

106. $C_2H_5Br \xrightarrow{AgCN} X \xrightarrow[Zn—Hg/HCl]{Reduction} Y$, Here Y is [Online May 7, 2012]

(1) Ethyl methyl amine $\quad$ (2) n-propylamine

(3) Isopropylamine $\quad$ (4) Ethylamine

107. Out of the following, the alkene that exhibits optical isomerism is **[2010]**
(1) 3-methyl-2-pentene (2) 4-methyl-1-pentene
(3) 3-methyl-1-pentene (4) 2-methyl-2-pentene

108. Consider the following bromides :

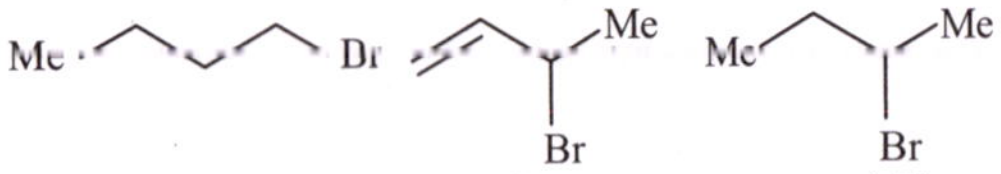

(A) (B) (C)

The correct order of S_N1 reactivity is
(1) B>C>A (2) B>A>C **[2010]**
(3) C>B>A (4) A>B>C

109. The number of stereoisomers possible for a compound of the molecular formula **[2009]**
$CH_3-CH=CH-CH(OH)-Me$ is:
(1) 2 (2) 4 (3) 6 (4) 3

110. The organic chloro compound, which shows complete stereochemical inversion during an S_N2 reaction, is **[2008]**
(1) $(C_2H_5)_2CHCl$ (2) $(CH_3)_3CCl$
(3) $(CH_3)_2CHCl$ (4) CH_3Cl

111. Which of the following molecules is expected to rotate the plane of plane-polarised light? **[2007]**

(1) 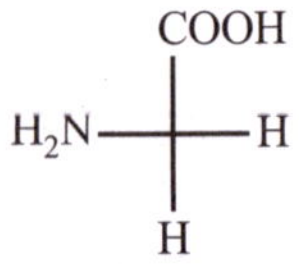 (2) 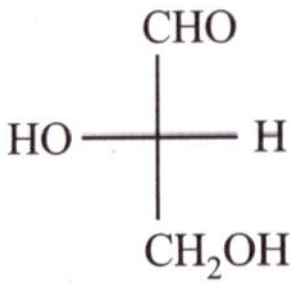

(3) 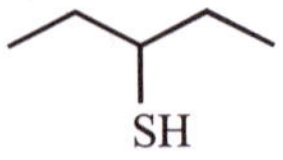 (4) 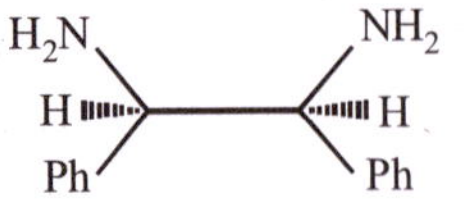

112. Tertiary alkyl halides are practically inert to substitution by S_N2 mechanism because of **[2005]**
(1) steric hindrance (2) inductive effect
(3) instability (4) insolubility

113. Alkyl halides react with dialkyl copper reagents to give
(1) alkenyl halides (2) alkanes **[2005]**
(3) alkyl copper halides (4) alkenes

114. Elimination of bromine from 2-bromobutane results in the formation of– **[2005]**
(1) Predominantly 2-butyne
(2) Predominantly 1-butene
(3) Predominantly 2-butene
(4) Equimolar mixture of 1 and 2-butene

115. Which type of isomerism is shown by 2, 3-dichlorobutane?
(1) Structural (2) Geometric **[2005]**
(3) Optical (4) Diastereo

116. Which of the following will have a mesoisomer also?
(1) 2, 3- Dichloropentane **[2004]**
(2) 2, 3-Dichlorobutane
(3) 2-Chlorobutane
(4) 2-Hydroxypropanoic acid

117. The reaction: **[2002]**

$(CH_3)_3C-Br \xrightarrow{H_2O} (CH_3)_3-C-OH$

(1) elimination reaction (2) substitution reaction
(3) free radical reaction (4) displacement reaction.

Numerical Value Questions (NVQs)

118. Consider all the structural isomers with molecular formula $C_5H_{11}Br$ are separately treated with KOH (aq) to give respective substitution products, without any rearrangement. The number of products which can exhibit optical isomerism from these is ________. **[Jan. 23, 2026 (I)]**

119. 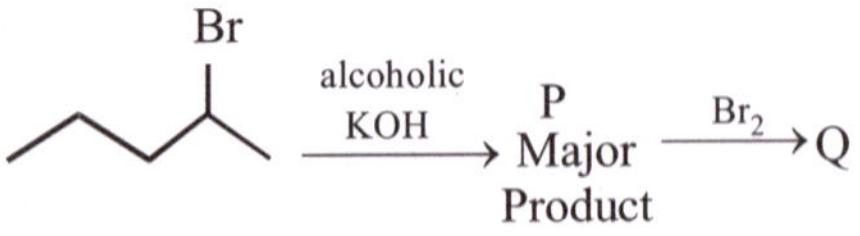

Consider the above sequence of reactions. 151 g of 2-bromopentane is made to react. Yield of major product P is 80% whereas Q is 100%.
Mass of product Q obtained is ________g.
(Given molar mass in g mol^{-1} H: 1, C : 12, O : 16, Br : 80) **[April 2, 2025 (II)]**

120. The number of optical isomers in following compound is : _____. **[April 8, 2024 (I)]**

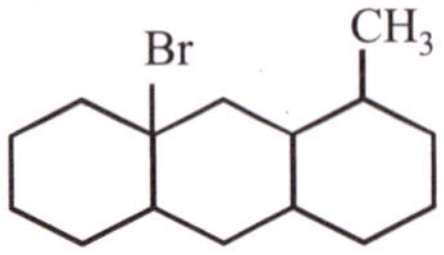

121. Total number of optically active compounds from the following is _______. **[April 8, 2024 (II)]**

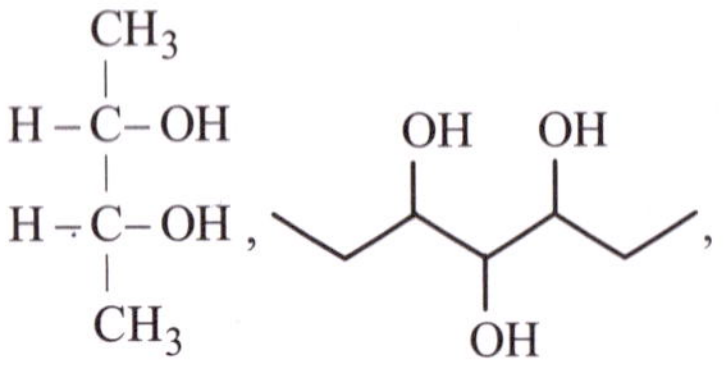

$CH_3-CH_2-CH_2-CH_2-OH$,
$CH_3-CH_2-CH(Cl)-CH_3$,
$CH_3-CH_2-CH_2-CH_2-Cl$,
$(CH_3)_2CH-CH_2-CH_2-Cl$

122. Number of optical isomers possible for 2 - chlorobutane _______ . **[Feb. 1, 2024 (I)]**

123. Total number of isomeric compounds (including stereoisomers) formed by monochlorination of 2- methylbutane is _____ . **[Feb. 1, 2024 (II)]**

124. $CH_3CH_2Br + NaOH \xrightarrow{C_2H_5OH}$ Product A
$CH_3CH_2Br + NaOH \xrightarrow{H_2O}$ Product B

The total number of hydrogen atoms in product A and product B is _____. **[Jan. 31, 2024 (I)]**

125. 2 - Chlorobutane $+ Cl_2 \rightarrow C_4H_8Cl_2$ (isomers)
Total number of optically active isomers shown by $C_4H_8Cl_2$, obtained in the above reaction is ________ . **[Jan. 30, 2024 (II)]**

126. 3-Methylhex-2-ene on reaction with HBr in presence of peroxide forms an addition product (A). The number of possible stereoisomers for 'A' is________. [Jan. 27, 2024 (I)]

127. Total number of compounds with Chiral carbon atoms from following is ________. [Jan. 27, 2024 (II)]

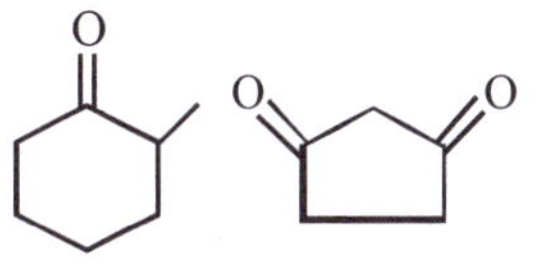

$CH_3 - CH_2 - CH(NO_2) - COOH$
$CH_3 - CH_2 - CHBr - CH_2 - CH_3$
$CH_3 - CH(I) - CH_2 - NO_2$
$CH_3 - CH_2 - CH(OH) - CH_2OH$
$CH_3 - CH - CH(I) - C_2H_5$
|
I

128. The total number of chiral compound/s from the following is _____. [Feb. 01, 2023 (I)]

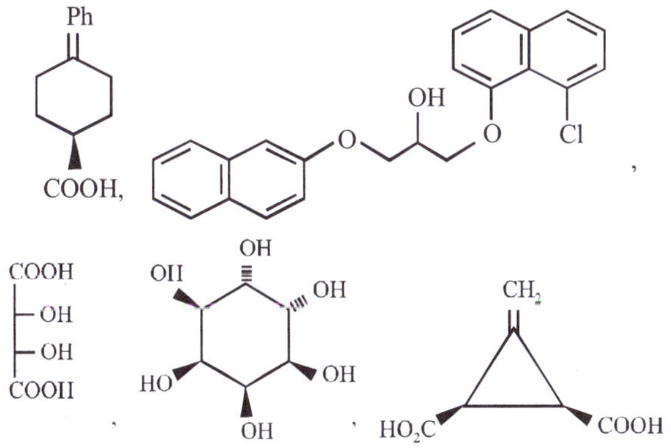

129. Testosterone, which is a steroidal hormone, has the following structure. [Feb. 01, 2023 (II)]

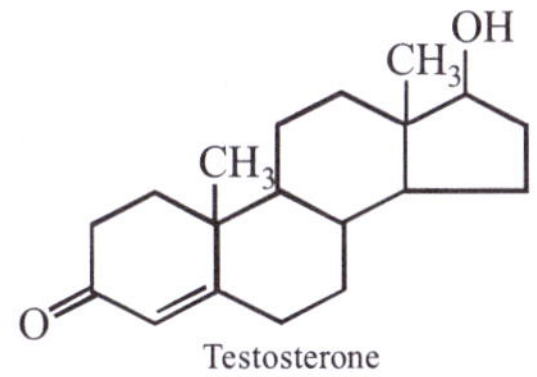

Testosterone

The total number of asymmetric carbon atom/s in testosterone is_______.

130. Optical activity of an enantiomeric mixture is +12.6° and the specific rotation of (+) isomer is +30°. The optical purity is___% [July 27, 2022 (I)]

131. Observe structures of the following compounds

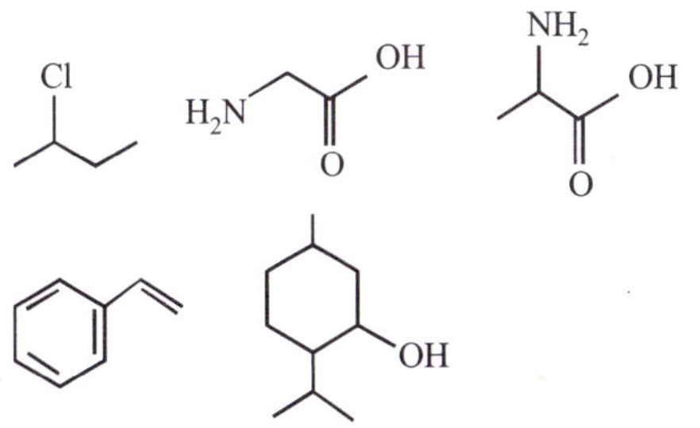

The total number of structures/compounds which possess asymmetric carbon atoms is __________. [June 29, 2022 (I)]

132. The major product of the following reaction contains _______ bromine atom(s). [June 28, 2022 (II)]

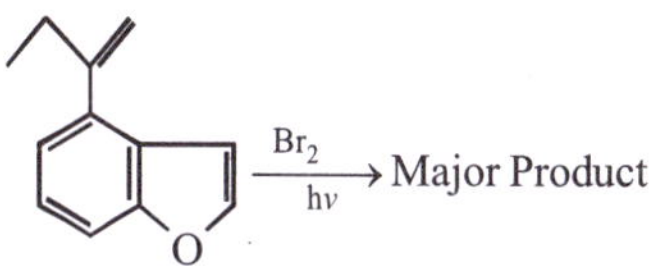

133.

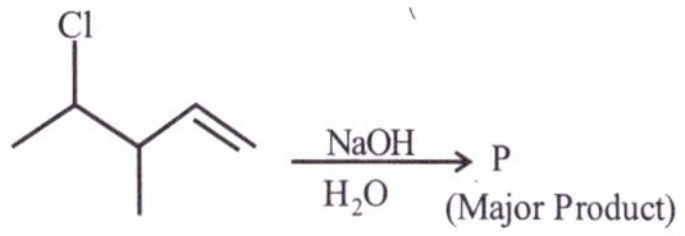

Consider the above reaction. The number of π electrons present in the product 'P' is____. [June 24, 2022 (II)]

134. A chloro compound "A".
(i) forms aldehydes on ozonolysis followed by the hydrolysis.
(ii) when vaporized completely 1.53 g of A, gives 448 mL of vapour at STP.
The number of carbon atoms in a molecule of compound A is _______. [Aug. 26, 2021 (II)]

135. The total number of monohalogenated organic products in the following (including stereoisomers) reaction is __________.

$$A \xrightarrow[\text{(ii) } X_2/\Delta]{\text{(i) } H_2/Ni/\Delta}$$

(Simplest optically active alkene) [Sep. 03, 2020 (I)]

Topic 2 - Preparation and Properties of Haloarenes

MULTIPLE CHOICE QUESTIONS (MCQs)

136. Given below are two statements : [April 08, 2026 (II)]
Statement I: Due to increase in van der Waals forces, the order of boiling points is $CH_3CH_2CH_2I > CH_3CH_2I > CH_3I$.
Statement II: As  is more symmetric, its melting point is higher than 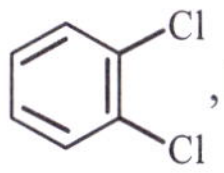, however its boiling point is lower than 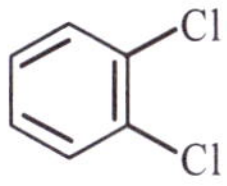
In the light of the above statements, choose the **correct** answer from the options given below :
(1) Both Statement-I and Statement-II are true.
(2) Both Statement-I and Statement-II are false.
(3) Statement-I is true but Statement-II is false
(4) Statement-I is false but Statement-II is true.

137. Given below are two statements : [April 06, 2026 (I)]
Statement-I: 3-phenylpropene reacts with HBr and gives secondary alkyl bromide having a chiral carbon atom as the major product.
Statement-II: Aryl chlorides and aryl cyanides can be prepared by Sandmeyer reaction as well as Gattermann reaction.
In the light of the above statements, choose the **correct** answer from the options given below.

(1) Both Statement-I and Statement-II are true.
(2) Both Statement-I and Statement-II are false.
(3) Statement-I is true but Statement-II is false.
(4) Statement-I is false but Statement-II is true.

138. Benzyl isocyanide can be obtained from

(A) 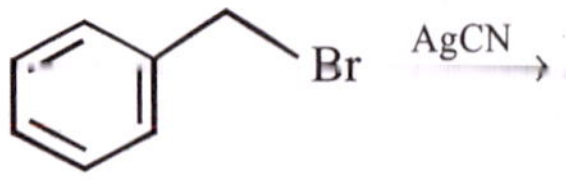

(B) 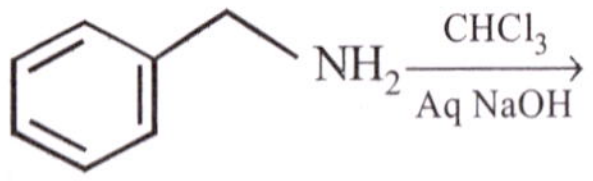

(C) 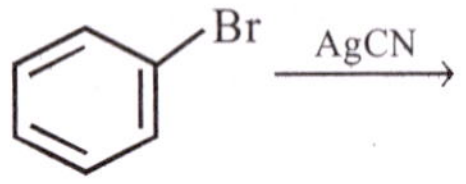

(D) 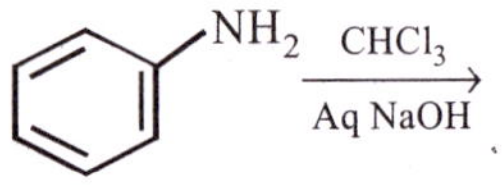

(E) 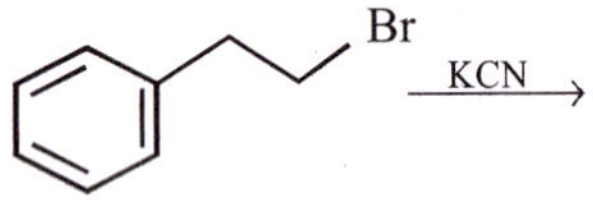

Choose the **correct** answer from the options given below: **[April 05, 2026 (I)]**

(1) A and B only (2) A and C only
(3) B and D only (4) D and E only

139. Match the **List-I** with **List-II** **[April 04, 2026 (I)]**

List-I Name of reaction		**List-II** Reagent or catalyst used	
A.	Finkelstein reaction	I.	SbF_3
B.	Swarts reaction	II.	Na, dry ether
C.	Sandmeyer's reaction	III.	NaI
D.	Fitting reaction	IV.	Cu_2Cl_2

Choose the **correct** answer from the options given below:
(1) A-I, B-IV, C-III, D-III (2) A-III, B-I, C-IV, D-II
(3) A-IV, B-II, C-I, D-III (4) A-I, B-III, C-II D-IV

140. 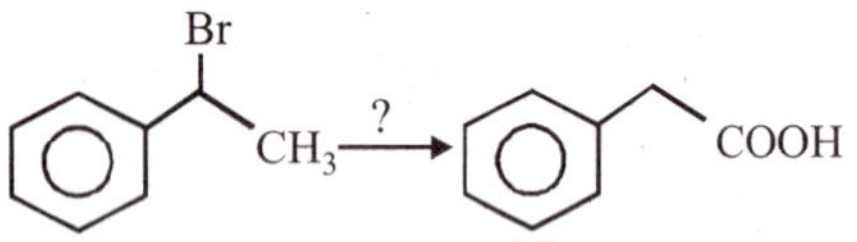

(X) (Y)

The correct sequence of reagent for the above conversion of X to Y is: **[Jan. 23, 2026 (I)]**
(1) (i) NaOH (aq) (ii) Jones reagent (iii) H_3O^+
(2) B_2H_6/H_2O_2 (ii) NaOEt (iii) Jones reagent
(3) (i) Jones reagent (ii) NaOEt (iii) Hot $KMnO_4$/KOH
(4) (i) NaOEt (ii) B_2H_6/H_2O_2 (iii) Jones reagent

141. As compared with chlorocyclohexane, which of the following statements correctly apply to chlorobenzene? **[Jan. 22, 2026 (I)]**
A. The magnitude of negative charge is more on chlorine atoms.
B. The C – Cl bond has partial double bond character.
C. C – Cl bond is less polar.
D. C – Cl bond is longer due to repulsion between delocalised electrons of the aromatic ring and lone pairs of electrons of chlorine.
E. The C–Cl bond is formed using sp^2 hybridised orbital of carbon.

Choose the correct answer from the options given below:
(1) A, C and E only (2) B, C and D only
(3) A, D and E only (4) B, C and E only

142. The correct order of reactivity of the following benzyl halides towards reaction with KCN is: **[Jan. 21, 2026 (II)]**

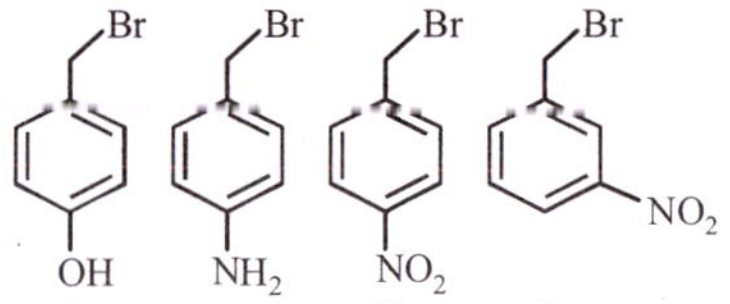

A B C D

(1) A > B > C > D (2) B > A > D > C
(3) B > A > C > D (4) A > B > D > C

143. Match **List-I** with **List-II** **[April 7, 2025 (II)]**

List-I (Conversion)	**List-II** (Reagents, Conditions used)
(A) 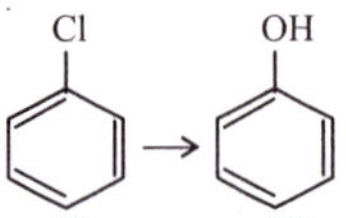	(I) Warm, H_2O
(B) 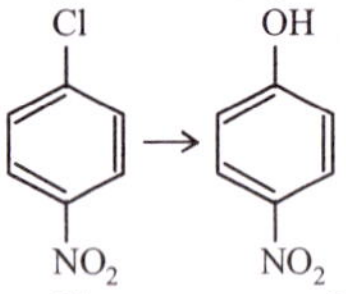	(II) (1) NaOH, 368 K; (2) H_3O^+
(C) 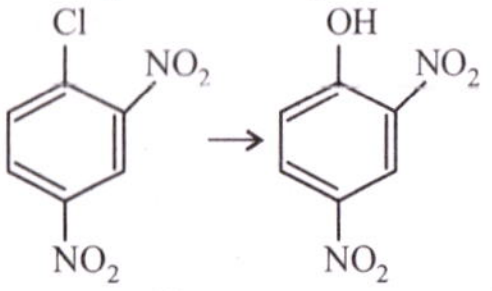	(III) (1) NaOH, 443 K; (2) H_3O^+
(D) 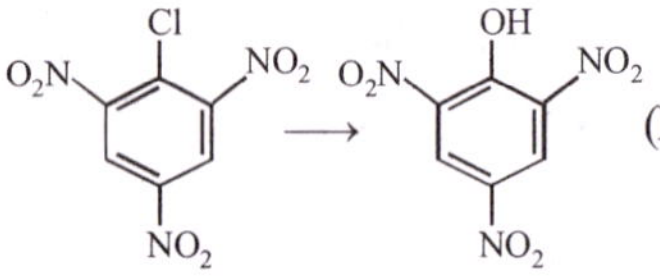	(IV) (1) NaOH, 623 K, 300 atm; (2) H_3O^+

Choose the **correct** answer from the options given below:
(1) (A)-(II), (B)-(III), (C)-(I), (D)-(IV)
(2) (A)-(III), (B)-(IV), (C)-(II), (D)-(I)
(3) (A)-(IV), (B)-(III), (C)-(II), (D)-(I)
(4) (A)-(IV), (B)-(III), (C)-(I), (D)-(II)

144. In the following series of reactions identify the major products A & B respectively. **[April 3, 2025 (II)]**

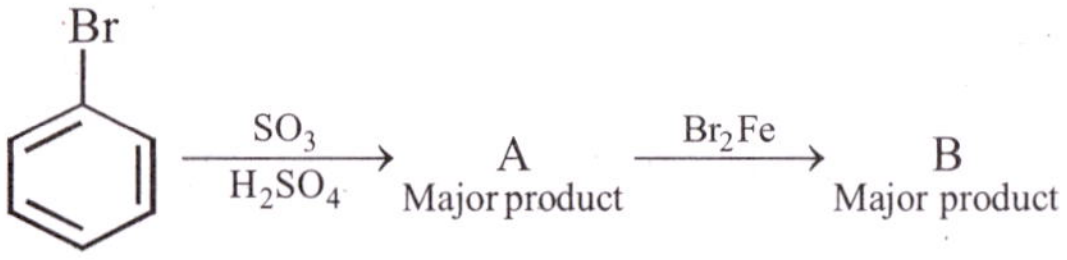

(1) 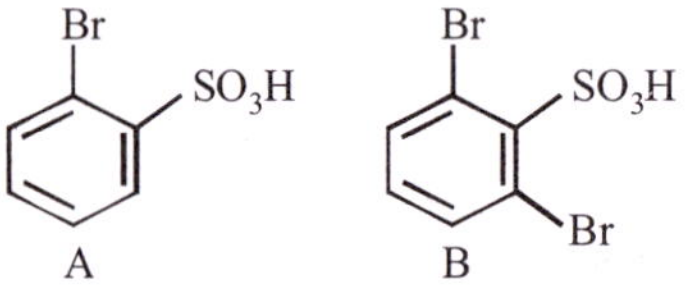

A B

(2) 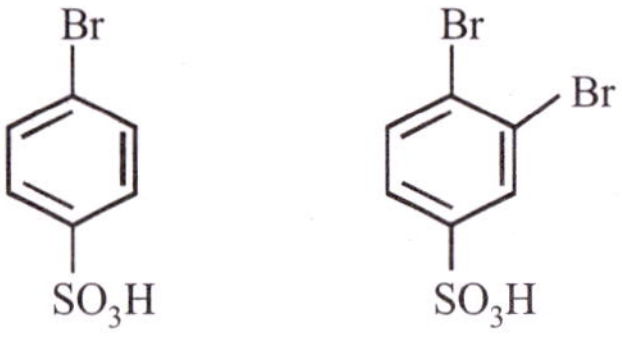

A B

(3) 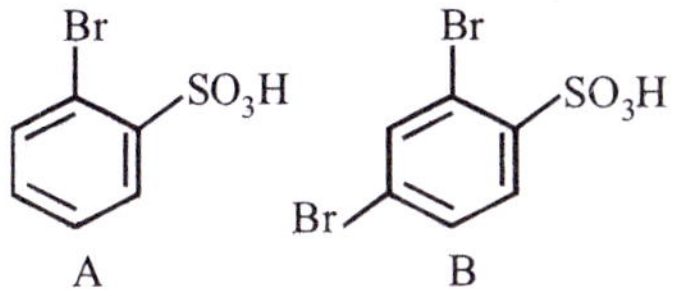

(4) 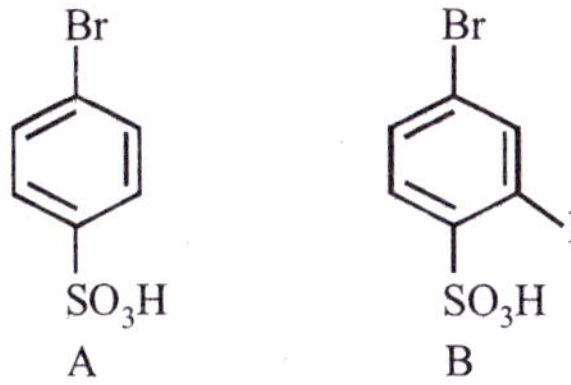

145. Match **List-I** with **List-II.** [April 2, 2025 (II)]

List-I (Reaction)	List-II (Name of reaction)
(A) 2 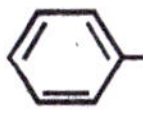 $-X - 2Na \xrightarrow[\text{Ether}]{\text{Dry}}$ 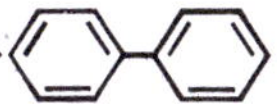 $+2Na$	(I) Lucas reaction
(B) $ArN_2^+X^- \xrightarrow[\text{HCl}]{\text{Cu}} ArCl + N_2\uparrow + CuX$	(II) Finkelstein reaction
(C) $C_2H_5Br + NaI \xrightarrow[\text{Acetone}]{\text{Dry}} C_2H_5I + NaBr$	(III) Fittig reaction
(D) $CH_3C(OH)(CH_3)CH_3 \xrightarrow[\text{ZnCl}_2]{\text{HCl}} CH_3C(Cl)(CH_3)CH_3$	(IV) Gatterman reaction

Choose the **correct** answer from the options given below:

(1) (A)-(III), (B)-(II), (C)-(IV), (D)-(I)
(2) (A)-(III), (B)-(IV), (C)-(II), (D)-(I)
(3) (A)-(IV), (B)-(III), (C)-(I), (D)-(II)
(4) (A)-(IV), (B)-(I), (C)-(II), (D)-(III)

146. In the following substitution reaction : [Jan. 29, 2025 (I)]

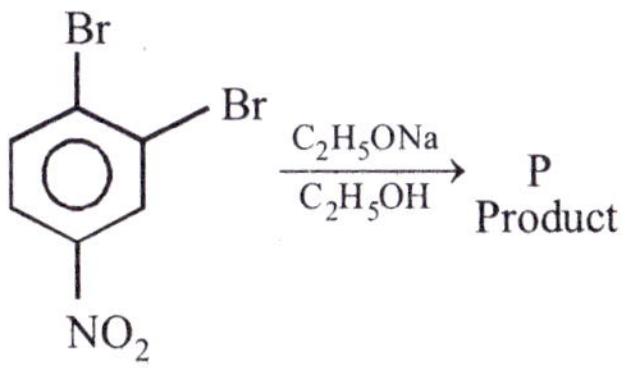

Product 'P' formed is

(1) 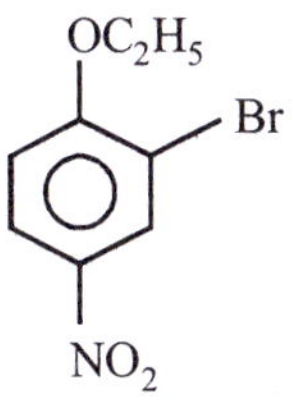 (2) 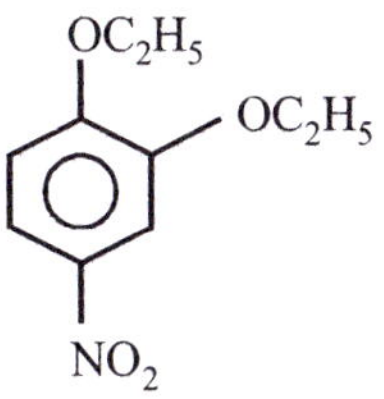

(3) 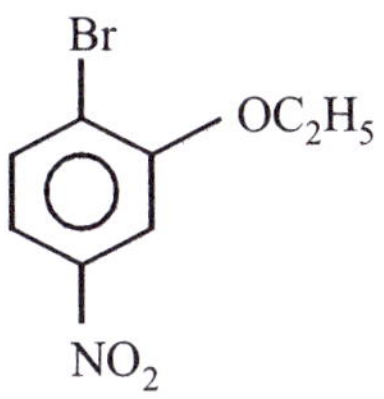 (4) 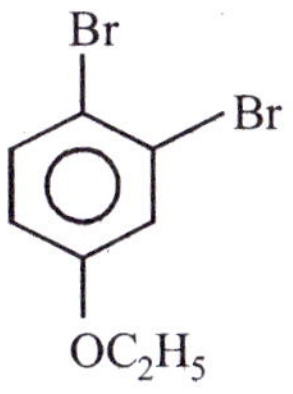

147. 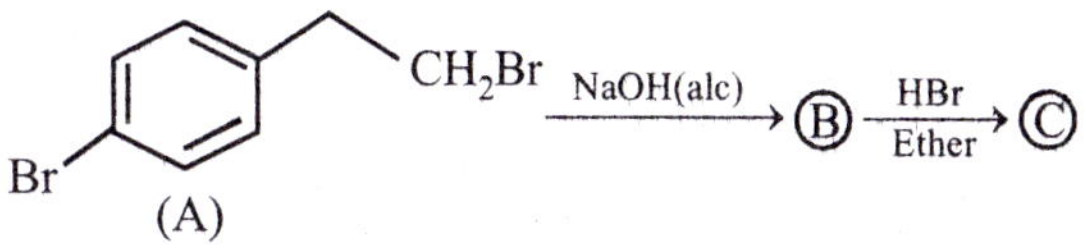

Identify (B) and (C) and how are (A) and (C) related?

(B) (C)

[April 4, 2024 (I)]

(1) 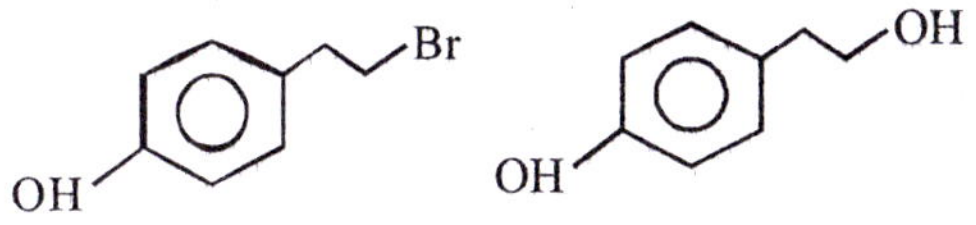

functional group isomers

(2) 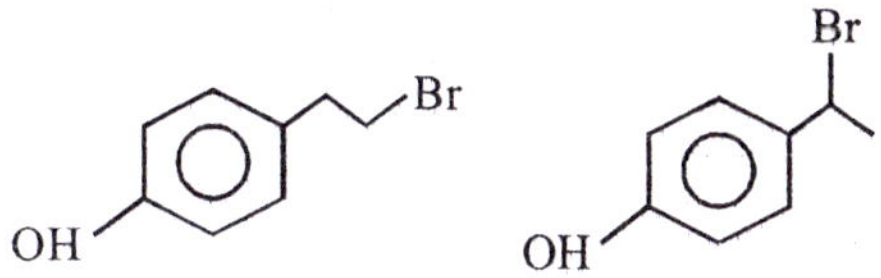

Derivative

(3) 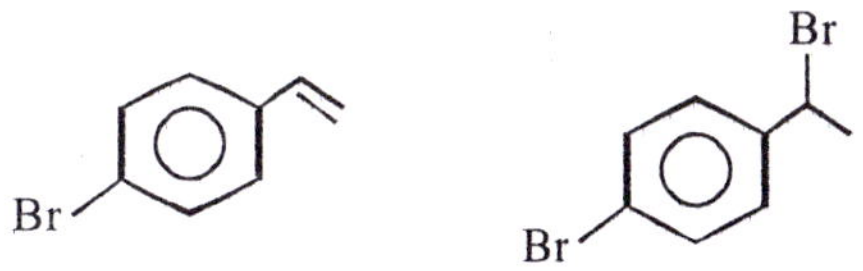

position isomers

(4) 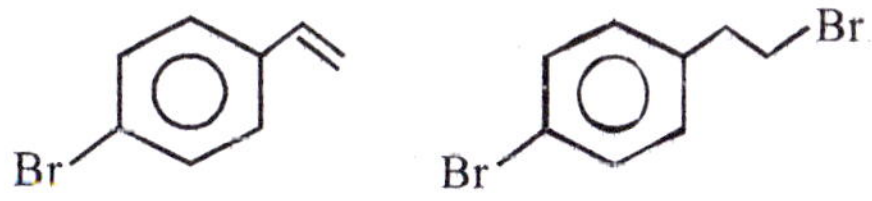

chain isomers

148. Identify A and B in the following reaction sequence. [Jan. 31, 2024 (II)]

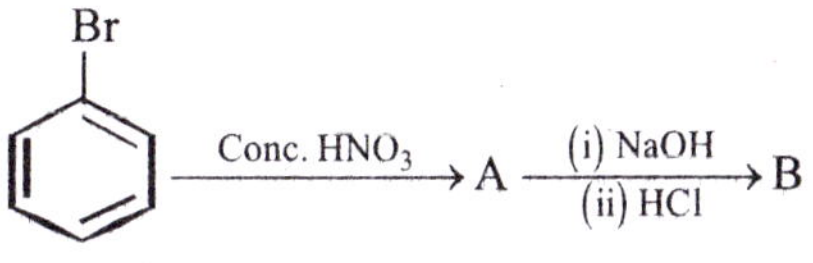

(1) A = 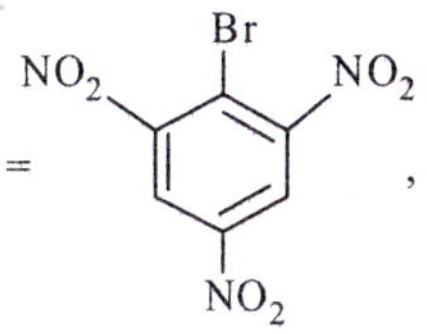 , B = 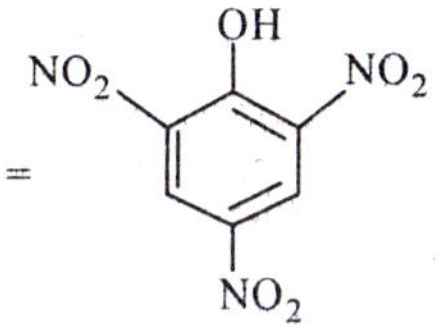

(2) A = 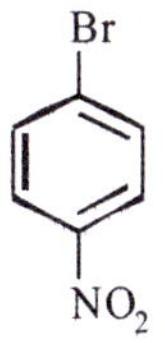 , B = 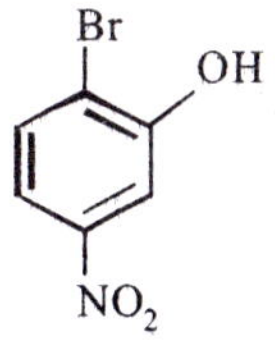

(3) A = 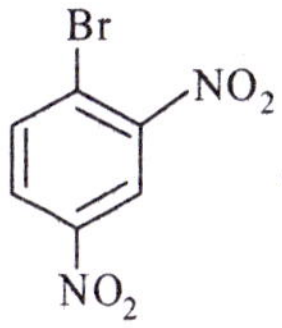 , B = 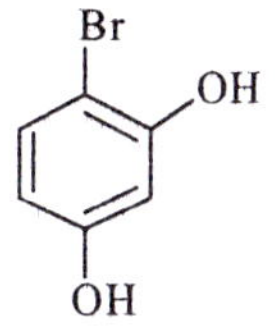

(4) A = 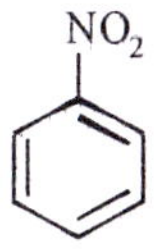 , B = 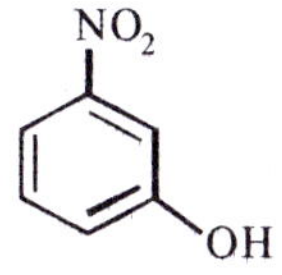

149. The major product formed in the Friedel-Craft acylation of chlorobenzene is: [April 15, 2023 (I)]

(1) 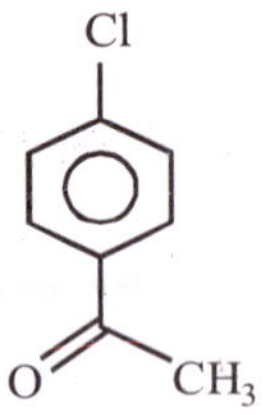
(2) 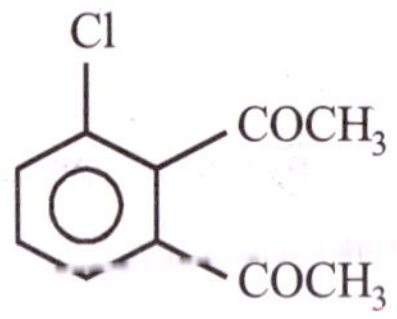
(3) 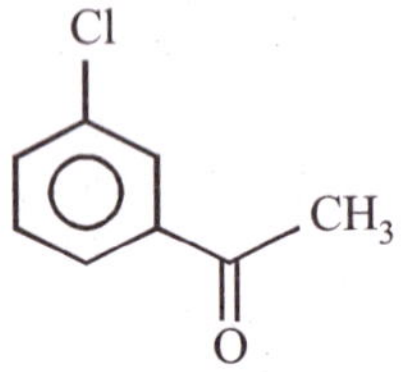
(4) 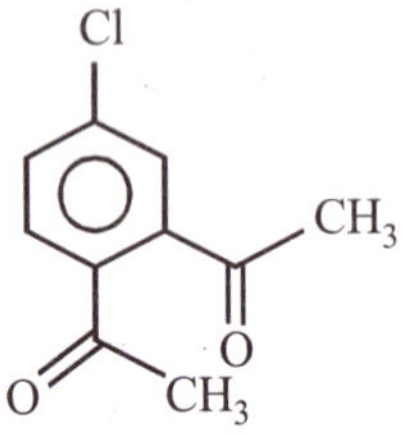

150. Identify the correct order of reactivity for the following pairs towards the respectively mechanism [April 10, 2023 (I)]

(A) S_N2 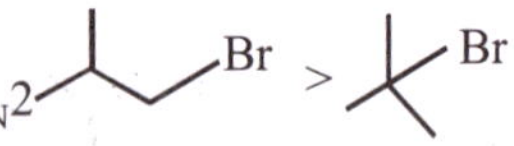

(B) S_N1 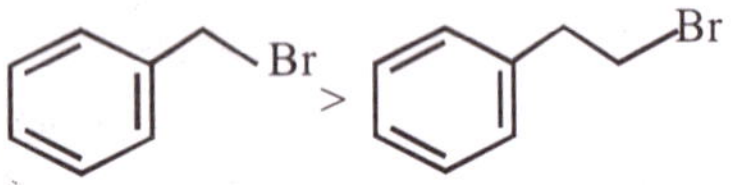

(C) Electrophilic substitution 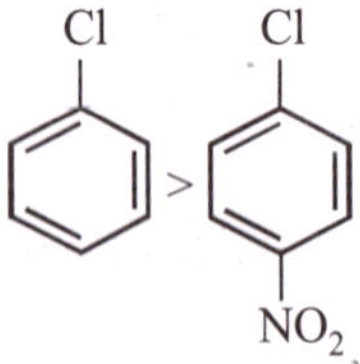

(D) Nucleophilic substitution 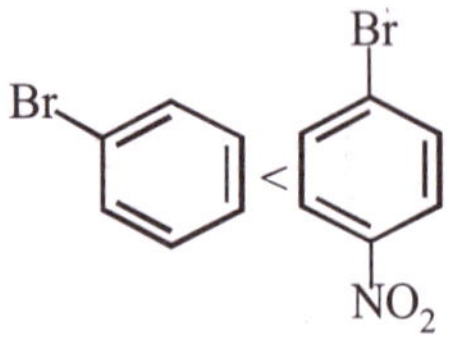

Choose the correct answer from the options given below:
(1) (A), (B) and (D) only (2) (A), (B), (C) and (D)
(3) (A), (C) and (D) only (4) (B), (C) and (D) only

151. The correct order of reactivity of following haloarenes towards nucleophilic substitution with aqueous NaOH is: [April 08, 2023 (II)]

A. 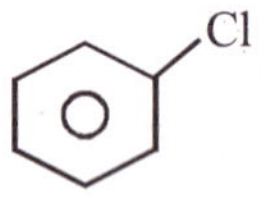
B. 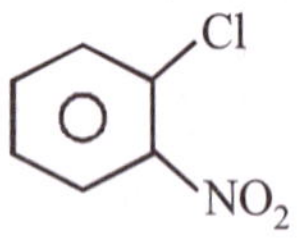
C. 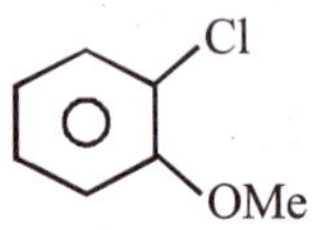
D. 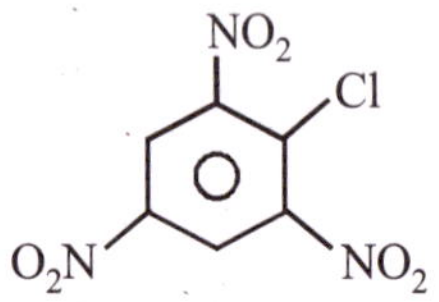

Choose the correct answer from the options given belowe:
(1) A>B>D>C (2) C>A>D>B
(3) D>C>B>A (4) D>B>A>C

152. Identify the incorrect option from the following: [Feb. 01, 2023 (I)]

(1) 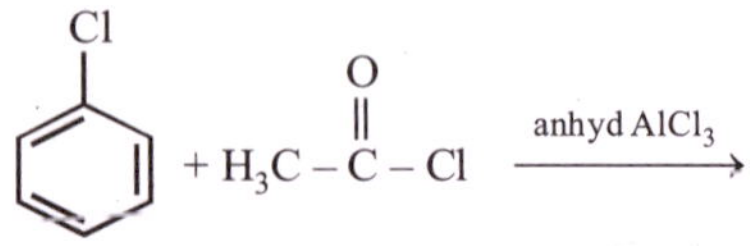
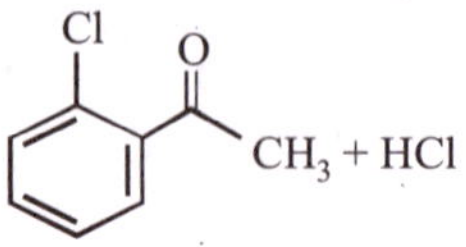

(2) 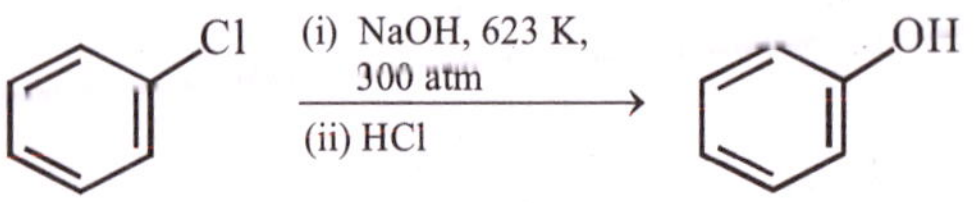
(3) 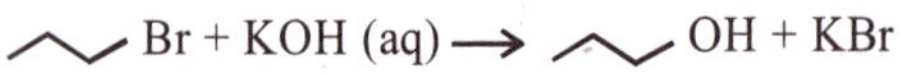
(4) 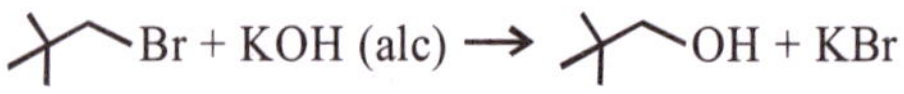

153. The correct order of melting point of dichlorobenzenes is [Jan. 31, 2023 (I)]

(1) 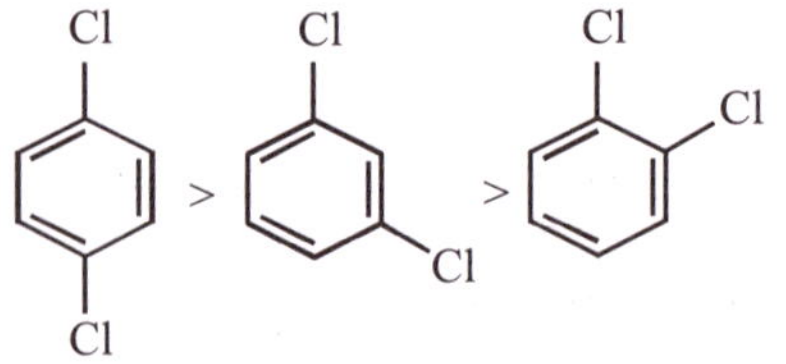
(2) 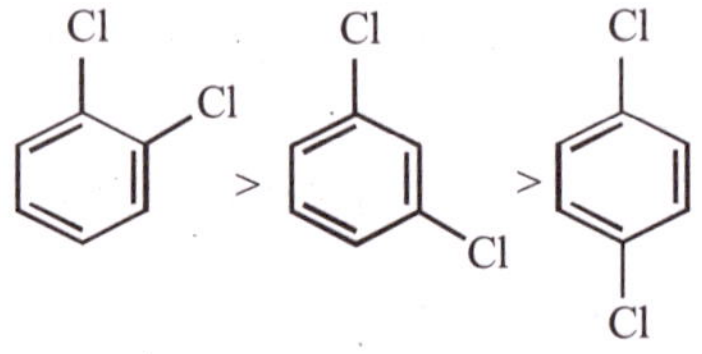
(3) 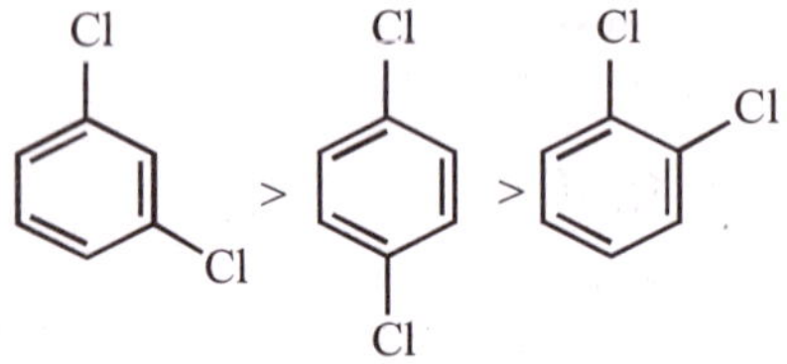
(4) 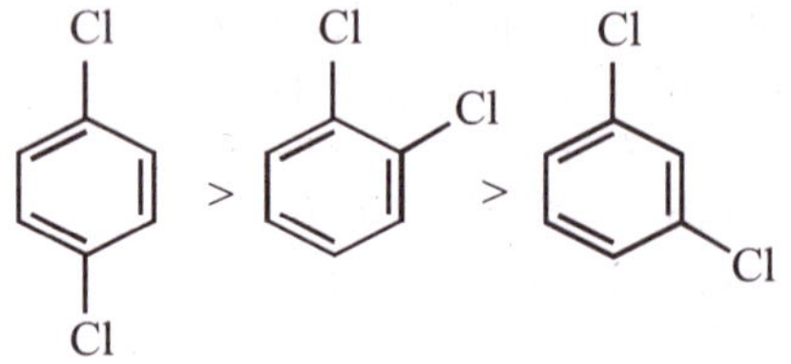

154. Match **List I** with **List II** [Jan. 30, 2023 (I)]

List I	List II
A. 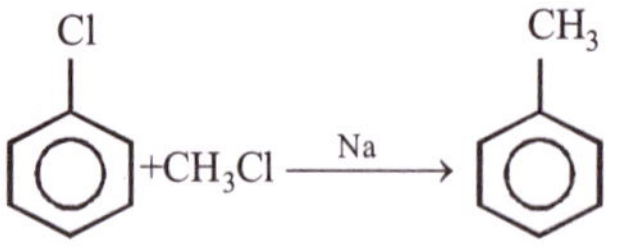	I. Fittig reaction
B. 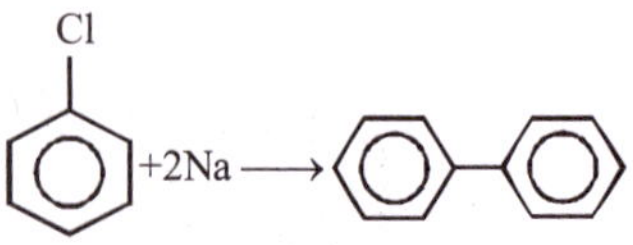	II. Wurtz Fittig reaction
C. 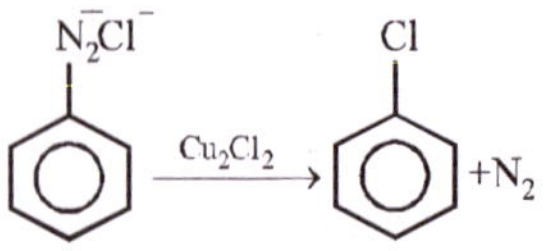	III. Finkelstein reaction
D. $C_2H_5Cl + NaI \rightarrow C_2H_5I + NaCl$	IV. Sandmeyer reaction

(1) A – II, B – I, C – III, D – IV
(2) A – III, B – II, C – IV, D – I

(3) A – IV, B – II, C – III, D – I
(4) A – II, B – I, C – IV, D – III

155. Decreasing order towards S_N1 reaction for the following compounds is: [Jan. 30, 2023 (II)]

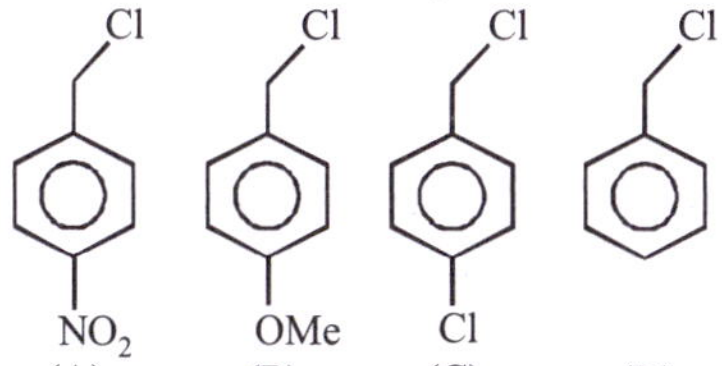

(A) (B) (C) (D)

(1) A > C > D > B (2) A > B > C > D
(3) B > D > C > A (4) D > B > C > A

156. The compound which will have the lowest rate towards nucleophilic aromatic substitution on treatment with OH^- is [Jan. 25, 2023 (I)]

(1) 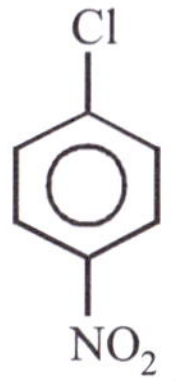

(2) 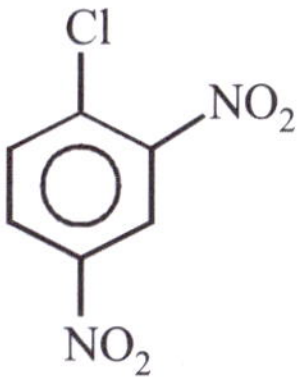

(3) 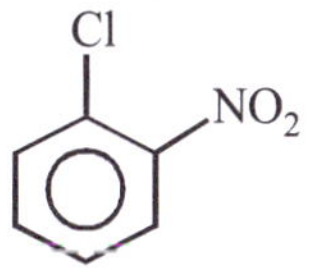

(4) 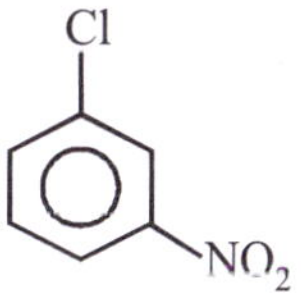

157. Given below are two statements, one is labelled as Assertion A and the other is labelled as Reason R.

Assertion A: Benzene is more stable than hypothetical cyclohexatriene.

Reason R: The delocalized π electron cloud is attracted more strongly by nuclei of carbon atoms.

In the light of the above statements, choose the correct answer from the options given below: [Jan. 24, 2023 (II)]

(1) A is true but R is false.
(2) A is false but R is true.
(3) Both A and R are correct and R is the correct explanation of A.
(4) Both A and R are correct but R is NOT the correct explanation of A.

158. Which among the following pairs of the structures will give different products on ozonolysis?

(Consider the double bonds in the structures are rigid and not delocalized.) [July 29, 2022 (I)]

(1) 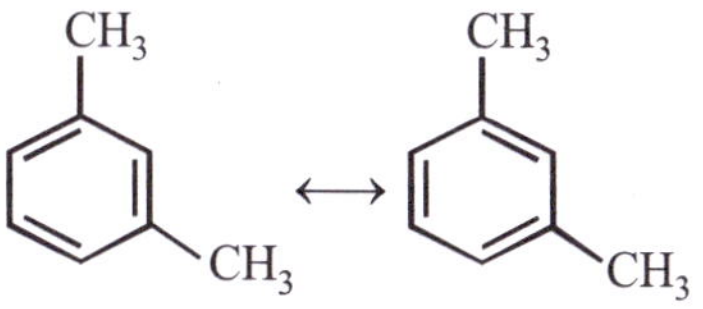

(2) 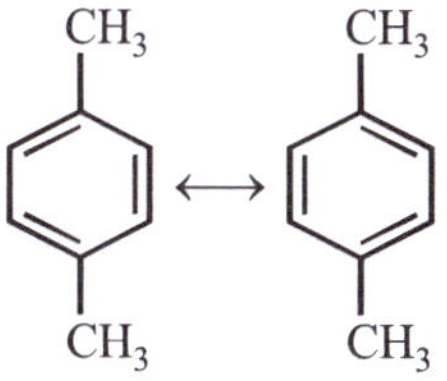

(3) 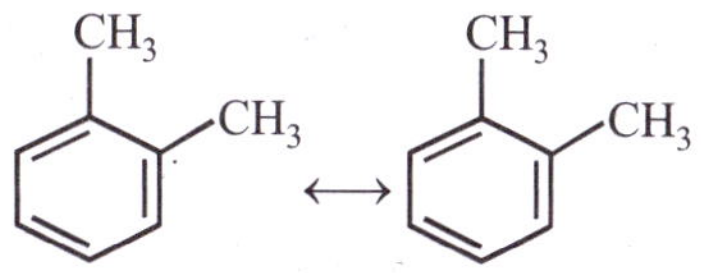

(4) 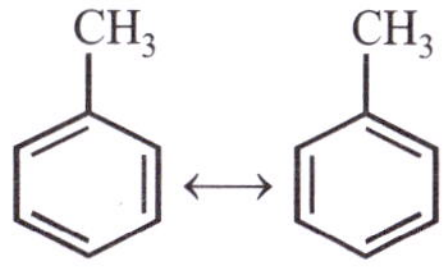

159. Arrange the following in increasing order of reactivity towards nitration [July 28, 2022 (II)]

(A) p-xylene (B) bromobenzene
(C) mesitylene (D) nitrobenzene
(E) benzene

Choose the correct answer from the options given below:

(1) C < D < E < A < B (2) D < B < E < A < C
(3) D < C < E < A < B (4) C < D < E < B < A

160. Product 'A' of following sequence of reactions is Ethylbenzene

$\xrightarrow[\text{(b)} Cl_2, \Delta \text{ (c) alc. KOH}]{\text{(a)} Br_2.Fe}$ 'A' (Major product) [June 27, 2022 (II)]

(1) 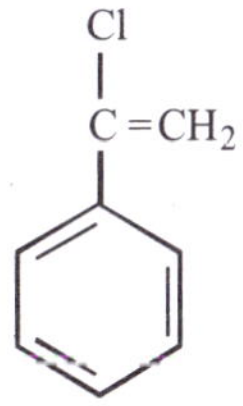

(2) 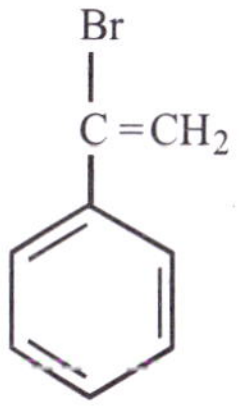

(3) 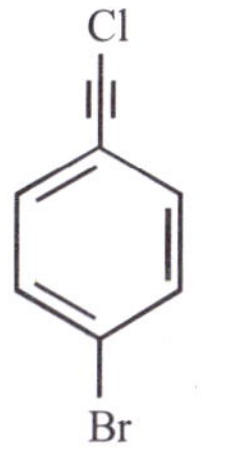

(4) 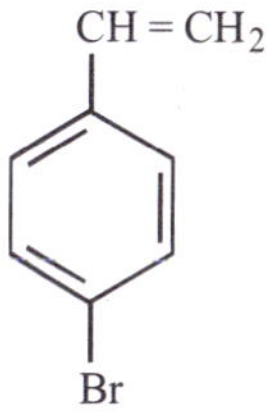

161. Halogenation of which one of the following will yield m-substituted product with respect to methyl group as a major product? [June 26, 2022 (II)]

(1) 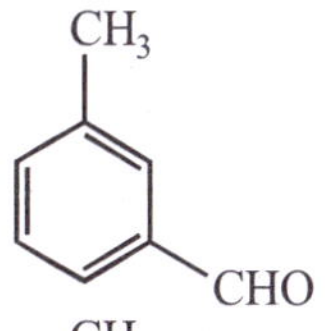

(2) 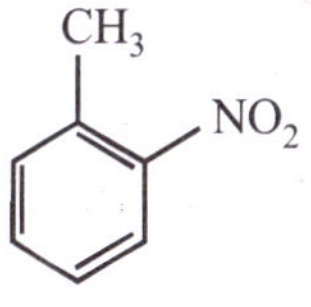

(3) 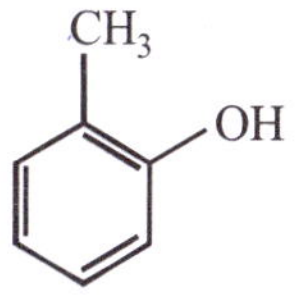

(4) 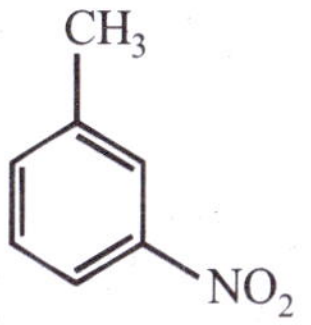

162. In the given reaction

$2A \xrightarrow[\text{(ii) Methyl benzoate (iii) } H_2O/H^+]{\text{(i) 2 Mg, THF}}$ 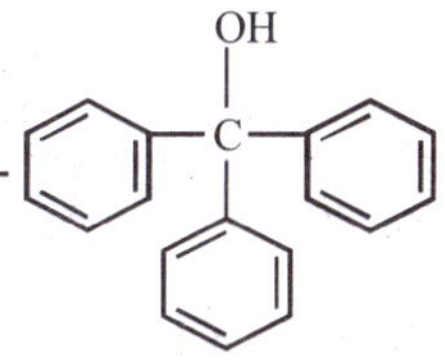

'A' can be [June 25, 2022 (II)]

(1) benzyl bromide (2) bromobenzene
(3) cyclohexyl bromide (4) methyl bromide

163. The correct order of reactivity of the given chlorides with acetate in acetic acid is : **[Aug. 31, 2021 (I)]**

(1) 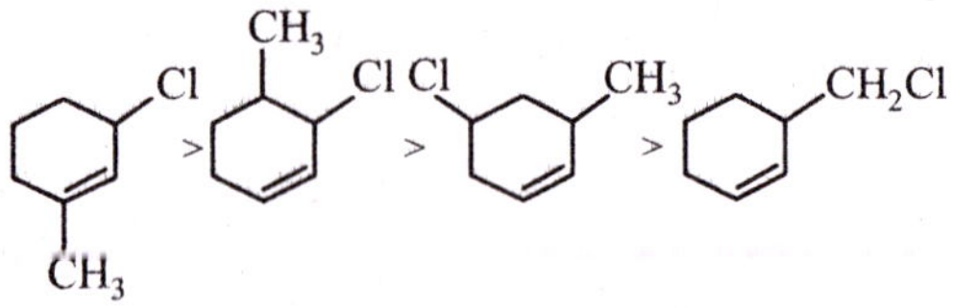

(2) 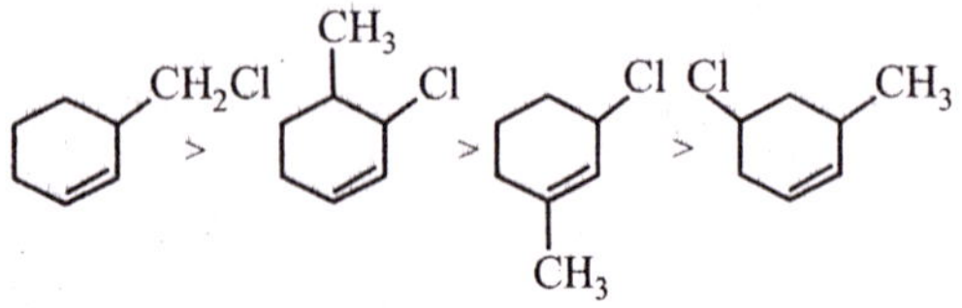

(3) 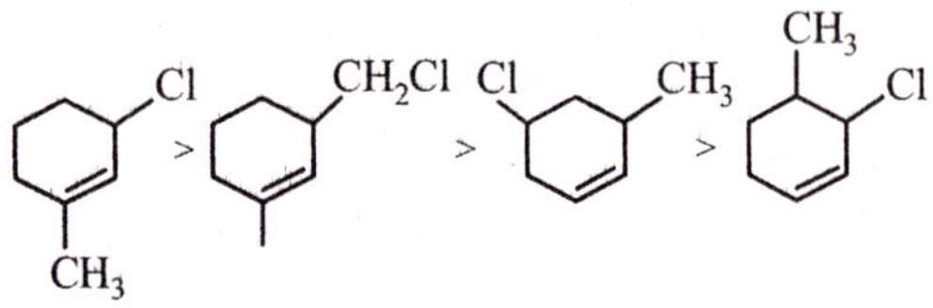

(4) 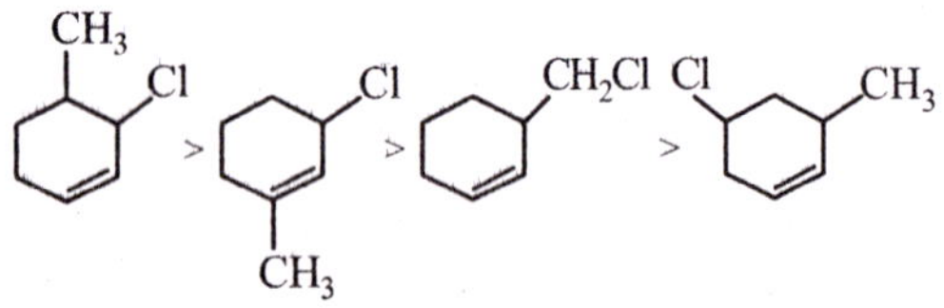

164. Among the following compounds I-IV, which one forms a yellow precipitate on reacting sequentially with (i) NaOH (ii) dil. HNO_3 (iii) $AgNO_3$? **[Aug. 26, 2021 (I)]**

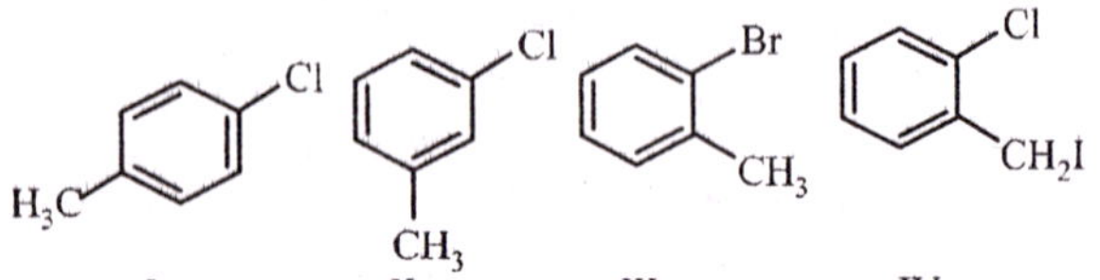

I II III IV

(1) II (2) IV (3) I (4) III

165. The correct decreasing order of densities of the following compounds is: **[July 25, 2021 (II)]**

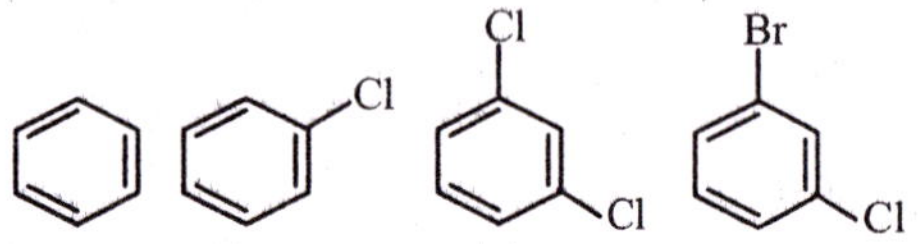

(A) (B) (C) (D)

(1) (D) > (C) > (B) > (A) (2) (C) > (D) > (A) > (B)
(3) (C) > (B) > (A) > (D) (4) (A) > (B) > (C) > (D)

166. Match List-I with List-II **[Feb. 26, 2021 (II)]**

List-I	List-II
(A) 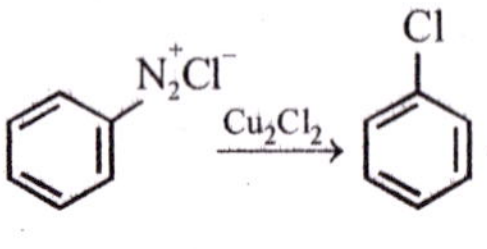 $+ N_2$	(I) Wurtz reaction
(B) 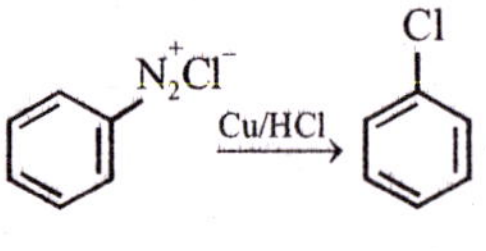 $+ N_2$	(II) Sandmeyer reaction
(C) $2CH_3CH_2Cl + 2Na \xrightarrow{\text{Ether}} C_2H_5-C_2H_5 + 2NaCl$	(III) Fittig reaction
(D) $2C_6H_5Cl + 2Na \xrightarrow{\text{Ether}} C_6H_5-C_6H_5 + 2NaCl$	(IV) Gattermann reaction

Choose the correct answer from the options given below

(1) (A)-(III); (B)-(IV); (C)-(I); (D)-(II)
(2) (A)-(II); (B)-(IV); (C)-(I); (D)-(III)
(3) (A)-(III); (B)-(I); (C)-(IV); (D)-(II)
(4) (A)-(II); (B)-(I); (C)-(IV); (D)-(III)

167. The correct order of the following compounds showing increasing tendency towards nucleophilic substitution reaction is : **[Feb. 24, 2021 (II)]**

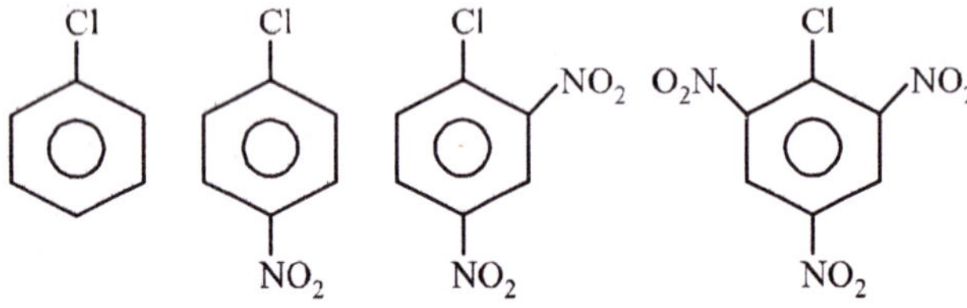

(i) (ii) (iii) (iv)

(1) (iv) < (iii) < (ii) < (i) (2) (iv) < (i) < (ii) < (iii)
(3) (iv) < (i) < (iii) < (ii) (4) (i) < (ii) < (iii) < (iv)

168. The decreasing order of reactivity of the following compounds towards nucleophilic substitution (S_N2) is : **[Sep. 03, 2020 (II)]**

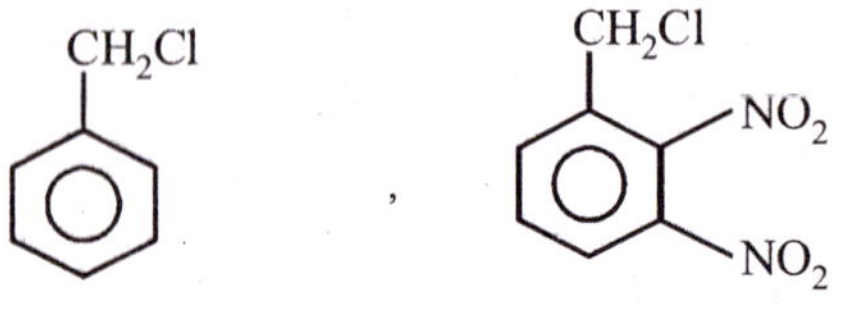

(I) (II)

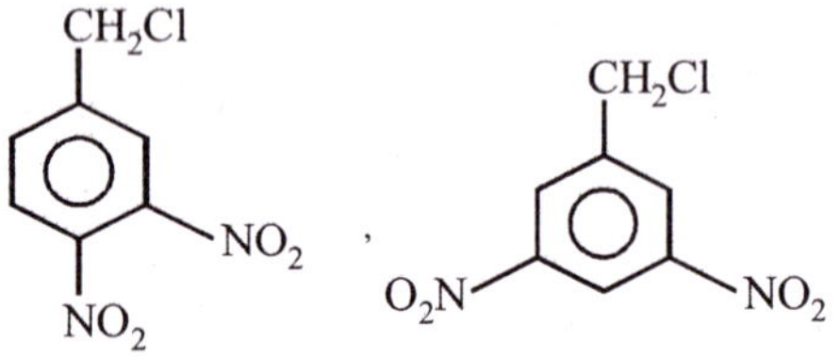

(III) (IV)

(1) (II) > (III) > (I) > (IV) (2) (II) > (III) > (IV) > (I)
(3) (III) > (II) > (IV) > (I) (4) (IV) > (II) > (III) > (I)

169. The major product of the following reaction is : **[April 8, 2019 (I)]**

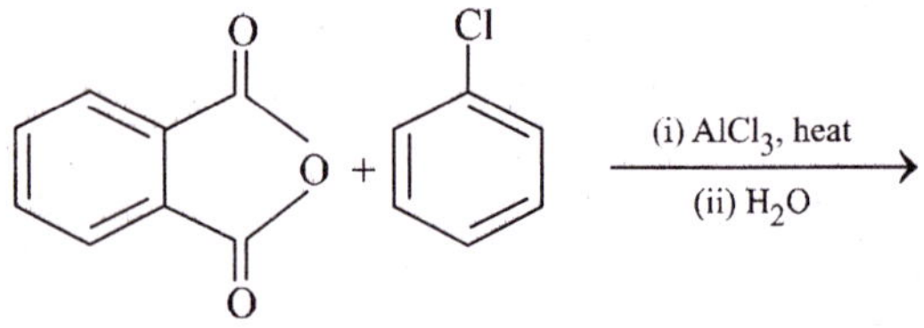

(1) 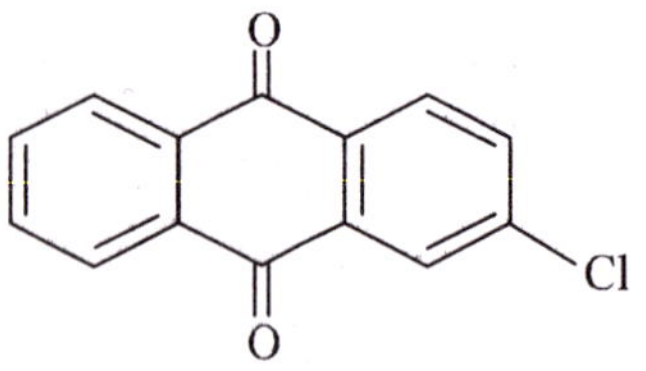

(2) 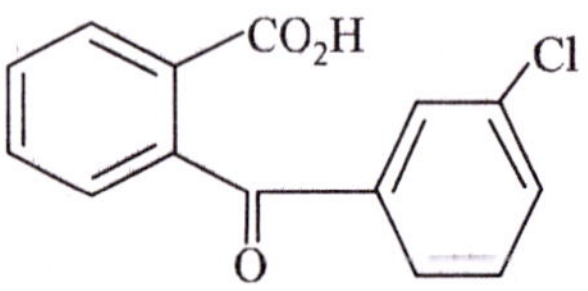

(3) 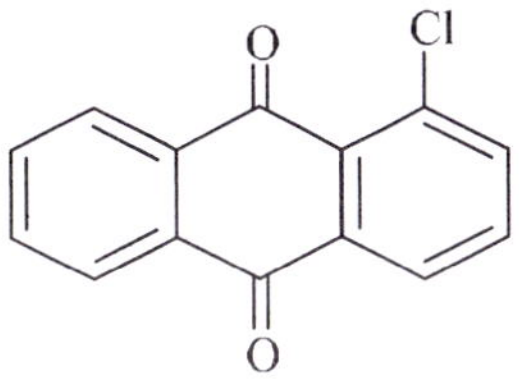

(4) 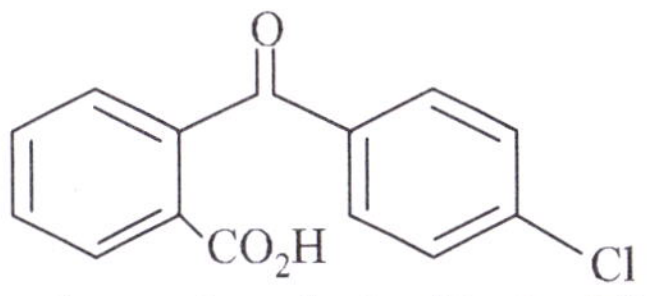

170. The major product obtained in the following reaction is :

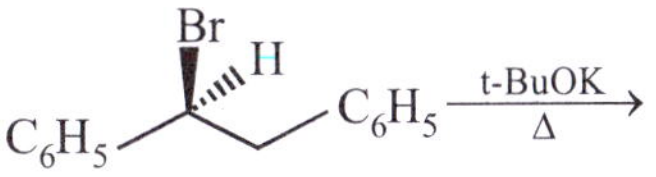

[2017]

(1) $(\pm)C_6H_5CH(O^tBu)CH_2C_6H_5$

(2) $C_6H_5CH=CHC_6H_5$

(3) $(+)C_6H_5CH(O^tBu)CH_2C_6H_5$

(4) $(-)C_6H_5CH(O^tBu)CH_2C_6H_5$

171. In a nucleophilic substitution reaction:

$$R-Br+Cl^- \xrightarrow{DMF} R-Cl+Br^-,$$

which one of the following undergoes complete inversion of configuration? [Online April 9, 2014]

(1) $C_6H_5CHC_6H_5Br$ (2) $C_6H_5CH_2Br$

(3) $C_6H_5CHCH_3Br$ (4) $C_6H_5CCH_3C_6H_5Br$

172. Compound (A), C_8H_9Br, gives a yellow precipitate when warmed with alcoholic $AgNO_3$. Oxidation of (A) gives an acid (B), $C_8H_6O_4$. (B) easily forms anhydride on heating. Identify the compound (A). [2013]

(1) 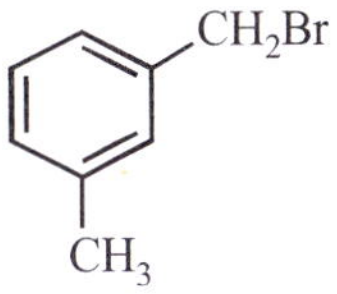 (2) 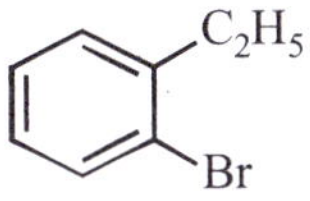

(3) 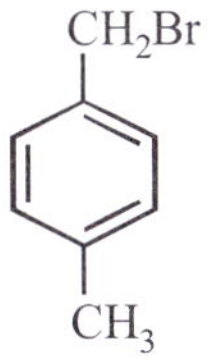 (4) 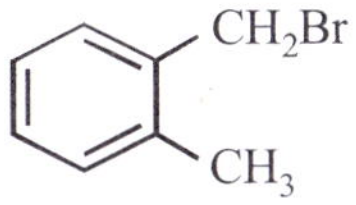

173. The Wurtz-Fittig reaction involves condensation of :

(1) two molecules of aryl halides [Online April 22, 2013]

(2) one molecule of each of aryl halide and alkyl halide.

(3) one molecule of each of aryl halide and phenol.

(4) two molecules of alkyl halides.

174. A major component of Borsch reagent is obtained by reacting hydrazine hydrate with which of the following ? [Online April 22, 2013]

(1) 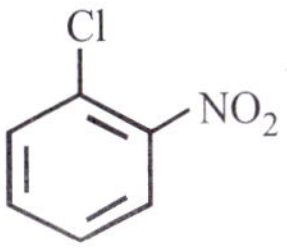 (2) 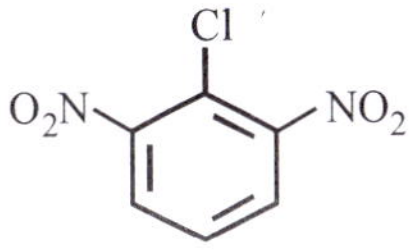

(3) 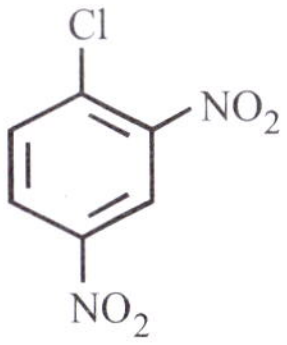 (4) 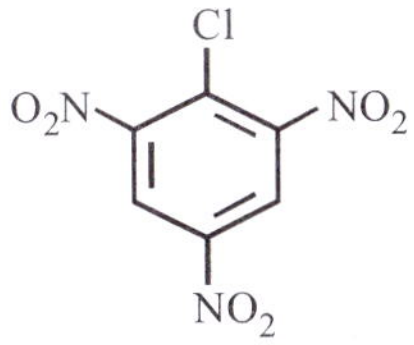

175. Aryl fluoride may be prepared from arene diazonium chloride using : [Online April 9, 2013]

(1) HBF_4/Δ (2) $HBF_4/NaNO_2,Cu,\Delta$

(3) CuF/HF (4) Cu/HF

176. Reaction of *trans* 2-phenyl-1-bromocyclopentane on reaction with alcoholic KOH produces [2006]

(1) 1-phenylcyclopentene (2) 3-phenylcyclopentene

(3) 4-phenylcyclopentene (4) 2-phenylcyclopentene

177. The structure of the major product formed in the following reaction is [2006]

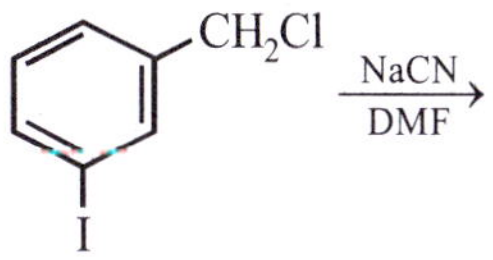

(1) 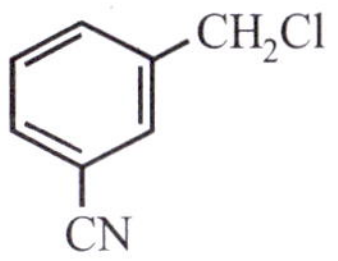 (2) 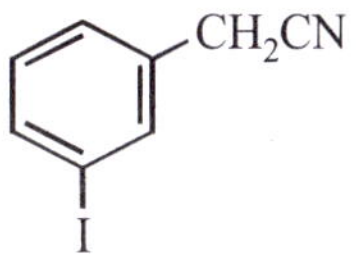

(3) 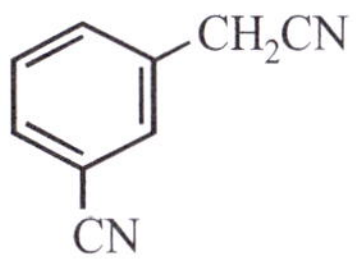 (4) 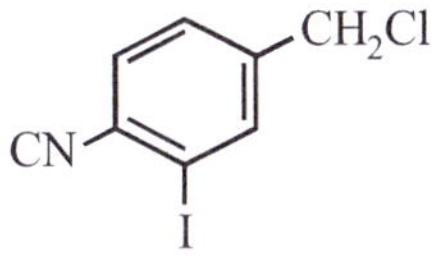

178. Fluorobenzene (C_6H_5F) can be synthesized in the laboratory

(1) by direct fluorination of benzene with F_2 gas [2006]

(2) by reacting bromobenzene with NaF solution

(3) by heating phenol with HF and KF

(4) from aniline by diazotisation followed by heating the diazonium salt with HBF_4

179. Phenyl magnesium bromide reacts with methanol to give [2005]

(1) a mixture of toluene and Mg(OH)Br

(2) a mixture of phenol and Mg(Me)Br

(3) a mixture of anisole and Mg(OH)Br

(4) a mixture of benzene and Mg(OMe)Br

180. Bottles containing C_6H_5I and $C_6H_5CH_2I$ lost their original labels. They were labelled A and B for testing. A and B were separately taken in test tubes and boiled with NaOH solution. The end solution in each tube was made acidic with dilute HNO_3 and then some $AgNO_3$ solution was added. Substance B gave a yellow precipitate. Which one of the following statements is true for this experiment ? [2003]

(1) A is $C_6H_5CH_2I$

(2) B is C_6H_5I

(3) Addition of HNO_3 was unnecessary

(4) A is C_6H_5I

NUMERICAL VALUE QUESTIONS (NVQs)

181. The number of halobenzenes from the following that can be prepared by Sandmeyer's reaction is .... [April 5, 2024 (I)]

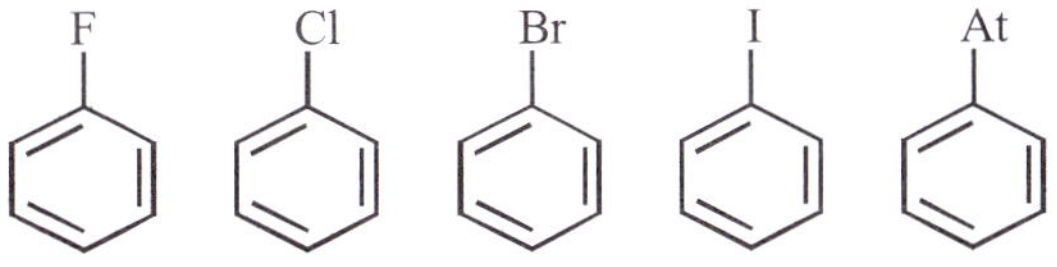

I II III IV V

Topic 3 - Some Important Polyhalogen Compounds

MULTIPLE CHOICE QUESTIONS (MCQs)

182. In the following sequence of reaction, the major products B and C respectively are : [April 9, 2024 (I)]

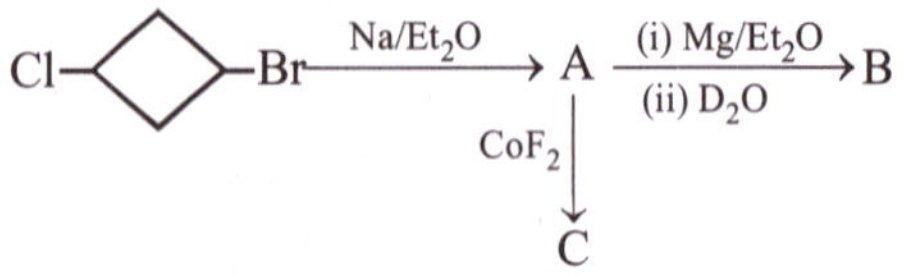

(1) 

(2) 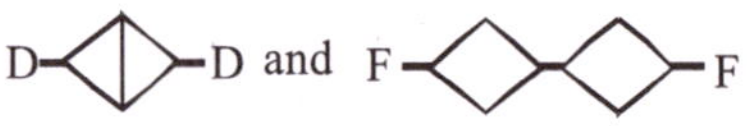

(3) 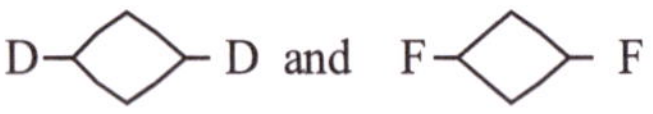

(4) 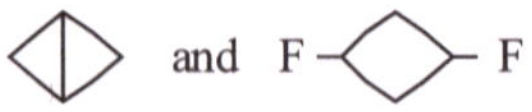

183. Match **List-I** with **List-II** : [April 6, 2024 (I)]

List-I (Compound)	List-II (Uses)
(A) Iodoform	(I) Fire extinguisher
(B) Carbon tetrachloride	(II) Insecticide
(C) CFC	(III) Antiseptic
(D) DDT	(IV) Refrigerants

Choose the correct answer from the options given below:

(1) A-(I), B-(II), C-(III), D-(IV)
(2) A-(III), B-(II), C-(IV), D-(I)
(3) A-(III), B-(I), C-(IV), D-(II)
(4) A-(II), B-(IV), C-(I), D-(III)

184. Match **List - I** with **List - II.** [Feb. 1, 2024 (II)]

List - I (Compound)	List - II (Use)
(A) Carbon tetrachloride	(I) Paint remover
(B) Methylene chloride	(II) Refrigerators and air conditioners
(C) DDT	(III) Fire extinguisher
(D) Freons	(IV) Non Biodegradable insecticide

Choose the correct answer from the options given below:

(1) (A)-(II), (B) - (III), (C) - (I), (D) - (IV)
(2) (A) - (III), (B) - (I), (C) - (IV), (D) - (II)
(3) (A) - (I), (B) - (II), (C) - (III), (D) - (IV)
(4) (1) - (IV), (B) - (III), (C) - (II), (D) - (I)

185. Which of the following compounds is an example of Freon? [April 11, 2023 (II)]

(1) $C_2Cl_2F_2$ (2) C_2HF_3
(3) $C_2H_2F_2$ (4) C_2F_4

186. In the following halogenated organic compounds the one with maximum number of chlorine atoms in its structure is: [Jan. 31, 2023 (II)]

(1) Chloral (2) Gammaxene
(3) Chloropicrin (4) Freon -12

187. Which of the following is true about freons ? [Jan. 24, 2023 (I)]

(1) These are chlorofluoro carbon compounds
(2) These are chemicals causing skin cancer
(3) These are radicals of chlorine and chlorine monoxide
(4) All radicals are called freons

188. Which among the following pairs has only herbicides ? [July 29, 2022 (I)]

(1) Aldrin and Dieldrin
(2) Sodium chlorate and Aldrin
(3) Sodium arsinate and Dieldrin
(4) Sodium chlorate and sodium arsenite.

189. The major organic compound formed by the reaction of 1, 1, 1-trichloroethane with silver powder is: [2014]

(1) Acetylene (2) Ethene
(3) 2 - Butyne (4) 2 - Butene

190. Chlorobenzne reacts with trichloroacetaldehyde in the presence of H_2SO_4.

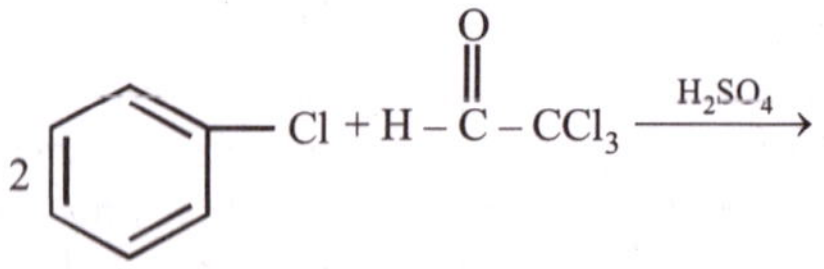

The major product formed is: [Online April 11, 2014]

(1) 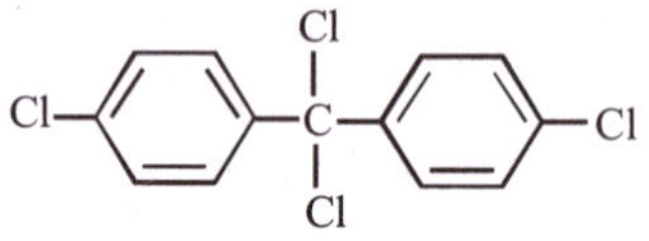

(2) 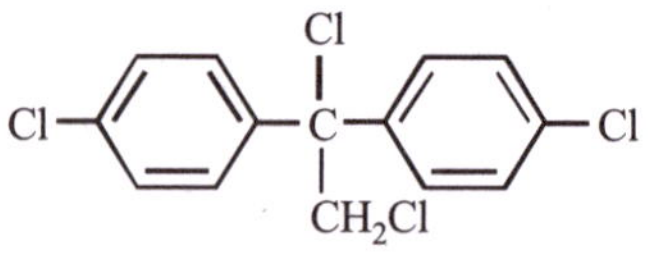

(3) 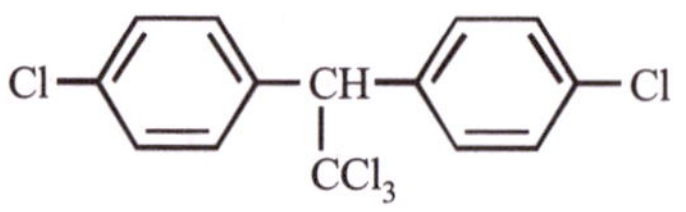

(4) 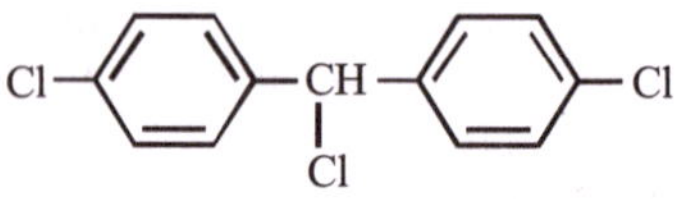

191. The major product formed when 1, 1, 1-trichloropropane is treated with aqueous potassium hydroxide, is: [Online April 19, 2014]

(1) propyne (2) 1-propanol
(3) 2-propanol (4) propionic acid

192. The compound formed on heating chlorobenzene with chloral in the presence of concentrated sulphuric acid, is [2004]

(1) freon (2) DDT
(3) gammexene (4) hexachloroethane

Part-B (JEE Advanced)

Topic 1- Preparation and Properties of Haloalkanes

1 MCQs with One Correct Answer

1. In the following reactions, P, Q, R, and S are the major products. [Adv. 2023]

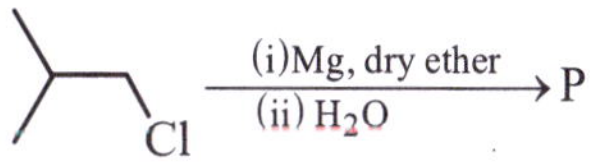

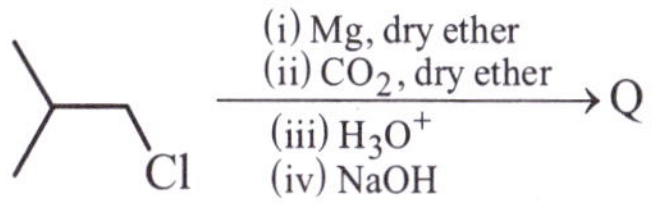

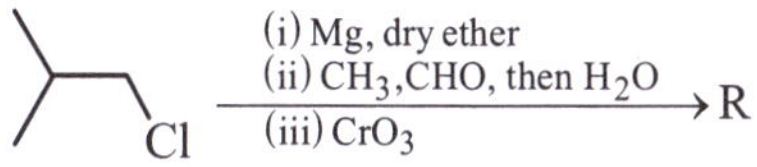

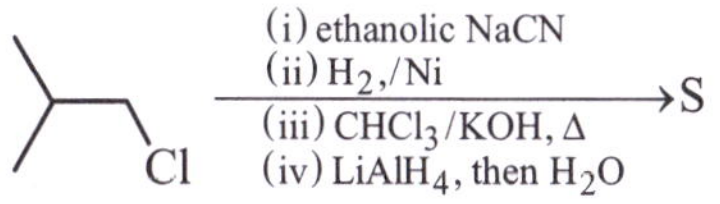

The correct statement about P, Q, R, and S is
(a) P is a primary alcohol with four carbons.
(b) Q undergoes Kolbe's electrolysis to give an eight-carbon product.
(c) R has six carbons and it undergoes Cannizzaro reaction.
(d) S is a primary amine with six carbons.

2. KI in acetone, undergoes S_N2 reaction with each of P, Q, R and S. The rates of the reaction vary as [Adv. 2013]

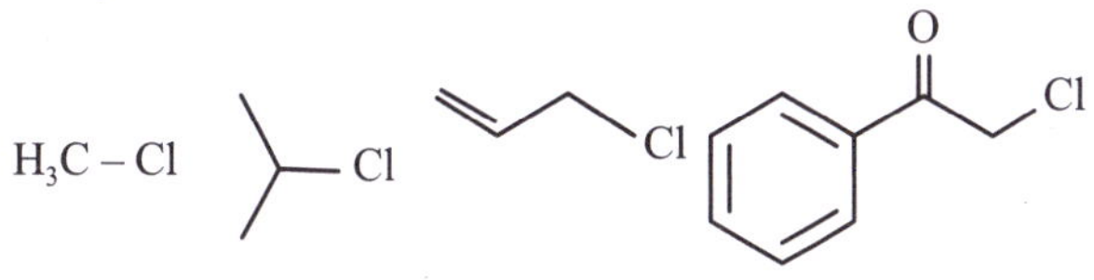

P Q R S

(a) P > Q > R > S (b) S > P > R > Q
(c) P > R > Q > S (d) R > P > S > Q

3. The reagent(s) for the following conversion,

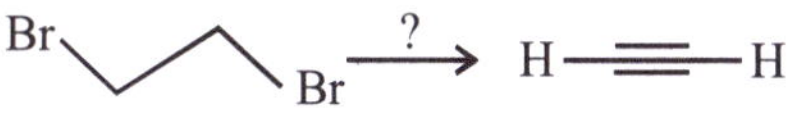

is/are [2007]
(a) alcoholic KOH
(b) alcoholic KOH followed by $NaNH_2$
(c) aqueous KOH followed by $NaNH_2$
(d) Zn/CH_3OH

4. What would be the product formed when 1–bromo–3–chlorocyclobutane reacts with two equivalents of metallic sodium in ether? [2005S]

(a) 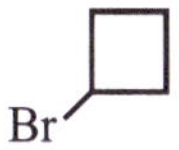

(b) 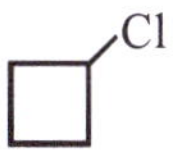

(c) 
(d) 

5. 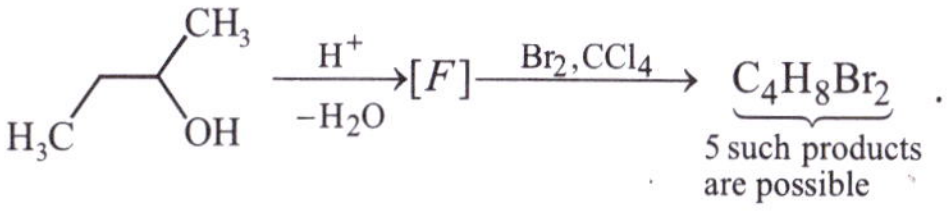

How many structures for F are possible? [2003S]
(a) 2 (b) 5 (c) 6 (d) 3

6. Identify the set of reagent / reaction conditions 'X' and 'Y' in the following set of transformations [2002S]

$$CH_3-CH_2-CH_2Br \xrightarrow{X} \text{Product} \xrightarrow{Y} CH_3-\underset{\underset{Br}{|}}{CH}-CH_3$$

(a) X = dilute aqueous NaOH, 20°C; Y = HBr/acetic acid, 20°C
(b) X = concentrated alcoholic NaOH, 80°C; Y = HBr/acetic acid, 20°C
(c) X = dilute aqueous NaOH, 20°C; Y = $Br_2/CHCl_3$, 0°C
(d) X = concentrated alcoholic NaOH, 80°C; Y = $Br_2/CHCl_3$, 0°C

7. A solution of (+) –2–chloro–2–phenylethane in toluene racemises slowly in the presence of small amount of $SbCl_5$, due to the formation of [1999 - 2 Marks]
(a) carbanion (b) carbene
(c) free-radical (d) carbocation

8. $(CH_3)_3CMgCl$ on reaction with D_2O produces : [1997 - 1 Mark]
(a) $(CH_3)_3CD$ (b) $(CH_3)_3OD$
(c) $(CD_3)_3CD$ (d) $(CD_3)_3OD$

9. Isobutyl magnesium bromide with dry ether and ethyl alcohol gives : [1995S]

(a) $CH_3\underset{\underset{CH_3}{|}}{CH}CH_2OH$ & CH_3CH_2MgBr

(b) $CH_3\underset{\underset{CH_3}{|}}{CH}CH_3$ & $MgBr(OC_2H_5)$

(c) $CH_3\underset{\underset{CH_3}{|}}{CH}, CH=CH_2$ & Mg(OH)Br

(d) $CH_3\underset{\underset{CH_3}{|}}{CH}CH_3$ & CH_3CH_2OMgBr

10. The chief reaction product of reaction between n-butane and bromine at 130°C is : [1995S]
(a) $CH_3CH_2CH_2CH_2Br$ (b) $CH_3CH_2CH(Br)CH_3$
(c) $CH_3-CH_2CH(Br)CH_2Br$ (d) $CH_3CH_2C(Br_2)CH_3$

11. 1-Chlorobutane on reaction with alcoholic potash gives [1991 - 1 Mark]
(a) 1-butene (b) 1-butanol
(c) 2-butene (d) 2-butanol

12. The number of structural and configurational isomers of a bromo compound, C_5H_9Br, formed by the addition of HBr to 2-pentyne respectively are [1988 - 1 Mark]
(a) 1 and 2 (b) 2 and 4
(c) 4 and 2 (d) 2 and 1

13. n-Propyl bromide on treatment with ethanolic potassium hydroxide produces **[1987 - 1 Mark]**

(a) Propane (b) Propene
(c) Propyne (d) Propanol

14. The reaction conditions leading to the best yields of C_2H_5Cl are : **[1986 - 1 Mark]**

(a) $C_2H_6\,(\text{excess}) + Cl_2 \xrightarrow{\text{uv light}}$

(b) $C_2H_6 + Cl_2 \xrightarrow[\text{room temperature}]{\text{dark}}$

(c) $C_2H_6 + Cl_2\,(\text{excess}) \xrightarrow{\text{uv light}}$

(d) $C_2H_6 + Cl_2 \xrightarrow{\text{uv light}}$

15. Among the following compounds, the most acidic is **[2011]**

(a) *p*-nitrophenol (b) *p*-hydroxybenzoic acid
(c) *o*-hydroxybenzoic acid (d) *p*-toluic acid

16. The number of stereoisomers obtained by bromination of *trans-2-butene* is **[2007]**

(a) 1 (b) 2 (c) 3 (d) 4

17. 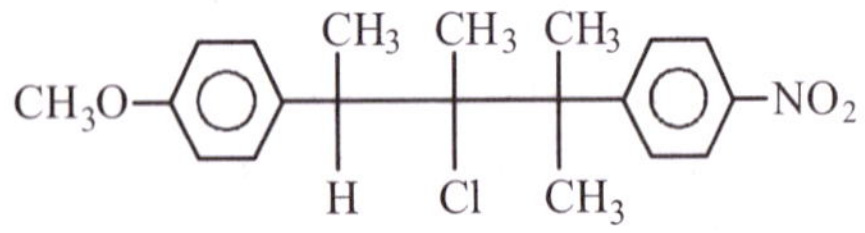

compound on hydrolysis in aqueous acetone will give **[2005S]**

(i) 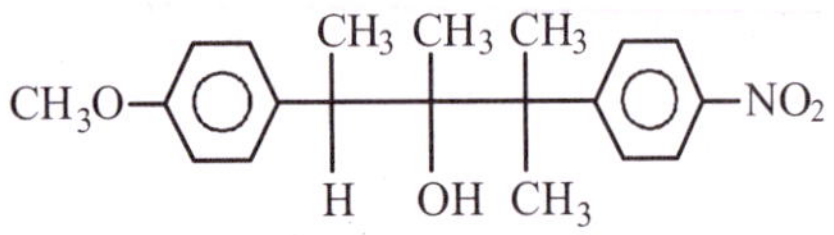

(ii) 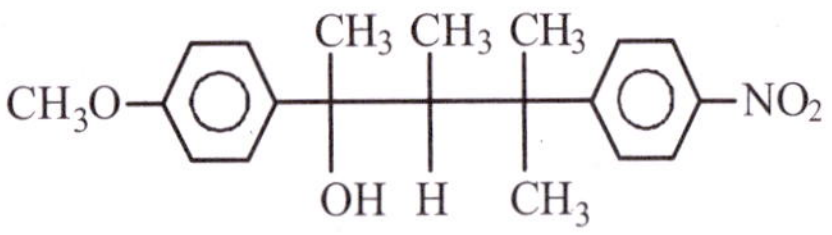

(iii) 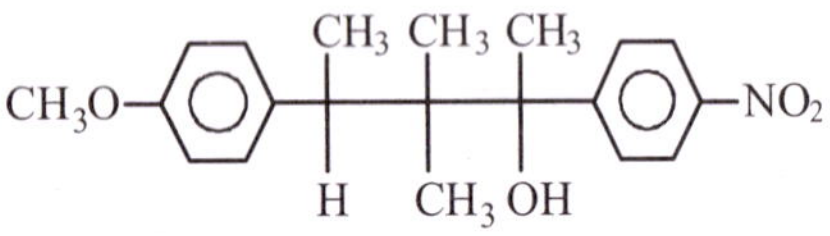

(a) Mixture of (i) and (ii) (b) Mixture of (i) and (iii)
(c) Only (iii) (d) Only (i)

18. Which of the following compounds exhibits stereoisomerism? **[2002S]**

(a) 2-methylbutene-1 (b) 3-methylbutyne-1
(c) 3-methylbutanoic acid (d) 2-methylbutanoic acid

19. An S_N2 reaction at an asymmetric carbon of a compound always gives **[2001S]**

(a) an enantiomer of the substrate
(b) a product with opposite optical rotation
(c) a mixture of diastereomers
(d) a single stereoisomer

20. The order of reactivity of the following alkyl halides for a S_N2 reaction is **[2000S]**

(a) RF > RCl > RBr > RI (b) RF > RBr > RCl > RI
(c) RCl > RBr > RF > RI (d) RI > RBr > RCl > RF

21. Which of the following has the highest nucleophilicity? **[2000S]**

(a) F^- (b) OH^- (c) CH_3^- (d) NH_2^-

22. How many optically active stereoisomers are possible for butane-2, 3-diol? **[1997]**

(a) 1 (b) 2 (c) 3 (d) 4

23. The Cl—C—Cl angle in 1,1,2,2-tetrachloroethene and tetrachloromethane respectively will be about **[1988]**

(a) 120° and 109.5° (b) 90° and 109.5°
(c) 109.5° and 90° (d) 109.5° and 120°

2 Integer Value Answer

24. For the given compound **X**, the total number of optically active stereoisomers is ______. **[Adv. 2018]**

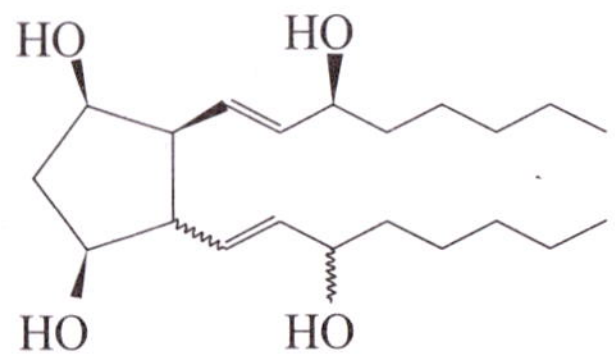

X

◄ This type of bond indicates that the configuration at the specific carbon and the geometry of the double bond is fixed

〰 This type of bond indicates that the configuration at the specific carbon and the geometry of the double bond is **NOT** fixed

25. In the following monobromination reaction, the number of possible chiral products is **[Adv. 2016]**

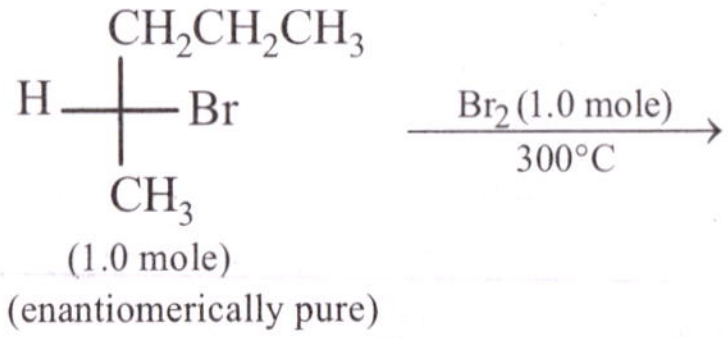

26. The total number of alkenes possible by dehydrobromination of 3-bromo-3-cyclopentylhexane using alcoholic KOH is **[2011]**

4 Fill in the Blanks

27. Vinyl chloride on reaction with dimethyl copper gives .... **[1997 - 1 Mark]**

28. The interaction of elemental sulphur with Grignard reagent gives .... **[1991 - 1 Mark]**

29. Isomers which are .... mirror images are known as .... (superimposable, non-superimposable, enantiomers, diastereomers, epimers) **[1988]**

30. The compound prepared by the action of magnesium on dry ethyl bromide in ether is known as .... reagent. **[1982 - 1 Mark]**

31. The halogen which is most reactive in the halogenation of alkanes under sunlight is .... (chlorine, bromine, iodine) **[1981 - 1 Mark]**

5 True / False

32. 2, 3, 4-Trichloropentane has three asymmetric carbon atoms. **[1990]**

33. During S_N1 reaction, the leaving group leaves the molecule before the incoming group is attached to the molecule. **[1990]**

34. The reaction of vinyl chloride with hydrogen iodide to give 1-chloro-1-iodoethane is an example of anti-Markovnikov's rule. **[1989 - 2 Marks]**

35. Iodide is a better nucleophile than bromide. **[1985 - ½ Mark]**

6 MCQs with One or More than One Correct Answer

36. For the reaction sequence given below, the correct statement(s) is(are) **[Adv. 2025]**

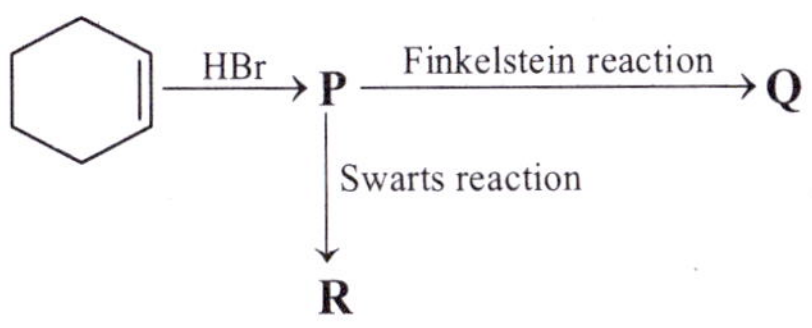

(In the options, X is any atom other than carbon and hydrogen, and it is different in **P**, **Q** and **R**)

(a) C – X bond length in **P**, **Q** and **R** follows the order **Q** > **R** > **P**.

(b) C – X bond enthalpy in **P**, **Q** and **R** follows the order **R** > **P** > **Q**.

(c) Relative reactivity toward S_N2 reaction in **P**, **Q** and **R** follows the order **P** > **R** > **Q**.

(d) pK_a value of the conjugate acids of the leaving groups in **P**, **Q** and **R** follows the order **R** > **Q** > **P**.

37. For the following compounds, the correct statement(s) with respect to nucleophilic substitution reaction is(are) **[Adv. 2017]**

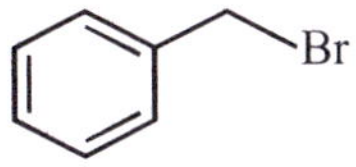

(I)

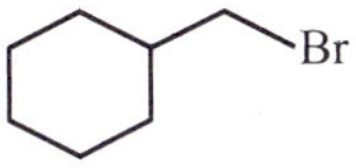

(II)

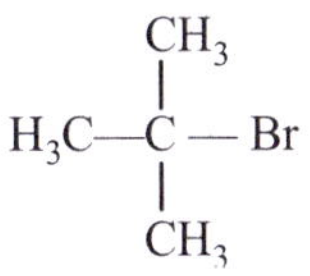

(III)

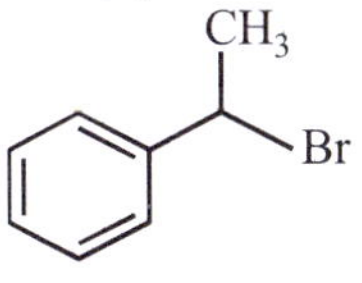

(IV)

(a) I and III follow S_N1 mechanism

(b) I and II follow S_N2 mechanism

(c) Compound IV undergoes inversion of configuration

(d) The order of reactivity for I, III and IV is : IV > I > III

38. The correct statement(s) about the compound given below is (are) **[2008]**

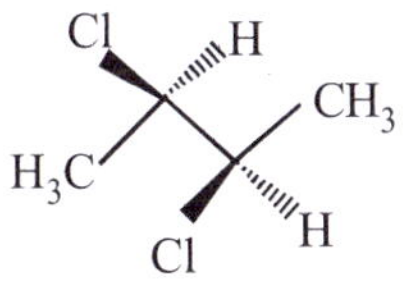

(a) The compound is optically active

(b) The compound possesses centre of symmetry

(c) The compound possesses plane of symmetry

(d) The compound possesses axis of symmetry

39. Which of the following have asymmetric carbon atom? **[1989]**

(a) 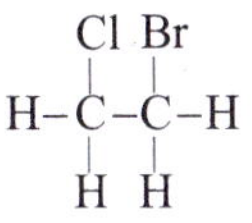

(b) 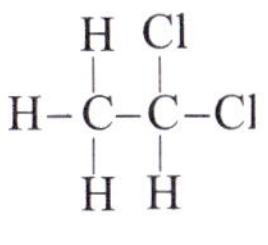

(c) 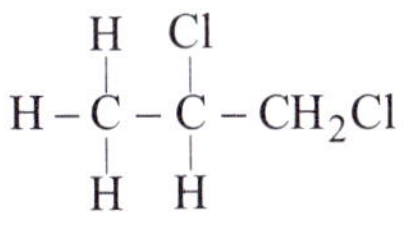

(d) 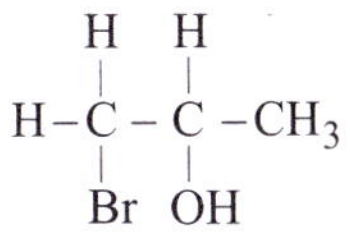

7 Match the Following

40. Match the reactions in List I with the features of their products in List–II and choose the correct option.

List–I	List–II
(P) (–)–1–Bromo–2–ethylpentane (single enantiomer) $\xrightarrow[S_N2\ \text{reaction}]{\text{aq.NaOH}}$	(1) Inversion of configuration
(Q) (–)–2–Bromopentane (single enantiomer) $\xrightarrow[S_N2\ \text{reaction}]{\text{aq.NaOH}}$	(2) Retention of configuration
(R) (–)–3–Bromo–3–methylhexane (single enantiomer) $\xrightarrow[S_N1\ \text{reaction}]{\text{aq.NaOH}}$	(3) Mixture of enantiomers
(S) 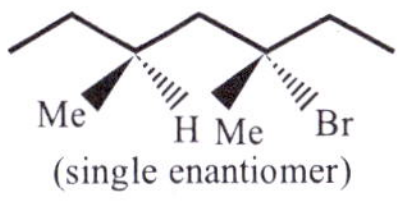 (single enantiomer) $\xrightarrow[S_N1\ \text{reaction}]{\text{aq.NaOH}}$	(4) Mixture of structural isomers
	(5) Mixture of diastereomers

[Adv. 2023]

(a) P → 1; Q → 2; R → 5; S → 3

(b) P → 2; Q → 1; R → 3; S → 5

(c) P → 1; Q → 2; R → 5; S → 4

(d) P → 2; Q → 4; R → 3; S → 5

41. The Fischer projection of D-erythrose is shown below.

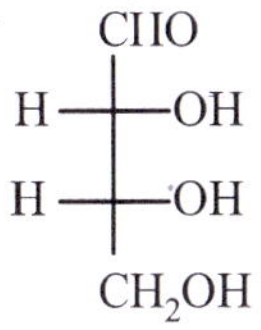

D-Erythrose

D-Erythrose and its isomers are listed as **P**, **Q**, **R**, and **S** in **Column-I**. Choose the correct relationship of **P**, **Q**, **R**, and **S** with D-erythrose from **Column II**. **[Adv. 2020]**

Column-I	Column-II
P. 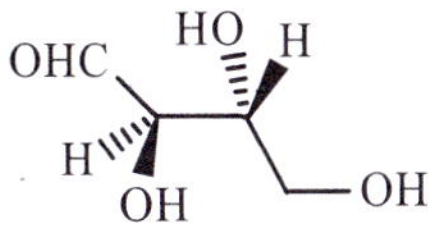	1. Diastereomer
Q. 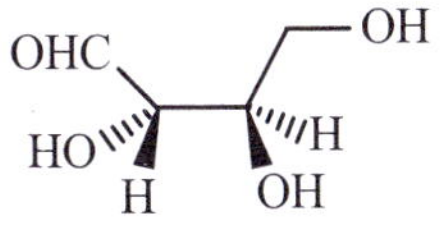	2. Identical
R. 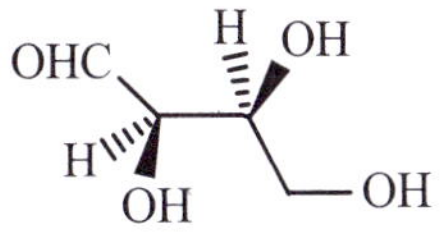	3. Enantiomer
S. 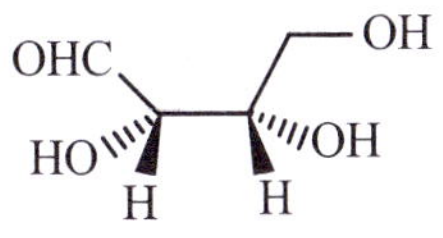	

(a) P → 2, Q → 3, R → 2, S → 2

(b) P → 3, Q → 1, R → 1, S → 2

(c) P → 2, Q → 1, R → 1, S → 3

(d) P → 2, Q → 3, R → 3, S → 1

9 Assertion and Reason/Statement Type Questions

Each question contains STATEMENT-1 (Assertion) and STATEMENT-2 (Reason). Each question has 4 choices (a), (b), (c) and (d) out of which ONLY ONE is correct. Mark your answer as

(a) If both Statement -1 and Statement -2 are correct, and Statement -2 is the correct explanation of the Statement -2.

(b) If both Statement-1 and Statement-2 are correct, but Statement -2 is not the correct explanation of the Statement -1.

(c) If Statement -1 is correct but Statement -2 is incorrect.

(d) If Statement -1 is incorrect but Statement -2 is correct.

42. **Statement - 1 :** Molecules that are not superimpossable on their mirror images are chiral. [2007]

Statement - 2 : All chiral molecules have chiral centres.

10 Subjective Problems

43. Cyclobutyl bromide on treatment with magnesium in dry ether forms an organometallic (A). The organometallic reacts with ethanal to give an alcohol (B) after mild acidification. Prolonged treatment of alcohol (B) with an equivalent amount of HBr gives 1–bromo–1–methylcyclopentane (C). Write the structures of (A), (B) and explain how (C) is obtained from (B). [2001 - 5 Marks]

44. Identify the pairs of enantiomers and diastereomers from the following compounds I, II and III [2000]

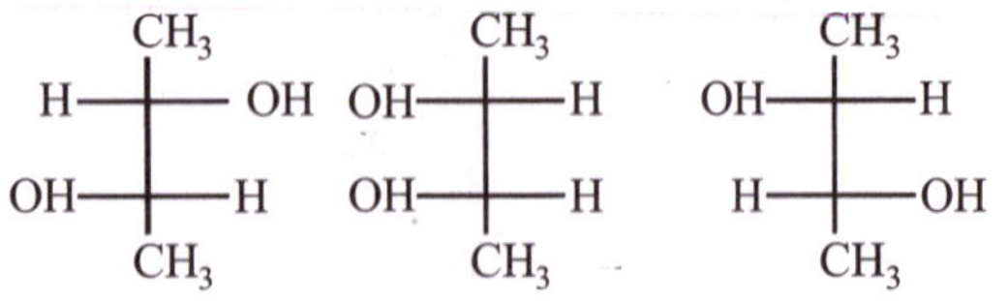

I II III

45. Write the structural formula of the major product in each of the following cases :

(*i*) $CH_3-\underset{CH_3}{\overset{CH_3}{\underset{|}{\overset{|}{C}}}}-CH_2Br \xrightarrow[\Delta]{C_2H_5OH}$ [2000 - 1 Mark]

(*ii*) $C_6H_5CH_2CHClC_6H_5 \xrightarrow[\text{heat}]{\text{alcoholic KOH}}$ 2 Products [1998 - 2 Marks]

(*iii*) $C_6H_5-CH_2-\underset{Br}{\underset{|}{CH}}-CH_3 \xrightarrow[\text{KOH, }\Delta]{\text{alcoholic}} ? \xrightarrow{HBr} ?$ [1993 - 1 Mark]

(*iv*) $CH_3CH_2CHCl_2 \xrightarrow[\text{alkali}]{\text{boil}}$ [1992 - 1 Mark]

(*v*) $(CH_3)_2\underset{Cl}{\underset{|}{C}}-CH_2CH_3 \xrightarrow[\text{alc. KOH}]{}$ [1992 - 1 Mark]

(*vi*) bromoethane reacts with one-half of the molar quantity of silver carbonate. [1981 - ½ Mark] [1992 - 1 Mark]

(*vii*) chloroform reacts with aniline in the presence of excess alkali [1981 - ½ Mark]

46. An alkyl halide, *X*, of formula $C_6H_{13}Cl$ on treatment with potassium tertiary butoxide gives two isomeric alkenes *Y* and *Z* (C_6H_{12}). Both alkenes on hydrogenation give 2, 3-dimethylbutane. Predict the structures of *X*, *Y* and *Z*. [1996 - 3 Marks]

47. Draw the stereochemical structures of the products in the following reaction : [1994 - 4 Marks]

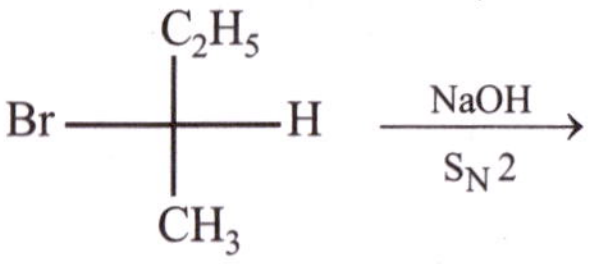

48. Arrange the following in :

Increasing reactivity in nucleophilic substitution reactions

CH_3F, CH_3I, CH_3Br, CH_3Cl [1992]

49. An organic compound *X*, on analysis gives 24.24 per cent carbon and 4.04 per cent hydrogen. Further, sodium extract of 1.0 g of *X* gives 2.90 g of silver chloride with acidified silver nitrate solution. The compound *X* may be represented by two isomeric structures, *Y* and *Z*. *Y* on treatment with aqueous potassium hydroxide solution gives a dihydroxy compound while *Z* on similar treatment gives ethanal. Find out the molecular formula of *X* and give the structures of *Y* and *Z*. [1989 - 4 Marks]

50. What effect should the following resonance of vinyl chloride have on its dipole moment? [1987 - 1 Mark]

$$CH_2=CH-Cl \longleftrightarrow \overset{-}{C}H_2-CH=Cl^+$$

51. Write the structure of all the possible isomers of dichloroethene. Which of them will have zero dipole moment? [1985 - 2 Marks]

52. State the conditions under which the following preparation are carried out. Give the necessary equations which need not be balanced :

(*i*) Lead tetraethyl from sodium-lead alloy [1983 - 1 Mark]

(*ii*) Methyl chloride from aluminium carbide [1983 - 1 Mark]

53. (a) Show by chemical equations only, how you would prepare the following from the indicated starting materials. Specify the reagents in each step of the synthesis.

(i) Hexachlorethane, C_2Cl_6, from calcium carbide.

(ii) Chloroform from carbon disulphide.

(b) Give one chemical test which would distinguish between C_2H_5OH from $CHCl_3$. [1979]

Topic 2 - Preparation and Properties of Haloarenes

1 MCQs with One Correct Answer

54. The major product of the following reaction is – [2008]

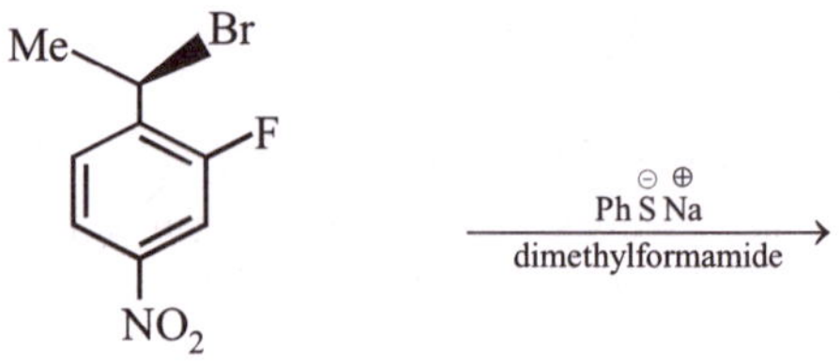

(a) 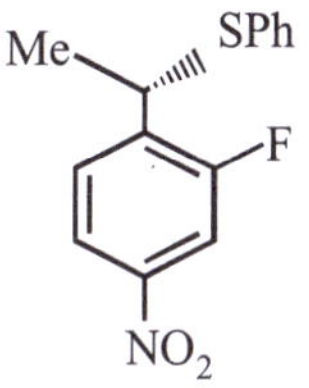

(b) 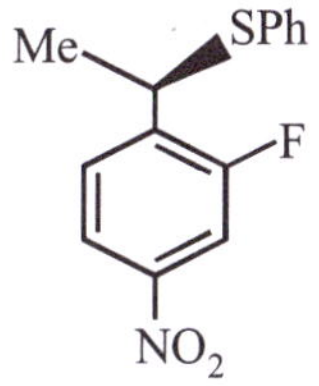

(c) 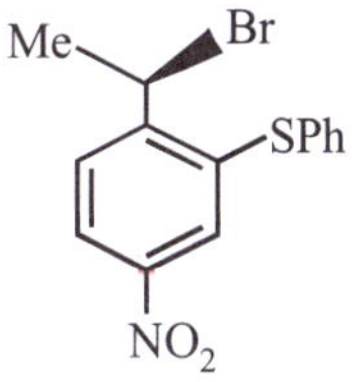

(d) 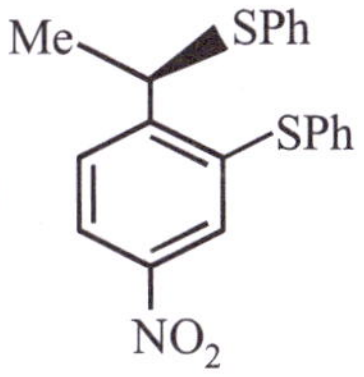

55. When phenyl magnesium bromide reacts with *tert* –butanol, the product would be [2005S]
    (a) Benzene (b) Phenol
    (c) *ter*–butylbenzene (d) *ter*–butyl phenyl ether

56. The reaction of toluene with chlorine in presence of ferric chloride gives predominantly : [1986 - 1 Mark]
    (a) benzoyl chloride (b) *m*-chlorotoluene
    (c) benzyl chloride (d) *o*- and *p*-chlorotoluene

57. Chlorobenzene can be prepared by reacting aniline with : [1984 - 1 Mark]
    (a) hydrochloric acid
    (b) cuprous chloride
    (c) chlorine in presence of anhydrous aluminium chloride
    (d) nitrous acid followed by heating with cuprous chloride

3 Numeric / New Stem Based Questions

58. The weight percentage of hydrogen in **Q**, formed in the following reaction sequence, is ________. [Adv. 2022]

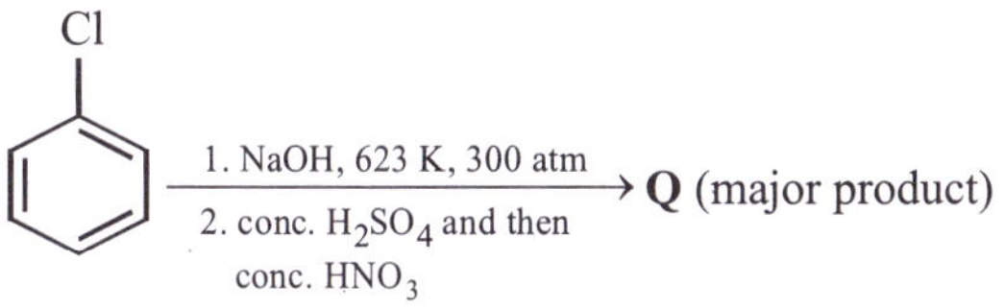

[Given: Atomic mass of H = 1, C = 12, N = 14, O = 16, S = 32, Cl = 35]

59. If the reaction sequence given below is carried out with 15 moles of acetylene, the amount of the product **D** formed (in g) is ________. [Adv. 2022]

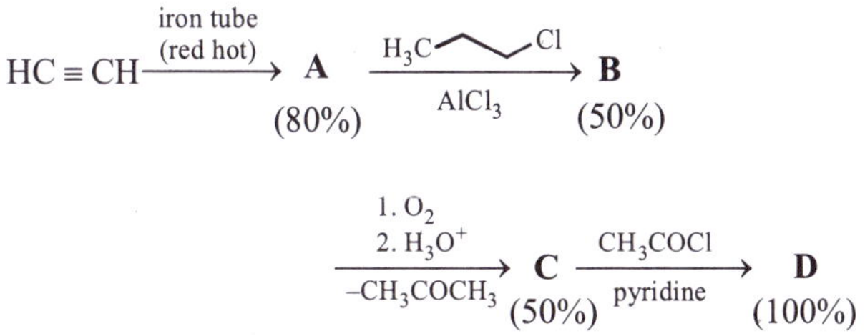

The yields of **A**, **B**, **C** and **D** are given in parentheses.
[Given: Atomic mass of H = 1, C = 12, O = 16, Cl = 35]

5 True / False

60. *m*-Chlorobromobenzene is an isomer of *m*-bromochlorobenzene. [1985 - ½ Mark]

6 MCQs with One or More than One Correct Answer

61. Choose the correct option(s) for the following set of reactions [Adv. 2019]

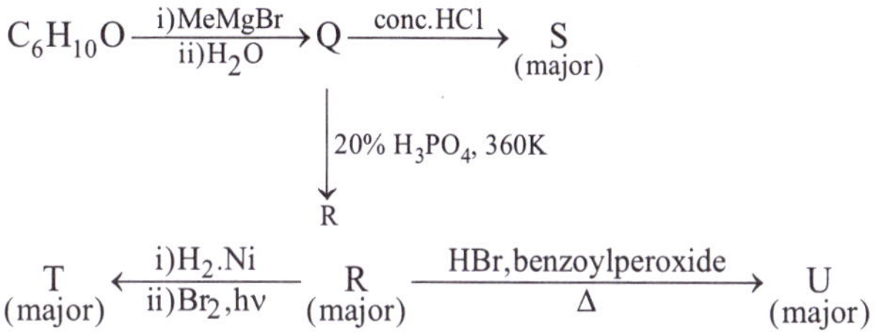

(a) 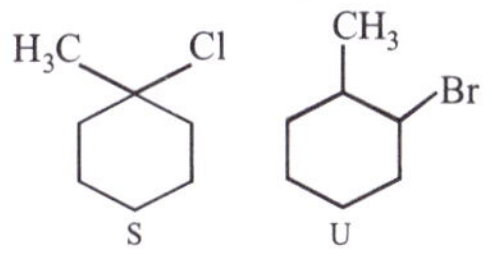

(b) 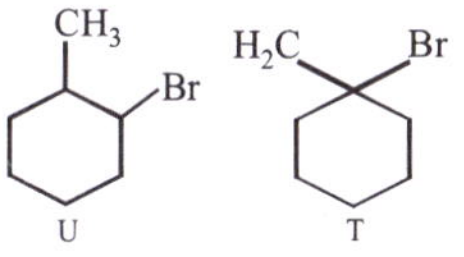

(c) 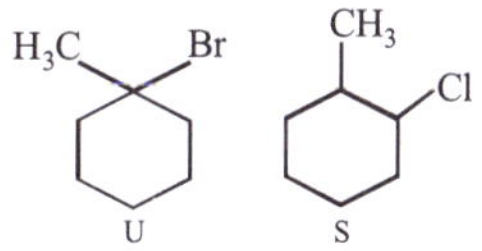

(d) 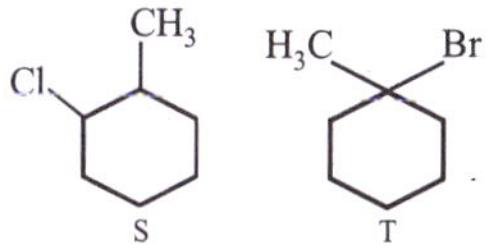

62. Benzyl chloride ($C_6H_5CH_2Cl$) can be prepared from toluene by chlorination with [1998 - 2 Marks]
    (a) SO_2Cl_2 (b) $SOCl_2$
    (c) Cl_2 (d) NaOCl

63. Aryl halides are less reactive towards nucleophilic substitution reaction as compared to alkyl halides due to: [1990 - 1 Mark]
    (a) The formation of less stable carbonium ion
    (b) Resonance stabilization
    (c) Longer carbon-halogen bond
    (d) The inductive effect
    (e) sp^2 hybridized carbon attached to the halogen.

7 Match the Following

64. Match the major products obtained in the reactions given in **List-I** with the corresponding structures in **List-II** and choose the correct option. [Adv. 2026]

List-I

(P) 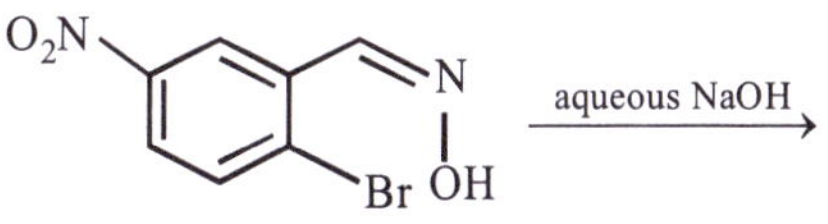

(Q) 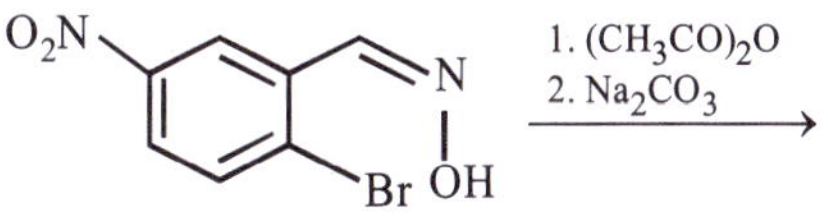

(R) 

(S) 

List-II

(1) 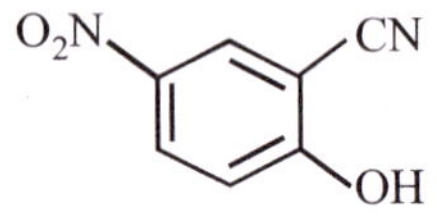

(2) 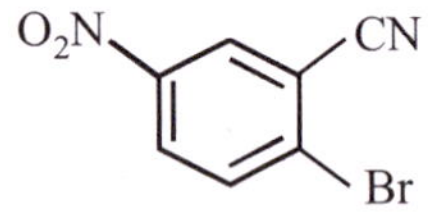

(3) 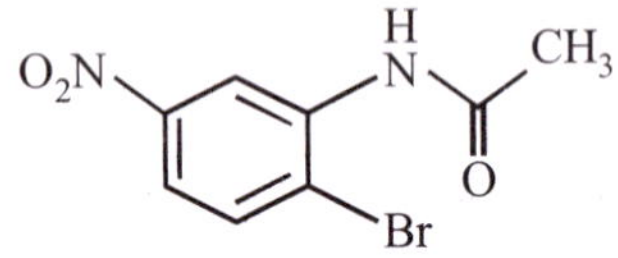

(4) 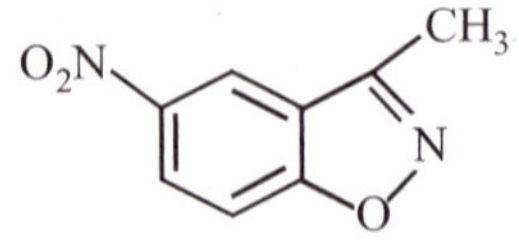

(5) 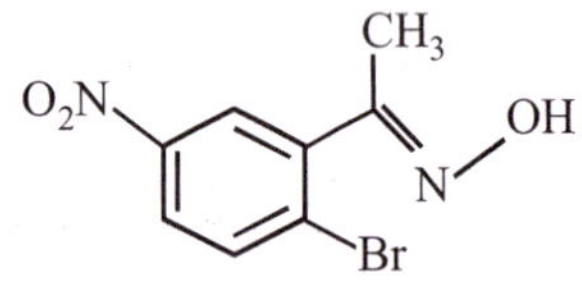

(a) $P \rightarrow 2, Q \rightarrow 1, R \rightarrow 5, S \rightarrow 4$
(b) $P \rightarrow 1, Q \rightarrow 2, R \rightarrow 4, S \rightarrow 5$
(c) $P \rightarrow 1, Q \rightarrow 2, R \rightarrow 3, S \rightarrow 4$
(d) $P \rightarrow 2, Q \rightarrow 1, R \rightarrow 3, S \rightarrow 5$

65. Given below are certain matching type questions, where two columns (each having 4 items) are given. Immediately after the columns the matching grid is given, where each item of **Column I** has to be matched with the items of **Column II**, by encircling the correct match(es). Note that an item of **Column I** can match with more than one item of **Column II**. All the items of **Column II** must be matched. Match the following : **[2006 - 6M]**

Column I	Column II
(A) $C_6H_5CH_2CD_2Br$ on reaction with $C_2H_5O^-$ gives C_6H_5–$CH{=}CD_2$	(p) E1 reaction
(B) $PhCHBrCH_3$ and $PhCHBrCD_3$, both react with the same rate	(q) E2 reaction
(C) $C_6H_5CH_2CH_2Br$ on treatment with $C_2H_5O^-$ and C_2H_5OD gives $C_6H_5CD{=}CH_2$	(r) E1cB reaction
(D) $C_6H_5CH_2CH_2Br$ reacts faster than $C_6H_5CD_2CH_2Br$ on reaction with $C_2H_5O^-$ in ethanol	(s) First order reaction

9 Assertion and Reason/Statement Type Questions

Each question contains **STATEMENT-1 (Assertion)** and **STATEMENT-2 (Reason)**. Each question has 4 choices (a), (b), (c) and (d) out of which ONLY ONE is correct. Mark your answer as

(a) If both Statement -1 and Statement -2 are correct, and Statement -2 is the correct explanation of the Statement -2.
(b) If both Statement -1 and Statement -2 are correct, but Statement -2 is not the correct explanation of the Statement -1.
(c) If Statement -1 is correct but Statement -2 is incorrect.
(d) If Statement -1 is incorrect but Statement -2 is correct.

66. Read the following Statement-1 (Assertion) and Statement -2 (Reason) and answer as per the options given below :

Statement-1 : Bromobenzene upon reaction with Br_2/Fe gives 1,4-dibromobenzene as the major product. **[2008S]**

Statement-2 : In bromobenzene, the inductive effect of the bromo group is more dominant than the mesomeric effect in directing the incoming electrophile.

10 Subjective Problems

67. Give reasons for the following :

(*i*) 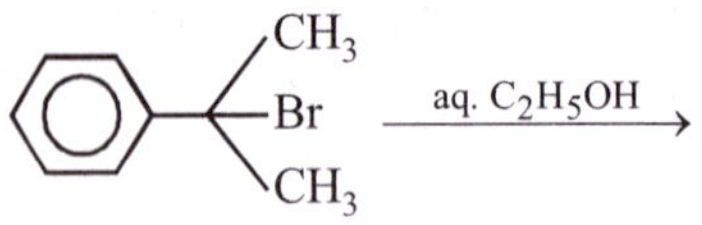 Acidic solution

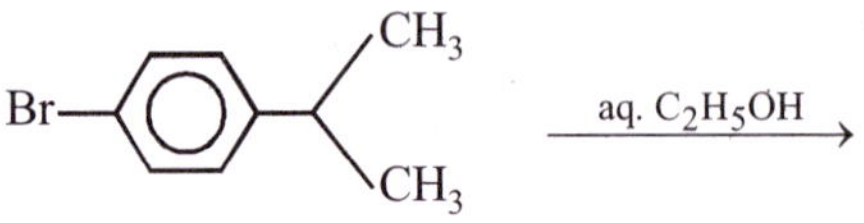 Neutral

solution. Explain. **[2005 - 1 Mark]**

(*ii*) 7-Bromo-1, 3, 5-cycloheptatriene exists as ionic compound, while 5-bromo-1, 3-cyclopentadiene does not ionise even in presence of Ag^+ ion. Explain. **[2004 - 2 Marks]**

68. Write the structural formula of the major product in the following case :

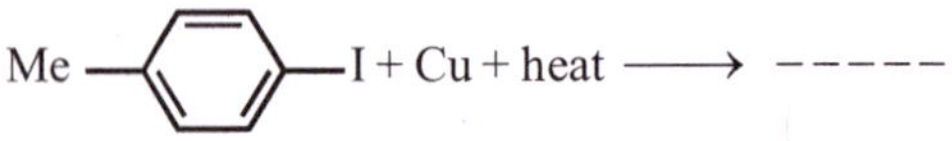

[1997 - 1 Mark]

69. How will you prepare *m*-bromoiodobenzene from benzene (in not more than 5-7 steps)? **[1996 - 2 Marks]**

70. Give reasons for the following :

Aryl halides are less reactive than alkyl halides towards nucleophilic reagents **[1994]**

71. What happens when excess chlorine is passed through boiling toluene in the presence of sunlight? **[1987 - 1 Mark]**

Answer Key

Part-A (JEE Main)																			
1	(1)	**21**	(2)	**41**	(4)	**61**	(1)	**81**	(2)	**101**	(1)	**121**	(1)	**141**	(4)	**161**	(3)	**181**	(2)
2	(2)	**22**	(1)	**42**	(2)	**62**	(4)	**82**	(3)	**102**	(2)	**122**	(2)	**142**	(2)	**162**	(2)	**182**	(1)
3	(3)	**23**	(1)	**43**	(4)	**63**	(1)	**83**	(3)	**103**	(1)	**123**	(6)	**143**	(3)	**163**	(1)	**183**	(3)
4	(1)	**24**	(3)	**44**	(4)	**64**	(2)	**84**	(3)	**104**	(2)	**124**	(10)	**144**	(2)	**164**	(2)	**184**	(2)
5	(4)	**25**	(1)	**45**	(1)	**65**	(4)	**85**	(1)	**105**	(2)	**125**	(6)	**145**	(2)	**165**	(1)	**185**	(1)
6	(3)	**26**	(4)	**46**	(2)	**66**	(3)	**86**	(4)	**106**	(1)	**126**	(4)	**146**	(1)	**166**	(2)	**186**	(2)
7	(3)	**27**	(3)	**47**	(3)	**67**	(4)	**87**	(1)	**107**	(3)	**127**	(5)	**147**	(3)	**167**	(4)	**187**	(1)
8	(3)	**28**	(1)	**48**	(2)	**68**	(2)	**88**	(4)	**108**	(1)	**128**	(2)	**148**	(1)	**168**	(2)	**188**	(4)
9	(2)	**29**	(3)	**49**	(2)	**69**	(4)	**89**	(2)	**109**	(2)	**129**	(6)	**149**	(1)	**169**	(4)	**189**	(3)
10	(4)	**30**	(3)	**50**	(1)	**70**	(4)	**90**	(1)	**110**	(4)	**130**	(42)	**150**	(2)	**170**	(2)	**190**	(3)
11	(1)	**31**	(2)	**51**	(1)	**71**	(4)	**91**	(1)	**111**	(2)	**131**	(3)	**151**	(4)	**171**	(2)	**191**	(4)
12	(2)	**32**	(3)	**52**	(2)	**72**	(3)	**92**	(2)	**112**	(1)	**132**	(1)	**152**	(4)	**172**	(4)	**192**	(2)
13	(2)	**33**	(3)	**53**	(3)	**73**	(3)	**93**	(4)	**113**	(2)	**133**	(2)	**153**	(4)	**173**	(2)		
14	(2)	**34**	(2)	**54**	(3)	**74**	(4)	**94**	(4)	**114**	(3)	**134**	(3)	**154**	(4)	**174**	(3)		
15	(3)	**35**	(2)	**55**	(3)	**75**	(2)	**95**	(2)	**115**	(3)	**135**	(8)	**155**	(3)	**175**	(1)		
16	(4)	**36**	(2)	**56**	(3)	**76**	(2)	**96**	(2)	**116**	(2)	**136**	(1)	**156**	(4)	**176**	(1)		
17	(4)	**37**	(1)	**57**	(3)	**77**	(1)	**97**	(4)	**117**	(2)	**137**	(3)	**157**	(3)	**177**	(2)		
18	(4)	**38**	(2)	**58**	(4)	**78**	(3)	**98**	(1)	**118**	(3)	**138**	(1)	**158**	(3)	**178**	(4)		
19	(4)	**39**	(1)	**59**	(3)	**79**	(3)	**99**	(2)	**119**	(184)	**139**	(2)	**159**	(2)	**179**	(4)		
20	(3)	**40**	(1)	**60**	(3)	**80**	(4)	**100**	(3)	**120**	(32)	**140**	(4)	**160**	(4)	**180**	(4)		
Part-B (JEE Advanced)																			
1	(b)	**8**	(a)	**15**	(c)	**22**	(c)	**34**	(False)	**41**	(c)	**59**	(136)	**65**	(A)-q; (B)-p, s; (C)-r, s; (D) - q				
2	(b)	**9**	(b)	**16**	(a)	**23**	(a)	**35**	(True)	**42**	(c)	**60**	(False)						
3	(b)	**10**	(b)	**17**	(a)	**24**	(7)	**36**	(b)	**54**	(a)	**61**	(a, b)	**66**	(c)				
4	(d)	**11**	(a)	**18**	(d)	**25**	(5)	**37**	(a, b,c)	**55**	(a)	**62**	(c)						
5	(d)	**12**	(b)	**19**	(d)	**26**	(5)	**38**	(a, d)	**56**	(d)	**63**	(b, e)						
6	(b)	**13**	(b)	**20**	(d)	**32**	(False)	**39**	(c, b)	**57**	(d)	**64**	(b)						
7	(d)	**14**	(a)	**21**	(c)	**33**	(True)	**40**	(b)	**58**	(1.31)								

Chapter-11

Alcohols, Phenols and Ethers

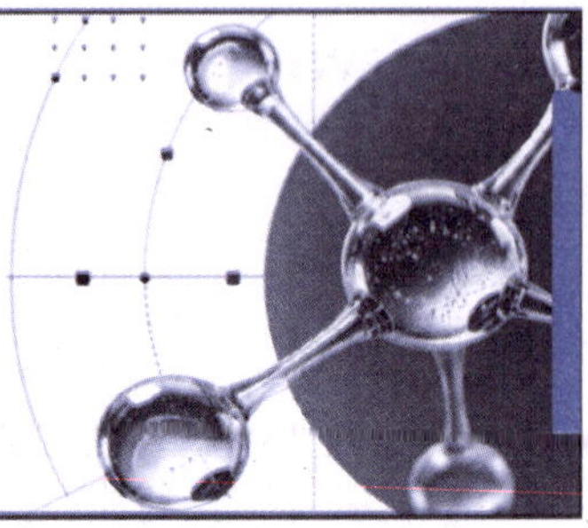

Part-A (JEE Main)

Topic 1- Preparation and Properties of Alcohols

MULTIPLE CHOICE QUESTIONS (MCQs)

1. Consider compounds A, B and C with following structural formulae **[April 05, 2026 (I)]**

$A = CH_3-CH_2-CH_2-CH_2-CH_2-OH$

$B = CH_2=CH-CH_2-CH_2-CH_3$

$C = HO-CH_2\,CH_2-CH(OH)-CH_3$

For the conversion of B from A, reagent (D) required is ____ and structural formula of product (E) obtained when C undergoes same reaction using excess reagent (D) is ______.

(1)	D	E
	Conc. H_2SO_4	$CH_2=CH-CH(OH/CH_3$

(2)	D	E
	PCC	$HO-CH_2-CH_2-CH=CH_2$

(3)	D	E
	PCC	$CH_2=CH-CH=CH_2$

(4)	D	E
	Conc. H_2SO_4 or H_3PO_4	$CH_2=CH-CH=CH_2$

2. From the following, how many compounds contain at least one secondary alcohol ? **[Jan. 24, 2026 (II)]**

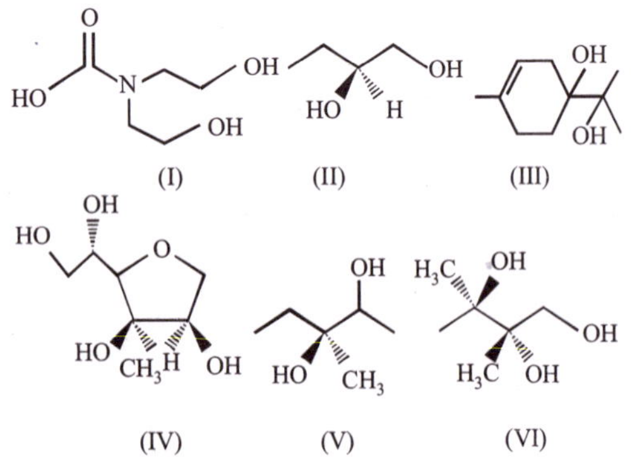

Choose the correct answer from the options given below:

(1) Five (2) Three

(3) Four (4) two

3. 3, 3-Dimethyl-2-butanol **cannot** be prepared by : **[Jan. 22, 2026 (II)]**

(A) 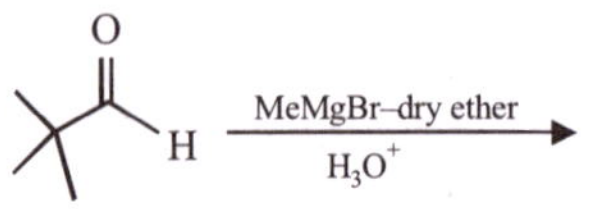

(B) 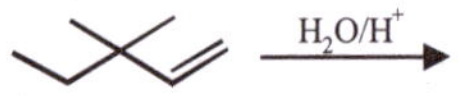

(C) 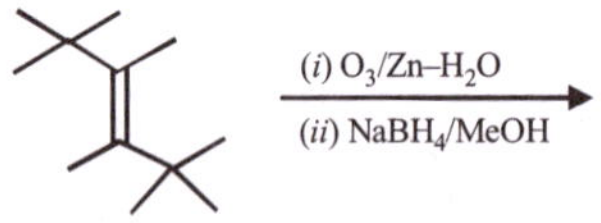

(D) 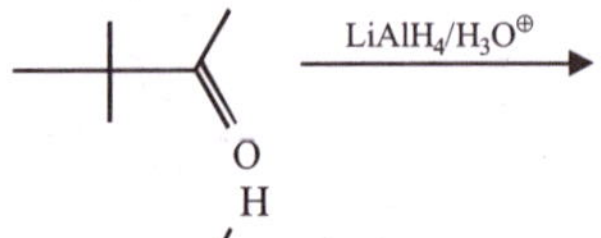

(E) 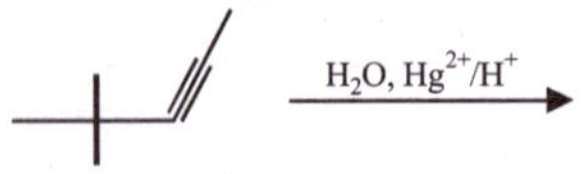

Choose the **correct** answer from the **options** given below:

(1) B only (2) B and E only

(3) B and C only (4) B, C and E only

4. A toxic compound "*A*" when reacted with NaCN in aqueous acidic medium yields an edible cooking component and food preservative '*B*'. "*B*" is converted to "*C*" by diborane and can be used as an additive to petrol to reduce emission. "*C*" upon reaction with oleum at 140°C yields an inhalable anesthetic "*D*". Identify "*A*", "*B*", "*C*" and "*D*", respectively. **[April 4, 2025 (II)]**

(1) Methanol; formaldehyde; methyl chloride; chloroform

(2) Ethanol; acetonitrile; ethylamine; ethylene

(3) Methanol; acetic acid; ethanol; diethyl ether

(4) Acetaldehyde;2- hydroxypropanoic acid; propanoic acid; dipropyl ether

5. Given below are two statements: **[Jan. 23, 2025 (II)]**

Statement I: The boiling points of alcohols and phenols increase with increase in the number of C-atoms.

Statement II : The boiling points of alcohols and phenols are higher in comparison to other class of compounds such as ethers, haloalkanes.

In the light of the above statements, choose the correct answer from the options given below

(1) Both Statement I and Statement II are false

(2) Statement I is false but Statement II is true

(3) Statement I is true but Statement II is false

(4) Both Statement I and Statement II are true

6. Identify the product A and product B in the following set of reactions. [April 9, 2024 (I)]

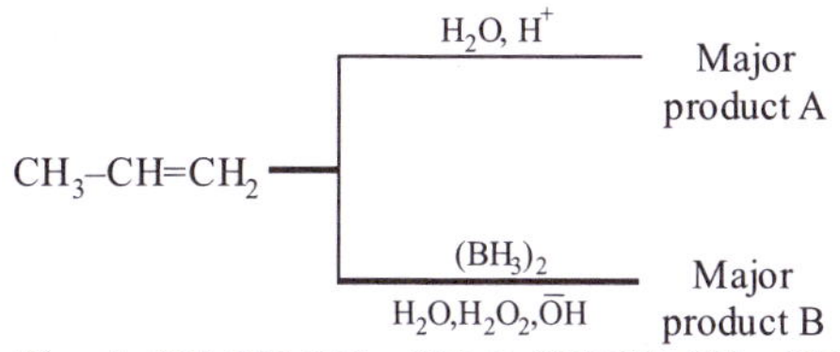

(1) A-$CH_3CH_2CH_2$–OH, B-$CH_3CH_2CH_2$–OH
(2) A – $CH_3CH_2CH_2$ – OH, B – CH_3 CH(OH) – CH_3
(3) A – CH_3 – CH(OH) – CH_3, B – $CH_3CH_2CH_2$ – OH
(4) A-$CH_3CH_2CH_3$, B-$CH_3CH_2CH_3$

7. Which one the following compounds will readily react with dilute NaOH? [April 8, 2024 (II)]
(1) $C_6H_5CH_2OH$ (2) C_2H_5OH
(3) $(CH_3)_3COH$ (4) C_6H_5OH

8. Identify the major products A and B respectively in the following set of reactions. [April 8, 2024 (I)]

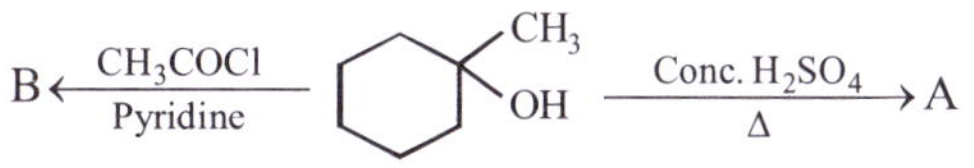

(1)
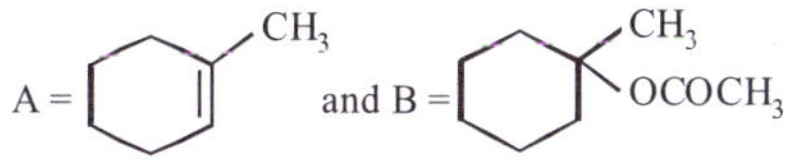

(2)
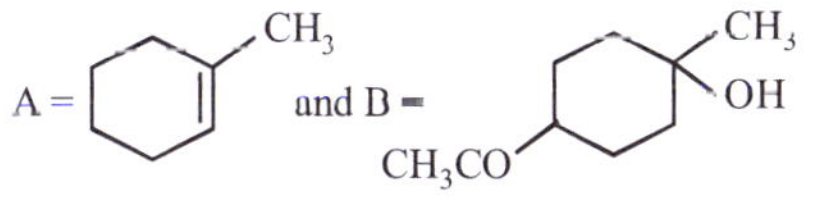

(3)
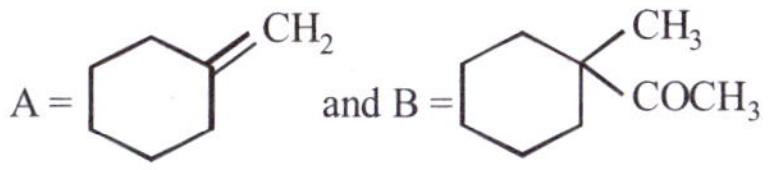

(4)
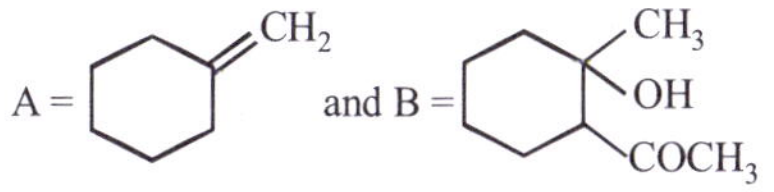

9. Given below are two statements : one is labelled as **Assertion A** and the other is labelled as **Reason R :**
Assertion A : Alcohols react both as nucleophiles and electrophiles. [Jan. 31, 2024 (I)]
Reason R : Alcohols react with active metals such as sodium, potassium and aluminum to yield corresponding alkoxides and liberate hydrogen.
In the light of the above statements, choose the **correct answer** from the options given below:
(1) Both A and R are true but R is NOT the correct explanation of A
(2) A is true but R is false
(3) Both A and R are true and R is the correct explanation of A
(4) A is false but R is true

10. The final product A, formed in the following reaction sequence is: [Jan. 27, 2024 (II)]

$$Ph-CH=CH_2 \xrightarrow[\text{(iii)HBr} \quad \text{(iv)Mg,ether,then } HCHO/H_3O^+]{\text{(i)}BH_3 \quad \text{(ii)}H_2O_2, {}^{\ominus}OH} A$$

(1) Ph – CH_2 – CH_2 – CH_3 (2) Ph – CH(CH_3) – CH_3
(3) Ph – CH(CH_2OH) – CH_3 (4) Ph – CH_2 – CH_2 – CH_2 – OH

11. Given below are two statements, one is labelled as **Assertion A** and the other is labelled as **Reason R**.
Assertion A : Order of acidic nature of the following compounds is A > B > C.

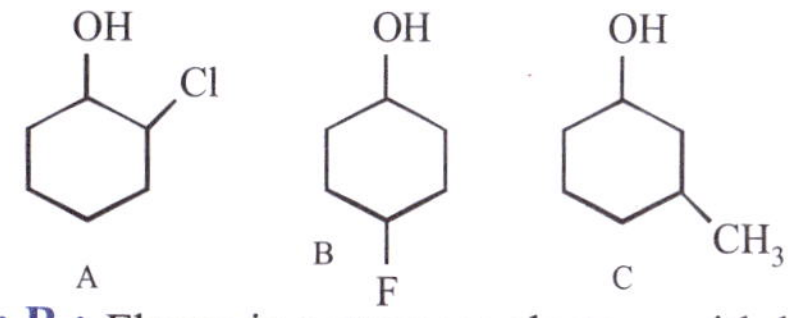

Reason R : Fluoro is a stronger electron withdrawing group than chloro group.
In the light of the above statements, choose the correct answer from the options given below: [April 13, 2023 (II)]
(1) A is false but R is true
(2) Both A and R are correct and R is the correct explanation of A
(3) Both A and R are correct but R is NOT the correct explanation of A
(4) A is true but R is false

12. $H_3C-CH_2-CH(OH)-CH_3 \xrightarrow[\text{(iii) } D_2O]{\text{(i) NaI, } H_3PO_4 \quad \text{(ii) Mg, Dry ether}}$ [X] Product

Product [X] formed in the above reaction is : [April 11, 2023 (II)]
(1) $H_3C-CH_2-CH(D)-CH_3$ (2) $H_3C-CH_2-C(H)(OH)-CH_3$
(3) $H_3C-CH_2-CH=CH_2$ (4) $H_3C-CH=CH-CH_3$

13. Incorrect method of preparation for alcohols from the following is: [April 10, 2023 (II)]
(1) Ozonolysis of alkene.
(2) Reaction of Ketone with RMgBr followed by hydrolysis.
(3) Hydroboration–oxidation of alkene.
(4) Reaction of alkyl halide with aqueous NaOH.

14. Given below are two statements: One is labelled as **Assertion A** and the other is labelled as **Reason R**.
Assertion A: Butan -1- ol has higher boiling point than ethoxyethane.
Reason R: Extensive hydrogen bonding leads to stronger association of molecules.
In the light of the above statements, choose the correct answer from the options given below: [April 08, 2023 (I)]
(1) Both **A** and **R** are true and **R** is the correct explanation of **A**.
(2) **A** is true but **R** is false.
(3) Both **A** and **R** are true but **R** is not the correct explanation of **A**.
(4) **A** is false but **R** is true.

15. In the following reaction, 'B' is [April 06, 2023 (II)]

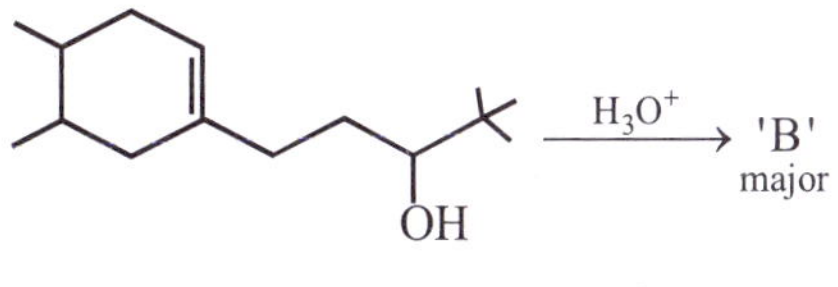

(1) 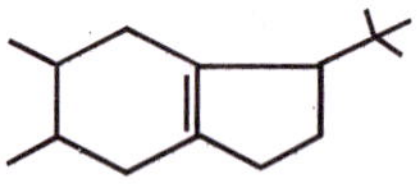 (2) 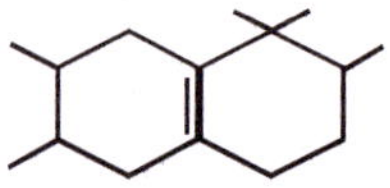

(3) 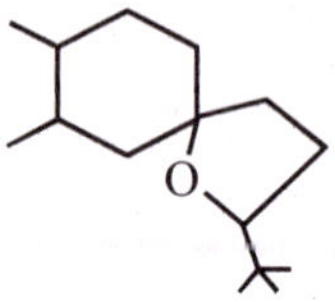 (4) 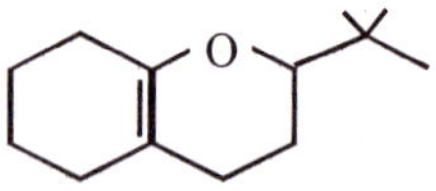

16. Compound A, $C_5H_{10}O_5$, gives a tetraacetate with Ac_2O and oxidation of A with $Br_2 - H_2O$ gives an acid, $C_5H_{10}O_6$. Reduction of A with HI gives isopentane. The possible strucutre of A is: **[Jan. 31, 2023 (II)]**

(1) 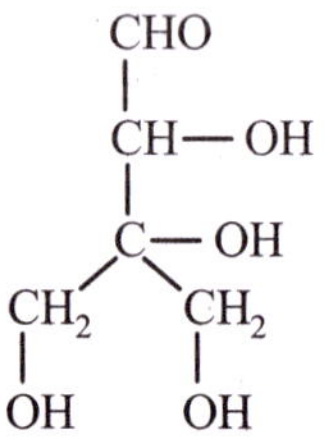

(2) 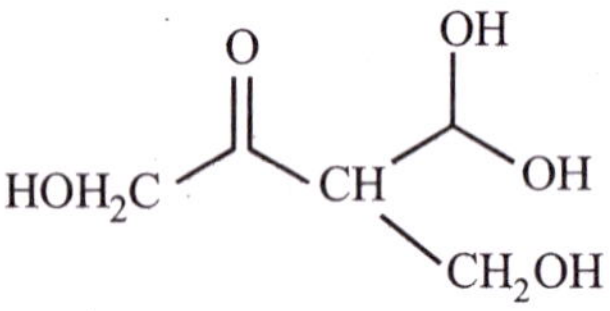

(3) 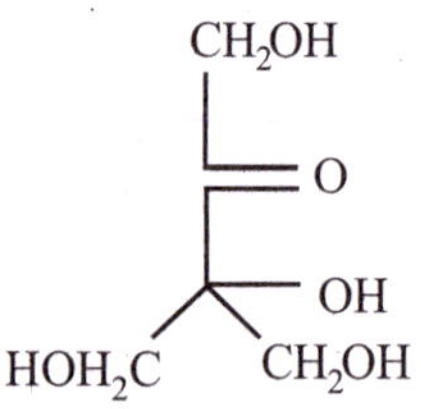

(4) 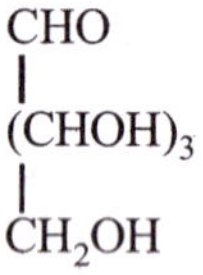

17. Find out the major product for the following reaction.

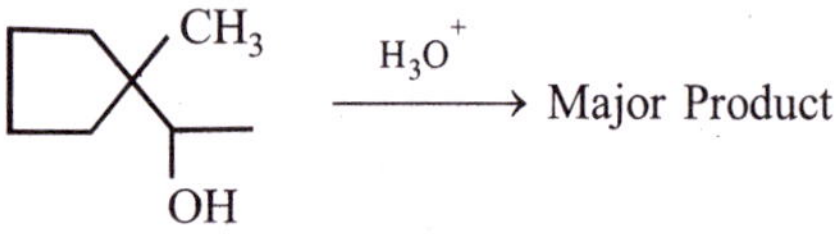

[Jan. 29, 2023 (II)]

(1) 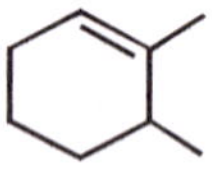 (2) 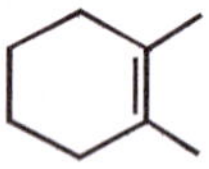

(3) 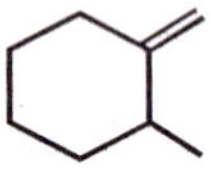 (4) 

18. Find out the major product from the following reaction.

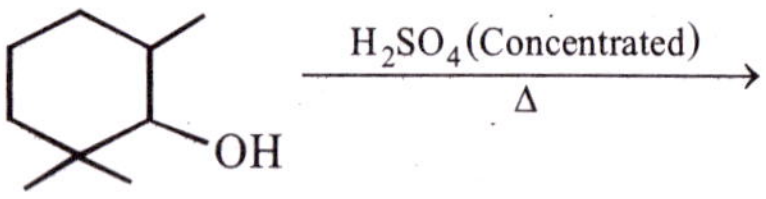

[Jan. 25, 2023 (II)]

(1) 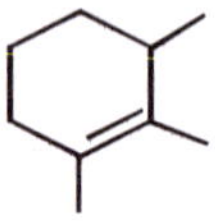 (2) 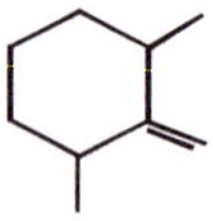

(3) 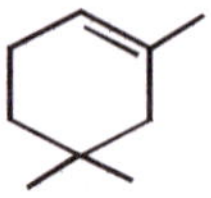 (4) 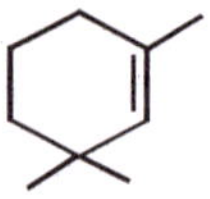

19. Find out the major products from the following reactions.

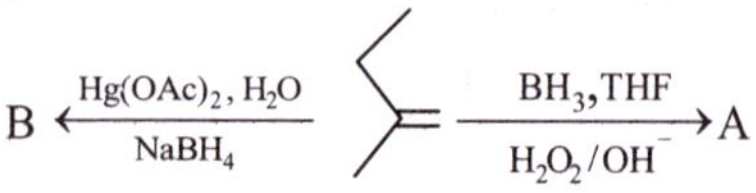

[Jan. 24, 2023 (II)]

(1) A = 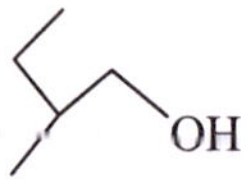 , B = 

(2) A =  , B = 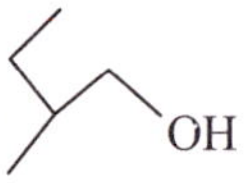

(3) A =  , B = 

(4) A = 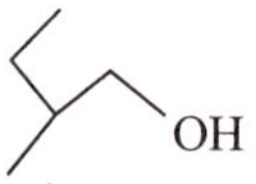 , B = 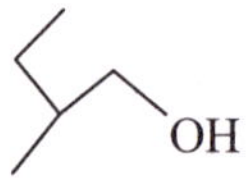

20. When enthanol is heated with conc. H_2SO_4, a gas is produced. The compound formed, when this gas is treated with cold dilute aqueous solution of Baeyer's reagent, is : **[July 29, 2022 (II)]**

(1) Formaldehyde (2) Formic acid

(3) Glycol (4) Ethanoic acid

21. The major product in the given reaction is

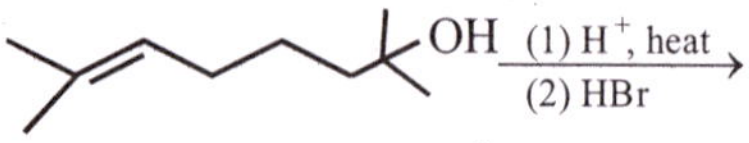

[July 28, 2022 (II)]

(1) 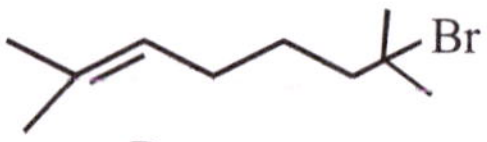

(2) 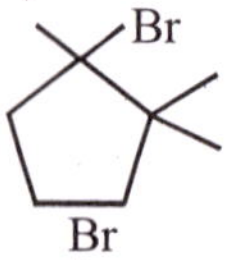

(3) 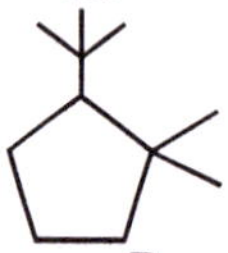

(4) 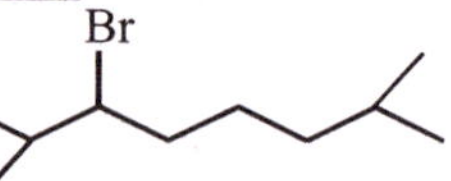

22. 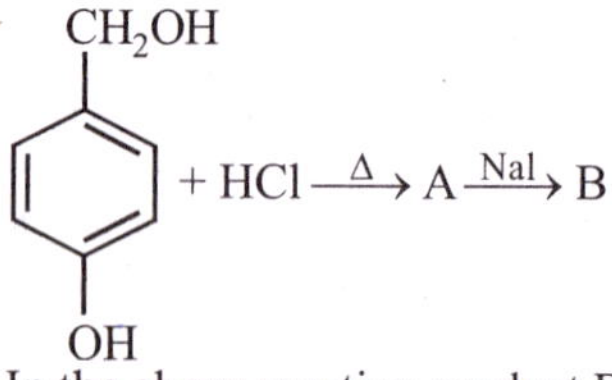

In the above reaction product B is: **[July 27, 2022 (I)]**

(1) 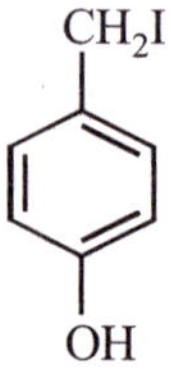 (2) 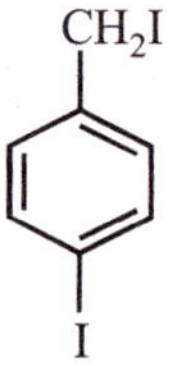

(3) 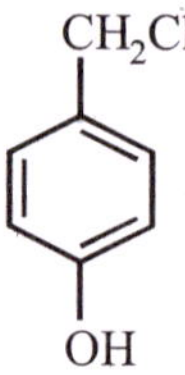 (4) 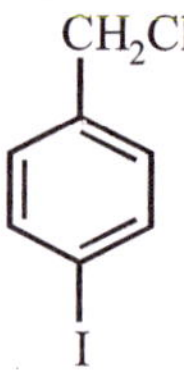

23. Most stable product of the following reaction is: [July 25, 2022 (I)]

(1) 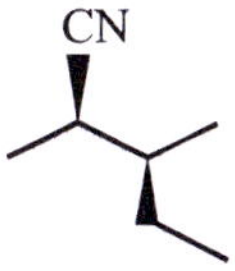 (2) 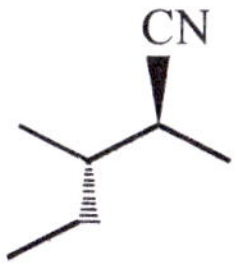

(3) 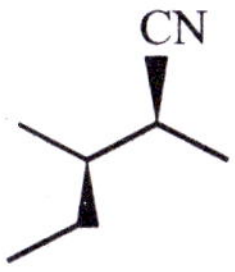 (4) 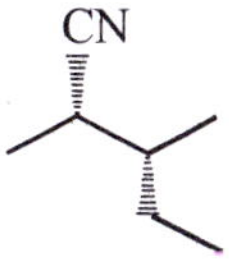

24. Given below are two statements : [July 25, 2022 (I)]

Statement I : On heating with $KHSO_4$, glycerol is dehydrated and acrolein is formed.

Statement II : Acrolein has fruity odour and can be used to test glycerol's presence.

Choose the **correct** option.

(1) Both Statement I and Statement II are correct.
(2) Both Statement I and Statement II are incorrect.
(3) Statement I is correct but Statement II is incorrect.
(4) Statement I is incorrect but Statement II is correct.

25. The major product (P) of the given reaction is (where, Me is $-CH_3$) [June 28, 2022 (II)]

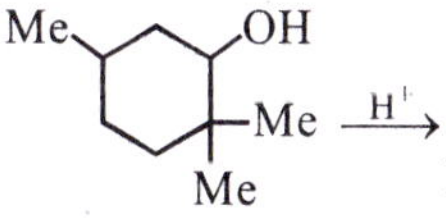

P
Major Product

(1) 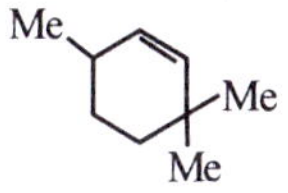 (2) 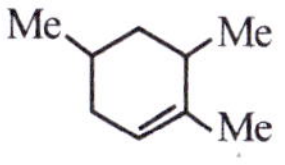

(3) 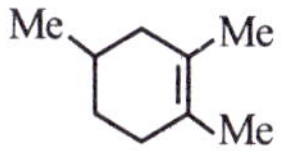 (4) 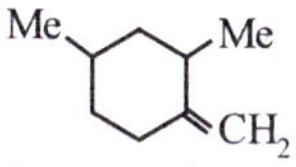

26. The major product in the following reaction [June 26, 2022 (II)]

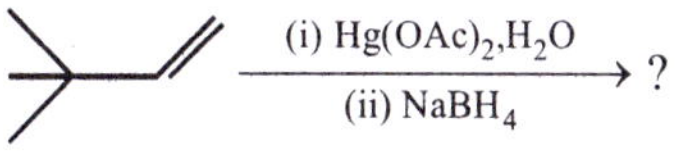

(1) 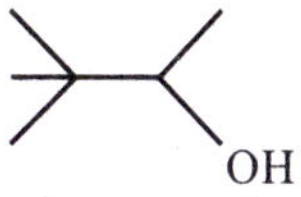 (2) 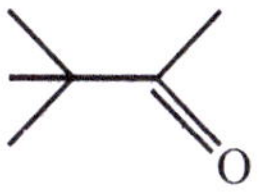

(3) 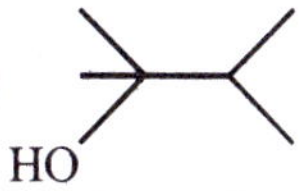 (4) 

27. The major product formed in the following reaction is : [Aug. 31, 2021 (I)]

$$CH_3-\underset{CH_3}{\overset{CH_3}{\underset{|}{\overset{|}{C}}}}-\underset{OH}{\underset{|}{CH}}-CH_3 \xrightarrow[\text{a few drops}]{\text{conc. } H_2SO_4} \text{Major product}$$

(1) $CH_3-C=CH-CH_2CH_3$ (2) 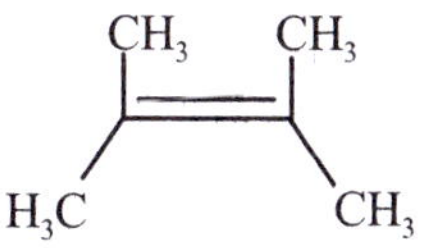

(3) $\begin{matrix}CH_3\\CH_3\end{matrix}\!\!>C=CH-CH_3$ (4) $CH_3-\underset{CH_3}{\underset{|}{C}}—CH=CH_2$

28. Ceric ammonium nitrate and $CHCl_3$/alc. KOH are used for the identification of functional groups present in _____ and _____ respectively. [Feb. 26, 2021 (II)]

(1) alcohol, phenol (2) amine, phenol
(3) amine, alcohol (4) alcohol, amine

29. What is 'X' in the given reaction? [Feb. 25, 2021 (II)]

$$\begin{matrix}CH_2OH\\|\\CH_2OH\end{matrix} + \text{oxalic acid} \xrightarrow{210°C} \underset{\text{major product}}{X}$$

(1) $\begin{matrix}CHO\\|\\CHO\end{matrix}$ (2) $\begin{matrix}CH_2\\||\\CH_2\end{matrix}$ (3) $\begin{matrix}CH-OH\\||\\CH_2\end{matrix}$ (4) $\begin{matrix}CH_2OH\\||\\CHO\end{matrix}$

30. What is the final product (major) 'A' in the given reaction? [Feb. 24, 2021 (I)]

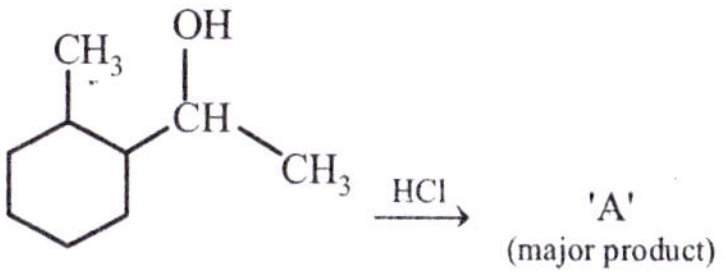

(1) 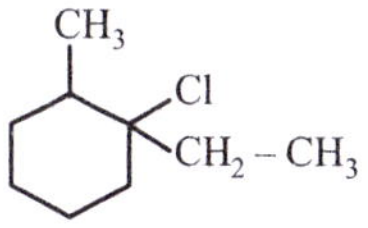 (2) 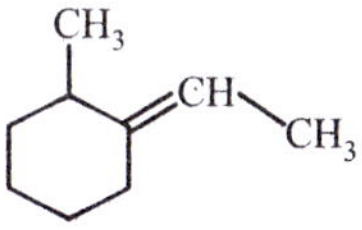

(3) 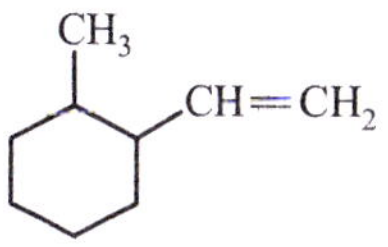 (4) 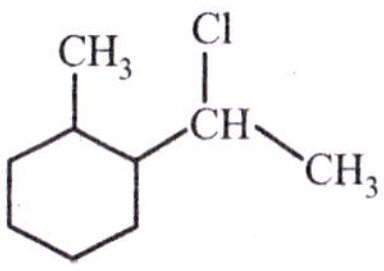

31. The major product of the following reaction is : [Sep. 05, 2020 (II)]

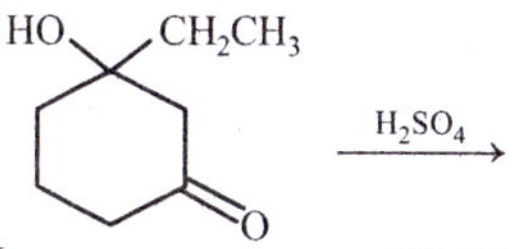

(1) 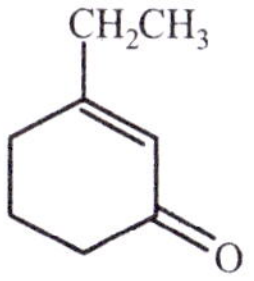 (2) 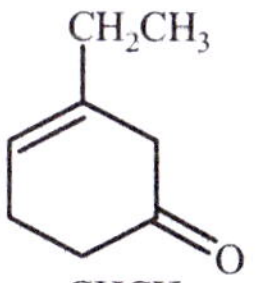

(3) 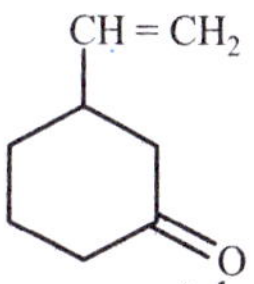 (4) 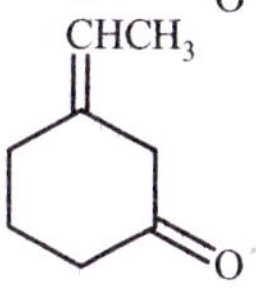

32. When neopentyl alcohol is heated with an acid, it is slowly converted into an 85 : 15 mixture of alkenes A and B, respectively. What are these alkenes? [Sep. 04, 2020 (I)]

(1) 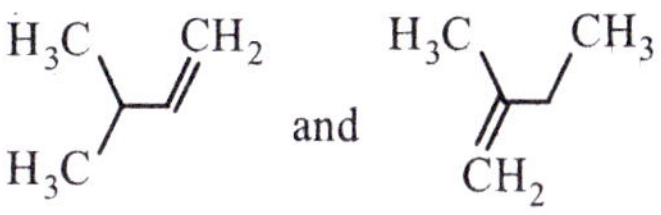

(2) 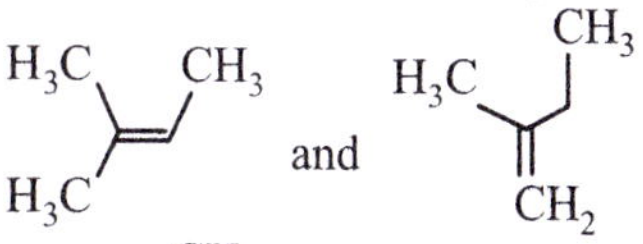

(3) 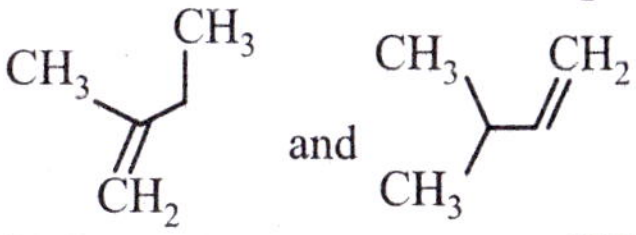

(4) 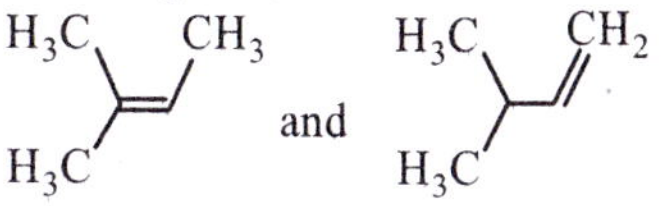

33. The major product [B] in the following sequence of reactions is: **[Jan. 08, 2020 (II)]**

$$CH_3-\underset{\underset{CH(CH_3)_2}{|}}{C}=CH-CH_2CH_3 \xrightarrow[\text{(ii) } H_2O_2,\ OH^{\ominus}]{\text{(i) } B_2H_6} [A] \xrightarrow[\Delta]{\text{dil. } H_2SO_4} [B]$$

(1) $CH_2=\underset{\underset{CH(CH_3)_2}{|}}{C}-CH_2CH_2CH_3$

(2) $CH_3-\underset{\underset{C(CH_3)_2}{||}}{C}-CH_2CH_2CH_3$

(3) $CH_3-\underset{\underset{CH(CH_3)_2}{|}}{C}=CH-CH_2CH_3$

(4) $CH_3-\underset{\underset{CH(CH_3)_2}{|}}{CH}-CH=CH-CH_3$

34. The major product of the following reaction is: **[Jan. 08, 2020 (I)]**

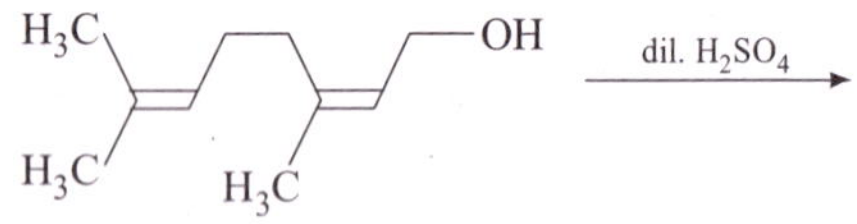

(1) 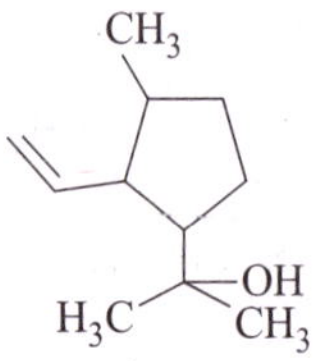 (2) 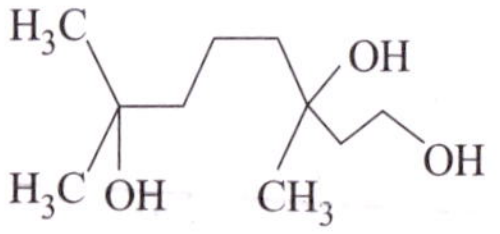

(3) 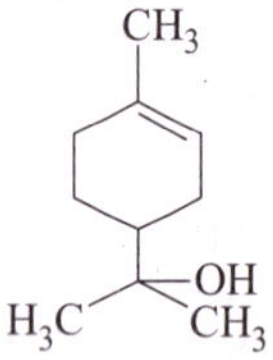 (4) 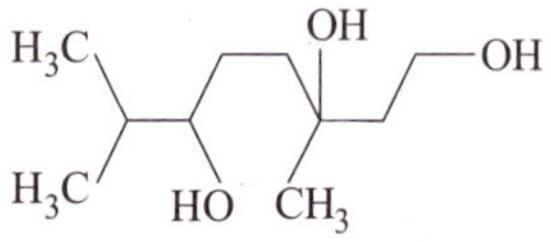

35. Heating of 2-chloro-1-phenylbutane with EtOK/EtOH gives X as the major product. Reaction of X with Hg $(OAc)_2/H_2O$ followed by $NaBH_4$ gives Y as the major product, Y is : **[April 12, 2019 (II)]**

(1) 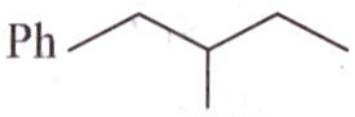 (2) 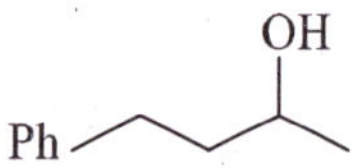

(3) 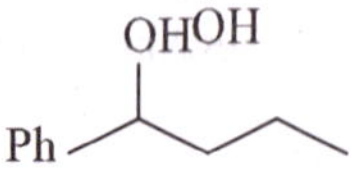 (4) 

36. $CH_3CH_2-\underset{\underset{Ph}{|}}{\overset{\overset{OH}{|}}{C}}-CH_3$ cannot be prepared by: **[Jan. 12, 2019 (I)]**

(1) $CH_3CH_2COCH_3 + PhMgX$
(2) $PhCOCH_2CH_3 + CH_3MgX$
(3) $PhCOCH_3 + CH_3CH_2MgX$
(4) $HCHO + PhCH(CH_3)CH_2MgX$

37. Which is the most suitable reagent for the following transformation? **[Jan. 10, 2019 (II)]**

$$CH_3-CH=CH-CH_2-\overset{\overset{OH}{|}}{CH}-CH_3 \longrightarrow CH_3-CH=CH-CH_2CO_2H$$

(1) Tollen's reagent (2) $I_2/NaOH$
(3) CrO_2Cl_2/CS_2 (4) alkaline $KMnO_4$

38. The major product 'X' formed in the following reaction is **[Jan. 10, 2019 (I)]**

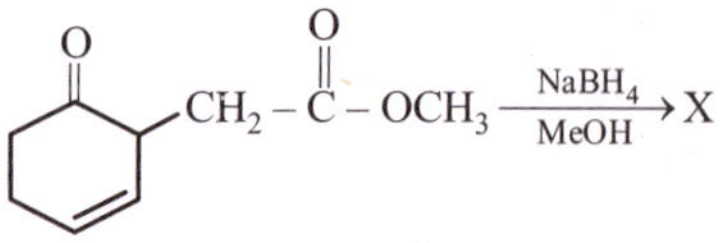

(1) 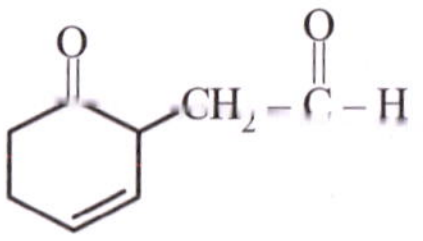 (2) 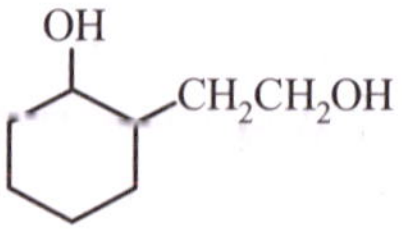

(3) 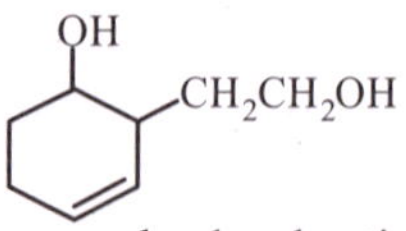 (4) 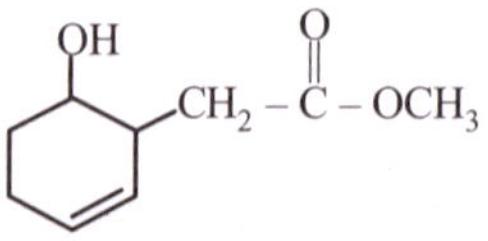

39. The gas evolved on heating CH_3MgBr in methanol is : **[Online April 9, 2016]**

(1) Methane (2) Ethane
(3) Propane (4) HBr

40. The most suitable reagent for the conversion of $R-CH_2-OH \rightarrow R-CHO$ is: **[2014]**

(1) $KMnO_4$
(2) $K_2Cr_2O_7$
(3) CrO_3
(4) PCC (Pyridinium chlorochromate)

41. In the Victor-Meyer's test, the colour given by 1°, 2° and 3° alcohols are respectively: **[Online April 12, 2014]**

(1) Red, colourless, blue (2) Red, blue, colourless
(3) Colourless, red, blue (4) Red, blue, violet

42. 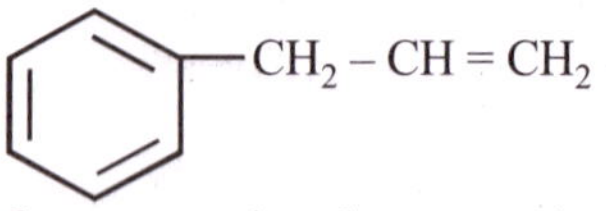

On mercuration-demercuration produces the major product: **[Online April 12, 2014]**

(1) 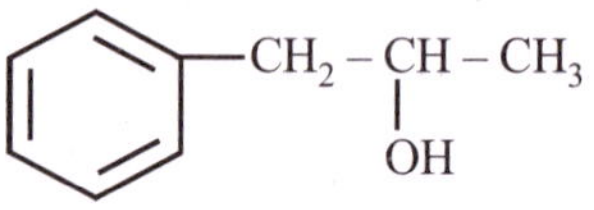

(2) 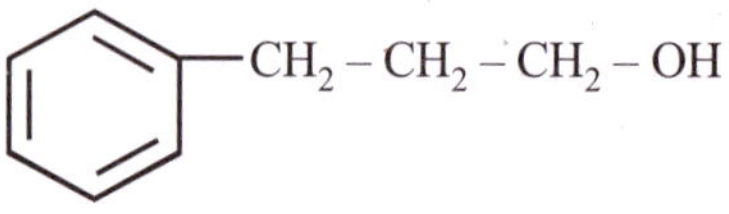

(3) 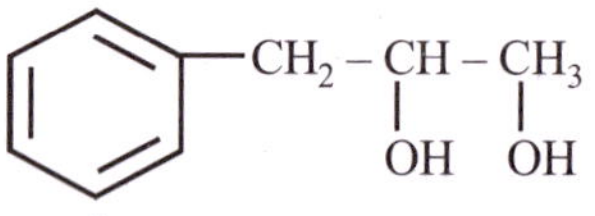

(4) 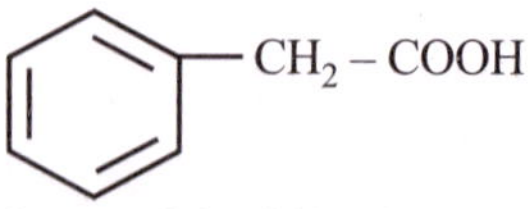

43. Which one of the following statements is not correct? **[Online April 11, 2014]**

(1) Alcohols are weaker acids than water
(2) Acid strength of alcohols decreases in the following $RCH_2OH > R_2CHOH > R_3COH$
(3) Carbon-oxygen bond length in methanol, CH_3OH is shorter than that of C – O bond length in phenol.
(4) The bond angle 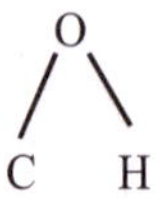 in methanol is 108.9°.

44. In the hydroboration - oxidation reaction of propene with diborane, H_2O_2 and NaOH, the organic compound formed is: [Online April 9, 2014]
(1) CH_3CH_2OH (2) $CH_3CHOHCH_3$
(3) $CH_3CH_2CH_2OH$ (4) $(CH_3)_3COH$

45. An unknown alochol is treated with the "Lucas reagent" to determine whether the alcohol is primary, secondary or tertiary. Which alcohol reacts fastest and by what mechanism : [2013]
(1) secondary alcohol by S_N1
(2) tertiary alcohol by S_N1
(3) secondary alcohol by S_N2
(4) tertiary alcohol by S_N2

46. Amongst the following alcohols which would react fastest with conc. HCl and $ZnCl_2$? [2010 (s), Online April 22, 2013]
(1) pentanol (2) 2-methyl butanol
(3) 2-pentanol (4) 2-methyl butan-2-ol

47. Rate of dehydration of alcohols follows the order: [Online April 9, 2013]
(1) $2° > 1° > CH_3OH > 3°$ (2) $3° > 2° > 1° > CH_3OH$
(3) $2° > 3° > 1° > CH_3OH$ (4) $CH_3OH > 1° > 2° > 3°$

48. Reagent used to convert allyl alcohol to acrolein is [Online May 19, 2012]
(1) MnO_2 (2) H_2O_2 (3) OsO_4 (4) $KMnO_4$

49. Consider thiol anion ($RS^{\ominus}$) and alkoxy anion ($RO^{\ominus}$). Which of the following statements is correct ? [2011RS]
(1) $RS^{\ominus}$ is less basic but more nucleophilic than $RO^{\ominus}$
(2) $RS^{\ominus}$ is more basic and more nucleophilic than $RO^{\ominus}$
(3) $RS^{\ominus}$ is more basic but less nucleophilic than $RO^{\ominus}$
(4) $RS^{\ominus}$ is less basic and less nucleophilic than $RO^{\ominus}$

50. The main product of the following reaction is

$$C_6H_5CH_2CH(OH)CH(CH_3)_2 \xrightarrow{conc.H_2SO_4} ?$$ [2010]

(1) $(H_5C_6)(H)C=C(H)(CH(CH_3)_2)$
(2) $(C_6H_5CH_2)(H)C=C(CH_3)(CH_3)$
(3) $(C_6H_5CH_2CH_2)(H_3C)C=CH_2$
(4) $(C_6H_5)(H)C=C(CH(CH_3)_2)(H)$

51. From amongst the following alcohols, the one that would react fastest with conc. HCl and anhydrous $ZnCl_2$, is [2010]
(1) 2-Butanol (2) 2- Methylpropan-2-ol
(3) 2-Methylpropanol (4) 1- Butanol

52. In the following sequence of reactions, [2007]

$$CH_3CH_2OH \xrightarrow{P+I_2} A \xrightarrow[ether]{Mg} B \xrightarrow{HCHO} C \xrightarrow{H_2O} D$$

the compound D is
(1) propanal (2) butanal
(3) *n*-butyl alcohol (4) *n*-propyl alcohol.

53. Among the following the one that gives positive iodoform test upon reaction with I_2 and NaOH is [2006]
(1) $CH_3-CH(CH_3)CH_2OH$
(2) $PhCHOHCH_3$
(3) $CH_3CH_2CH(OH)CH_2CH_3$
(4) $C_6H_5CH_2CH_2OH$

54. The best reagent to convert pent-3-en-2-ol into pent-3-en-2-one is [2005]
(1) Pyridinium chlorochromate
(2) Chromic anhydride in glacial acetic acid
(3) acidic dichromate
(4) Acidic permanganate

55. Among the following compounds which can be dehydrated very easily? [2004]
(1) $CH_3CH_2-C(CH_3)(OH)-CH_2CH_3$
(2) $CH_3CH_2CH_2CH(OH)CH_3$
(3) $CH_3CH_2CH_2CH_2CH_2OH$
(4) $CH_3CH_2CH(CH_3)CH_2CH_2OH$

56. During dehydration of alcohols to alkenes by heating with conc. H_2SO_4, the initiation step is [2003]
(1) formation of carbocation
(2) elimination of water
(3) formation of an ester
(4) protonation of alcohol molecule

Numerical Value Questions (NVQs)

57. One mole of phenol is treated with dilute HNO_3 at 298K to give a mixture of products. The mixture is separated by steam distillation. The steam volatile compound (X) is separated. The increase in percentage of oxygen in (X) with respect to phenol is ________ $10^{-1}\%$. [April 05, 2026 (I)]
(Given molar mass in g mol^{-1} H: 1, C:12, N:14, O:16)

58. Vanillin compound obtained from vanilla beans, has total sum of oxygen atoms and π electrons is _____. [April 4, 2024 (II)]

59. Number of compounds giving (i) red colouration with ceric ammonium nitrate and also (ii) positive iodoform test from the following is _______ [Jan. 25, 2023 (II)]

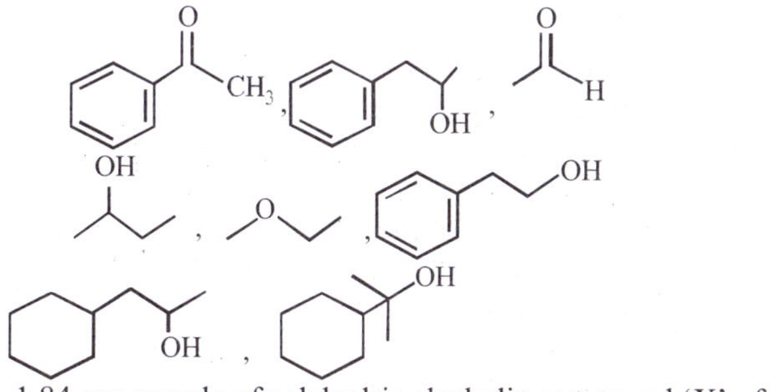

60. A 1.84 mg sample of polyhydric alcoholic compound 'X' of molar mass 92.0 g/mol gave 1.344 mL of H_2 gas at STP. The number of alcoholic hydrogens present in compound 'X' is____. [July 29, 2022 (II)]

61. A 100 mL solution of CH_3CH_2MgBr on treatment with methanol produces 2.24 mL of a gas at STP. The weight of gas produced is _______ mg. [nearest integer] [July 26, 2022 (II)]

62. A sample of 4.5 mg of an unknown monohydric alcohol, R–OH was added to methylmagnesium iodide. A gas is evolved and is collected and its volume measured to be 3.1 mL. The molecular weight of the unknown alcohol is _____ g/mol. [Nearest integer] [July 25, 2022 (II)]

63. The number of chiral alcohol(s) with molecular formula $C_4H_{10}O$ is______. [June 29, 2022 (II)]

64. To synthesise 1.0 mole of 2-methylpropan-2-ol from ethylethanoate ________ equivalents of CH_3MgBr reagent will be required. (Integer value) [July 20, 2021 (I)]

65. The number of chiral centres present in [B] is ________. [Sep. 04, 2020 (I)]

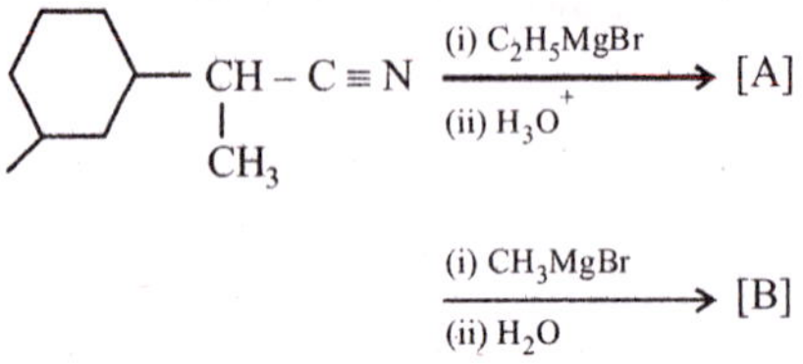

Topic 2 - Preparation and Properties of Phenols

MULTIPLE CHOICE QUESTIONS (MCQs)

66. Consider the following reaction sequence [Jan. 28, 2026 (I)]

Compound (X) [76.6%C, 6.38%H, vapour density 47] $\xrightarrow[\text{(ii) } H_3O^+]{\text{(i) } CO_2\text{, NaOH, 120°C, high pressure}}$ Compound (Y) (Major product)

Compound (Y) develops characteristic colour with neutral $FeCl_3$ solution.

Identify the **INCORRECT** statement from the following for the above sequence.

(1) Both compounds X and y will dissolve in NaOH.

(2) Compound Y will dissolve in $NaHCO_3$ and evolves a gas.

(3) Compound X is more acidic than compound Y.

(4) Both compounds X and Y will burn with sooty flame.

67. A hydroxy compound (X) with molar mass 122 g mol^{-1} is acetylated with acetic anhydride, using a large excess of the reagent ensuring complete acetylation of all hydroxyl groups. The product obtained has a molar mass of 290 g mol^{-1}. The number of hydroxyl groups present in compound (X) is : [Jan. 24, 2026 (I)]

(1) 3 (2) 5 (3) 2 (4) 4

68. Given below are two statements : [Jan. 22, 2026 (I)]

Statement I : Phenol on treatment with $CHCl_3$/aq. KOH under refluxing condition, followed by acidification produces *p*-hydroxy benzaldehyde as the major product and *o*-hydroxy benzaldehyde as the minor product.

Statement II : The mixture of *p*-hydroxybenzaldehyde and *o*-hydroxybenzaldehyde can be easily separated through steam distillation.

In the light of the above statements, choose the **correct** answer from the options given below :

(1) Both Statement I and Statement II are false

(2) Statement I is true but Statement II is false

(3) Both Statement I and Statement II are true

(4) Statement I is false but Statement II is true

69. Identify the products [A] and [B], respectively in the following reaction: [Jan. 23, 2025 (II)]

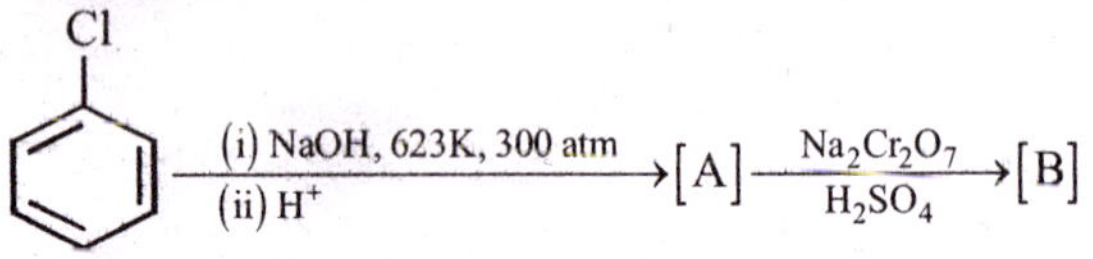

(1) [A] 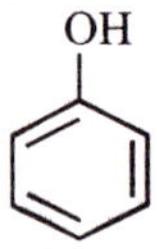 , [B] 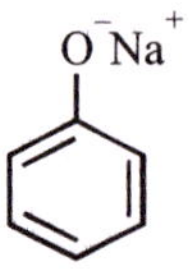

(2) [A] 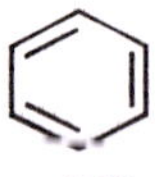 , [B] 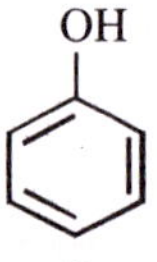

(3) [A] 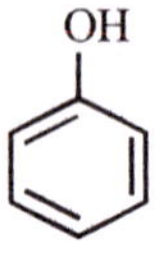 , [B] 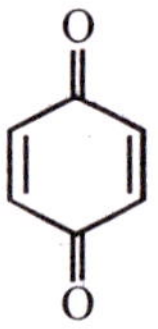

(4) [A]  , [B] 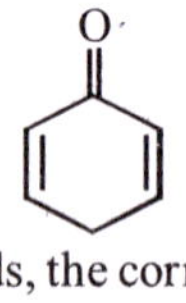

70. For the given compounds, the correct order of increasing pK_a value : [April 9, 2024 (I)]

(A) 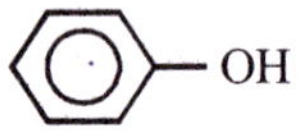 (B) 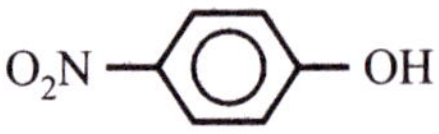

(C) 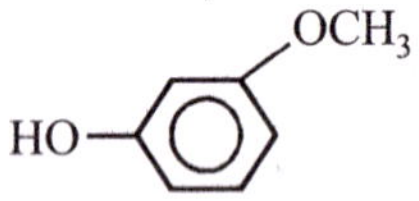 (D) 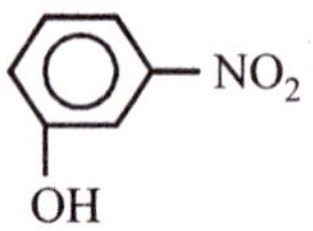

(E) 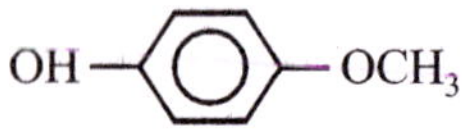

(1) (E) < (D) < (C) < (B) < (A)

(2) (D) < (E) < (C) < (B) < (A)

(3) (E) < (D) < (B) < (A) < (C)

(4) (B) < (D) < (A) < (C) < (E)

71. Match **List-I** with **List-II**. [April 8, 2024 (II)]

List-I (Reactions)		**List-II (Products)**	
(A)	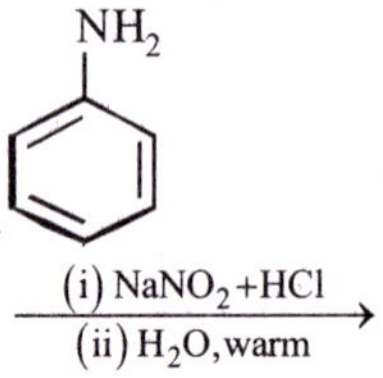	(I)	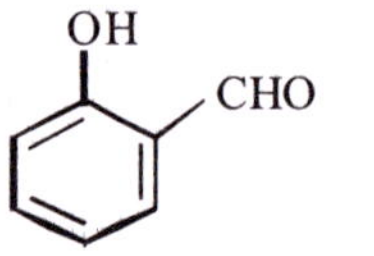
(B)	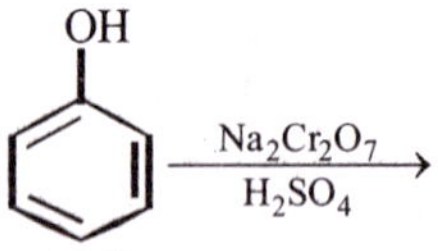	(II)	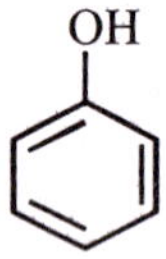
(C)	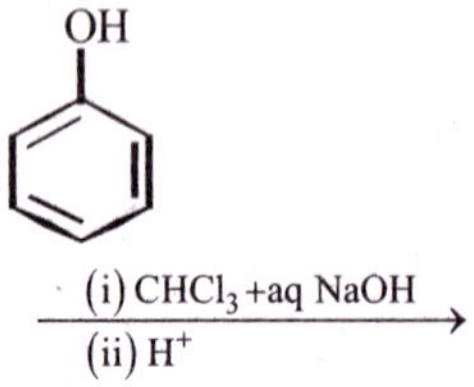	(III)	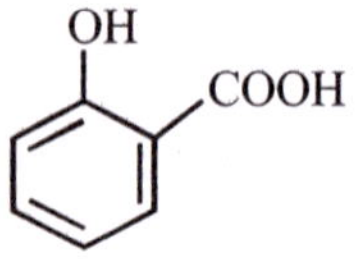
(D)	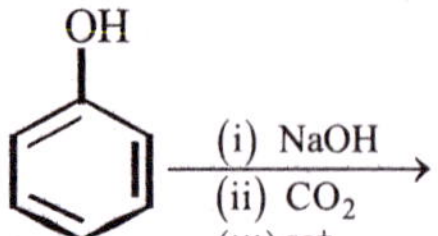	(IV)	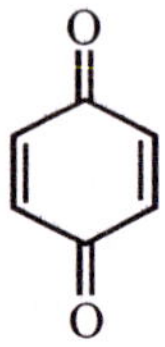

Choose the **correct** answer from the options given below:
(1) (A)-(III), (B)-(II), (C)-(I), (D)-(IV)
(2) (A)-(IV), (B)-(II), (C)-(III), (D)-(I)
(3) (A)-(I), (B)-(IV), (C)-(II), (D)-(III)
(4) (A)-(II), (B)-(IV), (C)-(I), (D)-(III)

72. Given below are two statements : **[April 6, 2024 (I)]**
Statement I : Picric acid is 2, 4, 6-trinitrotoluene.
Statement II : Phenol-2, 4-disulphuric acid is treated with conc. HNO_3 to get picric acid.
In the light of the above statement, choose the **most appropriate** answer from the options given below :
(1) Statement I is incorrect but Statement II is correct.
(2) Both Statement I and Statement II are incorrect.
(3) Statement I is correct but Statement II is incorrect.
(4) Both Statement I and Statement II are correct.

73. In Reimer - Tiemann reaction, phenol is converted into salicylaldehyde through an intermediate. The structure of intermediate is ______. **[April 6, 2024 (I)]**

(1) 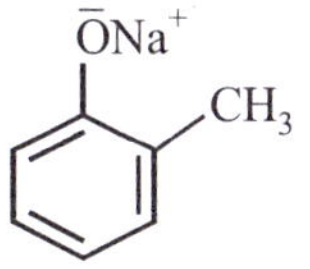
(2) 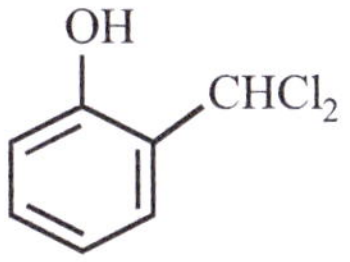
(3) 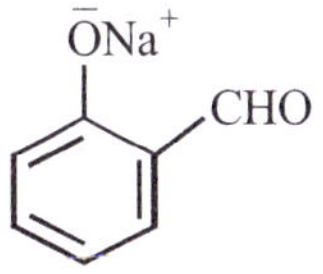
(4) 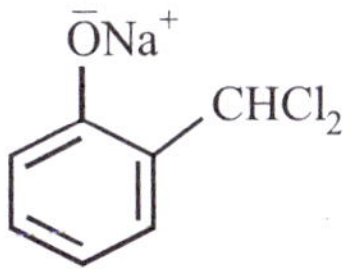

74. Identify compound (Z) in the following reaction sequence.

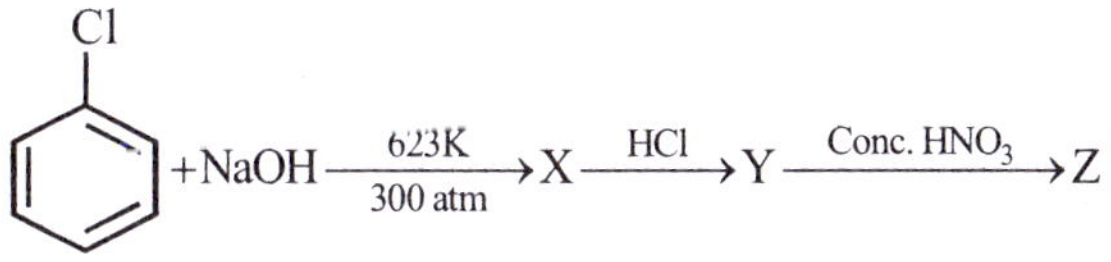

[April 5, 2024 (I)]

(1) 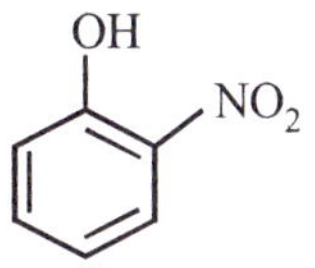
(2) 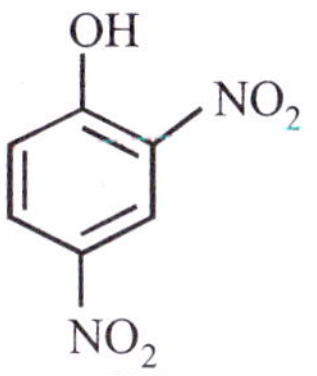
(3) 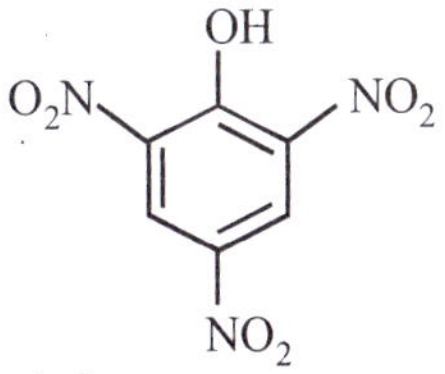
(4) 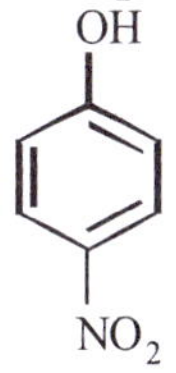

75. Given below are two statement : **[April 5, 2024 (I)]**
Statement I : Bromination of phenol in solvent with low polarity such as $CHCl_3$ or CS_2 requires Lewis acid catalyst.
Statement II : The Lewis acid catalyst polarises the bromine to generate Br^+.
In the light of the above statements, choose the **correct** answer from the options given below :
(1) Statement I is true but Statement II is false.
(2) Both Statement I and Statement II are true
(3) Both Statement I and Statement II are false.
(4) Statement I is false but Statement II is true.

76. Common name of benzene-1, 2-diol is **[April 4, 2024 (II)]**
(1) quinol (2) resorcinol
(3) catechol (4) *o*-cresol

77. Match List-I with List-II **[Feb. 1, 2024 (II)]**

List-I (Reactants)	**List-II (Product)**
(A) Phenol, Zn/ Δ	(I) Salicylaldehyde
(B) Phenol, $CHCl_3$, NaOH, HCl	(II) Salicylic acid
(C) Phenol, CO_2, NaOH, HCl	(III) Benzene
(D) Phenol, Conc. HNO_3	(IV) Picric acid

Choose the correct answer from the options given below:
(1) (A)- (IV), (B) - (I), (C) - (II), (D) - (III)
(2) (A) - (III), (B) - (I), (C) - (II), (D) - (IV)
(3) (A) - (IV), (B) - (II), (C) - (I), (D) - (III)
(4) (A) - (III), (B) - (IV), (C) - (I), (D) - (II)

78. Given below are two statements : one is labelled as **Assertion A** and the other is labelled as **Reason R :** **[Jan. 31, 2024 (I)]**
Assertion A : pK_a value of phenol is 10.0 while that of ethanol is 15.9.
Reason R : Ethanol is stronger acid than phenol.
In the light of the above statements, choose the **correct answer** from the options given below:
(1) A is true but R is false
(2) Both A and R are true but R is NOT the correct explanation of A
(3) Both A and R are true and R is the correct explanation of A
(4) A is false but R is true

79. Salicylaldehyde is synthesized from phenol, when reacted with **[Jan. 30, 2024 (II)]**

(1) 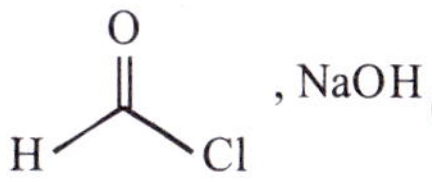 , NaOH (2) CO_2, NaOH
(3) CCl_4, NaOH (4) $HCCl_3$, NaOH

80. Phenol treated with chloroform in presence of sodium hydroxide, which further hydrolyzed in presence of an acid results **[Jan. 29, 2024 (II)]**
(1) Benzene- 1, 2-diol
(2) Benzene- 1, 3-diol
(3) 2-Hydroxybenzaldehyde
(4) Salicylic acid

81. Match List-I with List-II **[Jan. 29, 2024 (II)]**

LIST-I (Compound)	**LIST-II (pK_a value)**
A. Ethanol	I. 10.0
B. Phenol	II. 15.9
C. m-Nitrophenol	III. 7.1
D. p-Nitrophenol	IV. 8.3

Choose the correct answer from the options given below:
(1) A-I, B-II, C-III, D-IV (2) A-II, B-I C-IV, D-III
(3) A-III, B-IV, C-I, D-II (4) A-IV, B-I, C-II, D-III

82. Given below are two statements : one is labelled as **Assertion A** and the other is labelled as **Reason R :** **[Jan. 29, 2024 (I)]**
Assertion A : Aryl halides cannot be prepared by replacement of hydroxyl group of phenol by halogen atom.
Reason R : Phenols react with halogen acids violently.
In the light of the above statements, choose the **most appropriate** from the options given below:
(1) **A** is false but **R** is true
(2) Both **A** and **R** are true and **R** is the correct explanation of **A**
(3) **A** is true but **R** is false
(4) Both **A** and **R** are true but **R** is NOT the correct explanation of **A**

83. Match List-I with List-II. [Jan. 27, 2024 (II)]

List–I
(Reaction)

(A) 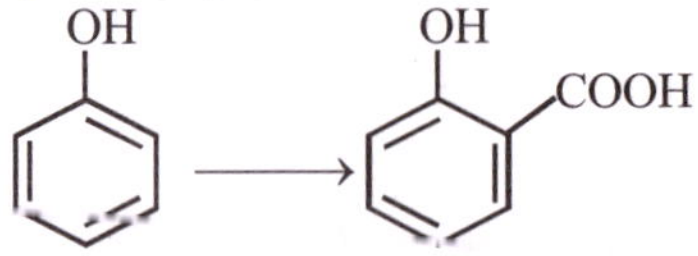

(B) 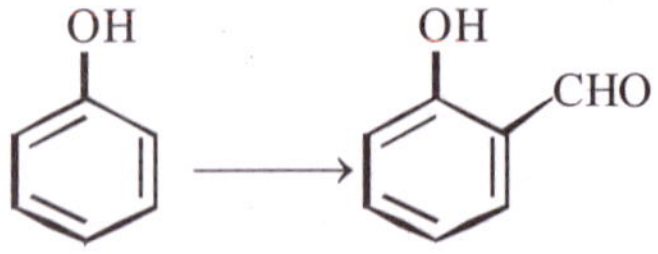

(C) 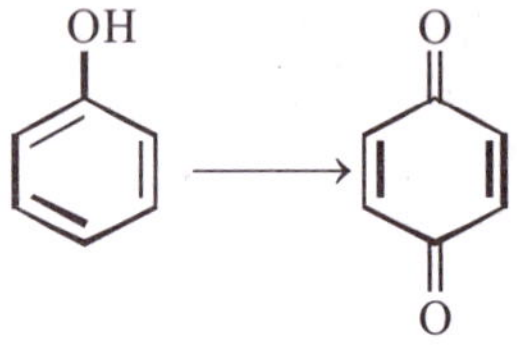

(D) 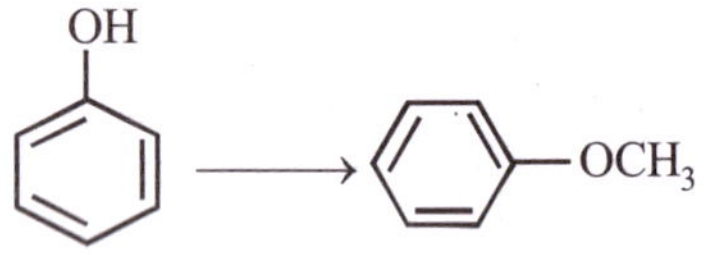

List–II
(Reagent(s))

(I) $Na_2Cr_2O_7$, H_2SO_4
(II) (i) NaOH (ii) CH_3Cl
(III) (i) NaOH, $CHCl_3$ (ii) NaOH (iii) HCl
(IV) (i) NaOH (ii) CO_2 (iii) HCl

Choose the correct answer from the options given below:
(1) (A)-(IV), (B)-(I), (C)-(III), (D)-(II)
(2) (A)-(II), (B)-(III), (C)-(I), (D)-(IV)
(3) (A)-(II), (B)-(I), (C)-(III), (D)-(IV)
(4) (A)-(IV), (B)-(III), (C)-(I), (D)-(II)

84. Given below are two statements : [Jan. 27, 2024 (I)]
Statement (I) : p-Nitrophenol is more acidic than m-nitrophenol and o-nitrophenol.
Statement (II) : Ethanol will give immediate turbidity with Lucas reagent.
In the light of the above statements, choose the correct answer from the options given below :
(1) Statement I is true but Statement II is false
(2) Both Statement I and Statement II are true
(3) Both Statement I and Statement II are false
(4) Statement I is false but Statement II is true

85. The ascending order of acidity of –OH group in the following compounds is : [Jan. 27, 2024 (I)]
(A) Bu – OH

(B) 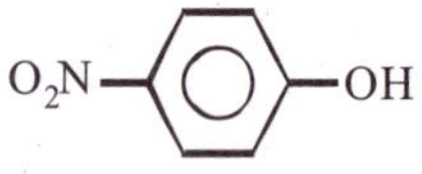 (C) 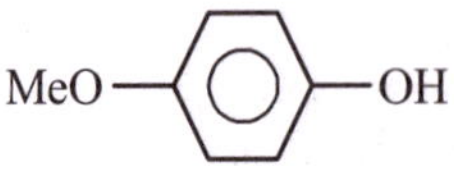

(D) 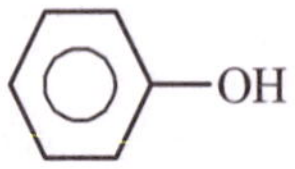 (E) 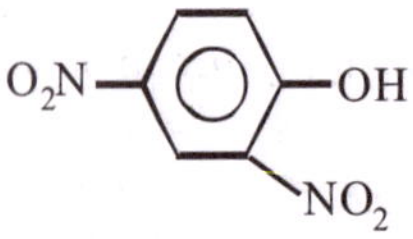

Choose the correct answer from the options given below:
(1) (A) < (D) < (C) < (B) < (E)
(2) (C) < (A) < (D) < (B) < (E)
(3) (C) < (D) < (B) < (A) < (E)
(4) (A) < (C) < (D) < (B) < (E)

86. Compound 'B' is [April 11, 2023 (II)]

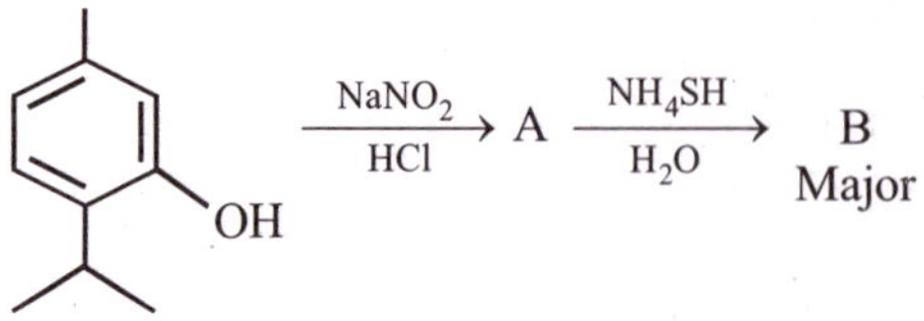

(1)  (2) 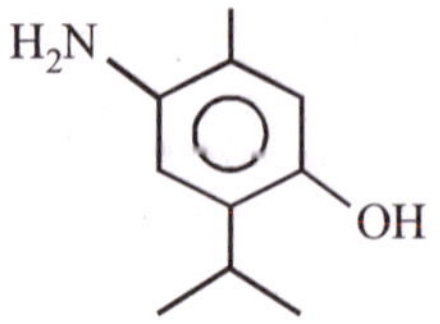

(3) 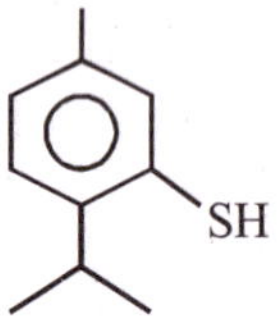 (4) 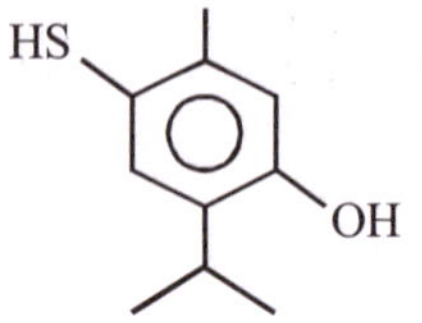

87. A compound 'X' when treated with phthalic anhydride in presence of concentrated H_2SO_4 yields 'Y'. 'Y' is used as an acid/base indicator. 'X' and 'Y' are respectively: [April 08, 2023 (II)]
(1) Carbolic acid, Phenolphthalein
(2) Anisole, Methyl orange
(3) Salicylaldehyde, Phenolphthalein
(4) Toludine, Phenolphthalein

88. The strongest acid from the following is [April 06, 2023 (II)]

(1) 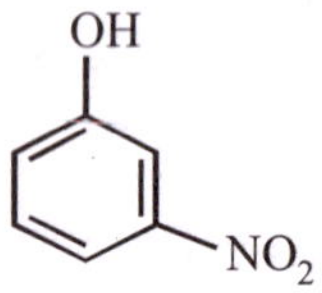 (2) 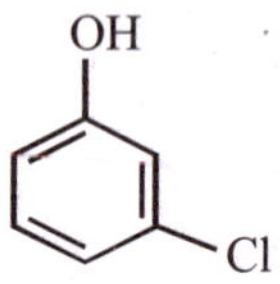

(3) 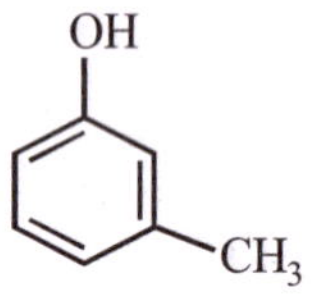 (4) 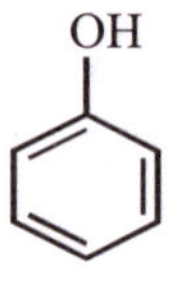

89. Decreasing order of dehydration of the following alcohols is [Feb. 01, 2023 (I)]

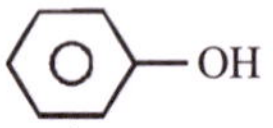 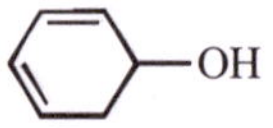 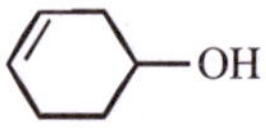 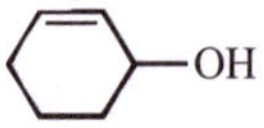

A B C D

(1) D > B > C > A (2) A > D > B > C
(3) B > A > D > C (4) B > D > C > A

90. An organic compound 'A' with empirical formula C_6H_6O gives sooty flame on burning. Its reaction with bromine solution in low polarity solvent results in high yield of B. B is [Jan. 31, 2023 (I)]

(1) 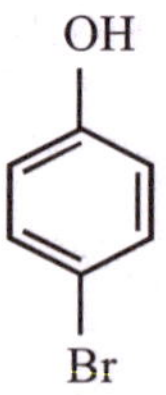 (2) 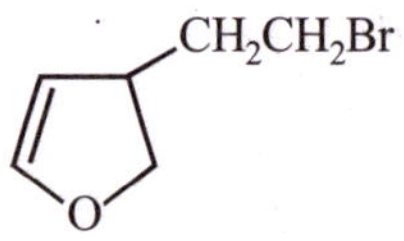

(3) 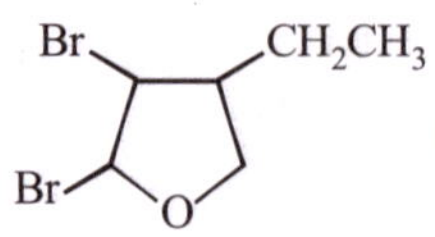 (4) 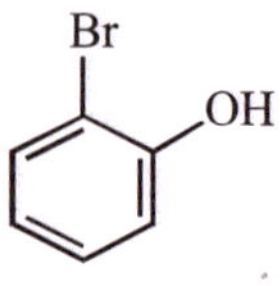

91. The correct order of pK_a values for the following compounds is: [Jan. 30, 2023 (II)]

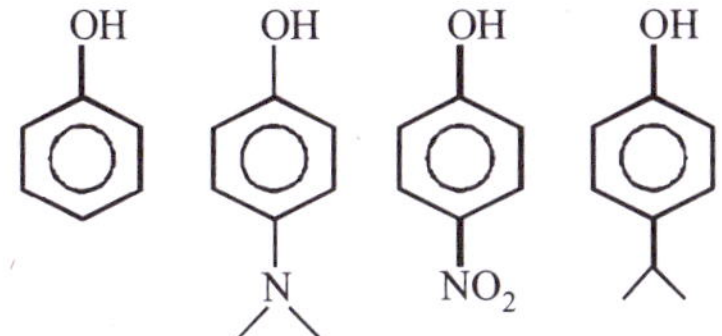

(A) (B) (C) (D)

(1) C > A > D > B (2) B > D > A > C
(3) B > A > D > C (4) A > B > C > D

92. The increasing order of pK_a for the following phenols is [Jan. 29, 2023 (I)]

(A) 2, 4-Dinitrophenol (B) 4 - Nitrophenol
(C) 2,4, 5-Trimethylphenol (D) Phenol
(E) 4-Chlorophenol

(1) A, E, B, D, C (2) A, B, E, D, C
(3) C, D, E, B, A (4) C, E, D, B, A

93. In the cumene to phenol preparation in presence of air, the intermediate is [Jan. 25, 2023 (I)]

(1) 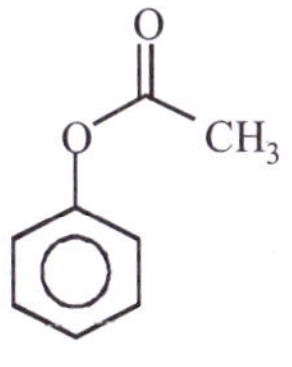 (2) 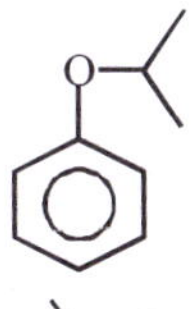

(3)  (4) 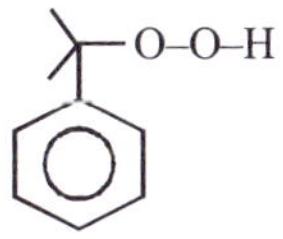

94. A compound 'X' is acidic and it is soluble in NaOH solution, but insoluble in $NaHCO_3$ solution. Compound 'X' also gives violet colour with neutral $FeCl_3$ solution. The compound 'X' is : [July 29, 2022 (I)]

(1) 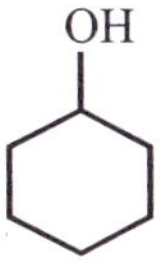 (2) 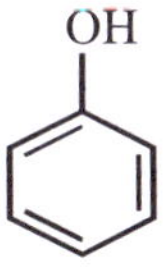

(3) 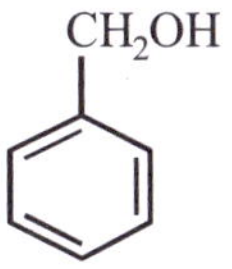 (4) 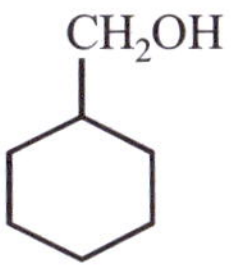

95. Identify the major product A and B for the below given reaction sequence. [July 28, 2022 (I)]

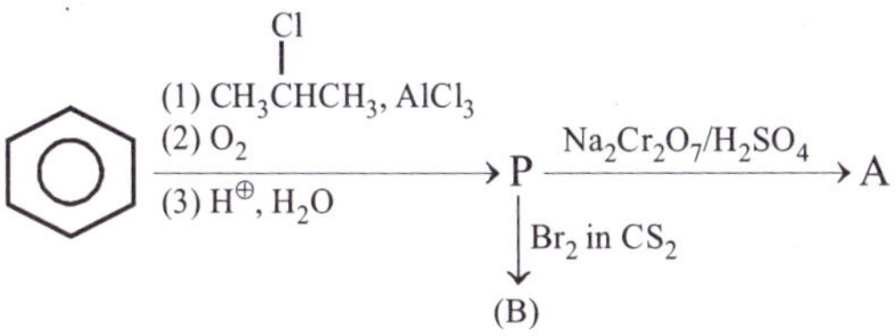

(A) (B)

(1) 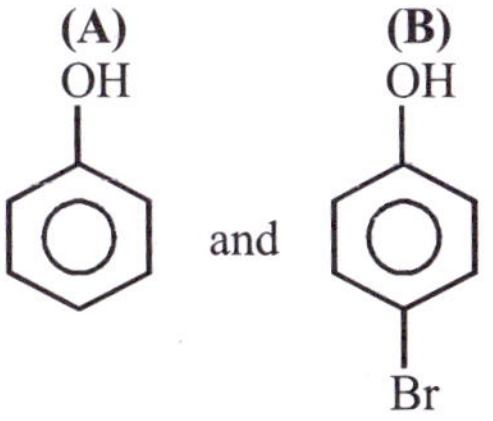

(2) 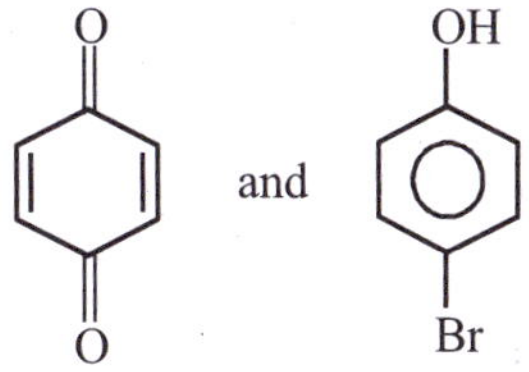

(3) 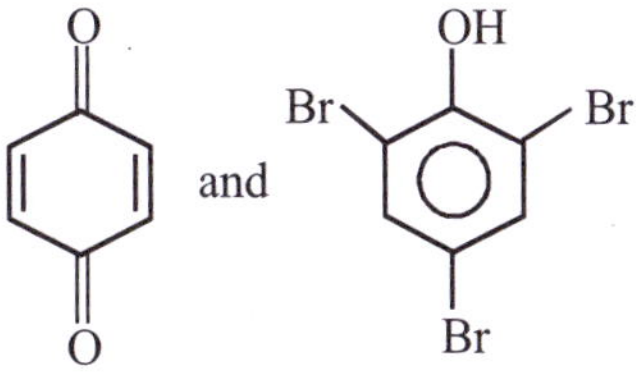

(4) 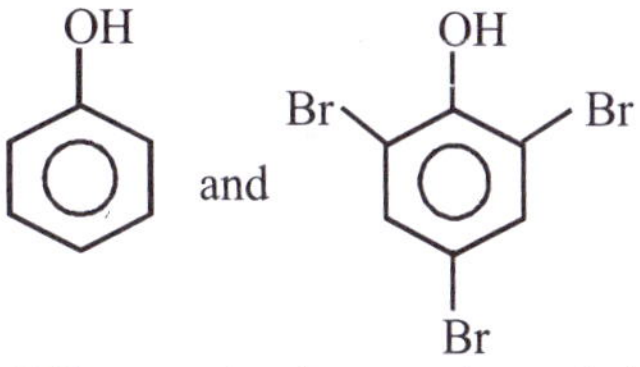

96. The difference in the reaction of phenol with bromine in chloroform and bromine in water medium is due to: [July 26, 2022 (I)]

(1) Hyperconjugation in substrate
(2) Polarity of solvent
(3) Free radical formation
(4) Electromeric effect the substrate

97. Arrange the following in decreasing acidic strength. [July 25, 2022 (II)]

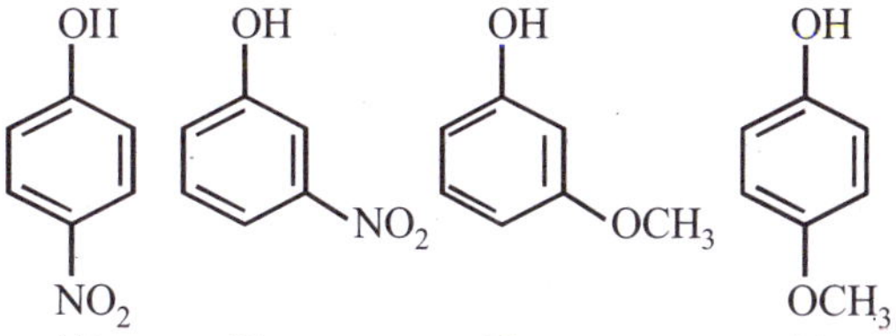

(A) (B) (C) (D)

(1) A > B > C > D (2) B > A > C > D
(3) D > C > A > B (4) D > C > B > A

98. 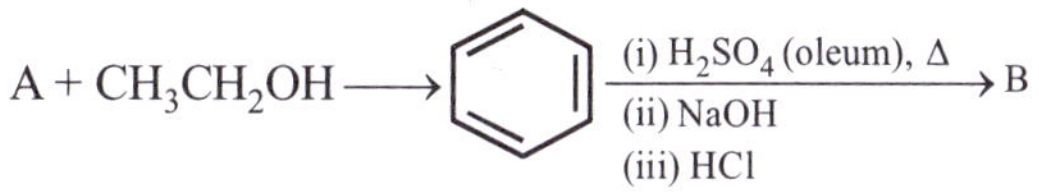

Consider the above reaction sequence. Identify the component A and component B: [June 30, 2022 (I)]

(1) A = $C_6H_5NH_2$ B = 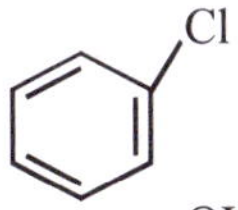

(2) A = $C_6H_5N_2^+Cl^-$ B = 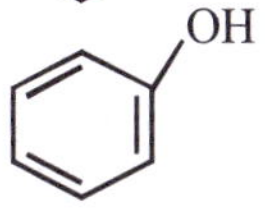

(3) A = 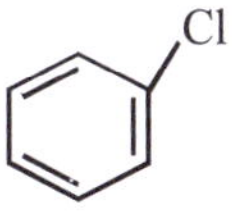 B = $C_6H_5NH_2$

(4) A = 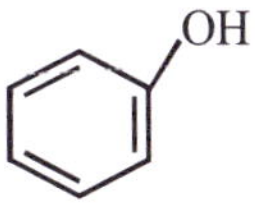 B = $C_6H_5N_2^+Cl^-$

99. Given below are two statements. [June 29, 2022 (I)]

Statement I : Phenols are weakly acidic.

Statement II : Therefore they are freely soluble in NaOH solution and are weaker acids than alcohols and water.

Choose the **most appropriate** option:

(1) Both Statement I and Statement II are correct.
(2) Both Statement I and Statement II are incorrect.
(3) Statement I is correct but Statement II is incorrect.
(4) Statement I is incorrect but Statement II is correct.

100. Match List-I with List-II [June 27, 2010, 2022 (II)]

List-I	List-II
(A) 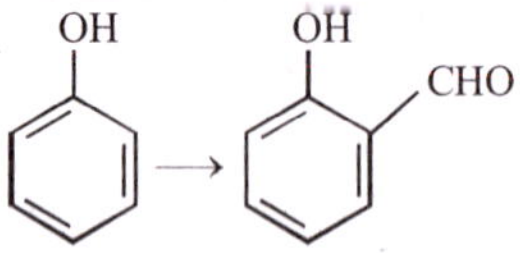	(I) Br_2 in CS_2
(B) 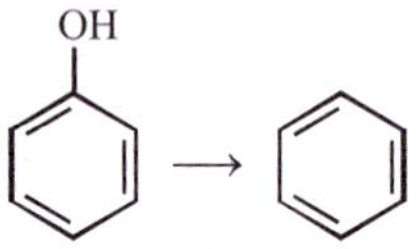	(II) $Na_2Cr_2O_7/H_2SO_4$
(C) 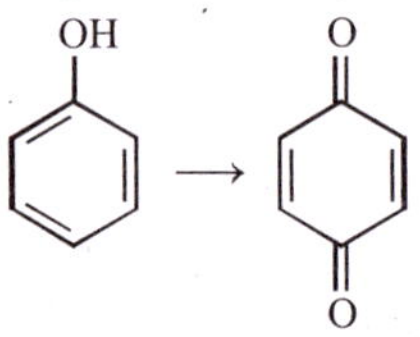	(III) Zn
(D) 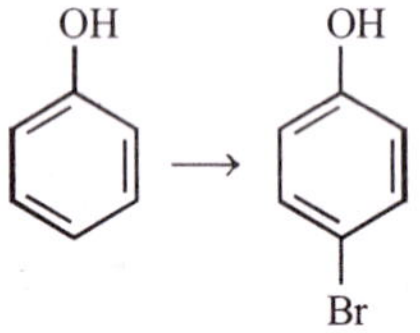	(IV) $CHCl_3/NaOH$

Choose the correct answer from the options given below:

(1) A-IV, B-III, C-II, D-I (2) A-IV, B-III, C-I, D-II
(3) A-II, B-III, C-I, D-IV (4) A-IV, B-II, C-III, D-I

101. Which of the following conditions or reaction sequence will NOT give acetophenone as the major product ? [June 25, 2022 (II)]

(1) (A) 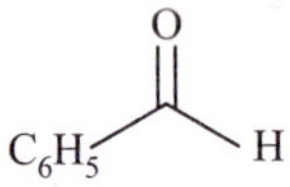 $H + CH_3MgBr$ (B) $Na_2Cr_2O_7$, H^+

(2) (A) 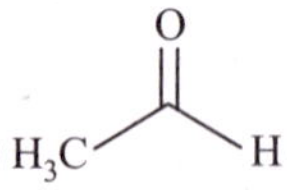 $H + C_6H_5MgBr$ (B) PCC, DCM

(3) 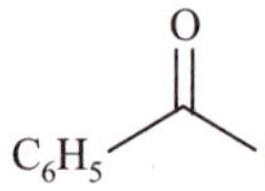 $OC_2H_5 + 2CH_3MgBr$

(4) 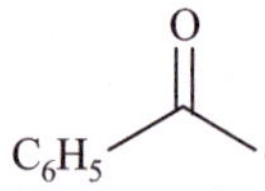 $Cl + CH_3MgBr + CdCl_2$

102. Phenol on reaction with dilute nitric acid, gives two products. Which method will be most effective for large scale separation ? [June 25, 2022 (I)]

(1) Chromatographic separation
(2) Fractional crystallisation
(3) Steam distillation
(4) Sublimation

103. The intermediate X, in the reaction [June 25, 2022 (I)]

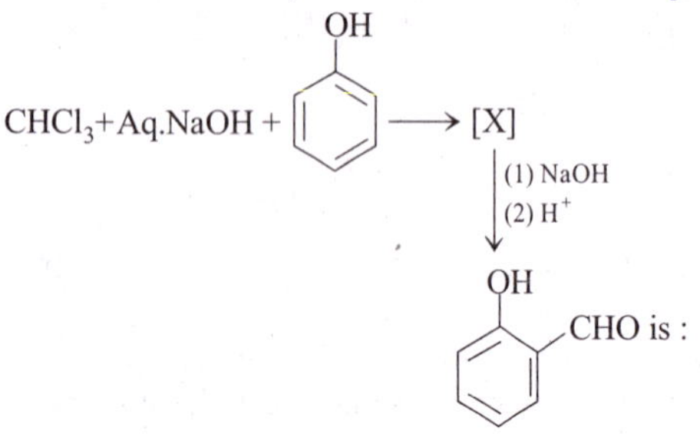

CHO is :

(1) 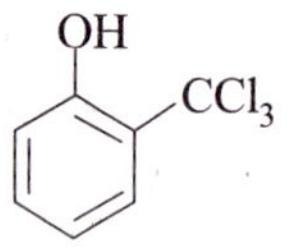 (2) 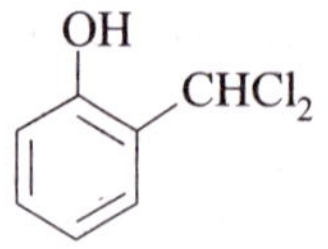

(3) 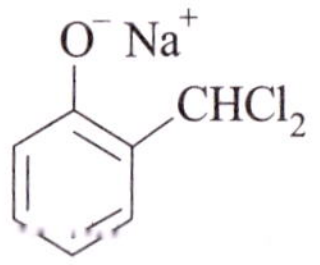 (4) 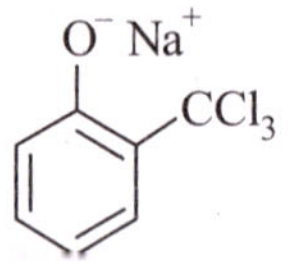

104. In the following reaction : [June 25, 2022 (I)]

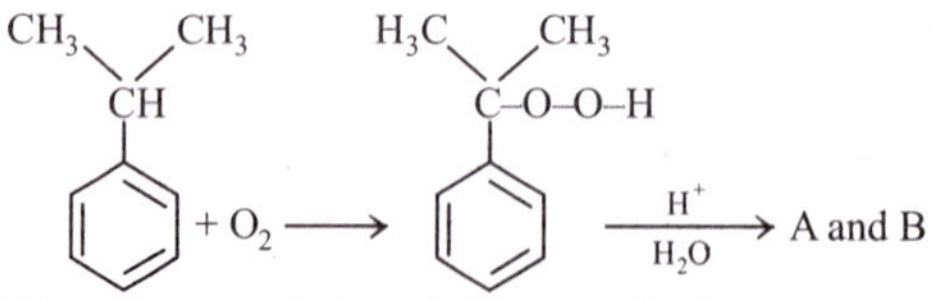

The compounds A and B respectively are:

(1) 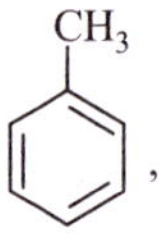, CH_3COOH (2) 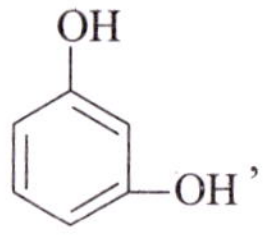, CH_3COOH

(3) 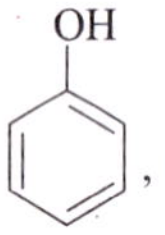, CH_3COCH_3 (4) 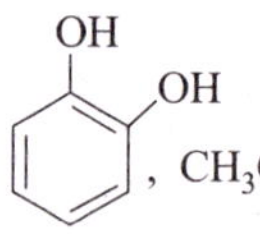, CH_3COCH_3

105. The correct structures of A and B formed in the following reactions are : [Aug. 27, 2021 (II)]

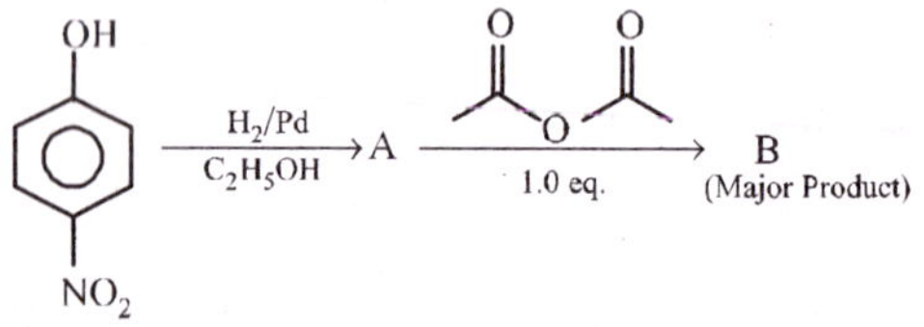

(1) A : 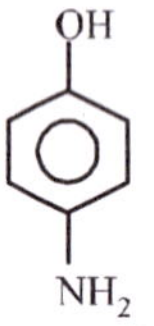, B : 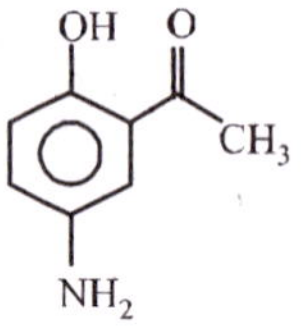

(2) A : 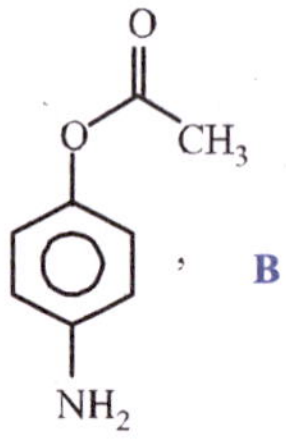, B : 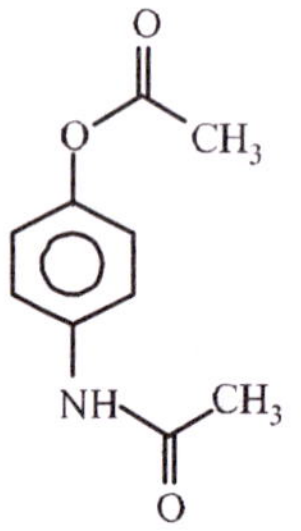

(3) A : , B : 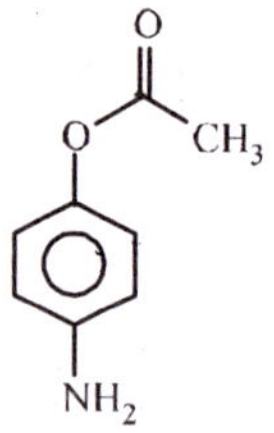

(4) A : , B : 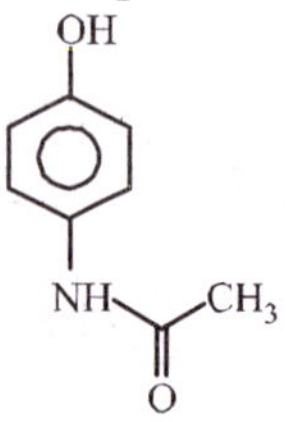

106. Which one of the following phenols does not give colour when condensed with phthalic anhydride in presence of conc. H_2SO_4? [Aug. 26, 2021 (II)]

(1) 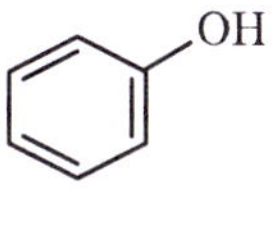 (2) 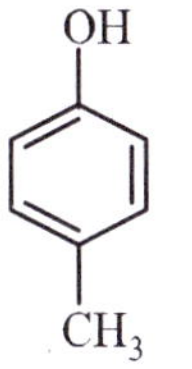

(3) 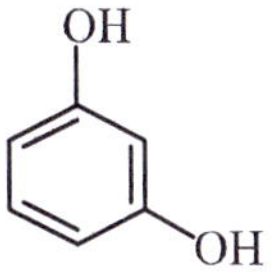 (4) 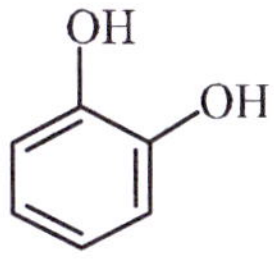

107. 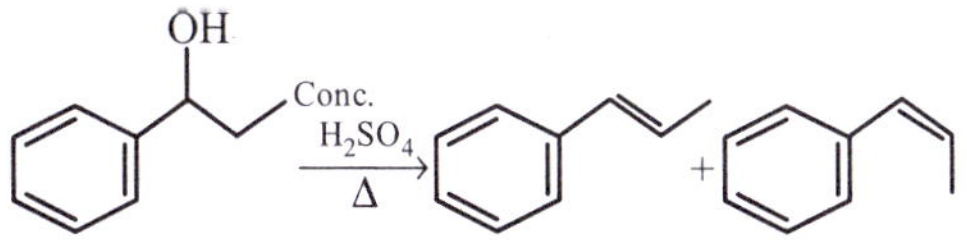

A B

Consider the above reaction, and choose the correct statement : [July 27, 2021 (II)]

(1) The reaction is not possible in acidic medium
(2) Both compounds A and B are formed equally
(3) Compound A will be the major product
(4) Compound B will be the major product

108. Reaction of Grignard reagent, C_2H_5MgBr with C_8H_8O followed by hydrolysis gives compound "A" which reacts instantly with Lucas reagent to give compound B, $C_{10}H_{13}Cl$. [March 18, 2021 (I)]

(1) 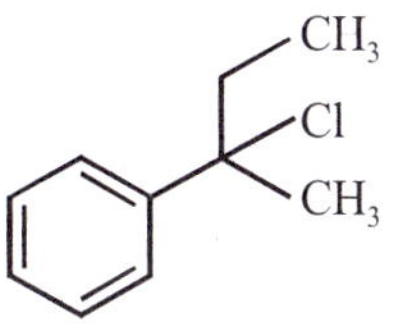 (2) 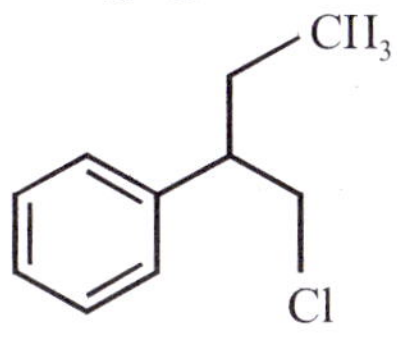

(3) 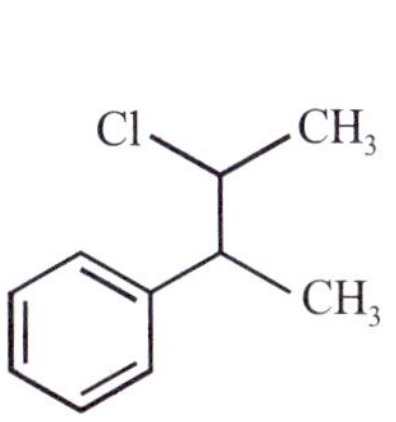 (4) 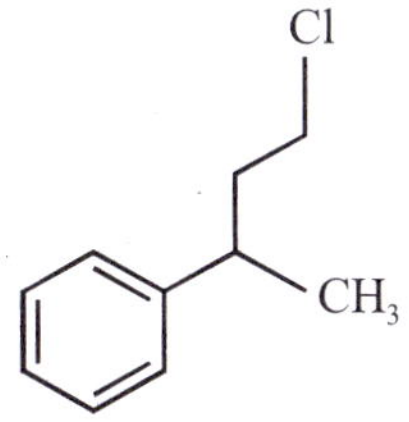

109. Identify A in the given reaction, [Feb. 26, 2021 (II)]

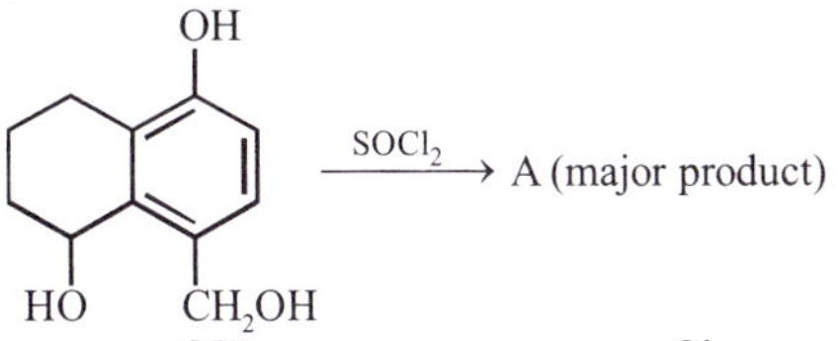

(1) 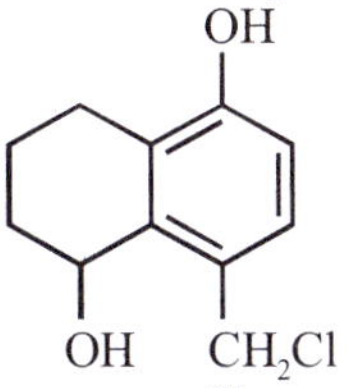 (2) 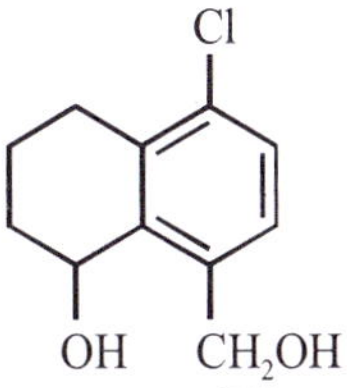

(3) 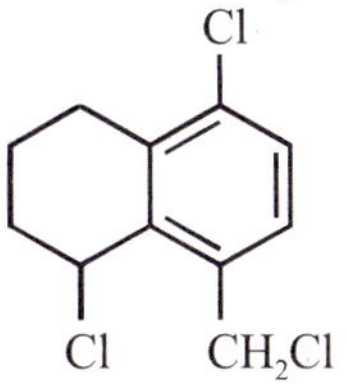 (4) 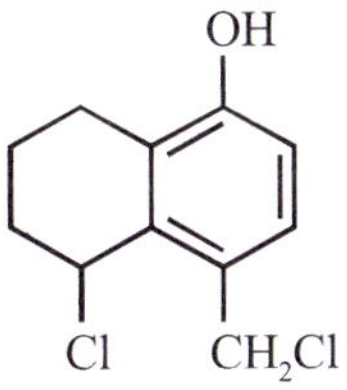

110. The increasing order of boiling points of the following compounds is : [Sep. 05, 2020 (II)]

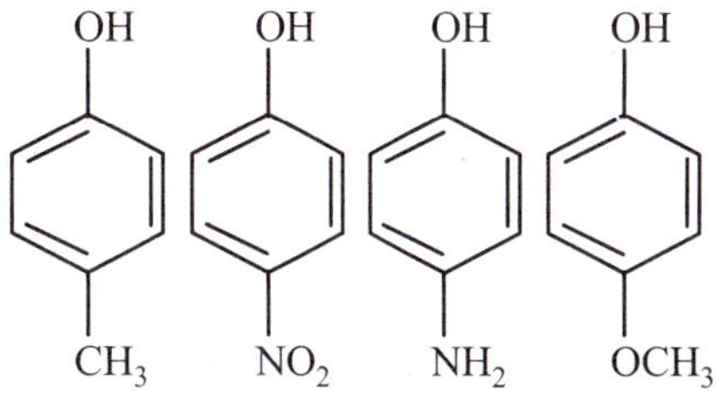

I II III IV

(1) I < III < IV < II (2) I < IV < III < II
(3) IV < I < II < III (4) III < I < II < IV

111. Consider the following reaction : [Sep. 03, 2020 (II)]

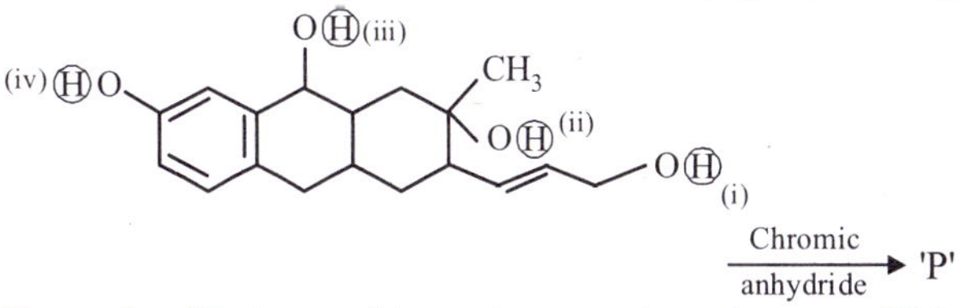

The product 'P' gives positive ceric ammonium nitrate test. This is because of the presence of which of these –OH group(s) ?

(1) (ii) only (2) (iii) and (iv)
(3) (iv) only (4) (ii) and (iv)

112. The major product of the following reaction is : [Sep. 02, 2020 (II)]

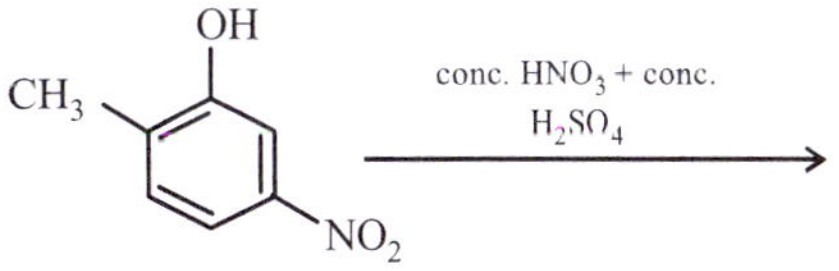

(1) 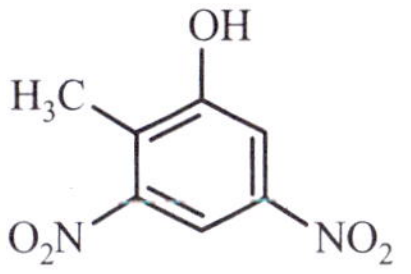 (2) 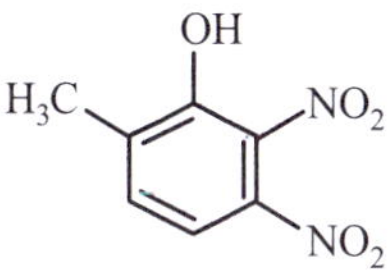

(3) 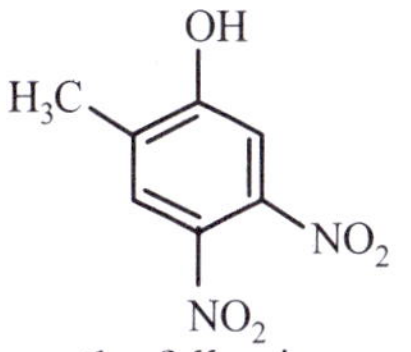 (4) 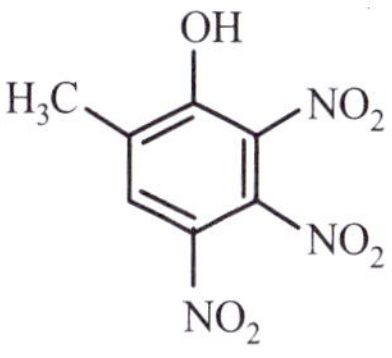

113. Arrange the following compounds in increasing order of C – OH bond length: methanol, phenol, *p*-ethoxyphenol [Jan. 08, 2020 (I)]

(1) methanol < *p*-ethoxyphenol < phenol
(2) phenol < methanol < *p*-ethoxyphenol
(3) phenol < *p*-ethoxyphenol < methanol
(4) methanol < phenol < *p*-ethoxyphenol

114. What will be the major product when *m*-cresol is reacted with propargyl bromide ($HC \equiv C—CH_2Br$) in presence of K_2CO_3 in acetone ? [April 12, 2019 (II)]

(1) 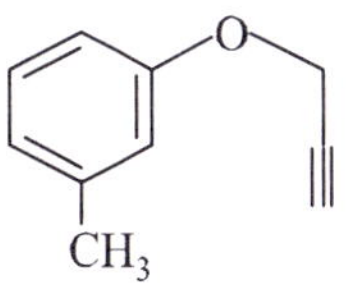 (2) 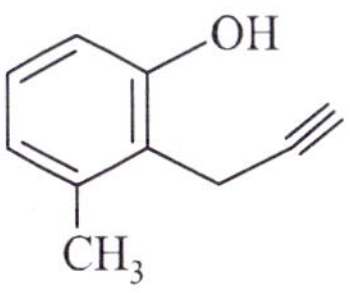

(3) 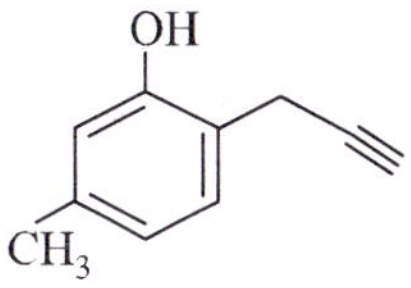 (4) 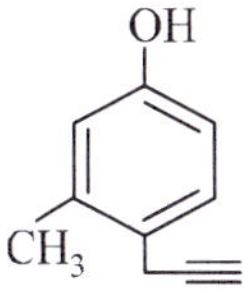

115. The major product of the following reaction is : **[Jan. 11, 2019 (II)]**

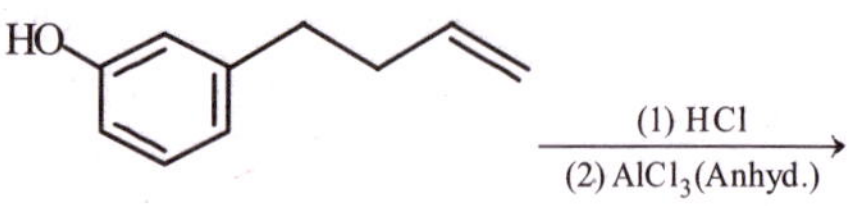

(1) 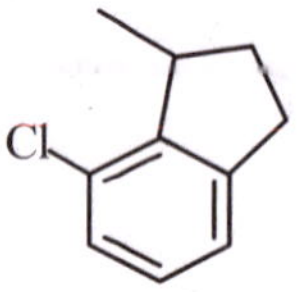

(2) 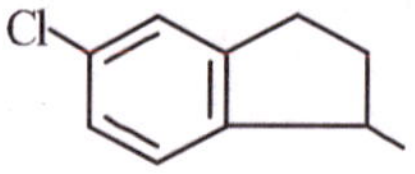

(3) 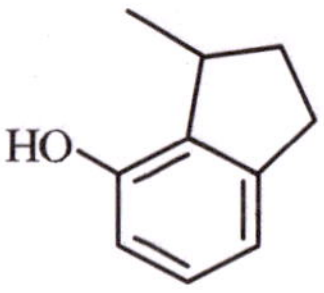

(4) 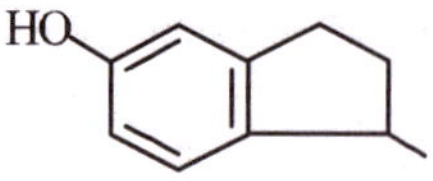

116. The major product of the following reaction is : **[Jan. 11, 2019 (I)]**

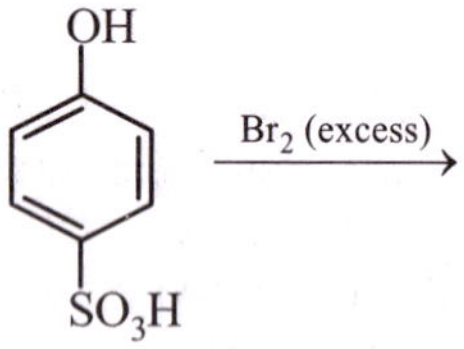

(1) 

(2) 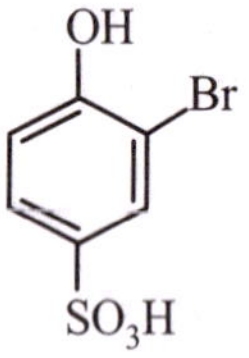

(3) 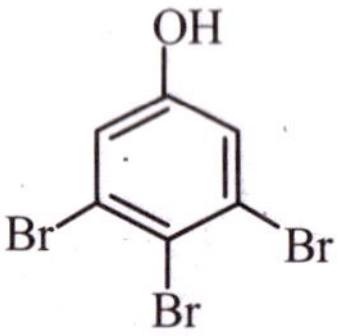

(4) 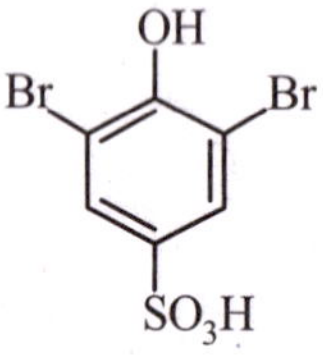

117. The major product of the following reaction is : **[Jan. 11, 2019 (I)]**

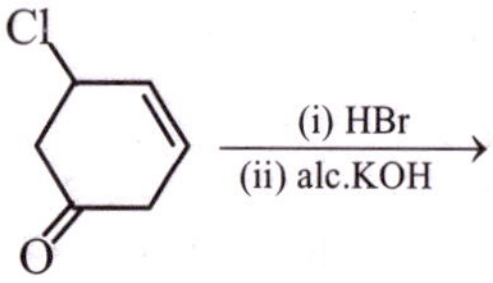

(1) 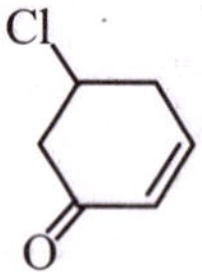

(2) 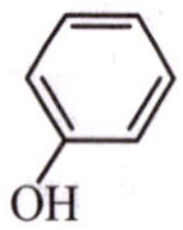

(3) 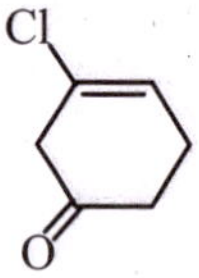

(4) 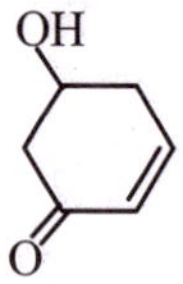

118. The major product of the following reaction is:

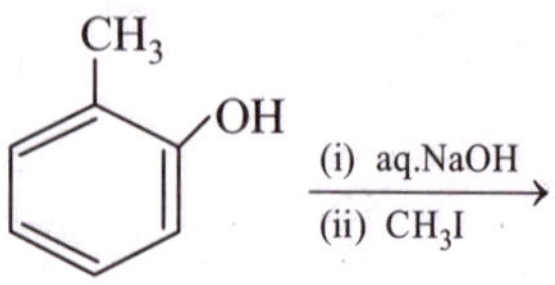

[Jan. 10, 2019 (II)]

(1) 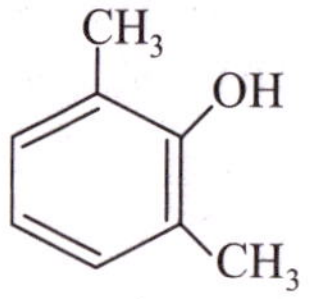

(2) 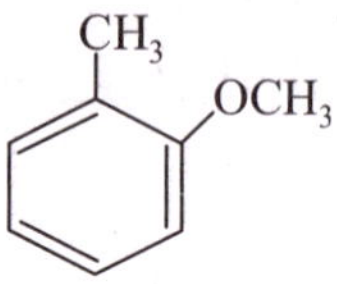

(3) 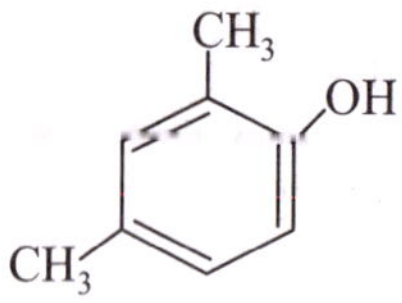

(4) 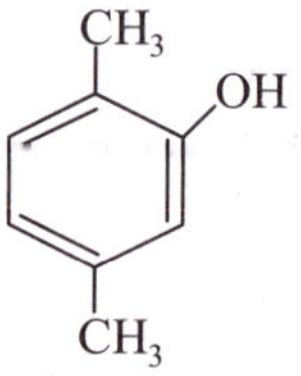

119. The increasing order of the pK_a values of the following compounds is: **[Jan. 10, 2019 (I)]**

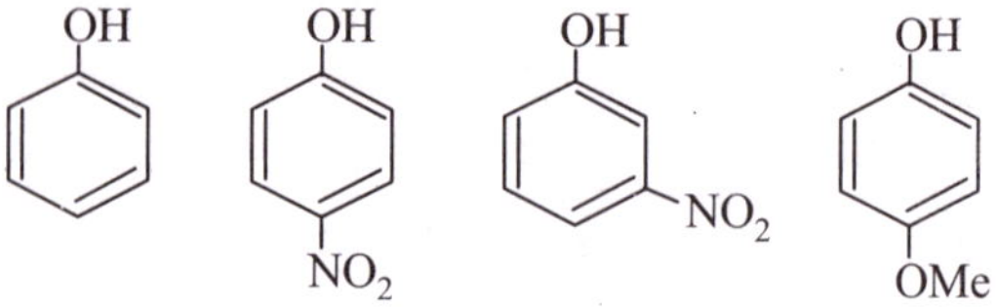

A B C D

(1) C<B<A<D (2) B<C<D<A

(3) D<A<C<B (4) B<C<A<D

120. The products formed in the reaction of cumene with O_2 followed by treatment with dil. HCl are: **[Jan. 9, 2019 (II)]**

(1)  and 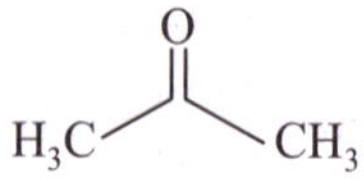

(2) 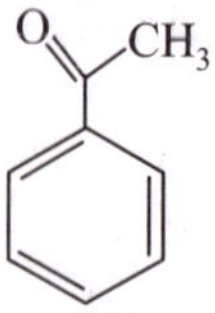

and $CH_3 - OH$

(3) 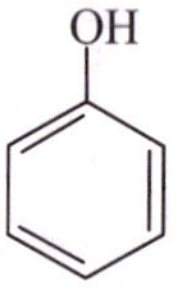

and 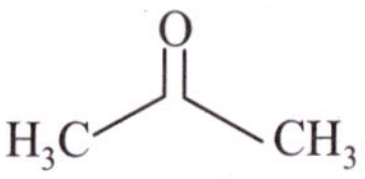

(4) 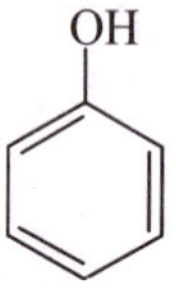

and 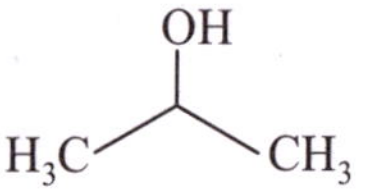

121. Phenol on treatment with CO_2 in the presence of NaOH followed by acidification produces compound X as the major product. X on treatment with $(CH_3CO)_2O$ in the presence of catalytic amount of H_2SO_4 produces: **[2018; Same 2014]**

(1) 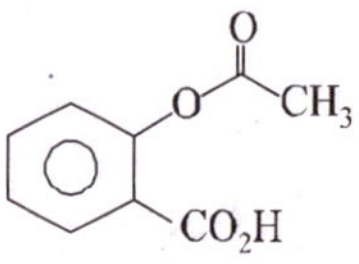

(2) 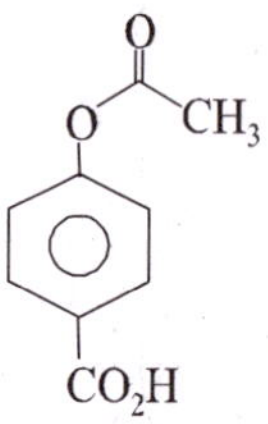

(3) 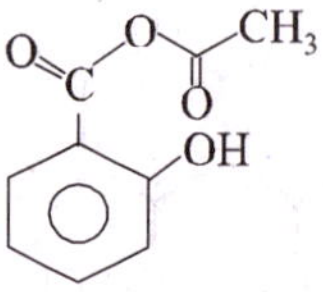

(4) 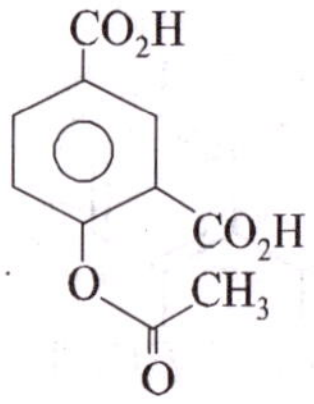

122. Phenol reacts with methyl chloroformate in the presence of NaOH to form product A. A reacts with Br_2 to form product B. A and B are respectively : **[2018]**

(1) 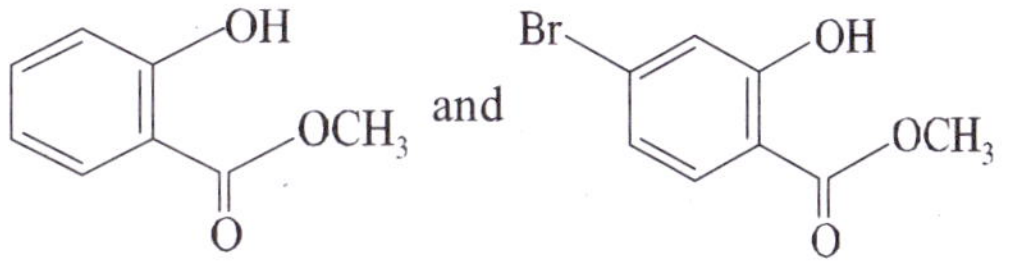

(2) 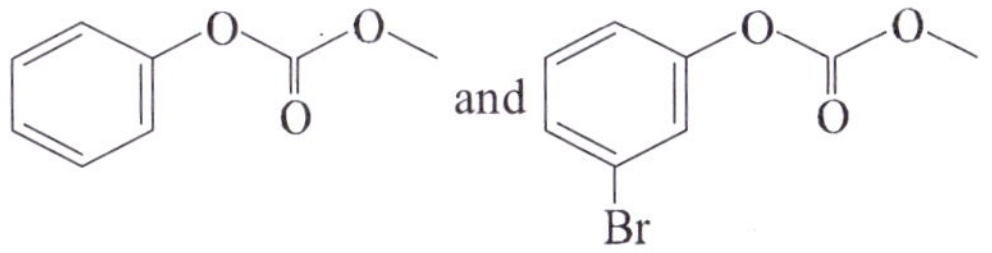

(3) 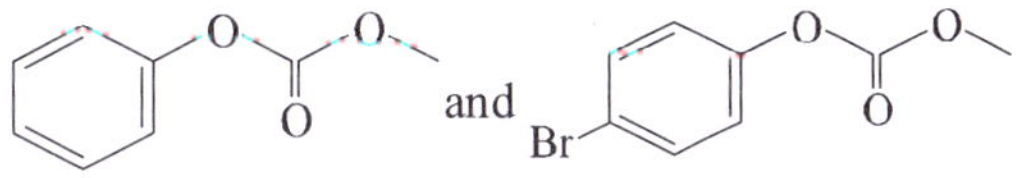

(4) 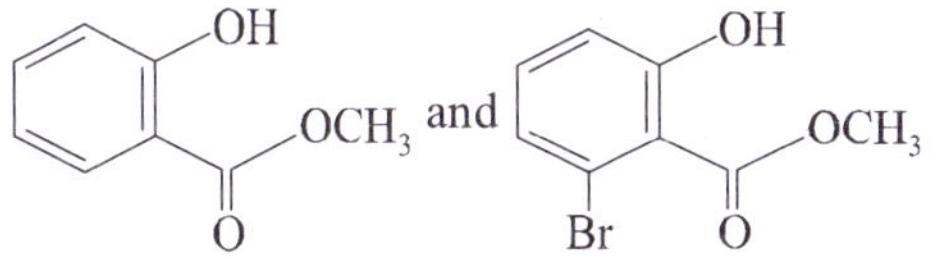

123. Which of the following , upon treatment with tert-BuONa followed by addition of bromine water, fails to decolourize the colour of bromine? **[2017]**

(1) 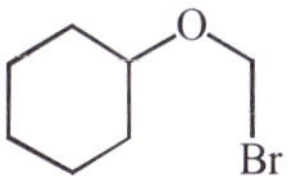 (2) 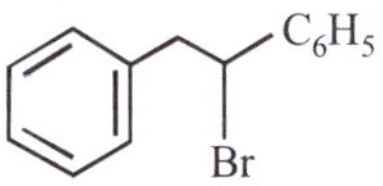

(3) 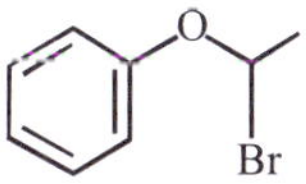 (4) 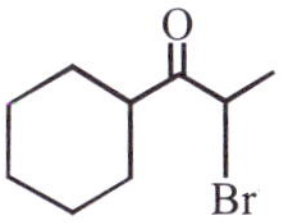

124. The major product of the following reaction is : **[Online April 8, 2017]**

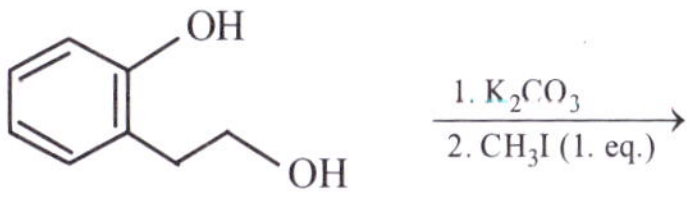

(1) 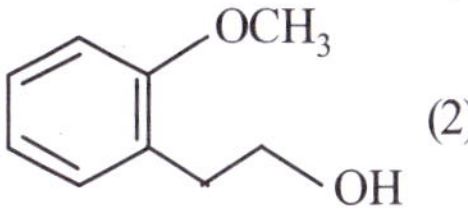 (2) 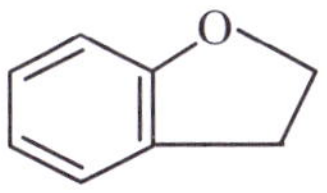

(3) 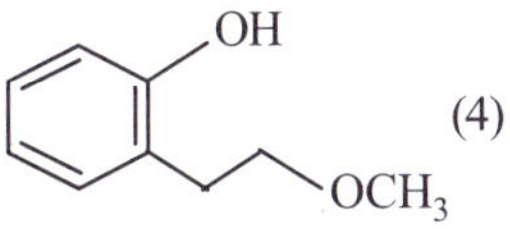 (4) 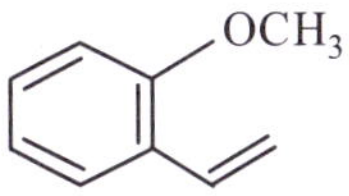

125. Which one of the following compounds will not be soluble in sodium bicarbonate? **[Online April 19, 2014]**

(1) 2, 4, 6 - Trinitrophenol (2) Benzoic acid

(3) *o* - Nitrophenol (4) Benzene sulphonic acid

126. Which one of the following substituents at *para*-position is most effective in stabilizing the phenoxide 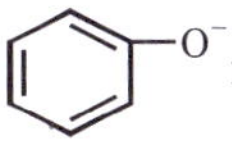 ion? **[Online April 19, 2014]**

(1) $-CH_3$ (2) $-OCH_3$

(3) $-COCH_3$ (4) $-CH_2OH$

127. The following reaction

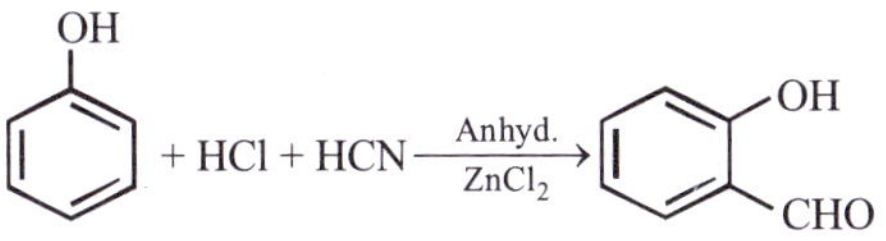

is known as: **[Online April 11, 2014]**

(1) Perkin reaction

(2) Gatterman-Koch formylation

(3) Kolbe's reaction

(4) Gattermann reaction

128. Sodium phenoxide when heated with CO_2 under pressure at 125 °C yields a product which on acetylation produces C **[2014]**

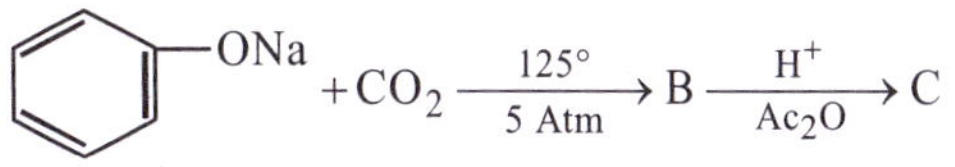

The major product C would be

(1) 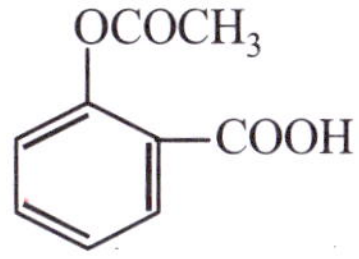 (2) 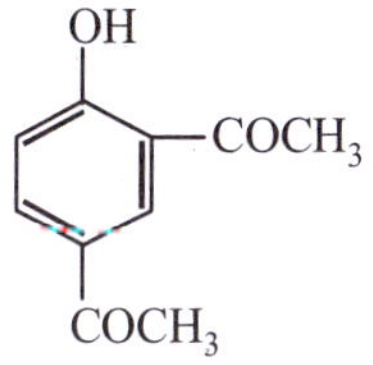

(3) 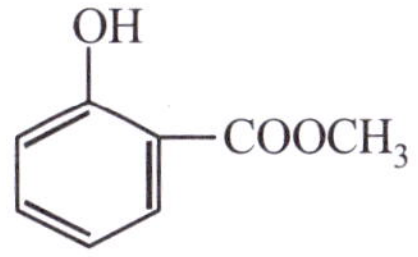 (4) 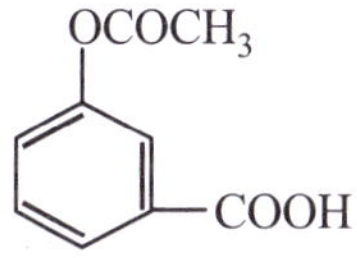

129. Arrange the following compounds in order of decreasing acidity : **[2013; Same 2011]**

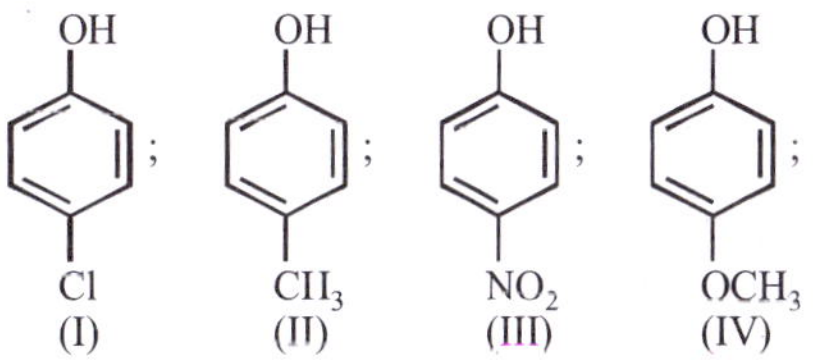

(1) II > IV > I > III (2) I > II > III > IV

(3) III > I > II > IV (4) IV > III > I > II

130. Phenol on heating with $CHCl_3$ and NaOH gives salicylaldehyde. The reaction is called: **[Online April 25, 2013]**

(1) Reimer - Tiemann reaction

(2) Claisen reaction

(3) Cannizzaro's reaction

(4) Hell - Volhard - Zelinsky reaction

131. The major product in the following reaction is :

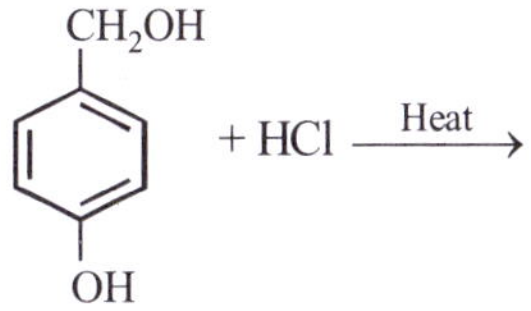

[Online April 25, 2013]

(1) 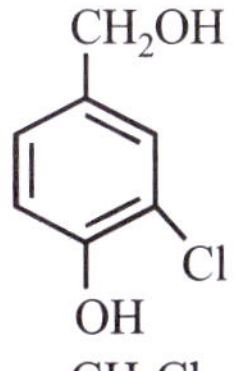 (2) 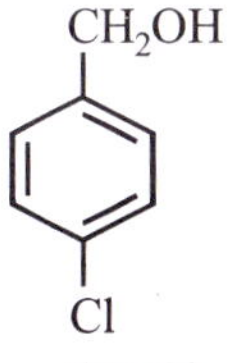

(3) 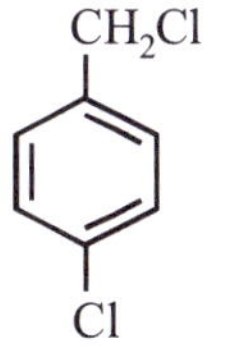 (4) 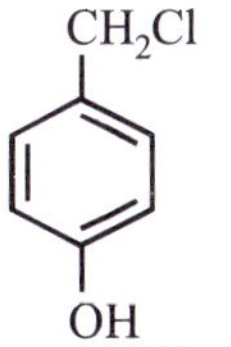

132. The reaction of phenol with benzoyl chloride to give phenyl benzoate is known as : **[Online April 23, 2013]**

(1) Claisen reaction

(2) Schotten-Baumann reaction

(3) Reimer-Tiemann reaction

(4) Gattermann-Koch reaction

133. What is the structure of the major product when phenol is treated with bromine water ? **[Online April 22, 2013]**

(1) 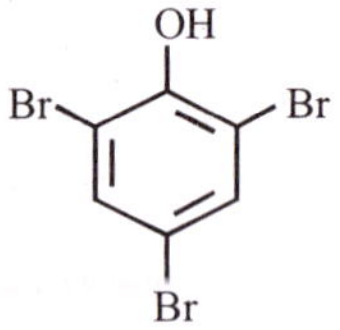

(2) 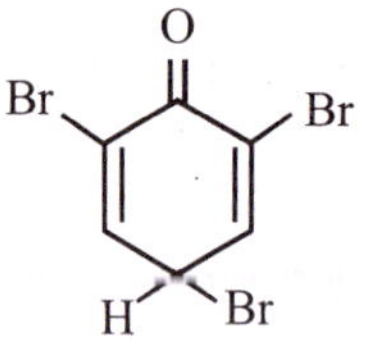

(3) 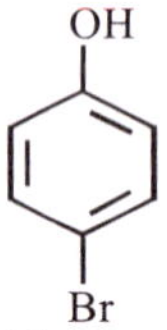

(4) 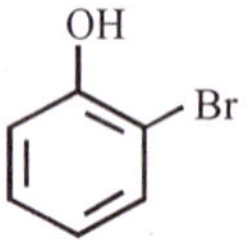

134. Consider the following reaction :

$C_2H_5OH + H_2SO_4 \longrightarrow$ Product

Among the following, which one cannot be formed as a product under any conditions ? **[2011RS]**

(1) Ethylene (2) Acetylene
(3) Diethyl ether (4) Ethylhydrogen sulphate

135. The major product obtained on interaction of phenol with sodium hydroxide and carbon dioxide is **[2009]**

(1) salicylaldehyde (2) salicylic acid
(3) phthalic acid (4) benzoic acid

136. Phenol, when it reacts first with concentrated sulphuric acid and then with concentrated nitric acid, gives **[2008]**

(1) 2, 4, 6-trinitrobenzene (2) *o*-nitrophenol
(3) *p*-nitrophenol (4) nitrobenzene

137. 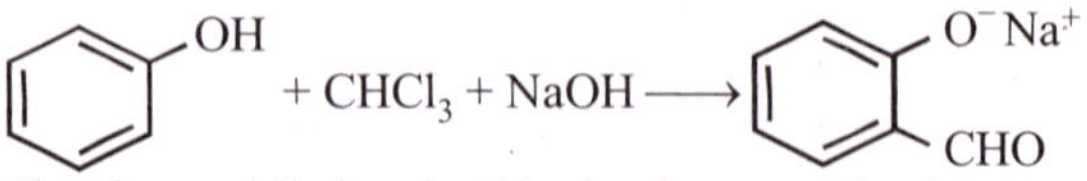

The electrophile involved in the above reaction is **[2006]**

(1) trichloromethyl anion ($\overset{\ominus}{C}Cl_3$)
(2) formyl cation
(3) phenoxide ion
(4) dichlorocarbene

138. The structure of the compound that gives a tribromo derivative on treatment with bromine water is **[2006]**

(1) 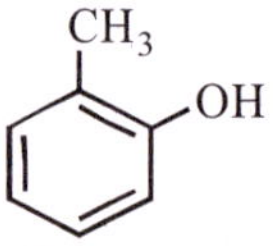

(2) 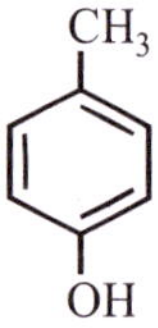

(3) 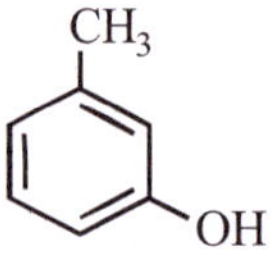

(4) 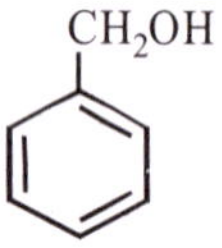

139. Picric acid is: **[2002]**

(1) 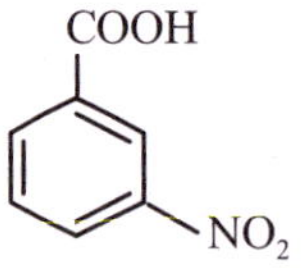

(2) 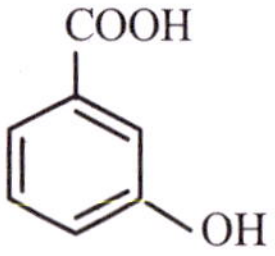

(3) 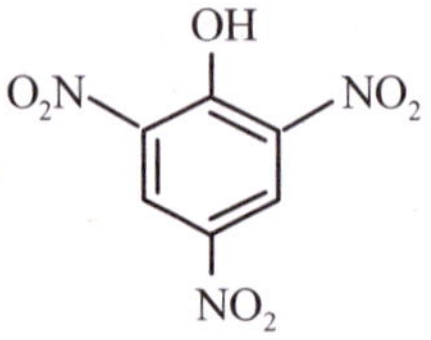

(4) 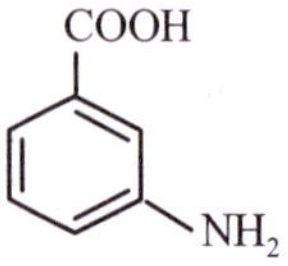

NUMERICAL VALUE QUESTIONS (NVQs)

140. 4.7 g of phenol is heated with Zn to give product X. If this reaction goes to 60% completion then the number of moles of compound X formed will be _____ $\times 10^{-2}$ (Nearest Integer) (Given molar mass in g mol^{-1} : H : 1, C : 12, O : 16) **[April 06, 2026 (I)]**

141. Consider the given chemical reaction sequence :

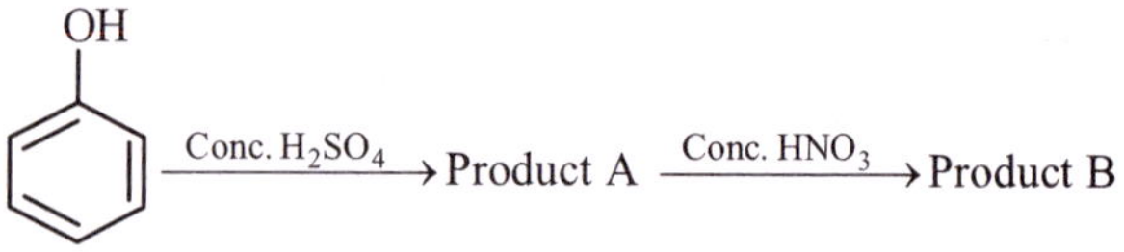

Total sum of oxygen atoms in product A and product B are .... **[April 5, 2024 (I)]**

142. A trisubstituted compound 'A', $C_{10}H_{12}O_2$ gives neutral $FeCl_3$ test positive. Treatment of compound 'A' with NaOH and CH_3Br gives $C_{11}H_{14}O_2$, with hydriodic acid it gives methyl iodide and with hot conc. NaOH gives a compound B, $C_{10}H_{12}O_2$. Compound 'A' also decolorises alkaline $KMnO_4$. The number of π bond/s present in the compound 'A' is _______. **[Jan. 30, 2023 (I)]**

143. In the following reaction

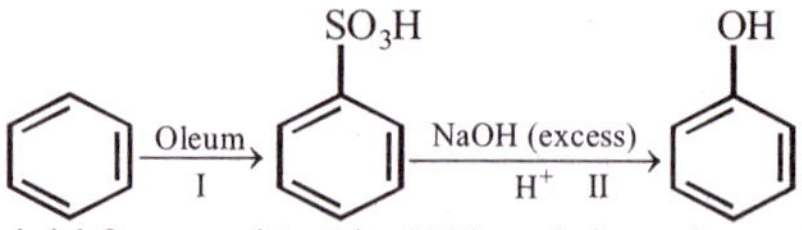

The % yield for reaction I is 60% and that of reaction II is 50%. The overall yield of the complete reaction is ___% [nearest integer] **[July 27, 2022 (I)]**

144. Compound 'P' on nitration with dil. HNO_3 yields two isomers (A) and (B). These isomers can be separated by steam distillation. Isomers (A) and (B) show the intramolecular and intermolecular hydrogen bonding respectively. Compound (P) on reaction with conc. HNO_3 yields a yellow compound 'C', a strong acid. The number of oxygen atoms present in compound 'C' is _______. **[June 26, 2022 (I)]**

145. A solution of phenol in chloroform when treated with aqueous NaOH gives compound P as a major product. The mass percentage of carbon in P is ______. (to the nearest integer) (Atomic mass: C = 12; H = 1; O = 16) **[Sep. 06, 2020 (II)]**

Topic 3 - Preparation and Properties of Ethers

MULTIPLE CHOICE QUESTIONS (MCQs)

146. Consider the following reaction.

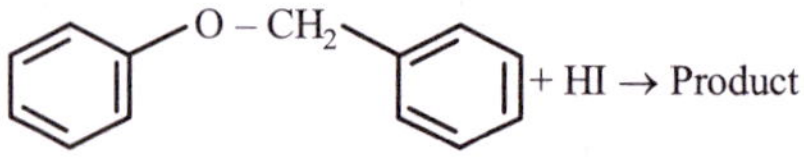

Statement I: In the above reaction, product formed will be a mixture of benzyl alcohol and iodobenzene.

Statement II: In the above reaction, the – O – CH_2 – bond is cleaved to give the product.

In the light of the above statements, choose the **correct** answer from the options given below: **[April 06, 2026 (II)]**

(1) Both Statement I and Statement II are true
(2) Both Statement I and Statement II are false
(3) Statement I is true but Statement II is false
(4) Statement I is false but Statement II is true

147. A mixed ether (P), when heated with excess of hot concentrated hydrogen iodide produces two different alkyl iodides which when treated with aq. NaOH give compounds (Q) and (R). Both (Q) and (R) give yellow precipitate with NaOI. Identify the mixed ether (P): [Jan. 23, 2026 (II)]

(1) 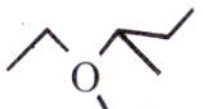 (2) 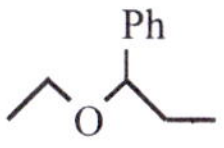

(3) 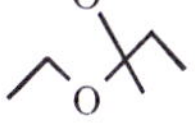 (4) 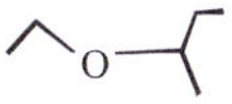

148. Which one of the following reactions will not lead to the desired ether formation in major proportion? [April 8, 2025 (II)]
(iso-Bu ⇒ isobutyl, sec-Bu ⇒ sec-butyl, nPr ⇒ n-propyl, ^tBu ⇒ tert-butyl, Et ⇒ ethyl)

(1) $^tBuO^{\ominus}Na^{\oplus} + EtBr \rightarrow {}^tBu\text{–}O\text{–}Et$

(2) 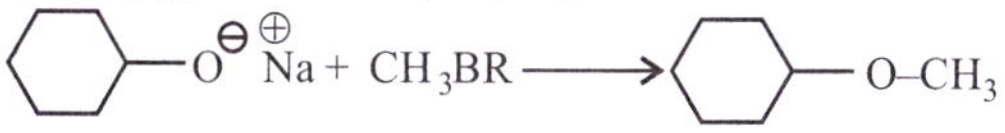

(3) 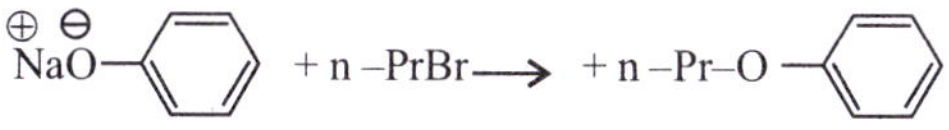

(4) iso-$BuO^{\ominus}Na^{\oplus}$ + sec – BuBr → Sec–Bu–O–iso–Bu

149. Given below are two statements : [April 7, 2025 (I)]
Statement I : Dimethyl ether is completely soluble in water. However, diethyl ether is soluble in water to a very small extent.
Statement II : Sodium metal can be used to dry diethyl ether and not ethyl alcohol.
In the light of given statements, choose the *correct* answer from the options given below
(1) Statement I is false but Statement II are true
(2) Both Statement I and Statement II are false
(3) Statement I is true but Statement II is false
(4) Both Statement I and Statement II are true

150. Which one of the following, with HBr will give a phenol ? [Jan. 29, 2025 (II)]

(1) 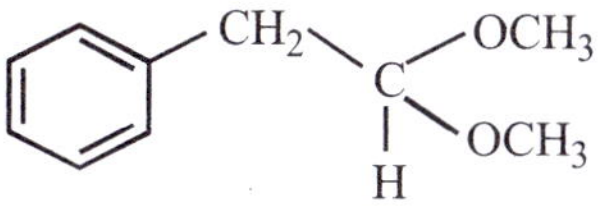

(2) 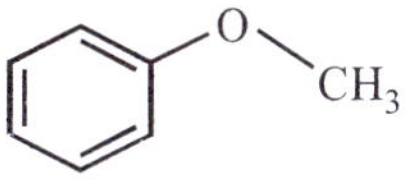

(3) 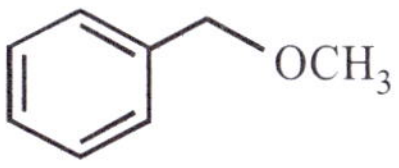

(4) 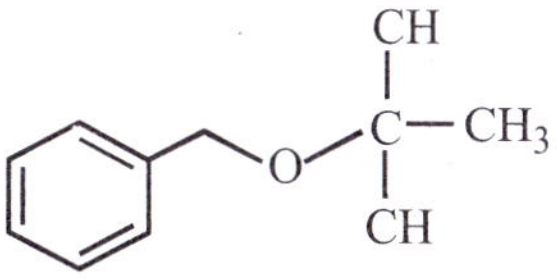

151. The major products formed: [April 6, 2024 (II)]

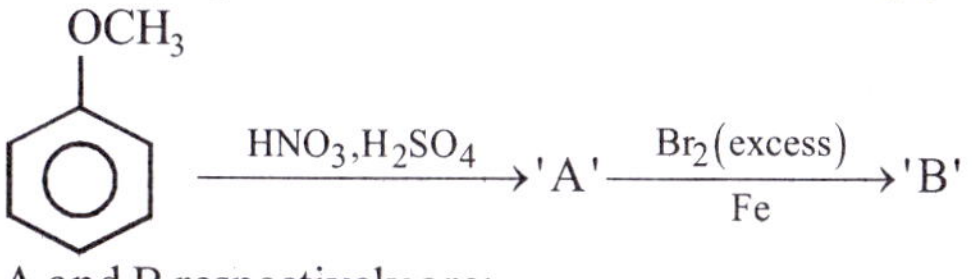

A and B respectively are:

(1) 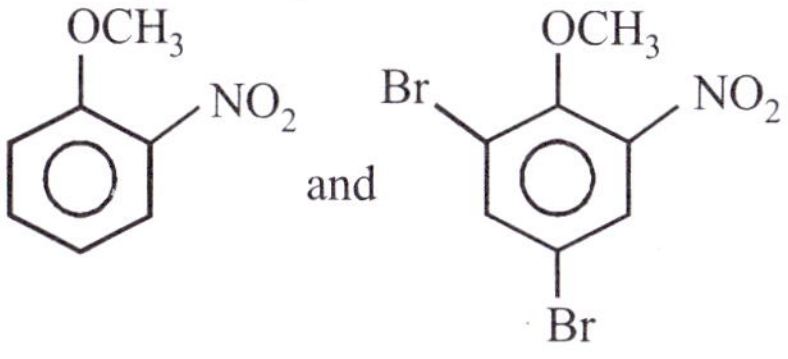

(2) 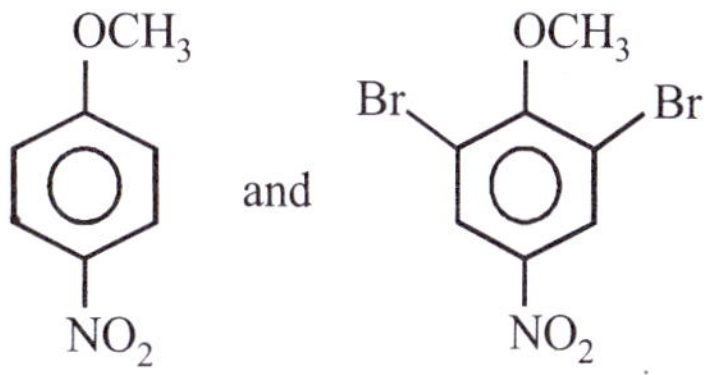

(3) 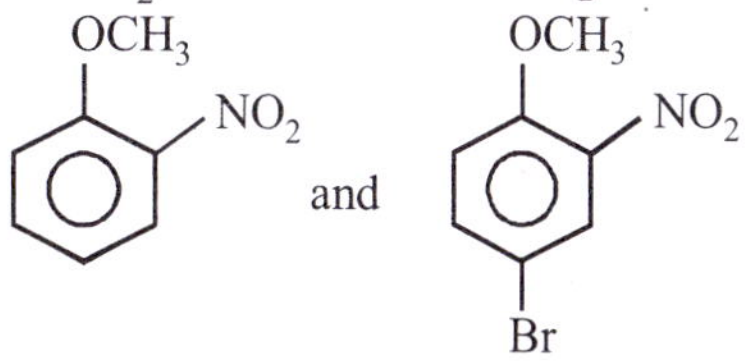

(4) 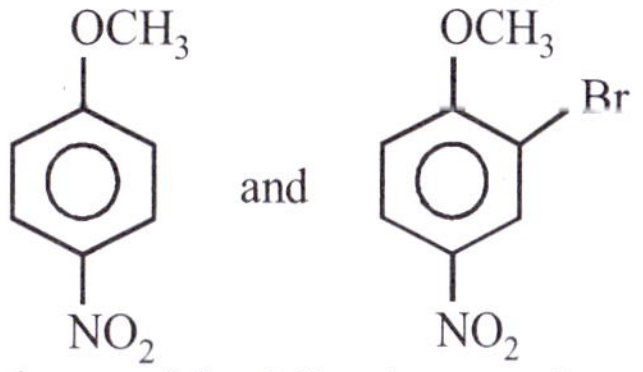

152. Which one of the following reactions is NOT possible? [April 5, 2024 (II)]

(1) 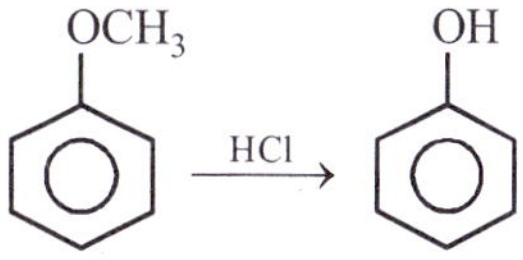

(2) 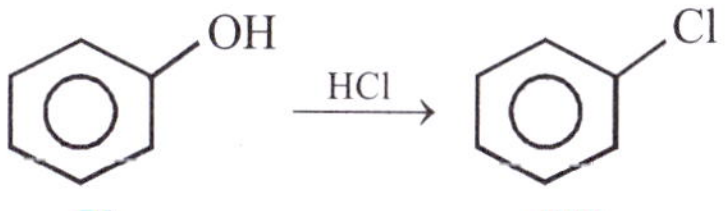

(3) 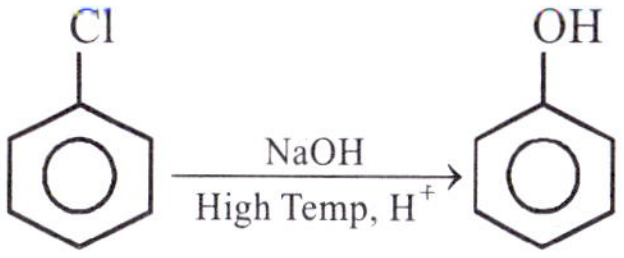

(4) 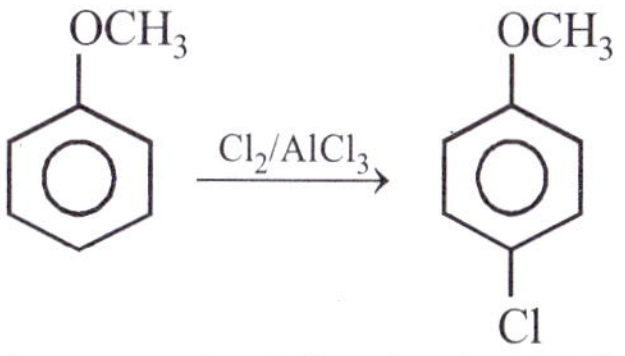

153. Which among the following has highest boiling point? [Feb. 1, 2024 (II)]
(1) $CH_3CH_2CH_2CH_2-OH$ (2) $CH_3CH_2CH_2CH_3$
(3) $CH_3CH_2CH_2CHO$ (4) $H_5C_2-O-C_2H_5$

154. The major product (P) in the following reaction is [Jan. 29, 2024 (I)]

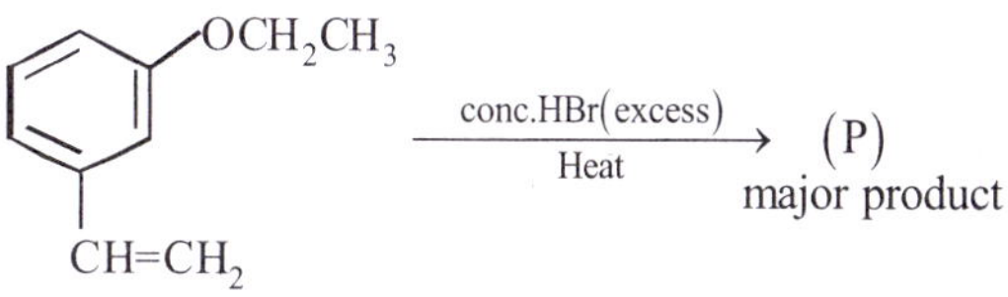

(1)  (2) 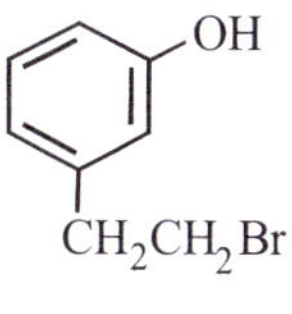

(3) 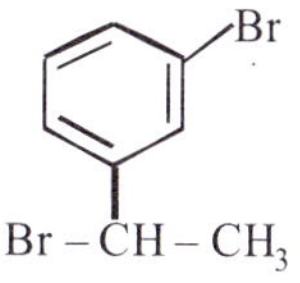 (4) 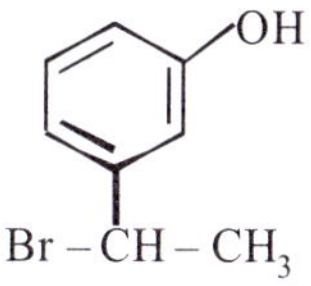

155. Major product formed in the following reaction is a mixture of: **[Jan. 27, 2024 (II)]**

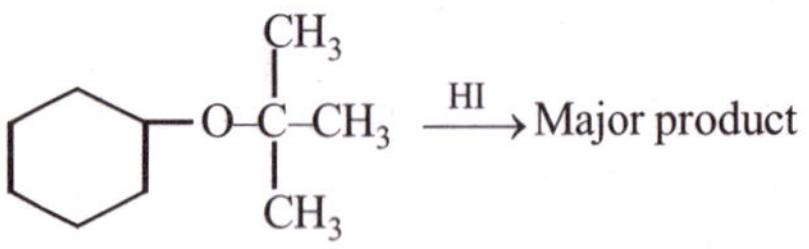

(1) 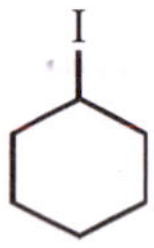 and $(CH_3)_3CI$ (2) 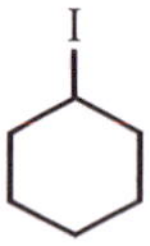 and $(CH_3)_3COH$

(3) 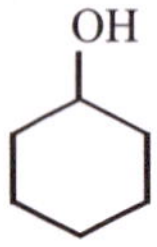 and $(CH_3)_3COH$ (4) 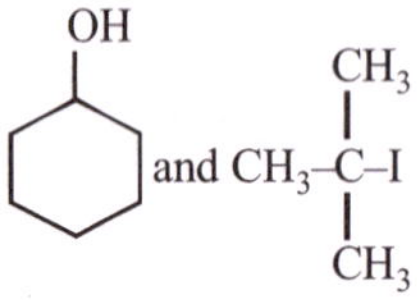

156. Suitable reaction condition for preparation of methyl phenyl ether is **[April 10, 2023 (I)]**

(1) $Ph-Br, MeO^-Na^+$ (2) PhO^-Na^+, MeOH

(3) PhO^-Na^+, MeBr (4) Benzene, MeBr

157. 'A' and 'B' formed in the following set of reactions are : **[Jan. 24, 2023 (I)]**

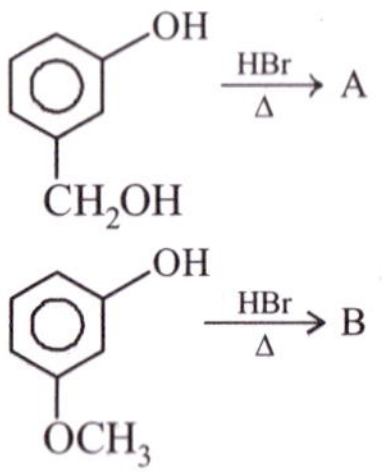

(1) A = 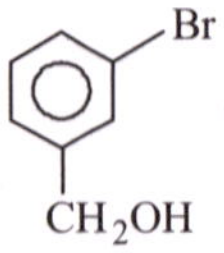, B = 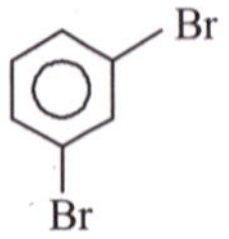

(2) A = 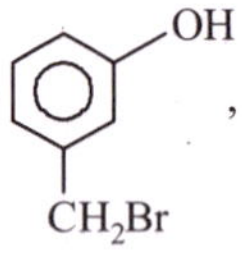, B = 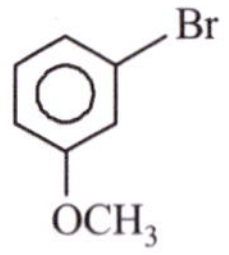

(3) A = 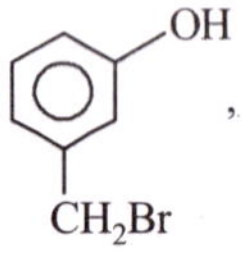, B = 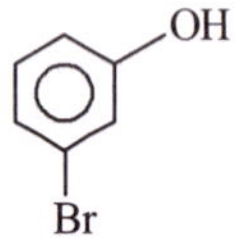

(4) A = 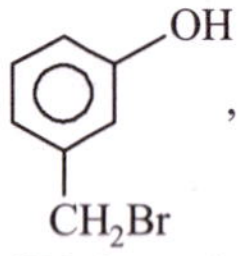, B = 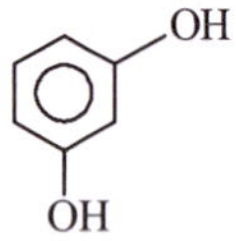

158. Compound I is heated with conc. HI to give a hydroxy compound A which is further heated with Zn dust to give compound B. Identify A and B. **[July 28, 2022 (II)]**

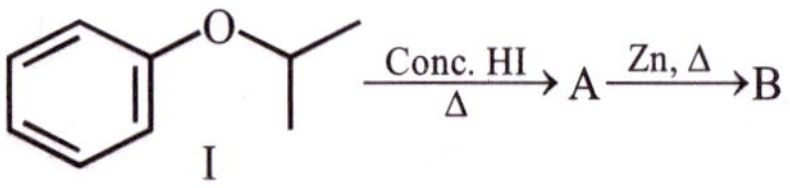

(1) A = 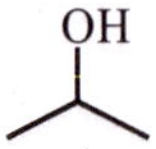, B = 

(2) A = 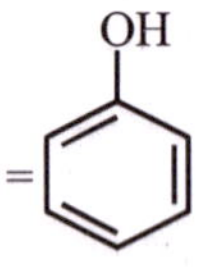, B = 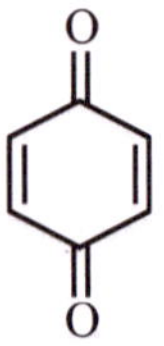

(3) A = 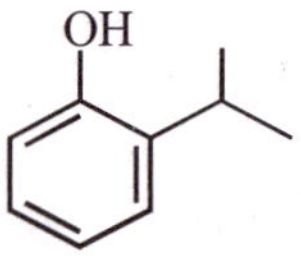, B = 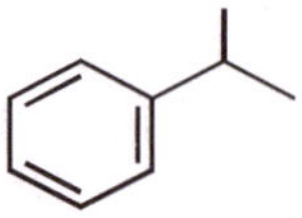

(4) A = 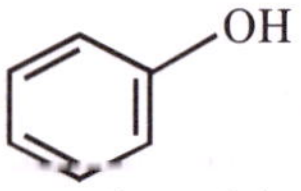, B = 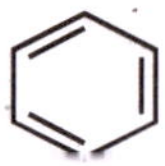

159. The major product of the following reaction is **[July 26, 2022 (I)]**

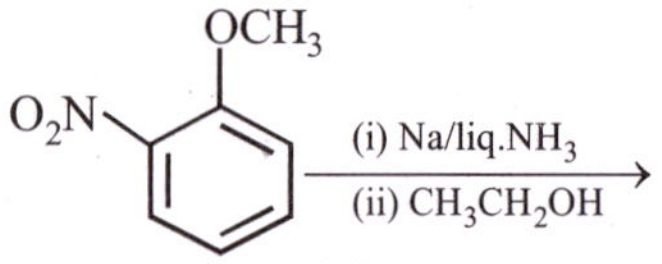

(1) 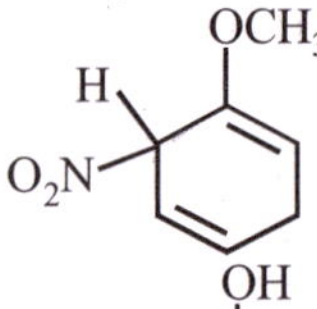 (2) 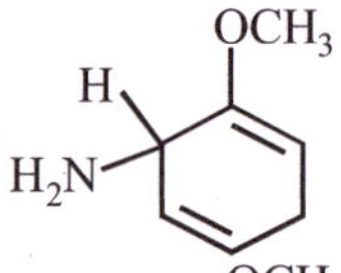

(3) 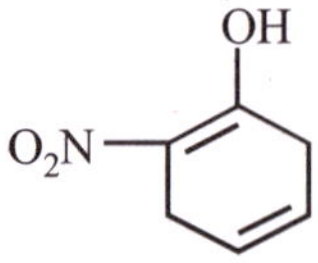 (4) 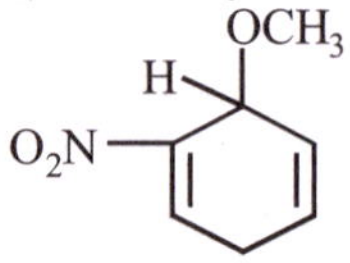

160. The major product formed in the following reaction, is **[June 25, 2022 (II)]**

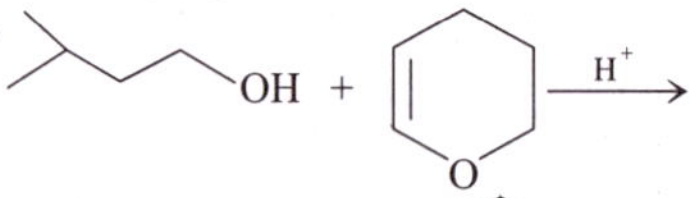

(1) 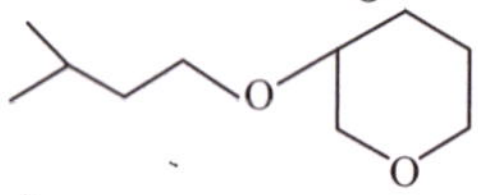

(2) 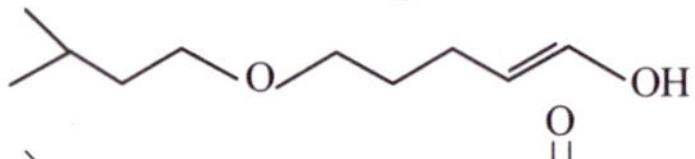

(3) 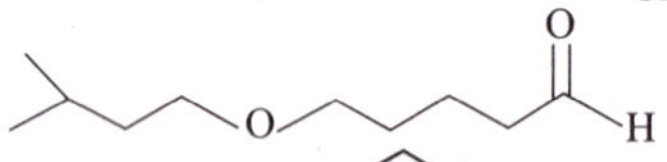

(4) 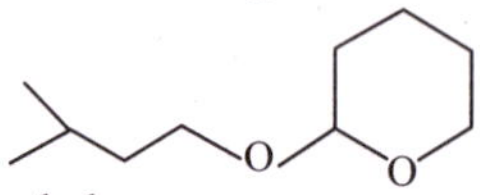

161. Given below are two statements, one is labelled as **Assertion (A)** and the other is labelled as **Reason (R)**. **[Aug. 27, 2021 (I)]**

Assertion (A) : Synthesis of ethyl phenyl ether may be achieved by Williamson synthesis.

Reason (R) : Reaction of bromobenzene with sodium ethoxide yields ethyl phenyl ether.

In the light of the above statements, choose the **most appropriate** answer from the options given below:

(1) Both (A) and (R) are correct and (R) is the correct explanation of (A)

(2) (A) is correct but (R) is not correct

(3) (A) is not correct but (R) is correct

(4) Both (A) and (R) are correct but (R) is NOT the correct explanation of (A)

162. The major product [B] in the following reactions is : **[Sep. 04, 2020 (II)]**

$$CH_3-CH_2-\underset{}{\overset{CH_3}{\overset{|}{C}H}}-CH_2-OCH_2-CH_3 \xrightarrow[\text{Heat}]{\text{HI}} [A]\text{alcohol} \xrightarrow[\Delta]{H_2SO_4} [B]$$

(1) $CH_2=CH_2$ (2) $CH_3-CH=\overset{CH_3}{\overset{|}{C}}-CH_3$

(3) $CH_3-CH_2-\overset{CH_3}{\overset{|}{C}}=CH_2$ (4) $CH_3-CH_2-CH=CH-CH_3$

163. An organic compound 'A' ($C_9H_{10}O$) when treated with conc. HI undergoes cleavage to yield compounds 'B' and 'C'. 'B' gives yellow precipitate with $AgNO_3$ whereas 'C' tautomerizes to 'D'. 'D' gives positive iodoform test. 'A' could be : **[Sep. 02, 2020 (II)]**

(1) 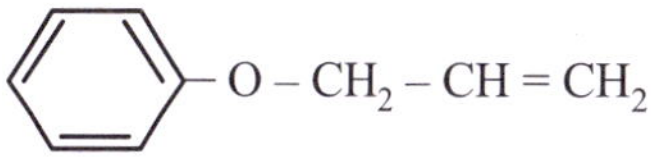

(2) 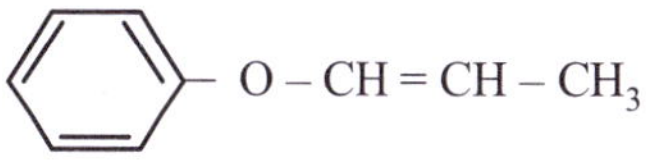

(3) 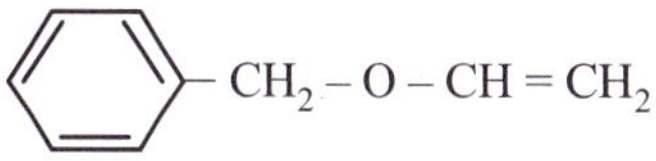

(4) 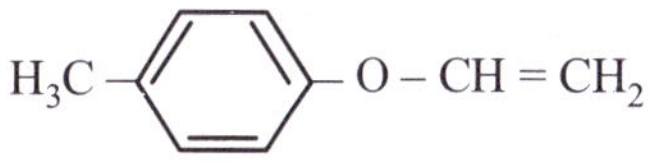

164. The major aromatic product C in the following reaction sequence will be : **[Sep. 02, 2020 (I)]**

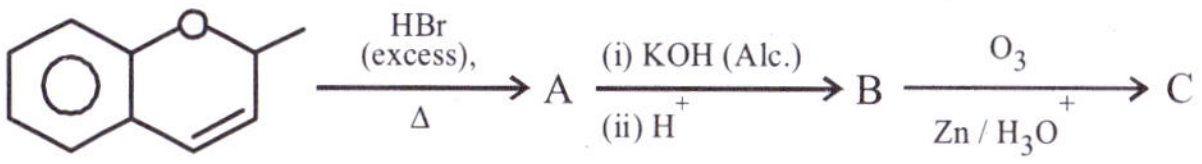

(1) 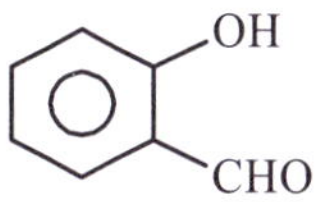

(2) 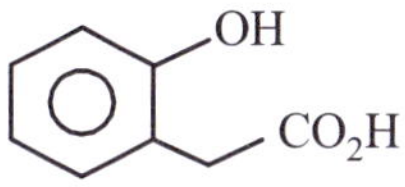

(3) 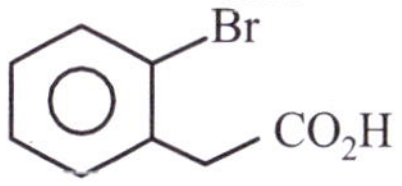

(4) 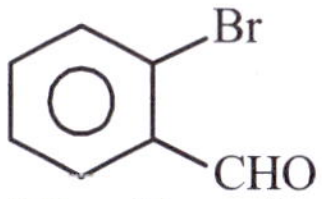

165. Among the compounds A and B with molecular formula $C_9H_{18}O_3$, A is having higher boiling point than B. The possible structures of A and B are: **[Jan. 08, 2020 (II)]**

(1) 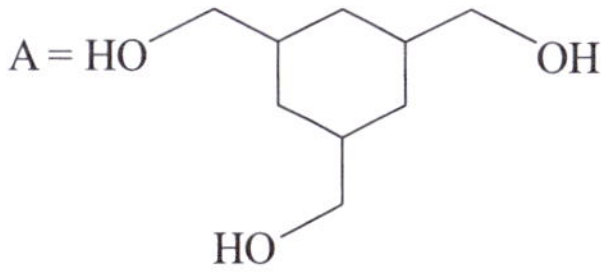

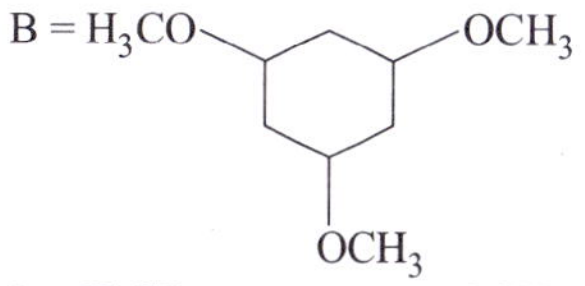

(2) 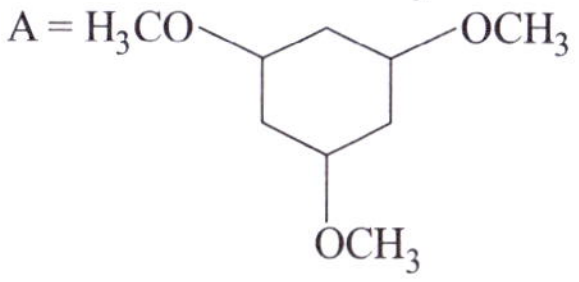

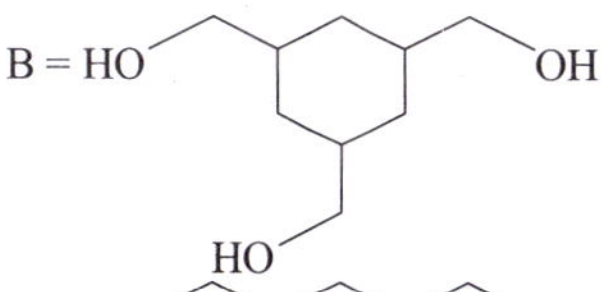

(3) 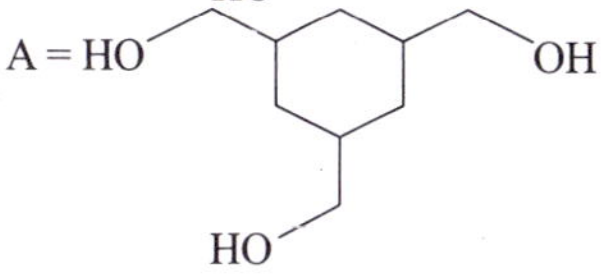

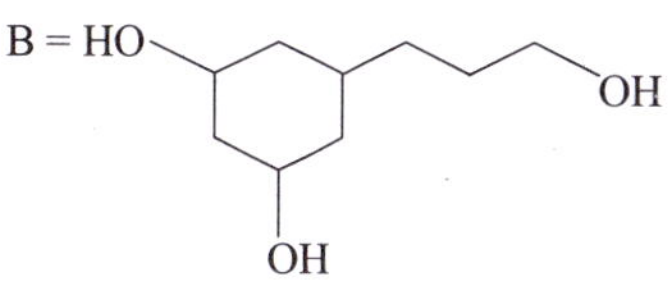

(4) 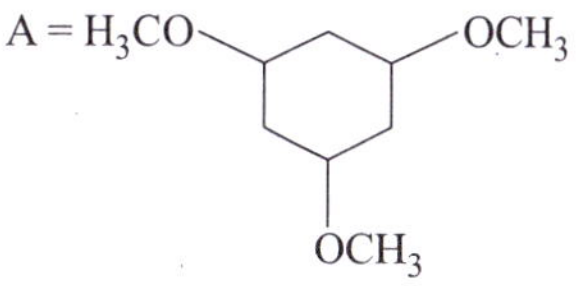

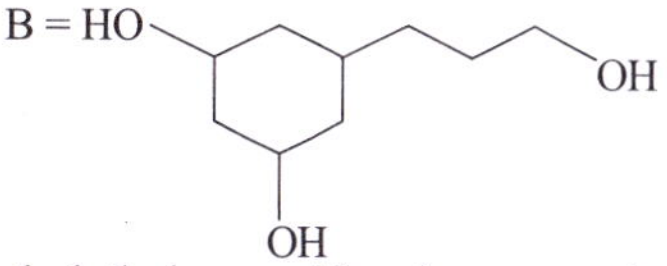

166. 1-Methylethylene oxide when treated with an excess of HBr produces: **[Jan. 07, 2020 (I)]**

(1) 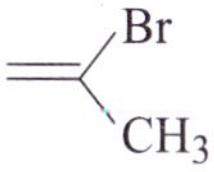

(2) 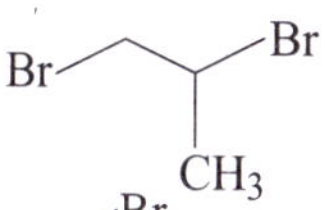

(3) 

(4) 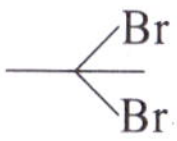

167. The major product obtained in the given reaction is : **[April 10, 2019 (II)]**

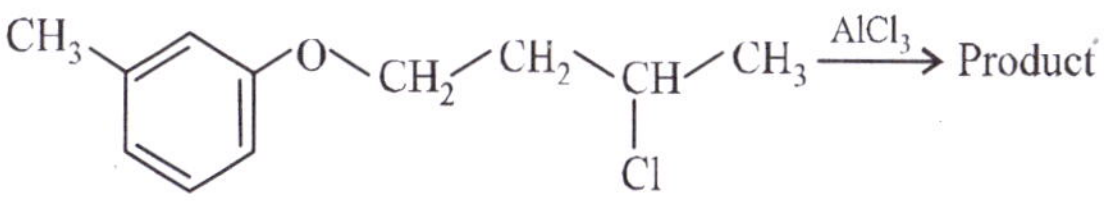

(1) 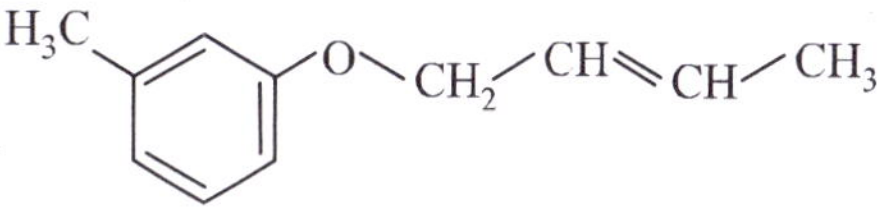

(2) 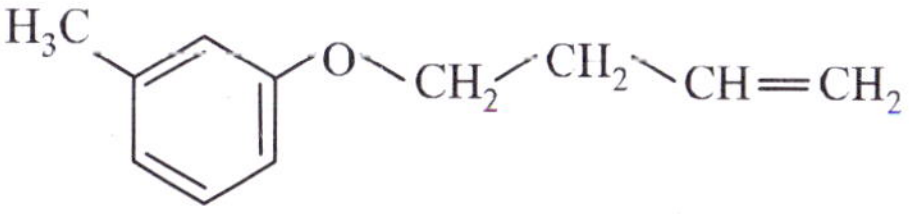

(3) 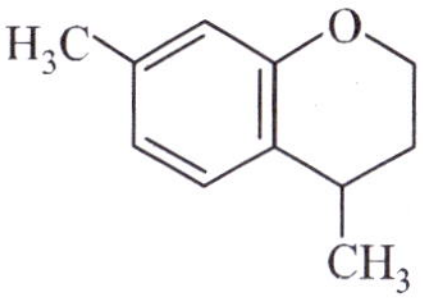

(4) 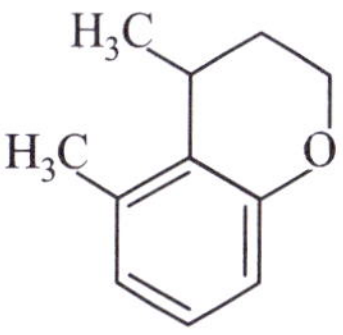

168. The major product of the following reaction is: **[April 10, 2019 (I)]**

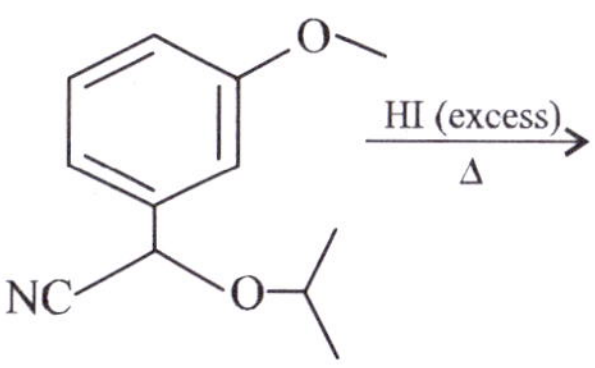

(1) 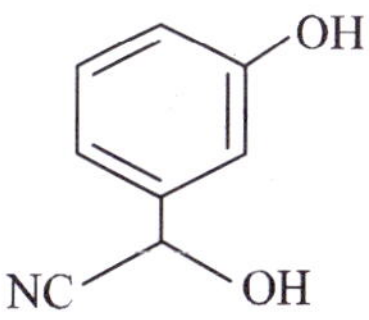

(2) 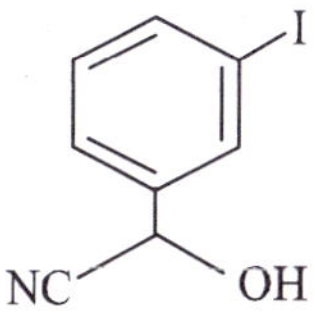

(3) 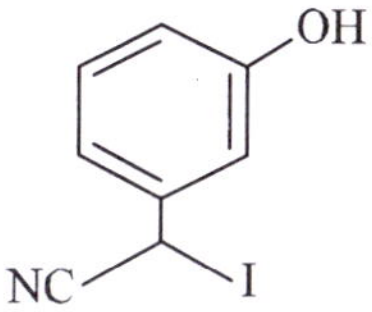

(4) 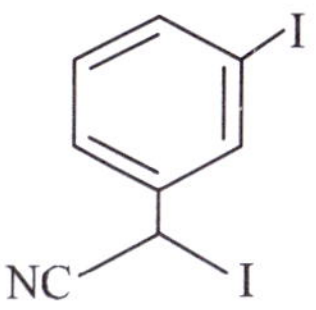

169. The major product of the following reaction is :

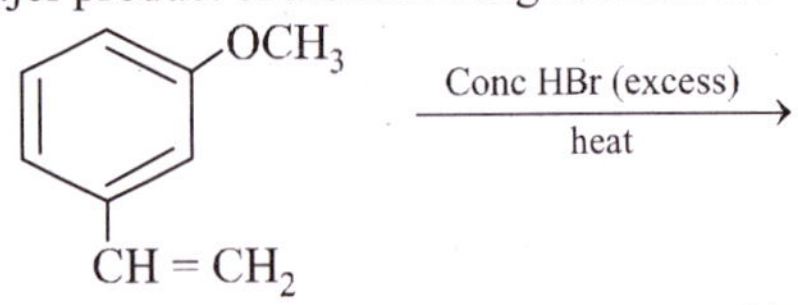

[April 8, 2019 (I)]

(1) 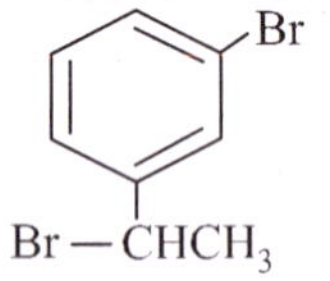

(2) 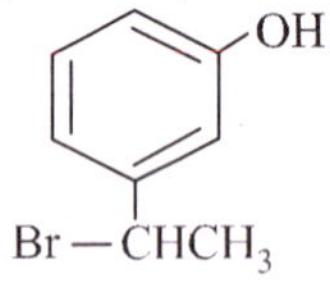

(3) 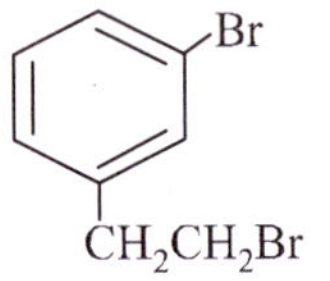

(4) 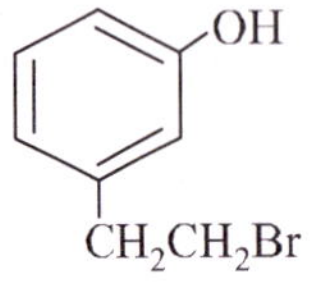

170. The major product in the following conversion is : [Jan. 12, 2019 (II)]

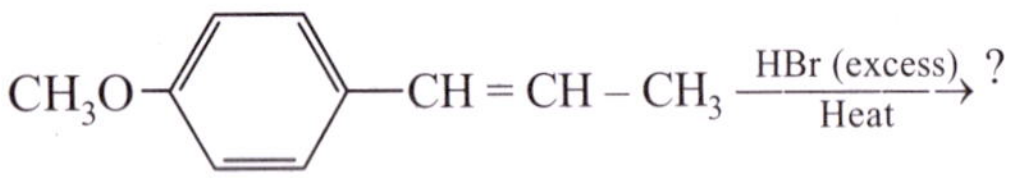

(1) 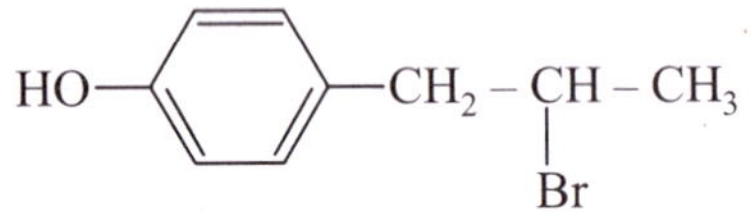

(2) 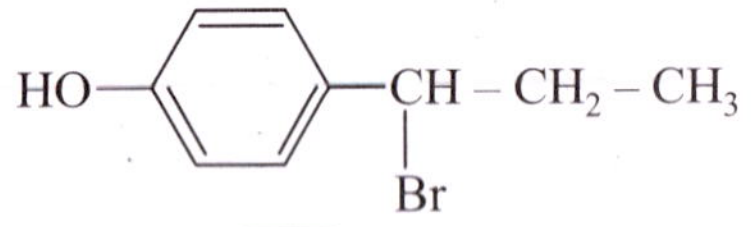

(3) 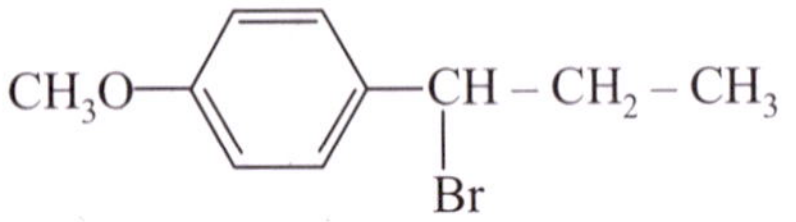

(4) 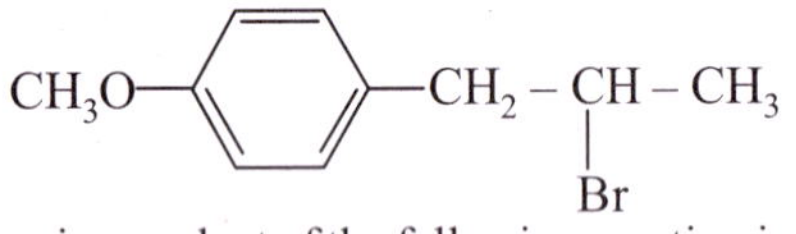

171. The major product of the following reaction is:

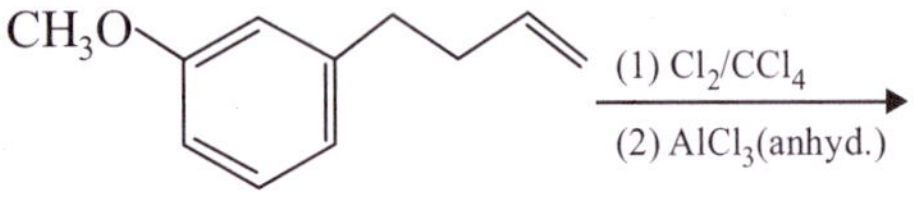

[Jan. 12, 2019 (I)]

(1) 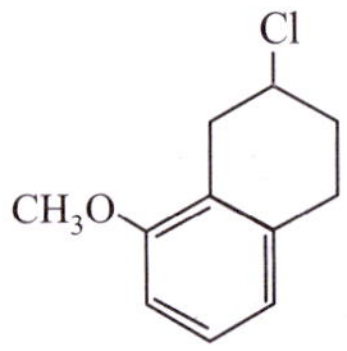

(2) 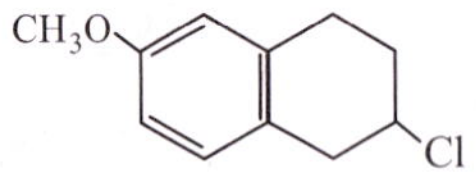

(3) 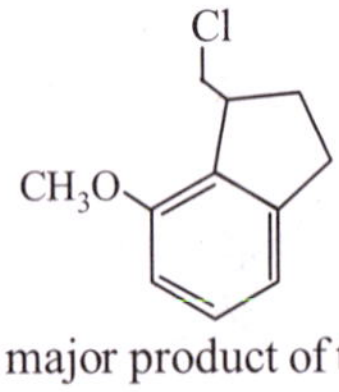

(4) 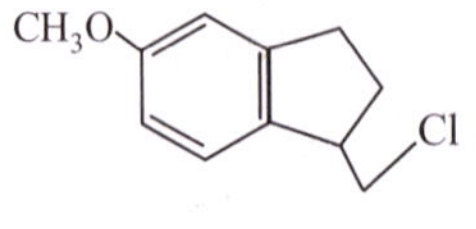

172. The major product of the following reaction is: [Jan. 10, 2019 (I)]

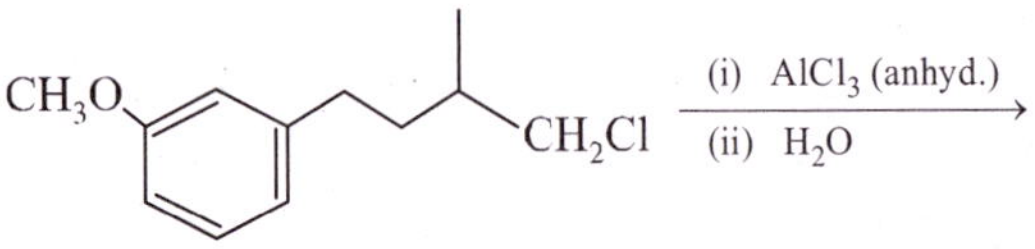

(1) 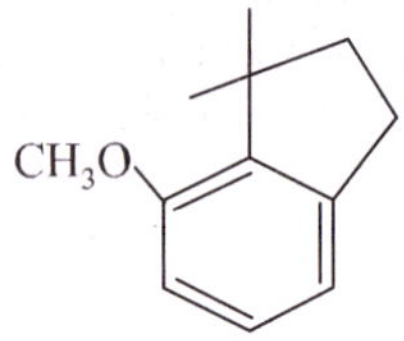

(2) 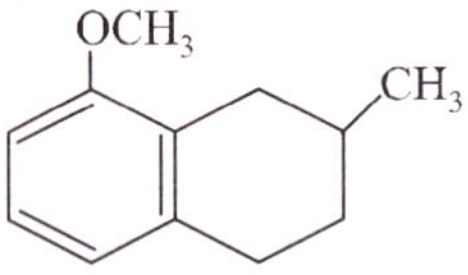

(3) 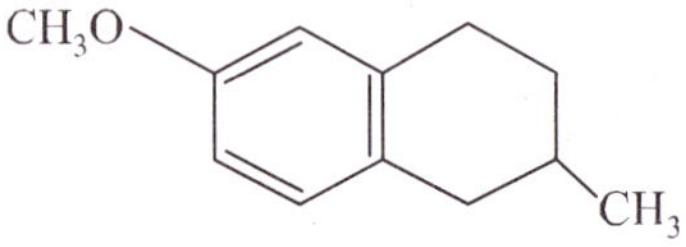

(4) 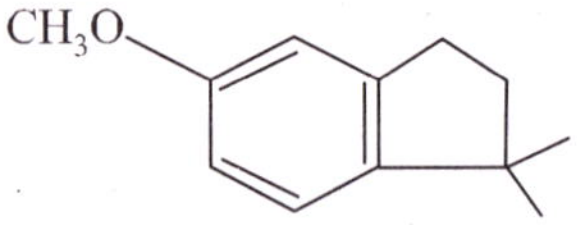

173. The major product formed in the following reaction is : [2018]

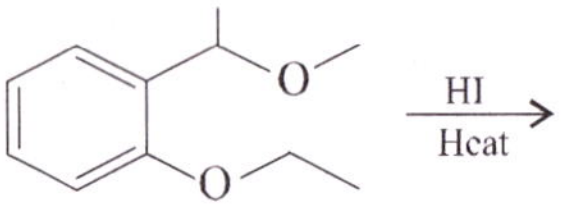

(1) 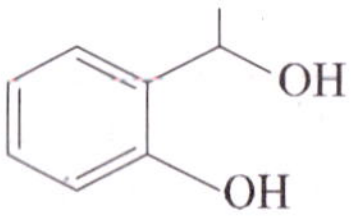

(2) 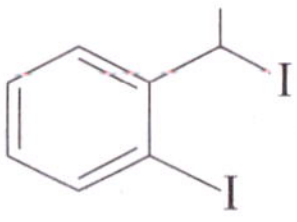

(3) 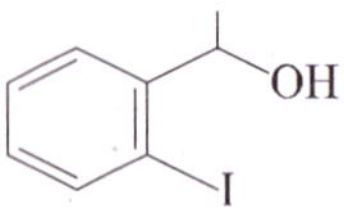

(4) 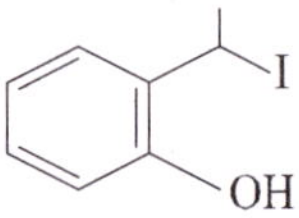

174. Consider the reaction sequence below : [Online April 10, 2016]

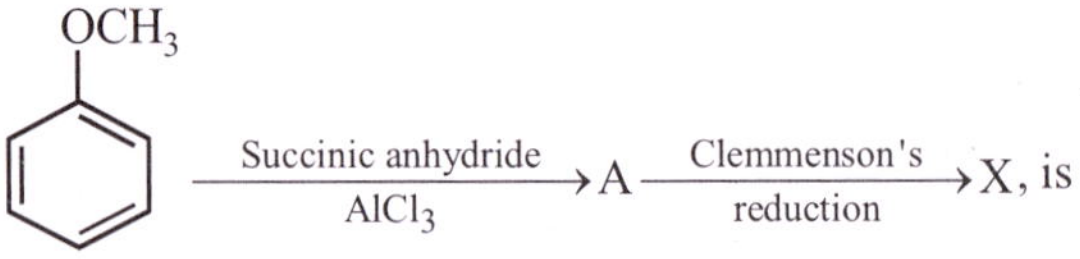

(1) 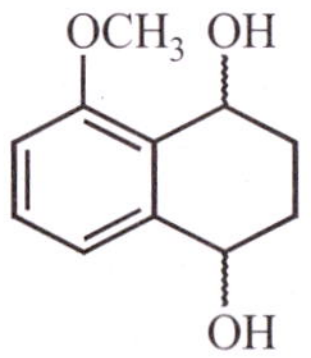

(2) 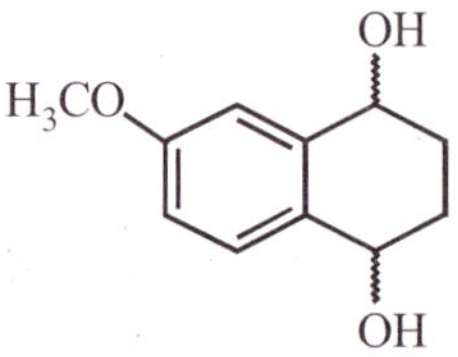

(3) 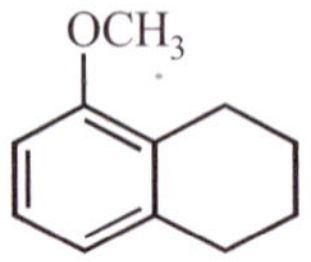

(4) 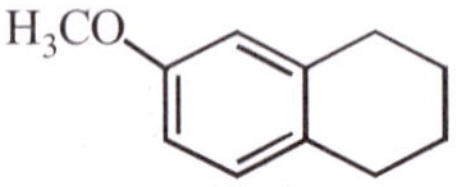

175. Williamson synthesis of ether is an example of: [Online April 19, 2014]

(1) Nucleophilic addition
(2) Electrophilic addition
(3) Electrophilic substitution
(4) Nucleophilic substitution

176. Allyl phenyl ether can be prepared by heating: [Online April 9, 2014]

(1) $C_6H_5Br + CH_2 = CH - CH_2 - ONa$
(2) $CH_2 = CH - CH_2 - Br + C_6H_5ONa$
(3) $C_6H_5 - CH = CH - Br + CH_3 - ONa$
(4) $CH_2 = CH - Br + C_6H_5 - CH_2 - ONa$

177. In Williamson synthesis of mixed ether having a primary and a tertiary alkyl group, if tertiary halide is used, then : **[Online April 22, 2013]**

(1) Rate of reaction will be slow due to slow cleavage of carbon-halogen bond.
(2) Alkene will be the main product.
(3) Simple ether will form instead of mixed ether.
(4) Expected mixed ether will be formed.

178. An ether (A), $C_5H_{12}O$, when heated with excess of hot concentrated HI produced two alkyl halides which when treated with NaOH yielded compounds (B) and (C). Oxidation of (B) and (C) gave propanone and ethanoic acid respectively. The name of the ether (A) is : **[Online April 9, 2013]**

(1) 2-ethoxypropane (2) ethoxypropane
(3) methoxybutane (4) 2-methoxybutane

179. HBr reacts with $CH_2 = CH - OCH_3$ under anhydrous conditions at room temperature to give **[2006]**

(1) $BrCH_2 - CH_2 - OCH_3$ (2) $H_3C - CHBr - OCH_3$
(3) CH_3CHO and CH_3Br (4) $BrCH_2CHO$ and CH_3OH

NUMERICAL VALUE QUESTIONS (NVQs)

180.

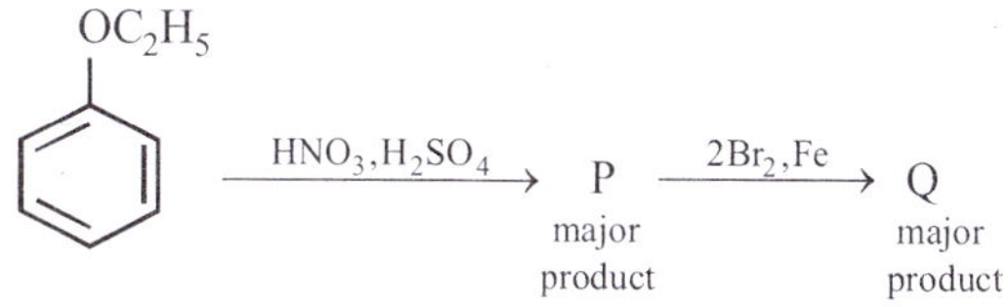

The ratio of number of oxygen atoms to bromine atoms in the product Q is________ $\times 10^{-1}$. **[April 6, 2024 (II)]**

Part-B (JEE Advanced)

Topic 1- Preparation and Properties of Alcohols

1 MCQs with One Correct Answer

1. The best method to prepare cyclohexene from cyclohexanol is by using **[2005S]**

(a) Conc. HCl + $ZnCl_2$ (b) Conc. H_3PO_4
(c) HBr (d) Conc. HCl

2. The product of acid catalyzed hydration of 2-phenylpropene is **[2004S]**

(a) 3-phenyl-2-propanol (b) 1-phenyl-2-propanol
(c) 2-phenyl-2-propanol (d) 2-phenyl-1-propanol

3. 1–propanol and 2–propanol can be best distinguished by **[2001S]**

(a) oxidation with alkaline $KMnO_4$ followed by reaction with Fehling solution
(b) oxidation with acidic dichromate followed by reaction with Fehling solution
(c) oxidation by heating with copper followed by reaction with Fehling solution
(d) oxidation with concentrated H_2SO_4 followed by reaction with Fehling solution

4. The compound that will react most readily with NaOH to form methanol is **[2001S]**

(a) $(CH_3)_4N^+I^-$ (b) CH_3OCH_3
(c) $(CH_3)_3S^+I^-$ (d) $(CH_3)_3CCl$

5. Which of the following compounds is oxidised to prepare methyl ethyl ketone? **[1987 - 1 Mark]**

(a) 2-Propanol (b) l-Butanol
(c) 2-Butanol (d) t-Butyl alcohol

6. HBr reacts fastest with : **[1986 - 1 Mark]**

(a) 2-methylpropan-2-ol (b) propan-1-ol
(c) propan-2-ol (d) 2-methylpropan-1-ol

7. An industrial method of preparation of methanol is : **[1984 - 1 Mark]**

(a) catalytic reduction of carbon monoxide in presence of $ZnO-Cr_2O_3$
(b) by reacting methane with steam at 900°C with a nickel catalyst
(c) by reducing formaldehyde with lithium aluminium hydride
(d) by reacting formaldehyde with aqueous sodium hydroxide solution

8. The compound which reacts fastest with Lucas reagent at room temperature is **[1981 - 1 Mark]**

(a) butan-1-ol (b) butan-2-ol
(c) 2-methylpropan-1-ol (d) 2-methylpropan-2-ol

9. Which of the following is basic **[1980]**

(a) $CH_3 - CH_2 - OH$ (b) $OH - CH_2 - CH_2 - OH$
(c) $H - O - O - H$ (d) $H_3C - \underset{\underset{O}{\|}}{C} - OH$

10. Ethyl alcohol is heated with conc H_2SO_4 the product formed is **[1980]**

(a) $H_3C - \underset{\underset{O}{\|}}{C} - OC_2H_5$ (b) C_2H_6
(c) C_2H_4 (d) C_2H_2

2 Integer Value Answer

11. Total number of hydroxyl groups present in a molecule of major product *P* is ___ **[Adv. 2019]**

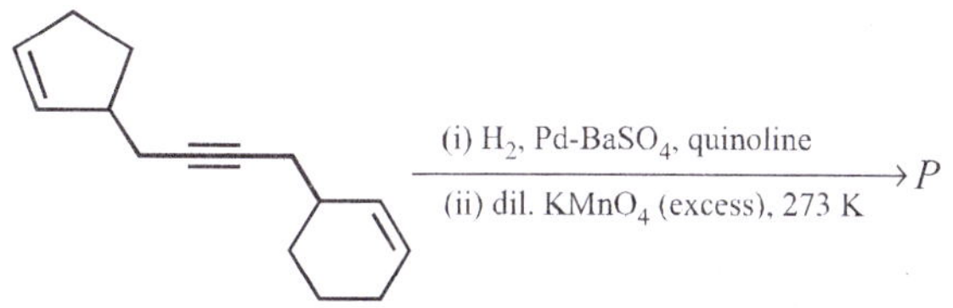

12. The number of hydroxyl group(s) in *Q* is **[Adv. 2015]**

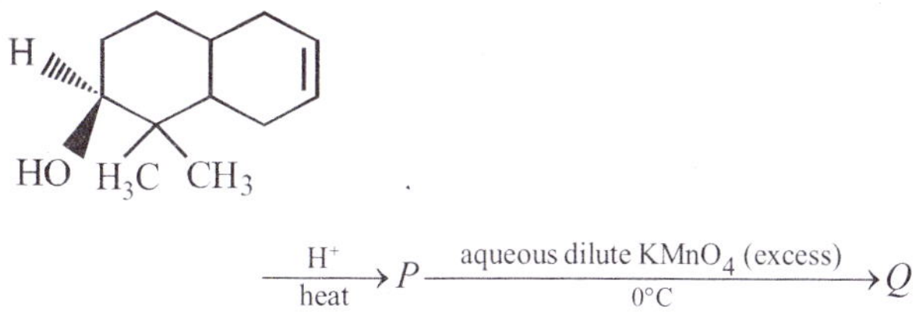

3 Numeric / New Stem Based Questions

13. Consider the following reaction. **[Adv. 2022]**

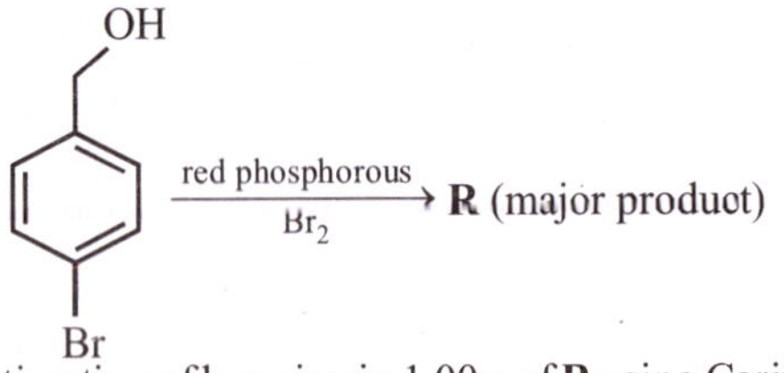

On estimation of bromine in 1.00 g of **R** using Carius method, the amount of AgBr formed (in g) is ________.
[Given: Atomic mass of H = 1, C = 12, O = 16, P = 31, Br = 80, Ag = 108]

4 Fill in the Blanks

14. Glycerine contains one .... hydroxy group. **[1997 - 1 Mark]**

15. Ethanol vapour is passed over heated copper and the product is treated with aqueous NaOH. The final product is .... **[1983 - 1 Mark]**

5 True / False

16. Sodium ethoxide is prepared by reacting ethanol with aqueous sodium hydroxide. **[1986 - 1 Mark]**

6 MCQs with One or More than One Correct Answer

17. The correct combination of names for isomeric alcohols with molecular formula $C_4H_{10}O$ is/are **[Adv. 2014]**

(a) tert-butanol and 2-methylpropan-2-ol
(b) tert-butanol and 1, 1-dimethylethan-1-ol
(c) *n*-butanol and butan-1-ol
(d) isobutyl alcohol and 2-methylpropan-1-ol

18. The reaction of 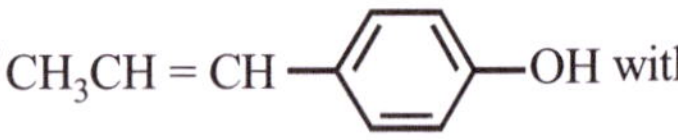 with HBr gives **[1998 - 2 Marks]**

(a) 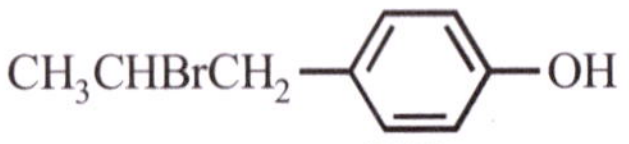
(b) 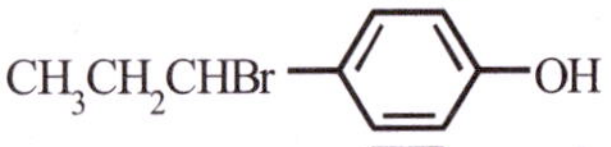
(c) 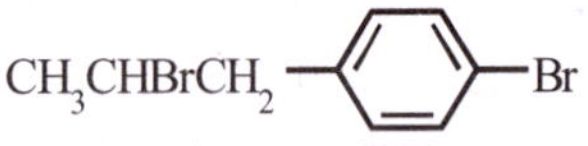
(d) 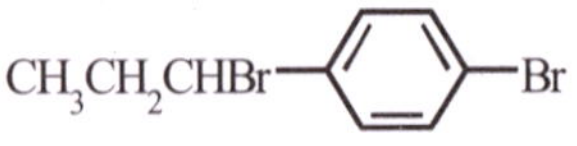

9 Assertion and Reason/Statement Type Questions

Each question contains **STATEMENT-1 (Assertion)** and **STATEMENT-2 (Reason)**. Each question has 4 choices (a), (b), (c) and (d) out of which ONLY ONE is correct. Mark your answer as

(a) If both Statement -1 and Statement -2 are correct, and Statement -2 is the correct explanation of the Statement -2.
(b) If both Statement -1 and Statement -2 are correct, but Statement -2 is not the correct explanation of the Statement -1.
(c) If Statement -1 is correct but Statement -2 is incorrect.
(d) If Statement -1 is incorrect but Statement -2 is correct.

19. Read the following statement and explanation and answer as per the options given below : **[1988 - 2 Marks]**

Statement : Solubility of *n*-alcohols in water decreases with increase in molecular weight.

Explanation : The relative proportion of the hydrocarbon part in alcohols increases with increasing molecular weight which permits enhanced hydrogen bonding with water.

10 Subjective Problems

20. Identify (*X*) and (*Y*) in the following reaction sequence. **[2005 - 2 Marks]**

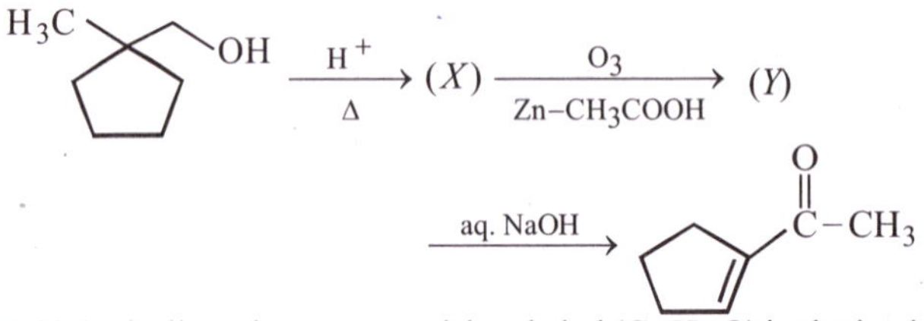

21. A biologically active compound, bombykol ($C_{16}H_{30}O$) is obtained from a natural source. The structure of the compound is determined by the following reactions. **[2002 - 5 Marks]**

(a) On hydrogenation, bombykol gives a compound **A**, $C_{16}H_{34}O$, which reacts with acetic anhydride to give an ester;
(b) Bombykol also reacts with acetic anhydride to give another ester, which on oxidative ozonolysis (O_3/H_2O_2) gives a mixture of butanoic acid, oxalic acid and 10-acetoxydecanoic acid.

Determine the number of double bonds in bombykol. Write the structures of compound *A* and bombykol. How many geometrical isomers are possible for bombykol?

22. Complete the following reaction with appropriate reagents : **[1999 - 3 Marks]**

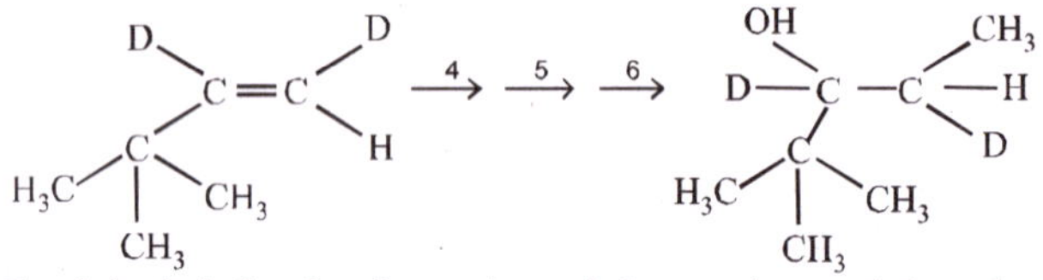

23. Explain briefly the formation of the products giving the structures of the intermediates. **[1999 - 3 Marks]**

(i) 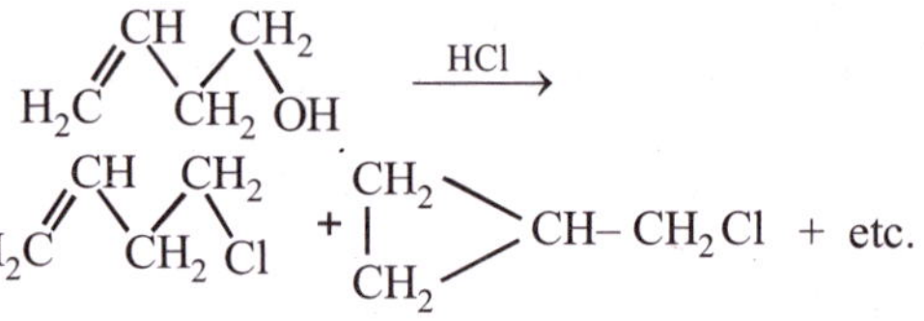

(ii) 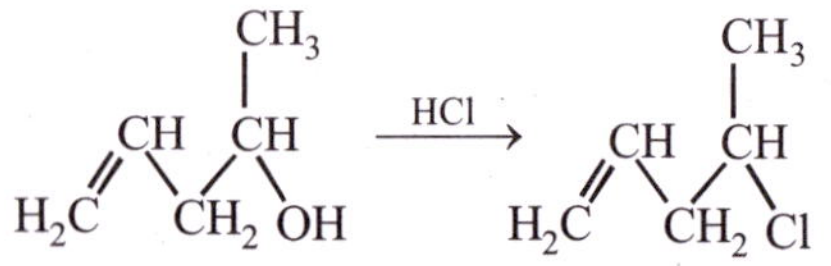 (Only)

24. Give reasons for the following:
Acid catalysed dehydration of *t*-butanol is faster than that of *n*-butanol. **[1998 - 2 Marks]**

25. An optically active alcohol *A* ($C_6H_{10}O$) absorbs two moles of hydrogen per mole of *A* upon catalytic hydrogenation and gives a product *B*. The compound *B* is resistant to oxidation by CrO_3 and does not show any optical activity. Deduce the structures of *A* and *B*. **[1996 - 2 Marks]**

26. 3,3-Dimethylbutan-2-ol loses a molecule of water in the presence of concentrated sulphuric acid to give tetramethylethylene as a major product. Suggest a suitable mechanism. **[1996 - 2 Marks]**

27. When *t*-butanol and *n*-butanol are separately treated with a few drops of dilute $KMnO_4$, in one case only the purple colour disappears and a brown precipitate is formed. Which of the two alcohols gives the above reaction and what is the brown precipitate? **[1994 - 2 Marks]**

28. Compound '*X*' (molecular formula, C_5H_8O) does not react appreciably with Lucas reagent at room temperature but gives a precipitate with ammonical silver nitrate. With excess of MeMgBr, 0.42 g of '*X*' gives 224 mL of CH_4 at STP. Treatment of '*X*' with H_2 in presence of Pt catalyst followed by boiling with excess HI, gives *n*-pentane. Suggest structure for '*X*' and write the equation involved. **[1992 - 5 Marks]**

29. State with balanced equations what happens when :
Ethylene glycol is obtained by the reaction of ethylene with potassium permanganate. **[1991 - 1 Mark]**

30. A ketone 'A' which undergoes haloform reaction gives compound B on reduction. B on heating with sulphuric acid gives compound C, which forms monoozonide D, D on hydrolysis in presence of zinc dust gives only acetaldehyde. Identify A, B and C. Write down the reactions involved. **[1989 - 4 Marks]**

31. A hydrocarbon A (molecular formula C_5H_{10}) yields 2-methylbutane on catalytic hydrogenation. A adds HBr (in accordance with Markownikoff's rule) to form a compound B which on reaction with silver hydroxide forms an alcohol C, $C_5H_{12}O$. Alcohol C on oxidation gives a ketone D. Deduce the structures of A, B, C and D and show the reactions involved. **[1988 - 5 Marks]**

32. Outline the reaction sequence for the conversion of
(ii) 1-propanol from 2-propanol (in three steps) **[1982 - 1 Mark]**
(i) ethyl alcohol to vinyl acetate. (in not more than 6 steps) **[1986 - 3 Marks]**

33. Give a chemical test/suggest a reagent to distinguish between methanol and ethanol. **[1985 - 1 Mark]**

34. An alcohol *A*, when heated with conc. H_2SO_4 gives an alkene *B*. When *B* is bubbled through bromine water and the product obtained is dehydrohalogenated with excess of sodamide, a new compound *C* is obtained. The compound *C* gives *D* when treated with warm dilute H_2SO_4 in presence of $HgSO_4$. *D* can also be obtained either by oxidizing *A* with $KMnO_4$ or from acetic acid through its calcium salt. Identify *A*, *B*, *C* and *D*. **[1983 - 4 Marks]**

35. An organic liquid (A), containing C, H and O with boiling point : 78°C, and possessing a rather pleasant odour, on heating with concentrated sulphuric acid gives a gaseous product (B) – with the empirical formula, CH_2. 'B' decolourises bromine water as well as alkaline $KMnO_4$ solution and takes up one mole of H_2 (per mole of 'B') in the presence of finely divided nickel at high temperature. Identify the substances 'A' and 'B'. **[1979]**

Topic 2 - Preparation and Properties of Phenols

1 MCQs with One Correct Answer

36. In the given reaction scheme, P is a phenyl alkyl ether, Q is an aromatic compound; R and S are the major products.

$$P \xrightarrow{HI} Q \xrightarrow[\text{(iii)}\,H_3O^+]{\text{(i)}\,NaOH,\ \text{(ii)}\,CO_2} R \xrightarrow[\text{(ii)}\,H_3O^+]{\text{(i)}\,(CH_3CO)_2O} S$$

The correct statement about S is **[Adv. 2023]**
(a) It primarily inhibits noradrenaline degrading enzymes.
(b) It inhibits the synthesis of prostaglandin.
(c) It is a narcotic drug.
(d) It is *ortho*–acetylbenzoic acid.

37. For the identification of β-naphthol using dye test, it is necessary to use **[Adv. 2014]**
(a) Dichloromethane solution of β-naphthol
(b) Acidic solution of β-naphthol
(c) Neutral solution of β-naphthol
(d) Alkaline solution of β-naphthol

38. The increasing order of boiling points of the below mentioned alcohols is **[2006 - 3M, –1]**
(I) 1,2-dihydroxybenzene (II) 1,3-dihydroxybenzene
(III) 1,4-dihydroxybenzene (IV) Hydroxybenzene
(a) I < II < IV < III (b) I < II < III < IV
(c) IV < II < I < III (d) IV < I < II < III

39. Which of the following acids has the smallest dissociation constant ? **[2002S]**
(a) $CH_3CHFCOOH$ (b) FCH_2CH_2COOH
(c) $BrCH_2CH_2COOH$ (d) $CH_3CHBrCOOH$

40. In the following compounds, **[1996]**

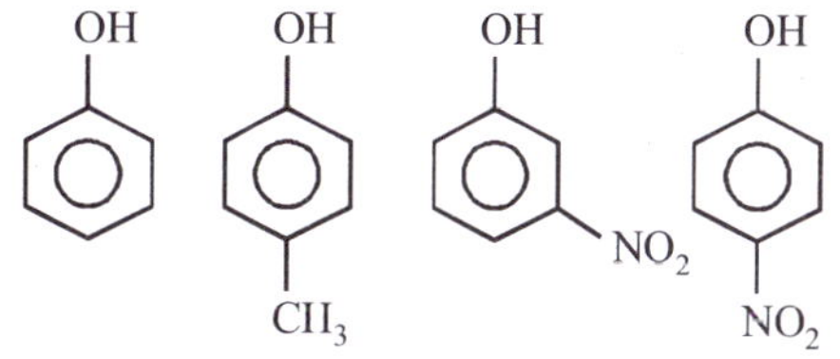

(I) (II) (III) (IV)

The order of acidity is :
(a) III > IV > I > II (b) I > IV > III > II
(c) II > I > III > IV (d) IV > III > I > II

41. When phenol is reacted with $CHCl_3$ and NaOH followed by acidification, salicyladehyde is obtained. Which of the following species are involved in the above mentioned reaction as intermediate? **[1995S]**

(a) 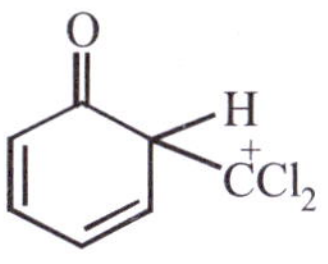

(b) 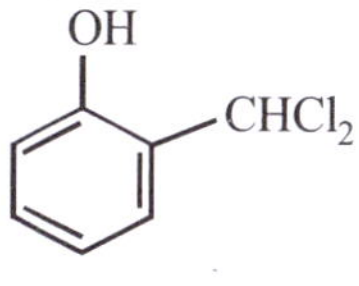

(c) 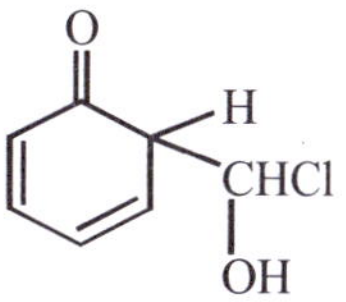

(d) 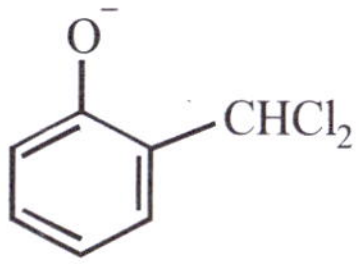

42. Chlorination of toluene in the presence of light and heat followed by treatment with aqueous NaOH gives **[1990 - 1 Mark]**
(a) *o*-Cresol (b) *p*-Cresol
(c) 2, 4-Dihydroxytoluene (d) Benzoic acid

43. Phenol reacts with bromine in carbon disulphide at low temperature to give **[1988 - 1 Mark]**
(a) *m*-bromophenol (b) *o*- and *p*-bromophenol
(c) *p*-bromophenol (d) 2, 4, 6-tribromophenol

44. When phenol is treated with excess bromine water, it gives: **[1984 - 1 Mark]**
(a) *m*-Bromophenol (b) *o*- and *p*-Bromophenol
(c) 2, 4-Dibromophenol (d) 2, 4, 6-Tribromophenol

2 Integer Value Answer

45. An organic compound ($C_8H_{10}O_2$) rotates plane-polarized light. It produces pink color with neutral $FeCl_3$ solution. What is the total number of all the possible isomers for this compound? **[Adv. 2020]**

46. The number of resonance structures for N is **[Adv. 2015]**

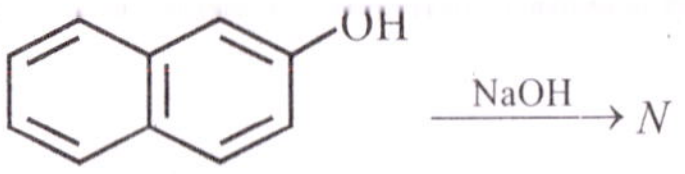

4 Fill in the Blanks

47. Phenol is acidic because of resonance stabilization of its conjugate base, namely .... **[1990 - 1 Mark]**

48. Formation of phenol from chlorobenzene is an example of .... aromatic substitution. **[1989 - 1 Mark]**

49. The acidity of phenol is due to the .... of its anion. **[1984 - 1 Mark]**

6 MCQs with One or More than One Correct Answer

50. The reactivity of compound Z with different halogens under appropriate conditions is given below: **[Adv. 2014]**

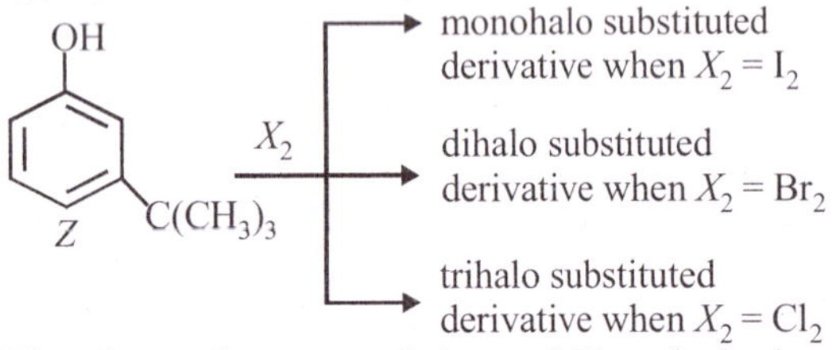

The observed pattern of electrophilic substitution can be explained by

(a) The steric effect of the halogen
(b) The steric effect of the *tert*-butyl group
(c) The electronic effect of the phenolic group
(d) The electronic effect of the *tert*-butyl group

51. The major product(s) of the following reaction is(are) **[Adv. 2013]**

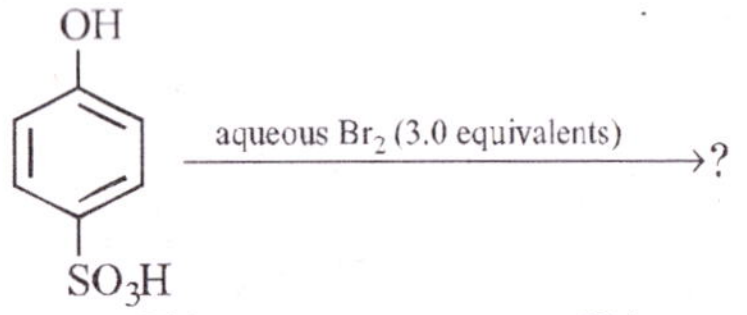

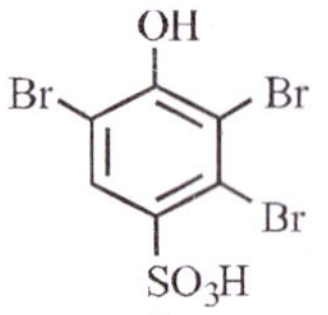

P

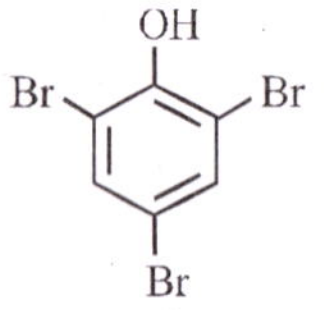

Q

R

S

(a) P (b) Q (c) R (d) S

52. In the reaction 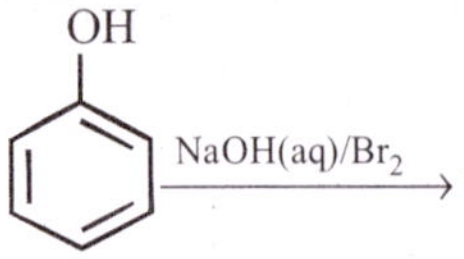 the intermediate (s) is (are) **[2010]**

(a) 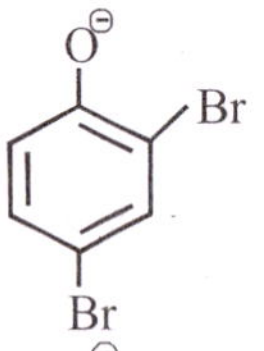
(b) 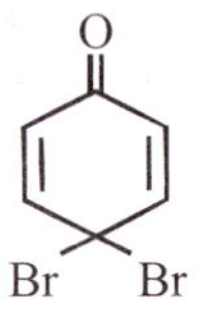
(c) 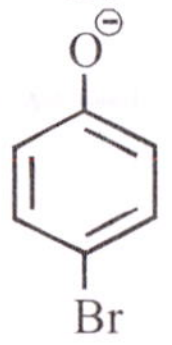
(d) 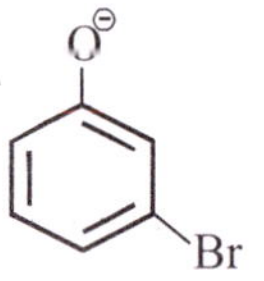

53. The reaction of 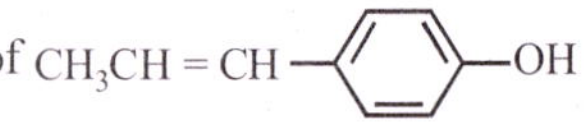 with HBr gives **[1998 - 2 Marks]**

(a) 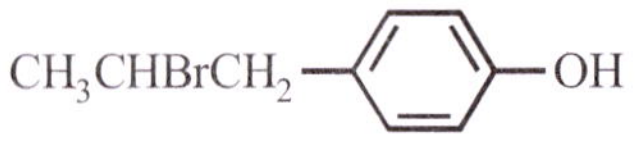
(b) 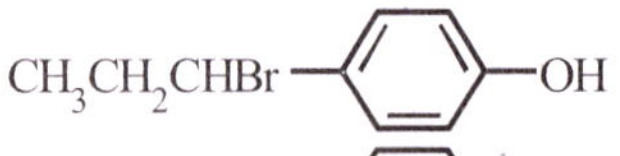
(c) 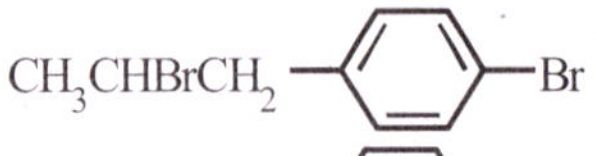
(d) 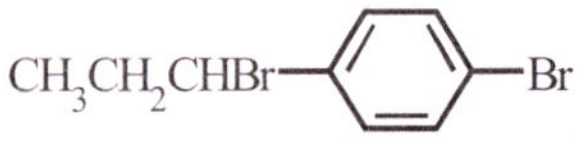

54. When phenol is reacted with $CHCl_3$ and NaOH followed by acidification, salicylaldehyde is obtained. Which of the following species are involved in the above mentioned reaction as intermediates? **[1995 - 2 Marks]**

(a) 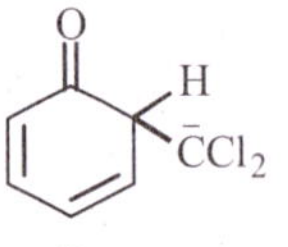
(b) 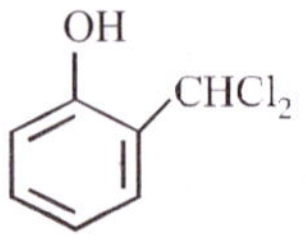
(c) 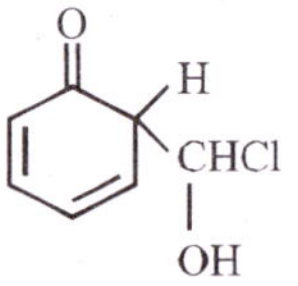
(d) 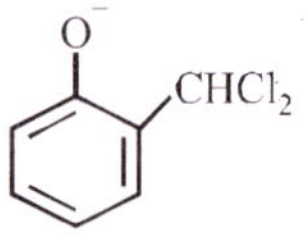

55. Phenol is less acidic than : **[1986]**

(a) acetic acid (b) p-methoxyphenol
(c) p-nitrophenol (d) ethanol

7 Match the Following

56. Match each item of the right hand column with an appropriate item in the left hand column for each of the following sections: **[1985 - 2½ × 4 = 10 Marks]**

A.	(i)	spinel	(a)	$MgAl_2O_4$
	(ii)	feldspar	(b)	PbCO
	(iii)	cerussite	(c)	$KAlSi_3O_8$
	(iv)	malachite	(d)	$MgSO_4.H_2O$
	(v)	kisserite	(e)	$Cu(OH)_2.CuCO_3$
B.	(vi)	liquid air	(f)	Deacon process
	(vii)	Na_2CO_3	(g)	Parke process
	(viii)	nitric oxide	(h)	Claude process
	(ix)	silver	(i)	Ostwald process
	(x)	chlorine	(j)	Solvay process
C.	(xi)	phenol	(k)	coloured glass
	(xii)	$Na_2S_2O_3$	(l)	antichlor
	(xiii)	salicylic acid	(m)	refractory material
	(xiv)	quick lime	(n)	antiseptic
	(xv)	CuO	(o)	analgesic

D. (xvi) Aston (p) radium
(xvii) Priestley (q) radioactivity
(xviii) Ramsay (r) oxygen
(xix) Marie Curie (s) inert gas
(xx) Bacquerel (t) mass spectrum

57. Match the following, choosing one item from column **X** and one from column **Y**. **[1982 - 3 Marks]**

	X		Y
(i)	pyrolysis of alkanes	(a)	elimination reaction
(ii)	benzene+chloroethane (+anhydrous $AlCl_3$)	(b)	saponification
(iii)	$CH_3COOC_2H_5$ + NaOH	(c)	Wurtz reaction
(iv)	preparation of alkanes	(d)	Friedel-Crafts reaction
(v)	phenol + $CHCl_3$ (NaOH)	(e)	Reimer-Tiemann reaction
(vi)	C_2H_5Br + alc. KOH	(f)	cracking

8 Comprehension Based Questions

Passage -I

Riemer-Tiemann reaction introduces an aldehyde group, on to the aromatic ring of phenol, ortho to the hydroxyl group. This reaction involves electrophilic aromatic substitution. This is a general method for the synthesis of substituted salicylaldehyde as depicted below.

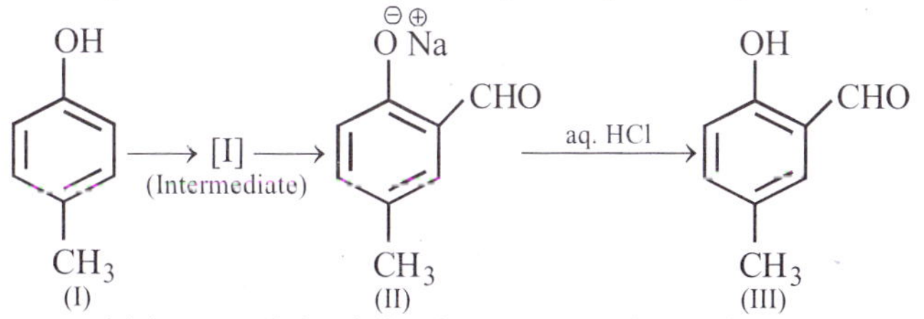

(I) (II) (III)

58. Which one of the following reagents is used in the above reaction ? **[2007]**
(a) aq.NaOH + CH_3Cl (b) aq.NaOH + CH_2Cl_2
(c) aq.NaOH + $CHCl_3$ (d) aq.NaOH + CCl_4

59. The electrophile in the reaction is **[2007]**
(a) :CHCl (b) $^+CHCl_2$ (c) :CCl_2 (d) CCl_3

60. The structure of the intermediate I is **[2007]**

(a) 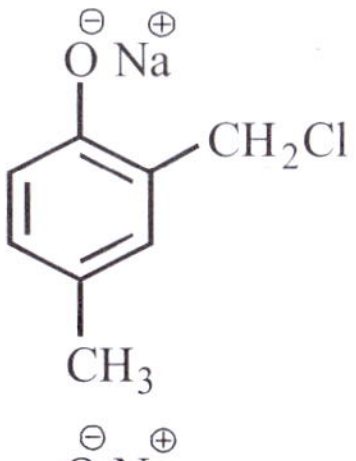

(b) 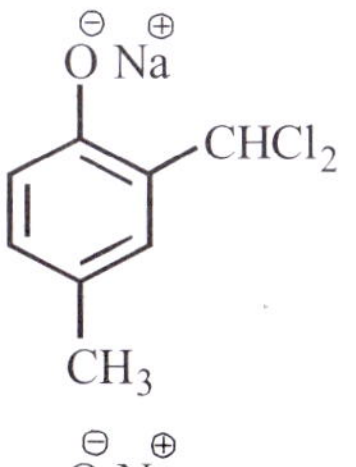

(c) 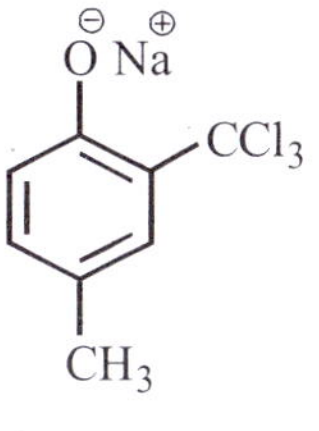

(d) 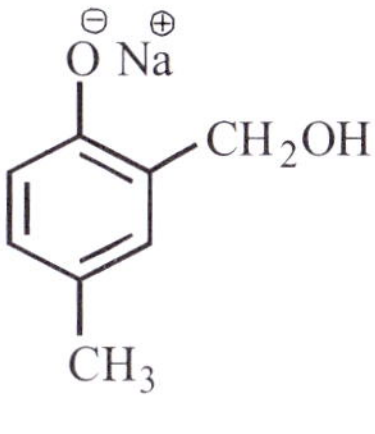

9 Assertion and Reason/Statement Type Questions

Each question contains **STATEMENT-1 (Assertion)** and **STATEMENT-2 (Reason)**. Each question has 4 choices (a), (b), (c) and (d) out of which ONLY ONE is correct. Mark your answer as

(a) If both Statement -1 and Statement -2 are correct, and Statement -2 is the correct explanation of the Statement -2.
(b) If both Statement -1 and Statement -2 are correct, but Statement -2 is not the correct explanation of the Statement -1.
(c) If Statement -1 is correct but Statement -2 is incorrect.
(d) If Statement -1 is incorrect but Statement -2 is correct.

61. **Statement - 1 :** Phenol is more reactive than benzene towards electrophilic substitution reactions.
Statement - 2 : In the case of phenol, the intermediate carbocation is more resonance stabilized. **[2000S]**

10 Subjective Problems

62. Outline the reaction sequence for the conversion of

(i) 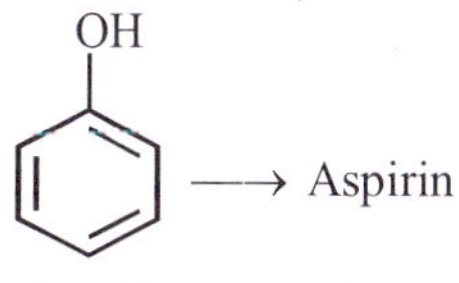

[2003 - 2 Marks]

(ii) phenol to acetophenone **[1989 - 1½ Marks]**

63. A compound D ($C_8H_{10}O$) upon treatment with alkaline solution of iodine gives a yellow precipitate. The filtrate on acidification gives a white solid E ($C_7H_6O_2$). Write the structures of D and E and explain the formation of E. **[1996 - 2 Marks]**

64. Complete the following with appropriate structures :

(i) 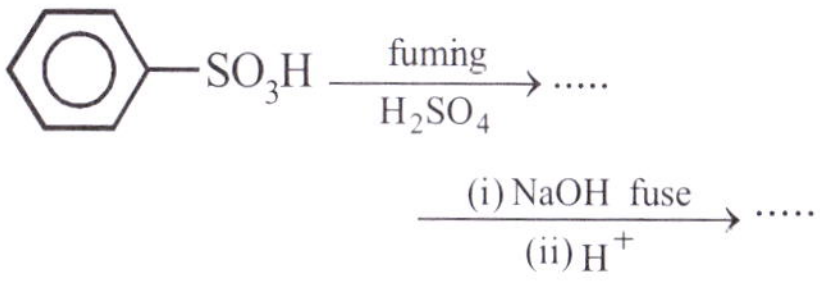

[1992 - 1 Mark]

(ii) 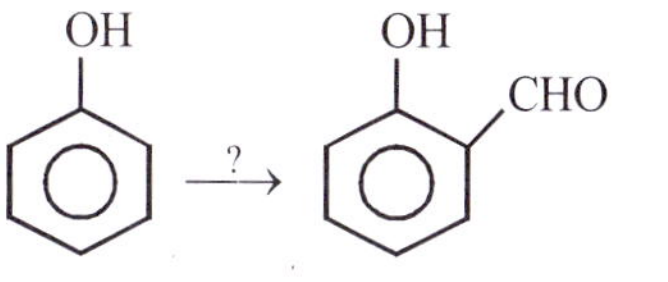

[1986 - 1 Mark]

65. Give reasons for the following :
Phenol is an acid but it does not react with sodium bicarbonate. **[1987 - 1 Mark]**

66. A compound of molecular formula C_7H_8O is insoluble in water and dilute sodium bicarbonate but dissolves in dilute aqueous sodium hydroxide. On treatment with bromine water, it readily gives a precipitate of $C_7H_5OBr_3$. Write down the structure of the compound. **[1985 - 2 Marks]**

67. What happens when p-xylene is reacted with concentrated sulphuric acid and the resultant product is fused with KOH. **[1984 - 2 Marks]**

68. State with balanced equations what happens when : acetic anhydride reacts with phenol in presence of a base. **[1982 - 1 Mark]**

Topic 3 - Preparation and Properties of Ethers

1 MCQs with One Correct Answer

69. The reaction products of $C_6H_5OCH_3 + HI \xrightarrow{\Delta}$ is : **[1995S]**
(a) $C_6H_5OH + CH_3I$ (b) $C_6H_5I + CH_3OH$
(c) $C_6H_5CH_3 + HOI$ (d) $C_6H_6 + CH_3OH$

2 Integer Value Answer

70. Total number of isomers, considering both structural and stereoisomers of cyclic ethers with the molecular formula C_4H_8O is_____ **[Adv. 2019]**

4 Fill in the Blanks

71. Aliphatic ethers are purified by shaking with a solution of ferrous salt to remove .... which are formed on prolonged standing in contact with air. **[1992 - 1 Mark]**

6 MCQs with One or More than One Correct Answer

72. The ether 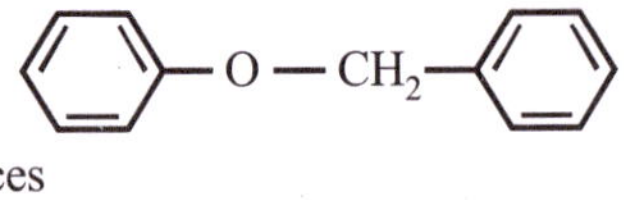 when treated with HI produces **[1999 - 3 Marks]**

(a) 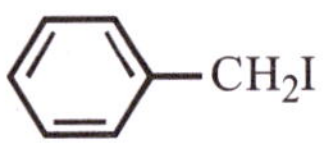 (b) 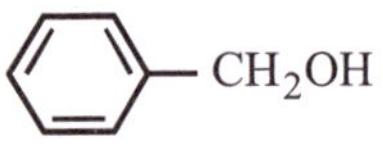

(c) 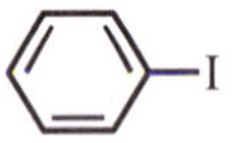 (d) 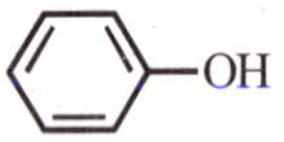

7 Match the Following

73. Match the chemical conversions in **List I** with the appropriate reagents in **List II** and select the correct answer using the code given below the lists : **[Adv. 2013]**

List I	List II
P 	1. (i) $Hg(OAc)_2$; (ii) $NaBH_4$
Q. 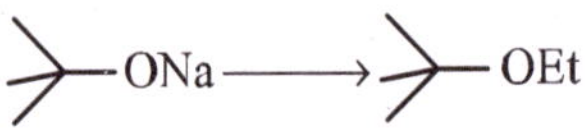	2. NaOEt
R. 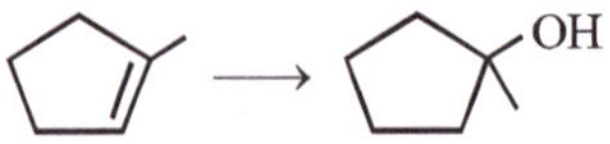	3. Et-Br
S. 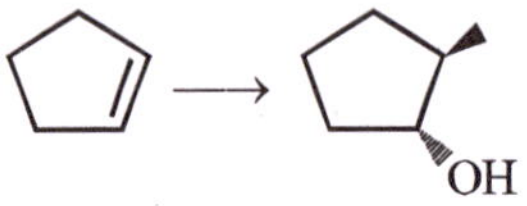	4. (i) BH_3; (ii) H_2O_2/NaOH

Codes :

	P	Q	R	S
(a)	2	3	1	4
(b)	3	2	1	4
(c)	2	3	4	1
(d)	3	2	4	1

74. Match the following, choosing one item from column **X** and the appropriate item from column **Y**. **[1983 - 2 Marks]**

X	Y
(i) Decarboxylation	(a) Addition reaction
(ii) Ozonolysis	(b) Soda lime
(iii) Williamson's synthesis	(c) Structure of alkene
(iv) Dichloroethylene	(d) Ether

10 Subjective Problems

75. An organic compound (*P*) of molecular formula $C_5H_{10}O$ is treated with dil. H_2SO_4 to give two compounds (*Q*) and (*R*) both of which respond iodoform test. The rate of reaction of (*P*) with dil. H_2SO_4 is 10^{10} faster than the reaction of ethylene with dil. H_2SO_4. Identify the organic compounds, (*P*), (*Q*) and (*R*) and explain the extra reactivity of (*P*). **[2004 - 4 Marks]**

76. Write the structural formula of the main organic product formed when :

$$(CH_3)_2CHOCH_3 \xrightarrow{\text{HI (excess), heat}} \text{2 Products.}$$

[1998 - 2 Marks]

77. Write the intermediate steps for the following reaction.

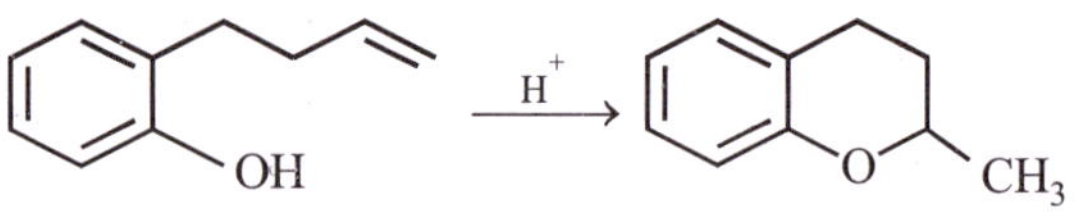

[1998 - 1 Mark]

78. 2, 2-Dimethyloxirane can be cleaved by acid (H^+). Write mechanism. **[1997 - 2 Marks]**

79. Which of the following is the correct method for synthesising methyl-t-butyl ether and why?

(i) $(CH_3)_3CBr + NaOMe \rightarrow$

(ii) $CH_3Br + NaO\text{-}t\text{-}Bu \rightarrow$ **[1997 - 2 Marks]**

80. An organic compound containing C, H and O exists in two isomeric forms A and B. An amount of 0.108 g of one of the isomers gives on combustion 0.308 g of CO_2 and 0.072 g of H_2O. A is insoluble in NaOH and $NaHCO_3$ while B is soluble in NaOH. A reacts with conc. HI to give compounds C and D. C can be separated from D by ethanolic $AgNO_3$ solution and D is soluble in NaOH. B reacts readily with bromine water to give compound E of molecular formula, $C_7H_5OBr_3$. Identify, A, B, C, D and E with justification and give their structures. **[1991 - 6 Marks]**

81. Give reasons for the following :

Sodium metal can be used for drying diethyl ether but not ethanol. **[1982 - 1 Mark]**

82. A compound (X) containing C, H and O is unreactive towards sodium. It does not add bromine. It also does not react with Schiff's reagent. On refluxing with an excess of hydriodic acid, (X) yields only one organic product (Y). On hydrolysis, (Y) yields a new compound (Z) which can be converted into (Y) by reaction with red phosphorus and iodine. The compound (Z) on oxidation with potassium permanganate gives a carboxylic acid. The equivalent weight of this acid is 60. What are the compounds (X), (Y) and (Z)? Write chemical equations leading to the conversion of (X) to (Y). **[1981 - 3 Marks]**

Answer Key

Part-A (JEE Main)																	
1	(4)	**19**	(1)	**37**	(2)	**55**	(1)	**73**	(4)	**91**	(2)	**109**	(4)	**127**	(2)	**145**	(69)	**163**	(3)
2	(2)	**20**	(3)	**38**	(4)	**56**	(4)	**74**	(3)	**92**	(2)	**110**	(2)	**128**	(1)	**146**	(4)	**164**	(1)
3	(2)	**21**	(3)	**39**	(1)	**57**	(175)	**75**	(4)	**93**	(4)	**111**	(1)	**129**	(3)	**147**	(1)	**165**	(1)
4	(3)	**22**	(1)	**40**	(4)	**58**	(11)	**76**	(3)	**94**	(2)	**112**	(3)	**130**	(1)	**148**	(4)	**166**	(2)
5	(4)	**23**	(2)	**41**	(2)	**59**	(3)	**77**	(2)	**95**	(2)	**113**	(3)	**131**	(4)	**149**	(4)	**167**	(3)
6	(3)	**24**	(3)	**42**	(1)	**60**	(6)	**78**	(1)	**96**	(2)	**114**	(1)	**132**	(2)	**150**	(2)	**168**	(3)
7	(4)	**25**	(3)	**43**	(3)	**61**	(3)	**79**	(4)	**97**	(1)	**115**	(4)	**133**	(1)	**151**	(2)	**169**	(2)
8	(1)	**26**	(1)	**44**	(3)	**62**	(33)	**80**	(3)	**98**	(2)	**116**	(1)	**134**	(2)	**152**	(2)	**170**	(2)
9	(1)	**27**	(2)	**45**	(2)	**63**	(1)	**81**	(2)	**99**	(3)	**117**	(2)	**135**	(2)	**153**	(1)	**171**	(4)
10	(4)	**28**	(4)	**46**	(4)	**64**	(2)	**82**	(3)	**100**	(1)	**118**	(2)	**136**	(N)	**154**	(4)	**172**	(4)
11	(3)	**29**	(2)	**47**	(2)	**65**	(4)	**83**	(4)	**101**	(3)	**119**	(4)	**137**	(4)	**155**	(4)	**173**	(4)
12	(1)	**30**	(1)	**48**	(1)	**66**	(3)	**84**	(1)	**102**	(3)	**120**	(3)	**138**	(3)	**156**	(3)	**174**	(4)
13	(1)	**31**	(1)	**49**	(1)	**67**	(4)	**85**	(4)	**103**	(3)	**121**	(1)	**139**	(3)	**157**	(4)	**175**	(4)
14	(1)	**32**	(2)	**50**	(1)	**68**	(4)	**86**	(2)	**104**	(3)	**122**	(3)	**140**	(3)	**158**	(4)	**176**	(2)
15	(2)	**33**	(2)	**51**	(2)	**69**	(3)	**87**	(1)	**105**	(4)	**123**	(1)	**141**	(14)	**159**	(1)	**177**	(2)
16	(1)	**34**	(3)	**52**	(4)	**70**	(N)	**88**	(1)	**106**	(2)	**124**	(1)	**142**	(4)	**160**	(4)	**178**	(1)
17	(2)	**35**	(3)	**53**	(2)	**71**	(4)	**89**	(4)	**107**	(3)	**125**	(3)	**143**	(30)	**161**	(2)	**179**	(2)
18	(1)	**36**	(4)	**54**	(1)	**72**	(1)	**90**	(1)	**108**	(1)	**126**	(3)	**144**	(7)	**162**	(2)	**180**	(15)

Part-B (JEE Advanced)																			
1	(b)	**6**	(a)	**11**	(6)	**18**	(b)	**39**	(c)	**44**	(d)	**52**	(a, c)	**58**	(c)	**70**	(10)		
2	(c)	**7**	(a)	**12**	(4)	**19**	(c)	**40**	(d)	**45**	(6)	**53**	(b)	**59**	(c)	**72**	(a, d)		
3	(c)	**8**	(d)	**13**	(1.50)	**36**	(b)	**41**	(d)	**46**	(9)	**54**	(a, d)	**60**	(b)	**73**	(a)		
4	(a)	**9**	(a)	**16**	False	**37**	(d)	**42**	(d)	**50**	(a,b,c)	**55**	(a, c)	**61**	(a)	**74**	(i)-(b); (ii)-(c); (iii)-(d); (iv)-(a)		
5	(c)	**10**	(c)	**17**	(a, c, d)	**38**	(d)	**43**	(b)	**51**	(b)	**57**	(i) (f) (ii) (d) (iii) (b) (iv) (c) (v) (e) (vi) (a)	**69**	(a)				

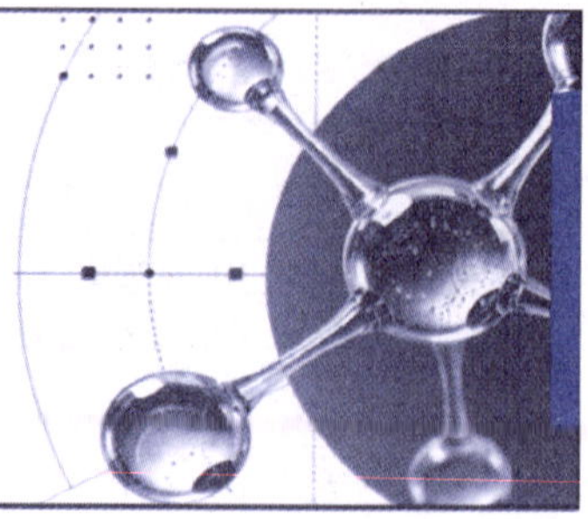

Chapter-12

Aldehydes, Ketones and Carboxylic Acids

Part-A (JEE Main)

Topic 1 – Methods of Preparation of Carbonyl Compounds

Multiple Choice Questions (MCQs)

1. Match List-I with List-II **[Jan. 24, 2025 (II)]**

List-I

(A) $RCN \xrightarrow[(ii)H_3O^+]{(i)SnCl_2,HCl} RCHO$

(B) 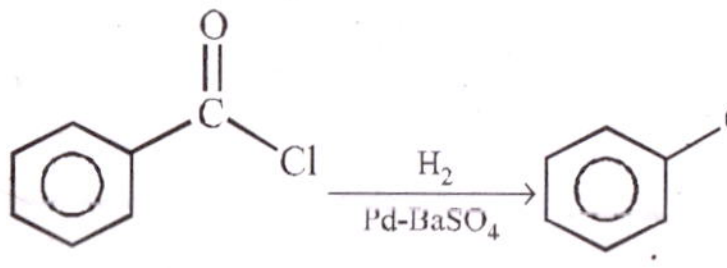

(C) 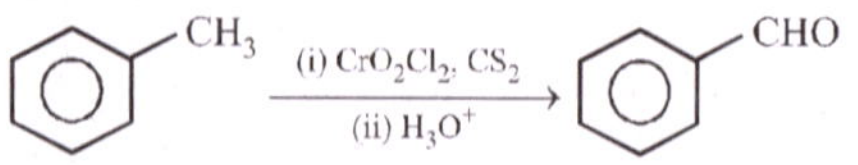

(D) 

List-II

(I) Etard reaction

(II) Gatterman – Koch reaction

(III) Rosenmund reduction

(IV) Stephen reaction

Choose the correct answer from the options given below :

(1) (A)-(IV), (B)-(III), (C)-(I), (D)-(II)

(2) (A)-(III), (B)-(IV), (C)-(II), (D)-(I)

(3) (A)-(I), (B)-(III), (C)-(II), (D)-(IV)

(4) (A)-(III), (B)-(IV), (C)-(I), (D)-(II)

2. 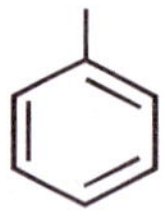 $\xrightarrow[\substack{(ii)\ H_3O^+ \\ (iii)\ NaHSO_3}]{(i)\ CrO_2Cl_2,\ CS_2}$ Filter ⟶ Residue (A)

Toluene (excess)

Residue (A) + HCl (dil.) → Compound (B) **[Jan. 22, 2025 (II)]**

Structure of residue (A) and compound (B)

Formed respectively is:

[A] **[B]**

(1) 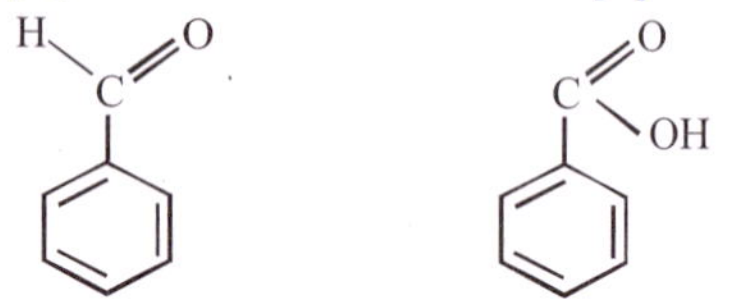

(2) 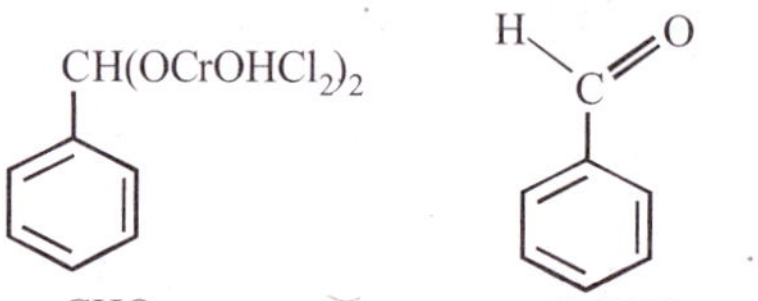

(3) 

(4) 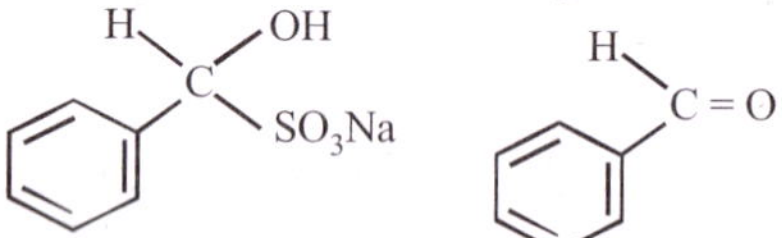

3. Major product of the following reaction is **[April 9, 2024 (II)]**

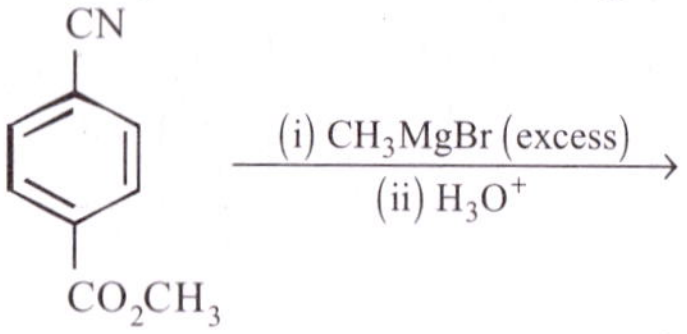

(1) 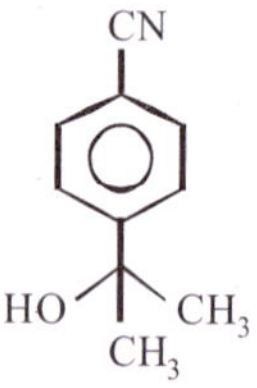

(2) 

(3) 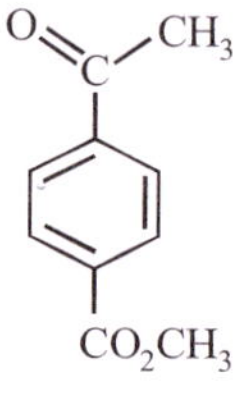

(4) 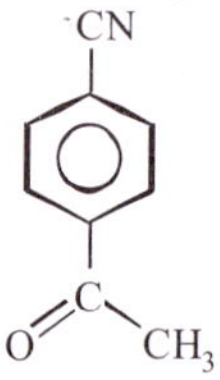

4. Identify major product "X" formed in the following reaction :

 $\xrightarrow[\text{Anhydrous } AlCl_3/CuCl]{CO,HCl}$ X (Major product)) **[April 9, 2024 (I)]**

(1) 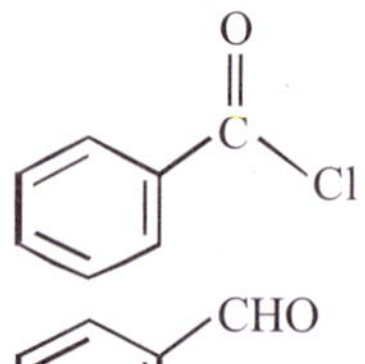

(2) 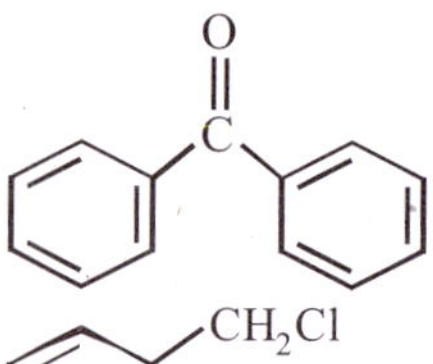

(3) 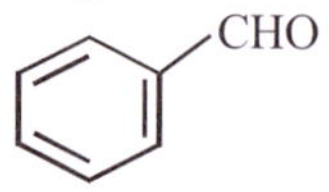

(4) 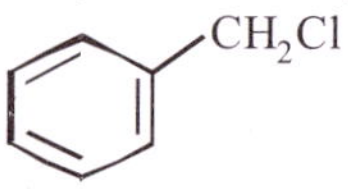

5. Identify A and B in the following sequence of reaction **[Feb. 1, 2024 (I)]**

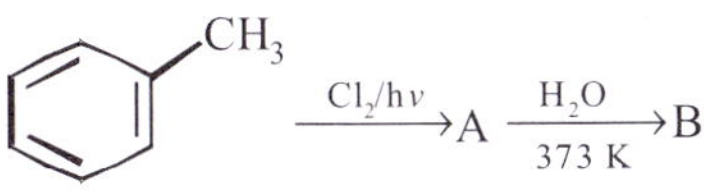

(1) A = 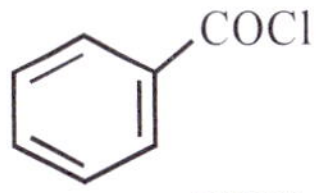 B = 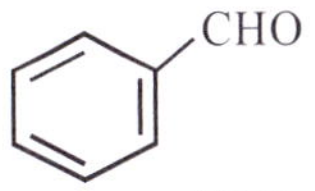

(2) A = 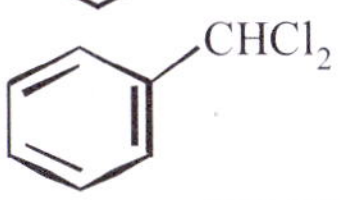 B = 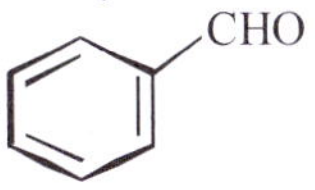

(3) A = 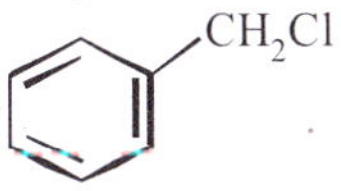 B = 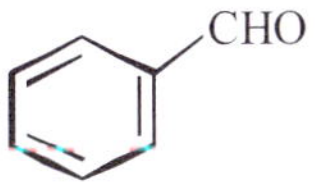

(4) A = 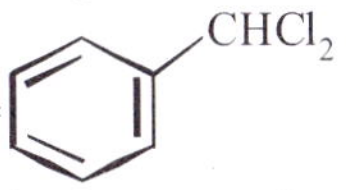 B = 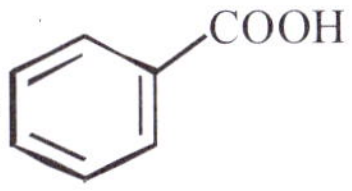

6. Identify the name reaction. **[Jan. 31, 2024 (II)]**

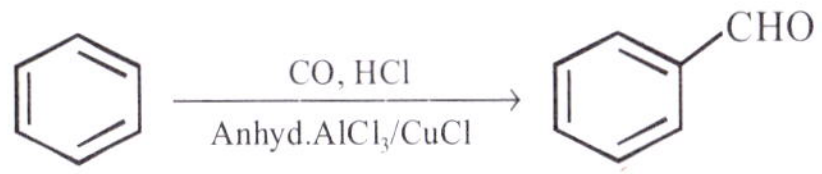

(1) Stephen reaction
(2) Etard reaction
(3) Gattermann-Koch reaction
(4) Rosenmund reduction

7. 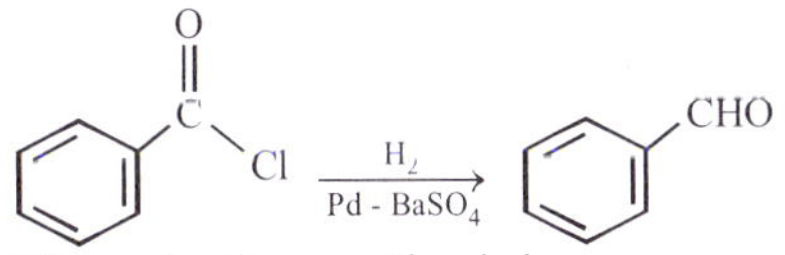

This reduction reaction is known as : **[Jan. 30, 2024 (I)]**
(1) Rosenmund reduction (2) Wolff - Kishner reduction
(3) Stephen reduction (4) Etard reduction

8. In the given reactions, identify the reagent A and reagent B. **[Jan. 30, 2024 (I)]**

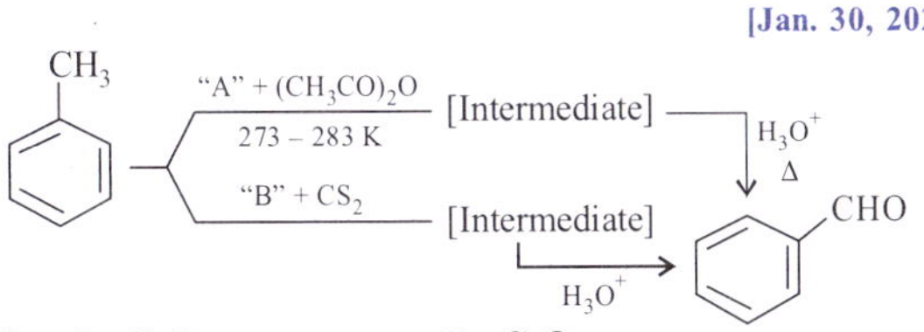

(1) A - CrO_3 B - CrO_3
(2) A - CrO_3 B - CrO_2Cl_2
(3) A - CrO_2Cl_2 B - CrO_2Cl_2
(4) A - CrO_2Cl_2 B - CrO_3

9. Correct structure of γ-methylcyclohexane carbaldehyde is : **[July 29, 2022 (II)]**

(1) 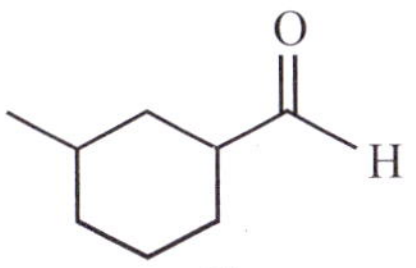 (2) 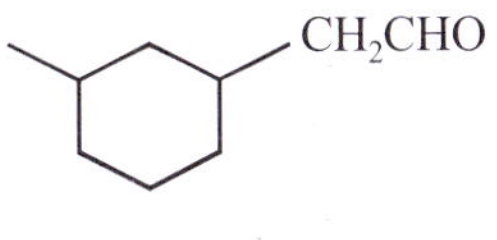

(3) 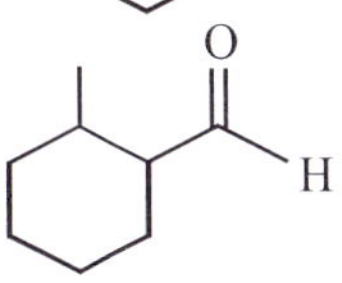 (4) 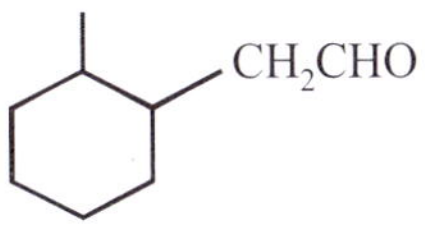

10. 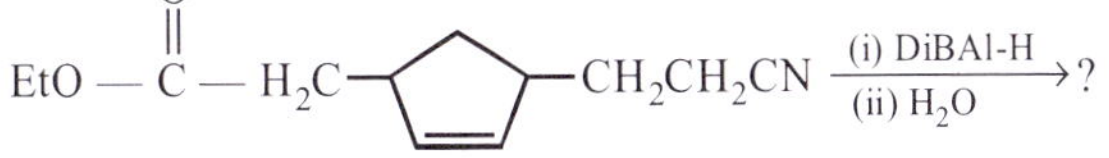

Consider the above reaction and predict the major product. **[July 26, 2022 (II)]**

(1) 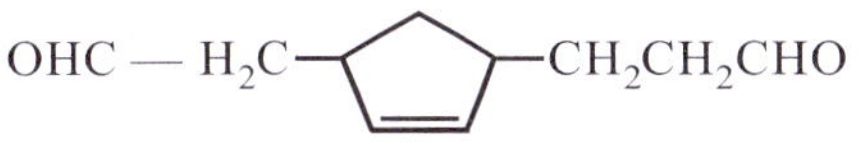

(2) 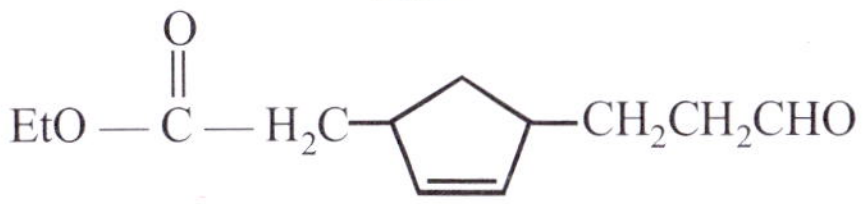

(3) 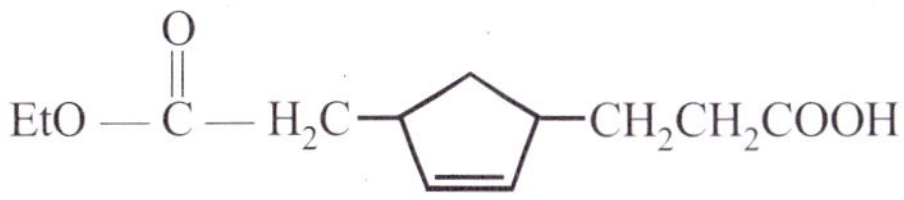

(4) 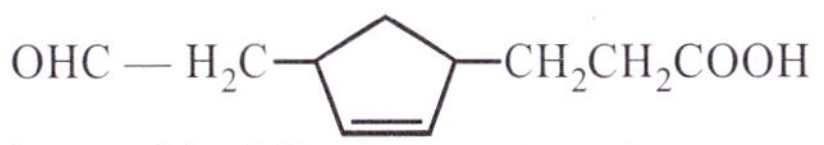

11. Which one of the following reactions does **not** represent correct combination of substrate and product under the given conditions? **[July 25, 2022 (I)]**

(1) 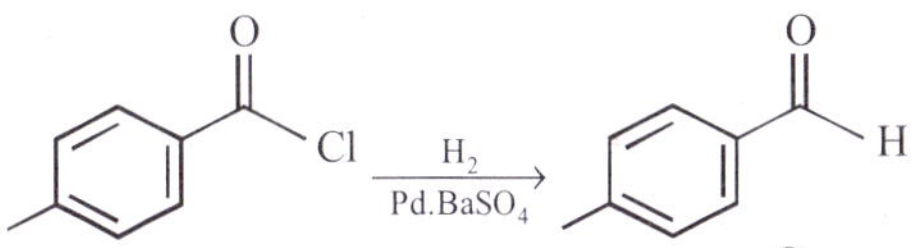

(2) 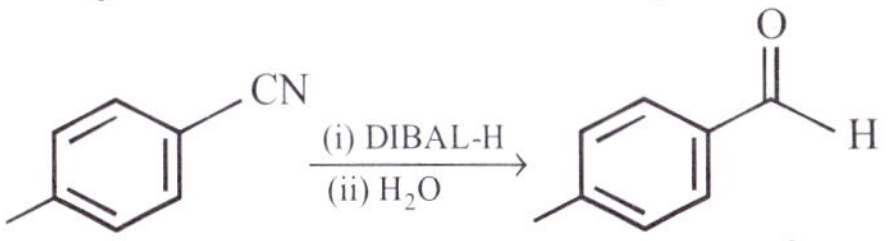

(3) 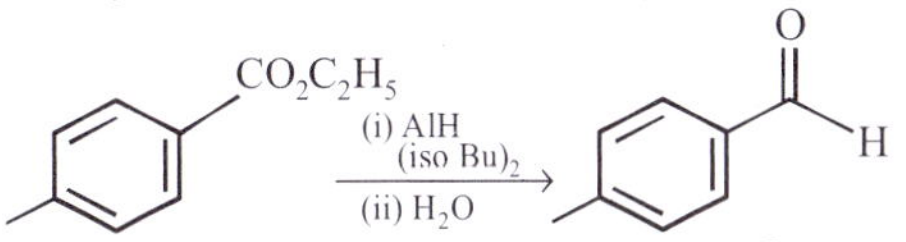

(4) 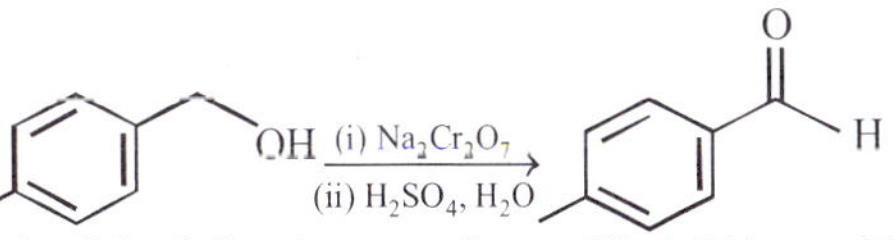

12. Which of the following reactions will yield benzaldehyde as a product? **[June 27, 2022 (I)]**

(A) 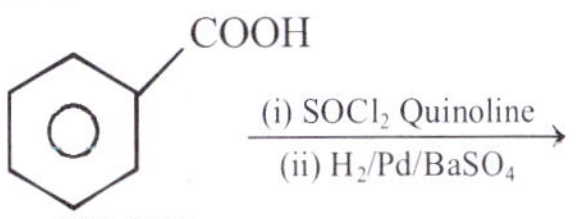

(B) 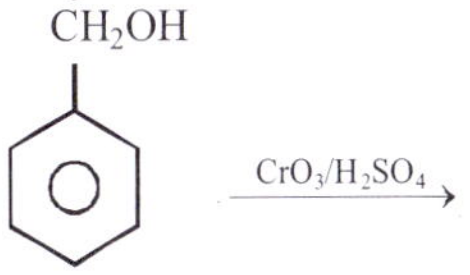

(C) 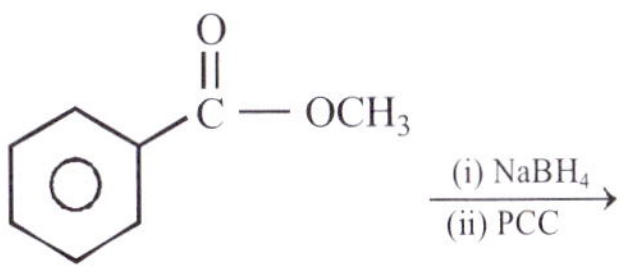

(D) 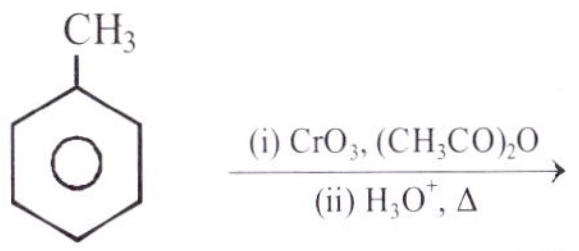

(1) (B) and (C) (2) (C) and (D)
(3) (A) and (D) (4) (A) and (C)

13. Which of the following reagents/ reactions will convert 'A' to 'B'? **[June 24, 2022 (II)]**

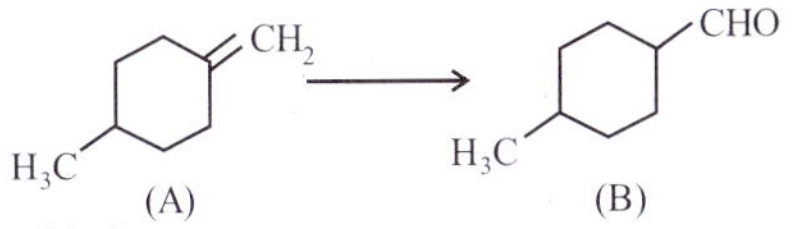

(1) PCC oxidation
(2) Ozonolysis
(3) BH_3, H_2O_2 / ^-OH followed by PCC oxidation
(4) HBr, hydrolysis followed by oxidation by $K_2Cr_2O_7$.

14. The major product of the following reaction is:

$$CH_3\underset{\displaystyle |}{\overset{CH_3}{\overset{|}{CH}}}CH_2CH_2-\overset{O}{\overset{\|}{C}}-Cl \xrightarrow[\text{(iii) NaNO}_2\text{,HCl} \quad \text{(vi) H}_2\text{O}]{\text{(i) alcoholic NH}_3 \quad \text{(ii) NaOH, Br}_2} \text{Major product}$$

[Aug. 27, 2021 (I)]

(1) $CH_3-CH(CH_3)-CH(Br)-CH_2OH$ (Br on the third carbon, CH_3 on the second carbon)

(2) $CH_3-CH(CH_3)-CH_2-CH_2-CH_2OH$

(3) $CH_3-CH(CH_3)-CH_2-CH_2OH$

(4) $CH_3-CH(CH_3)-CH_2-CH_2-Cl$

15. Which one of the following reactions will not form acetaldehyde? [Feb. 25, 2021 (I)]

(1) $CH_3CH_2OH \xrightarrow{CrO_3-H_2SO_4}$

(2) $CH_2=CH_2+O_2 \xrightarrow[H_2O]{Pd(II)/Cu(II)}$

(3) $CH_3CH_2OH \xrightarrow[573K]{Cu}$

(4) $CH_3CN \xrightarrow[(ii)H_2O]{(i)DIBAL-H}$

16. The correct match between Item - I (starting material) and Item - II (reagent) for the preparation of benzaldehyde is : [Sep. 06, 2020 (II)]

Item - I	Item - II
(I) Benzene	(P) HCl and $SnCl_2$, H_3O^+
(II) Benzonitrile	(Q) H_2, Pd-$BaSO_4$, S and quinoline
(III) Benzoyl chloride	(R) CO, HCl and $AlCl_3$

(1) (I) - (Q), (II) - (R) and (III) - (P)
(2) (I) - (P), (II) - (Q) and (III) - (R)
(3) (I) - (R), (II) - (P) and (III) - (Q)
(4) (I) - (R), (II) - (Q) and (III) - (P)

17. The major products of the following reaction are: [Sep. 06, 2020 (I)]

$$CH_3-\overset{CH_3}{\overset{|}{CH}}-\underset{OSO_2CH_3}{\underset{|}{CH}}-CH_3 \xrightarrow[(ii)O_3\,/\,H_2O_2]{(i)KO^tBu/\Delta}$$

(1) 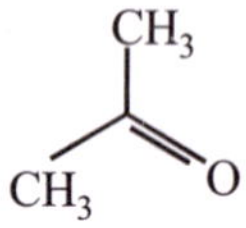 $+ CH_3CHO$

(2) 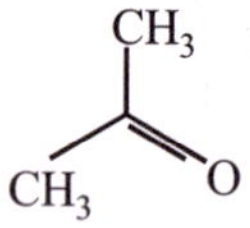 $+ CH_3COOH$

(3) 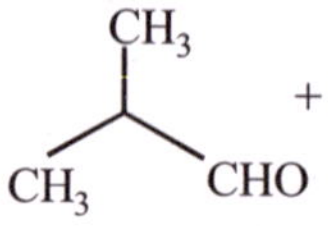 $+ HCHO$

(4) 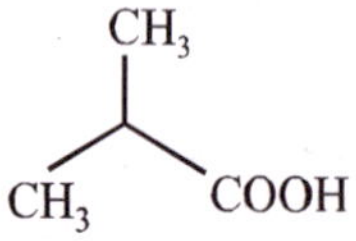 $+ HCOOH$

18. In the following reaction A is: [Jan. 09, 2020 (II)]

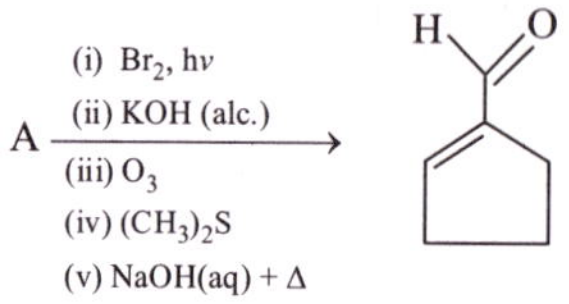

(1) 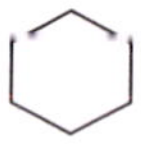 (2)  (3) 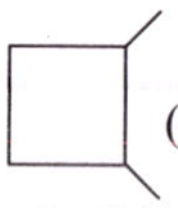 (4) 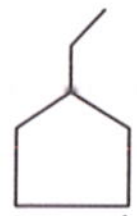

19. The major product(s) obtained in the following reaction is/are : [April 12, 2019 (I)]

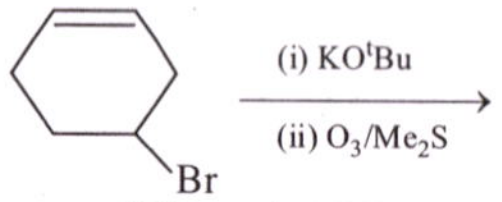

(1) 

(2) 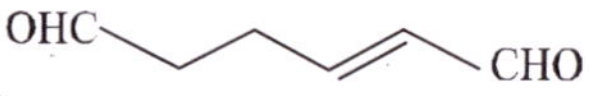

(3) 

(4) 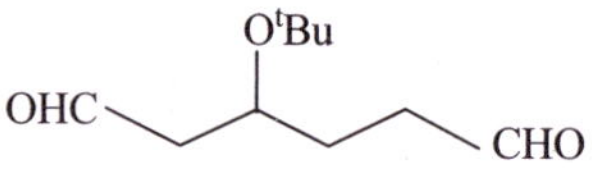

20. The major product of the following reaction is: [April 12, 2019 (I)]

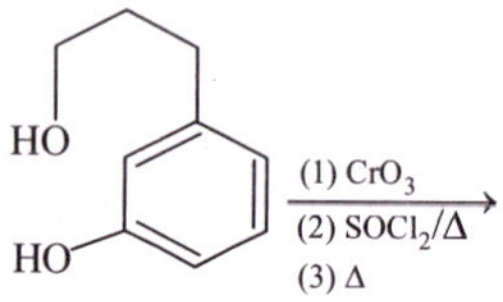

(1) 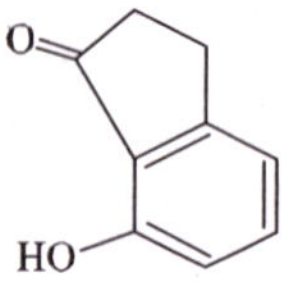

(2) 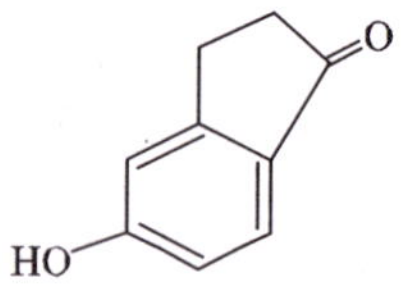

(3) 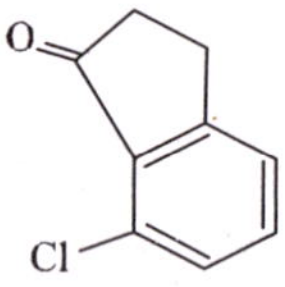

(4) 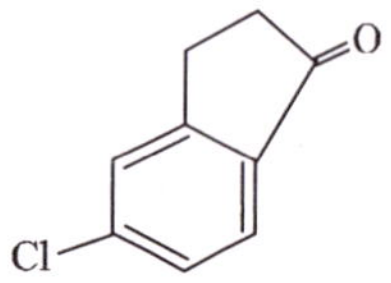

21. The major product of the following reaction is: [April 9, 2019 (I)]

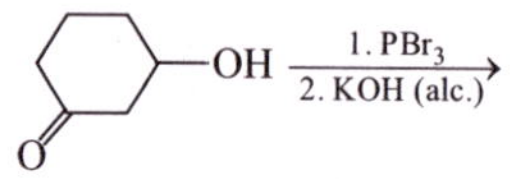

(1) 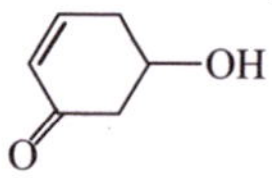

(2) 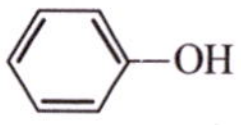

(3) 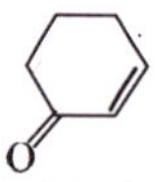

(4) 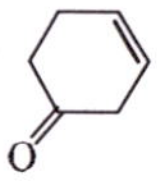

22. In the following reaction [April 9, 2019 (II)]

$$\text{Carbonyl compound} + \text{MeOH} \xrightleftharpoons{HCl} \text{Acetal}$$

Rate of the reaction is the highest for:

(1) Acetone as substrate and methanol in excess.
(2) Propanal as substrate and methanol in stoichiometric amount.
(3) Propanal as substrate and methanol in excess.
(4) Acetone as substrate and methanol in stoichiometric amount.

23. In the following reaction

$\text{Aldehyde} + \text{Alcohol} \xrightarrow{HCl} \text{Acetal}$

Aldehyde	Alcohol
HCHO	^tBuOH
CH_3CHO	MeOH

The best combination is: **[Jan. 12, 2019 (I)]**

(1) CH_3CHO and ^tBuOH (2) HCHO and MeOH
(3) CH_3CHO and MeOH (4) HCHO and ^tBuOH

24. The major product of the following reaction is:

$R-C\equiv N \xrightarrow[\text{(ii) } H_2O]{\text{(i) AlH (i-Bu)}_2}$ **[Jan. 9, 2019 (I)]**

(1) RCOOH (2) $RCONH_2$
(3) RCHO (4) RCH_2NH_2

25. The major product of the following reaction is: **[Jan. 9, 2019 (I)]**

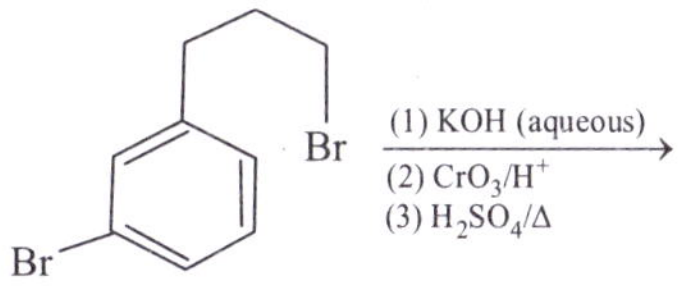

(1) 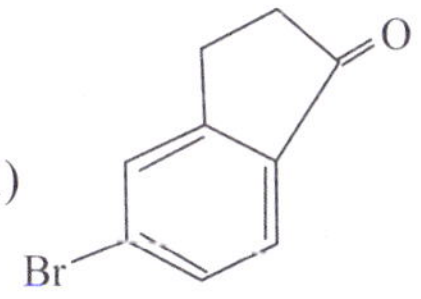 (2) 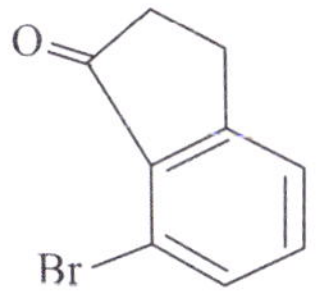

(3) 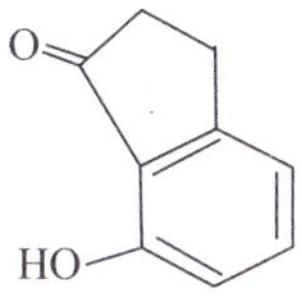 (4) 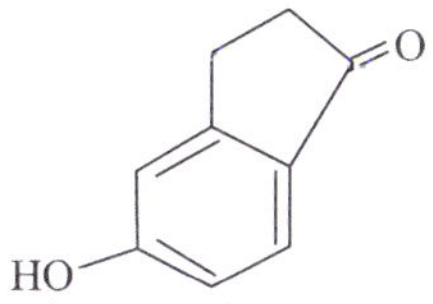

26. The major product of the following reaction is:

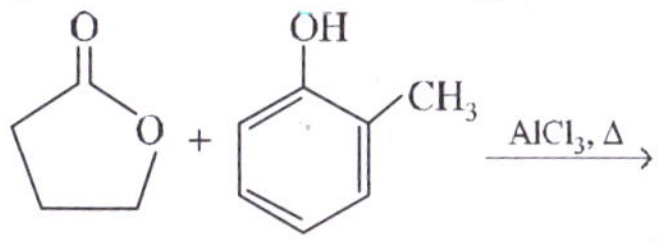

[Jan. 9, 2019 (II)]

(1) 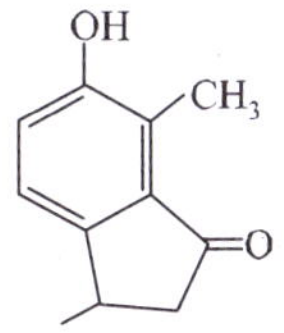 (2) 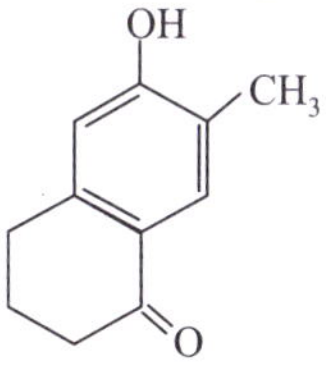

(3) 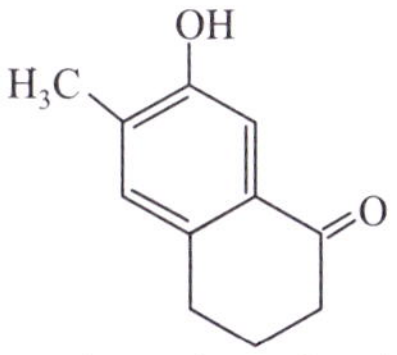 (4) 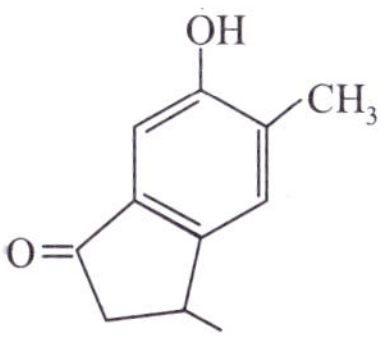

27. The total number of optically active compounds formed in the following reaction is: **[Online April 15, 2018 (II)]**

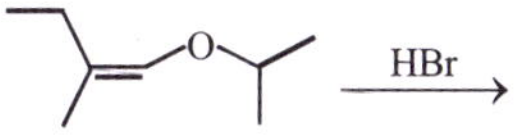

(1) Zero (2) Six (3) Four (4) Two

28. Which compound would give 5 - keto - 2 - methylhexanal upon ozonolysis ? **[2015]**

(1) 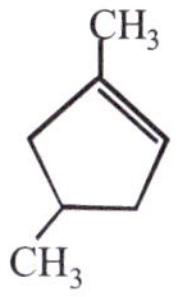 (2) 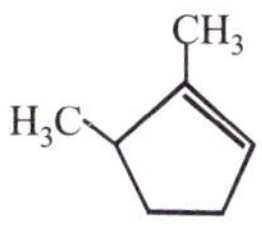

(3) 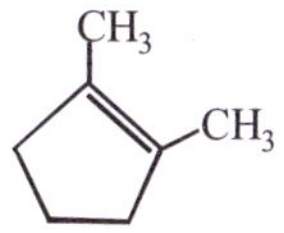 (4) 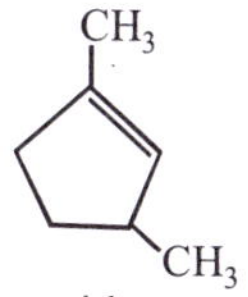

29. Which of the following on heating with aqueous KOH, produces acetaldehyde? **[2009]**

(1) CH_3CH_2Cl (2) CH_2ClCH_2Cl
(3) CH_3CHCl_2 (4) CH_3COCl

NUMERICAL VALUE QUESTIONS (NVQs)

30. Two moles of benzaldehyde and one mole of acetone under alkaline conditions using aqueous NaOH after heating gives X as the major product. The number of π bonds in the product X is **[April 8, 2024 (II)]**

Topic 2 - Properties of Carbonyl Compounds

MULTIPLE CHOICE QUESTIONS (MCQs)

31. Consider the following reaction

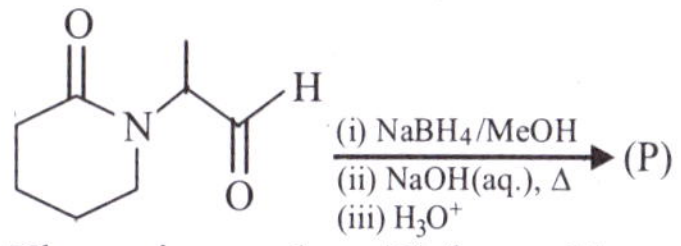

The major product (P) formed is : **[April 08, 2026 (II)]**

(1) 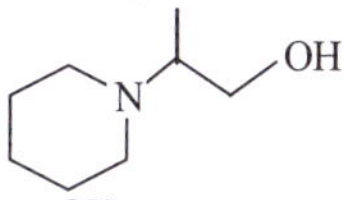

(2) 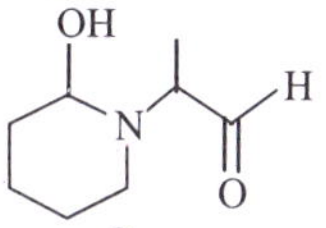

(3) 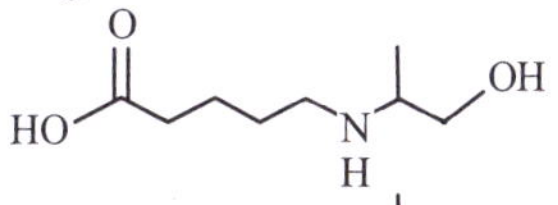

(4) 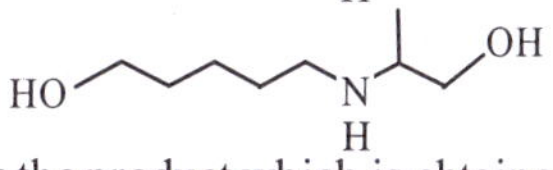

32. 'x' is the product which is obtained by the hydrolysis of prop-1-yne in the presence of mercuric sulphate under dilute acidic medium at 333 K. 'y' is the product which is obtained by the reaction of ethane nitrile with methyl magnesium bromide in dry ether followed by hydrolysis. IUPAC name of product obtained from 'x' and 'y' in the presence of barium hydroxide followed by heating is: **[April 06, 2026 (II)]**

(1) 2-Methylpent-4-en-3-one
(2) 4-Methylpent-3-en-2-one
(3) 4-Methylpent-1-ene
(4) 2-Methylpent-3-one

33. Shown below is the structure of methyl acetate with three different α, β and γ carbon-oxygen bonds.

$$H_3C-\overset{\overset{\displaystyle O}{\|\alpha}}{C}\underset{\beta}{-}O\underset{\gamma}{-}CH_3$$

The correct order of bond length of these bonds is : **[April 06, 2026 (II)]**

(1) $\alpha > \beta > \gamma$ (2) $\alpha < \beta < \gamma$
(3) $\alpha = \beta = \gamma$ (4) $\alpha < \beta = \gamma$

34. Given below are two statements : **[April 05, 2026 (II)]**

Statement-I: The condensation reaction between $CH_3-CH=O$ and $H_2N-\underset{H}{\underset{|}{N}}-\underset{O}{\underset{||}{C}}-NH_2$ under optimum pH will produce $CH_3-CH=N-\underset{O}{\underset{||}{C}}-\overset{H}{N}\;NH_2$

Statement-II: The molecule, $Ph-CH\begin{matrix} / O-H \\ \backslash O-CH_3 \end{matrix}$ will generate Ph–CH=O in the presence of dilute acid.

In the light of the above statements, choose the **correct answer** from the options given below.

(1) Both Statement I and Statement II are true
(2) Both Statement I and Statement II are false
(3) Statement I is false but Statement II is true
(4) Statement I is true but Statement II is false

35. Consider the following molecules/species:

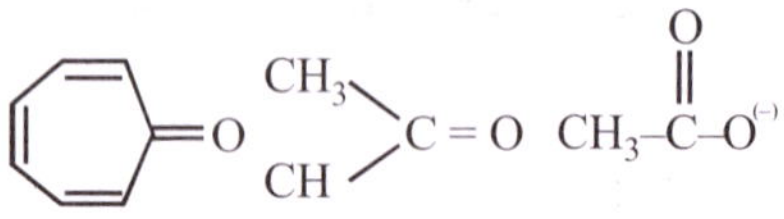

(x) (y) (z)

The correct order of carbon – oxygen double bond length is: **[April 04, 2026 (II)]**

(1) $x>y>z$ (2) $y>z>x$
(3) $z>x>y$ (4) $x>z>y$

36. A molecule (X) with following structure under mild acidic condition is hydrolysed to produce (Y) and (Z). Identify the correct statements about (Y) and (Z).

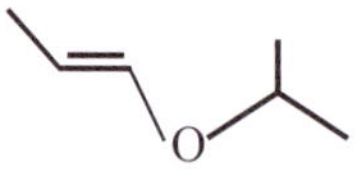

(X)

(A) Both (Y) and (Z) have same molar mass
(B) (Y) and (Z) can be distinguished from each other by $NaHCO_3$
(C) (Y) and (Z) react with HCN with same rates
(D) (Y) and (Z) undergo addition reaction with 2, 4-DNP

Choose the **correct** answer from the options given below: **[April 02, 2026 (II)]**

(1) A, B and C Only (2) B and C Only
(3) C and D Only (4) A and D Only

37. Given below are the four isomeric compounds (P, Q, R, S)

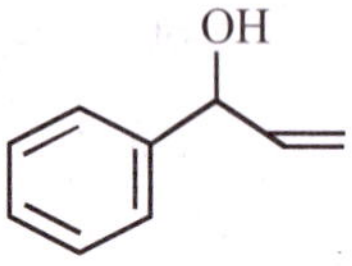

(P)

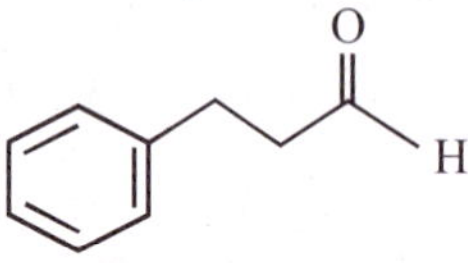

(Q)

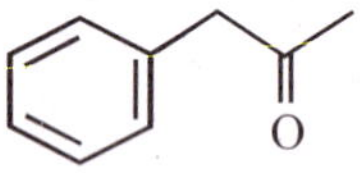

(R)

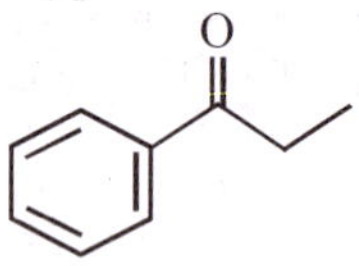

(S)

Identify **correct** statements from below. **[Jan. 28, 2026 (I)]**

A. Q, R and S will give precipitate with 2, 4 - DNP.
B. P and Q will give positive Baeyer's test.
C. Q and R will give sooty flame.
D. R and S will give yellow precipitate with I_2/NaOH.
E. Q alone will deposit silver with Tollen's reagent

Choose the correct option.

(1) A, C and E only (2) A and E only
(3) C and E only (4) A, B, D and E only

38. Given below are two statements: **[Jan. 24, 2026 (II)]**

Statement I: Cross aldol condensation between two different aldehydes will always produce four different products.

Statement II: When semicarbazide reacts with a mixture of benzaldehyde and acetophenone under optimum pH, it forms a condensation product with acetophenone only.

In the light of the above statements, choose the correct answer from the options given below:

(1) Both Statement I and Statement II are false.
(2) Statement I is false but Statement II is true.
(3) Both Statement I and Statement II are true.
(4) Statement I is true but Statement II is false.

39. The unsaturated ether on acidic hydrolysis produces carbonyl compounds as shown below:-

$$CH_3-CH=CH-O-CH=CH_2 \xrightarrow{H_3O^{\oplus}} CH_3-CH_2-CH=O \text{ and } O=CH-CH_3$$

Based on this, predict the solution/reagent that will to distinguish "P" and "Q" obtained in the following reaction. **[Jan. 24, 2026 (II)]**

$$CH_3-CH=CH-O-\overset{CH_3}{\overset{|}{C}}=CH_2 \xrightarrow[\Delta]{H_3O^{(+)}} P+Q$$

(1) Lucas reagent
(2) 2, 4-DNP reagent
(3) Saturated $NaHSO_3$ solution
(4) Fehling solution

40. A student is given one compound among the following compounds that gives positive test with Tollen's reagent.

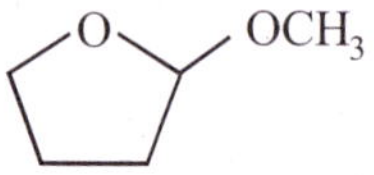

A

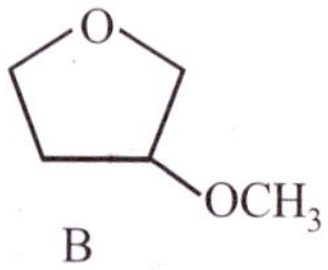

B

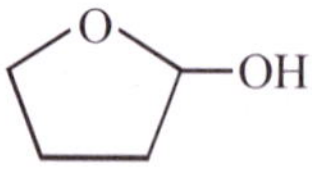

C

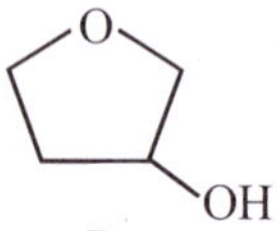

D

The compound is: **[Jan. 24, 2026 (I)]**

(1) D (2) A
(3) B (4) C

41. Which of the following statements are TRUE about Haloform reaction? **[Jan. 23, 2026 (II)]**

A. Sodium hypochlorite reacts with KI to give KOI.
B. KOI is a reducing agent.
C. α, β-unsaturated methyl ketone ($CH_3-CH=CH-\overset{O}{\overset{||}{C}}-CH_3$) will give iodoform reaction.
D. Isopropyl alcohol will not give iodoform test.
E. Methanoic acid will give positive iodoform test.

Choose the **correct** answer from the options given below:

(1) A, C & E only (2) A, B & C only
(3) A & C only (4) B, D & E only

42. Iodoform test can differentiate between **[Jan. 23, 2026 (II)]**

A. Methanol and Ethanol
B. CH_3COOH and CH_3CH_2COOH
C. Cyclohexene and cyclohexanone
D. Diethyl ether and Pentan-3-one
E. Anisole and acetone

Choose the **correct** answer from the options given below:

(1) A & E only (2) A & D only
(3) A, B & E only (4) B, C & E only

43. 'X' is the product which is obtained from propanenitrile and stannous chloride in the presence of hydrochloric acid followed by hydrolysis. 'y' is the product which is obtained from the but–2–ene by the ozonolysis followed by hydrolysis. From the following, which product is not obtained when one mole of 'x' and one mole of 'y' react with each other in the presence of alkali followed by heating? **[Jan. 23, 2026 (I)]**

(1) 2-Methylbut-2-enal (2) Pent-2-enal
(3) 2-Methylpent-2-enal (4) 3-Methylbut-2-enal

44. The dibromo compound [P] (molecular formula : $C_9H_{10}Br_2$) when heated with excess sodamide followed by treatment with dilute HCl gives [Q].
On warming [Q] with mercuric sulphate and dilute sulphuric acid yield [R] which gives positive. Iodoform test but negative Tollen's test. The compound [P] is : **[Jan. 22, 2026 (II)]**

(1) 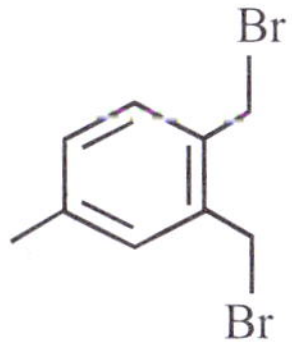 (2) 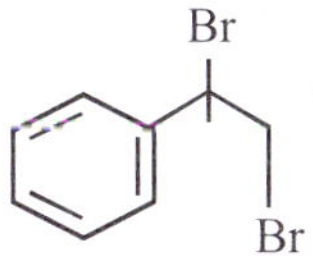

(3) 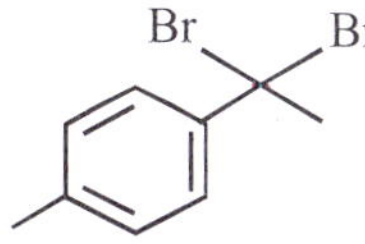 (4) 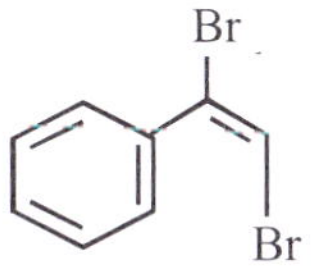

45. The compound A, $C_8H_8O_2$ reacts with acetophenone to form a single product via cross. Aldol condensation. The compound A on reaction with conc. NaOH forms a substituted benzyl alcohol as **[Jan. 22, 2026 (II)]**

(1) 2-hydroxy acetophenone
(2) 4-methyoxy benzaldehyde
(3) 4-hydroxy benzylaldehyde
(4) 4-methyl benzoic acid

46. Match the **LIST-I** with **LIST-II** **[Jan. 22, 2026 (I)]**

List-I (Reagents)	**List–II (Name of reactions involving carbonyl compound)**
A. NH_2-NH_2, KOH	I. Tollen's test
B. $Ag(NH_3)_2OH$	II. Clemmensen reduction
C. Aq. $CuSO_4$, Sodium potassium tartarate, KOH	III. Wolff-Kishner reduction
D. Zn – Hg, HCl	IV. Fehling's test

Choose the **correct** answer from the options given below:

(1) A-III, B-I, C-IV, D-II (2) A-II, B-I, C-IV, D-III
(3) A-IV, B-III, C-II, D-I (4) A-III, B-IV, C-I, D-II

47. Match **List-I** with **List-II**. **[Jan. 21, 2026 (II)]**

List-I (Reagents)	**List-II (Reaction Name) (Involving aldehydes))**
A. H_2, Pd-$BaSO_4$	I. Etard Reaction
B. $SnCl_2$, HCl	II. Rosenmund Reduction
C. CrO_2Cl_2, CS_2	III. Gatterman – Koch Reaction
D. CO, HCl, Anhyd. $AlCl_3$	IV. Stephen Reaction

Choose the **correct** answer from the options given below:

(1) A-II, B-III, C-IV, D-I (2) A-IV, B-III, C-I, D-II
(3) A-IV, B-I, C-II, D-III (4) A-II, B-IV, C-I, D-III

48. An organic compound "P" of molecular formula $C_6H_{12}O_3$ gives positive iodoform test but negative Tollen's test. When "P" is treated with dilute acid, it produces "Q". "Q" gives positive Tollen's test and also Iodoform test. The structure of "P" is: **[Jan. 21, 2026 (I)]**

(1) $CH_3-C(=O)-CH(OCH_3)-CH_2(OCH_3)$
(2) $CH_3-C(=O)-CH_2-CH(OCH_3)-OCH_3$
(3) $H-C(=O)-CH_2-CH_2-CH(OCH_3)-OCH_3$
(4) $CH_3-C(=O)-C(OCH_3)(OCH_3)-CH_3$

49. When 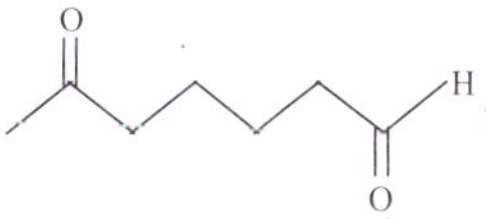 undergoes intramolecular aldol condensation, the major product formed is : **[April 8, 2025 (II)]**

(1) 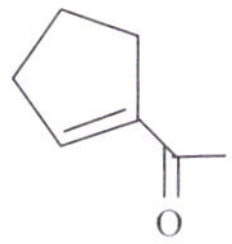 (2) 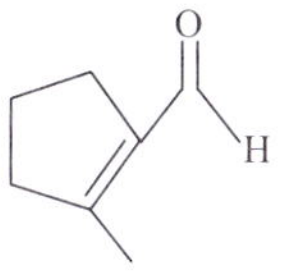

(3) 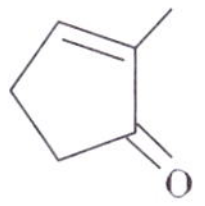 (4) 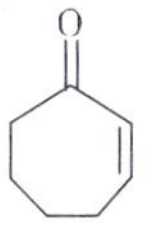

50. 1,2-dibromocyclooctane $\xrightarrow[\text{(ii) } NaNH_2 \text{ (iii) } Hg^{2+}/H^+ \text{ (iv) Zn–Hg}/H^+]{\text{(i) KOH (atc.)}}$ P (Major product)

'P' is **[April 8, 2025 (II)]**

(1) 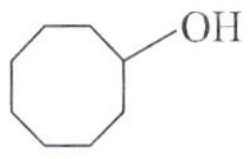 (2) 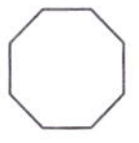

(3) 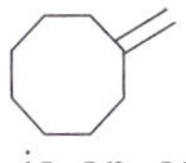 (4) 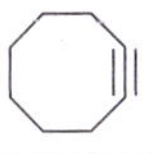

51. "*P*" is an optically active compound with molecular formula $C_6H_{12}O$. When "*P*" is treated with 2, 4-dinitrophenylhydrazine, it gives a positive test. However, in presence of Tollens reagent, "*P*" gives a negative test. Predict the structure of "*P*". **[April 7, 2025 (II)]**

(1) $CH_3-\overset{\overset{\displaystyle O}{\|}}{C}-CH_2-CH_2-CH_2-CH_3$

(2) $CH_3-\overset{\overset{\displaystyle O}{\|}}{C}-\underset{\underset{\displaystyle CH_2-CH_3}{|}}{CH}-CH_3$

(3) $H-\overset{\overset{\displaystyle O}{\|}}{C}-CH_2-\underset{\underset{\displaystyle CH_2-CH_3}{|}}{CH}-CH_3$

(4) $CH_3-\overset{\overset{\displaystyle O}{\|}}{C}-CH_2-\underset{\underset{\displaystyle CH_3}{|}}{CH}-CH_3$

52. Which among the following compounds give yellow solid when reacted with NaOI/NaOH? **[April 4, 2025 (II)]**

(A) $CH_3-\overset{\overset{\displaystyle OH}{|}}{CH}-C_2H_5$

(B) $CH_3-CH_2-CH_2-OH$

(C) 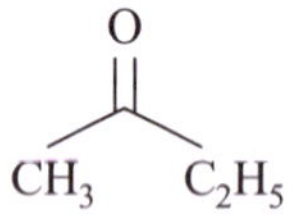

(D) 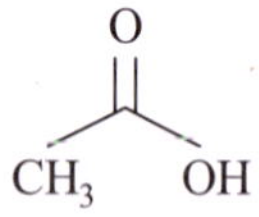

(E) 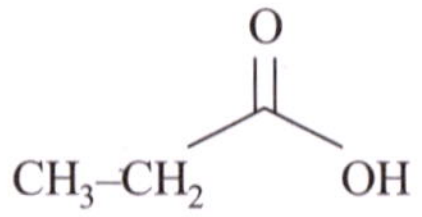

Choose the **correct** answer from the options given below :

(1) (B), (C) and (E) Only (2) (A) and (C) Only
(3) (C) and (D) Only (4) (A), (C) and (D) Only

53. An organic compound (*X*) with molecular formula C_3H_6O is not readily oxidised. On reduction it gives C_3H_8O (*Y*) which reacts with HBr to give a bromide (*Z*) which is converted to Grignard reagent. This Grinard reagent on reaction with (*X*) followed by hydrolysis give 2, 3-dimethylbutan-2-ol. **[April 4, 2025 (I)]**
Compounds (*X*), (*Y*) and (*Z*) respectively are :

(1) CH_3COCH_3, $CH_3CH_2CH_2OH$, $CH_3CH(Br)CH_3$
(2) CH_3COCH_3, $CH_3CH(OH)CH_3$, $CH_3CH(Br)CH_3$
(3) CH_3CH_2CHO, $CH_3CH_2CH_2OH$, $CH_3CH_2CH_2Br$
(4) CH_3CH_2CHO, $CH_3CH=CH_2$, $CH_3CH(Br)CH_3$

54. Aldol condensation is a popular and classical method to prepare α, β-unsaturated carbonyl compounds. This reaction can be both intermolecular and intramolecular. Predict which one of the following is not a product of intramolecular aldol condensation ? **[April 4, 2025 (I)]**

(1) 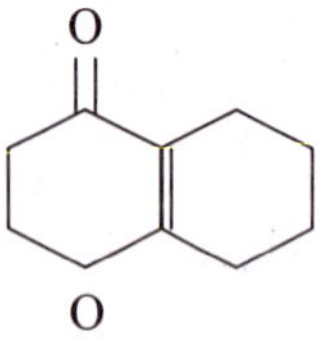

(2) 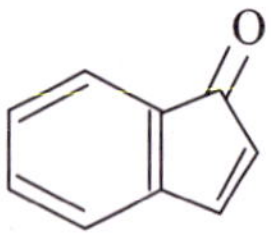

(3) 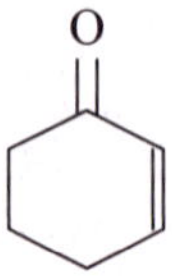

(4) 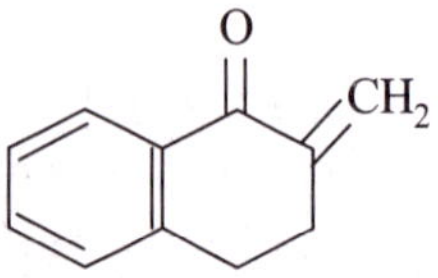

55. The major product (P) in the following reaction is :

$$PH-\underset{\underset{\displaystyle O}{\|}}{C}-\underset{\underset{\displaystyle O}{\|}}{C}-H \xrightarrow[\Delta]{KOH} \underset{\text{Major product}}{P}$$ **[April 3, 2025 (II)]**

(1) $Ph-\overset{\overset{\displaystyle OH}{|}}{CH}-CH_2OH$ (2) $Ph\ \ \underset{\underset{\displaystyle OH}{|}}{CH}-COO^-K^+$

(3) $Ph-\underset{\underset{\displaystyle O}{\|}}{C}-COO^-K^+$ (4) $Ph-\underset{\underset{\displaystyle O}{\|}}{C}-CH_2OH$

56. Number of molecules from below which cannot give iodoform reaction is :
Ethanol, Isopropyl alcohol, Bromoacetone, 2-Butanol, 2-Butanone, Butanal, 2-Pentanone, 3-Pentanone, Pentanal and 3-Pentanol **[April 3, 2025 (I)]**

(1) 5 (2) 4 (3) 3 (4) 2

57. Given below are two statements : **[April 2, 2025 (I)]**

Statement (I) : Vanillin 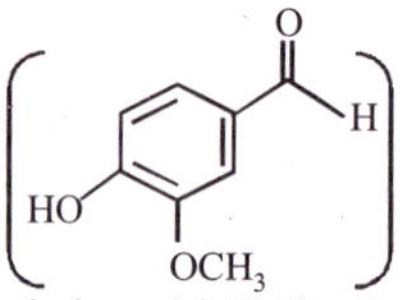

will react with NaOH and also with Tollen's reagent.

Statement (II) : Vanillin 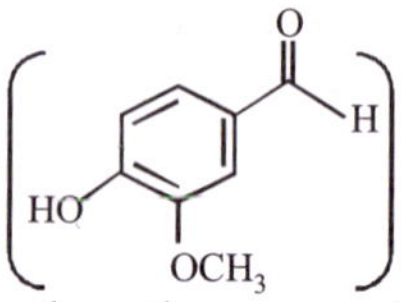

will undergo self aldol condensation very easily. In the light of the above statements, choose the **most appropriate answer** from the options given below :

(1) **Statement I** is incorrect but **Statement II** is correct
(2) **Statement I** is correct but **Statement II** is incorrect
(3) Both **Statement I** and **Statement II** are incorrect
(4) Both **Statement I** and **Statement II** are correct

58. The product (P) formed in the following reaction is: **[Jan. 29, 2025 (I)]**

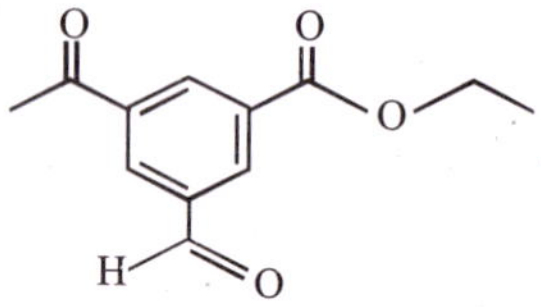

$\xrightarrow[HCl]{Zn-Hg}$ Product (P)

(1) 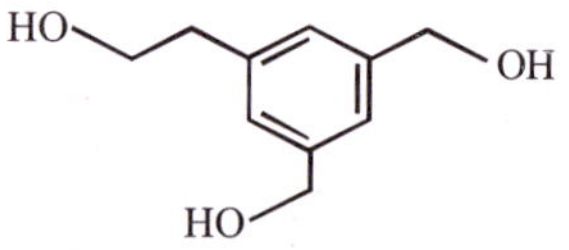

(2) 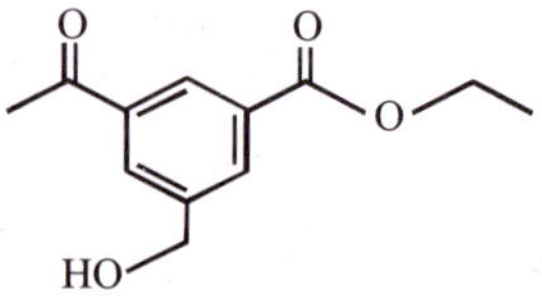

(3) 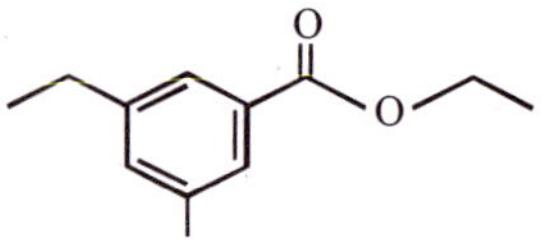

(4) 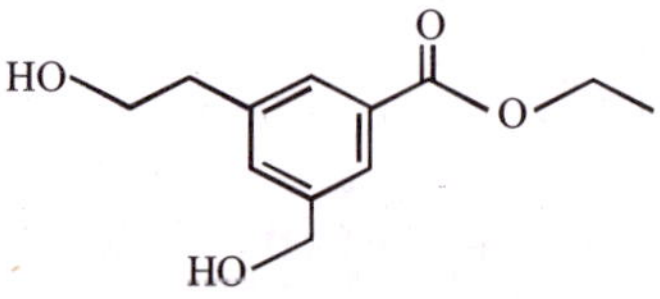

59. Both acetaldehyde and acetone (individually) undergo which of the following reactions? **[Jan. 28, 2025 (I)]**
(A) Iodoform Reaction (B) Cannizaro Reaction
(C) Aldol condensation (D) Tollen's Test
(E) Clemmensen Reduction
Choose the correct answer from the options given below :
(1) A, B and D only (2) A, C and E only
(3) C and E only (4) B, C and D only

60. The product (A) formed in the following reaction sequence is : **[Jan. 24, 2025 (I)]**

$$CH_3-C\equiv CH \xrightarrow[\text{(iii) } H_2/Ni]{\text{(i) } Hg^{2+}, H_2SO_4 \text{ (ii) } HCN} \underset{\text{Product}}{A}$$

(1) $CH_3-C(NH_2)(CH_3)-CH_2-OH$
(2) $CH_3-C(OH)(CH_3)-CH_2-NH_2$
(3) $CH_3-CH_2-CH(NH_2)-CH_2-OH$
(4) $CH_3-CH_2-CH(OH)-CH_2-NH_2$

61. Aman has been asked to synthesise the molecule 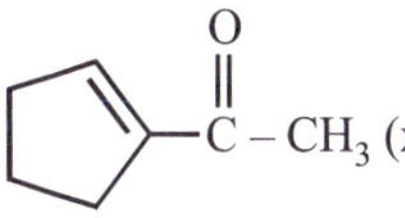 (x). He thought of preparing the molecule using an aldol condensation reaction. He found a few cyclic alkenes in his laboratory. He thought of performing ozonolysis reaction on alkene to produce a dicarbonyl compound followed by aldol reaction to prepare "x". Predict the suitable alkene that can lead to the formation of "x". **[Jan. 24, 2025 (I)]**

(1) 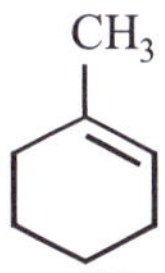 (2) 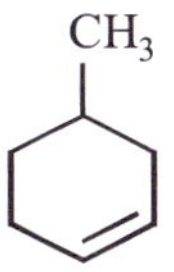
(3) 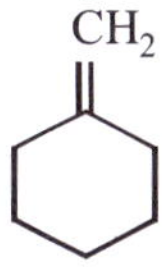 (4) 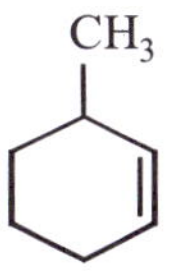

62. Which of the following arrangements with respect to their reactivity in nucleophilic addition reaction is correct? **[Jan. 24, 2025 (I)]**
(1) benzaldehyde < acetophenone < p-nitrobenzaldehyde < p-tolualdehyde
(2) acetophenone < benzaldehyde < p-tolualdehyde < p-nitrobenzaldehyde
(3) acetophenone < p-tolualdehyde < benzaldehyde < p-nitrobenzaldehyde
(4) p-nitrobenzaldehyde < benzaldehyde < p-tolualdehyde < acetophenone

63. Given below are two statements: **[Jan. 23, 2025 (II)]**
Consider the following reaction

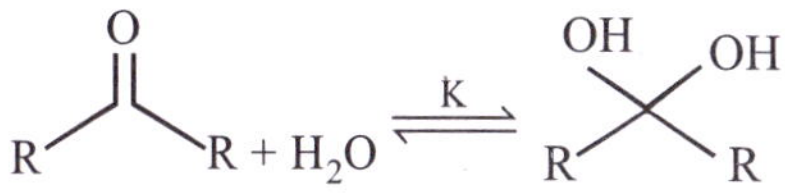

Statement (I): In the case of formaldehyde

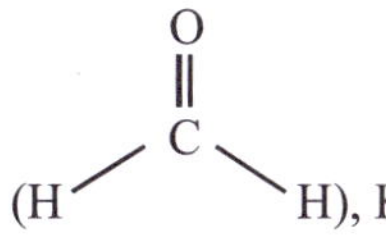

(H–CO–H), K is about 2280, due to small substituents, hydration is faster.
Statement (II) : In the case of trichloro acetaldehyde

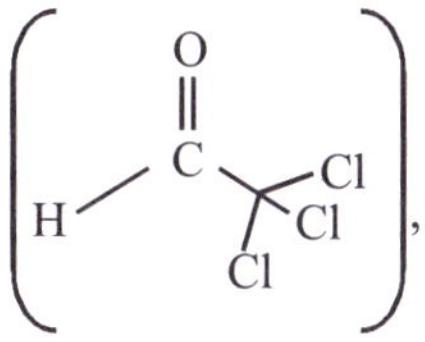

, K is about 2000 due to –I effect of –Cl.

In the light of the above statements, choose the correct answer from the options given below:
(1) Statement I true but Statement II is false
(2) Both Statement I and Statement II are true
(3) Statement I is false but Statement II is truc
(4) Both Statement I and Statement II are false

64. The major product of the following reaction is :

$$CH_3CH_2CH=O \xrightarrow[\text{reflux}]{\text{excess HCHO, alkali}} ?$$ **[Jan. 23, 2025 (I)]**

(1) $CH_3-CH_2-CH_2-OH$ (2) $CH_3-CH(CH_2-OH)-CH=O$
(3) $CH_3-C(CH_2-OH)_3$ (4) $CH_3-C(=CH_2)-CH=O$

65. The correct stability order of the following resonance structures of $CH_3-CH=CH-CHO$ is **[April 9, 2024 (II)]**

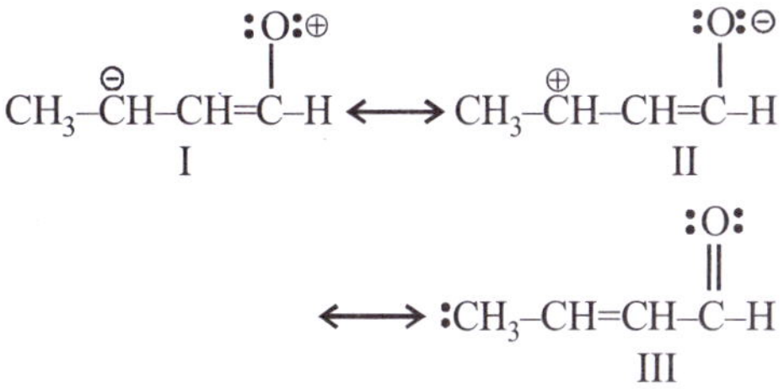

(1) II > III > I (2) III > II > I
(3) I > II > III (4) II > I > III

66. Which of the following compound can give positive iodoform test when treated with aqueous KOH solution followed by potassium hypoiodite. **[April 9, 2024 (II)]**

(1) $CH_3CH_2-CO-CH_2CH_3$ (2) $CH_3CH_2-CCl_2-CH_3$

(3) $CH_3CH_2CH_2CHO$ (4) 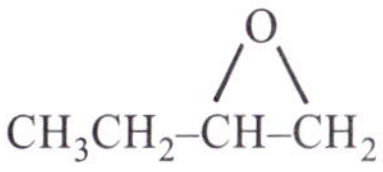

67. Which among the following aldehydes is most reactive towards nucleophilic addition reactions? **[April 6, 2024 (I)]**

(1) $H-\overset{O}{\overset{||}{C}}-H$ (2) $C_2H_5-\overset{O}{\overset{||}{C}}-H$

(3) $CH_3-\overset{O}{\overset{||}{C}}-H$ (4) $C_3H_7-\overset{O}{\overset{||}{C}}-H$

68. For the compounds: **[April 5, 2024 (I)]**

(A) $H_3C-CH_2-O-CH_2-CH_2-CH_3$

(B) $H_3C-CH_2-CH_2-CH_2-CH_3$

(C) $CH_3-CH_2-\underset{O}{\underset{||}{C}}-CH_2-CH_3$

(D) $H_3C-\underset{OH}{\underset{|}{C}H}-CH_2-CH_2-CH_3$

The increasing order of boiling point is :

Choose the correct answer from the options given below:

(1) (A) < (B) < (C) < (D) (2) (B) < (A) < (C) < (D)

(3) (D) < (C) < (A) < (B) (4) (B) < (A) < (D) < (C)

69. Identify 'A' in the following reaction : **[April 5, 2024 (I)]**

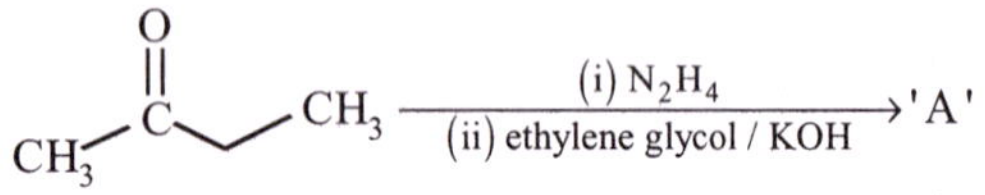

(1) 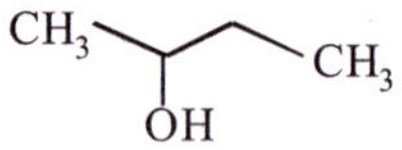 (2) 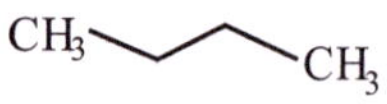

(3) 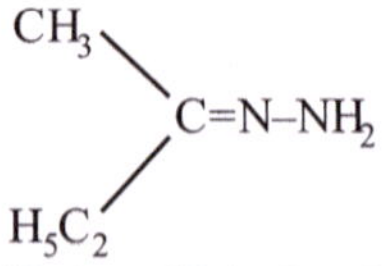 (4) 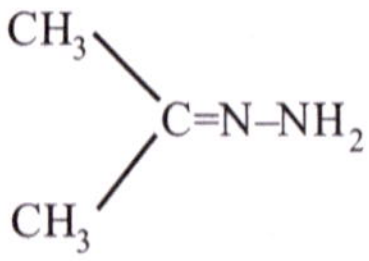

70. Identify A and B in the given chemical reaction sequence : -

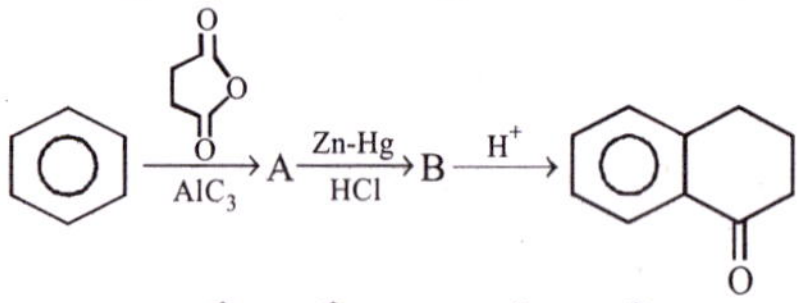

[April 5, 2024 (II)]

(1) 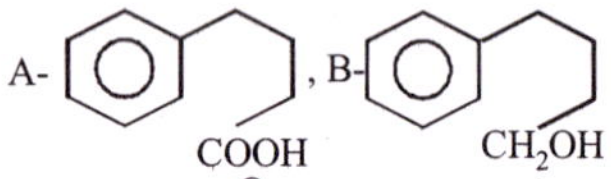

(2) 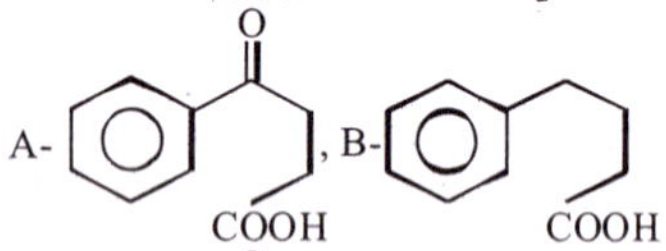

(3) 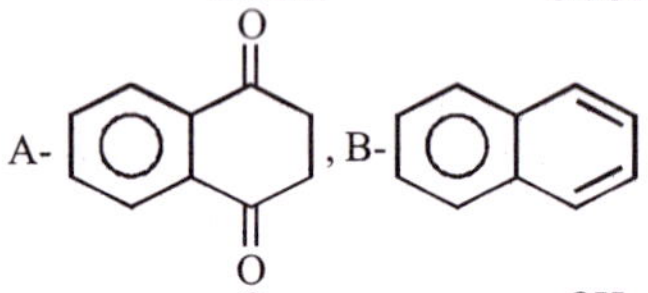

(4) 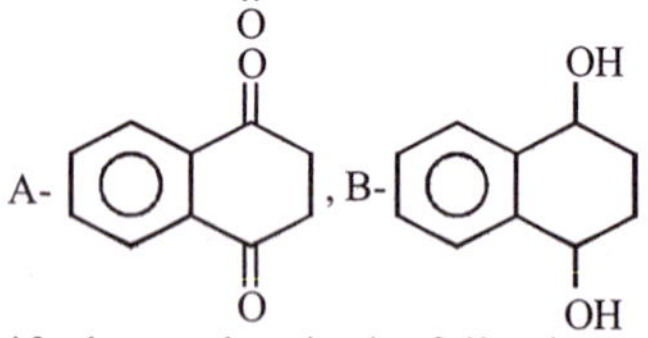

71. Identify the product in the following reaction : **[April 4, 2024 (I)]**

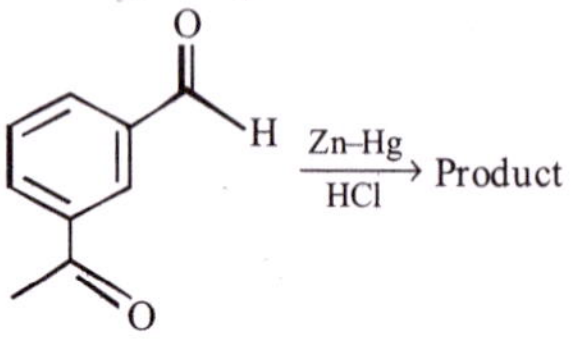

(1) 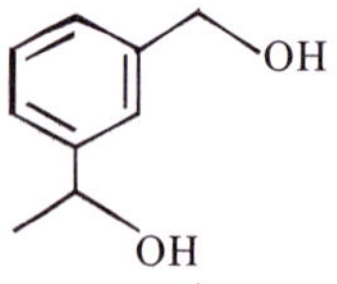 (2) 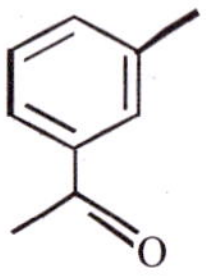

(3) 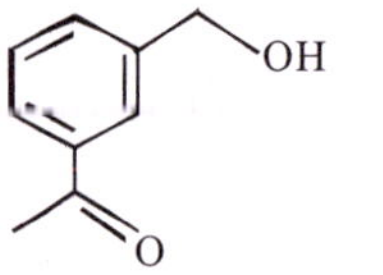 (4) 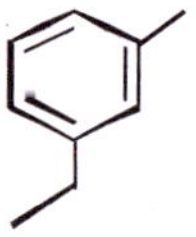

72. Given below are two statements : **[April. 4, 2024 (I)]**

Statement I : Acidity of α-hydrogens of aldehydes and ketones is responsible for aldol reaction.

Statement II : Reaction between benzaldehyde and ethanal will NOT give Cross – Aldol product.

In the light of above statements, choose the most appropriate answer from the options given below.

(1) Both Statement I and Statement II are correct.

(2) Both Statement I and Statement II are incorrect.

(3) Statement I is incorrect but Statement II is correct.

(4) Statement I is correct but Statement II is incorrect.

73. m-Chlorobenzaldehyde on treatment with 50% KOH solution yields **[Jan. 30, 2024 (II)]**

(1) 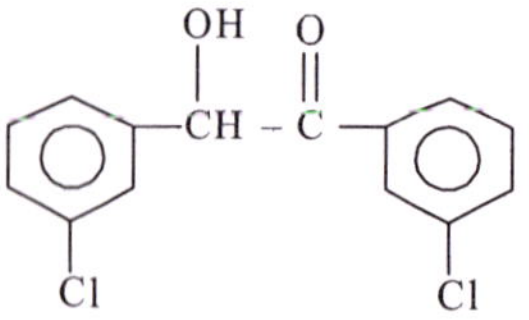

(2) 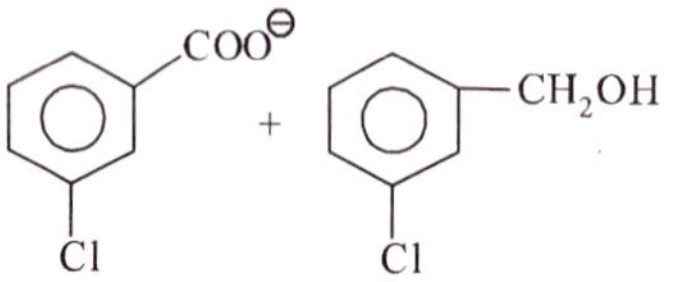

(3) 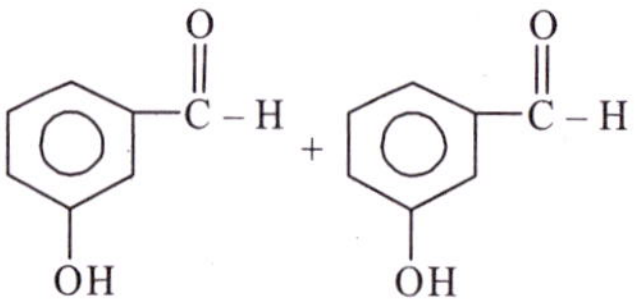

(4) 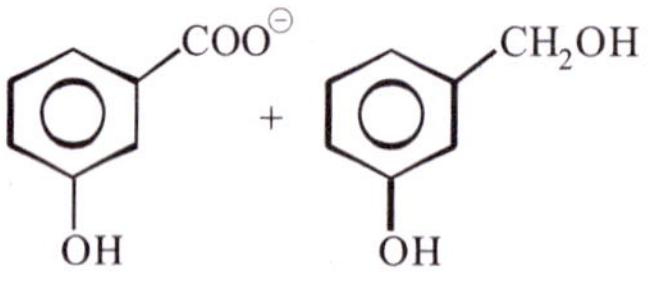

74. The final product A formed in the following multistep reaction sequence is **[Jan. 29, 2024 (I)]**

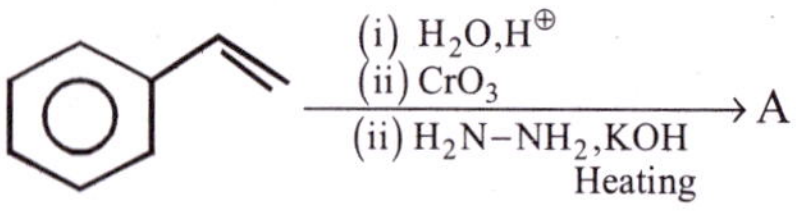

(1) 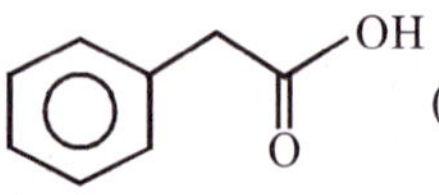 (2) 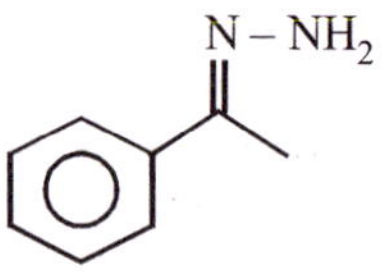

(3) 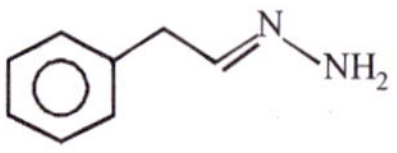 (4) 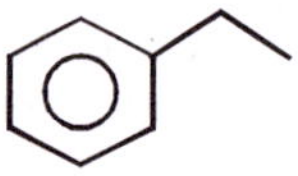

75. Identify the reagents used for the following conversion [Jan. 29, 2024 (II)]

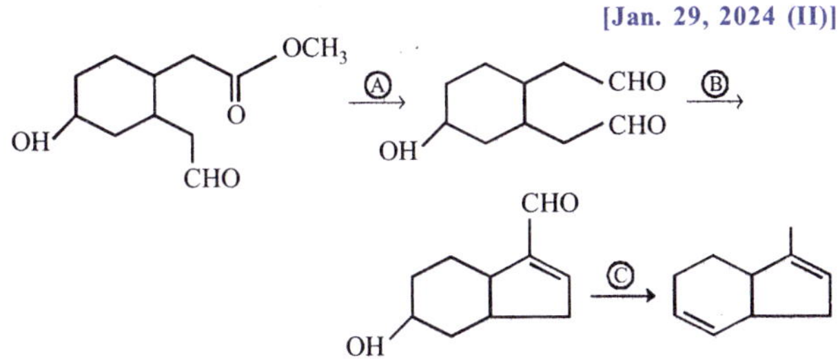

(1) A = $LiAlH_4$. B = $NaOH_{(aq)}$, C = NH_2-NH_2/KOH, ethylene glycol
(2) A = DiBAL-H, B = $NaOH_{(aq)}$, C = NH_2-NH_2/KOH, ethylene glycol
(3) A = DiBAL-H, B = $NaOH_{(alc)}$, C = Zn/HCl
(4) A = $LiAlH_4$, B = $NaOH_{(alc)}$, C = Zn/HCl

76. Highest enol content will be shown by : [Jan. 27, 2024 (I)]

(1) 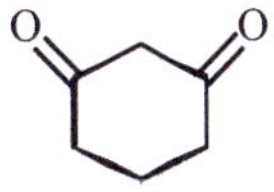
(2) 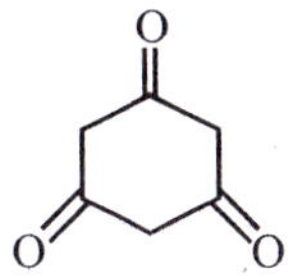
(3) 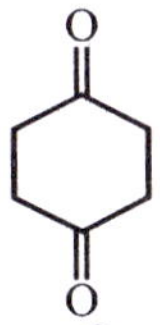
(4) 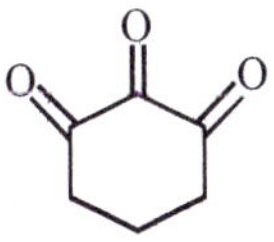

77. 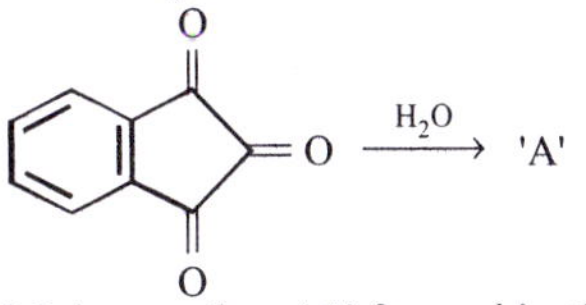

[April 15, 2023 (I)]

Major product 'A' formed in the above reaction is

(1) 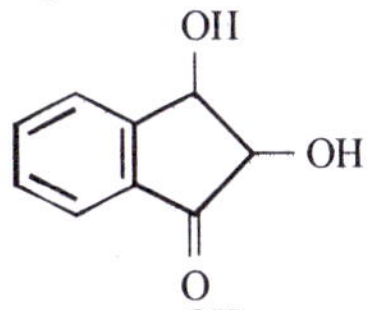
(2) 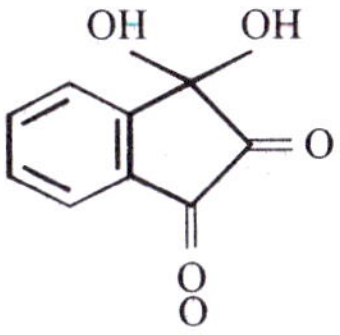
(3) 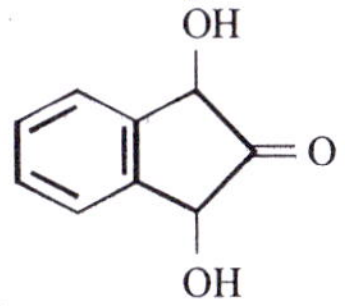
(4) 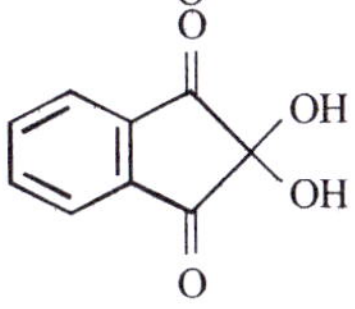

78. 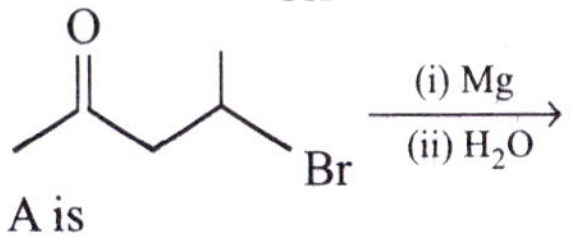

A is [April 12, 2023 (I)]

(1) 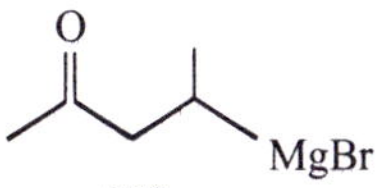

(2) 
(3) 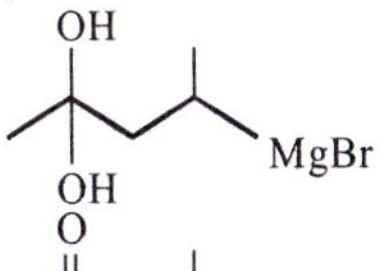

(4) 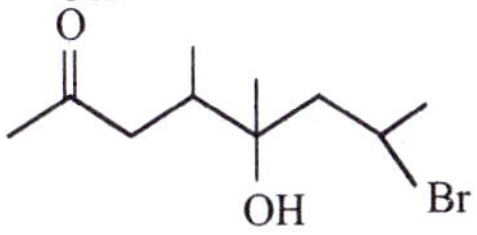

79. Correct statements for the given reaction are : [April 12, 2023 (I)]

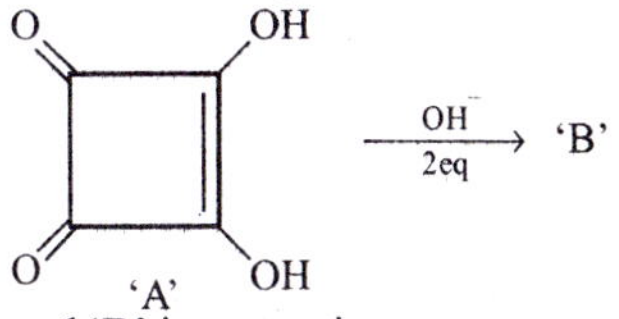

A. Compound 'B' is aromatic
B. The completion of above reaction is very slow
C. 'A' shows tautomerism
D. The bond lengths C-C in compound B are found to be same

Choose the correct answer from the options given below:
(1) A, B and D only (2) A, B and C only
(3) B, C and D only (4) A, C and D only

80. 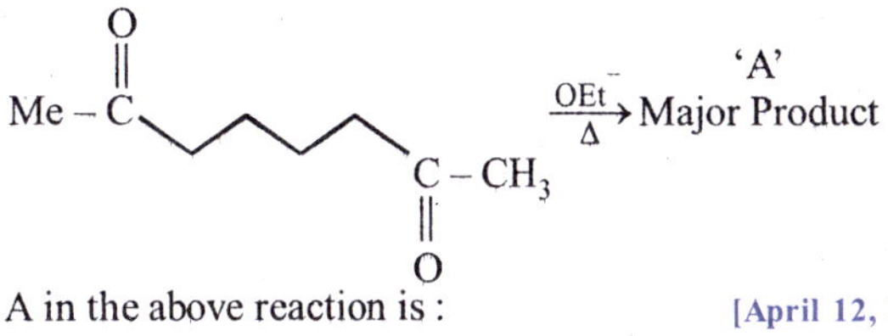

A in the above reaction is : [April 12, 2023 (I)]

(1) 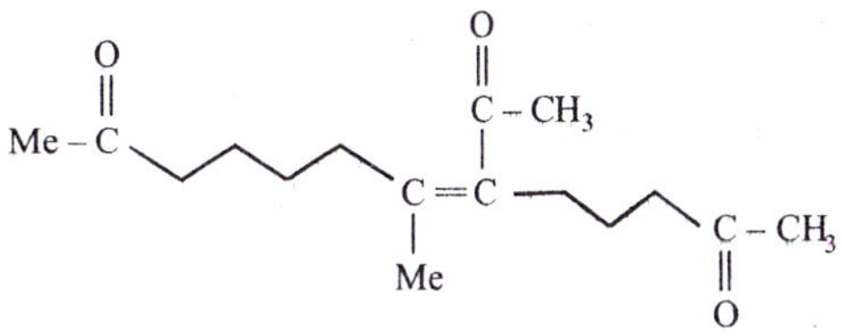
(2) 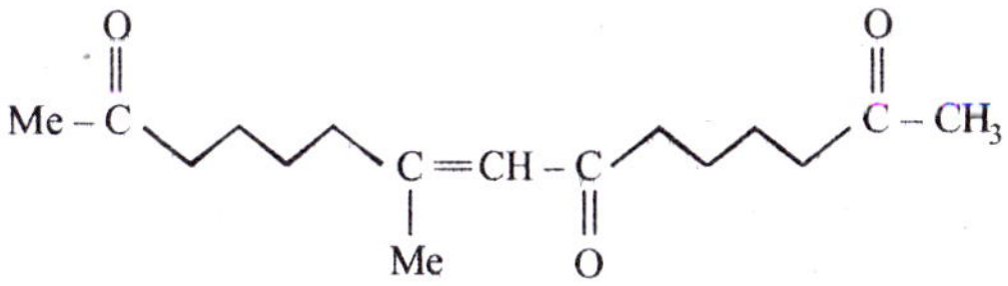
(3) 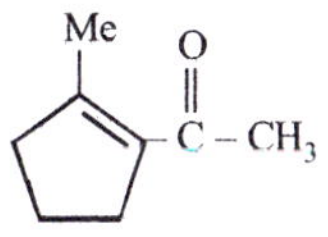
(4) 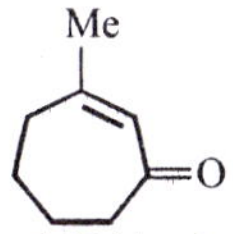

81. 'A' and 'B' in the below reactions are : [April 11, 2023 (I)]

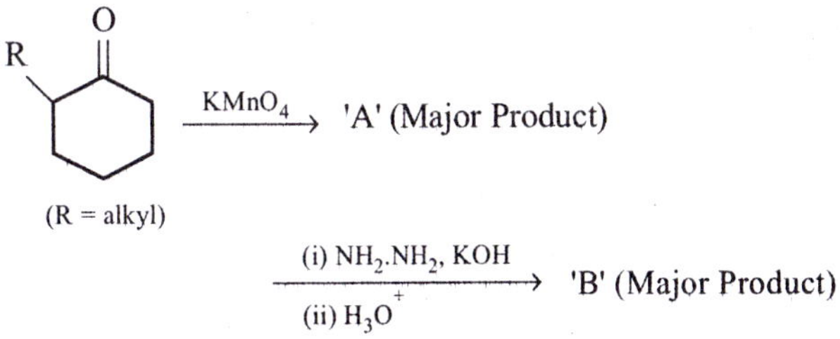

(1) 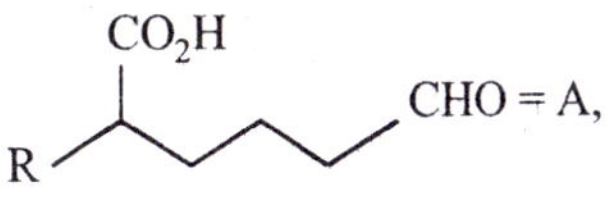

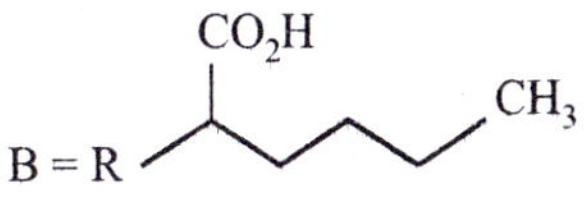

(2) 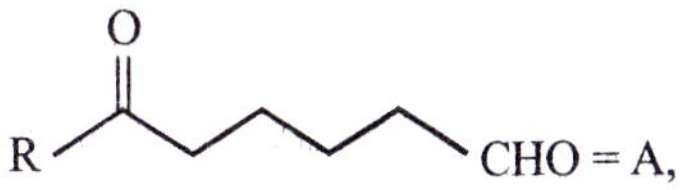

(3) 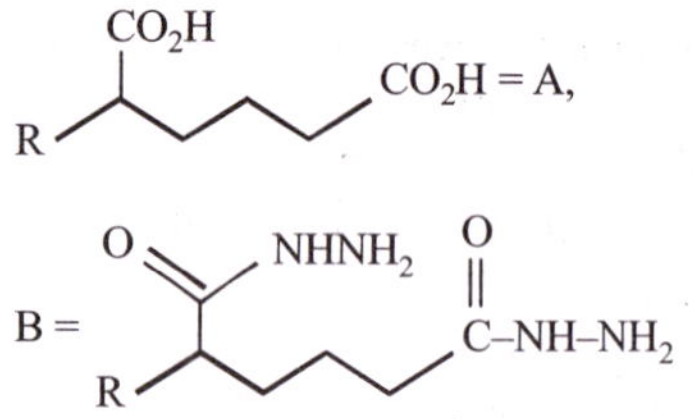

(4) 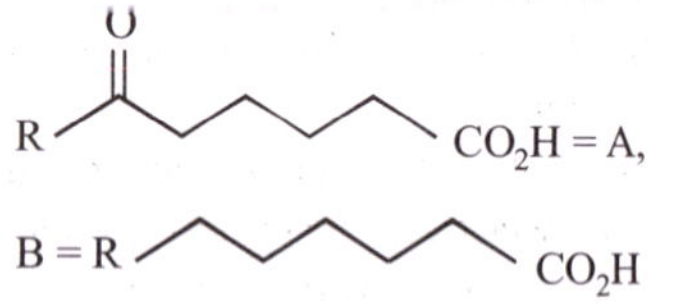

82. Given below are two statements, one is labelled as Assertion A and the other is labelled as Reason R. **[April 11, 2023 (II)]**

Assertion A : 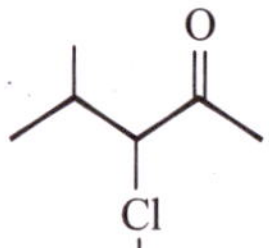 can be subjected to Wolff-Kishner reduction to give .

Reason R : Wolff-Kishner reduction is used to convert 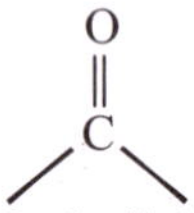 into 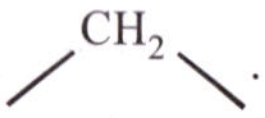.

In the light of the above statements, choose the correct answer from the options given below :

(1) Both A and R are true but R is NOT the correct explanation of A.
(2) A is true but R is false.
(3) A is false but R is true.
(4) Both A and R are true and R is the correct explanation of A.

83. The major product formed in the following reaction is:

$C_6H_5-CH(OH)-CH(CH_3)-CH_2-CHO \xrightarrow[\Delta]{Zn(Hg)/HCl}$ Major product

[April 11, 2023 (II)]

(A) $C_6H_5-CH(OH)-CH(CH_3)-C_2H_5$
(B) $C_6H_5-CH=C(CH_3)-C_2H_5$
(C) $C_6H_5-C(CH_3)=CH-C_2H_5$
(D) 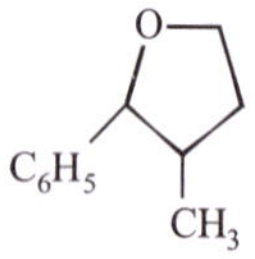

Choose the correct answer from the options given below:

(1) A only (2) B only
(3) C only (4) D only

84. In the reaction given below: **[April 10, 2023 (II)]**

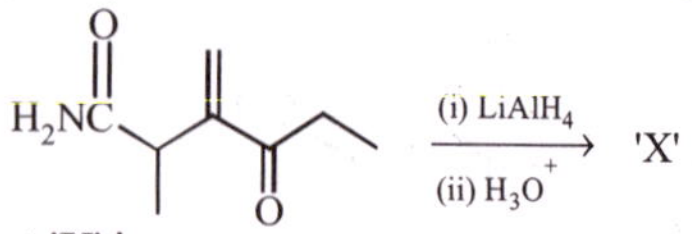

The product 'X' is:

(1) 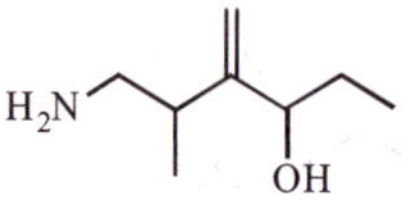 (2) 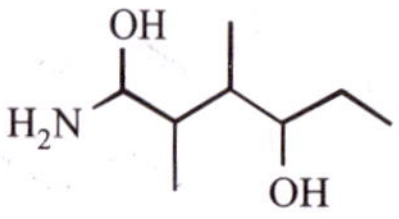

(3) 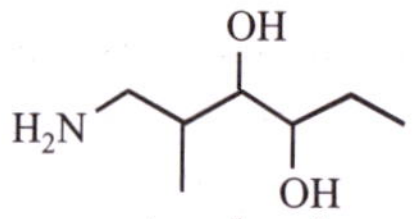 (4) 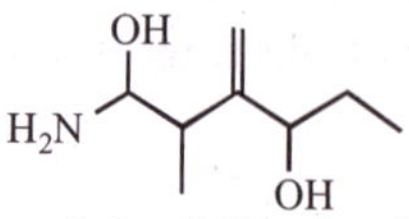

85. The correct order for acidity of the following hydroxyl compounds is: **[April 10, 2023 (II)]**

A. CH_3OH B. $(CH_3)_3COH$
C. 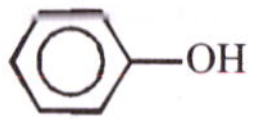 D. 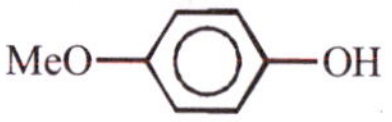
E. 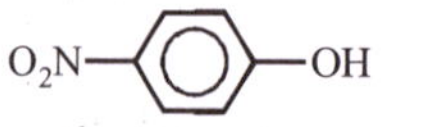

Choose the correct answer from the options given below:

(1) E > C > D > A > B (2) D > E > C > A > B
(3) C > E > D > B > A (4) E > D > C > B > A

86. Match List I with List II **[April 06, 2023 (I)]**

	List I (Name of reaction)		List II (Reagent used)
A	Hell–Volhard–Zelinsky reaction	I.	$NaOH + I_2$
B	Iodoform reaction	II.	(i) CrO_2Cl_2, CS_2 (ii) H_2O
C	Etard reaction	III.	(i) Br_2/red phosphorus (ii) H_2O
D	Gatterman–Koch reaction	IV.	CO, HCl, anhyd. $AlCl_3$

Choose the correct answer from the options given below:

(1) A–III, B–II, C–I, D–IV (2) A–III, B–I, C–IV, D–II
(3) A–I, B–II, C–III, D–IV (4) A–III, B–I, C–II, D–IV

87. Find out the major product from the following reaction

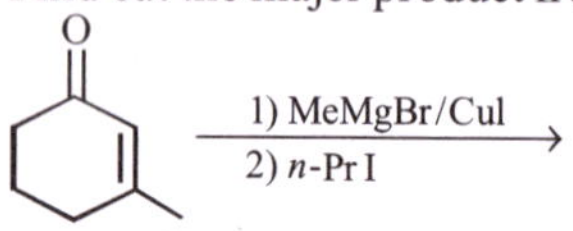

[April 06, 2023 (II)]

(1) 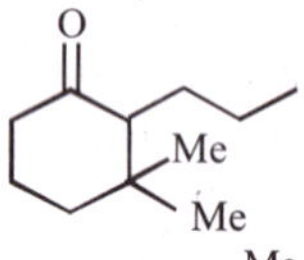 (2) 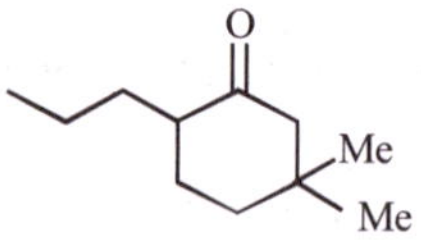

(3) 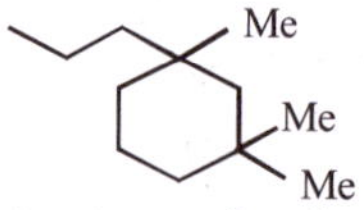 (4) 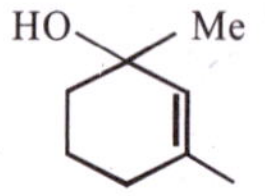

88. The structures of major products A, B and C in the following reaction sequence are. **[Feb. 01, 2023 (II)]**

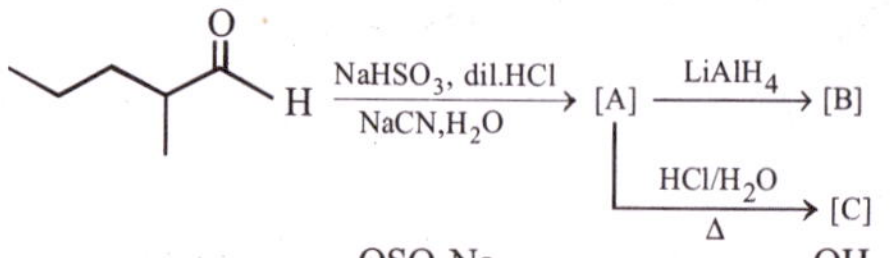

(1) 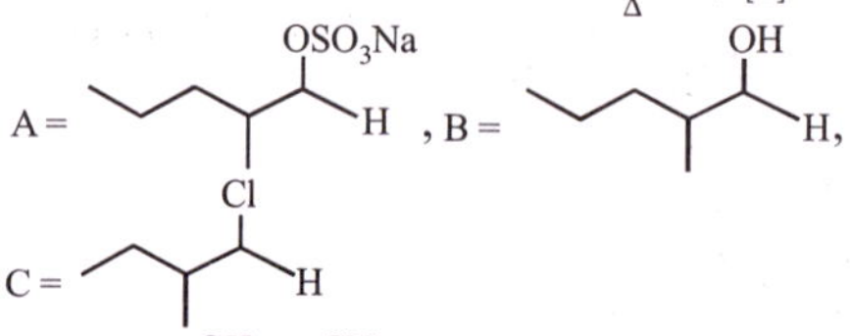

(2) 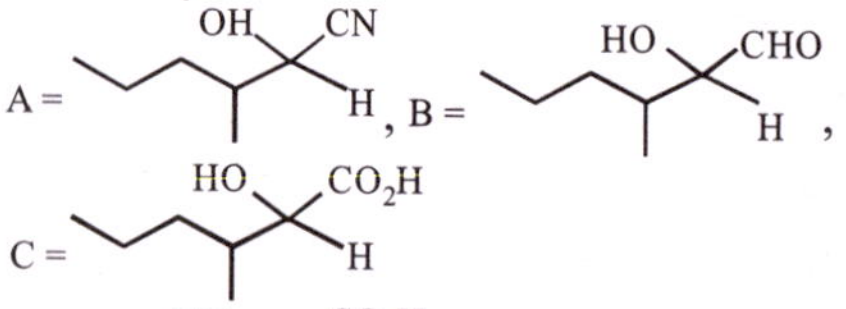

(3) 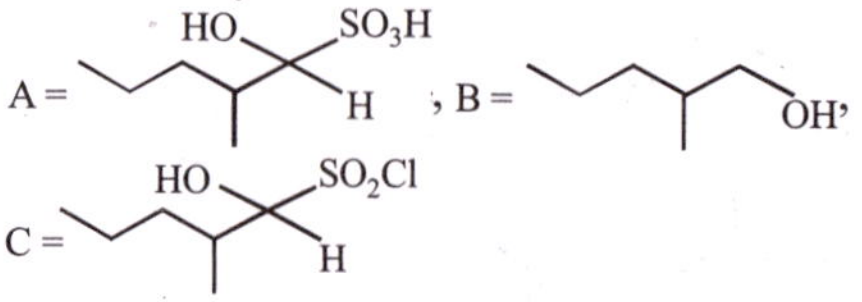

(4) A = 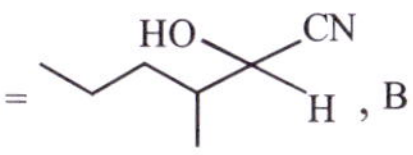 , B = 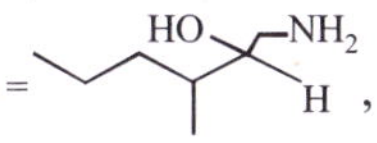 ,

C = 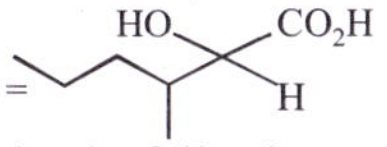

89. Consider the following reaction [Jan. 31, 2023 (I)]

Propanal + Methanal (X) $\xrightarrow[\text{(ii) }\Delta \text{ (iii) NaCN (iv) } H_3O^+]{\text{(i) dil.NaOH}}$ Product B ($C_5H_8O_3$)

The correct statement for product B is. It is

(1) optically active and adds one mole of bromine
(2) racemic mixture and is neutral
(3) racemic mixture and gives a gas with saturated $NaHCO_3$ solution
(4) optically active alcohol and is neutral

90. Given below are two statements: One is labelled as Assertion A and the other is labelled as Reason R.

Assertion A : 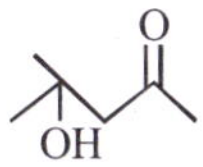 can be easily reduced using Zn-Hg/HCl to .

Reason R : Zn-Hg/HCl is used to reduce carbonyl group to $-CH_2-$ group.

In the light of the above statements, choose the correct answer from the options given below: [Jan. 30, 2023 (II)]

(1) A is false but R is true
(2) A is true but R is false
(3) Both A and R are true but R is not the correct explanation of A
(4) Both A and R are true and R is the correct explanation of A

91. Match List I with List II. [Jan. 29, 2023 (I)]

List-I (Reaction)		List-II (Reagents)
(A) Hoffmann degradation	(I)	Conc.KOH, Δ
(B) Clemmensen reduction	(II)	$CHCl_3$, $NaOH/H_3O^+$
(C) Cannizaro reaction	(III)	Br_2, NaOH
(D) Reimer-Tiemann reaction	(IV)	Zn-Hg/HCl

(1) (A) – III, (B) – IV, (C) – II, (D) - I
(2) (A) – II, (B) – IV, (C) – I, (D) - III
(3) (A) – III, (B) – IV, (C) – I, (D) - II
(4) (A) – II, (B) – I, (C) – III, (D) - IV

92. The major product 'P' for the following sequence of reactions is: [Jan. 29, 2023 (I)]

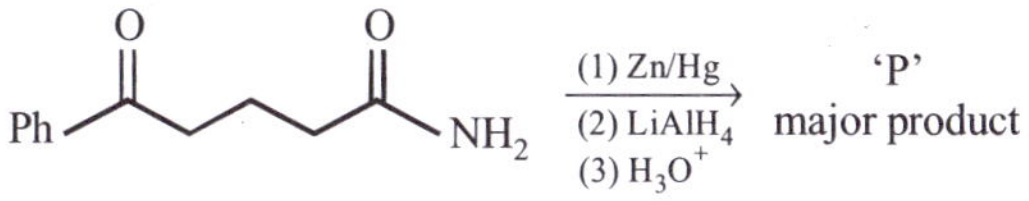

(1) 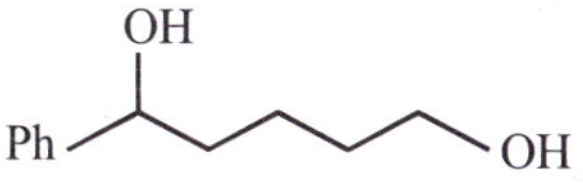

(2) 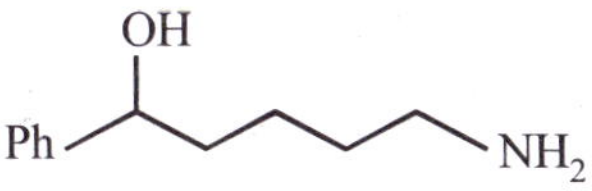

(3) 

(4) 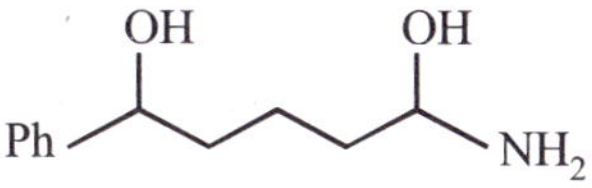

93. Find out the major products from the following reaction sequence. [Jan. 29, 2023 (II)]

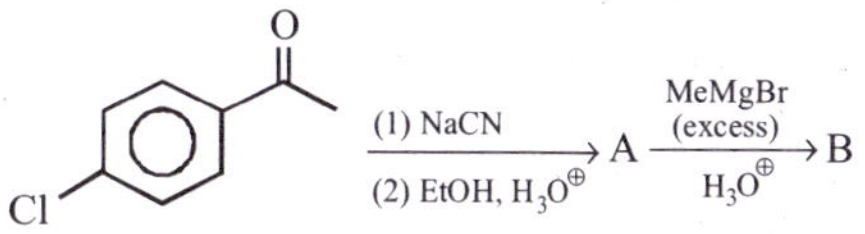

(1) A = 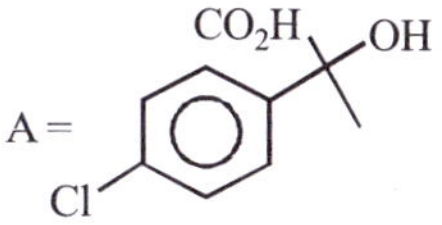

B = 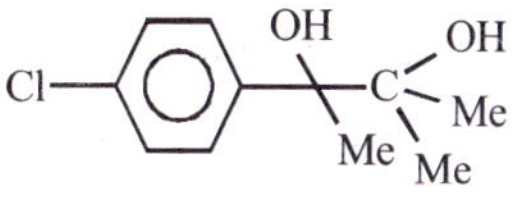

(2) A = 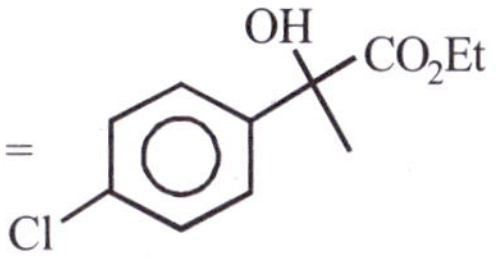

B = 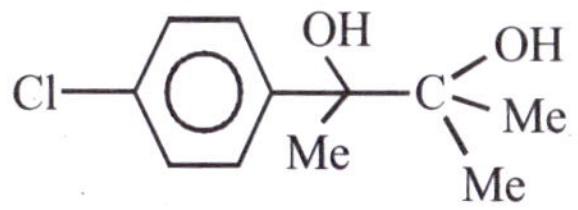

(3) A = 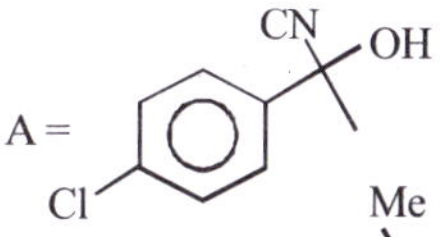

B = 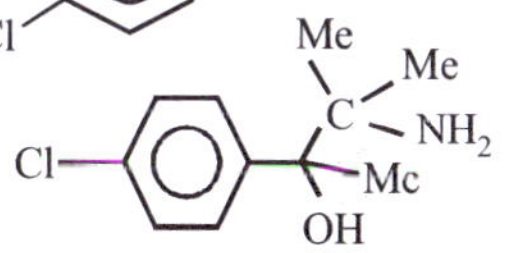

(4) A = 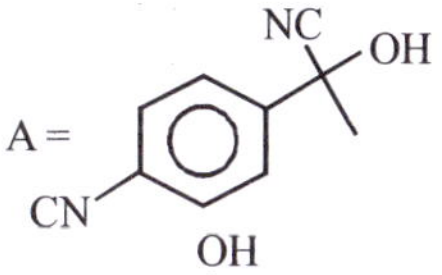

B = 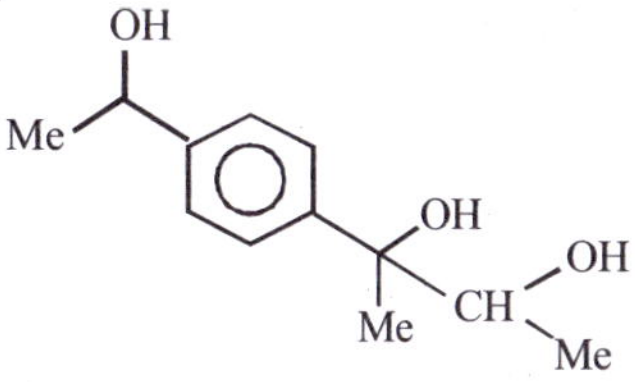

94. Given below are two statements : one is labelled as Assertion A and the other is labelled as Reason R :

Assertion A : Acetal/ketal is stable in basic medium.

Reason R : The high leaving tendency of alkoxide ion gives the stability to acetal/ketal in basic medium.

In the light of the above statements, choose the correct answer from the options given below: [Jan. 25, 2023 (I)]

(1) A is true but R is false
(2) A is false but R is true
(3) Both A and R are true and R is the correct explanation of A
(4) Both A and R are true but R is NOT the correct explanation of A

95.  $\xrightarrow{?}$ PhCOOH + $PhCH_2OH$

P Q R

The correct sequence of reagents for the preparation of Q and R is : [Jan. 25, 2023 (I)]

(1) (i) Cr_2O_3, 770 K, 20 atm ;
(ii) CrO_2Cl_2, H_3O^+ ;
(iii) NaOH ;
(iv) H_3O^+

(2) (i) CrO_2Cl_2, H_3O^+
(ii) Cr_2O_3, 770 K, 20 atm;
(iii) NaOH; (iv) H_3O^+

(3) (i) $KMnO_4, OH^-$;
(ii) Mo_2O_3, A;
(iii) NaOH; (iv) H_3O^+

(4) (i) Mo_2O_3, Δ;
(ii) CrO_2Cl_2, H_3O^+;
(iii) NaOH; (iv) H_3O^+

96. 'A' in the given reaction is [Jan. 25, 2023 (II)]

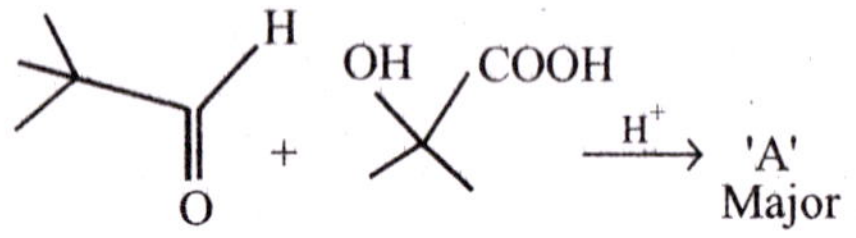

(1) 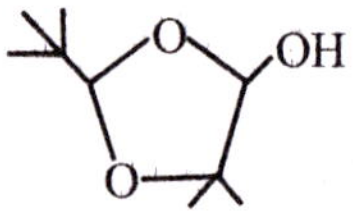 (2) 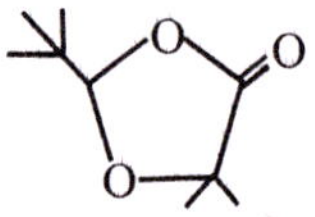

(3) 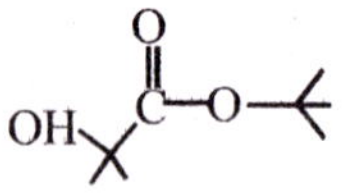 (4) 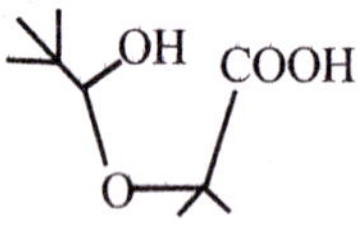

97. Compound (*X*) undergoes following sequence of reactions to give the lactone (*Y*), compound (*X*) is _____. [Jan. 24, 2023 (I)]

Compound (*X*) $\xrightarrow[\text{(iii) } H_3O^+]{\text{(i) HCHO, KOH; (ii) KCN (alc)}}$ 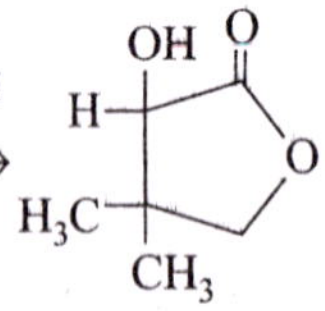 Lactone (*Y*)

(1) $H-C(CH_3)_2-CHO$

(2) $HOH_2C-C(CH_3)_2-CHO$

(3) $H_3C-CH(OH)-CH_2-CH_2-CHO$

(4) $H_3C-CH(CH_3)-CH_2-CHO$

98. 'R' formed in the following sequence of reactions is: [Jan. 24, 2023 (I)]

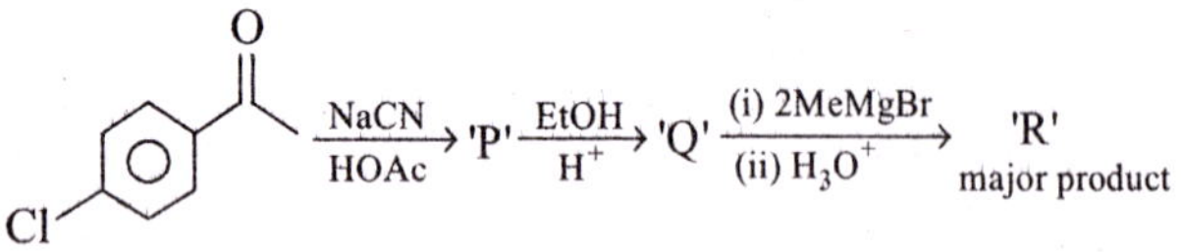

(1) 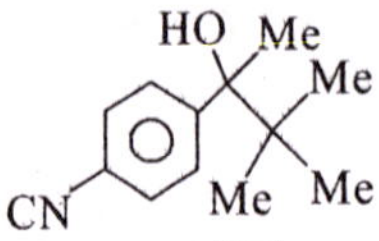 (2) 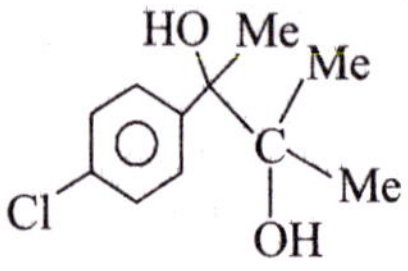

(3) 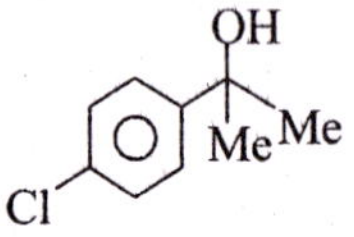 (4) 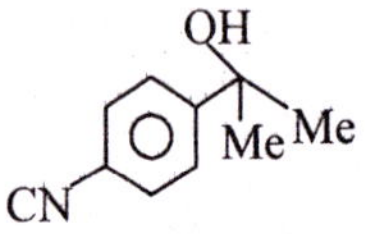

99. Given below are two statements: [Jan. 24, 2023 (II)]

Statement I: 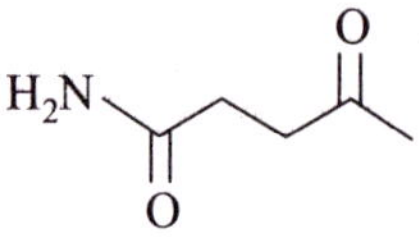 under Clemmensen reduction conditions will give 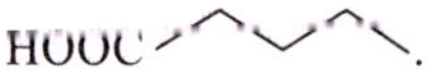.

Statement II: 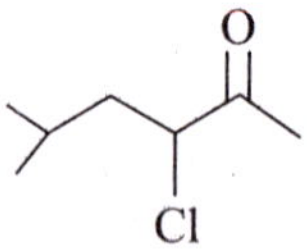 under Wolff-Kishner reduction condition will give 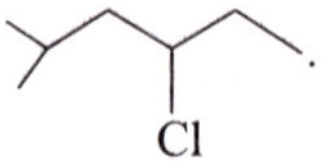.

In the light of the above statements, choose the correct answer from the options given below :

(1) Statement I is false but Statement II is true.
(2) Both Statement I and Statement II are false.
(3) Statement I is true but Statement II is false.
(4) Both Statement I and Statement II are true.

100. Which will undergo deprotonation most readily in basic medium? [Jan. 24, 2023 (II)]

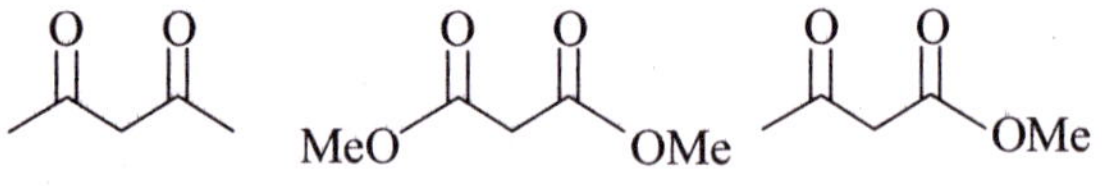

A B C

(1) A only (2) C only
(3) Both A and C (4) B only

101. 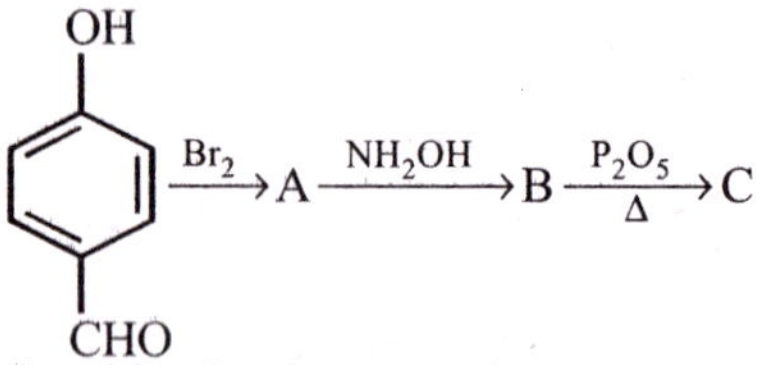

Consider the above reaction sequence, the product 'C' is: [July 29, 2022 (I)]

(1) 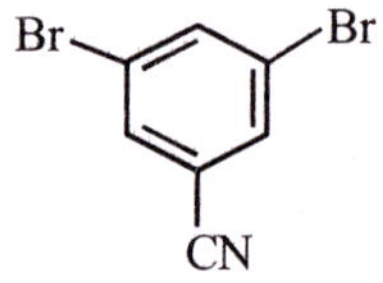 (2) 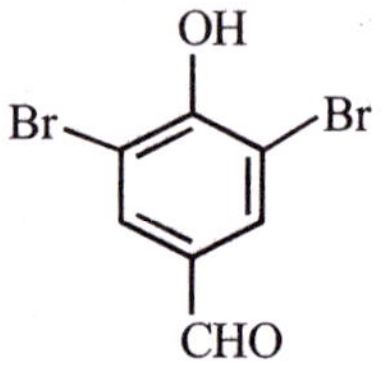

(3) 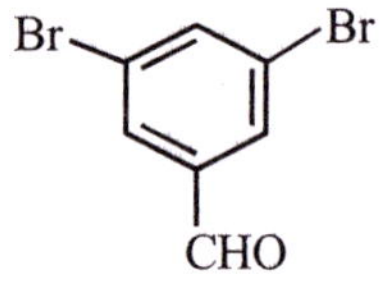 (4) 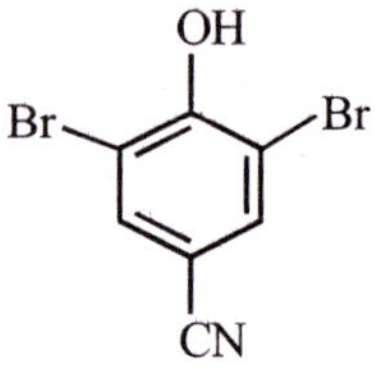

102. Match List-I with List-II.

List-I **List-II**

(A) 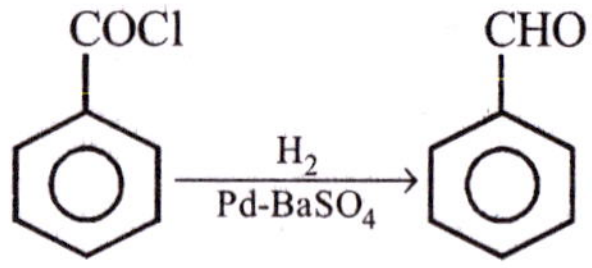 (I) Gatterman Koch reaction

(B) $CH_3-CN \xrightarrow[H_3O^+]{SnCl_2/HCl} CH_3-CHO$ (II) Etard reaction

(C) 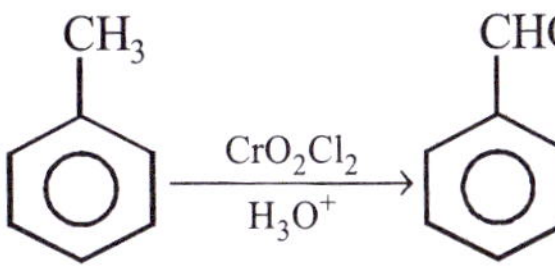 (III) Stephen reaction

(D) 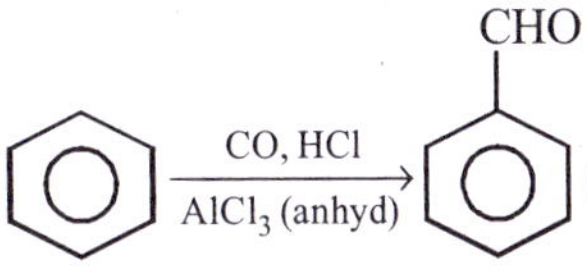 (IV) Rosenmund reaction

Choose the **correct** answer from the options given below:
(1) (A)-(IV), (B)-(III), (C)-(II), (D)-(I) [July 27, 2022 (II)]
(2) (A)-(I), (B)-(II), (C)-(III), (D)-(IV)
(3) (A)-(II), (B)-(III), (C)-(IV), (D)-(I)
(4) (A)-(III), (B)-(II), (C)-(I), (D)-(IV)

103. The structure of A in the given reaction is: [July 27, 2022 (II)]

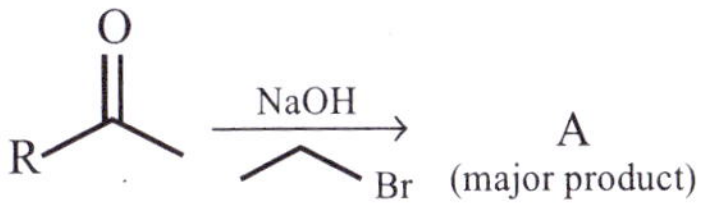

(1) 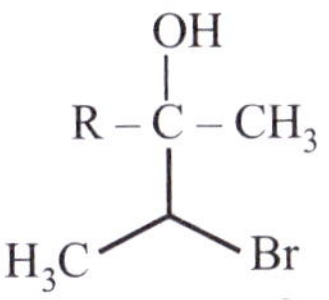 (2) 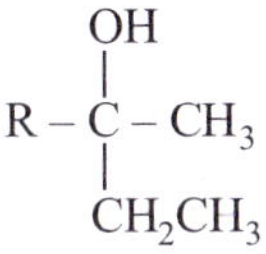

(3) 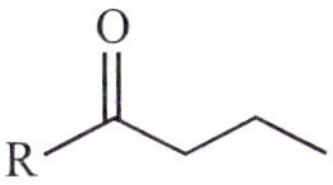 (4) 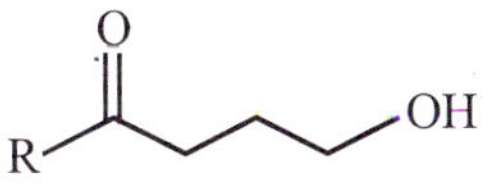

104. The products formed in the following reaction, **A** and **B** are [July 26, 2022 (I)]

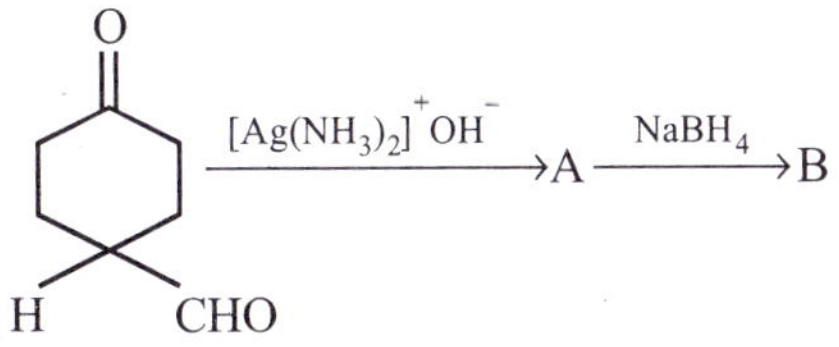

(1) A = 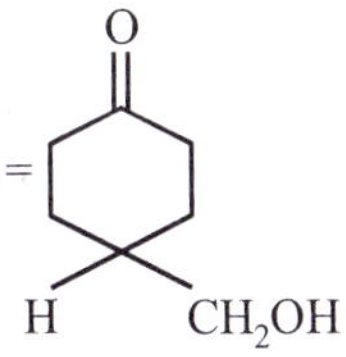, B = 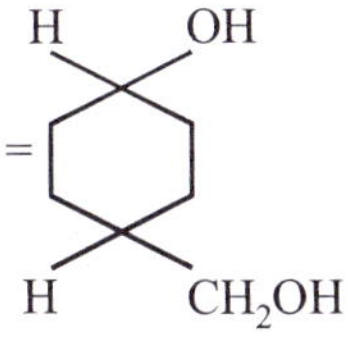

(2) A = 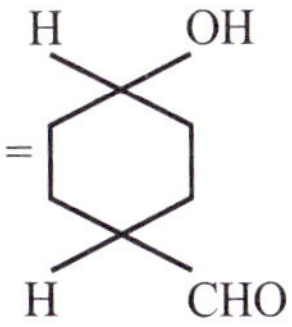, B = 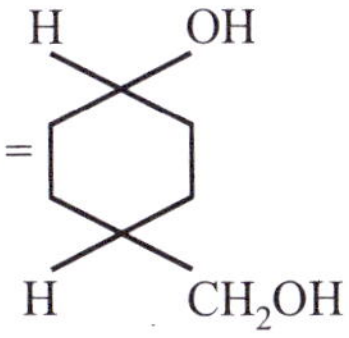

(3) A = 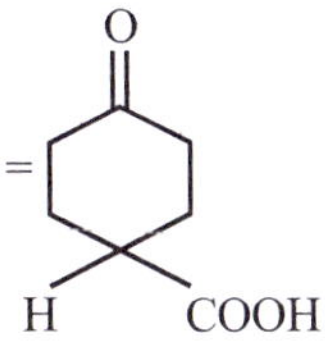, B = 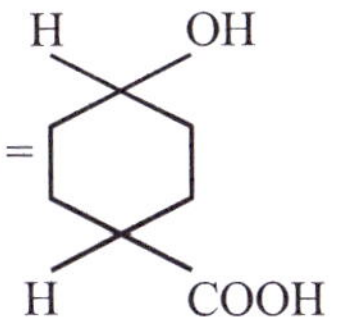

(4) A = 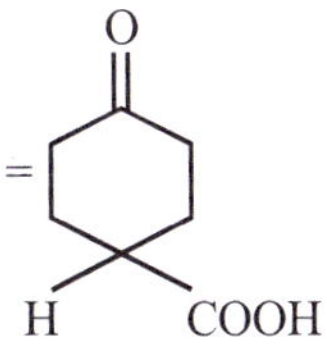, B = 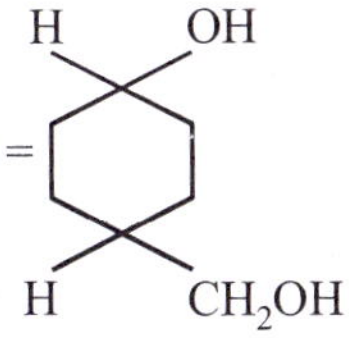

105. Which reactant will give the following alcohol on reaction with one mole of phenyl magnesium bromide (PhMgBr) followed by acidic hydrolysis? [July 26, 2022 (I)]

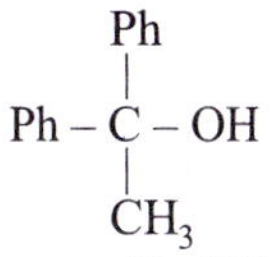

(1) $CH_3-C\equiv N$ (2) $PH-C\equiv N$

(3) $CH_3-\overset{O}{\overset{\|}{C}}-O-Ph$ (4) $Ph-\underset{O}{\underset{\|}{C}}-CH_3$

106. What is the major product of the following reaction? [July 25, 2022 (II)]

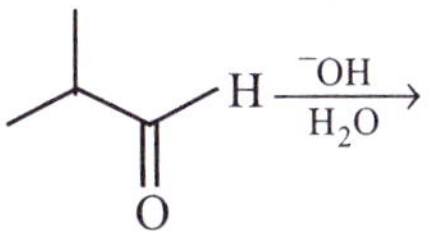

(1) 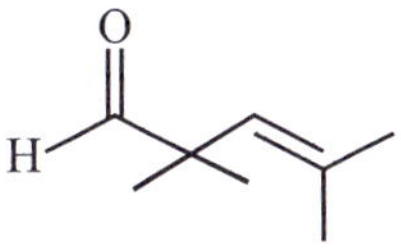 (2) 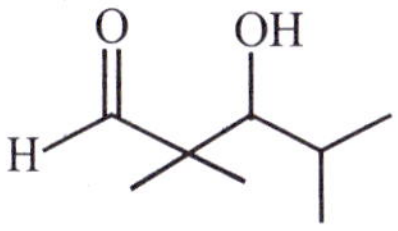

(3) 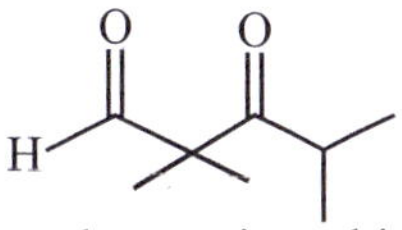 (4) 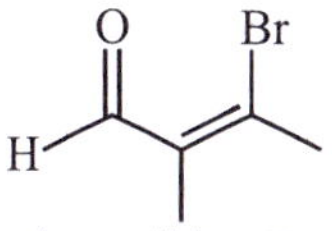

107. Choose the reaction which is **not** possible: [June 30, 2022 (I)]

(1) 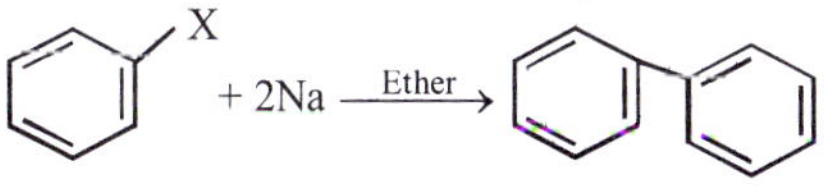

(2) 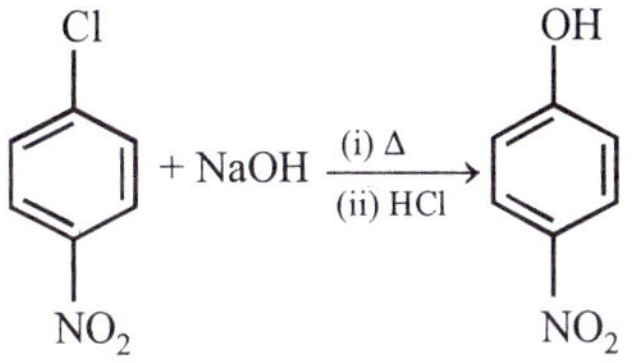

(3) 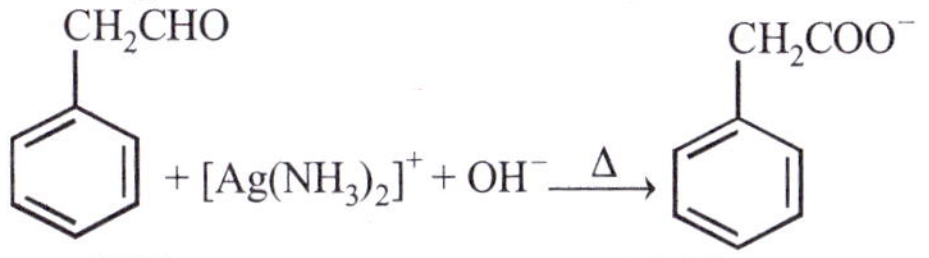

(4) 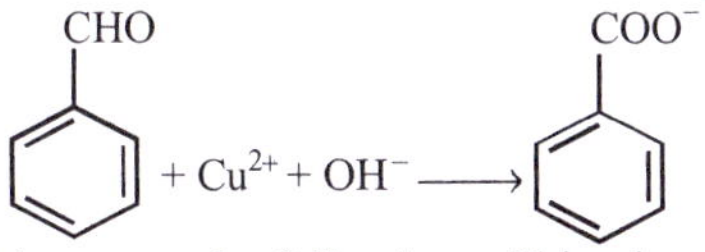

108. Which among the following will be the major product of the given reaction? [June 30, 2022 (I)]

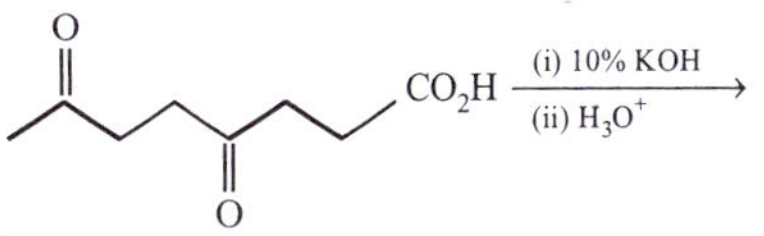

(1) 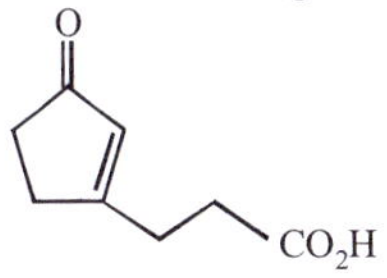 (2) 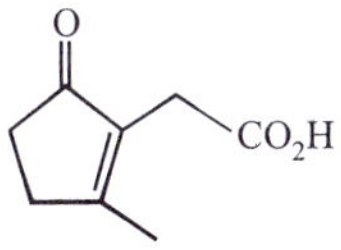

(3) 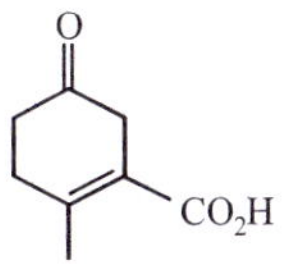 (4) 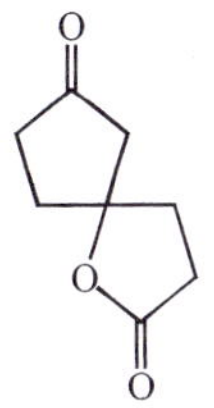

109. The correct structure of product 'A' formed in the following reaction.

$$PhCHO + PhCHO \xrightarrow[\text{in } D_2O]{NaOD} A + Ph-\overset{O}{\overset{\|}{C}}-O^-$$

(Ph is $-C_6H_5$) **[June 28, 2022 (I)]**

(1) 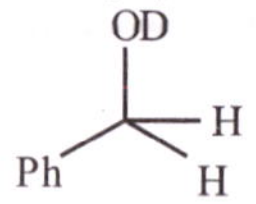 (2) 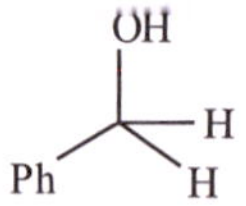

(3) 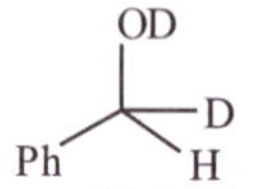 (4) 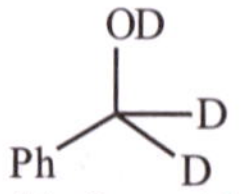

110. Isobutyraldehyde on reaction with formaldehyde and K_2CO_3 gives compound 'A'. Compound 'A' reacts with KCN and yields compound 'B', which on hydrolysis gives a stable compound 'C'. The compound 'C' is : **[June 28, 2022 (II)]**

(1) $HO-CH_2-\underset{CH_3}{\overset{CH_3}{\overset{|}{\underset{|}{C}}}}-\underset{OH}{\underset{|}{CH}}-COOH$

(2) $HO-CH_2CH_2-\underset{CH_3}{\underset{|}{CH}}-\underset{OH}{\underset{|}{CH}}-COOH$

(3) 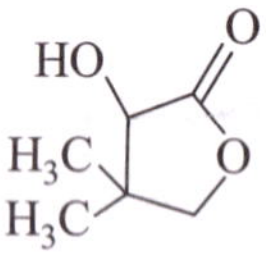 (4) 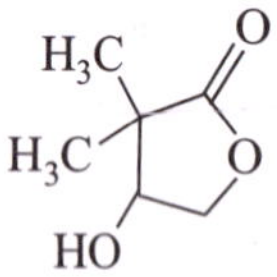

111. 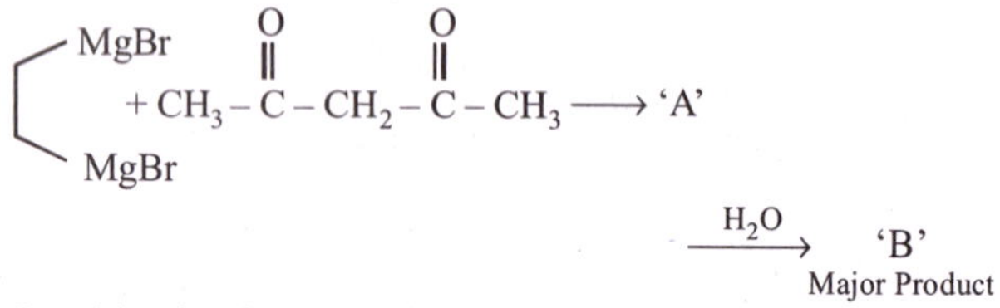

Consider the above reaction sequence and identify the product B. **[June 26, 2022 (I)]**

(1) 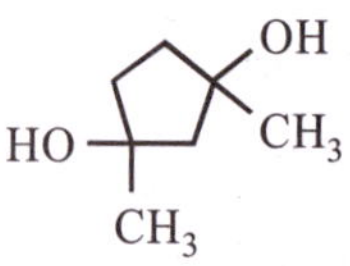 (2) 

(3) 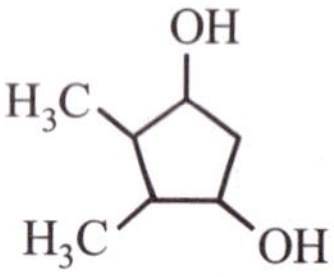 (4) 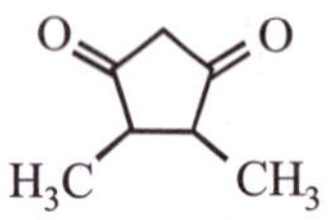

112. Which will have the highest enol content ? **[June 26, 2022 (I)]**

(1) 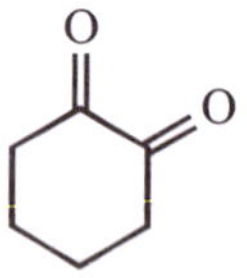 (2) 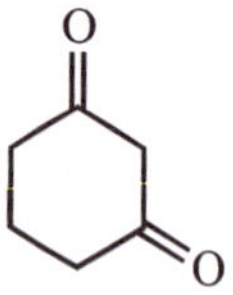

(3) 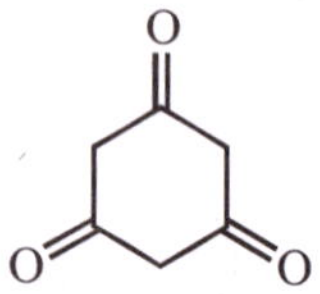 (4) 

113. The reagent, from the following, which converts benzoic acid to benzaldehyde in one step is **[June 26, 2022 (II)]**

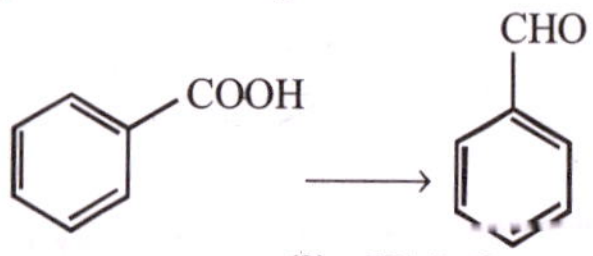

(1) $LiAlH_4$ (2) $KMnO_4$
(3) MnO (4) $NaBH_4$

114. The final product 'A' in the following reaction sequence

$$CH_3CH_2-\underset{O}{\underset{\|}{C}}-CH_3 \xrightarrow{HCN} ? \xrightarrow[\text{Heat}]{95\% H_2SO_4} A$$

[June 26, 2022 (II)]

(1) $CH_3-CH=\overset{CH_3}{\overset{|}{C}}-COOH$

(2) $CH_3-CH=\underset{CH_3}{\underset{|}{C}}-CN$

(3) $CH_3-CH-\overset{OH}{\overset{|}{\underset{CH_3}{\underset{|}{C}}}}-COOH$

(4) $CH_3-CH=\underset{CH_3}{\underset{|}{C}}-CONH_2$

115. Which of the following ketone will NOT give enamine on treatment with secondary amines?
[where t-Bu is –C(CH₃)₃] **[June 25, 2022 (II)]**

(1) 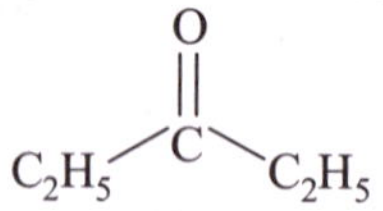 (2) 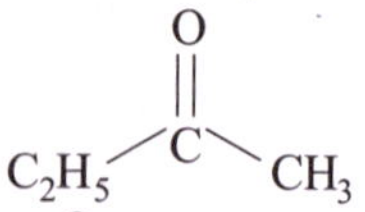

(3) 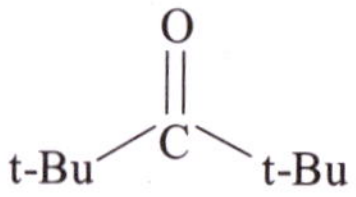 (4) 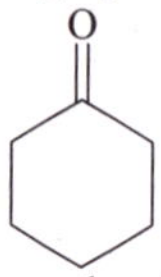

116. Which of the following is an example of conjugated diketone? **[June 24, 2022 (I)]**

(1) $CH_3-\overset{O}{\overset{\|}{C}}-CH_2-CH_2-CH_2-\overset{O}{\overset{\|}{C}}-CH_3$

(2) 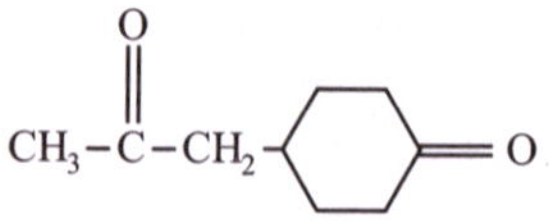

(3) 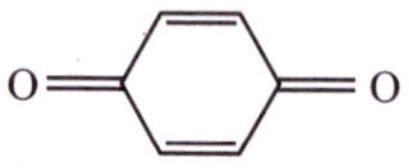

(4) 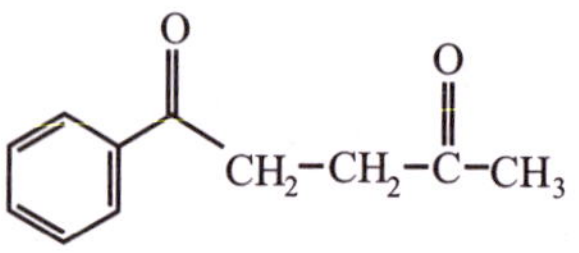

117. Hex-4-ene-2-ol on treatment with PCC gives '*A*'. '*A*' on reaction with sodium hypoiodite gives '*B*', which on further heating with soda lime gives '*C*'. The compound '*C*' is **[June 24, 2022 (II)]**

(1) 2- pentene (2) proponaldehyde
(3) 2-butene (4) 4-methylpent-2-ene

118. Given below are two statements :

Statement I : The nucleophilic addition of sodium hydrogen sulphite to an aldehyde or a ketone involves proton transfer to form a stable ion.

Statement II : The nucleophilic addition of hydrogen cyanide to an aldehyde or a ketone yields amine as final product.

In the light of the above statements, choose the **most appropriate** answer from the options given below : **[Sep. 1, 2021 (II)]**

(1) Both Statement I and Statement II are true.
(2) Statement I is true but Statement II is false.
(3) Statement I is false but Statement II is true.
(4) Both Statement I and Statement II are false.

119. For the reaction given below : **[Aug. 31, 2021 (II)]**

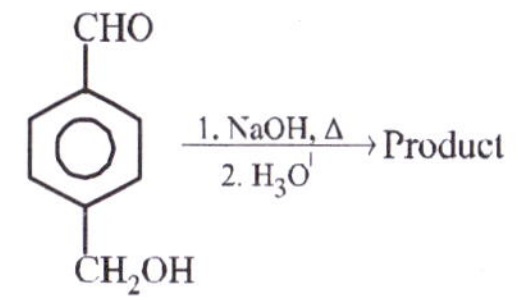

The compound which is **not** formed as a product in the reaction is a :

(1) compound with both alcohol and acid functional groups
(2) monocarboxylic acid
(3) dicarboxylic acid
(4) diol

120. The structure of the starting compound P used in the reaction given below is : **[Aug. 27, 2021 (I)]**

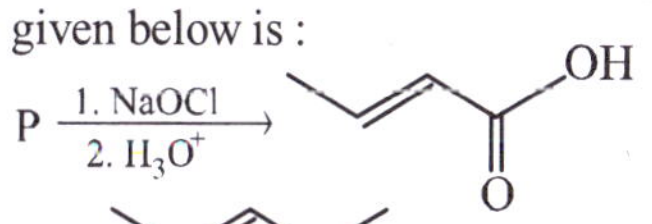

(1) 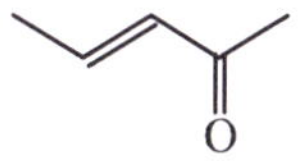 (2) 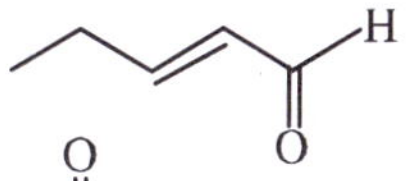

(3) 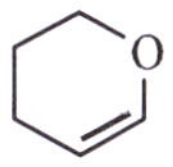 (4) 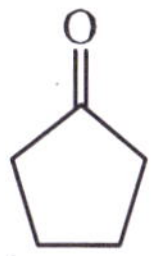

121. The major product of the following reaction is: **[Aug. 27, 2021 (I)]**

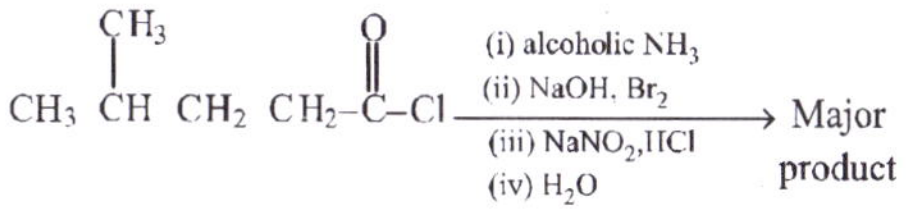

(1) $CH_3-CH(CH_3)-CH(Br)-CH_2OH$

(2) $CH_3-CH(CH_3)-CH_2-CH_2-CH_2OH$

(3) $CH_3-CH(CH_3)-CH_2-CH_2OH$

(4) $CH_3-CH(CH_3)-CH_2-CH_2-Cl$

122. Which one of the following is the major product of the given reaction? **[Aug. 27, 2021 (II)]**

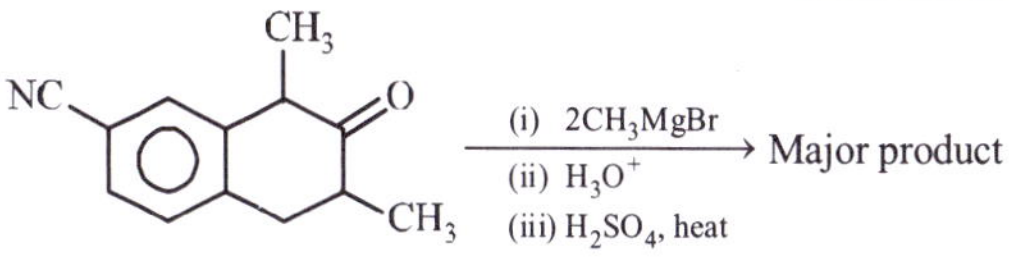

(1) 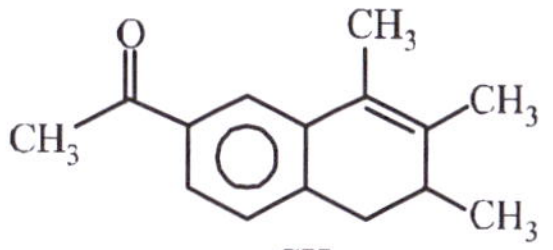

(2) 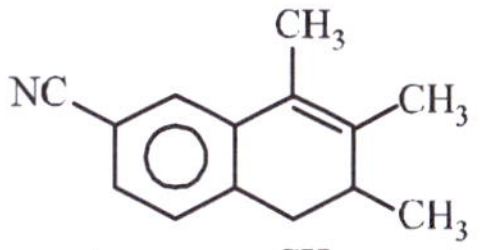

(3) 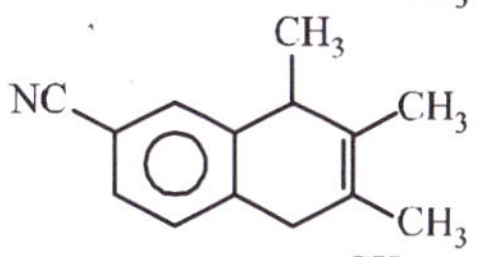

(4) 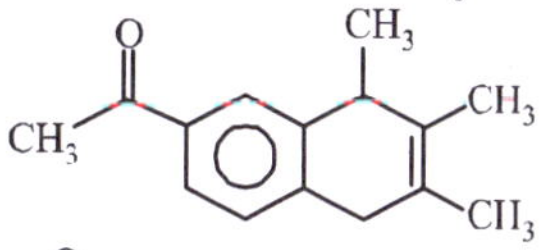

123.

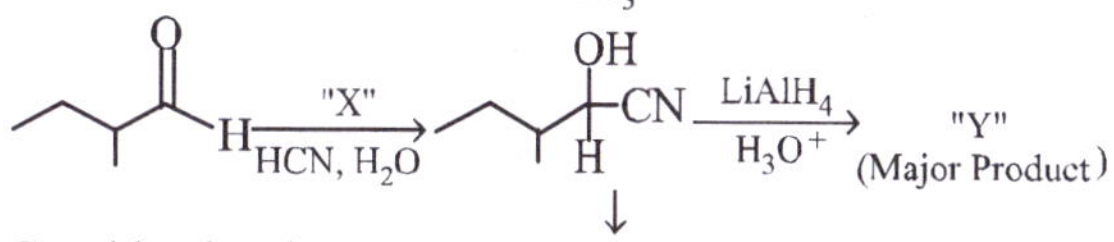

Consider the given reaction, Identify 'X' and 'Y' : **[Aug. 26, 2021 (II)]**

(1) X – NaOH Y – 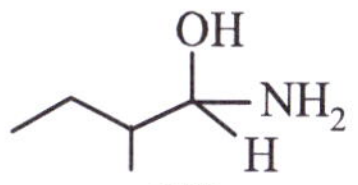

(2) X – HNO_3 Y – 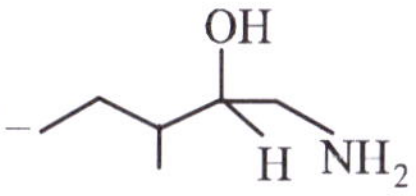

(3) X – NaOH Y – 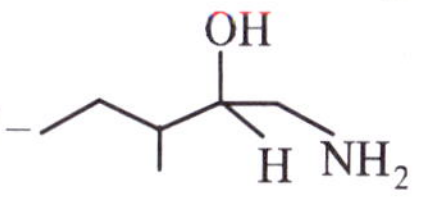

(4) X – HNO_3 Y – 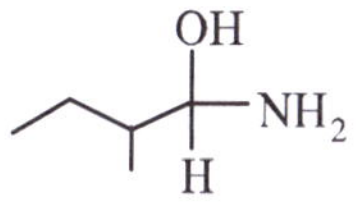

124.

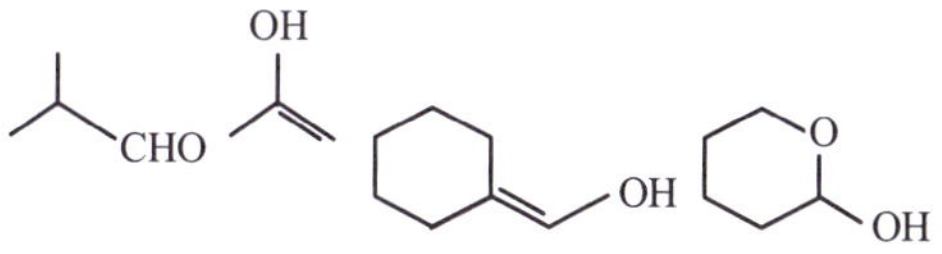

(I) (II) (III) (IV)

Which among the above compound/s does/do not form silver mirror when treated with Tollen's reagent? **[July 20, 2021 (I)]**

(1) (I), (III) and (IV) only (2) Only (IV)
(3) Only (II) (4) (III) and (IV) only

125.

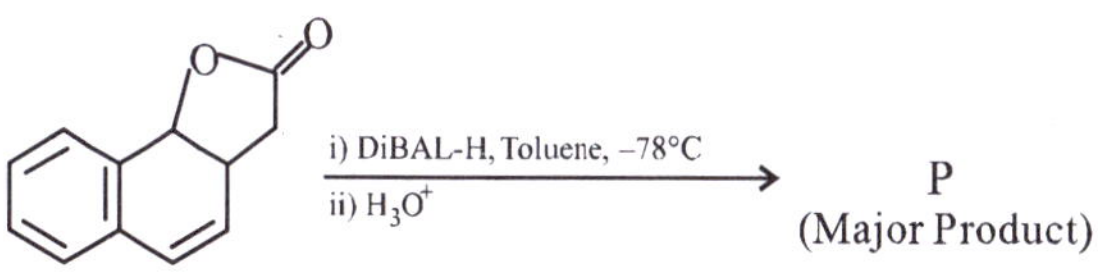

The product "P" in the above reaction is : **[March 16, 2021 (I)]**

(1) 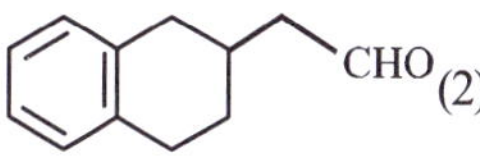 (2) 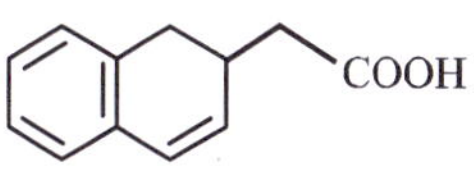

(3) 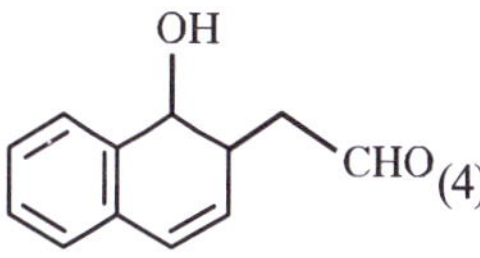 (4) 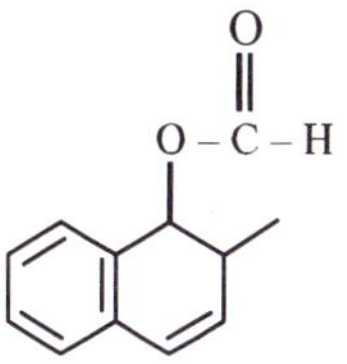

126. $\underset{(C_4H_8Cl_2)}{A} \xrightarrow[373K]{\text{Hydrolysis}} \underset{(C_4H_8O)}{B}$

B reacts with hydroxyl amine but does not give Tollen's test. Identify A and B. **[Feb. 26, 2021 (I)]**

(1) 2, 2-Dichlorobutane and Butanal
(2) 1, 1-Dichlorobutane and Butanal
(3) 1, 1 Dichlorobutane and 2-Butanone
(4) 2, 2-Dichlorobutane and Butan-2-one

127. 2,4-DNP test can be used to identify **[Feb. 26, 2021 (II)]**

(1) Aldehydes (2) Amines
(3) Ethers (4) Halogens

128. Which of the following reagent is used for the following reaction? **[Feb. 24, 2021 (I)]**

$$CH_3CH_2CH_3 \xrightarrow{?} CH_3CH_2CHO$$

(1) Manganese acetate
(2) Copper at high temperature and pressure
(3) Molybdenum oxide
(4) Potassium permanganate

129. Match list-I and list-II. **[Feb. 24, 2021 (II)]**

List-I	List-II
(A) $R-\overset{O}{\overset{\parallel}{C}}-Cl \rightarrow R-CHO$	(i) $Br_2/NaOH$
(B) $R-CH_2-COOH \rightarrow R-\underset{Cl}{\underset{\mid}{CH}}-COOH$	(ii) $H_2/Pd\text{-}BaSO_4$
(C) $R-\overset{O}{\overset{\parallel}{C}}-NH_2 \rightarrow R-NH_2$	(iii) Zn(Hg)/ Conc. HCl
(D) $R-\overset{O}{\overset{\parallel}{C}}-CH_3 \rightarrow R-CH_2-CH_3$	(iv) Cl_2/Red P, H_2O

Choose the correct answer from the options given below:

(1) (A)-(ii), (B)-(i), (C)-(iv), (D)-(iii)
(2) (A)-(iii), (B)-(iv), (C)-(i), (D)-(ii)
(3) (A)-(ii), (B)-(iv), (C)-(i), (D)-(iii)
(4) (A)-(iii), (B)-(i), (C)-(iv), (D)-(ii)

130. Which one of the following carbonyl compounds cannot be prepared by addition of water on an alkyne in the presence of $HgSO_4$ and H_2SO_4? **[Feb. 24, 2021 (II)]**

(1) $CH_3-\overset{O}{\overset{\parallel}{C}}-H$ (2) 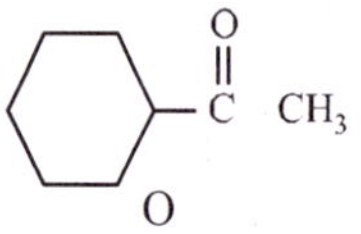

(3) $CH_3-CH_2-\overset{O}{\overset{\parallel}{C}}-H$ (4) $CH_3-\overset{O}{\overset{\parallel}{C}}-CH_2CH_3$

131. 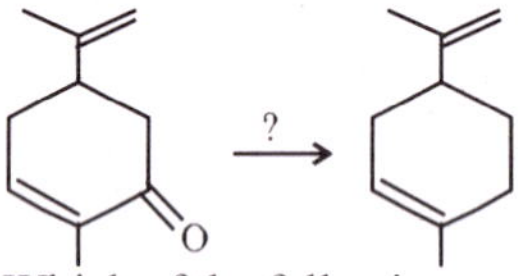 **[Feb. 24, 2021 (II)]**

Which of the following reagent is suitable for the preparation of the product in the above reaction?

(1) $NaBH_4$ (2) $NH_2-NH_2/C_2H_5\overset{\ominus}{O}\overset{\oplus}{Na}$
(3) Ni/H_2 (4) Red P + Cl_2

132. The major product [R] in the following sequence of reactions is : **[Sep. 04, 2020 (II)]**

$$HC \equiv CH \xrightarrow[\text{(ii) } (H_3C)((CH_3)_2CH)CH-Br]{\text{(i) } LiNH_2/\text{ether}} [P] \xrightarrow[\text{(ii) } NaBH_4]{\text{(i) } HgSO_4/H_2SO_4} [Q] \xrightarrow[\Delta]{\text{Conc. } H_2SO_4} [R]$$

(1) 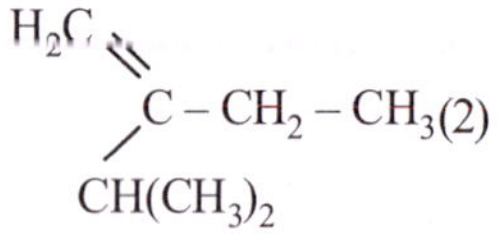 (2) 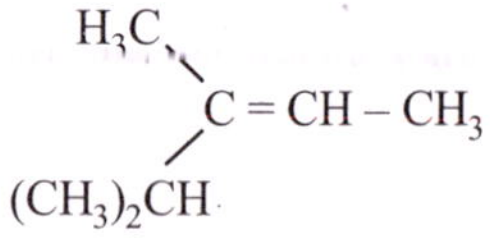

(3) 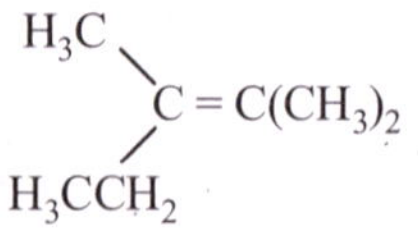 (4) 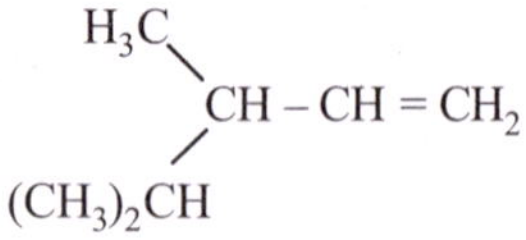

133. The major product [C] of the following reaction sequence will be : **[Sep. 04, 2020 (II)]**

$$CH_2=CH-CHO \xrightarrow[\text{(ii) } SOCl_2]{\text{(i) } NaBH_4} [A] \xrightarrow[\text{Anhy. } AlCl_3]{} [B] \xrightarrow{DBr} [C]$$

(1) 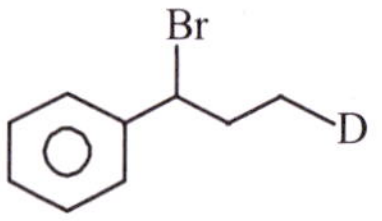 (2) 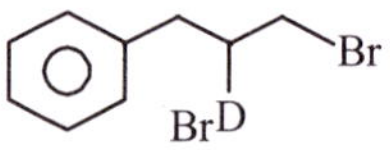

(3) 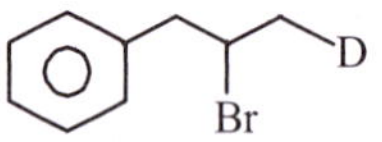 (4) 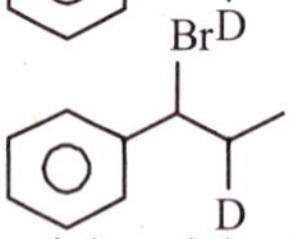

134. The increasing order of the reactivity of the following compounds in nucleophilic addition reaction is :
Propanal, Benzaldehyde, Propanone, Butanone **[Sep. 03, 2020 (II), Online May 7, 2012 (s)]**

(1) Benzaldehyde < Butanone < Propanone < Propanal
(2) Butanone < Propanone < Benzaldehyde < Propanal
(3) Propanal < Propanone < Butanone < Benzaldehyde
(4) Benzaldehyde < Propanal < Propanone < Butanone

135. The compound A in the following reactions is : **[Sep. 03, 2020 (II)]**

$$A \xrightarrow[\text{(ii) Conc. } H_2SO_4/\Delta]{\text{(i) } CH_3MgBr/H_2O} B \xrightarrow[\text{(ii) } Zn/H_2O]{\text{(i) } O_3} C + D$$

$C \xrightarrow[\text{(ii) } \Delta]{\text{(i) Conc. KOH}}$ 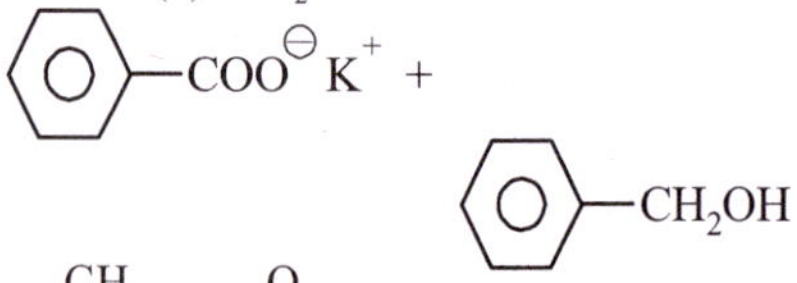

$$D \xrightarrow[\Delta]{Ba(OH)_2} H_3C-\overset{CH_3}{\overset{\mid}{C}}=CH-\overset{O}{\overset{\parallel}{C}}-CH_3$$

(1) $C_6H_5-\overset{O}{\overset{\parallel}{C}}-CH_3$ (2) $C_6H_5-\overset{O}{\overset{\parallel}{C}}-CH(CH_3)_2$

(3) $C_6H_5-CH_2-\overset{O}{\overset{\parallel}{C}}-CH_3$ (4) $C_6H_5-\overset{O}{\overset{\parallel}{C}}-CH_2CH_3$

136. The increasing order of the following compounds towards HCN addition is : **[Sep. 02, 2020 (I)]**

(i) 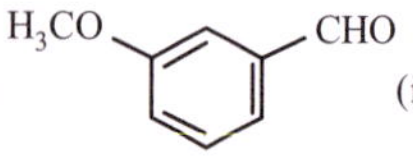 (ii) 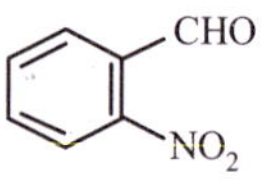

(iii) 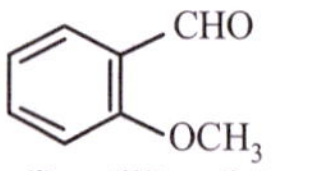 (iv) 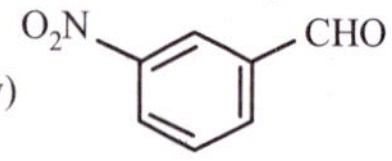

(1) (i) < (iii) < (iv) < (ii) (2) (iii) < (iv) < (i) < (ii)
(3) (iii) < (i) < (iv) < (ii) (4) (iii) < (iv) < (ii) < (i)

137. Identify (A) in the following reaction sequence: **[Jan. 09, 2020 (I)]**

(A) Gives positive iodoform test $\xrightarrow[\text{(ii) } H^+, H_2O \text{ (ii) Conc. } H_2SO_4/\Delta]{\text{(i) } CH_3MgBr}$ (B) $\xrightarrow{O_3/Zn,\ H_2O}$

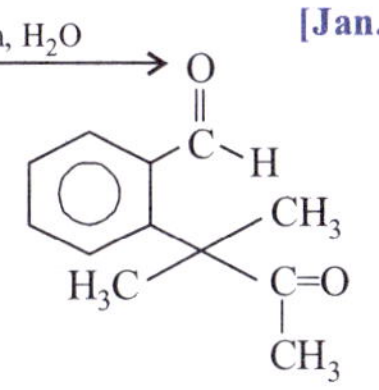

(1) 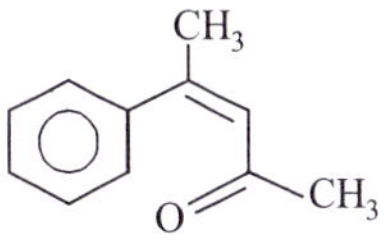 (2) 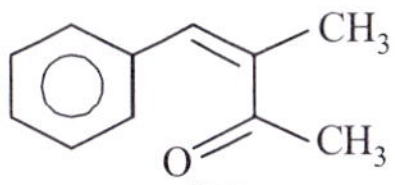

(3) 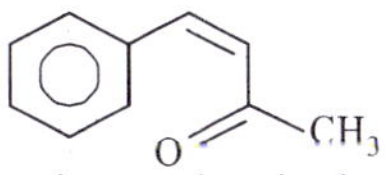 (4) 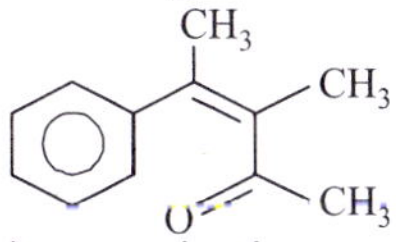

138. The major product in the following reaction is: **[Jan. 08, 2020 (II)]**

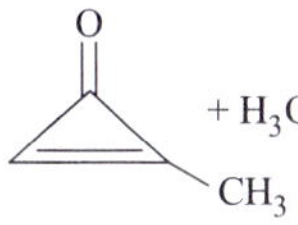 + $H_3O^{\oplus}$ $\longrightarrow$

(1) 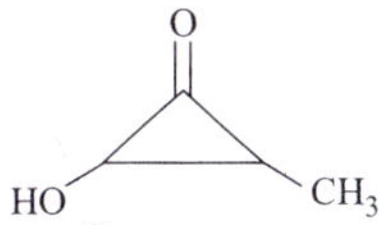 (2) 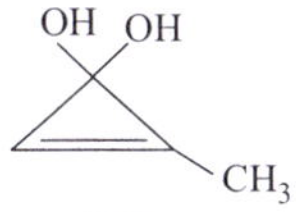

(3) 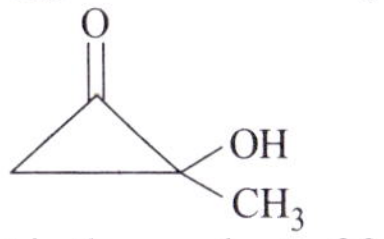 (4) 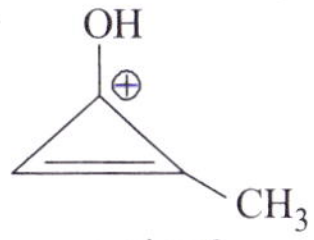

139. What is the product of following reaction?

Hex-3-ynal $\xrightarrow[\text{(ii) } PBr_3, \text{ (iii) Mg/ether, (iv) } CO_2/H_3O^+]{\text{(i) } NaBH_4}$?

[Jan. 07, 2020 (I)]

(1) 

(2) 

(3) 

(4) 

140. Major products of the following reaction are : **[April 10, 2019 (I)]**

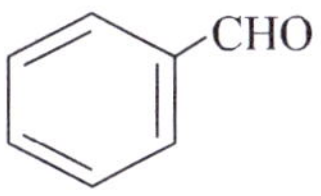 + HCHO $\xrightarrow[\text{(ii) } H_3O^+]{\text{(i) 50\% NaOH}}$

(1) 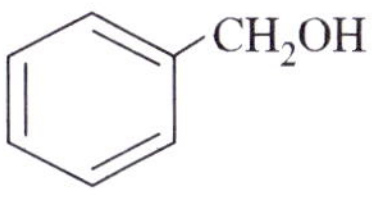 and 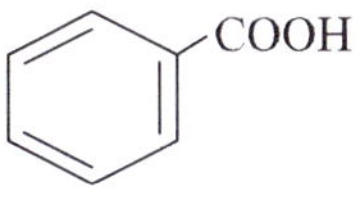

(2) CH_3OH and 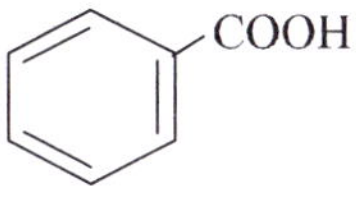

(3) HCOOH and 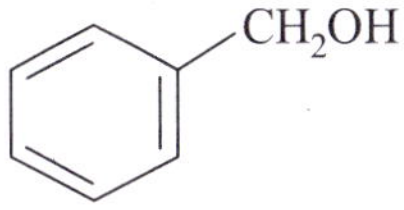

(4) CH_3OH and HCO_2H

141. Compound A($C_9H_{10}O$) shows positive iodoform test. Oxidation of A with $KMnO_4$/KOH gives acid B($C_8H_6O_4$). Anhydride of B is used for the preparation of phenolphthalein. Compound A is : **[April 10, 2019 (II)]**

(1) 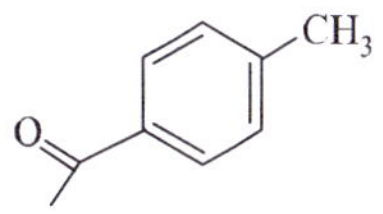 (2) 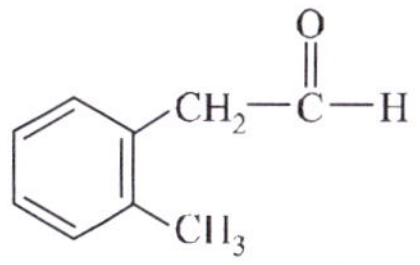

(3) 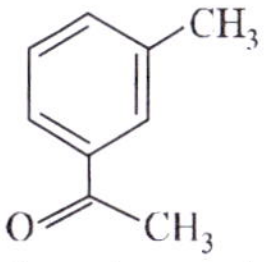 (4) 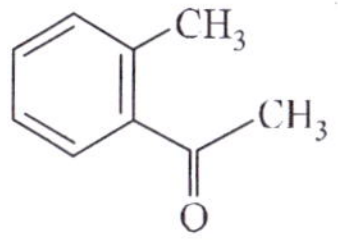

142. *p*-Hydroxybenzophenone upon reaction with bromine in carbon tetrachloride gives: **[April 9, 2019 (II)]**

(1) 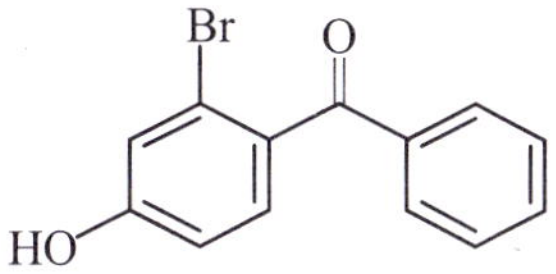

(2) 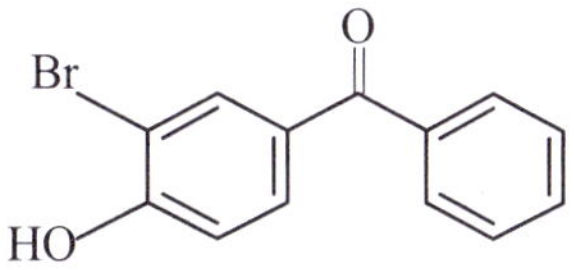

(3) 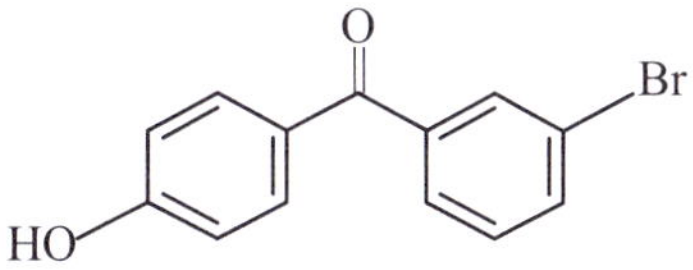

(4) 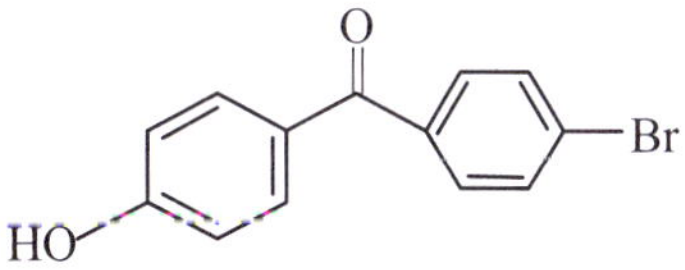

143. The major product of the following reaction is : **[April 8, 2019 (I)]**

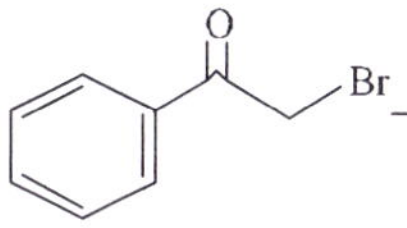 $\xrightarrow[\text{MeOH, } 25\ ^\circ C]{NaBH_4}$

(1) 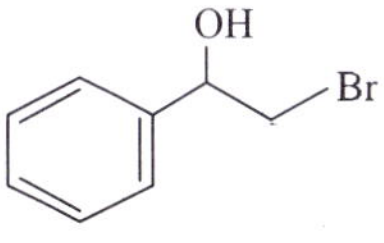 (2) 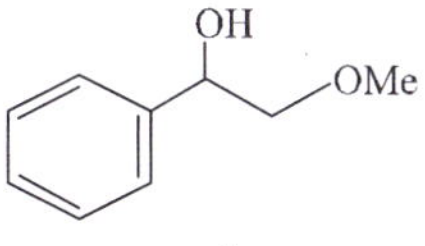

(3) 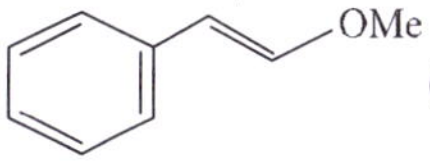 (4) 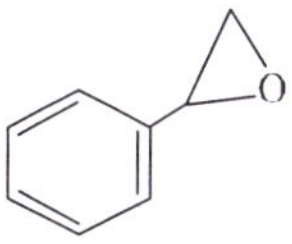

144. Which of the following compounds will show the maximum 'enol' content? **[April 8, 2019 (II)]**

(1) $CH_3COCH_2COOC_2H_5$ (2) $CH_3COCH_2COCH_3$

(3) CH_3COCH_3 (4) $CH_3COCH_2CONH_2$

145. An organic compound neither reacts with neutral ferric chloride solution nor with Fehling solution. It however, reacts with Grignard reagent and gives positive iodoform test. The compound is : **[April 8, 2019 (I)]**

(1) 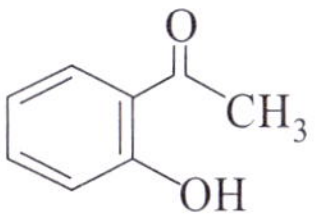 (2) 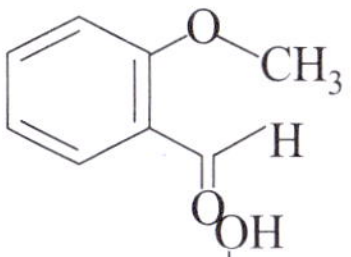

(3) 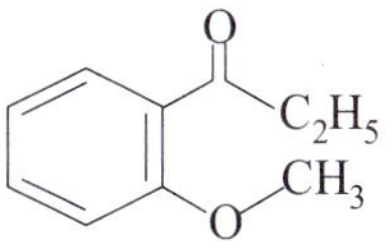 (4) 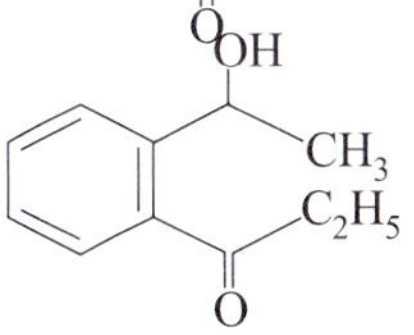

146. The major product obtained in the following reaction is : [April 8, 2019 (II)]

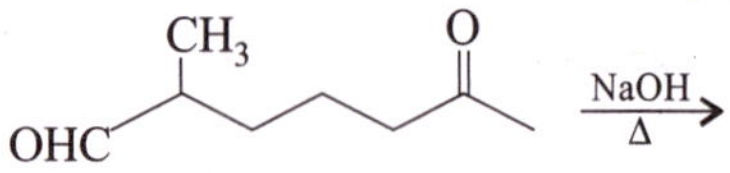

(1) 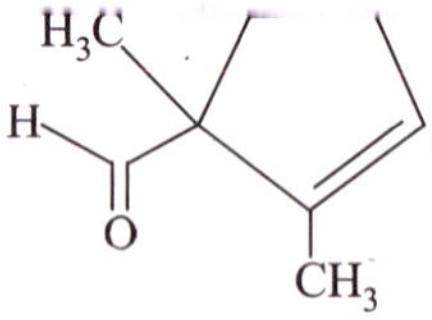 (2) 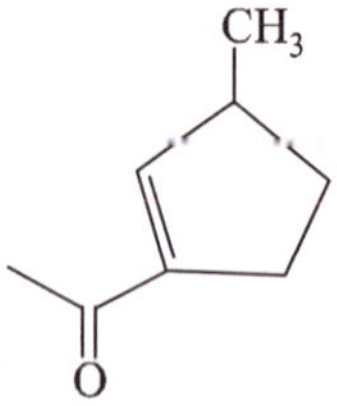

(3) 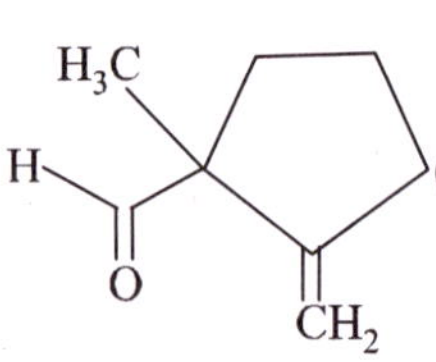 (4) 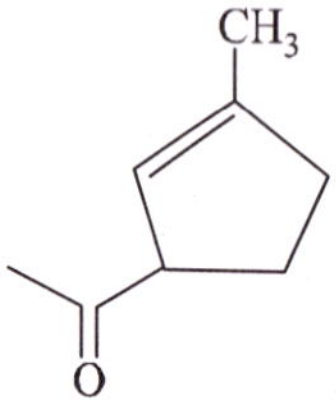

147. In the following reactions, products A and B are: [Jan. 12, 2019 (I)]

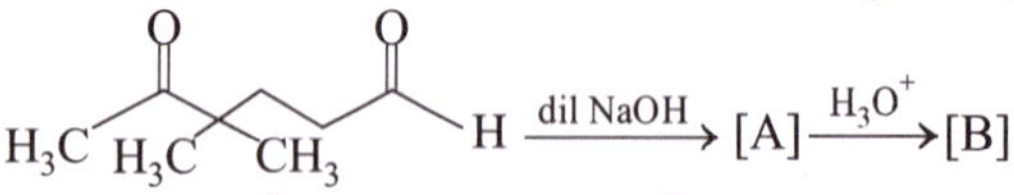

(1) 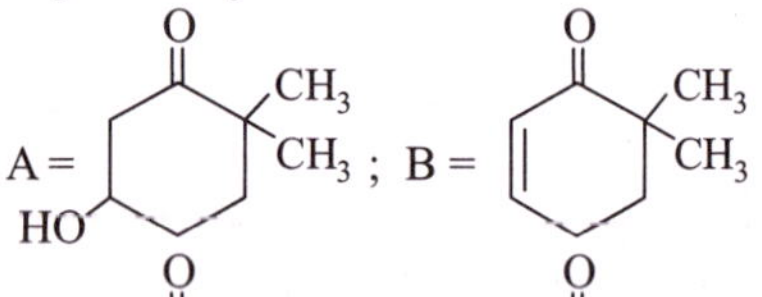

(2) 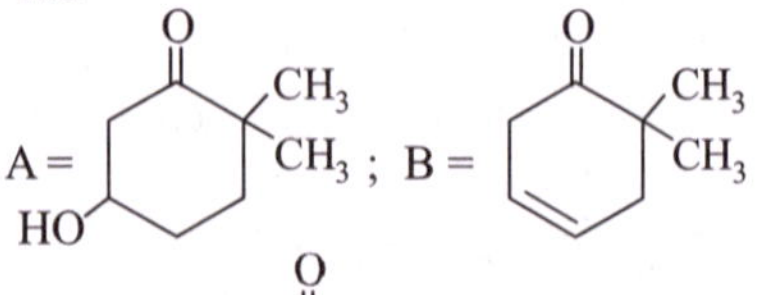

(3) 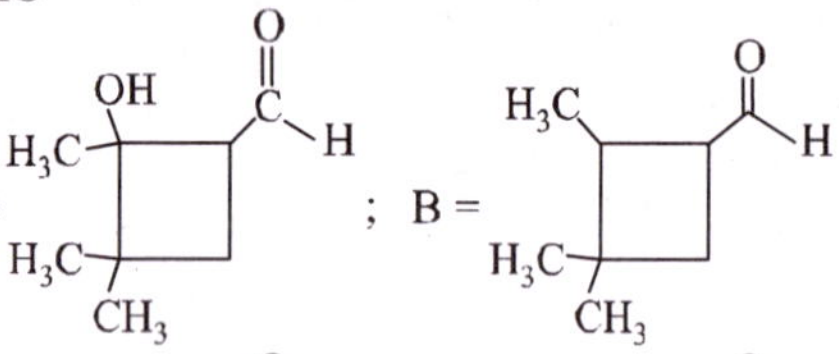

(4) 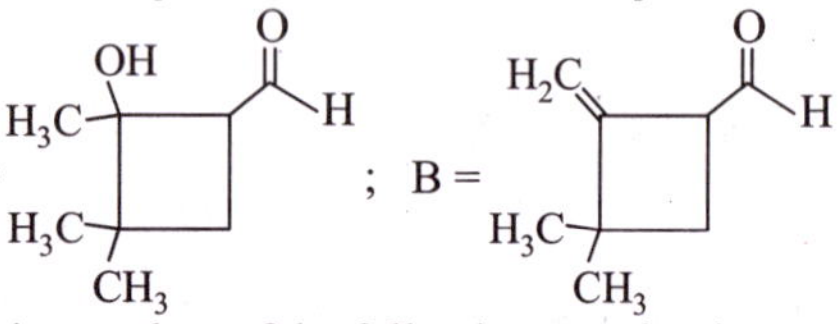

148. The major product of the following reaction is : [Jan. 12, 2019 (II)]

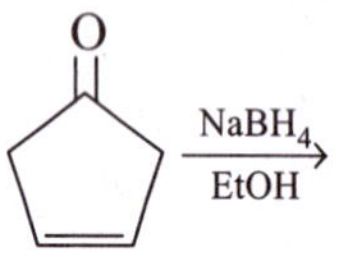

(1) 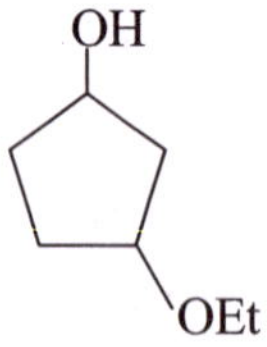 (2) 

(3) 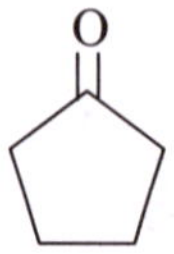 (4) 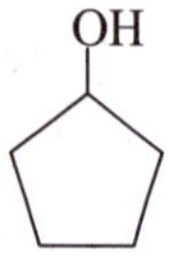

149. The aldehydes which will **not** form Grignard product with one equivalent of Grignard reagents are: [Jan. 12, 2019 (II)]

(A) 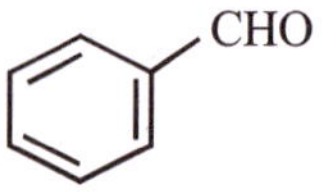

(B) 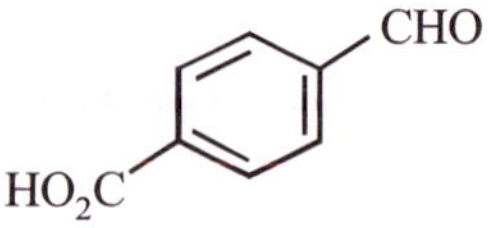

(C) 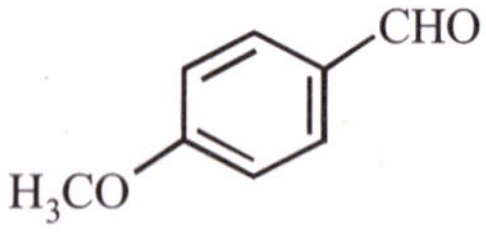

(D) 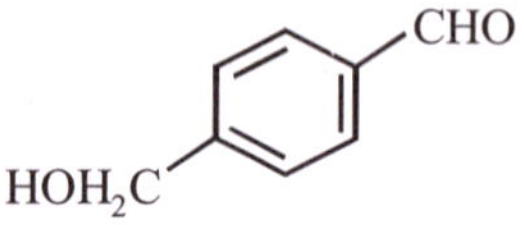

(1) (B), (D) (2) (B), (C)
(3) (B), (C), (D) (4) (C), (D)

150. The major product of the following reaction is: [Jan. 10, 2019 (II)]

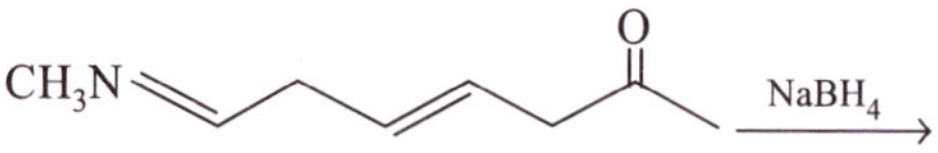

(1) 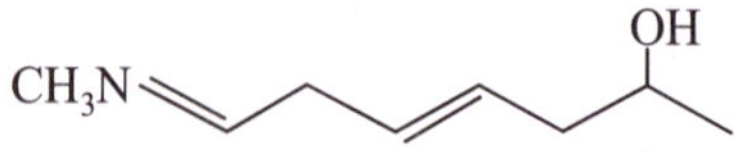

(2) 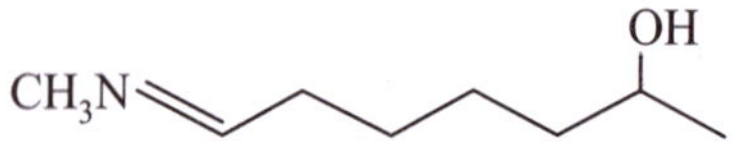

(3) 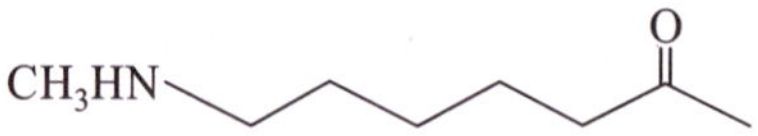

(4) 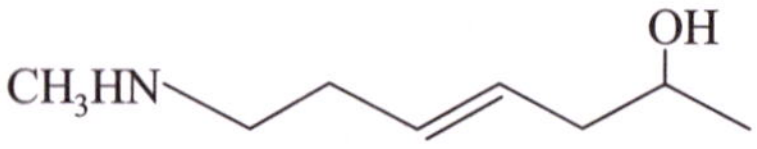

151. The major product formed in the following reaction is: [Jan. 9, 2019 (II)]

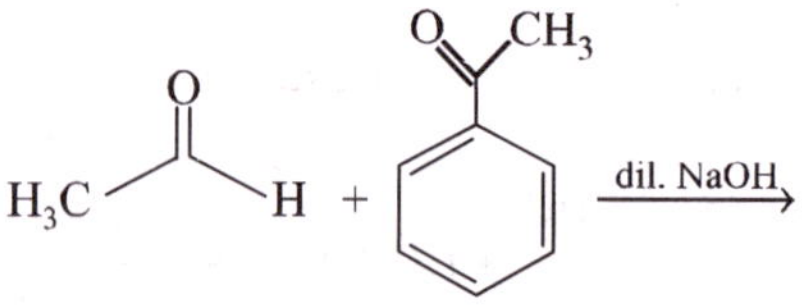

(1) 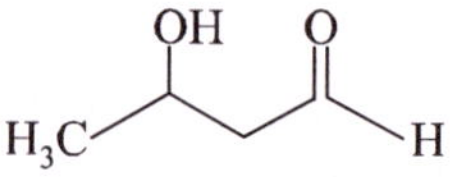

(2) 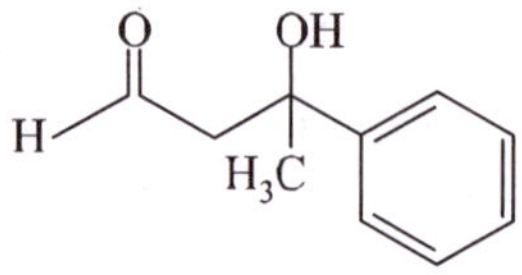

(3) 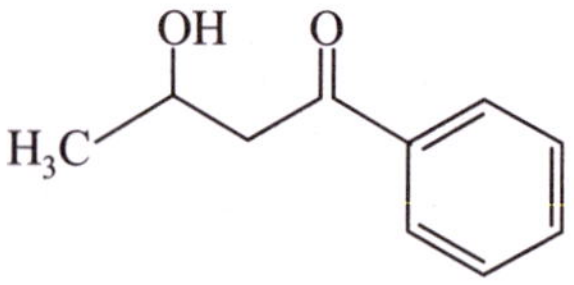

(4) 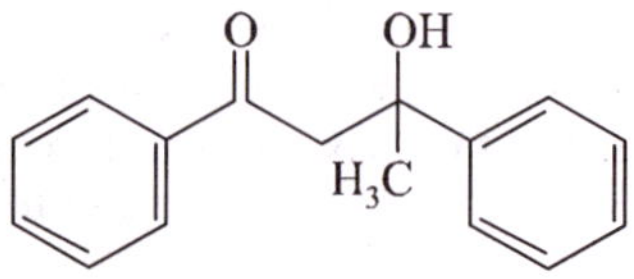

152. The major product B formed in the following reaction sequence is: [Online April 16, 2018]

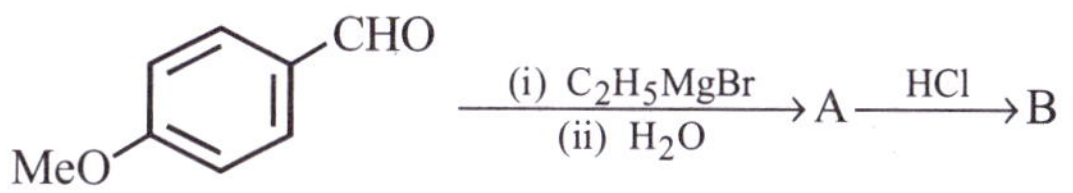

(1) 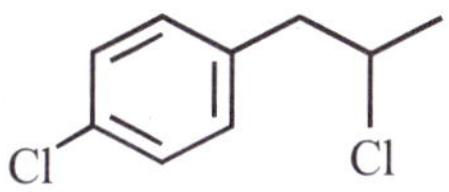

(2) 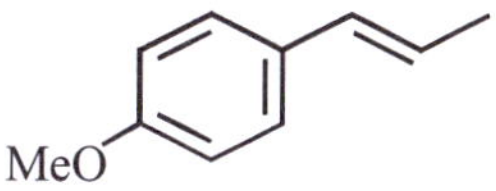

(3) 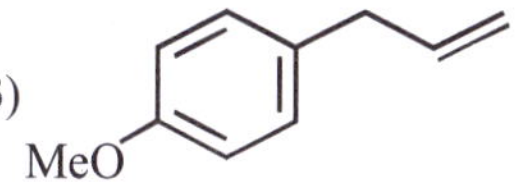

(4) 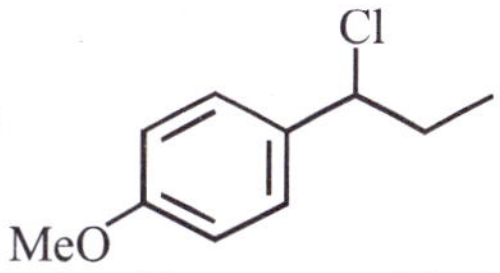

153. Which of the following compound will most readily be dehydrated to give alkene under acidic condition? [Online April 16, 2018]
(1) 4-Hydroxypentan-2-one
(2) 3-Hydroxypentan-2-one
(3) 1-Pentanol
(4) 2-Hydroxycyclopentanone

154. The correct sequence of reagents for the following conversion will be : [2017]

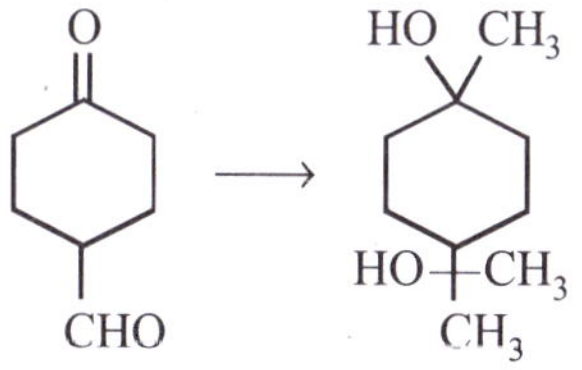

(1) $[Ag(NH_3)_2]^+ OH^-$, H^+/CH_3OH, CH_3MgBr
(2) CH_3MgBr, H^+/CH_3OH, $[Ag(NH_3)_2]^+ OH^-$
(3) CH_3MgBr, $[Ag(NH_3)_2]^+ OH^-$, H^+/CH_3OH
(4) $[Ag(NH_3)_2]^+ OH^-$, CH_3MgBr, H^+/CH_3OH

155. A compound of molecular formula $C_8H_8O_2$ reacts with acetophenone to form a single cross-aldol product in the presence of base. The same compound on reaction with conc. NaOH forms benzyl alcohol as one of the products. The structure of the compound is: [Online April 9, 2017]

(1) 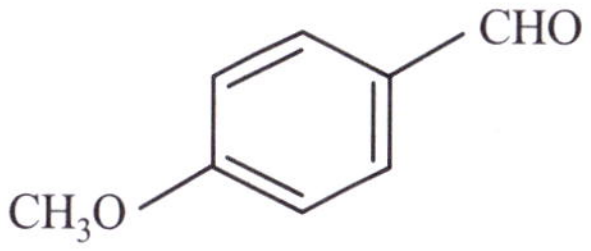

(2) 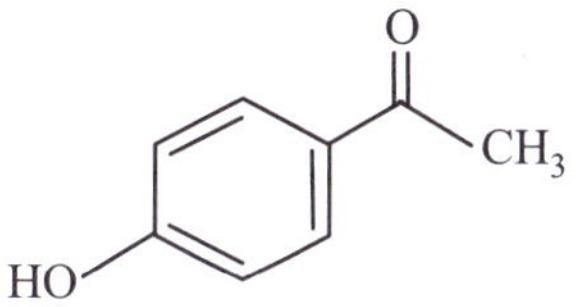

(3) 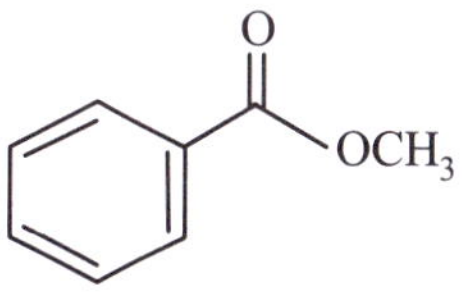

(4) 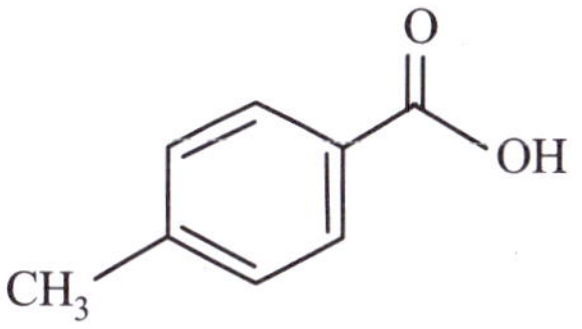

156. The correct statement about the synthesis of pentaerythritol ($C(CH_2OH)_4$) used in the preparation of PETN is : [Online April 10, 2016]
(1) Thy synthesis requires three aldol condensations and one Cannizzaro reaction.
(2) Alpha hydrogens of ethanol and methanol are involved in this reaction.
(3) The synthesis requires two aldol condensations and two Cannizzaro reactions.
(4) The synthesis requires four aldol condensations between methanol and ethanol.

157. In the reaction sequence [Online April 11, 2015]

$$2CH_3CHO \xrightarrow{OH^-} A \xrightarrow{\Delta} B$$; the product B is :

(1) $CH_3-CH_2-CH_2-CH_2-OH$
(2) $CH_3-CH=CH-CHO$
(3) $CH_3-CH_2-CH_2-CH_3$
(4) $CH_3-\overset{O}{\overset{\|}{C}}-CH_3$

158. Which one of the following reactions will not result in the formation of carbon-carbon bond? [Online April 9, 2014]
(1) Reimer-Tiemann reaction
(2) Friedel Craft's acylation
(3) Wurtz reaction
(4) Cannizzaro reaction

159. Which is major product formed when acetone is heated with iodine and potassium hydroxide? [Online April 9, 2014]
(1) Iodoacetone (2) Acetic acid
(3) Iodoform (4) Acetophenone

160. An organic compound A, C_5H_8O; reacts with H_2O, NH_3 and CH_3COOH as described below: [Online April 11, 2014]

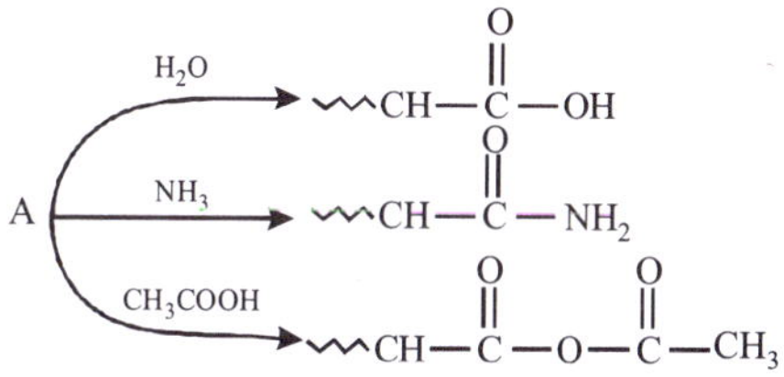

A is:
(1) $CH_3CH=\underset{CH_3}{\underset{|}{C}}-CHO$
(2) $CH_2=CH\,\underset{CH_3}{\underset{|}{CH}}-CHO$
(3) $CH_3-CH_2-\underset{CH_3}{\underset{|}{C}}=C=O$
(4) $CH_3-CH_2-\underset{CH_2H}{\underset{\|}{C}}-\underset{|}{C}=O$

161. Tischenko reaction is a modification of: [Online April 11, 2014]
(1) Aldol condensation (2) Claisen condensation
(3) Cannizzaro reaction (4) Pinacol-pinacolone reaction

162. Formaldehyde can be distinguished from acetaldehyde by the use of : [Online April 9, 2013]
(1) Schiff's reagent (2) Tollen's reagent
(3) I_2/Alkali (4) Fehling's solution

163. Clemmensen reduction of a ketone is carried out in the presence of : [Online April 22, 2013]
(1) $LiAlH_4$ (2) Zn-Hg with HCl
(3) Glycol with KOH (4) H_2 with Pt as catalyst

164. Which of the following is the product of aldol condensation? [Online April 23, 2013]

(1) 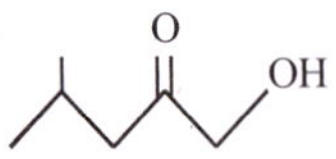 (2) 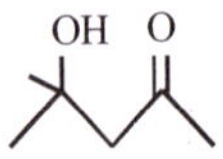

(3)  (4) 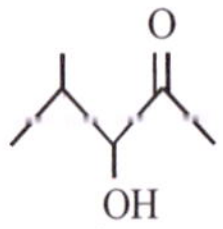

165. Cannizzaro's reaction is not given by : [Online April 25, 2013]

(1) 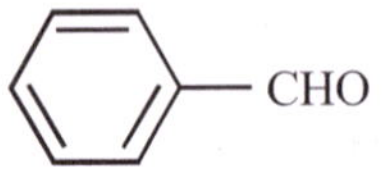 (2) 

(3) CH_3CHO (4) HCHO

166. Which of the following reagent(s) is/are used for the conversion? [Online April 25, 2013]

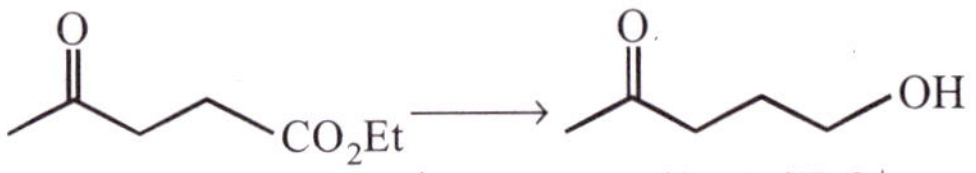

(1) glycol/$LiAlH_4$/H_3O^+ (2) glycol/NaH/H_3O^+

(3) $LiAlH_4$ (4) $NaBH_4$

167. Iodoform can be prepared from all except : [2012]

(1) Ethyl methyl ketone (2) Isopropyl alcohol

(3) 3-Methyl 2-butanone (4) Isobutyl alcohol

168. In the given transformation, which of the following is the most appropriate reagent ? [2012]

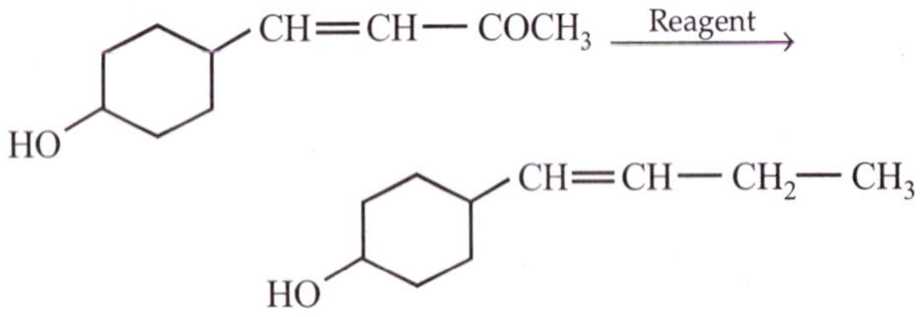

(1) $NH_2NH_2, \bar{O}H$ (2) Zn – Hg/ HCl

(3) Na, Liq. NH_3 (4) $NaBH_4$

169. Among the following the order of reactivity towards nucleophilic addition is [Online May 7, 2012]

(1) $CH_3CHO > CH_3COCH_3 > HCHO$

(2) $HCHO > CH_3CHO > CH_3COCH_3$

(3) $CH_3CHO > HCHO > CH_3COCH_3$

(4) $CH_3COCH_3 > CH_3CHO > HCHO$

170. The reaction, [Online May 12, 2012]

$$CH_3CHO \xrightarrow{Zn(Hg)/Conc.\ HCl[H]} CH_3CH_3 \text{ is}$$

(1) Cannizzaro's reaction (2) Rosenmund reduction

(3) Wolf-Kishner reduction (4) Clemmenson reduction

171. Tollen's reagent and Fehling solutions are used to distinguish between [Online May 26, 2012]

(1) acids and alcohols

(2) alkanes and alcohols

(3) ketones and aldehydes

(4) *n*-alkanes and branched alkanes

172. In Cannizzaro reaction given below

$$2PhCHO \xrightarrow{:\bar{O}H} PhCH_2OH + PhC\bar{O}_2$$

the slowest step is : [2009]

(1) the transfer of hydride to the carbonyl group

(2) the abstraction of proton from the carboxylic group

(3) the deprotonation of Ph CH_2OH

(4) the attack of : $\bar{O}H$ at the carboxyl group

173. The increasing order of the rate of HCN addition to compound A to D is [2006]

(A) HCHO (B) CH_3COCH_3

(C) $PhCOCH_3$ (D) PhCOPh

(1) $D < C < B < A$ (2) $C < D < B < A$

(3) $A < B < C < D$ (4) $D < B < C < A$

174. Reaction of cyclohexanone with dimethylamine in the presence of catalytic amount of an acid forms a compound if water during the reaction is continuously removed. The compound formed is generally known as [2005]

(1) an amine (2) an imine

(3) an enamine (4) a Schiff's base

175. On mixing ethyl acetate with aqueous sodium chloride, the composition of the resultant solution is [2004]

(1) $CH_3COCl + C_2H_5OH + NaOH$

(2) $CH_3COONa + C_2H_5OH$

(3) $CH_3COOC_2H_5 + NaCl$

(4) $CH_3Cl + C_2H_5COONa$

176. Which one of the following is reduced with zinc amalgam and hydrochloric acid to give the corresponding hydrocarbon? [2004]

(1) Acetamide (2) Acetic acid

(3) Ethyl acetate (4) Butan-2-one

177. Which one of the following undergoes reaction with 50% sodium hyroxide solution to give the corresponding alcohol and acid? [2004]

(1) Butanal (2) Benzaldehyde

(3) Phenol (4) Benzoic acid

178. Acetyl bromide reacts with excess of CH_3MgI followed by treatment with a saturated solution of NH_4Cl gives [2004]

(1) 2-methyl-2-propanol (2) acetamide

(3) acetone (4) acetyl iodide

179. On vigorous oxidation by permanganate solution. [2002]

$(CH_3)_2C = CH - CH_2 - CHO$ gives

(1) $CH_3 - \underset{CH_3}{\overset{OH}{C}} - \overset{OH}{CH} - CH_2CHO$

(2) $(CH_3)_2C = O + HOOCCH_2COOH$

(3) $(CH_3)_2C = O + OHCCH_2COOH$

(4) $(CH_3)_2C = O + OHCCH_2CHO$

NUMERICAL VALUE QUESTIONS (NVQs)

180. 'x' is the product which is obtained from benzene by reacting it with carbon monoxide and hydrogen chloride in the presence of cuprous chloride. 'y' is the major product obtained from the benzene by reacting it with ethanoyl chloride in the presence of anhydrous $AlCl_3$. Product (major) obtained by heating x and y in the presence of alkali is z. Total number of π (pi) electrons in z is _______. [April 08, 2026 (II)]

181. 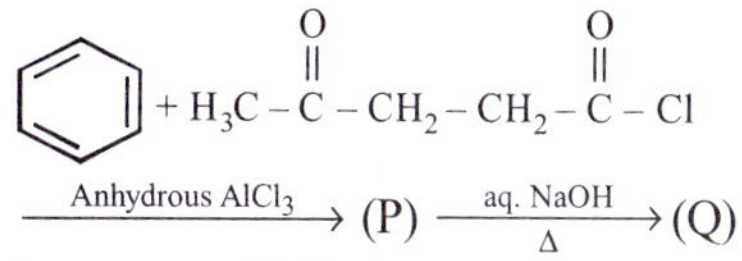

In compound (Q), the percentage of oxygen is _____%. (Nearest integer) [Jan. 23, 2026 (II)]

182. Identify the structure of the final product (D) in the following sequence of the reactions : [April 7, 2025 (II)]

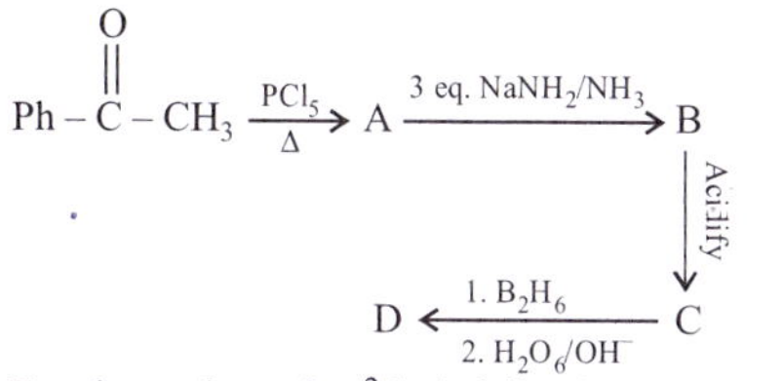

Total number of sp^2 hybridised carbon atoms in product D is.

183. In the Claisen-Schmidt reaction to prepare, dibenzalacetone from 5.3 g benzaldehyde, a total of 3.51 g of product was obtained. The percentage yield in this reaction was ______ %. [Jan. 29, 2025 (II)]

184. 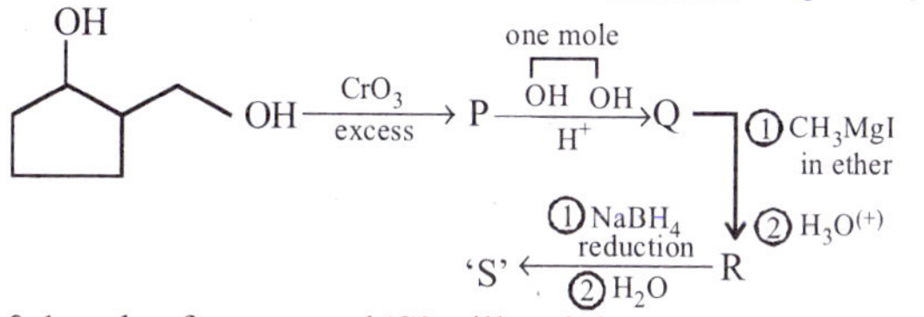

0.1 mole of compound 'S' will weigh_____g.
(Given molar mass in g mol^{-1} C:12, H:1, O:16) [Jan. 29, 2025 (I)]

185. In the Claisen-Schmidt reaction to prepare 351 g of dibenzalacetone using 87 g of acetone, the amount of benzaldehyde required is __________g. (Nearest integer) [April 5, 2024 (II)]

186. The product of the following reaction is P.

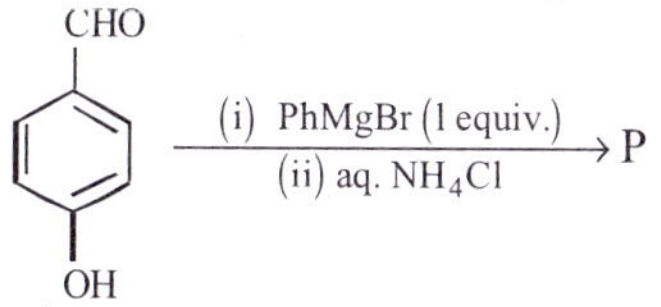

The number of hydroxyl groups present in the product P is ______. [Jan. 31, 2024 (I)]

187. The compound formed by the reaction of ethanal with semicarbazide contains ______ number of nitrogen atoms. [Jan. 30, 2024 (I)]

188. The mass of NH_3 produced when 131.8 kg of cyclohexanecarbaldehyde undergoes Tollen's test is ______ kg. (Nearest integer) [April 12, 2023 (I)]
Molar Mass of C = 12 g/mol, N = 14 g/mol, O = 16 g/mol

189. 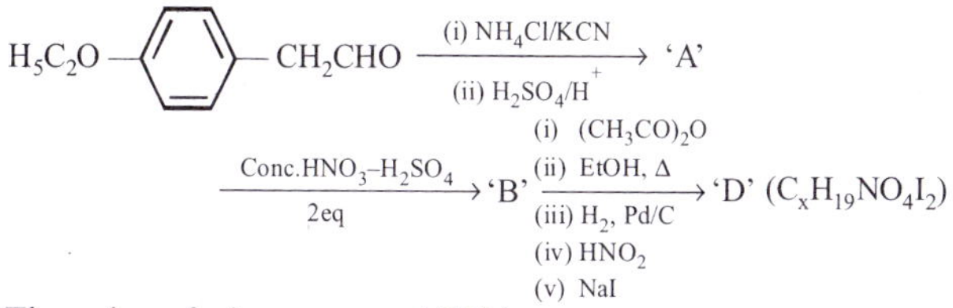

The value of x in compound 'D' is _____ [April 12, 2023 (I)]

190. 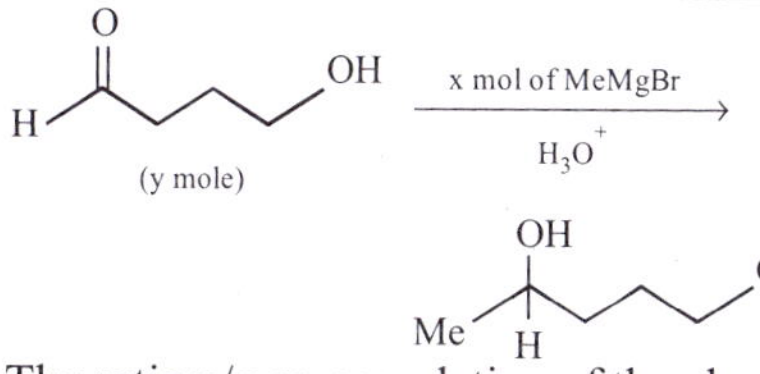

The ratio x/y on completion of the above reaction is ______. [April 11, 2023 (I)]

191. Among the following, the number of compounds which will give positive iodoform reaction is _______. [April 06, 2023 (II)]
(a) 1-Phenylbutan-2-one (b) 2-Methylbutan-2-ol
(c) 3-Methylbutan-2-ol (d) 1-Phenylethanol
(e) 3,3-dimethylbutan-2-one (f) 1-Phenylpropan-2-ol

192. Number of isomeric compounds with molecular formula $C_9H_{10}O$ which (i) do not dissolve in NaOH (ii) do not dissolve in HCl. (iii) do not give orange precipitate with 2,4-DNP (iv) on hydrogenation give identical compound with molecular formula $C_9H_{12}O$ is __________. [Feb. 01, 2023 (I)]

193. The number of molecules which gives haloform test among the following molecules is ____ . [Jan. 31, 2023 (II)]

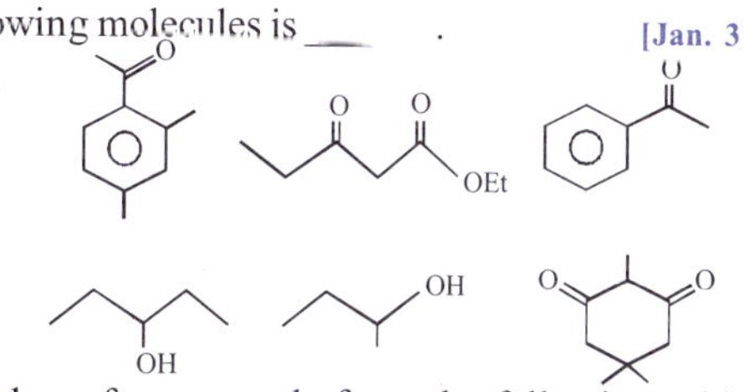

194. Number of compounds from the following which will not dissolve in cold $NaHCO_3$ and NaOH solutions but will dissolve in hot NaOH solution is ______ . [Jan. 30, 2023 (II)]

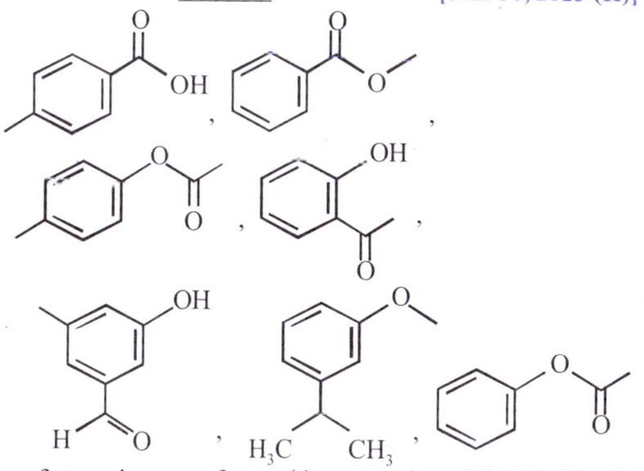

195. The number of stereoisomers formed in a reaction of (±) Ph(C=O)C(OH)(CN)Ph with HCN is_____. [July 29, 2022 (II)]

196. $C_6H_{12}O_6 \xrightarrow{\text{Zymase}} A \xrightarrow[\Delta]{\text{NaOI}} B + CHI_3$
The number of carbon atoms present in the product B is __________. [June 29, 2022 (I)]

197. In the given reaciton

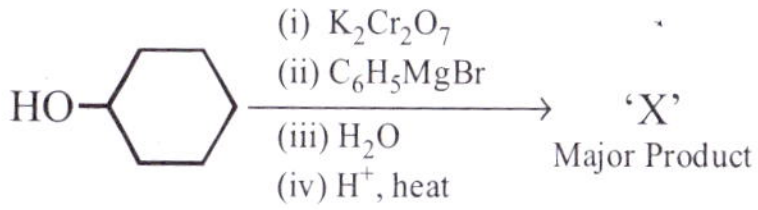

the number of sp^2 hybridised carbon(s) in compound 'X' is _________. [June 29, 2022 (II)]

198. In the given reaction

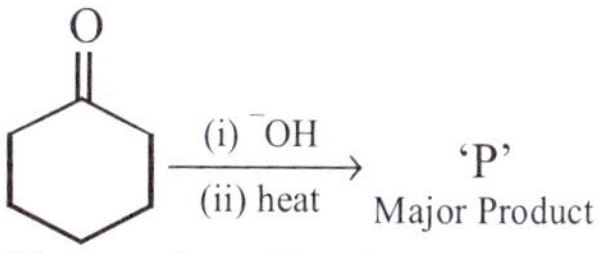

The number of π electrons present in the product 'P' is _____. [June 29, 2022 (II)]

199. Number of electrophilic centre in the given compound is ________ [June 24, 2022 (I)]

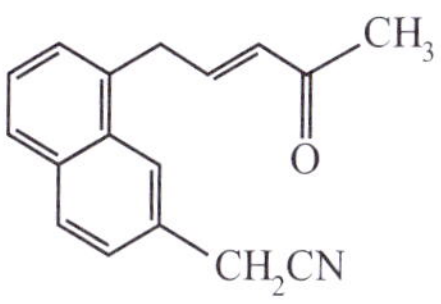

200. The number of nitrogen atoms in a semicarbazone molecule of acetone is ________. **[July 20, 2021 (I)]**

201. In Tollen's test for aldehyde, the overall number of electrons(s) transferred to the Tollen's reagent formula $[Ag(NH_3)_2]^+$ per aldehyde group to form silver mirror is ________. (Round off to the nearest integer). **[March 18, 2021 (II)]**

202. The total number of C – C sigma bonds in mesityl oxide ($C_6H_{10}O$) is ________. (Round off to the nearest integer). **[March 17, 2021 (II)]**

203. Consider the following reactions

$$A \xrightarrow[\text{(ii)}H_3O^+]{\text{(i)}CH_3MgBr} B \xrightarrow[573\text{ K}]{Cu} \text{2-Methyl-2-butene}$$

The mass percentage of carbon in A is ______. **[Jan. 09, 2020 (II)]**

Topic 3 - Preparation and Properties of Carboxylic Acids and their Derivatives

MULTIPLE CHOICE QUESTIONS (MCQs)

204. Consider the following sequence of reactions

 $-OH \xrightarrow[\text{(ii)}H^+,\ PhCOOH]{\text{(i)}Cu/573K} P$

The major product P is : **[April 06, 2026 (I)]**

(1) 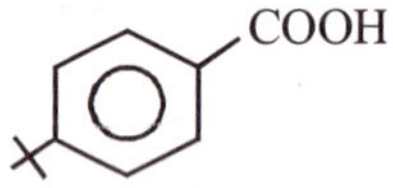

(2) 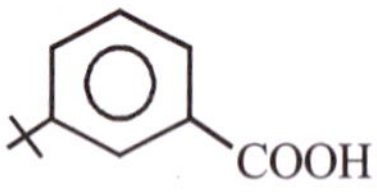

(3) 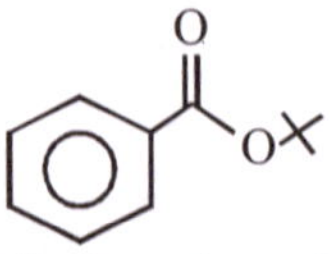

(4) 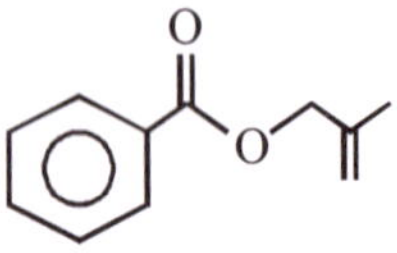

205. P" is hydrocarbon of molecular formula : C_8H_{14}.
On ozonolysis, "P" forms "Q". "Q" on treatment with alkali under reflux condition produces. "R", which on treatment with I_2/NaOH gives a yellow precipitate. Acidification of the solution gives "S".
The structure of "S" given below : **[April 05, 2026 (I)]**

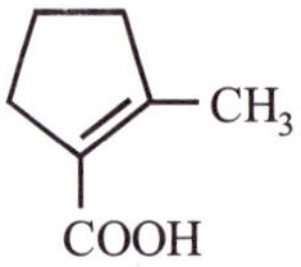

(1) 

(2) 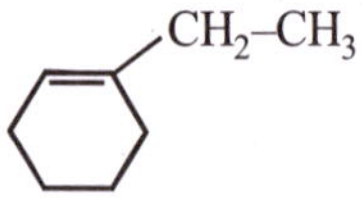

(3) 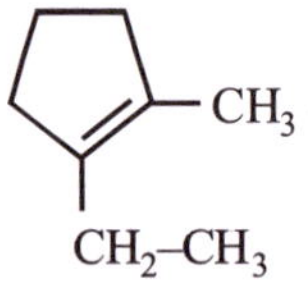

(4) 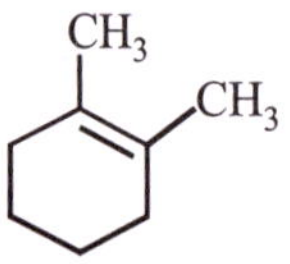

206. The descending order of acidity among the following compounds is

A. 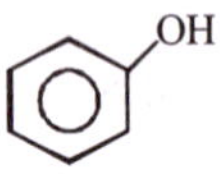

B. 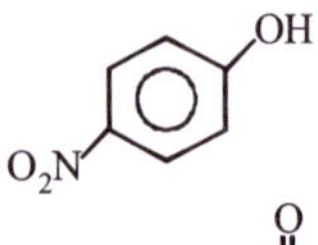

C. 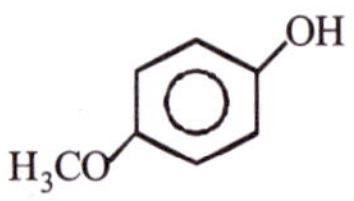

D. 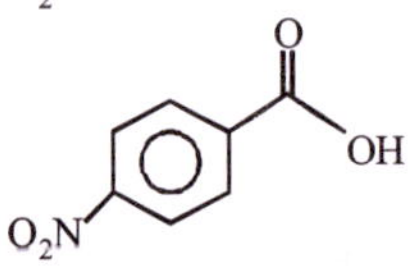

E. 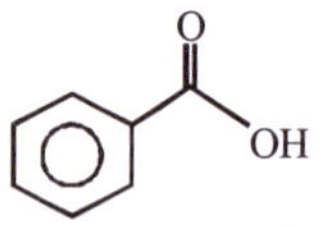

Choose the correct answer from the options given below. **[April 04, 2026 (II)]**

(1) B > D > E > A > C (2) D > B > E > A > C
(3) C > A > B > D > E (4) D > E > B > A > C

207. Amongst the following, the total number of compounds soluble in aqueous NaOH at room temperature is: **[April 04, 2026 (I)]**

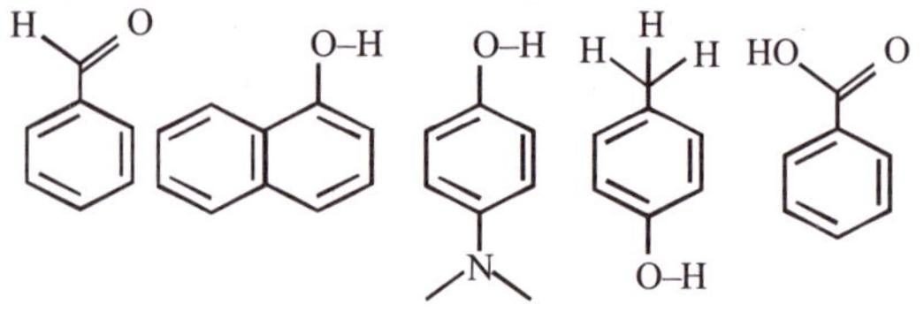

(I) (II) (III) (IV) (V)

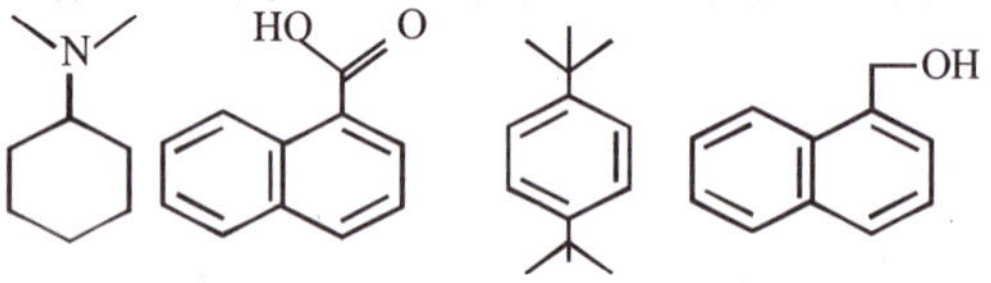

(VI) (VII) (VIII) (IX)

(1) 5 (2) 4 (3) 6 (4) 3

208. An organic compound "x" where molar ratio of C, O and H are equal, on treatment with 50% KOH under reflux followed by acidification produced "y". The most likely structure of "y" is: [Molar mass of 'x' is 58 g mol^{-1}] **[April 02, 2026 (II)]**

(1) $CH_2 = CH - \overset{O}{\overset{\|}{C}} - OH$ (2) $CH_3-CH=CH-CH=O$

(3) $O = \underset{CH_2-OH}{\underset{|}{C}} - OH$ (4) $CH_3 - \overset{O}{\overset{\|}{C}} - OH$

209. The correct order of acidic strength of the major products formed in the given reactions, is : **[Jan. 28, 2026 (II)]**

A. $PhNH_2 \xrightarrow[\text{(2) CuCN} \quad \text{(3) } H_3O^+/\Delta]{\text{(1) }NaNO_2+HCl\,(<5°C)} [A]$

B. $CH_3CH_2CHO \xrightarrow[\Delta]{[Ag(NH_3)_2]^+, OH^-} [B]$

C. $CH_4 + O_2 \xrightarrow[\text{(ii) } Na_2Cr_2O_7/H^+]{\text{(i) } Mo_2O_3} [C]$

D. $PhCH_2MgBr + CO_2 \xrightarrow[H_3O^+]{\text{Dry ether}} [D]$

Choose the **correct** answer from the options given below:

(1) C > B > A > D (2) A > D > C > B
(3) A > D > B > C (4) C > A > D > B

210. Consider the following two reactions A and B. **[Jan. 24, 2026 (I)]**

(A) 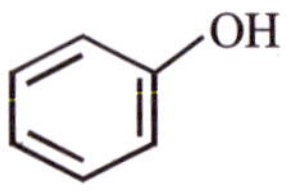

$\xrightarrow{Na}$ Main Product + gas(x)

(B) 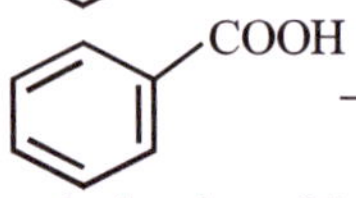

$\xrightarrow{NaHCO_3}$ Main Product + gas(y)

Numerical value of [molar mass of x + molar mass of y] is ______.

(1) 4 (2) 88 (3) 46 (4) 160

211. Given below are two statements: **[Jan. 21, 2026 (II)]**

Statement-I: Compound (X), shown below, dissolves in $NaHCO_3$ solution and has two chiral carbon atoms

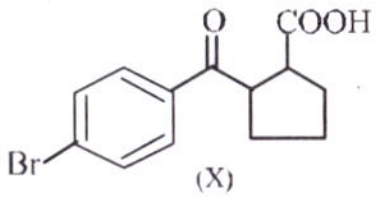

(X)

Statement-II: Compound (Y), shown below, has two carbons with sp^3 hybridization, one carbon with sp^2 and one carbon with sp hybridization

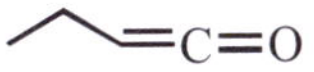

(Y)

In the light of the above statements, choose the **correct** answer from the options given below:

(1) Statement I is true but Statement II is false.
(2) Statement I is false but Statement II is true.
(3) Both Statement I and Statement II are true.
(4) Both Statement I and Statement II are false.

212. Mixture of 1 g each of chlorobenzene, aniline and benzoic acid is dissolved in 50 mL ethyl acetate and placed in a separating funnel, 5 M NaOH (30 mL) was added in the same funnel. The funnel was shaken vigorously and then kept aside. The ethyl acetate layer in the funnel contains : **[April 7, 2025 (II)]**

(1) benzoic acid
(2) benzoic acid and aniline
(3) benzoic acid and chlorobenzene
(4) chlorobenzene and aniline

213. Which of the following compounds is least likely to give effervescence of CO_2 in presence of aq. $NaHCO_3$? **[April 7, 2025 (I)]**

(1) 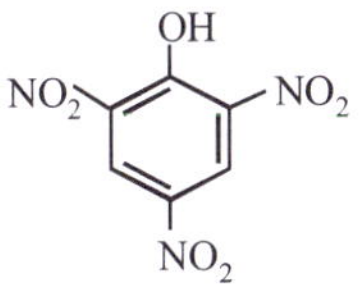

(2) 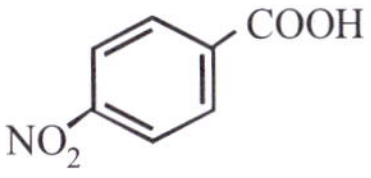

(3) $\overset{(+)\,(-)}{Ph-NH_3Cl}$

(4) 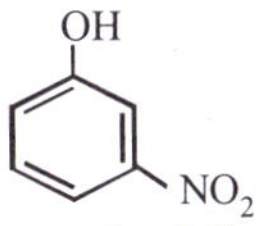

214. The least acidic compound, among the following is: **[April 3, 2025 (I)]**

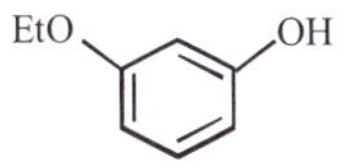

(A)

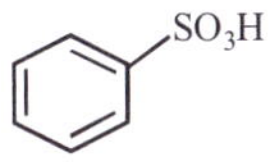

(B)

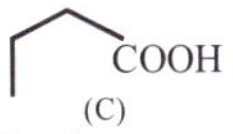

(C)

$EtO_2C-C\equiv C-H$

(D)

(1) D (2) A (3) B (4) C

215. Consider the following reactions. From these reactions which reaction will give carboxylic acid as a major product ? **[April 2, 2025 (II)]**

(A) $R-C\equiv N \xrightarrow[\text{mild condition}]{(i)\,H^+/H_2O}$

(B) $R-MgX \xrightarrow[(ii)\,H_3O^+]{(i)\,CO_2}$

(C) $R-C\equiv N \xrightarrow[(ii)\,H_3O^+]{(i)\,SnCl_2/HCl}$

(D) $R\cdot CH_2\cdot OH \xrightarrow{PCC}$

(E) 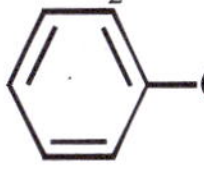 $-COCl \xrightarrow[(ii)\,Br_2\,water]{(i)\,H_2|Pd-BaSO_4}$

Choose the **correct** answer from the options given below :

(1) A and D only (2) A, B and E only
(3) B, C and E only (4) B and E only

216. Consider the following molecules : **[April 2, 2025 (I)]**

$CH_3-CH_2-\overset{O}{\overset{\|}{C}}-Cl$ (p)

$CH_3-CH_2-\overset{O}{\overset{\|}{C}}-O-\overset{O}{\overset{\|}{C}}-CH_3$ (q)

$CH_3-CH_2-\overset{O}{\overset{\|}{C}}-O-CH_2-CH_3$ (r)

$CH_3-CH_2-\overset{O}{\overset{\|}{C}}-NH_2$ (s)

The correct order of rate of hydrolysis is :

(1) $r>q>p>s$ (2) $q>p>r>s$
(3) $p>r>q>s$ (4) $p>q>r>s$

217. The total number of compounds from below when treated with hot $KMnO_4$ giving benzoic acid is : **[Jan. 28, 2025 (II)]**

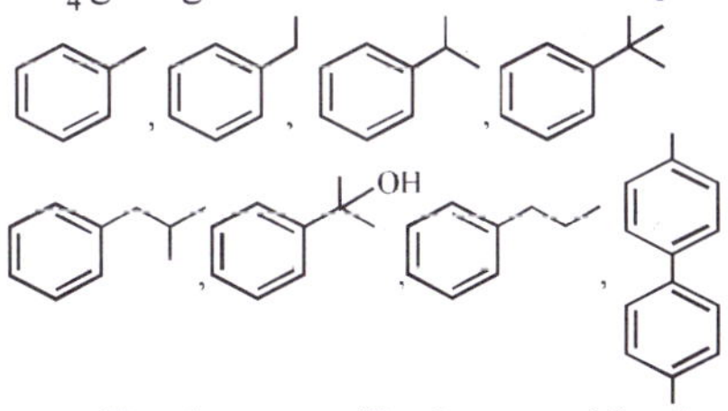

(1) 3 (2) 4 (3) 6 (4) 5

218. Identify the product (P) in the following reaction: **[April 8, 2024 (I)]**

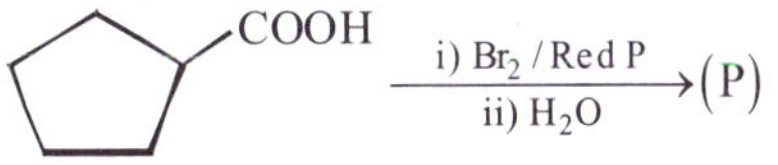

(1) 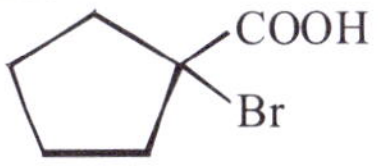 (2) 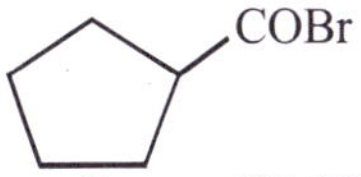

(3)  (4) 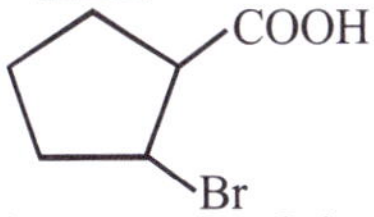

219. The correct sequence of acidic strength of the following aliphatic acids in their decreasing order is : **[April 8, 2024 (II)]**

$CH_3CH_2COOH, CH_3COOH, CH_3CH_2CH_2COOH, HCOOH$

(1) $HCOOH > CH_3COOH > CH_3CH_2COOH > CH_3CH_2CH_2COOH$
(2) $HCOOH > CH_3CH_2CH_2COOH > CH_3CH_2COOH > CH_3COOH$
(3) $CH_3CH_2CH_2COOH > CH_3CH_2COOH > CH_3COOH > HCOOH$
(4) $CH_3COOH > CH_3CH_2COOH > CH_3CH_2CH_2COOH > HCOOH$

220. Consider the given reaction, identify the major product P.

$$CH_3-COOH \xrightarrow[(iv)\,H_2O/\bar{O}H,\Delta]{(i)\,LiAlH_4\,(ii)\,PCC\,(iii)\,HCN/\bar{O}H} \text{"P"}$$

[April 6, 2024 (II)]

(1) $CH_3-CH_2-CH_2-OH$

(2) $CH_3-CH_2-\overset{O}{\overset{\|}{C}}-NH_2$

(3) $CH_3-\overset{\overset{O}{||}}{C}-CH_2CH_3$

(4) $CH_3-\underset{}{\overset{\overset{OH}{|}}{C}}H-COOH$

221. $CH_3CH_2-OH \xrightarrow[\text{(ii) } KMnO_4 \text{ (iii) NaOH, CaO, } \Delta]{\text{(i) Jone's Reagent}}$

Consider the above reaction sequence and identify the major product P. [April 5, 2024 (II)]

(1) Methane (2) Methanal
(3) Methoxymethane (4) Methanoic acid

222. Consider the given chemical reaction :

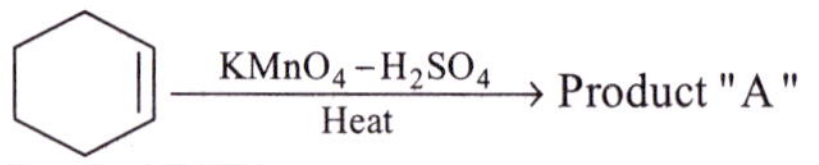

Product "A" is : [April 5, 2024 (II)]

(1) picric acid (2) oxalic acid
(3) acetic acid (4) adipic acid

223. Given below are two statements : [April 4, 2024 (I)]

Statement I : Acidity of α-hydrogens of aldehydes and ketones is responsible for aldol reaction.

Statement II : Reaction between benzaldehyde and ethanal will NOT give cross – aldol product.

In the light of above statements, choose the **most appropriate** answer from the options given below.

(1) Both **Statement I** and **Statement II** are correct.
(2) Both **Statement I** and **Statement II** are incorrect.
(3) **Statement I** is incorrect but **Statement II** is correct.
(4) **Statement I** is correct but **Statement II** is incorrect.

224. Match List - I with List - II. [Feb. 1, 2024 (I)]

List-I
(Reaction)

(A) $CH_3(CH_2)_5-\underset{\underset{O}{||}}{C}-OC_2H_5 \longrightarrow CH_3(CH_2)_5CHO$

(B) $C_6H_5COC_6H_5 \rightarrow C_6H_5CH_2C_6H_5$

(C) $C_6H_5CHO \rightarrow C_6H_5CH(OH)CH_3$

(D) $CH_3COCH_2COOC_2H_5 \rightarrow CH_3\underset{\underset{H}{|}}{C}(OH)CH_2COOC_2H_5$

List-II
(Reagents)

(I) CH_3MgBr, H_2O
(II) Zn(Hg) and conc. HCl
(III) $NaBH_4, H^+$
(IV) DiBAL - H, H_2O

Choose the correct answer from the option given below:

(1) (A)-(III), (B)-(IV), (C)-(I), (D)-(II)
(2) (A)-(IV), (B)-(II), (C)-(I), (D)-(III)
(3) (A)-(IV), (B)-(II), (C)-(III), (D)-(I)
(4) (A)-(III), (B)-(IV), (C)-(II), (D)-(I)

225. $C_2H_5Br \xrightarrow{\text{alc. KOH}} A \xrightarrow[CCl_4]{Br_2} B \xrightarrow[\text{Excess}]{KCN} C \xrightarrow[\text{Excess}]{H_3O^+} D$

Acid D formed in above reaction is: [Feb. 1, 2024 (II)]

(1) Malonic acid (2) Oxalic acid
(3) Succinic acid (4) Gluconic acid

226. The molecular formula of second homologue in the homologous series of mono carboxylic acids is________. [Jan. 27, 2024 (II)]

(1) $C_3H_6O_2$ (2) $C_2H_4O_2$
(3) CH_2O (4) $C_2H_2O_2$

227. The major product 'P' formed in the following sequence of reactions is [April 12, 2023 (I)]

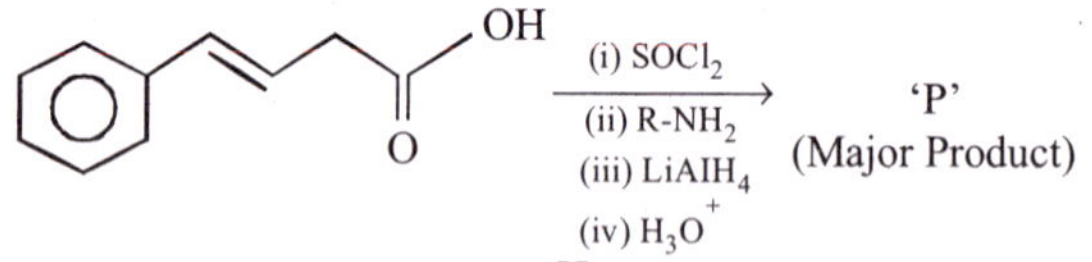

(1) 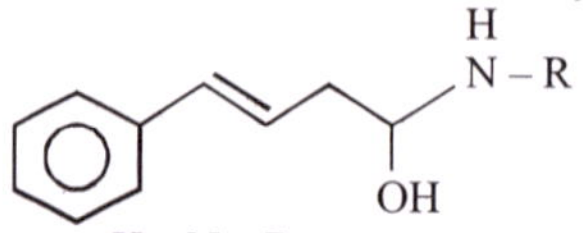

(2) 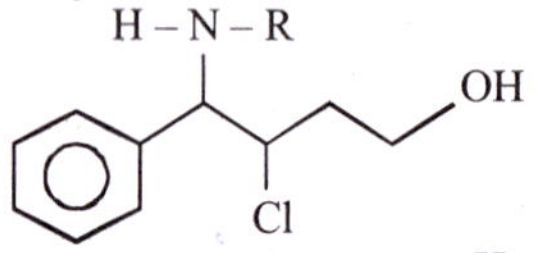

(3) 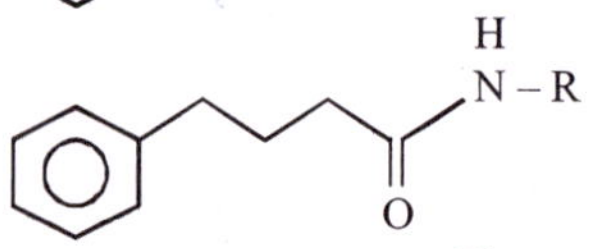

(4) 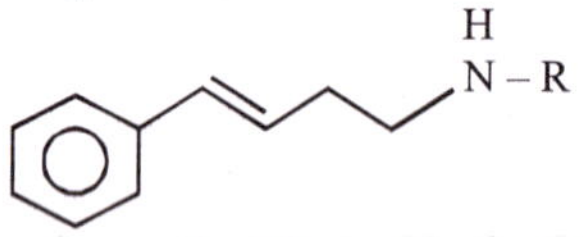

228. The major product formed in the following reaction is:

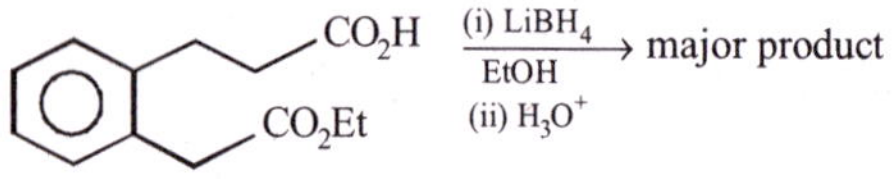

[April 08, 2023 (I)]

(1) 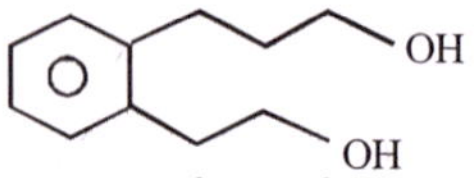

(2) 

(3) 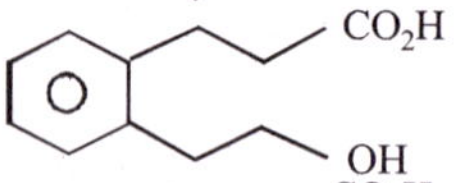

(4) 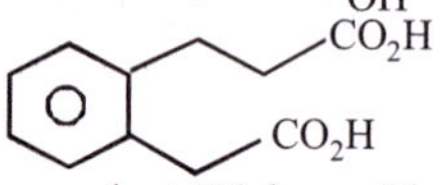

229. Major product 'P' formed in the following reaction is :-

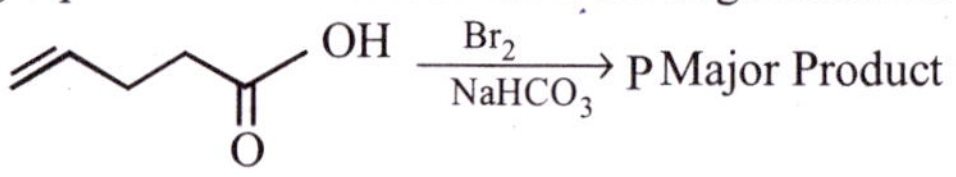

[April 08, 2023 (II)]

(1) 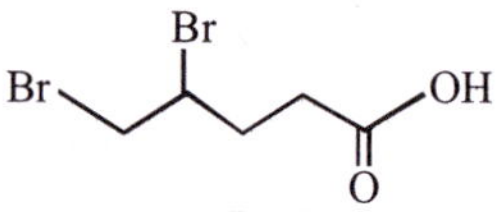

(2) 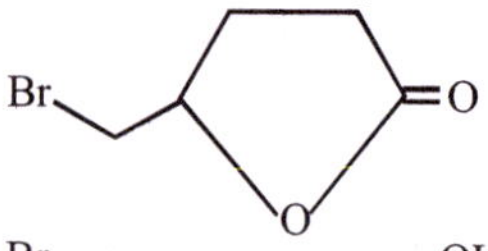

(3) 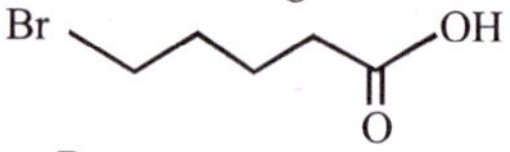

(4) 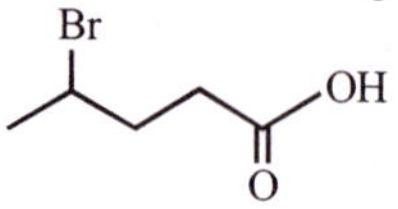

230. The descending order of acidity for the following carboxylic acid is : [April 08, 2023 (II)]

A. CH_3COOH B. $F_3C–COOH$
C. $ClCH_2–COOH$ D. $FCH_2–COOH$
E. $BrCH_2–COOH$

Choose the correct answer from the options given below:

(1) D > B > A > E > C (2) E > D > B > A > C
(3) B > C > D > E > A (4) B > D > C > E > A

231. In a reaction, [Feb. 01, 2023 (II)]

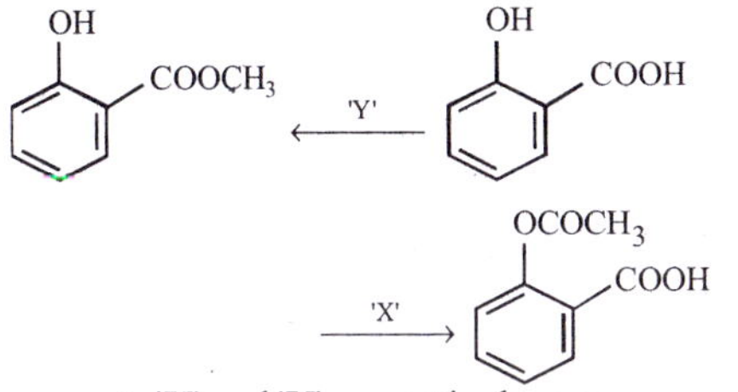

reagents 'X' and 'Y' respectively are :

(1) $CH_3OH/H^+, \Delta$ and $(CH_3CO)_2O/H^+$
(2) $(CH_3CO)_2O/H^+$ and $CH_3OH/H^+, \Delta$
(3) $CH_3OH/H^+, \Delta$ and $CH_3OH/H^+, \Delta$
(4) $(CH_3CO)_2O/H^+$ and $(CH_3CO)_2O/H^+$

232. Among the following marked proton of which compound shows lowest pK_a value? [July 28, 2022 (I)]

(1) $H_2\overset{H}{\overset{|}{C}}-COOH$ (2) $H_2\overset{H}{\overset{|}{C}}-\overset{O}{\overset{||}{C}}-CH_3$

(3) 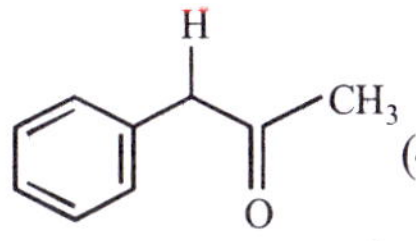 (4) 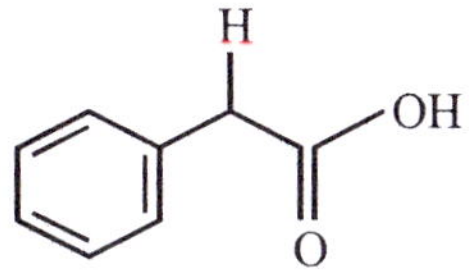

233. 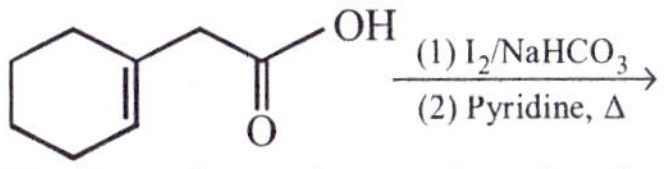

Find out the major product for the above reaction. [July 28, 2022 (II)]

(1) 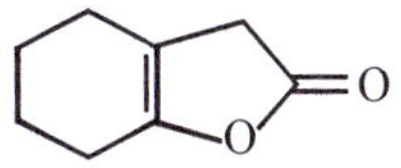 (2) 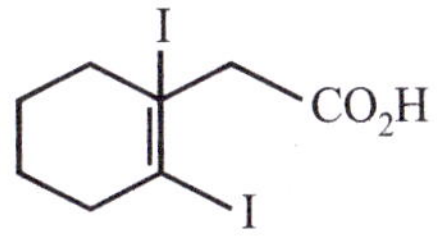

(3) 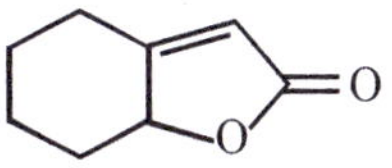 (4) 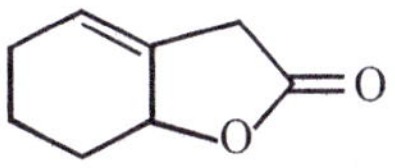

234. Consider the given chemical reaction. [June 30, 2022 (I)]

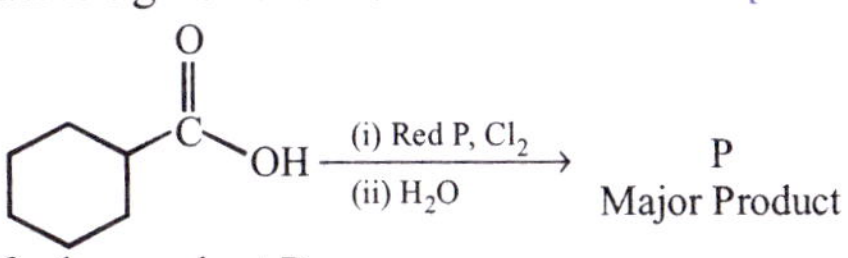

Identify the product P.

(1) 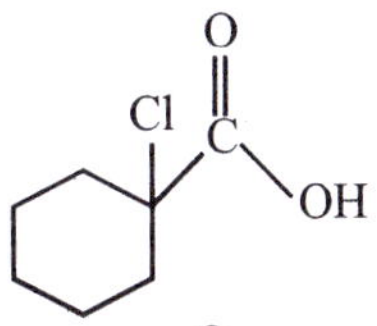 (2) 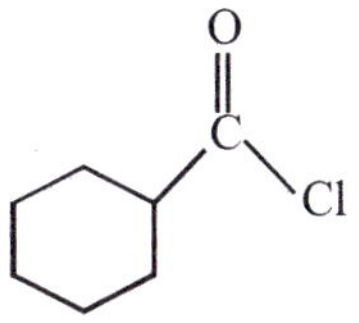

(3) 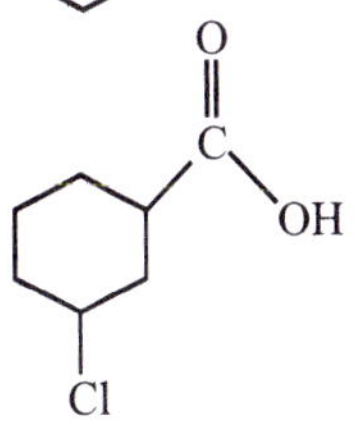 (4) 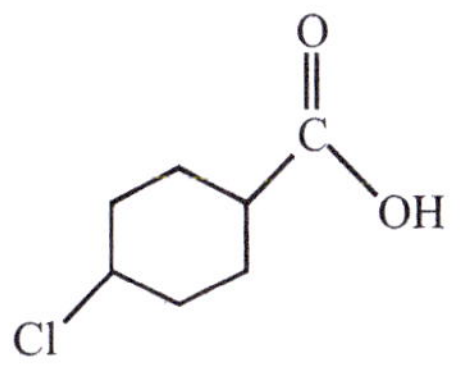

235. 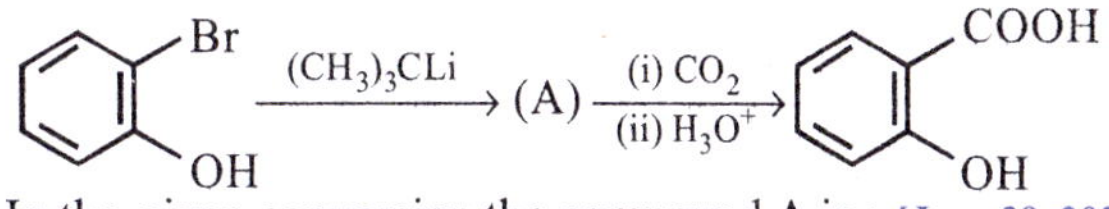

In the given conversion the compound A is : [June 29, 2022 (I)]

(1) 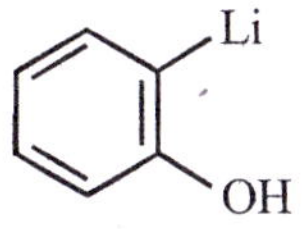 (2) 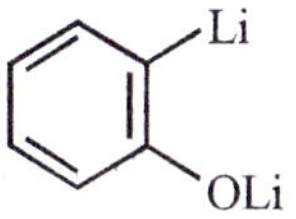

(3) 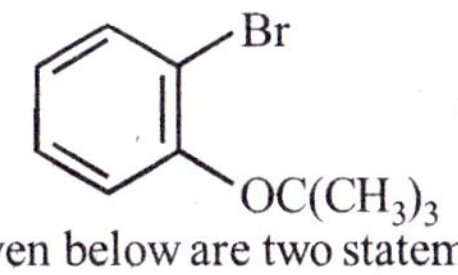 (4) 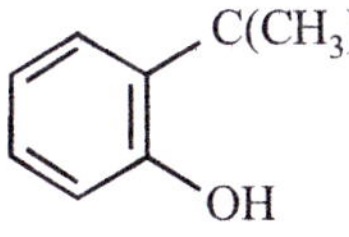

236. Given below are two statements : [June 29, 2022 (I)]

Statement I : The esterification of carboxylic acid with an alcohol is a nucleophilic acyl substitution.

Statement II : Electron withdrawing groups in the carboxylic acid will increase the rate of esterification reaction.

Choose the **most appropriate** option :

(1) Both Statement I and Statement II are correct.
(2) Both Statement I and Statement II are incorrect.
(3) Statement I is correct but Statement II is incorrect.
(4) Statement I is incorrect but Statement II is correct.

237. $A \xrightarrow[\text{(ii) } CN^- \text{ (iii) } H_2O/H^+]{\text{(i) } Cl_2, \Delta}$ 4-Bromophenyl acetic acid.

In the above reaction 'A' is [June 28, 2022 (II)]

(1) 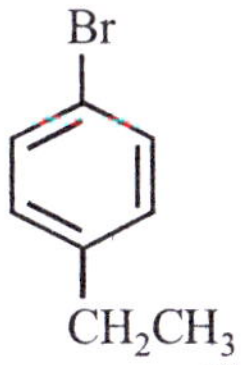 (2) 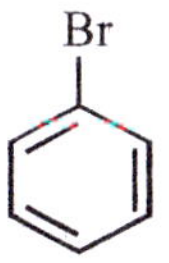

(3) 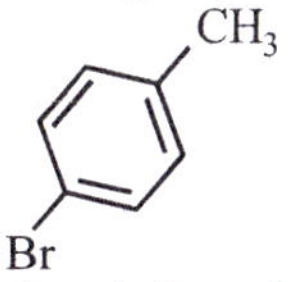 (4) 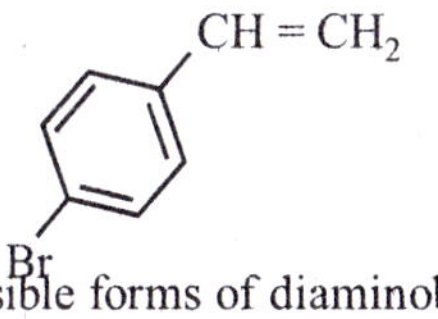

238. Decarboxylation of all six possible forms of diaminobenzoic acids $C_6H_3(NH_2)_2COOH$ yields three products A, B and C. Three acids give a product 'A', two acids give a product 'B' and one acid gives a product 'C'. The melting point of product 'C' is [June 27, 2022 (II)]

(1) 63°C (2) 90°C (3) 104°C (4) 142°C

239. Two statements are given below :

Statement I: The melting point of monocarboxylic acid with even number of carbon atoms is higher than that of with odd number of carbon atoms acid immediately below and above it in the series.

Statement II : The solubility of monocarboxylic acids in water decreases with increase in molar mass.

Choose the **most appropriate** option: [June 24, 2022 (I)]

(1) Both Statement I and Statement II are correct.
(2) Both Statement I and Statement II are incorrect.
(3) Statement I is correct but Statement II is incorrect.
(4) Statement I is incorrect but Statement II is correct.

240. In the following sequence of reactions a compound A, (molecular formula $C_6H_{12}O_2$) with a straight chain structure gives a C_4 carboxylic acid. A is : [Sep. 1, 2021 (II)]

$$A \xrightarrow[H_3O^+]{LiAlH_4} B \xrightarrow{\text{Oxidation}} C_4-\text{carboxylic acid}$$

(1) $CH_3-CH_2-COO-CH_2-CH_2-CH_3$

(2) $CH_3-CH_2-CH(OH)-CH_2-O-CH=CH_2$

(3) $CH_3-CH_2-CH_2-COO-CH_2-CH_3$

(4) $CH_3-CH_2-CH_2-O-CH=CH-CH_2-OH$

241. Given below are two statements : [Aug. 27, 2021 (II)]

Statement I : Ethyl pent–4–yn–oate on reaction with CH_3MgBr gives a 3°–alcohol.

Statement II : In this reaction one mole of ethyl pent–4–yn–oate utilizes two moles of CH_3MgBr.

In the light of the above statements, choose the **most appropriate** answer from the options given below :

(1) Both Statement I and Statement II are false.
(2) Statement I is false but Statement II is true.
(3) Statement I is true but Statement II is false.
(4) Both Statement I and Statement II are true.

242. 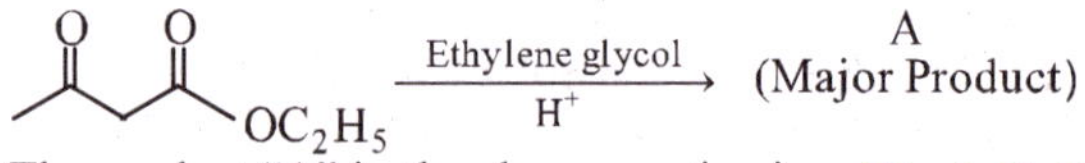

The product "A" in the above reaction is [March 17, 2021 (I)]

(1) 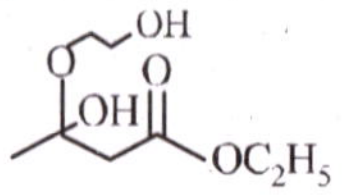 (2) 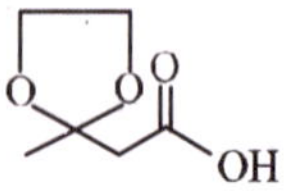

(3) 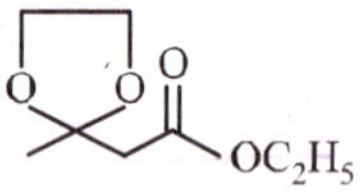 (4) 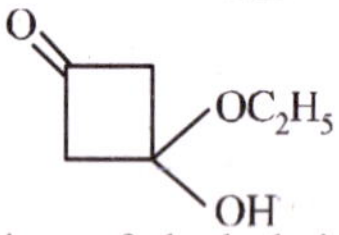

243. Which of the following derivatives of alcohols is unstable in an aqueous base? [Sep. 05, 2020 (I)]

(1) 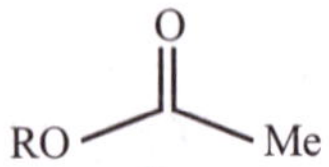 (2) 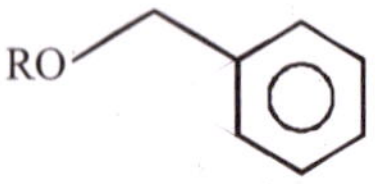

(3) 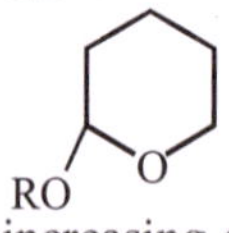 (4) $RO-CMe_3$

244. The increasing order of the acidity of the α-hydrogen of the following compounds is : [Sep. 05, 2020 (I)]

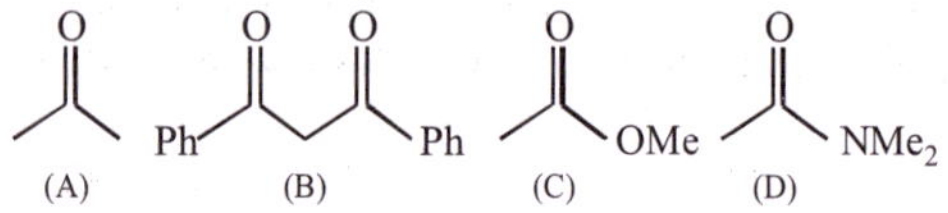

(1) (D) < (C) < (A) < (B) (2) (B) < (C) < (A) < (D)
(3) (A) < (C) < (D) < (B) (4) (C) < (A) < (B) < (D)

245. In the following reaction sequence the major products A and B are : [Sep. 05, 2020 (I)]

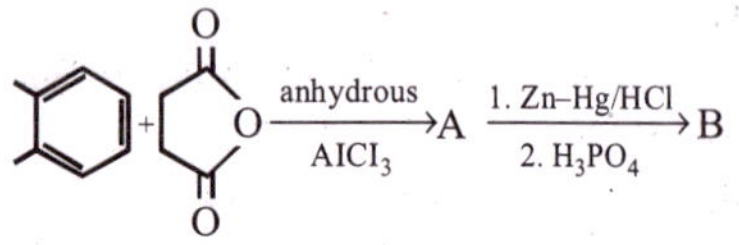

(1) 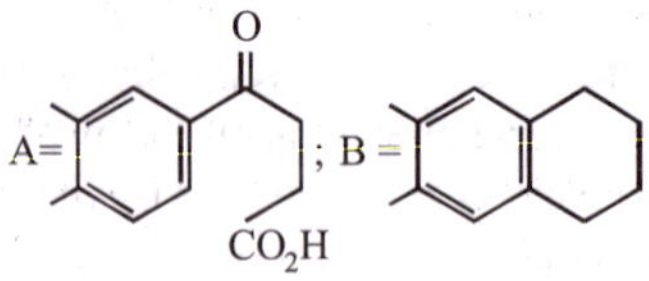

(2) 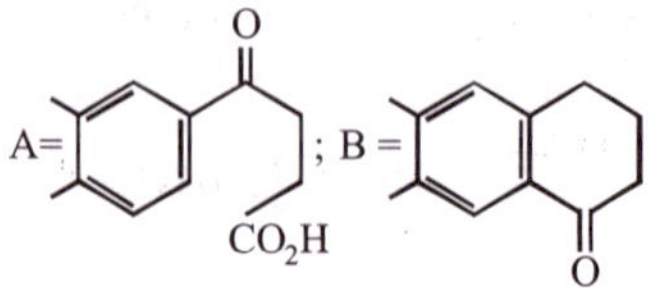

(3) 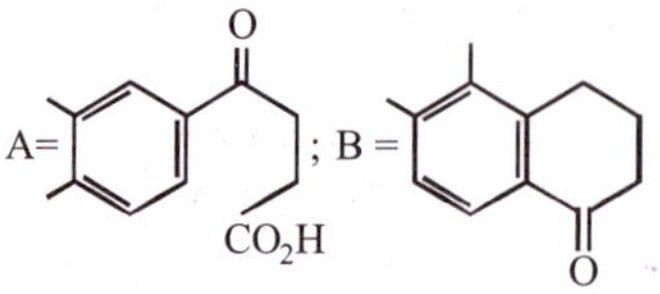

(4) 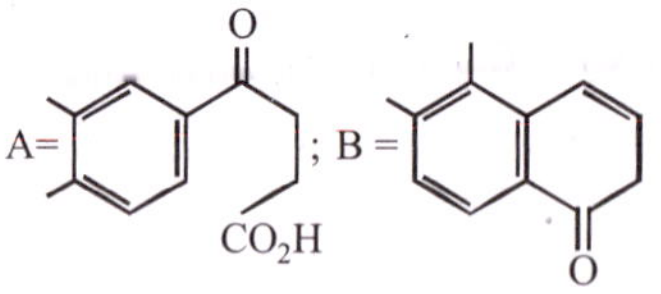

246. An organic compound (A) (molecular formula $C_6H_{12}O_2$) was hydrolysed with dil. H_2SO_4 to give a carboxylic acid (B) and an alcohol (C). 'C' gives white turbidity immediately when treated with anhydrous $ZnCl_2$ and conc. HCl. The organic compound (A) is : [Sep. 04, 2020 (I)]

(1) 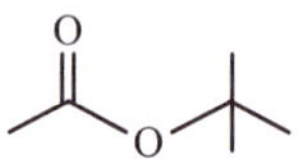 (2) 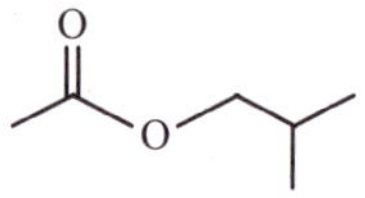

(3) 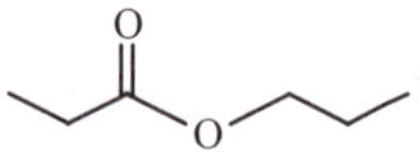 (4) 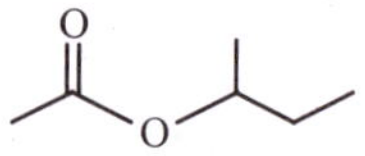

247. [P] on treatment with $Br_2/FeBr_3$ in CCl_4 produced a single isomer $C_8H_7O_2Br$ while heating [P] with soda-lime gave toluene. The compound [P] is : [Sep. 04, 2020 (I)]

(1) 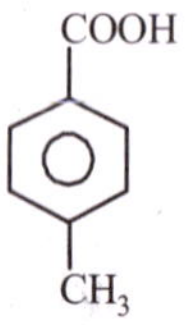 (2) 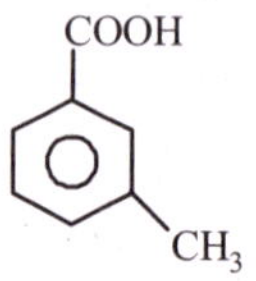

(3) 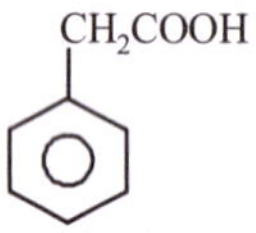 (4) 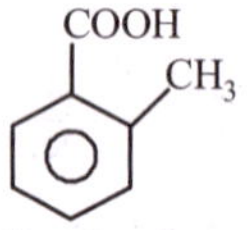

248. An organic compound [A], molecular formula $C_{10}H_{20}O_2$ was hydrolyzed with dilute sulphuric acid to give a carboxylic acid [B] and an alcohol [C]. Oxidation of [C] with $CrO_3-H_2SO_4$ produced [B]. Which of the following structures are not possible for [A]? [Sep. 03, 2020 (I)]

(1) $CH_3-CH_2-CH(CH_3)-OCOCH_2CH(CH_3)-CH_2CH_3$

(2) $CH_3-CH_2-CH(CH_3)-COOCH_2-CH(CH_3)-CH_2CH_3$

(3) $CH_3CH_2CH_2COOCH_2CH_2CH_2CH_3$

(4) $(CH_3)_3-C-COOCH_2C(CH_3)_3$

249. Consider the following molecules and statements related to them : [Sep. 03, 2020 (II)]

(A) 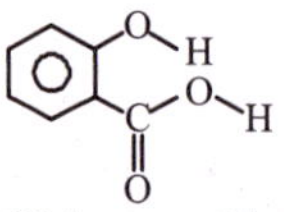 (B) 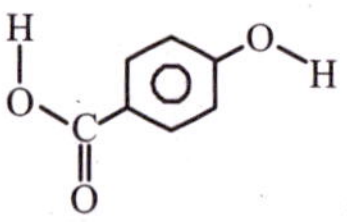

(a) (2) is more likely to be crystalline than (A)
(b) (2) has higher boiling point than (A)
(c) (2) dissolves more readily than (A) in water

Identify the **correct** option from below :
(1) (a) and (b) are true (2) (a) and (c) are true
(3) only (a) is true (4) (b) and (c) are true

250. Arrange the following labelled hydrogens in decreasing order of acidity : [Sep. 02, 2020 (II)]

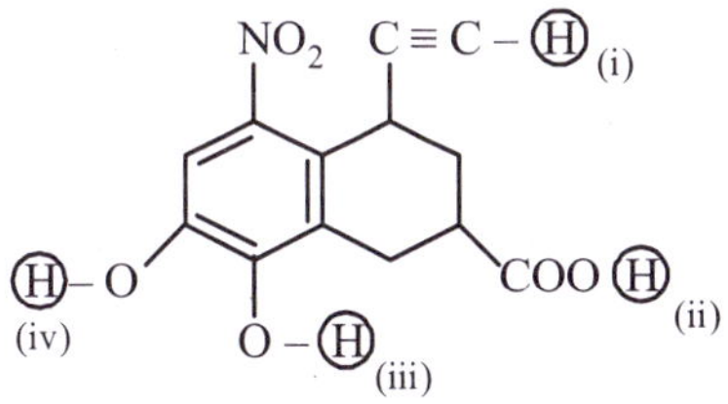

(1) (ii) > (i) > (iii) > (iv) (2) (iii) > (ii) > (iv) > (i)
(3) (ii) > (iii) > (iv) > (i) (4) (iii) > (ii) > (i) > (iv)

251. The most suitable reagent for the given conversion is: [Jan. 08, 2020 (I)]

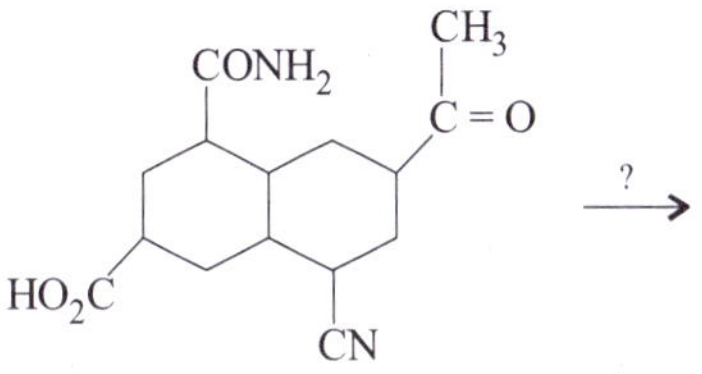

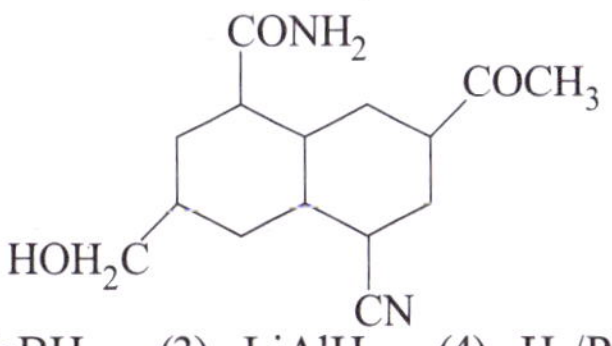

(1) B_2H_6 (2) $NaBH_4$ (3) $LiAlH_4$ (4) H_2/Pd

252. An unsaturated hydrocarbon *X* absorbs two hydrogen molecules on catalytic hydrogenation, and also gives following reaction:

$$X \xrightarrow[Zn/H_2O]{O_3} A \xrightarrow{[Ag(NH_3)_2]^+}$$

B (3-oxo-hexanedicarboxylic acid)
X will be : [Jan. 08, 2020 (II)]

(1) 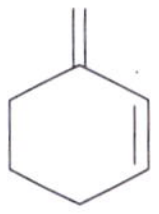 (2) 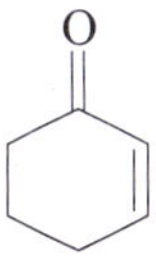

(3) 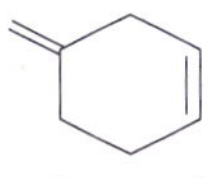 (4) 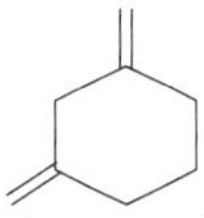

253. The major products of the following reaction are : [April 12, 2019 (I)]

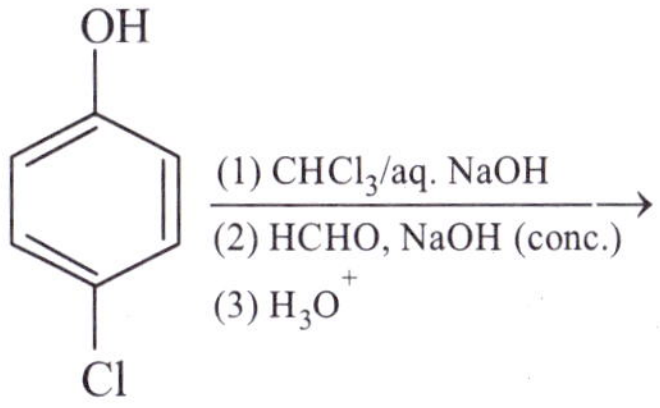

(1) 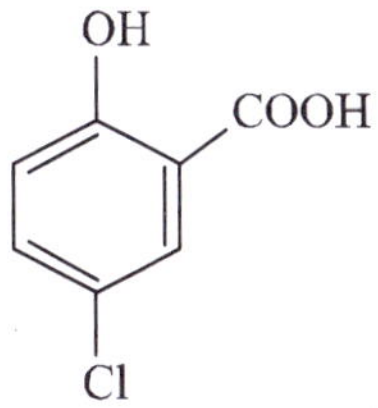 and Methanol

(2) 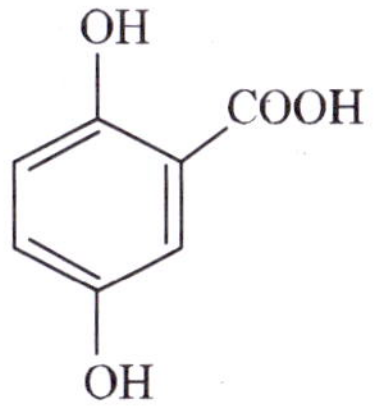 and Methanol

(3) 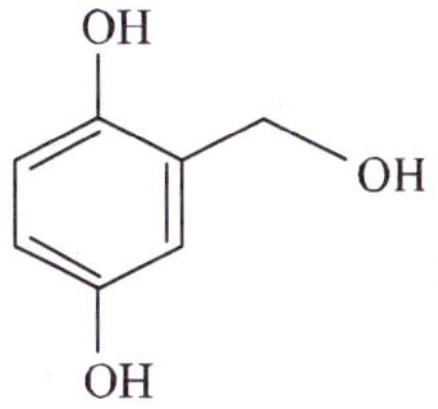 and Formic acid

(4) 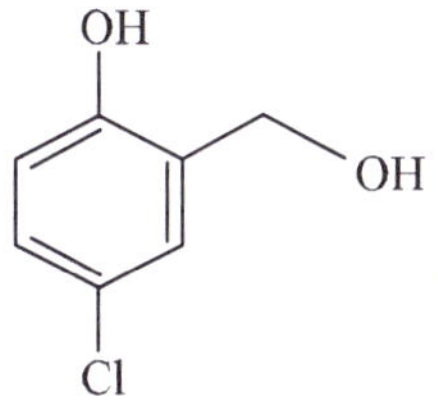 and Formic acid

254. The major product of the following reaction is:

$CH_3CH=CHCO_2CH_3 \xrightarrow{LiAlH_4}$ [April 9, 2019 (I)]

(1) $CH_3CH_2CH_2CO_2CH_3$ (2) $CH_3CH=CHCH_2OH$
(3) $CH_3CH_2CH_2CH_2OH$ (4) $CH_3CH_2CH_2CHO$

255. The major product of the following reaction is: [April 9, 2019 (I)]

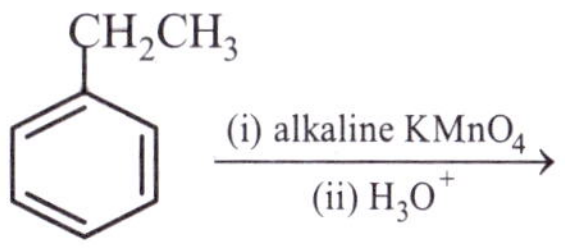

(1) 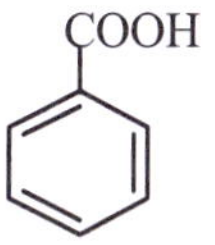 (2) 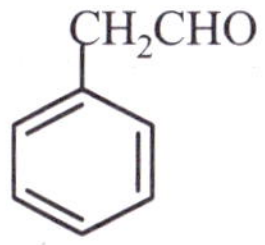

(3) 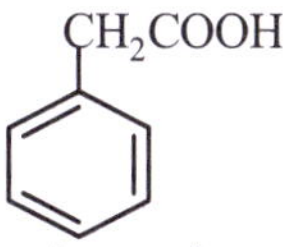 (4) 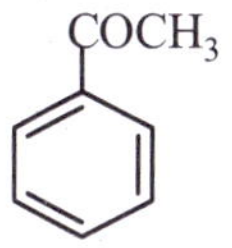

256. The major product of the following reaction is: [April 9, 2019 (II)]

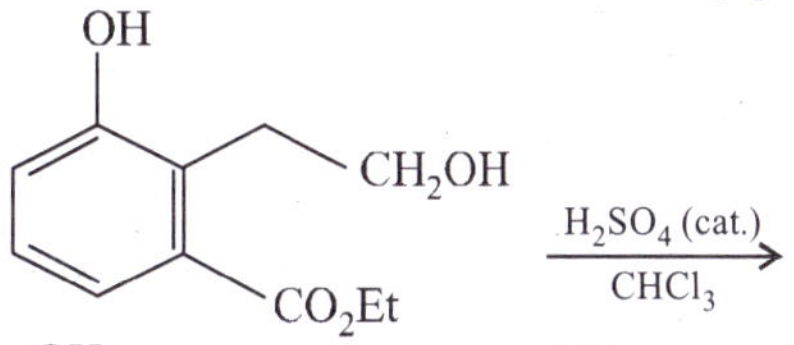

(1) 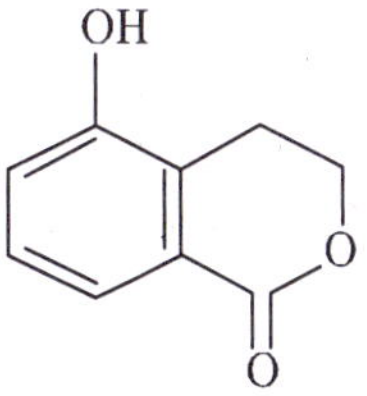 (2) 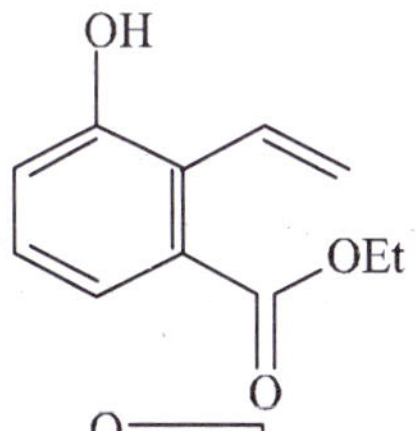

(3) 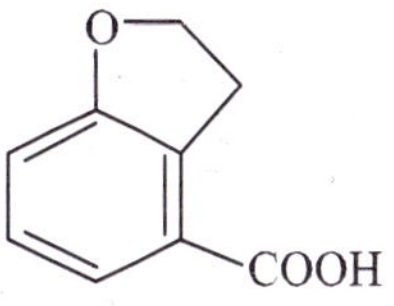 (4) 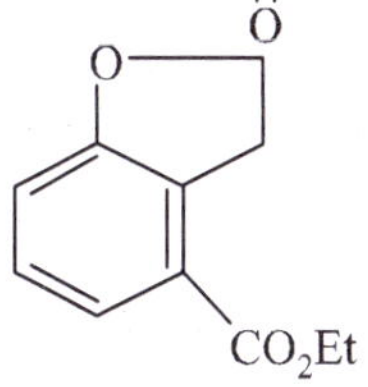

257. The major product of the following reaction is: [April 8, 2019 (II)]

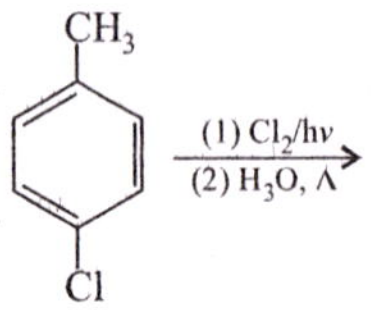

(1) 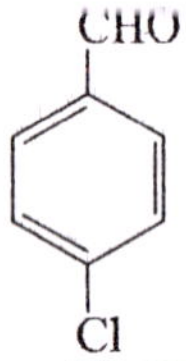

(2) 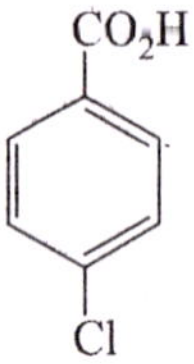

(3) 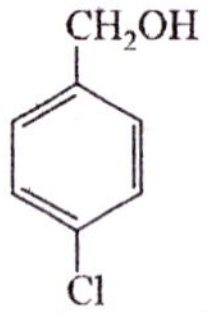

(4) 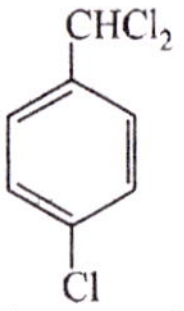

258. The major product of the following reaction is: [April 8, 2019 (II)]

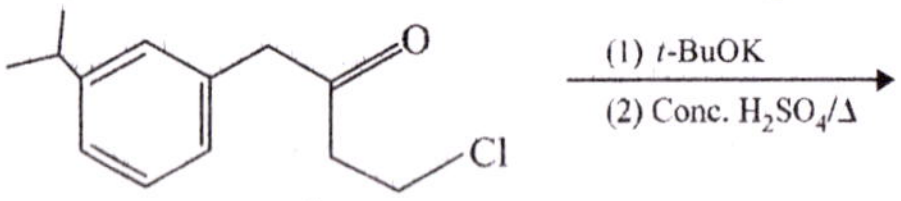

(1) 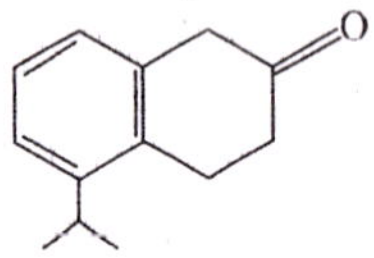
(2) 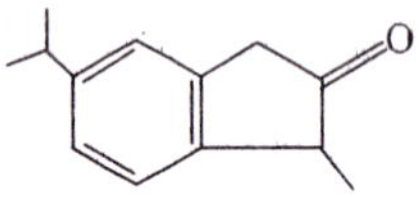

(3) 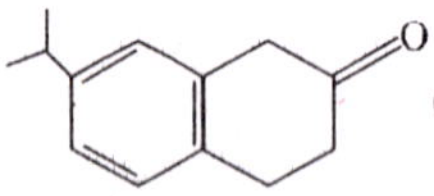
(4) 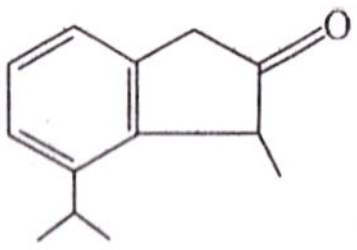

259. The major product of the following reaction is: [Jan. 12, 2019 (I)]

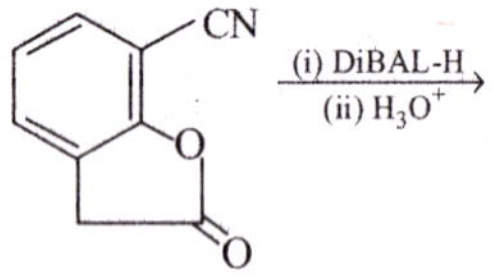

(1) 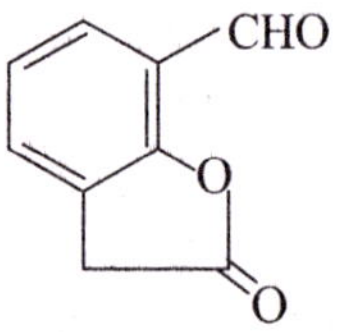

(2) 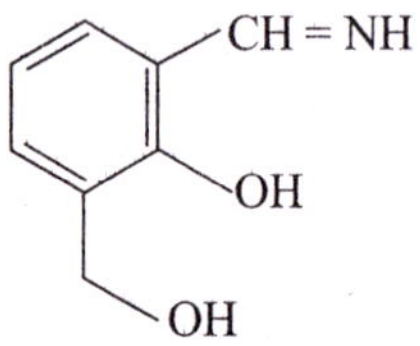

(3) 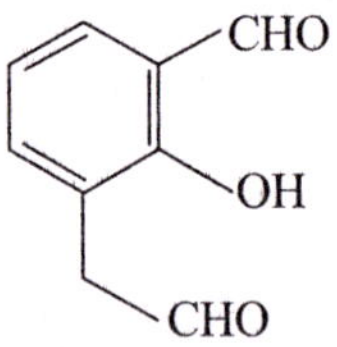

(4) 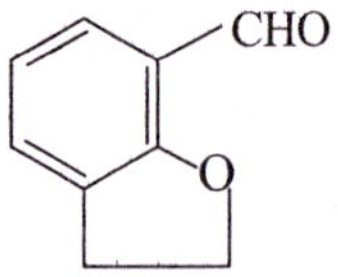

260. Among the following four aromatic compounds ,which one will have the lowest melting point? [Jan. 12, 2019 (I)]

(1) 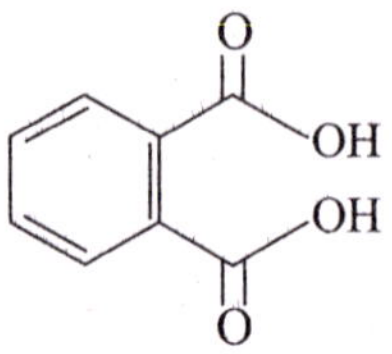

(2) 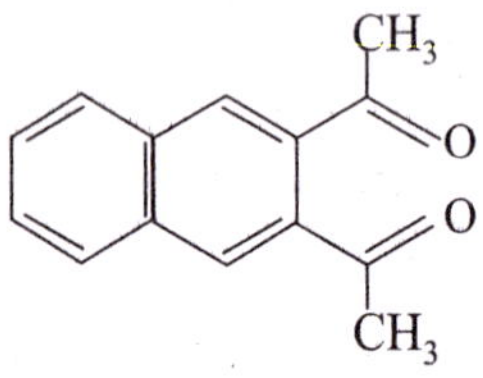

(3) 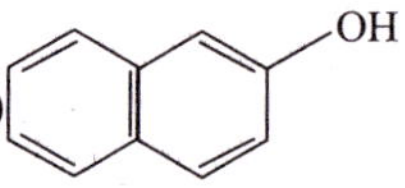

(4) 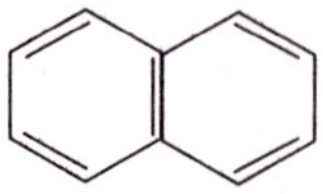

261. The major product of the following reaction is : [Jan. 12, 2019 (II)]

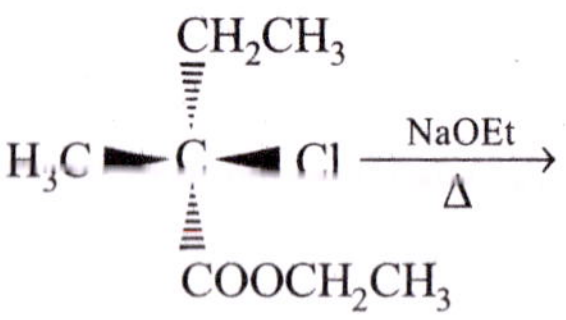

(1) 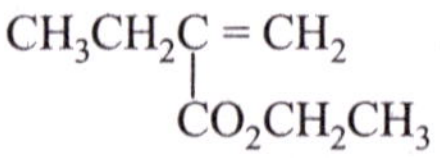

(2) 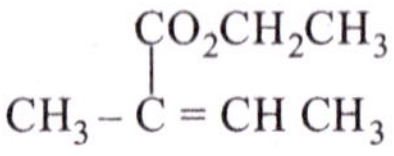

(3) 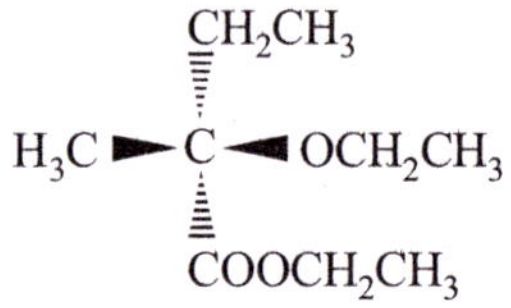

(4) 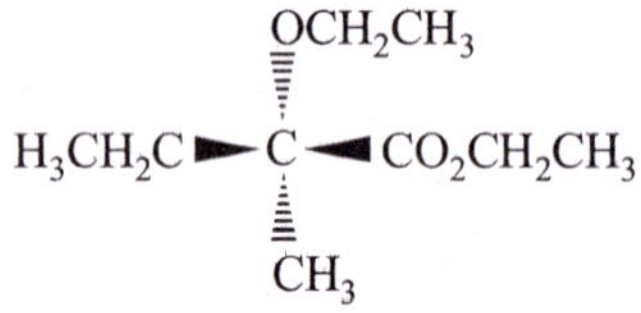

262. The major product of the following reaction is : [Jan. 11, 2019 (I)]

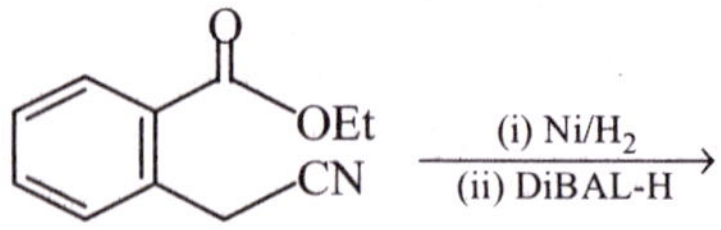

(1) 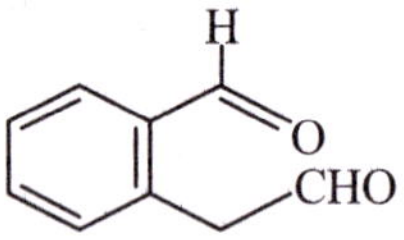

(2) 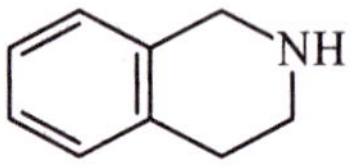

(3) 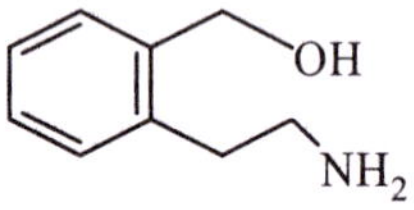

(4) 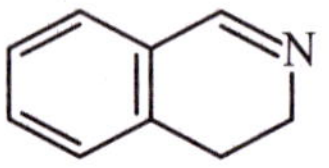

263. The major product of the following reaction is : [Jan. 11, 2019 (I)]

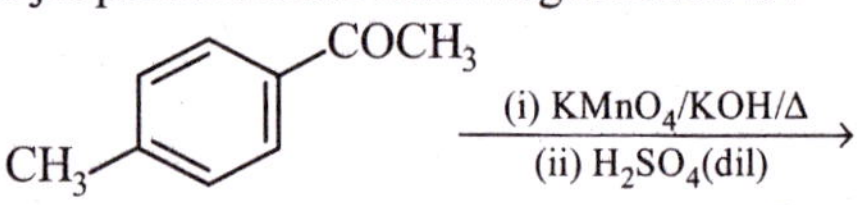

(1) 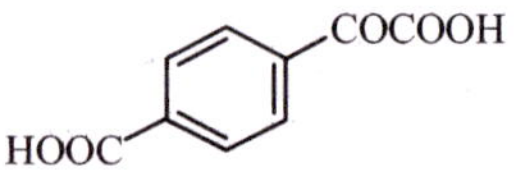

(2) 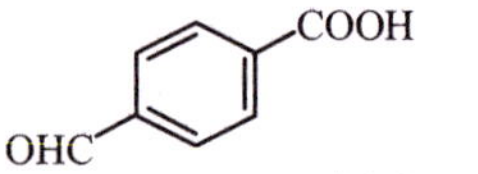

(3) 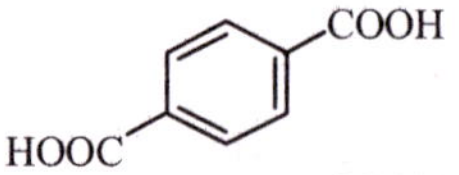

(4) 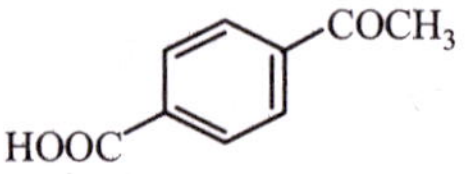

264. The major product obtained in the following reaction is: [Jan. 11, 2019 (II)]

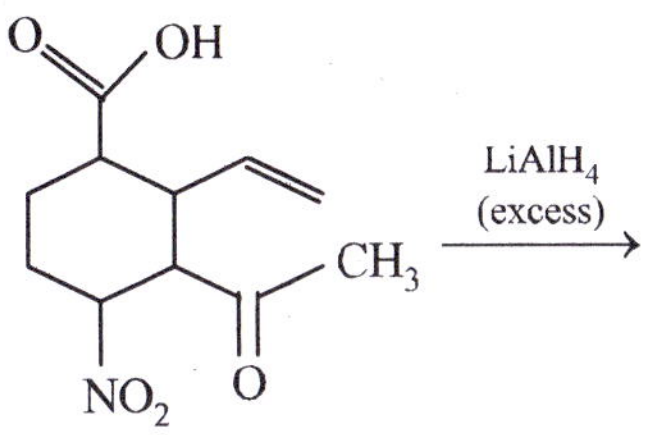

(1) 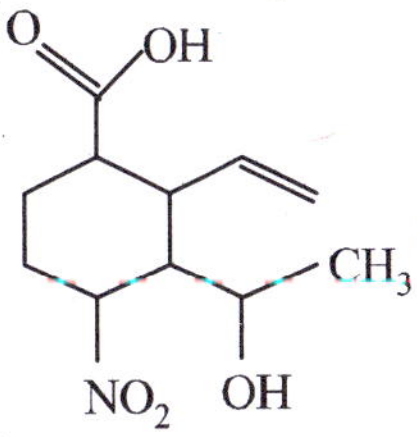 (2) 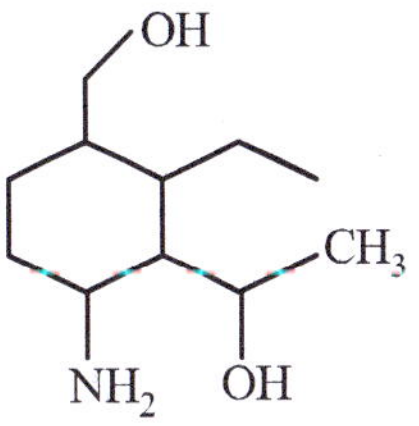

(3) 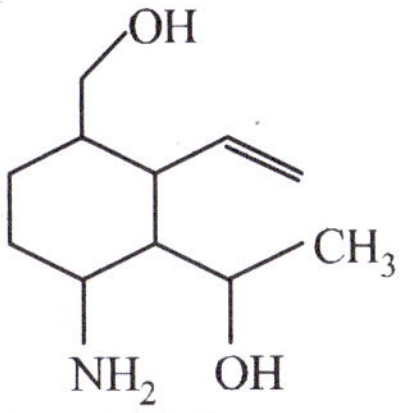 (4) 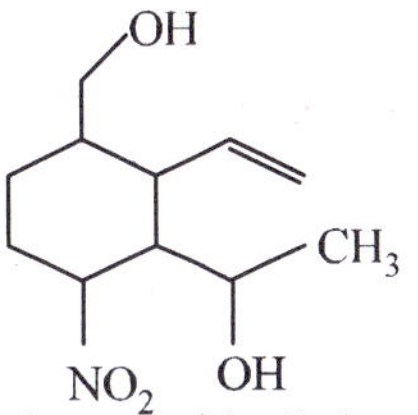

265. Which of the following compound reacts with ethyl magnesium bromide and also decolourizes bromine water solution? [Jan. 11, 2019 (II)]

(1) 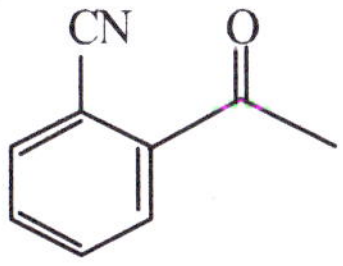 (2) 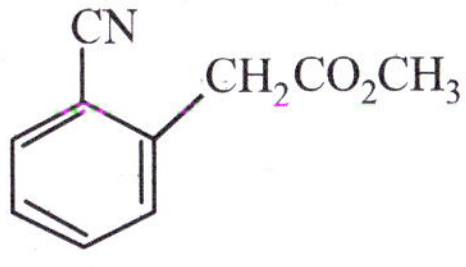

(3) 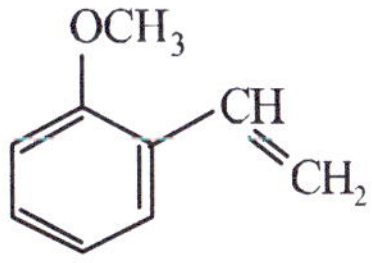 (4) 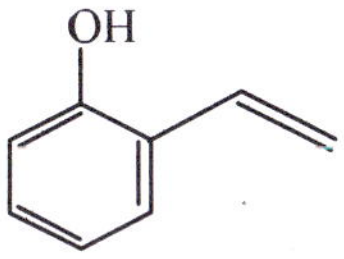

266. The major product obtained in the following conversion is: [Jan. 11, 2019 (II)]

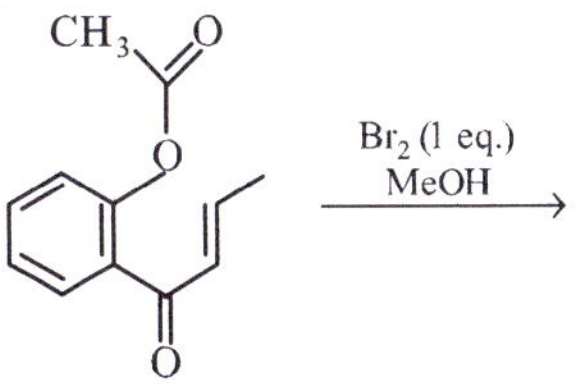

(1) 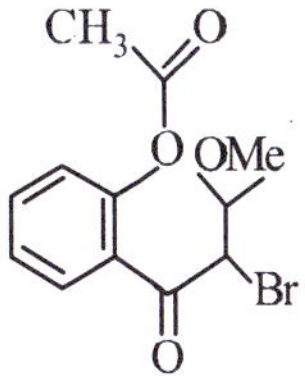 (2) 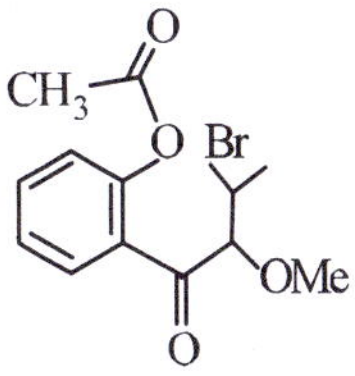

(3) 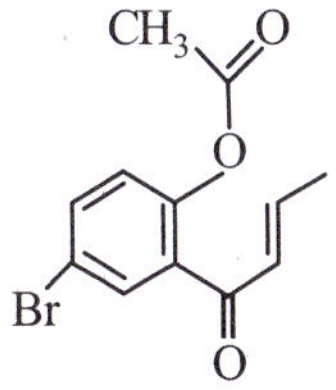 (4) 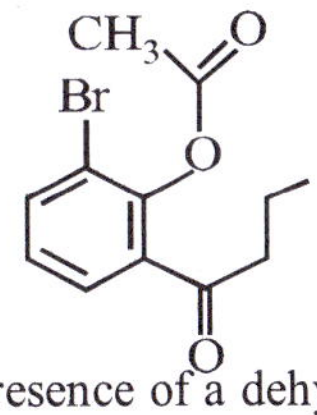

267. Which dicarboxylic acid in presence of a dehydrating agent is least reactive to give an anhydride? [Jan. 10, 2019 (I)]

(1) 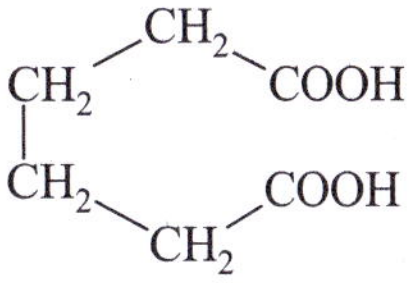 (2) 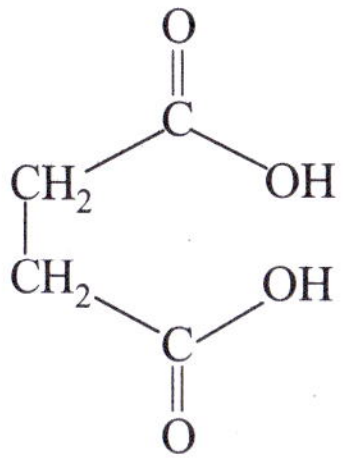

(3) 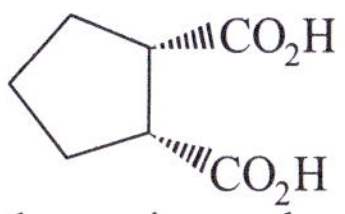 (4) 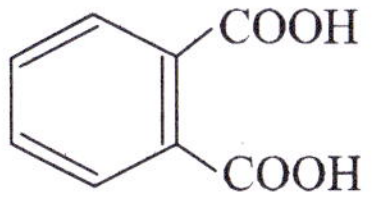

268. The decreasing order of ease of alkaline hydrolysis for the following esters is [Jan. 10, 2019 (I)]

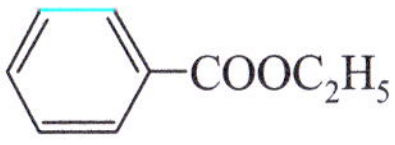
I

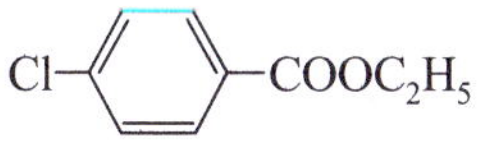
II

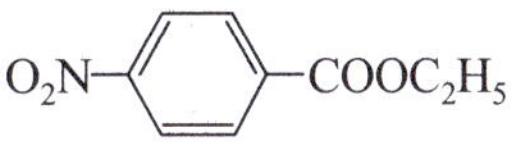
III

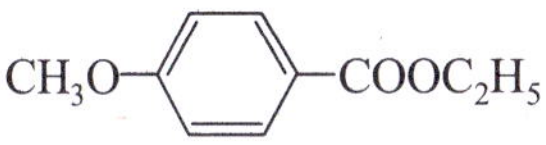
IV

(1) III > II > IV > I (2) III > II > I > IV
(3) IV > II > III > I (4) II > III > I > IV

269. An aromatic compound 'A' having molecular formula $C_7H_6O_2$ on treating with aqueous ammonia and heating forms compound 'B'. The compound 'B' on reaction with molecular bromine and potassium hydroxide provides compound 'C' having molecular formula C_6H_7N. The structure of 'A' is: [Jan. 10, 2019 (II)]

(1) 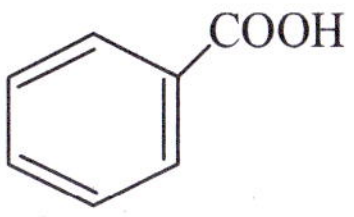 (2) 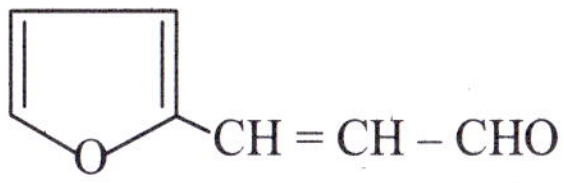

(3) 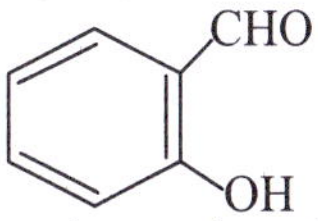 (4) 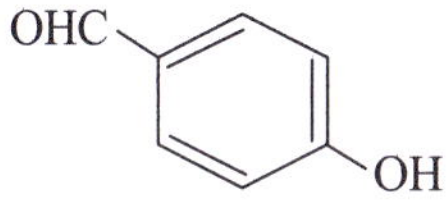

270. The major product obtained in the following reaction is: [Jan. 10, 2019 (II)]

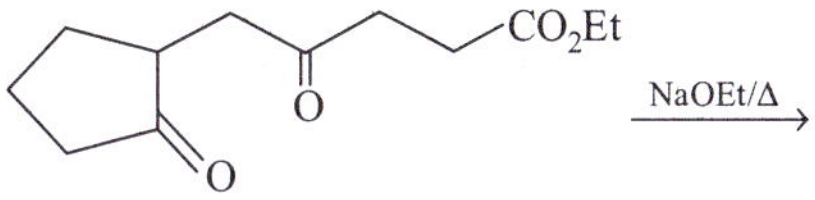

(1) 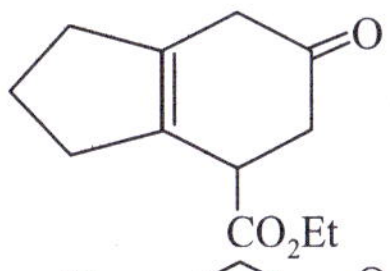 (2) 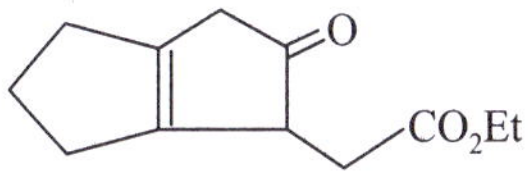

(3) 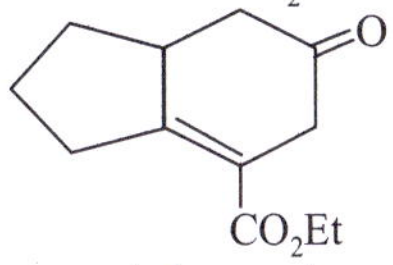 (4) 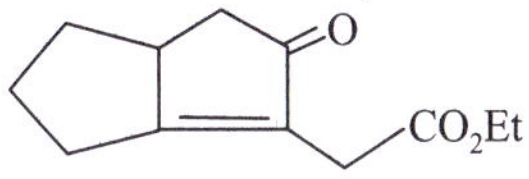

271. The correct decreasing order for acid strength is: [Jan. 9, 2019 (I)]

(1) $NO_2CH_2COOH > FCH_2COOH > CNCH_2COOH > ClCH_2COOH$
(2) $FCH_2COOH > CNCH_2COOH > NO_2CH_2COOH > ClCH_2COOH$
(3) $CNCH_2COOH > NO_2CH_2COOH > FCH_2COOH > ClCH_2COOH$
(4) $NO_2CH_2COOH > CNCH_2COOH > FCH_2COOH > ClCH_2COOH$

272. The major product of the given reaction is: **[Online April 16, 2018]**

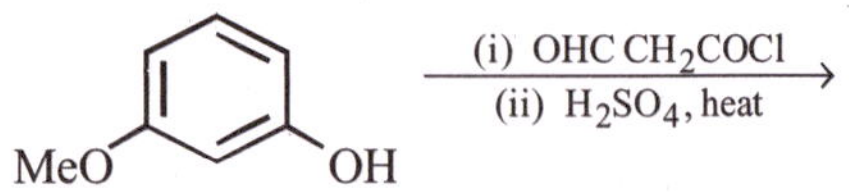

(1) 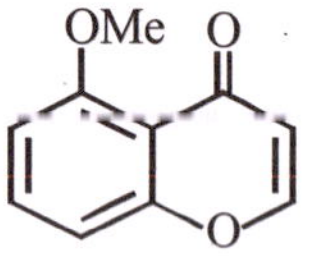 (2) 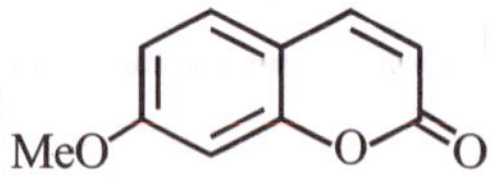

(3) 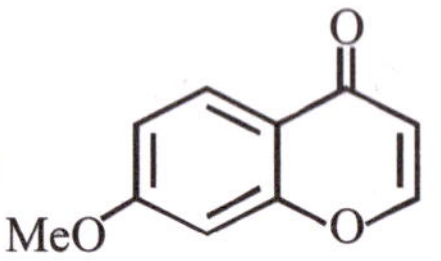 (4) 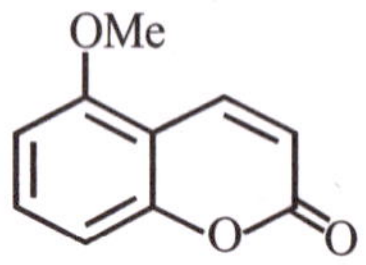

273. The major product formed in the following reaction is:

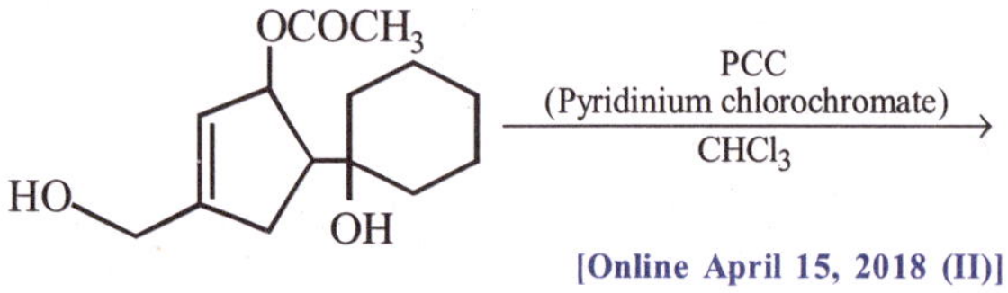

[Online April 15, 2018 (II)]

(1) 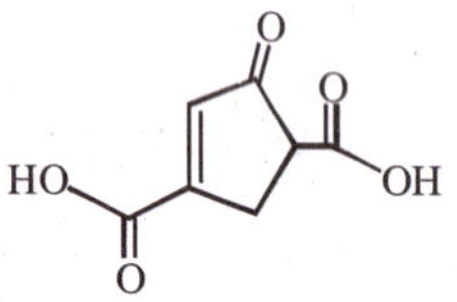 (2) 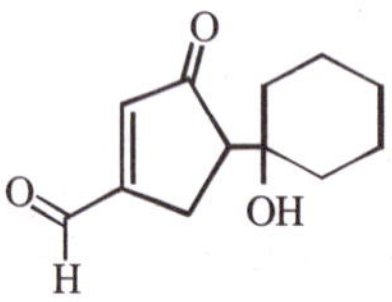

(3) 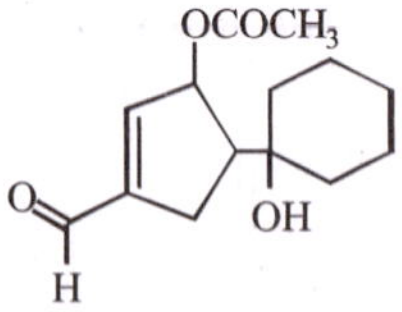 (4) 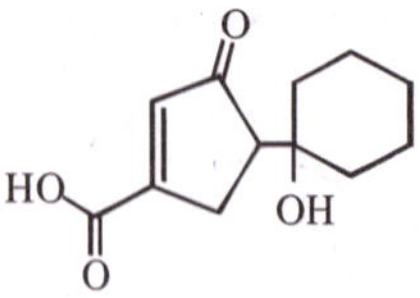

274. The increasing order of the acidity of the following carboxylic acids is: **[Online April 15, 2018 (II)]**

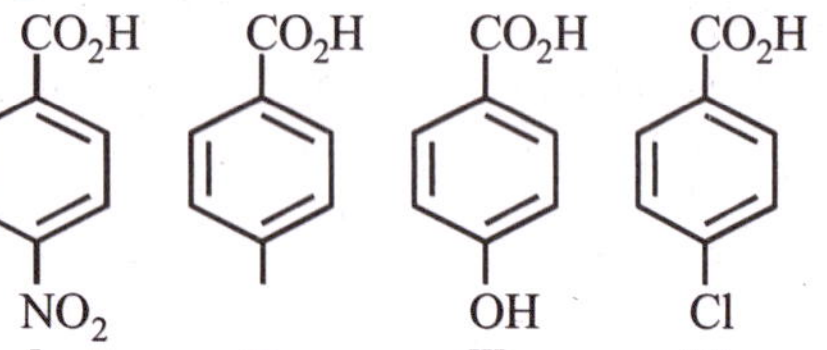

I II III IV

(1) III < II < IV < I (2) I < III < II < IV
(3) IV < II < III < I (4) II < IV < III < I

275. The major product of the following reaction is

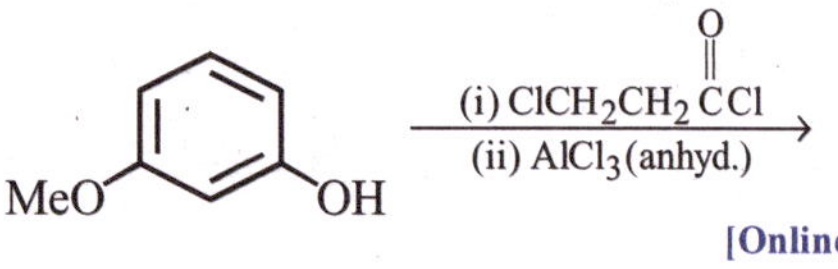

[Online April 15, 2018 (I)]

(1) 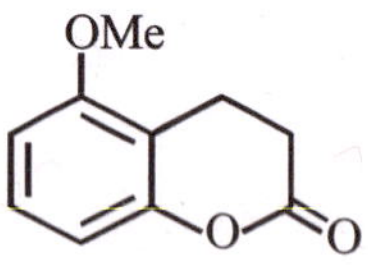 (2) 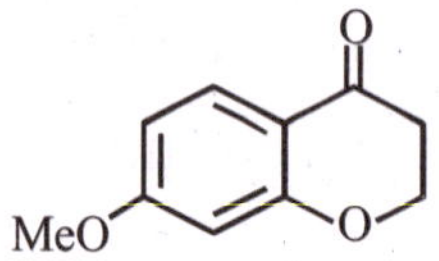

(3) 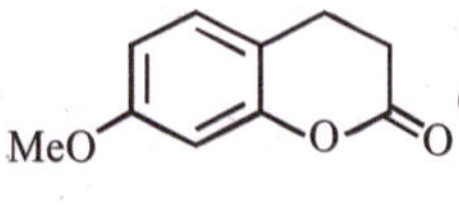 (4) 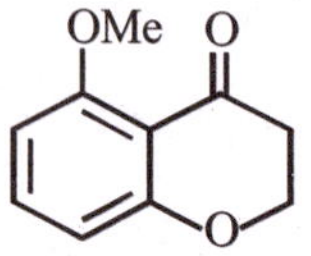

276. The reagent(s) required for the following conversion are :

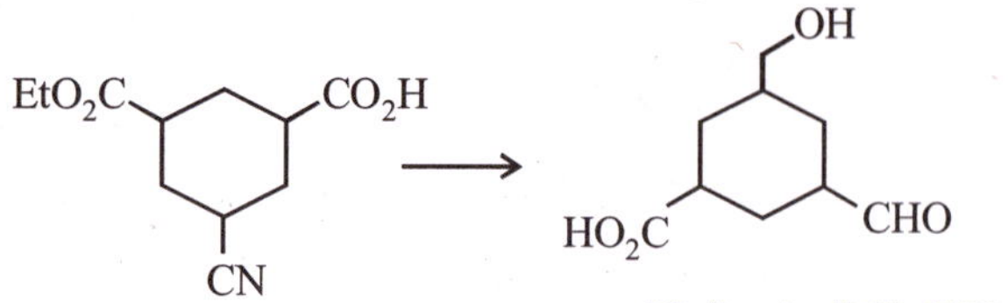

[Online April 15, 2018 (I)]

(1) (i) $NaBH_4$, (ii) Raney Ni/H_2, (iii) H_3O^+
(2) (i) $LiAlH_4$, (ii) H_3O^+
(3) (i) B_2H_6, (ii) DIBAL-H, (iii) H_3O^+
(4) (i) B_2H_6, (ii) $SnCl_2/HCl$, (iii) H_3O^+

277. The main reduction product of the following compound with $NaBH_4$ in methanol is :

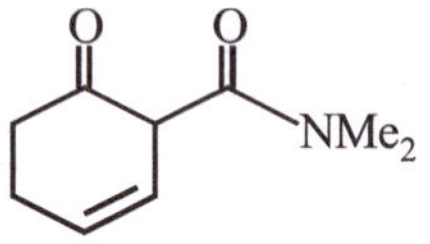

[Online April 15, 2018 (I)]

(1) 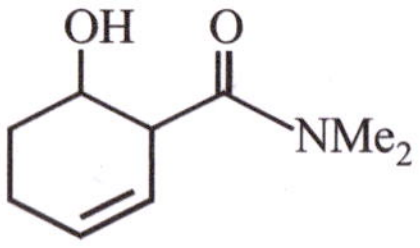 (2) 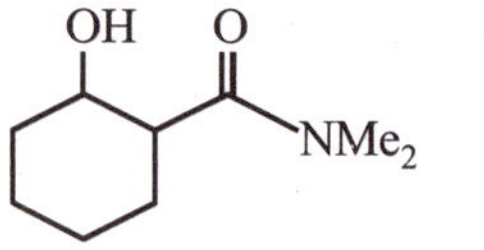

(3) 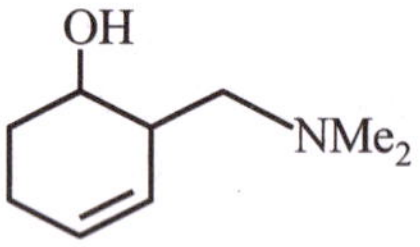 (4) 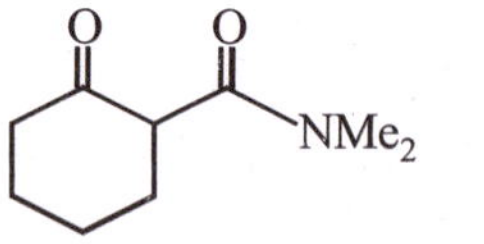

278. Sodium salt of an organic acid 'X' produces effervescences with conc. H_2SO_4. 'X' reacts with the acidified aqueous $CaCl_2$ solution to give a white precipitate which decolourises acidic solution of $KMnO_4$. 'X' is : **[2017]**

(1) C_6H_5COONa (2) HCOONa
(3) CH_3COONa (4) $Na_2C_2O_4$

279. The major product obtained in the following reaction is : **[2017]**

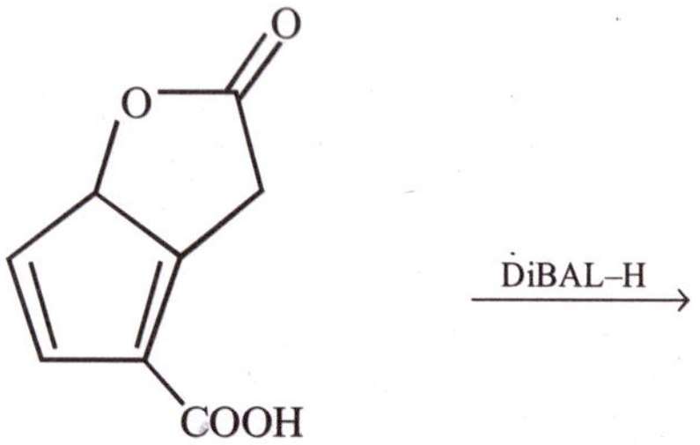

(1) 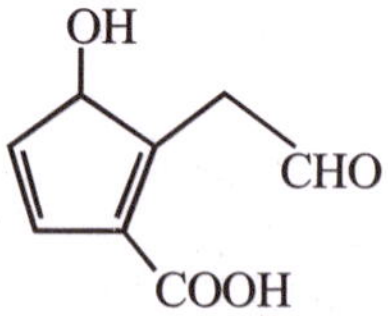 (2) 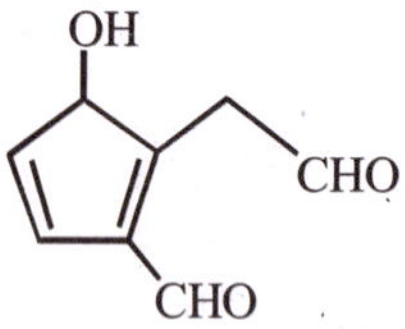

(3) 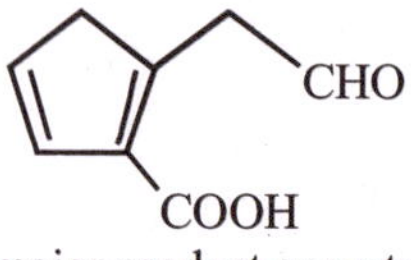 (4) 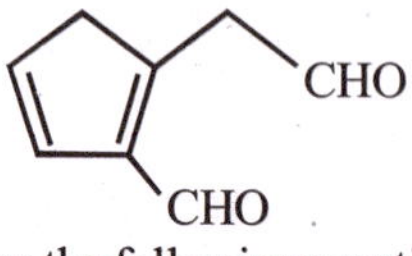

280. The major product expected from the following reaction is: **[Online April 8, 2017]**

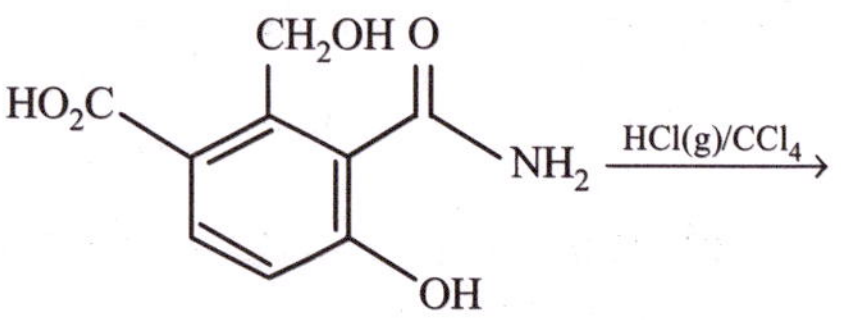

(1) 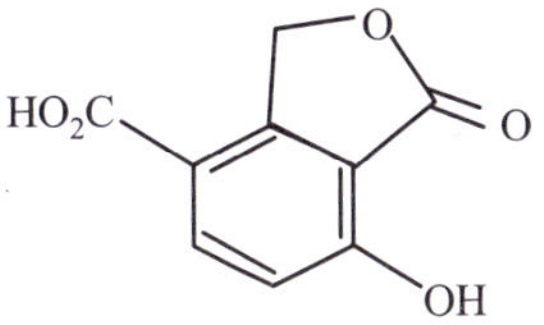

(2) 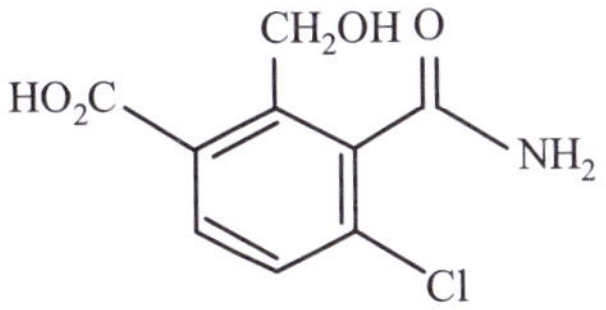

(3) 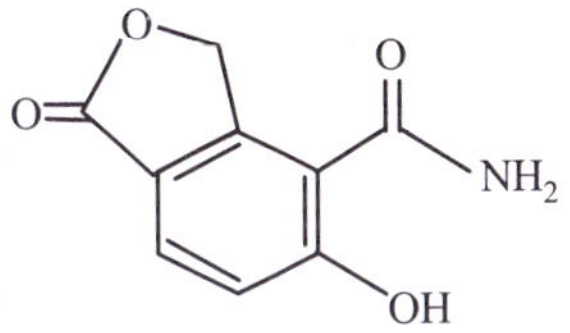

(4) 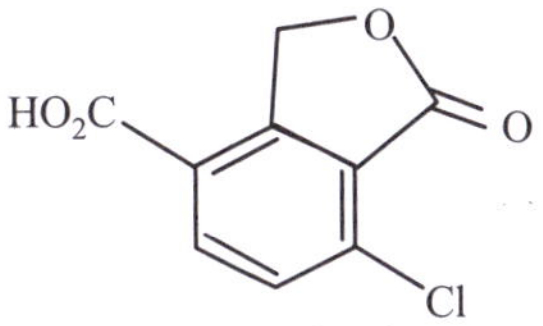

281. Bouveault–Blanc reduction reaction involves : **[Online April 9, 2016]**

(1) Reduction of an acyl halide with H_2/Pd
(2) Reduction of an anhydride with $LiAlH_4$
(3) Reduction of an ester with Na/C_2H_5OH
(4) Reduction of a carbonyl compound with Na/Hg and HCl.

282. 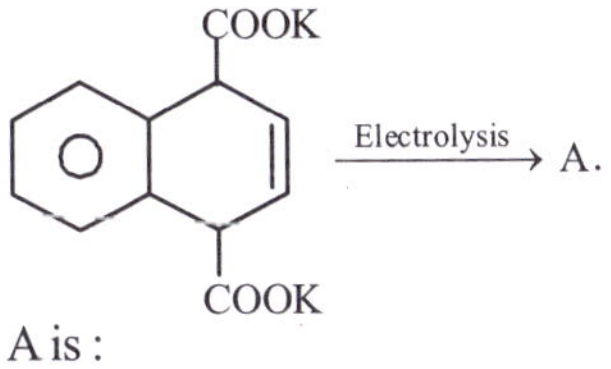

A is : **[Online April 10, 2015]**

(1) 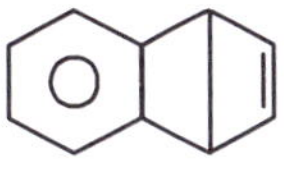 (2) 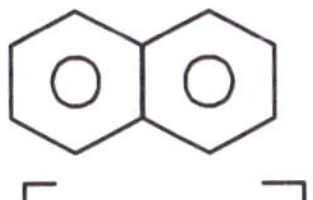

(3) 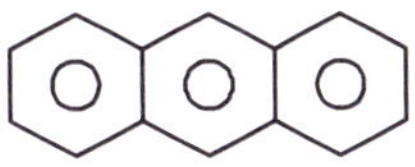 (4) 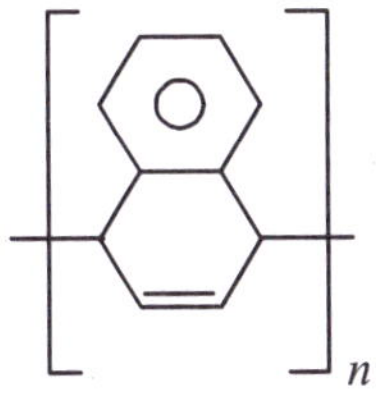

283. 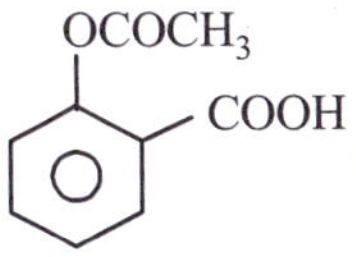 is used as : **[Online April 10, 2015]**

(1) Analgesic (2) Insecticide
(3) Antacid (4) Antihistamine

284. In the presence of a small amount of phosphorus, aliphatic carboxylic acids react with chlorine or bromine to yield a compound in which α-hydrogen has been replaced by halogen. This reaction is known as : **[Online April 10, 2015]**

(1) Wolff - Kishner reaction
(2) Rosenmund reaction
(3) Etard reaction
(4) Hell - Volhard - Zelinsky reaction

285. In the reaction,

$$CH_3COOH \xrightarrow{LiAlH_4} A \xrightarrow{PCl_5} B \xrightarrow{Alc.KOH} C,$$

the product C is: **[2014]**

(1) Acetaldehyde (2) Acetylene
(3) Ethylene (4) Acetyl chloride

286. Phthalic acid reacts with resorcinol in the presence of concentrated H_2SO_4 to give: **[Online April 12, 2014]**

(1) Phenolphthalein (2) Alizarin
(3) Coumarin (4) Fluorescein

287. Monocarboxylic acids are functional isomers of: **[Online April 23, 2013]**

(1) Ethers (2) Amines (3) Esters (4) Alcohols

288. Maleic acid and fumaric acids are **[Online May 26, 2012]**

(1) chain isomers (2) functional isomers
(3) tautomers (4) geometrical isomers

289. The correct order of increasing acid strength of the compounds **[2006]**

(A) CH_3CO_2H (B) $MeOCH_2CO_2H$

(C) CF_3CO_2H (D) 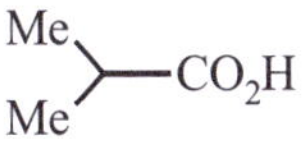

is

(1) D < A < B < C (2) A < D < B < C
(3) B < D < A < C (4) D < A < C < B

290. Among the following acids which has the lowest pK_a value? **[2005]**

(1) CH_3CH_2COOH (2) $(CH_3)_2CH-COOH$
(3) HCOOH (4) CH_3COOH

291. When $CH_2 = CH — COOH$ is reduced with $LiAlH_4$, the compound obtained will be **[2003]**

(1) $CH_2=CH—CH_2OH$ (2) $CH_3—CH_2—CH_2OH$
(3) $CH_3—CH_2—CHO$ (4) $CH_3—CH_2—COOH$

292. $CH_3CH_2COOH \xrightarrow[\text{red P}]{Cl_2} A \xrightarrow{\text{alc. KOH}} B$. What is *B*? **[2002]**

(1) CH_3CH_2COCl (2) CH_3CH_2CHO
(3) $CH_2=CHCOOH$ (4) $ClCH_2CH_2COOH$.

NUMERICAL VALUE QUESTIONS (NVQs)

293. Grignard reagent RMgBr(P) reacts with water and forms a gas (Q). One gram of Q occupies 1.4 dm^3 at STP. (P) on reaction with dry ice in dry ether followed by H_3O^+ forms a compound (Z). 0.1 mole of (Z) will weigh _______. (Nearest integer) **[Jan. 24, 2026 (II)]**

294. Major product B of the following reaction has _______ π-bond. **[April 8, 2024 (I)]**

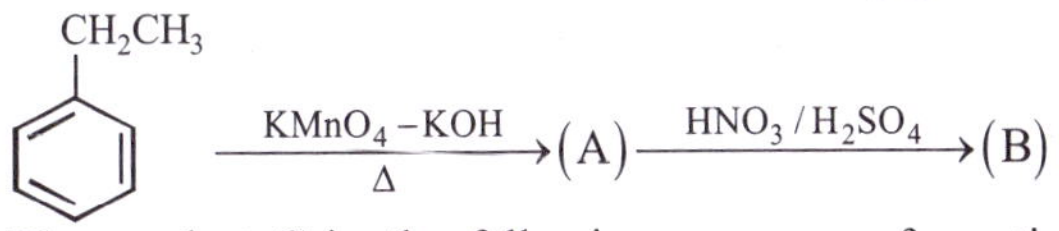

295. The product C in the following sequence of reactions has _______ π bonds. **[April 5, 2024 (II)]**

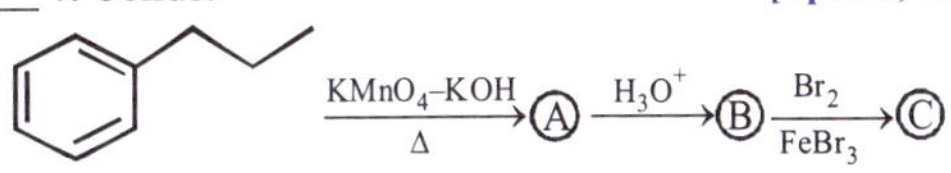

296. X g of ethylamine is subjected to reaction with $NaNO_2/HCl$ followed by water; evolved dinitrogen gas which occupied 2.24 L volume at STP.
X is ______ $\times 10^{-1}$ g. **[April 4, 2024 (I)]**

297. The total number of 'sigma' and 'Pi' bonds in 2-oxohex-4-ynoic acid is _____. **[April 4, 2024 (II)]**

298. In the given reaction

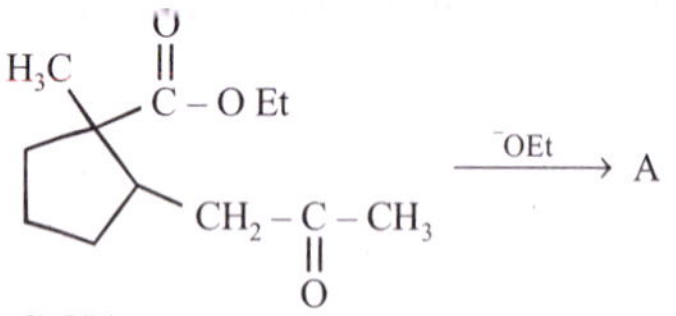

(Where Et is $-C_2H_5$)
The number of chiral carbon/s in product A is _________. **[July 25, 2022 (I)]**

299.

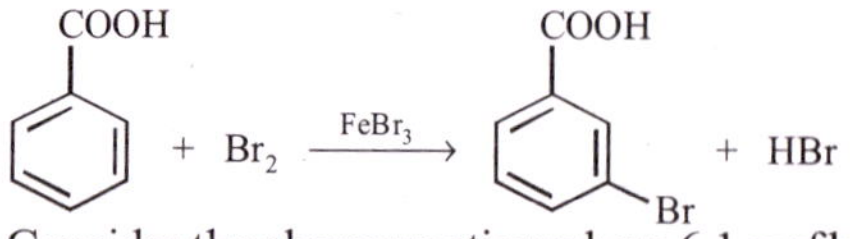

Consider the above reaction where 6.1 g of benzoic acid is used to get 7.8 g of *m*-bromobenzoic acid.
The precentage yield of the product is ________.
(Round off to nearest integer).
[Given : Atomic masses : C : 120 u, H : 1.0 u, O : 16.0 u, Br : 80.0 u] **[March 18, 2021 (II)]**

300. The number of compound/s given below which contain/s —COOH group is ____. (Integer answer) **[Feb. 25, 2021 (II)]**
(A) Sulphanilic acid (B) Picric acid
(C) Aspirin (D) Ascorbic acid

Part-B (JEE Advanced)

Topic 1 - Methods of Preparation of Carbonyl Compounds

2 Integer Value Answer

1. In the reaction given below, the total number of atoms having sp^2 hybridization in the major product P is _____

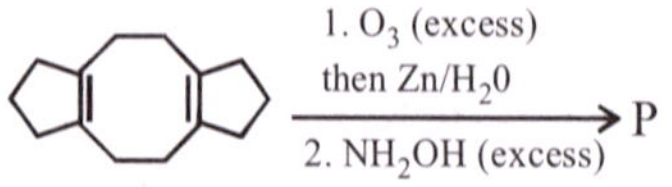

[Adv. 2021]

2. Among the following, the number of reaction(s) that produce(s) benzaldehyde is **[Adv. 2015]**

I. 

II. 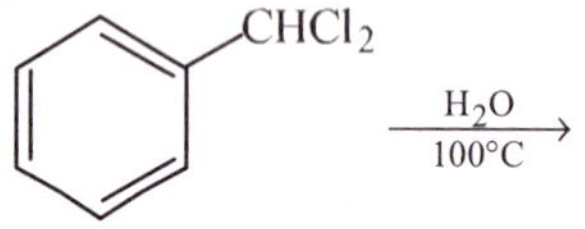

III. 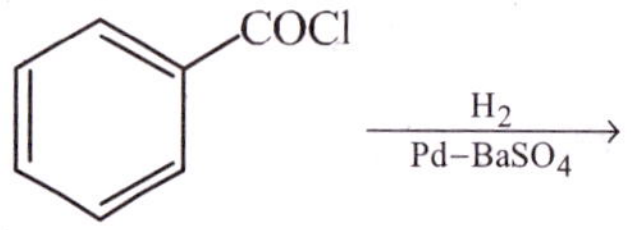

IV. 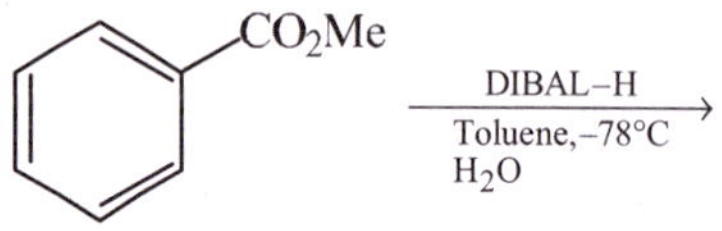

3. In the scheme given below, the total number of intramolecular aldol condensation products formed from '**Y**' is

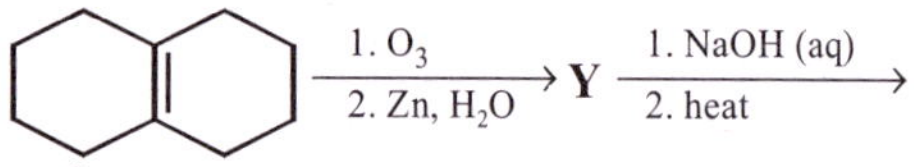

[2010]

4.  $+ Cl-CH_2CH_2CH_3$

$\xrightarrow{AlCl_3} P \xrightarrow[(ii)\ H_3O^+]{(i)\ O_2/\Delta} Q + \text{Phenol}$

The major products P and Q are **[2006 - 5M, –1]**

(a) 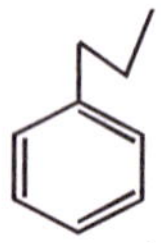 and CH_3CH_2CHO

(b) 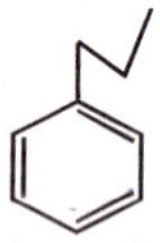 and CH_3COCH_3

(c) 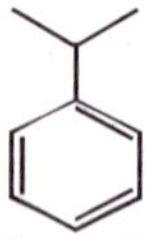 and CH_3COCH_3

(d) 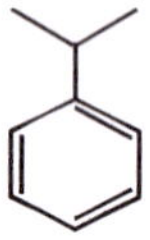 and CH_3CH_2CHO

4 Fill in the Blanks

5. The structure of the intermediate product, formed by the oxidation of toluene with CrO_3 and acetic anhydride, whose hydrolysis gives benzaldehyde is .... **[1992 - 1 Mark]**

6 MCQs with One or More than One Correct Answer

6. Reaction of *iso*-propylbenzene with O_2 followed by the treatment with H_3O^+ forms phenol and a by-product **P**. Reaction of **P** with 3 equivalents of Cl_2 gives compound **Q**. Treatment of **Q** with $Ca(OH)_2$ produces compound **R** and calcium salt **S**. The correct statement(s) regarding **P**, **Q**, **R** and **S** is(are) **[Adv. 2024]**

(a) Reaction of **P** with **R** in the presence of KOH followed by acidification gives

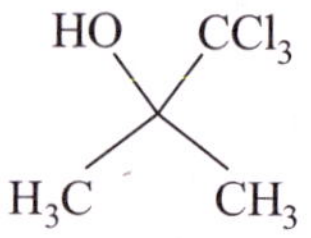

(b) Reaction of **R** with O_2 in the presence of light gives phosgene gas
(c) **Q** reacts with aqueous NaOH to produce Cl_3CCH_2OH and $Cl_3CCOONa$
(d) **S** on heating gives **P**

7. In the following reactions, P, Q, R and S are the major products. [Adv. 2023]

$$CH_3CH_2CH(CH_3)CH_2CN \xrightarrow[\text{(ii) PhMgBr, then } H_2O]{\text{(i) PhMgBr, then } H_3O^{\oplus}} P$$

$$Ph-H + CH_3\overset{O}{\overset{\|}{C}}Cl \xrightarrow[\text{(ii) PhMgBr, then } H_2O]{\text{(i) anhyd. } AlCl_3} Q$$

$$CH_3CH_2\overset{O}{\overset{\|}{C}}Cl \xrightarrow[\text{(ii) PhMgBr, then } H_2O]{\text{(i) } \frac{1}{2}(PhCH_2)_2Cd} R$$

$$PhCH_2CHO \xrightarrow[\text{(iii) HCN; (iv) } H_2SO_4, \Delta]{\text{(i) PhMgBr, then } H_2O\text{; (ii) } CrO_3, \text{dil. } H_2SO_4} S$$

The correct statement(s) about P, Q, R and S is(are)

(a) Both P and Q have asymmetric carbon(s).
(b) Both Q and R have asymmetric carbon(s).
(c) Both P and R have asymmetric carbon(s).
(d) P has asymmetric carbon(s), S does not have any asymmetric carbon.

8. Considering the following reaction sequence, the correct statement(s) is(are) [Adv. 2022]

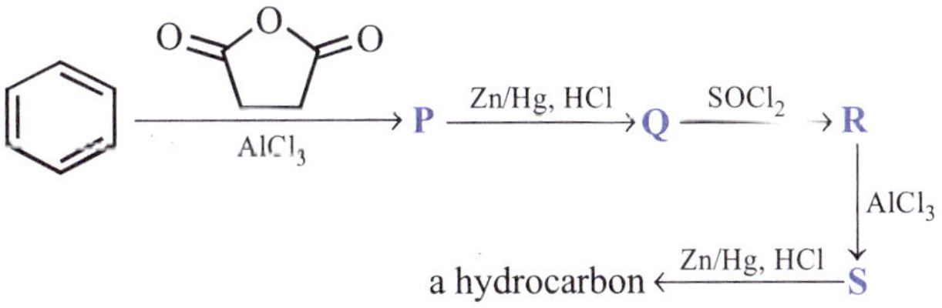

(a) Compounds P and Q are carboxylic acids.
(b) Compound S decolorizes bromine water.
(c) Compounds P and S react with hydroxylamine to give the corresponding oximes.
(d) Compound R reacts with dialkylcadmium to give the corresponding tertiary alcohol.

9. In the reaction scheme shown below, Q, R, and S are the major products. [Adv. 2020]

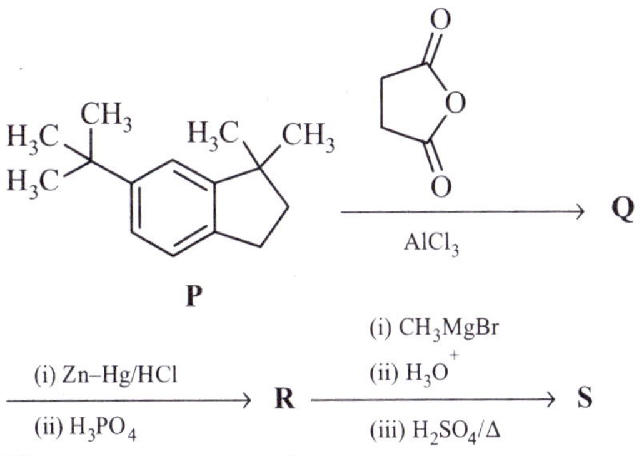

The correct structure of

(a) S is 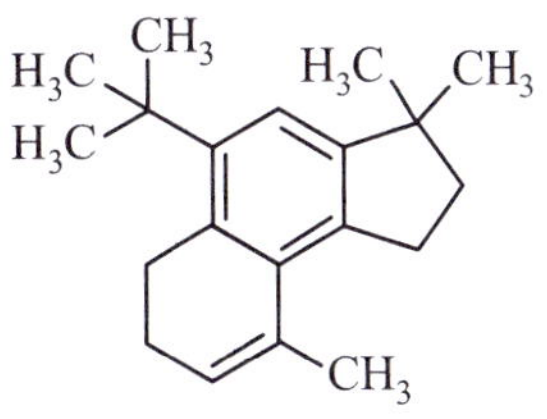

(b) Q is 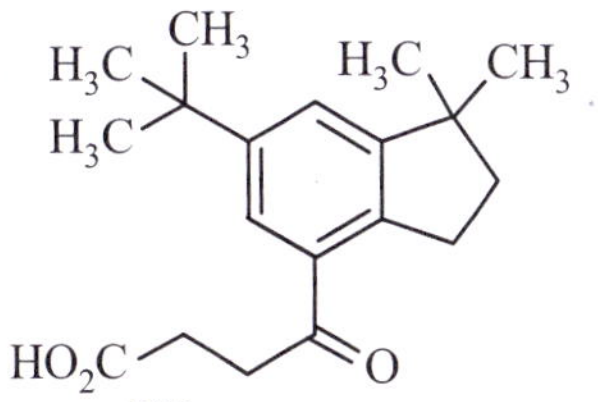

(c) R is 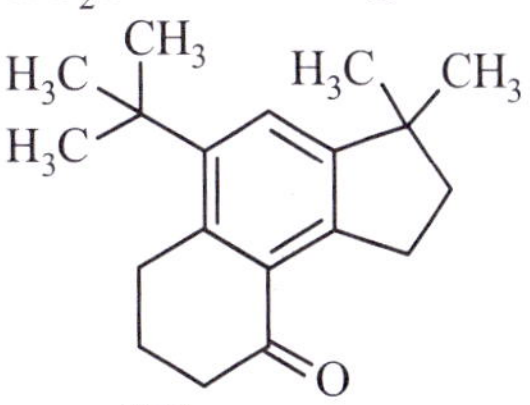

(d) S is 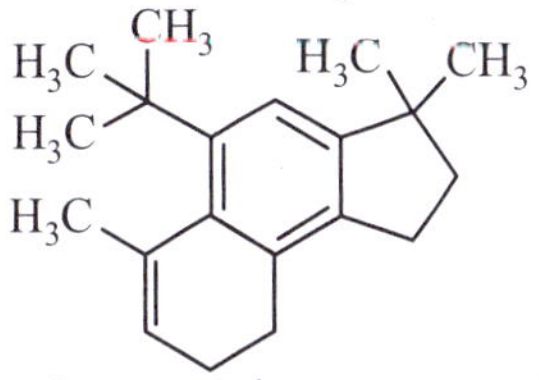

10. Compounds *P* and *R* upon ozonolysis produce *Q* and *S*, respectively. The molecular formula of *Q* and *S* is C_8H_8O. *Q* undergoes Cannizzaro reaction but not haloform reaction, whereas *S* undergoes haloform reaction but not Cannizzaro reaction

(i) $P \xrightarrow[\text{ii) } Zn/H_2O]{\text{i) } O_3/CH_2Cl_2} \underset{(C_8H_8O)}{Q}$

(ii) $R \xrightarrow[\text{ii) } Zn/H_2O]{\text{i) } O_3/CH_2Cl_2} \underset{(C_8H_8O)}{S}$

The option(s) with suitable combination of *P* and *R*, respectively, is (are) [Adv. 2017]

(a) 

(b) 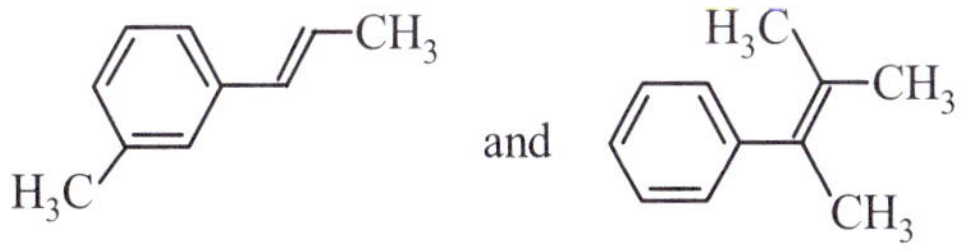

(c) 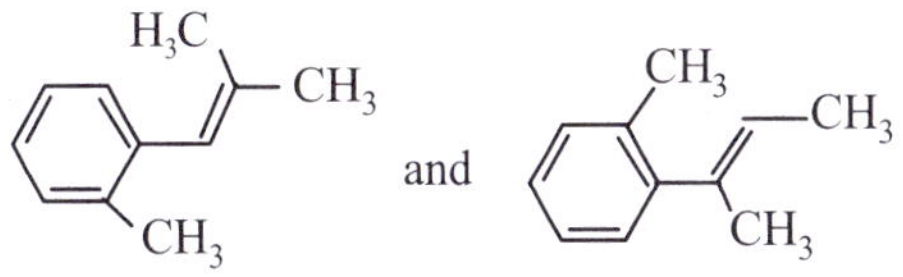

(d) 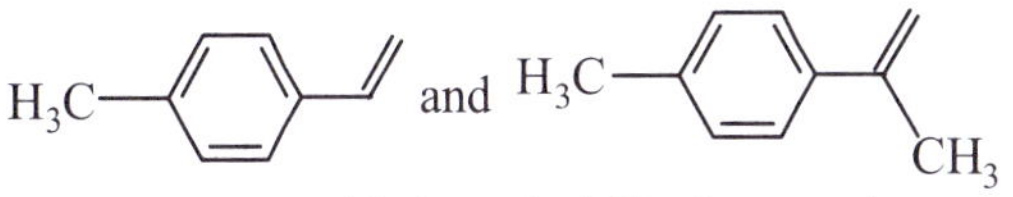

11. The correct statement(s) about the following reaction sequence is (are) [Adv. 2016]

$$\text{Cumene}(C_9H_{12}) \xrightarrow[\text{(ii) } H_3O^+]{\text{(i) } O_2} \mathbf{P} \xrightarrow{CHCl_3/NaOH} \mathbf{Q}\text{ (major)} + \mathbf{R}\text{ (minor)}$$

$$\mathbf{Q} \xrightarrow[PhCH_2Br]{NaOH} \mathbf{S}$$

(a) **R** is steam volatile
(b) **Q** gives dark violet coloration with 1% aqueous $FeCl_3$ solution
(c) **S** gives yellow precipitate with 2, 4–dinitrophenylhydrazine
(d) **S** gives dark violet coloration with 1% aqueous $FeCl_3$ solution

12. In the following reaction, the product(s) formed is (are) **[Adv. 2013]**

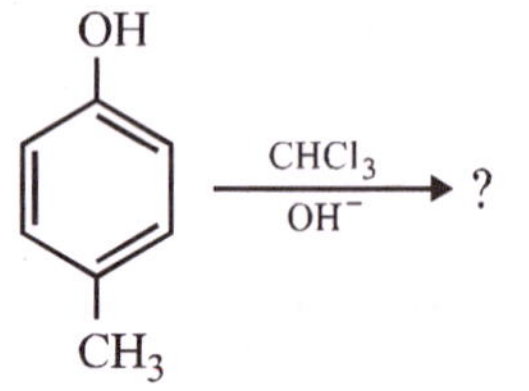

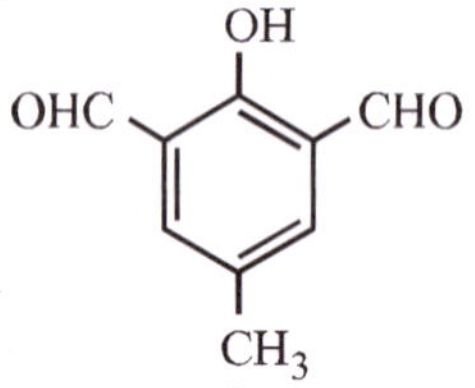

P

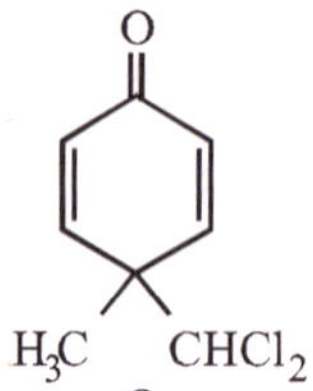

Q

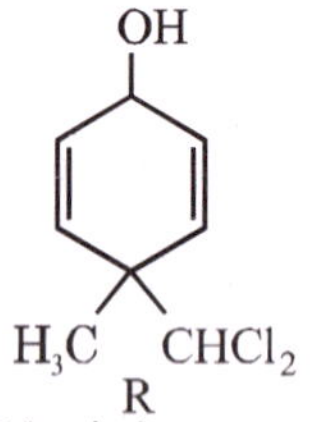

R

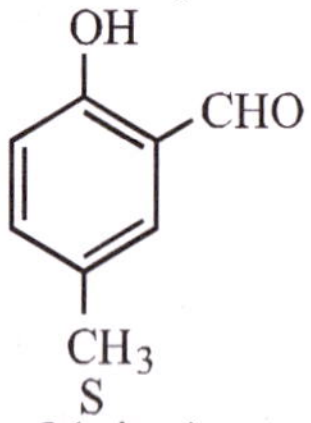

S

(a) *P* (major) (b) *Q* (minor)
(c) *R* (minor) (d) *S* (major)

8 Comprehension Based Questions

Passage-IV

A tertiary alcohol *H* upon acid catalysed dehydration gives a product I. Ozonolysis of *I* leads to compounds *J* and *K*. Compound *J* upon reaction with KOH gives benzyl alcohol and compound *L*, whereas *K* on reaction with KOH gives only *M*. **[2008]**

M = 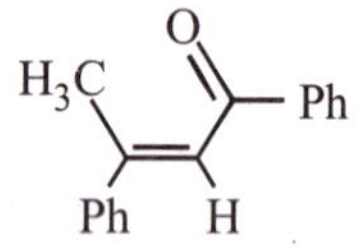

13. Compound *H* is formed by the reaction of

(a) 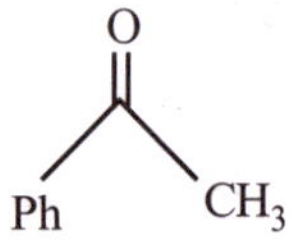 + PhMgBr

(b) 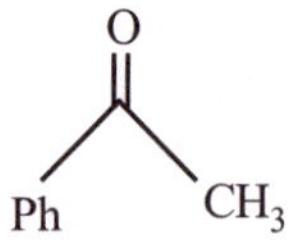 + $PhCH_2MgBr$

(c) 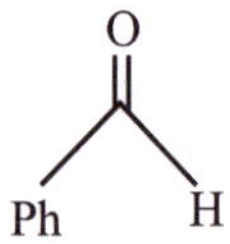 + $PhCH_2MgBr$

(d) 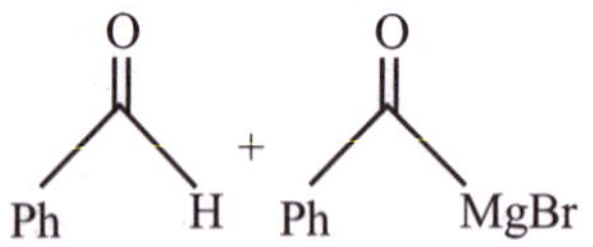

14. The structure of compound *I* is

(a) 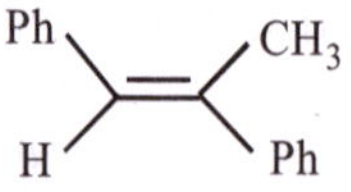 (b) 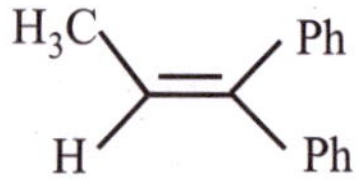

(c) 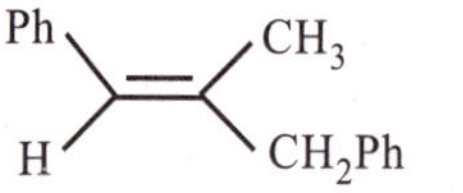 (d) 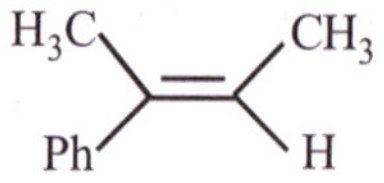

15. The structure of compounds *J*, *K* and *L* respecitvely, are –
(a) $PhCOCH_3$, $PhCH_2COCH_3$ and $PhCH_2COO^-K^+$
(b) PhCHO, $PhCH_2CHO$ and $PhCOO^-K^+$
(c) $PhCOCH_3$, $PhCH_2CHO$ and $CH_3COO^-K^+$
(d) PhCHO, $PhCOCH_3$ and $PhCOO^-K^+$

7 Match the Following

16. The **List-II** contains products obtained from the reaction of compounds in **List-I** with $O_3/Zn\text{-}H_2O$ followed by cyclization (via more stable enolate) in the presence of aqueous NaOH. Match each entry in **List-I** with appropriate entry in **List-II** and choose the correct option. **[Adv. 2026]**

List-I | **List-II**

(P) 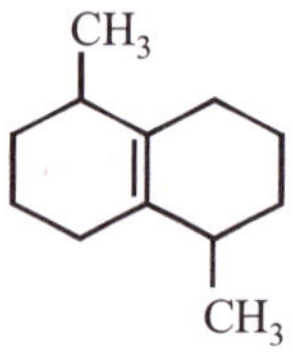 (1) 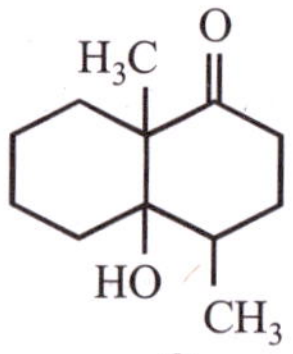

(Q) 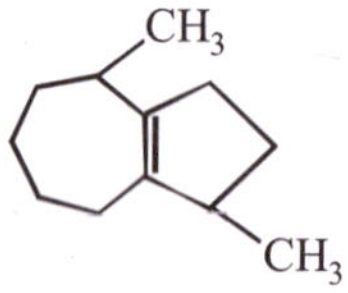 (2) 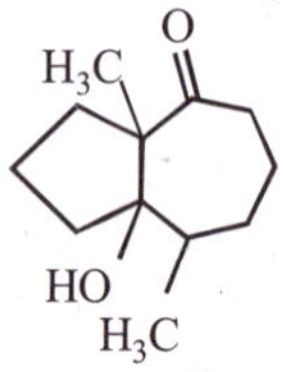

(R) 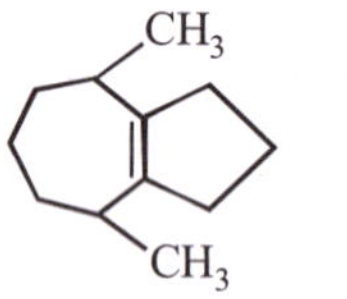 (3) 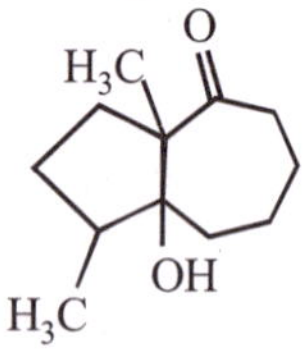

(S) 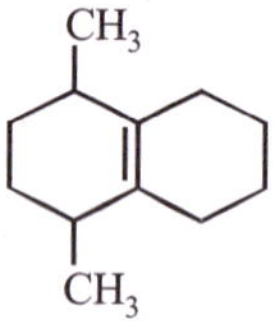 (4) 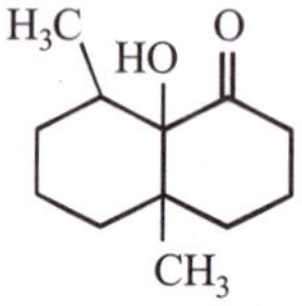

(5) 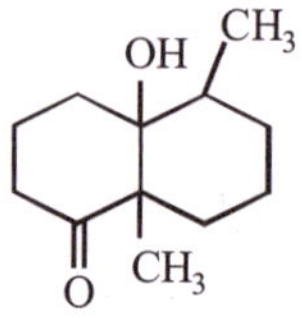

(a) P→2, Q→4, R→1, S→3
(b) P→3, Q→4, R→5, S→2
(c) P→2, Q→1, R→5, S→3
(d) P→3, Q→5, R→4, S→2

9 Assertion and Reason/Statement Type Questions

Each question contains **STATEMENT-1 (Assertion)** and **STATEMENT-2 (Reason)**. Each question has 4 choices (a), (b), (c) and (d) out of which ONLY ONE is correct. Mark your answer as
(a) If both Statement -1 and Statement -2 are correct, and Statement -2 is the correct explanation of the Statement -2.
(b) If both Statement -1 and Statement -2 are correct, but Statement -2 is not the correct explanation of the Statement -1.
(c) If Statement -1 is correct but Statement -2 is incorrect.
(d) If Statement -1 is incorrect but Statement -2 is correct.

17. **Statement-1 :** Dimethyl sulphide is commonly used for the reduction of an ozonide of an alkene to get the carbonyl compounds.
Statement-2 : It reduces the ozonide giving water soluble dimethyl sulphoxide and excess of it evaporates. **[2001S]**

10 Subjective Problems

18. A monomer of a polymer on ozonolysis gives two moles of CH_2O and one mole of CH_3COCHO. Write the structure of monomer and write all – '*cis*' configuration of polymer chain. **[2005 - 2 Marks]**

19. Identify (A), (B), (C), (D) and (E) in the following schemes and write their structures : **[2001 - 5 Marks]**

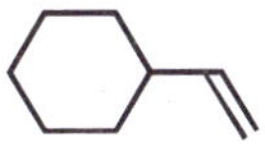 $\xrightarrow{Br_2/CCl_4}$ (A) $\xrightarrow{NaNH_2}$ (B)

$\xrightarrow{HgSO_4/H_2SO_4}$ (C) $\xrightarrow{NH_2NHCONH_2}$ (D)

[C] $\xrightarrow{NaOD/D_2O(excess)}$ (E)

20. Write the intermediate steps for the following reaction.

$$C_6H_5CH(OH)C \equiv CH \xrightarrow{H_3O^+} C_6H_5CH = CHCHO$$

[1998 - 2 Marks]

21. Suggest appropriate structures for the missing compounds. (The number of carbon atoms remains the same throughout the reactions.) **[1996 - 3 Marks]**

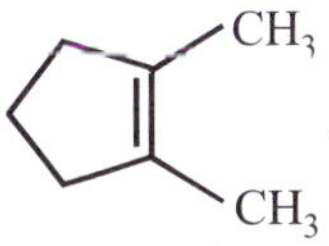 $\xrightarrow{\text{dil. } KMnO_4} A \xrightarrow{HIO_4} B \xrightarrow{OH^-} C$

22. An organic compound *A*, C_8H_6, on treatment with dilute sulphuric acid containing mercuric sulphate gives a compound *B*, which can also be obtained from a reaction of benzene with an acid chloride in the presence of anhydrous aluminium chloride. The compound *B*, when treated with iodine in aqueous KOH, yields *C* and a yellow compound *D*. Identify *A*, *B*, *C* and *D* with justification. Show how *B* is formed from *A*. **[1994 - 3 Marks]**

23. Outline the reaction sequence for the conversion of
(*i*) Carry out the following transformation in not more than three steps. **[1999 - 3 Marks]**

$$CH_3-CH_2-C\equiv C-H \rightarrow CH_3-CH_2-CH_2-\overset{\overset{\displaystyle O}{||}}{C}-CH_3$$

(*ii*) acetylene to acetone **[1985 - 1 Mark]**
(*iii*) methanal to ethanal (the number of steps should not be more than three). **[1981 - 2 Marks]**

Topic 2 - Properties of Carbonyl Compounds

1 MCQs with One Correct Answer

24. The correct reaction/reaction sequence that would produce a dicarboxylic acid as the major product is **[Adv. 2025]**

(a)  $\xrightarrow[\text{(iii) } H_3O^+]{\text{(i) NaCN, (ii) } HO^-, H_2O}$

(b) $\begin{matrix} CHO \\ | \\ (CHOH)_4 \\ | \\ CH_2OH \end{matrix} \xrightarrow{Br_2, H_2O}$

(c) 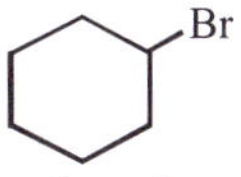 $\xrightarrow{\text{(i) KOH, EtOH; (ii) } KMnO_4, H_2SO_4, \Delta}$

(d) 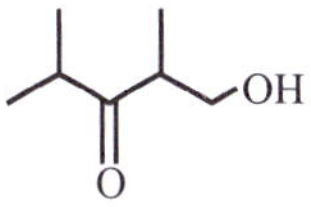 $\xrightarrow{H_2CrO_4}$

25. The major product of the following reaction sequence is **[Adv. 2016]**

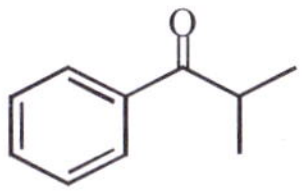 $\xrightarrow[\text{(ii) HCHO/}H^+\text{(catalytic amount)}]{\text{(i) HCHO (excess)/NaOH, heat}}$

(a) 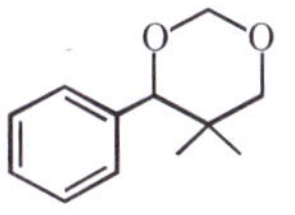
(b) 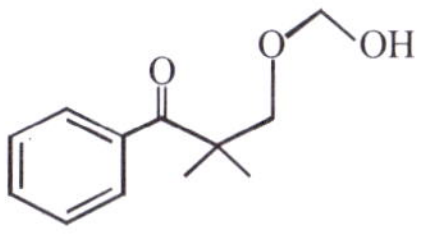
(c) 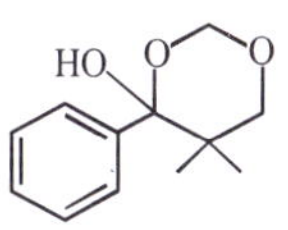
(d) 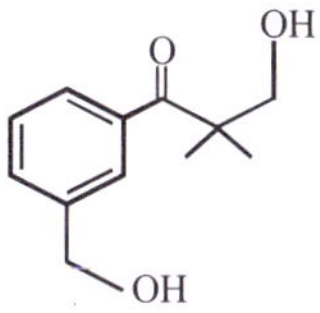

26. The major product of the following reaction is **[Adv. 2015]**

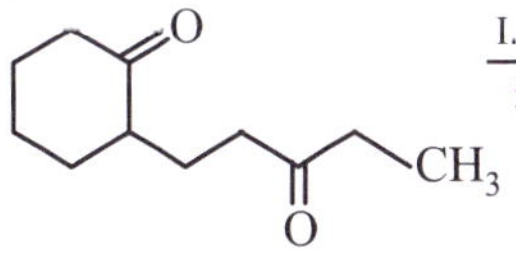 $\xrightarrow[\text{ii. } H^+\text{, heat}]{\text{I. KOH, } H_2O}$

(a) 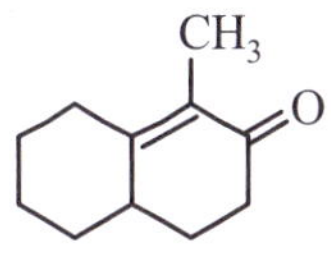
(b) 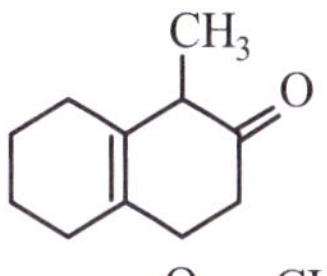
(c) 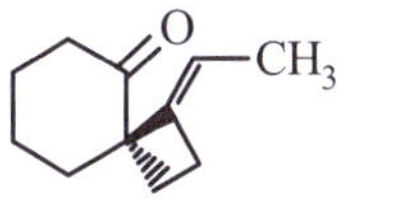
(d) 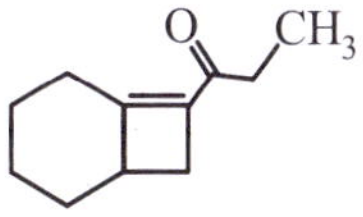

27. The major product in the following reaction is

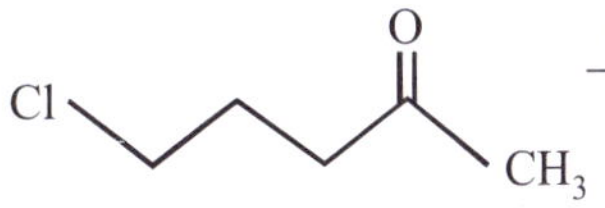 $\xrightarrow[\text{2. aq. acid}]{\text{1. } CH_3MgBr\text{, dry ether, 0°C}}$

[Adv. 2014]

(a) 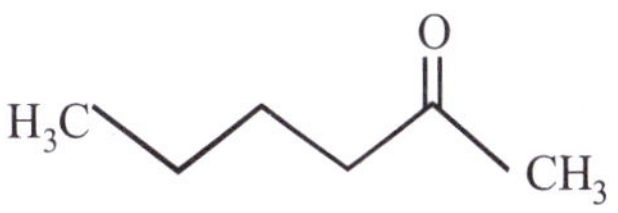
(b) 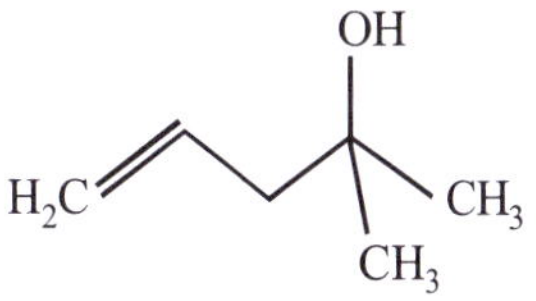
(c) 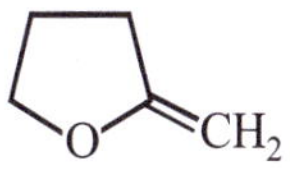
(d) 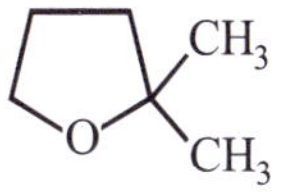

28. After completion of the reactions (I and II), the organic compound(s) in the reaction mixtures is (are) [Adv. 2013]

Reaction I : 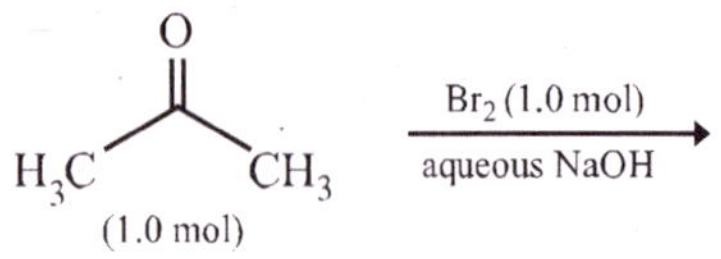

Reaction II : 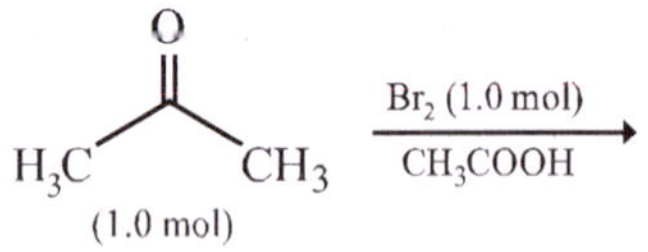

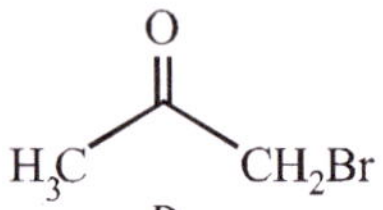
P

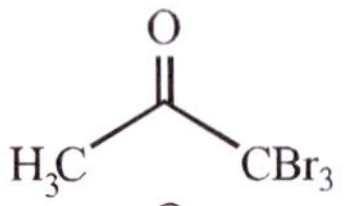
Q

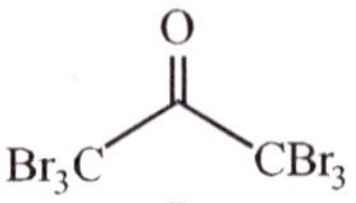
R

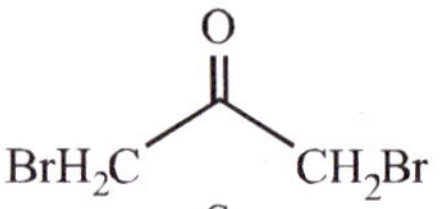
S

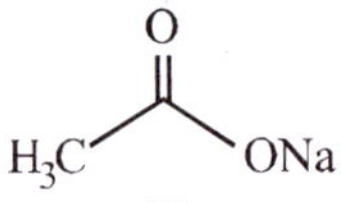
T

$CHBr_3$
U

(a) Reaction I : *P* and Reaction II : *P*
(b) Reaction I : *U*, acetone and Reaction II : *Q*, acetone
(c) Reaction I : *T*, *U*, acetone and Reaction II : *P*
(d) Reaction I : *R*, acetone and Reaction II : *S*, acetone

29. The major product *H* of the given reaction sequence is

$$CH_3—CH_2—CO—CH_3 \xrightarrow{CN^-} G \xrightarrow[\text{Heat}]{95\%\ H_2SO_4} H$$

[2012 - II]

(a) $CH_3—CH{=}C(CH_3)—COOH$
(b) $CH_3—CH{=}C(CH_3)—CN$
(c) $CH_3—CH_2—C(OH)(CH_3)—COOH$
(d) $CH_3—CH{=}C(CH_3)—CO—NH_2$

30. In the following reaction,

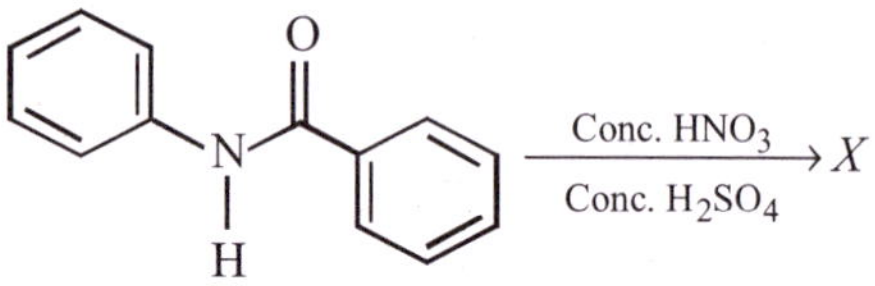

the structure of the major product '*X*' is [2007]

(a) 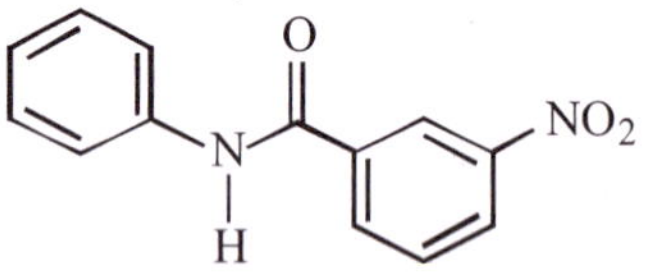

(b) 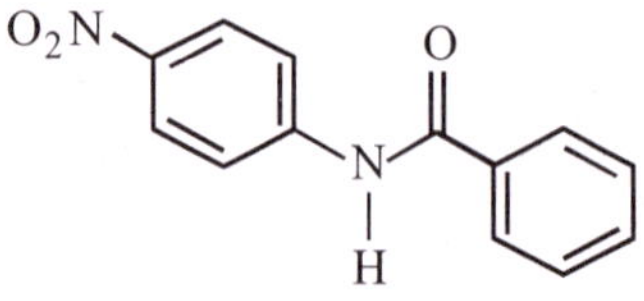

(c) 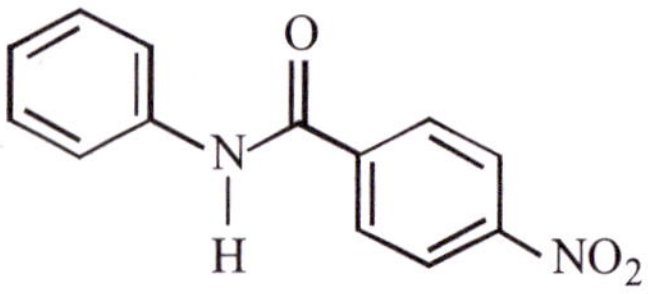

(d) 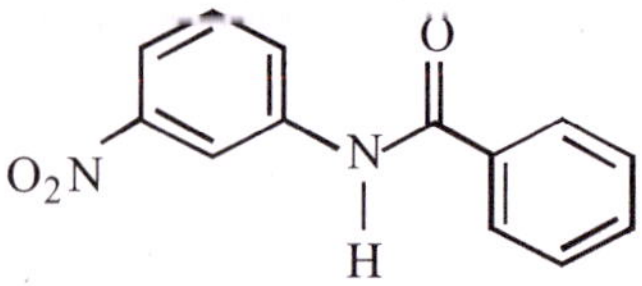

31. 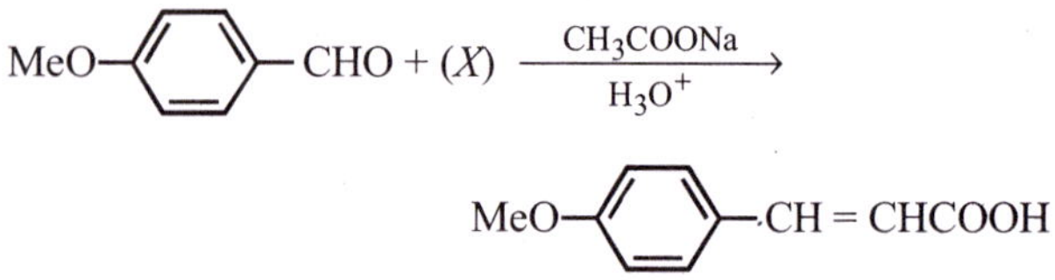

The compound (*X*) is [2005S]

(a) CH_3COOH (b) $BrCH_2-COOH$
(c) $(CH_3CO)_2O$ (d) $CHO-COOH$

32. How will you convert butan–2–one to propanoic acid? [2005S]

(a) Tollen's reagent (b) Fehling's solution
(c) $NaOH/I_2/H^+$ (d) $NaOH/NaI/H^+$

33. The correct order of reactivity of PhMgBr with [2004S]

$Ph-\overset{O}{\overset{\|}{C}}-Ph$ (I) $CH_3-\overset{O}{\overset{\|}{C}}-H$ (II) $CH_3-\overset{O}{\overset{\|}{C}}-CH_3$ (III) is

(a) (I) > (II) > (III) (b) (III) > (II) > (I)
(c) (II) > (III) > (I) (d) (I) > (III) > (II)

34. 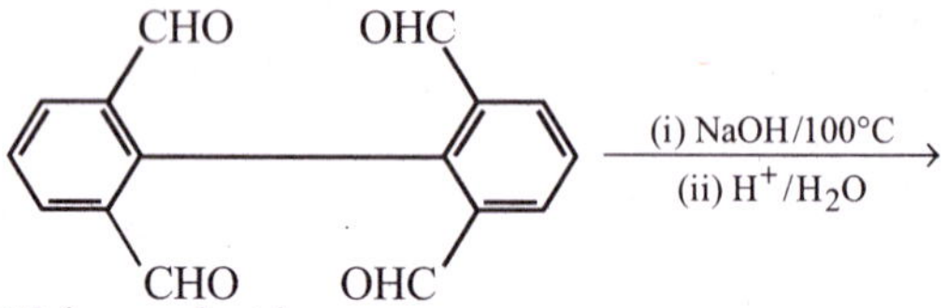

Major product is : [2003S]

(a) 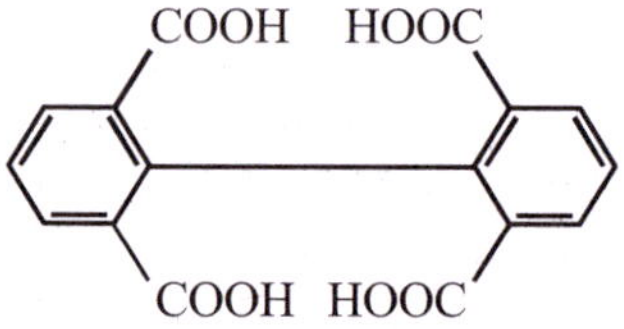

(b) 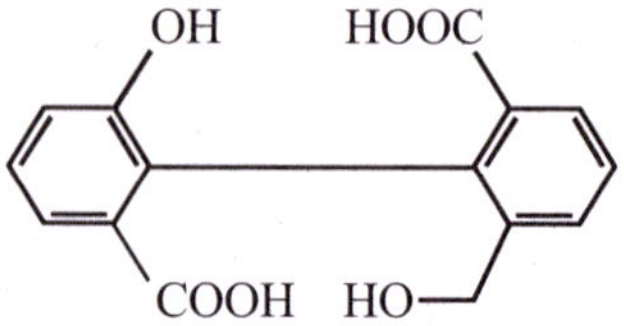

(c) 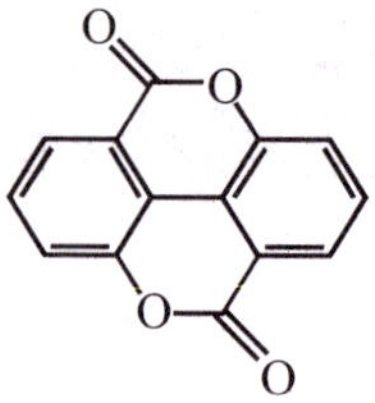

(d) 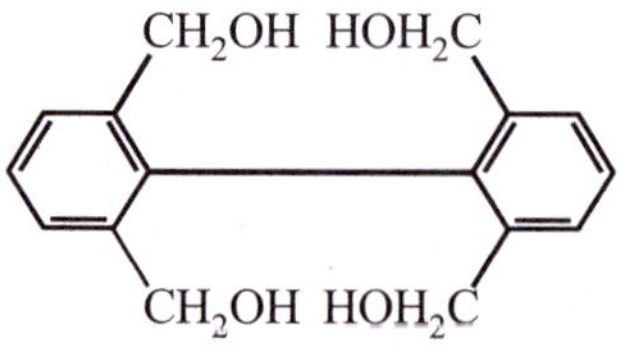

35. The product of acid hydrolysis of *P* and *Q* can be distinguished by [2003S]

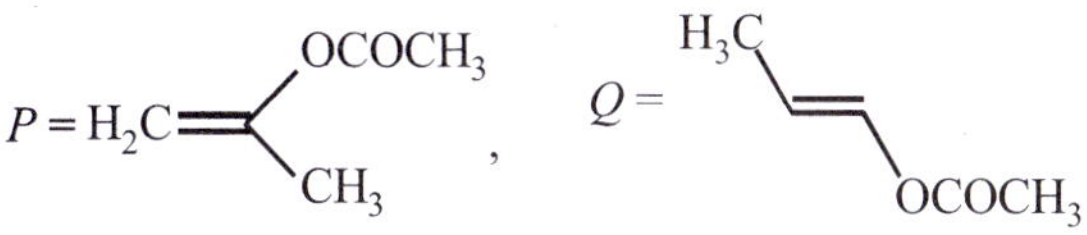

(a) Lucas Reagent (b) 2,4–DNP
(c) Fehling's Solution (d) $NaHSO_3$

36. A mixture of benzaldehyde and formaldehyde on heating with aqueous NaOH solution gives [2001S]
(a) benzyl alcohol and sodium formate
(b) sodium benzoate and methyl alcohol
(c) sodium benzoate and sodium formate
(d) benzyl alcohol and methyl alcohol

37. The appropriate reagent for the following transformation is [2000S]

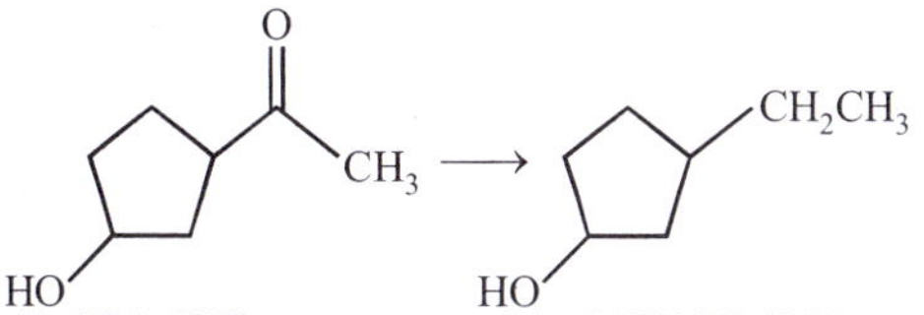

(a) Zn(Hg), HCl (b) NH_2NH_2/OH^-
(c) H_2/Ni (d) $NaBH_4$

38. Which one of the following will most readily be dehydrated in acidic condition ? [2000S]

(a) 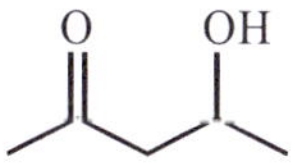

(b) 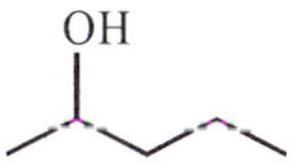

(c) 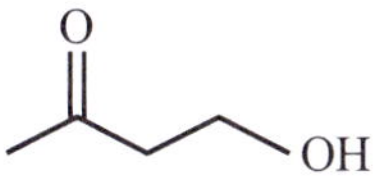

(d) 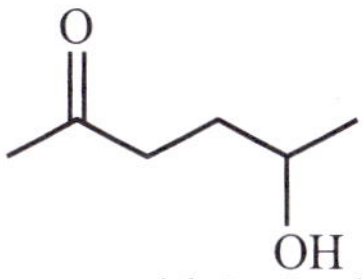

39. The enol form of acetone, after treatment with D_2O, gives. [1999 - 2 Marks]
(a) $CH_3-\overset{\overset{OD}{|}}{C}=CH_2$ (b) $CD_3-\overset{\overset{O}{||}}{C}-CD_3$
(c) $CH_2=\overset{\overset{OH}{|}}{C}-CH_2D$ (d) $CD_2=\overset{\overset{OD}{|}}{C}-CD_3$

40. In the Cannizzaro reaction given below, [1996 - 1 Mark]
$$2PhCHO \xrightarrow{OH^-} PhCH_2OH + PhCO_2^-,$$
the slowest step is
(a) the attack of ^-OH at the carbonyl group,
(b) the transfer of hydride to the carbonyl group,
(c) the abstraction of proton from the carboxylic acid,
(d) the deprotonation of $PhCH_2OH$.

41. The formation of cyanohydrin from a ketone is an example of : [1990 - 1 Mark]
(a) Electrophilic addition
(b) Nucleophilic addition
(c) Nucleophilic substitution
(d) Electrophilic substiution

42. *m*-Chlorobenzaldehyde on reaction with conc. KOH at room temperature gives [1991 - 1 Mark]
(a) potassium *m*-chlorobenzoate and *m*-hydroxybenzaldehyde
(b) *m*-hydroxybenzaldehyde and *m*-chlorobenzyl alcohol
(c) *m*-chlorobenzyl alcohol and *m*-hydroxybenzyl alcohol
(d) potassium *m*-chlorobenzoate and *m*-chlorobenzyl alcohol.

43. The enolic form of acetone contains [1990 - 1 Mark]
(a) 9 sigma bonds, 1 pi-bond and 2 lone pairs
(b) 8 sigma bonds, 2 pi-bonds and 2 lone pairs
(c) 10 sigma bonds, 1 pi-bond and 1 lone pair
(d) 9 sigma bonds, 2 pi-bonds and 1 lone pair

44. Polarisation of electrons in acrolein may be written as [1988 - 1 Mark]
(a) $\overset{\delta^-}{C}H_2=CH-\overset{\delta^+}{C}H=O$ (b) $\overset{\delta^-}{C}H_2=CH-CH=\overset{\delta^+}{O}$
(c) $\overset{\delta^-}{C}H_2=\overset{\delta^+}{C}H-CH=O$ (d) $\overset{\delta^+}{C}H_2=CH-CH=\overset{\delta^-}{O}$

45. The compound that will not give iodoform on treatment with alkali and iodine is : [1985 - 1 Mark]
(a) acetone (b) ethanol
(c) diethyl ketone (d) isopropyl alcohol

46. The Cannizzaro reaction is not given by [1983 - 1 Mark]
(a) trimethylacetaldehye (b) acetaldehyde
(c) benzaldehyde (d) formaldehyde

47. When acetaldehyde is heated with Fehling's solution it gives a precipitate of [1983 - 1 Mark]
(a) Cu (b) CuO
(c) Cu_2O (d) $Cu + Cu_2O + CuO$

48. A compound that gives a positive iodoform test is [1982 - 1 Mark]
(a) 1-pentanol (b) 2-pentanone
(c) 3-pentanone (d) pentanal

49. The reagent with which both acetaldehyde and acetone react easily is [1981 - 1 Mark]
(a) Fehling's reagent (b) Grignard reagent
(c) Schiff's reagent (d) Tollen's reagent

2 Integer Value Answer

50. Consider all possible isomeric ketones, including stereoisomers of MW = 100. All these isomers are independently reacted with $NaBH_4$ (**NOTE:** stereoisomers are also reacted separately). The total number of ketones that give a racemic product(s) is/are [Adv. 2014]

3 Numeric / New Stem Based Questions

51. An organic compound P having molecular formula $C_6H_6O_3$ gives ferric chloride test and does not have intramolecular hydrogen bond. The compound P reacts with 3 equivalents of NH_2OH to produce oxime Q. Treatment of P with excess methyl iodide in the presence of KOH produces compound R as the major product. Reaction of R with excess iso-butylmagnesium bromide followed by treatment with H_3O^+ gives compound S as the major product. [Adv. 2024]
The total number of methyl ($-CH_3$) group(s) in compound S is _____.

52. Complete reaction of acetaldehyde with excess formaldehyde, upon heating with conc. NaOH solution, gives **P** and **Q**. Compound **P** does not give Tollens' test, whereas **Q** on acidification gives positive Tollens' test. Treatment of **P** with excess cyclohexanone in the presence of catalytic amount of *p*-toluenesulfonic acid (PTSA) gives product **R**. Sum of the number of methylene groups ($-CH_2-$) and oxygen atoms in **R** is _____. [Adv. 2024]

4 Fill in the Blanks

53. The structure of the enol form of $CH_3–CO–CH_2–CO–CH_3$ with intramolecular hydrogen bonding is .... **[1993 - 1 Mark]**

54. Fehling's solution 'A' consists of an aqueous solution of copper sulphate, while Fehling's solution 'B' consists of an alkaline solution of .... **[1990 - 1 Mark]**

55. Each item from (i) to (x) given below indicates a reaction type, a process or a homologue. Match each of these items with the related phrase by writing the correct phrase in the corresponding vacant space given under each. The correct phrase must be picked only from those given below within brackets : **[1981 - 1 × 10 = 10 Marks]**

(Bayer's process, Nucleophilic addition, Free radical substitution, Ostwald's process, Homologous pair, Cyanamide process, Electrophilic substitution, Homolytic addition, Thermite process, Nucleophilic substitution)

(*i*) Cyclopropane, chlorine and light

....

(*ii*) Welding

....

(*iii*) Propanone and sodium bisulphite

....

(*iv*) Production of ammonia

....

(*v*) Chloromethane and methanol

....

(*vi*) Ore purification

....

(*vii*) Ethanal and methanal

....

(*viii*) Benzene, nitric acid and sulphuric acid

....

(*ix*) Production of nitric acid

....

(*x*) Propene, hydrogen bromide and a peroxide catalyst

....

5 True / False

56. The reaction of methyl magnesium iodide with acetone followed by hydrolysis gives secondary butanol. **[1987 - 1 Mark]**

57. The yield of ketone when a secondary alcohol is oxidized is more than the yield of aldehyde when a primary alcohol is oxidized. **[1983 - 1 Mark]**

58. Benzaldehyde undergoes aldol condensation in an alkaline medium. **[1982 - 1 Mark]**

6 MCQs with One or More than One Correct Answer

59. In the following reaction sequence, **Q**, **R**, **S** and **T** are the major products. **[Adv. 2026]**

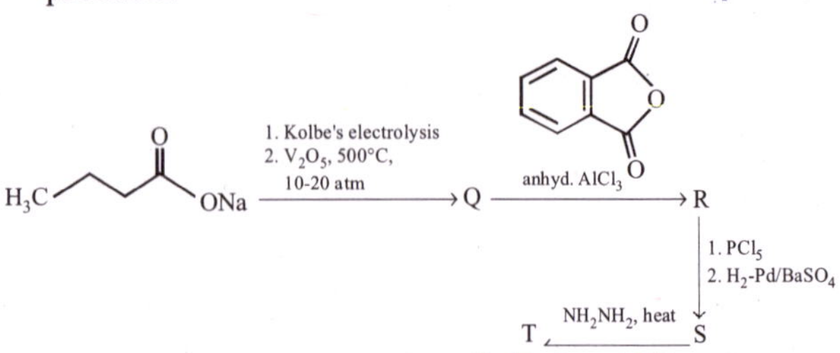

The correct statement(s) about **Q**, **R**, **S** and **T** is(are)

(a) **S** on warming with ammoniacal $AgNO_3$ results in the formation of silver mirror.

(b) **Q** on treatment with Cl_2(excess)/UV gives gammaxane.

(c) **T** is a heterocyclic compound.

(d) **R** on acid catalyzed intramolecular cyclization followed by treatment with Zn-Hg/HCl gives 9, 10-dihydroxyanthracene.

60. The option(s) with correct sequence of reagents for the conversion of P to Q is(are): **[Adv. 2024]**

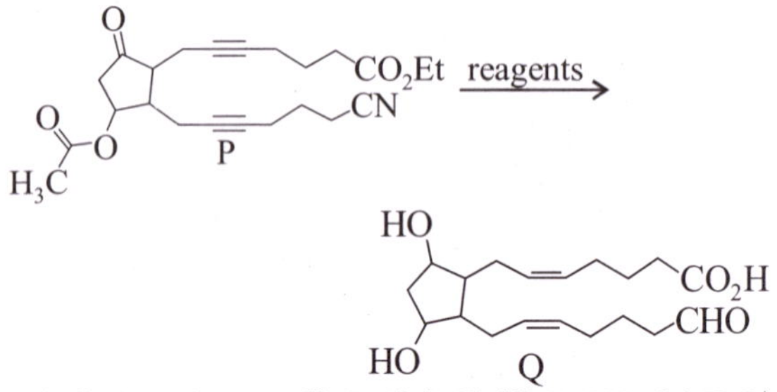

(a) (i) Lindlar's catalyst, H_2; (ii) $SnCl_2$/HCl; (iii) $NaBH_4$; (iv) H_3O^+

(b) (i) Lindlar's catalyst, H_2; (ii) H_3O^+; (iii) $SnCl_2$/HCl; (iv) $NaBH_4$

(c) (i) $NaBH_4$; (ii) $SnCl_2$/HCl; (iii) H_3O^+; (iv) Lindlar's catalyst, H_2

(d) (i) Lindlar's catalyst, H_2; (ii) $NaBH_4$; (iii) $SnCl_2$/HCl; (iv) H_3O^+

61. Positive Tollen's test is observed for **[Adv. 2016]**

(a) 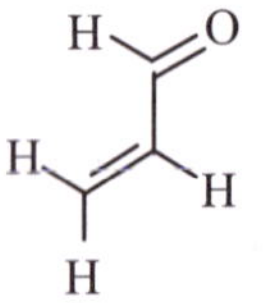

(b) 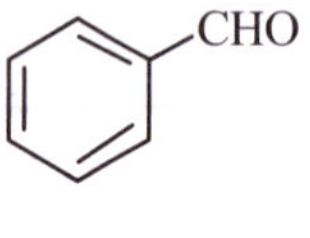

(c) 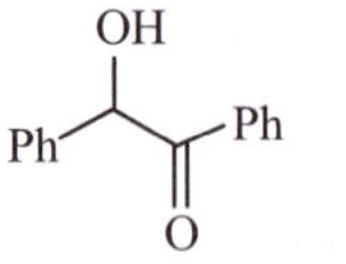

(d) 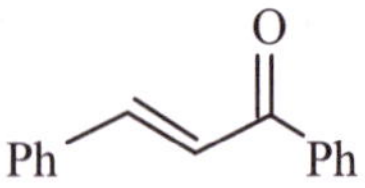

62. The smallest ketone and its next homologue are reacted with NH_2OH to form oxime **[2006 - 5M, –1]**

(a) Two different oximes are formed

(b) Three different oximes are formed

(c) Two oximes formed are optically active

(d) All oximes formed are optically active

63. Which of the following will undergo aldol condensation? **[1998 - 2 Marks]**

(a) acetaldehyde (b) propanaldehyde

(c) benzaldehyde (d) trideuteroacetaldehyde

64. Which of the following will react with water? **[1998 - 2 Marks]**

(a) $CHCl_3$ (b) Cl_3CCHO

(c) CCl_4 (d) $ClCH_2CH_2Cl$

65. A new carbon-carbon bond formation is possible in **[1998 - 2 Marks]**

(a) Cannizzaro reaction (b) Friedel–Craft alkylation

(c) Clemmensen reduction (d) Reimer–Tiemann reaction

66. Which of the following are examples of aldol condensation? **[1989 - 1 Mark]**

(a) $2CH_3CHO \xrightarrow{\text{dil. NaOH}} CH_3CHOHCH_2CHO$

(b) $2CH_3COCH_3 \xrightarrow{\text{dil. NaOH}} CH_3COH(CH_3)CH_2COCH_3$

(c) $2HCHO \xrightarrow{\text{dil. NaOH}} CH_3OH$

(d) $C_6H_5CHO + HCHO \xrightarrow{\text{dil. NaOH}} C_6H_5CH_2OH$

67. Keto-enol tautomerism is observed in **[1988 - 1 Mark]**

(a) $H_5C_6-\overset{O}{\overset{\|}{C}}-H$ (b) $H_5C_6-\overset{O}{\overset{\|}{C}}-CH_3$

(c) $H_5C_6-\overset{O}{\overset{\|}{C}}-C_6H_5$ (d) $H_5C_6-\overset{O}{\overset{\|}{C}}-CH_2-CH_3$

68. Base catalysed aldol condensation occurs with : **[1984 - 1 Mark]**

(a) propionaldehyde
(b) benzaldehyde
(c) 2-methylpropionaldehyde
(d) 2, 2-dimethylpropionaldehyde

7 Match the Following

69. The major products obtained from the reactions in List-II are the reactants for the named reactions mentioned in List-I. Match List-I with List-II and choose the correct option. **[Adv. 2023]**

List-I	List-II
(P) Etard reaction	(1) Acetophenone $\xrightarrow{Zn-Hg, HCl}$
(Q) Gattermann reaction	(2) Toluene $\xrightarrow[(ii) SOCl_2]{(i) KMnO_4, KOH, \Delta}$
(R) Gattermann–Koch reaction	(3) Benzene $\xrightarrow[\text{anhyd. } AlCl_3]{CH_3Cl}$
(S) Rosenmund reduction	(4) Aniline $\xrightarrow[273-278K]{NaNO_2 / HCl}$
	(5) Phenol $\xrightarrow{Zn, \Delta}$

(a) $P \to 2; Q \to 4; R \to 1; S \to 3$
(b) $P \to 1; Q \to 3; R \to 5; S \to 2$
(c) $P \to 3; Q \to 2; R \to 1; S \to 4$
(d) $P \to 3; Q \to 4; R \to 5; S \to 2$

70. Match the entries in Column I with the correctly related quantum number(s) in Column II. Indicate your answer by darkening the appropriate bubbles of the 4 × 4 matrix given in the ORS **[2008]**

Column I

(A) $H_2N-\overset{\oplus}{N}H_3\overset{\ominus}{Cl}$

(B) 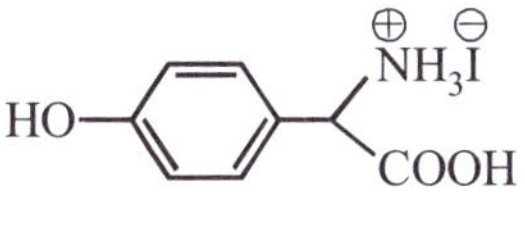

(C) 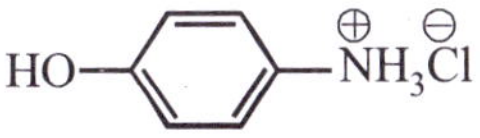

(D) 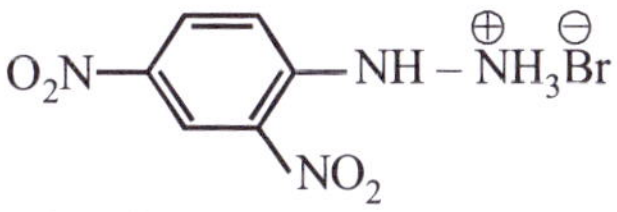

Column II

(p) sodium fusion extract of the compound gives Prussian blue colour with $FeSO_4$

(q) gives positive $FeCl_3$ test

(r) gives white precipitate with $AgNO_3$

(s) reacts with aldehydes to form the corresponding hydrazone derivative

8 Comprehension Based Questions

Passage - I

An organic compound P with molecular formula $C_9H_{18}O_2$ decolorizes bromine water and also shows positive iodoform test. P on ozonolysis followed by treatment with H_2O_2 gives Q and R. While compound Q shows positive iodoform test, compound R does not give positive iodoform test. Q and R on oxidation with pyridinium chlorochromate (PCC) followed by heating give S and T, respectively. Both S and T show positive iodoform test.

Complete copolymerization of 500 moles of Q and 500 moles of R gives one mole of a single acyclic copolymer U.

[Given, atomic mass: H=1, C=12, O=16]

71. Sum of number of oxygen atoms in S and T is _____. **[Adv. 2024]**

72. The molecular weight of U is _____. **[Adv. 2024]**

Passage-II

In the following reactions **[Adv. 2015]**

$$C_8H_6 \xrightarrow[H_2]{Pd-BaSO_4} C_8H_8 \xrightarrow[(ii)\ H_2O_2, NaOH, H_2O]{(i)\ B_2H_6} X$$

$$C_8H_6 \xrightarrow{H_2O,\ HgSO_4,\ H_2SO_4} C_8H_8O \xrightarrow[(ii)\ H^+, \text{heat}]{(i)\ EtMgBr, H_2O} Y$$

73. Compound *X* is

(a) 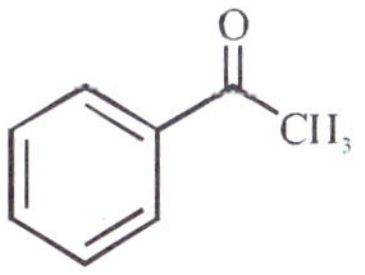 (b) 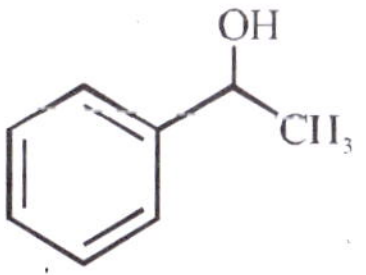

(c) 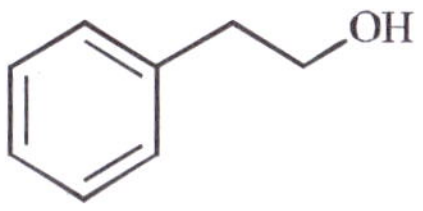 (d) 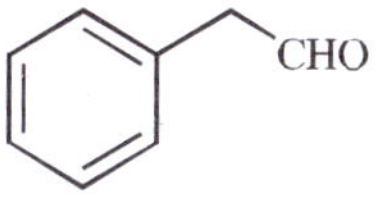

74. The major compound *Y* is

(a) 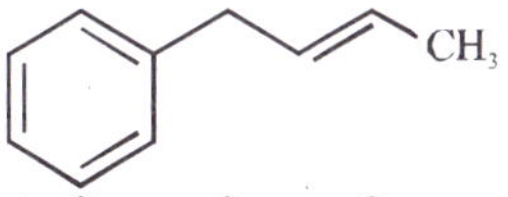

(b) 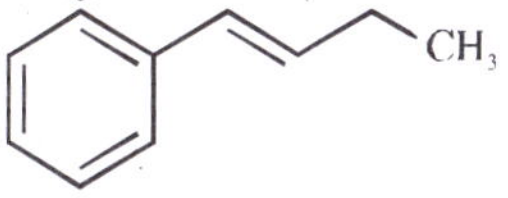

(c) 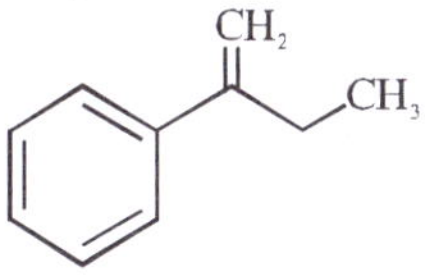

(d) 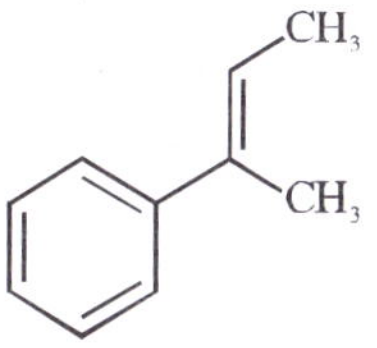

Passage-III

Two aliphatic aldehydes *P* and *Q* react in the presence of aqueous K_2CO_3 to give compound *R*, which upon treatment with HCN provides compound *S*. On acidification and heating, *S* gives the product shown below. **[2010]**

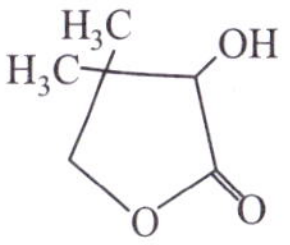

75. The compounds *P* and *Q* respectively are :

(a) 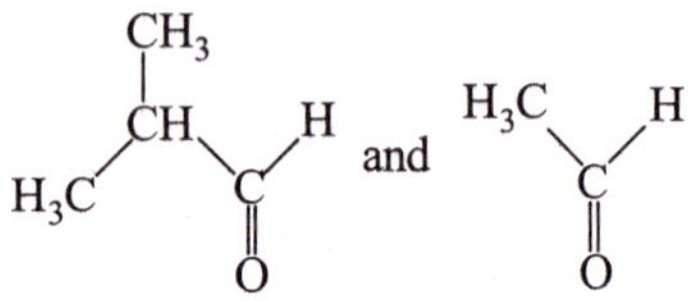

(b) 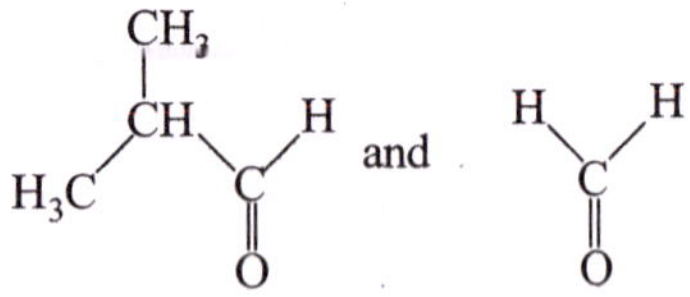

(c) 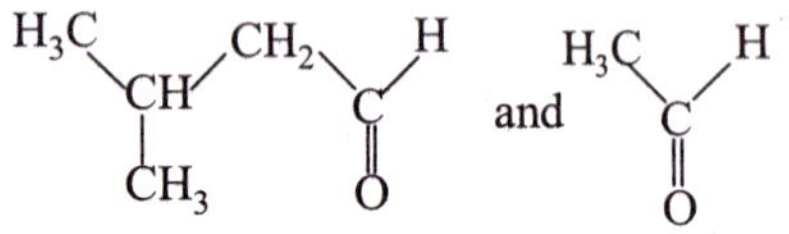

(d) 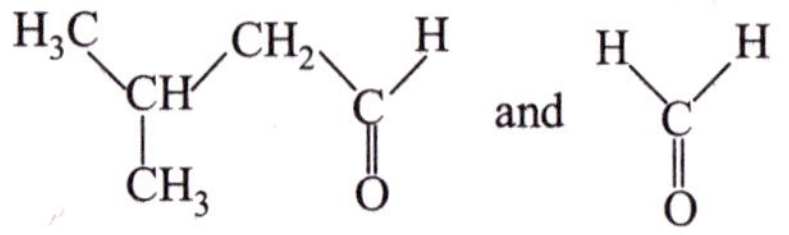

76. The compound *R* is :

(a) 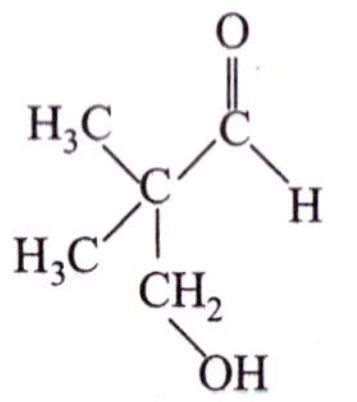

(b) 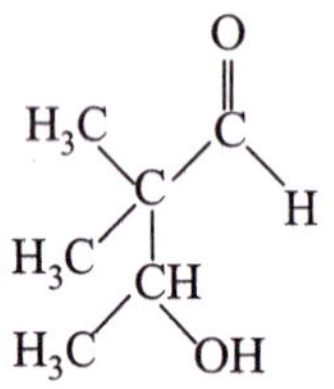

(c) 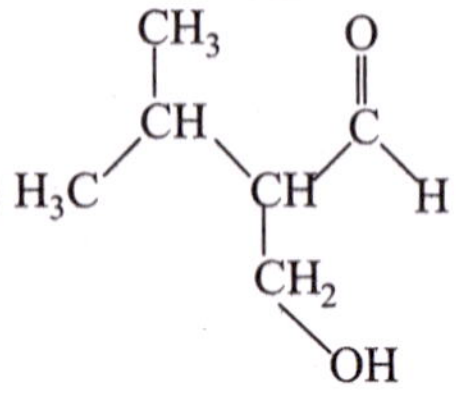

(d) 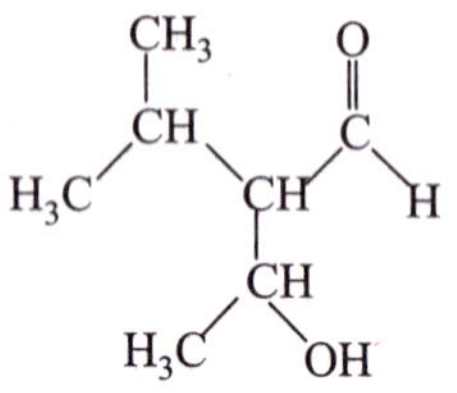

77. The compound *S* is :

(a) 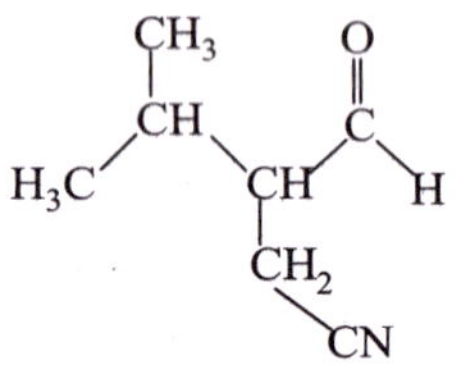

(b) 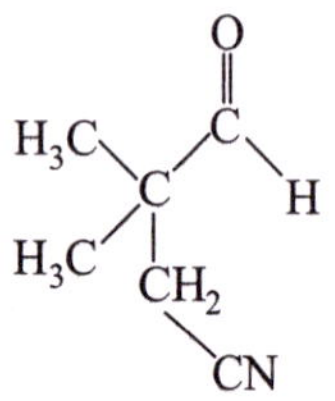

(c) 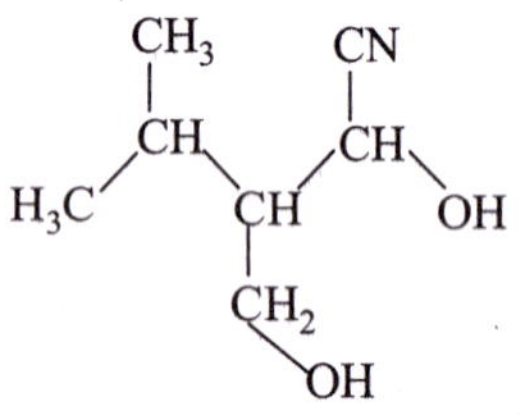

(d) 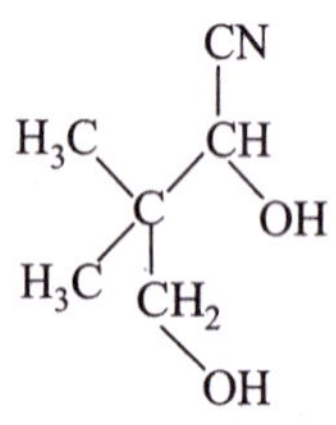

Passage-IV

In the following reaction sequence, products *I*, *J* and *L* are formed. *K* represents a reagent. **[2008]**

$$\text{Hex}-3-\text{ynal}\xrightarrow[\text{2.PBr}_3]{\text{1.NaBH}_4} I \xrightarrow[\text{3.H}_3\text{O}^+]{\text{1.Mg/ether, 2.CO}_2} J \xrightarrow{K}$$

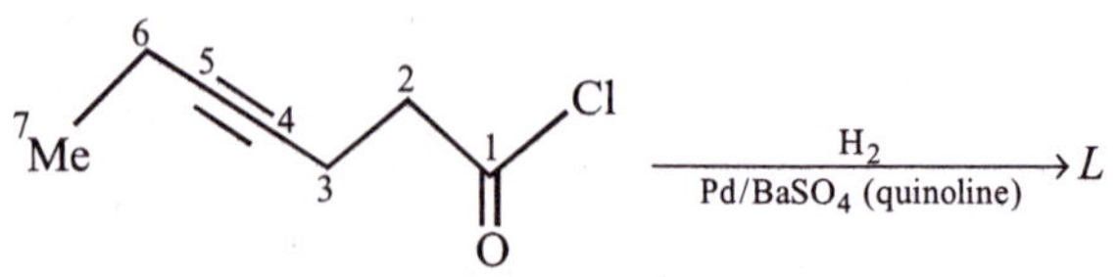

78. The structure of the product *I* is –

(a) 

(b) 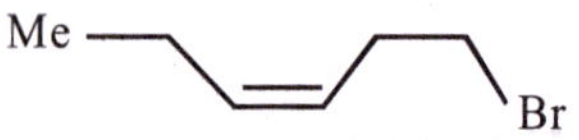

(c) 

(d) 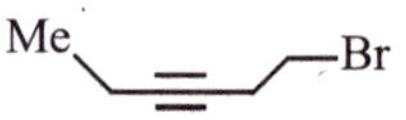

79. The structures of compounds *J* and *K*, respectively, are

(a) 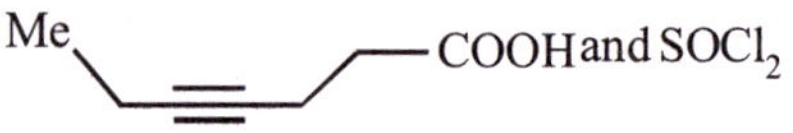

(b) 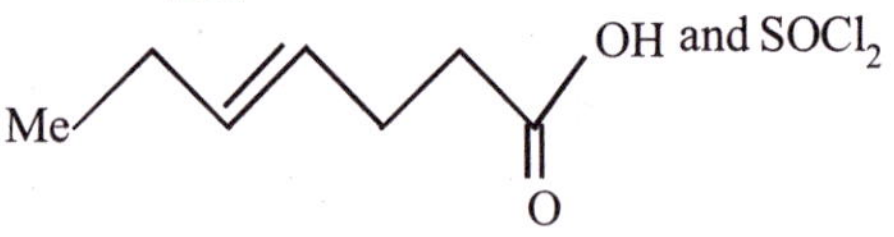

(c) 

(d) 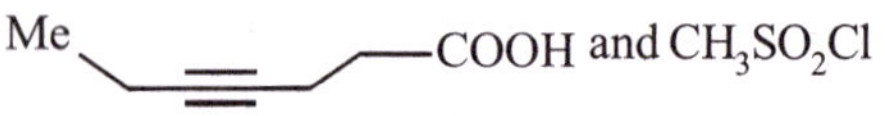

80. The structure of product *L* is

(a) 

(b) 

(c) 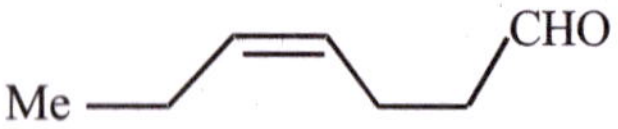

(d) 

10 Subjective Problems

81. $A(C_6H_{12}) \xrightarrow{HCl} \underset{(C_6H_{13}Cl)}{B + C}$ **[2003 - 4 Marks]**

$B \xrightarrow{\text{alc. KOH}} D$ (isomer of *A*)

$D \xrightarrow{\text{ozonolysis}} E$ (it gives negative test with Fehling solution but responds to iodoform test).

$A \xrightarrow{\text{Ozonolysis}} F + G$ (both gives positive Tollen's test but do not give iodoform test).

$F + G \xrightarrow{\text{conc. NaOH}} \text{HCOONa} + \text{A}$ primary alochol.

Identify from *A* to *G*.

82. Compound '*A*' of molecular formula $C_9H_7O_2Cl$ exists in keto form and predominantly in enolic form '*B*'. On oxidation with $KMnO_4$, '*A*' gives *m*–chlorobenzoic acid. Identify '*A*' and '*B*'. **[2003 - 2 Marks]**

83. An organic compound **A**, $C_6H_{10}O$ on reaction with CH_3MgBr followed by acid treatment gives compound **B**. The compound **B** on ozonolysis gives compound **C**, which in presence of a base gives 1-acetylcyclopentene **D**. The compound **B** on reaction with HBr gives compound **E**. Write the structures of **A, B, C** and **E**. Show how **D** is formed from **C** ? **[2000 - 5 Marks]**

84. Write the structural formula of the main organic product formed when :

(*i*) Identify A, B, C and give their structures.

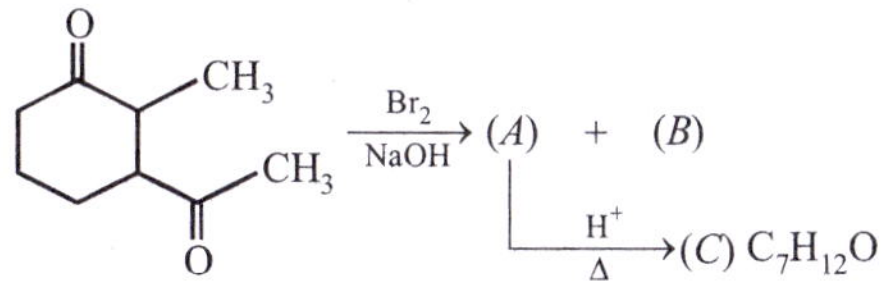

[2000 - 3 Marks]

(*ii*) 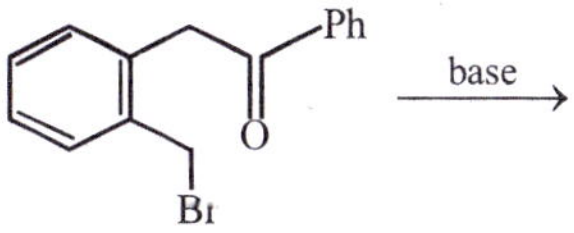 [2000 - 1 Mark]

(*iii*) $H_3CCOCOC_6H_5 + NaOH/H_3O^{\oplus} \longrightarrow$ ----

[1997 - 1 Mark]

(*iv*) $ClCH_2CH_2CH_2COPh + KOH + MeOH \longrightarrow$ ----

[1997 - 1 Mark]

(*v*) 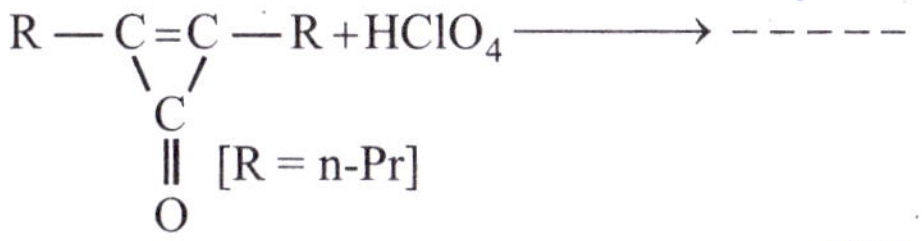

[1997 - 1 Mark]

(*vi*) 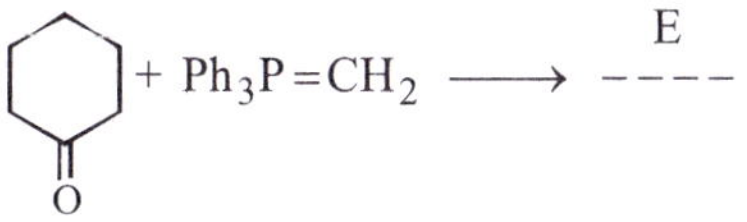

[1997 - 1 Mark]

(*vii*) Complete the following reaction with appropriate structure. [1996 - 1 Mark]

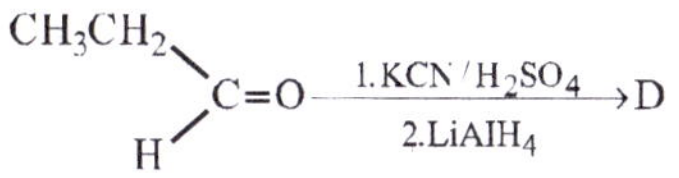

(*viii*) 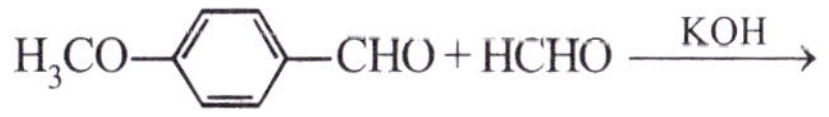

[1992 - 1 Mark]

(*ix*) propanal $\xrightarrow[\text{heat}]{\text{NaOH}}$ [1985 - 1 Mark]

(*x*) methanal reacts with ammonia [1981 - ½ Mark]

85. Complete the following reaction with appropriate structures of products/reagents : [1998 - 2 + 2 Marks]

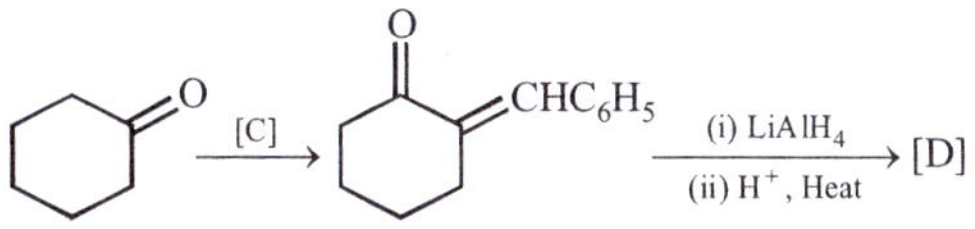

86. An aldehyde *A* ($C_{11}H_8O$), which does not undergo self aldol condensation, gives benzaldehyde and two moles of *B* on ozonolysis. Compound *B*, on oxidation with silver ion gives oxalic acid. Identify the compounds *A* and *B*. [1998 - 2 Marks]

87. Arrange the following in increasing order of expected enol content [1992 - 1 Mark]

$CH_3COCH_2CHO, CH_3COCH_3, CH_3CHO, CH_3COCH_2COCH_3$

88. Complete the following reactions :

(*i*) $H_3C-CHO \xrightarrow{CH_3MgBr} ? \xrightarrow{?} H_3C-\underset{H}{\overset{OH}{C}}-CH_3$

[1988 - 1 Mark]

(*ii*) $2CH_3\overset{O}{\overset{\|}{C}}CH_3 \xrightarrow{\text{base catalyst}} ? \xrightarrow{\text{acid catalyst}} ?$

[1988 - 1 Mark]

(*iii*) 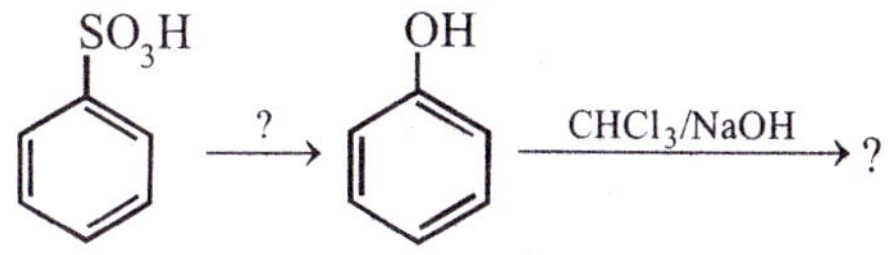

[1988 - 1 Mark]

89. An unknown compound of carbon, hydrogen and oxygen contains 69.77% carbon and 11.63% hydrogen and has a molecular weight of 86. It does not reduce Fehling solution, but forms a bisulphite addition compound and gives a positive iodoform test. What are the possible structures for the unknown compound? [1987 - 5 Marks]

90. A white precipitate was formed slowly when silver nitrate was added to a compound (A) with molecular formula $C_6H_{13}Cl$. Compound (A) on treatment with hot alcoholic potassium hydroxide gave a mixture of two isomeric alkenes (B) and (C), having formula C_6H_{12}. The mixture of (B) and (C), on ozonolysis, furnished four compounds: [1986 - 4 Marks]

(*i*) CH_3CHO; (*ii*) C_2H_5CHO;

(*iii*) CH_3COCH_3 and (*iv*) $H_3C-\underset{CH_3}{\underset{|}{CH}}-CHO$

What are the structures of (A), (B) and (C)?

91. Give a chemical test/suggest a reagent to distinguish between acetaldehyde from acetone. [1987 - 1 Mark]

92. Arrange the following in :

Increasing reactivity towards HCN [1985]

CH_3CHO, CH_3COCH_3, $HCHO$, $C_2H_5COCH_3$

93. Give reasons for the following :

(*i*) Explain why *o*-hydroxybenzaldehyde is a liquid at room temperature while *p*-hydroxybenzaldehyde is a high melting solid. [1999 - 2 Marks]

(*ii*) In acylium ion, the structure $R-C \equiv O^+$ is more stable than $R-C^+=O$. [1994 - 1 Mark]

(*iii*) Iodoform is obtained by the reaction of acetone with hypoiodite but not with iodide ion. [1991 - 1 Mark]

(*iv*) Hydrazones of aldehydes and ketones are not prepared in highly acidic medium. [1986 - 1 Mark]

(*v*) Suggest a reason for the large difference between the boiling points of butanol and butanal, although they have almost the same solubility in water. [1985 - 2 Marks]

Topic 3 - Preparation and Properties of Carboxylic Acids and their Derivatives

1 MCQs with One Correct Answer

94. Considering $LiBH_4$ reduces an ester group to the corresponding alcohol and does not reduce a carboxylic acid group, the correct statement about the major products P, Q, R and S is [Adv. 2026]

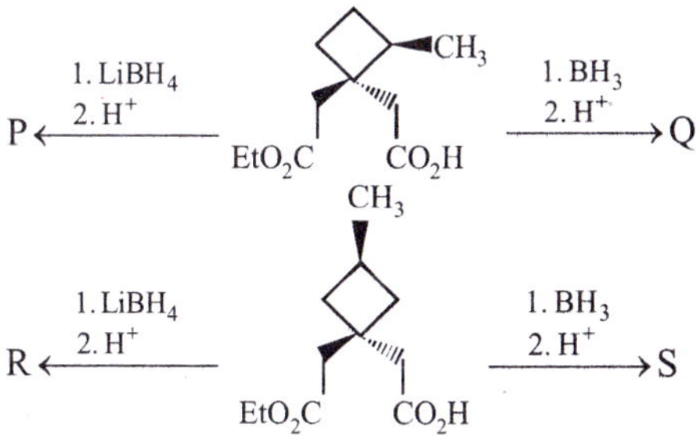

(a) P & Q are identical, and R & S are diastereomers.
(b) P & Q are diastereomers, and R & S are identical.
(c) P & Q are diastereomers, and R & S are diastereomers.
(d) P & Q are identical, and R & S are identical.

95. Monocyclic compounds P, Q, R and S are the major products formed in the reaction sequences given below.

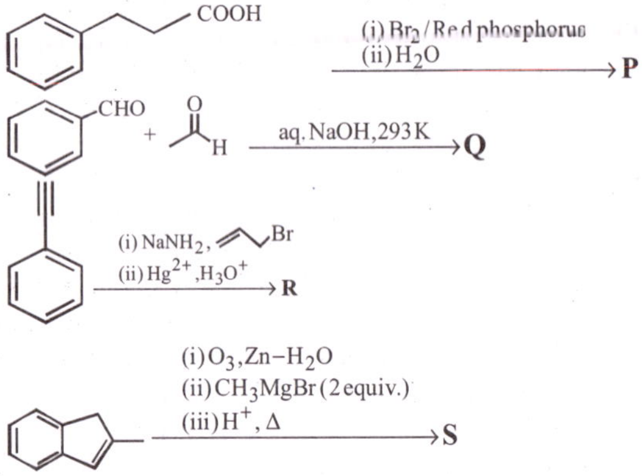

The product having the highest number of unsaturated carbon atom(s) is [Adv. 2025]
(a) P (b) Q
(c) R (d) S

96. The correct order of acid strength of the following carboxylic acid is [Adv. 2019]

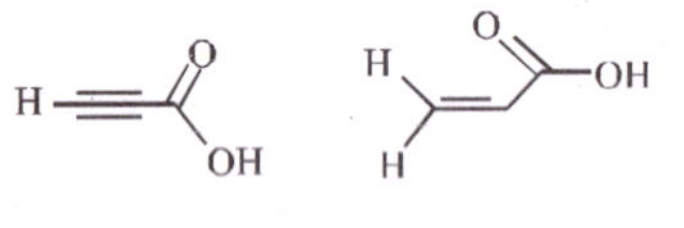

I II

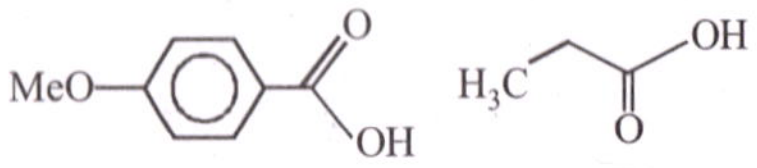

III IV

(a) I > II > III > IV (b) II > I > IV > III
(c) I > III > II > IV (d) III > II > I > IV

97. The correct order of acidity for the following compounds is

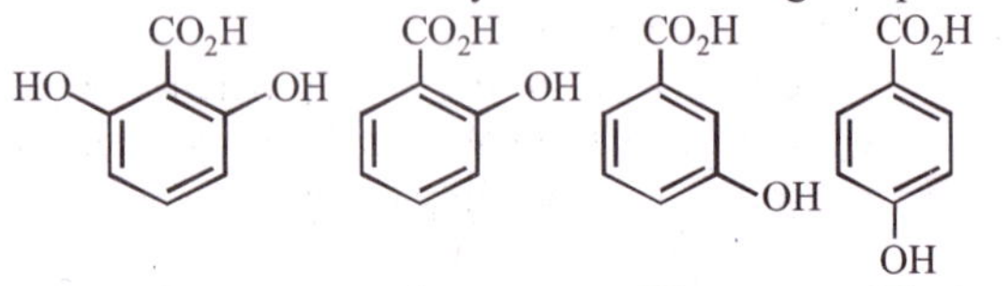

I II III IV

[Adv. 2016]

(a) I > II > III > IV (b) III > I > II > IV
(c) III > IV > II > I (d) I > III > IV > II

98. The compound that does NOT liberate CO_2, on treatment with aqueous sodium bicarbonate solution, is [Adv. 2013]
(a) Benzoic acid (b) Benzenesulphonic acid
(c) Salicylic acid (d) Carbolic acid (Phenol)

99. The compound that undergoes decarboxylation most readily under mild condition is [2012]

(a) 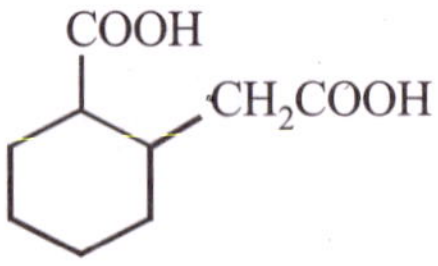

(b) 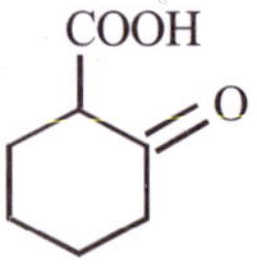

(c) 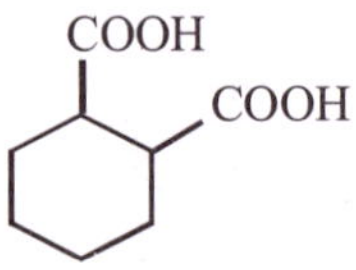

(d) 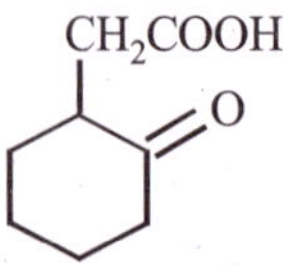

100. The carboxyl functional group (– COOH) is present in [2012]
(a) picric acid (b) barbituric acid
(c) ascorbic acid (d) aspirin

101. The major product of the following reaction is [2011]

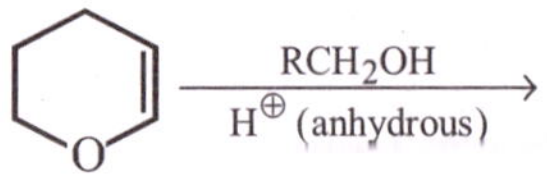

(a) a hemiacetal (b) an acetal
(c) an ether (d) an ester

102. In the reaction

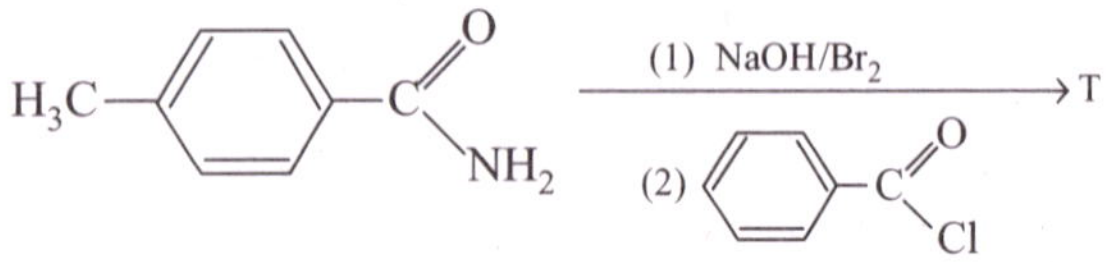

the structure of the product T is: [2010]

(a) 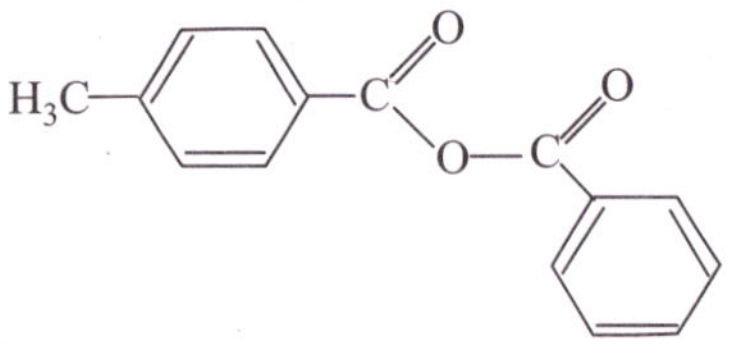

(b) 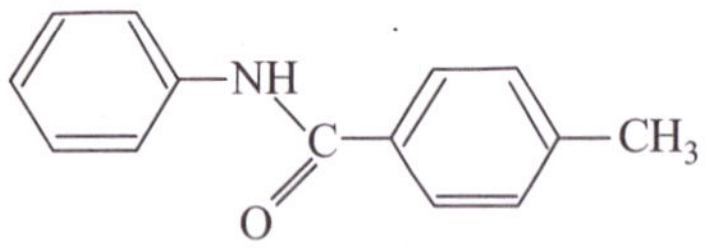

(c) 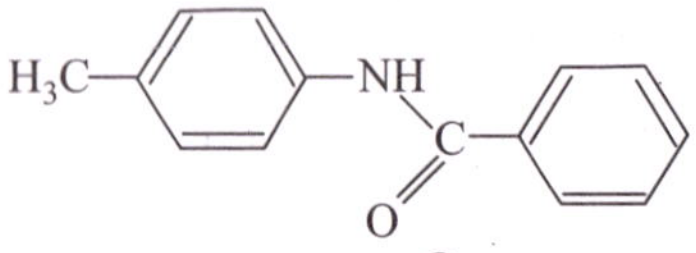

(d) 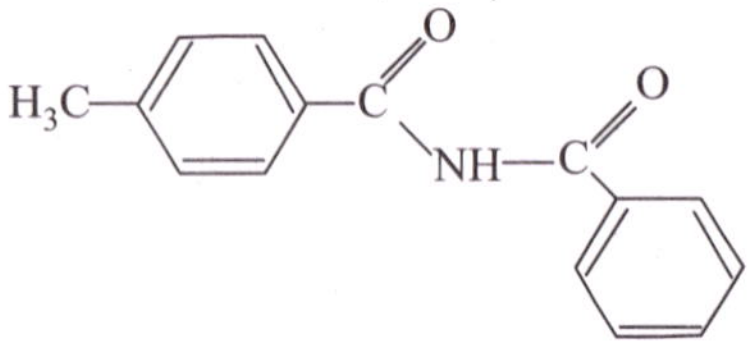

103. The compounds *P*, *Q* and *S*

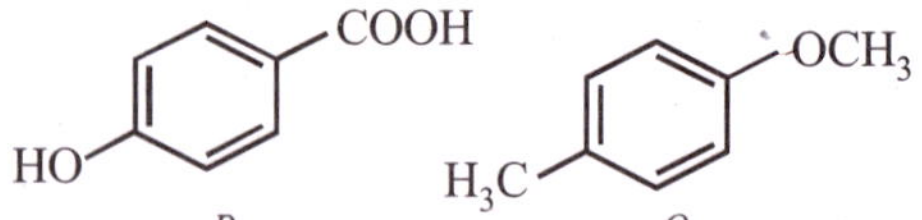

P *Q*

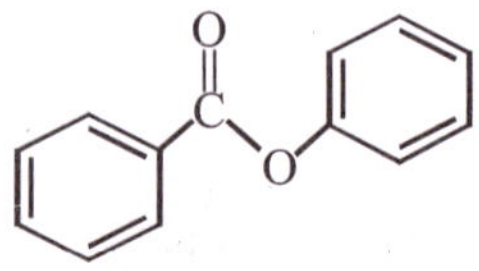

S

were separately subjected to nitration using HNO_3/H_2SO_4 mixture. The major product formed in each case respectively, is : [2010]

(a) 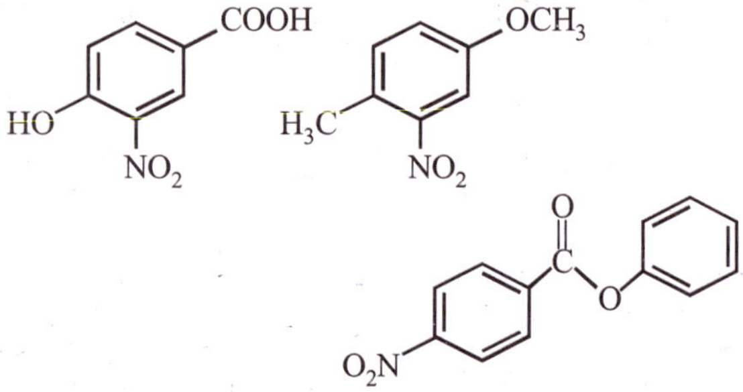

(b) 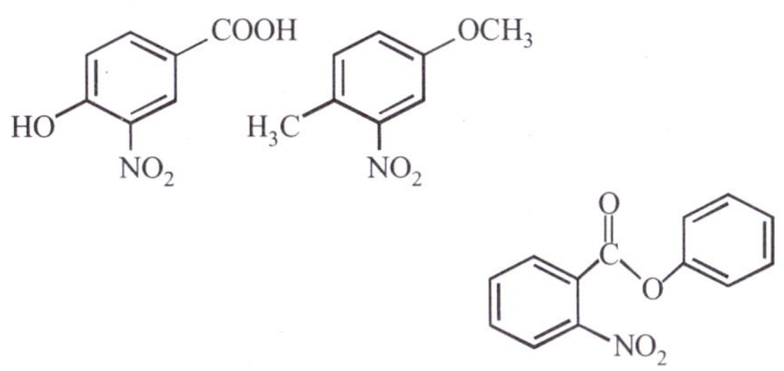

(c) 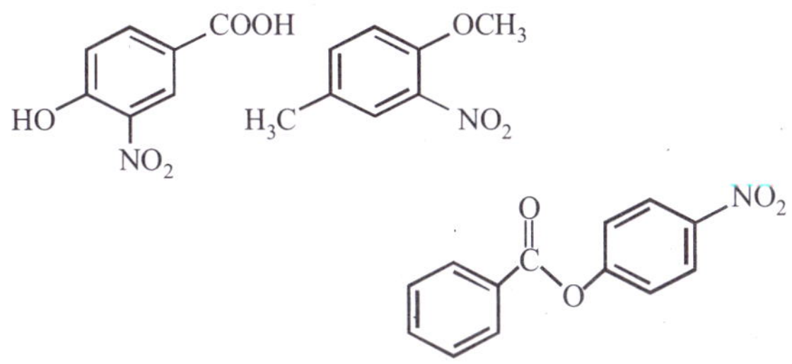

(d) 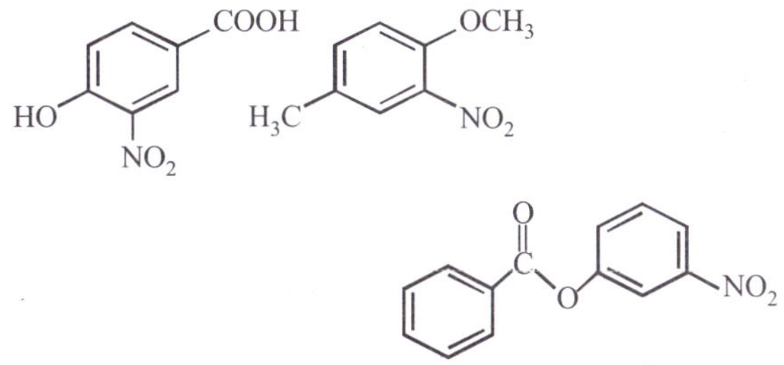

104. The correct acidity order of the following is **[2009S]**

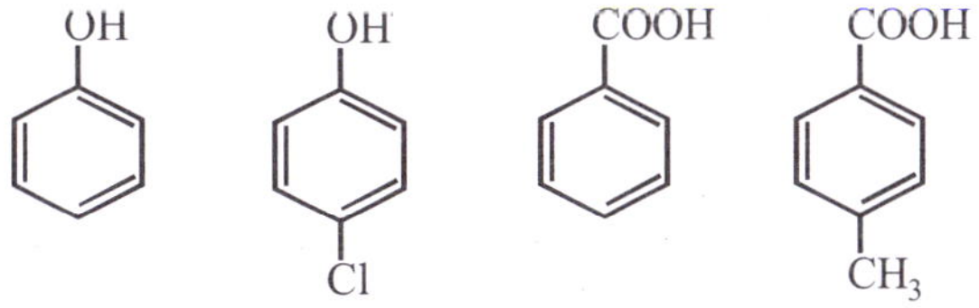

(I) (II) (III) (IV)

(a) (III) > (IV) > (II) > (I) (b) (IV) > (III) > (I) > (II)
(c) (III) > (II) > (I) > (IV) (d) (II) > (III) > (IV) > (I)

105. In the following reaction sequence, the correct structures of *E*, *F* and *G* are

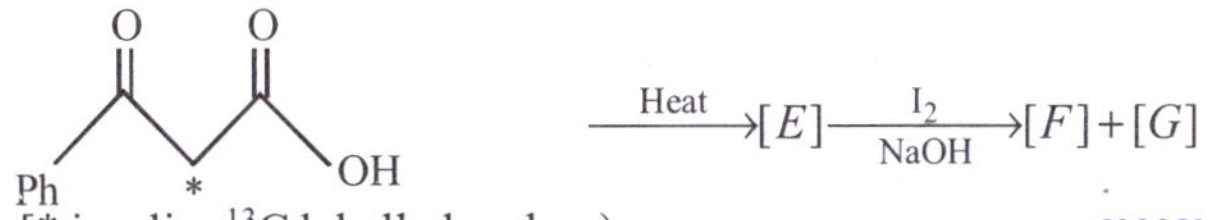

[* implies ^{13}C labelled carbon) **[2008]**

(a) 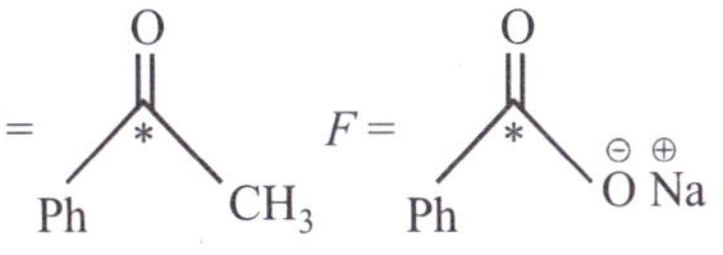 $G = CHI_3$

(b) 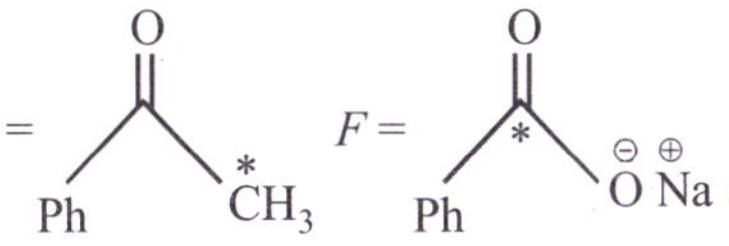 $G = CHI_3$

(c) 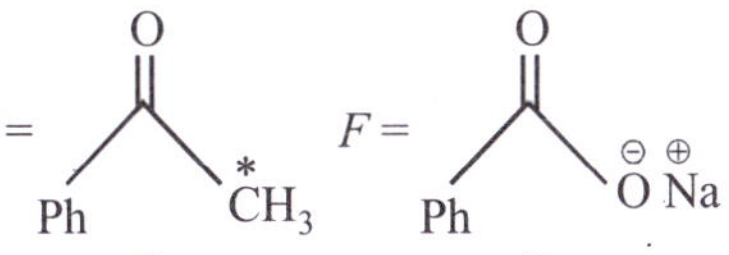 $G = \overset{*}{C}HI_3$

(d) 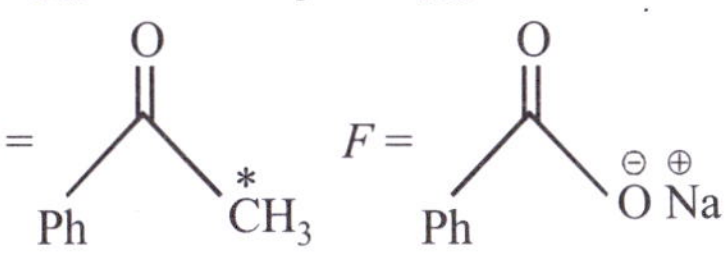 $G \overset{*}{=} CH_3I$

106. Which of the following reactants on reaction with conc. NaOH followed by acidification gives following lactone as the product? **[2006 - 5M, –1]**

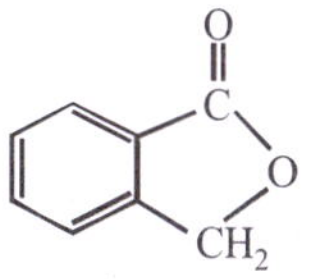

(a) 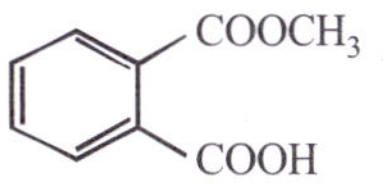 (b) 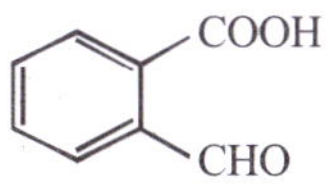

(c) 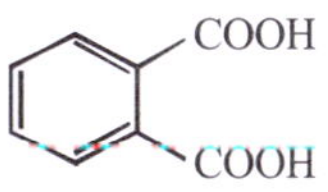 (d) 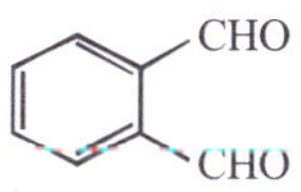

107. 4–Methylbenzenesulphonic acid reacts with sodium acetate to give **[2005S]**

(a) 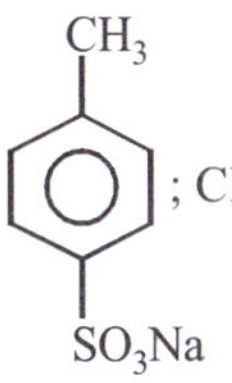 ; CH_3COOH (b) 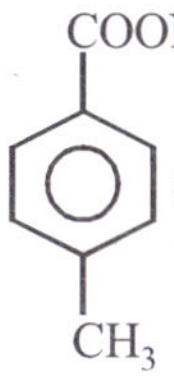 ; SO_3

(c) 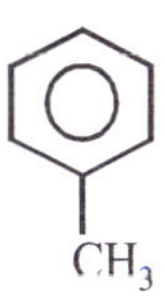 ; SO_3 (d) 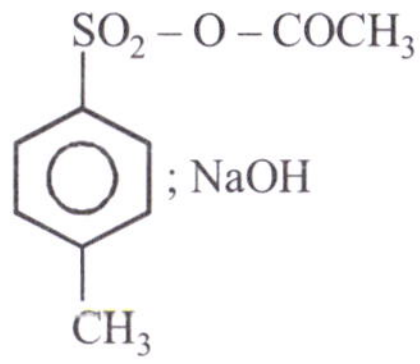 ; NaOH

108. An enantiomerically pure acid is treated with a racemic mixture of an alcohol having one chiral carbon. The ester formed will be **[2003S]**

(a) Optically active mixture (b) Pure enantiomer
(c) Meso compound (d) Racemic mixture

109. Ethyl ester $\xrightarrow[\text{excess}]{CH_3MgBr} P$. The product *P* will be **[2003S]**

(a) 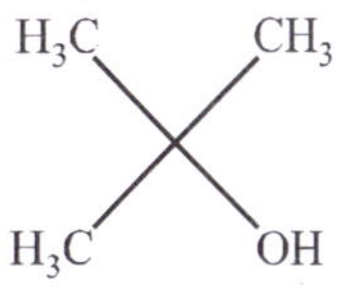 (b) 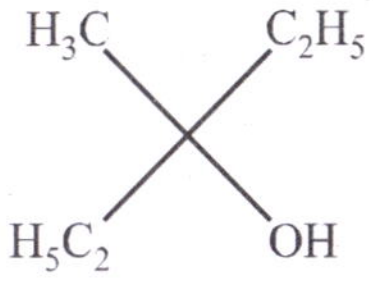

(c) 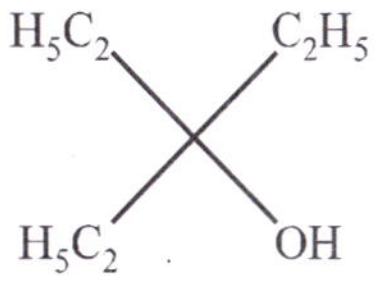 (d) 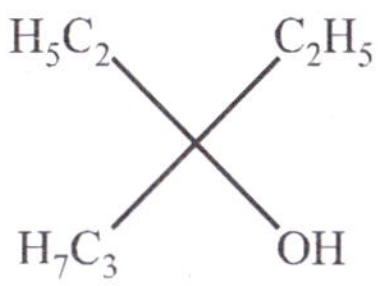

110. Benzoyl chloride is prepared from benzoic acid by **[2000S]**

(a) $Cl_2, h\nu$ (b) SO_2Cl_2
(c) $SOCl_2$ (d) Cl_2, H_2O

111. When propionic acid is treated with aqueous sodium bicarbonate, CO_2 is liberated. The 'C' of CO_2 comes from **[1999 - 2 Marks]**

(a) methyl group (b) carboxylic acid group
(c) methylene group (d) bicarbonate

112. The organic product formed in the reaction **[1995S]**

$$C_6H_5COOH \xrightarrow[\text{II } H_3O^+]{\text{I } LiAlH_4}$$

(a) $C_6H_5CH_2OH$ (b) C_6H_5COOH & CH_4
(c) $C_6H_5CH_3$ & CH_3OH (d) $C_6H_5CH_3$ & CH_4

2 Integer Value Answer

113. The total number of carboxylic acid groups in the product *P* is **[Adv. 2013]**

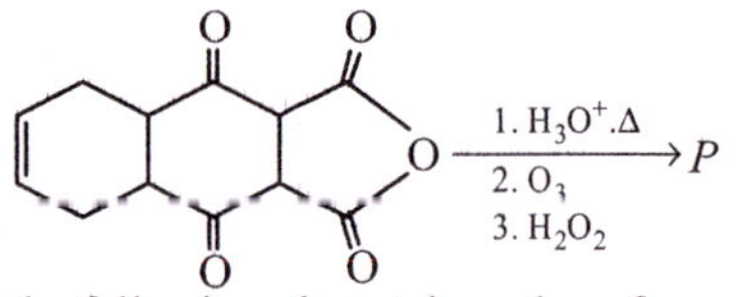

114. Amongst the following, the total number of compounds soluble in aqueous NaOH is **[2010]**

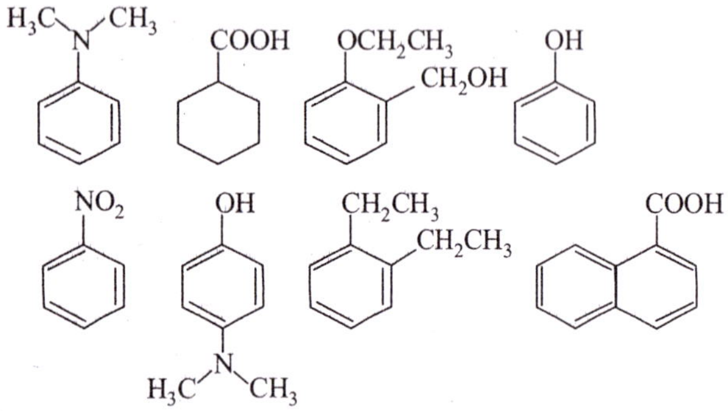

3 Numeric / New Stem Based Questions

115. The sum of total number of carbonyl groups (>C=O) present in the major products X and Y in the following reactions is ______. **[Adv. 2026]**

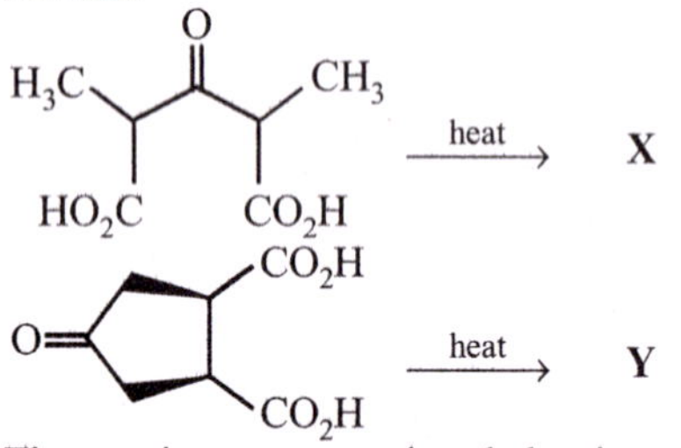

116. The reaction sequence given below is carried out with 16 moles of X. The yield of the major product in each step is given below the product in parentheses. The amount (in grams) of S produced is______. **[Adv. 2025]**

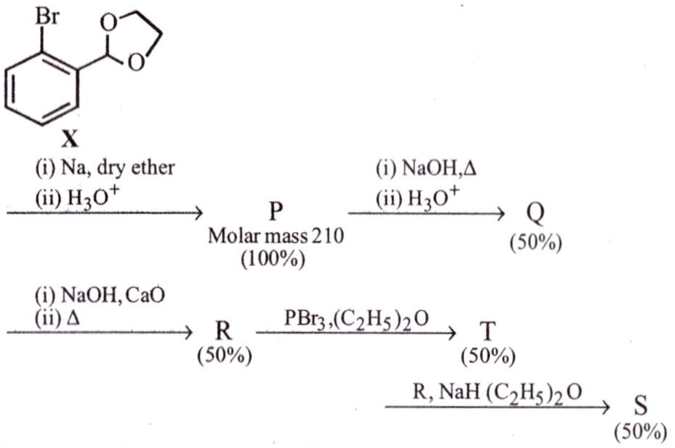

Use: Atomic mass (in amu): H = 1, C = 12, O = 16, Br = 80

117. In the following reaction sequence, the major product P is formed. **[Adv. 2024]**

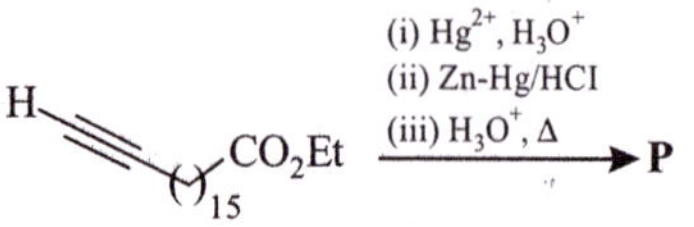

Glycerol reacts completely with excess P in the presence of an acid catalyst to form Q. Reaction of Q with excess NaOH followed by the treatment with $CaCl_2$ yields Ca-soap R, quantitatively. Starting with one mole of Q, the amount of R produced in gram is ______.
[Given, atomic weight: H = 1, C = 12, N = 14, O = 16, Na = 23, Cl = 35, Ca = 40]

118. In the following reaction, compound Q is obtained from compound **P** via an ionic intermediate. **[Adv. 2020]**

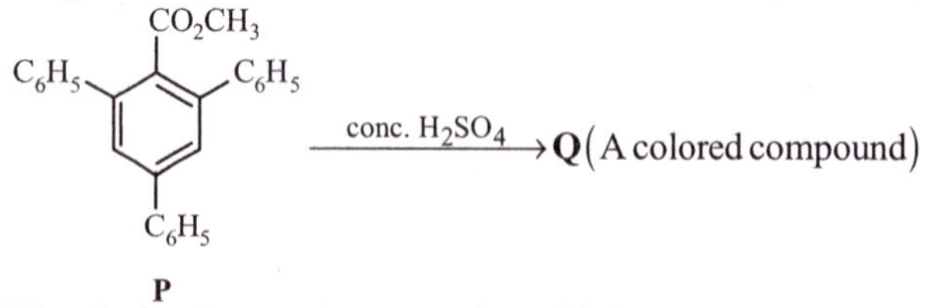

What is the degree of unsaturation of **Q**?

4 Fill in the Blanks

119. Formic acid when heated with conc. H_2SO_4 produces .... **[1983 - 1 Mark]**

5 True / False

120. The boiling point of propionic acid is less than that of *n*-butyl alcohol, an alcohol of comparable molecular weight. **[1991 - 1 Mark]**

121. Hydrolysis of an ester in presence of a dilute acid is known as saponification. **[1983 - 1 Mark]**

6 MCQs with One or More than One Correct Answer

122. For the reaction sequence given below, the correct statement(s) is(are) **[Adv. 2025]**

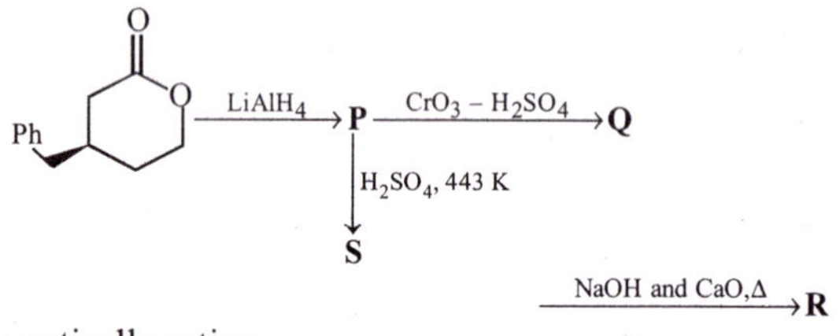

(a) P is optically active.
(b) S gives Bayer's test.
(c) Q gives effervescence with aq. $NaHCO_3$.
(d) R is an alkyne

123. In the following reactions, P, Q, R, and S are the major products. **[Adv. 2023]**

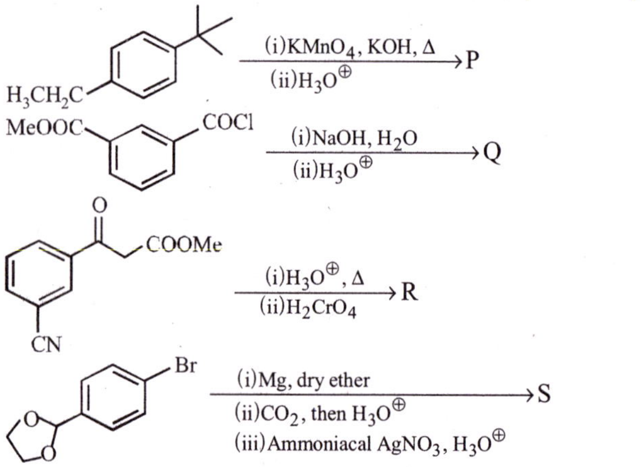

The correct statement(s) about P, Q, R, and S is(are)

(a) P and Q are monomers of polymers dacron and glyptal, respectively.

(b) P, Q, and R are dicarboxylic acids.

(c) Compounds Q and R are the same.

(d) R does not undergo aldol condensation and S does not undergo Cannizzaro reaction.

124. Consider the following reaction scheme and choose the correct option(s) for the major products Q, R and S.

$$\text{Styrene} \xrightarrow[\text{(ii) NaOH, } H_2O_2, H_2O]{\text{(i) } B_2H_6} P \xrightarrow[\text{(ii) } Cl_2\text{, Red phosphorus; (iii) } H_2O]{\text{(i) } CrO_3, H_2SO_4} Q$$

$$P \xrightarrow[\text{(ii) NaCN; (iii) } H_3O^+, \Delta]{\text{(i) } SOCl_2} R \xrightarrow{\text{conc. } H_2SO_4} S$$

[Adv. 2023]

(a)

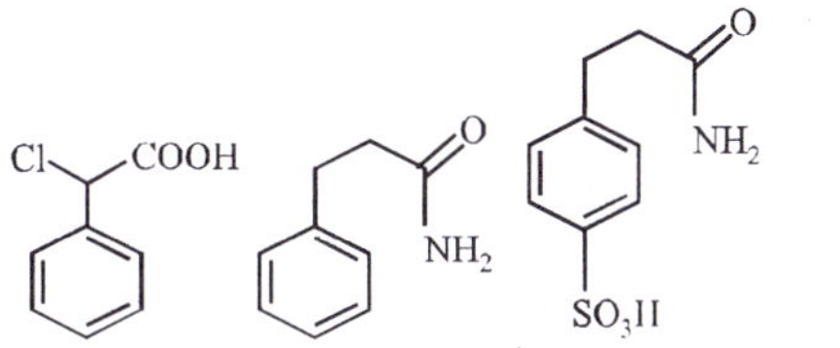

Q R S

(b)

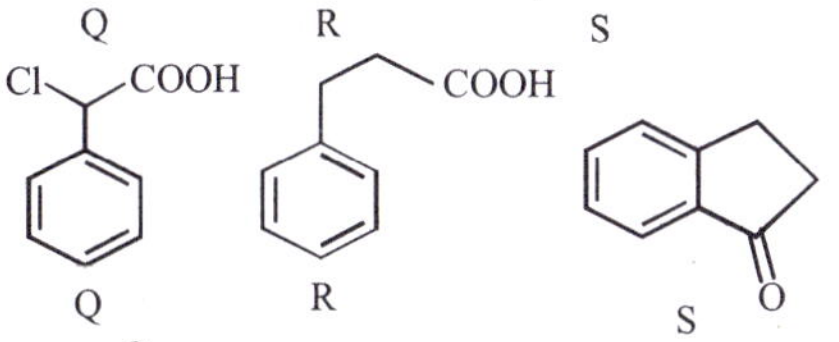

Q R S

(c)

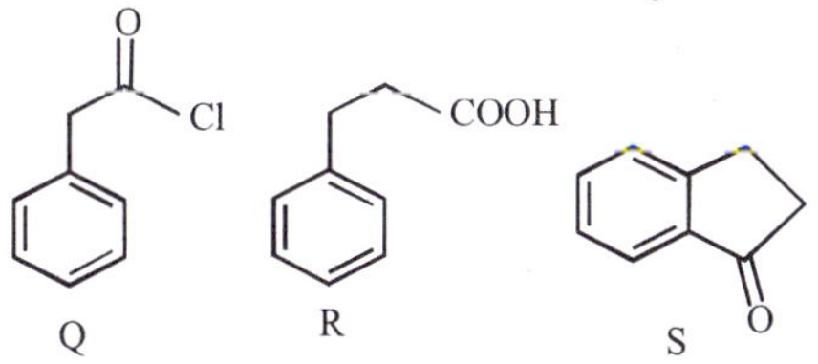

Q R S

(d)

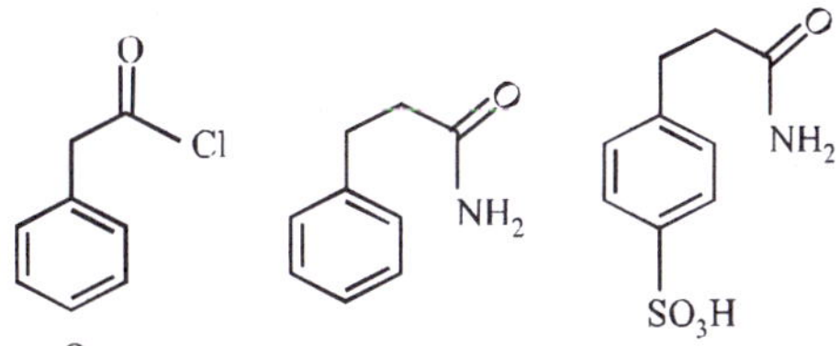

Q R S

125. Choose the correct option(s) for the following reaction sequence [Adv. 2019]

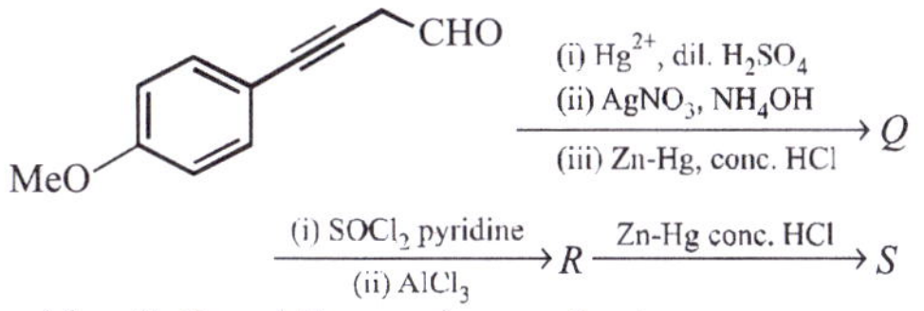

Consider Q, R and S as major products.

(a)

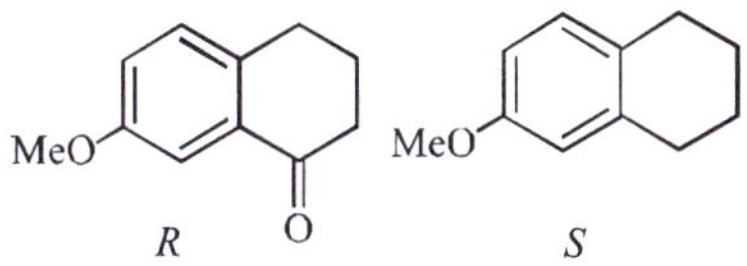

(b)

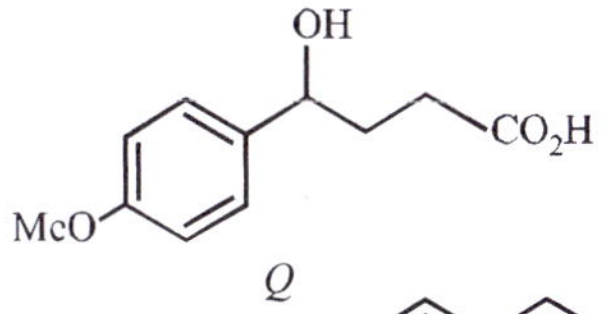

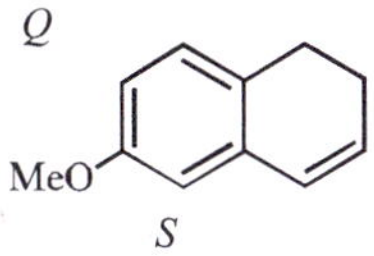

(c)

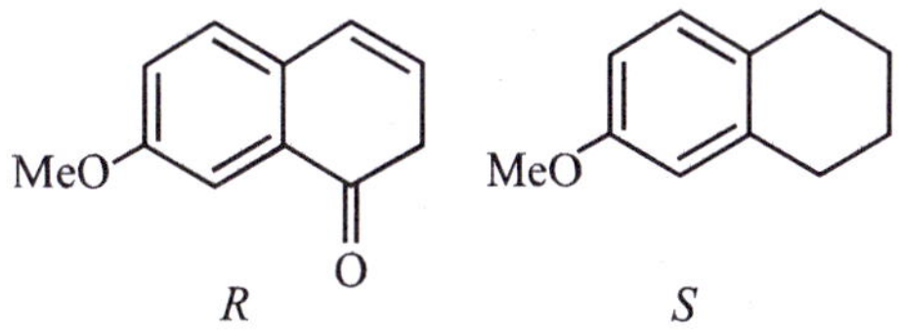

(d)

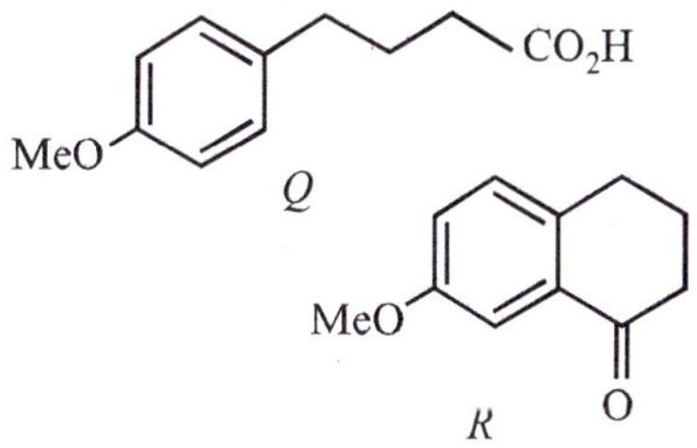

126. The reaction(s) leading to the formation of 1,3,5-trimethylbenzene is (are) [Adv. 2018]

(a)

(b) $Me-\!\!\equiv\!\!-H \xrightarrow[\text{873 K}]{\text{heated iron tube}}$

(c)

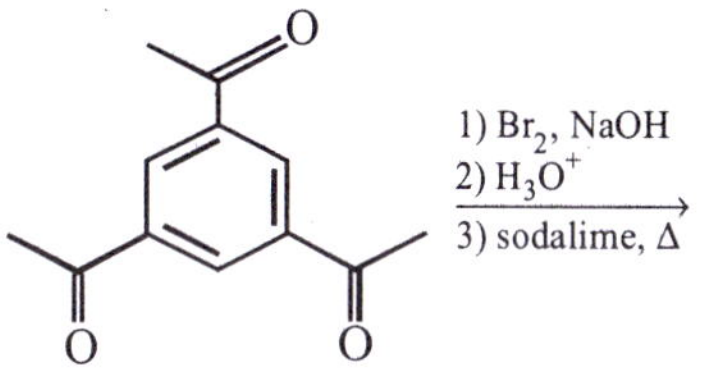

(d)

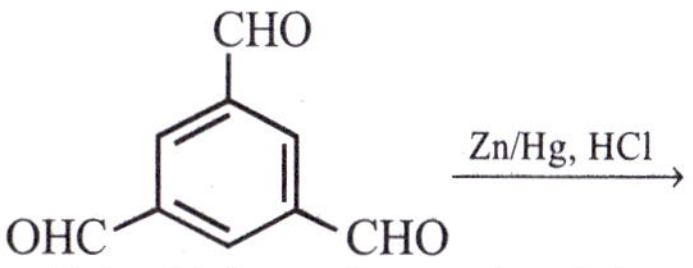

127. Reagent(s) which can be used to bring about the following transformation is (are) [Adv. 2016]

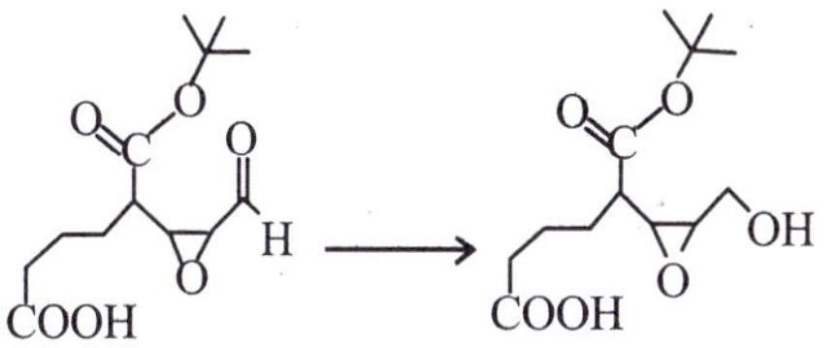

(a) $LiAlH_4$ in $(C_2H_5)_2O$ (b) BH_3 in THF

(c) $NaBH_4$ in C_2H_5OH (d) Raney Ni/H_2 in THF

128. With reference to the scheme given below, which of the given statement(s) about T, U, V and W is (are) correct ? [2012]

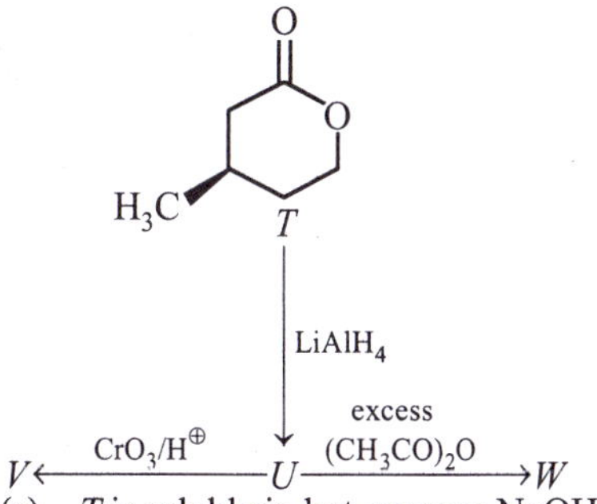

(a) T is soluble in hot aqueous NaOH

(b) U is optically active

(c) Molecular formula of W is $C_{10}H_{18}O_4$

(d) V gives effervescences on treatment with aqueous $NaHCO_3$.

129. Identify the binary mixture(s) that can be separated into individual compounds, by differential extraction as shown in the given scheme. **[2012]**

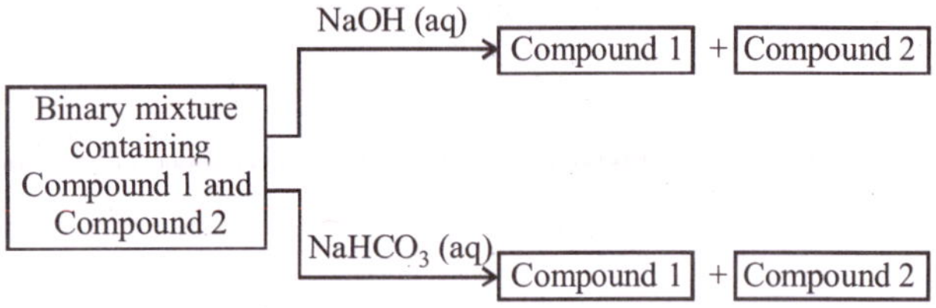

(a) C_6H_5OH and C_6H_5COOH
(b) C_6H_5COOH and $C_6H_5CH_2OH$
(c) $C_6H_5CH_2OH$ and C_6H_5OH
(d) $C_6H_5CH_2OH$ and $C_6H_5CH_2COOH$

130. Which of the following compounds will react with ethanolic KCN? **[1984 - 1 Mark]**

(a) ethyl chloride (b) acetyl chloride
(c) chlorobenzene (d) benzaldehyde

131. Which of the following compounds will give a yellow precipitate with iodine and alkali? **[1984 - 1 Mark]**

(a) 2-Hydroxypropane (b) acetophenone
(c) methyl acetate (d) acetamide

7 Match the Following

Directions [Qs. 132-133] : Answer the following by appropriately matching the lists based on the information given in the paragraph. List–I includes starting materials and reagents of selected chemical reactions. List–II gives structures of compounds that may be formed as intermediate products and/ or final products from the reactions of List–I. **[Adv. 2019]**

List–I **List–II**

(I)
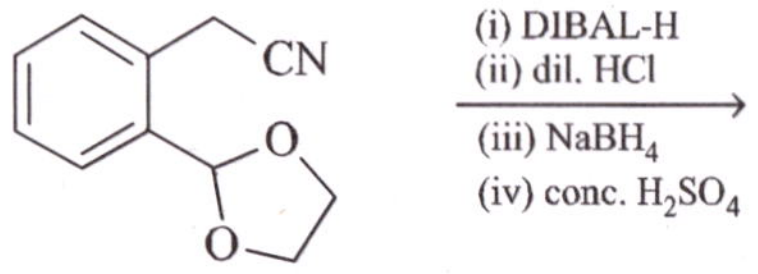

(P)
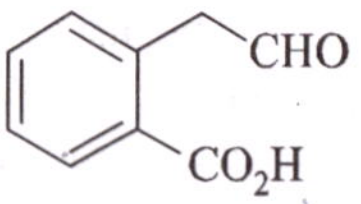

(II)
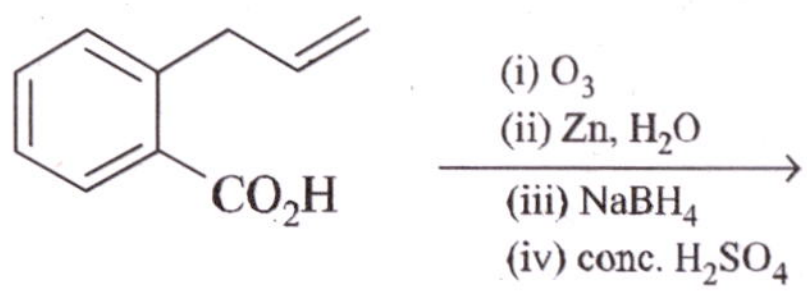

(Q)
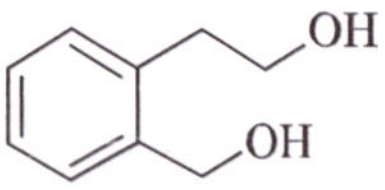

(III)
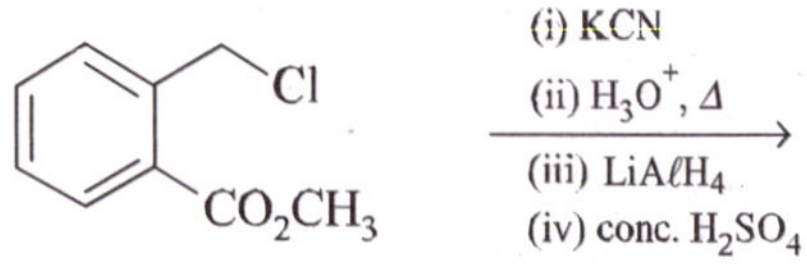

(R)
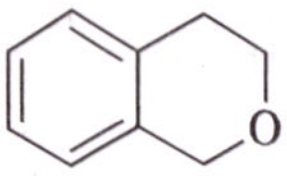

(IV)
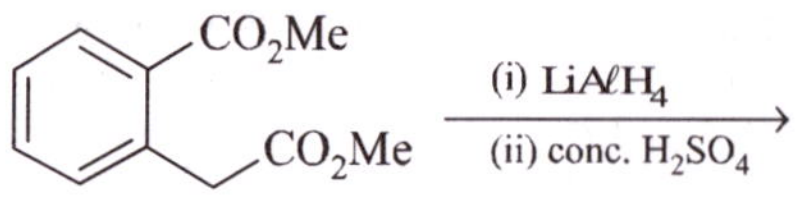

(S)
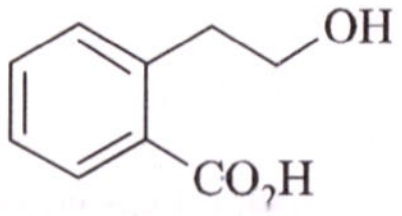

(T)
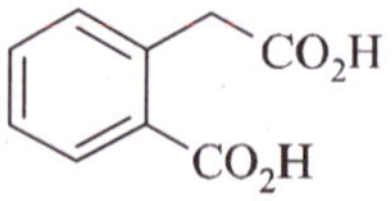

(U)
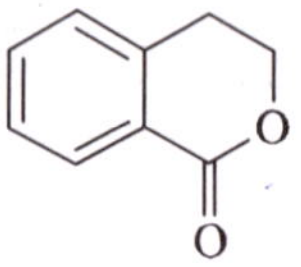

132. Which of the following options has correct combination considering List–I and List–II?

(a) (I), (S), (Q), (R) (b) (II), (P), (S), (U)
(c) (I), (Q), (T), (U) (d) (II), (P), (S), (T)

133. Which of the following options has correct combination considering List – I and List –II?

(a) (IV), (Q), (R) (b) (IV), (Q), (U)
(c) (III), (S), (R) (d) (III), (T), (U)

Directions [Qs. 134-136] : By appropriately matching the information given in the three columns of the following table. Columns 1, 2 and 3 contain starting materials, reaction conditions, and type of reactions, respectively. **[Adv. 2017]**

	Column 1	Column 2	Column 3
(I)	Toluene	(i) $NaOH/Br_2$	(P) Condensation
(II)	Acetophenone	(ii) Br_2/hv	(Q) Carboxylation
(III)	Benzaldehyde	(iii) $(CH_3CO)_2O/CH_3COOK$	(R) Substitution
(IV)	Phenol	(iv) $NaOH/CO_2$	(S) Haloform

134. For the synthesis of benzoic acid, the only CORRECT combination is

(a) (II) (i) (S) (b) (IV) (ii) (P)
(c) (I) (iv) (Q) (d) (III) (iv) (R)

135. The only CORRECT combination that gives two different carboxylic acids is

(a) (II) (iv) (R) (b) (IV) (iii) (Q)
(c) (III) (iii) (P) (d) (I) (i) (S)

136. The only CORRECT combination in which the reaction proceeds through radical mechanism is

(a) (III) (ii) (P) (b) (IV) (i) (Q)
(c) (II) (iii) (R) (d) (I) (ii) (R)

137. Different possible **thermal** decomposition pathways for peroxy esters are shown below. Match each pathway from **List-I** with an appropriate structure from **List-II** and select the correct answer using the code given below the lists. **[Adv. 2014]**

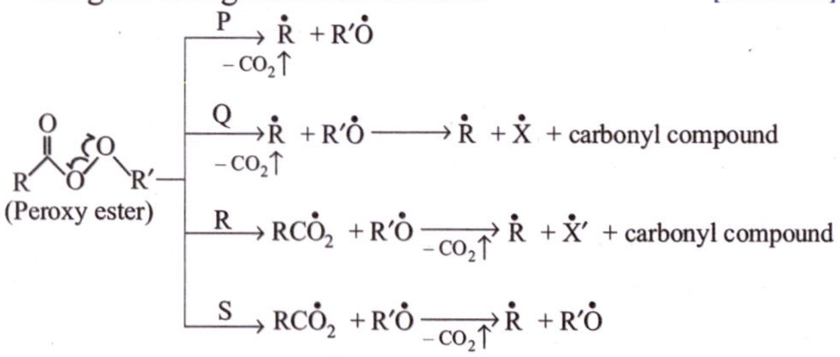

List-I		List-II
P. Pathway P	**1.**	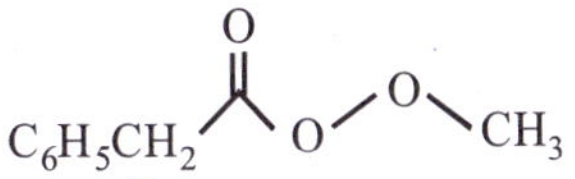
Q. Pathway Q	**2.**	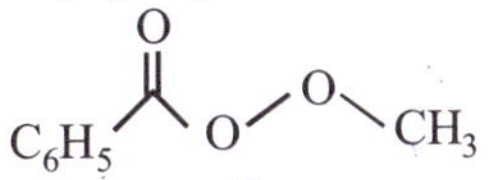
R. Pathway R	**3.**	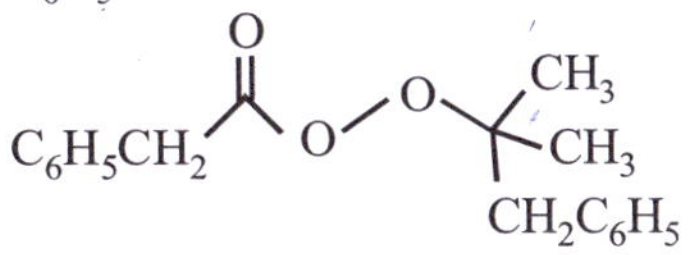
S. Pathway S	**4.**	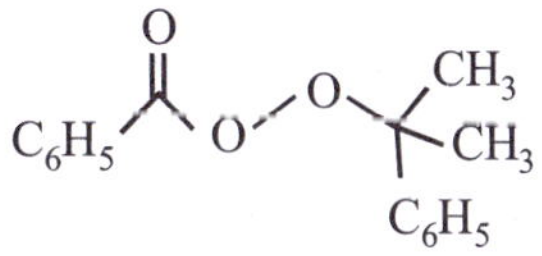

Code :

	P	Q	R	S
(a)	1	3	4	2
(c)	4	1	2	3
(b)	2	4	3	1
(d)	3	2	1	4

138. Match the reactions in **Column-I** with appropriate types of steps/reactive intermediate involved in these reactions as given in **Column-II.** **[2011]**

Column-I **Column-II**

(A) 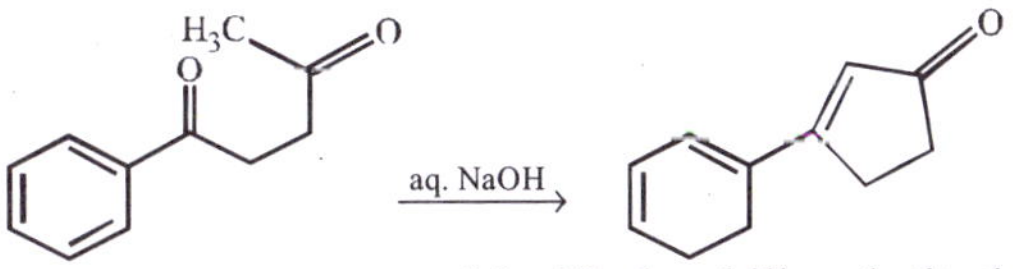

(p) Nucleophilic substitution

(B) 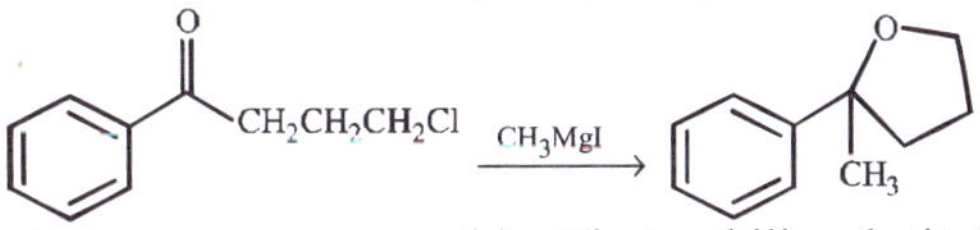

(q) Electrophilic substitution

(C) 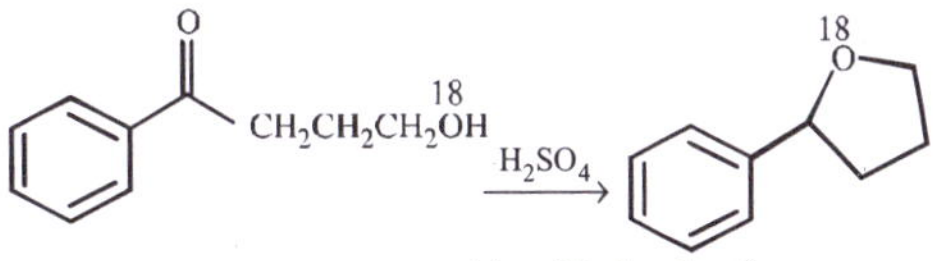

(r) Dehydration

(D) 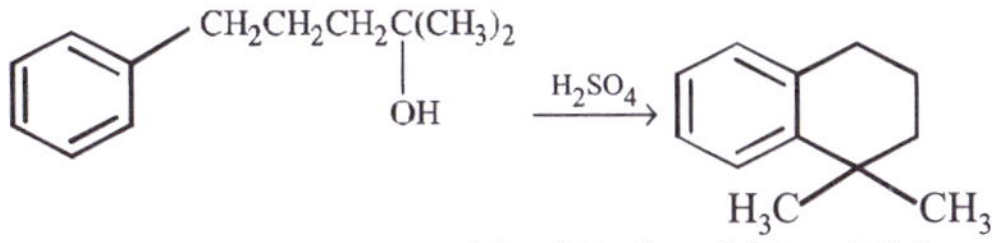

(s) Nucleophilic addition

(t) Carbanion

139. Match each of the compounds given in **Column-I** with the reaction(s), that they can undergo, given in **Column-II.** **[2009]**

Column-I **Column-II**

(A) 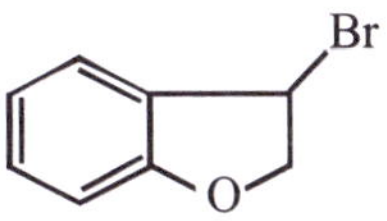 (p) Nucleophilic substitution

(B) 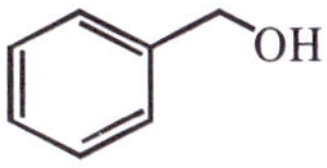 (q) Elimination

(C) 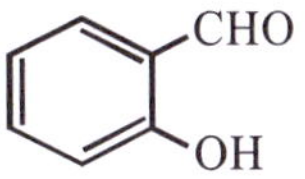 (r) Nucleophilic addition

(D) 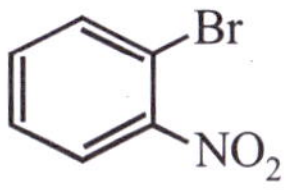 (s) Esterification with acetic anhydride

(t) Dehydrogenation

140. Match the compounds/ions in **Column-I** with their properties/ reactions in **Column-II**. Indicate your answer by darkening the appropriate bubbles of the 4 × 4 matrix given in the ORS. **[2007]**

Column-I	Column-II
(A) C_6H_5CHO	(p) gives precipitate with 2, 4-dinitrophenylhydrazine
(B) $CH_3C{\equiv}CH$	(q) gives precipitate with $AgNO_3$
(C) CN^-	(r) is a nucleophile
(D) I^-	(s) is involved in cyanohydrin formation

8 Comprehension Based Questions

Passage-I

An organic acid **P** ($C_{11}H_{12}O_2$) can easily be oxidized to a dibasic acid which reacts with ethylene glycol to produce a polymer dacron. Upon ozonolysis, **P** gives an aliphatic ketone as one of the products. **P** undergoes the following reaction sequences to furnish **R** *via* **Q**. The compound **P** also undergoes another set of reactions to produce **S**. **[Adv. 2018]**

S ← [1) H_2/Pd-C; 2) NH_3/Δ; 3) Br_2/NaOH; 4) $CHCl_3$, KOH, Δ; 5) H_2/ Pd-C] — **P** → [1) H_2/Pd-C; 2) $SOCl_2$; 3) MeMgBr, $CdCl_2$; 4) $NaBH_4$] → **Q** → [1) HCl; 2) Mg/Et_2O; 3) CO_2 (dry ice); 4) H_3O^+] → **R**

141. The compound **R** is

(a) 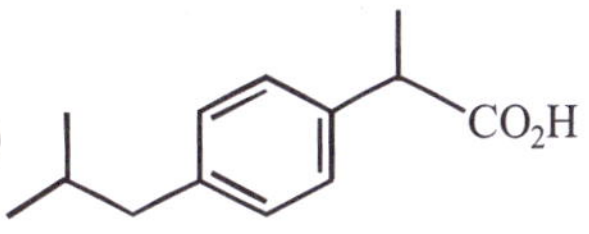 (b) 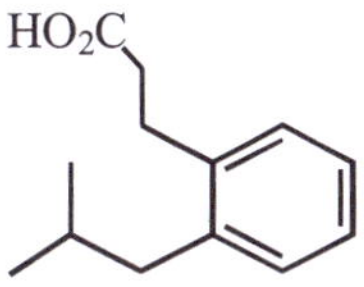

(c) 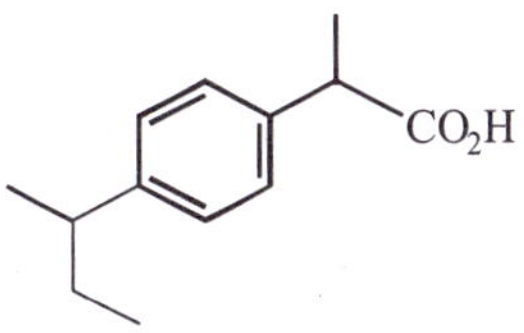 (d) 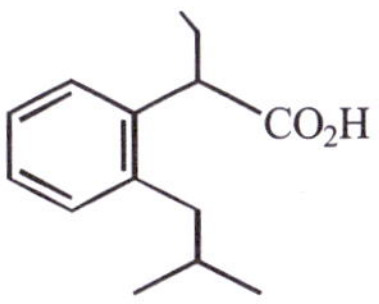

142. The compound **S** is **[Adv. 2018]**

(a) 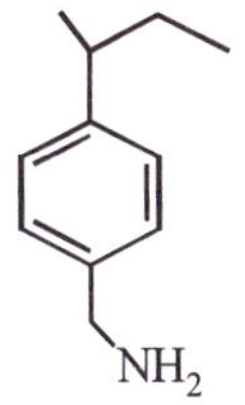 (b) 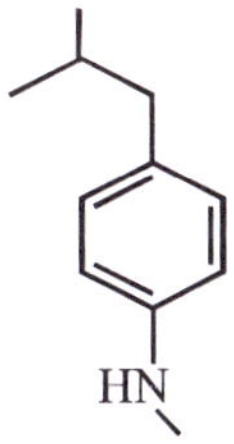

(c) 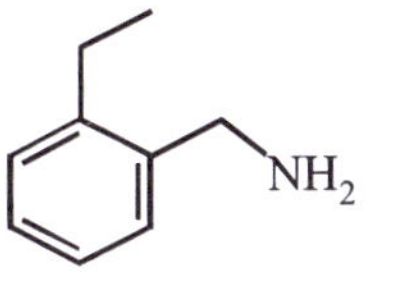 (d) 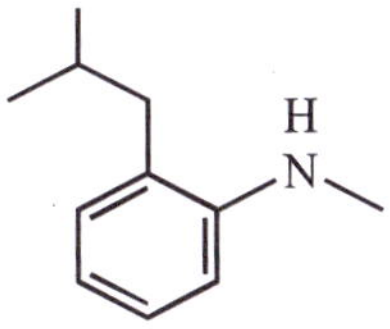

Passage-II

Treatment of benzene with CO/HCl in the presence of anhydrous $AlCl_3$/CuCl followed by reaction with Ac_2O/NaOAc gives compound **X** as the major product. Compound **X** upon reaction with Br_2/Na_2CO_3, followed by heating at 473 K with alc. KOH furnishes **Y** as the major product. Reaction of **X** with H_2/Pd-C, followed by H_3PO_4 treatment gives **Z** as the major product.

[Adv. 2018]

143. The compound **Y** is

(a) 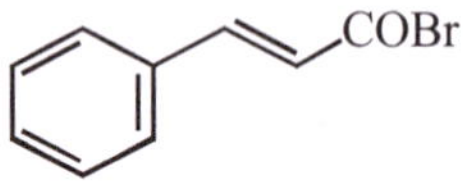

(b) 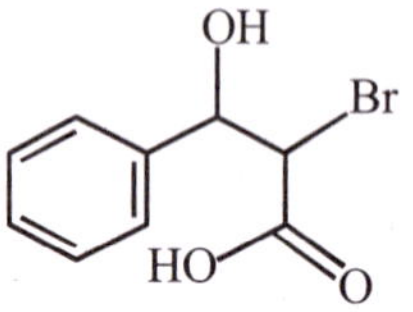

(c) 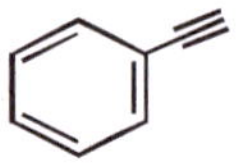

(d) 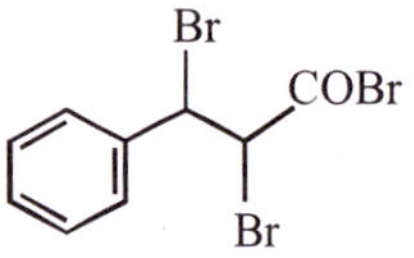

144. The compound **Z** is

(a) 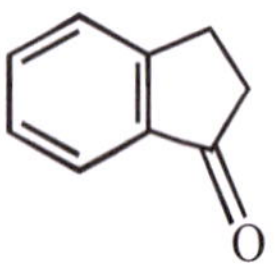

(b) 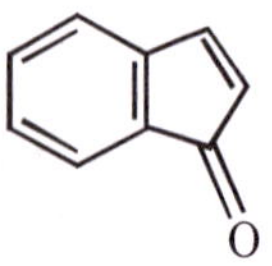

(c) 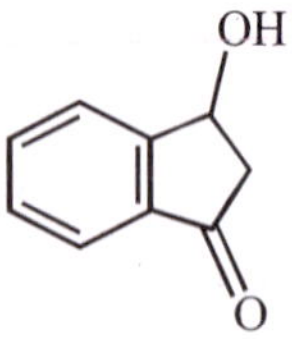

(d) 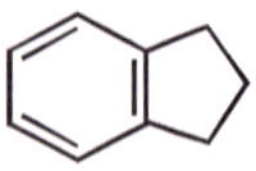

Passage-III

The reaction of compound *P* with CH_3MgBr (excess) in $(C_2H_5)_2O$ followed by addition of H_2O gives *Q*. The compound *Q* on treatment with H_2SO_4 at 0°C gives *R*. The reaction of R with CH_3COCl in the presence of anhydrous $AlCl_3$ in CH_2Cl_2 followed by treatment with H_2O produces compound *S*. [Et in compound *P* is ethyl group]

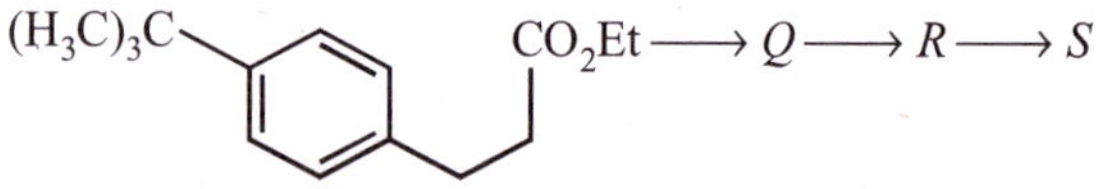

P

[Adv. 2017]

145. The product *S* is

(a) 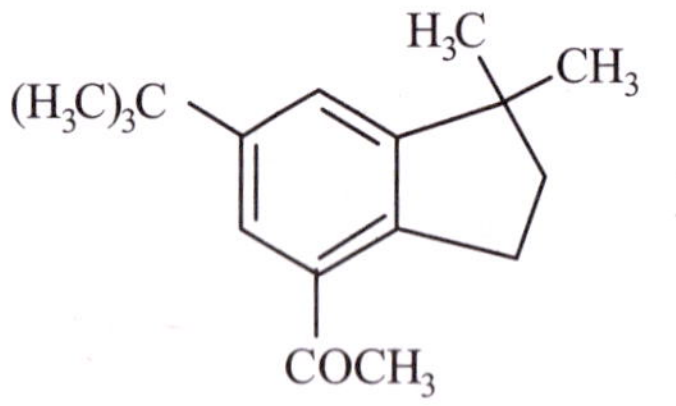

(b) 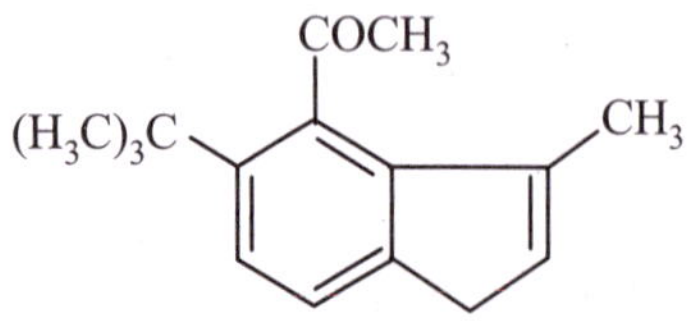

(c) 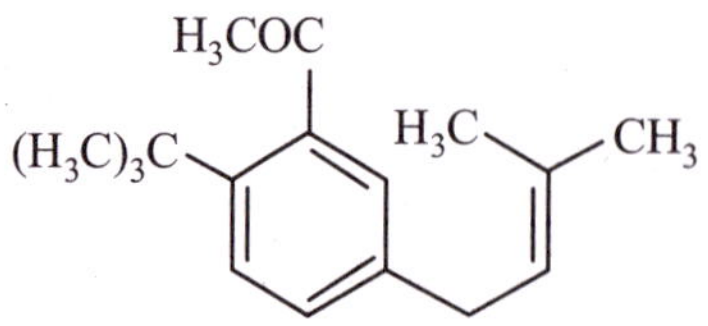

(d) 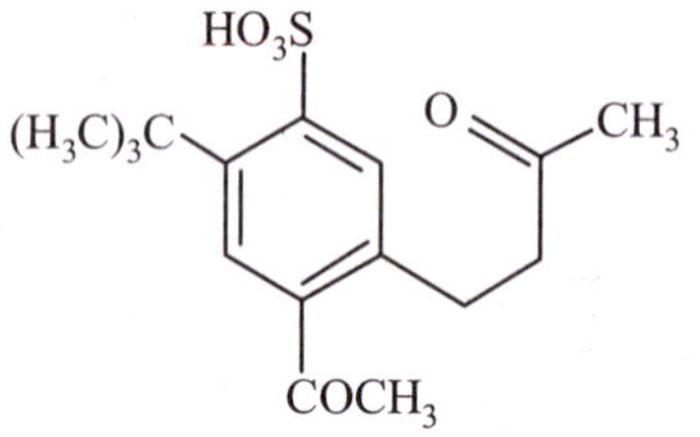

146. The reactions, *Q* to *R* and *R* to *S*, are
(a) Dehydration and Friedel–Craft's acylation
(b) Aromatic sulfonation and Friedel–Craft's acylation
(c) Friedel–Craft's alkylation, dehydration and Friedel–Craft's acylation
(d) Friedel–Craft's alkylation and Friedel–Craft's acylation

Passage-IV

P and **Q** are isomers of dicarboxylic acid $C_4H_4O_4$. Both decolorize Br_2/H_2O. On heating, **P** forms the cyclic anhydride. Upon treatment with dilute alkaline $KMnO_4$, **P** as well as **Q** could produce one or more than one from **S**, **T** and **U**.

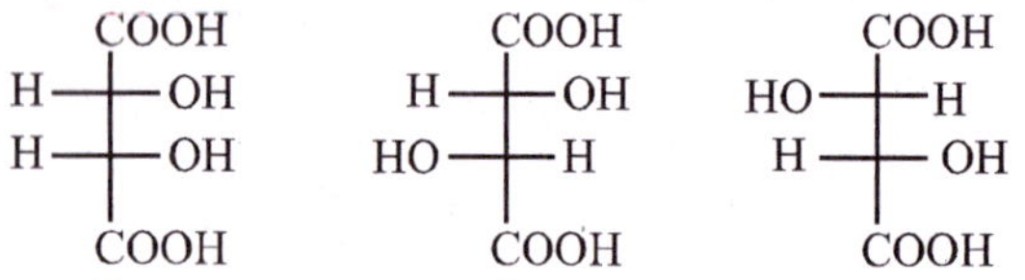

S **T** **U**

[Adv. 2013]

147. Compounds formed from **P** and **Q** are, respectively
(a) Optically active **S** and optically active pair **[T, U]**
(b) Optically inactive **S** and optically inactive pair **[T, U]**
(c) Optically active pair **[T, U]** and optically active **S**
(d) Optically inactive pair **[T, U]** and optically inactive **S**

148. In the following reaction sequences **V** and **W** are respectively

$$\mathbf{Q} \xrightarrow[\Delta]{H_2/Ni} \mathbf{V}$$

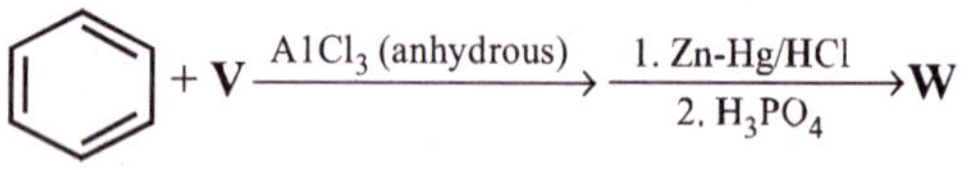

(a) 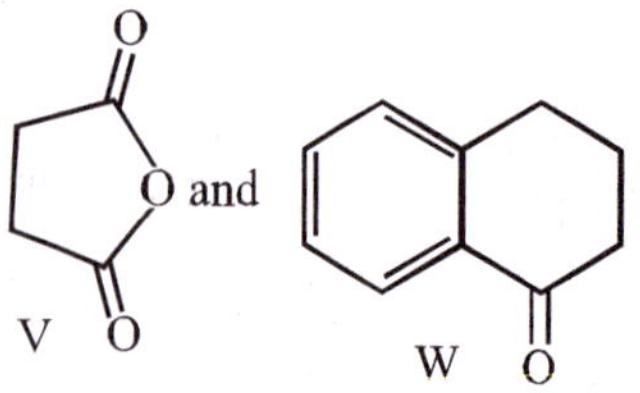

(b) 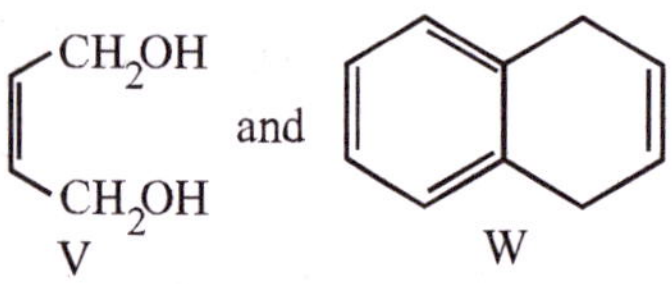

(c) 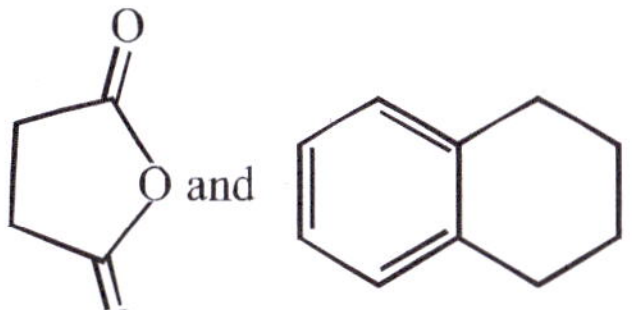

V and W

(d) 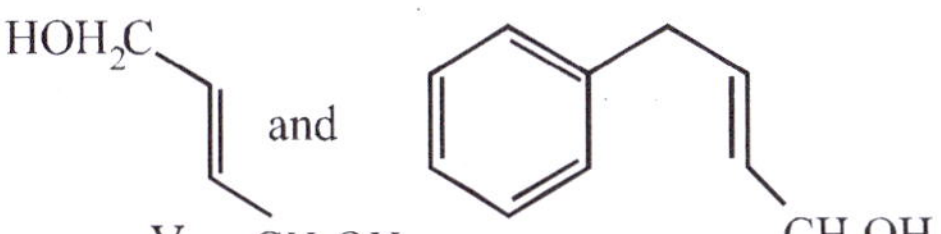

V and W

Passage-V

In the following reaction sequence, the compound **J** is an intermediate.

$$\mathbf{I} \xrightarrow[CH_3COONa]{(CH_3CO)_2O} \mathbf{J} \xrightarrow[\text{(ii) } SOCl_2 \text{ (iii) anhyd. } AlCl_3]{\text{(i) } H_2, Pd/C} \mathbf{K}$$

J ($C_9H_8O_2$) gives effervescences on treatment with $NaHCO_3$ and also give positive Baeyer's test. [2012]

149. The compound **I** is

(a) 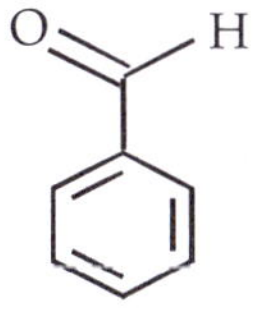

(b) 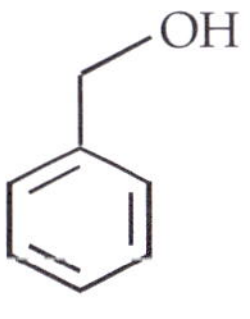

(c) 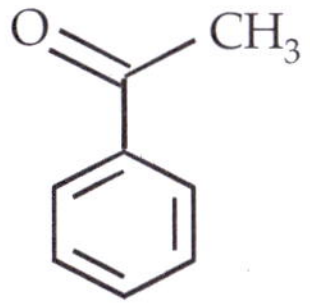

(d) 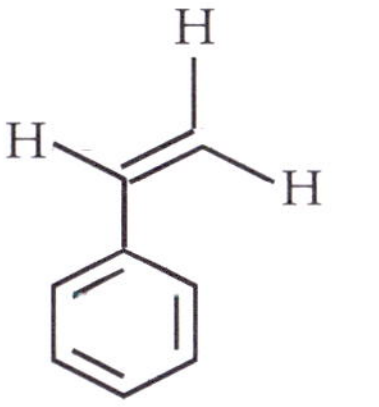

150. The compound **K** is

(a) 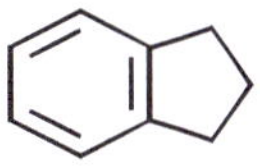

(c) 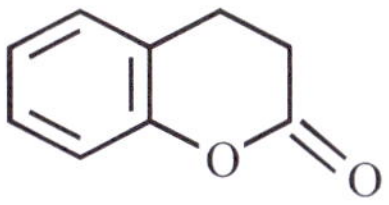

(b) 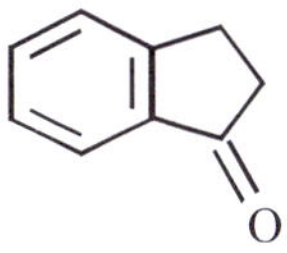

(d) 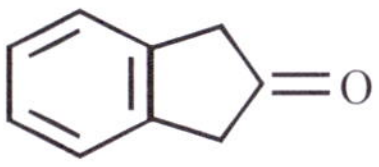

9 Assertion and Reason/Statement Type Questions

Each question contains STATEMENT-1 (Assertion) and STATEMENT-2 (Reason). Each question has 4 choices (a), (b), (c) and (d) out of which ONLY ONE is correct. Mark your answer as

(a) If both Statement -1 and Statement -2 are correct, and Statement -2 is the correct explanation of the Statement -2.

(b) If both Statement -1 and Statement -2 are correct, but Statement -2 is not the correct explanation of the Statement -1.

(c) If Statement -1 is correct but Statement -2 is incorrect.

(d) If Statement -1 is incorrect but Statement -2 is correct.

151. **Statement-1 :** *p*-Hydroxybenzoic acid has a lower boiling point than *o*-hydroxybenzoic acid.

Statement-2 : *o*-Hydroxybenzoic acid has intramolecular hydrogen bonding. [2007]

152. **Statement-1 :** Acetic acid does not undergo haloform reaction.

Statement-2 : Acetic acid has no alpha hydrogens. [1998 - 2 Marks]

153. **Statement-1 :** Acetate ion is more basic than the methoxide ion.

Statement-2 : The acetate ion is resonance stabilized [1994 - 2 Marks]

10 Subjective Problems

154. Match the K_a values [2003]

		Ka
(a)	Benzoic acid	6.4×10^{-5}
(b)	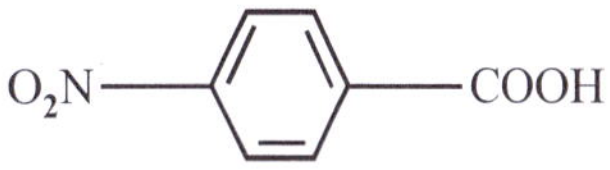	30.6×10^{-5}
(c)	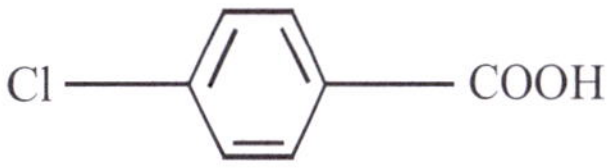	10.2×10^{-5}
(d)	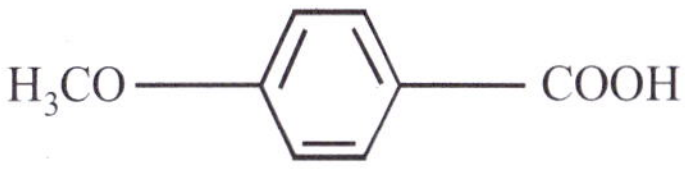	3.3×10^{-5}
(e)	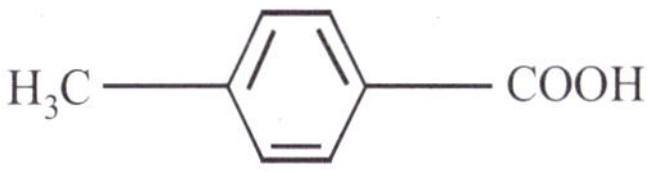	4.2×10^{-5}

155. A racemic mixture of (±) 2–phenylpropanoic acid on esterification with (+) 2–butanol gives two esters. Mention the stereochemistry of the two esters produced. [2003 - 2 Marks]

156. Identify (X), (Y) and (Z) in the following synthetic scheme and write their structures. [2001 - 5 Marks]

$$Ba\overset{*}{C}O_3 + H_2SO_4 \rightarrow (X) \text{ gas} \qquad [C^* \text{ denotes } C^{14}]$$

$$CH_2 = CH - Br \xrightarrow[\text{(ii) X (iii) } H_3O^+]{\text{(i) Mg/ether}} (Y) \xrightarrow{LiAlH_4} (Z)$$

Explain the formation of labelled formaldehyde (H_2C^*O) as one of the products when compound (Z) is treated with HBr and subsequently ozonolysed. Mark the C* carbon in the entire scheme.

157. An organic compound **A**, $C_8H_4O_3$, in dry benzene in the presence of anhydrous $AlCl_3$ gives compound **B**. The compound **B** on treatment with PCl_5, followed by reaction with H_2/Pd ($BaSO_4$) gives compound **C**, which on reaction with hydrazine gives a cyclic compound **D** ($C_{14}H_{10}N_2$). Identify **A, B, C** and **D**. Explain the formation of **D** from **C**. [2000 - 5 Marks]

158. Write the structural formula of the main organic product formed when :

(*i*) Write the structures of the products A and B.

$$CH_3 - \overset{O}{\overset{\|}{C}} - {}^{18}OC_2H_5 \xrightarrow{H_3O^+} A + B$$

[2000 - 2 Marks]

(*ii*) $(COOH)_2 + (CH_2OH)_2 + \text{Conc.}H_2SO_4 \xrightarrow{F} ----$ [1997 - 1 Mark] (*iii*) $o-HOOC-C_6H_4-CH_2-C_6H_5$

$\xrightarrow{SOCl_2} \overset{G}{------} \xrightarrow[AlCl_3]{anhydrous} \overset{H}{------}$

$H \xrightarrow[HCl]{Zn-Hg} \overset{I}{------}.$ **[1995 - 2 Marks]**

(*iv*) $C_6H_5-CHO + CH_3-COOC_2H_5$

$\xrightarrow[C_2H_5OH \text{ and heat}]{NaOC_2H_5 \text{ in absolute}} \overset{D}{------}$ **1995 - 1 Mark]**

(*v*) $CH_3-CH_2-COOH \xrightarrow[bromine]{P \text{ and}} \overset{A}{-----}$

$\xrightarrow[2.\ H^+]{1.\ alcoholic\ KOH (excess)} \overset{B}{-----}$ **[1995 - 2 Marks]**

(*vi*) $C_6H_5CH_2CO_2CH_3 \xrightarrow[(ii)\ H^+]{(i)\ CH_3MgBr(excess)}$

[1994 - 1 Mark]

(*vii*) $C_6H_5COOH + CH_3MgI \rightarrow ? + ?$ **[1993 - 2 Marks]**

(*viii*) benzene **[1985 - 1 Mark]**

(*ix*) ethyl acetate is treated with double the molar quantity of ethyl magnesium bromide and the reaction mixture poured into water. **[1981 - ½ Mark]**

159. Explain briefly the formation of the products giving the structures of the intermediates.

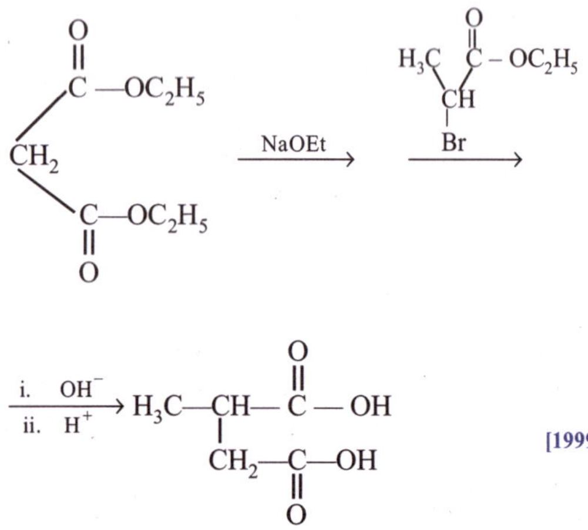

[1999 - 5 Marks]

160. An ester *A* ($C_4H_8O_2$), on treatment with excess methyl magnesium chloride followed by acidification, gives an alcohol *B* as the sole organic product. Alcohol *B*, on oxidation with NaOCl followed by acidification, gives acetic acid. Deduce the structures of *A* and *B*. Show the reactions involved.

[1998 - 6 Marks]

161. Give reasons for the following :

(*i*) Although phenoxide ion has more number of resonating structures than benzoate ion, benzoic acid is a stronger acid than phenol. Why? **[1997 - 2 Marks]**

(*ii*) Formic acid is a stronger acid than acetic acid; **[1985 - 1 Mark]**

(*iii*) Acetic acid can be halogenated in the presence of red P and Cl_2 but formic acid cannot be halogenated in the same way. **[1983 - 1 Mark]**

162. Which of the following carboxylic acids undergoes decarboxylation easily? Explain briefly. **[1995 - 2 Marks]**

(*i*) $C_6H_5-CO-CH_2-COOH$

(*ii*) $C_6H_5-CO-COOH$

(*iii*) $C_6H_5-\underset{\underset{OH}{|}}{CH}-COOH$

(*iv*) $C_6H_5-\underset{\underset{NH_2}{|}}{CH}-COOH$

163. An organic compound *A*, C_8H_6, on treatment with dilute sulphuric acid containing mercuric sulphate gives a compound *B*, which can also be obtained from a reaction of benzene with an acid chloride in the presence of anhydrous aluminium chloride. The compound *B*, when treated with iodine in aqueous KOH, yields *C* and a yellow compound *D*. Identify *A*, *B*, *C* and *D* with justification. Show how *B* is formed from *A*. **[1994 - 3 Marks]**

164. In the following reactions identify the compounds *A*, *B*, *C* and *D*. **[1994 - 1 × 4 = 4 Marks]**

$PCl_5 + SO_2 \longrightarrow A + B$

$A + CH_3COOH \longrightarrow C + SO_2 + HCl$

$2C + (CH_3)_2Cd \longrightarrow 2D + CdCl_2$

165. An organic compound 'A' on treatment with ethyl alcohol gives a carboxylic acid 'B' and compound 'C'. Hydrolysis of 'C' under acidic conditions gives 'B' and 'D'. Oxidation of 'D' with $KMnO_4$ also gives 'B'. 'B' on heating with $Ca(OH)_2$ gives 'E' (molecular formula, C_3H_6O). 'E' does not give Tollen's test and does not reduce Fehling's solution but forms a 2,4-dinitrophenylhydrazone. Identify 'A', 'B', 'C', 'D' and 'E'.

[1992 - 3 Marks]

166. Compound 'X', containing chlorine on treatment with strong ammonia gives a solid 'Y' which is free from chlorine. 'Y' analysed as C = 49.31%, H = 9.59% and N = 19.18% and reacts with Br_2 and caustic soda to give a basic compound 'Z'. 'Z' reacts with HNO_2 to give ethanol. Suggest structures for 'X', 'Y' and 'Z'. **[1992 - 1 Mark]**

167. Arrange the following in :

Increasing order of acid strength : **[1991]**

(I) $ClCH_2COOH$, (II) CH_3CH_2COOH,
(III) $ClCH_2CH_2COOH$, (IV) $(CH_3)_2CHCOOH$,
(V) CH_3COOH

168. (*i*) $C_6H_5COOH \xrightarrow{PCl_5} C \xrightarrow{NH_3}$

$D \xrightarrow{P_2O_5} C_6H_5CN \xrightarrow{H_2/Ni} E;$

Identify *C*, *D* and *E*. **[1991 - 2 Marks]**

(*ii*) $H_3C-CH=CH-CHO \xrightarrow{NaBH_4} F$

$\xrightarrow[ZnCl_2]{HCl} G \xrightarrow[H^+]{KCN} H;$

Identify *F*, *G* and *H*. **[1991 - 2 Marks]**

169. Compound *A* ($C_6H_{12}O_2$) on reduction with $LiAlH_4$ yielded two compounds *B* and *C*. The compound *B* on oxidation gave *D*, which on treatment with aqueous alkali and subsequent heating furnished *E*. The latter on catalytic hydrogenation gave *C*. The compound *D* was oxidized further to give *F* which was found to be a monobasic acid (molecular weight = 60.0). Deduce the structures of *A*, *B*, *C*, *D* and *E*. **[1990 - 4 Marks]**

170. The sodium salt of a carboxylic acid, *A*, was produced by passing a gas, *B*, into an aqueous solution of caustic alkali at an elevated temperature and pressure. *A*, on heating in presence of sodium hydroxide followed by treatment with sulphuric acid gave a dibasic acid, *C*. A sample of 0.4 g of acid *C*, on combustion gave 0.08 g of water and 0.39 g of carbon dioxide. The silver salt of the acid *C* weighing 1.0 g on ignition yielded 0.71 g of silver as residue. Identify *A*, *B* and *C*. **[1990 - 5 Marks]**

171. Outline the reaction sequence for the conversion of

(*i*) Ethanoic acid to a mixture of methanoic acid and diphenyl ketone. **[1990 - 2Marks]**

(*ii*) Ethanal to 2-hydroxy-3-butenoic acid. **[1990 - 2 Marks]**

(*iii*) acetic acid to tertiary-butyl alcohol. **[1989 - 1½ Marks]**

172. Complete the following reactions :

$$CH_3COOH \xrightarrow{?} ClCH_2COOH \xrightarrow{\text{excess ammonia}} ?$$

[1988 - 1 Mark]

173. An organic compound (A) on treatment with acetic acid in the presence of sulphuric acid produces an ester (B), (A) on mild oxidation gives (C), (C) with 50% potassium hydroxide followed by acidification with dilute hydrochloric acid generates (A) and (D), (D) with phosphorus pentachloride followed by reaction with ammonia gives (E), (E) on dehydration produces hydrocyanic acid.

Identify the compounds A, B, C, D and E. **[1987 - 5 Marks]**

174. Complete the following with appropriate structures :

(*i*) $(CH_3CO)_2O \xrightarrow{C_2H_5OH} CH_3COOH + ?$ **[1986 - 1 Mark]**

(*ii*) 

[1986 - 1 Mark]

175. A liquid (X), having a molecular formula $C_6H_{12}O_2$ is hydrolysed with water in the presence of an acid to give a carboxylic acid (Y) and an alcohol (Z). Oxidation of (Z) with chromic acid gives (Y). What are the structures of (X), (Y) and (Z)?

[1986 - 3 Marks]

176. Arrange the following in increasing ease of hydrolysis

$CH_3COOC_2H_5$, CH_3COCl, $(CH_3CO)_2O$, CH_3CONH_2.

[1986 - 1 Mark]

177. Write down the reactions involved in the preparation of the following using the reagents indicated against it in parenthesis:

Propionic anhydride from propionaldehyde

[$AgNO_3/NH_4OH$, P_2O_5]. **[1984 - 2 Marks]**

178. State the conditions under which the following preparations are carried out. Give the necessary equations which need not be balanced.

(*i*) Ethanol from acetylene **[1983 - 1 Mark]**

(*ii*) Acetic acid from methyl iodide **[1983 - 1 Mark]**

179. An alkene (*A*) on ozonolysis yields acetone and an aldehyde. The aldehyde is easily oxidized to an acid (*B*). When *B* is treated with bromine in presence of phosphorus, it yields a compound (*C*) which on hydrolysis gives a hydroxy acid (*D*). This acid can also be obtained from acetone by reaction with hydrogen cyanide followed by hydrolysis. Identify the compounds *A*, *B*, *C* and *D*.

[1982 - 2 Marks]

180. Outline the accepted mechanism of the following reaction. Show the various steps including the charged intermediates.

[1981 - 3 Marks]

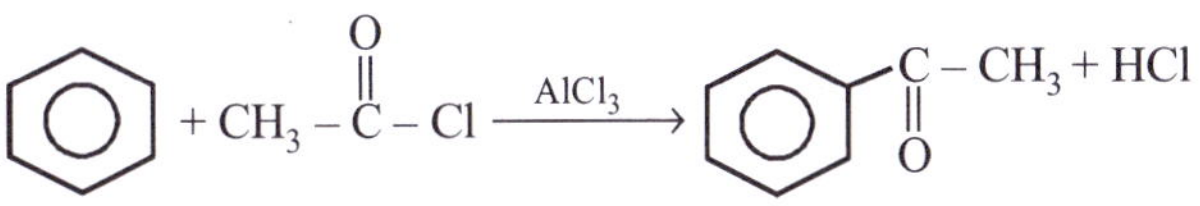

181. Write the chemical equation to show what happens when ethyl acetate is treated with sodium ethoxide in ethanol and the reaction mixture is acidified. **[1981 - 2 Marks]**

Answer Key

Part-A (JEE Main)

1	(1)	31	(3)	61	(1)	91	(3)	121	(3)	151	(1)	181	(10.1)	211	(3)	241	(3)	271	(4)
2	(4)	32	(2)	62	(3)	92	(3)	122	(1)	152	(4)	182	(7)	212	(4)	242	(3)	272	(2)
3	(2)	33	(2)	63	(2)	93	(2)	123	(3)	153	(1)	183	(60)	213	(4)	243	(1)	273	(2)
4	(3)	34	(3)	64	(3)	94	(1)	124	(3)	154	(1)	184	(13)	214	(1)	244	(1)	274	(1)
5	(2)	35	(4)	65	(2)	95	(1)	125	(3)	155	(1)	185	(318)	215	(4)	245	(2)	275	(3)
6	(3)	36	(4)	66	(2)	96	(2)	126	(4)	156	(1)	186	(0)	216	(4)	246	(1)	276	(4)
7	(1)	37	(1)	67	(1)	97	(1)	127	(1)	157	(2)	187	(3)	217	(4)	247	(1)	277	(1)
8	(2)	38	(1)	68	(2)	98	(2)	128	(3)	158	(4)	188	(60)	218	(1)	248	(1)	278	(4)
9	(1)	39	(4)	69	(2)	99	(3)	129	(3)	159	(3)	189	(15)	219	(1)	249	(4)	279	(2)
10	(1)	40	(4)	70	(2)	100	(1)	130	(3)	160	(3)	190	(2)	220	(4)	250	(3)	280	(3)
11	(4)	41	(3)	71	(4)	101	(4)	131	(2)	161	(3)	191	(4)	221	(1)	251	(1)	281	(3)
12	(3)	42	(1)	72	(4)	102	(1)	132	(3)	162	(3)	192	(2)	222	(4)	252	(3)	282	(2)
13	(3)	43	(4)	73	(2)	103	(3)	133	(1)	163	(2)	193	(3)	223	(3)	253	(4)	283	(1)
14	(3)	44	(3)	74	(4)	104	(3)	134	(2)	164	(2)	194	(3)	224	(2)	254	(2)	284	(4)
15	(1)	45	(2)	75	(2)	105	(4)	135	(3)	165	(3)	195	(3)	225	(3)	255	(1)	285	(3)
16	(3)	46	(1)	76	(2)	106	(2)	136	(3)	166	(1)	196	(1)	226	(2)	256	(1)	286	(4)
17	(4)	47	(4)	77	(4)	107	(4)	137	(2)	167	(4)	197	(8)	227	(4)	257	(1)	287	(3)
18	(1)	48	(2)	78	(4)	108	(2)	138	(4)	168	(1)	198	(4)	228	(3)	258	(3)	288	(4)
19	(1)	49	(1)	79	(4)	109	(1)	139	(4)	169	(2)	199	(3)	229	(2)	259	(3)	289	(1)
20	(2)	50	(2)	80	(3)	110	(3)	140	(3)	170	(4)	200	(3)	230	(4)	260	(4)	290	(3)
21	(3)	51	(2)	81	(4)	111	(1)	141	(4)	171	(3)	201	(2)	231	(1)	261	(2)	291	(1)
22	(3)	52	(2)	82	(3)	112	(3)	142	(2)	172	(1)	202	(5)	232	(3)	262	(4)	292	(3)
23	(2)	53	(2)	83	(2)	113	(3)	143	(4)	173	(1)	203	(66.67)	233	(3)	263	(3)	293	(6)
24	(3)	54	(4)	84	(1)	114	(1)	144	(2)	174	(3)	204	(3)	234	(1)	264	(3)	294	(5)
25	(1)	55	(2)	85	(1)	115	(3)	145	(4)	175	(3)	205	(4)	235	(2)	265	(4)	295	(4)
26	(3)	56	(2)	86	(4)	116	(3)	146	(2)	176	(4)	206	(4)	236	(1)	266	(1)	296	(45)
27	(3)	57	(2)	87	(1)	117	(3)	147	(1)	177	(2)	207	(1)	237	(3)	267	(1)	297	(18)
28	(4)	58	(3)	88	(4)	118	(2)	148	(2)	178	(1)	208	(3)	238	(4)	268	(2)	298	(2)
29	(3)	59	(2)	89	(3)	119	(3)	149	(1)	179	(2)	209	(4)	239	(1)	269	(1)	299	(78)
30	(9)	60	(2)	90	(1)	120	(1)	150	(4)	180	(16)	210	(3)	240	(3)	270	(4)	300	(1)

Part-B (JEE Advanced)

1	(12)	25	(a)	42	(d)	62	(b, c)	79	(a)	112	(a)	131	(a, b)	144	(a)				
2	(4)	26	(a)	43	(a)	63	(a, b, d)	80	(c)	113	(2)	132	(b)	145	(a)				
3	(1)	27	(d)	44	(d)	64	(b)	94	(c)	114	(4)	133	(a)	146	(a)				
4	(c)	28	(c)	45	(c)	65	(b, d)	95	(d)	115	(4.00)	134	(a)	147	(b)				
6	(a, b, d)	29	(a)	46	(b)	66	(a, b)	96	(a)	116	(175)	135	(c)	148	(a)				
7	(c, d)	30	(b)	47	(c)	67	(b, d)	97	(a)	117	(909)	136	(d)	149	(a)				
8	(a, c)	31	(c)	48	(b)	68	(a, c)	98	(d)	118	(18.00)	137	(a)	150	(b)				
9	(b, d)	32	(c)	49	(b)	69	(d)	99	(b)	120	False	138	A – r, s, t ; B – p, s ;C - r, s; D – q, r	151	(d)				
10	(a, b)	33	(c)	50	(5)	70	(A) - r, s; (B) - p, q; (C) - p, q, r; (D) - p, s	100	(d)	121	False	139	(A) - p, q, t ; (B) - p, s, t ; (C) – r, s ; (D) – p	152	(c)				
11	(b, c)	34	(b)	51	(12)	71	(2)	101	(b)	122	(b, c)	140	(A) - p, q, s; (B) - q; (C) - q, r, s; (D) - q, r	153	(d)				
12	(b, d)	35	(c)	52	(18)	72	(102018)	102	(c)	109	(a)	127	(c, d)						
13	(b)	36	(a)	56	False	73	(c)	103	(c)	110	(c)	128	(a, c, d)						
14	(a)	37	(b)	57	True	74	(d)	104	(a)	111	(d)	129	(b, d)						
15	(d)	38	(a)	58	False	75	(b)	105	(c)	123	(c, d)	130	(a, b, d)						
16	(c)	39	(b)	59	(a, b, c)	76	(a)	106	(d)	124	(b)	141	(a)						
17	(a)	40	(b)	60	(a, c, d)	77	(d)	107	(a)	125	(a, d)	142	(b)						
24	(c)	41	(b)	61	(a, b, c)	78	(d)	108	(a)	126	(a, b, d)	143	(c)						

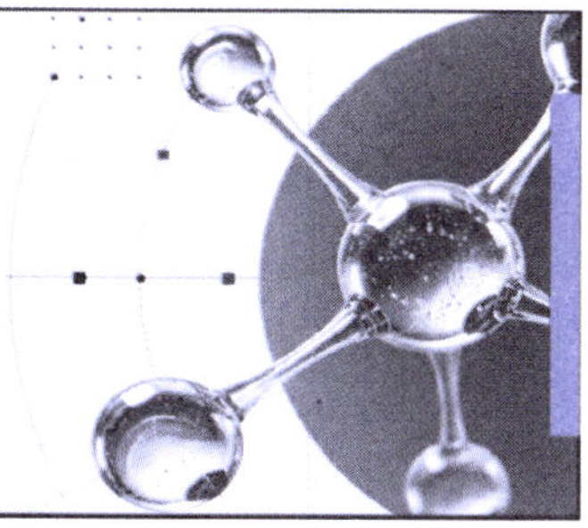

Chapter-13

Amines

Part-A (JEE Main)

Topic 1 - Aliphatic and Aromatic Amines

Multiple Choice Questions (MCQs)

1. Which statements are True? [April 08, 2026 (II)]
   A. In Hoffmann bromamide degradation, 4 moles of NaOH and 2 moles of Br_2 are consumed per mole, of an amide
   B. Hoffmann bromamide reaction is not given by alkyl amides.
   C. Primary amines can be synthesized by Hoffmann bromamide degradation.
   D. Secondary amide on reaction with Br_2 and NaOH will give secondary amine.
   E The by-products of Hoffmann degradation are Na_2CO_3, NaBr and H_2O.
   Choose the correct answer from the options given below:
   (1) A, C and E only (2) B, C and D only
   (3) C and E only (4) C, D and E only

2. The number of compounds from the following which can undergo reaction with Br_2/KOH/(alcoholic) to give respective products and these respective products can also be obtained separately by Gabriel phthalimide reaction is: [April 06, 2026 (II)]

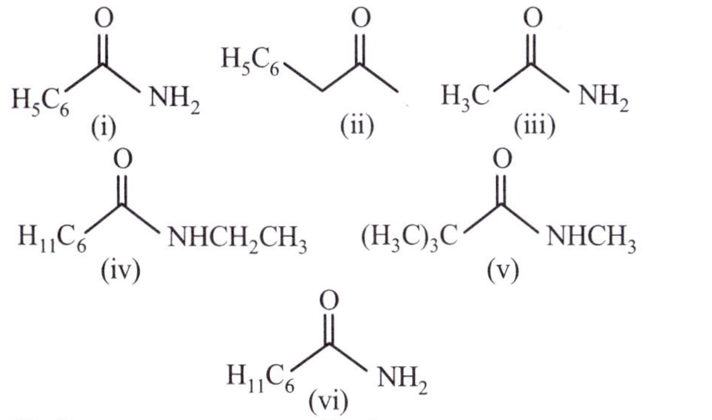

   (1) 5 (2) 4
   (3) 3 (4) 6

3. Arrange the following compounds according to increasing order of boiling points.
   n-C_4H_9OH (A), n-$C_4H_9NH_2$ (B), n-C_4H_{10} (C) and $C_2H_5NHC_2H_5$(D). [April 06, 2026 (I)]
   (1) C<B<A<D (2) D<C<B<A
   (3) C<D<B<A (4) D<B<A<C

4. Given below are two statements : [April 05, 2026 (II)]
   Statement-I: Heating benzamide with bormine in an ethanolic solution of sodium hydroxide will give benzylamine.
   Statement-II: Nitration of aniline with HNO_3/H_2SO_4 at 288K produces m-nitroaniline in higher amount than o-nitroaniline (pH adjusted).
   In the light of the above statements, choose the correct answer from the options given below.
   (1) Both Statement I and Statement II are true
   (2) Both Statement I and Statement II are false
   (3) Statement I is false but Statement II is true
   (4) Statement I is true but Statement II is false

5. Complete the following reaction sequence and give the name of major product 'P'. [April 05, 2026 (II)]

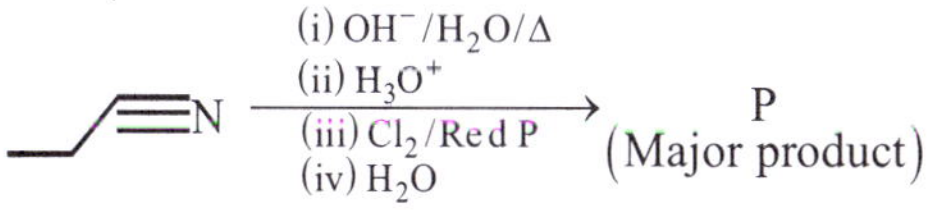

   (1) 2-Chloropropanoic acid (2) 3-Chloropropanoic acid
   (3) 1-Chloropropane (4) 2-Chloropropane

6. Identify the incorrect statements: [April 05, 2026 (I)]
   A. 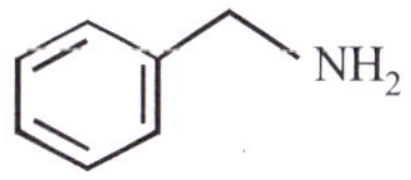 is a stronger base than 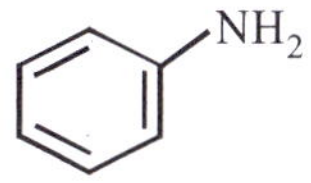
   B. 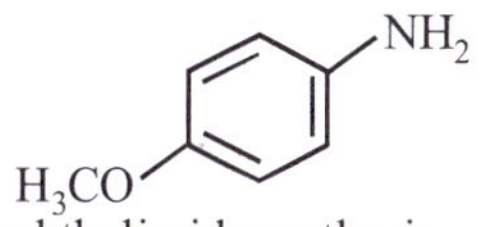 can be synthesized by Gabriel phthalimide synthesis.
   C. 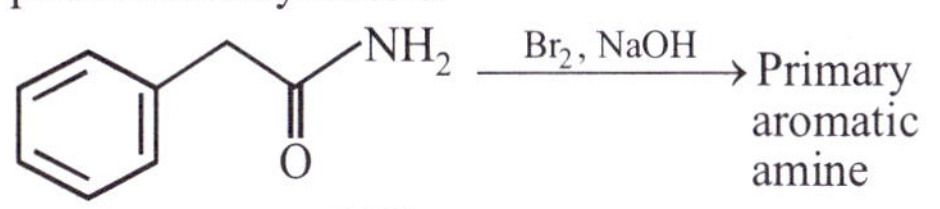
   D. 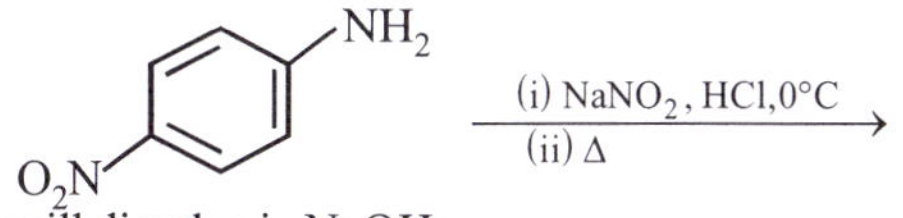 will dissolve in NaOH

   Choose the correct answer from the options given below:
   (1) A and D only (2) A and C only
   (3) B and C only (4) A and B only

7. The strongest conjugate acid will result from [April 04, 2026 (II)]

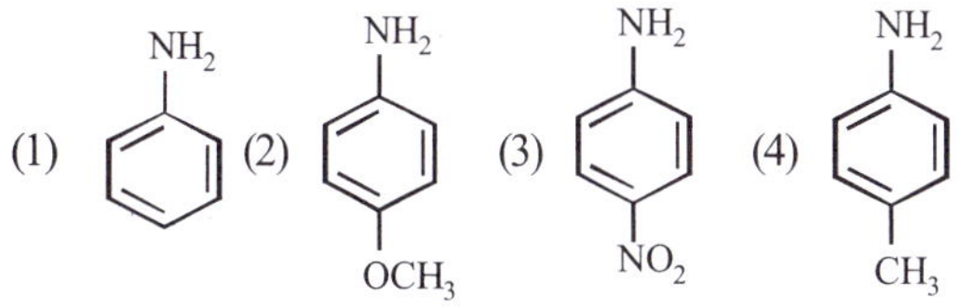

8. Consider the three aromatic molecules (P, Q and R) whose structures have been given below : [April 02, 2026 (I)]

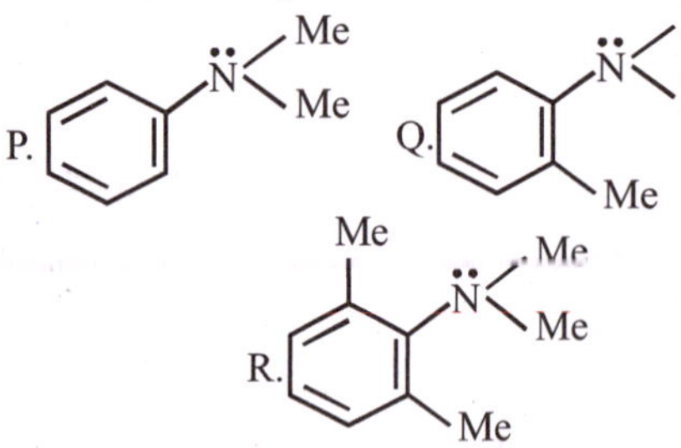

The correct order regarding the reactivity of these compounds with $Ph-N\equiv\overset{+}{N}Cl^{(-)}$ under optimum but slightly acidic medium is :

(1) P > Q > R (2) R > P > Q
(3) R > Q > P (4) P > R > Q

9. Total number of alkali insoluble solid sulphonamides obtained by reaction of given amines with Hinsberg's reagent is ....
Aniline, N-Methylaniline, Methanamine,
N, N-Dimethylmethanamine,
N-Methyl methanamine, Phenylmethanamine,
N-Propylaniline, N-Phenylaniline,
N, N-Dimethylaniline, Allyl amine,
Isopropyl amine [Jan. 28, 2026 (II)]

(1) 4 (2) 2 (3) 8 (4) 5

10. A student performed analysis of aliphatic organic compound 'X' which on analysis gave C = 61.01%, H = 15.25%, N = 23.74%. This compound, on treatment with HNO_2/H_2O produced another compound 'Y' which did not contain any nitrogen atom. However, the compound 'Y' upon controlled oxidation produced another compound 'Z' that responded to iodoform test. The structure of 'X' is: [Jan. 28, 2026 (II)]

(1) $CH_3CH_2CH_2NH_2$ (2) $Ph-\underset{CH_3}{\underset{|}{CH}}-NH_2$

(3) $\overset{CH_3}{\underset{CH_3}{>}}CH-NH_2$ (4) $CH_3-CH_2\underset{NH_2}{\underset{|}{CH}}-CH_3$

11. Consider the following reactions giving major product. Identify the correct reaction. [Jan. 28, 2026 (I)]

(1) 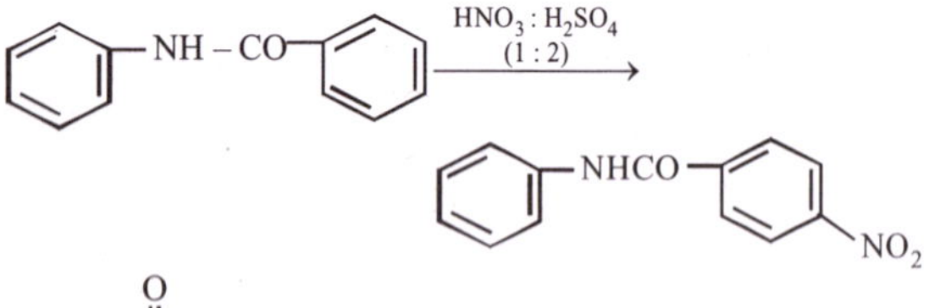

(2) 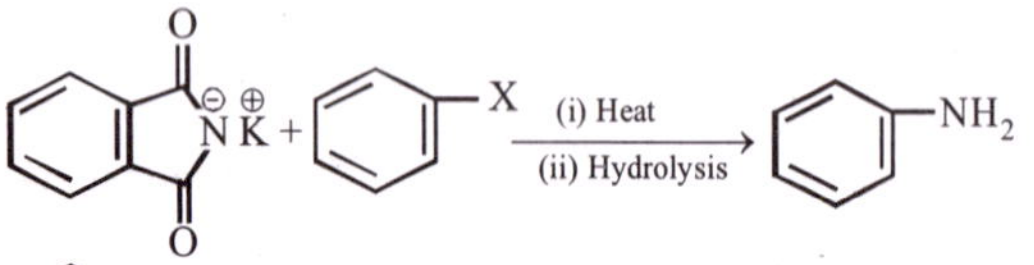

(3) 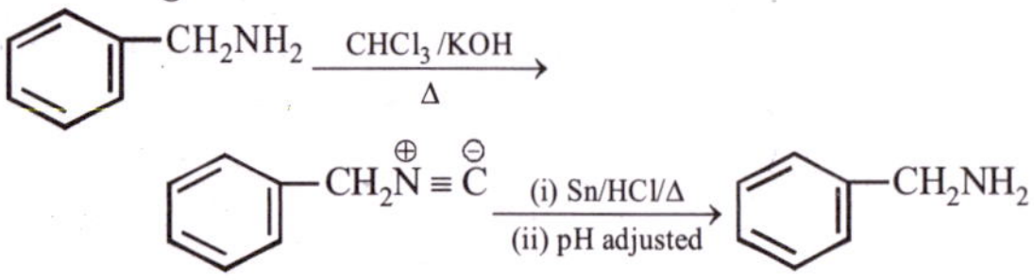

(4) $H_3CCH_2-\overset{O}{\overset{\|}{C}}-NH_2+Br_2+4KOH(alc.)\xrightarrow{\Delta}$
$H_3CCH_2NH_2+2KBr+K_2CO_3+2H_2O$

12. A student has planned to prepare acetanilide from aniline using acetic anhydride. The student has started from 9.3g of aniline. However, the student has managed to obtain 11 g of dry acetanilide.
The % yield of this reaction is :- [Jan. 24, 2026 (II)]

(1) 81.5% (2) 97.5% (3) 59.5% (4) 72.5%

13. A student has been given a compound "x" of molecular formula C_6H_7N. 'x' is sparingly soluble in water. However, on addition of dilute mineral acid, 'x' becomes soluble in water. 'x' when treated with $CHCl_3$ and KOH (alc.) 'y' is produced. 'y' has a specific unpleasant smell. On treatment with benzenesulphonyl chloride, 'x' gives a compound 'z' which is soluble in alkali. The number of different "H" atoms present in 'z' is: [Jan. 23, 2026 (II)]

(1) 5 (2) 8 (3) 4 (4) 7

14. 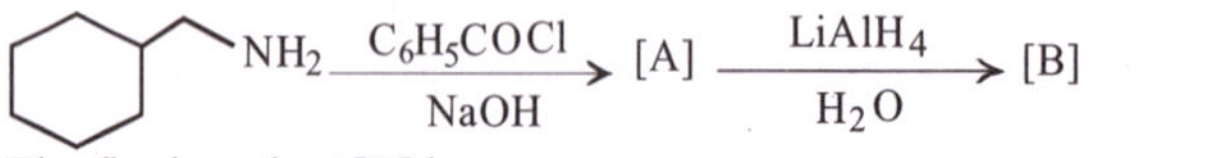

The final product [B] is : [Jan. 22, 2026 (II)]

(1) 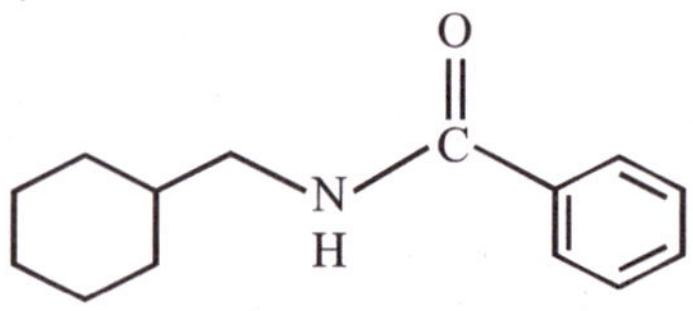

(2) 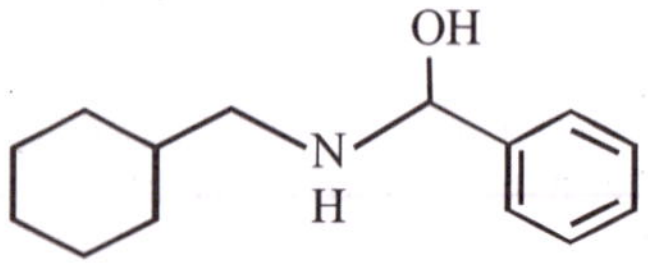

(3) 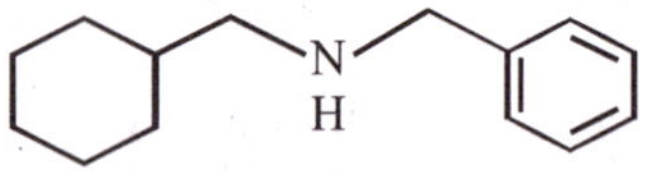

(4) 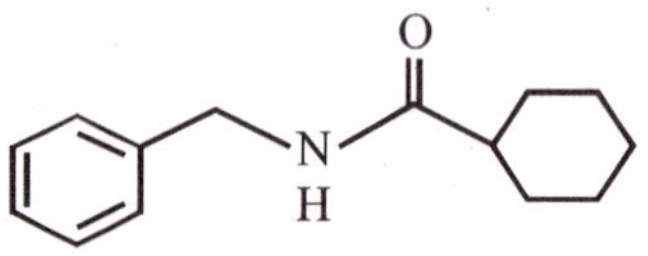

15. An organic compound (P) on treatment with aqueous ammonia under hot condition forms compound (Q) which on heating with Br_2 and KOH forms compound (R) having molecular formula C_9H_7N. Names of P, Q and R respectively are [Jan. 21, 2026 (I)]

(1) Benzoic acid, benzamide, aniline
(2) Toluic acid, methylbenzamide, 2-methylaniline
(3) Benzoic acid, 4-methylbenzamide, 4-methylaniline
(4) Phenylethanoic acid, phenylethanamide, benzamine

16. $A\xrightarrow[\text{(ii)}H_3O^+]{\text{(i)NaOH}}B\xrightarrow[\text{(ii)}H_2SO_4\Delta]{\text{(i)EtOH}}C$

'A' shows positive Lassaign's test for N and its molar mass is 121.
'B' gives effervescence with aq. $NaHCO_3$.
'C' gives fruity smell.
Identify A, B and C from the following. [April 8, 2025 (II)]

(1) 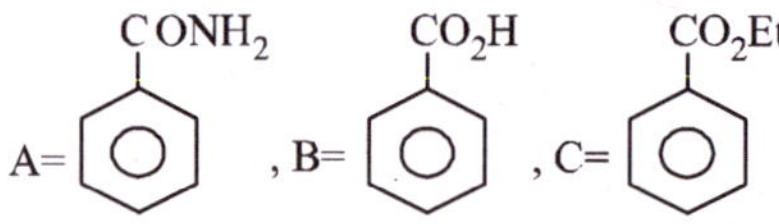

(2) 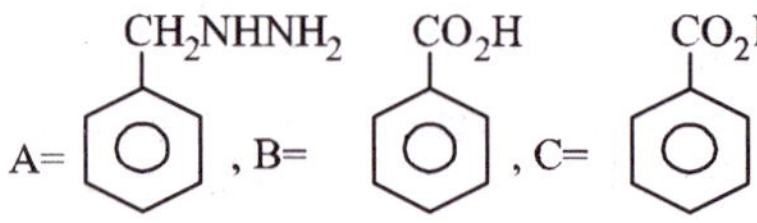

(3) A= , B= , C =

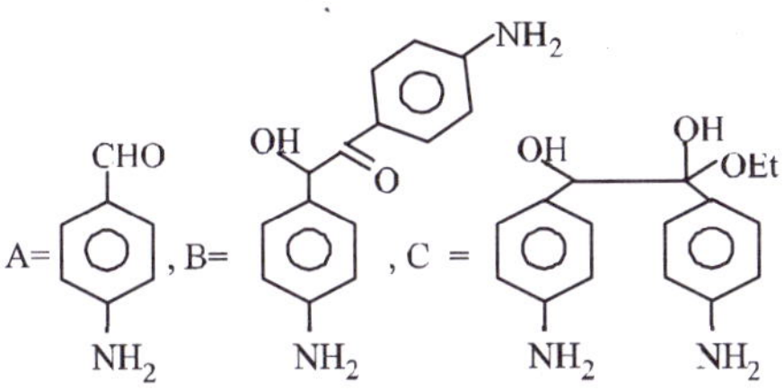

(4) A= , B= , C=

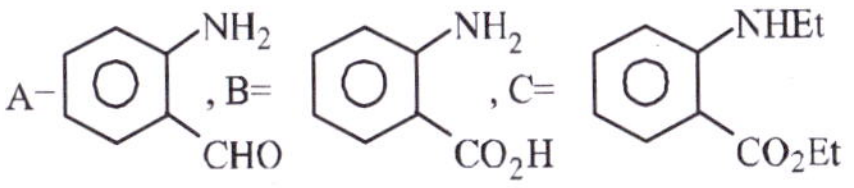

17. The descending order of basicity of following amines is: **[April 7, 2025 (II)]**

(A) 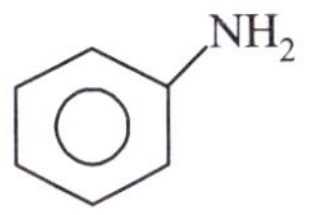

(B) 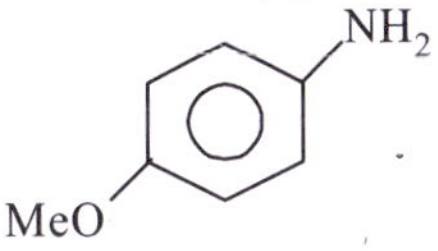

(C) 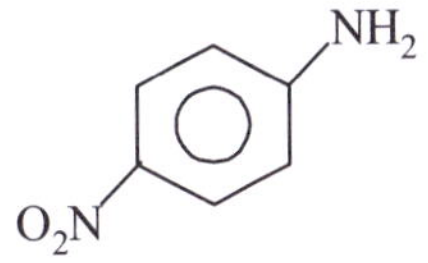

(D) CH_3NH_2

(E) $(CH_3)_2NH$

Choose the **correct** answer from the options given below:

(1) B > E > D > A > C (2) E > D > B > A > C
(3) E > D > A > B > C (4) E > A > D > C > B

18. Which of the following amine (s) show (s) positive carbylamines test ? **[April 7, 2025 (I)]**

(A) 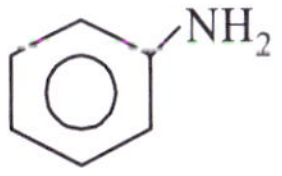

(B) $(CH_3)_2NH$

(C) CH_3NH_2

(D) $(CH_3)_3N$

(E) 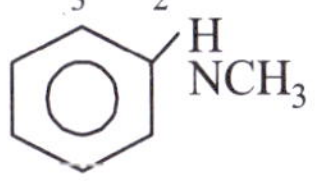

Choose the **correct** answer from the options given below :

(1) A and E Only (2) C Only
(3) A and C Only (4) B, C and D Only

19. The correct order of basicity for the following molecules is:

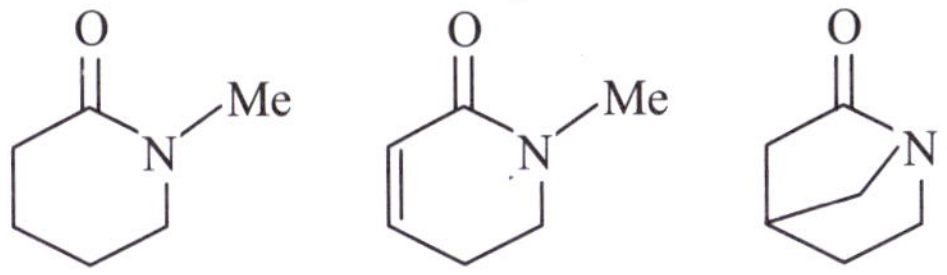

(P) (Q) (R)

[April 4, 2025 (II)]

(1) $P > Q > R$ (2) $R > P > Q$
(3) $Q > P > R$ (4) $R > Q > P$

20. The sequence from the following that would result in giving predominantly 3, 4, 5 –Tribromoaniline is : **[April 3, 2025 (II)]**

(1) 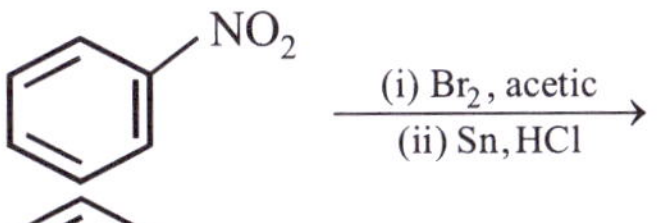

(2) 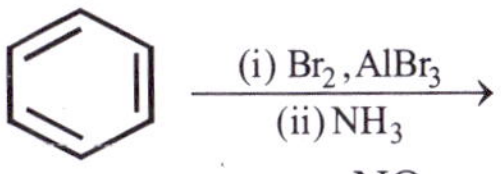

(3) 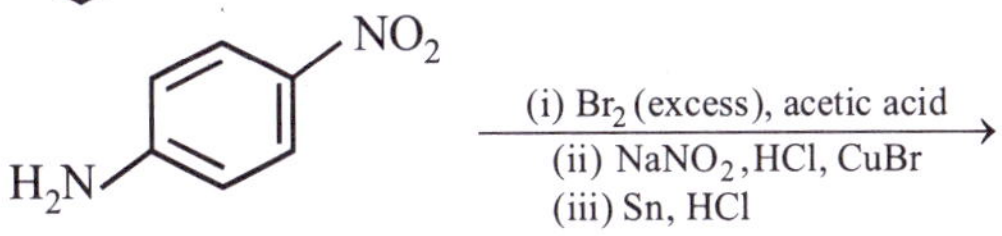

(4) 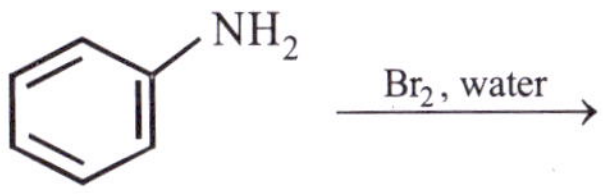

21. The correct order of basic nature on aqueous solution for the bases NH_3, $H_2N–NH_2$, $CH_3CH_2NH_2$, $(CH_3CH_2)_2NH$ and $(CH_3CH_2)_3N$ is : **[April 2, 2025 (I)]**

(1) $NH_3 < H_2N–NH_2 < (CH_3CH_2)_3N < CH_3CH_2NH_2 < (CH_3CH_2)_2NH$
(2) $NH_3 < H_2N–NH_2 < CH_3CH_2NH_2 < (CH_3CH_2)_2NH < (CH_3CH_2)_3N$
(3) $H_2N–NH_2 < NH_3 < (CH_3CH_2)_3N < CH_3CH_2NH_2 < (CH_3CH_2)_2NH$
(4) $NH_2–NH_2 < NH_3 < CH_3CH_2NH_2 < (CH_3CH_2)_3N < (CH_3CH_2)_2NH$

22. Identify correct statements : **[Jan. 28, 2025 (II)]**

(A) Primary amines do not give diazonium salts when treated with $NaNO_2$ in acidic condition.
(B) Aliphatic and aromatic primary amines on heating with $CHCl_3$ and ethanolic KOH form carbylamines.
(C) Secondary and tertiary amines also give carbylamine test.
(D) Benzenesulfonyl chloride is known as Hinsberg's reagent.
(E) Tertiary amines reacts with benzenesulfonyl chloride very easily.

Choose the correct answer from the options given below :

(1) (B) and (D) only (2) (A) and (B) only
(3) (D) and (E) only (4) (B) and (C) only

23. For reaction **[Jan. 24, 2025 (II)]**

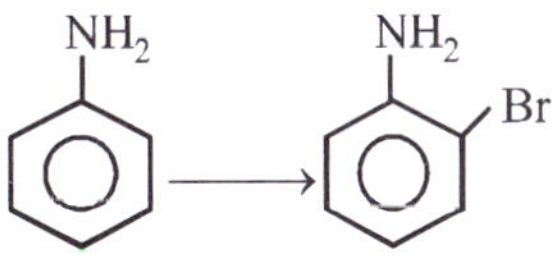

(Major)

The correct order of set of reagents for the above conversion is:

(1) $Br_2 | FeBr_3$, $H_2O(\Delta)$, NaOH
(2) H_2SO_4, Ac_2O, Br_2, $H_2O(\Delta)$, NaOH
(3) Ac_2O, Br_2, $H_2O(\Delta)$, NaOH
(4) Ac_2O, H_2SO_4, Br_2, NaOH

24. Which among the following react with Hinsberg's reagent? **[Jan. 23, 2025 (I)]**

(A) 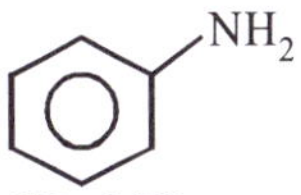

(B) 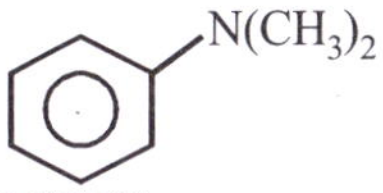

(C) $CH_3–NH_2$

(D) $N(CH_3)_3$

(E) 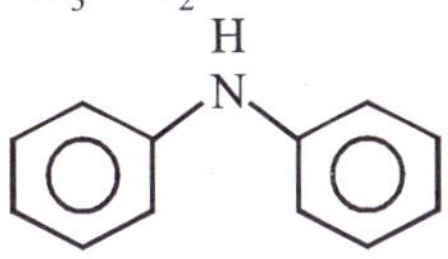

Choose the correct answer from the options given below:

(1) B and D only (2) C and D only
(3) A, B and E only (4) A, C and E only

25. Match the Compounds (List-I) with the appropriate Catalyst/Reagents (List-II) for their reduction into corresponding amines. **[Jan. 22, 2025 (II)]**

List-I (Compounds)	List-II (Catalyst/Reagents)
(A) $R-\overset{\displaystyle O}{\overset{\|\|}{C}}-NH_2$	(I) NaOH (aqueous)
(B) 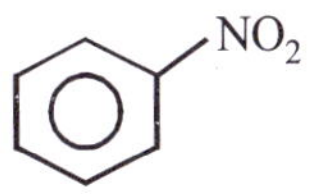	(II) H_2/Ni

(C) $R-C\equiv N$ (III) $LiAlH_4, H_2O$

(D) 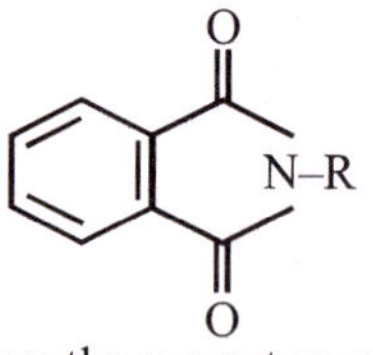 (IV) Sn, HCl

Choose the correct answer from the options given below.
(1) (A)-(III), (B)-(II), (C)-(IV), (D)-(I)
(2) (A)-(II), (B)-(IV), (C)-(III), (D)-(I)
(3) (A)-(II), (B)-(I), (C)-(III), (D)-(IV)
(4) (A)-(III), (B)-(IV), (C)-(II), (D)-(I)

26. Correct order of basic strength of Pyrrole 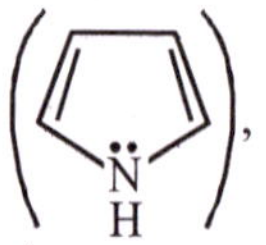, Pyridine  and Piperidine 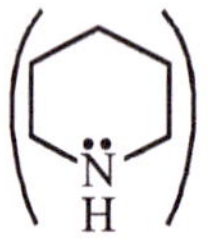 is: [April 9, 2024 (I)]
(1) Piperidine > Pyridine > Pyrrole
(2) Pyrrole > Pyridine > Piperidine
(3) Pyridine > Piperidine > Pyrrole
(4) Pyrrole > Piperidine > Pyridine

27. Given below are two statements: [April 8, 2024 (II)]
Statement I: All the following compounds react with p-toluenesulfonyl chloride.
$C_6H_5NH_2$, $(C_6H_5)_2NH$, $(C_6H_5)_3N$
Statement II: Their products in the above reaction are soluble in aqueous NaOH.
In the light of the above statements, choose the **correct** answer from the options given below:
(1) Both Statement I and Statement II is false
(2) Statement I is true but Statement II is false
(3) Statement I is false but Statement II is true
(4) Both Statement I and Statement II is true

28. Given below are two statements: [Feb. 1, 2024 (I)]
Statement I : Aminobenzene and aniline are same organic compounds.
Statement II : Aminobenzene and aniline are different organic compounds.
In the light of the above statements, choose the most appropriate answer from the options given below:
(1) Both Statement I and Statement II are correct
(2) Statement I is correct but Statement II is incorrect
(3) Statement I is incorrect but Statement II in correct
(4) Both Statement I and Statement II are incorrect.

29. Given below are two statements: [Feb. 1, 2024 (I)]
Statement I : The NH_2 group in aniline is ortho and para directing and a powerful activating group.
Statement II : Aniline does not undergo Friedel - Craft's reaction (alkylation and acylation).
In the light of the above statements, choose the most appropriate answer from the options given below:
(1) Both Statement I and Statement II are correct
(2) Both Statement I and Statement II are incorrect.
(3) Statement I is incorrect but Statement II in correct
(4) Statement I is incorrect but Statement II in incorrect

30. The final product A, formed in the following multi step reaction sequence is: [Jan. 30, 2024 (I)]

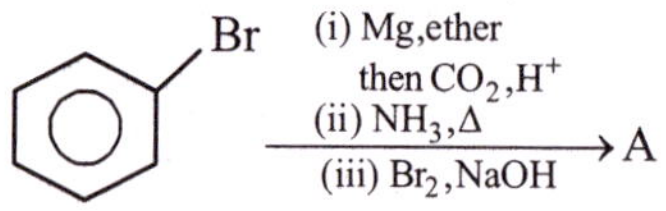

(1) 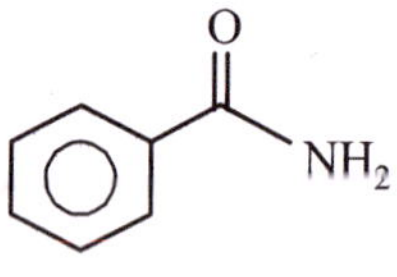 (2) 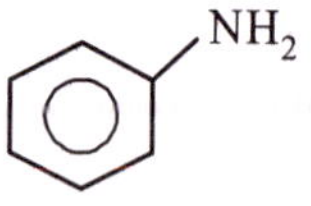

(3) 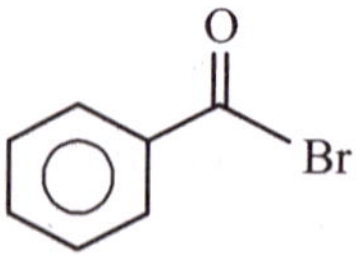 (4) 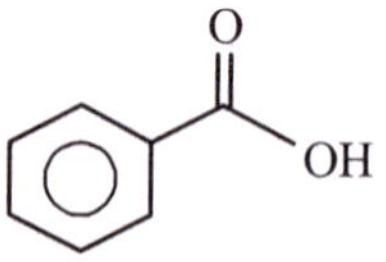

31. Which of the following reaction is correct? [Jan. 29, 2024 (II)]
(1) 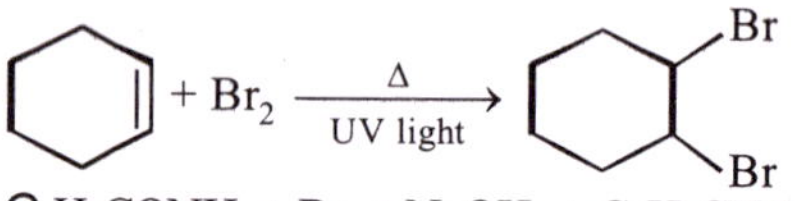
(2) $C_2H_5CONH_2 + Br_2 + NaOH \rightarrow C_2H_5CH_2NH_2 + Na_2CO_3 + NaBr + H_2O$
(3) $CH_3CH_2CH_2NH_2 \xrightarrow[H_2O]{HNO_2, 0°C} CH_3CH_2OH + N_2 + HCl$
(4) 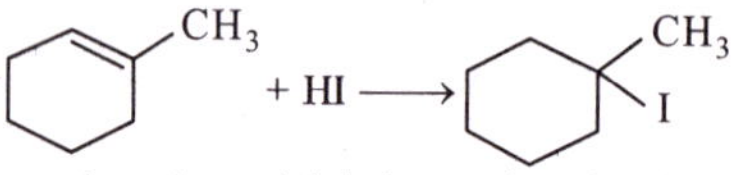

32. The arenium ion which is not involved in the bromination of aniline is _______. [Jan. 29, 2024 (I)]

(1) 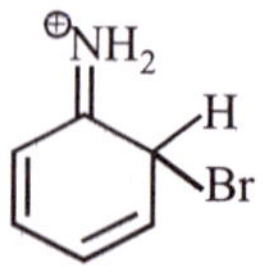 (2) 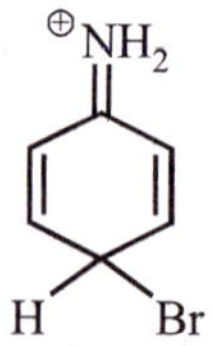

(3) 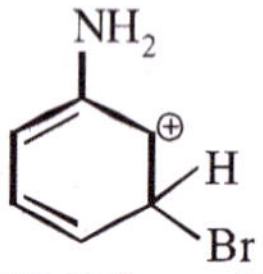 (4) 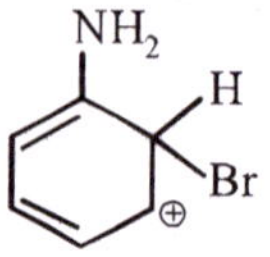

33. Identify B formed in the reaction. [Jan. 27, 2024 (II)]
$$Cl-(CH_2)_4-Cl \xrightarrow{\text{excess } NH_3} A \xrightarrow{NaOH} B + H_2O + NaCl$$
(1) 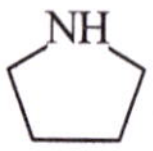
(2) $H_2N-(CH_2)_4-NH_2$
(3) $Cl\overset{+}{N}H_3-(CH_2)_4-\overset{+}{N}H_3Cl^-$
(4) 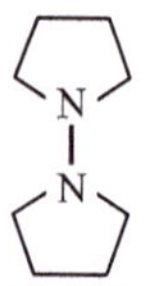

34. Which of the following is strongest Bronsted base? [Jan. 27, 2024 (I)]

(1) 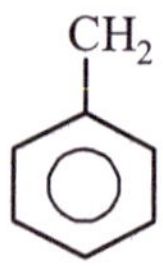 (2) 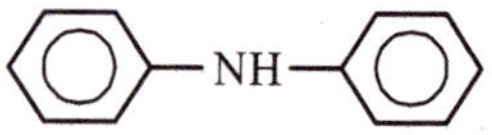

(3) 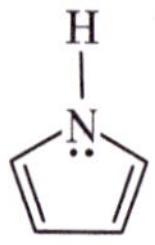 (4) 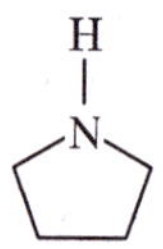

35. In the reaction given below [April 13, 2023 (I)]

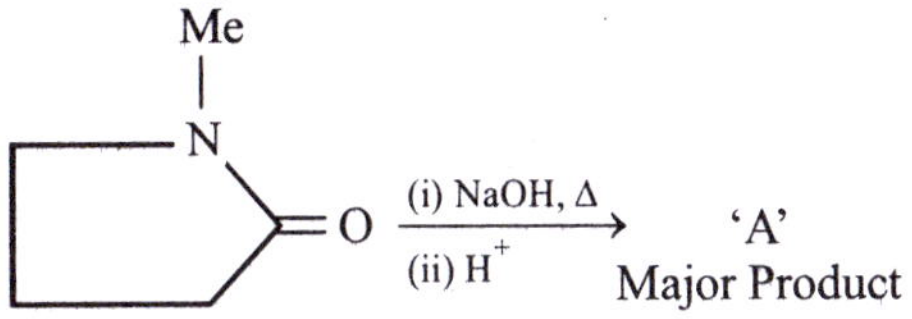

'A' is

(1) 

(2) 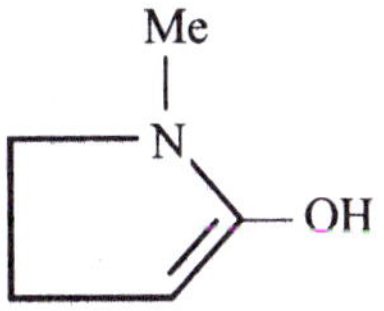

(3) 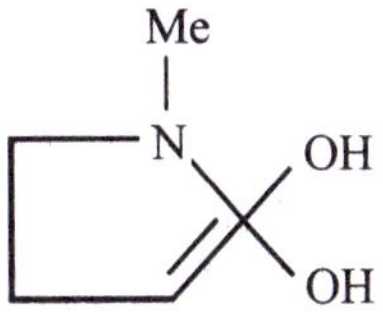

(4) 

36. The major product for the following reaction is: [April 13, 2023 (II)]

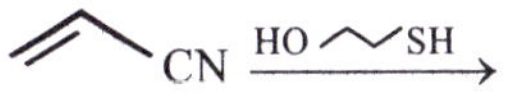

(1) 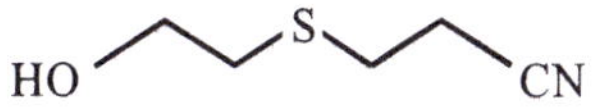

(2) 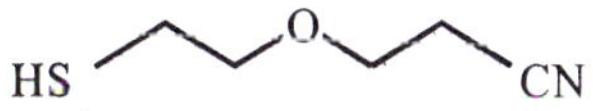

(3) 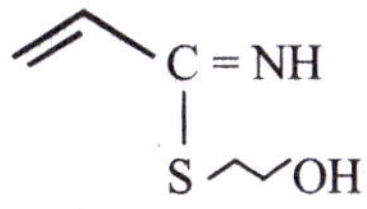

(4) 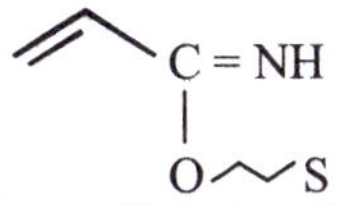

37. In the reaction given below [April 13, 2023 (I)]

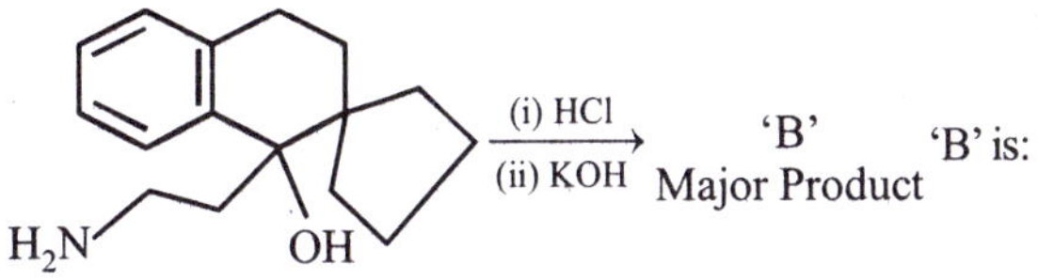

'B' is:

(1) 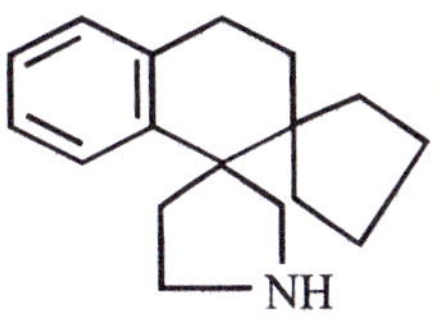

(2) 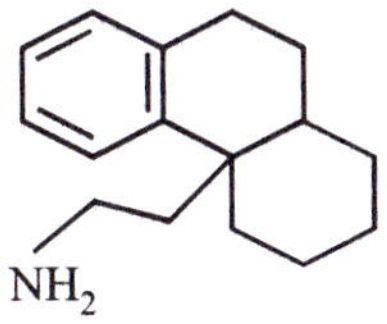

(3) 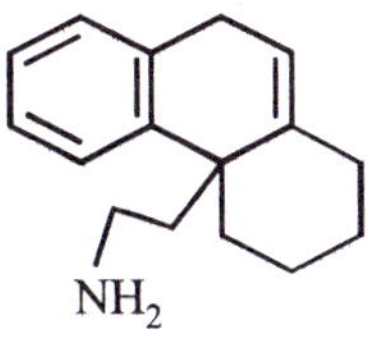

(4) 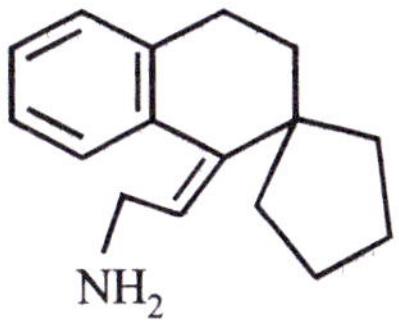

38. o-Phenylenediamine $\xrightarrow{HNO_2}$ 'X' Major Product

'X' is [April 11, 2023 (I)]

(1) 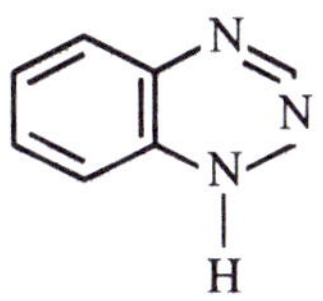

(2) 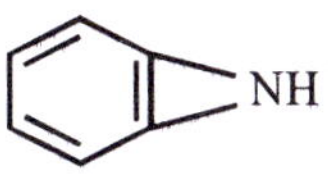

(3) 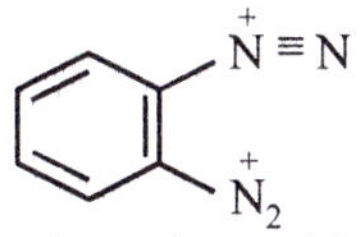

(4) 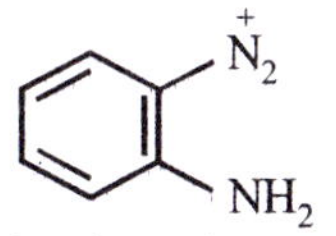

39. Isomeric amines with molecular formula $C_8H_{11}N$ give the following tests [April 10, 2023 (I)]

Isomer (P) ⇒ Can be prepared by Gabriel phthalimide synthesis

Isomer (Q) ⇒ Reacts with Hinsberg's reagent to give solid insoluble in NaOH

Isomer (R) ⇒ Reacts with HONO followed by β-naphthol in NaOH to give red dye.

Isomers (P), (Q) and (R) respectively are

(1) 

(2) 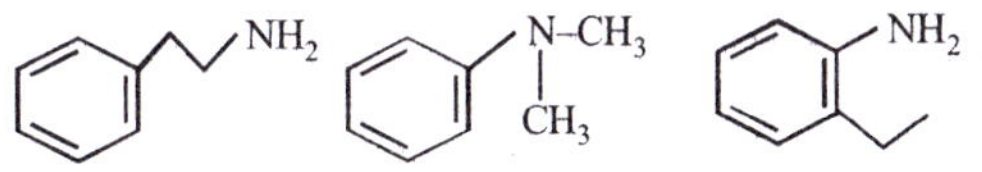

(3) 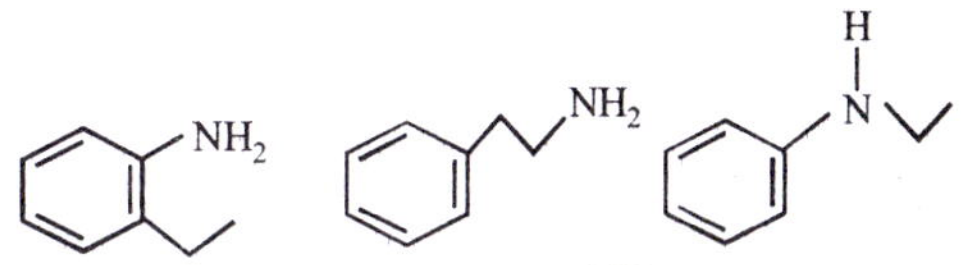

(4) 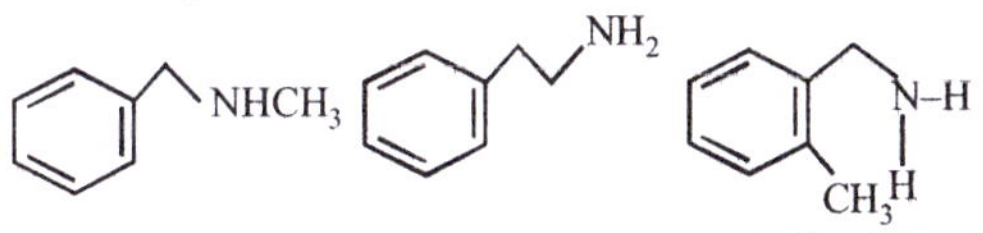

40. Compound P $\xrightarrow{HCl, \Delta}$ Filter → Residue Q; Filtrate $\xrightarrow{NaOH}$ Oily liquid R.

(M.F. $C_{14}H_{13}ON$)

M.F = Molecular formula

Compound P is neutral. Q gives effervescences with $NaHCO_3$ while R reacts with Hinsbergs reagent to give solid soluble in NaOH. Compound P is [April 06, 2023 (I)]

(1) 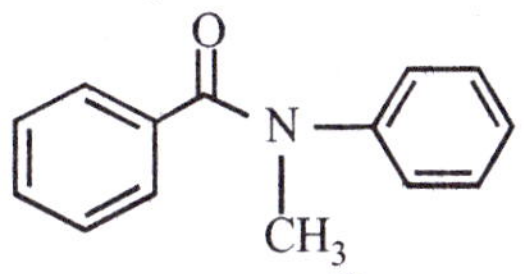

(2) 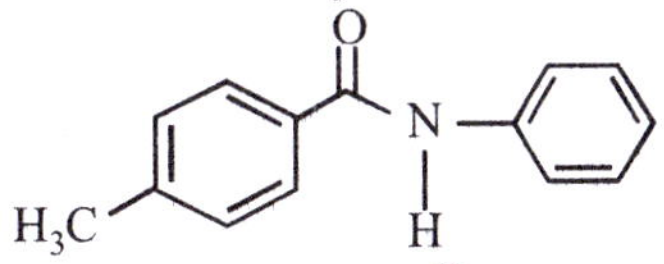

(3) 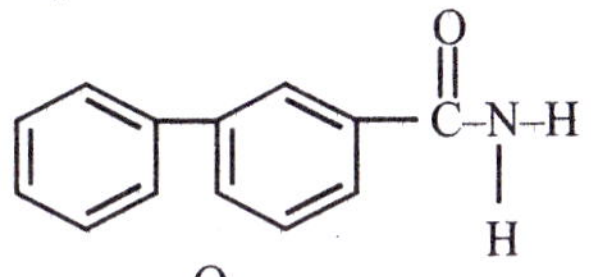

(4) 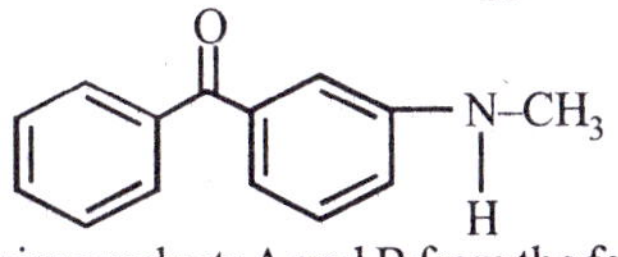

41. A major products A and B from the following reactions are: [April 06, 2023 (I)]

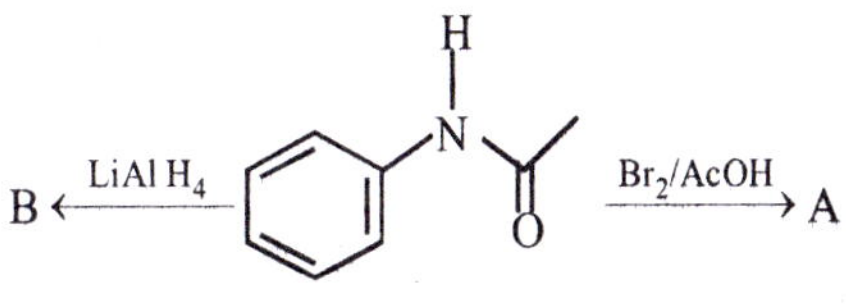

(1) A = 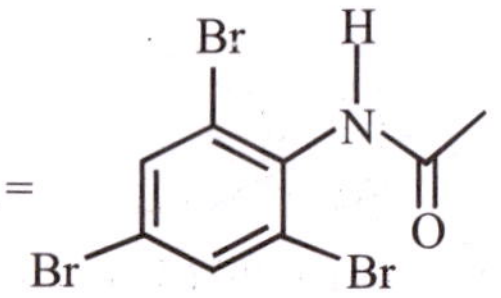 , B = 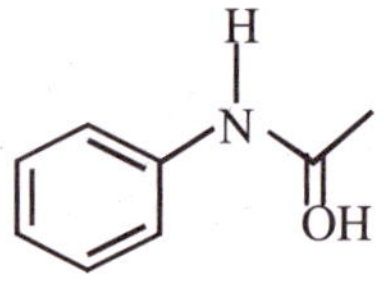

(2) A = 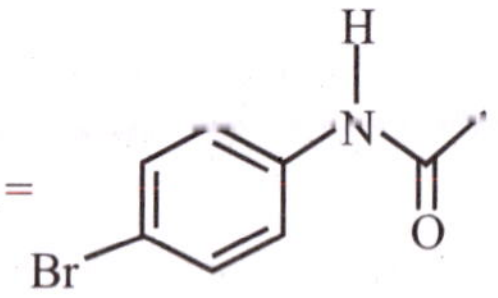 , B = 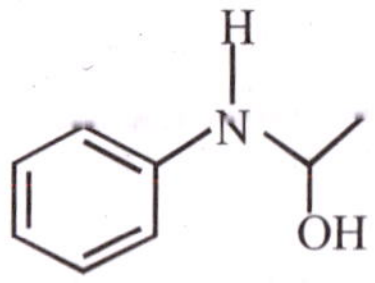

(3) A = 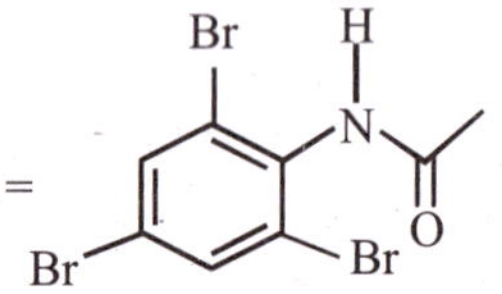 , B = 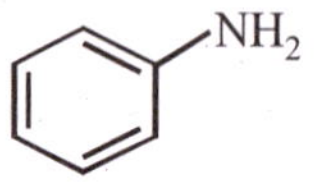

(4) A = 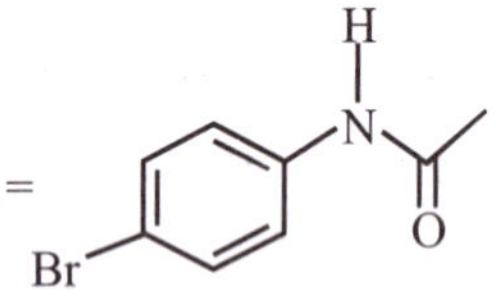 , B = 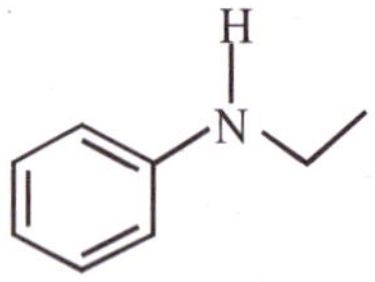

42. The major product formed in the following reaction is

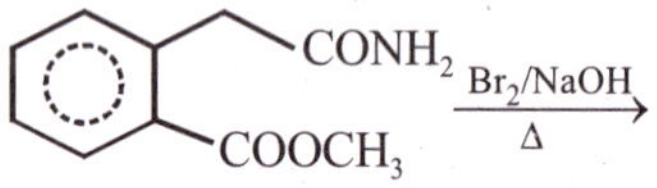

[April 06, 2023 (I)]

(1) 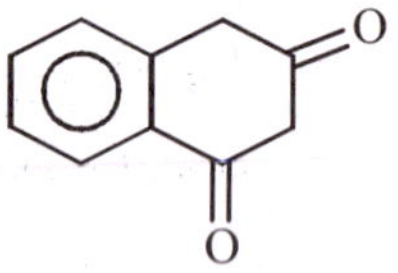 (2) 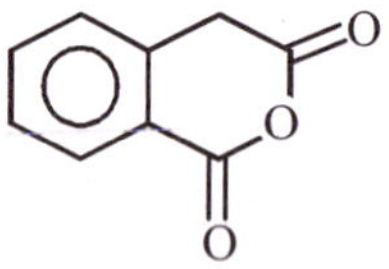

(3) 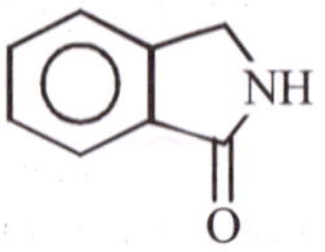 (4) 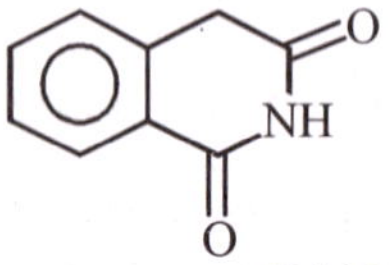

43. In the following reaction, 'A' is [Feb. 01, 2023 (I)]

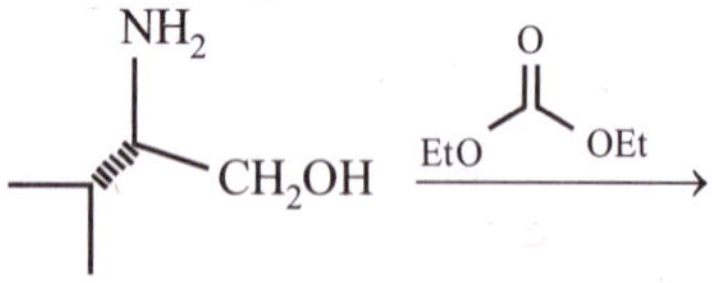 'A' Major product

(1) 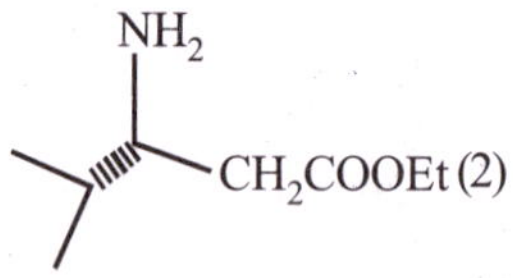 (2) 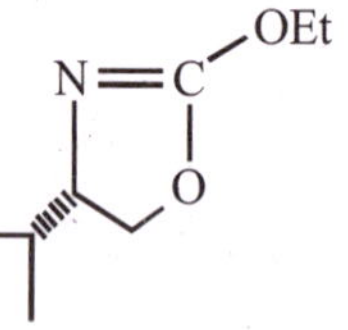

(3) 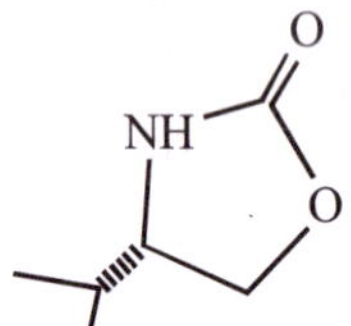 (4) 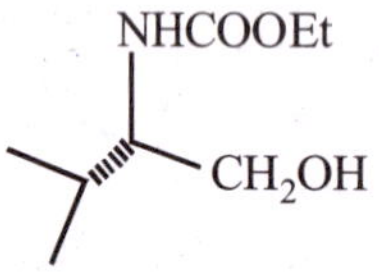

44. Given below are two statements: one is labelled as **Assertion (A)** and the other is labelled as **Reason (R)**.

Assertion (A): α-Halocarboxylic acid on reaction with dil NH_3 gives good yield of α-amino carboxylic acid whereas the yield of amine is very low when prepared from alkyl halides.

Reason (R): Amino acids exist in zwitter ion form in aqueous medium.

In the light of the above statements, choose the correct answer from the options given below [Feb. 01, 2023 (II)]

(1) Both **(A)** and **(R)** are correct but **(R)** is not the correct explanation of **(A)**
(2) **(A)** is correct but **(R)** is not correct
(3) Both **(A)** and **(R)** are correct and **(R)** is the correct explanation of **(A)**
(4) **(A)** is not correct but **(R)** is correct

45. Cyclohexylamine when treated with nitrous acid yields (P). On treating (P) with PCC results in (Q). When (Q) is heated with dil. NaOH we get (R) The final product (R) is: [Jan. 31, 2023 (II)]

(1) 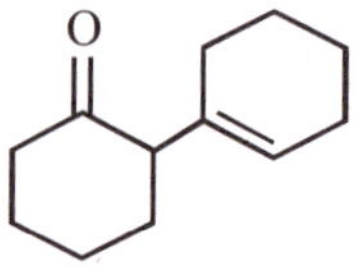 (2) 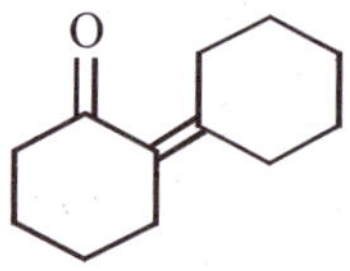

(3) 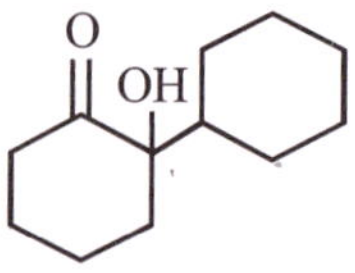 (4) 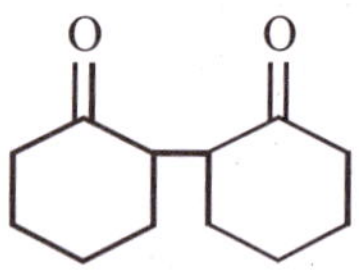

46. An organic compound [A]($C_4H_{11}N$), shows optical activity and gives N_2 gas on treatment with HNO_2. The compound [A] reacts with $PhSO_2Cl$ producing a compound which is soluble in KOH. The structure of A is: [Jan. 31, 2023 (II)]

(1)  (2) 

(3) 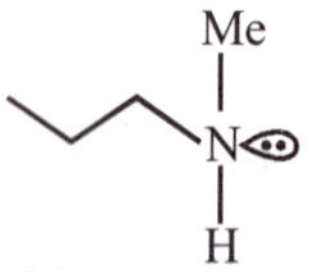 (4) 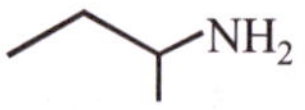

47. Benzyl isocyanide can be obtained by : [Jan. 30, 2023 (I)]

(A) 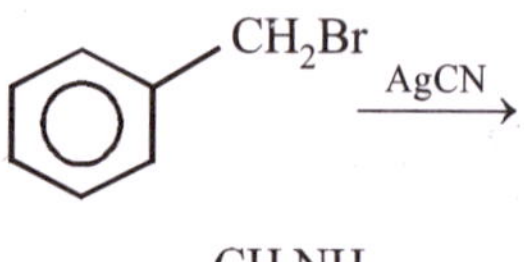

(B) 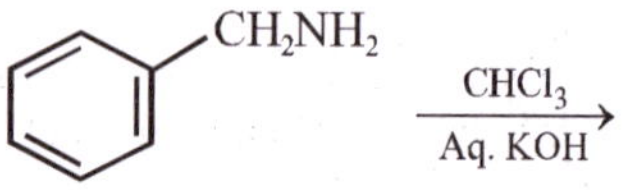

(C) 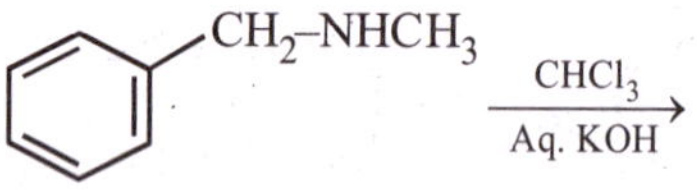

(D) 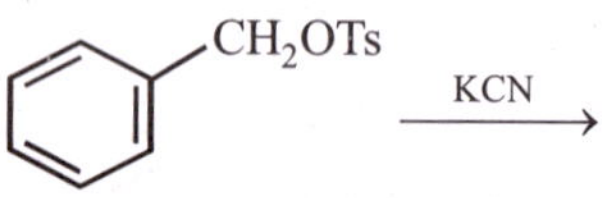

Choose the correct answer from the options given below:
(1) A and D (2) Only B
(3) A and B (4) B and C

48. Reaction of propanamide with Br_2 / KOH (aq) produces : [Jan. 29, 2023 (II)]
(1) Ethyl nitrile (2) Propyl amine
(3) Propane nitrile (4) Ethyl amine

49. The correct order in aqueous medium of basic strength in case of methyl substituted amines is : [Jan. 25, 2023 (I)]
(1) $Me_2NH > MeNH_2 > Me_3N > NH_3$
(2) $Me_2NH > Me_3N > MeNH_2 > NH_3$
(3) $NH_3 > Me_3N > MeNH_2 > Me_2NH$
(4) $Me_3N > Me_2NH > MeNH_2 > NH_3$

50. Match **List I** with **List II** [Jan. 25, 2023 (II)]

List I (Amines)	List II (pK_b)
A. Aniline	I. 3.25
B. Ethanamine	II. 3.00
C. N-Ethylethanamine	III. 9.38
D. N, N-Diethylethanamine	IV. 3.29

Choose the correct answer from the options given below:-

(1) A-I, B-IV, C-II, D-III (2) A-III, B-II, C-I, D-IV
(3) A-III, B-II, C-IV, D-I (4) A-III, B-IV, C-II, D-I

51. Given below are two statements : [Jan. 24, 2023 (II)]

Statement I: Pure aniline and other aryl amines are usually colourless.

Statement II: Aryl amines get coloured on storage due to atmospheric reduction.

In the light of the above statements, choose the most appropriate answer from the options given below :

(1) Both Statement I and Statement II are incorrect.
(2) Both Statement I and Statement II are correct.
(3) Statement I is correct but Statement II is incorrect.
(4) Statement I is incorrect but Statement II is correct.

52. Which among the following is the strongest Bronsted base? [July 29, 2022 (I)]

(1) 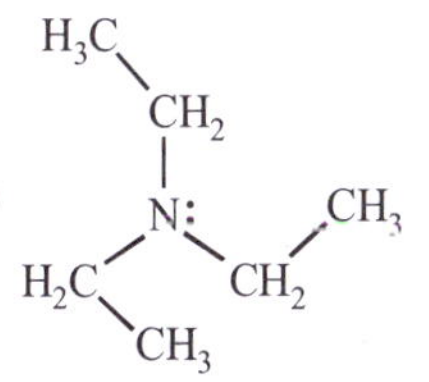 (2) 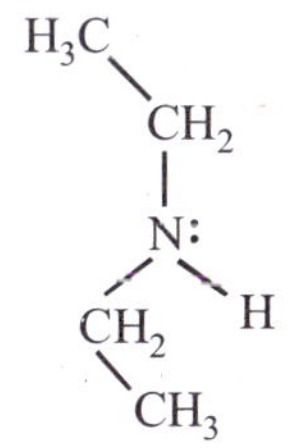

(3)  (4) 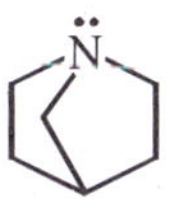

53. Consider the following reaction sequence :

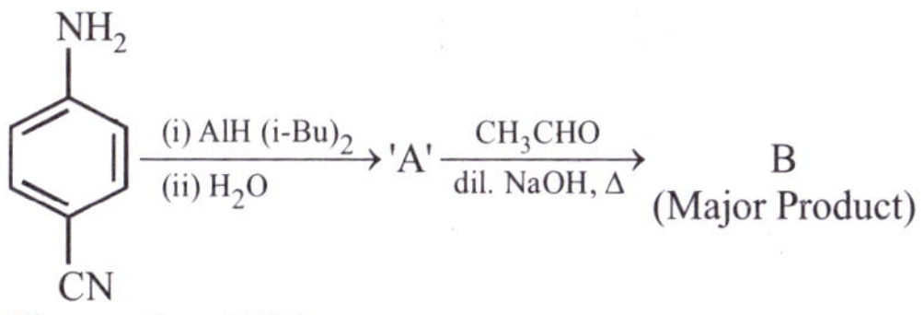

The product 'B' is : [July 29, 2022 (I)]

(1) 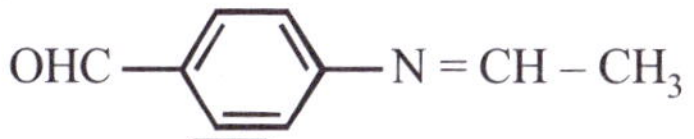

(2) 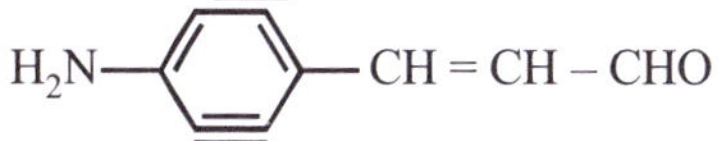

(3) 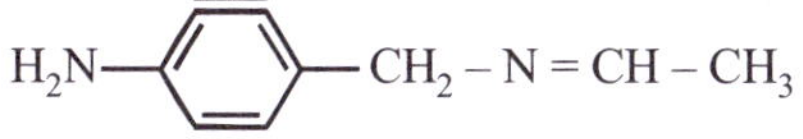

(4) 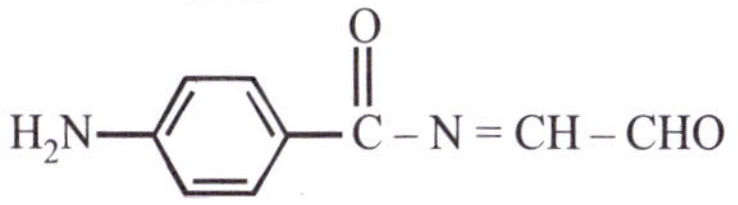

54. 'A' ($C_8H_6Cl_2O$) $\xrightarrow{NH_3}$ C_8H_8ClNO $\xrightarrow[NaOH]{Br_2}$

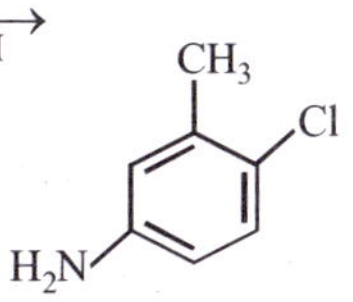

Consider the above reaction, the compound 'A' is : [July 29, 2022 (I)]

(1) 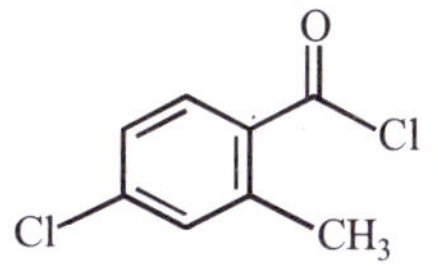 (2) 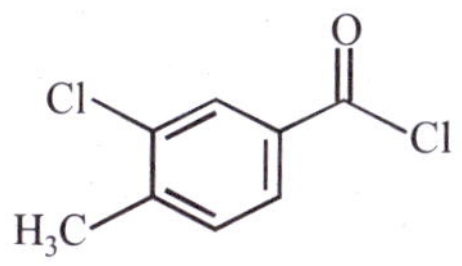

(3) 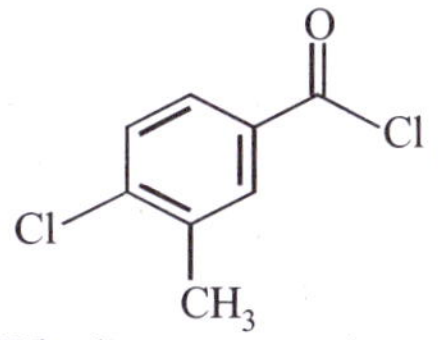 (4) 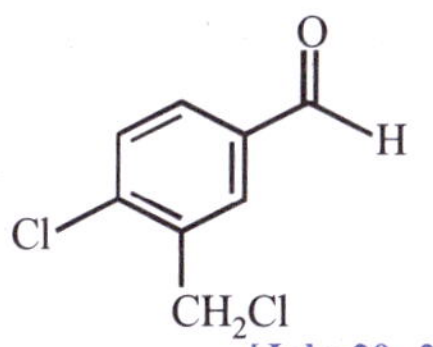

55. The Hinsberg reagent is : [July 29, 2022 (II)]

(1) 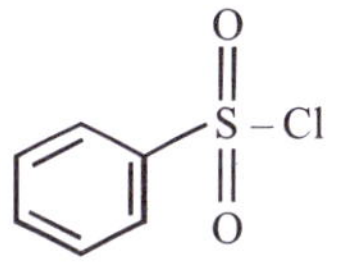 (2) 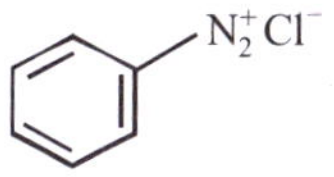

(3) 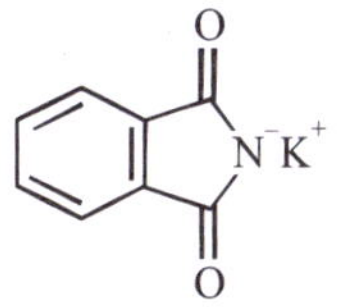 (4) 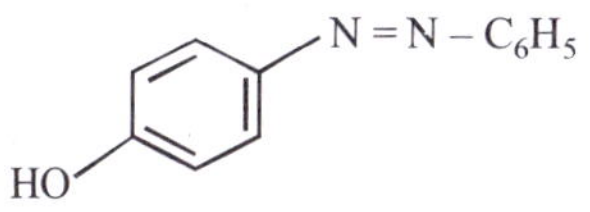

56. Identify the correct statement for the below given transformation. [July 28, 2022 (I)]

$$CH_3-CH_2-CH_2-\underset{\overset{|}{\oplus N(CH_3)_3}}{CH}-CH_3 \xrightarrow[C_2H_5OH]{C_2H_5ONa} \underset{(Major)}{A} + \underset{(Minor)}{B}$$

(1) $A-CH_3CH_2CH=CH-CH_3$, B $CH_3CH_2CH_2CH=CH_2$, Saytzeff products
(2) $A-CH_3CH_2CH=CH-CH_3$, $B-CH_3CH_2CH_2CH=CH_2$, = Hofmann products
(3) $A-CH_3CH_2CH_2CH=CH_2$, $B-CH_3CH_2CH=CHCH_3$, Hofmann products
(4) $A-CH_3CH_2CH_2CH=CH_2$, $B-CH_3CH_2CH=CHCH_3$, Saytzeff products

57. Given below are two statements: one is labelled as **Assertion A** and the other is labelled as **Reason R**.

Assertion A : Aniline on nitration yields ortho, meta and para nitro derivatives of aniline.

Reason R : Nitrating mixture is a strong acidic mixture.

In the light of the above statements, choose the correct answer from the options given below: [July 28, 2022 (II)]

(1) Both A and R are true and R is the correct explanation of A.
(2) Both A and R are true but R is NOT the correct explanation of A.
(3) A is true but R is false.
(4) A is false but R is true.

58. Match **List-I** with **List-II** [July 27, 2022 (I)]

List-I	List-II
(A) Benzene sulphonyl chloride	(I) Test for primary amines
(B) Hoffmann bromamide reaction	(II) *anti*-Saytzeff
(C) Carbylamine reaction	(III) Hinsberg reagent
(D) Hoffmann orientation	(IV) Known reaction of isocyanates.

Choose the **correct** answer from the options given below:

(1) A–IV, B–III, C–II, D–I (2) A–IV, B–II, C–I, D–III
(3) A–III, B–IV, C–I, D–II (4) A–IV, B–III, C–I, D–II

59. Given below are two statements: one is labelled as **Assertion (A)** and the other is labelled as **Reason (R).** [July 26, 2022 (I)]

Assertion (A): Experimental reaction of CH_3Cl with aniline and anhydrous $AlCl_3$ does **not** give *o* and *p*-methylaniline.

Reason (R): The $-NH_2$ group of aniline becomes deactivating because of salt formation with anhydrous $AlCl_3$ and hence yields *m*-methyl aniline as the product.

In the light of the above statements, choose the **most appropriate** answer from the options given below:

(1) Both (A) and (R) are true and (R) is the correct explanation of (A).
(2) Both (A) and (R) are true but (R) is not the correct explanation of (A).
(3) (A) is true, but (R) is false.
(4) (A) is false, but (R) is true.

60. $CH_3-CH_2-CN \xrightarrow[\text{Ether}]{CH_3MgBr} A \xrightarrow{H_3O^+} B \xrightarrow[\text{HCl}]{\text{Zn-Hg}} C$

The correct structure of C is [July 25, 2022 (II)]

(1) $CH_3-CH_2-CH_2-CH_3$
(2) $CH_3-CH_2-\overset{\overset{\displaystyle O}{\|}}{C}-CH_3$
(3) $CH_3-CH_2-\overset{\overset{\displaystyle OH}{|}}{CH}-CH_3$
(4) $CH_3-CH_2-CH=CH_2$

61. An organic compound 'A' on reaction with NH_3 followed by heating gives compound B. Which on further strong heating gives compound C ($C_8H_5NO_2$). Compound C on sequential reaction with ethanolic KOH, alkyl chloride and hydrolysis with alkali gives a primary amine. The compound A is: [July 25, 2022 (I)]

(1) 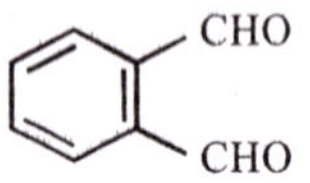
(2) 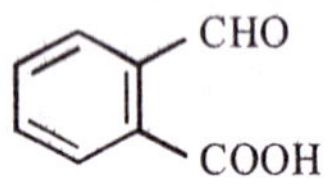
(3) 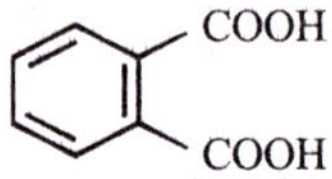
(4) 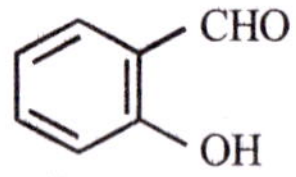

62. In Friedel-Crafts alkylation of aniline, one gets : [June 29, 2022 (II)]

(1) alkylated product with ortho and para substitution.
(2) secondary amine after acidic treatment.
(3) an amide product.
(4) positively charged nitrogen at benzene ring.

63. 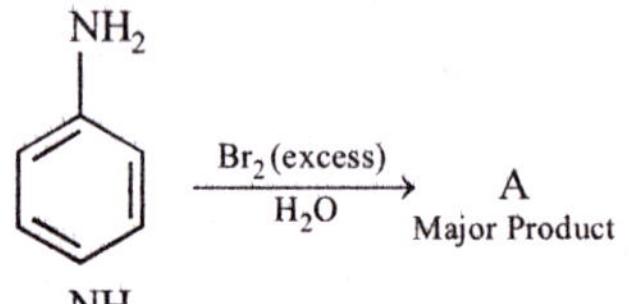

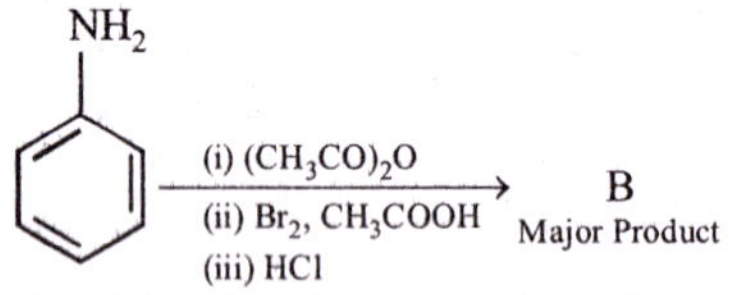

Consider the above reaction, the product A and product B respectively are [June 29, 2022 (I)]

(1) 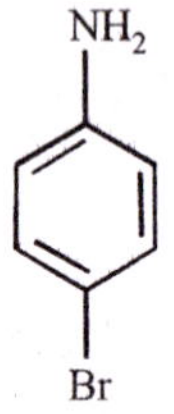 and 

(2) 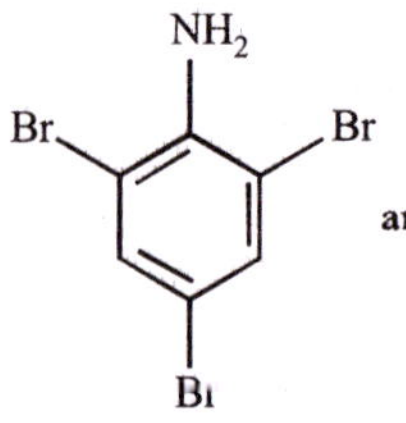 and 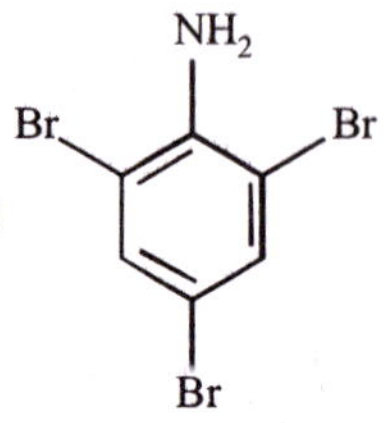

(3) 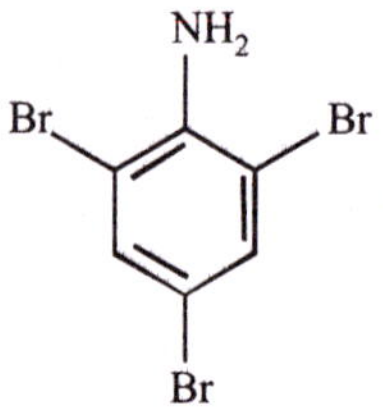 and 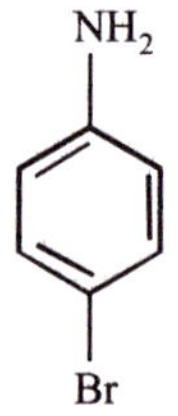

(4) 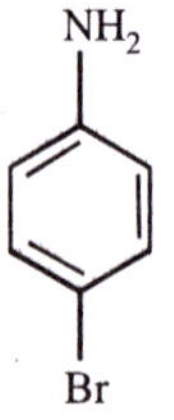 and 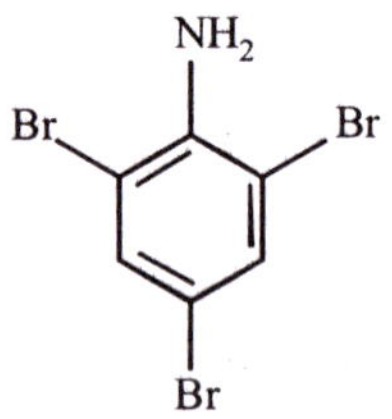

64. With respect to the following reaction, consider the given statements : [June 28, 2022 (II)]

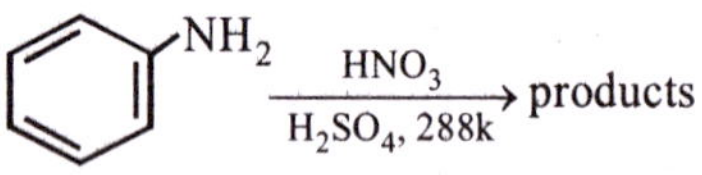

(A) o-Nitroaniline and p-nitroaniline are the predominant products
(B) p-Nitroaniline and m-nitroaniline are the predominant products
(C) HNO_3 acts as an acid
(D) H_2SO_4 acts as an acid

(1) (A) and (C) are correct statements.
(2) (A) and (D) are correct statements.
(3) (B) and (D) are correct statements.
(4) (B) and (C) are correct statements.

65. Identify the major product formed in the following sequence of reactions : [June 28, 2022 (I)]

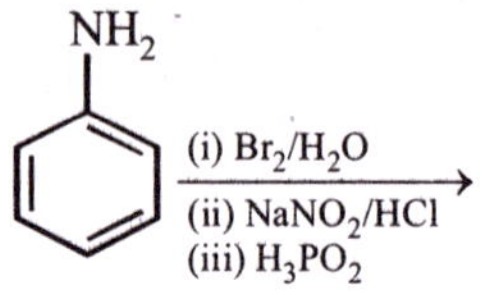

(1) 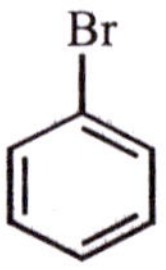
(2) 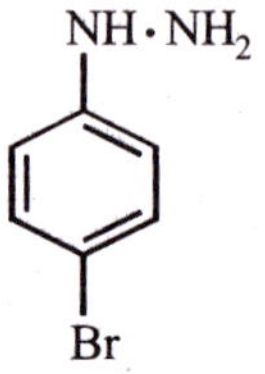
(3) 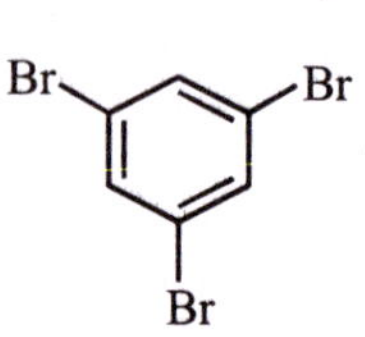
(4) 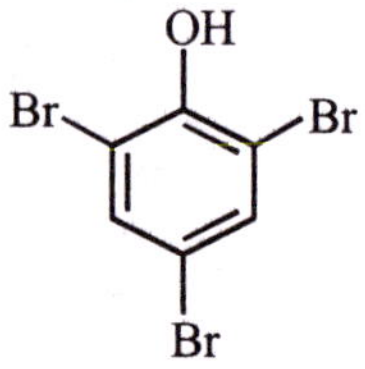

66. Given below are two statements:

Statement I : In Hofmann degradation reaction, the migration of only an alkyl group takes place from carbonyl carbon of the amide to the nitrogen atom.

Statement II : The group is migrated in Hofmann degradation reaction to electron deficient atom. [June 27, 2022 (I)]
In the light of the above statement, choose the **most appropriate** answer from the options given below:
(1) Both Statement I and Statement II are correct.
(2) Both Statement I and Statement II are incorrect.
(3) Statement I is correct but Statement II is incorrect.
(4) Statement I is incorrect but Statement II is correct.

67. Which statement is NOT correct for *p*-toluene sulphonyl chloride? [June 26, 2022 (II)]
(1) It is known as Hinsberg's reagent.
(2) It is used to distinguish primary and secondary amines.
(3) On treatment with secondary amine, it leads to a product, that is soluble in alkali.
(4) It doesn't react with tertiary amines.

68. Among the following structures, which will show the most stable enamine formation ? (Where Me is – CH_3) [June 26, 2022 (I)]
(1) 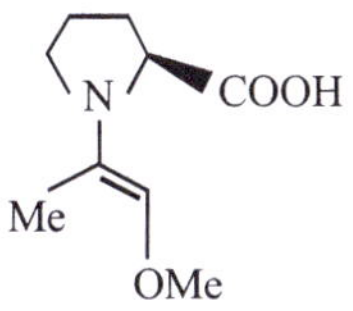
(2) 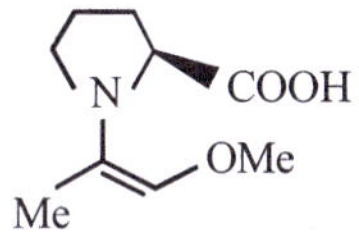
(3) 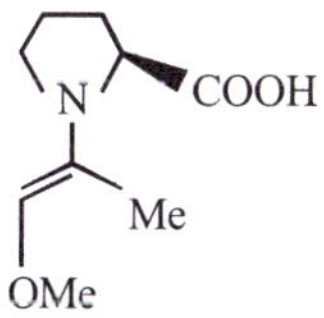
(4) 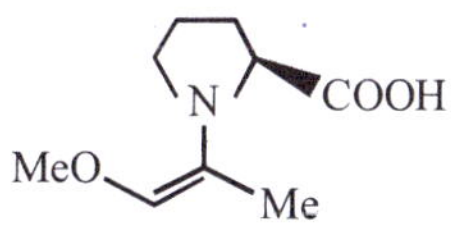

69. The reaction of $R-\underset{\underset{O}{||}}{C}-NH_2$ with bromine and KOH gives RNH_2 as the end product. Which one of the following is the intermediate product formed in this reaction ? [June 25, 2022 (I)]
(1) $R-\underset{\underset{O}{||}}{C}-NH-Br$ (2) $R-NH-Br$
(3) $R-N=C=O$ (4) $R-\underset{\underset{O}{||}}{C}-NBr_2$

70. The conversion of propan-1-ol to *n*-butylamine involves the sequential addition of reagents. The correct sequential order of reagents is. [June 24, 2022 (II)]
(1) (i) $SOCl_2$ (ii) KCN (iii) H_2/Ni, Na(Hg)/C_2H_5OH
(2) (i) HCl (ii) H_2/Ni, Na(Hg)/C_2H_5OH
(3) (i) $SOCl_2$ (ii) KCN (iii) CH_3NH_2
(4) (i) HCl (ii) CH_3NH_2

71. 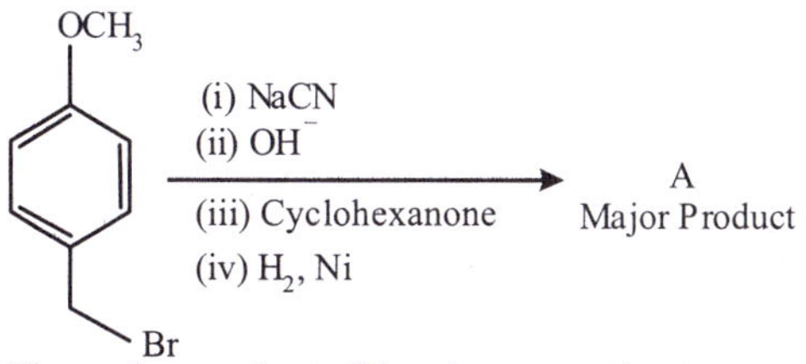
The major product of the above reaction is [June 24, 2022 (I)]
(1) 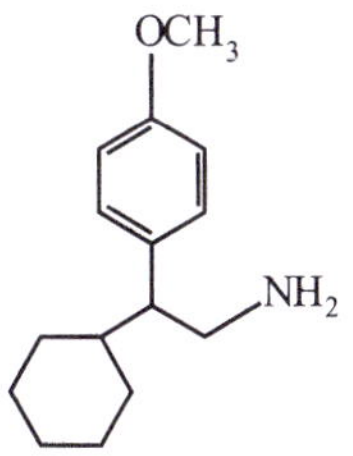
(2) 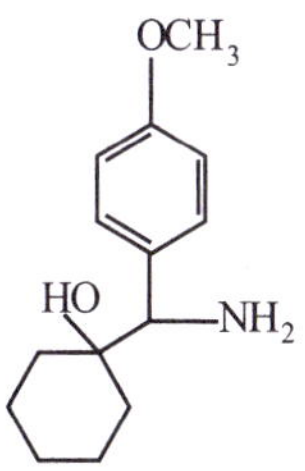
(3) 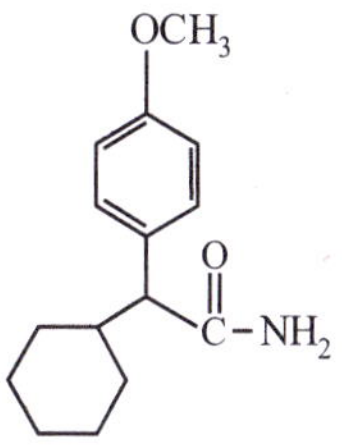
(4) 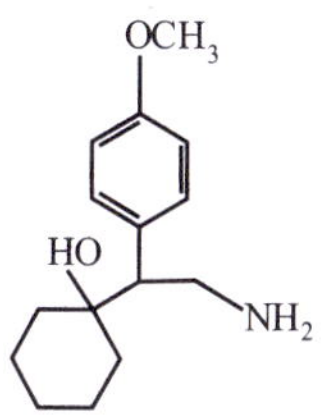

72. Which of the following is **not** a correct statement for primary aliphatic amines? [Aug. 27, 2021 (I)]
(1) The intermolecular association in primary amines is less than the intermolecular association in secondary amines.
(2) Primary amines on treating with nitrous acid solution form corresponding alcohols except methyl amine.
(3) Primary amines are less basic than the secondary amines.
(4) Primary amines can be prepared by the Gabriel phthalimide synthesis.

73. $R-CN \xrightarrow[(ii)H_2O]{(i)\ DiBAL-H} R-Y$ [July 27, 2021 (II)]
Consider the above reaction and identify "Y"
(1) $-CH_2NH_2$ (2) $-CONH_2$
(3) $-CHO$ (4) $-COOH$

74. Given below are two statements : [July 27, 2021 (I)]
Statement I : Aniline is less basic than acetamide.
Statement II : In aniline, the lone pair of electrons on nitrogen atom is delocalised over benzene ring due to resonance and hence less available to a proton.
Choose the **most appropriate** option :
(1) Statement I is true but statement II is false.
(2) Statement I is false but statement II is true.
(3) Both statement I and statement II are true.
(4) Both statement I and statement II are false.

75. A reaction of benzonitrile with one equivalent CH_3MgBr followed by hydrolysis produces a yellow liquid "P". The compound "P" will give positive____. [July 25, 2021 (II)]
(1) Iodoform test (2) Schiff's test
(3) Ninhydrin's test (4) Tollen's test

76. Which one of the following compounds will liberate CO_2, when treated with $NaHCO_3$? [July 25, 2021 (I)]
(1) $(CH_3)_3\overset{\oplus}{N}H\overset{\ominus}{Cl}$ (2) $(CH_3)_4\overset{\oplus}{N}\overset{\ominus}{O}H$
(3) $CH_3-\overset{\overset{O}{||}}{C}-NH_2$ (4) CH_3NH_2

77. 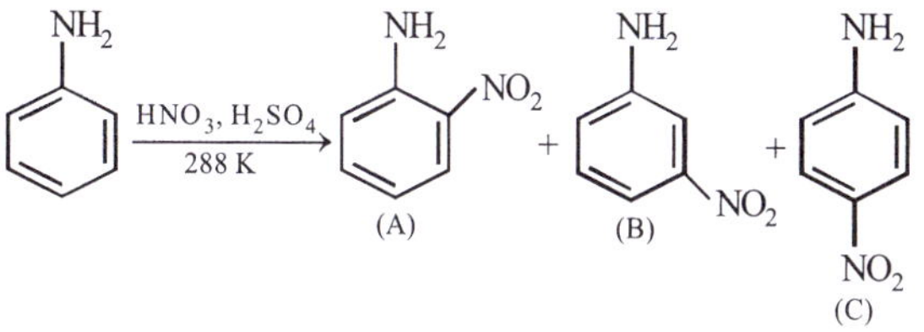
Consider the given reaction, precentage yield of : [March 18, 2021 (II)]
(1) C>B>A (2) C>A>B
(3) B>C>A (4) A>C>B

78. An organic compound "A" on treatment with benzene sulphonyl chloride gives compound B. B is soluble in dil. NaOH solution. Compound A is [March 18, 2021 (II)]
(1) $C_6H_5-N-(CH_3)_2$ (2) $C_6H_5-CH_2NHCH_3$
(3) $C_6H_5-\underset{\underset{CH_3}{|}}{CH}-NH_2$ (4) $C_6H_5-NHCH_2CH_3$

79. 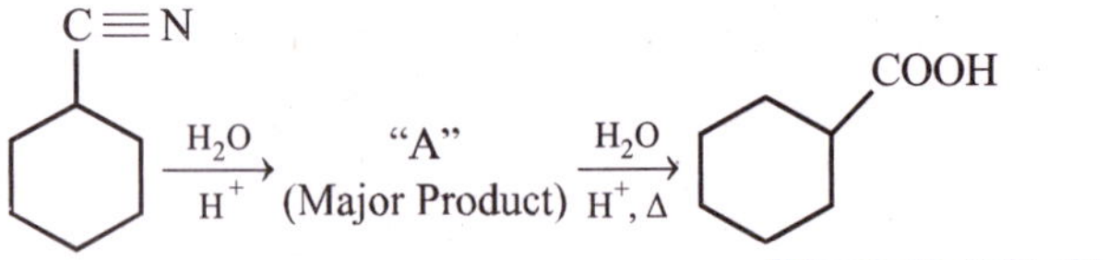

[March 18, 2021 (I)]

Consider the above chemical reaction and identify product "A"

(1) 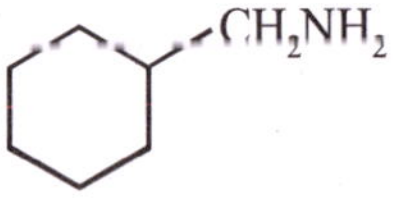 (2) 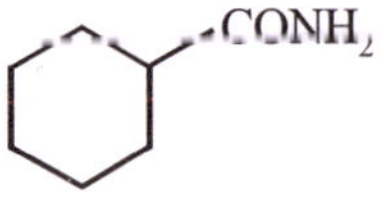

(3) 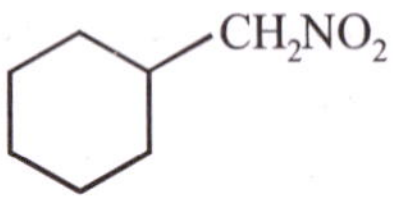 (4) 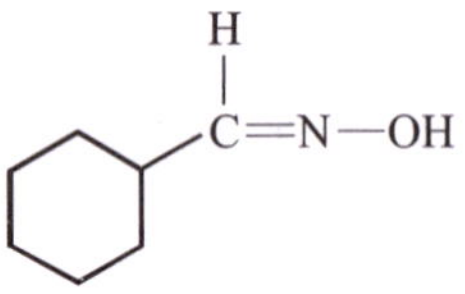

80. Primary, secondary and tertiary amines can be separated using
(1) para-Toluene sulphonyl chloride [March 17, 2021 (II)]
(2) Acetyl amide
(3) Chloroform and KOH
(4) Benzene sulphonic acid

81. Which of the following is least basic? [March 16, 2021 (II)]

(1) $(CH_3CO)\ddot{N}HC_2H_5$ (2) $(CH_3CO)_2\ddot{N}H$

(3) $(C_2H_5)_2\ddot{N}H$ (4) $(C_2H_5)_3\ddot{N}$

82. An amine on reaction with benzene sulphonyl chloride produces a compound insoluble in alkaline solution. This amine can be prepared by ammonolysis of ethyl chloride. The correct structure of amine is : [Feb. 26, 2021 (I)]

(1) $CH_3CH_2CH_2-\underset{}{\overset{H}{\overset{|}{N}}}-CH_2CH_3$
(2) $CH_3CH_2NH_2$
(3) $CH_3CH_2CH_2NHCH_3$
(4) 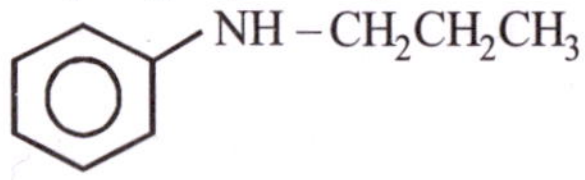

83. A. Phenyl methanamine B. N,N-Dimethylaniline
C. N-Methyl aniline D. Benzenamine
Choose the correct order of basic nature of the above amines. [Feb. 26, 2021 (II)]
(1) A > C > B > D (2) D > B > C > A
(3) D > C > B > A (4) A > B > C > D

84. For the given reaction : [Feb. 26, 2021 (I)]

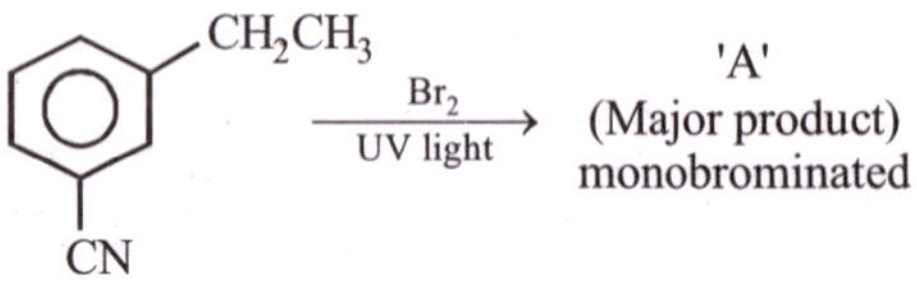

What is 'A'?

(1) 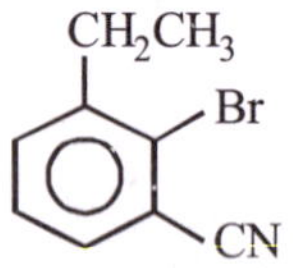 (2) 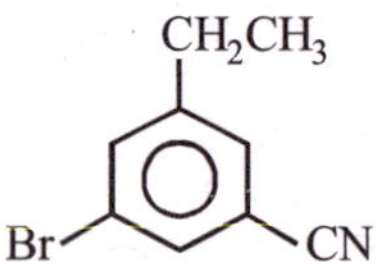

(3) 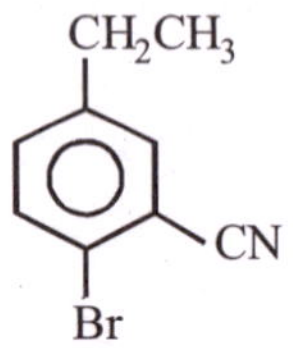 (4) 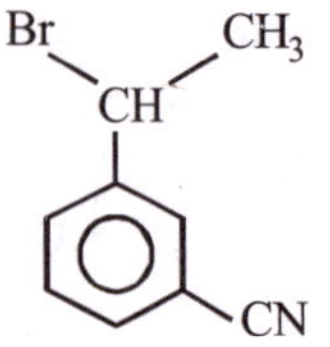

85. The major product of the following chemical reaction is:

$$CH_3CH_2CN \xrightarrow[(3)\ Pd/BaSO_4,\ H_2]{(1)\ H_3O^+,\Delta \quad (2)\ SOCl_2}$$ [Feb. 25, 2021 (I)]

(1) $CH_3CH_2CH_3$ (2) $CH_3CH_2CH_2OH$
(3) CH_3CH_2CHO (4) $(CH_3CH_2CO)_2O$

86. Carbylamine test is used to detect the presence of primary amino group in an organic compound. Which of the following compound is formed when this test is performed with aniline? [Feb. 25, 2021 (II)]

(1) 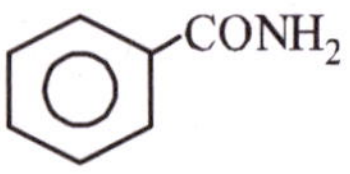 (2) 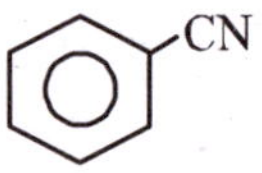

(3) 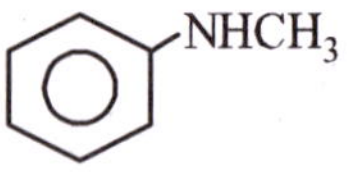 (4) 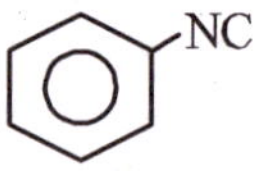

87. In the following reaction, the reason why meta-nitro product also formed is [Feb. 24, 2021 (I)]

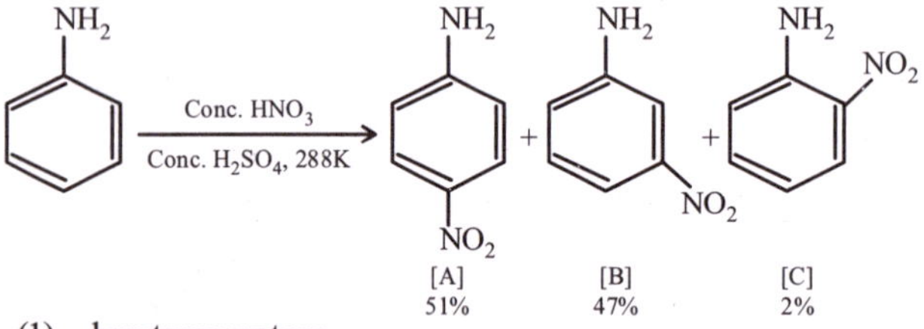

(1) low temperature
(2) $-NH_2$ group is highly meta-directive
(3) Formation of anilinium ion
(4) $-NO_2$ substitution always takes place at meta-position

88. Which of the following compounds can be prepared in good yield by Gabriel phthalimide synthesis? [Sep. 06, 2020 (II)]

(1) 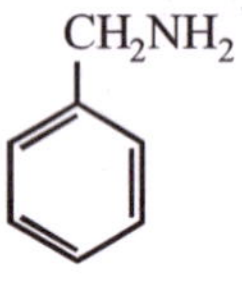 (2) $CH_3-CH_2-NHCH_3$

(3) 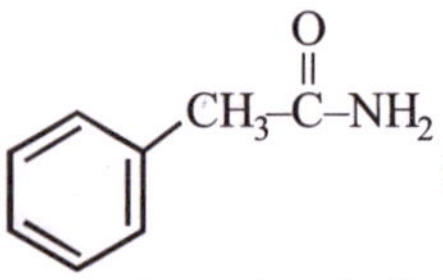 (4) 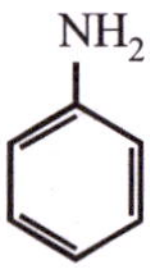

89. The increasing order of pK_b values of the following compounds is : [Sep. 06, 2020 (I)]

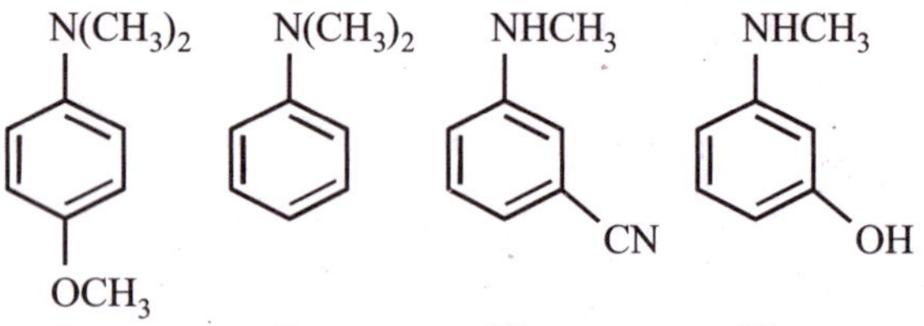

I II III IV

(1) II < IV < III < I (2) I < II < IV < III
(3) II < I < III < IV (4) I < II < III < IV

90. The final major product of the following reaction is :

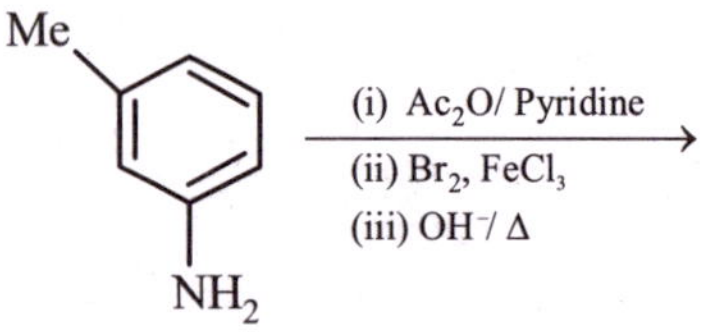

[Sep. 05, 2020 (II)]

(1) 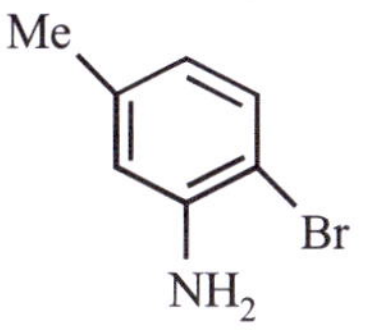 (2) 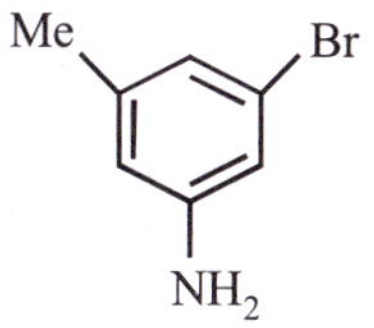

(3) 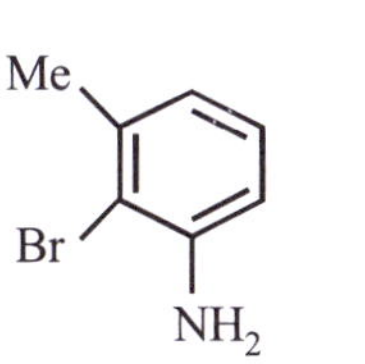 (4) 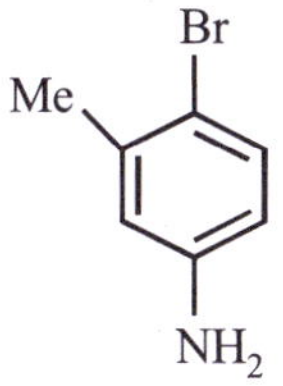

91. The most appropriate reagent for conversion of C_2H_5CN into $CH_3CH_2CH_2NH_2$ is : **[Sep. 05, 2020 (I)]**
(1) $NaBH_4$ (2) CaH_2
(3) $LiAlH_4$ (4) $Na(CN)BH_3$

92. The increasing order of basicity of the following compounds is: **[Sep. 05, 2020 (I)]**

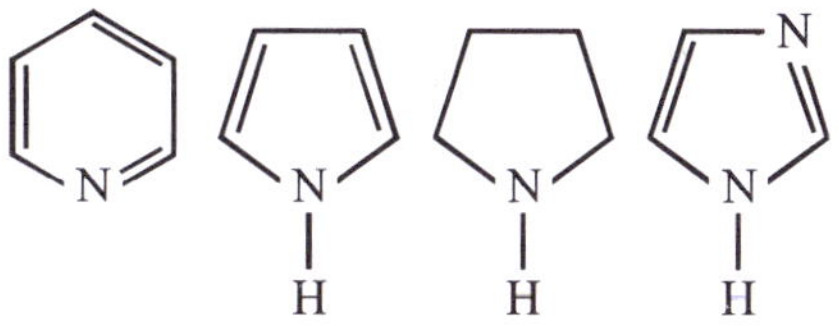

(A) (B) (C) (D)

(1) (A) < (B) < (C) < (D) (2) (B) < (A) < (D) < (C)
(3) (D) < (A) < (B) < (C) (4) (B) < (A) < (C) < (D)

93. The decreasing order of basicity of the following amines is: **[Jan. 09, 2020 (II)]**

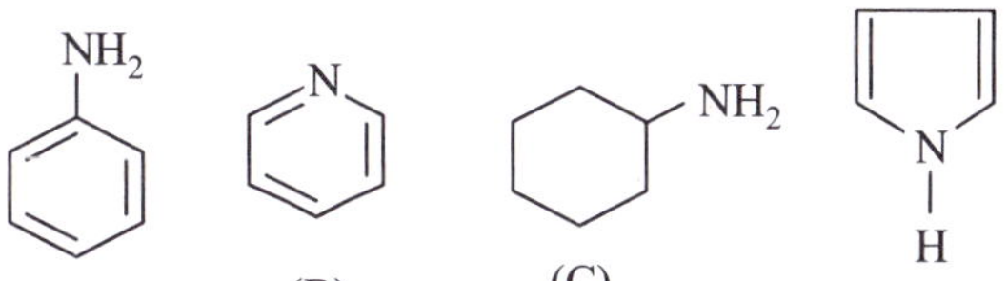

(A) (B) (C) (D)

(1) (A) > (C) > (D) > (B) (2) (C) > (A) > (B) > (D)
(3) (B) > (C) > (D) > (A) (4) (C) > (B) > (A) > (D)

94. Which of these will produce the highest yield in Friedel Crafts reaction? **[Jan. 09, 2020 (I)]**

(1) 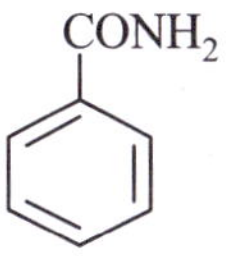 (2) 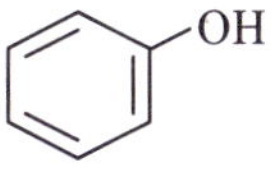

(3) 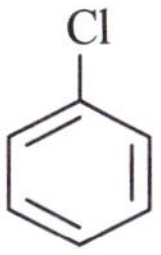 (4) 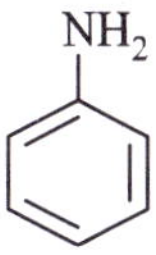

95. In the following reaction sequence,

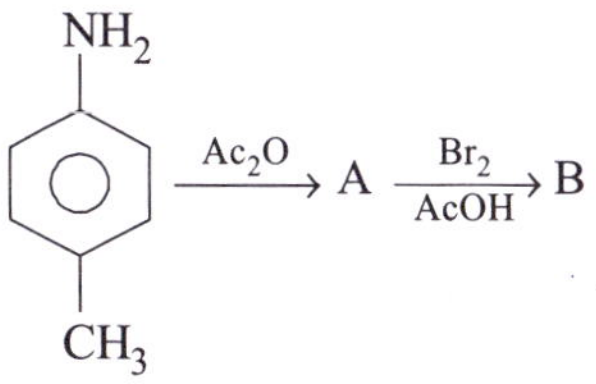

the major product B is: **[Jan. 07, 2020 (II)]**

(1) 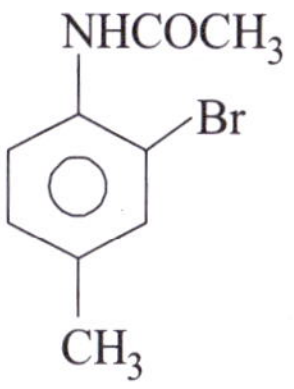 (2) 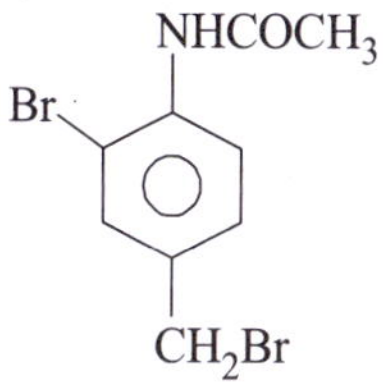

(3) 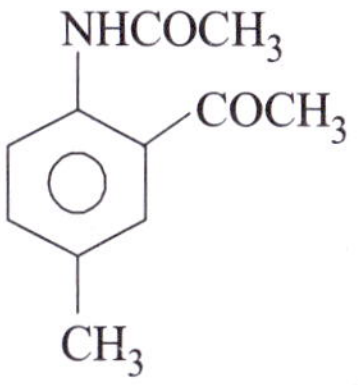 (4) 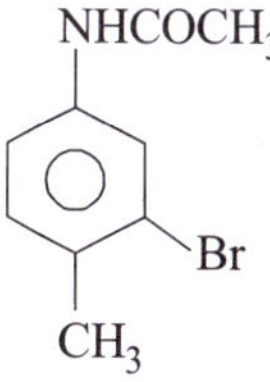

96. The increasing order of pK_b for the following compounds will be: **[Jan. 07, 2020 (I)]**

(A) $NH_2-CH=NH$, 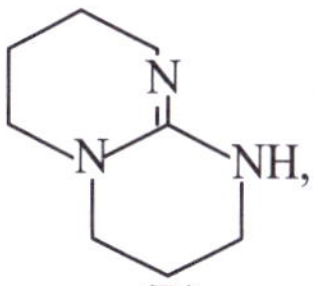 (B), CH_3NHCH_3 (C)

(1) (B) < (C) < (A) (2) (A) < (B) < (C)
(3) (C) < (A) < (B) (4) (B) < (A) < (C)

97. The increasing order of the pK_b of the following compounds is : **[April 12, 2019 (I)]**

(A) 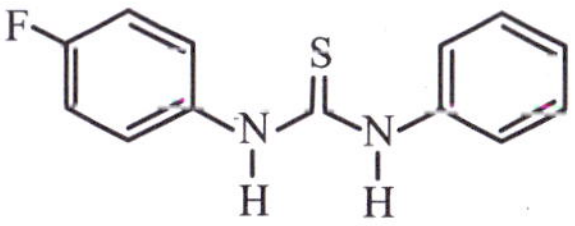

(B) 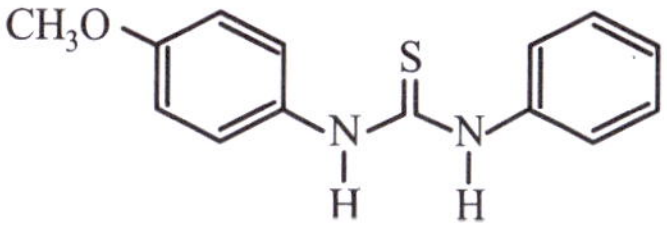

(C) 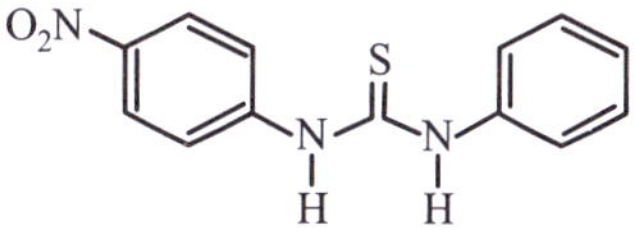

(D) 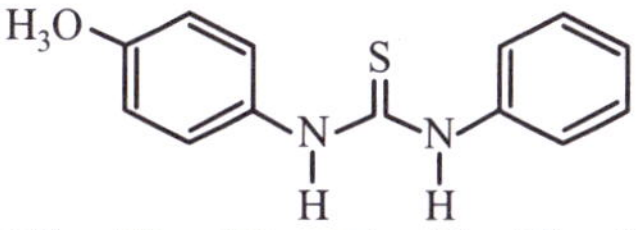

(1) (A) < (C) < (D) < (B) (2) (C) < (A) < (D) < (B)
(3) (B) < (D) < (A) < (C) (4) (B) < (D) < (C) < (A)

98. The major product 'Y' in the following reaction is: **[April 10, 2019 (II)]**

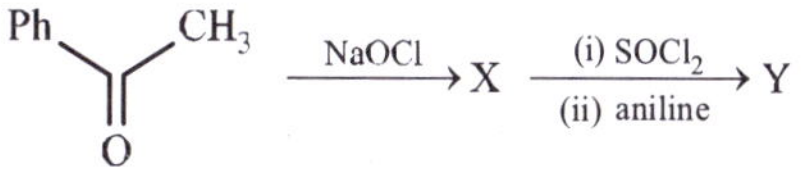

(1) 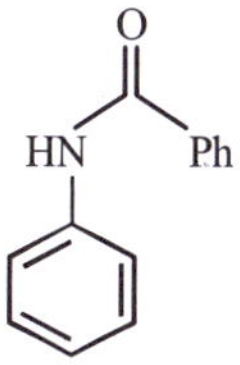 (2) 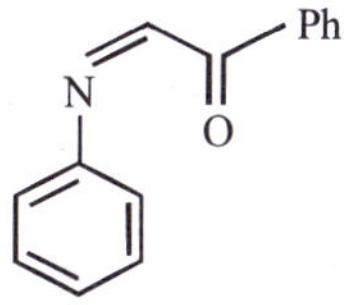

(3) 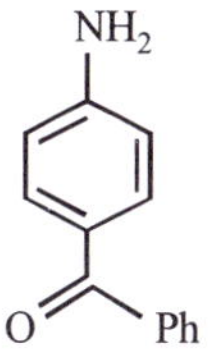 (4) 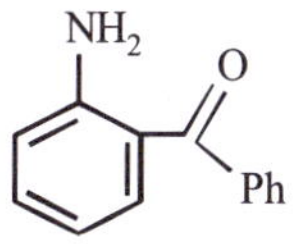

99. Which of the following is NOT a correct method of the preparation of benzylamine from cyanobenzene ? [April 10, 2019 (II)]

(1) H_2/Ni
(2) (i) $LiAlH_4$ (ii) H_3O^+
(3) (i) $SnCl_2 + HCl(gas)$ (ii) $NaBH_4$
(4) (i) HCl/H_2O (ii) $NaBH_4$

100. Ethylamine ($C_2H_5NH_2$) can be obtained from N-ethylphthalimide on treatment with . [April 10, 2019 (I)]

(1) NH_2NH_2 (2) CaH_2
(3) $NaBH_4$ (4) H_2O

101. The major product of the following reaction is: [April 10, 2019 (I)]

$$CH_3\overset{OH}{\overset{|}{C}}HCH_2CH_2NH_2 \xrightarrow[\text{triethylamine}]{\text{ethyl formate (1 equiv.)}}$$

(1) $CH_3-\overset{OH}{\overset{|}{C}}H-CH=CH_2$

(2) $CH_3\overset{OH}{\overset{|}{C}}HCH_2CH_2NHCHO$

(3) 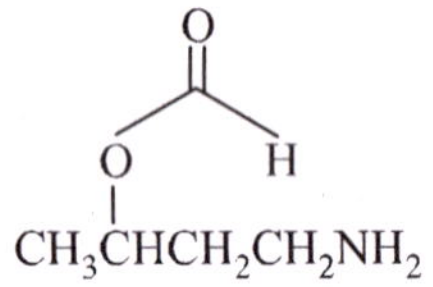

(4) $CH_3CH=CH-CH_2NH_2$

102. The major products A and B for the following reactions are, respectively: [April 9, 2019 (II)]

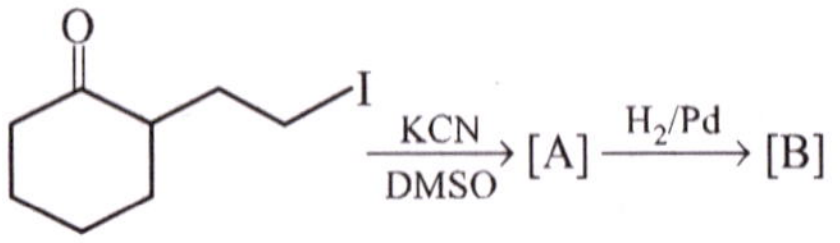

(1) 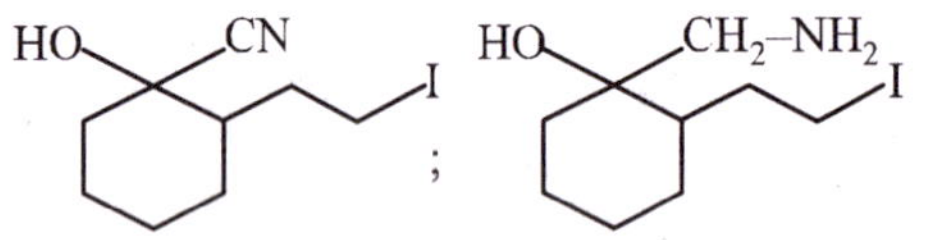

(2) 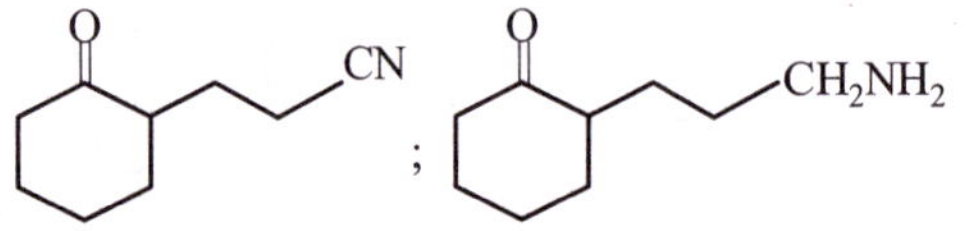

(3) 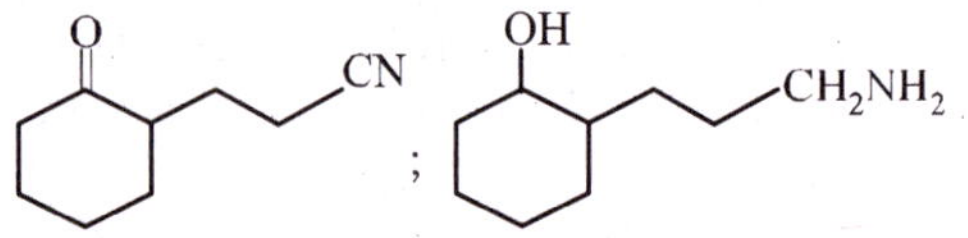

(4) 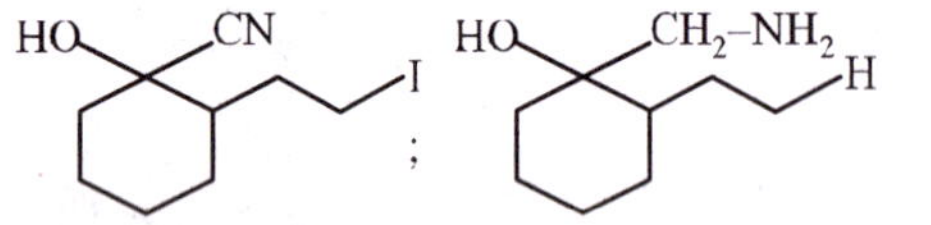

103. Hinsberg's reagent is: [April 9, 2019 (II)]

(1) C_6H_5COCl (2) $SOCl_2$
(3) $C_6H_5SO_2Cl$ (4) $(COCl)_2$

104. Polysubstitution is a major drawback in : [April 8, 2019 (II)]

(1) Friedel Craft's alkylation
(2) Reimer Tiemann reaction
(3) Acetylation of aniline
(4) Friedel Craft's acylation

105. The major product in the following reaction is: [April 8, 2019 (II)]

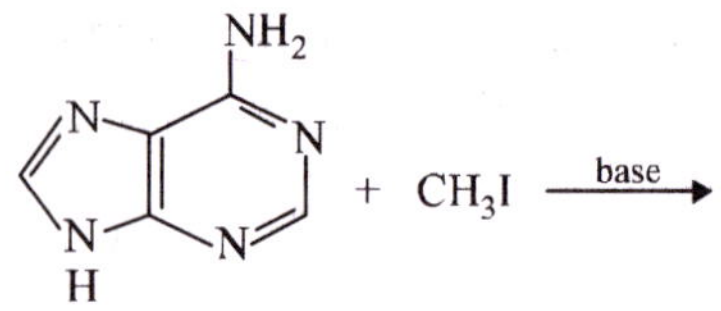

(1) 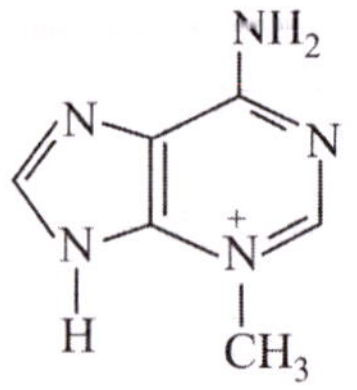

(2) 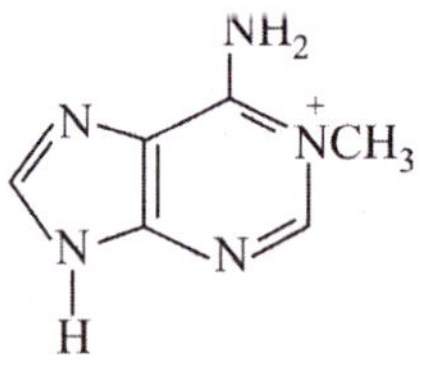

(3) 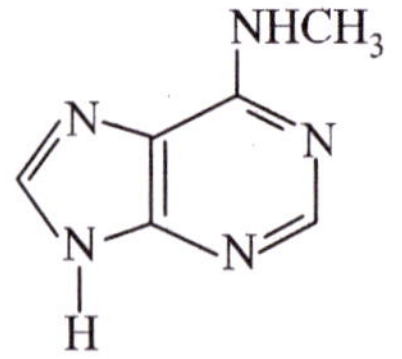

(4) 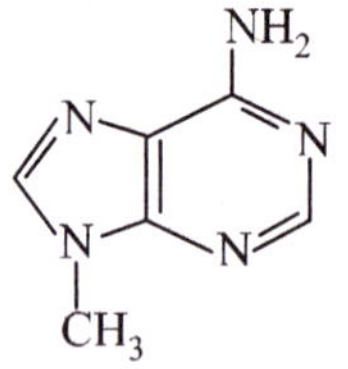

106. The major product obtained in the following reaction is : [April 8, 2019 (II)]

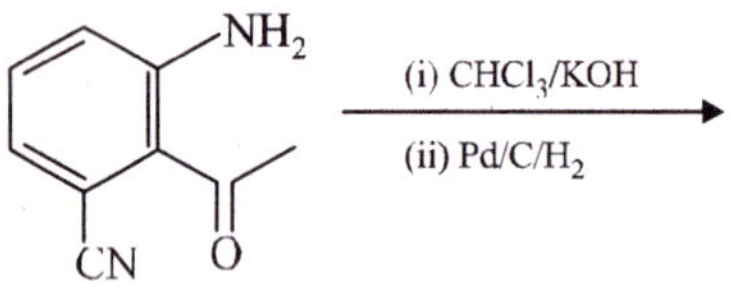

(1) 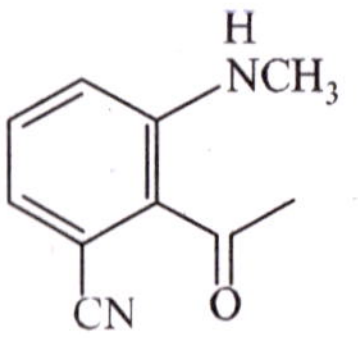

(2) 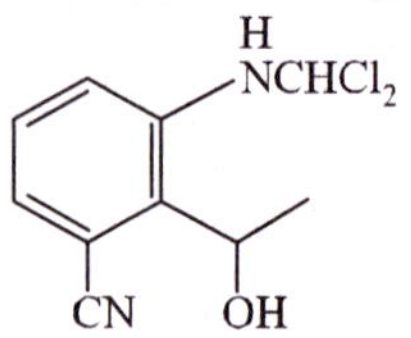

(3) 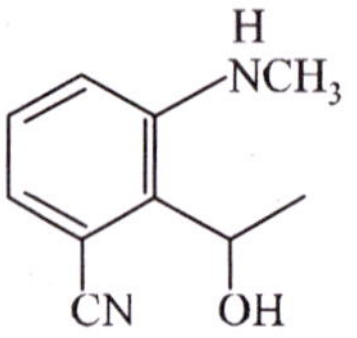

(4) 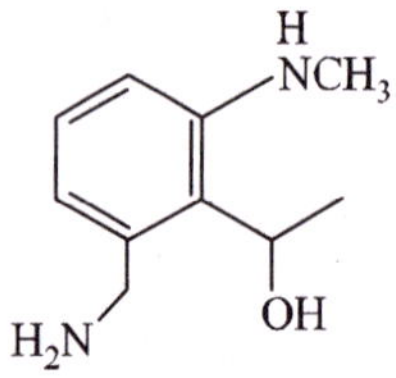

107. Which of the following amines can be prepared by Gabriel phthalimide reaction ? [April 8, 2019 (I)]

(1) *n*-butylamine (2) triethylamine
(3) *t*-butylamine (4) neo-pentylamine

108. In the following compounds, the decreasing order of basic strength will be: [April 8, 2019 (I)]

(1) $C_2H_5NH_2 > NH_3 > (C_2H_5)_2NH$
(2) $(C_2H_5)_2NH > NH_3 > C_2H_5NH_2$
(3) $(C_2H_5)_2NH > C_2H_5NH_2 > NH_3$
(4) $NH_3 > C_2H_5NH_2 > (C_2H_5)_2NH$

109. The major product of the following reaction is : [Jan. 12, 2019 (II)]

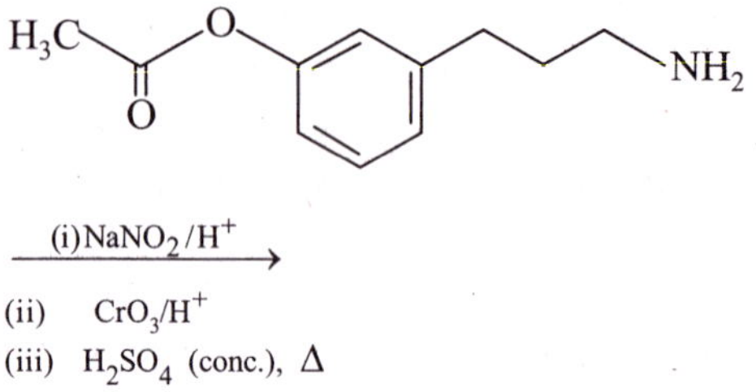

(1) 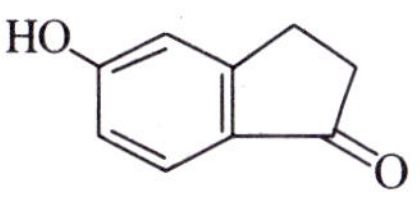

(2) 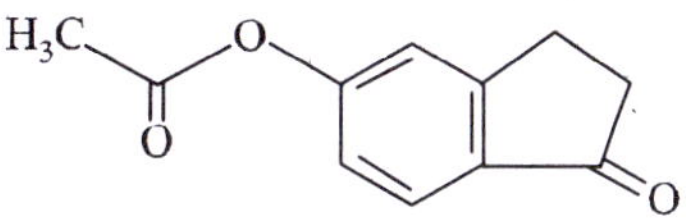

(3) 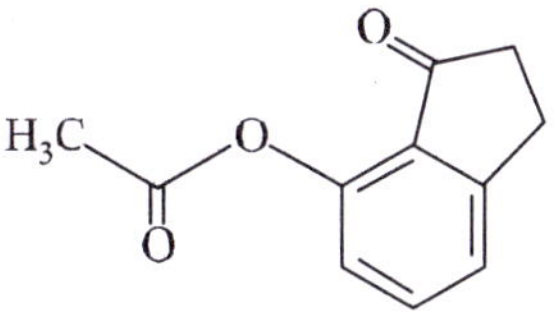

(4) 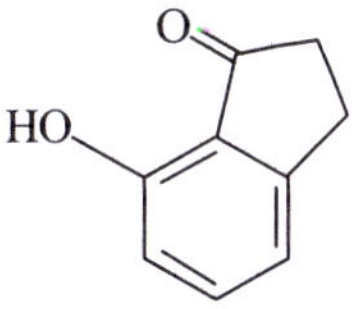

110. The increasing order of the reactivity of the following with $LiAlH_4$ is : **[Jan. 12, 2019 (II)]**

(A) 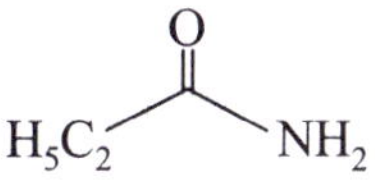

(B) 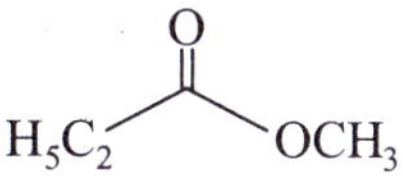

(C) 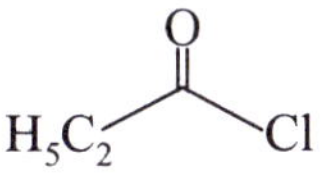

(D) 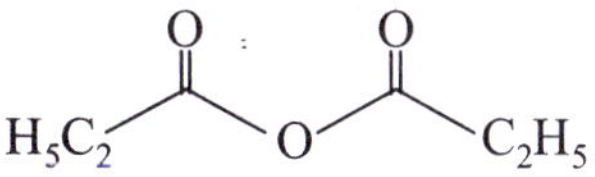

(1) (B) < (A) < (C) < (D) (2) (B) < (A) < (D) < (C)
(3) (A) < (B) < (D) < (C) (4) (A) < (B) < (C) < (D)

111. The increasing order of reactivity of the following compound towards reaction with alkyl halides directly is: **[Jan. 12, 2019 (I)]**

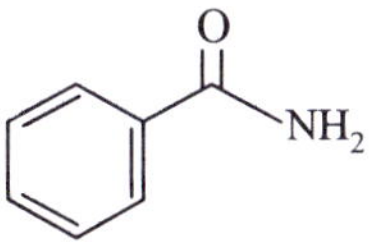
(A)

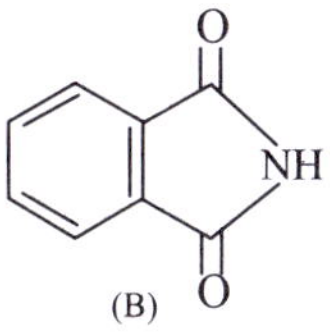
(B)

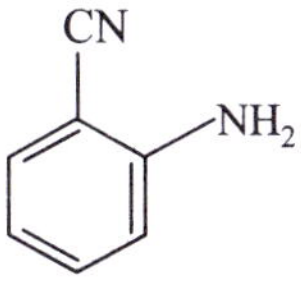
(C)

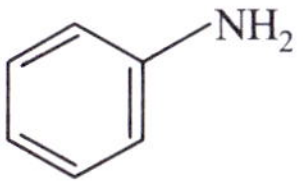
(D)

(1) (B) < (A) < (C) < (D) (2) (A) < (B) < (C) < (D)
(3) (B) < (A) < (D) < (C) (4) (A) < (C) < (D) < (B)

112. In the following compound,

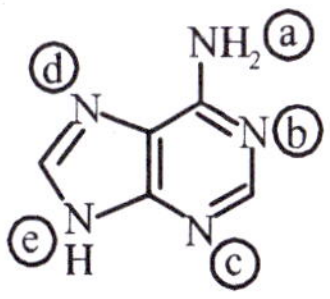

the favourable site/s for protonation is/are : **[Jan. 11, 2019 (II)]**

(1) (a) and (e) (2) (b), (c) and (d)
(3) (a) and (d) (4) (a)

113. A compound 'X' on treatment with Br_2/NaOH, provided C_3H_9N, which gives positive carbylamine test. Compound 'X' is : **[Jan. 11, 2019 (II)]**

(1) $CH_3COCH_2NHCH_3$ (2) $CH_3CH_2COCH_2NH_2$
(3) $CH_3CH_2CH_2CONH_2$ (4) $CH_3CON(CH_3)_2$

114. What will be the major product in the following mononitration reaction? **[Jan. 10, 2019 (II)]**

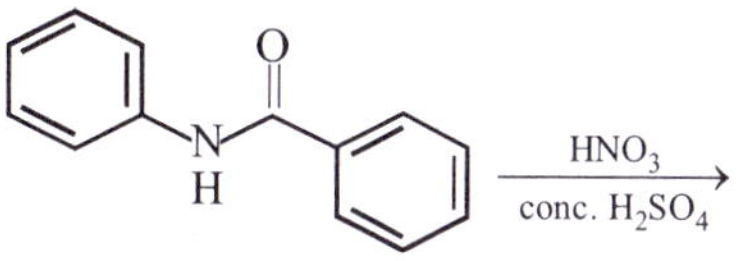

(1) 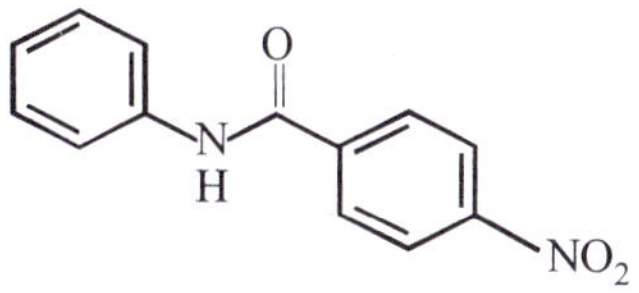

(2) 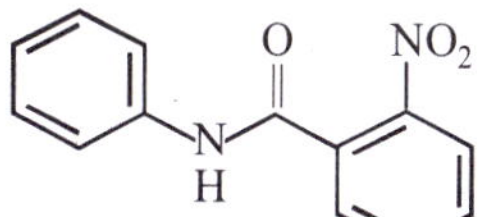

(3) 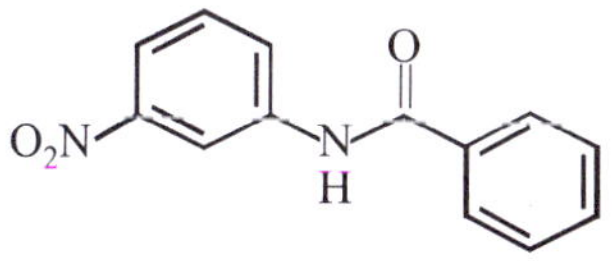

(4) 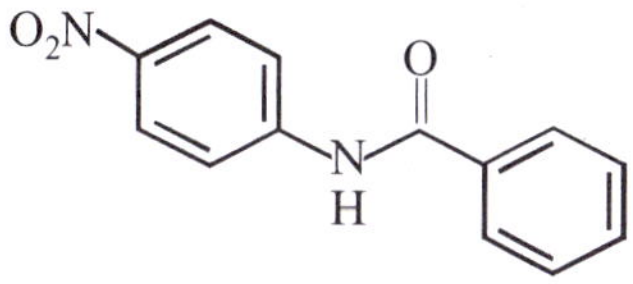

115. The major product formed in the reaction given below will be: **[Jan. 10, 2019 (I)]**

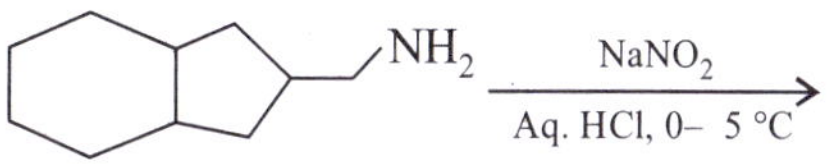

(1) 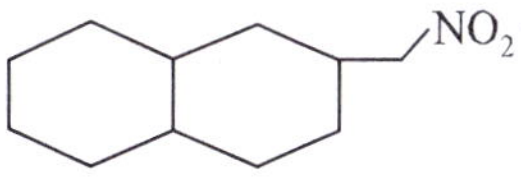

(2) 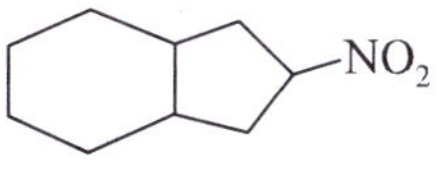

(3) 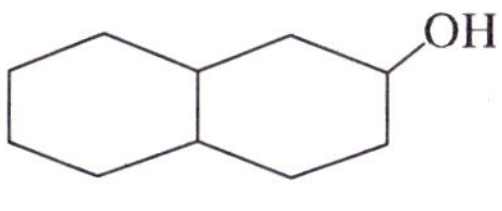

(4) 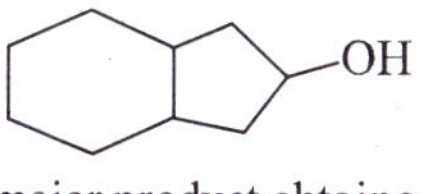

116. The major product obtained in the following reaction is:

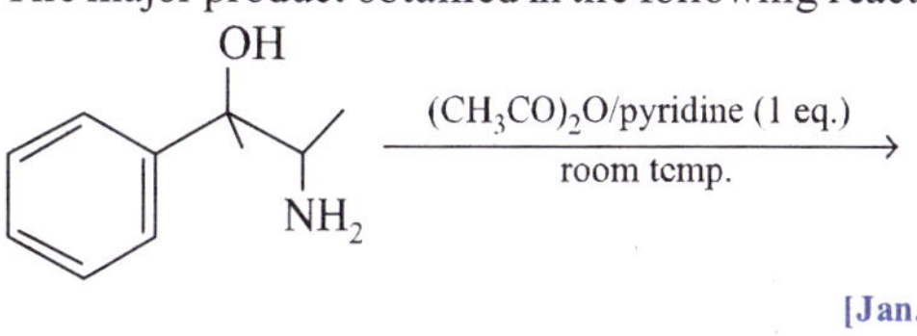

[Jan. 9, 2019 (II)]

(1) 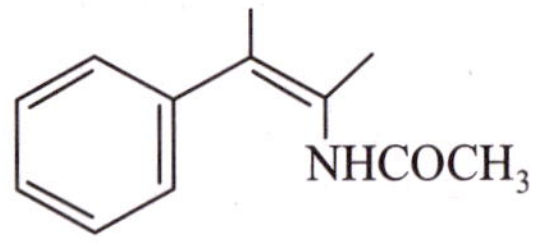

(2) 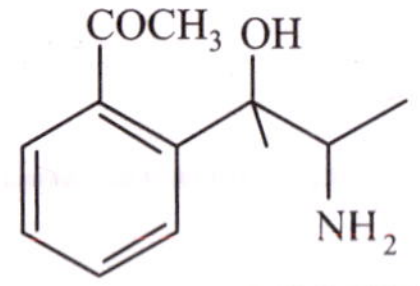

(3) 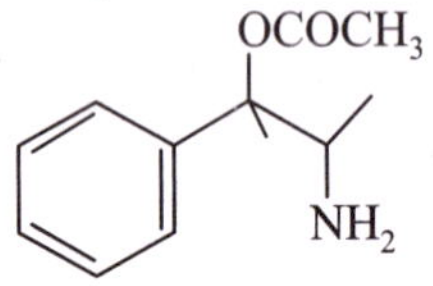

(4) 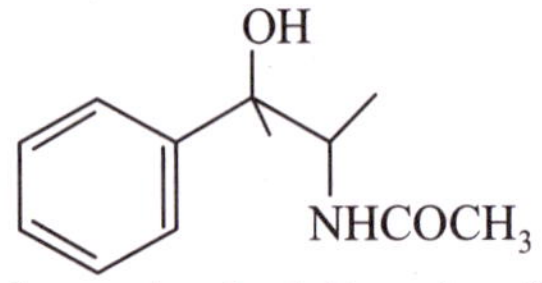

117. The increasing basicity order of the following compounds is: **[Jan. 9, 2019 (II)]**

(A) $CH_3CH_2NH_2$ (B) CH_3CH_2NH with CH_2CH_3 on N

(C) $H_3C-N-CH_3$ with CH_3 on N (D) $Ph-N-H$ with CH_3 on N

(1) (D) < (C) < (B) < (A) (2) (D) < (C) < (A) < (B)
(3) (A) < (B) < (C) < (D) (4) (A) < (B) < (D) < (C)

118. Arrange the following amines in the decreasing order of basicity: **[Jan. 9, 2019 (I)]**

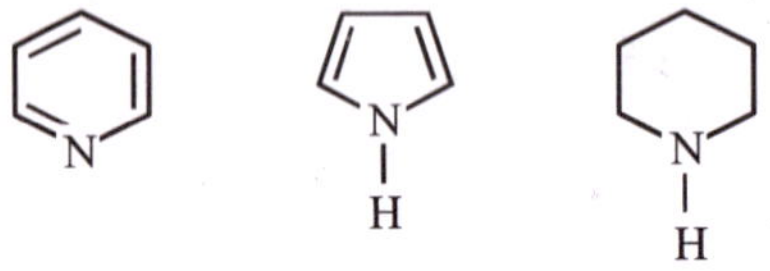

I II III

(1) I > II > III (2) III > I > II
(3) III > II > I (4) I > III > II

119. The major product of the following reaction is: **[Jan. 9, 2019 (II)]**

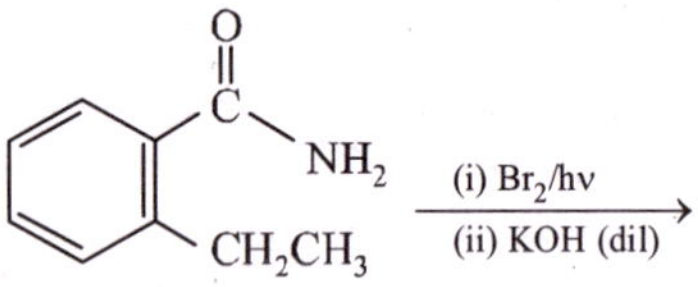

(1) 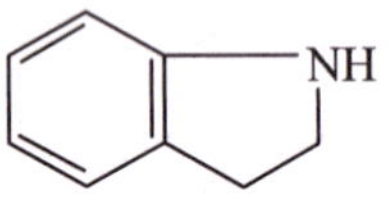 (2) 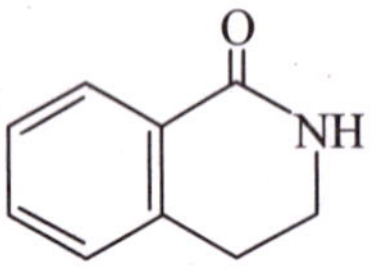

(3) 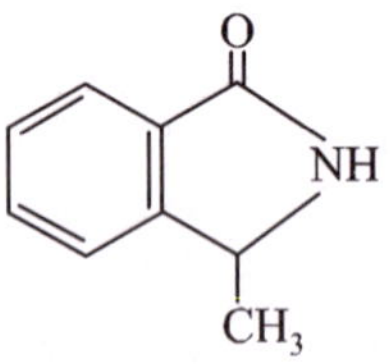 (4) 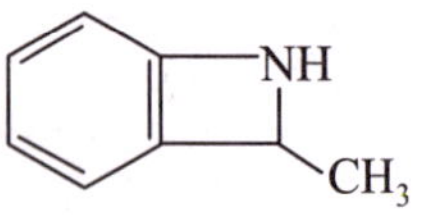

120. The compounds A and B in the following reaction are, respectively : **[Jan. 9, 2019 (I)]**

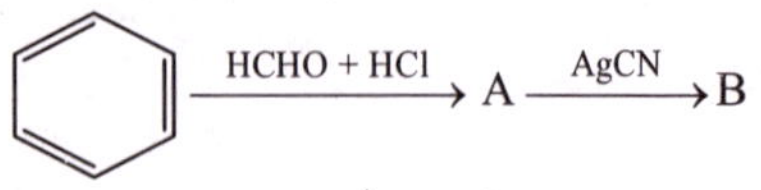

(1) A = Benzyl alcohol, B = Benzyl cyanide
(2) A = Benzyl chloride, B = Benzyl cyanide
(3) A = Benzyl alcohol, B = Benzyl isocyanide
(4) A = Benzyl chloride, B = Benzyl isocyanide

121. Products A and B formed in the following reactions are respectively : **[Online April 16, 2018]**

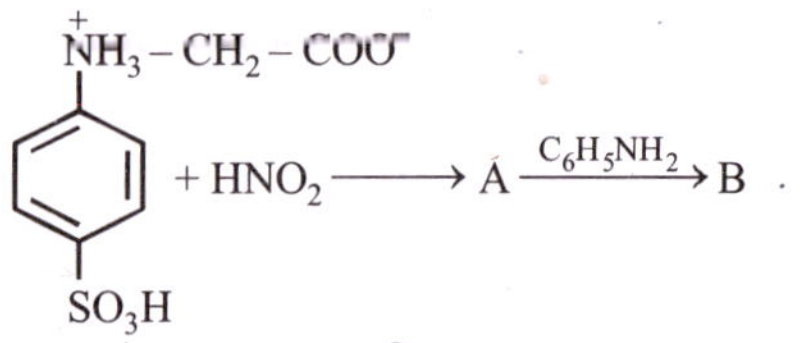

(1) 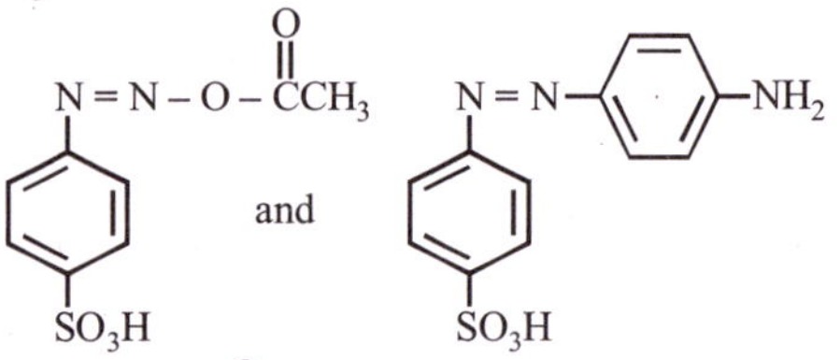

(2) 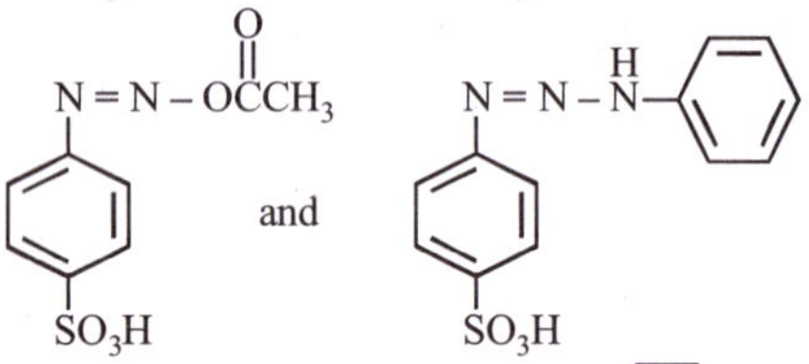

(3) 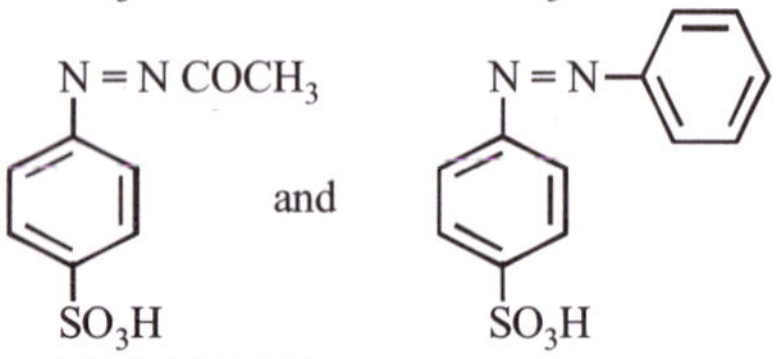

(4) 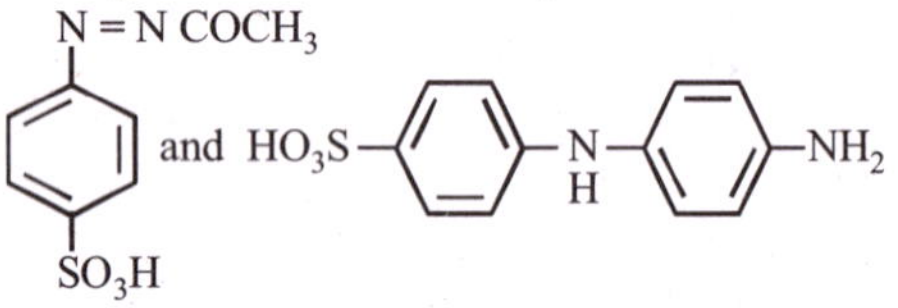

122. The increasing order of diazotisation of the following compounds is: **[Online April 15, 2018 (II)]**

(A)  (B) 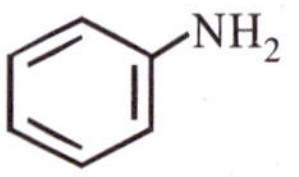

(C) 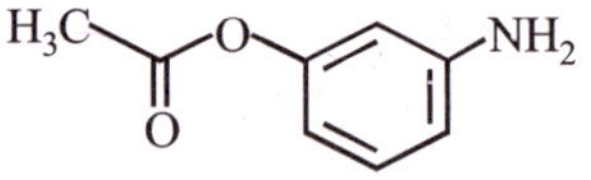

(D) 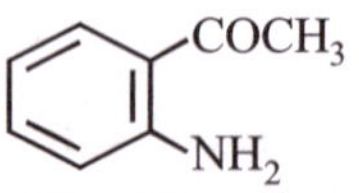

(1) (D) < (C) < (B) < (A) (2) (A) < (D) < (B) < (C)
(3) (A) < (B) < (C) < (D) (4) (A) < (D) < (C) < (B)

123. Among the following compounds, the increasing order of their basic strength is : **[Online April 9, 2017]**

(I) 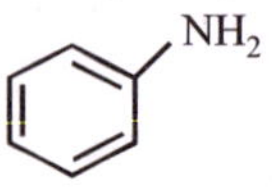 (II) 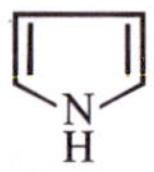

(III) 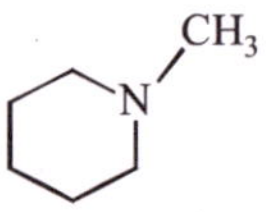 (IV) 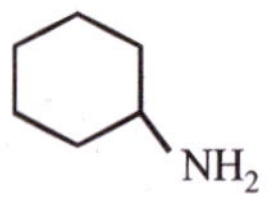

(1) (I) < (II) < (IV) < (III) (2) (I) < (II) < (III) < (IV)
(3) (II) < (I) < (IV) < (III) (4) (II) < (I) < (III) < (IV)

124. Which of the following compounds will form significant amount of meta product during mono-nitration reaction ? **[2017]**

(1) 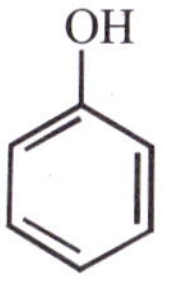

(2) 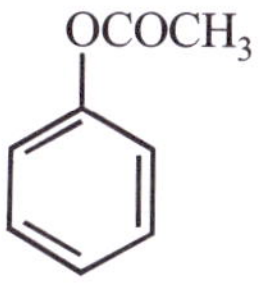

(3) 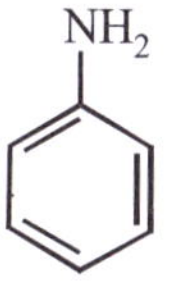

(4) 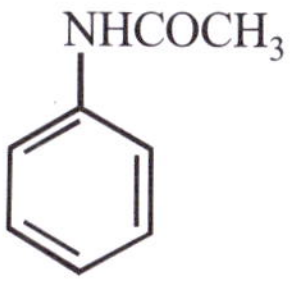

125. In the Hoffmann bromamide degradation reaction, the number of moles of NaOH and Br_2 used per mole of amine produced are : **[2016]**

(1) Two moles of NaOH and two moles of Br_2.

(2) Four moles of NaOH and one mole of Br_2.

(3) One mole of NaOH and one mole of Br_2.

(4) Four moles of NaOH and two moles of Br_2.

126. The test to distinguish primary, secondary and tertiary amines is : **[Online April 9, 2016]**

(1) Sandmeyer's reaction (2) Carbylamine reaction

(3) Mustard oil test (4) $C_6H_5SO_2Cl$

127. The "N" which does not contribute to the basicity for the compound is : **[Online April 10, 2016]**

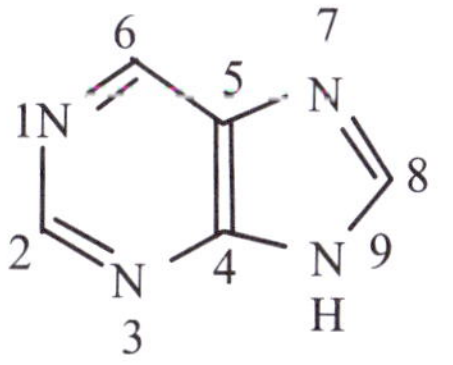

(1) N9 (2) N3 (3) N1 (4) N7

128. Arrange the following amines in the order of increasing basicity. **[Online April 10, 2015]**

(1) 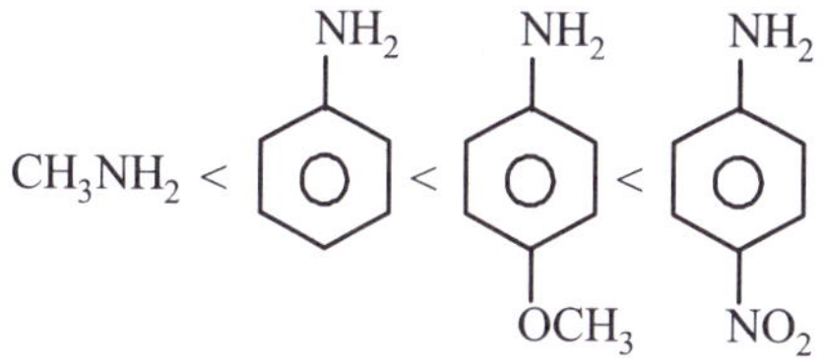

(2) 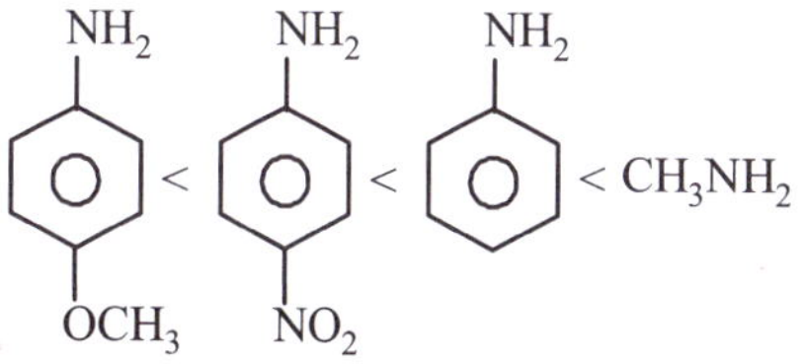

(3) 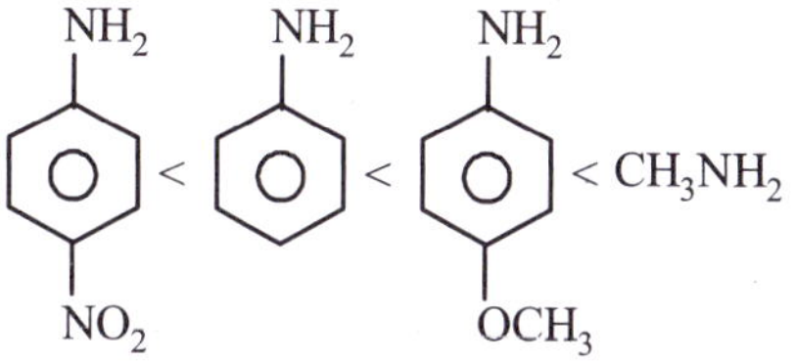

(4) 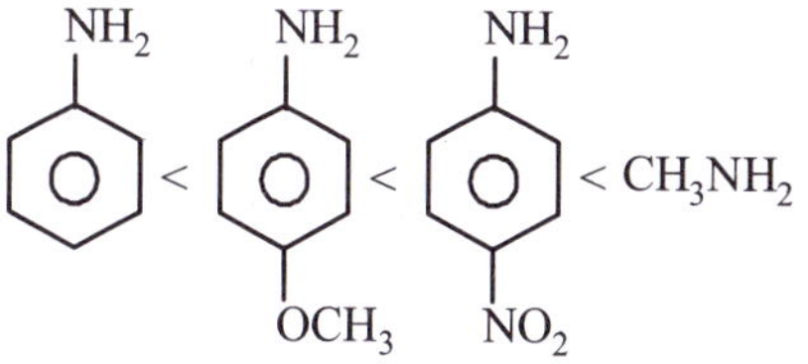

129. On heating an aliphatic primary amine with chloroform and ethanolic potassium hydroxide, the organic compound formed is: **[2014]**

(1) an alkanol (2) an alkanediol

(3) an alkyl cyanide (4) an alkyl isocyanide

130. Considering the basic strength of amines in aqueous solution, which one has the smallest pK_b value? **[2014; Same 2007]**

(1) $(CH_3)_2NH$ (2) CH_3NH_2

(3) $(CH_3)_3N$ (4) $C_6H_5NH_2$

131. The final product formed when methyl amine is treated with $NaNO_2$ and HCl is: **[Online April 19, 2014]**

(1) Diazomethane (2) Methyl alcohol

(3) Methylcyanide (4) Nitromethane

132. The major product of the reaction

[Online April 9, 2014]

(1) 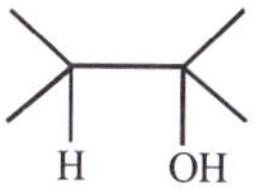

(2) 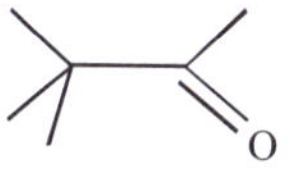

(3) 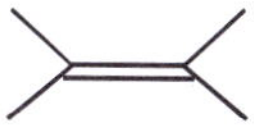

(4) 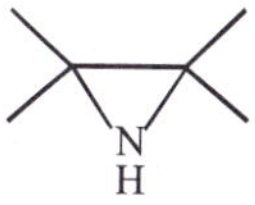

133. A compound with molecular mass 180 is acylated with CH_3COCl to get a compound with molecular mass 390. The number of amino groups present per molecule of the former compound is : **[2013]**

(1) 2 (2) 5 (3) 4 (4) 6

134. The order of basicity of amines in gaseous state is : **[Online April 23, 2013]**

(1) $1° > 2° > 3° > NH_3$ (2) $3° > 2° > NH_3 > 1°$

(3) $3° > 2° > 1° > NH_3$ (4) $NH_3 > 1° > 2° > 3°$

135. Carbylamine is formed from aliphatic or aromatic primary amine via which of the following intermediates ? **[Online April 25, 2013]**

(1) Carbanions (2) Carbenes

(3) Carbocations (4) Carbon radicals

136. The most basic compound among the following is **[2005, Online May 7, 2012]**

(1) Acetanilide (2) Benzyl amine

(3) *p*-Nitro aniline (4) Aniline

137. The order of basicity of the compounds is **[Online May 19, 2012]**

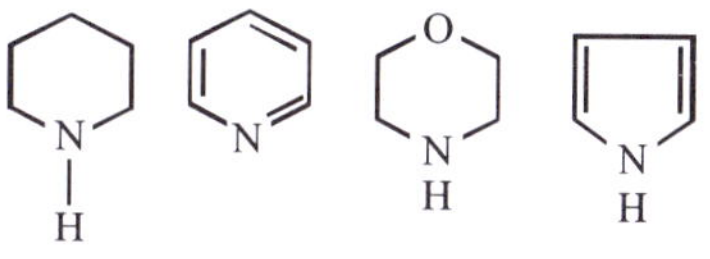

(I) (II) (III) (IV)

(1) IV > I > III > II (2) I > III > II > IV

(3) III > I > IV > II (4) II > I > III > IV

138. Which one of the following is the strongest base in aqueous solution ? **[2007]**

(1) Methylamine (2) Trimethylamine

(3) Aniline (4) Dimethylamine.

139. In the chemical reaction,

$CH_3CH_2NH_2 + CHCl_3 + 3KOH \rightarrow$

(A) + (B) + $3H_2O$, the compounds (A) and (B) are respectively **[2007]**

(1) C_2H_5NC and 3KCl

(2) C_2H_5CN and 3KCl

(3) $CH_3CH_2CONH_2$ and 3KCl

(4) C_2H_5NC and K_2CO_3.

140. Which one of the following methods is neither meant for the synthesis nor for separation of amines? [2005]
(1) Curtius reaction (2) Wurtz reaction
(3) Hofmann method (4) Hinsberg method

141. The correct order of increasing basic nature for the bases NH_3, CH_3NH_2 and $(CH_3)_2NH$ is [2003]
(1) $(CH_3)_2NH < NH_3 < CH_3NH_2$
(2) $NH_3 < CH_3NH_2 < (CH_3)_2NH$
(3) $CH_3NH_2 < (CH_3)_2NH < NH_3$
(4) $CH_3NH_2 < NH_3 < (CH_3)_2NH$

142. The reaction of chloroform with alcoholic KOH and *p*-toluidine forms [2003]

(1) 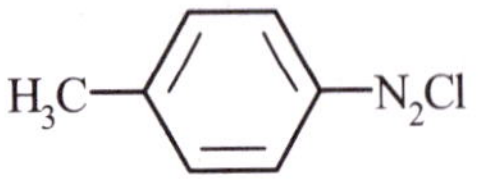

(2) 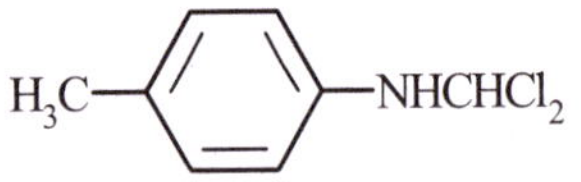

(3) 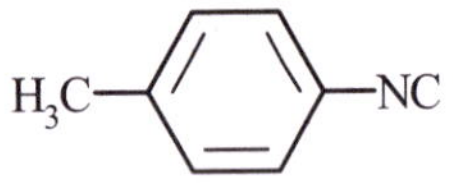

(4) 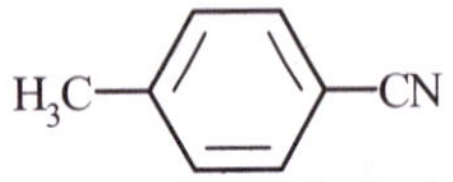

143. Ethyl isocyanide on hydrolysis in acidic medium generates [2003]
(1) propanoic acid and ammonium salt
(2) ethanoic acid and ammonium salt
(3) methylamine salt and ethanoic acid
(4) ethylamine salt and methanoic acid

144. When a primary amine reacts with chloroform in ethanolic KOH then the product is [2002]
(1) an isocyanide (2) an aldehyde
(3) a cyanide (4) an alcohol.

NUMERICAL VALUE QUESTIONS (NVQs)

145. Consider the following sequence of reactions.

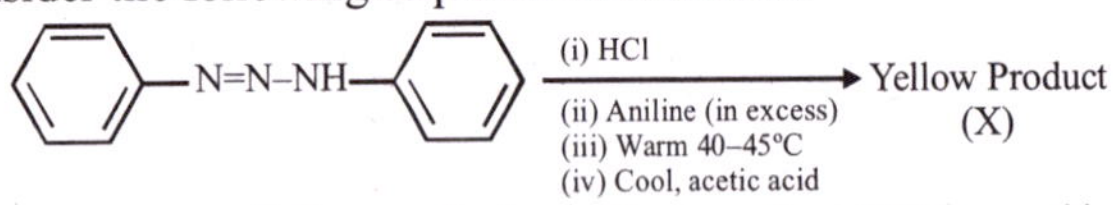

The percentage of nitrogen in the yellow product (X) formed is ______ %.
(Nearest Integer)
(Given Molar mass in g mol^{-1} H : 1, C : 12, N : 14)
[April 06, 2026 (I)]

146. Consider the following reactions sequence [April 02, 2026 (II)]

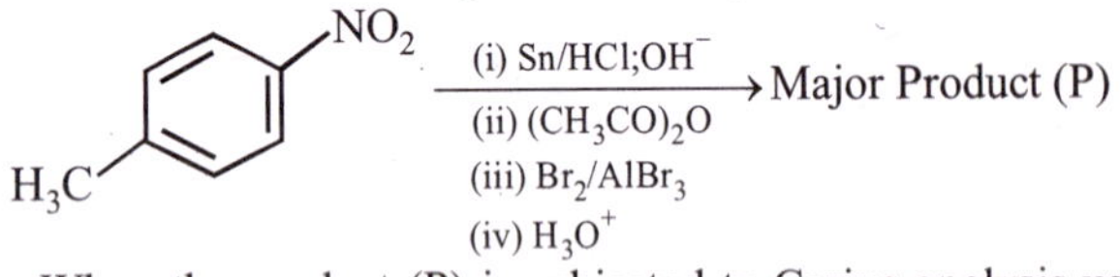

When the product (P) is subjected to Carius analysis using $AgNO_3$, 1.0 g of the product (P) will produce _______ g of the precipitate of AgBr.
(Nearest interger)
(Given : molar mass in g mol^{-1} C : 12, H : 1, O : 16, N : 14, Br : 80, Ag : 108)

147. X g of nitrobenzene on nitration gave 4.2 g of m-dinitrobenzene.
X = ________ g. (nearest integer)
[Given : molar mass (in g mol^{-1}) C : 12, H : 1, O : 16, N : 14]
[April 3, 2025 (II)]

148. Given below are some nitrogen containing compounds. [Jan. 29, 2025 (I)]

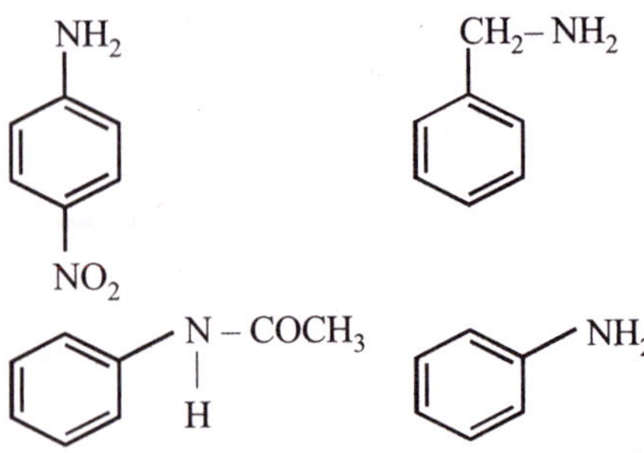

Each of them is treated with HCl separately. 1.0 g of the most basic compound will consume _______ mg of HCl.
(Given molar mass in g mol^{-1} C:12, H : 1, O : 16, Cl : 35.5)

149. Consider the following sequence of reactions : [Jan. 28, 2025 (I)]

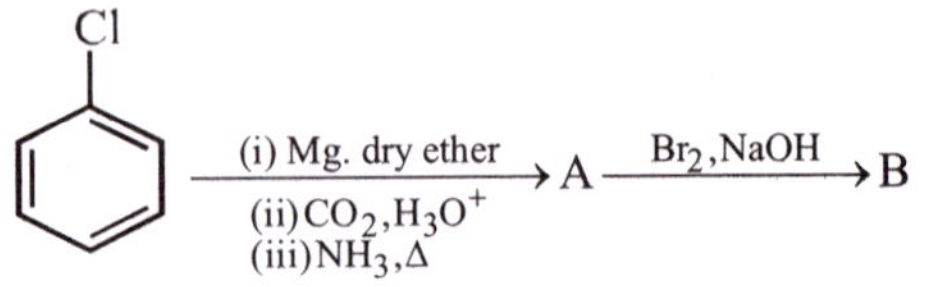

Chlorobenzene
11.25 mg of chlorobenzene will produce _____ $\times 10^{-1}$ mg of product B.
(Consider the reactions result in complete conversion.)
[Given molar mass of C, H, O, N and Cl as 12, 1, 16, 14 and 35.5 g mol^{-1} respectively]

150. Number of compounds from the following which **cannot** undergo Friedel-Crafts reactions is :____ toluene, nitrobenzene, xylene, cumene, aniline, chlorobenzene, m-nitroaniline, m-dinitrobenzene [April 9, 2024 (II)]

151. Number of amine compounds from the following giving solids which are soluble in NaOH upon reaction with Hinsberg's reagent is _____. [April 8, 2024 (I)]

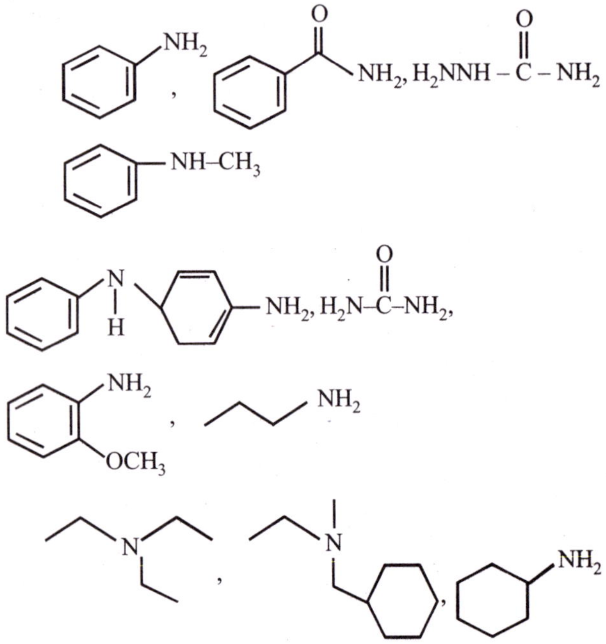

152. An amine (X) is prepared by ammonolysis of benzyl chloride. On adding p-toluenesulphonyl chloride to it the solution remains clear. Molar mass of the amine (X) formed is_____________ g mol^{-1}. [April 6, 2024 (II)]
(Given molar mass in gmol^{-1} C : 12, H : 1, O : 16, N : 14)

153. 9.3 g of pure aniline is treated with bromine water at room temperature to give a white precipitate of the product 'P'. The mass of product 'P' obtained is 26.4 g. The percentage yield is ......%. [April 5, 2024 (I)]

154. Phthalimide is made to undergo following sequence of reactions.

Phthalimide $\xrightarrow[\text{(ii) Benzyl chloride}]{\text{(i) KOH}}$ 'P' [April 4, 2024 (II)]

Total number of π bonds present in product 'P' is/are

155. The number of correct reaction(s) among the following is ______. [April 4, 2024 (I)]

(A) 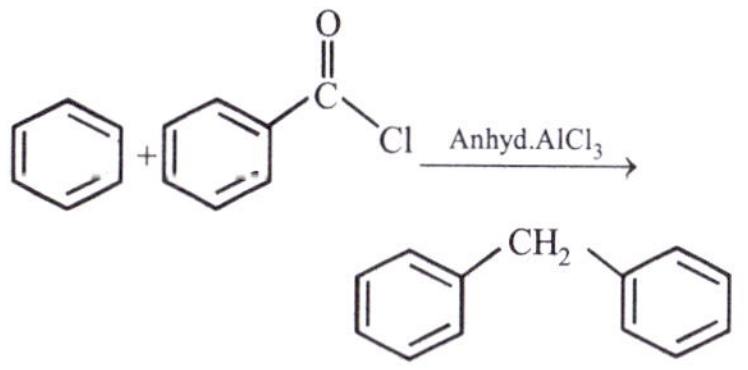

(B) 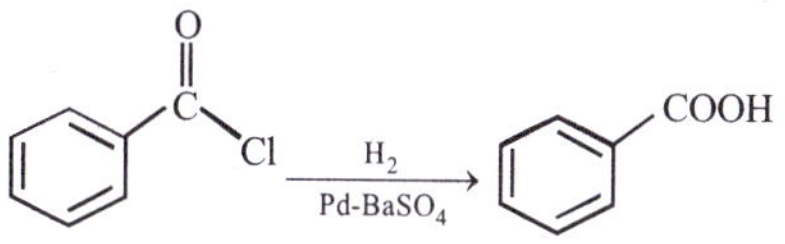

(C) 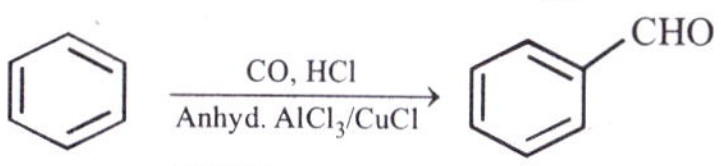

(D) 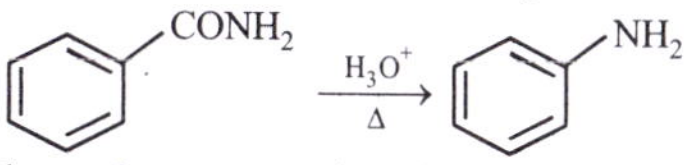

156. Number of compounds which give reaction with Hinsberg's reagent is ________ . [Feb. 1, 2024 (II)]

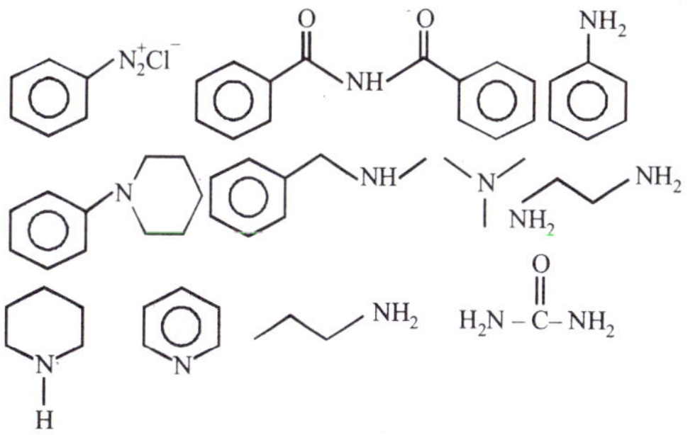

157. A compound (X) with molar mass 108 g mol^{-1} undergoes acetylation to give product with molar mass 192 g mol^{-1}. The number of amino groups in the compound (X) is ______. [Jan. 31, 2024 (II)]

158. Number of isomeric aromatic amines with molecular formula $C_8H_{11}N$, which can be synthesized by Gabriel phthalimide synthesis is _______ [April 06, 2023 (II)]

159. How many of the transformation given below would result in aromatic amines? [Jan. 31, 2023 (I)]

(A) 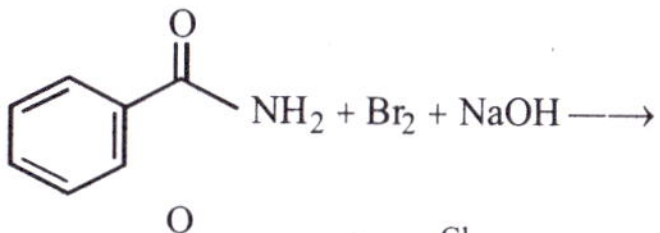

(B) 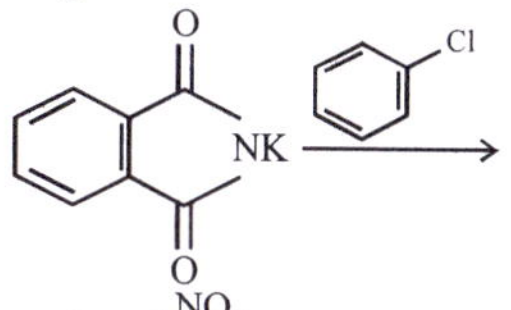

(C) 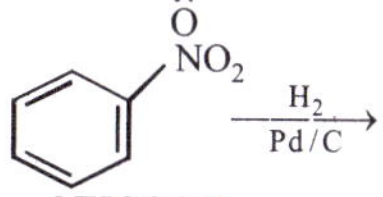

(D) 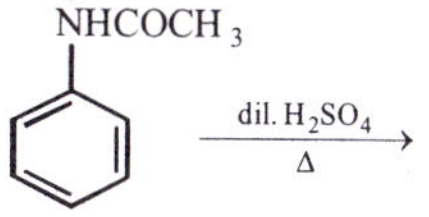

160. The total number of reagents from those given below, that can convert nitrobenzene into aniline is ______. (Integer answer) [Aug. 31, 2021 (I)]

I. Sn – HCl	II. Sn – NH_4OH
III. Fe – HCl	IV. Zn – HCl
V. H_2 – Pd	VI. H_2 – Raney nickel

161. In gaseous triethyl amine, the "–C–N–C–" bond angle is ________ degree. [July 27, 2021 (I)]

162. 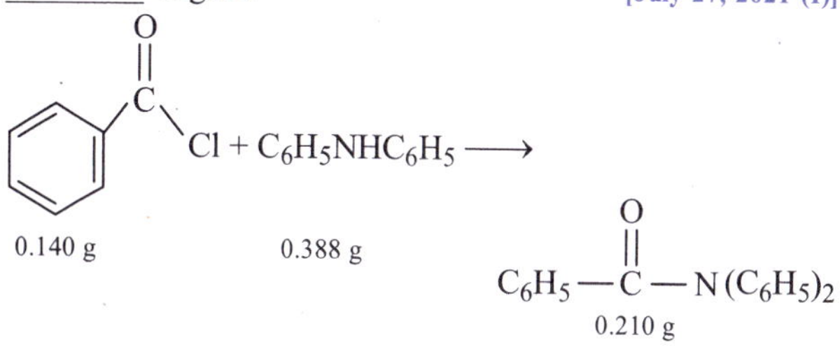

Consider the above reaction. The percentage yield of amide product is ________. (Round off to the nearest integer). (Given : Atomic mass : C : 12.0 u, H : 1.0 u, N : 14. 0 u, O : 16.0 u, Cl : 35.5 u) [March 17, 2021 (II)]

163. The total number of amines among the following which can be synthesized by Gabriel synthesis is ________. [Feb. 24, 2021 (II)]

(A) $(CH_3)_2CH-CH_2-NH_2$

(B) $CH_3CH_2NH_2$

(C) 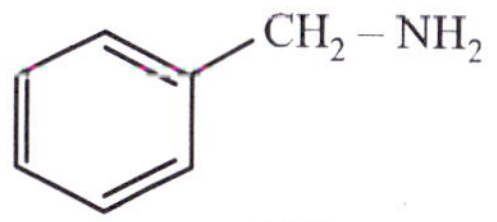

(D) 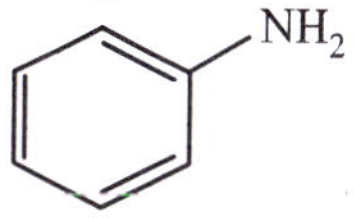

Topic 2 - Preparation, Chemical Reactions and Importance of Diazonium Salts

MULTIPLE CHOICE QUESTIONS (MCQs)

164. Consider the following organic reaction sequence. Choose the final product (X) from the following (consider the major product in all intermediate reactions) [April 06, 2026 (II)]

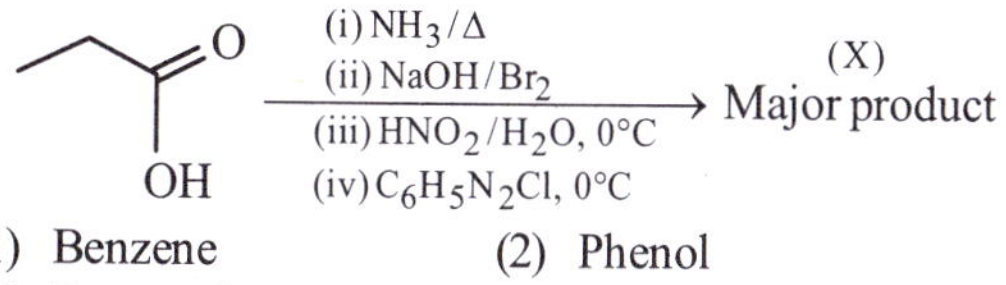

(1) Benzene (2) Phenol

(3) Propanol (4) Chlorobenzene

165. Product C of the following reaction sequence will be [April 04, 2026 (I)]

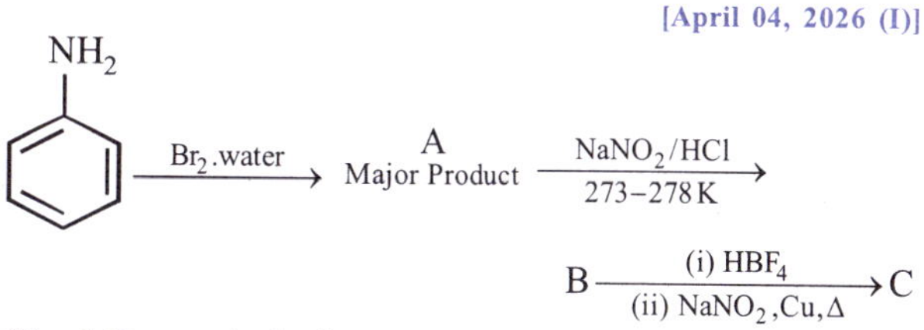

(1) 1-Bromo-4-nitrobenzene

(2) 1,3,5-Tribromo-2-nitrobenzene

(3) 4-Bromo-1-nitrobenzene

(4) 1,3,5-Tribromobenzene

166. Identify compounds A and E in the following reaction sequence. **[April 02, 2026 (II)]**

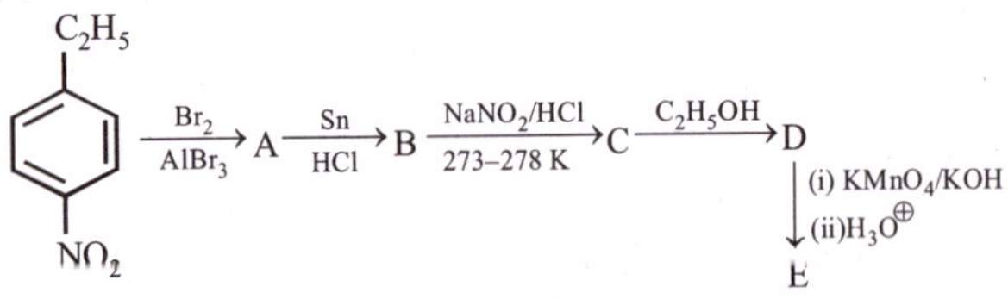

(1) A = 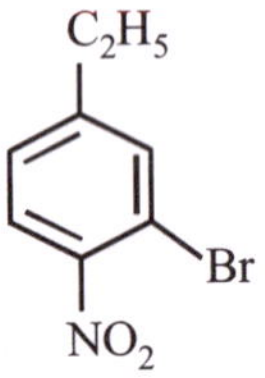 E = 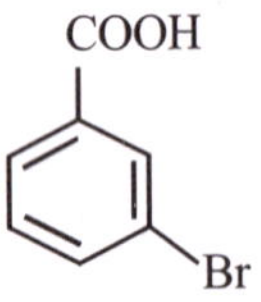

(2) A = 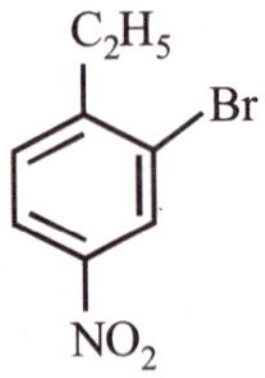 E = 

(3) A = 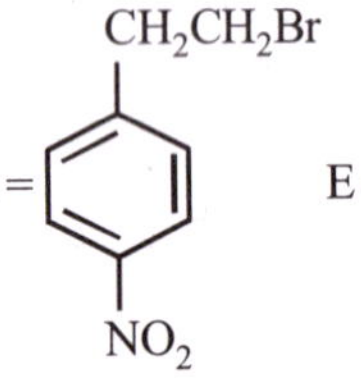 E = 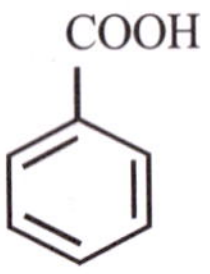

(4) A = 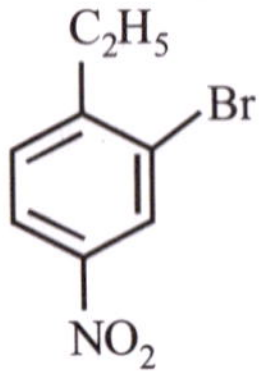 E = 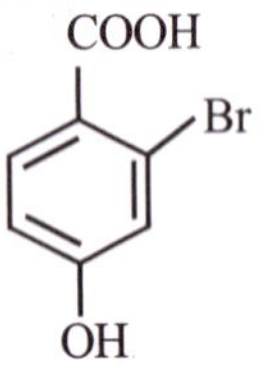

167. Given below are two statements : **[Jan. 24, 2026 (II)]**

Statement I: The dipole moment of R-CN is greater than R-NC and R-NC can undergo hydrolysis under acidic medium to produce $R-\overset{\overset{O}{\|}}{C}-OH$.

Statement II: R–CN hydrolyses under acidic medium to produce a compound which on treatment with $SOCl_2$, followed by the addition of NH_3 gives another compound(x). This compound (x) on treatment with NaOCl/NaOH gives a product, that on treatment with $CHCl_3$/KOH/Δ produces R-NC

In the light of the above statements, choose the correct answer from the options below:

(1) Both Statement I and Statement II are false.
(2) Both Statement I and Statement II are true.
(3) Statement I is true but Statement II is false.
(4) Statement I is false but Statement II is true.

168. Given below are two statements: **[Jan. 23, 2026 (II)]**

Statement-I: 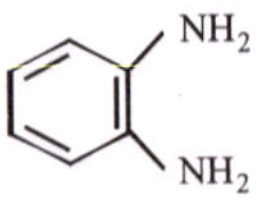 can be synthesized from

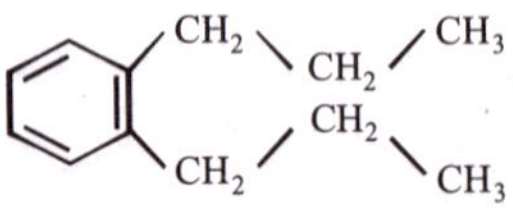 using simpler reagents in the order (i) Acidic $KMnO_4$, (ii) Ammonia, (iii) Bromine and alkali.

Statement-II:  can be converted into

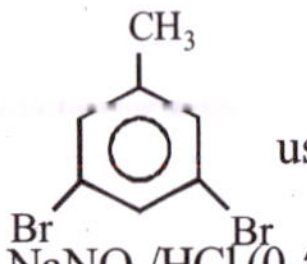 using reagents in the order (i) Bromine-H_2O (ii) $NaNO_2$/HCl (0-5 °C) (iii) aq. H_3PO_2.

In the light of the above statements, choose the **correct** answer from the options given below:

(1) Both Statement-I and Statement-II are false.
(2) Statement-I is true but Statement-II is false.
(3) Both Statement-I and Statement-II are true.
(4) Statement-I is false but Statement-II is true.

169. Compound 'P' undergoes the following sequence of reactions: **[Jan. 23, 2026 (I)]**

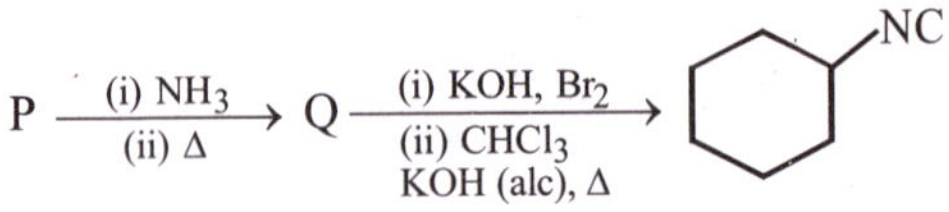

'P' is :

(1) 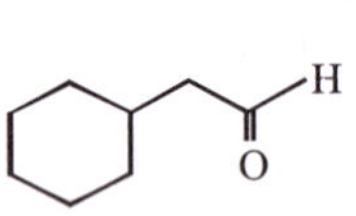 (2) 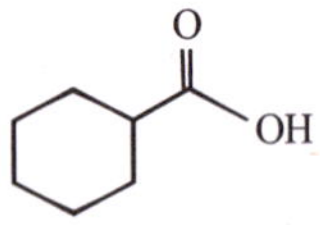

(3) 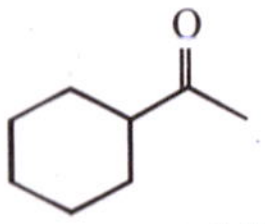 (4) 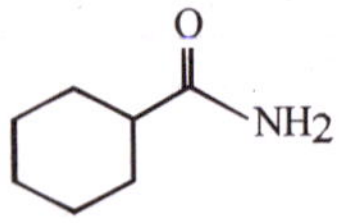

170. The correct stability order of the following diazonium salts is **[Jan. 24, 2026 (I)]**

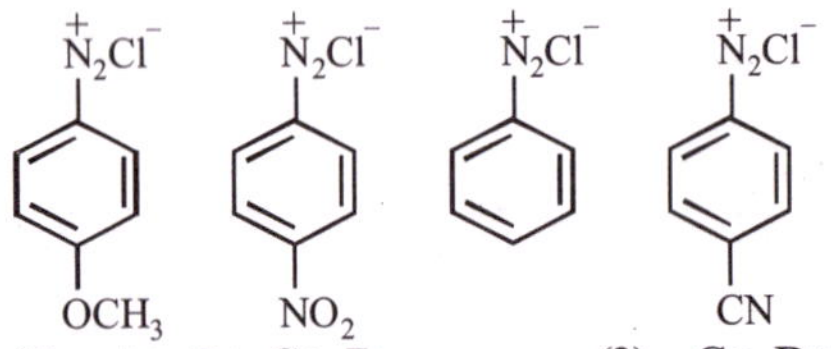

(1) A > B > C > D (2) C > D > B > A
(3) A > C > D > B (4) C > A > D > B

171. Consider the following sequence of reactions.

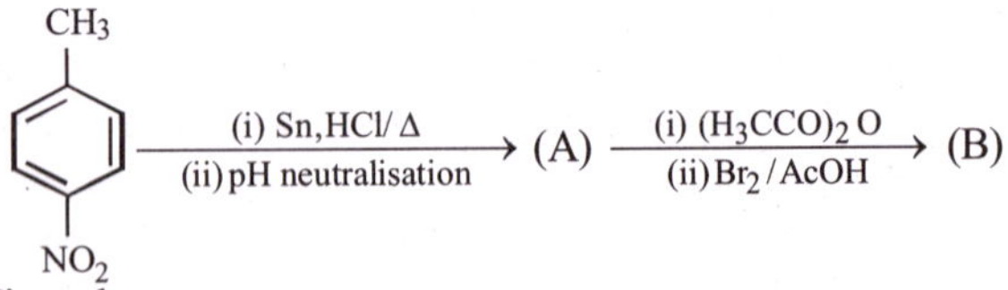

4–Nitrotoluene

Assuming that the reaction proceeds to completion, then 137 mg of 4-nitrotoluene will produce ________ mg of B.

(Given molar mass in g mol^{-1} H : 1, C : 12, N : 14, O : 16, Br : 80) **[Jan. 23, 2026 (I)]**

(1) 301 (2) 146 (3) 228 (4) 208

172. 'A' is a neutral organic compound (M. F : C_8H_9ON). On treatment with aqueous $Br_2/HO^{(-)}$, 'A' forms a compound 'B' which is soluble in dilute acid. 'B' on treatment with aqueous $NaNO_2$/HCl(0–5°C) produces a compound 'C' which on treatment with CuCN/NaCN produces 'D' Hydrolysis of 'D' produces 'E' which is also obtainable from the hydrolysis of 'A'. 'E' on treatment with acidified $KMnO_4$ produces 'F'. 'F' contains two different types of hydrogen atoms. The structure of 'A' is **[Jan. 22, 2026 (I)]**

(1) 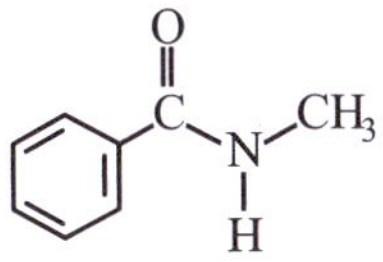 (2) 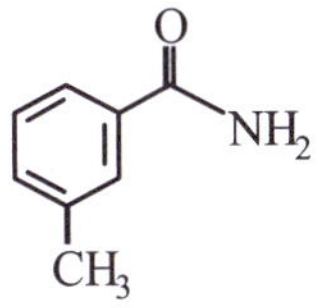

(3) 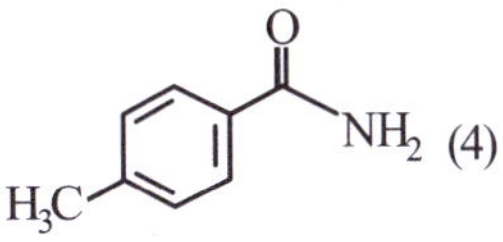 (4) 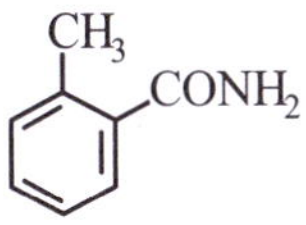

173. 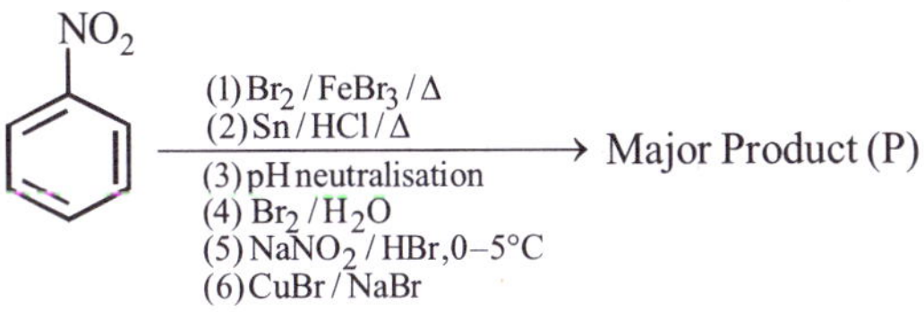

Consider the above sequence of reactions. The number of bromine atom(s) in the final product (P) will be: [Jan. 21, 2026 (II)]

(1) 1 (2) 6 (3) 5 (4) 3

174. A hydrocarbon 'P' (C_4H_8) on reaction with HCl gives an optically active compound 'Q' (C_4H_9Cl) which on reaction with one mole of ammonia gives compound 'R' ($C_4H_{11}N$). 'R' on diazotization followed by hydrolysis gives 'S'. Identify P, Q, R and S. [Jan. 21, 2026 (I)]

(1) $P=CH_3-CH_2-CH=CH_2$, $Q=CH_3-CH_2-CH_2-CH_2(Cl)$

$R=CH_3-CH_2-CH_2-NH(CH_3)$, $S=CH_3-CH_2-CH(OH)(CH_3)$

(2) 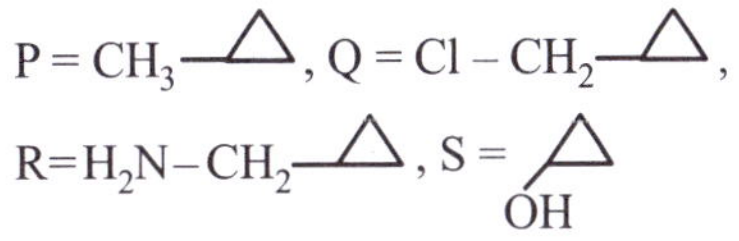

(3) $P=CH_3-CH=CH-CH_3$, $Q=CH_3-CH_2-CH(Cl)-CH_3$,

$R=CH_3-CH_2-CH(NH_2)-CH_3$, $S=CH_3-CH_2-CH(OH)-CH_3$

(4) $P=CH_3-CH=CH-CH_3$, $Q=CH_3-CH_2-CH_2-CH_2-Cl$

$R=CH_3-CH_2-CH_2-CH_2(NH_2)$, $S=CH_3-CH_2-CH_2(NH_2-OH)$

175. The major product (*A*) formed in the following reaction sequence is [April 4, 2025 (I)]

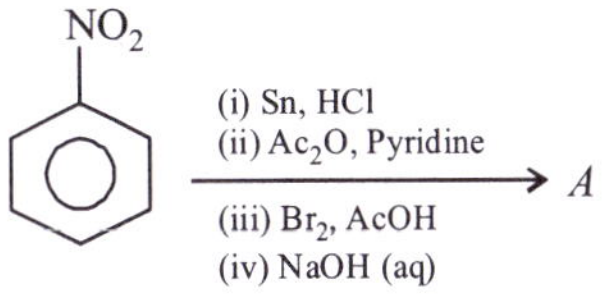

(1) 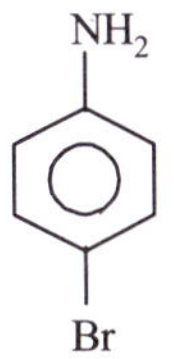 (2) 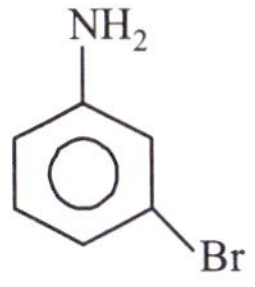

(3) 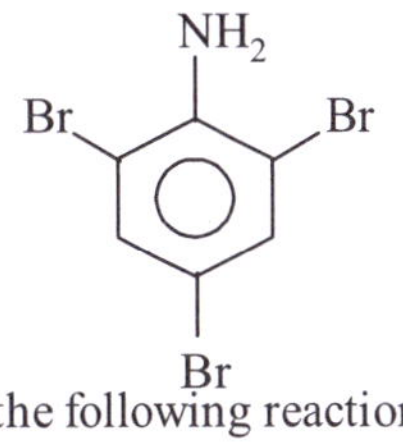 (4) 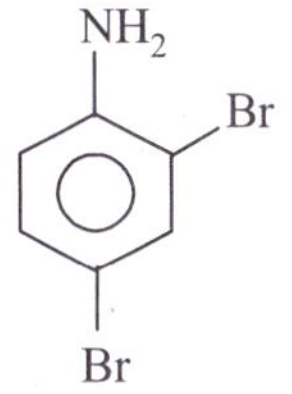

176. In the following reactions, which one is NOT correct? [April 3, 2025 (I)]

(1) 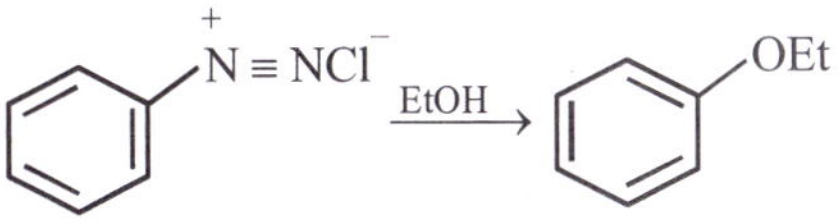

(2) 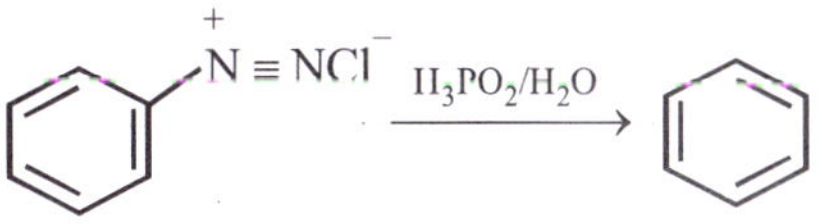

(3) 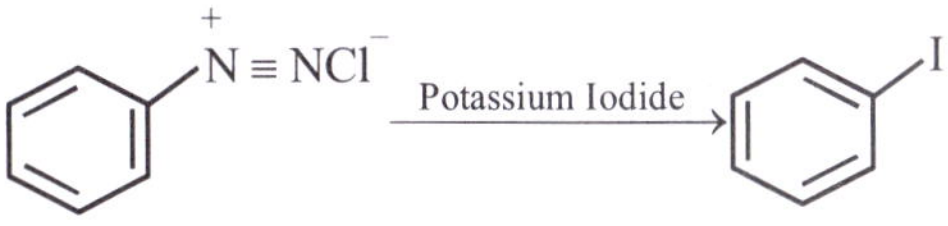

(4) 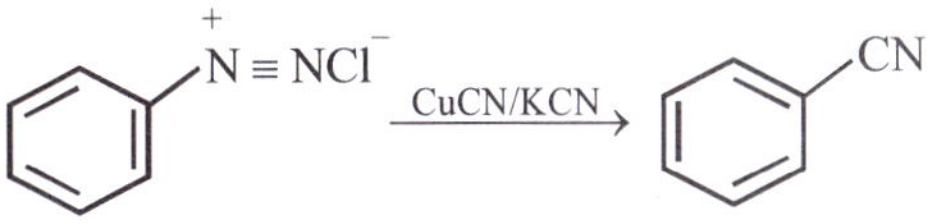

177. Identify [A], [B], and [C], respectively in the following reaction sequence : [April 3, 2025 (I)]

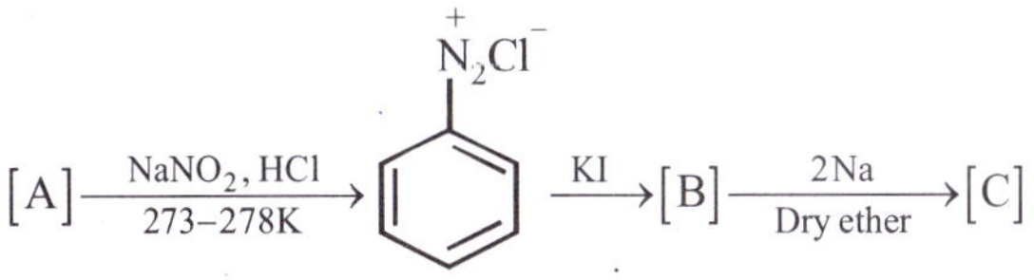

(1) 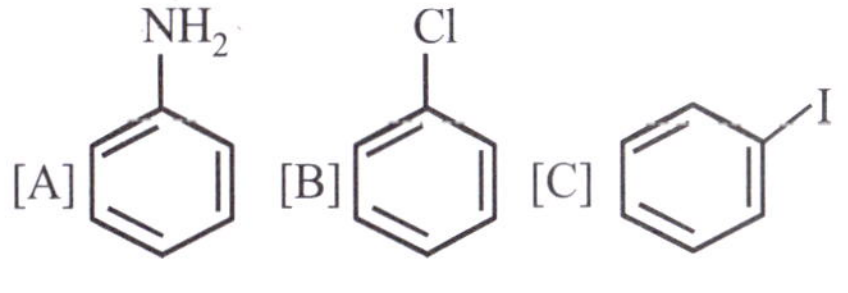

(2) 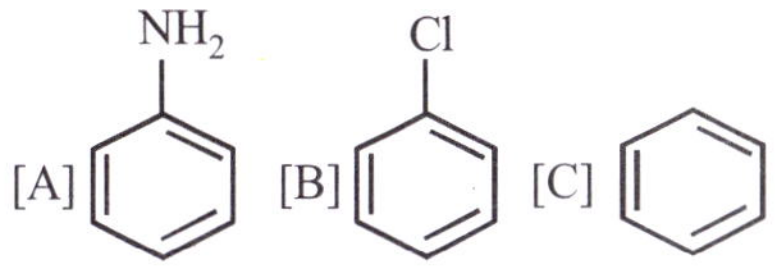

(3) 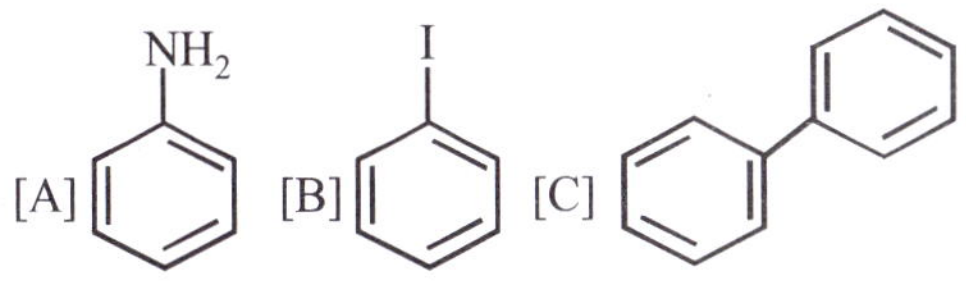

(4) 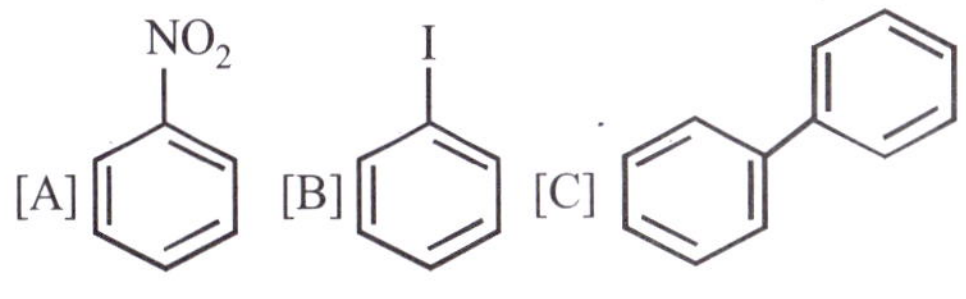

178. When a concentrated solution of sulphanilic acid and 1-naphthylamine is treated with nitrous acid (273 K) and acidified with acetic acid, the mass (g) of 0.1 mole of product formed is : [April 2, 2025 (II)]

(Given molar mass in g mol^{-1} H : 1, C : 12, N : 14, O : 16, S : 32)

(1) 343 (2) 330
(3) 33 (4) 66

179. Which one of the following reaction sequences will give an azo dye ? [Jan. 29, 2025 (II)]

(1) 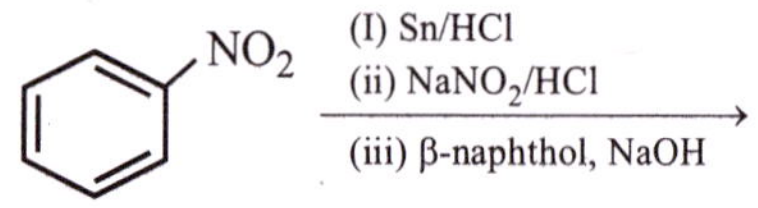

(2) 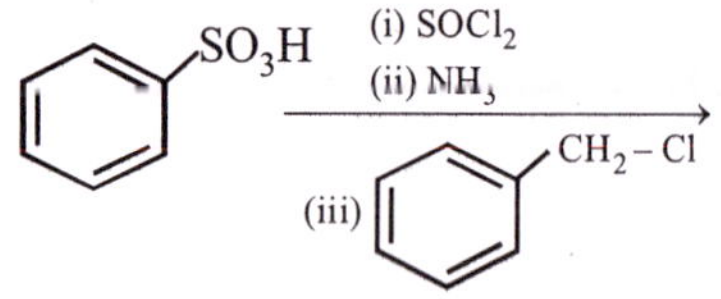

(3) 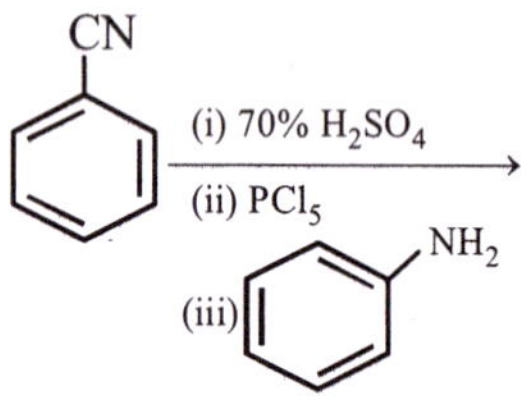

(4) 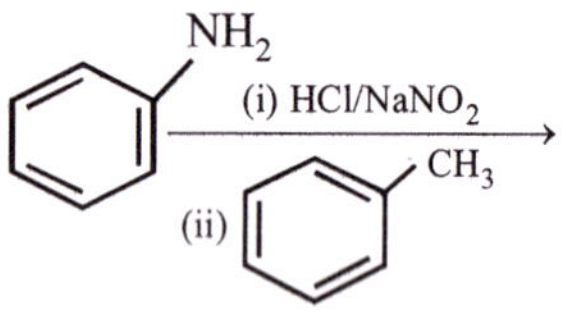

180. The products formed in the following reaction sequence are : [Jan. 22, 2025 (I)]

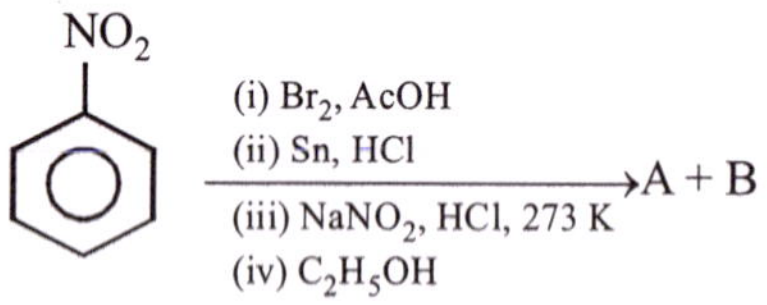

(1) 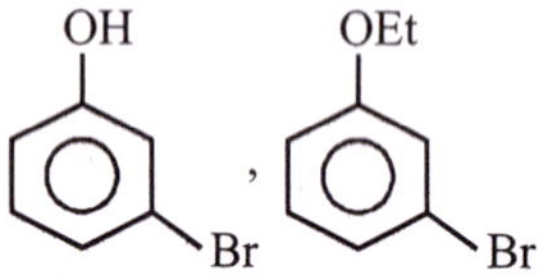

(2) 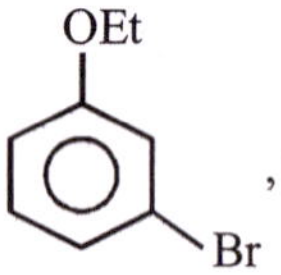

, CH_3–COOH

(3) 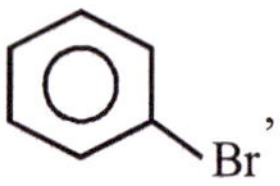

, CH_3CHO

(4) 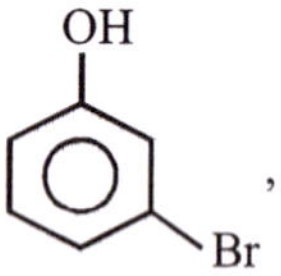

, CH_3–CHO

181. Identify the product Ⓐ in the following reaction. [April 6, 2024 (II)]

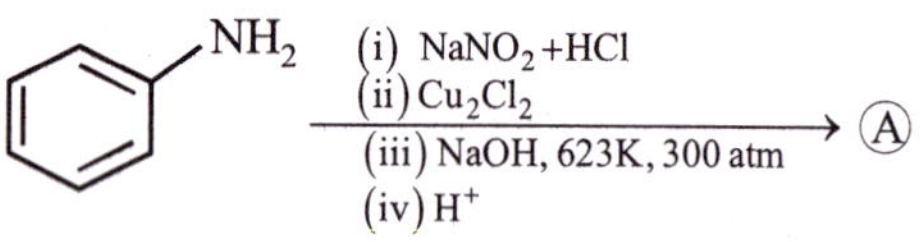

(1) 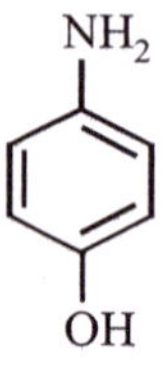

(2) 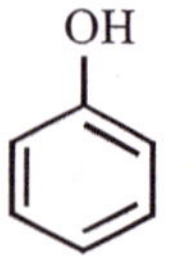

(3) 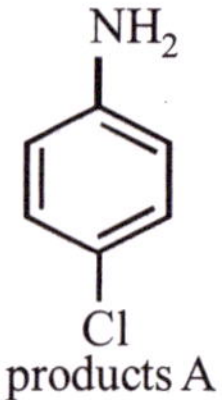

(4) 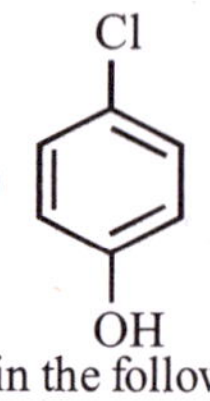

182. The products A and B formed in the following reaction scheme are respectively [Jan. 30, 2024 (II)]

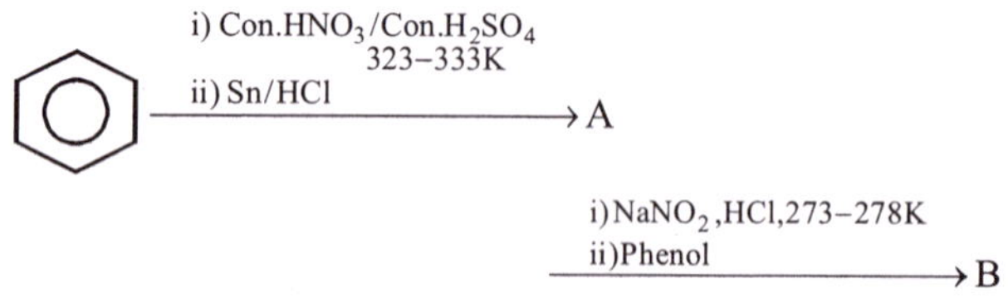

(1) 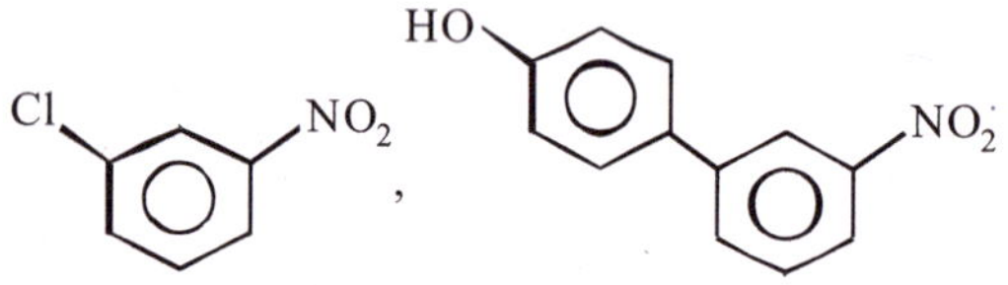

(2) 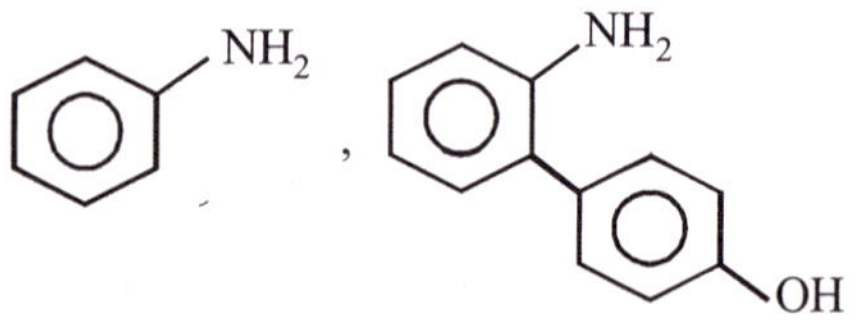

(3) 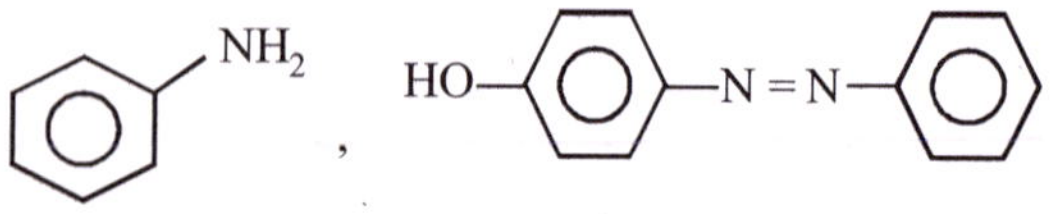

(4) 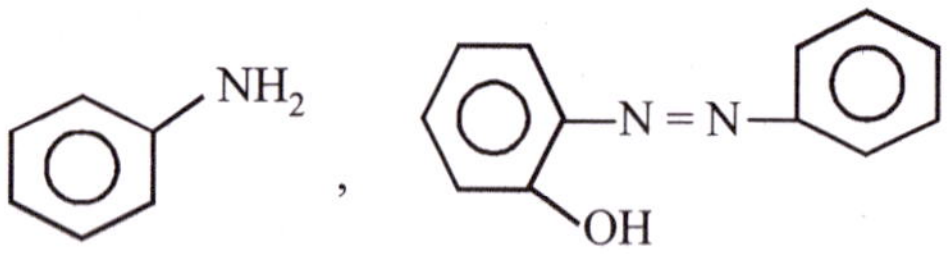

183. The product A formed in the following reaction is [Jan. 29, 2024 (II)]

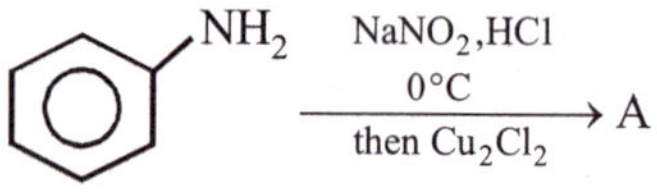

(1) 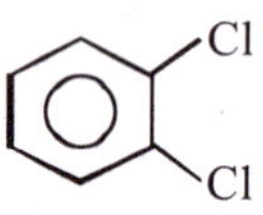

(2) 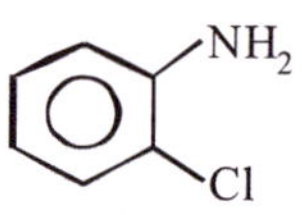

(3) 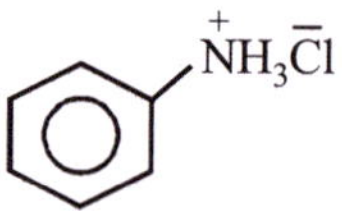

(4) 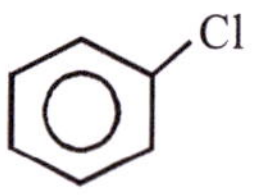

184. Consider the following sequence of reactions:

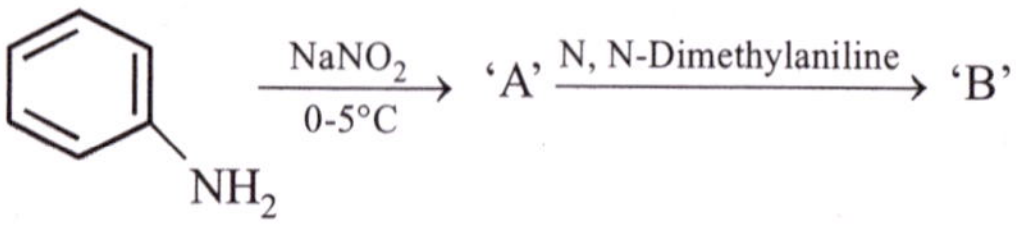

The product 'B' is [April 15, 2023 (I)]

(1) 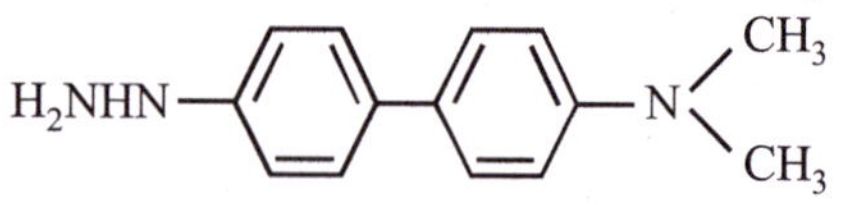

(2) 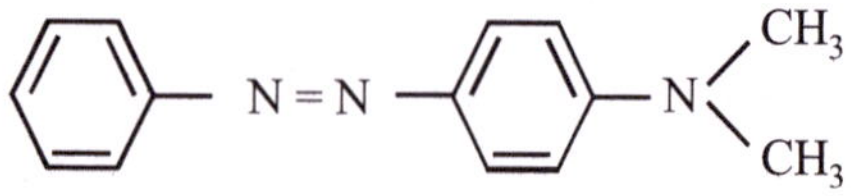

(3) 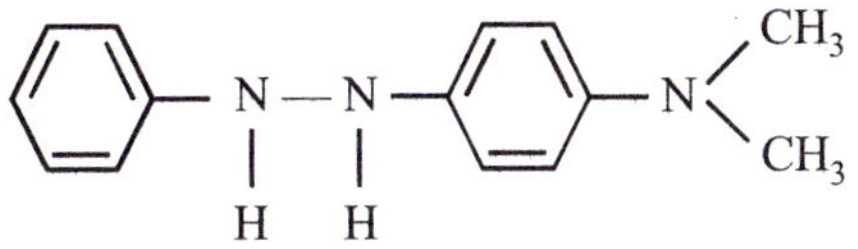

(4) 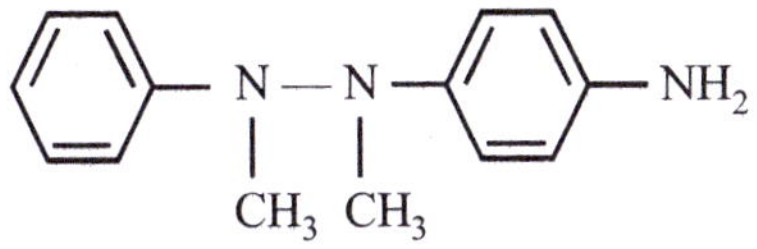

185. The incorrect statement regarding the reaction given below is **[April 12, 2023 (I)]**

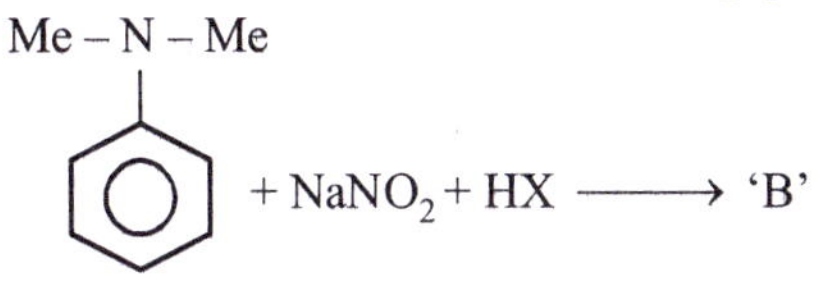

'A'

(1) The electrophile involved in the reaction is NO^+
(2) 'B' is N-nitroso ammonium compound
(3) The reaction occurs at low temperature
(4) The product 'B' formed in the above reaction is p-nitroso compound at low temperature

186. 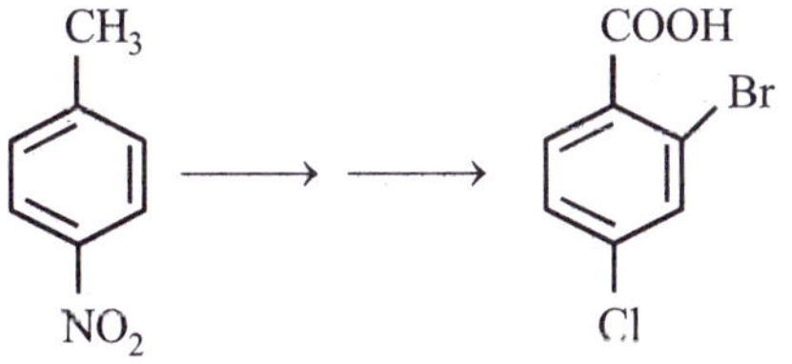

In the above conversion the correct sequence of reagents to be added is **[April 15, 2023 (I)]**

(1) (i) Fe/H^+, (ii) HONO, (iii) CuCl, (iv) $KMnO_4$, (v) Br_2
(2) (i) $KMnO_4$, (ii) Br_2/Fe, (iii) Fe/H^+, (iv) Cl_2
(3) (i) Br_2/Fe, (ii) Fe/H^+, (iii) HONO, (iv) CuCl, (v) $KMnO_4$
(4) (i) Br_2/Fe, (ii) Fe/H^+, (iii) $KMnO_4$, (iv) Cl_2

187. Compound A from the following reaction sequence is:

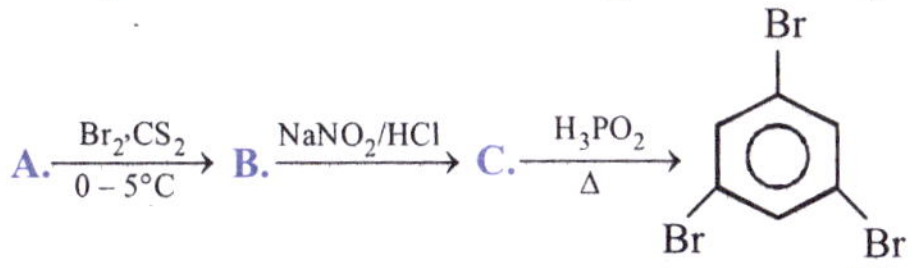

[April 13, 2023 (II)]

(1) Benzoic Acid (2) Phenol
(3) Salicylic Acid (4) Aniline

188. Match List I with List II: **[April 08, 2023 (I)]**

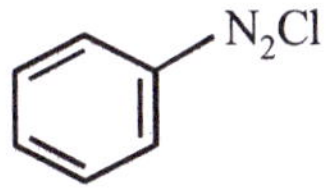 is reacted with reagents in List I to form products in List II.

List-I (Reagent)	**List-II (Product)**
A. 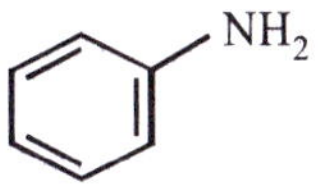	I. 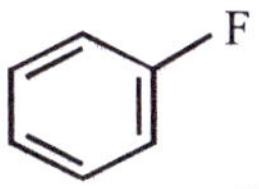
B. HBF_4, Δ	II. 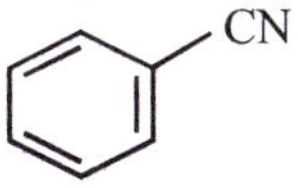
C. Cu, HCl	III. 
D. CuCN/KCN	IV. 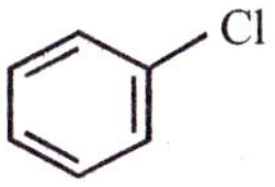

Choose the correct answer from the options given below:
(1) A-IV, B-III, C-II, D-I (2) A-I, B-III, C-IV, D-II
(3) A-III, B-I, C-II, D-IV (4) A-III, B-I, C-IV, D-II

189. The product (P) formed from the following multistep reaction is : **[April 08, 2023 (II)]**

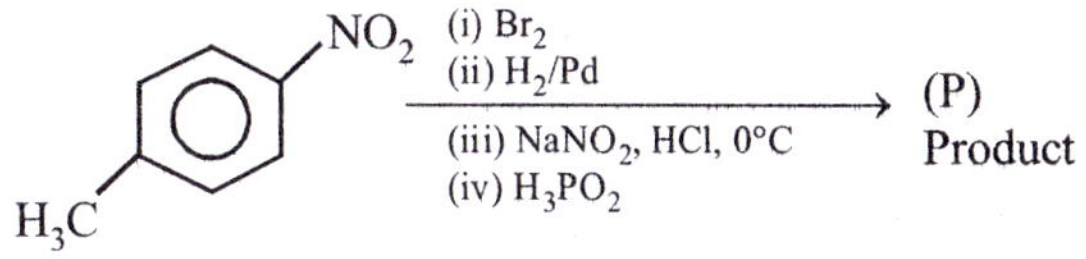

(1) 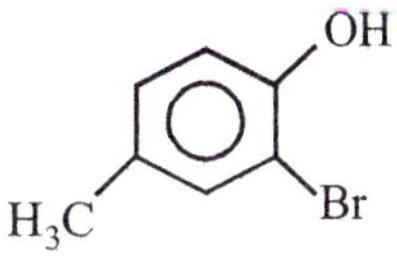 (2) 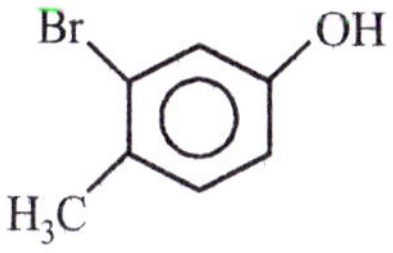

(3) 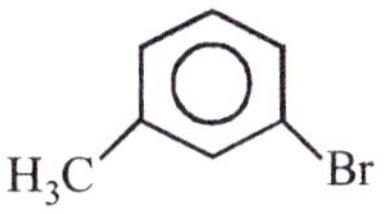 (4) 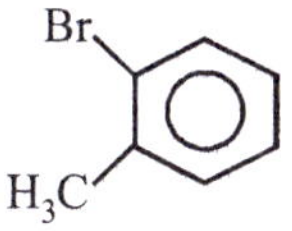

190. 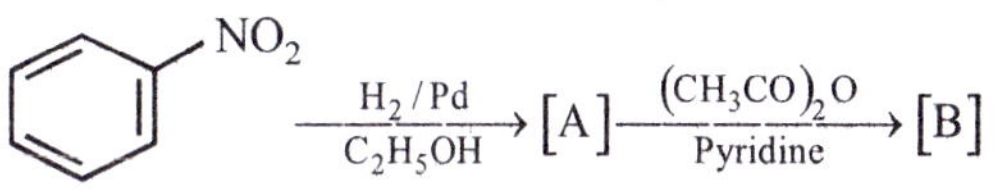

[Jan. 31, 2023 (I)]

(1) 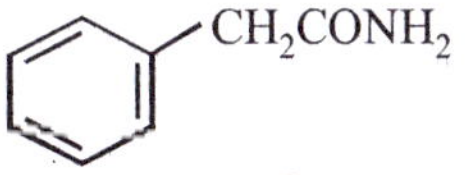

(2) 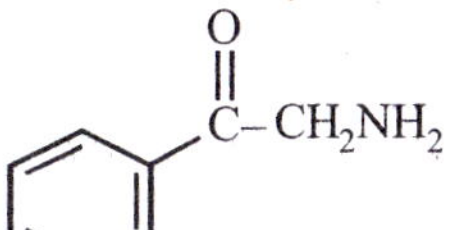

(3) 

(4) 

191 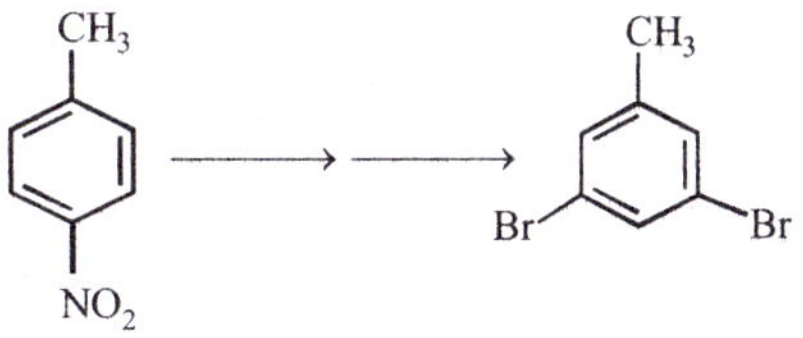

(X) (Y)

In the above conversion of compound (X) to product (Y), the sequence of reagents to be used will be: **[Jan. 30, 2023 (II)]**

(1) (i) Br_2, Fe (ii) Fe, H^+ (iii) $LiAlH_4$
(2) (i) Br_2(aq) (ii) $LiAlH_4$ (iii) H_3O^+
(3) (i) Fe, H^+ (ii) Br_2(aq) (iii) HNO_2 (iv) CuBr
(4) (i) Fe, H^+ (ii) Br_2(aq) (iii) HNO_2 (iv) H_3PO_2.

192. Identify the products formed (A and E) **[Jan. 25, 2023 (I)]**

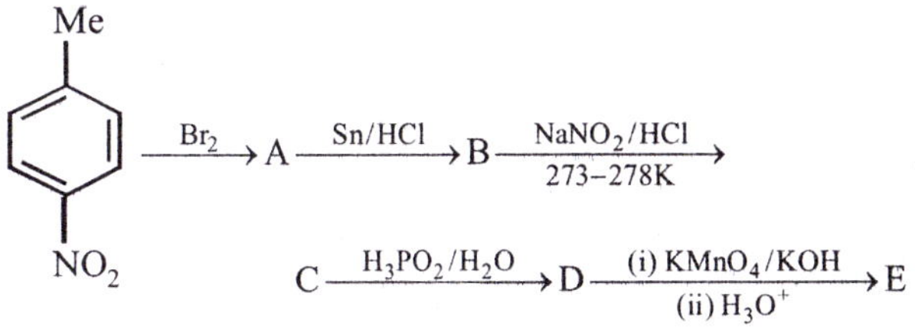

(1) A = 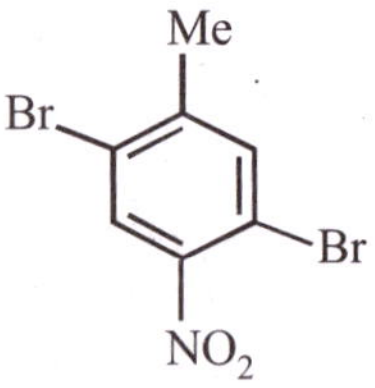 , E = 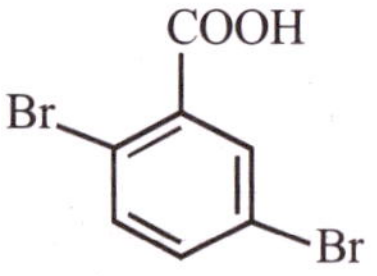

(2) A = 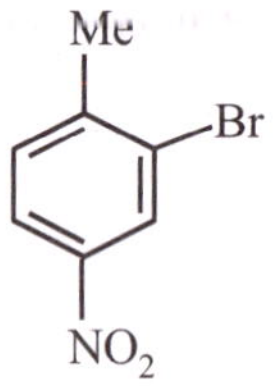 , E = 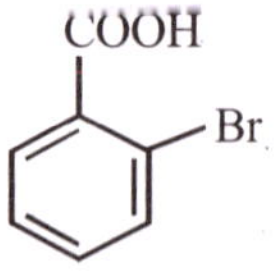

(3) A = 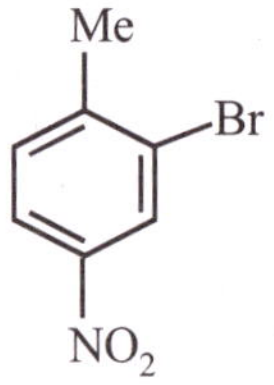 , E = 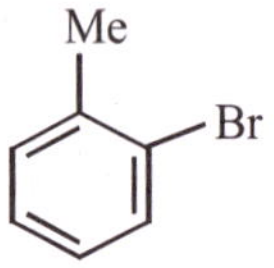

(4) A = 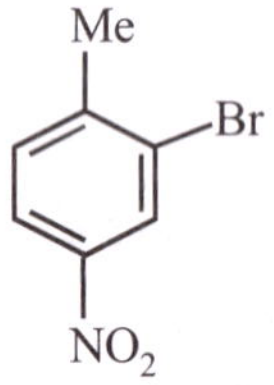 , E = 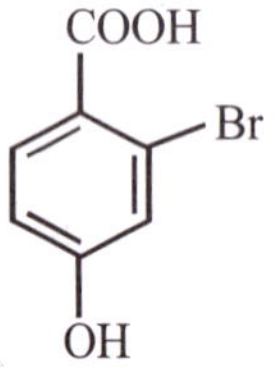

193. Choose the correct colour of the product for the following reaction. **[Jan. 24, 2023 (II)]**

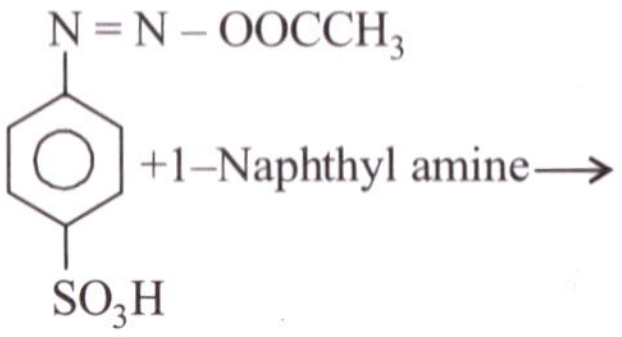

(1) Yellow (2) White (3) Red (4) Blue

194. 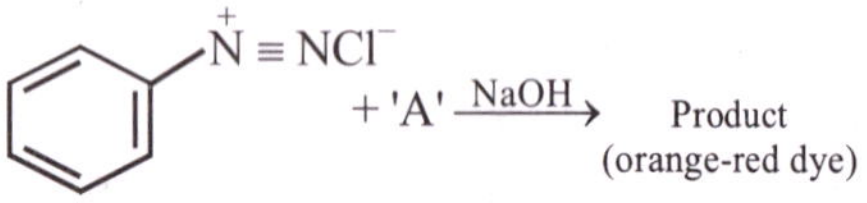

Which among the following represents reagent 'A'? **[July 29, 2022 (I)]**

(1) 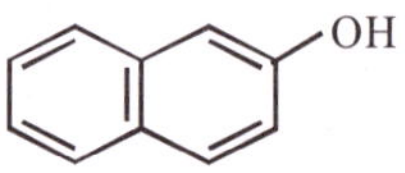 (2) 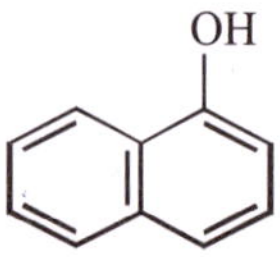

(3) 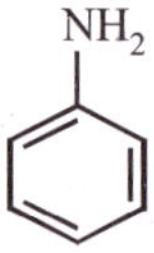 (4) 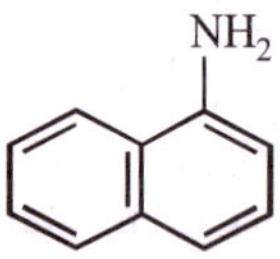

195. An organic compound 'A' contains nitrogen and chlorine. It dissolves readily in water to give a solution that turns blue litmus red. Titration of compound 'A' with standard base indicates that the molecular weight of 'A' is 131 ± 2. When a sample of 'A' is treated with aq. NaOH, a liquid separates which contains N but not Cl. Treatment of the obtained liquid with nitrous acid followed by phenol gives orange precipitate. The compound 'A' is : **[July 27, 2022 (II)]**

(1) 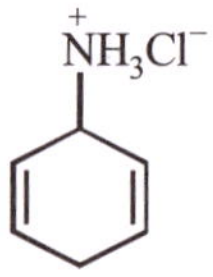 (2) 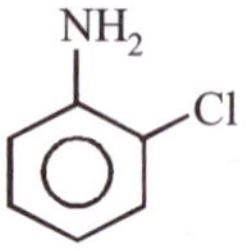

(3) 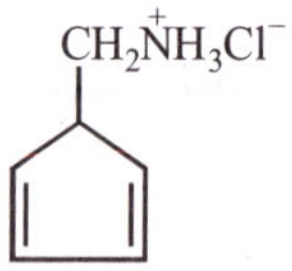 (4) 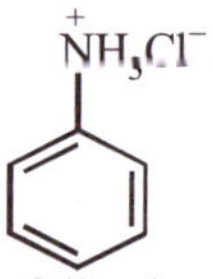

196. The correct stability order of the following diazonium salt is **[July 26, 2022 (I)]**

(A) 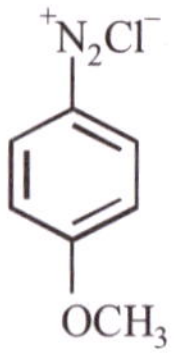 (B) 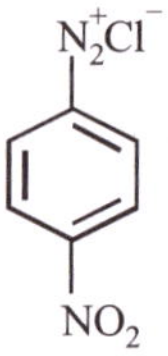

(C) 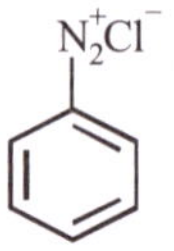 (D) 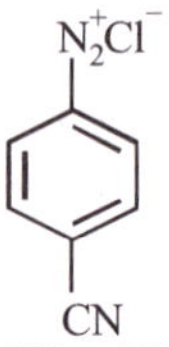

(1) (A) > (B) > (C) > (D) (2) (A) > (C) > (D) > (B)
(3) (C) > (A) > (D) > (B) (4) (C) > (D) > (B) > (A)

197. Hydrolysis of which compound will give carbolic acid ? **[July 26, 2022 (II)]**
(1) Cumene
(2) Benzene diazonium chloride
(3) Benzal chloride
(4) Ethylene glycol ketal

198. The correct sequential order of the reagents for the given reaction is : **[July 26, 2022 (II)]**

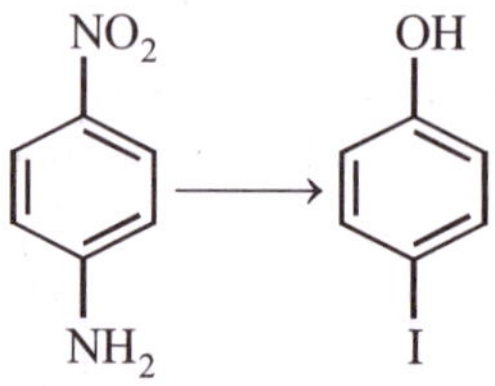

(1) HNO_2, Fe/H^+, HNO_2, KI, H_2O/H^+
(2) HNO_2, KI, Fe/H^+, HNO_2, H_2O/warm
(3) HNO_2, KI, HNO_2, Fe/H^+, H_2O/H^+
(4) HNO_2, Fe/H^+, KI, HNO_2, H_2O/warm

199. A primary aliphatic amine on reaction with nitrous acid in cold (273 K) and there after raising temperature of reaction mixture to room temperature (298 K). Gives a/an **[June 28, 2022 (I)]**
(1) nitrile (2) alcohol
(3) diazonium salt (4) secondary amine

200. The final product 'C' in the following series of reactions is **[June 26, 2022 (II)]**

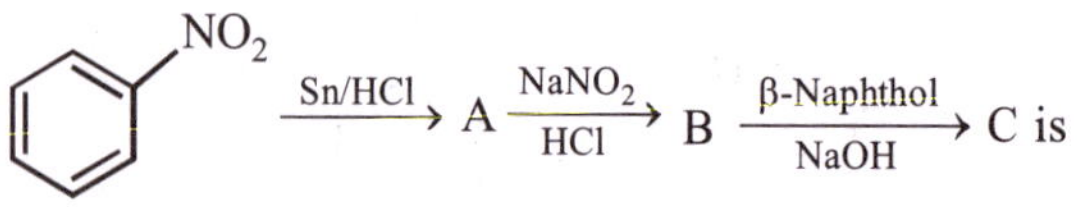

(1) 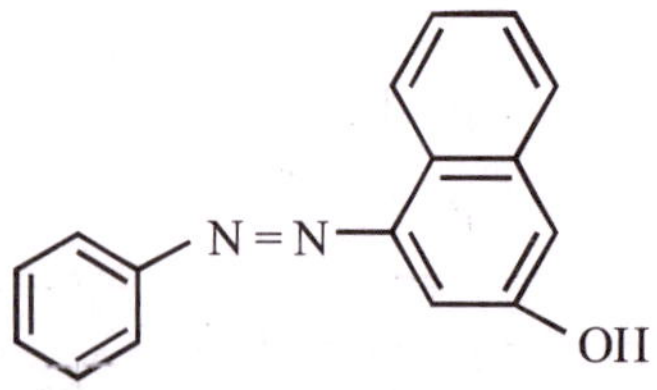

(2) 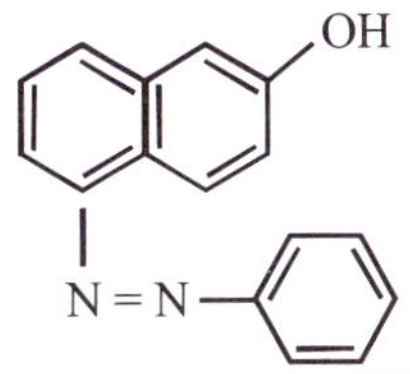

(3) 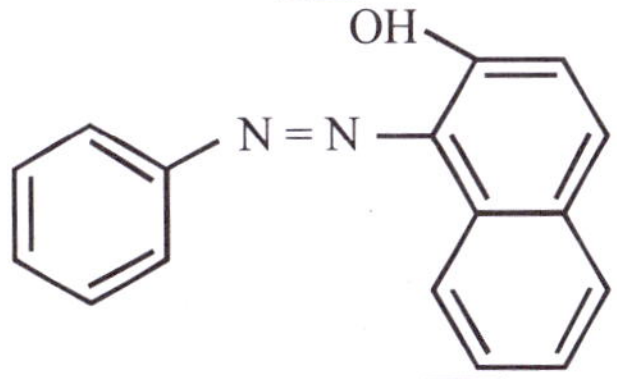

(4) 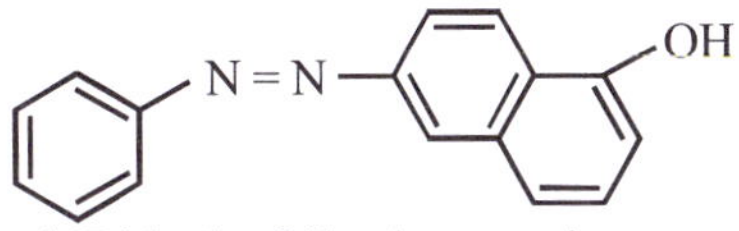

201. 'A' and 'B' in the following reactions are : [Feb. 24, 2021 (I)]

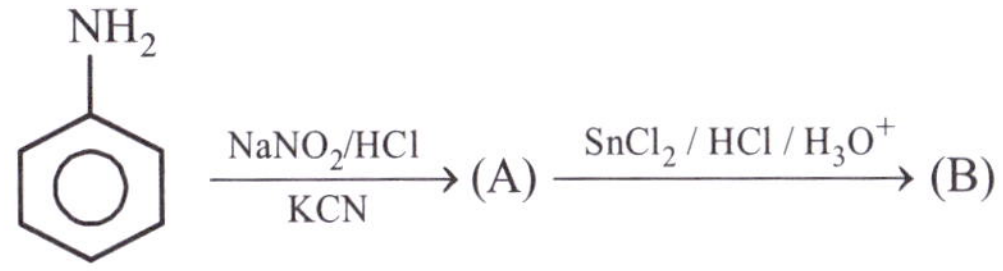

(1) (A) : 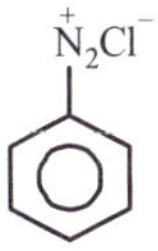 (B) : 

(2) (A) : 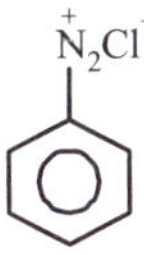 (B) : 

(3) (A) : 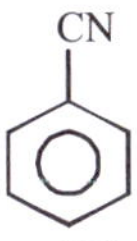 (B) : 

(4) (A) : 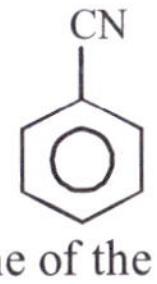 (B) : 

202. Which one of the following gives the most stable diazonium salt ? [Sep. 1, 2021 (II)]

(1) $CH_3–CH_2–CH_2–NH_2$ (2) 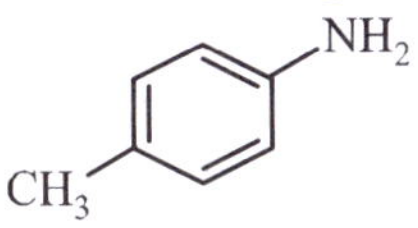

(3) $CH_3–\underset{H}{\overset{CH_3}{C}}–NH_2$ (4) 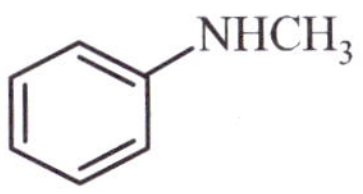

203. What is the correct sequence of reagents used for converting nitrobenzene into *m*-dibromobenzene?

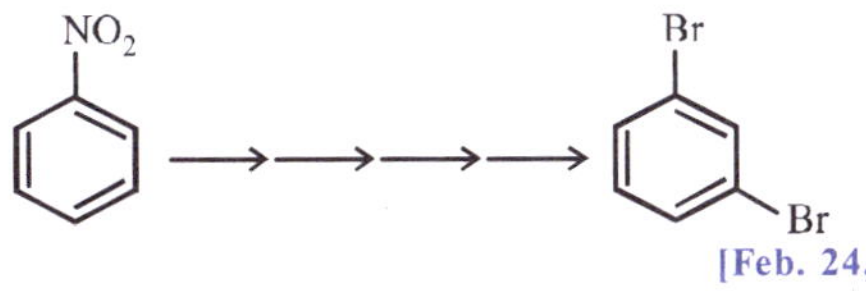

[Feb. 24, 2021 (II)]

(1) $\xrightarrow{NaNO_2}$ / $\xrightarrow{HCl}$ / $\xrightarrow{KBr}$ / $\xrightarrow{H^+}$

(2) $\xrightarrow{Br_2/Fe}$ / $\xrightarrow{Sn/HCl}$ / $\xrightarrow{NaNO_2/HCl}$ / $\xrightarrow{CuBr/HBr}$

(3) $\xrightarrow{Sn/HCl}$ / $\xrightarrow{KBr}$ / $\xrightarrow{Br_2}$ / $\xrightarrow{H^+}$

(4) $\xrightarrow{Sn/HCl}$ / $\xrightarrow{Br_2}$ / $\xrightarrow{NaNO_2}$ / $\xrightarrow{NaBr}$

204. In the following reaction sequence, [C] is : [Sep. 04, 2020 (II)]

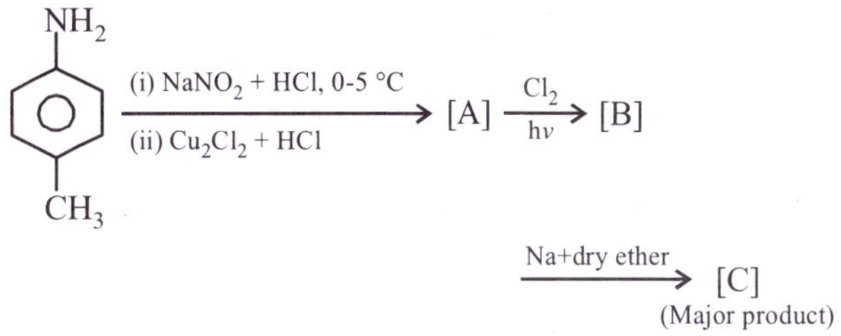

(1) 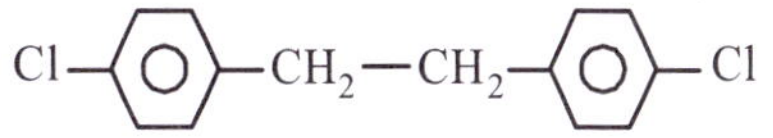

(2) 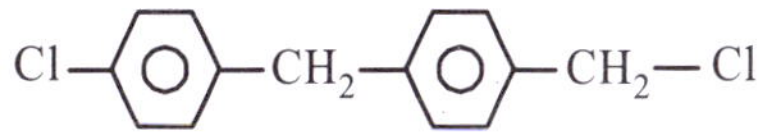

(3) 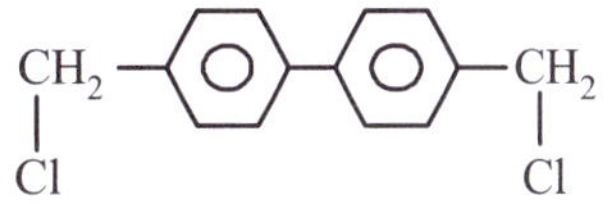

(4) 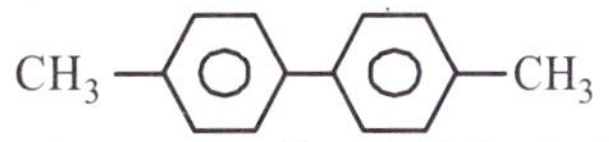

205. Three isomers A, B and C (mol. formula $C_8H_{11}N$) give the following results :

A and C $\xrightarrow{\text{Diazotization}}$ P + Q $\xrightarrow[\text{(ii) Oxidation (KMnO}_4+H^+)]{\text{(i) Hydrolysis}}$

R (product of A) + S (product of C)

R has lower boiling point than S

B $\xrightarrow{C_6H_5SO_2Cl}$ Alkali-insoluble product

A, B and C, respectively are : [Sep. 03, 2020 (II)]

(1) 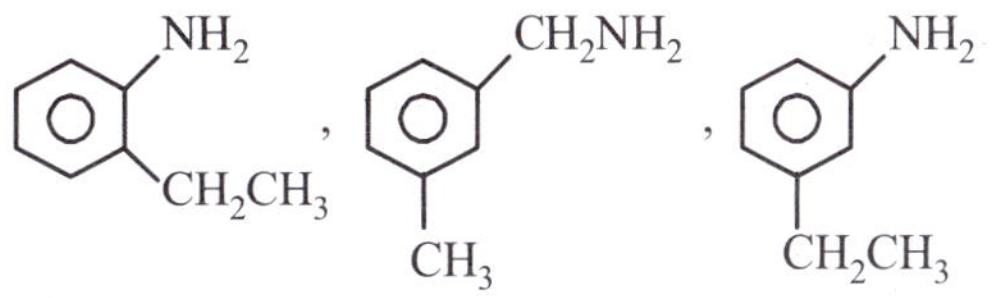

(2) 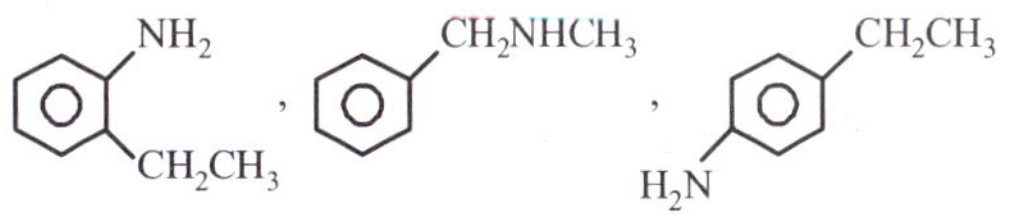

(3) 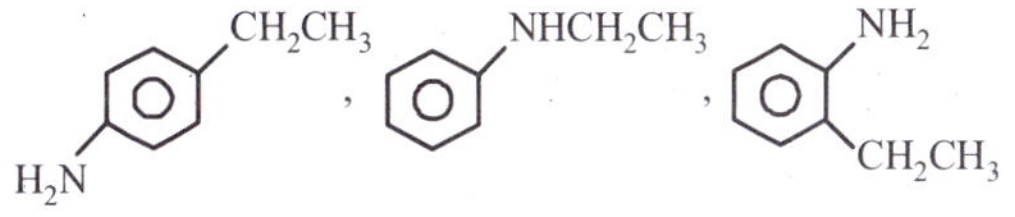

(4) 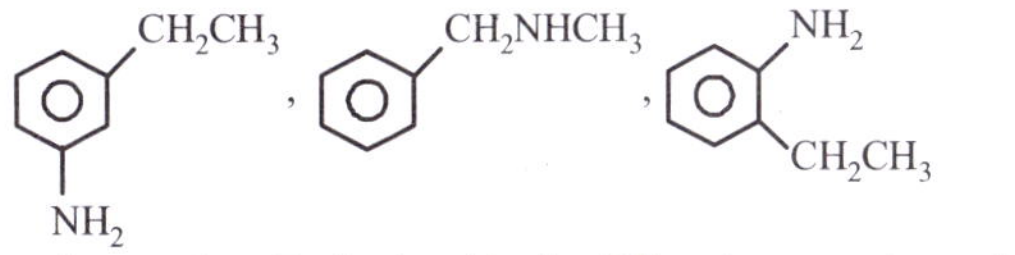

206. The major product Z obtained in the following reaction scheme is: [Jan. 09, 2020 (I)]

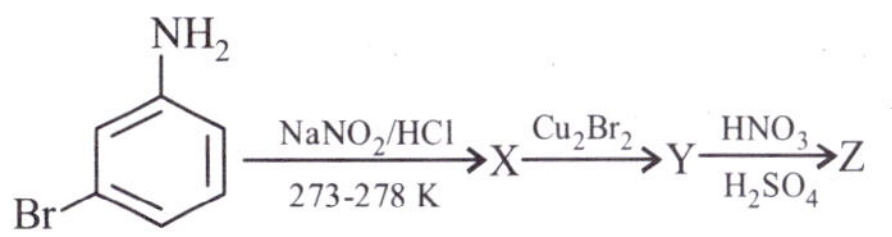

(1) 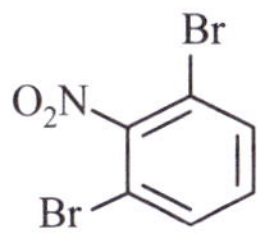 (2) 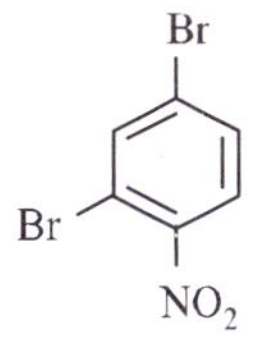

(3) 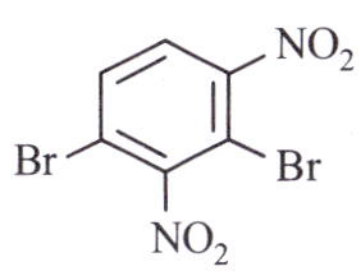 (4) 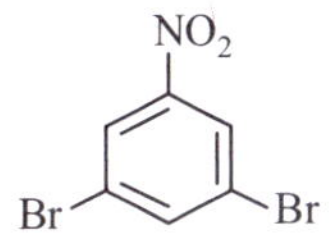

207. Consider the following reactions,

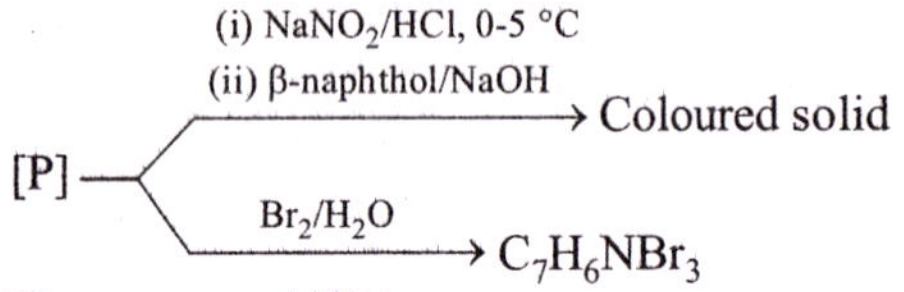

The compound [P] is: [Jan. 09, 2020 (II)]

(1) 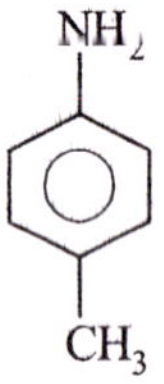

(2) 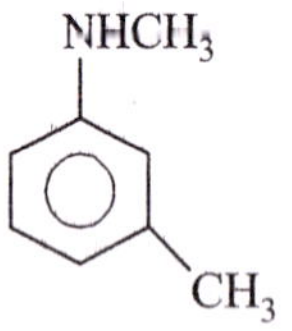

(3) 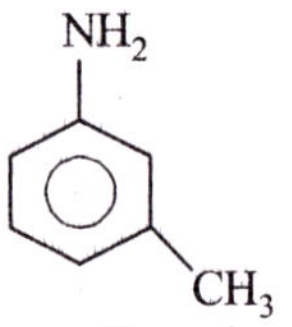

(4) 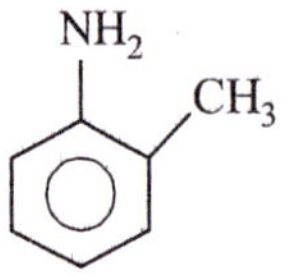

208. Benzene diazonium chloride on reaction with aniline in the presence of dilute hydrochloric acid gives : [April 12, 2019 (II)]

(1) 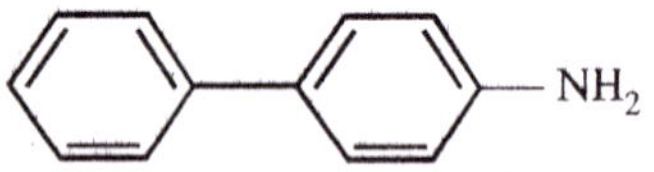

(2) 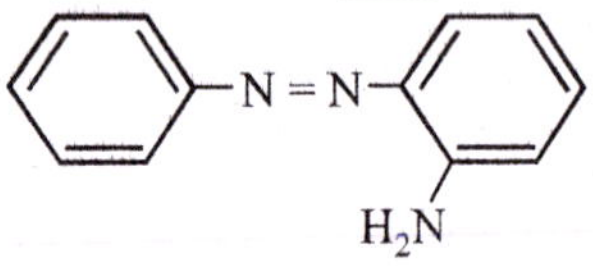

(3) 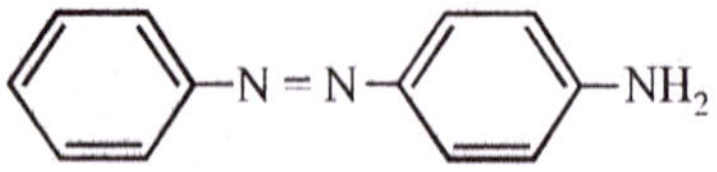

(4) 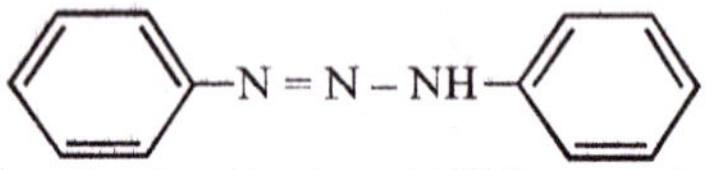

209. Aniline dissolved in dilute HCl is reacted with sodium nitrite at 0 °C. This solution was added dropwise to a solution containing equimolar mixture of aniline and phenol in dil. HCl. The structure of the major product is: [April 9, 2019 (I)]

(1) 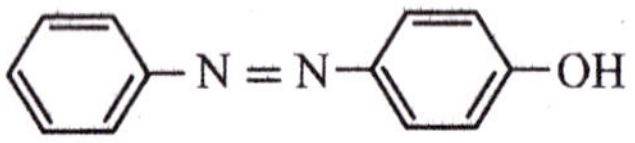

(2) 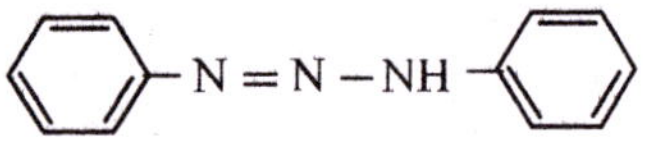

(3) 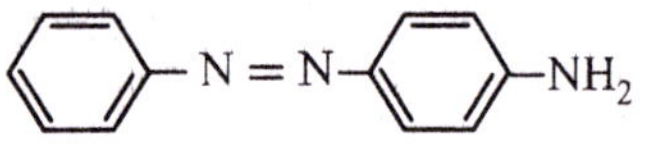

(4) 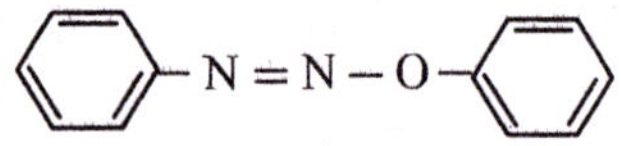

210. Coupling of benzene diazonium chloride with 1-naphthol in alkaline medium will give : [April 8, 2019 (I)]

(1) 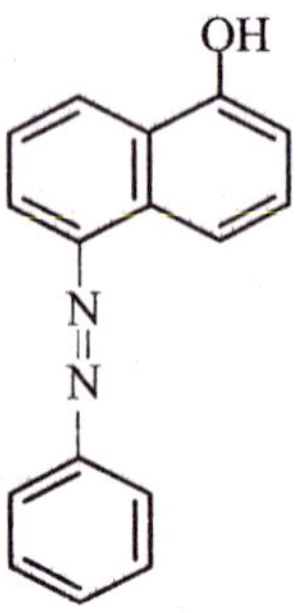

(2) 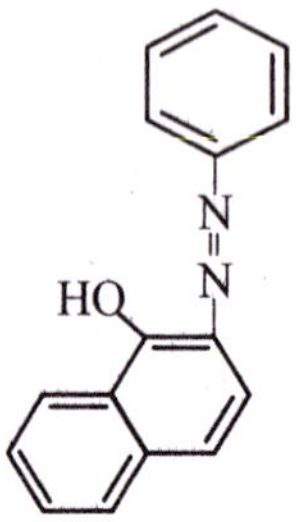

(3) 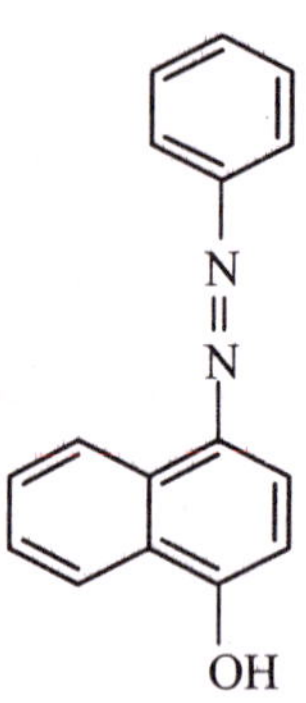

(4) 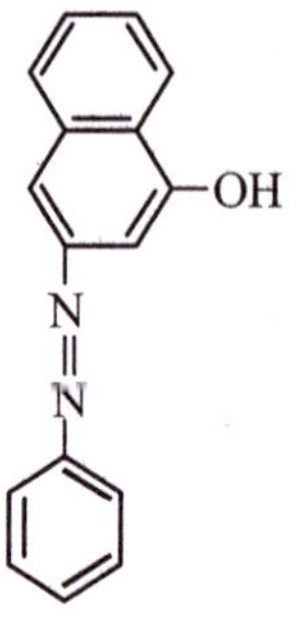

211. The major product formed in the following reaction is:

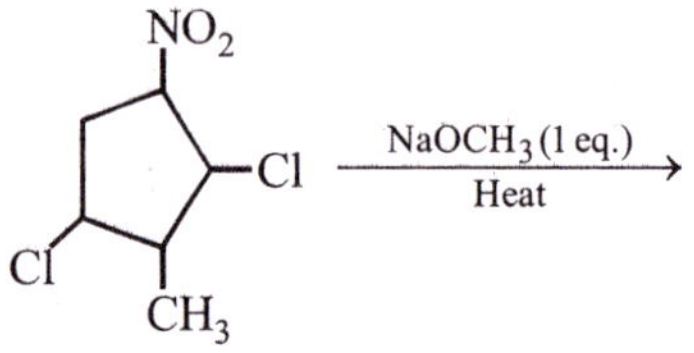

[Online April 15, 2018 (II)]

(1) 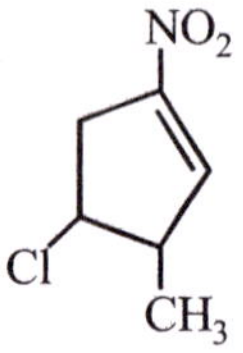

(2) 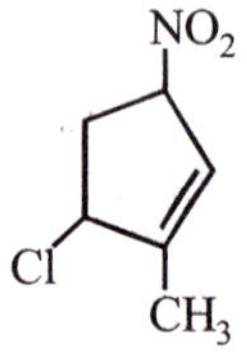

(3) 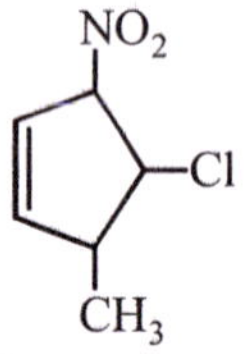

(4) 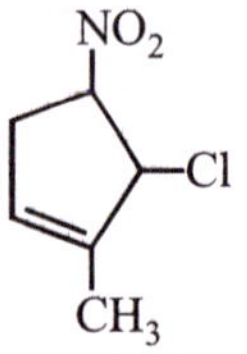

212. The increasing order of nitration of the following compounds is : [Online April 15, 2018 (I)]

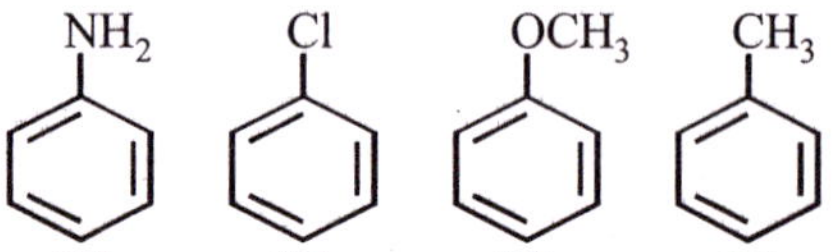

(A) (B) (C) (D)

(1) (A) < (B) < (D) < (C)
(2) (A) < (B) < (C) < (D)
(3) (B) < (A) < (C) < (D)
(4) (B) < (A) < (D) < (C)

213. Fluorination of an aromatic ring is easily accomplished by treating a diazonium salt with HBF$_4$. Which of the following conditions is correct about this reaction? [Online April 10, 2016]

(1) NaF/Cu (2) Cu$_2$O/H$_2$O
(3) Only heat (4) NaNO$_2$/Cu

214. In the reaction

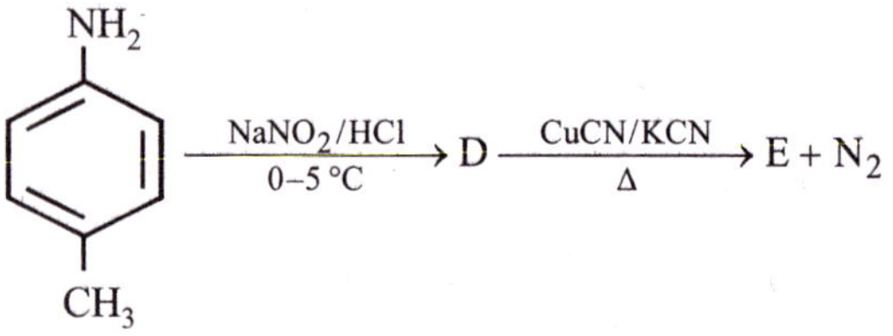

the product E is : [2015]

(1) 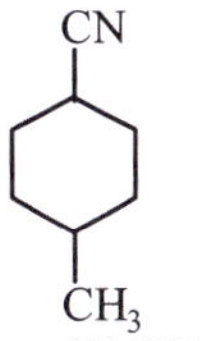

(2) 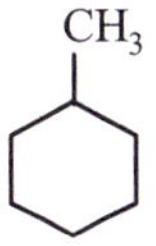

(3) 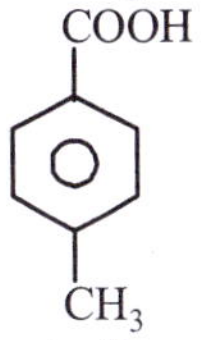

(4) 

215. In a set of reactions *p*-nitrotoluene yielded a product E.

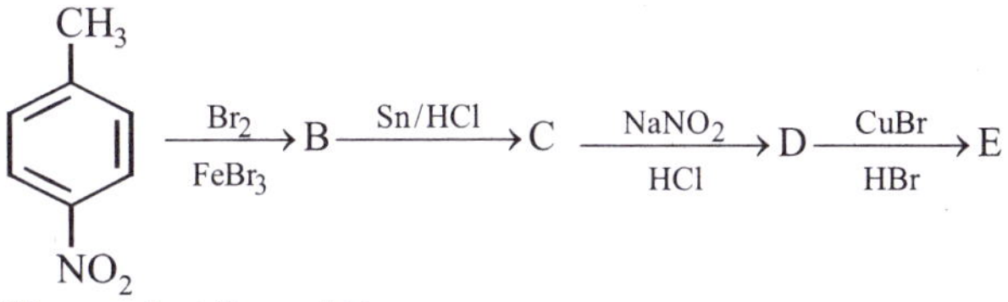

The product E would be: **[Online April 19, 2014]**

(1) 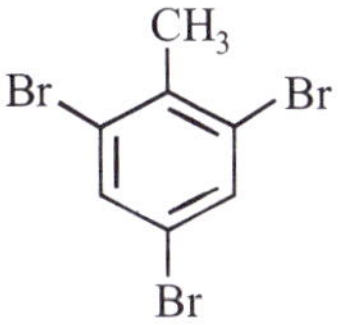

(2) 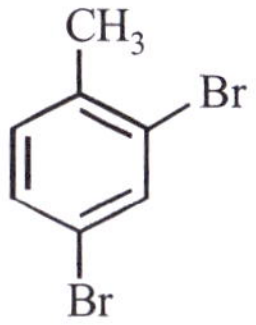

(3) 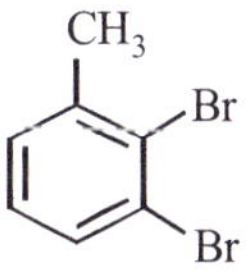

(4) 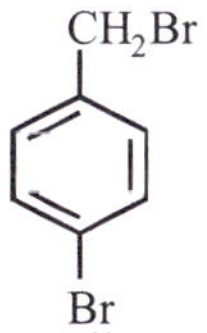

216. Complete reduction of benzene diazonium chloride with Zn/HCl gives: **[Online April 11, 2014]**

(1) Aniline (2) Phenylhydrazine
(3) Azobenzene (4) Hydrazobenzene

217. Conversion of benzene diazonium chloride to chlorobenzene is an example of which of the following reactions? **[Online April 12, 2014]**

(1) Claisen (2) Friedel-craft
(3) Sandmeyer (4) Wurtz

218. The conversion of benzene diazonium chloride to bromobenzene can be accomplished by **[Online May 12, 2012]**

(1) Reimer-Tiemann reaction
(2) Friedel-Crafts reaction
(3) Gattermann reaction
(4) Azo-Coupling reaction

219. In the chemical reactions : **[2012)]**

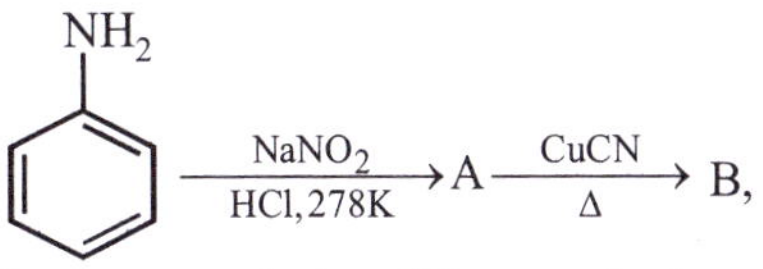

the compounds A and B respectively are :

(1) Benzene diazonium chloride and benzonitrile
(2) Nitrobenzene and chlorobenzene
(3) Phenol and bromobenzene
(4) Fluorobenzene and phenol

220. In the chemical reactions,

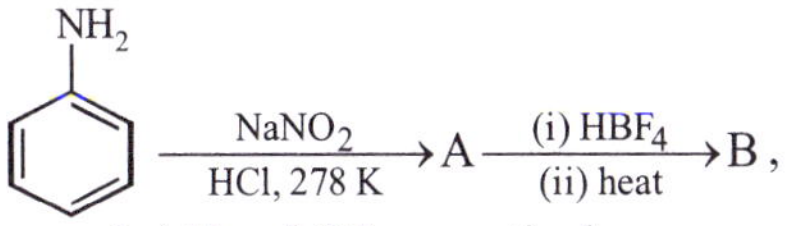

the compounds 'A' and 'B' respectively are **[2010]**

(1) nitrobenzene and fluorobenzene
(2) phenol and benzene
(3) benzene diazonium chloride and fluorobenzene
(4) nitrobenzene and chlorobenzene

NUMERICAL VALUE QUESTIONS (NVQs)

221. Consider the following sequence of reactions. **[Jan. 23, 2025 (II)]**

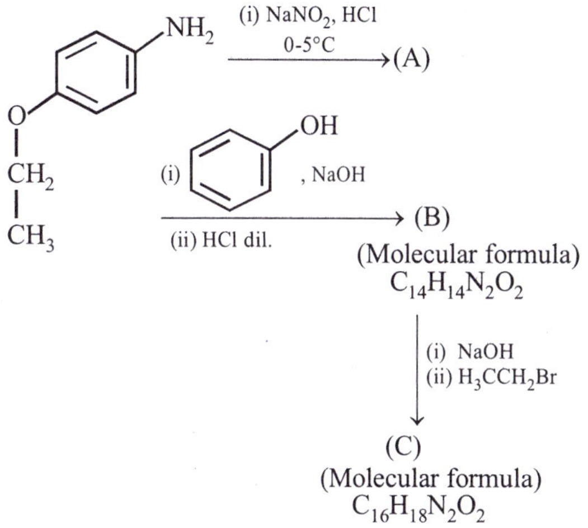

Total number of sp^3 hybridised carbon atoms in the major product C formed is ____.

222. Consider the following sequence of reactions to produce major product (A) **[Jan. 23, 2025 (I)]**

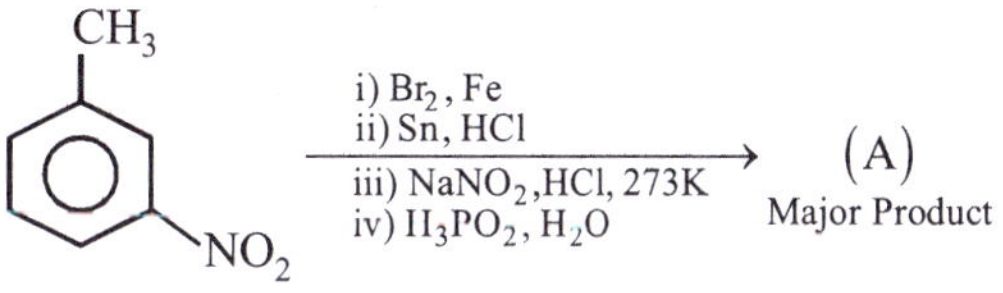

Molar mass of product (A) is ______ g mol^{-1}.
(Given molar mass in g mol–1 of C : 12, H : 1, O : 16, Br : 80, N : 14, P : 31)

223. Consider the following sequence of reactions :

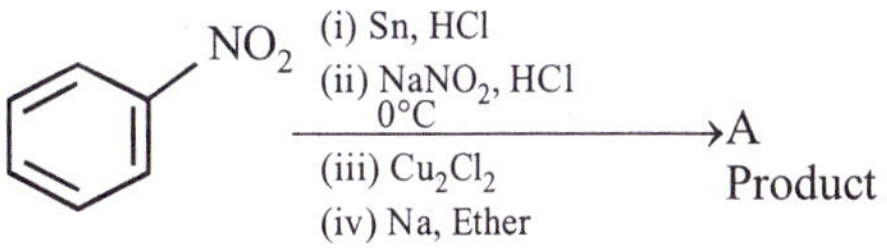

Molar mass of the product formed (A) is ______g mol^{-1}. **[Jan. 22, 2025 (I)]**

224. If 279 g of aniline is reacted with one equivalent of benzenediazonium chloride, the maximum amount of aniline yellow formed will be _____ g. (nearest integer)
(consider complete conversion) **[April 8, 2024 (I)]**

225. 9.3 g of pure aniline upon diazotisation followed by coupling with phenol gives an orange dye. The mass of orange dye produced (assume 100% yield/ conversion) is ________g. (nearest integer) **[April 6, 2024 (I)]**

226. X of enthanamine was subjected to reaction with $NaNO_2$/HCl followed by hydrolysis to liberate N_2 and HCl. The HCl generated was completely neutralised by 0.2 moles of NaOH. X is ____ g. **[April 5, 2024 (II)]**

Part-B (JEE Advanced)

Topic 1 - Aliphatic and Aromatic Amines

1 MCQs with One Correct Answer

1. The order of basicity among the following compounds is [Adv. 2017]

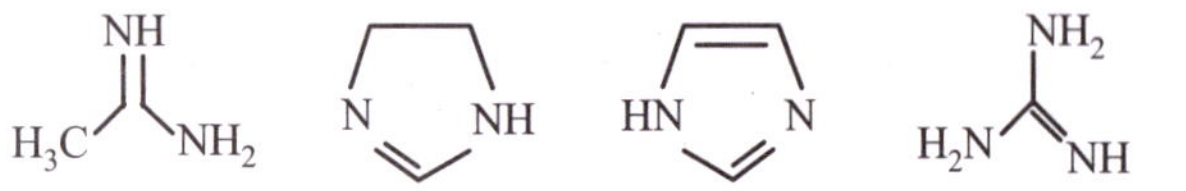

I II III IV

(a) II > I > IV > III (b) IV > II > III > I
(c) IV > I > II > III (d) I > IV > III > II

2. The major product of the following reaction is [2011 - I]

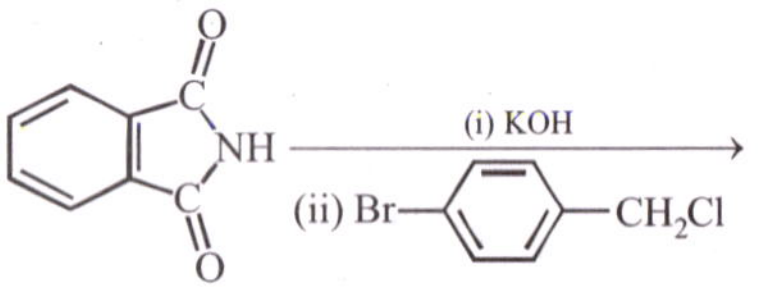

(a)
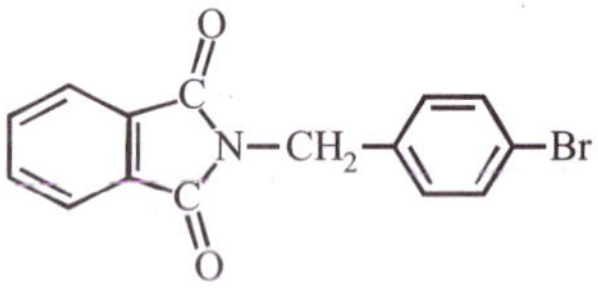

(b)
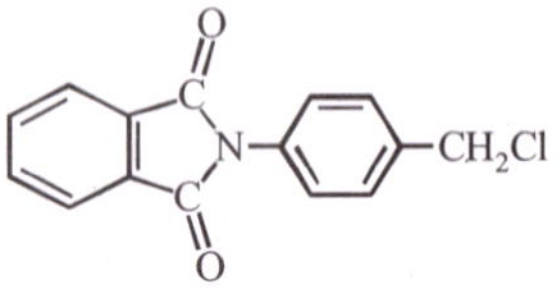

(c)
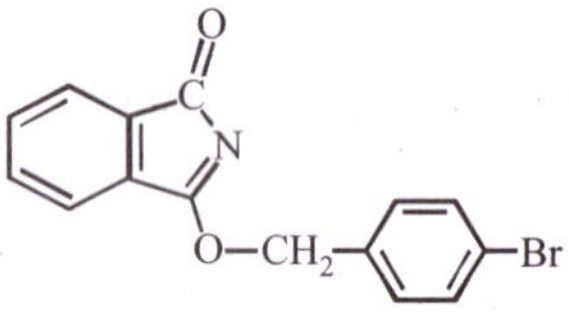

(d)
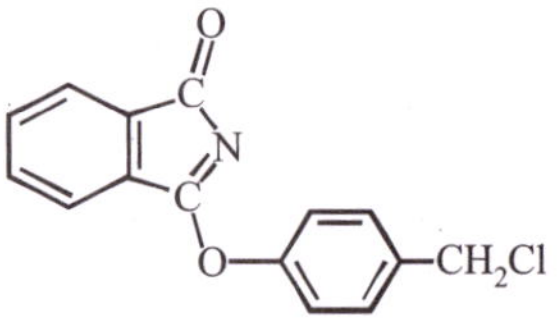

3. In the compound given below, the correct order of the acidity of the positions X, Y and Z is [2004S]

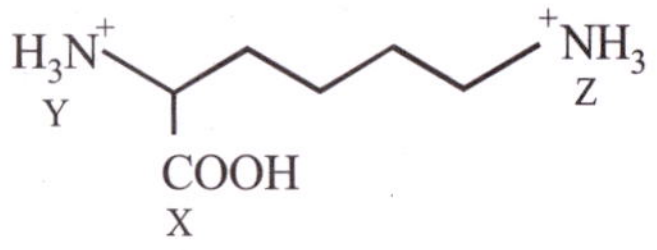

(a) Z > X > Y (b) X > Y > Z
(c) X > Z > Y (d) Y > X > Z

4. The major product obtained when Br_2/Fe is treated with

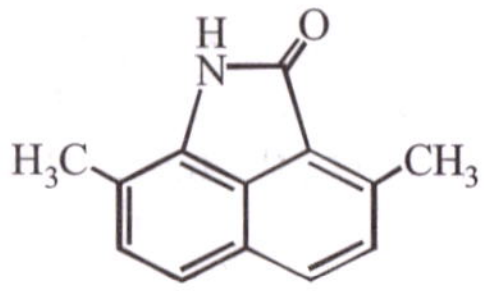

[2004S]

(a)
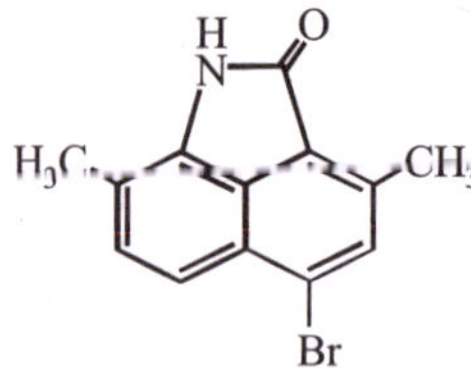

(b)
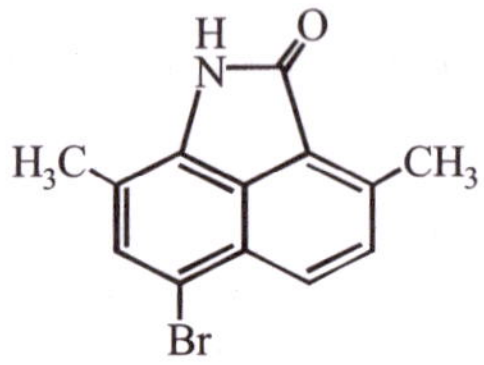

(c)
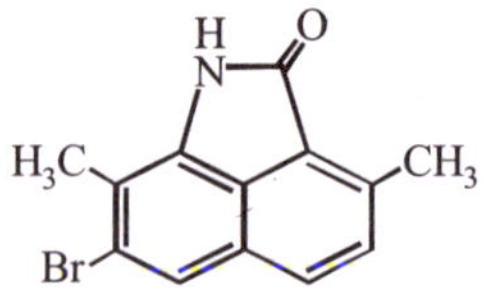

(d)
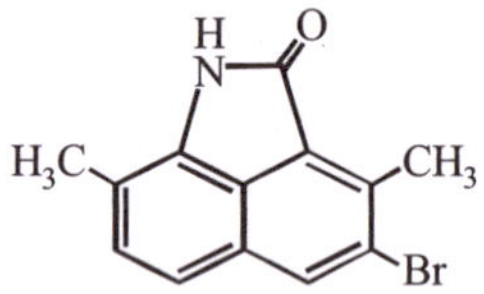

5. Benzamide on reaction with $POCl_3$ gives [2004S]
(a) aniline (b) chlorobenzene
(c) benzylamine (d) benzonitrile

6. Compound 'A' (molecular formula C_3H_8O) is treated with acidified potassium dichromate to form a product 'B' (molecular formula C_3H_6O). 'B' forms a shining silver mirror on warming with ammonical silver nitrate. 'B' when treated with an aqueous solution of $H_2NCONHNH_2.HCl$ and sodium acetate gives a product 'C'. Identify the structure of 'C'.
(a) $CH_3CH_2CH{=}NNHCONH_2$ [2002S]
(b) $CH_3-\underset{\underset{CH_3}{|}}{C}=NNHCONH_2$
(c) $CH_3-\underset{\underset{CH_3}{|}}{C}=NCONHNH_2$
(d) $CH_3CH_2CH{=}NCONHNH_2$

7. The correct order of basicities of the following compounds is [2001S]

1. $CH_3-C(=NH)-NH_2$ 2. $CH_3-CH_2-NH_2$

3. $(CH_3)_2NH$ 4. $CH_3-\overset{\overset{O}{||}}{C}-NH_2$

(a) 2 > 1 > 3 > 4 (b) 1 > 3 > 2 > 4
(c) 3 > 1 > 2 > 4 (d) 1 > 2 > 3 > 4

8. Among the following, the strongest base is [2000S]
(a) $C_6H_5NH_2$ (b) p-$NO_2.C_6H_4NH_2$
(c) m-$NO_2.C_6H_4.NH_2$ (d) $C_6H_5CH_2NH_2$

9. In the reaction p-chlorotoluene with KNH_2 in liq. NH_3, the major product is: [1997 - 1 Mark]
(a) o-toluidine (b) m-toluidine
(c) p-toluidine (d) p-chloroaniline.

10. Butanenitrile may be prepared by heating : **[1992 - 1 Mark]**
(a) Propyl alcohol with KCN
(b) Butyl alcohol with KCN
(c) Butyl chloride with KCN
(d) Propyl chloride with KCN

11. Amongst the following, the most basic compound is : **[1990 - 1 Mark]**
(a) Benzylamine (b) Aniline
(c) Acetanilide (d) *p*-Nitroaniline

12. The compound that is most reactive towards electrophilic nitration is : **[1985 - 1 Mark]**
(a) toluene (b) benzene
(c) benzoic acid (d) nitrobenzene

13. Carbylamine test is performed in alcoholic KOH by heating a mixture of : **[1984 - 1 Mark]**
(a) chloroform and silver powder
(b) trihalogenatedmethane and a primary amine
(c) an alkyl halide and a primary amine
(d) an alkyl cyanide and a primary amine

14. Acetamide is treated separately with the following reagents. Which one of these would give methylamine? **[1983 - 1 Mark]**
(a) PCl_5 (b) $NaOH + Br_2$
(c) soda lime (d) hot conc. H_2SO_4

15. The compound which on reaction with aqueous nitrous acid at low temperature produces an oily nitrosoamine is **[1981 - 1 Mark]**
(a) methylamine (b) ethylamine
(c) diethylamine (d) triethylamine

2 Integer Value Answer

16. The total number of sp^2 hybridised carbon atoms in the major product P (a non–heterocyclic compound) of the following reaction is _____. **[Adv. 2023]**

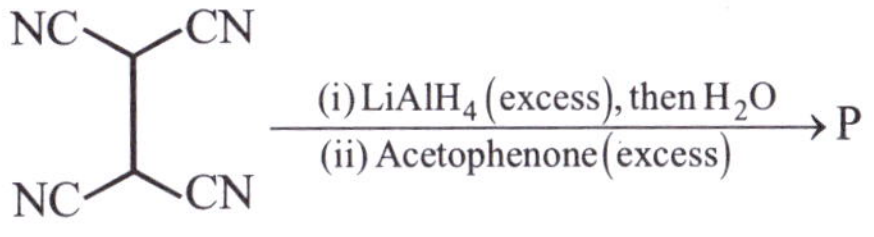

3 Numeric / New Stem Based Questions

17. In the following reaction sequence, the amount of **D** (in g) formed from 10 moles of acetophenone is ______.
(Atomic weights in g mol^{-1}: H = 1, C = 12, N = 14, O = 16, Br = 80. The yield (%) corresponding to the product in each step is given in the parenthesis) **[Adv. 2018]**

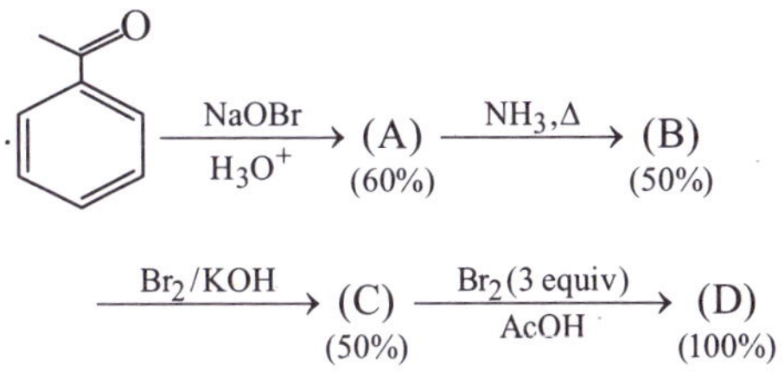

4 Fill in the Blanks

18. The high melting point and insolubility in organic solvents of sulphanilic acid are due to its........structure. **[1994 - 1 Mark]**

19. In an acidic medium, ... behaves as the strongest base. (nitrobenzene, aniline, phenol) **[1981 - 1 Mark]**

6 MCQs with One or More than One Correct Answer

20. For the reaction sequence given below, the correct statement(s) is(are) **[Adv. 2025]**

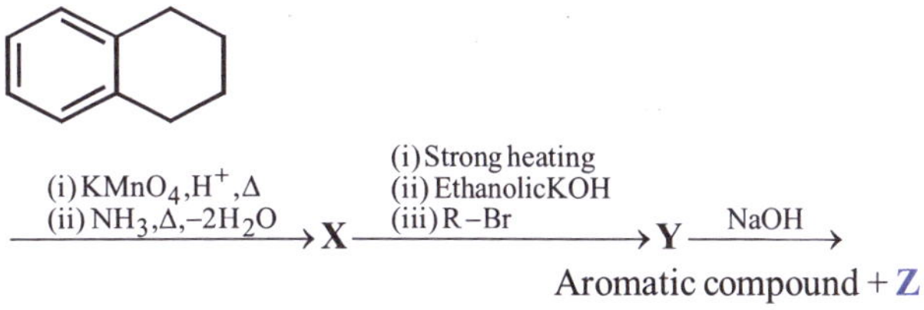

(a) Both **X** and **Y** are oxygen containing compounds.
(b) **Y** on heating with $CHCl_3$/KOH forms isocyanide.
(c) **Z** reacts with Hinsberg's reagent.
(d) **Z** is an aromatic primary amine.

21. Considering the reaction sequence given below, the correct statement(s) is(are) **[Adv. 2022]**

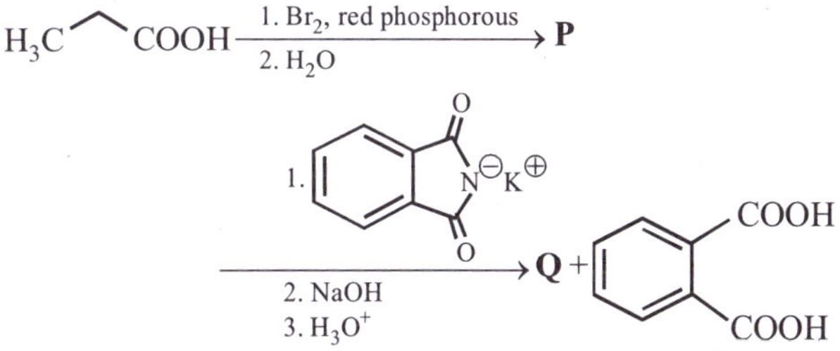

(a) **P** can be reduced to a primary alcohol using $NaBH_4$.
(b) Treating **P** with conc. NH_4OH solution followed by acidification gives **Q**.
(c) Treating **Q** with a solution of $NaNO_2$ in aq. HCl liberates N_2.
(d) **P** is more acidic than CH_3CH_2COOH.

22. Correct options for the following sequence of reactions is (are)

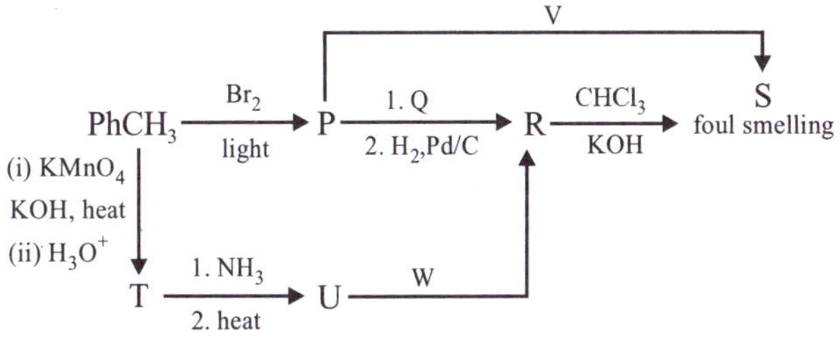

(a) $Q = KNO_2$, $W = LiAlH_4$ **[Adv. 2021]**
(b) R = benzenamine, V = KCN
(c) $Q = AgNO_2$, R = phenylmethanamine
(d) $W = LiAlH_4$, V = AgCN

23 Consider the following four compounds **I**, **II**, **III**, and **IV**.

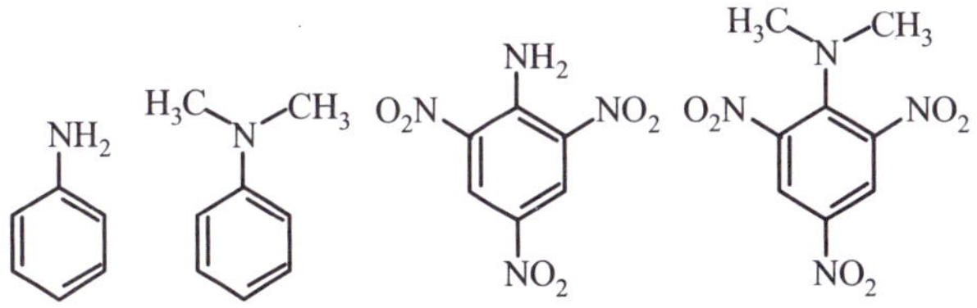

I **II** **III** **IV**

Choose the correct statement(s). **[Adv. 2020]**
(a) The order of basicity is **II** > **I** > **III** > **IV**.
(b) The magnitude of pK_b difference between **I** and **II** is more than that between **III** and **IV**.
(c) Resonance effect is more in **III** than in **IV**.
(d) Steric effect makes compound **IV** more basic than **III**.

24. In the following reactions, the product S is [Adv. 2015]

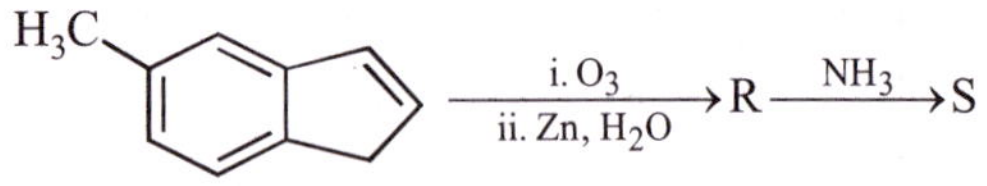

(a) 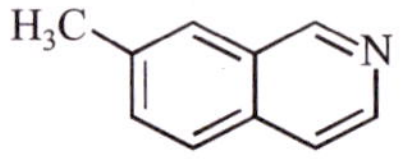 (b) 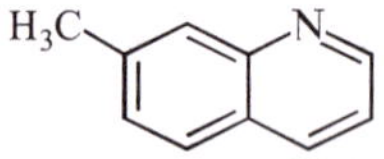

(c) 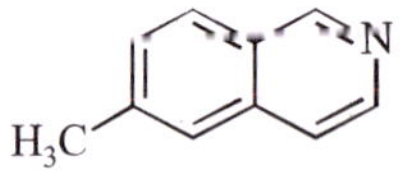 (d) 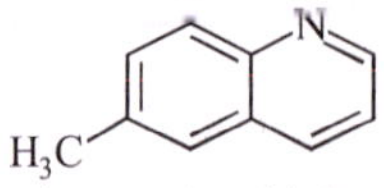

25. In the reaction shown below, the major product(s) formed is/are [Adv. 2014]

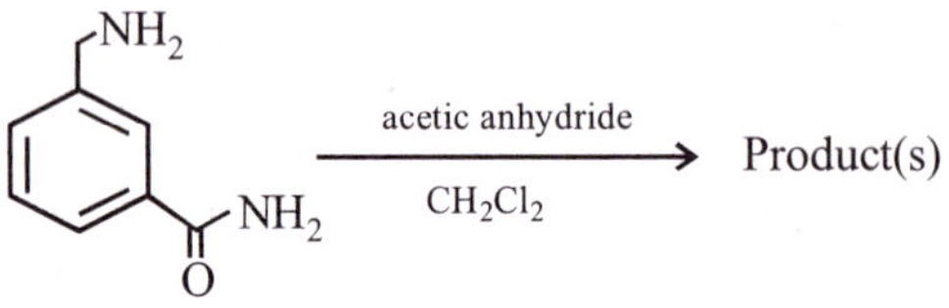

(a) 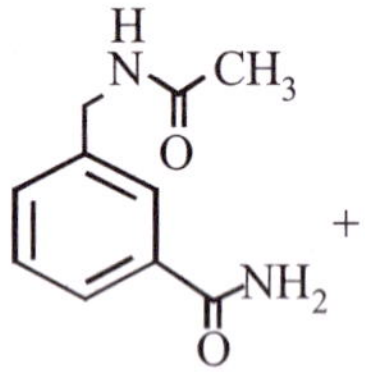 $+ CH_3COOH$

(b) 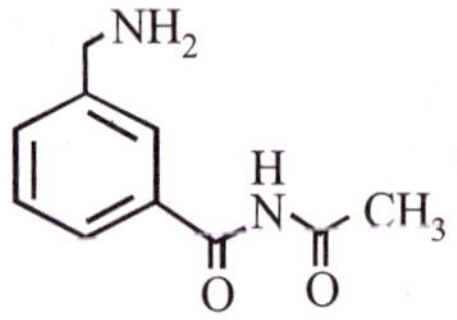 $+ CH_3COOH$

(c) 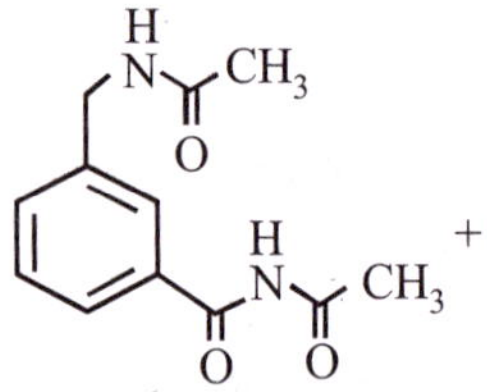 $+ H_2O$

(d) 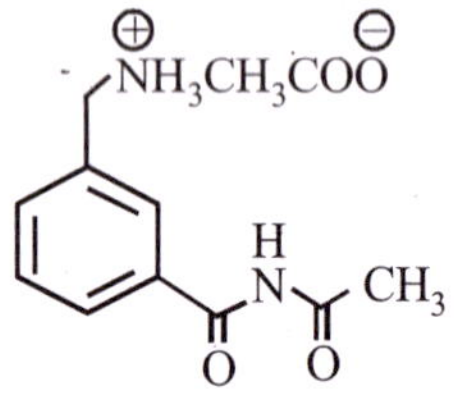

26. Hydrogen bonding plays a central role in the following phenomena [Adv. 2014]
(a) Ice floats in water
(b) Higher Lewis basicity of primary amines than tertiary amines in aqueous solutions
(c) Formic acid is more acidic than acetic acid
(d) Dimerisation of acetic acid in benzene

27. In the reaction $2\mathbf{X} + B_2H_6 \rightarrow [BH_2(\mathbf{X})_2]^+[BH_4]^-$ the amine(s) **X** is (are) [2009]
(a) NH_3 (b) CH_3NH_2
(c) $(CH_3)_2NH$ (d) $(CH_3)_3N$

28. A positive carbylamine test is given by [1999 - 2 Marks]
(a) N, N—dimethylaniline
(b) 2, 4–dimethylaniline
(c) N–methyl-*o*-methylaniline
(d) *p*-methylbenzylamine

29. Among the following compounds, which will react with acetone to give a product containing $> C = N$–bond ? [1998 - 2 Marks]
(a) $C_6H_5NH_2$ (b) $(CH_3)_3N$
(c) $C_6H_5NHC_6H_5$ (d) $C_6H_5NHNH_2$.

30. *p*-Chloroaniline and anilinium hydrochloride can be distinguished by [1998 - 2 Marks]
(a) Sandmeyer reaction (b) $NaHCO_3$
(c) $AgNO_3$ (d) Carbylamine test

31. Examine the following two structures for the anilinium ion and choose the correct statement from the ones given below [1993 - 1 Mark]

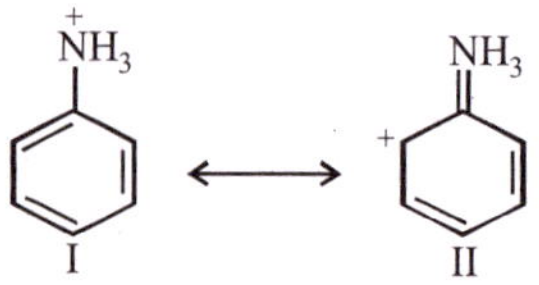

(a) II is not an acceptable canonical structure because carbonium ions are less stable than ammonium ions.
(b) II is not an acceptable canonical structure because it is non-aromatic.
(c) II is not an acceptable canonical structure because its nitrogen has 10 valence electrons.
(d) II is an acceptable canonical structure.

32. Reaction of $R-\overset{O}{\overset{\|}{C}}-NH_2$ with a mixture of Br_2 and KOH gives R-NH_2 as the main product. The intermediates involved in this reaction are : [1992 - 1 Mark]
(a) $R-\overset{O}{\overset{\|}{C}}-NHBr$ (b) $R-NHBr$
(c) $R-N=C=O$ (d) $R-\overset{O}{\overset{\|}{C}}-N\begin{matrix}Br\\Br\end{matrix}$

7 Match the Following

33. The major products obtained from the reactions in **List-II** are the reactants for the named reactions mentioned in **List-I**. Match each entry in **List-I** with the appropriate entry in **List-II** and choose the correct option. [Adv. 2025]

List-I	List-II
(P) Stephen reaction	(1) Tolune $\xrightarrow[(ii)\,H_3O^+]{(i)\,CrO_2Cl_2/CS_2}$
(Q) Sandmeyer reaction	(2) Benzoic acid $\xrightarrow{(i)\,PCl_5\ (ii)\,NH_3\ (iii)\,P_4O_{10},\Delta}$
(R) Hoffmann bromamide degradation reaction	(3) Nitrobenzene $\xrightarrow{(i)\,Fe,HCl\ (ii)\,HCl,NaNO_2\ (273-278\,K),H_2O}$
(S) Cannizzaro reaction	(4) Toluene $\xrightarrow{(i)\,Cl_2/hv,H_2O\ (ii)\,Tollen's\ reagent\ (iii)\,SO_2Cl_2\ (iv)\,NH_3}$
	(5) Aniline $\xrightarrow{(i)\,(CH_3CO)_2O,\,Pyridine\ (ii)\,HNO_3,H_2SO_4.288K\ (iii)\,aq.NaOH}$

(a) $P\rightarrow 2; Q\rightarrow 4; R\rightarrow 1; S\rightarrow 3$
(b) $P\rightarrow 2; Q\rightarrow 3; R\rightarrow 4; S\rightarrow 1$
(c) $P\rightarrow 5; Q\rightarrow 3; R\rightarrow 4; S\rightarrow 2$
(d) $P\rightarrow 5; Q\rightarrow 4; R\rightarrow 2; S\rightarrow 1$

34. List - I | List - II [Adv. 2014]

P. $H-C\equiv C-H$ 1. **Scheme I**

$$? \xrightarrow[\text{(iv) } NH_3]{\text{(i) } KMnO_4, HO^-, \text{ heat; (ii) } H^+, H_2O; \text{ (iii) } SOCl_2} C_7H_6N_2O_3$$

Q. 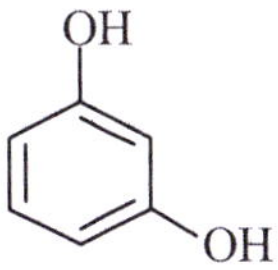 2. **Scheme II**

$$? \xrightarrow[\text{(vi) } HO^-]{\text{(i) Sn/HCl; (ii) } CH_3COCl; \text{ (iii) conc. } H_2SO_4; \text{ (iv) } HNO_3; \text{ (v) dil. } H_2SO_4, \text{ heat}} C_6H_6N_2O_2$$

R. 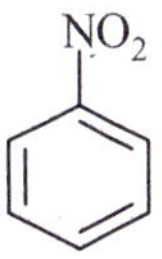 3. **Scheme III**

$$? \xrightarrow[\text{(v) hydrolysis}]{\text{(i) red hot iron, 873 K; (ii) fuming } HNO_3, H_2SO_4, \text{ heat; (iii) } H_2S.NH_3; \text{ (iv) } NaNO_2, H_2SO_4} C_6H_5NO_2$$

S. 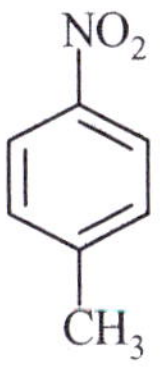 4. **Scheme IV**

$$? \xrightarrow[\text{heat}]{\text{(i) conc. } H_2SO_4, 60°C; \text{ (ii) conc. } HNO_3, \text{ conc. } H_2SO_4; \text{ (iii) dil. } H_2SO_4} C_6H_5NO_4$$

Code:

	P	Q	R	S
(a)	1	4	2	3
(b)	3	1	4	2
(c)	3	4	2	1
(d)	4	1	3	2

35. Match each of the compounds in **Column-I** with its characteristic reaction(s) in **Column-II**. [2009]

Column-I	Column-II
(A) $CH_3CH_2CH_2CN$	(p) Reduction with Pd–C/H_2
(B) $CH_3CH_2OCOCH_3$	(q) Reduction with $SnCl_2$/HCl
(C) CH_3–CH=CH–CH_2OH	(r) Development of foul smell on treatment with chloroform and alcoholic KOH
(D) $CH_3CH_2CH_2CH_2NH_2$	(s) Reduction with diisobutylaluminium hydride(DIBAL-H)
	(t) Alkaline hydrolysis

8 Comprehension Based Questions

Passage - I

Treatment of compound (O) with $KMnO_4/H^+$ gave (P), which on heating with ammonia gave (Q). The compound (Q) on treatment with Br_2/NaOH produced (R). On strong heating, (Q) gave (S), which on further treatment with ethyl 2-bromopropanoate in the presence of KOH followed by acidification, gave a compound (T). [Adv. 2016]

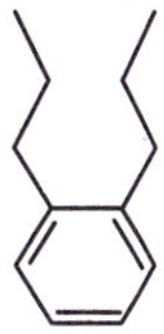

(O)

36. The compound (R) is

(a) 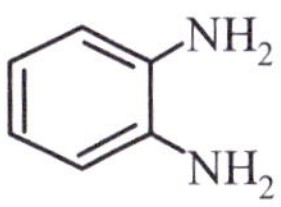 (b) 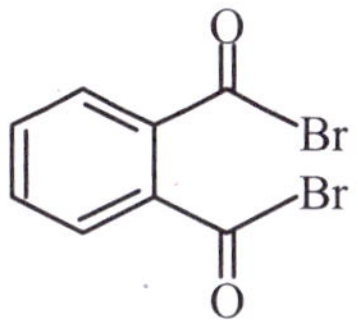

(c) 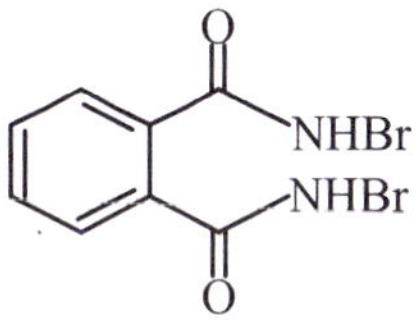 (d) 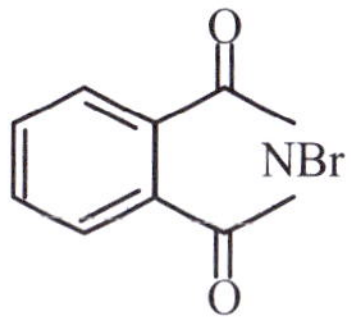

37. The compound (T) is

(a) glycine (b) alanine
(c) valine (d) serine

Passage - II

The conversion of an amide to an amine with one carbon atom less by the action of alkaline hypohalite is known as Hofmann bromamide degradation.

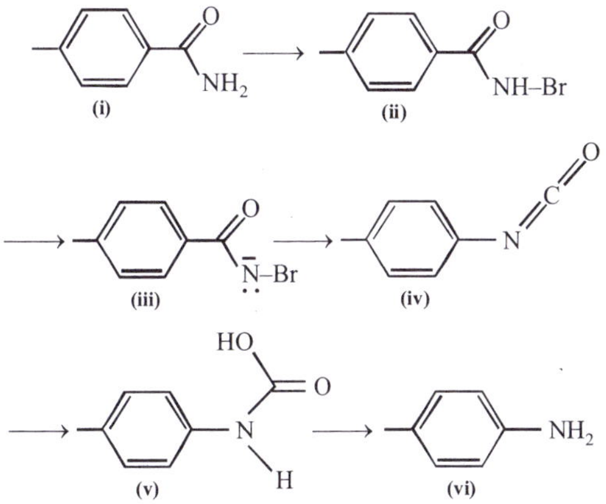

In this reaction, RCONHBr is formed from which the reaction has derived its name. Hofmann reaction is accelerated if the migrating group is more electron-releasing. Hofmann degradation reaction is an intramolecular reaction.

38. How can the conversion of (i) to (ii) be brought about? [2006 - 5M, –2]

(a) KBr (b) KBr + CH_3ONa
(c) KBr + KOH (d) Br_2 + KOH

39. Which is the rate determining step in Hofmann bromamide degradation? [2006 - 5M, –2]

(a) Formation of (i) (b) Formation of (ii)
(c) Formation of (iii) (d) Formation of (iv)

40. What are the constituent amines formed when the mixture of (i) and (ii) undergoes Hofmann bromamide degradation? [2006 - 5M, –2]

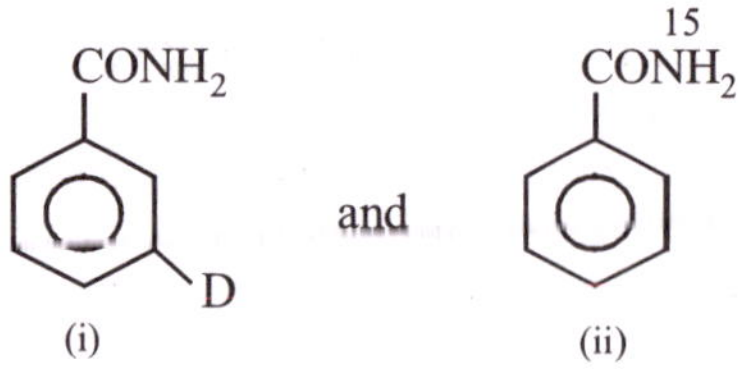

(a) 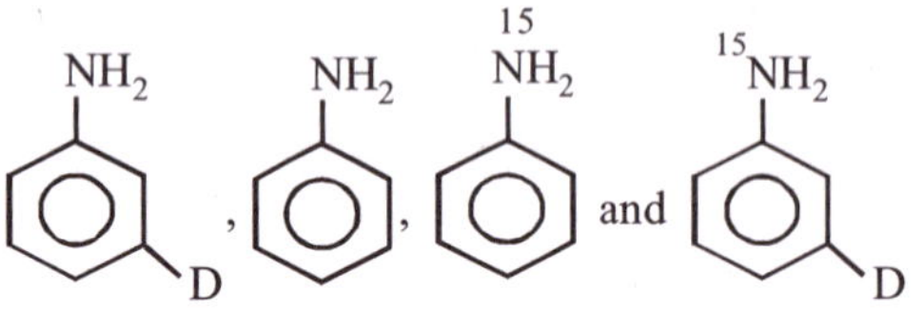

(b) 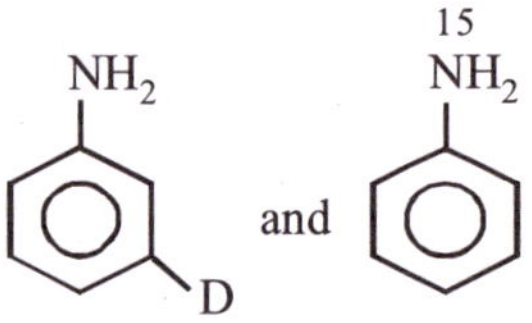

(c) 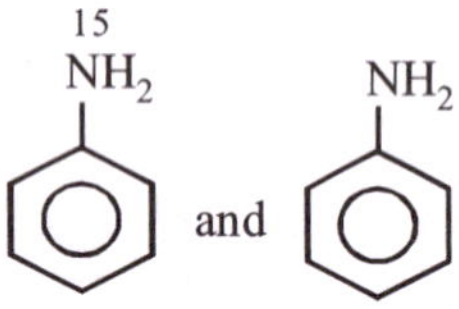

(d) 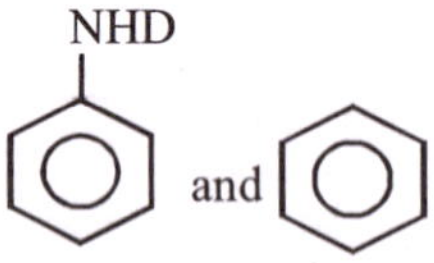

9 Assertion and Reason/Statement Type Questions

Each question contains **STATEMENT-1 (Assertion)** and **STATEMENT-2 (Reason)**. Each question has 4 choices (a), (b), (c) and (d) out of which ONLY ONE is correct. Mark your answer as

(a) If both Statement -1 and Statement -2 are correct, and Statement -2 is the correct explanation of the Statement -2.
(b) If both Statement -1 and Statement -2 are correct, but Statement -2 is not the correct explanation of the Statement -1.
(c) If Statement -1 is correct but Statement -2 is incorrect.
(d) If Statement -1 is incorrect but Statement -2 is correct.

41. **Statement-1 :** In strongly acidic solutions, aniline becomes more reactive towards electrophilic reagents.
Statement-2 : The amino group being completely protonated in strongly acidic solution, the lone pair of electrons on the nitrogen is no longer available for resonance. [2001S]

10 Subjective Problems

42. $\underset{\text{(Optical active) (X)}}{C_5H_{13}N} \xrightarrow[-N_2]{aq.NaNO_2/HCl} \underset{\text{Tertiary alcohol}}{(Y)}$ + Some other products [2005 - 4 Marks]

(i) Identify (X) and (Y)
(ii) Is (Y) optically active?
(iii) Give structure(s) of intermediate(s), if any, in the formation of (Y) from (X).

43. Give reasons for the following :

(i) 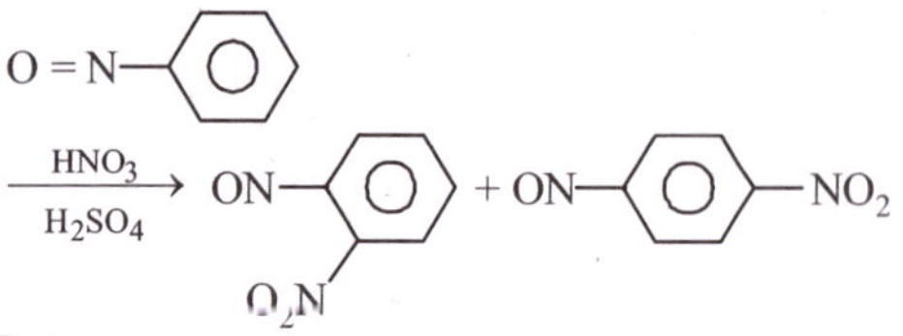

but

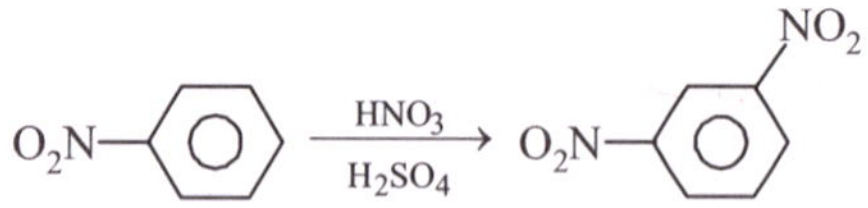

[2005 - 1 Mark]

(ii) (a) 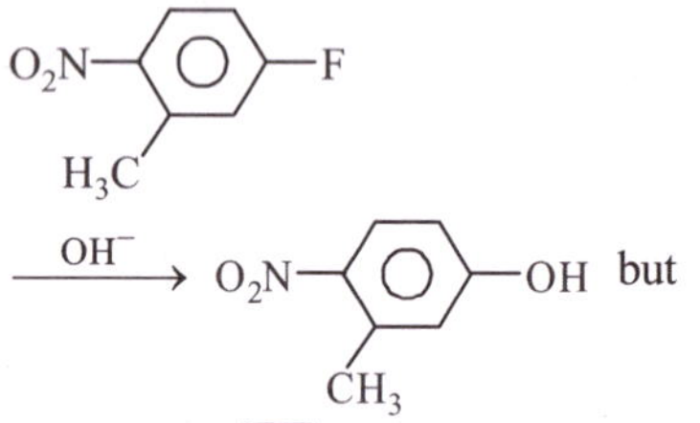

but

(b) 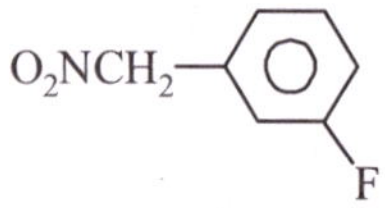

$\xrightarrow{OH^-}$ No release of F^- [2005 - 1 Mark]

(iii) Nitrobenzene does not undergo Friedel-Craft's alkylation [1998 - 2 Marks]
(iv) *o*-Nitrophenol is steam volatile whereas *p*-nitrophenol is not; [1985 - 1 Mark]

44. Identify (A) to (D) in the following series of reactions. [2004 - 4 Marks]

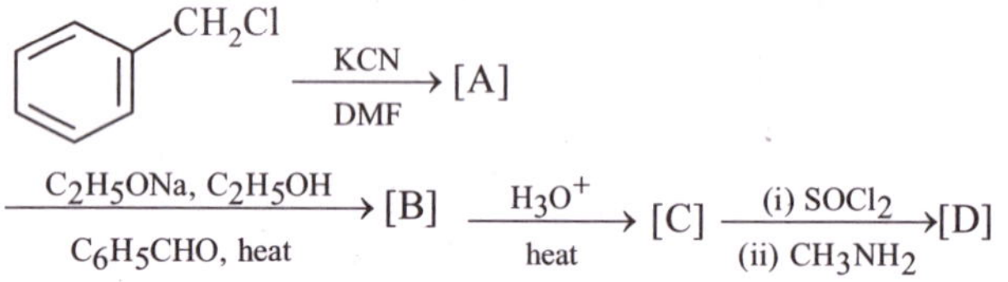

45. There is a solution of *p*–hydroxybenzoic acid and *p*-aminobenzoic acid. Discuss one method by which we can separate them and also write down the confirmatory tests of the functional groups present. [2003 - 4 Marks]

46. Write structures of the products **A, B, C, D** and **E** in the following scheme. [2002 - 5 Marks]

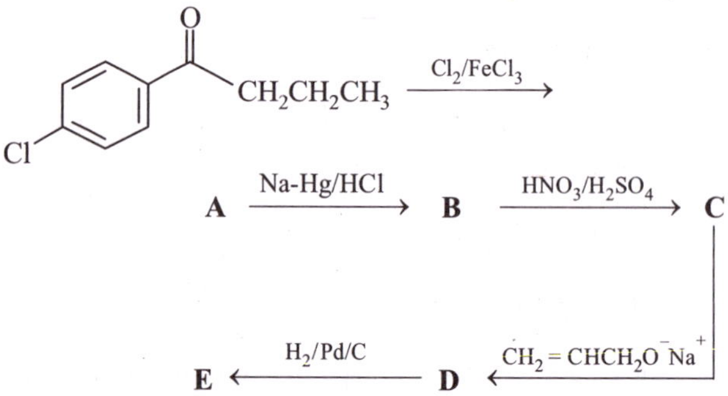

47. Explain briefly the formation of the products giving the structures of the intermediates. [1999 - 2 Marks]

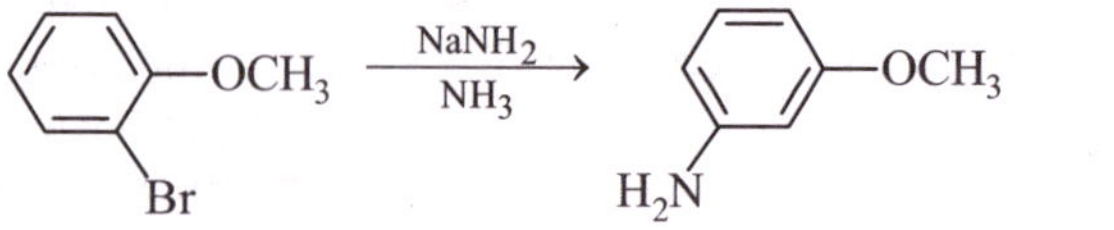

48. Compound *A* (C_8H_8O) on treatment with NH_2OH. HCl gives *B* and *C*. *B* and *C* rearrange to give *D* and *E*, respectively, on treatment with acid. *B*, *C*, *D* and *E* are all isomers of molecular formula (C_8H_9NO). When *D* is boiled with alcoholic KOH an oil F (C_6H_7N) separates out. F reacts rapidly with CH_3COCl to give back *D*. On the other hand, *E* on boiling with alkali followed by acidification gives a white solid *G* ($C_7H_6O_2$). Identify *A-G*. **[1999 - 7 Marks]**

49. Complete the following with appropriate structures :

(i) $CH_3CH_2NH_2 \xrightarrow{(CH_3CO)_2O,\text{heat}}$ 2 products **[1998 - 2 Marks]**

(ii) $CH_3CONHC_6H_5 \xrightarrow{Br_2,\ Fe}$ 2 products **[1998 - 2 Marks]**

(iii) 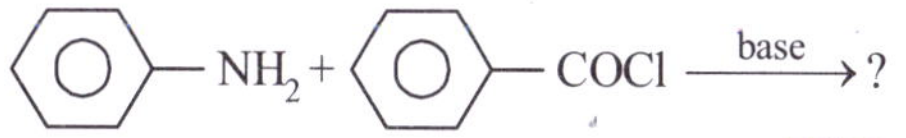

[1986 - 1 Mark]

50. Give reasons for the following :
   (i) Dimethylamine is a stronger base than trimethylamine. **[1998 - 2 Marks]**
   (ii) Cyclohexylamine is a stronger base than aniline. **[1982 - 1 Mark]**

51. Acetophenone on reaction with hydroxylamine hydrochloride can produce two isomeric oximes. Write structures of the oximes. **[1997 - 2 Marks]**

52. Write the structure of the foul-smelling compound obtained when aniline is treated with chloroform in the presence of KOH. **[1996 - 1 Mark]**

53. Identify, **A** (C_3H_9N) which reacts with benzensulphonyl chloride to give a solid, insoluble in alkali. **[1993 - 1 Mark]**

54. A basic, volatile nitrogen compound gave a foul smelling gas when treated with chloroform and alcoholic potash. A 0.295 g sample of the substance. dissolved in aq. HCl and treated with $NaNO_2$ solution at 0°C, liberated a colorless, odourless gas whose volume corresponded to 112 mL at STP. After the evolution of the gas was complete, the aqueous solution was distilled to give an organic liquid which did not contain nitrogen and which on warming with alkali and iodine gave a yellow precipitate. Identify the original substance. Assume that it contains one N atom per molecule. **[1993 - 4 Marks]**

55. A mixture of two aromatic compounds *A* and *B* was separated by dissolving a chloroform followed by extraction with aqueous KOH solution. The organic layer containing compound *A*, when heated with alcoholic solution of KOH produced a compound *C*(C_7H_5N) associated with an unpleasant odour. The alkaline aqueous layer on the other hand, when heated with chloroform and then acidified gave a mixture of two isomeric compounds *D* and *E* of molecular formula $C_7H_6O_2$. Identify the compounds *A*, *B*, *C*, *D*, *E* and write their structures. **[1990 - 4 Marks]**

56. Give a chemical test and the reagents used to distinguish between the following pair of compounds :
Ethylamine and diethylamine. **[1988 - 1 Mark]**

57. An organic compound *A*, containing C, H, N and O, on analysis gives 49.32% carbon, 9.59% hydrogen and 19.18% nitrogen. *A* on boiling with NaOH gives off NH_3 and a salt which on acidification gives a monobasic nitrogen free acid *B*. The silver salt of *B* contains 59.67% silver. Deduce the structures of *A* and *B*. **[1988 - 3 Marks]**

58. Arrange the following :
   (i) methylamine, dimethylamine, aniline, N-methylaniline in incerasing order of base strength. **[1988 - 1 Mark]**
   (ii) *p*-toluidine, N, N-dimethyl-*p*-toluidine, *p*-nitroaniline, aniline in increasing basicity. **[1986 - 1 Mark]**

59. Write balanced equations for the following reaction :
Acetamide is reacted with bromine in the presence of potassium hydroxide. **[1987 - 1 Mark]**

60. Show with equations how the following compounds are prepared (equations need not be balanced) :
   (i) aniline to chlorobenzene **[1985 - 1 Mark]**
   (ii) Acetoxime from acetaldehyde using the reagents, [$K_2Cr_2O_7/H^+$, $Ca(OH)_2$ and $NH_2OH.HCl$]. **[1984 - 2 Marks]**
   (iii) Aniline from benzene **[1983 - 1 Mark]**
   (iv) chlorobenzene from aniline (in two steps). **[1982 - 1 Mark]**
   (v) *n*-propyl amine from ethyl chloride (in two steps) **[1982 - 1 Mark]**

Topic 2 - Preparation, Chemical Reactions and Importance of Diazonium Salts

1 MCQs with One Correct Answer

61. The major product of the following reaction is

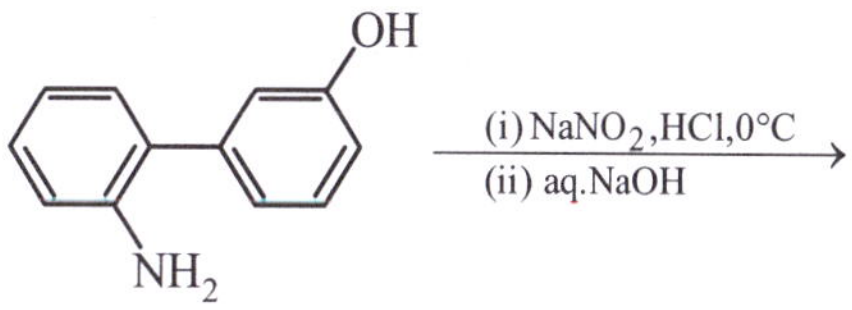

[Adv. 2017]

(a) 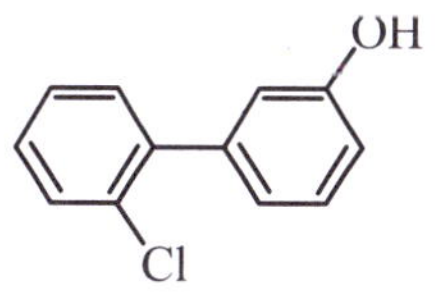

(b) 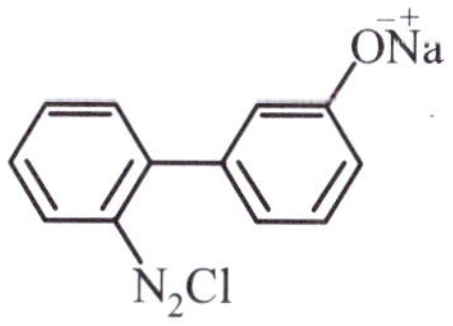

(c) 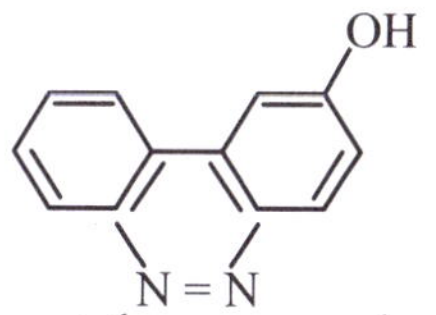

(d) 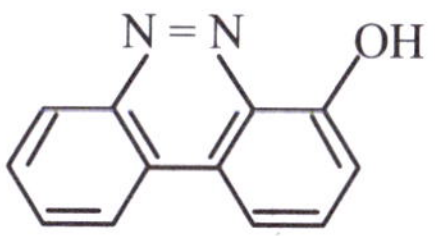

62. Amongst the compounds given, the one that would form a brilliant colored dye on treatment with $NaNO_2$ in dil. HCl followed by addition to an alkaline solution of β-naphthol is **[2011 - II]**

(a) 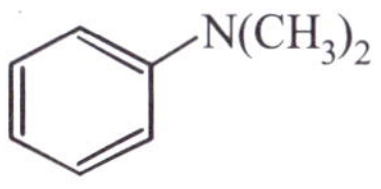

(b) 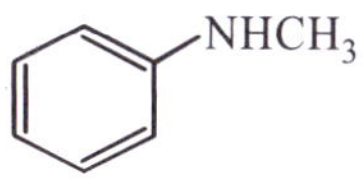

(c) 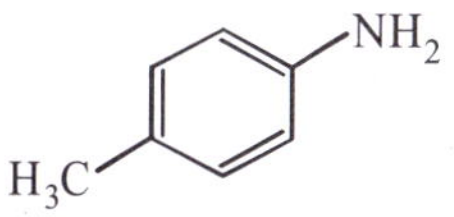

(d) 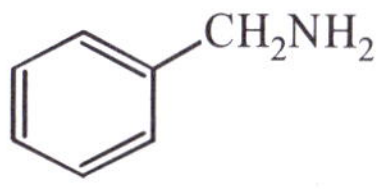

63. The correct stability order of the following resonance structures is **[2009]**

$H_2C=\overset{+}{N}=\overset{-}{N}$ (I)

$H_2\overset{+}{C}-N=\overset{-}{N}$ (II)

$H_2\overset{-}{C}-\overset{+}{N}\equiv N$ (III)

$H_2\overset{-}{C}-N=\overset{+}{N}$ (IV)

(a) (I) > (II) > (IV) > (III)
(b) (I) > (III) > (II) > (IV)
(c) (II) > (I) > (III) > (IV)
(d) (III) > (I) > (IV) > (II)

64. In the following reaction, [2006]

$CH_3NH_2 + CHCl_3 + KOH \longrightarrow$ Nitrogen containing compound $+ KCl + H_2O$.

The nitrogen containing compound is

(a) $CH_3-NH-CH_3$ (b) $CH_3-C\equiv N$

(c) $CH_3-\overset{+}{N}\equiv\overset{-}{C}$ (d) $CH_3-\overset{-}{N}\equiv\overset{+}{C}$

65. When benzenesulfonic acid and *p*-nitrophenol are treated with $NaHCO_3$, the gases released respectively are [2006]

(a) SO_2, NO (b) SO_2, NO_2

(c) CO_2, CO_2 (d) SO_2, CO_2

66.

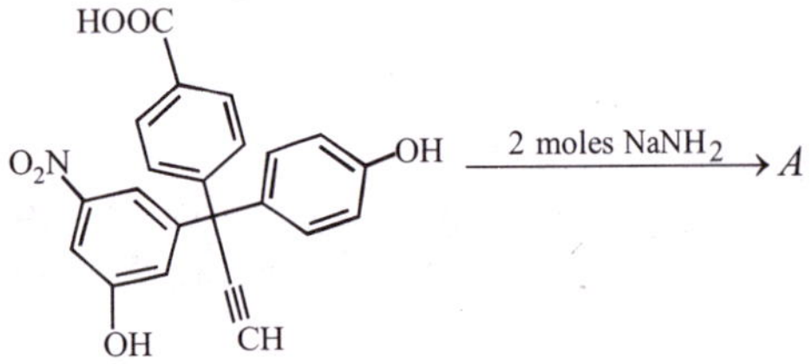

The product A will be [2003S]

(a)

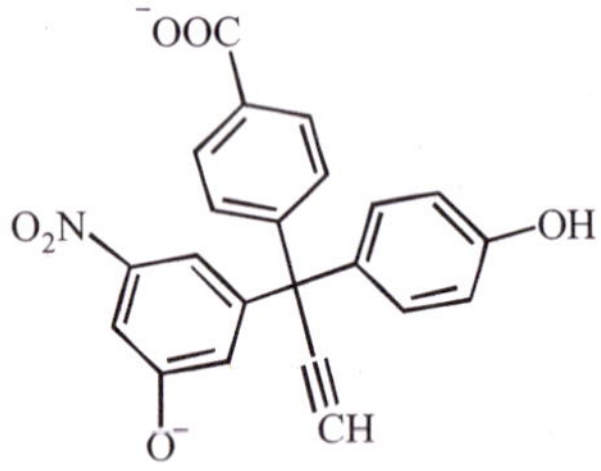

(b)

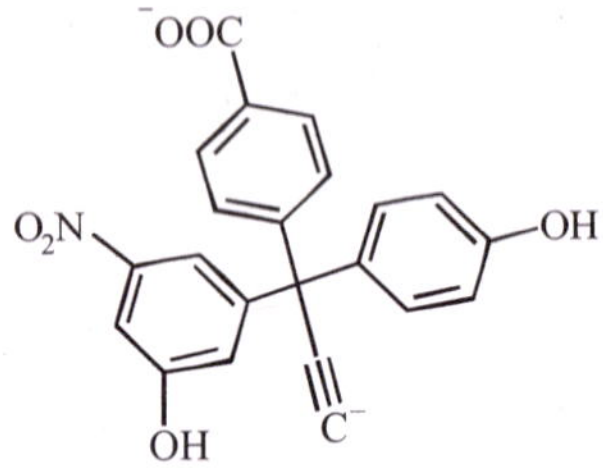

(c)

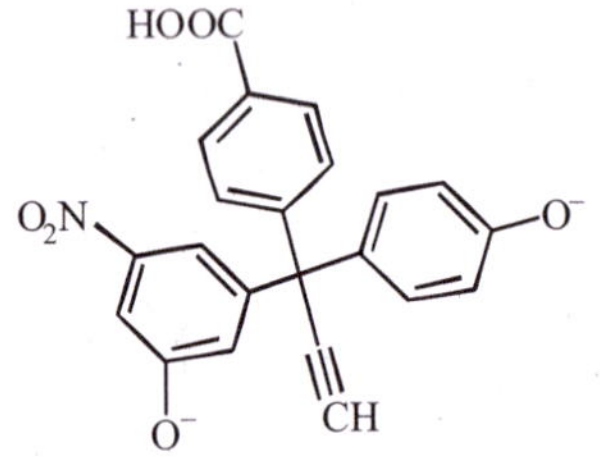

(d)

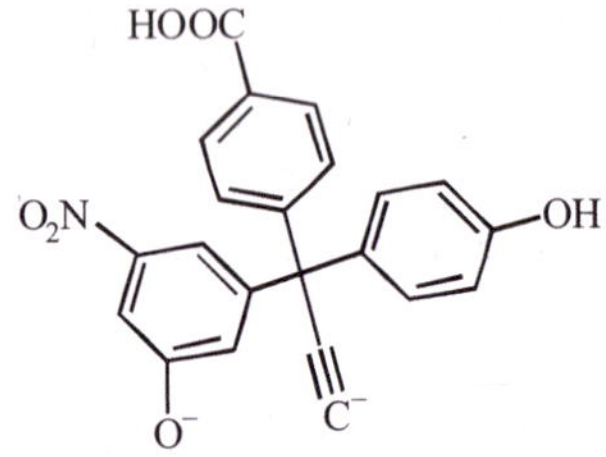

67.

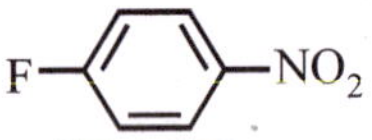

$\xrightarrow[\text{DMF, }\Delta]{(CH_3)_2NH} (A) \xrightarrow[\text{ii) } H_2/Ni]{\text{i) } NaNO_2/HCl,\ 0^\circ\text{-}5^\circ C} (B)$. (B) is: [2003S]

(a)

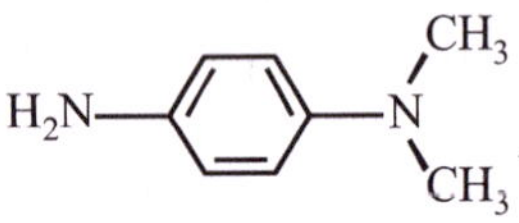

(b)

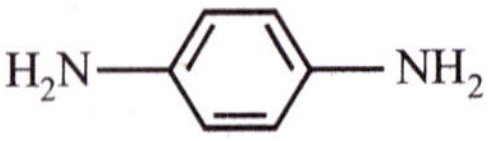

(c)

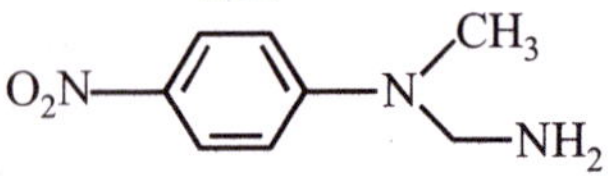

(d)

68.

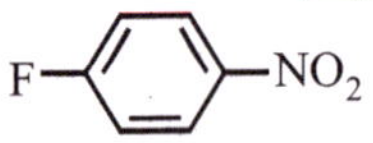

$\xrightarrow[\text{DMF, }\Delta]{(CH_3)_2NH} (A) \xrightarrow[\text{ii) } H_2/Ni]{\text{i) } NaNO_2/HCl,\ 0^\circ\text{-}5^\circ C} (B)$. (B) is: [2003S]

(a)

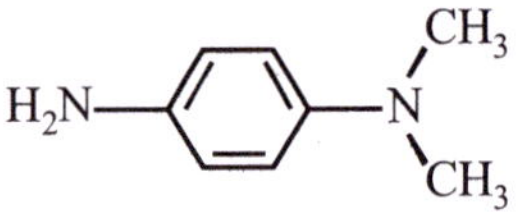

(b)

(c)

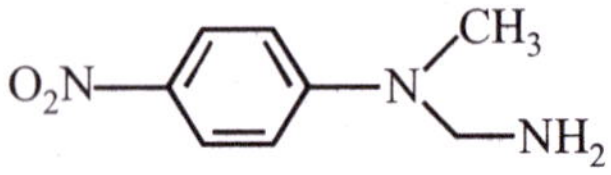

(d)

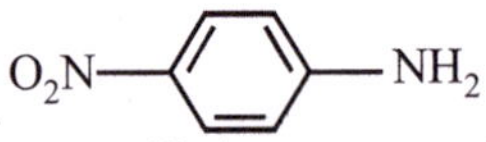

69. The most unlikely representation of resonance structures of *p*-nitrophenoxide ion is [1999 - 2 Marks]

(a)

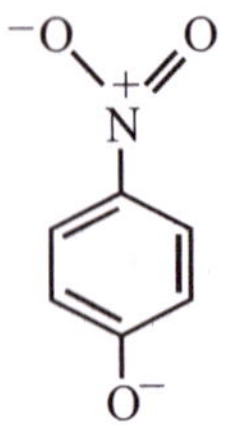

(b)

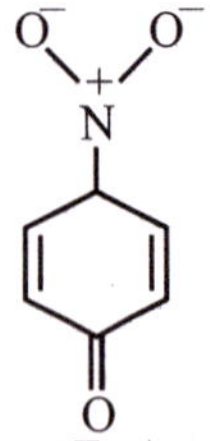

(c)

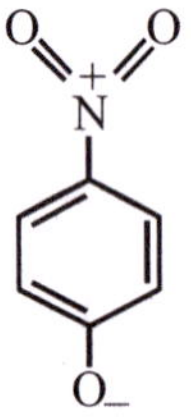

(d)

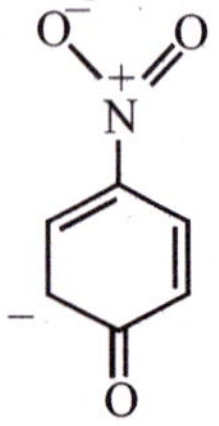

3 Numeric / New Stem Based Questions

70. Schemes 1 and 2 describes the conversion of *P* to *Q* and *R* to *S*, respectively, scheme 3 describes the synthesis of *T* from *Q* and *S*. The total number of Br atoms in a molecule of *T* is ____ [Adv. 2019]

Scheme 1 :

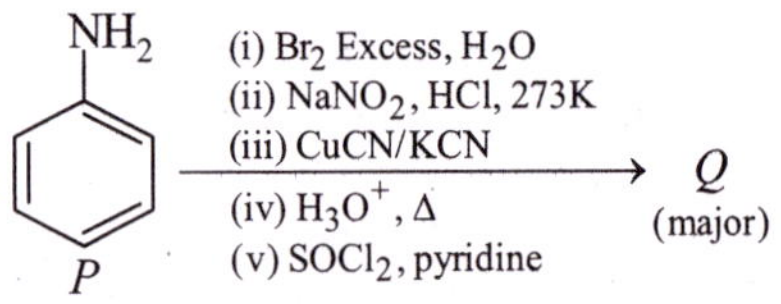

Scheme 2 :

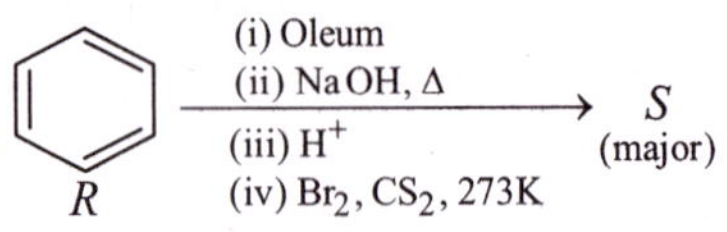

Scheme 3 : $S \xrightarrow[\text{(ii) Q}]{\text{(i) NaOH}} T$ (major)

71. Consider the reaction sequence from **P** to **Q** shown below. The overall yield of the major product **Q** from **P** is 75%. What is the amount in grams of **Q** obtained from 9.3 mL of **P**? (Use density of **P** = 1.00 g mL^{-1}; Molar mass of C = 12.0, H = 1.0, O = 16.0 and N = 14.0 g mol^{-1}) **[Adv. 2020]**

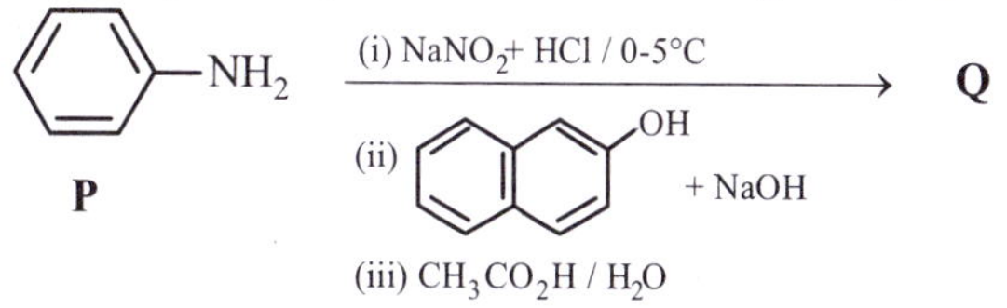

8 Comprehension Based Questions

Questions Stem

A trinitro compound, 1,3,5-tris-(4-nitrophenyl)benzene, on complete reaction with an excess of Sn/HCl gives a major product, which on treatment with an excess of $NaNO_2$/HCl at 0 °C provides P as the product. P, upon treatment with excess of H_2O at room temperature, gives the product Q. Bromination of Q in aqueous medium furnishes the product R. The compound P upon treatment with an excess of phenol under basic conditions gives the product S.

The molar mass difference between compounds Q and R is 474 g mol^{-1} and between compounds P and S is 172.5 g mol^{-1}.

72. The total number of carbon atoms and heteroatoms present in one molecule of S is ______ . **[Adv. 2023]**
[**Use:** Molar mass (in g mol^{-1}): H = 1, C = 12, N = 14, O = 16, Br = 80, Cl = 35.5
Atoms other than C and H are considered as heteroatoms]

73. The number of heteroatoms present in one molecule of R is ______ . **[Adv. 2023]**
[**Use:** Molar mass (in g mol^{-1}): H = 1, C = 12, N = 14, O = 16, Br = 80, Cl = 35.5
Atoms other than C and H are considered as heteroatoms]

4 Fill in the Blanks

74. Amongst the three isomers of nitrophenol, the one that is least soluble in water is .... **[1992 - 1 Mark]**

6 MCQs with One or More than One Correct Answer

75. In the following reaction sequence, **P**, **Q**, **S** and **T** are the major products. **[Adv. 2026]**

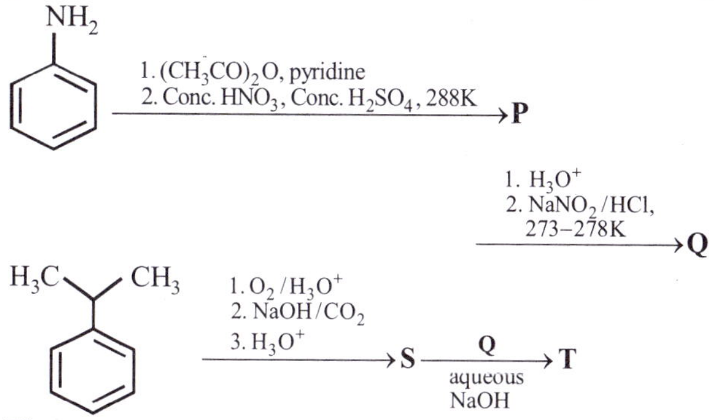

The correct statement(s) about **P**, **Q**, **S** and **T** is(are)

(a) **Q** on treatment with ethanol generates an aromatic aldehyde.
(b) **S** gives positive phthalein dye test.
(c) **P** is a dinitro compound.
(d) **T** is a coloured compound.

76. Considering the following reaction sequence, **[Adv. 2022]**

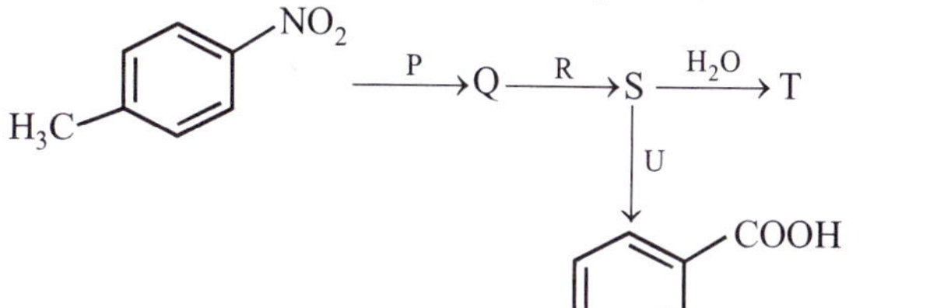

the correct option(s) is(are)

(a) **P** = H_2/Pd, ethanol **R** = $NaNO_2$/HCl
U = 1. H_3PO_2 2. $KMnO_4$ - KOH, heat
(b) **P** = Sn/HCl **R** = HNO_2

S =

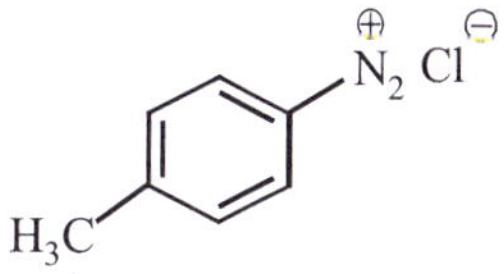

(c) **S** =

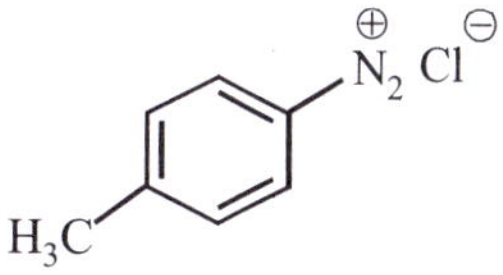

T =

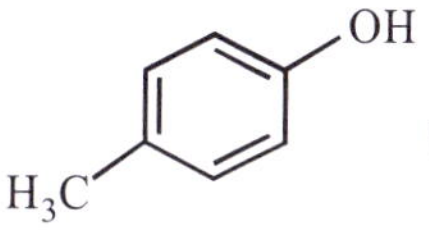

U = 1. CH_3CH_2OH
2. $KMnO_4$ - KOH, heat

(d) **Q** =

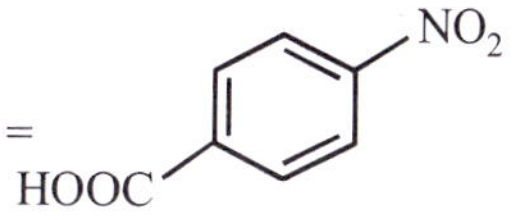

R = H_2/Pd, ethanol

T =

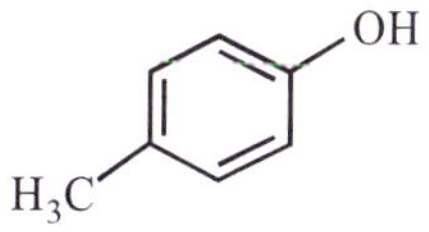

77. Aniline reacts with mixed acid (conc. HNO_3 and conc. H_2SO_4) at 288 K to give **P** (51 %), **Q** (47%) and **R** (2%). The major product(s) of the following reaction sequence is (are) **[Adv. 2018]**

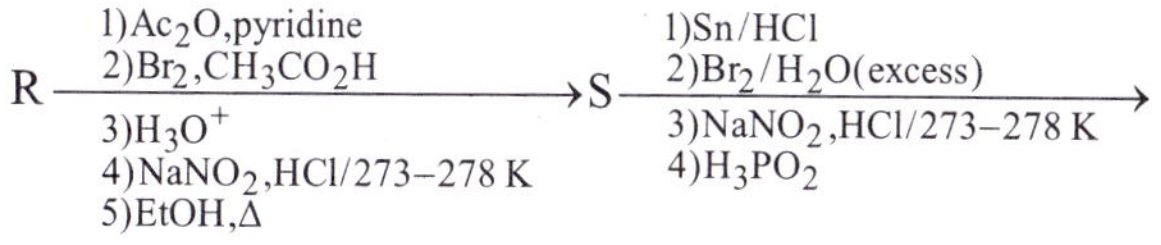

major product(s)

(a)

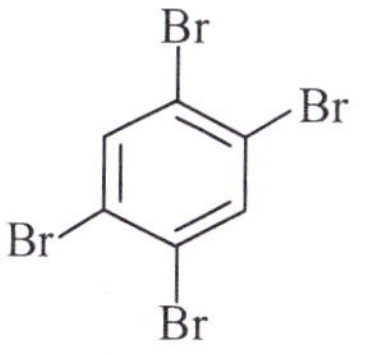

(b)

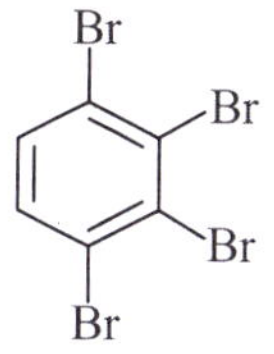

(c)

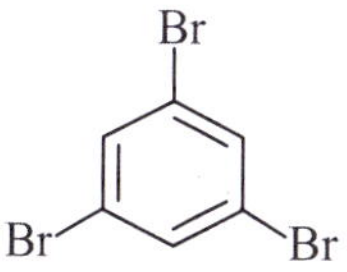

(d)

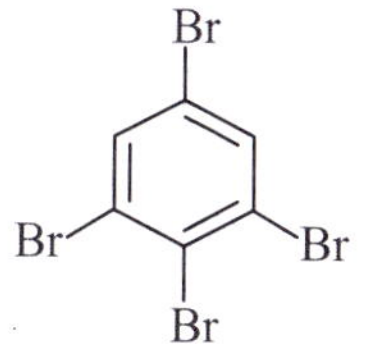

78. The product(s) of the following reaction sequence is (are) **[Adv. 2016]**

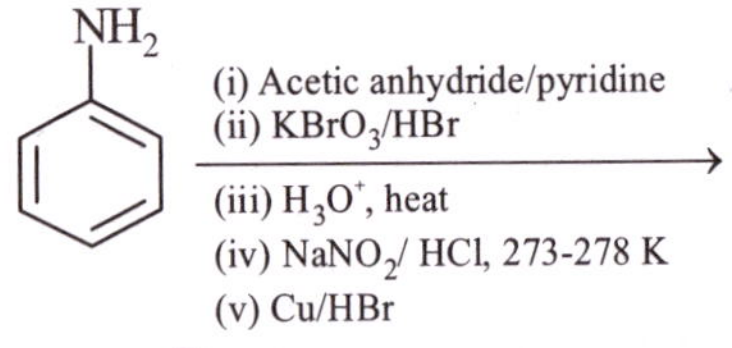

(a) 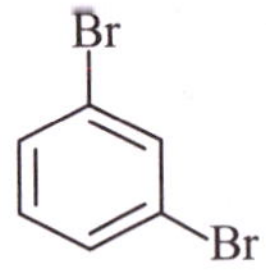

(b) 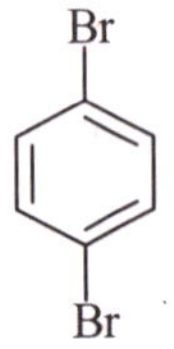

(c) 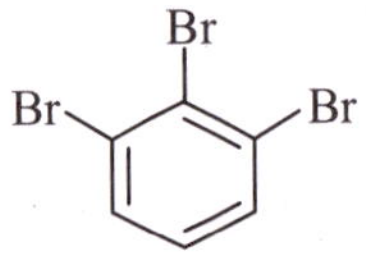

(d) 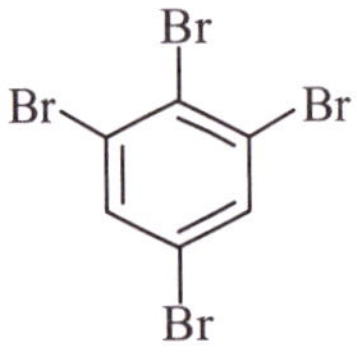

79. In the following reactions, the major product *W* is **[Adv. 2015]**

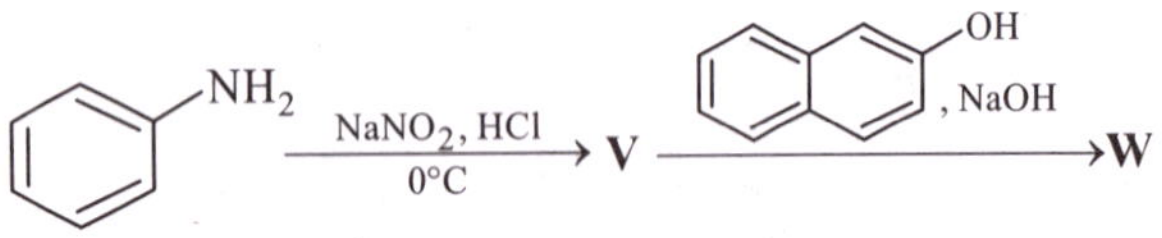

(a) 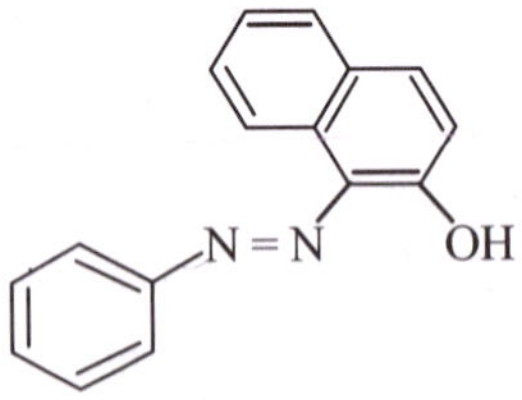

(b) 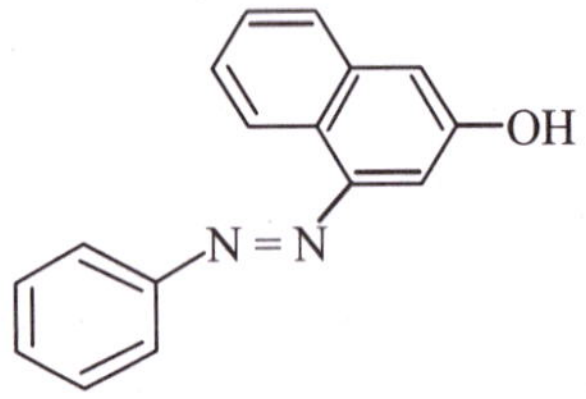

(c) 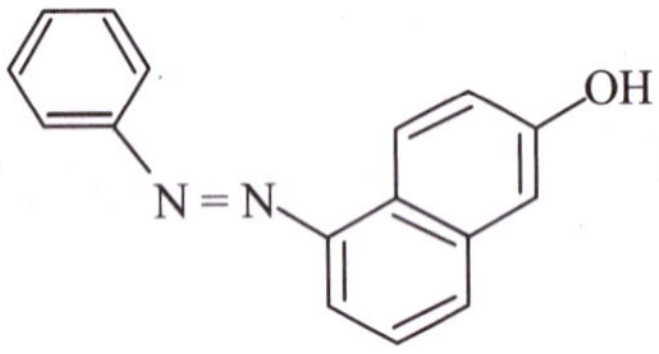

(d) 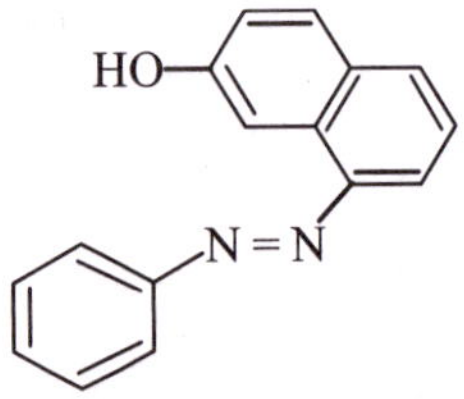

80. The major product of the reaction is **[Adv. 2015]**

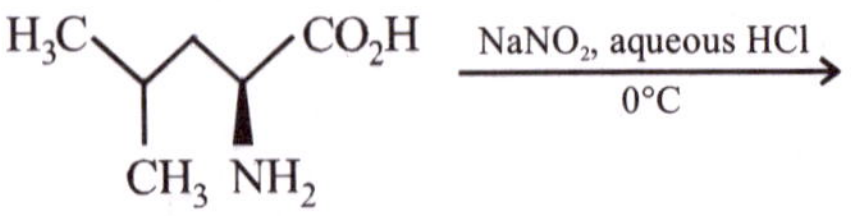

(a) 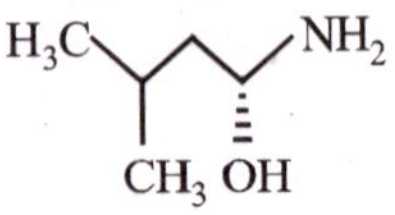

(b) 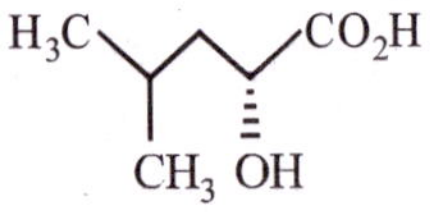

(c) 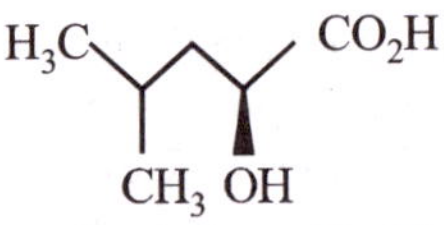

(d) 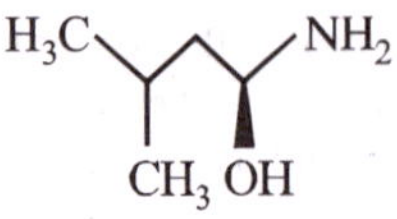

81. Benzenediazonium chloride on reaction with phenol in weakly basic medium gives **[1998 - 2 Marks]**

(a) diphenyl ether
(b) *p*-hydroxyazobenzene
(c) chlorobenzene
(d) benzene

82. When nitrobenzene is treated with Br_2 in presence of $FeBr_3$, the major product formed is *m*-bromonitrobenzene. Statements which are related to obtain the *m*-isomer are **[1992 - 1 Mark]**

(a) The electron density on meta carbon is more than that on ortho and para positions
(b) The intermediate carbonium ion formed after initial attack of Br^+ at the meta position is least destabilised
(c) Loss of aromaticity when Br^+ attacks at the ortho and para positions and not at meta position
(d) Easier loss of H^+ to regain aromaticity from the meta position than from ortho and para positions.

83. The products of reaction of alcoholic silver nitrite with ethyl bromide are **[1991 - 1 Mark]**

(a) ethane (b) ethene
(c) nitroethane (d) ethyl alcohol
(e) ethyl nitrite

7 Match the Following

84. **List-I** contains various reaction sequences and **List-II** contains different phenolic compounds. Match each entry in **List-I** with the appropriate entry in **List-II** and choose the correct option. **[Adv. 2024]**

List-I | **List-II**

(P) 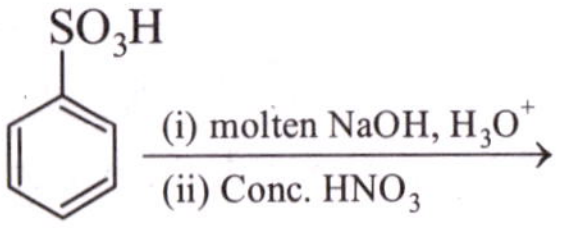

(1) 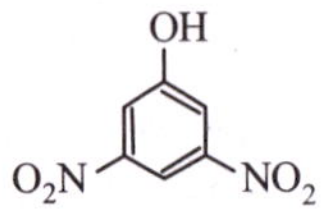

(Q) 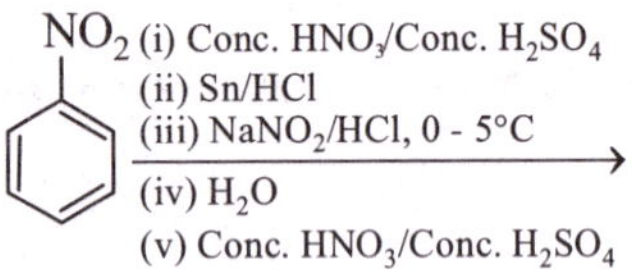

(2) 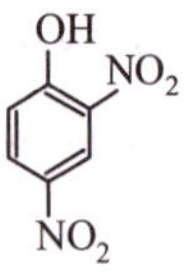

(R) 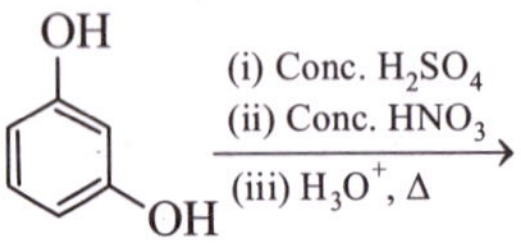

(3) 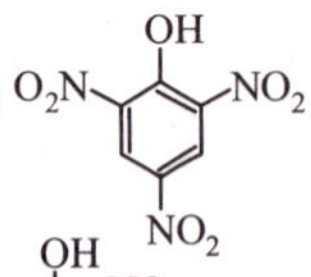

(S) 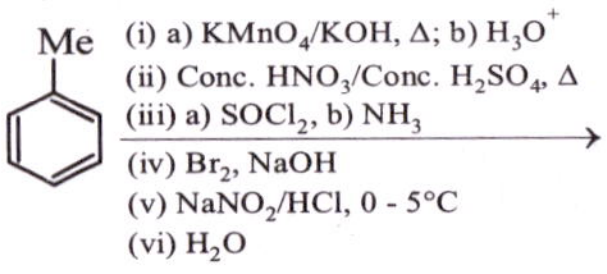

(4) 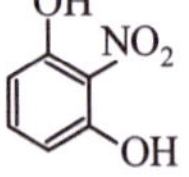

(5) 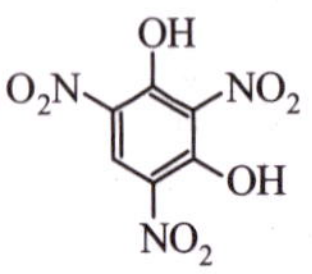

(a) P-2, Q-3, R-4, S-5 (b) P-2, Q-3, R-5, S-1
(c) P-3, Q-5, R-4, S-1 (d) P-3, Q-2, R-5, S-4

85. Match the reaction in Column-I with appropriate options in Column-II. **[2010]**

Column-I

(A) 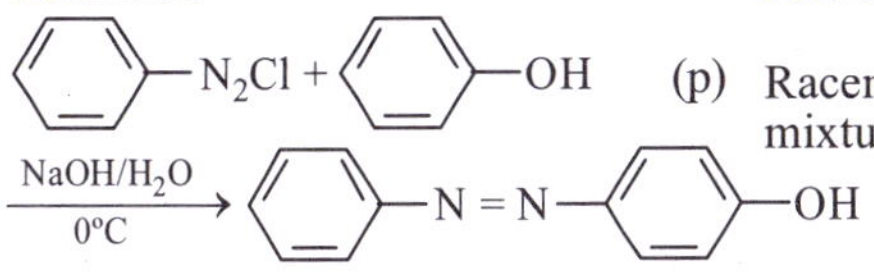

(B) 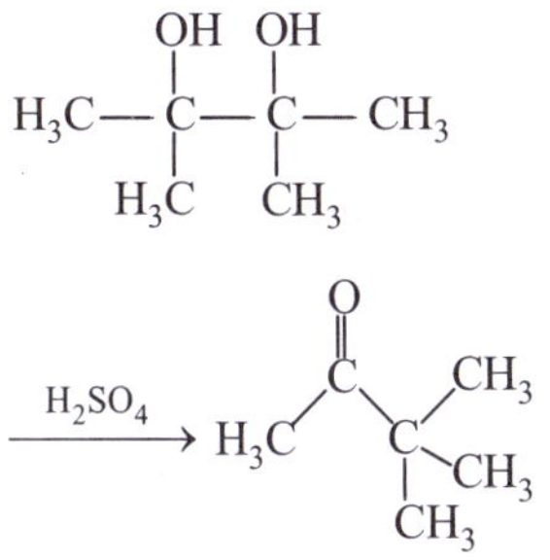

(C) 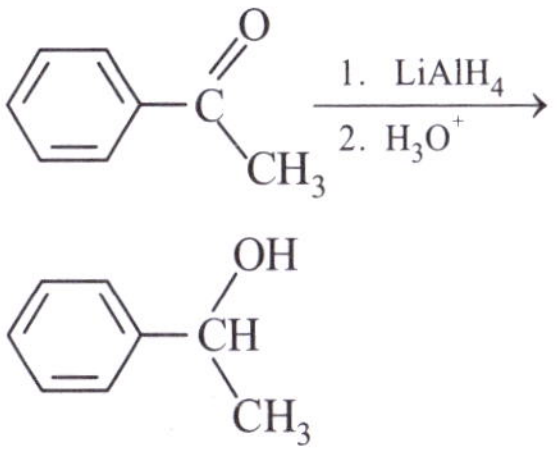

(D) 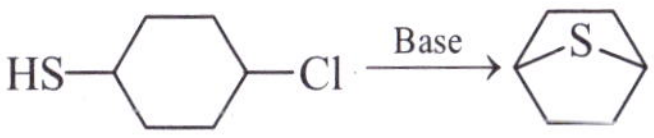

Column-II

(p) Racemic mixture

(q) Addition reaction

(r) Substitution reaction

(s) Coupling reaction

(t) Carbocation intermediate

9 Assertion and Reason/Statement Type Questions

Each question contains **STATEMENT-1 (Assertion)** and **STATEMENT-2 (Reason)**. Each question has 4 choices (a), (b), (c) and (d) out of which ONLY ONE is correct. Mark your answer as

(a) If both Statement -1 and Statement -2 are correct, and Statement -2 is the correct explanation of the Statement -2.

(b) If both Statement -1 and Statement -2 are correct, but Statement -2 is not the correct explanation of the Statement -1.

(c) If Statement -1 is correct but Statement -2 is incorrect.

(d) If Statement -1 is incorrect but Statement -2 is correct.

86. **Statement - 1:** Benzonitrile is prepared by the reaction of chlorobenzene with potassium cyanide.

Statement - 2 : Cyanide (CN^-) is a strong nucleophile. **[1998 - 2 Marks]**

87. **Statement - 1:** *p*-Nitrophenol is a stronger acid than *o*-nitrophenol.

Statement - 2 : Intramolecular hydrogen bonding makes the *o*-isomer weaker than the *p*-isomer. **[1989 - 2 Marks]**

10 Subjective Problems

88. Show with equations how the following compounds are prepared (equations need not be balanced) :

(i) Convert 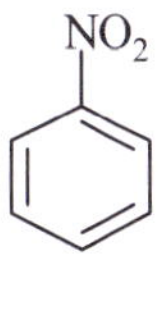 to 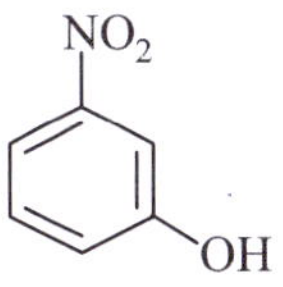 in not more than four steps. **[2004 - 4 Marks]**

(ii) 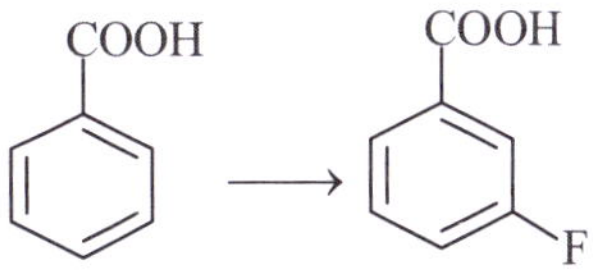 (in not more than 3 step) **[2003 - 2 Marks]**

(iii) Aniline ⟶ Benzylamine (in 3 steps) **[2000 - 3 Marks]**

(iv) benzamide from nitrobenzene **[1994 - 2 Marks]**

(v) 4-nitroaniline to 1, 2, 3-tribromobenzene. **[1990 - 2 Marks]**

(vi) benzaldehyde to cyanobenzene. (in not more than 6 steps) **[1986 - 2 Marks]**

89. How would you synthesise 4–methoxyphenol from bromobenzene in NOT more than five steps? State clearly the reagents used in each step and show the structures of the intermediate compounds in your synthetic scheme. **[2001 - 5 Marks]**

90. Identify the major product in the following reactions :

(i) 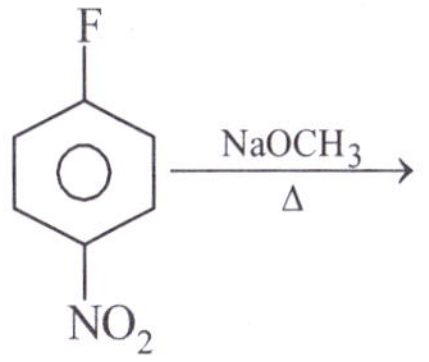 **[2000 - 1 Mark]**

(ii) 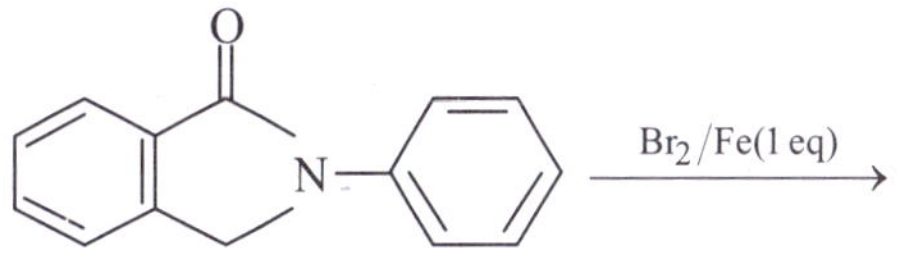 **[2000 - 1 Mark]**

(iii)  **[1993 - 1 Mark]**

91. Complete the following reaction with appropriate reagents:

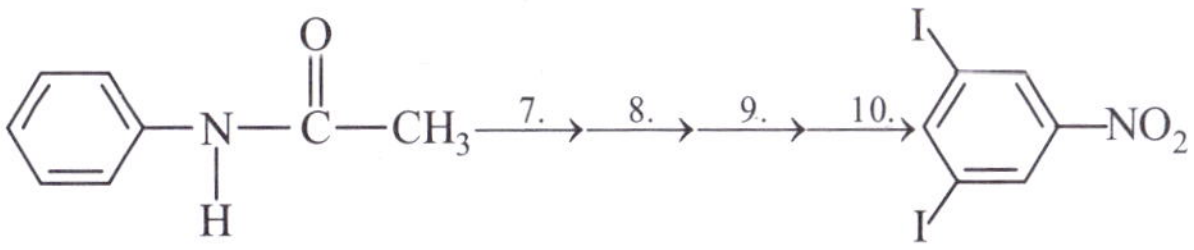

[1999 - 4 Marks]

92. Complete the following with appropriate structures :

(i) 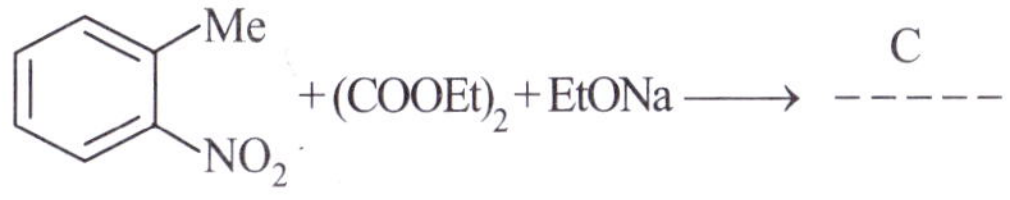 **[1997 - 1 Mark]**

(ii) 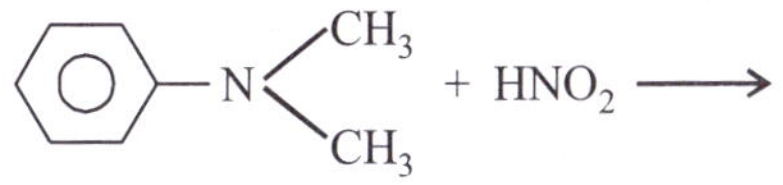 **[1992 - 1 Mark]**

(iii) 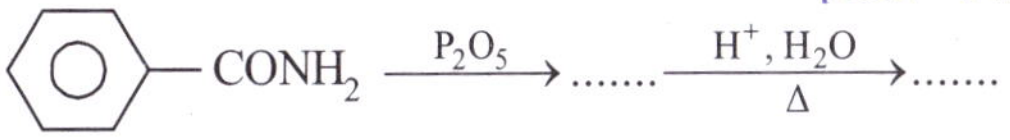 **[1992 - 1 Mark]**

93. Complete the following with appropriate structures :

$$\text{2, 4-Dinitroaniline} \xrightarrow[\text{2. anisole}]{\text{1. } NaNO_2 \text{ and HCl at } 5°C} \overset{\textbf{C}}{-----}$$

[1995 - 1 Mark]

94. Show with equations how the following compounds are prepared (equations need not be balanced) :

(i) 4-nitrobenzaldehyde from benzene. **[1994 - 2 Marks]**

(ii) *p*-bromonitrobenzene from benzene in two steps. **[1993 - 2 Marks]**

(iii) toluene to *m*-nitrobenzoic acid? **[1987 - 1 Mark]**

Answer Key

Part-A (JEE Main)																			
1	(3)	24	(4)	47	(3)	70	(1)	93	(4)	116	(4)	139	(1)	162	(77)	185	(2)	208	(3)
2	(3)	25	(4)	48	(4)	71	(4)	94	(3)	117	(2)	140	(2)	163	(3)	186	(3)	209	(3)
3	(3)	26	(1)	49	(1)	72	(1)	95	(1)	118	(2)	141	(2)	164	(1)	187	(4)	210	(3)
4	(3)	27	(1)	50	(4)	73	(3)	96	(4)	119	(3)	142	(3)	165	(2)	188	(4)	211	(1)
5	(1)	28	(2)	51	(3)	74	(2)	97	(3)	120	(4)	143	(4)	166	(2)	189	(4)	212	(1)
6	(3)	29	(1)	52	(4)	75	(1)	98	(1)	121	(1)	144	(1)	167	(4)	190	(4)	213	(3)
7	(3)	30	(2)	53	(2)	76	(1)	99	(4)	122	(4)	145	(21)	168	(3)	191	(4)	214	(1)
8	(1)	31	(4)	54	(3)	77	(1)	100	(1)	123	(3)	146	(1)	169	(2)	192	(2)	215	(2)
9	(1)	32	(3)	55	(1)	78	(3)	101	(2)	124	(3)	147	(3)	170	(3)	193	(3)	216	(1)
10	(3)	33	(2)	56	(3)	79	(2)	102	(3)	125	(2)	148	(340)	171	(3)	194	(1)	217	(3)
11	(4)	34	(4)	57	(1)	80	(1)	103	(3)	126	(4)	149	(93)	172	(3)	195	(4)	218	(3)
12	(1)	35	(1)	58	(3)	81	(2)	104	(1)	127	(1)	150	(4)	173	(3)	196	(2)	219	(1)
13	(4)	36	(1)	59	(3)	82	(1)	105	(4)	128	(3)	151	(5)	174	(3)	197	(2)	220	(3)
14	(3)	37	(3)	60	(1)	83	(4)	106	(4)	129	(4)	152	(287)	175	(1)	198	(2)	221	(4)
15	(1)	38	(1)	61	(3)	84	(4)	107	(1)	130	(1)	153	(80)	176	(1)	199	(2)	222	(171)
16	(1)	39	(1)	62	(4)	85	(3)	108	(3)	131	(2)	154	(8)	177	(3)	200	(3)	223	(154)
17	(2)	40	(2)	63	(3)	86	(4)	109	(1)	132	(2)	155	(1)	178	(3)	201	(3)	224	(591)
18	(3)	41	(4)	64	(3)	87	(3)	110	(3)	133	(2)	156	(5)	179	(1)	202	(2)	225	(20)
19	(4)	42	(3)	65	(3)	88	(1)	111	(1)	134	(3)	157	(2)	180	(3)	203	(2)	226	(9)
20	(3)	43	(3)	66	(4)	89	(2)	112	(2)	135	(2)	158	(6)	181	(2)	204	(1)		
21	(4)	44	(1)	67	(3)	90	(4)	113	(3)	136	(2)	159	(3)	182	(3)	205	(2)		
22	(1)	45	(2)	68	(3)	91	(3)	114	(4)	137	(2)	160	(5)	183	(4)	206	(2)		
23	(2)	46	(4)	69	(1,3)	92	(2)	115	(3)	138	(4)	161	(108°)	184	(2)	207	(3)		
Part-B (JEE Advanced)																			
1	(c)	9	(b)	17	(495)	27	(a,b,c)	35	(A)– p, q, s, t; (B) – s, t; (C) – p; (D) – r	62	(c)	70	(4.00)	79	(a)	87	(a)		
2	(a)	10	(d)	20	(a, c)	28	(b, d)	36	(a)	63	(b)	71	(18.60)	80	(c)				
3	(b)	11	(a)	21	(b, c, d)	29	(a, d)	37	(b)	64	(c)	72	(51)	81	(b)				
4	(b)	12	(a)	22	(c) (d)	30	(c)	38	(d)	65	(c)	73	(9)	82	(a, b, d)				
5	(d)	13	(b)	23	(c, d)	31	(c)	39	(d)	66	(a)	75	(b, d)	83	(c, e)				
6	(a)	14	(b)	24	(a)	32	(a, c)	40	(b)	67	(a)	76	(a, b, c)	84	(c)				
7	(b)	15	(c)	25	(a)	33	(b)	41	(d)	68	(a)	77	(d)	85	(A) – (r), (s) ; (B) – (t) ; (C) – (p), (q) ; (D) – (r)				
8	(d)	16	(28)	26	(a, b, d)	34	(c)	61	(c)	69	(c)	78	(b)	86	(d)				

Chapter-14

Biomolecules

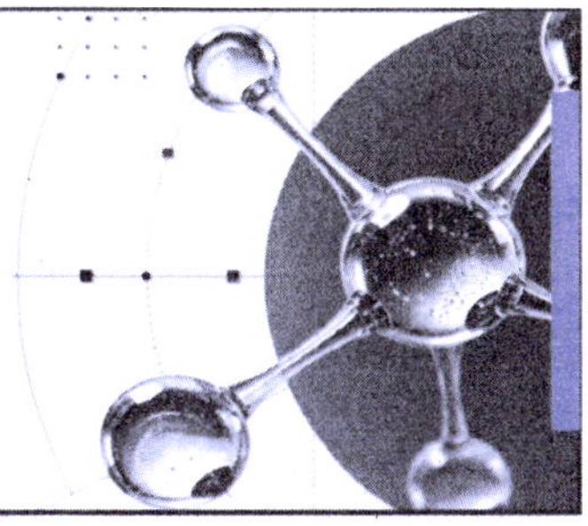

Part-A (JEE Main)

Topic 1 - Carbohydrates

MULTIPLE CHOICE QUESTIONS (MCQs)

1. The **incorrect** statement from the following with respect to carbohydrates is: [April 08, 2026 (II)]
(1) All monosaccharides are reducing sugars
(2) The monosaccharide units obtained from hydrolysis of oligosaccharides are always the same.
(3) Starch and cellulose are typical examples of polysaccharides, which are very high molecular weight compounds of more than ten monosaccharide units.
(4) Open chain and cyclic structures co-exist at equilibrium that are responsible for certain properties as in the case of D-(+)-glucose.

2. Consider the following reactions. Total number of electrons in the π bonds and lone pair of electrons in the product (X) is: [April 06, 2026 (II)]

$$\begin{matrix} CHO \\ | \\ (CHOH)_4 \\ | \\ CH_2OH \end{matrix} \xrightarrow[\text{(iii) Benzoyl chloride/Anhyd. } AlCl_3]{\text{(i) HI, } \Delta \quad \text{(ii) } V_2O_5/10-20\,atm/773\,K} \begin{matrix} (X) \\ \text{Major product} \end{matrix}$$

(1) 12 (2) 16
(3) 14 (4) 18

3. Given below are two statements : [April 05, 2026 (II)]
Statement-I:

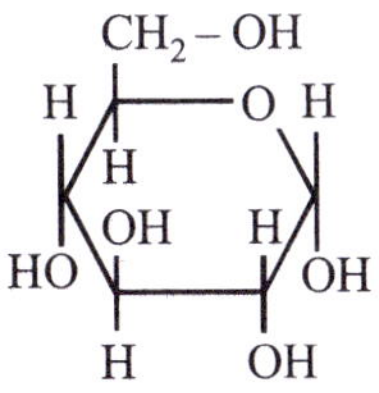 and 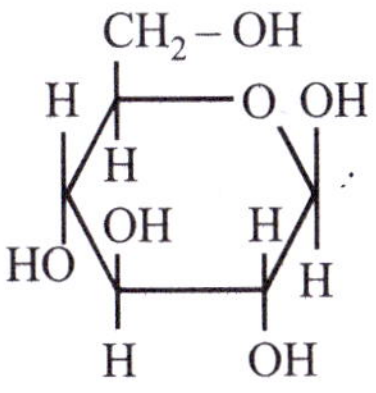

are two anomers of D-(+)-glucose.
Statement-II: The open chain forms of D-glucose and D-fructose contain three similar chiral carbons at C_3, C_4 and C_5.
In the light of the above statements, choose the **correct answer** from the options given below.
(1) Both Statement I and Statement II are true
(2) Both Statement I and Statement II are false
(3) Statement I is false but Statement II is true
(4) Statement I is true but Statement II is false

4. Identify the correct statements [April 05, 2026 (I)]
A. Glucose exist in two anomeric forms.
B. Anomers of glucose differ in configuration at C-1 in in cyclic hemiacetal structure.
C. Melting point of α-anomer of glucose is greater than β-anomer.
D. Specific rotation of α-anomer is +19° while for β-anomer is +112°.
E. α and β-anomers of glucose are prepared by crystallization of saturated glucose solution at 303 K and 371 K respectively.
Choose the correct answer from the options given below:
(1) A and B only (2) B and C only
(3) A, B and D only (4) A, B and E only

5. A D-aldotetrose on oxidation with concentrated HNO_3 resulted in optically inactive dicarboxylic acid. The structure of the D-aldotetrose is : [April 04, 2026 (II)]

(1) 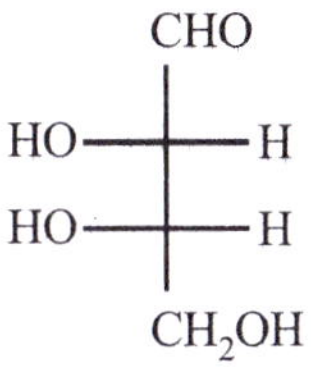 (2) 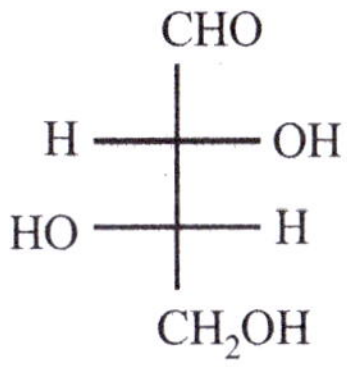

(3) 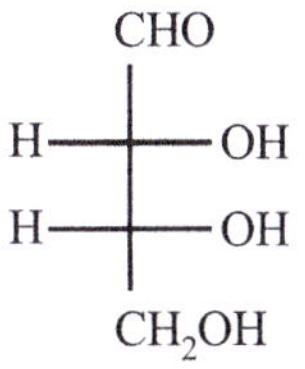 (4) 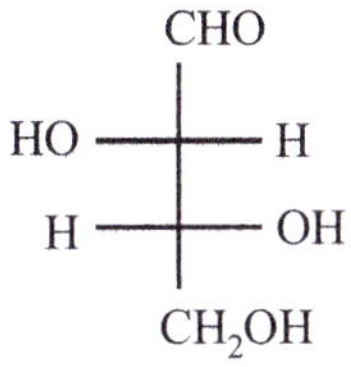

6. **Statement-I :** The structure of Maltose is given below :

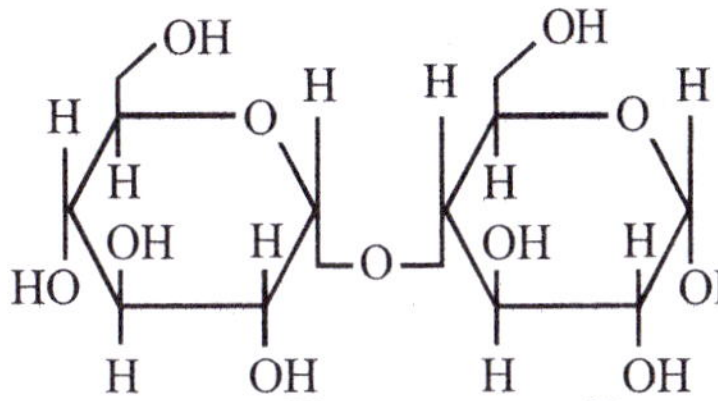

Maltose is a non reducing sugar

Statement-II : The structure of Lactose is given below

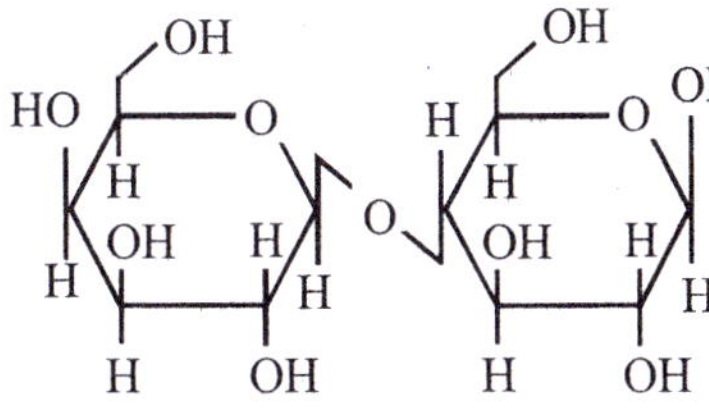

Lactose is a reducing sugar

In the light of the above statements, choose the **correct** answer from the options given below [April 04, 2026 (I)]

(1) Both Statement I and Statement II are true
(2) Both Statement I and Statement II are false
(3) Statement I is true but Statement II is false
(4) Statement I is false but Statement II is true

7. Structures of four disaccharides are given below. Among the given disaccharides, the non-reducing sugar is: [Jan. 28, 2026 (II)]

(1) 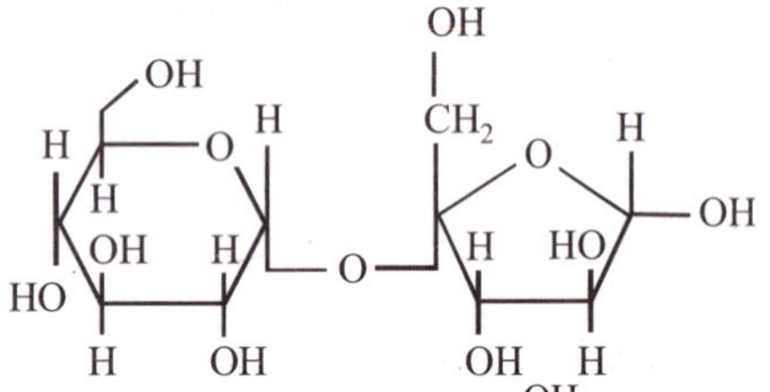

(2) 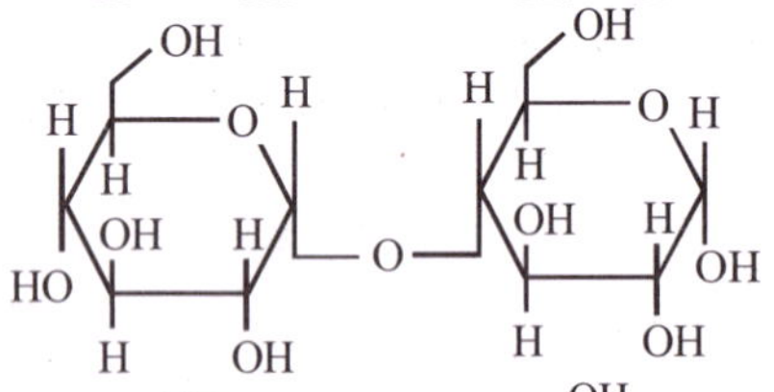

(3) 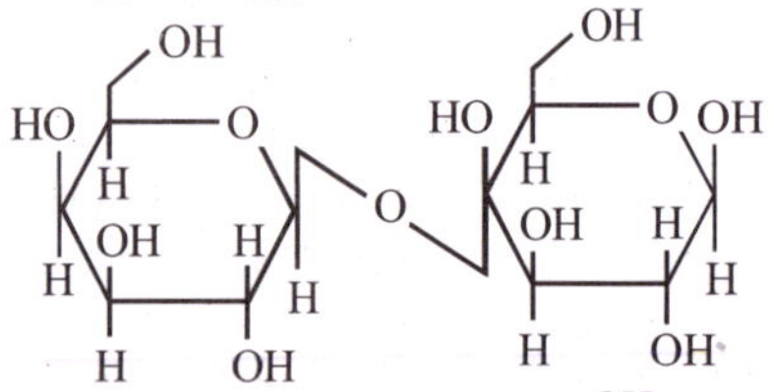

(4) 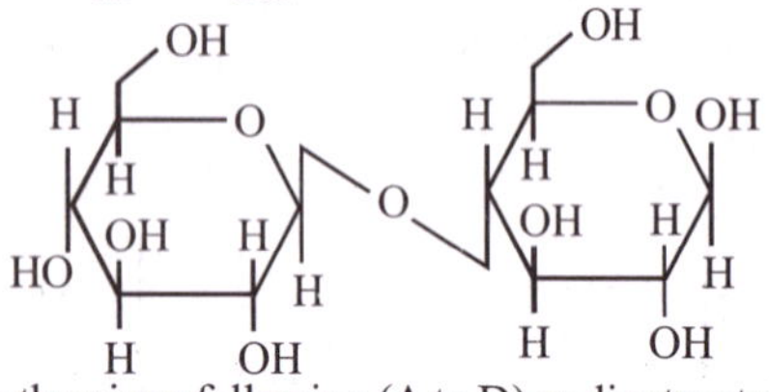

8. From the given following (A to D) cyclic structures, those which will not react with Tollen's reagent are: [Jan. 23, 2026 (I)]

(A) 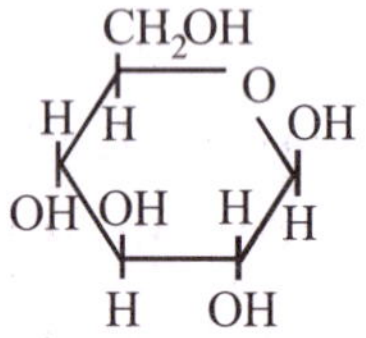

(B) 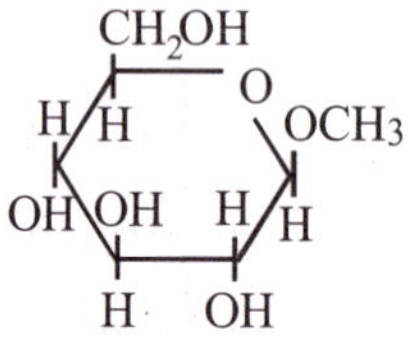

(C) 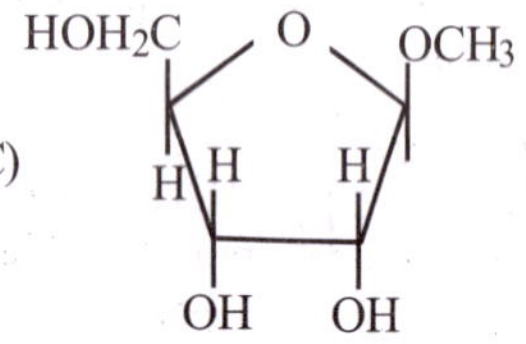

(D) 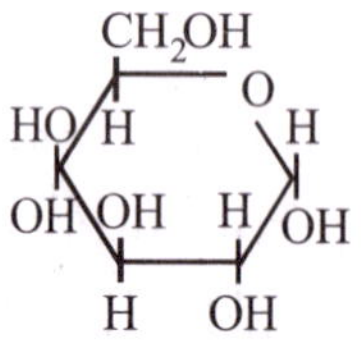

(1) B and D (2) A and D
(3) A and B (4) B and C

9. Match List-I with List-II. [Jan. 22, 2026 (II)]

List-I (Reaction of glucose with)	List-II (Product)
A. Hydroxylamine	I. Glouconic acid
B. Br_2 water	II. Glucose pentacetate
C. Excess acetic anhydride	III. Saccharic acid
D. Concentrated HNO_3	IV. Glucoxime

Choose the **correct** answer from the options given below:

(1) A-I, B-III, C-IV, D-II (2) A-IV, B-I, C-II, D-III
(3) A-III, B-I, C-IV, D-II (4) A-IV, B-III, C-II, D-I

10. Given below are two statements:

Statement I : Sucrose is dextrorotatory. However sucrose upon hydrolysis gives a solution having mixture of products. This solution shows laevorotation.

Statement II : Hydrolysis of sucrose gives glucose and fructose. Since the laevorotation of glucose is more than the dextrorotation of fructose the resulting solution becomes laevorotatory. [Jan. 22, 2026 (I)]

In the light of the above statements, choose the correct answer from the options given below.

(1) Statement I is false but Statement II is true.
(2) Both Statement I and Statement II are false.
(3) Both Statement I and Statement II are true.
(4) Statement I is true but Statement II is false.

11. Given below are two statements : [April 7, 2025 (I)]

Statement I: D – (+) – glucose + D – (+) fructose $\xrightarrow{-H_2O}$ sucrose

sucrose $\xrightarrow{\text{Hydrolysis}}$ D – (+) – glucose + D – (+) fructose

Statement II : Invert sugar is formed during sucrose hydrolysis.

In the light of the above statements, choose the correct answer from the options given below -

(1) Both Statement I and Statement II are true.
(2) Statement I is false but Statement II are true.
(3) Statement I is true but Statement II is false.
(4) Both Statement I and Statement II are false.

12. Given below are two statements : [April 3, 2025 (II)]

Statement I : Wet cotton clothes made of cellulose based carbohydrate takes comparatively longer time to get dried than wet nylon polymer based clothes.

Statement II : Intermolecular hydrogen bonding with water molecule is more in nylon-based clothes than in the case of cotton clothes.

In the light of above statements, choose the **Correct** answer from the options given below

(1) Statement I is false but Statement II is true
(2) Statement I is true but Statement II is false
(3) Both Statement I and Statement II are true
(4) Both Statement I and Statement II are false

13. Which of the following is the correct structure of L-fructose? [April 3, 2025 (I)]

(1) 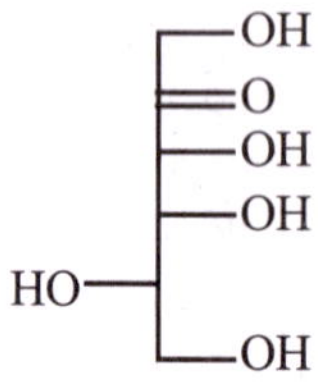

(2) 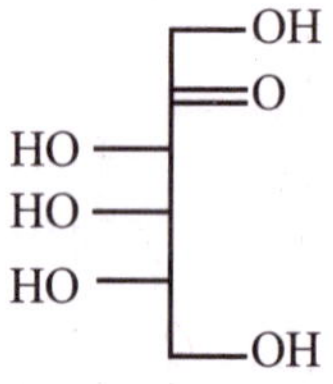

(3) 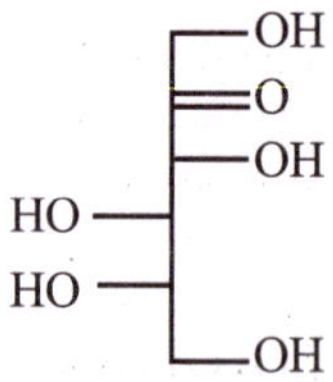

(4) 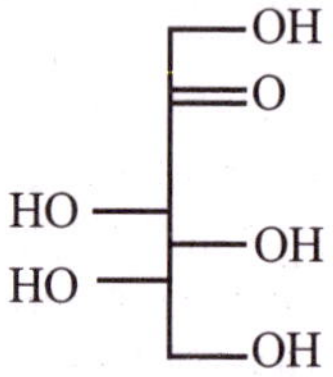

14. Match List – I with List – II. [Jan. 29, 2025 (I)]

List-I (Carbohydrate)	List-II (Linkage Source)
(A) Amylose	(I) β-C_1-C_4, plant
(B) Cellulose	(II) α-C_1-C_4, animal
(C) Glycogen	(III) α-C_1-C_4, α-C_1-C_6, plant
(D) Amylopectin	(IV) α-C_1-C_4, plant

Choose the correct answer form the options given below :
(1) (A)-(III), (B)-(II), (C)-(I), (D)-(IV)
(2) (A)-(IV), (B)-(I), (C)-(II), (D)-(III)
(3) (A)-(II), (B)-(III), (C)-(I), (D)-(IV)
(4) (A)-(IV), (B)-(I), (C)-(III), (D)-(II)

15. Identify correct conversion during acidic hydrolysis from the following : [Jan. 28, 2025 (II)]
(A) starch gives galactose.
(B) cane sugar gives equal amount of glucose and fructose.
(C) milk sugar gives glucose and galactose.
(D) amylopectin gives glucose and fructose.
(E) amylose gives only glucose.
Choose the correct answer from the options given below :
(1) (C), (D) and (E) only (2) (A), (B) and (C) only
(3) (B), (C) and (E) only (4) (B), (C) and (D) only

16. Match List-I with List-II [Jan. 28, 2025 (II)]

List-I (Saccharides)	List-II (Glycosidic-linkages found)
(A) Sucrose	(I) α 1 – 4
(B) Maltose	(II) α 1 – 4 and α 1 – 6
(C) Lactose	(III) α 1 – β 2
(D) Amylopectin	(IV) β 1 – 4

Choose the correct answer from the options given below :
(1) (A)-(III), (B)-(I), (C)-(IV), (D)-(II)
(2) (A)-(IV), (B)-(II), (C)-(I), (D)-(III)
(3) (A)-(II), (B)-(IV), (C)-(III), (D)-(I)
(4) (A)-(I), (B)-(II), (C)-(III), (D)-(IV)

17. Given below are two statements : [Jan. 28, 2025 (I)]
Statement I : D-glucose pentaacetate reacts with 2, 4-dinitrophenylhydrazine.
Statement II : Starch, on heating with concentrated sulfuric acid at 100°C and 2-3 atmosphere pressure produces glucose.
In the light of the above statements, choose the correct answer from the options given below
(1) Both Statement I and Statement II are false
(2) Statement I is false but Statement II is true
(3) Statement I is true but Statement II is false
(4) Both Statement I and Statement II are true

18. The carbohydrates "Ribose" present in DNA, is [Jan. 24, 2025 (I)]
A. A pentose sugar
B. present in pyranose from
C. in "D" configuration
D. a reducing sugar, when free
E. in α-anomeric form
Choose the correct answer from the options given below:
(1) A, C and D Only (2) A, B and E Only
(3) B, D and E Only (4) A, D and E Only

19. Given below are two statements: [Jan. 23, 2025 (I)]
Statement I: Fructose does not contain an aldehydic group but still reduces Tollen's reagent
Statement II : In the presence of base, fructose undergoes rearrangement to give glucose.
In the light of the above statements, choose the correct answer from the options given below
(1) Statement I is false but Statement II is true
(2) Both Statement I and Statement II are true
(3) Both Statement I and Statement II are false
(4) Statement I is true but Statement II is false

20. Identify the number of structure/s from the following which can be correlated to D-glyceraldehyde. [Jan. 22, 2025 (II)]

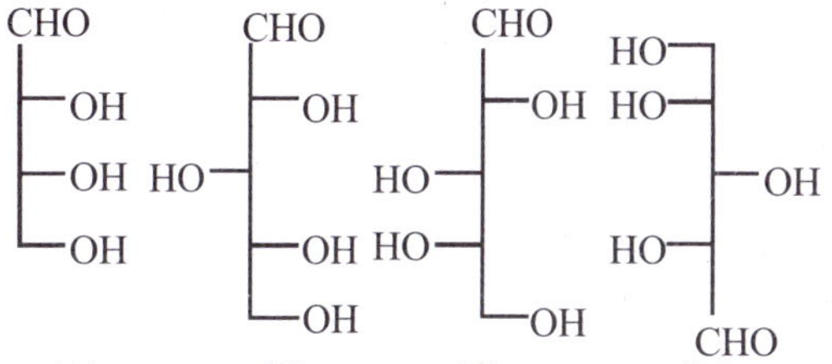

(A) (B) (C) (D)

(1) three (2) two (3) four (4) one

21. The **incorrect** statement about Glucose is : [April 9, 2024 (II)]
(1) Glucose is soluble in water because of having aldehyde functional group
(2) Glucose remains in multiple isomeric form in its aqueous solution
(3) Glucose is an aldohexose
(4) Glucose is one of the monomer unit in sucrose

22.

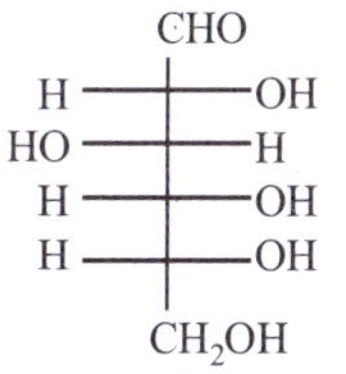

The **incorrect** statement regarding the given structure is [April 8, 2024 (I)]
(1) can be oxidized to a dicarboxylic acid with Br_2 water
(2) despite the presence of – CHO does not give Schiff's test
(3) has 4-asymmetric carbon atoms
(4) will coexist in equilibrium with 2 other cyclic structures

23. Which of the following is the correct structure of L-Glucose ? [April 4, 2024 (I)]

(1)
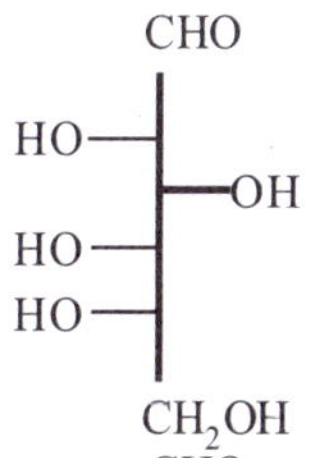

(2)
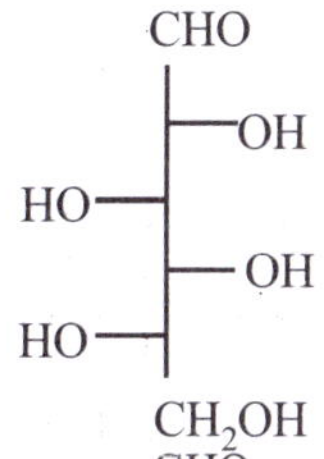

(3)
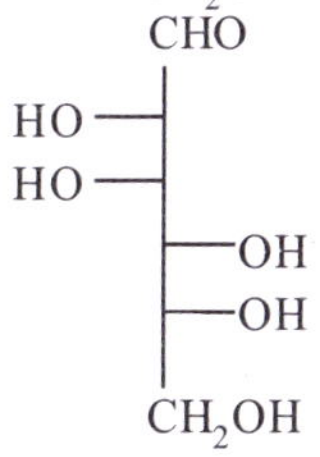

(4)
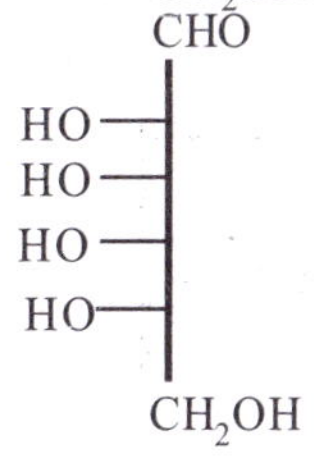

24. Match List-I with List-II [April 4, 2024 (II)]

List-I	List-II
(A) α- Glucose and α-Galactose	I. Functional isomers
(B) α- Glucose and β-Glucose	II. Homologues
(C) α- Glucose and α-Fructose	III. Anomers
(D) α- Glucose and α-Ribose	IV. Epimers

Choose the correct answer from the options given below:
(1) A-(III), B-(IV), C-(II), D-(I)
(2) A-(III), B-(IV), C-(I), D-(II)
(3) A-(IV), B-(III), C-(I), D-(II)
(4) A-(IV), B-(III), C-(II), D-(I)

25. Match List I with List II [Jan. 31, 2024 (I)]

LIST I	LIST II
A. Glucose/$NaHCO_3/\Delta$	I. Gluconic acid
B Glucose/HNO_3	II. No reaction
C Glucose/HI/Δ	III. n-Hexane
D. Glucose/Bromine water	IV. Saccharic acid

Choose the correct answer from the options given below:
(1) A-IV, B-I, C-III, D-II (2) A-I, B-IV, C-III, D-II
(3) A-II, B-IV, C-III, D-I (4) A-III, B-II, C-I, D-IV

26. Sugar which does not give reddish brown precipitate with Fehling's reagent, is : [Jan. 30, 2024 (I)]
(1) Sucrose (2) Lactose (3) Glucose (4) Maltose

27. D- (+)- Glyceraldehyde $\xrightarrow[\text{(ii) } H_2O/H^+ \text{ (iii) } HNO_3]{\text{(i) HCN}}$

The products formed in the above reaction are [April 13, 2023 (I)]
(1) Two optically active products
(2) One optically active and one meso product
(3) One optically inactive and one meso product
(4) Two optically inactive products

28. L - Isomer of tetrose X ($C_4H_8O_4$) gives positive Schiff's test and has two chiral carbons. On acetylation. 'X' yields triacetate. 'X' also undergoes following reactions

$$'A' \xleftarrow{HNO_3} 'X' \xrightarrow{NaBH_4} 'B'$$
Chiral compound

'X' is [April 11, 2023 (I)]

(1)
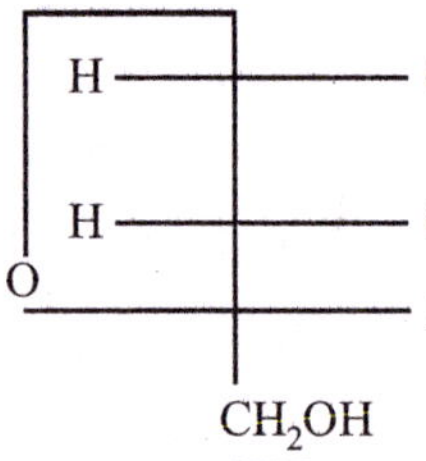

(2)
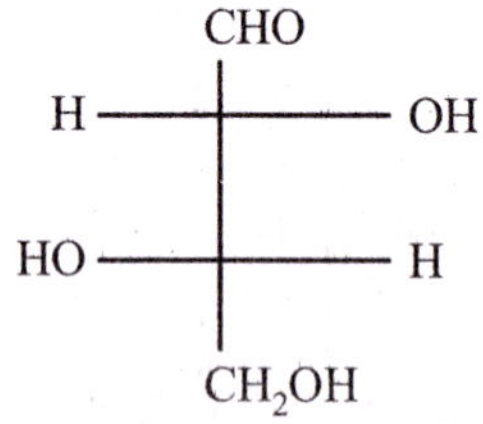

(3)
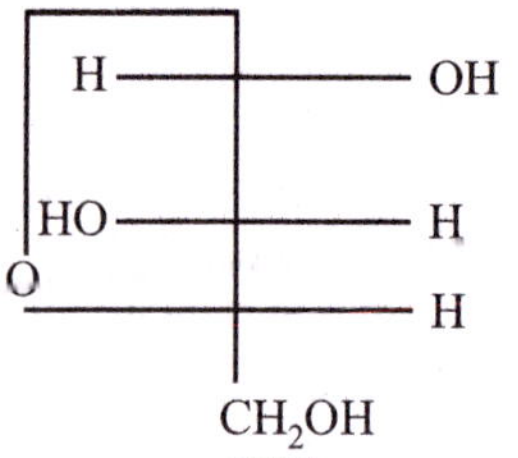

(4)
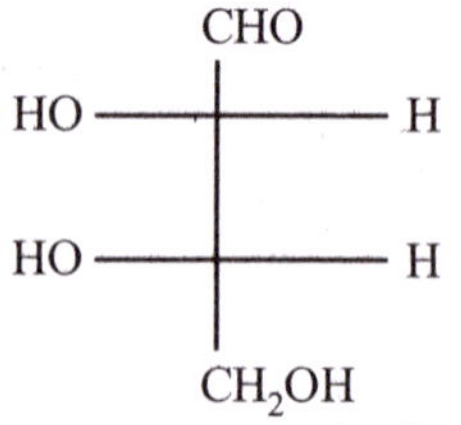

29. The correct representation in six membered pyranose form for the following sugar [X] is [Feb. 01, 2023 (I)]

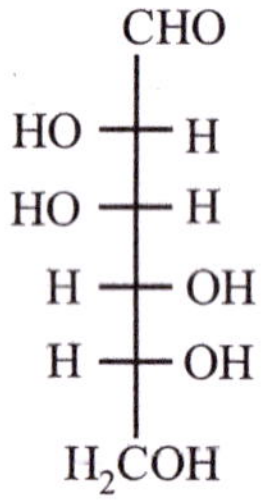
Sugar [X]

(1)
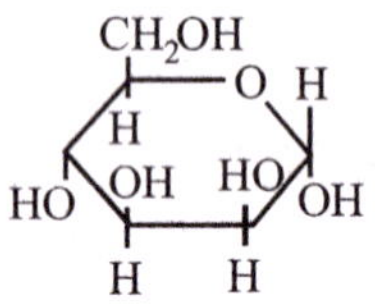
(2)
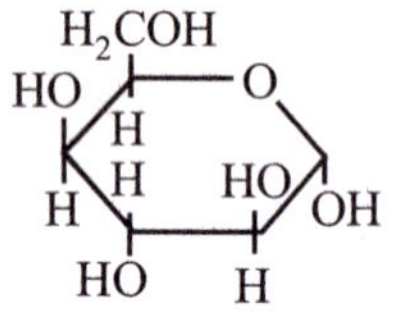
(3)
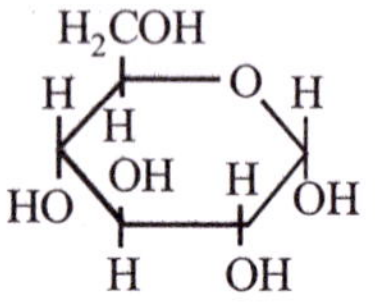
(4)
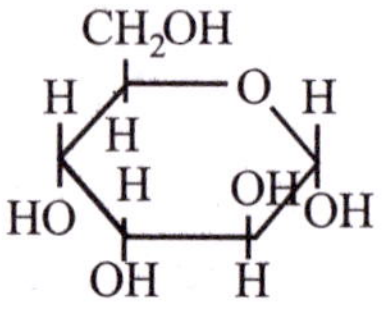

30. Given below are two statements : one is labelled as Assertion (A) and the other is labelled as Reason (R).
Assertion (A) : Ketoses give Seliwanoff's test faster than aldoses.
Reason (R) : Ketoses undergo β-elimination followed by formation of furfural.
In the light of the above statements, choose the correct answer from the options given below : [Jan. 30, 2023 (I)]
(1) (A) is false but (R) is true
(2) Both (A) and (R) are true and (R) is the correct explanation of (A)
(3) (A) is true but (R) is false
(4) Both (A) and (R) are true but (R) is not the correct explanation of (A)

31. Match items of Row I with those of Row II.
Row I : [Jan. 25, 2023 (I)]

(P)
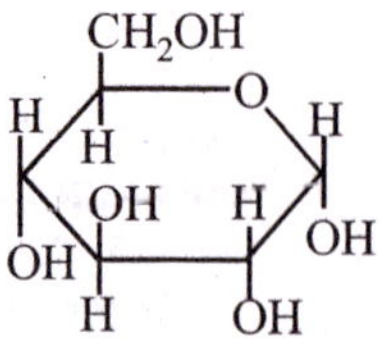
(Q)
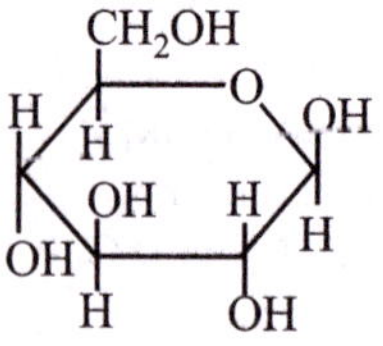

(R) 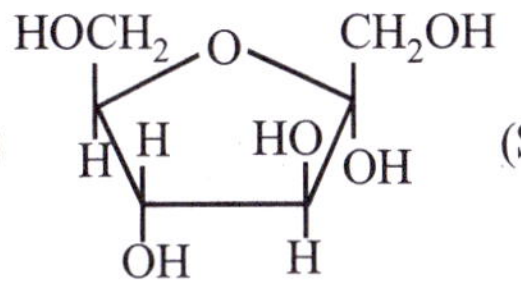 (S) 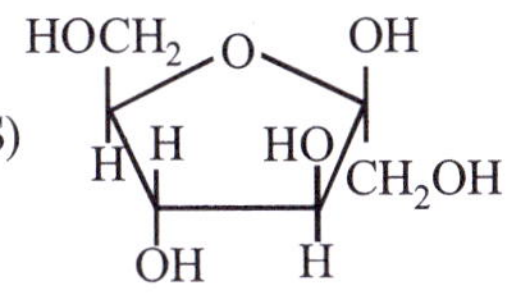

Row II :

(i) α-D-(–) Fructofuranose.

(ii) β-D-(–) Fructofuranose

(iii) α-D-(–) Glucopyranose.

(iv) β-D-(–) Glucopyranose

Correct match is

(1) P→iv, Q→iii, R→i, S→ii

(2) P→i, Q→ii, R→iii, S→iv

(3) P→iii, Q→iv, R→ii, S→i

(4) P→iii, Q→iv, R→i, S→ii

32. For the below given cyclic hemiacetal (X), the correct pyranose structure is : [July 28, 2022 (I)]

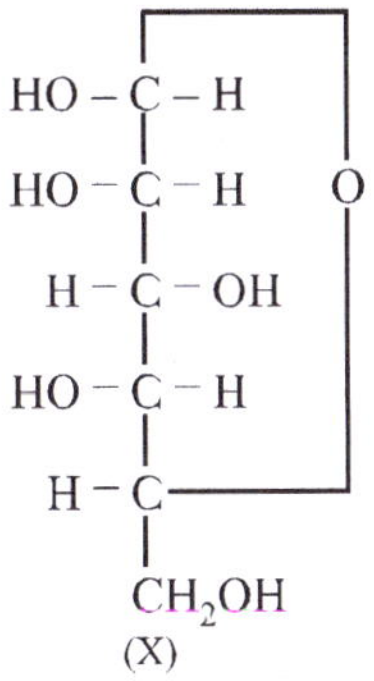

(X)

(1) 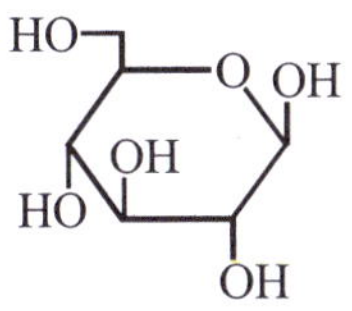 (2) 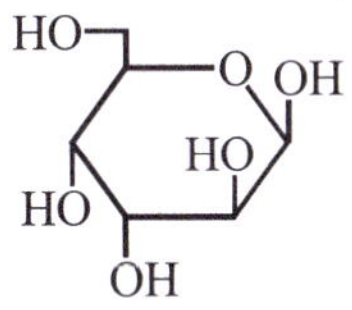

(3) 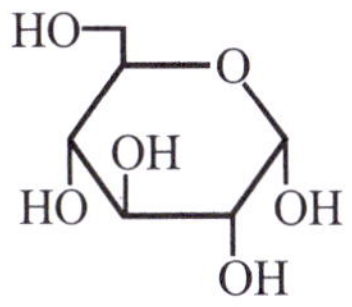 (4) 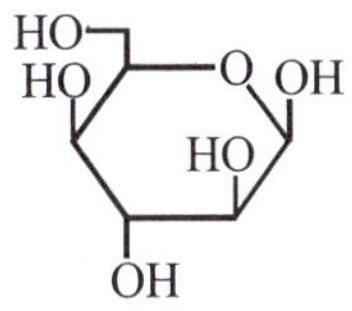

33. The formulas of A and B for the following reaction sequence

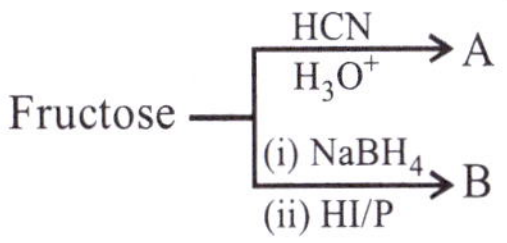

(1) $A = C_7H_{14}O_8, B = C_6H_{14}$ [July 28, 2022 (II)]

(2) $A = C_7H_{13}O_7, B = C_7H_{14}O$

(3) $A = C_7H_{12}O_8, B = C_6H_{14}$

(4) $A = C_7H_{14}O_8, B = C_6H_{14}O_6$

34. A sugar 'X' dehydrates very slowly under acidic condition to give furfural which on further reaction with resorcinol gives the coloured product after sometime. Sugar 'X' is [July 27, 2022 (I)]

(1) Aldopentose (2) Aldotetrose

(3) Oxalic acid (4) Ketotetrose

35. Match List-I with List-II [July 27, 2022 (II)]

List-I	List-II
(A) Glucose + HI	(I) Gluconic acid
(B) Glucose + Br_2 water	(II) Glucose pentacetate
(C) Glucose + acetic anhydride	(III) Saccharic acid
(D) Glucose + HNO_3	(IV) Hexane

Choose the correct answer from the options given below:

(1) (A)-(IV), (B)-(I), (C)-(II), (D)-(III)

(2) (A)-(IV), (B)-(III), (C)-(II), (D)-(I)

(3) (A)-(III), (B)-(I), (C)-(IV), (D)-(II)

(4) (A)-(I), (B)-(III), (C)-(IV), (D)-(II)

36. Which one of the following is a reducing sugar? [July 26, 2022 (I)]

(1) 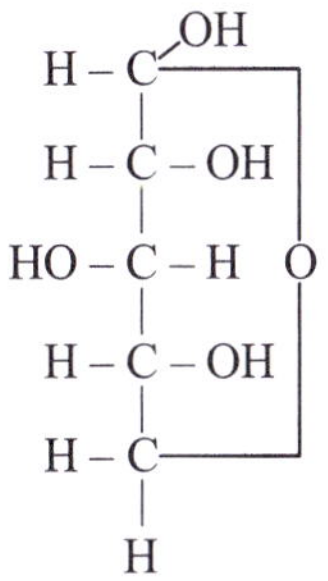 (2) 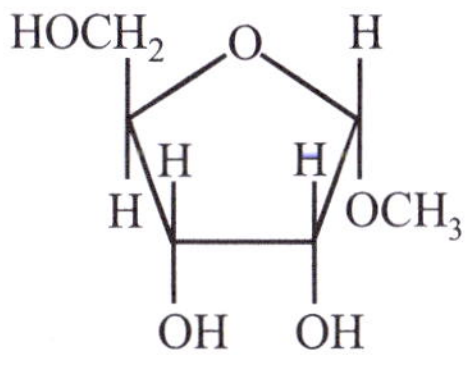

(3) 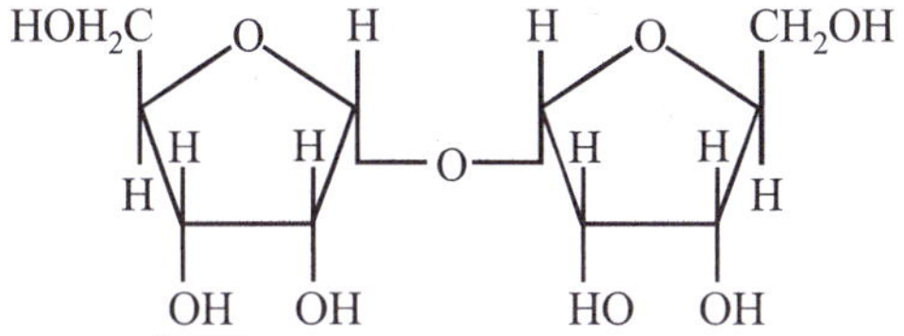

(4) 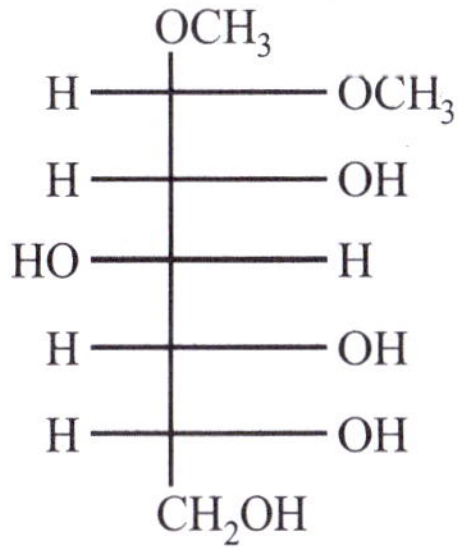

37. Glycosidic linkage between C_1 of α-glucose and C_2 of β-fructose is found in [July 25, 2022 (II)]

(1) maltose (2) sucrose

(3) lactose (4) amylose

38. When sugar 'X' is boiled with dilute H_2SO_4 in alcoholic solution, two isomers 'A' and 'B' are formed. 'A' on oxidation with HNO_3 yields saccharic acid whereas 'B' is laevorotatory. The compound 'X' is : [June 28, 2022 (II)]

(1) Maltose (2) Sucrose

(3) Lactose (4) Starch

39. L-Isomer of a compound 'A' ($C_4H_8O_4$) gives a positive test with $[Ag(NH_3)_2]^+$. Treatment of 'A' with acetic anhydride yield triacetate derivative. Compound 'A' produces an optically active compound (B) and an optically inactive compound (C) on treatment with bromine water and HNO_3 respectively, compound (A) is: [June 27, 2022 (I)]

(1) 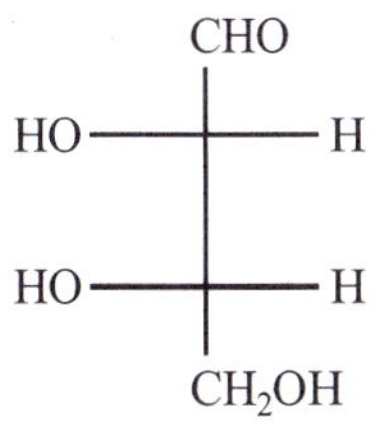 (2) 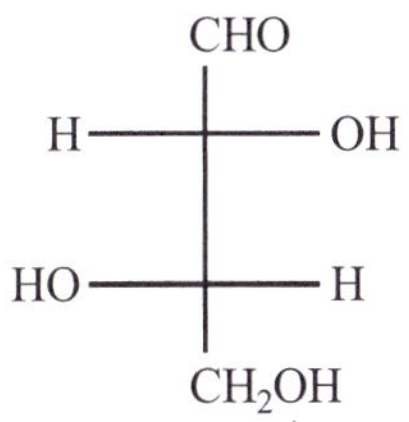

(3) 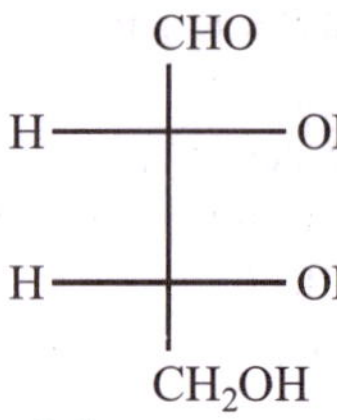 (4) 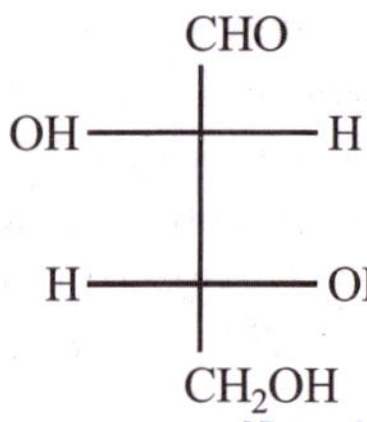

40. Given below are two statements. **[June 27, 2022 (II)]**

Statement I: Maltose has two α-D-glucose units linked at C_1 and C_4 and is a reducing sugar.

Statement II: Maltose has two monosaccharides: α-D-glucose and β-D-glucose linked at C_1 and C_6 and it is a non-reducing sugar.

In the light of the above statements, choose the correct answer from the options given below.

(1) Both Statement I and Statement II are true.
(2) Both Statement I and Statement II are false.
(3) Statement I is true but Statement II is false.
(4) Statement I is false but Statement II is true.

41. Which one of the following compounds contains β-C_1-C_4 glycosidic linkage? **[Aug. 31, 2021 (I)]**

(1) Lactose (2) Sucrose
(3) Maltose (4) Amylose

42. Given below are two statements, one is labelled as **Assertion (A)** and other is labelled as **Reason (R).**

Assertion (A) : Sucrose is a disaccharide and a non-reducing sugar.

Reason (R) : Sucrose involves glycosidic linkage between C_1 of β-glucose and C_2 of β-fructose.

Choose the **most appropriate** answer from the options given below: **[Aug. 26, 2021 (II)]**

(1) Both (A) and (R) are true but (R) is not the true explanation of (A)
(2) (A) is false but (R) is true.
(3) (A) is true but (R) is false
(4) Both (A) and (R) are true and (R) is the true explanation of (A)

43. Compound A gives D-galactose and D-glucose on hydrolysis. The compound A is : **[July 27, 2021 (II)]**

(1) Amylose (2) Sucrose
(3) Maltose (4) Lactose

44. Fructose is an example of **[March 17, 2021 (II)]**

(1) Pyranose (2) Heptose
(3) Aldohexose (4) Ketohexose

45. Which one of the following statements is not true? **[Sep. 06, 2020 (II)]**

(1) Lactose contains α-glycosidic linkage between C_1 of galactose and C_4 of glucose.
(2) Lactose is a reducing sugar and it gives Fehling's test.
(3) Lactose ($C_{11}H_{22}O_{11}$) is a disaccharide and it contains 8 hydroxyl groups.
(4) On acid hydrolysis, lactose gives one molecule of D(+)-glucose and one molecule of D(+)-galactose.

46. What are the functional groups present in the structure of maltose? **[Sep. 04, 2020 (I)]**

(1) One ketal and hemiketal
(2) Two acetals
(3) One acetal and one hemiacetal
(4) One acetal and one ketal

47. Consider the following reactions : **[Sep. 02, 2020 (I)]**

(i) Glucose + ROH $\xrightarrow{\text{dry HCl}}$ Acetal $\xrightarrow[(CH_3CO)_2O]{x\text{ eq. of}}$ Acetyl derivative

(ii) Glucose $\xrightarrow{Ni/H_2}$ A $\xrightarrow[(CH_3CO)_2O]{y\text{ eq. of}}$ Acetyl derivative

(iii) Glucose $\xrightarrow[(CH_3CO)_2O]{z\text{ eq. of}}$ Acetyl derivative

'x', 'y' and 'z' in these reactions are respectively.

(1) 5, 4 & 5 (2) 4, 6 & 5
(3) 4, 5 & 5 (4) 5, 6 & 5

48. The correct observation in the following reactions is :

$$\text{Sucrose} \xrightarrow[\substack{\text{cleavage}\\\text{(Hydrolysis)}}]{\text{Glycosidic bond}} A + B \xrightarrow[\text{reagent}]{\text{Seliwanoff's}} ?$$

[Sep. 02, 2020 (II)]

(1) Formation of blue colour
(2) Gives no colour
(3) Formation of red colour
(4) Formation of violet colour

49. Which of the following statement is not true for glucose? **[Jan. 08, 2020 (I)]**

(1) Glucose exists in two crystalline forms α and β
(2) Glucose gives Schiff's test for aldehyde
(3) Glucose reacts with hydroxylamine to form oxime
(4) The pentaacetate of glucose does not react with hydroxylamine to give oxime

50. Two monomers in maltose are: **[Jan. 08, 2020 (II)]**

(1) α-D-glucose and β-D-glucose
(2) α-D-glucose and α-D-galactose
(3) α-D-glucose and α-D-fructose
(4) α-D-glucose and α-D-glucose

51. Which of the following statements is correct? **[Jan. 07, 2020 (II)]**

(1) Gluconic acid can form cyclic (acetal/hemiacetal) structure
(2) Gluconic acid is a dicarboxylic acid
(3) Gluconic acid is a partial oxidation product of glucose
(4) Gluconic acid is obtained by oxidation of glucose with HNO_3

52. Glucose and galactose are having identical configuration in all the positions except position. **[April 12, 2019 (I)]**

(1) C – 3 (2) C – 4
(3) C – 2 (4) C – 5

53. Number of stereo centers present in linear and cyclic structures of glucose are respectively: **[April 10, 2019 (II)]**

(1) 5 & 4 (2) 4 & 4 (3) 5 & 5 (4) 4 & 5

54. Which of the following statements is not true about sucrose?

(1) It is a non reducing sugar **[April 9, 2019 (I)]**
(2) The glycosidic linkage is present between C_1 of α-glucose and C_1 of β-fructose
(3) It is also named as invert sugar
(4) On hydrolysis, it produces glucose and fructose

55. Maltose on treatment with dilute HCl gives : **[April 8, 2019 (I)]**

(1) D-Glucose and D-Fructose
(2) D-Fructose
(3) D-Galactose
(4) D-Glucose

56. Fructose and glucose can be distinguished by: **[April 8, 2019 (II)]**

(1) Benedict's test (2) Fehling's test
(3) Barfoed's test (4) Seliwanoff's test

57. Glucose on prolonged heating with HI gives : [2018]
(1) *n*-Hexane (2) 1- Hexene
(3) Hexanoic acid (4) 6-iodohexanal

58. Among the following, the incorrect statement is: [Online April 16, 2018]
(1) Cellulose and amylose have 1,4-glycosidic linkage
(2) Lactose contains β-D-galactose and β-D-glucose
(3) Maltose and lactose have 1,4-glycosidic linkage
(4) Sucrose and amylose have 1,2-glycosidic linkage

59. Which of the following compounds will behave as a reducing sugar in an aqueous KOH solution? [2017]

(1) 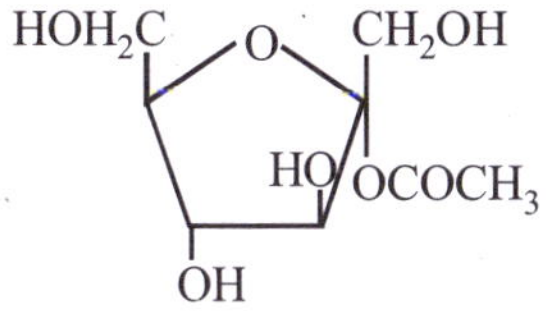

(2) 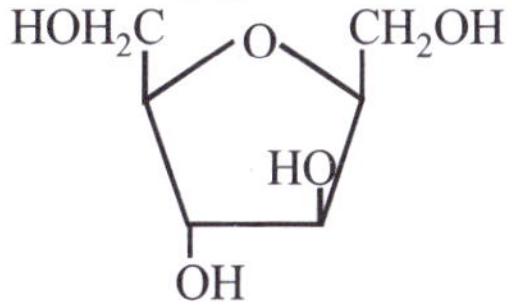

(3) 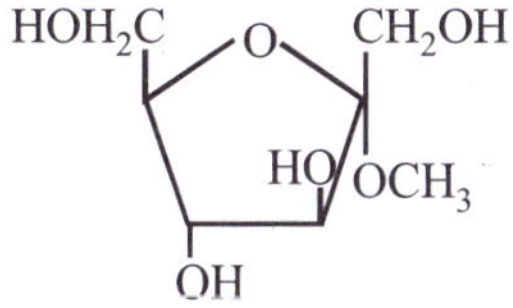

(4) 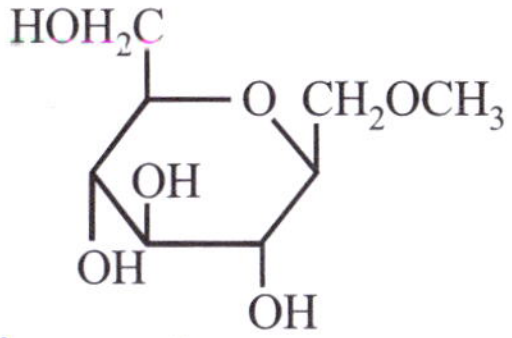

60. The incorrect statement among the following is : [Online April 9, 2017]
(1) α-D-glucose and β-D-glucose are anomers.
(2) α-D-glucose and β-D-glucose are enantiomers.
(3) Cellulose is a straight chain polysaccharide made up of only β-D-glucose units.
(4) The penta-acetate of glucose does not react with hydroxyl amine.

61. Accumulation of which of the following molecules in the muscles occurs as a result of vigorous exercise ? [Online April 11, 2015]
(1) Glycogen (2) Glucose
(3) Pyruvic acid (4) L-lactic acid

62. Which of the following will not show mutarotation? [Online April 12, 2014]
(1) Maltose (2) Lactose
(3) Glucose (4) Sucrose

63. Glycosidic linkage is actually a (an) : [Online April 23, 2013]
(1) Carbonyl bond (2) Ether bond
(3) Ester bond (4) Amide bond

64. Synthesis of each molecule of glucose in photosynthesis involves : [2013]
(1) 18 molecules of ATP (2) 10 molecules of ATP
(3) 8 molecules of ATP (4) 6 molecules of ATP

65. Natural glucose is termed D-glucose because : [Online April 23, 2013]
(1) – OH on the second carbon is on the right side in Fischer projection
(2) – OH on the sixth carbon is on the right side in Fischer projection.
(3) – OH on the fifth carbon is on the right side in Fischer projection.
(4) It is dextrorotatory.

66. Which of the following statement is not correct? [Online April 25, 2013]
(1) Amylopectin is a branched polymer of α - glucose.
(2) Cellulose is a linear polymer of β-glucose
(3) Glycogen is the food reserve of plants
(4) All proteins are polymers of α - amino acids.

67. Which of the following compounds can be detected by Molisch's test ? [2012]
(1) Nitro compounds (2) Sugars
(3) Amines (4) Primary alcohols

68. Which of the following is a non-reducing sugar? [Online May 19, 2012]
(1) Lactose (2) Fructose
(3) Sucrose (4) Maltose

69. The change in the optical rotation of freshly prepared solution of glucose is known as: [2011RS]
(1) racemisation (2) specific rotation
(3) mutarotation (4) tautomerism

70. The two functional groups present in a typical carbohydrate are: [2009]
(1) – CHO and – COOH (2) > C = O and – OH
(3) – OH and – CHO (4) – OH and – COOH

71. α - D-(+)-glucose and β-D-(+)-glucose are [2008]
(1) conformers (2) epimers
(3) anomers (4) enatiomers

72. The term anomers of glucose refers to [2006]
(1) enantiomers of glucose
(2) isomers of glucose that differ in configuration at carbon one (C-1)
(3) isomers of glucose that differ in configurations at carbons one and four (C-1 and C-4)
(4) a mixture of (D)-glucose and (L)-glucose

NUMERICAL VALUE QUESTIONS (NVQs)

73. Number of compounds from the following which will not produce orange red precipitate with Benedict solution is ______.
Glucose, maltose, sucrose, ribose, 2-deoxyribose, amylose, lactose. [April 11, 2023 (II)]

74. The number of chiral carbons present in sucrose is ____________. [Sep. 05, 2020 (II)]

Topic 2 - Proteins and Amino Acids

MULTIPLE CHOICE QUESTIONS (MCQs)

75. Which of the following amino acid will give violet coloured complex with neutral ferric chloride solution? [April 08, 2026 (II)]
(1) Threonine (2) Serine
(3) Tyrosine (4) Cysteine

76. Match the List-I with List-II [April 06, 2026 (I)]

List-I Amino acid		List-II Positive reaction/Test for functional group present in side chain of amino acid
A. Glutamine	I.	Hinsberg's test
B. Lysine	II.	Neutral $FeCl_3$ test
C. Tyrosine	III.	Ceric ammonium nitrate test
D. Serine	IV.	Hoffman bromamide degradation

Choose the correct answer from the options given below:

(1) A-IV, B-II, C-I, D-III (2) A-IV, B-I, C-II, D-III
(3) A-III, B-II, C-I, D-IV (4) A-IV, B-I, C-III, D-II

77. Identify the incorrect statement about tertiary structure of proteins. [April 05, 2026 (II)]
(1) They can be fibrous or globular in structure.
(2) The main forces that stabilize the structure are hydrogen bonding, disulphide links, van der Waals and electrostatic forces of attraction.
(3) The structure remains intact when exposed to pH changes.
(4) A linear polypeptide chain will convert to a secondary structure and then further folding of the secondary structure will convert to tertiary structure.

78. Match the List-I with List-II. [April 04, 2026 (I)]

List-I Name of amino acid		List-II One letter symbol/ type)
A. Arginine	I.	D / Non-essential
B. Aspartic acid	II.	R / Essential
C. Lysine	III.	E / Non-essential
D. Glutamic acid	IV.	K/Essential

Choose the correct answer from the options given below:

(1) A-II, B-I, C-IV, D-III (2) A-IV, B-III, C-II, D-I
(3) A-III, B-IV, C-I, D-II (4) A-II, B-IV, C-I, D-III

79. Identify the correct pair having amino acid (A) and the hormone (B) that is iodinated derivative of the amino acid (A). [April 02, 2026 (II)]
(T and Y represent one letter code for amino acids)

	Amino acid (A)	Hormone (B)
(1)	T	Insuline
(2)	T	Thyroxine
(3)	Y	Thyroxine
(4)	Y	Insuline

80. In the given pentapeptide, find out an essential amino acid (Y) and the sequence present in the pentapeptide:

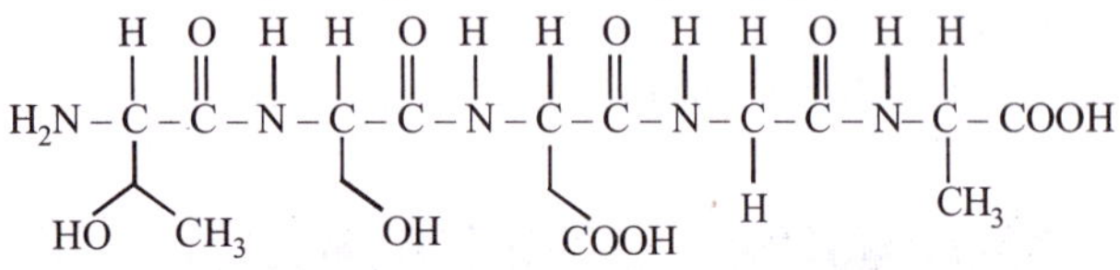

Choose the correct answer from the options given below: [Jan. 28, 2026 (I)]

(1)

(Y)	(Sequence)
Threonine	Ser – Thr – Asp – Gly – Ala

(2)

(Y)	(Sequence)
Serine	Thr – Ser – Asp – Ala – Gly

(3)

(Y)	(Sequence)
Threonine	Thr – Ser – Asp – Gly – Ala

(4)

(Y)	(Sequence)
Serine	Ser – Asp – Thr – Ala – Gly

81. The number of possible tripeptides formed involving alanine (ala), glycine (gly) and valine (val), where no amino acid has been used more than once is: [Jan. 24, 2026 (II)]
(1) 6 (2) 3
(3) 4 (4) 8

82. Identify the correct statements. [Jan. 21, 2026 (I)]
A. Arginine and Tryptophan are essential amino acids.
B. Histidine does not contain heterocyclic ring in its structure.
C. Proline is a six membered cyclic ring amino acid.
D. Glycine does not have chiral centre.
E. Cysteine has characteristic feature of side chain as MeS–CH_2–CH_2–.

Choose the correct answer from the options given below:
(1) C and E only (2) B and E only
(3) C and D only (4) A and D only

83. 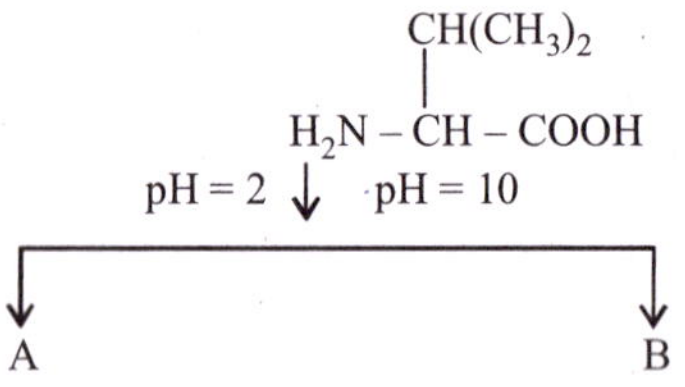

Choose the correct option for structures of A and B, respectively. [April 8, 2025 (II)]

(1) $H_3\overset{+}{N}-\underset{CH(CH_3)_2}{CH}-COOH$ and $H_2N-\underset{CH(CH_3)_2}{CH}-COOH$
(2) $H_2N-\underset{CH(CH_3)_2}{CH}-COO^-$ and $H_3\overset{+}{N}-\underset{CH(CH_3)_2}{CH}-COOH$
(3) $H_2N-\underset{CH(CH_3)_2}{CH}-COO^{\ominus}$ and $H_3\overset{+}{N}-\underset{CH(CH_3)_2}{CH}-COO^-$
(4) $H_3\overset{+}{N}-\underset{CH(CH_3)_2}{CH}-COO^-$ and $H_3\overset{+}{N}-\underset{CH(CH_3)_2}{CH}-COOH$

84. Given below are two statements : [April 7, 2025 (II)]
Statement (I) : On hydrolysis, oligo peptides give rise to fewer number of α-amino acids while proteins give rise to a large number of β-amino acids.
Statement (II) : Natural proteins are denatured by acids which convert the water soluble form of fibrous proteins to their water insoluble form.
In the light of the above statements, choose the most appropriate answer from the options given below :
(1) Both statement I and statement II are correct
(2) Statement I is incorrect but Statement II is correct
(3) Both statement I and statement II are incorrect
(4) Statement I is correct but Statement II is incorrect

85. A dipeptide, "x" on complete hydrolysis gives "y" and "z". "y" on treatment with aq. HNO_2 produces lactic acid. On the other hand "z" on heating gives the following cyclic molecule.

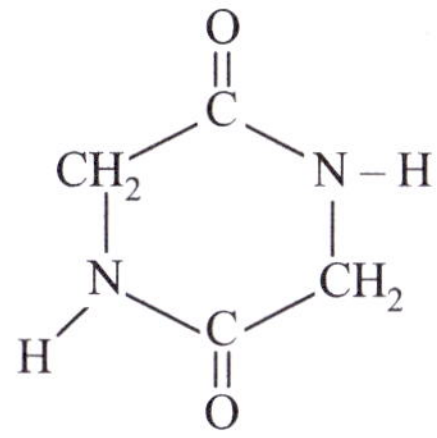

Based on the information given, the dipeptide x is: [April 4, 2025 (II)]

(1) valine-glycine (2) alanine-glycine
(3) valine-leucine (4) alanine-alanine

86. Identify the pair of reactants that upon reaction, with elimination of HCl will give rise to the dipeptide Gly-Ala. [April 4, 2025 (I)]

(1) NH_2-CH_2-COCl and $NH_2-\overset{\overset{CH_3}{|}}{CH}-COOH$

(2) NH_2-CH_2-COCl and $\overset{\oplus}{N}H_2-\overset{\overset{CH_3}{|}}{CH}-COCl$

(3) NH_2-CH_2-COOH and $NH_2-\overset{\overset{CH_3}{|}}{CH}-COCl$

(4) NH_2-CH_2-COOH and $NH_2-\overset{\overset{CH_3}{|}}{CH}-COOH$

87. A tetrapeptide "x" on complete hydrolysis produced glycine (Gly), alanine (Ala), valine (Val), leucine (Leu) in equimolar proportion each. The number of tetrapeptides (sequences) possible involving each of these amino acids is [April 2, 2025 (II)]

(1) 16 (2) 32 (3) 8 (4) 24

88. Identify the correct statement among the following: [April 2, 2025 (I)]

(1) All naturally occurring amino acids except glycine contain one chiral centre.
(2) All naturally occurring amino acids are optically active.
(3) Glutamic acid is the only amino acid that contains a –COOH group at the side chain.
(4) Amino acid, cysteine easily undergo dimerization due to the presence of free SH group.

89. Identify the essential amino acids from below : [Jan. 29, 2025 (II)]

(A) Valine (B) Proline
(C) Lysine (D) Threonine
(E) Tyrosine

Choose the correct answer from the options given below :

(1) (A), (C) and (D) only (2) (A), (C) and (E) only
(3) (B), (C) and (E) only (4) (C), (D) and (E) only

90. The α- Helix and β- Pleated sheet structures of protein are associated with its: [Jan. 23, 2025 (II)]

(1) quaternary structure (2) primary structure
(3) secondary structure (4) tertiary structure

91. Coagulation of egg, on heating is because of : [April 5, 2024 (II)]

(1) Denaturation of protein occurs
(2) The secondary structure of protein remains unchanged
(3) Breaking of the peptide linkage in the primary structure of protein occurs
(4) Biological property of protein remains unchanged

92. Type of amino acids obtained by hydrolysis of proteins is: [Jan. 29, 2024 (I)]

(1) γ (2) δ (3) β (4) α

93. Which structure of protein remains intact after coagulation of egg white on boiling? [Jan. 27, 2024 (II)]

(1) Primary (2) Tertiary
(3) Secondary (4) Quaternary

94. Which is not true for arginine? [April 15, 2023 (I)]

(1) It is a crystalline solid.
(2) It is associated with more than one pK_a values.
(3) It has a fairly high melting point.
(4) It has high solubility in benzene.

95. The naturally occurring amino acid that contains only one basic functional group in its chemical structure is [April 13, 2023 (II)]

(1) arginine (2) lysine
(3) asparagine (4) histidine

96. The one that does not stabilize 2° and 3° structures of proteins is [April 10, 2023 (I)]

(1) H-bonding (2) –S-S-linkage
(3) –O-O-linkage (4) van der Waals forces

97. Sulphur (S) containing amino acids from the following are: [April 08, 2023 (I)]

(A) isoleucine (B) cysteine
(C) lysine (D) methionine
(E) glutamic acid

(1) A, D (2) B, D (3) B, C, E (4) A, B, C

98. Match List I with List II

Choose the correct answer from the options given below: [April 08, 2023 (II)]

	List I (Natural amino acid)		List II (One letter code)
A.	Glutamic acid	I.	Q
B.	Glutamine	II.	W
C.	Tyrosine	III.	E
D.	Tryptophan	IV.	Y

(1) A-II, B-I, C-IV, D-III (2) A-IV, B-III, C-I, D-II
(3) A-III, B-I, C-IV, D-II (4) A-III, B-IV, C-I, D-II

99. Match List I with List II [April 06, 2023 (II)]

List I		List II	
Natural Amino acid		One Letter Code	
(A)	Arginine	(I)	D
(B)	Aspartic acid	(II)	N
(C)	Asparagine	(III)	A
(D)	Alanine	(IV)	R

Choose the correct answer from the options given below:-

(1) (A)-IV, (B)-I, (C)-III, (D)-II
(2) (A)-I, (B)-III, (C)-IV, (D)-II
(3) (A)-III, (B)-I, (C)-II, (D)-IV
(4) (A)-IV, (B)-I, (C)-II, (D)-III

100. A protein 'X' with molecular weight of 70,000 u, on hydrolysis gives amino acids. One of these amino acid is [Jan. 31, 2023 (I)]

(1) $NH_2-CH_2-\underset{\underset{CH_3}{|}}{CH}-CH_2CH_2COOH$

(2) 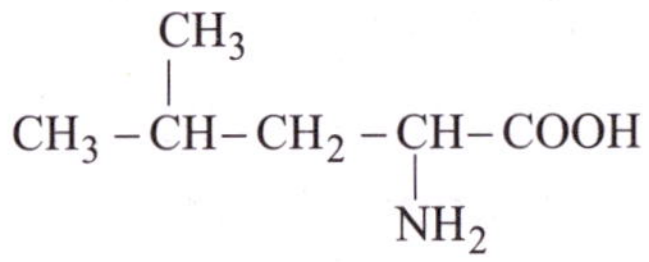

(3) 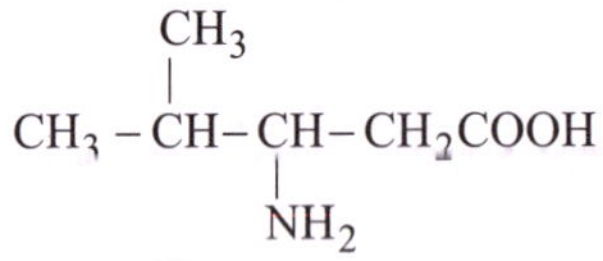

(4) 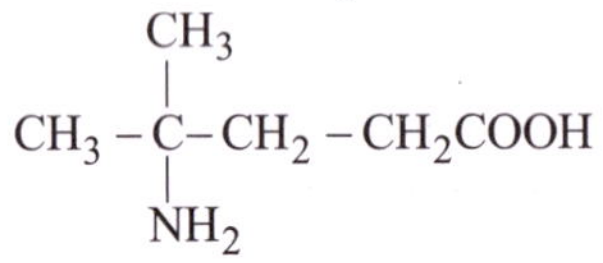

101. Number of cyclic tripeptides formed with 2 amino acids A and B is: [Jan. 29, 2023 (I)]

(1) 2 (2) 3
(3) 5 (4) 4

102. Following tetrapeptide can be represented as [Jan. 29, 2023 (II)]

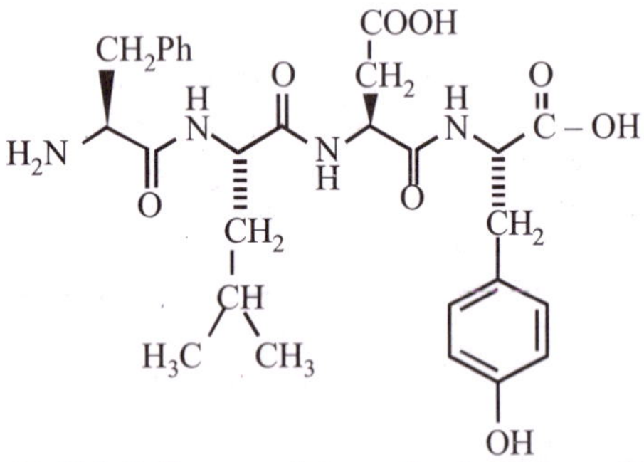

(F, L, D, Y, I, Q, P are one letter codes for amino acids)

(1) FIQY (2) FLDY (3) YQLF (4) PLDY

103. During the denaturation of proteins, which of these structures will remain intact? [July 25, 2022 (I)]

(1) Primary (2) Secondary
(3) Tertiary (4) Quaternary

104. The structure of protein that is unaffected by heating is: [June 29, 2022 (II)]

(1) secondary structure (2) tertiary structure
(3) primary structure (4) quaternary structure

105. Stability of α -Helix structure of proteins depends upon [June 28, 2022 (I)]

(1) dipolar interaction (2) H-bonding interaction
(3) van der Waals forces (4) π -stacking interaction

106. Which of the following is NOT an example of fibrous protein ? [Aug. 31, 2021 (II)]

(1) Keratin (2) Albumin
(3) Collagen (4) Myosin

107. The water soluble protein is : [July 25, 2021 (I)]

(1) Fibrin (2) Albumin (3) Myosin (4) Collagen

108. Seliwanoff test and xanthoproteic test are used for the identification of _______ and _____ respectively. [Feb. 26, 2021 (II)]

(1) ketoses, aldoses (2) proteins, ketoses
(3) ketoses, proteins (4) aldoses, ketoses

109. Out of the following, which type of interaction is responsible for the stabilisation of α-helix structure of proteins? [February 24, 2021 (I)]

(1) Ionic bonding (2) Hydrogen bonding
(3) Covalent bonding (4) van der Waals' forces

110. Which of the following is not an essential amino acid? [Sep. 05, 2020 (I)]

(1) Tyrosine (2) Leucine
(3) Valine (4) Lysine

111. Which of the following will react with $CHCl_3$ + alc. KOH? [Sep. 04, 2020 (I)]

(1) Adenine and proline (2) Thymine and proline
(3) Adenine and lysine (4) Adenine and thymine

112. The correct structure of product 'P' in the following reaction is: [Jan. 10, 2019 (I)]

$$\text{Asn} - \text{Ser} + \underset{\text{(excess)}}{(CH_3CO)_2O} \xrightarrow{NEt_3} P$$

(1) 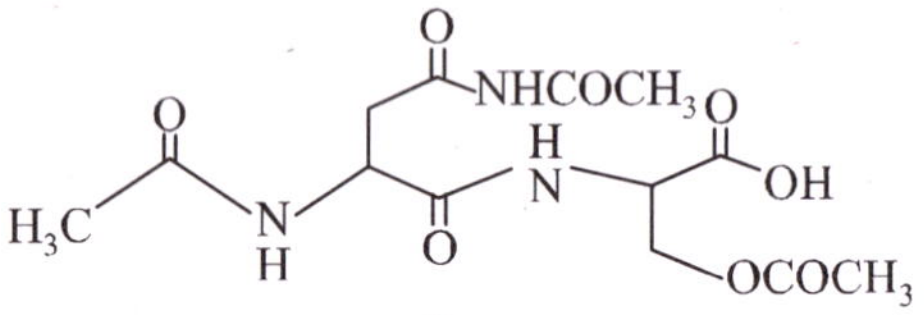

(2) 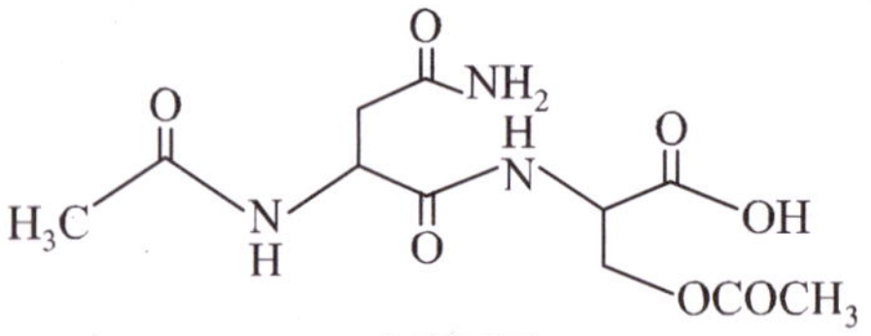

(3) 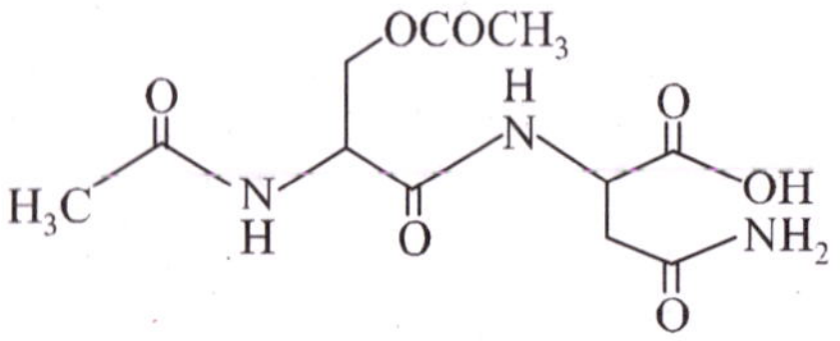

(4) 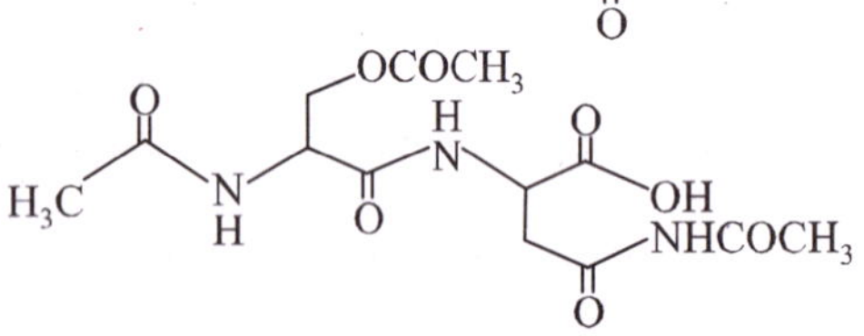

113. The peptide that gives positive ceric ammonium nitrate and carbylamine tests is: [April 9, 2019 (II)]

(1) Ser – Lys (2) Gln – Asp
(3) Lys – Asp (4) Asp – Gln

114. Among the following compounds most basic amino acid is: [Jan. 12, 2019 (I)]

(1) Asparagine (2) Lysine
(3) Serine (4) Histidine

115. The correct structure of histidine in a strongly acidic solution (pH = 2) is : [Jan. 12, 2019 (II)]

(1) 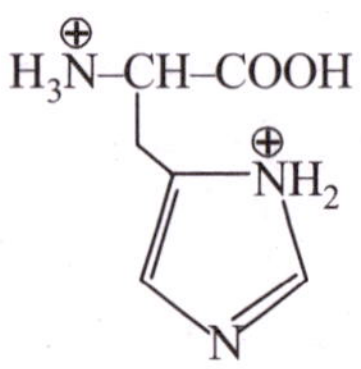 (2) 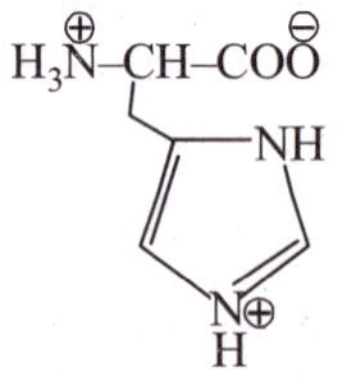

(3) 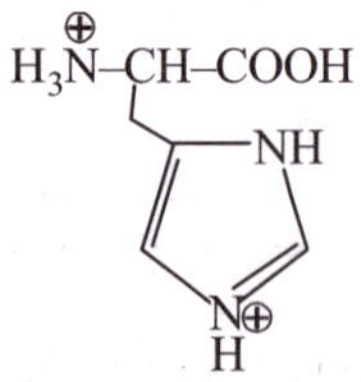 (4) 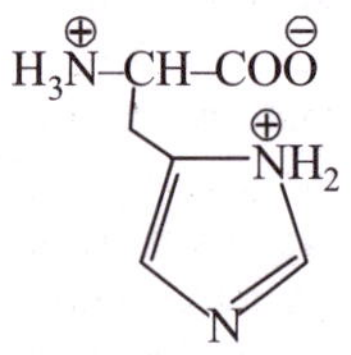

116. The increasing order of pK_a of the following amino acids in aqueous solution is: [Jan. 9, 2019 (I)]
Gly Asp Lys Arg
(1) Asp < Gly < Arg < Lys (2) Gly < Asp < Arg < Lys
(3) Asp < Gly < Lys < Arg (4) Arg < Lys < Gly < Asp

117. The correct sequence of amino acids present in the tripeptide given below is: [Jan. 9, 2019 (II)]

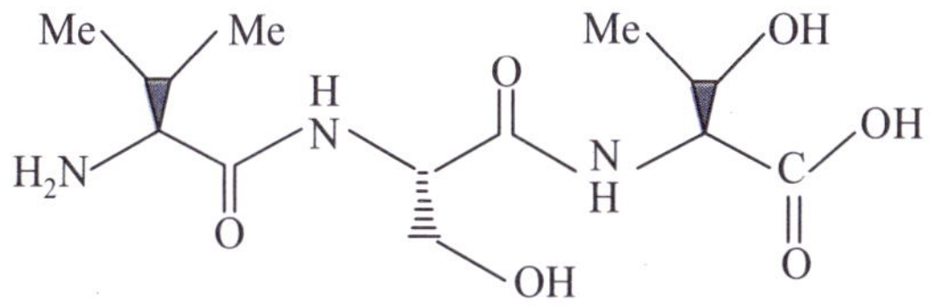

(1) Val - Ser - Thr (2) Thr - Ser - Val
(3) Leu - Ser - Val (4) Thr - Ser - Leu

118. Which of the following will not exist in zwitter ionic form at pH = 7 ? [Online April 15, 2018 (I)]

(1) 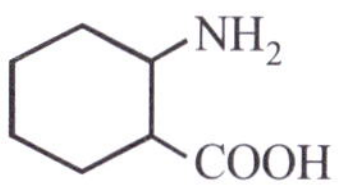 (2) 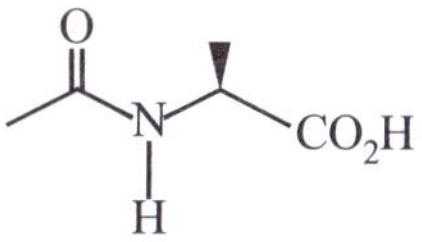

(3)  (4) 

119. Among the following, the essential amino acid is : [Online April 8, 2017]
(1) Alanine (2) Valine
(3) Aspartic acid (4) Serine

120. Observation of "Ruhemann's purple" is a confirmatory test for the presence of : [Online April 10, 2016]
(1) Starch (2) Reducing sugar
(3) Protein (4) Cupric ion

121. Which one of the following statements is correct? [2012]
(1) All amino acids except lysine are optically active
(2) All amino acids are optically active
(3) All amino acids except glycine are optically active
(4) All amino acids except glutamic acids are optically active

122. All of the following statements apply to proteins except [Online May 7, 2012]
(1) Proteins generally have no definite melting point
(2) Proteins contain the grouping —CONH—
(3) Proteins have high molecular weight
(4) Proteins can only contain the elements C, H, O and N.

123. Biuret test is **not** given by [2010]
(1) carbohydrates (2) polypeptides
(3) urea (4) proteins

124. The secondary structure of a protein refers to [2007]
(1) fixed configuration of the polypeptide backbone
(2) α–helical backbone
(3) hydrophobic interactions
(4) sequence of α –amino acids.

NUMERICAL VALUE QUESTIONS (NVQs)

125. Total number of essential amino acids among the given list of amino acids is _______. [April 9, 2024 (I)]
Arginine, Phenylalanine, Aspartic acid, Cysteine, Histidine, Valine, Proline.

126. The total number of carbon atoms present in tyrosine, an amino acid, is _______. [April 8, 2024 (II)]

127. The number of tripeptides formed by three different amino acids using each amino acid once is _________ . [Feb. 1, 2024 (II)]

128. In an oligopeptide named Alanylglycylphenylalanyl-isoleucine, the number of sp^2 hybridised carbons is _______. [April 12, 2023 (I)]

129. A short peptide on complete hydrolysis produces 3 moles of glycine (G), two moles of leucine (L) and two moles of valine (V) per mole of peptide. The number of peptide linkages in it are _______. [Jan. 30, 2023 (II)]

130. Total number of tripeptides possible by mixing of valine and proline is _________ [Jan. 24, 2023 (II)]

131. In a linear tetrapeptide (Constituted with different amino acids), (number of amino acids) - (number of peptide bonds) is______. [July 29, 2022 (I)]

132. How many of the given compounds will give a positive biuret test _________? Glycine, Glycylalanine, Tripeptide, Biuret [June 25, 2022 (II)]

133. In alanylglycylleucylalanylvaline, the number of peptide linkages is_________. [June 24, 2022 (II)]

134. A peptide synthesized by the reactions of one molecule each of Glycine, Leucine, Aspartic acid and Histidine will have ________ peptide linkages. [Sep. 1, 2021 (II)]

135. The total number of negative charge in the tetrapeptide, Gly-Glu-Asp-Tyr, at pH 12.5 will be_______. (Integer answer) [Aug. 26, 2021 (I)]

136. The number of chiral carbon(s) present in peptide, Ile-Arg-Pro, is ______. [Sep. 05, 2020 (I)]

137. The number of chiral centres present in threonine is ____________. [Sep. 04, 2020 (II)]

138. The number of $\rangle C=O$ groups present in a tripeptide Asp - Glu - Lys is ______________. [Sep. 03, 2020 (II)]

139. The mass percentage of nitrogen in histamine is ______. [Jan. 09, 2020 (I)]

Topic 3 - Enzymes and Nucleic Acid

MULTIPLE CHOICE QUESTIONS (MCQs)

140. Both human DNA and RNA are chiral molecules. The chirality in DNA and RNA arises due to the presence of [Jan. 23, 2026 (II)]
(1) Base unit (2) Chiral phosphate unit
(3) D-sugar component (4) L-sugar component

141. The **correct** statements are: [Jan. 21, 2026 (II)]
A. Activation energy for enzyme catalysed hydrolysis of sucrose is lower than that of acid catalysed hydrolysis.
B. During denaturation, secondary and tertiary structures of a protein are destroyed but primary structure remains intact.
C. Nucleotides are joined together by glycosidic linkage between C_1 and C_4 carbons of the pentose sugar
D. Quaternary structure of proteins represents overall folding of the polypeptide chain.
Choose the **correct** answer from the options given below:
(1) A, C and D Only (2) A, B and D Only
(3) A and B Only (4) B and C Only

142. Match List-I with List-II. [Jan. 24, 2025 (II)]

List-I		List-II
(A) Adenine	(I)	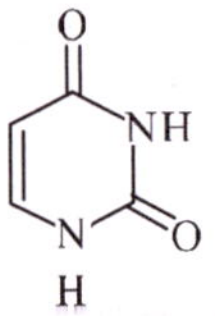
(B) Cytosine	(II)	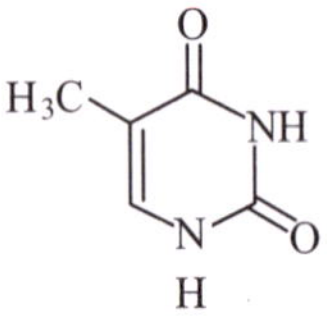
(C) Thymine	(III)	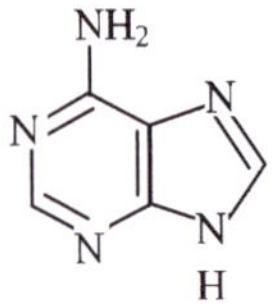
(D) Uracil	(IV)	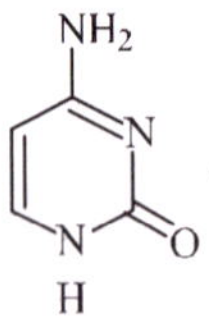

Choose the correct answer from the options given below :

(1) (A)-(III), (B)-(IV), (C)-(II), (D)-(I)
(2) (A)-(III), (B)-(I), (C)-(IV), (D)-(II)
(3) (A)-(IV), (B)-(III), (C)-(II), (D)-(I)
(4) (A)-(III), (B)-(IV), (C)-(I), (D)-(II)

143. DNA molecule contains 4 bases whoes structure are shown below. One of the structure is not correct, identify the **incorrect** base structure. [April 6, 2024 (I)]

(1) 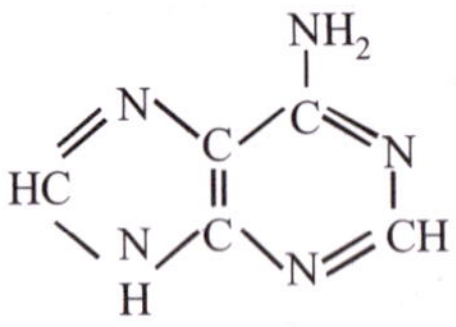

(2) 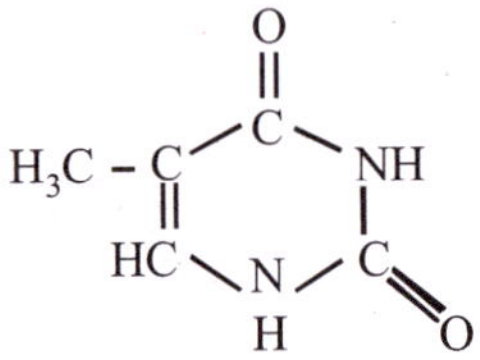

(3) 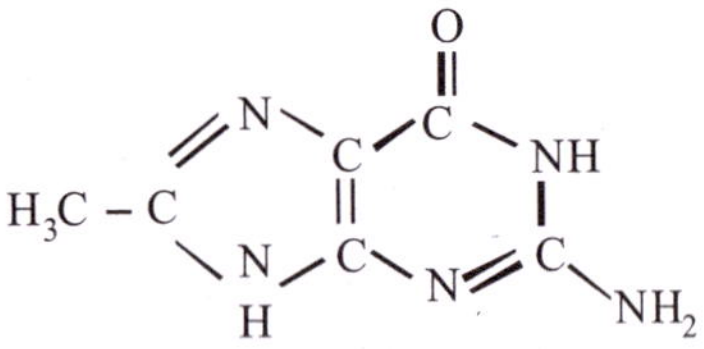

(4) 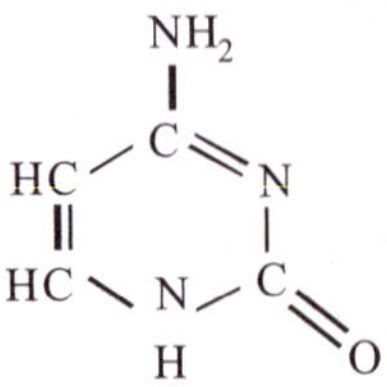

144. The incorrect statements regarding enzymes are: [April 6, 2024 (II)]

(A) Enzymes are biocatalysts.
(B) Enzymes are non-specific and can catalyse different kinds of reactions.
(C) Most Enzymes are globular proteins.
(D) Enzyme - oxidase catalyses the hydrolysis of maltose into glucose.

Choose the correct answer from the option given below:

(1) (B) and (C) (2) (B), (C) and (D)
(3) (C) and (D) (4) (A), (B) and (C)

145. If one strand of a DNA has the sequence ATGCTTCA, sequence of the bases in complementary strand is: [Feb. 1, 2024 (I)]

(1) CATTAGCT (2) TACGAAGT
(3) GTACTTAC (4) ATGCGACT

146. Match List I with List II [Jan. 29, 2024 (II)]

LIST I (Bio polymer)	LIST II (Monomer)
A. Starch	I. nucleotide
B. Cellulose	II. α-glucose
C. Nucleic acid	III. β-glucose
D. Protein	IV. α-amino acid

Choose the correct answer from the options given below:

(1) A-II, B-III, C-I, D-IV (2) A-II, B-I C-III, D-IV
(3) A-IV, B-II, C-I, D-III (4) A-I, B-III, C-IV, D-II

147. Two nucleotides are joined together by a linkage known as : [Jan. 27, 2024 (I)]

(1) Phosphodiester linkage (2) Glycosidic linkage
(3) Disulphide linkage (4) Peptide linkage

148. The sugar produced after complete hydrolysis of DNA is [June 30, 2022 (I)]

(1) a pentose sugar. (2) a hexose sugar.
(3) a polysaccharide. (4) a disaccharide.

149. Sugar moiety in DNA and RNA molecules respectively are [June 29, 2022 (I)]

(1) β-D-2-deoxyribose, β-D-deoxyribose
(2) β-D-2-deoxyribose, β-D-ribose
(3) β-D-ribose, β -D-2-deoxyribose
(4) β-D-deoxyribose, β -D-2-deoxyribose

150. Out of following isomeric forms of uracil, which one is present in RNA? [Aug. 27, 2021 (I)]

(1) 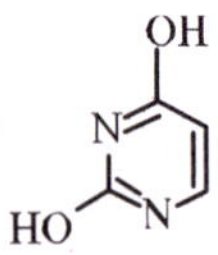 (2) 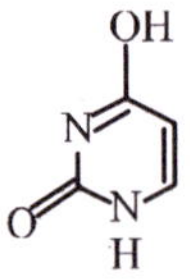

(3) 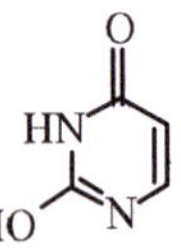 (4) 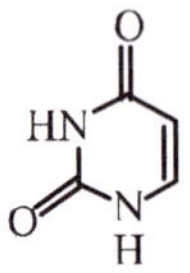

151. Which one of the following is correct structure for cytosine? [July 25, 2021 (II)]

(1) 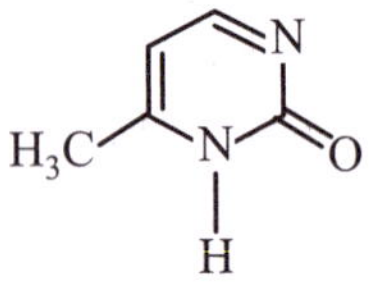 (2) 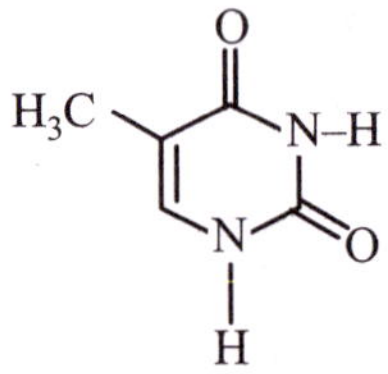

(3) 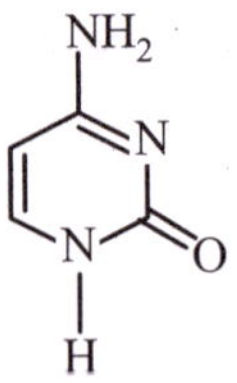 (4) 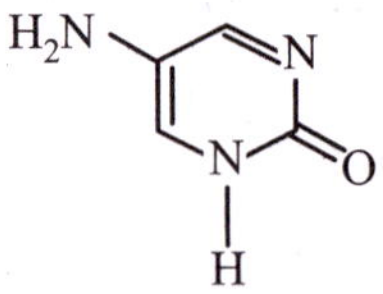

152. $C_{12}H_{22}O_{11} + H_2O \xrightarrow{\text{Enzyme B}} \underset{\text{Glucose}}{C_6H_{12}O_6} + \underset{\text{Fructose}}{C_6H_{12}O_6}$

$\underset{\text{Glucose}}{C_6H_{12}O_6} \xrightarrow{\text{Enzyme B}} 2C_2H_5OH + 2CO_2$

In the above reactions, the enzyme A and enzyme B respectively are [March 17, 2021 (II)]

(1) Amylase and Invertase (2) Invertase and Zymase
(3) Zymase and Invertase (4) Invertase and Amylase

153. Which of the following statements is not true about RNA?
(1) It controls the synthesis of protein. [April 12, 2019 (I)]
(2) It has always double stranded helix structure.
(3) It usually does not replicate.
(4) It is present in the nucleus of the cell.

154. Among the following compounds, which one is found in RNA only? [Jan. 11, 2019 (I)]

(1) 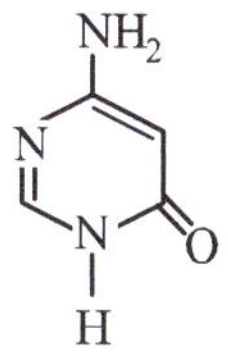
(2) 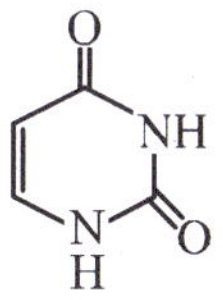
(3) 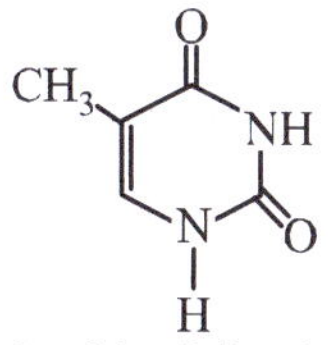
(4) 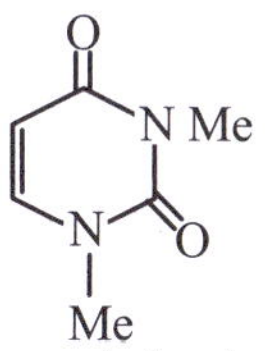

155. Which of the following is the correct structure of adenosine? [Online April 15, 2018 (I)]

(1) 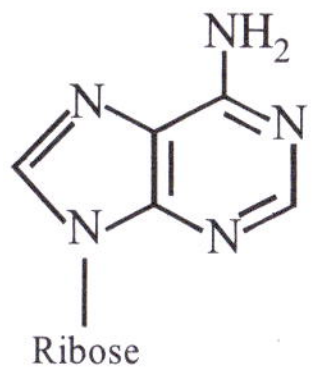
(2) 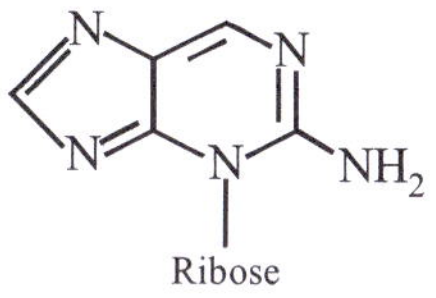
(3) 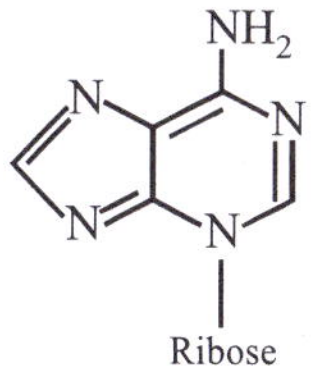
(4) 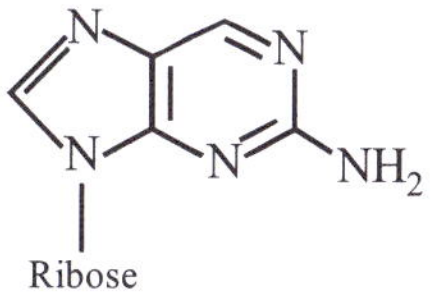

156. Thiol group is present in : [2016]
(1) Cysteine (2) Methionine
(3) Cytosine (4) Cystine

157. The reason for double helical structure of DNA is the operation of: [Online April 19, 2014; Same, 2003]
(1) Electrostatic attractions
(2) van der Waals forces
(3) Dipole - Dipole interactions
(4) Hydrogen bonding

158. Which one of the following bases is **not** present in DNA? [2014]
(1) Quinoline (2) Adenine
(3) Cytosine (4) Thymine

159. Among the following organic acids, the acid present in rancid butter is: [Online April 19, 2014]
(1) Pyruvic acid (2) Lactic acid
(3) Butyric acid (4) Acetic acid

160. Which of the following enzyme converts starch into maltose? [Online April 9, 2013]
(1) Diastase (2) Maltase
(3) Zymase (4) Invertase

161. Which of the following structures represents thymine ? [Online April 22, 2013]

(1) 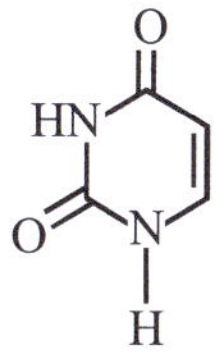
(2) 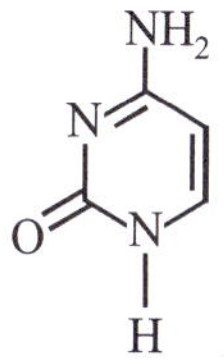
(3) 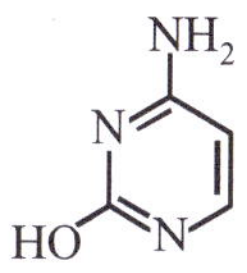
(4) 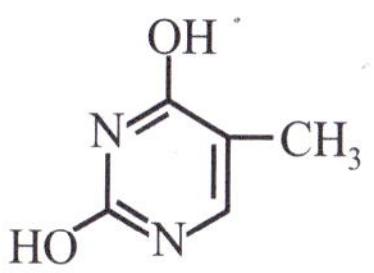

162. Which of the following statements is correct? [Online May 26, 2012]
(1) RNA controls the synthesis of proteins.
(2) The sugar present in DNA is D-(–)-ribose.
(3) RNA has double stranded α-helix structure.
(4) DNA mainly occurs in the cytoplasm of the cell.

163. The pyrimidine bases present in DNA are [2006]
(1) cytosine and thymine (2) cytosine and uracil
(3) cytosine and adenine (4) cytosine and guanine

164. In both DNA and RNA, heterocyclic base and phosphate ester linkages are at – [2005]
(1) C_5' and C_1' respectively of the sugar molecule
(2) C_1' and C_5' respectively of the sugar molecule
(3) C_2' and C_5' respectively of the sugar molecule
(4) C_5' and C_2' respectively of the sugar molecule

165. Which base is present in RNA but not in DNA ? [2004]
(1) Guanine (2) Cytosine
(3) Uracil (4) Thymine

166. RNA is different from DNA because RNA contains [2002]
(1) ribose sugar and thymine
(2) ribose sugar and uracil
(3) deoxyribose sugar and thymine
(4) deoxyribose sugar and uracil.

Numerical Value Questions (NVQs)

167. The total number of hydrogen bonds of a DNA-double Helix strand whose one strand has the following sequence of bases is _______.
5′–G–G–C–A–A–A–T–C–G–G–C–T–A–3′. [April 4, 2025 (I)]

168. The total number of correct statements, regarding the nucleic acids is _______. [Jan. 30, 2024 (II)]
A. RNA is regarded as the reserve of genetic information
B. DNA molecule self - duplicates during cell division
C. DNA synthesizes proteins in the cell

D. The message for the synthesis of particular proteins is present in DNA

E. Identical DNA strands are transferred to daughter cells.

169. Uracil is base present in RNA with the following structure. % of N in uracil is _______. [Jan. 24, 2023 (I)]

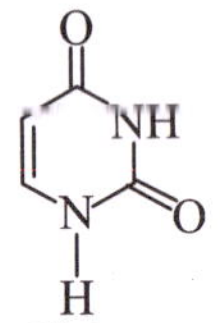

Given :

Molar mass N = 14 g mol^{-1}; O = 16 g mol^{-1};
C = 12 g mol^{-1}; H = 1 g mol^{-1}

170. The number of oxygen present in a nucleotide formed from a base, that is present only in RNA is ________. [June 26, 2022 (I)]

Topic 4 - Vitamins and Hormones

Multiple Choice Questions (MCQs)

171. Match the List-I with List-II : [April 06, 2026 (I)]

List-I Deficiency Disease		List-II Vitamin	
A.	Scurvy	I.	Pyridoxine
B.	Convulsions	II.	Vitamin A
C.	Cheilosis	III.	Ascorbic Acid
D.	Xerophthalmia	IV.	Riboflavin

Choose the correct answer from the options given below:
(1) A-I, B-III, C-II, D-IV (2) A-I, B-III, C-IV, D-II
(3) A-III, B-I, C-IV, D-II (4) A-III, B-I, C-II, D-IV

172. Match List-I with List-II. [April 02, 2026 (I)]

List-I Vitamin		List-II Name	
A.	Vitamin B_1	I.	Pyridoxine
B.	Vitamin B_2	II.	Ascorbic acid
C.	Vitamin B_6	III.	Thiamine
D.	Vitamin C	IV.	Riboflavin

Choose the correct answer from the options given below:
(1) A-II, B-I, C-III, D-IV (2) A-IV, B-III, C-II, D-I
(3) A-III, B-IV, C-I, D-II (4) A-I, B-III, C-II, D-IV

173. Fat soluble vitamins are : [April 3, 2025 (II)]
A. Vitamin B_1 B. Vitamin C
C. Vitamin E D. Vitamin B_{12}
E. Vitamin K
Choose the correct answer from the options given below:
(1) C & D Only (2) A & B Only
(3) B & C Only (4) C & E Only

174. Which of the following acids is a vitamin ? [Jan. 22, 2025 (I)]
(1) Adipic acid (2) Aspartic acid
(3) Ascorbic acid (4) Saccharic acid

175. Match List I with List II [April 06, 2023 (I)]

List I (Vitamin)		List II (Deficiency disease)	
A	Vitamin A	I.	Beri–beri
B	Thiamine	II.	Cheilosis
C	Ascorbic acid	III.	Xerophthalmia
D	Riboflavin	IV.	Scurvy

Choose the correct answer from the options given below:
(1) A–IV, B–II, C–III, D–I (2) A–III, B–II, C–IV, D–I
(3) A–IV, B–I, C–III, D–II (4) A–III, B–I, C–IV, D–II

176. Given below are two statements : [Jan. 24, 2023 (I)]
Statement I : Noradrenaline is a neurotransmitter.
Statement II : Low level of noradrenaline is not the cause of depression in human.
In the light of the above statements, choose the correct answer from the options given below
(1) Statement I is correct but Statement II is incorrect
(2) Statement I is incorrect but Statement II is correct
(3) Both Statement I and Statement II are correct
(4) Both Statement I and Statement II are incorrect

177. Which one of the following is a water soluble vitamin, that is not excreted easily? [June 26, 2022 (II)]
(1) Vitamin B_2 (2) Vitamin B_1
(3) Vitamin B_6 (4) Vitamin B_{12}

178. Thiamine and pyridoxine are also known respectively as : [July 22, 2021 (II)]
(1) Vitamin B_2 and Vitamin E
(2) Vitamin E and Vitamin B_2
(3) Vitamin B_6 and Vitamin B_2
(4) Vitamin B_1 and Vitamin B_6

179. Deficiency of vitamin K causes : [March 18, 2021 (II)]
(1) Cheilosis
(2) Increase in blood clotting time
(3) Increase in fragility of RBC's
(4) Decrease in blood clotting time

180. Match the following: [Jan. 07, 2020 (I)]

(i)	Riboflavin	(a)	Beriberi
(ii)	Thiamine	(b)	Scurvy
(iii)	Pyridoxine	(c)	Cheilosis
(iv)	Ascorbic acid	(d)	Convulsions

(1) (i) – (a), (ii) – (d), (iii) – (c), (iv) – (b)
(2) (i) – (c), (ii) – (d), (iii) – (a), (iv) – (b)
(3) (i) – (c), (ii) – (a), (iii) – (d), (iv) – (b)
(4) (i) – (d), (ii) – (b), (iii) – (a), (iv) – (c)

181. Which of the vitamins given below is water soluble ? [2015]
(1) Vitamin E (2) Vitamin K
(3) Vitamin C (4) Vitamin D

182. Among the following vitamins the one whose deficiency causes rickets (bone deficiency) is [Online April 25, 2013]
(1) Vitamin A (2) Vitamin B
(3) Vitamin D (4) Vitamin C

183. The change in the optical rotation of freshly prepared solution of glucose is known as: [2011RS]
(1) racemisation (2) specific rotation
(3) mutarotation (4) tautomerism

184. Insulin production and its action in human body are responsible for the level of diabetes. This compound belongs to which of the following categories ? [2004]
(1) An enzyme (2) A hormone
(3) A co-enzyme (4) An antibiotic

Numerical Value Questions (NVQs)

185. From the vitamins A, B_1, B_6, B_{12}, C, D, E and K, the number of vitamins that can be stored in our body is ______. [Jan. 31, 2024 (II)]

Part-B (JEE Advanced)

Topic 1 - Carbohydrates

1 MCQs with One Correct Answer

1. A known artificial sweetener **X** is composed of 4-chloro-4-deoxy-α-D-galactose and 1,6-dichloro-1,6-dideoxy-β-D-fructose joined by a glycosidic linkage. Structure of D-galactose is given below: [Adv. 2026]

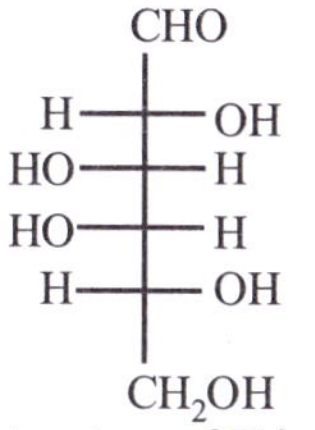

= D-galactose

The correct structure of **X** is

(a) 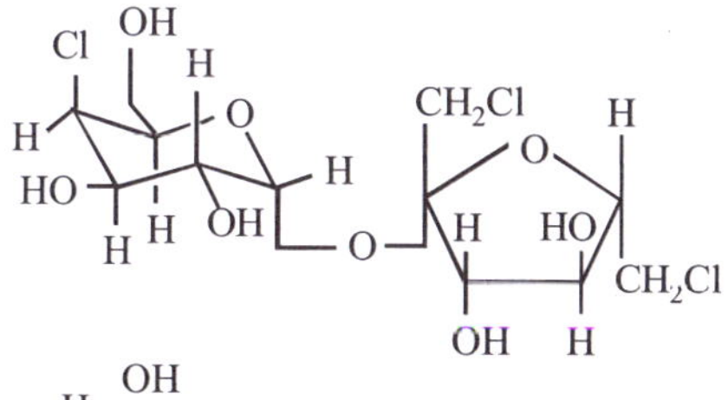

(b) 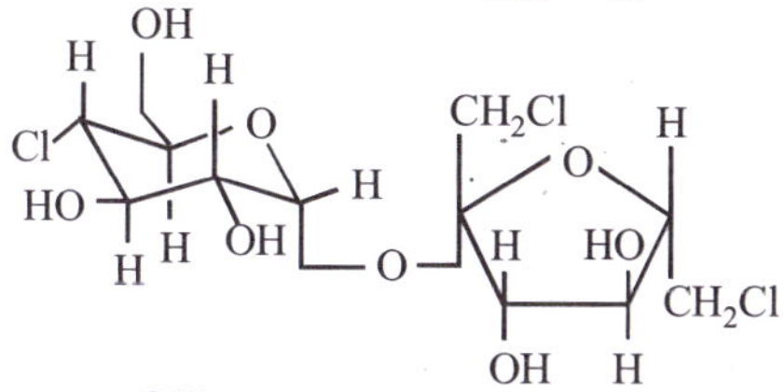

(c) 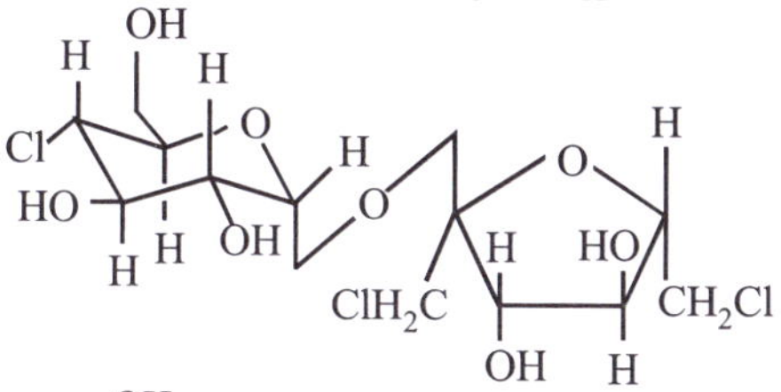

(d) 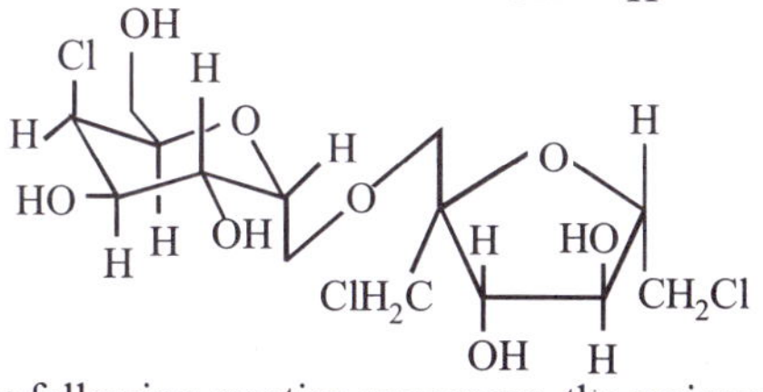

2. In the following reaction sequences, the major product Q is [Adv. 2024]

$$\text{L-Glucose} \xrightarrow[\text{(ii) } Cr_2O_3,\, 775\,K,\, 10\text{-}20\,atm]{\text{(i) HI},\,\Delta} P \xrightarrow[UV]{Cl_2\text{(excess)}} Q$$

(a) 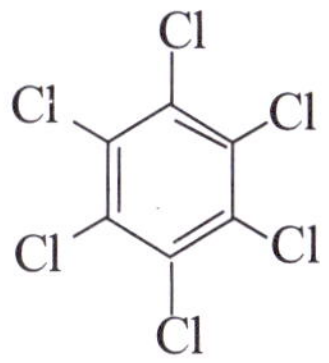

(b) 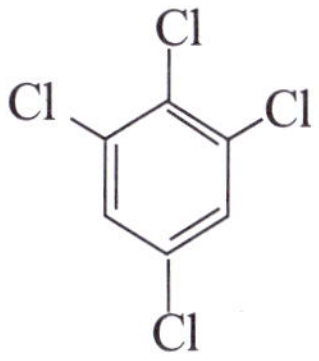

(c) 

(d) 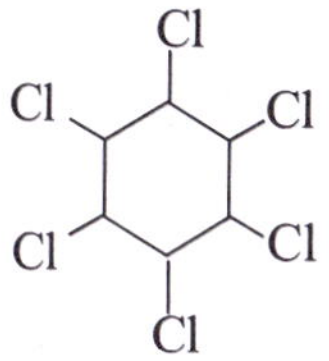

3. A disaccharide X cannot be oxidised by bromine water. The acid hydrolysis of X leads to a laevorotatory solution. The disaccharide X is [Adv. 2023]

(a) 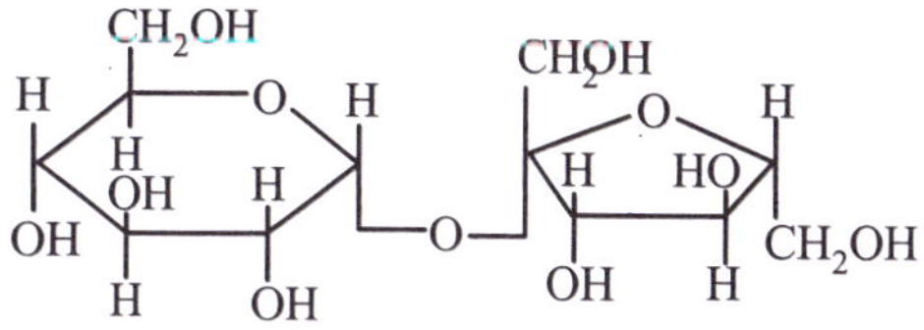

(b) 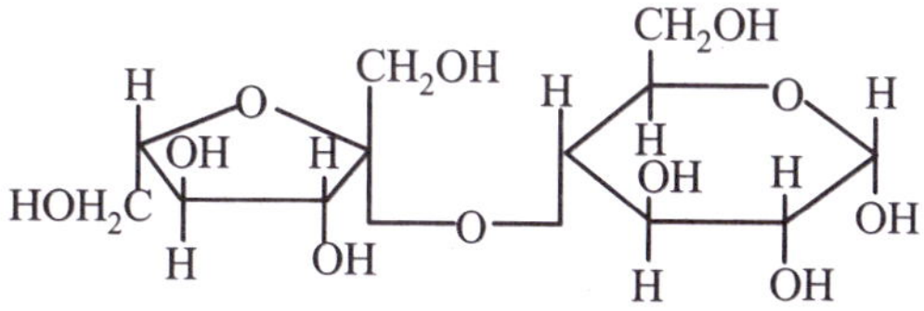

(c) 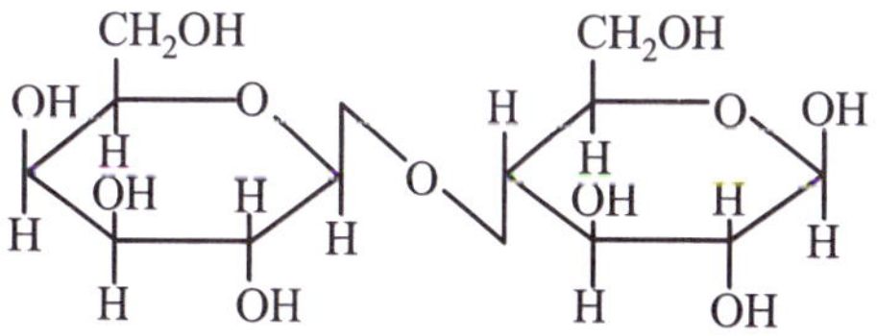

(d) 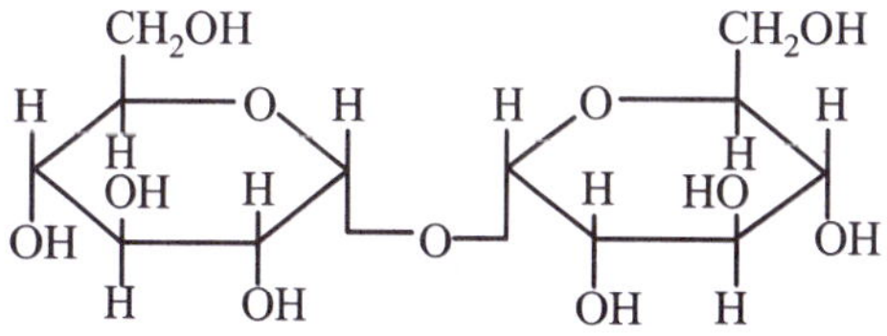

4. Treatment of D-glucose with aqueous NaOH results in a mixture of monosaccharides, which are [Adv. 2022]

(a) 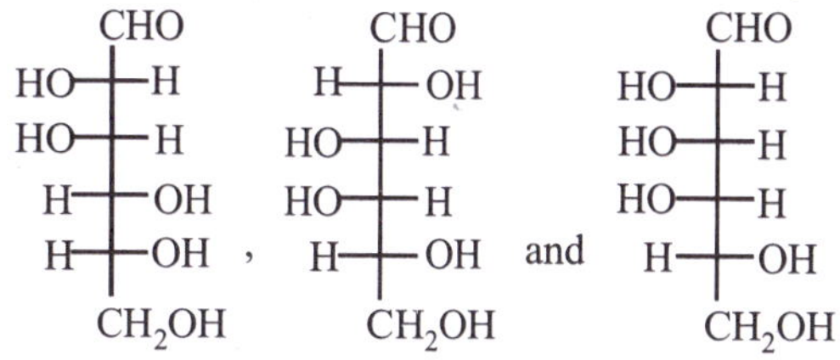

(b) 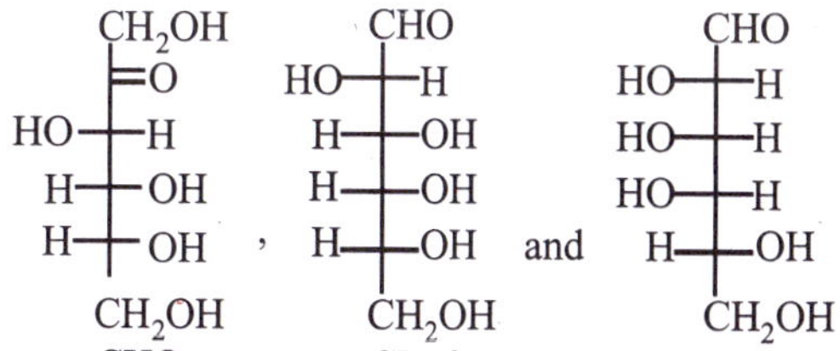

(c) 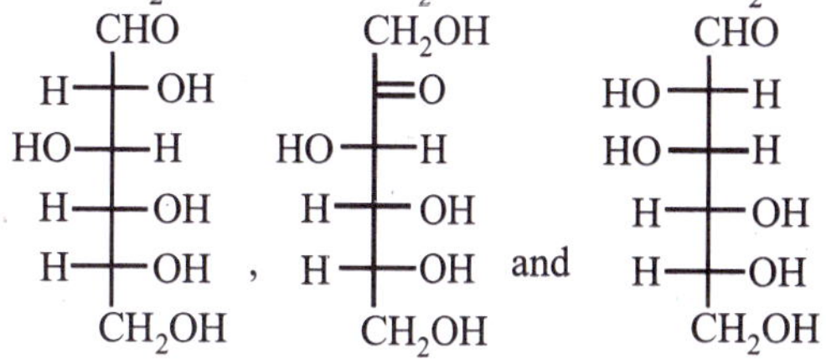

(d) 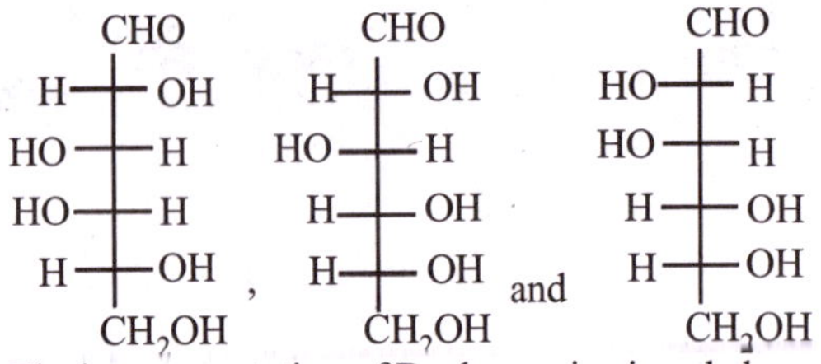

5. The Fischer presentation of D-glucose is given below. **[Adv. 2018]**

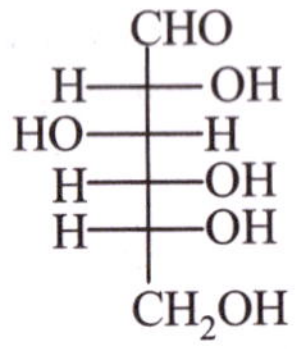

D-glucose

The correct structure(s) of β-L-glucopyranose is (are)

(a) 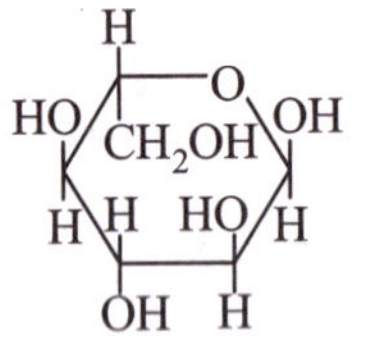

(b) 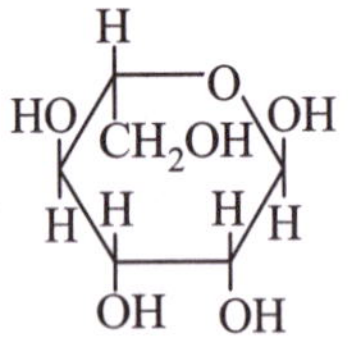

(c) 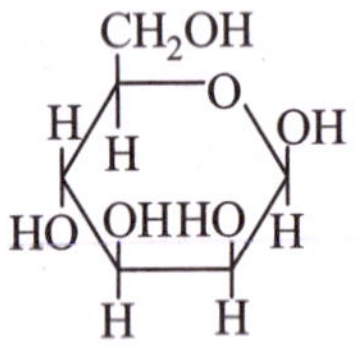

(d) 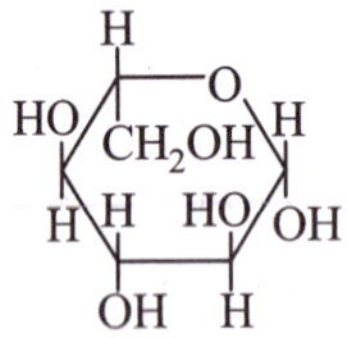

6. The structure of D-(+)-glucose is **[Adv. 2015]**

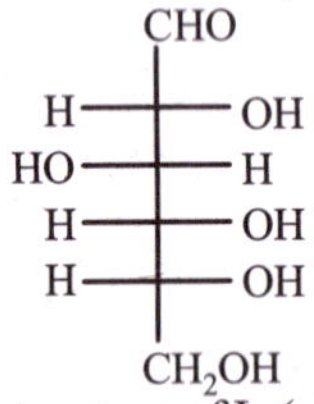

The structure of L-(–)-glucose is

(a) 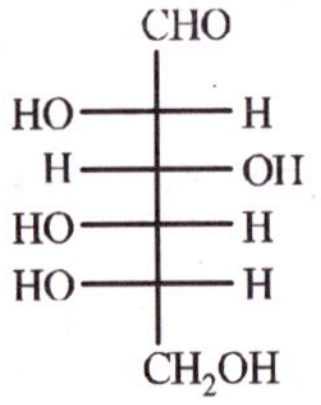

(b) 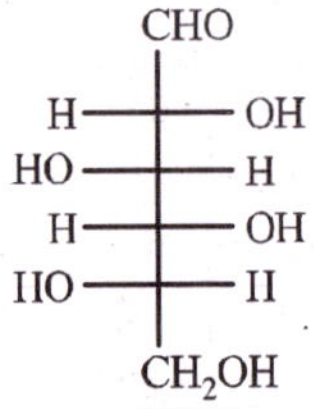

(c) 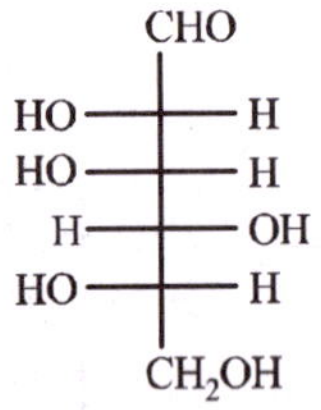

(d) 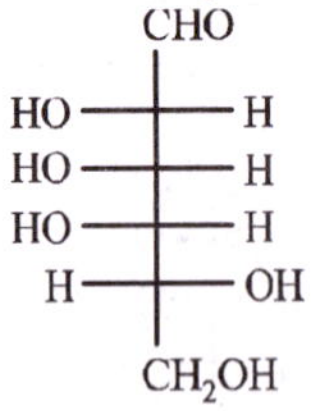

7. The following carbohydrate is **[2011 - II]**

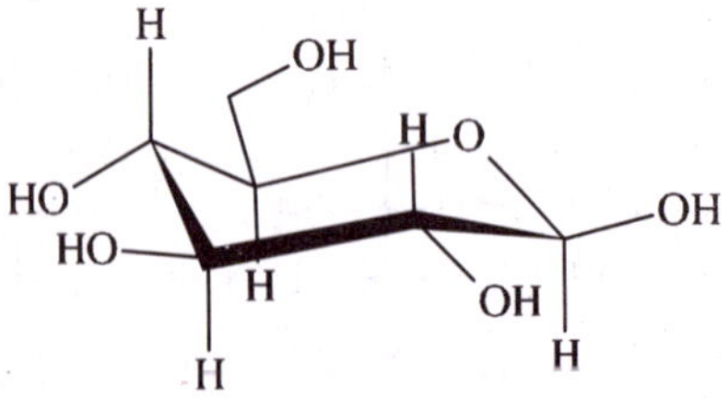

(a) a ketohexose (b) an aldohexose
(c) an α-furanose (d) an α-pyranose

8. The correct statement about the following disaccharide is

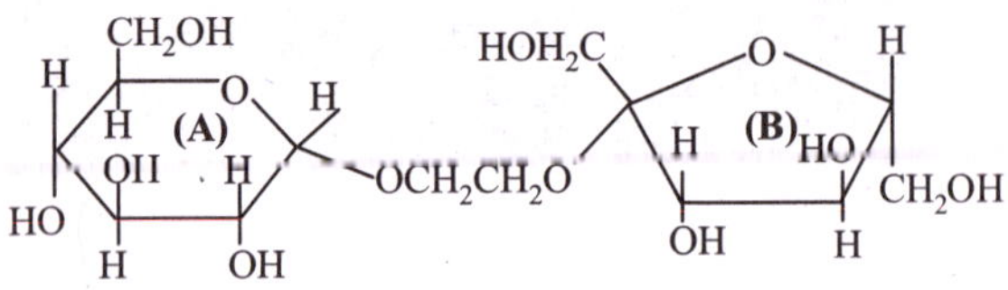

[2010]

(a) Ring (**A**) is pyranose with α - glycosidic link
(b) Ring (**A**) is furanose with α - glycosidic link
(c) Ring (**B**) is furanose with α - glycosidic link
(d) Ring (**B**) is pyranose with β - glycosidic link

9. Cellulose upon acetylation with excess acetic anhydride/H_2SO_4 (catalytic) gives cellulose triacetate whose structure is **[2008S]**

(a) 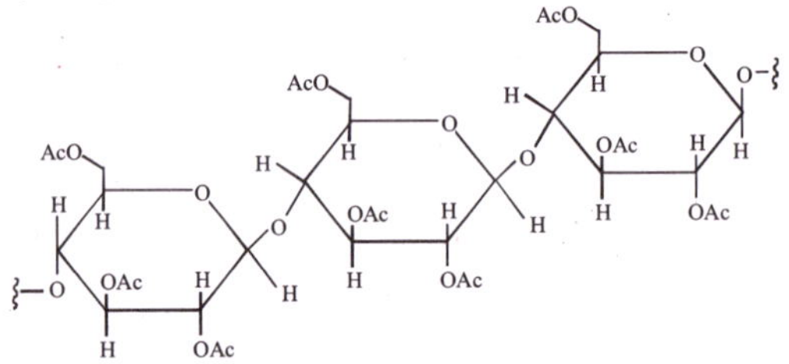

(b) 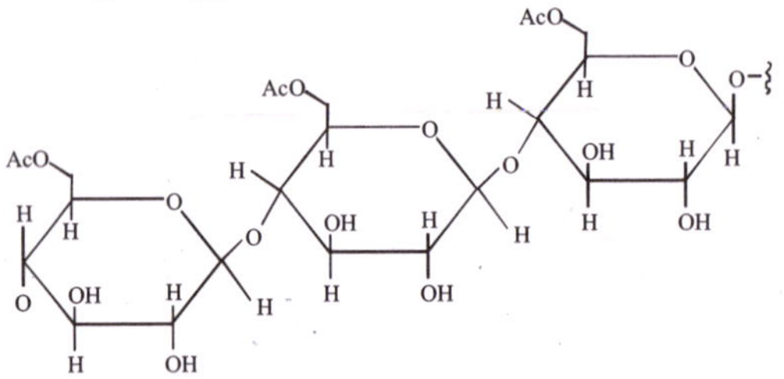

(c) 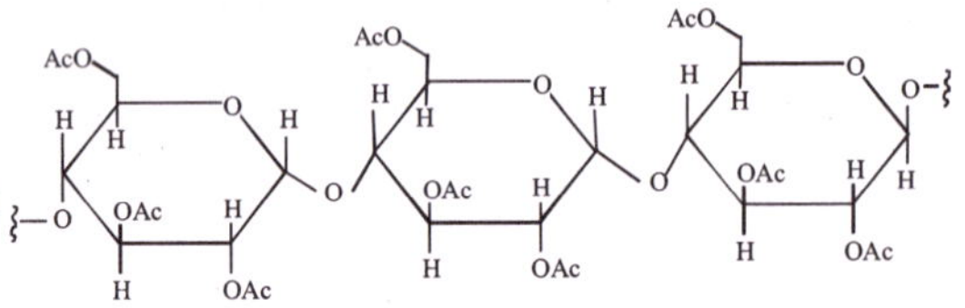

(d) 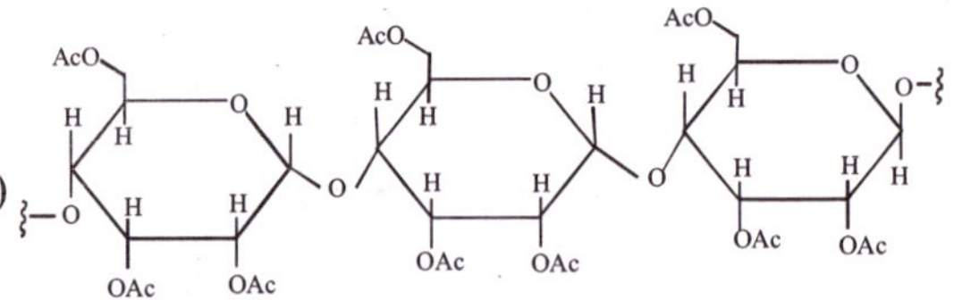

10. The two forms of D–glucopyranose obtained from the solution of D–glucose are called **[2005S]**
(a) Isomers (b) Anomers
(c) Epimers (d) Enantiomers

11. The pair of compounds in which both the compounds give positive test with Tollen's reagent is **[2004S]**
(a) Glucose and Sucrose
(b) Fructose and Sucrose
(c) Acetophenone and Hexanal
(d) Glucose and Fructose

2 Integer Value Answer

12. When the following aldohexose exists in its D-configuration, the total number of stereoisomers in its pyranose form is : **[2012]**

$CHO—CH_2—CHOH—CHOH—CHOH—CH_2OH$

3 Numeric / New Stem Based Questions

13. A linear octasaccharide (molar mass = 1024 g mol^{-1}) on complete hydrolysis produces three monosaccharides: ribose, 2-deoxyribose and glucose. The amount of 2-deoxyribose formed is 58.26 % (w/w) of the total amount of the monosaccharides produced in the hydrolyzed products. The number of ribose unit(s) present in one molecule of octasaccharide is _____. **[Adv. 2025]**

Use: Molar mass (in g mol^{-1}): ribose = 150, 2-deoxyribose = 134, glucose = 180; Atomic mass (in amu): H = 1, O = 16

6 MCQs with One or More than One Correct Answer

14. The correct statement(s) regarding sugars is(are)

Given: Specific rotations of L-(–)-glucose and L-(+)-fructose are –52.5° and +92.5°, respectively. **[Adv. 2026]**

(a) On treatment with HNO_3, gluconic acid is oxidized to saccharic acid, whereas glucose is not oxidized to saccharic acid.

(b) Fructose gives a positive Fehling's test because it isomerises to glucose and another aldohexose in the presence of Fehling's reagent.

(c) Invert sugar is an equimolar mixture of D-glucose and D-fructose formed after hydrolysis of the corresponding disaccharide.

(d) Specific rotation of invert sugar is –40°.

15. Given

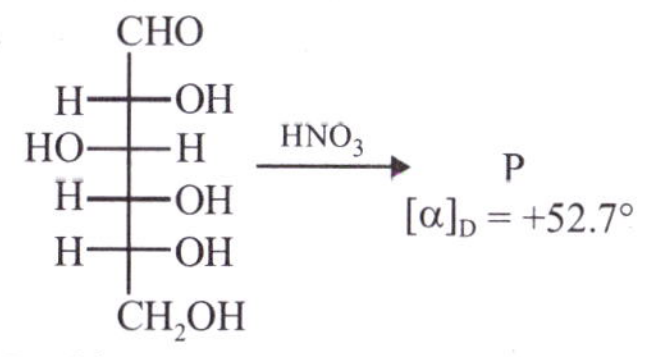

D-Glucose

The compound(s), which on reaction with HNO_3 will give the product having degree of rotation, $[\alpha]_D = -52.7°$ is (are) **[Adv. 2021]**

(a) 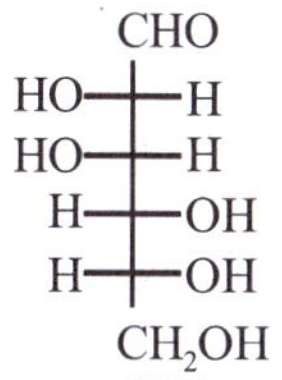

(b) 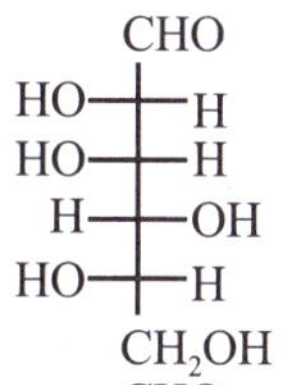

(c) 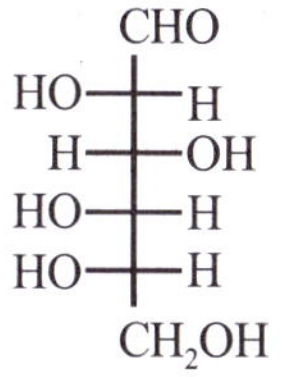

(d) 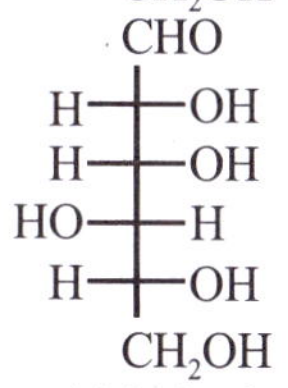

16. Which of the following statement(s) is (are) true? **[Adv. 2019]**

(a) Oxidation of glucose with bromine water gives glutamic acid.

(b) Hydrolysis of sucrose gives dextrorotatory glucose and laevorotatory fructose.

(c) The two six-membered cyclic hemiacetal forms of D-(+)-glucose are called anomers.

(d) Monosaccharides cannot be hydrolysed to give polyhydroxy aldehydes and ketones

17. For 'invert sugar', the correct statement(s) is(are)

(Given : specific rotations of (+) -sucrose, (+)-maltose, L-(-)-glucose and L-(+) fructose in aqueous solution are + 66°, +140°, –52° and +92°, respectively) **[Adv. 2016]**

(a) 'invert sugar' is prepared by acid catalyzed hydrolysis of maltose

(b) 'invert sugar' is an equimolar mixture of D-(+)-glucose and D-(-)-fructose

(c) specific rotation of 'invert sugar' is –20°

(d) on reaction with Br_2 water, 'invert sugar' forms saccharic acid as one of the products

18. The correct statement(s) about the following sugars **X** and **Y** is (are) **[2009S]**

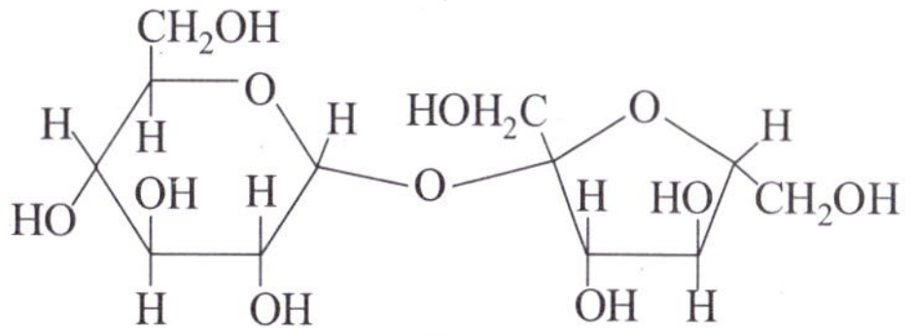

X

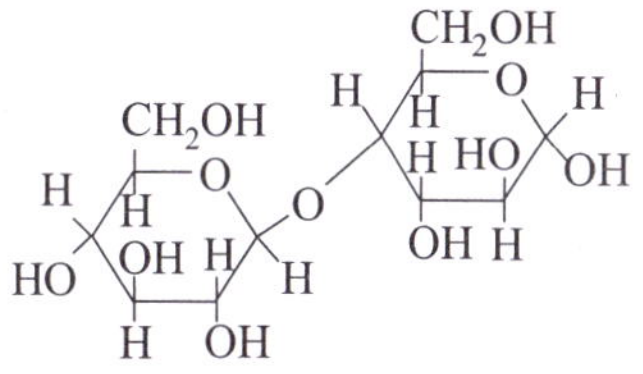

Y

(a) **X** is a reducing sugar and **Y** is a non-reducing sugar

(b) **X** is a non-reducing sugar and **Y** is a reducing sugar

(c) The glucosidic linkages in **X** and **Y** are α and β, respectively

(d) The glucosidic linkages in **X** and **Y** are β and α, respectively

7 Match the Following

19. Match the following, choosing one item from column **X** and the appropriate item from column **Y**. **[1983 - 2 Marks]**

(*i*) Lucas test	(a) Phenol
(*ii*) Neutral $FeCl_3$ test	(b) Glucose
(*iii*) Dye test	(c) Tertiary alcohol
(*iv*) Tollen's test	(d) Aniline

9 Assertion and Reason/Statement Type Questions

Each question contains **STATEMENT-1 (Assertion)** and **STATEMENT-2 (Reason)**. Each question has 4 choices (a), (b), (c) and (d) out of which ONLY ONE is correct. Mark your answer as

(a) If both Statement -1 and Statement -2 are correct, and Statement -2 is the correct explanation of the Statement -2.

(b) If both Statement -1 and Statement -2 are correct, but Statement -2 is not the correct explanation of the Statement -1.

(c) If Statement -1 is correct but Statement -2 is incorrect.

(d) If Statement -1 is incorrect but Statement -2 is correct.

20. **Statement-1 :** Glucose gives a reddish-brown precipitate with Fehling's solution.

Statement-2 : Reaction of glucose with Fehling's solution give CuO and gluconic acid. **[2007]**

10 Subjective Problems

21. Which of the following will reduce Tollen's reagent? Explain. **[2005 - 2 Marks]**

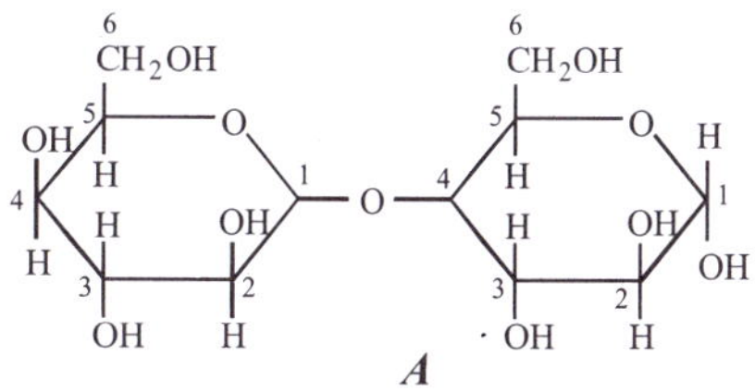

A

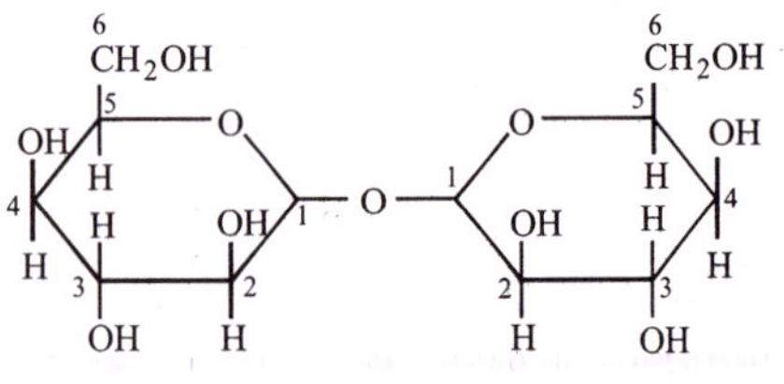

B

22. The Fisher projection of D-glucose is drawn below. [2004 - 2 Marks]

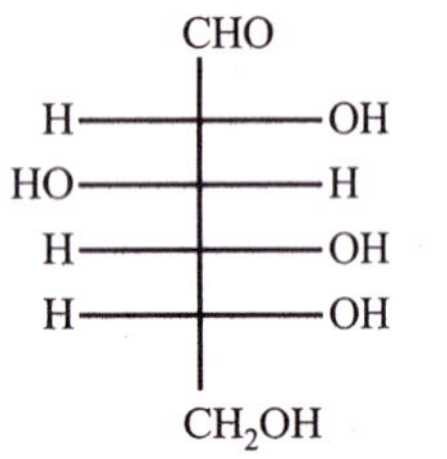

(i) Draw the Fisher projection of L-glucose.
(ii) Give the reaction of L-glucose with Tollen's reagent

Topic 2 - Proteins and Amino Acids

2 Integer Value Answer

23. The structure of a peptide is given below. [Adv. 2020]

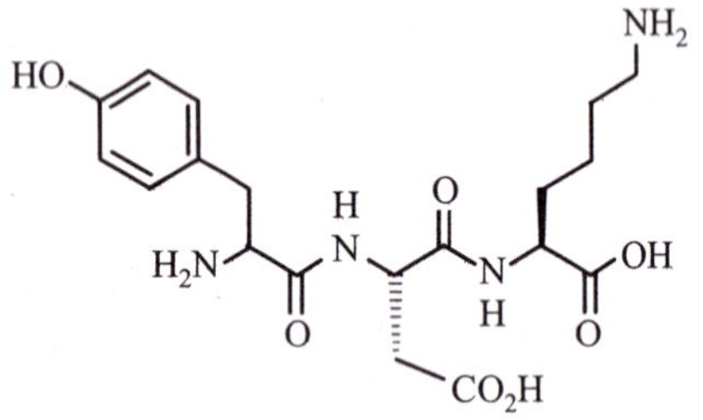

If the absolute values of the net charge of the peptide at pH = 2, pH = 6, and pH = 11 are $|z_1|$, $|z_2|$, and $|z_3|$, respectively, then what is $|z_1| + |z_2| + |z_3|$?

24. The total number of distinct naturally occurring amino acids obtained by complete acidic hydrolysis of the peptide shown below is [Adv. 2014]

25. A tetrapeptide has —COOH group on alanine. This produces glycine (Gly), valine (Val), phenyl alanine (Phe) and alanine (Ala), on complete hydrolysis. For this tetrapeptide, the number of possible sequences (primary structures) with —NH_2 group attached to a chiral center is [Adv. 2013]

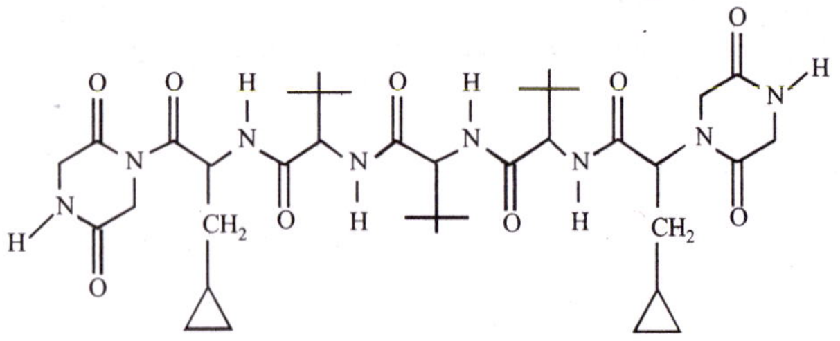

26. The substituents R_1 and R_2 for nine peptides are listed in the table given below. How many of these peptides are positively charged at pH = 7.0 ? [2012]

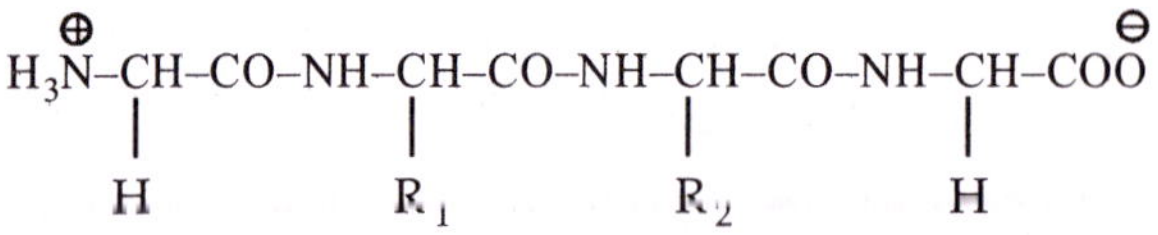

Peptide	R_1	R_2
I	H	H
II	H	CH_3
III	CH_2COOH	H
IV	CH_2CONH_2	$(CH_2)_4NH_2$
V	CH_2CONH_2	CH_2CONH_2
VI	$(CH_2)_4NH_2$	$(CH_2)_4NH_2$
VII	CH_2COOH	CH_2CONH_2
VIII	CH_2OH	$(CH_2)_4NH_2$
IX	$(CH_2)_4NH_2$	CH_3

27. A decapeptide (Mol. wt. 796) on complete hydrolysis gives glycine (Mol. wt. 75), alanine and phenylalanine. Glycine contributes 47.0% to the total weight of the hydrolysed products. The number of glycine units present in the decapeptide is [2011]

28. The total number of basic groups in the following form of lysine is [2010]

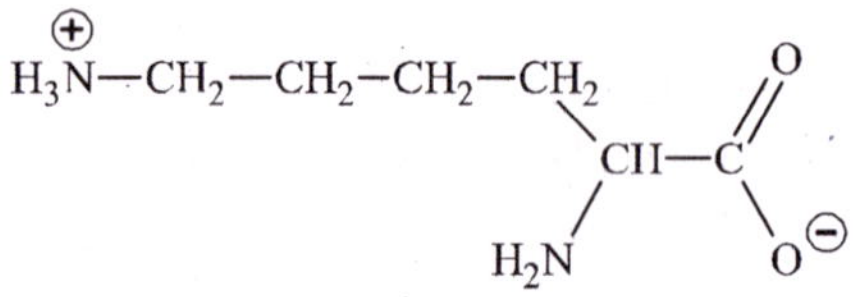

10 Subjective Problems

29. Following two amino acids lysine and glutamine form dipeptide linkage. What are two possible dipeptides? [2003 - 2 Marks]

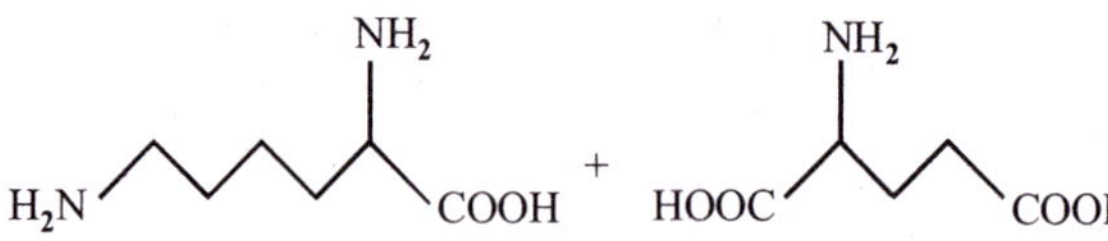

30. Aspartame, an artificial sweetener, is a peptide and has the following structure : [2001 - 5 Marks]

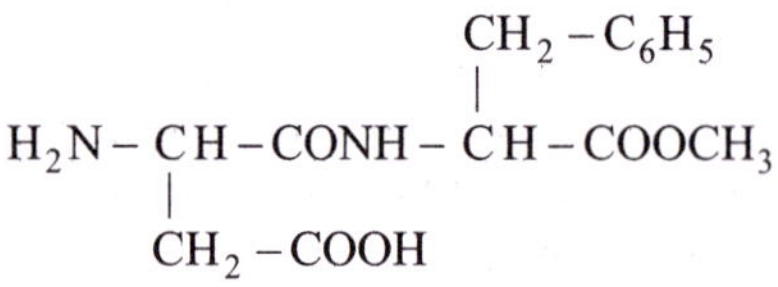

(i) Identify the four functional groups.
(ii) Write the zwitterionic structure.
(iii) Write the structures of the amino acids obtained from the hydrolysis of aspartame.
(iv) Which of the two amino acids is more hydrophobic?

31. Write the structures of alanine at pH = 2 and pH = 10. [2000 - 2 Marks]

Topic 3 - Enzymes and Nucleic Acid

2 Integer Value Answer

32. F or a double strand DNA, one strand is given below: [Adv. 2024]

5' ─────────── 3'
A G T C A C G T A A G T C

The amount of energy required to split the double strand DNA into two single strands is _____ kcal mol^{-1}.

[Given: Average energy per H-bond for A-T base pair = 1.0 kcal mol^{-1}, G-C base pair = 1.5 kcal mol^{-1}, and A-U base pair = 1.25 kcal mol^{-1}. Ignore electrostatic repulsion between the phosphate groups.]

Answer Key

Part-A (JEE Main)																			
1	(2)	21	(1)	41	(1)	61	(4)	81	(1)	101	(4)	121	(3)	141	(3)	161	(4)	181	(3)
2	(4)	22	(1)	42	(3)	62	(4)	82	(4)	102	(2)	122	(4)	142	(1)	162	(1)	182	(3)
3	(1)	23	(1)	43	(4)	63	(2)	83	(1)	103	(1)	123	(1)	143	(3)	163	(1)	183	(3)
4	(4)	24	(3)	44	(4)	64	(1)	84	(3)	104	(3)	124	(2)	144	(3)	164	(2)	184	(2)
5	(3)	25	(3)	45	(1)	65	(3)	85	(2)	105	(2)	125	(4)	145	(2)	165	(3)	185	(5)
6	(4)	26	(1)	46	(3)	66	(3)	86	(1)	106	(2)	126	(9)	146	(1)	166	(2)		
7	(1)	27	(2)	47	(2)	67	(2)	87	(4)	107	(2)	127	(6)	147	(1)	167	(33)		
8	(4)	28	(2)	48	(3)	68	(3)	88	(4)	108	(3)	128	(10)	148	(1)	168	(3)		
9	(2)	29	(1)	49	(2)	69	(3)	89	(1)	109	(2)	129	(6)	149	(2)	169	(25)		
10	(4)	30	(2)	50	(4)	70	(3)	90	(3)	110	(1)	130	(8)	150	(4)	170	(9)		
11	(2)	31	(4)	51	(3)	71	(3)	91	(1)	111	(3)	131	(1)	151	(3)	171	(3)		
12	(2)	32	(4)	52	(2)	72	(2)	92	(4)	112	(2)	132	(2)	152	(2)	172	(3)		
13	(3)	33	(1)	53	(4)	73	(2)	93	(1)	113	(1)	133	(4)	153	(2)	173	(4)		
14	(2)	34	(1)	54	(2)	74	(9)	94	(4)	114	(2)	134	(3)	154	(2)	174	(3)		
15	(3)	35	(1)	55	(4)	75	(3)	95	(3)	115	(3)	135	(4)	155	(1)	175	(4)		
16	(1)	36	(1)	56	(4)	76	(2)	96	(3)	116	(3)	136	(4)	156	(1)	176	(1)		
17	(2)	37	(2)	57	(1)	77	(3)	97	(2)	117	(1)	137	(2)	157	(4)	177	(4)		
18	(1)	38	(2)	58	(4)	78	(1)	98	(3)	118	(2)	138	(5)	158	(1)	178	(4)		
19	(2)	39	(1)	59	(1)	79	(3)	99	(4)	119	(2)	139	(37.84)	159	(3)	179	(2)		
20	(1)	40	(3)	60	(2)	80	(3)	100	(2)	120	(3)	140	(3)	160	(1)	180	(3)		
Part-B (JEE Advanced)																			
1	(a)	4	(c)	7	(b)	10	(b)	13	(2)	16	(b, c, d)	19	(i)–(c), (ii)–(a), (iii)–(d), (iv)–(b)	24	(1)	27	(6)		
2	(d)	5	(d)	8	(a)	11	(d)	14	(b, c)	17	(b, c)	20	(c)	25	(4)	28	(2)		
3	(a)	6	(a)	9	(a)	12	(8)	15	(c, d)	18	(b, c)	23	(5)	26	(4)	32	(41)		

Chapter-15

Polymers

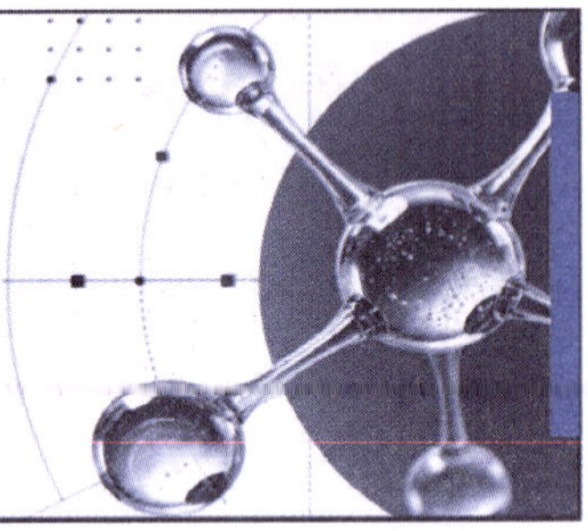

Part-A (JEE Main)

Topic 1 - Classification of Polymers

MULTIPLE CHOICE QUESTIONS (MCQs)

1. Which of the following is **NOT** a natural polymer ? **[July 29, 2022 (II)]**
   (1) Protein (2) Starch
   (3) Rubber (4) Rayon

2. Match List-I with List-II **[July 28, 2022 (II)]**

List-I (Polymer)	List-II (Nature)
(A) $\left[CH_2-\underset{\underset{Cl}{|}}{C}=CH-CH_2\right]_n$	(I) Thermosetting polymer
(B) $\left[\overset{H}{\overset{|}{N}}-(CH_2)_6-\overset{H}{\overset{|}{N}}-\overset{O}{\overset{\|}{C}}-(CH_2)_4-\overset{O}{\overset{\|}{C}}\right]_n$	(II) Fibers
(C) $\left[CH_2-\overset{Cl}{\overset{|}{CH}}\right]_n$	(III) Elastomer
(D) 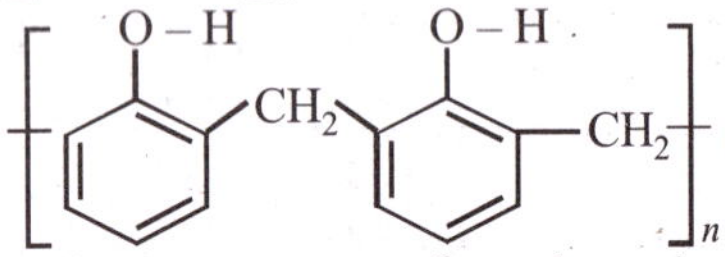	(IV) Thermo-plastic polymer

Choose the correct answer from the options given below:
(1) (A) - II, (B) - III, (C) - IV, (D) - I
(2) (A) - III, (B) - II, (C) - IV, (D) - I
(3) (A) - III, (B) - I, (C) - IV, (D) - II
(4) (A) - I, (B) - III, (C) - IV, (D) - II

3. Which one of the following is **NOT** a copolymer ? **[June 28, 2022 (I)]**
   (1) Buna-S (2) Neoprene
   (3) PHBV (4) Butadiene-styrene

4. Which of the following sets are **correct** regarding polymer ? **[June 26, 2022 (I)]**
   (A) Copolymer : Buna-S
   (B) Condensation polymer : Nylon-6,6
   (C) Fibre : Nylon-6,6
   (D) Thermosetting polymer : Terylene
   (E) Homopolymer : Buna-N
   Choose the **correct** answer from the options given below:
   (1) (A), (B) and (C) are correct
   (2) (B), (C) and (D) are correct
   (3) (A), (C) and (E) are correct
   (4) (A), (B) and (D) are correct

5. Which of the following is an example of polyester? **[June 24, 2022 (I)]**
   (1) Butadiene-styrene copolymer
   (2) Melamine polymer
   (3) Neoprene
   (4) Poly- β -hydroxybutyrate-co- β -hydroxy valerate

6. Which of the following is not an example of a condensation polymer? **[June 24, 2022 (II)]**
   (1) Nylon 6, 6 (2) Decron
   (3) Buna-N (4) Silicone

7. Which one of the following polymers is not obtained by condensation polymerisation? **[Sep. 05, 2020 (II)]**
   (1) Nylon 6,6 (2) Buna - N
   (3) Bakelite (4) Nylon 6

8. The correct name of the following polymer is: **[April 12, 2019 (II)]**

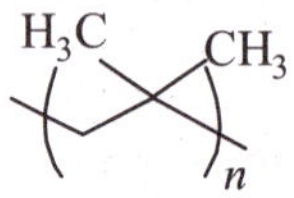

   (1) Polyisobutane (2) Polytert-butylene
   (3) Polyisoprene (4) Polyisobutylene

9. Which of the following is a condensation polymer? **[April 10, 2019 (I)]**
   (1) Buna - S (2) Neoprene
   (3) Teflon (4) Nylon 6, 6

10. The structure of Nylon-6 is: **[April 8, 2019 (II)]**
   (1) $\left[(CH_2)_4-\overset{O}{\overset{\|}{C}}-\overset{H}{\overset{|}{N}}\right]_n$ (2) $\left[\overset{O}{\overset{\|}{C}}-(CH_2)_6-\overset{H}{\overset{|}{N}}\right]_n$
   (3) $\left[\overset{O}{\overset{\|}{C}}-(CH_2)_5-\overset{H}{\overset{|}{N}}\right]_n$ (4) $\left[(CH_2)_6-\overset{O}{\overset{\|}{C}}-\overset{H}{\overset{|}{N}}\right]_n$

11. Which of the following polymers is synthesized using a free radical polymerisation technique? **[Online April 10, 2016]**
   (1) Terylene (2) Melamine polymer
   (3) Nylon 6, 6 (4) Teflon

12. Which one is classified as a condensation polymer? **[2014]**
   (1) Dacron (2) Neoprene
   (3) Teflon (4) Acrylonitrile

13. Which one of the following is an example of thermosetting polymers? **[Online April 19, 2014; April 12, 2019 (I) (S)]**
   (1) Neoprene (2) Buna-N
   (3) Nylon 6, 6 (4) Bakelite

14. Which of the following is a polyamide? [Online May 19, 2012, Online April 23, 2013; Similar 2005]
(1) Teflon (2) Orlon (3) Nylon (4) Terylene

15. Which one of the following is a chain growth polymerisation? [Online May 7, 2012]
(1) Nucleic acid (2) Polystyrene
(3) Protein (4) Starch

16. The polymer containing strong intermolecular forces, e.g., hydrogen bonding, is [2010]
(1) teflon (2) nylon 6, 6
(3) polystyrene (4) natural rubber

Topic 2 - Preparation and Properties of Polymers

Multiple Choice Questions (MCQs)

17. Terylene polymer is obtained by condensation of : [July 28, 2022 (I)]
(1) Ethane-1, 2-diol and Benzene-1, 3 dicarboxylic acid
(2) Propane-1, 2-diol and Benzene-1, 4 dicarboxylic acid
(3) Ethane-1, 2-diol and Benzene-1, 4 dicarboxylic acid
(4) Ethane-1, 2-diol and Benzene-1, 2 dicarboxylic acid

18. Match **List-I** with **List-II**

List-I Polymers	List-II Commenrcial names
(A) Phenolformaldehyde resin	(I) Glyptal
(B) Copolymer of 1,3-butadiene and styrene	(II) Novolac
(C) Polyester of glycol and phthalic acid	(III) Buna-S
(D) Polyester of glycol and terephthalic acid	(IV) Dacron

Choose the correct answer from the options given below:
(1) A–II, B–III, C–IV, D–I [July 27, 2022 (I)]
(2) A–II, B–III, C–I, D–IV
(3) A–II, B–I, C–III, D–IV
(4) A–III, B–II, C–IV, D–I

19. Match List-I with List-II.

List-I (Polymer)	List-II (Monomer)
(A) Neoprene	(I) Acrylonitrile
(B) Teflon	(II) Chloroprene
(C) Acrilan	(III) Tetrafluoroethene
(D) Natural rubber	(IV) Isoprene

Choose the correct answer from the option given below:
(1) (A)-(II), (B)-(III), (C)-(I), (D-(IV) [July 27, 2022 (II)]
(2) (A)-(II), (B)-(I), (C)-(III), (D-(IV)
(3) (A)-(II), (B)-(I), (C)-(IV), (D-(III)
(4) (A)-(I), (B)-(II), (C)-(III), (D-(IV)

20. Vulcanization of rubber is carried out by heating a mixture of : [July 26, 2022 (II)]
(1) isoprene and styrene (2) neoprene and sulphur
(3) isoprene and sulphur (4) neoprene and styrene

21. Melamine polymer is formed by the condensation of: [July 25, 2022 (I)]

(1) 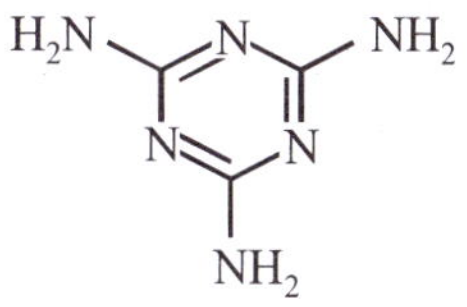 + HCHO

(2) 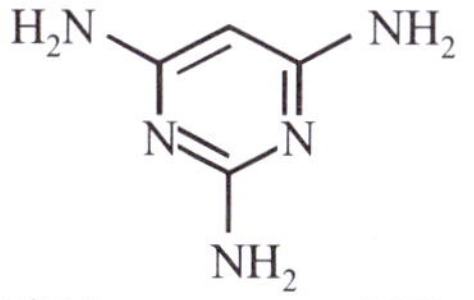 + HCHO

(3) 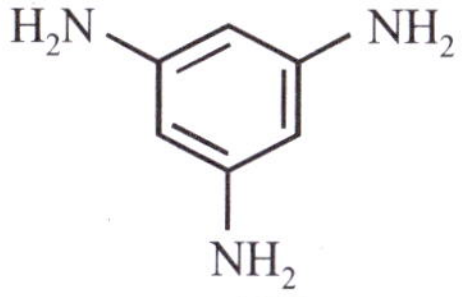 + HCHO

(4) 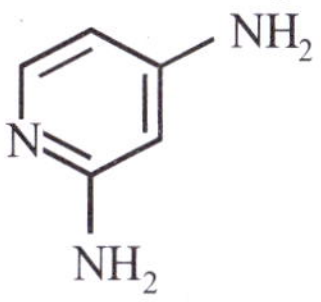 + HCHO

22. The polymer, which can be stretched and retains its original status on releasing the force is [June 29, 2022 (I)]
(1) Bakelite (2) Nylon 6,6
(3) Buna-N (4) Terylene

23. Given below are two statements : one is labelled as **Assertion A** and the other is labelled as **Reason R**. **Assertion A:** Dacron is an example of polyester polymer.
Reason R: Dacron is made up of ethylene glycol and terephthalic acid monomers.
In the light of the above statements, choose the **most appropriate** answer from the options given below. [June 29, 2022 (II)]
(1) Both A and B are correct and R is the correct explanation of A.
(2) Both A and B are correct but R is NOT the correct explanation of A.
(3) A is correct but R is not correct.
(4) A is not correct but R is correct.

24. Given below are two statements, one is **Assertion (A)** and other is **Reason (R)**. [June 28, 2022 (II)]
Assertion (A) : Natural rubber is a linear polymer of isoprene called cis-polyisoprene with elastic properties.
Reason (R) : The cis-polyisoprene molecules consist of various chains held together by strong polar interactions with coiled structure.
In the light of the above statements, choose the correct one from the options given below :
(1) Both (A) and (R) are true and (R) is the correct explanation of (A).
(2) Both (A) and (R) are true but (R) is not the correct explanation of (A).
(3) (A) is true but (R) is false.
(4) (A) is false but (R) is true.

25. Match List-I with List-II [June 27, 2022 (I)]

List-I (Polymer)	List-II (Used in)
(A) Bakelite	(I) Radio and television Cabinets
(B) Glyptal	(II) Electrical switches
(C) PVC	(III) Paints and Lacquers
(D) Polystyrene	(IV) Water Pipes

Choose the correct answer from the options given below:
(1) (A) – (II), (B) – (III), (C) – (IV), (D) – (I)
(2) (A) – (I), (B) – (II), (C) – (III), (D) – (IV)
(3) (A) – (IV), (B) – (III), (C) – (II), (D) – (I)
(4) (A) – (II), (B) – (III), (C) – (I), (D) – (IV)

26. Match List - I with List - II. [June 30, 2022 (I)]

List - I (Polymer)	List - II (Repeat Unit)
(A) Acrilan	(i) $-\!\!\left(CH_2-\underset{}{\overset{Cl}{\overset{\vert}{C}}}=CH-CH_2\right)\!\!-_n$
(B) Neoprene	(ii) $-\!\!\left(CH_2-\underset{C_6H_5}{\underset{\vert}{CH_2}}\right)\!\!-_n$
(C) Polystyrene	(iii) $-\!\!\left(CH_2-\underset{CN}{\underset{\vert}{CH}}\right)\!\!-_n$
(D) Buna-N	(iv) $-\!\!\left(CH_2-CH=CH-CH_2-CH_2-\underset{}{\overset{CN}{\overset{\vert}{CH}}}\right)\!\!-_n$

Choose the **correct** match from the options given below :
(1) (A)-(iii), (B)-(iv), (C)-(ii), (D)-(i)
(2) (A)-(iii), (B)-(ii), (C)-(i), (D)-(iv)
(3) (A)-(iii), (B)-(i), (C)-(ii), (D)-(iv)
(4) (A)-(ii), (B)-(iii), (C)-(i), (D)-(iv)

27. Which is true about Buna-*N*? [June 27, 2022 (II)]
(1) It is a linear polymer of 1, 3-butadiene.
(2) It is obtained by copolymerization of 1, 3-butadiene and styrene.
(3) It is obtained by copolymerization of 1, 3-butadiene and acrylonitrile.
(4) The suffix *N* in Buna-*N* stands for its natural occurrence.

28. Monomer of Novolac is : [Aug. 31, 2021 (I)]
(1) 3-Hydroxybutanoic acid
(2) phenol and melamine
(3) *o*-Hydroxymethylphenol
(4) 1,3-Butadiene and styrene

29. The polymer formed on heating Novolac with formaldehyde is : [Aug. 26, 2021 (I); Similar July 20, 2021 (II)]
(1) Bakelite (2) Polyester
(3) Melamine (4) Nylon 6,6

30. 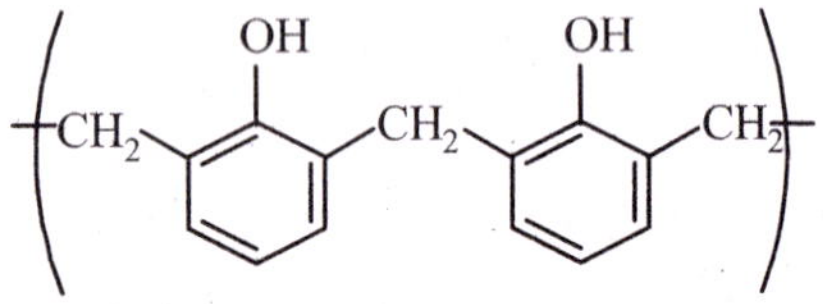

is a repeating unit for : [July 25, 2021 (I)]
(1) Novolac (2) Buna–N
(3) Acrilan (4) Neoprene

31. A biodegradable polyamide can be made from : [July 25, 2021 (II)]
(1) Glycine and isoprene
(2) Hexamethylene diamine and adipic acid
(3) Glycine and aminocaproic acid
(4) Styrene and caproic acid

32. Orlon fibers are made up of : [July 20, 2021 (I)]
(1) Polyacrylonitrile (2) Polyesters
(3) Polyamide (4) Cellulose

33. The structure of Neoprene is : [Feb. 26, 2021 (I)]

(1) 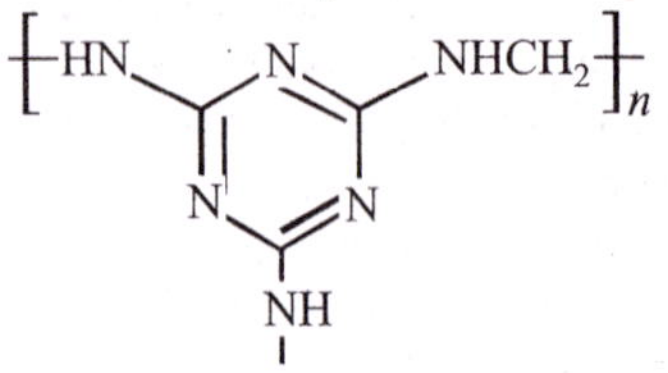

(2) $\left[CH_2-\underset{CN}{\underset{\vert}{CH}}\right]_n$

(3) $\left[CH_2CH=CH-CH_2-CH_2-\underset{CN}{\underset{\vert}{CH}}\right]_n$

(4) $\left[CH_2-\overset{Cl}{\overset{\vert}{C}}=CH-CH_2\right]_n$

34. In polymer Buna-S: 'S' stands for : [February 24, 2021 (II)]
(1) Sulphonation (2) Strength
(3) Sulphur (4) Styrene

35. Consider the **Assertion** and **Reason** given below.
Assertion (A) : Ethene polymerized in the presence of Ziegler Natta catalyst at high temperature and pressure is used to make buckets and dustbins.
Reason (R) : High density polymers are closely packed and are chemically inert. [Sep. 06, 2020 (I)]
Choose the correct answer from the following :
(1) (A) is correct but (R) is wrong.
(2) Both (A) and (R) are correct but (R) is not the correct explanation of (A).
(3) Both (A) and (R) are correct but (R) is the correct explanation of (A).
(4) (A) and (R) both are wrong.

36. The correct match between Item-I and Item - II is: [Sep. 06, 2020 (II)]

Item - I	Item - II
(A) Natural rubber	(I) 1, 3-butadiene + styrene
(B) Neoprene	(II) 1, 3-butadiene + acrylonitrile
(C) Buna-N	(III) Chloroprene
(D) Buna-S	(IV) Isoprene

(1) (A) - (III), (B) - (IV), (C) - (I), (D) - (II)
(2) (A) - (III), (B) - (IV), (C) - (II), (D) - (I)
(3) (A) - (IV), (B) - (III), (C) - (II), (D) - (I)
(4) (A) - (IV), (B) - (III), (C) - (I), (D) - (II)

37. Which polymer has 'chiral' monomer(s)? [Jan. 09, 2020 (II)]
(1) Neoprene (2) Buna-N
(3) Nylon 6, 6 (4) PHBV

38. Preparation of Bakelite proceeds via reactions: [Jan. 08, 2020 (II)]
(1) Electrophilic addition and dehydration
(2) Condensation and elimination
(3) Electrophilic substitution and dehydration
(4) Nucleophilic addition and dehydration

39. The correct match between **Item – I** and **Item- II** is : [April 10, 2019 (II)]

Item-I	Item-II
(A) High density polythene	(I) Peroxide catalyst
(B) Polyacrylonitrile	(II) Condensation at high temperature & pressure
(C) Novolac	(III) Ziegler-Natta catalyst
(D) Nylon 6	(IV) Acid or base catalyst

(1) (A)$\rightarrow$(IV), (B)$\rightarrow$(II), (C)$\rightarrow$(I), (D)$\rightarrow$(III)
(2) (A)$\rightarrow$(II), (B)$\rightarrow$(IV), (C)$\rightarrow$(I), (D)$\rightarrow$(III)
(3) (A)$\rightarrow$(III), (B)$\rightarrow$(I), (C)$\rightarrow$(II), (D)$\rightarrow$(IV)
(4) (A)$\rightarrow$(III), (B)$\rightarrow$(I), (C)$\rightarrow$(IV), (D)$\rightarrow$(II)

40. The major product of the following reaction is:

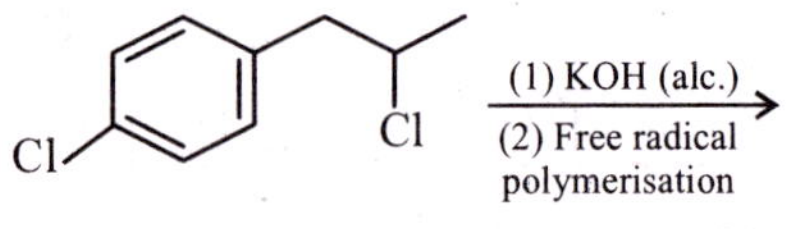

[April 9, 2019 (I)]

(1) 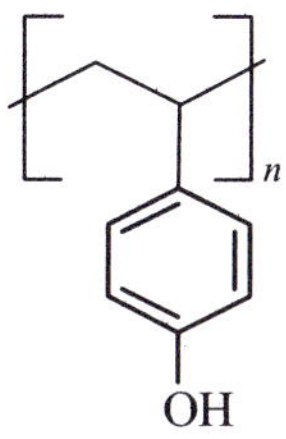 (2) 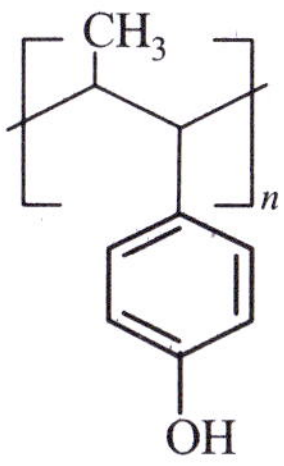

(3) 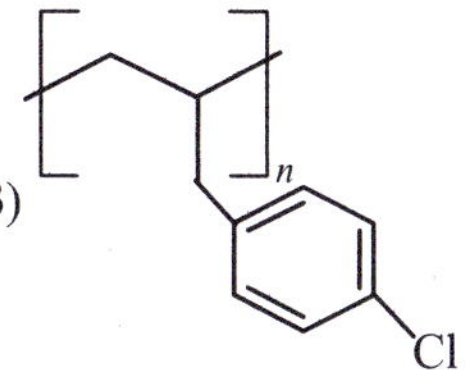 (4) 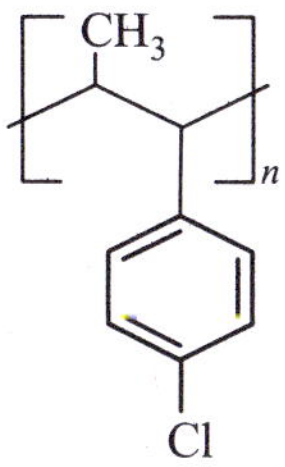

41. Which of the following compounds is a constituent of the polymer $-\!\!\left[HN-\overset{O}{\overset{\|}{C}}-NH-CH_2\right]\!\!-_n$? [April 9, 2019 (II)]

(1) N-Methyl urea (2) Formaldehyde
(3) Methylamine (4) Ammonia

42. Poly-β-hydroxybutyrate-co-β-hydroxy valerate (PHBV) is a copolymer of __________. [Jan. 12, 2019 (I)]

(1) 3-Hydroxybutanoic acid and 4-Hydroxypentanoic acid
(2) 2-Hydroxybutanoic acid and 3-Hydroxypentanoic acid
(3) 3-Hydroxybutanoic acid and 2-Hydroxypentanoic acid
(4) 3-Hydroxybutanoic acid and 3-Hydroxypentanoic acid

43. The two monomers for the synthesis of nylon 6, 6 are : [Jan. 12, 2019 (II)]

(1) $HOOC(CH_2)_4COOH, H_2N(CH_2)_6NH_2$
(2) $HOOC(CH_2)_6COOH, H_2N(CH_2)_6NH_2$
(3) $HOOC(CH_2)_4COOH, H_2N(CH_2)_4NH_2$
(4) $HOOC(CH_2)_6COOH, H_2N(CH_2)_4NH_2$

44. The polymer obtained from the following reactions is [Jan. 11, 2019 (I)]

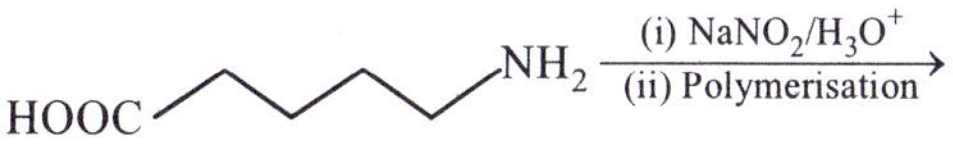

(1) $\left[\overset{O}{\overset{\|}{C}}-(CH_2)_4-\overset{H}{N}\right]_n$ (2) $\left[HN-\overset{O}{\overset{\|}{C}}-(CH_2)_4-\overset{O}{\overset{\|}{C}}-\overset{H}{N}\right]_n$

(3) $\left[O-(CH_2)_4-\overset{O}{\overset{\|}{C}}\right]_n$ (4) $\left[O-\overset{O}{\overset{\|}{C}}-(CH_2)_4-O\right]_n$

45. The homopolymer formed from 4-hydroxy-butanoic acids is : [Jan. 11, 2019 (II)]

(1) $\left[O-\overset{O}{\overset{\|}{C}}(CH_2)_3-O\right]_n$

(2) $\left[O-\overset{O}{\overset{\|}{C}}(CH_2)_2-\overset{O}{\overset{\|}{C}}\right]_n$

(3) $\left[O-\overset{O}{\overset{\|}{C}}-(CH_2)_2-\overset{O}{\overset{\|}{C}}-O\right]_n$

(4) $\left[\overset{O}{\overset{\|}{C}}(CH_2)_3-O\right]_n$

46. The major product of the following reaction is: [Jan. 10, 2019 (II)]

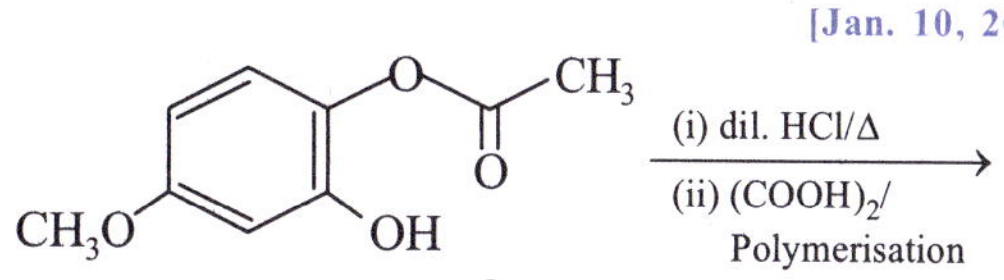

(1) 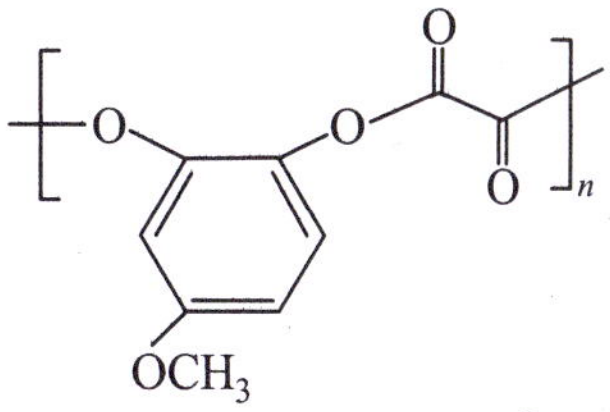

(2) 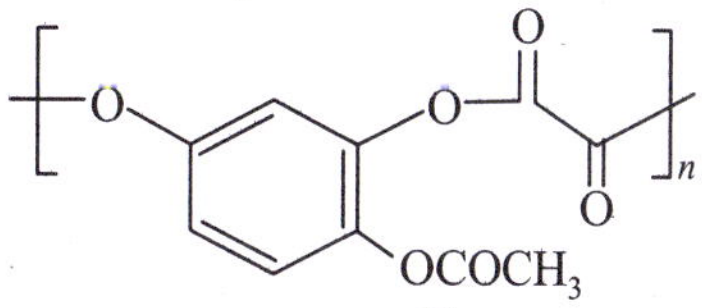

(3) 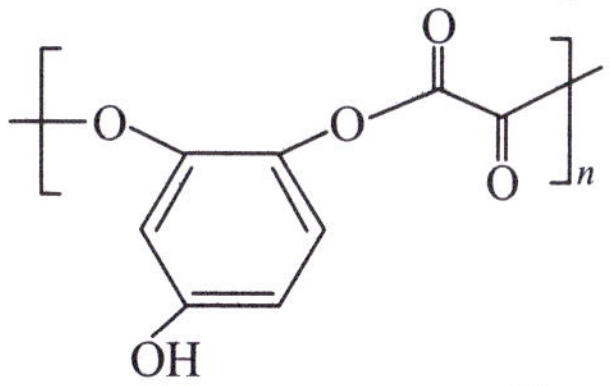

(4) 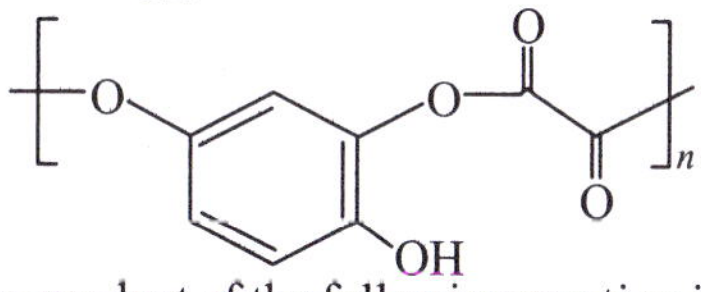

47. Major product of the following reaction is: [Jan. 9, 2019 (I)]

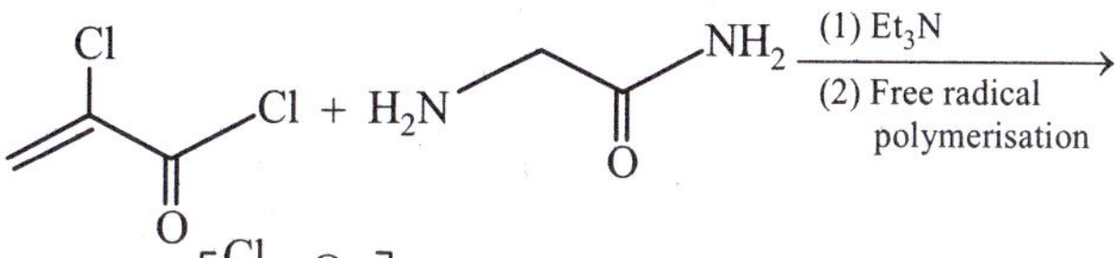

(1) 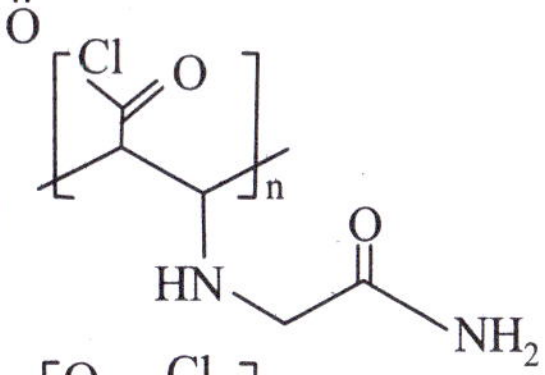

(2) 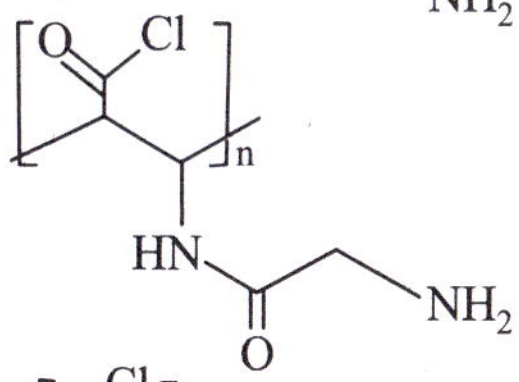

(3) 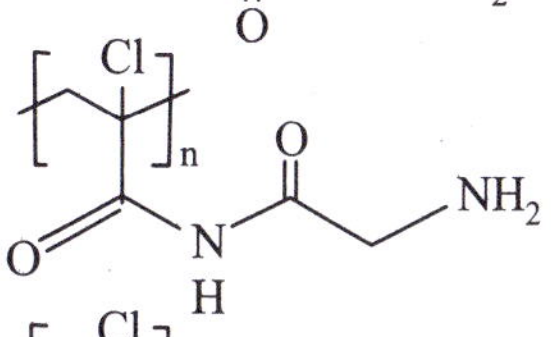

(4) 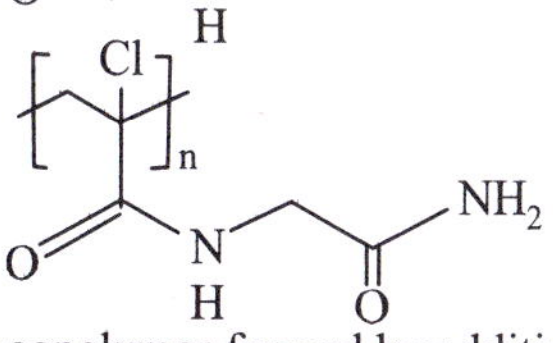

48. The copolymer formed by addition polymerization of styrene and acrylonitrile in the presence of peroxide is: [Online April 15, 2018 (I)]

(1) $\left[\overset{C_6H_5}{\underset{CH_3}{C}}-\overset{CN}{CH}-CH_2\right]_n$

(2) $\left[CH_2-CH(C_6H_5)-CH_2-CH(CN) \right]_n$

(3) $\left[CH_2-CH(C_6H_5)-CH(CN)-CH_2 \right]_n$

(4) $\left[CH(C_6H_5)-CH_2-CH_2-CH(CN) \right]_n$

49. Which of the following statements is not true? [Online April 15, 2018 (II)]
(1) Chain growth polymerisation involves homopolymerisation only
(2) Chain growth polymerisation includes both homopolymerisation and copolymerisation
(3) Nylon 6 is an example of step-growth polymerisation
(4) Step growth polymerisation requires a bifunctional monomer

50. The formation of which of the following polymers involves hydrolysis reaction? [2017]
(1) Nylon 6 (2) Bakelite
(3) Nylon 6, 6 (4) Terylene

51. Which of the following is a biodegradable polymer ? [Online April 9, 2017]

(1) $\left[HN-(CH_2)_5-\overset{O}{\overset{\|}{C}}-NH-CH_2-\overset{O}{\overset{\|}{C}} \right]_n$

(2) $\left[HN-(CH_2)_5-\overset{O}{\overset{\|}{C}} \right]_n$

(3) $\left[HN-(CH_2)_6\,NHCO-(CH_2)_4-\overset{O}{\overset{\|}{C}} \right]_n$

(4) 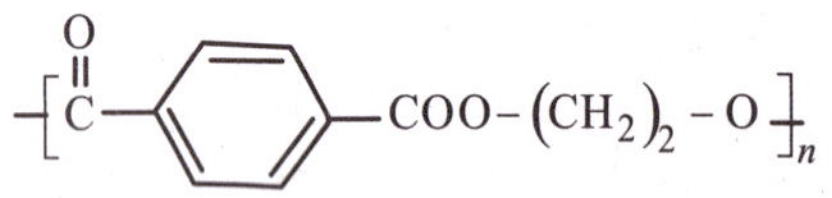

52. Which of the following statements about low density polythene is **FALSE**? [2016]
(1) Its synthesis requires dioxygen or a peroxide initiator as a catalyst.
(2) It is used in the manufacture of buckets, dust-bins etc.
(3) Its synthesis requires high pressure.
(4) It is a poor conductor of electricity.

53. **Assertion :** Rayon is a semisynthetic polymer whose properties are better than natural cotton.
Reason : Mechanical and aesthetic properties of cellulose can be improved by acetylation. [Online April 9, 2016]
(1) Both assertion and reason are correct, but the reason is not the correct explantion for the assertion
(2) Both assertion and reason are correct, and the reason is the correct explantion for the assertion
(3) Assertion is incorrect statement, but the reason is correct.
(4) Both assertion and reason are incorrect.

54. Which one of the following structures represents the neoprene polymer ? [Online April 11, 2015]
(1) $\left[CH(C_6H_5)-CH_2 \right]_n$
(2) $\left[CH_2-CH(CN) \right]_n$
(3) $\left[CH_2-C(Cl)=CH-CH_2 \right]_n$
(4) $\left[CH_2-CH(Cl) \right]_n$

55. Structure of some important polymers are given. Which one represents Buna-S? [Online April 9, 2014]
(1) $\left[CH_2-C(CH_3)=CH-CH_2 \right]_n$
(2) $\left[CH_2-CH=CH-CH_2-CH(C_6H_5)-CH_2 \right]_n$
(3) $\left[CH_2-CH=CH-CH_2-CH(CN)-CH_2 \right]_n$
(4) $\left[CH_2-C(Cl)=CH-CH_2 \right]_n$

56. If a polythene sample contains two monodisperse fractions in the ratio 2 : 3 with degree of polymerization 100 and 200, respectively, then its weight average molecular weight will be : [Online April 9, 2013]
(1) 4900 (2) 4600 (3) 4300 (4) 5200

57. Bakelite is obtained from phenol by reacting with: [Online April 25, 2013; Similar May 12, 2012; May 26, 2012; 2011 RS; 2008]
(1) Acetal (2) CH_3CHO
(3) HCHO (4) Chlorobenzene

58. The species which can best serve as an initiator for the cationic polymerisation is : [2012]
(1) $LiAlH_4$ (2) HNO_3 (3) $AlCl_3$ (4) BaLi

59. Buna-N synthetic rubber is a copolymer of : [2009]
(1) $H_2C=CH-CH=CH_2$ and $H_5C_6-CH=CH_2$
(2) $H_2C=CH-CN$ and $H_2C=CH-CH=CH_2$
(3) $H_2C=CH-CN$ and $H_2C=CH-C(CH_3)=CH_2$
(4) $H_2C=CH-C(Cl)=CH_2$ and $H_2C=CH-CH=CH_2$

60. Which of the following is fully fluorinated polymer? [2005]
(1) PVC (2) Thiokol
(3) Teflon (4) Neoprene

61. Nylon threads are made of [2003]
(1) polyester polymer (2) polyamide polymer
(3) polyethylene polymer (4) polyvinyl polymer

62. Polymer formation from monomers starts by [2002]
(1) condensation reaction between monomers
(2) coordinate reaction between monomers
(3) conversion of monomer to monomer ions by protons
(4) hydrolysis of monomers.

NUMERICAL VALUE QUESTIONS (NVQs)

63. The Novolac polymer has mass of 963 g. The number of monomer units present in it are [June 25, 2022 (II)]

Topic 3 - Uses of Polymers

MULTIPLE CHOICE QUESTIONS (MCQs)

64. Match List-I with List-II : [July 25, 2022 (II)]

List-I (Polymer)	List-II (used for items)
(A) Nylon 6, 6	(I) Buckets
(B) Low density polythene	(II) Non-stick utensils
(C) High density polythene	(III) Bristles of brushes
(D) Teflon	(IV) Toys

Choose the correct answer from the options given below:

(1) A-III, B-I, C-IV, D-II (2) A-III, B-IV, C-I, D-II

(3) A-II, B-I, C-IV, D-III (4) A-II, B-IV, C-I, D-III

65. Which of the following polymer is used in the manufacture of wood laminates? [March 16, 2021 (II)]

(1) Urea formaldehyde resin

(2) Phenol and formaldehyde resin

(3) Melamine formaldehyde resin

(4) *cis*-poly-isoprene

66. Match the polymers in column - A with their main uses in column - B and choose the correct answer: [Online April 10, 2015]

Column - A	Column - B
(A) Polystrene	(i) Paints and lacquers
(B) Glyptal	(ii) Rain coats
(C) Polyvinyl chloride	(iii) Manufacture of toys
(D) Bakelite	(iv) Computer discs

(1) (A) - (ii), (B) - (i), (C) - (iii), (D) - (iv)

(2) (A) (ii), (B) (iv), (C) - (iii), (D) - (i)

(3) (A) - (iii), (B) - (iv), (C) - (ii), (D) - (i)

(4) (A) - (iii), (B) - (i), (C) - (ii), (D) - (iv)

67. The polymer used for optical lenses is : [Online April 22, 2013]

(1) polypropylene (2) polyvinyl chloride

(3) polythene (4) polymethyl methacrylate

Part-B (JEE Advanced)

Topic 1 - Classification of Polymers

10 Subjective Problems

1. Write the matched set (of three) for each entry in column A: [Multiple Concepts, 1984 - 1 × 5 = Mark]

A	B	C
(i) Asbestos	(a) molecular sieve	(1) air pollutant
(ii) Fluorocarbons	(b) paramagnetic	(2) carcinogen
(iii) Lithium metal	(c) refrigeration	(3) fluorescent paint
(iv) Nitric oxide	(d) reducing agent	(4) electron donor
(v) Zeolites	(e) semi-conductor	(5) ion exchanger
(vi) Zinc oxide	(f) silicates of (Ca + Mg)	(6) propellant

Topic 2 - Preparation and Properties of Polymers

1 MCQs with One Correct Answer

2. Natural rubber on complete ozonolysis (O_3/Zn-H_2O) gives compound X as the major product. X gives positive iodoform and Tollen's tests. X on heating with aqueous NaOH gives Y as the major product. Y is [Adv. 2026]

(a) 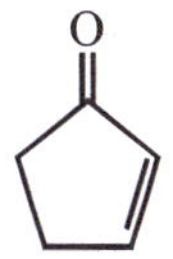

(b) 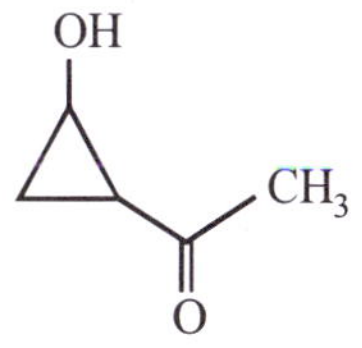

(c) 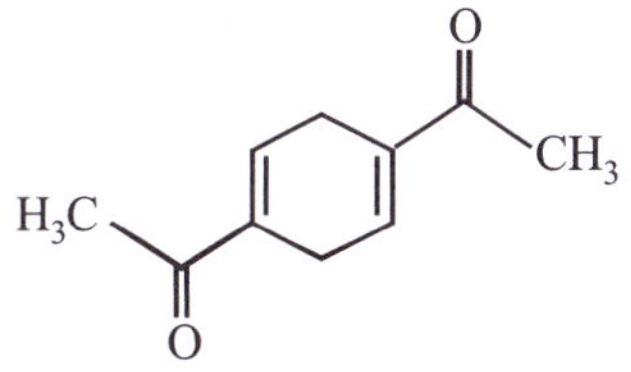

(d) 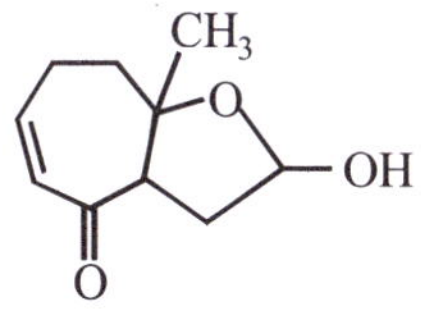

3. On complete hydrogenation, natural rubber produces [Adv. 2016]

(a) ethylene–propylene copolymer

(b) vulcanised rubber

(c) polypropylene

(d) polybutylene

4. Among cellulose, poly (vinyl chloride), nylon and natural rubber, the polymer in which the intermolecular force of attraction is weakest is [2009S]

(a) Nylon (b) Poly (vinyl chloride)

(c) Cellulose (d) Natural rubber

3 Numeric / New Stem Based Questions

5. In the following reaction sequence, major products X and Y are acyclic monomers. **[Adv. 2026]**

$$CH_3I \xrightarrow[4.\ NH_3\ (excess)]{1.\ KCN;\ 2.\ H_3O^+,\ \Delta;\ 3.\ Red\ P,\ Br_2} X$$

$$\text{Caprolactam} \xrightarrow[\Delta]{H_3O^+} Y$$

500 mol of X completely reacts with 500 mol of Y to give 1 mol of a single biodegradable acyclic copolymer Z as the only product. The amount of Z formed in grams is ____.

Given:

Atomic mass (in amu): H : 1, C : 12, N : 14, O : 16, Br : 80

6. The total number of lone-pairs of electrons in melamine is **[2013]**

4 Fill in the Blanks

7. Sulphur acts as .... agent in vulcanization of rubber. **[1989 - 1 Mark]**

6 MCQs with One or More than One Correct Answer

8. Among the following, the correct statement(s) about polymers is(are) **[Adv. 2022]**

(a) The polymerization of chloroprene gives natural rubber.

(b) Teflon is prepared from tetrafluoroethene by heating it with persulphate catalyst at high pressures.

(c) PVC are thermoplastic polymers.

(d) Ethene at 350-570 K temperature and 1000-2000 atm pressure in the presence of a peroxide initiator yields high density polythene.

9. Choose the correct option(s) from the following.

(a) Nylon-6 has amide linkages **[Adv. 2019]**

(b) Cellulose has only α-D-glucose units that are joined by glycosidic linkages

(c) Teflon is prepared by heating tetrafluoroethene in presence of a persulphate catalyst at high pressure

(d) Natural rubber is polyisoprene containing trans alkene units

10. The correct functional group X and the reagent/reaction conditions Y in the following scheme are **[2011 - II]**

$$X-(CH_2)_4-X \xrightarrow[HOOC-(CH_2)_4-COOH,\ heat]{(i)\ Y} \text{Condensation polymer}$$

(a) $X = COOCH_3$, $Y = H_2/Ni/heat$

(b) $X = CONH_2$, $Y = H_2/Ni/heat$

(c) $X = CONH_2$, $Y = Br_2/NaOH$

(d) $X = CN$, $Y = H_2/Ni/heat$

7 Match the Following

11. Match the chemical substances in Column I with type of polymers/type of bonds in Column II. **[2007]**

Column I	Column II
(A) Cellulose	(p) Natural polymer
(B) Nylon-6, 6	(q) Synthetic polymer
(C) Protein	(r) Amide linkage
(D) Sucrose	(s) Glycoside linkage

10 Subjective Problems

12. Give the structures of the products in each of the following reactions. **[2000 - 2 Marks]**

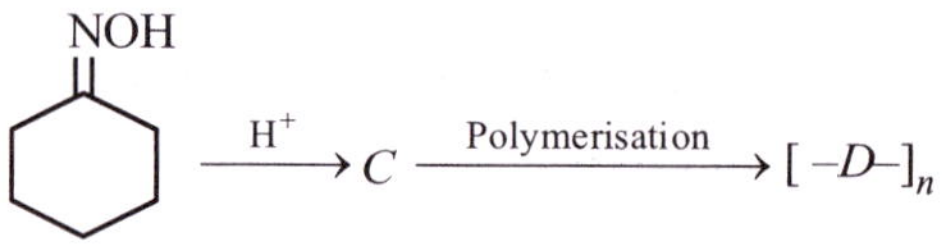

Answer Key

Part-A (JEE Main)																			
1	(4)	**8**	(4)	**15**	(2)	**22**	(3)	**29**	(1)	**36**	(3)	**43**	(1)	**50**	(1)	**57**	(3)	**64**	(2)
2	(2)	**9**	(4)	**16**	(2)	**23**	(1)	**30**	(1)	**37**	(4)	**44**	(3)	**51**	(1)	**58**	(3)	**65**	(1)
3	(2)	**10**	(3)	**17**	(3)	**24**	(3)	**31**	(3)	**38**	(3)	**45**	(4)	**52**	(2)	**59**	(2)	**66**	(4)
4	(1)	**11**	(4)	**18**	(2)	**25**	(1)	**32**	(1)	**39**	(4)	**46**	(1)	**53**	(1)	**60**	(3)	**67**	(4)
5	(4)	**12**	(1)	**19**	(1)	**26**	(3)	**33**	(4)	**40**	(4)	**47**	(4)	**54**	(3)	**61**	(2)		
6	(3)	**13**	(4)	**20**	(3)	**27**	(3)	**34**	(4)	**41**	(2)	**48**	(2)	**55**	(2)	**62**	(1)		
7	(2)	**14**	(3)	**21**	(1)	**28**	(3)	**35**	(3)	**42**	(4)	**49**	(2)	**56**	(1)	**63**	(9)		
Part-B (JEE Advanced)																			
2	(a)	**4**	(d)	**6**	(6)	**9**	(a, c)												
3	(a)	**5**	(85018.00)	**8**	(b, c)	**10**	(a,b,c,d)												

Chapter-16

Chemistry in Every Day Life

Part-A (JEE Main)

Topic 1 - Drugs Classification, Therapeutic Action of Drugs and Chemicals in Food

MULTIPLE CHOICE QUESTIONS (MCQs)

1. Which of the following compounds is an example of hypnotic drug ? [July 29, 2022 (I)]
(1) Seldane (2) Amytal
(3) Aspartame (4) Prontosil

2. Statements about Enzyme Inhibitor Drugs are given below : [July 28, 2022 (I)]
(A) There are Competitive and Non-competitive inhibitor drugs.
(B) Thcsc can bind at the active sites and allosteric sites.
(C) Competitive Drugs are allosteric site blocking drugs.
(D) Non-competitive Drugs are active site blocking drugs.
Choose the correct answer from the options given below :
(1) (A), (D) only (2) (A), (C) only
(3) (A), (B) only (4) (A), (B), (C) only

3. Towstatements in respect of drug-enzyme interaction are given below:
Statement I: Action of an enzyme can be blocked only when an inhibitor blocks the acitve site of the enzyme.
Statement II: An inhibitor can form a strong covalent bond with the enzyme.
In the light of the above statements, choose the correct answer from the options given below: [July 28, 2022 (II)]
(1) Both Statement I and Statement II are true.
(2) Both statement I and statement II are false.
(3) Statement I true but Statement II is false.
(4) Statement I is false but Statement II is true.

4. Match **List-I** with **List-II**

List-I	List-II
(A) 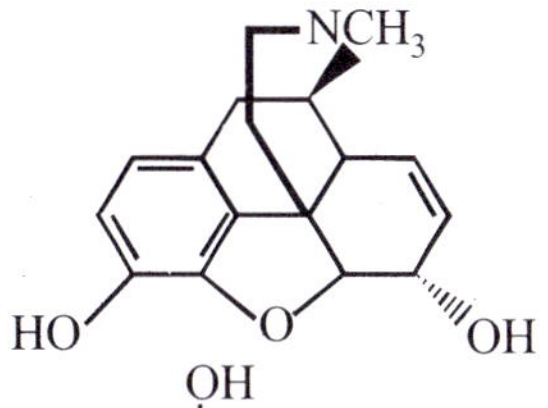	(I) Anti-depressant
(B) 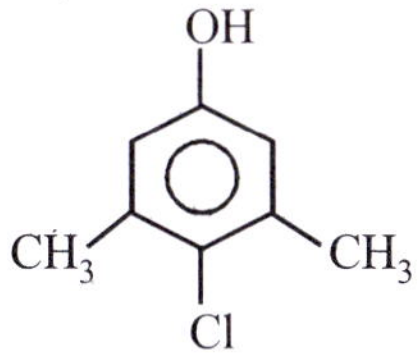	(II) 550 times sweeter
(C) 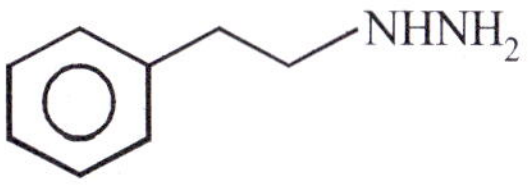	(III) Narcotic analgesic
(D) 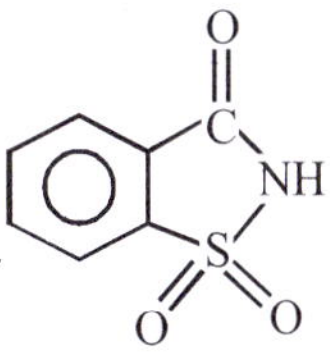	(IV) Antiseptic

Choose the correct answer from the options given below:
(1) A–IV, B–III, C–II, D–I [July 27, 2022 (I)]
(2) A–III, B–I, C–II, D–IV
(3) A–III, B–IV, C–I, D–II
(4) A–III, B–I, C–IV, D–II

5. Which of the following enhances the lathering property of soap? [July 27, 2022 (II)]
(1) Sodium stearate
(2) Sodium carbonate
(3) Sodium rosinate
(4) Trisodium phosphate

6. Stearic acid and polyethylene glycol react to form which one of the following soap/s detergents? [July 26, 2022 (I)]
(1) Cationic detergent (2) Soap
(3) Anionic detergent (4) Non-ionic detergent

7. Drugs used to bind to receptors, inhibiting its natural function and blocking a message are called : [July 25, 2022 (I)]
(1) Agonists (2) Antagonists
(3) Allosterists (4) Anti histaminists

8. Some drugs bind to a site other than the active site of an enzyme. This site is known as [July 25, 2022 (II)]
(1) non-active site (2) allosteric site
(3) competitive site (4) therapeutic site

9. Which of the following compound does not contain sulfur atom ? [June 29, 2022 (I)]
(1) Cimetidine (2) Ranitidine
(3) Histamine (4) Saccharin

10. The mixture of chloroxylenol and terpineol is an example of : [June 29, 2022 (II)]
(1) antiseptic (2) pesticide
(3) disinfectant (4) narcotic analgesic

11. Which amongst the following is not a pesticide ? [June 28, 2022 (I)]
(1) DDT (2) Organophosphates
(3) Dieldrin (4) Sodium arsenite

12. The drug tegamet is : [June 28, 2022 (II)]

(1) 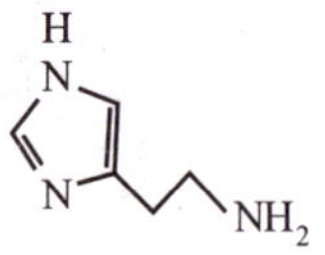

(2) 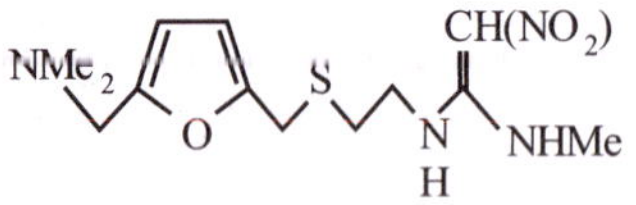

(3) 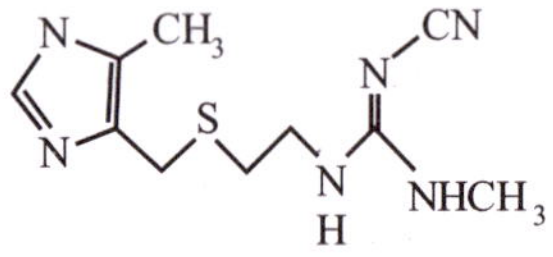

(4) 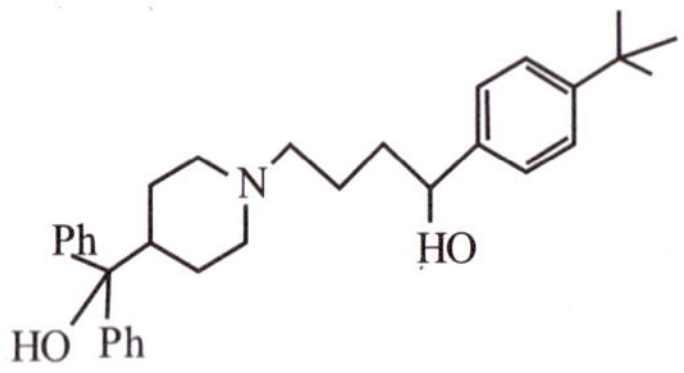

13. Match **List-I** with **List-II**

List-I

(A) $[CH_3(CH_2)_{15}-N(CH_3)_3]^+ Br^-$

(B) 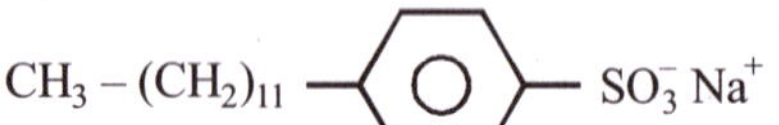

(C) $C_{17}H_{35}COO^-Na^+ + Na_2CO_3$ + Rosinate

(D) $CH_3(CH_2)_{16}COO(CH_2CH_2O)_nCH_2CH_2OH$

List-II

(I) Dishwashing powder

(II) Toothpaste

(III) Laundry soap

(IV) Hair conditionar [June 27, 2022 (I)]

(1) (A) – (III), (B) – (II), (C) – (IV), (D) – (I)
(2) (A) – (IV), (B) – (II), (C) – (III), (D) – (I)
(3) (A) – (IV), (B) – (III), (C) – (II), (D) – (I)
(4) (A) – (III), (B) – (IV), (C) – (I), (D) – (II)

14. Match **List-I** with **List-II**. [June 27, 2022 (II)]

List-I	**List-II**
(A) Antipyretic	(I) Reduces pain
(B) Analgesic	(II) Reduces stress
(C) Tranquilizer	(III) Reduces fever
(D) Antacid	(IV) Reduces acidity (Stomach)

Choose the correct answer from the options given below:
(1) A-III, B-I, C-II, D-IV (2) A-III, B-I, C-IV, D-II
(3) A-I, B-IV, C-II, D-III (4) A-I, B-III, C-II, D-IV

15. A chemical which stimulates the secretion of pepsin is : [June 26, 2022 (I)]
(1) Anti histamine (2) Cimetidine
(3) Histamine (4) Zantac

16. Which of the following is NOT an example of synthetic detergent? [June 26, 2022 (II)]

(1) 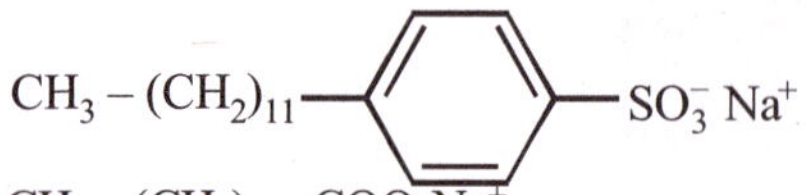

(2) $CH_3-(CH_2)_{16}-COO^-Na^+$

(3) $[CH_3-(CH_2)_{15}-N(CH_3)_3]^+ Br^-$

(4) $CH_3(CH_2)_{16}COO(CH_2CH_2O)_nCH_2CH_2OH$

17. Which one of the following is an example of artificial sweetner? [June 25, 2022 (I)]
(1) Bithional (2) Alitame
(3) Salvarsan (4) Lactose

18. An antiseptic dettol is a mixture of two compounds 'A' and 'B' where A has 6π electrons and B has 2π electrons. What is 'B'? [June 25, 2022 (II)]
(1) Bithionol (2) Terpineol
(3) Chloroxylenol (4) Chloramphenicol

19. Which of the following is not a broad spectrum antibiotic? [June 24, 2022 (I)]
(1) Vancomycin (2) Ampicillin
(3) Ofloxacin (4) Penicillin G

20. The structure shown below is of which well-known drug molecule [June 24, 2022 (II)]

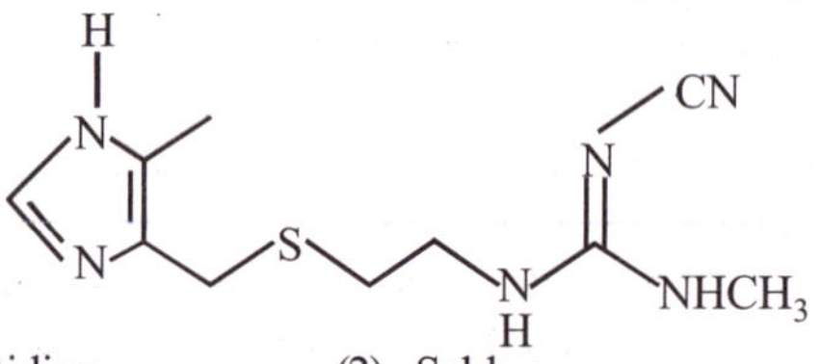

(1) Ranitidine (2) Seldane
(3) Cimetidine (4) Codeine

21. Which one of the following chemicals is responsible for the production of HCl in the stomach leading to irritation and pain? [Aug. 27, 2021 (II)]

(1) 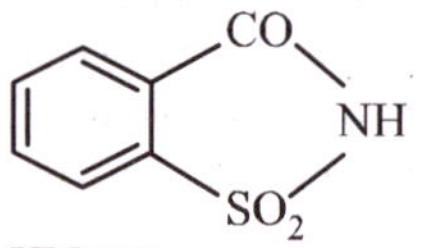

(2) 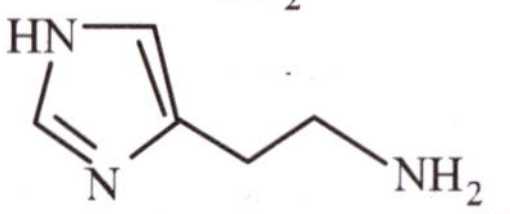

(3) 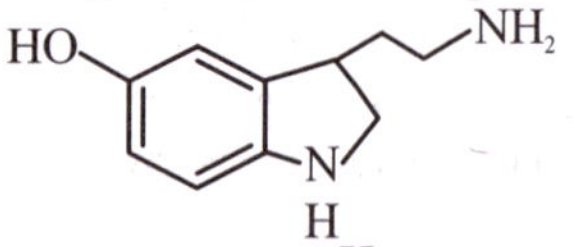

(4) 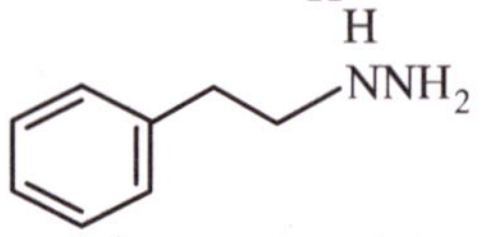

22. Given below are two statements :

Statement I : Penicillin is a bacteriostatic type antibiotic.

Statement II : The general structure of Penicillin is:

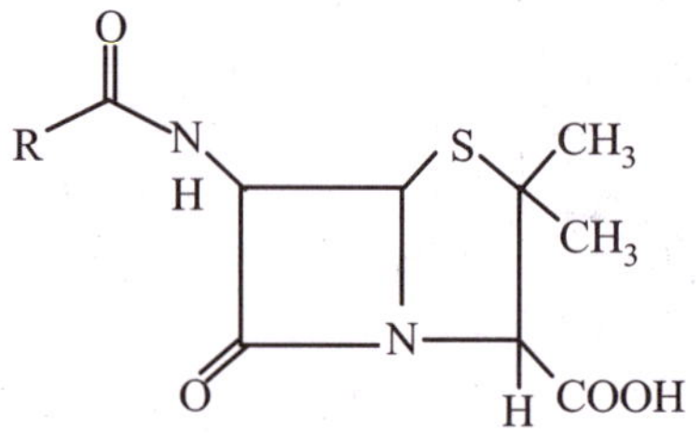

Choose the correct option : [July 27, 2021 (II)]
(1) Both statement I and statement II are false.
(2) Statement I is incorrect but statement II is true.
(3) Both statement I and statement II are true.
(4) Statement I is correct but statement II is false.

23. Sodium stearate $CH_3(CH_2)_{16}COO^-Na^+$ is an anionic surfactant which forms micelles in oil. [July 25, 2021 (I)]
Choose the correct statement for it from the following:
(1) It forms spherical micelles with $CH_3(CH_2)_{16}$ – group pointing towards the centre of sphere.
(2) It forms non–spherical micelles with $-COO^-$ group pointing outwards on the surface.
(3) It forms spherical micelles with $CH_3(CH_2)_{16}$ – group pointing outwards on the surface of sphere.
(4) It forms non–spherical micelles with $CH_3(CH_2)_{16}$ –group pointing towards the centre.

24. Which one of the following chemical agent is not being used for dry-clothes? [July 25, 2021 (I)]
(1) H_2O_2 (2) CCl_4
(3) Liquid CO_2 (4) $Cl_2C=CCl_2$

25. Match **List-I** with **List-II** [March 17, 2021 (II)]

List-I (Chemical Compound)	List-II (Used as)
(A) Sucralose	(i) Synthetic detergent
(B) Glyceryl ester of stearic acid	(ii) Artificial sweetener
(C) Sodium bezoate	(iii) Antiseptic
(D) Bithionol	(iv) Food preservative

(1) A–(iv), B–(iii), C–(ii), D–(i)
(2) A–(iii), B–(ii), C–(iv), D–(i)
(3) A–(ii), B–(i), C–(iv), D–(iii)
(4) A–(i), B–(ii), C–(iv), D–(iii)

26. If a person is suffering from the deficiency of nor-adrenaline, what kind of drug can be suggested? [Sep. 05, 2020 (I)]
(1) Anti-inflammatory (2) Antidepressant
(3) Antihistamine (4) Analgesic

27. The following molecule acts as an : [Sep. 05, 2020 (II)]

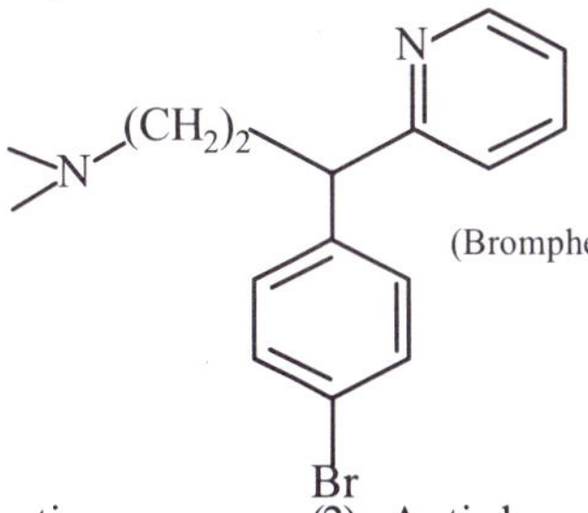

(Brompheniramine)

(1) Antiseptic (2) Anti-depressant
(3) Anti-bacterial (4) Anti-histamine

28. The mechanism of action of "Terfenadine" (Seldane) is : [Sep. 04, 2020 (II)]
(1) Activates the histamine receptor
(2) Inhibits the secretion of histamine
(3) Helps in the secretion of histamine
(4) Inhibits the action of histamine receptor

29. The antifertility drug "Novestrol" can react with : [Sep. 03, 2020 (I)]
(1) $ZnCl_2/HCl$; $FeCl_3$; Alcoholic HCN
(2) Br_2/ water; $ZnCl_2/HCl$; $FeCl_3$
(3) Alcoholic HCN; NaOCl; $ZnCl_2$/ HCl
(4) Br_2/ water; $ZnCl_2$/ HCl; NaOCl

30. Match the following drugs with their therapeutic actions : [Sep. 03, 2020 (II)]

(i) Ranitidine	(A) Antidepressant
(ii) Nardil (Phenelzine)	(B) Antibiotic
(iii) Chloramphenicol	(C) Antihistamine
(iv) Dimetane (Brompheniramine)	(D) Antacid
	(E) Analgesic

(1) (i)-(A); (ii)-(C); (iii)-(B); (iv)-(E)
(2) (i)-(D); (ii)-(A); (iii)-(B); (iv)-(C)
(3) (i)-(E); (ii)-(A); (iii)-(C); (iv)-(D)
(4) (i)-(D); (ii)-(C); (iii)-(A); (iv)-(E)

31. A chemist has 4 samples of artificial sweetener A, B, C and D. To identify these samples, he performed certain experiments and noted the following observations: [Jan. 09, 2020 (I)]
(I) A and D both form blue-violet colour with ninhydrin.
(II) Lassaigne extract of C gives positive $AgNO_3$ test and negative $Fe_4[Fe(CN)_6]_3$ test.
(III) Lassaigne extract of B and D gives positive sodium nitroprusside test.
Based on these observations which option is correct?
(1) A : Aspartame; B : Saccharin; C : Sucralose; D : Alitame
(2) A : Alitame; B : Saccharin; C : Aspartame; D : Sucralose
(3) A : Saccharin; B : Alitame; C : Sucralose; D : Aspartame
(4) A : Aspartame; B : Alitame; C : Saccharin; D : Sucralose

32. Noradrenaline is a/an: [April 9, 2019 (II)]
(1) Antacid (2) Neurotransmitter
(3) Antidepressant (4) Antihistamine

33. The correct match between **Item-I** and **Item-II** is : [Jan. 11, 2019 (I)]

Item - I	Item - II
(A) Norethindrone	(P) Antibiotic
(B) Ofloxacin	(Q) Antifertility
(C) Equanil	(R) Hypertension
	(S) Analgesics

(1) (A) → (Q); (B) → (R); (C) → (S)
(2) (A) → (Q); (B) → (P); (C) → (R)
(3) (A) → (R); (B) → (P); (C) → (S)
(4) (A) → (R); (B) → (P); (C) → (R)

34. The correct match between **Item-I** and **Item-II** is: [Jan. 11, 2019 (II)]

Item-I	Item-II
(A) Allosteric effect	(P) Molecule binding to the active site of enzyme
(B) Competitive inhibitor	(Q) Molecule crucial for communication in the body
(C) Receptor	(R) Molecule binding to a site other than the active site of enzyme
(D) Poison	(S) Molecule binding to the enzyme covalently

(1) (A) → (R); (B) → (P); (C) → (Q); (D) → (S)
(2) (A) → (P); (B) → (R); (C) → (Q); (D) → (S)
(3) (A) → (R); (B) → (P); (C) → (S); (D) → (Q)
(4) (A) → (P); (B) → (R); (C) → (S); (D) → (Q)

35. The predominant form of histamine present in human blood is (pK_a, Histidine – 6.0) [2018]

(1) 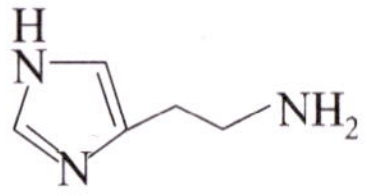
(2) 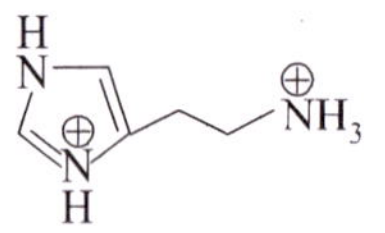
(3) 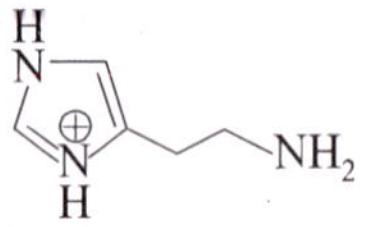
(4) 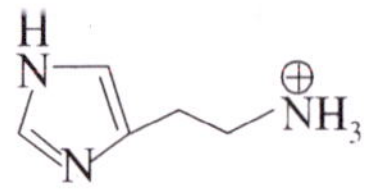

36. The correct match between items of List-I and List-II is: [Online April 16, 2018]

List-I	List-II
(A) Phenelzine	(p) Pyrimidine
(B) Chloroxylenol	(q) Furan
(C) Uracil	(r) Hydrazine
(D) Ranitidine	(s) Phenol

(1) (A)-(s); (B)-(r); (C)-(q); (D)-(p)
(2) (A)-(r); (B)-(s); (C)-(p); (D)-(q)
(3) (A)-(r); (B)-(s); (C)-(q); (D)-(p)
(4) (A)-(s); (B)-(r); (C)-(p); (D)-(q)

37. The reason for "drug induced poisoning" is : [Online April 8, 2017]
(1) Binding reversibly at the active site of the enzyme
(2) Bringing conformational change in the binding site of enzyme
(3) Binding irreversibly to the active site of the enzyme
(4) Binding at the allosteric sites of the enzyme

38. Which of the following is an anionic detergent? [2016]
(1) Cetyltrimethyl ammonium bromide.
(2) Glyceryl oleate.
(3) Sodium stearate.
(4) Sodium lauryl sulphate.

39. Which of the following is a bactericidal antibiotic ? [Online April 10, 2016]
(1) Ofloxacin (2) Tetracycline
(3) Chloramphenicol (4) Erythromycin

40. The artificial sweetener that has the highest sweetness value in comparison to cane sugar is : [Online April 9, 2016]
(1) Sucralose (2) Aspartane
(3) Saccharin (4) Alitame

41. Which of the following compounds is not an antacid ? [2015]
(1) Phenelzine (2) Ranitidine
(3) Aluminium hydroxide (4) Cimetidine

42. Which artificial sweetener contains chlorine ? [Online April 11, 2015]
(1) Sucralose (2) Alitame
(3) Aspartame (4) Saccharin

43. 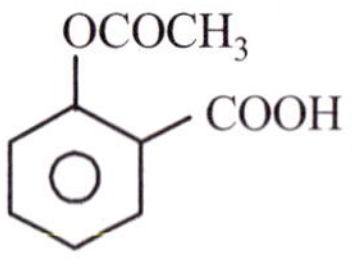
is used as : [Online April 10, 2015]
(1) Analgesic (2) Insecticide
(3) Antacid (4) Antihistamine

44. Which one of the following is used as Antihistamine? [Online April 11, 2014]
(1) Omeprazole (2) Chloranphenicol
(3) Diphenhydramine (4) Norethindrone

45. Aminoglycosides are usually used as: [Online April 12, 2014]
(1) antibiotic (2) analgesic
(3) hypnotic (4) antifertility

46. Which of the following statements about aspirin is **not** true? [Online April 22, 2013]
(1) It is effective in relieving pain.
(2) It is a neurologically active drug.
(3) It has antiblood clotting action.
(4) It belongs to narcotic analgesics.

47. H_1 – Receptor antagonists is a term associated with : [Online April 23, 2013]
(1) Antiseptics (2) Antihistamins
(3) Antacids (4) Analgesics

48. Which one of the following compounds is an anti-fertility drug? [Online May 7, 2012]
(1) Aspirin (2) Chloromycetin
(3) Saheli (4) Penicillin

49. Chemically heroin is [Online May 12, 2012]
(1) morphine monoacetate (2) morphine dibenzoate
(3) morphine diacetate (4) morphine monobenzoate

50. Aspirin can be prepared by the reaction of [Online May 19, 2012]
(1) Salicyldehyde with acetic anhydride in presence of H_2SO_4
(2) Salicylic acid with methanol in presence of H_2SO_4
(3) Salicylic acid with acetic anhydride in presence of H_2SO_4
(4) Cinnamic acid with acetic anhydride in presence of H_2SO_4

51. Sulphonamides act as [Online May 26, 2012]
(1) Antiseptic (2) Analgesic
(3) Antimicrobials (4) Antipyretic

52. Which one of the following types of drugs reduces fever ? [2005]
(1) Tranquiliser (2) Antibiotic
(3) Antipyretic (4) Analgesic

53. Which of the following could act as a propellant for rockets? [2003]
(1) Liquid oxygen + liquid argon
(2) Liquid hydrogen + liquid oxygen
(3) Liquid nitrogen + liquid oxygen
(4) Liquid hydrogen + liquid nitrogen

54. The compound 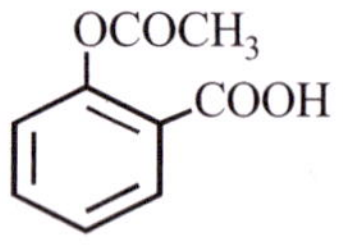 is used as [2002]
(1) antiseptic (2) antibiotic
(3) analgesic (4) pesticide.

NUMERICAL VALUE QUESTIONS (NVQs)

55. The number of chlorine atoms in bithionol is____. [July 29, 2022 (II)]

56. How many of the following drugs is/are example(s) of broad spectrum antibiotic ? Ofloxacin, Penicillin G, Terpineol, Salvarsan [July 26, 2022 (II)]

57. How many of the following drugs belong to the tranquilizer class _______ ? [June 30, 2022 (I)]
Veronal, Luminal, Propanal, Seconal

58. The number of chiral carbons present in the molecule given below is ________. [Sep. 02, 2020 (I)]

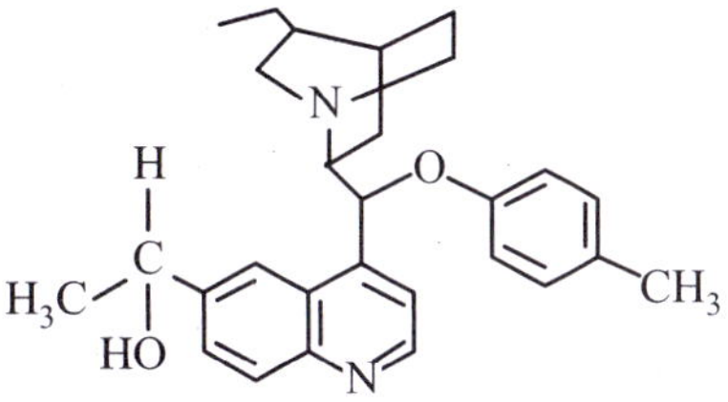

59. The number of chiral centres in penicillin is ________. [Jan. 08, 2020 (I)]

60. The number of chiral carbons in chloramphenicol is ________. [Jan. 07, 2020 (I)]

61. The number of sp^2 hybridised carbons present in "Aspartame" is ________. [Jan. 07, 2020 (II)]

Part-B (JEE Advanced)

Topic 1 - Drugs Classification, Therapeutic Action of Drugs and Chemicals in Food

1 MCQs with One Correct Answer

1. Aspartame, an artificial sweetener, is a dipeptide aspartyl phenylalanine methyl ester. The structure of aspartame is

Structures of phenylalanine and aspartic acid are given below.

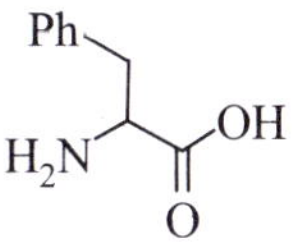

Phenylalanine

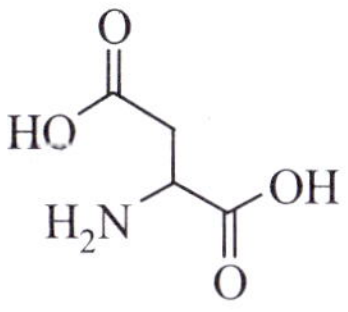

Aspartic acid

[Adv. 2024]

(a) 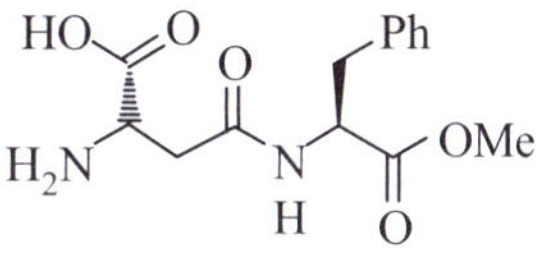

(b) 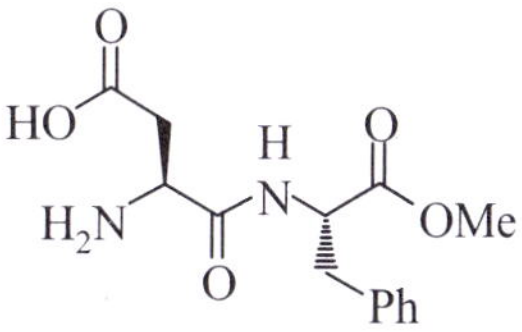

(c) 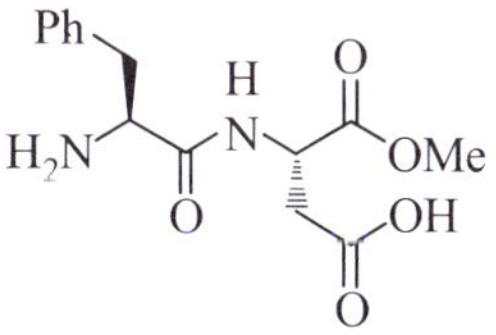

(d) 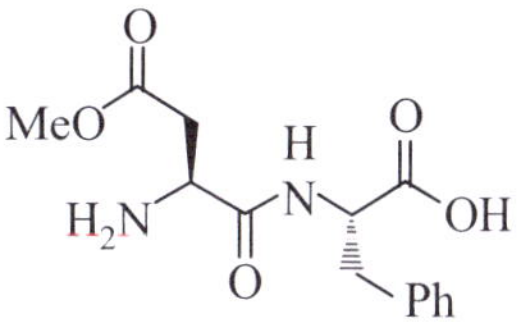

Answer Key

Part-A (JEE Main)																			
1	(2)	8	(2)	15	(3)	22	(2)	29	(4)	36	(2)	43	(1)	50	(3)	57	(3)		
2	(3)	9	(3)	16	(2)	23	(1)	30	(2)	37	(3)	44	(3)	51	(3)	58	(5)		
3	(4)	10	(1)	17	(2)	24	(1)	31	(1)	38	(4)	45	(1)	52	(3)	59	(3)		
4	(3)	11	(N)	18	(2)	25	(3)	32	(2)	39	(1)	46	(4)	53	(2)	60	(2)		
5	(3)	12	(3)	19	(4)	26	(2)	33	(2)	40	(4)	47	(2)	54	(3)	61	(9)		
6	(4)	13	(2)	20	(3)	27	(4)	34	(1)	41	(1)	48	(3)	55	(4)				
7	(2)	14	(1)	21	(2)	28	(4)	35	(4)	42	(1)	49	(3)	56	(1)				
Part-B (JEE Advanced)																			
1	(b)																		

Chapter-17

Principles Related to Practical Chemistry

Part-A (JEE Main)

Topic 1– Detection of Extra Elements, Detection of Organic Functional Groups and Preparation of Organic Compounds

MULTIPLE CHOICE QUESTIONS (MCQs)

1. Match the LIST-I with LIST-II [April 04, 2026 (II)]

List-I (Compound)	List-II (Test)
A. 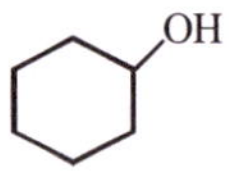	I. Hinsberg's reagent test
B. 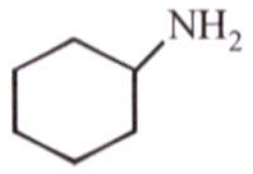	II. Phthalein dye test
C. 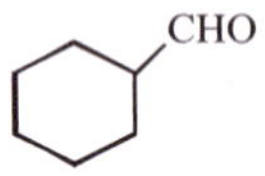	III. Lucas test
D. 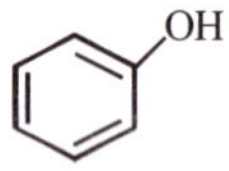	IV. Tollen's test

Choose the correct answer from the options given below:

(1) A-III, B-I, C-IV, D-II (2) A-III, B-IV, C-I, D-II
(3) A-I, B-III, C-II, D-IV (4) A-I, B-II, C-III, D-IV

2. Identify the colour of compound 'X' in the sequence of the reaction. [April 04, 2026 (I)]

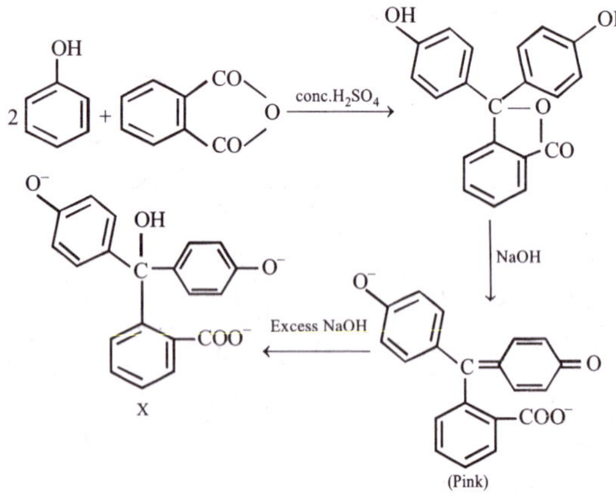

(1) Violet (2) Green
(3) Red (4) Colourless

3. Match List-I with List-II. [April 02, 2026 (I)]

List-I Mixture of Compounds	List-II Reagent used to distinguish
A. Diethyl amine + Ethyl amine	I. Bromine water
B. Acetaldehyde + Acetone	II. $CHCl_3 + KOH, \Delta$
C. Ethanol + Phenol	III. Neutral $FeCl_3$
D. Benzoic acid + Cinnamic acid	IV. Ammonical silver nitrate

Choose the **correct** answer from the options given below:

(1) A-IV, B-II, C-I, D-III (2) A-IV, B-II, C-III, D-I
(3) A-II, B-IV, C-I, D-III (4) A-II, B-IV, C-III, D-I

4. Match List-I with List-II. [Jan. 23, 2026 (I)]

List-I (Functional group detection)	List-II (Change observed during detection)
(A) Unsaturation (Baeyer's test)	(I) Red colour appears
(B) Alcoholic group (Ceric ammonium nitrate test)	(II) Silver mirror appears
(C) Aldehyde group (Tollen's reagent)	(III) Violet colour appears
(D) Phenolic group ($FeCl_3$ test)	(IV) Discharge of pink colour

(1) A-III, B-IV, C-II, D-I (2) A-III, B-IV, C-I, D-II
(3) A-IV, B-I, C-II, D-III (4) A-IV, B-III, C-II, D-I

5. Match the LIST-I with LIST-II [April 8, 2025 (II)]

LIST-I (Reagent)	LIST-II (Functional Group detected)
A. Sodium bicarbonate solution	I. double bond/unsaturation
B. Neutral ferric chloride	II. carboxylic acid
C. Ceric ammonium nitrate	III. phenolic - OH
D. Alkaline $KMnO_4$	IV. alcoholic - OH

Choose the **correct** answer from the options given below:

(1) A-II, B-III, C-IV, D-I (2) A-II, B-III, C-I, D-IV
(3) A-III, B-II, C-IV, D-I (4) A-II, B-IV, C-III, D-I

6. The compounds that produce CO_2 with aqueous $NaHCO_3$ solution are : **[Jan. 28, 2025 (I)]**

(A) 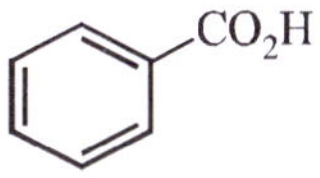

(B) 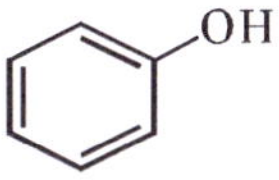

(C) 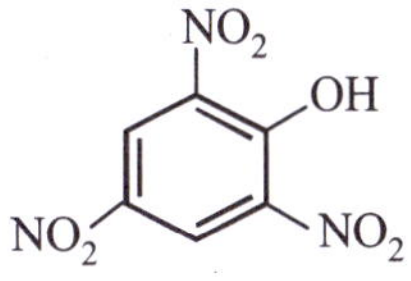

(D) 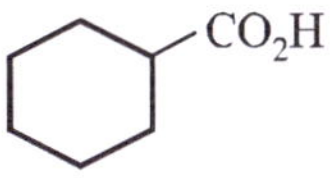

(E) 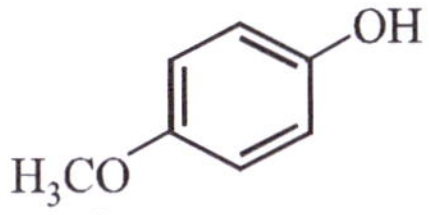

Choose the correct answer from the options given below :

(1) A and C only (2) A, B and E only

(3) A, C and D only (4) A and B only

7. Given below are two statements : **[Jan. 23, 2025 (I)]**

Statement I : In Lassaigne's test, the covalent organic molecules are transformed into ionic compounds.

Statement II : The sodium fusion extract of an organic compound having N and S gives prussian blue colour with $FeSO_4$ and $Na_4[Fe(CN)_6]$

In the light of the above statements, choose the correct answer from the options given below

(1) Both Statement I and Statement II are true

(2) Both Statement I and Statement II are false

(3) Statement I is false but Statement II is true

(4) Statement I is true but Statement II is false

8. Given below are two statement: **[Jan. 22, 2025 (II)]**

Statement I: Nitrogen, sulphur, halogen and phosphorus present in a organic compound are detected by Lassaigne's Test.

Statement II: The elements present in the compound are converted from covalent form into ionic form by fusing the compound with Magnesium in Lassaigne's test.

In the light of the above statements, choose the correct answer from the options given below :

(1) Both Statement I and Statement II are true

(2) Both Statement I and Statement II are false

(3) Statement I is true but Statement II is false

(4) Statement I is false but Statement II is true

9. The compounds which give positive Fehling's test are : **[Jan. 22, 2025 (I)]**

(A) 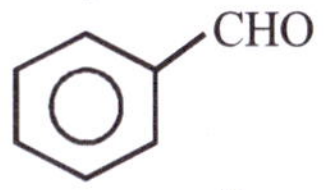

(B) 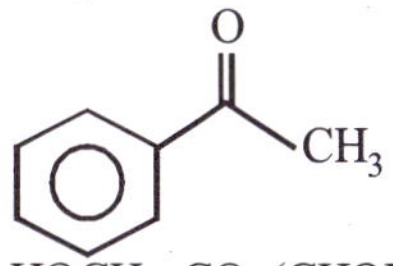

(C) $HOCH_2–CO–(CHOH)_3–CH_2–OH$

(D) $CH_3-\overset{O}{\overset{\|}{C}}-H$

(E) 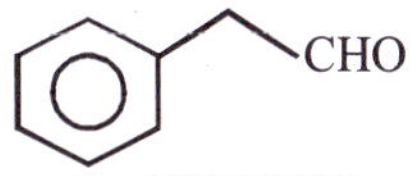

Choose the CORRECT answer from the options given below :

(1) (A),(C) and (D) Only (2) (A),(D) and (E) Only

(3) (C), (D) and (E) Only (4) (A), (B) and (C) Only

10. Match **List-I** with **List-II**. **[April 9, 2024 (II)]**

List-I (Test)	List-II (Observation)
(A) Br_2 water test	(I) Yellow orange or orange red precipitate formed
(B) Ceric ammonium nitrate test	(II) Reddish orange colour disappears
(C) Ferric chloride test	(III) Red colour appears
(D) 2, 4-DNP test	(IV) Blue, Green, Violet or Red colour appear

Choose the correct answer from the options given below:

(1) A-(I), B-(II), C-(III), D-(IV)

(2) A-(II), B-(III), C-(IV), D-(I)

(3) A-(III), B-(IV), C-(I), D-(II)

(4) A-(IV), B-(I), C-(II), D-(III)

11. Which of the following compounds will give silver mirror with ammoniacal silver nitrate? **[April 9, 2024 (II)]**

(A) Formic acid (B) Formaldehyde

(C) Benzaldehyde (D) Acetone

Choose the correct answer from the options given below:

(1) C and D only (2) A, B and C only

(3) A only (4) B and C only

12. Match **List-I** with **List-II**. **[April 8, 2024 (II)]**

List-I (Test)	List-II (Identification)
(A) Bayer's test	(I) Phenol
(B) Ceric ammonium nitrate test	(II) Aldehyde
(C) Phthalein dye test	(III) Alcoholic-OH group
(D) Schiff's test	(IV) Unsaturation

Choose the **correct** answer from the options given below:

(1) (A)-(III), (B)-(I), (C)-(IV), (D)-(II)

(2) (A)-(II), (B)-(III), (C)-(IV), (D)-(I)

(3) (A)-(IV), (B)-(I), (C)-(II), (D)-(III)

(4) (A)-(IV), (B)-(III), (C)-(I), (D)-(II)

13. Which of the following gives a positive test with ninhydrin ? **[April 5, 2024 (I)]**

(1) Cellulose (2) Starch

(3) Polyvinyl chloride (4) Egg albumin

14. Which of the following nitrogen containing compound does not give Lassaigne's test ? **[April 4, 2024 (I)]**

(1) Phenyl hydrazine (2) Glycene

(3) Urea (4) Hydrazine

15. Lassaigne's test is used for detection of: **[Feb. 1, 2024 (II)]**

(1) Phosphorous and halogens only

(2) Nitrogen, Sulphur and Phosphorous only

(3) Nitrogen, Sulphur and phosphorous and halogens

(4) Nitrogen and Sulphur only

16. The azo-dye (Y) formed in the following reactions is Sulphanilic acid + $NaNO_2 + CH_3COOH \rightarrow X$ **[Jan. 31, 2024 (II)]**

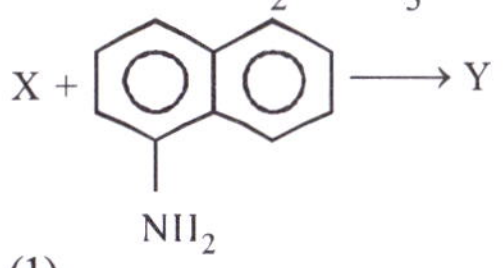

(1) 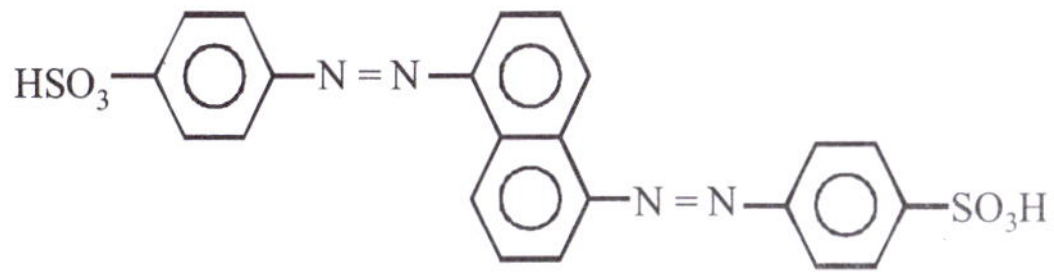

(2) 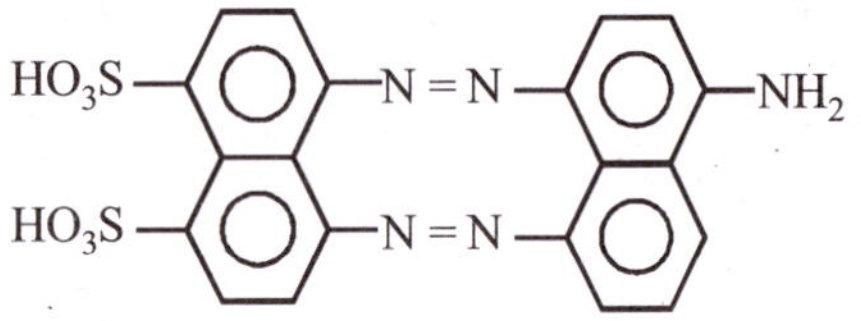

(3) 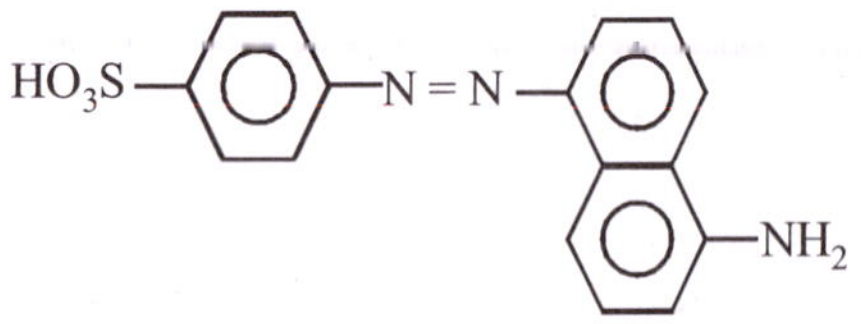

(4) 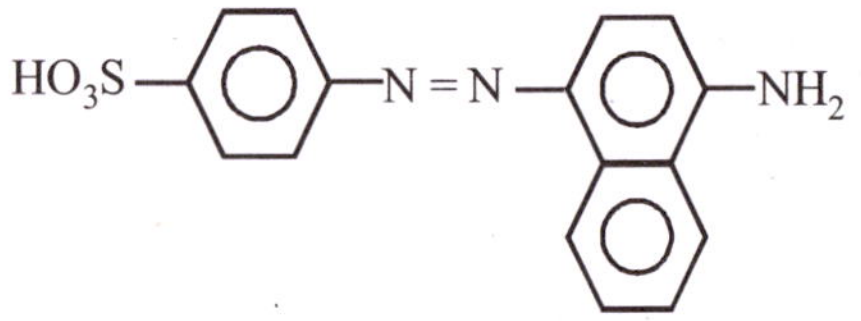

17. Given below are two statements : **[Jan. 31, 2024 (II)]**

Statement I : Aniline reacts with con. H_2SO_4 followed by heating at 453-473 K gives p-aminobenzene sulphonic acid, which gives blood red colour in the 'Lassaigne's test'.

Statement II : In Friedel - Craft's alkylation and acylation reactions, aniline forms salt with the $AlCl_3$ catalyst. Due to this, nitrogen of aniline aquires a positive charge and acts as deactivating group.

In the light of the above statements, choose the **correct answer** from the options given below :

(1) Statement I is false but Statement II is true.
(2) Both Statement I and Statement II are false.
(3) Statement I is true but Statement II is false.
(4) Both Statement I and Statement II are true.

18. Following is a confirmatory test for aromatic primary amines. Identify reagent (A) and (B). **[Jan. 30, 2024 (I)]**

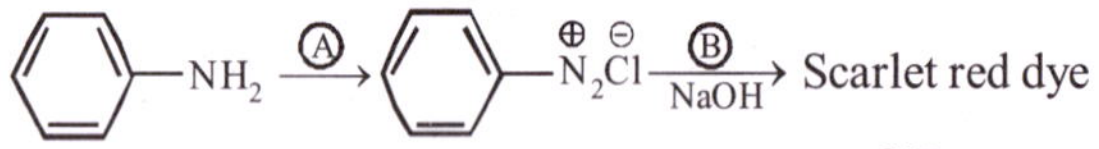

(1) $A = HNO_3/H_2SO_4$; B = 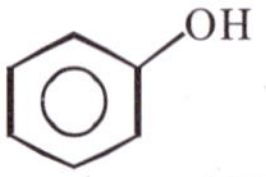

(2) $A = NaNO_2 + HCl$, 0-5°C; B = 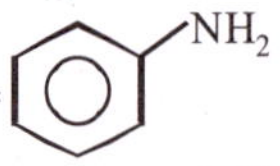

(3) $A = NaNO_2 + HCl$, 0 – 5°C; B = 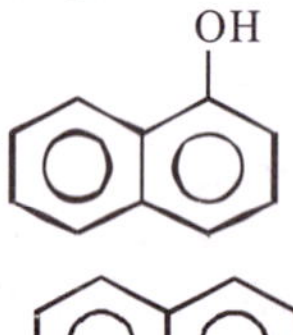

(4) $A = NaNO_2 + HCl$, 0– 5°C; B = 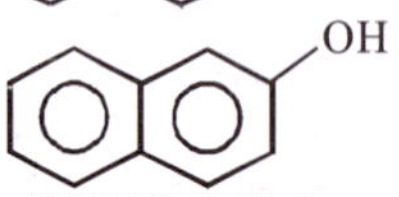

19. The Lassiagne's extract is boiled with dil HNO_3 before testing for halogens because, **[Jan. 30, 2024 (I)]**

(1) AgCN is soluble in HNO_3.
(2) Silver halides are soluble in HNO_3.
(3) Ag_2S is soluble in HNO_3.
(4) Na_2S and NaCN are decomposed by HNO_3.

20. Appearance of blood red colour, on treatment of the sodium fusion extract of an organic compound with $FeSO_4$ in presence of concentrated H_2SO_4 indicates the present of element/s **[Jan. 29, 2024 (I)]**

(1) N and S (2) N (3) S (4) Br

21. Phenolic group can be identified by a positive: **[Jan. 27, 2024 (II)]**

(1) Phthalein dye test (2) Lucas test
(3) Tollen's test (4) Carbylamine test

22. Match List-I with List-II : **[April 08, 2023 (I)]**

List-I (Reagents used)	List-II (Compound with functional group detected)
A. Alkaline solution of copper sulphate and sodium citrate	I. 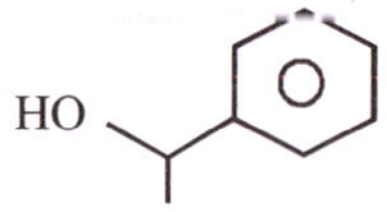
B. Neutral $FeCl_3$ solution	II. 
C. Alkaline chloroform solution	III. 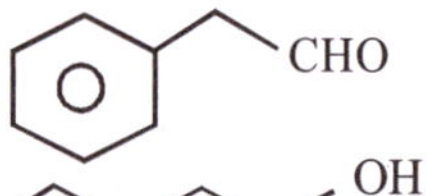
D. Potassium iodide and sodium hypochlorite	IV. 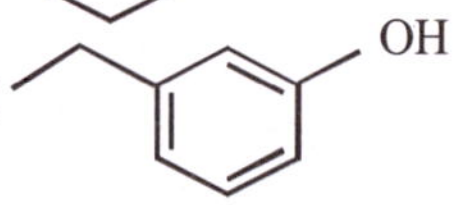

Choose the correct answer from the options given below:

(1) A-II, B-IV, C-III, D-I (2) A-IV, B-I, C-II, D-III
(3) A-III, B-IV, C-I, D-II (4) A-III, B-IV, C-II, D-I

23. Match List I with List II: **[Feb. 01, 2023 (I)]**

List I Test	List II Functional group / Class of Compound
(A) Molisch's Test	(I) Peptide
(B) Biuret Test	(II) Carbohydrate
(C) Carbylamine Test	(III) Primary amine
(D) Schiff's Test	(IV) Aldehyde

Choose the **correct** answer from the options given below:

(1) (A) – (I), (B) – (II), (C) – (III), (D) – (IV)
(2) (A) – (II), (B) – (I), (C) – (III), (D) – (IV)
(3) (A) – (III), (B) – (IV), (C) – (II), (D) – (I)
(4) (A) – (III), (B) – (IV), (C) – (I), (D) – (II)

24. Given below are two statements: **[Feb. 01, 2023 (II)]**

Statement I : Sulphanilic acid gives esterification test for carboxyl group.

Statement II: Sulphanilic acid gives red colour in Lassigne's test for extra element detection.

In the light of the above statements, choose the **most appropriate** answer from the options given below.

(1) **Statement I** is incorrect but **Statement II** is correct
(2) **Statement I** is correct but **Statement II** is incorrect
(3) Both **Statement I** and **Statement II** are correct
(4) Both **Statement I** and **Statement II** are incorrect

25. Which of the following compounds would give the following set of qualitative analysis ? **[Jan. 30, 2023 (I)]**

(i) Fehling's Test : Positive
(ii) Na fusion extract upon treatment with sodium nitroprusside gives a blood red colour.

(1) 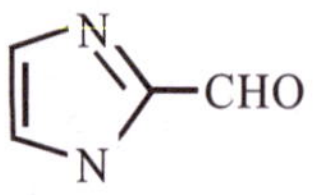 (2) 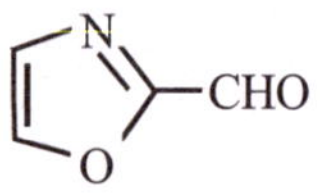

(3) 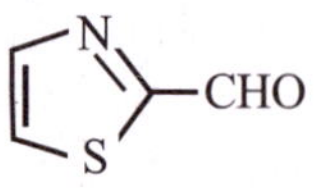 (4) 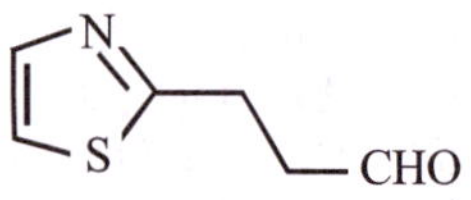

26. Compound that will give positive Lassaigne's test for both nitrogen and halogen is [Jan. 29, 2023 (I)]
(1) $N_2H_4.HCl$ (2) $CH_3NH_2.HCl$
(3) NH_4Cl (4) $NH_2OH.HCl$

27. The reagent neutral ferric chloride is used to detect the presence of ______ . [June 30, 2022 (I)]
(1) sulphide ion and alcoholic – OH group.
(2) acetate ion and phenolic – OH group.
(3) sulphide ion and phenolic – OH group.
(4) acetate ion and alcoholic – OH group.

28. The formula of the purple colour formed in Laissaigne's test for sulphur using sodium nitroprusside is [June 28, 2022 (I)]
(1) $NaFe[Fe(CN)_6]$ (2) $Na[Cr(NH_3)_2(NCS)_4]$
(3) $Na_2[Fe(CN)_5(NO)]$ (4) $Na_4[Fe(CN)_5(NOS)]$

29. Given below are two statements : [June 26, 2022 (I)]
Statement I : In 'Lassaigne's test, when both nitrogen and sulphur are present in an organic compound, sodium thiocyanate is formed.
Statement II : If both nitrogen and sulphur are present in an organic compound, then the excess of sodium used in sodium fusion will decompose the sodium thiocyanate formed to give NaCN and Na_2S.
In the light of the above statements, choose the **most appropriate** answer from the options given below :
(1) Both Statement I and Statement II are correct.
(2) Both Statement I and Statement II are incorrect.
(3) Statement I is correct but Statement II is incorrect.
(4) Statement I is incorrect but Statement II is correct.

30. During halogen test, sodium fusion extract is boiled with concentrated HNO_3 to [June 25, 2022 (II)]
(1) remove unreacted sodium
(2) decompose cyanide or sulphide of sodium
(3) extract halogen from organic compound
(4) maintain the pH of extract

31. Which one of the following set of elements can be detected using sodium fusion extract? [July 27, 2021 (II)]
(1) Sulfur, Nitrogen, Phosphorus, Halogens
(2) Phosphorus, Oxygen, Nitrogen, Halogens
(3) Nitrogen, Phosphorus, Carbon, Sulfur
(4) Halogens, Nitrogen, Oxygen, Sulfur

32. Which of the following compound is added to the sodium extract before addition of silver nitrate for testing of halogens?
(1) Hydrochloric acid (2) Sodium hydroxide
(3) Ammonia (4) Nitric acid [Feb. 25, 2021 (II)]

33. Consider the following reactions:

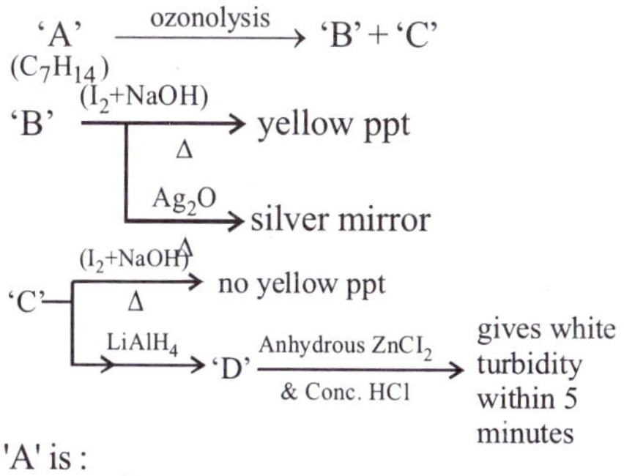

'A' is : [Sep. 06, 2020 (I)]

(1) 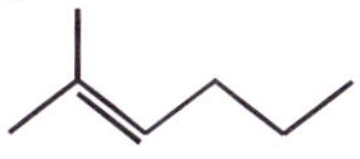 (2) 

(3) 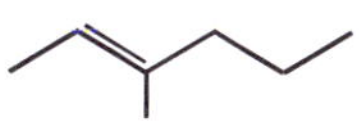 (4) 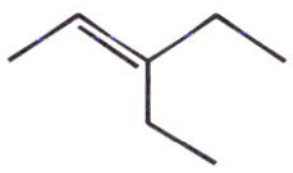

34. Match the following : [Sep. 06, 2020 (II)]

Column-I (Test / Method)	Column-II (Reagent)
(I) Lucas Test	(A) $C_6H_5SO_2Cl$ / aq. KOH
(II) Dumas method	(B) HNO_3 / $AgNO_3$
(III) Kjeldahl's method	(C) CuO / CO_2
(IV) Hinsberg test	(D) Conc. HCl and $ZnCl_2$
	(E) H_2SO_4

(1) (I)-(D), (II)-(C), (III)-(B), (IV)-(A)
(2) (I)-(B), (II)-(D), (III)-(E), (IV)-(A)
(3) (I)-(D), (II)-(C), (III)-(E), (IV)-(A)
(4) (I)-(B), (II)-(A), (III)-(C), (IV)-(D)

35. Two compounds A and B with same molecular formula (C_3H_6O) undergo Grignard's reaction with methyl-magnesium bromide to give products C and D. Products C and D show following chemical tests.

Test	C	D
Ceric ammonium nitrate test	Positive	Positive
Lucas test	Turbidity obtained after five minutes	Turbidity obtained immediately
Iodoform test	Positive	Negative

C and D respectively are : [Sep. 02, 2020 (II)]

(1) $C = H_3C - CH_2 - \overset{\overset{OH}{|}}{CH} - CH_3$; $D = H_3C - \underset{\underset{CH_3}{|}}{\overset{\overset{CH_3}{|}}{C}} - OH$

(2) $C = H_3C - CH_2 - CH_2 - CH_2 - OH$; $D = H_3C - CH_2 - \underset{\underset{OH}{|}}{CH} - CH_3$

(3) $C = H_3C - CH_2 - CH_2 - CH_2 - OH$; $D = H_3C - \underset{\underset{CH_3}{|}}{\overset{\overset{CH_3}{|}}{C}} - OH$

(4) $C = H_3C - \underset{\underset{CH_3}{|}}{\overset{\overset{CH_3}{|}}{C}} - OH$; $D = H_3C - CH_2 - \underset{\underset{OH}{|}}{CH} - CH_3$

36. A, B and C are three biomolecules. The results of the tests performed on them are given below:

	Molisch's Test	Barfoed Test	Biuret Test
A	Positive	Negative	Negative
B	Positive	Positive	Negative
C	Negative	Negative	Positive

A, B and C are respectively: [Jan. 09, 2020 (II)]
(1) A = Glucose, B = Fructose, C = Albumin
(2) A = Lactose, B = Glucose, C = Albumin
(3) A = Lactose, B = Glucose, C = Alanine
(4) A = Lactose, B = Fructose, C = Alanine

37. A solution of *m*-chloroaniline, *m*-chlorophenol and *m*-chlorobenzoic acid in ethyl acetate was extracted initially with a saturated solution of $NaHCO_3$ to give fraction *A*. The left over organic phase was extracted with dilute NaOH solution to

give fraction *B*. The final organic layer was labelled as fraction *C*. Fractions *A*, *B* and *C*, contain respectively: [Jan. 07, 2020 (I)]

(1) *m*-chlorobenzoic acid, *m*-chloroaniline and *m*-chlorophenol
(2) *m*-chlorobenzoic acid, *m*-chlorophenol and *m*-chloroaniline
(3) *m*-chlorophenol, *m*-chlorobenzoic acid and *m*-chloroaniline
(4) *m*-chloroaniline, *m*-chlorobenzoic acid and *m*-chlorophenol

38. Consider the following reaction: [Jan. 07, 2020 (I)]

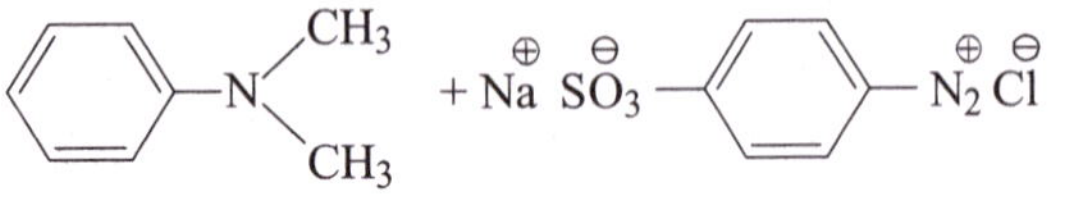

$\xrightarrow{OH^-}$ '*X*'

The product '*X*' is used:
(1) in protein estimation as an alternative to ninhydrin
(2) in acid base titration as an indicator
(3) as food grade colourant
(4) in laboratory test for phenols

39. An organic compound 'A' is oxidized with Na_2O_2 followed by boiling with HNO_3. The resultant solution is then treated with ammonium molybdate to yield a yellow precipitate.
Based on above observation, the element present in the given compound is: [April 12, 2019 (I)]
(1) Nitrogen (2) Phosphorus
(3) Fluorine (4) Sulphur

40. Which one of the following is likely to give a precipitate with $AgNO_3$ solution ? [April 12, 2019 (II)]
(1) $CH_2{=}CH{-}Cl$ (2) CCl_4
(3) $CHCl_3$ (4) $(CH_3)_3CCl$

41. The organic compound that gives following qualitative analysis is: [April 9, 2019 (I)]

Test	Inference
(A) Dil. HCl	Insoluble
(B) NaOH solution	Soluble
(C) Br_2/water	Decolourization

(1) 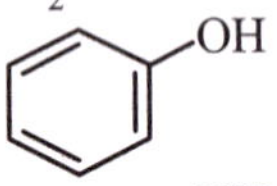

(2) 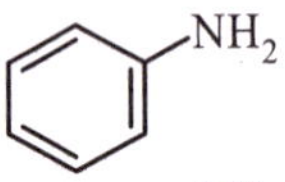

(3) 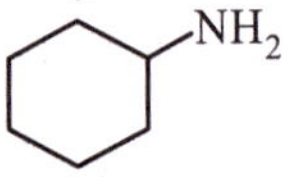

(4) 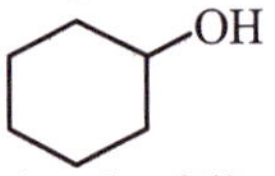

42. An organic compound 'X' showing the following solubility profile is: [April 8, 2019 (I)]

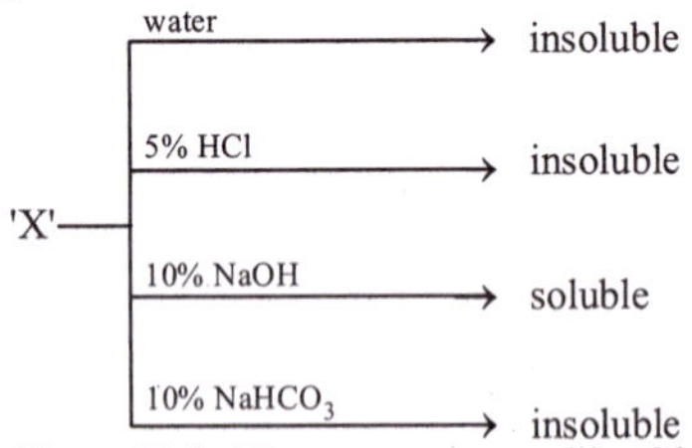

(1) *o*-Toluidine (2) Oleic acid
(3) *m*-Cresol (4) Benzamide

43. The correct match between Item-I and Item-II is: [Jan. 11, 2019 (II)]

Item-I	Item-II
(A) Ester test	(P) Tyr
(B) Carbylamine test	(Q) Asp
(C) Phthalein dye test	(R) Ser
	(S) Lys

(1) (A)→(Q, R); (B)→(S); (C)→(P)
(2) (A)→(R); (B)→(Q); (C)→(P)
(3) (A)→(R); (B)→(S); (C)→(Q)
(4) (A)→(Q); (B)→(S); (C)→(R)

44. Which of the following compounds will produce a precipitate with $AgNO_3$? [Jan. 11, 2019 (II)]

(1) 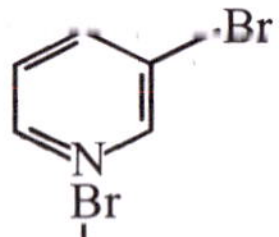

(2) 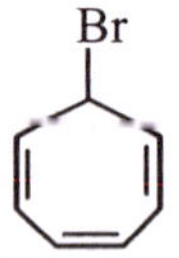

(3) 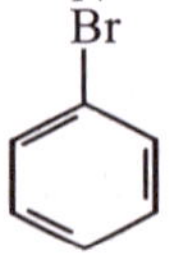

(4) 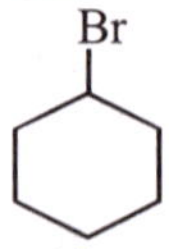

45. The correct match between Item-I and Item-II is: [Jan. 10, 2019 (II)]

Item-I (Compound)	Item-II (Reagent)
(A) Lysine	(P) 1-Naphthol
(B) Furfural	(Q) Ninhydrin
(C) Benzyl alcohol	(R) $KMnO_4$
(D) Styrene	(S) Ceric ammonium nitrate

(1) (A)→(Q); (B)→(P); (C)→(S); (D)→(R)
(2) (A)→(Q); (B)→(P); (C)→(R); (D)→(S)
(3) (A)→ (R); (B)→(P); (C)→(Q) : (D)→(S)
(4) (A)→(Q); (B)→(R); (C)→(S); (D)→(P)

46. Which of the following tests cannot be used for identifying amino acids? [Jan. 10, 2019 (II)]
(1) Biuret test (2) Barfoed test
(3) Ninhydrin test (4) Xanthoproteic test

47. The correct match between Item-I and Item-II is: [Jan. 9, 2019 (I)]

Item-I (Drug)	Item-II (Test)
(A) Chloroxylenol	(P) Carbylamine test
(B) Norethindrone	(Q) Sodium hydrogen carbonate test
(C) Sulphapyridine	(R) Ferric chloride test
(D) Penicillin	(S) Baeyer's test

(1) A→R ; B→P ; C→S ; D→Q
(2) A→Q ; B→S ; C→P ; D→R
(3) A→R ; B→S ; C→P ; D→Q
(4) A→Q ; B→P ; C→S ; D→R

48. The tests performed on compound X and their inferences are:

Test	Inference
(A) 2, 4 - DNP test	Coloured precipitate
(B) Iodoform test	Yellow precipitate
(C) Azodye test	No dye formation

Compound 'X' is: [Jan. 9, 2019 (II)]

(1) 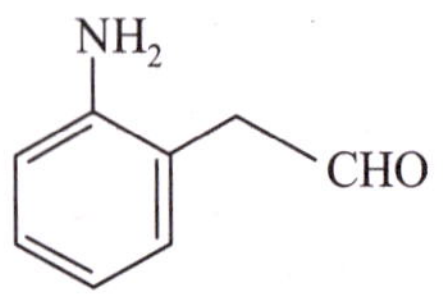

(2) 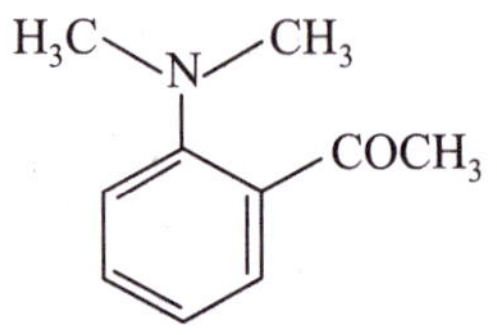

(3) 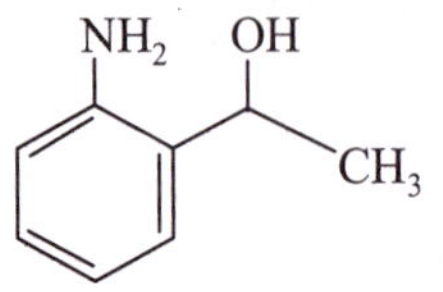

(4) 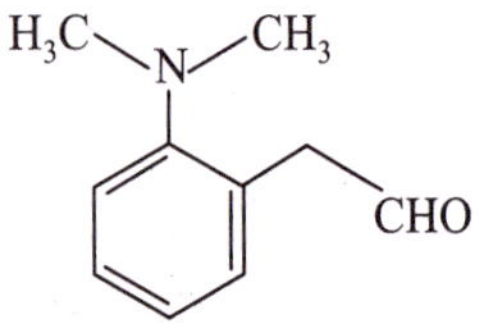

49. Match the organic compounds in Column-I with the Lassaigne's test results in Column-II appropriately : [Online April 11, 2015]

Column-I	Column-II
(A) Aniline	(i) Red colour with $FeCl_3$
(B) Benzene sulfonic acid	(ii) Violet colour with sodium nitroprusside
(C) Thiourea	(iii) Blue colour with hot and acidic solution of $FeSO_4$

(1) A-(ii); B-(iii); C-(i) (2) A-(iii); B-(i); C-(ii)
(3) A-(iii); B-(ii); C-(i) (4) A-(ii); B-(i); C-(iii)

50. Copper wire test for halogens is known as [Online May 7, 2012]
(1) Duma's Test (2) Beilstein's Test
(3) Liebig's Test (4) Lassigne's Test

51. Beilstein test is used for the estimation of which one of the following elements? [Online May 19, 2012]
(1) S (2) Cl (3) C and H (4) N

52. 29.5 mg of an organic compound containing nitrogen was digested according to Kjeldahl's method and the evolved ammonia was absorbed in 20 mL of 0.1 M HCl solution. The excess of the acid required 15 mL of 0.1 M NaOH solution for complete neutralization. The percentage of nitrogen in the compound is [2010]
(1) 59.0 (2) 47.4 (3) 23.7 (4) 29.5

53. An organic compound having molecular mass 60 is found to contain C = 20%, H = 6.67% and N = 46.67% while rest is oxygen. On heating it gives NH_3 along with a solid residue. The solid residue gives violet colour with alkaline copper sulphate solution. The compound is [2005]
(1) $CH_3CH_2CONH_2$ (2) $(NH_2)_2CO$
(3) CH_3CONH_2 (4) CH_3NCO

54. The compound formed in the positive test for nitrogen with the Lassaigne solution of an organic compound is [2004]
(1) $Fe_4[Fe(CN)_6]_3$ (2) $Na_3[Fe(CN)_6]$
(3) $Fe(CN)_3$ (4) $Na_4[Fe(CN)_5NOS]$

NUMERICAL VALUE QUESTIONS (NVQs)

55. From 6.55 g of aniline, the maximum amount of acetanilide that can be prepared will be _____ $\times 10^{-1}$ g. [April 4, 2024 (II)]

56. From the compounds given below, number of compounds which give positive Fehling's test is ______. [Jan. 29, 2024 (I)]
Benzaldehyde, Acetaldehyde, Acetone, Acetophenone, Methanal, 4-nitrobenzaldehyde, cyclohexane carbaldehyde.

57. 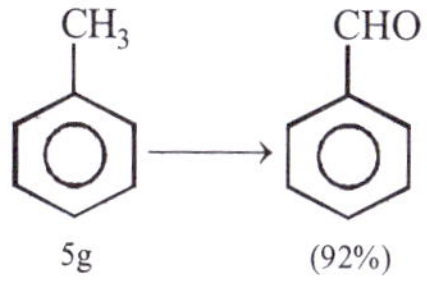

In the above reaction, 5 g of toluene is converted into benzaldehyde with 92% yield. The amount of benzaldehyde produced is_____ $\times 10^{-2}$ g. (Nearest integer) [July 27, 2022 (II)]

58. An organic compound with 51.6% sulfur is heated in a Carius tube. The amount of this compound which will form 0.752 g of barium sulphate is ________ $\times 10^{-1}$g.
(Given molar mass of barium sulphate 233 g mol^{-1}) (Nearest integer). [June 30, 2022 (I)]

59. In the estimation of bromine, 0.5 g of an organic compound gave 0.40 g of silver bromide. The percentage of bromine in the given compound is _________% (nearest integer)
(Relative atomic masses of Ag and Br are 108u and 80u, respectively). [June 28, 2022 (I)]

60. 0.2 g of an organic compound was subjected to estimation of nitrogen by Dumas method in which volume of N_2 evolved (at STP) was found to be 22.400 mL. The percentage of nitrogen in the compound is____.[nearest integer]
(Given: Molar mass of N_2 is 28 mol^{-1}. Molar volume of N_2 at STP : 22.4 L) [June 24, 2022 (II)]

61. In Carius method for estimation of halogens, 0.2 g of an organic compound gave 0.188 g of AgBr.
The percentage of bromine in the compound is _______ . (Nearest integer)
[Atomic mass : Ag = 108, Br = 80] [Aug. 27, 2021 (I)]

62. 0.8 g of an organic compound was analysed by Kjeldahl's method for the estimation of nitrogen. If the percentage of nitrogen in the compound was found to be 42%, then ______ mL of 1 M H_2SO_4 would have been neutralized by the ammonia evolved during the analysis. [July 25, 2021 (II)]

63. Methylation of 10 g of benzene gave 9.2 g of toluene. Calculate the percentage yield of toluene ______.(Nearest integer) [July 22, 2021 (II)]

64. When 0.15 g of an organic compound was analyzed using Carius method for estimation of bromine, 0.2397 g of AgBr was obtained. The percentage of bromine in the organic compound is _______. (Nearest integer) [July 20, 2021 (II)]
[Atomic mass : Silver = 108, Bromine = 80]

65. 1.86 g of aniline completely reacts to form acetanilide. 10% of the product is lost during purification. Amount of acetanilide obtained after purification (in g) is ____ $\times 10^{-2}$. [Feb. 24, 2021 (II)]

Topic 2 - Qualitative Salt Analysis and Preparation of Inorganic Compounds

MULTIPLE CHOICE QUESTIONS (MCQs)

66. Treatment of a gas 'X' with a freshly prepared ferrous sulphate solution gives a compound 'Y' as a brown ring. The compounds X and Y are. [April 06, 2026 (II)]
(1) NO and $[Fe(NO)]SO_4$
(2) NO_2 and $[Fe(NO_2)]SO_4$
(3) N_2O and $[Fe(N_2O)]SO_4$
(4) N_2O_4 and $[Fe(N_2O_4)]SO_4$

67. A paper dipped in a dil. H_2SO_4 solution of 'X' upon treatment with SO_2 gas turns into green. The compound 'X' is :
(1) KI-starch (2) $KMnO_4$ [April 05, 2026 (II)]
(3) $Pb(CH_3COO)_2$ (4) $K_2Cr_2O_7$

68. Statement I : Sodium dichromate and potassium dichromate are classified as primary standard in titrimetric analysis .
Statement II : Phenolphthalein is a weak base, therefore it dissociates in acidic medium. [April 05, 2026 (I)]
In the light of the above statements, choose the correct answer from the options given below.
(1) Both Statement I and Statement II are true
(2) Both Statement I and Statement II are false
(3) Statement I is false but Statement II is true
(4) Statement I is true but Statement II is false

69. Among Fe^{3+}, Pb^{2+}, Cu^{2+} and Mn^{2+} , identify the one that gets precipitated out while passing H_2S in presence of NH_4OH as group reagent. The highest possible oxidation state of the corresponding metal is [April 04, 2026 (II)]
(1) +3 (2) +4 (3) +2 (4) +7

70. In order to oxidise a mixture of 1 mole each of FeC_2O_4, $Fe_2(C_2O_4)_3$, $FeSO_4$ and $Fe_2(SO_4)_3$ in acidic medium, the number of moles of $KMnO_4$ required is : [April 04, 2026 (I)]

(1) 3 (2) 2 (3) 5 (4) 7

71. Among Fe^{2+}, Fe^{3+}, Cr^{2+} and Zn^{2+}, the ion that shows positive borax bead test and with highest ionisation enthalpy is: [April 02, 2026 (II)]

(1) Fe^{2+} (2) Zn^{2+} (3) Cr^{2+} (4) Fe^{3+}

72. A salt with few drops of conc. HCl gives apple green colour in flame test. The group precipitate of the salt is dissolved in acetic acid and treated with K_2CrO_4 to give yellow precipitate. When the sodium carbonate extract of the salt solution is heated with conc. HNO_3 and ammonium molybdate, it resulted a canary yellow precipitate. The cation and anion present in the salt are respectively. [April 02, 2026 (I)]

(1) Ca^{2+} and SO_4^{2-} (2) Ba^{2+} and PO_4^{3-}
(3) Mn^{2+} and PO_4^{3-} (4) Ba^{2+} and SO_4^{2-}

73. Consider the following reactions.

$$Na_2B_4O_7 \xrightarrow{\Delta} 2X + Y$$

$$CuSO_4 + Y \xrightarrow{\text{Non-luminous flame}} Z + SO_3$$

$$2Z + 2X + \text{Carbon} \xrightarrow{\text{Luminous flame}} 2Q + Na_2B_4O_7 + CO$$

The oxidation states of Cu in Z and Q, respectively are : [Jan. 28, 2026 (II)]

(1) +2 and +2 (2) +2 and +1
(3) +1 and +2 (4) +1 and +1

74. Given below are two statements: [Jan. 28, 2026 (I)]

Statement I: Griss-Ilosvay test is used for the detection of nitrite ion, which involves the use of sulphanilic acid and α-naphthylamine reagent.

Statement II: In the above test, sulphanilic acid is diazotized by the acidified nitrite ion, which on further coupling with α-naphthylamine forms an azo-dye.

In the light of the above statements, choose the **correct** answer from the options given below

(1) Statement I is false but Statement II is true
(2) Both Statement I and Statement II are true
(3) Statement I is true but Statement II is false
(4) Both Statement I and Statement II are false

75. In the Group analysis of cations, Ba^{2+} & Ca^{2+} are precipitated respectively as [Jan. 24, 2026 (II)]

(1) sulphide & sulphide (2) hydroxide & carbonate
(3) carbonate & carbonate (4) chromate & sulphide

76. When a salt is treated with sodium hydroxide solution it gives gas X. On passing gas X through reagent Y a brown coloured precipitate is formed. [April 7, 2025 (I)]

X and Y respectively, are

(1) $X = NH_3$ and $Y = HgO$
(2) $X = NH_3$ and $Y = K_2HgI_4 + KOH$
(3) $X = NH_4Cl$ and $Y = KOH$
(4) $X = HCl$ and $Y = NH_4Cl$

77. Formation of $Na_4[Fe(CN)_5NOS]$, a purple coloured complex formed by addition of sodium nitroprusside in sodium carbonate extract of salt indicates the presence of : [April 2, 2025 (II)]

(1) Sodium ion (2) Sulphate ion
(3) Sulphide ion (4) Sulphite ion

78. Choose the correct tests with respective observations. [April 2, 2025 (I)]

(A) $CuSO_4$ (acidified with acetic acid) + $K_4[Fe(CN)_6] \rightarrow$ Chocolate brown precipitate.
(B) $FeCl_3 + K_4[Fe(CN)_6] \rightarrow$ Prussian blue precipitate.
(C) $ZnCl_2 + K_4[Fe(CN)_6]$, neutralised with $NH_4OH \rightarrow$ White or bluish white precipitate.
(D) $MgCl_2 + K_4[Fe(CN)_6] \rightarrow$ Blue precipitate.
(E) $BaCl_2 + K_4[Fe(CN)_6]$, neutralised with NaOH $\rightarrow$ White precipitate.

Choose the **correct** answer from the options given below :

(1) A, D and E only (2) B, D and E only
(3) A, B and C only (4) C, D and E only

79. Identify the inorganic sulphides that are yellow in colour :

(A) $(NH_4)_2S$ (B) PbS (C) CuS
(D) As_2S_3 (E) As_2S_5 [Jan. 28, 2025 (II)]

Choose the correct answer from the options given below :

(1) (A) and (C) only (2) (A), (D) and (E) only
(3) (A) and (B) only (4) (D) and (E) only

80. The metal ion whose electronic configuration is not affected by the nature of the ligand and which gives a violet colour in non-luminous flame under hot condition in borax bead test is [Jan. 28, 2025 (I)]

(1) Ti^{3+} (2) Ni^{2+} (3) Mn^{2+} (4) Cr^{3+}

81. Find the compound 'A' from the following reaction sequences. [Jan. 24, 2025 (II)]

$$A \xrightarrow{\text{aqua-regia}} B \xrightarrow[\text{(2) AcOH}]{\text{(1) } KNO_2|NH_4OH} \text{yellow ppt}$$

(1) ZnS (2) CoS (3) MnS (4) NiS

82. Identify A, B and C in the given below reaction sequence [Jan. 23, 2025 (II)]

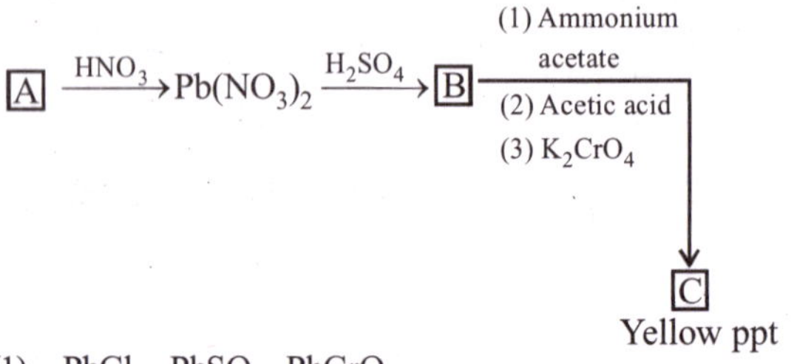

(1) $PbCl_2$, $PbSO_4$, $PbCrO_4$
(2) PbS, $PbSO_4$, $PbCrO_4$
(3) PbS, $PbSO_4$, $Pb(CH_3COO)_2$
(4) $PbCl_2$, $Pb(SO_4)_2$, $PbCrO_4$

83. On reaction of Lead Sulphide with dilute nitric acid which of the following is **not** formed ? [April 9, 2024 (I)]

(1) Lead nitrate (2) Sulphur
(3) Nitric oxide (4) Nitrous oxide

84. Match **List-I** with **List-II** : [April 8, 2024 (I)]

List-I (Name of the test)	List-II (Reaction sequence involved [M is metal])
(A) Borax bead test	(I) $MCO_3 \rightarrow MO \xrightarrow[\Delta]{Co(NO_3)_2} CoO.MO$
(B) Charcoal cavity test	(II) $MCO_3 \rightarrow MCl_2 \rightarrow M^{2+}$
(C) Cobalt nitrate test	(III) $MSO_4 \xrightarrow[\Delta]{Na_2B_4O_7} M(BO_2)_2 \rightarrow MBO_2 \rightarrow M$
(D) Flame test	(IV) $MSO_4 \xrightarrow[\Delta]{Na_2CO_3} MCO_3 \rightarrow MO \rightarrow M$

Choose the correct answer from the options below:
(1) A-(III), B-(I), C-(IV), D-(II)
(2) A-(III), B-(II), C-(IV), D-(I)
(3) A-(III), B-(I), C-(II), D-(IV)
(4) A-(III), B-(IV), C-(I), D-(II)

85. Match List-I with List-II : [April 8, 2024 (I)]

List-I (Compound)	List-II (Colour)
(A) $Fe_4[Fe(CN)_6]_3.xH_2O$	(I) Violet
(B) $[Fe(CN)_5NOS]^{4-}$	(II) Blood Red
(C) $[Fe(SCN)]^{2+}$	(III) Prussian Blue
(D) $(NH_4)_3PO_4.12MoO_3$	(IV) Yellow

Choose the correct answer from the options given below:
(1) A-(III), B-(I), C-(II), D-(IV)
(2) A-(IV), B-(I), C-(II), D-(III)
(3) A-(II), B-(III), C-(IV), D-(I)
(4) A-(I), B-(II), C-(III), D-(IV)

86. In qualitative test for identification of presence of phosphorous, the compound is heated with an oxidising agent. Which is further treated with nitric acid and ammonium molybdate respectively. The yellow coloured precipitate obtained is : [April 8, 2024 (II)]
(1) $Na_3PO_4 \cdot 12MoO_3$
(2) $(NH_4)_3PO_4 \cdot 12(NH_4)_2MoO_4$
(3) $(NH_4)_3PO_4 \cdot 12MoO_3$
(4) $MoPO_4 \cdot 21NH_4NO_3$

87. Match List-I with List-II : [April 6, 2024 (I)]

List-I (Precipitating reagent and conditions)	List-II (Cation)
(A) $NH_4Cl + NH_4OH$	(I) Mn^{2+}
(B) $NH_4OH + Na_2CO_3$	(II) Pb^{2+}
(C) $NH_4OH + NH_4Cl + H_2S$ gas	(III) Al^{3+}
(D) dilute HCl	(IV) Sr^{2+}

Choose the correct answer from the options given below:
(1) A-(IV), B-(III), C-(II), D-(I)
(2) A-(IV), B-(III), C-(I), D-(II)
(3) A-(III), B-(IV), C-(I), D-(II)
(4) A-(III), B-(IV), C-(II), D-(I)

88. During the detection of acidic radical present in a salt, a student gets a pale yellow precipitate soluble with difficulty in NH_4OH solution when sodium carbonate extract was first acidified with dil. HNO_3 and then $AgNO_3$ solution was added. This indicates presence of: [April 6, 2024 (II)]
(1) Br^- (2) CO_3^{2-} (3) I^- (4) Cl^-

89. While preparing crystals of Mohr's salt, dil. H_2SO_4 is added to a mixture of ferrous sulphate and ammonium sulphate, before dissolving this mixture in water, dil. H_2SO_4 is added here to: [April 5, 2024 (II)]
(1) prevent the hydrolysis of ferrous sulphate
(2) prevent the hydrolysis of ammonium sulphate
(3) make the medium strongly acidic
(4) increase the rate of formation of crystals

90. In the precipitation of the iron group (III) in qualitative analysis, ammonium chloride is added before adding ammonium hydroxide to : [April 4, 2024 (I)]
(1) prevent interference by phosphate ions
(2) decrease concentration of ^-OH ions
(3) increase concentration of Cl^- ions
(4) increase concentration of NH_4^+ ions

91. Given below are two statements: [Feb. 1, 2024 (II)]
Statement I : Dimethyl glyoxime forms a six - membered covalent chelate when treated with $NiCl_2$ solution in presence of NH_4OH.
Statement II : Prussian blue precipitate contains iron both in (+2) and (+3) oxidation states.
In the light of the above statements, choose the most appropriate answer from the options given below:
(1) Statement I is false but Statement II is true.
(2) Both Statement I and Statement II are true
(3) Statement I is true but Statement II is false
(4) Both Statement I and Statement II are false

92. The compound that is white in color is [Jan. 31, 2024 (I)]
(1) ammonium sulphide
(2) lead sulphate
(3) ammonium arsinomolybdate
(4) lead iodide

93. Given below are two statements: [Jan. 30, 2024 (I)]
Statement I : The gas liberated on warming a salt with dil H_2SO_4, turns a piece of paper dipped in lead acetate into black, it is a confirmatory test for sulphide ion.
Statement II : In statement - I the colour of paper turns black because of formation of lead sulphite.
In the light of the above statements, choose the most appropriate answer from the options given below:
(1) Both Statement I and Statement II are false.
(2) Statement I is false but Statement II is true.
(3) Statement I is true but Statement II is false.
(4) Both Statement I and Statement II are true.

94. A and B formed in the following reactions are:
$CrO_2Cl_2 + 4NaOH \rightarrow A + 2NaCl + 2H_2O,$
$A + 2HCl + 2H_2O_2 \rightarrow B + 3H_2O$ [Jan. 30, 2024 (II)]
(1) $A = Na_2CrO_4, B = CrO_5$ (2) $A = Na_2Cr_2O_4, B = CrO_4$
(3) $A = Na_2Cr_2O_7, B = CrO_3$ (4) $A = Na_2Cr_2O_7, B = CrO_5$

95. A reagent which gives brilliant red precipitate with Nickel ions in basic medium is [Jan. 29, 2024 (II)]
(1) dimethyl glyoxime (2) meta-dinitrobenzene
(3) sodium nitroprusside (4) neutral $FeCl_3$

96. On passing a gas, 'X', through Nessler's regent, a brown precipitate is obtained. The gas 'X' is [Jan. 29, 2024 (II)]
(1) Cl_2 (2) H_2S
(3) CO_2 (4) NH_3

97. NaCl reacts with conc. H_2SO_4 and $K_2Cr_2O_7$ to give reddish fumes (B), which react with NaOH to give yellow solution (C). (B) and (C) respectively are : [Jan. 27, 2024 (I)]
(1) CrO_2Cl_2, Na_2CrO_4 (2) Na_2CrO_4, CrO_2Cl_2
(3) $CrO_2Cl_2, KHSO_4$ (4) $CrO_2Cl_2, Na_2Cr_2O_7$

98. In the wet tests for detection of various cations by precipitation, Ba^{2+} cations are detected by obtaining precipitate of [April 13, 2023 (II)]
(1) Ba(ox) : Barium oxalate (2) $BaCO_3$
(3) $Ba(OAc)_2$ (4) $BaSO_4$

99. When a solution of mixture having two inorganic salts was treated with freshly prepared ferrous sulphate in acidic medium, a dark brown ring was formed whereas on treatment with neutral $FeCl_3$, it gave deep red colour which disappeared on boiling and a brown red ppt was formed. The mixture contains [April 11, 2023 (I)]
(1) CH_3COO^- & NO_3^- (2) $C_2O_4^{2}$ & NO_3^-
(3) SO_3^{2-} & CH_3COO^- (4) SO_3^{2-} & $C_2O_4^{2-}$

100. 25 mL of silver nitrate solution (1 M) is added dropwise to 25 mL of potassium iodide (1.05 M) solution. The ion(s) present in very small quantity in the solution is/are [April 11, 2023 (I)]

(1) NO_3^- only (2) K^+ only
(3) Ag^+ and I^- both (4) I^- only

101. Prolonged heating is avoided during the preparation of ferrous ammonium sulphate to [April 10, 2023 (I)]

(1) prevent oxidation (2) prevent reduction
(3) prevent hydrolysis (4) prevent breaking

102. Formation of which complex, among the following, is not a confirmatory test of Pb^{2+} ions [April 06, 2023 (II)]

(1) lead chromate (2) lead iodide
(3) lead nitrate (4) lead sulphate

103. Element not present in Nessler's reagent is:- [April 06, 2023 (II)]

(1) Hg (2) I (3) K (4) N

104. Which element is not present in Nessler's reagent?

(1) Mercury (2) Potassium [Feb. 01, 2023 (II)]
(3) Oxygen (4) Iodine

105. A solution of $FeCl_3$ when treated with $K_4[Fe(CN)_6]$ gives a prussian blue precipitate due to the formation of [Feb. 01, 2023 (I)]

(1) $Fe[Fe(CN)_6]$ (2) $Fe_3[Fe(CN)_6]_2$
(3) $Fe_4[Fe(CN)_6]_3$ (4) $K[Fe_2(CN)_6]$

106. During the qualitative analysis of SO_3^{2-} using dilute H_2SO_4, SO_2 gas is evolved which turns $K_2Cr_2O_7$ solution (acidified with dilute H_2SO_4): [Jan. 30, 2023 (I)]

(1) Black (2) Red (3) Green (4) Blue

107. In the wet tests for identification of various cations by precipitation, which transition element cation doesn't belong to group IV in qualitative inorganic analysis? [Jan. 30, 2023 (I)]

(1) Fe^{3+} (2) Zn^{2+} (3) Co^{2+} (4) Ni^{2+}

108. Formulae for Nessler's reagent is: [Jan. 30, 2023 (II)]

(1) KHg_2I_2 (2) $KHgI_3$ (3) K_2HgI_4 (4) HgI_2

109. During the borax bead test with $CuSO_4$, a blue green colour of the bead was observed in oxidising flame due to the formation of [Jan. 29, 2023 (I)]

(1) Cu_3B_2 (2) Cu
(3) $Cu(BO_2)_2$ (4) CuO

110. An ammoniacal metal salt solution gives a brilliant red precipitate on addition of dimethylglyoxime. The metal ion is : [Jan. 24, 2023 (I)]

(1) Cu^{2+} (2) Co^{2+} (3) Fe^{2+} (4) Ni^{2+}

111. Match the List-I with List-II : [Jan. 25, 2023 (I)]

Cations	Group reaction
$P \to Pb^{2+}, Cu^{2+}$	(i) H_2S gas in presence of dilute HCl
$Q \to Al^{3+}, Fe^{3+}$	(ii) $(NH_4)_2CO_3$ in presence of NH_4OH
$R \to Co^{2+}, Ni^{2+}$	(iii) NH_4OH in presence of NH_4Cl
$S \to Ba^{2+}, Ca^{2+}$	(iv) H_2S in presence of NH_4OH

(1) P→i, Q→iii, R→ii, S→iv (2) P→iv, Q→ii, R→iii, S→i
(3) P→iii, Q→i, R→iv, S→ii (4) P→i, Q→iii, R→iv, S→ii

112. When borax is heated with CoO on a platinum loop, blue coloured bead formed is largely due to : [July 29, 2022 (II)]

(1) B_2O_3 (2) $Co(BO_2)_2$
(3) CoB_4O_7 (4) $Co[B_4O_5(OH)_4]$

113. Fe^{3+} cation gives a prussian blue precipitate on addition of potassium ferrocyanide solution due to the formation of: [July 27, 2022 (II)]

(1) $[Fe(H_2O)_6]_2[Fe(CN)_6]$ (2) $Fe_2[Fe(CN)_6]_2$
(3) $Fe_3[Fe(OH)_2(CN)_4]_2$ (4) $Fe_4[Fe(CN)_6]_3$

114. A white precipitate was formed when $BaCl_2$ was added to water extract of an inorganic salt. Further, a gas '*X*' with characteristic odour was released when the formed white precipitate was dissolved in dilute HCl. The anion present in the inorganic salt is : [June 29, 2022 (II)]

(1) I^- (2) SO_3^{2-} (3) S^{2-} (4) NO_2^-

115. Match List-I with List-II. [June 27, 2022 (II)]

List-I (Anion)	List-II (Gas evolved on reaction with dil. H_2SO_4)
(A) CO_3^{2-}	(I) Colourless gas which turns lead acetate paper black
(B) S^{2-}	(II) Colourless gas which turns acidified potassium dichromate solution green.
(C) SO_3^{2-}	(III) Brown fumes which turns acidified KI solution containing starch blue.
(D) NO_2^-	(IV) Colourless gas evolved with brisk effervescences, which turns lime water milky.

Choose the correct answer from the options given below:

(1) A-III, B-I, C-II, D-IV (2) A-II, B-I, C-IV, D-III
(3) A-IV, B-I, C-III, D-II (4) A-IV, B-I, C-II, D-III

116. Which statement is **not** true with respect to nitrate ion test? [June 26, 2022 (I)]

(1) A dark brown ring is formed at the junction of two solutions.
(2) Ring is formed due to nitroferrous sulphate complex.
(3) The brown complex is $[Fe(H_2O)_5(NO)]SO_4$.
(4) Heating the nitrate salt with conc. H_2SO_4, light brown fumes are evolved.

117. During the qualitative analysis of salt with cation y^{2+}, addition of a reagent (X) to alkaline solution of the salt gives a bright red precipitate. The reagent (X) and the cation (y^{2+}) present respectively are: [June 24, 2022 (I)]

(1) Dimethylglyoxime and Ni^{2+}
(2) Dimethylglyoxime and Co^{2+}
(3) Nessler's reagent and Hg^{2+}
(4) Nessler's reagent and Ni^{2+}

118. In the flame test of a mixture of salts, a green flame with blue centre was observed. Which one of the following cations may be present? [June 24, 2022 (II)]

(1) Cu^{2+} (2) Sr^{2+} (3) Ba^{2+} (4) Ca^{2+}

119. Match List-I with List-II :

List-I (Metal Ion)	List-II (Group in Qualitative analysis)
(A) Mn^{2+}	(i) Group - III
(B) As^{3+}	(ii) Group - IIA
(C) Cu^{2+}	(iii) Group - IV
(D) Al^{3+}	(iv) Group - IIB

Choose the most appropriate answer from the options given below : [Aug. 31, 2021 (II)]
(1) (A)-(i), (B)-(ii), (C)-(iii), (D)-(iv)
(2) (A)-(iii), (B)-(iv), (C)-(ii), (D)-(i)
(3) (A)-(i), (B)-(iv), (C)-(ii), (D)-(iii)
(4) (A)-(iv), (B)-(ii), (C)-(iii), (D)-(i)

120. To an aqueous solution containing ions such as Al^{3+}, Zn^{2+}, Ca^{2+}, Fe^{3+}, Ni^{2+}, Ba^{2+} and Cu^{2+} conc. HCl was added, followed by H_2S.
The total number of cations precipitated during this reaction is/are : [July 27, 2021 (II)]
(1) 1 (2) 3 (3) 4 (4) 2

121. Reagent, 1-naphthylamine and sulphanilic acid in acetic acid is used for the detection of [March 18, 2021 (I)]
(1) NO_3^- (2) N_2O (3) NO_2^- (4) NO

122. On treating a compound with warm dil. H_2SO_4, gas X is evolved which turns $K_2Cr_2O_7$ paper acidified with dil. H_2SO_4 to a green compound Y. X and Y respectively are : [Feb. 26, 2021 (I)]
(1) $X = SO_3, Y = Cr_2O_3$ (2) $X = SO_3, Y = Cr_2(SO_4)_3$
(3) $X = SO_2, Y = Cr_2(SO_4)_3$ (4) $X = SO_2, Y = Cr_2O_3$

123. The incorrect statement out of the following is [Online April 16, 2018]
(1) Cu^{2+} ion gives chocolate coloured precipitate with potassium ferrocyanide solution
(2) Cu^{2+} and Ni^{2+} ions give black precipitate with H_2S in presence of HCl solution
(3) Ferric ion gives blood red colour with potassium thiocyanate
(4) Cu^{2+} salts give red coloured borax bead test in reducing flame.

124. A white sodium salt dissolves readily in water to give a solution which is neutral to litmus. When silver nitrate solution is added to the afore-mentioned solution, a white precipitate is obtained which does not dissolve in dil. nitric acid. The anion is : [Online April 15, 2018 (I)]
(1) CO_3^{2-} (2) SO_4^{2-} (3) S^{2-} (4) Cl^-

125. A solution containing a group–IV cation gives a precipitate on passing H_2S. A solution of this precipitate in dil.HCl produces a white precipitate with NaOH solution and bluish–white precipitate with basic potassium ferrocyanide. The cation is : [Online April 8, 2017]
(1) Co^{2+} (2) Ni^{2+} (3) Mn^{2+} (4) Zn^{2+}

126. The hottest region of Bunsen flame shown in the figure below is : [2016]

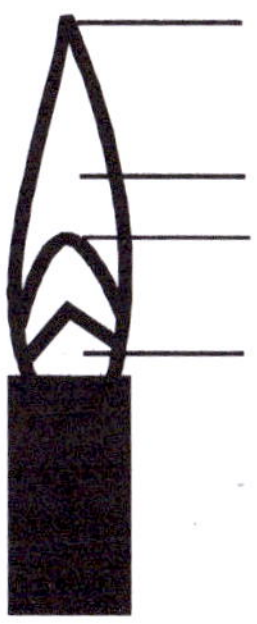

(1) region 3 (2) region 4 (3) region 1 (4) region 2

127. An aqueous solution of a salt X turns blood red on treatment with CNS^- and blue on treatment with $K_4[Fe(CN)_6]$. X also gives a positive chromyl chloride test. The salt X is :
(1) $CuCl_2$ (2) $FeCl_3$ [Online April 10, 2015]
(3) $Cu(NO_3)_2$ (4) $Fe(NO_3)_3$

128. The cation that will not be precipitated by H_2S in the presence of dil. HCl is : [Online April 10, 2015]
(1) Pb^{2+} (2) Cu^{2+} (3) Co^{2+} (4) As^{3+}

129. A pink coloured salt turns blue on heating. The presence of which cation is most likely? [Online April 11, 2015]
(1) Co^{2+} (2) Cu^{2+} (3) Zn^{2+} (4) Fe^{2+}

130. Sodium carbonate cannot be used in place of $(NH_4)_2CO_3$ for the identification of Ca^{2+}, Ba^{2+} and Sr^{2+} ions (in group V) during mixture analysis because :
(1) Mg^{2+} ions will also be precipitated. [Online April 9, 2013]
(2) Concentration of CO_3^{2-} ions is very low.
(3) Sodium ions will react with acid radicals.
(4) Na^+ ions will interfere with the detection of Ca^{2+}, Ba^{2+}, Sr^{2+} ions.

131. Which of the following statements is incorrect ? [Online April 22, 2013]
(1) Fe^{2+} ion also gives blood red colour with SCN^- ion.
(2) Fe^{3+} ion also gives blood red colour with SCN^- ion.
(3) On passing H_2S into Na_2ZnO_2 solution a white ppt of ZnS is formed.
(4) Cupric ion reacts with excess of ammonia solution to give deep blue colour of $[Cu(NH_3)_4]^{2+}$ ion.

132. Potassium dichromate when heated with concentrated sulphuric acid and a soluble chloride, gives brown-red vapours of :
(1) CrO_3 (2) $CrCl_3$ [Online April 9, 2013]
(3) CrO_2Cl_2 (4) Cr_2O_3

133. Calomel (Hg_2Cl_2) on reaction with ammonium hydroxide gives
(1) HgO (2) Hg_2O [2005]
(3) $NH_2 - Hg - Hg - Cl$ (4) $HgNH_2Cl$

134. Which one of the following statements is correct ? [2003]
(1) From a mixed precipitate of AgCl and AgI, ammonia solution dissolves only AgCl
(2) Ferric ions give a deep green precipitate on adding potassium ferrocyanide solution
(3) On boiling a solution having K^+, Ca^{2+} and HCO_3^- ions, we get a precipitate of $K_2Ca(CO_3)_2$
(4) Manganese salts give a violet borax bead test in the reducing flame

135. When H_2S is passed through Hg_2S we get [2002]
(1) HgS (2) $HgS + Hg_2S$
(3) $Hg_2S + Hg$ (4) None of these.

136. How do we differentiate between Fe^{3+} and Cr^{3+} in group III?
(1) by taking excess of NH_4OH solution [2002]
(2) by increasing NH_4^+ ion concentration
(3) by decreasing OH^- ion concentration
(4) both (2) and (3)

Numerical Value Questions (NVQs)

137. 500 mL of 0.2 M MnO_4^- solution in basic medium when mixed with 500 mL of 1.5 M KI solution, oxidises iodide ions to liberate molecular iodine. This liberated iodine is then titrated with a standard x M thiosulphate solution in presence of starch till the end point. If 300 mL of thiosulphate was consumed, then the value of x is______. [April 06, 2026 (II)]

138. Among the following cations, the number of cations which will give characteristic precipitate in their identification tests with $K_4[Fe(CN)_6]$ is :
Cu^{2+}, Fe^{3+}, Ba^{2+}, Ca^{2+}, NH_4^+, Mg^{2+}, Zn^{2+}. [Jan. 24, 2025 (I)]

139. Consider the following test for a group-IV cation.
$M^{2+} + H_2S \rightarrow A$ (Black precipitate) + by product
A + aqua regia $\rightarrow B + NOCl + S + H_2O$
$B + KNO_2 + CH_3COOH \rightarrow C$ + by product
The spin only magnetic moment value of the metal complex C is ______ BM.
(Nearest integer) [April 9, 2024 (II)]

140. The number of white coloured salts, among the following is ______ . [Feb. 1, 2024 (I)]

(a) $SrSO_4$	(b) $Mg(NH_4)PO_4$
(c) $BaCrO_4$	(d) $Mn(OH)_2$
(e) $PbSO_4$	(f) $PbCrO_4$
(g) AgBr	(h) PbI_2
(i) CaC_2O_4	(j) $[Fe(OH)_2(CH_3COO)]$

141. Molar mass of the salt from NaBr, $NaNO_3$, KI and CaF_2 which does not evolve coloured vapours on heating with concentrated H_2SO_4 is _____ g mol^{-1}.
(Molar mass in g mol^{-1} : Na : 23, N : 14, K : 39, O : 16, Br : 80, I : 127, F : 19, Ca : 40) [Jan. 31, 2024 (I)]

142. Number of metal ions characterized by flame test among the following is _______ . [Jan. 30, 2024 (II)]
Sr^{2+}, Ba^{2+}, Ca^{2+}, Cu^{2+}, Zn^{2+}, Co^{2+}, Fe^{2+}

143. The oxidation number of iron in the compound formed during brown ring test for NO_3^- ion is _____. [Jan. 29, 2024 (II)]

144. Consider the sulphides HgS, PbS, CuS, Sb_2S_3, As_2S_3 and CdS. Number of these sulphides soluble in 50% HNO_3 is ________. [Aug. 31, 2021 (I)]

Topic 3 - Principles Involved in Titrimetric Exercises, Calorimetry, Preparation of Colloidal Sols and Kinetic Study of Reaction Between I^- and H_2O_2

MULTIPLE CHOICE QUESTIONS (MCQs)

145. Identify the **correct** statements : [Jan. 22, 2026 (II)]
A. Hydrated salts can be used as primary standard.
B. Primary standard should not undergo any reaction with air.
C. Reactions of primary standard with another substance should be instantaneous and stoichiometric.
D. Primary standard should not be soluble in water.
E. Primary standard should have low relative molar mass.
Choose the **correct** answer from the options given below:
(1) A, B, C and E only (2) A, B, and C only
(3) A, B and E only (4) D and E only

146. Compounds that should not be used as primary standards in titrimetric analysis are : [April 3, 2025 (II)]
A. $Na_2Cr_2O_7$ B. Oxalic acid
C. NaOH D. $FeSO_4 . 6H_2O$
E. Sodium tetraborate
Choose the **most appropriate** answer from the options given below:
(1) B and D Only (2) D and E Only
(3) C, D and E Only (4) A, C and D Only

147. 10 mL of 2 M NaOH solution is added to 20 mL of 1 M HCl solution kept in a beaker. Now, 10 mL of this mixture is poured into a volumetric flask of 100 mL containing 2 moles of HCl and made the volume upto the mark with distilled water. The solution in this flask is : [April 3, 2025 (II)]
(1) 0.2 M NaCl solution (2) 20 M HCl solution
(3) 10 M HCl solution (4) Neutral solution

148. 40 mL of a mixture of CH_3COOH and HCl (aqueous solution) is titrated against 0.1 M NaOH solution conductometrically. Which of the following statement is **correct**? [April 3, 2025 (II)]

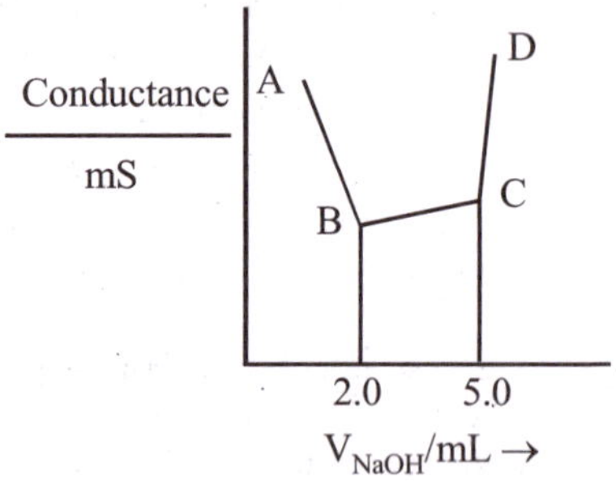

(1) The concentration of CH_3COOH in the original mixture is 0.005 M
(2) The concentration of HCl in the original mixture is 0.005 M
(3) CH_3COOH is neutralised first followed by neutralisation of HCl
(4) Point 'C' indicates the complete neutralisation HCl

149. Given below are two statements : [Jan. 28, 2025 (I)]
Statement I : In the oxalic acid vs $KMnO_4$ (in the presence of dil H_2SO_4) titration the solution needs to be heated initially to 60°C, but no heating is required in Ferrous ammonium sulphate (FAS) vs $KMnO_4$ titration (in the presence of dil H_2SO_4)
Statement II : In oxalic acid vs $KMnO_4$ titration, the initial formation of $MnSO_4$ takes place at high temperature, which then acts as catalyst for further reaction. In the case of FAS vs $KMnO_4$, heating oxidizes Fe^{2+} into Fe^{3+} by oxygen of air and error may be introduced in the experiment. In the light of the above statements, choose the correct answer from the options given below :
(1) Statement I is false but Statement II is true
(2) Both Statement I and Statement II are true
(3) Statement I is true but Statement II is false
(4) Both Statement I and Statement II are false

150. Which of the following mixing of 1M base and 1M acid leads to the largest increase in temperature? [Jan. 24, 2025 (II)]
(1) 30 mL HCl and 30 mL NaOH
(2) 30 mL CH_3COOH and 30 mL NaOH
(3) 50 mL HCl and 20 mL NaOH
(4) 45 mL CH_3COOH and 25 mL NaOH

151. Identify the **incorrect** statements regarding primary standard of titrimetric analysis [April 9, 2024 (I)]

(A) It should be purely available in dry form.
(B) It should not undergo chemical change in air.
(C) It should be hygroscopic and should react with another chemical instantaneously and stoichiometrically.
(D) It should be readily soluble in water.
(E) $KMnO_4$ & NaOH can be used as primary standard.

Choose the **correct** answer from the options given below:

(1) (C) and (D) only (2) (B) and (E) only
(3) (A) and (B) only (4) (C) and (E) only

152. Given below are two statements : One is labelled as **Assertion (A)** and the other is labelled as **Reason (R)** [April 5, 2024 (I)]

Assertion (A): Enthalpy of neutralisation of strong monobasic acid with strong monoacidic base is always -57 kJ mol^{-1}

Reason (R): Enthalpy of neutralisation is the amount of heat liberated when one mole of H^+ ions furnished by acid combine with one mole of ^-OH ions furnished by base to form one mole of water. In the light of the above statements, choose the **correct** answer from the options given below.

(1) (A) is true but (R) is false
(2) Both (A) and (R) are true and (R) is the correct explanation of (A)
(3) (A) is false but (R) is true
(4) Both (A) and (R) are true but (R) is not the correct explanation of (A)

153. Given below are two statements: [Feb. 1, 2024 (I)]

Statement I : Potassium hydrogen phthalate is a primary standard for standardisation of sodium hydroxide solution.

Statement II : In this titration phenophthalein can be used as indicator.

In the light of the above statements, choose the most appropriate answer from the options given below:

(1) Both Statement I and Statement II are correct
(2) Statement I is correct but Statement II is incorrect
(3) Statement I is incorrect but Statement II in correct
(4) Both Statement I and Statement II are incorrect.

154. Given below are two statements :- [April 08, 2023 (II)]

Statement I : Methyl orange is a weak acid.

Statement II : The benzenoid form of methyl orange is more intense/deeply coloured than the quinonoid form.

In the light of the above statement, choose the most appropriate answer from the options given below :-

(1) Statement I is correct but Statement II is incorrect.
(2) Statement I is incorrect but statement II is correct.
(3) Both Statement I and Statement II are incorrect.
(4) Both statement I and Statement II are correct.

155. Given below are two statements :- [April 08, 2023 (II)]

Statement I : In redox titration, the indicators used are sensitive to change in pH of the solution.

Statement II : In acid-base titration, the indicators used are sensitive to change in oxidation potential.

In the light of the above statements, choose the most appropriate answer from the options given below

(1) Both statement I and Statement II are correct.
(2) Statement I is incorrect but statement II is correct.
(3) Statement I is correct but Statement II is incorrect.
(4) Both Statement I and Statement II are incorrect.

156. The volume of 0.02 M aqueous HBr required to neutralize 10.0 mL of 0.01 M aqueous $Ba(OH)_2$ is (Assume complete neutralization) [April 06, 2023 (II)]

(1) 2.5 mL (2) 5.0 mL (3) 10.0 mL (4) 7.5 mL

157. Given below are two statements: one is labelled as **Assertion (A)** and the other is labelled as **Reason (R).**

Assertion (A): An aqueous solution of KOH when used for volumetric analysis, its concentration should be checked before the use.

Reason (R): On aging, KOH solution absorbs atmospheric CO_2.

In the light of the above statements, choose the **correct** answer from the options given below [Feb. 01, 2023 (II)]

(1) Both (A) and (R) are correct but (R) is not the correct explanation of (A)
(2) (A) is correct but (R) is not correct
(3) Both (A) and (R) are correct and (R) is the correct explanation of (A)
(4) (A) is not correct but (R) is correct

158. Incorrect statement for the use of indicators in acid-base titration is: [Jan. 31, 2023 (II)]

(1) Methyl orange may be used for a weak acid vs weak base titration.
(2) Methyl orange is a suitable indicator for a strong acid vs weak base titration
(3) Phenolphthalein is a suitable indicator for a weak acid vs strong base titration
(4) Phenolphthalein may be used for a strong acid vs strong base titration.

159. An indicator 'X' is used for studying the effect of variation in concentration of iodide on the rate of reaction of iodide ion with H_2O_2 at room temp. The indicator 'X' forms blue colored complex with compound 'A' present in the solution. The indicator 'X' and compound 'A' respectively are [Jan. 29, 2023 (II)]

(1) Starch and iodine (2) Methyl orange and H_2O_2
(3) Starch and H_2O_2 (4) Methyl orange and iodine

160. Choose the correct representation of conductometric titration of benzoic acid *vs* sodium hydroxide. [Jan. 24, 2023 (II)]

(1)
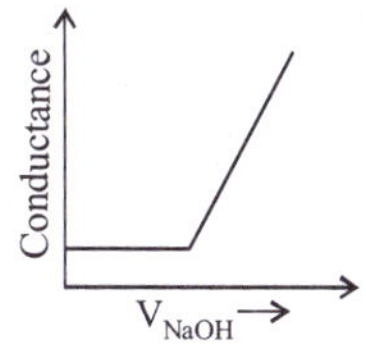

(2)
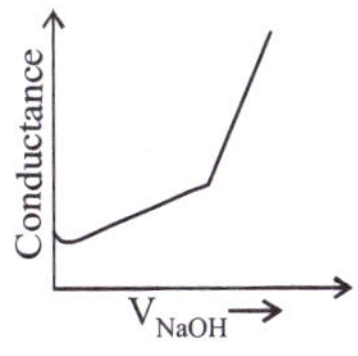

(3)
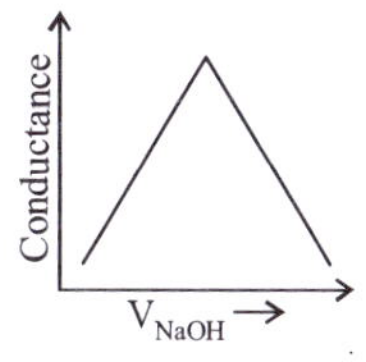

(4)
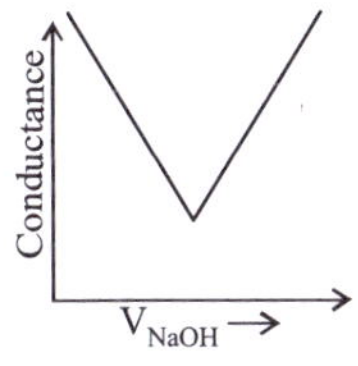

161. For kinetic study of the reaction of iodide ion with H_2O_2 at room temperature : [July 28, 2022 (I)]
(A) Always use freshly prepared starch solution.
(B) Always keep the concentration of sodium thiosulphate solution less than that of KI solution.
(C) Record the time immediately after the appearance of blue colour.
(D) Record the time immediately before the appearance of blue colour.
(E) Always keep the concentration of sodium thiosulphate solution more than that of KI solution.
Choose the correct answer from the options given below:
(1) (A), (B), (C) only (2) (A), (D), (E) only
(3) (D), (E) only (4) (A), (B), (E) only

162. Given below are two statements: One is labelled as **Assertion A** and the other is labelled as **Reason R**.
Assertion A: Permanganate titrations are not performed in presence of hydrochloric acid.
Reason R: Chlorine is formed as a consequence of oxidation of hydrochloric acid.
In the light of the above statements, choose the correct answer from the options given below: [July 28, 2022 (II)]
(1) Both A and R are true and R is the correct explanation of A.
(2) Both A and R are true but R is NOT the correct explanation of A.
(3) A is true but R is false.
(4) A is false but R is true.

163. The plot of pH-metric titration of weak base NH_4OH vs strong acid HCl looks like: [July 27, 2022 (II)]

(1)
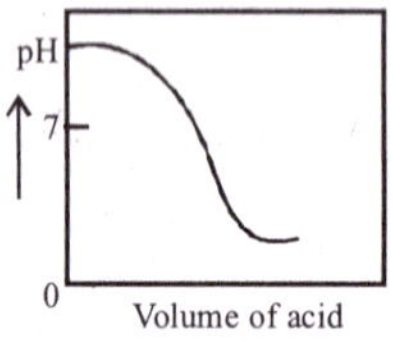

(2)
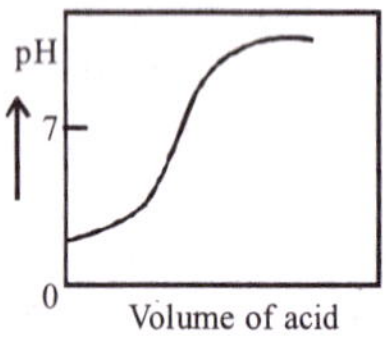

(3)
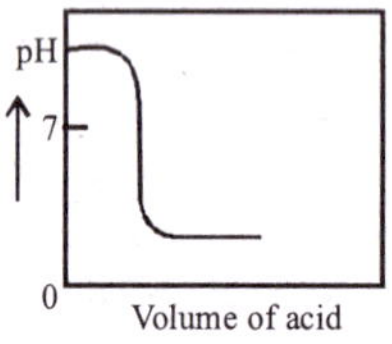

(4)
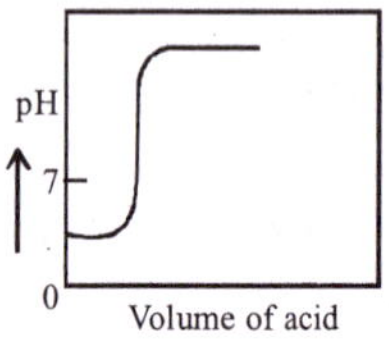

164. The dark purple colour of $KMnO_4$ disappears in the titration with oxalic acid in acidic medium. The overall change in the oxidation number of manganese in the reaction is: [July 26, 2022 (I)]
(1) 5 (2) 1 (3) 7 (4) 2

165. In base vs acid titration, at the end point methyl orange is present as [July 25, 2022 (II)]
(1) quinonoid form (2) heterocyclic form
(3) phenolic form (4) benzenoid form

166. If you spill a chemical toilet cleaning liquid on your hand, your first aid would be : [Sep. 02, 2020 (II)]
(1) vinegar (2) aqueous NaOH
(3) aqueous $NaHCO_3$ (4) aqueous NH_3

167. While titrating dilute HCl solution with aqueous NaOH, which of the following will **not** be required? [Sep. 02, 2020 (I)]
(1) Burette and porcelain tile
(2) Pipette and distilled water
(3) Clamp and phenophthalein
(4) Bunsen burner and measuring cylinder

168. The strength of an aqueous NaOH solution is *most accurately* determined by titrating: (Note: consider that an appropriate indicator is used) [Jan. 08, 2020 (I)]
(1) Aq. NaOH in a pipette and aqueous oxalic acid in a burette
(2) Aq. NaOH in a burette and aqueous oxalic acid in a conical flask
(3) Aq. NaOH in a burette and concentrated H_2SO_4 in a conical flask
(4) Aq. NaOH in a volumetric flask and concentrated H_2SO_4 in a conical flask

169. An alkali is titrated against an acid with methyl orange as indicator, which of the following is a correct combination? [2018]

	Base	Acid	End point
(1)	Weak	Strong	Colourless to pink
(2)	Strong	Strong	Pinkish red to yellow
(3)	Weak	Strong	Yellow to Pinkish red
(4)	Strong	Strong	Pink to colourless

170. For standardizing NaOH solution, which of the following is used as a primary standard? [Online April 16, 2018]
(1) Sodium tetraborate
(2) Ferrous ammonium sulfate
(3) Oxalic acid
(4) dil. HCl

171. In a complexometric titration of metal ion with ligand M(Metal ion) + L(Ligand) $\longrightarrow$ C(Complex)
end point is estimated spectrophotometrically (through light absorption) If 'M' and 'C' do not absorb light and only 'L' absorbs, then the titration plot between absorbed light (A) versus volume of ligand 'L' (V) would look like: [Online April 16, 2018]

(1)
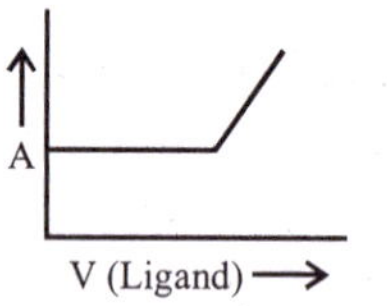

(2)
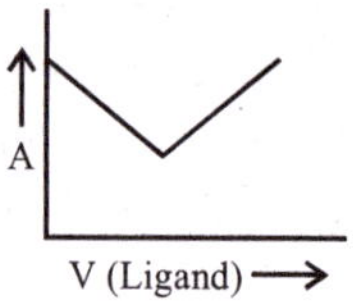

(3)
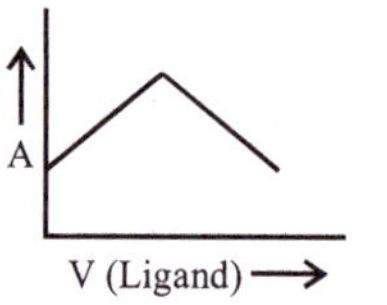

(4)
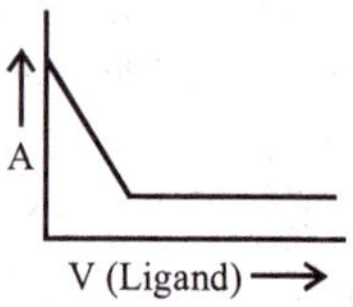

172. Amount of oxalic acid present in a solution can be determined by its titration with $KMnO_4$ solution in the presence of H_2SO_4. The titration gives unsatisfactory result when carried out in the presence of HCl, because HCl [2008]
(1) gets oxidised by oxalic acid to chlorine
(2) furnishes H^+ ions in addition to those from oxalic acid
(3) reduces permanganate to Mn^{2+}
(4) oxidises oxalic acid to carbon doxide and water

173. Which of the following compounds can be detected by Molisch's test ? [2012]
(1) Nitro compounds (2) Sugars
(3) Amines (4) Primary alcohols

NUMERICAL VALUE QUESTIONS (NVQs)

174. 500 mL of 1.2 M KI solution is mixed with 500 mL of 0.2 M $KMnO_4$ solution in basic medium. The liberated iodine was titrated with standard 0.1 M $Na_2S_2O_3$ solution in the presence of starch indicator till the blue color disappeared. The volume (in L) of $Na_2S_2O_3$ consumed is_______. (Nearest integer) [Jan. 28, 2026 (I)]

175. 200 cc of $x \times 10^{-3}$ M potassium dichromate is required to oxidise 750 cc of 0.6 M. Mohr's salt solution in acidic medium. Here x = [Jan. 23, 2026 (II)]

176. x mg of pure HCl was used to make an aqueous solution. 25.0 mL of 0.1M $Ba(OH)_2$ solution is used when the HCl solution was titrated against it.

The numerical value of x is ____ $\times 10^{-1}$. (Nearest integer)

Given : Molar mass of HCl and $Ba(OH)_2$ are 36.5 and 171.0 g mol^{-1} respectively. [Jan. 23, 2026 (I)]

177. A sample of n-octane (1.14 g) was completely burnt in excess of oxygen in a bomb calorimeter, whose heat capacity is 5 kJ K^{-1}. As a result of combustion reaction, the temperature of the calorimeter is increased by 5 K. The magnitude of the heat of combustion of octane at constant volume is _____ kJ mol^{-1} (nearest integer). [April 3, 2025 (II)]

178. The heat of solution of anhydrous $CuSO_4$ and $CuSO_4 \cdot 5H_2O$ are –70 kJ mol^{-1} and +12 kJ mol^{-1} respectively.

The heat of hydration of $CuSO_4$ to $CuSO_4 \cdot 5H_2O$ is –x kJ. The value of x is_____. [April 9, 2024 (I)]

179. When equal volume of 1M HCl and 1M H_2SO_4 are separately neutralised by excess volume of 1M NaOH solution. X and y kJ of heat is liberated respectively. The value of y/x is _______. [April 9, 2024 (I)]

180. Only 2 mL of $KMnO_4$ solution of unknown molarity is required to reach the end point of a titration of 20 mL of oxalic acid (2 M) in acidic medium. The molarity of $KMnO_4$ solution should be ______ M. [April 4, 2024 (I)]

181. 20 mL of calcium hydroxide was consumed when it was reacted with 10 mL of unknown solution of H_2SO_4. Also 20 mL standard solution of 0.5 M HCl containing 2 drops of phenolphthalein was titrated with calcium hydroxide the mixture showed pink colour when burette displaced the value of 35.5 mL whereas the burette showed 25.5 mL initially. The concentration of H_2SO_4 is _____ M (Nearest integer) [April 13, 2023 (I)]

182. $KMnO_4$ is titrated with ferrous ammonium sulphate hexahydrate in presence of dilute H_2SO_4. Number of water molecules produced for 2 molecules of $KMnO_4$ is ___. [April 13, 2023 (I)]

183. The titration curve of weak acid vs. strong base with phenolphthalein as indictor) is shown below. The $K_{phenolphthalein} = 4 \times 10^{-10}$. [April 08, 2023 (I)]

Given: log2 = 0.3

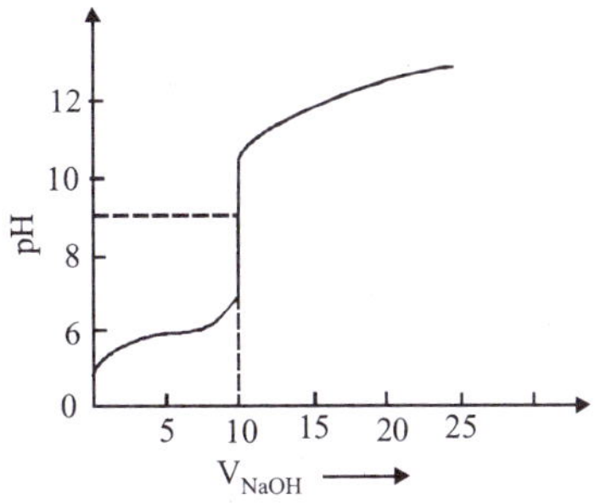

The number of following statements which is/are correct about phenolphthalein is_________

(A) It can be used as an indicator for the titration of weak acid with weak base.

(B) It begins to change colour at pH = 8.4.

(C) It is a weak organic base.

(D) It is colourless in acidic medium.

184. 20 mL of 0.02 M $K_2Cr_2O_7$ solution is used for the titration of 10 mL of Fe^{2+} solution in the acidic medium.

The molarity of Fe^{2+} solution is ______ $\times 10^{-2}$ M. (Nearest Integer) [July 27, 2022 (I)]

185. The normality of H_2SO_4 in the solution obtained on mixing 100 mL of 0.1 M H_2SO_4 with 50 mL of 0.1 M NaOH is____$\times 10^{-1}$ N. (Nearest integer) [July 27, 2022 (II)]

186. 20 mL of 0.02 M hypo solution is used for the titration of 10 mL of copper sulphate solution, in the presence of excess of KI using starch as an indicator. The molarity of Cu^{2+} is found to be _____ $\times 10^{-2}$ M [nearest integer]

Given : $2Cu^{2+} + 4I^- \rightarrow Cu_2I_2 + I_2$

$I_2 + 2S_2O_3^{2-} \rightarrow 2I^- + S_4O_6^{2-}$ [July 26, 2022 (II)]

187. A 2.0 g sample containing MnO_2 is treated with HCl liberating Cl_2. The Cl_2 gas is passed into a solution of KI and 60.0 mL of 0.1 M $Na_2S_2O_3$ is required to titrate the liberated iodine. The percentage of MnO_2 in the sample is ________. (Nearest integer)

[Atomic masses (in u) Mn = 55; Cl = 35.5; O = 16, I = 127, Na – 23, K – 39, S = 32] [June 28, 2022 (I)]

188. 0.01 M $KMnO_4$ solution was added to 20.0 mL of 0.05 M Mohr's salt solution through a burette. The initial reading of 50 mL burette is zero. The volume of $KMnO_4$ solution left in the burette after the end point is ___ mL. (nearest integer) [June 28, 2022 (II)]

189. A 0.166 g sample of an organic compound was digested with conc. H_2SO_4 and then distilled with NaOH. The ammonia gas evolved was passed through 50.0 mL of 0.5 N H_2SO_4. The used acid required 30.0 mL of 0.25 N NaOH for complete neutralization. The mass percentage of nitrogen in the organic compound is____. [June 24, 2022 (I)]

190. When 10 mL an aqueous solution of $KMnO_4$ was titrated in acidic medium, equal volume of 0.1 M of an aqueous solution of ferrous sulphate was required for complete discharge of colour. The strength of $KMnO_4$ in grams per litre is ________ $\times 10^{-2}$. (Nearest integer)

[Atomic mass of K = 39, Mn = 55, O = 16] [Aug. 27, 2021 (I)]

191. 10.0 mL of 0.05 M $KMnO_4$ solution was consumed in a titration with 10.0 mL of given oxalic acid dihydrate solution. The strength of given oxalic acid solution is _______ $\times 10^{-2}$ g/L. [July 27, 2021 (II)]

192. 10 mL of Na_2CO_3 solution is titrated against 0.2 M HCl solution. The following titre values were obtained in 5 readings.

4.8 mL, 4.9 mL, 5.0 mL, 5.0 mL and 5.0 mL

Based on these readings and convention of titrimetric estimation, the concentration of Na_2CO_3 solution is _______mM (Round off the nearest integer). [March 18, 2021 (II)]

193. 0.4 g mixture of NaOH, Na_2CO_3 and some inert impurities was first titrated with $\frac{N}{10}$ HCl using phenolphthalein as an indicator, 17.5 mL of HCl was required at the end point. After this, methyl orange was added and titrated. 1.5 mL of same HCl was required for the next end point. The weight percentage of Na_2CO_3 in the mixture is _______ . (Rounded-off to the nearest integer) [Feb. 25, 2021 (I)]

194. Consider titration of NaOH solution versus 1.25 M oxalic acid solution. At the end point, following burette readings were obtained.

(i) 4.5 mL (ii) 4.5 mL (iii) 4.4 mL

(iv) 4.4 mL (v) 4.4 mL

If the volume of oxalic acid taken was 10.0 mL then the molarity of the NaOH solution is ____ M. (Rounded-off to the nearest integer) [Feb. 25, 2021 (II)]

195. The volume, in mL, of 0.02 M $K_2Cr_2O_7$ solution required to react with 0.288 g of ferrous oxalate in acidic medium is _________. (Molar mass of Fe = 56 g mol^{-1}) [Sep. 05, 2020 (II)]

196. A 20.0 mL solution containing 0.2 g impure H_2O_2 reacts completely with 0.316 g of $KMnO_4$ in acid solution. The purity of H_2O_2 (in %) is __________ (mol. wt. of H_2O_2 = 34; mol. wt. of $KMnO_4$ = 158) [Sep. 04, 2020 (I)]

Part-B (JEE Advanced)

Topic 1- Detection of Extra Elements, Detection of Organic Functional Groups and Preparation of Organic Compounds

1 MCQs with One Correct Answer

1. During sodium nitroprusside test of sulphide ion in an aqueous solution, one of the ligands coordinated to the metal ion is converted to [Adv. 2025]

(a) NOS^- (b) SCN^- (c) SNO^- (d) NCS^-

2 Integer Value Answer

2. The reaction of 4-methyloct-1-ene (P, 2.52 g) with HBr in the presence of $(C_6H_5CO)_2O_2$ gives two isomeric bromides in a 9 : 1 ratio, with a combined yield of 50%. Of these, the entire amount of the primary alkyl bromide was reacted with an appropriate amount of diethylamine followed by treatment with aq. K_2CO_3 to give a non-ionic product S in 100% yield.
The mass (in mg) of S obtained is ___. [Adv. 2023]
[Use molar mass (in g mol^{-1}): H = 1, C = 12, N = 14, Br = 80]

3 Numeric / New Stem Based Questions

Question Stem for Question Nos. 3 and 4

For the following reaction scheme, percentage yields are given along the arrow :

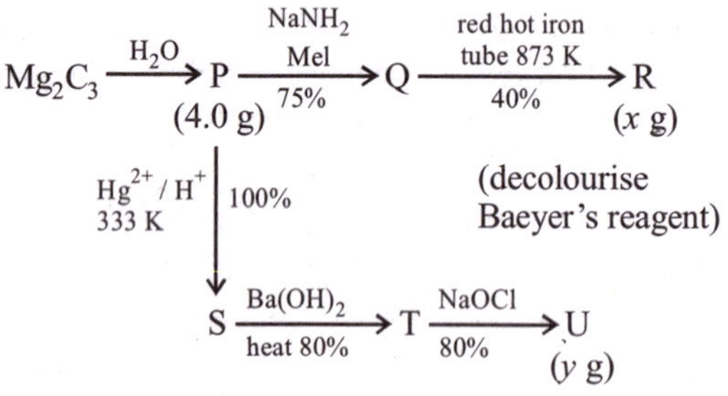

x g and y g are mass of R and U, respectively.
(Use : Molar mass (in g mol^{-1}) of H, C and O as 1, 12 and 16, respectively)

3. The value of x is_____.

4. The value of y is_____.

Question Stem for Question Nos. 5 and 6

Reaction of x g of Sn with HCl quantitatively produced a salt. Entire amount of the salt reacted with y g of nitrobenzene in the presence of required amount of HCl to produce 1.29 g of an organic salt (quantitatively).
(Use Molar masses (in g mol^{-1}) of H, C, N, O, Cl and Sn as 1, 12, 14, 16, 35 and 119, respectively).

5. The value of x is ____. [Adv. 2021]

6. The value of y is ____. [Adv. 2021]

6 MCQs with One or More than One Correct Answer

7. The reaction of Q with PhSNa yields an organic compound (major product) that gives positive Carius test on treatment with Na_2O_2 followed by addition of $BaCl_2$. The correct option(s) for Q is (are). [Adv. 2021]

(a) 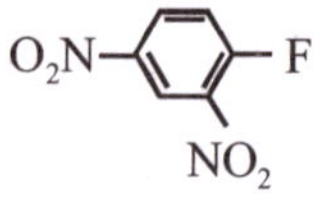

(b) 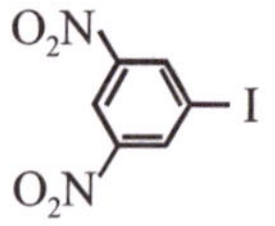

(c) 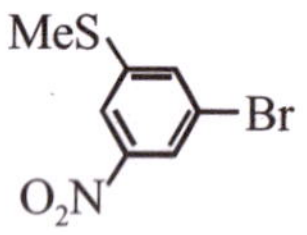

(d) 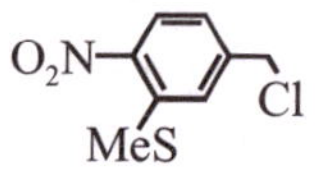

7 Match the Following

8. Match the compounds in List-I with the appropriate observations in List-II and choose the correct option. [Adv. 2025]

List-I	List-II
(P) 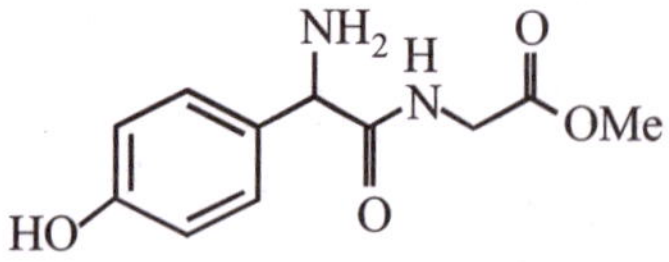	(1) Reaction with phenyl diazonium salt gives yellow dye.

(Q) 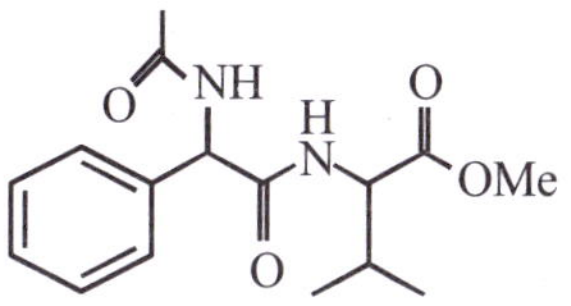

(2) Reaction with ninhydrin gives purple color and it also reacts with $FeCl_3$ to give violet color.

(R) 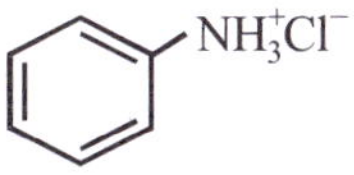

(3) Reaction with glucose will give corresponding hydrazone.

(S) 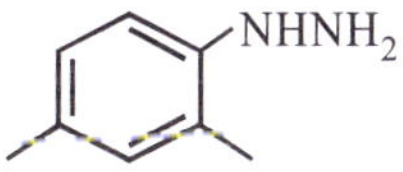

(4) Lassaigne extract of the compound treated with dilute HCl followed by addition of aqueous $FeCl_3$ gives blood red color.

(5) After complete hydrolysis, it will give ninhydrin test and it **DOES NOT** give positive phthalein dye test.

(a) P → 1; Q → 5; R → 4; S → 2
(b) P → 2; Q → 5; R → 1; S → 3
(c) P → 5; Q → 2; R → 1; S → 4
(d) P → 2; Q → 1; R → 5; S → 3

9. Match the compounds in LIST-I with the observations in LIST-II, and choose the correct option. [Adv. 2022]

List - I	List - II
(I) Aniline	(P) Sodium fusion extract of the compound on boiling with $FeSO_4$, followed by acidification with conc. H_2SO_4, gives Prussian blue color.
(II) *o*-Cresol	(Q) Sodium fusion extract of the compound on treatment with sodium nitroprusside gives blood red color.
(III) Cysteine	(R) Addition of the compound to a saturated solution of $NaHCO_3$ results in effervescence.
(IV) Caprolactam	(S) The compound reacts with bromine water to give a white precipitate.
	(T) Treating the compound with neutral $FeCl_3$ solution produces violet color.

(a) I → P,Q; II → S; III → Q,R; IV → P
(b) I → P; II → R,S; III → R; IV → Q,S
(c) I → Q,S; II → P,T; III → P; IV → S
(d) I → P,S; II → T; III → Q,R; IV → P

10. The desired product **X** can be prepared by reacting the major product of the reactions in **LIST-I** with one or more appropriate reagents in **LIST-II**. (given, order of migratory aptitude: aryl > alkyl > hydrogen) [Adv. 2018]

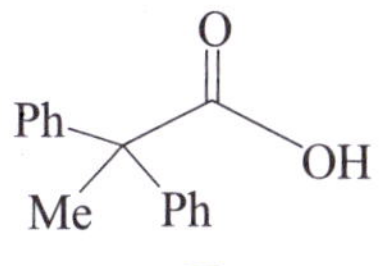

X

List - I	List - II
(P) 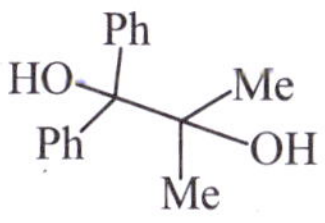 + H_2SO_4	(1) I_2, NaOH
(Q) 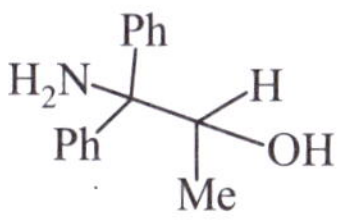 + HNO_2	(2) $[Ag(NH_3)_2]OH$
(R) 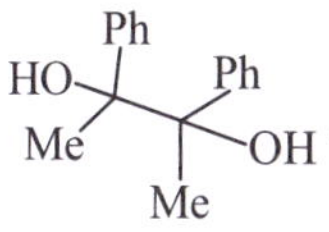 + H_2SO_4	(3) Fehling solution
(S) 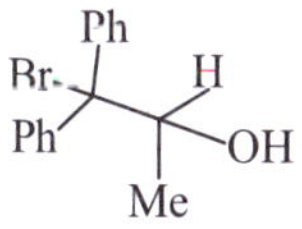 + $AgNO_3$	(4) HCHO, NaOH
	(5) NaOBr

The correct option is
(a) P-1; Q-2, 3 ; R-1, 4; S-2, 4
(b) P-1, 5 ; Q-3, 4 ; R-4, 5; S-3
(c) P-1, 5; Q-3, 4 ; R-5; S-2, 4
(d) P-1, 5; Q-2, 3 ; R-1, 5; S-2, 3

Topic 2 - Qualitative Salt Analysis and Preparation of Inorganic Compounds

1 MCQs with One Correct Answer

11. The green colour produced in the borax bead test of a chromium (III) salt is due to [Adv. 2019]
(a) $Cr_2(B_4O_7)_3$ (b) Cr_2O_3
(c) $Cr(BO_2)_3$ (d) CrB

12. The reagent(s) that can selectively precipitate S^{2-} from a mixture of S^{2-} and SO_4^{2-} in aqueous solution is (are) [Adv. 2016]
(a) $CuCl_2$ (b) $BaCl_2$
(c) $Pb(OOCCH_3)_2$ (d) $Na_2[Fe(CN)_5NO]$

13. Upon treatment with ammoniacal H_2S, the metal ion that precipitates as a sulfide is [Adv. 2013]
(a) Fe(III) (b) Al(III) (c) Mg(II) (d) Zn(II)

2 Integer Value Answer

14. Among PbS, CuS, HgS, MnS, Ag_2S, NiS, CoS, Bi_2S_3 and SnS_2, the total number of **BLACK** coloured sulphides is [Adv. 2014]

15. The treatment of an aqueous solution of 3.74 g of $Cu(NO_3)_2$ with excess KI results in a brown solution along with the formation of a precipitate. Passing H_2S through this brown solution gives another precipitate X. The amount of X (in g) is ________.
[Given: Atomic mass of H = 1, N = 14, O = 16, S = 32, K = 39, Cu = 63, I = 127] [Adv. 2022]

6 MCQs with One or More than One Correct Answer

16. A mixture of two salts is used to prepare a solution S, which gives the following results : [Adv. 2021]

$$\text{white precipitate(s) only} \xleftarrow[\text{Room temperature}]{\text{Dilute NaOH(aq.)}} \text{S (aq.solution of the salts)} \xrightarrow[\text{Room temperature}]{\text{Dilute HCl(aq.)}} \text{White precipitate(s) only}$$

The correct option(s) for the salt mixture is(are)
(a) $Pb(NO_3)_2$ and $Zn(NO_3)_2$ (b) $Pb(NO_3)_2$ and $Bi(NO_3)_3$
(c) $AgNO_3$ and $Bi(NO_3)_3$ (d) $Pb(NO_3)_2$ and $Hg(NO_3)_2$

17. With reference to *aqua regia*, choose the correct option(s)
(a) Reaction of gold with *aqua regia* produces NO_2 in the absence of air [Adv. 2019]
(b) Reaction of gold with *aqua regia* produces an anion having Au in +3 oxidation state
(c) *Aqua regia* is prepared by mixing conc. HCl and conc. HNO_3 in 3 : 1 (*v/v*) ratio
(d) The yellow colour of *aqua regia* is due to the presence of NOCl and Cl_2

18. The correct option(s) to distinguish nitrate salts of Mn^{2+} and Cu^{2+} taken separately is (are) [Adv. 2018]
(a) Mn^{2+} shows the characteristic green colour in the flame test
(b) Only Cu^{2+} shows the formation of precipitate by passing H_2S in acidic medium
(c) Only Mn^{2+} shows the formation of precipitate by passing H_2S in faintly basic medium
(d) Cu^{2+}/Cu has higher reduction potential than Mn^{2+}/Mn (measured under similar conditions)

19. The pair(s) of ions where BOTH the ions are precipitated upon passing H_2S gas in presence of dilute HCl, is (are) [Adv. 2015]
(a) Ba^{2+}, Zn^{2+} (b) Bi^{3+}, Fe^{3+}
(c) Cu^{2+}, Pb^{2+} (d) Hg^{2+}, Bi^{3+}

7 Match the Following

20. The correct match of the group reagents in List-I for precipitating the metal ion given in List-II from solutions, is [Adv. 2025]

	List-I		List-II
(P)	Passing H_2S in the presence of NH_4OH	(1)	Cu^{2+}
(Q)	$(NH_4)_2CO_3$ in the presence of NH_4OH	(2)	Al^{3+}
(R)	NH_4OH in the presence of NH_4Cl	(3)	Mn^{2+}
(S)	Passing H_2S in the presence of dilute HCl	(4)	Ba^{2+}
		(5)	Mg^{2+}

(a) $P \rightarrow 3; Q \rightarrow 4; R \rightarrow 2; S \rightarrow 1$
(b) $P \rightarrow 4; Q \rightarrow 2; R \rightarrow 3; S \rightarrow 1$
(c) $P \rightarrow 3; Q \rightarrow 4; R \rightarrow 1; S \rightarrow 5$
(d) $P \rightarrow 5; Q \rightarrow 3; R \rightarrow 2; S \rightarrow 4$

8 Comprehension Based Questions

Passage-I

The reaction of $K_3[Fe(CN)_6]$ with freshly prepared $FeSO_4$ solution produces a dark blue precipitate called Turnbull's blue. Reaction of $K_4[Fe(CN)_6]$ with the $FeSO_4$ solution in complete absence of air produces a white precipitate X, which turns blue in air. Mixing the $FeSO_4$ solution with $NaNO_3$, followed by a slow addition of concentrated H_2SO_4 through the side of the test tube produces a brown ring.

21. Precipitate X is [Adv. 2021]
(a) $Fe_4[Fe(CN)_6]_3$ (b) $Fe[Fe(CN)_6]$
(c) $K_2Fe[Fe(CN)_6]$ (d) $KFe[Fe(CN)_6]$

22. Among the following, the brown ring is due to the formation of [Adv. 2021]
(a) $[Fe(NO)_2(SO_4)_2]^{2-}$ (b) $[Fe(NO)_2(H_2O)_4]^{3+}$
(c) $[Fe(NO)_4(SO_4)_2]$ (d) $[Fe(NO)(H_2O)_5]^{2+}$

Passage-II

An aqueous solution of metal ion *M*1 reacts separately with reagents *Q* and *R* in excess to give tetrahedral and square planar complexes, respectively. An aqueous solution of another metal ion *M*2 always forms tetrahedral complexes with these reagents. Aqueous solution of *M*2 on reaction with reagent *S* gives white precipitate which dissolves in excess of *S*. The reactions are summarized in the scheme given below:

Scheme:

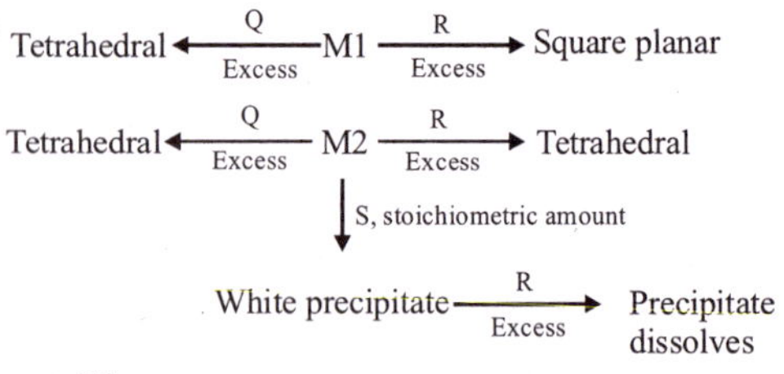

23. Reagent *S* is [Adv. 2014]
(a) $K_4[Fe(CN)_6]$ (b) Na_2HPO_4
(c) K_2CrO_4 (d) KOH

24. $M1$, Q and R, respectively are [Adv. 2014]

(a) Zn^{2+}, KCN and HCl (b) Ni^{2+}, HCl and KCN

(c) Cd^{2+}, KCN and HCl (d) Co^{2+}, HCl and KCN

Passage-III

An aqueous solution of a mixture of two inorganic salts, when treated with dilute HCl, gave a precipitate (P) and a filtrate (Q). The precipitate (P) was found to dissolve in hot water. The filtrate (Q) remained unchanged, when treated with H_2S in a dilute mineral acid medium. However, it gave a precipitate (R) with H_2S in an ammoniacal medium. The precipitate (R) gave a coloured solution (S), when treated with H_2O_2 in an aqueous NaOH medium. [Adv. 2013-II]

25. The coloured solution (S) contains

(a) $Fe_2(SO_4)_3$ (b) $CuSO_4$

(c) $ZnSO_4$ (d) Na_2CrO_4

26. The precipitate (P) contains

(a) Pb^{2+} (b) Hg_2^{2+}

(c) Ag^+ (d) Hg^{2+}

Topic 3 - Principles Involved in Titrimetric Exercises, Calorimetry, Preparation of Colloidal Sols and Kinetic Study of Reaction Between I^- and H_2O_2

3 Numeric / New Stem Based Questions

27. 5.00 mL of 0.10 M oxalic acid solution taken in a conical flask is titrated against NaOH from a burette using phenolphthalein indicator. The volume of NaOH required for the appearance of permanent faint pink color is tabulated below for five experiments. What is the concentration, in molarity, of the NaOH solution? [Adv. 2020]

Exp. No.	Vol. of NaOH(mL)
1	12.5
2	10.5
3	9.0
4	9.0
5	9.0

28. To measure the quantity of $MnCl_2$ dissolved in an aqueous solution, it was completely converted to $KMnO_4$ using the reaction,

$$MnCl_2 + K_2S_2O_8 + H_2O \rightarrow KMnO_4 + H_2SO_4 + HCl$$

(equation not balanced).

Few drops of concentrated HCl were added to this solution and gently warmed. Further, oxalic acid (225 mg) was added in portions till the colour of the permanganate ion disappeared. The quantity of $MnCl_2$ (in mg) present in the initial solution is ______.

(Atomic weights in g mol^{-1}: Mn = 55, Cl = 35.5) [Adv. 2018]

Answer Key

Part-A (JEE Main)																			
1	(1)	**21**	(1)	**41**	(1)	**61**	(40)	**81**	(2)	**101**	(1)	**121**	(3)	**141**	(78)	**161**	(1)	**181**	(1)
2	(4)	**22**	(4)	**42**	(3)	**62**	(12)	**82**	(2)	**102**	(3)	**122**	(2)	**142**	(4)	**162**	(1)	**182**	(68)
3	(4)	**23**	(2)	**43**	(1)	**63**	(78)	**83**	(4)	**103**	(4)	**123**	(2)	**143**	(1)	**163**	(1)	**183**	(2)
4	(3)	**24**	(1)	**44**	(2)	**64**	(68)	**84**	(4)	**104**	(3)	**124**	(4)	**144**	(4)	**164**	(1)	**184**	(24)
5	(1)	**25**	(4)	**45**	(1)	**65**	(243)	**85**	(1)	**105**	(3)	**125**	(4)	**145**	(2)	**165**	(1)	**185**	(1)
6	(3)	**26**	(2)	**46**	(2)	**66**	(1)	**86**	(3)	**106**	(3)	**126**	(4)	**146**	(4)	**166**	(3)	**186**	(4)
7	(4)	**27**	(2)	**47**	(3)	**67**	(4)	**87**	(3)	**107**	(1)	**127**	(2)	**147**	(2)	**167**	(4)	**187**	(13)
8	(3)	**28**	(4)	**48**	(2)	**68**	(2)	**88**	(1)	**108**	(3)	**128**	(3)	**148**	(2)	**168**	(2)	**188**	(30)
9	(3)	**29**	(1)	**49**	(2)	**69**	(4)	**89**	(1)	**109**	(3)	**129**	(1)	**149**	(2)	**169**	(3)	**189**	(148)
10	(2)	**30**	(2)	**50**	(2)	**70**	(2)	**90**	(2)	**110**	(4)	**130**	(1)	**150**	(1)	**170**	(3)	**190**	(316)
11	(2)	**31**	(1)	**51**	(2)	**71**	(4)	**91**	(1)	**111**	(4)	**131**	(1)	**151**	(4)	**171**	(1)	**191**	(1575)
12	(4)	**32**	(4)	**52**	(3)	**72**	(2)	**92**	(2)	**112**	(2)	**132**	(3)	**152**	(2)	**172**	(3)	**192**	(50)
13	(4)	**33**	(4)	**53**	(2)	**73**	(2)	**93**	(3)	**113**	(4)	**133**	(4)	**153**	(1)	**173**	(2)	**193**	(4)
14	(4)	**34**	(3)	**54**	(1)	**74**	(2)	**94**	(1)	**114**	(2)	**134**	(1)	**154**	(3)	**174**	(3)	**194**	(6)
15	(3)	**35**	(1)	**55**	(95)	**75**	(3)	**95**	(1)	**115**	(4)	**135**	(3)	**155**	(4)	**175**	(375)	**195**	(50)
16	(4)	**36**	(2)	**56**	(3)	**76**	(2)	**96**	(4)	**116**	(2)	**136**	(4)	**156**	(3)	**176**	(1825)	**196**	(85)
17	(4)	**37**	(2)	**57**	(530)	**77**	(3)	**97**	(1)	**117**	(1)	**137**	(1)	**157**	(3)	**177**	(2500)		
18	(4)	**38**	(2)	**58**	(2)	**78**	(3)	**98**	(2)	**118**	(1)	**138**	(4)	**158**	(1)	**178**	(82)		
19	(4)	**39**	(2)	**59**	(34)	**79**	(2)	**99**	(1)	**119**	(2)	**139**	(0)	**159**	(1)	**179**	(2)		
20	(1)	**40**	(4)	**60**	(14)	**80**	(2)	**100**	(3)	**120**	(1)	**140**	(5)	**160**	(2)	**180**	(8)		
Part-B (JEE Advanced)																			
1	(a)	**4**	(3.2)	**7**	(a, d)	**10**	(d)	**13**	(d)	**16**	(a, b)	**19**	(c, d)	**22**	(d)	**25**	(d)	**28**	(126)
2	(1791)	**5**	(3.57)	**8**	(b)	**11**	(c)	**14**	(7)	**17**	(b, c, d)	**20**	(a)	**23**	(d)	**26**	(a)		
3	(1.62)	**6**	(1.23)	**9**	(d)	**12**	(a)	**15**	(0.32)	**18**	(b, d)	**21**	(c)	**24**	(b)	**27**	(0.10)		

Chapter-1

The Solid State

HINTS & SOLUTIONS

Part-A (JEE Main)

1. **(4)** Covalent solid have very high melting point due to network like structure and they are insulator in both solid and in molten state, e.g., diamond, SiO_2.

2. **(4)** $CCl_4 \rightarrow$ Non conductor in solid and liquid phase.

3. **(3)**

4. **(1)** For triclinic crystal
$a \neq b \neq c$ Edge lengths
$\alpha \neq \beta \neq \gamma$ Axial angle

5. **(3)** Quartz exhibits piezoelectricity and thus, can be used as a piezoelectric material.

6. **(3)** RbCl, NaCl and LiCl share the same crystal structure with 6 : 6 coordination no. ratio. CsCl crystallises with 8:8 C.N. ratio.

7. **(2)**

8. **(3)** Radius ratio

$$= \frac{\text{Radius of cation}}{\text{Radius of anion}} = \frac{94}{146} = 0.643$$

Since the value is between 0.414 – 0.732, the coordination no. will be 6 and geometry will be octahedral.

9. **(2)** Molecular solid is the best description of water in the solid phase, i.e., ice is hydrogen bonded molecular solid.

10. **(4)** $d = \frac{Z \times M}{a^3 \times N_A} \Rightarrow Z = 2$ for bcc

$$5\text{ g/cm}^3 = \frac{2 \times M}{(200 \times 10^{-10}\text{ cm})^3 \times 6.0 \times 10^{23}}$$

$\Rightarrow \quad M = 12$ g

12 g of element contain $= N_A$ atoms

$$300\text{ g of element contains} = N_A \times \frac{300}{12} = 25\,N_A$$

11. **(4)** Frenkel defect occurs when constituent particle occupy the interstitial site. It create vacancy defect at original and interstitial defect at new location. It does not change density of crystal.

12. **(3)**

13. **(3)**

14. **(1)** AgBr shows both Schottky as well as Frenkel defects.

15. **(1)**

	M_1	M_2	O
	50% octahedral void	12.5% tetrahedral void	ccp
	$\frac{50}{100} \times 4$	$\frac{12.5}{100} \times 8$	4
	2	1	4
Charge	$2x$	$1y$	4×2

So, $2x + y = 8$

$x = 2$; $y = 4$ (among given options, only option (a) is possible)

16. **(1)**

17. **(4)**

18. **(2)**

19. **(3)**

20. **(2)** No. of lattice points in ccp structure = 4

$\therefore$ No. of atoms of B = 4

No. of octahedral voids = 4

$$\text{No. of atoms of A} = \frac{1}{2} \times \text{No. of } Oh \text{ voids}$$

$$= \frac{1}{2} \times 4 = 2$$

No. of atoms of O = No. of all Td voids
= 2 × No. of lattice points
= 2 × 4 = 8

Hence, A : B : O = 2 : 4 : 8 = 1 : 2 : 4

Therefore, the formula of the compound is AB_2O_4.

21. **(3)**

22. **(4)** For bcc

$$\Rightarrow R = \frac{\sqrt{3}\,a}{4}$$

$$\therefore \text{Empty space at edge} = a - 2R = a - \frac{\sqrt{3}\,a}{2}$$

= diameter of sphere.

$$\therefore \quad r_{\text{sphere}} = \frac{a - \frac{\sqrt{3}}{2}a}{2} = \left(\frac{2-\sqrt{3}}{4}\right)a = 0.067\,a$$

23. **(4)** Here, A_2B_3 can also be written as A_4B_6.

No. of effective atoms in hcp lattice = 6.

So, 'B' forms hcp lattice and 'A' is present in tetrahedral voids.

Total tetrahedral voids = 12

$\therefore$ Fraction of tetrahedral voids occupied by A

$$= 4/12 = \frac{1}{3}$$

24. **(4)** We know that, density $= \frac{Z(M)}{N_A \times a^3}$

Given that: $Z = 4$ (fcc)
$M = 63.5$ g
$N_A = 6 \times 10^{23}$
$a = x \times 10^{-8}$cm

After putting the values, we get

$$d = \frac{4 \times 63.5}{6.022 \times 10^{23} \times x^3 \times 10^{-24}} = \frac{422\text{ g cm}^{-3}}{x^3}$$

25. **(3)** In Frenkel defect some of ions (usually cations due to their small size) missing from their normal position, occupy position in interstitial sites.

26. **(4)**

27. **(4)** For an fcc unit cell

$$r = \frac{\sqrt{2}\,a}{4}$$

$$\therefore \text{ closest distance } (2r) = \frac{2\sqrt{2}\,a}{4} = \frac{a}{\sqrt{2}}$$

28. **(3)** In bcc, the atoms touch along body diagonal.

$$\therefore 2r + 2r = \sqrt{3}a$$

$$\therefore r = \frac{\sqrt{3}a}{4} = \frac{\sqrt{3} \times 4.29}{4} = 1.857\text{ Å}$$

29. **(3)** $2r_{Cs^+} + 2r_{Cl^-} = \sqrt{3}a$

$$r_{Cs^+} + r_{Cl^-} = \frac{\sqrt{3}a}{2}$$

30. **(2)** For fcc,

$$r = \frac{\sqrt{2}a}{4} = \frac{a}{2\sqrt{2}} = 0.3535a$$

$r = 0.3535 \times 361 = 128$ pm

31. **(4)** Number of atoms, $A = \frac{1}{8} \times 8 = 1$

Number of atoms, $B = \frac{1}{2} \times (6-1) = \frac{5}{2}$

(Since atom B is missing from one of the face centered point)

$\therefore$ Formula of ionic compound $= AB_{5/2} = A_2B_5$

32. **(3)** The hcp arrangement of atoms occupies 74% of the available space and thus has 26% vacant space.

33. **(4)** When the magnetic moments align in a regular pattern with neighbouring spins pointing in opposite directions the substance shows **antiferromagnetism.**

34. **(3)** For bcc structure,

$$2r = d = \frac{\sqrt{3}\,a}{2}$$

where d = distance between two atoms
a = edge length

$$1.73 = \frac{\sqrt{3}}{2}a$$

$$a = \frac{2 \times 1.73}{\sqrt{3}} = 2\text{Å} = 200\text{ pm}$$

35. **(2)**

36. **(4)** For an fcc crystal,

$$r_{\text{cation}} + r_{\text{anion}} = \frac{\text{Edge length}}{2}$$

$$110 + r_{\text{anion}} = \frac{508}{2}$$

$r_{\text{anion}} = 254 - 110 = 144$ pm

37. **(2)**

38. **(1)**

39. **(2)** The face centered cubic unit cell contains 4 atoms

$$\therefore \text{ Total volume of atoms } = 4 \times \frac{4}{3}\pi r^3 = \frac{16}{3}\pi r^3$$

40. **(2)**

41. **(4)** Since in NaCl type of structure 4 formula units form a cell.

No. of unit cells present in a cubic crystal

$$= \frac{\text{Mass of all unit cells}}{\text{Mass of one unit cell}}$$

$$= \frac{d \times a^3}{M / N_A \times Z} = \frac{m \times N_A}{M \times Z}$$

$\therefore$ units cells $= \dfrac{1.0 \times 6.02 \times 10^{23}}{58.5 \times 4}$

$= 2.57 \times 10^{21}$ unit cells.

42. (4) In bcc - points are at corners and one in the centre of the unit cell.

Number of atoms per unit cell

$= 8 \times \frac{1}{8} + 1 = 2$

In fcc - points are at the corners and also centre of the six faces of each cell.

Number of atoms per unit cell

$= 8 \times \frac{1}{8} + 6 \times \frac{1}{2} = 4$

43. (85)

44. (566) $a = 2(r_+ + r_-) \Rightarrow a = 566$ pm

45. (87)

46. (22)

$\therefore 6 g/cm^3 = \dfrac{Z \times A}{N_A (300 \times 10^{-10})}$

$= \dfrac{2 \times A}{27 \times 10^{-27} \times 6.023 \times 10^{23}}$

$\therefore$ Atomic mass = 48.6g

Atoms of M = $3.7 \times 6 \times 10^{23} = 22.22 \times 10^{23}$ atoms

47. (3) atoms of A (left) in the unit cell after the removal of two opposite face atoms.

$A \rightarrow 4 - \left(2 \times \frac{1}{2}\right) = 3$

atoms of B present in the lattice is

$B \rightarrow 12 \times \frac{1}{4} + 1 \times 1 = 4$

So, Compound is A_3B_4

The value of x is 3.

48. (1)

49. (43) No. of tetrahedral voids = 2 × No. of atoms

$X \rightarrow 6Y \rightarrow \frac{2}{3} \times 2 \times 6 = 8$

$\% X = \frac{16}{14} \times 100 = 42.8 \simeq 43\%$

50. (1) Anions occupy ccp or fcc (B^-) = 4 B^- per unit cell

Cations occupy all octahedral voids (A^+) = 4A^+ per unit cell.

$\therefore$ Cell formula $\Rightarrow A_4B_4$.

Empirical formula $\Rightarrow$ AB

$\Rightarrow x = 1$

51. (106) $d = \dfrac{Z \times M}{a \times N_A}$

$7.62 = \dfrac{4 \times M}{(0.4518 \times 10^{-7})^3 \times 6.022 \times 10^{23}}$

$M = 105.8$ g/mol = 106 g/mol

52. (8) No. of carbon atoms in fcc lattice

$= 8 \times \frac{1}{8} + 6 \times \frac{1}{2} = 4$

No. of tetrahedral voids = 4 × 2 = 8

No. of carbon atoms in tetrahedral voids

$= 8 \times \frac{1}{2} = 4$

$\therefore$ Total atoms = 4 + 4 = 8

53. (5) One Sr^{2+} replaces two K^+. It occupies one of the position and one void is created.

Moles of KBr given $= \frac{1}{119}$

Number of vacancies in 1 mole of KBr

$= \frac{10^{-5}}{100} \times 6.023 \times 10^{23}$

$\therefore$ Total vacancies in KBr

$= \dfrac{10^{-7} \times 6.023 \times 10^{23}}{119} = 5.06 \times 10^{14} \approx 5 \times 10^{14}$.

54. (33) For bcc,

$4r = \sqrt{3}a \Rightarrow r = \frac{\sqrt{3}}{4} \times 27$

For fcc,

$\Rightarrow a = 2\sqrt{2}r$

$= 2\sqrt{2} \times \frac{\sqrt{3}}{4} \times 27$

$= \dfrac{2 \times 1.41 \times 1.73 \times 27}{4} = 33$ Å

55. (15)

56. (1) If number of lattice point = N then effective octahedral voids = N

So, octahedral voids/lattice site = 1

57. (9076) Copper crystallises is fcc unit cell with edge length,

$a = 3.569$ Å

$d = \dfrac{z \times M}{N_A (a)^3}$

$\Rightarrow d = \dfrac{4 \times 63.54}{1000 \times 6.022 \times 10^{23} (3.596 \times 10^{-10})^3}$

$\simeq 9077 \text{ kg m}^{-3}$

58. (8) The coordination number of an atom in a body centered cubic structure is 8.

59. (143) $d = \dfrac{Z \times M}{N_A \times \text{Volume}}$

$2.7 = \dfrac{Z \times 27}{6.02 \times 10^{23} \times [4.05 \times 10^{-8}]^3}$

$Z = 4 \Rightarrow 4r = \sqrt{2}a$

$r = \dfrac{1.414 \times 405}{4} = 143.1675 \text{ pm} = 143.17 \text{ pm}$

Part-B (JEE Advanced)

1. A-p, s; B-p,q; C-q; D-q, r.

A→ p, s; Parameters of a cubic system are $a = b = c$ and $\alpha = \beta = \gamma = 90°$

There are three types of lattices in cubic system. These are simple, face centred and body centred.

B→ p, q ; The parameters of a Rhombohedral system are $a = b = c$; $\alpha = \beta = \gamma \neq 90°$

Cubic and Rhombohedral are two crystal systems. There are seven crystal systems in all.

C→ q; These are two crystal systems.

D→ q, r; Hexagonal and monoclinic are two crystal systems. The parameters of these are

Hexagonal; $a = b \neq c$ and $\alpha = \beta = 90°, \gamma = 120°$

Monoclinic; $a \neq b \neq c$ and $\alpha = \gamma = 90°, \beta \neq 90°$

2. (b) Atom X in fcc lattice sites

$= 8 \times \frac{1}{8} + 6 \times \frac{1}{2} = 4$

Atom X in alternate tetrahedral voids

$= \frac{1}{2} \times 8 = 4$

Total number of atom X in one unit cell = 4 + 4 = 8

T.V. forms at 1/4th of body diagonal, thus,

$\frac{1}{4} \times \sqrt{3}a = 2r \Rightarrow a = \frac{8r}{\sqrt{3}}$

Packing efficiency = $\dfrac{\text{Volume of atoms}}{\text{Volume of unit cell}}$

$= \dfrac{8 \times \frac{4}{3}\pi r^3}{a^3} \times 100 = \dfrac{8 \times \frac{4}{3}\pi r^3}{\left(\frac{8r}{\sqrt{3}}\right)^3} \times 100 \approx 35\%$

3. (a) The arrangement given, shows octahedral void arrangement; limiting radius ratio for octahedral void is

$\dfrac{r_{A^+}}{r_{X^-}} = 0.414$

$r_{A^+} = 0.414 \times r_{X^-} = 0.414 \times 250 = 103.5 \approx 104$ pm.

4. (b) No. of M atoms $= \frac{1}{4} \times 4 + 1 = 1 + 1 = 2$

No. of X atoms $= \frac{1}{2} \times 6 + \frac{1}{8} \times 8 = 3 + 1 = 4$

So, formula $= M_2X_4 = MX_2$

5. (d) Packing effieincy

$= \dfrac{\text{Area occupied by effective circles}}{\text{Area of square}}$

Effective no. of atoms in 2D $= \frac{1}{4} \times 4 + 1 = 2$

$\therefore$ Area occupied by effective circles = $2\pi r^2$

$= \dfrac{2\pi r^2}{L^2} \times 100 = \dfrac{2\pi r^2}{(2\sqrt{2}r)^2} \times 100$

$= \frac{\pi}{4} \times 100 = 78.54\%$

6. (a) Effective number of atoms at corners,

$A = 8 \times \frac{1}{8} = 1$

Effective number of atoms at centre,

$B = \frac{1}{2} \times 6 = 3$

Thus, the composition will be AB_3.

7. (d) In the given solid AB, A atoms occupy the corners and face centres of cubic unit cell (NaCl structure).

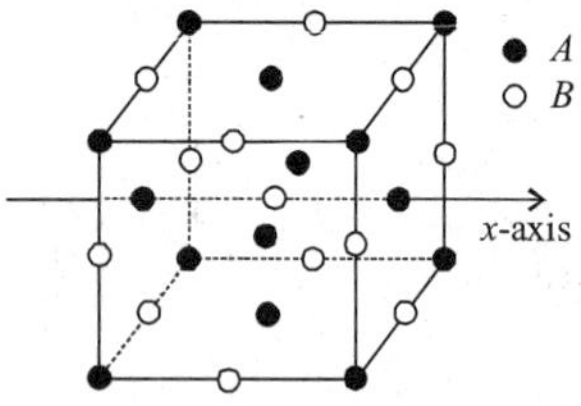

NaCl type AB solid

B atoms occupy the body and edge-centres.

If all the face-centred atoms along x-axis are removed *i.e.* removal of two A atoms then effective number of A atoms in a unit cell

$$= \left(8 \times \frac{1}{8}\right) + \left(4 \times \frac{1}{2}\right) = 3$$

Effective number of B atoms

$$= (1) + \left(12 \times \frac{1}{4}\right) = 4$$

Hence, the resultant stoichiometry of solid $= A_3B_4$

8. **(a)** In an *hcp* structure, atoms are located at the corners and at centres of two hexagons placed parallel to each other; three more atoms are placed in a parallel plane midway between these two planes. Here, each atom is surrounded by 12 others and is said to have co-ordination number of 12.

9. **(a)** For *bcc* structure, atomic radius,

$$r = \frac{\sqrt{3}}{4}a = \frac{\sqrt{3}}{4} \times 4.3 = 1.86$$

Since, r = half the distance between two nearest neighbouring atoms.

$\therefore$ Shortest inter ionic distance $= 2 \times 1.86 = 3.72$

10. **(3)**

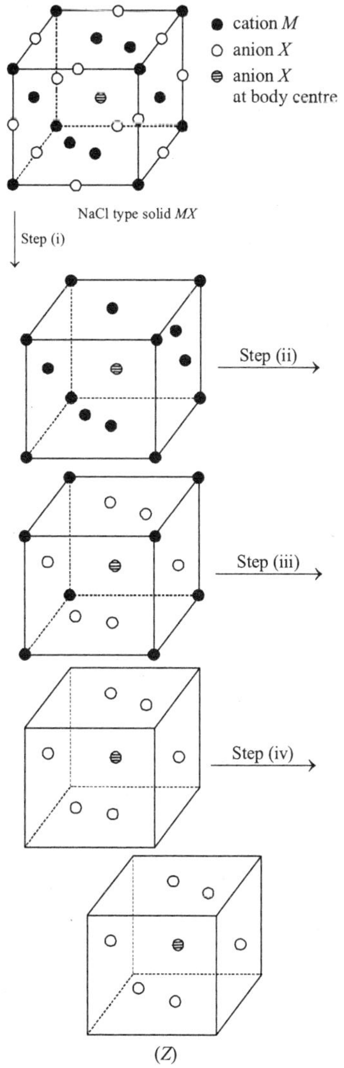

(Z)

Number of cation = 1

Number of anions $= 6 \times \frac{1}{2} = 3$

$$\frac{\text{Number of anions}}{\text{Number of cations}} = \frac{3}{1} = 3$$

11. **(2)** Density $(d) = \dfrac{Z \times M}{a^3 \times N_A}$

For *fcc*, $Z = 4$

Given $a = 4 \times 10^{-8}$ cm

$$8 = \frac{4 \times M}{(4 \times 10^{-8})^3 \times N_A}$$

$$M = \frac{8 \times (4 \times 10^{-8})^3 \times N_A}{4}$$

$= 2 \times (4 \times 10^{-8})^3 \times N_A = 128 \times 10^{-24} N_A$

No. of atoms $= \dfrac{\text{Given mass}}{\text{Molar mass}} \times N_A$

$$= \frac{256 \times N_A}{128 \times 10^{-24} \times N_A} = 2 \times 10^{24}$$

$\therefore$ Value of $N = 2$

12. **(8)** A truncated octahedron is constructed by removing 6 square pyramids from its 6 vertices. In the remaining structure there are 6 square faces and 8 hexagonal faces.

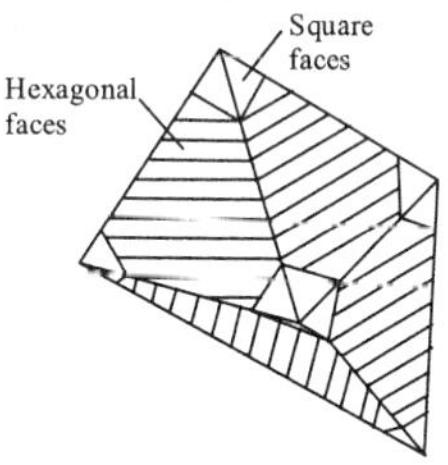

13. **(11)**

$$\rho = \left(\frac{Z \times M}{N_A \times a^3}\right) = \frac{4 \times 105.6}{6 \times 10^{23} \times (4 \times 10^{-8})^3}$$

$$= \frac{4 \times 105.6}{6 \times 10^{23} \times 64 \times 10^{-24}}$$

$$= \frac{422.4 \times 10}{6 \times 64} = 11 \text{ g cm}^{-3}.$$

14. **(1.259)** *fcc* unit cell length = 3.5Å; *bcc* unit cell length = 3.0 Å

Density in *fcc* $= \dfrac{z_1 \times \text{at.wt.}}{V_1 \times N_A}$

Density in *bcc* $= \dfrac{z_2 \times \text{at.wt.}}{V_2 \times N_A}$

$$\therefore \frac{d_{fcc}}{d_{bcc}} = \frac{z_1}{z_2} \times \frac{V_2}{V_1}$$

z_1 for *fcc* = 4; Also $V_1 = a^3 = (3.5 \times 10^{-8})^3$ cm^3

z_2 for *bcc* = 2; Also $V_2 = a^3 = (3.0 \times 10^{-8})^3$ cm^3

$$\therefore \frac{D_{fcc}}{D_{bcc}} = \frac{4 \times (3.0 \times 10^{-8})^3}{2 \times (3.5 \times 10^{-8})^3} = \mathbf{1.259}$$

15. **(124.27)** For *bcc* lattice,

$$r = \frac{\sqrt{3} \times a}{4} = \frac{\sqrt{3}}{4} \times 287 = \mathbf{124.27\,pm}$$

Now, Density $= \dfrac{z \times \text{at.wt.}}{V \times N_A} = \dfrac{z \times \text{at. wt.}}{a^3 \times N_A}$

$z = 2$ for *bcc*; $a = 287 \times 10^{-10}$ cm

$$\therefore \text{Density} = \frac{2 \times 51.99}{(287 \times 10^{-10})^3 \times 6.023 \times 10^{23}}$$

$= \mathbf{7.30\ g/cm^3}$

16. **(1.86)** For *bcc* lattice, (radius), $r = \dfrac{\sqrt{3}a}{4}$

Solution

$$\therefore r = \frac{\sqrt{3} \times 4.29\text{Å}}{4} = \frac{1.73 \times 4.29\text{Å}}{4} = \mathbf{1.86\ Å}$$

17. **(a, b, d)**

Element	X	Y	Z
Packing	FCC	BCC	Primitive
Edge	L_x	L_y	L_z
Relation between edge length and radius	$L_x = 2\sqrt{2}r_x$	$L_y = \frac{4}{\sqrt{3}}r_y$	$L_z = 2r_z$
Packing fraction	$\frac{\pi}{3\sqrt{2}}$	$\frac{\sqrt{3}\pi}{8}$	$\frac{\pi}{6}$

Now, $r_y = \dfrac{8}{\sqrt{3}}r_x$ & $r_z = \dfrac{\sqrt{3}}{2}r_y$

$$= \frac{\sqrt{3}}{2} \times \frac{8}{\sqrt{3}}r_x \Rightarrow r_z = 4r_x$$

So, $L_x = 2\sqrt{2}r_x$, $L_y = \dfrac{4}{\sqrt{3}} \times \dfrac{8}{\sqrt{3}}r_x$, $L_z = 8r_x$

$L_x = 2\sqrt{2}r_x$, $L_y = \dfrac{32}{3}r_x$, $L_z = 8r_x$

So, $L_y > L_z > L_x$

Density $= \dfrac{4M_x}{L_x^3}, \dfrac{2xM_y}{L_y^3}$

[density of X and Y $= \dfrac{\text{Mass}}{\text{Volume}}$, n (per atom) = 4 and 2 for FCC and BCC respectively]

Now, $3M_x = \dfrac{3M_y}{2}$ or $M_x \times 2 = M_y$ (Given)

$$\frac{\text{density (x)}}{\text{density (y)}} = \frac{4M_x}{2M_y} \times \frac{L_y^3}{L_x^3} = \frac{4M_x}{4M_x} \times \frac{\left(\frac{32}{3}\right)^3}{(2\sqrt{2})^3}$$

Hence d(x) > d(y)

18. **(a, c)**

(a) Contribution of M $= 2 \times \frac{1}{2} = 1$

Contribution of X $= 4 \times \frac{1}{4} = 1$

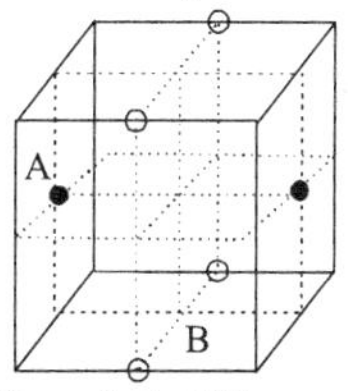

$\therefore$ Empirical formula is MX

(b) Coordination numbers of both M and X is 8.

(c) Bond length of M – X bond

$$= AB = \sqrt{3}.\frac{a}{2} = 0.866a$$

(d) Assuming anions are in contact, the ratio of ionic radii of cation to anion is 0.732 which is the radius ratio of cubical void.

19. **(b, c, d)** ccp is ABC ABC ...... type packing

(a) In topmost layer, each atom is in contact with 6 atoms in same layer and 3 atoms below this layer. Thus, the number, of nearest neighbours is 9.

(b) Packing fraction $= \dfrac{4 \times \frac{4}{3}\pi r^3}{\left(\frac{4r}{\sqrt{2}}\right)^3} = (0.74)$

(c) Each fcc unit has effective no. of atoms = 4
Octahedral void = 4
Tetrahedral void = 8

(d) $4r = a\sqrt{2} \Rightarrow a = 2\sqrt{2}r$

20. **(a)** In *ccp*, number of O^{2-} ions are 4.
Hence, total negative charge = –8
Let number of Al^{3+} ions be *x*, and Mg^{2+} ions be *y*.
Total positive charge = $3x + 2y$
$\Rightarrow 3x + 2y = 8$
This relation is satisfied only by $x = 2$ and $y = 1$.
Hence, number of $Al^{3+} = 2$.
and number of $Mg^{2+} = 1$.
$\Rightarrow n$ = fraction of octahedral holes occupied by Al^{3+}
$= \dfrac{2}{4} = \dfrac{1}{2}$
and *m* = fraction of tetrahedral holes occupied by $Mg^{2+} = \dfrac{1}{8}$

21. **(b, c)** Frenkel defect is a dislocation effect, observed when the size of the cation and anion differ largely. F-center is created when an anion is lost from the lattice and vacancy is filled by trapping of an electron. Schottky defect changes the density of a crystalline solid.

22. **(a, c, d)** The crystals of CsCl has body - centred cubic unit cell. Hence, each ion in this structure has coordination number of eight.
In case of crystals of NaCl, each Na^+ ion is located half way between two Cl^- ions and vice versa. A unit cell of NaCl crystal has Cl^- ions at the corners as well as at the face centres and Na^+ ions are located at body and edge-centers. On each edge of cubic unit cell, there are two Cl^- ions and one Na^+ ions. Hence,

$a = 2\left(r_{Na^+} + r_{Cl^-}\right)$
$= 2(95\text{ pm} + 181\text{ pm}) = 552\text{ pm}$.

23. **(b)** In 1 unit cell of *hcp*, the number of atoms can be calculated as follows
Number of atoms in a unit cell of *hcp*
$= 12 \times \dfrac{1}{6} + 2 \times \dfrac{1}{2} + 3 = 6$
i.e. the correct answer is option (b).

24. **(a)** The volume of hcp unit cell is given by the formula:-
Volume of hexagon = Area of base × height
$= 6 \times \dfrac{\sqrt{3}}{4}(2r)^2 \times 4r\sqrt{\dfrac{2}{3}} = 24\sqrt{2}\,r^3$
i.e. the correct answer is option (a).

25. **(d)** In an *hcp* unit cell, the space occupied is 74%, as calculated below:

$$\text{Packing fraction} = \frac{\text{Volume of the atoms in a unit cell}}{\text{Volume of a unit cell}}$$

$$= \frac{6 \times \frac{4}{3}\pi r^3}{24\sqrt{2}\,r^3} = \frac{\pi}{3\sqrt{2}} = \frac{22}{7} \times \frac{1}{3\sqrt{2}} = 0.74 \text{ or } 74\%$$

∴ Empty space is HCP unit cell = (100–74)% = 26%
i.e. the correct answer is option (d).

26. **(a)** Schottky defect is defined as a vacancy developed for anion and cation site, so cation and anion vacancy will be same in number. Therefore, an ionic solid *MX* with Schottky defects will still have the same number of anions and cations.

27. $d = \dfrac{n \times M}{N_A \times a^3}$ or $n = \dfrac{d \times N_A \times a^3}{M}$

$\Rightarrow n = \dfrac{2 \times 6 \times 10^{23} \times (5 \times 10^{-8})^3}{75} = 2$

Therefore, metal crystallizes in BCC structure and for a BCC lattice $\sqrt{3}a = 4r$

$r = \dfrac{\sqrt{3}}{4}a = \dfrac{\sqrt{3} \times 5}{4} = 2.165\text{Å} = 216.5\text{ pm}$

So, the required answer is **217** pm.

28. For an octhedral void $a = 2(r + R)$
In *fcc* lattice, the largest void present is octahedral void. If the radius of void sphere is *R* and of lattice sphere is *r*.

Then, $r = \dfrac{\sqrt{2} \times 400}{4} = 141.12\text{ pm}(a = 400\text{ pm})$

Applying condition for octahedral void,
$2(r + R) = a$
$\therefore 2R = a - 2r = 400 - (2 \times 141.12) = 117.16\text{ pm}$
∴ Diameter of greatest sphere = **117.16 pm**

29. (a) Density of $AB = \dfrac{Z \times M}{N_A \times a^3}$

Here, $Z = 4$ (for *fcc*), $M = 6.023\,Y$,
$a = 2\,Y^{1/3}\text{ nm} = 2\,Y^{1/3} \times 10^{-9}\text{m}$
Thus,

$\text{Density} = \dfrac{4 \times 6.023Y}{6.023 \times 10^{23} \times (2Y^{1/3} \times 10^{-9})^3}$

= 5.0 kgm^{-3}

(b) Since the observed density (20 kg m^{-3}) of AB is higher than the calculated (5 kg m^{-3}), the compound must have **metal excess defect.** (non-stoichiometric defect).

30. The area of square = 4 × 4 = 16 cm^2
Again to have the maximum number of spheres the packing must be *hcp*.
Maximum number of spheres
= 14 + 8 = 14 + 4 = 18.
full half

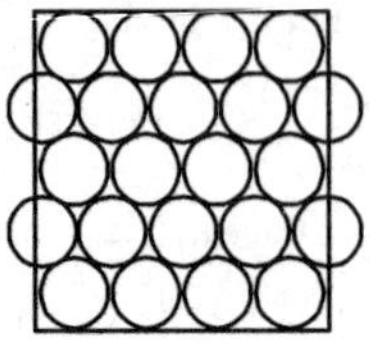

Area = 16 cm^2

∴ Number of spheres per cm^2 = $\dfrac{18}{16} = 1.126$

31.

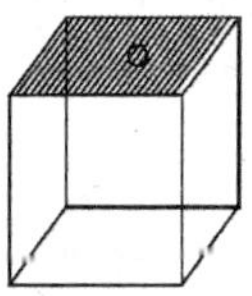

(I)
Face plane

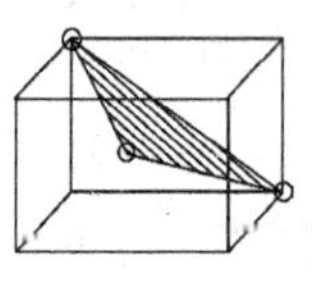

(II)
(1, 1, 1) plane

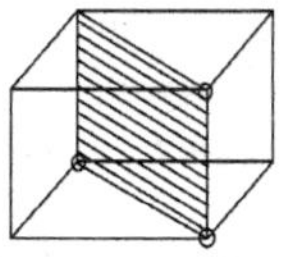

(III)
Diagonal plane

32. For an *hcp* unit cell, there are 6 atoms per unit cell. If r is the radius of the metal atoms, volume occupied by the metallic atoms

$= 6 \times \dfrac{4}{3} \times \pi \times r^3$

$= 6 \times 1.33 \times \dfrac{22}{7} \times r^3 = 25.08 \times r^3$

Geometrically it has been shown that the base area of hcp unit cell

$= 6 \times \dfrac{\sqrt{3}}{4} \times 4r^2$ and the height $= 4r \times \sqrt{2/3}$

∴ Volume of the unit cell

$= \text{Area} \times \text{height} = 6 \times \dfrac{\sqrt{3}}{4} \times 4r^2 \times 4r \times \sqrt{\dfrac{2}{3}}$

$= 33.94\,r^3$

∴ Volume of the empty space of one unit cell
$= 33.94\,r^3 - 25.08\,r^3 = 8.86\,r^3$

∴ Percentage void $= \dfrac{8.816r^3}{33.94r^3} \times 100 = $ **26.1%**

33. Avogadro's number = 6.023×10^{23}
At. wt. of mercury(Hg) = 200 g/mol
∵ In 1 g of Hg, the total number of atom

$= \dfrac{6.023 \times 10^{23}}{200} = \dfrac{6.023 \times 10^{23}}{2 \times 10^2}$

$= 3.0115 \times 10^{21} = 3.012 \times 10^{21}$
Density of Mercury (Hg) = 13.6 g/c.c.
∵ mass of 3.012×10^{21} atoms = 1g

∴ mass of 1 atom $= \dfrac{1}{3.012 \times 10^{21}}$ g

Now volume of 1 atom of mercury (Hg)

$= \dfrac{1}{3.012 \times 10^{21} \times 13.6}$ c.c.

$= 2.44 \times 10^{-23}$ c.c.

Since, each mercury atom occupies a cube of edge length equal to its diameter, therefore,

diameter of one Hg atom $= (2.44 \times 10^{-23})^{\frac{1}{3}}$ cm

$= (24.4 \times 10^{-24})^{\frac{1}{3}}$ cm.
$= 2.905 \times 10^{-8}\text{ cm} \cong$ **2.91 Å**

Chapter-2

Solutions

Part-A (JEE Main)

1 (1)

I. Let wt. of solution = 100 g
Wt. of methanol = 30 g
Wt. of $CCl_4 = 100 - 30 = 70$ g

Mole fraction of $CCl_4 = \dfrac{n_{CCl_4}}{n_{CCl_4} + n_{CH_3OH}}$

$$= \frac{\frac{70}{154}}{\frac{70}{154}+\frac{30}{32}} = 0.3267 \approx 0.33$$

II. CCl_4 & methanol is mixture of positive deviation from Raoult's law.

So, statement (I) & statement (II) are true.

2. (4) 10% by weight of aq. urea means 10 g urea present in 100 g solution

∴ water = 90 g

Molar mass of urea = 12 + 16 + 28 + 4 = 60 g/mol.

Moles of urea $= \dfrac{10}{60} = 0.1667$ mol

Molar mass of water = 18 g/mol

Moles of water $= \dfrac{90}{18} = 5$ mol.

Mole fraction of water $= \dfrac{5}{5+0.1667} \approx 0.967$

3. (4) $P_{N_2} = K_H \cdot X_{N_2}$

$P_{N_2} = 0.8 \times 10 = 8$ atm $= 8 \times 760$ mm Hg

$\Rightarrow 8 \times 760 = 6.5 \times 10^7 \times X_{N_2}$

$$X_{N_2} = \frac{8 \times 760}{6.5 \times 10^7} = 9.35 \times 10^{-5}.$$

4. (3) Total weight of H_2SO_4

$$= \left(100 \times \frac{98}{100}\right) + \left(100 \times \frac{49}{100}\right) = 147 \text{ g}$$

Total weight of solution = 200 g

Total weight of $H_2O = 200 - 147 = 53$ g

Mole fraction of H_2SO_4

$$= \frac{\text{Moles of } H_2SO_4}{\text{Moles of } H_2SO_4 + \text{Moles of } H_2O}$$

Mole fraction of $H_2SO_4 = \dfrac{\frac{147}{98}}{\left(\frac{147}{98}+\frac{53}{18}\right)} = 0.337$

5. (2) K_H depends on the nature of gas and solvent.

6. (3) From Henry's law:

$P(g) = K_H.X$

$\log P(g) = \log K_H + \log X(g)$

7. (4)

$$\frac{P_{N_2}}{P_{O_2}} = \frac{X_{N_2}.P_T}{X_{O_2}.P_T} = \frac{n_{N_2}}{n_{O_2}} = \frac{70/28}{27/32} = 2.96$$

$$\frac{P_{O_2}}{P_{Ar}} = \frac{n_{O_2}}{n_{Ar}} = \frac{27/32}{3/40} = 11.25$$

8. (2) The Volume of water is minimum at 4°C, hence molarity will be maximum at 4°C.

9. (4) For ideal permanent gases, as temperature increases, solubility first decreases then increases hence K_H first increases then decreases in aqueous solutions. The solubility order is He < N_2 < CH_4, So the K_H order will be $CH_4 < N_2 <$ He.

10. (2)

$$\text{Molality} = \frac{1000 \times \text{Molarity}}{1000 \times \text{density} - \text{molarity} \times \text{molar mass of solute}}$$

$$3 = \frac{1000 \times x}{1000 \times 1.12 - (x \times 40)} \Rightarrow x = 3$$

11. (2) Molality $= \dfrac{M \times 1000}{d \times 1000 - (M \times MM)}$

$$= \frac{3 \times 100}{1.25 \times 1000 - (3 \times 58.5)} = 2.79 \text{ m}$$

12. (2) Molarity (M) $= \dfrac{n_{NaCl}}{V_{sol}(\text{in L})}$

$$\Rightarrow \quad M = \frac{\frac{5.85}{58.5}}{0.5} = 0.2\,M$$

13. (2) Mole fraction of 'C'

$$= \frac{n_C}{n_A + n_B + n_C} = \frac{\text{Moles of C}}{\text{Total number of moles}}$$

14. (1)

$$\text{Molarity} = \frac{\text{Moles of solute}}{\text{Volume of solution in 'L'}}$$

Since volume depends on temperature, molarity will change upon change in temperature.

15. (4) Solute (X) = 2 g

Solvent (H_2O) = 1 mole = 18 g

Total mass = 2 + 18 = 20 g

$$\%\text{ mass of X} = \frac{2}{20} \times 100 = 10\%$$

16. (4) $M = \dfrac{3.15 \times 1000}{126 \times 250} = \dfrac{12.6}{126} = 0.1$

17. (1) Molality of solution does not depend on the temperature as it depends on the moles of solute and mass of solvent.

Hence, both assertion and reason are true.

18. (3) Let, 1 mole be present in the solution,

$n_{solute} = 0.2$ mol; $n_{solvent} = 0.8$ mol $= 0.8 \times 18$ g

$$\text{molality} = \frac{\text{Moles of solute}}{\text{Kilogram of solvent}}$$

$$\therefore \quad m = \frac{0.2 \times 1000}{0.8 \times 18} = \frac{1000}{4 \times 18} \approx 13.88$$

19. (4) 20% W/W KI solution (Given)

i.e., 100 g solution contains 20 g KI

∴ Mass of solvent = 100 – 20 = 80 g

No. of moles of KI is $\dfrac{20}{166} = 0.121$ mol

$$\therefore \text{Molality} = \frac{0.121 \times 1000}{80} = 1.51 \text{ m}$$

20. (2) Number of moles in 92 g of Na^+

$$= \frac{92}{23} = 4 \text{ mol}$$

$$\text{Molality } (m) = \frac{\text{Number of moles}}{\text{Mass of solvent (in kg)}}$$

$$\therefore m = \frac{4}{1} = 4 \text{ mol kg}^{-1}$$

21. (4) According to Henry's law

$$\frac{P_1}{P_2} = \frac{S_1}{S_2} \Rightarrow \frac{500}{750} = \frac{0.01}{S_2}$$

$$\therefore S_2 = \frac{750 \times 0.01}{500} = 0.015 \text{ g/L}$$

22. (1) From molarity equation :

$M_1V_1 + M_2V_2 = M \times V$

$$M = \frac{M_1V_1 + M_2V_2}{V}$$

$$= \frac{750 \times 0.5 + 250 \times 2}{750 + 250} = 0.875 \text{ M}$$

23. (4)

$$\text{Molarity} = \frac{\text{Moles of solute}}{\text{Volume of solution in L}}$$

Mass of solution = 1000 + 120 = 1120

$$V = \frac{M}{d} = \frac{1120}{1.15} \text{ mL}$$

$$\text{Molarity} = \frac{120 \times 1.15}{60 \times 1120} \times 1000 = 2.05 \text{ M}$$

24. (2) Sea water is 6 Molar in NaCl, So 1000 mL of sea water contains 6 mol of NaCl.

mass of solution = Volume × density

= 1000 × 2 = 2000 g

$$\text{ppm} = \frac{\text{mass of } O_2}{2000 \text{ g}} \times 10^6$$

$$\Rightarrow \text{mass of } O_2 = 5.8 \times 2000 \times \frac{1}{10^6}$$

$$= 5.8 \times 2 \times 10^{-3} = 1.16 \times 10^{-2} \text{ g}$$

$$\text{molality for } O_2 = \frac{1.16 \times 10^{-2}/32}{(2000 - 6 \times 58.5)} \times 1000$$

$$= \frac{1.16 \times 10}{32 \times 1649} = 0.000219 = 2.19 \times 10^{-4}$$

∴ Correct answer is 2.

25. (9) Let mass of iron = w g

$\Rightarrow \frac{w}{150\times10^3}\times10^6=12$

$\Rightarrow w=150\times12\times10^{-3}=1.8$ g

Let mass of $FeSO_4.7H_2O=w_1$ g

molar mass of $FeSO_4.7H_2O$
$=56+32+64+126=278$ g/mol.

$\Rightarrow$ Moles of Fe $=\frac{1.8}{56}$

Moles of $FeSO_4.7H_2O=\frac{w_1}{278}$

Each mole of $FeSO_4.7H_2O$ provides 1 mole of Fe

$\Rightarrow \frac{w_1}{278}=\frac{1.8}{56}=w_1=\frac{278\times1.8}{56}$

$\Rightarrow w_1=8.935\text{ g}\approx9$ g.

26. (164) Let's take 500 mL of the solution.

$Mass_{sol^n}=V_{sol^n}\times d_{sol^n}=500\times1.25=625g$

Mass of solute (x) = Molarity × Volume × Molar mass of $CuSO_4=0.2\times0.5\times159.5$
$=15.95g$

$n_{solute}=0.1$ mol.

Mass of solvent = Mass of solution – Mass of solute $=625-15.95=609.05g$

$$\text{Molality}=\frac{\text{No. of moles}}{\text{mass of solvent in kg}}=\frac{0.1}{\frac{609.05}{1000}}=0.164$$

$m=164\times10^{-3}$

27. (74) Molality of urea is 4.44 m, that means 4.44 moles of urea present in 1000 gm of water.

$$\frac{X_B}{X_A}=\frac{n_B}{n_A}\Rightarrow\frac{X_B}{1-X_B}=\frac{4.44}{\frac{1000}{18}}\ (n_B=m)$$

$X_B=0.07992(1-X_B)$

$\Rightarrow X_B=0.07992-0.07992X_B$

$1.07992X_B=0.07992$

$\Rightarrow X_B=\frac{0.07992}{1.07992}=0.0740$

$X_B=74\times10^{-3}$

28. (22) Mass percent of Alcohol

$=\frac{\text{Mass of ethyl alcohol}}{\text{Total mass of solution}}\times100$

$=\frac{1\times46}{1\times46+9\times18}\times100=\frac{4600}{208}=22.11\approx22$

29. (11) Given : V = 1L = 1000 mL

Specific gravity $=1.54\text{ g/cm}^3$

M = 98 g/mol

Mass of solution = 1.54 × 1000 = 1540 g

% purity of $H_3PO_4=70\%$

Weight of $H_3PO_4=\frac{70}{100}\times1540=1078$ g

No. of moles of $H_3PO_3=\frac{1078}{98}=11$

Molarity $=\frac{11}{1}=11M$

30. (7)

$$\text{Molarity}=\frac{\text{Mass}\times1000}{\text{Molar mass}\times\text{Volume (mL)}}$$

$\Rightarrow 0.35=\frac{\text{Mass}\times1000}{82.02\times250}\Rightarrow\text{mass}=7.176$ g

31. (4) 31.4% H_2SO_4 by mass

$\Rightarrow$ 31.4 g of H_2SO_4 in 100 g of water

density $=1.25\text{ g/mL}=\frac{100}{V}\Rightarrow V=\frac{100}{1.25}$ mL

Molarity $=\frac{m}{M}\times\frac{1000}{100}\times1.25$

$=\frac{31.4}{98}\times\frac{1000}{100}\times1.25=4.0051\text{ M}\approx4\text{ M}$

32. (815) Let's take 1000 mL of this solution (1000 cm^{-3})

$\Rightarrow$ Mass of this solution = density × volume
$=1.06\text{ g cm}^{-3}\times1000\text{ cm}^{-3}=1.06\times10^3$ g

$\Rightarrow$ Mass of H_2SO_4 present = n × molar mass
$=0.8\times98=78.4$ g

$\Rightarrow$ Mass of solvent = Mass of solution – Mass of solute
$=1060\text{ g}-78.4\text{ g}=981.6\text{ g}=0.982$ kg

$\Rightarrow \text{Molality}=\frac{n}{\text{mass of solvent in Kg}}=\frac{0.8}{0.982}$
$=0.815\text{ mol Kg}^{-1}=815\times10^{-3}\text{ mol Kg}^{-1}$

33. (3.64) 3M NaCl solution = 3 mol of NaCl $=(3\times58.5)$g = 175.5 g of NaCl is present in 1000 ml solution = 1000 g of solution. (as density = 1g/ml)

$\therefore$ Mass of solvent = 1000g – 175.5 g = 824.5g

$\therefore$ Molality of the solution $=\frac{3\times1000}{824.5}=3.64$

34. (139) 10 mL solute in 90 mL solvent

$\therefore$ mass of solute = 10 × 3.2 = 32g

mass of solvent = 90 × 1.6g

$m=\frac{32\times1000}{160\times90\times1.6}=1.388$

$m=138.8\times10^{-2}\approx139\times10^{-2}$ molal

35. (148)

$$\text{Molarity}=\frac{\text{mole of solute}}{\text{volume of the solution (L)}}$$

$2.6\times10^{-3}=\frac{X/85}{0.67141}\Rightarrow X=0.148$ g

Conc. of DCM in ppm $=\frac{0.148}{1.49\times671.141}\times10^6$

= 148 ppm.

36. (180) $M=\frac{5}{40}\times\frac{1000}{450}$

$M_1V_1=M_2V_2$

$\left(\frac{5}{40}\times\frac{1000}{450}\right)\times V_1=0.1\times500\Rightarrow V_1=180$

37. (5) Mass percent, mole fraction, molarity, ppm, molality are used for measuring concentration terms.

38. (54) Moles of nitric acid
$=800\times10^{-3}\times0.5=0.4$

Moles of nitric acid evaporated $=\frac{11.5}{63}=0.182$

Moles left = 0.4 – 0.182 = 0.218

Molarity $=\frac{0.218\times10^3}{400}=0.545\text{ M}=54.5\times10^{-2}$ M

39. (64) Moles of $CuSO_4.5H_2O$

$=\frac{80}{249.54}$ Molarity

$=\frac{80}{249.54}\times\frac{1}{5}=64.117\times10^{-3}\simeq64\times10^{-3}$

40. (20) $[H_2C_2O_4\cdot2H_2O]=\frac{\text{weight / M.W.}}{V(L)}$

$\Rightarrow x\times10^{-2}=\frac{6.3/126}{250/1000}\Rightarrow x=20$

41. (3) Molality

$\Rightarrow\frac{3.30\text{ mol}}{\text{kg}}=\frac{3.30\text{ mol solute}}{1000\text{ g solvent}}$

Mass of solution (solute + solvent)
$=(3.3\times74.5)+1000=1245.85$ g

Volume of solution $=\frac{1245.85}{1.2}$ mL

So Molarity $=\frac{3.3\times1.2}{1245.85}\times1000=3.17\simeq3.0$ M

42. (50) Mole of $Na^+=\frac{3.45}{23}=0.15$ mol

3 mol of Na^+ is present in 1 mole of Na_3PO_4.

0.15 mol of Na^+ is present in $\frac{0.15}{3}$ mole of Na_3PO_4

Molarity $=\frac{0.15\times1000}{3\times100}=0.5$ mol / L
$=50\times10^{-2}$ mol/L

43. (64) Molality of an aqueous solution of a solute = 100 m

Let the mole fraction of solute be x_a

$$\text{Molality}=\frac{\text{Mole fraction of solute}\times1000}{\text{Mole fraction of solvent}\times\text{Mol. mass of solvent}}$$

Then, $\frac{x_a\times1000}{(1-x_a)\times18}=100$ m

$\Rightarrow x_a=\frac{18}{28}=0.6428=64.28\times10^{-2}\simeq64\times10^{-2}$

44. (9) Molality $=\frac{\text{Moles of solute}}{\text{Mass of solvent (in kg)}}$

6.50 molal solution of KOH means

6.50 moles of KOH in 1000 g of water (solvent)

So, weight of solute = 6.5 × 56 = 364 g

Weight of solution = 1000 + 364 = 1364 g

Volume of solution

$=\frac{\text{Mass of solution}}{\text{Density of solution}}=\frac{1364}{1.89}$

Molarity $=\frac{\text{Moles of solute}}{\text{Volume of solution (in L)}}$

$=\frac{6.50\times1.89\times1000}{1364}=9.00$ M

45. (4) According to Raoult's Law

$P_{solution}=P^\circ_{solvent}\,\chi_{solvent}$

So, $\frac{P_{solution}}{P^\circ_{solvent}}=\chi_{solvent}$

$\chi_{solvent}=\frac{1}{1+0.25}=\frac{1}{1.25}=0.8$

Hence x% = 0.8 × 100 = 80%

46. (4) 2 moles of A + 3 moles of B

$\downarrow$

$X_A=2/5,\ X_B=3/5$

$P_s = X_A P^o_A + X_B P^o_B$

$320 = P^o_A\left(\frac{2}{5}\right) + P^o_B\left(\frac{3}{5}\right)$

$2P^o_A + 3P^o_B = 1600$ ...(i)

Now 1 mole of A & 1 mole of B is added

$X'_A = \frac{3}{7},\ X'_B = \frac{4}{7}$

$P'_S = 328.6 = P^o_A\left(\frac{3}{7}\right) + P^o_B\left(\frac{4}{7}\right)$

$3P^o_A + 4P^o_B = 2300.2$ ...(ii)

Now eq (i) × 3 – eq (ii) × 2

$6P^o_A + 9P^o_B = 4800 \Rightarrow 6P^o_A + 8P^o_B = 4600.4$

$P^o_B \simeq 200$ mm of Hg, $P^o_A \simeq 500$ mm of Hg

47. (1) Let X_A and X_B = mole fraction of A and B in solution respectively.

Y_A and Y_B = mole fraction of A and B in vapour respectively.

$\Rightarrow \frac{Y_A}{Y_B} = \frac{P^0_A}{P^0_B}\cdot\frac{X_A}{X_B}$

Given : $Y_A = 0.8; Y_B = 0.2$

$P^0_A = 55\ \text{k Nm}^{-2}, P^0_B = 15\ \text{k Nm}^{-2}$

$\therefore \frac{0.8}{0.2} = \frac{55}{15}\times\frac{X_A}{X_B} \Rightarrow \frac{X_A}{X_B} = \frac{60}{55} = \frac{12}{11}$

For a particular composition, n_A (solution) = 12 and n_B (solution) = 11.

Thus, $X_A = \frac{n_A}{n_A + n_B} = \frac{12}{12+11} = 0.5217.$

48. (1) $P^o = 640$ mm Hg, $P_S = 600$ mm Hg

$\therefore \Delta P = 40$ mm Hg, $\Delta T_b = 375 - 373 = 2K$

moles of solute $= \frac{W}{M}$

$\frac{\Delta P}{P^o} = i.X_{solute}$

Again : $\Delta T_b = i \times k_b \times m$

$2 = i \times 0.52 \times \frac{W/M}{100} \times 1000 \Rightarrow i = \frac{2}{5.2}\times\frac{M}{W}$

$X_{solute} = \frac{\Delta P}{P^o}\times\frac{1}{i} = \frac{40}{640}\times\frac{1}{i} = \frac{1}{16}\times\frac{5.2}{2}\times\frac{W}{M}$

$X_{solute} = \frac{1.3}{8}\times\frac{W}{M}$

49. (1) Mixture of CS_2 and $CH_3.CO.CH_3$ show positive deviation from Raoult's law. Therefore, the partial pressures of both CS_2 and acetone will be higher than expected ideal values.

50. (2) Binary mixture of C_6H_5OH and $C_6H_5NH_2$ shows negative deviation from Raoult's law. Vapour pressure of solution is less than vapour pressure of pure C_6H_5OH & $C_6H_5NH_2$. So boiling point of solution is greater than boiling point of pure C_6H_5OH & $C_6H_5NH_2$. Hence it shows maximum boiling azeotrope.

51. (1)

52. (3) $P^0_A = 350$ mm Hg, $P^0_B = 750$ mmHg

$\Rightarrow P^0_A < P^0_B \Rightarrow \frac{P^0_A}{P^0_B} < 1 \Rightarrow \frac{350}{750} < 1$

$\frac{y_A}{y_B} = \frac{P^0_A\, x_A}{P^0_B\, x_B} = \frac{350\, x_A}{750\, x_B}$

$\Rightarrow \frac{y_A/y_B}{x_A/x_B} < 1 \Rightarrow \frac{y_A}{y_B} < \frac{x_A}{x_B}$

53. (4)

$P_S = P^o_A.X_A + P^o_B.X_B$

$500 = 200 \times \frac{1}{4} + P^o_B.\frac{3}{4} \Rightarrow P^o_B = 600$ mm Hg

As $P^o_A < P^o_B \Rightarrow$ A is least volatile.

54. (1) $\because$ For liquid solution of two liquids '1' and '2'

$P_1 = P_T y_1 = P^0{}_1 x_1$ (From Dalton's Law and Raoult's Law)

$\therefore \frac{P_T}{x_1} = \frac{P^0_1}{y_1}$

$\therefore \frac{P^0_2 + x_1(P^0_1 - P^0_2)}{x_1} = \frac{P^0_1}{y_1}$

$\therefore \frac{P^0_2}{x_1} + (P^0_1 - P^0_2) = \frac{P^0_1}{y_1}$

$\therefore \frac{P^0_2}{x_1} = \frac{P^0_1}{y_1} - (P^0_1 - P^0_2)$

$\therefore \frac{P^0_2}{x_1} = \frac{P^0_1}{y_1} + (P^0_2 - P^0_1)$

$\therefore \frac{1}{x_1} = \left(\frac{P^0_1}{P^0_2}\right)\left(\frac{1}{y_1}\right) + \left(\frac{P^0_2 - P^0_1}{P^0_2}\right)$

Since $y = mx + c$

$\therefore$ Slope $= \left(\frac{P^0_1}{P^0_2}\right)$ $\therefore$ Intercept $= \left(\frac{P^0_2 - P^0_1}{P^0_2}\right)$

55. (4) $(CH_3)_2CO + CS_2$ exibits positive deviations from Raoult's Law.

56. (4) Solution with negative deviation has $P_T < P_{A^o}X_A + P_{B^o}X_B$ (Raoult's Law) $\Rightarrow P_A < P_{A^o}X_A$ and

$P_B < P_{B^o}X_B$

If vapour pressure decreases so boiling point increases, due to stronger intermolecular forces.

57. (1) According to Henry's law

$p = K_H\chi$

for χ constant : $p \propto K_H$; p has max partial pressures.

So it should have maximum K_H and s have minimum partial pressure, So K_H should be minimum.

58. (4)

59. (1, 3) Initial : $H_2O(l) \rightleftharpoons H_2O(g)$

On addition of solute :

$H_2O(l)$ + glucose $\rightleftharpoons H_2O(g)$

Hence, the rate at which water molecules leaves the vapour increases and that leaves the solution decreases.

60. (3) At a particular temperature, as intermolecular forces of attraction increases, vapour pressure decreases.

Thus, intermolecular forces are inversely proportional to vapour pressure and directly proportional to temperature.

Therefore, X has lower intermolecular interactions compared to Y.

61. (2) (1) Mixture of carbon disulphide and acetone will show positive deviation from Raoult's Law.

(3) The dipolar interaction between solute (CS_2) and solvent (acetone) molecules in solution are weaker than that between solute-solute and solvent-solvent. So the vapour pressure of solution will be greater than the individual vapour pressure of pure components.

(2) For positive deviation, $\Delta_{mix}V > 0$ and (4) $\Delta_{mix}H > 0$.

62. (1) When these two beakers are sealed together, then there will be natural flow of solvent from low concentration to high concentration. Thus, solvent will flow from the beaker with pure solvent towards beaker with solution.

Hence, the volume of solution increases and the volume of the solvent decreases.

63. (4) $P^0_M = 450$ mmHg, $P^0_N = 700$ mmHg

$P_M = P^0_M x_M = y_M P_T \Rightarrow P^o{}_M = \frac{y_M}{x_M}(P_T)$

Similarly, $P^0_N = \frac{y_N}{x_N}(P_T)$; given, $P^0_M < P^o{}_N$

$\Rightarrow \frac{y_M}{x_M} < \frac{y_N}{x_N} \Rightarrow \frac{y_M}{y_N} < \frac{x_M}{x_N}$

64. (4) $P = x_B p^o{}_B + x_A p^o{}_A$

$= (0.5 \times 600) + (0.5 \times 400) = 300 + 200 = 500$ mm Hg

Using the relation $p_i = y_i P_{Total}$, we can calculate the mole fractions of the components in vapour phase.

$p_B = y_B P_{total}$

$y_B = \frac{p_B}{P_{Total}} = \frac{300}{500} = \frac{3}{5} = 0.6$

$y_A = \frac{p_A}{P_{Total}} = \frac{200}{500} = \frac{2}{5} = 0.4$

65. (1) According to Henry's law,

$P = K_H . X_{gas}$

$\because X_{gas} + X_{H_2O} = 1 \Rightarrow X_{gas} = 1 - X_{H_2O}$

$\therefore P = K_H(1 - X_{H_2O}) \Rightarrow P = K_H - K_H . X_{H_2O}$

$y = c + mx$; $m = -K_H \Rightarrow$ slope is negative and $c = K_H$

Therefore, the intercept on Y axis increases from w to z.

66. (2) $P^o{}_A = 7 \times 10^3$; $P^o{}_B = 12 \times 10^3$

$x'_A = 0.4$; $x'_B = 1 - 0.4 = 0.6$

$P_{total} = P^o{}_A x'_A + P^o{}_B x'_B$

$= 7 \times 10^3 \times 0.4 + 12 \times 10^3 \times 0.6$

$= (7 \times 0.4 + 12 \times 0.6) \times 10^3 = 10^4$

$x_A = \frac{P^o_A x'_A}{P_{total}} = \frac{7\times10^3\times0.4}{10^4}$

$\therefore x_A = 0.28, x_B = 1 - 0.28 = 0.72$

67. (1) The solubility of the gas in liquids decreases with the increase in value of K_H at a given pressure.

68. (3)

69. (4) According to Henry's law,

$\frac{P_1}{P_2} = \frac{S_1}{S_2} \Rightarrow \frac{500}{750} = \frac{0.01}{S_2}$

$\therefore S_2 = \frac{750\times0.01}{500} = 0.015$ g/L

70. (2) Total V.P. of solution

$= P_A^\circ X_A + P_B^\circ X_B$

Given, $P_A^\circ = 74.7$ torr, $P_B^\circ = 22.3$ torr

$n_{benzene} = 1.5$ mol, $n_{toluene} = 3.5$ mol

$n_{solution} = 1.5 + 3.5 = 5$ mol

$$x_A = \frac{n_{benzene}}{n_{solution}} = \frac{1.5}{5} = 0.3$$

$$x_B = \frac{n_{toluene}}{n_{solution}} = \frac{3.5}{5} = 0.7$$

Total V.P. of solution $= (0.3 \times 74.7 + 0.7 \times 22.3)$ torr $= (22.4 + 15.6)$ torr $= 38$ torr

Mole fraction of benzene in vapour form

$$= \frac{22.4}{38} = 0.589$$

71. (2) If we plot a graph between the vapour pressure and the temperature, we would get a curve that rises faster as T increases, giving a curved line.

72. (4) Solutions in which solute - solute and solvent-solvent interactions are almost similar to solute-solvent interactions are known as ideal solution.

73. (2) $P_A = P_A^\circ \times X_A =$ total pressure $\times y_A$

$P_B = P_B^\circ \times X_B =$ total pressure $\times y_B$

where x and y represents mole fraction in liquid and vapour phase respectively.

$$\frac{P_B^\circ X_B}{P_A^\circ X_A} = \frac{y_B}{y_A}; \frac{P_B^\circ(1-X_A)}{P_A^\circ X_A} = \frac{1-y_A}{y_A}$$

on putting values $\frac{119(1-0.50)}{37 \times 0.50} = \frac{1-y_A}{y_A}$

on solving $y_A = 0.237$

74. (3) Let vapour pressure of $A = P_A^\circ$

Vapour pressure of $B = P_B^\circ$

In first solution,

Mole fraction of $A(X_A) = \frac{1}{1+2} = \frac{1}{3}$

Mole fraction of $B(X_B) = \frac{2}{1+2} = \frac{2}{3}$

According to Raoult's law,

Total vapour pressure $= 250 = P_A^\circ X_A + P_B^\circ X_B$

$$250 = \frac{1}{3}P_A^\circ + \frac{2}{3}P_B^\circ \quad ...(i)$$

In second solution

Mole fraction of $A(X_A) = \frac{2}{2+2} = \frac{2}{4} = \frac{1}{2}$

Mole fraction of $B(X_B) = \frac{2}{4} = \frac{1}{2}$

$\therefore$ Total vapour pressure $= 300 = P_A^\circ X_A + P_B^\circ X_B$

$$300 = \frac{1}{2}P_A^\circ + \frac{1}{2}P_B^\circ \quad ...(ii)$$

Multiplying equation (i) by $\frac{1}{2}$ and equation (ii) by $\frac{1}{3}$

$$\frac{1}{6}P_A^\circ + \frac{2}{6}P_B^\circ = 125 \Rightarrow \frac{\frac{1}{6}P_A^\circ + \frac{1}{6}P_B^\circ = 100}{\frac{1}{6}P_B^\circ = 25}$$

$P_B^\circ = 25 \times 6 = 150$ mm Hg

On substituting value of P_B° in equation (ii) we get

$$300 = P_A^\circ \times \frac{1}{2} + 150 \times \frac{1}{2}$$

$P_A^\circ = 450$ mm Hg

75. (1) $P_{Total} = (P^\circ_A X_A) + (P^\circ_B X_B)$

$= (P^\circ_{Heptane} X_{Heptane}) + (P^\circ_{Octane} X_{Octane})$

$$= \left(105 \times \frac{25/100}{\frac{25}{100} + \frac{35}{114}}\right) + \left(45 \times \frac{35/114}{\frac{25}{100} + \frac{35}{114}}\right)$$

$$= \left(105 \times \frac{0.25}{0.25 + 0.3}\right) + \left(45 \times \frac{0.3}{0.25 + 0.3}\right)$$

$$= \frac{105 \times 0.25}{0.55} + \frac{45 \times 0.3}{0.55} = \frac{26.25 + 13.5}{0.55} = 72 \text{ kPa}$$

76. (2) For this solution, intermolecular interactions between n-heptane and ethanol are weaker than n-heptane - n-heptane and ethanol-ethanol interactions. Hence, the solution of n-heptane and ethanol is non-ideal and shows positive deviation from Raoult's law.

77. (2) $P_{total} = P_x^\circ X_x + P_y^\circ X_y$

$$550 = P_x^\circ \times \frac{1}{4} + P_y^\circ \times \frac{3}{4}$$

$$P_x^\circ + 3P_y^\circ = 550 \times 4 \quad ...(i)$$

In second case, $P_{total} = P_x^\circ \times \frac{1}{5} + P_y^\circ \times \frac{4}{5}$

$$P_x^\circ + 4P_y^\circ = 560 \times 5 \quad ...(ii)$$

Subtract (i) from (ii)

$\therefore P_y^\circ = 560 \times 5 - 550 \times 4 = 600$

$\because P_x^\circ = 400$

78. (4) At 1 atmospheric pressure, the boiling point of mixture is 80 °C.

At boiling point, the vapour pressure of mixture, $P_T = 1$ atmosphere $= 760$ mm Hg.

Using the relation,

$P_T = P_A^\circ X_A + P_B^\circ X_B$, we get

$\{P_B^\circ = 1000$ mm Hg, $X_A + X_B = 1\}$

or $760 = (520X_A) + (1000(1 - X_A))$

or $X_A = \frac{240}{480} = \frac{1}{2}$ or 50 mol. percent

79. (2) $P_A^\circ = ?$, Given $P_B^\circ = 200$ mm, $X_A = 0.6$, $X_B = 1 - 0.6 = 0.4$, $P = 290$

$P = P_A + P_B = (P_A^\circ X_A) + (P_B^\circ X_B)$

$290 = (P_A^\circ \times 0.6) + (200 \times 0.4)$

$\therefore P_A^\circ = 350$ mm

80. (4) Let the mass of methane and oxygen = m g.

Mole fraction of O_2

$$= \frac{\text{Moles of } O_2}{\text{Moles of } O_2 + \text{Moles of } CH_4}$$

$$= \frac{m/32}{(m/32) + (m/16)} = \frac{m/32}{3m/32} = \frac{1}{3}$$

Partial pressure of O_2 = Total pressure × mole fraction of O_2

$$\Rightarrow P_{O_2} = P \times \frac{1}{3} = \frac{1}{3}P$$

81. (4) Vapour pressure of benzene = 75 torr

Vapour pressure of toluene = 22 torr

Moles of benzene $= \frac{78}{78} = 1$ mol

Moles of toluene $= \frac{46}{92} = 0.5$ mol

Now partial pressure of benzene

$$= P^\circ_b \cdot X_b = 75 \times \frac{1}{1 + 0.5} = 75 \times \frac{1}{1.5} = 75 \times \frac{2}{3} = 50$$

82. (2) Equimolar solutions of normal solutes in the same solvent will have the same b. pts and same f. pts. As these are colligative properties.

83. (4) Gaseous densities of ethanol and dimethyl ether would be same at same temperature and pressure. The heat of vaporisation, V.P. and b.pts will differ due to H-bonding in ethanol.

84. (2)

85. (4) When A and B form an ideal solution, $\Delta H_{mix} = 0$

86. (4) In solution containing A and B component showing negative deviation A–A and B–B interactions are weaker than that of A–B interactions. For such solutions,

$\Delta H = -$ve and $\Delta V = -$ve

87. (5) As we know,

$$\frac{P^o - Ps}{Ps} = \text{Molality} \times \frac{(\text{M.solvent})}{1000} \quad ...(1)$$

$\Delta T_b = 320 - 319.5 = 0.5$ K

$$\Delta T_b = K_b \cdot \text{molality} \Rightarrow \text{molality} = \frac{0.5}{0.3} \quad ...(2)$$

$$\frac{P^o - Ps}{Ps} = \frac{760 - 750}{750} = \frac{10}{750} = \frac{1}{75} \quad ...(3)$$

Putting (2) and (3) in (1), we get

(Molecular mass) $= \frac{600}{75}$ g

Moles of solvent $= \frac{40}{600/75} = 5.$

88. (200) $X_A = \frac{3}{4}, X_B = \frac{1}{4}$

$P_S = P_A^o X_A + P_B^o X_B$

$$500 = P_A^o \times \frac{3}{4} + P_B^o \times \frac{1}{4}$$

$$3P_A^o + P_B^o = 2000 \quad ...(1)$$

Now 1 moles of A is further added,

So $n_A = 4$ mole, and $n_B = 1$ mole

$$X'_A = \frac{4}{5}, X'_B = \frac{1}{5}$$

$$P_S = 520 = P_A^o \times \frac{4}{5} + P_B^o \times \frac{1}{5}$$

$$4P_A^o + P_B^o = 2600 \quad ...(2)$$

From equation (1) and (2) we get,

$P_A^o = 600$ mm Hg, $P_B^o = 200$ mm Hg

89. (23) wt of solute = 30 g
Volume of solution = 100 mL
wt. of solution = 1.2 × 100 = 120 g
wt. of solvent = 120 – 30 = 90 g

(n_1) mole of glucose $= \frac{30}{180} = 0.1666$

(n_2) mole of water $= \frac{90}{18} = 5$

$$\frac{P^\circ - P}{P^\circ} = \frac{n_1}{n_1 + n_2} = \frac{0.1666}{5 + 0.166} = \frac{0.1666}{5.166} = 0.32$$

$$\frac{24 - P}{24} = 0.32 \Rightarrow 24 - P = 0.77 \Rightarrow P = 23.23$$

90. (82) Boiling point is the temperature at which its vapour pressure is equal to the atmospheric pressure (1 atm or 1.013 bar). Therefore Boiling point of solvent is 82°C.
Boiling point of solution is 83°C

91. (555) $P_X = \chi_x P_T = \dfrac{\frac{0.6}{20}}{\frac{0.6}{20} + \frac{0.45}{45}} \times 740$

$P_X = 555$ mm Hg

92. (314) Let V.P. of pure A be $P_A^{\,0}$ Let V.P. of pure B be $P_B^{\,0}$
When $X_A = 0.7$ & $X_B = 0.3$; $P_s = 350$ mm Hg

$\Rightarrow P_A^0 \times 0.7 + P_B^0 \times 0.3 = 350$ ...(i)

When $X_A = 0.2$ & $X_B = 0.8$

$P_s = 410$ mm Hg $\Rightarrow P_A^0 \times 0.2 + P_B^0 \times 0.8 = 410$ ...(ii)
Solving (i) and (ii)
$P^0_A = 314$ mm Hg; $P^0_B = 434$ mm Hg

93. (1) $p = K_H \times x$

$$0.920 \text{ bar} = 46.82 \times 10^3 \text{ bar} \times \frac{\text{mol of } O_2}{\text{mol of } H_2O}$$

$$0.920 = 46.82 \times 10^3 \times \frac{\text{mol of } O_2}{1000/18}$$

$$0.920 = 46.82 \times n_{O_2} \times 18 \Rightarrow n_{O_2} = 1.09 \times 10^{-3}$$

$\Rightarrow$ mmol of $O_2 = 1$

94. (2) mole fraction of substance A in liquid phase = 0.2
$P_A = Y_A \times P_T$
$P_A = 0.5 \times 0.8 = 0.4$
The Raoult's law equation is: $P_A = \left(P_A^\circ\right) \times X_A$

$$\Rightarrow 0.4 \text{ atm} = P_A^\circ \times 0.2 \Rightarrow P_A^\circ = \frac{0.4}{0.2} = 2 \text{ atm}$$

95. (2) 40% Hydrogen contains 40g Hydrogen gas and 60g of Oxygen gas.

$$n_{H_2} = \frac{40}{2} = 20;\ n_{O_2} = \frac{60}{32} = 1.8$$

$$\chi_{H_2} = \frac{n_{H_2}}{n_{Total}} = \frac{20}{20 + 1.8} = \frac{20}{21.8} = 0.92$$

$$p_{H_2} = \chi_{H_2} \cdot P_{Total} = 0.92 \times 2.2 = 2.02 \text{ bar}$$

96. (14) $P_{Total} = P_A^0 x_A + P_B^0 x_B$
$= (50 \times 0.3) + (100 \times 0.7) = 85$ torr
$P_B = y_B \times P_{Total} \Rightarrow 70 = y_B \times 85$

$$y_B = \frac{70}{85} = \frac{14}{17} \Rightarrow x = 14$$

97. (1222) $P = K_H X_{CO_2}$

$$0.835 = 1.67 \times 10^3 \times \frac{n_{CO_2}}{n_{CO_2} + n_{H_2O}}$$

$$\approx 1.67 \times 10^3 \times \frac{n_{CO_2}}{n_{H_2O}} \Rightarrow 0.5 \times 10^{-3} = \frac{n_{CO_2}}{1000/18}$$

$$\Rightarrow n_{CO_2} = 27.775 \times 10^{-3}$$

$$w_{CO_2} = 44 \times 27.775 \times 10^{-3} = 1222.1 \times 10^{-3} \text{ g}$$

98. (25) From Henry's law,
$P_{Gas} = K_H . X_{gas}$

$$X_{gas} = \frac{n_{gas}}{n_{gas} + n_{H_2O}} \approx \frac{n_{gas}}{n_{H_2O}}$$

$$0.835 = 1.67 \times 10^3 \times \frac{n_{(CO_2)}}{\frac{0.9 \times 1000}{18}}$$

$n_{(CO_2)} = 0.025$ mole or 25 m mol

99. (77) $P^o_B = 70$ torr and $P^o_{MB} = 20$ torr
$P_B = P^o_B X_B$ and $P_{MB} = P^o_{MB} X_{MB}$
$P_{total} = P_B + P_{MB}$ $X_B = X_{MB} = 0.5$
[∵ mixture is equimolar]
$\therefore P_{total} = (70 \times 0.5) + (20 \times 0.5) = 45$ torr
Mole fraction of benzene in the vapour phase above the equimolar mixture

$$= \frac{P_B^0 X_B}{P_{total}} = \frac{70 \times 0.5}{45} = 0.77 = 77 \times 10^{-2}$$

100. (1) Given $P_A° = 90$ mm Hg, at 25 °C
$P_B° = 15$ mm Hg

$$\left.\begin{matrix} X_A = 0.6 \\ X_B = 0.4 \end{matrix}\right\} P_T = X_A P_A^\circ + X_B P_B^\circ$$

$= (0.6 \times 90) + (0.4 \times 15) = 54 + 6 = 60$ mm
Now mol fraction of B in the vapour phase

$$= \frac{P_B}{P_T} = \frac{X_B P_B^\circ}{60} = \frac{15 \times 0.4}{60} = 0.1 = 1 \times 10^{-1}$$

101. (1389) $P_{O_2} = KH.\chi_{O_2}$

$$\chi_{O_2} = \frac{20}{8 \times 10^4} = \chi_{O_2} 25 \times 10^{-5}$$

$$\chi O_2 = \frac{n_{O_2}}{n_{O_2} + nH_2O} = 25 \times 10^{-5}$$

$\therefore n_{O_2} = 25 \times 10^{-5} \times nH_2O$ $[nH_2O >> n_{O_2}]$

$$n_{O_2} = 25 \times 10^{-5} \times \frac{1000}{18} = 1388.88 \times 10^{-5} \text{ moles}$$

$$\text{Molar Solubility} = \frac{\text{Moles of } O_2}{\text{Volume of the Solution}}$$

$$= \frac{1388.88 \times 10^{-5}}{1} = 1388.88 \times 10^{-5}$$

102. (19) Using Raoult's law,

$$P_s = \chi_A P_A^o + \chi_B P_B^o$$

$$= \left(\frac{1}{3} \times 21\right) + \left(\frac{2}{3} \times 18\right) = 19 \text{ kPa}$$

103. (600)

$$P_{total} = \left(P^o_{hexane} \cdot X_{hexane}\right) + \left(P^o_{heptane} \cdot X_{heptane}\right)$$

$$550 = \left([P^o_{hexane}] \times \frac{1}{4}\right) + \left([P^o_{heptane}] \times \frac{3}{4}\right)$$

$$\Rightarrow 2200 = P^o_{hexane} + 3P^o_{heptane} \text{ ...(i)}$$

$$560 = \left([P^o_{hexane}] \times \frac{1}{5}\right) + \left([P^o_{heptane}] \times \frac{4}{5}\right)$$

$$\Rightarrow 2800 = P^o_{hexane} + 4P^o_{heptane} \text{ ...(ii)}$$

From equation (i) and (ii),

$P^o_{heptane} = 600$ mm of Hg

104. (3)
(A) K_b for water = 0.512 K Kg mol^{-1}
K_f for water = 1.86 K Kg mol^{-1}
So, $K_b < K_f$ for water.
(B) Since, $K_b < K_f$ so $\Delta T_b < \Delta T_f$ [same equimolal solution]
(C) Osmotic pressure is used to determine molar mass of protiens & polymers
(D) dimerised form of benzoic acid in benzene

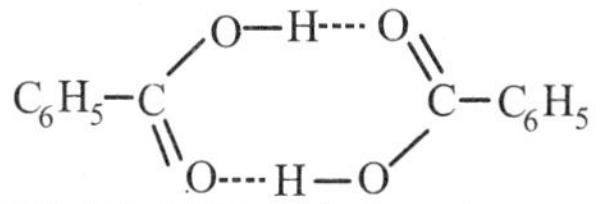

Option (A), (B) & (D) are incorrect.

105. (4) On side (I) $C_1 = \frac{18/180}{0.1} = 1$M

On side (II)

$$C_2 = \frac{30/180}{250/1000} = \frac{2}{3} \text{M};\ \{\pi_1 > \pi_2\}$$

Flow of water from diluted to concentrated solution thus water flow from chamber 2 to chamber 1.
Statement I is false.
Calculation of osmotic pressure,
$\pi = iCRT$; i = 3 for K_2SO_4

$$\text{And } \pi = 3 \times \left(\frac{50 \times 10^{-3}/174}{2}\right) \times 0.083 \times 300$$

$= 0.0107$ bar
Thus statement II is true.

106. (3) For X,

$$K_4[Fe(CN)_6] \longrightarrow 4K^+ + [Fe(CN)_6]^{4-}$$

It contains 5 moles of ions.
So, $\pi = iCRT = 5 \times 0.1 \times RT = 0.5$ RT

For Y, $FeCl_3 \longrightarrow Fe^{+3} + 3Cl^-$

It contains 4 moles of ions.
So, $\pi = iCRT = 4 \times 0.1 \times RT = 0.4$ RT
Hence, solution 'Y' is hypotonic w.r.t. 'X'.

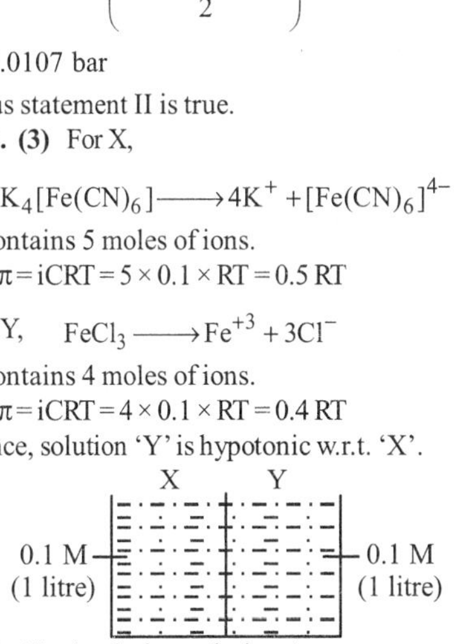

107. (1) Analysing solution A,
$m_A = 1$ g, $V_A = 0.5$ L

$$C_A = \frac{1/50000}{0.5} = 4 \times 10^{-5} \text{ mol/L}$$

$$\pi_A = x = C_A RT = 4 \times 10^{-5} \times 0.083 \times 300$$
$= 9.96 \times 10^{-4}$ bar

Analysing solution B,

$m_B = 2g, V_B = 1\,L$

$$C_B = \frac{2/50000}{1} = 4\times10^{-5}\ \text{mol/L}$$

Since, $C_A = C_B$ and temperature is the same, osmotic pressure is identical :

$\pi_B = y = 9.96\times10^{-4}$ bar

Analysing the mixed solution,

Total mass = 1 g + 2 g = 3 g

Total volume = 0.5 L + 1 L = 1.5 L

$$C_{mix} = \frac{3/50000}{1.5} = 4\times10^{-5}\ \text{mol/L}$$

Since, the final concentration remains identical, the final osmotic pressure (z) is also identical.

$z = \pi_{mix} = 9.96 \times 10^{-4}$ bar.

108. (4) Finding van't Hoff factor

$\Delta T_f = iK_f m$

$$\text{No. of moles} = \frac{19.5}{78} = \frac{1}{4}\ \text{mol}$$

$$1 = i \times 1.86 \times \frac{1/4}{1/2} \Rightarrow i = \frac{2}{1.86}$$

Finding degree of dissociation

$i = 1 + (n-1)\alpha \Rightarrow i = 1 + \alpha$

$$\alpha = \frac{2}{1.86} - 1 = 0.075$$

Calculating K_a, Given C = 0.5

$$K_a = \frac{C\alpha^2}{1-\alpha} = \frac{1}{2}\times\frac{(0.075)^2}{(1-0.075)} = 3\times10^{-3}$$

109. (1) $\Delta T_b = i.k_b.m$

For dilute solutions (M = m)

Molarity, m	m × i
(I) $M_{glucose} = \frac{2.2}{180}\times\frac{1000}{125} = 0.098$	0.098 × 1
(II) $M_{CaCl_2} = \frac{1.9}{111}\times\frac{1000}{250} = 0.068$	0.068 × 3
(III) $M_{urea} = \frac{9}{60}\times\frac{1000}{500} = 0.3$	0.3 × 1
(IV) $M_{Al_2(SO_4)_3} = \frac{20.5}{342}\times\frac{1000}{750} \simeq 0.08$	0.08 × 5

Order of $\Delta T_b = Al_2(SO_4)_3 >$ Urea $> CaCl_2 >$ Glucose

So order of BP $= Al_2(SO_4)_3 >$ Urea $> CaCl_2 >$ Glucose

(IV) (III) (II) (I)

110. (2) Mass of solute 'A' = 0.3 g

$$\text{Moles of solute 'A'} = \frac{0.3g}{60g/mol} = \frac{1}{200}\ \text{mol}$$

Mass of solute 'B' = 0.9 g

$$\text{Moles of solute 'B'} = \frac{0.9g}{180g/mol} = \frac{1}{200}\ \text{mol}$$

Total molarity of all solutes in solution (C)

$$= \frac{2/200}{100}\times1000 = \frac{1}{10}\ \text{M}$$

$$\therefore \pi = CRT = \frac{1}{10\ \text{mol L}^{-1}}\times 0.082\ \text{L atm mol}^{-1}\text{K}^{-1}\times300\ \text{K}$$

$\pi = 2.46$ atm.

111. (4) $(\Delta T_b)_{PQ} = K_b m$

$$1.176 = 5\times\frac{1}{M_1}\times\frac{1000}{50}$$

$M_1 = 85.03$

$$(\Delta T_b)_{PQ2} = 5\times\frac{1}{M_2}\times\frac{1000}{50} = 0.689$$

$M_2 = 145.13$

Let the atomic mass of P be M_P and element Q be M_Q respectively

For PQ,

$M_P + M_Q = 85.03$ ... (1)

For PQ,

$= M_P + 2M_Q = 145.13$ ... (2)

Subtract eq. (1) from the eq. (2) to find M_Q

$M_Q = 60.1$ g/mol

$M_P = 24.93 \approx 25$ g/mol.

For determining MP, take eq. (1)

$M_P + M_Q = 85.03$

$M_P = 85.03 - 60.1 = 24.93$

112. (1) $\Delta T_f = ik_f m$

$$i = \frac{\Delta T_f}{k_f m} \Rightarrow i = \frac{0.20}{1.8\times0.1} = 1.11$$

$$\alpha = \frac{i-1}{n-1}\ \text{for HA, n = 2}\ \therefore\ \alpha = \frac{1.11-1}{1} = 0.11$$

$$K_a = \frac{[H^+][A^-]}{[HA]} = \frac{c\alpha^2}{1-\alpha} = \frac{0.1\times(0.11)^2}{1-0.11} = 1.38\times10^{-3}$$

113. (4) Statement-I: Molar depression constant

$$K_f = \frac{M_1RT_f^2}{\Delta H_{fus}} = \frac{M_1R\cdot T^2}{\Delta S_{fus}\cdot T_f} = \frac{M_1RT_f}{\Delta S_{fus}}$$

Hence statement-I is correct.

Statement-II: K_f for benzene = 5.12 K kg mol^{-1}; K_f for water = 1.86 K kg mol^{-1}.

Hence statement- II is incorrect.

114. (3)

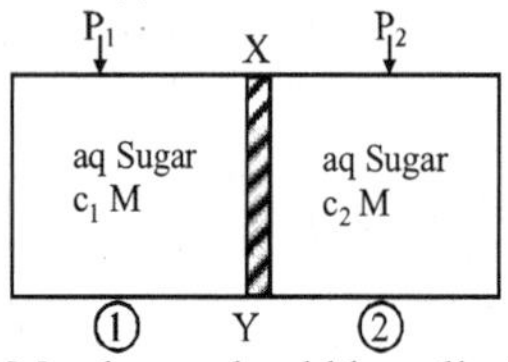

Given $C_2 > C_2$

Membrane should be cellophane or parchment paper

Normal osmosis occurs from (2) to (1)

For reverse osmosis from (1) to (2)

Pressure : $P_1 >$ Osmotic pressure (π)

∴ Correct options are A and C.

115. (2) $\Delta T_b = i_1 m_1 k_b + i_2 m_2 k_b$

$$= 1\times\frac{2}{0.5}\times0.52 + \frac{1\times2}{0.5}\times0.52 = 4.16$$

$(T_b)_{solution} = 373.16 + 4.16 = 377.3$K

116. (2) Salt and ice mixture is colder so it prevents early melting due to depression in freezing point

117. (4) For AX_2:- $\Delta T_b = K_b \times m \times i$

$$0.0156 = 0.52\times\frac{0.01}{1}\times i_{AX_2}$$

$\Rightarrow i_{AX_2} = 3 \Rightarrow$ complete ionisation

For AY_2:- $\Delta T_b = K_b \times m \times i$

$$0.026 = 0.52\times\frac{25.4}{250\times2}\times i_{AY_2}$$

$\Rightarrow i_{AY_2} = 0.98 \simeq 1$ ∴ completely unionised

118. (N)

NTA Ans. (4) Living cell = 0.9 gm in 100 gm of solution % w/w = 0.9

Solution has equal moles of glucose and water = 0.5 each Weight of solution = 0.5 × 180 + 0.5 × 18 = 99 g ⇒ w/w % = (90/99) × 100 = 90.9%

Concentrated solution ⇒ Cell will shrink.

119. (1) $\Delta T_f = K_b.m$

$$0.4 = K_b\frac{\frac{1}{256}}{50\times10^{-3}} \Rightarrow K_b = 5.12\ \text{Kg/mol}$$

120. (3) A solution will freeze when its vapour pressure equal the vapour pressure of the pure solid solvent.

On adding non-volatile solute in a solvent, the freezing point of solution decreases.

$T_f < T_f^0$ ∴ $\Delta T_f = T_f^0 - T_f$

F.P. of solution < F.P. of pure solvent

Also V.P. of solution decreases on adding non-volatile solute in a solvent.

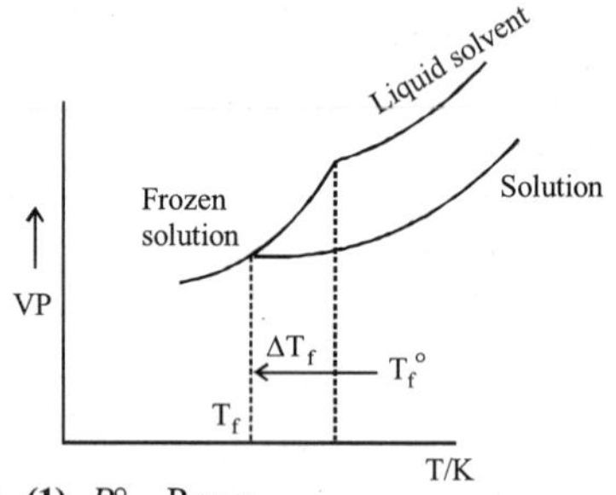

121. (1) $P^\circ - P \propto x_{solute}$

$P^\circ - P = 10$ mm Hg and $x_{solute} = 0.2$

If $P^\circ - P = 20$ mm Hg, then $x_{solute} = 0.4$

$\therefore x_{solvent} = 1 - x_{solute} = 1 - 0.4 = 0.6$

122. (3) Given : $\Delta T_f = 0.558°C$, $k_f = 1.86\ \frac{K\times kg}{mol}$,

Molality = 0.1 m $\Rightarrow \Delta T_f = i\times k_f\times m$

$\Rightarrow 0.558 = i\times1.86\times0.1 \Rightarrow i = 3$

When 100% ionisation takes place

$$[Cr(NH_3)_5Cl]Cl_2 \xrightarrow{\text{aqueous solution}} [Cr(NH_3)_5Cl]^{2+} + 2Cl^-$$

number of ions = 3

123. (1) $\Delta T_b \propto$ i.C. Where C = concentration and i = van't Hoff factor.

(i) 2×10^{-4} (ii) 1×10^{-4}

(iii) 2×10^{-3} (iv) 2×10^{-2}

B.P. order:

(ii) < (i) < (iii) < (iv)

124. (4) Only solvent Molecules are allowed to pass through the SPM. Water moves from $K_2Cr_2O_7$ to $CuSO_4$ solution causing molarity of $CuSO_4$ to decrease.

125. (1) A = 0.1 M, B = 0.01 M, C = 0.001 M

Value of i increases as solution becomes dilute so the correct order is: $i_A < i_B < i_C$

126. (4) Depression in freezing point of benzene takes place.

127. (1)

Freezing point depression (Colligative property)

$\propto$ number of solute particles (Van't Hoff factor, i)

Acetic acid dissociates in water so 2i will be the van't Hoff factor. Acetic acid and benzoic acid will dimense in benzene so $\frac{i}{\alpha}$ will be the actual van't Hoff factor.

Glucose does not dissociate in water so i will be the actual van't Hoff factor.

Thus, Acetic acid in water will have the highest van't Hoff factor and thus highest depression in the freezing point.

128. (4) $\frac{P^o - P_s}{P^o} = \frac{n}{N}$ (for dilute solution)

$\frac{0.2}{54.2} = \frac{n \times 18}{100} \Rightarrow n = \frac{100}{271 \times 18}$

$w = \frac{100 \times 180}{271 \times 18} \Rightarrow w = 3.69$ g

129. (4) $\Delta T_b \propto i \times C$

Osmosis always takes place from hypotonic to hypertonic solution.

When KI solutions is added to $AgNO_3$ solution, positively charged solution results due to adsorption of Ag^+ ions from dispersion medium.

130. (1) A-III, B-I, C-II, D-IV

131. (1) Vapour pressure of solvent is greater than vapour pressure of solution.

Only solvent molecules is solidify at the freezing point.

132. (2) $\Delta_A T_b = \frac{2}{M_A \times 98} \times 1000 \times K_b$

$\Delta_B T_b = \frac{8}{M_B \times 92} \times 1000 \times K_b$

Given, $\Delta_A T_b = \Delta_B T_b$

$\Rightarrow \frac{2}{98 \times M_A} = \frac{8}{92 \times M_B} \Rightarrow M_B = 4\text{–}261\ M_A$

133. (3) $\Delta T_f = 0.0405$ K

Concentration of formic acid (HCOOH) solution = 0.5 ml/L

$= 0.5 \times 1.05$ g/L $= 0.525$ g ($\because d_{HCOOH} = 1.05$ g/ml)

No. of moles of HCOOH in 1 L solution

$= \frac{0.525}{46} = 0.0114 (\because Mw_{HCOOH} = 46)$

Molality (m) of HCOOH = 0.0114 mol/kg (considering d_{H2O} = 1kg/L)

$\Delta T_f = i K_f m$

$\therefore i = \frac{\Delta T_f}{K_f m} = \frac{0.0405}{1.86 \times 0.0114} = 1.91 \approx 1.9$

$\therefore$ The van't Hoff factor of the formic acid solution is nearly 1.9.

134. (2) Depression in freezing point is a colligative property and depends only on the amount of solute.

$\Delta T_f = k_f . m$

$$\frac{\Delta T_{fx}}{\Delta T_{fy}} = \frac{k_f \cdot m_x}{k_f \cdot m_y} = \frac{\frac{1/M_x}{1}}{\frac{1/M_y}{1}} \Rightarrow \frac{1}{4} = \frac{M_y}{M_x}$$

$\Rightarrow M_x : M_y = 1 : 0.25$

135. (4) $\Delta T_f = i \times K_f \times m$

$0.2 = i \times 1.86 \times \frac{0.7}{93} \times \frac{1000}{42} \Rightarrow i = 0.60$

$$\begin{array}{lll} 2A & \rightleftharpoons & A_2 \\ 1-\alpha & & \frac{\alpha}{2} \end{array}$$

$i = 1 - \alpha + \frac{\alpha}{2} \Rightarrow 1 - \frac{\alpha}{2} = 0.60 \Rightarrow \alpha = 0.80$

$\Rightarrow$% association = 80%

136. (3) Relative lowering in vapour pressure

$= \frac{P - P_s}{P} = \frac{n}{n+N}$.

$n \rightarrow$ moles of solute; $N \rightarrow$ moles of solvent

RLVP $\propto n$ or RLVP $\propto \frac{1}{\text{Mol. Mass}}$

$n_A = \frac{10}{100}, n_B = \frac{10}{200}, n_C = \frac{10}{10000}$

From the above relation,

RLVP(A) > RLVP(B) > RLVP(C)

137. (3) Osmotic pressure (π) = CRT

Since, there are two solutes i.e., urea and glucose.

$\therefore \pi = (C_1 + C_1)RT$

$= \left(\frac{0.6}{60 \times 0.1} + \frac{1.8}{180 \times 0.1}\right) \times 0.0821 \times 300$

$= 0.2 \times 0.0821 \times 300 = 4.926$ atm

138. (3) Relative lowering of vapour pressure, is given by,

$\frac{p^\circ - p}{p^\circ} = x_A = \frac{n_A}{n_A + n_B} \simeq \frac{n_A}{n_B}$

Given, $p^\circ = 35$ mmHg, $n_{urea} = \frac{0.60}{60}, n_{water} = \frac{360}{18}$

$\frac{p^\circ - p}{35} = \frac{0.6 \times 18}{60 \times 360} = \frac{1}{2000}$

$\Rightarrow \Delta p = p^\circ - p = 0.017$

139. (3) Ebullioscopic constant (molal elevation const.) is given by,

$K_b = \frac{\Delta T_b}{m}$ and it is given that, $\frac{(K_b)_A}{(K_b)_B} = \frac{1}{5}$

$\therefore \frac{(\Delta T_b)_A}{(\Delta T_b)_B} = \frac{(k_b)_A}{(k_b)_B} = \frac{1}{5}$

140. (2) We know, $\pi = iCRT$; $\pi_{XY} = 4\pi_{BaCl_2}$

Since both are ionic solute, i.e.,

$XY \rightarrow X^+ + Y^- \Rightarrow i = 2$

$BaCl_2 \rightarrow Ba^{2+} + 2Cl^- \Rightarrow i = 3$

Now, 2[XY]RT = 4 × 3 × [0.01]RT

$[XY] = 0.06 = 6 \times 10^{-2}$ mol/L

141. (4) Dissociation of potassium sulphate (K_2SO_4),

$[K_2SO_4 \longrightarrow 2K^+ + SO_4^{2-}]$

i (van't Hoff factor) = 3, We know that, $\Delta T_f = iK_f m$ where, K_f is molal depression constant and m is molality.

$= 3 \times 4 \times 0.03 = 0.36$ K

142. (1) $(\Delta T_f)_X = (\Delta T_f)_Y$

$K_f . m_X = K_f . m_Y \Rightarrow m_X = m_Y$

$\frac{\text{No. of moles of } X}{\text{wt. of solvent in kg}} = \frac{\text{No. of moles of } Y}{\text{wt. of solvent in kg}}$

Given, 4% aqueous solution of X = 4 g of solute X present in 100 g of H_2O.

12% aqueous solution of Y = 12 g of solute Y present in 100 g of H_2O.

Now, $\frac{4\text{ g}}{m} \times \frac{1000}{100} = \frac{12\text{ g}}{m} \times \frac{1000}{100}$

$\frac{4}{A} = \frac{12}{m}$ where $[m = A]$

$m = 3A$

143. (1) $2C_6H_5COOH \longrightarrow (C_6H_5COOH)_2$

Initial	1	0
Final	$1-\alpha$	$\frac{\alpha}{2}$

Moles at equilibrium $= 1 - \alpha + \frac{\alpha}{2} = 1 - \frac{\alpha}{2}$

$i = \frac{\text{Total moles at equilibrium}}{\text{Initial moles}} = \frac{1 - \frac{\alpha}{2}}{1}$

$i = 1 - \frac{0.8}{2} = 0.6 \Rightarrow \Delta T_f = iK_f m$

$2 = 0.6 \times 5 \times \left(\frac{\frac{w}{122}}{30}\right) \times 1000 \Rightarrow w = 2.4$ g

144. (2) Freezing point of diluted milk = –0.2 °C

$\Rightarrow \Delta T'_f = 0.2$ °C

Freezing point of pure milk = –0.5 °C

$\Rightarrow \Delta T_f = 0.5$ °C

$\frac{\Delta T_f}{\Delta T'_f} = \frac{K_f \times m}{K_f \times m'}$; $m = \frac{\text{mole of solute}}{\text{mass of solvent (kg)}}$

Moles of solute are same in both samples.

$\therefore \frac{0.5}{0.2} = \frac{W'}{W} \Rightarrow \frac{W'}{W} = \frac{5}{2} \Rightarrow W' = \frac{5}{2}W$

2 cups of pure milk is mixed with 3 cups of water to make 5 cups of diluted milk.

145. (2) $K_2HgI_4 \rightleftharpoons 2K^+ + [HgI_4]^{2-}$; $n = 3$

$\because \alpha = \frac{i-1}{n-1} \Rightarrow 0.4 = \frac{i-1}{3-1} \Rightarrow i = 1.8$

146. (4) According to the question we can write,

$\Delta T_b = K_b m \Rightarrow K_b(1) = 2 \Rightarrow K_b = 2$ Km^{-1}

$\Delta T_f = K_f m \Rightarrow K_f(2) = 2 \Rightarrow K_f = 1$ Km^{-1}

So, $\frac{K_b}{K_f} = \frac{2}{1} \Rightarrow K_b = 2K_f$

147. (3) As we know, $\Delta T_f = K_f . m$

$\Rightarrow 10 = 1.86 \times \frac{62 \times 1000}{62 \times w}$

$\Rightarrow w = \frac{1.86 \times 1000}{10} = 186$ g

Total amount of water = 250 g

$\therefore$ The amount of water separated as ice = 250 – 186 = 64 g

148. (4)

		Number of particles (i)
(1)	$[Co(H_2O)_6]Cl_3$	4
(2)	$[Co(H_2O)_5Cl]Cl_2 \cdot H_2O$	3
(3)	$[Co(H_2O)_4Cl_2]Cl \cdot 2H_2O$	2
(4)	$[Co(H_2O)_3Cl_3] \cdot 3H_2O$	1

Remember, greater the no. of particles, lower will be the freezing point. Compound (d) will have the highest freezing point due to least number of particles.

149. (3) Molar mass of octane = 114 g/mol
From the lowering of vapour pressure we have,

$$\frac{\Delta P}{P} = \frac{\frac{W_2}{M_2}}{\frac{W_2}{M_2} + \frac{W_1}{M_1}} \Rightarrow \frac{75}{100} = \frac{\frac{W_2}{50\text{ g/mol}}}{\frac{W_2}{50\text{ g/mol}} + \frac{114\text{ g}}{114\text{ g/mol}}}$$

$W_2 = 150$ g

150. (1) The relationship between molar masses of the two solvents is

$$M_X = \frac{3}{4} M_Y \quad \ldots (i)$$

The relative lowering of vapour pressure of the two solutions is

$$\left(\frac{\Delta P}{P}\right)_X = m\left(\frac{\Delta P}{P}\right)_Y$$

But, the relative lowering of vapour pressure of solutions is directly proportional to the mole fraction of solute.
Given 5 molal solution, means 5 moles of solute are dissolved in 1 kg (or 1000 g) of solvent.

The number of moles of solvent $= \frac{1000\text{ g}}{M}$

The mole fraction of solute

$$= \frac{5}{1000/M} = M \times \frac{5}{1000}$$

Hence, $M_X \times \frac{5}{1000} = m \times M_Y \times \frac{5}{1000} \quad \ldots (ii)$

Substitute equation (i) in equation (ii),

$$\frac{3}{4} \times M_Y \times \frac{5}{1000} = m \times M_Y \times \frac{5}{1000} \Rightarrow m = \frac{3}{4}$$

151. (3) $Na_2SO_4 \rightleftharpoons 2Na^+ + SO_4^{2-}$

$1-\alpha \quad 2\alpha \quad \alpha$

$i = 1 - \alpha + 2\alpha + \alpha = 1 + 2\alpha$

Given, $\alpha = 81.5\% \Rightarrow i = 1 + (2 \times 0.815)$

$\Rightarrow i = 2.63$

$\Delta T_f = i \times K_f \times m$

$$\Rightarrow 3.82 = 2.63 \times 1.86 \times \frac{5/142}{x/1000} \Rightarrow x = 45.$$

152. (4) In benzene,

$2CH_3COOH \rightleftharpoons (CH_3COOH)_2$

$1-\alpha \quad \alpha/2$

$i = 1 - \alpha + \alpha/2 = 1 - \alpha/2$

Here α is degree of association

$\Delta T_f = i \times K_f \times m$

$$0.45 = \left(1 - \frac{\alpha}{2}\right)(5.12)\frac{\left(\frac{0.2}{60}\right)}{\frac{20}{1000}}$$

$$1 - \frac{\alpha}{2} = 0.527 \Rightarrow \alpha = 0.945$$

% degree of association = 94.6%

153. (1) van't Hoff factor (i)

$$= \frac{\text{No. of particles after association or dissociation}}{\text{No. of particles before association or dissociation}}$$

For $MX_2 \rightleftharpoons M^{2+} + 2X^-$

Initial	1	0	0
Final	$1-\alpha$	α	2α

Total no. of particles $= 1 - \alpha + \alpha + 2\alpha = 1 + 2\alpha$

$$\therefore i = \frac{1+2\alpha}{1} = 2 \Rightarrow \alpha = 0.50 \text{ or } 50\%.$$

154. (1) According to Raoult's Law

$$\frac{P^\circ - P_s}{P_s^\circ} = \frac{W_B \times M_A}{M_B \times W_A} \quad \ldots (i)$$

Here P° = Vapour pressure of pure solvent,
P_s = Vapour pressure of solution
W_B = Mass of solute, W_A = Mass of solvent
M_B = Molar mass of solute, M_A = Molar Mass of solvent
Vapour pressure of pure water at 100 °C (by assumption = 760 torr)
By substituting values in equation (i) we get,

$$\frac{760 - P_s}{P_s^\circ} = \frac{18 \times 18}{180 \times 178.2} \quad \ldots (ii)$$

On solving (ii) we get, $P_s = 752.4$ torr

155. (4) Using relation, $\frac{P^\circ - P_s}{P^\circ} = \frac{w_2 M_1}{w_1 M_2}$

where w_1, M_1 = mass in g and mol. mass of solvent
w_2, M_2 = mass in g and mol. mass of solute
$P^\circ = 185$ torr, $P_s = 183$ torr

$$\frac{185 - 183}{185} = \frac{1.2 \times 58}{100 \times M_2}$$

(Mol. mass of acetone = 58)
$M_2 = 64.68 \approx 64$
$\therefore$ Molar mass of substance = 64

156. (4) Acetic acid contain carboxylic group – COOH which can form H-bonding, so acetic acid dimerises.

157. (1) $\pi = iCRT$

$\pi_{C_2H_5OH} = 1 \times 0.500 \times R \times T = 0.5RT$

$\pi_{Mg_3(PO_4)_2} = 5 \times 0.100 \times R \times T = 0.5RT$

$\pi_{KBr} = 2 \times 0.250 \times R \times T = 0.5RT$

$\pi_{Na_3PO_4} = 4 \times 0.125 \times RT = 0.5RT$

Hence, the osmotic pressures of all the given solutions are equal.

158. (1) Given, $\pi_{ob} = 10.8$ atm
$\pi_{nor} = CRT = 0.10 \times 0.082 \times 298 = 2.446$ atm
Now experimental value of (i)

$$= \frac{\text{Observed osmotic pressure}}{\text{Normal osmotic pressure}} = \frac{10.8}{2.446} = 4.42$$

$Fe(NH_4)_2(SO_4)_2 \rightleftharpoons Fe^{2+} + 2NH_4^+ + 2SO_4^{2-}$
Expected $i = 5$.

159. (3) $\frac{P^\circ - P_s}{P^\circ} = \frac{W_B \times M_A}{M_B \times W_A}$

$$0.1 = \frac{12}{M_B} \times \frac{18}{108} \Rightarrow M_B = \frac{12 \times 18}{0.1 \times 108} = 20$$

160. (3) $nM \rightleftharpoons (M)_n$

initial	1	0
final	$1-\alpha$	α/n

$$i = \frac{(1-\alpha) + (\alpha/n)}{1}$$

$\Rightarrow 0.9 = (1 - 0.2) + (0.2/n) \Rightarrow n = 2.$

161. (3) $\Delta T_f = K_f m$
where m = molality

$$\Delta T_f = T_f - T_f' = K_f \times m = K_f \times \frac{w}{M \times \text{Mass or Volume}}$$

Exact frazing point of water.

$$273.15 - 268 = 1.86 \times \frac{w}{M \times V}$$

$$5.15 = 1.86 \times \frac{w}{32 \times 10} \Rightarrow w = \frac{5.15 \times 32 \times 10}{1.86} =$$

$886.02 \approx 886.02$ g

162. (2) $\Delta T_f = i \times K_f \times m$
Given $\Delta T_f = 2.8$, $K_f = 1.86$ K kg mol^{-1}, $i = 1$
(ethylene glygol is a non- electrolyte)
Wt. of solvent = 1 kg; Let of wt of solute = x
Mol. wt of ethylene glycol = 62

$$2.8 = 1 \times 1.86 \times \frac{x}{62 \times 1} \Rightarrow x = \frac{2.8 \times 62}{1.86} = 93 \text{ g}$$

163. (2) $\text{Mol.wt.}(M_B) = \frac{K_f \times W_B \times 1000}{\Delta T_f \times W_A}$

$$= \frac{1.86 \times 0.85 \times 1000}{0.23 \times 125} \approx 55 \text{ g}$$

Where, $W_B = 0.85$ g; $W_A = 125$ g
$\Delta T_f = 0\ °C - (-23\ °C) = 23\ °C$

Now, $i = \frac{M_{\text{normal}}}{M_{\text{observed}}} = \frac{136.3}{55} = 2.47$

$ZnCl_2 \rightleftharpoons Zn^{2+} + 2Cl^-$

$1-\alpha \quad \alpha \quad 2\alpha$

van't Hoff factor (i) $= \frac{1 - \alpha + \alpha + 2\alpha}{1} = 2.47$

$\Rightarrow \alpha = 0.735 = 73.5\%$

164. (4) $\Delta T_f = i \times K_f \times m$

$$i = \frac{\Delta T_f}{K_f \times m} = \frac{1.91}{1.86 \times 1} = 1.027$$

For $HF \rightleftharpoons H^+ + F^-$
$(1-\alpha) \quad \alpha \quad \alpha$
$1 - \alpha + \alpha + \alpha = i = 1.027$
$\Rightarrow \alpha = 0.027$ or 2.7%

165. (2) For isotonic solutions
$\pi_1 = \pi_2$

$$\frac{5/342}{0.1} = \frac{1/M}{0.1} \Rightarrow M = \frac{342}{5} = 68.4 \text{ g/mol}$$

166. (2) Sodium sulphate dissociates as

$Na_2SO_4(s) \longrightarrow 2Na^+ + SO_4^{2-}$

Hence van't Hoff factor, i = 3

Now $\Delta T_f = iK_f.m = 3 \times 1.86 \times 0.01 = 0.0558$ K

167. (1) Note: On addition of glucose to water, vapour pressure of water will decrease. The vapour pressure of a solution of glucose in water can be calculated using the relation

$$\frac{P^\circ - P_s}{P^\circ} = \frac{\text{Moles of glucose in solution}}{\text{Moles of water in solution}}$$

or $\frac{17.5 - P_s}{17.5} = \frac{18/180}{178.2/18} \quad [\because P^\circ = 17.5]$

or $17.5 - P_s = \frac{0.1 \times 17.5}{9.9}$ or $P_s = 17.325$ mm Hg.

168. (1) Osmotic pressure (π) of isotonic solutions are equal.
For solution of unknown substance, ($\pi = CRT$)

$$C_1 = \frac{5.25/M}{V}$$

For solution of urea, $C_2 = \frac{1.5/60}{V}$

Given, $\pi_1 = \pi_2 \because \pi = CRT$

$\therefore\ C_1RT = C_2RT$ or $C_1 = C_2$ or $\frac{5.25/M}{V} = \frac{1.5/60}{V}$

$\therefore\ M = 210$ g/mol

169. (3)

170. (4) $\Delta T_f = i \times K_f \times m; \Delta T_f = 1.85 \times 0.2 \times 1.3 = 0.480°C$

$(HX \rightleftharpoons H^+ + X^-,\ i = 1.3)$
$1-0.3 \quad 0.3 \quad 0.3$

$\Delta T_f = T_f^\circ - T_f$

$\therefore\ T_f = 0 - 0.480°C = -0.480°C$

171. (4) $\Delta T_f = i \times K_f \times m$. Since K_f has different values for different solvents, even if m is same, ΔT_f will be different.

172. (1) On increasing pressure, the temperature is also increased. Thus in pressure cooker due to increase in pressure the b.p. of water increases.

173. (2)

$\Delta T_b = K_b \times m \times 1000;\ \Delta T_f = K_f \times m \times 1000;$

$\frac{\Delta T_b}{\Delta T_f} = \frac{K_b}{K_f} \Rightarrow \frac{\Delta T_b}{0.186} = \frac{0.512}{1.86} \Rightarrow \Delta T_b = 0.0512°C.$

174. (62) Osmotic pressure, $\pi = \rho gh$
$= 1000 \times 10 \times 80 \times 10^{-3} = 800$ Pa

$\pi = CRT = \frac{n}{V}RT = \frac{W}{MV}RT,\ M = \frac{W}{\pi V}RT$

Now putting the values in the formula :

$M = \frac{20 \times 8.3 \times 300}{0.8 \times 1} = 62250 \text{ g/mol} = 62.25 \text{ Kg/mol} \approx 62.$

175. (3) Mole of Y $- \frac{150}{300} - 0.5$

$\Delta T_b = i \times K_b \times m$

$0.5 = i \times m \times 5$

$i \times m = \frac{0.5}{5} = 0.1$

$i \times y = \frac{15}{1000}$ (where y = moles of solute)

Now,

$\frac{P_o - P_s}{P^o} = i$

$X_{solute} = i \times \frac{y}{y + \frac{15}{300}}$

$= i \times \frac{y}{\frac{1}{2}} = \frac{\frac{15}{1000}}{\frac{1}{2}} = \frac{30}{1000} = 3 \times 10^{-2}$

176. (15) $\pi = iCRT$

$12 - 2 \times C \times 0.08 \times 300 \Rightarrow 12 = 2 \times C \times 24$

$C = \frac{1}{4}$ mol/L

Then g/L strength of NaCl solution

$= \frac{1}{4} \times 58.5$ g/L $= 14.625$ g/L $= 15$ g/L

177. (25) For AB

$\Delta T_b = 2.7, K_b = 0.5$ K kg mol^{-1}

$m = \frac{\Delta T_b}{K_b} = 5.4m$

For AB_2, $\Delta T_b = 1.5$ K, $m = \frac{1.5}{0.5} = 3m$

Moles of AB = 0.081 mol

Moles of AB_2 = 0.045 mol

Molar mass of AB $= \frac{\text{mass}}{\text{moles}} = \frac{1}{0.081} = 12.35$ g/mol

Molar mass of $AB_2 = \frac{1}{0.045} = 22.22$ g/mol

Let atomic mass of A = a, B = b

$\therefore\ a + b = 12.35 \Rightarrow a + 2b = 22.22$

On solving we get $a = 2.48 = 24.8 \times 10^{-1}$

178. (33) $MX_3 \to M^{3+} + 3X^-$

$i = 1 + (n-1)\alpha \Rightarrow 2 = 1 + (4-1)\alpha \Rightarrow 2 - 1 = 3\alpha$

$\Rightarrow \alpha = \frac{1}{3} = 33.33\% \approx 33\%$

179. (16) $A_2B \to 2A^+ + B^{-2}; n = 3$

$\alpha = 0.3; i = 1 + (n-1)\alpha$

$= 1 + (3-1)(0.3) = 1.6 = 16 \times 10^{-1}$

180. (75) $MX_2 \to M^{+2} + 2X^-$

$n = 3$

$i = \frac{\text{normal molar mass}}{\text{abnormal molar mass}};\ i = \frac{164}{65.6}$

$1 + (n-1)\alpha = \frac{164}{65.6};\ 1 + (3-1)\alpha = \frac{164}{65.6}$

$2\alpha = \frac{98.4}{65.6};\ \alpha = 0.75$

Percent dissociation = 75%

181. (23) $X_{\text{methylbenzene}} = 0.5$

$Y_{\text{methylbenzene}} = \frac{P_{\text{methylbenzene}}}{P_{\text{total}}}$

$= \frac{X_{\text{methylbenzene}} \times P^\circ_{\text{methylbenzene}}}{P_{\text{total}}}$

$= \frac{0.5 \times 24}{(0.5 \times 80) + (0.5 \times 24)}$

$= \frac{12}{40 + 12} = 0.23 = 23 \times 10^{-2}$

182. (5) Ionisation of AB_2 occurs as:

$AB_2 \to A^{2+} + 2B^-$
$1-\alpha \quad \alpha \quad 2\alpha$

$i = 1 + 2\alpha$

$\Delta T_b = k_b\, im \Rightarrow 0.52 = 0.52\,(1 + 2\alpha)\frac{10}{20}$

$1 = (1 + 2\alpha)\frac{10}{20} \Rightarrow 2 = 1 + 2\alpha \Rightarrow \alpha = 0.5$

Hence, $\alpha = 5 \times 10^{-1}$

183. (76) $HX \rightleftharpoons H^+ + X^-;\ K_a = 1.2 \times 10^{-5}$

Initial conc. 0.03M

Conc. at eq. 0.03–x x x

$K_a = 1.2 \times 10^{-5} = \frac{x^2}{0.03 - x}$

$0.03 - x \approx 0.03$ (K_a is very small)

$\Rightarrow \frac{x^2}{0.03} = 1.2 \times 10^{-5} \Rightarrow x = 6 \times 10^{-4}$

Conc. of final solution $= 0.03 - x + x + x$
$= 0.03 + x = 0.03 + 6 \times 10^{-4}$

osmotic pressure, $\pi = CRT$

$\pi = (0.03 + (6 \times 10^{-4})) \times 0.083 \times 300$
$= 76.19 \times 10^{-2} \approx 76 \times 10^{-2}$ bar

184. (543) $\Delta T_f = 2.5°C$

$\Delta T_f = i \times k_f \times m$

$2.5 = 1 \times 1.86 \times \frac{n_B}{0.1}$

$n_B = \frac{2.5 \times 0.1}{1.86} = 0.1344$ mol

mass of methanol $= 0.1344 \times 32 = 4.3$ g

$d = \frac{m}{v} \Rightarrow v = \frac{m}{d}$

$v = \frac{4.3}{0.792}$ mL $= 5.43$ mL $= 543 \times 10^{-2}$ mL

$x = 543$

185. (25)

$NaCl \longrightarrow Na^+ + Cl^-$
0.05M 0.05M 0.05M

Total $C_1 = 0.05 + 0.05 = 0.1$ M (NaCl)

$C_2 = 0.2$ M (glucose)

$\pi = (C_2 - C_1)RT$

$= (0.2 - 0.1) \times 0.083 \times 300 = 2.49$ bar

$= 24.9 \times 10^{-1}$ bar

186. (19) $CH_3COOH(aq) \rightleftharpoons CH_3COO^-(aq) + H^+(aq)$

$K_a = 6.25 \times 10^{-5}$

mass of acetic acid $= 1.2 \times 5$ g

concentration $= \frac{1.2 \times 5}{60 \times 1}$ M

$\alpha = \sqrt{\frac{K_a}{c}} = \sqrt{\frac{6.25 \times 10^{-5}}{(1.2 \times 5)/60}}$

$= \sqrt{625 \times 10^{-6}} = 25 \times 10^{-3}$

$i = \alpha n + (1 - \alpha) = 1 - \alpha + n\alpha$

$= 1 + (n-1)\alpha$

$= 1 + (2-1)\,0.025 = 1.025$

$\Delta T_f = 1.025 \times 1.86 \times 0.1 = 0.19065$

$= -19.065 \times 10^{-2} = 19$

187. (707) $\Delta T_b = k_b \times m$

$\Rightarrow 2 = 0.52 \times m;\ m = \frac{2}{0.52}$

$\frac{\Delta P}{P^\circ} = \frac{n_{solute}}{n_{solvent}} \Rightarrow \frac{\Delta P}{P^\circ} = \frac{m}{1000} \times M_{solvent}$

$\Delta P = P^\circ \times \frac{m}{1000} \times M_{solvent} = 760 \times \frac{\frac{2}{0.52}}{1000} \times 18 = 52.615$

$P^0 = 760 - 52.615 = 707.385$ mm of Hg

188. (31) As moles of water > moles of CH_3COOH

Molality of acetic acid

$= \frac{2700}{60} \times \frac{1}{2.7} = 16.66$

$\Delta T_f = k_f \times 16.66 = 1.86 \times 16.66 = 31K.$

189. (15) Given: T = (–24)°C

$K_f = 1.86$ K kgmol^{-1}

$M_2 = 62$ g/mol $= 0.062$ Kg/mol^{-1}

$W_1 = 18.6$ Kg

$W_2 = ?$

$\Delta T_f = T^\circ - T = (273 - 249)$ K $= 24$ K

$\therefore \Delta T_f = \frac{K_f \times W_2}{M_2\, W_1} \Rightarrow W_2 = \frac{\Delta T_f \times M_2 W_1}{K_f}$

$= \frac{24 \times 0.062 \times 18.6}{1.86} \Rightarrow W_2 = 14.88$ Kg ≈ 15 Kg

190. (72.56) Osmotic pressure $\pi_1 = 7 \times 10^5$ Pa

Temperature $T_1 = 273$ K

Osmotic pressure at 283 K, $\pi_2 = ?$

Osmotic pressure is a colligative property.

$\therefore \pi_1 = C RT_1 ; \pi_2 = C RT_2$

$$\frac{\pi_1}{\pi_2} = \frac{T_1}{T_2} \Rightarrow \frac{7\times10^5}{\pi_2} = \frac{273}{283} \Rightarrow \pi_2$$

$$= \frac{283}{273}\times 7\times10^5 = 7.25\times10^5 = 72.56\times10^4 \text{Nm}^{-2}$$

191. (240) For Isotonic solutions,

$\pi_1 = \pi_2 \Rightarrow C_1 = C_2$

$\frac{12}{x} = 0.05 \Rightarrow x = 240$ [$x \rightarrow$ Molar Mass of A]

192. (116) Amount of solvent

$= 100 - (29.25 + 19) = 51.75$ g

$$\Delta T_b = \left[\frac{2\times29.25\times1000}{58.5\times51.75} + \frac{3\times19\times1000}{95\times51.75}\right]\times 0.52$$

(For NaCl and $MgCl_2$)

$\Delta T_b = 16.075$

$\Delta T_b = (T_b)_{solution} - (T_b)_{solvent}$

$\Rightarrow (T_b)_{solution} = 100 + 16.07 = 116.07°C$

193. (48) $MgCl_2 \rightarrow Mg^{+2} + 2Cl^-$

$1-\alpha \quad \alpha \quad 2\alpha$

$i = 1 + 2\alpha\ (\alpha = 0.8) \Rightarrow i = 2.6$

$$\frac{\Delta p}{p^o} = \frac{i\times n_2}{n_1} \Rightarrow \Delta p = 2.34$$

$p_s = 47.66 \approx 48$

194. (75) $\pi_{K_2SO_4} = \pi_{Glucose}$

(Isotonic solutions)

$i\times0.004\times RT = 0.01\times RT \Rightarrow i = 2.5$

For K_2SO_4 {for dissociation $i = 1 + (n-1)\alpha$}

$$(\alpha) = \frac{i-1}{n-1} = \frac{2.5-1}{3-1} = 0.75\ \%\ \text{dissociation}$$

195. (30) $i = 1 + \alpha$ (for HA) $= 1.3$

$$\%\text{increase} = \frac{(\Delta T_f)_{obs} - (\Delta T_f)_{cal}}{(\Delta T_f)_{cal}}\times100$$

$$= \frac{(K_f\times i\times m) - (K_f\times m)}{K_f\times m}\times100$$

$$= \frac{i-1}{1}\times100 = 30\%$$

196. (40535) $\because \pi = CRT$

$$\pi = \frac{n}{V}\frac{RT}{M} \Rightarrow \pi = \frac{w}{V}\frac{RT}{M}$$

$$\Rightarrow M = \frac{wRT}{\pi\times V}\quad M = \frac{0.63\times0.083\times300}{1.29\times10^{-3}\times300\times10^{-3}}$$

$M = 40535$ g/mol

197. (8) $\frac{(T_B)_x}{(T_B)_y} = \frac{2}{1}; \frac{(\Delta H)_x}{(\Delta H)_y} = \frac{1}{2}$(given)

$$\frac{(\Delta T_B)_x}{(\Delta T_B)_y} = \frac{(K_B)_x\times\text{molality}}{(K_B)_y\times\text{molality}}$$

$(K_B)_x/(K_B)_y = m$

$$K_B = \frac{R\times M_1\times T_b^2}{1000\times\Delta_{vap}H} \Rightarrow K_B\ \alpha\ \frac{T_b^2}{\Delta_{vap}H}$$

$$\therefore \frac{(K_B)_x}{(K_B)_y} = \frac{(T_b)_x^2}{(T_b)_y^2}\times\frac{\Delta H_y}{\Delta H_x} = m$$

$$\therefore m = \frac{2^2}{1^2}\times\frac{2}{1} = 4\times2 = 8$$

198. (1111) $\frac{p^o - p}{p^o} = \frac{n_{solute}}{n_{solute} + n_{solvent}}$

$$\Rightarrow \frac{100-75}{100} = \frac{\frac{x}{60}}{\frac{x}{60}+\frac{1000}{18}} = \frac{1}{4} = \frac{\frac{x}{60}}{\frac{x}{60}+55.55}$$

$$\therefore 4 = 1 + \frac{55.55\times60}{x} \Rightarrow x = \frac{55.55\times60}{3} \Rightarrow x = 1111\text{g}$$

199. (4) $\pi = icRT \Rightarrow \pi \propto i\times c$

$\therefore$ (A) $\pi_{NaCl} \propto 2\times1; \pi_{Urea} \propto 1\times2$

(B) $\pi_{CaCl_2} \propto 3\times1; \pi_{KCl} \propto 2\times1.5$

(C) $\pi_{AlCl_3} \propto 4\times1.5; \pi_{Na_2SO_4} \propto 3\times2$

(D) $\pi_{KCl} \propto 2\times2.5; \pi_{Al_2(SO_4)_3} \propto 5\times1$

$\therefore$ A, B. C and D are isotonic pairs.

200. (3.4) $KCl + AgNO_3 \rightarrow AgCl\downarrow + KNO_3$

1 mol 1 mol 1 mol 1 mol

At equivalence point, 1 mmol KCl

= 1 mmol $AgNO_3$

$\therefore$ 20 ml of 1 M $AgNO_3$ = 20 mmol of $AgNO_3$

= 20 mmol of KCl

$= 20\times10^{-3}\times74.5$ g of KCl = 1.49 g of KCl

25 ml of an aqueous solution of KCl

= 25g of aqueous solution of KCl.

($\because$ density of solution = 1g/ml)

$\therefore$ Mass of solvent = 25 – 1.49 = 23·51g

$$\therefore \text{Molality of the solution} = \frac{20\times10^{-3}\times10^3}{23\cdot51} = 0.85$$

$\therefore \Delta T_f = iK_f m$ [i=2 considering 100% ionization]

$= 2\times2\times0.85 = 3.4$

201. (372) 20% of acetic acid is dissociated in the solution.

$\therefore i = 1 + (n-1)\alpha$

$i = 1 + 0.2(2-1) = 1.2$

$\Delta T_f = iK_f m$

$$\Delta T_f = 1.2\times1.86\times\frac{5\times1000}{60\times500} = 0.372$$

$\Delta T_f = 372\times10^{-3}$°C.

202. (62250) $\pi = CRT$

$$\frac{400}{10^5} = \frac{2.5}{M_o}\times\frac{1000}{250}\times0.083\times300 \Rightarrow M_o = 62250$$

203. (100) $\Delta T_b = 373.52 - 373 = 0.52$ K

$\Delta T_b = K_b\times$ molality

$$0.52 = 0.52\times\frac{2}{\text{molecular mass}}\times\frac{1000}{20}$$

Molecular Mass of solute = 100g/mol

204. (243) $\Delta T_f = i.K_f.m$

$$\Rightarrow \Delta T_f = 2.67\times1.8\times\frac{38}{98}\times\frac{1000}{62}$$

$\Rightarrow \Delta T_f = 30.05$ K

$\therefore$ Freezing temperature = 273 – 30 = 243 K.

205. (13) Let a mole of $Pb(NO_3)_2$ be added

$Pb(NO_3)_2 \rightarrow Pb^{2+} + 2NO_3^-$

a a 2a

$\Delta T_b = 0.15 = 0.5[3a] \Rightarrow a = 0.1$

$Pb^{2+}(aq) + 2Cl^-(aq) \rightarrow PbCl_2(s)$

t = 0: 0.1, 0.2

t = ∞: (0.1 – x), (0.2 – 2x)

In final solution; ions present

$= Pb^{2+}, Na^+, NO_3^-, Cl^-$

$$\Delta T_f = 0.8 = 1.8\left[\frac{0.3 - 3x + 0.2 + 0.2}{1}\right]$$

$$\Rightarrow x = \frac{2.3}{7}\quad K_{sp} = [Pb^{2+}][Cl^-]^2$$

$$= \left(0.1 - \frac{2.3}{27}\right)\left(0.2 - \frac{4.6}{27}\right)^2 = 13\times10^{-6}$$

206. (41,500) The relation related to the osmotic pressure:

$$\pi = \left(\frac{W/M}{V}\right)RT$$

(W = mass of PVC, M = molecular weight of PVC and V = the volume of the solution)

$$\Rightarrow \pi = \left(\frac{W}{V}\right)\left(\frac{1}{M}\right)RT = C\left(\frac{RT}{M}\right)$$

$$\Rightarrow \frac{\pi}{C} = \frac{RT}{M}$$

As per the given conditions:

$$\text{Slope} = \frac{RT}{M} = \frac{0.083\times300}{M} = 6\times10^{-4}$$

$$\Rightarrow M = \frac{0.083\times300}{6\times10^{-4}} = \frac{830\times300}{6} = 41{,}500\ \text{g/mole}$$

207. (4) $\pi = iCRT \Rightarrow \pi \propto iC$

A, B, D and E have same value of osmotic pressure.

208. (80) Mass of C_2H_5OH = density × volume

$= 62.5\times0.8 = 50$ g

$\Delta T_f = K_f\times m$ [m = molality]

$$0.9 = 2\times\frac{1.8\times1000}{M_w\times50} \Rightarrow M_w = \frac{2\times1.8\times1000}{0.9\times50} = 80$$

209. (15) Moles of ascorbic acid

$$= \frac{10.2\text{g}}{176\text{g mol}^{-1}} = 0.058\ \text{mol}$$

$\Delta T_f = (x\times10^{-1})$°C $= K_f$. molality

$$= 3.9\times\frac{0.058}{150}\times1000 = 1.5\text{°C} = 15\times10^{-1}\text{°C}$$

210. (6) $\frac{p^0 - p_s}{p^o} = \frac{n_{solvent}}{n_{solute}}$

$$\Rightarrow \frac{P^0 - \frac{P^0}{2}}{P^0} = \frac{n_{solute}}{n_{solvent} + n_{solute}}$$

$$\frac{1}{2} = \frac{n_{solute}}{n_{solute} + n_{solvent}}$$

$$2 = 1 + \frac{5.55}{n_{solute}}$$

$n_{solute} = 5.55 \approx 6$ mol

211. (1) $\Delta T_b = iK_b m_1; \Delta T_f = iK_f m_2$

$$\frac{\Delta T_b}{\Delta T_f} = \frac{K_b \times 1}{K_f \times 2} \Rightarrow \frac{3}{6} = \frac{1}{2} = \frac{K_b}{K_f} \times \frac{1}{2}$$

$$\Rightarrow \frac{K_b}{K_f} = \frac{1}{1} \Rightarrow x = 1$$

212. (5) $d = \frac{\text{Mass}}{\text{volume}}$

$\therefore$ Mass of acetic acid = 1.02 × 1.2 = 1.224 g

Moles of acetic acid = $\frac{\text{Mass}}{\text{Molar mass}}$

$$= \frac{1.224}{60} = 0.0204$$

Mass of water = 2 kg [1L ≈ 1kg for water]

Molality (m) = $\frac{0.0204}{2} = 0.0102$ mol/kg

Now $\Delta T_f = i \times K_f \times m$

$\Rightarrow i = \frac{\Delta T_f}{K_f \times m}$ and $i = 1 + \alpha$ for acetic acid

$\therefore 0.0198 = (1 + \alpha) \times 1.85 \times 0.0102$

$\alpha = 0.04928 \cong 5\%$

213. (3) Elevation in boiling point and depression in freezing point are colligative properties and their value depends only on the amount of solute present in the solution.

$\Delta T_b = iK_b m$; $\Delta T_f = iK_f m$

$$\frac{\Delta T_b}{\Delta T_f} = \frac{4}{4} = \frac{K_b 1.5}{K_f 4.5} \Rightarrow \frac{K_b}{K_f} = 3$$

214. (330) $\pi = CRT$

$5.03 \times 10^{-3} = C \times 0.083 \times 300$

$C = 0.202 \times 10^{-3}$ M; $C = \frac{n}{\text{volume}}$

Moles of protein = $0.202 \times 10^{-3} \times 0.5$
$= 10^{-4} \times 1.01$

moles = $\frac{\text{given weight}}{\text{molar mass}}$

$1.01 \times 10^{-4} = \frac{2.5}{M}$; M(molar mass of protein) = 24752

$\therefore$ No. of glycine units

$$= \frac{\text{molar mass of protein}}{\text{Normal molar mass of Glycine}} = \frac{24752}{75} = 330.03$$

215. (8) Given : $\frac{(K_b)_A}{(K_b)_B} = \frac{1}{8}$

$$\frac{(\Delta T_b)_A}{(\Delta T_b)_B} = \frac{(K_b)_A \cdot m}{(K_b)_B \cdot m} = \frac{1}{8} \Rightarrow \frac{x}{y} = \frac{1}{8}$$

$\therefore y = 8$ (nearest integer)

216. (45) $w_{Solute} = 2.5\,g K_b = 0.52$ K kg mol^{-1}
$w_{solvent} = 75$gM = Mol. Wt. of solute

$T_b^{\circ} = 373.15$K ; $\Delta T_b = 0.385$ K

$$\Delta T_b = \frac{K_b \times 10^3 w_{solute}}{M_{solute} \times w_{solvent}}$$

$$\to 0.385 - 0.52 \times \left(\frac{2.5}{M} \times \frac{1000}{75}\right)$$

M = 45 g mol^{-1}.

217. (98) 0.5% solution of KCl

Molality, $m = \frac{0.5}{74.6} \times \frac{1000}{99 \times 5}$

$\Delta T_f = i \times m \times K_f$

$$0.24 = i \times \frac{0.5}{74.6} \times \frac{1000}{100} \times 1.8 \Rightarrow i = 1.979$$

$$\underset{1-\alpha}{KCl} \rightleftharpoons \underset{\alpha}{K^+} + \underset{\alpha}{Cl^-} \Rightarrow 1.979 = 1 + \alpha$$

$\Rightarrow \alpha = 0.979$

$\% \alpha = 79.9\% \approx 98\%$

218. (415) $\pi = iCRT$; i = 1 here

$$\left(C = \frac{n}{v} = \frac{\text{weight}}{\text{Molar mass} \times \text{volume (1 kg = 1000 g)}}\right)$$

$$= \frac{1 \times 2}{60000 \times 0.2} \times 0.083 \times 300 = 0.00415 \text{ bar}$$

($\because$ 1 bar = 10^5 Pa)

So, 0.00415 × 10^5 Pa = 415 Pa

219. (54) π = C.R.T

$7.47 = C \times 0.083 \times 300 \Rightarrow C = 0.3\,M = 0.3 \times 180$ gL^{-1}
$= 54$ gL^{-1}

220. (13) K_b (benzene) = 2.6 K kg mol^{-1}
K_b (acetone) = 1.7 K kg mol^{-1}

With benzene as solvent, $\Delta T_b = iK_b m$

$$\Delta T_b = \frac{1}{2} \times 2.6 \times \frac{1.22/M_w}{100/1000} \quad ...(i)$$

With Acetone as solvent, $\Delta T_b = iK_b m$

$$0.17 = 1 \times 1.7 \times \frac{1.22/M_w}{100/1000} \quad ...(ii)$$

Dividing equation (i) by (ii)

$$\frac{\Delta T_b}{0.17} = \frac{\frac{1}{2} \times 2.6}{1 \times 1.7} \Rightarrow \Delta T_b = 13 \times 10^{-2} \Rightarrow x = 13$$

221. (518) Molality of sucrose sol. = 0.75 m, i.e., 0.75 mol/kg (solvent)

Mass of sucrose = 0.75 × 342g = 256.5 g

Mass of solution = mass of sucrose + mass of H_2O
= 1000 + 256.5 = 1256.5 g

Mass of solution in 1 kg (given) solution

$$= \frac{256.5}{1256.5} \times 1000 = 204.1 \text{ g}$$

Mass of H_2O in 1 kg solution
= 1000 – 204.1 = 795.9 g

After cooling the solution to – 4°C,
$\Delta T_f = iK_f m$

$$4 = 1 \times 1.86 \times \frac{204.1 \times 1000}{342 \times w_B}$$

[where w_B = Mass of H_2O left]

$\Rightarrow w_B = 277.5$ g

Mass of ice separated = 795.9 – 277.5
= 518.4 ≃ 518 g

222. (271)

$$\text{molality} = \frac{\left(\frac{40}{180}\right)\text{mol}}{0.2 \text{ kg}} = \left(\frac{10}{9}\right)\text{molal}$$

$$\Rightarrow \Delta T_f = K_f \times m = T_f - T_f' = 1.86 \times \frac{10}{9}$$

$$\Rightarrow T_f' = 273.15 - \left(1.86 \times \frac{10}{9}\right)$$

= 271.08 K ≃ 271K (nearest-integer)

223. (269) k_f = 1.86 k. kg/mol

Solute : $C_2H_6O_2$ (83 g) (Non dissociative solute)

$$\Rightarrow \Delta T_f = k_f \times m = 1.86 \times \frac{83/62}{624/1000}$$

$$= \frac{1.86 \times 83 \times 1000}{62 \times 625} = \frac{154380}{38750}$$

$\Delta T_f = 4$K

$\because \Delta T_f = T^o - T_s$

$\therefore T_s = T^o - \Delta T_f = 273 - 4 = 269$K

224. (4) $\Delta T_f = iK_f m$

$\Delta T_f \propto i$

Compound	**i**
C_2H_5OH	1
$Ba_3(PO_4)_2$	5
Na_2SO_4	3
KCl	2
Li_3PO_4	4

So, all compounds will exhibit lower freezing point as compared to C_2H_5OH.

225. (15) π = CRT

$$\pi = \frac{W_B \times R \times T}{M_B \times \text{Vol (in L)}}$$

$$\Rightarrow 2.42 \times 10^{-3} = \frac{1.46 \times 0.083 \times 300}{M_B \times 0.1}$$

$\Rightarrow M_B = 15.02 \times 10^4$ g/mol

226. (125) 50% i.e., 0.5 moles of HA dimerizes :

	2HA	→	$(HA)_2$
Initial	a		0
Final	$(1-0.5)$a		$\frac{0.5a}{2}$

Remaining 50% of HA dissociates :

	HA	→	H^+	+	A^-
Initial	0.5a		0		0
Final	0		0.5a		0.5a

van't Hoff factor (i)

$$= \frac{\text{Total no. of particles dissociated or associated}}{\text{Initial total no. of particles}}$$

$$= \frac{0.25a + 0.5a + 0.5a}{a} = 1.25 = 125 \times 10^{-2}$$

227. (250) Let molar mass of 'X' be M.

$\Delta T_b = K_b$ m.

$$\Rightarrow 0.6 = 5 \times \left(\frac{\text{Wt.} \times 1000}{M \times \text{Mass of solvent (g)}}\right)$$

$$\Rightarrow 0.6 = 5 \times \frac{3 \times 1000}{M \times 100}$$

$$\Rightarrow M = \frac{150}{0.6} = 250 \text{ g/mol.}$$

228. (50) Given freezing point = – 3.885 °C.

$\Delta T_f = 0 - (-3.885) = 3.885$

$\Delta T_f = iK_f \cdot m$

$\Rightarrow 3.885 = i \times 1.85 \times 2 \Rightarrow i = 1.05$

$$\underset{1-\alpha}{HA} \rightleftharpoons \underset{\alpha}{H^+} + \underset{\alpha}{A^-}$$

(α = degree of dissociation)
van't Hoff factor (i) = $1 + \alpha \Rightarrow 1.05 = 1 + \alpha$
$\therefore \alpha = 0.05 = 50 \times 10^{-3}$

229. (100) $\underset{1-\alpha}{2A} \rightarrow \underset{\frac{\alpha}{2}}{A_2}$

$i = 1 - \alpha + \frac{\alpha}{2}$; Also, $\Delta T_b = iK_b m$

$\Delta T_b = T_b - T_b^0 = 100.52 - 100 = 0.52$

$\Rightarrow 0.52 = i \times 0.52 \times 2 \Rightarrow i = \frac{1}{2}$

$\Rightarrow \frac{1}{2} = 1 - \alpha + \frac{\alpha}{2} \Rightarrow 1 - \frac{\alpha}{2} = \frac{1}{2} \Rightarrow \alpha = 1$

So, % association = 100%.

230. (85)

	$K_4Fe(CN)_6 \rightleftharpoons$	$4K^+$	$+ Fe(CN)_6^{4-}$
Initial	1	0	0
Final	$1-\alpha$	4α	α

Total moles = $1 - \alpha + 4\alpha + \alpha$

$i = \frac{1+4\alpha}{1} = 1 + (4 \times 0.4) = 2.6$

$\Delta T_{b_1} = 2.6 \times K_b \times 1$ ...(i)

For a non-electrolyte solute (A), $i = 1$
18.1 weight percent solution means 18.1 g solute is present in 100 g solution, then weight of solvent $= 100 - 18.1 = 81.9$g

$(\Delta T_b)_A = 1 \times \frac{18.1 \times 1000}{M_A \times 81.9} \times K_b$ ...(ii)

Given, $\Delta T_b = (\Delta T_b)_A$

$2.6 \times K_b \times 1 = \frac{1 \times 18.1 \times 1000}{M_A \times 81.9} \times K_b \Rightarrow M_A = 85.$

231. (106)

	$AB_2 \rightleftharpoons$	A^{2+}	$+ 2B^-$
$t = 0$	1	0	0
$t = t$	$1-\alpha$	α	2α

$i = 1 + \frac{2\alpha}{1} = 1 + 2 \times 0.1 \Rightarrow i = 1.2$

$\Delta T_b = i \times k_b \times m = 1.2 \times 0.5 \times 10 = 6$

$\Delta T_b = T_s - T^\circ$

$T_s = \Delta T_b + T^\circ = 6 + 100 = 106°C$

232. (2) $\Delta T_f = iK_f m$

$\Delta T_f = T_f^0 - T_f = 0 - (-0.93) = 0.93$

$0.93 = i \times 1.86 \times \frac{12.2 \times 1000}{122 \times 100} \Rightarrow i = 0.5$

$n(\text{Benzoic acid}) \longrightarrow (\text{Benzoic acid})_n$

$i = \frac{\text{Total number of particles after association}}{\text{Number of particles before association}}$

$0.5 = \frac{1}{n} \Rightarrow n = 2$

233. (24)
234. (3)

235. (375) $\underset{1-0.6}{A_2B_3} \longrightarrow \underset{2\times0.6}{2A^{3+}} + \underset{3\times0.6}{3B^{2-}}$

$\therefore$ van't Hoff factor $= 1 - 0.6 + 1.2 + 1.8 = 3.4$

$\Delta T_b = i \cdot k_b \cdot m = 3.4 \times 0.52 \times 1 = 1.768$

$\because \Delta T_b = T_b - T_b^0$

$\therefore T_b = 373 + 1.768 = 374.77 \approx 375$

236. (35)

$ClCH_2COOH \rightleftharpoons ClCH_2COO^- + H^+$

Degree of dissociation = α

$\Rightarrow i = 1 + \alpha$ and $\Delta T_f = iK_f m$

$\Rightarrow 0.5 = (1+\alpha)(1.86) \times \frac{9.45}{94.5} \times \frac{1000}{500}$

$\Rightarrow 0.5 = (1+\alpha)(1.86)(0.2)$

$\Rightarrow 1 + \alpha = 1.34 \Rightarrow \alpha = 0.34$

	$ClCH_2COOH \rightleftharpoons$	$ClCH_2COO^-$	$+ H^+$
Initial	α	0	0
Final	$C - C\alpha$	$C\alpha$	$C\alpha$

$K_a = \frac{(C\alpha)^2}{C - C\alpha} = \frac{C\alpha^2}{1-\alpha}$ $[C \approx m]$

$\Rightarrow K_a = \frac{0.2 \times (0.34)^2}{1 - 0.34} = 0.035$

$\therefore K_a = 35 \times 10^{-3}$.

237. (1) $\Delta T_f = K_f \times m$

(Solute is non-dissociating)

$\Rightarrow (T_f^0 - T'_f) = 5.12 \times \frac{10 \times 1000}{58 \times 200}$

$\Rightarrow 5.5 - T'_f = \frac{5.12 \times 5 \times 10}{58} \Rightarrow T'_f = 1.086\,°C \approx 1\,°C.$

238. (5)
Molality of $CaCl_2$ solution = 0.05 m

$\Delta T_b = iK_b m = 3 \times K_b \times 0.05 = 0.15 K_b$

Molality of $CrCl_3 . xNH_3$ = 0.10 m

$\Delta T_b' = iK_b \times 0.10 \Rightarrow \Delta T_b' = 2\Delta T_b$

$iK_b \times 0.10 = 2 \times 0.15 K_b \Rightarrow i = 3$

Since, co-ordination number of Cr is 6.
$\therefore$ The complex is $[Cr(NH_3)_5Cl]Cl_2$.
$\therefore x = 5$.

239. (167) For NaCl :

$\pi_1 = iC_1RT \Rightarrow C_1 = \frac{0.10}{2RT}$

For Glucose : $\pi_2 = C_2RT \Rightarrow C_2 = \frac{0.20}{RT}$

When 1 L of NaCl solution and 2 L glucose solution are Mixed.

$\therefore C'_1 = \frac{0.1}{2RT} \times \frac{1}{(1+2)} = \frac{0.10}{6RT}$ and

$C'_2 = \frac{0.2}{RT} \times \frac{2}{(1+2)} = \frac{0.40}{3RT}$

$\therefore \pi_{Total} = iC'_1 RT + C'_2 RT$

$= \left(2 \times \frac{0.1}{6RT} + \frac{0.4}{3RT}\right) RT = \frac{0.50}{3}$

$\pi_{Total} = 167 \times 10^{-3}$ atm

240. (177)

$\pi_A = C_A RT,\ \pi_B = C_B RT$

For isotonic solution, $\pi_A = \pi_B$

$\Rightarrow C_1 = C_2$ (For protein $i = 1$)

$\Rightarrow \frac{0.73 \times 1000}{M_A \times 250} = \frac{1.65}{M_B \times 1}$

$\therefore \frac{M_A}{M_B} = \frac{0.73 \times 4}{1.65} = 1.77 = 177 \times 10^{-2}$

241. (1.75) $\Delta T_f = iK_f m$; i(NaCl) = 2

$0.2 = 2 \times 2 \times \frac{w}{58.5} \times \frac{1000}{600}$

$w = \frac{0.2 \times 58.5 \times 600}{1000 \times 4} = 1.75$ g

242. (2.18)

$K_f = 2.0$ K kg mol^{-1}; m = 0.5

$\Delta T_f = K_f m = 0.5 \times 2 = 1$

$T_{initial} = 273 - 1 = 272$ K; $n = 0.1$ mol; V = 1 dm^3

$P_{gas} = \frac{nRT}{V} = \frac{0.1 \times 0.08 \times 272}{1} = 2.176$ atm

After releasing piston, $P_1V_1 = P_2V_2$

$2.176 \times 1 = 1 \times V_2 \Rightarrow V_2 = 2.18$ dm^3

Part-B (JEE Advanced)

1. **(c)** Number of moles of urea $= \frac{120}{60} = 2$

Total mass of solution = 1000 + 120 = 1120 g
Total volume of solution (in L)

$= \frac{\text{Mass}}{\text{Density}} = \frac{1120}{1.15 \times 10^3} = \frac{112}{115}$ L

Molarity of the solution

$= \frac{\text{Number of moles}}{\text{Volume in L}} = \frac{2 \times 115}{112} = 2.05$ mol L^{-1}

2. **(a)** $p_{N_2} = \kappa_H \chi_{N_2}$; $0.8 \times 5 = 1 \times 10^5 \times \chi_{N_2}$

$\therefore \chi_{N_2} = 4 \times 10^{-5}$; Solubility in 10 moles $= 4 \times 10^{-4}$.

3. **(19)** $P_T = p^\circ_A X_A + p^\circ_B X_B$

$\Rightarrow 45 = 20(0.5) + p_B^o(0.5) \Rightarrow p_B^o = 70$

$22.5 = 20X_A + 70(1 - X_A)$

$\Rightarrow X_A = \frac{4.75}{5} = 0.95 \Rightarrow X_B = 0.05$

$\Rightarrow \frac{X_A}{X_B} = 19$

4. **(9)** 1 mole solution has 0.1 mole solute and 0.9 mole solvent.
Let M_1 = Molar mass solute
M_2 = Molar mass solvent

Molality, $m = \frac{0.1}{0.9M_2} \times 1000$ .... (1)

Molarity, $M = \frac{0.1}{0.1M_1 + 0.9M_2} \times 2 \times 1000$ .... (2)

Molarity = $\frac{\text{Moles of solute}}{(\text{Mass of solution/density})}$ $\because m = M$

$\Rightarrow \frac{0.1 \times 1000}{0.9M_2} = \frac{200}{0.1M_1 + 0.9M_2} \Rightarrow \frac{M_1}{M_2} = 9$

5. (8)

$$\text{Molality} = \frac{\text{wt. of solute in 1 L of solution} \times 1000}{\text{wt. of solvent in 1 L of solution} \times \text{mol. wt. of solute}}$$

Calculation of wt. of solvent
1 mL of solvent = 0.4 g
1000 mL of solvent = 400 g
Calculation of wt. of solute
1000 mL of solution contain = 3.2 × 80 g solute = 256 g

$\therefore$ Molality $= \frac{256 \times 1000}{400 \times 80} = 8$

6. (8) Molarity of stock solution of HCl

$= \frac{29.2 \times 1000 \times 1.25}{100 \times 36.5}$

Let the volume of stock solution required = V mL

Thus, $V \times \frac{29.2 \times 1000 \times 1.25}{100 \times 36.5} = 200 \times 0.4$

$\Rightarrow V = 8$ mL

7. (2.98) $X_{urea} = 0.05$
Mass of water = 900 g

Number of moles of water $= n_{H_2O} = \frac{900}{18} = 50$

Let moles of urea = n_U

$\frac{n_U}{n_U + 50} = 0.05 \Rightarrow n_U = \frac{50}{19} = 2.63$

Mass of urea = $2.63 \times 60 = 157.8$ g
Given density (d) of solution = 1.2 g/mL.

$V_{solution} = \frac{\text{mass of solution}}{\text{density (g/cc)}} = \frac{900 + 157.8}{1.2}$

= 881.5 mL

$M = \frac{n_V}{V_{mL}} \times 1000 = \frac{2.63}{881.5} \times 1000 = 2.98$

8. (3.84) $N_1 = 1, V_1 = ?, N_2 = 26.7, V_2 = 0.4$

$N_1V_1 = N_2V_2;\ 1 \times V_1 = 26.7 \times 0.4$

$\Rightarrow V_1 = \frac{26.7 \times 0.4}{1} = 10.68$

49g ($\because$ eq wt of $H_2SO_4 = 49$) of H_2SO_4 will be neutralised by = 1N 1000 mL NaOH

$\therefore$ 0.5g of H_2SO_4 will be neutralised by

$= \frac{1000}{49} \times 0.5 = 10.20$ mL 1N NaOH

Volume of 1 N NaOH used by dissolved $SO_3 = 10.68 - 10.20 = 0.48$ mL

$SO_3 + 2NaOH \longrightarrow Na_2SO_4 + H_2O$

$\therefore$ Eq wt of $SO_3 = \frac{\text{Mol wt}}{2} = \frac{80}{2} = 40$

Wt of SO_3 in 0.48 mL of 1 M solution

$= \frac{40}{1000} \times 0.48 = 0.0192$ g

% of $SO_3 = \frac{0.0192}{0.5} \times 100 = $ **3.84%**

9. Molality, $M = \frac{\text{moles of solute}}{\text{kg of solvent}}$

Mole fraction, $x_1 = \frac{n_1}{n_1 + n_2}$, $x_2 = \frac{n_2}{n_1 + n_2}$

$p_1 = x_1 p^\circ_1$

$\therefore x_1 = \frac{p_1}{p_1^\circ} = \frac{750}{760} = $ **0.9868**

x_2 (solute) $= 1 - 0.9868 = 0.0132$

$$\text{Molality} = \frac{\text{Mole fraction of solute} \times 1000}{\text{Mole fraction of solvent} \times \text{Mol. mass of solvent}}$$

Molality, $m = \frac{x_2}{x_1 M_1} \times 1000 = \frac{0.0132 \times 1000}{0.9868 \times 18}$

= **0.7503 mol kg^{-1}**

10. $M = \frac{86/98}{100/1.787} \times 1000$

$= \frac{0.8775}{55.5} \times 1000 = $ **15.81 M**

$M_1V_1 = M_2V_2$
$M_1 = 15.81, V_1 = ?$
$M_2 = 0.2, V_2 = 1$ L $= 1000$ mL
$\therefore 15.81 \times V_1 = 0.2 \times 1000$ or V_1

$= \frac{0.2 \times 1000}{15.81} = 12.65$ mL

$\therefore$ Amount of acid to be used to make 1 L of 0.2 M H_2SO_4 = **12.65 mL.**

11. (i)

$\text{Molarity} = \frac{\text{Moles of solute}}{\text{Volume of solution in L}}$

(ii) $N_1V_1 = N_2V_2$

A 13% solution (by weight) contains 13g of solute (*i.e.* H_2SO_4) per 100 g of solution.

Moles of solute $= \frac{\text{Mass of } H_2SO_4}{\text{M. wt. of } H_2SO_4}$

$= \frac{13}{98} = 0.1326$

Volume of solution in L

$= \frac{\text{Mass of solution}}{\text{density of solution} \times 1000}$

$= \frac{100}{1.02 \times 1000} = 0.0980$ Litre

$\therefore$ Molarity of solution $= \frac{0.1326}{0.0980} = $ **1.35 M**

Again, Molality $= \frac{\text{Moles of solute}}{\text{Mass of solvent in kg}}$

Mass of solute in 100 mL of solution = 13 g [13 % solution]
Mass of solvent = Mass of solution – Mass of solvent
= 100 – 13 = 87 g

$\therefore$ Molality $= \frac{13/98}{87/1000} = $ **1.57 m**

Normality = Molarity × $\frac{\text{Mol. wt.}}{\text{Eq. wt.}}$ or

$1.35 \times \frac{98}{49} = 2.70$ N

$N_1 = 2.70,\ V_1 = 100$ mL, $N_2 = 1.5,\ V_2 = ?$

$\left[\because \text{Eq. wt} = \frac{98}{2} = 49\right]$

$N_1V_1 = N_2V_2;\ 2.70 \times 100 = 1.5 \times V_2$

or $V_2 = \frac{2.70 \times 100}{1.5} = 180$ mL.

$\therefore$ 100 mL of this acid should be diluted to 180 mL to prepare 1.5 N solution.

12. (b) $\frac{p^o - p}{p^o} = \frac{w/m}{w/m + w/m}$ = mole fraction of solute = x_B

[Mathematical statement of Raoult's law]

13. (c) Lower the B. pt., higher will be the V.P. The V.P. of the mixture is greater than either of the two liquids.
In case of positive deviation from Raoult's law, the partial vapour pressure of each liquid and total vapour pressure of solution will be greater as compared to initial solution.

14. (33.16) $B \rightleftharpoons \underbrace{2C + 2D}_{n=4}$

Let molar mass of S = x molar mass of B = 10x.

$K_b = \left(\frac{R \times M \times T_b^{o2}}{1000 \times \Delta H_{vap}}\right)$

$K_b = \left\{\frac{R \times x \times (400)^2}{1000 \times 10R}\right\}$

$K_b = 16x$
$\Delta T_b = i \times K_b \times m$

$T_b - T_b^\circ = i \times 16x \times \frac{0.25 \times 1000}{10x \times 99.75}$

$408 - 400 = i \times 4 \Rightarrow 8 = i \times 4 \therefore i = 2$
$i = 1 + (n-1)\alpha \Rightarrow 2 = 1 + 3\alpha \Rightarrow \alpha = 0.3333$

Dissociated mol % of B = 0.3333 × 100 = 33.33%

15. (0.02) Using Raoult's law

$P_T = P_A^o X_A + P_B^o X_B$

$0.3 = P_A^o \times 0.25 + P_B^o \times 0.75$ ...(i)

$0.4 = P_A^o \times 0.5 + P_B^o \times 0.5$

on solving eq^n (i) and (ii)

$P_A^o = 0.6,\ P_B^o = 0.2$

16. (0.735)

(i) Volume $= \frac{\text{No. of moles} \times \text{molar mass}}{\text{density}}$

(ii) $PV = nRT$ or $P = \frac{nRT}{V}$

Volume of 1 mole of liq. benzene $= \frac{78}{0.877}$

Volume of 1 mole of toluene $= \frac{92}{0.867}$

In vapour phase,
At 20°C, for 1 mole of benzene,

Volume $= \frac{1 \times 78 \times 2750}{0.877} = 244583.80$ mL
= 244.58 L

Similarly for 1 mole of toluene,

Volume $= \frac{1 \times 92}{0.867} \times 7720 = 819192.61$ mL
= 819.19 L

As we know that, $PV = nRT$

For benzene, $P_B^o = \frac{nRT}{V}$

$= \frac{1 \times 0.0821 \times 293}{244.58}$ atm $= 0.098$ atm

For toluene, $P_T^o = \frac{nRT}{V}$

$= \frac{1 \times 0.0825 \times 293}{819.19}$ atm $= 0.029$ atm

$P = P_B^o \cdot X_B + P_T^o \cdot X_T$

$P = P_B^o \cdot X_B + P_T^o(1 - X_B)$

Total vapour-pressure = 46 torr

$= \frac{46}{760} = 0.060$ atm

Thus, $0.060 = 0.098\,X_B + 0.029\,(1 - X_B)$

$\therefore \quad X_B = \frac{0.031}{0.069} = 0.45$ (in liquid phase)

$X_B + X_T = 1 \Rightarrow X_T = 1 - 0.45 = 0.55$ (in liquid phase)

Also, $P_B' = P_B^o X_B = P X_B'$

So, $0.098 \times 0.045 = 0.060 \times X'_B$

$X'_B = \frac{0.098 \times 0.45}{0.060}$ = **0.735 (in gas phase)**

17. (0.156) $\frac{P^o - P}{P^o} = \frac{n}{N}$

[Raoult's Equation]

Let the molality of the solution = m

Now, the solution contains 'm' moles of solute per 1000 g of benzene.

Vapour pressure of benzene, $P^o = 639.7$ mm

Vapour pressure of solution, $P = 631.9$ mm

Moles of benzene (Mol. wt. 78), $N = \frac{1000}{78}$

Moles of solute, $n = ?$

Substitute these values in the Raoult's equation

$\frac{P^o - P}{P^o} = \frac{n}{N}$ or $\frac{639.7 - 631.9}{639.7} = \frac{n \times 78}{1000}$

or $\frac{7.8}{639.7} = \frac{78\,n}{1000}$ $\quad \therefore n = \frac{1000 \times 7.8}{78 \times 639.7} = \mathbf{0.156}$

Hence, molality of solution = **0.156 m**

18. (b, d)

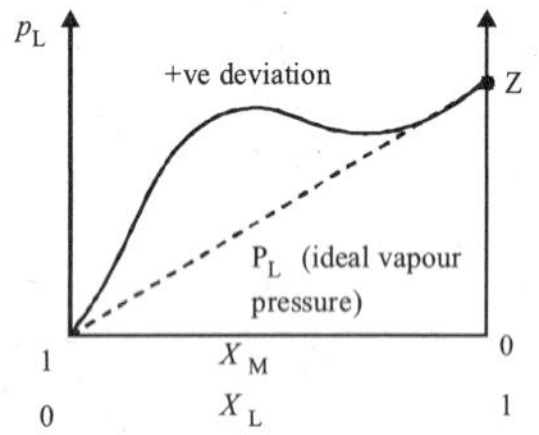

(b) From the given figure it is clear that at point Z, the mole fraction of M is zero, in that case 'Z' represents the vapour pressure of pure liquid L and Raoult's law is obeyed when $x_L \to 1$.

(d) The solution formed by mixing two liquids L and M shows positive deviation from Raoult's law. Thus, intermolecular forces of attraction between L – L in pure liquid L and M – M in pure liquid M are stronger than that between L – M in the solution.

19. (a, b)

(a) H-bonding of methanol breaks when CCl_4 is added so bonds become weaker, resulting positive deviation.

(b) Mixing of polar and non-polar liquids will produce a solution of weaker interaction, resulting positive deviation.

(c) Ideal solution

(d) –ve deviation, because stronger H-bond is formed.

20. (b, d) For ideal solution, $\Delta S_{system} > 0$; $\Delta H_{mixing} = 0$

Since, there is no heat exchange between system and surrounding;

$\Delta S_{surroundings} = 0 \times \Delta G = \Delta H - T\Delta S_{system} < 0$

21. Molecular weight of CH_3OH

$= 12 + 3 + 16 + 1 = 32$

Molecular weight of $C_2H_5OH = 24 + 5 + 16 + 1 = 46$

According to Raoult's law, $P_{total} = p_1 + p_2$

where P_{total} = Total vapour pressure of the solution

p_1 = Partial vapour pressure of one component

p_2 = Partial vapour pressure of other component

Again, p_1 = Vapour pressure $(p°_1)$ × mole fraction

Similarly, p_2 = Vapour pressure $(p°_2)$ × mole fraction

Mole fraction of $CH_3OH = \frac{40/32}{40/32 + 60/46} = 0.49$

Mole fraction of ethanol $= 1 - 0.49 = 0.51$

Thus now let us first calculate the partial vapour pressures, *i.e.*, p_1 and p_2 of the two components.

Partial vapour pressure of $CH_3OH(p_1)$

$= 88.7 \times 0.49 = 43.48$ mm

Partial vapour pressure of $C_2H_5OH(p_2)$

$= 44.5 \times 0.51 = 22.69$ mm

$\therefore$ Total vapour pressure of the solution

$= 43.48 + 22.69$ mm $=$ **66.17 mm**

Mole fraction of CH_3OH in vapour

$= \frac{43.48}{66.17} = \mathbf{0.65}$

22. If they form an ideal solution which obeys Raoult's Law and for which

$\Delta H_{mixing} = 0$ and $\Delta V_{mixing} = 0$

Thus, we can separate two volatile and miscible liquids by fractional distillation if they should not form azeotropic solutions.

23. The chemical equation for the combustion of organic compound $C_xH_{2y}O_y$ can be represented as :

$C_xH_{2y}O_y + 2xO_2 = x\,CO_2 + y\,H_2O + x\,O_2$

The gases obtained after cooling $= x + x = 2x$

$\therefore 2x = 2.24$ litres $\quad [\because H_2O$ is in liquid state$]$

or $x = \frac{2.24}{2} = 1.12$ litres

Number of moles of $CO_2 = \frac{1.12 \text{ litres}}{22.4 \text{ litres / mole}}$

$[\because 22.4$ L at NTP = 1 mole$]$

$\Rightarrow \quad x = \frac{1}{20}$ mole = 0.05 mole

Moles of H_2O formed $(y) = \frac{0.9}{18} = 0.05$;

$x : y = 0.05 : 0.05 = 1 : 1$

The empirical formula of the organic compound is

CH_2O ...(1)

The mole fraction of the solute (A)

= relative decrease in vapour pressure of the solvent (B)

$\frac{p^o - p}{p^o} = \frac{W_A / M_A}{W_B / M_B + W_B / M_B}$

or $\frac{0.104}{17.5} = \frac{50 / M_A}{50 / M_A + 1000/18}$

$[M_A$ = mol.wt. of $A]$

or $\frac{0.104}{17.5} = \frac{50}{M_A\left(\frac{50 \times 18 + 1000 M_A}{18 M_A}\right)}$

or $\frac{104}{17500} = \frac{50 \times 18}{900 + 1000 M_A}$ or $M_A = 150.6$

Molecular wt. of the organic compound, $(CH_2O)_n = 150$

Molecular wt. of $CH_2O = 12 + 2 + 16 = 30$

$\therefore 30 \times n = 150$ or $n = \frac{150}{30} = 5$

$\therefore$ Molecular formula of the given organic compound is $\mathbf{(CH_2O)_5}$ or $\mathbf{C_5H_{10}O_5}$.

24. (c) As T increase, V.P. increases

$\Delta T_f = K_f \times m$ (m = Molality of solute)

$273 - T_f' = 2 \times \frac{34.5 \times 1000}{46 \times 500}$

$\therefore \quad T_f = 270$ K

Thus, freezing point of solution = 270K. Further, as T increases, vapour pressure increases. Hence, these facts coincide with the curve given in (c).

25. (a) From elevation in boiling point, $\Delta T_b = K_b \times m$

$2 = 0.76 \times m$

$m = \frac{2}{0.76}$ ...(ii)

From Raoult's law,

$\frac{p° - p}{p°} = \frac{\text{No. of moles of solute}}{\text{No. of moles of solvent} + \text{No. of moles of solute}}$

When the concentration of solute is much lower than the concentration of solvent,

$\frac{p° - p}{p°} = \frac{\text{No. of moles of solute}}{\text{No. of moles of solvent}}$

$\frac{760 - p}{760} = \frac{2.5/m}{100/18}$ ...(i)

From (i) and (ii), $p = 724$ mm

26. (a) $\Delta T_f = i \times K_f \times m$

Where m = Molality of the solution

(*i.e.* number of moles of solute per 1000 g of the solvent)

Here, $m = \frac{0.1}{329} \times \frac{1000}{100} = \frac{1}{329}$

Assuming 100% $K_3[Fe(CN)_6]$ dissociates in 4 ions;

$\therefore$ i = 4 ionization.

Thus, $\Delta T_f = 4 \times 1.86 \times \frac{1}{329} = 2.3 \times 10^{-2}$

$\therefore T_f = 0 - 2.3 \times 10^{-2} = -2.3 \times 10^{-2}\ °C$

27. (a) Molecular weight of naphthoic acid $C_{11}H_8O_2 = 172\ gmol^{-1}$.

The theoretical value of depression in freezing point

$= K_f \times \text{ molality} = 1.72 \times \frac{20 \times 1000}{172 \times 50} = 4K$

van't Hoff factor,

$i = \frac{\text{Observed value of colligative property}}{\text{Theoretical value of colligative property}}$

$= \frac{2}{4} = 0.5$.

28. (a)

(i) $i = \frac{\text{No. of particles after ionisation}}{\text{No. of particles before ionisation}}$

(ii) $\Delta T_b = i \times K_b \times m$

$$\begin{array}{lccc} CuCl_2 & \longrightarrow & Cu^{2+} & + 2Cl^- \\ 1 & & 0 & 0 \\ (1-\alpha) & & \alpha & 2\alpha \end{array}$$

$i = \frac{1+2\alpha}{1}$ $\quad i = 1 + 2\alpha$

Assuming 100% ionization

So, $i = 1 + 2 = 3$

$\Delta T_b = 3 \times 0.52 \times 0.1 = 0.156 \approx 0.16$

$\left[m = \frac{13.44}{134.4} = 0.1\right]$

29. (a) At the freezing point liquid and solid remain in equilibrium. If a solution of a non-volatile solute is cooled to a temperature below the freezing point of solution, some of liquid solvent will separate as a solid solvent and thus the concentration of solution will increase.

30. (b) Benzoic acid exists as dimer in benzene.

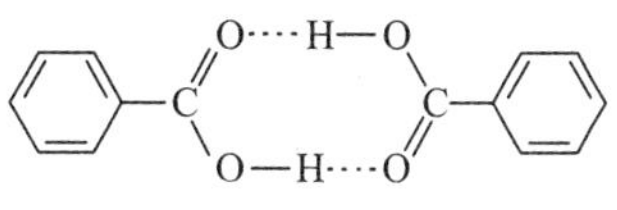

[Normal molecular mass = 122 amu and observed molecular mass = 244 amu, in case of complete association]

31. (a) Depression in freezing point, $\Delta T_f = i \times K_f \times m$

van't Hoff factor, $i = \frac{1 - \alpha + n\alpha}{1}$, where n = no. of ions produced by complete dissociation of 1 mole of HX.

$HX \rightleftharpoons H^+ + X^- \quad \Rightarrow n = 2$

$\therefore i = \frac{1 - 0.2 + 2 \times 0.2}{1} = 1.2$

$\left[\text{For 20\% ionisation, } \alpha = \frac{20}{100} = 0.2\right]$

$\therefore \Delta T_f = 1.2 \times 1.86 \times 0.2 = 0.45\ [\because m = 0.2]$

Hence, freezing point of solution is 0 – 0.45

= –0.45°C

32. (d) The salt that ionises to least extent will have highest freezing point. [*i.e.*, minimum ΔT_f] Glucose, being non electrolyte, gives minimum no. of particles and hence, minimum ΔT_f or maximum F. pt.

33. (a) The salt producing highest number of ions will have lowest freezing point.

$K_2SO_4 \rightarrow 2K^+ + SO_4^{2-}$; K_2SO_4 gives highest number of particles (2 + 1 = 3).

Glucose, being non-electrolyte gives minimum no. of particles and hence minimum ΔT_f or maximum F. pt.

34. (a) Added HgI_2 forms a complex with KI in the solution as follows

$2KI + HgI_2 \rightarrow K_2[HgI_4]$

As a result, number of particles decreases and so ΔT_f increases.

Depression in freezing point is a colligative property which is not observed here.

35. (150) Vessel-1

$(\Delta T_b)_1 = i_X \frac{w_2}{M_X} \cdot \frac{1}{w_1} \times 1000 \times k_b$

M_X = Molar mass of 'X'

Vessel-2

$(\Delta T_b)_2 = i_Y \frac{w_2}{M_Y} \cdot \frac{1}{w_1} \times 1000 \times k_b$

M_Y = Molar mass of 'Y'

$\frac{(\Delta T_b)_1}{(\Delta T_b)_2} \times 100 = \frac{i_X}{i_Y} \cdot \frac{M_X}{M_Y} \times 100$

$= 1.2 \times \frac{100}{80} \times 100 = 150\%$

36. (682)

Weight of 50 mL 0.2 molal urea

$= V \times d = 50 \times 1.012$

= 50.6 gm

Given 0.2 molal implies

1000 gm solvent has 0.2 moles urea

So weight of solution = 1000 + (0.2 × 60) = 1012 gm.

So wt. of urea in 50.6 gm solution

$= \frac{12 \times 50.6}{1012} = 0.6\text{gm}$

Total urea = 0.6 + 0.06 = 0.66 gm

Total volume = 300 mL = 0.3 L

Now, osmotic pressure $\pi = C \times R \times T$

$= \frac{0.66 \times 62 \times 300}{60 \times 0.3} = 682$ Torr.

37. (5) Vapour pressure of water,

$P^o = 60$ mm Hg.

Vapour pressure of solution,

$P_s = 59.724$ mm Hg

Let the number of ions present per formula unit of the ionic salt S be n.

$$\begin{array}{ccc} S_n & \longrightarrow & nS \\ 0.1 & & 0 \\ 0.1(1-0.9) & & 0.1 \times 0.9 \times n \end{array}$$

$i = \frac{0.1 \times 0.9 \times n + 0.1(1-0.9)}{0.1} \Rightarrow i = 0.1 + 0.9n$

$\frac{P^o - P_s}{P^o} = i \times \text{mole fraction of solute}$

$\Rightarrow \frac{60 - 59.724}{60} = i \times \frac{0.1}{\frac{1800}{18} + 0.1}$

$\Rightarrow i = 4.6 \Rightarrow 0.1 + 0.9n = 4.6 \Rightarrow n = 5.$

38. (1) Given $\Delta T_f = 0.0558°C$

as we know, $\Delta T_f = i \times K_f \times m$

$\Rightarrow 0.0558 = i \times 1.86 \times 0.01 \Rightarrow \boxed{i = 3}$

Therefore, the complex will be $[Co(NH_3)_5Cl]Cl_2$.

$[Co(NH_3)_5Cl]Cl_2 \rightarrow [Co(NH_3)_5Cl]^+ + 2Cl^-$

Hence, number of chloride in co-ordination sphere is 1.

39. (2)

$$\begin{array}{ccccc} MX_2 & \rightleftharpoons & M^{2+} & + & 2X^- \\ 1-\alpha & & \alpha & & 2\alpha \end{array}$$

$i = \frac{\text{obser. depression in f.pt of aq. solution}}{\text{depression of f.p.t. in absence of ionic dissociation}}$

$= \frac{1 - \alpha + \alpha + 2\alpha}{1} = 1 + 2\alpha = 1 + 2 \times 0.5 = 2$

40. (2000.00)

$P_T = X_A P_A^o + X_B P_B^o$

$100 = \frac{1000/50}{\frac{1000}{50} + 5}(105) + \frac{5}{\frac{1000}{50} + 5}\left(P_B^o\right)$

$P_B^o = 80$ mm of Hg

Molar volume of vapour of B

$$V_m = \frac{RT}{P} = \frac{0.08 \times 300}{80/760} \Rightarrow V_m = 228 \text{ lit/mol}$$

Molar volume of B in liquid phase

$$V'_m = \frac{57}{0.5} = 114 \text{ ml/mol}$$

$$\text{Ratio} = \frac{\text{molar volume in vapour}}{\text{molar volume in liquid}}$$

$$= \frac{228 \times 1000}{114} = 2000$$

41. (0.16)

$$P_B = x_B P_B^\circ \Rightarrow P_B = 0.2 \times 80 = 16$$

$$y_B = \frac{P_B}{P_T} = \frac{16}{100} = 0.16$$

42. (2.49) $\pi = \rho gh = CRT$

$\rho gh = 10^3 \times 10 \times 2 \times 10^{-2}$ pascal = 200 pascal

$$\therefore 200 = \frac{2}{M} \times 1000 \times 8.3 \times 300$$

$M = 24900 = 2.49 \times 10^4$ g/mol

$\therefore$ X = 2.49.

43. (100.1) In solution-A

$$AgNO_3(aq) \longrightarrow Ag^+(aq) + NO_3^-(aq)$$

$\because \Delta T_b = iK_b m$

$\Delta T_b = 2 \times 0.5 \times 0.1 [i = 2]$

$(T_s - T^0) = 0.1$ K

$T_s = 373 + 0.1 = 373.1$ K = 100.1°C.

44. (2.5) In solution – B

$$\underset{0.05\,m}{AgNO_3(aq)} \longrightarrow \underset{0.05\,m}{Ag^+(aq)} + \underset{0.05\,m}{NO_3^-(aq)}$$

$$\underset{0.05\,m}{BaCl_2(aq)} \longrightarrow \underset{0.05\,m}{Ba^{2+}(aq)} + \underset{0.1\,m}{2Cl^-(aq)}$$

Ag^+ and Cl^- combine to form AgCl precipitate.

	$Ag^+(aq)$	+ $Cl^-(aq)$	$\longrightarrow AgCl(s)$
t = 0	0.05 m	0.1 m	
t = ∞	0	0.05 m	

In final solution (B) total concentration of all ions is

$[Cl^-] + [NO_3^-] + [Ba^{2+}] = 0.05 + 0.05 + 0.05 = 0.15$ m

$\Delta T_b = 0.5 \times 0.15 = 0.075$ °C

B.P. of solution 'B' = 100.075 °C

B.P. of solution 'A' = 100.1 °C

$|y| = 100.1 - 100.075 = 0.025 = 2.5 \times 10^{-2}$

45. (1.03)

We know that,

$$\frac{P^0 - P}{P^0} = \frac{w_2 \times m_1}{m_2 \times w_1}; \quad \frac{10}{640} = \frac{0.5 \times 78}{m_2 \times 39}$$

$m_2 = 64$ g

$$\Delta T_f = \frac{K_f \times w_2 \times 1000}{m_2 \times w_1}$$

$$\Delta T_f = \frac{5.12 \times 0.5 \times 1000}{64 \times 39} = 1.0256 \approx 1.03$$

46. (0.05) From graph

For solvent 'X' $\Delta T_{b(x)} = 362 - 360 = 2$ K

$\Delta T_{b(x)} = m_{NaCl} \times K_{b(x)}$ ...(1)

For solvent 'Y' $\Delta T_{b(y)} = 368 - 367 = 1$ K

$\Delta T_{b(y)} = m_{NaCl} \times K_{b(y)}$ ...(2)

Dividing equation (1) by (2)

$$\Rightarrow \frac{K_{b(x)}}{K_{b(y)}} = 2$$

For solute 'S'

Given solute S dimerizes in solvent. Hence,

2(S)	→	S_2	
1			initial
(1 – α)		α/2	after dimerisation

$$i = 1 - \alpha + \frac{\alpha}{2} = 1 - \frac{\alpha}{2}$$

For solvent 'Y', $\alpha_Y = 0.7$

$$i_Y = 1 - \frac{0.7}{2} = 0.65$$

For solute 'S', it is given that

$$\Delta T_{b(x)} = 3\Delta T_{b(y)}$$

$$i_x \times K_{b(x)} \times m_5 = i_Y \times K_{b(y)} \times m_5 \times 3$$

$$i_x \times \frac{K_{b(x)}}{K_{b(y)}} = i_Y \times 3; \; i_x \times 2 = 3 \times 0.65; \; i_x = 0.975$$

$$1 - \frac{\alpha_x}{2} = 0.975; \; \alpha_x = 0.05$$

47. (75)

	$2C_6H_5OH$	⇌	$(C_6H_5OH)_2$
Initial no. of moles	1		0
No. of moles at equilibrium	1 – α		α / 2

Total number of moles at equilibrium

$$= 1 - \alpha + \frac{\alpha}{2} = 1 - \frac{\alpha}{2}$$

$\Delta T_f = iK_f \times$(molality)

$$\Rightarrow 7 = 14 \times \frac{75.2}{94} \times \left(1 - \frac{\alpha}{2}\right)$$

$$\left[\begin{array}{l}\text{weight of phenol} = 75.2\text{g} \\ \text{mol.wt of phenol} = 94\end{array}\right]$$

$\therefore \alpha = 0.75$

So, the percentage of phenol that dimerises = **75%.**

48. (0.228) $\Delta T_f = i \times K_f \times m$

Weight of water = 500 × 0.997 = 498.5 g

(weight = volume × density)

No. of moles of acetic acid

$$= \frac{\text{Wt. of } CH_3COOH \text{ in g}}{\text{Mol. wt. of } CH_3COOH} = \frac{3 \times 10^{-3} \times 10^3}{60} = 0.05$$

Since 498.5 g of water has 0.05 moles of CH_3COOH

$$1000 \text{ g of water has} = \frac{0.05 \times 1000}{498.5} = 0.1$$

Therefore, molality of the solution = 0.1

Determination of van't Hoff factor, i

	CH_3COOH	$\longrightarrow CH_3COO^-$	+ H^+
at start	1	0	0
at equb.	1 – 0.23	0.23	0.23

Therefore, van't Hoff factor

$$= \frac{\text{No. of particles after dissociation}}{\text{No. of particles before dissociation}}$$

$$= \frac{1 - 0.23 + 0.23 + 0.23}{1} = 1.23$$

Now, we know that,

$\Delta T_f = i \times K_f \times m = 1.23 \times 1.86 \times 0.1 =$ **0.228 K**

49. (23.44) Depression in freezing point,

$\Delta T_f = 0 - (-0.30) = 0.30$

Now we know that $\Delta T_f = K_f m$

$$\therefore \quad m = \frac{\Delta T_f}{K_f} = \frac{0.30}{1.86} = 0.161$$

According to Raoult's law

$$\frac{p^\circ - p}{p^\circ} = \frac{\text{No. of moles of solute}}{\text{No. of moles of solvent}}$$

$$\frac{23.51 - p}{23.51} = \frac{0.161}{1000/18} = \frac{0.161 \times 18}{1000}$$

$$\left(\because \text{No. of moles of } H_2O = \frac{1000}{18}\right)$$

$$\Rightarrow \frac{23.51 - p}{23.51} = 0.0020898$$

p **= 23.44 mm Hg**

50. (156.056) Given Wt. of benzene (solvent),

W = Volume × density = 50 × 0.879 = 43.95 g

Wt. of compound (solute), w = 0.643 g

Mol. wt. of benzene, M_1 = 78; Mol. wt. of solute, m_2 = ?

Depression in freezing point, $\Delta T_f = 5.51 - 5.03 = 0.48$

Molal freezing constant, $K_f = 5.12$

Now we know that,

$$m_2 = \frac{1000 \times K_f \times w}{W \times \Delta T_f} = \frac{1000 \times 5.12 \times 0.643}{43.95 \times 0.48}$$

= 156.056

51. (746.3) First find moles of $Ca(NO_3)_2$ and water. Then use the expression $\frac{p^\circ - p}{p^\circ} = \frac{n}{n + N}$ to find vapour pressure of solution.

Let initially 1 mole of $Ca(NO_3)_2$ is taken

$$\text{Degree of dissociation of } Ca(NO_3)_2 = \frac{70}{100} = 0.7$$

Ionisation of $Ca(NO_3)_2$ can be represented as

	$Ca(NO_3)_2$	⇌	Ca^{2+}	+	$2NO_3^-$
At start	1		0		0
At equilibrium	1 – 0.7		0.7		2 × 0.7

$\therefore$ Total number of moles in the solution at equilibrium

$= (1 - 0.7) + 0.7 + 2 \times 0.7 = 2.4$

No. of moles when the solution contains 1 g of calcium nitrate instead of 1 mole of the salt

$$= \frac{2.4}{164}$$ (164 is the mol. wt. of Cal. nitrate)

$\therefore$ No. of moles of the solute in the solution containing 7 g of salt, *i.e.*,

$$n = \frac{2.4}{164} \times 7 = 0.102$$

No. of moles of water (N)

$$= \frac{\text{Wt. of water}}{\text{Mol. wt. of water}} = \frac{100}{18} = 5.55$$

Applying Raoult's law, $\frac{p^0 - p}{p^0} = \frac{n}{n+N}$

$$\frac{760-p}{760} = \frac{0.102}{0.102+5.55}$$

$$\Rightarrow \frac{760}{760} - p = 0.0180$$

$\Rightarrow$ $p = 760 - (760 \times 0.0180) = $ **746.3 mm Hg**

52. **(65.25)** According to Raoult's law,

$$\frac{p^o - p}{p^o} = \frac{w/m}{w/m + W/M}$$

Here, $p^o = 640$ mm $\quad p = 600$ mm

$w = 2.175$ g $\quad W = 39.0$

$m = ?$ $M = 78$

Substituting the various values in the above equation for Raoult's law :

$$\frac{640-600}{640} = \frac{2.175/m}{2.175/m + 39/78}$$

$$\Rightarrow \frac{1}{16} = \frac{2.175}{2.175 + 0.5m} \Rightarrow m = \mathbf{65.25}$$

53. K_f; Depression in freezing point, $\Delta T_f = K_f$. m, where K_f is the molar depression constant or cryoscopic constant of the solvent.

54. **(a, d)** The freezing point of a solvent depresses as a non-volatile solute is added to a solvent. According to Raoult's law, when a non-volatile solute is added to a solvent the vapour pressure of the solvent decreases. At the freezing point, it will be only the solvent molecules which will solidify.

55. **(d)** The solution M is a mixture of ethanol and water. In it the mole fraction of ethanol is 0.9 and mole fraction of water is 0.1 (1.0 – 0.9 = 0.1). Here, solvent is ethanol and solute is water.

Also given are:-

Standard freezing point of ethanol = 155.7K.

Freezing point depression constant (K_f) for ethanol = 2.0 K kg mol^{-1}

Molecular weight of ethanol (C_2H_5 OH) = 46

now molality (m) of solution $= \frac{0.1 \times 1000}{0.9 \times 46}$

Using the formula :-

Depression in freezing point (ΔT_f) = $K_f \times m$,

Substituting various values, we get

$$\Delta T_f = \frac{2 \times 0.1 \times 1000}{0.9 \times 46} = \frac{2000}{414} = 4.83 \text{ Kmn.}$$

$\therefore$ Freezing point of solution 'M' = (155.7 – 4.83) K = 150.9 K

i.e. (d) is the correct answer.

56. **(b)** **Given :** vapour pressure of pure ethanol (P_A^o) = 40 mm Hg Mole fraction (X_A) of ethanol in solution = 0.9

Using the formula : Total pressure

$$(P) = P_A^o \, X_A^o$$

Substituting the given values, we get

$P = 40 \times 0.9 = 36.0$ mm Hg

i.e. (b) is the correct answer.

It is given that the solute is non-volatile, in the question.

57. **(b)** **Given:** Standard boiling point of water = 373K

Boiling point elevation constant of water (K_b) = 0.52 Kg mol^{-1}

Molecular weight of water (H_2O) = 18

Mole fraction of ethanol in solution = 0.1 (1.0 – 0.9 = 0.1)

Now, solvent is water and solute is ethanol

$$\text{molality } (m) = \frac{0.1 \times 1000}{0.9 \times 18}$$

using the relation $\Delta T_b = K_b \times m$

Substituting the given values, we get

$$\Delta T_b = 0.52 \times \frac{0.1 \times 1000}{0.9 \times 18} = \frac{520}{9 \times 18} = 3.20 \text{ K}$$

$\therefore$ Boiling point of solution = (373 + 3.20) K = 376.2 K

i.e. option (b) is correct answer.

58. (i) $\Delta T_b = K_b \times m$

In first case,

$$\Delta T_b = K_b \times m = K_b \times \frac{\left(\frac{\text{Wt. of solute}}{\text{Mol. wt. of solute}}\right)}{\text{wt. of solvent (in kg)}}$$

$$\text{or } 0.17 = 1.7 \times \frac{1.22}{M \times 100 \times 10^{-3}} \quad \text{or } \mathbf{M = 122}$$

Thus, benzoic acid exists as a monomer in acetone

(ii) In second case,

$$\Delta T_b = K_b \times \frac{\frac{\text{Wt. of solute}}{\text{M. W. of solute}}}{\text{wt. of solvent} \times 10^{-3}}$$

$$\text{or } 0.13 = 2.6 \times \frac{1.22}{M' \times 100 \times 10^{-3}} \Rightarrow \mathbf{M' = 224}$$

Double the expected molecular weight of benzoic acid (244) in benzene solution indicates that benzoic acid exists as a dimer in benzene.

59. Higher the value of dipole-dipole interaction higher is b.p.

Higher value of K_b of a solvent suggests larger polarity of solvent molecules which in turn leads to higher dipole – dipole interaction implies higher boiling point due to dipole – dipole interaction. Therefore, the correct order of K_b values of the three given solvents is

$$\text{Mathematically, } K_b = \frac{M_A R T_b^2}{\Delta H_{\text{vap}} \times 1000}$$

or $K_b \propto T_b$ (b.pt.)

Thus,

	Boiling point	K_b
x	100	0.63
y	27	0.53
z	253	0.98

60.

Element	%	Relative no.	Simplest ratio of atoms
C	42.86	$\frac{42.86}{12} = 3.57$	$\frac{3.57}{1.19} = 3$
H	2.40	$\frac{2.40}{1} = 2.40$	$\frac{2.40}{1.19} = 2$
N	16.67	$\frac{16.67}{14} = 1.19$	$\frac{1.19}{1.19} = 1$
O	38.07	$\frac{38.07}{16} = 2.38$	$\frac{2.38}{1.19} = 2$

$\therefore$ Empirical formula of the minor product is $C_3H_2NO_2$

Molar empirical formula mass of the minor product = 3 × 12 + 2 × 1 + 1 × 14 + 2 × 16 = 84 g mol^{-1}

Let M be the molar mass of the minor product. For 5.5 g of the minor product dissolved in 45 g benzene, the molality (m) of the solution

$$= \frac{5.5 \text{ g}/M}{0.045 \text{ kg}}$$

Substituting this in the expression of elevation of boiling point, $\Delta T_b = K_b m$

$$\Rightarrow \quad 1.84 \text{ K} = (2.53 \text{ K kg mol}^{-1}) \left(\frac{5.5 \text{ g}/M}{0.045 \text{ kg}}\right)$$

or $\mathbf{M = 168 \text{ g mol}^{-1}}$

No. of unit of empirical formula in molecular formula

$$= \frac{168 \text{ g mol}^{-1}}{84 \text{ g mol}^{-1}} = 2$$

Hence, the molecular formula of the minor product is **$2(C_3H_2NO_2)$, *i.e.*, $C_6H_4(NO_2)_2$.**

The product is ***m*-dinitrobenzene**

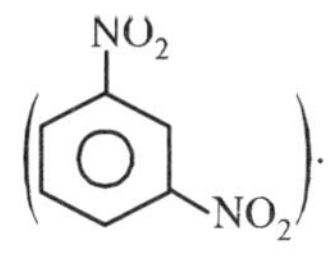

61. $\frac{p^o - p}{p^o} = \frac{\frac{w}{m}}{\frac{w}{m} + \frac{W}{M}}$

Here, w and m are wt. and molecular wt. of solute, W and M are wt. and molecular weight of solvent. p = Vap. Pressure of solution; p^o = Normal vapour pressure

Let the initial (normal) pressure (p^o) = p

$\therefore$ Pressure of solution $= \frac{75}{100} \times p = \frac{3}{4} p$

$m = 60$, $M = 18$, $W = 100$ g

$$\therefore \frac{p - \frac{3}{4}p}{p} = \frac{w/60}{\frac{w}{60} + \frac{100}{18}}$$

$$\Rightarrow \frac{1}{4} = \frac{w/60}{(w/60) + 5.55}$$

$$\frac{4w}{60} = \frac{w}{60} + 5.55$$

$$\Rightarrow \frac{3w}{60} = \frac{w}{20} = 5.55 \text{ or } w = \mathbf{111 \text{ g}}$$

$$\text{Molality} = \frac{\text{No. of moles of solute}}{\text{Wt. of solvent}} \times 1000$$

$$= \frac{111 \times 1000}{60 \times 100} = \mathbf{18.52 \text{ m}}$$

$\Rightarrow$ 4T_b = 0.0512 °C.

Electrochemistry

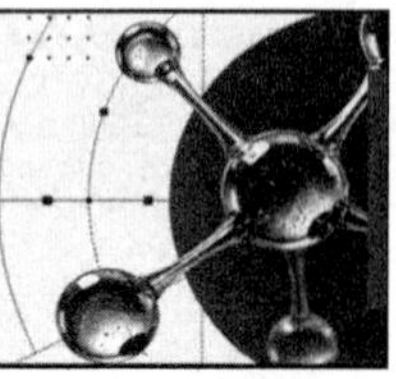

Part-A (JEE Main)

1. (2)

$Fe^{+2}(aq) + 2e^- \rightarrow Fe(s)$ .... ΔG_1° ...(i)

$Fe^{+3}(aq) + 3e^- \rightarrow Fe(s)$ .... ΔG_2° ...(ii)

$Fe^{+3}(aq) + e^- \rightarrow Fe^{+2}(aq)$ .... ΔG_3° ...(iii)

According to equations (i), (ii) & (iii),

Eq.(iii) = Eq.(ii) – Eq.(i)

$$-1 \times F \times E^o_{Fe^{3+}/Fe^{2+}} = -3 \times F \times Y - (-2 \times F \times X)$$

$$E^o_{Fe^{3+}/Fe^{2+}} = 3Y - 2X.$$

2. (4)

Anode:

$[Red_1 \rightarrow Ox_1^{n_1^+} + n_1e^-] \times n_2$ (Oxidation)

Cathode:

$[Ox_2 + n_2e^- \rightarrow Red_2^{n_2^-}] \times n_1$ (Reduction)

Overall reaction :

$$n_2Red_1 + n_1Ox_2 \rightarrow n_2Ox_1^{n_1^+} + n_1Red_2^{n_2^-}$$

$$E = E^o - 2.303\frac{RT}{nF}\log Q$$

$$\frac{(E - E^o)}{RT/F} = -\frac{2.303}{n}\log Q$$

$$\text{Slope} = \frac{-2.303}{n}$$

Straight line passing through origin.

Electrical work = Charge × Potential difference

Thus statement (4) is incorrect.

3. (2) Cell Reaction :

$$Zn(s) + 2Ag^{\oplus}(aq) \longrightarrow 2Ag(s) + Zn^{+2}(aq)$$

$$E_{cell} = E^0_{cell} - \frac{0.059}{n}\log\frac{[Zn^{2+}]}{[Ag^+]^2}$$

$$1.6 = 1.56 - \frac{0.059}{2}\log\frac{1}{(Ag^+)^2}$$

$$\Rightarrow 0.04 = 1.56 - \frac{0.059}{2}\log\frac{1}{(x)^2}$$

$$\Rightarrow \frac{0.04 \times 2}{0.059} = 2\log x$$

$$\Rightarrow \log x = \frac{4}{5.9}$$

4. (1) Overall reaction

$$Fe(s) + 2OH^-(aq) + 2AgBr(s) \rightleftharpoons Fe(OH)_2(s) + 2Ag(s) + 2Br^-(aq)$$

$$E^o_{cell} = E^o_{R(cathode)} - E^o_{R(anode)}$$

$$= 0.07 - (-0.88) = 0.95\text{ V}$$

Option (1) is correct.

Option (2) is incorrect as the calculated E^o_{cell} is positive.

Option (3) is incorrect as Fe is oxidised.

Option (4) is incorrect because E^o_{cell} is an intensive property.

5. (3)

At anode: $M \rightarrow M^+_{(a)} + e^-$

At cathode: $M^+_{(c)} + e^- \rightarrow M$

$$M^+_{(c)} \rightarrow M^+_{(a)}$$

$$E_{cell} = -\frac{RT}{F}\ln\frac{M^+_{(a)}}{M^+_{(c)}}$$

In order to have $E_{cell} > 0$

$$\left[M^+_{(c)}\right] > \left[M^+_{(a)}\right]$$

6. (4) C, D and E are correct

Anode: $Ag + Cl^- \rightarrow AgCl + e^-$

Cathode: $Fe^{3+} + e^- \rightarrow Fe^{2+}$

Net

$Ag + Cl^- + Fe^{3+} \rightarrow AgCl + Fe^{2+}$

$E_{cell}\uparrow$ if $Q\downarrow$

(C) $\Rightarrow$ Fe^{2+} is decreased $\Rightarrow Q\downarrow$

(D) $\Rightarrow$ Fe^{3+} is increased $\Rightarrow Q\downarrow$

(E) $\Rightarrow$ Cl^- is increased $\Rightarrow Q\downarrow$

7. (2) The standard cell potential E°_{cell} is fixed for a specific electrochemical reaction at a given temperature and pressure. It does not change over time. Thus the graph between E°_{cell} versus time should be a horizontal line.

8. (2) According to Nernst equation:-

$$E = E^\circ - \frac{RT}{nF}\ln Q$$

Cell reaction:-

$$Mg(s) + 2Ag^+(aq) \rightleftharpoons 2Ag(s) + Mg^{+2}(aq)$$

According to Nernst equation,

$$E_{cell} = E^\circ_{cell} - \frac{RT}{2F}\ln\frac{\left[Mg^{+2}\right]}{\left[Ag^+\right]^2}$$

$$\Rightarrow E_{cell} = E^\circ + \frac{RT}{2F}\ln\frac{\left[Ag^+\right]^2}{\left[Mg^{2+}\right]}$$

9. (4) Higher the standard reduction potential value (+ve) higher will be the oxidising capacity. Therefore option (d) is the corect answer.

10. (2) For strongest reducing agent Reduction potential should be lowest Hence Cr is the strongest reducing agent.

11. (2)

$$Fe^{2+}(aq) + Ag^+(aq) \rightarrow Fe^{3+}(aq) + Ag(s)$$

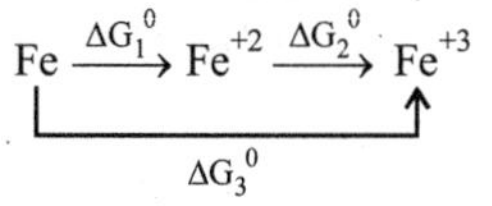

$$\Delta G_3^0 = \Delta G_1^0 + \Delta G_2^0$$

$$= -3F(-z) = -2F(-y) + \Delta G_2^0$$

$$\Delta G_2^0 = 3Fz - 2Fy\ ;$$

Also $\Delta G_2^0 = -nFE^0_{Fe^{+2}/Fe^{+3}}$

$$3Fz - 2Fy = -1F\left(E^0_{Fe^{+2}/Fe^{+3}}\right)$$

$E^0_{Fe^{+2}/Fe^{+3}} = 2y - 3z$; E^0_{cell} for reaction will be

$$E^0_{Ag^+/Ag} + E^0_{Fe^{+2}/Fe^{+3}} = x + 2y - 3z$$

12. (1) $\because \Delta G^\circ = -nFE^\circ$

Option (1) $E^\circ_{cell} = 0.8 + 0.76 = 1.56$ V

$\therefore\ \Delta G^\circ = -2 \times F \times 1.56 = -3.12$ V

Option (2) $E^\circ_{cell} = -2.37 + 0.76 = -1.61$ V

$\therefore\ \Delta G^\circ = -2 \times F \times (-1.61) = +3.22$ V

Option (3) $E^\circ_{cell} = -2.37 - 0.8 = -3.17$ V

$\therefore\ \Delta G^\circ = -2 \times F \times (-3.17) = +6.34$

Option (4) $E^\circ_{cell} = 0.8 - 0.34 = 0.46$ V

$\therefore\ \Delta G^\circ = -2 \times F \times 0.46 = -0.92$ V

13. (1)

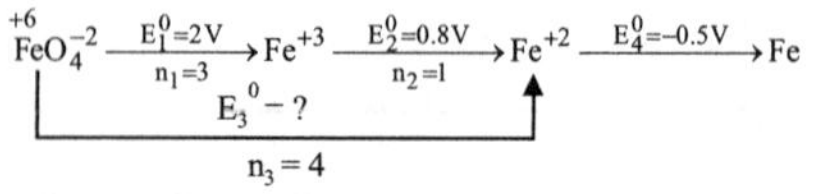

$$\Delta G_3^\circ = \Delta G_1^\circ + \Delta G_2^\circ \Rightarrow -n_3FE_3^\circ = -n_1FE_1^0 - n_2FE_2^\circ$$

$$\Rightarrow +4E_3^\circ = 3 \times 2 + (1 \times 0.8)$$

$$\Rightarrow E_3^\circ = \frac{6.8}{4}\text{V} \Rightarrow E_3^\circ = 1.7\text{ V}$$

14. (4)

15. (1)

16. (3) $E^\circ_{cell} = E^\circ_{cathode} - E^\circ_{anode}$

= 0.34 – 0.46

= –0.12V (negative value nonspontaneous)

Thus reverse reaction is spontaneous.

17. (1) $2H^+(aq) + 2e^- \rightarrow H_2(g)$

$$E = E^\circ - \frac{0.059}{n}\log\frac{P_{H_2}}{\left[H^+\right]^2}$$

For pure water, $[H^+] = 10^{-7}$M

$$0 = 0 - \frac{0.059}{2}\log\frac{P_{H_2}}{\left(10^{-7}\right)^2}$$

$$\log\frac{P_{H_2}}{\left(10^{-7}\right)^2} = 0 \Rightarrow \frac{P_{H_2}}{10^{-14}} = 1$$

$P_{H_2} = 10^{-14}$bar

18. (1) Calomel electrode is a metal-insoluble salt anion electrode.

19. (2) Anode : $\frac{1}{2}H_2(g) \rightleftharpoons H^+(aq) + e^-$

Cathode : $AgCl(s) + e^- \rightarrow Ag(s) + Cl^-(aq)$

20. (3) Metal cation with negative value of reduction potential (M^{+3}/M^{+2}) or with positive value of oxidation potential (M^{+2}/M^{+3}) will liberate H_2 from dilute acid.

21. (2)

22. (3) A cell with less variation in EMF with temperature is preferred as reference electrode because it can be used for wider range of temperature without much deviation from standard value. So a cell with less $\left(\frac{\partial E}{\partial T}\right)_p$ is preferred.

23. (1) According to standard electrode potential table, the order of electrode couple from top to bottom is :

$Cl_2/Cl^- > Ag^+/Ag > I_2/I^- > Na^+/Na > Li^+/Li$

24. (2) $E^o_{cell} = E^o_{Cu^{2+}|Cu} - E^o_{Zn^{2+}|Zn} = 1.1\,V$

So, if $E_{ext.} = 1.1$ V, no electron will flow.

At, $E_{ext.} > 1.1$ V, cell act as electrolytic cell and electron will flow from Cu to Zn.

At, $E_{ext.} < 1.1$ V, cell act as electrochemical cells so Zn dissolves at anode and Cu deposits at cathode.

25. (4)

26. (2) $Cu^{2+} + 2e^- \longrightarrow Cu, \Delta G_1^\circ = -2F(0.34)$ ...(i)

$Cu^+ + e^- \longrightarrow Cu, \Delta G_2^\circ = -F(0.522)$ ...(ii)

Subtract (ii) from (i)

$Cu^{2+} + e^- \longrightarrow Cu^+, \quad \Delta G_3^\circ = -F(E^0)$

$\therefore \quad \Delta G_1^\circ - \Delta G_2^\circ = \Delta G_3^\circ$

$\Rightarrow \quad -FE^\circ = -2F(0.34) + F(0.522)$

$\Rightarrow \quad E^\circ = 0.68 - 0.522 = 0.158\,V$

27. (1) $\Delta G^\circ = -nFE^\circ_{cell}$

$= -2 \times (96000) \times 2\,V = -384000$ J/mol

$= -384$ kJ/mol

28. (2) Higher the reduction potential, higher will be oxidising power.

So, $Bi^{3+} < Ce^{4+} < Pb^{4+} < Co^{3+}$

29. (3)

$Ag^+(aq) + Fe^{2+}(aq) \longrightarrow Fe^{3+}(aq) + Ag(s)$

$E^\circ_{cell} = E^\circ_{Ag^+/Ag} - E^\circ_{Fe^{3+}/Fe^{2+}}$

$E^\circ_{Fe^{3+}/Fe^{2+}} = 3z - 2y; \quad E^\circ_{Ag^+/Ag} = x \quad$ (given)

$\therefore \; E^\circ_{cell} = x - 3z + 2y$

30. (3) Reduction potential is maximum for $S_2O_8^{2-}$, therefore, it is the strongest oxidising agent amongst the given species.

31. (1) $\Delta_r H^\circ = \Delta G_r^\circ + T\Delta S^\circ$

$\Delta_r H^\circ = -nFE^\circ + TnF\frac{dE^\circ}{dT}$

Cell reaction: $Zn(s) + Cu^{2+}(aq) \longrightarrow Zn^{2+}(aq) + Cu(s)$

$\Delta_r H^\circ = -nF\left(E^\circ - \frac{TdE^\circ}{dT}\right)$

$\Delta_r H^\circ = -2 \times 96000[2 - 300 \times (-5 \times 10^{-4})]$

$= -412.8 \times 10^3$ J/mol $= -412.8$ kJ/mol

32. (3) $E^0_{cell} = \frac{2.303RT}{nF}\log K_C$

or $\quad E^0_{cell} = \frac{0.059V}{n}\log K_C$

$= \frac{0.059V}{2}\log 10^{16} = 0.4736\,V$

33. (1) $Zn + 2Ag^+ \longrightarrow 2Ag + Zn^{2+} \; (n = 2e^-)$

$E^\circ_{cell} = (E^\circ_{R.P.})_{cathode} - (E^\circ_{R.P.})_{anode}$

$= 0.80 - (-0.76) = 1.56$ V for $2e^-$

$\therefore \; E^\circ_{cell}$ for $1e^- = \frac{1.56}{2} = 0.78\,V$

34. (4)

35. (2) Higher the oxidation potential, higher will be the reducing power. So, the order of reducing behaviour is: Ca > Mg > Zn > Ni

36. (4) We know that, $\ln K = \frac{nFE^\circ_{cell}}{RT}$

After putting the given values, we get

$\ln K = \frac{2 \times 96000 \times 2}{8 \times 300} = 160 \quad \therefore K = e^{160}$

37. (2) Cell reaction :

$M(s) + 3Ag^+(aq) \longrightarrow M^{3+}(aq) + 3Ag(s)$

$E = E^0 - \frac{0.0591}{n}\log\frac{[\text{Reduced state}]}{[\text{Oxidised state}]}$

$0.421 = E^0 - \frac{0.0591}{3}\log\frac{0.001}{(0.01)^3}; \; E^0 = 0.48\,V$

$E^0 = E^0_{Ag^+/Ag} - E^0_{M^{3+}/M}$

$E^0_{M^{3+}/M} = 0.8V - 0.48V = 0.32$ Volt

38. (1) $E^\circ_{Mn^{2+}/MnO_4^-} = -1.51\,V$

$E^\circ_{Cl^-/Cl_2} = -1.36\,V; \; E^\circ_{Cr^{3+}/C_2O_7^{2-}} = -1.33\,V$

$E^\circ_{Cr/Cr^{3+}} = +0.74\,V$

Since Cr is having highest oxidation potential, Cr is the best reducing agent.

39. (3)

(i) $\underset{\text{Most stable}}{Al^{3+}} \xrightarrow{E^\circ = -1.66} Al \xleftarrow{E^\circ = +0.55} \underset{\text{Less stable}}{Al^+}$

(ii) $\underset{\text{Less stable}}{Tl^{3+}} \xrightarrow{E^\circ = +1.26} Tl \xleftarrow{E^\circ = -0.34} \underset{\text{More stable}}{Tl^+}$

Tl^+ has negative electrode potential ($E^\circ = -0.34$) means, it does not prefer to convert into Tl but reverse must be preferred that's why it is more stable than Tl^{3+} ($E^\circ = +1.26$). In Al, Al^{3+} is more stable ($E^\circ = -1.66$) than Al^+ ($E^\circ = +0.55$) and also from Tl^+ due to more negative value of E°. Therefore, by comparison it confirms that Tl^+ is more stable than Al^+.

40. (1) $\Delta G^\circ = -nFE^\circ$

(i) $Fe^{2+} + 2e^- \longrightarrow Fe; \quad E^0 = -0.47V;$

(ii) $Fe^{3+} + e^- \longrightarrow Fe^{2+}; \; E^0 = +0.77V;$

(iii) $Fe^{3+} + 3e^- \longrightarrow Fe$

(i) $\Delta G^0 = -nFE^0 = -2(-0.47)F = 0.94F$

(ii) $\Delta G^0 = -nFE^0 = -1(+0.77)F = -0.77F$

(iii) On adding : $\Delta G^0 = +0.17$ F

$\Delta G^0 = -nFE^0 \Rightarrow E^0$ for

$(Fe^{3+} \longrightarrow Fe) = \frac{\Delta G^0}{-nF} = \frac{0.17F}{-3F} = -0.057V$

41. (3) No reaction will occur as the Zn is placed above Cu in electrochemical series. Hence, there will be no displacement reaction.

42. (2) $MnO_4^- + 8H^+ + 5e^- \longrightarrow Mn^{2+} + 4H_2O$

$E = E^\circ - \frac{0.059}{n}\log\frac{[Mn^{2+}]}{[MnO_4^-][H^+]^8}$

Taking Mn^{2+} and MnO_4^- in standard state, i.e., 1 M,

$E = 1.51 - \frac{0.059}{5} \times 8\log\frac{1}{[H^+]}$

$= 1.51 - \frac{0.059}{5} \times 8 \times 3 = 1.2268\,V$

Hence at this pH, MnO_4^- will oxidise only Br^- and I^- as SRP of Cl_2/Cl^- is 1.36 V which is greater than that for MnO_4^-/Mn^{2+}.

43. (3) EMF of galvanic cell = 1.1 volt

If E_{ext} < EMF then electrons flow steadily from anode to cathode, while if E_{ext} > EMF then electrons flow from cathode to anode as polarity is changed.

44. (2) The order of electrode oxidation Potential : $Br^- < Fe^{2+} < Al$

The same order will be for their reducing power.

45. (3) As the value of reduction potential decreases, the sequence of deposition also decreases.

C > B > D < A

(0.85) (0.6) (–0.76) (–1.2)

46. (1) $Mn^{2+} + 2e^- \rightarrow Mn; E^\circ = -1.18\,V;$ ...(i)

$Mn^{3+} + e^- \rightarrow Mn^{2+}; E^\circ = +1.51\,V;$ ...(ii)

Now multiplying equation (ii) by two and subtracting from equation (i):

$3Mn^{2+} \rightarrow Mn^+ + 2Mn^{3+};$

$E^\circ = -1.18 - (+1.51) = -2.69\,V$

–ve value of EMF (i.e., ΔG = +ve) shows that the reaction is non-spontaneous.

47. (3) Higher the value of reduction potential stronger will be the oxidising agent, hence based on the given values Ag^+ will be strongest oxidizing agent.

48. (4) Higher the value of standard reduction potential, stronger is the oxidising agent, hence MnO_4^- is the strongest oxidising agent.

49. (4) For a spontaneous reaction, ΔG° must be –ve.

$\Delta G^\circ = -nFE^\circ$

Hence for ΔG° to be –ve, ΔE° has to be positive which is possible when X = Zn, Y = Ni

$Zn + Ni^{2+} \longrightarrow Zn^{2+} + Ni$

$E^\circ_{cell} = E^\circ_{Ni^{2+}/Ni} - E^\circ_{Zn^{2+}/Zn}$

$= -0.23 - (-0.76) = +0.53$ (positive)

50. (2) $Cu^{2+} + e^- \longrightarrow Cu^+;$

$E_1^\circ = 0.15V; \Delta G_1^\circ = -n_1E_1^\circ F$

$Cu^{2+} + 2e \longrightarrow Cu;$

$E_2^\circ = 0.34V; \Delta G_2^\circ = -n_2E_2^\circ F$

On subracting eq.(i) from eq. (ii) we get

$Cu^+ + e^- \longrightarrow Cu \; ; \; \Delta G^\circ = \Delta G_2^\circ - \Delta G_1^\circ$

$-nE^\circ F = -(n_2E_2^\circ F - n_1E_1^\circ F)$

$E^\circ = \frac{n_2E_2^\circ F - n_1E_1^\circ F}{nF}$

$= \frac{2 \times 0.34 - 0.15}{1} = 0.53V$

51. (3) $\frac{2}{3}Al_2O_3 \rightarrow \frac{4}{3}Al + O_2$

The ionic reactions are,

$\frac{2}{3} \times (2Al^{3+}) + 4e^- \rightarrow \frac{4}{3}Al$

$\frac{2}{3} \times (3O^{2-}) \rightarrow O_2 + 4e^-$

Thus, no. of electron transferred ⇒ n = 4

$\Delta G = -nFE°$

or $E° = \frac{\Delta G}{-nF} = \frac{966 \times 10^3}{4 \times 96500} = -2.5$ V

∴ The potential difference needed for the reduction = 2.5 V.

52. (2)

53. (4) From the given representation of the cell, E_{cell} can be found as follows.

$2Cr + 3Fe^{2+} \longrightarrow 2Cr^{3+} + 3Fe$

$$E_{cell} = E^o_{Fe^{2+}/Fe} - E^o_{Cr^{3+}/Cr} - \frac{0.059}{6}\log\frac{[Cr^{3+}]^2}{[Fe^{2+}]^3}$$

$$= -0.42 - (-0.72) - \frac{0.059}{6}\log\frac{(0.1)^2}{(0.01)^3}$$

$$= 0.3 - \frac{0.059}{6}\log\frac{10^{-2}}{10^{-6}} = 0.3 - \frac{0.059}{6} \times 4$$

$= 0.30 - 0.0393 = 0.26$ V

54. (4) $E_{cell} = 0$; when cell is completely discharged.

$$E_{cell} = E°_{cell} - \frac{0.059}{2}\log\left(\frac{[Zn^{2+}]}{[Cu^{2+}]}\right)$$

$$\text{or } 0 = 1.1 - \frac{0.059}{2}\log\left(\frac{[Zn^{2+}]}{[Cu^{2+}]}\right)$$

$$\log\left(\frac{[Zn^{2+}]}{[Cu^{2+}]}\right) = \frac{2 \times 1.1}{0.059} = 37.3$$

$$\therefore \left(\frac{[Zn^{2+}]}{[Cu^{2+}]}\right) = 10^{37.3}$$

55. (4) For spontaneous reaction, ΔG should be negative. Equilibrium constant should be more than one.

$\Delta G = -2.303\,RT \log K_c$, If $K_c = 1$ then $\Delta G = 0$; If $K_c > 1$ then ΔG = –ve.

Again, $\Delta G = -nFE°_{cell}$.

$E°_{cell}$ must be +ve to have ΔG –ve.

56. (1) $E_{cell} = E°_{cell} - \frac{0.059}{n}\log K_c$

or $0 = 0.591 - \frac{0.0591}{1}\log K_c$

or $\log K_c = \frac{0.591}{0.0591} = 10$ or $K_c = 1 \times 10^{10}$

57. (1) $Zn(s) + 2H^+(aq) \rightleftharpoons Zn^{2+}(aq) + H_2(g)$

$$E_{cell} = E°_{cell} - \frac{0.059}{2}\log\frac{[Zn^{2+}][H_2]}{[H^+]^2}$$

Addition of H_2SO_4 will increase $[H^+]$ and E_{cell} will also increase and thus, the equilibrium shifts towards RHS.

58. (1) $Sn(s) + 2Fe^{3+}(aq) \rightarrow 2Fe^{2+}(aq) + Sn^{2+}(aq)$

Standard potential for the given reaction

or $E°_{cell} = E°_{Fe^{3+}/Fe^{2+}} - E°_{Sn^{2+}/Sn}$

$= 0.77 - (-0.14) = 0.91$ V

59. (3) The given values show that Cr has maximum oxidation potental, therefore its oxidation will be easiest. (Change the sign to get the oxidation values)

60. (4)

A	B	C
+0.5V	–3.0V	–1.2V

Note: The higher the negative value of reduction potential, the more is the reducing power.

Hence B > C > A.

61. (3) The equilibrium constant is related to the standard emf of cell by the expression

$$\log K = E°_{cell} \times \frac{n}{0.059} = 0.295 \times \frac{2}{0.059}$$

$$\log K = \frac{590}{59} = 10 \text{ or } K = 1 \times 10^{10}$$

62. (2) $E_{cell} = E°_{cell} + \frac{0.059}{n}\log\frac{[Cu^{2+}]}{[Zn^{2+}]}$

$= 1.10 + \frac{0.059}{2}\log[0.1]$

$= 1.10 - 0.0295 = 1.07$ V

63. (4) Oxidation half cell:-

$H_2(g)_{P_1} \longrightarrow 2H^+(1M) + 2e^-$

Reduction half cell

$2H^+(1M) + 2e^- \longrightarrow H_2(g)_{P_2}$

The net cell reaction,

$H_2(g)_{P_1} \longrightarrow H_2(g)_{P_2}$

$E°_{cell} = 0.00$ V; n = 2

$$\therefore E_{cell} = E°_{cell} - \frac{RT}{nF}\log_e K = 0 - \frac{RT}{nF}\log_e\frac{P_2}{P_1}$$

$$\text{or } E_{cell} = \frac{RT}{2F}\log_e\frac{P_1}{P_2}$$

64. (1) $2Cr^{3+} + 7H_2O \rightarrow Cr_2O_7^{2-} + 14H^+$

O.S. of Cr changes from +3 to +6 by loss of electrons. At anode, oxidation takes place.

65. (2) At cathode :

$M^{x+}(aq) + xe^- \longrightarrow M(s)$

At anode :

$Fe(s) \longrightarrow Fe^{+3}(aq) + 3e^-$

$E^0_{cell} = E^0_C - E^0_A$

$= 0.15 - (-0.036) = 0.186$ V

$$E_{cell} = E^0_{cell} - \frac{0.059}{n}\log Q$$

$$0.2057 = 0.186 - \frac{0.059}{n}\log(10^{-2})$$

On solving, n = 5.98

$\boxed{n \simeq 6}$

$xFe(s) + 3M^{x+}(aq) \longrightarrow 3M(s) + xFe^{+3}(aq)$

So total electrons transfered (n) = 3 × x

$n = 3x$

$x = \frac{6}{3} \Rightarrow \boxed{x = 2}$

66. (78) We have considered

$$E°_{[Sn(OH)_6]^{2-}/HSnO_2^-} = -0.9V$$

$Pt|HSnO_2^-(aq), [Sn(OH)_6]^{2-}(aq), OH^-(aq)| Bi_2O_3(s)| Bi(s)|$

0.5 M 0.05 M

$E°_{cell} = +0.9 - 0.44 = 0.46$ V

Oxidation Half :

$HSnO_2^- + H_2O + 3OH^- \rightarrow [Sn(OH)_6]^{2-} + 2e^-$

Reduction Half :

$Bi_2O_3 + 3H_2O + 6e^- \rightarrow 2Bi + 6OH^-$

$3HSnO_2^-(aq) + Bi_2O_3(s) + 6H_2O + 3OH^-(aq) \rightarrow 3[Sn(OH)_6]^{2-}(aq) + 2Bi(s)$

$$E_{cell} = E°_{cell} - \frac{0.059}{6}\log\frac{(0.5)^3}{(0.05)^3 \times [OH^-]^3}$$

$$0.2353 = 0.46 - \frac{0.059}{6} \times 3\log\left[\frac{10}{OH^-}\right]$$

$$\log\left[\frac{10}{OH^-}\right] = \frac{2 \times 0.2247}{0.059} = 7.6$$

1 + pOH = 7.6

pOH = 6.6

pH = 14 – 6.6 = 7.4

$$pH = pK_{a_1} + \log\frac{[HCO_3^-]}{[H_2CO_3]}$$

$$7.4 = 6.11 + \log\frac{5x}{20}$$

$$1.29 = \log\frac{x}{4} \Rightarrow \frac{x}{4} = 19.5 \Rightarrow x = 78$$

Note : In question paper,

$$E^o_{HSnO_2^-/[Sn(OH)_6]^{2-}} = -0.9V$$

data is given, but NTA has given answer by considering $E^o_{[SnOH)_6]^{2-}/HSnO_2^-} = -0.9V$

67. (24) $3BH_4^- + 4ClO_3^- \rightarrow 4Cl^- + 3H_2BO_3^- + 3H_2O$

n-factor = 8

moles = 3

∴ n = 3 × 8 = 24

68. (4)

Anode : $H_2O \longrightarrow 2H^+ + \frac{1}{2}O_2 + 2e^-$

$E^0_{O_2/H_2O} = 1.23V$

Cathode : $M^{2+} + 2e^- \longrightarrow M(s)$

$E^0_{M^{2+}/M} = 0.994V$

$E^0 = E^{Cathode}_{Red} - E^{Anode}_{Red}$

$E^0 = 0.994 - 1.23 = -0.236$

For spontaneity $E_{cell} > 0$, at limiting condition $E_{cell} = 0$

$$E = E^0 - \frac{0.059}{n}\log\left[\frac{(H^+)^2 P_{O_2}^{1/2}}{1}\right]$$

$$0 = -0.236 - \frac{0.059}{2} \times 2\log[H^+]$$

$$\frac{0.236}{0.059} = -\log[H^+]$$

⇒ pH = 4.

69. (4) $E°_{cell} = -0.44 - (-0.90) = +0.46$ V

Applying Nernst equation:-

$$E_{cell} = E°_{cell} - \frac{0.059}{n}\log Q$$

$E_{cell} = 0.46 - \frac{0.06}{6}\log 10^6$

$E_{cell} = 0.46 - 0.06$

$E_{cell} = 4 \times 10^{-1}$

$x = 4$

70. (200)

$$E^0_{X^-/MX(s)/M} = E^0_{M^+/M} + \frac{0.0591}{n}\log K_{sp}$$

$$= 0.79 + \frac{0.059}{1}\log 10^{-10}$$

$$= 0.79 - 0.59 = 0.20\text{ V} = 200\text{ mV}$$

71. (10) $Cr_2O_7^{-2}(aq) + 14H^+(aq) + 6e^- \rightarrow 2Cr^{+3}(aq) + 7H_2O(aq)$

Using Nernst equation

$$E_R = E_R^\circ - \frac{0.059}{6}\log\frac{[Cr^{+3}]^2}{[Cr_2O_7^{-2}][H^+]^{14}}$$

$$0 = 1.33 - \frac{0.059}{6}\log\frac{10^{-6}}{[H^+]^{14}}$$

$$\Rightarrow \frac{1.33 \times 6}{0.059} = \log\frac{10^{-6}}{[H]^{14}}$$

$\Rightarrow$ $135.254 = -6 - 14\log[H^+]$

$\Rightarrow$ $141.254 = 14\,pH$

$$\Rightarrow pH = \frac{141.254}{14} - 10.08$$

72. (400)

	Cu^{2+} + $2e^-$	$\rightarrow$	Cu (1 Faraday = charge on 1 mole of electron)
t = 0	1.5	1 mole	
t = t	1	–	0.5 mole

$[Cu^{+2}] = 1$ M after electrolysis

	Ag^+ +	e^-	$\rightarrow$ Ag
t = 0	0.2	0.1 mole	
t = t	0.1	–	

$[Ag^+] = 0.1$ M after electrolysis

Cell reaction

$$Cu(s) + 2Ag^+(aq) \rightarrow Cu^{+2}(aq) + 2Ag(s)$$

$$E = E^\circ - \frac{0.06}{n}\log\frac{[Cu^{+2}]}{[Ag^+]^2}$$

$$E = (0.8 - 0.34) - \frac{0.06}{2}\log\frac{1}{(0.1)^2} = 0.4\text{V}$$

$\therefore$ Correct answer = 400 mV

73. (6)

$QH_2 + 2Ag^+ \rightarrow 2Ag + Q + 2H^+$ (Cell reaction)

$E = E^\circ - \frac{0.06}{2}\log[H^+]^2$ (as $[Ag^+] = 1M$)

$E = E^\circ - 0.06 \times \log[H^+]$

$$pH = -\log[H^+] = \frac{E - E^\circ}{0.06} = \frac{0.4 - 0.1}{0.06}$$

$$= \frac{0.3}{0.06} = 5$$

NH_4X is electrolyte here.

Now, $pH = 7 - \frac{1}{2}pK_b - \frac{1}{2}\log C$

$$= 7 - \frac{1}{2}pK_b - \frac{1}{2}\log[NH_4X]$$

$$\Rightarrow 5 = 7 - \frac{1}{2} \times pK_b - \frac{1}{2}\log(10^{-2}) \Rightarrow pK_b = 6$$

74. (3) Fe, Ni, Ag will be oxidized due to lower S.R.P than that of $Cr_2O_7^{2-}$.

75. (338.9)

$$2MnO_4^- + 6H^+ + 5H_2C_2O_4 \rightleftharpoons 2Mn^{2+} + 8H_2O + 10CO_2$$

$E^0_{cell} = E^0_{(R\cdot P)_C} - E^0_{(R\cdot P)_A} = 1.51 - (-0.49) = 2.0\text{V}$

At Equilibrium $E_{cell} = 0$

$$E^0_{cell} = \frac{0.059}{n}\log K \Rightarrow 2 = \frac{0.059}{10}\log K$$

$$2 = \frac{0.059}{10}\log 10^x$$

$\log 10^x = 338.9 \Rightarrow x = 338.9$

76. (1) $2H^+(aq) + 2e^- \longrightarrow H_2(g)$

$[H^+] = 1$ M, $P_{H_2} = 2$ atm, $n = 2$

$$E_{H^+/H_2} = E^o_{H^+/H_2} - \frac{2.303RT}{nF}\log\frac{(P_{H_2})}{([H^+]^2)}$$

$$= 0 - \frac{2.303RT}{2F}\log\frac{(2)}{(1)^2}$$

$\{E^\circ_{H^+/H_2} = 0\}$

$$= -\frac{0.06}{2}\log 2 = -9.03 \times 10^{-3} = -.9 \times 10^{-2}\text{ V}$$

77. (18) $2H^+(aq) + 2e^- \rightarrow H_2(g)$

$pH = -\log[H^+]$

$3 = -\log[H^+]$ $\therefore [H+] = 10^{-3}$

$$E_{cell} = E^0_{cell} - \frac{0.059}{2}\log\frac{P_{H_2}}{[H^+]^2}$$

$$\therefore E_{cell} = 0 - \frac{0.059}{2}\log\frac{P_{H_2}}{[10^{-3}]^2}$$

$$= 0 - \frac{0.059}{2} \times 6$$

$E_{cell} = 0 - 0.059 \times 3 = -0.177$ volts. $= -17.7 \times 10^{-2}$ V.

78. (25) $Cu(OH)_2(s) \rightleftharpoons Cu^{2+}(aq) + 2OH^-(aq)$

$K_{sp} = [Cu^{2+}][OH^-]^2$

$pH = 14$; $pOH = 0$; $[OH^-] = 1$ M

$$\therefore [Cu^{2+}] = \frac{K_{sp}}{[1]^2} = 10^{-20}\text{ M}$$

$Cu^{2+}(aq) + 2e^- \rightarrow Cu(s)$

$$E = E^o - \frac{0.059}{2}\log_{10}\frac{1}{[Cu^{2+}]}$$

$$= 0.34 - \frac{0.059}{2}\log_{10}\frac{1}{10^{-20}}$$

$= -0.25 = -25 \times 10^{-2}$

79. (4) Given statements A, B, C and D are correct. Statement 'A' and 'B' can be explained by the definition of standard electrode potential and EMF. Statement 'C' and 'D' can be explained using Faraday's Laws of electrolysis.

80. (2) $Pb^{2+} + 2e^- \rightarrow Pb$ $\quad \Delta G_1^o = -2FE_1^o$

$Pb^{4+} + 4e^- \rightarrow Pb$ $\quad \Delta G_2^o = -4FE_2^o$

$Pb^{2+} \rightarrow Pb^{4+} + 2e^-$ $\quad \Delta G_3^o = -2FE_3^o$

$\Delta G_3^o = \Delta G_1^o - \Delta G_2^o$

$\Rightarrow -2FE_3^o = 2F(2n - m)$

$\Rightarrow E_3^o = m - 2n = m - xn$

Hence x = 2.

81. (1825)

$$FeO_4^{2-} \xrightarrow[n=3]{2.2V} Fe^{3+} \xrightarrow[n=1]{0.7V} Fe^{2+} \xrightarrow{-0.45V} Fe$$

$FeO_4^{2-} \rightarrow Fe^{2+}$: E = ?, n = 4 $\quad [\Delta G = -nF\,E^\circ_{cell}]$

$4 \times E^\circ = (3 \times 2.2) + (1 \times 0.7)$

$$E^\Theta = \frac{7.3}{4} = 1.825\text{ V} = 1825 \times 10^{-3}\text{ V}$$

82. (1) Option B is incorrect, E° cell is independent from pressure. So, Answer is 1

83. (3) Metal having lower Standard Reduction potential than 0.97V will be oxidised by NO_3^-.

84. (3) $MnO_4^-(0.1\text{ M}) / Mn^{2+}(0.001\text{ M})$

Reaction at cathode : $MnO^-_4 + 8H^+ + 5e^- \rightarrow Mn^{2+} + 4H_2O$

$$E = E^o - \frac{0.059}{5}\log\frac{[Mn^{2+}]}{[MnO_4^-][H^+]^8}$$

$$1.282 = 1.54 - \frac{0.059}{5}\log\frac{10^{-3}}{10^{-1} \times [H^+]^8}$$

$$\frac{0.258 \times 5}{0.059} = \log\frac{10^{-2}}{[H^+]^8}$$

$21.86 = -2 + 8\,pH$

$pH = 2.98 \approx 3$

85. (6)

86. (10) $Pt(g)|H_2(g, 1\text{ atm})|\ H^+(aq, 1M)||\ Fe^{3+}(aq), Fe^{3+}(aq)|Pt(s)$ at anode $H_2 \longrightarrow 2H^+ + 2e^-$

At cathode $2Fe^{3+}(aq) + 2e^- \longrightarrow 2Fe^{2+}(aq)$

$$E^\circ = E^\circ_{Fe^{3+}|Fe^{2+}} - E^\circ_{H_2|H^+} = 0.771\text{ V}$$

$$E = E^o - \frac{0.06}{2}\log\frac{[Fe^{2+}]^2}{[Fe^{3+}]^2}[H^+]^2$$

$$0.712 = 0.771 - \frac{0.06}{2}\log\frac{[Fe^{2+}]^2}{[Fe^{3+}]^2}$$

$$\log\frac{Fe^{2+}}{Fe^{3+}} = \frac{0.059}{0.06} \approx 1 \Rightarrow \frac{Fe^{2+}}{Fe^{3+}} = 10$$

87. (275) $X + Y^{2+} \rightarrow Y + X^{2+}$

$E^0_{Cell} = 0.36 - (-2.36) = 2.72$ V

$$E_{Cell} = 2.72 - \frac{0.06}{2}\log\frac{0.001}{0.01}$$

$$= 2.72 + 0.03 = 2.75\text{ V} = 275 \times 10^{-2}\text{ V}$$

88. (17) $Zn(s) + Sn^{2+}(aq) \rightleftharpoons Zn^{2+}(aq) + Sn(s)$

$\Delta G^\circ = -2.303\,RT\log_{10}K_{eq}$

$-nF(E^0_{cell}) = -2.303RT\log_{10}K_{eq}$,

$$E^0_{Sn^{2+}/Sn} - E^0_{Zn^{2+}/Zn} = \frac{0.059}{2}\log_{10}K_{eq}$$

$$E^0_{Sn^{2+}/Sn} + 0.76 = \frac{0.059}{2}\log_{10}10^{20}$$

$$E^0_{Sn^{2+}/Sn} + 0.76 = \frac{0.059 \times 20}{2}$$

$E^0_{Sn^{2+}/Sn} = 0.59 - 0.76 = -0.17$

$E^0_{Sn/Sn^{2+}} = 17 \times 10^{-2}$ V

89. (3) At Cathode: $M^{3+} + 2e \rightarrow M^+, E^\circ_{M^{3+}/M^+} = 0.2V$

At Anode: $H_2 \rightarrow 2H^+ + 2e, E^\circ_{H_2/2H^+} = 0.0V$

Overall reaction:

$H_2(g) + M^{3+}(aq) \rightarrow M^+(aq) + 2H^+(aq)$

$E_{Cell} =$

$$E^\circ_{Cathode} - E^\circ_{anode} - \frac{0.059}{2}\log\frac{[M^+]\times 1^2}{[M^{+3}]1}$$

$$0.1115 = 0.2 - \frac{0.059}{2}\log\frac{[M^+]}{[M^{+3}]}$$

$$3 = \log\frac{[M^+]}{[M^{+3}]} \Rightarrow \therefore a = 3$$

90. (10)

91. (917) $Cr_2O_7^{2-} + 14H^+ + 6e^- \rightarrow 2Cr^{3+} + 7H_2O$

$$E = 1.33 - \frac{0.059}{6}\log\frac{(0.1)^2}{(10^{-2})(10^{-3})^{14}}$$

$$E = 1.33 - \frac{0.059}{6} \times 42 = 0.917 = 917 \times 10^{-3}$$

$\therefore x = 917$

92. (34) cell reaction $\rightarrow Cu + 2Ag^+ \rightarrow Cu^{2+} + 2Ag$

According to Nernst equation

$$E_{cell} = E^\circ_{cell} - \frac{0.06}{2}\log\frac{[Cu^{2+}]}{[Ag^+]^2}$$

$$0.43 = E^\circ_{cell} - \frac{0.06}{2}\log\frac{(0.001)}{(0.01)^2};\ E^\circ_{cell} = 0.46V$$

$$E^\circ_{cell} = E^\circ_{Ag^+/Ag} - E^\circ_{Cu^{2+}/Cu}$$

$$\Rightarrow 0.46 = 0.80 - E^\circ_{Cu^{2+}/Cu}$$

$$E^\circ_{Cu^{2+}/Cu} = 0.34 \text{ volt} \Rightarrow 34 \times 10^{-2}$$

93. (4) From Nernst Eq.

$$E_{cell} = E^\circ_{cell} - \frac{RT}{nF}\ln Q$$

$$\text{or, } E_{cell} = E^\circ_{cell} - \frac{2.303RT}{nF}\log Q$$

$$\text{or, } 0.801 = (0.008 + 0.763) - \frac{0.06 \times (-2)}{n} \quad \text{or, } n = 4$$

94. (7) The cell reaction is

$H_2(g) + Cu^{2+}(a\,q.) \rightarrow 2H^+(aq.) + Cu(s)$

$n = 2$

According to Nernst equation

$$0.31 = 0.34 - \frac{0.06}{2}\log\frac{[H^+]^2}{[Cu^{2+}]}$$

$$-0.03 = -0.03\log\frac{[H^+]^2}{Cu^{2+}} \Rightarrow 10^1 = \frac{[10^{-3}]^2}{Cu^{2+}}$$

$Cu^{2+} = 10^{-7} \Rightarrow x = 7$

95. (16) $\Delta G = -nFE^\circ_{cell}$

$Sn^{2+} + 2e^- \rightarrow Sn \quad \Delta G^\circ_1 = +2 \times 0.140 \times F$ ...(i)

$Sn^{4+} + 4e^- \rightarrow Sn \quad \Delta G^\circ_2 = -4 \times 0.01 \times F$ ...(ii)

Subtracting (i) from (ii), we get

$Sn^{4+} + 2e^- \rightarrow Sn^{2+} \quad \Delta G^\circ_3 = -2 \times E^\circ_{Sn^{4+}/Sn^{2+}} \times F$

$\Delta G^\circ_3 = \Delta G^\circ_2 - \Delta G^\circ_1$

$-2 \times E^\circ \times F = -(0.04 + 0.28) \times F$

$E^\circ = 0.16 \text{ volt} = 16 \times 10^{-2}\text{ V}$

96. (51) $\frac{1}{2}H_2 + Ag^+ \rightarrow H^+ + Ag;\ n = 1$

$\Delta G^\circ = -nE^\circ F = -1 \times 0.5332 \times 96500$ J

$= -51.45$ kJ/mol

9 volume is 1 or 2

Become according to question, the equastion given $H_2 \rightarrow H^+$ not ½ $H_2 \rightarrow H^+$ So n = 2, So ans, should be = 102.9 Kg/mol

97. (983)

98. (23) $Fe^{3+} + I^- \longrightarrow I_2 + Fe^{2+}$

The E° value for spontaneous reaction is positive.

$E^\circ_{Cell} = E^\circ_{cathode} - E^\circ_{anode} = (0.77 - 0.54)$ V

$= 0.23\text{ V} = 23 \times 10^{-2}$ V.

99. (5) Anode : $H_2 \rightarrow 2H^+ + 2e^-$

Cathode : $Cu^{2+} + 2e^- \rightarrow Cu$

$Cu^{2+} + H_2 \rightarrow 2H^+ + Cu$

$$E_{cell} = E^0_{cell} - \frac{0.06}{2}\log\frac{[H^+]^2}{[Cu^{2+}]}$$

$$0.576 = 0.34 - \frac{0.06}{2}\log\left\{\frac{[H^+]^2}{[0.01]}\right\}$$

$\Rightarrow 7.87 = -\log[H^+]^2 + \log 10^{-2}$

$\Rightarrow pH = 4.93 \simeq 5$

100. (25) $\Delta G^\circ = -nFE^\circ = \Delta H^\circ - T\Delta S^\circ$

$$\Delta S^\circ = \frac{\Delta H^\circ + nFE^\circ}{T}$$

$$= \frac{(-825.2 \times 10^3) + (2 \times 96487 \times 4.315)}{298}$$

$= 25.11\ JK^{-1}mol^{-1} \approx 25\ JK^{-1}mol^{-1}$

101. (109) Galvanic cell:

$Zn(s) + \underset{0.02M}{Cu^{2+}(aq)} \rightarrow \underset{0.04M}{Zn^{2+}(aq)} + Cu(s)$

Nernst equation

$$E_{cell} = E^\circ_{cell} - \frac{0.059}{2}\log\frac{[Zn^{2+}]}{[Cu^{2+}]}$$

$$\Rightarrow E_{cell} = \left[E^\circ_{Cu^{2+}/Cu} - E^\circ_{Zn^{+2}/Zn}\right] - \frac{0.059}{2}\log\frac{0.04}{0.02}$$

$$\Rightarrow E_{cell} = [0.34 - (-0.76)] - \frac{0.059}{2}\log 2$$

$\Rightarrow E_{cell} = 1.0911 = 109.11 \times 10^{-2}$

102. (28) Cell reaction is :

$Cu(s) + 2Ag^+(aq) \longrightarrow Cu^{2+}(aq) + 2Ag(s)$

$$\text{Now, } E_{cell} = E^0_{cell} - \frac{0.059}{2}\log\frac{[Cu^{2+}]}{[Ag^+]^2} \quad ...(1)$$

For cell 1,

$$0.3095 - E^0_{cell} \quad \frac{0.059}{2}\log\frac{(0.1)}{(0.01)^2};$$

$E^0_{cell} = 0.3981$

For cell 2,

$$E_{cell} = 0.3981 - \frac{0.059}{2}\log\frac{(0.01)}{(0.001)^2}\lim_{x\to\infty}$$

$= 0.28 = 28 \times 10^{-2}$

103. (24) $Zn \longrightarrow Zn^{2+} + 2e^-$

$2Fe^{3+} + 2e^- \longrightarrow 2Fe^{2+}$

$Zn + 2Fe^{3+} \longrightarrow Zn^{2+} + 2Fe^{2+}$

$E^0_{cell} = 0.77 - (-0.76) = 1.53$ V

$$1.50 = 1.53 - \frac{0.06}{2}\log\left(\frac{[Fe^{2+}]}{[Fe^{3+}]}\right)^2$$

$$\Rightarrow \log\left(\frac{[Fe^{2+}]}{[Fe^{3+}]}\right) = \frac{0.03}{0.06} = \frac{1}{2}$$

$$\Rightarrow \frac{[Fe^{2+}]}{[Fe^{3+}]} = 10^{\frac{1}{2}} = \sqrt{10}$$

$$\Rightarrow \frac{[Fe^{2+}] + [Fe^{3+}]}{[Fe^{3+}]} = \sqrt{10} + 1$$

$$\Rightarrow \frac{[Fe^{3+}]}{[Fe^{2+}] + [Fe^{3+}]} = \frac{1}{1 + \sqrt{10}} = \frac{1}{4.16}$$

$= 0.2402 = 24 \times 10^{-2}$.

104. (3) $Cu + \underset{10^{-3}M}{2Ag^+} \longrightarrow \underset{0.250\,M}{Cu^{2+}} + 2Ag$

$$E_{cell} = E^\circ_{cell} - \frac{0.059}{2}\log\frac{[Cu^{2+}]}{[Ag^+]^2}$$

$$= 2.97 - \frac{0.059}{2}\log\frac{(0.25)}{(10^{-3})^2} = 2.81\text{ V} \approx 3.0\text{V}$$

105. (147) $Zn(s) \longrightarrow Zn^{2+}(aq) + 2e^-$

$2Ag^+(aq) + 2e^- \longrightarrow 2Ag(s)$

$Zn(s) + 2Ag^+(aq) \longrightarrow Zn^{2+}(aq) + 2Ag(s)$

$E^0_{cell} = E^0_{Ag^+/Ag} - E^0_{Zn^{2+}/Zn}$

$= 0.80 - (-0.76) = 1.56V$

$$E^0_{cell} = E^0 - \frac{2.303RT}{nF}\log_{10}\frac{[Zn^{2+}]}{[Ag^+]^2}$$

$$= 1.56 - \frac{0.059}{2}\log\frac{0.1}{(0.01)^2}$$

$$= 1.56 - \frac{0.059}{2} \times 3 = 1.56 - 0.0885 = 1.4715$$

$= 147.15 \times 10^{-2}$

106. (4)

107. (3776) Equation is

$$MnO_4^-(aq) + 8H^+(aq) + 5e^- \rightarrow Mn^{2+}(aq) + 4H_2O(l)$$

Nernst equation :

$$E_{cell} = E^0_{cell} - \frac{0.059}{5}\log\frac{[Mn^{2+}]}{[MnO_4^-]}\frac{1}{[H^+]^8}$$

(I) Given $[H^+] = 1M$

$$E_1 = E^0 - \frac{0.059}{5}\log\frac{[Mn^{2+}]}{[MnO_4^-]}$$

(II) Now: $[H^+] = 10^{-4}$ M

$$E_2 = E^0 - \frac{0.059}{5}\log\frac{[Mn^{2+}]}{[MnO_4^-]}\times\frac{1}{(10^{-4})^8}$$

Therefore,

$$|E_1 - E_2| = \frac{0.059}{5}\times 32$$

$$= 0.3776\,V = 3776\times 10^{-4}$$

$\therefore x = 3776.$

108. (– 6)

$$\Delta G^o = -nFE^o_{cell}$$

$$17.37\times 10^3 = -3\times 96500\times E^o_{cell}$$

$$E^o_{cell} = -0.06\ V \simeq -6.0\times 10^{-2}\ V$$

109. (142)

110. (144)

$$2Cu^+(aq) \rightleftharpoons Cu(s) + Cu^{2+}(aq)$$

$$E^o_{cell} = E^o_{Cu^+/Cu} - E^o_{Cu^{2+}/Cu^+} = 0.52 - 0.16 = 0.36\ V$$

$$E^o_{cell} = \frac{RT}{nF}\ln K_{eq} \Rightarrow 0.36 = \frac{0.025}{1}\ln K_{eq}$$

$$\Rightarrow \ln K_{eq} = 14.4 = 144\times 10^{-1}$$

111. (96500)

$$E^0_{cell} = E^0_{Sn^{2+}/Sn} - E^0_{Cu^{2+}/Cu}$$

$$= -0.16 - 0.34 = -0.50\ V$$

$$\Delta G^0 = -nFE^0_{cell}$$

$$= -2\times 96500\times(-0.5) = 96500\ J$$

112. (2.15)

At equilibrium state, $E_{cell} = 0$; $E^0_{cell} = 0.01$ V

$$Sn(s) + Pb^{2+}(aq) \longrightarrow Sn^{2+}(aq) + Pb(s)$$

$$E = E^0_{cell} - \frac{0.06}{n}\log\frac{[P]}{[R]}$$

$$0 = 0.01 - \frac{0.06}{2}\log\frac{[Sn^{2+}]}{[Pb^{2+}]}$$

$$\frac{1}{3} = \log\frac{[Sn^{2+}]}{[Pb^{2+}]}$$

$$\frac{[Sn^{2+}]}{[Pb^{2+}]} = 10^{1/3} = 2.15$$

113. (1.52) $E^o_{cell} = -1.23$ (Because is Ques E^o_{cell} is given for reduction Rx4 But in reaction oxidation in going on)

$$E_{cell} = -1.23 - \frac{0.0591}{4}\log[H^+]^4$$

$$= -1.23 + 0.0591\times PH$$

$$= -1.23 + 0.0591\times 5 = -0.935\ V$$

$$E = 1.23 - \frac{0.0591}{4}\log[H^+]^4 = 1.23 + 0.0591\times pH$$

$$= 1.23 + 0.0591\times 5 = 1.23 + 0.2955 = 1.52\ V$$

114. (1)

$$BaSO_4(s) \rightleftharpoons \underset{s}{Ba^{+2}(aq)} + \underset{s}{SO_4^{2-}(aq)}$$

$K_{sp} = s^2$

k of $BaSO_4 = x$ S cm^{-1}

$$\overset{o}{\Lambda}_m(BaSO_4) = \overset{o}{\Lambda}_m(BaCl_2) + \overset{o}{\Lambda}_m(H_2SO_4) - 2\overset{o}{\Lambda}_m(HCl)$$

$\therefore \overset{o}{\Lambda}_m$ of $BaSO_4 = (x_1 + x_2 - 2x_3)$

$$\Lambda_m = \alpha\overset{o}{\Lambda}_m = \frac{k\times 1000}{s}$$

$$s = \frac{x\times 1000}{\alpha(x_1 + x_2 - 2x_3)}$$

$$K_{sp} = s^2 = \frac{1}{\alpha^2}\left[\frac{x\times 1000}{x_1 + x_2 - 2x_3}\right]^2$$

$$= \frac{x^2\times 10^6}{\alpha^2(x_1 + x_2 - 2x_3)^2}$$

115. (3) Mohr's salt : $FeSO_4.(NH_4)_2SO_4.6H_2O$

Using Kohlrousch law, $\lambda^\infty_m = x_1 + 2x_2 + 2x_3$

116. (2) Higher the charge of the metal cation higher is the molar conductivity. For same charge, higher is solvation lower will be the molar conductivity where $H^\oplus$ cation is a exceptional case.

117. (4)

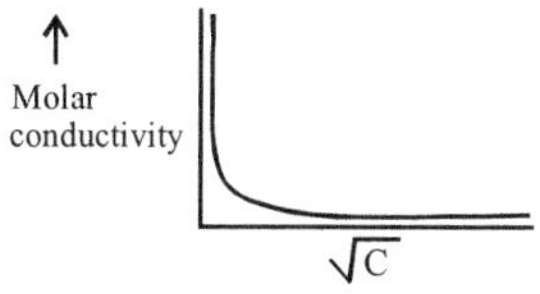

Molar conductivity decreases sharply with the increase in concentration because the total volume V of solution containing 1 mol of electrolyte decreases

118. (3) A → Weak electrolyte (greater change of Λ_m)

B → Strong electrolyte (less effect of C on Λ_m)

119. (1) $HA(aq) \rightleftharpoons H^+(aq) + A^-(aq)$

$$K_a = \frac{\alpha^2 C}{1-\alpha} \qquad \Rightarrow \alpha^2 C + K_a\alpha - K_a = 0$$

$$\Rightarrow \left(\frac{\lambda_m}{\lambda^\infty_m}\right)^2 C + K_a\frac{\lambda_m}{\lambda^\infty_m} - K_a = 0$$

$$\Rightarrow \lambda^2_m C + K_a\lambda_m\lambda^\infty_m - K_a\left(\lambda^\infty_m\right)^2 = 0$$

120. (2) Solution in the conductivity cell is infinitely dilute, hence on addition of water at constant temperature, the molar conductivity of the cell will remain same.

121. (2) Given : $\Lambda_+ = 57$ S cm^2mol^{-1}; $\Lambda_- = 73$ S cm^2mol^{-1}

Apply Kohlrausch's law

$$\Lambda_m = \lambda_+ + \lambda_- = 57 + 73 = 130\ S\,cm^2mol^{-1}$$

122. (1) $\Lambda_m = \Lambda^o_m - A\sqrt{C}$

Unit of $A\sqrt{C}$ = unit of Λ_m = S $cm^2\,mol^{-1}$

Unit of A = S cm^2 $mol^{-3/2}$ $L^{1/2}$.

123. (3) Conductivity of electrolytic cell is affected by concentration of electrolyte, nature of electrolyte and nature of solvent.

124. (2) **Statement I:** KI is strong electrolyte, it dissociates completely. So, λ_m increases slowly on dilution.

Statement II: In weak electrolyte λ_m increases, sharply as the dissociation increases.

125. (1) $\Lambda_m = \frac{\kappa}{C} \Rightarrow \kappa = \Lambda_m\times C$

from question, $\kappa_1 = \kappa_2$

$$\Lambda_{m_1}\times C_1 = \Lambda_{m_2}\times C_2$$

$$\Lambda_{m_1}\times\frac{10\times 10^3}{20} = \Lambda_{m_2}\times\frac{20\times 10^3}{80}$$

$$\Rightarrow 2\Lambda_{m_1} = \Lambda_{m_2}$$

126. (4)

Ion	H^+	K^+	Cl^-	CH_3COO^-
$\Lambda^\infty_{m\,Scm^2/mole}$	349.8	73.5	76.3	40.9

So $\Lambda^\infty_{m\,CH_3COOH} = \Lambda^\infty_{m(H^+)} + \Lambda^\infty_{m\,CH_3COO^-}$

$= 349.8 + 40.9 = 390.7$ Scm^2/mole

$\Lambda^\infty_{m\,KCl} = \Lambda^\infty_{m(K^+)} + \Lambda^\infty_{m(Cl^-)} = 73.5 + 76.3$

$= 149.8$ Scm^2/mole

So statement-I is wrong.

Molar conductivity of a solution increases with decrease in concentration because of the increase in the total volume containing one mole of electrolyte. Therefore statement II is also not correct.

127. (4)

128. (N) Ionic mobility increases with increase in temperature, which increases the conductance of the solution. $BaSO_4$ is a sparingly soluble salt, hence at any T, $C_{NaCl} >> C_{BaSO_4}$. Therefore, all options are true.

129. (4) $(\Lambda^0_m)_{NaBr} - (\Lambda^0_m)_{NaI}$

$= \Lambda^o_m\overset{+}{Na} + \Lambda^o_m Br^- - (\Lambda^o_m\overset{+}{Na} + \Lambda^o_m I^-)$

$= \Lambda^o_m Br^- - \Lambda^o_m I^-$ $(\Lambda^0_m)_{KBr} - (\Lambda^0_m)_{NaBr}$

$= \Lambda^o_m K^+ + \Lambda^o_m Br^- - (\Lambda^o_m Na^+ + \Lambda^o_m Br^-)$

$= \Lambda^o_m K^+ - \Lambda^o_m\overset{+}{Na}$

$\therefore (\Lambda^0_m)_{NaBr} - (\Lambda^0_m)_{NaI} \neq (\Lambda^0_m)_{KBr} - (\Lambda^0_m)_{NaBr}$

130. (1) Order of acidic strength is :

$HCOOH > C_6H_5COOH > CH_3COOH$

More the acidic strength, more will be the dissociation of acid into ions and more will be the conductivity.

Thus, order of conductivity will be,

$HCOOH > C_6H_5COOH > CH_3COOH$

(A) > (C) > (B)

131. (2) Conductivity of an electrolyte is the conductance of 1 cm^3 of the given electrolyte. It increases with the increase in concentration of electrolyte due to increase in the number of ions per unit volume.
Molar conductivity (Λ_m) is the conductance of a solution containing 1 mole of the electrolyte. It increases with the decrease of concentration due to increase in the total volume having one mole of electrolyte.
Therefore, (S_1) is wrong and (S_2) is correct.

132. (1) Since, KCl is more conducting than NaCl, therefore, graph (a) is correct.

133. (3) Given: $\Lambda^\circ_m(NaCl) = 126.4\ S\ cm^2\ mol^{-1}$, $\Lambda^\circ_m(HCl) = 425.9\ S\ cm^2\ mol^{-1}$ and $= 100.5\ S\ cm^2\ mol^{-1}$

$$\Lambda^\circ_m(HA) = \lambda^\circ_{H^+} + \lambda^\circ_{A^-}$$

$$= \lambda^\circ_{H^+} + \lambda^\circ_{Cl^-} + \lambda^\circ_{A^-} + \lambda^\circ_{Na^+} - \lambda^\circ_{Cl^-} - \lambda^\circ_{Na^+}$$

$$= \Lambda^\circ_m(HCl) + \Lambda^\circ_m(NaA) - \Lambda^\circ_m(NaCl)$$

$$\Lambda^\circ_m(HA) = 425.9 + 100.5 - 126.4$$
$$= 400\ S\ cm^2\ mol^{-1}$$

$\kappa(HA) = 5 \times 10^{-5}\ S\ cm^{-1}$

$$\Lambda_m(HA) = \frac{\kappa(HA) \times 1000}{\text{Molarity of HA}} = \frac{5 \times 10^{-5} \times 1000}{0.001}$$
$$= 50\ S\ cm^2\ mol^{-1}$$

$$\alpha = \frac{\Lambda_m(HA)}{\Lambda^\circ_m(HA)} = \frac{50}{400} = 0.125$$

134. (3) According to Debye – Huckle – Onsager equation,

$$\lambda_C = \lambda_\infty - B\sqrt{C}$$

135. (1) Given for 0.2 M solution,

$$R = \rho\frac{l}{A} = \frac{1}{\kappa} \times \frac{l}{A}$$

$$\Rightarrow \frac{l}{A} = R \times \kappa = 50 \times 1.4 \times 10^{-2}\ cm^{-1}$$

For 0.5 M solution,

$$\frac{l}{A} = 50 \times 1.4 \times 10^{-2}\ cm^{-1} \Rightarrow R = \rho\frac{l}{A} = \frac{1}{\kappa} \times \frac{l}{A}$$

$$\Rightarrow \kappa = \frac{1}{280} \times 50 \times 1.4 \times 10^{-2}$$

$$= \frac{1}{280} \times 70 \times 10^{-2}$$
$$= 2.5 \times 10^{-3}\ S\ cm^{-1}$$

$$\text{Now, } \Lambda_m = \frac{\kappa \times 1000}{M}$$

$$= \frac{2.5 \times 10^{-3}\ S\ cm^{-1} \times 1000\ cm^3/L}{0.5\ mol/L}$$

$= 5\ S\ cm^2\ mol^{-1} = 5 \times 10^{-4}\ S\ m^2\ mol^{-1}$

136. (2) According to Kohlrausch's law, equivalent conductivity of weak electrolyte, acetic acid (CH_3COOH) can be calculated as follows:

$$\Lambda^\circ_{CH_3COOH} = \left(\Lambda^\circ_{CH_3COONa} + \Lambda^\circ_{HCl}\right) - \Lambda^\circ_{NaCl}$$

$\therefore$ Value of Λ°_{NaCl} should also be known for calculating value of $\Lambda^\circ_{CH_3COOH}$.

137. (2) $\overset{\circ}{\Lambda}_{CH_3COOH}$ is given by the following equation

$$\overset{\circ}{\Lambda}_{CH_3COOH} = (\overset{\circ}{\Lambda}_{CH_3COONa} + \overset{\circ}{\Lambda}_{HCl}) - (\overset{\circ}{\Lambda}_{NaCl})$$

Hence $\overset{\circ}{\Lambda}_{NaCl}$ is required.

138. (2) $R = 100\,\Omega$, $\kappa = \frac{1}{R}\left(\frac{l}{a}\right)$,

$\frac{l}{a}$ (cell constant) $= 1.29 \times 100\ m^{-1}$
Given, $R = 520\Omega$,
$C = 0.2\ M = 0.2 \times 1000\ mol\ m^{-3}$
Λ_m (molar conductivity) = ?
$\Lambda_m = K \times V$ (K can be calculated as cell constant is known.)

$$K = \frac{1}{R}\left(\frac{l}{a}\right)$$

$$\text{Hence, } K = \frac{129 m^{-1}}{520\ \Omega} = 0.248\ Sm^{-1}$$

$$\Lambda_m = \frac{K}{C} = \frac{0.248\ Sm^{-1}}{0.2 \times 1000\ mol\ m^{-3}}$$
$$= 12.4 \times 10^{-4}\ Sm^2\ mol^{-1}$$

139. (1) Difluoro acetic acid being strongest acid will furnish maximum number of ions showing highest electrical conductivity. The decreasing acidic strength of the carboxylic acids given is difluoro acetic acid > fluoro acetic acid > chloro acetic acid > acetic acid.

140. (3)
$\Lambda^\circ_{NaCl} = \lambda^\circ_{Na^+} + \lambda_{Cl^-}$ ....(i)
$\Lambda^\circ_{KBr} = \lambda^\circ_{K^+} + \lambda^\circ_{Br^-}$ ....(ii)
$\Lambda^\circ_{KCl} = \lambda^\circ_{K^+} + \lambda_{Cl^-}$ ....(iii)
Operating (i) + (ii) – (iii)
$\Lambda^\circ_{NaBr} = \lambda^\circ_{Na^+} + \lambda^\circ_{Br^-}$
$= 126 + 152 - 150 = 128\ S\ cm^2\ mol^{-1}$

141. (2)
$\Lambda^\infty_{HCl} = 426.2$ ....(i)
$\Lambda^\infty_{AcONa} = 91.0$ ....(ii)
$\Lambda^\infty_{NaCl} = 126.5$ ....(iii)
$\Lambda^\infty_{AcOH} = (i) + (ii) - (iii)$
$= [426.2 + 91.0 - 126.5] = 390.7$

142. (2) Given conductivity $\propto \frac{\text{area} \times \text{conc.}}{\text{length}}$

$$= \frac{Km^2 mol}{m \times m^3}$$

$\therefore$ $K = Sm^2\ mol^{-1}$

143. (15) Let molarity of MX solution = y M

$$\Rightarrow \Lambda_m = C(l/A) \times \frac{1000}{y}$$

$$\Rightarrow 123.5 = 1.9 \times 10^{-3} \times 1.3 \times \frac{1000}{y}$$

$\Rightarrow y = 0.02\ M$
Let molar mass of MX = M g mol^{-1}
0.02 mol MX is dissolve in 1 L solution
(0.02×75) gm MX is dissolved in 1000 g solution

$$\%\ w/w = x\% = \frac{0.02 \times 75}{1000} \times 100 = 0.15\%$$

$= 15 \times 10^{-2}$

144. (4) Using equation : $\Lambda_m = \Lambda^0_m - A\sqrt{c}$

$96.1 = \Lambda^0_m - A\sqrt{0.04}$
$96.1 = \Lambda^0_m - A \times 0.2$ ... (1)
$95.7 = \Lambda^0_m - A \times \sqrt{0.09}$
$95.7 = \Lambda^0_m - A \times 0.3$ ... (2)
From eq. (1) and eq. (2)
$0.4 = A(0.3 - 0.2)$
$0.4 = 0.1A$
$A = 4$

145. (2) $K_a(HQ) = C_1\alpha_1^2$ $\alpha_1 = \frac{\lambda_m(HQ)}{\lambda^\infty_m(HQ)}$

$$K_a(HZ) = C_2\alpha_2^2 \quad \alpha_2 = \frac{\lambda_m(HZ)}{\lambda^\infty_m(HZ)}$$

$$\frac{K_a(HQ)}{K_a(HZ)} = \frac{C_1}{C_2} \cdot \left(\frac{\alpha_1}{\alpha_2}\right)^2 = \frac{0.18}{0.02} \cdot \left[\frac{\lambda_m(HQ)}{\lambda_m(HZ)}\right]^2$$

$$\frac{K_a(HQ)}{K_a(HZ)} = 9 \times \left(\frac{1}{30}\right)^2 = \frac{1}{100}$$

$\Rightarrow$ $pK_a(HQ) - pK_a(HZ) = 2$ (using logarithm property)

146. (6) pH = 5
$[H^+] = 10^{-5} = [HX].\alpha$

$$= [HX] \cdot \frac{\Lambda_m}{\Lambda^\infty_m}$$

$$\Lambda_m = \frac{k \times 1000}{[HX]}$$

$$K = G \cdot G^* = 4 \times 10^{-5} \times \frac{15}{1} = 6 \times 10^{-4}\ S.cm^{-1}$$

$$[H^*] = 10^{-5} = [HX] \times \frac{6 \times 10^{-4} \times 1000}{\Lambda^\infty_m \times [HX]}$$

$\Lambda^\infty_m = 60000\ S.cm^2 mol^{-1}$
$\Lambda^\infty_m = 6\ S.m^2 mol^{-1}$.

147. (3) $\lambda^\circ_m(NH_4Cl) = 185\ Scm^2\ mol^{-1}$

$$\Rightarrow \left(\lambda^\circ_m\right)_{NH_4^+} + \left(\lambda^\circ_m\right)_{Cl^-} = 185$$

$$\left(\lambda^\circ_m\right)_{NH_4^+} = \left(\lambda^\circ_m\right)_{NH_4Cl} - \left(\lambda^\circ_m\right)_{Cl^-}$$

$= 185 - 70 = 115\ Scm^2\ mol^{-1}$

$$\left(\lambda^\circ_m\right)_{NH_4OH} = \left(\lambda^\circ_m\right)_{NH_4^+} + \left(\lambda^\circ_m\right)_{OH^-}$$

$= 115 + 170 = 285\ Scm^2\ mol^{-1}$

$$\text{Degree of dissociation} = \frac{(\lambda_m)NH_4OH}{\left(\lambda^\circ_m\right)_{NH_4OH}}$$

$$= \frac{85.5}{285} = 0.3 = 3 \times 10^{-1}$$

148. (23) We have 0.2 g of NaOH in 100 mL solution

$$M = \frac{0.2}{40 \times 100} \times 1000 = 0.05\ M$$

$$\kappa = \frac{1}{0.870} = 1.15 Sm^{-1} = 1.15 \times 10^{-2}\ S\ cm^{-1}$$

$$\Lambda_m = \frac{\kappa \times 1000}{M} = \frac{1.15 \times 10^{-2} \times 1000}{0.05}$$
$$= 230\ S\ cm^2\ mol^{-1}$$

$= 23 \times 10^4\ m\ S\ cm^2\ mol^{-1} = 23 \times 10^2\ m\ S\ dm^2\ mol^{-1}$

149. (150) $R = \rho\frac{l}{A}$

$\kappa = G.G^* \quad G = \frac{1}{R}; \kappa = \frac{1}{\rho} \quad G^* = \frac{l}{A}$

R = Resistance

ρ = Resistivity

$\frac{l}{A}$ = cell constant (G*)

$\frac{\kappa_c}{\kappa_d} = \frac{R_d}{R_c}; \lambda_m = \frac{\kappa \times 1000}{C}$

$\frac{\kappa_c}{\kappa_d} = \frac{(\lambda_m.C)_c}{(\lambda_m.C)_d} = \frac{R_d}{R_c}$

c = concentrated sol.

d = diluted solution

$\frac{100.(0.15)^2}{150.(0.1)^2} = \frac{R_d}{100} \qquad \boxed{R_d = 150\Omega}$

150. (4) $2.1\times10^3, 1.2\times10, 3.91, 1.0\times10^3$

151. (3) (B) On dilution, the number of ions per unit volume that carry current in a solution decreases.

(E) On dilution degree of dissociation of weak electrolyte increases.

So answer is (A), (C) & (D).

152. (66) $\Lambda_m = \frac{k}{C} \times 1000$

Given $k = 5 \times 10^{-5}$ S cm^{-1}; C = 0.0025 M

$\Lambda_m = \frac{5\times10^{-5}\times10^3}{0.0025} = \frac{5\times10^{-2}}{2.5\times10^{-3}}$

$= 20$ S cm^2 mol^{-1}

$\alpha = \frac{20}{400} = \frac{1}{20} \quad \left(\alpha = \frac{\Lambda_m}{\Lambda_m^o}\right)$

$K_a = \frac{C\alpha^2}{1-\alpha} = \frac{0.0025 \times \frac{1}{20} \times \frac{1}{20}}{\frac{19}{20}}$

$= \frac{0.0025}{19 \times 20} = 6.6\times10^{-6} = 66 \times 10^{-7}$

153. (13.4)

$AgNO_3 \xrightarrow[\text{dissociation}]{\text{complete}} Ag^+ + NO_3^-$

10^{-5} M $\qquad 10^{-5} \quad 10^{-5}$

$AgBr \rightleftharpoons Ag^+ + Br^-$

$\qquad S \quad S$

$K_{sp} = [Ag^+][Br^-] = 4.9\times10^{-13}$

∴ $S^2 = 4.9 \times 10^{-13}$

∴ $S = 7 \times 10^{-7}$ M

∴ $[Ag^+] = (7 \times 10^{-7} + 100 \times 10^{-7}) = 107 \times 10^{-7}$ M

∴ $[Br^-] = \frac{K_{sp}}{[Ag^+]} = \frac{4.9\times10^{-13}}{107\times10^{-7}} = 0.0458\times10^{-6}$

$= 4.58\times10^{-8}$ M

Conductivity

$= \nu_{Ag^+}\lambda^o_{Ag^+} \nu_{Br^-}\lambda^o_{Br^-} + \nu_{NO_3^-}\lambda^o_{NO_3^-}$

$= (107\times10^{-7}\times6\times10^{-3}) + (4.58\times10^{-8}\times8\times10^{-3})$
$+ (10^{-5}\times7\times10^{-3})$

$= 642 \times 10^{-10} + 36.64 \times 10^{-11} + 7 \times 10^{-8}$

$= (642 + 3.664 + 700) \times 10^{-10} = 1346 \times 10^{-10}$ Sm^{-1}

$= 13.46 \times 10^{-8}$ Sm^{-1}

154. (25) $\Lambda_m = \frac{\kappa\times1000}{M}$

$\Rightarrow \Lambda_m = \frac{1}{\rho}\times\frac{1000}{M}$

$= \frac{1}{5\times10^{-3}}\times\frac{1000}{0.8} = 25 \times 10^4\,\Omega^{-1}$ cm^{-2} mol^{-1}

155. (2) Statement (A) and Statement (C) are incorrect.

(A) is weak electrocyle and (B) is strong electrolyte

156. (1000) Ratio of Molarity $= \frac{M_1}{M_2}$

$= \frac{W_1/mV}{W_2/mV} = \frac{W_1}{W_2} = \frac{ppm_1}{ppm_2} = \frac{74.5}{149} = \frac{1}{2}$

$\frac{\Lambda_1}{\Lambda_2} = \frac{\kappa_1 \times \frac{1000}{M_1}}{\kappa_2 \times \frac{1000}{M_2}} = \frac{\kappa_1}{\kappa_2}\times\frac{M_2}{M_1} = \frac{50}{100}\times2$

$= 1{,}000 \times 10^{-3}$

157. (3)

$A_2X_3(s) \rightleftharpoons 2A^{+3}(aq) + 3X^{-2}(aq)$

Solubility = s M $\qquad$ 2s $\qquad$ 3s

$(2s)^2(3s)^3 - 1.1 \times 10^{-23}$

$108\, s^5 = 1.1 \times 10^{-23}$

$s = 10^{-5}M = 10^{-5}\frac{mol}{L} = 0.01\frac{mol}{m^3}$

∴ $[1L = 0.001m^3]$

Now, $\Lambda_m \simeq \Lambda_m^\infty = \frac{k}{m} = \frac{k}{s}$

$\Rightarrow \Lambda_m^\infty = \frac{3\times10^{-5}}{0.01} = 3\times10^{-3}Sm^2/mol$

158. (14)

$\lambda_m^\infty(AgI) = \lambda_m^\infty(NaI) + \lambda_m^\infty(AgNO_3) - \lambda_m^\infty(NaNO_3)$

$= 12.7 + 13.3 - 12.0 = 14.0$

159. (266) $k = \frac{1}{R}\times$ cell constant

$0.152 \times 10^{-3} = \frac{1}{1750}\times$ cell constant

Cell constant $= 266 \times 10^{-3}$

160. (12) $\Lambda_m = 1000\times\frac{\kappa}{M}$

$= 1000\times\frac{2\times10^{-5}}{0.001} = 20$ S cm^2 mol^{-1}

$\Rightarrow \alpha = \frac{\Lambda_m}{\Lambda_m^o} = \frac{20}{190} = \left(\frac{2}{19}\right)$

$HA \rightleftharpoons H^+ + A^-$

$0.001(1-\alpha) \qquad 0.001\,\alpha \qquad 0.001\alpha$

$K_a = 0.001\left(\frac{\alpha^2}{1-\alpha}\right) = \frac{0.001\times\left(\frac{2}{19}\right)^2}{1-\left(\frac{2}{19}\right)}$

$= 12.3 \times 10^{-6}$

161. (760)

$\kappa = \frac{1}{R}\times{}^{l}/_{A} = \left(\left(\frac{1}{1500}\right)\times1.14\right)$ S cm^{-1}

$\Rightarrow \Lambda_m = \frac{1000\times\kappa}{C} = 1000\times\frac{\left(\frac{1.14}{1500}\right)}{0.001}$ S cm^{2}mol^{-1}

$= 760$ S cm^2 mol^{-1}

162. (288) From Kohlrausch's law

$\Lambda_m^\infty BaSO_4 = \Lambda_m^\infty Ba^{2+} + \Lambda_m^\infty SO_4^{2-}$

$\Lambda_m^\infty(BaSO_4) = \Lambda_m^\infty(BaCl_2) + \Lambda_m^\infty(H_2SO_4) - 2\Lambda_m^\infty(HCl)$

$= 280 + 860 - 2 \times 426 = 288$ S cm^2 mol^{-1}

163. (57) For KCl,

$\kappa_1 = 0.14$ Sm^{-1}; $R_1 = 4.19\Omega$

For HCl, $\kappa_2 = x$ Sm^{-1}; $R_2 = 1.03\Omega$

$\kappa = \frac{1}{R}\cdot\frac{l}{A}$

Hence κR = Constant (cell is same)

∴ $\kappa_1R_1 = \kappa_2R_2 \Rightarrow 0.14\times4.19 = 1.03\times\kappa_2$

$\Rightarrow \kappa_2 = 0.5695$ Sm$^{-1} \approx 57\times10^{-2}$ Sm^{-1}

164. (14) Conductance

$= \frac{\text{Conductivity}}{\text{Cell constant}}$

∴ Conductivity $= 0.55 \times 10^{-3} \times 1.3$ S cm^{-1}

Molar conductivity

$= \frac{\text{Conductivity (S cm}^{-1})\times1000}{\text{Molarity (mol/L)}}$ (cm^3/L)

$= \frac{0.55\times10^{-3}\times1.3\times1000}{5\times10^{-3}} = 143$ S cm^2 mol^{-1}

∴ 1 S cm^2 mol^{-1} = 10^{-4} S m^2 mol^{-1}

143 S cm^2 mol^{-1} = 143×10^{-4} S m^2 mol^{-1}

$= 143 \times 10^{-4} \times 10^3$ m S m^2 mol^{-1}

$= 14.3$ mS m^2 mol^{-1} ≈ 14 mS m^2 mol^{-1}

165. (2) Ag will deposit first due to higher SRP followed by Hg and Cu. Mg will not deposit. H_2O will get reduced and H_2 gas will evolve at cathode.

166. (1)

167. (3) At anode.

$2HSO_4^- \rightarrow H_2S_2O_8 + 2e^-$

168. (3) $Al^{3+} + 3e^- \longrightarrow Al(s)$

3F required for 27 gm

$W = \frac{E\times i\times t}{96500} = \frac{M\times i\times t}{n\times96500}$

$= \frac{27\times2\times30\times60}{3\times96500} = 0.336$ gm Al

169. (1) An electrochemical can be converted into an electrolyte cell by applying an external potential greater than E^o_{cell}. The entire cell reaction will be reversed. The oxidised species will be reduced and the reduced species will be oxidised.

170. (4) $Ag^+(aq) + e^- \rightarrow Ag(s)$

w = Zq; q = 1C

w = Z = 1 electrochemical equivalent

171. (1)

172. (2) Electrolysis of concentrated solution of acidified sulphate solution yields $H_2S_2O_8$.

Anode : $2H_2SO_4 \rightarrow H_2S_2O_8 + 2H^+ + 2e^-$

$2H_2O \rightarrow O_2 + 4H^+ + 4C^-$

Cathode : $H^+ + e^- \rightarrow \frac{1}{2}H_2$

173. (1) Millimoles of $Au^+ = 0.1 \times 250 = 25$

Mole of $Au^+ = \frac{25}{1000} = \frac{1}{40} = 0.025$

Similarly, moles of $Ag^+ = 0.025$

Charge passed $= I \times t = 1 \times 15 \times 60 = 900$ C

Moles of e^- passed $= \frac{900}{96500} = 0.0093$ mol.

Species with higher value of SRP will get deposited first at cathode.

$Au^+(aq) + e^- \longrightarrow Au(s)$

0.025 0.0093

So, only Au will get deposited.

174. (1) According to the Faraday's law of electrolysis, nF of current is required for the deposition of 1 mol.

According to the reaction,

$Ni(NO_3)_2 \longrightarrow Ni^{2+} + 2NO_3^-$

2 F of current deposits = 1 mol

$\therefore$ 0.1 F of current deposits $= \frac{0.1}{2} = 0.05$ mol

175. (3) The anodic half cell reaction when the battery is recharged :

$PbSO_4(s) + 2e^- \longrightarrow Pb(s) + SO_4^{2-}(aq)$

We require 2F for the electrolysis of 1 mol or 303 g of $PbSO_4$.

$\therefore$ Amount of $PbSO_4$ electrolysed by 0.05 F

$= \frac{303}{2} \times 0.05 = 7.575\ g \approx 7.6\ g$

176. (3)

177. (3) 9.65 ampere current was passed for 1.0 hour (3600 seconds)

Number of moles of electrons passed

$= \frac{I(A) \times t(s)}{96500} = \frac{9.65A \times 3600s}{96500} = 0.36$ moles

$C_6H_5NO_2 + 4e^- + 4H^+ \longrightarrow$ *p*-Aminophenol $+ H_2O$

$\because$ 4 moles of electrons reduces 1 mole of nitrobenzene to *p*-aminophenol.

$\therefore$ 0.36 moles of electrons will reduce $\frac{0.36}{4} = 0.09$ moles of nitrobenzene to *p*-aminophenol.

p-aminophenol molar mass = 109.14 g/mol

Mass of *p*-aminophenol obtained

= 109.14 g/mol × 0.09 mol = 9.81 g

178. (4) Reduction at cathode:

$2e^- + 2H_2O \rightarrow H_2 + 2OH^-$ (valence factor) $H_2 = 2$

At NTP, 22400 mL of H_2 = 1 mole of H_2

112 mL of H_2

$= \frac{1}{22400} \times 112 = 0.005$ mole of H_2

Mass of H_2 produced $= \frac{I \times t}{96500} E_{H_2}$

$2 \times 0.005 = \frac{I \times 965}{96500} \times \frac{2}{2}$; $\therefore I = 1.0$ A

179. (4) $Cu^{2+} + 2e^- \longrightarrow Cu$

2F i.e. 2 × 96500 C deposit Cu = 1 mol = 63.5g

180. (3) According to Faraday's first law of electrolysis,

$W = \frac{E \times I \times t}{96500}$

Where E = equivalent weight

$= \frac{\text{mol. mass of metal}(M)}{\text{oxidation state of metal }(x)}$

Substituting the value in the formula

$W = \frac{M}{x} \times \frac{I \times t}{96500}$

or $x = \frac{M}{W} \times \frac{I \times t}{96500} = \frac{10 \times 2 \times 60 \times 60}{96500 \times 0.250} = 3$

$\left[\text{Given : no. of moles} = \frac{W}{M} = 0.250\right]$

Hence, oxidation state of metal is (+3).

181. (3) $Cu \longrightarrow Cu^{2+} + 2e^-$

i.e, to deposit 1 mole of Cu at cathode from Cu^{2+} SO_4^{2-} solution = 2 moles of electrons are required

i.e, To deposit 6.35g $= \frac{6.35}{63.5} \times 2 = \frac{2}{10} = \frac{1}{5}$ mol

Thus, total no. of electrons required $= \frac{N_A}{5}$

182. (2) W = ZIt

where Z = Electrochemical equivalent

Eq. wt. of copper $= \frac{63}{2} = 31.5$

$Z = \frac{31.5}{96500}$

$W = ZIt = \frac{31.5}{96500} \times 1.5 \times 10 \times 60 = 0.2938$ g

183. (3) Reduction half reaction :

$Cr_2O_7^{2-} + 6e^- + 14H^+ \longrightarrow 2Cr^{3+} + 7H_2O$

Oxidation half reaction :

$Cr \longrightarrow Cr^{3+} + 3e^-$

Overall reaction :

$Cr_2O_7^{2-} + Cr + 14H^+ + 3e^- \longrightarrow 3Cr^{3+} + 7H_2O$

3F of electricity = 3 moles of Cr^{3+}

1F of electricity $= \frac{3}{3}$ moles of Cr^{3+}

184. (3) More the reduction potential, more is the deposition of metals at cathode. Cation having E° value less than – 0.83V (reduction potential of H_2O) will not deposit from aqueous solution. Hence, correct order of deposition of the metal at the cathode is:

Ag > Hg > Cu

185. (4) 1 mole of e^- = 1F = 96500 C

27g of Al is deposited by 3 × 96500 C

5120 g of Al will be deposited by

$= \frac{3 \times 96500 \times 5120}{27} = 5.49 \times 10^7$ C

186. (1) When 96500 coulomb of electricity is passed through the electroplating bath the amount of Ag deposited = 108g

$\therefore$ when 9650 coulomb of electricity is passed deposited Ag.

$= \frac{108}{96500} \times 9650 = 10.8$ g

187. (4) Pure metal always deposits at cathode.

188. (111) Eq. of Cu = Eq. of O_2

$\Rightarrow n_{Cu} \times 2 = n_{O_2} \times 4$

$\Rightarrow \frac{300 \times 10^{-3} \times 2}{63.54} = n_{O_2} \times 4$

$2.36 \times 10^{-3} = n_{O_2}$

When current is further passed ;

$n_{O_2} \times 4 = \frac{600 \times 28 \times 60}{96500 \times 1000}$

$n_{O_2} = 2.611 \times 10^{-3}$

Total O_2 released

$= [10^{-3} \times (2.36 + 2.611)] \times 22400$ mL

= 111.35 mL

189. (2)

190. (1.93)

$H_2O^{-2} \longrightarrow O_2^0 + H_2$; n = 2

$\therefore$ Amount of electricity in coulomb required is = 2F

$= 2 \times 96500 = 193000 = 1.93 \times 10^5$ C

191. (5) $Cu^{2+} + 2e^- \rightarrow Cu$

2 Faraday → 1 mol Cu

1 Faraday → 0.5 mol Cu deposit

0.5 mol = 0.5 g atom = 5×10^{-1}

$\therefore$ x = 5

192. (2) $Au^{3+} + 3e^- \longrightarrow Au(s)$

$\Rightarrow \frac{W}{E} = \frac{Q}{1F}$

$\Rightarrow Q = \frac{W \times F}{E} = \frac{W}{E}F = \frac{1.314}{\frac{197}{3}}F$

$= 0.02F = 2 \times 10^{-2}$ F

193. (45.75) I = 0.015 A

t = 15 min. = 15 × 60 sec.

$Zn^{2+} + 2e^- \longrightarrow Zn$

$W = Z \times I \times t \Rightarrow = \frac{65.4}{2 \times 96500} \times 0.015 \times 15 \times 60 =$

45.75×10^{-4} g

194. (108)

Eq. of Ag = Eq. of O_2 (given)

Let x gm silver be displaced,

$\Rightarrow \frac{x \times 1}{108} = \frac{5.6}{22.7} \times 4$

(Molar volume of gas at STP = 22.7 lit)

$\Rightarrow$ x = 106.57 gm

or, as per old STP data, molar volume = 22.4 lit

$\frac{x \times 1}{108} = \frac{5.6}{22.7} \times 4, \Rightarrow x = 108$ g.

195. (161) W = z × i × t.

Density × volume $= \frac{E \times i \times t}{96500}$

(10) × (100 × 0.0001) =

$\frac{\left(\frac{\text{atomic wt.}}{\text{v.f}}\right) \times 2 \times x}{96500}$ (v.f = 2)

$\therefore$ x = 161 sec.

196. (3) Fe_3O_4 is a mixture of FeO and Fe_2O_3. The oxidation state of Fe in FeO and Fe_2O_3 are +2 and +3 respectively.

$$\overset{(+2)}{FeO} \xrightarrow{2e^-} Fe$$

$$\overset{(+3)}{Fe_2O_3} \xrightarrow{6e^-} 2Fe$$

$$Fe_3O_4 \xrightarrow{8F} 3Fe$$

3 mole Fe requires = 8F

1 mole Fe requires $\frac{8}{3}F = 2.6\,F \approx 3$

197. (127) $2H_2O \rightarrow O_2(g) + 4H^+ + 4e^-$: At anode

$2H_2O + 2e^- \rightarrow H_2(g) + 2OH^-$: At cathode

Now number of gm eq. $= \frac{i \times t}{96500}$

$= \frac{0.1 \times 2 \times 60 \times 60}{96500} = 0.00746$

$V_{O_2} = \frac{0.00746}{4} \times 22.7 = 0.0423$

$V_{H_2} = \frac{0.00746}{2} \times 22.7 = 0.0846$

$V_{total} \approx 127$ mL or cc

198. (20) $Fe^{3+} + 3e^- \longrightarrow Fe$

$3F \equiv 1$ mole Fe is deposited

For 56 g $\equiv 3 \times 96500$ (required charge)

For 0.3482 g $\equiv \frac{3 \times 96500}{56} \times 0.3482 = 1800$ coulomb

$Q = it \Rightarrow 1800 = 1.5\,t \Rightarrow t = 1200$ s = 20 min

199. (25)

$MnO_4^- + 8H^+ + 5e^- \rightarrow Mn^{+2} + 4H_2O$, $E^o = 1.51$ V

$\because$ 1 mole of MnO_4^- required 5 moles of electrons or 5 F electricity.

$\therefore$ 5 moles of MnO_4^- required = 25 F electricity.

200. (11)

$$6OH^- + Cl^- \longrightarrow ClO_3^- + 3H_2O + 6e^-$$

For synthesis of 1 mole of ClO_3^-, 6 F of charge is required.

$\because$ Current efficiency = 60%

$\therefore$ To synthesis 1 mole of ClO_3^-, 10 F of charge is required.

To synthesis $\frac{10}{122}$ moles of $KClO_3$, charge

$= \frac{10 \times 10}{122}F$

$Q = I \cdot t$

$t = \frac{100 \times 96500}{122 \times 2} = 39549.18$ s $= 10.99$ h

$\therefore t = 11$ h.

201. (60) Charge $(Q) = It = 2 \times 8 \times 60 = 960$ C

$\Rightarrow \frac{960}{96000} = 0.01F$

$$\underset{\substack{6F \\ 0.01F}}{Cr_2O_7^{2-} + 14H^+ + 6e^-} \longrightarrow \underset{\substack{2\text{ moles} \\ \frac{1}{3}\times 0.01\text{ mole}}}{2Cr^{3+}} + 7H_2O$$

Theoritical mass of Cr^{3+}

$= \frac{1}{3} \times \frac{960}{96000} \times 52 = 0.173$ g

So, efficiency

$= \frac{W_{actual}}{W_{Theoritical}} \times 100 = \frac{0.104}{0.173} \times 100 = 60\%$

202. (5.68) No. of moles of silver deposited

$= \frac{108}{108} = 1$ mol

$Ag^+ + e^- \rightarrow Ag$

1 F charge is required to deposit 1 mole of Ag

$H_2O \rightarrow \frac{1}{2}O_2 + 2H^+ + 2e^-$

2 F charge deposit $\rightarrow \frac{1}{2}$ moles of oxygen

1 F charge will deposit $\rightarrow \frac{1}{4}$ moles of oxygen

$V_{O_2} = \frac{nRT}{P}$

$= \frac{1}{4} \times \frac{0.08314\,L\,bar\,K^{-1}mol^{-1} \times 273K}{1\,bar}$

$= 5.675$ L

203. (2) During the charging of lead storage battery, the following reaction takes place.

$2PbSO_4(s) + 2H_2O(l) \rightarrow Pb(s) + PbO_2(s) + 2H_2SO_4(aq)$

At anode $PbSO_4$ reduced back to metallic Pb. At cathode $PbSO_4$ oxidised back to PbO_2.

From the balanced reacation.

$\because x_1 = +2, y_1 = 0$

$x_2 = +2, y_2 = 4$

204. (1) $CH_3OH(\ell) + \frac{3}{2}O_2(g) \rightarrow CO_2(g) + 2H_2O(\ell)$

$\because E^o_{cell} = E^o_{cathode} - E^o_{Anode} = E^o_{O_2/H_2O} - E^o_{CO_2/CH_3OH}$

$1.21 = 1.229 - E^o_{CO_2/CH_3OH}$

$\Rightarrow E^o_{CO_2/CH_3OH} = 0.019\,V = 19\,mV$

Fuel cell involves oxidation of methanol which will occur at anode and reduction of O_2 will occur at cathode.

205. (1)

206. (4) Corrosion is an example of electrochemical reaction due to redox reaction in which pure metal act as anode and impure metal (rusted metal) act as cathode.

Corrosion is more favourable in acid medium than alkaline medium as H^+ ions S quickly accept electrons from the metal.

207. (3) Leclanche cell and Hg-cell are primary batteries while Ni – Cd is a secondary cell.

208. (1) Reaction occurs at cathode is :

$MnO_2 + NH_4^+ + e^- \longrightarrow MnO(OH) + NH_3$

In the cathode reaction manganese (Mn) is reduced from the +4 oxidation state to the +3 state.

209. (4) Fuel cell is used in spaceship and it is type of galvanic cell. It has efficiency of about 70%.

210. (2) Mn, Ni and Cd metals used in battery industries.

211. (1) As tin coating is peeled off, then iron is exposed to atmosphere and rusting of iron will take place.

212. (1) Lead storage battery consists of lead anode and a grid of lead packed with lead oxide (PbO_2) as cathode, a 38% solution of H_2SO_4 is used as an electrolyte.

On charging the battery the reaction is reversed and $PbSO_4(s)$ on anode and cathode is converted into Pb and PbO_2 respectively.

213. (3)

(1) $Cd(s) + 2Ni(OH)_3(s) \rightarrow CdO(s) + 2Ni(OH)_2(s) + H_2O(l)$

Discharge of secondary Battery

(2) $Zn(Hg) + HgO(s) \rightarrow ZnO(s) + Hg(l)$

(Primary Battery Mercury cell)

(3) $2PbSO_4(s) + 2H_2O(l) \rightarrow Pb(s) + PbO_2(s) + 2H_2SO_4(aq)$

Charging of secondary Battery

(4) $2H_2(g) + O_2(g) \rightarrow 2H_2O(l)$ – Fuel cell

214. (3) $CH_3OH\,(l) + \frac{3}{2}O_2\,(g) \rightarrow CO_2\,(g) + 2H_2O\,(l)$

$\Delta G_r = \Delta G_f(CO_2, g) + 2\Delta G_f(H_2O, l) - \Delta G_f(CH_3OH, l) - \frac{3}{2}\Delta G_f(O_2, g)$

$= -394.4 + 2(-237.2) - (-166.2) - 0$

$= -394.4 - 474.4 + 166.2 = -702.6$ kJ

% efficiency $= \frac{702.6}{726} \times 100 = 97\%$

215. (2)

Cathode: $O_2(g) + 2H_2O(l) + 4e^- \rightarrow 4OH^-(aq)$

Anode : $2H_2(g) + 4OH^-(aq) \rightarrow 4H_2O(l) + 2e^-$

Overall : $2H_2(g) + O_2(g) \rightarrow 2H_2O(l)$

Due to potential difference between cathode and anode, production of electricity is possible.

216. (560) $E^o_{cell} = E^o_{cathode} - E^o_{anode}$

$E^o_{cell} = [1.23 - 0.02] = 1.21$ V

$W_{max} = \Delta G^o = -nFE^o = -6 \times 96500 \times 1.21$ J

W_{actual} = Efficiency $\times W_{max}$

$= \frac{80}{100} \times \Delta G^o = -P\Delta V$

$\Rightarrow 0.8 \times 6 \times 96500 \times 1.21 = 1 \times 10^3(\Delta V)$

$\therefore \Delta V = 560.472\,m^3$.

Part-B (JEE Advanced)

1. **(b)** $Zn(s) + Cu^{2+}(aq) \rightarrow Zn^{2+}(aq) + Cu(s)$

$\Delta G = \Delta G^\circ + 2.303\,RT\log_{10}Q;\ Q = \dfrac{[Zn^{2+}]}{[Cu^{2+}]}$

$[\Delta G^\circ = -nFE^\circ] = -2 \times F \times 1.1$

Given $[Zn^{2+}] = 10[Cu^{2+}]$

$\therefore\ \Delta G = -2F(1.1) + 2.303\,RT\log_{10}10$

$= 2.303\,RT - 2.2F$

2. **(d)** At anode : $H_2(g) \rightleftharpoons 2H^+(aq) + 2e^-$

At cathode : $M^{4+}(aq) + 2e^- \rightleftharpoons M^{2+}(aq)$

Net cell reaction : $H_2(g) + M^{4+}(aq) \rightleftharpoons 2H^+(aq) + M^{2+}(aq)$

$$\text{Now, } E_{cell} = \left(E^\circ_{M^{4+}/M^{2+}} - E^\circ_{H^+/H_2}\right) - \frac{0.059}{n}\cdot\log\frac{[H^+]^2[M^{2+}]}{P_{H_2}\cdot[M^{4+}]}$$

$$\text{or, } 0.092 = (0.151 - 0) - \frac{0.059}{2}.\log\frac{1^2 \times [M^{2+}]}{1 \times [M^{4+}]}$$

$$\therefore\ \frac{[M^{2+}]}{[M^{4+}]} = 10^2 \Rightarrow x = 2$$

3. **(d)** Here $n = 4$, and $[H^+] = 10^{-pH} = 10^{-3}$

Applying Nernst equation

$$E = E^\circ - \frac{0.059}{n}\log\frac{[Fe^{2+}]^2}{[H^+]^4\,p_{O_2}}$$

$$= 1.67 - \frac{0.059}{4}\log\frac{(10^{-3})^2}{(10^{-3})^4 \times 0.1}$$

$$= 1.67 - \frac{0.03}{2}\log 10^7 = 1.67 - 0.105 = 1.565\text{ V}$$

4. **(a)** $Fe(s) \longrightarrow Fe^{2+} + 2e^-;\quad E^\circ = 0.44\text{ V}$

$2H^+ + 2e^- + \frac{1}{2}O_2 \longrightarrow H_2O(l);\quad E^\circ = +1.23\text{ V}$

$Fe(s) + 2H^+ + \frac{1}{2}O_2 \longrightarrow Fe^{2+} + H_2O;$

$E^\circ_{cell} = 0.44 + 1.23 = 1.67\text{ V}$

$\therefore\ \Delta G^\circ = -nFE^\circ_{cell} = -2 \times 96500 \times 1.67 = -322\text{ kJ}$

5. **(b)** Cell reaction :

$Zn + Fe^{2+} \longrightarrow Zn^{2+} + Fe$

Using Nernst equation

$$E_{cell} = E^\circ_{cell} - \frac{0.0591}{n}\log\left[\frac{Zn^{2+}}{Fe^{2+}}\right]$$

$$E_{cell} = E^\circ_{cell} - \frac{0.0591}{2}\log\frac{10^{-2}}{10^{-3}}$$

At 298 K, $E = 0.2905$

$$E^\circ_{cell} = 0.2905 + \frac{0.0591}{2} = 0.32$$

$$\text{or}\quad 0.32 = \frac{0.0591}{2}\log K_{eq.}$$

$$\text{or } K_{eq} = 10^{\frac{0.32}{0.0295}}.$$

6. **(c)** In an electrolytic cell, electrons do not flow themselves. It is the migration of ions towards oppositely charged electrodes that indirectly constitutes the flow of electrons from cathode to anode through internal supply.

7. **(a)** MnO_4^- will oxidise Cl^- ion according to the following equation:

$2MnO_4^- + 16H^+ + 10Cl^- \longrightarrow 2Mn^{2+} + 8H_2O + 5Cl_2\uparrow$

The cell corresponding to this reaction is as follows :

Pt, Cl_2 (1 atm) | Cl^- || MnO_4^-, Mn^{2+}, H^+ | Pt

$E^\circ_{cell} = 1.51 - 1.40 = 0.11\text{ V}$

E°_{cell} being +ve, ΔG° will be –ve and hence, the above reaction is feasible. MnO_4^- will not only oxidise Fe^{2+} ion but also Cl^- ion simultaneously. So, the quantitative estimation of aq $Fe(NO_3)_2$ cannot be done by this.

8. **(c)** The salt used to make 'salt–bridge' must be such that the ionic mobility of cation and anion are of comparable order so that they can keep the anode and cathode half cells neutral at all times. KNO_3 is used becasue velocities of K^+ and NO_3^- ions are nearly same.

9. **(b)** For $M^+ + X^- \longrightarrow M + X$, $E^\circ_{cell} = 0.44 - 0.33 = 0.11$V is positive, hence reaction is spontaneous.

10. **(a)** The given order of reduction potentials is $Z > Y > X$. A spontaneous reaction will have the following characteristics:

(i) Z reduced and Y oxidised,
(ii) Z reduced and X oxidised and
(iii) Y reduced and X oxidised

Hence, Y will oxidise X and not Z.

11. **(c)** We have

Half-cell	Half-cell reaction	$\Delta G^\circ = -nFE^\circ$
Cu^{2+} | Cu	$Cu^{2+} + 2e^- = Cu$	$\Delta G_1^\circ = -2FE^\circ_{Cu^{2+}\mid Cu}$
Cu^{2+} | Cu^+	$Cu^{2+} + e^- = Cu^+$	$\Delta G_2^\circ = -FE^\circ_{Cu^{2+}\mid Cu^+}$
Cu^+ | Cu	$Cu^+ + e^- = Cu$	$\Delta G_3^\circ = -FE^\circ_{Cu^+\mid Cu}$

From the half-cell reactions, it follows that

$$\Delta G_3^\circ = \Delta G_1^\circ - \Delta G_2^\circ$$

i.e.,

$$-FE^\circ_{Cu^+\mid Cu} = -2FE^\circ_{Cu^{2+}\mid Cu} - \left(-FE^\circ_{Cu^{2+}\mid Cu^+}\right)$$

$$\text{or } E^\circ_{Cu^+\mid Cu} = 2E^\circ_{Cu^{2+}\mid Cu} - E^\circ_{Cu^{2+}\mid Cu^+}$$

$= 2(0.337\text{ V}) - 0.153\text{ V} = 0.521\text{ V}$

12. **(a)** H_2O is more readily reduced at cathode than Na^+. It is also more readily oxidized at anode than SO_4^{2-}. Hence, the electrode reactions are

$2H_2O + 2e^- \longrightarrow H_2\uparrow + 2OH^-$ [at cathode]

$H_2O \longrightarrow \frac{1}{2}O_2\uparrow + 2H^+ + 2e^-$ [at anode]

13. **(b)**

(i) In a galvanic cell oxidation occurs at anode and reduction occurs at cathode.

(ii) Oxidation occurs at electrode having higher oxidation potential and it behaves as anode and other electrode acts as cathode.

(iii) $E_{Cell} = E_C - E_A$

(substitute reduction potential at both places).

$Fe^{2+} + Zn \longrightarrow Zn^{2+} + Fe$

$\because Zn \longrightarrow Zn^{++} + 2e^-$

and $Fe^{2+} + 2e^- \longrightarrow Fe$

$\therefore$ Zn is anode and Fe is cathode.

$E_{cell} = E_C - E_A = -0.41 - (-0.76) = 0.35$V.

14. **(a)** Water is reduced at the cathode and oxidized at the anode instead of Na^+ and SO_4^{2-}.

Cathode : $2H_2O + 2e^- \rightarrow H_2 + 2OH^-$; $E^\circ = -0.83$ V

Anode : $H_2O \rightarrow 2H^+ + \frac{1}{2}O_2 + 2e^-$;

$E^\circ = -1.23$ V

Note : The standard electrode, reduction potential of Na^+ is less than that of water.

$Na^+ + e^- \longrightarrow Na(s)\quad E^\circ = -2.71$ V
oxidation

The standard electrode, oxidation potential of SO_4^{2-} is less than that of water.

$2SO_4^{2-} \longrightarrow S_2O_8^{2-} + 2e^-\quad E^\circ = -2.01$ V

15. **(c)** Oxidation is loss of electron and in a galvanic cell it occurs at anode. Reduction is gain of electron and in a galvanic cell it occurs at cathode.

Cell representation :

Anode / Anodic electrolyte || Cathodic electrolyte / Cathode

Reaction at Anode : $H_2 \rightarrow 2H^+ + 2e^-$

Reaction at Cathode : $AgCl + e^- \rightarrow Ag + Cl^-$

16. **(c)** The reduction potentials (as given) of the ions are in the order :

$Ag^+ > Hg_2^{2+} > Cu^{2+} > Mg^{2+}$

Mg^{2+} (aq.) will not be reduced as its reduction potential is much lower than that of water (–0.83 V). Hence, the sequence of deposition of the metals will be Ag, Hg, Cu.

17. **(a)** More negative is the value of reduction potential, higher will be the reducing property, *i.e.*, the power to give up electrons.

18. **(10)**

$Mg(s) \longrightarrow Mg^{2+}(aq) + 2e^-$

$Cu^{2+}(aq) + 2e^- \longrightarrow Cu(s)$

$Mg(s) + Cu^{2+}(aq) \longrightarrow Mg^{2+}(aq) + Cu(s)$

$$E_{cell} = E^\circ_{cell} - \frac{RT}{nF}\ln x$$

$$E = 2.67 = 2.7 - \frac{RT}{nF}\ln\frac{x}{1}$$

$$0.03 = \frac{300}{2 \times 11500}\ln x$$

$2.3 = \ln x;\ x = 10$

19. (4) $X \longrightarrow Y; \quad \Delta G° = -193 \text{ kJ mol}^{-1}$

$M^{+} \longrightarrow M^{3+} + 2e^{-} \qquad E° = -0.25\text{V}$

Hence $\Delta G°$ for oxidation will be

$\Delta G° = -nFE°$

$= -2 \times 96500 \times (-0.25) = 48250 \text{ J} = 48.25 \text{ kJ}$

48.25 kJ energy oxidises one mole M^{+}

$\therefore$ 193 kJ energy oxidises $\frac{193}{48.25}$ mole M^{+}

$= 4$ mole M^{+}

20. (0.77)

$$\overset{\text{VII}}{MnO_4^-} \xrightarrow[E_1^\circ = 1.68\text{ V}]{+3e} \overset{\text{IV}}{MnO_2^-} \xrightarrow[E_2^\circ = 1.21\text{ V}]{+2e} Mn^{2+} \xrightarrow[E_3^\circ = -1.03\text{ V}]{+2e} Mn$$

$$MnO_4^- \xrightarrow[E_4^\circ = ?]{+7e} Mn$$

$$\Delta G_4^\circ = G_1^\circ + \Delta G_2^\circ + \Delta G_3^\circ$$

or $-n_4FE_4^\circ = n_1FE_1^\circ - n_2FE_2^\circ - n_3FE_3^\circ$

[Applying Nernst Eq.]

or $-7E_4^\circ = -3E_1^\circ - 2E_2^\circ - 2E_3^\circ$

or $E_4^\circ = \frac{(3 \times 1.68) + (2 \times 1.21) - (2 \times 1.03)}{7} \text{V}$

$= 0.77\text{V}$

21. (– 11.62)

$A\text{ (s)} \mid A^{n+}\text{ (aq, 2M)} \parallel B^{2n+}\text{ (aq, 1M)} \mid B\text{ (s)}$

Reactions

Anode $\quad (A \longrightarrow A^{n+} + ne^{-}) \times 2$

Cathode $\quad B^{2n+} + 2ne^{-} \longrightarrow B$

Overall reaction :

$$2A + B^{2n+} \longrightarrow 2A^{n+} + B$$

$$E = E° - \frac{RT}{2nF}\ln Q$$

$$0 = E° - \frac{RT}{2nF}\ln\frac{[A^{n+}]^2}{[B^{2n+}]}$$

$$E° = \frac{RT}{2nF}\ln\frac{2^2}{1}; \quad E° = \frac{RT}{2nF}\ln 4$$

Now,

$$\Delta G° = -2nFE° = -\frac{2nFRT}{2nF}\ln 4 = -RT\ln 4$$

$\Delta G° = \Delta H° - T\Delta S° = 2\Delta G° - T\Delta S°$ (Given $\Delta H° = 2\Delta G°$)

$T\Delta S° = \Delta G°$

$$\Delta S° = \frac{\Delta G°}{T} = \frac{-RT\ln 4}{T} = -R\ln 4 = -R \times 2\ln 2$$

$= -8.3 \times 2 \times 0.7 = -11.62 \text{ JK}^{-1}\text{mol}^{-1}$

22. (0.05) Daniel cell is : $Zn \mid Zn^{2+} \mid\mid Cu^{2+} \mid Cu$

Let there be two Daniel cells with their E_{cell} as given below:

$Zn \mid Zn^{2+}(C_1) \mid\mid Cu^{2+}(C = ?) \mid Cu$,

$E_{cell} = E_1$

$Zn \mid Zn^{2+}(C_2) \mid\mid Cu^{2+}(C = 0.5\text{ M}) \mid Cu$

$E_{cell} = E_2$ where $E_2 > E_1$

According to question, $E_2 - E_1 = 0.03$ and $C_2 = C_1$

The cell reaction is

$$Zn + Cu^{2+} \longrightarrow Zn^{2+} + Cu, Q = \frac{[Zn^{2+}]}{[Cu^{2+}]}.$$

Thus, $E_1 = E^o_{cell} - \frac{0.06}{2}\log\frac{C_1}{C}$;

and $E_2 = E^o_{cell} - \frac{0.06}{2}\log\frac{C_2}{0.5}$;

Since, same $ZnSO_4$ is used in both cells $C_1 = C_2$

So, $E_2 - E_1 = \frac{0.06}{2}\left[\log\frac{C_1}{C} \times \frac{0.5}{C_1}\right]$

$$\Rightarrow 0.03 = \frac{0.06}{2}\log\frac{0.5}{C} \Rightarrow \log\frac{0.5}{C} = 1 \text{ or } C$$

= 0.05 M

23. (–0.22) At pH = 14; $[H^+] = 1 \times 10^{-14}$ M; $[OH^-] = 10^0 = 1$M

$(\because [H^+][OH^-] = 1 \times 10^{-14})$

$\therefore$ Cu $(OH)_2$ ionises as follows:

$$Cu(OH)_2 \rightleftharpoons Cu^{2+} + 2OH^-$$

$\therefore K_{sp}$ of $Cu(OH)_2 = [Cu^{2+}][OH^-]^2$

$1.0 \times 10^{-19} = [Cu^{2+}][1]^2$; $[Cu^{2+}] = 1.0 \times 10^{-19}$ M

The standard reduction potential of Cu^{2+}/Cu is represented in the form of following equation:

$$Cu^{2+}(aq) + 2e^- \rightarrow Cu(s)$$

On applying Nernst equation

$$E = E^o - \frac{0.0591}{n}\log\frac{1}{[Cu^{2+}]}$$

$$= +0.34 - \frac{0.0591}{2}\log\frac{1}{1 \times 10^{-19}}$$

$$= \left[0.34 - \frac{0.0591}{2} \times 19\right] = 0.34 - 0.56$$

= – 0.22V

24. (0.792)

	2Hg + $2Fe^{3+}$ $\longrightarrow$	Hg_2^{2+} +	$2Fe^{2+}$
Initial conc.	1.0×10^{-3}	0	0
Eqilb. conc.	0.05×10^{-3}	$\frac{0.95 \times 10^{-3}}{2}$	0.95×10^{-3}

$$E = E^o_{Fe^{3+}/Fe^{2+}} - E^o_{Hg_2^{2+}/Hg} - \frac{0.059}{n}\log\frac{[Fe^{2+}]^2[Hg_2^{2+}]}{[Fe^{3+}]^2}$$

At equilibrium, $E = 0$

$$\Rightarrow 0 = 0.77 - E^o_{Hg_2^{2+}/Hg} - \frac{0.059}{2}$$

$$\log\frac{(0.95 \times 10^{-3})^2(0.475 \times 10^{-3})}{(0.05 \times 10^{-3})^2}$$

On usual calculations, $E^o_{Hg_2^{2+}/Hg}$ = **0.792V**

25. negative, greater; Among the various metals, since sodium has the minimum reduction potential, it must be strongest reducing agent. In general, more the reduction potential lesser is its reducing action.

26. False : When the temperature is 273 K, the value of the factor will come out as 0.0541 instead of 0.0591. The value 0.0591 comes out at 298 K and not at 273 K.

27. (b, c, d)

(a) $M(s) + 2H^+(aq) \longrightarrow H_2(g) + M^{2+}(aq)$

No. of moles of exchanged e^-, $n = 2$

$$\Delta S = nF\frac{dE_{cell}}{dT} = 2F\left(\frac{R}{F}\right) = 2R$$

(b) $Pt(s) \mid H_2(g)$, 1 bar $\mid H^+$ (aq, 0.01 M) $\mid\mid H^+$ (aq, 0.1 M) $\mid H_2$(g, 1 bar) $\mid$ Pt(s)

$$E_{cell} = \frac{2.303RT}{nF}\log\frac{0.01}{0.1}$$

$$\Rightarrow E_{cell} = \frac{2.303RT}{F} > 0 \Rightarrow \Delta G < 0$$

For concentration cell, $\Delta H = 0$

$\Delta G = \Delta H - T\Delta S \Rightarrow \Delta S > 0$

Hence, it is an entropy driven process.

(c) Racemization of an optically active compound is a spontaneous process, i.e., $\Delta G < 0$.

As $\Delta H = 0 \Rightarrow \Delta S > 0$

(d) $[Ni(H_2O)_6]^{2+} + 3\text{en} \longrightarrow \underset{\text{more stable}}{[Ni(en)_3]^{2+}} + 6H_2O$

The no. of molecules in the product side is increasing hence, $\Delta S > 0$.

28. (a, b, c)

$$X(s) \rightarrow X^{+2}(0.001M) + 2e^-(\text{anode})$$

$$Y^{+2}(0.1M) + 2e^- \rightarrow Y(s)(\text{cathode})$$

$$E_{cell} = E°_{cell} - \frac{0.06}{2}\log\frac{X^{+2}}{Y^{+2}}$$

$E_{cell} = E°_{cell} + 0.06$

(a) Cd (X) and Ni (Y) $E°_{cell} = +0.4 - 0.24$; $E_{cell} = 0.22$

(b) Cd (X) and Fe (Y) $E°_{cell} = -0.04$; $E_{cell} = 0.02$

(c) Ni (X) and Pb (Y) $E°_{cell} = 0.11$; $E_{cell} = 0.17$

(d) Ni (X) and Fe (Y) $E°_{cell} = -0.2$; $E_{cell} = -0.14$

Since in (a) (b) (c), E_{cell} is positive, hence answer is (a) (b) (c).

29. (a) Salt bridge is introduced to keep the solutions of two electrodes separate, so that the ions in electrodes do not mix freely with each other. Salt bridge maintains the diffusion of ions from one electrode to another.

30. (a, b, d)

The species having less reduction potential with respect to NO_3^- ($E° = +0.96$ V) will be oxidised by NO_3^-. These species are V, Fe and Hg.

31. (a) More negative or lower is the reduction potential, more is the reducing property. Thus, the reducing power of the corresponding metal will follow the reverse order, *i.e.* $Y > Z > X$.

32. (d)

(P) $$Fe^{3+} \xrightarrow[n=1]{+0.77\text{V}} Fe^{2+} \xrightarrow[n=2]{-0.44\text{V}} Fe$$

$$Fe^{3+} \xrightarrow[n=3]{x\text{ V}} Fe$$

$$\Delta G^o_{Fe^{3+}/Fe} = \Delta G^o_{Fe^{3+}/Fe^{2+}} + \Delta G^o_{Fe^{2+}/Fe}$$

$$\Rightarrow -3 \times FE^o_{(Fe^{+3}/Fe)} = -1 \times FE^o_{(Fe^{+3}/Fe^{+2})} + \left(-2 \times FE^o_{Fe^{+2}/Fe}\right)$$

$$\Rightarrow 3 \times x = 1 \times 0.77 + 2 \times (-0.44)$$

$$\Rightarrow x = -\frac{0.11}{3}\text{V} \simeq -0.04\text{ V}.$$

(Q)

$2H_2O \longrightarrow O_2 + 4H^+ + 4e^-$	$E° = -1.23\,V$
$4e + O_2 + 2H_2O \longrightarrow 4OH^-$	$E° = +0.40\,V$
$4H_2O \longrightarrow 4H^+ + 4OH^-$	$E° = -0.83\,V$

(R)

$Cu^{2+} + 2e \longrightarrow Cu$	$E° = +0.34\,V$
$2Cu \longrightarrow 2Cu^+ + 2e$	$E° = -0.52\,V$
$Cu^{2+} + Cu \longrightarrow 2Cu^+$	$E° = -0.18\,V$

(S) $Cr^{3+} \xrightarrow[n=1]{x} Cr^{2+} \xrightarrow[n=2]{-0.91V} Cr$

$-0.74V, \quad n = 3$

$x \times 1 + 2 \times (-0.91) = 3 \times (-0.74)$

$x - 1.82 = -2.22 \Rightarrow x = -0.4V$

33. (d) At anode : $M(s) + 2X^-(aq) \longrightarrow MX_2(aq) + 2e^-$

At cathode : $M^{+2}(aq) + 2e^- \longrightarrow M(s)$

Thus, here $n = 2$

$\Delta G = -nFE_{cell}$

$= -2 \times 96500 \times 0.059 \times 10^{-3}$ kJ/mole

$= -11.4$ kJ/ mole

34. (b)

$M(s) + M^+(aq)1M \longrightarrow M^+(aq)(.05M) + M(s)$

According to Nernst equation,

$$E_{cell} = 0 - \frac{2.303RT}{F}\log\frac{M^+_{(.05M)}}{M^+_{(1M)}}$$

$$= 0 - \frac{2.303RT}{F}\log(5\times10^{-2}) = +\text{ve}$$

Hence, $|E_{cell}| = E_{cell} = 0.070$ V and $\Delta G < 0$ for the feasibility of the reaction.

For concentration cell, $E^o_{cell} = 0$

35. (c) From above equation

$$\frac{-2.303RT}{F}\log(5\times10^{-2}) = 0.07$$

$$\Rightarrow \frac{2.303RT}{F}\times1.3 = 0.07$$

$$\Rightarrow \frac{2.303RT}{F} = 0.0538$$

$$\text{So, } E_{cell} = E^{\circ}_{cell} - \frac{0.0538}{1}\log 0.0025$$

$$= 0 - \frac{0.0538}{1}\log 0.0025 \approx 0.13988V \approx 140mV$$

36. (c) $2I^- + Cl_2 \longrightarrow I_2 + 2Cl^-$

$E° = E^o_{I^-/I_2} + E^o_{Cl_2/Cl^-} = -0.54 + 1.36;$

$E° = 0.82V$

$E°$ is positive hence, iodide ion is oxidized by chlorine.

37. (d)

$4Mn^{3+} + 2H_2O \longrightarrow 4Mn^{2+} + O_2 + 4H^+$

$E^o_{Mn^{3+}/Mn^{2+}} + E^o_{H_2O/O_2} = 1.50 + (-1.23) = 0.27V$

Reaction is feasible. [$\therefore E°$ is positive]

38. (a) The precipitate formed in this reaction is of

III II

$Fe_4[Fe(CN)_6]_3$.

39. (d) In the given reaction,

Ag^+ ions are reduce to Ag and Glucose is oxidised to gluconic acid as per the given reactions,

$Ag^+ + e^- \longrightarrow Ag;\ E^o_{red} = +0.800\,V$ and

$C_6H_{12}O_6 + H_2O \longrightarrow \underset{\text{Gluconic acid}}{C_6H_{12}O_7} + 2H^+ + 2e^-;$

$E^o_{ox} = -0.05V$

Hence, $E^o_{cell} = 0.8 - 0.05 = 0.75\,V$

$\Delta G^o_{cell} = -nFE = 2F \times 0.75 = -RT\ln K$

$$\Rightarrow \ln K = \frac{2F}{RT}(0.75) = 2\times38.92\times0.75 = 58.38$$

40. (a) For the reaction,

$C_6H_{12}O_6 + H_2O \longrightarrow \underset{\text{Gluconic acid}}{C_6H_{12}O_7} + 2H^+ + 2e^-$

$$E = E° - \frac{0.0591}{n}\log\frac{[P]}{[R]} = E° - \frac{0.0591}{2}\log[H^+]^2$$

$$E - E° = -\frac{0.0591}{2}\times 2(-pH) = 0.0591\times11 = 0.65$$

So, $E_{oxidation}$ increases over $E^o_{oxidation}$ by 0.65 V.

41. (b) During Tollen's test, oxidation of silver ion requires an alkaline medium. Under these conditions it forms insoluble silver oxide, hence to dissolve this oxide a complexing agent, ammonia is added, which brings silver ion as diamminosilver (I) ion, $[Ag(NH_3)_2]^+$. It is a soluble complex.

42. (a) From the given details, the reactions can be written as:

At anode:	$Ag(s) + Cl^-(aq) \longrightarrow AgCl(s) + e^-$
At cathode:	$Ag^+(aq) + e^- \longrightarrow Ag(s)$
Complete reaction	$Ag^+(aq) + Cl^-(aq) \longrightarrow AgCl(s)$

Hence, cell representation is

Ag (s) | AgCl (s) | Cl^- (aq) | | Ag^+ (aq) | Ag (s)

$$\Delta G° = \Delta G^o_f(AgCl) - [\Delta G^o_f(Ag^+) + \Delta G^o_f(Cl^-)]$$

$= -109 - (-129 + 77) = -57$ kJ/mol $= -57000$ J/mol

We know that, $\Delta G° = -nFE^o_{cell}$

$-57000 = -1 \times 96500 \times E^o_{cell}$ ($\because n$ = electron transferred = 1)

$$E^o_{cell} = \frac{57000}{96500} = \textbf{0.59 volts}$$

$$\text{Again } E^o_{cell} = \frac{0.0591}{n}\log K_c$$

$$\text{or } E^o_{cell} = \frac{0.0591}{n}\log\frac{AgCl}{[Ag^+][Cl^-]}$$

$$E^o_{cell} = \frac{0.0591}{1}\log\left(\frac{1}{K_{sp}}\right)$$

or $0.59 = -0.059\log K_{sp}$

or $\log K_{sp} = -10 \Rightarrow K_{sp} = \mathbf{10^{-10}}$

(b) When Zn is added to 100 mL of saturated AgCl solution.

$2Ag^+ + Zn(s) \rightleftharpoons 2Ag(s) + Zn^{2+}$

$Ag^+ + e^- \rightleftharpoons Ag;\ E° = 0.80\,V$

$Zn^{2+} + 2e^- \rightleftharpoons Zn;\ E° = -0.76\,V$

$E^o_{cell} = E^o_{Ag^+|Ag(s)} - E^o_{Zn^{2+}|Zn(s)}$

$= 0.80 - (-0.76) = 1.56\,V$

$$E^o_{cell} = \frac{0.059}{n}\log_{10}\frac{[Zn^{2+}]}{[Ag^+]^2}$$

$$\Rightarrow 1.56 = \frac{0.059}{2}\log_{10}\frac{[Zn^{2+}]}{[Ag^+]^2}$$

$$\Rightarrow \log_{10}\frac{[Zn^{2+}]}{[Ag^+]^2} = \mathbf{52.9}$$

As the value of equilbrium constant is very high so the reaction moves in forward direction completely.

$[Ag^+]$ from (a) $= \sqrt{10^{-10}} = 10^{-5}$

$[(\because K_{sp} = 10^{-10} = [Ag^+][Cl^-]]$

$$\therefore Ag^+ \text{ in 100 mL of solution} = \frac{10^{-5}\times100}{1000}$$

$= \mathbf{10^{-6}}$ **mol.**

43. The required reaction can be obtained in the following way.

$Cu^{2+} + e^- \longrightarrow Cu^+$	$\Delta G° = -0.15\,F$
$In^{2+} + e^- \longrightarrow In^+,$	$\Delta G° = +0.40\,F$
$In^+ \longrightarrow In^{3+} + 2e^-,$	$\Delta G° = -0.84\,F$

On adding,

$Cu^{2+} + In^{2+} \longrightarrow In^{3+} + Cu^+,\ E° = -0.59\,F$

Now we know that $-nFE° = -0.59\,F$

or $-E^o_{cell} = -0.59V$ or $E^o_{cell} = 0.59V$

$$E_{cell} = E^o_{cell} - \frac{0.0591}{n}\log K_c;$$

$$E_{cell} = 0,\ 0.59 = \frac{0.0591}{1}\log K_c$$

$$\log K_c = \frac{0.59}{0.0591} = 10;\ \boldsymbol{K_c = 10^{10}}$$

44. (*i*) The half cell reactions are

At anode $\quad \frac{1}{2}H_2(g) \longrightarrow H^+(aq) + e^-$

At cathode

$AgCl(s) + e^- \longrightarrow Ag(s) + Cl^-(aq)$

The cell reaction $\frac{1}{2}H_2(g) + AgCl(s)$

$\rightleftharpoons H^+(aq) + Ag(s) + Cl^-(aq)$

(*ii*) We know that $\Delta S° = nF\dfrac{dE°}{dT}$

$n \rightarrow$ No. of transferred electrons = 1

$F \rightarrow$ faraday number = 96500 coulombs

$dE° \rightarrow$ Difference of electrode potential at two different temperatures $= (0.21 - 0.23) = -0.02V$

$dT \rightarrow$ Difference of two temperatures $=(35°C-15°C)=20°C$

$\therefore \Delta S° = 1\times 96500\times\frac{-0.02}{20} = -96.5 J/K\ mole$;

$\therefore \overset{o}{E}_{15} = 0.23V$; $\Delta G° = -nE°F$

so $\Delta G°_{15} = -1\times 0.23\times 96500 J$
$=$ **-22195 Jmole**

$\Delta H° = \Delta G° - T\Delta S° = -22195 - 288\times(-96.5)$
$=$ **-49987 J/mole.**

(*iii*) $\overset{o}{E}_{25°C}$ of cell

$= \overset{o}{E}_{15} - \frac{dE}{dT}\times\Delta T = \left(0.23 - \frac{0.02}{20}\times 10\right)V = 0.22V$

The corresponding cell is represented as :

$Ag_{(s)} \mid Ag^+_{(aq)} \parallel Cl^-_{(aq)} (AgCl_{(s)}) \mid Ag_{(s)}$

In form of oxidised electrode potential

$E°_{cell} = E°_{anode} - E°_{Cathode}$
$= E°_{Ag/Ag^+} - E°_{Ag/AgCl/Cl^-}$
$= -0.80 - (0.22) = 0.58V$

$E°_{cell} = \frac{0.0591}{n}\log_{10} K_{eq}$

$AgCl_{(s)} \rightleftharpoons Ag^+ + Cl^-$

$E°_{cell} =$
$\frac{0.0591}{n}\log_{10}[Ag^+][Cl^-] = \frac{0.0591}{n}\log_{10} K_{sp}$

Therefore $-0.58 = \frac{0.0591}{1}\log_{10} K_{sp}$

or $\log_{10}K_{sp} = -9.8139 = \overline{10}.1861$; K_{sp}
$= 1.54\times 10^{-10}$

K_{sp} of AgCl $= 1.54\times 10^{-10}$ (mol Litre^{-1})^2

Solubility of AgCl

$= \sqrt{K_{sp}} = \sqrt{1.54\times 10^{-10}} =$ **1.24×10^{-5} mol / L**

45. Given, $E°_{Ce^{4+}/Ce^{3+}} = 1.61V$;

$E°_{Fe^{3+}/Fe^{2+}} = 0.77V$

Thus for $E°_{cell}$ to be positive, following reaction should occur

$Ce^{4+} + Fe^{2+} \longrightarrow Fe^{3+} + Ce^{3+}$

Hence, Ce^{4+}/Ce^{3+} electrode will act as cathode and Fe^{3+}/Fe^{2+} electrode will act as anode.

Therefore, current will flow from Ce electrode to Fe electrode.

Current will **decrease** with time.

46. Note that the given cell will not work as electrochemical cell since $E°_{OP_{Cu}} > E°_{OP_{Ag+}}$

The equation for electro-chemical cells will be:

$Cu \rightarrow Cu^{2+} + 2e^-$

$2Ag^+ + 2e^- \rightarrow 2Ag$

Thus, e.m.f. of cell $Cu \mid Cu^{2+} \parallel Ag^+ \mid Ag$ will be

$E_{cell} = E°_{OP_{Cu}} + E°_{RPAg} + \frac{0.059}{2}\log\frac{[Ag^+]^2}{[Cu^{2+}]}$

$\because [Ag^+] = 1M$ and $[Cu^{2+}] = 1M$

$\therefore E_{cell} = E°_{OP_{Cu}} + E°_{RP_{Ag}}$

$\left(E°_{cell} = E°_{OP_{Cu}} + E°_{RP_{Ag}}\right) \Rightarrow E_{cell} = E°_{cell}$

After the passage of 9.65 ampere for 1 hr *i.e.* $9.65\times 60\times 60$ Coulomb charge, during which the cell reactions are reversed, the Ag metal passes in solution state and Cu^{2+} ions are discharged. The reactions during the passage of current are:

$2Ag \rightarrow 2Ag^+ + 2e^-$ and $Cu^{2+} + 2e^- \rightarrow Cu$

Thus, Ag^+ formed $= \frac{9.65\times 60\times 60}{96500} = 0.36$ eq.

$= 0.36$ mole

Cu^{2+} discharged $= \frac{9.65\times 60\times 60}{96500} = 0.36$ eq.

$= 0.18$ mole

Thus, $[Ag^+]$ left $= 1 + 0.36 = 1.36$ mole

$[Cu^{2+}]$ left $= 1-0.18 = 0.82$ mole.

Now e.m.f. can be given as:

$E_{cell} = E°_{cell} + \frac{0.059}{2}\log\frac{(1.36)^2}{0.82}$

$= E°_{cell} + 0.010 V$

Thus, E_{cell} increases by **0.010 V.**

47. The cell reaction can be written as:

$Ag \mid Ag^+ (Ag_2CrO_4 Sat.) \parallel Ag^+ (0.1M) \mid Ag$;
$E = 0.164 V$

At cathode : $Ag^+_{cathode} + e^- \longrightarrow Ag$

At anode : $Ag \longrightarrow Ag^+_{anode} + e^-$

Net reaction : $Ag^+_{cathode} \longrightarrow Ag^+_{anode}$;
$E = 0.164 V$

Thus here, $n = 1$, $E = 0.164$ V, $[Ag^+]_{cathode} = 0.1$ M

Let the solubility of Ag_2CrO_4 be S M

Since, Ag_2CrO_4 gives 2 Ag^+.

$\therefore$ Here, concentration of $[Ag^+]_{anode} = 2S$ M

$E = E° - \frac{0.059}{n}\log\frac{[Product]}{[Reactant]}$

(for a concentration cell, $E° = 0$)

$\therefore 0.164 = -\frac{0.059}{1}\log\frac{[Ag^+]_{anode}}{[Ag^+]_{cathode}}$

$0.164 = -\frac{0.059}{1}\log\frac{2S}{0.1}$

or $0.164 = \frac{0.059}{1}\log\frac{0.1}{2S}$ $\quad\therefore 2S = 1.697\times 10^{-4}$

Hence, $S = 0.8485\times 10^{-4}$ M

For Ag_2CrO_4; $Ag_2CrO_4 \rightleftharpoons 2Ag^+ + CrO_4^{2-}$

$K_{sp} = (2S)^2(S) = 4S^3$

$\therefore K_{sp} = 4\times(0.8485\times 10^{-4})^3 =$ **2.44×10^{-12}**

48. For the change

$2Fe^{3+} + 3I^- \rightleftharpoons 2Fe^{2+} + I_3^-$,

$E°_{cell} = E°_{RP_{Fe^{3+}/Fe^{2+}}} + E°_{OP_{I^-/I_3^-}}$

$= 0.77 - 0.54 = 0.23$ V

$E_{cell} = E°_{cell} - \frac{0.059}{2}\log K_c$

At equilibrium, $E_{cell} = 0$ (Using Nernst equation)

Thus, $0.23 = \frac{0.059}{2}\log K_c$ $\quad\therefore K_C =$ **6.26×10^7**

49. $Fe^{2+}(aq) \longrightarrow Fe^{3+}(aq) + e^-$; $E° = 0.68V$

$Ce^{4+}(aq) + e^- \longrightarrow Ce^{3+}(aq)$; $E° = 1.44V$

$E°_{cell} = 1.44 - 0.68 = +0.76$ V

At equilibrium, $E_{cell} = 0$

$E°_{cell} = \frac{0.0591}{n}\log_{10} K_c$; 0.76

$= \frac{0.0591}{1}\log_{10} K_c$

or $\log_{10} K_c = \frac{0.76}{0.0591} = 12.859$

$\therefore$ **$K_c = 7.6\times 10^{12}$**

50. The thin protective layer of oxides of aluminium is formed which protects the metal from further attack of water and air and make it stable.

51. (*i*) Given $E°_{Ni_2O_3/NiO} = +0.40V$; $E°_{FeO/Fe} = -0.87$ V

$E°_{NiO/Ni_2O_3} = -0.40V$; $E°_{Fe/FeO} = +0.87V$

Since $E°_{ox.}$ for Fe/FeO $> E°_{ox.}$ for NiO/Ni_2O_3.

Redox changes can be written as

At anode : $Fe(s) + 2OH^- \rightarrow FeO(s) + H_2O(l) + 2e^-$

At cathode : $Ni_2O_3(s) + H_2O(l) + 2e^- \rightarrow 2NiO(s) + 2OH^-$

Cell reaction : $Fe(s) + Ni_2O_3(s) \rightarrow FeO(s) + 2NiO(s)$

(*ii*) $E_{cell} = E°_{OP\ Fe/FeO} + E°_{RP\ Ni_2O_3/NiO}$

$= 0.87 + 0.40 =$ **1.27 V**

It is independent of conc. of KOH.

(*iii*) Electrical energy $= nFE_{cell}$
$= 2\times 96500\ J\ V^{-1}\times 1.27\ V =$ **2.45×10^5 J**

52. $E°$ = Standard reduction potential of the Ag^+/Ag electrode $= 0.799$ V

$AgI(s) \rightleftharpoons Ag^+ + I^-$

$K_{sp} = [Ag^+][I^-] = 8.7\times 10^{-17}$ (given)

If 'S' is the solubility of AgI, then $K_{sp} = S^2$

$\therefore S = \sqrt{K_{sp}} = \sqrt{8.7\times 10^{-17}} = 9.327\times 10^{-9}$ mol L^{-1}

$\therefore [Ag^+] = [I^-] = 9.327\times 10^{-9}$ M

Reaction : $Ag^+ + e^- \longrightarrow Ag$

$\therefore E = E° - \frac{0.059}{n}\log\frac{[Ag]}{[Ag^+]}$

$= 0.799\ V - \frac{0.059}{1}\log\frac{1}{9.327\times 10^{-9}}$ V

[$\because$ Activity of the electrode material in pure solid state is taken as one]

$= 0.799 - 0.059\log 0.1072\times 10^9$ V

$= 0.799 - 0.474 =$ **0.325 V**

Again,

L.H.S. electrode reaction :

$Ag \rightarrow Ag^+ + e^-$

R.H.S. electrode reaction :

$AgI(s) + e^- \rightarrow Ag + I^-$

Cell reaction : $AgI(s) \rightarrow Ag^+ + I^-$

K = Equilibrium constant $= [Ag^+][I^-] = 8.7\times 10^{-17}$

The standard cell emf $E°$ and the equilibrium constant K are related by the expression:

$E°_{cell} = \frac{0.059}{n}\log K$ at 298 K, Here, $n = 1$,

$K = 8.7\times 10^{-17}$

$E^o_{cell} = 0.059 \log 8.7 \times 10^{-17} = 0.059\,[0.9395 - 17] = -0.948\text{ V}$

But $E^o_{cell} = E^o_{R.H.S.} - E^o_{L.H.S.}$

$\therefore E^o_{R.H.S.} = E^o_{cell} + E^o_{L.H.S.} = -0.948 + 0.799$
$= \mathbf{-0.149\ V}$

53. For the half-cell reaction
$NO_3^-(aq) + 2H^+(aq) + e^- \longrightarrow NO_2(g) + H_2O(l)$
The Nernst equation is

$$E = E^o - \frac{0.059}{n}\log\frac{[\text{Products}]}{[\text{Reactants}]}$$

Substituting the values in case of (i)

$$E = 0.78 - \frac{0.059}{1}\log\frac{1}{(8)^2} = 0.78 + 0.059\log 64$$

= **0.887 V**

Substituting the value in the Nernst equation in case (ii)

$$E = 0.78 - \frac{0.059}{1}\log\frac{1}{(10^{-7})^2}$$

$= 0.78 - 0.059\log 10^{-14}$
$= 0.78 - (0.059)\times(14) = \mathbf{-0.046\ V}$

54. $2Cl^-(aq) + 2H_2O = 2OH^-(aq) + H_2(g) + Cl_2(g)$
Reaction at anode : $2Cl^- \to Cl_2 + 2e^-$
Reaction at cathode : $2H_2O + 2e^- \to H_2 + 2OH^-$

$$i = \frac{62}{100}\times 25 = 15.4\text{ amperes}$$

Weight of Cl_2 deposited = 1 kg or 1000 g

We know that $\frac{W}{E} = \frac{Q}{F} = \frac{it}{F}$; $\frac{1000}{35.5} = \frac{15.4\times t}{96500}$

$t = 175300$ sec. or **48.69 hours**

No. of moles of Cl_2 thus produced $= \frac{1000}{71}$
$= 14.08$

Amount of OH^- released in the electrolysis
$= 2\times 14.08$ moles $= 28.16$ moles

$\therefore$ Molarity with respect to $OH^- = \frac{28.16\text{ mol}}{20\text{ L}}$

= 1.408 M

55. Ag | AgCl(s), KCl(0.2M) || KBr(0.001M), AgBr(s) | Ag
AnodeCathode
$K_{sp}(AgCl) = 2.8\times 10^{-10}$
$K_{sp}(AgBr) = 3.3\times 10^{-13}$
At anode, ${}_1Ag \to {}_1Ag^+ + e^-$
At cathode, ${}_2Ag^+ + e^- \to {}_2Ag$
$\therefore$ Cell reaction ${}_1Ag + {}_2Ag^+ \to {}_2Ag + {}_1Ag^+$
The subscripts 1 and 2 on Ag denote the species concerned with anode and cathode respectively.
Applying Nernst equation

$$E = E^o - \frac{0.059}{n}\log\left[\frac{\text{Products}}{\text{Reactants}}\right]$$

$$= 0 - \frac{0.059}{1}\log\left[\frac{{}_2Ag\times {}_1Ag^+}{{}_1Ag\times {}_2Ag^+}\right]$$

$[{}_1Ag] = [{}_2Ag] = 1$ ($\because$ these are in solid state)

$K_{sp}(AgCl) = 2.8\times 10^{-10}$
or $[{}_1Ag^+][Cl^-] = 2.8\times 10^{-10}$

$$[{}_1Ag^+] = \frac{2.8\times 10^{-10}}{0.2} = 14\times 10^{-10}$$

($\because [Cl^-] = 0.2$)

$K_{sp}(AgBr) = 3.3\times 10^{-13}$ or
$[{}_2Ag^+][Br^-] = 3.3\times 10^{-13}$

$$[{}_2Ag^+] = \frac{3.3\times 10^{-13}}{0.001} = 3.3\times 10^{-10}\ (\because [Br^-] = 0.001)$$

$$\therefore E = -\frac{0.059}{1}\log\left[\frac{14\times 10^{-10}}{3.3\times 10^{-10}}\right]$$

$$= -0.059\log\left[\frac{14}{3.3}\right] = -0.059\times 0.6276 = \mathbf{-0.037\ V}$$

Since, emf is negative this shows that the reaction is non-spontaneous.
For the reaction to be spontaneous, its emf should be positive *i.e.* **E = 0.037 V** and its polarities should be reversed *i.e.* anode should be made cathode and *vice-versa.*
So, the galvanic cell is : Ag | AgBr(s), KBr || AgCl(s), KCl | Ag
In other words, Ag | AgBr acts as anode and AgCl | Ag acts as cathode.

56. The following chemical cell sets up :
$Zn\,|\,Zn^{2+}\,||\,Ni^{2+}\,|\,Ni$
The net cell reaction is : $Zn + Ni^{2+} \rightleftharpoons Zn^{2+} + Ni$

The e.m.f. is given by

$$E_{cell} = E^o_{Ni^{2+}/Ni} - E^o_{Zn^{2+}/Zn} - \frac{0.059}{2}\log\frac{[Zn^{2+}]}{[Ni^{2+}]}$$

$$= -0.24 - (-0.75) - 0.0295\log\frac{[Zn^{2+}]}{[Ni^{2+}]}$$

$$= 0.51 - 0.0295\log\frac{[Zn^{2+}]}{[Ni^{2+}]}$$

At equilibrium $E_{cell} = 0$
Let x mol L^{-1} be the concentration of Ni^{2+} at equilibrium.
Then $[Zn^{2+}] = 1 - x$ [$\because$ 1 mole of Ni^{2+} gives 1 mole of Zn^{2+}]

$$\therefore 0.0295\log\frac{1-x}{x} = 0.51$$

$$\text{or}\log\frac{1-x}{x} = \frac{0.51}{0.0295} = 17.29\ \text{or}\ \frac{1-x}{x}$$

$= 1.95\times 10^{17}$

$$\text{or}\ x = \frac{1}{1.95\times 10^{17}} = \mathbf{5.128\times 10^{-18}\ mol^{-1}}$$

57. $E^o_{Cu^{2+}/Cu} = 0.337$ and $E^o_{Ag^+/Ag} = 0.799$ V

$E^o_{Ag^+/Ag} + E^o_{Cu/Cu^{2+}} = 0.799 - 0.337 = 0.462$ V

$\therefore Cu + 2Ag^+ \to Cu^{2+} + 2Ag$; $E^o_{cell} = 0.462$ V
Hence, the galvanic cell in question will consist of **anode of copper and cathode of silver.**
Calculation of concentration :

$$E^o_{cell} = \frac{0.059}{n}\log\frac{[\text{Products}]}{[\text{Reactants}]}\quad [\because E_{cell} = 0]$$

$$0.462 = \frac{0.059}{2}\log\frac{0.01}{[Ag^+]^2}\quad [n = 2]$$

$$\frac{462\times 2}{59} = \log(10^{-2}) - \log[Ag^+]^2$$

$$\frac{924}{59} = -2 - 2\log[Ag^+] \Rightarrow [Ag^+]$$

$= \mathbf{1.48\times 10^{-9}\ M}$

58. $E^o_{cell} = \frac{0.0591}{n}\log K_c$ **or** $\frac{RT}{nF}\log K_c$

Let us split the desired reaction into two half cell reactions:

Oxidation half reaction :

$H_2O + \frac{1}{2}H_2(g) \rightleftharpoons H_3O^+ + e^-\quad E^o = 0.00$ V

Reduction half reaction :

$H_2O + e^- \longrightarrow \frac{1}{2}H_2 + OH^-\quad E^o = -08.277$ V

Net reaction :
$2H_2O \rightleftharpoons H_3O^+ + OH^-$
$E^o_{cell} = -0.8277$ V
So, the number of electrons involved in redox reaction,
$(n) = 1$

We know that $E^o_{cell} = \frac{0.0591}{n}\log K_c$

$$\log K_c = \frac{E^o_{cell}\times n}{0.0591} = \frac{(-0.8277)\times 1}{0.0591} = -14.005$$

$K_c = \text{Antilog}[-14.005] = \mathbf{9.88\times 10^{-15}}$

59. For a concentration cell

$$E_{cell} = \frac{0.059}{n}\log\frac{C_1}{C_2}$$

It is a concentration cell as both the electrodes are made of same element. Negative electrode acts as anode in a galvanic cell.

At anode; $H_2 \longrightarrow 2H^+ + 2e^-\quad [H^+] = 10^{-6}M$

At cothode; $2H^+ + 2e^- \longrightarrow H_2\qquad [H^+] = ?$

$$E_{cell} = \frac{0.059}{1}\log\left[\frac{C_{H^+}}{10^{-6}}\right]\ \text{or}\ 0.118$$

$$= \frac{0.059}{1}\log\left(\frac{C_{H^+}}{10^{-6}}\right)$$

$$\log\frac{C_{H^+}}{10^{-6}} = \frac{0.118}{0.059} = 2 \Rightarrow C_{H^+} = \mathbf{10^{-4}M}$$

60. Half cell reactions will be
$Zn^{2+} + 2e^- \rightleftharpoons Zn$ ......(*i*)
$H^+ + e^- \rightleftharpoons \frac{1}{2}H_2$ or $2H^+ + 2e^- \rightleftharpoons H_2$ ......(*ii*)

We know that $E_{Zn/Zn^{2+}}$

$$= E_{Zn/Zn^{2+}} - \frac{RT}{nF}\ln\frac{[Zn^{2+}]}{[Zn]}$$

Here, $R = 8.314\ Jmol^{-1}K^{-1}$, $T = 298\ K$, $F = 96{,}500\ c/mol$

$n = 2$, $E^\circ_{Zn/Zn^{2+}} = 0.76\ V$.

Substituting the values in the above equation

$$E_{Zn/Zn^{2+}} = 0.76 - \frac{8.314\times 298}{2\times 96500}\ln\frac{0.1}{1} = 0.79\ V$$

Similarly, $E_{H^+/H_2} = E^\circ_{H^+/H_2} - \frac{RT}{nF}\ln\frac{[H_2]}{[H^+]^2}$

$$= 0 - \frac{8.314\times 298}{2\times 96500}\ln\frac{[1]}{[H^+]^2}$$

$= 0.05915\log_{10}[H^+] = -0.05915pH$

($\because -\log_{10}[H^+] = pH$)

Now since $E = E_{Zn/Zn^{2+}} + E_{H^+/H_2}$

$0.28 = 0.79 - 0.05915\,pH$

$\Rightarrow pH = \frac{0.51}{0.05915} = \mathbf{8.62}$

61. (i) The two half cell reactions can be written as below :

Oxidation half reaction : $Zn \rightarrow Zn^{2+} + 2e^-$

Reduction half reaction : $Cu^{2+} + 2e^- \rightarrow Cu$

Thus the cell reaction will be : $Zn + Cu^{2+} \rightarrow Zn^{2+} + Cu$

(ii) EMF of cell, $E^\circ_{cell} = E^\circ_{cathode} - E^\circ_{anode}$

$E^\circ_{cell} = 0.350 - (-0.763)$

$= 0.350 + 0.763$ volts $=$ **1.113 volts**

(iii) Since emf of the cell is **positive**, the reaction as written is **spontaneous**.

62. (a)

$$\underset{\substack{2[119+(2\times 35.5)]\\ =380}}{2SnCl_2} \longrightarrow \underset{119}{Sn} + \underset{\substack{119+(4\times 35.5)\\ =261}}{SnCl_4}$$

$\because$ 119g Sn deposits from = 380g $SnCl_2$

$\therefore$ 0.119g Sn deposits from $= \frac{380}{119}\times 0.119$

$= 0.380g\ SnCl_2$

$\because$ 380g $SnCl_2$ gives = 261g $SnCl_4$

$\therefore$ 0.380 g $SnCl_2$ gives $= \frac{261}{380}\times 0.380$

$= 0.261g\ SnCl_4$

$\therefore$ Wt of $SnCl_2$ left after decomposition

$= 19.00 - 0.380 = 18.620$ g.

Ratio $SnCl_2 : SnCl_4$

$\Rightarrow$ 18.620 : 0.261 $\Rightarrow$ **71.34 : 1**

(b) $NaCl \xrightarrow{Electricity} Na^+ + Cl^-$

At Cathode; $Na^+ + e^- \longrightarrow Na$

$2Na + H_2O \longrightarrow 2NaOH + H_2$

At anode; $Cl^- \longrightarrow Cl + e^-$

$Cl + Cl \longrightarrow Cl_2$

$2OH^- + Cl_2 \longrightarrow Cl^- + OCl^- + H_2O$

$OCl^- + 2HOCl \longrightarrow ClO_3^- + 2Cl^- + 2H^+$

$Na^+ + ClO_3^- \longrightarrow \underset{\text{Sod.Chlorate}}{NaClO_3}$

On prolonged electrolysis

$ClO_3^- + ClO^- \longrightarrow Cl^- + ClO_4^-$

$Na^+ + ClO_4^- \longrightarrow \underset{\text{Sod. perchlorate}}{NaClO_4}$

(c) Charge on $N^{3-} = 3$

No. of ions in 14 g of $N^{3-} = 6.02\times 10^{23}$

No. of ions in 1g of $N^{3-} = \frac{6.02\times 10^{23}}{14}$

No. of electronic charges on 1 g N^{3-}

$= \frac{6.02\times 10^{23}}{14}\times 3$

Charge on 1 g of N^{3-}

$= \frac{6.023\times 10^{23}\times 3\times 1.6\times 10^{-19}}{14}$ Coulombs

($\because$ Charge on one electron is 1.6×10^{-19} Coulombs)

= 2.06 × 10⁴ Coulombs

63. **(c)** $\Lambda_M = \Lambda^\circ_M - b\sqrt{c}$

$\Lambda_{M.1} - \Lambda_{M.2} = b\left[\sqrt{c_2} - \sqrt{c_1}\right]$

$$b = \frac{\Lambda_{M.1} - \Lambda_{m.2}}{\sqrt{0.04} - \sqrt{0.01}} = \frac{\Lambda_{M.1} - \Lambda_{M.2}}{0.1}$$

For $NaNO_3$, $b = \frac{111-101}{0.1} = 100$

$\left[\Lambda^\circ_M\right]_{NaNO_3} = \Lambda_M + b\sqrt{c} = 111 + 100(0.1)$

$= 121\ S\ cm^2/mol$

For NaCl, $b = \frac{117-107}{0.1} = \frac{10}{0.1} = 100$

$\left[\Lambda^\circ_M\right]_{NaCl} = 117 + 100\times 0.1 = 127\ S\ cm^2/mol$

For $AgNO_3$, $b = \frac{125-116}{0.1} = 90$

$\left[\Lambda^\circ_M\right]_{AgNO_3} = 125 + 90(0.1) = 134\ S cm^2/mol$

$\left[\Lambda^\circ_M\right]_{AgCl} = \left[\Lambda^\circ_M\right]_{AgNO_3} + \left[\Lambda^\circ_M\right]_{NaCl} - \left[\Lambda^\circ_M\right]_{NaNO_3}$

$= 134 + 127 - 121 = 140\ S\ cm^2/mol$

Now, $\Lambda_M = \Lambda^\circ_M = \frac{k\times 1000}{S}$

$= \frac{1.4\times 10^{-6}\times 10^{-3}}{S} = 140$

$S = \frac{1.4\times 10^{-3}}{140} = 10^{-5}M = X$

$\log(X^{-1}) = \log(10^{-5})^{-1} = 5$

64. **(a)** For weak acid, $\alpha = \frac{\Lambda_m}{\Lambda_0}$

$K_a\frac{c\alpha^2}{1-\alpha} \Rightarrow K_a(1-\alpha) = c\alpha^2$

$\Rightarrow K_a\left(1 - \frac{\Lambda_m}{\Lambda_0}\right) = c\left(\frac{\Lambda_m}{\Lambda_0}\right)^2$

$\Rightarrow K_a - \frac{\Lambda_m K_a}{\Lambda_0} = \frac{c\Lambda_m^2}{(\Lambda_0)^2}$

$\Rightarrow K_a = \frac{c\Lambda_m^2}{(\Lambda_0)^2} + \frac{\Lambda_m K_a}{\Lambda_0}$

Divide by 'Λ_m'

$\Rightarrow \frac{K_a}{\Lambda_m} = \frac{c\Lambda_m}{(\Lambda_0)^2} + \frac{K_a}{\Lambda_0}$

$\Rightarrow \frac{1}{\Lambda_m} = \frac{c\Lambda_m}{K_a(\Lambda_0)^2} + \frac{1}{\Lambda_0}$

For the Plot $\frac{1}{\Lambda_m}$ vs $c\Lambda_m$,

Slope $= \frac{1}{K_a(\Lambda_0)^2} = S$

y–intercept $= \frac{1}{\Lambda_0} = P$

$$\therefore \frac{P}{S} = \frac{\frac{1}{\Lambda_0}}{\frac{1}{K_a(\Lambda_0)^2}} = K_a\Lambda_0$$

65. **(d)** Sodium stearate at low concentration (*i.e.*, below CMC) behaves as normal strong electrolyte, but at higher concetration (*i.e.* above CMC) exhibits colloidal behaviour due to the formation of micelles. Thus, plot (d) correctly represents relation between Λ_m and $\sqrt{C}$ for sodium strearate.

66. **(d)** $AgNO_3(aq) + KCl(aq) \longrightarrow AgCl(s) + KNO_3(aq)$

Conductivity of the solution is almost compensated due to formation of $KNO_3(aq)$. However, after end point, conductivity increases more rapidly due to addition of excess $AgNO_3$ solution.

67. **(b)** As we go down the group 1 (*i.e.* from Li^+ to K^+), the ionic radius increases, degree of solvation decreases and hence effective size decreases resulting in increase in ionic mobility. Hence, equivalent conductance at infinite dilution increases in the same order.

68. **(7)**

$\Lambda^\circ(U_mY_p) = m\lambda^\circ(U^{P+}) + p\lambda^\circ(Y^{m-})$

$\Rightarrow$ $25m + 100p = 250$

$\Rightarrow$ $m + 4p = 10$ ...(1)

$\Lambda^\circ(V_mX_n) = m\lambda^\circ(V^{n+}) + n\lambda^\circ(X^{m-})$

$\Rightarrow$ $100m + 80n = 440$

$\Rightarrow$ $5m + 4n = 22$ ...(2)

For electrolyte Z_mX_n from the given curve,

$\Lambda(Z_mX_n) = \Lambda^\circ(Z_mX_n) - A\sqrt{C}$

Slope, $m = -A = \frac{339-336}{0.01-0.04}$

$\Rightarrow$ $A = 100$

For $\lambda_m = 339\ S\,cm^2\,mol^{-1}$, $\sqrt{C} = 0.01\ (mol\ L^{-1})^{1/2}$

$339 = \Lambda^\circ(Z_mX_n) - 100\times 0.01$

$\Rightarrow$ $\Lambda^\circ(Z_mX_n) = 340\ S\ cm^2\ mol^{-1}$

$\Rightarrow$ $m\lambda^\circ(Z^{n+}) + n\lambda^\circ(Z^{m-}) = 340$

$\Rightarrow \quad 50m + 80n = 340$

$\Rightarrow \quad 5m + 8n = 34$ ...(3)

From eq. (2) and (3),

n = 3 and m = 2

Putting value of m in eq. (1),

p = 2

Therefore, m + n + p = 2 + 3 + 2 = 7.

69. (6) The formula for conductance is

$$G = \kappa \times \frac{a}{l}$$

$$5\times10^{-7} = \kappa \times \frac{1}{120} = 6\times10^{-5}\text{S cm}^{-1}$$

$$\Lambda_m^c = \frac{\kappa\times1000}{M} = \frac{6\times10^{-5}\times1000}{0.0015} = 40$$

$\because$ pH = 4

$$\therefore [H^+] = 10^{-4} = c\alpha = 0.0015\,\alpha;\ \alpha = \frac{10^{-4}}{0.0015}$$

$$\text{Also, } \alpha = \frac{\Lambda_m^c}{\Lambda_m^o} \Rightarrow \frac{10^{-4}}{0.0015} = \frac{40}{\Lambda_m^o}$$

$$\Lambda_m^o = \frac{40\times0.0015}{10^{-4}} = 6\times10^2\text{S cm}^2\text{ mol}^{-1};$$

$Z \approx 6$

70. (3)

$1 \rightarrow HX \qquad 2 \rightarrow HY$

$$\alpha_1 = \frac{(\lambda_m)_{HX}}{\lambda_m^o} \qquad \alpha_2 = \frac{(\lambda_m)_{HY}}{\lambda_m^o}$$

HX	$\rightleftharpoons$	H^+	+	X^-
1		0		0
$(1-\alpha)C_1$		$\alpha_1 C_1$		$\alpha_1 C_1$

$$K_{a_1} = \frac{\alpha_1^2 C_1}{1-\alpha_1}\ [\because \alpha << 1 \therefore 1-\alpha_1 \approx 1]$$

$$\therefore K_{a_1} = \alpha_1^2 C_1^2;\ K_{a_2} = \alpha_2 C_2$$

$$K_{a_1} = C_1\alpha_1^2;\ K_{a_2} = C_2\alpha_2^2$$

$$= 0.01\frac{(\lambda_m)_{HX}^2}{(\lambda_m^o)^2} = 0.1\frac{(\lambda_m)_{HY}^2}{(\lambda_m^o)^2}$$

$$\therefore \frac{K_{a_1}}{K_{a_2}} = \frac{0.01(\lambda_m)_{HX}^2}{0.1(\lambda_m)_{HY}^2} = 0.1\left(\frac{(\lambda_m)_{HX}}{(\lambda_m)_{HY}}\right)^2$$

$$= 0.1\left(\frac{1}{10}\right)^2 = 10^{-3}$$

$$pK_a(HX) - pK_a(HY) = -\log\frac{K_{a_1}}{K_{a_2}} = -\log10^{-3} = 3$$

71. (0.22)

$$K_a = \frac{\Lambda_m^2 C}{\Lambda_m^o(\Lambda_m^o - \Lambda_m)}$$

$$= \frac{(y\times10^2)^2\times C}{4\times10^2(4\times10^2 - y\times10^2)}$$

$$= \frac{(3y\times10^2)^2\times\frac{C}{20}}{4\times10^2(4\times10^2 - 3y\times10^2)} \Rightarrow y = \frac{44}{51}$$

$$\therefore \Lambda_m = \frac{44}{51}\times10^2\text{ S cm}^2\text{ mol}^{-1}$$

$$\alpha = \frac{\frac{44}{51}\times10^2}{4\times10^2} = 0.2156\ (\alpha = 0.22 \text{ or } 0.21)$$

72. (0.86)

73. increased; Formed salt will be a strong electrolyte.

74. (a, c)

$$\overset{-2}{N_2H_4} + \overset{2}{O_2} \longrightarrow \overset{0}{N_2} + \overset{-2}{H_2O}$$

Oxidation (anode) Reduction (cathode)

At anode : $N_2H_4 + 4OH^- \rightarrow N_2 + 4H_2O + 4e^-$

At cathode : $O_2 + 2H_2O + 4e^- \rightarrow 4OH^-$

Complete reaction : $N_2H_4 + O_2 \rightarrow N_2 + 2H_2O$

Thus, statements (a) and (c) are correct.

75. (c)

(P) $KCl + AgNO_3 \rightarrow AgCl\downarrow + KNO_3$

Cl^- ions is replaced by NO_3^- ions

Conductance will first decrease and then after equivalence point, it will increase P $\rightarrow$ (3)

(Q) $AgNO_3 + KCl \rightarrow AgCl + KNO_3$

Ag^+ ions is replaced by K^+ ions

Conductance will first increase slightly and then will increase further

(R) $NaOH + HCl \rightarrow NaCl + H_2O$

OH^- ions is replaced by Cl^- ions

(S) $NaOH + CH_3COOH \rightarrow CH_3COONa + H_2O$, OH^- ions is replaced by CH_3COO^-, ions conductance will first decrease and them become almost constant due to buffer formation.

76. (a) (P) $\underset{X}{(C_2H_5)_3N} + \underset{Y}{CH_3COOH} \longrightarrow$

$$(C_2H_5)_3NH^+CH_3COO^-$$

Initially conductivity increases because on neutralisation ions are created. After that it becomes practically constant because X alone cannot form ions.

(Q)

$$\underset{X}{KI(0.1M)} + \underset{Y}{AgNO_3(0.01M)} \longrightarrow AgI\downarrow + KNO_3$$

Number of ions in the solution remains constant as only $AgNO_3$ precipitated as AgI. Thereafter, conductance increases due to increase in number of ions.

(R) Initially conductance decreases due to the decrease in the number of $\bar{O}H$ ions as OH^- is getting replaced by CH_3COO^- which has poorer conductivity. Thereafter, it slowly increases due to the increase in number of H^+ ions.

(S) Initially it decreases due to decrease in H^+ ions and then increases due to the increase in OH^- ions.

77. (i)-(s); (ii)-(r); (iii)-(q); (iv)-(p)

78. Given :

$\Lambda_m^\infty(Ag^+) = 6\times10^{-3};\ \Lambda_m^\infty(Br^-) = 8\times10^{-3};$

$\Lambda_m^\infty(NO_3^-) = 7\times10^{-3}$ and $K_{sp}(AgBr) = 12\times10^{-14}$

To find the specific conductivity (κ) of the final solution of AgBr in which $AgNO_3$ (10^{-7} M) is mixed, we must find the individual κ of the ions.

or $\kappa_{soln} = \kappa_{Ag^+} + \kappa_{Br^-} + \kappa_{NO_3^-}$

Again, $\kappa = \Lambda_m^\infty \times$ molar concentration

Calculation of molar concentration of ions :

Concentration,

$[NO_3^-] = 10^{-7}$ moles / L $\equiv 10^{-4}$ moles / m^3

Let x be the molar concentration of Ag^+ from AgBr

$\Rightarrow (x + 10^{-7})x = 12\times10^{-14}$

or $x^2 + 10^{-7}x - 12\times10^{-14} = 0;$

$\Rightarrow x = 3\times10^{-7}$ M

$\Rightarrow [Br^-] = 3\times10^{-7}$ M $\equiv 3\times10^{-4}$ moles / m^3

and $[Ag^+] = 3\times10^{-7} + 10^{-7} = 4\times10^{-7}$ M

$= 4\times10^{-4}$ moles / m^3

$\kappa_{Ag^+} = \mathbf{6\times10^{-3}\times4\times10^{-4}}$

$= 24\times10^{-7}\ (Sm^2mol^{-1}\times mol/m^3) = 24\times10^{-7}$ S/m

Similarly,

$\kappa_{Br^-} = 8\times10^{-3}\times3\times10^{-4} = \mathbf{24\times10^{-7}\ S/m}$

and $\kappa_{NO_3^-} = 7\times10^{-3}\times10^{-4} = 7\times10^{-7}$ S/m

$\Rightarrow \kappa = (24+24+7)\times10^{-7}$ S/m $= 55\times10^{-7}$ S/m

So, the correct answer is **55.**

79. (b) Give : I = 10 milliamperes ; F = 96500 C mol^{-1}

t = ? ; Moles of H_2 produces = 0.01 mol

From the law of electrolysis, we have

$$\text{Equivalents of } H_2 \text{ produces} = \frac{I\times t(\text{sec})}{96500}$$

Substituting given values, we get

$$0.01\times2 = \frac{10\times10^{-3}(\text{amperes})\times t(\text{sec})}{96500}$$

$= 19.3\times10^4$ sec.

80. (d) Charge of one mole of electrons = 96500 C $\therefore$ 1 gram equivalent of substance will be deposited by one mole of electrons.

81. (c) $\frac{W_1}{W_2} = \frac{E_1}{E_2} = \frac{Z_1 it}{Z_2 it} \quad \therefore \frac{Z_1}{Z_2} = \frac{E_1}{E_2}$

Here E_1 & E_2 are equivalent weights of the ions.

82. (105.5)

$$C_4H_{10} + \frac{13}{2}O_2 \longrightarrow 4CO_2 + 5H_2O$$

$$\Delta_rG^\circ = 4\Delta G_f^\circ(CO_2) + 5\Delta G_f^\circ(H_2O) - \Delta G_f^\circ(C_4H_{10})$$

$$-26 \times F \times E^\circ_{cell} = [4\times(-394) + 5(-237) - (-18)] \times 10^3$$

$$\therefore E^\circ_{cell} = \frac{2743\times10^3}{26F} = 105.5\times\frac{10^3}{F}.$$

83. (100) As per the question, the reaction is :-

$$Cr_2O_7^{2-} + 14H^+ + 6e^- \rightarrow 2Cr^{+3} + 7H_2O$$

- From the above balanced reaction, 2 moles of Cr^{+3} are produced by 6 moles of electrons.

$\therefore$ 1 mol of Cr^{+3} is produced by $\frac{6}{2} = 3$ moles of electrons.

Hence, n = 3 moles of electrons

- As Q = it & Q = nF where, Q = charge

$\therefore$ nF = it, n = No. of moles of electrons transferred in a reaction

as, t = 48.25 × 60 s

n = 3 mol

F = 96500 C/mol, F = Faraday's constant

i = current

t = time

So, $i \times 48.25 \times 60 = 3 \times 96500$

$$\therefore i = \frac{3\times96500}{48.25\times60} = 100$$

84. (13.32)

$$H_2(g) + \frac{1}{2}O_2(g) \longrightarrow H_2O(l)$$

$E^0_{cell} = 1.23 - 0.00 = 1.23$ V

$\Delta G^0_{cell} = -nFE^0_{cell} = -2 \times 96500 \times 1.23$ J

$\therefore$ Work derived from this fuel cell using 70% efficiency

$$= \frac{70}{100}\times\left(-\Delta G^0_{cell}\right)\times10^{-3}$$

$= 0.7 \times 2 \times 96500 \times 1.23 \times 10^{-3} = 166.17$ J

For insulated vessel, q = 0

Therefore for monoatomic gas,

w = ΔU

$166.17 = nC_{v,m}\Delta T$; $\therefore$ $\Delta T = 13.32$K

85. (0.154) $i = \frac{1.70\times90}{100}$ ampere

No. of equivalents of Zn^{2+} which are lost

$$= \frac{i\times t}{96500} = \frac{1.70\times90\times230}{100\times96500} = 3.646\times10^{-3}$$

$\therefore$ Milli equivalents of Zn^{2+} which are lost 3.646

$\therefore$ Initial value of $Zn^{2+} = 300 \times 0.160 \times 2 = 96$

$\therefore$ Mili equivalents of Zn^{2+} left in solution $= 96 - 3.646 = 92.354$

$$[ZnSO_4] = \frac{92.354}{2\times300} = 0.154 \text{ M}$$

86. (19.06)

Watt = Volt × Current ⇒ 100 = 110 × Current

or Current $= \frac{100}{110} = \frac{10}{11}$ amp.

Now we know that,

$$Q = i\times t = \frac{10}{11}\times10\times3600\times\frac{1}{96500} = 0.339 \text{ F}$$

$$\text{Wt. of cadmium deposited} = \frac{0.339\times112.4}{2}$$

= 19.06 g

87. (125.09) Volume of the surface = area × thickness

$$= 80 \text{ cm}^2 \times \frac{0.005}{10}\text{ cm} = \frac{1}{25}\text{ cm}^3$$

Mass of Ag deposited = Volume × Density

$$= \frac{1}{25}\times10.5 \text{ g/cm}^3 = \frac{21}{50}\text{ g}$$

Cell reaction : $Ag^+ + e^- \rightarrow Ag$

We know that, $\frac{W}{E} = \frac{Q}{F} = \frac{it}{F}$

E = Eq. wt. of Ag = 108 $\therefore$ $\frac{21/50}{108} = \frac{i\times t}{96500}$

$$\frac{21}{50\times108} = \frac{3\times t}{96500} \quad \therefore \ t = \mathbf{125.09 \text{ sec}}$$

88. (27171.96) Wt. of Cu deposited = Zit

Electrochemical equivalent of Cu $= \frac{63.5}{2}$

= 31.75

Volume of surface = area × thickness

$= 10 \times 10 \times 10^{-2} = 1$ cc

Weight of Cu = density × volume = 8.94 × 1

= 8.94 g

According to Faraday's laws of electrolysis

31.75 g of Cu is deposited by = 96500 coulombs of electricity

$$\therefore 8.94 \text{ g of Cu is deposited by} = \frac{96500}{31.75}\times8.94$$

= 27171.96 coulombs

89. (b) Reaction at anode: $2Cl^- \longrightarrow Cl_2 + 2e^-$

moles of $Cl^- = 4 \times 500 \times 10^{-3} = 2$

moles $Cl_2 = \frac{1}{2}\times2 = 1$

90. (d) 500 mL of 4.0 molar NaCl has 2 mole of NaCl.

By electrolysis, we can get a maximum of 2 moles of sodium which can combine with exactly 2 moles of mercury to give amalgam.

$\therefore$ The maximum weight of amalgam which can be formed from this solution

= weight of 2 mole of sodium + weight of 2 mole of mercury

= 2 × 23 + 2 × 200 = 446g

91. (d) $Na^+ + e^- \longrightarrow Na$

Total number of moles of Na^+ discharged at cathode = 2mole

$\therefore$ The number of electron required for this purpose = 2 mole

$\therefore$ Total charge required

= 2 faraday = 2 × 96500 = 193000 coulombs.

92. $w = Zit$

Z for Cu $= \frac{63.5/2}{96500}$; $t = 16 \times 60$ sec

$$\therefore w = \frac{63.5}{2\times96500}\times2\times10^{-3}\times16\times60$$

$$= \frac{63.5\times16\times60\times10^{-3}}{96500}\text{ g}$$

Wt. of Cu at 50% electrolysis of $CuSO_4$

$$= \frac{63.5\times16\times60\times10^{-3}}{96500}\text{ g}$$

Wt. of Cu at 100% electrolysis of $CuSO_4$

$$= \frac{63.5\times2\times16\times60\times10^{-3}}{96500}\text{ g}$$

$= 0.198 \times 63.5 \times 10^{-4}$g

159.5 g of $CuSO_4 \equiv 63.5$g Cu

Weight of $CuSO_4$ in 250 mL $= \frac{0.198\times63.5\times10^{-4}}{63.5}$

$= 0.198 \times 10^{-4}$ mol.

$\therefore$ Conc. of $CuSO_4 = 0.198 \times 10^{-4} \times \frac{1000}{250}$

= 7.95 × 10⁻⁵ mol/L

93. $W_{Ag} = \frac{E.i.t}{96500}$

$$= \frac{107.8\times8.46\times8\times60\times60}{96500} = \mathbf{34.02g}$$

Volume of Ag $= \frac{34.02}{10.5} = 3.24$ cm³

$\therefore$ Surface area $= \frac{3.24}{0.00254} = \mathbf{1275.6\ cm^2}$

94. $CrO_3 + 6H^+ + 6e^- \longrightarrow Cr + 3H_2O$

Eq. wt. of Cr

$$= \frac{\text{At. wt.}}{\text{No. of Electrons lost or gained by one molecule of Cr}} = \frac{52}{6}$$

(i) $\because$ 96500 coulomb deposit $= \left(\frac{52}{6}\right)$ g Cr

$\therefore$ 24000 coulomb deposit $= \frac{52}{6} \times \frac{24000}{96500}$

= 2.1554 g of Cr

(ii) Also given, $w_{Cr} = 1.5$ g, $i = 12.5$ ampere,

$t = ?, E_{Cr} = \frac{52}{6}$

$\therefore w = \frac{Eit}{96500}$ or $1.5 = \frac{52 \times 12.5 \times t}{6 \times 96500}$

$\therefore$ **t = 1336.15 second**

95. $C_6H_5NO_2 + 6H^+ + 6e^- \rightarrow C_6H_5NH_2 + 2H_2O$

Eq. wt of $C_6H_5NO_2 = \frac{\text{M.wt.}}{6} = \frac{123}{6}$

$w = \frac{Eit}{96500}$; $i = \frac{50 i_0}{100}$

$\because$ current efficiency = 50%

$\therefore 12.3 = \frac{123 \times i \times t \times 50}{6 \times 100 \times 96500}$; $i \times t = Q$

= 115800 Coulomb

Energy used = 115800 × 3 = **347.4 kJ.**

96. The chemical reactions taking place at the two electrodes are

At cathode : $Cu^{2+} + 2e^- \rightarrow Cu$

$H_2O \rightleftharpoons H^+ + OH^-$

Only Cu^{2+} ions will be discharged so as these are present in solution and H^+ ions will be discharged only when all the Cu^{2+} ions have been deposited.

At anode : $2OH^- \rightarrow H_2O + O + 2e^-$

$O + O \rightarrow O_2$

Thus in first case, Cu^{2+} ion will be discharged at the cathode and O_2 gas at the anode. Let us calculate the volume of gas (O_2) discharged during electrolysis.

According to Faraday's second law

31.75 g Cu ≡ 8 g of oxygen ≡ 5.6 litres of O_2 at NTP

0.4 g Cu $= \frac{5.6}{31.75} \times 0.4$ litres of O_2 at NTP

= 0.07055 litres = 70.55 mL

As explained earlier, when all the Cu^{2+} ion will be deposited at cathode, H^+ ions will start going to cathode liberating hydrogen (H_2) gas, *i.e.*

$H^+ + e^- \rightleftharpoons H \rightleftharpoons H + H \rightarrow H_2$

However, the anode reaction remains same as previous. Thus in the second (latter) case, amount of H_2 collected at cathode should be calculated.

8 g of O_2 ≡ 1 g of H_2

5.6 litres of O_2 at NTP = 11.2 litres of hydrogen

Quantity of electricity passed after 1st electrolysis,

i.e. Q = i × t = 1.2 × 7 × 60 = 504 coulombs

504 coulombs will liberate $= \frac{5.6 \times 504}{96500} = 29.24$ mL of O_2.

Similarly, H_2 liberated by 504 coulombs

$= 11.2 \times \frac{504}{96500} = 58.48$ mL

(Twice the volume of O_2 liberated in latter phase = 2 × 29.24 = 58.48 mL)

Total volume of O_2 liberated = 70.55 + 29.24

= 99.79 mL

Vol. of H_2 liberated = **58.48 mL**

Alternate Solution:

Weight of solution before discharge

= 3500 × 1.294 = 4529 g

Weight of H_2SO_4 before discharge

$= \frac{39}{100} \times 4529 = 1766.31$ g.

Weight of solution after discharge

= 3500×1.139 = 3986.5 g

Weight of H_2SO_4 after after $= \frac{20}{100} \times 3986.5$

= 797.3 g

Loss of mass of H_2SO_4 during discharge

= 1766.31 – 797.3 = 969.01 g

From, first law of electrolysis (W) $= \frac{Q \times E}{96500}$

$\Rightarrow 969.01 = \frac{Q \times 98}{96500}$

Q = 954178.21 Coloumb

Ampere hour $= \frac{\text{Coloumb}}{3600} = \frac{954178.21}{3600}$

= 265.04 Ampere - hour

97. For the given reactions, it is obvious that 22.4 litres of H_2 gas require 2 Faraday electricity.

$\therefore$ 67.2 litres of H_2 will produce = 6 Faraday electricity

$Q = i \times t$; $6 \times 96500 = i \times 15 \times 60$

$i = \frac{6 \times 96500}{15 \times 60}$ = **643.3 ampere**

Calculation of amount of Cu deposited by 6 F

Since 1 F deposits $= \frac{63.5}{2} = 31.75$ g of Cu

6 F will deposit = 31.75 × 6 g = **190.50 g**

98. In lead storage battery, the anodic and cathodic reactions during discharge (or operation or working) are as :

(i) Anodic reaction :

$Pb(s) + SO_4^{2-}(aq) \longrightarrow PbSO_4(s) + 2e^-$

(ii) Cathodic reaction :

$PbO_2(s) + SO_4^{2-}(aq) + 4H^+(aq) + 2e^- \longrightarrow PbSO_4(s) + 2H_2O(l)$

In both the half cell reactions, H_2SO_4 is consumed and hence, conc. of H_2SO_4 decreases during the working (discharging, of the battery. For the withdrawl of 2F = 2 × 96500 C of electric charge, 2 mol of H_2SO_4 are consumed. Density of H_2SO_4 solution (used as electrolyte) falls during working of the cell.

Both reactions get reversed on charging the battery, leading to regeneration of H_2SO_4 as :

Formerly anode but now cathode (recharging)

$PbSO_4(s) + 2e^- \longrightarrow Pb(s) + SO_4^{2-}(aq)$

Formerly cathode but now anode :

$PbSO_4(s) + 2H_2O(l) \rightarrow PbO_2(s) + SO_4^{2-}(aq) + 4H^+(aq) + 2e^-$

Molarity of H_2SO_4 before electrolysis

$= \frac{39 \times 1.294 \times 1000}{98 \times 100} = 5.15$ M

Moles of H_2SO_4 before electrolysis

= 5.15 × 3.5 = 18.025

Molarity of H_2SO_4 after electrolysis

$= \frac{20 \times 1.139 \times 1000}{98 \times 100} = 2.32$ M

Moles of H_2SO_4 after electrolysis

= 2.32 × 3.5 = 8.12

The overall discharging reaction is :

$$Pb(s) + PbO_2(s) + 2SO_4^{2-} + 4H^+ \longrightarrow 2PbSO_4 + 2H_2O$$

Here, $2SO_4^{2-}$ requires $2e^-$ hence *n*-factor = 1

i.e. $N_{H_2SO_4} = M_{H_2SO_4}$

Equivalent mass of H_2SO_4 = 98/1 = 98

Moles or equivalents of H_2SO_4 used

= 18.025 – 8.12 = 9.905

Number of coulomb required = 9.905 × 96500

$i \times t$ = 955350 A-s = 265.375 A-h

99. Gold deposited in the first cell = 9.85 g

At. wt. of Gold = 197, Oxidation number of gold = +3

Eq. Wt. of Gold $= \frac{197}{3}$

$W = Zit$

$\because$ Charge required to deposit 1 g eq. of gold = 1*F* = 96,500 C

$\therefore$ Charge required to deposit 9.85 g of gold or

$\frac{9.85}{197/3}$ g eq. of gold $= \frac{96{,}500 \times 9.85 \times 3}{197}$ C

= 965 × 5 × 3 C = 14475 C

According to Faraday's second law,

$$\frac{\text{Wt. of Cu}}{\text{Eq. wt. of Cu}} = \frac{\text{Wt. of Gold}}{\text{Eq. wt. of Gold}}$$

$\Rightarrow$ Wt. of Cu deposited $= \frac{9.85 \times 3}{197} \times \frac{63.5}{2}$

= 4.7625 g

Current $= \frac{Q}{t} = \frac{14475}{5 \times 3600}$ A $= \frac{193}{240}$ A

= 0.8042 A

Chapter-4

Chemical Kinetics and Nuclear Chemistry

Part-A (JEE Main)

1. **(4)** At, t = 10 minutes

For reaction II,

Rate of reaction $= -\frac{1}{5}\frac{\Delta[Br^-]}{\Delta t}$

$= \frac{1}{5} \times (2 \times 10^{-4}) = 4 \times 10^{-5}$

For reaction A → P at t = 10 minutes

Rate of reaction (I) = Rate of reaction (II) (given)

∴ Rate of reaction (I) $= 4 \times 10^{-5}$

Rate of reaction $= 4 \times 10^{-5} = k[A]$

$k = 4 \times 10^{-5} \times 10^{-2} = 4 \times 10^{-3}\ min^{-1}$.

2. **(3)** For I[st] order reaction

Rate ∝ [A]; Rate = k[A]

Run1 = 10 M; Run2 = 20 M

Run3 = 5 M

With decrease in concentration of A, rate of reaction decreases.

3. **(2)** $A(g) \rightarrow B(g) + C(g)$

	A(g)	B(g)	C(g)
initial	1	0	0
after 100 min	1 – P	P	P

$P_{total} = 1 + P$

$1.5 = 1 + P$

$P = 0.5$

$$k = \frac{2.303}{100}\log\frac{1}{0.5} = \frac{0.693}{100} = 6.9 \times 10^{-3}\ mn^{-1}$$

4. **(4)** According to given conditions t_1 and t_2 can be given as

$$t_1 = t_{\frac{1}{4}} = \frac{1}{k}\ln\frac{A_0}{A_0} = \frac{1}{k}\ln 4$$

$$t_2 = t_{\frac{1}{8}} = \frac{1}{k}\ln\frac{A_0}{A_0} = \frac{1}{k}\ln 8$$

∴ Ratio t_1/t_2 will be given by

$$\frac{t_1}{t_2} = \frac{\ln 4}{\ln 8} = \frac{2\ln 2}{3\ln 2} = \frac{2}{3}$$

5. **(3)**

	A(g) →	B(g) +	C(g)
t = 0	P^o	0	0
t = t	P^o - x	x	x
t = ∞	0	P^o	P^o

$$P_t = P^o + x \Rightarrow x = P_t - P^o = P_t - \frac{P_\infty}{2}$$

$$P_\infty = 2P^o \Rightarrow P^o = \frac{P_\infty}{2}$$

$$k = \frac{1}{t}\ell n\frac{P^o}{P^o - x}$$

$$k = \frac{1}{t}\ell n\frac{P_\infty}{2(P_\infty - P_t)}$$

6. **(3)** $A(g) \longrightarrow 2B(g) + C(g)$

	A(g)	2B(g)	C(g)
t = 0	P_0		
t → ∞	0	$2P_0$	P_0

$P_\infty = 3P_0 = 240$

$P_0 = 80$ mm of Hg

$$kt = \ln\left(\frac{P_\infty - P_0}{P_\infty - Pt}\right)$$

$$k \times 10 = \ln\left(\frac{240-80}{240-160}\right)$$

$$k = \frac{\ln 2}{10} = 0.0693\ min^{-1}$$

7. **(1)** $r_1 = k[A]^n[B]^m$

Now A is doubled & B is halved in concentration

$$\Rightarrow r_2 = k2^n[A]^n.\frac{[B]^m}{2^m}$$

$$r_2 = k[A]^n[B]^m.2^n.\left(\frac{1}{2}\right)^m$$

Now $\frac{r_2}{r_1} = 2^n\left(\frac{1}{2}\right)^m \Rightarrow \frac{r_2}{r_1} = 2^{(n-m)}$

8. **(4)** Rate of formation of B is

$$\frac{d[B]}{dt} = K_1[A] - K_2[B] = 0$$

$$K_1[A] = K_2[B] \Rightarrow [B] = \frac{K_1[A]}{K_2}$$

9. **(3)** For a given reaction,

$$rate = -\frac{1}{2}\frac{dn_A}{dt} = -\frac{1}{3}\frac{dn_B}{dt} = -\frac{2}{3}\frac{dn_C}{dt}$$

$$rate = \frac{dn_A}{dt} = \frac{2}{3}\frac{dn_B}{dt} = \frac{4}{3}\frac{dn_C}{dt}$$

10. **(4)** $2H_2(g) + 2NO(g) \rightarrow N_2(g) + 2H_2O(g)$

$R_f = K_f[NO]^2[H_2]$

This means that, for forward reaction the order of reaction w.r.t. [NO] is 2 and that w.r.t. $[H_2]$ is 1.

Assuming the equilibrium expression,

$$K_{eq} = \frac{[N_2][H_2O]^2}{[H_2]^2[NO]^2} = \frac{K_f}{K_b}$$

$$\Rightarrow K_f[NO]^2[H_2] = K_b\frac{[N_2][H_2O]^2}{[H_2]}$$

At equilibrium, $R_f = R_b$

$$\Rightarrow R_b = K_b\frac{[N_2][H_2O]^2}{[H_2]}$$

Note: In this question, the order of reaction for backward reaction is not given.

11. **(3)** $xA \rightarrow yB$

$$\therefore \frac{-dA}{xdt} = \frac{1}{y}\frac{dB}{dt} \Rightarrow \frac{-dA}{dt} = \frac{dB}{dt} \times \frac{x}{y}$$

$$\log\left[\frac{-dA}{dt}\right] = \log\left[\frac{dB}{dt}\right] + \log\left(\frac{x}{y}\right)$$

Comparing this equation with the equation given in question. We get,

$\log\frac{x}{y} = 0.3010$ or $\log\frac{x}{y} = \log 2$ ∴ $\frac{x}{y} = 2$

∴ The reaction is of type 2A ® B.

Hence, option (c) is correct.

12. **(2)** $2N_2O_5(g) \rightarrow 4NO_2(g) + O_2(g)$

Rate of reaction

$$= \frac{-1}{2}\frac{d[N_2O_5]}{dt} = \frac{1}{4}\frac{d[NO_2]}{dt} = \frac{d[O_2]}{dt}$$

According to the question

$$\frac{-d[N_2O_5]}{dt} = -\frac{(2.75-3)}{30} = \frac{0.25}{30}\ M\ min^{-1}$$

$$\therefore \frac{d[NO_2]}{dt} = 2 \times \frac{-d[N_2O_5]}{dt} = 2 \times \frac{0.25}{30}$$

$= 1.67 \times 10^{-2}\ M\ min^{-1}$

13. **(4)** $A \xrightarrow{k_1} B \xrightarrow{k_2} C$ [It is a consecutive reaction]

$$\frac{d[B]}{dt} = k_1[A] - k_2[B] = 0 \Rightarrow [B] = \frac{k_1[A]}{k_2}$$

14. **(1)** $2A + B \longrightarrow C$

Rate $= k[A]^x[B]^y$

Exp-1, $0.045 = k[0.05]^x[0.05]^y$ ...(i)

Exp-2, $0.090 = k[0.1]^x[0.05]^y$ ...(ii)

Exp-3, $0.72 = k[0.2]^x[0.1]^y$ ...(iii)

Divide equation (i) by equation (ii)

$$\frac{0.045}{0.090} = \left(\frac{1}{2}\right)^x \Rightarrow x = 1$$

Divide equation (i) by equation (iii)

$$\frac{0.045}{0.72} = \left(\frac{0.05}{0.1}\right)^y\left(\frac{0.05}{0.2}\right)^1$$

$$\frac{0.045}{0.72} \times \frac{0.2}{0.05} = \left(\frac{0.05}{0.1}\right)^y$$

$$\left(\frac{1}{2}\right)^2 = \left(\frac{1}{2}\right)^y \Rightarrow y = 2$$

Rate law $= k[A]^1[B]^2$.

15. **(4)** $A_2 \underset{k_{-1}}{\overset{k_1}{\rightleftharpoons}} 2A$

$$-\frac{1}{2}\frac{d[A]}{dt} = k_1[A_2] - k_{-1}[A]^2$$

$$\Rightarrow \frac{d[A]}{dt} = 2k_1[A_2] - 2k_{-1}[A]^2$$

16. **(1)**

$O_3(g) + Cl^\bullet(g) \longrightarrow O_2(g) + ClO^\bullet(g),\ k_i$

$ClO^\bullet(g) + O^\bullet(g) \longrightarrow O_2(g) + Cl^\bullet(g),\ k_{ii}$

$O_3(g) + O^\bullet(g) \longrightarrow 2O_2(g),\ k_{Rate} = k_i \times k_{ii}$

$= 5.2 \times 10^9 \times 2.6 \times 10^{10} \approx 1.4 \times 10^{20}\ mol^{-1}\ L\ s^{-1}$

Note: When reaction added, the rate of the reaction are multiply

17. (4) Rate constant is independent of concentration.

18. (3) Rate $= k[A][B] = R$

$R' = k[A][2B]$

$$\frac{R}{R'} = \frac{k[A][B]}{k[A][2B]} = \frac{k[A][B]}{2k[A][B]}$$

$\Rightarrow 2R = R'$, i.e., rate becomes double.

19. (4) Let rate of reaction $= \frac{d[C]}{dt} = k[A]^x[B]^y$

k is rate constant, for formation of C.

Now from the given data,

$1.2\times10^{-3} = k[0.1]^x[0.1]^y$ ...(i)

$1.2\times10^{-3} = k[0.1]^x[0.2]^y$ ...(ii)

$2.4\times10^{-3} = k[0.2]^x[0.1]^y$ ...(iii)

Dividing equation (i) by (ii),

$$\frac{1.2\times10^{-3}}{1.2\times10^{-3}} = \frac{k[0.1]^x[0.1]^y}{k[0.1]^x[0.2]^y}$$

We find, $y = 0$

Now dividing equation (i) by (iii),

$$\frac{1.2\times10^{-3}}{2.4\times10^{-3}} = \frac{k[0.1]^x[0.1]^y}{k[0.2]^x[0.1]^y}$$

We find, $x = 1$

Hence, $\frac{d[C]}{dt} = k[A]^1[B]^0$

Note: In this type of questions, either Divide in such away that, x on y cancel with another equation.

20. (4) From rate law

$$-\frac{1}{2}\frac{d[SO_2]}{dt} = -\frac{d[O_2]}{dt} = \frac{1}{2}\frac{d[SO_3]}{dt}$$

$$\therefore -\frac{d[SO_2]}{dt} = -2\times\frac{d[O_2]}{dt} = -2\times2.5\times10^{-4}$$

$= -5\times10^{-4}\ \text{mol L}^{-1}\text{s}^{-1}$

21. (2) Rate of disappearance of reactant = Rate of appearance of products

$$-\frac{1}{2}\frac{d[N_2O_5]}{dt} = \frac{1}{4}\frac{d[NO_2]}{dt};$$

$$\frac{1}{2}k[N_2O_5] = \frac{1}{4}k'[N_2O_5] \Rightarrow \frac{k}{2} = \frac{k'}{4} \therefore k' = 2k$$

22. (4) For the reaction

$3A + 2B \longrightarrow C + D$

From the rate law,

$$= -\frac{1}{3}\frac{d[A]}{dt} = \frac{d[C]}{dt} = k[A]^n[B]^m$$

23. (2) Given $-\frac{d[MnO_4^-]}{dt} = 4.56\times10^{-3}\ \text{Ms}^{-1}$

From the reaction given,

$$-\frac{1}{2}\frac{d[MnO_4^-]}{dt} = \frac{4.56\times10^{-3}}{2}\text{Ms}^{-1}$$

$$-\frac{1}{2}\frac{d[MnO_4^-]}{dt} = \frac{1}{5}\frac{dI_2}{dt}$$

$$\Rightarrow -\frac{5}{2}\frac{d[MnO_4^-]}{dt} = \frac{dI_2}{dt}$$

On substituting the given value

$$\therefore \frac{dI_2}{dt} = \frac{4.56\times10^{-3}\times5}{2} = 1.14\times10^{-2}\ \text{M/s}$$

24. (4) $\text{Rate}_1 = k[A]^n[B]^m$

$\text{Rate}_2 = k[2A]^n\left[\frac{1}{2}B\right]^m$

$$\therefore \frac{\text{Rate}_2}{\text{Rate}_1} = \frac{k[2A]^n\left[\frac{1}{2}B\right]^m}{k[A]^n[B]^m} = (2)^n\left(\frac{1}{2}\right)^m$$

$= 2^n\,.\,(2)^{-m} = 2^{n-m}$

25. (4)

26. (2) The rate of reaction for the given reaction, $\frac{1}{2}A \longrightarrow 2B$ can be written either as $-2\frac{d}{dt}[A]$ with respect to 'A' or $\frac{1}{2}\frac{d}{dt}[B]$ with respect to 'B'

From the above, we have

$$-2\frac{d}{dt}[A] = \frac{1}{2}\frac{d}{dt}[B] \quad\text{or}\quad -\frac{d}{dt}[A] = \frac{1}{4}\frac{d}{dt}[B]$$

27. (4)

28. (1) For a zero order reaction.

rate $= k[A]^o$ i.e. rate $= k$

Hence unit of $k = \text{M.sec}^{-1}$

For a first order reaction.

Rate $= k[A]$

$\therefore\ k = \text{M.sec}^{-1}/\text{M} = \text{sec}^{-1}$

29. (47) The graph given is for 1st order reaction

$$k = \frac{2.303}{t}\log\frac{[A_0]}{[A_t]} \Rightarrow k = \frac{2.303}{25}\log\frac{[50]}{[10]}$$

$$k = \frac{2.303}{25}\log5 = \frac{2.303\times0.7}{25}s^{-1}$$

$$= \frac{2.303\times0.7}{25} = \frac{2.303}{t}\log\frac{50}{2.5}$$

$$\Rightarrow \frac{1.699}{25} = \frac{2.303}{t}\log20$$

$$t = \frac{1.301\times25\times2.303}{0.699} = 46.46\,s \approx 47\,s$$

30. (2435)

$t_{1/2} \propto [A]_0 \Rightarrow$ Order = zero

$$t_{1/2} = \frac{A_0}{2k} \Rightarrow \text{Slope} = \frac{1}{2k} = 76.92$$

$$k = \frac{1}{2\times76.92}$$

$$[A]_{10} = -kt + A_0 = -\frac{1}{2\times76.92}\times10 + 2.5$$

$= 2.435 = 2435\times10^{-3}$ mol/L

31. (8) $r = k[A]^2|B|$; if conc. are doubled

$r' = k[2A]^2[2B]^1 \Rightarrow r' = 8r \Rightarrow x = 8$

For a zero order reaction

$$k = \frac{[R]_0 - [R]}{t} \Rightarrow \text{Zero order, } y = 0$$

Thus, $x + y = 8$

32. (3) $A(g) \rightarrow 2B(g) + C(g)$.

$P_t = P_0 + 2x = 200$; $P_\infty = 3P_0 = 300$; $P_0 = 100$

$$K = \frac{1}{t}\ln\frac{P_\infty - P_0}{P_\infty - P_t} = \frac{2.303}{t}\log\frac{P_\infty - P_0}{P_\infty - P_t}$$

$$= \frac{2.303}{23}\log\frac{300-100}{300-200}$$

$$= \frac{2.3\times0.301}{23} = 0.0301 = 3.01\times10^{-2}\ \text{sec}^{-1}$$

33. (17) The given reaction is a first order reaction. (from its mechanism, we know that it is a first-order reaction).

Rate of NO_2 formation

$$= \frac{d[NO_2]}{dt} = 2\times\frac{d[N_2O_5]}{dt}$$

(from stoichiometry)

$$\Rightarrow \frac{d[NO_2]}{dt} = 2\times\frac{d[N_2O_5]}{dt}$$

$$= 2\times\frac{\Delta[N_2O_5]}{\Delta t} = 2\times\frac{(3-2.75)}{30}$$

$= 1.67\times10^{-2} = 16.7\times10^{-3}\ \text{mol L}^{-1}\text{min}^{-1}$

$= x\times10^{-3}\ \text{mol L}^{-1}\text{min}^{-1}$ Thus, $x = 16.7$ or 17.

34. (2200) $t_{1/2} = 10$ minutes

$$(P_A)_{30\,min} = (P_A)_0\left(\frac{1}{2}\right)^{30/10} = 100\ \text{mm Hg}$$

	A (g) →	2B (g) +	C (g)
at t = 0,	800 mm	0	0
at t = 30,	100 mm	1400 mm	700 mm

Total pressure after 30 minutes = 2200 mm Hg

35. (333) $KClO_3 + 6FeSO_4 + 3H_2SO_4 \longrightarrow KCl + 3Fe_2(SO_4)_3 + 3H_2O$

$$\text{Rate} = \frac{-\Delta[KClO_3]}{\Delta t} = \frac{-1}{6}\frac{\Delta[FeSO_4]}{\Delta t}$$

$$= \frac{+1}{3}\frac{\Delta[Fe_2(SO_4)_3]}{\Delta t}$$

$$\Rightarrow \frac{\Delta[Fe_2(SO_4)_3]}{\Delta t} = \frac{-1}{2}\frac{\Delta[FeSO_4]}{\Delta t}$$

$$= \frac{1}{2}\frac{(10-8.8)}{30\times60}$$

$= 0.333\times10^{-3} = 333\times10^{-6}\ \text{mol litre}^{-1}\text{sec}^{-1}$

36. (7) $t_{10\%} = \frac{1}{K}\ln\left(\frac{a}{a-x}\right) = \frac{1}{K}\ln\left(\frac{100}{90}\right)$

$$t_{10\%} = \frac{2.303}{K}(\log10 - \log9)$$

$$\Rightarrow t_{10\%} = \frac{2.303}{K}(0.04)$$

$$t_{90\%} = \frac{1}{K}\ln\left(\frac{100}{10}\right) \Rightarrow t_{90\%} = \frac{2.303}{K}$$

$$\frac{t_{90\%}}{t_{10\%}} = \frac{1}{0.046} = 21.7$$

From the rate law of 1st order reaction

$$e^{kt} = \frac{a}{a-x} \Rightarrow \frac{a-x}{a} = e^{-kt}$$

$$1 - \frac{x}{a} = e^{-kt} \Rightarrow x = a(1 - e^{-kt})$$

$$\Rightarrow \alpha = \frac{x}{a} = (1 - e^{-kt})$$

Only statement D is correct

37. (40) $r = k[X][Y]^0 = k[X]$

Using I and II:

$$\frac{4\times10^{-3}}{2\times10^{-3}} = \left(\frac{L}{0.1}\right) \Rightarrow L = 0.2$$

Using I and III:

$$\frac{M\times10^{-3}}{2\times10^{-3}}=\frac{0.4}{0.1}\Rightarrow M=8\Rightarrow\frac{M}{L}=\frac{8}{0.2}=40$$

38. (1)

39. (4)

$$\text{Rate}=-\frac{1}{3}\frac{d[C_2H_6O]}{dt}=\frac{1}{2}\frac{d[Cr_2(SO_4)_3]}{dt}$$

$$-\frac{d[C_2H_6O]}{dt}=\frac{3}{2}\times2.67=4\text{ mol min}^{-1}$$

40. (4) $\text{Rate}=\frac{d[B]}{dt}=\frac{0.2}{0.5}=0.4=4\times10^{-1}$

41. (106) As the unit of rate constant is min^{-1}, so it must be a first order reaction: $k\times t=2.303\log A_0/A_t$

In 1 min, 10% is inactivated so:

$A_0=100,\ A_t=90$

$$k\times1=2.303\times\log\frac{100}{90}=2.303\times(\log10-2\log3)$$

$=2.303\times(1-2\times0.477)=0.10593$

$=105.93\times10^{-3}\approx106\times10^{-3}$

42. (27) $2A+B_2\longrightarrow2AB$

Rate of reaction is $r=k[A]^2[B_2]$

The conc. of A and B_2 will become 3 times as the volume of reaction vessel is reduced by a factor of 3 and hence, the rate becomes $3^2\times3=27$ times of initial rate.

43. (1) $A(g)\longrightarrow B(g)+C(g)$

	A(g)	B(g)	C(g)
$t=0$	P_i	–	–
$t=t$	P_i-x	x	x

As given $(P_i-x)+x+x=P_t$

$x=(P_t-P_i)$

$$k=\frac{1}{t}\ln\frac{P_i}{P_i-x}=\frac{1}{t}\ln\frac{P_i}{P_i-(P_t-P_i)}$$

$$\Rightarrow k=\frac{1}{t}\ln\frac{P_i}{2P_i-P_t}$$

44. (2) Curve 1st represent zero order kinetics

$[R_t]=[R]_0-Kt$

Concentration decreases linearly with time in curve 1 whereas curve 2 and 3 decreases exponentially, so order is not same for all curves.

$\Rightarrow\ K_3>K_2$

$\Rightarrow$ Reaction-1 is zero order having unit of rate constant same as rate of reaction i.e., molarity/second.

Thermal decomposition of HI on Au surface is zero order reaction.

45. (4) $R\longrightarrow P$

	R	P
$t=0$	a	0
$t=t$	$a-x$	x

From integrated rate law expression for 1st order reaction.

$$a-x=ae^{-k_1t}$$

$$\frac{x}{a}=1-e^{-k_1t}.$$

46. (1) Rate of constant $(k)=10^{-3}\ M^{-1}\ sec^{-1}$

$\because\ 1-n=-1$ [unit of $k=M^{1-n}\ sec^{-1}$]

$n=2\Rightarrow$ 2nd order reaction

rate $=k[A]^2$

Now, evaluating the statements,

(A) Rate $=k[X]^2$,

If $[X]=4[X]$

Rate $\longrightarrow k(4[X])^2=16k[X]^2$

If concentration of A is 4 times then reaction will become 16 times.

(B) Order of reaction is 2.

(C) Half-life of the reaction is independent of the concentration of the reactant for 1st order, not for 2nd order reaction.

$$t_{\frac{1}{2}}=\frac{1}{k[X]_0}\ \text{(For 2nd order)}$$

(D) Decomposition of N_2O_5 is an example of 1st order reaction.

(E) A plot of $\ln\frac{[R]_0}{[R]_t}$ vs time (t) is straight line for 1st order not for 2nd order reaction. For a 2nd order reaction, the linear plot is $\frac{1}{[R]}$ vs. time.

So, statement (A) and (B) are correct.

47. (2) Zero-Order:

$$t_{100\%}=\frac{a}{k};\ t_{1/2}=\frac{a}{2k};\ t_{100\%}=2t_{1/2}$$

1st order

$t_{100\%}=\infty$

$$t_{1/2}=\frac{\ln2}{k}$$

$$t_{100\%}=(t_{1/2})^{\infty}$$

48. (1) $t=\frac{1}{k}\ell n\frac{A_0}{A_t}$

$$t_{1/8}=\frac{1}{k}\ell n\frac{A_0}{A_0/8}=\frac{1}{k}\ell n8$$

$$t_{1/10}=\frac{1}{k}\ell n\frac{A_0}{A_0/10}=\frac{1}{k}\ell n10$$

$$\frac{t_{1/8}}{t_{1/10}}=\frac{\ell n8}{\ell n10}=\frac{\log8}{\log10}$$

$$\frac{t_{1/8}}{t_{1/10}}=\log8=3\log2=0.9$$

$$\frac{t_{1/8}}{t_{1/10}}\times10=9$$

49. (2) $A\rightarrow nB$

	A	nB
Initial	0.05	0
at 10 min.	0.04	0.01 × n

$\Rightarrow\ 0.01\times n=0.03$ (from the graph)

$\Rightarrow\ n=3$

50. (2) Before applying medicine

$$\frac{dA}{dt}=k[A]\ \text{(First order Rate law)}$$

$$\frac{A}{A_0}=\frac{N}{N_0}=e^{kt}$$

After applying medicine

Active Bacteria $\rightarrow$ Inactive Bacteria

(A) (I)

$$r=-\frac{dA}{dt}=k[A]^2\quad\text{(Rate law)}$$

$y=kx^2$ Parabola

51. (3) For zero order reaction :

$$\text{Half life}=\frac{[A]_o}{2k}\Rightarrow60\text{min}=\frac{2}{2k}$$

$$\Rightarrow k=\frac{1}{60}\text{M/min}$$

Now, $[A]_t=[A]_o-kt\Rightarrow t=\frac{[A]_o-[A]_t}{k}$

$$=\frac{0.5-0.25}{1/60}=15\text{ min}$$

52. (4) Half life of A = 10 min

Half life of B = 40 min

$[A]_0=8[B]_0$ (given)

$$\Rightarrow\ [A]_0=8[B]_0\xrightarrow{10\text{ min}}4[B]_0\xrightarrow{10\text{ min}}2$$

$$[B_0]\xrightarrow{10\text{ min}}[B_0]\xrightarrow{10\text{ min}}\frac{[B]_0}{2}$$

$$\Rightarrow\ [B_0]\xrightarrow{40\text{ min}}\frac{[B]_0}{2}$$

Therefore, 40 min. is the time required for concentration to be same for both the reactants A and B.

53. (4) $\frac{t_2}{t_1}=\left(\frac{C_1}{C_2}\right)^n$ and $n=1$

$C_o=16$ mg/mL

$t_1=?,\ C_1=8$ mg/mL

$t_2=12,\ C_2=4$ mg/mL

$$\Rightarrow t_1=\frac{t_2\times C_2}{C_1}=\frac{12\times4}{8}=6$$

54. (1) $\underset{k_{-1}}{\overset{k_1}{\rightleftharpoons}}A+A$

$\text{Rate}=k_2A_2[A][B_2]$ ... (1)

(from slow step)

$$\Rightarrow\ \left(\frac{k_1}{k_{-1}}\right)=K_{eq}=\left(\frac{[A]^2}{[A_2]}\right)$$

$$\Rightarrow\ [A]=\sqrt{\frac{k_1}{k_{-1}}}\cdot\sqrt{[A_2]}$$

Substituting in (1); we get

$$\text{Rate}=k_2\sqrt{\frac{k_1}{k_{-1}}}\cdot[A_2]^{1/2}\cdot[B_2]$$

$$\therefore\ \text{Order}=1+\frac{1}{2}=\left(\frac{3}{2}\right)=1.5$$

55. (4) Because no. of bacteria initial $=N_0$ and No. of bacteria at any time $t=N$

Since bacterial growth is given as $N=N_0e^{Kt}$

Where K = growth constant for bac terial growth

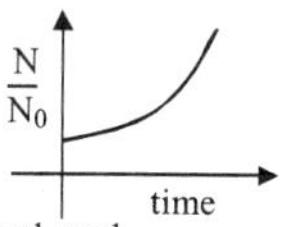

56. (2) $r_1=k[A]^1[B]^1$

$$r_1=k\left[\frac{n_A}{V}\right]^1\left[\frac{n_B}{V}\right]^1$$

By Reducing the volume to 1/3 of its original volume, the concentration of rectants will becomes 3 times.

$$r_2=k\left[\frac{3n_A}{V}\right]^1\left[\frac{3n_B}{V}\right]^1$$

$\boxed{r_2=9r_1}$ Rate will become 9 times to its initial rate.

57. (3) (A) $t_{1/2} \propto \frac{1}{A_0^{n-1}}$

$$\frac{(t_{1/2})_1}{(t_{1/2})_2} = \frac{(A_0)_2^{n-1}}{(A_0)_1^{n-1}}$$

$$\frac{200}{100} = \left(\frac{0.025}{0.100}\right)^{n-1} \Rightarrow 2 = \left(\frac{1}{4}\right)^{n-1}$$

$$\Rightarrow \frac{1}{4} \times 2 = \left(\frac{1}{4}\right)^n \Rightarrow 2 = 4^n \Rightarrow n = \frac{1}{2} \text{ (order)}$$

(B) $t_{1/2} \propto \sqrt{A_0}$

$$\frac{200}{t_{1/2}} = \frac{(1)^{-1/2}}{(0.1)^{-1/2}} = \frac{(0.1)^{1/2}}{(1)^{1/2}} \text{ when } A_0 = 1\text{ M}$$

$t_{1/2} = 200\sqrt{10}$ min

(C) The order would change to 0.769 if the concentration was 0.500 instead of 0.100. Thus, C is not correct.

(D) $\frac{200}{t_{1/2}} = \frac{(0.1)^{1/2}}{(1.6)^{1/2}}$ when $A_0 = 1.6$ M

$t_{1/2} = 800$ min

58. (4) For first order reaction $t_{1/2} = \frac{\ln 2}{k}$

$t_{1/2}$ is independent of $[R]_0$

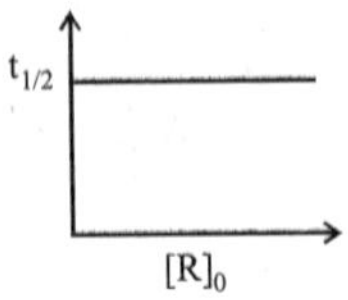

For first order reaction

$$k = \frac{2.303}{t}\log\frac{[R_0]}{[R]}$$

$$\log\frac{[R]_0}{[R]} = \frac{1}{2.303}kt\ ;\ \log\frac{[R]_0}{[R]} = \left(\frac{k}{2.303}\right) \times t$$

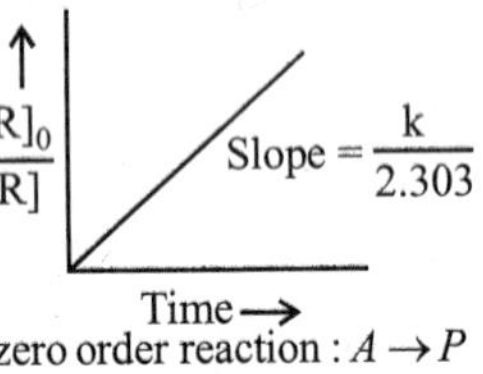

59. (2) For zero order reaction : $A \rightarrow P$

$$\text{Rate} = k \Rightarrow \frac{[A]_t - [A]_0}{t} = k$$

Thus, concentration $\propto t$ and rate is constant.

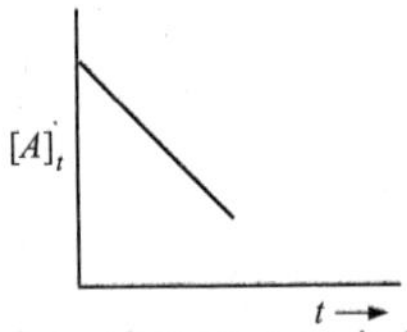

60. (3) Nuclear phenomenon independent of temperature. Decay constant is a nuclear property.

61. (4)

	A(g)	→	B(g)	+	C(g)
	P_i		0		0
	$P_i - x$		x		x

$P_t = P_i + x$

$P_i - x = P_i - P_t + P_i = 2P_i - P_t$

$$k = \frac{2.303}{t}\log\frac{P_i}{2P_i - P_t}$$

62. (1) $r = K[A]^1[B]^1$

$0.1 = K(20)^1(0.5)^1$ ...(i)

$0.40 = K(x)^1(0.5)^1$ ...(ii)

$0.80 = K(40)^1(y)^1$ ...(iii)

From (i) and (ii); $x = 80$

From (i) and (iii) ; $y = 2$

63. (2) $t_{1/2} \propto (P_0)^{1-n}$

$$\frac{(t_{1/2})_1}{(t_{1/2})_2} = \frac{(P_0)_1^{1-n}}{(P_0)_2^{1-n}} \Rightarrow \left(\frac{4}{2}\right) = \left(\frac{50}{100}\right)^{1-n}$$

$$\Rightarrow 2 = \left(\frac{1}{2}\right)^{1-n} \Rightarrow 2 = (2)^{n-1}$$

$\Rightarrow n - 1 = 1 \quad \Rightarrow n = 2$

64. (3) For Ist order reaction

$$k = \frac{2.303}{t}\log\frac{a_0}{0.2a_0} \text{ also } t_{1/2} = \frac{0.693}{k}$$

$$k = \frac{0.693}{200} \Rightarrow \frac{0.693}{200} = \frac{2.303}{t}\log\frac{1}{0.2}$$

$$t = \frac{2.303}{0.693} \times 200\log\frac{1}{0.2} = 466.675 \approx 467 \text{ sec}$$

65. (3) Given $t_{0.90} = xt_{1/2}$

First order rate constant,

$$k = \frac{\ln 2}{t_{1/2}} = \frac{1}{xt_{1/2}}\ln\frac{A_0}{A_0 - A_0 \times \frac{90}{100}}$$

$$\frac{\ln 2}{t_{1/2}} = \frac{\ln 10}{xt_{1/2}} \Rightarrow x$$

$$= \frac{\ln 10}{\ln 2} = \frac{2.303}{2.303 \times 0.3010} = 3.32$$

66. (3) Rate $= k[A]^n$

Comparing units

$$\frac{(\text{mol/L})}{\text{sec}} = k\left(\frac{\text{mol}}{\text{L}}\right)^n \Rightarrow k = \text{mol}^{(1-n)}\,\text{L}^{(n-1)}\,\text{s}^{-1}$$

67. (4) Rate $= k[A]^n$

$\log[\text{Rate}] = \log k + n\log[A]$

Slope $= n$ [n is order of the reaction]

∴ Correct sequence for the order of the reaction is (iv) > (ii) > (i) > (iii)

68. (2) $C_t = C_0 e^{-kt}$; $k = \frac{\ln 2}{t_{1/2}}$

$$(C_t)_A = (C_0)_A e^{-k_A t};\ k_A = \frac{\ln 2}{300}$$

$$(C_t)_B = (C_0)_B e^{-k_B t};\ k_B = \frac{\ln 2}{180}$$

$$\frac{(C_t)_B}{(C_t)_A} = \frac{(C_0)_B}{(C_0)_A} \times e^{(k_B - k_A)t} \Rightarrow 4 = e^{(k_B - k_A)t}$$

$$\Rightarrow 2\ln 2 = \left[\frac{\ln 2}{180} - \frac{\ln 2}{300}\right]t$$

$$\Rightarrow 2\ln 2 = \ln 2\left[\frac{1}{180} - \frac{1}{300}\right]t$$

$$\Rightarrow 2 = \left(\frac{120}{180 \times 300}\right)t \Rightarrow t = \frac{2 \times 180 \times 300}{120} = 900 \text{ sec}$$

69. (2) For an elementary or single step reaction, order and molecularity are same. But for a reaction, molecularity ≠ 0. Hence, the zero order reaction must be a multi-step so that overall order of reaction becomes zero.

70. (3) Rate (R) $= k[A]^a[B]^b$

Exp. I $\Rightarrow 6.0 \times 10^{-3} = k[0.1]^a[0.1]^b$

Exp. II $\Rightarrow 24.0 \times 10^{-3} = k[0.1]^a[0.2]^b$

Exp. III $\Rightarrow 12.0 \times 10^{-3} = k[0.2]^a[0.1]^b$

Exp. IV $\Rightarrow 72 \times 10^{-3} = k[X]^a[0.2]^b$

Exp. V $\Rightarrow 288 \times 10^{-3} = k[0.3]^a[Y]^b$

From Exp. I and II,

$$\Rightarrow \frac{6.0 \times 10^{-3}}{24 \times 10^{-3}} = \frac{k[0.1]^a[0.1]^b}{k[0.1]^a[0.2]^b}$$

$$\Rightarrow \frac{1}{4} = \left(\frac{0.1}{0.2}\right)^b \Rightarrow \left(\frac{1}{2}\right)^2 = \left(\frac{1}{2}\right)^b \quad \therefore b = 2$$

Similarly, from exp. I and III we get,

$$\frac{1}{2} = \left(\frac{1}{2}\right)^a \Rightarrow a = 1$$

From Exp. II and IV,

$$\Rightarrow \frac{24 \times 10^{-3}}{72 \times 10^{-3}} = \frac{k[0.1]^a[0.2]^b}{k[X]^a[0.2]^b}$$

$$\Rightarrow \frac{1}{3} = \left(\frac{0.1}{X}\right)^a \Rightarrow \frac{1}{3} = \left(\frac{0.1}{X}\right)^1 \quad \therefore X = 0.3$$

From Exp. I and V,

$$\Rightarrow \frac{6.0 \times 10^{-3}}{288 \times 10^{-3}} = \left(\frac{0.1}{0.3}\right)^a\left(\frac{0.1}{Y}\right)^b \Rightarrow \frac{1}{48} = \left(\frac{1}{3}\right)^1\left(\frac{0.1}{Y}\right)^2$$

$$\Rightarrow \left(\frac{1}{4}\right)^2 = \left(\frac{0.1}{Y}\right)^2 \Rightarrow \frac{1}{4} = \frac{0.1}{Y} \quad \therefore Y = 0.4$$

71. (3)

72. (1) In graph (i), ln [Reactant] vs time is linear with positive intercept and negative slope. Hence it is 1st order. In graph (ii), [Reactant] vs time is linear with positive intercept and negative slope. Hence, it is zero order.

73. (1) Rate constant of decomposition of X = 0.05 μg/year.

Unit of rate constant confirms that the decomposition of X is a zero order reaction.

For zero order kinetics,

$[X] = [X]_0 - kt$

$$kt = [X]_0 - [X],\ t = \frac{[X]_0 - [X]}{k}$$

$$t = \frac{5 - 2.5}{0.05} = \frac{2.5}{0.05} = 50 \text{ years}$$

74. (3) For the reaction 2X → B, follow zeroth order

Rate equation is $kt = [A]_0 - [A]$

For the half-life; $t = t_{1/2}$ and $[A] = 0.1$

$kt_{1/2} = 0.2 - 0.1$

$$k = \frac{0.2 - 0.1}{6} = \frac{0.1}{6}\text{M hr}^{-1}$$

∴ Time required to reach from 0.5 M to 0.2 M = t

$kt = [A]_0 - [A]$

$$\frac{0.1}{6} \times t = (0.5 - 0.2) \Rightarrow t = 18 \text{ hour}$$

75. (2) From experiment I and II, it is observed that order of reaction w.r.t. (B) is zero. From experiment II and III, x can be calculated as:

$$\frac{6.93\times10^{-3}}{13.86\times10^{-3}}=\left(\frac{0.1}{0.2}\right)^x\left(\frac{0.25}{0.3}\right)^y$$

$\because$ Order of reaction w.r.t. B = 0, so y = 0

$$\therefore \frac{6.93\times10^{-3}}{13.86\times10^{-3}}=\left(\frac{0.1}{0.2}\right)^x \text{ or } \left(\frac{1}{2}\right)^1=\left(\frac{1}{2}\right)^x$$

$x = 1$, Now, Rate = $k[A]^1[B]^o$

or $6.93\times10^{-3}=k[0.1]^1[0.2]^o$

$k = 6.93\times10^{-2}$

$$\therefore t_{1/2}=\frac{0.693}{k}=\frac{0.693}{6.93\times10^{-2}}=10$$

76. (2) $\frac{r_2}{r_1}=2^x.2^y=8$ ...(i)

$\frac{r_2}{r_1}=2^x=2\Rightarrow x=1$

Put $x = 1$ in eq(i), $2^y=4\Rightarrow y=2$

77. (1) $CH_3CHO\longrightarrow CH_4+CO$

From rate law, $r\propto(a-x)^m$

m = order of reaction, a – x = unreacted

$r_1 = 1$ torr s^{-1}, when 5% reacted

$r_2 = 0.5$ torr s^{-1}, when 33 % reacted

$(a-x_1)=0.95$(unreacted)

$(a-x_2)=0.67$(unreacted)

$$\frac{r_1}{r_2}=\left[\frac{(a-x_1)}{(a-x_2)}\right]^m\Rightarrow\frac{1}{0.5}=\left(\frac{0.95}{0.67}\right)^m$$

$2=(1.41)^m\Rightarrow2=(\sqrt{2})^m\Rightarrow m=2$

78. (4) This is first order reaction as half life is constant.

79. (2) $N_2O_5 \rightarrow 2NO_2+\frac{1}{2}O_2$

	N_2O_5	$2NO_2$	$\frac{1}{2}O_2$
At t = 0	50	0	0
At t = 50 min	$50-p_1$	$2p_1$	$\frac{p_1}{2}$

Total pressure at 50 minutes

$$=50-p_1+2p_1+\frac{p_1}{2}=87.5$$

$$\therefore p_1=\frac{37.5\times2}{3}=25=\frac{1}{2}\times50\text{ mm Hg}$$

This means, 50 min. is half-life.

For 100 minutes, i.e., 2 half lives, $50-p_2=12.5$

$\therefore p_2=37.5$ mm of Hg

Total pressure at 100 minutes

$$=50-p_2+2p_2+\frac{p_2}{2}$$

$$=50+\frac{3p_2}{2}=50+\frac{3}{2}\times37.5=106.25\text{ mm of Hg}$$

80. (3) The half life $t_{1/2}=10$ days

The decay constant,

$$k=\frac{0.693}{t_{1/2}}=\frac{0.693}{10\text{ days}}=0.0693\text{ days}^{-1}$$

The time required for one fourth conversion

$$t=\frac{2.303}{k}\log_{10}\frac{a}{a-x}$$

$$=\frac{2.303}{0.0693\text{ day}^{-1}}\log_{10}\frac{1}{1-(1/4)}=4.1\text{ days}$$

81. (4) $H_2O_2(aq)\rightarrow H_2O(aq)+\frac{1}{2}O_2(g)$

For a first order reaction,

$$k=\frac{2.303}{t}\log\frac{a}{(a-x)}$$

Given a = 0.5, (a – x) = 0.125, t = 50 min

$$\therefore k=\frac{2.303}{50}\log\frac{0.5}{0.125}=2.78\times10^{-2}\text{ min}^{-1}$$

$r=k[H_2O_2]=2.78\times10^{-2}\times0.05$

$=1.386\times10^{-3}$ mol min^{-1}

Now

$$-\frac{d[H_2O_2]}{dt}=\frac{d[H_2O]}{dt}=\frac{2d[O_2]}{dt}=k[H_2O_2]$$

$$\therefore \frac{d[O_2]}{dt}=-\frac{1}{2}\times\frac{d[H_2O_2]}{dt}$$

$$=\frac{1.386\times10^{-3}}{2}=6.93\times10^{-4}\text{ mol min}^{-1}$$

82. (1) Rate law for first order reaction $=k[N_2O_5]$

	$2N_2O_5(g)\longrightarrow$	$4NO_2(g)$	$+O_2(g)$
t = 0 min (Pressure in mm Hg)	50	0	0
t = 30 min (Pressure in mm Hg)	$50-2p$	$4p$	p

Total pressure $50-2p+4p+p=50+3p$

$=87.5$ mm Hg

$\therefore p=12.5$ mm Hg

$\therefore p_0=50$ & p (t = 30 min) = 25 for N_2O_5 reactant

$$\therefore k=\frac{2.303}{30\text{ min}}\times\log\left(\frac{50}{25}\right)=\frac{2.303}{60\text{ min}}\times\log\left(\frac{50}{x}\right)$$

On solving $x=12.5$ mm Hg $=50-2p$

$\therefore p=18.75$ mm Hg

$\therefore$ Total pressure $=50+3p=106.25$ mm Hg

83. (3) Reactions of higher order (>3) are very rare due to very less chances of many molecules to undergo effective collisions.

84. (3) Given $t_{1/2}=15$ minutes

Total time $(T)=1$ hr = 60 min

From $T=n\times t_{1/2}\Rightarrow n=\frac{60}{15}=4$

Now, from the formula,

$$\frac{N}{N_0}=\left(\frac{1}{2}\right)^n=\left(\frac{1}{2}\right)^4=\frac{1}{16}$$

85. (4) For a zero order reaction,

$[A_t]=[A_0]-k.t$

$\Rightarrow 0.5=[A_0]-2\times10^{-2}\times25\Rightarrow[A_0]=1$ M

86. (3) Given $t_{1/2}=3$; Total time T = 12

$$\text{No. of half lives }(n)=\frac{12}{3}=4\Rightarrow\left(\frac{1}{2}\right)^n=\frac{N}{N_o}$$

$$\therefore \left(\frac{1}{2}\right)^4=\frac{3}{N}\Rightarrow N=48\text{ g}$$

87. (2) For a first order reaction

$$k=\frac{2.303}{t}\log\frac{a}{a-x}=\frac{2.303}{40}\log\frac{0.1}{0.025}$$

$=3.47\times10^{-2}$

$R=k[A]^1=3.47\times10^{-2}\times0.01=3.47\times10^{-4}$

88. (4) $A\longrightarrow B$

Initial concentration	Rate of reaction
2×10^{-3} M	2.40×10^{-4} Ms^{-1}
1×10^{-3} M	0.60×10^{-4} Ms^{-1}

$r=k[A]^x$

where x = order of reaction, hence,

$2.40\times10^{-4}=k[2\times10^{-3}]^x$ ......(i)

$0.60\times10^{-4}=k[1\times10^{-3}]^x$ ......(ii)

On dividing eqn.(i) from eqn. (ii) we get,

$4=(2)^x$

$\therefore x=2$, i.e., order of reaction = 2

89. (2) Plot given is for zero order reaction. From a zero order reaction,

$$[A_t]=[A_0]-k_t;\ t_{1/2}=\frac{A_0}{2k}$$

$\Rightarrow \log t_{1/2}=-\log2k+\log[A_0]$

90. (3) For the reaction

A $\rightarrow$ Product

Given $t_{1/2}=1$ hour

$$\text{For a zero order reaction; }\therefore t_{1/2}=\frac{[A_0]}{2k}$$

$$\text{or } k=\frac{[A_0]}{2t_{1/2}}=\frac{2}{2\times1}=1\text{ mol L}^{-1}\text{hr}^{-1}$$

Further for a zero order reaction,

$$k=\frac{dx}{dt}=\frac{\text{change in concentration}}{\text{time}}$$

$$1=\frac{0.50-0.25}{\text{time}}\Rightarrow\text{time}=0.25\text{ hr.}$$

91. (2) For first order reaction,

$$k=\frac{2.303}{t}\log\frac{100}{100-99}\quad\frac{0.693}{6.93}=\frac{2.303}{t}\log\frac{100}{1}$$

$$\frac{0.693}{6.93}=\frac{2.303\times2}{t};\ t=46.06\text{ min}$$

92. (2) For a first order reaction, $t_{1/2}=\frac{0.693}{K}$

i.e. for a first order reaction $t_{1/2}$ does not depend upon the concentration. From the given data, we can say that order of reaction with respect to B = 1 because change in concentration of B alone does not change half life.

Order of reaction with respect to A = 1 because rate of reaction doubles when concentration of A is doubled keeping concentration of B constant.

$\therefore$ Order of reaction = 1 + 1 = 2 and units of second order reaction are L mol^{-1} sec^{-1}.

93. (1) Suppose activity of safe working = A

Given $A_0=10A$

$$\lambda=\frac{0.693}{t_{1/2}}=\frac{0.693}{30}$$

$$t_{½}=\frac{2.303}{\lambda}\log\frac{A_0}{A}=\frac{2.303}{0.693/30}\log\frac{10A}{A}$$

$$=\frac{2.303\times30}{0.693}\times\log10=100\text{ days.}$$

94. (2) (i) $NO(g)+Br_2(g)\rightleftharpoons NOBr_2(g)$

(ii) $NOBr_2(g)+NO(g)\longrightarrow2NOBr(g)$

Rate law equation $=k[NOBr_2][NO]$

But $NOBr_2$ is intermediate and must not appear in the rate law equation.

$$\text{From Ist step, }K_c=\frac{[NOBr_2]}{[NO][Br_2]}$$

$\therefore [NOBr_2]=K_c[NO][Br_2]$

$\therefore$ Rate law equation $=k.K_c[NO]^2[Br_2]$

Hence, order of reaction w.r.t. NO is 2.

95. (1) Since the reaction is 2nd order w.r.t CO. Thus, rate law is given as.
$r = k[CO]^2$
Let initial concentration of CO is a
i.e. $[CO] = a$
$\therefore \quad r_1 = k(a)^2 = ka^2$
When concentration becomes doubled, i.e. $[CO] = 2a$
$\therefore \quad r_2 = k(2a)^2 = 4ka^2 \quad \therefore \ r_2 = 4r_1$
So, the rate of reaction becomes 4 times.

96. (4) The molecularity of a reaction is the number of reactant molecules taking part in a single step of the reaction.
Therefore, the reaction involving two different reactants can never be unimolecular.

97. (3) $t_{1/4} = \dfrac{2.303}{k}\log\dfrac{1}{3/4} = \dfrac{2.303}{k}\log\dfrac{4}{3}$

$= \dfrac{2.303}{k}(2 \times 0.301 - 0.4771) = \dfrac{0.29}{k}$

98. (4) The velocity constant depends on temperature only. It is independent of concentration of reactants.

99. (2) $r = k[O_2][NO]^2$. When the volume is reduced to 1/2, the conc. will double.
$\therefore$ New rate $= k[2O_2][2NO]^2 = 8k[O_2][NO]^2$
The new rate increases to eight times of its initial value.

100. (4) $t_{1/2} = 3$ h, $T = 18$ h

$\because T = n \times t_{1/2} \ \therefore n = \dfrac{18}{3} = 6$

Initial mass $(C_0) = 256$ g

$\therefore C_n = \dfrac{C_0}{2^n} = \dfrac{256}{(2)^6} = \dfrac{256}{64} = 4$ g

101. (1) $Rt = \log C_0 - \log C_t$

$\Rightarrow t = \dfrac{-\log C_t}{R} + \dfrac{\log C_0}{R}$

It is clear from the equation that if we plot a graph between $\log C_t$ and time, a straight line with a slope equal to $-1/R$ and intercept equal to $\log[C_0]/R$ will be obtained.

102. (3) As the concentration of reactant decreases from 0.8 to 0.4 in 15 minutes, hence the $t_{1/2}$ is 15 minutes. To fall the concentration from 0.1 to 0.025, we need two half lives, i.e., 30 minutes.

103. (1) $N_t = N_0\left(\dfrac{1}{2}\right)^n$ where n is number of half life periods.

$n = \dfrac{\text{Total time}}{\text{half life}} = \dfrac{24}{4} = 6$

$\therefore N_t = 200\left(\dfrac{1}{2}\right)^6 = 3.125$ g

104. (1) Order is the sum of the power of the concentration terms in rate law expression.
Hence the order of reaction is $= 1 + 2 = 3$

105. (4) $t_{1/2} = 5$ years, $T = 15$ years.

Hence total number of half life periods $= \dfrac{15}{5} = 3$

$\therefore$ Amount left $= \dfrac{64}{(2)^3} = 8$ g

106. (25) The hydrolysis of sucrose is given by reaction:-
Sucrose + $H_2O \rightarrow$ glucose + fructose
For first order reaction, the amount of substance remaining after n half lives is given by the formula.

$[R] = \dfrac{[R_0]}{2^n}$ ...1

$\because$ [R] is the concentration remaining at time t
$\because$ $[R_0]$ is the initial concentration
$\because$ n is the number of half lives $\left(n = \dfrac{t}{t_{1/2}}\right)$

Given: $t_{1/2} = 3$ hours t = 6 hours

$\therefore n = \dfrac{6 \text{ hours}}{3 \text{ hours}} = 2$

Now, substituting the value of n in (1) we get:

$[R] = \dfrac{[R_0]}{2^2} = \dfrac{[R]}{[R_0]} = \dfrac{1}{4}$

$\therefore$ Percentage of sucrose remaining after 6 hours is $\dfrac{1}{4} \times 100 = 25\%$

107. (90) Straight line starting from origin. The given graph follow zero order kinetics.

For zero order $t_{1/2} = \dfrac{A_0}{2K}$

$y = mx$
$t_{1/2} \propto A_0$

$\dfrac{(t_{1/2})_1}{(t_{1/2})_2} = \dfrac{(A_0)_1}{(A_0)_2}$

$\Rightarrow \dfrac{240}{x} = \dfrac{4 \times 10^{-3}}{1.5 \times 10^{-3}}$

$\Rightarrow x = \dfrac{240 \times 1.5}{4} = 90.$

108. (46) $t_{1/2} = 6.93$ min

$\Rightarrow K_1 = \dfrac{0.693}{6.93} \Rightarrow 0.1 \text{ min}^{-1}$

$t_{99\%} = \dfrac{1}{K}\ln\left[\dfrac{A_0}{A_t}\right]$

$= \dfrac{2.303}{0.1}\log\left[\dfrac{100}{1}\right]$ [log(100) = 2]

$= \dfrac{2 \times 2.303}{0.1} \Rightarrow 46.06 \text{ min} \approx 46$

109. (7) $k = \dfrac{1}{t}\ln\left[\dfrac{A_0}{A_t}\right]$

$k = \dfrac{1}{20}\ln\left[\dfrac{0.6500}{0.00065}\right] = \dfrac{3\ln 10}{20} \text{ min}^{-1}$

Also, $x = \dfrac{1}{k}\ln\left[\dfrac{A_0}{A_x}\right] = \dfrac{20}{3\ln 10}\ln\left(\dfrac{0.6500}{0.0650}\right)$

$= \dfrac{20}{3} \text{ min} \Rightarrow 6.7 \text{ min}$

110. (102) $t_{25\%} = \dfrac{1}{k}\ln\dfrac{100}{75}$

$= \dfrac{245}{\ln 2} \times (\ln 4 - \ln 3) \quad \left(k = \dfrac{\ln 2}{t_{1/2}}\right)$

$= \dfrac{245 \times (2 \times \log 2 - \log 3)}{\log 2}$

$t_{25\%} = 101.66$ days

111. (10) For the thermal decomposition of reaction AB which follows zero order reaction.
$[AB]_0 - [AB]_t = kt$
$0.60 - 0.55 = k(100)$
$k = 5 \times 10^{-4}$

Half life $(t_{1/2}) = \dfrac{[AB]_0}{2k}$

$= \dfrac{0.60}{2 \times 5 \times 10^{-4}} = 600 \text{ sec} = 10 \text{ min}$

112. (897)

Given : $2N_2O_5(g) \rightarrow 2N_2O_4(g) + O_2(g)$

	$2N_2O_5(g)$	$2N_2O_4(g)$	$O_2(g)$
t = 0	0.6	0	0
t = 100s	0.6 – y	y	y/2

$P_{Total} = (0.6 - y + y + \dfrac{y}{2}) = 0.6 + \dfrac{y}{2}$

As given in question,
$k = 4.606 \times 10^{-2} \text{ sec}^{-1}$

$kt = \ln\dfrac{0.6}{0.6 - y} = 2.303\log\dfrac{0.6}{0.6 - y}$

$4.606 \times 10^{-2} \times 100 = 2.303 \log\dfrac{0.6}{0.6 - y}$

$\Rightarrow y = 0.594$

$\Rightarrow P_{Total} = 0.6 + \dfrac{0.594}{2} = 0.897$ atm

$= 897 \times 10^{-3}$ atm

Thus, $x = 897 \times 10^{-3}$ and answer is 897.

113. (1) For 1st order reaction

$A \xrightarrow{t=x} \dfrac{A}{2} \xrightarrow{t=x} \dfrac{A}{4}$

t = 2x; 75% life = 2 × 50% life
So, order with respect to A will be first order.
So, order with respect to B will be zero.
Overall order of reaction = 1 + 0 = 1

114. (3) $k = \dfrac{1}{t_{99.9\%}}\ell n\left(\dfrac{100}{0.1}\right) = \dfrac{1}{t_{90\%}}\ell n\left(\dfrac{100}{10}\right)$

$t_{99.9\%} = t_{90\%}\dfrac{\ell n(10^3)}{\ell n 10}$ [in $a^b = b \ln a$]

$t_{99.9\%} = t_{90\%} \times 3$

115. (17) $A + B \rightarrow C$ Reaction ...(1)
$P \rightarrow Q$ Reaction ...(2)

$K_1 = \dfrac{0.693}{(t_{1/2})_1}$; $K_2 = \dfrac{0.693}{(t_{1/2})_2}$

$\dfrac{K_2}{K_1} = \dfrac{(t_{1/2})_1}{(t_{1/2})_2} = \dfrac{5}{2}$

Time required

$\dfrac{(t_{2/3})_1}{(t_{4/5})} = \dfrac{K_2 \log 3}{K_1 \log 5}$

$\Rightarrow = \dfrac{5}{2}\dfrac{0.477}{0.699} = 1.7 = 17 \times 10^{-1}$

116. (3) $r = k[A]^x[B]^y$

From (I) $6\times10^{-3} = k[0.1]^x[0.1]^y$ ...(1)

From (IV) $2.4\times10^{-2} = k[0.4]^x[0.1]^y$ ...(2)

On dividing (2) by (1) we get

$\Rightarrow 4 = (4)^x \Rightarrow x = 1$

$r = k[A]^x[B]^y$

From (III) $2.88\times10^{-1} = k[0.3]^x[0.4]^y$ ...(3)

From (II) $7.2\times10^{-2} = k[0.3]^x[0.2]^y$ ...(4)

On dividing (3) by (4) we get

$\Rightarrow 4 = 2^y \Rightarrow y = 2$

Overall order $= x + y = 1 + 2 = 3$

117. (315) $2A(g) + B(g) \rightarrow C(g)$

As this is single step reaction,

$r = k_f[A]^2[B]$

	2A(g)	+ B(g)	→ C(g)
t = 0	1.5	0.7	0
t = 't'	1.5 – 2x	0.7 – x	x

$\Rightarrow x = 0.5$ atm

$r_1 = k_r(0.5)^2(0.7)$; $r_2 = k_r(0.5)^2(0.2)$

$$\frac{r_1}{r_2} = \frac{(1.5)^2(0.7)}{(0.5)^2(0.2)} = \frac{1.5\times1.5\times0.7}{0.5\times0.5\times0.2}\ \frac{r_1}{r_2}$$

$$= \frac{15\times15\times7}{5\times5\times2} = \frac{63}{2} = 31.5 = 315\times10^{-1}$$

118. (100) $k = \dfrac{k_1k_2}{k_3}$

$$Ae^{\frac{-E_a}{RT}} = \frac{A_1e^{\frac{-E_{a_1}}{RT}}A_2e^{\frac{-E_{a_2}}{RT}}}{A_3e^{\frac{-E_{a_3}}{RT}}}$$

$$Ae^{\frac{-E_a}{RT}} = \frac{A_1A_2}{A_3}e^{\frac{-(E_{a_1}+E_{a_2}+E_{a_3})}{RT}}$$

$E_a = E_{a_1} + E_{a_2} - E_{a_3} \Rightarrow 400 = 300 + 200 - Ea_3$

$E_{a_3} = 100$ kJ/mol

Note: Direct trick: $k_1 \times k_2 \rightarrow$ Multiplication factor will act added and on division k_3 will act substracted

$E_a = E_{a_1} + E_{a_2} - E_{a_3}$

119. (50) $A + B \rightarrow C$

Since, $[A] = [B] \Rightarrow r = k[A]$

Apply first order rate equation.

$$k = \frac{2.303}{t}\log\frac{1}{0.1}$$

$$\Rightarrow 4.6\times10^{-2} = \frac{2.303}{t} \Rightarrow t = 50 \text{ sec.}$$

120. (17190)

$$\frac{^{14}C}{^{12}C} = \frac{1}{8}\ ;\ t_{1/2}(^{14}C) = 5730 \text{ years}$$

$$\frac{N}{N_o} = \frac{1}{8} = \frac{1}{2^{t/t_{1/2}}} \Rightarrow \frac{N_t}{N_o} = \frac{1}{2^{t/t_{1/2}}}$$

Age of wood = t = ?

$$\frac{1}{8} = \frac{1}{2^{t/5730}} \Rightarrow \frac{1}{2^3} = \frac{1}{2^{t/5730}} \Rightarrow \frac{t}{5730} = 3$$

t = 17190 years

121. (2)

	A	⟶ 2B	+ C
t = 0	0.1	–	–
t = 115 sec	0.1 – x	2x	x

$0.1 - x + 2x + x = 0.28$

$0.1 + 2x = 0.28 \Rightarrow 2x = 0.28 - 0.1 \Rightarrow x = 0.09$

$$K = \frac{1}{t}\ln\left(\frac{0.1}{0.1-0.09}\right)$$

$$= \frac{2.303}{115}\log\left(\frac{0.1}{0.01}\right) = \frac{2.303}{115}\times1$$

$K = 0.0200 = 2\times10^{-2}\ s^{-1}$

122. (399)

$r = k[A]$

So this is first order reaction.

For 50% decomposition, time req is 120 min

$\therefore\ t_{1/2} = 120$ min

$$\therefore\quad k = \frac{0.693}{t_{1/2}} = \frac{0.693}{120}$$

$k = 0.00577 = 5.77\times10^{-3}\ \text{min}^{-1}$

For 90% decomposition

$$k = \frac{2.303}{t}\log\frac{a}{(a-x)}$$

$$\Rightarrow t = \frac{2.303}{0.00577}\log\left(\frac{100}{10}\right)$$

t = 399 min

123. (24)

124. (63) Radioactive decays are first order reactions.

$$\text{Thus, } K = \frac{0.693}{t_{1/2}} = \frac{0.693}{36\ h} = 0.01925\ h^{-1}$$

$$\Rightarrow t = \frac{2.303}{K}\log\frac{a}{a-x}$$

$$\Rightarrow 24\ h = \frac{2.303}{0.01925\ h^{-1}}\log\frac{a}{a-x}$$

$$\Rightarrow \frac{a}{a-x} = \frac{24\times0.01925}{2.303} = 0.2006$$

$$\rightarrow \frac{a}{a-x} = \text{antilog}(0.2006) = 1.587.$$

$$\Rightarrow \text{Fraction left} = \frac{a-x}{a} = \frac{1}{1.587} = 0.630 = 63.0\times10^{-2}$$

125. (2) Let, $R = k[HI]^n$

Using any two of given data,

$$\frac{R_2}{R_1} = \left[\frac{[HI]_2}{[HI]_1}\right]^n \Rightarrow \frac{3\times10^{-3}}{7.5\times10^{-4}} = \left(\frac{0.01}{0.005}\right)^n$$

$\Rightarrow n = 2$

126. (10)

$$\frac{t_{99.9\%}}{t_{1/2}} = \frac{\frac{2.303}{k}\left(\frac{a}{a-x}\right)}{\frac{2.303}{k}\log 2}$$

$$= \frac{\log\left(\frac{100}{100-99.9}\right)}{\log 2} = \frac{\log 10^3}{\log 2} = \frac{3}{0.3} = 10$$

$t_{99.9\%} = 10t_{1/2}$

127. (3) $A_t = A_0 \times \dfrac{12.5}{100} = \dfrac{A_0}{8}$

[87.5% complete] (1st order rate law)

$$A_0 \xrightarrow{t_{1/2}} \frac{A_0}{2} \xrightarrow{t_{1/2}} \frac{A_0}{4} \xrightarrow{t_{1/2}} \frac{A_0}{8}$$

$t_{87.5} = 3t_{1/2}\ \therefore\ x = 3$

128. (3) Rate determing step : $NOBr_2 + NO \rightarrow 2NOBr$

$r = K[NOBr_2][NO]$ ...(i)

$$K_{eq} = \frac{[NOBr_2]}{[NO][Br_2]} \quad ...(ii)$$

From (i) & (ii)

$r = K.K_{eq.}[NO][Br_2][NO]$

$r = K'[NO]^2[Br_2]$; Overall order = 3

129. (2) $\dfrac{1}{t_{1/2}} = \dfrac{1}{3} + \dfrac{1}{12} = \dfrac{4+1}{12} = \dfrac{5}{12}$

$$\Rightarrow t_{1/2} = \frac{12}{5}\text{ min} = 2.4 \sim 2$$

130. (17) $K = 20\ \text{min}^{-1}$

$$t = \frac{2.303}{K}\log\frac{[a_0]}{a_0\times\frac{1}{32}}$$

$$t = \frac{2.303}{20}\log 32 = \frac{2.303}{20}\times\log 2^5$$

$$= \frac{2.303}{20}\times5\times0.30 = 0.1727 \approx 17\times10^{-2}\ \text{min}$$

131. (623) For first order section

$$t = \frac{2.303}{k}\log\frac{[A]_0}{[A]_t} = \frac{2.303}{2.011\times10^{-3}}\log\frac{7}{2}$$

$$= \frac{2.303\times0.544}{2.011\times10^{-3}} = 622.989 \approx 623$$

132. (1350)

$$\frac{t_1}{t_2} = \frac{\frac{1}{K}\ln\frac{a_0}{0.4\ a_0}}{\frac{1}{K}\ln\frac{a_0}{0.1\ a_0}}$$

$$\frac{540}{t_2} = \frac{\ln\frac{10}{4}}{\ln 10} \Rightarrow \frac{540}{t_2} = \frac{\log10 - \log4}{\log10}$$

$$\frac{540}{t_2} = \frac{1-0.6}{1} \Rightarrow \frac{540}{t_2} = 0.4$$

$$\Rightarrow t_2 = \frac{540}{0.4} = 1350 \text{ sec}$$

133. (1) (A) For zero order $t_{1/2} = \dfrac{[A]_0}{2K}$ as concentration decreases half life decreases. (Correct statement)

134. (15) Suppose, at time 't', the concentration of both A and B will be same.

$[A]_t = [A]_o e^{-k_A t}$

$$k_A = \frac{0.693}{t_{1/2}} = \frac{0.693}{15};\ \therefore [A]_t = [A]_0 e^{\left(-\frac{0.693}{15}\right)t}$$

and $[B]_t = [B]_o e^{-k_B t} = 4[A]_o e^{\left(-\frac{0.693}{5}\right)t}$

At time t, $\therefore\ [A]_t = [B]_t$

$$\text{or, } [A]_o e^{\left(-\frac{0.693}{15}\right)t} = 4[A]_o e^{\left(-\frac{0.693}{5}\right)t}$$

$$\text{or } -\frac{0.693}{15}t = \ln 4 - \frac{0.693}{5}t$$

$$\text{or } 0.693t\left(\frac{1}{5} - \frac{1}{15}\right) = 2\ln 2 \text{ or } t = 15 \text{ min.}$$

135. (75) Assume reaction starts with 1 mole A

$$t_{1/2} = \frac{a}{2k},\ k = \frac{1}{2\times50}$$

For zero order reaction. $[R_0] - [R] = Kt$

For 75% completion

$$a - \frac{a}{4} = kt \Rightarrow t = \frac{3a}{4k} = \frac{3}{4}\times\frac{100}{a} = 75$$

136. (2) (B) and (C) are correct.

137. (2) As unit of rate constant is $(\text{conc.})^{1-n}\ \text{time}^{-1}$

$\Rightarrow (L\ mol^{-1}) \Rightarrow 1 - n = -1\ \therefore n = 2$

138. (60) $t_{1/2} = 30$ min

Time taken for $\frac{3}{4}$th completion of the reaction

$= 2t_{1/2} = 30 \times 2 = 60$ min

139. (200) $k_A = \frac{\ln 2}{100}; k_B = \frac{\ln 2}{50}$

$A_t = A_0 \times e^{-k_A t}$

$A_t = A_0 \times e^{\left(\frac{-\ln 2}{100}\times t\right)}; \; B_t = B_0 \times e^{\left(\frac{-\ln 2}{50}\times t\right)}$

Given, $A_0 = B_0$ and $A_t = 4B_t$

$$e^{\frac{-\ln 2}{100}\times t} = 4 \times e^{\frac{-\ln 2}{50}\times t} \Rightarrow e^{\frac{\ln 2}{100}\times t} = 4$$

$$\Rightarrow \frac{\ln 2}{100}\times t = \ln 4 = 2\ln 2 \Rightarrow t = 200\text{sec.}$$

140. (1) $t = \frac{1}{k}\ln\left(\frac{a}{a-x}\right)$

$$k = \frac{\ln 2}{t_{1/2}} = \frac{\ln 2}{30} \Rightarrow 100 = \frac{30}{\ln 2}\ln\left(\frac{1}{w}\right)$$

$\frac{1}{w} = 10 \Rightarrow w = 1 \times 10^{-1}$ µg

141. (100) $A \to B$

For first order reaction; $k = \frac{2.303}{t}\log\left[\frac{A_0}{A_t}\right]$

$$\Rightarrow \frac{0.693}{t_{1/2}} = \frac{2.303}{2}\log\left(\frac{A_0}{A_t}\right)$$

$$\Rightarrow \frac{0.693}{0.3010} = \frac{2.303}{2}\log\frac{A_0}{A_t}$$

$$\log\frac{A_0}{A_t} = 2 \; \therefore \; \frac{A_0}{A_t} = 10^2 = 100$$

142. (2) When initial pressure of NO increases by a factor of 2 ($2 \times 20.1 \approx 40$) then initial rate increases by a factor of 4 ($4 \times 0.033 \approx 0.135$). Therefore, order of reaction w.r.t. NO = 2

143. (165) $t_{1/2} = 70$ min; for Ist order reaction,

$$k = \frac{0.693}{t_{1/2}} = \frac{0.693}{70\times 60}s^{-1} = \frac{6930}{7\times 6}\times 10^{-6}s^{-1}$$

$= 165 \times 10^{-6}\,s^{-1}$

144. (2)

- For zero order, $t_{1/2}$ decrease with decrease in concentration of reactions.
- For first order, $t_{1/2}$ is independent of concentration of reaction
- For second order, $t_{1/2}$ increases with decrease in concentration of reactant.

$$\frac{(t_{1/2})_1}{(t_{1/2})_2} = \left(\frac{[A_0]_2}{[A_0]_1}\right)^{1-n} \Rightarrow \frac{100}{50} = \left(\frac{0.5}{1.0}\right)^{1-n}$$

$$\Rightarrow 2 = \left(\frac{1}{2}\right)^{1-n} \Rightarrow 2 = 2^{n-1}$$

$\Rightarrow n - 1 = 1 \quad \Rightarrow \quad n = 2$

145. (1) Since the half life for the decomposition of gas is not dependent on initial concentration, the order of the reaction is one.

146. (2) For first order reaction

$$k = \frac{1}{t}\ln\left(\frac{P_\circ}{P}\right)$$

$$\ln\left(\frac{P_\circ}{P}\right) = kt \Rightarrow \ln\left(\frac{P}{P_\circ}\right) = -kt$$

On comparing with straight line equation

$y = mx$; k = slope $= 3.465 \times 10^4$

$$t_{1/2} = \frac{\ln 2}{k} = \frac{0.693}{3.465\times 10^4} = 2\times 10^{-5}\,s$$

147. (16) $t = \frac{1}{k}\ln\left(\frac{a_0}{a_0 - n}\right)$ or $k = \frac{\ln(2)}{t_{1/2}}$

$$t_{67\%} = \frac{1}{k}\ln\left(\frac{1}{1-0.67}\right) = \frac{t_{1/2}}{\ln 2}\times\ln\left(\frac{1}{1-\frac{2}{3}}\right)$$

$$t_{67\%} = \frac{t_{1/2}}{\log 2}\times\log 3 = \frac{t_{1/2}\times 0.4771}{0.301}$$

$\Rightarrow t_{67\%} = 1.585 \times t_{1/2}$ or $x \times 10^{-1} = 1.585$

$\Rightarrow x = 15.85 \simeq 16$

148. (75) $t_{1/2} = \frac{0.693}{k}$ or $k = \frac{0.693}{t_{1/2}}$

$$t = \frac{2.303}{k}\log\frac{[A_0]}{[A]_t} = \frac{t_{1/2}}{0.3}\log\frac{[A_0]}{[A]_t}$$

$$83 = \frac{200}{0.3}\log\frac{[A_0]}{[A]_t} \text{ or } 0.125 = \log\frac{[A_0]}{[A]_t}$$

$$\frac{[A_0]}{[A]_t} = 1.333 \cong \frac{4}{3}$$

$$\therefore \frac{[A]_t}{[A]_0}\times 100 = \frac{3}{4}\times 100 = 75\%$$

149. (0) $t_{1/2} \propto \frac{1}{[P_0]^{n-1}}$

$$\frac{(t_{1/2})_1}{(t_{1/2})_2} = \frac{(P_2)^{n-1}}{(P_1)^{n-1}} \Rightarrow \frac{340}{170} = \left(\frac{27.8}{55.5}\right)^{n-1}$$

$$\Rightarrow 2 = \frac{1}{(2)^{n-1}} \Rightarrow n = 0$$

150. (1) Rate $= k[NO]^x[H_2]^y$

$7 \times 10^{-9} = k \times (8 \times 10^{-5})^x (8 \times 10^{-5})^y$ ...(i)

$2.1 \times 10^{-8} = k \times (24 \times 10^{-5})^x (8 \times 10^{-5})^y$ ...(ii)

Now dividing equation (i) by equation (ii):

$$\frac{7\times 10^{-9}}{2.1\times 10^{-8}} = \left(\frac{8\times 10^{-5}}{24\times 10^{-5}}\right)^x$$

$$\text{or} \quad \frac{1}{3} = \left(\frac{1}{3}\right)^x \Rightarrow x = 1$$

First check the unit of rate constant

151. (2) $k = \frac{2.303}{t}\log\frac{a}{a-x}$

$$\frac{2.303}{t_{50\%}}\log\frac{100}{100-50} = \frac{2.303}{t_{75\%}}\log\frac{100}{100-75}$$

$$t_{75\%} = 2t_{50\%}.$$

152. (654)

	A ⟶	2B
t = 0	1	0
t = 100 min	1 − x	2x
	= 0.9mol	= 0.2mol

$$2x = 0.2 \Rightarrow x = \frac{0.2}{2} = 0.1$$

$$k = \frac{2.303}{t}\log\frac{[A_\circ]}{[A_t]}$$

$$\Rightarrow k = \frac{2.303}{100}\log\frac{1}{0.9} \Rightarrow k = 1.059\times 10^{-3}$$

$$t_{1/2} = \frac{0.693}{k} = \frac{0.693}{1.059\times 10^{-3}} = 654.3$$

153. (7) For the first order reaction $kt = \ln\frac{[R]_0}{[R]}$

$$k\times 60 = \ln\frac{(2.4\times 10^{-2})}{(1.6\times 10^{-2})}$$

$$= 2.303 \times (\log 3 - \log 2)\frac{1}{60}$$

$$= 2.303 \times (0.477 - 0.301)\frac{1}{60}$$

$\Rightarrow k = 6.7 \times 10^{-3}\,\text{min}^{-1} \approx 7 \times 10^{-3}\,\text{min}^{-1}$

154. (1)

	$PCl_5(g)$ $\xrightarrow[300K]{1^{st}\text{ order}}$	$PCl_3(g)$ +	$Cl_2(g)$
t = 0	50 M	0	0
t = 120 min	10 M	40 M	40 M

$$\Rightarrow k = \frac{2.303}{t}\log\frac{[A_0]}{[A_t]} \Rightarrow k = \frac{2.303}{120}\log\frac{50}{10}$$

$$\Rightarrow k = \frac{2.303}{120}\times 0.6989 = 0.013413\,\text{min}^{-1}$$

$= 1.3413 \times 10^{-2}\,\text{min}^{-1} \approx 1 \times 10^{-2}\,\text{min}^{-1}$.

155. (3)

156. (10) This question is incomplete, order is not given. If we consider, the given reaction to be first order, then

$$t_{1/2} = \frac{0.693}{k} = \frac{\ln 2}{k} \Rightarrow k = \frac{\ln 2}{1}$$

$$\therefore t = \frac{1}{k}\ln\frac{100}{100-99.9} \Rightarrow t = \frac{1}{\ln 2}\ln 10^3$$

$$\Rightarrow t = \frac{1\times 3\times\ln 10}{\ln 2} = \frac{3\times 2.3}{0.69} = 10\text{ min.}$$

157. (2) A → Products

Rate constant of a first order is given as,

$$k = \frac{2.303}{t}\log\frac{[A]_0}{[A]_t}$$

$$= \frac{2.303}{570}\log\frac{100}{32} = 2\times 10^{-3}\,s^{-1}$$

158. (108) Initially: $[A_0]=[B_0]=a$

After time 't' min.: $[A]=16[B]$

$[A]=[A_0]e^{-k_A t}$

$[B]=[B_0]e^{-k_B t} \Rightarrow a\cdot e^{-k_A t}=16a\cdot e^{-k_B t}$

$\Rightarrow e^{-(k_A-k_B)t}=16 \Rightarrow (k_B-k_A)t=\ln 16$

$\because k_A=\frac{\ln 2}{t_{1/2}}=\frac{\ln 2}{54};\ \therefore k_B=\frac{\ln 2}{18}$

$\Rightarrow \left(\frac{\ln 2}{18}-\frac{\ln 2}{54}\right)t=\ln 16$

$\Rightarrow \ln 2\left(\frac{1}{18}-\frac{1}{54}\right)t=4\ln 2$

$\Rightarrow t=\frac{54\times 18\times 4}{36}=108$ min

159. (81) Given:

$$\underset{\text{Sucrose}}{C_{12}H_{22}O_{11}}+H_2O\xrightarrow{\text{First order}}\underset{\text{Glucose}}{C_6H_{12}O_6}+\underset{\text{Fructose}}{C_6H_{12}O_6}$$

$t_{1/2}=3.33$ h

$t=0,\ a=[A]_0;\ t=9$ h., $(a-x)=[A]_t$

From first order kinetic: $\frac{k\times t}{2.303}=\log\frac{[A]_0}{[A]_t}$

$\Rightarrow \frac{0.693\times 9}{3.33\times 2.303}=\log\left(\frac{1}{f}\right)\ \left(\because t_{1/2}=\frac{0.693}{k}\right)$

$\Rightarrow \log\left(\frac{1}{f}\right)=0.813=81.3\times 10^{-2}\approx 81\times 10^{-2}$

160. (26) $k=\frac{1}{t}\cdot\ln\frac{[A]_0}{[A]_t}\Rightarrow t=\frac{1}{k}\ln\left(\frac{100}{60}\right)$

$\Rightarrow t=\frac{1}{3.3\times 10^{-4}}(\ln 10-\ln 6)$

$\Rightarrow t=1548.49$ sec $=25.81$ min ≈ 26 min.

161. (60) $t=\frac{2.303}{k}\log\left[\frac{100}{100-x\%}\right]$

$t_{75\%}=\frac{2.303}{k}\log\left[\frac{100}{25}\right]=90$

$t_{60\%}=\frac{2.303}{k}\log\left[\frac{100}{40}\right]$

$\frac{t_{75\%}}{t_{60\%}}=\frac{2\log 2}{\log 2.5}\Rightarrow\frac{90}{t_{60\%}}=\frac{2\times 0.3}{0.4}$

$\Rightarrow t_{60\%}=\frac{90\times 4}{6}=60$ min.

162. (23.03) $t_{1/2}=6.93$ years, $a=10^{-6}$ g

$t_{1/2}=\frac{0.693}{k}\Rightarrow k=\frac{0.693}{t_{1/2}}=\frac{0.693}{6.93}=0.1$

For 1st order reaction,

$t=\frac{2.303}{k}\log\frac{a}{a-x}$

$=\frac{2.303}{0.1}\log\frac{10^{-6}}{10^{-7}}=\frac{2.303}{0.1}=23.03$ years

163. (3) Statement-I: $\ln\left(\frac{K_2}{K_1}\right)=\frac{E_a}{R}\left[\frac{1}{T_1}-\frac{1}{T_2}\right]$

$\ln\left(\frac{K_2}{K_1}\right)=\frac{12.6\times 10^3\times 4.2}{8.314}\left[\frac{1}{298}-\frac{1}{308}\right]$

$\ln\left(\frac{K_2}{K_1}\right)=6.365\times 10^3\left[\frac{308-298}{298\times 308}\right]$

$\Rightarrow \ln\left(\frac{K_2}{K_1}\right)=0.693\Rightarrow\frac{K_2}{K_1}=2$

Statement-II: For Ist order

$t_{1/2}=\frac{\ln 2}{K};\ t_{1/2}\propto [A]^0=$ constant.

Statement-I is true but statement-II is false.

164. (2) $\Delta G°=\Delta H°-T\Delta S°$

$-2.303\,RT\log K=\Delta H°-T\Delta S°$

$\log_{10}K=-\frac{\Delta H°}{2.303RT}+\frac{\Delta S°}{2.303R}$

Intercept $=\frac{\Delta S°}{2.303R}$

Slope $=-\frac{\Delta H°}{2.303R}$

165. (2) $\frac{k_{catalysed}}{k_{uncatalysed}}=\frac{Ae^{-Ea_1/RT}}{Ae^{-Ea_2/RT}}=e^{\frac{\Delta E_a}{RT}}$

$\ell n\frac{k_{catalysed}}{k_{uncatalysed}}=\frac{\Delta E_a}{RT}$

$\log\frac{k_{catalysed}}{k_{uncatalysed}}=\frac{\Delta E_a}{2.303RT}$

$=\frac{10\times 1000}{2.303\times 8.314\times 300}$

$\log\frac{k_{catalysed}}{k_{uncatalysed}}=1.741$

166. (3) Given:

$A\rightarrow D;\ \Delta H=(+)ve=E_D-E_A\ \therefore E_D>E_A$

Steps:

$A\rightarrow B;\ \Delta_r H=(+)ve\Rightarrow E_B>E_A$

$B\rightarrow C;\ \Delta_r H=(-)ve\Rightarrow E_C<E_B$

$C\rightarrow D;\ \Delta_r H=(-)ve\Rightarrow E_D<E_C$

167. (4)

(A) In the Arrhenius equation, $K=A\times e^{-Ea/RT}$, the factor $e^{-Ea/RT}$ corresponds to the fraction of molecules that have kinetic energy greater than E_a.

(D) $\log k=\log A-\frac{E_a}{2.303RT}$

The plot of log k vs. $\frac{1}{T}$ gives straight line with slope

$=-\frac{E_a}{2.303R}$

So, 'A' and 'D' statements are incorrect.

168. (1) $k=A\,e^{-Ea/RT}$

$\log k=\log A-\frac{E_a}{2.303RT}$

For graph between log k with $\frac{1}{T}$

$|\text{Slope of curve}|=\frac{E_a}{2.303R}$

From given graph:

Magnitude of slope $\Rightarrow (2)>(1)>(3)$

Hence $Ea_2>Ea_1>Ea_3$

169. (1) $A_2+B_2\rightleftharpoons 2AB$

$E_f=180$ kJ mol^{-1}

$E_b=200$ kJ mol^{-1}

$\Delta H=E_f-E_b=-20$ kJ mol^{-1}

In presence of catalyst:

$E_f=180-100=80$ kJ mol^{-1}

$E_b=200-100=100$ kJ mol^{-1}

Catalyst does not change ΔH or ΔG of a reaction. Catalyst only reduces activation energy of the reaction.

170. (3) Arrhenius equation holds true for elementary as well as complex reactions.

Unit of A is same as unit of k. Rate of reaction is high if activation energy is low.

A and E_a are temperature independent.

171. (1) A net release of heat is observed in the process.

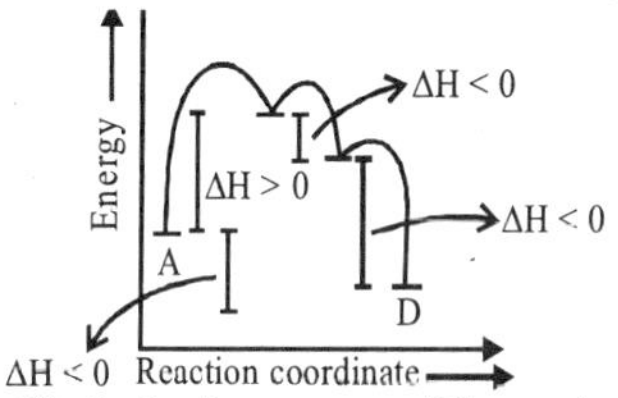

172. (3) Activation energy of forward reaction $=E_1+E_2$

Energy of product > Energy of reactant

Stability $\propto\frac{1}{\text{Energy}}$

173. (2) By using positive catalyst:

(i) ΔH does not change, (ii) Activation energy decreases

174. (2) Step II with highest activation energy is rate Determining Step.

No. of activated complex = 3

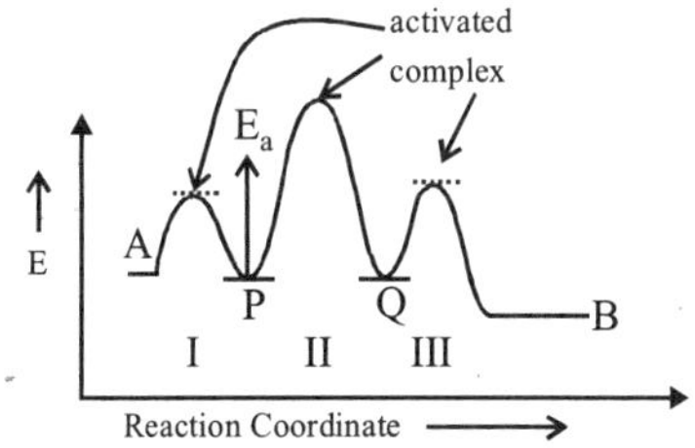

Number of intermediates = 2

175. (4) Arrhenius equation:

$k=Ae^{-E_a/RT}$

$\ln k=\ln A-\left(\frac{E_a}{R}\right)\frac{1}{T}$

$\Rightarrow \ln k=\ln A-\left(\frac{E_a}{R\times 10^3}\right)\times\frac{10^3}{T}$

Slope of graph $=\frac{-E_a}{R\times 10^3}=\frac{-10}{5}$

$E_a=2R\times 10^3$ J $=2R$ kJ mol^{-1}

176. (1) $A \xrightarrow{700K(k_1)} \text{Product}$

$A \xrightarrow[\text{Catalyst}]{500K(k_2)} \text{Product}$

Given: $k_1 = k_2$

$$Ae^{-\frac{E_{a_1}}{RT_1}} = Ae^{-\frac{E_{a_2}}{RT_2}}$$

$$\frac{E_{a_1}}{T_1} = \frac{E_{a_2}}{T_2}$$

$$E_{a_2} = E_{a_1} - 30 \Rightarrow E_{a1} = E_{a_2} + 30$$

$$\frac{E_{a_2}+30}{T_1} = \frac{E_{a_2}}{T_2}$$

$$\frac{E_{a_2}+30}{700} = \frac{E_{a_2}}{500} \Rightarrow 150 = 2E_{a_2}$$

$$E_{a_2} = 75 \text{kJ/mol}$$

177. (1) The rate constant of a reaction is given by, $k = Ae^{-E_a/RT}$

The rate constant in presence of catalyst is given by,

$$k' = Ae^{-E_a'/RT} \text{ or } \frac{k'}{k} = e^{-(E_a'-E_a)/RT}$$

$$10^6 = e^{-(E_a'-E_a)/RT};\ \ln 10^6 = -\frac{(E_a'-E_a)}{RT}$$

$$E_a' - E_a = -6(2.303)RT$$

178. (3) Arrhenius equation, $k = Ae^{-E/RT}$

$$\log k = \log A - \frac{E}{2.303\ RT};\ \text{slope} = -\frac{E}{2.303\ R}$$

∴ More negative the slope, greater will be the E.

So correct order is: $E_c > E_a > E_d > E_b$

179. (1) So, $\log\frac{k_2}{k_1} = \frac{E_a}{2.303R}\left(\frac{1}{T_1}-\frac{1}{T_2}\right)$

$$\log\frac{1}{2.5\times10^{-4}} = \frac{E_a}{8.314\times2.303}\left(\frac{1}{600}-\frac{1}{800}\right)$$

$$\Rightarrow 3.6 = \frac{E_a}{8.314\times2.303}\times\frac{200}{600\times800}$$

$\Rightarrow E_a = 165.4$ kJ/mol

180. (1) As we can see from the graph that activation enthalpy to form D from A+B is 15 – 5 = 10 kJ mol^{-1}, whereas, to form C from A+B is 20 – 5 = 15 kJ mol^{-1}. Therefore, activation enthalpy to form C is 5 kJ more than that to form D. Due to more E_a, C is thermodynamically stable and D is kinetically stable product.

181. (3) From Arrhenius equation,

$$\ln K = \ln A - \frac{E_a}{RT}$$

$$\text{Slope} = \frac{-E_a}{R} = -4606\ K$$

$$\log\left(\frac{K_2}{K_1}\right) = \frac{E_a}{2.303R}\left(\frac{1}{T_1}-\frac{1}{T_2}\right)$$

$$\log\left(\frac{K_2}{10^{-5}}\right) = \frac{1}{2.303}\times4606\left(\frac{1}{400}-\frac{1}{500}\right)$$

$$= \frac{4606\times(100)}{2.303\times400\times500} = 1$$

$$\Rightarrow \log\left(\frac{K_2}{10^{-5}}\right) = 1;\ \frac{K_2}{10^{-5}} = \text{Antilog}(1)$$

$$\frac{K_2}{10^{-5}} = 10 \Rightarrow K_2 = 10^{-5}\times10 = 10^{-4}\ s^{-1}$$

182. (2) From Arrhenius equation,

$$k = Ae^{-E_a/RT};\ \ln k = \ln A - \frac{E_a}{RT}$$

slope = $-y$ (given) $-y = -E_a \Rightarrow E_a = y$

183. (2) From Arrhenius equation,

$k = Ae^{-E_a/RT}$

So, as E_a increases, $e^{-E_a/RT}$ decreases, k decreases and as T increases, $\frac{E_a}{RT}$ decreases, $e^{-E_a/RT}$ increases.

184. (4) $k_1 = A\cdot e^{-E_{a_1}/RT}$ ...(i)

$k_2 = A\cdot e^{-E_{a_2}/RT}$ ...(ii)

On dividing equation (ii) by (i)

$$\Rightarrow \frac{k_2}{k_1} = e^{\frac{(E_{a_1}-E_{a_2})}{RT}}$$

$$\ln\left(\frac{k_2}{k_1}\right) = \frac{E_{a_1}-E_{a_2}}{RT} = \frac{10,000}{8.314\times300} = 4$$

Trick: Remember the formula (direct)

$$\ln\left(\frac{k_2}{k_1}\right) = \frac{E_{a_1}-E_{a_2}}{RT}$$

185. (2)

186. (1) $\ln\frac{k_1}{k_2} = \frac{E_a}{R}\left(\frac{1}{T_1}-\frac{1}{T_2}\right)$

$$\ln 4 = \frac{E_a}{8.314}\left(\frac{310-300}{310\times300}\right)$$

$$2\ln 2 = \frac{E_a}{8.314}\left(\frac{310-300}{310\times300}\right)$$

$$E_a = \frac{0.693\times2\times8.314\times300\times310}{10} = 107.2 \text{ kJ/mol}$$

187. (1) $A(g) \rightleftharpoons B(g)\ \Delta H = -40$ kJ

If activation energy of the reaction = x

Since, $\frac{E_f}{E_b} = \frac{2}{3}$, therefore,

$$E_f = \frac{2x}{5} \text{ and } E_b = \frac{3x}{5}$$

$$E_b - E_f = -\Delta H = +40$$

$$\Rightarrow \frac{3x}{5} - \frac{2x}{5} = 40 \Rightarrow x = 200$$

Therefore, $E_b = \frac{3x}{5} = \frac{3\times200}{5} = 120$ kJ mol^{-1}

$$E_f = \frac{2x}{5} = \frac{2\times200}{5} = 80 \text{ kJ mol}^{-1}$$

188. (2) According to Arrhenius equation

$$\log\frac{k_2}{k_1} = \frac{E_a}{2.303R}\left(\frac{1}{T_1}-\frac{1}{T_2}\right)$$

$$\log\frac{1.3\times10^{-3}}{1.3\times10^{-4}} = \frac{E_a}{2.303\times8.314}\left[\frac{1}{373}-\frac{1}{423}\right]$$

$$1 = \frac{E_a}{2.303\times8.314}\left[\frac{1}{373}-\frac{1}{423}\right]$$

$E_a = 60$ kJ / mol

189. (2) $X \longrightarrow Y$; $\Delta H = -135$ kJ/mol,

$E_a = 150$ kJ/mol

For an exothermic reaction,

$E_{a(F.R.)} = \Delta H + E'_{a(B.R.)}$

$\Rightarrow 150 = -135 + E'_{a(B.R.)}$

$\Rightarrow E'_{a(B.R.)} = 285$ kJ/mol

190. (1) Activation energy can be calculated from the equation

$$\frac{\log k_2}{\log k_1} = \frac{E_a}{2.303R}\left(\frac{1}{T_1}-\frac{1}{T_2}\right)$$

Given $\frac{k_2}{k_1} = 2$; $T_2 = 310$ K; $T_1 = 300$ K

$$\therefore\ \log 2 = \frac{E_a}{2.303\times8.314}\left(\frac{1}{300}-\frac{1}{310}\right)$$

$E_a = 53598.6$ J/mol $= 53.6$ kJ/mol.

191. (3) Activation energy can be calculated from the equation

$$\frac{\log k_2}{\log k_1} = \frac{E_a}{2.303\ R}\left(\frac{1}{T_1}-\frac{1}{T_2}\right)$$

Given, $\frac{\log k_2}{\log k_1} = 2$ $T_2 = 308$; $T_1 = 298$

$$\therefore\ \log 2 = \frac{E_a}{2.303\times8.314}\left(\frac{1}{298}-\frac{1}{308}\right)$$

$$E_a = 52.9 \text{ kJ mol}^{-1}$$

192. (3) $k_1 = A_1e^{-Ea_1/RT}$ ...(i)

$k_2 = A_2e^{-Ea_2/RT}$ ...(ii)

On dividing eqn. (i) by eqn. (ii),

$$\frac{k_1}{k_2} = \frac{A_1}{A_2}\exp.\{(E_{a_2}-E_{a_1})/RT\} \quad ...(iii)$$

Given $E_{a_2} = 2E_{a_1}$

On substituting this value in eqn. (iii),

$k_1 = k_2A\times e^{Ea_1/RT}$ $[A = A_1/A_2]$

193. (3) In Arrhenius equation, $k = Ae^{-Ea/RT}$, E_a is the energy of activation, which is required by the colliding molecules to react resulting in the formation of products.

194. (4)

195. (4) Enthalpy of reaction $(\Delta H) = E_f - E_b$

For an endothermic reaction, $\Delta H = +$ve, hence, $E_b < E_f$

196. (2) In equation $k = Ae^{-E_a/RT}$;
A = Frequency factor
k = velocity constant, R = gas constant and E_a = energy of activation

197. (232)

Given, $K = (5.5\times10^{11}s^{-1})e^{-\frac{28000K}{T}}$

Arrheneus equation is $K = Ae^{-\frac{E_a}{RT}}$
On comparing,

$\frac{E_a}{R} = 28000$

$\Rightarrow E_a = \frac{28000\times R}{1000} = 232.4$ kJ / mol.

198. (41) We know, $K = Ae^{-E_a/RT}$

$\therefore \ln k = \ln A - \frac{E_a}{RT}$

$\ln\left(\frac{K_2}{K_1}\right) = \frac{E_a}{R}\left[\frac{T_2 - T_1}{T_1T_2}\right]$

$\ln\left(\frac{4.5\times10^3}{1.5\times10^3}\right) = \frac{60\times10^3}{8.3}\left[\frac{T_2 - 300}{300.T_2}\right]$

$\ln 3 = \frac{60\times10^3}{8.3}\left[\frac{T_2 - 300}{300.T_2}\right]$

$\Rightarrow T_2 = 314.4$ K
$\Rightarrow t = 41.4°C$

199. (5) A ⟶ B
C ⟶ D
For 1st reaction
$K_{500} = 2k_{300}$

$\log\frac{k_{500}}{k_{300}} = \frac{E_a}{2.303R}\left[\frac{1}{300} - \frac{1}{500}\right]$

$\log\frac{\frac{0.693}{2}}{k_{300}} = \frac{(E_a)_1\times200}{2.303\times R\times300\times500}$ ...(i)

For 2nd reaction

$\log\frac{k_{500}}{k_{300}} = \frac{(E_a)_2\times200}{2.303\times R\times300\times500}$

$\log\frac{0.693}{k_{300}} = \frac{\frac{E_{a_1}}{2}\times200}{2.303\times R\times300\times500}$ ...(ii)

From (i) and (ii), $k_{300K} \cong 5\times10^{-1}$

200. (57) According to Arrhenius equation

$\log_{10} K = \log A - \frac{E_a}{2.303RT}$

$\frac{E_{a_1}}{2.303R} = 1.5\times10^4$

$E_{a_1} = 1.5\times10^4\times2.303\times8.314$

$E_{a_1} = 28.7207\times10^4$ J

$E_{a_1} = 287.207$ kJ

$\therefore E_{a_2} = \frac{E_{a_1}}{5} = \frac{287.207}{5} = 57.44$ kJ.

201. (1303) Given, $k_1 = k_2$

$10^4 e^{\frac{-24000}{T}} = 10^6 e^{\frac{-30000}{T}}$

$e^{\frac{6000}{T}} = 100$

Taking natural logarithm (ln) in both sides

$\frac{6000}{T} = 2\ln10$

$T = \frac{6000}{2\times2.303}$; $T = 1302.64$ K $\Rightarrow T \approx 1303$ K

202. (8) $\ln k_1 = \ln A - \frac{E_1}{300R}$

$\ln k_2 = \ln A - \frac{E_2}{300R}$

Assuming 'A' same for both reactions.

$\ln\left(\frac{k_2}{k_1}\right) = \frac{E_1 - E_2}{300R} = \frac{20\times1000}{300\times8.3} = 8.032.$

203. (20) $k = \sqrt{\frac{k_1k_3}{k_2}}$

$A.e^{-Ea/RT} = \sqrt{\frac{A_1e^{-Ea_1/RT}\times A_3e^{-Ea_3/RT}}{A_2e^{-Ea_2/RT}}}$

Taking ln and both sides and comparing exponential terms.

$\frac{E_a}{RT} = \frac{1}{2}\times\left(\frac{E_{a_1}}{RT} + \frac{E_{a_3}}{RT} - \frac{E_{a_2}}{RT}\right)$

$E_a = (E_{a_1} + E_{a_3} - E_{a_2})/2$

$E_a = (60 + 10 - 30)/2 = 20$ kJ/mol^{-1}

(**Note:** The values of A, A_1, A_2, and A_3 should also be considered here but their values are not given so we must ignore them to get overall E_a)

204. (69) $t_{1/2} = \frac{0.693}{k}$; $K - Ae^{-E_a/RT}$

$= 10^{20}\times e^{-\frac{191.48\times10^3}{8.314\times1000}}$

$= 10^{20}\times e^{-23.031} = 10^{20}\times - e^{\ln10\times10}$

$= \frac{10^{20}}{10^{10}} = 10^{10}$ sec.

$t_{1/2} = \frac{0.693}{10^{10}} = 6.93\times10^{-11} = 69.3\times10^{-12}$ sec.

= 69.3 pico second

205. (30) $k = Ae^{-Ea/RT}$

Overall rate constant $K = \frac{k_1k_2}{k_3}$

$E_a = E_{a_1} + E_{a_2} - E_{a_3}$

$\Rightarrow Ae^{-E_a/RT} = \frac{A_1\cdot e^{-E_{a_1}/RT}\cdot A_2\cdot e^{-E_{a_2}/RT}}{A_3\cdot e^{-E_{a_3}/RT}}$

$= \left(\frac{A_1\cdot A_2}{A_3}\right)\cdot e^{-\frac{(E_{a_1} + E_{a_2} - E_{a_3})}{RT}}$

$\therefore$ Overall $E_a = E_{a1} + E_{a2} - E_{a3} = 40 + 50 - 60$
$= 90 - 60 = 30$ kJ/mol

$\left[K = A.e^{-E_a/RT}\right]$

206. (2) $k = A.e^{-E_aRT}$

$\log k = \log A - \frac{E_a}{2.303R}\cdot\frac{1}{T}$

Higher is E_a, stronger is the temperature dependence of k (i.e. steeper the slope)

$\Rightarrow \frac{1}{k}\frac{dk}{dT} = \frac{E_a}{R}\frac{1}{T^2} \Rightarrow \frac{dk}{dT} = A\times e^{-\frac{E_a}{R}}\cdot\frac{E_a}{RT^2}$

Now, if $E_a = 0 \Rightarrow \frac{dk}{dT} = 0$

Thus, (A) and (B) are correct.

207. (2) $k = Ae^{-\frac{E_a}{RT}}$; $k_1 = Ae^{-\frac{(E_a)_1}{RT}}$

$k_2 = Ae^{-\frac{(E_a)_2}{RT}} \Rightarrow \frac{k_2}{k_1} = e^{\frac{(E_a)_1-(E_a)_2}{RT}}$

$\log\frac{k_2}{k_1} = \frac{(E_a)_1 - (E_a)_2}{2.3RT}$

$= \frac{(41.4 - 30)\times1000}{2.3\times8.3\times300} = 1.99 \cong 2$

208. (2520)

$\log\frac{K_{300}}{K_{200}} = \frac{E_a}{2.3\times8.314}\left(\frac{1}{T_1} - \frac{1}{T_2}\right)$

$\log\frac{0.05}{0.03} = \frac{E_a}{2.3\times8.3}\times\left[\frac{1}{200} - \frac{1}{300}\right]$

$E_a = 2519.88$ J $\Rightarrow E_a = 2520$ J

209. (3) (A) $k = Ae^{\frac{-E_a}{RT}}$
As E_a increases k decreases

(B) Temperature coefficient = $\frac{k_{T+10}}{k_T}$

(C) Option (C) is wrong. Δk may be greater or lesser depending on temperature.

(D) $\ln k = \ln A - \frac{E_a}{RT}$

210. (8) $k = Ae^{-E_a/RT}$
Taking ln both sides;

$\ln k = \frac{-E_a}{RT} + \ln A$

$y = mx + c$

Slope $= \frac{E_a}{R} = \frac{20}{5}$; $E_a = 4R = 8$ cal/mol

211. (225) $k = Ae^{\frac{-E_a}{RT}}$ or $\ln k = \ln A - \frac{E_a}{RT}$

$\Rightarrow \log(2\times10^{-6}) = \log(4\times10^{13}) - \frac{8.3\times10^3}{8.3\times T\times2.3}$

$\Rightarrow \log(2) - 6 = 2\times\log(2) + 13 - \frac{8.3\times10^3}{8.3\times T\times2.3}$

$\Rightarrow -19.3 = -\frac{10^3}{T\times2.3} \Rightarrow T = 225\times10^{-1}$ K

212. (1) $\ln\left(\frac{k_2}{k_1}\right)=\frac{E_a}{R}\left(\frac{1}{T_1}-\frac{1}{T_2}\right)$

$\ln\left(\frac{k_2}{k_1}\right)=\frac{532611}{8.3}\times\left(\frac{10}{300\times310}\right)$

where k_2 is at 310 K & k_1 is at 300 K

$\ln\left(\frac{k_2}{k_1}\right)=6.9$ or $=k_1=x\times10^{-3}$

$\frac{k_2}{k_1}=999.2$; $\frac{k_2}{x\times10^{-3}k_2}=999.2$

$x=1$

213. (216)

$k=Ae^{-E_a/RT}=(6.5\times10^{12}\,s^{-1})e^{-26000\,K/T}$

$\frac{E_a}{8.314}=26000\Rightarrow E_a=216.164$ kJ/mol.

214. (166) $\ln k=\ln A-\frac{E_a}{RT}$

Given : $\ln k=33.24-\frac{2.0\times10^4}{T}$

$\therefore$ on comparing $\frac{E_a}{R}=2.0\times10^4$

$\Rightarrow E_a=2.0\times10^4\times8.3\text{ J}=16.6\times10^4\text{ J}=166\text{ kJ}$

215. (59) $\log_{10}\frac{k_2}{k_1}=\frac{E_a}{2.303R}\left[\frac{1}{T_1}-\frac{1}{T_2}\right]$

$T_1=300\text{K}, T_2=309\text{K}$

$\log_{10}\frac{k_2}{k_1}=\frac{E_a}{2.303R}\left(\frac{1}{300}-\frac{1}{309}\right)$

$0.3=\frac{E_a}{2.303\times8.3}\left(\frac{9}{300\times309}\right)$

$E_a=\frac{0.3\times2.303\times8.3\times300\times309}{9}$

$=59065.04$ J/mol

$E_a=59.06$ kJ/mol

216. (4) According to the Arrhenius equation

$k=Ae^{\frac{-Ea}{RT}}$

$k_{cat}=Ae^{\frac{E_a^1}{RT}}$, $k_{uncat.}=Ae^{\frac{Ea}{RT}}$

$\frac{k_{cat}}{k_{uncat.}}=e^{\frac{Ea-E_a^1}{RT}}=e^{\frac{10\times1000}{8.3\times300}}=e^{4.009}=e^x$

$[E_a-E_a^1=10\text{ kJ}]$

$\therefore x=4$

217. (154) $\ln k=\ln A-\frac{E_a}{10^3RT}\times10^3$

$=\ln A+\frac{10^3}{T}\left[-\frac{E_a}{10^3R}\right]$

From the graph,

$\frac{-E_a}{10^3\times R}=-18.5$; $E_a=153.735=154$ kJ/mol

218. (45) ΔH = Energy of products – Energy of reactants

$=Z-(Z+Y)=15-(15+45)=-45$

$|\Delta H|=45$ kJ/mol

219. (47) Given $\log K=20.35-\frac{(2.47\times10^3)}{T}$ ...(i)

We know $\log K=\log A-\frac{E_a}{2.303\,RT}$ ...(ii)

comparing equation (i) with equation (ii)

$\frac{E_a}{2.303\,RT}=\frac{2.47\times10^3}{T}$

$E_a=2.47\times10^3\times2.303\times8.314$

$=4\,7293.44$ J/mol$=47.29=47$ kJ/mol

220. (16) $E_a=209$ kJ/mol

k_2 at 700 °C $=6.36\times10^{-3}\,s^{-1}$,

$k_1=x\times10^{-6}\,s^{-1}$

$\log\left(\frac{k_2}{k_1}\right)=\frac{-E_a}{2.303R}\left(\frac{1}{T_2}-\frac{1}{T_1}\right)$

$\log\left(\frac{6.36\times10^{-3}}{k_1}\right)=\frac{209\times1000}{2.303\times8.31}\left(\frac{100}{700\times600}\right)$

$\log(6.36\times10^{-3})-\log k_1=2.6$

$\Rightarrow\log k_1=-2.19-2.6=-4.79$

$\Rightarrow k_1=10^{-4.79}=1.62\times10^{-5}=16.2\times10^{-6}\Rightarrow x=16$

221. (10) $\log\frac{k_2}{k_1}=\frac{E_a}{2.303R}\left[\frac{1}{T_1}-\frac{1}{T_2}\right]$

k_1 (at 200 K) = ?

k_2 (at 300 K) $=1\times10^{-3}\,s^{-1}$

$\log\frac{1\times10^{-3}}{k_1}=\frac{11.488\times10^3}{2.303\times8.314}\left[\frac{1}{200}-\frac{1}{300}\right]$

$\log\frac{1\times10^{-3}}{k_1}=\frac{11.488\times10^3\times100}{2.303\times8.314\times200\times300}$

$\log\frac{1\times10^{-3}}{k_1}=1$

$\frac{1\times10^{-3}}{k_1}=10\Rightarrow k_1=10\times10^{-5}\,s^{-1}$

222. (50) $X\rightarrow Y$

$\Delta E=(E_a)_f-(E_a)_b$; $-20=30-(E_a)_b$

$\Rightarrow (E_a)_b=50$ kJ

223. (14) Energy of activation,

$E_a=80.9$ kJ mol^{-1} = 80900 J/mol.

Temperature of reaction, $T=700$ K

In the Arrhenius equation, $k=Ae^{-Ea/RT}$, A is the frequency factor and $e^{-Ea/RT}$ is the fraction of collision having enough energy to overcome the activation energy.

$\therefore$ Fraction of molecules having enough energy to react $=e^{-Ea/RT}=e^{-x}$ (given)

$\therefore\quad x=\frac{E_a}{RT}=\frac{80900}{8.31\times700}=13.9\simeq14$

224. (526) Arrhenius equation is,

$k=Ae^{-E_a/RT}$

$\log k=\log A-\frac{E_a}{2.303R}\times\frac{1}{T}$

Given : Slope $=-10000=\frac{-E_a}{2.303R}$

$\because\log\frac{k_2}{k_1}=\frac{E_a}{2.303R}\left[\frac{1}{T_1}-\frac{1}{T_2}\right]$

$\log\frac{10^{-4}}{10^{-5}}=10000\left[\frac{1}{500}-\frac{1}{T_2}\right]$

$\Rightarrow1=20-\frac{10000}{T_2}$

$\Rightarrow T_2=\frac{10000}{19}=526.3\approx526$ K

225. (52)

$\log\frac{k_2}{k_1}=\frac{E_a}{2.303R}\left[\frac{1}{T_1}-\frac{1}{T_2}\right]$

$\log5=\frac{E_a}{2.303\times8.314}\left[\frac{1}{300}-\frac{1}{325}\right]$

$\Rightarrow E_a=52271$ J mol^{-1} = 52.271 kJ mol^{-1}

226. (100) The Arrhenius equation is $k=Ae^{\frac{E_a}{RT}}$

Assuming A and E_a to be independent of temperature

$\ln\frac{k_2}{k_1}=\frac{E_a}{R}\left(\frac{1}{T_1}-\frac{1}{T_2}\right)$

$\ln3.555=\frac{E_a}{8.314}\left(\frac{1}{303}-\frac{1}{313}\right)$

$\Rightarrow E_a=\frac{1.268\times8.314\times303\times313}{10}$

$=99980.7=99.98$ kJ / mol

227. (84297.48) $\because k=Ae^{-E_a/RT}$.

$\ln\frac{k_2}{k_1}=\frac{E_a}{R}\left(\frac{1}{T_1}-\frac{1}{T_2}\right)$

$\Rightarrow\ln5=\frac{E_a}{R}\left(\frac{1}{300}-\frac{1}{315}\right)$

$\Rightarrow E_a=\frac{1.6094\times8.314\times300\times315}{15}=84297.48$ J/mol.

228. (3.98)

For a first order reaction, $kt=\ln\frac{[A]}{[A_0]}$

At 300 K, $k_1\times60=\ln\frac{[A]}{[A_0]}$ ...(1)

At 400 K, $k_2\times40=\ln\frac{[A]}{[A_0]}$ ...(2)

From equation (1) and (2), $\frac{k_2}{k_1}=\frac{60}{40}$

$\ln\frac{k_2}{k_1}=\frac{E_a}{R}\left[\frac{1}{T_1}-\frac{1}{T_2}\right]$; $\ln\left(\frac{60}{40}\right)$

$=\frac{E_a}{8.3}\times\frac{100}{400\times300}\Rightarrow E_a=0.4\times8.3\times1200$

$=3984$ J/mol. $=3.984$ kJ/mol.

Part-B (JEE Advanced)

1. **(b)** $M \rightarrow N$
$r = k[M]^x$
when concentration $= 2M$; rate $= 8r$, thus
$8r = k[2M]^x$
$8 = (2)^x$
$x = 3$

2. **(c)** For first order reaction,
Rate $= k$ [conc. of reactant]
Since 0.1 M of A changes to 0.025 M in 40 minutes,
$t_{1/2}$ of reaction $= 40/2 = 20$ minutes
Rate of reaction of A

$$= k[A] = \frac{0.693}{t_{1/2}} \times [A] = \frac{0.693}{20} \times 0.01$$
$$= 3.47 \times 10^{-4}\,\text{M min}^{-1}$$

3. **(c)** $k = \frac{2.303}{t}\log\frac{[A]_0}{[A]_t}$

$$\therefore\ k = \frac{2.303}{2 \times 10^4}\log\frac{800}{50} = 1.386 \times 10^{-4}\,\text{s}^{-1}$$

4. **(a)** Individual rates of reactants and products become equal when each of these is divided by their respective stoichiometric coefficient. With time, concentration of reactants decreases and is represented by negative sign whereas concentration of products increases and is represented by positive sign.
The given reaction is

$$N_2(g) + 3H_2(g) \longrightarrow 2NH_3(g)$$

$\therefore$ Correct relationship amongst the rate expression is shown in (a).

5. **(b)** The rate of photochemical process varies with the intensity of absorption.
Since, greater the intensity of absorbed light, more photons will fall at a point, and further, each photon causes one molecule to undergo reaction.

6. **(d)** Find the order of reaction and then use appropriate equation.
As unit of k is sec^{-1}, reaction is of **first order**,

$$r = k[N_2O_5];\ \therefore\ [N_2O_5] = \frac{2.4 \times 10^{-5}}{3 \times 10^{-5}} = 0.8\ \text{mol/L}$$

7. **(d)** It is a characteristic constant of a particular reaction at a given temperature. It does not depend upon initial concentration of the reactants, time of reaction and extent of reaction.

8. **(a)** It is a constant of a particular reaction at a given temperature. It does not depend upon initial concentration of the reactants, time of reaction and extent of reaction.

9. **(8)** $8H^+ + 5[Fe(H_2O)_2(Ox)_2]^{2-} + MnO_4^-$
$\rightarrow Mn^{2+} + 5[Fe(H_2O)_2(Ox)_2]^- + 4H_2O$

$$\text{Rate} = \frac{1}{8}\frac{d[H^+]}{dt} = -\frac{d[MnO_4^-]}{dt}$$

$$\text{Hence, } \frac{\text{rate of } [H^+]\text{ decay}}{\text{rate of } [MnO_4^-]\text{ decay}} = 8$$

10. **1.765×10^{-4} kg/hr;**
$N_2(g) + 3H_2(g) \longrightarrow 2NH_3(g)$

$$\text{Rate of reaction} = \frac{1}{3}\ [\text{Rate of disappearance of } H_2]$$
$$= \frac{1}{2}\ [\text{Rate of appearance of } NH_3]$$

or $\frac{1}{3}\frac{d[H_2]}{dt} = \frac{1}{2}\frac{d[NH_3]}{dt}$

$$\Rightarrow \frac{d[H_2]}{dt} = \frac{3}{2}\frac{d[NH_3]}{dt}$$

$$\frac{d[NH_3]}{dt} = 0.01\ \text{kg/hr}$$
$$= \frac{0.001}{17} \times 1000 = \frac{1}{17}\ \text{mole/hr}$$
$$\therefore \frac{d[H_2]}{dt} = \frac{1}{17} \times \frac{3}{2}$$
$$= \frac{3}{34}\ \text{mole/hr} = \frac{3}{34} \times \frac{2}{1000}\ \text{kg/hr}$$

$= 1.765 \times 10^{-4}$ kg/hr.

11. **Product of active masses of reactants at that time**

12. **(b,c, d)** Rate $= \frac{dP}{dt} = k[X]^1$

	$2X$ +	Y →	P
t = 0	2 mol	1 mol	
t=50s	1 mol	0.5 mol	0.5 mol

$$-\frac{1}{2}\frac{d[X]}{dt} = \frac{d[P]}{dt} = k[X]^1$$
$$-\frac{d[X]}{dt} = 2k[X]^1$$
$$2k = \frac{ln2}{50} \Rightarrow k = \frac{ln2}{100} = 6.93 \times 10^{-3}\,\text{s}^{-1}$$
$$t_{1/2} = \frac{ln2}{2k} = \frac{ln2 \times 50}{ln2} = 50\,\text{sec}$$

At 50 sec $\frac{-d[X]}{dt} = 2k \times (1)^1 = \frac{ln2}{50}$

$$= 13.86 \times 10^{-3}\,\text{mol L}^{-1}\text{s}^{-1}$$

At 100 sec $-\frac{1}{2}\frac{d[X]}{dt} = \frac{-d[Y]}{dt}$

$$\Rightarrow -\frac{d[Y]}{dt} = \frac{ln2}{100} \times \frac{1}{2}\left\{\frac{-d[Y]}{dt} = k[X]^1\right\}$$
$$\frac{d[Y]}{dt} = 3.46 \times 10^{-3}\,\text{mol L}^{-1}\,\text{s}^{-1}$$

13. **(b, c)** As rate $= k$ [RCl], so it is first order reaction. On decreasing the concentration of RCl to half, the rate will also be halved. Rate will also increase with temperature.

14. (a) From the rate law expression, $R_0 = k[A_0]^a[B_0]^b$ and from the table it is clear that :
(i) when the concentration of $[A_0]$ is doubled, keeping $[B_0]$ constant (see readings 1 and 2), the rate also doubles *i.e.* rate is directly proportional to $[A_0]$ or $a = 1$.
(ii) when the concentration of $[B_0]$ is reduced, keeping $[A_0]$ constant (see readings 1 and 3), the rate remains constant. *i.e.* rate is independent of $[B_0]$ or $b = 0$
Thus, rate equation becomes $\boldsymbol{R_0 = k[A_0]}$

(b) $k = \frac{R_0}{[A_0]} = \frac{0.05}{0.10} = \mathbf{0.5\ sec^{-1}}$

15. $A \rightarrow B; k = 4.5 \times 10^{-3}\,\text{min}^{-1}; [A]_0 = 1\text{M}$

For first order reaction, $k = \frac{2.303}{t}\log\frac{a}{(a-x)}$

Find $(a-x)$ at $t = 60$ min.

$$4.5 \times 10^{-3} = \frac{2.303}{60}\log\frac{1}{(a-x)}$$
$\therefore\ (a-x) = 0.7634$
Thus, rate after 60 minute $= k(a-x)$
$= 4.5 \times 10^{-3} \times 0.7634$
$= 3.4354 \times 10^{-3}$ M min^{-1}

16. (i) According to Fig. in the given time of 4 hours (1 to 5) concentration of A falls from 0.5 to 0.3 M, while in the same time concentration of B increases from 0.2 M to 0.6 M.
Decrease in concentration of A in 4 hours
$= 0.5 - 0.3 = 0.2$ M
Increase in concentration of B in 4 hours
$= 0.6 - 0.2 = 0.4$ M
Thus increase in concentration of B in a given time is twice the decrease in concentration of A.
Thus $\boldsymbol{n = 2}$.

(ii) $K = \frac{[B^2]_{eq.}}{[A]_{eq.}} = \frac{(0.6)^2}{0.3} = \mathbf{1.2\ M}$

(iii) Initial rate of conversion of A
= Change in conc. of A during 1 hour

$$= \frac{0.6 - 0.5}{1} = \mathbf{0.1\ mole\ litre^{-1}\ hour^{-1}}$$

17. **(d)** For P, if $t_{50\%} = x$ then $t_{75\%} = 2x$
This is true only for first order reaction.
So, order with respect to P is 1.
Further, the graph shows that concentration of Q decreases with time. So, rate with respect to Q, remains constant. Hence, it is zero order wrt Q.
So, overall order is $1 + 0 = 1$

18. **(a)** The values of rate constants k_0, k_1 for zero order and first order reaction, respectively, are given by the following equation:

$k_0 = \frac{A_{rs}}{2 \times t_{1/2}}$ [where A_{rs} = initial concentration, and $t_{1/2}$ = half-life period] and $k_1 = \frac{0.693}{t_{1/2}}$

substituting various given values, we get

$$k_0 = \frac{1.386\,\text{mol litre}^{-1}}{2 \times 20\ \text{sec}} \quad ...(i)$$
$$\text{and } k_1 = \frac{0.693}{40\,\text{sec}} \quad ...(ii)$$

Dividing (ii) by (i), we get

$$\frac{k_1}{k_0} = \frac{0.693}{40} \times \frac{2 \times 20}{1.386}\,\text{mol}^{-1}\,\text{litre}$$
$$= \frac{0.693}{1.386}\,\text{mol}^{-1}\text{ litre} = 0.5\,\text{mol}^{-1}\text{ litre}$$
$= 0.5\,\text{mol}^{-1}\,\text{dm}^3$ [1 litre = 1dm^3]
Thus, the correct answer is (a).

19. **(d)** Overall order = sum of orders w.r.t each reactant.
Let the order be x and y for G and H respectively

Exp.No.	[G]mole litre^{-1}	[H]mole litre^{-1}	rate (mole litre^{-}time^{-1})
1	a	b	r
2	$2a$	$2b$	$8r$
3	$2a$	b	$2r$

Applying $r = k[G]^x[H]^y$ we get, $x = 1, y = 2$

$\because$ For (1) and (3), the rate is doubled when conc. of G is doubled keeping that of H constant *i.e.*, rate $\propto [G]$ $\therefore x = 1$

From (2) and (3), $y = 2$ $\therefore$ Overall order is 3.

20. **(d)** Order of a reaction can be fractional. Rest of all are true.

Order of a reaction can be determined experimentally

21. **(143)** Life of sample $\rightarrow$ t year

$[A]_0 \propto$ Initial mole of U-238

$[A]_t \propto$ Final mole of U-238

$$\frac{[A]_0}{[A]_t} = \frac{\frac{1}{238} + \frac{7}{206}}{\frac{1}{238}} = \frac{0.0042 + 0.0340}{0.0042} = 9.1$$

$$\Rightarrow \frac{2.303}{t}\log\frac{[A]_0}{[A]_t} = \frac{2.303}{t}\log(9.1) = k$$

$$\text{Now, } k = \frac{\ln 2}{t_{1/2}} = \frac{0.693}{4.5\times10^9} = 1.54\times10^{-10}$$

$$\Rightarrow \frac{2.303}{t}\log(9.1) = 1.54\times10^{-10}$$

$\Rightarrow t = 1.434 \times 10^{10} = 143 \times 10^8$ years $\Rightarrow P = 143$

22. **(3)** Rate of reaction is determined by slowest step as:

$\Rightarrow r = k_2[N_2O_2][H_2]$...(1)

Now concentration for intermediate $[N_2O_2]$ can be written as:

$$\frac{k_1}{k_{-1}} = \frac{[N_2O_2]}{[NO]^2} \Rightarrow [N_2O_2] = \frac{k_1}{k_{-1}}[NO]^2 \text{ ...(2)}$$

From equation (1) and (2) we get,

$$r = \frac{k_2k_1}{k_{-1}}[NO]^2[H_2]$$

Overall order of reaction $= 2 + 1 + 3$

23. **(9)** $t_{1/8} = \frac{2.303\log 8}{k} = \frac{2.303\times3\log2}{k}$

$$t_{1/10} = \frac{2.303}{k}\log10 = \frac{2.303}{k}$$

$$\left[\frac{t_{1/8}}{t_{1/10}}\right]\times10 = \frac{\left(\frac{2.303\times3\log2}{k}\right)}{\left(\frac{2.303}{k}\right)}\times10 = 9$$

24. **(0)**

Δx	1 – 0.75	1 – 0.4	1 – 0.1
Δt	0.05	0.12	0.18
$\frac{\Delta x}{\Delta t}$	5	5	5

Since, rate $= \Delta x/\Delta t$ is constant for different concentration values, order of reaction is zero.

25. **(4.17)** Rate = k[A][R]

	A	+	R	$\rightarrow$	Product
$t = 0$	A_0		$10A_0$ (Excess)		
$t = t$	$0.6A_0$		$10A_0$		

At start, $R_1 = k[A]_0(10[A_0]) = 10k[A_0]^2$

When reaction is 40% complete

	A	+	R	$\rightarrow$	Product
$t = 0$	A_0		$10A_0$		
$t = t$	$0.6A_0$		$9.6A_0$		

$R_2 = k[0.6A]_0(9.6[A_0]) = k(0.6\times9.6)[A_0]^2$

For pseudo 1st order reaction

$R'_1 = k'[A_0]$

$R'_2 = k'\times0.6[A_0] = k\times10[A_0]\times0.6[A_0]$

$= k\times6[A_0]^2$

$$\%\text{ error} = 100\times\frac{\Delta\text{ Rate}}{\text{Rate}_1} = \left[\frac{6-0.6\times9.6}{0.6\times9.6}\right]100$$

$$= \frac{6-5.76}{5.76} = 4.17.$$

26. **(6.75)** Rate of the reaction $= K[A]^x[B]^y[C]^z$

Compairing experiment 1 with 2 we get that, $y = 0$

Compairing experiment 1 with 3 we get that, $z = 1$

Compairing experiment 1 with 4 we get that, $x = 1$

putting values of x, y, z in rate equation for experiment 1.

$6\times10^{-5} = K\times(0.2)^1\times(0.1)^1\times(0.1)^0$

$K = 3\times10^{-3}$

Now, for the given concentration of A, B and C, rate of reaction will be,

Rate $= 3\times10^{-3}\times0.15\times1\times0.15 = 6.75\times10^{-5}$

Therefore, value of Y = 6.75.

27. **(2.30)**

$$2N_2O_5(g) \xrightarrow{\Delta} 2N_2O_4(g) + O_2(g)$$

Given rate constant of the reaction $= 5\times10^{-4}\text{sec}^{-1}$

Thus, it is a first order reaction.

$$\frac{1}{2}\frac{dP_{N_2O_5}}{dt} = K_{overall}P_{N_2O_5}$$

$$\Rightarrow \frac{dP_{N_2O_5}}{dt} = 2K_{overall}P_{N_2O_5}$$

$$2N_2O_5(g) \rightarrow 2N_2O_4(g) + O_2(g)$$

$1 - x \quad x \quad x/2$

$t = Y\times10^3$ sec.

$$1 + \frac{x}{2} = 1.45$$

$x = 0.90$ atm

For a first order reaction,

$$t = \frac{2.303}{K}\log\frac{[P]_o}{[P]}$$

$$t = \frac{2.303}{2K}\log\frac{[P]_o}{[P]}$$

$$Y\times10^3 = \frac{2.303}{2\times5\times10^{-4}}\log\frac{1}{0.1}$$

$$Y = \frac{23.03}{10} = 2.30$$

28. **(24.14)** $r_1 = kc_1$ and $r_2 = kc_2$

Since, rate of first order reaction is directly proportional to the concentration of its reactant,

$$\therefore \frac{r_1}{r_2} = \frac{c_1}{c_2} = \frac{0.04}{0.03}$$

According to first order reaction

$$k = \frac{2.303}{t_{20}-t_{10}}\log\frac{c_1}{c_2}$$

On substituting the various values

$k = 0.0287\text{ min}^{-1}$

$$t_{1/2} = \frac{0.693}{k} = \frac{0.693}{0.0287} = 24.14\text{ min}$$

29. **(0.749)**

$CH_3-O-CH_3(g) \rightarrow CH_4(g) + CO(g) + H_2(g)$

Given $t_{1/2} = 14.5$ min, initial pressure = 0.40 atm, $t = 12$ min.

$$\text{Now, } k = \frac{0.693}{14.5} = 4.78\times10^{-2}\text{ min}^{-1}$$

Writing first order equation and substituting the given values, we get

$$4.78\times10^{-2} = \frac{2.303}{12\text{ min.}}\log\frac{0.4}{0.4-x}$$

which gives $x = 0.175$ atm

Since, volume and temp. are constant, final pressure :

$CH_3-O-CH_3(g) \rightarrow CH_4(g) + CO(g) + H_2(g)$

0.4 $\quad$ 0.4 – 0.175 $\quad$ 0.175 $\quad$ 0.175 $\quad$ 0.175

Hence, total pressure $= 0.4 - 0.175 + 3\times0.175$

$= 0.749$ atm

30. **acidic, Pseudo first**

31. **True :** The rate of reaction of first order is directly proportional to the concentration of reacting substance.

32. **(c, d)**

(a) $R \longrightarrow P$

$t = 0 \quad R_0$

$t = t \quad R_0 - P_t \quad P_t$

$[P]_t = [R]_0(1 - e^{-kt})$

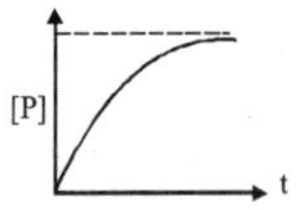

(b) $\text{Rate} = \frac{-d[R]}{dt} = k[R]$

$$\frac{d[R]}{dt} = -k[R]$$

$y = mx$

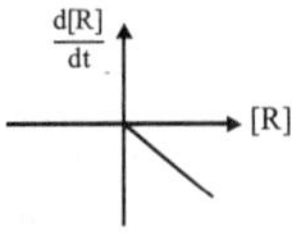

(c) $\text{Rate} = \frac{-d[R]}{dt} = \frac{d[P]}{dt} = k[R]$

$$\text{So, } \frac{d[P]}{dt} = k[R]_0e^{-kt}$$

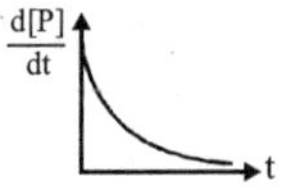

(d) k = constant (k only depends on temperature)

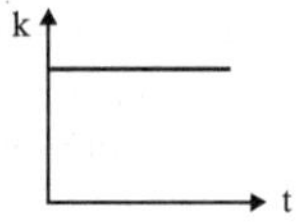

33. **(a)**

$(CH_3)_3C-Br + NaOH \xrightarrow[\text{(first order)}]{S_N1} (CH_3)_3C-OH + NaBr$

At time $t = 0 \quad P_0 \qquad\qquad 0$

At time $t \quad P \qquad\qquad P_0 - P$

rate $= k[(CH_3)_3C\text{Br}]$

$$\Rightarrow \ln\frac{P_0}{P} = kt;\ t_{1/2} = \frac{0.693}{k}\ \ln\frac{P}{P_0} = -kt;$$

$$\frac{[Q]}{[P]_0} = \frac{[P_0]-[P]}{[P_0]} = 1 - \frac{[P]}{[P_0]} = 1 - e^{-kt}$$

34. **(a, d)**

$$A(g) \xrightarrow{\text{First order}} 2B(g) + C(g)$$

V = constant, T = 300K

$t = 0 \quad P_0$

$t = t_{1/3}\left(P_0 - \frac{2P_0}{3}\right) \quad \frac{4P_0}{3} \quad \frac{2P_0}{3}$

$t = tP_0 - x2xx$

So, $P_t = P_0 - x + 2x + x = P_0 + 2x$

or $2x = P_t - P_0$

$t = \frac{1}{k}\ln\frac{P_0}{(P_0 - x)}$

or $t = \frac{1}{k}\ln\frac{P_0}{P_0 - \frac{(P_t - P_0)}{2}} = \frac{1}{k}\ln\frac{2P_0}{2P_0 - P_t + P_0}$

or $kt = \ln\frac{2P_0}{3P_0 - P_t}$, $kt = \ln 2P_0 - \ln(3P_0 - P_t)$

or $\ln(3P_0 - P_t) = -kt + \ln 2P_0$

Graph between $\ln(3P_0 - P_t)$ vs 't' is a straight line with negative slope.

Sincc, rate constant is a constant quantity and independent of initial concentration.

So, graph (a) and (d) are correct.

35. (a, b, d) For first order reaction

$[A] = [A]_0 e^{-kt}$

Hence, concentration of $[NO_2]$ decreases exponentially.

Also, $t_{1/2} = \frac{0.693}{K}$. Which is independent of concentration and $t_{1/2}$ decreases with the increase of temperature.

$t_{99.6} = \frac{2.303}{K}\log\left(\frac{100}{0.4}\right)$

$t_{99.6} = \frac{2.303}{K}(2.4) = 8 \times \frac{0.693}{K} = 8t_{1/2}$

36. (a, b, d) The relevant expressions are as follows.

Choice (a) $\log K_p = -\frac{\Delta H}{R}\frac{1}{T} + I$

Choice (b) $\log [X] = \log [X]_0 + kt$

Choice (c) P/T = constant (V constant)

Choice (d) PV = constant (T constant).

37. (a, d) In first order reaction, if α is the degree of dissociation then

$kt = \log_e\frac{1}{(1-\alpha)} = -\log_e(1-\alpha)$

or $e^{-kt} = 1 - \alpha$

$\therefore \alpha = 1 - e^{-kt}$

The Arrhenius equation is, $k = Ae^{-E_a/RT}$

Plot of reciprocal concentration of the reactant vs time is linear. Dimensions of pre-exponential factor 'A' are equivalent to dimensions of k, which is T^{-1} for a first order reaction.

38. (a)

$\text{Rate} = \frac{k[X]}{X_s + [X]} = \frac{k}{\frac{X_s}{[X]} + 1} = k\frac{1}{\frac{X_s}{[X]} + 1}$

(I) When $[X]_0 \longrightarrow \infty$; Rate = $k \Rightarrow$ Rate is independent of substrate's concentration.

$\therefore$ Order of reaction = 0

(I) → (P), (S)

(II) When $[X]_0 << X_s$ $\therefore$ Rate = k $\frac{[X]}{X_s}$

⇒ This implies Order of the reaction = 1.

(II) → (Q), (T)

(III) $[X]_0 >> X_s$; $[X] + X_s \approx [X]$

$\text{Rate} = \frac{k[X]}{[X]} = k$

⇒ This implies Order of the reaction = 0.

(III) → (P), (S)

(IV) $\text{Rate} = \frac{k[X]^2}{X_s + [X]}$

$[X] >> X_s$; Rate = $k[X] \Rightarrow$ This implies Order of the reaction = 1.

(IV) → (Q), (T)

39. (i) From the given data, it is evident that the $t_{1/2}$ (half-life period) for the decomposition of X (g) is constant (100 minutes) therefore, the order of reaction is **one**.

(ii) Rate constant, $k = \frac{0.693}{t_{1/2}} = \frac{0.693}{100}$

$= \mathbf{6.93 \times 10^{-3}\ min^{-1}}$

(iii) Time taken for 75% completion of reaction

$= 2t_{1/2} = 2 \times 100 = 200$ minutes

(iv)	$2X$	⟶	$3Y$	+	$2Z$
Initial pressure	800		0		0
After time t	(800 – 2P)		3 P		2 P

when the pressure of X is 700 mm of Hg then,

$800 - 2P = 700$

$2P = 100$; $P = 50$ mm of Hg

Total pressure $= 800 - 2P + 3P + 2P = 800 + 150$

= 950 mm of Hg

40. Let the number of moles of A left after 100 min = x

Total number of moles after 100 min

$= x + 12 + 0.525$

$P_{mix} = p_A + p_B$

$= \left(\frac{x}{12+x} \times 300\right) + \left(\frac{12}{12+x} \times 500\right)$

$= \frac{6000 + 300x}{12 + x}$

According to Raoult's law

$\frac{\left(\frac{6000+300x}{12+x}\right) - 400}{\frac{6000+300x}{12+x}} = \frac{0.525}{x + 12 + 0.525}$

On solving we get, $x = 9.9$

Now according to first order kinetics,

$k = \frac{2.303}{100}\log\frac{10}{9.9} = \mathbf{1.005 \times 10^{-4}\ min^{-1}}$

41.

$2N_2O_5(g)$	⟶ $4NO_2(g)$	+ $O_2(g)$	
a	0	0	$t = 0$
$a - x$	$2x$	$x/2$	$t = 30$ min
0	$2a$	$a/2$	final

$2a + a/2 = 584.5$ mm Hg

$a = 584.5 \times \frac{2}{5}$ mm Hg

$a = 233.8$ mm Hg

Therefore, initial pressure of $N_2O_5 = 233.8$ mm Hg

Total pressure after 30 min

$= 284.5 \text{ mm Hg} = 233.8 - x + 2x + \frac{x}{2}$

or $233.8 + \frac{3x}{2} = 284.5$ or $x = 33.8$ mm Hg

Hence pressure of N_2O_5 after 30 min.

$= 233.8 - 33.8 = 200$ mm Hg

$k = \frac{2.303}{t}\log\frac{a}{(a-x)} = \frac{2.303}{30}\log\frac{233.8}{200}$

$k = \frac{2.303}{30} \times 0.0679 = \mathbf{5.2 \times 10^{-3}\ min^{-1}}$

42. For a first order reaction we know that

$k = \frac{2.303}{t}\log\frac{a}{(a-x)}$

Here, $t = 10 \times 60 \times 60$ sec. and let $a = 1$, then substituting the values, we get

$1.5 \times 10^{-6} = \frac{2.303}{10 \times 3600}\log\frac{1}{(1-x)}$

$\frac{1.5 \times 10^{-6} \times 10 \times 3600}{2.303} = \log\frac{1}{(1-x)}$

$0.0234 = \log\frac{1}{(1-x)}$

Taking antilog, $1.055 = \frac{1}{(1-x)}$

or $1.055 - 1.055x = 1$

$\Rightarrow x = \frac{(1.055 - 1)}{1.055} = 0.052$

Thus, **5.2%** of the initial concentration has changed into product.

Again we know that

$t_{1/2} = \frac{0.693}{k} = \frac{0.693}{1.5 \times 10^{-6}}$

$= 462000$ second $=$ **128.33 hours**

43. Assuming that the decomposition of N_2O_5 is a **first order reaction,** then

$k = \frac{2.303}{t}\log\frac{a}{(a-x)} = \frac{2.303}{t}\log\frac{P_0}{P}$

or $\log P = \frac{-kt}{2.303} + \log P_0$

Thus $\log P$ vs time graph is linear with slope $= \frac{-k}{2.303}$ if the given reaction is of first order which is in accordance with the given statement. Thus, the reaction obeys first order reaction.

44. From data (i) and (ii), it is obvious that when the concentration of B is kept constant (0.01 mol litre^{-1}) and the concentration of A is doubled (0.01 to 0.02 mol litre^{-1}), the rate of reaction is also doubled (0.005 to 0.010 mol litre^{-1} min^{-1}). Hence, **the order of reaction with respect to A is 1**.

Similarly, from data (i) and (iii) it is obvious that when the concentration of A is kept constant (0.01 mol litre^{-1}) and the concentration of B doubled (0.01 to 0.02 mol litre^{-1}), the rate of reaction remains constant (0.005 mol litre^{-1} min^{-1}). This shows that the **order of reaction with respect to B is zero**.

Now we know that the rate of reaction, $A + B \rightarrow$ Products, is given by

Rate $r = k[A]^1[B]^0 \quad \Rightarrow r = k[A]$

$\therefore k = \frac{r}{[A]} = \frac{0.005}{0.01} = 0.5\ \text{min}^{-1}$

[$\because r = 0.005$ mol/l/m and $[A] = 0.01$ mol/l]

For a first order reaction :

We know that $t_{1/2} = \frac{0.693}{k}$

$= \frac{0.693}{0.5}$ [$\because k = 0.5\ \text{min}^{-1}$]

= 1.386 minutes

Alternate Solution:

Rate = k $[A]^m [B]^n$

$0.005 = k[0.01]^m[0.01]^n$

$0.010 = k[0.02]^m[0.01]^n$

$0.005 = k[0.01]^m[0.02]^n$

On solving

m = 1, n = 0

$$k = \frac{[r]}{[A]} = \frac{0.005}{0.01} = 0.5\ \text{Min}^{-1}$$

$$t_{1/2} = \frac{0.693}{k} = \frac{0.693}{0.5} = 1.38\ \text{min}$$

$t_{1/2} = 1.386$ min

45. (a) As per Arrhenius equation $(k = Ae^{-E_a/RT})$, the rate constant increases exponentially with temperature.

46. (d) $\log k = \log A - \frac{E_a}{2.303RT}$ ....(1)

Also given $\log k = 6.0 - (2000)\frac{1}{T}$ ....(2)

On comparing equations, (1) and (2)

$\log A = 6.0 \Rightarrow A = 10^6 s^{-1}$ and $\frac{E_a}{2.303\,R} = 2000$;

$\Rightarrow E_a = 2000 \times 2.303 \times 8.314 = 38.29$ kJ mol^{-1}

47. (b) The Arrhenius equation is : $k = A \exp(-E_a/RT)$

As $T \to \infty$, $\exp(-E_a/RT) \to 1$. Hence, $k = A$ where A, the Arrhenius parameter, is $6.0 \times 10^{14} s^{-1}$

'A' is also known as frequency factor

48. (c) A catalyst decreases the activation energy of the reactants and thus shortens time of reaction. So (c) is the correct option.

49. (5) Log $k_f = 9$ at 500K (given)

$\therefore\ k_f = 10^9$

$$\log k_b = \frac{(E_a)_b}{2.303R}\cdot\frac{1}{T} + \log A_b$$

$$k_{eq} = 10^6 = \frac{10^9}{k_b} \Rightarrow k_b = 10^3$$

$$\therefore 3 = \frac{-(E_a)_b}{2.303R}\cdot\frac{1}{T} + 11$$

$$\therefore 3 = \frac{-(E_a)_b}{2.303R}\cdot\frac{1}{T} + 11$$

$$\frac{(E_a)_b}{2.303R} = 8 \times 500 = 4000$$

$$\log k_b = -4000 \times \frac{1}{250} + 11 = -5$$

$(\log k_b) = 5$

50. (– 8500) $A(g) + B(g) \rightleftharpoons AB(g)$

$E_{a_b} = E_{a_f} + 2RT$ and $A_f = 4A_b$

Now, Rate constant of forward reaction

$$k_f = A_f e^{-Ea_f/RT}$$

Rate constant of reverse reaction

$$k_b = A_b e^{-E_{ab}/RT}$$

Equilibrium constant

$$K_{eq} = \frac{k_f}{k_b} = \frac{A_f}{A_b} e^{-(E_{af} - E_{ab})/RT}$$

$$K_{eq} = 4e^{2RT/RT} = 4e^2$$

Now, $\Delta G° = -RT \ln K_{eq} = -2500 \ln(4e^2)$

$= -2500 (\ln 4 + \ln e^2) = -2500 (1.4 + 2)$

$= -2500 \times 3.4 = -8500$ J/mol.

51. (100) According to Arrhenius equation

$k = Ae^{-E_a/RT}$

Let E_a of the reaction in absence of catalyst = x kJ mol^{-1}

Therefore E_a of the reaction in presence of catalyst

$= x - 20$ kJ mol^{-1}

The Arrhenius equations in the two conditions can thus be written as

$$k = Ae^{-\frac{x}{R \times 500}} \quad ...(i)$$

$$k = Ae^{-\frac{x-20}{R \times 400}} \quad ...(ii)$$

Dividing equation (i) by (ii), we get

$$e^{-\frac{x}{500R}} = e^{-\frac{x-20}{400R}}$$

$$\Rightarrow \frac{x}{500} = \frac{x-20}{400} \text{ or } x = 100 \text{ kJ mol}^{-1}$$

52. (20.34)

$$k_{653\,K} = \frac{0.693}{360} = 1.925 \times 10^{-3}\ \text{min}^{-1}$$

Calculation of $k_{723\,K}$

$E_a = 200$ kJ mol^{-1} = 200×10^3 J mol^{-1}

$T_2 = 723$ K, $T_1 = 653$ K

$k_{653\,K} = 1.925 \times 10^{-3}$min^{-1}

We know that, $\frac{k_2}{k_1} = e^{-\frac{E_a}{R}\left(\frac{1}{T_2} - \frac{1}{T_1}\right)}$

or $$\frac{k_{723K}}{1.925 \times 10^{-3}} = e^{-\frac{200 \times 10^3}{8.314}\left[\frac{1}{723} - \frac{1}{653}\right]}$$

On usual calculations, $k_{723\,K} = 6.81 \times 10^{-2}$ min^{-1}

Calculation of time for 75% decomposition at 723 K

Let the initial amount of H_2O_2, $a = 1$

$\therefore$ Amount at the required time, $(a - x) = 0.25$

Substituting the values in the given relation,

$$t = \frac{2.303}{k_{723K}} \log \frac{a}{a-x} = \frac{2.303}{6.81 \times 10^{-2}} \log \frac{1}{0.25}$$

$= 20.34$ min

53. (311.35) According to Arrhenius equation

$$\log k = \log A - \frac{E_a}{2.303\,RT}$$

We know that

$$k = \frac{0.693}{t_{½}} = \frac{0.693}{10 \times 60} \quad (t_{½} = 10 \times 60 \text{ sec.})$$

$= 1.555 \times 10^{-3}$

Substituting the various values in the above equation, we get

$$\log 1.155 \times 10^{-3} = \log 4 \times 10^{13} - \frac{98.6}{2.303 \times 8.314 \times 10^{-3} \times T}$$

On usual calculations, $T = 311.35$ K

54. very high temperature ($T = \infty$) or zero activation energy.

55. True : The rate of a reaction increases with increase in temperature because at higher temperature more number of molecules attain the activation energy.

Note : For a reversible exothermic reaction, the rate of both forward and backward reactions will increase with increase in temperature. But according to Le Chatelier's principle, increased temperature favour the backward reaction.

56. False : Catalyst lowers the energy of activation and therefore influences the rate as well as rate constant of the reaction.

57. False : Catalyst does not make a reaction more exothermic, but decreases the activation energy and hence, increase the rate of reaction.

58. (a, c) On increasing temperature, concentration of product decreases.

Hence, reaction is exothermic $\Rightarrow \Delta H° < 0$

$$\frac{\ln K_{T_1}}{\ln K_{T_2}} > 1 \Rightarrow \ln K_{T_1} > \ln K_{T_2} \text{ so, } K_{T_1} > K_{T_2}$$

Also, $\frac{\ln K_{T_1}}{\ln K_{T_2}} > \frac{T_2}{T_1}$

or $T_1 \ln K_{T_1} > T_2 \ln K_{T_2}$

$\Rightarrow -RT_1 \ln K_{T_1} > -RT_2 \ln K_{T_2}$

or $\Delta G^\circ_{T_1} < \Delta G^\circ_{T_2}$ ($\because \Delta G = -RT \ln K$)

or $\Delta H° - T_1 \Delta S° < \Delta H° - T_2 \Delta S°$

(Also $\Delta G° = \Delta H° - T\Delta S°$: Gibbs Helmholtz equation)

As $T_1 < T_2$, this is possible only when $\Delta S° < 0$

59. (a, b) According to Arrhenius equations

$$k = Ae^{-Ea/RT}$$

where, A = Frequency factor

Taking into account orientation factor,

$P = \frac{A}{Z}$ or $A = PZ$

$k = PZe^{-Ea/RT}$

where, P = steric factor, Z = collision frequency

The value of steric factor lies between 0 and 1 predicted by Arrhenius equation. Thus, the experimentally determined value of frequency factor is higher than that predicted by Arrhenius equation. Thus, (b) is correct.

60. (b, c, d)

(a) High activation energy usually implies a **slow reaction**.

(b) Rate constant of a reaction increases with increase in temperature due to increase in number of collisions whose energy exceeds the activation energy.

(c) $k = P \times Z \times e^{-E_a/RT}$

(d) So, pre-exponential factor $(A) = P \times Z$ and it is independent of activation energy or energy of molecules.

61. (b, c) A catalyst provides a new path of lower activation energy. The catalyst reacts with the reactants to form an activated complex of low activation energy. The activated complex then decomposes to form the products along with regeneration of catalyst. Thus, the reaction mechanism changes completely.

62. (a) Assertion is correct as for every 10ºC raise in temperature, the specific rate constant, K nearly doubles. (Although, it is not correct for all reactions. For some reactions, K even gets tripled for 10ºC raise).

[The value of temperature coefficient is the ratio of rate constants at two different temperature (which differ by 10°C) generally lies between 2 and 3]

The statement is clearly true and it explains the assertion, as the rate of collision among the molecules doubles for 10° rise in temperature.

63. $\log\frac{k_2}{k_1} = \frac{E_a}{2.303 \times R}\left[\frac{T_2 - T_1}{T_1 T_2}\right]$

$\therefore \log\frac{4.5\times10^7}{1.5\times10^7} = \frac{E_a}{8.314\times2.303}\left[\frac{373-323}{373\times323}\right]$

$\therefore E_a = \mathbf{2.2\times10^4\,J\,mol^{-1}}$

Now $k = Ae^{-E_a/RT}$

$\therefore 4.5\times10^7\,J\,mol^{-1} = A.e^{-\frac{2.2\times10^4}{8.314\times373}}$

$\therefore A = \mathbf{5.42\times10^{10}\,sec^{-1}}$

64. (i) The Arrhenius equation is $k = A\exp(-E_a/RT)$

Taking natural logarithm, we get

$\ln k = \ln A - E_a/RT$

Thus $\log k = \log A - \frac{Ea}{2.303R}.\frac{1}{T}$

Comparing this expression with the given one, we get

$\frac{E_a}{2.303R} = 1.25\times10^4\,K$

Hence, $E_a = (1.25\times10^4\,K)(2.303)(8.314\,JK^{-1}mol^{-1})$
$= 2.39\times10^5\,J\,mol^{-1} = \mathbf{239\,kJ\,mol^{-1}}$

(ii) The reaction is first order as the unit of rate constant is s^{-1}. For a first order reaction,

$t_{1/2} = \frac{0.693}{k}$

Hence, $k = \frac{0.693}{256\times60\,s} = 4.51\times10^{-5}\,s^{-1}$

Substituting this in the given expression, we get

$\log\left(4.51\times10^{-5}\right) = 14.34 - \frac{1.25\times10^{-5}K}{T}$

or $-4.346 = 14.34 - \frac{1.25\times10^{-4}K}{T}$

or $T = \frac{1.25\times10^{-4}K}{18.686} = \mathbf{669\,K}$

65. Let the order with respect to A is x and the order with respect to B is y

Then, Rate $= k[A]^x[B]^y$

$5.0\times10^{-4} = k[2.5\times10^{-4}]^x[3.0\times10^{-5}]^y$ ...(*i*)

$4.0\times10^{-3} = k[5.0\times10^{-4}]^x[6.0\times10^{-5}]^y$ ...(*ii*)

$1.6\times10^{-2} = k[1.0\times10^{-3}]^x[6.0\times10^{-5}]^y$ ...(*iii*)

From (*ii*) and (*iii*), we get $2^x = 4;\ x = 2$

From (*i*) and (*ii*), we get $2^{x+y} = 8;\ x+y = 3; y = 1$

$\therefore$ The rate equation for the reaction is

Rate $= k[A]^2[B]$

(i) Thus, **order of reaction with respect to $A = 2$ and order of reaction with respect to $B = 1$.**

(ii) Rate constant (k_1) at 300 K

$k_1 = \frac{\text{Rate}}{[A]^2[B]}$

$= \frac{5.0\times10^{-4}\,mol\,L^{-1}\,s^{-1}}{(2.5\times10^{-4}\,mol\,L^{-1})^2\,(3.0\times10^{-5}\,mol\,L^{-1})}$

$= \mathbf{2.66\times10^8\,mol^{-2}\,L^2\,s^{-1}}$

(iii) Determination of energy of activation : Rate constant (k_2) at 320 K

$k_2 = \frac{\text{Rate}}{[A]^2[B]}$

$k_2 = \frac{2.0\times10^{-3}\,mol\,L^{-1}\,s^{-1}}{(2.5\times10^{-4}\,mol\,L^{-1})^2\,(3.0\times10^{-5}\,mol\,L^{-1})}$

$= 1.066\times10^9\,mol^{-2}\,L^2\,s^{-1}$

$\log\frac{k_2}{k_1} = \frac{E_a}{2.303\,R}\left[\frac{T_2 - T_1}{T_1\,T_2}\right]$

or $E_a = 2.303\,R\left[\log\frac{k_2}{k_1}\right]\left[\frac{T_1\,T_2}{T_2 - T_1}\right]$

$= 2.303\times8.314\left[\log\frac{1.066\times10^9}{2.66\times10^8}\right]\left[\frac{320\times300}{20}\right]$

$= \mathbf{5.54\times10^4\,J}$

(iv) Determination of the pre-exponential factor

$\log k = -\frac{E_a}{2.303\,R}\times\frac{1}{T} + \log A$

$\log 2.66\times10^8 = \frac{5.54\times10^4}{2.303\times8.314}\times\frac{1}{300} + \log A$

or $8.425 + 9.645 = \log A;\ \log A = 18.07$

$\therefore\ A = \mathbf{1.175\times10^{18}\,L^2\,mol^{-2}\,sec}$

66. According to Arrhenius equation

$\log\frac{k_2}{k_1} = \frac{E_a}{2.303\,R}\left(\frac{1}{T_1} - \frac{1}{T_2}\right)$

Calculation of k at 25°C (298 K), i.e. k_1

Here $a = 100$, $a - x = 100 - 25$, $t = 20$ mts.

Thus $k_1 = \frac{2.303}{20}\log\frac{100}{75} = 0.014386\,min^{-1}$

Calculation of k at 40°C (313 K) i.e., k_2

Substituting various values, we get

$\log\frac{k_2}{0.014386} = \frac{70\times10^3}{2.303\times8.314}\times\left(\frac{1}{298} - \frac{1}{313}\right)$

$= \frac{70\times10^3\times15}{2.303\times8.314\times298\times313} = 0.587$

$\log k_2 = 0.014386\times0.587;\ k_2 = 0.05570\,min^{-1}$

Calculation of % decomposition at 40°C

Thus here $a = 100$, $a - x = 100 - x$, $t = 20$ min, $k_2 = 0.05570$

Substituting the values in the first order reaction equation :

$k_2 = 0.05570 = \frac{2.303}{20}\log\frac{100}{100 - x}$

On usual calculations, $x = 67.169 = \mathbf{67.17\%}$

Value of rate constant k does not depend on the initial concentration (20% or 30% solution).

67. $A \rightarrow$ Products...(i)

$B \rightarrow$ Product...(ii)

Half-life of (i) reaction at 310 K = 30 min.

$\therefore {}_Ak_{310} = \frac{0.693}{30}\,min^{-1}$...(1)

Also given, $\frac{{}_Ak_{310}}{{}_Ak_{300}} = 2$ ...(2)

Also at 310 k, ${}_Bk_{310} = 2\,{}_Ak_{310}$ ...(3)

Also $E_B = \frac{1}{2}E_A$

According to Arrhenius equation,

$k = Ae^{-E/RT}$ ...(4)

or $\log\frac{k_2}{k_1} = \frac{E}{2.303R}\left[\frac{T_2 - T_1}{T_1\,T_2}\right]$

For reaction (*i*),

$\log\frac{{}_Ak_{310}}{{}_Ak_{300}} = \frac{E_A}{2.303R}\left[\frac{10}{300\times310}\right]$ ...(5)

For reaction (*ii*),

$\log\frac{{}_Bk_{310}}{{}_Bk_{300}} = \frac{E_B}{2.303R}\left[\frac{10}{300\times310}\right]$ ...(6)

Dividing (6) by (5),

$\frac{\log\frac{{}_Bk_{310}}{{}_Bk_{300}}}{\log\frac{{}_Ak_{310}}{{}_Ak_{300}}} = \frac{E_B}{E_A} = \frac{1}{2}\left(\because E_B = \frac{1}{2}E_A\right)$

or $\log\frac{{}_Ak_{310}}{{}_Ak_{300}} = 2\log\frac{{}_Bk_{310}}{{}_Bk_{300}} = \log\left(\frac{{}_Bk_{310}}{{}_Bk_{300}}\right)^2$

or $\frac{{}_Ak_{310}}{{}_Ak_{300}} = \left(\frac{{}_Bk_{310}}{{}_Bk_{300}}\right)^2$ ...(7)

Combining (7), (2) and (3), we get $\left(\frac{2\,{}_Ak_{310}}{{}_Bk_{300}}\right)^2 = 2$

$2\,{}_Ak_{310} = \sqrt{2}\times{}_Bk_{300};\ {}_Bk_{300} = \sqrt{2}\times{}_Ak_{310}$

${}_Bk_{300} = 1.414\times\frac{0.693}{30} = \mathbf{3.267\times10^{-2}\,min^{-1}}$

68. $k = \frac{0.693}{t_{1/2}};\ \log\frac{k_2}{k_1} = \frac{E_a}{2.303\,R}\left[\frac{T_2 - T_1}{T_1.T_2}\right]$

$T_1 = 300K$, $T_2 = 320K$

Substituting the value at the two given conditions

$k_{27°} = \frac{0.693}{30} = \mathbf{0.0231\,min^{-1}}$; $k_{47°}$

$= \frac{0.693}{10} = 0.0693\,min^{-1}$

We also know that

$\log\frac{k_{47}}{k_{27}} = \frac{E_a}{2.303R}\times\frac{T_2 - T_1}{T_1\times T_2}$

or $E_a = \frac{2.303R\times T_1\times T_2}{T_2 - T_1}\log\frac{k_{47}}{k_{27}}$

$= \frac{2.303\times8.314\times10^{-3}\times300\times320}{320 - 300}\times\log\frac{0.0693}{0.0231}$

$= \mathbf{43.848\,kJ\,mol^{-1}}$

$t_{1/2} = 1.386$ min

69. **(a)** ${}_{13}Al^{27} + {}_2\alpha^4 \longrightarrow {}_{14}Si^{30} + {}_1p^1$ (X)

${}_{13}Al^{27} + {}_2\alpha^4 \longrightarrow {}_{15}P^{30} + {}_0n^1$ (Y)

${}_{15}P^{30} \longrightarrow {}_{14}Si^{30} + {}_1\beta^0$ (Z)

70. **(c)** ${}^{23}_{11}Na \longrightarrow {}^{23}_{10}X + {}^{0}_{1}\beta$

71. **(a)** *n*/*p* ratio of ^{24}Na nuclide is 13/11 i.e. greater than unity and hence ^{24}Na is radioactive. To achieve stability, it would tend to adjust its n/p ratio to the proper value of unity. This can be done by breaking a neutron into proton and electron.

${}_0n^1 \longrightarrow {}_{+1}p^1 + {}_{-1}e^0$ or β^-

The proton will stay inside the nucleus whereas electron which cannot exist in the nucleus, will be emitted out as β-ray.

72. (d) The sum of mass number and atomic numbers of reactants = The sum of mass number and atomic no. of products in a nuclear reaction. The given nuclear fission reaction is

$$^{235}_{92}U + ^{1}_{0}n \to ^{139}_{54}Xe + ^{94}_{38}Sr + 3\,^{1}_{0}n$$

73. (b) The species $_{13}Al^{29}$ (No. of neutrons = 16) contains more neutrons than the stable isotope $_{13}Al^{27}$ (No. of neutrons = 14).

Neutron on decomposition shows β -emission.

$$_0n^1 \to {}_{+1}p^1 + {}_{-1}e^0 \quad (\beta\text{–particle})$$

74. (d) $N = N_0\left(\frac{1}{2}\right)^n$

where, N = Amount of radioactive substance which is left after certain number of half-life periods (n)

N_0 = Initial amount of radioactive substance.

$$\text{No. of half-lives} = \frac{\text{total time}}{\text{half life period}} = \frac{560}{140} = 4$$

In 'n' half-lives, the element will reduce to

$$\left(\frac{1}{2}\right)^n \times \text{Initial wt.} = \left(\frac{1}{2}\right)^4 \times 1 = \frac{1}{16}\text{g}$$

75. (d) α- and β-rays, made up of positively and negatively charged particles and are deflected by a magnetic field in opposite directions; γ-rays remain undeflected. (as they do not have charge).

76. (b) $_{92}U^{238} \to {}_{90}U^{234} + {}^{4}_{2}He$ (α-particle)

77. (9) Number of moles in gas phase, at start $(n_i) = 1$

$$^{238}_{92}U \to ^{206}_{82}Pb + 8\,^{4}_{2}He + 6\,^{0}_{-}\beta$$

Now number of moles in gas phase, after decomposition $(n_F) = 1 + 8 = 9$ mole

At constant temperature and pressure

$$\frac{P_F}{P_{in}} = \frac{n_F}{n_{in}} = \frac{9}{1} = 9$$

78. (8)

$$^{63}_{29}Cu + ^{1}_{1}H \to 6\,^{1}_{0}n + ^{4}_{2}He + 2\,^{1}_{1}H + ^{A}_{Z}X$$

Balancing the atomic mass and atomic number

$63 + 1 = (6 \times 1) + 4 + 2 + A \Rightarrow A = 52$

$29 + 1 = (6 \times 0) + 2 + 2 + Z \Rightarrow Z = 26$

Thus $^{A}_{Z}X = ^{52}_{26}X$ or $^{52}_{26}Fe$

Hence, X belongs to group 8 in the periodic table.

79. (3) $_{92}U^{235} \to {}_{54}Xe^{142} + {}_{38}Sr^{90} + y\,_0n^1$

$235 = 142 + 90 + y \Rightarrow y = 3.$

The number of neutrons emitted are 3.

80. (8)

$$_{92}U^{238} \xrightarrow{-6\alpha} {}_{80}X^{214} \xrightarrow{-2\beta} {}_{82}Pb^{214}$$

Hence, total number of particles emitted are 2 + 6 = 8

81. (1.20) $^{238}_{92}U \longrightarrow ^{206}_{82}Pb$

$$\text{Initial moles of } U^{238} = \frac{68\times10^{-6}}{238} = x$$

Number of α-particles emitted per nuclei of

$$^{238}_{92}U = \frac{238-206}{4} = 8$$

Moles of U^{238} decayed in three half-lives $= \frac{7}{8}x$

No. of α-particles emitted $= \left(\frac{7}{8}x\right) \times 8 \times N_A$

$$= 7 \times \frac{68\times10^{-6}}{238} \times 6.022\times10^{23} = 1.204 \times 10^{18}$$

∴ Z = 1.20

82. (0.061) Wt. of ^{90}Sr at start, $N_0 = 1\,\mu g$, Wt. of ^{90}Sr after 20 years,

N_t = ?, Time t = 20 years

Half-life of ^{90}Sr, $T_{1/2}$ = 28.1 years

Calculation of decay constant,

$$\lambda = \frac{0.693}{T_{1/2}} = \frac{0.693}{28.1} = 0.0247\ \text{years}^{-1}$$

Substituting the values in the relation

$N_t = N_0 e^{-\lambda t} = 1\,e^{-0.0247\times20} = 0.061\ \mu g$

83. Isotope; (because new atom has same atomic number but different atomic mass). [Refer to Q. 1 above]

84. 8; $^{14}_{6}C \to ^{14}_{7}N + ^{0}_{-1}e$

Due to emission of the β particle, atomic number increases by 1.

85. $^{A-4}_{Z}M$

When an element emits α -particle, atomic mass decreases by four and atomic number decreases by two. Loss of β -particle results in increase in atomic number by 1 and no change in atomic mass.

$$^{A}_{Z}M \xrightarrow{-\alpha} {}^{A-4}_{Z-2}X \xrightarrow{-\beta} {}^{A-4}_{Z-1}Y \xrightarrow{-\beta} {}^{A-4}_{Z}M$$

86. False : In β-emission ($_{-1}e^0$) the atomic number of the daughter nuclei increases by 1.

87. (a, b, d)

$^{238}_{92}U \longrightarrow ^{234}_{90}Th + ^{4}_{2}He;\ x_1 = \alpha$

$^{234}_{90}Th \longrightarrow ^{234}_{91}Pa + ^{0}_{-1}e;\ x_2 = \beta$

$^{234}_{91}Pa \longrightarrow ^{234}_{92}Z + ^{0}_{-1}e;\ x_3 = \beta$

$^{234}_{92}Z \longrightarrow ^{230}_{90}Th + ^{4}_{2}He;\ x_4 = \alpha$

88. (b, d) When N/P ratio is less than one, then proton changes into neutron to increase the ratio.

Position emission : $^{1}_{1}H \longrightarrow ^{1}_{0}n + ^{0}_{+1}\beta$

K-electron capture :

$^{1}_{1}H + ^{0}_{-1}e \longrightarrow ^{1}_{0}n + \text{X-rays}$

89. (a, b) The reaction can occur by following two ways.

$^{9}_{4}Be + \gamma \to ^{8}_{4}Be + ^{1}_{0}n\ ;\ ^{9}_{4}Be + ^{1}_{1}H \to ^{2}_{1}H + ^{8}_{4}Be$

90. (a,c,d)

(a) $_zX^A \xrightarrow{-\alpha} {}_{z-2}Y^{A-4}$ (α - emission) ;

(b) $_zX^A \xrightarrow{-\beta} {}_{z+1}Z^A$ (β - emission);

(c) $_zX^A \xrightarrow{-{}_{+1}e^0} {}_{z-1}D^A$ (positron - emission);

(d) $_zX^A + {}_{-1}e^0 \to {}_{z-1}D^A$ (electron - capture)

Atomic number increases during β-emission

91. (a, d) Balance various given nuclear reactions.

(a) $_{13}Al^{27} + {}_2He^4 \to {}_{15}P^{30} + {}_0n^1$

(d) $_{96}Am^{241} + {}_2He^4 \to {}_{97}Bk^{244} + {}_1e^0 + {}_0n^1$

92. (c) It is clear from the 3rd paragraph, which states that in living organisms a dynamic equilibrium is established whereby the ratio of C^{14} to C^{12} remains constant. The C^{14} which decays into N^{14} is replenished by the production of new isotopes.

93. (b) As the half-life of C^{14} is 5760 years, so a 6 year old fossil's age can't be determined. Further this technique cannot be used to date objects older than 30,000 years. After this length of time the radioactivity is too low to be measured.

94. (a) $T_1 = \frac{1}{\lambda}\ln C_1,\ T_2 = \frac{1}{\lambda}\ln C_2$

Let the concentration of C^{14} in the fossil be C. In nearby areas concentration of C^{14} in living beings will be C_1 and in far off places C_2, obviously $C_1 > C_2$.

Hence, age of fossil in nearby areas,

$$T_1 = \frac{1}{\lambda}\ln\frac{C_1}{C} \quad \text{....... (i)}$$

And age in far off places, $T_2 = \frac{1}{\lambda}\ln\frac{C_2}{C}$ ..... (ii)

From (i) and (ii), $T_1 - T_2 = \frac{1}{\lambda}\ln\frac{C_1}{C_2}$

Since, $C_1 > C_2$, R.H.S. is positive *i.e.*, $T_1 > T_2$.

95. (a)

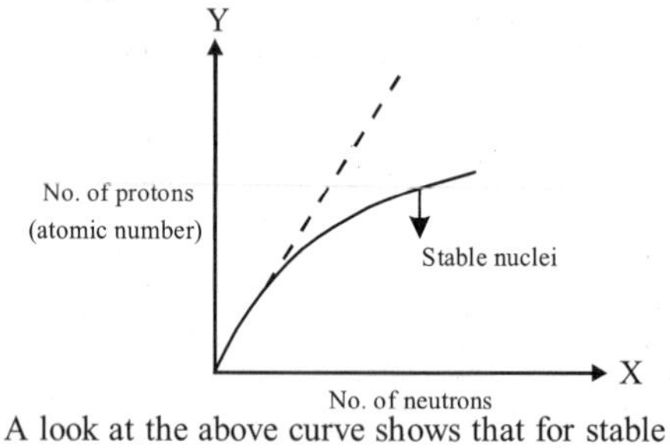

A look at the above curve shows that for stable nuclei it shows a curvature towards x-axis from the line of 45° slope (dotted line) as the atomic number (*i.e.* number of protons) increases. So statement 1 is true.

The proton - proton repulsion would overcome the attractive force of proton and neutron. Thus, statement 2 in True. Also this statement 2 is a correct explanation for statement 1. Therefore the correct answer is option (a).

96. (c) Nuclides having both even number of protons and neutrons have maximum stability. So the reason is incorrect. But the assertion is correct as $^{40}Ca_{20}$ has even number of neutrons and protons as compared to $^{30}Al_{13}$, which has odd neutrons and protons.

97. (i) Sum of atomic masses of reactant = sum of atomic masses of products

(ii) Sum of atomic numbers of reactant = sum of atomic numbers of products

(*i*) The atomic number of the final stable product = 90 – 7 × 2 + 1 × 6 = 82 and the mass number of the final stable product = 234 – 7 × 4 + 0 = 206. Thus, the element should be $_{82}Pb^{206}$.

(*ii*) $_{92}U^{235} + {}_0n^1 \longrightarrow {}_{52}Te^{137} + {}_{40}Zr^{97} + 2\,_0n^1$

(*iii*) $_{34}Se^{86} \longrightarrow 2\,_{-1}e^0 + {}_{36}Kr^{86}$

98. 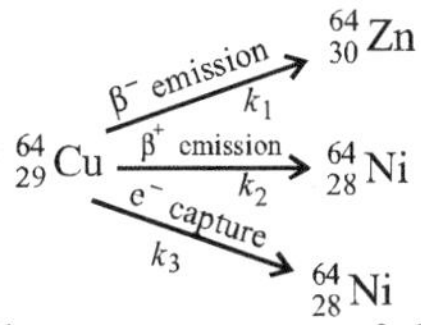

Let the rate constants of the above emission processes be k_1, k_2 and k_3 respectively and the overall rate constant be k. Then,

$$k = k_1 + k_2 + k_3 = \frac{0.693}{t_{1/2}} = \frac{0.693}{12.8} h^{-1}.$$

Also, $k_1 = 0.38 k = 0.38 \times \frac{0.693}{12.8} h^{-1}$

$$t_1 = \frac{0.693 \times 12.8}{0.38 \times 0.693} = \mathbf{33.68h}$$

Similarly,

$$t_2 = \frac{0.693}{k_2} = \frac{0.693}{0.19k} = \frac{0.693}{0.19 \times 0.693} \times 12.8$$

= 67.36 h

$$t_3 = \frac{0.693}{k_3} = \frac{0.693}{0.43k} = \frac{12.8}{0.43} = \mathbf{29.76h}$$

where t_1, t_2 and t_3 are the partial half-lives for β^- emission, β^+ emission and electron capture processes, respectively.

Alternate Solution:

Let β^- emission $= \lambda_1, \beta^-$ emission $= \lambda_2$, r electron emission $= \lambda_3$

$\lambda_{net} = \lambda_1 + \lambda_2 + \lambda_3$

$$\text{Fractional yield of path (I)} = \frac{\lambda 1}{\lambda net} = \frac{(t_{1/2})net}{(t_{1/2})_1}$$

$$\text{Half life of path I} = \frac{(t_{1/2})}{fractional\ yield}$$

$$= \frac{12.8}{0.38} = 33.684 \text{ hr}$$

$$\text{Half life of path II} = \frac{12.8}{0.19}, = 67.368 \text{ hr}$$

$$\text{Half life of path III} = \frac{12.8}{0.43}, = 29.767 \text{ hr}$$

99. Let the number of α-particles emitted $= a$ and number of β-particles emitted $= b$

$$_{92}U^{238} \longrightarrow a\,_2\alpha^4 + b\,_{-1}\beta^0 + _{82}Pb^{206}$$

Thus $4a + 206 = 238$; Therefore $a = 8$

Further $2a - b + 82 = 92$; Therefore $b = 6$

Composition of the ore indicates that it has 1 g of U and 0.1 gm of Pb; thus here $N_t = 1$ g

Determination of N_0

206 g Pb is obtained from 238 g of U

0.1 g Pb is obtained from

$$= \frac{238}{206} \times 0.1g = 0.1155g$$

Therefore, initial amount of U (N_0) = 1 + 0.1155 = 1.1155

Now we know that

$$t = \frac{2.303}{k} \log \frac{N_0}{N_t} = \frac{2.303}{0.693/4.5 \times 10^9} \log \frac{1.1155}{1}$$

By usual calculations, $\mathbf{t = 7.097 \times 10^8}$ **years.**

100. $_7N^{14} + _2He^4 \rightarrow [_9F^{18}] \rightarrow _8O^{17} + _1H^1$

101. $\lambda_{Ac} = \frac{0.693}{21.8} = 3.18 \times 10^{-2}$ year^{-1}

Since, the decay involves two parallel paths

$Th^{227} \longleftarrow Ac^{227} \longrightarrow Fr^{223}$

$\lambda_{Ac} = \lambda_{Th} + \lambda_{Fr}$

Thus, Fractional yield of Th $= \frac{\lambda_{Th}}{\lambda_{Ac}}$

or $\lambda_{Th} = 3.18 \times 10^{-2} \times \frac{1.2}{100} = \mathbf{3.81 \times 10^{-4} yr^{-1}}$

Similarly, Fractional yield of Fr $= \frac{\lambda_{Fr}}{\lambda_{Ac}}$

$\therefore \lambda_{Fr} = 3.18 \times 10^{-2} \times \frac{98.8}{100} = \mathbf{3.14 \times 10^{-2} yr^{-1}}$

102. The ratio of $H^3 : H^1 :: 8 \times 10^{-18} : 1$

$\because$ No. of H atoms in 18 g $H_2O = 2N_A$

$\therefore$ No. of H^3 atoms in 18 g of H_2O

$= 2N_A \times 8 \times 10^{-18}$

$= 2 \times 6.023 \times 10^{23} \times 8 \times 10^{-18}$ atoms

$\therefore$ No. of H^3 atoms in 10 g H_2O

$$= \frac{2 \times 6.023 \times 10^{23} \times 8 \times 10^{-18} \times 10}{18} \text{ atoms}$$

$= 5.354 \times 10^6$ atoms

No. of atoms left after 40 years are derived as follows using the relation

$$t = \frac{2.303}{\lambda} \log \frac{N_0}{N}$$

$$40 = \frac{2.303 \times 12.3}{0.693} \log \frac{5.354 \times 10^6}{N}$$

$\therefore$ $\mathbf{N = 5.624 \times 10^5}$ **atoms**

103. Minimum number of β-particles required in one minute = 346

No. of β-particles required for carrying out the experiment for 6.909×60 minutes $= 346 \times 6.909 \times 60 = 143431$

$\therefore$ Amount of β-particles required

$$= \frac{143431}{6.023 \times 10^{23}} = 2.3814 \times 10^{-19} \text{mol}$$

Now we know that,

$$\lambda = \frac{0.693}{t_{½}} = \frac{0.693}{66.6} = 0.0104 \text{ hr}^{-1}$$

Further we know that, $\lambda = \frac{2.303}{t} \log \frac{a}{a-x}$

where a = Initial concentration of β-particles

x = Consumed concentration of β-particles

$$\log \frac{a-x}{a} = -\frac{\lambda t}{2.303}$$

$$= -\frac{0.0104 \times 6.909}{2.303} = -0.0312 = \bar{1}.9688$$

or $\frac{a-x}{a} = 0.931$ [Taking antilog]

or $\frac{a - 2.3814 \times 10^{-19}}{a} = 0.931$

$[\because x = 2.3814 \times 10^{-19}]$

On usual calculations, $a = \mathbf{3.451 \times 10^{-18}}$ **mol**

Alternate Solution:

The Minimum rate of decay required after 6.909 h is 346 particle min^{-1}

Rate = kN

$$N = \frac{Rate}{k} = \frac{346 \times 66.6 \times 60}{0.693} = 1.995 \times 10^6 \text{ atom}$$

$$kt = \ln \frac{N_o}{N}$$

$$\frac{\ln 2}{66.6} \times 6.909 \Rightarrow \frac{\ln No}{N} = 0.0715$$

$$= \frac{N_o}{N} 1.074$$

$\Rightarrow$ $N_o = 1.074 \times N = 1.074 \times 1.995 \times 10^6$

$\Rightarrow$ 2.14×10^6 atom of N_o

$\Rightarrow$ Mass of N_o required $= \frac{2.14 \times 10^6 \times 99}{6.023 \times 10^{23}}$

$= 3.56 \times 10^{-16}$ g.

104. $^{234}_{90}Th \rightarrow ^{206}_{82}Pb$

For emission of one α-particle, atomic mass decreases by 4 and atomic number by 2. Further, for the emission of one β-particle, the atomic mass does not change but the atomic number increases by 1.

So, we first find the α-particles :

Decrease in atomic mass = 234 – 206 = 28

No. of α-particles emitted $= \frac{28}{4} = 7$

Hence, atomic number should have decrease to $90 - (7 \times 2) = 76$

Now, atomic number of Pb = 82,

which is more by (82 – 76) = 6

This increase is due to emission of β-particles.

Therefore, β-particles emitted= **6**

105. $N = N_0 \left(\frac{1}{2}\right)^n$

Half life, $t_{1/2}$ = 5770 years

Let the original sample be 1 g.

$\therefore$ After every 5770 years one-half of radioactive carbon would decay or disintegrate.

Thus, 1 g sample becomes ½ g after 5770 years

and $\frac{1}{2} \times \frac{1}{2} = \mathbf{\frac{1}{4}}$ left after 11,540 years.

$\therefore$ 25% of radioactive carbon remains after 11540 years.

Rate constant, k for first order reaction,

$$k = \frac{0.693}{t_{½}} = \frac{0.693}{5770} = \mathbf{1.2 \times 10^{-4}\ year^{-1}}.$$

Chapter-5

Surface Chemistry

Part-A (JEE Main)

1. **(3)** Due to polar nature of SO_2, it has higher critical temperature than CH_4. Activated charcoal adsorbs SO_2 more efficiently than CH_4 due to more intermolecular attraction forces in SO_2. Hence (A) is correct but (R) is not correct.

2. **(4)** $\frac{x}{m} \alpha P^{1/n}\left(0<\frac{1}{n}<1\right)$

On Increasing temperature, extent of $\left(\frac{x}{m}\right)$ decreases.

3. **(3)**

(i) Adsorption leads to a decrease in the residual forces on the surface adsorbent. This causes a decrease in surface energy of the adsorbent. Therefore, adsorption is always exothermic, i.e., $\Delta H < 0$.

(ii) When a gas is adsorbed, randomness of units molecules become restricted due to association between adsorbate and adsorbent. Therefore, entropy decreases, i.e., $\Delta S < 0$.

4. **(4)** Freundlich adsorption isotherm can be plotted using $\frac{x}{m} = kP^{1/n}$ ...(i)

Given : At moderate pressure, $\frac{x}{m} \propto P^{\chi}$ ...(ii)

Comparing equations (i) and (ii), we get $\chi = \frac{1}{n}$

5. **(4)** Freundlich adsorption isotherm,

$\frac{x}{m} = kp^{1/n}$

Taking logarithm, $\log\left(\frac{x}{m}\right) = \log k + \frac{1}{n}\log p$

$\Rightarrow y = c + mx$

$m = 1/n$ so slope will be equal to $1/n$.

$\frac{1}{n}$ can have value between 0 and 1.

6. **(3)**

(A) When gas is adsorbed on metal surface, ΔH becomes less negative with progress of adsorption.

(B) The gas having greater value of critical temperature (T_C) is adsorbed more compared to lower T_C containing gases. As $T_C(NH_3) > T_C(N_2)$, so NH_3 is adsorbed more than N_2 gas.

7. **(2)** Initially, adsorption of gases at the surface of charcoal occurs rapidly which results in a sudden decrease in pressure. As the number of vacant sites at the surface of adsorbent decreases with the passage of time, rate of adsorption decreases. Therefore, pressure tends to be constant.

8. **(3)** Freundlich adsorption is applicable for physical adsorption. The variation of extent of adsorption with (i) Pressure and (ii) Temperature is given by the following curves.

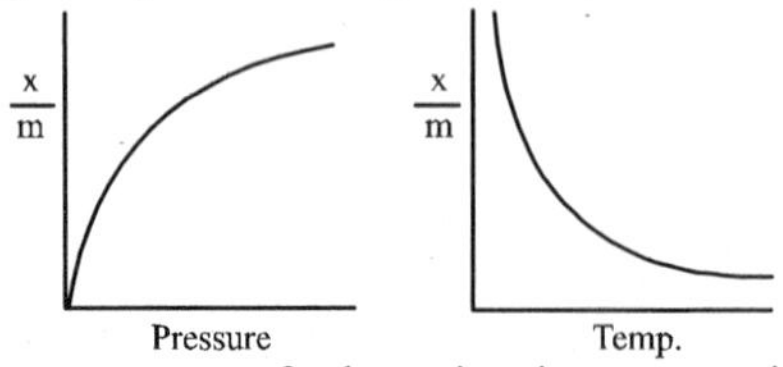

Hence, extent of adsorption increases with increase in pressure and decrease in temperature.

9. **(2)** No. of surfactant molecule

$= 6 \times 10^{23} \times 10 \times 10^{-3} \times 1 \times 10^{-3} = 6 \times 10^{18}$ molecule

Let edge length = a cm

Total surface area of surfactant = $6 \times 10^{18} \times a^2$

$= 0.24\ cm^2$

$\Rightarrow a = 2 \times 10^{-10}\ cm = 2$ pm

10. **(1)** According to Freundlich adsorption isotherm,

$\log\frac{x}{m} = \log k + \frac{1}{n}\log p$

Slope $= \frac{1}{n} = \frac{2}{3} \Rightarrow \frac{x}{m} \propto p^{\frac{2}{3}}$

11. **(4)** Smaller the value of critical temperature of gas, lesser is the extent of adsorption.

$\therefore$ Least adsorption is shown by H_2 (least critical temperature).

12. **(4)** The physical adsorption is temperature and pressure dependent. It increases with increase in pressure and decreases with increase in temperature. Also, when there is more surface area then physical adsorption will be more. Physical adsorption forms multiple layer of adsorption. Thus, statement given in option (4) is incorrect.

13. **(2)** According to Freundlich adsorption isotherm.

$\frac{x}{m} = kP^{\frac{1}{n}} \Rightarrow \log_{10}\frac{x}{m} = \frac{1}{n}\log_{10}P + \log_{10}k$

This is the equation of straight line of type $y = mx + c$.

Hence slope is $1/n$ (m) and intercept is $\log_{10}k$.

14. **(3)**

According to Freundlich adsorption isotherm

$\log\frac{x}{m} = \log k + \frac{1}{n}\log p$

$\frac{x}{m} = k \cdot p^{1/n}$

$\Rightarrow \boxed{\frac{x}{m} \propto p^{1/2}}$ $\left(\text{Slope } \frac{1}{n} = 0.5\right)$

15. **(4)** The characteristics given suggests that this should be physical adsorption.

Physical adsorption usually takes place at low temperature and decreases with increase in temperature. The force of attraction holding the adsorbate are van der Waals forces. Heat of adsorption is low. It is reversible and forms multimolecular layer. It does not acquire any activation energy.

16. **(3)** Adsorption is a phenomenon of attracting and retaining the molecules of a substance on the surface of a liquid or a solid resulting into a higher concentration of the molecules on the surface.

After adsorption, there is a decrease in the residual forces due to bond formation. ΔG, ΔH and ΔS all are negative in the case of adsorption.

17. **(3)** Let the weight of acetic acid initially be w_1 in 50 mL of 0.060 N solution.

$N_1 = \frac{w_1 \times 1000}{\text{M. wt.} \times 50} \Rightarrow 0.06 = \frac{w_1 \times 1000}{60 \times 50}$

$\Rightarrow w_1 = \frac{0.06 \times 60 \times 50}{1000} = 0.18\ g = 180$ mg.

After an hour, the strength of acetic acid = 0.042 N

Let the weight of acetic acid be w_2

$N_2 = \frac{w_2 \times 1000}{60 \times 50}$

$0.042 = \frac{w_2 \times 1000}{3000} \Rightarrow w_2 = 0.126\ g = 126$ mg

So amount of acetic acid adsorbed per 3g

= 180 – 126 mg = 54 mg

$\therefore$ Amount of acetic acid adsorbed per g

= 54/3 = 18 mg

18. **(4)** The Freundlich adsorption isotherm is mathematically represented as, $\frac{x}{m} = kp^{1/n}$

At high pressure, $1/n = 0$, Hence, $x/m \propto p^{\circ}$

At low pressure, $1/n = 1$, Hence, $x/m \propto p^1$

19. **(3)** Both are exothermic.

20. **(3)** Physiosorption is an exothermic process, so ΔH will always be negative.

21. **(1)** According to Langmuir's adsorption isotherm,

$\frac{x}{m} = \frac{kp}{1+kp} \Rightarrow \frac{x}{m} = \frac{1}{(1/kp)+1}$

As we increase p, $\{(1/kp) + 1\}$ decreases and $\frac{1}{(1/kp)+1}$ increases or we can say that x/m increases.

We cannot say that, the mass of gas striking a given area of surface (x/m) is proportional to the pressure of gas (p) but we can say that on increasing p, x/m also increases.

22. **(1)** Considering high pressure, adsorption on metal surface do not depend on concentration of gas, i.e., zero order reaction.

23. (3) The freundlich adsorption isotherm equation is:

$$\frac{x}{m} = kP^{\frac{1}{n}}$$

taking log both sides; $\log\frac{x}{m} = \frac{1}{n}\log P + \log k$

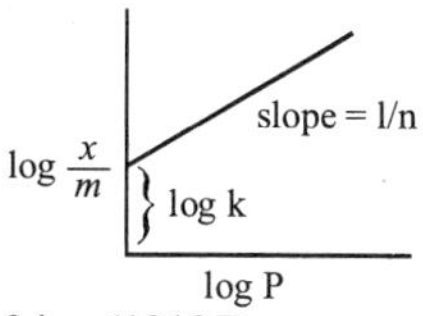

24. (13125)

Initial moles of acetic acid $= M_1 \times V_1$

$= 100 \times 0.3 = 30$ m mol.

moles after adsorption $= 100 \times 0.125 = 12.5$ m mol.

moles adsorbed $= 30 - 12.5 = 17.5$ m mol.

for acetic acid; weight adsorbed.

$$\text{moles} = \frac{\text{Weight}}{\text{Molar Mass}}$$

weight $= 17.5 \times 10^{-3} \times 60 = 1050 \times 10^{-3}$g

for 0.8g charcoal mass of acetic acid adsorbed $= 1050 \times 10^{-3}$ g

$$1\text{g charcoal} = \frac{1050\times10^{-3}}{0.8}$$

$= 1312.5 \times 10^{-3}$ g $= 13125 \times 10^{-4}$ g

25. (12) Using Freundlich adsorption isotherm equation.

$$\frac{x}{m} = kP^{\frac{1}{n}};\ \log\frac{x}{m} = \log k + \frac{1}{n}\log P$$

From graph

$$\text{Slope} = \frac{1}{n} = 1 \Rightarrow n = 1$$

Intercept $= \log k = 0.602$; $k = 4$

$$\frac{x}{m} = 4\times(0.03)^{\frac{1}{1}} \Rightarrow \frac{x}{m} = 12\times10^{-2}$$

So mass of gas adsorbed per gram of the adsorbent is 12×10^{-2} g

26. (9960)

$$\text{Volume of } H_2 = \frac{nRT}{p} = \frac{2}{2}\times\frac{0.083\times300}{1}$$

$= 24.9$ L $= 24900$ mL

$$\text{Vol. of } H_2 \text{ adsorbed per g of Pt} = \frac{249000}{2.5}\text{mL}$$

$= 9960$ mL.

27. (2) Weight of wood charcoal = 0.6 g

Mass of acetic acid adsorbed

$$= \frac{M_1V_1 - M_2V_2}{1000}\times60$$

$$= \frac{0.2\times200 - 0.1\times200}{1000}\times60 = 1.2\text{g}$$

Mass of acetic acid adsorbed per gram of carbon

$$= \frac{1.2}{0.6} = 2$$

28. (128) We know $\frac{x}{m} = kp^{1/n}$

$$\Rightarrow \frac{10}{1} = k\times(100)^{1/n} \quad \text{...(i)}$$

$$\frac{15}{1} = k\times(200)^{1/n} \quad \text{...(ii)}$$

$$\frac{V}{1} = k\times(300)^{1/n} \quad \text{...(iii)}$$

Dividing equation (i) by equation (ii) :

$$\frac{15}{10} = 2^{1/n} \Rightarrow \log\left(\frac{3}{2}\right) = \frac{1}{n}\log 2$$

$$\frac{1}{n} = \frac{\log3-\log2}{\log2} = \frac{0.4771-0.3010}{0.3010}\quad \frac{1}{n} = 0.585$$

Dividing equation (iii) by (i) :

$$\frac{V}{10} = 3^{1/n} \Rightarrow \log\left(\frac{V}{10}\right) = \frac{1}{n}\log 3$$

$$\log\left(\frac{V}{10}\right) = 0.585\times0.4771 = 0.2791$$

$$\frac{V}{10} = 10^{0.28} \Rightarrow V = 10\times10^{0.28}$$

$\Rightarrow V = 10^{1.28} = 10^x \Rightarrow x = 1.278 = 128 \times 10^{-2}$

29. (17) $\frac{x}{m} = k\cdot p^{\frac{1}{n}}$ ...(i)

When pressure is doubled,

$$64\frac{x}{m} = k(2p)^{\frac{1}{n}} \quad \text{...(ii)}$$

Divide (ii) by (i) :

$$\left(\frac{64}{1}\right) = (2)^{\frac{1}{n}} \Rightarrow \frac{1}{n} = 6 \Rightarrow n = 0.166 \cong 17\times10^{-2}$$

30. (2) Using ideal gas equation :

$PV = nRT$

$$V = \frac{3.12\times0.0821\times300}{32\times1} = 2.40\text{L}$$

∴ Volume of O_2(g) adsorbed per gram of the adsorbent $= \frac{2.4}{1.2} = 2$

31. (48) Freundlich adsorption isotherm :

$$\log\frac{x}{m} = \log k + \frac{1}{n}\log p$$

Slope $= \left(\frac{1}{n}\right) = 2$, Intercept $= \log k = 0.4771$,

so k = Antilog (0.4771) = 3

$$\text{So, } \left(\frac{x}{m}\right) = k(p)^{1/n}$$

$$\frac{x}{m} = 3\cdot p^2 = 3\times(0.04)^2 = 48\times10^{-4}$$

32. (3) (A) – (III), (B) – (IV), (C) – (II), (D) – (I)

33. (3) (A) - (III), (B) - (IV), (C) - (I), (D) - (II)

34. (2) (A) – (III), (B) – (II), (C) – (I), (D) – (IV)

35. (3) A-IV, B-III, C-I, D-II

36. (2) A-II, B-III, C-IV, D-I

37. (1) Enzymes are mostly proteins. They function as catalysts in biochemical reaction by lowering the energy of activation and are also highly specific w.r.t. temp. and pH.

38. (2) Partially deactivated palladised charcoal is called as Lindlar's catalyst.

39. (2) $C + O_2 \rightarrow CO_2$ (No catalyst is required) In Ostwald process, Haber's process and hydrogenation of vegetable oils, catalyst (solid) such as Pt/Rh, Fe and Ni are used.

40. (4) (A) - (ii), (B) - (i), (C) - (iv), (D) - (iii)

41. (4) Enzymes are very specific biological catalysts possessing well - defined active sites

42. (1) It is the standard method for the preparation of lyophilic sol.

43. (3) Particles of true solution pass through parchment paper but not the colloidal particles (size 10^{-9} to 10^{-6} m).

44. (1) The colour of the colloidal solution depends on the wavelength of the light scattered by the dispersed particles.

45. (3) For micelle formation, $\Delta S > 0$ (hydrophobic effect) This is possible because, the decrease in entropy due to clustering is offset by increase in entropy due to desolvation of the surfactant, Also $\Delta H > 0$.

46. (4) As equal & similar charged particles will repel each other, so precipitate never formed.

47. (N) The minimum concentration of electrolyte required to cause coagulation in milli-moles per litre is called coagulating value of sol. In the question volume term is not given hence data is insufficient.

48. (3) Colloids particles have two layer in solution, one is fixed layer and other is diffused layer. The potential difference between the fixed and diffused layer of charges in a colloidal particle is called zeta potential.

49. (3) The individual particles in macromolecular colloids have high molecular weight

Negative charged sol → CdS

Macromolecular colloid → starch

Positively charged sol → $Fe_2O_3.xH_2O$

Cheese → gel

50. (1)

(A) Lyophilic colloid act as protective colloid as the lyophilic particles form a layer around lyophobic particles and thus protect the latter form electrolytes.

(B) Emulsions are liquid in liquid colloidal solutions.

(C) $FeCl_3$ + hot water forms positively charged colloidal solution of hydrated ferric oxide.

(D) $FeCl_3$ + NaOH forms negatively charged colloidal solution due to preferential adsoption of OH^- ions.

51. (2) The diameter of dispersed particle should be near the wavelength of light.

52. (3) Micelle formation only takes place above CMC.

53. (3) Statement I : Correct.

Statement II : The principle emulsifying agents for O/W emulsions are proteins, gum natural synthetic soaps, etc.

54. (4) Lyophilic sols are those colloids which have strong affinity between particles of dispersed phase and dispersion medium. In the lyophilic colloids, the colloidal particles are extensively solvated.

55. (3) When silver nitrate solution is added to KI solution, AgI gets precipitated which absorbs iodide ions from the dispersion medium and gives negatively charged colloidal solution (AgI/I^-).

56. (1)

$$AgNO_3 + \underset{\text{(excess)}}{KI} \rightarrow \underset{\text{colloid}}{AgI/I^-} + KNO_3$$

So, negative colloid (AgI/I^-) will be formed.

57. **(3)** CdS is negatively charged sol whereas TiO_2 is positively charged sol.

58. **(3)** A colloidal system consisting of a gas dispersed in a solid is called a 'solid sol'.

59. **(1)** For flocculation of anion, cations are used. According to Hardy Schulze rule, greater the charge on cation, higher its flocculating power. So, Ba^{2+} has the highest flocculating power.

60. **(3)** Ordinary filter paper cannot stop the flow of colloidal particles. An ultra-filter paper is required to stop the flow of colloidal particles.

61. **(4)** When $FeCl_3$ is added in excess of hot water positively charged sol of hydrated ferric oxide is formed.

62. **(4)** According to Hardy-Schulz rule, for the coagulation of negatively charged sol, (+)ve ion with more charge is most efficient.
Blood is negatively charged sol, therefore, Fe^{3+} will be best for coagulation.

63. **(4)** Above Kraft temperature, the formation of micelles takes place and the conc. above which micelle formation become appreciable is called critical micelles concentration.

64. **(1)** In micelle formation, above "CMC", hydrocarbon chains are pointing towards the centre of sphere with COO^- part remaining outward on the surface.

65. **(1)** If the surface tension and vapour pressure are low and the substance solidifies on the surface then stability will be enhanced. Egg white possess all these properties.

66. **(4)** Tyndall effect is observed only when the following two conditions are satisfied :
(i) The diameter of the dispersed particles is not much smaller than the wavelength of the light used.
(ii) The refractive indices of the dispersed phase and the dispersion medium differ greatly in magnitude.

67. **(3)** Excess water to liquid

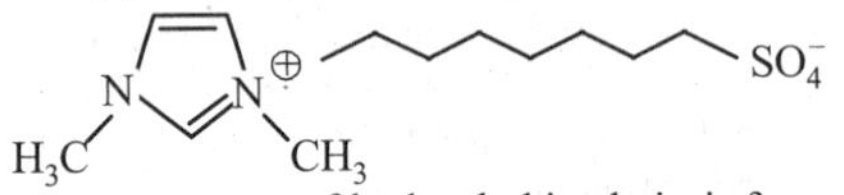

Due to presence of hydrophobic chain it forms micelle.

68. **(3)** Bredig's Arc method is used for preparation of colloidal sol's of less reactive metal like Au, Ag, Pt.

69. **(1)** It is an example of osmosis. Osmosis is the movement of solvent across a semipermeable membrane towards a higher concentration of solute (concentrated solution).

70. **(1)** According to Hardy-Schulze,
Coagulation value or flocculation value

$$\propto \frac{1}{\text{Coagulation power}}$$

For a positively charged sol $Fe(OH)_3/Fe^{3+}$, order of coagulation power:
$[Fe(CN)_6]^{3-} > CrO_4^{2-} > Br^- = NO_3^- = Cl^-$
$\therefore$ order of flocculation value :
$K_3[Fe(CN)_6] < K_2CrO_4 < KBr = KNO_3 = AlCl_3$

71. **(2)** Peptization is the process of converting a precipitate into a colloidal sol by shaking it with dispersion medium in the presence of small amount of electrolyte. It is also known as deflocculation.

72. **(3)** Due to the association of solute molecule till they acquire colloidal dimensions, the osmotic pressure of a colloidal solution is of lower order than that of true solution at the same concentation. $\pi = iCRT$
i (van't Hoff factor) is less in colloidal solution than true solution. So, π (osmotic pressures) will also be less in colloidal solution.

73. **(3)** The process of electrophoresis is used to precipitate colloidal particles in lyophobic sols.

74. **(1)** In aerosol, the dispersion medium is gas while the dispersed phase can be both solid or liquid.

75. **(4)** Latex is a colloidal solution of negatively charged rubber particles.

76. **(2)** Gem stones are solid sol.

77. **(2)**

Colloids phase	Dispersed medium	Dispersion
Cheese (C)	liquid	solid
Milk (M)	liquid	liquid
Smoke (S)	solid	gas

78. **(1)** Haemoglobin and gold sol (metal) are examples of positive and negative sols, respectively.

79. **(2)** For coagulation of a negatively charged arsenious sulphide sol, the cation which is in higher oxidation state will be most effective.

80. **(2)** The depression in freezing point is a colligative property and depends on the number of solute particles and independent of size or shape of solute particles. In colloids, due to association, the number of effective particles decreases. Thus, the depression in freezing point will be less or we can say that the freezing point will be high for colloidal solution.

81. **(1)** Brownian movement is the random motion of particles suspended in a fluid (a liquid or gas) resulting from their collision with the fast moving atom or molecules in the liquid or gaseous state of the matter. That means smaller particles are responsible for the Brownian movement than for bigger particles.

82. **(1)** Preparation of egg albumin sol. is done in following two steps.
(i) Prepare 100 mL of 5% (w/V) solution of NaCl in water in a 250 mL beaker.
(ii) Break one egg in a porcelain dish and pipette out the albumin and pour it in sodium chloride solution. Stir well to ensure the preparation of sol.

83. **(1)** Gold number $\propto \frac{1}{\text{Protective power}}$
i.e., The smaller the value of gold number of lyophilic sol, the greater is the protective action. Hence, gelatin will be better protective colloid.

84. **(4)** Greater the surface area, greater the van der Waals forces of attraction and therefore, at lesser concentration, micelle formation will take place. In case of $CH_3(CH_2)_{15}\overset{\oplus}{N}(CH_3)_3Br^-$ due to greater chain length, greater will be van der Waals forces.

85. **(2)** The motion of a liquid through a membrane under the influence of an applied electric field is known as electroosmosis.

86. **(4)** In first case, the given compounds have same anion but different cations having different charge. Hence, they will precipitate negatively charged sol., i.e., 'A'.
In second case, the given compounds have same cation but different anions with different charge. Hence, they will precipitate positively charged sol., i.e., 'B'.

87. **(3)** Smoke is an example of solid particles dispersed in gas.

88. **(2)** Gold Number is the minimum amount of lyophilic colloid in miligrams (starch in given case) which prevents the coagulation of 10 mL gold sol by the addition of 1 mL of 10% NaCl solution.
$\therefore$ Gold number = miligram of starch = 0.025×10^3 mg = 25

89. **(3)** Fog has liquid as dispersed phase and gas as dispersion medium.

90. **(1)** Particle size of colloidal particle is 10 Å to 1000 Å.
For a true solution, diameter range is 1Å – 10Å.
Taking lower limits :

$$V_C = \frac{4}{3}\pi r^3;\ V_C = \frac{4}{3}\pi(10/2)^3$$

$$V_S = \frac{4}{3}\pi(1/2)^3;\ \therefore \frac{V_C}{V_S} = 10^3$$

91. **(2)** When oppositely charged electrolytes are mixed, the colloidal particles come closer to each other and form large aggregates and precipitate out. This phenomenon is also termed as coagulations.
Mg^{2+} can coagulate negative charged sol (gold) more than Cl^- can coagulate positive charged sol (iron (III) hydroxide).

92. **(0.37)** For 1 L sol, 30 m mol of HCl is required
$\therefore$ For 1 L sol, 15 m mol H_2SO_4 is required
For 250 mL of sol, H_2SO_4 required

$$= \frac{15}{1000} \times 250 \text{ m mol } H_2SO_4 = 3.75 \text{ m mol of } H_2SO_4$$

1 mol H_2SO_4 = 98 g $\Rightarrow$ 3.75 m mol of H_2SO_4
$= 3.75 \times 98 \times 10^{-3} = 0.3675$ g H_2SO_4

Part-B (JEE Advanced)

1. **(b)** The adsorption of methylene blue on activated charcoal is an example of physisorption which is exothermic, multilayer and does not have energy barrier.

2. **(b)** When a gas is adsorbed on the surface, the freedom of movement of its molecules becomes restricted. This causes decrease in the entropy of the gas after adsorbtion, *i.e.* ΔS becomes negative. Thus, adsorption to be sontaneous, ΔH has to be negative from the equation $\Delta G = \Delta H - T\Delta S$.

3. **(a)** Rate of physisorption increases with decrease of temperature

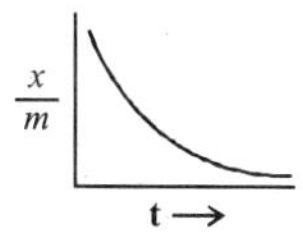

$\frac{x}{m}$ = amt of gas adsorbed per unit mass of adsorbent

t = temperature

4. (2500) Number of moles of unadsorbed CH_3COOH

$= \frac{V_{NaOH} \times M_{NaOH}}{1000} = \frac{40 \times 1}{1000} = 4 \times 10^{-2}$ mol

Number of moles adsorbed CH_3COOH

$= \frac{V_{CH_3COOH} \times M_{CH_3COOH}}{1000}$

$= \frac{100 \times 0.5}{1000} - (4 \times 10^{-2}) = 10^{-2}$ mol

Surface area occupied by one molecular of CH_3COOH

$= \frac{\text{Total surface area}}{\text{Number of molecules of } CH_3COOH \text{ adsorbed}}$

$= \frac{1.5 \times 10^2}{10^{-2} \times 6 \times 10^{23}} = \frac{150 \times 10^2 \times 10^{-23}}{6}$

$= 2500 \times 10^{-23}\ m^2 \Rightarrow P = 2500$

5. (2) $P_{N_2} = 0.001$ atm, $T = 300$ K,

$V = 2.46\ cm^2$

$\therefore$ Number of N_2 molecules

$= \frac{PV}{RT} \times N_A = \frac{0.001 \times 2.46 \times 10^{-3}}{0.0821 \times 300} \times 6.023 \times 10^{23}$

$= 6.016 \times 10^{16}$

Now, total number of surface sites = Density × Total surface area

$= 6.023 \times 10^{14} \times 1000 = 6.023 \times 10^{17}$

Sites occupied by N_2 molecules

$= \frac{20}{100} \times 6.023 \times 10^{17} = 12.04 \times 10^{16}$

$\therefore$ No. of sites occupied by each N_2 molecule

$= \frac{12.04 \times 10^{16}}{6.016 \times 10^{16}} \approx \mathbf{2}$

6. (15.62)

From Freundlich isotherm

$\frac{x}{m} = K(C)^{1/n}$

$\frac{4}{m} = K(10)^{1/n}$ ...(1)

$\frac{10}{m} = K(16)^{1/n}$ ...(2)

On doing (2) / (1)

$\Rightarrow \frac{10}{4} = \left(\frac{16}{10}\right)^{1/n} \Rightarrow \frac{5}{2} = \left(\frac{8}{5}\right)^{1/n}$

$\Rightarrow \log 5 - \log 2 = [\log 8 - \log 5]\frac{1}{n}$

$\Rightarrow 0.4 = [3 \times 0.3 - 0.7]\frac{1}{n} \Rightarrow n = \frac{1}{2}$

$= \left(\frac{y}{m}\right) = K(20)^2$ ...(3)

From equation (3) and (2)

$\frac{y}{10} = \left(\frac{20}{16}\right)^2 \Rightarrow \frac{y}{10} = \left(\frac{5}{4}\right)^2 \Rightarrow \frac{y}{10} = \frac{25}{16}$

$\therefore$ $y = 15.625$ mg g^{-1}.

7. (1.50)

Given, $K_a = 1 \times 10^{-5}$

After shaking pH = 3

$[H^+] = 10^{-3} = \sqrt{K_a + C}$

$10^{-3} = \sqrt{10^{-5} \times C} \Rightarrow C = 0.1$ M

Let mass left after shaking is 'w'.

$10^{-1} = \frac{\frac{w}{60}}{50} \times 1000 \Rightarrow w = 0.3$ g

mass absorbed = 0.45 – 0.30 = 0.15 g

$\frac{x}{m} = kC^{1/n}$

$\log\frac{x}{m} = \log k + \frac{1}{n}\log C$

Given, Slope = 1 $= \frac{1}{n} \Rightarrow n = 1$

$\therefore \frac{x}{m} = kC^1 \Rightarrow 0.15 = k \times 0.1 \Rightarrow \therefore k = 1.5$

8. (a, d)

(a) Chemisorption occurs due to the chemical bond (covalent or ionic) formation between adsorbate and adsorbent. Thus, it results in a unimolecular layer.

(b) In physisorption, the adsorbate is adsorbed on the solid surface by weak van der Waals force. So, the enthalpy of adsorption is very low, ~ 20–40 kJ/mol.

(c) Chemisorption is exothermic, its enthalpy of adsorption is ~ 80–240 kJ/mol.

(d) Lowering of temperature favours the physisorption over chemisorption.

9. (b, c, d) Reaction on metal surface

$M \longrightarrow M^+ + e^-$; $O_2 + e^- \longrightarrow O_2^-$

This is an example of chemisorption.

(c) There are 2 unpaired electrons in π^*2p orbital of O_2 molecule : In O_2^- the extra electron will go to π^*2p orbital.

(d) This will decrease the bond order and increase the bond length.

10. (a, c) Graph (I) and (III) represent physisorption because, in physisorption, the amount of adsorption decreases with the increase of temperature and increases with the increase of pressure.

Graph (II) represent chemisorption, because in chemisorption amount of adsorption increases with the increase of temperature. Graph (IV) is showing the formation of a chemical bond, hence chemisorption.

11. (a, b, d)

(a) ΔH is negative for adsorption

(b) Fact based

(d) Chemical bonds are stronger than van der Waal's forces, so chemical adsorption is more exothermic.

12. Number of moles of acetic acid in 100 mL before adding charcoal = 0.05

Number of moles of acetic acid in 100 mL after adding charcoal = 0.049

Number of moles of acetic acid adsorbed on the surface of charcoal = 0.001

Number of molecules of acetic acid adsorbed on the surface of charcoal = 0.001 × 6.02 × 10^{23} = 6.02×10^{20}

Surface area of charcoal = 3.01 × $10^2\ m^2$ (given)

Area occupied by single acetic acid molecule on the surface of charcoal

$\frac{3.01 \times 10^2}{6.02 \times 10^{20}} = 5 \times 10^{-19}\ m^2$

13. (d) Enzymes are very specific biological catalysts possessing well - defined active sites

14. (a) (III) Dissolving surfactant in a proper solvent will only form micelles at temperature above Kraft's temperature.

(IV) For Tyndall effect there must be a large difference in refractive index between dispersed phase and dispersion medium in order to have diffraction of light.

15. (c) As Sb_2S_3 is a negative sol, so $Al_2(SO_4)_3$ will be the most effective coagulant due to higher positive charge on Al (Al^{3+}) – **Hardy-Schulze rule**.

16. (a) Larger the hydrophobic fragment of surfactant easier will be the micellisation, smaller tha critical micelle concentration. Therefore $CH_3(CH_2)_{15}N^+(CH_3)_3Br^-$ will have forest critical micelle concentration.

17. (d) Lyophilic sols are self stabilizing because these sols are reversible and are highly hydrated in the solution.

Lyophilic sols can be coagulated by adding a suitable solvent like alcohol and an electrolyte both.

18. (b, c)

(a) Process of precipitating colloidal solution is called coagulation. Hence, this statement is wrong.

(b) Colloidal solutions exhibit colligative properties. Hence, this statement is correct.

(c) At CMC surfactant form micelles. Hence, this statement is correct.

(d) Micelles are associated colloids, not macromolecular colloids. Hence, this statement is incorrect.

19. (a, b)

(a) Since adsorption is an exothermic process, hence, enthalpy decreases during this process. On adsorption, the randomness of the adsorbate molecules decreases, thus, entropy decreases.

(b) Higher the critical temperature of a gas, greater the amount of gas adsorbed thus, the extent of adsorption ethane will be more than nitrogen.

(c) Cloud is an aerosol in which liquid is dispersed phase and gas is dispersion medium, Whereas, emulsion is liquid in liquid colloidal system.

(d) Brownian motion of colloidal particles depends on the size of the particles as well as on viscosity of the solution.

20. (a, d)

(a) When two or more ions are present in the dispersion medium, preferential adsorption of the ion common to the colloidal particle takes place.

(d) This is called zeta potential which helps in increasing the stability of the lyophobic sol.

21. (b) Statement-1 is correct because the surfactant molecules aggregate to form micelles only at or above the critical micellar concentration (CMC). Although, statement-2 is also correct, *i.e.*, the conductivity of the solution having surfactant molecules decreases sharply at CMC. However, statement-2 is not the explanation for statement-1. Each micelle contains at least 100 molecules and thus, with the formation of micelles, the number of ions in solution decreases and mobility of the bulkier micelle particles decreases. This, finally leads to decrease in conductivity of the solution.

Chapter-6

General Principles and Process of Isolation of Elements

Part-A (JEE Main)

1. (1) For the reaction,
$2Mg(s) + O_2(g) \longrightarrow 2MgO(s)$
$\Delta S^\circ = -ve$; $\Delta G^\circ = \Delta H^\circ - T\Delta S^\circ$
$-T\Delta S^\circ$ is positive,
Hence, ΔG° increase with increase in temperature. Also, when there is phase change there is an abrupt increase in the slope.

2. (1)

3. (4) For Zn $\longrightarrow$ KCN is used Ag/Au leached by cyanide process
$4Au + 8NaCN + O_2 + 2H_2O \longrightarrow 4Na[Au(CN)g] + 4NaOH$

4. (4)

5. (2) From Ellingham diagram given in NCERT, it can be seen that Mg, MgO line crosses Al, Al_2O_3 line after 1350°C i.e., Al can reduce MgO hence assertion is true.
Yes, Mg have lower MP and BP than aluminium but reduction of one compound by other is govern by the change in Gibbs free energy (ΔG°) of the process.

6. (4) In leaching of gold Au is oxidised to Au^+
$4Au(s) + 8CN^-(aq) + 2H_2O + O_2(g) \longrightarrow 4[Au(CN)_2]^-(aq) + 4OH(aq)$
Zinc is oxidised to Zn^{2+}
$2[Au(CN)_2]^-(aq) + Zn(s) \longrightarrow 2Au(s) + [Zn(CN)_4]^{2-}(aq)$

7. (2)

8. (4) Metal oxide with lower (more negative) ΔG° is more stable as this is equal to enthalpy of formation of that oxide.
Hence, statement II is correct.

9. (1) Depressant helps in separation of two sulphide are by selectively preventing one are from foaming.

10. (3) $Au + NaCN \rightarrow Na[Au(CN)_2]$
$Zn + Na[Au(CN)_2] \rightarrow Na_2[Zn(CN)_4] + Au$

11. (1) (A) - (IV), (B) - (III), (C) - (II), (D) - (I)

12. (1) Earthy and undesired materials present in the ore, other then the desired metal, is known as gangue.

13. (3) In this method tin, a low melting point metal, is made to flow on a sloping surface in liquid form. In this way it is separated from higher melting impurities.

14. (3)
$Al_2O_3 + 2NaOH + 3H_2O \rightarrow 2Na[Al(OH)_4]$

15. (1) Al is extracted from $Al_2O_3 . 2H_2O$ i.e., Bauxite ore

16. (4) The main impurity in the copper ores is iron oxide (FeO) which reacts with silica (SiO_2) to form Fe SiO_3 as slag.

17. (2) Calamine $\Rightarrow ZnCO_3$
Malachite $\Rightarrow Cu(OH)_2 \cdot CuCO_3$

18. (4) The value of ΔS decreases from left to right due to consumption of oxygen and formation of solid oxide.

19. (1)

20. (1) Ellingham diagram is a plot between ΔG° and T. It does not give any information about kinetics of reaction.

21. (4) $\underset{\text{Acidic gangue}}{SiO_2} + \underset{\text{basic flux}}{2NaOH} \rightarrow \underset{\text{slag}}{Na_2SiO_2} + H_2O$

22. (1) A–(iii), B–(ii), C–(i), D–(iv)

23. (1) Sulphide ion being a soft base will prefer to bind with soft acids (Pb^{2+} and Ag^+).

24. (1)
(A) $ZnCO_3(s) \xrightarrow{\Delta} ZnO(s) + CO_2(g)$
Heating in absence of oxygen is calcination.
(B) $2ZnS(s) + 3O_2(g) \rightarrow 2ZnO(s) + 2SO_2(g)$
Heating in presence of oxygen is roasting.

25. (1) Cryolite is used to lower the M.P. of alumina.

26. (3) Elingham diagram is a plot between Gibbs free energy (ΔG°) and temperature.

27. (1) Sphalerite ore : ZnS

28. (4) Brass contains 70% Cu and 30% Zn.

29. (2)

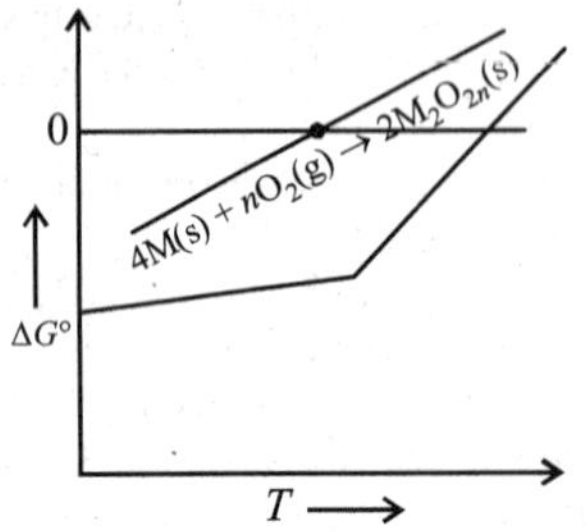

Ellingham Diagram

From graph it is evident that the temperature below which the oxide is stable, is the point at which free energy change shows a change from negative to positive.

30. (2) Ellingham diagram is the graph of ΔG^0 vs T of any metal/ element oxide. Since
$$\Delta G^\circ = \Delta H^\circ - T\Delta S^\circ$$

31. (2)
Calcination : Metal carbonate $\xrightarrow{\Delta}$ Metal oxide + CO_2 (causes global warming)
Roasting : Metal sulphide + $O_2 \longrightarrow$ Metal oxide + SO_2 (causes acid rain)

32. (1) All statements are correct.

33. (2) $A + BO_2 \longrightarrow B + AO_2$
$\Delta G = -ve$
When the temperature is above 1400 °C, A reduces BO_2.

34. (3) During the reduction of Fe_2O_3 to FeO, Fe_2O_3 is first reduced to Fe_3O_4 and then to FeO, thus follows reaction (B).
In order to remove impurity of the ore, limestone is added along with ore which is decomposed to CaO.
The resulted CaO reacts with SiO_2 to form $CaSiO_3$, i.e., follows reaction (A) and is removed as a slag.

35. (2) Froth floatation was discovered by washer woman. It is a method of concentration of ores.

36. (1)
(c) Cast iron is obtained from pig iron.
(d) Blister of copper is due to evolution of SO_2.
(b) Hall-Heroult process is used for the production of aluminium.

37. (1) Ti is refined by van Arkel method. Silver is leached by dilute solution of NaCN. Zincite is oxide ore of zinc, i.e., ZnO. Aniline is a froth stabilizer.

38. (4) Bauxite$\rightarrow AlO_x(OH)_{3-2x}$

39. (1) Cryolite is Na_3AlF_6.

40. (4) For the extraction of iron, haematite ore (Fe_2O_3) is used.

41. (4) Ellingham diagram helps to predict feasibility of its thermal reduction.

42. (2) In Hall-Heroult process, steel vessel with lining of carbon acts as cathode and graphite is used as anode.

43. (1) Calcination involves the conversion of metal carbonates or hydrated oxides into metal oxides. ZnO and MgO are oxides, therefore, do not require calcination.

44. (2) Calcination involves heating of carbonates or hydrated oxides in absence of air.
$\therefore\ 2Cu_2S + 3O_2 \xrightarrow{\Delta} 2Cu_2O + 2SO_2$; is roasting (heating in a regular supply of air).

45. (2)

46. (3) Hall-Heroult's process is an electrochemical process used in extraction of Al from alumina.

47. (1) The electrolytes used in the electroplating of Au and Ag are $[Au(CN)_2]^-$ and $[Ag(CN)_2]^-$ respectively.

48. (1) In the given Ellingham diagram, the metal which has a lower value of ΔG° (more negative) can reduce a metal oxide whose curve lies above it. So, Al can reduce ZnO at 1400 °C.

49. (1) Amongst the given ores; copper pyrite ($CuFeS_2$), dolomite ($MgCO_3 \cdot CaCO_3$), malachite $[CuCO_3 \cdot Cu(OH)_2]$ and azurite $[2CuCO_3 \cdot Cu(OH)_2]$ only copper pyrite contains both copper and iron.

50. (3) $Cu_2S + 2Cu_2O \longrightarrow 6Cu + SO_2$
It is an example of auto-reduction, Cu_2O is reduced to copper metal.

51. (4) $\underset{\text{Alumina}}{Al_2O_3(s)} + 2NaOH(aq) + 3H_2O(l) \xrightarrow[35-36\text{ bar}]{473-523\text{ K}} \underset{\text{Sodium aluminate (X)}}{2Na[Al(OH)_4](aq)}$
$2Na[Al(OH)_4](aq) + CO_2(g) \longrightarrow \underset{\text{Hydrated alumina (Y)}}{Al_2O_3 \cdot xH_2O(s)} + 2NaHCO_3(aq)$

52. (1) Froth floatation method is mainly applicable for sulphide ores.
Galena ore : PbS (Sulphide ore)

53. (3) FeO is gangue and SiO_2 is flux to form slag $FeSiO_3$.

$$\underset{\text{Basic impurity}}{FeO} + \underset{\text{Acidic flux}}{SiO_2} \longrightarrow \underset{\text{Slag}}{FeSiO_3}$$

54. (2) In the metallurgy of aluminium, purified Al_2O_3 is mixed with Na_3AlF_6 or CaF_2 which lowers the melting point of the mix and brings conductivity.

55. (3) $\underset{\text{Limestone}}{CaCO_3} \xrightarrow{\Delta} \underset{\text{Calcium oxide}}{CaO} + CO_2 \uparrow$

$\underset{\text{Malachite}}{CuCO_3 \cdot Cu(OH)_2} \xrightarrow{\Delta} \underset{\text{Copper Oxide}}{2CuO} + H_2O \uparrow + CO_2 \uparrow$

56. (3) The iron obtained from blast furnace contains about 4% C and impurities (eg. S, P, Si, Mn). This is known as pig iron.

57. (4)

58. (3) Reduction by powdered aluminium is known as Gold-Schmidt aluminothermic process.

59. (2) Ca and K are strong reducing agents, hence their oxides can not be reduced with carbon.

60. (3) Extraction of Zn from zinc blende is achieved by roasting followed by reduction with carbon.

61. (2) Froth stabilizers like cresols, aniline are added to stabilize the froth and enhance the non-wettability of the mineral particles.

62. (3)

63. (3) Pure aluminium can be obtained by electrolysis of a mixture containing alumina, crayolite and fluorspar in the ratio 20 : 24 : 20.

64. (2) Silver ore forms a soluble complex with NaCN from which silver is precipitated using scrap zinc.

$$Ag_2S + 2NaCN \rightarrow \underset{\substack{\text{sodium argentocynanide} \\ \text{(soluble)}}}{Na[Ag(CN)_2]} \xrightarrow{Zn} Na_2[Zn(CN)_4] + Ag \downarrow$$

65. (4)
Bauxite — $AlO_x(OH)_{3-2x}$ (where $0 < x < 1$)
Siderite — $FeCO_3$
Cuprite — Cu_2O
Calamine — $ZnCO_3$
Haematite — Fe_2O_3
Kaolinite — $Al_2(OH)_4Si_2O_5$
Malachite — $CuCO_3 . Cu(OH)_2$
Magnetite — Fe_3O_4
Sphalerite — ZnS
Limonite — $Fe_2O_3.3H_2O$
Cryolite — Na_3AlF_6

66. (3)

67. (1) Calcination and leaching are methods used for concentration of ores.

68. (1) Liquation is used for the purification of metals having low melting point.

69. (2) Iron obtained from blast furnance is called pig iron. It contains about 4% carbon content.
Cast iron is made by melting pig iron with scrap iron and coke. It contains 3% carbon content.

70. (1)

71. (2) Hall Heroult process is the major industrial process for extraction of aluminium.

72. (4) Aluminium is reactive metal so aluminium is extracted by electrolysis of alumina with molten mixture of Cryolite. The O.S. of Al in cryolite (Na_3AlF_6) is +3.

73. (2) Liquation method is used to purify lower melting point metals.

74. (4)

75. (2) Zn, Cd and Hg are purified by fractional distillation process.

76. (2) Germanium (Ge), silicon (Si), boron (B), gallium (Ga) and indium (In) can be obtained in pure state by zone refining process.

77. (4) Cast iron is made from pig iron and is used for production of wrought iron and steel.

78. (2) Wrought iron is purest form of commercial iron.

79. (3)

80. (2) Ca is s-block metal and is highly electropositive. It cannot be obtained by electrolysis of its salt as cations.

81. (4) van Arkel : In this process, an impure metal is converted into metal iodide which is easily decomposed to pure metal and free I_2.

82. (6) Anode mud contains Sb, Se, Te, Ag, Au and Pt.

Part-B (JEE Advanced)

1. (a) Calamine → $ZnCO_3$
Malachite → $CuCO_3.Cu(OH)_2$
Magnetite → Fe_3O_4
Cryolite → Na_3AlF_6

2. (b) The reactions involved in cyanide extraction process are :

$$\underset{\text{(argentite)}}{Ag_2S} + 4NaCN \rightarrow 2Na[Ag(CN)_2] + Na_2S$$

$$4Na_2S + \underset{\substack{\text{Oxiding} \\ \text{agent}}}{5O_2} + 2H_2O \rightarrow 2Na_2SO_4 + 4NaOH + 2S$$

$$2Na[Ag(CN)_2] + \underset{\substack{\text{(reducing} \\ \text{agent)}}}{Zn} \rightarrow Na_2[Zn(CN)_4] + 2Ag \downarrow$$

3. (d) (i) Haematite is Fe_2O_3 in which Fe is present in III oxidation state.
(ii) Magnetite (Fe_3O_4) is an equimolar mixture of FeO and Fe_2O_3.
Oxidation state of Fe in FeO is II.
Oxidation state of Fe in Fe_2O_3 is III.

4. (b) Extraction of Zn from ZnS (Zinc blende) is achieved by roasting followed by reduction with carbon.

$$2ZnS + 3O_2 \longrightarrow 2ZnO + 2SO_2$$
$$ZnO + C \longrightarrow Zn + CO$$

5. (c) Cuprite : Cu_2O; Chalcocite : Cu_2S; Chalcopyrite : $CuFeS_2$; Malachite: $Cu(OH)_2.CuCO_3$. We see that $CuFeS_2$ contains both Cu and Fe.

6. (a) $2Au + 4CN^- + H_2O + \frac{1}{2}O_2 \longrightarrow \underset{'X'}{2[Au(CN)_2]^-} + 2OH^-$

$$2[Au(CN)_2]^- + Zn \longrightarrow \underset{'Y'}{[Zn(CN)_4]^{2-}} + 2Au$$

7. (a) $\underset{\text{(fused anhydrous)}}{MgCl_2} \rightarrow Mg^{+2} + 2Cl^-$

At cathode : $Mg^{+2} + 2e^- \rightarrow Mg$;

At anode : $2Cl^- - 2e^- \rightarrow Cl_2 \uparrow$

8. (c) Al_2O_3 is mixed with cryolite (Na_3AlF_6) which lowers the melting point of the mixture and brings conductivity.

9. (c) Haematite ore (Fe_2O_3) is first reduced to cast iron which is then oxidised for removing carbon (impurity) as CO_2.

10. (b) During the extraction of copper, iron is present in the ore as impurity (FeS).
The ore together with a little coke and silica is smelted; FeS present as impurity in the ore is oxidized to iron oxide, which then reacts with silica to form fusible ferrous silicate which is removed as slag.

$$2FeS + 3O_2 \longrightarrow 2FeO + 2SO_2 \uparrow;$$

$$FeO + SiO_2 \longrightarrow \underset{\text{(Slag)}}{FeSiO_3}$$

11. (c) Al_2O_3 is electrolyte, while Na_3AlF_6 is used to decrease the melting point of Al_2O_3 and to increase the conductivity.

12. (c) Al reduces Fe_2O_3 or Cr_2O_3 to respective metals and acts as a reducing agent.

$$Fe_2O_3 + 2Al \rightarrow Al_2O_3 + 2Fe$$

13. (a) $CaCO_3 \xrightarrow{\text{Heat}} CaO + CO_2$;

$$\underset{\text{Flux}}{CaO} + \underset{\text{impurity}}{SiO_2} \longrightarrow CaSiO_3 \text{ (slag)}.$$

14. (c) Malachite is $CuCO_3 . Cu(OH)_2$ which is ore of copper.

15. sintering, smelting.

16. Magnesia and lime; calcium silicate
The lining of corverter is made of magnesia & lime. Slag formed consists of $CaSiO_3$.

17. Aluminium

$$Fe_2O_3 + 2Al \longrightarrow Al_2O_3 + 2Fe + \text{Heat}$$

[Thermite reaction]

18. Tin. It is SnO_2.

19. (a, d)
The treatment of galena with HNO_3 produces nitric oxide gas which is paramagnetic.

$$PbS + HNO_3 \longrightarrow Pb(NO_3)_2 + S\downarrow + NO\uparrow + 4H_2O$$

NO is colourless, neutral, paramagnetic gas.
NO is of linear geometry : $:\dot{N} = \ddot{O}: \longleftrightarrow :\dot{N} = \ddot{O}:$

20. (a, b, c) Exhaust gases do not cantain NO_2.

21. (a, c, d)

(a) $PbS + 2PbO \rightarrow 3Pb + SO_2$ (self reduction)

(b) Silica is added to remove impurity of Fe in the form of slag $FeSiO_3$. Hence, this statement is wrong.

(c) Sulphide ore is partially oxidized first by roasting and then self reduction of Cu takes place to produce blister copper.

$$Cu_2S + \frac{3}{2}O_2 \rightarrow Cu_2O + SO_2 \uparrow$$

$$Cu_2S + 2Cu_2O \rightarrow 6Cu + SO_2 \uparrow$$

(self reduction)

The molten copper obtained is poured into large container and allowed to cool and during cooling the dissolved SO_2 comes up to the surface and forms blisters. It is known as blister copper.

(d) $2Na[\overset{+1}{Au}(CN)_2] + \underset{\text{Reducing Agent}}{Zn} \rightarrow Na_2[\overset{+2}{Zn}(CN)_4] + 2\overset{0}{Au}$

22. (a, b, c, d)

(a) $2Na[Al(OH)_4] + CO_2 \longrightarrow NaCO_3 + H_2O + 2Al(OH)_3(\downarrow)$ or $Al_2O_3.2H_2O$

(b) Function of Na_3AlF_6 is to lower the melting point of electrolyte.

(c) During electrolysis of Al_2O_3, the reactions at anode are:

$$2Al^{3+}(\ell) + 3O^{2-}(\ell) \xrightarrow{\text{Al anode}} O_2(g) + 2e^-$$

$$C(graphite) + O_2 \longrightarrow CO(g) + CO_2(g)$$

(d) The steel vessel with a lining of carbon acts as cathode.

23. (a, b, c) Gold extraction:

$$4Au + 8NaCN + 2H_2O + \underset{(Q)}{O_2} \rightarrow \underset{(R)}{4Na[Au(CN)_2]} + 4NaOH$$

$$\underset{(R)}{4Na[Au(CN)_2]} + \underset{(T)}{2Zn} \rightarrow \underset{(Z)}{2Na_2[Zn(CN)_4]} + 4Au$$

24. (a, b, c)

Copper pyrite [$CuFeS_2$]

↓ Crushing into fine powder

Concentrated by froth floatation process

↓

Roasting take place in reverberatory furnace

$$\left\{\begin{array}{l} 2CuFeS_2 + \underset{(air)}{O_2} \xrightarrow{\Delta} Cu_2S + 2FeS + SO_2 \\ FeS + O_2 \xrightarrow{\Delta} FeO + SO_2 \end{array}\right.$$

↓ Smelting $\left[\begin{array}{l} \text{Iron is removed as slag of } FeSiO_3 \\ FeO + SiO_2\text{ (flux)} \xrightarrow{\Delta} FeSiO_3\text{(l)(slag)} \end{array}\right]$

Copper matte ($Cu_2S + FeS$)

↓ Self reduction $\left[\begin{array}{l} \text{Self reduction occurs in Bessemer converter} \\ 2Cu_2S + 3O_2 \longrightarrow 2Cu_2S + 2SO_2(\uparrow) \\ Cu_2S + 2Cu_2O \longrightarrow 6Cu + SO_2(\uparrow) \end{array}\right]$

(Blister copper)

↓

Refining of blister copper is done by poling followed by electrorefining but not by carbon reduction method.

25. (b, c, d)

(a) $2CuFeS_2 + O_2 \xrightarrow{\Delta} Cu_2S + 2FeS + SO_2$

(b) $4CuO \xrightarrow{\Delta} 2Cu_2O_2 + O_2$

$$2Cu_2O + Cu_2S \xrightarrow{\Delta} 6Cu + SO_2$$

(c) From above, Cu_2O gives copper metal.

(d) $CuSO_4 \xrightarrow{\Delta} CuO + SO_2 + \frac{1}{2}O_2$

From above, CuO also gives copper metal.

26. (c, d) Al from Al_2O_3 and Mg from $MgCO_3.CaCO_3$ are separately extracted by electrolytic reduction.

27. (a, c, d) Cassiterite (SnO_2) contains impurity of S, As, Fe, Cu *etc.*

Roasting : Concentrated ore is heated in a current of air impurities of iron and copper changed to their oxides and sulphates.

SnO_2 is treated with coke to reduce it to Sn.

$$SnO_2 + 2C \longrightarrow Sn + 2CO$$

28. (b, c) To make the fused mixture very conducting and to reduce the temperature of the melt.

29. (b, d) Both Mg and Al have their reduction potentials less than that of water [$E^\circ = -0.83$ V]. Hence, their ions in the aqueous solution can not be reduced. Instead water will be reduced : $2H_2O + 2e^- \rightarrow H_2 + 2OH^-$

30. (a, b) Because of high melting point (2050ºC), pure alumina cannot be electrolysed. Hence a mixture of alumina, cryolite (m.p. 1000ºC) and calcium fluoride (to lower the temperature of the melt) is electrolysed at about 900ºC.

The function of cryolite is to increase the electrical conductivity of the electrolyte, and to lower the temperature of the melt.

31. A-(p, q, s), B-(t), C-(q, r), D-(r)

Siderite	→	$FeCO_3$
Malachite	→	$CuCO_3 \cdot Cu(OH)_2$
Bauxite	→	$Al_2O_3.xH_2O$ or $AlO_x(OH)_{3-2x}$ $0 < x < 1$
Calamine	→	$ZnCO_3$
Argentite	→	Ag_2S

32. (A) - p; (B) - q; (C) - p, r; (D) - p, s

33. (A)- p, r; (B) - p, r; (C) - q; (D) - s

The oxides and sulphides of less active metals like Hg, Cu & Pb are unstable to heat and hence no reducing agent is required. They undergo self reduction.

(A) $2Cu_2S + 3O_2 \longrightarrow 2Cu_2O + 2SO_2$;

$$Cu_2S + 2Cu_2O \longrightarrow 6Cu + SO_2$$

$2PbS + 3O_2 \longrightarrow 2PbO + 2SO_2$;

$$PbS + 2PbO \longrightarrow 3Pb + SO_2$$

Hence, (A) $\rightarrow$ (p), (r)

The oxides of less electropositive metals like Pb, Zn, Fe, Sn, Cu, etc. are reduced by strongly heating them with coke or coal.

(B) $PbO + C \longrightarrow Pb + CO_2$;

$$2Cu_2O + C \longrightarrow 4Cu + CO_2$$

Hence, (B) $\rightarrow$ (p), (r)

(C) Extraction from argentite (Ag_2S)

$Ag_2S + 2NaCN \rightleftharpoons Na_2S + 2AgCN$;

$AgCN + NaCN \longrightarrow \underset{\text{Sod. argentocyanide (soluble)}}{Na[Ag(CN)_2]}$

Zn, being more electropositive than Ag, displaces Ag from the complex.

$$2Na[Ag(CN)_2] + Zn \longrightarrow Na_2[Zn(CN)_4] + 2Ag \downarrow$$

Hence, (C) $\rightarrow$ (q)

(D) Among the halides of boron, BI_3 is unstable because of the large size of Iodine and small size of boron atom. Hence, it decomposes to give boron. Thus, (D) $\rightarrow$ (s).

34. (*i*) (g) (*ii*) (h) (*iii*) (f) (*iv*) (c) (*v*) (e) (*vi*) (d) (*vii*) (b) (*viii*)(a)

35. (c) (*i*) (b) (*ii*) (c) (*iii*) (d) (*iv*) (a)

36. (b) $2CuFeS_2 + O_2 \longrightarrow Cu_2S + SO_2\uparrow + 2FeS$

$2FeS + 3O_2 \longrightarrow 2FeO + 2SO_2\uparrow$

$2Cu_2S + 3O_2 \longrightarrow 2Cu_2O + 2SO_2\uparrow$

37. (d) $FeO + SiO_2 \longrightarrow FeSiO_3$

38. (c) $Cu_2S + 2Cu_2O \longrightarrow 6Cu + SO_2$

The reducing species is the one which gets oxidised. So, it is S^{2-} ion getting oxidised to S^{4+}.

39. Calcination of the ore A_1 to form CO_2 indicates that A_1 should be a carbonate. Further, reaction of A_1 with HCl and KI to evolve I_2 indicates that A_1 would also be hydroxide. So the possible formula for the ore, should be $CuCO_3.Cu(OH)_2$ which explains all the given reactions

$$\underset{(A_1)}{CuCO_3.Cu(OH)_2} \xrightarrow{\text{calcination}} \underset{(C)\text{ black}}{2CuO} \downarrow + CO_2 + H_2O$$

$$\underset{(A_1)}{CuCO_3.Cu(OH)_2} + 4HCl \longrightarrow 2CuCl_2 + CO_2 + 3H_2O$$

$$2CuCl_2 + 4KI \longrightarrow \underset{(D)}{Cu_2I_2} \downarrow + 4KCl + I_2$$

Roasting of A_2 gives gas G whose nature is identified as SO_2 as it gives green colour with acidified $K_2Cr_2O_7$. So A_2 should be sulphide of copper.

$$\underset{(A_2)}{2Cu_2S} + 3O_2 \xrightarrow{\text{roasting}} 2Cu_2O + \underset{(G)}{2SO_2}$$

$$Cu_2S + 2Cu_2O \xrightarrow{\text{self reduction}} \underset{(M)}{6Cu} + SO_2$$

$$\underset{(G)}{3SO_2} + K_2Cr_2O_7 + H_2SO_4 \longrightarrow K_2SO_4 + \underset{\text{(Green colour)}}{Cr_2(SO_4)_3} + 4H_2O$$

40. $2PbS + 3O_2 \longrightarrow 2PbO + 2SO_2$

$PbS + 2O_2 \longrightarrow PbSO_4$

$PbS + 2PbO \longrightarrow 3Pb + SO_2$

$PbS + PbSO_4 \longrightarrow 2Pb + 2SO_2$

Oxidation number of Pb in litharge (PbO) is +2

41. Haematite (Fe_2O_3) on burning with coke and lime at 2000 °C results in the following reactions.

$C + O_2 \rightarrow CO_2$

$CO_2 + O \rightarrow 2CO$

$3CO + Fe_2O_3 \rightarrow 2Fe + 3CO_2$ (Reduction of Fe_2O_3 to form steel)

$\underset{\text{(Lime)}}{SiO_2 + CaO} \rightarrow \underset{\text{(Slag)}}{CaSiO_3}$

(Slag, $CaSiO_3$ is used as building material)

42. (*i*)

Roasting:

$\underset{\text{(copper pyrites)}}{2CuFeS_2} + O_2 \xrightarrow{\Delta} Cu_2S + 2FeS + SO_2$

$2Cu_2S + 3O_2 \longrightarrow 2Cu_2O + 2SO_2$

$2FeS + 3O_2 \longrightarrow 2FeO + 2SO_2$

Smelting with coke and sand:

$Cu_2O + FeS \longrightarrow Cu_2S + FeO$

$FeO + SiO_2 \longrightarrow FeSiO_3$ (*slag*)

Bessemerization:

$2Cu_2S + 3O_2 \longrightarrow 2Cu_2O + 2SO_2$

$Cu_2S + 2Cu_2O \longrightarrow 6Cu + SO_2 \uparrow$

(*ii*) $\underset{\text{(silver glance)}}{Ag_2S} + 4NaCN \rightleftharpoons 2\,Na[Ag(CN)_2] + Na_2S$

$4Na_2S + SO_2 + 2H_2O \rightarrow 2\,Na_2SO_4 + 4\,NaOH + 2S$

[Na_2S is converted into Na_2SO_4 to avoid reversibility of first reaction]

$2NaAg(CN)_2 + Zn \rightarrow \underset{\text{Sod. Zincocyanide}}{Na_2[Zn(CN)_4]} + 2Ag$

43. Recovery of Pb from galena :

$2PbS + 3O_2 \rightarrow 2PbO + 2SO_2 \uparrow$

$PbS + 2PbO \rightarrow 3Pb + SO_2$

44. Carbon monoxide is the actual reducing agent of haematite in blast furnace.

$3Fe_3O_4 + CO \longrightarrow 2Fe_3O_4 + CO_2$

$Fe_3O_4 + CO \longrightarrow 3FeO + CO_2$

45. (*i*) Excess of Air (used during roasting) is necessary for converting chalcocite (a sulphide ore) to oxide. Calcination does not convert it to oxide.

(*ii*) Metals can be recovered by chemical methods because they occur as oxides, carbonates, sulphides which have to be calcined or roasted.

46. Equations for extraction of silver from its sulphide ore.

Cyanide Process :

$\underset{\text{(sulphide ore)}}{Ag_2S} + 2NaCN \rightarrow Na_2S + 2AgCN$

$AgCN + NaCN \rightarrow Na[Ag(CN)_2]$ Sod. argentocyanide (soluble)

$2Na[Ag(CN)_2] + Zn \rightarrow Na_2[Zn(CN)_4] + 2Ag \downarrow$

[Zn is more electropositive than Ag.]

47. Extraction of aluminium from bauxite :

At cathode : $[Al^{3+}(melt) + 3e^- \longrightarrow Al(l)] \times 2$

At anode :

$C(s) + O^{2-}(melt) \longrightarrow CO(g) + 2e^-$

$C(s) + 2O^{2-}(melt) \longrightarrow CO_2(g) + 4e^-$

Net reaction :

$2Al^{3+} + 2C + 3O^{2-} \longrightarrow 2Al + CO + CO_2$

$Al_2O_3 + 2C \longrightarrow Al + CO + CO_2$

48.

Bleaching agent	⟶	Chlorine
Smelling salt	⟶	Ammonium carbonate
Cryolite	⟶	Aluminium
Bell metal	⟶	Tin
Fluorspar	⟶	Calcium
Fertilizer	⟶	Ammonium phosphate
Anthracite	⟶	Carbon

49. (a) Galena is roasted in excess of air in a reverberatory furnace

$2PbS + \underset{\text{(air)}}{3O_2} \longrightarrow 2PbO + 2SO_2$

$PbS + 2O_2 \longrightarrow PbSO_4$

(b) It is followed by self reduction

$PbS + PbSO_4 \longrightarrow 2Pb + 2SO_2$

$PbS + 2PbO \longrightarrow 3Pb + SO_2$

(i) Silver ⟶ (D) Amalgamation

(ii) Calcium ⟶ (A) Fused salt electrolysis

(iii) Zinc ⟶ (B) Carbon reduction

(iv) Iron ⟶ (C) Carbon monoxide reduction

(v) Copper ⟶ (E) Self reduction

50. (b, c, d)

(a) Under roasting condition,

$\underset{\text{malachite}}{CuCO_3 \cdot Cu(OH)_2} \rightarrow 2CuO + CO_2 + H_2O$

(b) $\underset{\text{(Calamine)}}{ZnCO_3} \rightarrow \underset{\text{(Zincite)}}{ZnO} + CO_2 \uparrow$

(c) Copper pyrites is heated in a reverberatory furnace after mixing with silica. Which results into iron silicate as slag from iron oxide and copper is produced in the form of copper matte.

$FeO + SiO_2 \rightarrow \underset{\text{(Slag)}}{FeSiO_3}$

(d)

$Ag + KCN + O_2 + H_2O \rightarrow [Ag(CN)_2]^- + KOH$

$\xrightarrow{Zn} Ag \downarrow + [Zn(CN)_4]^{2-}$

51. (b, c, d)

(a) In electrochemical extraction of aluminium from bauxite ore does not involve the reduction of Al_2O_3 with (C)

(b) The neutralization of aluminate $[Al(OH)_4]^-$ solution is performed by passing CO_2 gas –

$[Al(OH)_4]^- \xrightarrow{CO_2} \underset{\text{ppt.}[Al_2O_3 \cdot 3H_2O]}{Al(OH)_3} \downarrow + OH^-$

$CO_2 + OH^- \longrightarrow CO_3^{2-} + H_2O$

(c) Concentrated bauxite is carried out by heating the powdered Al_2O_3 in concentrated NaOH solution at 473 K – 523 K and 35 – 36 bar pressure.

$Al_2O_3(s) + 2NaOH(aq) \longrightarrow 2Na[Al(OH)_4](aq)$

(d) In electrolysis method Al_2O_3 is mixed with Na_3AlF_6 or CaF_2 to reduce the melting temperature of the mixture from 2050 °C to 900 °C.

Cathode : $Al^{3+}(melt) + 3e^- \longrightarrow Al(l)$

Anode : $C(s) + 2O^{2-}(melt) \longrightarrow CO_2(g) + 4e^-$

52. (b, c, d) In electrolytic refining of blister Cu, acidified $CuSO_4$ is used as electrolyte, pure Cu deposits at cathode and impurities settle as anode mud. Impure metal act as an anode.

53. MgO is used for the lining of steel making furnace because it acts as basic flux and facilitates the removal of acidic impurities of Si, P and S from steel through slag formation.

54. *Zone refining* is based on the difference in solubility of impurities in molten and solid state of the metal. This method is used for obtaining metals of very high purity.

Ge, Si and Ga used as semi-conductors are refined in this manner. These metals can be easily melted and can easily crystallise out from the melt form.

Chapter-7

The *p*-Block Elements (Group 15, 16, 17 & 18)

Part-A (JEE Main)

1. (3)

(A) Reducing nature:

$NH_3 < PH_3 < AsH_3 < SbH_3 < BiH_3$

On moving down the group, E–H bond gets weaker, making it easy to release Hydrogen, hence reducing strength increases. [E = central element]

(B) Lone pair donation tendency:

$NH_3 > PH_3 > AsH_3 > SbH_3 > BiH_3$

Size of central atom increases down the group due to which lone pair dispersed over larger volume making it less available for donation.

(C) Stability of Hydrides:

$NH_3 > PH_3 > AsH_3 > SbH_3 > BiH_3$

Bond dissociation enthalpy decreases down the group leading to lower thermal stability of hydrides.

(D) HEH bond angle:

$NH_3 > PH_3 > AsH_3 > SbH_3 > BiH_3$

Down the group, electronegativity of central atom decreases so electron pairs gets further away from nucleus reducing the repulsions and hence small bond angle.

2. (4)

(A) In the case of nitrogen, oxidation states ranging from + 1 to + 4 tend to undergo disproportionation in acidic solution. For example:

$$3HNO_2 \rightarrow HNO_3 + H_2O + 2NO$$

(B) Nitrogen form $p_\pi - p_\pi$ bonds with itself and other elements with small size and high electronegativity. It does not form $d_\pi - p_\pi$ bond because it lacks *d*-orbitals.

(C) The single N-N bond is weaker than the single P-P bond due to greater lone pair-lone pair repulsion in nitrogen which arises from its small atomic size and short bond length.

(D) Nitrogen has lowest density in its group.

(E) The maximum covalency of nitrogen is four as it can utilize its valence *s* and *p*-orbital to form upto four bonds.

Correct answer are (A) and (E).

3. (1)

(A) Stability : $NH_3 > PH_3 > AsH_3 > SbH_3 > BiH_3$

(B) Basicity : $NH_3 > PH_3 > AsH_3 > SbH_3 > BiH_3$

(C) Reducing character : $NH_3 < PH_3 < AsH_3 < SbH_3 < BiH_3$

(D) Boiling point : $PH_3 < AsH_3 < NH_3 < SbH_3 < BiH_3$

Thus statements-A, B and C are true while D is false.

4. (1)

Element	Electronegativity value
N(X)	3.0
P	2.1
As (Y)	2.0
Sb	1.9
Bi	1.9

Generally electronegativity value decreases down the group.

5. (4) Nitrogen can exhibit +5 oxidation state in its oxides because it can form pπ–pπ bond with oxygen which helps to stabilize the higher oxidation state. e.g. N_2O_5

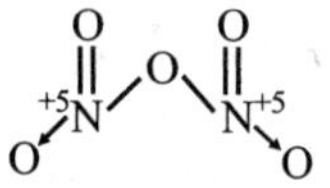

Nitrogen cannot form halide in +5 oxidation state because it does not contain d-orbital. Being 2nd period element N has only s and p orbitals which cannot expand its octet. e.g. NX_5 does not exist where X = halide

6. (2) $\ddot{N} - \ddot{N}$ single bond is weaker than $\ddot{P} - \ddot{P}$ bond due to more ℓp-ℓp repulsion. Bond length of P-P bond is greater than N-N bond.

In group 15 elements only nitrogen and phosphorus show disproportionation in +3 oxidation state. As, Sb and Bi are almost inert for disproportionation in +3 oxidation state. So both statements I and II are false.

7. (1) Group of phosphate mineral are called apatite family phosphorous is main component of phosphate mineral which have ionisation enthalpy 1012 kJ mol^{-1}.

8. (4) SO_2 can oxidise as well as reduce as it is in + 4 state.

Hence it can act as both oxidising and reducing agent.

9. (4) N – N : Single (σ) bond strength is lowest due to L.P. - L.P. repulsion. For N, the maximum possible accomodation is $8e^-$ thus covalency of nitrogen is 4.

10. (4) (D) Due to inert pair effect lower oxidation state is more stable.

(E) Nitrogen belongs to 2nd period and cannot expand its octet due to unavailability of d-orbital

11. (4) PH_3 has lower boiling point than NH_3 because NH_3 in liquid state have hydrogen bonding between molecules. PH_3 does not form hydrogen bonds.

12. (3) The stability of hydrides decreases from NH_3 to BiH_3 but their reducing character increases down the group.

13. (3) The E–H bond strength in group 15 follows the order:

$N-H > P-H > As-H > Sb-H > Bi-H$

Thus, the thermal stability will follow:

$NH_3 > PH_3 > AsH_3 > SbH_3 > BiH_3$

The reducing power will follow:

$NH_3 < PH_3 < AsH_3 < SbH_3 < BiH_3$

The basicity follows the oder:

$NH_3 > PH_3 > AsH_3 > SbH_3 > BiH_3$

This is due to N being unable to easily accommodate a lone pair due to small size while the larger atoms being able to do so.

14. (1)

$$\underset{\text{Steam}}{CH_4(g) + H_2O(g)} \xrightarrow[1270K]{Ni} CO(g) + 3H_2(g)$$

$$NaNO_2(aq) + NH_4Cl(aq) \longrightarrow N_2(g) + NaCl(aq) + 2H_2O(\ell)$$

15. (3) $N_2(g) + O_2(g) \rightarrow 2NO(g)$

$2NO(g) + O_2(g) \rightarrow 2NO_2(g)$

Higher concentration of NO_2 damage the leaves and retard the rate of photosynthesis.

16. (4) The reaction between dinitrogen and dioxygen is an endothermic process and feasible at high temperature.

$$N_2 + O_2 \xrightleftharpoons{(1483–2000K)} 2NO$$

17. (4) Red phosphrous is obtained by heating white phosphorous at 573 K in inert atmosphere. When red phosphorous is heated in a sealed tube at 803 K, we get α-black phosphorous.

β-Black phosphorous is prepared by heating white phosphorous at 473 K under high pressure.

18. (2) $Ba(N_3)_2 \longrightarrow Ba + 3N_2 \uparrow$

$Ba(NO_3)_2 \longrightarrow 2BaO + 4NO_2 + O_2 \uparrow$

$2NaNO_3 \longrightarrow 2NaNO_2 + O_2 \uparrow$

$2NaNO_2 \longrightarrow Na_2O + NO \uparrow + NO_2 \uparrow$

19. (4) As positive oxidation state increase, the acidic character increases.

Pentavalent oxides, E_2O_5 has E in +5 oxidation state.

Trivalent oxides; E_2O_3 has E in +3 oxidation state.

So, Acidic character $E_2O_5 > E_2O_3$

Down the group non-metallic/acidic character decreases.

20. (2) The reaction result in formation of nitrogen gas.

$$NH_4Cl(aq) + NaNO_2(aq) \rightarrow NH_4NO_2(aq) + NaCl(aq)$$
$$NH_4NO_2 \rightarrow N_2\uparrow + 2H_2O(l)$$

21. (2) PCl_5 forms five bonds by using the d-orbitals to "expand the octet". But NCl_5 does not exist because there are no d-orbitals in the valence shell (2nd shell). Therefore, there is no way to expand the octet.

22. (1) Due to increase in bond length, order of stability :

$$NF_3 > NCl_3 > NBr_3 > NI_3$$

23. (4) When red phosphorus is heated in a sealed tube at 803 K, α-black phosphorus is formed.

24. (2) NH_3 is only a mild reducing agent while BiH_3 is the strongest reducing agent amongst all the hydrides due to the longest bond length.

25. (3) The reducing character of the hydrides of group 15 elements increases from NH_3 to BiH_3 because the reducing character depends upon the stability of hydrides. The greater the unstability of the hydrides, the greater is its reducing character, the BiH_3 is the least stable in the series, hence it has strongest reducing power.

26. (4) (1) N_2 is diamagnetic in nature.

(2) $N_2 + O_2 \xrightleftharpoons{2000K} NO(g)$

(3) Liquid N_2 is used in cryosurgery.

(4) Because of its inertness, it is used where an inert atmosphere is required.

27. (4) $NH_3 + \underset{(excess)}{3Cl_2} \longrightarrow NCl_3 + 3HCl$

If NH_3 is used in excess then NH_4Cl is formed instead of NCl_3.

$$\underset{(excess)}{8NH_3} + 3Cl_2 \longrightarrow 6NH_4Cl + N_2$$

28. (2) No. of pentagons in C_{60} (Buckminsterfullerene) = 12

No. of triangles in P_4 (White phosphorous) = 4

29. (4) (1) $Ba(N_3)_2 \xrightarrow{\Delta} Ba + 3N_2$

(2) $(NH_4)_2Cr_2O_7 \xrightarrow{\Delta} Cr_2O_3 + N_2 + 4H_2O$

(3) $NH_4NO_2 \xrightarrow{\Delta} N_2 + 2H_2O$

(4) $(NH_4)_2SO_4 \xrightarrow{\Delta} 2NH_3 + H_2SO_4$

30. (3) Hint: Number of moles $= \dfrac{\text{Weight}}{\text{Molecular mass}}$

$$\underset{\substack{221g \\ 1g}}{Ba(N_3)_2(s)} \rightarrow Ba(s) + \underset{\substack{3\ moles \\ 3/221\ moles}}{3N_2(g)}$$

$$\underset{1g}{(NH_4)_2Cr_2O_7(s)} \rightarrow \underset{1/252\ moles}{N_2(g)} + 4H_2O(g) + Cr_2O_3(s)$$

$$\underset{1g}{2NH_3(g)} \rightarrow \underset{1/34\ moles}{N_2(g)} + 3H_2(g)$$

$$\underset{1g}{2NH_4NO_3(s)} \rightarrow \underset{1/80\ moles}{2N_2(g)} + 4H_2O(g) + O_2(g)$$

Hence, thermal decomposition of NH_3 will produce maximum amount of N_2.

31. (1)

$$Li(s) + N_2(g) \xrightleftharpoons[\text{high temperature}]{} 2Li_3N(s)$$

$$Li_3N(s) + 3H_2O(l) \longrightarrow 3LiOH(aq) + NH_3(g)$$

$$CuSO_4(aq) + 4NH_3(g) + 2H_2O(l) \longrightarrow \underset{\text{deep blue compound}}{[Cu(NH_3)_4(OH_2)_2]SO_4(aq)}$$

32. (3) All the members form volatile halides of the type AX_3. All halides are pyramidal in shape. The bond angle decreases on moving down the group (from NCl_3 to $SbCl_3$) due to decrease in bond pair-bond pair repulsion or increase in lone pair-bond pair repulsion.

33. (1) $(NH_4)_2SO_4 + 2H_2O \longrightarrow 2H_2SO_4 + NH_4OH$

H_2SO_4 is strong acid and increases the acidity of soil.

34. (4) Acetonitrile ($\overset{\delta+}{CH_3} - C \equiv \overset{\delta-}{N}$) and acetone both are polar molecules, hence dipole-dipole interaction exists

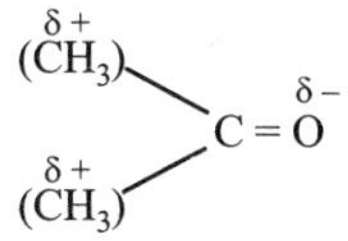

between them. Between KCl and water, ion-dipole interaction is found and in benzene-ethanol and benzene–carbon tetrachloride, dispersion force is present.

35. (2) Among the given compounds, $\ddot{N}H_3$ is most basic. Hence it has highest proton affinity.

36. (1) $_7N = 1s^2 2s^2 2p^3$; $_{15}P = 1s^2 2s^2 2p^6 3s^2 3p^3$

Note: In phosphorus the 3*d*-orbitals are available. Hence phosphorus can form pentahalides but nitrogen can not form pentahalide due to absence of *d*-orbitals.

37. (15) Phosphorus belongs to 15th group and forms dπ–dπ bond with transition metal and PH_3 is strongest base among the other group members.

38. (1) Statement-I: Flourine has the highest electro-negativity in the periodic table.

Across a period (N → O → F) electronegativity increases due to increasing nuclear charge and decreasing atomic size.

Statement-II: Oxidation state of oxygen in OF_2

$OF_2 \Rightarrow x - 1 \times 2 = 0 \Rightarrow x = +2$

Oxidation state of oxygen in Na_2O

$Na_2O \Rightarrow +1 \times 2 + x = 0 \Rightarrow x = -2$

Both the statement I and statement II are correct.

39. (1) Oxygen can have covalency 4.

In SO_2 ; O.S. of O = –2

In OF_2 ; O.S. of O = +2

(F is more electronegative than Oxygen)

Anomalous behaviour of oxygen is due to high EN and small size.

40. (4)

Acidic strength : $H_2Se < H_2Te$

$\Delta_{dis}H$	:	H_2Se	> H_2Te
		[276 kJ/mol]	[238 kJ/mol]

41. (2) Oxygen exists as O_2 (Atomicity = 2)

Sulphur exists as S_8 (Atomicity = 8)

Hence, Melting point & Boiling point of sulphur are significantly large compared to oxygen.

42. (3) Oxygen can from 2pπ-2pπ multiple bond with itself due to its small size while sulphur cannot form multiple bond with itself as 3pπ-3pπ bond will be unstable due to large size of sulphur, but sulphur can form multiple bond with small size atom like C and N.

eg. S = C = S

Thus, (A) is correct but (R) is incorrect

43. (2) H_2Te is more acidic than H_2S as H – Te bond is weaker than H – S bond due to a greater size difference between H and Te atom.

Thus, both A and R are true and R explains A.

44. (1) The first member of a group exhibits anomalous behaviour and shows different physical and chemical properties than the rest of the elements due to smallest size and highest electronegativity.

E.g. Oxygen is diatomic while other members are polyatomic.

45. (4) A and B only

All group 16 element form oxides of genral formula EO_2 and EO_3, Both are acidic in nature.

The reducing property increases from H_2S to H_2Te down the group.

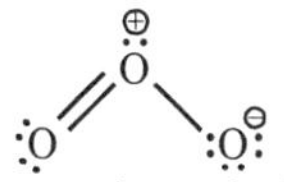

Ozone molecule contain total of six lone pair of electrons.

46. (3) Statement (I) : Oxygen can have oxidation state from –2 to +2

0 oxidation state in elemental form, –2 oxidation state in normal oxide, –1 oxide state in peroxide, +2 → oxidation state in OF_2.

So statement I is incorrect

Statement (II) : On moving down the group stability of +4 oxidation state increases whereas stability of +6 oxidation state decreases down the group, due to inert pair effect.

So both statements are incorrect.

47. (1) Bond dissociation energy of E–H bond in hydrides of group 16 follows the order:

$H_2O > H_2S > H_2Se > H_2Te$

48. (1) Water molecule is significantly more polar than the other group 16 hydride molecules so the attraction between $H^{\delta+}$ and $O^{\delta-}$ between water molecules is stronger than the attraction of $H^{\delta+}$ and $E^{\delta-}$ between other group 16 hydride molecules.

49. (3) $X = Na_2SO_3$

$$\underset{(X)}{SO_3^{2-}} + H_2SO_4 \rightarrow [H_2SO_3] \rightarrow H_2O + \underset{(Y)}{SO_2}$$

$$SO_2 + 2NaOH \rightarrow \underset{(X)}{Na_2SO_3} + H_2O$$

$$\underset{(X)}{Na_2SO_3} + \underset{(Y)}{SO_2} + H_2O \rightarrow \underset{(Z)}{2NaHSO_3}$$

50. (1) At room temperature, water is liquid and has boiling point 373 K due to hydrogen bonding. Whereas H_2S is a gas and it has no hydrogen bonding. Hence boiling point of H_2S is less than 300 K (boiling point of H_2S is –60°C).

51. (1) The S_6 molecule has a chair-form hexagon ring with the approx same bond length as that in S_8, but with some what smaller bond angles, i.e., bond lengths are approx. same but bond angles are different.

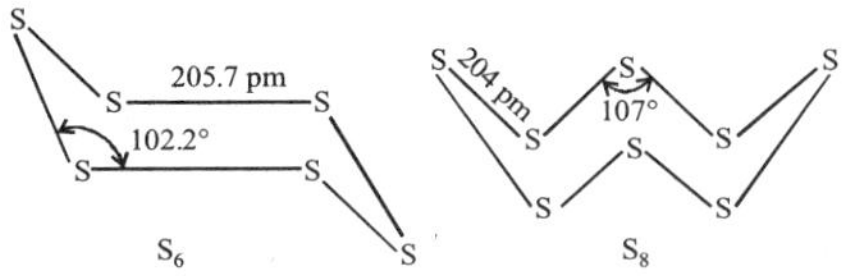

52. (7) Fuming sulphuric acid is a mixture of conc. $H_2SO_4 + SO_3$ or $H_2S_2O_7$

So, Number of Oxygen atoms = 7

53. (1) α-sulphur and β-sulphur are diamagnetic. S_2-form is paramagnetic similar to O_2.

54. (1)

(I) Oxidising power refers to the ability to accept electrons.

$$F_2 > Cl_2 > Br_2 > I_2$$

(II) $NaBr + Cl_2\text{ - water} \longrightarrow Br_2 + NaCl$
(Orange/Brownlayer in CCl_4)
$NaI + Cl_2\text{ - water} \longrightarrow I_2 + NaCl$
(Violet/Purple layer in CCl_4)

55. (1) Correct order of
(i) Boiling point : HF > HI > HBr > HCl
(ii) Melting point : HI > HF > HBr > HCl

56. (3)
HF < HCl < HBr < HI (bond length order)
F < Cl < Br < I (covalent radius order)
$PH_3 < AsH_3 < NH_3 < SbH_3 < BiH_3$ (Boiling point order)
Maximum possible covalency of phosphorus is 6.

57. (3) F < Cl < Br < I (Covalent radius)
Cl > F > Br > I (Electron affinity) (Irregularity)
$F^- < Cl^- < Br^- < I^-$ (Ionic radius)
F > Cl > Br > I (Ist ionization energy)

58. (1) Fluoroapatite ⇒ $[3(Ca_3(PO_4)_2).\,CaF_2]$
Hydroxyapatite ⇒ $[3(Ca_3(PO_4)_2).\,Ca(OH)_2]$

59. (4) N,O, F can't extend their valencies upto their group number due to the non-availability of vacant d orbitals.

60. (4) Fluorospar is CaF_2. Therefore option (4) is incorrect pair.

61. (1)

62. (4) Fluorine does not show variable oxidation state as it is the most electronegative element and shows only –1 state.

63. (4) Only I^- among halides can be oxidised to Iodine by oxygen in acidic medium due to its reduction potential.
$4I^-(aq) + 4H^+(aq) + O_2(g) \rightarrow 2I_2(s) + 2H_2O(l)$

64. (2)
(1) $\Delta_{eg}H(Cl) < \Delta_{eg}H(F)$
(–349) (–333) Correct
(2) $\Delta_{eg}H(Se) < \Delta_{eg}H(S)$
(–195) (–200) Incorrect
(3) $\Delta_{eg}H(I) < \Delta_{eg}H(At)$
(–296) (–270) Correct
(4) $\Delta_{eg}H(Te) < \Delta_{eg}H(Po)$
(–190) (–174) Correct

65. (1) Bond energy of F_2 less than Cl_2 due to lone pair-lone pair repulsions and small size of fluorine.
∴ Bond energy order $Cl_2 > Br_2 > F_2 > I_2$

66. (1) Due to small size and high electronegativity of fluorine it can not expand its oxidation state and show only one oxoacid i.e., HOF.

67. (1) Bond dissociation energy refers to the energy required to break bond between two atoms. Its value depends on the strength of the bonds. As size of atom increases, its bond strength and dissociation energy decreases. So, dissociation energy decreases down the group as size increases. In the halogen family, size of F is very small, so it faces a strong electron repulsion between its electrons, which make its bond weak, thus, reducing its dissociation energy. Therefore, the correct order is: $Cl_2 > Br_2 > F_2 > I_2$

68. (1) Generally, bond energy
$$\propto \frac{1}{\text{Bond length}}$$
So, bond energy order is: C–F > C–Cl > C–Br > C–I

69. (3) Due to strong H-bonding between HF molecules. HF has highest boiling point among the hydrogen halides.

70. (3) Fluorine is most electronegative element in periodic table and exhibits O.S. value of –1 only.

71. (1) NaF is composed of Na^+ and F^-.
$Na^+ \longrightarrow [He]2s^2 2p^6$
$F^- \longrightarrow [He]2s^2 2p^6$
Hence, configuration of Na^+ and F^- do not match with the configuration given in the question.

72. (2) Order of reactivity of halogens: $Cl_2 > Br_2 > I_2$
But the interhalogen compounds are generally more reactive than halogens (except F_2), since the bond between two dissimilar electronegative elements is weaker than the bond between two similar atoms, i.e, X – X.

73. (2) The molecular geometry of BrF_5 is square pyramidal with asymmetric charge distribution on the central atom.

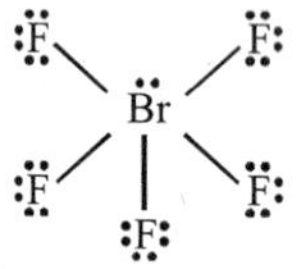

74. (3) The electron gain enthalpy order for halogens is Cl > F > Br > I
Due to small size of fluorine the extra electron to be added feels more electron-electron repulsion. Therefore fluorine has less value for electron affinity than chlorine.

75. (3) The H–X bond strength decreases from HF to HI, i.e., HF > HCl > HBr > HI. Thus HF is most stable while HI is least stable. The decreasing stability of the hydrogen halide is also reflected in the values of dissociation energy of the H–X bond:

H – F	H – Cl	H – Br	H – I
$135 kcal\ mol^{-1}$	$103 kcal\ mol^{-1}$	$87 kcal\ mol^{-1}$	$71 kcal\ mol^{-1}$

76. (1) The oxidising nature of fluorine is due to its small size that result in high enthalpy of hydration

77. (1) IF_5 has 1 lone pair and IF_7 has 0 lone pair
∴ 1 + 0 = 1

78. (3) AB_5 species have square pyramidal structures.
Hence, BrF_5, IF_5 and ClF_5.

79. (4)
(i) Due to unpaired e^- in O_2^-, KO_2 is paramagnetic.
(ii) $^{\ominus}O - \overset{+}{N} = O$ is paramagnetic due to unpaired e^- with N atom.
(iii) $O = \ddot{Cl} = O$ is paramagnetic due to unpaired e^- with Cl atom.
(iv) $\cdot\ddot{N} = O$ is paramagnetic.

80. (1)
(A)
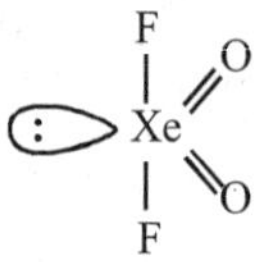
See-saw
So (A) is true (T)
(B) Xe has 7 (not 5) electron pairs in XeO_2F_2 valence shell (F)
(C) O–Xe–O Bond angle is close to 120° (not 180°) (F)
(D) F–Xe–F Bond angle is close to 180° (T)
(E) Xe has 14 (not 16) valence electrons in XeO_2F_2 (F)

81. (4) Statement I and II are False.
Noble gases have low boiling points and they are held together by weak dispersion forces.

82. (3)

Element	$\Delta_{eg}H$[kJ / mol]
He	+48
Ne	+116
Kr	+96
Xe	+77

83. (1) A-II, B-I, C-IV, D-III

84. (4) Number of lone pair of electrons on the central atom
$$= \frac{\text{Number of valence electrons on central atom 2(Number of bivalent atoms) (Number of monovalent atoms)}}{2}$$

Species	No. of lone pair
XeF_2	$n = (8-2)/2 = 3$
XeO_2F_2	$n = \frac{8-(2\times 2)-2}{2} = 1$
XeO_3F_2	$n = \frac{8-(2\times 3)-2}{2} = 0$
XeF_4	$n = \frac{8-4}{2} = 2$

85. (4) Radon is radioactive element and not present in atmosphere.

86. (1) Structure of XeO_3F_2

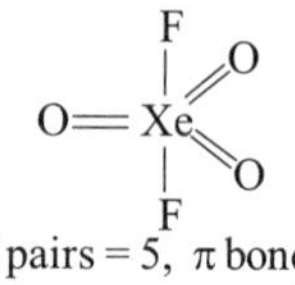

So, bond pairs = 5, π bonds = 3 lone pairs = 0

87. (1) Instantaneous dipole-induced dipole forces are most responsible in allowing xenon gas to liquify.

88. (3) In $XeOF_4$, Xe is sp^3d^2, hybridised having 5 bond pairs and 1 lone pair respectively.

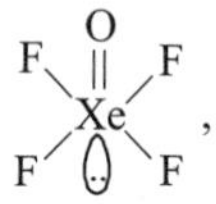
$XeOF_4$
Square pyramidal
(sp^3d^2)

89. (2) As we move down the group, the melting and boiling points show a regular increase due to corresponding increase in the magnitude of their van der Waals forces of attraction as the size of the atom increases.

90. (2) The shape of XeO_3F_2 is trigonal Pyramidal.

91. (2) The hybridization of XeO_3F_2 is sp^3d and its structure is trigonal bipyramidal in which oxygen atoms are situated on the plane and the fluoride atoms are on the top and bottom.

92. (2) XeF_4 has zero dipole moment. It has square planar structure due to which the bond moments of Xe – F cancel each other.

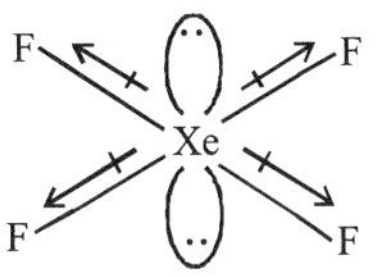

93. (3) XeF_2:

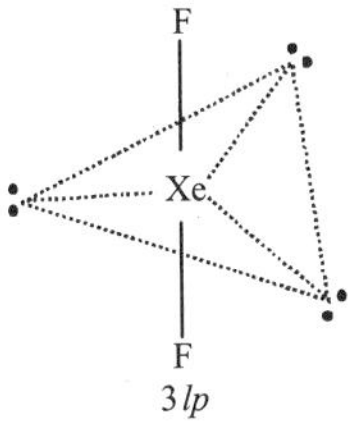

3 *lp*

XeF_4:

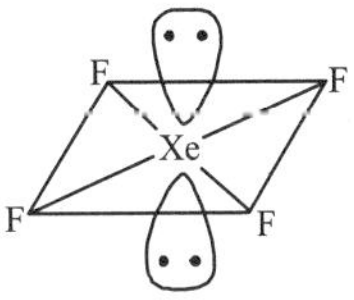

2 *lp*

XeF_6:

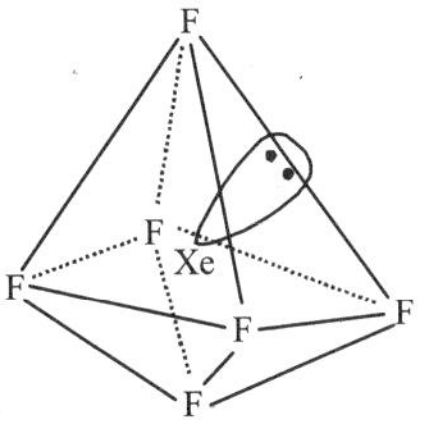

1 *lp*

XeO_3:

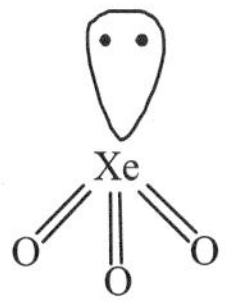

1 *lp*

94. (4)

95. (3) Helium is heavier than hydrogen although it is non-inflammable.

96. (4) XeF_5^+,

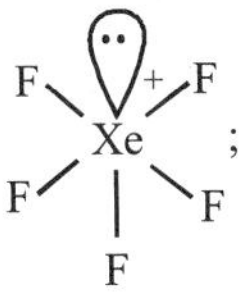

;

XeO_3,

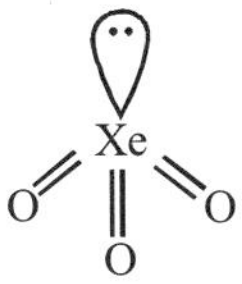

XeO_2F_2,

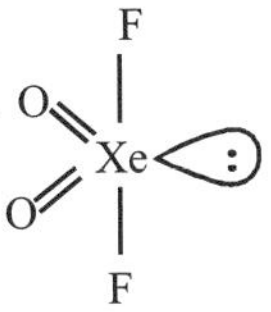

$XeOF_4$,

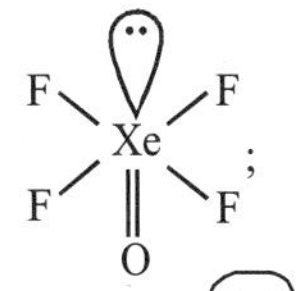

;

XeF_4,

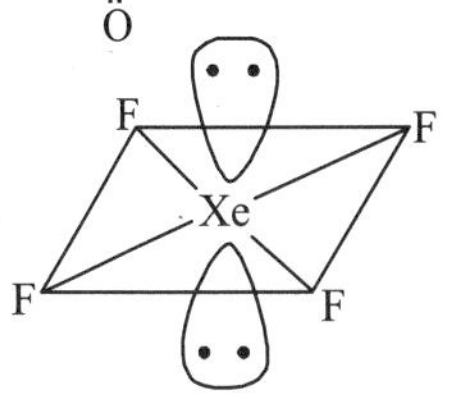

2 *lp*

So, Answer is 4

97. (3)

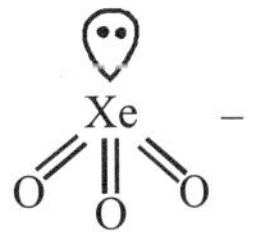

S.N of Xe = 4 (3B.P + 1ℓ.P)

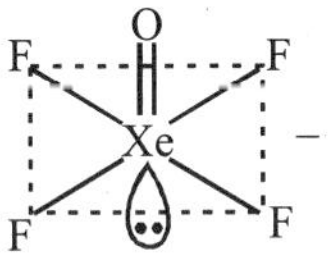

S.N of Xe = 6 (5B.P + 1ℓ.P)

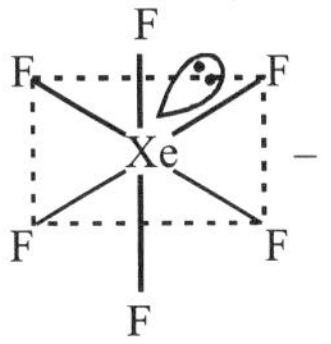

S.N of Xe = 7 (6B.P + 1ℓ.P)

Total lone pair = 1 + 1 + 1 = 3

Part-B (JEE Advanced)

1. (b) **Bond Angle**

$NO_2^+ \longrightarrow O=\overset{+}{N}=O$ (sp) 180°

$NO_2 \longrightarrow$

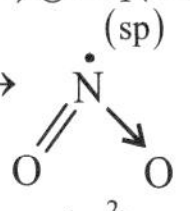

(sp²) 134°

$NO_3^- \longrightarrow$

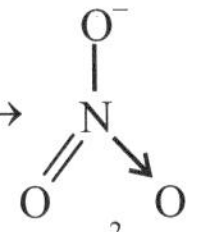

(sp²) 120°

$NO_2^- \longrightarrow$

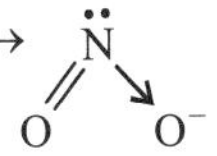

(sp²) 115°

So, order of bond angle is

$NO_2^- < NO_3^- < NO_2 < NO_2^+$

2. (a) Ammonium nitrite

$NH_4NO_2 \xrightarrow{\Delta} N_2 + 2H_2O$

Ammonium nitrate

$NH_4NO_3 \xrightarrow{\Delta} N_2O + 2H_2O$

3. (b) If PCl_5 is fluorinated in a polar solvent, ionic isomers are formed e.g.:-

$[PCl_4]^+[PCl_4F_2]^-$ (colourless crystals)

and $[PCl_4]^+[PF_6]^-$ (white crystals)

4. (a) Disproporationation of HNO_2 in aqueous solution takes place as:

$\underset{+3}{3HNO_2} \rightleftharpoons \underset{+5}{HNO_3} + \underset{+2}{2NO} + H_2O$

Hence, the products formed are H_3O^+, NO_3^- and NO.

5. (c) Let oxidation states of phosphorus in H_3PO_2, H_3PO_4, H_3PO_3 and $H_4P_2O_6$ be *w*, *x*, *y* and *z* respectively.

Thus, in H_3PO_2:

$3 \times (+1) + w + 2 \times (-2) = 0 \quad \therefore w = +1$

In H_3PO_4:

$3 \times (+1) + x + 4 \times (-2) = 0 \quad \therefore x = +5$

In H_3PO_3:

$3 \times (+1) + y + 3 \times (-2) = 0 \quad \therefore y = +3$

In $H_4P_2O_6$:

$4 \times (+1) + 2z + 6 \times (-2) = 0 \quad \therefore z = +4$

Thus, the order of oxidation state is :

$H_3PO_4 > H_4P_2O_6 > H_3PO_3 > H_3PO_2$

6. (a)

$\underset{\text{white phosphorus}}{P_4} + 8SOCl_2 \rightarrow 4PCl_3 + 4SO_2 + 2S_2Cl_2$

7. (b) The slow decomposition of HNO_3 is represented by the eqn.

$4HNO_3 \rightarrow \underset{\text{(yellow-brown)}}{4NO_2} + 2H_2O + O_2$

8. (c)

$\overset{0}{P_4} + 3NaOH + 3H_2O \xrightarrow[\text{inert atmosphere}]{\Delta} \overset{-3}{PH_3} + 3Na\overset{+1}{H_2PO_2}$ (sodium hypophosphite)

(oxidation: $P_4 \rightarrow NaH_2PO_2$; reduction: $P_4 \rightarrow PH_3$)

9. (b) $\overset{+5}{HNO_3}, \overset{+2}{NO}, \overset{0}{N_2}, \overset{-3}{NH_4Cl}$

10. (d) $Ba(N_3)_2 \longrightarrow Ba + 3N_2$

11. (b) $P_4 + 3O_2 \xrightarrow{\text{in presence of } N_2} P_4O_6$

Here N_2 acts as a diluent and thus retards further oxidation of P_4O_6 to P_4O_{10}. Reaction of P_4 under other three conditions.

(a) $P_4 + 3O_2 \longrightarrow P_4O_6 \xrightarrow{2O_2} P_4O_{10}$

(c) In moist air, P_4O_6 is hydrolysed to form H_3PO_3

$P_4O_6 + 6H_2O \longrightarrow 4H_3PO_3$

(d) In presence of NaOH,

$P_4 + 3OH^- + 3H_2O \longrightarrow PH_3 + 3H_2PO_2^-$

12. (d) In P_4, the P–P linkage is formed by sp^3–sp^3 hybridised orbital overlapping. So, the percentage of π-character will be 75%.

13. (b) PbO_2 is a powerful oxidizing agent and liberate O_2 when treated with acids.

$2PbO_2 + 4HNO_3 \longrightarrow 2Pb(NO_3)_2 + 2H_2O + O_2 \uparrow$

14. (b)

$NO\,(g) + NO_2\,(g) \xrightarrow{-30°C} \underset{\text{(blue liquid)}}{N_2O_3}$

15. (c) The ignition temperature of black phosphorus is highest among all its allotropes, hence is most stable.

16. (a) The structure of H_3PO_3 is as follows: There are only two –OH groups and hence **dibasic**. The oxidation number of P in this acid is +3. Whereas P can have +5 oxidation state also. Therefore, H_3PO_3 can be oxidised which means H_3PO_3 is a **reducing agent**.

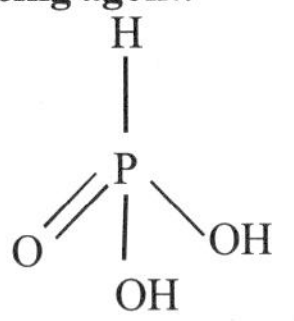

17. (c) NH_3 does not react with basic oxide, CaO, while other reacts with NH_3

$2NH_3 + H_2SO_4 \rightarrow (NH_4)_2SO_4$

$P_4O_{10} + 12NH_3 \rightarrow 4(NH_4)_3PO_4 + 6H_2O$

$CaCl_2 + 8NH_3 \rightarrow CaCl_2.8NH_3$

18. (c) In cyclic metaphosphoric acid number of P–O–P bonds is three.

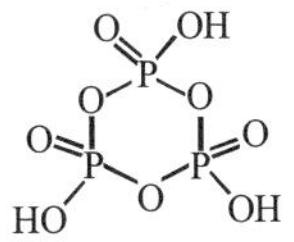

19. (b) In BCl_3, $H = \frac{1}{2}(3+3+0-0) = 3$; sp^2 hybridization (bond angle = 120°). Similarly PCl_3, $AsCl_3$ and $BiCl_3$ are found to have sp^3 hybridized central atom with one lone pair of electrons on the central atom. The bond angle ≤ 109°28′, since the central atoms belong to the same group. The bond angle of the chlorides decreases as we go down the group. Thus, the order of bond angle is, $BCl_3 > PCl_3 > AsCl_3 > BiCl_3$.

20. (d) $(NH_4)_2Cr_2O_7 \rightarrow N_2 + Cr_2O_3 + 4H_2O$

21. (c) $Ca_3P_2 + 6H_2O \rightarrow 3Ca(OH)_2 + 2PH_3$; *i.e.* 2 moles of phosphine are produced.

22. (a) Least basic trihalogen of nitrogen is NF_3 because of the highest electronegativity of fluorine.

23. (d) NO_2 is reddish brown coloured gas. Rest of the oxides are colourless.

24. (b)

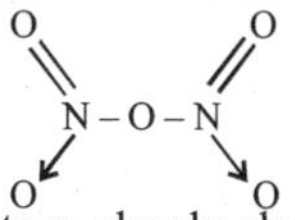

The structure clearly shows the presence of covalent and co-ordinate bonds.

25. (a) Only nitrates of heavy metals and lithium decompose on heating to produce NO_2.

26. (d) None; it reacts with all given compounds. It forms addition compounds with them because it is basic in nature and given compounds are acidic in nature.

It can be dried over any metal oxide.

27. (c)

$$\underset{\substack{\text{Nitric oxide}\\\text{(colourless)}}}{2NO} + O_2 \longrightarrow \underset{\substack{\text{Nitrogen dioxide}\\\text{(reddish brown coloured)}}}{2NO_2}$$

28. (8)

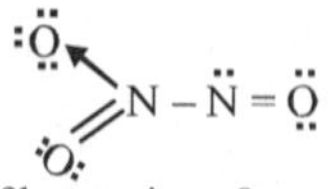

Number of lone pairs = 8

29. (4) $PCl_5 + SO_2 \longrightarrow POCl_3 + \underset{\text{(Thionyl chloride)}}{SOCl_2}$

$6PCl_5 + P_4O_{10} \longrightarrow 10POCl_3$

When equimolar amounts of PCl_5 and H_2O are used, the reaction is gentle: $PCl_5 + H_2O \rightarrow POCl_3 + 2HCl$

[**Note:** PCl_5 reacts violently with H_2O on complete hydrolysis to produce phosphoric acid.

$PCl_5 + 4H_2O \rightarrow H_3PO_4 + 5HCl$

$2PCl_5 + H_2SO_4 \rightarrow 2POCl_3 + \underset{\text{(sulphuryl chloride)}}{SO_2Cl_2} + 2HCl$

30. (2.38)

$$\underset{\text{(White phosphorous)}}{P_4} + 3NaOH + 3H_2O \xrightarrow[\text{atmosphere}]{\text{Inert}} \underset{\text{Phosphine}}{PH_3\uparrow} + 3NaH_2PO_2$$

PH_3 : a non-inflammable gas in its pure form; slightly soluble in water.

When PH_3 is absorbed in $CuSO_4$ solution cupric phosphide forms:

$2PH_3 + 3CuSO_4 \longrightarrow CuP_2 + 3H_2SO_4$

1 mol of $P_4 = 31 \times 4 = 124$ g

∴ 1.24 g of white phosphorous = 0.01 mol

∴ 0.01 mol of P_4 forms 0.01 mol of PH_3

No. of moles of $CuSO_4$ is required for complete consumption of 0.01 mol $= 0.01 \times \frac{3}{2} = 15 \times 10^{-3}$

M.W. of $CuSO_4$ = 159 g/mol

∴ Amount of $CuSO_4$ required $= 15 \times 10^{-3} \times 159 =$ 2.38g

31. four.

Each 'P' atom is linked to 4 'O' atoms as confirmed by its structure. It is linked to three 'O' atoms by single bond and one 'O' atom by double bond. Each P atom is linked to 4 oxygen atoms

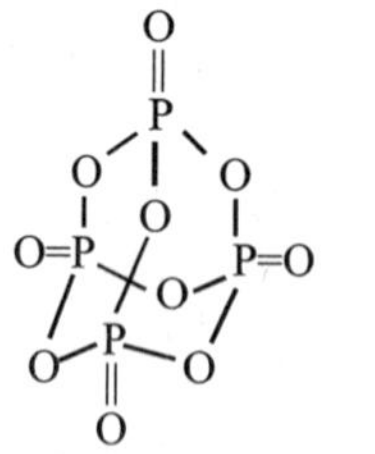

32. two;

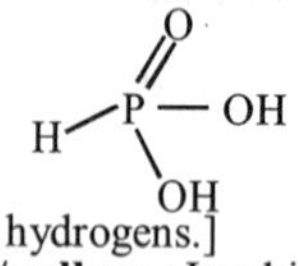

[It contains two replaceable hydrogens.]

33. white/ yellow; : In white phosphorus, each phosphorus atom is linked to the other three atoms by covelent bonds. P–P–P bond angle is 60°, due to which the molecule remains under strain and hence is active in nature.

34. –3; Nitrogen has 5 electrons in its valence shell. Thus, it can accept maximum 3 electrons.

35. True : The molecule of NO has eleven valence electrons (5 due to N and 6 due to O). It is impossible for all of them to be paired, hence the nitric oxide molecule contains an *odd electron* which makes gaseous nitric oxide *paramagnetic*.

$:N \overset{\bullet\bullet\bullet}{=\!=} O:$

In the liquid and solid states, nitric oxide is polymerised to a dimer which is *diamagnetic*.

36. True :

The central element in the metal hydrides of group 15 elements is although in sp^3 hybrid state, the H – M – H bond angle is less than the normal tetrahedral bond angle of 109° 28'; *e.g.* the bond angle, H – N – H in NH_3 is 106° 45'. This is due to greater repulsion between a lone pair and a bond pair of electrons than between the two bond pairs of electrons.

The decrease in bond angle from 107.8° in ammonia to about 90° in AsH_3 can be explained by the fact that in the latter case sp^3 hybridisation becomes less and less distinct with the increasing size of their electron clouds, *i.e.*, pure *p* orbitals (instaed of sp^3 hybrid orbitals) are used for M – H bonding and the lone pair of electrons is present in spherical *s*-orbital.

37. False : Red phosphorus is ***polymeric substance***. It exist as chains of P_4 tetrahedra linked together. Therefore, it is less volatile than white phosphorus.

38. (b,d)

H_3PO_4 (Phosphoric acid)

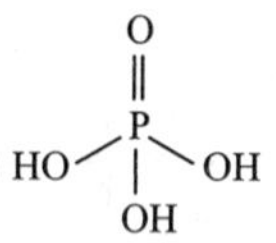

H_3PO_3 (Phosphorous acid)

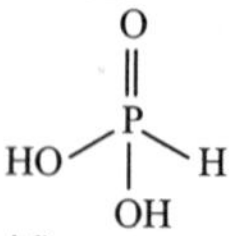

$H_4P_2O_7$ (Pyrophosphoric acid)

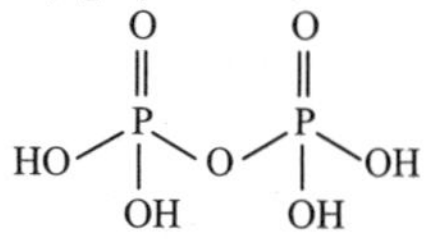

H_3PO_2 (Hypophosphorous acid)

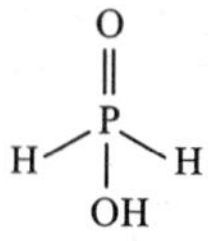

39. (a, b, d)

(a) $4H_3PO_3 \xrightarrow{\Delta} 3H_3PO_4 + PH_3$ (correct)

(b) H_3PO_4 has "P" in its highest oxidation state, hence cannot act as a reducing agent (correct)

(c)

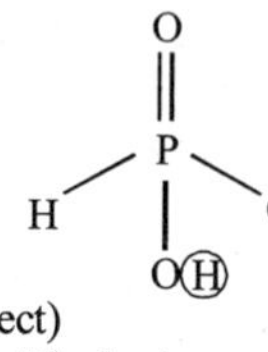

So, it is dibasic acid (incorrect)

(d) The hydrogen which is directly attached to phosphorous does not ionized in water.

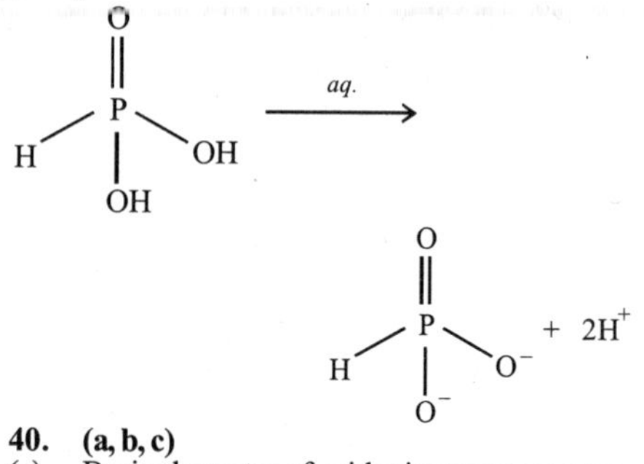

40. (a, b, c)

(a) Basic character of oxides increases on moving down the group therefore, Bi_2O_5 is more basic than N_2O_5.

(b) Covalent nature depends on electronegativity difference between bonded atoms. In NF_3, N and F are non-metals but in BiF_3, Bi is metal while F is non metal therefore, NF_3 is more covalent than BiF_3.

(c) In PH_3, hydrogen bonding is absent but in NH_3, hydrogen bonding is present, therefore PH_3 boils at lower temperature than NH_3.

(d) Due to small size in N–N single bond, l.p. – l.p. repulsion is more than P–P single bond therefore, N–N single bond is weaker than the P–P single bond.

41. (b, c)

(a) $NH_4NO_3 \xrightarrow[\text{below } 300°C]{} N_2O + 2H_2O$

$2N_2O \xrightarrow[\text{above } 600°C]{} 2N_2 + O_2$

(b)

$$(NH_4)_2Cr_2O_7 \xrightarrow{\Delta} \underset{\text{(colourless}}{N_2} + \underset{\text{(Green colour)}}{Cr_2O_3} + 4H_2O$$

(c) $Ba(N_3)_2 \xrightarrow[180°C]{\Delta} Ba + 3N_2$

(d) $Mg_3N_2 \xrightarrow[\text{above } 700°C]{} 3Mg + N_2$

Hence, only $(NH_4)_2Cr_2O_7$ and $Ba(N_3)_2$ can provide N_2 gas on heating below 300°C

42. (b, d)

$$P_4O_{10} + 4HNO_3 \xrightarrow{\text{dehydration of } HNO_3} \underset{\text{(required product)}}{4(HPO_3) + 2N_2O_5}$$

(a) $P_4 + 20HNO_3 \rightarrow 4H_3PO_4 + 20NO_2\uparrow + 4H_2O$

(b) N_2O_5 is diamagnetic in nature

(c) $N_2O_5 \rightarrow$

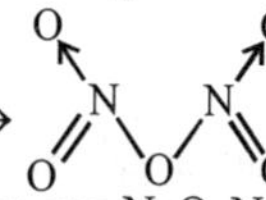

N_2O_5 contains one N–O–N bond but not N–N bond.

(d) $Na + N_2O_5 \rightarrow NaNO_3 + \underset{\text{(Brown gas)}}{NO_2\uparrow}$

43. (a, b, c)

$N \equiv N \rightarrow O$

(a)

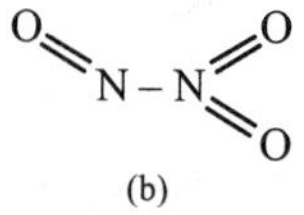

(b)

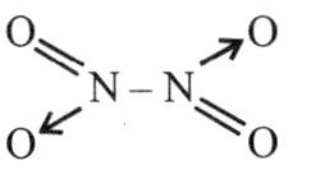
(c)

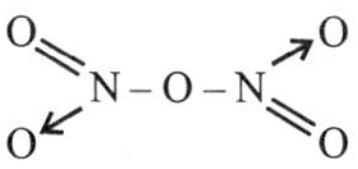
(d)

44. (a,b) When ammonium salt NH_4NO_3 or NH_4NO_2 (ammonium salts are colourless) is boiled with excess of NaOH, ammonia (NH_3) gas is evolved as follows:

$NH_4NO_2 + NaOH \longrightarrow NaNO_2 + NH_3 + H_2O$
$NH_4NO_3 + NaOH \longrightarrow NaNO_3 + NH_3 + H_2O$

The NH_3 gas evolved is non-flammable gas.
When the gas evolution ceases we are left with $NaNO_2$ or $NaNO_3$ in solution.
These salts get reduced when Zn is added to this solution containing salt ($NaNO_2$ or $NaNO_3$). Again NH_3 gas evolves.

$$NaNO_2 + 6(H) \xrightarrow{Zn / NaOH} NaOH + NH_3 + H_2O$$

$$NaNO_3 + 8(H) \xrightarrow{Zn / NaOH} NaOH + NH_3 + 2H_2O$$

Thus, the colourless salt [H] is either NH_4NO_2 or NH_4NO_3.
Thus, (a) and (b) are correct answers.

45. (c) $2NH_3 + OCl^- \rightarrow NH_2.NH_2 + H_2O + Cl^-$

46. (a, c, d) The four atoms in a P_4 molecule are situated at the corners of a tetrahedron. There are six P - P single bonds with PPP bond angle equal to 60°. Each phosphorus has a lone pair of electrons.

47. (a, b, d) Sodium nitrate on decomposition upto 500°C gives $NaNO_2$ and oxygen.

$2NaNO_3 \xrightarrow{\Delta} 2NaNO_2 + O_2\uparrow$

While at higher temperature (*i.e.* above 800°C), $NaNO_2$ further decomposes into Na_2O, N_2 and O_2.

$2NaNO_2 \xrightarrow{800°C} Na_2O + 3/2O_2\uparrow + N_2\uparrow$

48. (a, d) $NH_4NO_3 \longrightarrow N_2O + 2H_2O$
$NH_2OH + HNO_2 \longrightarrow N_2O + 2H_2O$

49. (d) (P) $P_2O_3 + 3H_2O \rightarrow 2H_3PO_3$
(Q) $P_4 + 3NaOH + 3H_2O \rightarrow 3NaH_2PO_2 + PH_3$
(R) $PCl_5 + CH_3COOH \rightarrow CH_3COCl + POCl_3 + HCl$
(S) $H_3PO_2 + 2H_2O + 4AgNO_3 \rightarrow 4Ag + 4HNO_3 + H_3PO_4$

50. (i) (d) (ii) (b) (iii) (a) (iv) (c)

51. (c) We know that phosphates have a biological significance in human, therefore statement (a) is not correct.
Since nitrates are more soluble in water so they are less abundant in earth's crust where as phosphates are less soluble in water and so they are more abundant in earth's crust. Thus, statement (b) is False and statement (c) is correct.

In nitrates (NO_3^-) nitrogen is in +5 oxidation state which is the highest oxidation state exhibited by nitrogen. Because of this, nitrates can not be oxidized (oxidation means increase in oxidation state). Hence, statement (d) is not correct.
The correct answer is (c).

52. (c) In case of group 15 (nitrogen group), on moving down the group, there occurs a decrease in bond angle of metal hydrides. This decrease in bond angle of metal hydrides of this group may be attributed to the increased *p*-character in the bond pair which results in more *s*-character in lone pair orbital.
The directional character is more for sp^3 hybrid orbital than on *s*- orbital.
Thus, the correct answer is (c).

53. (b) The reaction between NaOH and white phosphorus (P_4) can be represented as follows:

$$\underset{\text{(white Phosphorus)}}{\overset{0}{P_4}} + 3NaOH + 3H_2O \longrightarrow 3NaH_2\overset{+1}{P}O_2 + \underset{\text{(Phosphine)}}{\overset{-3}{P}H_3}$$

In this reaction, Phosphorus is oxidised as well as reduced, so it is a disproportionation reaction.
∴ The correct answer is (b).

54. (b) Nitrogen cannot form pentahalides because it cannot expand its octet due to non-availability of *d*-orbitals. So E is not correct explanation of S.

55. $(CH_3)_3N$ and $(Me_3Si)_3N$ are not isostructural, the former is pyramidal while the latter is trigonal planar. Silicon has vacant *d* orbitals which can accommodate lone pair of electrons from N (back bonding) leading to planar shape.

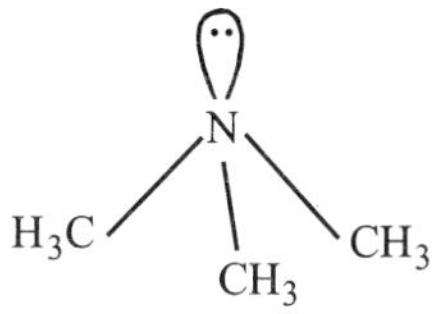

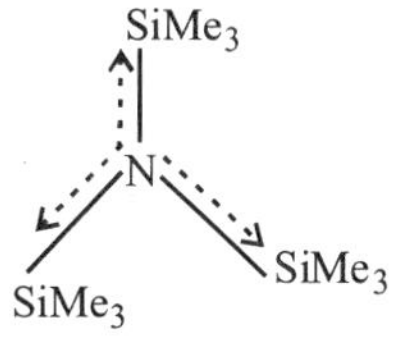

56. $6CaO + P_4O_{10} \longrightarrow 2Ca_3(PO_4)_2$

Moles of $P_4O_{10} = \frac{852}{284} = 3$

Moles of CaO = 3 × 6 = 18; wt. of CaO = 18 × 56 = 1008 g

57. Elemental nitrogen exists as a diatomic molecule because nitrogen can form $p\pi$ - $p\pi$ multiple bonds which is not possible in case of phosphorus due to repulsion between non-bonded electrons of the inner core. There is no such repulsion in case of smaller nitrogen atoms as they have only $1s^2$ electrons in their inner core.

58. $A = Ca(OH)_2$, $B = NH_4HCO_3$, $C = Na_2CO_3$, $D = NH_4Cl$ and $E = CaCl_2$

$$CaO + H_2O \longrightarrow \underset{(A)}{Ca(OH)_2}$$

$$NH_3 + CO_2 + H_2O \longrightarrow \underset{\text{Sod. Bicarbonate}(B)}{NH_4HCO_3}$$

$$NH_4HCO_3 + NaCl \longrightarrow NaHCO_3 + \underset{\text{Amm. chloride}(D)}{NH_4Cl}$$

$$2NaHCO_3 \xrightarrow{\Delta} \underset{\text{Sod. carbonate }(C)}{Na_2CO_3} + H_2O + CO_2$$

$$Ca(OH)_2 + 2NH_4Cl \longrightarrow \underset{(E)}{CaCl_2} + \underset{\text{can be used again}}{2NH_3} + 2H_2O$$

59. $7H_3PO_4 + Ca_5(PO_4)_3F \rightarrow \underset{\text{Triple superphosphate}}{5Ca(H_2PO_4)_2} + HF$

60. The poisonous element *M* may be As. So, on the basis of the given facts,

$$AsCl_3 + 6H \xrightarrow{Zn/H_2SO_4} \underset{(N)}{AsH_3} + 3HCl$$

$$2AsH_3 \xrightarrow{\Delta} \underset{(M)}{2As} + 3H_2\uparrow$$

Hence, *M* = As; *N* = As H_3

61.

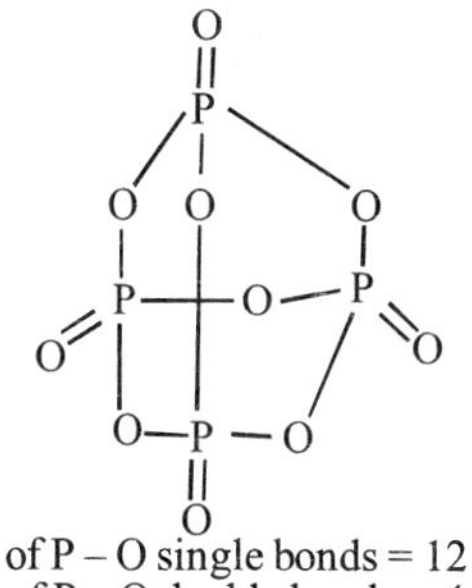

No. of P – O single bonds = 12
No. of P – O double bonds = 4

62. (*i*) $2P + 3I_2 + 6H_2O \rightarrow 2H_3PO_4 + 6HI$
(*ii*) $2KNO_3 + 10K \rightarrow 6K_2O + N_2$

63. Since *carbon* has no *d*-orbital, it cannot extend its coordination number beyond four, its halides are not attacked (hydrolysed) by water. On the other hand, silicon have vacant *d*-orbitals to which water molecules can coordinate and hence their halides are hydrolysed by water. As the charge on central atom increases, the tendency of attack of a nucleophile (OH^-) increases.
Increasing order of extent of hydrolysis
$CCl_4 < MgCl_2 < AlCl_3 < SiCl_4 < PCl_5$

64. (*i*) $KClO_3 + 2H_2C_2O_4 + H_2SO_4 \rightarrow KHSO_4 + HCl + 6CO_2 + 3H_2O$
(*ii*) $(NH_4)_2SO_4 + NO + NO_2 \rightarrow 2N_2 + 3H_2O + H_2SO_4$

65. N_2O has two principal resonance structures :

$$:\overset{-}{\ddot{N}} = \overset{+}{N} = \ddot{O}: \longleftrightarrow :N \equiv \overset{+}{N} - \overset{-}{\ddot{\underset{..}{O}}}:$$

66. (*i*) $P_4O_{10} + 6PCl_5 \rightarrow 10POCl_3$
(*ii*) $[2HNO_3 \rightarrow H_2O + 2NO_2 + [O]] \times 10$

$P_4 + 10[O] \rightarrow P_4O_{10}$
$P_4O_{10} + 6H_2O \rightarrow 4H_3PO_4$

$P_4 + 20HNO_3 \rightarrow 4H_3PO_4 + 20NO_2 + 4H_2O$

(*iii*) $P_4 + 20HNO_3 \xrightarrow[\text{catalyst}]{I_2 \text{ as}} 4H_3PO_4 + 20NO_2 + 4H_2O$
(*iv*) $2KMnO_4 + 2NH_4OH \rightarrow 2MnO_2 + N_2 + 2KOH + 4H_2O$
(*v*) $Na_2CO_3 + NO + NO_2 \rightarrow 2NaNO_2 + CO_2$
(*vi*) $NaNO_2 + 6H \xrightarrow{Zn / NaOH} NaOH + NH_3 + H_2O$
(*vii*) $CaSO_4 + 2NH_3 + CO_2 + H_2O \rightarrow CaCO_3\downarrow + (NH_4)_2SO_4$
(*viii*) $15CaO + 4P_4 \xrightarrow{\Delta} 5Ca_3P_2 + 3P_2O_5\uparrow$
$[Ca_3P_2 + 6H_2O \rightarrow 3Ca(OH)_2 + 2PH_3\uparrow] \times 5$

$15CaO + 4P_4 + 30H_2O \rightarrow 15Ca(OH)_2 + 3P_2O_5\uparrow + 10PH_3\uparrow$

(*ix*) $2H_3PO_2 \xrightarrow{Heat} \underset{\text{Phosphine}}{PH_3} + H_3PO_4$
(*x*) $4P + 10HNO_3 + H_2O \rightarrow 5NO + 5NO_2 + 4H_3PO_4$
(*xi*) $4Sn + \underset{\text{(dilute)}}{10HNO_3} \rightarrow 4Sn(NO_3)_2 + NH_4NO_3 + 3H_2O$
(*xii*) Phosphine gas (PH_3) is evolved when white phosphorous is boiled with aqueous NaOH or alcoholic solution of potassium hydroxide.
$P_4 + 3NaOH + 3H_2O \rightarrow \underset{\text{Sod. Hypophosphite}}{NaH_2PO_2} + PH_3\uparrow$

67. (*i*) Nitrogen and fluorine both are small and have high electron density, they repel the bonded pair of electrons leading to larger bond length than expected.
(*ii*) It is due to self ionization of NH_3, the reaction is

$2NH_3 \rightarrow NH_4^+ + NH_2^-$

Thus, on addition of NH_4Cl the concentration of NH_4^+ radical increases and therefore, NH_4Cl acts as an acid in liquid NH_3.

(*iii*) As compared to P, N atom has higher electronegativity and small size and shows H-bonding.

Thus, ammonia molecule show association where as phosphine does not.

(*iv*) H_3PO_3 is a dibasic acid because it contains two OH groups in its molecule.

In the two P–OH bonds, the hydrogen is ionisable.

(*v*) Orthophosphorus acid is a dibasic acid as it has 2–OH groups in its formula : whereas orthophoric acid has 3 – OH groups.

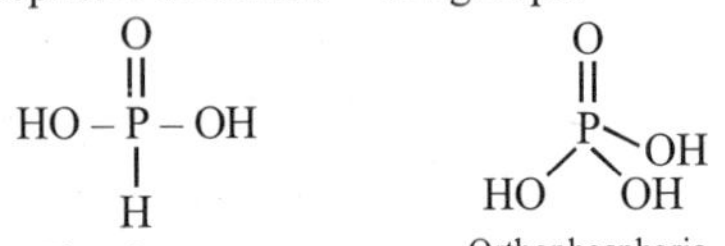

Phosphorous acid　　Orthophosphoric acid

(*vi*) Liquor ammonia possesses high vapour pressure at room temperature and thus before opening a bottle of liquor ammonia, it should be cooled to lower the pressure of NH_3 inside the bottle, otherwise NH_3 will bump out of the bottle.

(*vii*) In H_3PO_4 and H_3PO_3, the P atom is attached to 3 and 2 OH groups respectively. The H atom of these P – OH bonds are ionisable. This clearly shows that H_3PO_4 is tribasic and H_3PO_3 is dibasic.

68. 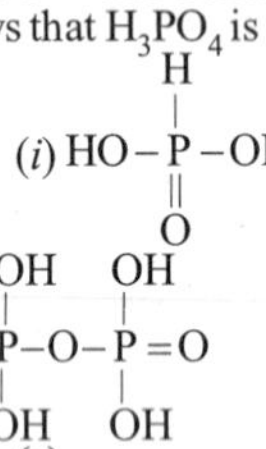

69. (a)

$$S_2O_3^{2-} \xrightarrow{Ag^+} \underset{\text{clear solution}}{\underset{(X)}{[Ag(S_2O_3)_2]^{3-}}}$$

$$\xrightarrow{Ag^+} \underset{\text{white precipitate}}{\underset{(Y)}{Ag_2S_2O_3\downarrow}}$$

$$\xrightarrow{\text{on standing}} \underset{\text{black precipitate}}{\underset{(Z)}{Ag_2S\downarrow}}$$

70. (b) The following reaction occurs $Na_2S_2O_3 + 4Cl_2 + 5H_2O \rightarrow 2\,NaHSO_4 + 8HCl$.

71. (c) In $KMnO_4$, manganese is already present in its highest possible oxidation state *i.e.* +7. So, no further oxidation is possible.

72. (c) Among oxyacids of sulphur, only Caro's acid (H_2SO_5) and Marshall's acid ($H_2S_2O_8$) have the O – O linkage.

73. (d) In sulphur trioxide trimer S_3O_9 (also called γ–sulphur trioxide), two sulphur atoms are linked to each other via O atoms, hence there is no S –S bond.

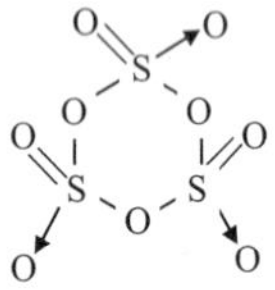

74. (b) $Na_2SO_3 + S \xrightarrow[\text{medium}]{\text{In alkaline}} Na_2S_2O_3$

75. (c)

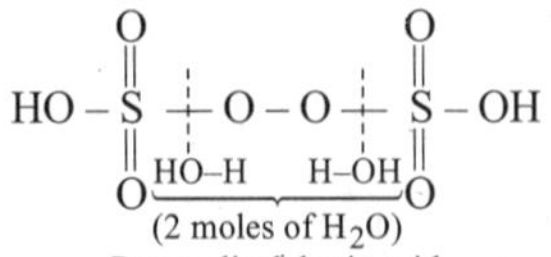

(2 moles of H_2O)
Peroxodisulphuric acid

complete hydrolysis →

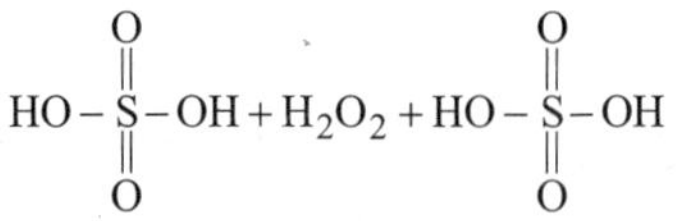

(2 moles of sulphuric acid and 1 mole of H_2O_2)

But this is not one of the options.

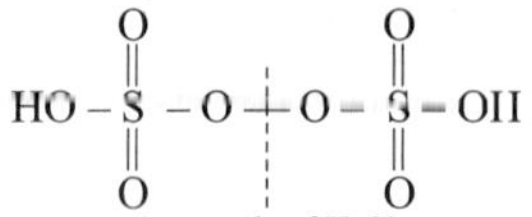

(one mole of H_2O)
Peroxodisulphuric acid

partial hydrolysis →

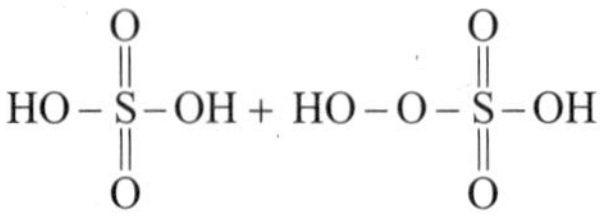

Sulphuric acid　　Peroxomonosulphuric acid

This matches to option (c).

76. (b) HBr is not prepared by heating NaBr with conc. H_2SO_4 because HBr is a strong reducing agent and reduce H_2SO_4 to SO_2 and is itself oxidised to bromine.

$$NaBr + H_2SO_4 \longrightarrow NaHSO_4 + HBr$$

$$H_2SO_4 + 2H\overset{-1}{Br} \longrightarrow SO_2 + \overset{0}{Br_2} + 2H_2O$$

77. (b) $2HgO \xrightarrow{\Delta} 2Hg + O_2$

78. (c) SO_2 is highly soluble in water and therefore, cannot be collected over water.

79. (d) All are colourless gases.

80. (6)

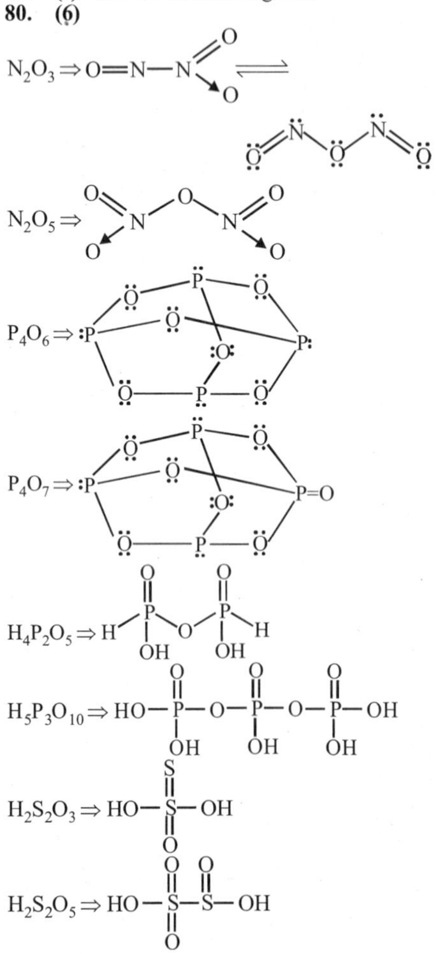

81. (288) Concentrated nitric acid oxidises sulphur to sulphuric acid (+6 oxidation state).

$$\underset{\text{Rhombic sulphur}}{S_8} + 48HNO_3 \rightarrow 8H_2SO_4 + 48NO_2 + 16H_2O$$

1 mole S_8 produces 16 × 18 g of H_2O *i.e.* 288 g of H_2O.

82. Nitrogen dioxide, [NO_2]

In lead chamber process, SO_2 is oxidised by atomic oxygen under the influence of NO_2 catalyst.

Lead chamber process

$$2SO_2 + O_2 \xrightarrow[NO_2]{\text{Oxidation}} 2SO_3$$

83. (b, d)

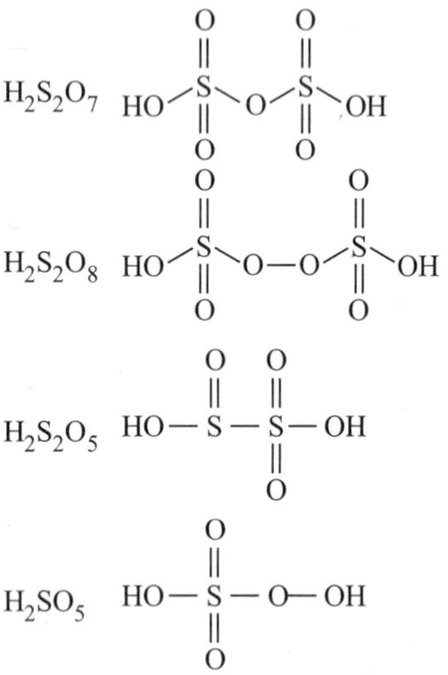

84. (a, c, d)

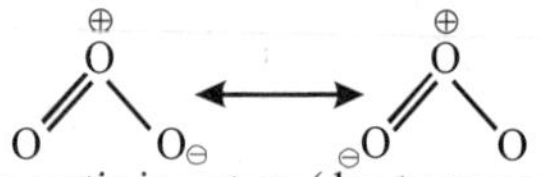

Ozone is diamagnetic in nature (due to presence of paired electron) and both the O – O bond length are equal. It has a bent structure.

85. (a, c, d)

$$HX + AgNO_3 \longrightarrow AgX\downarrow + HNO_3\ (X = Cl, Br, I)$$

$$AgX + 2Na_2S_2O_3 \longrightarrow Na_3[Ag(S_2O_3)_2] + NaX$$

86. (c) $2KClO_3 \xrightarrow[MnO_2]{\Delta} 2KCl + \underset{(W)}{3O_2\uparrow}$

$$\underset{\text{(white)}}{P_4} + \underset{\underset{(W)}{\text{(excess)}}}{O_2} \longrightarrow \underset{(X)}{P_4O_{10}}$$

87. (b) $\underset{(X)}{P_4O_{10}} + 4HNO_3 \rightarrow \underset{(Y)}{2N_2O_5} + \underset{(Z)}{4HPO_3}$

88. (a)

$$Cl_2 + 2NaOH(\text{dil.}) \xrightarrow{\text{cold}} \underset{(P)}{NaOCl} + NaCl + H_2O$$

$$3Cl_2 + 6NaOH(\text{conc.}) \xrightarrow{\text{hot}} \underset{(Q)}{NaClO_3} + 5NaCl + 3H_2O$$

(*P*) and (*Q*) are salts of HOCl and $HClO_3$ respectively.

89. (a) $SO_2 + Cl_2 \xrightarrow[\text{Catalyst}]{\text{Charcoal}} \underset{(R)}{SO_2Cl_2}$

$$\underset{(R)}{10SO_2Cl_2} + P_4 \rightarrow \underset{(S)}{4PCl_5} + 10SO_2$$

$$\underset{(S)}{PCl_5} + 4H_2O \rightarrow \underset{(T)}{H_3PO_4} + 5HCl$$

90. $\underset{[A]}{2\,NaBr} + 2H_2SO_4 + MnO_2 \rightarrow$

$$Na_2SO_4 + MnSO_4 + \underset{\underset{[B]}{\text{Brown fumes and pungent smell}}}{Br_2\uparrow} + 2H_2O$$

$2H_2SO_4 + HNO_3 \rightarrow 2HSO_4^- + NO_2^+ + H_3O^+$
[C]

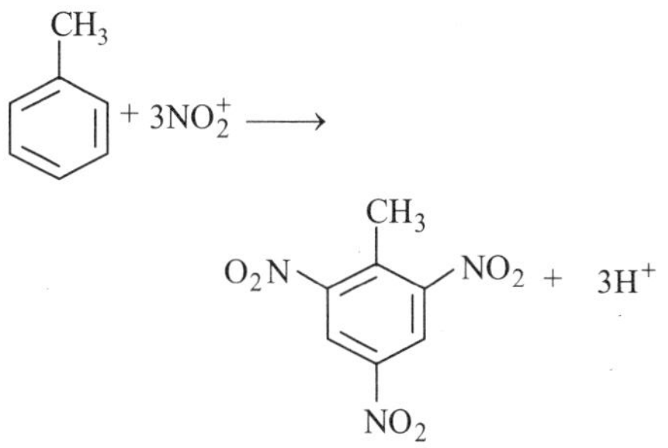

[D]
Trinitrotoluene (TNT)
(Explosive)

91.

$Na_2CO_3 + 2SO_2 + H_2O \rightarrow 2NaH\overset{+4}{S}O_3 + CO_2$;
(A)

$2NaHSO_3 + Na_2CO_3 \rightarrow 2Na_2\overset{+4}{S}O_3 + H_2O + CO_2$
(B)

$Na_2SO_3 + S \rightarrow 2Na_2\overset{+2}{S_2}O_3$
(C)

$2Na_2S_2O_3 + I_2 \rightarrow 2Na_2\overset{+2.5}{S_4}O_6 + 2NaI$
(D)

Oxidation states of 'S' are : +4 in (A), (+4) in B and +2 in (C), +2.5 in (D)

92. Sulphur trioxide produced in the contact process is passed into 98% sulphuric acid forming $H_2S_2O_7$. It is not dissolved in water as the reaction is violent and it gives a dense chemical mist of sulphuric acid particles.
The catalyst used in the contact process is vanadium pentoxide.

93. The reaction is
$HNO_2 + 2H_2SO_3 + H_2O \rightarrow NH_2OH + 2H_2SO_4$
(A) (B) (C) (D)
The structures of A, B, C and D are as follows.

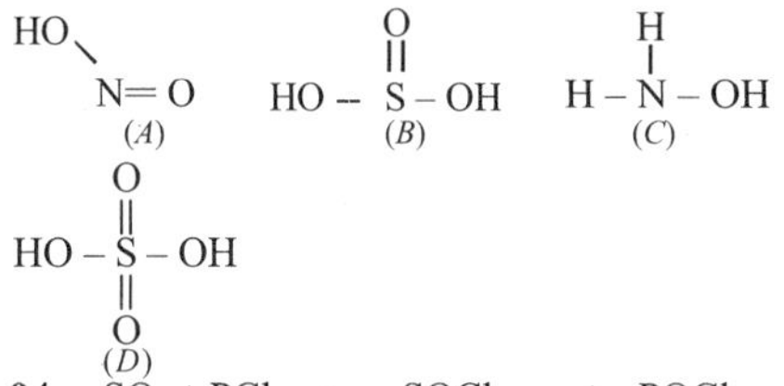

94. $SO_2 + PCl_5 \rightarrow SOCl_2 + POCl_3$
Thionyl chloride
$FeCl_3.6H_2O + 6SOCl_2 \rightarrow FeCl_3 + 12HCl + 6SO_2$
$FeCl_3.6H_2O + 6CH_3-C(OCH_3)_2-CH_3 \rightarrow FeCl_3 + 12CH_3OH + 6CH_3COCH_3$

95. $3PbS \xrightarrow[\text{air}]{\text{heat in}} 2PbO + PbS$
(A)
$\xrightarrow{\text{heat in the absence of air (B)}} 3Pb + SO_2$

96. The two resonating structures of ozone are :

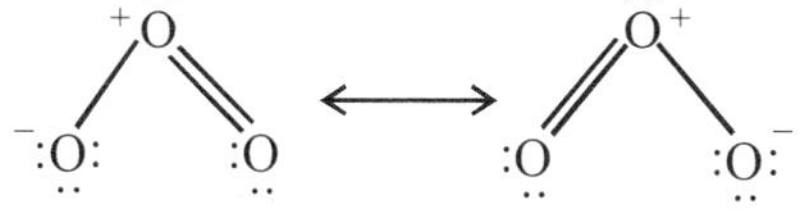

97. $SO_2 + H_2O + \frac{1}{2}O_2 \rightarrow H_2SO_4$
$H_2SO_4 + 2NaCl \rightarrow Na_2SO_4 + 2HCl$

98. $SiO_2 < CO_2 < N_2O_5 < SO_3$.
Among oxides of the non-metals, the acidic strength increases with oxidation state. Hence SO_3 (O.S. of S = +6) is most acidic followed by N_2O_5 (O.S. of N = +5) and CO_2 and SiO_2 (O.S. of C and Si = +4). Further CO_2 is more acidic than SiO_2 because of small size of C-atom.

99. (i) H_2S oxidises into S,
$SO_2 + 2H_2S \rightarrow 3S + 2H_2O$
(ii) $Sn + 4HNO_3 \rightarrow$
conc.
$H_2SnO_3 + 4NO_2 + H_2O$
meta stannic acid
(iii) $Pb_3O_4 + 4HNO_3 \rightarrow$
$2Pb(NO_3)_2 + 2H_2O + PbO_2 \downarrow$

100. By boiling Na_2SO_3 solution with powder of sulphur in absence of air, sodium thiosulphate is prepared. Unreacted S is removed, filtrate is evaporated to give crystals of sod. thiosulphate.
$Na_2SO_3 + S \rightarrow Na_2S_2O_3$

101. $2H_2S + NaHSO_3 + H^+ \rightarrow 3S \downarrow + 3H_2O + Na^+$

102. (i) The reducing nature of SO_2 is represented as
$$SO_2 + 2OH^- \rightarrow SO_4^{2-} + 2H^+ + 2e^-$$
Hence, with the increase of OH^- (alkalinity) the forward reaction is favoured.
(ii) Oxygen is the 2nd most electronegative element after the fluorine and thus invariably show negative oxidation state.
Furthermore, it has $2s^2 2p^4$ configuration and thus requires only two electrons to complete its octet to show –2 oxidation state. Although sulphur also possess $3s^2 3p^4$ configuration but due to availability of d-orbitals in their outermost shell – 2, +2, +4, +6 oxidation state are also shown. Oxygen, however, cannot exceed –2 oxidation state due to non-availability of d-orbitals in its outermost shell.
(iii) Sulphur consists of S_8 rings held together by weak van der Waal's forces. As sulphur melts at 119°C, these van der Waal's forces are overcome and S_8 rings slip and roll over one another giving rise to a clear mobile liquid. Above 160°C, the S_8 rings begin to open up and form long chains which gets tangled with each other, thereby gradually increasing the viscosity.

103. (a) $ICl + H_2O \longrightarrow H^+ + OI^- + HCl$
$BrF_5 + 3H_2O \longrightarrow H^+ + BrO_3^- + 5HF$
$ClF_3 + 2H_2O \longrightarrow H^+ + ClO_2^- + 3HF$

104. (c)
$2HClO_3 + 2HCl \longrightarrow 2ClO_2 + Cl_2 + 2H_2O$
(paramegnetic)
$2ClO_2 + 2O_3 \longrightarrow Cl_2O_6 + 2O_2$

105. (a) X : Ag , Y : Pb
$AgNO_3 \xrightarrow{NaCl} AgCl(s)$;
$PbNO_3 \xrightarrow{NaCl} PbCl_2(s)$
$AgCl(s) + NH_3(aq) \longrightarrow [Ag(NH_3)_2]^+$
(P) soluble
$PbCl_2(s) + KI \longrightarrow PbI_2$
(Q) yellow ppt.
$AgCl(s) + Na_2S_2O_3 \rightarrow Na_3[Ag(S_2O_3)_2] + NaCl$

106. (c) $Cl_2 + NaOH \rightarrow NaCl + NaClO + H_2O$
[cold and dilute]

107. (b) The ions consisting of two or more atoms of which at least one is 'N' atom is known as pseudohalide ions. These are univalent, have properties similar to those of halide ions. They combine to form dimers comparable with halogen molecules, X_2. These dimers are known as pseudohalogens.

108. (c) $KF + HF \rightarrow KHF_2 \rightleftharpoons K^+ + (HF_2)^-$

109. (c) $CsBr_3$ may be represented as $Cs^+Br_3^-$

110. (b) Chlorine is stronger oxidising agent than bromine therefore, chlorine water will liberate bromine from KBr solution.
$2KBr + Cl_2 \rightarrow 2KCl + Br_2$

111. (b) Bleaching action of chlorine is only in presence of moisture where nascent oxygen is displaced from H_2O.
$Cl_2 + H_2O \rightarrow HCl + HClO$ (unstable)
$HClO \rightarrow HCl + |O|$

112. (d) HI and HBr (in that order) are the strongest reducing hydracids and hence they reduce H_2SO_4. HCl is quite stable and hence is oxidised by strong oxidising agent like $KMnO_4$. HF is not a reducing agent. In the smallest F^- ion, the electron which is to be removed during oxidation is closest to the nucleus and therefore most difficult to be removed. Therefore, HF is a poor reducing agent.

113. (a) ∵ Cl_2 is more reactive than bromine.

114. (a) Due to highest bond dissociation energy.

115. (6) $2ClO_2 + 2O_3 \rightarrow Cl_2O_6 + 2O_2$
$Cl_2O_6 \Rightarrow 2x + 6(-2) = 0$
$x = +6$
Average oxidation state of Cl in Cl_2O_6 is (+6).

116. (5)
$3Br_2 + 3Na_2CO_3 \longrightarrow 5NaBr + NaBrO_3 + 3CO_2$

117. I_3^- **complex ion;** $I_2 + I^- \longrightarrow I_3^-$

118. Hypobromous; bromite.
$HBrO \rightleftharpoons H^+ + BrO^-$

119. KI_3; complexes are more soluble in water as compared to normal salts. $[KI + I_2 \longrightarrow KI_3]$

120. HF; HF is the weakest of the three, because the ionisation (i.e. acidic character) of HX is a multistep process and when its ΔH, heat of ionisation, is calculated it comes out to be the minimum. This is due to the strong H – F bond, large heat of hydration (because of H-bonding) and low value of electron affinity of F-atom.

121. $NaIO_3$
$3I_2 + 6NaOH \longrightarrow 5NaI + NaIO_3 + 3H_2O$
(hot)

122. False : None amongst HBr and HI, exhibit hydrogen bonding. HI is a stronger acid than HBr. Although the electronegativity and electron affinity of Br is greater than that of I, the bond dissociation energy of H–I (+295 kJ/mol) is lower than that of H–Br (+363 kJ/mol). Also the enthalpy of hydration of $I^{\ominus}$ is –394kJ/mol, whereas for $Br^{\ominus}$ is –339 kJ/mol, which indicates the bettter stability of $I^{\ominus}$ in aqueous medium than $Br^{\ominus}$. Thus, HI has a stronger tendency to release protons to water molecules and hence, is a stronger acid.

123. False : The reason of accepting electrons by fluorine more readily than chlorine is as follows:
(i) F_2 has lower enthalpy of dissociation (+159 kJ mol^{-1}) due to weak F–F bond than Cl_2(+243 kJ mol^{-1}).
(ii) F_2 has higher enthalpy of hydration due to the smaller size of F^- ion (–513 kJ/mol) than that of Cl^- ion (–370 kJ mol^{-1}). Thus, F_2 is stronger reductant than Cl_2 both in solution and when dry.
Similarly, Cl_2 will displace Br^- and I^- ions from their solutions and Br_2 will displace I^- ions.
In general, *a halogen of low atomic number will oxidise the halide ion of higher atomic number.*

124. True : $Fe + 2HCl \rightarrow$
$\overset{+2}{Fe}Cl_2 + H_2$; $E^{\circ}_{Fe^{2}/Fe} = -044V$

125. (a, c)

$MnO_2 + \text{conc. } HCl \rightarrow MnCl_2 + \underset{\substack{(X) \\ \text{(greenish yellow gas)}}}{Cl_2} + H_2O$

$NH_3 + \underset{(\text{excess})}{Cl_2} \rightarrow \underset{(Y)}{NCl_3} + HCl$

$Cl_2 + \underset{(\text{excess})}{F_2} \xrightarrow{573K} \underset{(Z)}{ClF_3}$

(a) Cl_2 is used for sterilizing drinking water

(b) NCl_3 is non-planar as hybridization of nitrogen is sp^3.

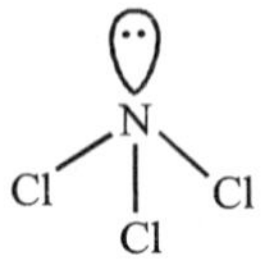

sp^3

(Pyramidal)

(c) ClF_3 is used in enrichment of ^{235}U to produce volatile UF_6

$U + 3ClF_3 \rightarrow UF_6 + 3ClF$

(d) NCl_3 is weaker Lewis base than NH_3 because chlorine is highly electronegative so its withdraw electron density from nitrogen.

126. (a, b, d)

(a) Acidic strength order :

$\overset{+1}{HClO} < \overset{+5}{HClO_3} < \overset{+7}{HClO_4}$

Conjugate base order :

$ClO^- > ClO_3^- > ClO_4^-$

(b) Hypochlorite ion (ClO^-) : :C̈l –Ö: Linear

Chlorate ion : (ClO_3^-)

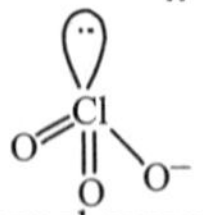

Trigonal pyramidal

Perchlorate ion : (ClO_4^-)

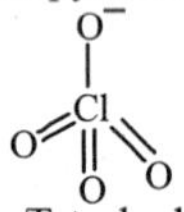

Tetrahedral

In chlorate ion bond angle changes due to presence of lone pair on chlorine atom. While there is no effect of lone pair on hypochlorite ion and perchlorate ion.

(c) Disproportionation reaction of

Hypochlorite ion : $3\overset{+1}{Cl}O^- \rightarrow 2\overset{-1}{Cl^-} + \overset{+5}{Cl}O_3^-$

Chlorate ion : $4\overset{+5}{Cl}O_3^- \rightarrow 3\overset{+7}{Cl}O_4^- + \overset{-1}{Cl^-}$

(d) $ClO^- + \overset{+4}{S}O_3^{2-} \rightarrow \overset{+6}{S}O_4^{2-} + Cl^-$

127. (c, d) Energy, $E = \dfrac{hc}{\lambda}$

On moving down the group, the colour of the X_2 molecule of group 17 elements changes gradually from yellow to violet. This happens because the amount of energy required for the excitation of the halogen atom decreases down the group. HOMO (π^*)- LUMO (σ^*) gap decreases down the group that makes π^* to σ^* excitation easier. Lesser the energy gap, more is the wavelength of light absorbed and hence, lesser is the wavelength of light emitted.

128. (a, b, d)

(a) In both the acids central atom is sp^3 hybridized.

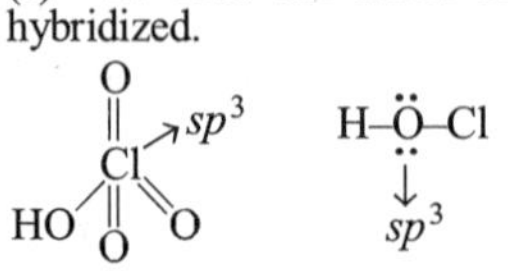

(b) $HClO_4$ is more acidic than HClO because ClO_4^- is more stable than ClO^- due to resonance.

(c) $Cl_2 + H_2O \longrightarrow HCl + HClO$

(d) $HClO_4$ is a stronger acid than H_3O^+, hence conjugate base of $HClO_4$, *i.e.* ClO_4^- is weaker base than H_2O.

129. (a, d) NH_3 and CF_2Cl_2 (freon-12) are used as refrigerants.

130. (a) $Ca(OCl)Cl \longrightarrow Ca^{2+} + {}^-OCl + Cl^-$

^-OCl (Hypochlorite ion) is anion of the acid HOCl which on dehydration gives Cl_2O.

$2HOCl \longrightarrow H_2O + Cl_2O$

131. (c) Bleach + 2KI $\longrightarrow I_2$ + Products

$I_2 + 2Na_2S_2O_3 \longrightarrow Na_2S_4O_6 + 2NaI$

Number of millimole of hypo = 0.25×48

= 2 × millimole of I_2

$\therefore$ Number of millimole of $I_2 = \dfrac{0.25 \times 48}{2} = 6$

millimole of I_2 = millimole of bleach

Molarity of bleaching solution

$\dfrac{\text{Millimoles of bleach}}{\text{Vol. (in mL) of bleach}} = \dfrac{6}{25} = 0.24$

132. (c) F has slightly less electron affinity than chlorine because F has very small atomic size (only two shells). Hence, there is a tendency of electron-electron repulsion, which results in less evolution of energy in the formation of F^- ion. Assertion is correct but reason is incorrect.

133. More electronegative halogen displaces lesser electronegative halogen from its halide. Thus,

$Cl_2 + 2KBr \text{ (or } 2KI) \longrightarrow 2KCl + Br_2 \text{ (or } I_2)$

134. (a) $2KI + Cl_2 \rightarrow 2KCl + I_2$

Since, Cl_2 is more powerful oxidising agent than I_2, Cl_2 is able to displace I^- to form I_2.

$2I^- \rightarrow I_2 + 2e^-, E° = +0.54 \text{ V}$ ...(i)

$Cl_2 + 2e^- \rightarrow 2Cl^- \; E° = 1.36V$ ...(ii)

On subtracting eq. (i) from eq. (ii), we get

$Cl_2(g) + 2I^-(aq) + \rightarrow 2Cl^-(aq) + I_2(s)$,

$E° = 0.82V$

(b) $2KClO_3 + I_2 \rightarrow 2KIO_3 + Cl_2$.

Here, ClO_3^- is more powerful oxidising agent than IO_3^-, so Cl is displaced by I.

$2IO_3^- + 12H^+ + 10e^- \rightarrow I_2 + 6H_2O, \; E° = 1.195V$ ...(i)

$2ClO_3^- + 12H^+ + 10e^- \rightarrow Cl_2 + 6H_2O, E° = 1.47V$ ..(ii)

On subtracting eq. (i) from eq. (ii), we get

$2ClO_3^- + I_2 \rightarrow 2IO_3^- + Cl_2, \; E° = 0.275V$

135. At first, $Bi(NO_3)_3$ hydrolyses to give nitric acid which being an oxidising agent, oxidises potassium iodide liberating free iodine responsible for dark brown precipitate. Iodine dissolves in excess of potassium iodide forming soluble KI_3 imparting yellow colour to solution.

$Bi(NO_3)_3 + H_2O \longrightarrow [Bi(OH)(NO_3)_2] + HNO_3$

$NO_3^- + 4H^+ + 3e^- \longrightarrow NO_2 + 2H_2O\,] \times 2$

$2I^- \longrightarrow I_2 + 2e^-\,] \times 3$

$2NO_3^- + 8H^+ + 6I^- \longrightarrow 2NO_2 + 4H_2O + \underset{\text{(dark brown ppt)}}{3I_2 \downarrow}$

$KI + I_2 \longrightarrow \underset{\text{(Yellow solution)}}{KI_3}$

136. $2NH_3 + NaOCl \rightarrow \underset{\text{Hydrazine}}{H_2N.NH_2} + NaCl + H_2O$

137. (*i*) $2\overset{+2}{Fe}\overset{-1}{Br_2} + 3\overset{0}{Cl_2} \rightarrow 2\overset{+3}{Fe}\overset{-1}{Cl_3} + 2\overset{0}{Br_2}$

(*ii*) $2SnCl_2 + 2I_2 \rightarrow SnCl_4 + SnI_4$

138. (*i*) $HOCl < HOClO < HOClO_2 < HOClO_3$

As the number of oxygen atoms increase, the -ve charge dispersal becomes more and more from Cl atom due to more electronegativity of oxygen atom. Due to more double bond character between Cl and O atoms, the bond length decreases and thus, bond strength increases. Hence, thermal stability increases.

(*ii*) $HI < HBr < HCl < HF$

The strength of H–*X* bond decreases from HF to HI. The larger is H–*X* bond length, lower is the bond energy, lesser is the bond strength.

139. (*i*) This is a method used to prepare I_2.

$5NaHSO_3 + 2NaIO_3 \rightarrow 3NaHSO_4 + 2Na_2SO_4 + I_2 + H_2O$

(*ii*) $NaClO_3 + SO_2 + 10H^+ \rightarrow NaCl + S + 5H_2O$

(*iii*) $NaBrO_3 + F_2 + 2NaOH \rightarrow NaBrO_4 + 2NaF + H_2O$

(*iv*) $2IO_3^- + 5HSO_3^- \rightarrow I_2 + H_2O + 3HSO_4^- + 2SO_4^{2-}$

140. (*i*) The repulsive forces between fluorine atoms are high due to its small size and high electronegativity. It makes dissociation of F – F bond easy. So, bond dissociation energy of F_2 is less than Cl_2

(*ii*) The standard reduction potential of fluorine is highest and thus, it cannot be oxidized by any reagent.

$\left(\dfrac{1}{2}\right) F_2 + e^- \longrightarrow F^-; \; E^°_{RP}$ = maximum

(*iii*) Anhydrous HCl, being a covalent compound, is a bad conductor however, an aqueous solution of HCl is ionised to give H^+ and Cl^- ions and is a good conductor.

(*iii*) HI cannot be prepared by heating hydrogen iodide with conc. H_2SO_4 because it is a strong oxidising agent and oxidises HI to I_2.

$H_2SO_4 + 2HI \rightarrow SO_2 + I_2 + 2H_2O$

Hence, HI is prepared by heating iodides with conc. phosphoric acid.

$3KI + H_3PO_4 \rightarrow K_3PO_4 + 3HI$

H_3PO_4 is not a strong oxidising agent.

141 (*i*)

$\underset{\text{Alumina}}{Al_2O_3} + 3C + 3Cl_2 \longrightarrow \underset{\text{Aluminium chloride}}{2AlCl_3} + 3CO$

(*ii*) $\underset{\text{slaked lime}}{3Ca(OH)_2} + 2Cl_2 \longrightarrow \underset{\text{bleaching powder}}{Ca(OCl)_2 \cdot CaCl_2 \cdot Ca(OH)_2 \cdot 2H_2O}$

(*iii*) $\underset{\text{Cassiterite}}{SnO_2} + 2C \longrightarrow \underset{\text{Tin}}{Sn} + 2CO$

(*iv*) $\underset{\text{Sodium chloride}}{2NaCl} + MnO_2 + 2H_2SO_4 \xrightarrow{\Delta} Na_2SO_4 + MnSO_4 + 2H_2O + Cl_2$

142. (*i*) HBr is a reducing agent and it reduces H_2SO_4 to SO_2.

(*ii*) Acids turn blue litmus red, so HClO also turns blue litmus red. The colour of litmus is decolourised because HClO is also a strong oxidising agent.

143. (c)

$$XeF_6 \xrightarrow[\text{Hydrolysis}]{\text{Complete}} \underset{(P)}{XeO_3} + HF \quad \text{highly explosive}$$

$$XeO_3 \xrightarrow{OH^-/H_2O} \underset{\substack{(\text{Xenate})\\(Q)}}{HXeO_4^-} \xrightarrow[\text{(Slow disproportionation)}]{OH^-/H_2O} \underset{\text{perxenate ion}}{[XeO_6]^{4-}} + Xe(g) + H_2O + O_2(g)$$

144. (d) XeO_2F_2 has trigonal bipyramidal geometry, but due to presence of lone pair of electrons on equitorial position,

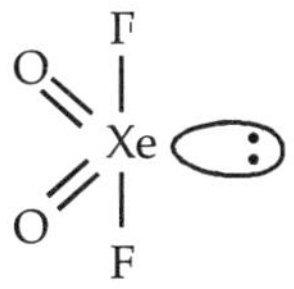

its shape is *see-saw.*

145. (b) In $XeOF_4$, Xenon is sp^3d^2 hybridised and has one lone pair.

146. (4)

$$Xe + 2O_2F_2 \longrightarrow \underset{(P)}{XeF_4} + 2O_2$$

$$6XeF_4 + 12H_2O \longrightarrow 4Xe + 2XeO_3 + 24HF + 3O_2$$

Number of moles of HF produced by the complete hydrolysis of 1 mol of P = 4.

147. (19) $XeF_4 + O_2F_2 \rightarrow XeF_6 + O_2$

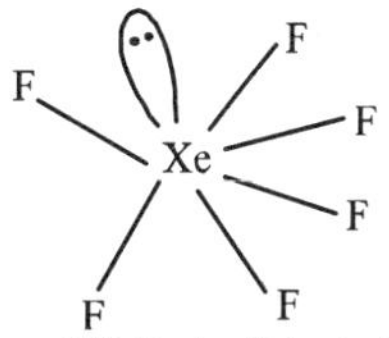

Shape of XeF_6 is distorted octahedral contains one lone pair e^-s on central atom 3 lone pair e^-s on each F atom surrounded by Xe.

Total no. of lone pairs: 1 + 18 = 19

148. (a, c, d) $Xe(g) + F_2(g) \xrightarrow[7\text{ bar}]{873K} \underset{(P)}{XeF_4}$

1 : 5

$$XeF_4 + O_2F_2 \xrightarrow{143\text{ K}} \underset{(Q)}{XeF_6} + O_2$$

$$XeF_6 + H_2O \xrightarrow[\text{hydrolysis}]{\text{complete}} \underset{(R)}{XeO_3} + HF$$

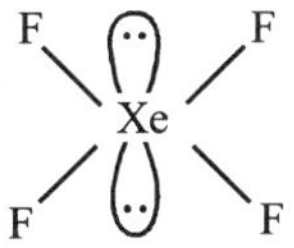

2 lone pair

(P)

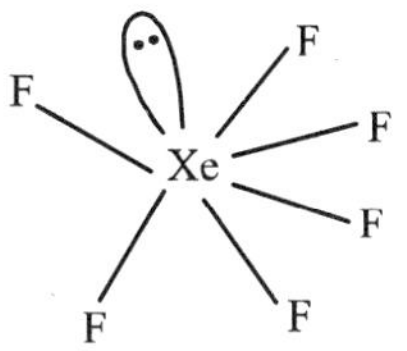

Distorted octahedral & Fluorinating agent

(Q)

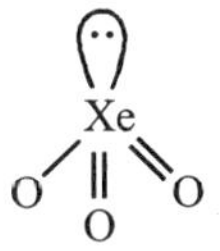

Trigonal Pyramidal

(R)

149. (b, c)

$PtF_6 + O_2 \rightarrow O_2^+[PtF_6]^-$

$X^+ = O_2^+$, $Y^\ominus = [PtF_6]^{-1}$

(a) Bond order of O_2^+ is 2.5

(b) Electronic configuration of Pt is $[Xe]4f^{14}5d^96s^1$

$[PtF_6]^{-1} \Rightarrow Pt^{+5}$:- $4f^{14}5d^5$

5 electron in its valence d orbital.

(c) PtF_6 oxidizes O_2 into O_2^+ hence act as oxidant.

(d) In this reaction no 'F' atom is transferred from PtF_6.

So it does not act as a fluorinating agent.

150. (d) (P)

$$2PbO_2 + 2H_2SO_4 \xrightarrow{\text{Warm}} 2PbSO_4 + 2H_2O + O_2$$

(Q)

$$Na_2S_2O_3 + 5H_2O + 4Cl_2 \longrightarrow 2NaHSO_4 + 8HCl$$

(R) $N_2H_4 + 2I_2 \longrightarrow N_2 + 4HI$

(S) $XeF_2 + 2NO \longrightarrow Xe + 2NOF$

151. (A) – (p), (s)

$$(CH_3)_2SiCl_2 + 2H_2O \xrightarrow[-2HCl]{} (CH_3)_2Si(OH)_2 \xrightarrow{\text{Polymerisation}} [(CH_3)_2Si{-}O]_n.$$

(B) – (p), (q), (r), (t)

$XeF_4 + H_2O \longrightarrow Xe + XeO_3 + H_2F_2 + O_2$

$SiO_2 + 4HF \longrightarrow SiF_4 + 2H_2O$

glass

$SiF_4 + 2HF \longrightarrow H_2[SiF_6]$

Soluble hexafluorosilicic(IV) acid

(C) – (p), (q) $Cl_2 + H_2O \longrightarrow HCl + HOCl$

(D) – (p), (q) $VCl_5 + 7H_2O \longrightarrow [V(H_2O)_6]^{3+} + 3Cl^- + HCl + HOCl$

152. (a) Argon, being a noble gas, will not react with the metals, thus, can be used in arc welding.

153. (c) In XeO_3 there are total of 4 electron pairs around central atom. Out of which 3 are bonding electron pair and one is non-bonding electron pair. This combination provides sp^3-hybridization and pyramidal shape.

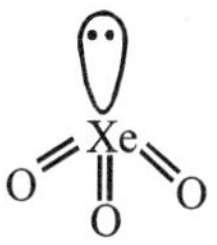

154. (a) All xenon fluorides are strongly oxidizing, XeF_4 can act as reducing agent (with F_2) as well as oxidizing agent but XeF_6 can only function as an oxidizing agent.

$$6XeF_4 + 12H_2O \longrightarrow 4Xe + 2XeO_3 + 24HF + 3O_2$$

$$XeF_6 + 3H_2O \longrightarrow XeO_3 + 6HF$$

155. (*i*)

$Al_4C_3 + 12H_2O \longrightarrow 4Al(OH)_3 + 3CH_4 \uparrow$

(*ii*) $CaNCN + 3H_2O \longrightarrow CaCO_3 \downarrow + 2NH_3$

Ammonia formed dissolves in water to form NH_4OH

$CaNCN + 5H_2O \longrightarrow 2NH_4OH + CaCO_3 \downarrow$

(*iii*) $4BF_3 + 3H_2O \longrightarrow 3HBF_4 + B(OH)_3$

(*iv*) $NCl_3 + 3H_2O \longrightarrow NH_3 + 3HOCl$

(*v*) $3XeF_4 + 6H_2O \longrightarrow XeO_3 + 2Xe + \frac{3}{2}O_2 + 12HF$

156. Use the formula

H (hydridisation),

$H = \frac{1}{2}(V + M - C + A)$ where

V = number of electron in valence shell of central atom

M = number of monovalent atoms surrounding the central atom

C = Charge on cation

A = Charge on anion

XeF_2: $H = \frac{1}{2}(8 + 2 - 0 + 0) = 5$

Hence, hybridisation is sp^3d, and thus its shape is linear.

XeF_4: $H = \frac{1}{2}(8 + 4 - 0 + 0) = 6$,

Hence, hybridisation is sp^3d^2. and thus its shape is square planar.

XeO_2F_2: $H = \frac{1}{2}(8 + 2 - 0 + 0) = 5$,

Hence, hybridisation is sp^3d. and shape is see saw.

Chapter-8

The *d*-and *f*-Block Elements

Part-A (JEE Main)

1. (2) The Statement I is incorrect because first row transition metals utilize 3*d* and 4*s* electrons for bonding on the catalyst surface.

The Statement II is incorrect because adsorption increases reactant concentration on the catalyst surface and weakens bonds facilitating formation of activated complex and lowering activation energy.

2. (2) $Cr - [Ar]\,3d^5\,4s^1$ $Cr \rightarrow Cr^+ ; \Delta_i H = 653$

$Mn - [Ar]\,3d^5\,4s^2$ $Mn \rightarrow Mn^+ ; \Delta_i H = 717$

$Cr^+ \rightarrow Cr^{2+} ; \Delta_i H = 1592$

$M^+ \rightarrow Mn^{2+} ; \Delta_i H = 1509$

$IE_1 - Mn > Cr$

$IE_2 - Mn < Cr$

3. (3)

A. The compound V_2O_5 is amphoteric which is used in manufacture of H_2SO_4.

B. Ziegler Natta catalyst is $\overset{+4}{TiCl_4} + \overset{+3}{AlEt_3}$ is used in

Ti^{4+} : $3d^0\,4s^0$ so, number of unpaired electrons are zero.

Hence, magnetic moment is zero.

C. Oxidation state of Al is +3

D. $\overset{+2}{Pd}Cl_2$ is used in wacker process

$Pd^{2+} : 4d^8\,5s^0$

C and D are correct.

4. (2) They are very hard and interstitial hydrides are not ionic in nature.

5. (3)

Oxide of Mn in its highest O.S. = Mn_2O_7 $[Mn^{+7}]$

Fluoride of Mn in its highest O.S. = MnF_4 $[Mn^{+4}]$

$\Rightarrow$ Difference in O.S. = $|x| = 3$

$Sc^{3+} \Rightarrow [Ar]$; 0 unpaired e^-

$Zn^{2+} \Rightarrow [Ar]\,3d^{10}$; 0 unpaired electrons

$Fe^{2+} \Rightarrow [Ar]\,3d^6$; 4 unpaired electrons

$Co^{2+} \Rightarrow [Ar]\,3d^7$; 3 unpaired electrons

$V^{2+} \Rightarrow [Ar]\,3d^3$; 3 unpaired electrons

6. (2) $Cr = (Ar)\,3d^5\,4s^1$

$Mn = (Ar)\,3d^5\,4s^2$

Thus $IE_1(Cr) < IE_1(Mn)$

Statement I is correct.

$IE_2(Cr) > IE_2(Mn)$

$IE_3(Cr) < IE_3(Mn)$

Statement II is false.

7. (3) In 3d series Vanadium has highest enthalpy of atomization and colour of V^{3+} is green. Oxide form by V^{3+} is V_2O_3 which is a basic oxide.

8. (3) From the given elements, Chromium has lowest enthalpy of atomisation.

$Cr = [Ar]3d^5 4s^1$

$\therefore$ Total valence electrons in Cr = 6

9. (4) $Mn^{2+} : [Ar]3d^5$ and $Fe^{2+} : [Ar]\,3d^6$

Due to greater stability of half-filled shell, first ionization energy of Mn^{2+} is higher than that of Fe^{2+}.

10. (1)

Configuration		No. of unpaired e^-
(1) $V^{3+} \Rightarrow [Ar]3d^3 4s^0$	1 | 1 | 1 | |	3
$Co^{2+} \Rightarrow [Ar]3d^7 4s^0$	1l | 1l | 1 | 1 | 1	3

So V^{2+} and Co^{2+} have same number of unpaired electrons.

11. (2) Cr (VI) is less stable than Mo(VI)

Hence, CrO_3 easily reduce into Cr^{+3} as compared to MoO_3 and show stronger oxidizing nature.

$E^o_{Cr^{6+}/Cr^{3+}} = +0.77V$; $E^o_{Mo^{6+}/Mo^{3+}} - 0.19V$

12. (3) In any row the melting points of transition metals rise to maximum at d^5 except for anomalous values of Mn and Tc and fall regularly as the atomic number increases.

This trend is not followed by elements of 5d series due to relativistic effects and number of unpaired electrons, leading to stronger metallic bonding in Re

Therefore melting point order is:

Mn < Fe, Tc < Ru, Os < Re

13. (2) ClF_3

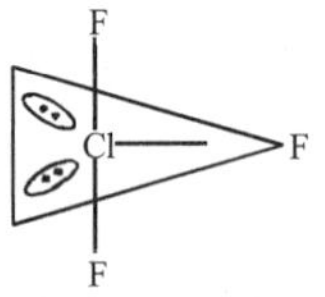

n = 2 (No of lone pair present in equatorial position)

		(Unpaired e^-)
(A)	$V^{+3} : [Ar]3d^2$	2
(B)	$Ti^{3+} : [Ar]3d^1$	1
(C)	$Cu^{+2} : [Ar]3d^9$	1
(D)	$Ni^{+2} : [Ar]3d^8$	2
(E)	$Ti^{+2} : [Ar]3d^2$	2

14. (2) Spin only magnetic moment

$(\mu) = \sqrt{n(n+2)}$

$Sc^{+3} = 3d^0 \therefore$ $\mu_{spin} = 0$

$V^{+2} = 3d^3 \therefore$ $\mu_{spin} = 3.87$ B.M.

$Ni^{+2} = 3d^8 \therefore$ $\mu_{spin} = 2.84$ B.M.

$Ti^{+3} = 3d^1 \therefore$ $\mu_{spin} = 1.73$ B.M.

15. (2)

16. (1)

(a) V^{2+}(Violet), Cr^{3+}(Violet), Mn^{3+}(Violet)

(b) Zn^{2+}(Colourless), V^{3+}(Green), Fe^{3+}(Yellow)

(c) Ti^{4+}(Colourless), V^{4+}(Blue), Mn^{2+}(Pink)

(d) Sc^{3+}(Colourless), Ti^{3+}(Purple), Cr^{2+}(Blue)

17. (3) Cu(II) is more stable than Cu(I) because hydration energy of Cu^{+2} ion compensate for high IE_2 of Cu.

18. (3)

19. (4) Reactions involve are:

$2Fe^{3+} + 2I^- \longrightarrow 2Fe^{2+} + I_2$

$2Fe^{2+} + S_2O_8^{2-} \longrightarrow 2Fe^{3+} + 2SO_4^{2-}$

Fe^{3+} oxidises I^- to I_2 and convert itself into Fe^{2+}.

This Fe^{2+} reduces $S_2O_8^{2-}$ to SO_4^{2-} and converts itself into Fe^{3+}.

20. (4) Electronic configuration of the given metals

(A) $Sc \Rightarrow 3d^1\,4s^2 \Rightarrow n = 1$

(B) $Cr \Rightarrow 3d^5\,4s^1 \Rightarrow n = 6$

(C) $V \Rightarrow 3d^3\,4s^2 \Rightarrow n = 3$

(D) $Ti \Rightarrow 3d^2\,4s^2 \Rightarrow n = 2$

(E) $Mn \Rightarrow 3d^5\,4s^2 \Rightarrow n = 5$

$\therefore$ Correct order of increasing number of unpaired electrons.

(A) < (D) < (C) < (E) < (B)

21. (2)

22. (3) Ions with negative SRP can liberate H_2 from a dilute acid.

Ti^{2+}, Cr^{2+}, V^{2+} all can liberate H_2 gas from a dilute acid.

23. (2) Co, Ti, Ni can show +2, +3 and +4 oxidation state, But 'Sc' only shows +3 stable oxidation state.

24. (4) $M = 3.86 = \sqrt{n(n+2)}$

$\Rightarrow n = 3$ unpaired electrons.

$_{22}Ti^{2+} \Rightarrow [Ar]3d^2$

$_{23}V^{2+} \Rightarrow [Ar]3d^3$

$_{25}Mn^{2+} \Rightarrow [Ar]3d^5$

$_{26}Fe^{2+} \Rightarrow [Ar]3d^6$

25. (1) $_{25}Mn = [Ar]\,3d^5\,4s^2$

3rd Ionization enthalpy will be highest for Mn because when 3rd electron is removed it will loose its stability of half filled electron system.

26. (2) Both (A) and (R) are true and (R) is the correct explanation of (A)

Cr^{2+} is reducing as it configuration changes from d^4 to d^3 due to formation of Cr^{+3}, which has half filled t_{2g} level, on other hand the change Mn^{3+} to Mn^{2+} result half filled d^5 configuration which has extra stability.

27. (2) (A) – (III), (B) – (IV), (C) – (I), (D) – (II)

28. (2)

29. (4)

30. (4) $\mu = \sqrt{n(n+2)}$ B.M.

	$3d^7$	$3d^8$	$3d^3$	$3d^6$
No. of unpaired e^-	3	2	3	4
Spin only magnetic moment	$\sqrt{15}$ BM	$\sqrt{8}$ BM	$\sqrt{15}$ BM	$\sqrt{24}$ BM

31. (2)

32. (4) The catalyst used in Wacker's process is $PdCl_2$
The process of oxidation of terminal alkenes to aldehydes and ketones by using palladium and copper compounds as catalysts. Palladium chloride and copper chloride are used as catalyst.

33. (1) (A) The M^{3+}/M^{2+} reduction potential for manganese is greater than iron

$$E^\circ_{Fe^{+3}/Fe^{+2}} = +0.77,\ E^\circ_{Mn^{+3}/Mn^{+2}} = +1.57$$

(C) $E^\circ_{Cr^{+3}/Cr^{+2}} = -0.26$

$$\therefore Cr^{2+} + H^+ \rightarrow Cr^{3+} + \frac{1}{2}H_2$$

(D) $V^{2+} \Rightarrow 3d^3 4s^0$, unpaired electron

Magnetic Moment = 3.87 B.M

34. (4) In CrO_2Cl_2 oxidation state of Cr is +6
Cr (VI) = $[Ar]^{18}\,3d^0$, Mn(VII) = $[Ar]^{18}\,3d^0$, Fe(III) = $[Ar]^{18}\,3d^5$, Ti(III) = $[Ar]^{18}\,3d^1$, V(IV) = $[Ar]^{18}\,3d^1$
Hence Cr (VI) and Mn (VII) have same d^0 configuration.

35. (1) $2Cu^+ \rightarrow Cu^{2+} + Cu$
Cu^{2+}(aq) is more stable than Cu^+(aq). $\Delta_{hyd}H$ of Cu^{2+} (aq) is much more negative than Cu^+(aq), which compensates for the second ionisation enthalpy of Cu.

36. (2) Sc does not show +4 oxidation state.

37. (1) Sc^{3+} is with d^0 and Zn^{2+} is with d^{10} configuration.

38. (2)

39. (1) In reactivity series, Cu lies below Hydrogen and it is least electropositive among the given metals.

40. (1) ${}_{78}Pt$ = [Xe] $4f^{14}\,5d^9\,6s^1$ (Exceptional electronic configuration)

41. (3) Zinc dissolves in excess of aqueous alkali

$$Zn(s) + 2OH^-(aq) + 2H_2O(l) \rightarrow \underset{\text{Tetrahydroxozincate (II) ion}}{[Zn(OH)_4]^{2-}(aq)} + H_2(g)$$

ZnO_2^{2-} is anhydrous form of $[Zn(OH)_4]^{2-}$.
So in aqueous medium best answer of this question is $[Zn(OH)_4]^{2-}$.

42. (3) $Cr^{2+}/Cr \rightarrow -0.90$ V ; $Fe^{2+}/Fe \rightarrow -0.44$ V
$Cu^{2+}/Cu \rightarrow +0.34$ V ; $Zn^{2+}/Zn \rightarrow -0.76$ V
For $Cu(s) \rightarrow Cu^{2+}(aq)$
Copper has high atomization and low hydration enthalpy.
So, energy required for the conversion is not balanced by hydration entropy.

43. (1) Strongest oxidising agent have highest reduction potential value.

$$E^\circ_{Mn^{3+}/Mn^{2+}} = 1.51\text{ V (highest)}$$

44. (4)

45. (1) Such ions, which have unpaired electrons exhibit colour and paramagnetic behaviour.
$Cu^{2+} = 3d^9$ (one unpaired e^-)
$Cr^{3+} = 3d^3$ (three unpaired e^-)
$Sc^+ = 3d^1\,4s^1$ (two unpaired e^-)

46. (3) Chromium Cr ($Z = 24$) $\Rightarrow$ [Ar] $3d^4 4s^2$
$\therefore$ Common positive oxidation state of Cr are from +2 to +6.
Where +3 and +6 are the most stable oxidation states.

47. (1) The element having atomic number 25 is manganese.
Mn (25) = $3d^5 4s^2$; $Mn^{2+} \Rightarrow 3d^5$
In aqueous solution, it exists as $[Mn(H_2O)_6]^{2+}$ which is high spin complex due to weak field nature of H_2O ligand.
So, total no. of unpaired electrons = 5

$$\mu = \sqrt{n(n+2)}\text{ BM}$$

$$\mu = \sqrt{5(5+2)} = \sqrt{35} = 5.92\text{ BM}$$

48. (3) $Cr^+ - 4s^0\,3d^5$; $Mn^{2+} - 4s^0\,3d^5$
So, the outermost electronic configuration will be same in case of Cr^+ and Mn^{2+}.

49. (1) German silver contains Cu (50%), Zn (30%) and Ni (20%) respectively. It does not contain Ag.

50. (2) ${}_{26}Fe$ = [Ar] $3d^6 4s^2$. Third ionisation results into stable d^5 configuration.

51. (1) Reactant should not be adsorbed strongly which might result into immobilisation that inhibit further adsorption on the catalyst's surface.

52. (1) Atomic size of elements of 4*d* and 5*d* transition series are nearly same due to Lanthanoid contraction.

53. (4) Mo and W belong to group-6 and period 5 (4*d* series) and 6 (5*d* series) respectively.
Due to lanthanoid contraction, radius of Mo and W are almost same, i.e., 0.140 nm and 0.141 nm respectively.

54. (2)

55. (2) I.E. increases on moving left to right in a period.
$\therefore$ Ti < Mn < Ni < Zn

56. (3)

57. (1) Interstitial compounds are inert, i.e., they are chemically non-reactive.

58. (3) Sc shows oxidation state of +3 only.

59. (1) $\sqrt{5(5+2)} = \sqrt{35} = 5.92$ BM

60. (4) As zinc has no unpaired electrons to take part in the bond, it has least enthalpy of atomisation amongst the given transition elements.

61. (4) $(M^{2+} + 2H^+ - e^- \longrightarrow M^{3+} + H_2)$

62. (2) In MnO_4^{2-}, manganese is in +6 oxidation state which is having highest stability. Stability increase with increase in Solvation energy.
Solvation $\propto$ Effective nuclear charge
Since MnO_4^{2-} have +6 oxidation number and high effective nuclear charge so attracted towords H_2O and thus acquire high Solvation energy.

63. (3)

64. (2) $\mu = \sqrt{n(n+2)}$ Þ $1.73 = \sqrt{n(n+2)}$
No. of unpaired electrons, n = 1
V(23) = [Ar] $3d^3 4s^2$; For one unpaired e^-, V^{4+} = [Ar]$3d^1$
$\therefore$ Its chloride has the formula is : $V^{4+}Cl^- \equiv VCl_4$

65. (4) Maganese exhibits the most number of oxidation states. The most common oxidation states of Mn are +2, +3, + 4, + 6 and +7.

66. (1) Cu_2O is yellow in colour.

67. (1)

68. (4)

69. (1) The element with outer electron configuration $3d^5 4s^2$ is Mn which exhibits oxidation states from + 2 to + 7.

70. (4) Fe^{3+} is easily hydrolysed than Fe^{2+} due to more positive charge.

71. (1) Zr (40), Hf (72) have similar properties as they belong to same group and have same valence shell electronic configuration.

72. (1) Across the first transition series, the negative values for standard electrode potential decrease (or the value increases) except for Mn due to the stable d^5 configuration.
Thus, correct order is :

$$Mn^{2+} > Cr^{2+} > Fe^{2+} > Co^{2+}.$$

73. (4) $(n-1)d^5 ns^2$ attains the maximum O.S. of +7.

74. (1) Mn^{2+} = 5 unpaired electrons
Fe^{2+} = 4 unpaired electrons
Ti^{2+} = 2 unpaired electrons
Cr^{2+} = 4 unpaired electrons
Magnetic moment $\propto$ Number of unpaired electrons

75. (3) Mn^{+2} = [Ar] $3d^5 \rightarrow$
5 unpaired electrons : Paramagnetic
Cu^{+2} = [Ar] $3d^9 \rightarrow$
1 unpaired electron : aramagnetic
Zn^{+2} = [Ar] $3d^{10} \rightarrow$
0 unpaired electron : Diamagnetic
Yb^{+2} = [Xe] $4f^{14} \rightarrow$
0 unpaired electron : Diamagnetic
Sc^{+3} = [Ar] $3d^0 \rightarrow$
0 unpaired electron : Diamagnetic
La^{+3} = [Xe] $4f^0 5d^0 \rightarrow$
0 unpaired electron : Diamagnetic
Gd^{+3} = [Xe] $4f^7 \rightarrow$
7 unpaired electrons : Paramagnetic
Lu^{+3} = [Xe] $4f^{14} \rightarrow$
0 unpaired electron : Diamagnetic
Ti^{+4} = [Ar] $3d^0 \rightarrow$
0 unpaired electron : Diamagnetic
Ce^{+4} = [Xe] $4f^0 \rightarrow$
0 unpaired electron : Diamagnetic.
Therefore, 3 paramagnetic ions are present here (Mn^{+2}, Cu^{+2} and Gd^{+3}).

76. (3) Among the given oxides, three oxides are mixed oxides.
$Mn_3O_4 = MnO\,.\,Mn_2O_3$
$Fe_3O_4 = FeO\,.\,Fe_2O_3$
$Co_3O_4 = CoO\,.\,Co_2O_3$.

77. (4)

	Sc	Mn	Co	Cu
Enthalpy of Atomisation (kJ/mol)	326	281	425	339

The element with highest $\Delta_a H$ is Co.
Co^{2+} : [Ar] $3d^7$; [↿⇂] [↿⇂] [↿] [↿] [↿]
$\Rightarrow$ n = 3

$$\mu = \sqrt{n(n+2)} = \sqrt{15} = 3.87 \approx 4$$

78. (0) $K_2[NiCl_4] \Rightarrow Ni^{2+} \Rightarrow 3d^8$
$\Rightarrow$ Paramagnetic
$[Zn(H_2O)_6]Cl_2 \Rightarrow Zn^{2+} \Rightarrow 3d^{10} \Rightarrow$ diamagnetic
$K_3[Mn(CN)_6] \Rightarrow Mn^{3+} \Rightarrow 3d^4 \Rightarrow$ paramagnetic

$[Cu(PPh_3)I] \Rightarrow Cu^{+1} \Rightarrow 3d^{10} \Rightarrow$ diamagnetic
Here two diamagnetic complexes
$\therefore n = 2$
Zn has least enthalpy of atomisation
$Zn^{2+} \Rightarrow \mu = 0$

79. (11) $z = 41 \rightarrow$ Nb (Niobium) : $[Kr]_{36} 4d^4 5s^1$
Number of electron in 4d = 4 = x
$Z = 44 \rightarrow$ Ru (Ruthenium) $[Kr]_{36} 4d^7 5s^1$
Number of electron in 4d = 7 = y
x + y = 11

80. (6) Among given metals, Cr has maximum IE_2 because second electron is removed from stable configuration $3d^5$.
Cr^+: [ar] $3d^5 4s^0$.
$\therefore$ No of unpaired e^- in Cr^+ is 5, n = 5
So, Magnetic moment $= \sqrt{n(n+2)}$ B.M.
$= \sqrt{5(5+2)} = 5.92$ BM ≈ 6

81. (0) Metal having least metallic radii among Sc, Ti, V, Cr, Mn and Zn is Cr.
Spin only magnetic moment of CrO_4^{2-}.
Here Cr^{+6} is in d^0 configuration (diamagnetic).
$\mu = \sqrt{n(n+2)}$ BM
$n = 0 \Rightarrow \mu = 0$

82. (0) VO_2^+ has least oxidising ability due to more stability of their octate in penultimate shell.
$VO_2^+ \Rightarrow V^{+5} \Rightarrow 3d^0 4s^0$
There is no unpaired electron present. So, magnetic moment = 0

83. (5) Among the given species
Strongest oxidising agent is Co^{3+}
No. of unpaired e^- in $Co^{3+}[3d^6] = 4$
Hence $\mu = \sqrt{n(n+2)} = \sqrt{24}$ BM = 4.89 BM
Nearest integer = 5 BM

84. (6) The metaborates of transition metals have characteristic colours.
Fe^{3+} will give green coloured bead when heated at point B.
Number of unpaired electrons in $Fe^{3+} = 5$
$\mu = \sqrt{n(n+2)}$ BM $= \sqrt{5(5+2)}$ BM = 5.91 BM
Nearest integer = 6.

85. (0) 'V' has highest enthalpy of atomisation (515 kJ/mol) among first row transition elements.
V_2O_5 is amphoteric oxide.
Here 'V' is in +5 oxidation state
$V^{+5} \Rightarrow 1s^2 2s^2 2p^6 3s^2 3p^6$ (no unpaired electrons)

86. (5)
$MnF_4 \Rightarrow$ +4 state $\Rightarrow$ $[Ar]3d^3$
$MnF_3 \Rightarrow$ +3 state $\Rightarrow$ $[Ar]3d^4$ (Strongest O.A)
$MnF_2 \Rightarrow$ +2 state $\Rightarrow$ $[Ar]3d^5$
For n = 4,
$\therefore \mu = \sqrt{4(4+2)} = \sqrt{24} = 4.89 = 5$

87. (5) Co^{3+} cannot liberate H_2 from dilute mineral acid solution.
Co^{3+} (d^6) in gas phase : | ⇅ | 1 | 1 | 1 | 1 |
Unpaired electrons (n) = 4

$M_{s.o.} = \sqrt{4(4+2)}$ B.M. = 4.89 B.M. $\approx$ 5 B.M.

88. (4) $E^\circ_{Cr^{+3}/Cr^{+2}} = -0.41V$; $[Cr^{+3}] = 4s^0 3d^3$
$\mu = \sqrt{n(n+2)}$B.M $= \sqrt{15}$ B.M ~ 4 B.M

89. (3) A, B, C are correct and D is incorrect because Fehling solution has Cu(II).

90. (12) Vanadium (Z = 23); V = $1s^2 2s^2 2p^6 3s^2 3p^6 3d^3 4s^2$
Number of electrons in *p*-orbital are 12.

91. (49) Fe (Z = 26) $\Rightarrow [Ar] 3d^6 4s^2$
Number of unpaired electrons = 4
$\because \mu = \sqrt{n(n+2)}$ BM
$\therefore \mu = \sqrt{4(4+2)} = \sqrt{24}$ BM $= 4.89 \approx 49 \times 10^{-1}$ BM

92. (2)
(A) Potassium dichromate is an oxidising agent and its oxidising nature can be explained by the reaction given below:
$\overset{+6}{Cr_2O_7^{2-}} + 14H^+ + 6Fe^{+2} \rightarrow 2Cr^{+3} + 6Fe^{+3} + 7H_2O$
(B) $Na_2Cr_2O_7$ generally not used as a primary standard in volumetric analysis becuase it is deliquescent which results in wrong calculation.
(C) In acidic medium $Cr_2O_7^{2-}$ exists and it get converted into CrO_4^{2-} in basic medium.
$2CrO_4^{2-} + 2H^+ \longrightarrow Cr_2O_7^{2-} + H_2O$
$Cr_2O_7^{2-} + 2OH^- \longrightarrow 2CrO_4^{2-} + H_2O$
(D) Bond angle of Cr–O–Cr bond in $Cr_2O_7^{2-}$ is 126°. It consists of two tetrahedra sharing one one corner with Cr–O–Cr.

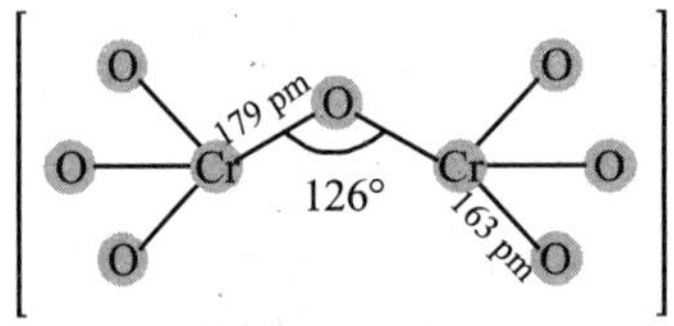

Dichromate ion

93. (1)
Reducing agent + $K_2Cr_2O_7 \xrightarrow{\text{acidic medium}} Cr^{3+}$ + Product
Given Reducing Agent
$Fe^{+2} \rightarrow Fe^{+3}$
$Sn^{+2} \rightarrow Sn^{+4}$
$I^- \rightarrow I_2$
$S^{-2} \rightarrow S$
$NO_2^- \rightarrow NO_3^-$
$SO_2 \rightarrow SO_4^{2-}$
$C_2O_4^{2-} \rightarrow CO_2$
Fe^{+3}, Sn^{+4}, SO_4^{2-} cannot be oxidised as they are already present in their highest oxidation states.

94. (4) Manganate ion = MnO_4^{-2}
Permanganate ion = MnO_4^-
(A) Both are tetrahedral (d^3s Hybridisation)
(B) Permanganate (MnO_4^-) = +7 oxidation state
Manganate (MnO_4^{2-}) = +6 oxidation state
(C) $2Mn^{2+} + 5S_2O_8^{2-} + 8H_2O \longrightarrow 2MnO_4^- + 10SO_4^{2-} + 16H^+$
(D) MnO_4^- = Diamagnetic
MnO_4^{2-} = Paramagnetic
(E) Acidified permanganate is an oxidising (not reducing) agent. It oxidises oxalates to carbon dioxide, nitrites to nitrates and iodides to free iodine.

95. (1) Lightest element of Group 7 is Mn
$K_2\overset{+6}{Mn}O_4 \Rightarrow$ Green

96. (2) $K_2Cr_2O_7 + 6KI + 7H_2SO_4 \longrightarrow 4K_2SO_4 + \overset{+3}{Cr_2}(SO_4)_3 + 3I_2 + 7H_2O$

97. (2) $4NaCl + K_2Cr_2O_7 + 6H_2SO_4 \longrightarrow 2KHSO_4 + 2\overset{+6}{Cr}O_2Cl_2 + 4NaHSO_4 + 3H_2O$
(Chromyl chloride)
In chromyl chloride, Cr is in +6 oxidation state.

98. (2)
- Mn_2O_7 : Mn is in +7 (highest) oxidation state.
- Multiple bond between Mn and O stabilizes Mn.
- Mn_2O_7 is a covalent oxide. Oxygen is connected with two Mn atom.

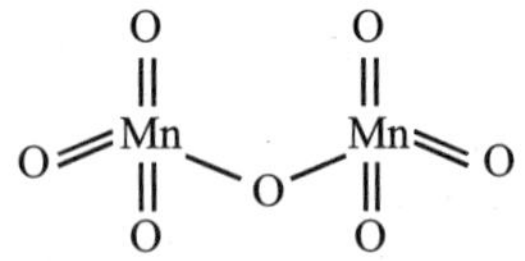

99. (3)
$3MnO_4^{2-} + 4H^+ \rightarrow 2MnO_4^- + MnO_2 + 2H_2O$

100. (4)
$PbCl_2$ (Hot solution) + $K_2CrO_4 \rightarrow PbCrO_4$ (A) + 2KCl
$PbCrO_4 + 4NaOH$ (excess) $\rightarrow Na_2[Pb(OH)_4] + Na_2CrO_4$ (B)
$PbSO_4 + 4CH_3COONH_4 \rightarrow (NH_4)_2[Pb(CH_3COO)_4]$ (X) $+ (NH_4)_2SO_4$

101. (3) V_2O_5 + alkali $\rightarrow VO_4^{3-}$
$VO_4^{-3} = x + 4(-2) = -3 \Rightarrow x = 8 - 3 \Rightarrow x = +5$

102. (3) Both $K_2Cr_2O_7$ and $KMnO_4$ act as oxidizing agent in an acidic medium.
$I^- \xrightarrow{H^+} I_2$; $I^- \xrightarrow{OH^-} IO_3^-$
$S^{-2} \xrightarrow{H^+} S$; $S_2O_3^{2-} \xrightarrow{OH^-} SO_4^{2-}$
$Fe^{+2} \xrightarrow{H^+} Fe^{+3}$; $S_2O_3^{2-} \xrightarrow{H^+} S\downarrow + SO_4^{2-}$

103. (4) $MnO_2 \xrightarrow[KNO_3, \Delta]{KOH} K_2MnO_4$

104. (3)
$K_2Cr_2O_7 \xrightarrow[-H_2O]{KOH} K_2CrO_4$ [A] $\xrightarrow[-H_2O]{H_2SO_4} K_2Cr_2O_7 + K_2SO_4$ [B]

$Cr_2O_7^{2-}$ and CrO_4^{2-} exist in equilibrium in aqueous solutions and are interconvertible depending on the pH of the solution.

105. (2) Chromate and dichromate ions both are interconvertible due to pH change.

$$Cr_2O_7^{2-} \underset{H^+}{\overset{OH^-}{\rightleftharpoons}} 2CrO_4^{2-}$$

106. (4) $MnO_2 + 4KOH + O_2 \xrightarrow{\text{fused}} \underset{\substack{\text{Potassium manganate} \\ \text{Dark green}}}{2K_2MnO_4} + 2H_2O$

Electrolytic oxidation in alkaline medium :

At anode :

Manganate change in permanganate ion.

$$MnO_4^{2-} \rightarrow MnO_4^- + e^-$$

107. (2) $CuSO_4.5H_2O$

In $K_2Cr_2O_7$ and K_2CrO_4, the Cr is in +6 oxidation state and hence contain no '*d*' electron.

In $KMnO_4$, Mn is in +7 oxidation state and this also do not contain any '*d*' electrons.

They exhibit colour due to charge transfer spectra and not due to d-d transistion.

108. (4) A. CrO_4^{2-} is tetrahedral

B. $2Na_2CrO_4 + 2H^+ \rightarrow Na_2Cr_2O_7 + 2Na^+ + H_2O$

C. Green manganate is paramagnetic with 1 unpaired electron.

D. $3MnO_4^{2-} + 4H^+ \rightarrow 2MnO_4^- + MnO_2 + 2H_2O$

$3MnO_4^{2-} + 2H_2O \rightarrow 2MnO_4^- + MnO_2 + 4OH^-$

E. As the oxidation state increases ionic character decreases.

109. (2) A and C only

V_2O_4 react with acid to give VO^{2+} V_2O_5 is amphoteric in nature and it react with acid as well as with base.

110. (1) Cr and Mn in $K_2Cr_2O_7$ and $KMnO_4$ are in +6 and +7 states which are highest in them and results in d^0 configurations.

Thus, their colours are due to charge transfer transitions and not d–d transitions.

111. (2)

$$MnO_2 \xrightarrow[\text{air / }KNO_3]{\text{Fusion with KOH,}} \underset{(A)}{K_2MnO_4}$$

$$MnO_2 \xrightarrow[\text{oxidation in alkaline medium}]{\text{electrolytic}} \underset{(B)}{KMnO_4}$$

112. (4) When the yellow solution (sodium chromate) is acidified with dilute sulphuric acid and adding small amount of amyl alcohol and 10%, H_2O_2, the organic layer turns blue and confirms the formation of chromium pentoxide. The oxidation state of Cr in chromium pentoxide is +6.

$$CrO_4^{2-} + 2H^+ + 2H_2O \longrightarrow \underset{\substack{\text{Chromium} \\ \text{pentoxide}}}{CrO_5} + 3H_2O$$

113. (4) In faintly alkaline solution, $KMnO_4$ oxidises iodide to iodate.

$$2MnO_4^- + H_2O + I^- \longrightarrow 2MnO_2 + 2OH^- + IO_3^-$$

114. (2) $KMnO_4 \xrightarrow[\Delta]{513\text{ K}} K_2MnO_4 + MnO_2 + O_2$

115. (4) $PbCrO_4 + NaOH\text{ (hot excess)} \rightarrow [Pb(OH)_4]^{-2} + Na_2CrO_4$

The product is dianionic complex with coordination number four.

116. (3) $K_2Cr_2O_7$ is used as primary standard. The concentration of $Na_2Cr_2O_7$ changes in aq. solution.

$K_2Cr_2O_7$ is less soluble than $Na_2Cr_2O_7$.

117. (1) $2Cu^{2+} + 4I^- \rightarrow Cu_2I_2(s) + I_2$

118. (2) In neutral or weakly alkaline solution oxidation state of Mn changes by 3 unit

$$\overset{+7}{Mn}O_4^{-1} \rightarrow \overset{+4}{Mn}O_2$$

119. (2)

120. (1) With increase in oxygen acidic nature of oxide of an element increase and basic nature decreases.

121. (2)

122. (1) In acidic medium

$2MnO_4^- + 10I^- + 16H^+ \rightarrow 2MnO^{2+} + 5I_2 + 8H_2O$

In neutral/faintly alkaline solution

$2MnO_4^- + I^- + H_2O \rightarrow 2MnO_2 + 2OH^- + IO_3^-$

123. (2)

$$\overset{+6}{Cr_2}O_7^{2-} + 14H^+ + 6e \longrightarrow 2Cr^{3+} + 7H_2O \quad (ox)$$

124. (4) $3SO_2 + Cr_2O_7^{2-} + 2H^+ \rightarrow 3SO_4^{2-} + \underset{\text{green}}{2Cr^{+3}} + H_2O$

125. (4) $8\overset{+7}{Mn}O_4^- + 3S_2O_3^{2-} + H_2O \rightarrow 8\overset{+4}{Mn}O_2 + 6SO_4^{2-} + 2OH^-$

Change in oxidation state of Mn is from +7 to +4 i.e., 3.

126. (2) Copper get reduced and hydrogen get oxidised CuO + H_2 → Cu + H_2O (under hot conditions)

127. (4) $8MnO_4^- + 3S_2O_3^{2-} + H_2O \xrightarrow[\text{alk.solution}]{\text{neutral or}} 8MnO_2 + 6SO_4^{2-} + 2OH^-$

128. (2)

129. (3)

130. (4) Potassium permangnate on heating at 513 K gives potassium manganate (K_2MnO_4) and its colour is green.

$$2KMnO_4 \xrightarrow[513K]{\Delta} \underset{\text{Green}}{K_2MnO_4} + \underset{\text{Black}}{MnO_2} + O_2$$

In K_2MnO_4, manganese oxidation state is +6 and hence, it has one unpaired electron and paramagnetic.

131. (4) $KMnO_4$ on heating dissociates as

$$\underset{\text{Permanganate}}{2K\overset{+7}{Mn}O_4} \xrightarrow{\Delta} \underset{\text{Manganate}}{K_2\overset{+6}{Mn}O_4} + MnO_2 + O_2$$

$Mn^{+7} \Rightarrow 3d^04s^0$ (Diamagnetic)

$Mn^{+6} \Rightarrow 3d^14s^0$ (Paramagnetic)

Both permanganate and manganate are tetrahedral but only manganate is paramagnetic.

132. (3)

133. (3)

Manganate :	**Permanganate :**
Paramagnetic, green in colour, Tetrahedral & contains *pπ-dπ* bond	Diamagnetic, purple in colour, Tetrahedral & contains *pπ-dπ* bond

Manganate ion is paramagnetic while permanganate ion is diamagnetic.

134. (1)

$$\underset{(X)}{2KMnO_4} \xrightarrow{513K} \underset{(Y)}{K_2MnO_4} + MnO_2 + \underset{\text{gas}}{O_2}$$

$$MnO_2 + 4NaCl + \underset{\text{Conc.}}{4H_2SO_4} \longrightarrow MnCl_2 + 4NaHSO_4 + \underset{\substack{(Z) \\ \text{Pungent gas}}}{Cl_2} + 2H_2O$$

135. (2)

$$\underset{(A)}{2MnO_2} \xrightarrow{KOH,\ O_2} \underset{\substack{\text{(Green)} \\ (B)}}{2K_2MnO_4} + 2H_2O$$

$$K_2MnO_4 \xrightarrow{4\ HCl} \underset{\substack{\text{(Purple)} \\ (C)}}{2KMnO_4} + MnO_2 + 2H_2O$$

$$2KMnO_4 \xrightarrow{H_2O,\ KI} \underset{(A)}{2MnO_2} + 2KOH + \underset{(D)}{KIO_3}$$

136. (1) $MnO_2 + KOH + KNO_3 \longrightarrow \underset{\text{(dark green)}}{K_2MnO_4} + KNO_2 + H_2O$

$$MnO_4^{2-} + 4H^+ \xrightarrow[\text{tionation}]{\text{dispropor-}} \underset{\text{(dark purple)}}{2MnO_4^-} + MnO_2 + 2H_2O$$

137. (4)

138. (1, 2) Hydration energy of $Na^{\oplus}$ is greater than $K^{\oplus}$. Because of smaller size of $Na^{\oplus}$, $Na_2Cr_2O_7$ is more soluble than $K_2Cr_2O_7$.

139. (2)

140. (4) In equation (1) $Fe_2(SO_4)_3$, and in equation (2) $FeSO_4$, on decomposition will form oxide instead of Fe. In equation (3) $FeCl_3$ cannot be reduced when heated in air. Hence equation (4) is correct.

141. (1) $K_2Cr_2O_7 + 4H_2SO_4 + 3H_2S \longrightarrow K_2SO_4 + Cr_2(SO_4)_3 + 7H_2O + 3S$

142. (2) $KMnO_4$ reacts with H_2SO_4 to form Mn_2O_7 which is highly explosive substance.

$2KMnO_4 + H_2SO_4 \longrightarrow K_2SO_4 + Mn_2O_7 + H_2O$

143. (1) $Cr_2O_7^{2-} + 6I^- + 14H^+ \longrightarrow 2Cr^{3+} + 7H_2O + 3I_2$

144. (4) $2Cu_2O + Cu_2S \longrightarrow 6Cu + SO_2$

self reduction.

145. (1) $\ddot{N}H_3 + H^+ \text{ (acid medium)} \rightleftharpoons \overset{+}{N}H_4$

146. (1)

147. (3) $AgNO_3$ on heating till red **hot** decomposes as follows:

$$AgNO_3 \rightarrow Ag + NO_2 + \frac{1}{2}O_2$$

148. (1) $2K_2CrO_4 + 2HNO_3 \longrightarrow K_2Cr_2O_7 + 2KNO_3 + H_2O$

149. (3)

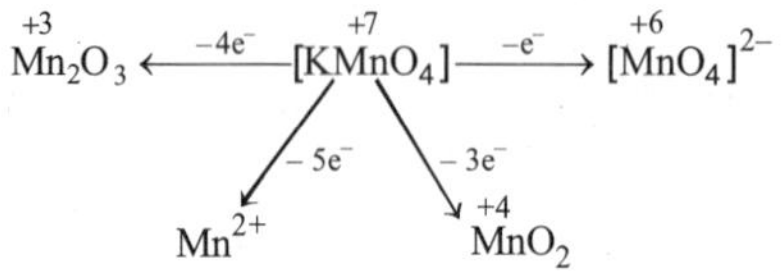

150. (12)

$Cr_2O_7^{2-} + 14H^+ + 6I^- \rightarrow 2Cr^{3+} + 3I_2 + 7H_2O$

No. of moles of e^- involved = x = 6

$Cr_2O_7^{2-} + 3S^{2-} + 14H^+ \rightarrow 3S + 2Cr^{3+} + 7H_2O$

No. of moles e^- involved = y = 6

Sum of x + y = 6 + 6 = 12

151. (10) $\overset{+7}{K\,MnO_4}\ (O.A) \xrightarrow{\text{Acidic medium}} Mn^{2+}$

X is difference in oxidation state. 7 – 2 = 5

So X = 5

$6CH_3COO^- + Fe^{3+} + H_2O \rightarrow [Fe_3(OH_2)(CH_3COO)_6]^{\oplus} + 2H^{\oplus}$

$[Fe_3(OH)_2(CH_3COO)_6]^{\oplus} + 4H_2O \rightarrow [Fe(OH)_2(CH_3COO)] + CH_3COOH + H^{\oplus}$

Brown red ppt

$Fe^{3+} \Rightarrow 3d^5 4s^0$ contains 5d electrons

So Y = 5

$\therefore$ X + Y = 5 + 5 = 10

152. (6) $Cr_2O_7^{2-} + NaCl + H_2SO_4 \rightarrow \underset{(A)}{CrO_2Cl_2}$

$CrO_2Cl_2\,(\text{Vapour}) + NaOH \rightarrow \underset{(B)}{Na_2CrO_4} + NaCl + H_2O$

$\underset{(B)}{Na_2CrO_4} + H^{\oplus} \rightarrow \underset{(C)}{Na_2Cr_2O_7} + H_2O$

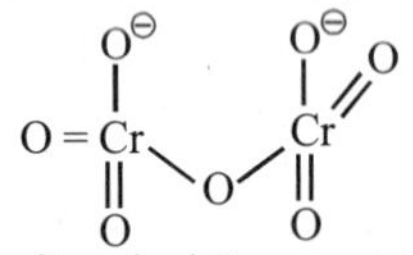

Number of terminal Oxygen = 6

153. (160) $4FeCr_2O_4 + 8Na_2CO_3 + 7O_2 \rightarrow 8Na_2CrO_4 + 2Fe_2O_3 + 8CO_2$

Fe_2O_3 is water insoluble, so its molar mass

$\Rightarrow$ [2 × 56 + 3 × 16] $\Rightarrow$ 160 g/mol

154. (5) Strong oxidising power among the option is Mn_2O_3.

$E^{\circ}_{Mn^{3+}/Mn^{2+}} = +1.57\,V$

$Mn^{+3} \rightarrow d^4$ configuration

$\mu = \sqrt{24}$BM = 4.89 BM $\Rightarrow$ 5

155. (6) Spin only magnetic moment of Mn^{+7} in $KMnO_4$ is 0. Spin only value of manganese product (Mn^{+2}) formed during titration of $KMnO_4$ against oxalic acid in acidic medium is 6

$\therefore$ Difference = 6 – 0 = 6

156. (877) CrO is Basic oxide

Cr_2O_3 is Amphoteric oxide

In CrO, Cr exist as Cr^{+2} and μ spin only = 4.90 BM

In Cr_2O_3, Cr exist as Cr^{+3} and μ spin only = 3.87 BM

Sum of spin only magnetic moment = 4.90 + 3.87 = 8.77 = 877 × 10^{-2} BM.

157. (6)

158. (4) $MnO_2 + KOH + O_2 \rightarrow \underset{(A)}{K_2MnO_4} + H_2O$

$K_2MnO_4 \xrightarrow{\text{Neutral/acidic solution}} \underset{(B)}{\overset{+7}{KMnO_4}} + \underset{(C)}{\overset{+4}{MnO_2}}$

Mn^{+4} : [Ar]$3d^3$; Mn^{+7} : [Ar]$3d^0 \Rightarrow \mu = 0$

n = 3, $\mu = \sqrt{3(3+2)}$ = 3.87 B.M

Nearest integer is (4)

159. (6) The reaction will be as follows

$K_2Cr_2O_7 + 4KCl + \underset{\text{conc.}}{6H_2SO_4}$

$\downarrow$

$2CrO_2Cl_2 + 6KHSO_4 + 3H_2O$

Oxidation state of Cr in CrO_2Cl_2 is

$x - 4 - 2 = 0 \Rightarrow x = +6$

160. (5) $\overset{(+7)}{KMnO_4} \rightarrow \overset{(+2)}{Mn^{2+}}$ (In acidic medium)

Change in oxidation state of Mn = 5

161. (3) In faintly alkaline medium,

$\overset{+7}{MnO_4^-} + 3e^- + 2H_2O \rightarrow \overset{+4}{MnO_2} + 4OH^-$

No. of electrons gained = 3

162. (10) $\overset{(0)}{Fe(CO)_5}\quad \overset{(+4)}{VO^{2+}}\quad \overset{(+6)}{WO_3}$

So, Sum of oxidation state = 0 + 4 + 6 = 10

163. (3)

$\overset{+7}{MnO_4^-} + 4H^+ + 3e^- \longrightarrow \overset{+4}{MnO_2} + 2H_2O$

164. (4) $MnO_4^{2-} \xrightarrow{H^+} MnO_4^- + MnO_2$

No. of unpaired e^- (n) = 3

$\therefore\ \mu = \sqrt{n(n+2)} = \sqrt{15}$ = 3.877B.M.

165. (1) $\overset{(+3)}{(COOH)_2} \longrightarrow \overset{(+4)}{CO_2}$

Change in O.N. = 4 – 3 = 1

166. (4) $2\overset{+7}{KMnO_4} + 3H_2O_2 \xrightarrow{\text{basic medium}} 2\overset{+4}{MnO_2} + 3O_2 + 2H_2O + 2KOH$

167. (6) $4FeCr_2O_4 + 8Na_2CO_3 + 7O_2 \rightarrow \underset{(A)}{\underbrace{8Na_2CrO_4}} + 2Fe_2O_3 + 8CO_2$

$2Na_2CrO_4 + 2H^+ \rightarrow \underset{(B)}{\underbrace{Na_2Cr_2O_7}} + 2Na^+ + H_2O$

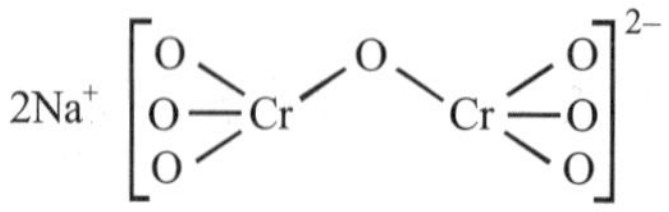

168. (0)

$3\overset{+6}{MnO_4^{2-}} + 4H^+ \longrightarrow 2\overset{+7}{MnO_4^-} + \overset{+4}{MnO_2} + 2H_2O$

In the product MnO4⁻ has highest oxidation state i.e + 7. For which number of unpaired electors will be zero.

Hence $\mu = \sqrt{n(n+1)}$; n = number of unpaired electrons.

for n = 0 $\Rightarrow$ μ = 0 B.M.

169. (6) $2KMnO_4 + 5H_2C_2O_4 + 3H_2SO_4 \rightarrow K_2SO_4 + 2MnSO_4 + 10CO_2 + 8H_2O.$

Mn^{2+} has 5 unpaired electrons, therefore, the magnetic moment $= \sqrt{5(5+2)} = \sqrt{35}$ BM. = 5.91B.M ≈ 6

170. (0)

171. (3)

172. (3) Most basic oxide is V_2O_3.

$V^{+3} \rightarrow$ [Ar] $3d^2$

No. of unpaired e^-, n = 2

$\mu = \sqrt{n(n+2)} = \sqrt{2(2+2)}$ = 2.84 BM ≈ 3

173. (6) $Cr_2O_7^{2-} \xrightarrow{OH^-} CrO_4^{2-}$

$\therefore$ Oxidation state of Cr in CrO_4^{2-} is +6.

174. (6)

175. (173)

$\overset{+6}{CrO_4^{2-}} + \overset{+2}{S_2O_3^{2-}} \longrightarrow \overset{+3}{Cr(OH)_4^-} + \overset{+6}{SO_4^{2-}}$

Using law of equivalence,

m. equivalents of CrO_4^{2-} used = m. equivalents of $S_2O_3^{2-}$ used

n-factor of CrO_4^{2-} = 3;

n-factor of $S_2O_3^{2-}$ = 4 × 2 = 8

$\therefore$ 0.154 × V × 3 = 0.25 × 40 × 8

$\Rightarrow$ V = 173.16 mL ≈ 173 mL

176. (18)

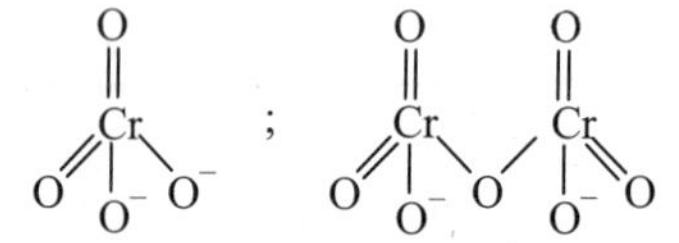

Chromate ion ; Dichromate ion

$\Rightarrow$ Total number of Cr and O bonds is 6 + 12 = 18.

177. (18.00)

$4NaCl + K_2Cr_2O_7 + \underset{\text{(Conc.)}}{3H_2SO_4} \longrightarrow \underset{(A)}{2CrO_2Cl_2} + KHSO_4 + 2NaHSO_4 + 3H_2O$

$\underset{(A)}{CrO_2Cl_2} + NaOH \longrightarrow \underset{(B)}{Na_2CrO_4} + H_2O + NaCl$

$\underset{(B)}{Na_2CrO_4} + \underset{\text{(dilute)}}{H_2SO_4} + H_2O_2 \longrightarrow \underset{(C)}{CrO_5} + Na_2SO_4 + H_2O$

The sum of total no. of atoms in one molecule each of A, B and C = 5 + 7 + 6 = 18.0

178. (1) Ti^{+4} is colourless due to d^0 configuration

$V^{2+} = (3d^3)$

$Mn^{2+} = (3d^5)$

$Fe^{3+} = (3d^4)$

$Cr^{2+} = (3d^4)$

These all are coloured due to unpaired electrons.

$\therefore$ Statement I is correct.

La^{+3}, Yb^{+2}, Lu^{+3}, Ce^{+4}, Al^{+3} & Lr^{+3} all are diamagnetic because they have paired electrons.

$\therefore$ Statement II is correct.

179. (4) Gd – [Xe] $4f^7 5d^1 6s^2$

Eu – [Xe] $4f^7 6s^2$

Dy – [Xe] $4f^{10} 6s^2$

Ho – [Xe] $4f^{11} 6s^2$

Hf – [Xe] $4f^{14} 5d^2 6s^2$

Yb – [Xe] $4f^{14} 6s^2$

Lu – [Xe] $4f^{14} 5d^2 6s^2$

Tm – [Xe] $4f^{13} 6s^2$

180. (3)

(A) Mo(VI) and W(VI) more stable than Cr(VI).

(B) In lanthanoids + 3 oxidation state is more stable hence +4 ions act as oxidising agents and +2 ions act as reducing agents.

(C) Cm – [Rn] $5f^7 6d^1 7s^2$ = 8 unpaired electrons. Am – [Rn] $5f^7 7s^2$ = 7 unpaired electrons

(D) Actinoid contraction is more than lanthanoid contraction due to poor shielding effect of 5-*f* electrons.

181. (3) Sc^{3+}, Ti^{4+} and Zn^{2+} are colourless. Th^{4+} cannot act as a reducing agent. Thus both statements are false.

182. (3) Tb^{4+} is strongest oxidising agent due to tendency to accept electron because of its very high standard reduction potential of +3.1V.

183. (1) ${}_{63}Eu^{2+}$ – [Xe] $4f^7 6s^0$

${}_{64}Gd^{3+}$ – [Xe] $4f^7 5d^0 6s^0$

${}_{63}Eu^{3+}$ – [Xe] $4f^6 6s^0$

${}_{65}Tb^{3+}$ – [Xe] $4f^8 6s^0$

${}_{62}Sm^{2+}$ – [Xe] $4f^6 6s^0$

Eu^{2+} & Gd^{3+}

184. (2) Einsteinium (atomic No = 99) : [Rn]$5f^{11} 6d^0 7s^2$

185. (3) Cm is Actinoid

Eu, Er, Tb, Yb, Lu are lanthanoid.

186. (2) La^{3+} and Ce^{4+} are diamagnetic

187. (4) The electronic configurations of the given species are:-

Gd^{3+} = [Xe]$4f^7$ (Stable half-filled configuration)

Lu^{3+} = [Xe]$4f^{14}$ (Stable fully-filled configuration)

Ce^{4+} = [Xe]$4f^0$ (Acts as oxidant, reverts back to Ce^{3+} due to high reduction potential of Ce^{4+}/Ce^{3+})

Eu^{2+} = [Xe]$4f^7$ (Strong reducing agent despite stable configuration due to low reduction potential of Eu^{3+}/Eu^{2+} = –0.36V)

188. (1) Electronic configuration of Nd(Z = 60) is; [Xe]$4f^4 6s^2$ (*f*-block)

189. (2) Statement (I) is true, Ce^{+4} has noble electronic configuration. as Ce^{+4} = [Xe] $4f^0 5d^0 6s^0$.

Statement (II) is also true due to high reduction potential for Ce^{4+}/Ce^{3+} (+1.74V), and stability of Ce^{3+}, Ce^{4+} acts as strong oxidizing agent.

190. (4) Eu^{2+} can donate an electron to become Eu^{3+} and Ce^{4+} will become Ce^{3+}.

191. (2)

192. (2) Nd(60) = [Xe] $4f^4 5d^0 6s^2$

Nd^{2+} = [Xe] $4f^4 5d^0 6s^0$

193. (1)

${}_{62}Sm : 4f^6 6s^2$; ${}_{64}Gd : 4f^7 5d^1 6s^2$; ${}_{63}Eu : 4f^7 6s^2$; ${}_{65}Tb : 4f^9 6s^2$; ${}_{61}Pm : 4f^5 6s^2$

194. (4) Ce^{+4} and Tb^{+4} act as oxidising agent.

195. (1, 4)

196. (3)

197. (3)

198. (1) Ce = [Xe] $4f^1 5d^1 6s^2$

Ce^{3+} = [Xe] $4f^1 5d^0$

Ce^{4+} = [Xe] $4f^0 5d^0$ (Noble gas configuration)

Ce^{3+} get easily oxidised to Ce^{4+} by acheiving noble gas configuration.

199. (2) Cerium exists in two different oxidation states +3 and +4.

$Ce^{4+} + e^- \rightarrow Ce^{3+}$ $E^0 = +1.61$ V

It shows, Ce^{4+} prefer to gain electron and acts as a strong oxidising agent.

200. (2) Ytterbium shows +2 oxidation state with diamagnetic nature due to $4f^{14}6s^2$ configuration.

201. (3) Eu → [Xe] $4f^7 6s^2$.

Eu^{2+} → [Xe] $4f^7$

202. (2) $Ce^{4+} \xrightarrow{e^-} Ce^{3+}$ $E^o = +1.74$ V

Positive SRP means greater oxidising power. So, Ce^{4+} can be easily reduced to Ce^{3+}. Hence, Ce^{4+} is less stable than Ce^{3+}.

203. (1)

204. (3) Europium (Eu)

Atomic No. - 63

Electronic configuration - [Xe]$4f^7 6s^2$

It shows only +2 and +3 oxidation state.

205. (1) Mischmetal is an alloy consisting mainly of lanthanoid metals.

Lan. metal ⇒ ≈ 95% ; Iron ⇒ ≈ 5%

S, C, Ca, Al ⇒ traces

206. (2)

207. (2) Eu^{2+} : [Xe]$4f^7$; Ce^{3+} : [Xe]$4f^1$

208. (1) Maximum oxidation state shown by uranium is + 6 and plutonium is + 7.

209. (4) Maximum oxidation state is shown by Np and Pu.

210. (2) Sm^{3+} has unpaired electron

211. (4) Among Lanthanides, radii decrease due to lanthanoid contraction as atomic number increase.

212. (3) Due to lanthanoid contraction, size of atoms as well as ions of lanthanoid decreases.

213. (1)

214. (2) Gd (64) = [Xe]$4f^7 5d^1 6s^2$

∴ No. of unpaired electrons = 8

215. (3) Gd (64) = [Xe] $4f^7 5d^1 6s^2$

Gd^{3+} = [Xe] $4f^7 5d^0 6s^0$

i.e., no. of unpaired electrons = 7

$$\mu = \sqrt{n(n+2)} = \sqrt{7(7+2)} = \sqrt{63} = 7.93 \text{ BM}$$

216. (3) Only Ce (Z = 58) shows stable (+4) oxidation state.

217. (2) Most of the Ln^{3+} compounds except La^{3+} and Lu^{3+} are coloured due to the presence of unpaired *f*-electrons.

218. (2) There is lesser energy difference between 5*f* and 6*d* orbitals as compared to that between 4*f* and 5*d* orbitals.

In case of actinoids, we can remove electrons from 5*f* as well as from 6*d* and due to this, actinoids exhibit larger number of oxidation states than lanthanoids.

219. (1) 4*f* orbital is nearer to nucleus as compared to 5*f* orbital, therefore, shielding of 4*f* is more than 5*f*.

220. (1) More the distance between nucleus and outer orbitals, lesser will be force of attraction on them. Distance between nucleus and 5 *f* orbitals is more as compared to distance between 4*f* orbitals and nucleus. Therefore, there is lesser energy difference between 5*f* and 6*d* than between 4*f* and 5d orbitals.

221. (2) The imperfect shielding of *f* electrons in lanthanoids is due to the shape of *f* orbitals which is very much diffused. Thus, as the atomic number increases, the nuclear charge increases by unity at each step, while no comparable increase in the mutual shielding effect of 4*f* occurs. This causes a contraction in the size of the 4*f* subshell as a result of which atomic and ionic radii decrease gradually from La to Lu.

222. (4) Lanthanide contraction is associated with the intervention of the 4*f* orbitals which must be filled before the 5*d* series of elements begin. The filling of 7*f* before 5*d* orbitals results in a regular decrease in atomic radii called lanthanoid contraction which essentially compensates for the expected increase in atomic size with increasing atomic number. The factor responsible for the lanthanoid contraction is the imperfect shelding of one electron by another in the same set of orbitals.

223. (2) Note: In vertical columns of transition elements, there is an increase in size from first member to second member as expected but from second member to third member, there is very small chang in size and some times sizes are same. This is due to lanthanide contraction this is the reason for Zr and Hf to have same radius.

224. (3) Due to lanthanoid contraction, the ionic radii decreases from La^{3+} to Lu^{3+}. Thus, only possible option is (3).

225. (2) *f*-Block elements show a regular decrease in atomic size due to lanthanide/actinide contraction.

226. (3) Common oxidation states of Ce(cerium) are + 3 and + 4.

227. (1) In lanthanides there is a regular decrease in the atomic radii as well as ionic radii of trivalent ions as the atomic number increases from Ce to Lu. This decrease in size of atoms and ions is known as **lanthanide contraction**. Although the atomic radii do show some irregularities but ionic radii decrease from La to Lu. Thus the correct order is.

$Yb^{3+} < Pm^{3+} < Ce^{3+} < La^{3+}$

86.8pm 97pm 102pm 103pm

228. (1) The +4 oxidation state of cerium is also known in solution due to the configuration [Xe]$4f^1 5d^1 6s^2$

229. (2) La^{+3} and Lu^{+3} do not show any colour because no unpaired electron is present for the f–f transition.

230. (2) Noble gas configuration = $ns^2 np^6$

[Sr^{2+}] = [Kr]

[Cs^+] = [Xe]

[Yb^{2+}] = [Xe] $4f^{14}$

[La^{2+}] = [Xe] $5d^1$

[Pb^{2+}] = [Xe] $4f^{14} 5d^{10} 6s^2$

[Fe^{2+}] = [Ar] $3d^6$

231. (18) Np [93] = [Rn] $5f^4 6d^1 7s^2$

So, no. of *f* electrons in 5*f* = 4

No. of *f* electrons in 4f = 14

Total no. of *f* electrons = 14 + 4 = 18

232. (7) Gd(64) ⇒ [Xe] $4f^7 5d^1 6s^2$ ⇒ Gd^{2+} ⇒ [Xe] $4f^7 5d^1 6s^0$

So, the number of 4*f* electrons in the ground state electronic configuration of Gd^{2+} is 7.

233. (10) Ho(67) = [Xe]$4f^{11} 6s^2$

Ho^{3+} = [Xe]$4f^{10}$

Part-B (JEE Advanced)

1. (c)

$$MnCl_2 + NaOH \rightarrow \underset{\substack{(P)\\(\text{white ppt.})}}{Mn(OH)_2} \downarrow + \underset{\substack{(Q)\\(\text{Filterate})}}{NaCl}$$

$$Mn(OH)_2 \xrightarrow[\text{heat}]{PbO_2 + H^+(H_2SO_4)} \underset{(X)}{MnO_4^-} + Pb^{2+}$$

$$Cl^- \xrightarrow{MnO(OH)_2/\text{conc. } H_2SO_4/\text{warm}} MnCl_2 + H_2O + NaHSO_4 + Cl_2$$

$$Cl_2 + KI \rightarrow I_2 \xrightarrow{\text{Starch}} \text{blue colouration}$$

2. (c) (Fe becomes passive on reaction with concentrated HNO_3). However, cold relatively conc. HNO_3 reacts with Fe as below.

$Fe + 6HNO_3 \rightarrow Fe(NO_3)_3 + 3NO_2 + 3H_2O$

$Cu + 4HNO_3\text{(conc.)} \rightarrow Cu(NO_3)_2 + 2NO_2 + 2H_2O$

$4Au + 8NaCN + O_2 + 2H_2O \rightarrow 4Na[Au(CN)_2] + 4NaOH$

$Zn + 2NaOH \rightarrow Na_2ZnO_2 + H_2$

3. (a) $CuSO_4$ will be absorbing orange-red colour and hence will be of blue colour.

4. (b) In the presence of oxygen, Ag metal forms a water soluble complex Na [Ag $(CN)_2$] with dilute solution of NaCN

$$4Ag + 8NaCN + 2H_2O + O_2 \longrightarrow \underset{(\text{Soluble})}{4Na[Ag(CN)_2]} + 4NaOH$$

5. (a)

$$CuSO_4 + 2KCN \longrightarrow Cu(CN)_2 + K_2SO_4$$

$$\underset{\text{unstable}}{Cu(CN)_2} \rightleftharpoons CuCN + \frac{1}{2}(CN)_2$$

$$CuCN + 3KCN \rightleftharpoons \underset{\text{colourless}}{K_3[Cu(CN)_4]^{3-}}$$

6. (b)

$$2Fe + 3Cl_2\text{ (dry)} \rightarrow 2FeCl_3\text{(anhydrous)}$$

7. (c) Chrome alum is $K_2SO_4.Cr_2(SO_4)_3.24H_2O$

Sodium peroxide (Na_2O_2) will act as an oxidizing agent. It will oxidise Cr^{3+} to Cr^{6+} and Fe^{2+} to Fe^{3+}.

$$Cr^{3+} \longrightarrow \underset{\text{yellow}}{CrO_4^{2+}}$$

$$Fe^{2+} \longrightarrow \underset{\text{brown residue}}{Fe(OH)_3 \downarrow}$$

Hence, the filtrate will be yellow in colour and the residue will be brown in colour.

8. (a) HgS does not dissolved in hot dil. HNO_3.

9. (a) $2KMnO_4 + 16HCl \longrightarrow 2KCl + 2MnCl_2 + 8H_2O + 3Cl_2$

10. (c) Solder is an alloy containing Sn – 67% and Pb – 33%.

11. (b) In Zn–Cu couple, Zn is activated by Cu. It is used as a reducing agent in organic synthesis.

The proportion of Zn is about 90% and it can be prepared by coating Zn with copper.

12. (d) Conc. HNO_3 renders iron passive by forming a thin protective film of Fe_3O_4 on its surface.

13. (c) Hypo solution ($Na_2S_2O_3$) is used in photography to remove the unaffected AgBr in the form of soluble complex.

$$AgBr + 2Na_2S_2O_3 \longrightarrow \underset{\text{Sod. argentothiosulphate}}{Na_3[Ag(S_2O_3)_2]} + NaBr$$

14. (b) ↑↓ | ↑↓ | ↑↓ | ↑ | ↑

It has 2 unpaired electrons.

3*d* orbital of Ni^{2+} ion. At No. of Ni = 28.

15. (b)

$$Zn + 2NaOH \longrightarrow \underset{\text{Sod.meta zincate(soluble)}}{Na_2[ZnO_2]} + H_2$$

16. (b) German silver is alloy of Cu + Zn + Ni

17. (d) ∵ Basicity of hydroxides decreases on moving left to right in a period.

18. (a) $Zn + H_2SO_4 \longrightarrow ZnSO_4 + H_2$

$Zn + 2NaOH \longrightarrow Na_2[ZnO_2] + H_2$

∴ Ratio of H_2 evolved is 1 : 1.

19. (7) $K_2Cr_2O_7 + KI + H_2SO_4 \rightarrow K_2SO_4 + Cr_2(SO_4)_3 + I_2 + H_2O$

$2CuSO_4 + KI \rightarrow 2CuI + I_2 + 2K_2SO_4$

$H_2O_2 + 2KI \rightarrow 2KOH + I_2$

$Cl_2 + 2KI \rightarrow 2KCl + I_2$

$O_3 + H_2O + 2KI \rightarrow 2KOH + O_2 + I_2$

$FeCl_3 + 2KI \rightarrow 2KCl + FeCl_2 + I_2$

$HNO_3 + KI \rightarrow KNO_3 + I_2 + NO$

20. (6)

$$2MnO_2 + 4KOH + O_2 \longrightarrow 2K_2MnO_4 + 2H_2O$$

Oxidation number of Mn in K_2MnO_4 is 6

K_2MnO_4; $2 + x - 8 = 0$

$x = 6$

21. H_2S; It is due to formation of sulphide of silver (Ag_2S) which is black.

22. $FeSO_4.7H_2O$, $ZnSO_4.7H_2O$;

23. Hydration/solvation; [A substance dissolves when its $\Delta H_{\text{hydration}}$ > lattice energy].

24. Zinc;

25. PbO_2; Pb^{4+} can be easily reduced to Pb^{2+}.

26. $K[Ag(CN)_2]$

27. False : Dipositive zinc exhibits diamagnetism (and not paramagnetism) because it has no unpaired electron.

28. True : Insolubility of AgCl in H_2O is due to its high lattice energy. Further, AgCl forms a complex with conc. NaCl solution and is therefore soluble.

29. True : Hydration energy of AgF is appreciably higher than its lattice energy because of smaller F^- ion and thus AgF is soluble in water. In rest of the halides, lattice energy is more than hydration energy to make them insoluble.

30. (b, c, d)

$$Zn + \underset{\text{Hot \& Conc.}}{H_2SO_4} \rightarrow G + R + X$$

$$Zn + H_2SO_4 \rightarrow \underset{(G)}{ZnSO_4} + \underset{(R)}{SO_2} + \underset{(X)}{2H_2O}$$

$$Zn + \text{Conc. } NaOH \rightarrow \underset{(T)}{Na_2ZnO_2} + \underset{(Q)}{H_2}$$

$$\underset{(G)}{ZnSO_4} + H_2S + NH_4OH \rightarrow \underset{(Z)}{ZnS} + \underset{(Y)}{(NH_4)_2SO_4} + \underset{(X)}{H_2O}$$

(a) Oxidation state of Zn in Na_2ZnO_2 is +2

(b) Bond order of *Q* is one for H_2.

(c) ZnS is white in colour

(d) SO_2 is angular in shape

31. (a, b)

(a) $2Fe^{3+} + H_2\overset{-1}{O_2} + 2OH^- \longrightarrow 2Fe^{2+} + 2H_2O + \overset{0}{O_2}$ ([R]: $Fe^{3+} \rightarrow Fe^{2+}$; [O]: $H_2O_2 \rightarrow O_2$)

(b) $Na_2O_2 + 2H_2O \longrightarrow H_2O_2 + 2NaOH$

The formed H_2O_2 will reduce Fe^{3+} to Fe^{2+}.

32. (a, b, c) Cr^{2+} is a reducing agent and Mn^{3+} is an oxidizing agent and both have electronic configuration d^4.

$\overset{\circ}{E}_{Cr^{3+}/Cr^{2+}} = -0.41V$ $\overset{\circ}{E}_{Mn^{3+}/Mn^{2+}} = 1.51V$

Above E° values explains reducing nature of Cr^{2+} and oxidizing behaviour of Mn^{3+}.

33. (a, b, c) $Na + NH_3\text{ (excess)} \rightarrow$ Dilute solution of Na in liq. NH_3 → Paramagnetic due to ammoniated e^-.

$K + O_2\text{ (excess)} \rightarrow KO_2$ (O_2^- is paramagnetic)

$Cu + HNO_3\text{ (dil.)} \rightarrow Cu(NO_3)_2 + NO$ (NO is paramagnetic)

2-Ethylanthraquinol + O_2 → 2-Ethylanthraquinone + H_2O_2

(H_2O_2 is diamagnetic)

34. (b, c, d) Cu^{2+} ions will react with CN^- and SCN^- forming $[Cu(CN)_4]^{3-}$ and $[Cu(SCN)_4]^{3-}$ leading the reaction in the backward direction.

$$Cu^{2+} + 2CN^- \rightarrow Cu(CN)_2$$

$$2Cu(CN)_2 \rightarrow 2CuCN + (CN)_2$$

$$\overset{+1}{Cu}CN + 3CN^- \rightarrow [\overset{+1}{Cu}(CN)_4]^{3-}$$

$$Cu^{2+} + 4SCN^- \rightarrow [\overset{+1}{Cu}(SCN)_4]^{3-}$$

Cu^{2+} also combines with $CuCl_2$ which reacts with Cu to produce CuCl pushing the reaction in the backward direction.

$$CuCl_2 + Cu \rightarrow 2CuCl \downarrow$$

35. (a, d) Potassium permanganate will oxidize itself in this process.

In acidic medium

$$MnO_4^- \longrightarrow Mn^{2+}$$

Change in oxidation state of Mn = 7 – 2 = 5

Thus electrons lost = 5

In neutral medium

$$MnO_4^- \longrightarrow MnO_2$$

Change in oxidation state of Mn = 7 – 4 = 3

∴ Electrons lost = 3

36. (a, c) Mn makes steel harder and increases its elasticity and tensile strength. Further Mn acts as deoxidiser. MnO reacts with S present in cast iron, gets oxidised and then combine to form slag.

37. (b, c) Brass : Cu (60-80%), Zn (40-20%); Gun Metal : Cu (87%), Sn (10%), Zn (3%); Bronze : Cu, Sn; Type metal : Pb, Sb, Sn

38. (c, d) Aqueous solution of $Co(NO_3)_2$ and $CrCl_3$ in which Co^{2+} (d^7) and Cr^{3+} (d^3) contains incompletely filled *d*-orbitals are coloured.
$Zn(NO_3)_2$, $LiNO_3$, $KAl(SO_4)_2 \cdot 12H_2O$ (potash alum) do not have unpaired electrons in *d*-orbital.

39. (A) –p, s ; (B) –q, s ; (C) –r, t ; (D) –q, t

(A) $3Cu + 8\,HNO_3(\text{dil.}) \longrightarrow \underset{p}{2NO} + \underset{s}{Cu(NO_3)_2} + 4H_2O$

(B) $Cu + 4HNO_3(\text{conc.}) \longrightarrow \underset{q}{2NO_2} + \underset{s}{Cu(NO_3)_2} + 2H_2O$

(C) $4Zn + 10HNO_3(\text{dil.}) \longrightarrow \underset{t}{4Zn(NO_3)_2} + \underset{r}{N_2O} + 5H_2O$

(D) $Zn + 4HNO_3(\text{conc.}) \longrightarrow \underset{t}{Zn(NO_3)_2} + \underset{q}{2NO_2} + 2H_2O$

40. *(i)* (g) *(ii)* (d)*(iii)*(e) *(iv)*(a) *(v)* (f) *(vi)* (b)*(vii)*(c)

41. (b) **42.** (a)

43. (c)

$\underset{(M)}{Cu} + \underset{(N)}{2AgNO_3} \longrightarrow \underset{\text{Blue}}{Cu(NO_3)_2} + 2Ag$

$NaCl + \underset{(\text{remaining})}{AgNO_3} \longrightarrow \underset{(O)}{AgCl \downarrow} + NaNO_3$

$2AgCl + 2NH_3(aq) \longrightarrow \underset{\text{Soluble}}{[Ag(NH_3)_2]^+}$

$\underset{(\text{remaining})}{Cu(NO_3)_2} + 4NH_3(aq) \rightarrow \underset{(\text{Deep blue colour})}{[Cu(NH_3)_4](NO_3)_2}$

44. (b) 3*d* 4*s*

Zn [↑↓|↑↓|↑↓|↑↓|↑↓] [↑↓]

Zn^{2+} [↑↓|↑↓|↑↓|↑↓|↑↓] [ ]

Zn^{2+} is diamagnetic because of absence of unpaired electrons.

45. (c) The statement-1 is correct

$\underset{(\text{yellow})}{2K_2CrO_4} + H_2SO_4 \rightarrow \underset{(\text{orange})}{K_2Cr_2O_7} + Na_2SO_4 + H_2O$

Oxidation state of Cr in K_2CrO_4 and $K_2Cr_2O_7$ is +6, *i.e.* no change in O.S. So explanation is wrong.

46. Reaction involved in developing of a black and white photographic film.

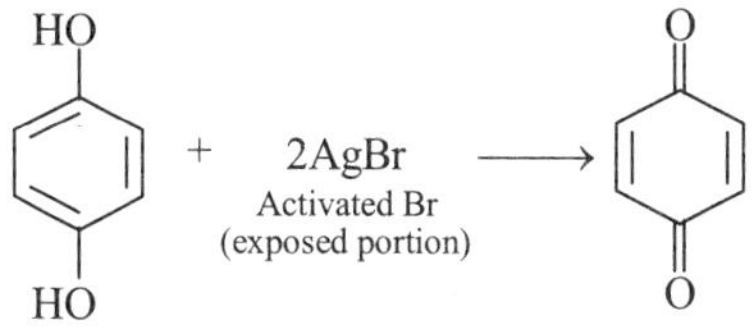

Hydroquinone
(developer)

$+ 2Ag + 2HBr$

$\underset{\text{unexposed portion}}{AgBr} + 2Na_2S_2O_3 \longrightarrow Na_3[Ag(S_2O_3)_2] + NaBr$

$Na_2S_2O_3 + 2H^+ \rightarrow 2Na^+ + H_2SO_3 + \underset{\text{colloidal sulphur}}{S \downarrow}$

47. *(i)* Argentite is Ag_2S. Silver is extracted from its ore argentite (silver glance, Ag_2S) as follows :

(1) Silver glance is concentrated by froth flotation.

(2) Leaching : The concentrated ore is ground to fine powder and dissolved in dilute solution of sodium cyanide.

$Ag_2S + 4NaCN \longrightarrow 2NaAg(CN)_2 + Na_2S$

Oxygen of air converts Na_2S to Na_2SO_4 thereby preventing reaction to take place in the reversible direction.

(3) Recovery of silver.

Silver is precipitated out by adding electropositive metal, Zn.

$2Na[Ag(CN)_2] + Zn \rightarrow Na_2[Zn(CN)_4] + 2Ag$

(ii) For development, activated grains are preferentially reduced by mild reducing agents like hydroquinone

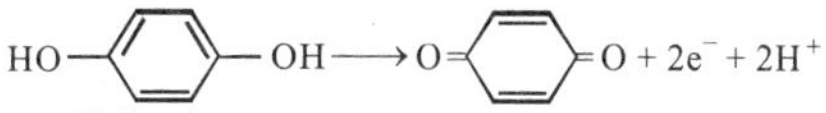

Hydroquinone Quinone

$AgBr(s) + e^- \longrightarrow Ag(s) + Br^-$

(Reduction of activated AgBr to elemental silver.)

The photographic film is permanently fixed by immediately washing out any non activated AgBr grains in hypo emulsion.

$AgBr(s) + \underset{\text{hypo}}{2Na_2S_2O_3} \longrightarrow \underset{\text{soluble}}{Na_3Ag(S_2O_3)_2} + NaBr$

48. $NaNO_3 + H_2SO_4 \rightarrow NaHSO_4 + HNO_3$

$6FeSO_4 + 2HNO_3 + 3H_2SO_4 \longrightarrow 3Fe_2(SO_4)_3 + 4H_2O + 2NO$

$[Fe(H_2O)_6]SO_4.H_2O + NO \longrightarrow \underset{\text{brown ring}}{[Fe(H_2O)_5\,NO].SO_4} + 2H_2O$

49. $2Cu + H_2O + CO_2 + O_2 \rightarrow \underset{\text{Green basic copper carbonate}}{CuCO_3.Cu(OH)_2}$

50. $_{29}Cu = 1s^2, 2s^2\,2p^6, 3s^2 3p^6 3d^{10}, 4s^1$
$_{30}Zn = 1s^2, 2s^2\,2p^6, 3s^2 3p^6 3d^{10}, 4s^2$

On the basis of configuration of Cu and Zn, first ionisation potential of Zn is greater than that of copper because in zinc the electron is removed from $4s^2$ configuration while in copper it is removed from $4s^1$ configuration. So more amount of energy is required for the removal of electron of $4s^2$ (completely filled orbital) than that of $4s^1$ while the second ionisation potential of Cu is higher than that of zinc because Cu^+ has $3d^{10}$ (stable configuration) in comparison to Zn^+ ($4s^1$ configuration).

51. $3[Fe(H_2O)_6]^{2+} + NO_3^- + 4H^+ \rightarrow NO + 3[Fe(H_2O)_6]^{3+} + 2H_2O$

$[Fe(H_2O)_6]^{2+} + NO \rightarrow [Fe(H_2O)_5NO]^{2+} + H_2O$

52. $ZnO + 2NaOH \rightarrow \underset{\text{Sod. Zincate}}{Na_2ZnO_2} + H_2O$

53. $2Mn(OH)_2 + 5NaBiO_3 + 18H^+ \rightarrow 2MnO_4^- + 5Bi^{3+} + 5Na^+ + 11H_2O$

54. *(i)* CrO_3 is acid anhydride of H_2CrO_4 (Chromic acid)

[Anhydride are formed by loss of water from acid]

$H_2CrO_4 \xrightarrow{\text{Dehydration}} CrO_3 + H_2O$

In $H_2Cr_2O_4$, Cr is present in + 6 oxidation state.

(ii) Cu^{2+} is reduced to Cu^+ by I^-, hence cupric iodide is converted into cuprous iodide so $[CuI_4]^{2-}$ does not exist, Cl^- cannot effect this change and thus $[CuCl_4]^{2-}$ exists.

(iii) Mercurous chloride changes from white to black when treated with ammonia due to the formation of finely divided mercury.

*(iv)*Zinc is cheaper and stronger reducing agent than copper and zinc is volatile

(v) The transition metals form **coloured compounds** and **coloured complexes**. They have vacant *d*-orbitals. Electrons take up energy from the visible region and move to higher energy levels. The visible colour of the substance is the complementary colour of the absorbed light.

[The colour is due to *d*-*d* transitions]

(vi) It is because silver bromide being sensitive to light, reduces into metallic silver grains when light fall on it.

55. $SO_2 + MnO_4^- + H^+ \rightarrow SO_4^{2-} + Mn^{2+} + H_2O$

56. On standing, $FeCl_3$ is hydrolysed and produces colloidal solution of $Fe(OH)_3$ which is in form of brown precipitate.

$FeCl_3 + 3H_2O \rightleftharpoons \underset{\text{Brown ppt}}{Fe(OH)_3 \downarrow} + 3HCl$

57. *(i)* $2Ag + 2H_2SO_4 \longrightarrow \underset{(\text{white})}{Ag_2SO_4} + SO_2 + 2H_2O$

(ii) $(NH_4)_2Cr_2O_7 \xrightarrow{\Delta} \underset{(\text{green})}{Cr_2O_3} + N_2 + 4H_2O$

(iii) $2KMnO_4 + 3H_2SO_4 + 5H_2S \xrightarrow{\Delta} K_2SO_4 + 2MnSO_4 + \underset{(\text{turbidity})}{5S \downarrow} + 8H_2O$

58. (b) In neutral or faintly alkaline solutions a notable reaction is the oxidation of iodide to iodate:

$2MnO_4^- + H_2O + \underset{\text{iodide}}{I^-} \rightarrow 2MnO_2 + 2OH^- + \underset{\text{iodate}}{IO_3^-}$

59. (c) Colour is due to *d* – *d* transitions. Coloured compounds contain partly filled *d*-orbital. The oxidation state of copper in various compounds is + 1 and + 2. In CuF_2 it is in + 2 oxidation state. In + 2 state its configuration is

$Cu^{2+} = 1s^2 2s^2 2p^6 3s^2 3p^6 3d^9$

3*d*

i.e. [↑↓|↑↓|↑↓|↑↓|↑]

It has one unpaired electron due to which it is coloured.

(CuF_2 possesses blue colour in crystalline form)

60. (b) Colour of transition metal ion salt is due to *d*-*d* transition of unpaired electrons of *d*-orbital. Metal ion salt having similar number of unpaired electrons in *d*-orbitals shows similar colour in aqueous medium.

V^{4+}: $[Ar]3d^1$ | 1 | | | | |

Cu^{2+}: $[Ar]3d^9$ | ↿⇂ | ↿⇂ | ↿⇂ | ↿⇂ | 1 |

Number of unpaired electrons = 1

61. (a)

$$(NH_4)_2Cr_2O_7 \xrightarrow{heat} N_2 + Cr_2O_3 + 4H_2O$$

$NH_4NO_2 \rightarrow N_2 + 2H_2O$ (Same gas i.e., N_2)

$$NH_4NO_3 \longrightarrow N_2O + 2H_2O$$

$$Mg_3N_2 + 3H_2O \longrightarrow 3Mg(OH)_2 + 2NH_3$$

$$2NaNO_2 + H_2O_2 \longrightarrow Na_2O_2 + 2HNO_2$$

62. (a) $6MnO_4^- + I^- + 6OH^- \longrightarrow 6MnO_4^{2-} + IO_3^- + 3H_2O$

63. (a) Stable oxidation state of Mn in alkaline medium is +6. So, MnO_2 is oxidised to K_2MnO_4 (purple green) by atmospheric oxygen in KOH medium.

$$2MnO_2 + 4KOH + O_2 \rightarrow \underset{\text{(Purple)}}{2K_2MnO_4} + 2H_2O$$

64. (b) The structure of $Cr_2O_7^{2-}$

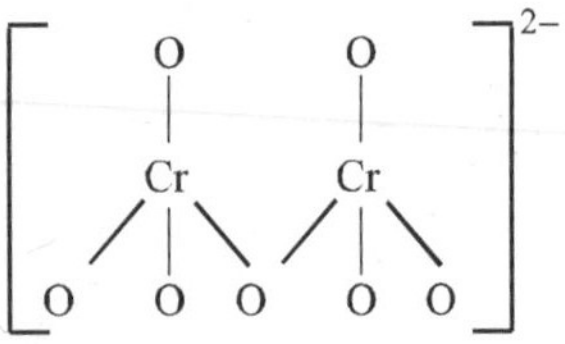

There are six normal Cr – O bonds and two bridged Cr – O bonds.
The six normal Cr – O bonds are expected to be equivalent and different from those of the bridged Cr – O bonds.

65. (b) The electronic configurations of cations in the given salts are
$Ag^+(4d^{10})$, $Cu^{2+}(3d^9)$, $Mg^{2+}(2s^2,2p^6)$, $Cu^+(3d^{10})$
Only Cu^{2+} ion has one unpaired electron in 3d orbital and so, its salt is expected to be coloured.

66. (a) The reaction of MnO_4^- and SO_3^{2-} in acidic medium is derived as follows:

$$[MnO_4^- + 8H^+ + 5e^- \rightarrow Mn^{2+} + 4H_2O] \times 2$$

$$[SO_3^{2-} + H_2O \rightarrow SO_4^{2-} + 2H^+ + 2e^-] \times 5$$

$$2MnO_4^- + 5SO_3^{2-} + 6H^+ \rightarrow 2Mn^{2+} + 5SO_4^{2-} + 3H_2O$$

Hence, 2 mole $2MnO_4^- \equiv 5$ mol SO_3^{2-}

i.e., $\frac{2}{5}$ mol $MnO_4^- \equiv 1$ mol SO_3^{2-}

67. (b)

$$(NH_4)_2Cr_2O_7 \xrightarrow{\Delta} \underset{\text{(green)}}{Cr_2O_3} + N_2 + 4H_2O$$

Hence, green coloured powder blown in the air is Cr_2O_3.

68. (d) $CuSO_4 + 2KCN \longrightarrow Cu(CN)_2 + K_2SO_4$

$2Cu(CN)_2 \longrightarrow Cu_2(CN)_2 + (CN)_2$ (Cyanogen)

$Cu_2(CN)_2 + 6KCN \longrightarrow 2K_3[Cu(CN)_4]$

69. (6) $8\overset{+7}{Mn}O_4^- + 3\overset{+2}{S_2}O_3^{2-} + H_2O \longrightarrow 8\overset{+4}{Mn}O_2 + 6\overset{+6}{S}O_4^{2-} + 2OH^-$

$\therefore$ 8 moles of MnO_4^- form 6 moles of SO_4^{2-}.

70. (a, b, c)

$$2MnO_2 + 4KOH + O_2 \xrightarrow{\Delta} \underset{(W)}{2K_2MnO_4} + 2H_2O$$

$$\underset{(W)}{2K_2MnO_4} + 2H_2O \rightarrow \underset{(X)}{2KMnO_4} + 2KOH(g) + H_2$$

$$\underset{(Y)}{K_2MnO_4} \longrightarrow 2K^+ + MnO_4^{2-} \text{ (mangnate ion)}$$

$$\underset{(Z)}{KMnO_4} \longrightarrow K^+ + MnO_4^- \text{ (permangnate ion)}$$

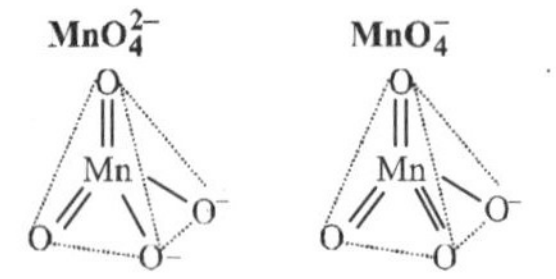

MnO_4^{2-}	MnO_4^-
sp^3, Tetrahedral	sp^3, Tetrahedral
Green colour•	Purple colour
$Mn^{6+}: [Ar]4s^0 3d^1$	$Mn^{7+}: [Ar]4s^0 3d^0$
Paramagnatic	Diamagnetic

Disproportionation of MnO_4^{2-} undergoes in acidic medium but not in base, concerned reaction is as under :

$$3\overset{+6}{Mn}O_4^{2-} + 4H^+ \rightarrow \overset{+4}{Mn}O_2 + 2\overset{+7}{Mn}O_4^- + 2H_2O$$

71. (a, c, d)

(a) $K\overset{-1}{I}(aq) + K_3[\overset{+3}{Fe}(CN)_6](aq) \longrightarrow K\overset{-3}{I_3}(aq) + K_4[\overset{+2}{Fe}(CN)_6](aq)$ (Brownish-yellow)

$\downarrow ZnSO_4(aq)$

(b) $K_2Zn_3[Fe(CN)_6]_2$ (white ppt) or $\{K_2Zn[Fe(CN)_6]\}\downarrow$ (white ppt) + $KI_3(aq)$

$\downarrow Na_2S_2O_3$

$I^-(aq) + S_4O_6^{2-}(aq)$ (Colourless)

(c) When the filterate containing KI_3 add to start solution, the dissolved I_2 will produce a blue colour solution.

(d) $\underset{\text{white ppt.}}{K_2Zn[Fe(CN)_6]} + NaOH \longrightarrow [Zn(OH)_4]^{2-}(aq) + [Fe(CN)_6]^{4-}(aq)$

72. (b, c) $2MnO_2 + 4KOH + O_2 \xrightarrow{heat} 2K_2MnO_4 + 2H_2O$;

$2KOH + HCHO + 2KMnO_4 \longrightarrow 2K_2MnO_4 + H_2O + HCOOH$

73. (2) From this equation we need 2 mol of KI

$$2KI + 2K_3[Fe(CN)_6] \rightleftharpoons I_2 + \underset{(P)}{2K_4[Fe(CN)_6]}$$

74. (3 or 2)

$$\underset{(P)}{2K_4[Fe(CN)_6]} + 3ZnCl_2 \rightarrow \underset{(Q)}{K_2Zn_3[Fe(CN)_6]_2} + 6KCl$$

OR

$$K_4[Fe(CN)_6] \rightarrow Zn_2[Fe(CN)_6]$$

75. $[A] = [Ti(H_2O)_6]Cl_3$ $[B] = HCl$

$$TiCl_4 \xrightarrow[heat]{Zn} TiCl_3 \xrightarrow{H_2O} \underset{\text{purple [A]}}{[Ti(H_2O)_6]Cl_3}$$

$$\underset{\text{(moist air)}}{TiCl_4 + (n+2)H_2O} \rightarrow \underset{\text{white}}{TiO_2(H_2O)_n} + \underset{\text{fumes [B]}}{4HCl\uparrow}$$

$Ti^{4+} = [Ar]3d^0$; $Ti^{3+} = [Ar]3d^1$

$TiCl_4$ is colourless since Ti^{4+} has no *d* electrons, hence *d-d* transition is impossible. On the other hand, Ti^{3+} is coloured due to *d-d* transition. Ti^{3+} absorbs greenish yellow compound of white light, hence its aqueous solution is purple which is complementary colour of greenish yellow in white light.

76. (*i*) $CuSO_4 + 4NH_4OH \rightarrow [Cu(NH_3)_4]SO_4 + 4H_2O$

or $CuSO_4 + 2NH_4OH \rightarrow Cu(OH)_2 + (NH_4)_2SO_4$

$Cu(OH)_2 + 2(NH_4)_2SO_4 \rightarrow \underset{\text{deep blue complex}}{[Cu(NH_3)_4]SO_4} + 2H_2O + H_2SO_4$

(*ii*) $H_2O_2 \rightarrow H_2O + [O] \times 3$ $\underset{\text{green}}{CrCl_3} + 3NaOH \rightarrow Cr(OH)_3\downarrow + 3NaCl$

$2Cr(OH)_3 + 4NaOH + 3[O] \rightarrow 2Na_2CrO_4 + 5H_2O$

$2CrCl_3 + 10NaOH + 3H_2O_2 \rightarrow \underset{\text{yellow}}{2Na_2CrO_4} + 6NaCl + 8H_2O$

77. (*i*) $[Zn + 2HNO_3(dil) \rightarrow Zn(NO_3)_2 + 2[H]] \times 4$

$2HNO_3 + 8[H] \rightarrow N_2O + 5H_2O$

$4Zn + 10HNO_3 \rightarrow 4Zn(NO_3)_2 + N_2O + 5H_2O$

(*ii*) $\underset{\text{(Argentite)}}{Ag_2S} + 4KCN \longrightarrow \underset{\text{potassium Argentocyanide}}{2K[Ag(CN)_2]} + K_2S$

(*iii*) $3[MnO_4]^{2-} + 4H^+ \rightarrow MnO_2 + 2[MnO_4]^- + 2H_2O$

(*iv*) $3SO_2 + \overset{+6}{Cr_2}O_7^{2-} + 2H^+ \rightarrow 2\overset{+3}{Cr^{3+}} + 3SO_4^{2-} + H_2O$

(*v*) $5(NH_4)_2S_2O_8 + 2MnSO_4 + 8H_2O \longrightarrow 2MnO_4^- + 10SO_4^{2-} + 16H^+$

(*vi*) $AgBr + 2Na_2S_2O_3 \rightarrow \underset{\text{Sod. argento thiosulphate}}{Na_3[Ag(S_2O_3)_2]} + NaBr$

(*vii*) $K_2Cr_2O_7 + 14HCl \xrightarrow{heat} 2KCl + 2CrCl_3 + 7H_2O + 3Cl_2\uparrow$

(*viii*) $2CuSO_4 + 2Na_2CO_3 + H_2O \rightarrow CuCO_3.Cu(OH)_2 + 2Na_2SO_4 + CO_2$

(*ix*) The individual reactions are

$3Cu + 8HNO_3 \text{ (dil.)} \rightarrow 2NO + 3Cu(NO_3)_2 + 4H_2O$

$Cu + 4HNO_3 \text{ (conc.)} \rightarrow Cu(NO_3)_2 + 2NO_2 + 2H_2O$

For the molar ratio of 2 : 1 of NO and NO_2, we will have

$7Cu + 20HNO_3 \rightarrow 7Cu(NO_3)_2 + 4NO + 2NO_2 + 10H_2O$

(*x*) $2KMnO_4 + 3MnSO_4 + 2H_2O \rightarrow 5MnO_2 + K_2SO_4 + 2H_2SO_4$

This is known as **Volhard method** for estimation of manganese.

(*xi*)(a) $4Fe + 10HNO_3(6\% \text{ dilute}) \rightarrow 4Fe(NO_3)_2 + NH_4NO_3 + 3H_2O$

(b) $4Fe + 10HNO_3(20\% \text{ dilute}) \rightarrow 4Fe(NO_3)_2 + N_2O + 5H_2O$

(*xii*) $K_2Cr_2O_7 + 4NaCl + 6H_2SO_4 \rightarrow \underset{\text{chromyl chloride (orange)}}{2CrO_2Cl_2} \uparrow + 4NaHSO_4 + 2KHSO_4 + 3H_2O$

(*xiii*)

Roasting:

$\underset{\text{(copper pyrites)}}{2CuFeS_2} + O_2 \xrightarrow{\Delta} Cu_2S + 2FeS + SO_2$

$2Cu_2S + 3O_2 \longrightarrow 2Cu_2O + 2SO_2$

$2FeS + 3O_2 \longrightarrow 2FeO + 2SO_2$

Smelting with coke and sand:

$Cu_2O + FeS \longrightarrow Cu_2S + FeO$

$FeO + SiO_2 \longrightarrow FeSiO_3 \text{ (slag)}$

Bessemerization:

$2Cu_2S + 3O_2 \longrightarrow 2Cu_2O + 2SO_2$

$Cu_2S + 2Cu_2O \longrightarrow 6Cu + SO_2 \uparrow$

(*xiv*) $CoCl_2 + 2KNO_2 \rightarrow Co(NO_2)_2 + 2KCl$

$KNO_2 + CH_3COOH \rightarrow CH_3COOK + HNO_2$

$Co(NO_2)_2 + 3KNO_2 + 2HNO_2 \rightarrow \underset{\text{Pot. cobaltinitrite (yellow ppt.)}}{K_3[Co(NO_2)_6]} \downarrow + NO + H_2O$

(*xv*) $AgCl + 2NaCN \rightleftharpoons NaCl + Na[Ag(CN)_2]$

$\underset{\text{soluble}}{2Na[Ag(CN)_2]} + Zn \rightarrow Na_2[Zn(CN)_4] + 2Ag\downarrow$

(*xvi*) $\underset{\text{(silver glance)}}{Ag_2S} + 4NaCN \rightleftharpoons 2Na[Ag(CN)_2] + Na_2S$

$4Na_2S + 5O_2 + 2H_2O \rightarrow 2Na_2SO_4 + 4NaOH + 2S$

[Na_2S is converted into Na_2SO_4 to avoid reversibility of first reaction]

$2NaAg(CN)_2 + Zn \rightarrow \underset{\text{Sod. Zincocyanide}}{Na_2[Zn(CN)_4]} + 2Ag$

(*xvii*) $\underbrace{3HCl + HNO_3}_{\text{(aqua regia)}} \rightarrow NOCl + 2H_2O + 2(Cl)$

$Au + 2[Cl] \xrightarrow{HCl}$

$AuCl_3 \xrightarrow{HCl} \underset{\text{Chloroauric acid}}{HAuCl_4}$

(*xviii*) $2KMnO_4 + 3H_2SO_4 \rightarrow K_2SO_4 + 2MnSO_4 + 3H_2O + 5[O]$

$5H_2C_2O_4 + 5[O] \rightarrow 10CO_2 \uparrow + 5H_2O$

$2KMnO_4 + 3H_2SO_4 + 5(COOH)_2 \rightarrow K_2SO_4 + 2MnSO_4 + 10CO_2 + 8H_2O$

(*xix*) $K_4[Fe(CN)_6] + 6H_2SO_4 + 6H_2O \xrightarrow{\text{heat}} 2K_2SO_4 + FeSO_4 + 3(NH_4)_2SO_4 + 6CO$

(*xx*) $2KMnO_4 + 4KOH + MnO_2 \rightarrow 3K_2MnO_4 + 2H_2O$

(*xxi*) $\underset{\text{(yellow)}}{2K_2CrO_4} + H_2SO_4 \rightarrow \underset{\text{(orange red)}}{K_2Cr_2O_7} + K_2SO_4 + H_2O$

(*xxii*) $2MnO_4^{2-} + 4H^+ \rightarrow MnO_2 + MnO_4^- + 2H_2O$

(*xxiii*) $Fe_2(SO_4)_3 + 2KI \rightarrow 2FeSO_4 + K_2SO_4 + I_2$

(*xxiv*) $2CuSO_4 + SO_2 + 2H_2O + 2KCNS \rightarrow \underset{\text{Cuprous thiocyanate (white)}}{2CuCNS} \downarrow + K_2SO_4 + 2H_2SO_4$

78. (*i*) $\underset{(A)}{ZnCO_3} \xrightarrow{\Delta} \underset{(C)}{ZnO} + \underset{(B)}{CO_2}$

(*ii*) $\underset{(C)}{ZnO} + 2HCl \longrightarrow H_2O + \underset{\text{(soluble)}}{ZnCl_2}$

(*iii*) $2ZnCl_2 + K_4[Fe(CN)_6] \longrightarrow 4KCl + \underset{\text{(white ppt)}}{Zn_2[Fe(CN)_6]} \downarrow$

(*iv*) $\underset{(A)}{ZnCO_3} + HCl \longrightarrow \underset{\text{(soluble)}}{CO_2 + ZnCl_2}$

(*v*) $\underset{(B)}{CO_2} + Ca(OH)_2 \longrightarrow \underset{\text{(Milky)}}{CaCO_3} + H_2O$

(*vi*) $CaCO_3 + CO_2 + H_2O \longrightarrow \underset{\text{(soluble)}}{Ca(HCO_3)_2}$

(*vii*) $ZnCl_2 + H_2S \xrightarrow{NH_4OH} 2HCl + \underset{\text{(white) (D)}}{ZnS} \downarrow$

(*viii*) $ZnCl_2 + 2NaOH \longrightarrow 2NaCl + \underset{\text{(white) (E)}}{Zn(OH)_2} \downarrow$

(*ix*) $Zn(OH)_2 + 2NaOH \longrightarrow \underset{\text{sod. zincate (soluble)}}{Na_2ZnO_2} + H_2$

79. (*i*) Since the compound (A) on strong heating gives two oxides of sulphur (C and D) which might be SO_2 and SO_3, it must be a **sulphate.**

(*ii*) The reaction of compound (E) with thiocyanate to give blood red coloured compound (H) indicates that (E) must have **Fe^{3+} ion.**

Thus the compound (A) must be ferrous sulphate, $FeSO_4.7H_2O$, which explains all given reactions as below (Fe^{2+} ion of $FeSO_4$ is changed to Fe^{3+} during heating).

$\underset{\text{Ferrous sulphate (A)}}{FeSO_4.7H_2O} \xrightarrow{\text{heat}} \underset{\text{Ferrous sulphate}}{FeSO_4} + 7H_2O\uparrow$

$\underset{\text{Ferrous sulphate}}{2FeSO_4} \xrightarrow{\text{heat}} \underset{\text{Ferric oxide, (Blackish brown powder) (B)}}{Fe_2O_3} + \underset{(C)}{SO_2} + \underset{(D)}{SO_3}$

$\underset{(B)}{Fe_2O_3} + 6HCl \longrightarrow \underset{\text{Ferric chloride (Yellow solution), (E)}}{2FeCl_3} + 3H_2O$

$\underset{(E)}{2FeCl_3} + H_2S \longrightarrow \underset{\text{Apple green (G)}}{2FeCl_2} + 2HCl + \underset{\text{White turbidity (F)}}{S}$

$\underset{(E)}{FeCl_3} + 3NH_4CNS \longrightarrow \underset{\text{Ferric thiocyanate (blood red colour)(H)}}{Fe(CNS)_3} + 3NH_4Cl$

80. **(a, b)**

	No. of Unpaired e^-	Magnetic Nature
$_{57}La^{+3}$: $[Xe]_{54}\ 4f^0\ 5d^0\ 6s^0$	0	Diamagnetic
$_{58}Ce^{+4}$: $[Xe]_{54}\ 4f^0\ 5d^0\ 6s^0$	0	Diamagnetic
$_{70}Yb^{+2}$: $[Xe]_{54}\ 4f^{14}\ 5d^0\ 6s^0$	0	Diamagnetic
$_{71}Lu^{+3}$: $[Xe]_{54}\ 4f^{14}\ 5d^0\ 6s^0$	0	Diamagnetic
$_{57}La^{+2}$: $[Xe]_{54}\ 4f^0\ 5d^1\ 6s^0$	1	Paramagnetic
$_{58}Ce^{+3}$: $[Xe]_{54}\ 4f^1\ 5d^0\ 6s^0$	1	Paramagnetic
$_{70}Yb^{+3}$: $[Xe]_{54}\ 4f^{13}\ 5d^0\ 6s^0$	1	Paramagnetic
$_{71}Lu^{+2}$: $[Xe]_{54}\ 4f^{14}\ 5d^1\ 6s^0$	1	Paramagnetic

Chapter-9

Co-ordination Compounds

Part-A (JEE Main)

1. **(2)** Coordination isomerism is shown by compounds in with both cationic and anionic parts are complex ions, and there is interchange of ligands between the two metal centres.

So, we check each compound:

A. $[Ag(NH_3)_2][Ag(CN)_2]$

Here, both complex ions contain the same metal, Ag.

If we try to interchange ligands, we still get the same compound becuase both metal centres are identical.

So, coordination isomerism is not possible.

B. Here, cationic complex is $[Co(NH_3)_6]^{3+}$ and anionic complex is $[Cr(CN)_6]^{3+}$. Ligands can interchange to give: $[Cr(NH_3)_6][Co(CN)_6]$. This is a different compound. So, coordination isomerism is possible.

C. $[Co(NH_3)_6][Co(CN)_6]$

Both complexes contain the same metal, Co. On interchanging ligands, no new compound is formed. So, coordination isomerism is not possible.

D. $[Fe(NH_3)_6][Co(CN)_6]$. Ligand interchange gives :

$[Co(NH_3)_6][Fe(CN)_6]$. This is a different compound. So, coordination isomerism is possible.

E. $[Co(NH_3)_6][Fe(CN)_6]$. This is the coordination isomer of option D. So, coordination isomerism is possible. Hence, the complexes which show coordination isomerism are : B, D and E

2. **(1)**

$$\text{Molarity of } Br^- \text{ in solution} = \frac{0.4\times1}{4} = 0.1$$

$$\text{Molarity of } SO_4^{2-} \text{ in solution} = \frac{0.4\times1}{4} = 0.1$$

mol of Br^- in 2L of solution 'X' $= 0.1\times2 = 0.2$

mol of SO_4^{2-} in 2L of solution 'X' $= 0.1\times2 = 0.2$

$$AgNO_3 + Br^- \longrightarrow \underset{'Y'}{AgBr} + NO_3^-$$

$$BaCl_2 + SO_4^{2-} \longrightarrow \underset{'Z'}{BaSO_4}$$

mol of AgBr (Y) = mol of $Br^- = 0.2$

mol of $BaSO_4$(Z) = mol of $SO_4^{2-} = 0.2$

'Y' is AgBr and 'Z' is $BaSO_4$.

3. **(4)** $Cu \xrightarrow{\text{dil. HCl}}$ no reaction

$$CuSO_4 \xrightarrow{KCN} \underset{\text{Complex}}{[Cu(CN)_4]^{3-}} \xrightarrow{H_2S} \text{no sulphide ppt}$$

4. **(4)** Homoleptic complex of type $[Ma_6]$ cannot show geometrical as well as optical isomerism. (Where a $\Rightarrow$ monodentate ligand) *Cis*-platin and *trans*-platin has formula $[Pt(NH_3)_2Cl_2]$ which is a heteroleptic complex of platinum.

5. **(2)** Primary valency = Oxidation state

Secondary valency = Co-ordination number

6. **(1)** Among VO_2, V_2O_3, V_2O_5, CrO_3 and Mn_2O_7.

CrO_3 = Acidic; Mn_2O_7 = Acidic

$\therefore$ x = 2

$$[Co(H_2NCH_2CH_2NH_2)_3]_2(SO_4)_3 \rightleftharpoons 2[Co(H_2NCH_2CH_2NH_2)_3]^{3+} + 3SO_4^{2-}$$

Co have +3 charge so the primary valency = 3

$\therefore$ x + y = 5

7. **(3)**

$Co.5NH_3.ClSO_4$

$\underset{(A)}{[Co(NH_3)_5SO_4]Cl} \xrightarrow{AgNO_3} \underset{\text{(white ppt)}}{AgCl}$

$\underset{(B)}{[Co(NH_3)_5Cl]SO_4} \xrightarrow{BaCl_2} \underset{\text{(white ppt)}}{BaSO_4}$

(A)complex is $[Co(NH_3)_5(SO_4)]Cl$

(B)complex is $[Co(NH_3)_5Cl]SO_4$

Both (A) and (B) are Ionisation isomers.

8. **(3)** $[CrCl_3(Py)_3]$

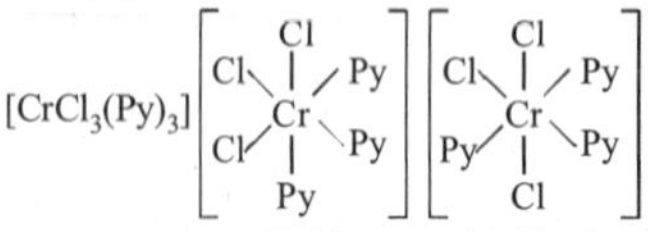

Facial Meridional

So total stereo isomer = 2

$[CrCl_2(ox)_2]^{3-}$

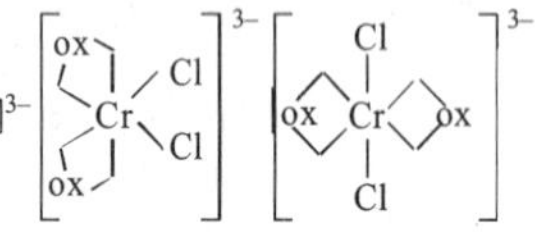

Cis-isomer (Optically active) *Trans*-isomer

Geometrical isomer = 2 (1 cis +1 trans)

Optical isomer = 3 (2 optically active + 1 optically inactive)

Stereoisomer = 3

9. **(1)**

(A) $Fe^{+6} \Rightarrow d^2$ (Homoleptic)

(B) $Fe^{+3} \Rightarrow d^5$ (Homoleptic)

(C) $Fe^{+2} \Rightarrow d^6$

(D) $Co^{+2} \Rightarrow d^7$ (Homoleptic)

(E) $Co^{+3} \Rightarrow d^6$

10. **(1)** $[Co(NH_3)_5Cl]Cl_2 \rightarrow$

$$\underbrace{[Co(NH_3)_5Cl]^{2+}(aq) + 2Cl^-(aq)}_{\text{3 ions in water}}$$

$$[Co(NH_3)_5Cl]Cl_2(aq) + 2AgNO_3(aq) \rightarrow [Co(NH_3)_5Cl](NO_3)_2(aq) + 2AgCl(s)$$

11. **(1)** Ma_3b_3 type complexes show Facial - Meridional isomerism

(1) $[Co(NH_3)_3Cl_3] \Rightarrow Ma_3b_3$

(2) $[Co(NH_3)_4Cl_2]^+ \Rightarrow Ma_4b_2$

(3) $[Co(en)_3]^{3+} \Rightarrow M(AA)_3$

(4) $[Co(en_2)Cl_2]^+ \Rightarrow M(AA)_2b_2$

a, b = NH_3, Cl^- (Monodentate ligands)

AA = en (Bidentate ligand)

12. **(3)** (A) $[Fe(CN)_5NO]^{-2} \rightarrow$ Heteroleptic, Fe^{+2}, $3d^6$, $t_{2g}^6 e_g^0$, d^2sp^3, Low spin

(3d series + SFL) (0) unpaired electron

(B) $[CoF_6]^{-3} \rightarrow$ Homoleptic, sp^3d^2, High spin, Co^{+3}, $3d^6$ (3d series + WFL) (4) unpaired e^-

(C) $[Fe(CN)_6]^{-4} \rightarrow$ Homoleptic Fe^{+2}, $3d^6$, d^2sp^3, $t_{2g}^6 e_g^0$ Low spin (3d series + SFL) (0) unpaired e^-

(D) $[Co(NH_3)_6]^{+3} \rightarrow$ Homoleptic, Co^{+3}, $3d^6$, d^2sp^3, $t_{2g}^6 e_g^0$, Low spin (3d series + SFL) (0) unpaired e^-

(E) $[Cr(H_2O)_6]^{+2} \rightarrow$ Homoleptic, Cr^{+2}, $3d^4$, sp^3d^2, High spin $t_{2g}^3 e_g^1$, (3d series + WFL) (4) unpaired

13. **(3)** As on reaction with excess of $AgNO_3$ solution 2 moles of AgCl formed.

So $2Cl^-$ ions should be ionisable.

$$[\overset{+3}{Co}(NH_3)_nCl]Cl_2 + \text{excess } AgNO_3 \rightarrow 2AgCl$$

oxidation state of Co : (2 moles)

$x + n(0) - 1 = +2 \Rightarrow x = +3$

Complex is octahedral so, n = 5

$\therefore$ x + n = 8

14. **(3)** The correct IUPAC name of $[PtBr_2(PMe_3)_2]$ is dibromobis (trimethyl phosphine) platinum (II).

15. **(2)** $[MABXL] \rightarrow sp^3 \rightarrow$ Tetrahedral

Tetrahedral complexes do not show geometrical isomerism as they have same relative position.

16. **(1)** Complex having all ligands same are called homoleptic complex. $[Ni(CN)_4]^{2-}$ has only CN^- ligands.

17. **(2)** Here, Mn is in +6 state and the suffix-ate is used as $[MnO_4]^{2-}$ is an anionic entity.

Thus, following the IUPAC convention, the correct IUPAC name will be Potassium tetraoxidomanganate (VI).

The term 'oxo' is a common name while 'oxido' is the IUPAC term for oxygen here.

18. **(2)** $[Cr(ox)_2ClBr]^{-3}$

- Trans structure has plane of symmetry. So no optical isomerism will be shown.
- Cis structure does not contain plane of symmetry. So two forms d as well as l will be shown.

19. **(1)** Ambidentate ligand are the ligand which has two different donor atoms. For example NO_2^- and NCS^-.

20. **(1)** $[MA_3B_3]$ type of compound exists are facial and meridonial isomer.

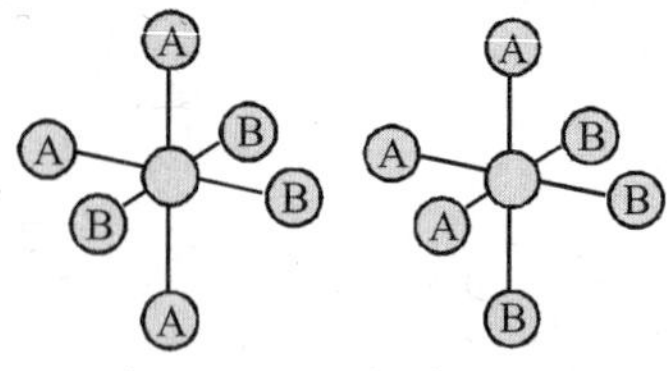

mer-isomer fac-isomer

21. (1) IUPAC name of $K_3[Co(C_2O_4)_3]$ is Potassium trioxalatocobaltate(III)

22. (3) $[Co(NH_3)_5NO_2]^{2+}$; $[Co(NH_3)_5ONO]^{2+}$
Two linkage isomers possible as NO_2 is an ambidentate ligand.

23. (4) The double salts dissociates comple-tely into the constituting ions on dissolution. This is possible for $FeSO_4 \cdot (NH_4)_2SO_4 \cdot 6H_2O$ and $K_2SO_4 \cdot Al_2(SO_4)_3 \cdot 24H_2O$.

24. (2) $\underset{0.01\ mol}{[Co(NH_3)_5SO_4]Br} + \underset{excess}{AgNO_3} \rightarrow \underset{0.01\ Mol}{AgBr\downarrow}$

$\underset{0.01\ mol}{[Co(NH_3)_5Br]SO_4} + \underset{excess}{BaCl_2} \rightarrow \underset{0.01\ Mol}{BaSO_4\downarrow}$

25. (1) Absence of symmetry elements (plane of symmetry centre of symmetry)

26. (4) Oxidation number of Co is + 3 in the given complex. So primary valency is 3.
It is an octahedral complex so secondary valency is 6.

27. (1) Only the counter ions are replaced easily by other reagents.

$[Co(NH_3)_5Br]SO_4 + BaCl_2 \rightarrow [Co(NH_3)_5Br]Cl_2 + BaSO_4\downarrow$ White ppt.

It does not have chloride ion as counter ion so it will not give white ppt with $AgNO_3$.

28. (3)

29. (3) $AgCl + 2NH_3 \rightarrow \underset{soluble}{[Ag(NH_3)_2]^+Cl^-}$

30. (1)

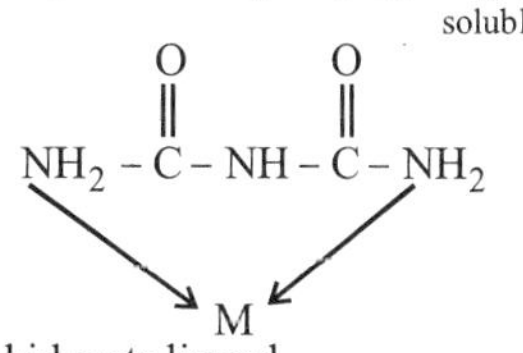

Biuret is bidenate ligand

31. (3) $[PtCl_2(NH_3)_2]$

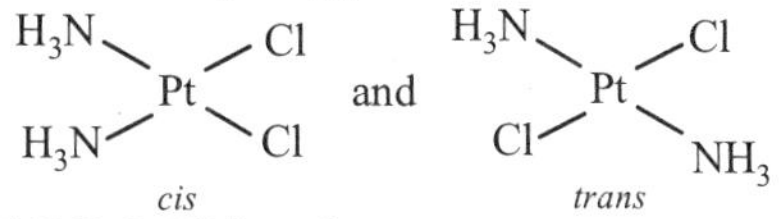

cis *trans*

$[Ni(CO)_4]$: All ligands are same.
Hence, zero geometrical isomers.
$[Ru(H_2O)_3Cl_3]$

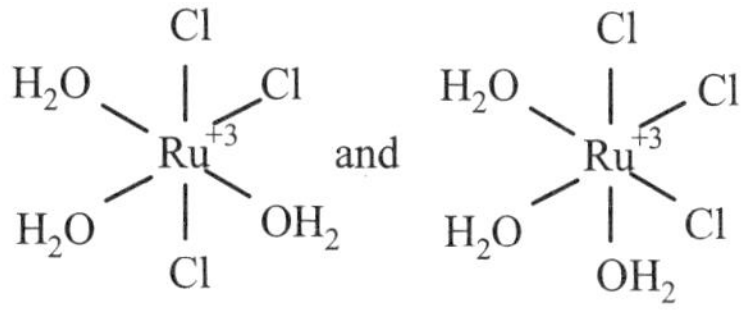

mer– *fac–*

$[CoCl_2(NH_3)_4]^+$

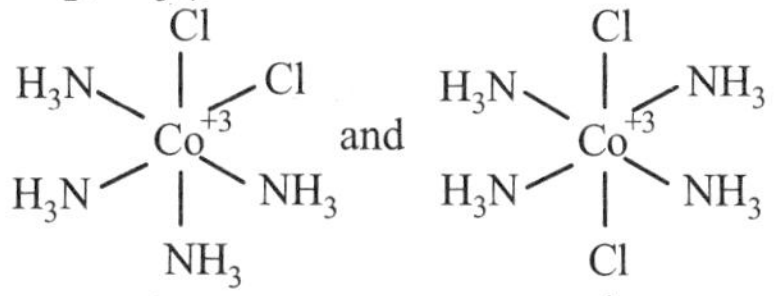

trans *cis*

32. (2) Ethylene diamine (en) is a chelating ligand. In $[Co(en)_3]Cl_3$, three rings are formed as chelating ring by ethylene diamine, so it will be most stable.

33. (4) *trans*-$[Co(en)_2Cl_2]^+$ (A) contains a plane of symmetry so (A) cannot be optically active, whereas, (B) *cis* $[Co(en)_2Cl_2]^+$ does not contain any plane of symmetry, so (B) will be optically active.

34. (4) $[Pt(en)(NO_2)_2]$

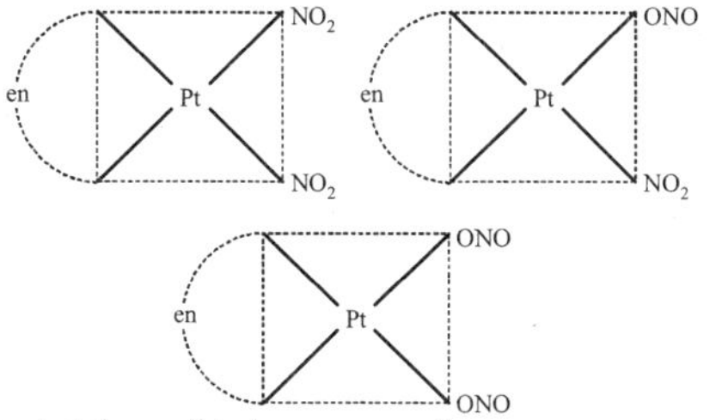

So, total possible isomers are 3.

35. (4) Only *cis*-$[CrCl_2(ox)_2]^{3-}$ shows optical isomerism while its *trans* form does not show optical isomerism due to presence of plane of symmetry.

36. (4) Conc. H_2SO_4 acts as dehydrating agent.
Molar mass of given complex = 266.5 g/mol.
On treating with conc. H_2SO_4 the mass lost by the

complex $= \frac{13.5}{100}(266.5) \approx 36\,g = 2$ moles of H_2O

Formula of the complex $= [Cr(H_2O)_4Cl_2]Cl \cdot 2H_2O$

37. (3)
(1) $[Ni(NH_3)_4(H_2O)_2]^{2+}$ shows geometrical isomerism.
(2) $[Ni(en)_3]^{2+}$ shows optical isomerism.
(3) $[Ni^{2+}(NH_3)_2Cl_2] \Rightarrow Ni^{2+} \Rightarrow 3d^8 4s^0$
$\Rightarrow sp^3$ hybridisation $\Rightarrow$ tetrahedral
So, $[Ni(NH_3)_2Cl_2]$ does not show isomerism.
(4) $[Pt(NH_3)_2Cl_2]$ shows geometrical isomerism.

38. (4) $\mu = \sqrt{n(n+2)}$ B.M. = 3.83 B.M.
$n = 3$ (n = No. of unpaired e^-)
Therefore, oxidation number of Cr should be +3.
Hence complex is $Cr(H_2O)_6Cl_3$.
Complex shows geometrical isomerism, therefore formula of complex is $[Cr(H_2O)_4Cl_2]Cl \cdot 2H_2O$.
IUPAC Name:
Tetraaquadichloridochromium(III) chloride dihydrate

39. (4)

40. (4) $[Ma_3b_3]$ type complex shows facial and meridional isomerism.
So, the complex $[Co(NH_3)_3(NO_2)_3]$ will show fac and mer-isomers.

41. (3) $[Pt(NH_3)_2Cl(NO_2)]$ and $[Pt(NH_3)_4ClBr]^{2+}$ are the Ma_4bc and Ma_2bc type of complexes.
Each shows two geometrical isomers, i.e., *cis* and *trans* isomers.

42. (1) $[Pt^{(+2)}(N^{(0)}H_3)_2Cl^{(-1)}(NH_2C^{(0)}H_3)]^+Cl^-$
Diamminechlorido (methanamine) platinum (II) chloride

43. (3) Tetrahedral (sp^3) and square planar (dsp^2) complexes do not show optical isomerism. $[MA_2B_2]$ will not exhibit optical isomerism in both conditions.

44. (4) $[Co(en)_2Cl]Cl$
Cl^- - monodentate ligand
en - bidentate ligand
$\therefore$ Co-ordination Number of Co = (2 × 2) + 1 = 5
$K_3[Al(C_2O_4)_3]$
$C_2O_4^{2-}$ - bidentate ligand
$\therefore$ Co-ordination Number of Al = 2 × 3 = 6

45. (4)

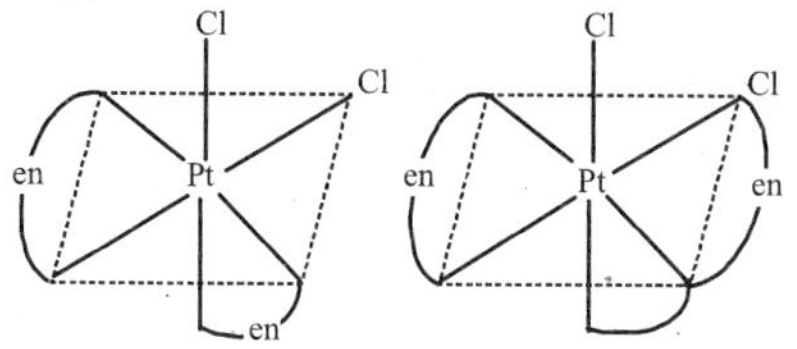

cis *trans*

cis-trans isomerism is possible with $[Pt(en)_2Cl_2]^{2+}$.
$[Cr(en)_2Ox]^+$ shows only optical isomerism. The other two complexes, i.e., $[Pt(en)Cl_2]$ and $[Zn(en)Cl_2]$ do not show stereoisomerism.

46. (2) The maximum possible denticites of the given ligand towards transition metal ion is 6 and towards inner transition metal ion (due to greater ionic radii and more atomic orbitals) is 8.

47. (3)

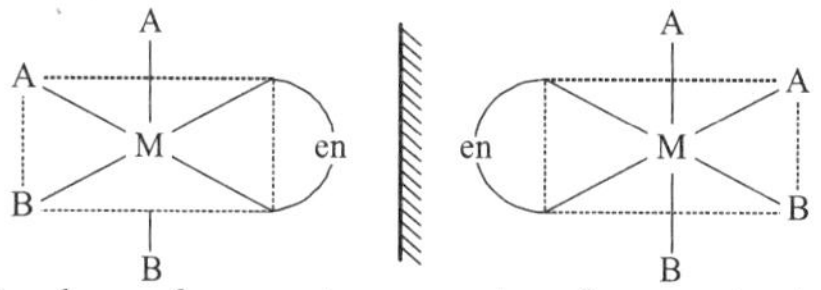

No plane of symmetry or centre of symmetry is present.
Hence, it will be optically active.

48. (2) It has four atoms (2 O and 2 N) containing lone pair of e^-, therefore, it will be able to donate these lone pairs and acts as a tetradentate ligand.

49. (4) $K_4[Th(C_2O_4)_4(H_2O)_2]$
$C_2O_4^{2-}$ (oxalato) : bidentate ligand
H_2O (aqua): monodentate
$\therefore$ Co-ordination no. of Th = 2 × 4 + 2 = 10

50. (4)

Possible complexes	No. of isomers
$[M(F)(Cl)(SCN)(NO_2)] \rightarrow$	3
(Two *cis* and one *trans*)	
$[M(F)(Cl)(SCN)(ONO)] \rightarrow$	3
$[M(F)(Cl)(NCS)(NO_2)] \rightarrow$	3
$[M(F)(Cl)(NCS)(ONO)] \rightarrow$	3
Total =	12

51. (1) Reaction for the given condition can be written as:
$[CoCl_6]^{3-} + 2(en) \longrightarrow [CoCl_2(en)_2]^+$
(1:2 mole ratio)(*cis-trans*-isomer)
A = optically active (*cis*-isomer), violet
B = optically inactive (*trans*-isomer), green

52. (2)

53. (3) $[Co(H_2O)Cl(en)_2]Cl$

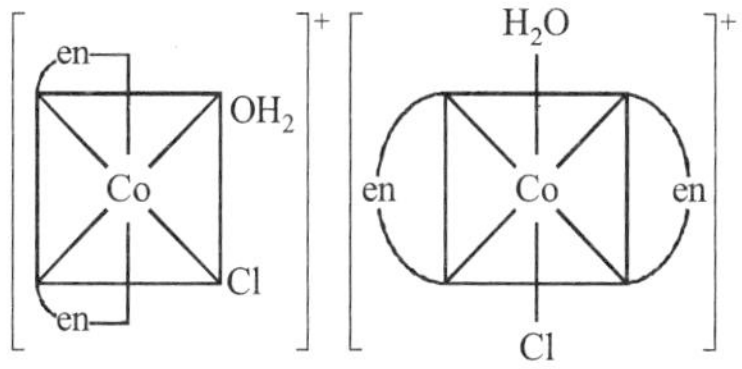

(Geometrical isomers)

54. (2)

55. (4) Moles of complex
$= \frac{Molarity \times Volume(mL)}{1000} = \frac{100 \times 0.1}{1000}$
= 0.01 mole

Moles of ions precipitated with excess of

$AgNO_3 = \frac{1.2 \times 10^{22}}{6.02 \times 10^{23}} = 0.02$ moles

$\therefore$ Number of Cl^- present in ionisation sphere

$= \dfrac{\text{Number of moles of ions precipitated}}{\text{Number of moles of complex}}$

$= \dfrac{0.02}{0.01} = 2$

It means $2Cl^-$ ions present in ionization sphere. Thus, formula of the complex is,

$[Co(H_2O)_5Cl]Cl_2 \cdot H_2O$

56. (1)

(1) $Cr\overset{2-}{O_2}\overset{-1}{Cl_2}$: Cr = +6; MnO_2 : Mn = + 4 (Highest oxidation states)

(2) $[Ni\overset{-1}{Cl_4}]^{2-}$: Ni = +2; $[Co\overset{-1}{Cl_4}]^{2-}$: Co = +2

(3) $[Fe\overset{-1}{(CN)_6}]^{3-}$: Fe = +3; $[Cu\overset{-1}{(CN)_4}]^{2-}$: Cu = +2

(4) $[Fe\overset{-1}{Cl_4}]^-$: Fe = +3; $Co_2\overset{-2}{O_3}$:Co = +3

57. (4) More equivalents of $AgNO_3$ aqueous solution will be consumed if complex will furnish more Cl^- ions in solution. Hence, complex $[Cr(H_2O)_6]Cl_3$ will consume more equivalents of aqueous solution of $AgNO_3$.

58. (1) Complexes having only one type of ligands are examples of homoleptic complex.

59. (4) Square planar complexes of type M[ABCD] form three isomers. Their position may be obtained by fixing the position of one ligand and placing at the *trans* position any one of the remaining three ligands one by one.

60. (4) (B) and (C) show only geometrical isomerism.

61. (2) Pentacyanomangnate (III) is the correct IUPAC name of $[Mn(CN)_5]^{2-}$.

62. (2) The two possible isomers for the given octahedral complex are $[M(NH_3)_5\ SO_4]$ Cl and $[M(NH_3)_5\ Cl]\ SO_4$. They respectively give chloride ion (indicated by precipitation with $AgNO_3$) and sulphate ion (indicated by precipitation with $BaCl_2$). Hence, the type of isomerism exhibited by the complex is ionisation isomerism.

63. (3) Octahedral coordination entities of the type Ma_3b_3 exhibit geometrical isomerism. The compound exists as facial and meridional isomers, both contain plane of symmetry.

64. (1) The compound shows linkage isomerism because the ligand in the compound is an ambidenate ligand that can bond at more than one atomic site. i.e., $:NCS^-$ and $:SCN^-$

65. (2) $[Cr(en)_2Br_2]Br$

dibromidobis(ethylenediamine) chromium (III) bromide.

66. (2) Shows cis and trans isomer.

67. (1) $\underset{2.675\ g}{CoCl_3 \cdot 6NH_3} \longrightarrow xCl^-$

$xCl^- + AgNO_3(\text{excess}) \longrightarrow \underset{4.78\ g}{x\,AgCl\downarrow}$

Number of moles of the complex $= \dfrac{2.675}{267.5} = 0.01$ mol

Number of moles of AgCl obtained

$= \dfrac{4.78}{143.5} = 0.03$ mol

$\therefore$ No. of moles of AgCl obtained

= 3 × No. of moles of complex

Hence, number of Cl atoms in ionisation sphere = 3

68. (2) Complexes of Zn^{2+} cannot show optical isomerism as they are tetrahedral complexes with plane of symmetry.

$[Co(H_2O)_4(en)]^{3+}$ has two planes of symmetry hence it is also optically inactive.

69. (3)

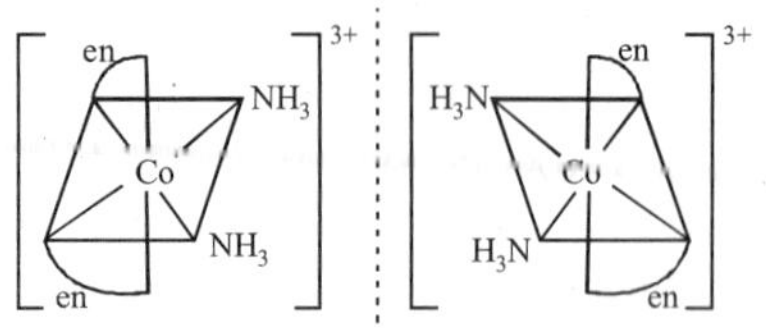

Enantiomers of *cis*-$[Co(en)_2(NH_3)_2]^{3+}$

70. (1) The SCN^- ion can coordinate through S or N atom giving rise to linkage isomerism

M ← SCN thiocyanato;

M ← NCS isothiocyanato.

71. (4) In the given complex we have two bidentate ligands (i.e, en and C_2O_4), so coordination number of E is 6

$(2 \times 2 + 1 \times 2 = 6)$

Let the oxidation state of E in complex be x, then $x - 2 = 1$ or $x = +3$, so its oxidation state is + 3.

72. (2) $[Co(NO_2)(NH_3)_5]Cl_2$

Pentaamminenitrito-N-cobalt (III) chloride

73. (4) Let, oxidation state of Cr i

$[Cr(NH_3)_4Cl_2]^+$ be x, $1 \times x + 4 \times 0 + 2(-1) = 1$. Therefore, $x = 3$.

74. (3) $K_3[Fe(CN)_6]$ is potassium hexacyanoferrate (III).

75. (3)

76. (4)

77. (1) CN^- ion acts good complexing as well as reducing agent.

78. (4) $Co(NH_3)_5Cl_3 \rightleftharpoons [Co(NH_3)_5Cl]^{2+} + 2Cl^-$

$\therefore$ Structure is $[Co(NH_3)_5Cl]Cl_2$.

Now $[Co(NH_3)_5Cl]Cl_2 + 2AgNO_3$

$\longrightarrow [Co(NH_3)_5Cl](NO_3)_2 + 2AgCl$

79. (1) Let the O. N. of Ni in $K_4[Ni(CN)_4]$ be x then

$4(+1) + x + (-1) \times 4 = 0 \Rightarrow 4 + x - 4 = 0; x = 0$

80. (2)

81. (5) Mole of $[Cr(H_2O)_4Cl_2]$ Cl in 100 ml 0.05 M solution = 0.005 mole

One mole of AgCl precipitated by one mole $[Cr(H_2O)_4Cl_2]Cl$

$[Cr(H_2O)_4Cl_2]Cl + AgNO_3$

$\longrightarrow AgCl\downarrow + [Cr(H_2O)_4Cl_2]NO_3$

Thus mole of AgCl precipitated = 0.005

$= 5 \times 10^{-3}$ moles $\therefore$ x = 5

82. (33) $[Cr(H_2O)_6]Cl_3 + \underset{(\text{excess})}{AgNO_3} \rightarrow \underset{(\text{ppt})}{3AgCl(s)}$

$n_{complex} = \dfrac{5.33}{266.5} = 0.02$ mol.

$n_{AgCl} = \dfrac{8.61}{143.5} = 0.06$ mol

$\dfrac{\text{no. of moles of complex reacted}}{\text{no. of moles of AgCl precipitated}} \times 100$

$= \dfrac{1}{3} \times 100 = 33.3$

83. (1) $[Pt(NH_3)(H_2O)BrCl] \Rightarrow$ [M a b c d]

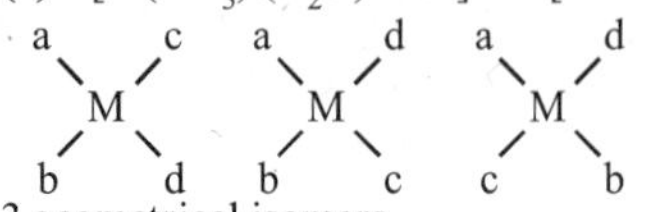

⇒ 3 geometrical isomers.

X = 3

$[CrCl_2(ox)_2]^{3-} \Rightarrow [M(AA)_2a_2]^{\pm n}$

⇒ Optically inactive

Y = 1

$[Co(NH_3)_3(NO_2)_3] \Rightarrow [M\ a_3\ b_3]$

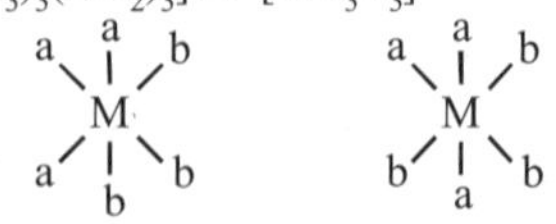

fac mer

Z = 2

X + Y + Z = 6.

84. (3) $\mu = 3.87$ BM

⇒ Number of unpaired e^- = 3

and it corresponds to 1 : 2 in electrolyte means the complex is : $[Cr(H_2O)_5Cl]Cl_2.H_2O$

Moles of complex $= \dfrac{2.75}{266.5} = 0.01$

Mass of AgCl formed = 0.01 × 2 × 143.5 = 2.87 g

85. (2) 1 lone pair hence 1 mole of AgCl.

$\therefore$ Complex is $[M(NH_3)_3Cl_3]$ Cl.

It shows 2 geometrical isomers (Ma_3b_3 type); facial (fac) and meridional (Mer)

86. (2) $FeCl_3 + KOH + H_2SO_4 \rightarrow \underset{(A)}{K_3[Fe(C_2O_4)_3]}$

$\Rightarrow [Fe(C_2O_4)_3]^{3-}$ is $[M(AA)_3]$ type complex. So total optical isomers = 2

87. (14)

88. (3) Ligands which have two diffeernt donor sites but at a time connects with only one donor site to central metal are ambidentate ligands.

Ambidentate ligands are NO_2^-; SCN^-; CN^-

89. (4) trans-$[Pt(en)_2Cl_2]^{2+}$ and trans-$[Cr(ox)_2Cl_2]^{3-}$ do not show optical isomerism as they are symmetrical in nature.

90. (4) $\underset{0.01M,\ 20\ mL}{[Cr(H_2O)_5Cl]Cl_2} + \underset{0.1\ M}{2AgNO_3} \rightarrow$

⇒ For 0.2 milli mole $AgNO_3$ required = 0.4 milli mole

$0.4 = 0.1 \times V \Rightarrow V = 4$ mL

91. (4) Ambidentate ligand has two donor site, so ambidentate ligand is SCN.

92. (4) Denticity of the ligand is 4.

93. (5) $[Co(NH_3)_4Cl_2]Cl \Rightarrow$ Gives 1 mole AgCl

$[Ni(H_2O)_6]Cl_2 \Rightarrow$ Gives 2 moles AgCl

$[Pt(NH_3)_2]Cl_2] \Rightarrow$ Gives No AgCl

$[Pd(NH_3)_4]Cl_2 \Rightarrow$ Gives 2 moles AgCl

Total number of moles of AgCl = 5 mole.

94. (9) Coordination no. = 6

Oxidation state is : $x - 4 = -1$

$x = +3$; Sum = 3 + 6 = 9.

95. (3) The isomers of $[Cu(en)_2(SCN)_2]$ are :

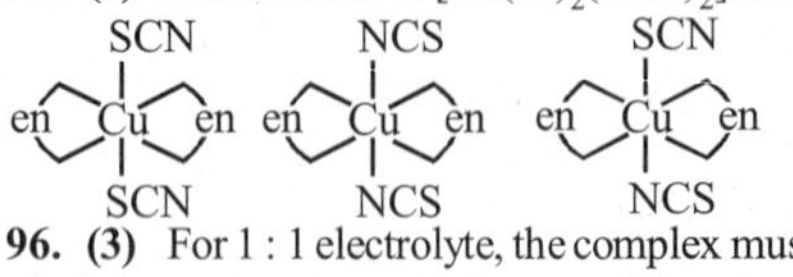

96. (3) For 1 : 1 electrolyte, the complex must be $[Co(NH_3)_4Cl_2]$ Cl. Oxidation number of Co in $[Co(NH_3)_4Cl_2]^+$ is +3.

97. (1) $[Fe(H_2O)_6]^{2+} + NO \longrightarrow [Fe(H_2O)_5(NO)]^{2+} + H_2O$

Reaction for brown ring test

NO ligand exists as NO^+

98. (5) $CoCl_3.xNH_3 + AgNO_3 \rightarrow AgCl\downarrow$ (2 mol)

2 moles of AgCl are ppt out that means 2 Cl^- ions are present as counter ions. Considering it as octahedral complex we have

$[Co(NH_3)_5Cl]Cl_2 + AgNO_3 \rightarrow AgCl\downarrow$

$x = 5$ (2 mol)

99. (1) Geometrical isomers have different spatial arrangement of atoms

(A) $CoCl_3.4NH_3 = [Co(NH_3)_4Cl_2]Cl$ can exhibit Geometrical Isomerism (GI).

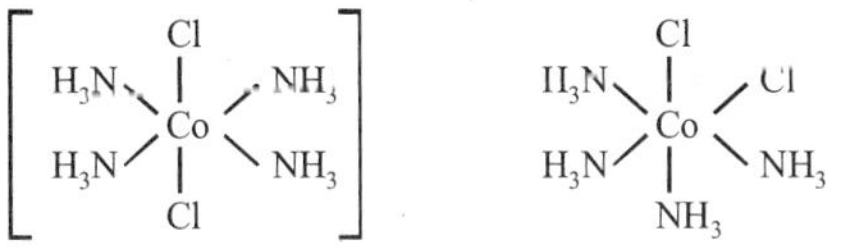

trans-isomer *cis*-isomer

100. (2) $\left[\overset{+1}{Ag}(NH_3)_2\right]^+ \left[\overset{+1}{Ag}(CN)_2\right]^-$

101. (1) One effective hydrogen bonded H_2O molecule.

102. (2) $MCl_3 \cdot 2L \Rightarrow$ octahedral complex (Given)

$MCl_3 \cdot 2L \xrightarrow{Ex.\ AgNO_3} 1$ mole of AgCl

1mole (Given)

Its means that one Cl^- ion is present in ionization sphere.

$\therefore$ Formula = $[MCl_2L_2]Cl$

For octahedral complex, coordination no. is 6.

$\therefore$ L acts as bidentate ligand.

103. (2) The number of optical isomers for $[Cr(C_2O_4)_3]^{3-}$ is two.

104. (5) Mohr's salt $\Rightarrow (NH_4)_2Fe(SO_4)_2 \cdot 6H_2O$

$K_2SO_4.Al_2(SO_4) \cdot 24H_2O$ Potash alum

$\Rightarrow KAl(SO_4)_2 \cdot 12H_2O$

Required ratio

$= \frac{6}{12} = 0.5 = 5\times10^{-1}, \frac{6}{24} = \frac{1}{4} = 2.5\times10^{-1}$

105. (2) $[Co(CN)_5(NC)]^{3-}$ cannot show geometrical isomerism.

106. (2) Triamminetrinitrocobalt (III) $\rightarrow [Co(NO_2)_3(NH_3)_3]$

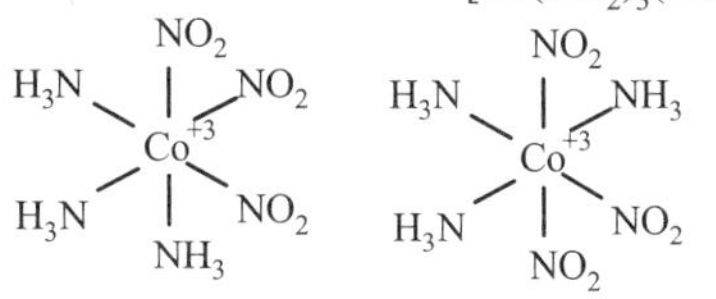

Two geometrical isomers (X = 2)

trioxalatochromate (III) ion $\rightarrow [Cr(C_2O_4)_3]^{3-}$

Zero geometrical isomer (Y = 0)

X + Y = 2 + 0 = 2

107. (6)

$3\left[Co(en)_2Cl_2\right]Cl + 3AgNO_3 \rightarrow 3AgCl$

(excess) (white ppt.)

Secondary valency (C.N.) of Co = 6

108. (0)

$[Cr(H_2O)_3(NH_3)_3]Cl_3 \longrightarrow [Cr(H_2O)_3(NH_3)_3]^{3+} + 3Cl^-$

(1 mole) (1 mole) (3mole)

$\xrightarrow{AgNO_3\ (Excess)} 3AgCl$

None of the chloride ion is directly bonded to metal ion. Hence, number of chloride ions satisfying the secondary valency of the metal ion is zero.

109. (3) $K_3[Cr(oxalate)_3] \equiv 3K^+[Cr(C_2O_4)_3]^{3-}$

Cr (24) $\Rightarrow 3d^54s^1$ or $Cr^{3+} \Rightarrow 3d^3$

$\therefore$ No. of unpaired electrons = 3

110. (6)

$$Fe^{3+} + 3K^+ + 3C_2O_4^{2-} \longrightarrow \underset{(A)}{K_3[Fe(C_2O_4)_3]}$$

Secondary valency of Fe in (A) is 6.

111. (2) In given complex, NH_3 is a neutral ligand. Each ethylene diamine replaces two NH_3 ligands. So, two ethylene diamine are required to replace all neutral monodentate (NH_3) ligands.

112. (3) Total number of stereoisomers possible for $[Co(ox)_2Br(NH_3)]^{2-}$ is 3.

113. (6)

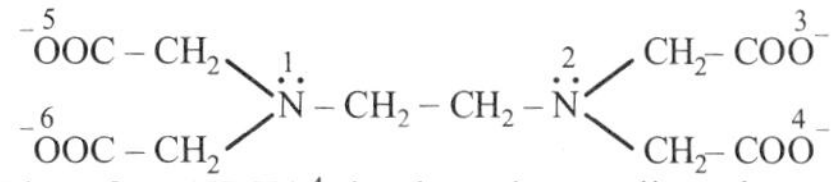

Therefore, $EDTA^{4-}$ is a hexadentate ligand.

114. (4) $[Ni(CO)_4]$ is sp^3, tetrahedral and diamagnetic.

115. (1) In a tetrahedral crystal field, the d orbitals split into two sets: e set contains the d_{z^2} and d_{x2-y2} obitals. These point between the ligands and experience less replusion hence lower in energy. t_2 set contains the d_{xy}, d_{yz} and d_{zx} orbitals. These point closer to the ligands and experience more repulsion hence higher in energy. so, the energy order is:-

$$d_{z^2} = d_{x^2-y^2} < d_{xy} = d_{yz} = d_{zx}$$

Hence, based on the energy order A, B and D are correct.

116. (3) In a tetrahedral complex, CFSE is calculated for $e^pt_2^q$ configuration by the formula.

$$CFSE = \left[-\frac{3}{5}p + \frac{2}{5}q\right]\Delta_t$$

For

$d^2 : \underbrace{e^{1.1}}_{p=2} \underbrace{t_2^{0,0,0}}_{q=0}$; C.F.S.E $= \left[\frac{-3}{5}(2)\right]\Delta t = -1.2\Delta t$

For

$d^4 : \underbrace{e^{1.1}}_{p=2} \underbrace{t_2^{1,1,0}}_{q=2}$; C.F.S.E $= \left[\frac{-3}{5}(2) + \frac{2}{5}(2)\right]\Delta t = -0.4\Delta t$

For

$d^6 : \underbrace{e^{2.1}}_{p=3} \underbrace{t_2^{1,1,1}}_{q=3}$; C.F.S.E $= \left[\frac{-3}{5}(3) + \frac{2}{5}(3)\right]\Delta t = -0.6\Delta t$

For

$d^8 : \underbrace{e^{2.2}}_{p=4} \underbrace{t_2^{2,1,1}}_{q=4}$; C.F.S.E $= \left[\frac{-3}{5}(4) + \frac{2}{5}(4)\right]\Delta t = -0.8\Delta t$

A - III, B - IV, C - I, D - II

117. (3)

(A) en is a bidentate ligand while NH_3 is a monodentate ligand. Complex with en form chelate rings which gives extra stability.

(B) So stability : $[Cu(NH_3)_4]^{2+} < [Cu(en)_2]^{2+}$

(E) NO_2^- and SCN^- both are ambidentate ligands. NO_2^- can bind via N (Nitro) or O (nitrito) and SCN^- can bind via S (thiocyanato) or N (isothiocyanato).

118. (1) Statement I : Transition metal have unpaired d-electrons that participate in metallic bonding in solid state.

More are the number of unpaired electrons. stronger is metal-metal bonding so more energy require to break them thus higher enthalpy of atomisation.

Statement II :

$[Fe(H_2O)_6]^{3+}$ is octahedral complex.

Ligands approach along x, y, z axes.

Thus,

$[t_{2g}(d_{xy} = d_{yz} = d_{zx}) < e_g(d_{x^2-y^2} = d_{x^2})]$

$[NiCl_4]^{2-}$ is tetrahedral complex.

Ligands approach between axes.

Thus,

$[e(d_{x^2-y^2} = d_{z^2}) < t_2(d_{xy} = d_{yz} = d_{zx})]$.

119. (1) Statement-I:

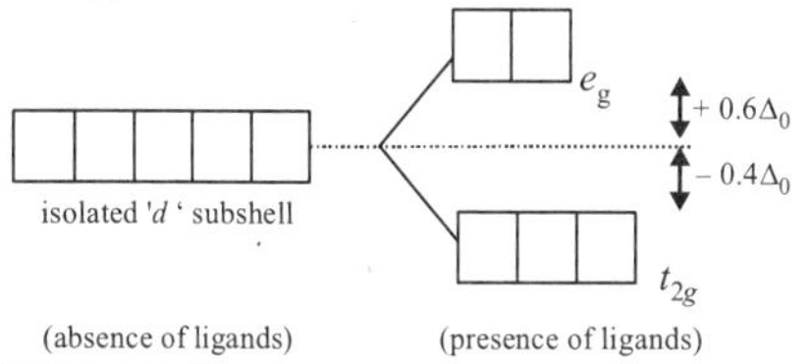

Statement -II:

All five *d*-orbital have equal energy (degenrate state) when ligand approach the electrostatic field of ligands cause splitting of *d*-orbital. This remove degenracy.

120. (1)

(A) $Ni^{2+} \Rightarrow 3d^8$, SFL, $t_{2g}^{2.2.2}\ e_g^{1.1}$

$\Rightarrow$ unpaired $e^- = 2$

(B) $Ni^{2+} \Rightarrow 3d^8$, WFL, tetrahedral, $e^{2.2}\ t_2^{2.1.1}$

$\Rightarrow$ unpaired $e^- = 2$

(C) $Ni^{2+} \Rightarrow 3d^8$, SFL, $t_{2g}^{2.2.2}\ e_g^{1.1}$

$\Rightarrow$ unpaired $e^- = 2$,

As $\Delta_0 > \Delta_t$ and Ligand strength : en > NH_3 > Cl^-

Strength of ligand increases, Δ increases, hence more is the frequency required for absorption.

121. (4)

Spin only magnetic moment, $\mu = \sqrt{n(n+2)}$

where n = no. of unpaired e^-.

H_2O is weak field ligand.

$\therefore [Cr(H_2O)_6]^{+2} \Rightarrow Cr^{+2}(3d^4)$; $n = 4$

$\mu = \sqrt{4(4+2)} \simeq 4.90$ BM(III)

Similarly,

$Co^{+2}(3d^7)$; $n = 3$ $\mu = \sqrt{3(3+2)} \simeq 3.87$ BM (I)

$Cu^{+2}(3d^9)$; $n = 1$ $\mu = \sqrt{1(1+2)} \simeq 1.73$ BM (IV)

$Mn^{+2}(3d^5)$; $n = 5$ $\mu = \sqrt{5(5+2)} \simeq 5.92$ BM (II)

122. (1) Evaluating statement I:

Sc : [Ar] $3d^14s^2 \Rightarrow Sc^{+3}$ is [Ar]

Removal of 3rd electron gives highly stable noble gas configuration.

Hence, it will have least ionisation energy.

Mn : [Ar] $3d^5\,4s^2 \Rightarrow Mn^{+3}$ is [Ar] $3d^4$

Cu : [Ar] $3d^9\,4s^2 \Rightarrow Cu^{+3}$ is [Ar] $3d^8$

Zn : [Ar] $3d^{10}\,4s^2 \Rightarrow Zn^{+3}$ is [Ar] $3d^9$

Removal of 4th electron is difficult because it has to be removed from fully filled *d*-orbital which requires very high ionisation energy.

Sc < Mn < Cu < Zn

Evaluating statement II:
CFSE $\Rightarrow$ SFL > WFL.
(en > H_2O)
CFSE $\propto$ charge on metal
$Co^{+2} < Co^{+3}$
$[Co(H_2O_6]^{2+} < [Co(H_2O_6]^{3+} < [Co(en)_3]^{3+}$

123. (4) All complexes contain Cr^{+3}, the only variation is ligand.
CFSE is proportional to strength of ligand followed by chelation.
CN^- is stronger ligand than en
$CN^- > en > H_2O > F^-$

124. (4) Magnetic moment depends upon the number of unpaired electron present in the molecule.
(A) $[MnBr_4]^{2-}$
$Mn^{2+} \rightarrow 3d^5\,4s^0$; Br is a weak field ligand
Number of unpaired electrons = 5.
(B) $[Cu(H_2O)_6]^{2+}$
$Cu^{2+} \rightarrow 3d^9\,4s^0$ (H_2O) is a weak field ligand
Number of unpaired electron = 1.
(C) $[Ni(CN)_4]^{2-}$
$Ni^{2+} \rightarrow 3d^8\,4s^0$; CN is a strong field lignad
Number of unpaired electron = 0
(D) $[Ni(H_2O)_6]^{2+}$
$Ni^{2+} \rightarrow 3d^8\,4s^0$; H_2O is a weak field ligand
Number of unpaired electrons = 2.
Thus increasing order of magnetic order is
$$\underset{0}{C} < \underset{1}{B} < \underset{2}{D} < \underset{5}{A}$$

125. (3)
$[Ni(CN)_4]^{2-} \rightarrow 3d^8 \rightarrow$ diamagnctic $\rightarrow dsp^2$
$[Ni(CO)_4] \rightarrow 3d^{10} \rightarrow$ diamagnetic $\rightarrow sp^3$
$[NiCl_4]^{2-} \rightarrow 3d^8 \rightarrow e^{2,2}\,t_2^{2,1,1} \rightarrow$ paramagnetic $\rightarrow sp^3$.

126. (4) Wavelength of light absorbed increases as C.F.S.E of complex decreases with decreasing strength of the ligand.
Order of ligand strength : $F^- < H_2O < NH_3 < CN^-$
$\Rightarrow$ Correct wavelength order
V > II > IV > I > III

127. (1) Paramagnetic species :
$[CoF_6]^{3-}$, $[TiF_6]^{3-}$, $[Fe(CN)_6]^{3-}$
Diamagnetic species : V_2O_5 (V = + 5 State)
In $K_4[Fe(CN)_6] \Rightarrow$ No. of unpaired electron = 0
$K_3[Fe(CN_6)] \Rightarrow$ No. of unpaired electron = 1
$[Fe(H_2O)_6]SO_4.H_2O$
$\Rightarrow$ No. of unpaired electron = 4
$[Fe(H_2O)_6]Cl_3 \Rightarrow$ No. of unpaired electron = 5

128. (3) In $K_3[Co(CO_3)_3]$
Co^{3+}, $3d^6$, CO_3^{2-} is WFL, hence pairing of *d* electrons is not possible.
$\therefore$ $3d^6$ = 2, 1, 1, 1, 1, sp^3d^2, octahedral
$\therefore$ n = 4
Hence $\mu = \sqrt{n(n+2)} = \sqrt{24} = 4.9$ BM
In $[Ni(CN)_4]^{2-}$
Ni^{2+}; $3d^8$, CN^{2-} is SFL, pairing of *d* electrons is possible
$\therefore$ $3d^8$ = 2, 2, 2, 2, 0, dsp^2, square planar
$\therefore$ n = 0; μ = 0
In $[MnBr_4]^{2-}$
Mn^{2+}; $3d^5$, Br^- is WFL
So $3d^5$ = 1, 1, 1, 1, 1; sp^3, tetrahedral
$\therefore$ n = 5; μ = 5.9 BM; In $[CoF_6]^{3-}$
Co^{3+}; $3d^6$; F^- is WFL
So $3d^6$ = 2, 1, 1, 1, 1; sp^3d^2, octahedral
$\therefore$ n = 4; μ = 4.9 BM

129. (2)
(A) Co^{3+} : - $3d^6$, $NH_3 \Rightarrow$ strong field ligand
Hybridisation $\Rightarrow d^2sp^3$, Inner orbital complex
(B) Mn^{3+} : - $3d^4$, $Cl^- \Rightarrow$ weak field ligand
Hybridisation $\Rightarrow sp^3d^2$, Outer orbital complex
(C) Co^{3+} : - $3d^6$, $F^- \Rightarrow$ weak field ligand
Hybridisation $\Rightarrow sp^3d^2$, Outer orbital complex
(D) Fe^{3+} : - $3d^5$, $F^- \Rightarrow$ weak field ligand
Hybridisation $\Rightarrow sp^3d^2$, Outer orbital complex
(E) Ni^{2+} : - $3d^8$, $CN^- \Rightarrow$ strong field ligand
Hybridisation $\Rightarrow dsp^2$, Inner orbital complex

130. (3)

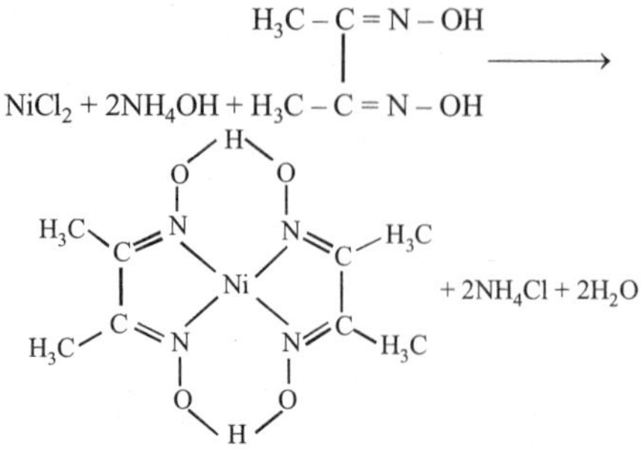

Complex of red colour
(Stable form of complex)

In the above complex, Ni is present in +2 oxidation number.
(A) It is rosy red ppt
(B) It is precipitated in basic medium
(C) Ni^{+2} : $3d^8$; Hybridisation : dsp^2; Unpaired e^- = 0; Geometry : Square planar
(D) N–Ni–N Bond angle is close to 90°
(E) 2 five membered metal containing rings are present.

131. (2) Statement 1: True
Strength of ligand : $Cl^- > Br^-$
$\Delta_t : [CoCl_4]^{2-} > [CoBr_4]^{2-}$
$E_{absorbed} : [CoCl_4]^{2-} > [CoBr_4]^{2-}$
Statement 2: False
Strength of ligand : $I^- < Cl^-$
$\Delta_t : [CoI_4]^{2-} < [CoCl_4]^{2-}$

132. (2) $[Ni(PPh_3)_2Cl_2]$
Ni^{2+} : $4s^0 3d^8$
Given : Paramagnetic complex hence it must be tetrahedral.
(A) Tetrahedral complex does not show geometrical isomerism.
(B) Complex is blue in colour.
(C) Calculated spin only magnetic moment of the complex is 2.84 B.M.
(D) Crystal field splitting of Ni^{2+} is $e^4t_2^{\,4}$.
C.F.S.E. = $-0.6\,\Delta_t(4) + 0.4\,\Delta_t(4)$
= $-0.8\,\Delta_t$ (not $-0.8\Delta_0$)
(E) $Ni(CO)_4$ also tetrahedral
Hence only A, B, D correct.

133. (4)

Metal	No. of e^- in *d* orbitals	Electronic distribution in e_g and t_{2g} orbital	Unpaired electron
Co^{3+}	$3d^6$	$t_{2g}^{2,2,2}e_g^{0,0}$	0
Fe^{3+}	$3d^5$	$t_{2g}^{2,2,1}e_g^{0,0}$	1
Cr^{3+}	$3d^3$	$t_{2g}^{1,1,1}e_g^{0,0}$	3
Mn^{3+}	$3d^4$	$t_{2g}^{2,1,1}e_g^{0,0}$	2

$\therefore$ Correct decreasing order of unpaired electron is
$Cr^{3+} > Mn^{3+} > Fe^{3+} > Co^{3+}$

134. (1) $[Cr(H_2O)_6]^{2+} \Rightarrow t_{2g}^3e_g^1$
CFSE = $(-0.4 \times 3 + 0.6 \times 1)\,\Delta_0$
= $(-1.2 + 0.6)\ \Delta_0 = -0.6\ \Delta_0$
$[Mn(H_2O)_6]^{2+} \Rightarrow t_{2g}^3e_g^2$
CFSE = $(-0.4 \times 3 + 0.6 \times 2)\,\Delta_0 = 0$
Statement I is correct
I. $K_3[Fe(CN)_6] \Rightarrow Fe^{3+} \Rightarrow 3d^5$
Unpaired e^- (n) = 1
II. $Na_4[Fe(CN)_6] \Rightarrow Fe^{2+} \Rightarrow 3d^6$
n = 0; $\mu_I > \mu_{II}$
Statement II is correct.

135. (4) In $[Cu(NH_3)_4]^{2+} \Rightarrow Cu^{+2}$ show d^9 electronic configuration and has dsp^2 hybridisation, one unpaired electron.
In $[Ni(en)_3]^{2+} \Rightarrow Ni^{+2}$ has d^8 electronic configuration and has sp^3d^2 hybridisation, two unpaired electrons.
In $[Ni(NH_3)_6]^{2+} \Rightarrow Ni^{+2}$ has d^8 electronic configuration and has sp^3d^2 hybridisation, two unpaired electrons.
In $[Mn(H_2O)_6]^{2+} \Rightarrow Mn^{+2}$ has d^5 electronic configuration and has sp^3d^2 hybridisation, five unpaired electrons.
So statement I is true.
$[Ni(CO)_4]$ (diamagnetic)
$[NiCl_4]^{-2}$ (paramagnetic)
Statement II is false.
$[Ni(CN)_4]^{-2}$ (diamagnetic)

136. (3)
(A) $[Ni(CO)_4]$, Ni^0 : $[Ar]3d^8\,4s^2$
Valence orbitals of Ni^0 in pre-hybridisation state :

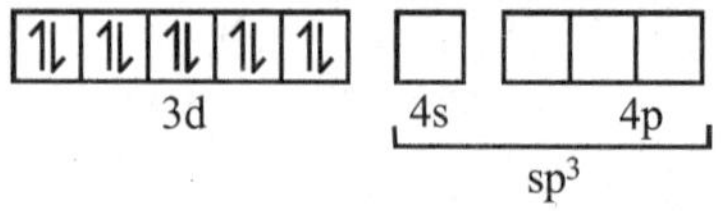

Tetrahedral, Diamagnetic, μ = 0 B.M.
(B) $[Ni(CN)_4]^{2-}$, Ni^{+2} : $[Ar]3d^84s^0$
Valence orbitals of Ni^{+2} in pre-hybridisation state :

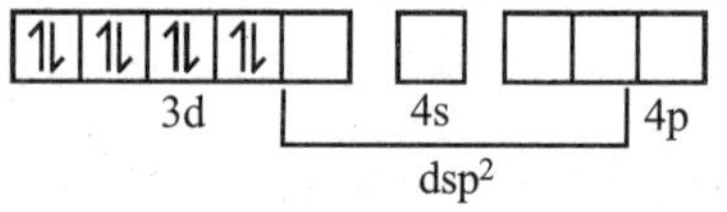

Square planar, Diamagnetic, μ = 0 B.M.
(C)$[NiCl_4]^{2-}$, Ni^{+2}, $[Ar]3d^84s^0$
Valence orbitals of Ni^{+2} in ground state :

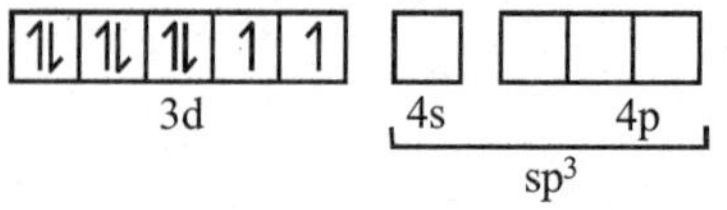

Tetrahedral, paramagnetic, $\mu = \sqrt{8}$ = 2.8 B.M.
(D)$[MnBr_4]^{2-}$, Mn^{+2} : $[Ar]3d^5$
Valence orbitals of Mn^{+2} in ground state :

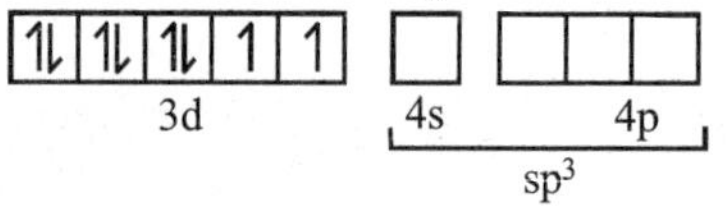

Tetrahedral, paramagnetic, $\mu = \sqrt{35}$ = 5.9 B.M.

137. (4)
(i)$[Co(NH_3)_6]^{3+}$
$Co^{3+} \Rightarrow 3d^6$; $t_{2g}^6e_g^0$
strong field ligand, d^2sp^3 hybridisation

(ii) SF_6

[1] [1|1|1] [1|1| | |]

3s 3p 3d

sp^3d^2 hybridisation

(iii) $[CrF_6]^{3+}$

$Cr^{3+} \Rightarrow 3d^3$; $t_{2g}^3e_g^0$, d^2sp^3 hybridisation

(iv) $[CoF_6]^{3-}$

$Co^{3+} \Rightarrow 3d^6$; $t_{2g}^4e_g^2$, sp^3d^2 hybridisation

(v) $[Mn(CN)_6]^{3-}$

$Mn^{3+} \Rightarrow 3d^4$; $t_{2g}^4e_g^0$,

strong field ligand, d^2sp^3 hybridisation

(vi) $[MnCl_6]^{3-}$

$Mn^{3+} \Rightarrow 3d^4$; $t_{2g}^4e_g^1$, sp^3d^2 hybridisation

Thus total number of sp^3d^2 hybridised molecule is 3.

138. (3)

$Cu^+ \Rightarrow 3d^{10} \Rightarrow$ [⇅|⇅|⇅|⇅|⇅] $\Rightarrow$ 0 BM

$Cu^{2+} \Rightarrow 3d^9 \Rightarrow$ [⇅|⇅|⇅|⇅|1] $\Rightarrow$ 1.73 BM

$Cr^{2+} \Rightarrow 3d^4 \Rightarrow$ [1|1|1|1|] $\Rightarrow$ 4.89 BM

$Cr^{3+} \Rightarrow 3d^3 \Rightarrow$ [1|1|1| |] $\Rightarrow$ 3.87 BM

So order :

$Cr^{2+} > Cr^{3+} > Cu^{2+} > Cu^+$

139. (1)

			Ligand field		Unpaired electrons
$[Fe(CN)_6]^{3-}$	Fe^{3+}	$3d^5$	strong	$t_{2g}^5\ e_g^0$	1
$[FeF_6]^{3-}$	Fe^{3+}	$3d^5$	weak	$t_{2g}^3\ e_g^2$	5
$[CoF_6]^{3-}$	Co^{3+}	$3d^6$	strong	$t_{2g}^4\ e_g^2$	4
$[Mn(CN)_6]^{3-}$	Mn^{3+}	$3d^4$	strong	$t_{2g}^4\ e_g^0$	2

140. (4) Most stable is $[Fe(C_2O_4)_3]^{3-}$ due to Chelation effect.

$[Fe(C_2O_4)_3]^{3-}$: [Ar] $t_{2g}^5e_g^0 \Rightarrow X = 5$ So compound is V_2O_5.

V_2O_5 is amphoteric.

141. (4) F^- is a weak-field ligand while CN^- and CO are strong-field ligands.

$[FeF_6]^{3-}$: [Ar]

3d [1|1|1|1|1] 4s [ ] 4p [| |]

5 unpaired electrons

$[CoF_6]^{3-}$: [Ar]

3d [⇅|1|1|1|1] 4s [ ] 4p [| |]

4 unpaired electrons

$[Ni(CN)_4]^{2-}$: [Ar]

3d [⇅|⇅|⇅|⇅|] 4s [ ] 4p [| |]

Zero unpaired electrons

$[Ni(CO)_4]$: [Ar]

3d [⇅|⇅|⇅|⇅|⇅] 4s [ ] 4p [| |]

Zero unpaired electrons

142. (4)

(1) $[Fe(CN)_6]^{-4}$ $Fe^{2+} \Rightarrow 3d^64s^0$

$\overset{\ominus}{CN}$ is SFL — e_g [|] ; t_{2g} [⇅|⇅|⇅] $\mu = 0$

CFSE $= -2.4\Delta_0$

(2) $[Co(NH_3)_6]^{3+}$ $Co^{3+} \Rightarrow 3d^64s^0$

NH_3 is strong field ligand — e_g [1|1] ; t_{2g} [⇅|1|1]

$= [-0.4 \times 6 + 0.6 \times (0)]\Delta_0 = -2.4\,\Delta_0$

(3) $[FeF_6]^{4-}$ $Fe^{2+} \Rightarrow 3d^64s^0$

e_g [1|1] ; t_{2g} [⇅|1|1]

$\mu = \sqrt{24}$ B.M. = 4.89B.M.

CFSE $= (-0.4 \times 4 + 0.6 \times 2)\Delta_0 = -1.2\Delta_0$

(4) $[Mn(SCN)_6]^{4-}$

$Mn^{2+} \Rightarrow 3d^5\,4s^0$

$\overset{\ominus}{SCN}$ is weak field — [1|1] ; [1|1|1]

$\mu = \sqrt{35}$ B.M. = 5.96 B.M.

CFSE $= (-0.4 \times 3 + 0.6 \times 2)\Delta_0$

So $\Delta_0 = 0$

143. (4)

(A) $[Mn(CN)_6]^{3-}$; Mn^{+3}

e_g^0 [|]

↕ $\Delta_0 > P$ Paramagnetic

t_{2g}^4 [⇅|1|1]

3d

(B) $[Co(NH_3)_6]^{3+}$; Co^{+3}

e_g^0 [|]

↕ $\Delta_0 > P$ Diamagnetic

t_{2g}^4 [⇅|⇅|⇅]

3d

(C) $[Fe(CN)_6]^{4-}$; Fe^{+2}

e_g^0 [|]

↕ $\Delta_0 > P$ Diamagnetic

t_{2g}^4 [⇅|⇅|⇅]

3d

(D) $[Co(H_2O)_3F_3]$; Co^{+3}

e_g^2 [1|1]

↕ $\Delta_0 > P$ Paramagnetic

t_{2g}^4 [⇅|1|1]

3d

144. (1) $E = h\nu = \dfrac{hc}{\lambda} \Rightarrow E \propto \dfrac{1}{\lambda}$

Here all Co in +3 oxidation state. So, as the ligand field strength ↑, CFSE ↑

Order of field strength of ligand : $CN^- > NH_3 > H_2O > Cl^-$

$\therefore$ CFSE order : C > B > A > D

$\Rightarrow$ Wavelength order : D > A > B > C

145. (4) Given magnetic moment = 4.9 B.M.

Magnetic moment $= \sqrt{n(n+2)}$B.M.

Magnetic moment = 4.9 BM (given)

$\therefore 4.9 = \sqrt{n(n+2)} \Rightarrow n = 4$

(A) $_{24}Cr^{2+} \Rightarrow [Ar]3d^4$(4 unpaired e^-)

(B) $_{26}Fe^{2+} \Rightarrow [Ar]3d^6$(4 unpaired e^-)

(C) $_{26}Fe^{3+} \Rightarrow [Ar]3d^5$(5 unpaired e^-)

(D) $_{27}Co^{2+} \Rightarrow [Ar]3d^7$(3 unpaired e^-)

(E) $_{25}Mn^{3+} \Rightarrow [Ar]3d^4$(4 unpaired e^-)

146. (2) $Mn^{3+} \Rightarrow 3d^4 \Rightarrow t_{2g}^3e_g^1$

Cl^- is a weakfield ligand, so electron pairing occurs.

Hybridisation is sp^3d^2 with 4 unpaired electron and paramagnetic.

147. (1) Among the given complexes $[Co(en)_3]^{3+}$ has highest CFSE

Co^{3+} has $3d^6$ configuration.

en is a strong field ligand so the complex formed with low spin configuration in octahedral field.

$= t_{2g}^6e_g^0$

CFSE $= (6 \times (-0.4) + 0 \times 0.6)\,\Delta_0 = -2.4\,\Delta_0$

148. (4) For octahedral complex :

If $\Delta_0 < P$, then high spin complexes are formed due to weaker field strength of the ligand.

If $\Delta_0 > P$, then low spin complexes are formed. But in tetrahedral complex $\Delta_t = \dfrac{4}{9}\,\Delta_0$.

Thus, $\Delta_t < P$ and mainly high spin complexes are formed and low spin complexes are rarely formed.

149. (2)

$K_3\left[Fe(OH)_6\right]; Fe^{+3} \Rightarrow d^5 \Rightarrow t_{2g}^3\ e_g^2$

$\Rightarrow$ 5 unpaired electron

$\Rightarrow \mu = \sqrt{n(n+2)} = \sqrt{35} = 5.92$

$K_4\left[Fe(OH)_6\right]; Fe^{2+} \Rightarrow d^6 \Rightarrow t_{2g}^4\ e_g^2$

$\Rightarrow$ 4 unpaired electron

$\mu = \sqrt{4(4+2)} = \sqrt{24} = 4.90$

150. (4)

(A) **$[MnBr_4]^{2-}$**

$Mn^{+2} \Rightarrow [Ar]\,3d^5$

In presence of weak ligand field

$\Rightarrow$ [Ar] 3d [1|1|1|1|1] 4s [ ] 4p [| |]

$\Rightarrow$ sp^3 hybridization, paramagnetic in nature

(B) **$[FeF_6]^{3-}$**

$Fe^{+3} \Rightarrow [Ar]\,3d^5$

In presence of weak ligand field

$\Rightarrow$ [Ar] 3d [1|1|1|1|1] 4s [ ] 4p [| |] 4d [| | | |]

$\Rightarrow$ sp^3d^2 hybridizations, & paramagnetic

(C) **$[Co(C_2O_4)_3]^{3-}$**

$Co^{+3} \Rightarrow [Ar]\,3d^6$

In presence of strong ligand field

$\Rightarrow$ [Ar] 3d [⇅|⇅|⇅| |] 4s [ ] 4p [| |]

$\Rightarrow$ d^2sp^3 hybridization, & diamagnetic

(D) **$[Ni(CO)_4]$**

$Ni^0 \Rightarrow [Ar]\,3d^84s^2$

In presence of strong ligand field

$\Rightarrow$ 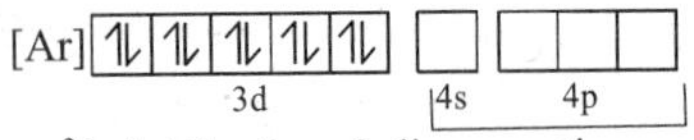

$\Rightarrow$ sp^3 hybridization, & diamagnetic

151. (3)

(I) $[Mn(CN)_6]^{3-}$; $Mn^{3+} \Rightarrow d^4 \Rightarrow t_{2g}^4 e_g^0$

$\Rightarrow (-0.4 \times 4 + 0 \times 0.6)\Delta_0 = -1.6\Delta_0$

(II) $[Co(CN)_6]^{4-}$, $Co^{+2} \Rightarrow d^7 \Rightarrow t_{2g}^6 e_g^1$

$\Rightarrow (-0.4 \times 6 + 1 \times 0.6)\Delta_0 = -1.80\Delta_0$

(III) $[Fe(CN)_6]^{4-}$; $Fe^{+2} \Rightarrow d^6 \Rightarrow t_{2g}^6 e_g^0$

$\Rightarrow (-0.4 \times 6 + 0 \times 0.6)\Delta_0 = -2.4\Delta_0$

(IV) $[Fe(CN)_6]^{3-}$; $Fe^{+3} \Rightarrow d^5 \Rightarrow t_{2g}^5 e_g^0$

$\Rightarrow (-0.4 \times 5 + 0 \times 0.6)\Delta_0 = -2.0\Delta_0$

Hence correct increasing order is:

(I) < (II) < (IV) < (III)

152. (2)

(A) $[CoF_6]^{-3}$

$Co^{3+} \rightarrow 3d^6$

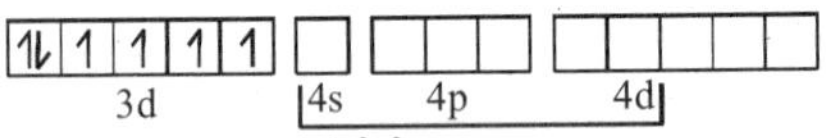

sp^3d^2 hybridisation

(B) $[NiCl_4]^{2-}$

$Ni^{2+} \rightarrow 3d^8$

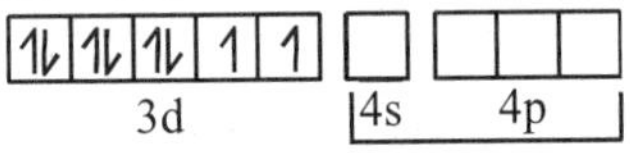

sp^3 hybridisation

(C) $[Co(NH_3)_6]^{3+}$

$Co^{3+} \rightarrow 3d^6$

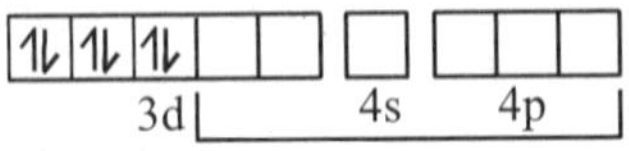

d^2sp^3 hybridisation

(D) $[Ni(CN)_4]^{2-}$

$Ni^{2+} \rightarrow 3d^8$

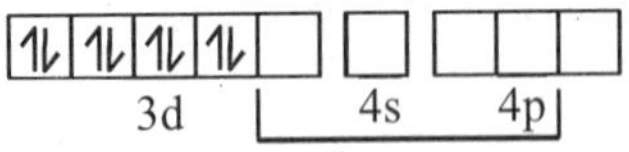

dsp^2 hybridisation

153. (1) For $3d^4$

If ligand is strong field ligand : $t_{2g}^4 e_g^0$ (low spin)

If ligand is weak field ligand : $t_{2g}^3 e_g^1$ (high spin)

154. (3)

$[Ni(H_2O)_6]^{+2}(aq) + en(aq) \rightarrow$
$[Ni(H_2O)_4(en)]^{+2}(aq) + 2H_2O$ Green

$[Ni(H_2O)_4(en)]^{+2}(aq) + en(aq) \rightarrow$
$[Ni(H_2O)_2(en)_2]^{+2}(aq) + 2H_2O$ Pale blue

$[Ni(H_2O)_2(en)_2]^{+2}(aq) + en(aq) \rightarrow$
$[Ni(en)_3]^{+2}(aq) + 2H_2O$ Blue/purple Violet

155. (3)

$Fe^{+6} = [Ar]3d^2$

$Mn^{+3} = [Ar]3d^4$

$Fe^{+3} = [Ar]3d^5$

$Cr^{+2} = [Ar]3d^4$

$Ni^{+4} = [Ar]3d^6$

156. (3) Co^{2+} complex having $\mu = 3.95$ BM

$\therefore$ Number of unpaired electrons = 3

$Co^{2+} \Rightarrow (Ar)_{18} 3d^7 4s^0$

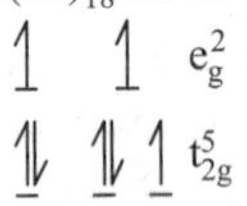

157. (2) Crystal field splitting energy $(\Delta) \propto$ charge or oxidation state of central metal atom.

Crystal field splitting energy $(\Delta) \propto$ field strength of ligand (and chelation)

Crystal field stabilisation energy (CFSE)

$= [-0.4\, t_{2g} + 0.6\, e_g]\, \Delta_0$ (for octahedral)

For,

$[Co(en)_3]^{3+} : Co^{3+} : t_{2g}^6 e_g^0$; CFSE = $-2.4\,(\Delta_0)_1$

$[Co(NH_3)_6]^{3+} : Co^{3+} : t_{2g}^6 e_g^0$; CFSE = $-2.4\,(\Delta_0)_2$

$[Co(NH_3)_6]^{2+} : Co^{2+} : t_{2g}^5 e_g^2$; CFSE = $-0.8\,(\Delta_0)_3$

$[Co(NH_3)_4]^{2+} : Co^{2+} : e^4 t_2^3$; CFSE = $-1.2\,\Delta_t$

$\left(\text{as} : (\Delta_t) = \frac{4}{9}(\Delta_0)_3\right)$

$\because \Delta_t < (\Delta_0)_3 < (\Delta_0)_2 < (\Delta_0)_1$

158. (2) $[NiCl_4]^{2-}$

Ni^{+2} – [Ar] $3d^8 4s^0 \rightarrow sp^3$, Tetrahedral

Number of unpaired electron = 2 paramagnetic

$[Ni(CO)_4]$,

$Ni(0) \rightarrow [Ar]\, 3d^{10} 4s^0$ (After rearrangement)

No unpaired electron sp^3, Tetrahedral, Diamagnetic

159. (4) For complex $K_3[Fe(SCN)_6]$

electron donating by sulphur in SCN^-, it is a weak field ligand.

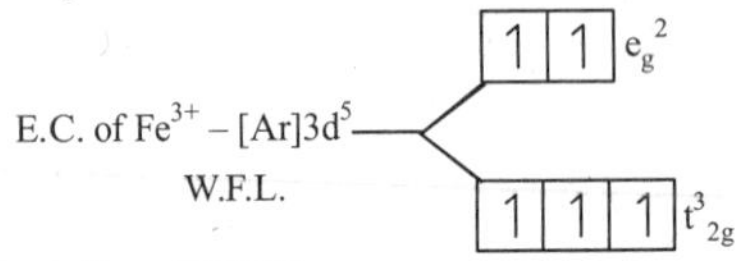

Calculation of CFSE

$= (-0.4 \times 3 + 0.6 \times 2)\, \Delta_0 = 0.\Delta_0$.

160. (3) $[Co(en)_2Cl_2]^+$ has octahedral geometry due to d^2sp^3 hybridization with two geometrical isomers (cis and trans).

161. (4)

(A) $K_2\left[\overset{+2}{Ni}(CN)_4\right]$

Ni^{2+}; $[Ar]3d^8 4s^0$, (CN^- is S.F.L)

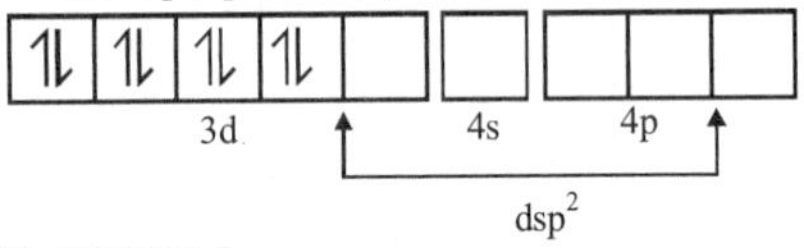

(B) $[Ni(CO)_4]$

Ni : $[Ar]\, 3d^8 4s^2$

CO is S.F.L, so pairing occur

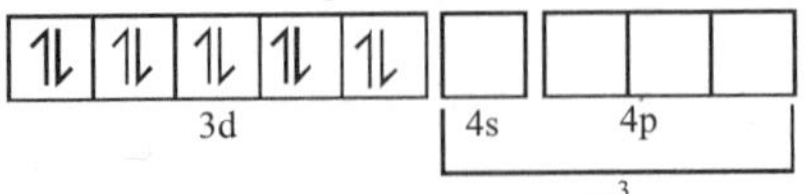

(C) $[Co(NH_3)_6]Cl_3$

Co^{+3} : $[Ar]3d^6 4s^0$

With Co^{3+}, NH_3 act as S.F.L

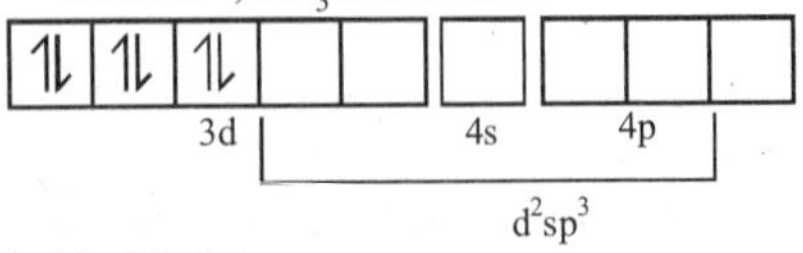

(D) $Na_3[CoF_6]$

Co^{3+} : $[Ar]3d^6$($F^{\ominus}$; W.F.L)

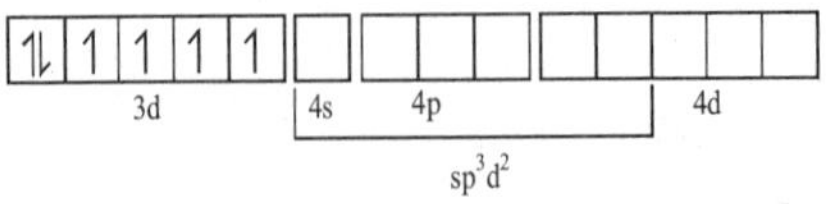

162. (2) $EDTA^{4-} \rightarrow$ Hexadentate ligand

$[Ca(EDTA)]^{2-}$

So Coordination environment is octahedral

163. (3) $N(CH_3)_3$ and $P(CH_3)_3$ both are Lewis base and act as ligand, however, $P(CH_3)_3$ has π-acceptor character due to presence of vacant d-orbitals but $N(CH_3)_3$ is not a π-acceptor due to absence of d-orbitals.

164. (2) Consider H_2O as a weak field ligand.

$[Fe(H_2O)_6]^{2+}$

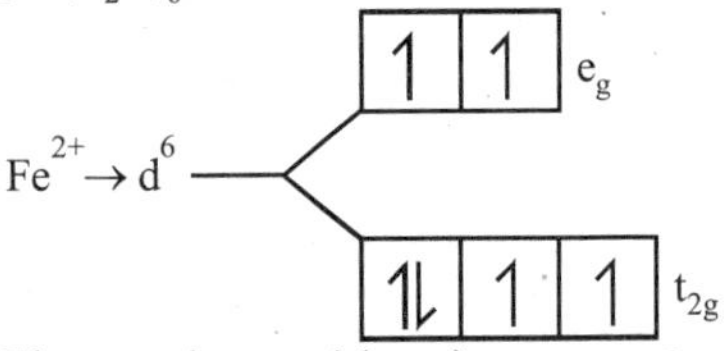

Electrons in t_{2g} = 4 (even)

$[Co(H_2O)_6]^{2+}$

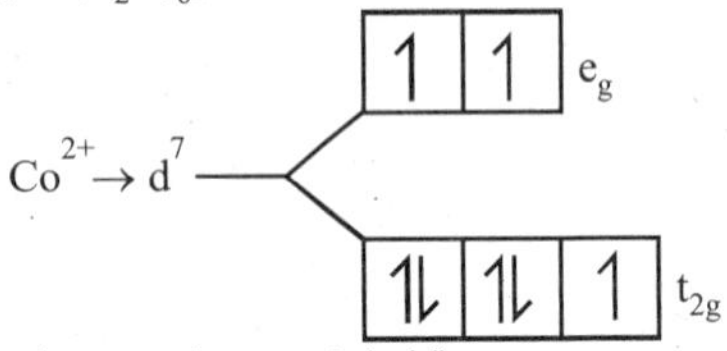

Electrons in t_{2g} = 5 (odd)

$[Co(H_2O)_6]^{3+}$

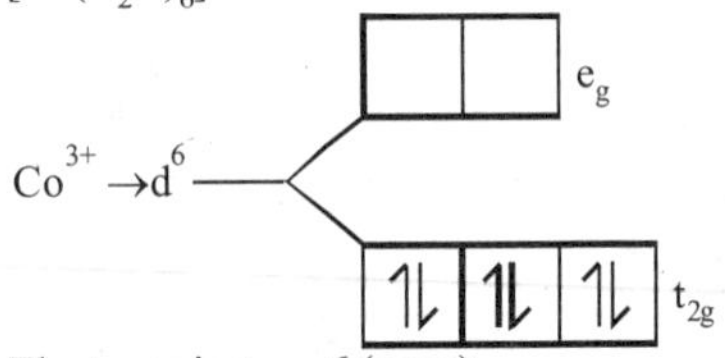

Electrons in t_{2g} = 6 (even)

Note : With Co^{3+}, H_2O acts as a strong field ligand.

$[Cu(H_2O)_6]^{2+}$

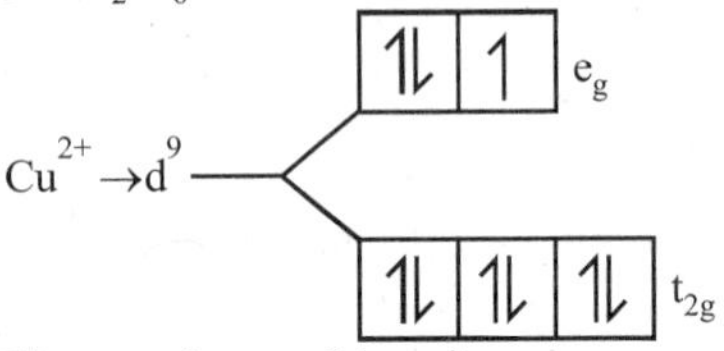

Electrons in t_{2g} = 6 (even)

$[Cr(H_2O)_6]^{2+}$

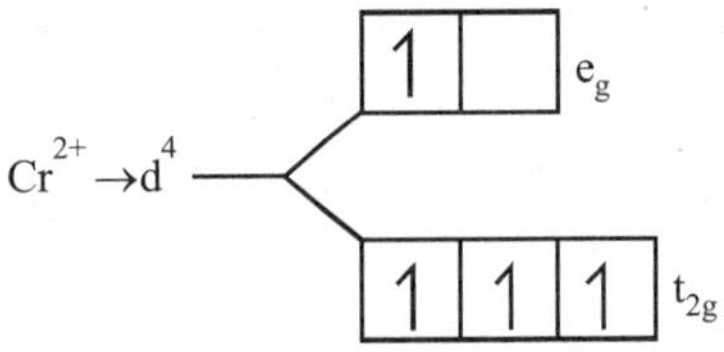

Electrons in t_{2g} = 3 (odd)

165. (3)

(A) $[Cr(NH_3)_6]^{3+}$ Strong ligand Cr^{3+} : $3d^3$

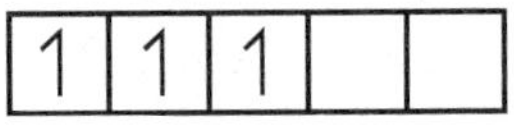

$\mu = \sqrt{n(n+2)}$

n = 3 (unpaired electrons)

$\mu \simeq 3.87$ B.M. (II)

(B) $[NiCl_4]^{2-}$ Weak ligand

Ni^{2+} : $3d^8$ 

n = 2

$\mu \simeq 2.83$ B.M. (IV)

(C) $[CoF_6]^{3-}$ Weak ligand

$Co^{3+}: 3d^6$ [⇅ | ↑ | ↑ | ↑ | ↑]

n = 4

$\mu \simeq 4.90$ B.M. (I)

(D) $[Ni(CN)_4]^{2-}$ Strong ligand

$Ni^{2+}: 3d^8$ [⇅ | ⇅ | ⇅ | ⇅ |]

n = 0

$\mu \simeq 0$ B.M. (III)

166. (2) As ligand field increases, radiation of high energy is absorbed. Energy of radiation and wave number is the / directly proportional. Therefore strong field ligand absorb light of higher wave number.

167. (4) $TiCl_4$; Ti^{+4}; $3d^04s^0$, $e^0t_2^0$

$[FeO_4]^{2-}$; Fe^{+6}; $3d^24s^0$; $e^2t_2^0$

$[FeCl_4]^-$; Fe^{+3}; $3d^54s^0$; $e^2t_2^3$

$[COCl_4]^{2-}$; Co^{2+}; $3d^74s^0$; $e^4t_2^3$

168. (1) H_2O is a weak field ligand. Unpaired e^- and is given in the table as :

Complex	Number of unpaired e^-	$M = \sqrt{n(n+2)}$ B.M.
$[Co(H_2O)_6]^{2+}$	3	3.87
$[Fe(H_2O)_6]^{2+}$	4	4.89
$[Mn(H_2O)_6]^{2+}$	5	5.92
$[Cr(H_2O)_6]^{2+}$	4	4.89

More the unpaired electrons more is the paramagnetic nature.

169. (3) According to spectrochemical series correct order of ligand field strength is :

$Br^- < F^- < H_2O < NH_3$

170. (1) Filling of electrons in tetrahedral complex as

$TiCl_4 \rightarrow Ti^{+4} \rightarrow [Ar]4s^03d^0 \rightarrow e^0t_2^0$

$[MnO_4]^- \rightarrow Mn^{+7} \rightarrow [Ar]4s^03d^0 \rightarrow e^0t_2^0$

$[FeO_4]^{2-} \rightarrow Fe^{+6} \rightarrow [Ar]4s^03d^2 \rightarrow e^2t_2^0$

$[FeCl_4]^- \rightarrow Fe^{3+} \rightarrow [Ar]4s^03d^5 \rightarrow e^2t_2^3$

$[CoCl_4]^{2-} \rightarrow Co^{+2} \rightarrow [Ar]4s^03d^7 \rightarrow e^4t_2^3$

Three complexes have no electrons in t_2 orbital.

171. (1) According to spectrochemical series ligand field strength is $CO > H_2O > F^- > S^{2-}$

172. (1) $[V(H_2O)_6]^{3+}$

$_{23}V$:- $[Ar]3d^34s^2$

V^{+3}:- $[Ar]3d^2$, n = 2 (even number of unpaired e^-)

$[Cr(H_2O)_6]^{2+}$

$_{24}Cr$:- $[Ar]3d^54s^1$

Cr^{2+}:- $[Ar]3d^4$, n = 4 (even number of unpaired e^-)

$[Fe(H_2O)_6]^{3+}$

Fe^{3+} :- $[Ar]3d^54s^0$

n = 5 (odd number of unpaired e^-)

$[Ni(H_2O)_6]^{3+}$

Ni:- $[Ar]3d^84s^2$

Ni^{3+}:- $[Ar]3d^7$, n = 3 (odd number of unpaired e^-)

$[Cu(H_2O)_6]^{2+}$

Cu: $[Ar]\ 3d^{10}4s^1$

Cu^{2+} :- $[Ar]3d^94s^0$

n = 1 (odd number of unpaired e^-)

173. (2) Balancing charge

$3 + x(0) + y(-1) = -1 \Rightarrow y = 4$

As the complex would be a octahedral complex, hence x will be 2.

x = 2, y = 4

so x + y = 6

174. (2) H_2O is a weak field ligand so pairing of electrons does not take place

$[Ni(H_2O)_6]^{2+}$ (d^8):

3d [⇅ | ⇅ | ⇅ | ↑ | ↑] 4s [××] 4p [×× | ×× | ××] 5d [×× | ×× | | |]

(4s, 4p, 5d ×× : sp^3d^2)

There is two unpaired electrons present, so complex is paramagnetic and have green colour.

$[Ni(CN)_4]^{2+}$ (d^8):

3d [⇅ | ⇅ | ⇅ | ⇅ | ××] 4s [××] 4p [×× | ×× |]

(dsp^2)

CN^- ligands cause pairing of electrons.

There is no unpaired electron present, so complex is diamagnetic and colourless.

175. (2) $[Co(NH_3)_6]^{+3}$

Co^{+3} = [⇅ | ↑ | ↑ | ↑ | ↑] (3d) [ ] (4s) [| |] (4p)

[⇅ | ⇅ | ⇅ | |] [ ] [| |] (d^2sp^3 Hybridization)

$[Co(NH_3)_6]^{+3}$ = [⇅ | ⇅ | ⇅] [⇅ | ⇅ | ⇅ | ⇅ | ⇅ | ⇅]

$NH_3\ NH_3\ NH_3\ NH_3\ NH_3\ NH_3$

The complex has octahedral geometry and is diamagnetic. Since the inner *d* orbital is involved in bonding is called an inner orbital or low spin or spin paired complex.

For $[CoF_6]^{3-}$

Co^{+3} = [⇅ | ↑ | ↑ | ↑ | ↑] (3d) [ ] (4s) [| |] (4p)

[| | | |] (4d)

Co^{+3} = [⇅ | ↑ | ↑ | ↑ | ↑]

[ ] [| |] [| | | |] (sp^3d^2 Hybridization)

$[CoF_6]^{3-}$ = [⇅ | ↑ | ↑ | ↑ | ↑] [⇅ | ⇅ | ⇅ | ⇅ | ⇅ | ⇅] [| |]

F F F F F F

The complex has octahedral geometry and is paramagnetic. Since the outer d-orbitals is involved in bonding is called outer orbital or high spin or spin free complex.

176. (1) A.Field strength of ligands are explained by LFT and MOT.

B.VBT does not explain stability of complex.

C.Hybridisation of $[Ni(CN)_4]^{-2}$ is dsp^2.

D.Cis-$[PtCl_2(en)_2]$ also has optical isomers.

177. (4) A-II, B-III, C-IV, D-I

$[Cr(H_2O)_6]^{3+}$	Cr^{+3} :	$[Ar]3d^3 : t_{2g}^3e_g^0$
$[Fe(H_2O)_6]^{3+}$	Fe^{+3} :	$[Ar]3d^5 : t_{2g}^3e_g^2$
$[Ni(H_2O)_6]^{2+}$	Ni^{+2} :	$[Ar]3d^8 : t_{2g}^6e_g^2$
$[V(H_2O)_6]^{3+}$	V^{+3} :	$[Ar]3d^2 : t_{2g}^2e_g^0$

178. (4) $[Ni(CO)_4]$ diamagnetic,

$[NiCl_4]^{2-}$ paramagnetic

In $[Ni(CO)_4]$, Ni = $[Ar]\ 3d^8\ 4s^2$

[⇅ | ⇅ | ⇅ | ↑ | ↑] (3d) [⇅] (4s) [| |] (4p)

CO is strong field ligand so pairing will take place.

Ni = [⇅ | ⇅ | ⇅ | ⇅ | ⇅] [ ] [| |]

sp^3 hybridization

$[Ni(CO)_4]$, [⇅ | ⇅ | ⇅ | ⇅ | ⇅] [⇅] [⇅ | ⇅ | ⇅]

CO CO CO CO

Its a diamagnetic complex.

In $[NiCl_4]^{2-}$, Ni = +2 oxidation state Cl is weak field ligand so no pairing will take place.

∴ [⇅ | ⇅ | ⇅ | ↑ | ↑] [⇅] [⇅ | ⇅ | ⇅]

Cl Cl Cl Cl

sp^3 Hybridization

Its a paramagnetic complex.

179. (3) $[FeF_6]^{3-}$: Fe^{+3} : $[Ar]d^5$

F : Weak field Ligand [↑ | ↑ | ↑ | ↑ | ↑]

No. of unpaired electrons = 5

$\mu = \sqrt{5(5+2)} \Rightarrow \mu = \sqrt{35}$ BM

$[V(H_2O)_6]^{+2}$: V^{+2} : $3d^3$ [↑ | ↑ | ↑ | |]

No. of unpaired electrons – 3

$\mu = \sqrt{3(3+2)} \Rightarrow \mu = \sqrt{15}$ BM

$[Fe(H_2O)_6]^{+2}$: Fe^{+2} : $3d^6$

H_2O : Weak field Lignad [⇅ | ↑ | ↑ | ↑ | ↑]

No. of unpaired electrons = 4

$\mu = \sqrt{4(4+2)} \Rightarrow \mu = \sqrt{24}$ BM

180. (1) Cr^3 + (H.S) is d^3 configuration

So its CFSE is t_{2g}^3 i.e., $3 \times (-0.4) = -1.2\,\Delta_0$

181. (2) Complex with maximum number of unpaired electron will exhibit maximum attraction to an applied magnetic field.

$[Zn(H_2O)_6]^{2+} \rightarrow d^{10}$ system $\rightarrow t_{2g}^6\ e_g^4$, 0 unpaired e^-

$[Co(H_2O)_6]^{2+} \rightarrow d^7$ system $\rightarrow t_{2g}^5\ e_g^2$, 3 unpaired e^-

$[Co(en)_3]^{3+} \rightarrow d^6$ system $\rightarrow t_{2g}^6\ e_g^0$, 0 unpaired e^-

$[Ni(H_2O)_6]^{2+} \rightarrow d^8$ system $\rightarrow t_{2g}^6\ e_g^2$, 2 unpaired e^-

182. (1) CFSE = $(-0.4\,nt_{2g} + 0.6\,ne_g)\Delta_o$

nt_{2g} = number of electrons in t_{2g} orbital

ne_g = number of electrons in e_g orbital

Complex	No. of at electrons	CFSE (Δ_0)
$[Cu(NH_3)_6]^{+2}$	d^9 (S.L.) $t_{2g}^{2,2,2}e_g^{2,1}$	–0.6
$[Ti(H_2O)_6]^{+3}$	d^1 (W.L.) $t_{2g}^{1,0,0}e_g^{0,0}$	–0.4
$[Fe(CN)_6]^{3-}$	d^5 (S.L.) $t_{2g}^{2,2,1}e_g^{0,0}$	–2.0
$[NiF_6]^{4-}$	d^8 (W.L.) $t_{2g}^{2,2,2}e_g^{1,1}$	–1.2

183. (2) $Fe_4[Fe(CN)_6]_3$ Prussian Blue–water insoluble $K_3[Co(NO_2)_6]$ very poorly water soluble $(NH_4)_3[As(Mo_3O_{10})_4]$ water insoluble ammonium arseno molybdate, $[Fe_3(OH)_2(OAc)_6]Cl$ is water soluble.

184. (1) Since H_2O is strong field ligand compared to chloride and Co^{3+} ion is present.
$\therefore$ CFSE is higher for $[Co(NH_3)_5H_2O]^{+3}$, hence it will absorb at lower wavelength. Thus, 'A' is false.

185. (4) $[Fe(H_2O)_6]^{3+}$
$Fe^{3+} \Rightarrow [Ar]\,3d^5\,4s^0$,

1	1	1	1	1

$\mu = \sqrt{n(n+2)} = \sqrt{5(5+2)}$

$\mu = \sqrt{35} = 5.92$ B.M.

$[Fe(CN)_6]^{-3}$
$Fe^{3+} \Rightarrow [Ar]\,3d^5\,4s^0$,

1	1	1	1	1

Pairing occur due to strong field ligand CN^-
$\therefore$ Unpaired $e^- \Rightarrow 1$

$\mu = \sqrt{n(n+2)} = \sqrt{1(1+2)} = \sqrt{3} = 1.732$ B.M.

186. (1) $Mg(NH_4)PO_4 \Rightarrow$ White,
$K_3[Co(NO_2)_6] \Rightarrow$ Yellow
$MnO(OH)_2 \Rightarrow$ Brown, $Fe_4[Fe(CN)_6]_3 \Rightarrow$ Blue

187. (2) $[NiBr_2Cl_2]^{2-} \rightarrow$ This complex species is tetrahedral as Br^- & Cl^- are weak field ligands.
$[PtBr_2Cl_2]^{2-} \rightarrow$ As Pt belongs to 5d series. This complex species is square planar.
Td complex are optically active and square plannar complex are optically inactive.
$[NiBr_2Cl_2]^{2-}$, being tetrahedral does not show Geometrical Isomerism.
$[PtBr_2Cl_2]^{2-}$ shows two geometrical isomers.
Ni^{2+} is paramagnetic while pt^{2+} is diamagnetic.

188. (4) $[Co(NH_3)_6]^{3+}$,
$NH_3 \Rightarrow$ SFL, $Co^{+3} = [Ar]3d^6$, $t_{2g}^6\,e_g^0$

1l	1l	1l	••	••		••		••	••	••

$\therefore$ Diagmagnetic & Low spin complex.

189. (2) A. $[Fe(CN)_6]^{3-}$; n = 1 (CN^- = Strong Field Ligand)
B. $[FeF_6]^{3-}$; n = 5 (F^- = Weak Field Ligand)
C. $[CoF_6]^{3-}$; n = 4 (F^- = WFL)
D. $[Cr(oxalate)_3]^{3-}$; n = 3 (oxalate = SFL)
E. $[Ni(CO)_4]$; n = 0 (CO = SFL)

190. (2) A–IV, B–I, C–II, D–III
(A) $[Ti(H_2O)_6]^{2+}$
$Ti^{2+} \Rightarrow 3d^2\,4s^0,\ t_{2g}^2 e_g^0$
$CFSE = [-0.4 \times 2 + 0.6 \times 0]\Delta_o = -0.8\,\Delta_o$
(B) $[V(H_2O)_6]^{2+}$
$V^{2+} \Rightarrow 3d^3\,4s^0,\ t_{2g}^3 e_g^0$
$CFSE = [-0.4 \times 3 + 0.6 \times 0]\Delta_o = -1.2\,\Delta_o$
(C) $[Mn(H_2O)_6]^{3+}$
$Mn^{3+} \Rightarrow 3d^4\,4s^0,\ t_{2g}^3 e_g^1$
$CFSE = [-0.4 \times 3 + 0.6 \times 1]\Delta_o = -0.6\,\Delta_o$
(D) $[Fe(H_2O)_6]^{3+}$
$Fe^{3+} \Rightarrow 3d^5\,4s^0,\ t_{2g}^3 e_g^2$
$CFSE = [-0.4 \times 3 + 0.6 \times 2]\Delta_o = 0\,\Delta_o$

191. (3) $Co^{+3} = t_{2g}^{\ 6}\,e_g^{\ 0}$
CN^-–strong field ligand
All d–electrons should be paired ($\mu_s = 0$) Hence diamagnetic.

192. (2)

		Unpaired e^-
$[Fe(CN)_6]^{3-}$	$Fe^{+3} \Rightarrow t_{2g}^{\ 5}\,e_g^{\ 0}$,	1 (Strong field ligand)
$[Mn(CN)_6]^{3-}$	$Mn^{+3} \Rightarrow t_{2g}^{\ 4}\,e_g^{\ 0}$,	2 (Strong field ligand)
$[CoF_6]^{3-}$	$Co^{+3} \Rightarrow t_{2g}^{\ 4}\,e_g^{\ 2}$,	4 (Weak field ligand)
$[MnBr_4]^{2-}$	$Mn^{+2} \Rightarrow e_g^{\ 2}\,t_{2g}^{\ 3}$,	5 (Weak field ligand)

Spin magnetic moment $\mu = \sqrt{n(n+2)}$ B.M
$\therefore$ Correct option is (b).

193. (1) For option (A), $Cr^{+3}: 3d^3$; CN^-: SFL
$\Rightarrow$ No. of unpaired electrons = 3
For option (B), $Fe^{+2}: 3d^6$; H_2O: WFL
$\Rightarrow$ No. of unpaired electrons = 4
For option (C), $Co^{+3}: 3d^6$; NH_3: SFL
$\Rightarrow$ No. of unpaired electrons = 0
For option (D), $Ni^{+2}: 3d^8$; NH_3: SFL
$\Rightarrow$ No. of unpaired electrons = 2

194. (1) $[Fe(CN)_6]^{3-}$

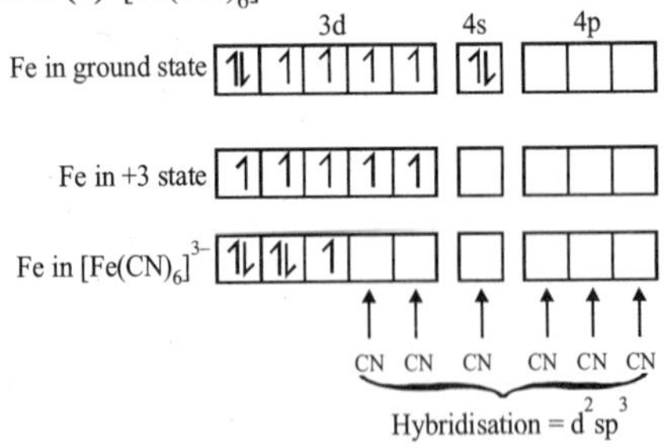

Unpaired electron = 1
$\mu = \sqrt{n(n+2)} = \sqrt{1 \times 3} = 1.74$ B.M.
$[Fe(H_2O)_6]^{3+}$ No pairing because H_2O is WFL
Number of unpaired electrons = 5, $\mu = 5.92$ B.M.
Assertion is true, Reason is true but not correct explanation.

195. (1) CN^- is the strong field ligand among the given ligands. So, the Δ_0 value of $[Fe(CN)_6]^{3-}$ will be the highest one among all. Here, the central atom is present in Fe^{3+} state in all cases.

196. (1) $CoCl_2 + 6H_2O \longrightarrow [Co(H_2O)_6]Cl_2$
Pink(X) octahedral
$\downarrow$ +HCl(conc.)
$[CoCl_4]^{2-}$
(Y) Blue solution
(Z) Tetrahedral

197. (2) The increasing order of field strength of ligands (according to spectrochemical series)
$S^{2-} < C_2O_4^{2-} < NH_3 < en < CO$

198. (4) For $[Fe(NH_3)_6]^{+2}$, $\Delta_0 < P$, hence the pairing of electrons does not occur in t_{2g}. Therefore complex is outer orbital and its hybridisation is $sp^3\,d^2$.

199. (2) The Cl – Co – Cl bond angle in above octahedral complex is 90°.

200. (1) $[FeF_6]^{3-}: Fe^{3+}: 3d^5$, $F^{\ominus}$ is a weak field ligand.

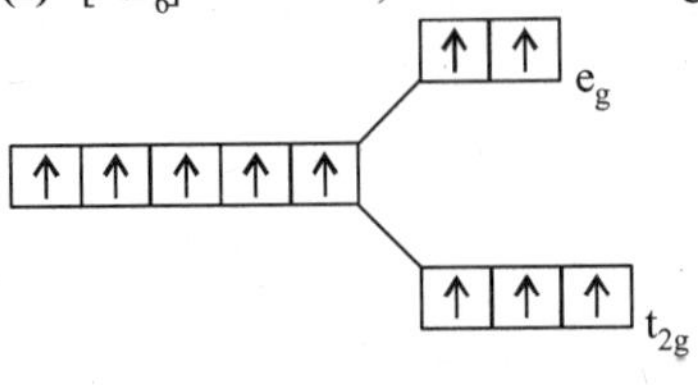

Number of unpaired electrons = 5
$\therefore \mu = \sqrt{35}$ BM
$[CoF_6]^{3-}: Co^{3+}: 3d^6$

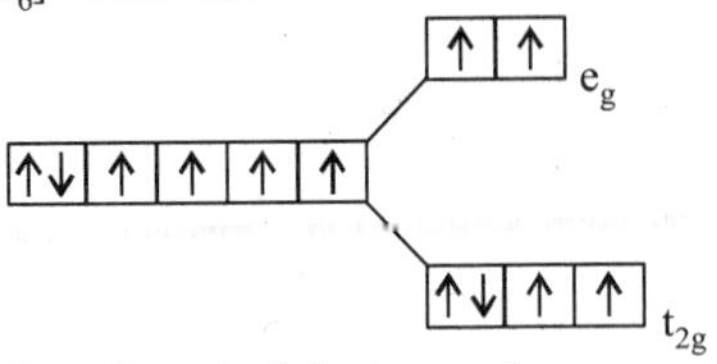

Number of unpaired electrons = 4
$\therefore \mu = \sqrt{24}$ BM
$[Co(C_2O_4)_3]^{3-}: Co^{3+}: 3d^6$ Oxalate is a strong field ligand

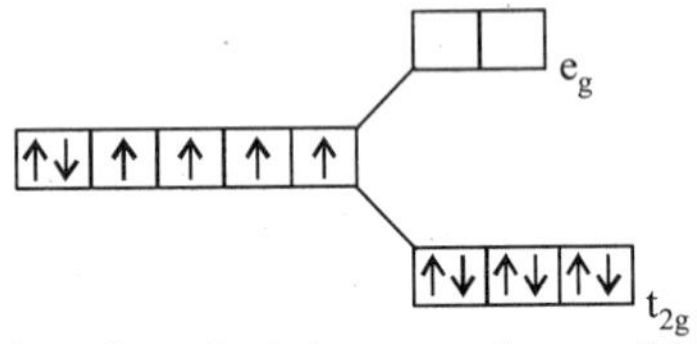

Number of unpaired electrons = 0 $\therefore \mu = 0$ BM

201. (2) Lower the wavelength absorbed = High the energy of the absorbed light $\Rightarrow$ Larger is the splitting of d – orbitals $\Rightarrow$ caused by high O.S. of central metal atom and strong field ligand.

$E = \dfrac{hc}{\lambda} \Rightarrow E \propto \dfrac{1}{\lambda}$

$\Rightarrow \Delta(CFSE) \propto \dfrac{1}{\lambda_{absorb}} \propto$ strength of ligand.

202. (3) $Cr^{+2}: [Ar]\,3d^4\,4s^0$;
$n = 4,\ \mu = \sqrt{4(4+2)} = 4.89$ BM
$Mn^{+2}: [Ar]\,3d^5\,4s^0$; n = 5,
$\mu = \sqrt{5(5+2)} = 5.91$ BM
$V^{+2}: [Ar]\,3d^3\,4s^0$; n = 3,
$\mu = \sqrt{3(3+2)} = 3.87$ BM
$Ti^{+2}: [Ar]\,3d^2\,4s^0$; n = 2,
$\mu = \sqrt{2(2+2)} = 2.82$ BM

203. (1) Spectro-chemical series is based on the absorption of light by complexes with different ligands and determined experimentally. Hence, Crystal field theory cannot be explained the relative strength of ligands as given in spectrochemical series.

204. (3) $[Co(NH_3)_6]^{3+}$
$Co^{+3} \Rightarrow 3d^6$
NH_3 behaves as strong field ligand in case of Co^{+3} and pairing will takes place.

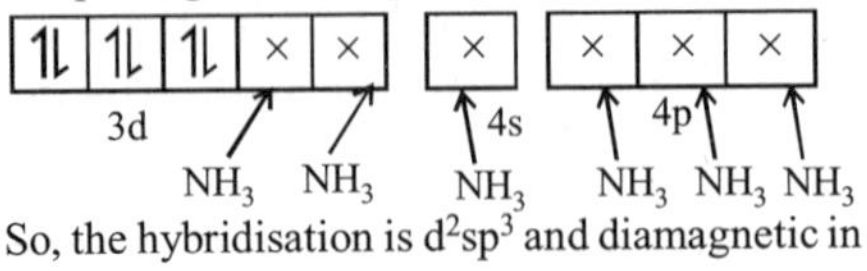

So, the hybridisation is d^2sp^3 and diamagnetic in nature.

205. (1) Strong John - Teller distortion is observed in case of unsymmetrical filling of e_g orbitals.
When the ligands are same, repulsion from all side are symmetrical and maximum distortion is observed.

206. (2) CO and CN behaves as SFL i.e., they have tendency to pair up the electrons while F behaves as WFL i.e. no tendency to pair up the electrons. Therefore; A–I, B–IV, C–III, D–II.

207. (1) Metals of low oxidation state forms complexes with ligands which have π-accepting character.

208. (1) $K_3[Cu(CN)_4]$
O.N. of copper is Cu^+
$Cu^+ \Rightarrow [Ar]3d^{10} \Rightarrow$ Diamagnetic

209. (1) Energy of absorption depends on the splitting of d-orbital in crystal field splitting.
A strong field ligand have high crystal field splitting. Order of strength of ligands is

$$en > NH_3 > H_2O$$

Hence, the order of energy is

$$\underset{(A)}{[Ni(en)_3]^{2+}} > \underset{(B)}{[Ni(NH_3)_6]^{2+}} > \underset{(C)}{[Ni(H_2O)_6]^{2+}}$$

210. (2) Ni in zero oxidation state with any ligand always form tetrahedral complex with sp^3 hybridization.
Ni^{2+} ion with strong field ligand form square planar complex with dsp^2 hybridisation.
Ni^{2+} ion with weak field ligand such as Cl^- form tetrahedral complex.
$[Ni(CN)_4]^{2-}$: d^8 configuration, SFL, sq. planar splitting (dsp^2), diamagnetic.
$[Ni(CO)_4]$: d^{10} configuration (after excitation), SFL, tetrahedral splitting (sp^3), diamagnetic.
$[NiCl_4]^{2-}$: d^8 configuration, WFL, tetrahedral splitting (sp^3), paramagnetic (2 unpaired e^-).

211. (3) CN^- act as strong field ligand and it has maximum tendency to split the energy cause the splitting of d-orbital.
Therefore, $[Co(CN)_6]^{3-}$ have maximum stabilization due to crystal field.

212. (2) Magnetic moment depends upon the number of unpaired electron present in the molecule.
(A) $[FeF_6]^{3-}$
$Fe^{3+} \to 3d^5 4s^0$; F^- is a weak field ligand.
Number of unpaired electron = 5.
(B) $[Fe(CN)_6]^{3-}$
$Fe^{3+} \to 3d^5 4s^0$; (CN^-) is a strong field ligand.
$n = 1$.
(C) $[MnCl_6]^{3-}$
$Mn^{3+} \to 3d^4 4s^0$; Cl^- is a weak field ligand.
$n = 4$.
(D) $[Mn(CN)_6]^{3-}$
$Mn^{3+} \to 3d^4 4s^0$; CN^- is a strong field ligand.
$n = 2$
$\mu \Rightarrow A > C > D > B$.

213. (2) $V^{2+}: 1s^2 2s^2 2p^6 3s^2 3p^6 3d^3$
[↑ | ↑ | ↑ | |] (3d) (unpaired e^- = 3)
$Ni^{2+}: 1s^2 2s^2 2p^6 3s^2 3p^6 3d^8$
[↑↓ | ↑↓ | ↑↓ | ↑ | ↑] (3d) (unpaired e^- = 2)
$Cr^{2+}: 1s^2 2s^2 2p^6 3s^2 3p^6 3d^4$
[↑ | ↑ | ↑ | ↑ |] (3d) (unpaired e^- = 4)
$Fe^{2+}: 1s^2 2s^2 2p^6 3s^2 3p^6 3d^6$
[↑↓ | ↑ | ↑ | ↑ | ↑] (3d) (unpaired e^- = 4)

214. (4) CFSE of octahedral complexes with water is greater for 5d series metal centre ion as compared to 3d and 4d series metal centre.

215. (4) Given, $\mu = 3.87 \Rightarrow n = 3$
$CFSE = -0.8\Delta_0$
$CFSE = [-0.4n_{t_{2g}} + 0.6n_{e_g}]\Delta_0$

M^{z+}	n	CFSE (Δ_0)	d-orbital configuration				
V^{3+}	2	– 0.8	↑	↑			
Cr^{3+}	3	– 1.2	↑	↑	↑		
Mn^{4+}	3	– 1.2	↑	↑	↑		
Co^{2+}	3	– 0.8	↑↓	↑↓	↑	↑	↑
			t_{2g}			e_g	

Thus, only Co^{2+} can exhibit both properties.

216. (4) $[Fe(CO)_4(C_2O_4)]^+$
CO is strong field ligand, so pairing will take place.
∴ Spin only magnetic moment is:
$\mu = \sqrt{n(n+2)} = \sqrt{1(1+2)} = 1.73$ B.M.

217. (4)
(1) $[Fe(CN)_6]^{4-} \to$ Pale yellow solution
(2) $[Fe(SCN)_6]^{4-} \to$ Blood red colour
(3) $Fe_4[Fe(CN_6)]_3 \cdot H_2O \to$ Prussian blue
(4) $[Fe(CN)_5NOS]^{4-} \to$ Violet colour

218. (2) In $[Co(H_2O)_6]^{2+}$, Co is in +2 as but in $[CoF_6]^{3-}$ Co in +3. So order would be $B < A < C < D$
Note: Cation with a higher oxidation state has a larger value of Δ CFSE.
(A) $[\overset{III}{Co}F_6]^{3-}$; $F \to$ weak field ligand.
$CFSE = [-0.4 \times 4 + 0.6 \times 2]\Delta_0 = -0.4\Delta_0$.
(B) $[\overset{II}{Co}(H_2O)_6]^{2+}$; $H_2O \to$ weak field ligand.
$CFSE = [-0.4 \times 5 + 0.6 \times 2]\Delta_0 = -0.8\Delta_0$.
(C) $[\overset{III}{Co}(NH_3)_6]^{3+}$; $NH_3 \to$ strong field ligand.
$CFSE = [-0.4 \times 6 + 0.6 \times 0]\Delta_0$
$= -2.4\Delta_0 + 2P$ (P is pairing energy)
(D) $[\overset{III}{Co}(en)_3]^{3+}$;
$CFSE = [-0.4 \times 6 + 0.6 \times 0]\Delta_0 = -2.4\Delta_0 + 2P$
en is the strongest field ligand among the given species. Hence, $[Co(en)_3]^{3+}$ will have highest CFSE.
Thus, order is: (A) < (B) < (C) < (D).

219. (4) $[MnCl_6]^{3-}$
$Mn^{3+} \to [Ar]$ 3d [↑ | ↑ | ↑ | ↑ |] 4s [ ] 4p [| |] 4d [| | | |]
sp^3d^2
Cl^- is a weak field ligand, so Mn^{3+} with weak field ligand forms high spin complex or outer orbital complex. Hence, the complex will be sp^3d^2 hybridised.

220. (1) $[Fe(H_2O)_6]^{3+}$
Fe^{3+} : $[Ar]3d^5$; Hybridisation : sp^3d^2
Magnetic nature : Paramagnetic (so this complex responds to external magnetic field)

221. (2) $Ti(22) \Rightarrow 3d^2 4s^2$; $Ti^{3+} \Rightarrow 3d^1 4s^0 \Rightarrow n = 1$
$Sc(21) \Rightarrow 3d^1 4s^2$; $Sc^{3+} \Rightarrow 3d^0 4s^0 \Rightarrow n = 0$
$V(23) \Rightarrow 3d^3 4s^2$; $V^{2+} \Rightarrow 3d^3 \Rightarrow n = 3$
Hence, the spin only magnetic moment will increase in the order; $Sc^{3+} < Ti^{3+} < V^{2+}$

222. (2)
(1) $NiCl_2 \cdot 6H_2O : Ni^{+2} \to [Ar]3d^8 4s^0$
Cl and $H_2O \to$ weak field ligand
sp^3d^2 Hybridisation
C.N. = 6 [↑↓ | ↑↓ | ↑↓ | ↑ | ↑] 3d
(2) $K_2[Ni(CN)_4]$
$Ni^{+2} \to [Ar]3d^8 4s^0$
C.N. = 4
$CN^- \to$ Strong field ligand
dsp^2 Hybridisation
3d [↑↓ | ↑↓ | ↑↓ | ↑ | ↑] 4s [ ] 4p [↑↓]
3d [↑↓ | ↑↓ | ↑↓ | ↑↓ |] 4s [ ] 4p [| |]
$CN^⊖$ $CN^⊖$ $CN^⊖$ $CN^⊖$
dsp^2 Hybridisation
(3) $[Ni(CO)_4]$
CO - Strong field ligand
$Ni \to [Ar]\,3d^{10}$
$3d^{10}$ [↑↓ | ↑↓ | ↑↓ | ↑↓ | ↑↓] 4s [ ] 4p [| |]
sp^3 Hybridisation
(4) $Na_2[NiCl_4] : Ni^{2+} \to [Ar]3d^8 4s^0$
C.N. = 4
sp^3 Hybridisation
$Cl^⊖ \to$ weak field ligand
[↑↓ | ↑ | ↑] t_{2g}
[↑↓ | ↑↓] e_g
tetrahedral splitting

223. (3)
$\Delta_t < \Delta_o < \Delta_{sq}$
$[NiCl_4]^{2-} < [Ni(H_2O)_6]^{2+} < [Ni(CN)_4]^{2-}$
Splitting/absorbed energy order
$[NiCl_4]^{2-} > [Ni(H_2O)_6]^{2+} > [Ni(CN)_4]^{2-}$
Intensity of (complementary) colour of compound

224. (3) Fe^{2+} : $[Ar]3d^6 4s^0$:
[↑↓ | ↑ | ↑ | ↑ | ↑]
In the presence of strong field ligand, pairing occurs and there will be no unpaired electron.
[↑↓ | ↑↓ | ↑↓ | |]

225. (2) Among given species, only CuI does not contain any unpaired electron because it furnishes Cu^+ ion. So, for Cu^+, $\mu = 0$.

226. (1) $[Mn(CN)_6]^{4-}$
$Mn^{2+} = 3d^5 4s^0$
$[Fe(CN)_6]^{3-}$
$Fe^{3+} = 3d^5 4s^0$
CN^- is a strong field ligand, so pairing will occur in both complexes.
[↑↓ | ↑↓ | ↑ | : | :] [:] [: | : | :]
d^2sp^3
Paramagnetic

227. (3)

Complex ions	Unpaired e^- (n)	Magnetic moment
(i) $[FeF_6]^{3-}$	5	$\sqrt{35}$
(ii) $[Co(NH_3)_6]^{3+}$	0	0
(iii) $[NiCl_4]^{2-}$	2	$\sqrt{8}$
(iv) $[Cu(NH_3)_4]^{2+}$	1	$\sqrt{3}$

So, correct order of spin only magnetic moment is: (ii) < (iv) < (iii) < (i)

228. (2)

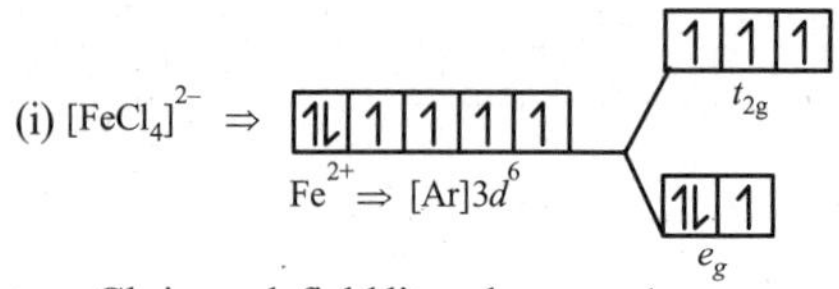

Cl^- is weak field ligand. ∴ $n = 4$.

$\mu = \sqrt{n(n+2)}$ BM

$= \sqrt{4(4+2)} = \sqrt{24} = 4.90$ BM

(ii) $[Co(C_2O_4)_3]^{3-}$

$C_2O_4^{2-}$ is strong field ligand.

∴ $n = 0$.

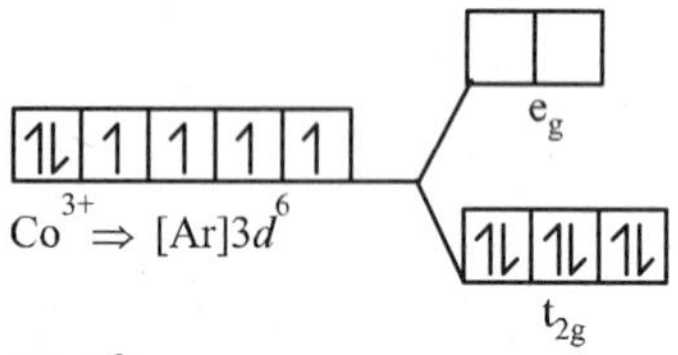

∵ $\mu = 0$

(iii) MnO_4^{2-}

$Mn^{+6} \Rightarrow [Ar]3d^1$

$\mu = \sqrt{1(1+2)}$ BM $= \sqrt{3}$ BM $= 1.73$ BM

229. (4) $[MnBr_4]^{2-} \Rightarrow Mn^{2+} \Rightarrow d^5$(Td)

[∵ Br is weak field ligand]

d^5(Td) is high spin complex.

So, $\mu = \sqrt{5(5+2)} = 5.91$ B.M.

230. (1) If $\Delta_0 > P$ then for $d^4 \Rightarrow t_{2g}^4 e_g^0$

If $\Delta_0 < P$ then for $d^4 \Rightarrow t_{2g}^3 e_g^1$

231. (1) For d^6 configuration, high spin complex.

(i) In case of octahedral field, $t_{2g}^4 e_g^2$

$CFSE = [-0.4p + 0.6q]\Delta_0 + n(P)$

$= [-0.4 \times 4 + 0.6 \times 2]\Delta_0 + 0 = -0.4\Delta_0$

(ii) In case of tetrahedral field, $e^3 t_2^3$

$CFSE = [-0.6p + 0.4q]\Delta_t$

$= [-0.6 \times 3 + 0.4 \times 3]\Delta_t = -0.6\Delta_t$

232. (1)

$[Cr(H_2O)_6]^{2+} \Rightarrow Cr^{2+} \Rightarrow 3d^4 \Rightarrow t_{2g}^{1,1,1}, e_g^{1,0}$

No. of unpaired e^- : 4

$[Fe(H_2O)_6]^{2+} \Rightarrow Fe^{2+} \Rightarrow 3d^6 \Rightarrow t_{2g}^{2,1,1}, e_g^{1,1}$

No. of unpaired e^- : 4

$[Fe(NH_3)_6]^{2+} \Rightarrow Fe^{2+} \Rightarrow 3d^6 \Rightarrow t_{2g}^{2,1,1}, e_g^{1,1}$

No. of unpaired e^- : 4

$[Co(OH)_4]^{2-} \Rightarrow Co^{2+} \Rightarrow 3d^7 \Rightarrow e^{2,2}, t_2^{1,1,1}$

No. of unpaired e^- : 3

$[CoCl_4]^{2-} \Rightarrow Co^{2+} \Rightarrow 3d^7 \Rightarrow e^{2,2}, t_2^{1,1,1}$

No. of unpaired e^- : 3

$[Mn(H_2O)_6]^{2+} \Rightarrow Mn^{2+} \Rightarrow 3d^5 \Rightarrow t_{2g}^{1,1,1}, e_g^{1,1}$

No. of unpaired e^- : 5

So $[Cr(H_2O)_6]^{2+}$ and $[Fe(H_2O)_6]^{2+}$ have same magnetic moment (spin only).

233. (2) $\Delta_0 < P$, hence there will be no pairing of e^-.

$[CoF_3(H_2O)_3] \Rightarrow Co^{3+} \Rightarrow d^6$ or $t_{2g}^4 e_g^2$

$CFSE = -[0.4p + 0.6q]\Delta_0 + n(P)$

$= [-4 \times 0.4 + 2 \times 0.6]\Delta_0 + 0 = -0.4\,\Delta_0$.

234. (3) Complex No. of unpaired e^-

$[Fe(en)(bpy)(NH_3)_2]^{2+}$: $Fe^{2+} - 3d^6$, 0

$[Pd(gly)_2]$: $Pd^{2+} - 3d^8$, 0

$[Co(ox)_2(OH)_2]^-$: $Co^{5+} - 3d^4$, 2

$[Ti(NH_3)_6]^{3+}$: $Ti^{3+} - 3d^1$, 1

Thus, $[Co(ox)_2(OH)_2]^-$ exhibits highest paramagnetic behaviour due to highest number of unpaired electrons.

235. (4) $[Ti(H_2O)_6]^{3+} \Rightarrow Ti^{3+} = 3d^1 4s^0$

∴ Electronic configuration is $t_{2g}^1 e_g^0$

$CFSE = [-0.4nt_{2g} + 0.6ne_g]\Delta_0$

$= [(-0.4) \times 1 + 0]\,20300 = -8120\ cm^{-1}$

$= \frac{-8120}{83.7}$ kJ/mol $= -97$ kJ/mol

236. (3) $[Ru(en)_3]^{2+} \Rightarrow Ru^{2+} = 4d^6 = t_{2g}^6, e_g^0$

$[Fe(H_2O)_6]^{2+} \Rightarrow Fe^{2+} = 3d^6 = t_{2g}^4, e_g^2$

237. (2) Spin only magnetic moment $= 4 \cdot 9 = \sqrt{n(n+2)}$

∴ Number of unpaired electrons = 4

Two possible arrangements are

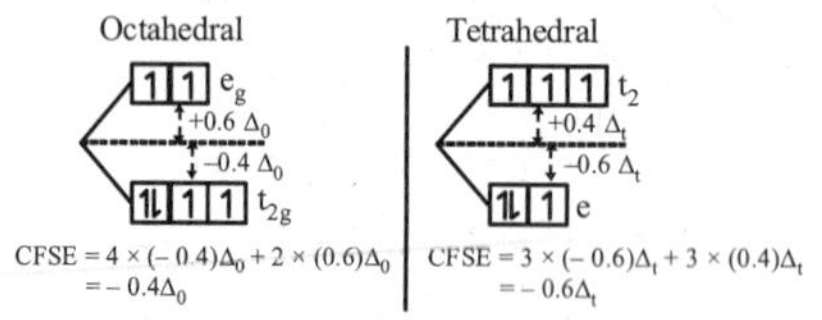

238. (3) With weak field ligands Mn(II) will be of high spin and with strong field ligands it will be of low spin. Ni(II) tetrahedral complexes will be generally of high spin due to sp^3 hybridisation. Mn(II) is of light pink colour in aqueous solution.

239. (1) Stronger the ligand, greater is splitting of d-orbitals and smaller will be wavelength of light absorbed.

According to spectrochemical series, the splitting power of ligands is: $NH_3 > NCS^- > F^-$.

Thus, order of wavelength of light absorbed is:

$\lambda_{NH_3} < \lambda_{NCS^-} < \lambda_{F^-}$.

240. (3) Number of Geometrical isomers (n) in square planar

$[Pd(F)(Cl)(Br)(I)]^{2-} = 3$

$[Fe(CN)_6]^{3-6} = [Fe(CN)_6]^{3-}$

$Fe^{3+} = 3d^5$

$\mu = \sqrt{n(n+2)} = 1.73$ B.M.

$CFSE = -0.4\,\Delta_0 \times nt_{2g} + 0.6\Delta_0 \times ne_g$

$= -0.4\,\Delta_0 \times 5 = -2.0\,\Delta_0$

241. (3) As, $\mu_s = \sqrt{n(n+2)}$

$Na_4[Fe(CN)_6] \rightarrow Fe^{2+}$

$\Rightarrow$ S.F.L. $\Rightarrow t_{2g}^{\ 6} e_g^{\ 0} \Rightarrow \mu_s = 0$

$[Cr(H_2O)_6]\,Br_2 \rightarrow Cr^{2+} \Rightarrow$ W.F.L. $\Rightarrow t_{2g}^{\ 3} e_g^{\ 1}$

$\Rightarrow \mu_s = \sqrt{4(4+2)} = \sqrt{24}$ B.M.

$(Et_4N)_2[CoCl_4] \equiv 2(Et_4N)^+[CoCl_4]^{2-} \rightarrow Co^{2+}$

$\Rightarrow$ W.F.L. $\Rightarrow e^4 t_2^{\ 3} \Rightarrow \mu_s = \sqrt{3(3+2)} = \sqrt{15}$ B.M.

$Na_3[Fe(C_2O_4)_3] \rightarrow Fe^{3+} \Rightarrow$ S.F.L. $\Rightarrow t_{2g}^{\ 5} e_g^{\ 0}$

$\Rightarrow \mu_s = \sqrt{1(1+2)} = \sqrt{3}$ B.M.

242. (4) No. of unpaired e^- μ

(A) $Ni(CO)_4$; $Ni = 3d^8 4s^2$(SFL) 0 0

(B) $[Ni(H_2O)_6]$; $Cl_2 Ni^{2+} = 3d^8$(WFL) 2 $\sqrt{8}$ BM

(C) $Na_2[Ni(CN)_4]$; $Ni^{2+} = 3d^8$(SFL) 0 0

(D) $PdCl_2(PPh_3)_2$; $Pd^{2+} = 4d^8$ 0 0

Correct order of the calculated spin only magnetic moments of complexes A to D is (A) ≈ (C) ≈ (D) < B

243. (1)

(A) Co^{3+} with strong field ligand forms complex of zero magnetic moment.

t_{2g}^6 e_g^0 when $\Delta_0 > p$

(D) $\Delta_t = \left(\frac{4}{9}\right)\Delta_0 \Rightarrow \Delta_t = \frac{4}{9} \times 18000 = 8000\ cm^{-1}$

244. (4) $CFSE = [-0.4\,p + 0.6\,q]\Delta_0$

$Ni^{2+}(d^8) \longrightarrow Ni^{3+}(d^7)$; $t_{2g}^6 e_g^2$ → $t_{2g}^6 e_g^1$; CFSE = −1.2 → −1.8

$Co^{2+}(d^7) \longrightarrow Co^{3+}(d^6)$; $t_{2g}^6 e_g^1$ → $t_{2g}^6 e_g^0$; CFSE = −1.8 → −2.4

$Zn^{2+}(d^{10}) \longrightarrow Zn^{3+}(d^9)$; $t_{2g}^6 e_g^4$ → $t_{2g}^6 e_g^3$; CFSE = 0.0 → −0.6

$Fe^{2+}(d^6) \longrightarrow Fe^{3+}(d^5)$; $t_{2g}^6 e_g^0$ → $t_{2g}^5 e_g^0$; CFSE = −2.4 → −2.0

So, only Fe^{2+} will lose crystal field stabilisation energy upon oxidation to +3 state, others will gain crystal field stabilisation energy.

245. (4) After removal of both axial ligands from octahedral complex the field becomes square planar and leads to following energy patterns.

$$d_{x^2-y^2} > d_{xy} > d_{z^2} > d_{xz} = d_{yz}$$

246. (4) Wavelength of the energy absorbed by the co-ordination compound is inversely proportional to ligand field strength of the given co-ordination compound. The strong field ligand causes higher splitting of the d-orbitals. The decreasing order of ligand field strength is $NH_3 > H_2O > Cl$. Therefore, decreasing order of absorbed wavelength is:

(I) > (II) > (III).

247. (2) $[Fe(H_2O)_6]^{2+} \longrightarrow t_{2g}^4 e_g^2$

$CFSE_0 = (-4 \times 0.4 + 2 \times 0.6)\,\Delta_0 = -0.4\,\Delta_0$

$[NiCl_4]^{2-} \longrightarrow e_g^4 t_{2g}^4$

$CFSE_t = (-4 \times 0.6 + 4 \times 0.4)\,\Delta_t = -0.8\,\Delta_t$

248. (1) Ruby is aluminium oxide (Al_2O_3) containing about 0.5 – 1% Cr^{3+} ions, which are randomly distributed in place of Al^{3+} ions.

249. (1) Cr^{3+} has d^3 configuration and forms an octahedral inner orbitals complex, therefore the set of degenerate orbitals are (d_{xy}, d_{yz} and d_{xz}) and ($d_{x^2-y^2}$ and d_{z^2}).

250. (3) Valence bond theory cannot distinguish between weak field ligands and strong field ligands. Therefore, it cannot predict quantitatively the magnetic properties of transition metal complexes.

251. (1) $[Fe(H_2O)_6]^{2+}$: cationic species and $[Fe(CN)_6]^{4-}$: anionic species
For the complex, $[Fe(H_2O)_6]_2[Fe(CN)_6]$
Now, $[Fe(H_2O)_6]^{2+}[Fe(CN)_6]^{4-}$
n = 4 n = 0
$\therefore\ \mu = \sqrt{4(4+2)}\ ;\ \mu = \sqrt{0(0+2)}$
$\mu = 4.9;\quad \mu = 0$

252. (2) $K_3[Co(CN)_6]$ is an octahedral complex. During splitting of *d*-orbitals in octahederal complexes, $d_{x^2-y^2}$ and d_{z^2} orbitals point towards the direction of ligands (i.e., they experience more repulsion and will be raised in energy by $\frac{3}{5}\Delta_0$).

253. (1) Given $\mu = 3.9$ BM
$\mu = \sqrt{n(n+2)}$ B.M. ; $3.9 = \sqrt{n(n+2)}$; $n = 3$
So, the central metal ion has 3 unpaired electrons.
∴ Configuration is either d^3 or d^7
As H_2O is a weak field ligand. V^{2+} and Co^{2+} will have 3 unpaired electrons.
V^{2+} has d^3 configuration; Co^{2+} has d^7 configuration.

254. (3) Electronic configuration of Mn^{2+} is,
$Mn^{2+} : 3d^5$ [↿ | ↿ | ↿ | ↿ | ↿] ($3d^5$)

$\mu = \sqrt{n(n+2)} = 5.9 \Rightarrow n = 5$
Presence of 5 unpaired e^- show that the complex of Mn^{2+} has only weak field ligand (NCS^-).

255. (1) Wilkinson catalyst : $[Rh(PPh_3)_3Cl]$
Chlorophyll : $C_{55}H_{72}O_5N_4Mg$
Vitamin B_{12} contains Co.
Carbonic anhydrase contains a Zn ion.

256. (3)

Metal ion	No. of unpaired elec.		Difference in the unpaired electrons
	High spin	Low spin	
Ni^{2+} ($3d^8$)	2	2	0
Mn^{2+} ($3d^5$)	5	1	4
Fe^{2+} ($3d^6$)	4	0	4
Co^{2+} ($3d^7$)	3	1	2

257. (1) Here, both the complexes (A) and (B) are paramagnetic with 3 unpaired electrons each. H_2O is also a weak field ligand, which causes lesser splitting than NH_3, which is comparatively stronger field ligand. Hence, the (Δ_0) value of (A) and (B) are calculated from the wavelengths of light absorbed and not from the wavelengths of light emitted.

258. (4) In case of similar metal atom or ion, the value of co-ordination number and the strength of the ligands determine the value of crystal field splitting energy.
Strength of ligands : $CN^- > NH_3 > H_2O > Cl^-$
∴ $K_3[Co(CN)_6]$ has the highest crystal field splitting energy.

259. (1) Lesser the wavelength of light absorbed (more energy) greater will be ligand strength.
Energy: Blue > Green > Red
Ligand strength: $L_2 > L_1 > L_3$

260. (3) Wilkinson catalyst is $[RhCl(PPh_3)_3]$.
$[RhCl(PPh_3)_3] \equiv [\overset{(+1)}{Rh}\ \overset{(-1)}{Cl}(\overset{(0)}{PPh_3})_3]$
$Rh^+ \to [Kr]$
4d: [↿⇂ | ↿ | ↿ | ↿ | ↿] 5s: [ ] 4p: [| |]
PPh_3 is a strong field ligand
[↿⇂ | ↿⇂ | ↿ | ↿ | ××] [××] [×× | ×× |]
dsp^2
Hence, Rh will undergo dsp^2 hybridisation to form square planar complex.

261. (3) $[Ni(CN)_4]^{2-}$ is square planar, diamagnetic (0 unpaired electrons) with dsp^2 hybridisation.
$[Ni(CO)_4]$ is tetrahedral, diamagnetic (0 unpaired electrons) with sp^3 hybridisation.
$[NiCl_4]^{2-}$ is tetrahedral, paramagnetic (2 unpaired electrons) with sp^3 hybridisation.
Hence, the option (c) is the correct answer.

262. (4) The complex having higher number of unpaired electrons will have higher value of spin only magnetic moment.
In all these complexes, the central metal ion is in +2 oxidation state.
Zn^{2+} has $3d^{10}$ outer electronic configuration with 0 unpaired electron.
Ni^{2+} has $3d^8$ outer electronic configuration with 2 unpaired electrons.
Co^{2+} has $3d^7$ outer electronic configuration with 3 unpaired electrons.
Mn^{2+} has $3d^5$ outer electronic configuration with 5 unpaired electrons
Hence, the correct order of spin-only magnetic moments is:
$[MnCl_4]^{2-} > [CoCl_4]^{2-} > [NiCl_4]^{2-} > [ZnCl_4]^{2-}$.

263. (3)

264. (3) The splitting is affected by the oxidation state of the central metal ion. A higher oxidation state leads to larger splitting, hence,
Δ_o of $[Ti(H_2O)_6]^{3+} > \Delta_o$ of $[Ti(H_2O)_6]^{2+}$
Further Δ_o also depends of $Z_{eff.}$ and $Z_{eff.}$ of 4*d* series is more than 3*d* series.
Hence,
Δ_o of $[Cr(H_2O)_6]^{2+} < \Delta_o$ of $[Mo(H_2O)_6]^{2+}$

265. (4)

	Metal ion	Unpaired electron	Magnetic moment
(i)	Cr^{2+}	4	$\sqrt{24}$ B.M.
(ii)	Fe^{2+}	4	$\sqrt{24}$ B.M.
(iii)	Co^{2+}	3	$\sqrt{15}$ B.M.
(iv)	Mn^{2+}	5	$\sqrt{35}$ B.M.

Since (i) and (ii), each has 4 unpaired electron, so they will exhibit same magnetic moment. Thus option (d) is correct.

266. (3) $Zn_2[Fe(CN)_6] \equiv 2Zn^{2+}[Fe(CN)_6]^{4-}$
$[Fe(CN)_6]^{4-} \equiv \left[Fe^{2+}(\overset{-1}{CN})_6\right]$;
Fe^{2+} [↿⇂ | ↿⇂ | ↿⇂ | |]
In presence of strong field ligand CN^-, e^- will pair and there will be no unpaired e^-. Hence the complex will not be yellow.

267. (1) $KMnO_4$ is colored because it absorbs light in the visible range of electromagnetic radiation. The permanganate ion is the source of color, as a ligand to metal (L → M) charge transfer takes place between oxygen's *p*-orbitals and the empty *d*-orbitals on the metal. This charge transfer takes place when a photon of light is absorbed which leads to the purple color of the compound.

268. (1) Symmetrically filled t_{2g} and e_g are those, which contain equal distribution of electrons.
$[FeF_6]^{3-}$ [↑ | ↑] e_g ; [↑ | ↑ | ↑] t_{2g} — {equal distribution of electrons in orbitals t_{2g} and e_g}

269. (1) Aqueous solution of $CoCl_2$ contains $[Co(H_2O)_6]^{2+}$ which is pinkish in colour so option (4) is incorrect.
Reduction potential of $Co^{3+} \longrightarrow Co^{2+}$ is high so option (2) is incorrect because Co^{2+} does not oxidises easily to Co^{3+}.
It is general case that symmetrical substituted octahedral complexes are less deeper in colour than tetrahedral complexes. So $[CoCl_4]^{2-}$ is deep blue in colour.

270. (3) $[Co(NH_3)_6]^{3+}$
[↿⇂ | ↿⇂ | ↿⇂ | ×× | ××] [××] [×× | ×× | ××]
(d^2sp^3)
Octahedral and diamagnetic.

271. (1) Both are paramagnetic, but CN^- will result in one unpaired e^- and F^- will result in 5 unpaired electrons.

272. (3) $\underset{\text{Lewis acid}}{Co^{3+}} + \underset{\text{Lewis base}}{6NH_3} \longrightarrow \underset{\text{adduct}}{\left[Co(NH_3)_6\right]^{3+}}$

273. (4) Following is the order of increasing value of Δ_o for the ligands:
$I^- < Br^- < Cl^- < F^- < H_2O < Py = NH_3 < en < NO_2^- << CN^- < CO$

274. (3) The absorption of visible light is responsible for coloured nature of the transition metal cation due to the promotion of one or more unpaired - *d* - electron from a lower to higher level within same *d*-subshell. Hence, higher the number of unpaired electron higher will be the absorption in visible light.
The electronic configuration of the given elements is
$Sc^{3+}(18) = 1s^2 2s^2 2p^6 3s^2 3p^6 3d^0 4s^0$
- no unpaired e^-.
$Ti^{4+}(18) = 1s^2 2s^2 2p^6 3s^2 3p^6 3d^0 4s^0$
- no unpaired e^-.
$V^{3+}(20) = 1s^2 2s^2 2p^6 3s^2 3p^6 3d^2 4s^0$
- Two unpaired e^-.
$Zn^{2+}(28) = 1s^2 2s^2 2p^6 3s^2 3p^6 3d^{10} 4s^0$
- no unpaired e.
Hence, $[V(NH_3)_6]^{3+}$ will most likely absorb visible light.

275. (2) $[NiL_4]^{2-} \Rightarrow$ Diamagnetic in nature (given)
$Ni^{2+}(d^8)$ 3d: [↿⇂ | ↿⇂ | ↿⇂ | ↿ | ↿] 4s: [ ] 4p: [| |]
$[NiL_4]^{2-}$ [↿⇂ | ↿⇂ | ↿⇂ | ↿⇂ | ××] [××] [×× | ×× | ××]
dsp^2
So, no. of unpaired electron = 0

276. (2) For a given metal ion, weak field ligands create a complex with smaller Δ, which will absorbs light of longer λ and thus lower frequency. Conservely, stronger field ligands create a larger Δ, absorb light of shorter λ and thus higher ν i.e. higher energy.

Red	< Yellow	< Green	< Blue
$\lambda = 650$ nm	570 nm	490 nm	450 nm

So order of ligand strength is

$L_1 < L_3 < L_2 < L_4$

277. (2) Among four ligands, CN^- and ox^{2-} are strong field ligands. So pairing takes places in $[(FeCN)_6]^{3-}$ and $[Co(ox)_3]^{3-}$ complexes. Fe^{3+} having one unpaired electron in its complex while Co^{3+} does not have unpaired electron in its complex.

278. (4) $[Cr(NO)(NH_3)(CN)_4]^{2-}$

Let O.S. of NO be +1, if x is the O.S. of Cr.

$\therefore x + (+1) + (0) + 4(-1) = -2 \Rightarrow x = +1$

But Cr does not exist in +1 O.S. Now, let O.S. of NO be zero.

$\therefore x + (0) + (0) + 4(-1) = -2 \Rightarrow x = +2$

$Cr^{2+} (d^4) \longrightarrow$ [↑|↑|↑|↑|]

Due to strong field ligands, pairing of e^- will take place, i.e.,

[|] e_g

[↑↓|↑|↑] t_{2g}

No. of unpaired e^-, n = 2

$\mu = \sqrt{n(n+2)} = \sqrt{2(2+2)} = \sqrt{8} \approx 2.82$ BM

279. (1) $[NiCl_4]^{2-}$: $Ni^{2+} (d^8)$

[↑↓|↑↓|↑↓|↑|↑] 3d [ ] 4s [| |] 4p

Due to weak field ligand Cl^-, pairing of e^- will not take place and the hybridisation will be sp^3.

[↑↓|↑↓|↑↓|↑|↑] [:] [:|:|:]

sp^3 Hybridisation

$[PdCl_4]^{2-}$: $Pd^{2+} (d^8)$

[↑↓|↑↓|↑↓|↑|↑] 4d [ ] 5s [| |] 5p

In case of heavier elements like Pd (4d series), the CFSE for square planar complex (Δ_s) increases such that a weak field ligand can also force the e^- from highest energy orbital ($d_{x^2-y^2}$) will pair with e^- of lower energy orbital (d_{xy}).

[↑↓|↑↓|↑↓|↑↓|:] [:] [:|:|]

d_{yz} d_{zx} d_{z^2} d_{xy} $d_{x^2-y^2}$ 5s 5p

dsp^2 hybridisation

280. (1) $[PtCl_2(NH_3)_2]$ has square planar geometry.

281. (1)

	Electronic configuration		No. of unpaired electrons
Co^{2+}	d^7	[↑↓|↑↓|↑|↑|↑]	3
Cr^{2+}	d^4	[↑|↑|↑|↑|]	4
Mn^{2+}	d^5	[↑|↑|↑|↑|↑]	5
Fe^{2+}	d^6	[↑↓|↑|↑|↑|↑]	4

$\therefore Co^{2+}$ has lowest no. of unpaired electrons, hence, lowest paramagnetic behaviour is shown by $[Co(H_2O)_6]^{2+}$.

282. (2) Correct order is : $CN^- > en > NCS^- > Cl^-$

283. (4)

Ion	Electronic Configuration		Magnetic moment $\left(\sqrt{n(n+2)}\right)$
Cr^{2+}	d^4	[↑|↑|↑|↑|]	$\sqrt{4(4+2)} = 4.9$
Mn^{2+}	d^5	[↑|↑|↑|↑|↑]	$\sqrt{5(5+2)} = 5.8$
Ti^{2+}	d^2	[↑|↑|		]	$\sqrt{2(2+2)} = 2.4$
V^{2+}	d^3	[↑|↑|↑|	]	$\sqrt{3(3+2)} = 3.8$

Cr^{2+} has magnetic moment of the order of 5 B.M.

284. (1) In octahedral complex, the magnitude of Δ_0 will be highest in a complex having strongest ligand. Of the given ligands, CN^- is strongest, so Δ_0 will be highest for $[Co(CN)_6]^{3-}$ Thus option (1) is correct.

285. (1) Complexes with dsp^2 hybridisation are square planar. All the complexes of Pt^{2+} are square planar including those with weak field ligand such as halide ions. Thus option (1) is correct.

286. (4) $[Ni X_4]^{2-}$, the electronic configuration of Ni^{2+} is

[↑↓|↑↓|↑↓|↑|↑] 3d [××] 4s [××|××|××] 4p

sp^3

It contains two unpaired electrons and the hybridisation is sp^3 (tetrahedral).

287. (3) The number of unpaired electrons in $Ni^{2+}(aq) = 2$. Water is a weak ligand, hence no pairing will take place.

$\therefore$ spin magnetic moment

$= \sqrt{n(n+2)} = \sqrt{2(2+2)} = \sqrt{8} = 2.82$ B.M

288. (4) d^5 —— strong ligand field

[↑↓|↑↓|↑| |]

$\mu = \sqrt{n(n+2)} = \sqrt{1(1+2)} = \sqrt{3} = 1.73$BM

d^3— in weak as well as in strong field

[↑|↑|↑| |]

$\mu = \sqrt{3(3+2)} = \sqrt{15} = 3.87$ B.M.

d^4– in weak ligand field

[↑|↑|↑|↑|]

$\mu = \sqrt{4(4+2)} = \sqrt{24} = 4.89$ B.M

d^4– in strong ligand field

[↑↓|↑|↑| |]

$\mu = \sqrt{2(2+2)} = \sqrt{8} = 2.82$ B.M

289. (2)

(1) Co^{3+}; n = 0 (2) Fe^{3+}; n = 1

(3) Mn^{3+}; n = 2 (4) Cr^{3+}; n = 3

Where n = number of unpaired electrons.

Due to strong field ligand CN^-, pairing of e^- will occur.

290. (3)

$[Fe(CN)_6]^{4-} \rightarrow$ [↑↓|↑↓|↑↓|:|:] [:] [:|:|:]

No of unpaired electron = 0

$[MnCl_4]^{2-} \rightarrow$ [↑|↑|↑|↑|↑] [:] [:|:|:]

No of unpaired electrons = 5

$[CoCl_4]^{2-} \rightarrow$ [↑↓|↑↓|↑|↑|↑] [:] [:|:|:]

No of unpaired electrons = 3

The greater the number of unpaired electrons, greater the magnitude of magnetic moment. Hence, the correct order will be :

$[MnCl_4]^{2-} > [CoCl_4]^{2-} > [Fe(CN)_6]^{4-}$

291. (4) $[Fe(CN)_6]^{4-}$ (d^2sp^3), $[Mn(CN)_6]^{4-}$ (d^2sp^3),

$[Co(NH_3)_6]^{3+}$ (d^2sp^3), $[Ni(NH_3)_6]^{2+}$ (sp^3d^2)

Hence, $[Ni(NH_3)_6]^{2+}$ is outer orbital complex.

292. (3) A square planar complex is formed by hybridisation of s, p_x, p_y and $d_{x^2-y^2}$ atomic orbitals.

[↑↓|↑↓|↑↓|↑↓|:] [:] [:|:|]

d_{yz} d_{zx} d_{z^2} d_{xy} $d_{x^2-y^2}$ s p_x p_y p_z

dsp^2

293. (3) The cyano and hydroxo complexes are far more stable than those formed by halide ion. This is due to the fact that CN^- and OH^- are strong Lewis bases (nucleophiles). Further, $[Fe(OH)_5]^{3-}$ is not possible because usual C.N. of Fe^{2+} is 6.

294. (1) It is general case that symmetrical substituted octahedral complexes are less deeper in colour than tetrahedral complexes. So $[CoCl_4]^{2-}$ is deep blue in colour.

295. (6) $[MnBr_4]^{2-}$: Br^- is weak field ligand

Mn^{+2} ($3d^5$), so high spin complex (paramagnetic)

$[NiCl_4]^{2-}$: Ni^{+2} ($3d^8$), and Cl^- is weak field ligand, so tetrahedral geometry (paramagnetic)

$[CoF_6]^{3-}$: Co^{+3} ($3d^6$) and F^- is weak field ligand,

$\therefore$ high spin complex (paramagnetic)

$[Mn(CN)_6]^{3-}$: Mn^{+3} ($3d^4$), CN^- is strong field ligand, $\therefore$ low spin complex but still has unpaired e^-, (paramagnetic)

$[Ti(CN)_6]^{3-}$: Ti^{+3} ($3d^1$), CN^- is strong field ligand, but has unpaired e^-, so (paramagnetic)

$[Cu(H_2O)]^{+2}$: Cu^{+2} ($3d^9$), it has 1 unpaired e^-, so paramagnetic.

$[MnBr_4]^{2-}$, $[NiCl_4]^{2-}$, $[CoF_6]^{3-}$, $[Mn(CN)_6]^{3-}$, $[Ti(CN)_6]^{3-}$, $[Cu(H_2O)_6]^{2+}$ are paramagnetic.

296. (15) $d^3 \rightarrow [t_{2g}^3, e_g^0]$, 3 unpaired electrons

d^4 (low spin) $\rightarrow [t_{2g}^4, e_g^0]$, 2 unpaired electrons

d^5 (high spin) $\rightarrow [t_{2g}^3, e_g^2]$, 5 unpaired electrons

d^6 (high spin) $\rightarrow [t_{2g}^4, e_g^2]$, 4 unpaired electrons

d^7 (low spin) $\rightarrow [t_{2g}^6, e_g^1]$, 1 unpaired electrons.

297. (6)

Ion	No. of e^-	No. of e^- in d orbital
Sc^{3+}	18	$3d^0$
Cr^{2+}	22	$3d^4$
Mn^{3+}	22	$3d^4$
Co^{3+}	24	$3d^6$
Fe^{3+}	23	$3d^5$

Thus Cr^{2+} and Mn^{3+} are isoelectronic.

Hence n = 2

Expressing the concerned reaction

$[CoCl_3(en)_2]Cl_2 + 2AgNO_3 \rightarrow [CoCl_3(en)_2](NO_3)_2 + 2AgCl$

∴ Moles of AgCl formed = 2

Hence formula of complex = $[Co(en)_2NH_3Cl]Cl_2$

For Co^{3+} $(3d^6) = t_{2g}^{2.2.2}\ e_g^{0.0}$

298. (2) In $[Ni(CO)_4]$, $Ni^0 : 3d^8\,4s^2$

Hybridisation state : sp^3; Unpaired electron = 0

In $[NiCl_4]^{2-}$, Ni^{2+} : $3d^8$; Hybridisation state : sp^3

Unpaired electron = 2 ; In $[PtCl_4]^{2-}$, $Pt^{2+} : 5d^8$

Hybridisation state : dsp^2; Unpaired electron = 0

In $[Ni(CN)_4]^{2-}$, $Ni^{2+} : 3d^8$; Hybridisation state : dsp^2

Unpaired electron = 0; In $[Pt(NH_3)_2Cl_2]$, $Pt^{2+} : 5d^8$

Hybridisation state : dsp^2; Unpaired electron = 0

299. (2) Pairing energy neglected w.r.t. Δ_0

$[Co(ox)_3]^{3-} ; Co^{+3} : d^6;\ t_{2g}^6 e_g^0$

$CFSE = 6 \times (-0.4\Delta_0) = -2.4\Delta_0$

$[Cr(ox_3)]^{3-} : Cr^{+3} : d^3;\ t_{2g}^3 e_g^0$

$CFSE = 3 \times (-0.4\Delta_0) = -1.2\Delta_0$

$$\frac{(CFSE)_{Co^{+3}}}{(CFSE)_{Cr^{+3}}} = 2$$

300. (7) $V^{3+} = [Ar_{18}]\,3d^2, n = 2$

$Ti^{2+} = [Ar_{18}]\,3d^2, n = 2$

$Ni^{2+} = [Ar_{18}]\,3d^8, n = 2$

$Fe^{2+} = [Ar_{18}]\,3d^6, n = 4$

$Co^{2+} = [Ar_{18}]\,3d^7, n = 3$

Only for Fe^{2+} and Co^{2+}, μ is more than 3.0 B.M.

[↿|↿]

Fe^{2+}, $3d^6$, [↿⇂|↿|↿], $n = 4, \mu > 3$

[↿|↿]

Co^{2+} $\cdot\ 3d^7$, [↿⇂|↿⇂|↿], $n = 3, \mu > 3$

∴ Number of unpaired electrons = 4 + 3 = 7

301. (2)

				Unpaired	Nature
$[Co(NH_3)_6]^{3+}$	Co^{3+}	$3d^6$	$t_{2g}^6\ e_g^0$	0	Diamagnetic
$[Co(C_2O_4)_3]^{3-}$	Co^{3+}	$3d^6$	$t_{2g}^6\ e_g^0$	0	Diamagnetic
$[MnCl_6]^{3-}$	Mn^{3+}	$3d^4$	$t_{2g}^3\ e_g^1$	4	Paramagnetic
$[Mn(Cl)_6]^{3-}$	Mn^{3+}	$3d^4$	$t_{2g}^4\ e_g^0$	2	Paramagnetic
$[CoF_6]^{3-}$	Co^{3+}	$3d^6$	$t_{2g}^4\ e_g^2$	4	Paramagnetic
$[Fe(CN)_6]^{3-}$	Fe^{3+}	$3d^5$	$t_{2g}^5\ e_g$	1	Paramagnetic
$[FeF_6]^{3-}$	Fe^{3+}	$3d^5$	$t_{2g}^3\ e_g$	5	Paramagnetic

302. (2)

$[FeF_6]^{3-}$ Paramagnetic (Weak field ligand) sp^3d^2

$[Fe(CN)_6]^{3-}$ Paramagnetic (Strong field ligand) d^2sp^3

$[Mn(CN)_6]^{3-}$ Paramagnetic (Strong field ligand) d^2sp^3

$[Co(C_2O_4)_3]^{3-}$ Diamagnetic (Strong field ligand) d^2sp^3

$[MnCl_6]^{3-}$ Paramagnetic (Weak field ligand) sp^3d^2

$[CoF_6]^{3-}$ Paramagnetic (Weak field ligand) sp^3d^2

Only $[Fe(CN)_6]^{3-}$ and $[Mn(CN)_6]^{3-}$ are paramagnetic and d^2sp^3 hybridised.

303. (1) Co has highest standard electrode potential (M^{+3}/M^{+2}) among Mn, Cr, Co, Fe

∴ Complex is $[Co(CN)_6]^{4-}$ and its splitting is as follows.

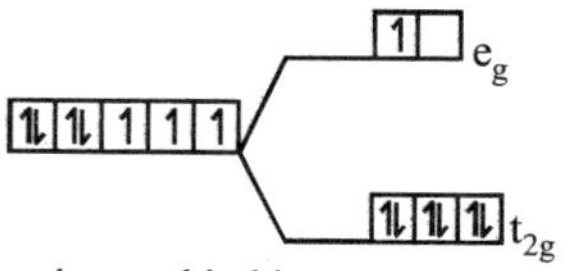

∴ electron in e_g orbital is one.

304. (0)

(O)	Compound	Colour	Unpaired e^-
A	$K_3[Co(NO_2)_6]$	Yellow	0
B	$K_4[Fe(CN)_6]$	Light Yellow	0
C	$K_3[Fe(CN)_6]$	Bright red	1
D	$Cu_2[Fe(CN)_6]$	Reddish Brown	0
E	$Zn_2[Fe(CN)_6]$	Bluish – white	0

305. (2) NH_3 is a strong field ligands

For octahedral symmetry

$Co^{+3} : 3d^6;\ t_{2g}^{2,2,2}\ e_g^{0,0}$

Unpaired electrons = 0

For weak field ligand, tetrahedral symmetry

$Ni^{+2} : 3d^8\ t_{2g}^{2,2,2}\ eg^{1,1}$

Unpaired electron = 2

306. (3) $\Rightarrow [Co(H_2O)_6]^{3+}$

With Co^{3+} H_2O acts as a strong field ligand.

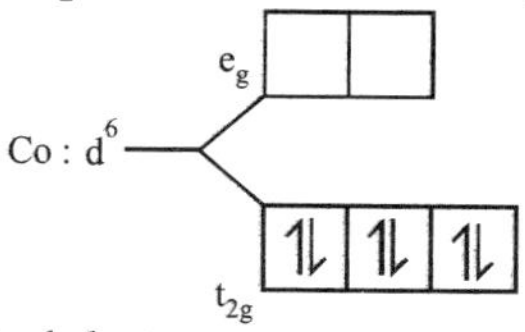

No unpaired electrons.

307. (28) NH_3 act as WFL with Ni^{2+}

$Ni^{2+} = 3d^8$

[↿⇂|↿⇂|↿⇂|↿|↿]

No. of unpaired electron = n = 2

$\mu = \sqrt{n(n+2)} = \sqrt{8} = 2.82$ BM

$= 28.2 \times 10^{-1}$ BM

308. (0) Pt^{2+} (d^8)

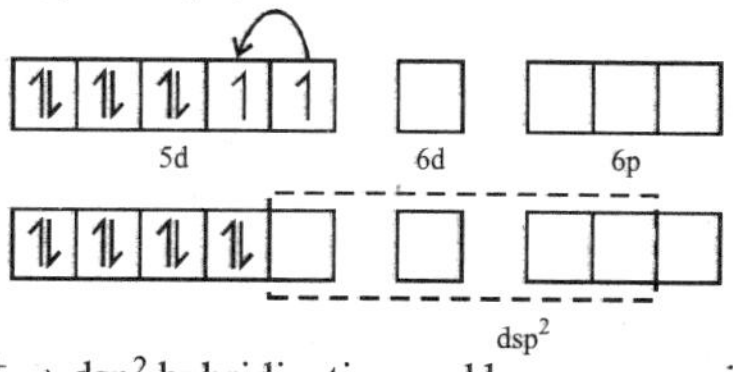

$Pt^{2+} \rightarrow dsp^2$ hybridization and have no unpaired electron.

∴ Magnetic moment = 0

309. (1) Co^{2+} : $3d^7$ configuration

$t_{2g}^{2,2,1}\ e_g^{1,1}$

310. (1) Spin magnetic moment of

$[Cr(CN)_6]^{3-}\ (t_{2g}^3 e_g^0)$

$\mu_1 = \sqrt{3(3+2)} = \sqrt{15}$ BM (Considering it as an outer orbital complex)

Spin magnetic moment of

$[Cr(H_2O)_6]^{3+}\ (t_{2g}^3 e_g^0)$

$\mu_2' = \sqrt{3(3+2)} = \sqrt{15}$ BM

$$\frac{\mu_1}{\mu_2} = \frac{\sqrt{15}}{\sqrt{15}} = 1$$

311. (3) $K_4[Fe(CN)_6] \Rightarrow Fe^{+2} = [Ar]3d^6$, CN

$\Rightarrow$ SFL, $t_{2g}^6\ e_g^0$

t_{2g} contain 6 electron so it become 3 pairs.

312. (4) $\mu = \sqrt{n(n+2)}$BM

$4 = \sqrt{n(n+2)} \Rightarrow n = 4$

$= \sqrt{4(4+2)} = \sqrt{24} = 4.90$

313. (4) $[Mn(NCS)_6]^{x-}$

Number of unpaired electron = 5

So, Mn must be in +2 oxidation state (Mn^{+2})

$\Rightarrow\ 2 + (-6) = -x\ \Rightarrow\ -4 = -x\ \Rightarrow x = 4$

314. (6) $[Mn(H_2O)_6]^{2+}$, $Mn^{2+} \Rightarrow 3d^5$

$\mu = \sqrt{5(5+2)} = 5.91$ BM

315. (480) $[Ti^{+3}(H_2O)_6]^{3+}$; $Ti^{+3} : 3d^1 4s^0 + 2g^1 e_g$

C.F.S.E. $= -0.4 \times \Delta_0 + 0$

$$\Rightarrow -\frac{96 \times 10^3}{N_A} J = -0.4 \times \Delta_0$$

$$\Rightarrow \Delta_0 = \frac{96 \times 10^3}{0.4 \times 6 \times 10^{23}} \Rightarrow \frac{hc}{\lambda} = \frac{96 \times 10^3}{0.4 \times 6 \times 10^{23}}$$

$$\lambda = \frac{0.4 \times 6 \times 10^{23} \times 6.4 \times 10^{-34} \times 3 \times 10^8}{96 \times 10^3}$$

$= 0.48 \times 10^{-6}$ m $= 480 \times 10^{-9}$ m $= 480$ nm

316. (2)

Ions:	V^{3+}	Cr^{3+}	Fe^{2+}	Ni^{3+}
Electronic Configuration:	$3d^2 4s^0$	$3d^3 4s^0$	$3d^6 4s^0$	$3d^7 4s^0$
Electrons(n):	2	3	4	3

Cr^{3+} and Ni^{3+} has the same number of unpaired electron. Hence, same magnetic moment.

317. (4) $[Ni(CN_4]^{2-} : Ni^{+2}$

$3d^8$

$\Rightarrow$ [↿⇂|↿⇂|↿⇂|↿⇂|] diamagnetic

CN^- : strong field ligand

$3d^{10}$

$[Ni(CO)_4] : Ni \Rightarrow$ [↿⇂|↿⇂|↿⇂|↿⇂|↿⇂] diamagnetic

$[NiCl_4]^{2-} : Ni^{2+}$

$3d^8$

$\Rightarrow$ [↿⇂|↿⇂|↿⇂|↿|↿] paramagnetic

Cl^- : weak field ligand

$[Fe(CN)_6]^{4-} : Fe^{2+}$

$3d^6$

$\Rightarrow$ [↿⇂|↿⇂|↿⇂| |] diamagnetic

CN^- : strong field ligand

$[Cu(NH_3)_4]^{+2} : Cu^{+2}$

$\Rightarrow$ [↿⇂|↿⇂|↿⇂|↿⇂|↿] paramagnetic

$3d^5$

$[Fe(CN)_6]^{-3} : Fe^{+3} \Rightarrow$ [↿⇂|↿⇂|↿| |]

paramagnetic, CN^- : strong field ligand

$3d^6$

$[Fe(H_2O)_6]^{2+} : Fe^{2+} \Rightarrow$ [↿⇂|↿|↿|↿|↿]

paramagnetic, H_2O : Weak field ligand

318. (7) $Co^{2+}: 3d^7 4s^0$, Cl^- is a weak field legand.

1 1 1 t_2

1l 1l e

Configuration $e^4 t_2^3 \Rightarrow m = 4$

Number of unpaired electrons = 3

$\therefore$ Sum of m and unpaired electrons = 4 + 3 = 7

319. (2) $[Fe(CN)_6]^{3-}$

CN^- is a strong field ligand, thus valence shell electronic configuration of Fe^{3+} will be :

$Fe^{3+} \equiv 3d^5 (t^5_{2g} e^0_g)$

$CFSE = 5(-0.4\Delta_0) = -2.0\Delta_0$

320. (2) Fehling solution is a complex of Cu^{2+}

$Cu^{2+} = 3d^9$; No. of unpaired $e^- = 1$

$\mu_{S.} = \sqrt{1(1+2)} = \sqrt{3} = 1.73 BM$

321. (0) $[Co(H_2O)_6]Cl_2$

Oxidation state of Co is :

$x + 6(0) = +2 \Rightarrow x = +2$.

H_2O is a weak field ligand thus, Co^{2+} electronic configuration = $3d^7 4s^0$

energy ↑ 1 1 e_g ; 1l 1l 1 t_{2g}

Number of unpaired $e^- = 3$

$\mu_s = \sqrt{n(n+2)} = \sqrt{3\times5} = \sqrt{15}$ B.M

For $[Cr(H_2O)_6]Cl_3$

Oxidation state of Cr = +3

Electronic configuration = $3d^3 4s^0$

energy ↑ – – e_g ; 1 1 1 t_{2g}

Number of unpaired electrons = 3

$\mu_s = \sqrt{3\times5} = \sqrt{15}$ B.M

Difference in spin only magnetic moment is $= \sqrt{15} - \sqrt{15} = 0$

322. (0) In the given complexes, Co present in +3 O.S. and all ligands occupy inner orbitals.

Shortest wavelength is shown by complex having maximum crystal field splitting.

$[Co(CN)_6]^{3-}$ energy ↑ — — e_g ; 1l 1l 1l t_{2g}

$M_s = \sqrt{0(0+2)} = 0$ B.M.

323. (2) $K_3[Fe(CN)_6]$

CN^- have –1 oxidation State.

Let oxidation state of Fe in the complex is x.

So, $x - 6 = -3 \Rightarrow x = +3$

Now, CN^- ion is a strong field ligand.

For Fe^{3+}: $-3d^5 4s^0$ Energy ↑ — — e_g ; 1l 1l 1 t_{2g}

Unpaired electron = 1

$\mu_B = \sqrt{n(n+2)} = \sqrt{1\times(1+2)} = 1.7$

$\mu_B = 2$ B.M

324. (6) Oxidation state of Mn

$x - 6 = -4 \Rightarrow x = +2$

Br is weak field ligand

$Mn^{2+} \rightarrow t_{2g}^3 e_g^2$; n = 5

$\mu_s = \sqrt{n(n+2)} = \sqrt{35} = 5.91 = 6$

325. (2) CN^- is a strong field ligand.

$[Fe(CN)_6]^{4-}$ Diamagnetic

$[Fe(CN)_6]^{3-}$ Paramagnetic (1 unpaired electron)

$[Ti(CN)_6]^{3-}$ Paramagnetic (1 unpaired electron)

$[Ni(CN)_4]^{2-}$ Square planar complex diamagnetic in nature

$[Co(CN)_6]^{3-}$ Diamagnetic

326. (3) $CoCl_3 \cdot 4NH_3 \rightarrow [Co(NH_3)_4Cl_2]Cl$

$NiCl_2 \cdot 6H_2O \rightarrow [Ni(H_2O)_6]Cl_2$

$PtCl_4 \cdot 2HCl \rightarrow H_2[PtCl_6]$

$[Ni(H_2O)_6]Cl_2 \xrightarrow{2AgNO_3} 2AgCl\downarrow + [Ni(H_2O)_6](NO_3)_2$

| 1l | 1l | 1l | 1 | 1 | $n = 2$
3d

$\mu = \sqrt{2(2+2)}$ B.M = 2.84BM ≈ 3

327. (6) $[Co(H_2O)_6]^{2+} + NH_3$ (excess) $\rightarrow [Co(NH_3)_6]^{3+} + 6H_2O$

for the Co^{3+}; NH_3 ligands are strong field ligand. Hence complex will be inner orbital or low spin complex.

The configuration of Co^{3+} is $\rightarrow 3d^6 \Rightarrow t_{2g}^6 e_g^0$

Total number electrons = 6.

328. (766) $\Delta_t = \dfrac{hc}{\lambda} = \dfrac{6.63\times10^{-34}\times3.08\times10^8}{600\times10^{-9}}$

$= 340.34 \times 10^{-21}$ J

$\Delta_0 = \dfrac{9}{4}\Delta_t = \dfrac{9}{4} \times 340.34 \times 10^{-21}$ J

$\approx 766 \times 10^{-21}$ J

329. (2)

$[Fe(H_2O)_3Cl_3]$, $\underbrace{K_3[Fe(CN)_6], [Co(NH_3)_6]Cl_3}_{\text{inner orbital complexes}}$

$K_3[Fe(CN)_6]$ has more value of Δ_0 than that of $[Co(NH_3)_6]Cl_3$; as CN^- is a strong ligand.

More $\Delta_0 \Rightarrow$ smaller value of absorbed λ

$K_3[Fe(CN)_6]$: Fe^{3+} : $3d^5$ $4s^0$ $4p^0$

| 1l | 1l | 1 | | | | | | | (d²sp³)

$n = 1$, $(\mu) = \sqrt{3}$BM = 1.732 BM ≈ 2 BM.

330. (173)

$B_2^+ \Rightarrow \sigma 1s^2 \sigma^* 1s^2 \sigma 2s^2 \sigma^* 2s^2 \pi 2p_x^1 = \pi 2p_y^0$

No. of unpaired electrons = 1

$\therefore \mu = \sqrt{n(n+2)}$ B.M $= \sqrt{1(1+2)}$ B.M

$= 1.73 \simeq 173 \times 10^{-2}$

331. (1) $[Co(NH_3)_6]Cl_2$

NH_3 is strong field ligand. Hence, pairing will occur.

$\Rightarrow Co^{2+} = 3d^7; t_{2g}^6 e_g^1$

Hybridisation will be d^2sp^3 with one unpaired e^-.

$[Co(NH_3)_6]Cl_3$ – d^2sp^3 hybridised with no unpaired e^-.

$Co^{+3} \Rightarrow 3d^6$; $t_{2g}^6 e_g^0$

332. (2) $[Co(CN)_6]^{4-}$

Co^{2+} : [Ar] $3d^7$

and CN^- is a strong field ligand which can pair electrons of central atom. After pairing of electron, one electron will remain unpaired.

Co^{2+} : [Ar] $t_{2g}^6 e_g^1$

It has one unpaired electron (n). So spin only magnetic moment, $\mu = \sqrt{n(n+2)}$B.M.

$\mu = \sqrt{3}$ B.M = 1.73 B.M. ≈ 2 B.M.

333. (2) $[Ni(CN)_6]^{2-}$

$Ni^{+4} \rightarrow d^6$

| ↑↓ | ↑↓ | ↑↓ | | |

CN^- : strong field ligand

Pairing will be there, hence zero unpaired electron.

$NiCl_2(aq) \rightarrow [Ni(H_2O)_6]Cl_2$

H_2O : Weak field ligand

| ↑↓ | ↑↓ | ↑↓ | ↑ | ↑ | → two unpaired e^-

$\therefore$ Total change in number of unpaired electrons = 2

334. (4) Octahedral splitting energy

$= \dfrac{hc}{\lambda} = \dfrac{6.626\times10^{-34}\times3\times10^8}{498\times10^{-9}}$

$= 0.0399\times10^{-17}$ J $= 3.99\times10^{-19}$ J $\approx 4\times10^{-19}$ J

335. (2) $Cu^{2+} : 3d^9$ | 1l | 1l | 1l | 1l | 1 |

No. of unpaired electron = 1

$\mu = \sqrt{1(1+2)} = \sqrt{3} = 1.73$ BM ≃ 2 BM

336. (0) $Ru^{2+} = 4d^6 = t_{2g}^6 e_g^0$ since $\Delta_0 > P$

No. of unpaired electrons = 0

$\therefore$ Magnetic moment = 0 B.M.

337. (26.92)

Number of moles of the complex $= \dfrac{0.3}{267.46}$

Number of moles of $AgNO_3 = 0.125 \times V \times 10^{-3}$

From stoichiometry:

$\dfrac{0.3}{267.46} \times 3 = 0.125 \times V \times 10^{-3}$

or, $V = \dfrac{0.3\times3\times1000}{267.46\times0.125} = 26.92$ mL

338. (20.0)

$\angle120° = 3$; $\angle90° = 6$; $\angle90° = 8$; $\angle180° = 2$

$\angle180° = 1 \Rightarrow$ Total = 10 $\Rightarrow$ Total = 10

Total number of 180°, 90° and 120° L–M–L bond angles = 10 + 10 = 20

339. (2) The metal to ligand bonding creates a synergic effect which strengthens the bond between CO and the metal i.e. metal carbon bond becomes strong.

Corrrect answer are A, C, D

340. (3) (D)Wilkinson catalyst have triphenyl phosphine as a ligand, $[(Ph_3P)_3RhCl]$

(E)Ca^{2+} and Mg^{2+} ions from more stable complex with EDTA.

341. (1) $Mn_2(CO)_{10}$ = decacarbonyl dimanganese (0)

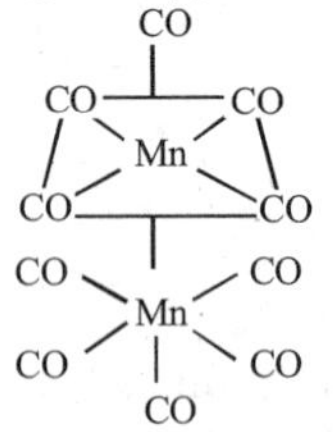

octahedral geometry around Mn atom

342. (3) $[Co_2(CO)_8]$

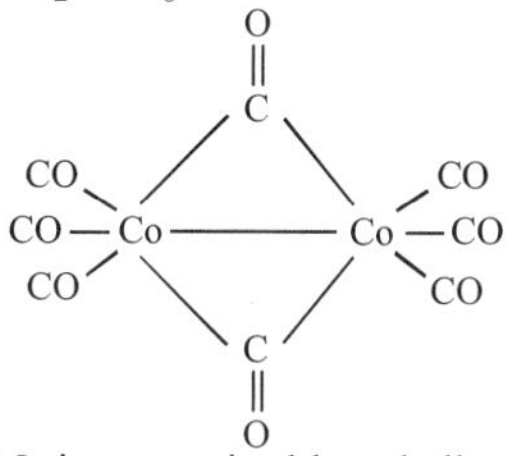

343. (1) Mg is present in chlorophyll and in black and white photography the developed film is fixed by washing with hypo solution which dissolves the undecomposed AgBr to form a complex ion $[Ag(S_2O_3)_2]^{3-}$

344. (1) *Cis* - Platin is used in chemotherapy to inhibits the growth of tumors.

345. (2) The covalant character of the bonding (M – C σ and M – C π bonding) which exists between the metal and the carbon atom of the CO can only be explained by the molecular orbital theory.

346. (4) EDTA is used in the treatment of lead poisoning. Deferrioxime B is used in treatment of iron poisoning and D-penicillamine is used in treatment of heavy metal poisoning, while *cis*-platin is used for treating cancer.

347. (3) *cis*-Platin is used as an anti-cancer drug.

348. (1) It contains two bridging CO ligands and one metal – metal (Co – Co) bond.

349. (3) Wilkinson's catalyst is $[Rh(PPh_3)_3Cl]$

350. (1)

351. (2) The donor atoms, molecules or anions which donate a pair of electrons to the metal atom or ion and form a coordinate bond with it are called ligands. In methane, there are no electrons for donation to central metal atom/ion, it is stable with complete octet configuration.

352. (4) Due to some backbonding by sideways overlapping between *d*-orbitals of metal and *p*-orbital of carbon, the Fe–C bond in $Fe(CO)_5$ has both σ and π character.

353. (1) EDTA is hexadentate, four donor O atoms and 2 donor N atoms, and for the formation of octahedral complex one molecule is required.

354. (3) The chlorophyll molecule plays an important role in photosynthesis. It has porphyrin rings and the metal Mg, not Ca.

355. (2)

Part-B (JEE Advanced)

1. (c) Complexes of type $[Ma_3b_3]^{n+}$ form facial and meridional isomers.

Set-I:

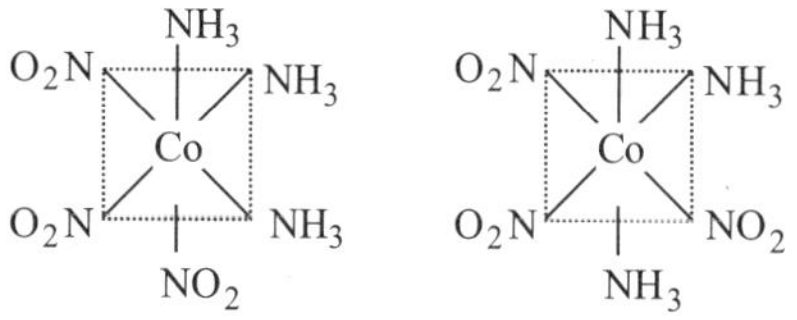

(Facial) (Meridional)

Complexes of type $[MX_2(L-L)_2]^{\pm}$ give *cis* and *trans* isomers.

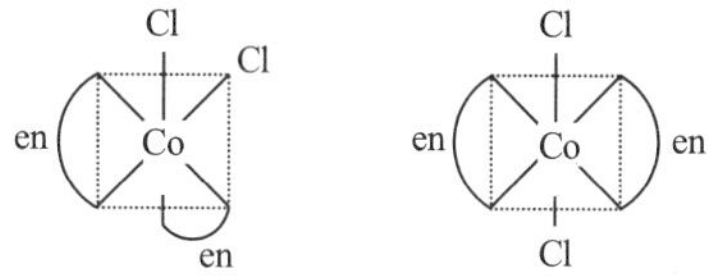

(*cis*) (*trans*)

Set-II: $[Co(NH_3)_5Cl]SO_4$ and $[Co(NH_3)_5SO_4]Cl$ are ionisation isomers.

2. (d) $[Co(H_2O)_4(NH_3)_2]Cl_3$ = Diamminetetraaquacobalt (III) chloride.

3. (c) The correct structure of EDTA is

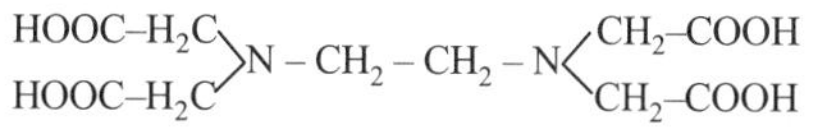

4. (b) Ionisation isomer of $[Cr(H_2O)_4Cl(NO_2)]Cl$ is $[Cr(H_2O)_4Cl_2]NO_2$.

5. (c) The correct IUPAC name of the given compound is tetraamminenickel (II) - tetrachloronickelate (II) thus (c) is the correct answer.

6. (a) $Co(NH_3)_4Br_2Cl$ will show both geometrical and ionization isomerism. $[Co(NH_3)_4Br_2]Cl$ and $[Co(NH_3)_4BrCl]Br$ are ionization isomers and geometrical isomers are

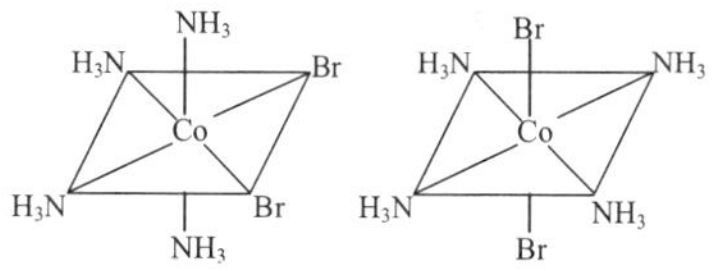

cis *trans*

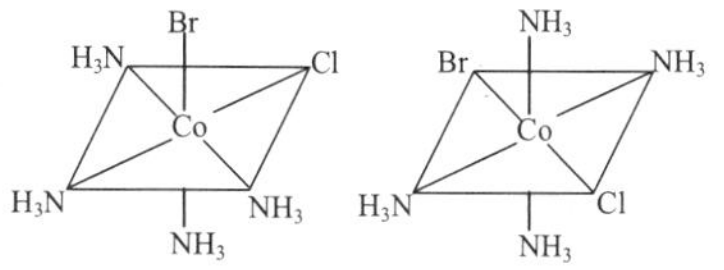

cis *trans*

7. (a) In $[MnO_4]^-$, Mn is in +7 oxidation state.
Electronic configuration of Mn (Z = 25) : $[Ar]\,3d^5 4s^2$
Electronic configuration of Mn^{7+} : $[Ar]\,3d^0 4s^0$
Central atom in other ions have definite number of *d* electrons.
No. of electrons

$[Co(NH_3)_6]^{3+}$ in Co^{3+} = Six; $[Fe(CN)_6]^{3-}$ in Fe^{3+} = Five; $[Cr(H_2O)_6]^{3+}$ in Cr^{3+} = three

8. (d) Organometallic compounds are those compounds in which metal atom is directly bonded with C-atom. H_3C-Li.

9. (3) Species having the equal number of valence shell electron are considered as isoelectonic species.
Valence electron in Ni $(CO)_4$ = 18

Species		Total valence electrons in species
$V(CO)_6$	–	17
$Cr(CO)_5$	–	16
$Cu(CO)_3$	–	17
$Mn(CO)_5$	–	17
$Fe(CO)_5$	–	18
$[Co(CO)_3]^{3-}$	–	18
$[Cr(CO)_4]^{4-}$	–	18
$Ir(CO)_3$	–	15

Note: Total valence electrons = Valence electron of metal + Electron donated by ligand.

10. (6) Possible isomers are
(I) $[Pt(NH_3)_4Cl_2]Br_2 \Rightarrow$ G.I. = 2
(II) $[Pt(NH_3)_4Br_2]Cl_2 \Rightarrow$ G.I. = 2
(III) $[Pt(NH_3)_4BrCl]Br.Cl \Rightarrow$ G.I. = 2
I, II, III are ionisation isomers of given complex and each exhibits two geometrical isomers. Therefore total possible isomers will be 6.

11. (6)

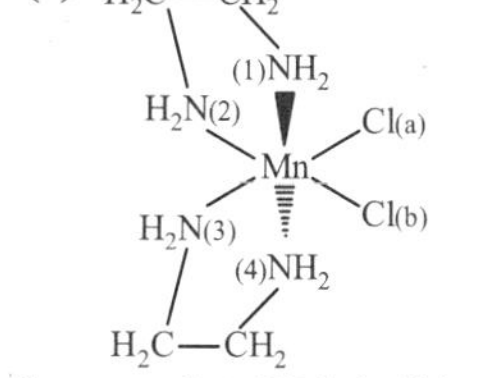

Structure of *cis*-$[Mn(en)_2Cl_2]$

Number of *cis* N – Mn – Cl bonds, *i.e.* Mn – N and Mn – Cl bonds are in adjacent positions.
1 – Cl (a) – Mn – N (1)2 – Cl (a) – Mn – N (2)
3 – Cl (a) – Mn – N (4)4 – Cl (b) – Mn – N (4)
5 – Cl (b) – Mn – N (3)6 – Cl (b) – Mn – N (1)
Hence, the answer is 6.
Note : All adjacent positions (at 90°) are *cis* to each other.

12.(5) $[CoL_2Cl_2]^-$ ($L = H_2NCH_2CH_2O^-$)
L is unsymmetrical didentate ligand.
So, the complete is equivalent to $[M(AB)_2a_2]$ Possible G.I. are

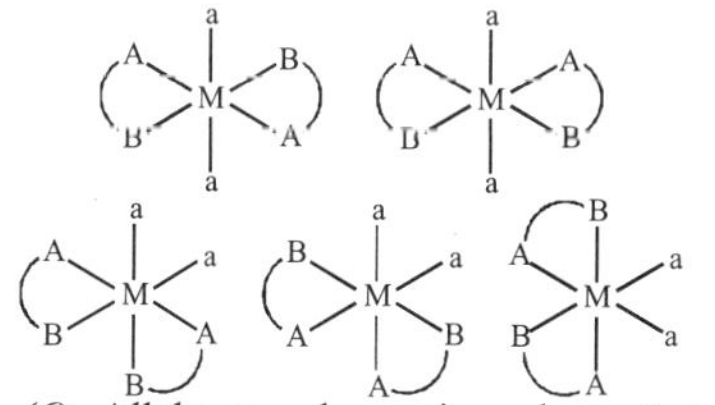

13. (6) All the complexes given show *cis-trans* isomerism
$[Co(NH_2-CH_2-CH_2-NH_2)_2Cl_2]^+, [CrCl_2(C_2O_4)_2]^{3-}$

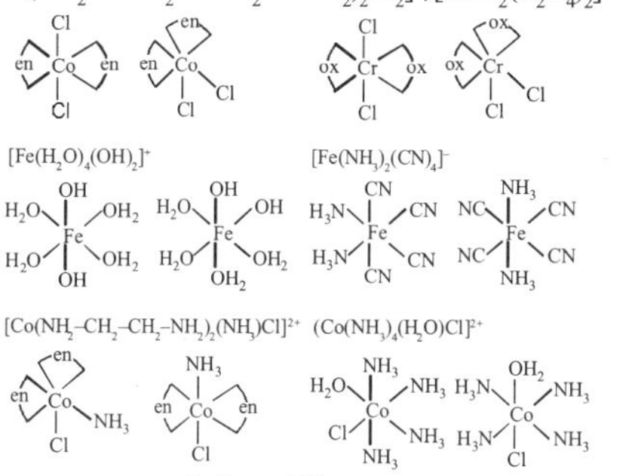

14. (3)

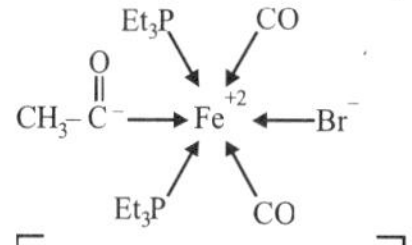

15. (8)

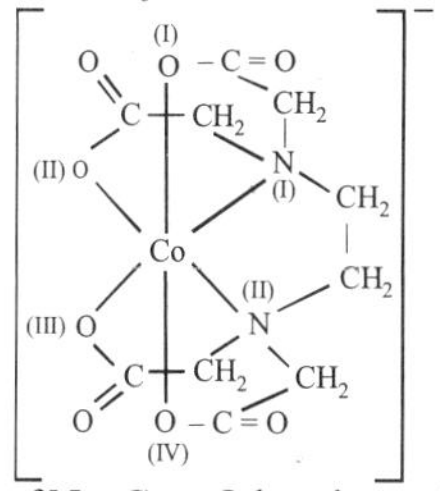

Total no. of N – Co – O bonds are 8.

16. (6) *m* moles of $[Cr(H_2O)_5Cl]Cl_2 = 0.01 \times 30 = 0.3$
m moles of $Cl^- = 0.3 \times 2 = 0.6$

[1 mole of complex gives 2 Cl^- ions]

m moles of $Ag^+ = m$ moles of Cl^-

$0.1 \times V = 0.6 \Rightarrow V = 6$ mL

17. (3)

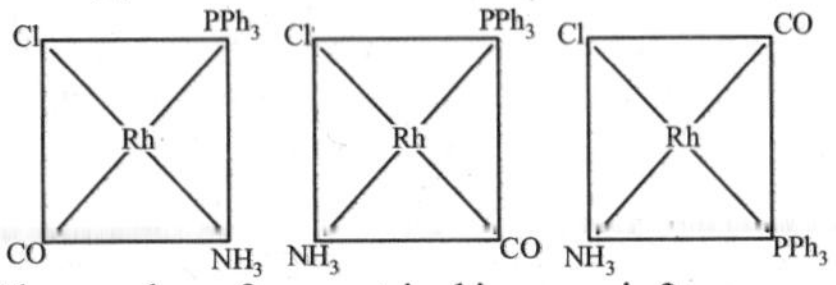

The number of geometrical isomers is 3.

18. (4) The number of water molecules directly bonded to the metal centre in $CuSO_4.5H_2O$ is 4.

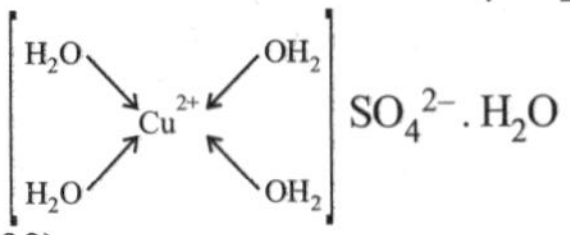

19. (6.00)

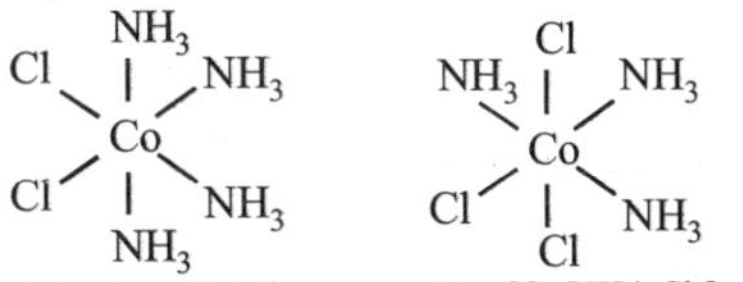

Cis-$[Co(NH_3)_4Cl_2]Cl$ — Two triangular faces having one N and two Cl toms.

Mer-$[Co(NH_3)_3Cl_3]$ — 4 triangular faces having one N and two Cl atoms.

$\therefore$ Sum = 2 + 4 = 6.

20. (8.00)

Total geometrical isomers are 8 which are given below

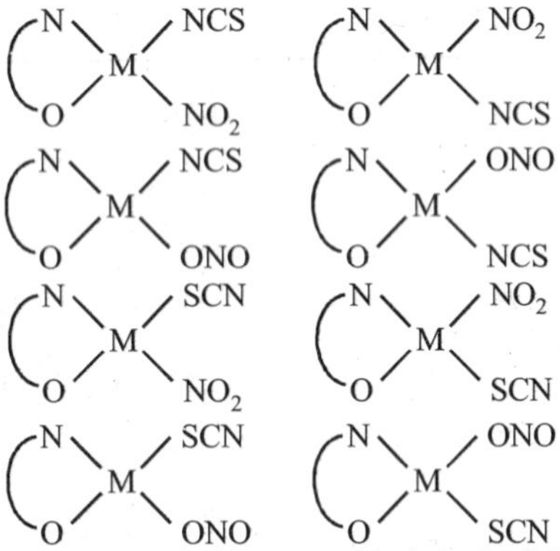

Note: 2 Geometrical isomers with gly, both of NO_2 and SCN which shows linkage isomerism.

So, total isomers = 2 × 2 × 2 = 8

21. Hexammine cobalt (III) chloride

22. (c, d) $\overset{+2}{\left[Pt(NH_3)_2 Br_2\right]}$

Hybridisation : dsp^2, geometry : square planar, show geometrical isomerism

(A) $[Pt(en)(SCN)_2]$: square planar, geometrical not possible

(B) $[Zn(NH_3)_2Cl_2]$: tetrahedral, geometrical not possible

(C) $[Pt(NH_3)_2Cl_4]$: octahedral, geometrical possible

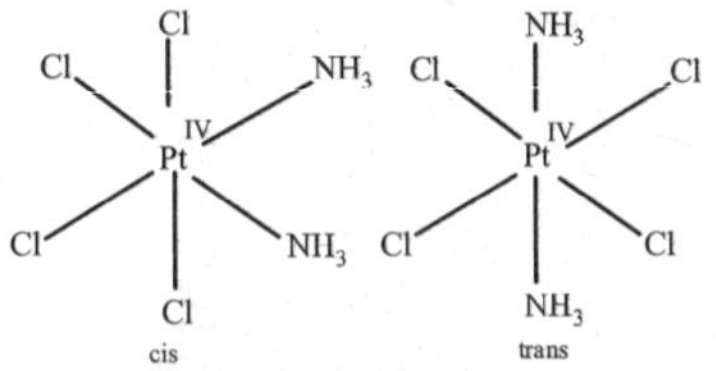

(D) $[Cr(en)_2(H_2O)SO_4]^+$: Octahedral, geometrical isomerism possible.

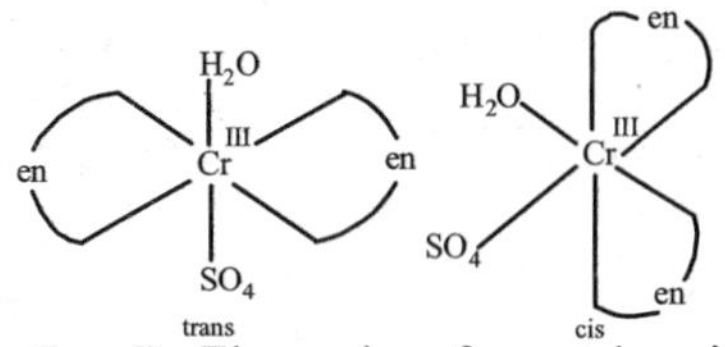

23. (b, d) The pair of complex ions $[Co(NH_3)_4Cl_2]^+$ and $[Pt(NH_3)_2(H_2O)Cl]^+$ show geometrical isomerism.

The pair of complexes $[Pt(NH_3)_3(NO_3)]Cl$ and $[Pt(NH_3)_3Cl]Br$ show ionisation isomerism. The other pairs given do not have same type of isomerism.

The complex $[CoBr_2Cl_2]^{2-}$ will show sp^3 hybridisation. The tetrahedral shape will not show any kind of isomerism.

(c) The complex $[PtBr_2Cl_2]^{2-}$ will show dsp^2 hybridisation. The square planar shape will show geometrical isomerism.

Henc, (c) is not an answer.

Note : Br and Cl are weak field ligand but in this case they will cause pairing of electrons. Most of the platinum complexes with four ligand are square planar in shape.

24. (c, d)

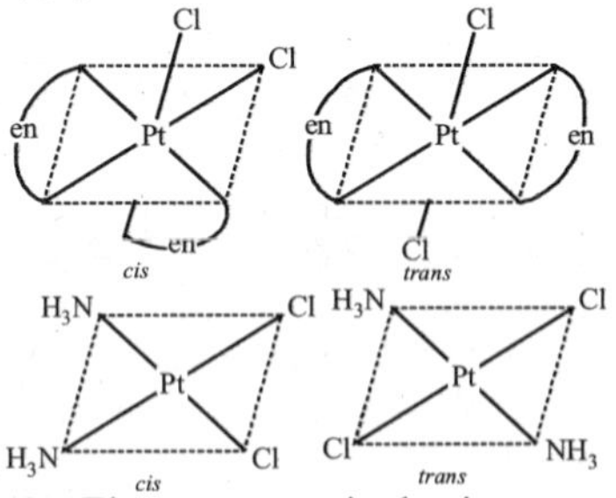

25. (b) The geometrical isomers of $[M(NH_3)_4Cl_2]$ can be represented as follows:-

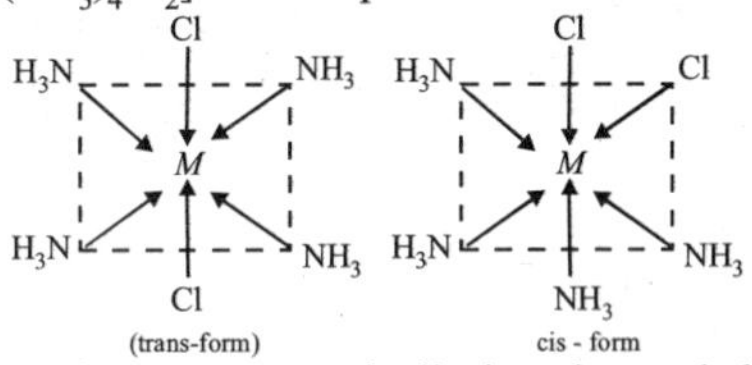

These isomers are optically inactive and they posses plane of symmetry in both and axis of symmetry in trans.

For a molecule to be optically active it should not possess axis of symmetry, centre of symmetry, or plane of symmetry. Therefore, statement I is correct and statement II is partially correct but not the correct explanation of statement I.

26. The complex A does not react with concentrated H_2SO_4 implying that all water molecules are coordinated with Cr^{3+} ion. Hence, its structure would be $[Cr(H_2O)_6]Cl_3$.

The compound B loses 6.75% of its original mass when treated with concentrated H_2SO_4. This loss is due to the removal of water molecules which is/are not directly coordinated to Cr^{3+} ion.

The mass of water molecules removed per mole of the complex = $\frac{6.75}{100} \times$ molar mass of the complex

$= \frac{6.75}{100} \times 266.5\text{ g} = 17.98\text{ g}$

This corresponds to one mole of water. Hence, the structure of the compound B will be $[Cr(H_2O)_5Cl](H_2O)Cl_2$

The compound C loses 13.5% of its mass when treated with concentrated H_2SO_4 which is twice of the mass lost by the compound B. Hence, the structure of the compound C will be $[Cr(H_2O)_4Cl_2](H_2O)_2Cl$.

27. (*i*) $[CoCl(NH_3)_5]^{2+}$ Formula of pentaamminechlorocobalt (III)

(*ii*) $LiAlH_4$ Formula of lithium tetrahydroa luminate (III)

28. (*i*) Pentaamminecarbonatochromium (III) chloride.

(*ii*) Potassium hexacyanochromate (III)

(*iii*) Pentaamminenitrito-o-cobalt (III) chloride

29. (a)

- All the given compounds are Co^{+3} Octahedral complexes consisting of strong & weak field ligands. So, the crystal field splitting energy depends upon the ligand strength. According to spectrochemical series, the increasing order of ligand strength is :-

$Cl^- < H_2O < NH_3 < CN^-$

- The crystal field splitting energy increases with increasing the ligand strength.
- So, the energy of absorption is more for stronger field ligands.
- As, $E = \frac{hc}{\lambda}$ where, E = Energy of absorption

$\therefore E \propto \frac{1}{\lambda}$ h = Planck's constant

c = speed of light

λ = wavelength

Thus, wavelength of absorption is less for stronger ligands

- Hence, the correct order is :-

$[Co(CN)_6]^{3-} < [Co(NH_3)_6]^{+3} < [Co(NH_3)_5(H_2O)]^{+3} < [Co(NH_3)_5Cl]^{+2}$

30. (c)

$Pb(NO_3)_2 + 2NaCl \longrightarrow PbCl_2 \downarrow + 2NaNO_3$

$PbCl_2 + 2Cl^-(\text{excess}) \longrightarrow [PbCl_4]^{2-}$ tetrachloroplumbate(II)ion

31. (a) $[Cr(NH_3)_6]^{3+}$

$Cr^{3+} \Rightarrow [Ar]\ 3d^3$

In presence of NH_3 ligand,

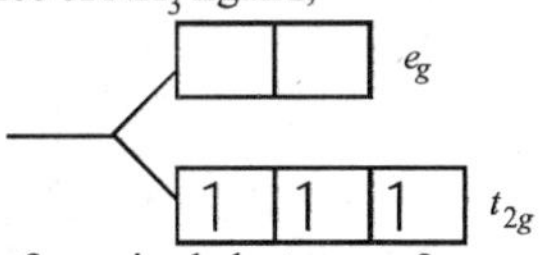

Number of unpaired electrons = 3

$\mu = \sqrt{n(n+2)}$ B.M.

$\mu = \sqrt{3(3+2)}$ B.M. = 3.87 B.M.

$[CuF_6]^{3-}$

$Cu^{3+} \Rightarrow [Ar]\ 3d^8$

In presence of F^- Ligand

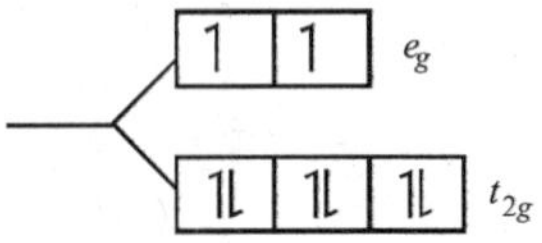

Number of unpaired electrons = 2

$\mu = \sqrt{n(n+2)}$ B.M.

$\mu = \sqrt{2(2+2)}$ B.M. = 2.84 B.M.

32. (a) Ni^{2+} with NH_3 shows CN = 6 forming $[Ni(NH_3)_6]^{2+}$ (sp^3d^2 hybridisation, Octahedral)
Pt^{2+} with NH_3 shows CN = 4 forming $[Pt(NH_3)_4]^{2+}$ (d^2sp hybridisation, square planner)
Zn^{2+} with NH_3 shows CN = 4 forming $[Zn(NH_3)_4]^{2+}$ (sp^3 hybridisation, tetrahedral)

33. (b) The electronic configuration of central metal ion in complex ions P, Q and R are

$P = [FeF_6]^{3-}$; Fe^{3+} : 

$Q = [V(H_2O)_6]^{2+}$; V^{2+} 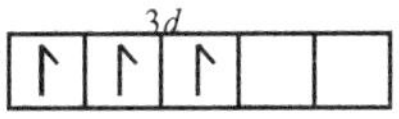

3d

$R = [Fe(H_2O)_6]^{2+}$; Fe^{2+} 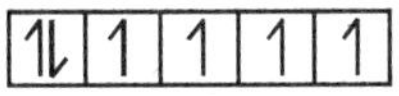

3d

Higher the no. of unpaired electron(s), higher will be magnetic moment.
Thus, the correct order of spin only magnetic moment is $Q < R < P$

34. (c) In both states (paramagnetic and diamagnetic) of the given complex, Ni exists as Ni^{2+} whose electronic configuration is [Ar] $3d^8 4s^0$.

Ni^{2+} :

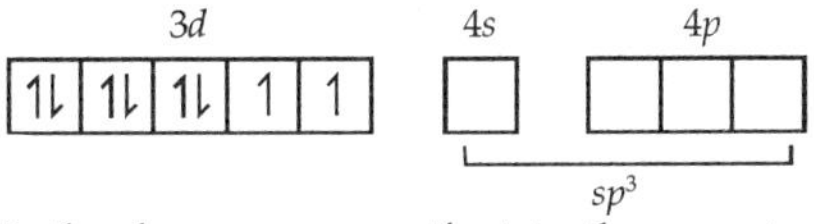

In the above paramagnetic state, the geometry of the complex is sp^3 giving tetrahedral geometry.
The diamagnetic state is achieved by pairing of electrons in 3*d* orbital.

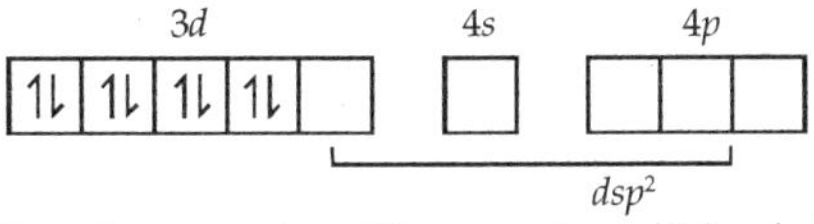

Thus, the geometry of the complex will be dsp^2 giving square planar geometry.

35. (c)

Complex	No. of electrons in outer *d* orbital	No. of unpaired electron (s)
$[Fe(CN)_6]^{3-}$ (K)	$3d^5$	1 (CN^- causes pairing of electrons)
$[Co(NH_3)_6]^{3+}$ (L)	$3d^6$	0(NH_3 causes pairing of electrons)
$[Co(oxal.)_3]^{3-}$ (M)	$3d^6$	0(oxalate causes pairing of electrons)
$[Ni(H_2O)_6]^{2+}$ (N)	$3d^8$	2
$[Pt(CN)_4]^{2-}$ (O)	$5d^8$	0(CN^- causes pairing of electrons)
$[Zn(H_2O)_6]^{2+}$ (P)	$3d^{10}$	–

Thus L, M, O and P are diamagnetic.

36. (b) $Ni^{+2} + 4Cl^- \xrightarrow{} \underset{sp^3}{[NiCl_4]^{2-}}$

$[NiCl_4]^{2-}$. = $3d^8$ configuration with nickel in + 2 oxidation state, Cl^- being weak field ligand does not compel for pairing of electrons.

So,

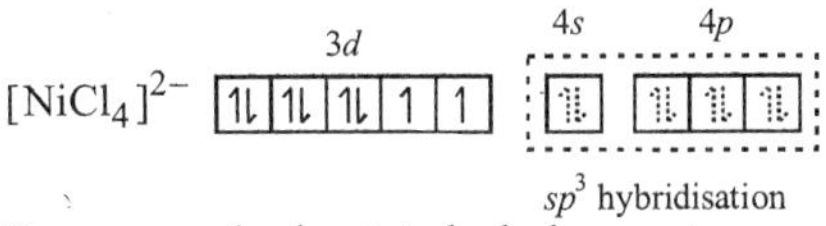

Hence, complex has tetrahedral geometry

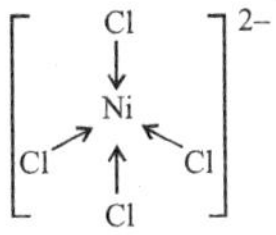

$Ni^{+2} + 4CN^- \longrightarrow [Ni(CN)_4]^{2-}$

$[Ni(CN)_4]^{2-} = 3d^8$ configuration with nickel in + 2 oxidation state, CN^- being strong field ligand compels for pairing of electrons.

$[NiCN_4]^{-2}$

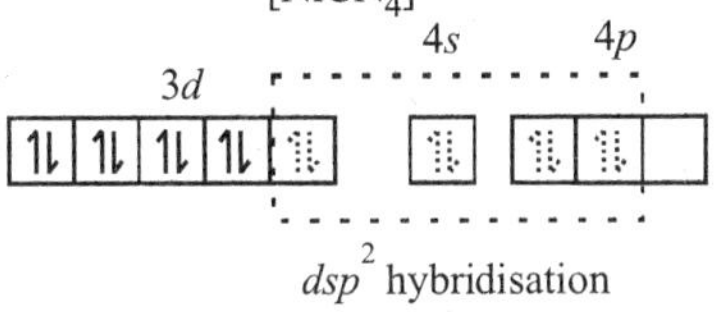

Hence, complex has square planar geometry.

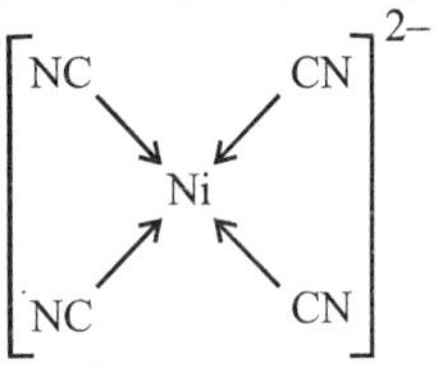

$Ni^{+2} + 6H_2O \longrightarrow [Ni(H_2O)_6]^{2+}$

$[Ni(H_2O)_6] = 3d^8$ configuration with nickel in + 2 oxidation state. As with $3d^8$ configuration two *d*-orbitals are not available for d^2sp^3 hybridisation.
So, hybridisation of Ni (II) is sp^3d^2 and Ni (II) with six co-ordination will have octahedral geometry.

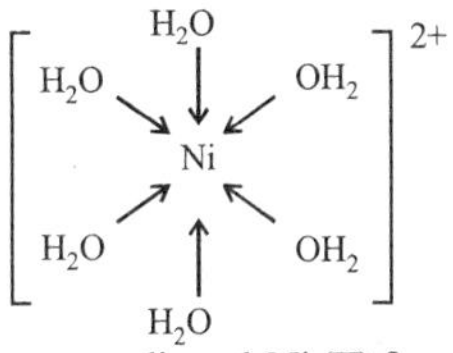

Note : With water as ligand, Ni (II) forms octahedral complexes.

37. (b) $[NiCl_4]^{2-}$, O.S. of Ni = +2
Ni(28) = $3d^8 4s^2$
Ni^{2+}

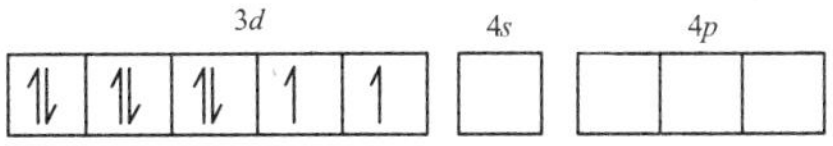

Cl^- being weak ligand it cannot pair up the two electrons present in 3*d* orbital

$[NiCl_4]^{2-}$

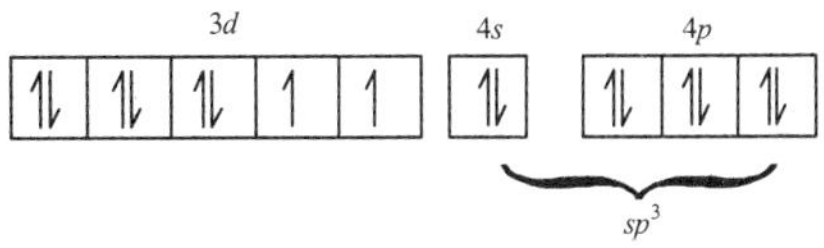

No. of unpaired electrons = 2
Magnetic moment, μ = 2.82 BM.

38. (a) Chromium in $Cr(CO)_6$ is in zero oxidation state and has $[Ar]^{18}\ 3d^5 4s^1$ as the electronic configuration. However, CO is a strong ligand, hence pairing up of electrons takes place leading to following configuration in $Cr(CO)_6$.

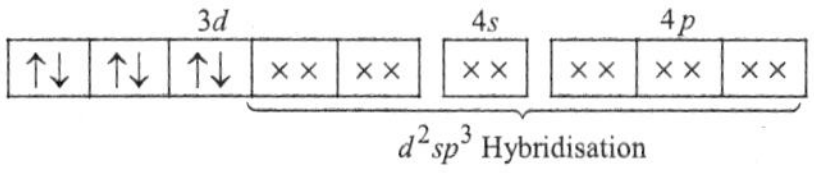

Since, the complex has no unpaired electron, its magnetic moment is zero.

39. (b) In carbonyls, O.S. of metal is zero.
In $[Ni(CO)_4]$, the oxidation state of nickel is zero.
Its configuration in $Ni(CO)_4$ is

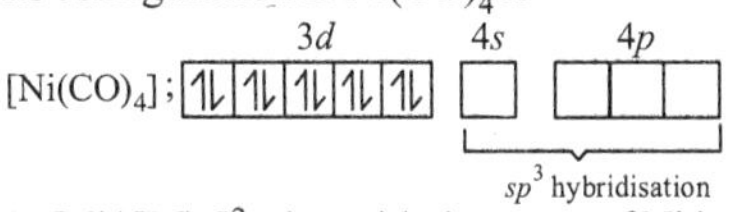

In $[Ni(CN)_4]^{2-}$ the oxidation state of Ni is 2+ and its configuration is

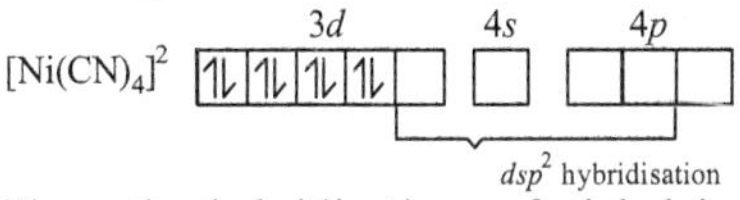

Thus, the hybridisations of nickel in these compounds are sp^3 and dsp^2 respectively.

40. (d) An anionic carbonyl complex can delocalise more electron density to antibonding pi-orbital of CO and hence, lowers the bond order.

41. (b) The required reaction is

$Ag^+ + 2NH_3 \rightleftharpoons [Ag(NH_3)_2]^+$; $K = ?$

From the given equations, we have

$$k_1 = \frac{[Ag(NH_3)]^+}{[Ag^+][NH_3]};\ k_2 = \frac{[Ag(NH_3)_2]^+}{[Ag(NH_3)]^+[NH_3]}$$

∴ The value of K is given by
$K = k_1 \times k_2 = 6.8 \times 10^{-3} \times 1.6 \times 10^{-3} = 1.08 \times 10^{-5}$.

42. (c) $\mu = \sqrt{n(n+2)}$ BM

(μ = spin magnetic moment)

Here Co is present as Co^{2+} ion which has 3 unpaired electrons. So the spin magnetic moment will be $\sqrt{3(3+2)}$, *i.e.* $\sqrt{15}$ BM.

43. (d) The configuration of Ni^{2+} is $3d^8$. For the elements of the first transition series, Cl^- behaves as a weak field/high spin ligand. Hence, Ni in $[NiCl_4]^{2-}$ is sp^3 hybridised leading to tetrahedral shape.

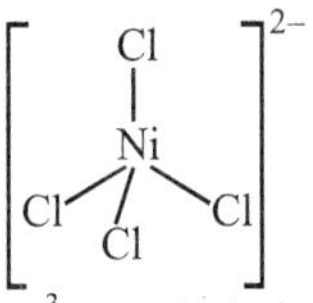

sp^3, Tetrahedral

44. (c) In metal carbonyl the metal is in **zero** oxidation state.
In $Ni(CO)_4$, O.N. of Ni = 0
For Ni (Z = 28)

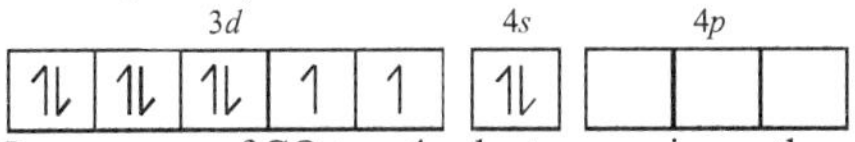

In presence of CO two 4*s* electrons pair up, thus

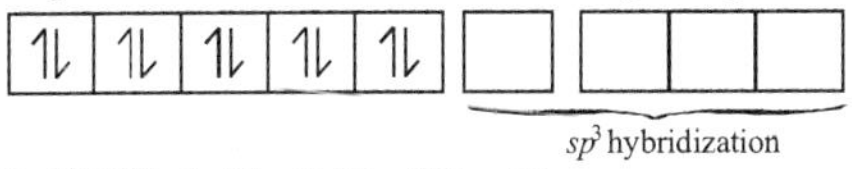

In $Ni(PPh_3)_2Cl_2$, O.N. of Ni = +2
For Ni^{2+}

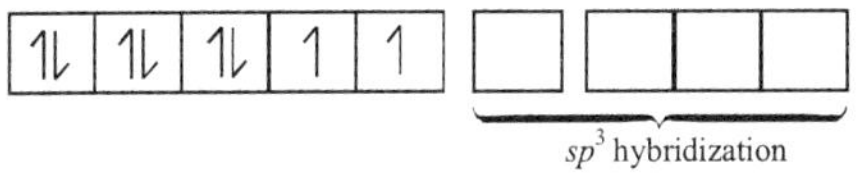

PPh_3 and Cl^- can't pair up d-electrons, leading to sp^3 hybridization and tetrahedral geometry

45. (c) The magnetic moment (μ) of a species is related to its number of unpaired electrons (n) in form of following expressions.

$\mu = \sqrt{n(n+2)}$ B.M

The number of unpaired electrons in the given pairs are as follows:

$Fe^{2+} = 1s^2, 2s^2 2p^6, 3s^2 3p^6 3d^6$

or 3d: ↑↓ | ↑ | ↑ | ↑ | ↑ Thus, here $n = 4$.

$Fe^{3+} = 1s^2, 2s^2 2p^6, 3s^2 3p^6 3d^5$

or 3d: ↑ | ↑ | ↑ | ↑ | ↑ Thus, here $n = 5$

$\overset{+}{N}O$ or $\overset{+}{N} = \ddot{O}$: $n = 0$; NO or $\times N = O$: $n = 1$

The given combinations differ in the number of unpaired electrons. Hence, these can be differentiated by the measurement on the solid state magnetic moment of nitroprusside ion.

46. (b) Highest parmagnetic character will be shown by the ion having maximum number of unpaired electrons in their d-subshells.

${}_{25}Cr^{3+}$ has 3 unpaired electrons ;
${}_{26}Fe^{2+}$ has 4 unpaired electrons
${}_{29}Cu^{2+}$ has 1 unpaired electrons ;
${}_{30}Zn^{2+}$ has no unpaired electrons

So (a), (b) & (c) show paramagnetism. Out of which (b) has the highest paramagnetism.

47. (c)

Atom/Ion Complex	Configuration (3d / 4s / 4p)	No. of unpaired electrons	Magnetic nature
Ni^{2+} (d^8)	3d: ↑↓ ↑↓ ↑↓ ↑ ↑; 4s: □; 4p: □ □ □	2	Paramagnetic
$[NiCl_4]^{2-}$	3d: ↑↓ ↑↓ ↑↓ ↑ ↑; 4s: :; 4p: : : : (sp^3)	2	Paramagnetic
$[Ni(CN)_4]^{2-}$	3d: ↑↓ ↑↓ ↑↓ ↑↓ :; 4s: :; 4p: : : □ (Rearrangement, dsp^2)	0	Diamagnetic
Ni (d^8s^2)	3d: ↑↓ ↑↓ ↑↓ ↑ ↑; 4s: ↑↓; 4p: □ □ □	2	Paramagnetic
$[Ni(CO)_4]$	3d: ↑↓ ↑↓ ↑↓ ↑↓ ↑↓; 4s: :; 4p: : : : (Rearrangement, sp^3)	0	Diamagnetic

48. (b) Mn^{2+} in $MnSO_4.4H_2O$ has d^5 configuration (five unpaired electrons); Cu^{2+} in $CuSO_4.5H_2O$ has d^9 configuration (one unpaired electron); Fe^{2+} in $FeSO_4.6H_2O$ has d^6 configuration (four unpaired electron); and Ni^{2+} in $NiSO_4.6H_2O$ has d^8 configuration (two unpaired electron). Thus, $CuSO_4.5H_2O$ has lowest degree of paramagnetism.

49. (5) $[Co(CN)_4]^{4-} \Rightarrow Co^0 \; 3d^7 4s^2 \Rightarrow 3d^9$

Due to SFL, CN– pairing and transference of electron takes place and hybridisation is sp^3.

Geometry ⇒ Tetrahedral 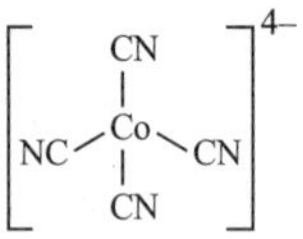

$[Co(CO)_3NO]$
$Co^0 \Rightarrow 3d^7 4s^2 \Rightarrow 3d^9$ due to SFL, CO and NO $\Rightarrow sp^3$ hybridisation

Geometry = Tetrahedral 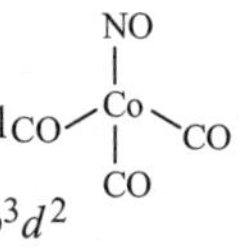

$XeF_4 \Rightarrow 4bp + 2lp = sp^3d^2$

Square planar 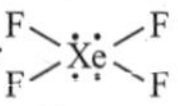

$PCl_4^+ \Rightarrow 4pb + 0lp$

$sp^3 \Rightarrow$ tetrahedral 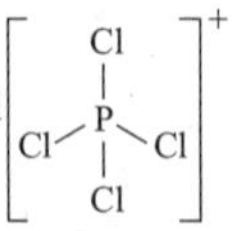

$[PdCl_4]^{2-} \Rightarrow Pd^{2+}$, Cl^- behaves as SFL

$Pd^{2+} \Rightarrow 4d^8 \Rightarrow dsp^2 \Rightarrow$ square planar

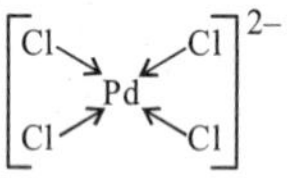

$ICl_4^- \Rightarrow 4bp + 2lp$

sp^3d^2 square planar 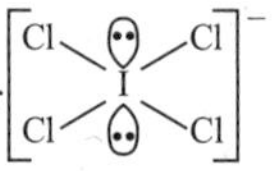

$[Cu(CN)_4]^{3-} \Rightarrow Cu^{+1} \Rightarrow 3d^{10} \Rightarrow sp^3$

Tetrahedral 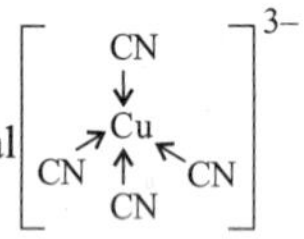

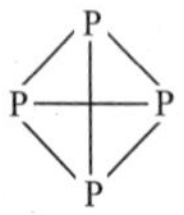 Tetrahedral

50. (1) $[Mn(NH_3)_6]^{3+} \Rightarrow d^4$ system : ↑↓ | 1 | 1 | ×× | ××, NH_3 strong field, ligand. Paramagnetic

$[MnCl_6]^{3-} \Rightarrow d^4$ system : 1 | 1 | 1 | 1 | □, Cl^- a weak field ligand. Paramagnetic

$[FeF_6]^{3-} \Rightarrow d^5$ system : 1 | 1 | 1 | 1 | 1, F^- a weak field ligand. Paramagnetic

$[CoF_6]^{3-} \Rightarrow d^6$ system : ↑↓ | 1 | 1 | 1 | 1, F^- a weak field ligand. Paramagnetic

$[Fe(NH_3)_6]^{3+} \Rightarrow d^5$ system : ↑↓ | 1 | 1 | 1 | ××, NH_3 a strong field ligand. Paramagnetic

$[Co(en)_3]^{3+} \Rightarrow d^6$ system : ↑↓ | ↑↓ | ↑↓ | ×× | ××, en a strong field ligand. Diamagnetic

Only 1 complex is diamagnetic.

51. (1)

- H-atom – $1s^1$ Paramagnetic
- NO_2 = 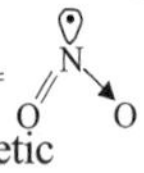 odd electron species Paramagnetic
- O_2^- (superoxide) = One unpaired electron in π^* M.O. Paramagnetic
- S_2 (in vapour phase) = same as O_2, two unpaired e^-s are present in π^* M.O. Paramagnetic
- $Mn_3O_4 = \overset{+2}{2MnO} . \overset{+4}{MnO_2}$ Paramagnetic
- $(NH_4)_2[FeCl_4] = Fe^{2+} = 3d^6 4s^0$ Paramagnetic
- $(NH_4)_2[NiCl_4] = Ni = 3d^8 4s^2$

$Ni^{2+} = 3d^8 4s^0$ Paramagnetic

- $K_2MnO_4 = 2K^+$ 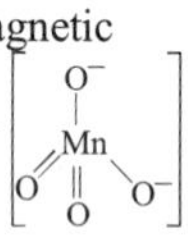, $Mn^{6+} = [Ar] 3d^1$

Paramagnetic

- $K_2CrO_4 = 2K^+$ 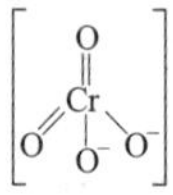, $Cr^{6+} = [Ar] 3d^0$ Diamagnetic

52. (4) $Fe(26) \to [Ar]_{18} 3d^6 4s^2$

$Fe^{3+} \to [Ar]_{18} 3d^5 4s^0$

SCN^- is weak field ligand hence pairing will not occur.

$\therefore$ $Fe^{3+} \to$ 1 | 1 | 1 | 1 | 1

Unpaired electrons = 5

Magnetic moment = $\sqrt{5(5+2)}$ B.M.

$= \sqrt{35}$ B.M. = 5.92 B.M.

CN^- is strong field ligand hence pairing will take place.

$\therefore$ $Fe^{3+} \to$ ↑↓ | ↑↓ | 1 | □ | □

Unpaired electrons = 1

Magnetic moment = $\sqrt{1(1+2)}$ B.M.

$= \sqrt{3}$ B.M. = 1.732

Difference = 5.92 – 1.732 = 4.188

Hence, answer is (4).

53. (7.73)

$[MnBr_6]^{3-}$	$[Mn(CN)_6]^{3-}$
Ligand nature = Weak field ligand	Strong field ligand
Oxidation state = Mn^{3+}	Mn^{3+}
Outer most electron = $3d^4$	$3d^4$
1 – e_g; 1 1 1 – t_{2g}	— — e_g; ↑↓ 1 1 t_{2g}
Unpaired electron = 4	2
Magnetic moment ($\sqrt{n(n+2)}$) = 4.89	2.84
Sum = 4.89 + 2.84 = 7.73	

54. Paramagnetism;

$[Mn(H_2O)_6]^{2+}$ shows paramagnetism because of presence of 5 unpaired electrons in the outer most orbital ($3d^5$) of Mn^{2+}.

55. False : Octahedral complexes of Fe(III) like $[Fe(CN)_6]^{3-}$ are low spin (d^2sp^3 hybridization) with one unpaired electron and have magnetic moment of about 1.9 BM. On the other hand, complexes of Fe(II) like $[Fe(CN)_6]^{2-}$ are low spin complex (d^2sp^3) with no unpaired electron and thus diamagnetic.

56. (a, b, d)

(a) **$[FeCl_4]^-$**

$Fe \Rightarrow [Ar]3d^6\,4s^2$

$Fe^{+3} \Rightarrow [Ar]\,3d^5\,4s^0$

Cl^- is weak field ligand, So pairing does not take place.

$\therefore$ In $[FeCl_4]^-$

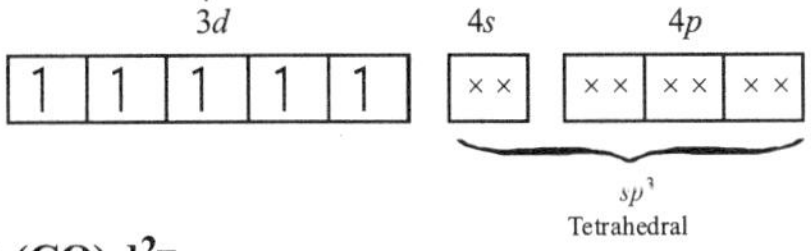

$[Fe(CO)_4]^{2-}$

$Fe \Rightarrow [Ar]3d^6\,4s^2$

$Fe^{-2} \Rightarrow [Ar]\,3d^8\,4s^2$

$\therefore$ In $[Fe(CO)_4]^{2-}$

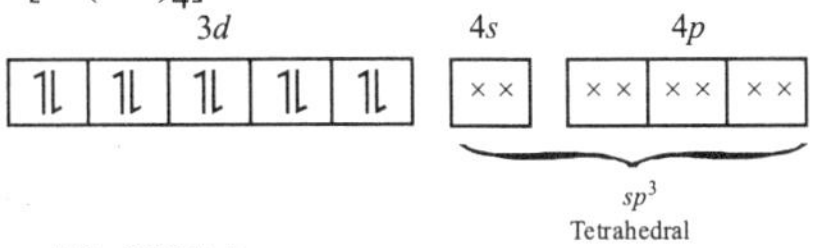

(b) **$[Co(CO)_4]^-$**

$Co \Rightarrow [Ar]3d^7\,4s^2$

$Co^{-1} \Rightarrow [Ar]\,3d^8\,4s^2$

$\therefore$ In $[Co(CO)_4]^-$

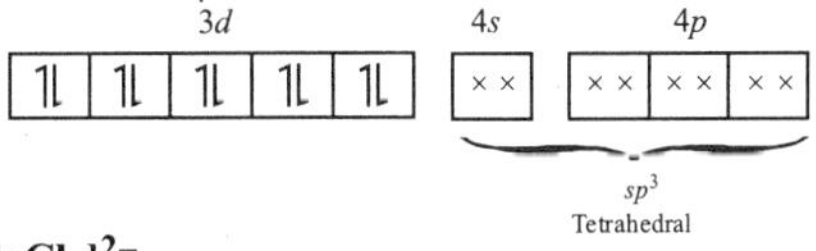

$[CoCl_4]^{2-}$

$Co \Rightarrow [Ar]3d^7\,4s^2$

$Co^{+2} \Rightarrow [Ar]\,3d^7\,4s^0$

$\therefore$ In $[CoCl_4]^{2-}$

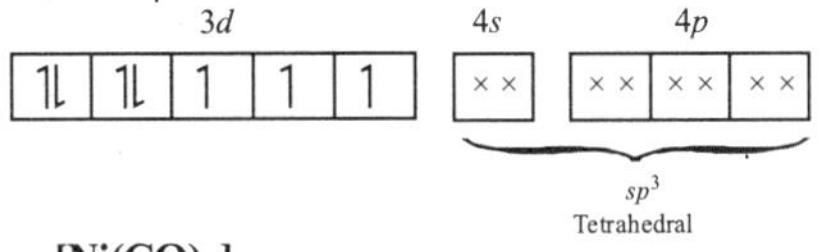

(c) **$[Ni(CO)_4]$**

$Ni \Rightarrow [Ar]3d^8\,4s^2$

$Ni^0 \Rightarrow [Ar]\,3d^8\,4s^2$

$\therefore$ In $[Ni(CO)_4]$

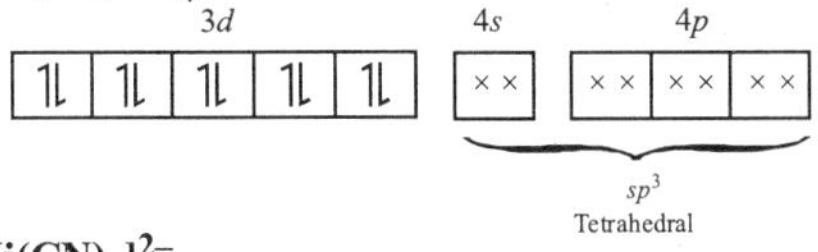

$[Ni(CN)_4]^{2-}$

$Ni \Rightarrow [Ar]3d^8\,4s^2$

$Ni^{+2} \Rightarrow [Ar]\,3d^8\,4s^0$

$\therefore$ In $[Ni(CN)_4]^{2-}$

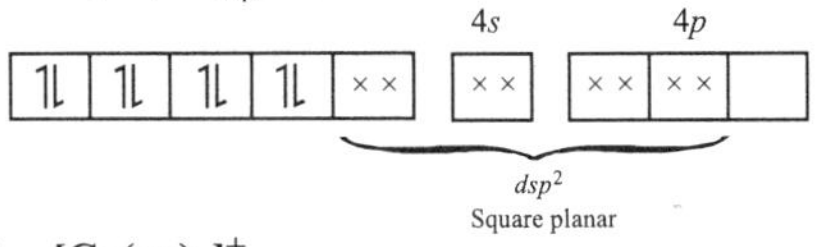

(d) **$[Cu(py)_4]^+$**

$Cu \Rightarrow [Ar]3d^{10}\,4s^1$

$Cu^+ \Rightarrow [Ar]\,3d^{10}\,4s^0$

$\therefore$ In $[Cu(py)_4]^+$

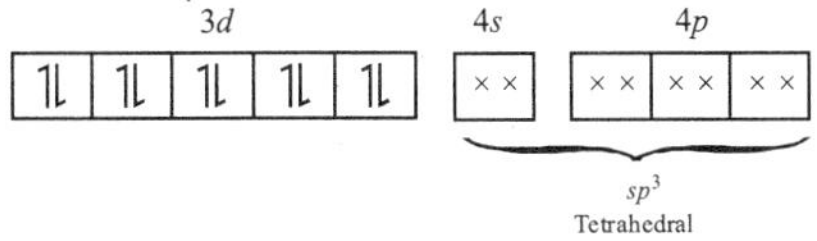

$[Cu(CN)_4]^{3-}$

$Cu \Rightarrow [Ar]3d^{10}\,4s^1$

$Cu^+ \Rightarrow [Ar]\,3d^{10}\,4s^0$

CN^- is so pairing occurs.

$\therefore$ In $[Cu(CN)_4]^{3-}$

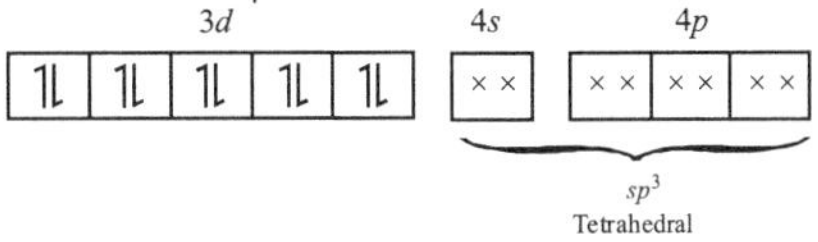

57. (a, c)

(a) $[FeCl_4]^-$

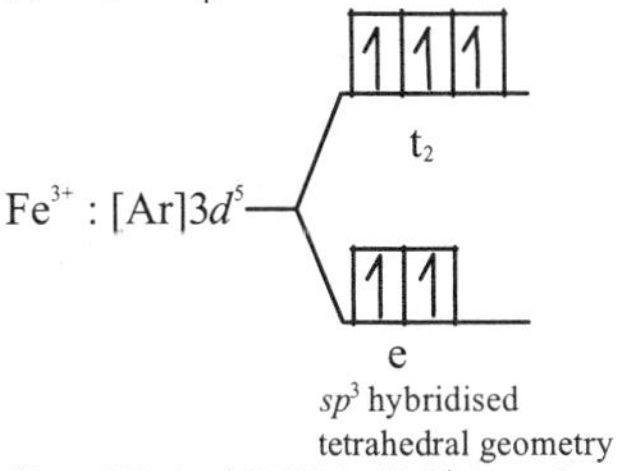

(b) $[Co(en)(NH_3)_2Cl_2]^+$ has three geometrical isomers.

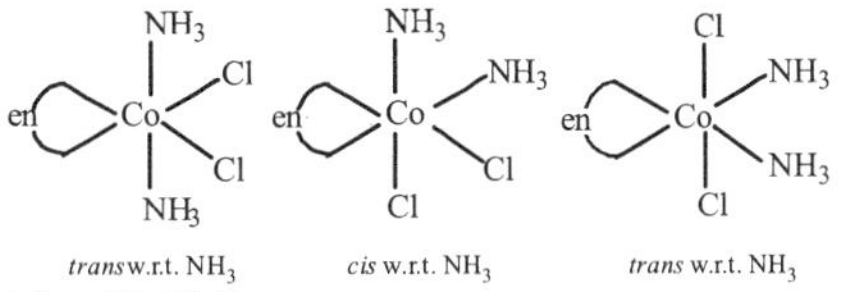

(c) $[FeCl_4]^-$

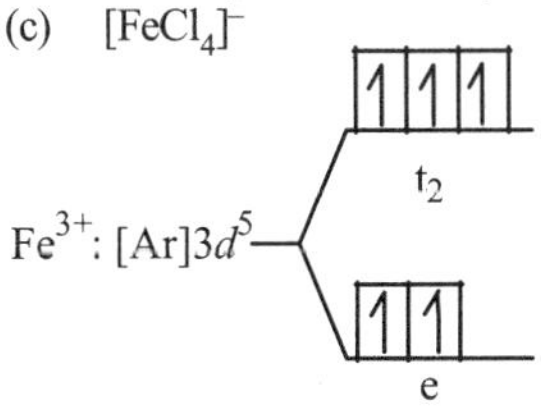

Number of unpaired electrons (n) = 5

Spin only magnetic moment $= \sqrt{n(n+2)}$ B.M.

= 5.92 B.M.

$[Co(en)(NH_3)_2Cl_2]^+$

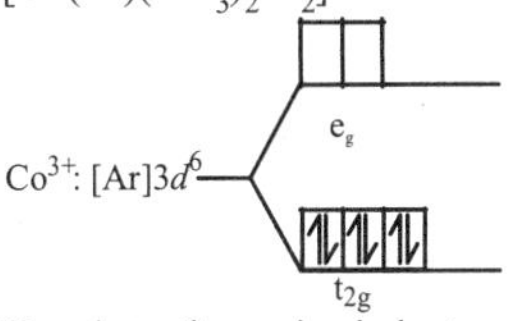

Number of unpaired electrons (n) = 0

Spin only magnetic moment $= \sqrt{n(n+2)}$ B.M. = 0

(d) $[Co(en)(NH_3)_2Cl_2]^+$

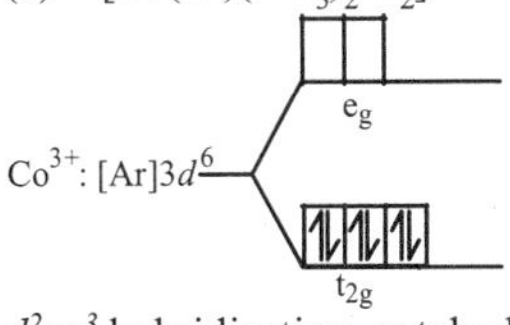

d^2sp^3 hybridisation, octahedral geometry

58. (a, b, d)

(a) $[Co(en)(NH_3)_3(H_2O)]^{3+}$ has 2 geometrical isomers.

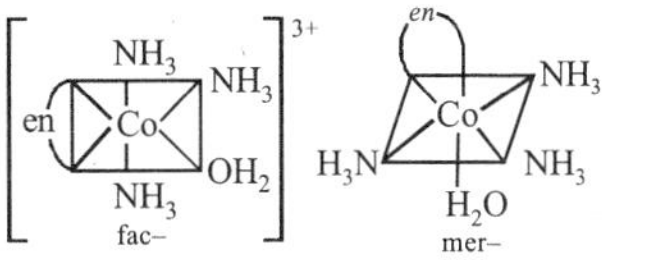

(b) Compound $[Co(CN)_2(NH_3)_3(H_2O)]^+$ will have three geometrical isomers.

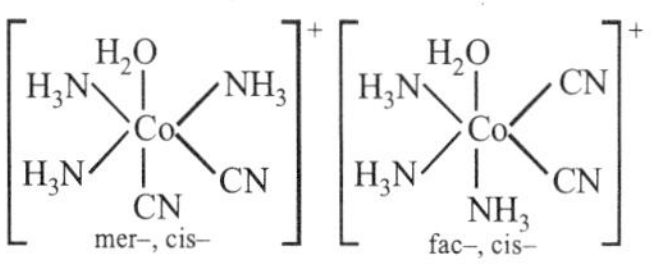

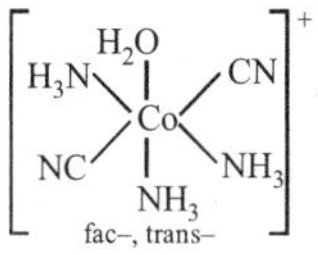

[Note: fac– and mer– w.r.t. NH_3, cis– and trans– w.r.t. –CN]

(c) $[Co(en)(NH_3)_3(H_2O)]^{3+}$ is diamagnetic

Due to the presence of strong field ligand 'en' d^6 system (Co^{3+}) forms low spin (l.s) complex in Oh splitting of d orbitals.

(d) $[Co(en)(NH_3)_4]^{3+}$ has larger gap between e_g and t_{2g} than $[Co(en)(NH_3)_3(H_2O)]^{3+}$.

So, $[Co(en)(NH_3)_3(H_2O)]^{3+}$ absorbs light at longer wavelength as compared to $[Co(en)(NH_3)_4]^{3+}$.

59. (b, c)

(a) $[(Fe(CO_5)]$ & $[Ni(CO)_4]$ complexes have 18-electrons in their valence shell.

For transition elements of 3*d*-series, the valence shells are 3*d*- and 4*s*.

(b) Due to strong ligand field, carbonyl complexes are predominantly low spin complexes.

(c) As electron density increases on metals (with lowering oxidation state on metals), the extent of synergic bonding increases. Hence, M-C bond strength increases.

(d) While positive charge on metals increases and the extent of synergic bond decreases and hence, C–O bond becomes stronger.

60. (b, c, d) Magnetic moment,

$\mu = \sqrt{n(n+2)}$ B.M.

where, n = No. of unpaired electrons

Given for X and Z, $\mu = 3.87$ B.M.

i.e. $\sqrt{n(n+2)} = 3.87$

$n^2 + 2n - 15 = 0 \quad \therefore n = 3$

For complex Y (1 : 3 electrolyte) given $\mu = 0$

i.e. $\sqrt{n(n+2)} = 0 \Rightarrow n = 0$

$CoCl_2.6H_2O$ or $[Co(H_2O)_6)]\ Cl_2$ (X) is pink coloured compound and octahedral in structure. On adding excess of HCl at room temperature, $[Co(H_2O)_6]Cl_2$ (X) changes into tetrahedral, $[CoCl_4]^{2-}$ (Z) and on adding excess of NH_3 and NH_4Cl in the presence of air forms $[Co\,(NH_3)_6]Cl_3$ (Y).

$$[Co(H_2O)_6]Cl_2 \xrightarrow[NH_4Cl + air]{NH_3\,(excess)} [Co(NH_3)_6]Cl_3$$

(X) Pink (Y)

(μ= 3.87 B.M.)(μ = 0)

61. (a, c) The metal to ligand bonding creates a synergic effect which strengthens the bond between CO and the metal (Fe).

As the bond strength $Fe \leftarrow C \equiv O$

between Fe – C increases, the bond order between C – O decreases and the effective bonding can be represented as Fe = C = O.

This will increase the C – O bond length. Thus, the answers can be 1.13 Å or 1.15 Å.

62 (d) 1. $\left[\overset{II}{Fe}(H_2O)_6\right]^{+2}$ WFL

configuration $3d^6$
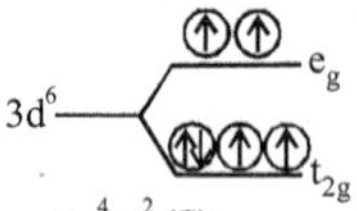
$t_{2g}^4 e_g^2$ (S)

2. $\left[\overset{II}{Mn}(H_2O)_6\right]^{+2}$ WFL

configuration $3d^5$
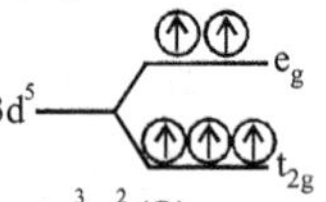
$t_{2g}^3 e_g^2$ (Q)

3. $\left[\overset{III}{Co}(NH_3)_6\right]^{+3}$ SFL

configuration $3d^6$
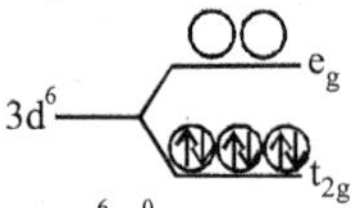
$t_{2g}^6 e_g^0$ (P)

4. $\left[\overset{III}{Fe}\,Cl_4\right]^{-}$ WFL

configuration $3d^5$
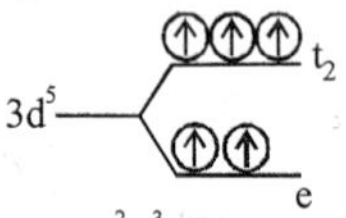
$e^2 t_2^3$ (R)

5. $\left[\overset{II}{Co}Cl_4\right]^{-2}$ WFL

configuration $3d^7$
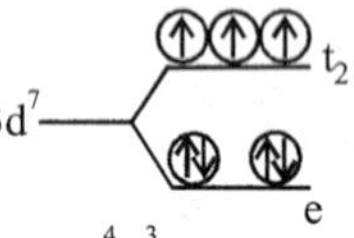
$e^4 t_2^3$ (None)

63. (a)

(I) $[Cr(CN)_6]^{4-} \Rightarrow Cr^{2+}$ (d^4) $\Rightarrow$ low spin Oh complex as CN^- is strong field ligand.

Electronic configuration = $t_{2g}^4 e_g^0$; $\Delta_0 > P$;

$\mu_{S.O.} = \sqrt{2(2+2)} = 2.82$ BM

(P), (R), (T)

(II) $[RuCl_6]^{2-} \Rightarrow Ru^{4+}$ (d^4) $\Rightarrow$ Low spin Oh complex as Ru is of large size.

Electronic configuration = $t_{2g}^4 e_g^0$; $\Delta_0 > P$;

$\mu_{S.O.} = \sqrt{2(2+2)} = 2.82$ BM

(P), (R), (S), (T)

(III) $[Cr(H_2O)_6]^{2+} \Rightarrow Cr^{2+}$ (d^4) $\Rightarrow$ high spin Oh complex as H_2O is weak field ligand.

Electronic configuration = $t_{2g}^3 e_g^1$; $\Delta_0 < P$;

$\mu_{S.O.} = \sqrt{4(4+2)} = 4.89$ BM

(Q), (T)

(IV) $[Fe(H_2O)_6]^{2+} \Rightarrow Fe^{2+}$ (d^6) $\Rightarrow$ high spin Oh complex

Electronic configuration = $t_{2g}^4 e_g^2$; $\Delta_0 < P$;

$\mu_{S.O.} = \sqrt{4(4+2)} = 4.89$ BM

(P), (Q)

64. (c)

(p) $[FeF_6]^{4-}$, $Fe^{2+} = 3d^6$ and F^- is weak field ligand

$\therefore$ Hybridization is sp^3d^2 (high spin complex)

(q) $[Ti(H_2O)_3Cl_3]$, $Ti^{3+} = 3d^1$ (No effect of ligand field strength)

$\therefore$ Hybridization is d^2sp^3

(r) $[Cr(NH_3)_6]^{3+}$, $Cr^{3+} = 3d^3$ (No effect of ligand field strength)

$\therefore$ Hybridization is d^2sp^3

(s) $[FeCl_4]^{2-}$, $3d^6$ and Cl^- is weak field ligand

$\therefore$ Hybridization is sp^3

(t) $[Ni(CO)_4]$, $Ni = 3d^{10}$ and CO is strong field ligand

$\therefore$ Hybridization is sp^3

(w) $[Ni(CN)_4]^{2-}$, $Ni^{2+} = 3d^8$ and CN is strong field ligand

$\therefore$ Hybridization is dsp^2

65. (b)

Complex	Magnetic character	Isomerism
A, $[Cr(NH_3)_4Cl_2]Cl$	Cr^{3+} is d^3, hence paramagnetic	cis-trans
B, $[Ti(H_2O)_5Cl](NO_3)_2$	Ti^{3+} is d^1, hence paramagnetic.	ionization
C, $[Pt(en)(NH_3)Cl]NO_3$	Pt^{2+} is d^8, complex is square planar, all electrons are paired, hence diamagnetic	ionization
D, $[Co(NH_3)_4(NO_3)_2]NO_3$	Co^{3+} is d^6, all electrons are paired due to strong ligands, hence diamagnetic	cis-trans

66. (A) : (p), (q) and (s)

In $[Co(NH_3)_4(H_2O)_2]^{2+}$, Co is in + 2 state having $3d^7$ configuration, which makes it paramagentic due to odd electrons. Moreover, it is an octahedral complex showing *cis-trans* isomerism w.r.t., H_2O.

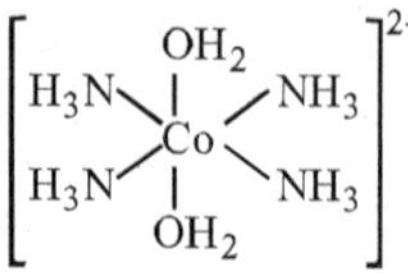

trans

$Co^{2+} = 3d^7$ (Paramagnetic)

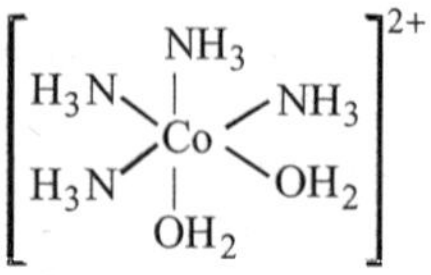

(*cis*)

(B) : (p), (r) and (s)

In $[Pt(NH_3)Cl_2]$, Pt is in + 2 state with configuration $5d^8$. Since NH_3 is a strong field ligand, it will pair all the electrons making the complex diamagnetic. Moreover, it is a square planar complex showing cis-trans isomerism.

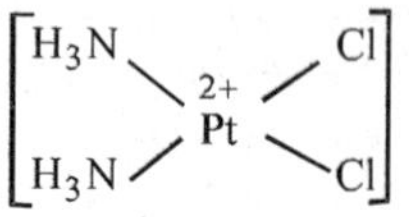

(*cis*)

$Pt^{2+} = 5d^8 4s^0$ (diamagnetic)

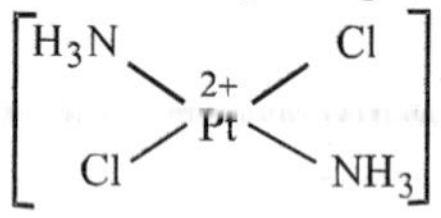

(*trans*)

(C) : (q) and (s)

In $[Co(H_2O)_5Cl]Cl$, Co is in + 2 state with $3d^7$ configuration making it paramagnetic.

(D) : (q) and (s)

In $[Ni(H_2O)_6]Cl_2$, Ni is in + 2 state with $3d^8$ configuration. It is attached with weak field ligands, therefore it is paramagnetic.

67. (a) $Ni^{2+} \xrightarrow[\text{excess}]{KCN} K_2[Ni(CN)_4]$;

potassium tetracyano nickelate (II)

$Ni^{2+} \xrightarrow[\text{(excess)}]{KCl} K_2[NiCl_4]$

potassium tetrachloro nichelate (II)

68. (c) For $[Ni(CN)_4]^{2-}$;

$Ni = [Ar]3d^8 4s^2$; $Ni^{2+} = [Ar]3d^8$

Ni^{2+}
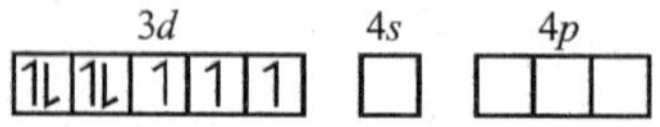

However, CN^- is a strong field ligand so it forces the 3*d* electrons to pair up and hence the effective configuration in this case will be:

Ni^{2+} in presence of CN^-

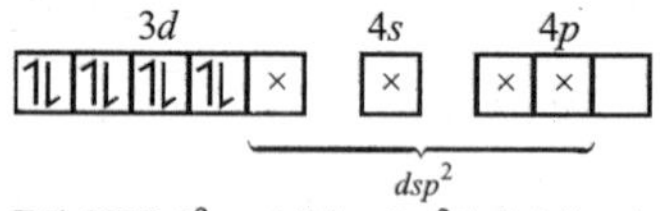

Thus, $[Ni(CN)_4]^{2-}$ exhibits dsp^2 hybridization and square planar shape. Since here, number of unpaired electrons is zero, the complex will be *diamagnetic*.

In case of $[NiCl_4]^{2-}$, Cl^- is a weak field ligand, so the effective configuration of Ni^{2+} in this complex will be as follows :

Ni^{2+} in presence of Cl^-

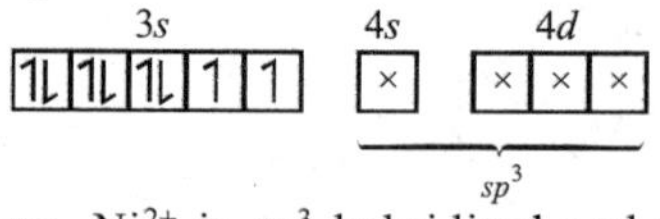

So here, Ni^{2+} is sp^3 hybridised and thus, tetrahedral in shape. Since, the complex has two unpaired electrons, it will be *paramagnetic*.

69. (a) Discussed above.

70. (a) In $[Fe(H_2O)_5NO]SO_4$, Let the oxidation state of Fe be *x*. Then for $[Fe(H_2O)_5NO]^{2+}$,

$x + 1 = +2$ or $x = +2 - 1 = +1$

Hence, in this complex the oxidation state of Fe is + 1

Electronic configuration of Fe^+ can be represented as $Fe^+ = 1s^2 2s^2 2p^6 3s^2 3p^6 3d^7$. This unexpected configuration is due to presence of NO^+. Due to which 1 electron from $4s^1$ gets shifted to 3*d*– orbitals. The $3d^7$ electrons in five 3*d*– orbitals can be shown as

$3d^7$;
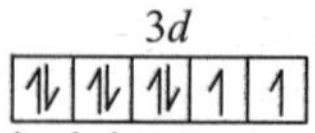

In it, we find 3 unpaired electrons.

Because of the presence of unpaired electrons the complex is paramagnetic *i.e.* statement 1 is true.

As is clear from above there are three unpaired electrons in this complex *i.e.* statement 2 is true. Since, paramagnetic behaviour is due to presence of unpaired electrons in it so statement 2 is correct explanation of statement 1.

71. $Fe^{3+} + \underset{\text{(excess)}}{SCN^-} \longrightarrow \underset{\substack{\text{(blood red)}\\ \text{(A)}}}{[Fe(SCN)(H_2O)_5]^{2+}}$

$$[Fe(SCN)(H_2O)_5]^{2+} + 6F^- \rightarrow \underset{\text{(B)}}{[FeF_6]^{3-}} + SCN^-$$

IUPAC name of (A) is pentaaquathiocyanatoferrate (III) ion

IUPAC name of (B) is hexafluoroferrate (III)

In $[FeF_6]^{3-}$ coordination no. of Fe = 6

In $[FeF_6]^{3-}$ oxidation state of Fe = + 3

∴ It has 5 unpaired electrons, $n = 5$, Fe^{3+} is $3d^5$

Magnetic moment $(\mu) = \sqrt{n(n+2)}$ (B.M.)

$\sqrt{5(5+2)} = \sqrt{35} = 5.92$ B.M.

72. (a)

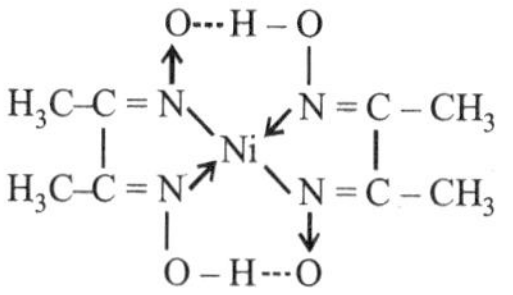

Bis(dimethylglyoximato) nickel (II)

(b) Charge on Ni in the complex is +2 and it is dsp^2 hybridised

(c) Since number of unpaired electrons in Ni^{2+} is zero, the complex is diamagnetic.

73. The spin magnetic moment, m of the complex is 1.73 BM.

$\mu = \sqrt{n(n+2)} = 1.73 \Rightarrow n = 1$

It means that nucleus of the complex, chromium ion has one unpaired electron. So the ligand NO is unit positively charged.

Cr (z = 24) = $3d^5 4s^2$, after pairing up of electrons, there shall be one unpaired electron left.

Thus, chromimum can be in Cr^+ state.

IUPAC name :

Potassium amminetetracyanonitrosochromate (I).

(a) Electronic configuration of Cr^+ :

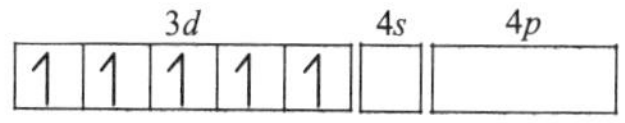

(b) Electronic configuration of Cr^+ under the influence of strong field ligand CN^-

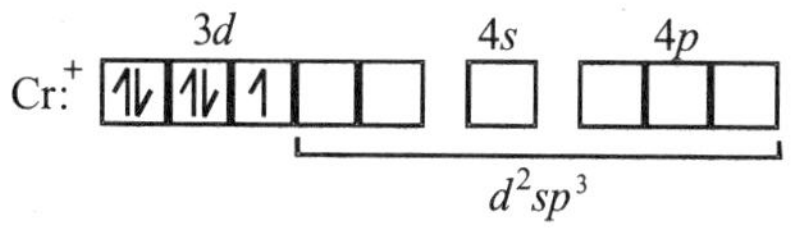

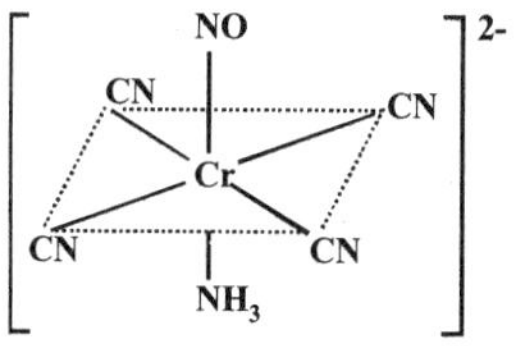

So, Hybridization : d^2sp^3; Shape : Octahedral

74. Cl^- is a weak ligand which is unable to pair the electrons of Ni^{2+}. Therefore, here hybridisation is sp^3 and shape will be **tetrahedral.**

Electronic configuration of Ni^{+2} (No. of electrons = 26) in presence of Cl^- ion, a weak ligand.

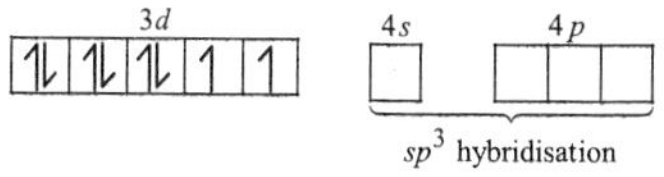

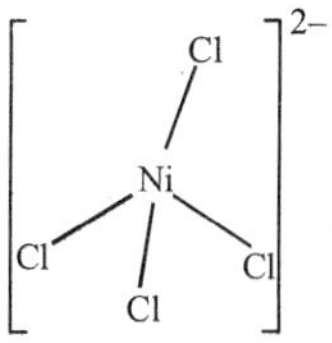

Magnetic moment of $[NiCl_4]^{2-}$

$= \sqrt{2(2+2)} = 2.82$ BM

On the other hand, CN^- is a stong ligand which pairs up the electrons of Ni^{2+}. Therefore, here hybridisation is dsp^2 and shape will be square planar.

Electronic configuration of Ni^{2+} in presence of CN^- ion, a strong ligand.

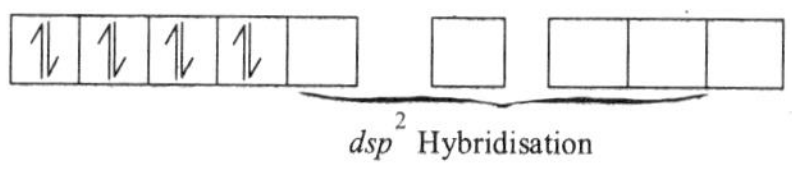

Magnetic moment of $[Ni(CN)_4]^{2-}$

$= \sqrt{0(0+2)} = 0.0$BM

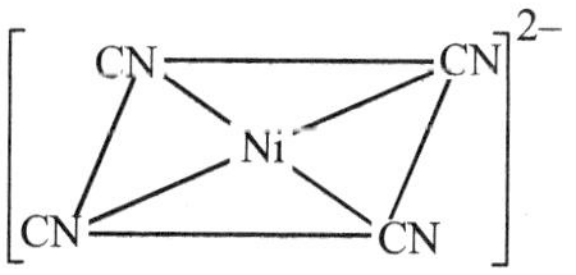

Square planar dsp^2 hybridisation

75. Compound (*A*) on treatment with $AgNO_3$ gives white precipitate of AgCl, which is readily soluble in dil.aq. NH_3. Therefore it has at least one Cl^- ion in the ionization sphere. Furthermore, chromium has coordination number equal to 6. So, its formula is **$[Cr(NH_3)_4BrCl]Cl$.**

Compound (*B*) on treatment with $AgNO_3$ gives pale yellow precipitate of AgBr soluble in conc. NH_3. Therefore it has Br^- in the ionization sphere. So, its formula is **$[Cr(NH_3)_4Cl_2]Br$.**

Cr^{3+} (*Z* = 24)

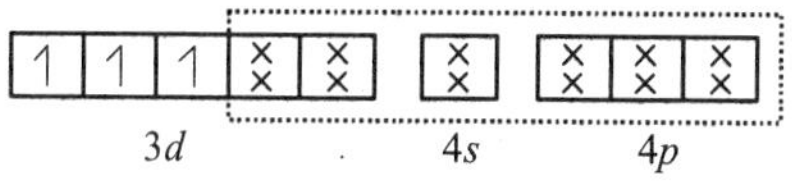

State of hybridization of chromium in both (*A*) and (*B*) is **d^2sp^3.**

Spin magnetic moment of (*A*) or (*B*),

$\mu_{spin} = \sqrt{n(n+2)}$

$= \sqrt{3(3+2)} = \sqrt{15} =$ **3.87 BM**

76. $[Co(NH_3)_6]^{3+}$

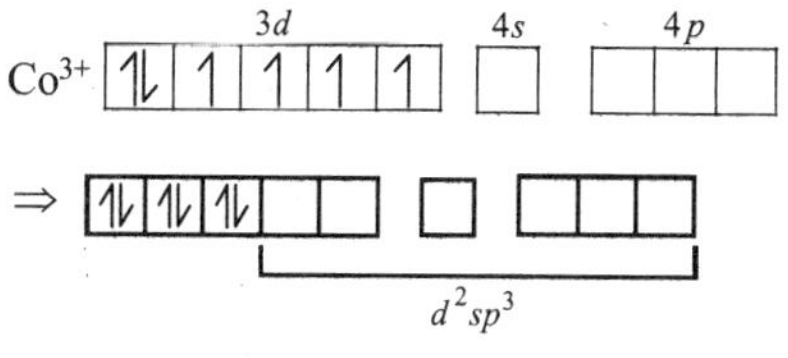

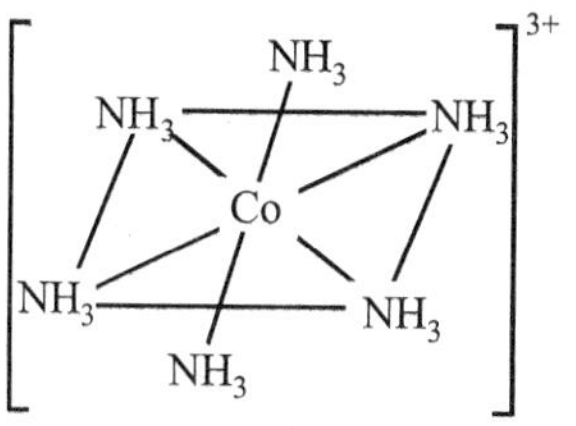

Octahedral complex, d^2sp^3 hybridisation

$[Ni(CN)_4]^{2-}$

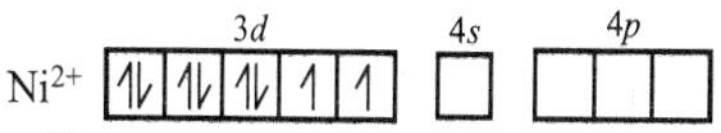

Ni^{2+} (after rearrangement)

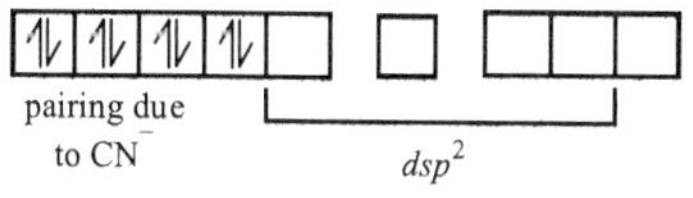

$[Ni(CO)_4]$

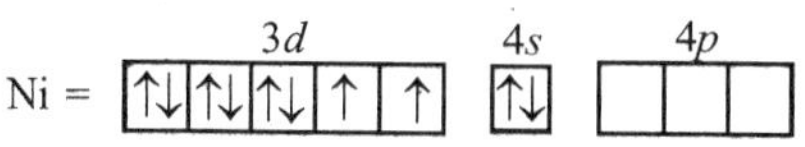

Ni (after rearrangement)

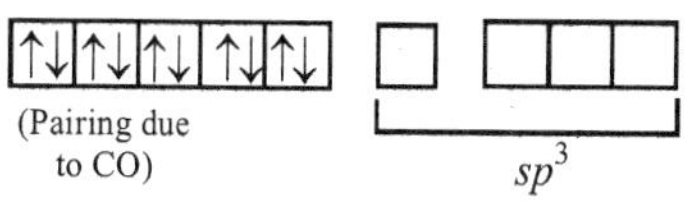

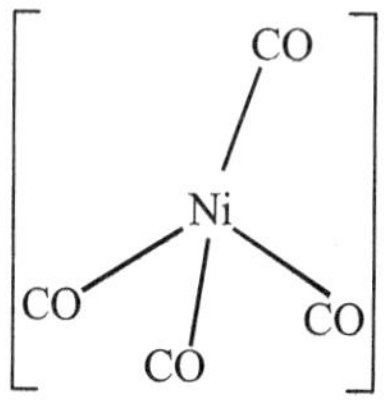

Tetrahedral (sp^3 hybridisation)

77. Magnetic moment $(\mu) = \sqrt{n(n+2)}$ BM

where $n \rightarrow$ number of unpaired electrons

$\mu = 1.73$ BM for vanadium ion

1.73 BM $= \sqrt{n(n+2)}$ So, $(1.73)^2 = n(n+2)$

$3.0 = n^2 + 2n$ or $n^2 + 2n - 3 = 0$

$n^2 + 3n - n - 3 = 0$ ∴ $n(n+3) - 1(n+3) = 0$

$(n-1)(n+3) = 0$ Correct value of $n = 1$

Thus, no. of unpaired electrons in vanadium ion = 1

${}_{23}V = 1s^2, 2s^2 2p^6, 3s^2 3p^6 3d^3, 4s^2$

It will have one unpaired electron if it will lose two electrons from 4*s* and two from 3*d*.

∴ Vanadium (IV) has one unpaired electron.

$V^{4+} = 1s^2, 2s^2 2p^6, 3s^2 3p^6 3d^1$

78. For the explanation of colouration of complexes, first of all find out the number of unpaired electrons present in outer available *d*-orbitals

(i) $[Ti(NO_3)_4]$; ${}_{22}Ti^{4+}$: [Ar] $3d^0 4s^0$

(ii) $[Cu(NCCH_3)_4]^+ BF_4^-$; ${}_{29}Cu^+$: [Ar] $3d^{10} 4s^0$

(iii) $[Cr(NH_3)_6]^{+3} 3Cl^-$; ${}_{24}Cr^{3+}$: [Ar] $3d^3 4s^0$

(iv) $K_3[VF_6]$; ${}_{23}V^{+3}$: [Ar] $3d^2 4s^0$

Due to the presence of unpaired electrons in *d*-orbitals, two complexes *i.e.*, $[Cr(NH_3)_6]^{3+} 3Cl^-$ and $K_3[VF_6]$ are coloured. Others having all paired electrons are colourless.

Chapter-10

Haloalkanes and Haloarenes

Part-A (JEE Main)

1. (1)

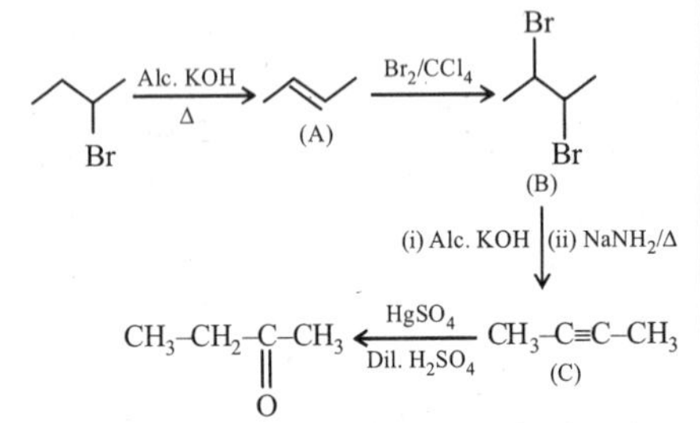

Due to α methyl group in compound, final product gives +ve haloform test.

2. (2) No chiral center (2 methyl groups)

3. (3) Statement (A) is true.

Statement (B) is false. The carbocation formed from $(C_6H_5)-\underset{|}{\underset{Cl}{CH}}-(C_6H_5)$ is $(C_6H_5)_2\overset{\oplus}{C}H$ which is more stable than the benzyl carbocation $(Ph\overset{\oplus}{C}H_2)$. Thus, order is reversed.

Statement (C) is false. Aryl Halides are highly unreactive towards nucleophilic substitution compared to alkyl halides because C-X bond have a slightly double bond character in aryl halides due to resonance with benzene ring.

Statement (D) is false. Vinyl chloride (CH_2=CHCl) is a haloalkene. However, allyl chloride (CH_2=CH–CH_2Cl) is also a haloalkene.

Statement (E) is true.

4. (1) Rate of $S_N1 \propto$ Stability of carbocation

Order of stability of carbocation:

$Ph-\overset{\oplus}{C}H_2 > CH_3-\overset{\oplus}{C}H_2$

Hyperconjugation in $CH_3-\overset{\oplus}{C}H_2$

Resonance in $Ph-\overset{\oplus}{C}H_2$

Resonance stabilization is generally a much more stabilizing effect than hyperconjugation.

5. (4) Statement-I : True

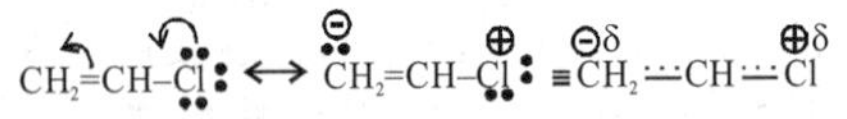

C–Cl bond is strong in vinyl chloride because of partial double bond character.

Statement-II: false

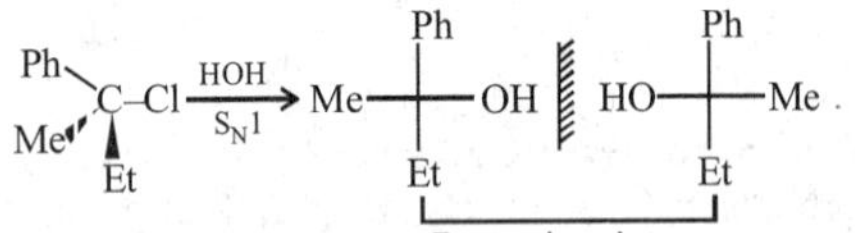

Racemic mixture

Racemic mixture is optically inactive, which can not rotate plane polarised light.

6. (3) A-III, B-IV, C-II, D-I

7. (3)

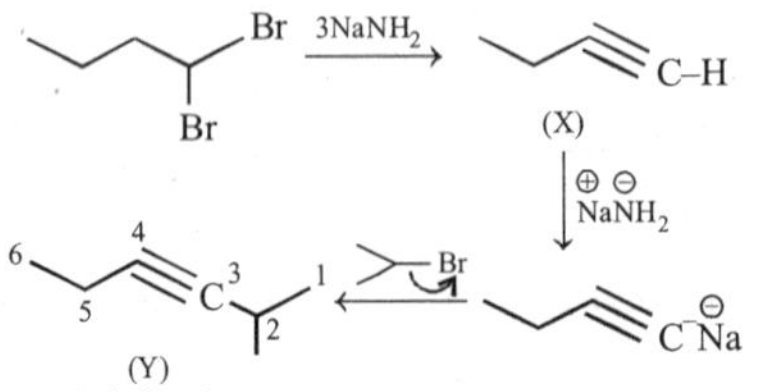

(Y)
2-Methylhex-3-yne

8. (3) Rate of $S_N1 \propto$ Stability of carbocation formed

(I) and (II) are unstable due to Bredt's rule, (I) has more +I effect.

$$(II) < (I) < (III) < (IV)$$

9. (2) CH_3Br in methanol follow S_N1 pathway due to polar protic solvent So order of nucleophilicity is

$$I^- > C_2H_5O^- > PhO^- > F^-$$

10. (4) Stability order of anion.

$$ClO_4^- > CH_3-\overset{O}{\overset{||}{C}}-O^- > Ph-O^- > OH^-$$ and nucleophlicity order (nucleophilic strength) is reverse of stability of anion, and greater the number of resonating structures of the anion, greater will be its stability.

$$OH^- > Ph-O^- > CH_3-\overset{O}{\overset{||}{C}}-O^- > ClO_4^-$$

11. (1) Statement - I: C—Cl bond is more polar than C—Br bond. Hence

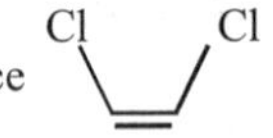

is more polar than

Statement - II:

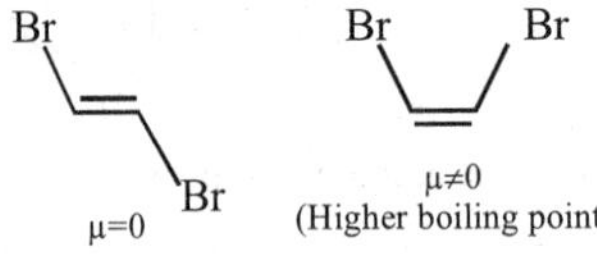

μ=0 μ≠0 (Higher boiling point)

12. (2)

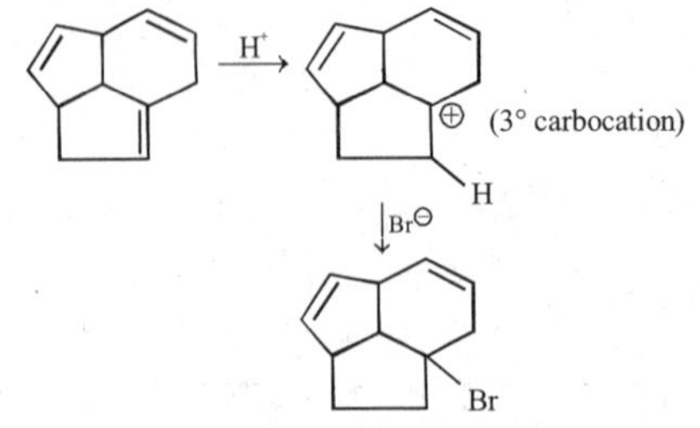

(Major Product)

13. (2) Statement-I:

$$R-Cl \xrightarrow{(aq.KOH)} R-OH(S_N \text{ reaction})$$

Statement-II:

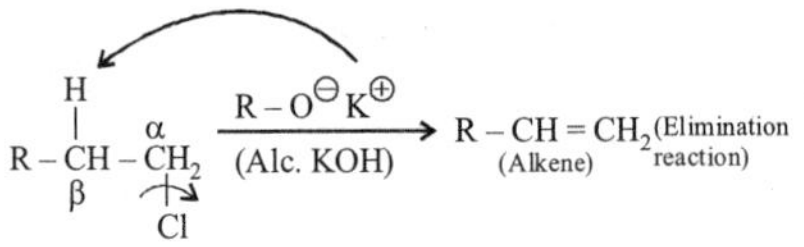

14. (2)

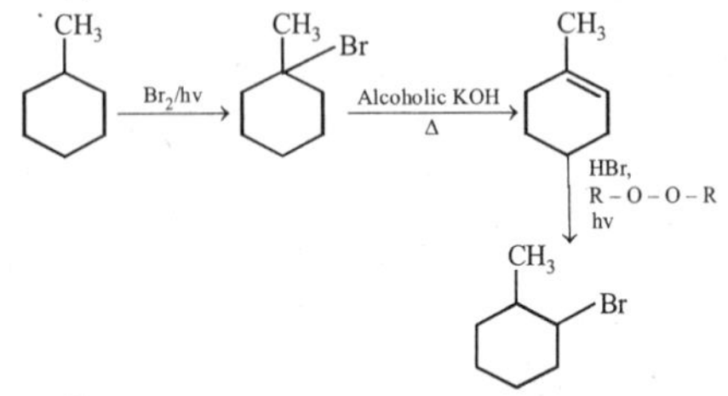

15. (3)

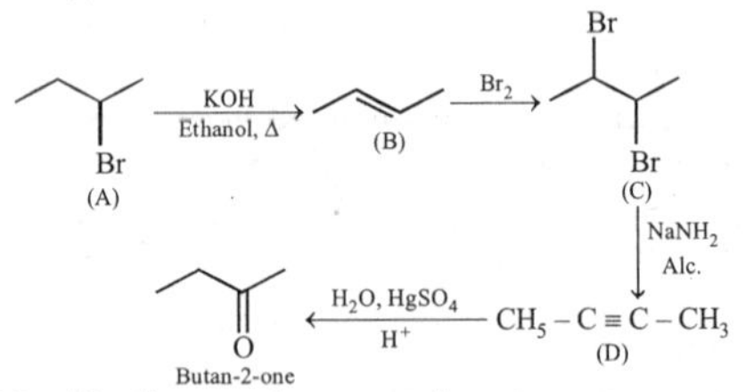

16. (4) Greater the resonating structure greater will be stability of carbocation.

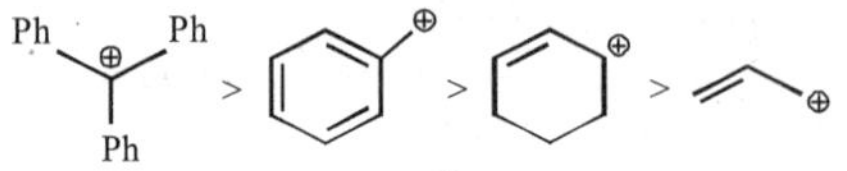

17. (4)

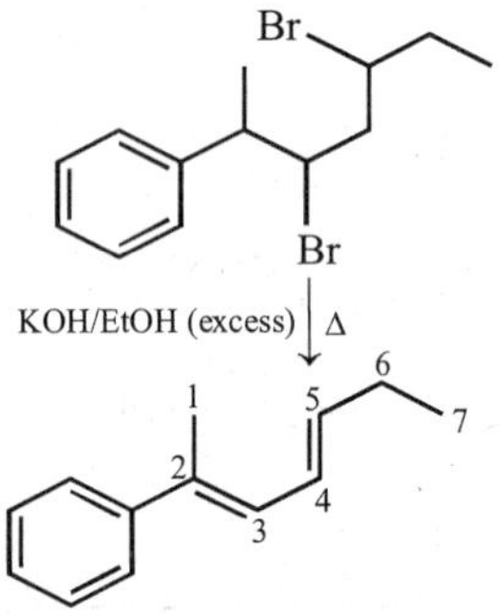

2-Phenylhepta-2,4-diene

18. (4)

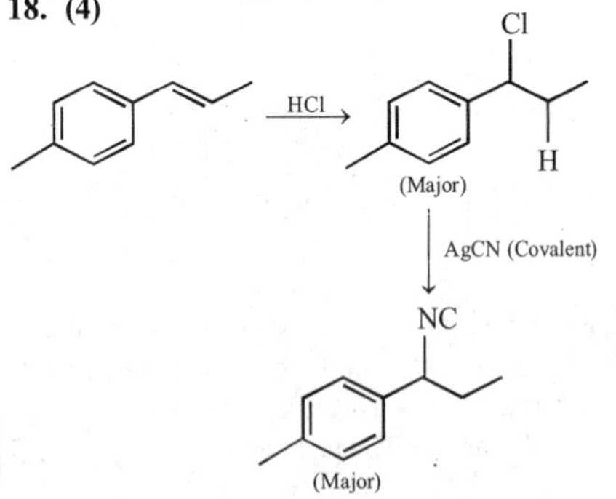

19. (4)

$CH_3-CH_2-CH_2-NO_2 \xleftarrow{Ag-NO_2}$ (A)

$CH_3-CH_2-CH_2-Br \xrightarrow{AgCN} CH_3-CH_2-CH_2-NC$ (B)

$AgNO_2$ and AgCN are covalent compounds in which there are free – N and – N terminals respectively.

20. (3)

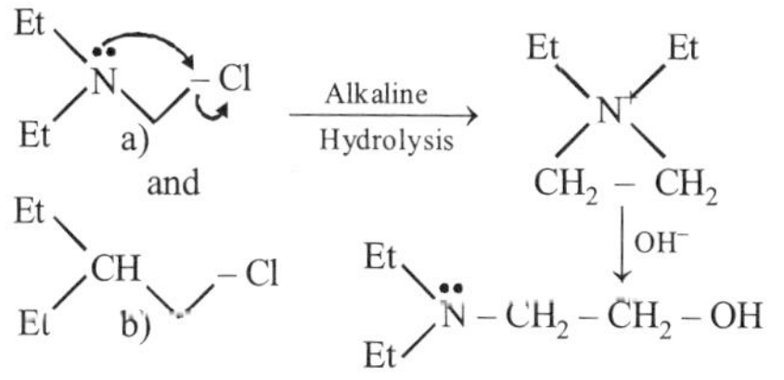

Due to intramolecular substitution rate of (a) is faster than rate of (b). (Neighbouring Group participation)

21. (2)

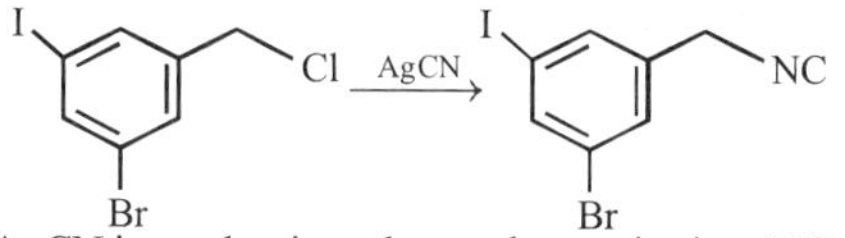

AgCN is predominantly covalent so it gives NC^- nucleophile and forms an isocyanide.

22. (1)

$CH_3-CH_2-CH_2-CH_2-Cl+\overset{\ominus}{O}H$

Nucleophile (higher charge density)

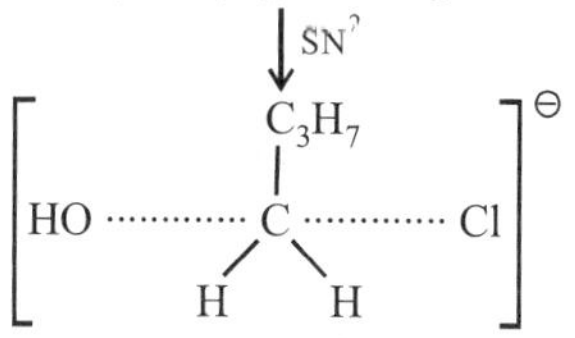

Transition state (less charge density as charge is dispersed) ⇒ This reaction will proceed faster in less polar medium which will not increase the activation energy value.

$CH_3-CH_2-CH_2-CH_2-Cl+R_3N$

Reactant (low charge density)

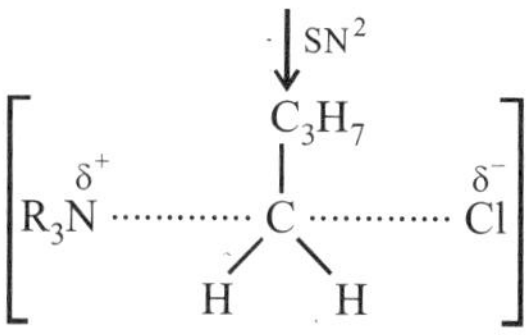

Transition state (higher charge density)

⇒ This reaction will proceed faster in more polar medium which will decrease the activation energy value.

23. (1) Solvolysis of $S_N1 \propto$ stability of carbocation stability order of carbocation and hence relative rate of solvolysis.

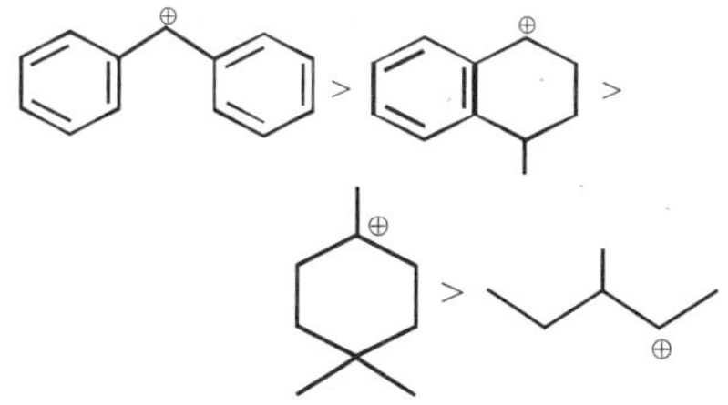

24. (3)

	List-I Name reaction	List-II Product obtainable
A.	Swarts reaction	I. $Et-I \xrightarrow[DMF]{AgF} Et-F$
B.	Sandmeyer's reaction	II. $PhN_2^{\oplus}Cl^- \xrightarrow{CuCN/KCN} PhCN+N_2$
C.	Wurtz Fittig reaction	III. $Ph-Cl+EtCl \xrightarrow[Dry\ ether]{Na} Ph-Et$
D.	Finkelstein reaction	IV. $Et-Cl \xrightarrow[acetone]{NaI}$ Et-I + NaCl

25. (1)

$RBr \xrightarrow{Mg,\ dry\ ether} RMgBr \xrightarrow{H_2O} RH+Mg(OH)$ Br. These four statements give 2– methylbutane upon the given reactions.

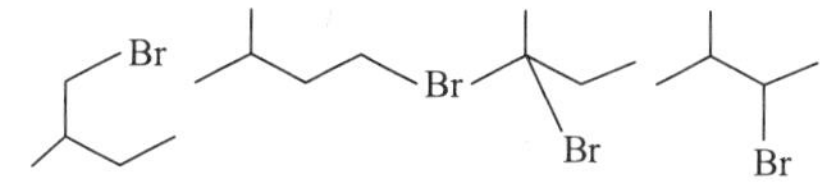

Total number of possible structures of RBr = 4

26. (4) CH_3-O-CH_2-Cl will undergo S_N1 mechanism because $CH_3-O-\overset{+}{C}H_2$ is highly stable.

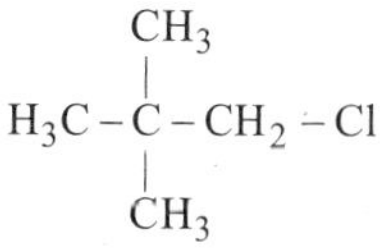

Neopentyl chloride will undergo S_N2 mechanism at a slow rate because it's sterically crowded. So, it prefers S_N1.

27. (3) The benzyl group acts by using the π-system of the benzene ring for conjugation with the p-orbital of the α-carbon in the transition state.

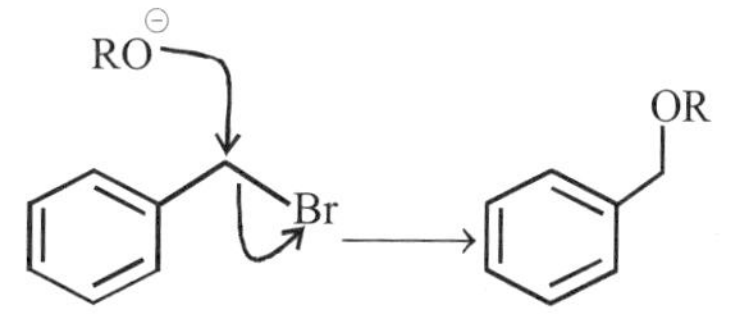

benzyl bromide

28. (1)

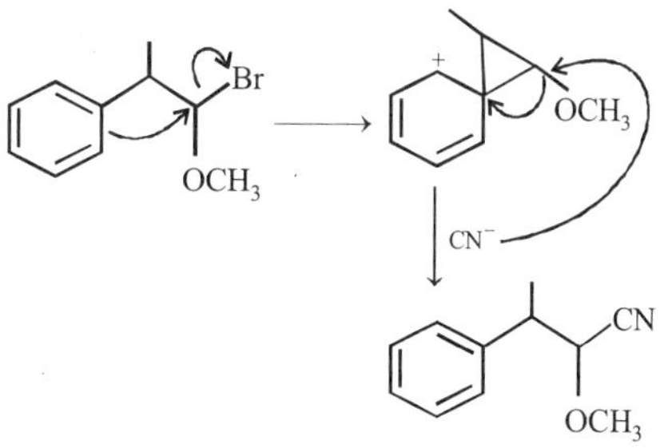

29. (3) Less the steric hindrance more will be here rate of S_N2 reaction. Order of rate of alkyl halide towards S_N2 reaction is $CH_3-X > 1° > 2° > 3°$

30. (3) S_N2 reactions are stereospecific and inversion of configurations observed.

S_N1 reaction generally racemisation is observed.

31. (2) Hint: Rearrangement of carbocation to gain stability. In aqueous medium the reaction follows S_N1 mechanism majority

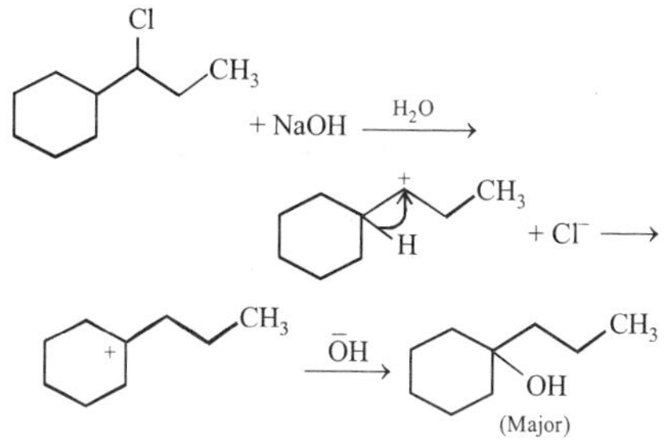

32. (3)

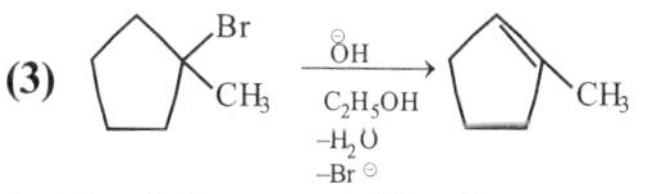

Elimination follow saytzeff's rule.

33. (3)

$CH_3-CH_2-CH_2-Br \xrightarrow[80°C]{X = conc.\ alc.\ NaOH}$

$CH_3-CH=CH_2 \xrightarrow{Y=HBr/Acetic\ acid}$

$CH_3-CHBr-CH_3$

34. (2)

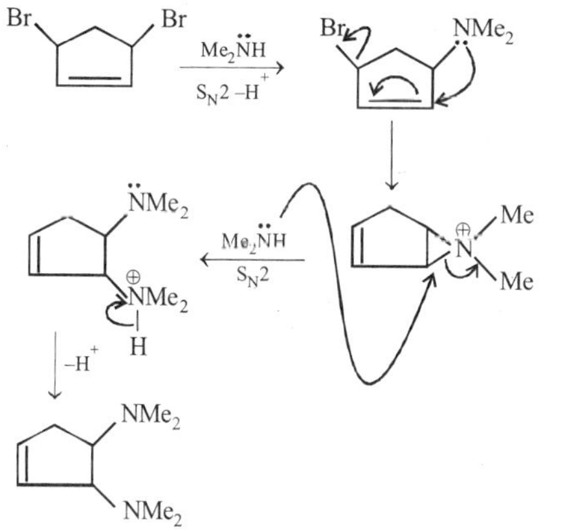

The above mechanism valid for both cis and trans isomers. So the products are same for both cis and trans isomers.

35. (2)

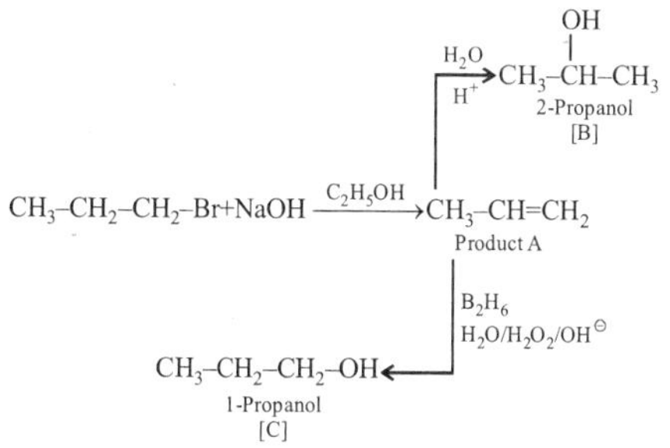

36. (2) Hint: Alc. KOH abstract most acidic hydrogen

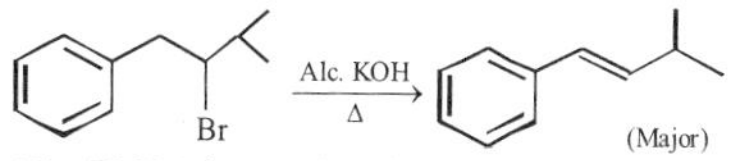

37. (1) KCN is predominantly ionic, it gives cyanide ions in solution. Although both C and N atoms can donate electrons, attack takes mainly through C because C – C bond formed in product is more stable than C – N bond and thus reaction mainly occurs through carbon of cyanide ion. But AgCN is mainly covalent in nature and nitrogen is free to donate electron pair foming isocyanide as the main product.

38. (2)

$$CH_3-CH_2-Br \xrightarrow{KOH(alc.)} CH_3-CH=CH_2$$

$$\downarrow HBr$$

$$\underset{\underset{OH}{|}}{CH_3-CH-CH_3} \xleftarrow[KOH(aq)]{\Delta} \underset{\underset{Br}{|}}{CH_3-CH-CH_3}$$

Propan-2-ol

39. (1) In vinylic halide halogen atom bonds with carbon atom of a C = C bond.

40. (1) In allylic halides halogen atom bonds with sp^3 hybridised carbon atom.

$$CH_2=CH-\underset{sp^3-C}{CH_2}-Cl$$

41. (4) S_N2 mechanism is preferred for non-bulky substituents and substrates that are substituted (1° or 2°).

Strong nucleophiles are required for S_N2 mechanisms.

Thus, Statement I is correct.

A secondary alkyl halide will be stable enough to undergo S_N1 and would require a weak nucleophile like ethanol for it to take place (polar protic solvents).

Thus, Statement II is also correct.

42. (2) KCN gives nitrile (cyanide) while AgCN gives isocyanide (isonitrile) as Ag and C are bonded covalently in AgCN and only N terminal is free that bonds to the C-atom in R – X.

43. (4) S_N1 – Racemisation due to two enantiomers in equal amount.

S_N2 – Inversion due to a concerted mechanism of nucleophilic attack.

44. (4) sp^2 hybridized carbocation [$CH_3-CH=\overset{+}{C}H$] is very unstable, therefore $CH_3-CH=CH-Cl$ cannot give S_N1 reaction.

45. (1)

(i) 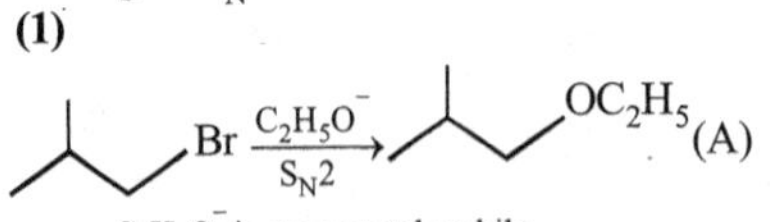

$C_2H_5O^-$ is strong nucleophile.

(ii) 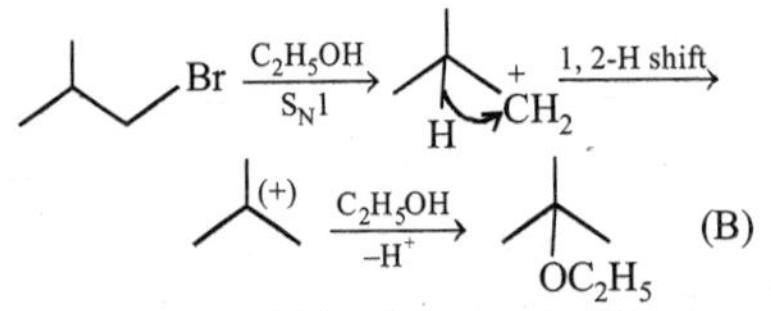

C_2H_5OH is weak nucleophile.

46. (2) $CH_3-CH_2-CH_2-Br + KOH\,(Alc) \rightarrow \underset{(Alkene)}{CH_3-CH=CH_2}$

$CH_3-CH_2-CH_2-Br + KCN\,(Alc) \rightarrow \underset{(Nitrile)}{CH_3-CH_2-CH_2-CN}$

$CH_3-CH_2-CH_2-Br + AgNO_2 \rightarrow \underset{(Nitroalkane)}{CH_3-CH_2-CH_2-NO_2} + AgBr\downarrow$

$CH_3-CH_2-CH_2-Br + CH_3-COOAg \rightarrow \underset{(Ester)}{CH_3-COO-CH_2-CH_2-CH_3} + AgBr\downarrow$

47. (3)

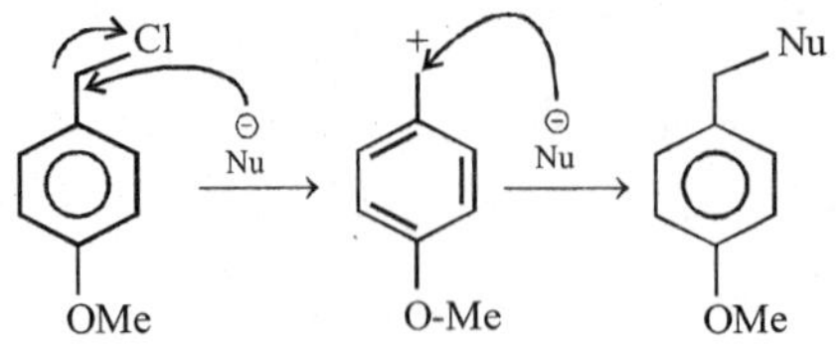

– OMe is Electron Donating group which stabilises the carbocation. The rate determining step is step one where formation of carbocation takes place.

⇒ S_N1 Mech., 1st order

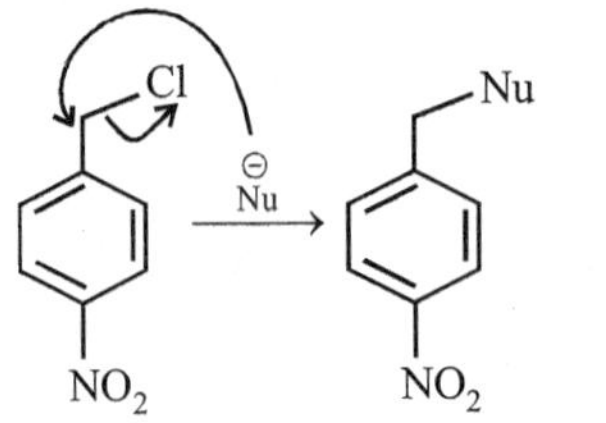

⇒ S_N2 Mech., 2nd order

$-NO_2$ is a Electron withdrawing group

48. (2) Assertion:

$$\underset{Glycine}{H_2N-CH_2-COOH} \xrightarrow[\substack{Cl_2\\(HVZ)}]{Red\ P} NH_2-\overset{*}{\underset{\underset{Cl}{|}}{CH}}-COOH$$

Optically active

Reason: A molecule with two chiral carbons is not always opticaly active since it may have some symmetry element present.

49. (2) Abstract the halogen where stability of carbocation moles

50. (1) This is finkelstein reaction

$$:\ddot{I}:^- \; R-\overset{\delta+}{CH_2}-\overset{\delta-}{Br} \xrightarrow[S_N2]{} R-CH_2-I + Br^-$$

Transition state 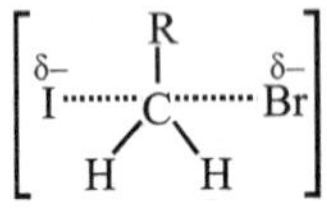

Clearly, the transition state is less polar than free anions. Br^- and I^-.

Acetic acid is protic which does not support S_N2. Acetone does not solvate anion.

Br^- gets precipitated and hence can not compete with I^- So only (a) is correct.

51. (1) It is S_N1 reaction so rate of reaction depends on the concentration of alkyl halide only.

52. (2) Boiling point of alkyl halides increase with increase in size, mass of halogen atom and size of alkyl group. Boiling point of isomeric alkyl halides decrease with increase in branching. Density increases with increase in atomic mass of halogen atom.

53. (3) $H_3C-\underset{\underset{Br}{|}}{\overset{\overset{H}{|}}{C^*}}-\underset{\underset{D}{|}}{\overset{\overset{H}{|}}{C^*}}-CH_3$

54. (3) The rate of hydrolysis of alkyl chloride improves because of better nucleophilicity of I^-.

55. (3) NaCl is ionic and gives CN^- which attacks through carbon atom and not through N atom. AgCN is covalent in nature where nitrogen is free to donate its electron pair giving isonitrile as the main product.

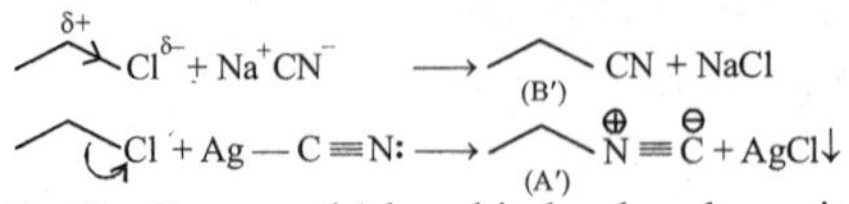

56. (3) Compound A has chiral carbon, hence it is optically active.

Compund B is not mirror image of compound A because both have same (*S*) configuration.

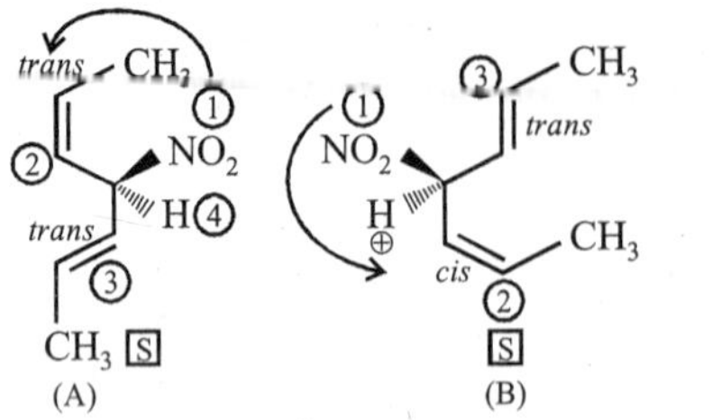

57. (3)

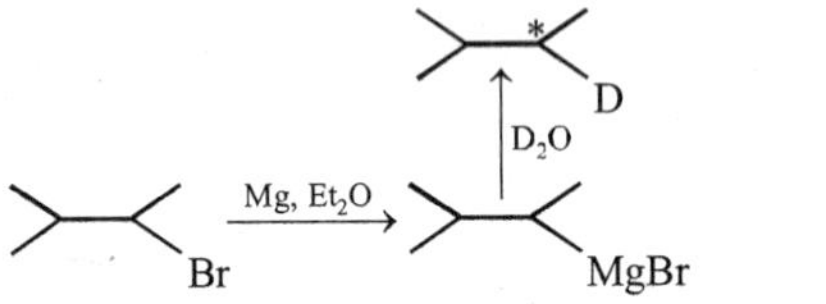

58. (4)

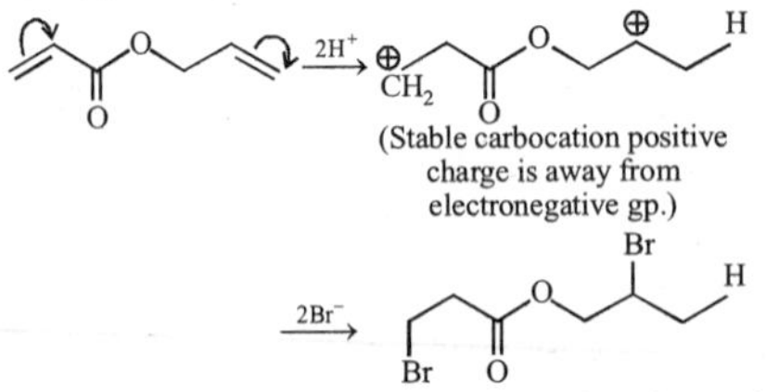

59. (3) In S_N1 reaction carbocation is formed. The carbocation formed at Bridge head in norbornane ring is very unstable due to restricted bond angle.

So it is inactive towards S_N1.

60. (3)

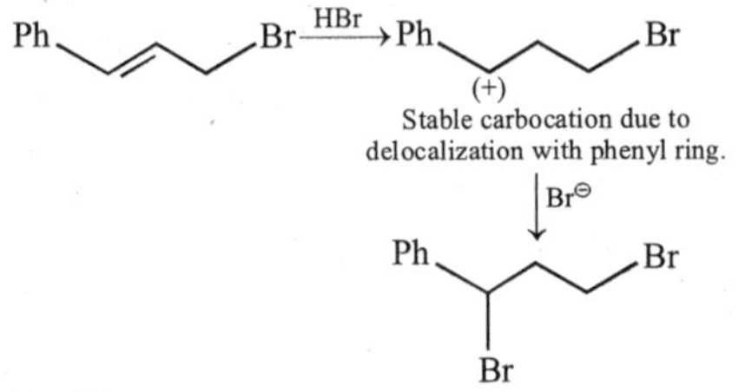

61. (1)

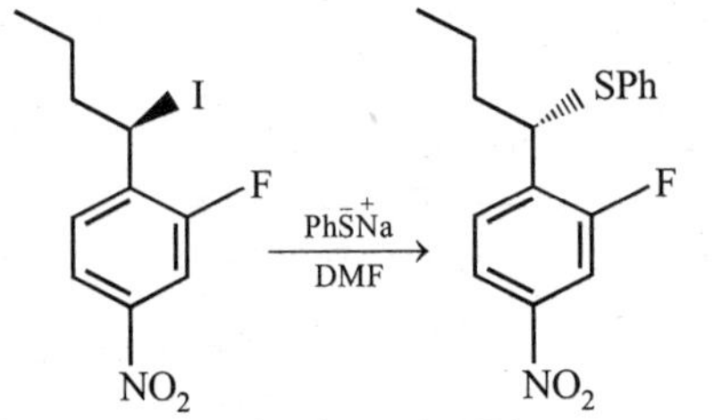

Due to polar aprotic solvent (DMF),

PhS (a good nucleophile) undergoes S_N2 type reaction which occurs at benzylic carbon by inversion in configuration.

62. (4) We are given a bulky base, hence elimination will take place & not substitution.

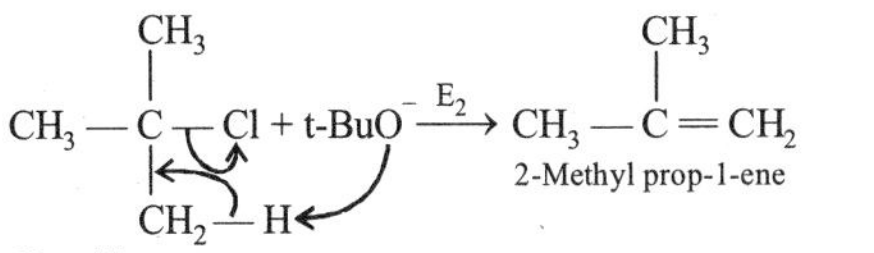

63. (1)

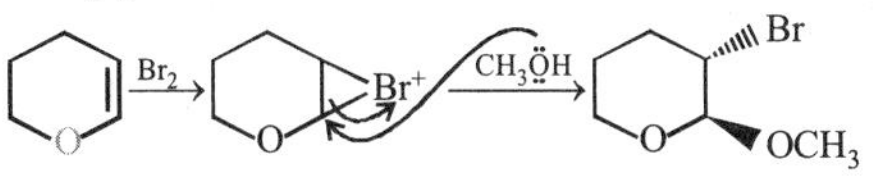

64. (2) Degree of unsaturation in C_8H_{10}
$= 9 - 5 = 4$

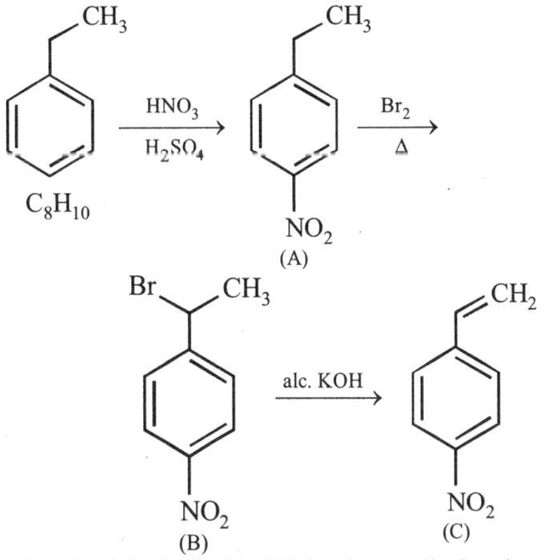

65. (4) Alcoholic-NaOH leads to elimination (E2) reaction.

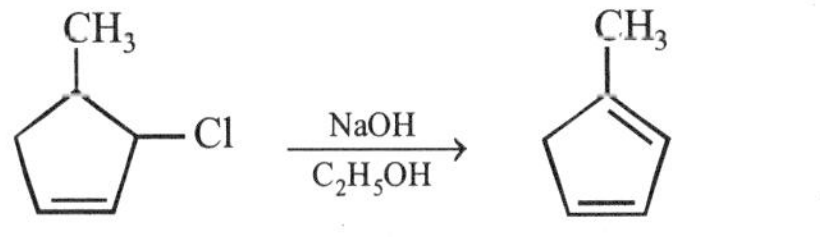

1-Methylcyclopent-1, 3-diene

66. (3) During ammonolysis of alkyl halide, acid liberated during the reaction combines with the amine and forms amine salt. For liberating the free amine from amine salt a base is needed.

$$R-X+NH_3 \longrightarrow RNH_3^+ + X^-$$

$$R-NH_3^+ + OH^- \longrightarrow RNH_2 + H_2O$$

So to remove acidic impurities, NaOH is added.

67. (4)

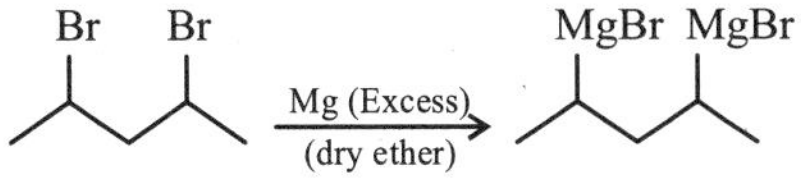

68. (2) Among the given bases (A) and (B), *t*-butoxide being bulky base favours elimination reaction and ethoxide favours substitution reaction.

$\therefore$ when $Z^- = CH_3CH_2O^-$,
(substitution reaction favoured)

and when Z^- =

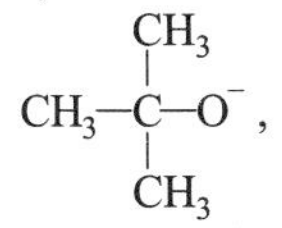

(elimination reaction favoured)

Hence, $\mu_A > \mu_B$ and $k_e(B) > k_e(A)$

69. (4) Given reaction is S_N1 reaction. In S_N1 reaction

Rate of reaction $\propto$ Stability of C^+

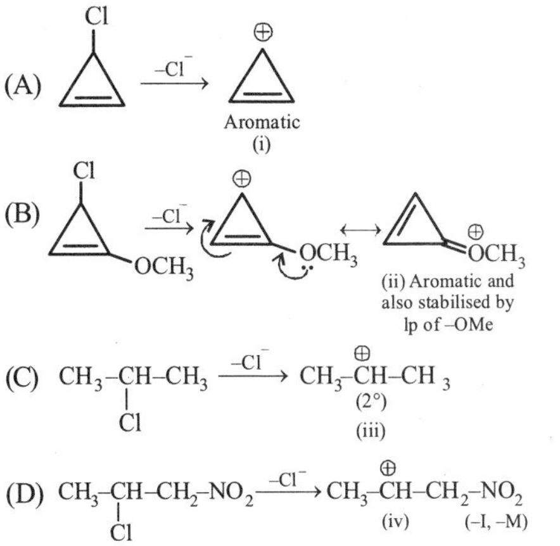

Stability of C^+ : ii > i > iii > iv

Reactivity order : B > A > C > D

70. (4)

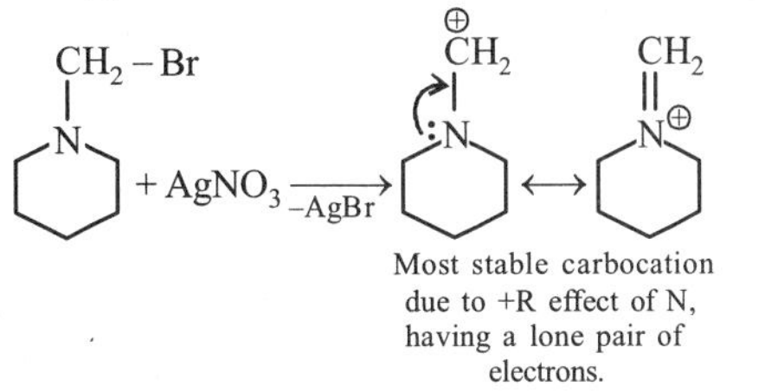

Most stable carbocation due to +R effect of N, having a lone pair of electrons.

Ease of precipitation of AgBr depends upon the rate of formation of carbocation.

71. (4) Due to conjugation of lone pair of Cl with π bond, partial double bond character decreases bond length. That's why compound (4) has shortest C–Cl bond length.

72. (3)

(A) The reaction is favoured by weak nucleophile as strong nucleophile will favour elimination over substitution reaction.

(B) If the substituents are bulky, then it will stabilise the intermediate $R^{\oplus}$.

(C) The nucleophile $Y^{\ominus}$ can attack the planar $R^{\oplus}$ from both the sides.

Hence, the reaction is accompanied by racemization.

(D) Polar solvent helps to separate the ion pair.

73. (3)

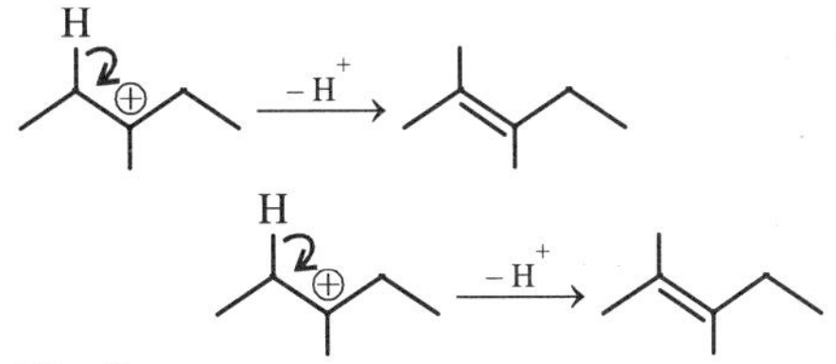

74. (4)

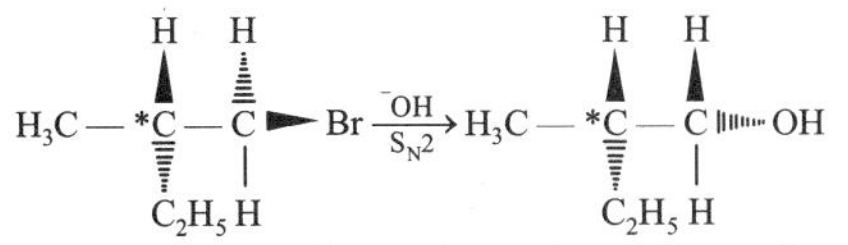

Inversion of configuration takes place when nucleophile attacks but at the stereogenic carbon.

75. (2)

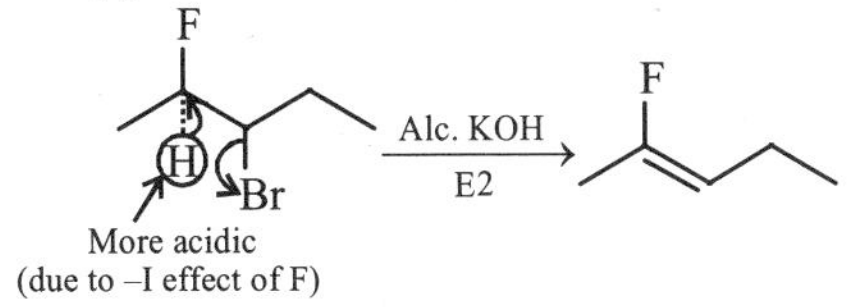

76. (2) First reaction is S_N1 in which rate does not depend on conc. of nucleophile but depends on reactant conc. Second reaction is E2 reaction in which rate depends on conc. of base as well as reactant conc. Therefore, changing in the concentration of base will have no effect on rate of reaction (1).

77. (1) E_1 reaction proceeds via carbocation formation, therefore greater the stability of carbocation, faster will be the E_1 reaction.

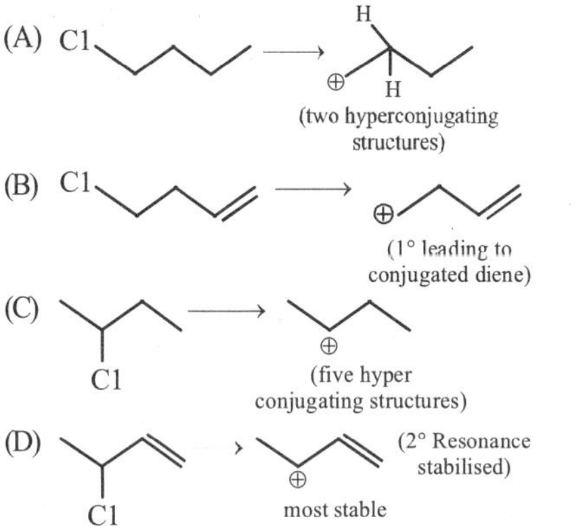

Thus correct decreasing order of the given halides towards dehydrohalogenation by E_1 is:
D > B > C > A

78. (3)

79. (3)

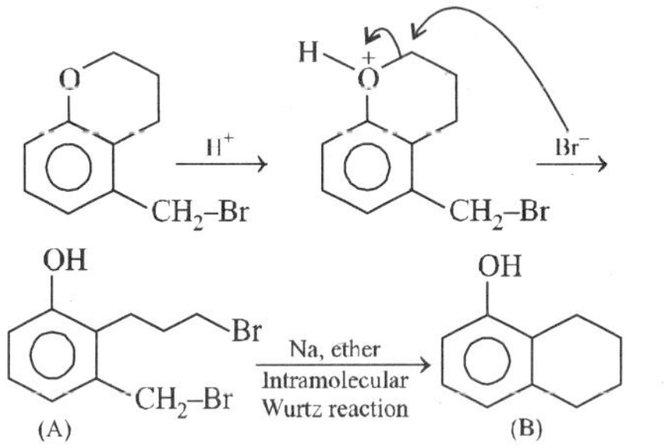

80. (4) $\overset{\curvearrowleft}{CH_2}=CH-\ddot{Cl} \longleftrightarrow \overset{\ominus}{CH_2}-CH=Cl^+$

Due to partial double bond character of C–halogen bond, halogen leaves with great difficulty, if at all it does. Hence, vinyl halides do not undergo nucleophilic substitution easily. So, assertion is correct.

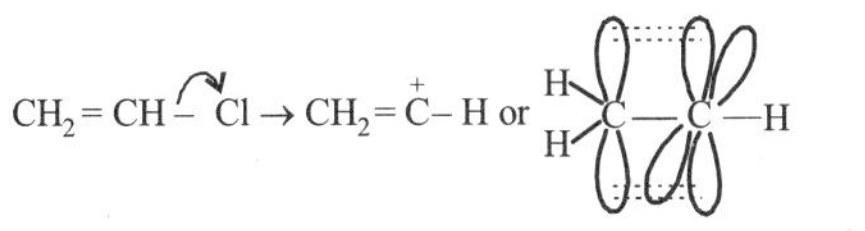

Intermediate carbocation is not stabilised by loosely held-π electrons because empty orbital, being at 90°, cannot overlap with *p*-orbitals of π bond. So, reason is wrong.

81. (2)

82. (3) CH_3OH is a protic solvent and poor nucleophile, so the reaction follows S_N1 pathway.

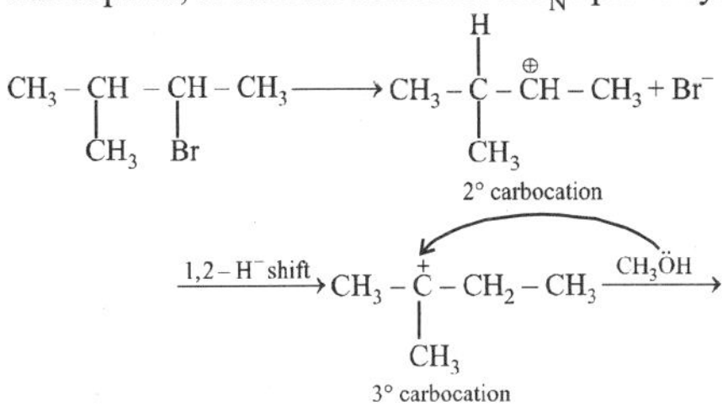

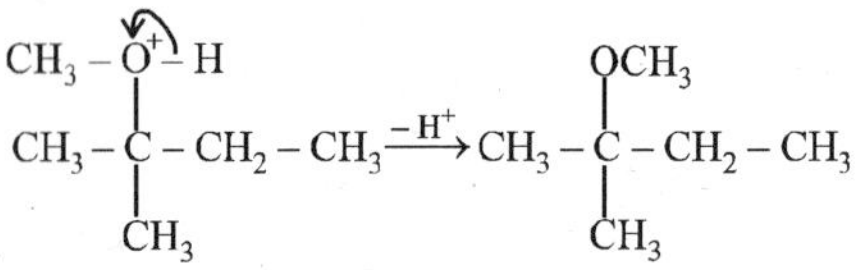

83. (3)

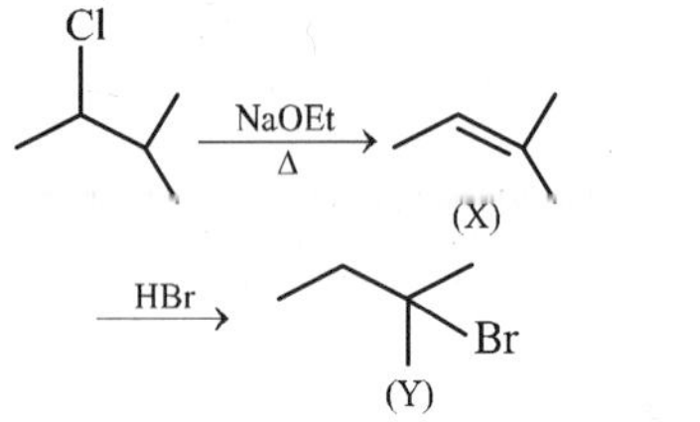

[Markovnikov product]

84. (3) In S_N1 reaction carbocation acts as an intermediate.

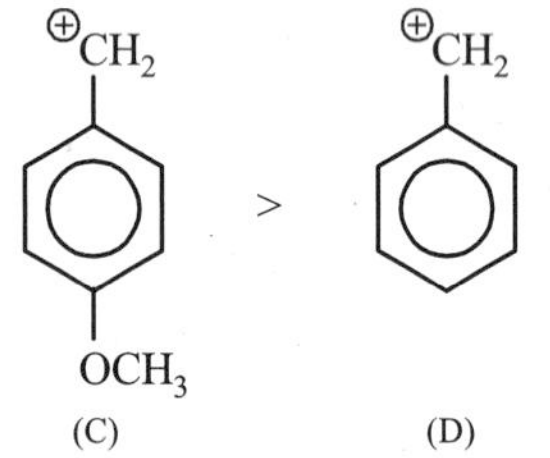

Carbocation produced by (C) is more stable than carbocation produced by (D) due to + M effect of $-OCH_3$ group.

Further in (A) there is formation of tertiary carbocation after rearrangement while (B) is primary carbocation.

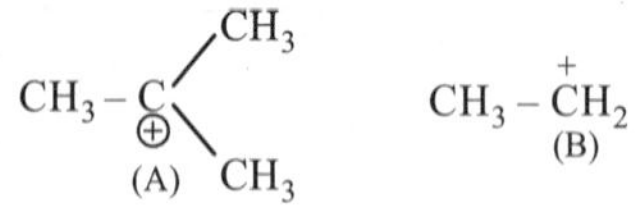

So, the correct order is (C) > (D) > (A) > (B).

85. (1) S_N1 reaction is a two step reaction, in which first step is rate determining step. Hence peak of the first step should be higher than second step.

86. (4)

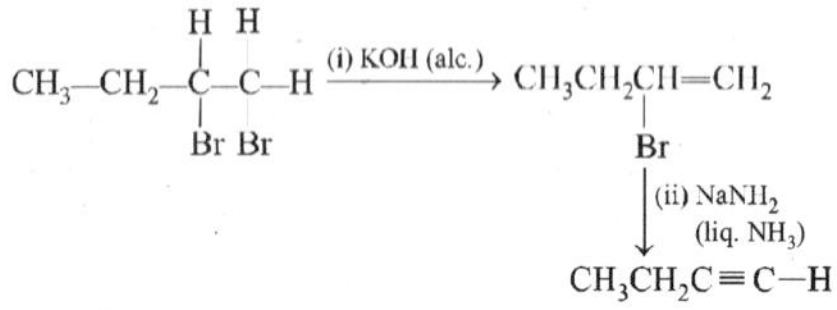

87. (1) Dehydrohalogenation (β–elimination) occurs as:

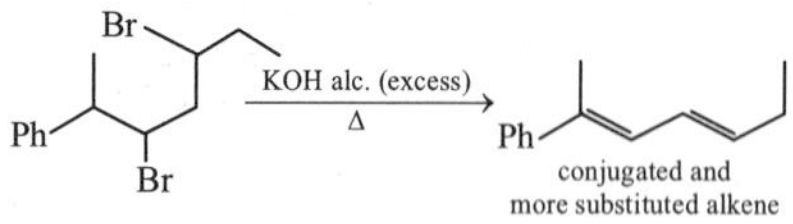

88. (4) Sodium borohydride is a selective reducing agent. It reduces carbonyl group to alcoholic group, N-methylimino group (MeN = CH –) to 2° amines, but does not reduce an isolated carbon-carbon double bond.

Reaction involved:

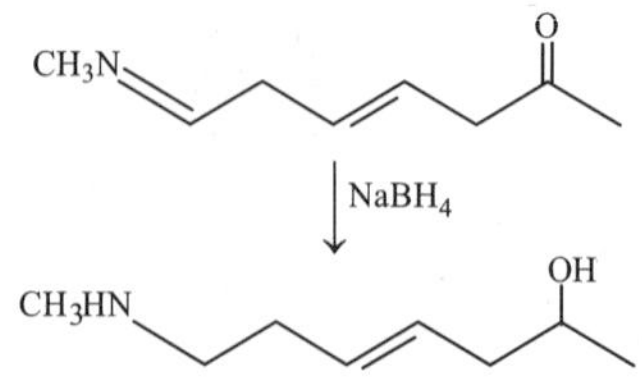

(major product)

89. (2) Allylic H is easily replaced due to the greater stability of allylic free radical.

$$CH_3-CH_2-CH=CH_2+Br_2 \xrightarrow{hv} CH_3-\underset{Br}{\underset{|}{CH}}-CH=CH_2$$

90. (1) Mechanism involved for the given reaction is:

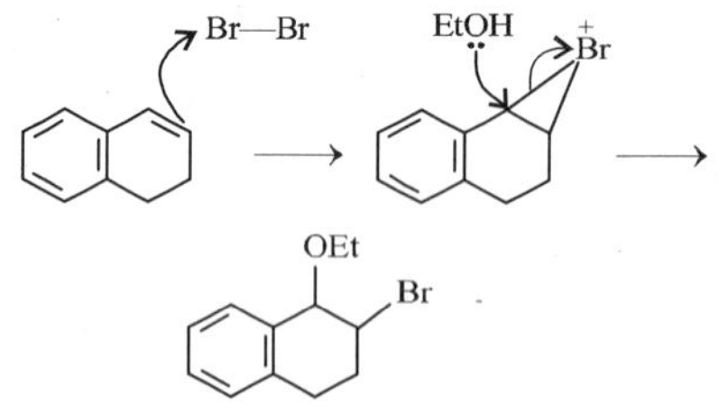

(Major product)

91. (1) Here dehydrohalogenation goes by E_1CB and most stable carbanion formation is favoured in (1).

92. (2) CH_3O^- is a strong base and strong nucleophile, so favourable condition is S_N2/E2. The given alkyl halide is 2° and β carbons are 4° and 2°, so sufficiently hindered, thus E2 dominates over S_N2.

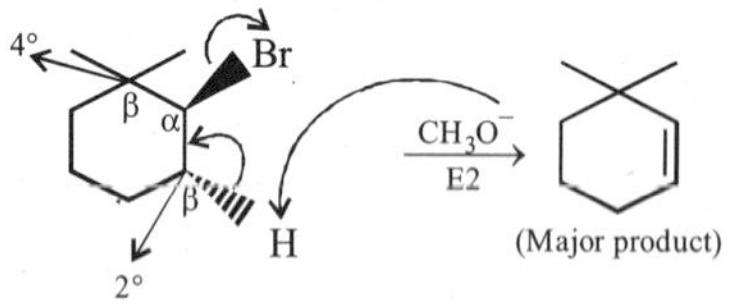

93. (4)

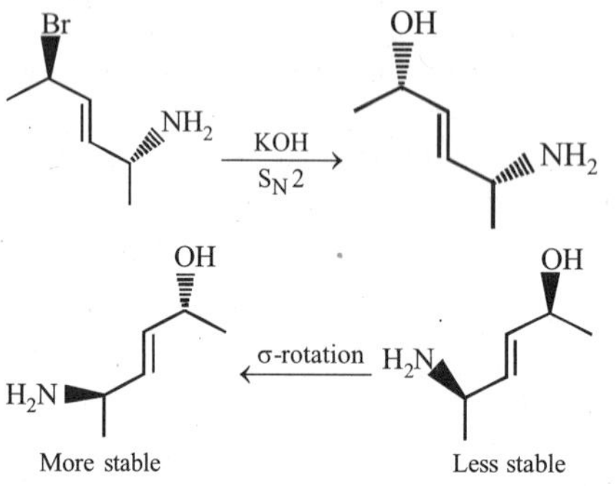

94. (4) $CH_3-\underset{Br}{\underset{|}{CH}}-CH_2-\underset{Br}{\underset{|}{CH}}-CH_2-CH_3 \xrightarrow[E2]{KOH,\ CH_3OH/\Delta} CH_3-CH=CH-CH=CH_2-CH_3$

(Saytzeff product)

95. (2) S_N1 reactions involve the formation of carbocation. Hence higher the stability of a carbocation more will be its reactivity towards S_N1 reactions. Hence, reactivity order is :

$$\underset{1°\text{ carbocation}}{CH_3-CH_2-\overset{\oplus}{C}H_2} < \underset{2°\text{ carbocation}}{CH_3-\overset{\oplus}{C}H-CH_2-CH_3}$$

$$< p-H_3CO-C_6H_4-\overset{\oplus}{C}H_2$$

Max. stable due to +M effect of $-OCH_3$ group.

96. (2)

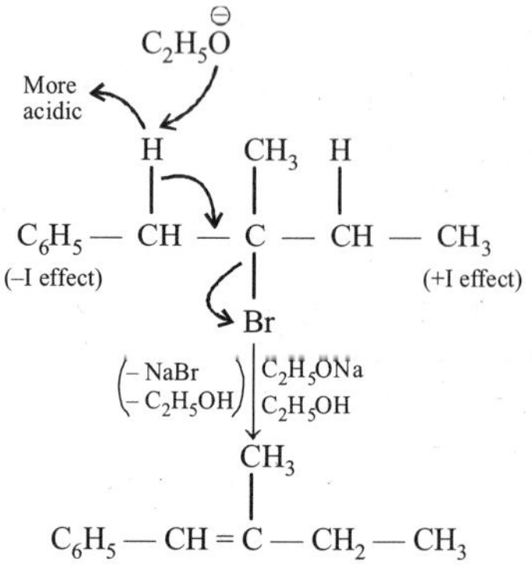

97. (4)

98. (1) Alkyl chloride or bromide undergo substitution and get converted to an alkyl iodide on treatment with a solution of sodium iodide in acetone. e.g.

$$CH_3CH_2CH_2Br+NaI \xrightarrow{\text{acetone}} CH_3CH_2CH_2I+NaBr$$

This reaction is also known as Finkelstein Reaction.

99. (2) Alkyl fluorides are more conveniently prepared by heating suitable chloro – or bromo-alkanes with inorganic fluorides such as AsF_3, SbF_3, CoF_2, AgF, Hg_2F_2 etc. This reaction is called **Swarts reaction**.

$$CH_3Br+AgF \longrightarrow CH_3F+AgBr$$

$$2CH_3CH_2Cl+Hg_2F_2 \longrightarrow 2CH_3CH_2F+Hg_2Cl_2$$

100. (3)

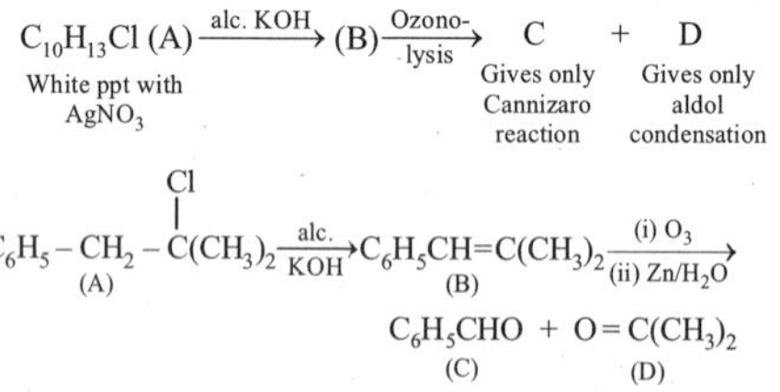

101. (1) The correct order of increasing bond length is $CH_3F < CH_3Cl < CH_3Br < CH_3I$

102. (2) Steric hindrance around the carbon atom having Cl will slow down the S_N2 reaction, hence lesser the hindrance, faster will be the reaction. So, the order of reactivity is

$CH_3Cl > (CH_3)CH_2 - Cl > (CH_3)_2CH - Cl > (CH_3)_3CCl$

103. (1) The order depends on the carbon-halogen bond energy; the carbon-fluorine bond energy is maximum and thus, fluorides are least reactive while carbon-iodine bond energy is minimum, hence iodides are most reactive.

104. (2)

105. (2)

106. (1) $C_2H_5Br \xrightarrow{AgCN} \underset{(X)}{C_2H_5NC} + AgBr$

$\downarrow$ Reduction Zn – Hg / HCl

$C_2H_5NHCH_3$

(Y) Ethyl methyl amine

107. (3) $H_2C=HC-\underset{CH_3}{\underset{|}{\overset{H}{\overset{|}{C^*}}}}-CH_2-CH_3$

3- Methyl-1-pentene

108. (1) S_N1 reactions involve the formation of carbocations, so higher the stability of the carbocation more will be reactivity of the corresponding bromide.

109. (2) $CH_3 - CH = CH - \overset{*}{C}HCH_3$ (with OH on the starred carbon)

It exhibits both geometrical as well as optical isomerism.

cis - R   *cis* - S
trans - R   *trans* - S

110. (4) S_N2 reaction is favoured by small groups on the carbon atom attached to halogen.
So, the order of reactivity is:

$$CH_3Cl > (CH_3)_2CHCl > (CH_3)_3CCl > (C_2H_5)_2CHCl$$

111. (2) The organic compounds which have chiral carbon atom and do not have plane of symmetry hence rotate plane polarised light.

$$\begin{array}{c} CHO \\ | \\ HO - \overset{*}{C} - H \\ | \\ CH_2OH \end{array}$$ (* is asymmetric carbon)

112. (1) Due to steric hindrance, tertiary alkyl halides do not react by S_N2 mechanism, they react by S_N1 mechanism. S_N2 mechanism is followed in case of primary and secondary alkyl halides.

113. (2) In Corey House synthesis of alkanes, alkyl halides react with lithium dialkyl cuprate.

$$R'X + LiR_2Cu \longrightarrow R' - R + RCu + LiX$$

114. (3)

$$CH_3 - \underset{Br}{CH} - CH_2 - CH_3 \xrightarrow{\text{Alc. KOH}} CH_3 - CH = CH - CH_3 + HBr$$

The formation of 2-butene is in accordance to **Saytzeff's rule** (more substituted alkene is formed).

115. (3) 2, 3-Dichlorobutane exhibits optical isomerism due to the presence of two asymmetric carbon atoms.

116. (2) Compounds containing two similar chiral C-atoms have plane of symmetry and can exist in *meso* form too.

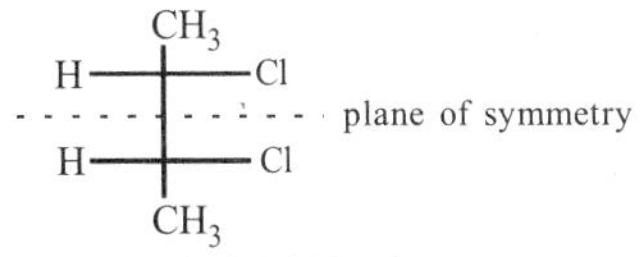

meso-2, 3-Dichlorobutane

117. (2) The hydrolysis of *t*-butyl bromide is an example of S_N1 reaction.

118. (3) $C_5H_{11}Br$

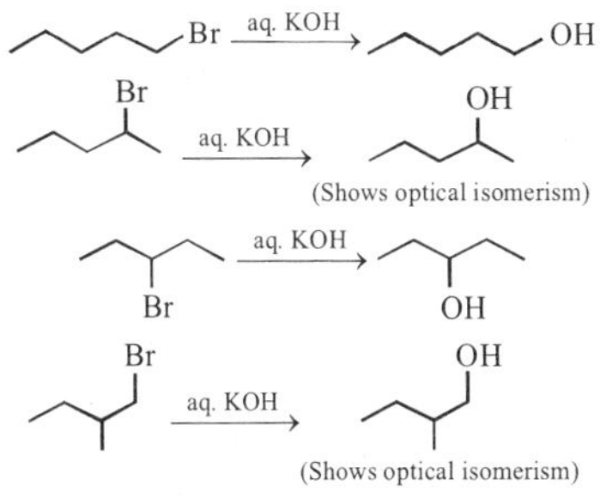

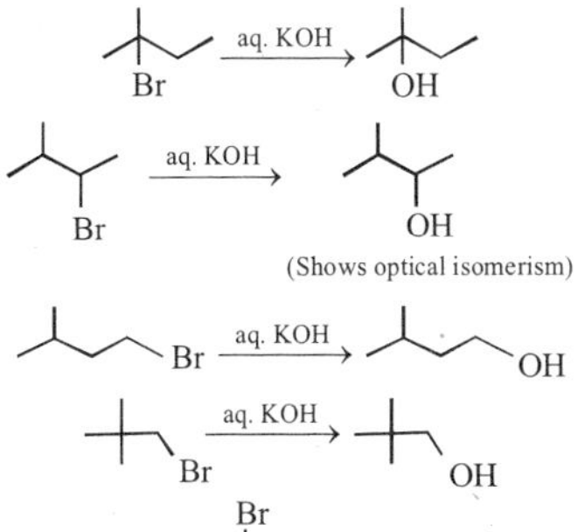

119. (184)

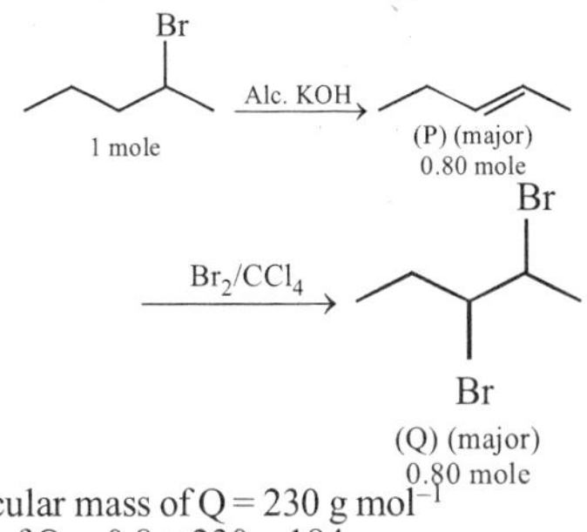

Molecular mass of Q = 230 g mol^{-1}
Mass of Q = 0.8 × 230 = 184 g

120. (32) Number of optical isomers is given by 2^n, n is the number of chiral centre.

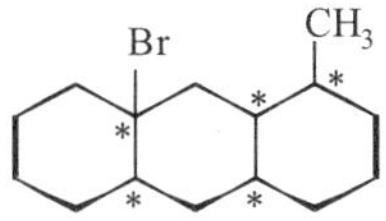

Total chiral centre = 5
No. of optical isomers = $2^n = 2^5 = 32$.

121. (1) Chiral compound is optically active.

$$CH_3 - CH_2 - \underset{Cl}{\overset{*}{C}H} - CH_3$$

122. (2) $CH_3 - \underset{*}{\overset{Cl}{CH}} - CH_2 - CH_3$

Optical isomer = 2^n, n = number of chiral centre
n = 1, Optical isomers = $2^1 = 2$

123. (6)

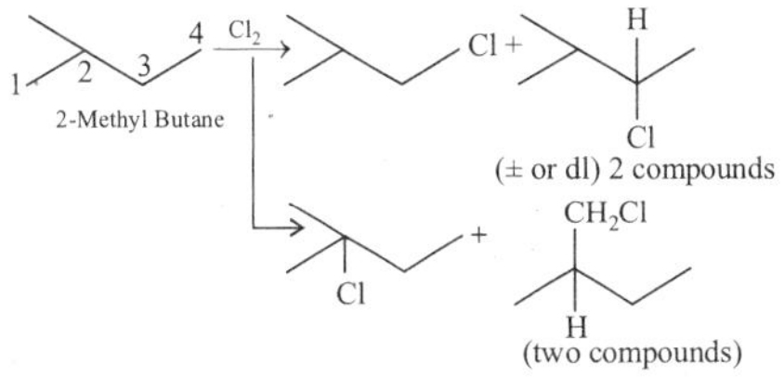

124. (10)

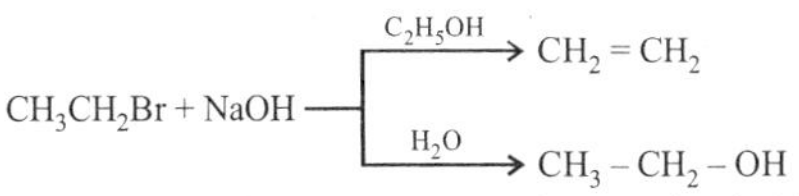

Total number of hydrogen atom in A and B is 10.

125. (6)

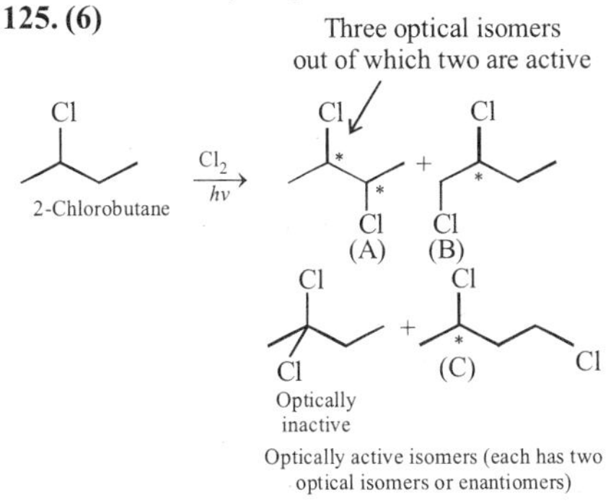

Total number of optically active isomers = 2(from A) + 2 (from B) + 2 (from C) = 6.

126. (4)

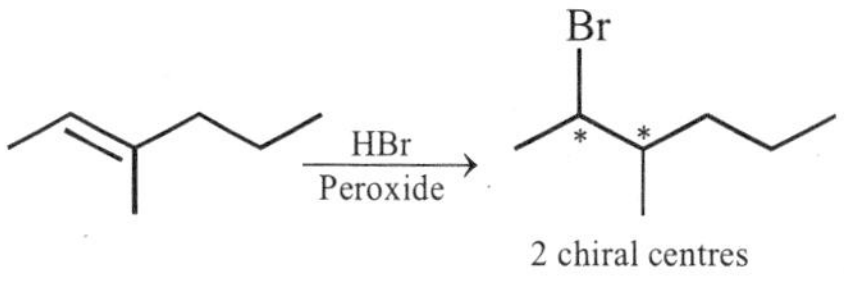

No. of stereoisomers = $2^n = 2^2 = 4$

127. (5) Chiral carbons are marked by.

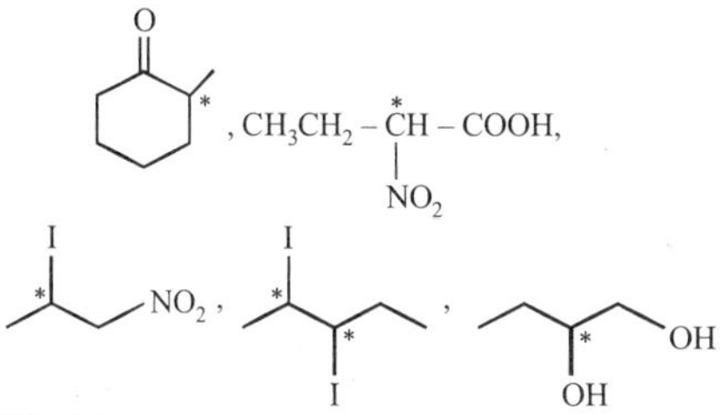

128. (2)

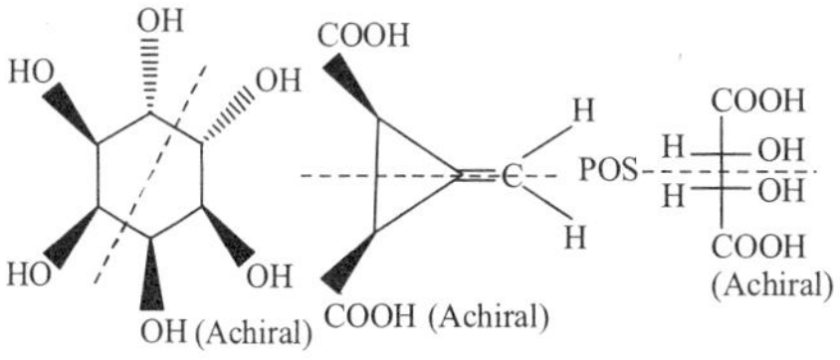

129. (6)

130. (42) % optical purity

$$= \frac{\text{optical rotation of mixture}}{\text{optical rotation of pure enantiomer}} \times 100$$

$$= \frac{12.6^\circ}{30^\circ} \times 100 = 42$$

131. (3) Carbon attached to four different group is called asymmetric carbon.

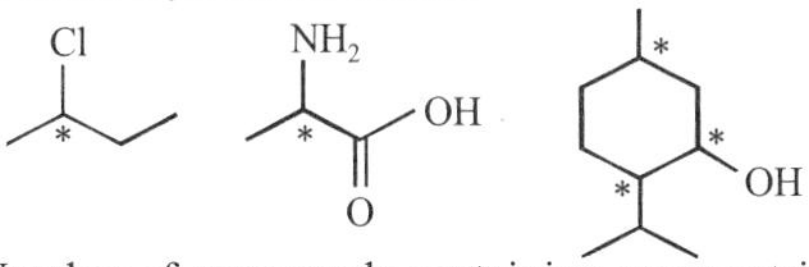

Number of compounds containing asymmetric carbons are three.

132. (1) Bromine in presence of light undergoes allylic bromination.

133. (2) 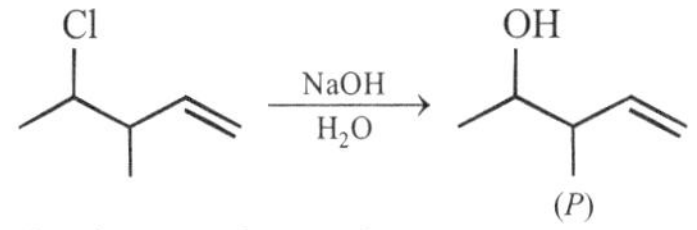

Number of π electrons in *P* = 2

134. (3)

$$\text{Mole} = \frac{\text{Given mass}}{\text{Molar mass}} = \frac{\text{Give volume(L)}}{22.4(L)}$$

$$\frac{1.53}{\text{Molar mass}} = \frac{448 \times 10^{-3}}{22.4}$$

Molar mass = 76.5 g mol^{-1}

Possible compound is $Cl - CH_2 - CH = CH_2$ (A)

$$\underset{(A)}{Cl - CH_2 - CH = CH_2} \xrightarrow[\text{(ii) Zn/AcOH}]{\text{(i) } O_3} ClCH_2 - CHO + HCHO$$

Thus, no. of carbon atom in (A) is 3.

135. (8) Hint: H_2|Ni used for hydrogenation of alkene.

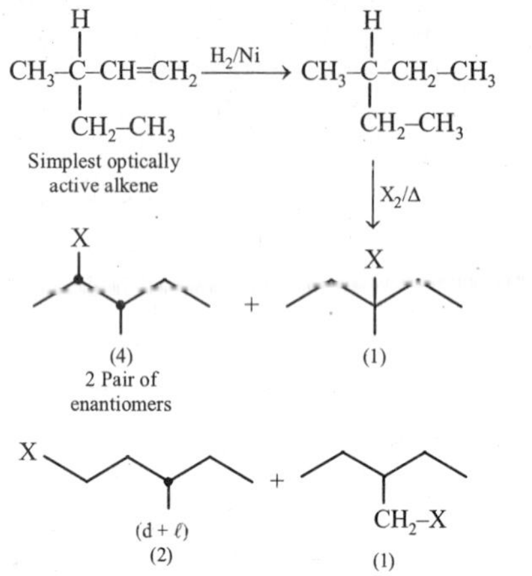

136. (1) Order of boiling point:

$CH_3CH_2CH_2I > CH_3CH_2I > CH_3I$

Van der Waals forces increases due to increase in number of carbon in alkyl group result increase in molecular weight, thus statement I is true.

Order of melting point:

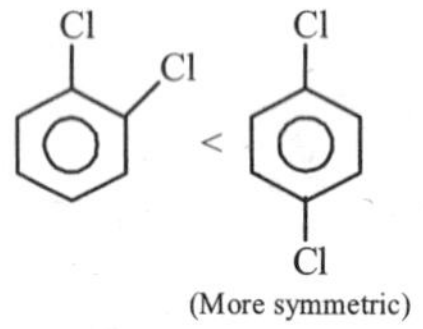

Boiling point order:

(Boiling point $\propto \mu$)

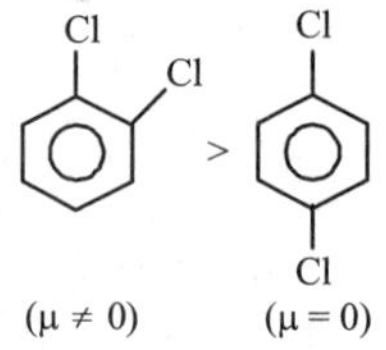

Thus statement II is true.

137. (3)

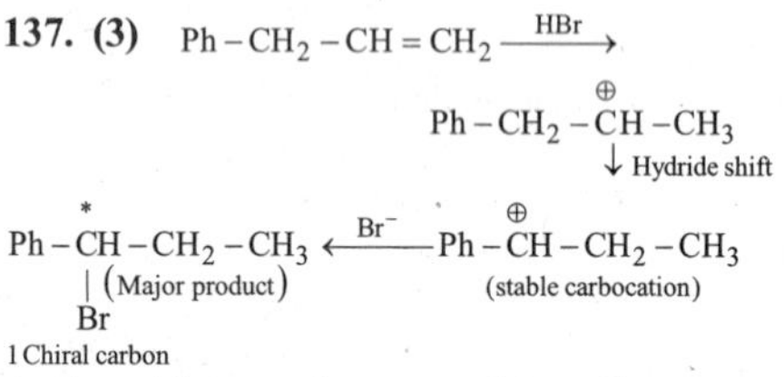

Aryl cyanide cannot be prepared from Gattermann reaction.

138. (1)

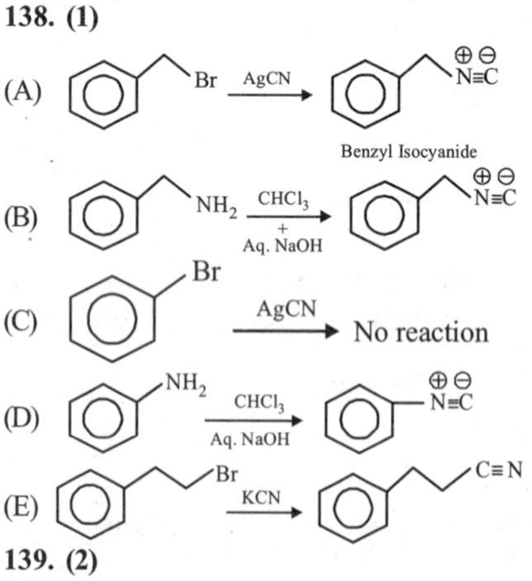

139. (2)

(A) Finkelstein reaction:

$$R-X + NaI \longrightarrow R-I + NaX$$

(B) Swarts reaction:

$$CH_3-Br + AgF \longrightarrow CH_3-F + AgBr$$

(C) Sandmeyer's reaction:

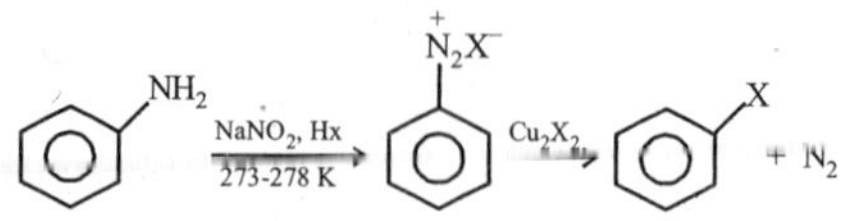

(D) Fittig reaction:

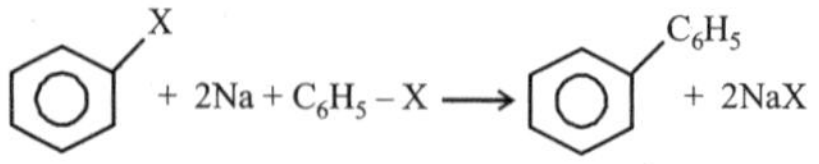

140. (4)

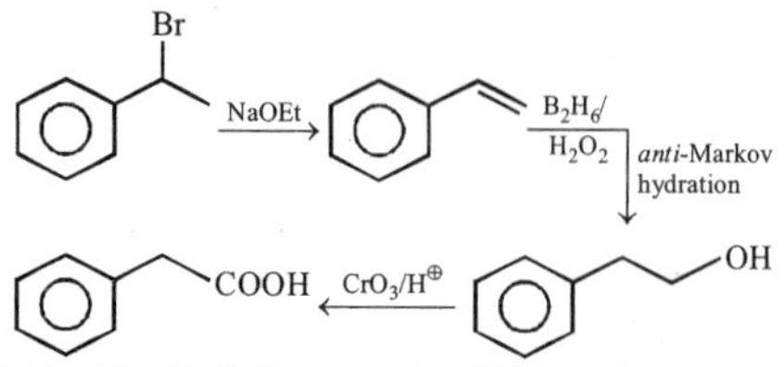

141. (4) In haloarene, C–Cl bond has partial double bond character due to resonance. Also in case of haloarene carbon atom attached to halogen is sp^2 hybridised which is more electronegative.

142. (2) The reaction with benzyl halide with KCN proceeds via S_N1. Rate of S_N1, depends on stability of carbocation formed.

- OH^- group is electron donating group but less than $-NH_2$ group
- NO_2 group is electron withdrawing, effect is more at *m*- than at *p*.

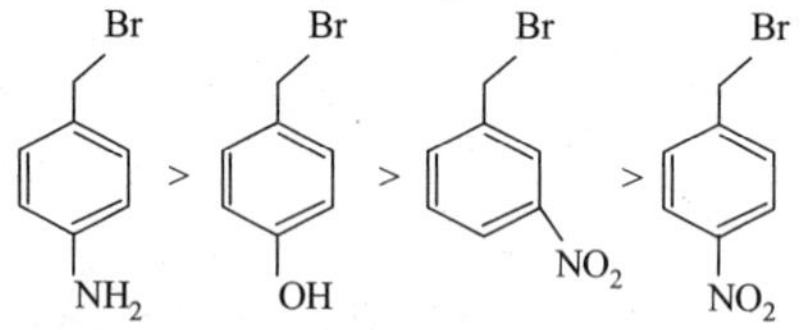

143. (3) Presence of electron withdrawing group at ortho and para position of chlorobenzene stabilize the intermediate in aromatic nucleophilic substitution and increases the rate of reaction.

144. (2)

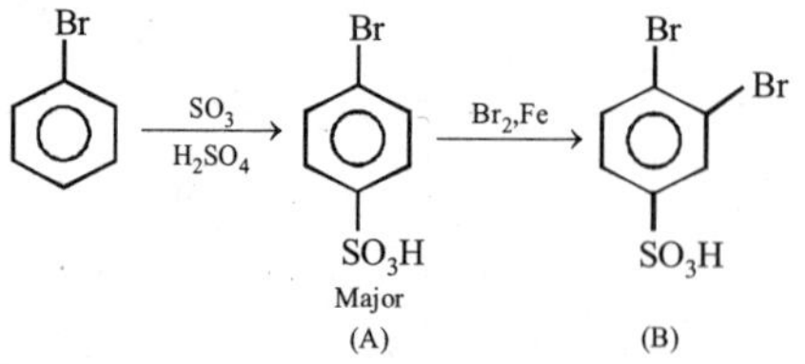

Both reactions are electrophilic substitution reaction first is sulphonation and second is halogenation. – Br is o- and p-directing and gives a para product (major) while $-SO_3H$ is meta directing and so cumulative directive influence of – Br and $-SO_3H$ give the final product (B).

145. (2)

146. (1)

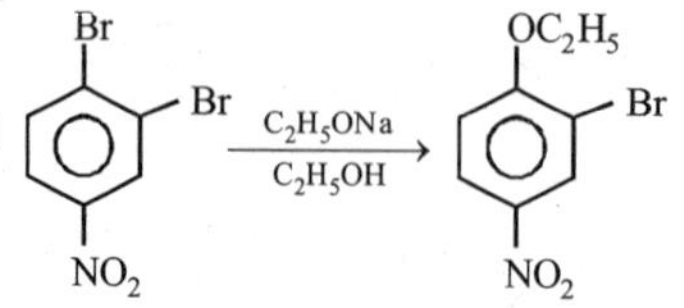

(Due to – I of – Br and – I and –M of NO_2 this Br is displaced)

147. (3)

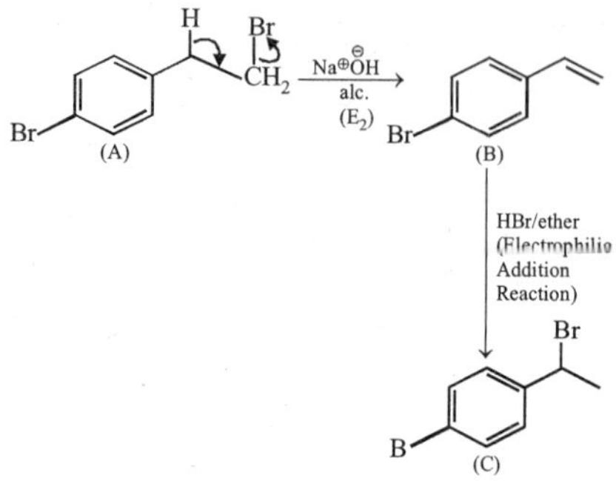

A and C are position isomer.

148. (1)

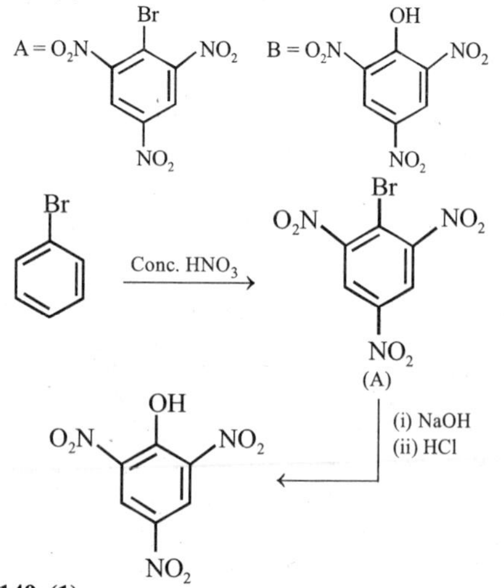

149. (1)

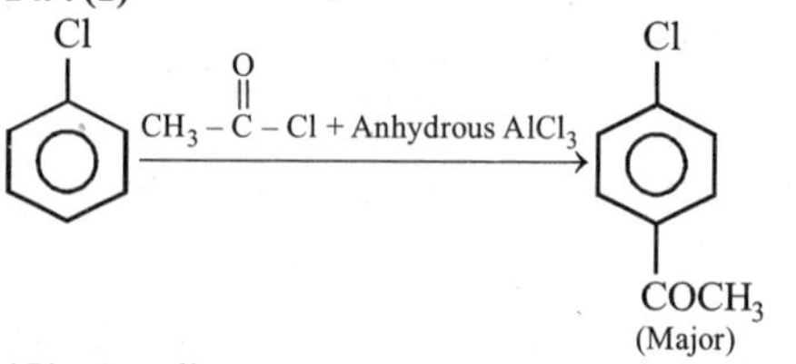

150. (2) All are correct

(A) S_N2 reaction decreases with increase in steric crowding.

(B) S_N1 reaction increases with stability of carbocation.

(C) EAS reaction decreases with decrease in electron density.

(D) Presence of electron withdrawing group at ortho and para–position to a halogen in haloarene increase nucleophilic aryl substitution.

151. (4) ∴ correct order is D > B > A > C

(– M) group increases reactivity where as (+M) group decreases reactivity of Halobenzene towards Nucleophilic substitution reaction.

152. (4) OH^- is a strong base. The substrate with the bulky tertiary group is present in less dielectric medium, alcohol than water. Hence, the elimination reaction is favourable than substitution.

153. (4) 1, 4-dichlorobenzene is more symmetrical and has better crystal fitting therefore its melting point is highest among the given options.

154. (4) A – II, B – I, C – IV, D – III

155. (3) The rate of S_N1 reaction depends upon stability of carbocation which follows the order.

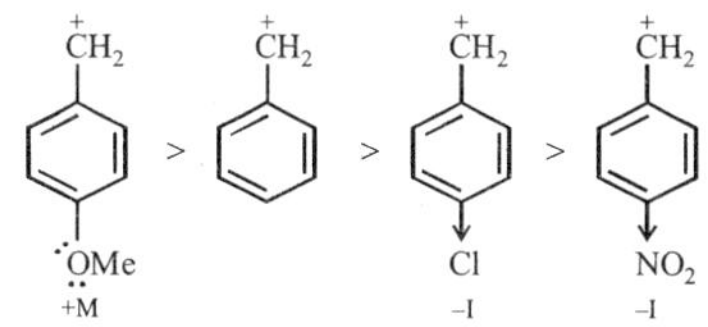

∴ Reactivity order : (B) > (D) > (C) > (A)

156. (4) Electron withdrawing groups are highly ineffective at meta position in nucleophilic aromatic substitution reactions. They have greater effect on ortho and para position.

157. (3) **Assertion A :** Benzene is more stable than cyclohexatriene.

Reason R : The delocalized π-electron cloud is attracted more strongly by nuclei of carbon atoms as the $\pi - e^-$ lies in bonding molecular orbital.

158. (3)

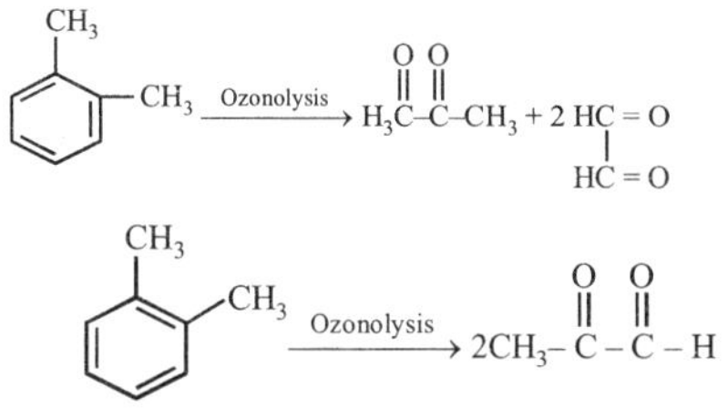

159. (2) Nitration of aromatic compounds is electrophilic substitution reaction. Sice electron donating groups (like $-CH_3$) increase electron density while electron withdrawing groups ($-NO_2$ > –Br) decrease electron density, the order of reactivity of nitration in the given compounds is

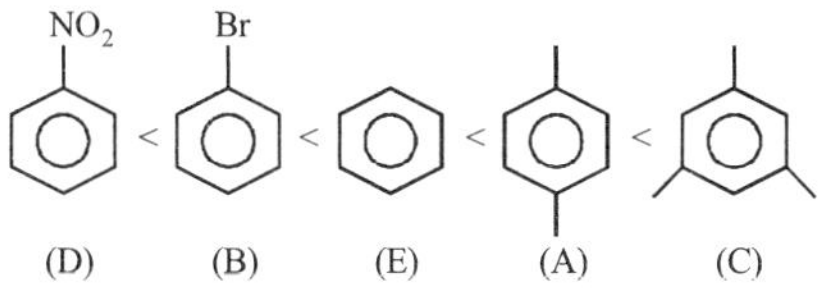

160. (4)

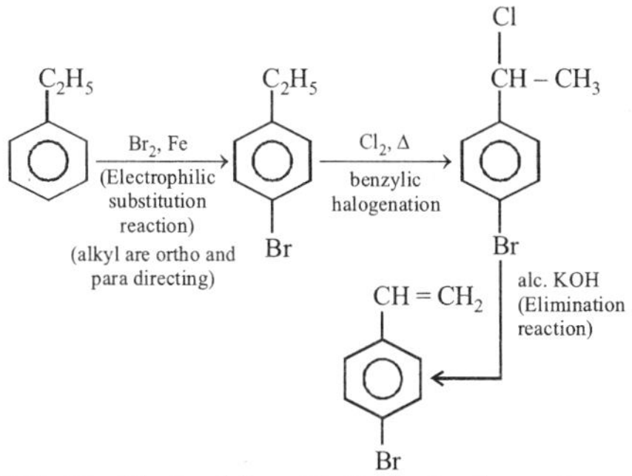

161. (3) Electrophile will attack at ortho and para position with respect to better electron releasing group (ERG)

–OH donates through +M and $-CH_3$ donates through +H

ERG : – OH > – CH_3

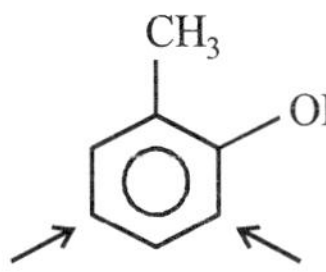

Para position with respect to – OH (+R) group and it will be meta position with respect to $-CH_3$ group.

162. (2)

163. (1) Acetate in acetic acid with R-X gives S_N1 reaction.

Allyl halide undergo faster S_N1 reaction, hence correct order is:

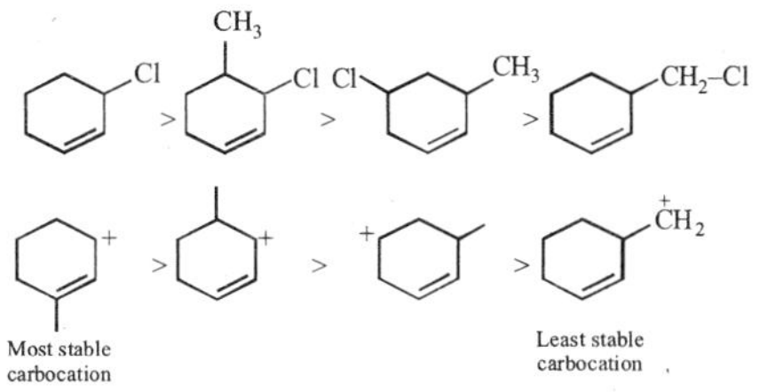

164. (2)

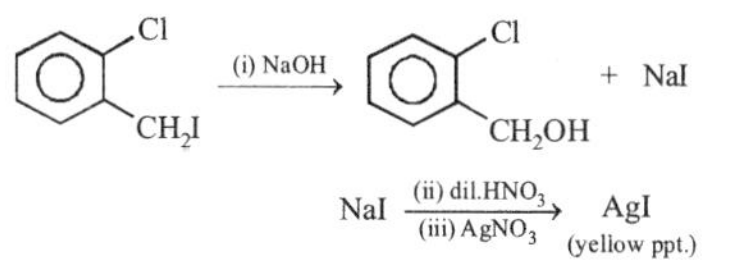

165. (1) Density is proportional to molar mass. So, correct relation is: D > C > B > A.

166. (2)

167. (4)

$-NO_2$ group is a strong electron withdrawing group and it activates the aromatic ring towards nucleophillic substituion.

168. (2) $\text{Rate of } S_N2 \text{ reaction} \propto \dfrac{1}{\text{Stability of carbocation}}$

$$\text{Stability of carbocation} \propto \frac{1}{-I, -M \text{ effect(s)}}$$

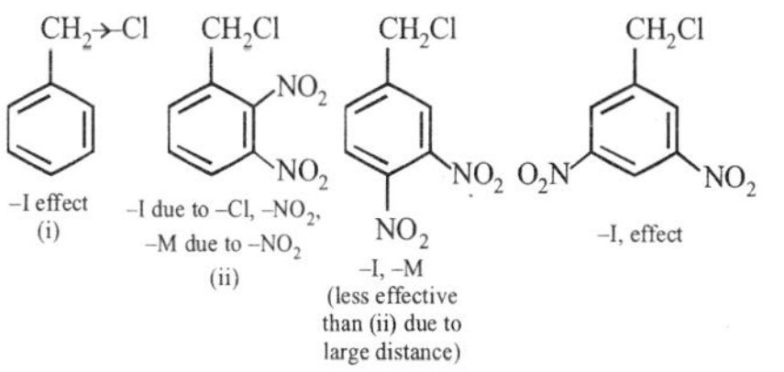

169. (4)

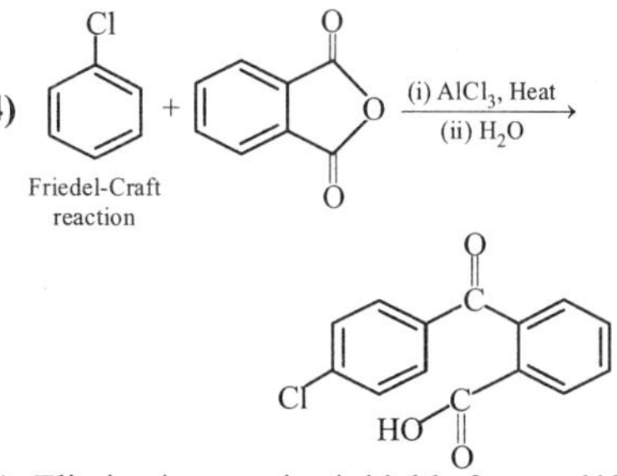

170. (2) Elimination reaction is highly favoured if

(1) Bulkier base is used

(2) Higher temperature is used

Hence, in given reaction, biomolecular elimination reaction provides major product.

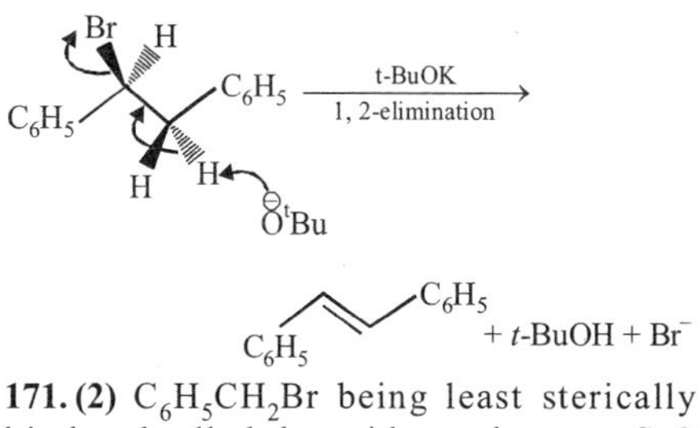

171. (2) $C_6H_5CH_2Br$ being least sterically hindered alkyl bromide undergoes S_N2 nucleophilic substitution reaction most easily. Hence, it undergoes complete inversion of configuration.

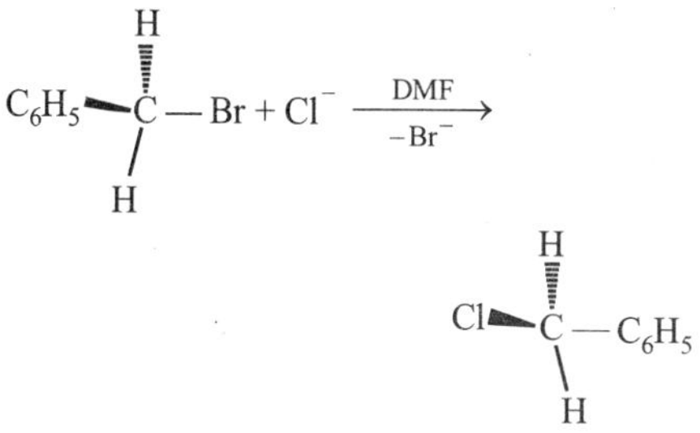

172. (4) $\underset{(A)}{C_8H_9Br} \xrightarrow{\text{Oxidation}} \underset{(B)}{\text{Acid}} \xrightarrow{\Delta}$ Anhydride

Formation of anhydride by acid B indicates the two – COOH groups in acid B are adjacent, i.e. 1, 2-positions and thus compound A has two oxidisable groups (side chains). Compound A gives yellow precipitate with $AgNO_3$ solution, hence Br must be present in side chain.

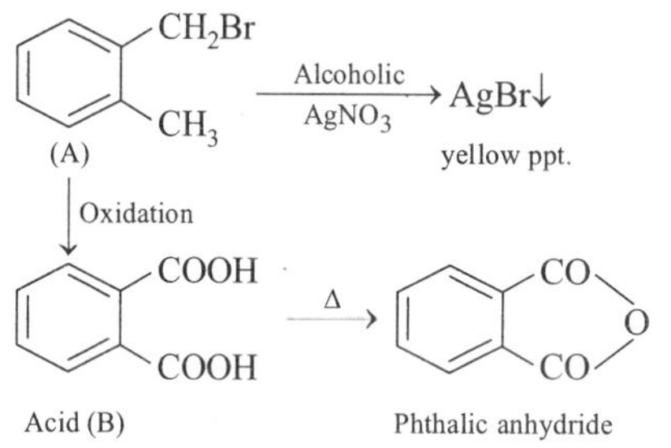

173. (2) Reaction between alkyl halides, aryl halides and sodium in presence of dry ether to give substituted aromatic compound is known as Wurtz-Fittig reaction:

$$C_6H_5Cl + 2Na + ClCH_3 \xrightarrow[\text{ether}]{\text{dry}} \underset{\text{Toluene}}{C_6H_5CH_3} + 2NaCl$$

174. (3) The major component of Borsch reagent (Brady's reagent) is 2,4-dinitrophenylhydrazine which can be obtained by reaction of 2,4-dinitrochlorobenzene and hydrazine sulphate.

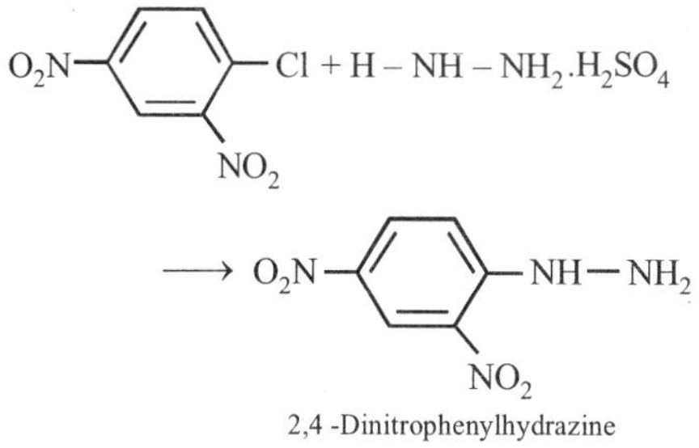

175. (1)

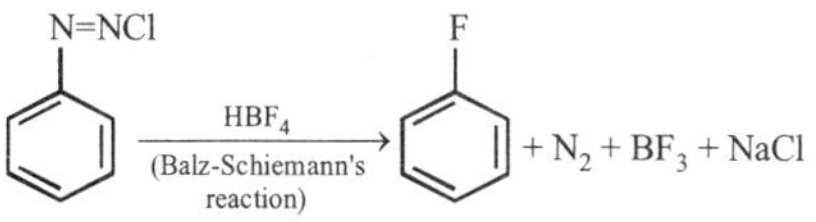

176. (1) The reaction is dehydrohalogenation

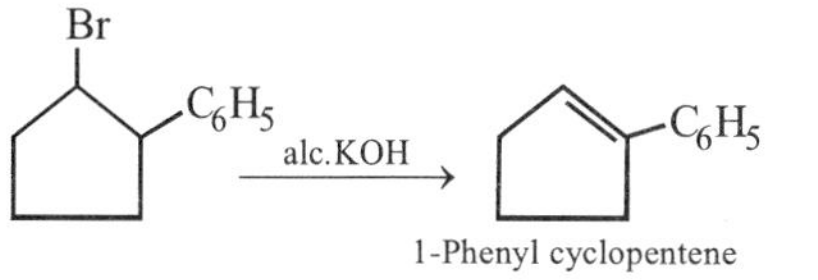

177. (2)

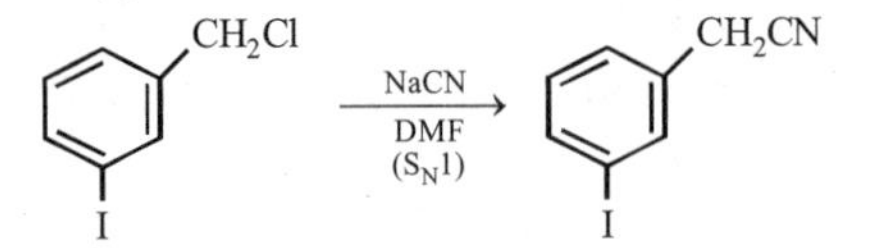

178. (4)

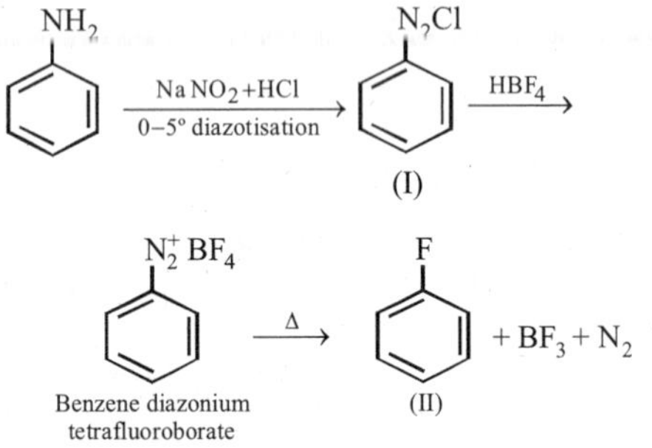

Conversion of (I) to (II) is known as Balz-schilmann reaction.

179. (4) $CH_3OH + C_6H_5MgBr \longrightarrow CH_3OMgBr + C_6H_6$

180. (4)

$C_6H_5I \xrightarrow{NaOH}$ No reaction $\xrightarrow[\text{(ii) AgNO}_3]{\text{(i) dil·HNO}_3}$ No yellow ppt.

$C_6H_5{-}CH_2I \xrightarrow{NaOH} C_6H_5CH_2OH + \mathbf{NaI}$

$\xrightarrow[\text{(ii) AgNO}_3]{\text{(i) dil·HNO}_3} \underset{\text{yellow}}{AgI\downarrow}$

So tube having $C_6H_5CH_2I$ is B as it responds $AgNO_3$ test and the other tube having C_6H_5I is A.

181. (2) In Sandmayer reaction only bromobenzene and chlorobenzene are prepared.

182. (1)

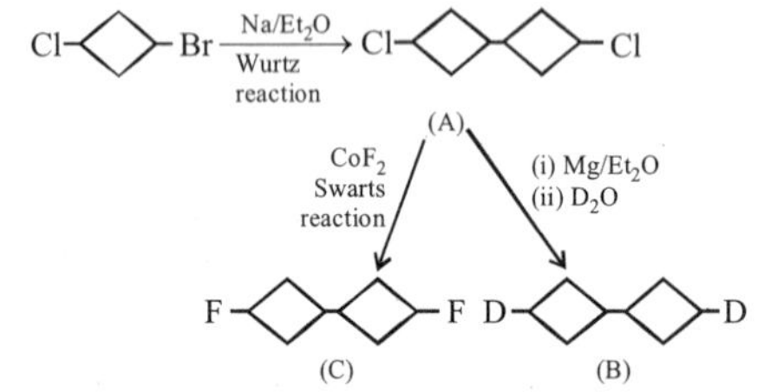

183. (3) A-III, B-I, C-IV, D-II

184. (2) (A) – (III), (B) – (I), (C) – (IV), (D) – (II)

185. (1) Freons are chlorofluorocarbons.

186. (2)

187. (1) Freons are chlorofluoro carbon compounds.

188. (4) Sodium chlorate and Sodium arsenite are herbicides.

189. (3)

190. (3) Chloral on reaction with chlorobenzene in the presence of a catalytic amount of sulphuric acid forms DDT (dichloro diphenyl trichloroethane).

191. (4) $\underset{\text{1, 1, 1–Trichloropropane}}{Cl_3C - CH_2CH_3} + KOH \xrightarrow[(-3\,KCl)]{\text{heat}}$

$(OH)_3C - CH_2CH_3 \longrightarrow \underset{\text{Propionic acid}}{CH_3CH_2\overset{O}{\overset{\|}{C}}{-}OH}$

192. (2) DDT is prepared by heating chlorbenzene and chloral with concentrated sulphuric acid

Part-B (JEE Advanced)

1. (b)

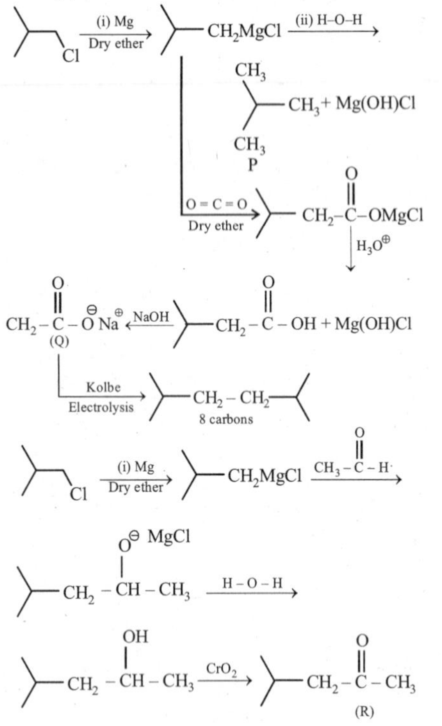

It does not give Cannizaro reaction.

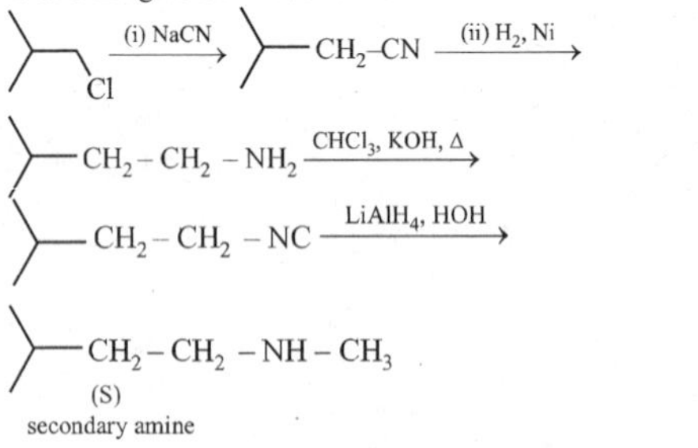

2. (b)

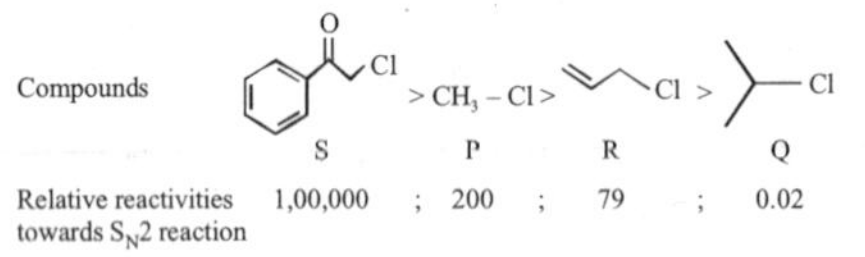

The electron withdrawing groups (–I effect) stabilise the transition state formed in the rate determining step of S_N2 reaction. Hence, the benzoylic, allylic system increases the reaction rate to more than 1° carbon system. But reactivity is higher for —CH_3 than that for allylic (CH_2=CH–) system.

3. (b)

$BrCH_2 - CH_2Br \xrightarrow{\text{Alc.KOH}} CH_2 = CHBr$

$\xrightarrow{NaNH_2} CH \equiv CH$

Elimination of HBr from CH_2= CHBr requires a stronger base because here, C – Br acquires partial double bond character due to resonance.

4. (d) It is an example of intramolecular Wurtz reaction.

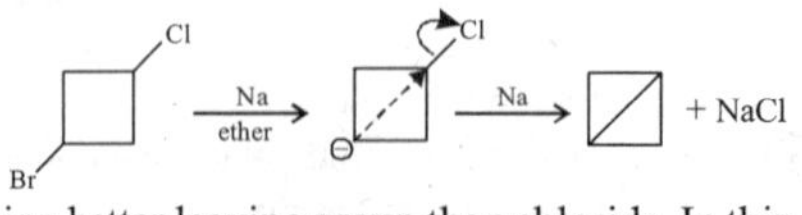

Br^- is a better leaving group than chloride. In this reaction alkali metal (Na) is electron donor.

5. (d)

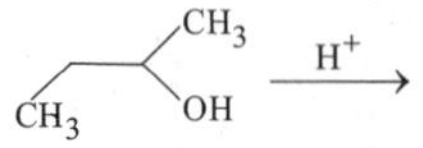

$\underset{\text{Butene-1}}{CH_3CH_2CH = CH_2} + \underset{\text{Butene-2 (cis- and trans)}}{CH_3CH = CHCH_3}$

F

$\underset{\text{Butene-1}}{CH_3CH_2CH = CH_2} \xrightarrow{Br_2} CH_3\,CH_2\overset{*}{C}HCH_2Br$ (with Br on the starred carbon)

(+) – and (–) enantiomer

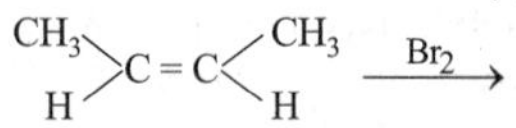

cis-Butene-2

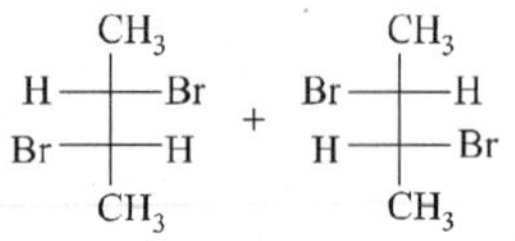

Enantiomers

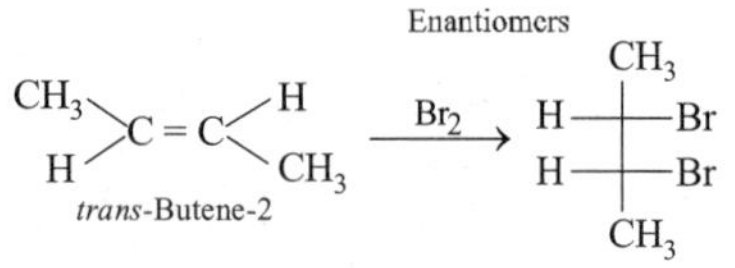

trans-Butene-2 *meso*

So *F* can have three possible structures.

6. (b) Dehydrobromination by strong base (alc. NaOH) followed by Markownikoff addition of HBr.

7. (d) Occurrence of racemization points out the formation of carbocation as intermediate, which being planar can be attacked from either side, which leads to racemisation.

$\underset{(+)}{C_6H_5 - \overset{Cl}{\overset{|}{\underset{H}{\underset{|}{C}}}} - CH_3} \xrightarrow[(SbCl_6^-)]{SbCl_5}$

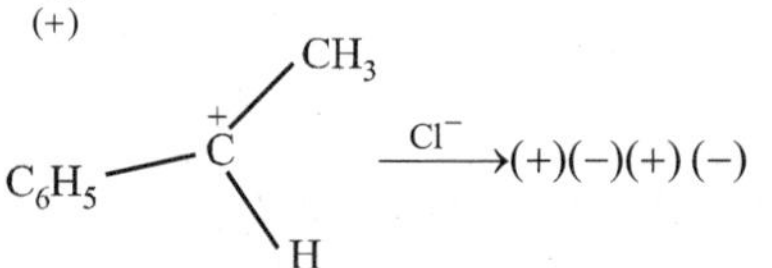

(carbocation intermediate)

8. (a) $(CH_3)_3C - MgCl + D_2O \longrightarrow (CH_3)_3C - D + Mg(OD)Cl$

9. (b) $(CH_3)_2CHCH_2MgBr \xrightarrow{C_2H_5OH} (CH_3)_2\,CHCH_3$

10. (b) The reaction proceeds via free radical mechanism.

As 2° free radical is more stable than 1°, so $CH_3CH_2CH(Br)CH_3$ would be formed.

11. (a)

$H_3C - H_2C - H_2C - CH_2Cl \xrightarrow{\text{alc. KOH}}$

$CH_3-CH_2CH=CH_2$ (Elimination reaction)
1-butene

Alkyl halides give alcohols with **aq.** KOH, which is a substitution reaction.

12. (b) Addition of HBr to 2-pentyne gives two structural isomers (I) and (II)

$$CH_3-C\equiv C-CH_2CH_3 \xrightarrow{HBr} \underset{(I)}{CH_3C(Br)=CHCH_2CH_3} + \underset{(II)}{CH_3CH=C(Br)CH_2CH_3}$$

Each one of these will exist as a pair of geometrical isomers. Thus, there are two structural (I and II) and four configurational isomers *cis*-and *trans*-of I and II.

13. (b)

$$\underset{n\text{-Propyl bromide}}{CH_3-CH_2-CH_2Br} \xrightarrow[-HBr]{\text{ethanolic KOH}} \underset{\text{Propene}}{CH_3-CH=CH_2}$$

14. (a) Chlorination beyond monochlorination during the preparation of alkyl halides in presence of UV light can be suppressed by taking **alkane in excess**.

15. (c) A mono-substituted benzoic acid is stronger than a mono-substituted phenol as former being a carboxylic acid. Among the given substituted benzoic acid, *ortho*-hydroxy acid is strongest acid although —OH causes electron donation by resonance effect which tends to decrease acid strength.

It is due to a very high stabilisation of congugate base by intramolecular H-bond which outweigh the electron donating resonance effect of —OH.

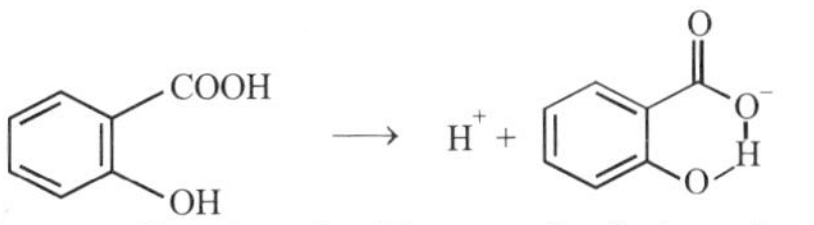

The overall order of acid-strength of given four acids is *ortho*-hydroxybenzoic acid ($pK_a = 2.98$). > Toluic acid ($pK_a = 4.37$) > *p*-hydroxybenzoic acid ($pK_a = 4.58$) > *p*-nitrophenol ($pK_a = 7.15$)

16. (a) Anti addition of Br_2 on *trans* alkene provides meso compound.

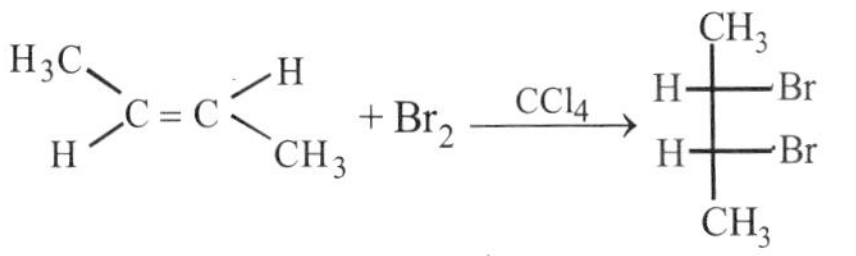

17. (a) This is an example of S_N1 reaction involving carbocation s intermediate.

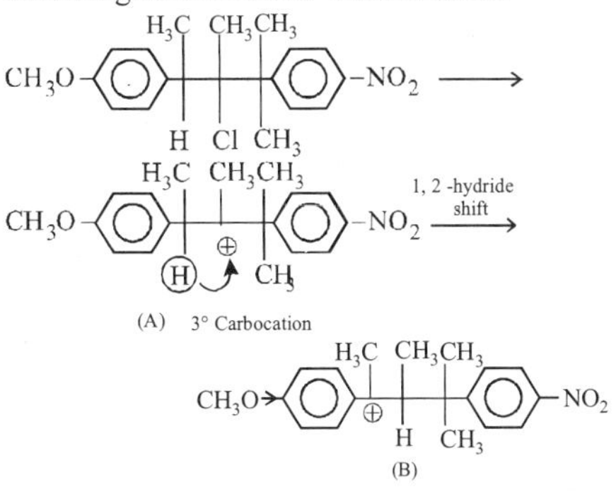

This carbocation is especially stabilised through resonance in which $-\ddot{O}-CH_3$ group acts as a good electron donor.

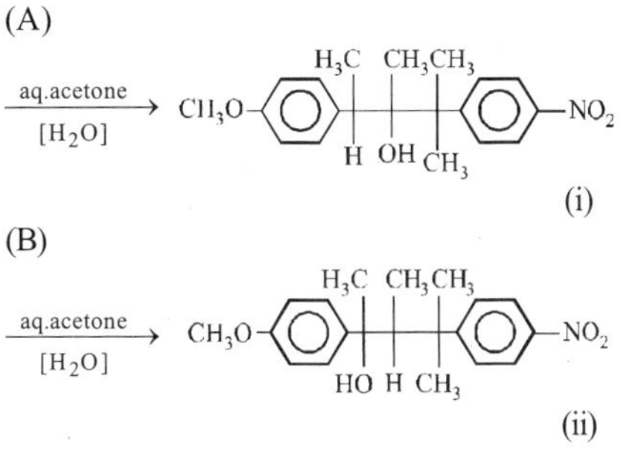

18. (d) 2-Methylbutanoic acid contains one asymmetric centre

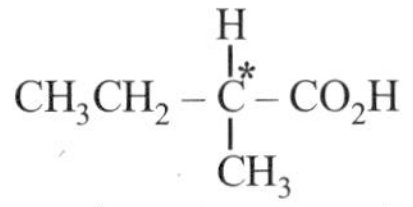

19. (d) S_N2 reaction at asymmetric carbon occur with inversion of configuration and a single steroisomer is formed because the reactant and product are not enantiomer. Therefore the sign of optical rotation may or may not change.

20. (d) Rate of reaction will be R – I > R – Br > R – Cl > R – F. because I^- is the best, while F^- is the poorest leaving group among halide ions.

21. (c) CH_3^- is the best nucleophile because carbon is least electronegative among the given options. The order is

$$H_3\bar{C} > \bar{N}H_2 > \bar{O}H > F^-$$

22. (c) The stereoisomers of butane -2,3-diol are

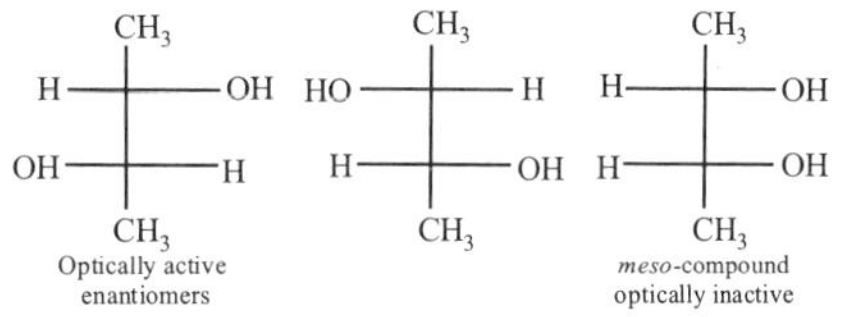

23. (a) The bond angle in sp^3, sp^2 and sp hybridisad carbon atoms is respectively 109.28', 120° and 180°.

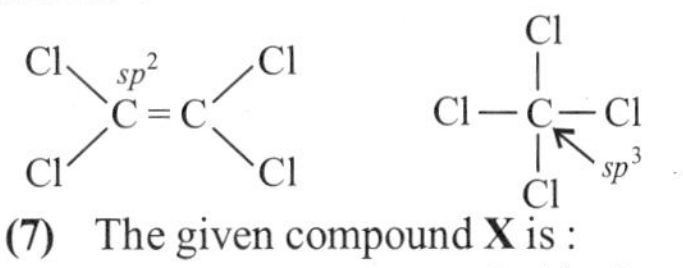

24. (7) The given compound **X** is :

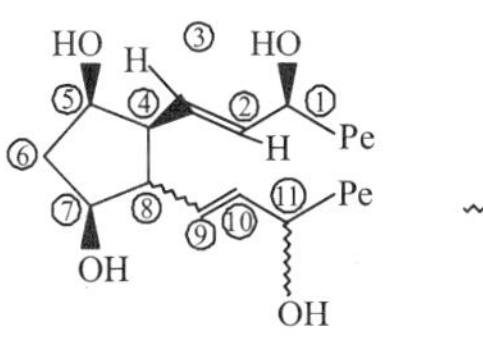

◄ Position in space and geometry (*cis-trans* configuration) is fixed.
〰 Position in space and geometry is not fixed.
Pe : *n*-pentane

Trick : If the bottom double bond C⑨ = C⑩ is in *trans* position then there can be distinct possibilities with the C⑧ – C⑨ and C⑪ – OH. Similarly, there will be additional 4 distinct possibilities when the double bond C⑨ = C⑩ is in *cis* configuration. Now, out of these 8 distinct possibilities, only one configuration has a plane of symmetry. This makes it an optically inactive compound. We can also solve this problem by considering the two chiral centers at C⑧ and C⑪.

Number of stereoisomers : : $= 2^2 = 4$

Now, each of these 4 stereoisomers can be in *cis* or *trans* configuration about C⑨ = C⑩ bond. This makes the total number of stereoisomers = 4 × 2 = 8. Total number of optically active stereoisomers = 8 – 1 = 7

25. (5)

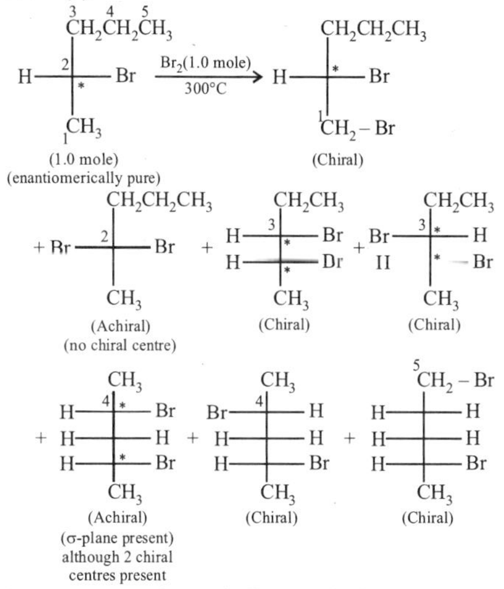

26. (5) Total no. of alkenes will be = 5

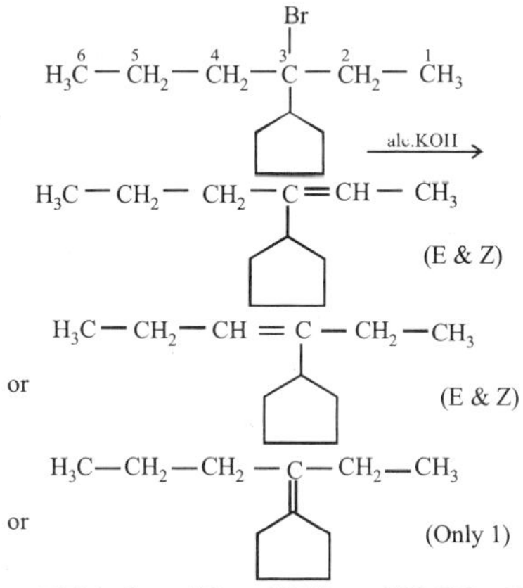

27. $(CH_3)_2Cu + CH_2=CHCl \rightarrow \underset{\text{Propene}}{CH_3CH=CH_2}$

28. Thioalcohol

29. non-superimposable, enantiomers;

30. Grignard (RMgX)

31. Chlorine; because rate of formation of $^\bullet CH_3$ (one of the propagating steps) is high when $X^\bullet$ is Cl. $CH_4 + X^\bullet \longrightarrow {}^\bullet CH_3 + H-X$

32. False : There are only two asymmetric (marked with *) carbon atoms.

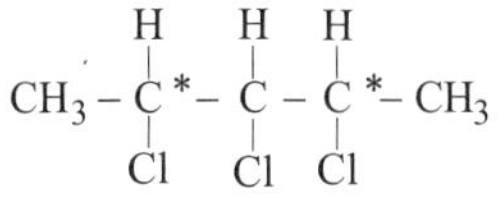

Note that C-3 has two similar groups, i.e., CH_3CHCl^-

33. True : S_N1 Reactions (unimolecular nucleophilic substitution reaction) take place in two steps.

(i) Formation of carbocation which is slow and rate determining step

(ii) Attack of nucleophile, a fast step.

34. False : $CH_2=CHCl + HI \rightarrow CH_3CHCl(I)$

Vinyl chloride 1- Chloro-1-iodoethane

This is an example of Markownikoff's rule as I^- is added at the C with less number. of H-atoms. ***anti*-Markovnikov's rule is applicable only to HBr, but not to HI and HCl.**

35. True: Iodide is bigger in size than bromide, hence its electrons are more dispersed than that of bromide, with the result, it is better nucleophile than bromide.

36. (b)

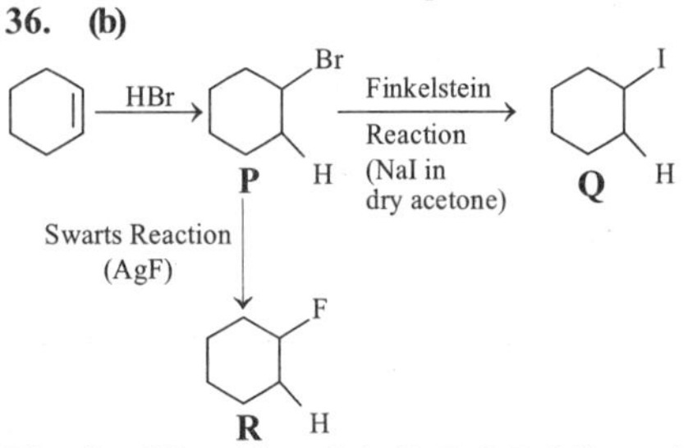

The C – X bond length in P, Q & R follows the order Q > P > R.

As, Bond length $\propto \frac{1}{\text{Bond Enthalpy}}$

Hence, the C – X bond enthalpy in *P*, *Q* & *R* follows the order Q < P < R

37. (a, b, c)

(a) I and III both forms stable carbocations

(I) 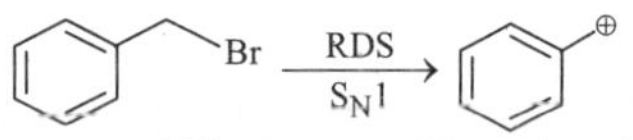

(Resonance stabilised very stable carbocation intermediate)

(III) 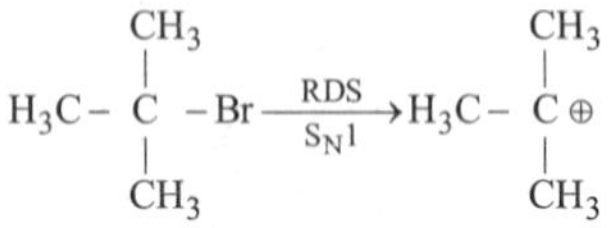

(Very stable tertiary carbocation intermediate)

Thus, they can follow S_N1 mechanism.

(b) I and II both are 1° alkyl halide and forms a stable transition state.

(I) 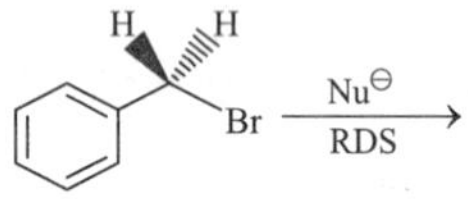

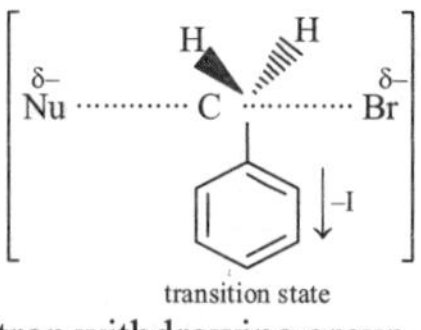

transition state

Due to an electron withdrawing group, it forms a very stable transition state.

(II) 1° alkyl halide always prefers S_N2 mechanism.

Thus, they can follow S_N2 mechanism.

(c) Compound (IV) can form a stable carbocation and also a stable transition state as :

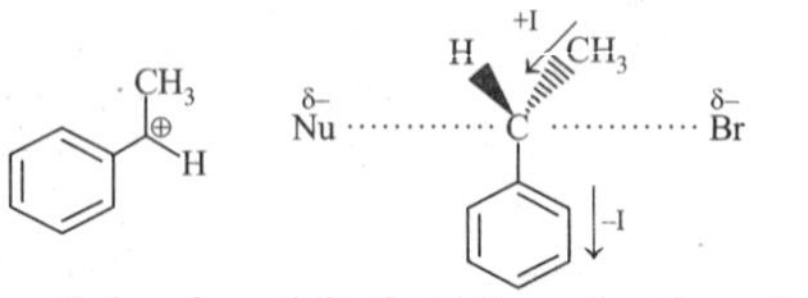

(d) Order of reactivity for S_N1 reaction depends on the stability of intermediate carbocation formed in the rate determining step:

Stability order of carbocations :

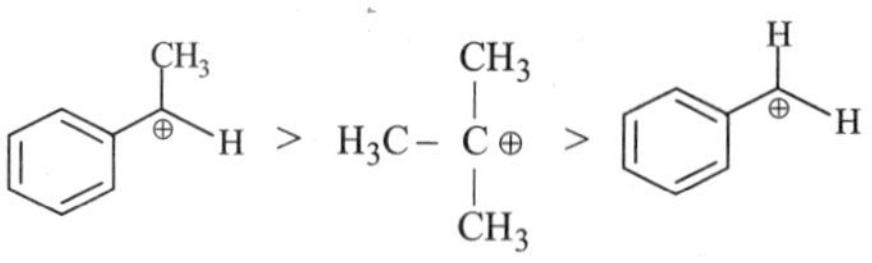

(IV) (III) (I)

Order of reactivity for S_N2 mechanism depends on the stability of T.S. and steric factor.

Stability order of transition states :

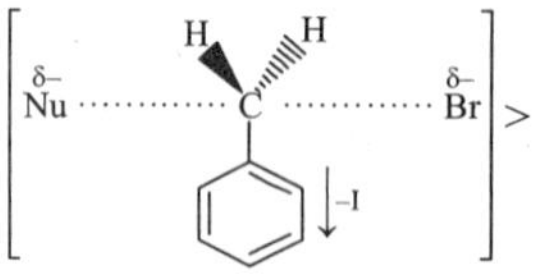

(I)

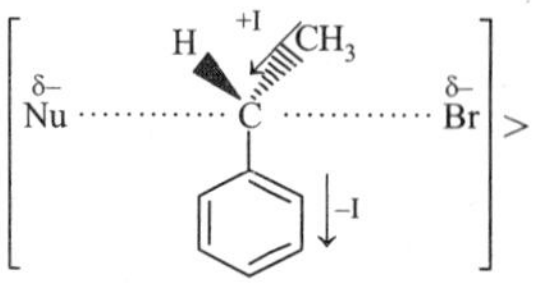

(IV)

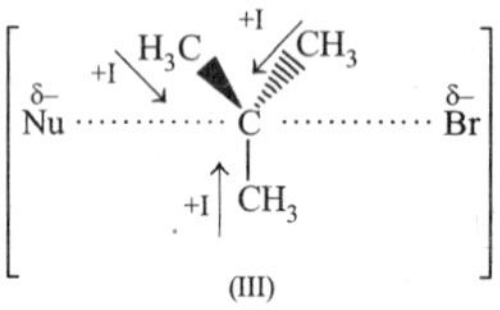

(III)

Hence, option (d) is wrong.

38. (a, d)

(a) 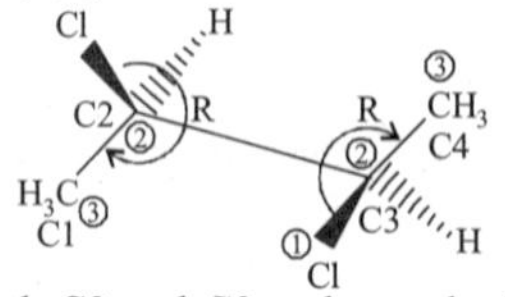

In both C2 and C3 carbons, the lowest priority group/atom (*i.e.* H) is in wedge position. Thus, we can directly assign the R–S configuration to the given compound.

The R–S configuration of the compound is (2R, 3R). Hence, the compound is opticaly active.

Option (b) and (c) can not be correct as the molecule is optically active.

(d) If we make a conformer of the given compound by taking the C2 – C3 axis and rotating the groups on C3 carbon by 180°, then the compound can be represented as:

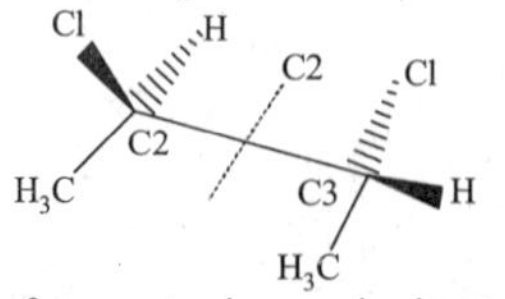

A C_n axis of symmetry is an axis about which the molecule can be rotated by 360°/*n* to produce a molecule indistinguishable from the original molecule. If we rotate the molecule by 180° along the axis which is perpendicular to C2 – C3 bond, then we will get the exactly same molecule. Hence, this molecule has C2 axis of symmetry.

Any optically active compound may or may not possess axis of symmetry but it cannot possess alternate axis of symmetry.

39. (c, d) An asymmetric carbon atom is one which is attached with 4 different groups. Hence, (c) & (d) are correct.

40. (b) $P \rightarrow 2, Q \rightarrow 1, R \rightarrow 3, S \rightarrow 5$

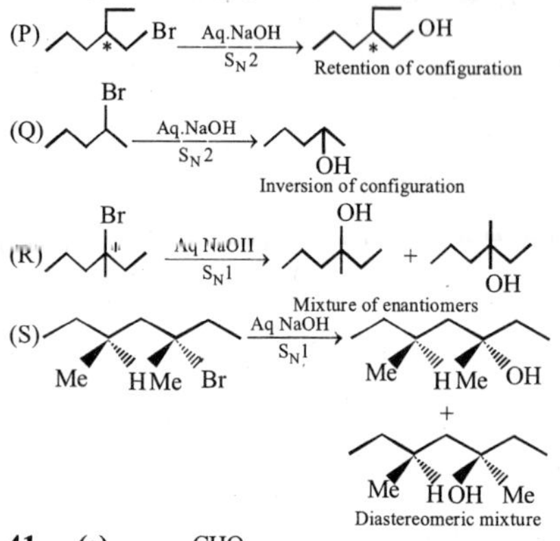

41. (c)

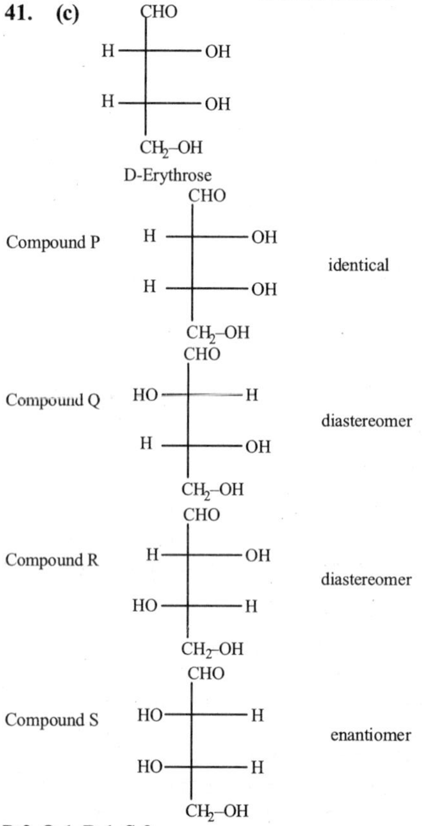

P-2, Q-1, R-1, S-3

42. (c) Statement-1 is correct. Statement-2 is incorrect because compound can be chiral even in the absence of chiral atoms.

43. 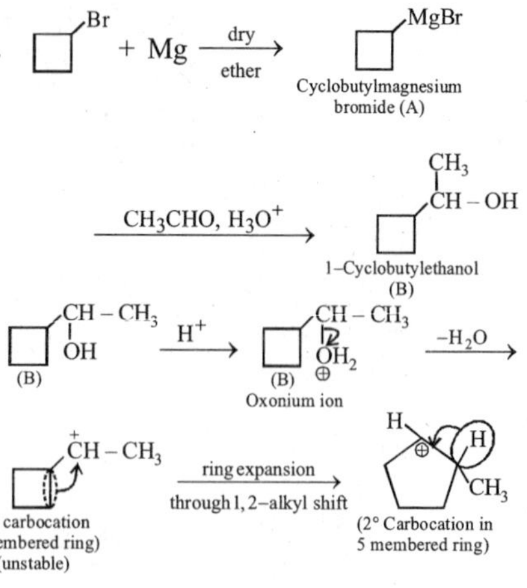

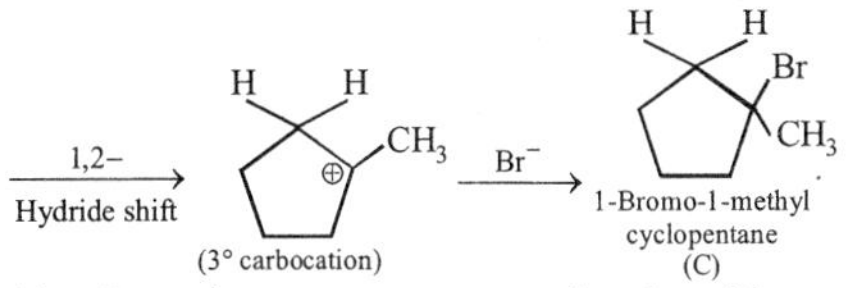

(3° carbocation) 1-Bromo-1-methyl cyclopentane (C)

44. In order to convert a molecule with two stereogenic centres to its enantiomer, the confuguration at both centres must be reversed. Reversing the configuration at only one stereogenic centre converts it to a distereomeric structure. Thus, structures I and III are enantiomers; while structures I and II as well as II and III are diastereomers.

45. (*i*) $CH_3-C(CH_3)_2-CH_2Br \xrightarrow[(S_N2\ reaction)]{C_2H_5OH} CH_3-C(CH_3)_2-CH_2OC_2H_5$

(*ii*) $C_6H_5-CH_2-CH(Cl)-C_6H_5 \xrightarrow[(E2)\Delta]{KOH\ (alcoholic)} C_6H_5(H)C=C(H)C_6H_5$

cis and *trans* - Stilbene (*trans* is major)

(*iii*) $C_6H_5CH_2-CH(Br)-CH_3 \xrightarrow[\Delta]{alc.\ KOH} \underset{(Saytzeff\ product)}{C_6H_5CH=CH-CH_3} \xrightarrow{HBr} C_6H_5-CH(Br)-CH_2CH_3$

(according to markovnikov rule)

[$C_6H_5CH_2\overset{+}{C}HCH_3$ and $C_6H_5\overset{+}{C}HCH_2CH_3$ carbocations are formed on addition of HBr on $C_6H_5CH{=}CHCH_3$, the latter being benzylic carbocation, is stabilised due to resonance and hence Br^- adds on it forming $C_6H_5CHBr.CH_2CH_3$ as the final product.]

(*iv*) $CH_3CH_2CHCl_2 \xrightarrow[alkali]{boil} CH_3CH_2CH(OH)_2 \xrightarrow[(-H_2O)]{} CH_3CH_2CHO$

(*v*) $(CH_3)_2C(Cl)-CH_2CH_3 \xrightarrow{alc.\ KOH} (CH_3)_2C=CHCH_3$

(*vi*) $2C_2\overset{\delta+}{H_5}-\overset{\delta-}{Br} + 2Ag^+ + {}^-O-C(=O)-O^- \xrightarrow{-2AgBr} (C_2H_5O)_2C=O$ Diethyl carbonate

(*vii*) $C_6H_5NH_2 + CHCl_3 \xrightarrow[KOH]{Alc.} C_6H_5-N\overset{\rightarrow}{=}C$ Phenyl isocyanide

46. Summary of the given facts

$\underset{X}{C_6H_{13}Cl} \xrightarrow{(CH_3)_3COK}$ Two isomeric alkenes Y and Z $\xrightarrow{H_2} CH_3-CH(CH_3)-CH(CH_3)-CH_3$

2,3–Dimethylbutane

The two isomeric precursors (*Y* and *Z*) of 2, 3-dimethylbutane are

$CH_3-C(CH_3)=C(CH_3)-CH_3 \quad CH_2=C(CH_3)-CH(CH_3)-CH_3$

Hence, the precursor of *Y* and *Z* should have following structure which explains all the given facts

$\underset{(X)}{CH_3-C(CH_3)(Cl)-CH(CH_3)-CH_3} \xrightarrow{(CH_3)_3COK} CH_2=C(CH_3)-CH(CH_3)-CH_3 + CH_3-C(CH_3)=C(CH_3)-CH_3$ (*Y*/*Z*)

both $\xrightarrow{H_2} CH_3-CH(CH_3)-CH(CH_3)-CH_3$

2, 3 - Dimethylbutane

47. S_N2 reaction leads to inversion in configuration.

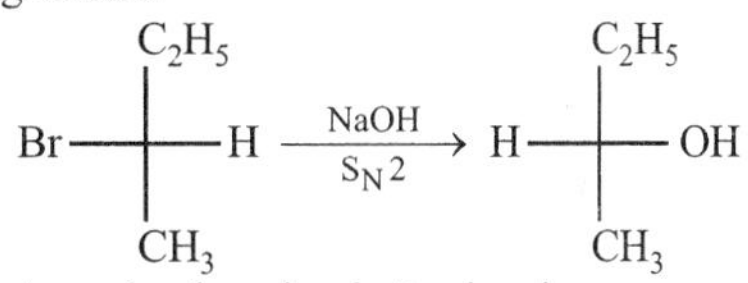

48. A weaker base is a better leaving group. Thus, rate of reaction will be R–I > R–Br > R–Cl > R–F. because I^- is the best, while F^- is the poorest leaving groups among halide ions.

49. $\%\ of\ Cl\ in\ X = \frac{35.5 \times 2.9}{143.5} \times 100 = 71.72\%$

Empirical formula of (*X*)

Element	%	Relative no. of atoms	Simplest ratio
C	24.24	2.02	1
H	4.04	4.04	2
Cl	71.72	2.02	1

∴ Empirical formula of (*X*) is $\mathbf{CH_2Cl}$

Since, *X* has two isomers *Y* and *Z*; both react with KOH(aq).

$Y \xrightarrow{KOH(aq.)}$ dihydroxy compound *i.e.* 2Cl atoms on adjacent carbon

$Z \xrightarrow{KOH(aq.)} CH_3CHO$ *i.e.* *Z* should have 2Cl atoms on one C atom

Thus *Z* should be $\mathbf{CH_3CHCl_2}$ (1, 1-dichlorethane) and *Y* should be $\mathbf{CH_2ClCH_2Cl}$ (1, 2-dichloroethane)

Reactions :

$\underset{(Z)}{CH_3CHCl_2} \xrightarrow{KOH(aq.)} CH_3CH(OH)_2 \xrightarrow{-H_2O} \underset{Ethanal}{CH_3CHO}$

$\underset{(Y)}{CH_2ClCH_2Cl} \xrightarrow{KOH(aq.)} \underset{ethane\text{-}1,\ 2\text{-}diol}{CH_2OHCH_2OH}$

50. Resonance decreases the dipole moment of vinyl chloride ($CH_2=CHCl$).

The positive charge on Cl and a negative charge on C (developed by resonance) oppose each other and hence diminish the electronegativity of Cl and thus polarity (and dipole moment) of the bond. The dipole moments of vinyl chloride and chlorobenzene are 1.4D and 1.7D respectively, while the dipole moment of alkyl halides is 2–2.2D.

51. Dichloroethene exists in three isomeric forms.

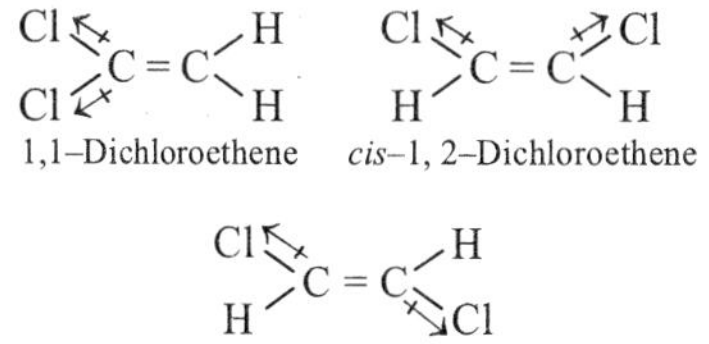

1,1–Dichloroethene *cis*–1, 2–Dichloroethene

trans–1, 2–Dichloroethene

trans-1, 2-Dichloroethene has zero dipole moment.

52. (*i*) $4C_2H_5Br + 4(Na\text{–}Pb) \xrightarrow{dry\ ether} Pb(C_2H_5)_4 + 4NaBr + 3Pb$

(*ii*) $Al_4C_3 \xrightarrow{H_2O} CH_4 \xrightarrow[sun\ light]{Cl_2} CH_3Cl$

53. (a) (i) $CaC_2 + 2H_2O \longrightarrow Ca(OH)_2 + C_2H_2$

$C_2H_2 \xrightarrow{H_2/Ni} C_2H_6 \xrightarrow[UV]{Cl_2(excess)} CCl_3-CCl_3$

(ii) $CS_2 + 3Cl_2 \longrightarrow CCl_4 + S_2Cl_2$

$CS_2 + 2S_2Cl_2 \longrightarrow CCl_4 + 6S$

$CCl_4 + 2[H] \xrightarrow{Fe/H_2O} CHCl_3 + HCl$

(b) Carbylamine test.

$CHCl_3$ + aq KOH + aniline (i.e. primary amine) → Bad smelling isocyanide

C_2H_5OH + aq KOH + aniline → No reaction

54. **(a)** The product (a) will be formed.

Nucleophilic substitution of an alkyl halide is easier as compared to that of an aryl halide.

PhS^- is a strong nucleophile and dimethyl formamide $HCONMe_2$ is a highly polar aprotic solvent. These reagents favour S_N2 reactions at 2° benzylic carbon.

In a S_N2 reaction, the major product formed is inversion product.

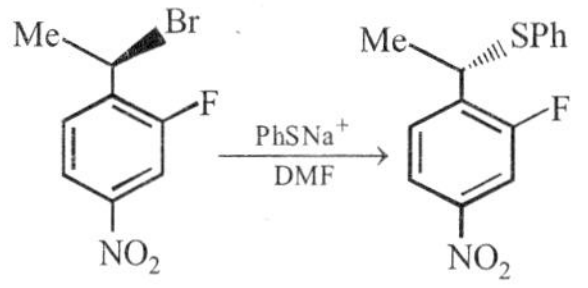

55. **(a)** Grignard reagents react with compounds containing active hydrogen to form hydrocarbons corresponding to alkyl (or aryl) part of the Grignard reagent.

$C_6H_5MgBr + Me_3COH \longrightarrow C_6H_6 + Me_3COMgBr$

56. **(d)** The given reaction is an example of electrophilic substitution. Further, CH_3 group in toluene is *o*, *p*-directing

57. **(d)**

$C_6H_5NH_2 \xrightarrow[HCl]{HONO} C_6H_5\overset{+}{N_2}\overset{-}{Cl} \xrightarrow{CuCl} C_6H_5Cl$

58. **(1.31)**

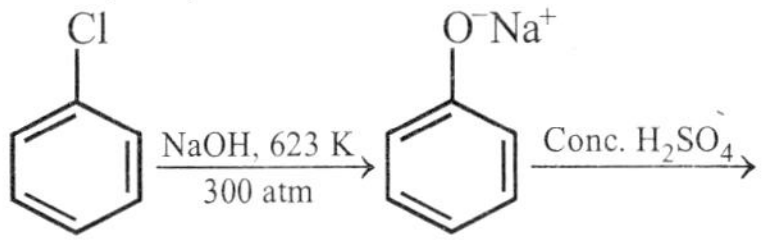

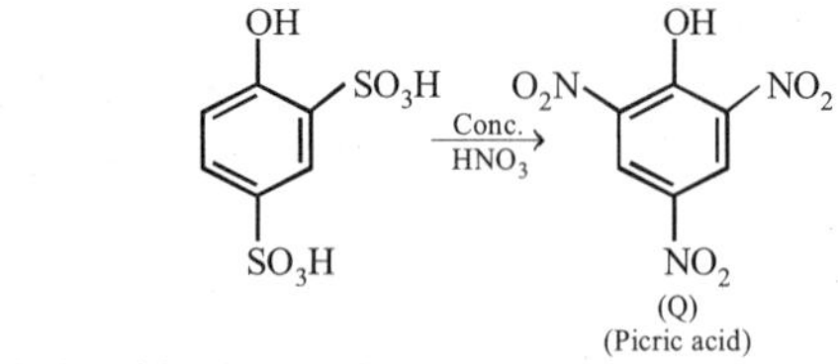

Picric acid = $C_6H_3N_3O_7$

∴ M.W. of picric acid

$= 6 \times 12 + 3 \times 1 + 3 \times 14 + 7 \times 16 = 229$

1 mol of picric acid contains 3-H atoms

∴ The wt.(%) of 'H' in (Q) $= \frac{3 \times 100}{229} = 1.31$

59. (136)

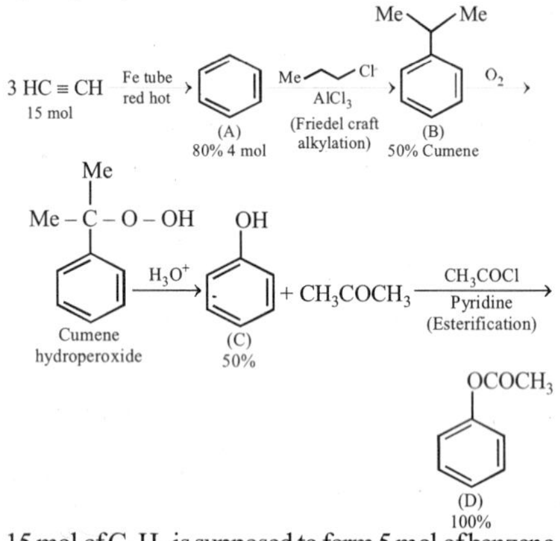

15 mol of C_2H_2 is supposed to form 5 mol of benzene (A) theoretically actual formation of (A) is 80%.

∴ No. of moles of (A) $= \frac{5 \times 80}{100} = 4$ mol

4 moles of (A) produces 50% of (B) = 2 moles of (B). Further, 2 moles of (B) produces 50% of (C) = 1 mol of (C)

∴ 1 mole of (C) produces 100% of (D) = 1 mol of (D).

M.W. of (D), 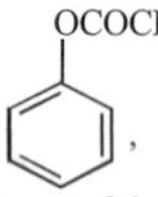 , acetophenone = 136

∴ The amount of the product D formed is 136 g.

60. False : *m*-Chlorobromobenzene and *m*-bromochloro-benzene is one and the identical compound.

61. (a, b)

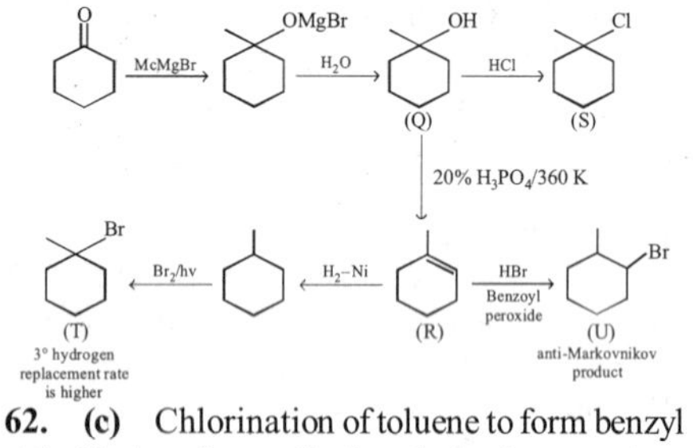

62. (c) Chlorination of toluene to form benzyl chloride is a free radical substitution reaction; only Cl_2 can give $\dot{C}l$ in presence of light.

63. (b, e)

Aryl halides are stable due to resonance stabilization. The resonating structures

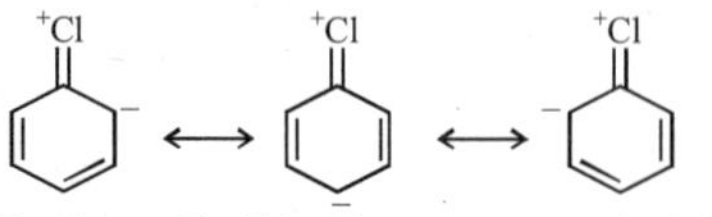

stabilise the aryl halide. These structures include a double bond between C and Cl which is shorter and thus stronger than the usual C – Cl single bond. The sp^2 hybridised carbon, being electronegative, makes the C – Cl bond shorter and stronger.

64. (b)

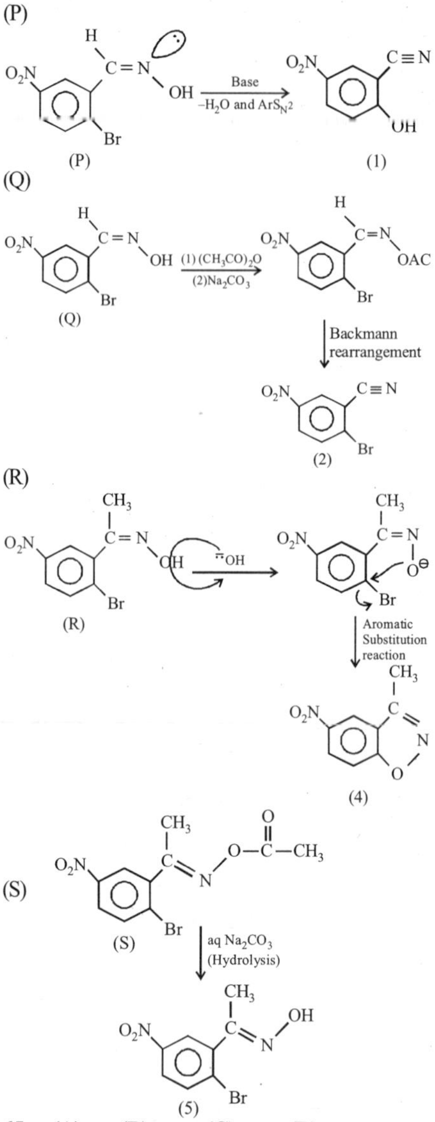

65. (A) - q; (B) - p, s; (C) - r, s; (D) - q

E1 mechanism is encountered only with tertiary or secondary substrates and in presence of either a weak base or a base in low concentration. So, primary substrates will follow E2 mechanism, *i.e.* (A)→E2 and (D)→E2. Further, E1 mechanism (similar to S_N1) proceeds by first order kinetics and is determined by the slower (first) step of the formation of carbanion. Hence, (B)→E1 and first order reaction.

Reaction of $C_6H_5CH_2CH_2Br$ on treatment with $C_2H_5O^-$ in presence of C_2H_5OD gives $C_6H_5CD{=}CH_2$. This reaction follows $E1_CB$ (Elimination unimolecular conjugate base) mechanism.

66. (c) When halogen is present directly on the benzene nucleus it produces two opposing effects namely + M (activating effect) and –I (deactivating) The inductive effect (–I) of bromo group is more than the mesomeric effect (+M).

This is the reason why bromo group deactivates the ring. But the directing influence with always governed by the mesomeric effect.

Hence, due to +M effect of –Br, the incoming electrophile attacks at *ortho* and *para* positions.

67. (i) The former halide is a 3° halide, hence it undergoes S_N1 reaction forming HBr, as one of the products, which make solution acidic.

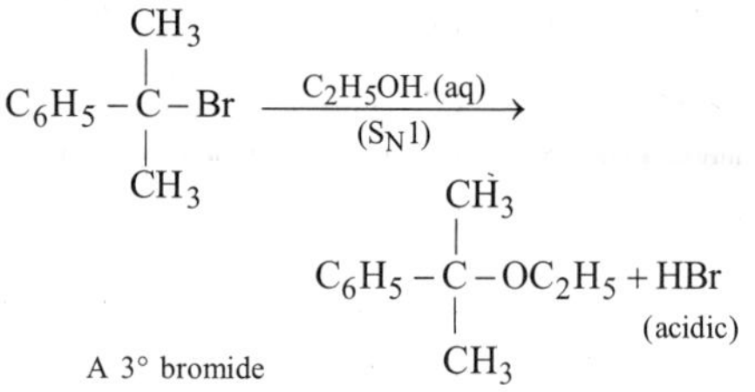

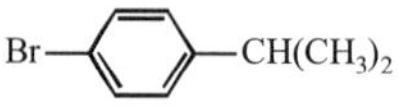 is an aryl halide so it does not undergo nucleophilic substitution reactions. Hence, the solution will remain neutral.

(ii) 7-Bromo-1,3,5-cycloheptatriene is aromatic whereas 5-Bromo-1,3-cycloheptadiene is non aromatic.

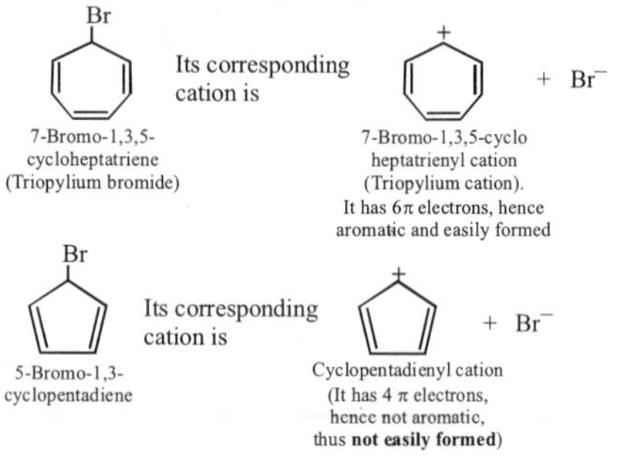

68. 2Me

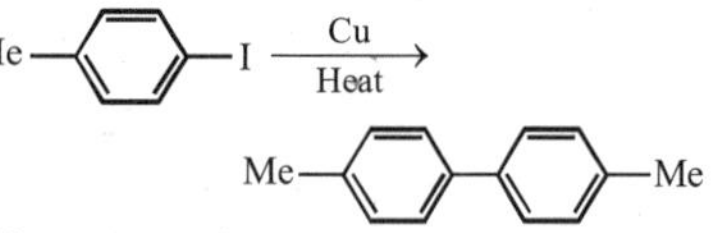

This is ullmann reaction

69.

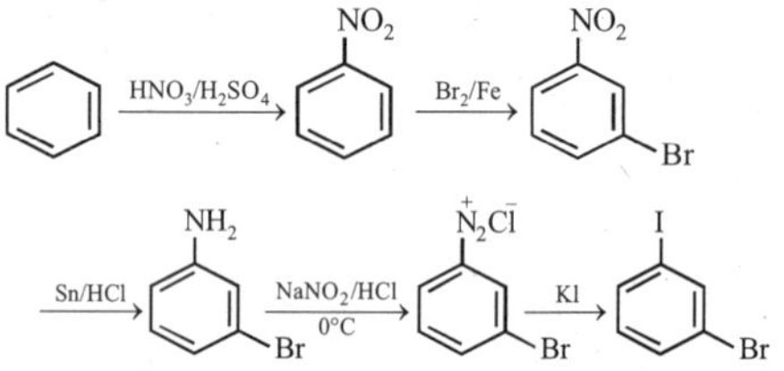

70. The low reactivity of halogen atom in aryl and vinyl halides towards nucleophiles is due to resonance.

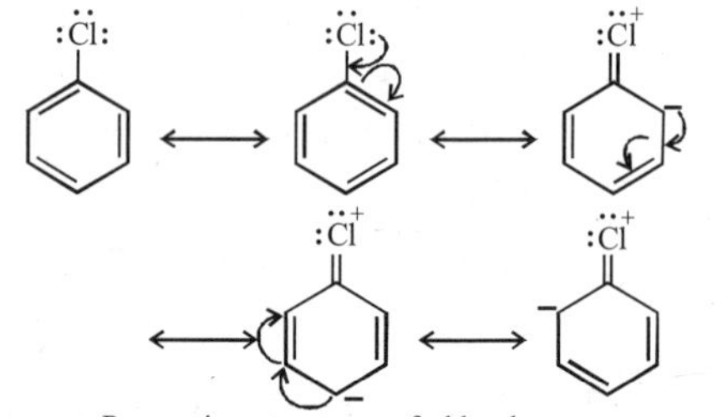

Resonating structures of chlorobenzene

Due to resonance, carbon-chlorine bond acquires partial double bond character, hence it becomes shorter and stronger and thus cannot be easily replaced by nucleophiles.

71. $C_6H_5CH_3 \xrightarrow[\text{UV Light}]{Cl_2} C_6H_5CH_2Cl$

$\xrightarrow{Cl_2} C_6H_5CHCl_2 \xrightarrow{Cl_2} C_6H_5CCl_3$ Benzotrichloride

[This follows free radical mechanism.]

Chapter-11

Alcohols, Phenols and Ethers

Part-A (JEE Main)

1. (4)

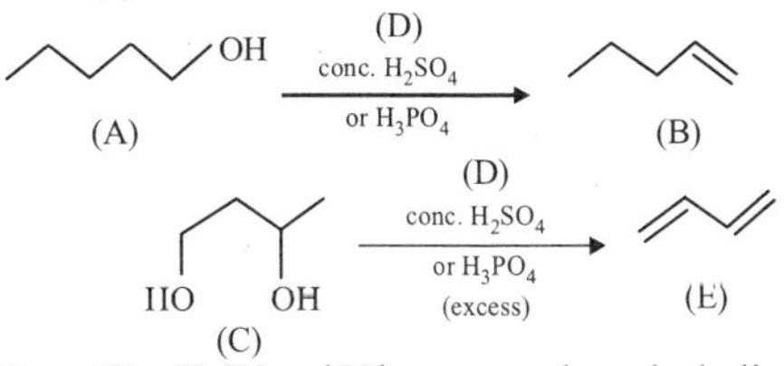

2. (2) II, IV and V have secondary alcoholic group.

3. (2)

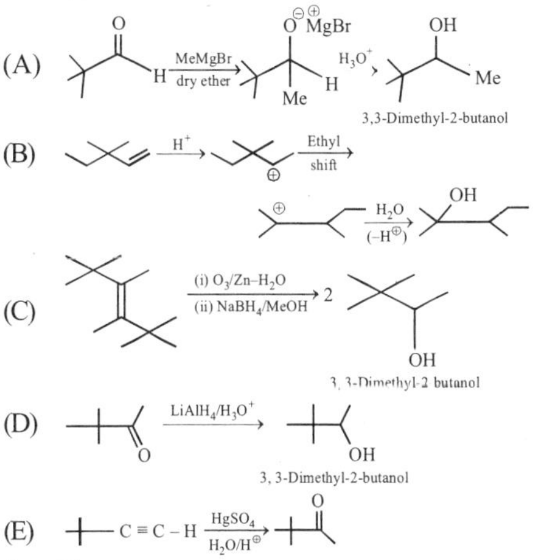

4. (3)

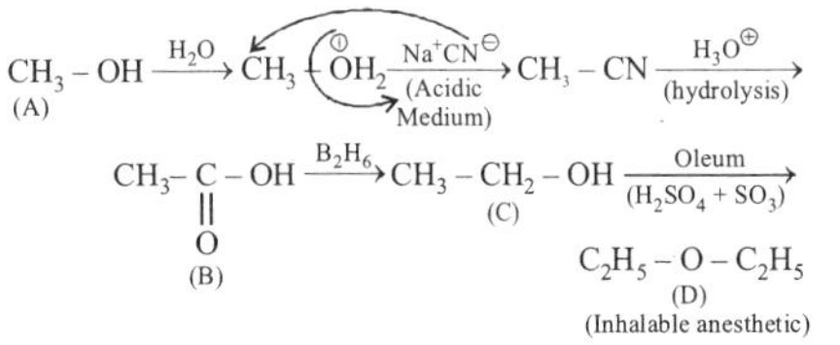

5. (4) B.P. ∝ M.W.
B.P. ∝ Intermolecular hydrogen bonding. Alcohols and phenols have intermolecular H-bonding so their boiling points are higher than species of comparable molecular mass.

6. (3) Hydration Reaction: (Markovnikoff)

$$CH_3-CH=CH_2+H^+ \longrightarrow \underset{\text{(More stable)}}{CH_3-\overset{+}{C}H-CH_3}$$

$$CH_3-\overset{+}{C}H-CH_3+H_2O \longrightarrow (CH_3-\underset{\underset{\text{(A)}}{OH}}{CH}-CH_3)+H^+$$

Hydroboration Oxidation Reaction:
(anti-Markovnikoff)

$$3CH_3-CH=CH_2+B_2H_6 \xrightarrow{THF} 2(CH_3CH_2CH_2)_3B$$

$$(CH_3CH_2CH_2)_3B+3H_2O_2 \xrightarrow{OH^-} \underset{\text{(B)}}{3CH_3CH_2CH_2OH+H_3BO_3}$$

7. (4)

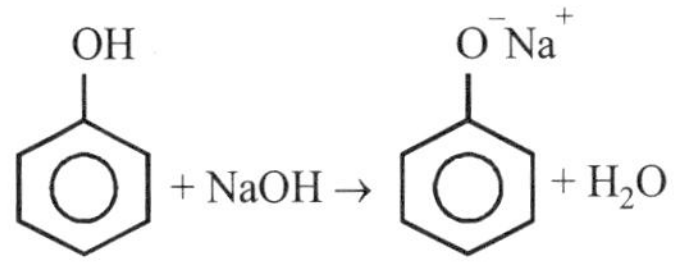

Stronger acid than H_2O due to more resonating structure of phenoxide ion.

8. (1) The major products (A) and (B) are formed as:

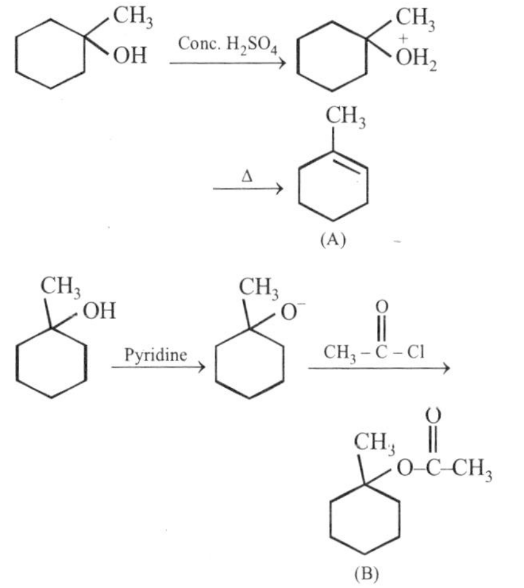

9. (1)

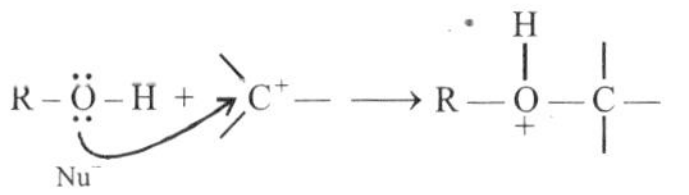

$$R-CH_2-OH+H^+ \longrightarrow R-CH_2-\overset{+}{O}H_2$$

$$Br^- + \underset{\underset{R}{|}}{CH_2}-OH_2^+ \longrightarrow Br-\underset{\underset{R}{|}}{CH_2}+H_2O$$

10. (4)

$$PhCH=CH_2 \xrightarrow[\text{(anti-Marko. addition)}]{BH_3/H_2O_2,\ OH^-} PhCH_2CH_2OH$$

$$PhCH_2CH_2OH+HBr \longrightarrow PhCH_2CH_2Br+H_2O$$

$$\downarrow \text{Mg/dry ether}$$

$$PhCH_2CH_2CH_2OH \xleftarrow[\text{(II) }H_3O^+]{\text{(I) }H-\overset{O}{\overset{\|}{C}}-H} PhCH_2CH_2MgBr$$

11. (3)

$$\text{Acidic strength} \propto -\text{I effect} \propto \frac{1}{+\text{I effect}}$$

F, Cl exerts –I effect, methyl exerts +I effect, compound C is least acidic.
Among A and B; since inductive effect is distance dependent too, extent of –I effect is higher in A followed by B even though F is stronger electron withdrawing group than Cl. Thus, A is more acidic than B.

12. (1)

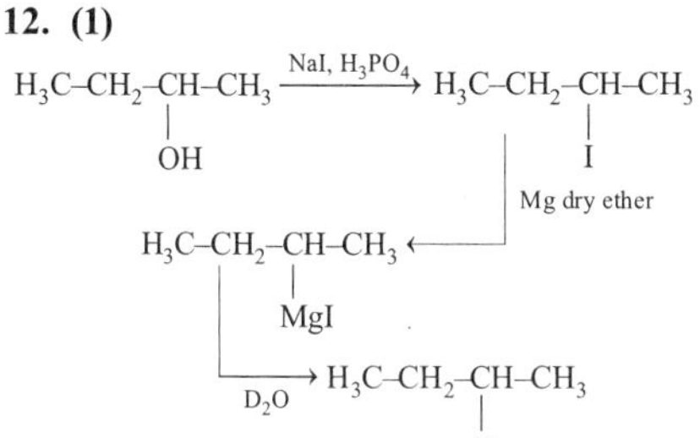

13. (1) Ozonolysis of alkene, gives aldehyde, ketone & carboxylic acid.

14. (1) Butan–1–ol ($CH_3CH_2CH_2CH_2OH$) can undergo hydrogen bonding. Ethoxyethane ($CH_3CH_2–O–CH_2CH_3$) has no hydrogen (attached with F, O, N) which can undergo hydrogen bonding. More is the extent of intermolecular H–bonding, more will be association of molecules. Thus leading to higher boiling point.
Hence both Assertion (A) and Reason(R) are true and (R) is the correct explanation of (A).

15. (2)

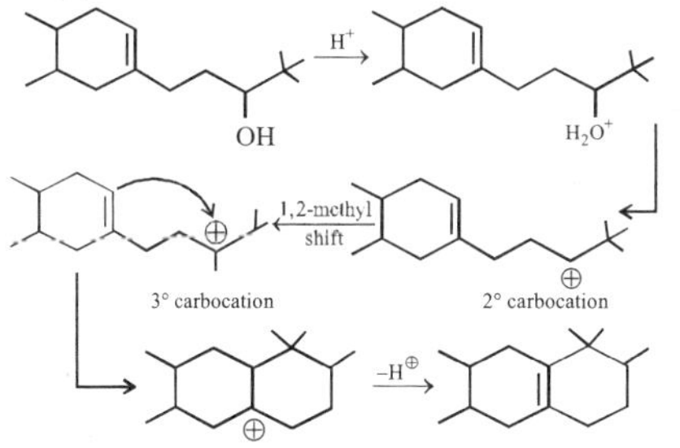

16. (1)

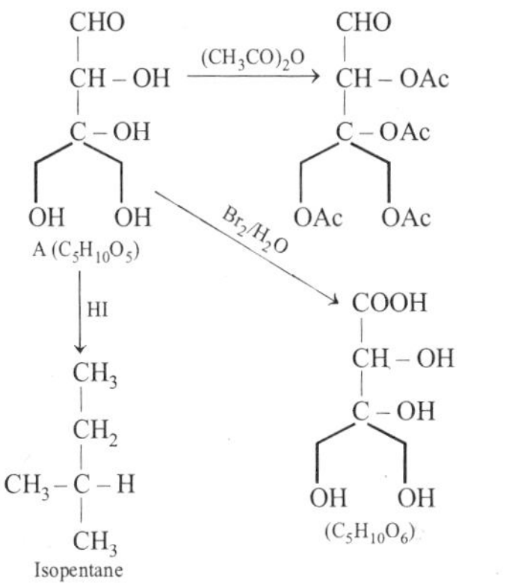

17. (2) Hint: Rearrangement of carbocation occur to gain stability.

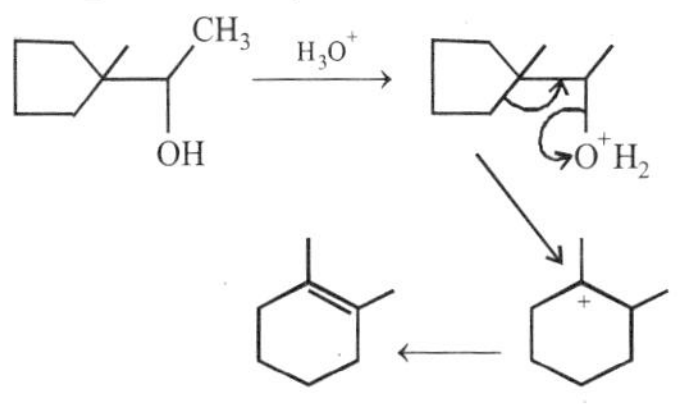

18. (1)

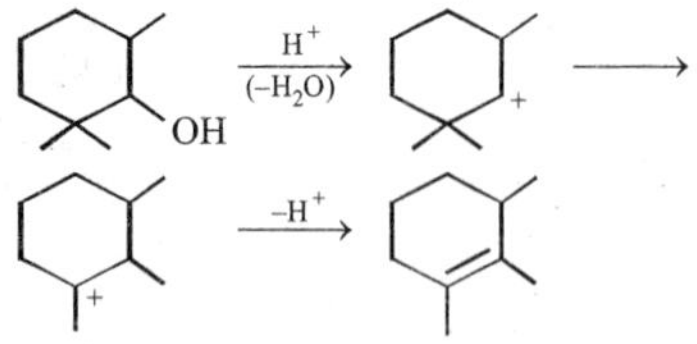

19. (1) Hint: In hydrocarbocation, hydrogen from BH_3 attached at stable carbocation position.

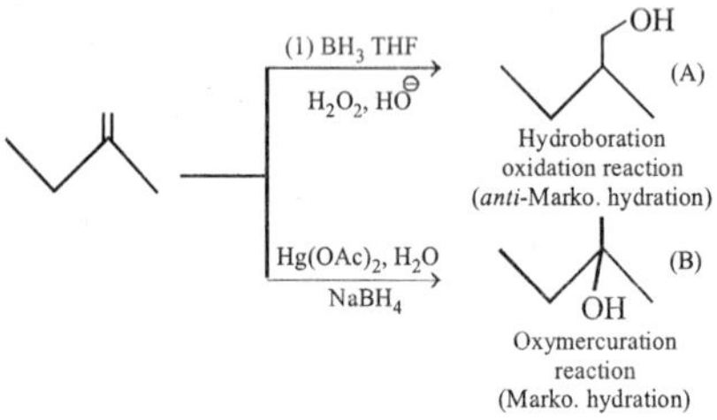

20. (3) $CH_3-CH_2-OH \xrightarrow[\text{(dehydration)}]{\text{conc. } H_2SO_4, \Delta} CH_2=CH_2\uparrow$ (Ethene) $\xrightarrow[\text{(alk-KMnO}_4)]{\text{Baeyer's reagent}} \underset{OH}{CH_2}-\underset{OH}{CH_2}$ (Glycol)

21. (3)

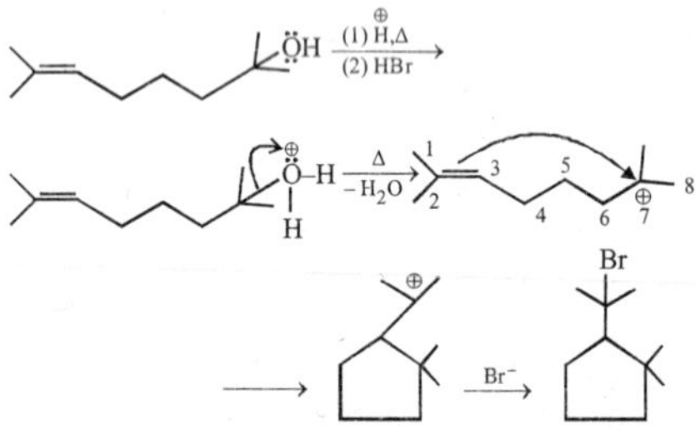

22. (1)

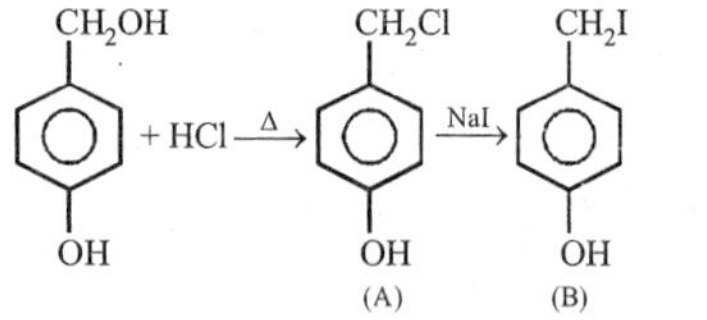

23. (2) Hint: In S_N2, there is an inversion attack by CN^-

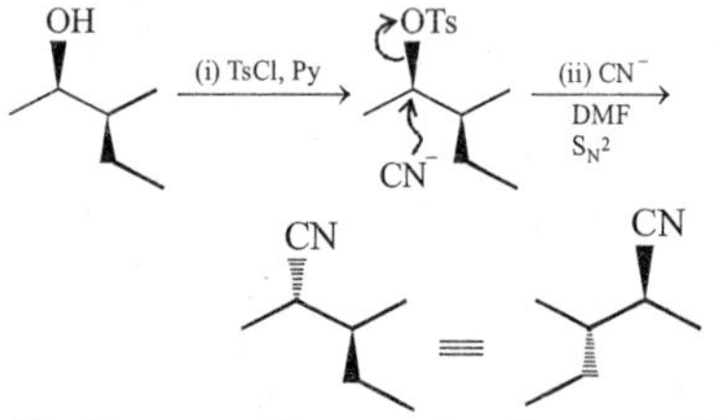

24. (3) Statement I is correct but Statement II is incorrect

25. (3)

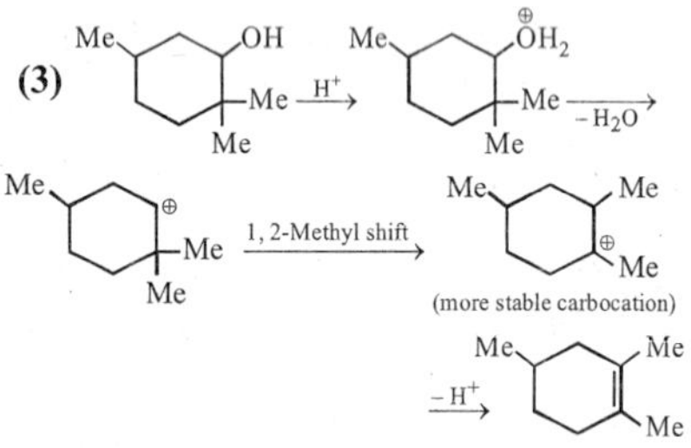

26. (1)

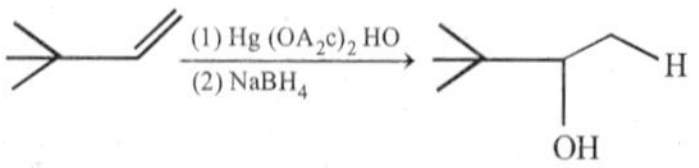

Oxymercuration–demercuration reaction follow Markovnikov's addition.

27. (2)

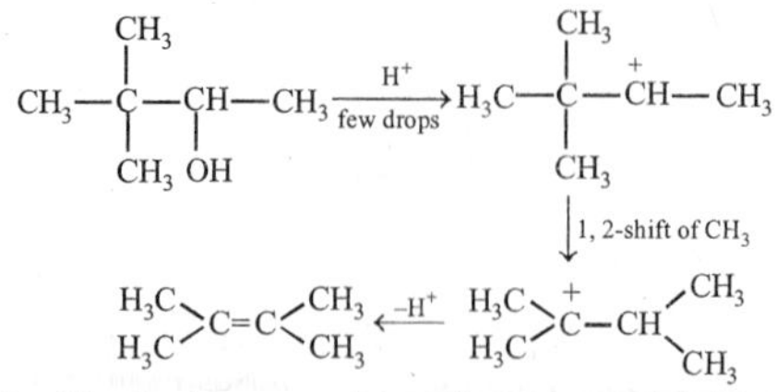

28. (4) Ceric ammonium nitrate is used for the identification of alcohol.
$CHCl_3/KOH$ is used for the identification of primary amines.

29. (2)

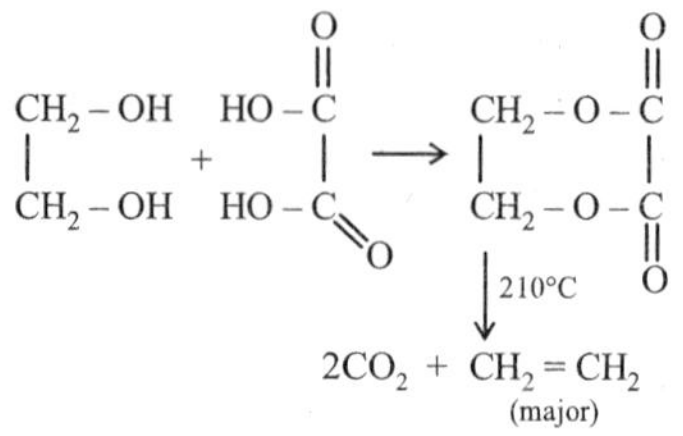

30. (1)
31. (1)

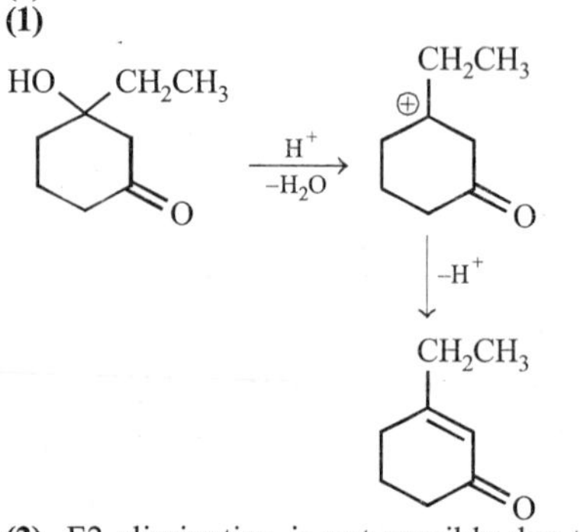

32. (2) E2 elimination is not possible due to absence of β-hydrogen.

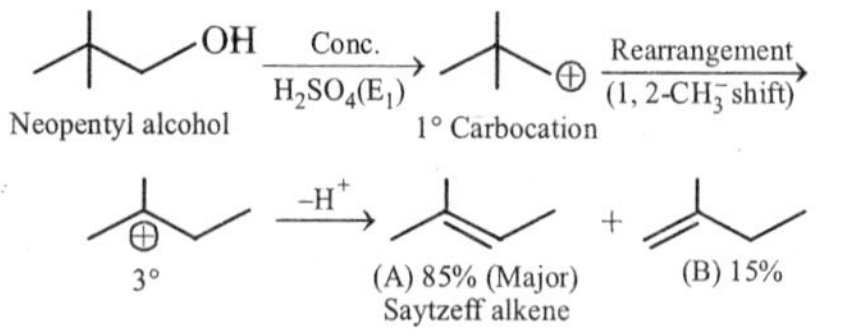

33. (2)

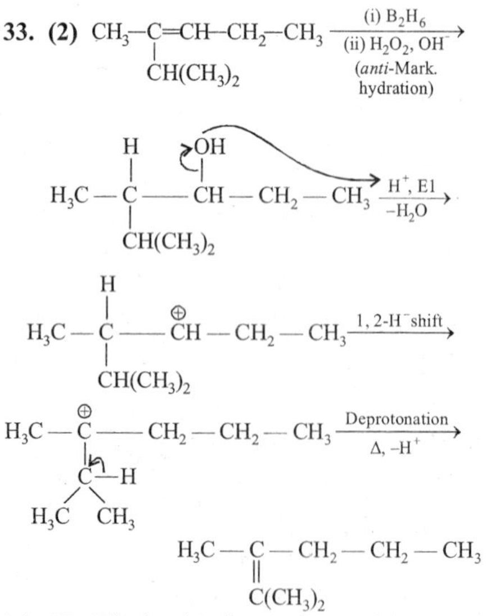

34. (3) Elimination is not favourable due to dilute acid and low temperature.

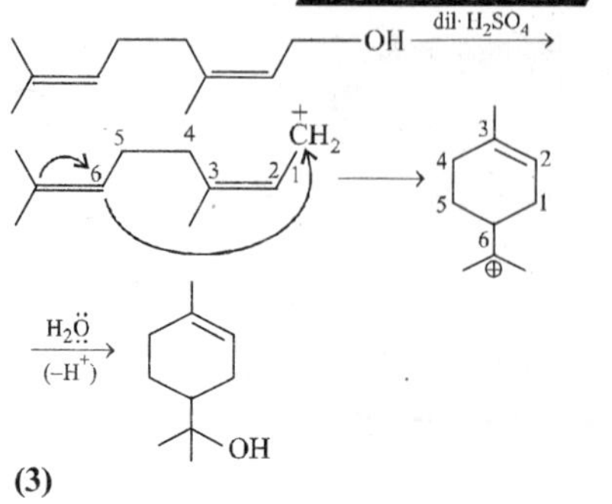

35. (3)

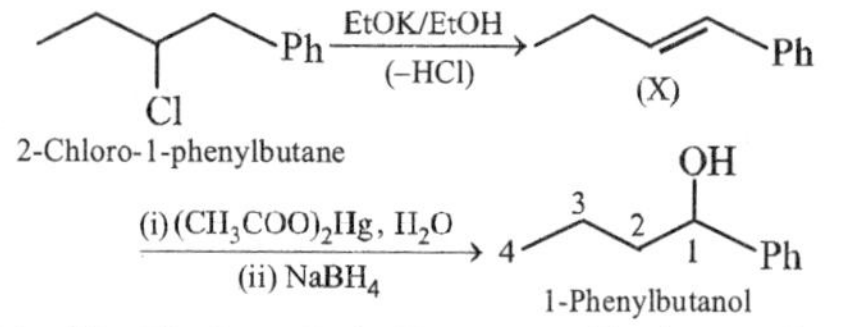

36. (4) Tertiary alcohol is prepared by the reaction of Grignard reagent with a ketone (formaldehyde is used to prepare primary alcohol).

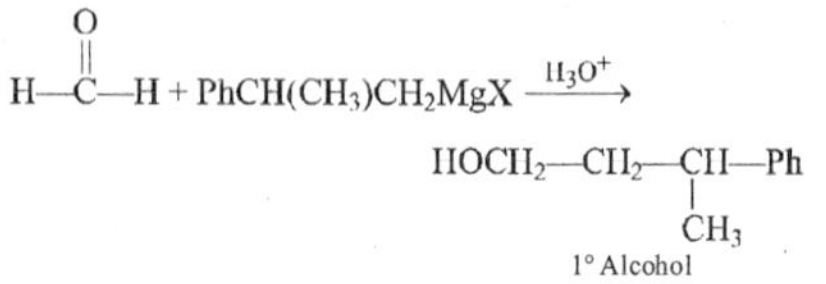

37. (2) The most suitable reagent for the given reaction is I_2/NaOH (Iodoform reaction).

38. (4) $NaBH_4$ is a selective reducing agent, used for the reduction of aldehydes and ketones, it does not affect alkene and ester.

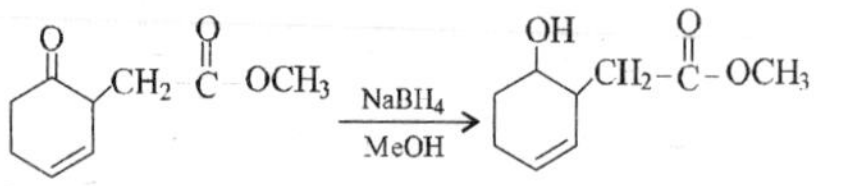

39. (1) $CH_3MgBr + CH_3OH \longrightarrow Mg(Br)OCH_3 + CH_4\uparrow$ (Methane)

40. (4) $R-CH_2-OH \xrightarrow{PCC} R-CHO$

41. (2)

42. (1)

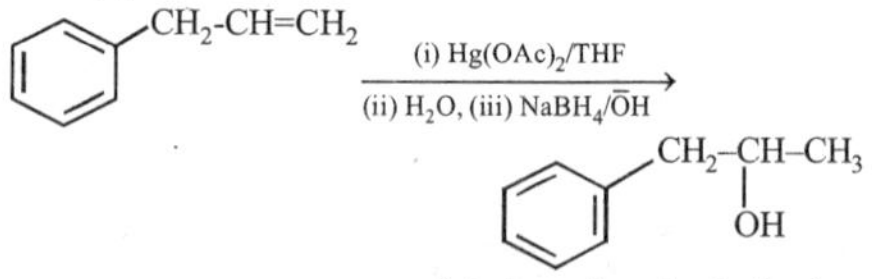

(Markovnikov hydration)

43. (3) The C – O bond length in alcohols is 142 pm and in phenol it is 136 pm. The C – O bond length in phenol is shorter than that in methanol due to the conjugation of unshared pair of electrons on oxygen with the ring, which imparts double bond character to the C – O bond (resonance).

44. (3)

$CH_3CH=CH_2 \xrightarrow[H_2O_2, NaOH]{B_2H_6} CH_3CH_2CH_2OH$

Anti-Markovnikov hydration

45. (2) Reaction of alcohols with Lucas reagent proceeds through carbocation formation (S_N1 reaction).
Further, 3° carbocations (from tertiary alcohols) are highly stable, thus reaction proceeds through S_N1 mechanism.

46. (4) $ZnCl_2$ + conc. HCl is Lucas reagent. Lucas reagent reacts fastest with tertiary alcohol because the corresponding carbocation formed is the most stable (3°).
3° Alcohol + Lucas reagent = Immediate turbidity

47. (2) The order of dehydration among three type of alcohols is 3° > 2° > 1° > CH_3OH. This behaviour is related to the relative stabilities of carbocations (3° > 2° > 1° > $\overset{+}{C}H_3$).

48. (1)

$$CH_2 = CH{\cdot}CH_2OH \xrightarrow{MnO_2} CH_2 = CH{\cdot}CHO$$

Allyl alcohol → Acraldehyde (Acrolein)

49. (1) On moving down a group, the basicity & nucleophilicity are inversely related i.e., nucleophilicity increases while basicity decreases, i.e., $RS^{\ominus}$ is more nucleophilic but less basic than $RO^{\ominus}$.

50. (1)

$$C_6H_5-CH_2-\underset{OH}{CH}-\underset{CH_3}{CH}-CH_3 \xrightarrow{Conc.\,H_2SO_4}$$

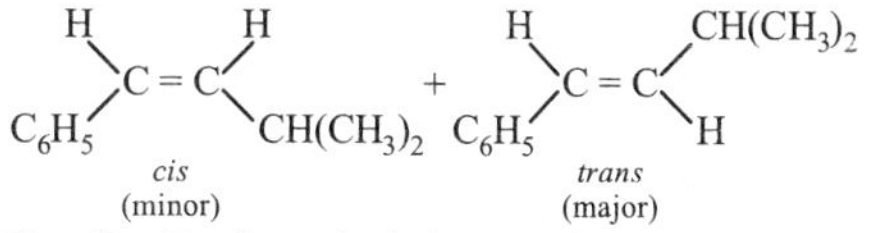

51. (2) Tertiary alcohols react fastest with conc. HCl and anhydrous $ZnCl_2$ (Lucas reagent) as its mechanism proceeds through the formation of stable tertiary carbocation.

52. (4) $CH_3CH_2OH \xrightarrow{P+I_2} \underset{(A)}{CH_3CH_2I} \xrightarrow[Ether]{Mg} \underset{(B)}{CH_3CH_2MgI} \xrightarrow{HCHO}$

$$\underset{(C)}{H-\overset{CH_2CH_3}{\underset{H}{C}}-OMgI} \xrightarrow{H_2O} \underset{(D)}{H-\overset{CH_2CH_3}{\underset{H}{C}}-OH}$$

n-Propyl alcohol

53. (2) Ethanol and only those 2° alcohols which contain $-CHOHCH_3$ group undergo haloform reaction.

54. (1) $CH_3-\overset{OH}{CH}-CH=CH-CH_3 \xrightarrow{PCC} CH_3-\overset{O}{\overset{\|}{C}}-CH=CH-CH_3$

Pent-3-en-2-ol → Pent-3-en-2-one

55. (1) 3-Methylpentan-3-ol will be dehydrated most readily since it produces a very stable, tertiary carbonium ion as intermediate.

56. (4) Initiation step is formation of protonated alcohol.

$$CH_3-CH_2-\ddot{O}-H + H^+ \rightleftharpoons CH_3CH_2-\overset{+}{O}{<}^{H}_{H}$$

(Protonated ethanol)

57. (175)

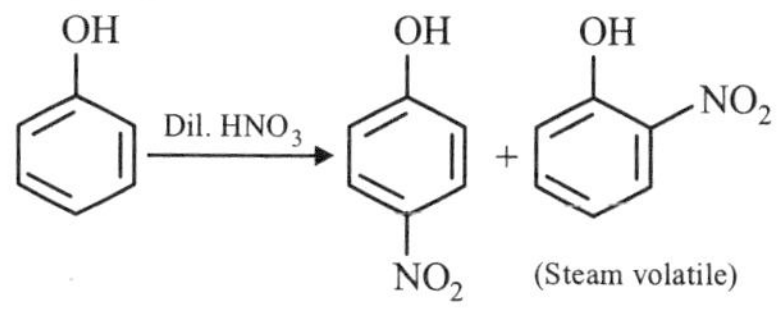

% oxygen in phenol = $\frac{16}{94} \times 100 = 17.02\%$

Molar mass of $C_6H_5NO_3$ = 139 g/mol

% oxygen in o-Nitrophenol ($C_6H_5NO_3$)

$= \frac{48}{139} \times 100 = 34.53\%$

% of increase = (34.53 – 17.02) = 17.5%

$= 175 \times 10^{-1}\%$

58. (11) Total sum of oxygen atoms and π-electrons is 3 + 8 = 11

Total number of oxygen atoms = 3

Total number of π-bonds = 4

∴ Total number of π-electrons = 8

59. (3)

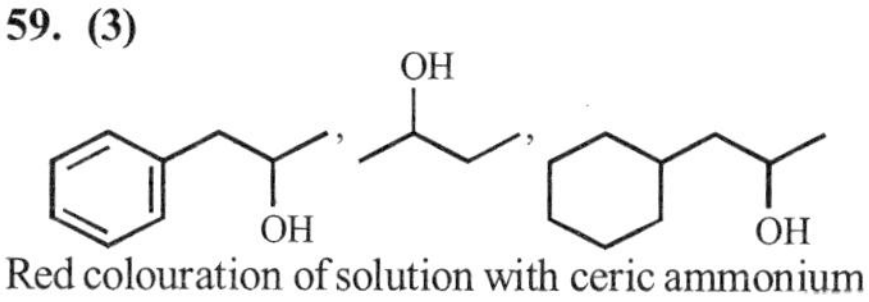

Red colouration of solution with ceric ammonium nitrate is the identification test for alcohol. Whereas the iodoform test is for the secondary alcohols having one alkyl group as $-CH_3$.

60. (6) $R(OH)x \longrightarrow H_2$

Alcoholic hydrogens are involved in hydrogen gas formation.

Number of alcoholic hydrogens present in x = (numbers of hydrogen atoms in H_2 gas)

$$x\left(\frac{1.84\times10^{-3}}{92}\right) = \left(\frac{1.344\times10^{-3}}{22.4}\right)\times 2 \Rightarrow x = 6$$

61. (3)

$$\underset{100\text{ mL}}{CH_3CH_2MgBr} + CH_3OH \longrightarrow \underset{2.24\text{ mL}}{CH_3CH_3} + Mg(Br)OCH_3$$

Moles of ethane formed = $\frac{2.24}{22400} = 10^{-4}$

Weight of ethane formed = Moles × Mol. mass of ethane

$= 10^{-4} \times 30$

$= 3 \times 10^{-3}$ g = 3 mg.

62. (33) 1 mole of alcohol gives 1 mole of methane

No. of moles in 3.1 mL = $\frac{1}{22400} \times 3.1$

$= 1.38 \times 10^{-4}$ moles

Thus 1.38×10^{-4} moles of alcohol are used

Also, No. of moles of alcohol (n)

$= \frac{\text{Wt. of alcohol taken}}{\text{Mol.wt. of alcohol}}$

$1.38 \times 10^{-4} = \frac{4.5\times10^{-3}\,g}{\text{Mol.wt}}$

Mol. wt. = $\frac{4.5\times10^{-3}}{1.38\times10^{-4}}$

$= 3.26 \times 10 = 32.6 \approx 33.$

63. (1) $CH_3-CH_2-CH_2-CH_2-OH$

$CH_3-\overset{*}{\underset{OH}{CH}}-CH_2-CH_3$ [Chiral]

$CH_3-\overset{CH_3}{CH}-CH_2-OH$

$CH_3-\overset{CH_3}{\underset{OH}{C}}-CH_3$

64. (2)

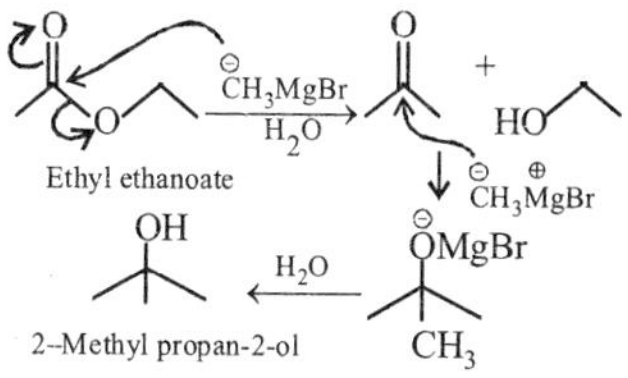

65. (4)

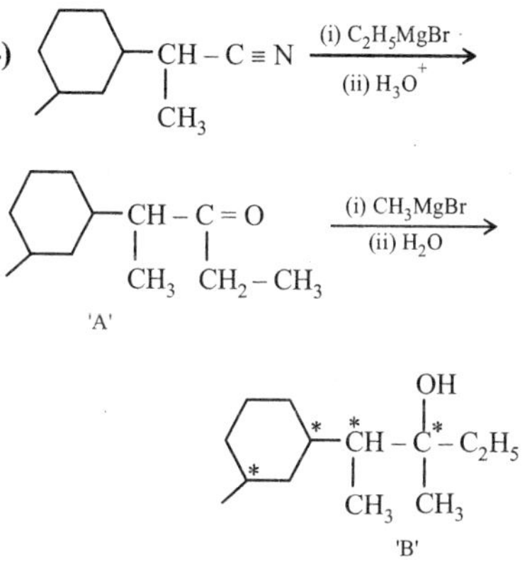

66. (3) Kolbe Schmidt reaction

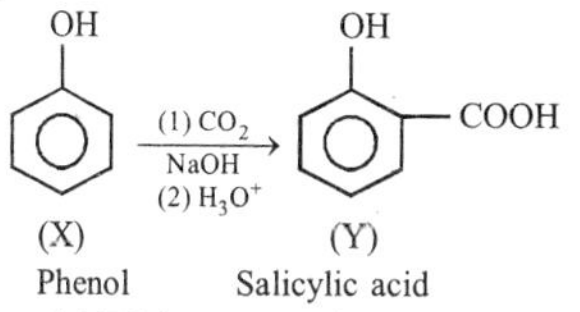

Salicylic acid (Y) is more acidic than phenol (X).

67. (4) $R-O-H + (CH_3CO)_2O \longrightarrow R-O-COCH_3 + CH_3COOH$

Mol. mass = 43

One –H is replaced by one CH_3CO- group on using one mole of $(CH_3CO)_2O$, resulting in an increase of molar mass by (43 – 1 = 42).

Increase in mass of X due to acetylation

= 290 – 122 = 168

Number of –OH group present in (X) = $\frac{168}{42}$ = 4.

68. (4) Reimer-Tiemann reaction:

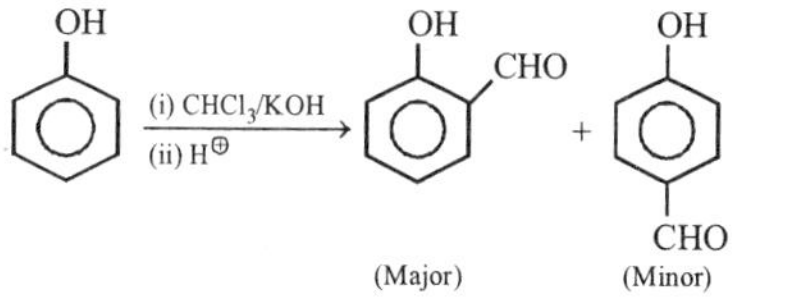

p-Hydroxybenzaldehyde forms inter-molecular hydrogen bonding and is less volatile and *o*-Hydroxy benzaldehyde is more volatile due to intramolecular hydrogen bonding. Therefore, the two can be separated by steam distillation.

69. (3) A is phenol and B is para benzoquinone. First step is nucleophilic substitution second is oxidation.

70. (N) Acidic strength order:- (Actual)

B > D > C > A > E

This is so because $-OCH_3$ acts as - I group at meta position while as an EDG at para position.

Thus, Correct pKa Order: B < D < C < A < E

All options are incorrect.

71. (4) (A)-(II), (B)-(IV), (C)-(I), (D)-(III)

72. (1)

73. (4)

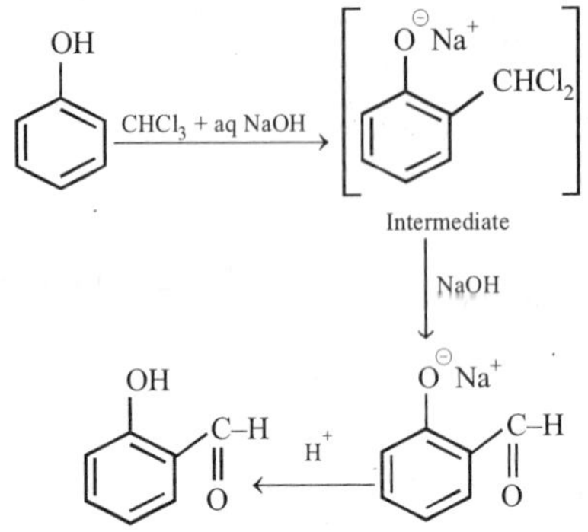

Intermediate in Reimer-Tiemann reaction is substituted benzal chloride.

74. (3) Reactions taking place as :

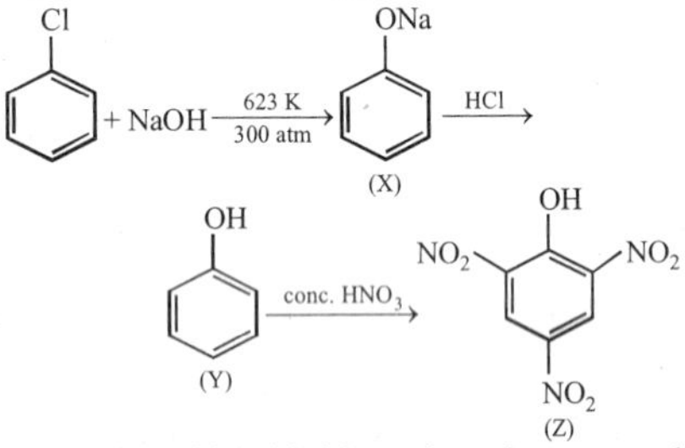

75. (4) Phenol is a highly activated compound due to – OH group, it can undergo bromination directly with bromine without any Lewis acid.

76. (3) IUPAC name : Benzene- 1,2-diol; Common name : Catechol.

77. (2) (A) - (III), (B) - (I), (C) - (II), (D) - (IV)

78. (1) Phenol is more acidic than ethanol because conjugate base of phenol is more stable than conjugate base of ethanol.

79. (4)

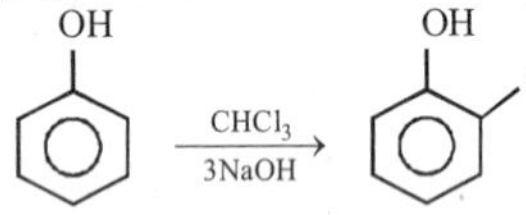

(Reimer-Tiemann reaction)

80. (3)

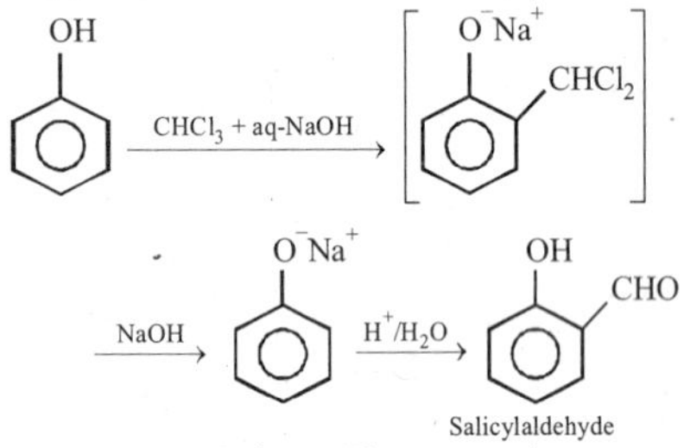

Reimer – Tiemann reaction

81. (2) The order of acidic strength of the given compounds is:

Ethanol < Phenol < m – Nitrophenol < p – Nitrophenol
(Neutral) (Acidic) (more acidic due to –I effect) (more acidic due to –I and –M effect)

Thus, the order of pK_a values will be:

D	<	C	<	B	<	A
(7.1)		(8.3)		(10.0)		(15.9)
(III)		(IV)		(I)		(II)

82. (3) Aryl halides cannot be prepared by replacement of hydroxyl group of phenol by halogen atom because the carbon-oxygen bond in phenols has a partial double bond character and is difficult to break being stronger than a single bond. Hence, phenols cannot react with halogen acids violently.

83. (4)

(A) $\rightarrow$ Kolbe Schmidt Reaction (IV)

(B) $\rightarrow$ Reimer Tiemann Reaction (III)

(C) $\rightarrow$ Oxidation of phenol to p-benzoquinone (I)

(D) $\rightarrow$ $PhOH + NaOH \rightarrow H_2O + PhO^-$

$PhO^- + CH_3 - Cl \longrightarrow PhOCH_3 + Cl^-$ (II)

84. (1) Acidic strength

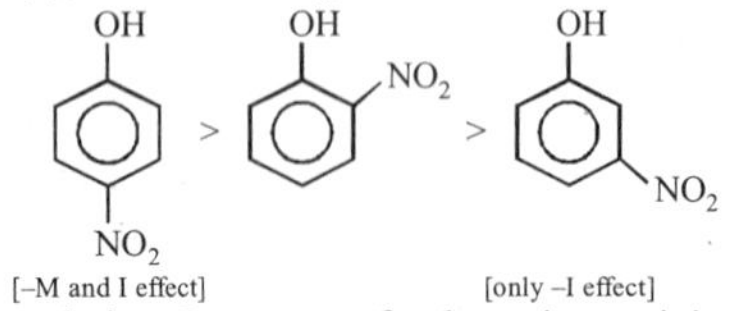

Ethanol gives Lucas test after long time as it is a 1° alcohol and Lucas test follows S_N1 mechanism. Thus, statement I is correct but II is incorrect.

85. (4)

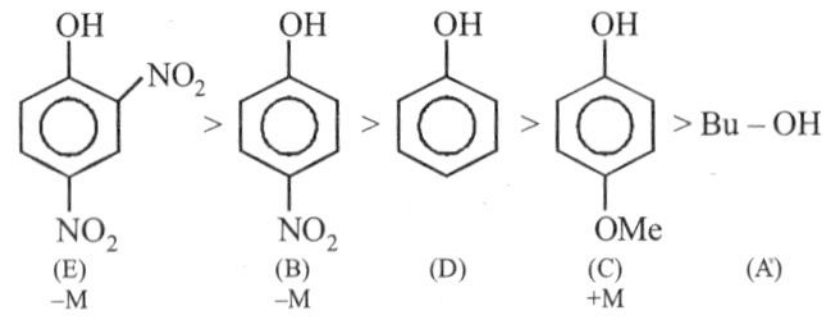

86. (2)

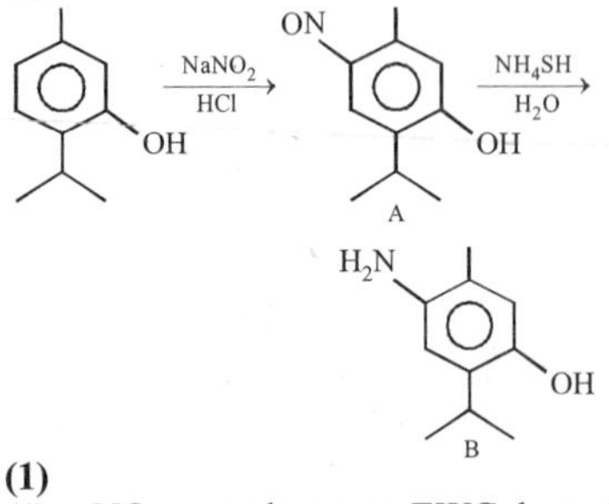

87. (1)

88. (1) $-NO_2$ group has more EWG than –Cl.

89. (4) The ease of dehydration of alcohols depends on the stability of the product formed during the reaction

B > D > C > A

90. (1) Aromatic compounds burns with sooty flame

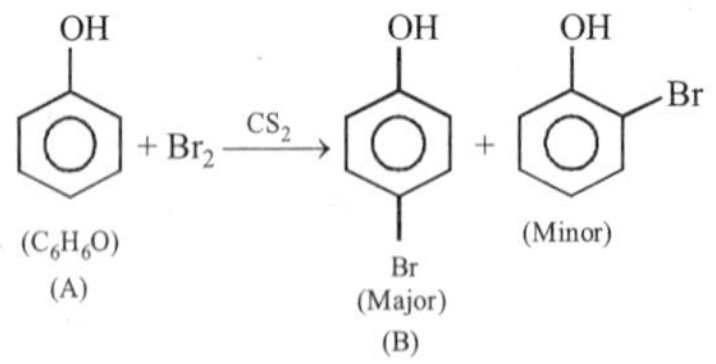

91. (2) Due to –M effect of $-NO_2$ group, it increases acidity +M effect of $N(CH_3)_2$ decreases acidity.

Hyperconjugation and +I effect of isopropyl decreases acidity.

∴ Order of acidic strength.

(C) > (A) > (D) > (B)

Hence order of pk_a will be (B) > (D) > (A) > (C).

92. (2) Order of acidity for following phenol is

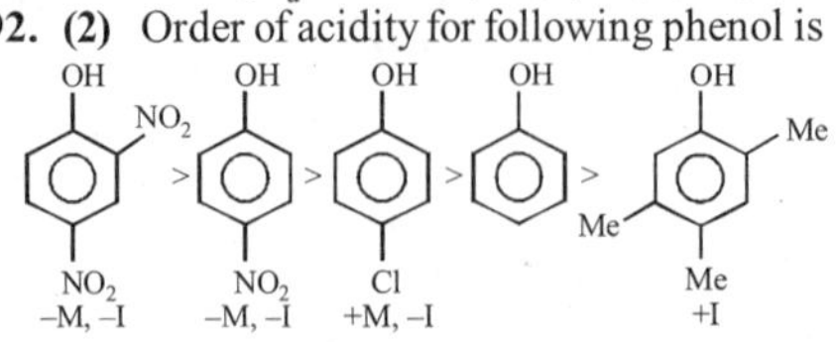

(A) (B) (E) (D) (C)

– M and –I effect of the substituents increase acidity + M and +I effect of the substituents decrease acidity.

∴ Increasing order of pKa

(A) < (B) < (E) < (D) < (C).

93. (4)

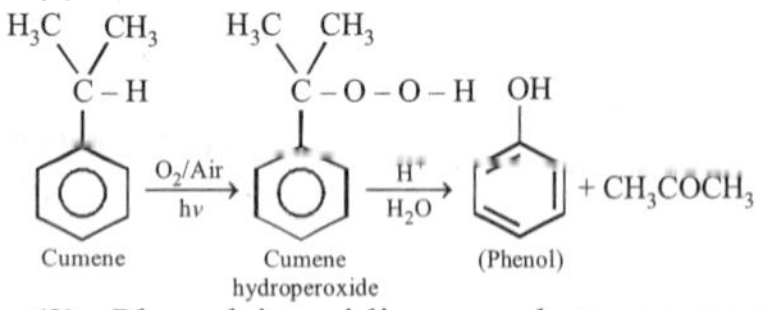

94. (2) Phenol is acidic enough to react with strong base NaOH but not with a weak base $NaHCO_3$ solution.

It gives violet colour with neutral $FeCl_3$ solution.

95. (2)

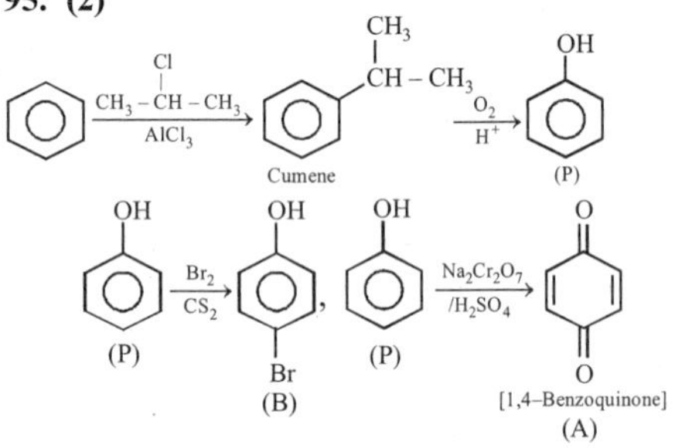

96. (2) Polarity of solvent

- Monosubstituted phenol is formed in case of low polarity solvent, viz. $CHCl_3$.

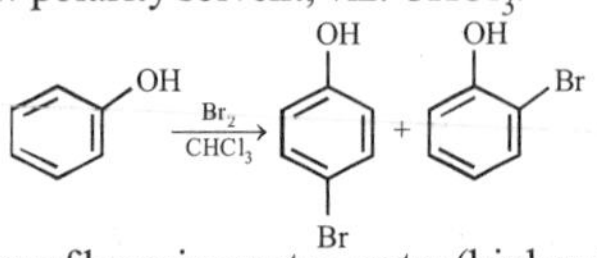

- In case of bromine water, water (high polarity solvent) ionises Br –Br easily.

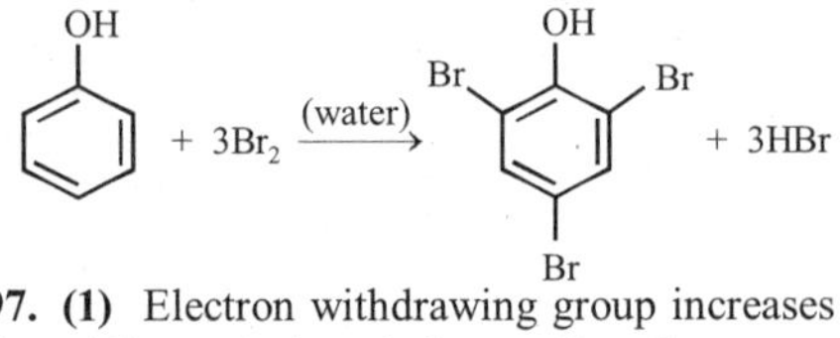

97. (1) Electron withdrawing group increases the acidic strength and electron donating group decreases the acidic strength. So the order is

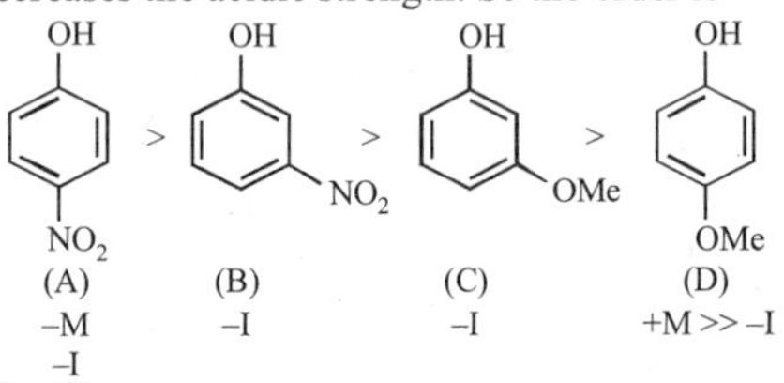

98. (2)

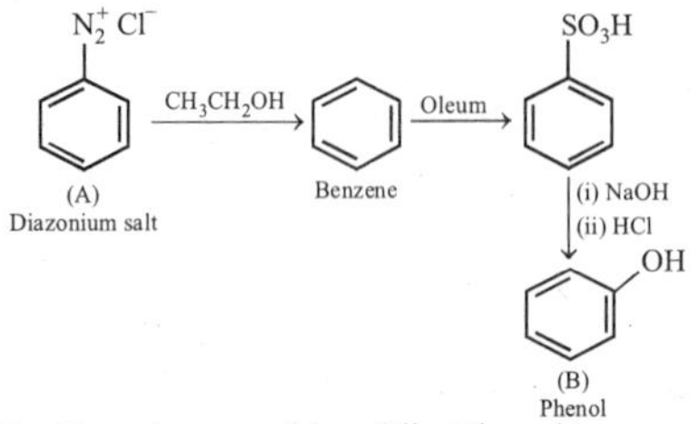

99. (3) Phenol are weakly acidic. Phenol is more acidic than alcohol and H_2O because phenoxide ion is more resonance stabilized therefore readily formed from phenol. Hence statement (I) is correct but (II) is incorrect.

100. (1)

101. (3)

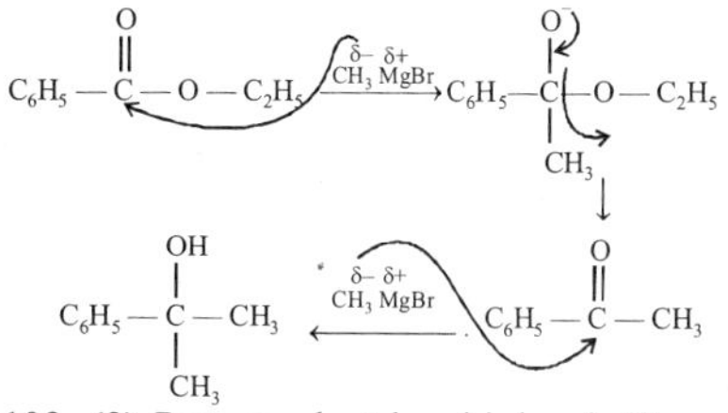

102. (3) Para product has higher boiling point than ortho as intermolecular H-bond is possible in para, whereas intramolecular H-bond is possible in ortho product. Steam distillation can separate them as ortho product is steam volatile.

103. (3)

104. (3) Given reaction is cumene-peroxide method for the preparation of phenol. Acetone is formed as by-product in this reaction.

105. (4)

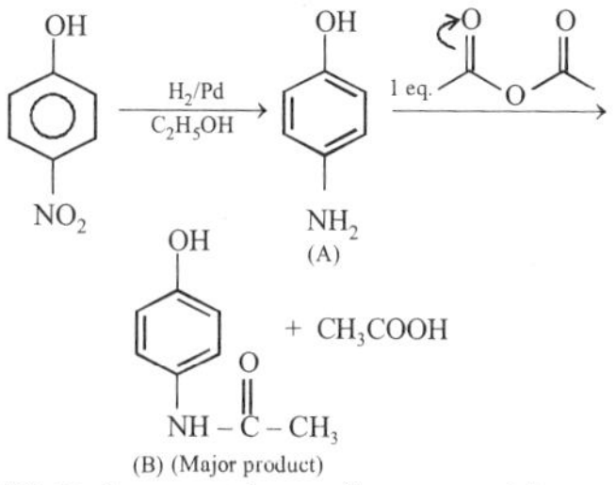

106. (2) It does not have free *p*-position.

107. (3)

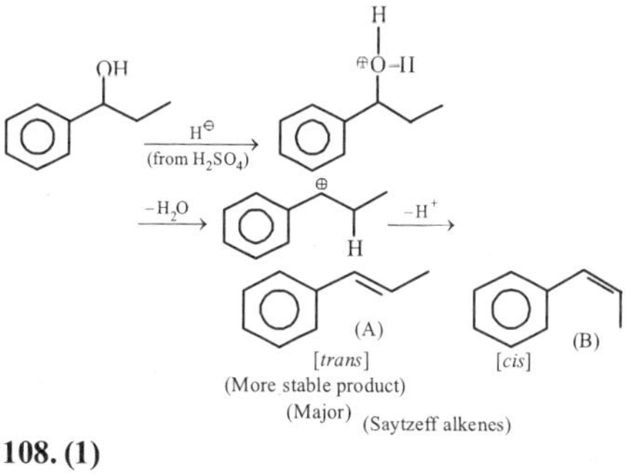

108. (1)

109. (4)

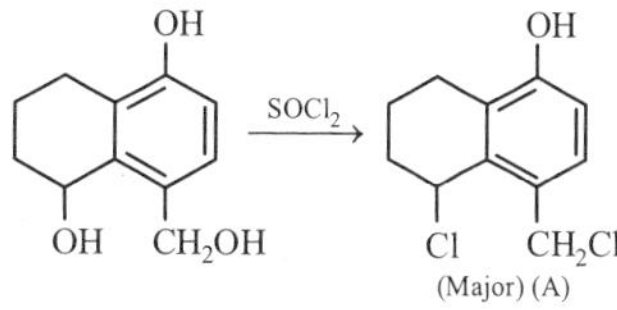

In case of – OH group present on the aromatic ring, carbon oxygen bond has partial double bond character. Hence, it is a strong bond and difficult to cleave.

110. (2)

111. (1)

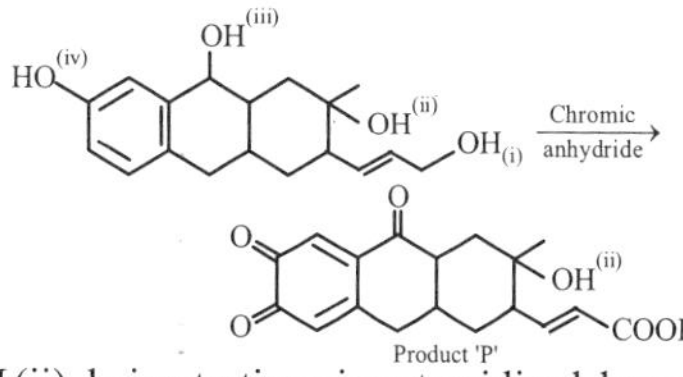

The –OH (ii), being tertiary, is not oxidised, hence it gives red colour with ceric ammonium nitrate.

112. (3)

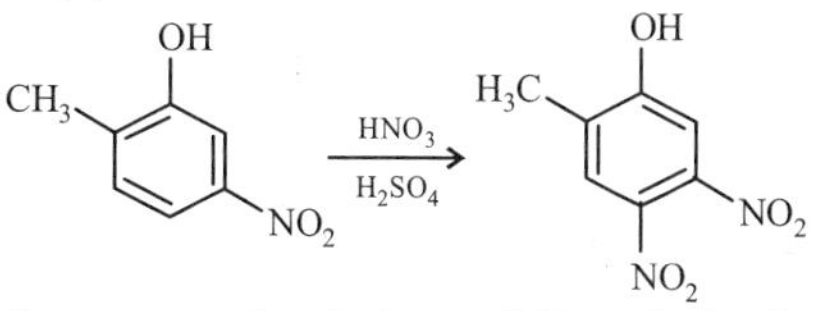

It is an example of electrophilic substitution reaction. Position of electrophile is directed by the strong ring activating group (–OH), present in the ring due to +M effect.

113. (3) Order of C–OH bond length:

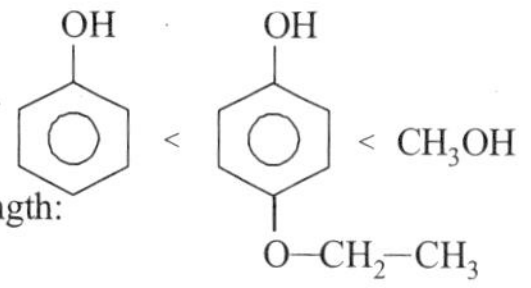

114. (1)

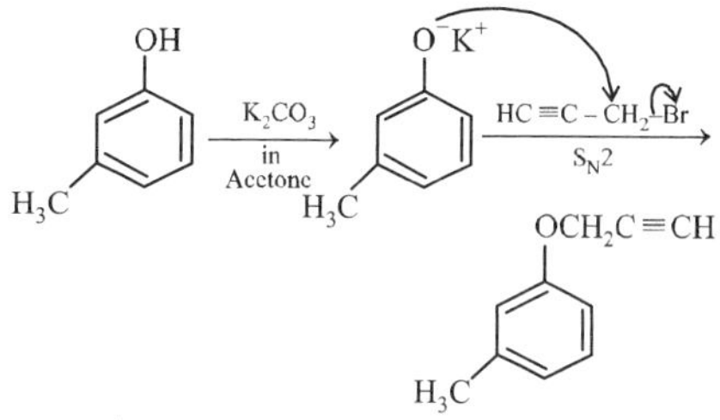

115. (4)

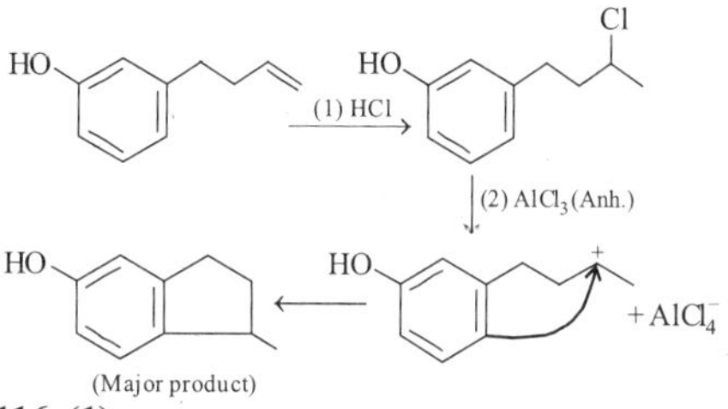

116. (1)

117. (2)

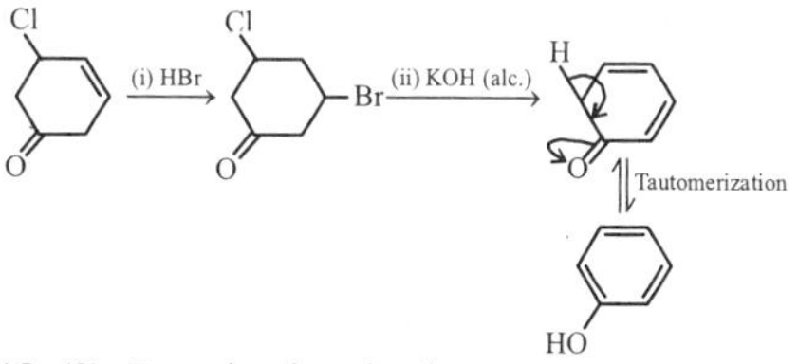

118. (2) Reaction involved:

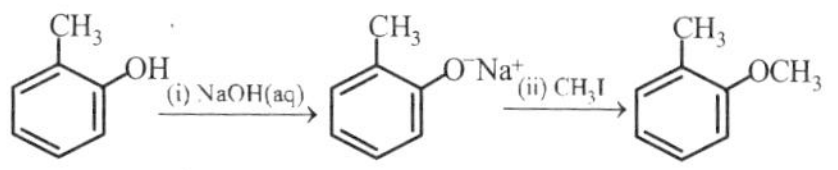

119. (4) Electron withdrawing substituents increase the acidic strength, while electron releasing groups decrease the acidic strength.

$\therefore$ Acidic strength $\propto K_a \propto \dfrac{1}{pK_a}$

$pK_a : B < C < A < D$

120. (3) Reaction involved:

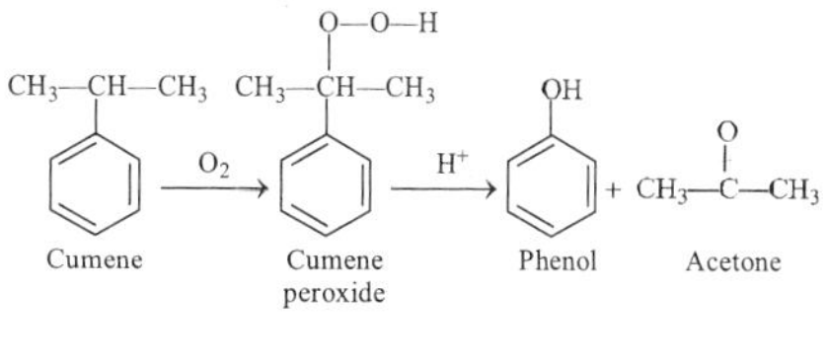

121. (1)

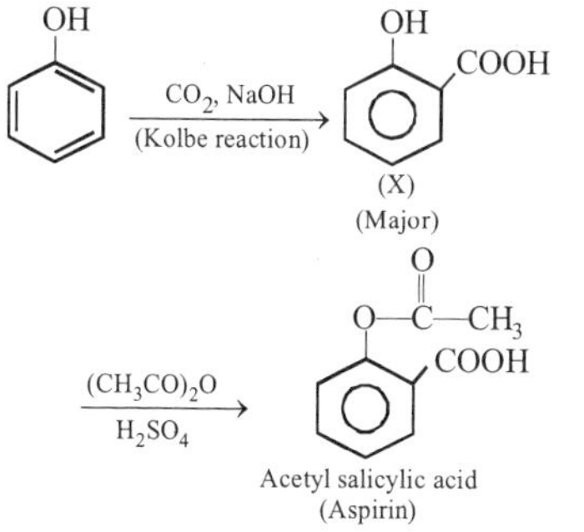

122. (3)

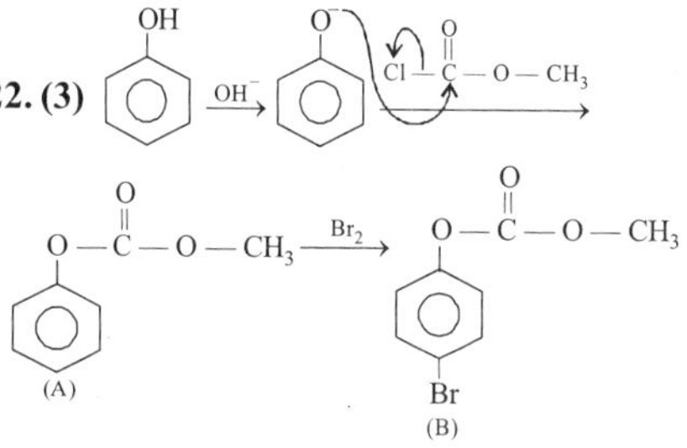

123. (1)

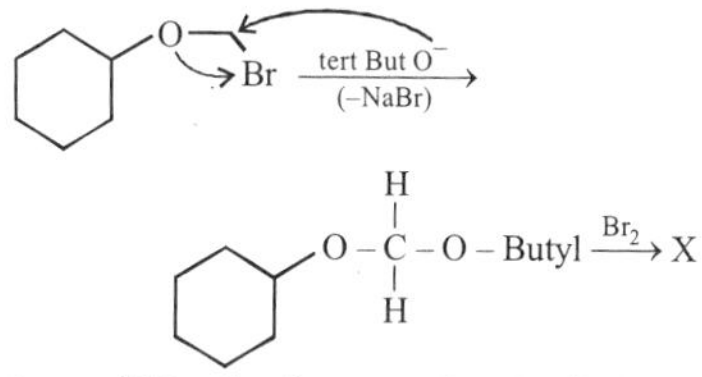

Other options will lead to form a – C = C – linkage which will decolorise Br_2 water color.

124. (1)

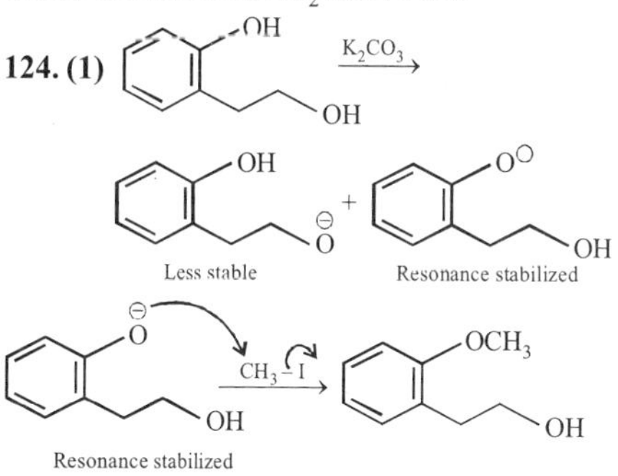

125. (3) Due to intramolecular hydrogen bonding this will not be soluble in sodium bicarbonate.

126. (3) 127. (2)

128. (1)

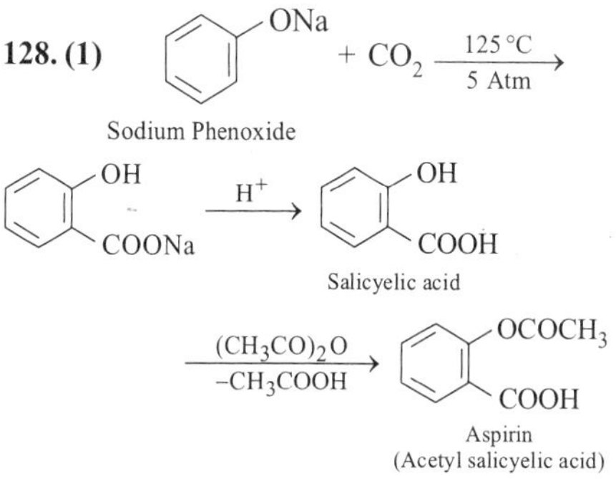

129. (3) Electron withdrawing substituents like $-NO_2$, –Cl increase the acidity of phenol while electron releasing substituents like $-CH_3$, $-OCH_3$ decrease acidity.

130. (1)

131. (4) Substitution of –OH group attached to benzene ring will not take place due to partial double bond character.

132. (2) Acetylation and benzoylation of phenol by acid chloride is known as Schotten-Baumann reaction.

133. (1)

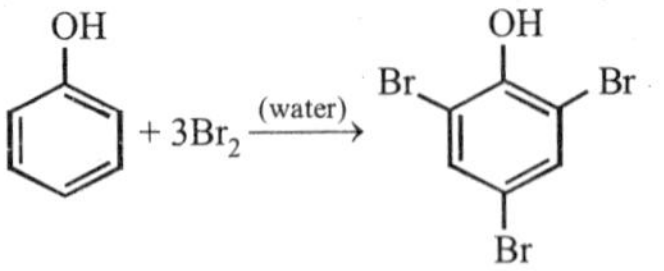

2, 4, 6-Tribromophenol

134. (2)

$C_2H_5-OH + H_2SO_4 \xrightarrow{433\ K} CH_2=CH_2$ (Ethylene)

$\xrightarrow{413\ K} CH_3-CH_2-O-CH_2-CH_3$ (Diethyl ether)

$\xrightarrow{383\ K} CH_3CH_2HSO_4 + H_2O$ (Ethyl hydrogen sulphate)

Acetylene is not formed under any condition.

135. (2)

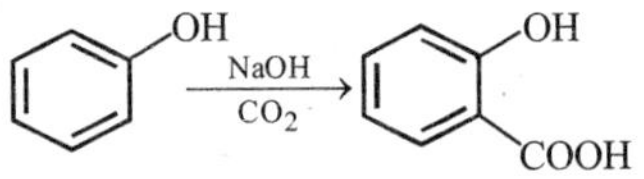

136. (N) Phenol on reaction with conc. H_2SO_4 gives a mixture of *o*- and *p*- products. Which on treatment with conc. HNO_3 will yield *picric acid*

137. (4) This is Riemer-Tiemann reaction and the electrophile is dichlorocarbene.

138. (3) In (3) both groups are activating and undergo electrophilic substitution in the same positions.

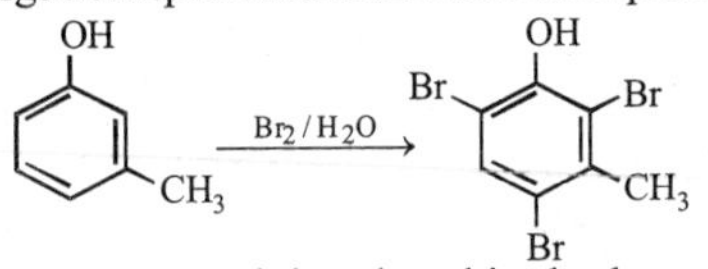

139. (3) 2,4,6 –Trinitrophenol is also known as picric acid.

140. (3)

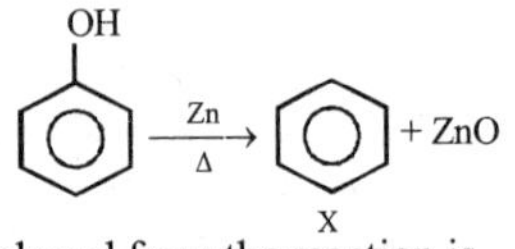

Moles of phenol from the reaction is

$= \frac{4.7}{94.0} = 0.05$ mol

The Reaction is 1 : 1 that means 0.05 mole of phenol will give 0.05 mole of benzene on 100% yield.

For 60% yield $= \frac{0.05}{100} \times 60 = 0.03$ mol.

mol of 'X' $= 3 \times 10^{-2}$ mol.

141. (14) Picric acid is prepared by treating phenol first with concentrated sulphuric acid which converts it to phenol-2,4-disulphonic acid and then with concentrated nitric acid to get 2, 4, 6-trinitrophenol.

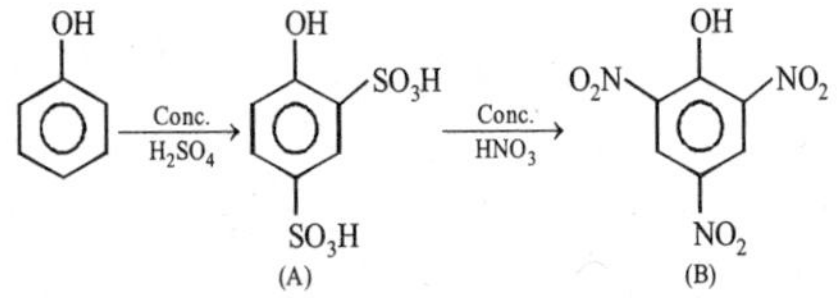

Total number of oxygen atoms in A and B = 7 + 7 = 14

142. (4) The compound must be

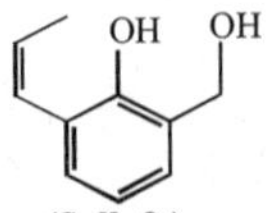

($C_{10}H_{12}O_2$)

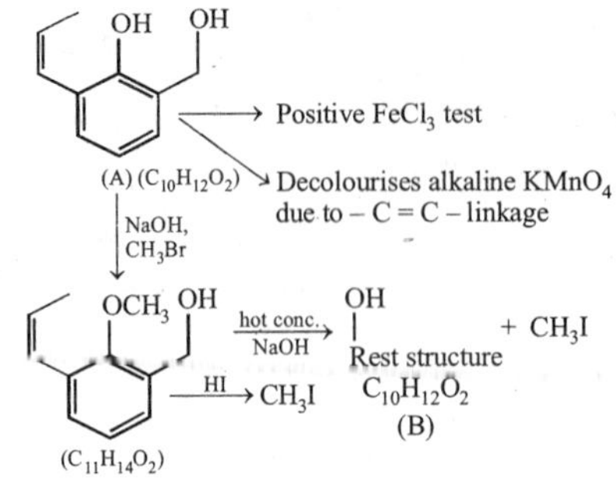

The number of π bonds in compound A are 4.

143. (30) Overall yield of complete reaction

$= \frac{60}{100} \times \frac{50}{100} = 0.3 \Rightarrow \%\ \text{yield} = 30\%$

144. (7)

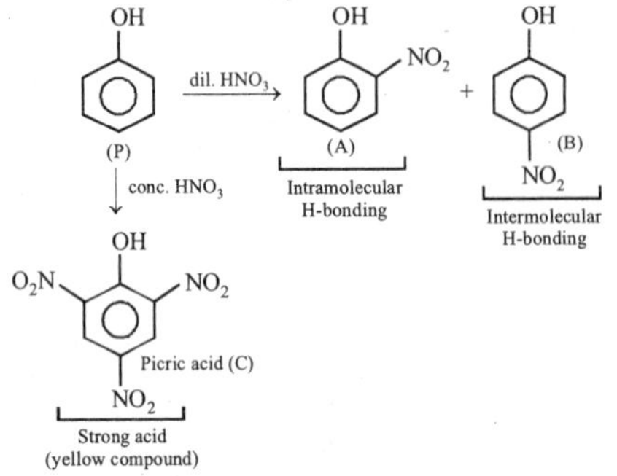

Number of oxygen atoms present in compound 'C' are 7.

145. (69)

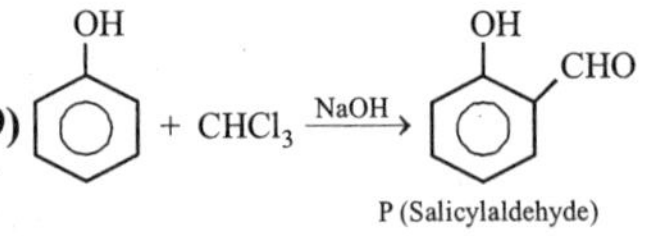

Molecular formula of product 'P' = $C_7H_6O_2$

So, mass % of C in 'P'

$= \frac{12 \times 7}{84 + 6 + 32} \times 100 = 68.85\% \approx 69\%$

146. (4) HI break phenylether as

$Ph-O-CH_2-Ph \xrightarrow{HI} Ph-OH + I-CH_2-Ph$ (Phenol, Benzyl Iodide)

Thus statement I false and statement II is true.

147. (1)

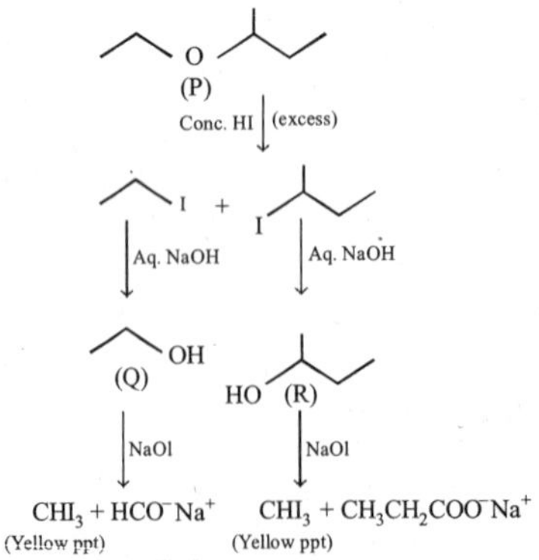

148. (4)

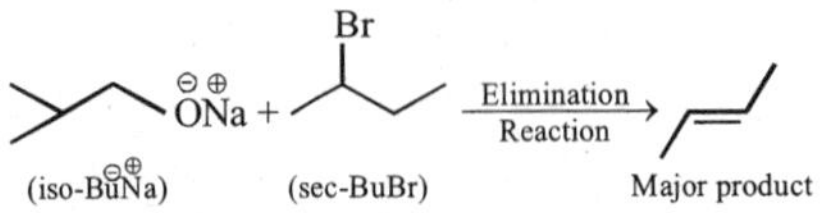

For secondary alkylhalide elimination would be more favourable than substitution.

149. (4)

150. (2)

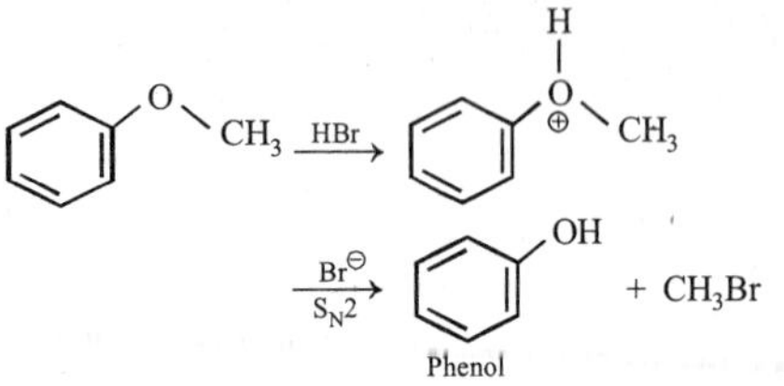

151. (2)

152. (2)

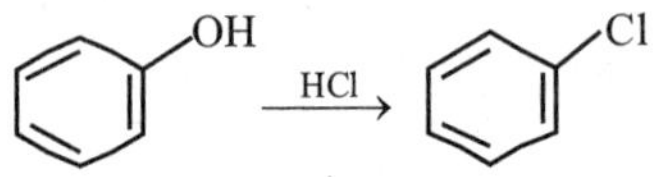

The above reaction is not possible as due to resonance C–O bond has partial double bond character. So, bond strength will increase and substitution will be less likely.

153. (1) $CH_3CH_2CH_2CH_2-OH$

Boiling point of alcohols are higher than the aldehydes having similar molecular masses due to H-bonding (intermolecular) in alcohols.

The boiling points of aldehydes are higher than hydrocarbons and ethers of comparable molecular masses.

154. (4)

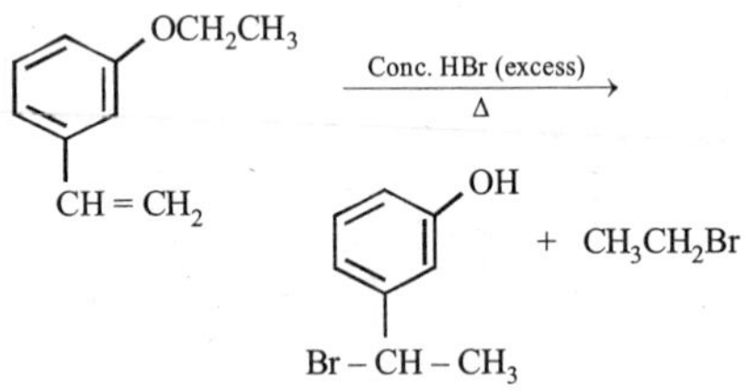

155. (4)

156. (3) $PhO^-Na^+ + Me-Br \xrightarrow{S_N2} Ph-O-Me$

157. (4)

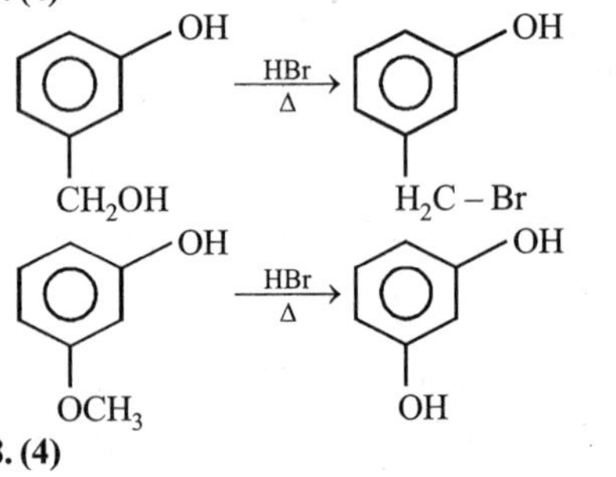

158. (4)

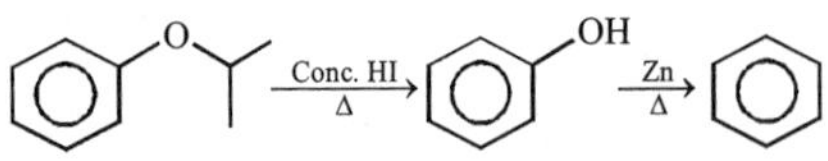

159. (1)

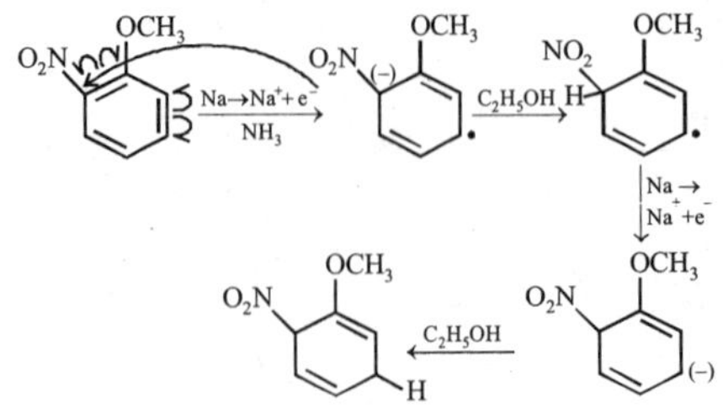

160. (4)

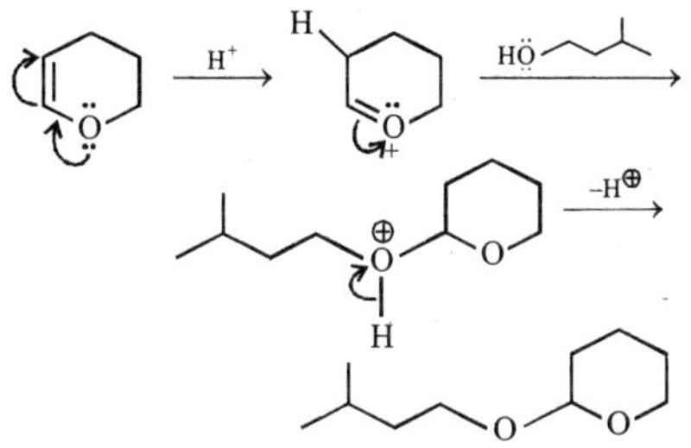

161. (2)

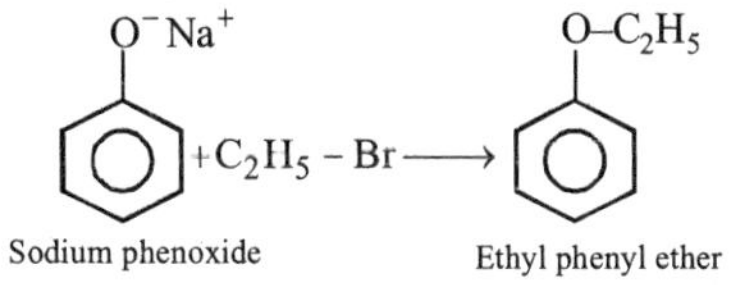

Bromobenzene does not undergo SN reactions due to resonance which leads to partial double bond to C – Br bond.

162. (2)

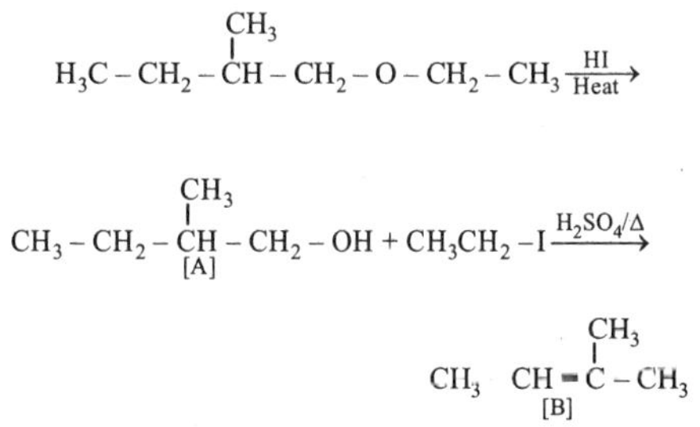

163. (3)

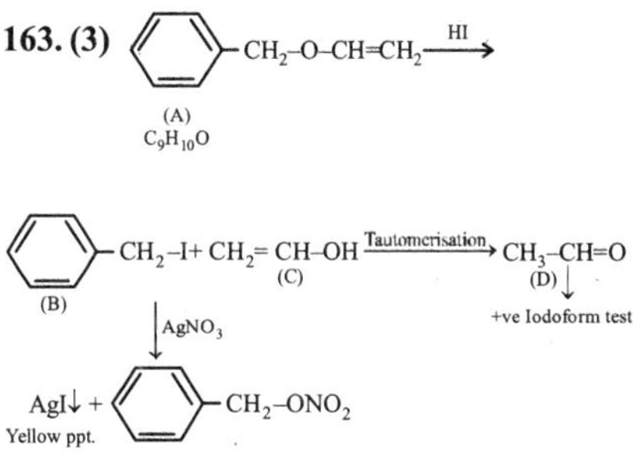

164. (1)

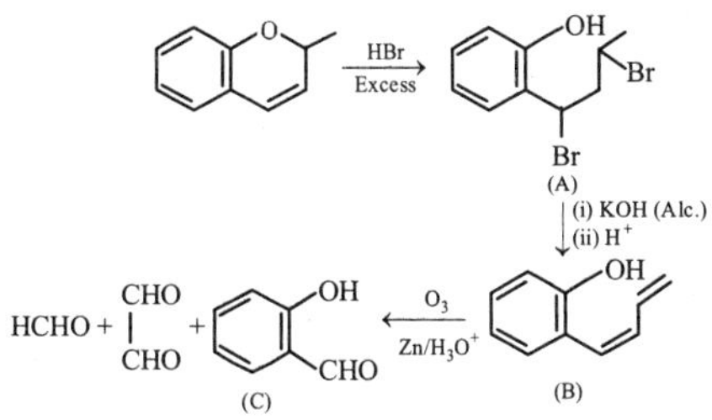

165. (1) In (A), –OH group is present, so inter-molecular H-bonding is possible while in (B), due to methoxy group there is no possibility of inter-molecular H-bonding. So, A is having higher boiling point than B.

166. (2)

167. (3)

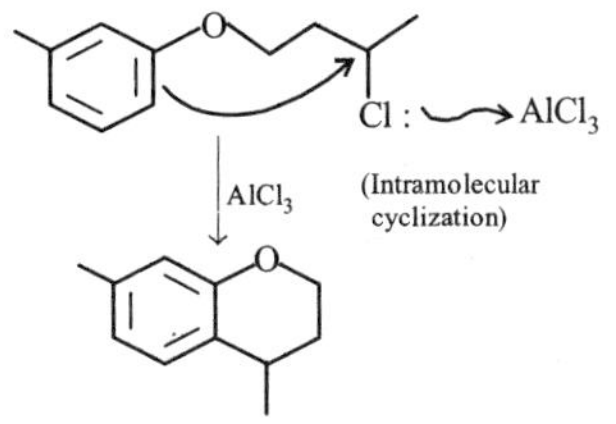

168. (3)

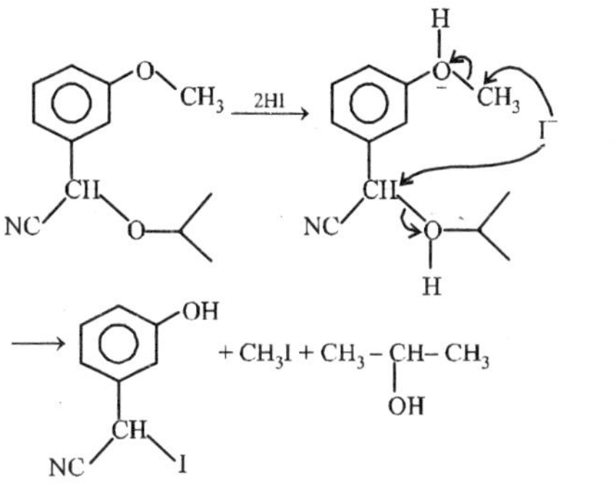

169. (2)

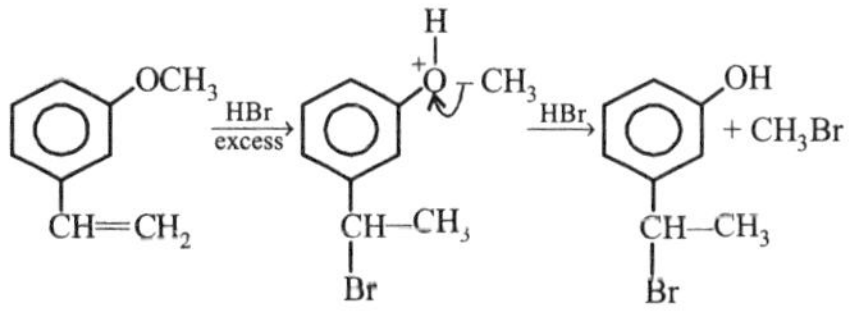

170. (2)

171. (4)

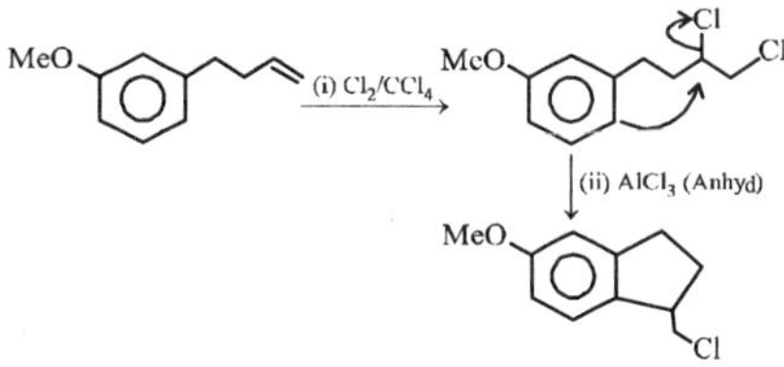

172. (4)

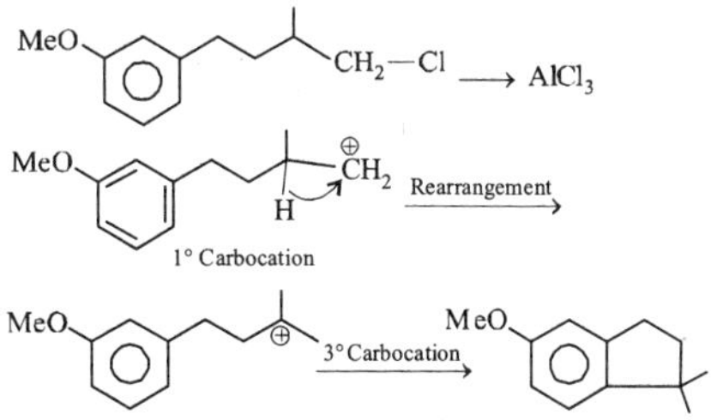

173. (4)

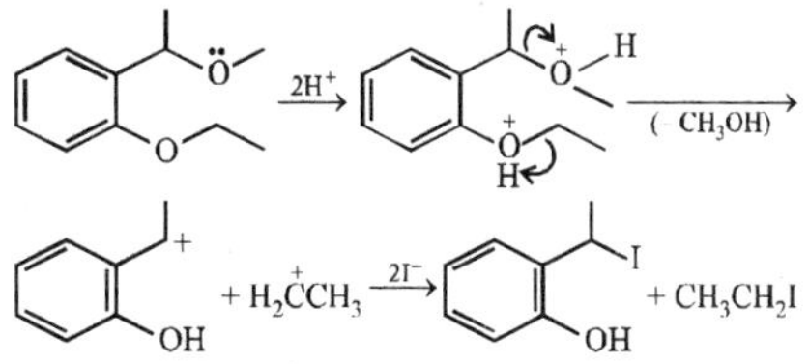

174. (4)

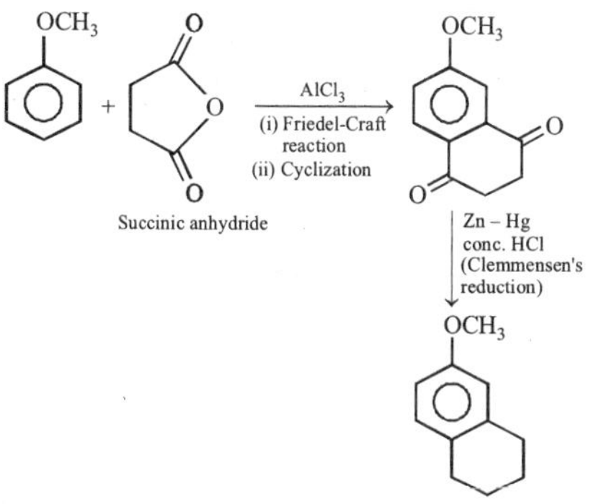

175. (4)

176. (2) $C_6H_5ONa + Br - CH_2 - CH = CH_2 \longrightarrow C_6H_5 - O - CH_2 - CH = CH_2 + NaBr$

Allyl phenyl ether

In other three options – Br is attached to $sp^2 - C$ atom.

177. (2) The tertiary alkyl halides undergo elimination reaction to give alkenes.

178. (1)

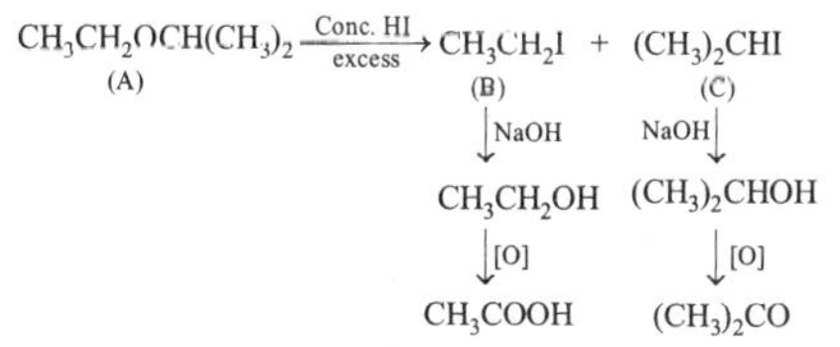

Hence the IUPAC name of compound (A) is

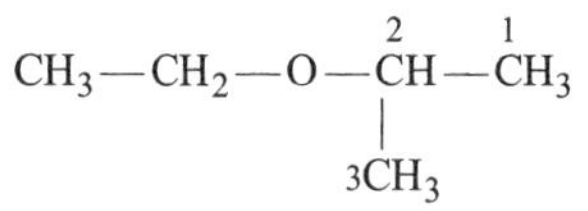

2–Ethoxypropane

179. (2)

$$CH_2 = CH - OCH_3 \xrightarrow{HBr} CH_3 - \underset{\displaystyle Br}{CH} - O - CH_3$$

180. (15)

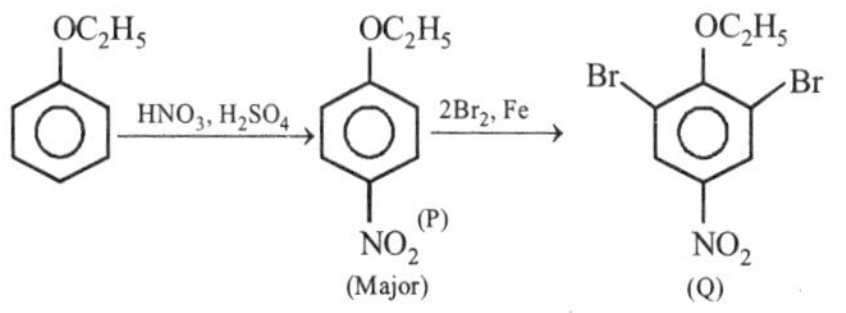

Number of O-atoms in Q = 3; Number of Br-atoms = 2

$$\text{Required ratio} = \frac{3}{2} = 1.5 = 15 \times 10^{-1}.$$

Part-B (JEE Advanced)

1. **(b)** Conc. HCl, HBr and conc. $HCl + ZnCl_2$ all are nucleophiles, thus convert alcohols to alkyl halides. However, conc. H_3PO_4 is a good dehydrating agent which converts an alcohol to an alkene.

2. **(c)** Addition of water to 2-phenylpropene follows Markovnikov's rule.

$$CH_3-\underset{}{\overset{Ph}{\overset{|}{C}}}=CH_2 \xrightarrow{H_3O^+} CH_3-\overset{Ph}{\overset{|}{\underset{OH}{\underset{|}{C}}}}-CH_3$$

2-Phenyl-2-propanol

3. **(c)**

$$\underset{\text{Propanol-1}}{CH_3CH_2CH_2OH} \xrightarrow[\text{heat}]{Cu} \underset{\text{Propanal (responds Fehling solution)}}{CH_3CH_2CHO}$$

$$\underset{\text{Propanol-2}}{CH_3CHOHCH_3} \longrightarrow \underset{\text{Propanone (negative to Fehling solution)}}{CH_3COCH_3}$$

4. **(a)** Compound $(CH_3)_4N^+I^-$ is most reactive due to (i) better leaving group, I^- and (ii) due to the fact that the methyl group, with positive N, is more electron deficient.

Hence this group is more reactive towards nucleophile, OH^-

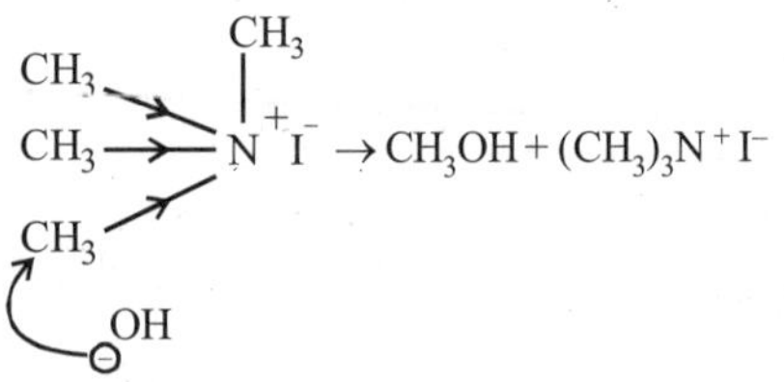

5. **(c)** Secondary alcohols oxidise to produce keone.

$$\underset{\text{2-Butanol}}{CH_3CHOHCH_2CH_3} \xrightarrow{(O)} \underset{\text{Ethyl methyl ketone}}{CH_3COCH_2CH_3}$$

6. **(a)** Reactions involving cleavage of carbon-oxygen bond, (C – OH) follows the following order :

Tertiary > Secondary > Primary

7. **(a)**

$$\underbrace{CO + H_2}_{\text{Water gas}} + H_2 \xrightarrow[300°C]{Cr_2O_3-ZnO} \underset{\text{Methanol}}{CH_3OH}$$

8. **(d)** **Lucas test** is based on the difference in the three types of alcohols (having 6 or less carbon) towads **Lucas reagent** (a mixture of conc. hydrochloric acid and anhydrous zinc chloride) at room temperature.

$$ROH + HCl \xrightarrow{ZnCl_2} RCl + H_2O$$

The tertiary alcohols produce turbidity immediately, the secondary alcohols give turbidity within 5 – 10 minutes, and the primary alcohols do not give turbidity at all, at room temperature. Thus, the order of reactivity of alcohol with Lucas reagent is *tert.* > *sec.* > *pri.*

Hence, 2-methylpropan-2-ol (a 3° alcohol) reacts fastest.

9. **(a)** Among the given options, ethyl alcohol is most basic.

10. **(c)** $C_2H_5OH \xrightarrow[\Delta]{H_2SO_4} C_2H_4 + H_2O$

11. **(6)**

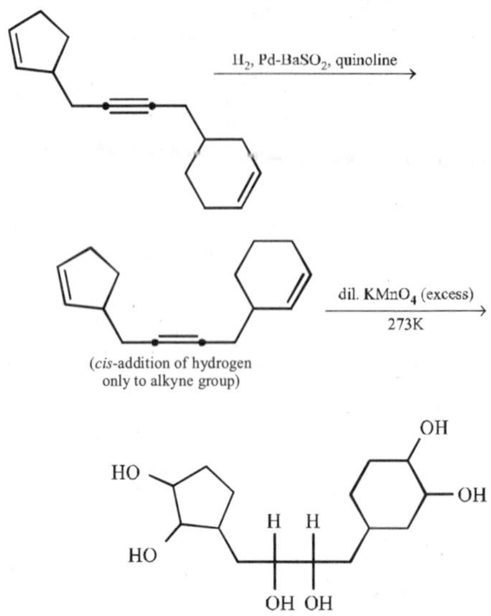

12. **(4)**

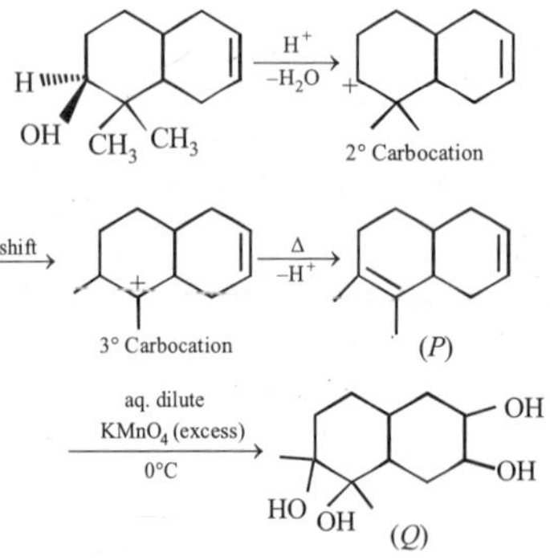

13. **(1.50)**

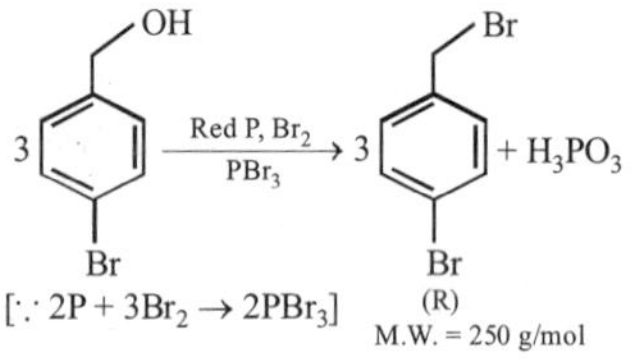

In Carius method, the halogen present in the organic compound completely forms the silver halide.

No. of moles of (R) $= \dfrac{1\text{ g}}{250\text{ g/mol}} = \dfrac{1}{250}$ mol

1 mol of (R) ≡ 2 mol of Br ≡ 2 mol of AgBr

$\therefore \dfrac{1}{250}$ mol of (R) $\equiv \dfrac{2}{250}$ mol of

Br $\equiv \dfrac{2}{250}$ mol of AgBr

M.W. of AgBr = 108 + 80 = 188 g/mol

$\therefore$ The amount of AgBr formed $= \dfrac{2\times 188}{250}$

$= 1.50$ g

14. **Secondary.**

15. **aldol (β-hydroxybutanal);**

$$CH_3CH_2OH \xrightarrow[\text{heat}]{Cu} CH_3CHO \xrightarrow{OH^-} CH_3CH(OH)CH_2CHO$$

16. **False :** Ethanol is not acidic enough to react with aq. NaOH. Thus, sod. ethoxide (C_2H_5ONa) is prepared by the reaction of Na metal with ethyl alcohol.

$2\,C_2H_5OH + 2Na \rightarrow 2\,C_2H_5ONa + H_2\uparrow$

17. **(a, c, d)** Isomeric alcohols with molecular formula $C_4H_{10}O$ are

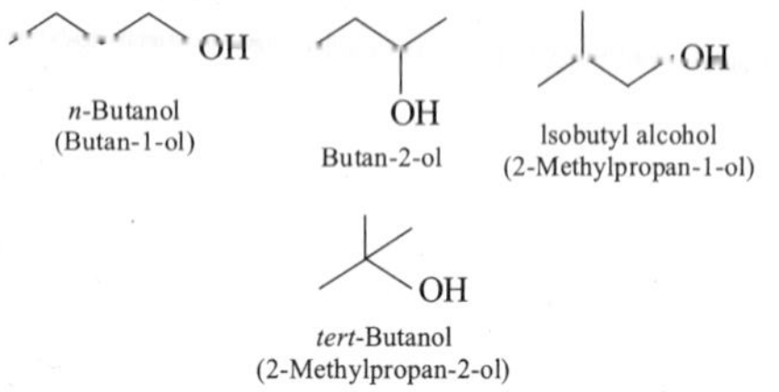

18. **(b)** The mechanism of this reaction is represented as follows.

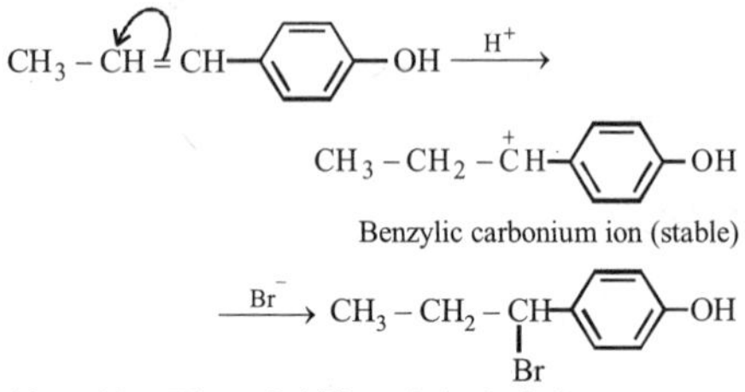

19. **(c)** The solubility of alcohols in water can be explained due to the formation of hydrogen bond between the highly polarised –OH groups present both in alcohol and water. However, in higher alcohols the hydrocarbon character (alkyl chain) of the molecule increases and thus, alcohols tend to resemble hydrocarbon (which are insoluble in water) and hence, the solubility in water decreases. When the ratio of C to OH is more than 4, alcohols have little solubility in water. So, statement is correct but explanation is not.

20.

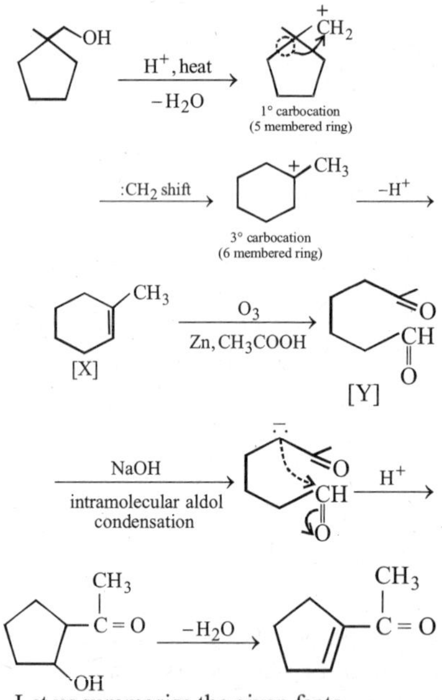

21. Let us summarise the given facts.

$$\underset{\text{Bombykol}}{C_{16}H_{30}O} \xrightarrow{\text{Hydrogenation}} \underset{(A)}{C_{16}H_{34}O} \xrightarrow{(CH_3CO)_2O} \text{Ester}$$

$\downarrow (CH_3CO)_2O$

Ester $\xrightarrow{\text{Oxidative ozonolysis}}$

$CH_3CH_2CH_2COOH$ + $\begin{matrix}COOH\\|\\COOH\end{matrix}$ + $AcOCH_2(CH_2)_8COOH$
Butanoic acid — Oxalic acid — 10 - Acetoxydecanoic acid

(i) Hydrogenation of bombykol ($C_{16}H_{30}O$) to $C_{16}H_{34}O$ (*A*) indicates the presence of two double bonds in bombykol.

(ii) Reaction of *A* with acetic anhydride to form ester indicates the presence of an alcoholic group in *A* and hence, also in bombykol.

(iii) Products of oxidative ozonolysis of bombykol ester suggests the structure of bombykol.

$CH_3CH_2CH_2COOH + HOOC.COOH + HOOC.(CH_2)_8.CH_2OAc$
Butanoic acid — Oxalic acid — 10 - Acetoxydecanoic acid

↑ oxidative ozonolysis

$CH_3CH_2CH_2CH = CH - CH = CH.(CH_2)_8.CH_2OAc$
Bombykol ester

The structure of the bombykol ester suggests that bombykol has the following structure :
$CH_3CH_2CH_2CH = CH - CH = CH{\cdot}(CH_2)_8{\cdot}CH_2OH(C_{16}H_{30}O)$
(Bombykol)
and the structure of *A* is
$CH_3CH_2CH_2CH_2CH_2CH_2CH_2(CH_2)_8.CH_2OH$ or $C_{16}H_{33}OH$.
Four geometrical isomers are possible for the above bombykol structure (as it has two double bonds).

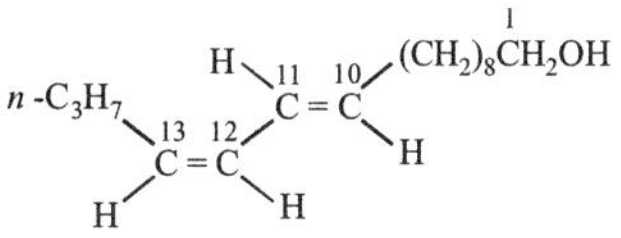
10 E, 12 Z

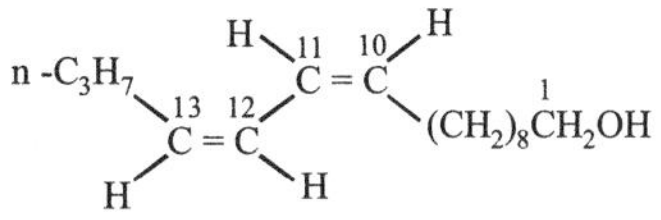
10 Z, 12 Z

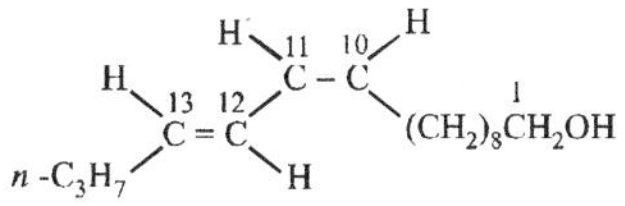
10 Z, 12 E

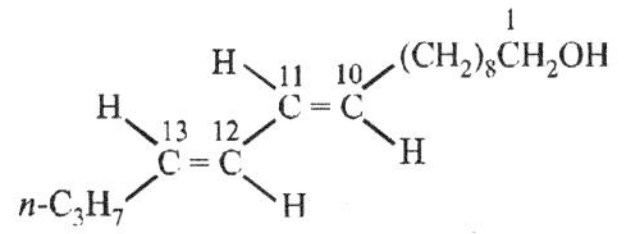
10E, 12E

22.

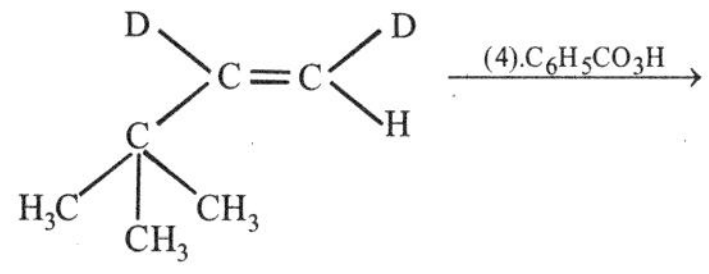

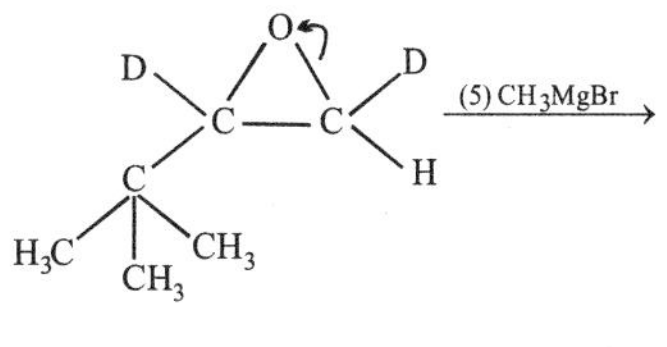

23. (i) Since the large propenyl group is attached to the carbon atom bearing the hydroxyl group, so the reaction is likely to occur via S_N1 mechanism.

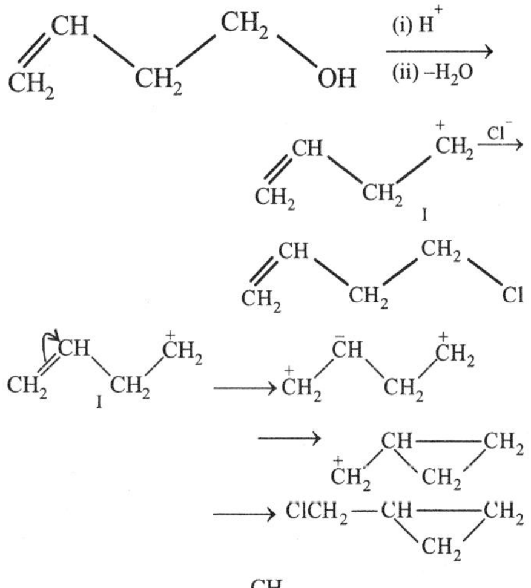

(ii)

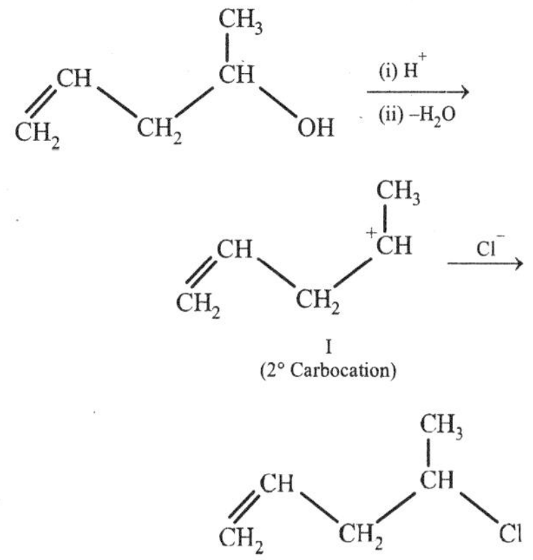

In the intermediate carbocation, I, carbon bearing positive charge has CH_3 group which decreases the positive charge and hence prevents cyclisation of the compound.

24. Since, 3° carbocation (formed in case of *t*-butanol) is more stable than 1° (formed in *n*-butanol), dehydration in the former proceeds faster than in the latter.

25. (a) Since (B, $C_6H_{14}O$) is resistant to oxidation, it must be *ter*-alcohol.

(b) Since (B) is optically inactive, it must have at least two similar alkyl groups.

$$\underset{(A)}{C_6H_{10}O} \xrightarrow{2H_2} \underset{(B)}{C_6H_{14}O}$$

Thus, the five carbon atoms can be adjusted into three alkyl groups (of which two are similar) either as $-CH_3$, $-CH_3$, and $-C_3H_7$, or as $-C_2H_5$, $-C_2H_5$ and $-CH_3$. Thus the possible structure of alcohol (*B*) is either

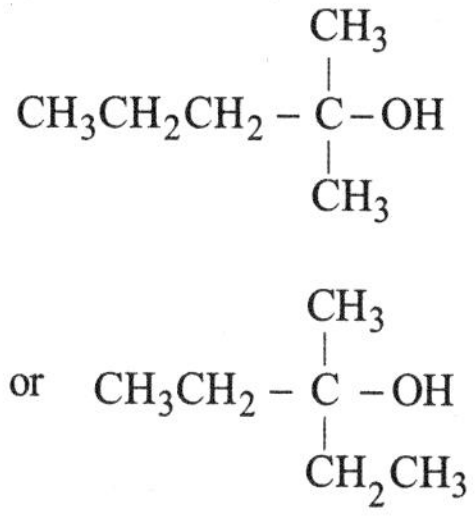

Hence the corresponding compound (*A*) is either

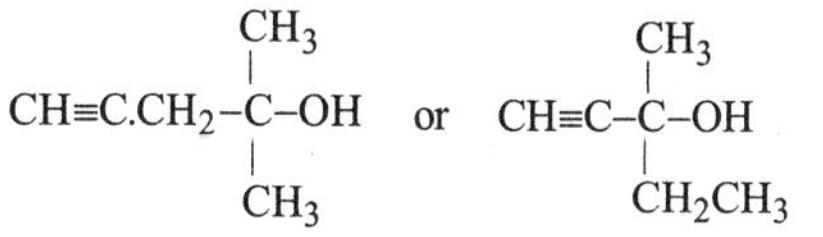
(Optically inactive) (Optically active)

However, the compound (*A*) is optically active, so (*A*) and hence also (*B*) should have right side structure.

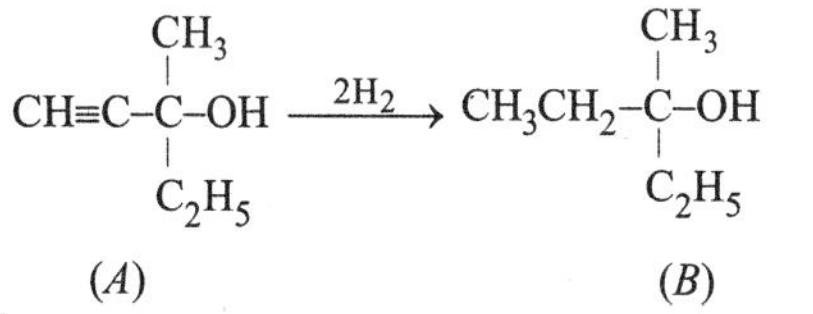

26.

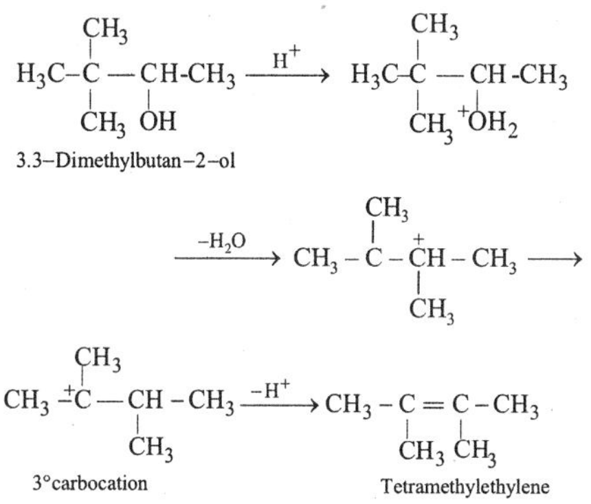

27. *n*-Butanol gives the following reaction in which the purple colour of $KMnO_4$ changes to brown. *tert*-Alcohols are not oxidisable easily, hence purple colour of $KMnO_4$ remains same.

$$\underset{n\text{-Butanol}}{CH_3CH_2CH_2CH_2OH} + \underset{\text{Purple}}{KMnO_4} \longrightarrow \underset{(\text{Soluble in } H_2O)}{CH_3CH_2CH_2COOK} + \underset{\text{Brown}}{MnO_2}\downarrow + KOH$$

The brown precipitate is of MnO_2.

28. (i) Since the compound *X* (C_5H_8O) does not react appreciably with Lucas reagent, it indicates that the compound has a primary alcoholic group ($-CH_2OH$).

(ii) Reaction of the compound *X* with ammonical silver nitrate to give a precipitate indicates that it has an acetylenic hydrogen atom, *i.e.*, $\equiv C - H$ grouping is present.

(iii) Treatment of *X* with H_2/Pt followed by boiling with excess of HI gives *n*-pentane. It indicates that the compound does not have any branch.

On the basis of the above points, compound *X* (C_5H_8O) may be assigned following structure.

$HC \equiv C - CH_2 - CH_2 - CH_2OH$
(*X*) 4-Pentyn-1-ol (Mol. wt. 84, Eq. wt. = 42)

The above structure for the compound X is in accordance with its equivalent weight obtained from the given data.

224 ml. of CH_4 at STP is obtained from 0.42 g

22400 ml. of CH_4 at STP $= \frac{0.42}{224} \times 22400 = 42$ g

$\therefore$ Eq. wt. of the compound X = 42

Reactions of the compound X;

(i) $HC \equiv C.CH_2\,CH_2.CH_2OH$ (X) $\xrightarrow[NH_4OH]{AgNO_3} AgC \equiv C.CH_2CH_2CH_2OH \downarrow$

(ii) $HC \equiv C.CH_2\,CH_2.CH_2OH$ (X) $\xrightarrow{2CH_3MgBr}$ $MgBrC \equiv C.CH_2CH_2CH_2OMgBr + 2CH_4$

(iii) $HC \equiv C.CH_2\,CH_2.CH_2OH$ (X) $\xrightarrow{H_2/Pt}$ $CH_3CH_2CH_2CH_2CH_2OH \xrightarrow{HI}$ $CH_3CH_2CH_2CH_2CH_3$ (n–Pentane)

29.

$CH_2{=}CH_2 + H_2O + [O]$ (from $KMnO_4$) $\longrightarrow$ $CH_2OH{-}CH_2OH$ (Ethylene glycol) $+ MnO_2$

30. The compound A, a ketone, undergoes haloform reaction. Thus, it must contain CH_3CO group.

The compound C gives mono-ozonide D, which shows that the compound C contains a double bond. Since the hydrolysis of D gives only acetaldehyde, the compound C would be an alkene having four carbon atoms, *i.e.* $CH_3-CH=CH-CH_3$ (butene-2).

The compound B is obtained by the reduction of compound A (which contains CH_3CO group). Hence, the compound B would be an alcohol, which on heating with H_2SO_4 gives (C). Hence B and A would be

$CH_3-CH(OH)-CH_2-CH_3$ butan–2–ol (B) $\quad CH_3-C(=O)-CH_2-CH_3$ butan–2–one (A)

The reactions involved :

$(A) \xrightarrow{reduction} (B) \xrightarrow[-H_2O]{conc.\ H_2SO_4} CH_3-CH=CH.CH_3\ (C) \xrightarrow{O_3}$

$CH_3-CH-O-CH-CH_3$ (with O—O bridge) mono-ozonide (D) $\xrightarrow{HOH} 2CH_3CHO$

31. For this type of problem, students are advised to summarise the whole problem in the form of reactions.

$H_3C.CH(CH_3).CH_2CH_3$ (2–Methylbutane) $\xleftarrow{H_2}$ A (C_5H_{10}) $\xrightarrow[Markow.\ addition]{HBr}$ B (having Br)

$B \xrightarrow{AgOH} C\ (C_5H_{12}O,\ alcohol) \xrightarrow{(O)} D$ (ketone)

Let us draw some conclusions from the above set of reactions.

(i) The molecular formula C_5H_{10} (C_nH_{2n}) for A indicates that it is an alkene having one double bond.

(ii) Since the alcohol C on oxidation gives a ketone D, C must be a secondary alcohol and hence B must be a secondary bromide.

(iii) The structure of 2-methylbutane, the hydrogenated product of A, indicates that the secondary bromide must have following structure.

$CH_3-CH(CH_3)-CH(Br)-CH_3$ (B)

(iv) Thus the corresponding olefin A must have structure A which on Markownikoff addition of HBr gives the bromide B, the other possible alkene A' will not give B when HBr is addd on it according to Markownikoff rule.

$CH_3-CH(CH_3).CH=CH_2$ (A); $\quad CH_3C(CH_3)=CH.CH_3$ (A')

Thus the reaction involved can be represented as below:

$CH_3-CH(CH_3)-CH=CH_2$ (3–Methylbutene–1) (A) $\xrightarrow{HBr}$ $CH_3-CH(CH_3)-CH(Br)-CH_3$ (2–Bromo–3–methylbutane) (B)

$\xrightarrow{AgOH}$ $CH_3-CH(CH_3)-CH(OH)-CH_3$ (3–Methylbutanol–2) (C)

$\xrightarrow{(O)}$ $CH_3-CH(CH_3)-C(=O)-CH_3$ (3–Methylbutanone–2) (D)

32. (i) CH_3CH_2OH (Ethanol) $\xrightarrow[350°C]{Al_2O_3}$ $CH_2=CH_2 \xrightarrow{Br_2} BrCH_2.CH_2Br \xrightarrow[(ii)\ NaNH_2]{(i)\ alc.KOH} CH \equiv CH$

$\xrightarrow{CH_3COOH} CH_2=CHOCOCH_3$ (Vinyl acetate)

or CH_3CH_2OH (Ethanol) $\xrightarrow{[O]} CH_3CHO \longrightarrow CH_3COOH \xrightarrow{CH\equiv CH}$ $CH_2=CHOCOCH_3$ (Vinyl acetate)

(ii) $CH_3CH(OH)CH_3$ (2-Propanol) $\xrightarrow[-H_2O]{Conc.\ H_2SO_4} CH_3.CH{=}CH_2$

$\xrightarrow[peroxide]{HBr} CH_3.CH_2.CH_2Br \xrightarrow{aq\ NaOH}$ $CH_3.CH_2.CH_2OH$ (1-Propanol)

33. Iodoform test is used to distinguish methanol and ethanol.

Ethanol gives iodoform test while methanol does not respond.

$C_2H_5OH + 4I_2 + 6NaOH \rightarrow CHI_3 \downarrow + 5NaI + HCOONa + 5H_2O$

34. The given problem can be sketched as below.

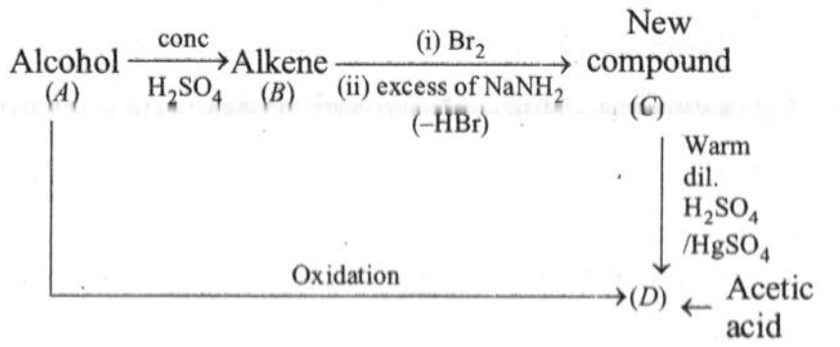

From the problem it appears that the compound C is an alkyne, hence D must be an aldehyde or ketone. Further, since D can be obtained from acetic acid through its calcium salt it may be either acetaldehyde or acetone. Hence going back, A must be ethyl alcohol, which explains the given set of reactions.

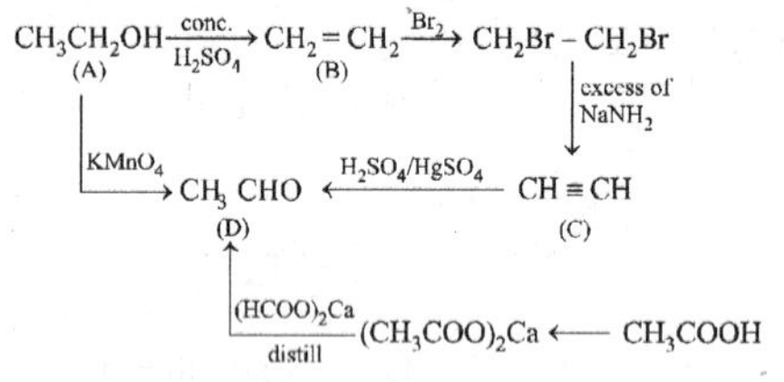

Hence,

A is ethyl alcohol, CH_3CH_2OH

B is ethylene, $CH_2=CH_2$

C is acetylene, $CH\equiv CH$

D is acetaldehyde, $CH_3.CHO$

35. 'A' is C_2H_5OH and 'B' is C_2H_4

C_2H_5OH (A) (Ethyl alcohol) $\xrightarrow[-H_2O]{Conc.\ H_2SO_4} C_2H_4$ (B) (Ethene)

$C_2H_4 + alk.\ KMnO_4 \longrightarrow CH_2OH{-}CH_2OH$;

$C_2H_4 + H_2 \xrightarrow{Ni} C_2H_6$ (Colourless)

36. (b)

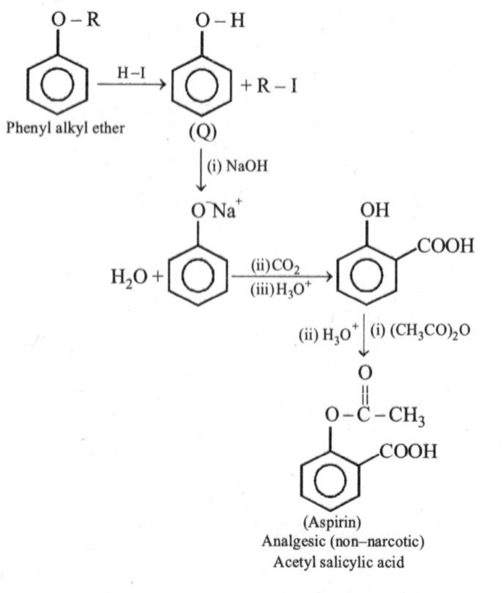

Aspirin inhibits the synthesis of prostaglandin.

37. (d) In dye test, phenolic — OH group of p-naphthol is converted to — O^- which activates the ring towards electrophilic aromatic substitution.

38. (d)

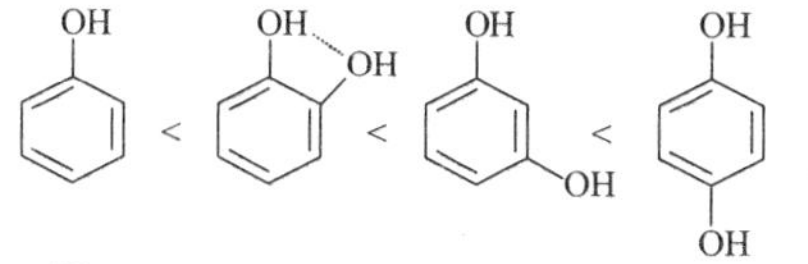

IV Lowest molar mass < I Chelation occurs (no intermolecular H-bonding) < II Weaker intermolecular H-bonding < III Stronger intermolecular H-bonding

39. (c) (i) The acidity increases with the increase in electronegativity of the halogen present.

(ii) The inductive effect decreases with increase in distance of halogen atom from the carboxylic group and hence, the strength of acid proportionally decreases.

Smallest dissociation constant means weakest acid, which is $BrCH_2CH_2COOH$ because here Br (less electronegative than F) is two carbon atoms away from – COOH

40. (d) $-NO_2$ is an electron-attracting group where as $-CH_3$ is an electron-releasing group.

An electron - attracting substituent tends to disperse the negative charge of the phenoxide ion and thus, makes it more stable. This, in turn, **increases the acid strength of phenol.** The substituent in para position is more effective than in the meta position as the former involves a resonating structure bearing negative charge on the carbon attached to the electron - withdrawing substituent.

An **electron - releasing substituent tends to** intensify the negative charge of the phenoxide ion and thus makes it more unstable. This, in turn, **decreases the acid strength of phenol.** Hence, the order of acid strength is

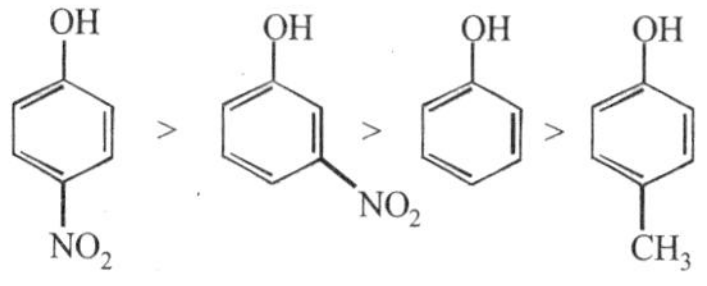

(IV) > (III) > (I) > (II)

41. (d) Riemer-Tiemann reaction is an electrophilic substitution on the highly reactive phenoxide ring.

$$HCCl_3 + OH^- \rightarrow H_2O + {}^-:CCl_3 : \bar{C}Cl_3 \rightarrow Cl^- + :CCl_2$$

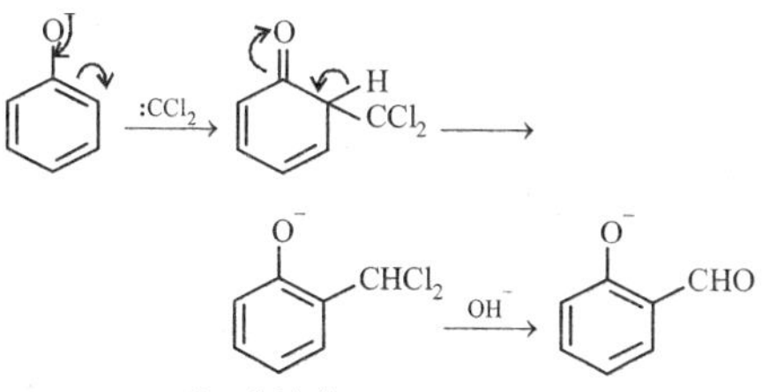

Benzal chloride (A)

42. (d)

$$C_6H_5CH_3 + Cl_2(exc.) \xrightarrow{\text{light, heat}} C_6H_5CCl_3$$

$$\xrightarrow{\text{aq. NaOH}} C_6H_5C(OH)_3 \xrightarrow{-H_2O} C_6H_5COOH$$

43. (b)

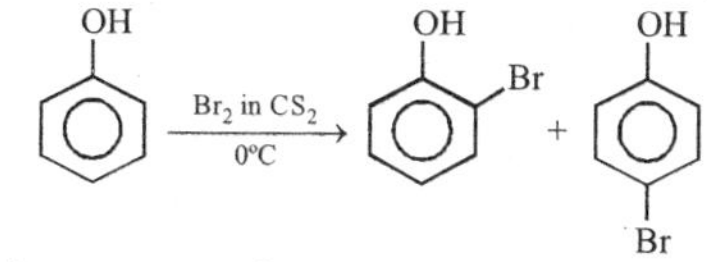

Bromine in presence of CS_2 (non-polar solvent) at low temperature ionises easily. Further in absence of CS_2, polyhalogenation in *o*- and *p*-positions takes place.

44. (d)

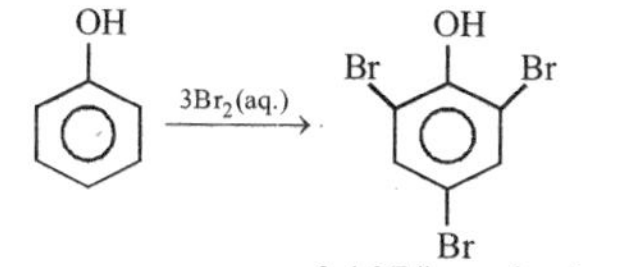

2, 4,6-Tribromophenol

Bromine in water (polar) solvent ionises earily to give Br^+ ions. Further the –OH group in phenol, being activating group, facilitates substitution in the *o*- and *p*-positions.

45. (6) $C_8H_{10}O_2$ gives $FeCl_3$ test means it has phenolic group.

It rotates plane polarized light means it is optically active.

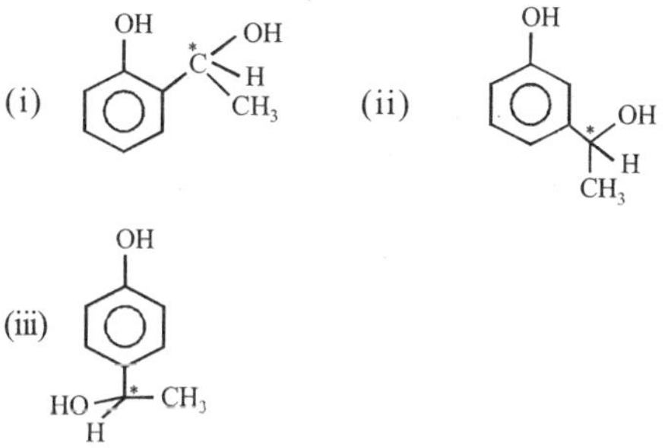

(i), (ii), (iii) are optically active and their enantiomers as well.

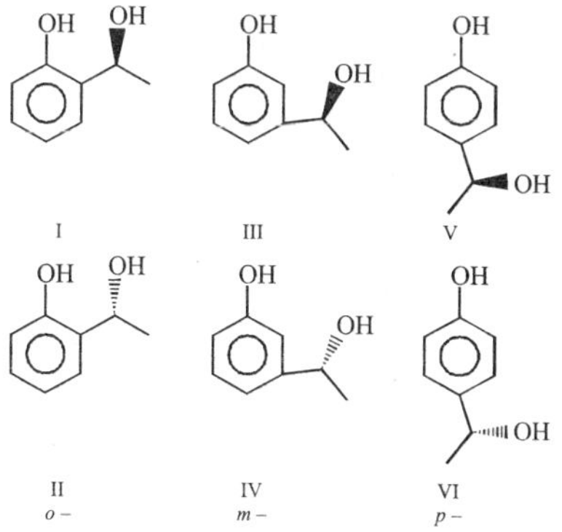

I ⋮ II
o ⋮ *o*
(Enantiomers)
III ⋮ IV
m ⋮ *m*
(Enantiomers)
V ⋮ VI
p ⋮ *p*
(Enantiomers)

46. (9)

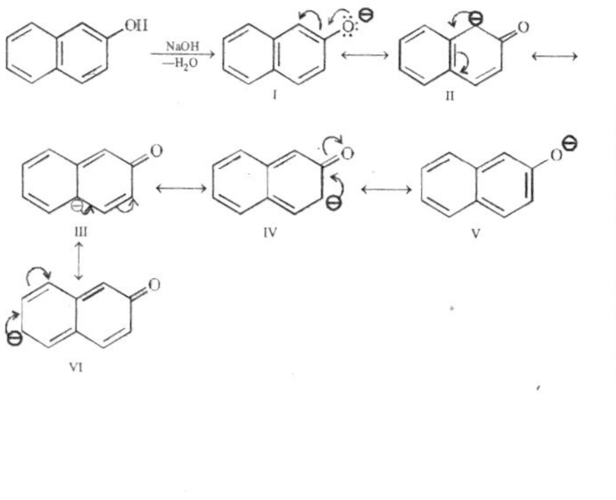

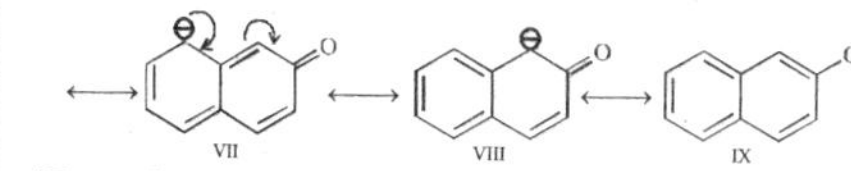

47. phenoxide ion

48. nucleophilic

49. resonance stabilization

50. (a, b, c) —OH group is strongly activating and *o*, *p*-directing due to +M effect. Thus positions *a*, *b* and *c* are the sites for attack by an electrophile. However, sites *b* and *c* are not preferred by bulky electrophile due to steric crowding. Thus, more bulky electrophile (like I_2) can attack only site *a*, which is least sterically hindered, a bit smaller electrophile (Br_2) can attack at sites *a* and also *b* (relatively less sterically hindered site) and the smallest electrophile (Cl_2) can attack all the three sites, *viz*., *a*, *b* and *c* (most sterically hindered site).

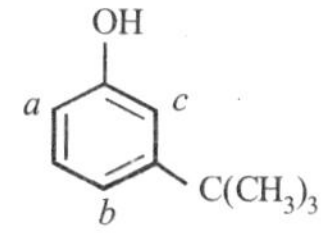

51. (b) Phenolic C—OH group is activating and *o-p-* directing group.

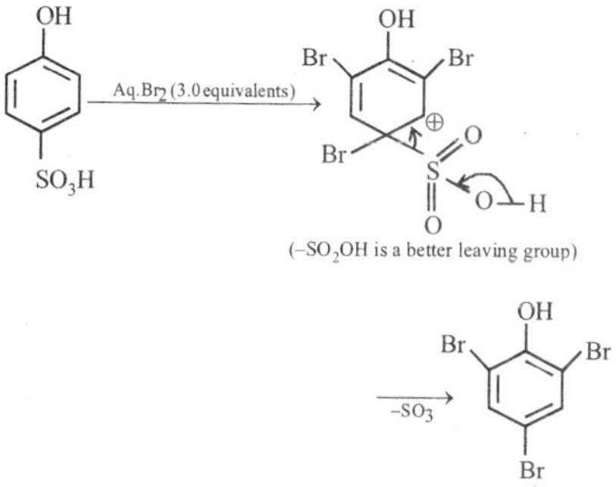

52. (a, c)

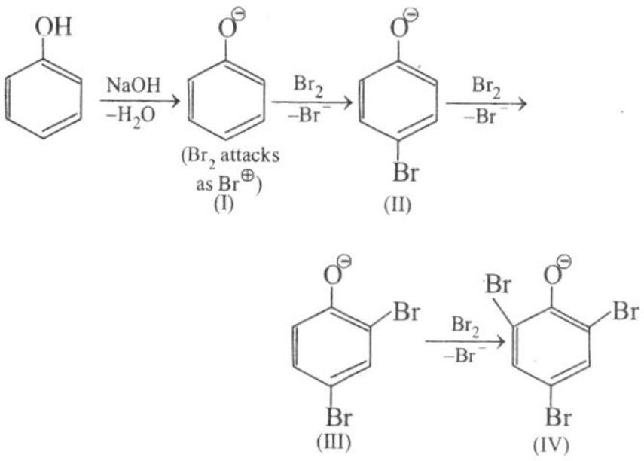

Product of reaction of phenol with NaOH/Br_2 is sodium salt of 2,4,6-tribromophenol. Hence, species (I), (II), (III) are formed as intermediate.

53. (b) The mechanism of this reaction is represented as follows.

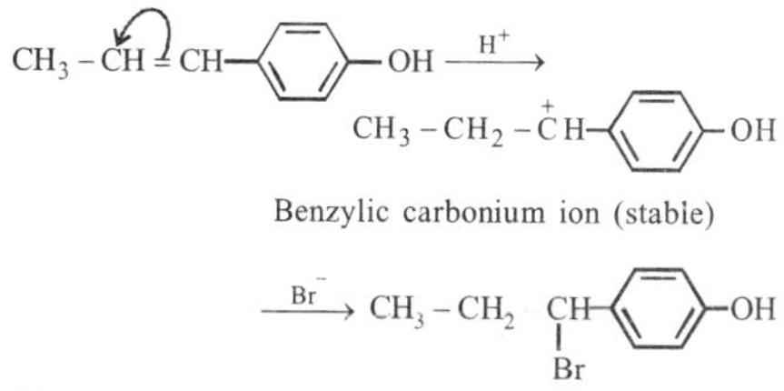

54. (a, d) The reaction involves electrophilic substitution on the highly reactive phenoxide ion.

Here, the electrophile is dichlorocarbene which is formed by the action of strong alkali on chloroform.

$$CHCl_3 + OH^- \rightleftharpoons CCl_3^- + H_2O$$
$$:CCl_3^- \rightleftharpoons :CCl_2 + Cl^-$$
Dichlorocarbene

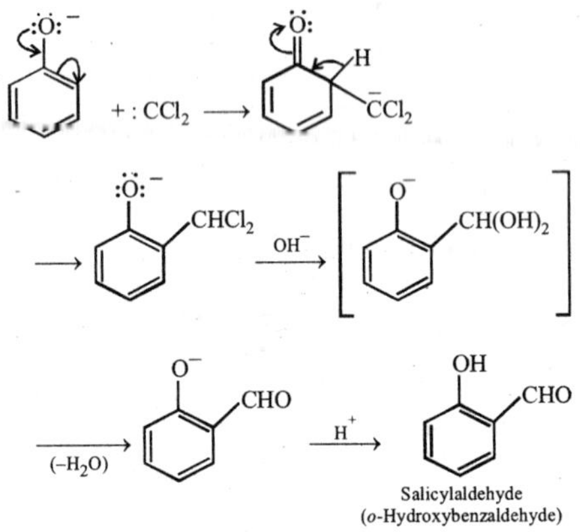

55. (a, c) Higher the stability of the corresponding anion, more will be the acidic character of the parent compound.

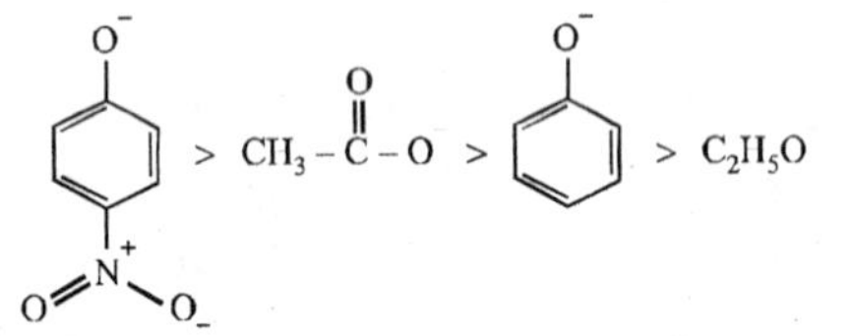

Higher stability of acetate ions than phenoxide ion is due to equivalent resonating structures in the former

56. A. (i) (a) (ii) (c) (iii) (b) (iv) (e) (v) (d)
B. (vi) (h) (vii) (j) (viii) (i) (ix) (g) (x) (f)
C. (xi) (n) (xii) (l) (xiii) (o) (xiv) (m) (xv) (k)
D. (xvi) (t) (xvii) (r) (xviii) (s) (xix) (p) (xx) (q)

57. (i) (f) (ii) (d) (iii) (b) (iv) (c) (v) (e) (vi) (a)

58. (c) Reagents for Reimer - Tiemann reaction are aq. $NaOH + CHCl_3$.

59. (c)

$$OH^- + CHCl_3 \rightleftharpoons :\bar{C}Cl_3 + H_2O$$
$$:\bar{C}Cl_3 \longrightarrow Cl^- + :CCl_2 \text{ (Dichlorocarbene)}$$

60. (b)

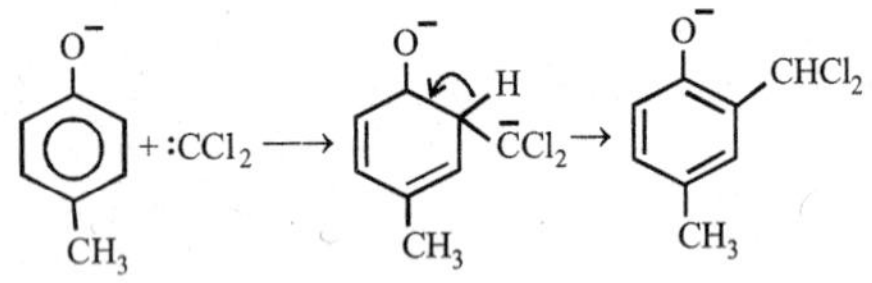

61. (a) Due to +M effect of $-\ddot{O}H$, its intermediate carbocation is more stable than the one in benzene.

62. (i)

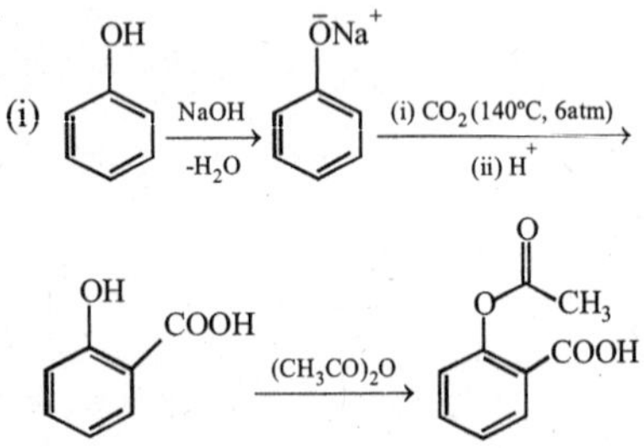

(ii) $C_6H_5OH \xrightarrow{\text{Zn dust distance}} C_6H_6$
Phenol

$$\xrightarrow[\text{(anhy. } AlCl_3)]{CH_3COCl} C_6H_5COCH_3$$
Acetophenone

63. The reaction of D ($C_8H_{10}O$) with alkaline solution of iodine is an iodoform reaction. This reaction is possible if the compound D has $-\underset{O}{\overset{\|}{C}}-CH_3$ or $-\underset{OH}{\overset{|}{C}H}-CH_3$ group.

The high carbon content in D indicates that D is an aromatic compound containing a benzene ring. To account for the given formula, the compound D may be $C_6H_5CH(OH)CH_3$.

The given reactions are

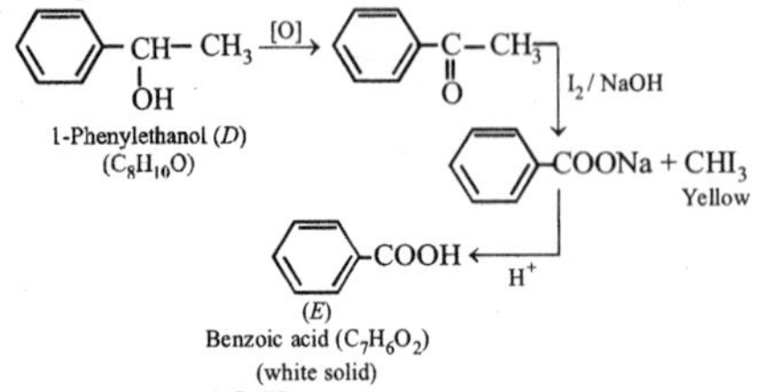

64. (i)

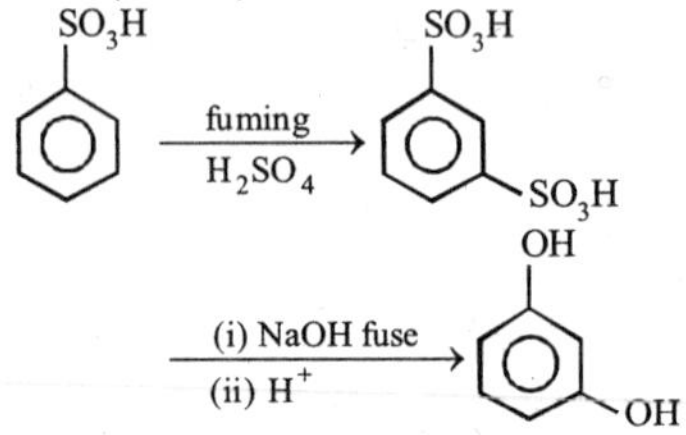

(ii)

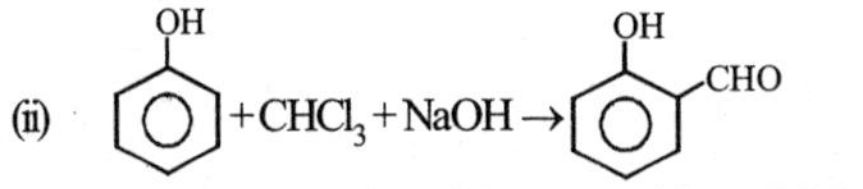

65. Phenol (a weaker acid) reacts with $NaHCO_3$ (a weaker base) to form phenoxide ion (a stronger base) and carbonic acid (a stronger acid).

$$C_6H_5OH + NaHCO_3 \rightleftharpoons C_6H_5ONa + H_2CO_3$$
Weaker acid   Weaker base   Stronger base   Stronger acid

Since, acid-base equilibria lies towards the weaker acid and weaker base, phenol does not decompose $NaHCO_3$ (*difference from carboxylic acids*).

$$RCOOH + NaHCO_3 \rightleftharpoons RCOONa + H_2CO_3$$
Stronger acid   Stronger base   Weaker base   Weaker acid

66. (i) The compound (C_7H_8O) is soluble in aq. NaOH but insoluble in $NaHCO_3$, indicating it to have a phenolic group.

(ii) The compound, on treatment with Br_2 water, gives $C_7H_5OBr_3$. Taking into account of molecular formulae of the two compounds, the parent compound seems to be cresol.

(iii) Bromination of the compound reveals that it is *m*-cresol as it forms tribromo derivative.

(iv) The reactions are

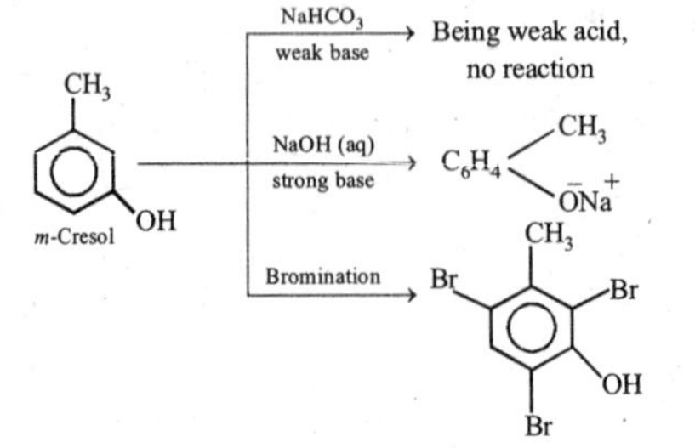

67.

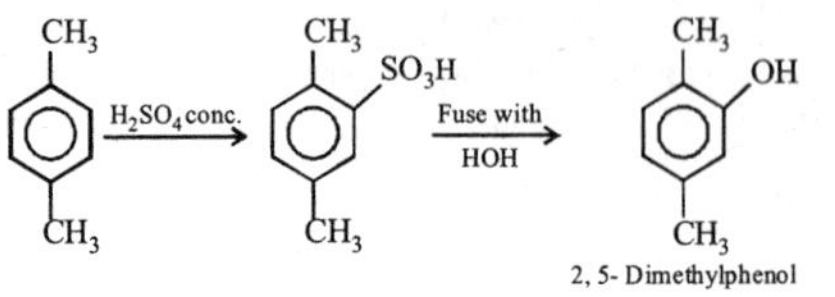

68.

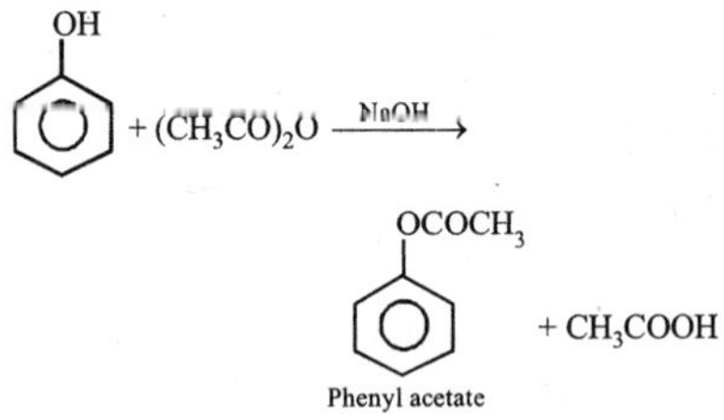

69. (a) $C_6H_5OCH_3 \xrightarrow{HI} C_6H_5OH + CH_3I$

70. (10)

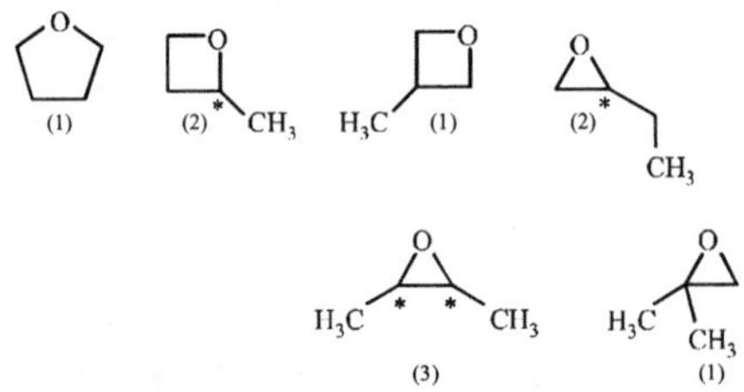

The ether with 2 chiral carbons has a plane of symmetry in *cis*-configuration. Therefore, it will have 3 stereo-isomers.

Total isomers = 1 + 2 + 1 + 2 + 3 + 1 = 10.

71. peroxides. On standing in contact with air, ethers are converted into unstable peroxides ($R_2O \rightarrow O$) *which are highly explosive* even in low concentrations. Hence, *ether is always purified before distillation*. Purification (removal of peroxides) can be done by washing ether with a solution of ferrous salt (which reduces peroxides to alcohols) or by distillation with conc. H_2SO_4 (which oxidises peroxides).

72. (a, d) The given ether is cleaved to give phenol as one of the products because benzyl (a stable carbocations) is formed as an intermediate.

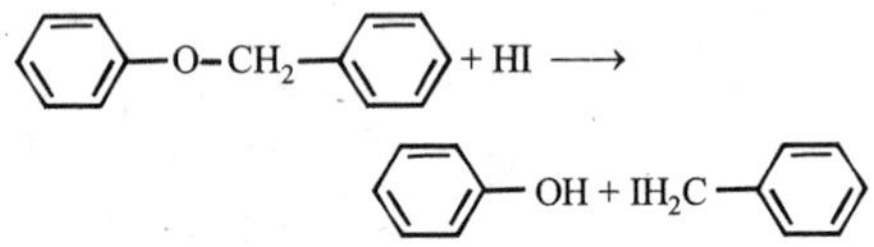

73. (a)

(P)

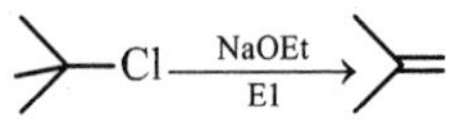

(Q)

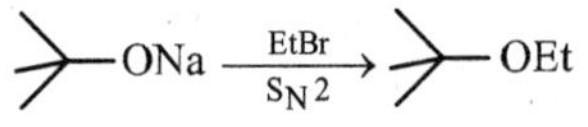

(R)

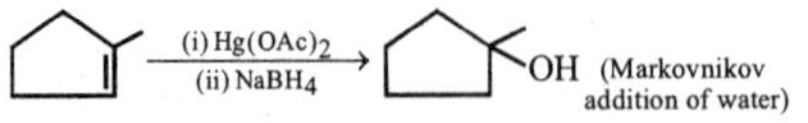

(S)

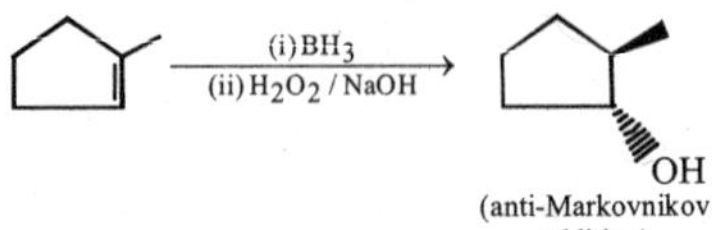

74. (i)-(b); (ii)-(c); (iii)-(d); (iv)-(a)

75. (i) Molecular formula of *P*, $C_5H_{10}O$ indicates 1° of unsaturation. So, it should have double bond.

(ii) Acidic hydrolysis of *P* to *Q* and *R*, both of which responds iodoform test, indicates that *Q* and *R* should have following structure.

CH_3CH_2OH, $(CH_3)_2CHOH$, CH_3CHO or CH_3COR

The only possible linkage that can explain such hydrolysis is ether. Hence *P* should have following type of structure.

C_2 – Component – O – C_3 – Component

Further either the C_2 – or the C_3 – component should have double bond, thus the possible structure for *P* should be either of the following two structures which explains all the given reactions.

$CH_2 = CH - O - CH(CH_3)_2$ (*P*)

↓

$[CH_3CHO \rightleftharpoons CH_2 = CHOH] + HOCH(CH_3)_2$

(*Q* and *R*), Both responds iodoform test

or $CH_3CH_2 - O - C(CH_3){=}CH_2$ (*P*)

↓

$CH_3CH_2OH + HO - C(CH_3) = CH_2 \rightleftharpoons O = C(CH_3) - CH_3]$

(*Q* and *R*), Both responds iodoform test

Extra reactivity of *P* toward dil. H_2SO_4 than ethylene is due to formation of highly stable carbocation.

$H_3C\text{-}\overset{+}{C}H\text{-}\ddot{O}\text{-}CH(CH_3)_2$

2°cabocation

or $CH_3CH_2\ \ddot{O}\ \overset{+}{C}(CH_3)\ CH_3$

3°carbocation

$CH_2 = CH_2 \xrightarrow{H^+} CH_3\overset{+}{C}H_2$

1° carbocation

76.

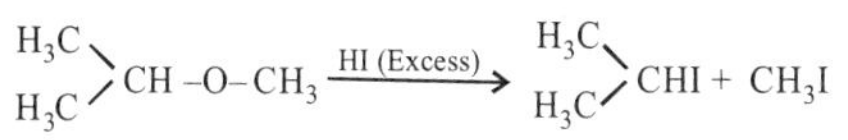

77.

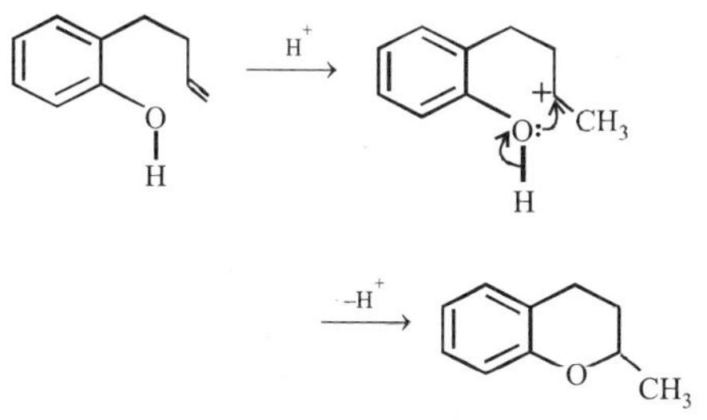

78. The oxirane ring is cleaved via S_N2 mechanism

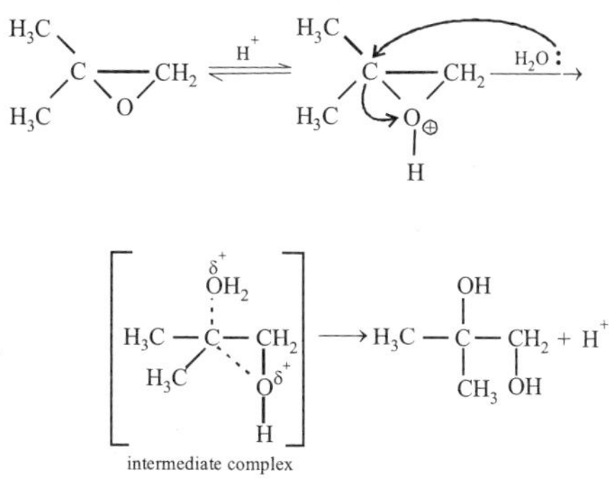

79. The method given in (ii) is the correct method for the formation of ether because 3° alkyl halides in method (i) leads alkene as the main product.

3° alkyl halides are easily dehydrohalogenated by base.

(i) $(CH_3)_3CBr \xrightarrow[E_2]{CH_3ONa} (CH_3)_2C = CH_2$ (Alkene)

(ii) $CH_3Br + NaOC(CH_3)_3 \xrightarrow[S_N2]{} CH_3 - O - C(CH_3)_3$ (Ether)

80. Empirical formula of A and B.

RelativeSimplest

$$\text{No. of atomsratio \% of C} = \frac{12}{44} \times \frac{0.308}{0.108} \times 100$$

$$= 77.77 \quad \frac{77.77}{12} = 6.48 \quad \frac{6.48}{0.92} = 7$$

$$\text{\% of H} = \frac{2}{18} \times \frac{0.072}{0.108} \times 100$$

$$= 7.40 \quad \frac{7.40}{1} = 7.40 \quad \frac{7.40}{0.92} = 8$$

$$\therefore \text{\% of O} = 100 - (77.77 + 7.40) \quad \frac{14.83}{16} = 0.92 \quad \frac{0.92}{0.92}$$

$$= 1 = 14.83$$

∴ Empirical formula of A and B = C_7H_8O

Nature of (A) : Since A is insoluble in NaOH and $NaHCO_3$, it can't have –OH and –COOH groups. Further the reaction of A with conc. HI to give compounds C and D separable by means of ammonical $AgNO_3$ and solubility of D in NaOH indicates that C and D are alkyl halide and phenol respectively. Thus A is an ether i.e. it is $C_6H_5.O.CH_3$ which explains all the given reactions.

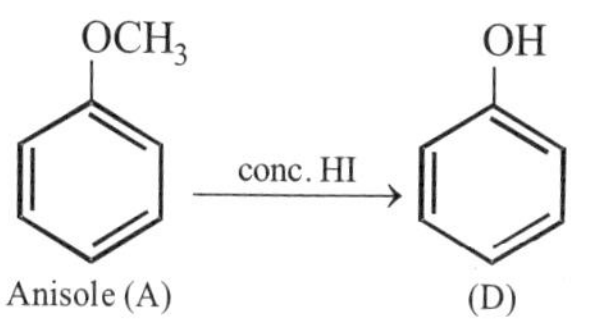

$+ CH_3I$ (C) $\xrightarrow[AgNO_3]{C_2H_5OH} AgI$

Nature of (B): Solubility of B (C_7H_8O) in NaOH indicates that it is a phenol which is further confirmed by its reaction with bromine water to give compound E of molecular formula, $C_7H_5OBr_3$. Further bromination of B to give tribromo product indicates that it is *m*-cresol.

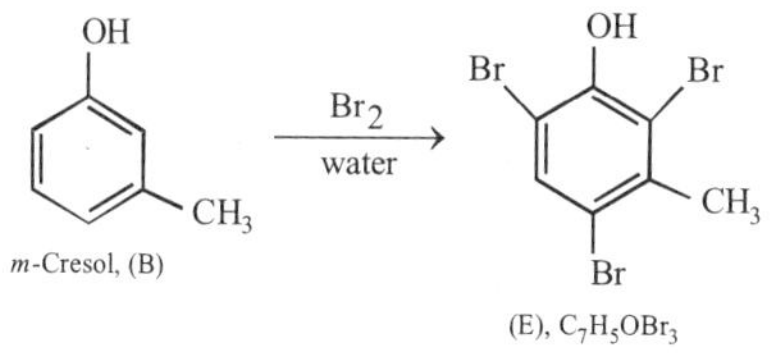

81. Ethanol (due to the presence of active hydrogen atom, $C_2H_5 - O - H$) reacts with sodium metal, while ether has does no such hydrogen atom and hence, does not react with sodium and thus, can be dried by metallic sodium.

82. The unreactivity of the compound (X) towards sodium indicates that it is neither an acid nor an alcohol, further its unreactivity towards Schiff's base indicates that it is not an aldehyde. The reaction of compound (X) with excess of HI to form only one product indicates that it should be a symmetric ether. Hence, its other reactions are sketched as below.

$R - O - R \xrightarrow[\text{excess of HI}]{\text{Reflux with}} 2RI \xrightarrow{\text{hydrolysis}} 2ROH$

(X) (Y) (Z)

$Z \xrightarrow{P+I_2} Y\ ROH\ (Z) \xrightarrow[(O)]{KMnO_4} -COOH$

eq wt 60

Since, the carboxylic acid has equivalent weight of 60, it must be acetic acid (CH_3COOH). Hence, Z must be ethyl alcohol, (Y) ethyl iodide and (X) diethyl ether.

$C_2H_5 - O - C_2H_5 + 2HI \xrightarrow{\text{reflux}} 2C_2H_5I$

Diethyl ether (X) Ethyl iodide (Y)

$\xrightarrow{OH^-} 2C_2H_5OH \xrightarrow{KMnO_4} CH_3COOH$

Ethyl alcohol (Z) Acetic acid

(Eq. wt. = 60)

Chapter-12

Aldehydes, Ketones and Carboxylic Acids

Part-A (JEE Main)

1. **(1)**

(A) $RCN \xrightarrow[\text{(ii) } H_3O^+]{\text{(i) } SnCl_2.HCl} RCHO$

Stephen reaction

(B) 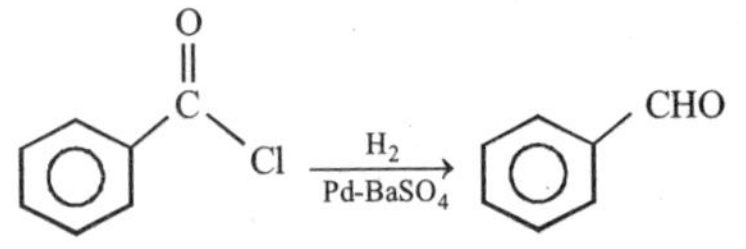

Rosenmund reduction

(C) 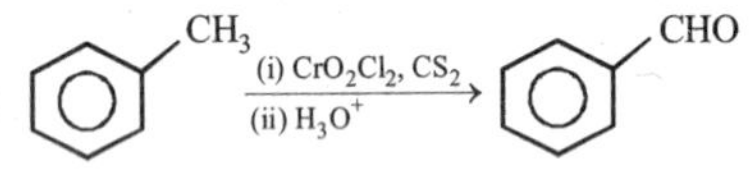

Etard reaction

(D) 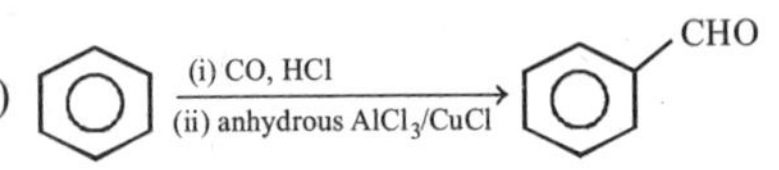

Gatterman –Koch reaction

2. **(4)**

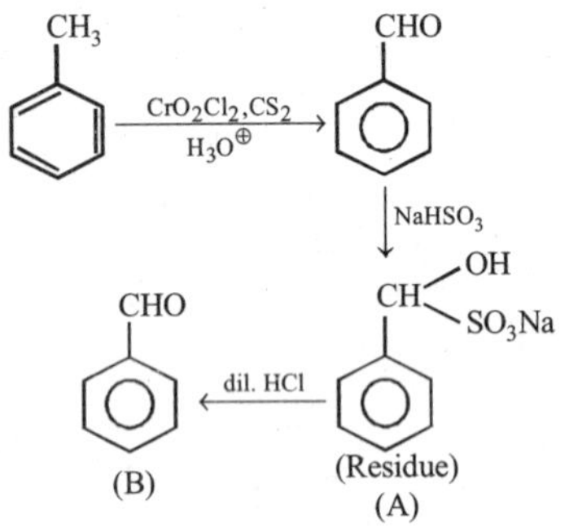

3. **(2)**

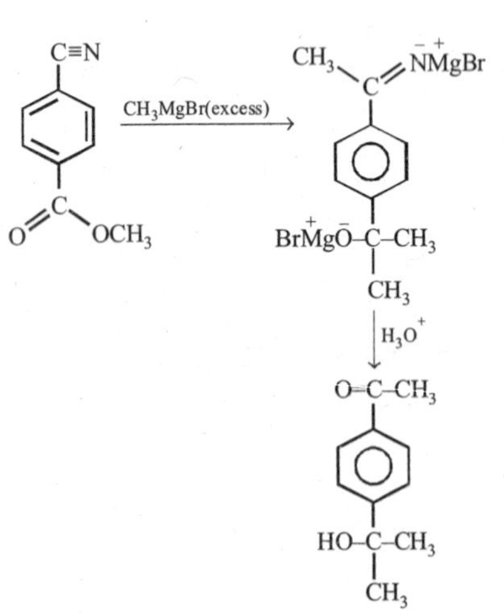

4. **(3)** This is Gattermann-Koch reaction.

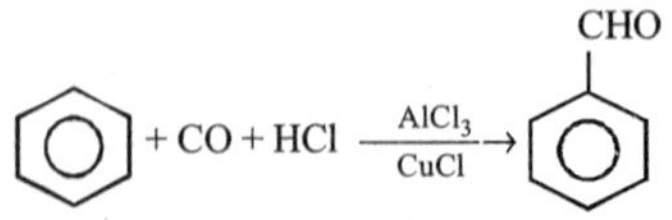

5. **(2)**

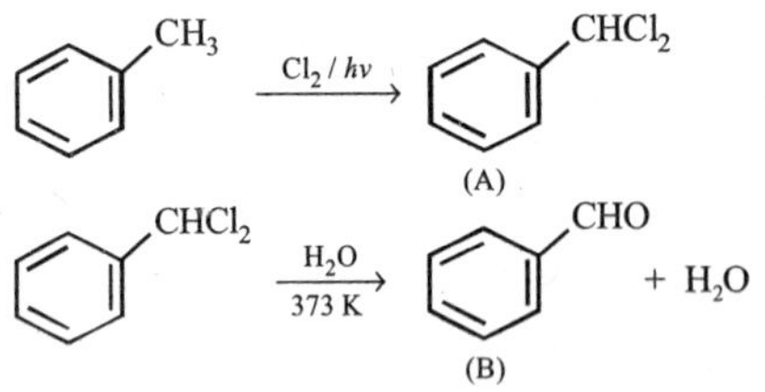

6. **(3)** Gatterman - Koch reaction.

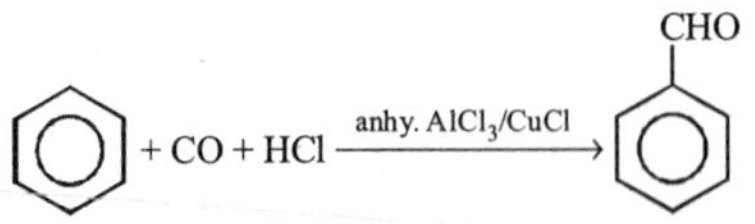

7. **(1)** This reaction is known as Rosenmund reduction.

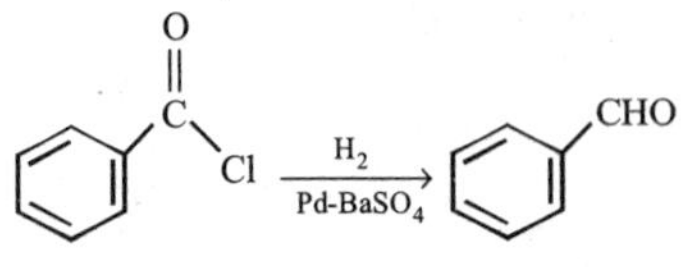

8. **(2)**

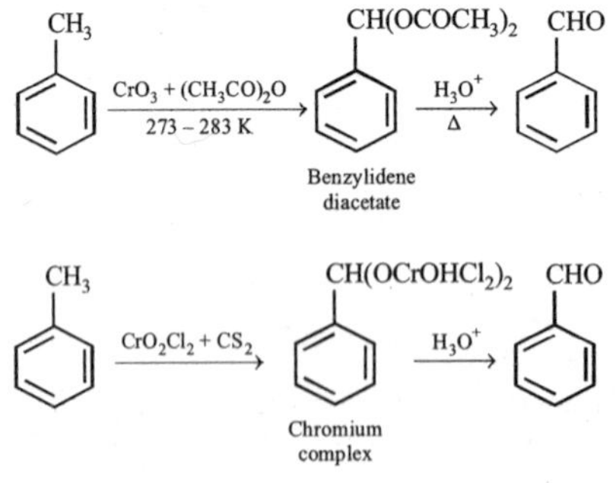

9. **(1)**

10. **(1)**

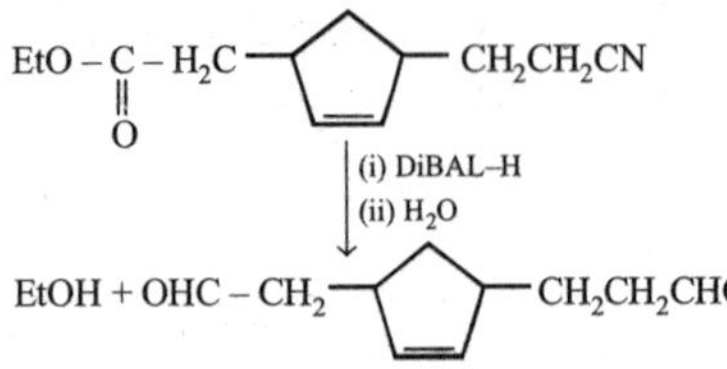

Diisobutylaluminium hydride [DiBALH or $(iBu_2)_2AlH$] reduces both cyanides as well as ester group without affecting double bond.

11. **(4)**

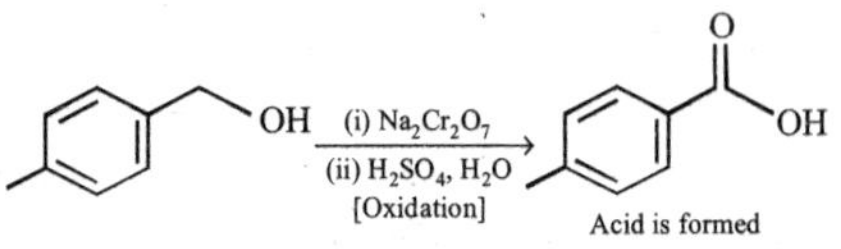

12. **(3)**

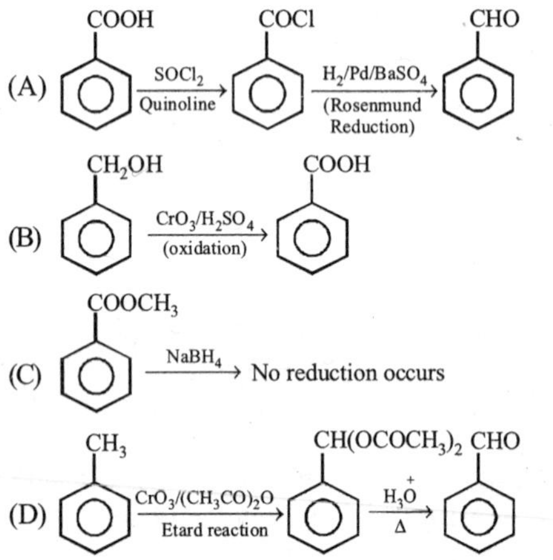

$NaBH_4$ is a mild reducing agent and lack, the reactivity needed to break the C = O bond in esters.

13. **(3)** BH_3, H_2O_2/OH^- followed by PCC oxidation.

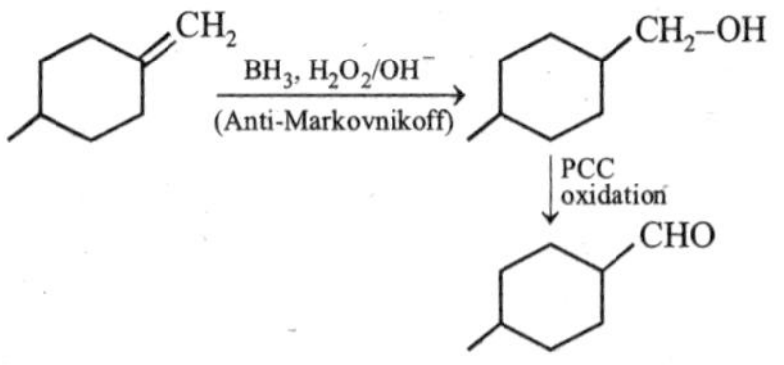

14. **(3)**

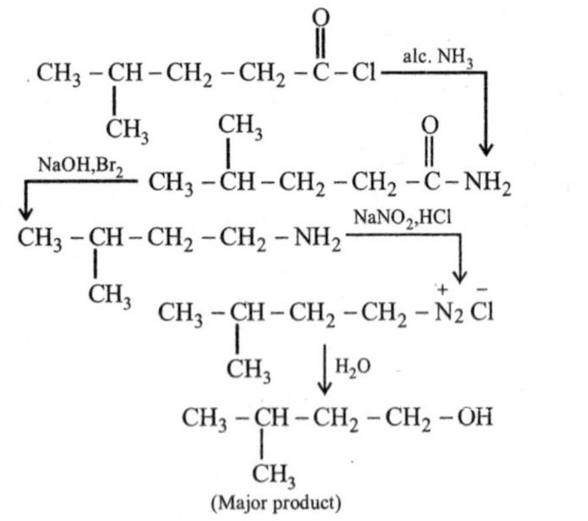

15. (1)

$$CH_3CH_2OH \xrightarrow[\text{Jones reagent}]{CrO_3-H_2SO_4} CH_3COOH$$

16. (3)

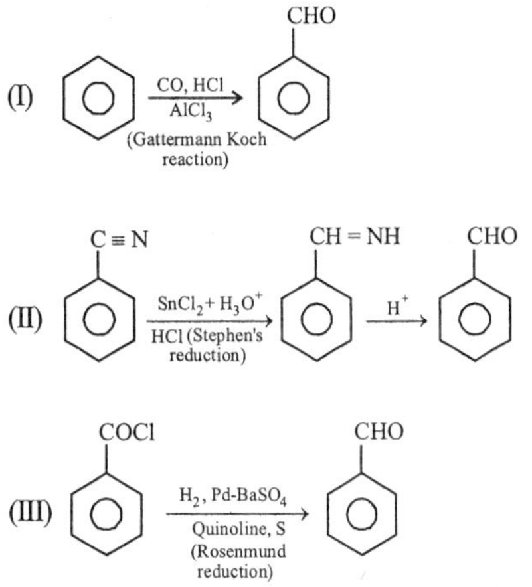

17. (4)

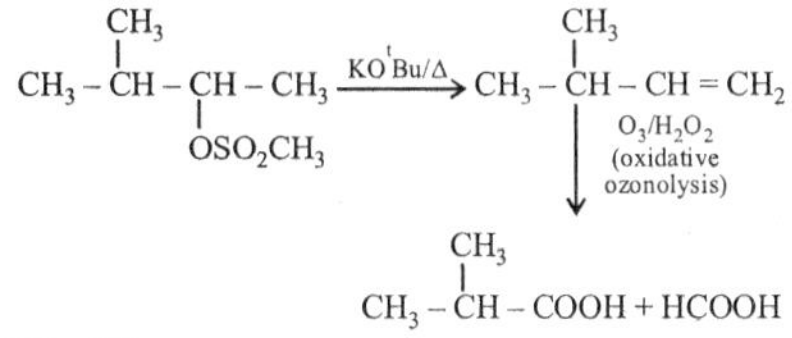

18. (1)

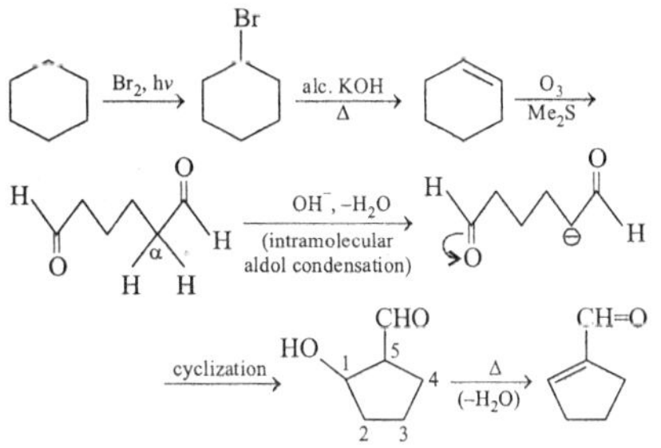

19. (1)

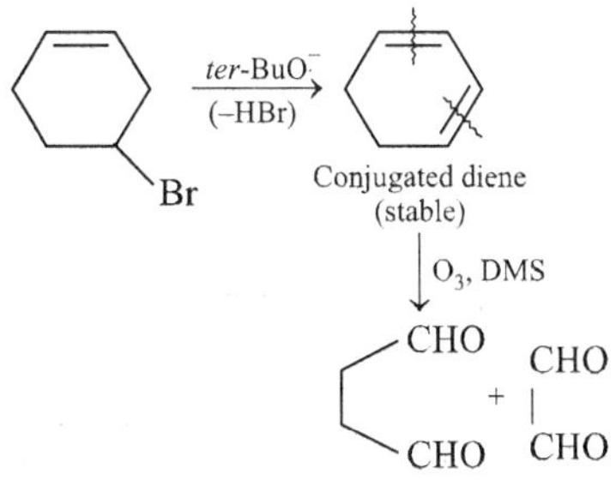

20. (2)

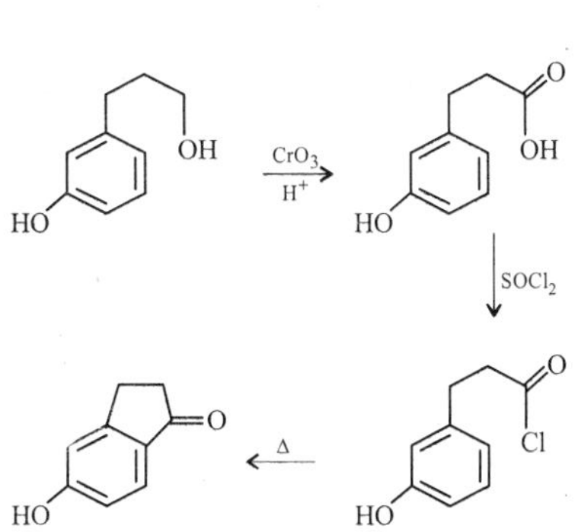

21. (3)

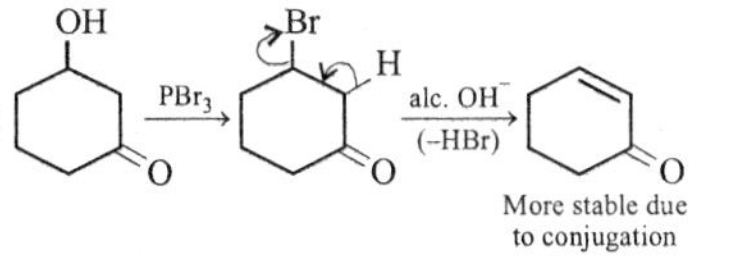

22. (3) $CH_3-\overset{O}{\overset{\|}{C}}-CH_3$ (Acetone) $\quad CH_3-CH_2-CHO$ (Propanal)

Generally, aldehydes are more reactive than ketones in nucleophilic addition reactions.

$$\underset{\text{(Propanal)}}{CH_3-CH_2-CHO} > \underset{\text{(Acetone)}}{CH_3-\overset{O}{\overset{\|}{C}}-CH_3}$$

∴ Rate of reaction with alcohol to form acetal and ketal is

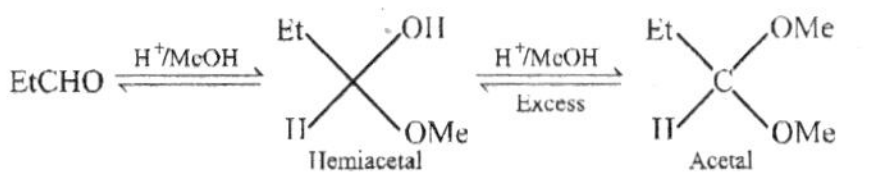

23. (2) Best combination is HCHO (more reactive aldehyde) and MeOH (less sterically hindered alcohol).

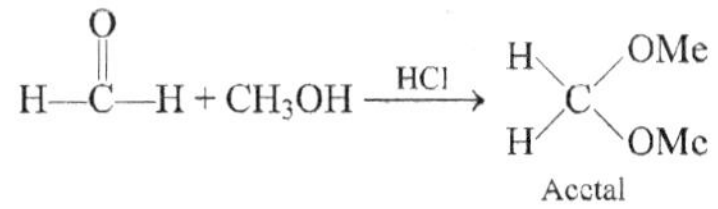

24. (3)

$$R-C\equiv N \xrightarrow[\text{(ii) } H_2O]{\text{(i) } AlH(i\text{-}Bu)_2} R-CHO$$

The reduction of nitriles to aldehydes can be done using DiBAL-H[AlH(i-Bu)$_2$].

25. (1) For the given reaction condition, the major product is:

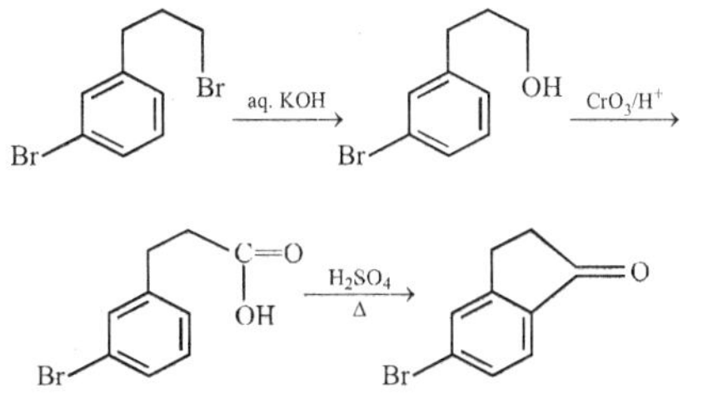

26. (3) Reaction involved:

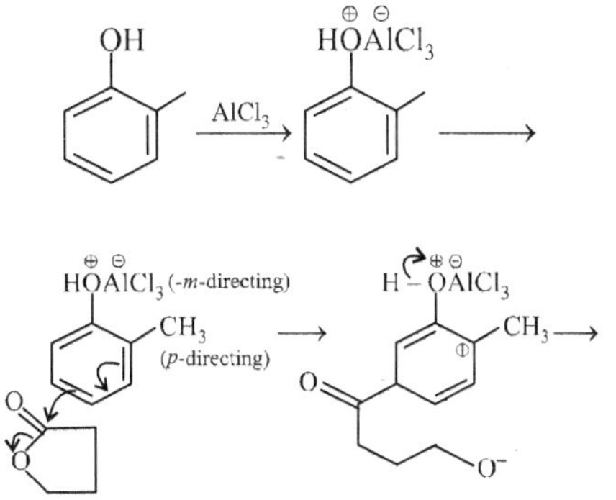

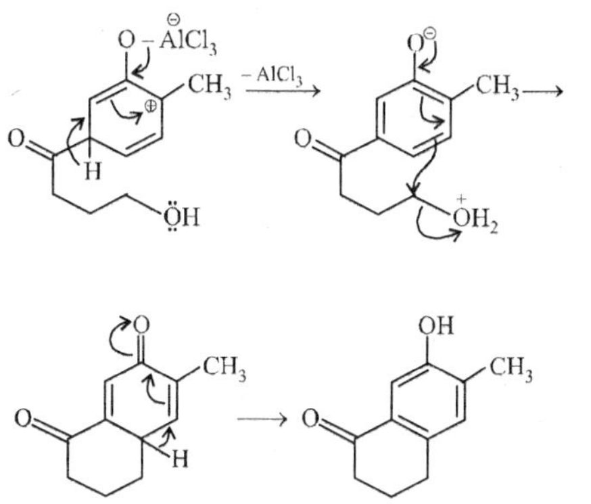

27. (3) The total number of optically active compounds formed is four. The product has two chiral C atoms. Thus, it has $2^n = 2^2 = 4$ stereoisomers.

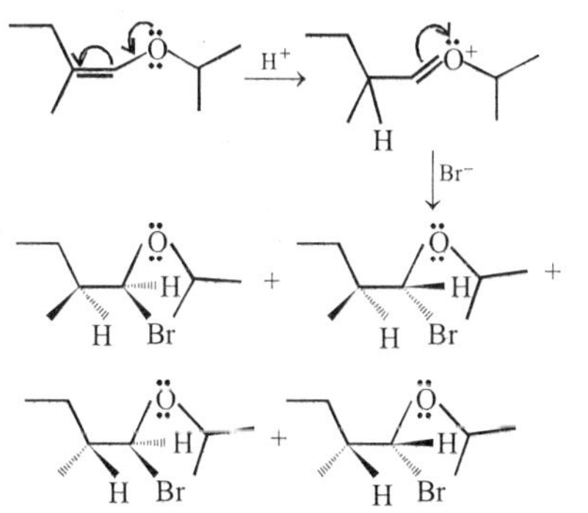

28. (4)

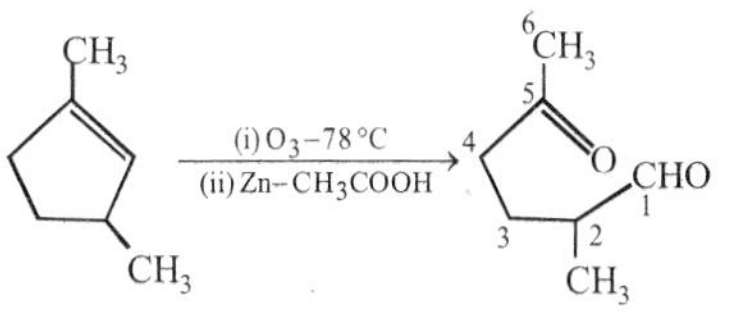

5-Keto-2-methylhexanal

29. (3) $CH_3CHCl_2 \xrightarrow{\text{aq. KOH}} CH_3CH(OH)_2 \xrightarrow{-H_2O} CH_3CHO$

30. (9) Aldol condensation reaction

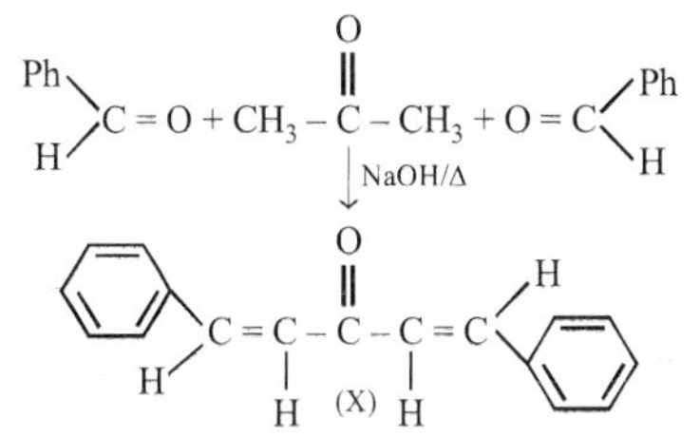

31. (3) $NaBH_4$ is a selective reducing agent which reduces aldehyde to alcohol.

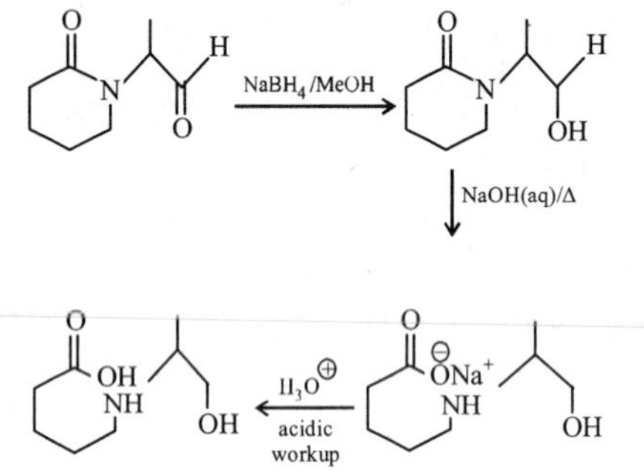

32. (2)

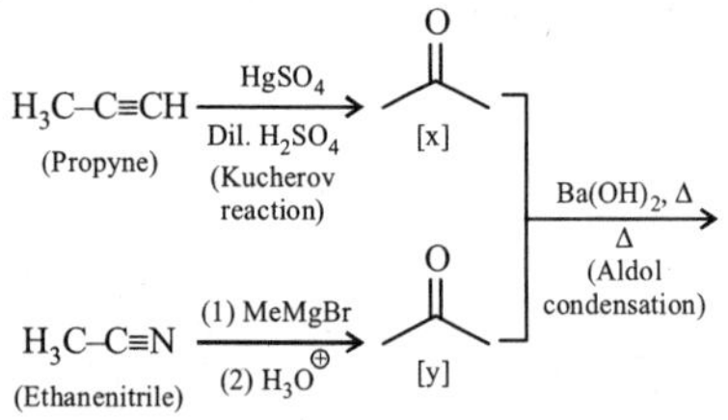

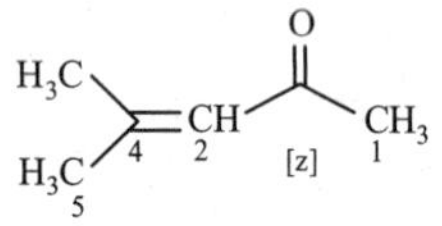

4-Methylpent-3-en-2-one

33. (2)

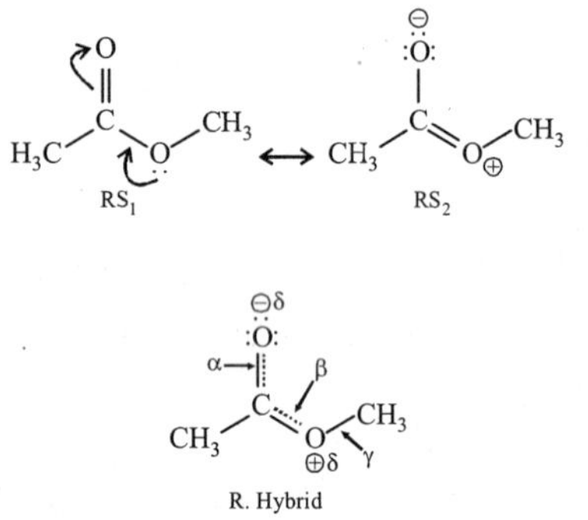

Positive charge on electronegative atom is less stable.

Thus, RS_1 is more stable than RS_2.

$\therefore$ Contribution of RS_1 is more than RS_2 in it's R.H.

$\therefore$ Order of bond length $\alpha < \beta < \gamma$.

34. (3) Statement-I :

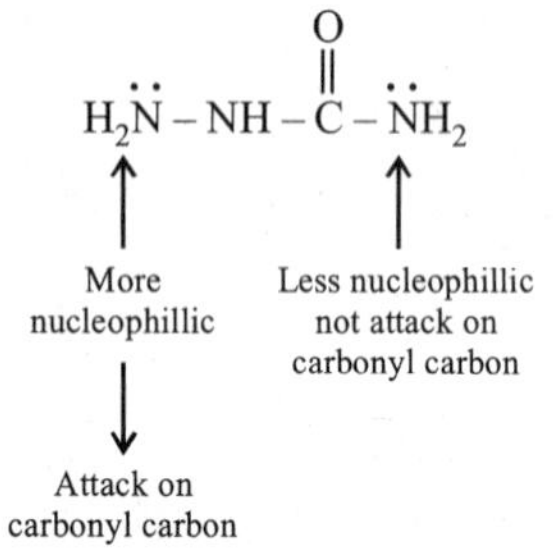

Hence Statement-I is false.

Statement-II

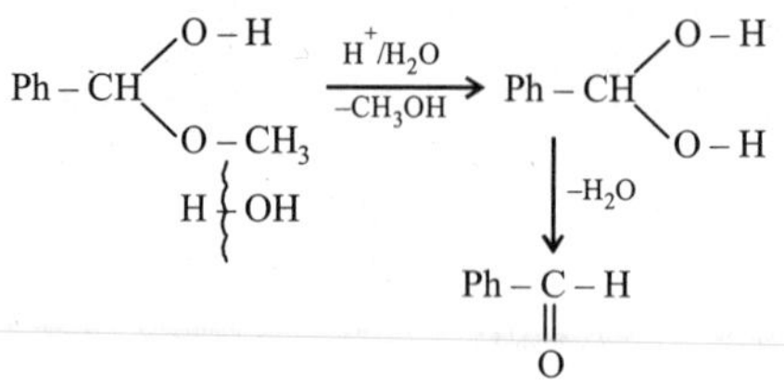

Statement-II is true.

35. (4)

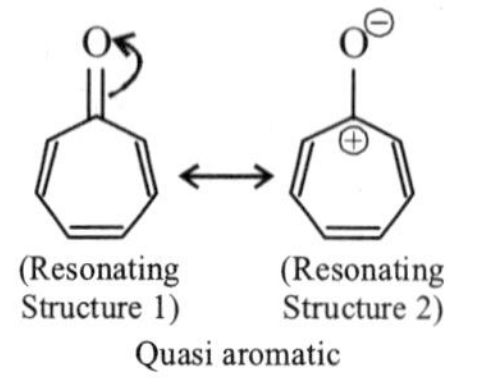

C-O bond length is maximum in compound (x) because RS2 of this compound has aromatic character which gives the C–O bond single bond character.

In (z), negative charge delocalises on both oxygen making it partial double bond, while in (y) C = O is purely double bond.

So, bond length : $x > z > y$.

36. (4)

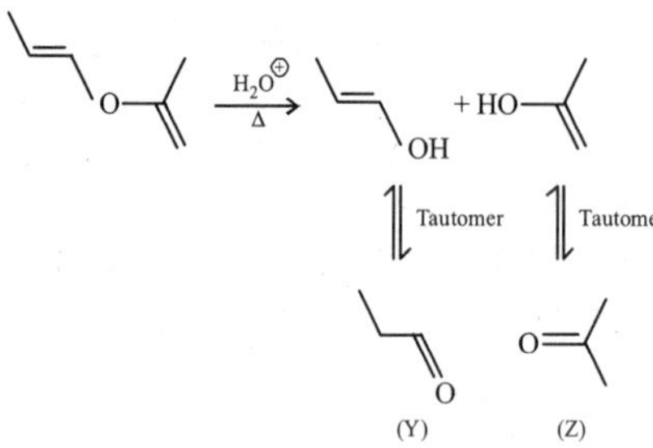

Molar mass of [Z] $(C_3H_6O) = 58$

Molar mass of [Y] $(C_3H_6O) = 58$.

Statement (D): 2,4-DNP is a test for aldehydes and ketones.

Statement (C): Aldehydes being more reactive, reacts faster than ketone.

37. (1)

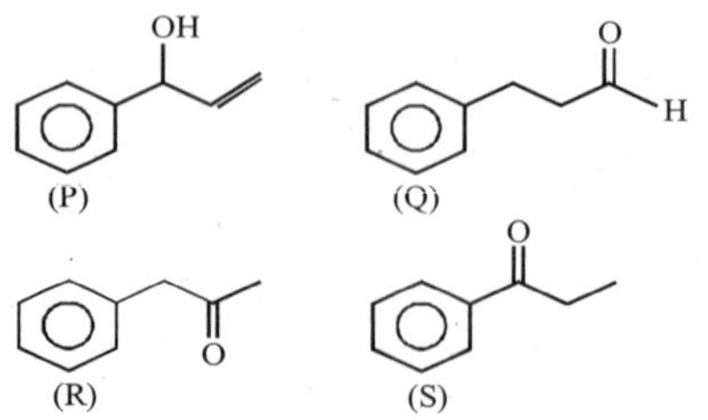

(A) Q, R, S all three give yellow crystalline product with 2, 4-DNP test as they have aldehyde/ketone group

(B) Only P gives Baeyer's test.

(C) Q & R give sooty flame

(D) Iodoform test positive only for $-COCH_3$

- R gives iodoform test
- S does not have $COCH_3$ group, it will not give iodoform test

(E) Q gives Tollen's reagent test

38. (1) Statement I : False

Cross aldol can give 2 or 4 products. It can give only two-products if one compound lacks $\alpha-H$

Statement II : False

Benzaldehyde and acetophenone both react with semicarbazide.

39. (4)

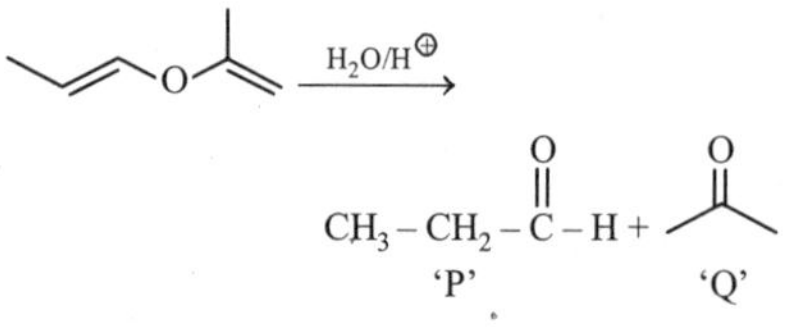

'P' (an aldehyde) and 'Q' (a ketone) can be differentiated by Fehling's test.

P gives positive Fehling test as it is an aldehyde

Q gives negative Fehling test as it is a ketone.

40. (4)

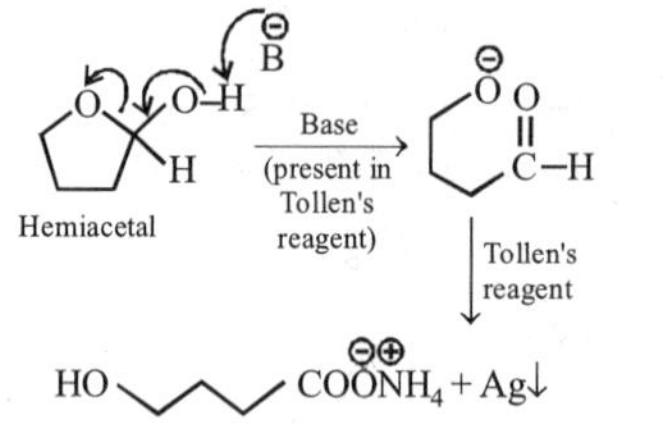

41. (3)

(A) NaOCl + KI $\rightarrow$ NaCl + KOI (Correct Statement)

(B) Incorrect statement

KOI is not a reducing agent.

(C) $CH_3-CH=CH-\overset{\overset{O}{||}}{C}-CH_3$ gives iodoform reaction.

(D) Isopropyl alcohol shows positive iodoform test.

(E) Methanoic acid will not give positive iodoform test.

So only (A) and (C) are correct.

42. (1) Compounds having

$$\overset{O}{\overset{||}{-C}}-CH_3,\ -\underset{OH}{\underset{|}{CH}}-CH_3 \text{ groups and } CH_3CH_2OH$$

show positive iodoform test.

$$R-\overset{O}{\overset{||}{C}}-CH_3 \text{ or } R-\underset{OH}{\underset{|}{CH}}-CH_3 \xrightarrow[+I_2]{NaOH} R-CO\overset{\ominus}{O}\overset{\oplus}{N}a + \underset{\text{Yellow ppt}}{CHI_3\downarrow}$$

(A) CH_3-OH and CH_3-CH_2OH

(–ve iodoform test) (shows +ve iodoform test)

(E) 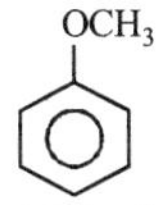 and $CH_3-\overset{O}{\overset{||}{C}}-CH_3$

(–ve iodoform test) (shows + ve idoform test)

Thus (A) and (E) can be differentiated through iodoform test.

43. (4)

$$\underset{\text{Propanenitrile}}{CH_3-CH_2-C\equiv N} \xrightarrow[H_3O^+]{SnCl_2/HCl} \underset{\underset{(X)}{\text{Propanal}}}{CH_3-CH_2-CHO}$$

$$\underset{\text{But-2-ene}}{CH_3-CH=CH-CH_3} \xrightarrow{\text{Ozonolysis}} \underset{\underset{(Y)}{\text{Ethanal}}}{CH_3-CHO}$$

$$(X)+(Y) \xrightarrow[\Delta]{NaOH} 3-\text{Methylbut}-2-\text{enal (d)}$$

(Not formed) (d)

$$CH_3-CH_2-CHO + CH_3-CHO \xrightarrow[\Delta]{NaOH}$$

$$CH_3-CH_2-CH=\underset{CH_3}{\underset{|}{C}}-CHO +$$

2-Methylpent-2-enal (c)

$$\underset{\text{But-2-enal}}{CH_3-CH=CH-CHO} + CH_3-CH=\underset{CH_3}{\underset{|}{C}}-CHO$$

2-Methyl but-2-enal (a)

$$+\ \underset{\text{Pent-2-enal (b)}}{CH_3-CH_2-CH=CH-CHO}$$

44. (3)

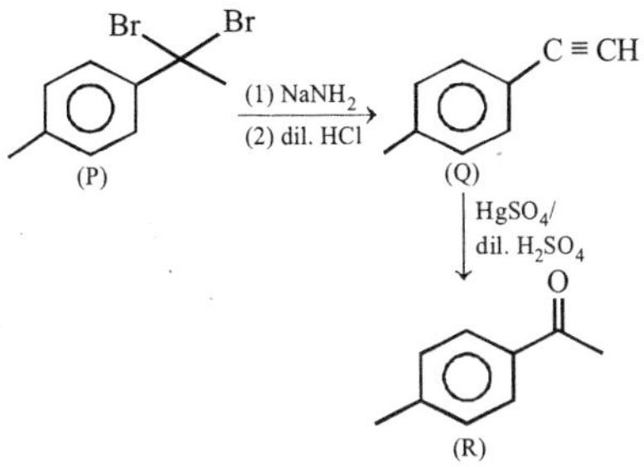

R gives positive iodoform test but negative Tollen's test.

45. (2) Degree of unsaturation in $C_8H_8O_2$ = 5

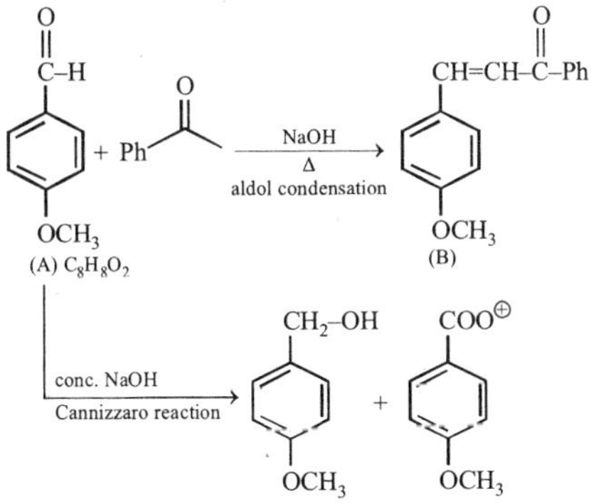

46. (1) A-IV, B-III, C-II, D-I

Wolf-Kishner reduction

$$>C=O \xrightarrow[-H_2O]{NH_2NH_2} >C=NNH_2 \xrightarrow[\text{heat}]{\text{KOH/ethylene glycol}} >CH_2 + N_2$$

Clemmensen reduction

$$>C=O \xrightarrow[HCl]{Zn-Hg} >CH_2 + H_2O$$

Tollen's test

$Ag(NH_3)_2OH$

Fehling's test

Aqueous copper sulphate–Fehling A

Sodium potassium tartarate–Fehling B (Rochelle salt)

47. (4) A-II, B-IV, C-I, D-III

Etard Reaction

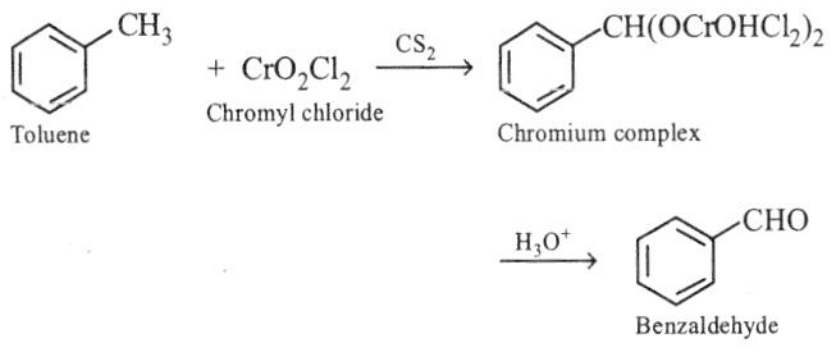

Rosenmund reduction

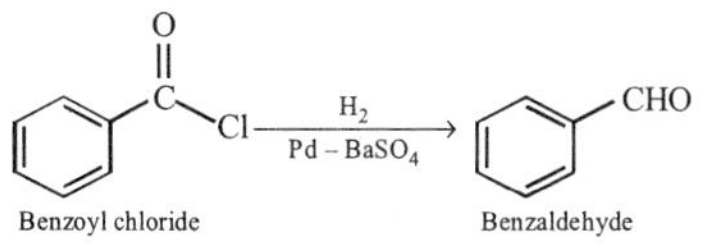

Gatterman–Koch Reaction

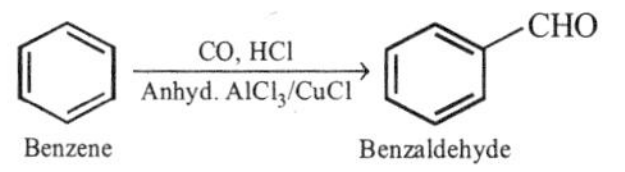

Stephen Reaction

$$RCN + SnCl_2 + HCl \longrightarrow$$

$$\underset{\text{Imine}}{RCH=NH} \xrightarrow{H_3O^+} RCHO$$

48. (2) Compound P gives a positive idoform test (indicating a CH_3CO or $CH_3CH(OH)$ group), but negative Tollen's Test (indicating the absence of aldehyde group)

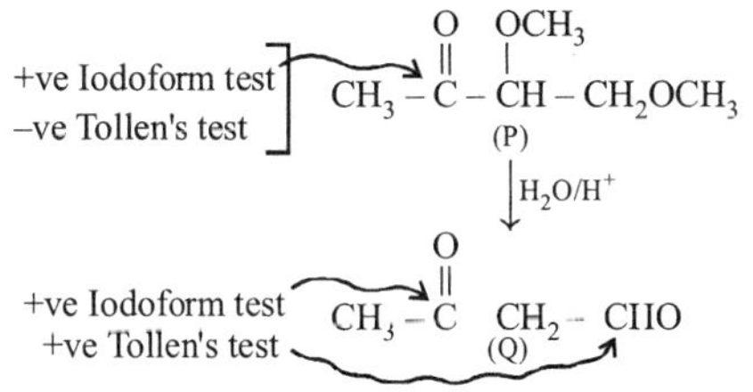

The compound Q gives both positive tests.

49. (1)

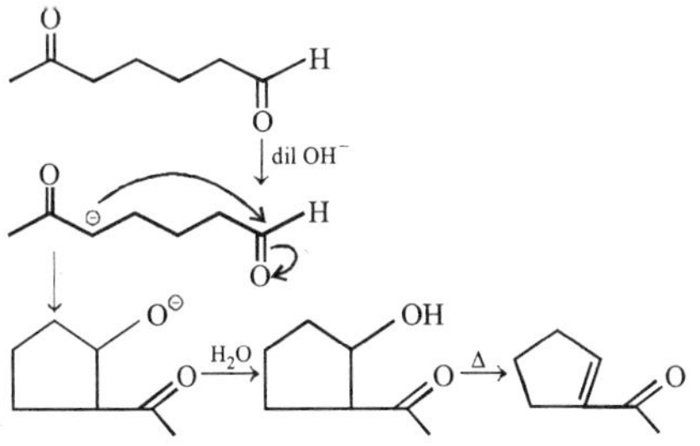

50 . (2)

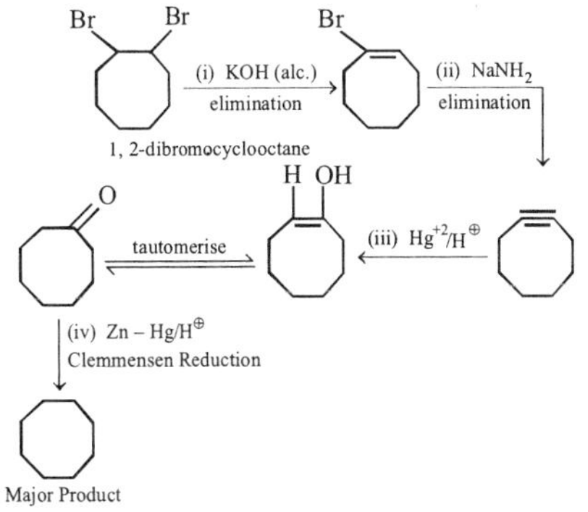

51. (2) P gives negative test with Tollen's reagent so it is not an aldehyde. It gives positive test with 2, 4-DNP.

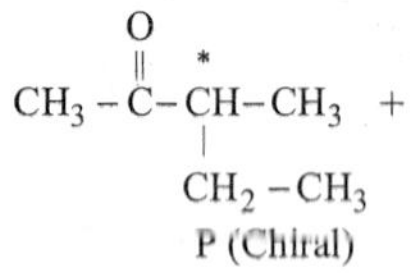

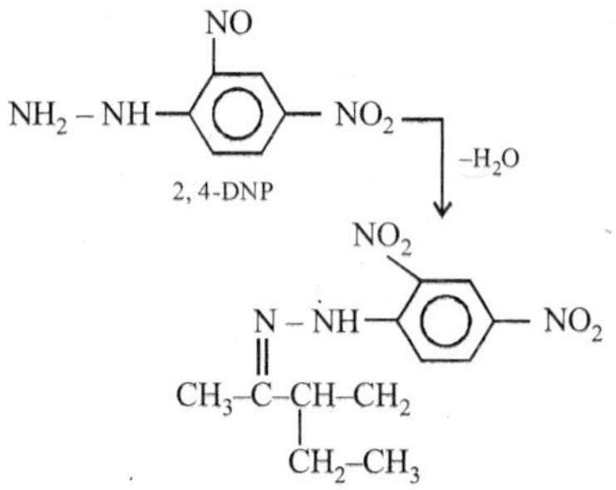

52. (2) Methyl ketones and methyl alcohols upon oxidation with NaOI/NaOH give iodoform.

53. (2)

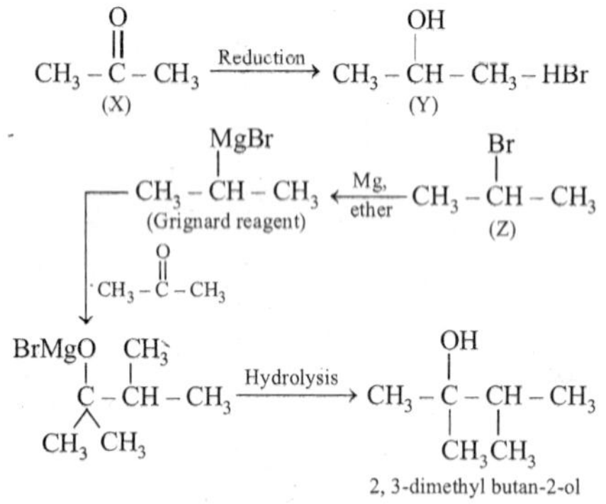

54. (4)

(1) 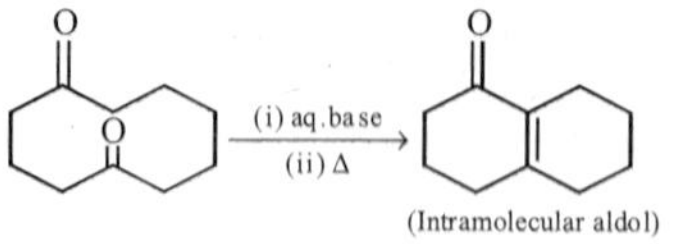

(2) 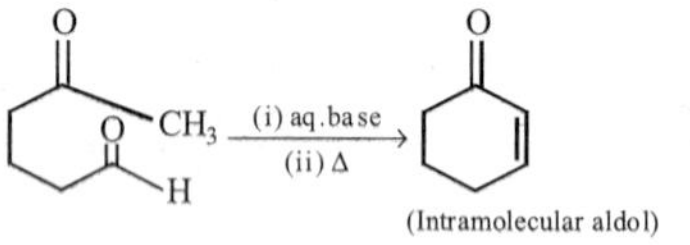

(3) 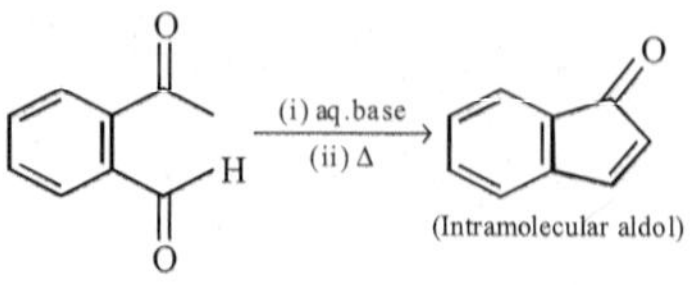

(4) 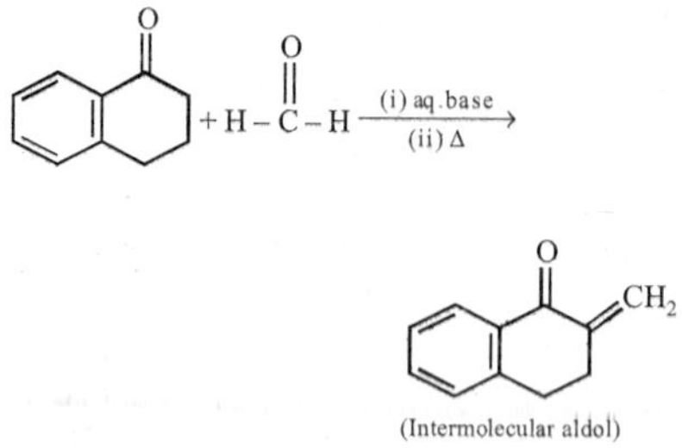

55. (2) Nucleophile OH^- attacks the more reactive aldehydic group.

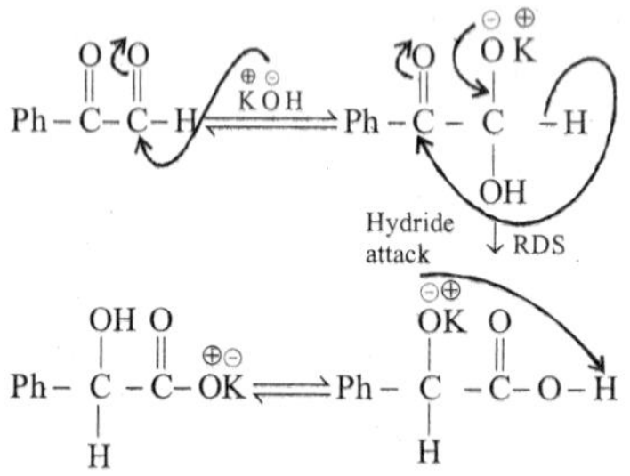

56. (2) Following compounds will not give iodoform reaction/test.

(1) Butanal $CH_3CH_2CH_2CHO$

(2) 3-Pentanone $CH_3CH_2-\overset{O}{\overset{\|}{C}}-CH_2-CH_3$

(3) Pentanal $CH_3CH_2CH_2CH_2CHO$

(4) 3-Pentanol $CH_3-CH_2-\underset{OH}{\underset{|}{CH}}-CH_2-CH_3$

57. (2)

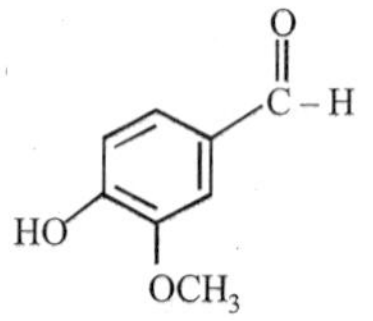

Phenolic group soluble in NaOH

Benzaldehyde derivative reacts with Tollen's reagent.

Vanillin does not give self-aldol reaction due to lack of acidic α-H for condensation.

58. (3)

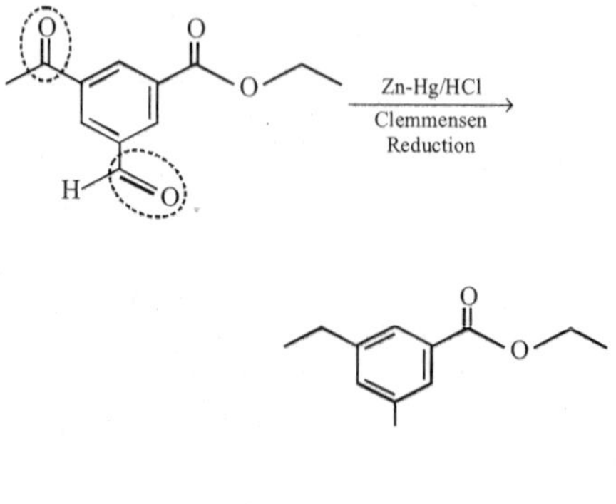

59. (2)

S. No.	Name of Reaction	Acetaldehyde $CH_3-\underset{\|\|}{C}-H$ O	Acetone $CH_3-\underset{\|\|}{C}-CH_3$ O
1	Iodoform reaction	Positive	Positive
2	Cannizaro	Negative	Negative
3	Aldol reaction	Positive	Positive
4	Tollen's test	Positive	Negative
5	Clemmensen reduction	Positive	Positive

Ans. (2) A, C and E only

60. (2)

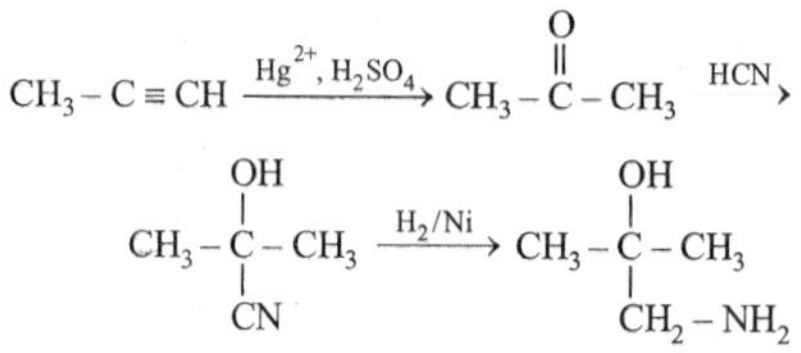

61. (1)

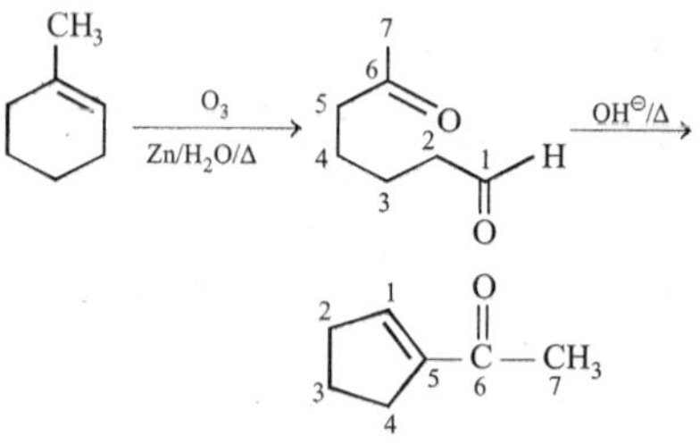

62. (3) The rate of nucleophilic addition is increased by electron withdrawing group so the reactivity towards nucleophilic addition will be

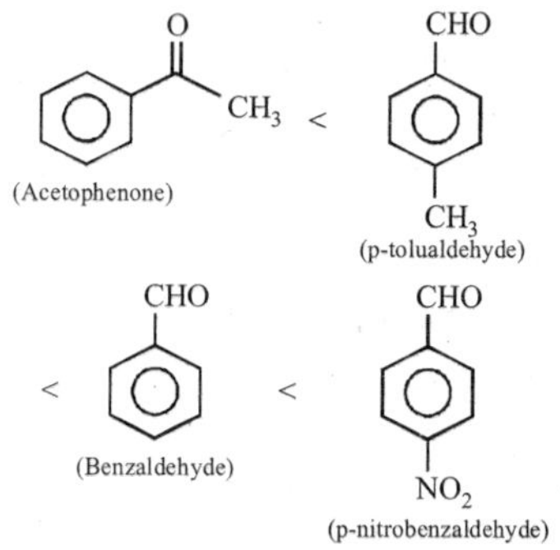

63. (2)

$k = 2280$ is for HCHO

$k = 2000$ is for chloral

k is much higher for both cases because HCHO and chloral are more electrophilic than R–CO–R

64. (3) Propanal undergoes aldol conden-sation with excess of HCHO in presence of alkali, it happens due to presence of two α-hydrogen in propanal which further react with HCHO and undergoes cannizzaro reaction.

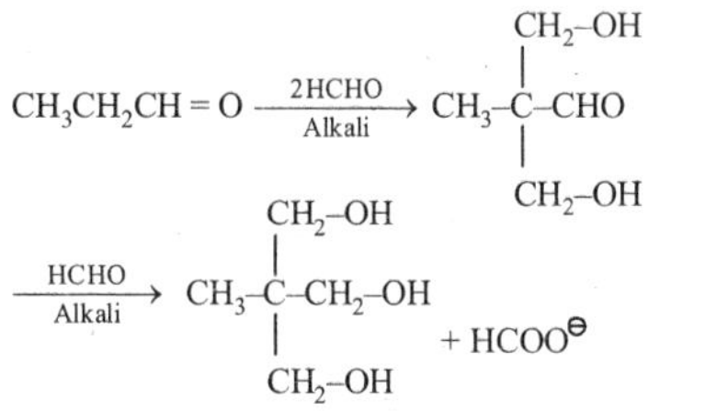

65. (2) $CH_3-CH=CH-\overset{O}{\overset{\|}{C}}H$ (III)
↓
Non polar R.S.
More No. of covalent bonds

$CH_3-\overset{\oplus}{C}H-CH=\overset{O^{\ominus}}{\overset{|}{C}}H$ (II)
↓
Having – ve charge on more electronegative atom

$CH_3-\overset{\ominus}{C}H-CH=\overset{O^{\oplus}}{\overset{|}{C}}H$ (I)
↓
Having – ve charge on less electronegative atom

∴ Stability order III > II > I

66. (2)

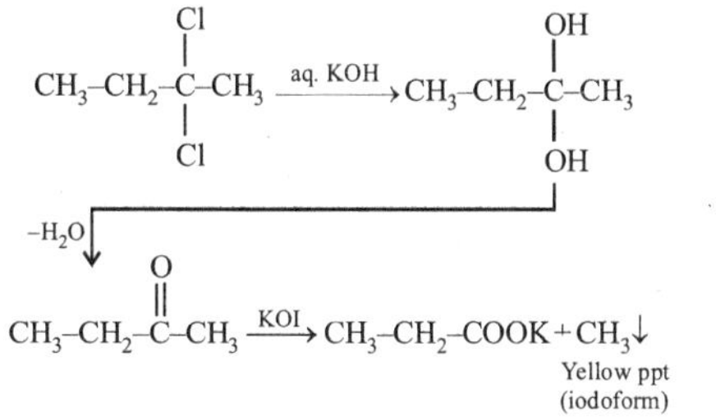

67. (1) $H-\overset{O}{\overset{\|}{C}}-H$ has low steric hindrance and high partial positive charge at carbonyl carbon. Therefore it is most reactive towards nucleophilic addition reactions.

68. (2) Compounds having same number of carbon atoms follow the boiling point order as:

$(\text{Boiling point})_{\text{Hydrogen bonding}} > (\text{Boiling point})_{\text{high polarity}} > (\text{Boiling point})_{\text{low polarity}} > (\text{Boiling point})_{\text{non polar}}$

69. (2) Reaction given is know as Wolff-Kishner reduction.

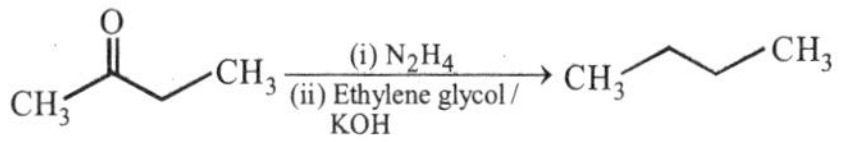

70. (2)

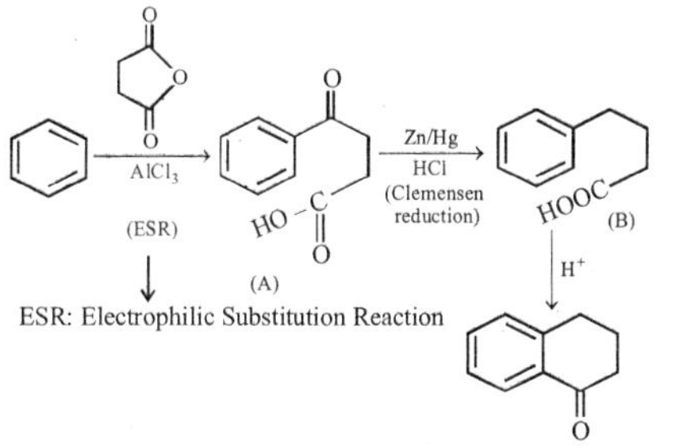

ESR: Electrophilic Substitution Reaction

71. (4) Zn–Hg reduces both aldehyde and ketone.

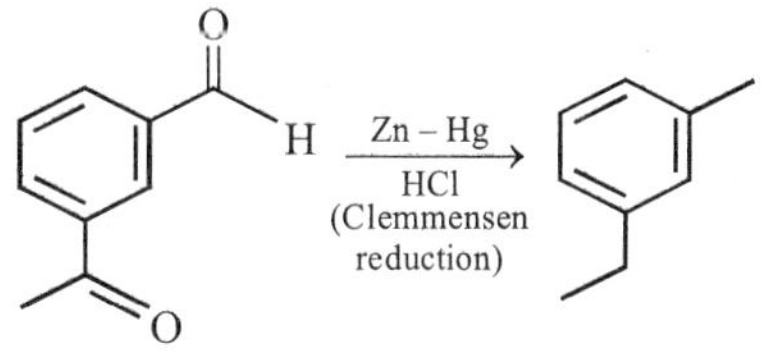

72. (4) Aldehyde and ketones having acidic α-hydrogen show aldol reaction. The acidity of α-H is due to the strong electron withdrawing effect of carbonyl group.

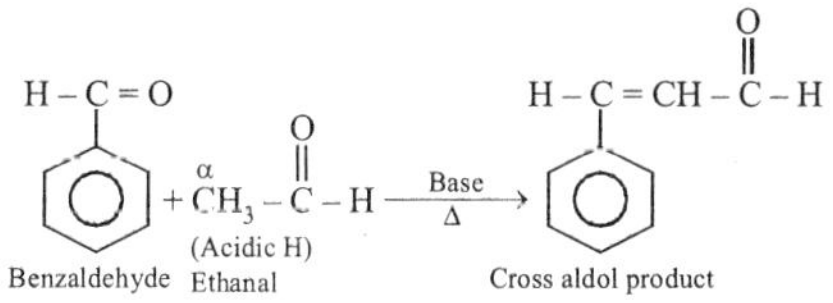

So statement II is incorrect.

73. (2)

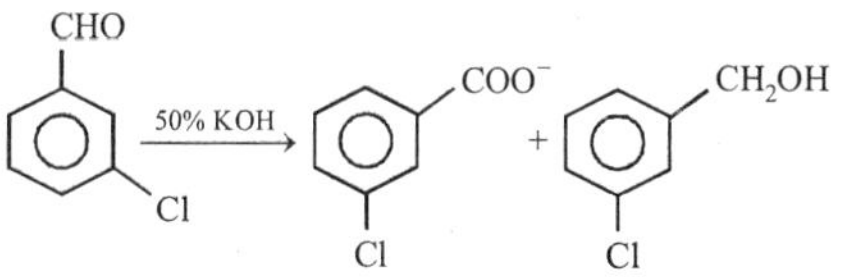

Since there are no α-hydrogens, it will undergo Cannizzaro reaction with 50% KOH and will give meta-substituted benzoic acid and benzyl alcohol.

The –Cl will not be substituted here as C – Cl bond of the ring is strong due to partial double bond character and thus will not undergo nucleophilic substitution.

74. (4)

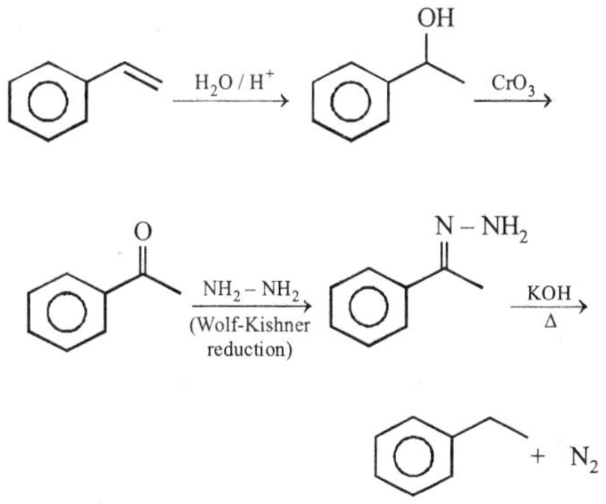

75. (2)

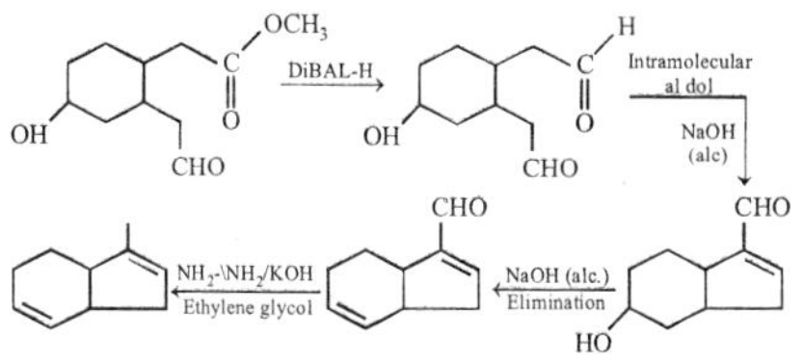

Thus, A = Di-BAL-H, B = NaOH(aq.), C = NH_2–NH_2/KOH, ethyleneglycol.

76. (2)

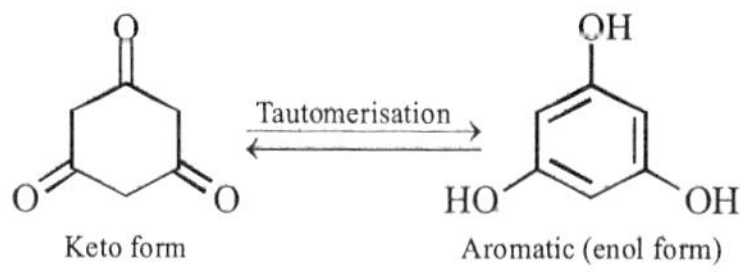

77. (4) The given reaction involves the conversion of a ketone to an acetal by hydrolysis.

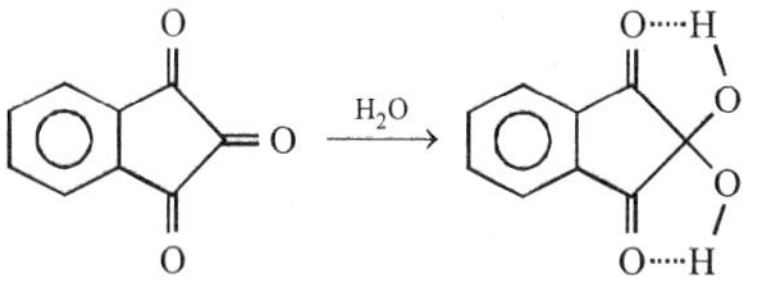

78. (4)

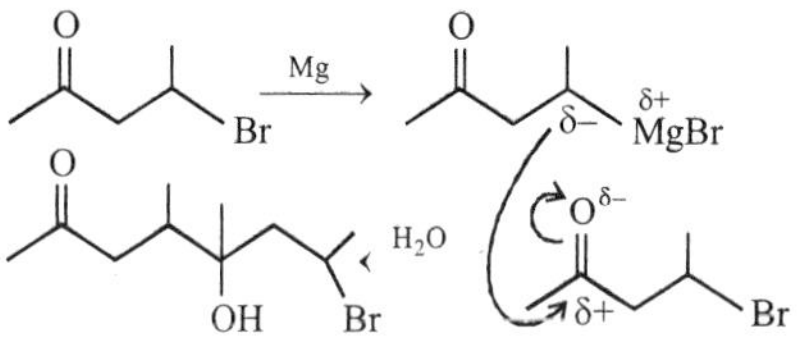

79. (4)

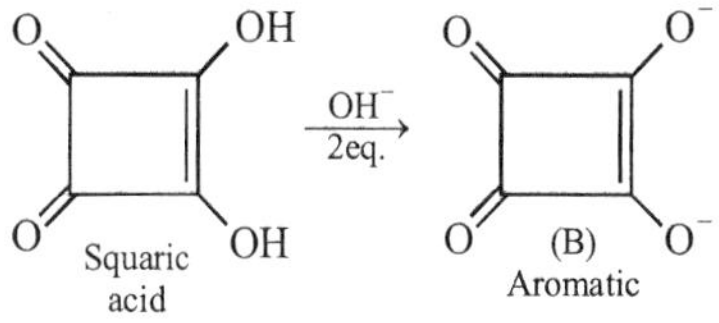

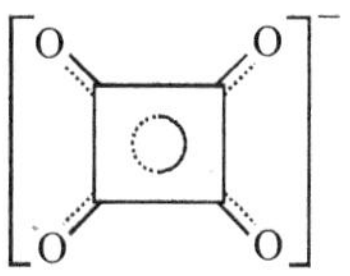

Resonance hybrid of B showing all C-C bond length same

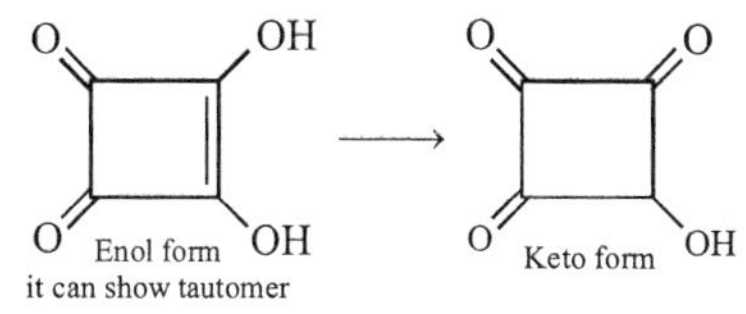

80. (3)

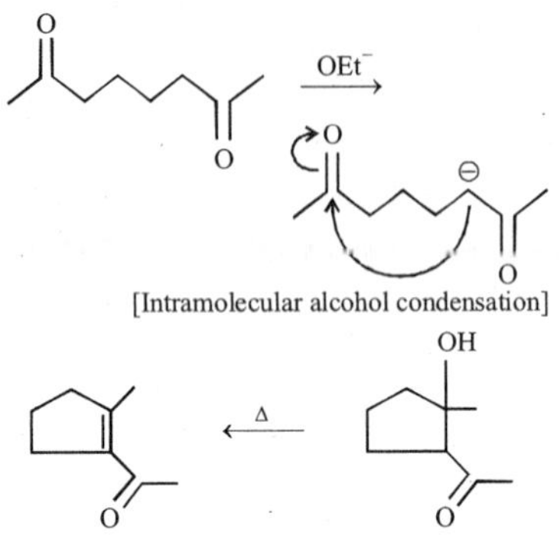

81. (4)

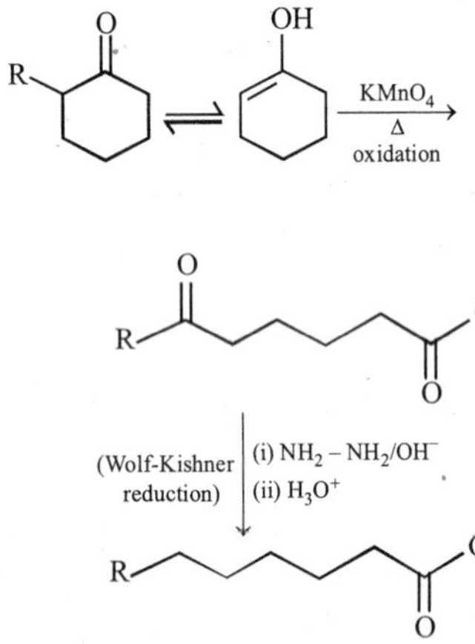

82. (3) Wolff-Kishner reduction is not suitable for base sensitive group.

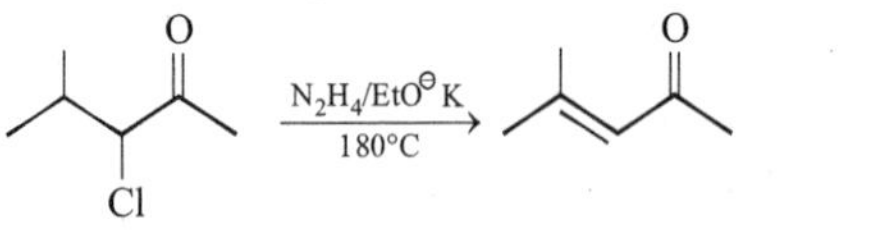

83. (2)

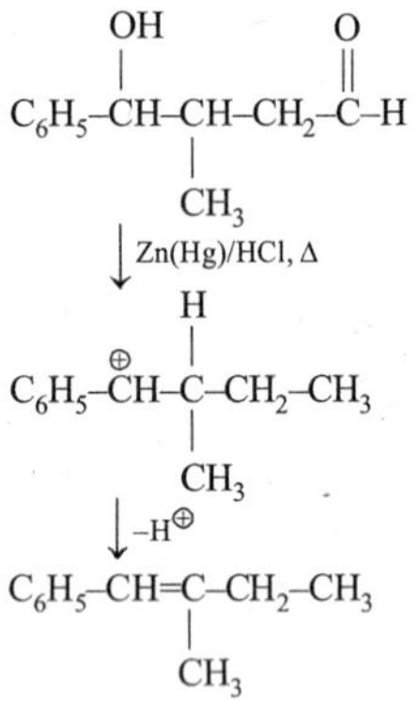

84. (1) $LiAlH_4$ is a strong reducing agent that reduces carbonyl and amide groups but not C = C bond.

85. (1) E > C > D > A > B [E is resonance stabilized with an EWG]

86. (4) HVZ reactions = Br_2 / red P

Iodoform reaction = NaOH + I_2

Etard reaction = (i) CrO_2Cl_2, CS_2 (ii) H_2O

Gattermann–Koch Reaction = CO, HCl, Anhydrous, $AlCl_3$

87. (1)

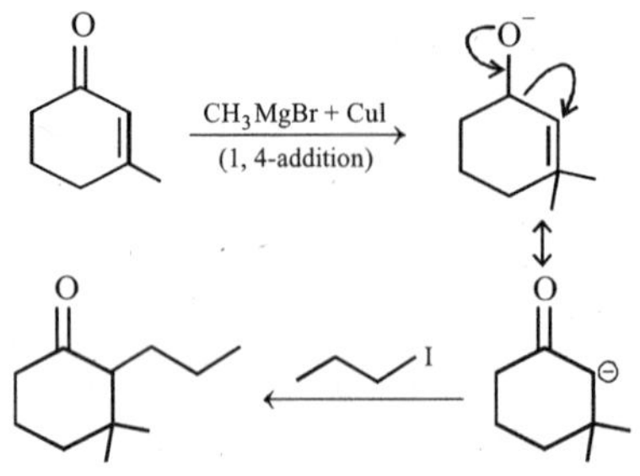

88. (4) Hint: $LiAlH_4$ reduces cyanide into amine.

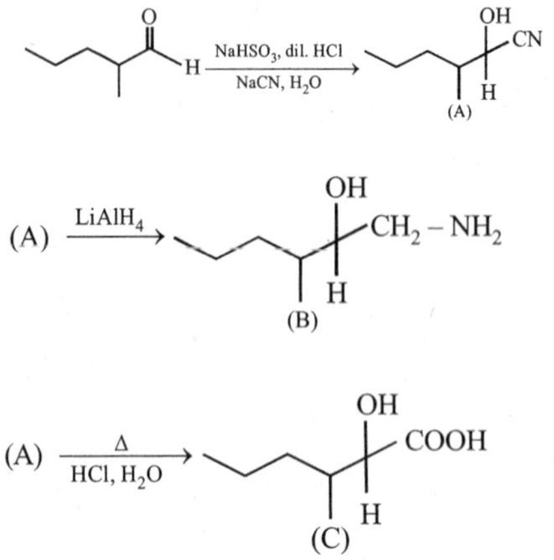

In the first reaction hydrogen sulphite addition compound converted back to original carbonyl compound by treating it with dil.HCl. which on further reaction with NaCN forms cyanohydrin.

89. (3)

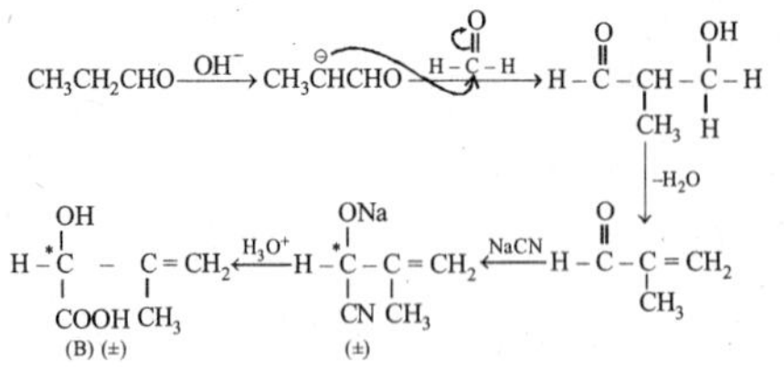

Compound B is a racemic mixture and gives off CO_2 gas with saturated solution of $NaHCO_3$.

90. (1)

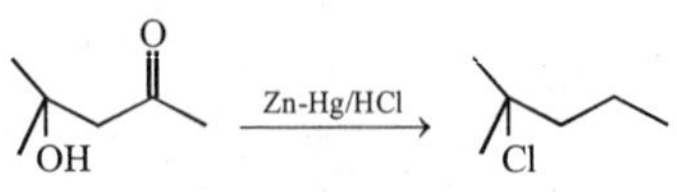

The acid sensitive alcoholic group reacts with HCl, hence Clemmensen reduction is not suitable for above conversion.

91. (3) A-III, B-IV, C-I, D-II

92. (3)

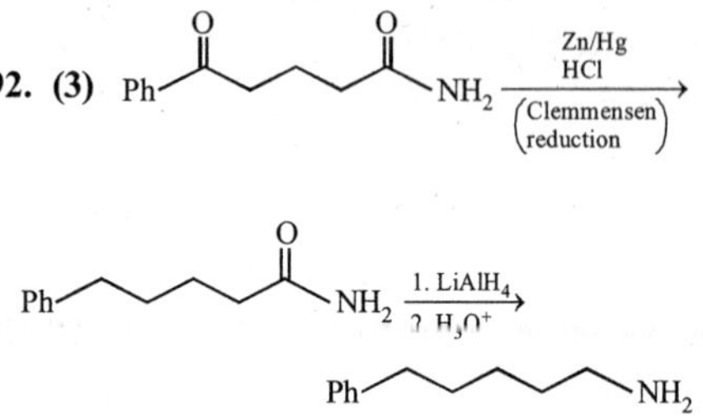

93. (2)

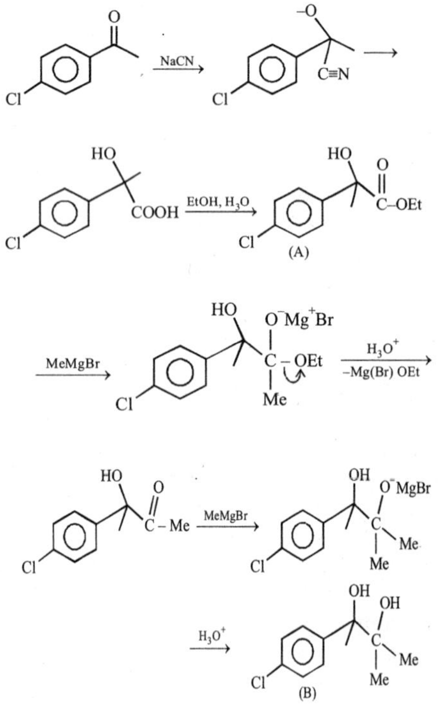

94. (1) Assertion : Acetals and ketals are cyclic ethers hence they are stable in basic medium but protonated and opened up in acidic medium.

Reason: Alkoxide ion (RO^-) is not considered as a good leaving group.

95. (1) Cannizzaro reaction is a redox process, where on aldehyde is oxidized to from the acid, and the other is reduced to form the alcohol.

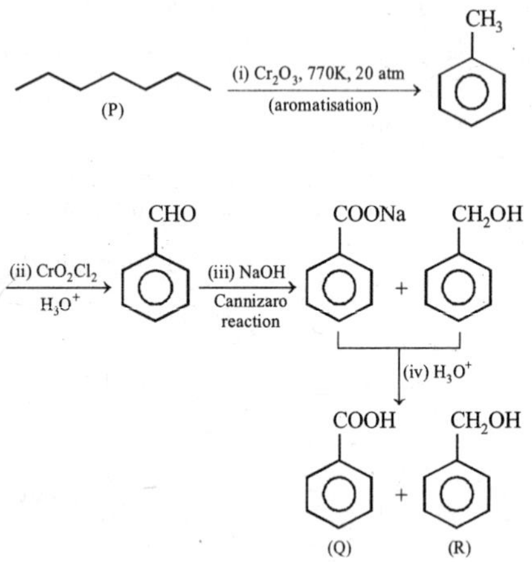

96. (2)

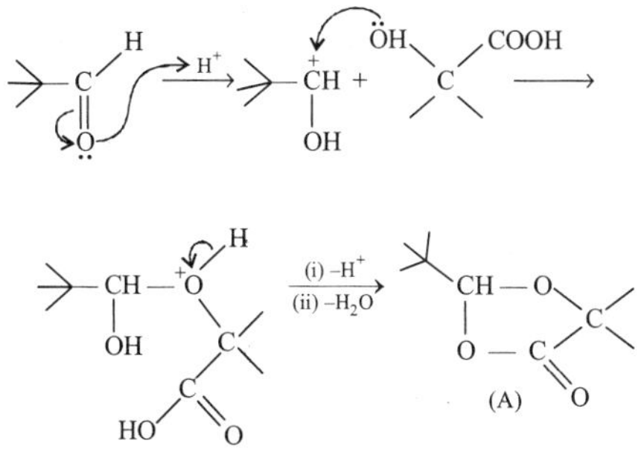

97. (1)

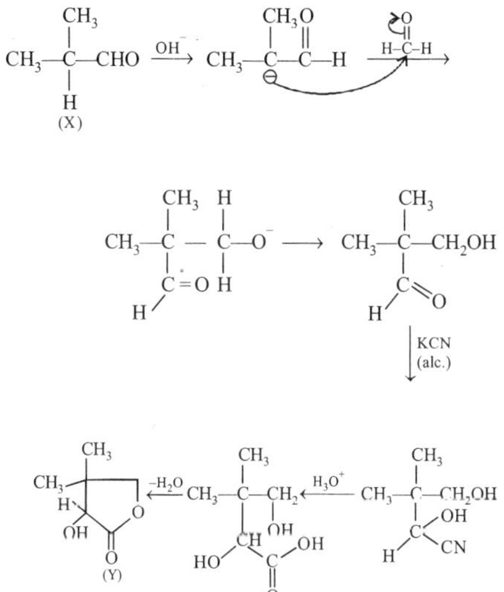

98. (2)

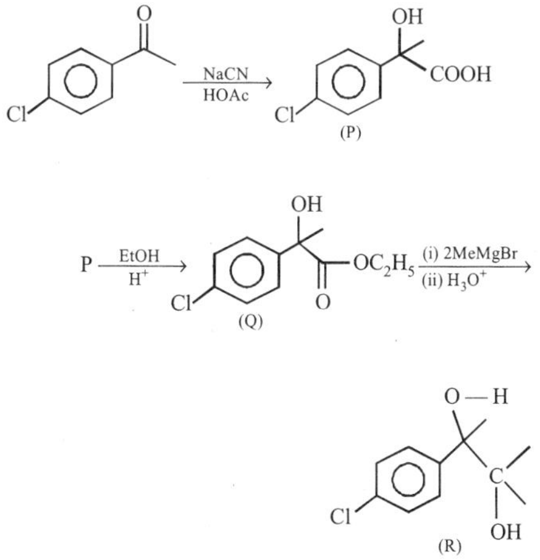

99. (3) Statement I:

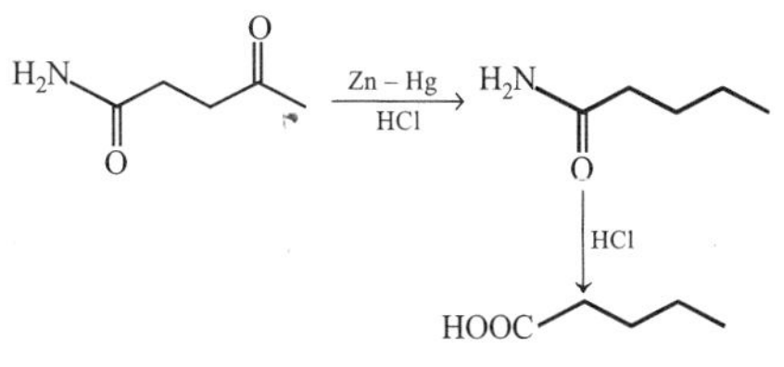

Statement II:

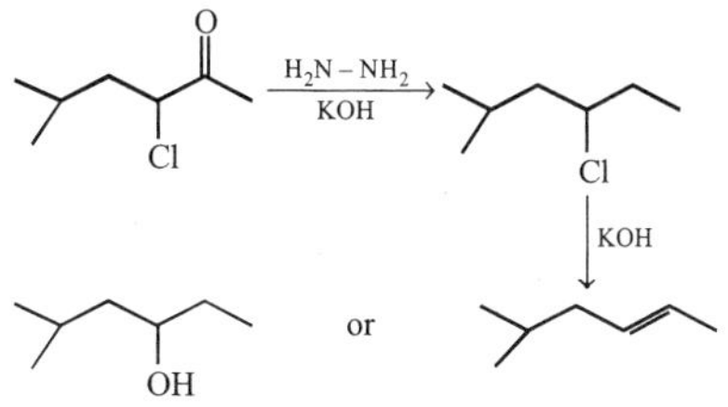

100. (1)

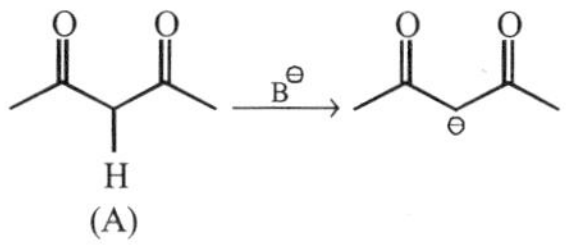

In structure (A), deprotonation occurs most readily because after the departure of proton from methylene, the product is resonance stabilised.

101. (4)

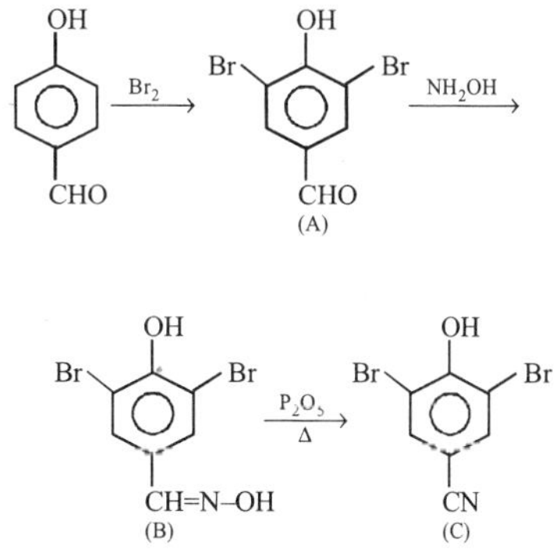

102. (1)

(A)

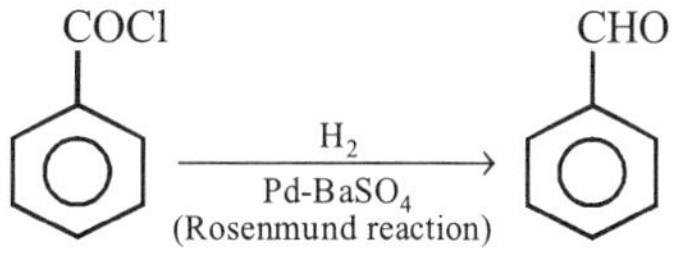

(B) $CH_3 - C \equiv N \xrightarrow[H_3O^+]{SnCl_2-HCl} CH_3 - CHO$

(Stephen reaction)

(C)

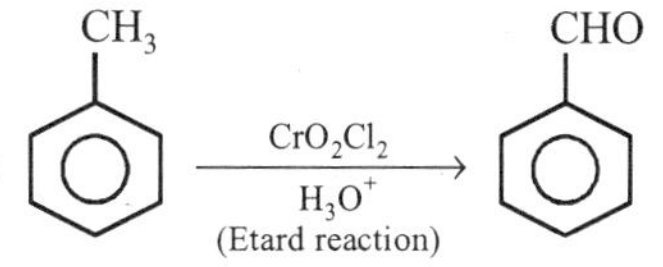

(D)

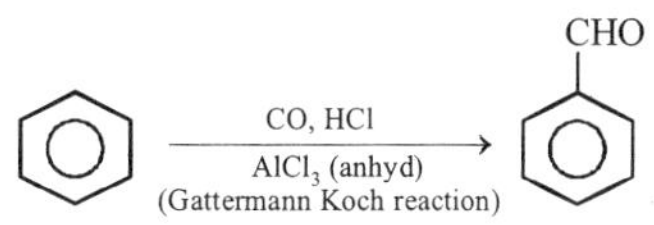

103. (3)

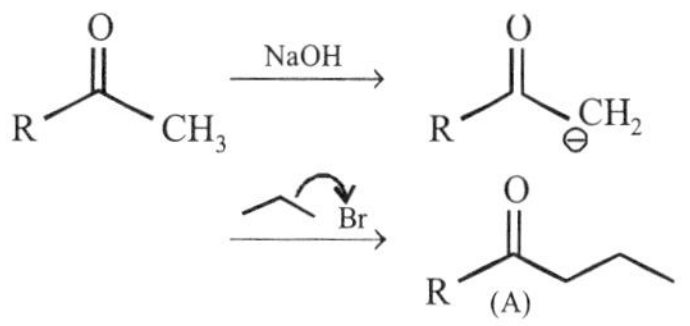

104. (3)

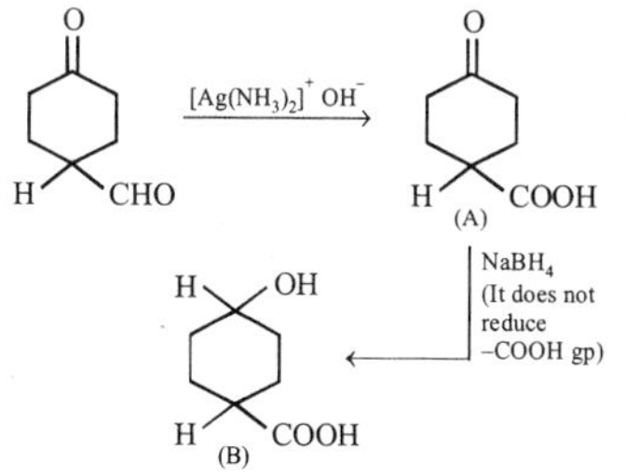

105. (4)

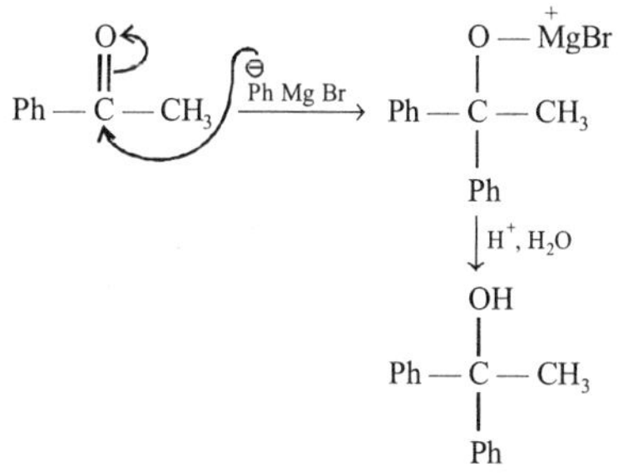

106. (2)

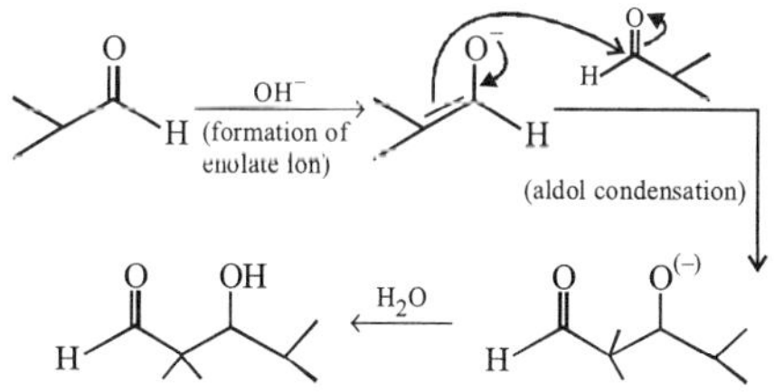

It does not contain α-hydrogen so no dehydration takes place.

107. (4) Fehling reagent (aq. $CuSO_4$ + alkaline sodium potassium tartarate) can not oxidise aromatic aldehydes.

108. (2)

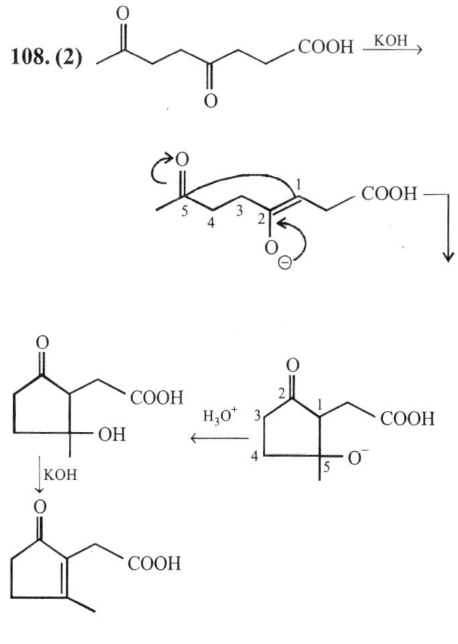

109. (1) C_6H_5CHO does not have any α-hydrogen. So it will give Cannizzaro reaction. One molecule is oxidised to acid and other is reduced to alcohol.

$$PhCH=O + PhCH=O \xrightarrow{OD/D_2O} PhCH_2OD + PhCO_2^-$$

110. (3)

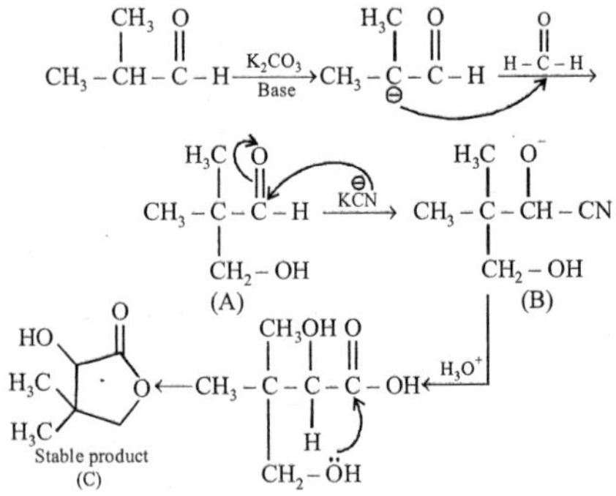

111. (1)

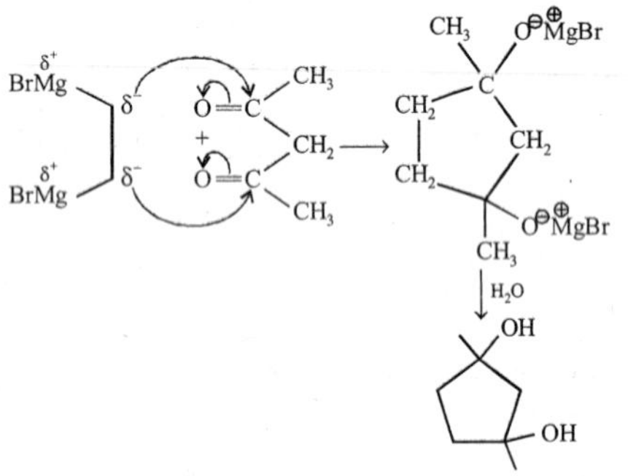

112. (3)

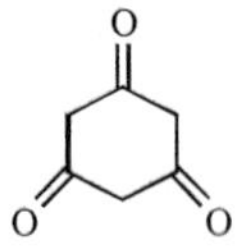 is tautomer of 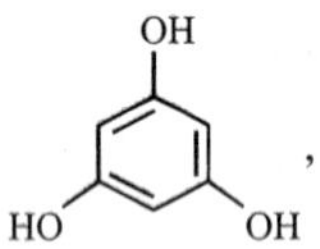,

(aromatic, hence stable)

Further it has three (max.) enolic group.

113. (3)

$$C_6H_5-\overset{O}{\overset{||}{C}}-OH + HO-\overset{O}{\overset{||}{C}}-H \xrightarrow[\Delta]{MnO} C_6H_5-\overset{O}{\overset{||}{C}}-H + CO_2 + H_2O$$

114. (1)

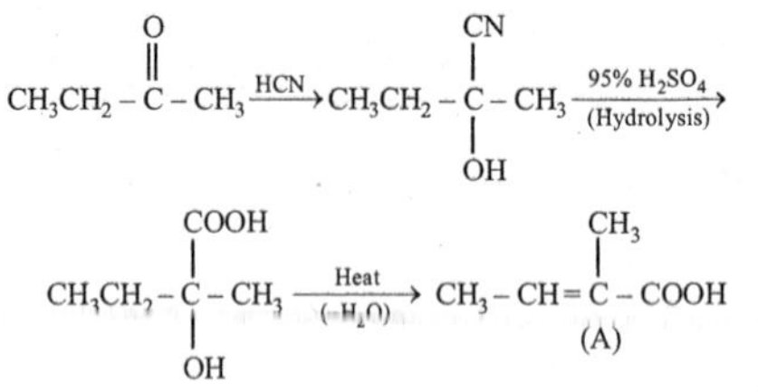

115. (3) Enamine formation is an example of nucleophilic addition elimination reaction.

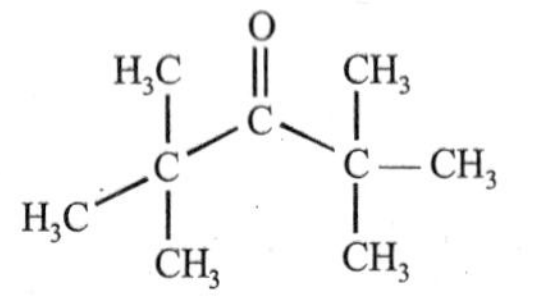

Carbonyl group is highly sterically hindered, hence attack of nuclephile will not be possible.

116. (3) 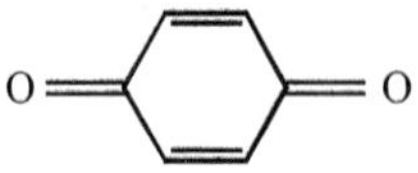

is a conjugated diketone.

117. (3)

$$CH_3-CH=CH-CH_2-\underset{OH}{\underset{|}{CH}}-CH_3$$

$$\downarrow PCC$$

$$CH_3-CH=CH-CH_2-\underset{O}{\underset{||}{C}}-CH_3 \ (A)$$

$$\downarrow NaOI \text{ (Haloform reaction)}$$

$$CH_3-CH=CH-CH_2-COOH + CHI_3$$
(B)

$$\downarrow NaOH + CaO, -CO_2$$

$$CH_3-CH=CH-CH_3$$
(C)
But-2-ene

118. (2) Statement I : Correct

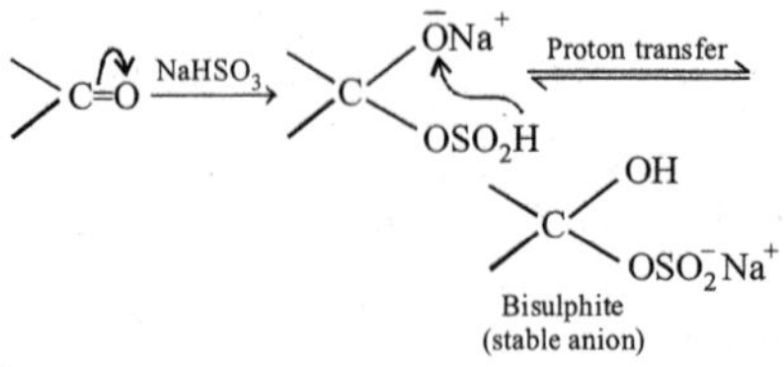

Statement II : Wrong

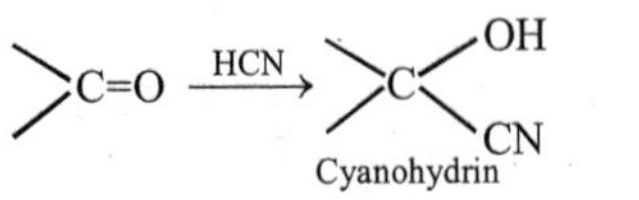

119. (3)

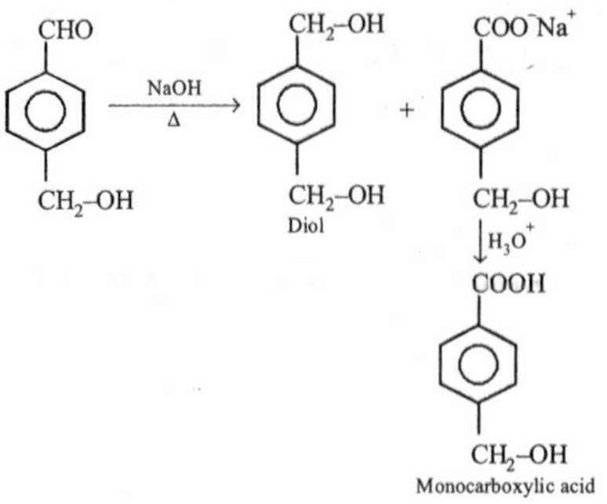

120. (1)

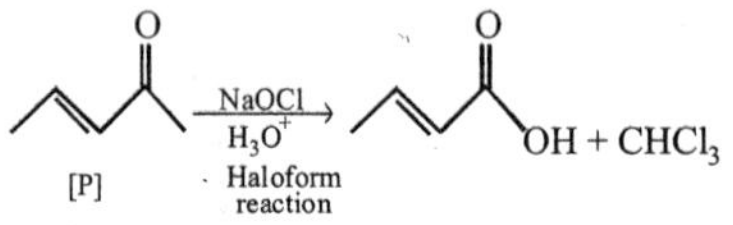

121. (3)

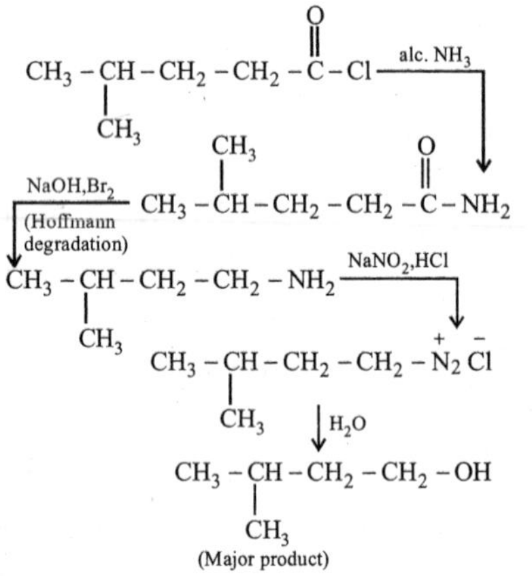

122. (1)

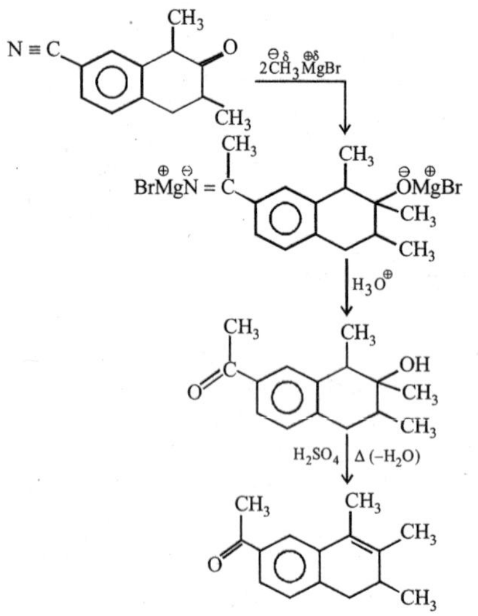

123. (3)

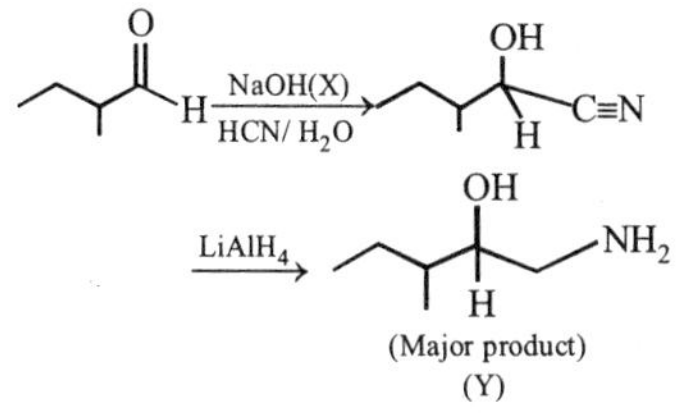

124. (3) Aldehydes give positive Tollen's test (Silver mirror test).

(I) 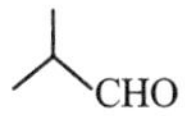 gives Tollen's test.

(II) 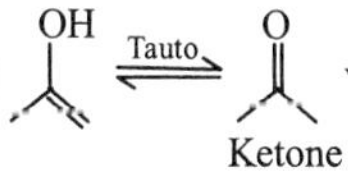 will not give Tollen's test

(III) 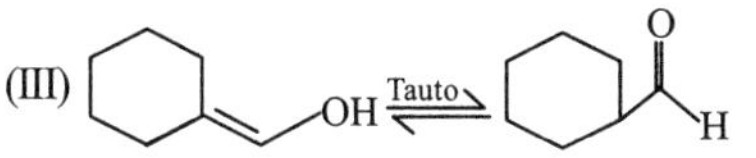

gives positive Tollen's test.

(IV) 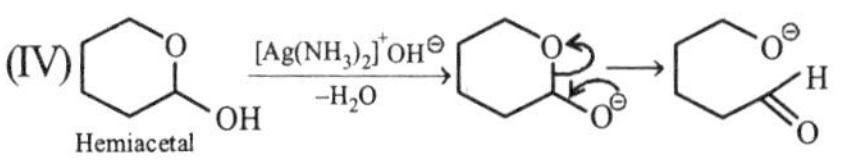

gives positive Tollen's test.

125. (3) Di-isobutyl aluminium hydride (DiBAL-H) selectively reduces nitriles and esters to aldehydes and it does not reduce double bond.

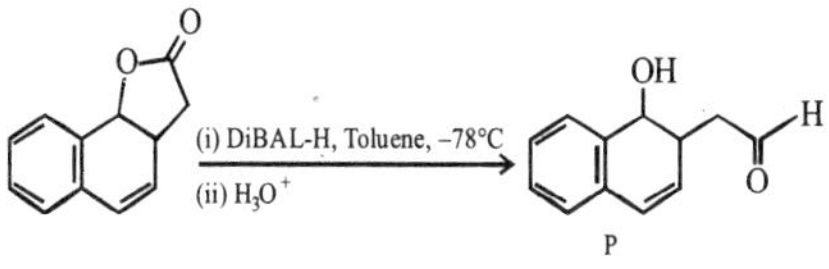

126. (4)

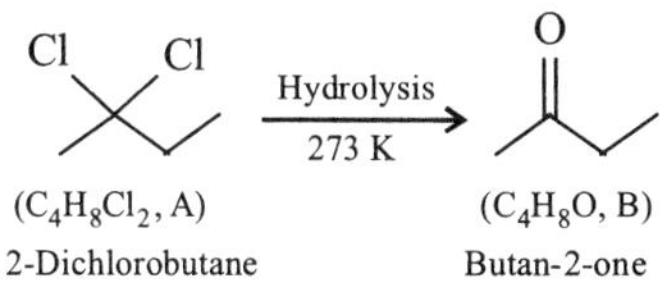

$(C_4H_8Cl_2, A)$ 2, 2-Dichlorobutane $\qquad$ (C_4H_8O, B) Butan-2-one

$\xrightarrow{Ag(NH_3)_2OH}$ No reaction

Ketones react with hydroxyl amine to give oxime.

127. (1) 2, 4-Dinitrophenylhydrazine reacts with aldehydes and ketones to form 2, 4-dinitrophenylhydrazone derivatives. It is a condensation reaction with elimination of water.

128. (3)

$$CH_3-CH_2-CH_3 \xrightarrow{Mo_2O_3} CH_3-CH_2-CHO$$

129. (3)

(A) $R-\overset{O}{\overset{||}{C}}-Cl \xrightarrow[Pd-BaSO_4]{H_2} R-\overset{O}{\overset{||}{C}}H$

(Rosenmund reduction)

(B) $R-CH_2-COOH \xrightarrow[H_2O]{Cl_2/P} R-\underset{Cl}{\underset{|}{C}H}-COOH$

(HVZ reaction)

(C) $R-\overset{O}{\overset{||}{C}}-NH_2 \xrightarrow[NaOH]{Br_2} R-NH_2$

(Hoffmann bromamide reaction)

(D) $R-\overset{O}{\overset{||}{C}}-CH_3 \xrightarrow[Conc.HCl]{Zn(Hg)} R-CH_2-CH_3$

(Clemmensen reduction)

130. (3) Reaction of $HgSO_4$/dil. H_2SO_4 with alkyne results in addition of water as per Markonikov's rule.

(1) $HC\equiv CH \xrightarrow[H_2SO_4]{HgSO_4} CH_2=\underset{OH}{\underset{|}{CH}} \rightleftharpoons CH_3-\overset{O}{\overset{||}{C}}H$

(2) 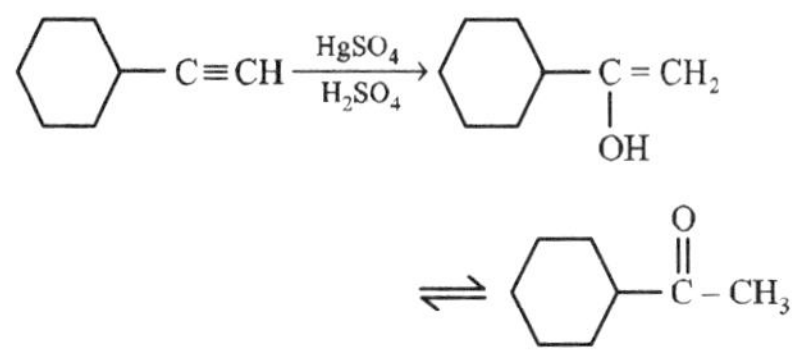

(3) $CH_3-C\equiv CH \xrightarrow[H_2SO_4]{HgSO_4} CH_3-\underset{OH}{\underset{|}{C}}=CH_2$

$\rightleftharpoons CH_3-\overset{O}{\overset{||}{C}}-CH_3$, but not CH_3CH_2CHO

(4) $CH_3-C\equiv C-CH_3 \xrightarrow[H_2SO_4]{HgSO_4} CH_3-\underset{OH}{\underset{|}{C}}=CH-CH_3 \rightleftharpoons CH_3-\underset{O}{\underset{||}{C}}-CH_2-CH_3$

131. (2) 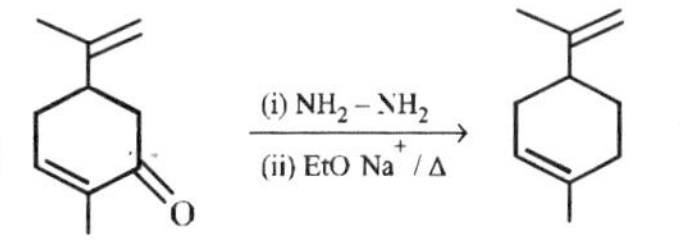

To reduce carbonyl group into alkane ($>C=O \rightarrow >CH_2$), Wolff-Kishner reduction is used, which does not affect the double bond.

132. (3)

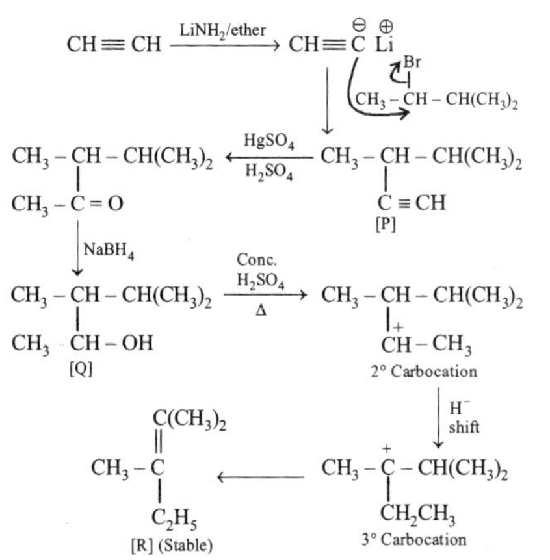

133. (1)

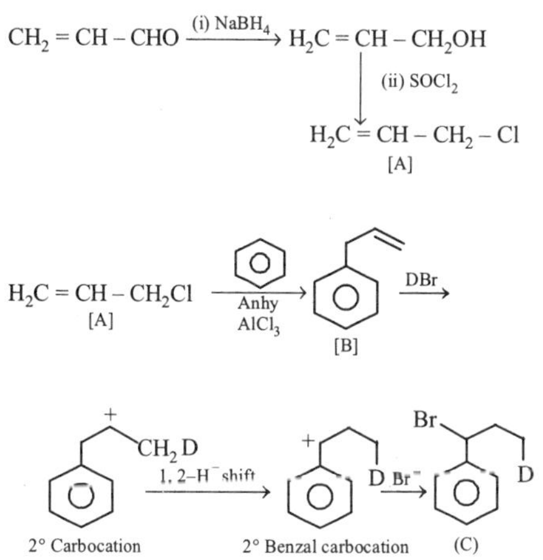

134. (2) Rate of nucleophillic addition reaction is directly proportional to the –I and –M effect of the substituents present in the substrate. Ketones are less susceptible to the nucleophillic addition, due to the presence of alkyl (R) group which has +I effect. Thus reactivity order is

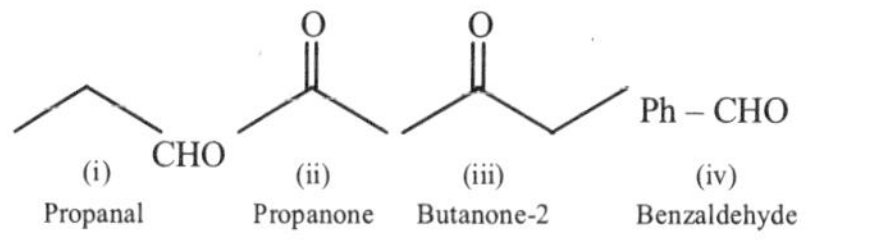

(i) > (iv) > (ii) > (iii)

135. (3) 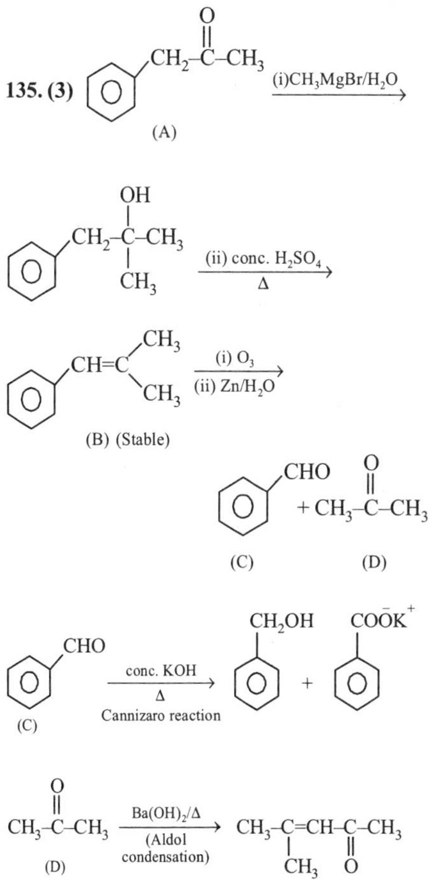

136. (3) –I effect of NO_2 increases reactivity towards nucleophilic addition reaction with HCN. – OCH_3 group is electron donating due to resonance effect which decreases the reactivity towards nucleophilic addition.

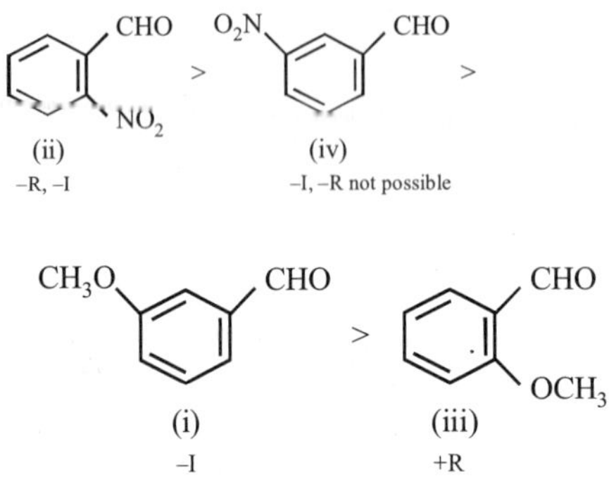

137. (2)

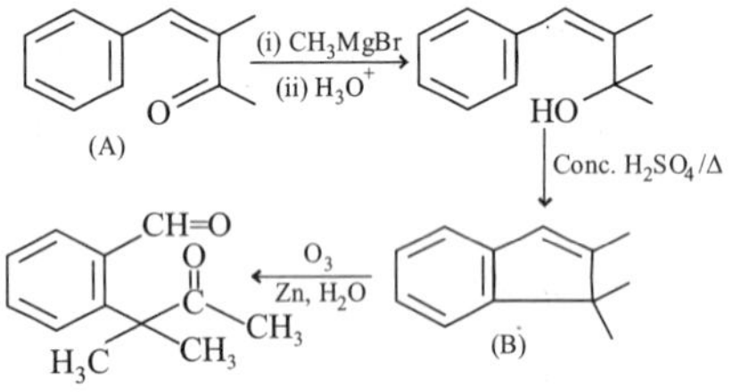

138. (4)

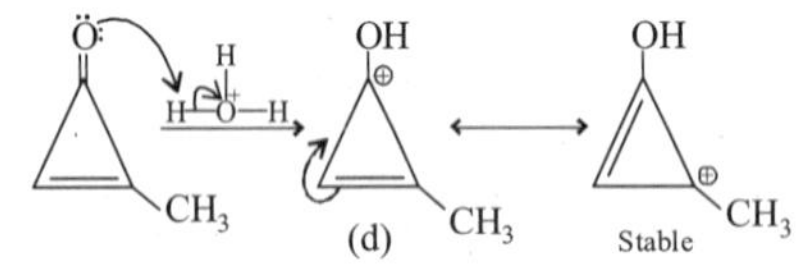

139. (4)

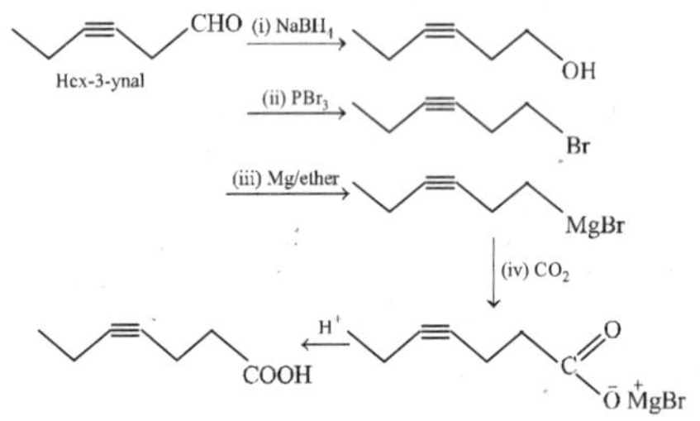

140. (3) Oxidation number of carbon in – CHO group of Ph – CHO and H– CHO are + 1 and zero respectively.

So, CHO is the stronger reducing agent than PhCHO. As a result, HCHO is oxidised to HCOONa (by donation of hydride, H^-) and PhCHO is reduced (by accepting H^-) to $PhCH_2OH$.

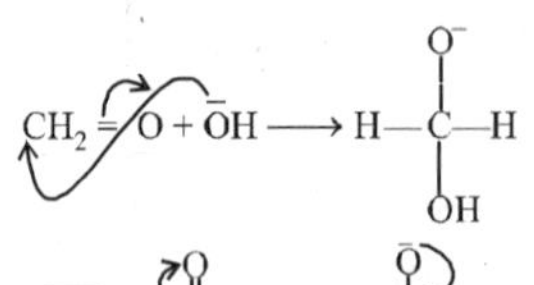

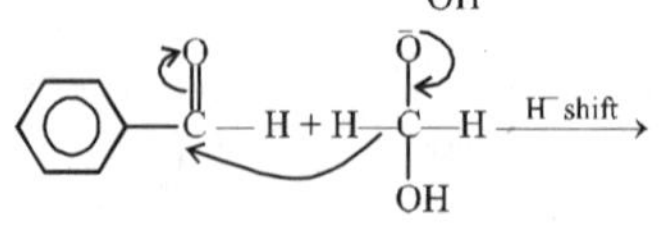

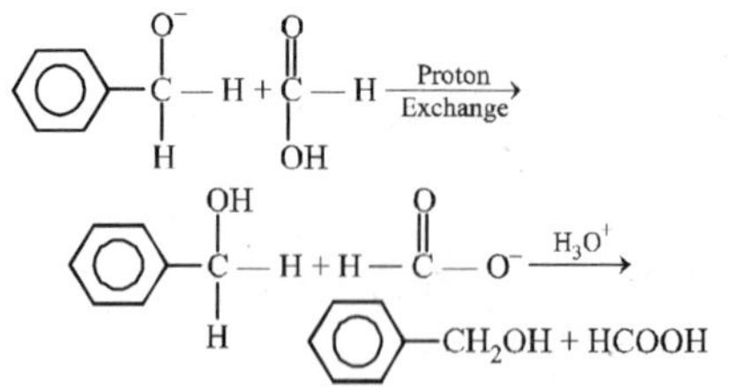

141. (4) (i) Since compound A ($C_9H_{10}O$) contains relatively very low H as compared to C, so it must contain a benzene ring.

(ii) Further, the oxidation product B ($C_8H_6O_4$) of A is a dicarboxylic acid which forms anhydride on heating, hence the acid is phthalic acid which is further confirmed by the fact that it is used in the preparation of phenolphthalein when condensed with phenol in presence of conc. H_2SO_4. So, the given compound A corresponds to option (d).

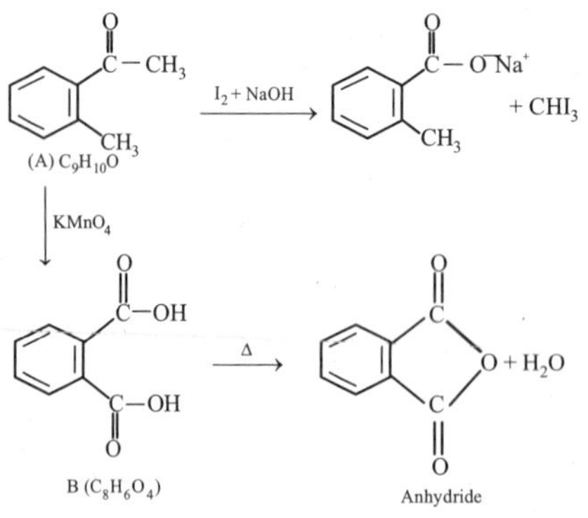

142. (2)

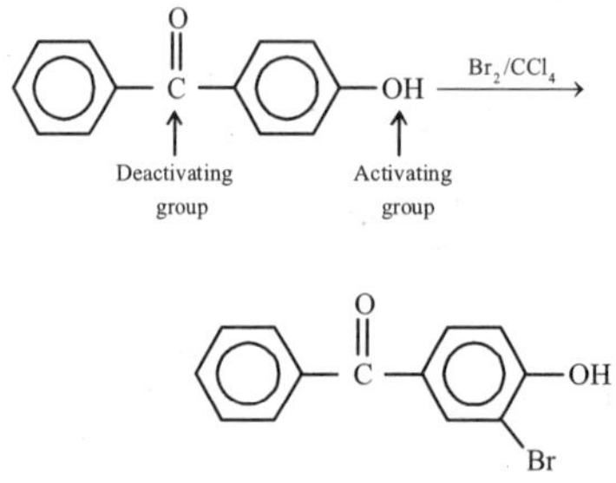

143. (4)

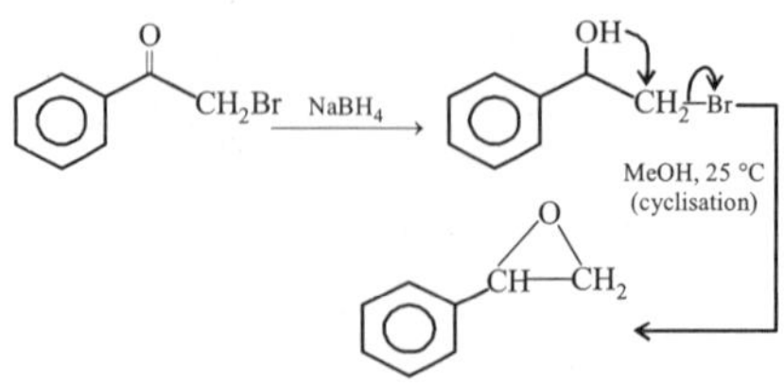

144. (2) Hint: Resonance and hydrogen bonding increases enol content.

$$H_3C-\overset{O}{\overset{\|}{C}}-CH_2-\overset{O}{\overset{\|}{C}}-CH_3 \rightleftharpoons H_3C-\overset{O-H\,......\,O}{C}=CH-\overset{\|}{C}-CH_3$$

Stable 6-membered ring due to H-bonding

145. (4)

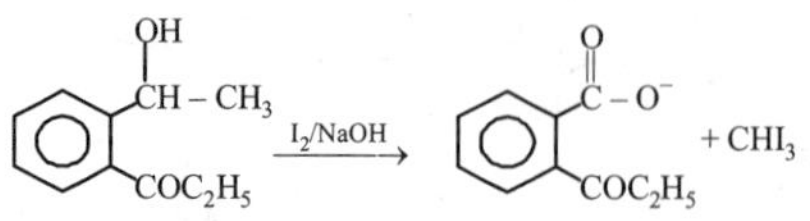

Test	Observation	Reason
Reaction with Grignard reagent,	+ve	Electrophilic centre or acidic hydrogen is present
Fehling solution test	–ve	CHO group is absent
Neutral $FeCl_3$ test	–ve	phenolic group is absent
Iodoform test	+ve	$-COCH_3$ or $-CH(OH)-CH_3$ is present

146. (2)

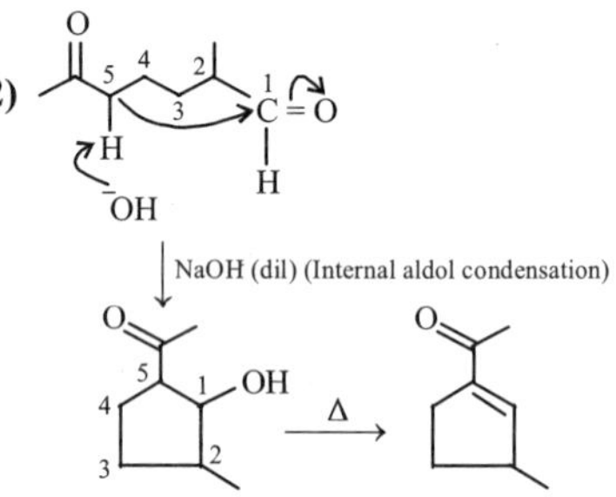

147. (1)

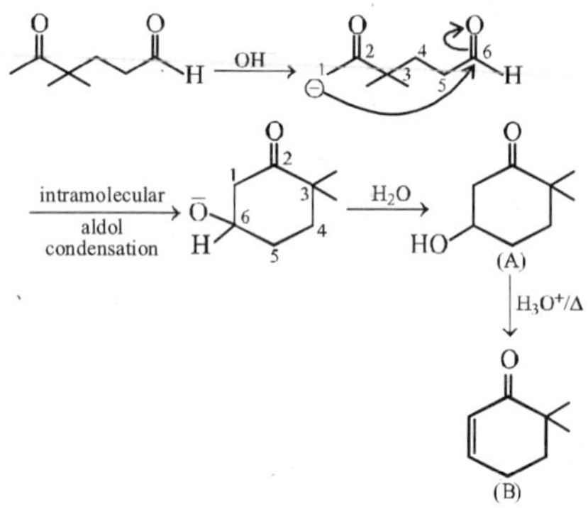

148. (2)

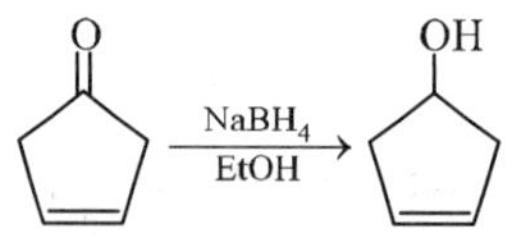

$NaBH_4$ does not reduce the double bond but can reduce keto group ($\rangle C=O$) to —CHOH group.

149. (1) The only one equivalent of Grignard reagent will not react with aldehydes (–CHO) if it has a functional group which has acidic hydrogen. In such cases RMgX (1 equi.) will react with the active hydrogen (acid-base reaction) Thus options (B) and (D) have —COOH and —CH_2OH respectively which has acidic H-atom. Therefore, acid-base reaction occurs.

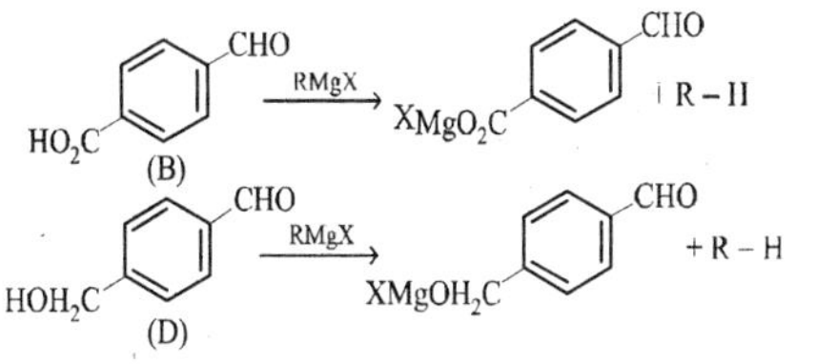

Other options (A) and (C) will react with one equi. of RMg as usual.

150. (4) Sodium borohydride is a selective reducing agent. It reduces carbonyl group to alcoholic group, N-methylimino group (MeN = CH –) to 2° amines, but does not reduce an isolated carbon-carbon double bond.

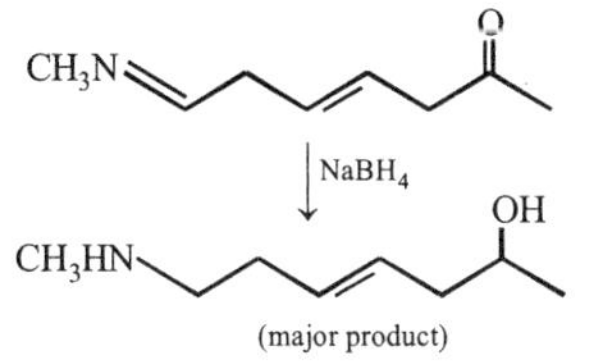

151. (1) Here four products are possible.

(i) From self aldol condensation of CH_3CHO.

(ii) From self aldol condensation of $C_6H_5COCH_3$

(iii) From cross aldol condensation between CH_3CHO and $C_6H_5–COCH_3$ when CH_3CHO acts as a carbanion.

(iv) From cross aldol condensation between CH_3CHO and $C_6H_5COCH_3$ when $C_6H_5COCH_3$ acts as a carbanion.

However, CH_3CHO is more reactive than ketone so it acts as a good donor, and its carbanion is also more stable, so it also acts as a good carbanion donor. Thus the major product should be (a).

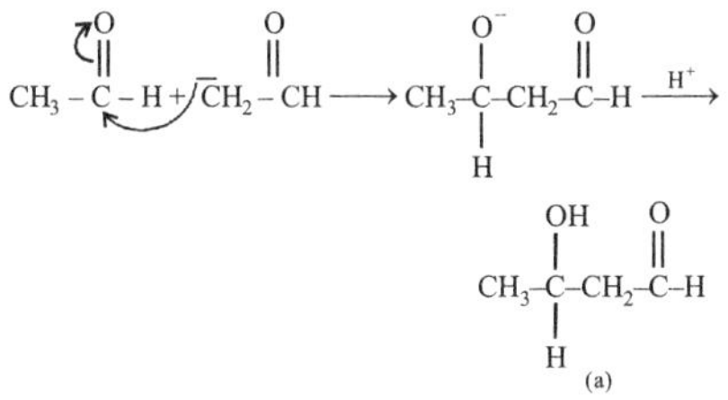

152. (4)

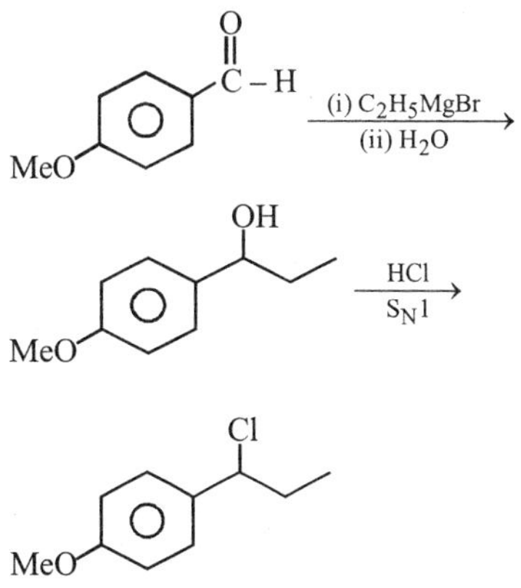

(A)
(B)

153. (1)

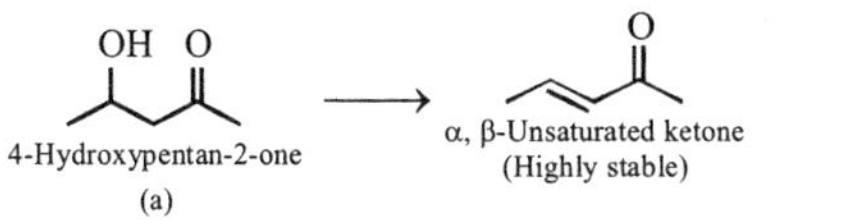

154. (1)

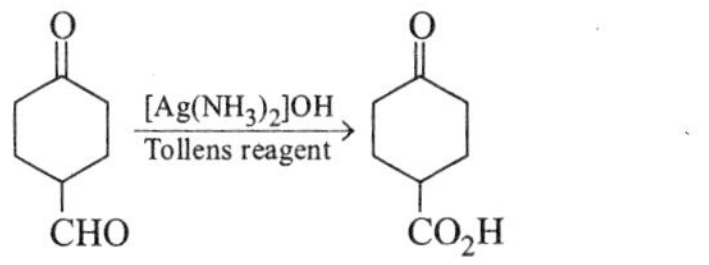

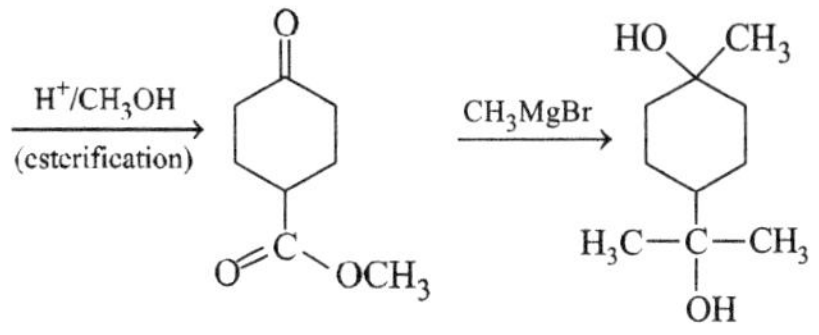

155. (1)

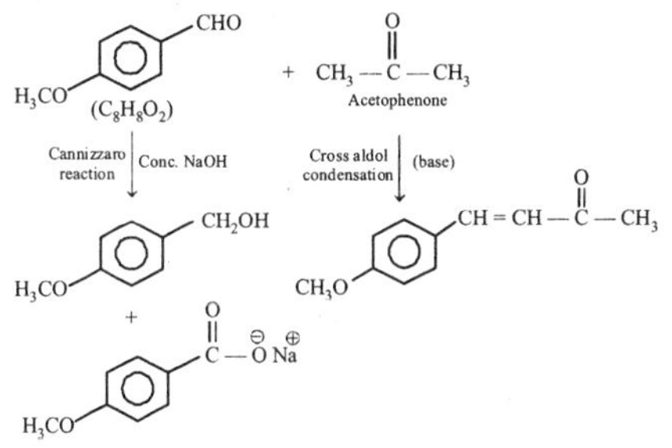

156. (1) The pentaerythritol is produced via a base-catalyzed reaction of acetaldehyde with excess formaldehyde. The aldol condensation of three moles of formaldehyde with one mole of acetaldehyde followed by a crossed Cannizzaro reaction with formladehyde gives the final product (pentaerythritol) and sodium formate as a by-product. These reactions are shown below

$$CH_3CHO \xrightarrow[\text{(three aldol condensation)}]{3HCHO/OH^-} \underset{\text{Pentaerythrose}}{C(CH_2OH)_3CHO}$$

The resulting compound does not have any α–H, hence it can undergo Cannizzaro's reaction.

$$C(CH_2OH)_3CHO \xrightarrow[\text{OH}^-\ \text{(Cannizzaro's reaction)}]{HCHO} \underset{\text{Pentaerythritol}}{C(CH_2OH)_4} + HCOONa$$

It is used in the preparation of pentaerythritol tetranitrate (PETN), an important explosive.

157. (2) $CH_3-\overset{O}{\overset{\|}{C}}-H \xrightarrow{OH^-}$

$$\left[\overset{\ominus}{C}H_2-\overset{O}{\overset{\|}{C}}-H \longleftrightarrow H_2C=\overset{O^{\ominus}}{\overset{|}{C}}-H\right]$$

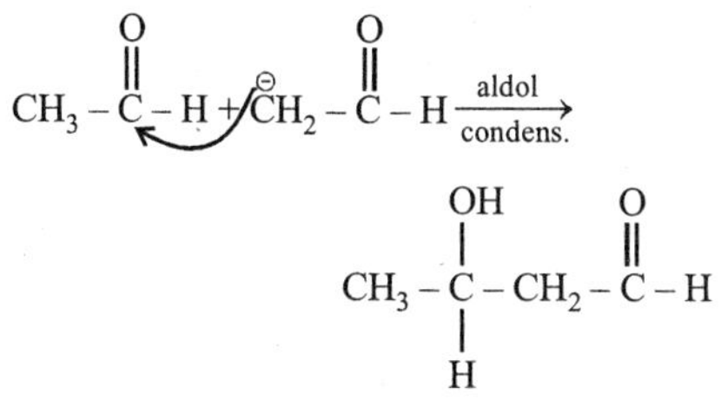

$$\xrightarrow{\Delta} CH_3-CH=CH-\overset{O}{\overset{\|}{C}}-H$$

158. (4) Cannizzaro's reaction is a disproportionation reaction of aldehyde in which one molecule of aldehyde is reduced to alcohol whereas other oxidised to salt of carboxylic acid

$$C_6H_5CHO + KOH \longrightarrow C_6H_5CH_2OH + C_6H_5COOK$$

159. (3) $CH_3COCH_3 + 3I_2 + 4KOH \longrightarrow CHI_3 + CH_3COOK + 3HI + 3H_2O$

160. (3) Given compound A is

$$CH_3-CH_2-\underset{CH_3}{\underset{|}{C}}=C=O$$

Reactions are as below:

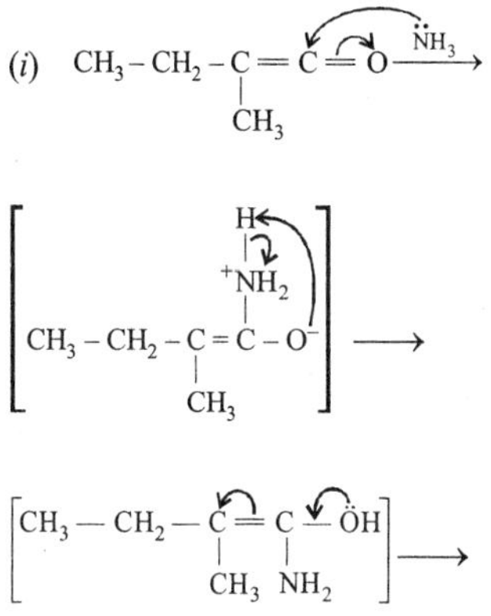

$$CH_3-CH_2-\underset{CH_3}{\underset{|}{CH}}-\overset{O}{\overset{\|}{C}}-NH_2$$

(*ii*) $CH_3-CH_2-\underset{CH_3}{\underset{|}{C}}=C=O \xrightarrow{H_2O} CH_3-CH_2-\underset{CH_3}{\underset{|}{CH}}-COOH$

(*iii*) $CH_3-CH_2-\underset{CH_3}{\underset{|}{C}}=C=O \xrightarrow{CH_3COOH} CH_3-CH_2-\underset{CH_3}{\underset{|}{CH}}-\overset{O}{\overset{\|}{C}}-O-\overset{O}{\overset{\|}{C}}-CH_3$

161. (3) Tishchenko reaction is a modification of Cannizzaro reaction. This reaction involves disproportionation of an aldehyde, lacking a hydrogen atom in the alpha position, in the presence of an alkoxide. The reaction product is an ester. Catalysts are aluminium alkoxide or sodium alkoxide.

In Cannizzaro's reaction the base is sodium hydroxide and the oxidation product is a carboxylic acid and the reduction product is an alcohol.

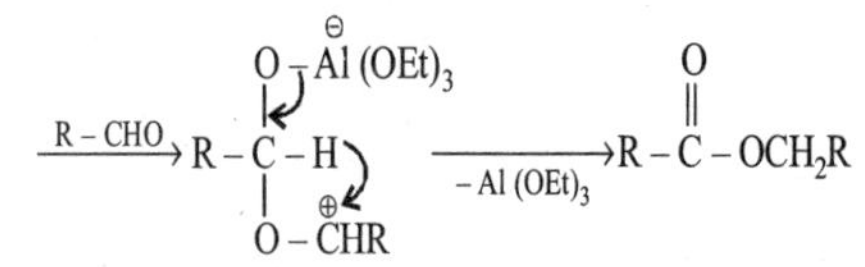

162. (3) Only acetaldehyde and methyl ketones give iodoform test.

163. (2) 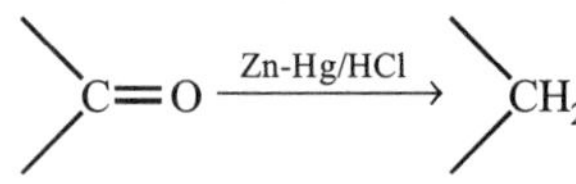

Ketone Clemmensen reduction

164. (2)

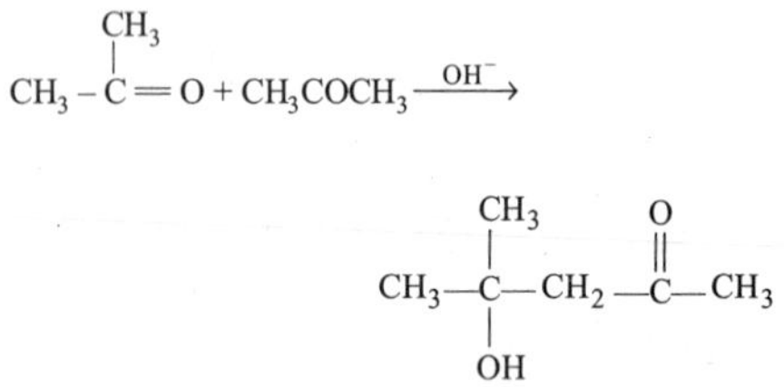

4–Hydroxy–4–methylpentan-2-one

165. (3) Only those aldehydes which do not have α-H atom undergo Cannizzaro's reaction. Hence CH_3CHO will not undergo Cannizzaro's reaction as it has 3 α H atoms.

166. (1)

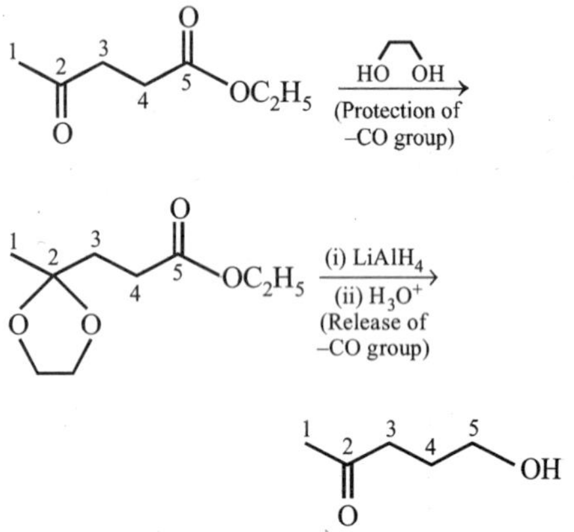

167. (4) Iodoform test is given by methyl ketones, acetaldehyde, ethyl alcohol and methyl secondary alcohols.

Isobutyl alcohol is a primary alcohol; except ethanol, primary alcohols do not give haloform test.

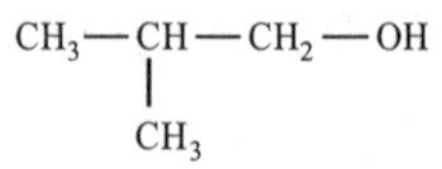

168. (1) Aldehydes and ketones can be reduced to hydrocarbons by the action (i) of amalgamated zinc and concentrated hydrochloric acid (Clemmensen reduction), or (ii) of hydrazine (NH_2NH_2) and a strong base like NaOH, KOH or potassium *tert*-butoxide in a high-boiling alcohol like ethylene glycol or triethylene glycol (Wolf-Kishner reduction). But –OH group is acid-sensitive, so Clemmensen reduction can not be used.

169. (2) Presence of alkyl group in carbonyl compounds decreases their reactivity towards nucleophilic addition. Further greater the number of such groups lesser will be the reactivity towards nucleophilic addition, thus correct order is

$$HCHO > CH_3CHO > CH_3COCH_3$$

170. (4) The reaction given is a Clemmensen reduction.

171. (3) Aldehydes (aliphatic and aromatic) show reaction with Tollen's reagent and aliphatic aldehydes show Fehling solution, but ketones do not show these reactions.

172. (1)

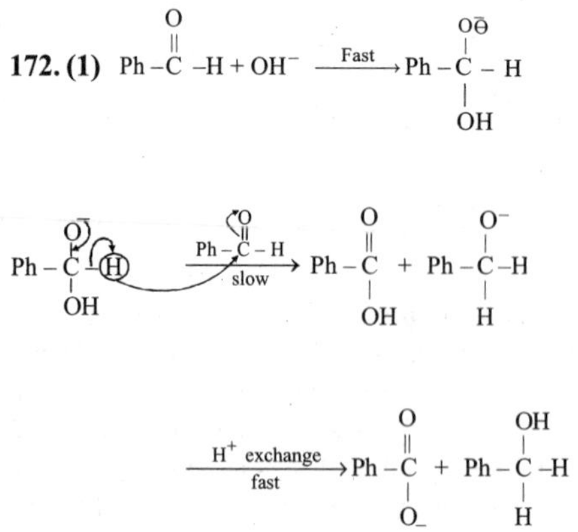

173. (1) Addition of HCN to carbonyl compounds is nucleophilic addition reaction. The order of reactivity of carbonyl compounds is

Aldehydes (smaller to higher) > Ketones (smaller to higher). Therefore,

$$HCHO > CH_3COCH_3 > PhCOCH_3 > PhCOPh$$

(A) (B) (C) (D)

174. (3)

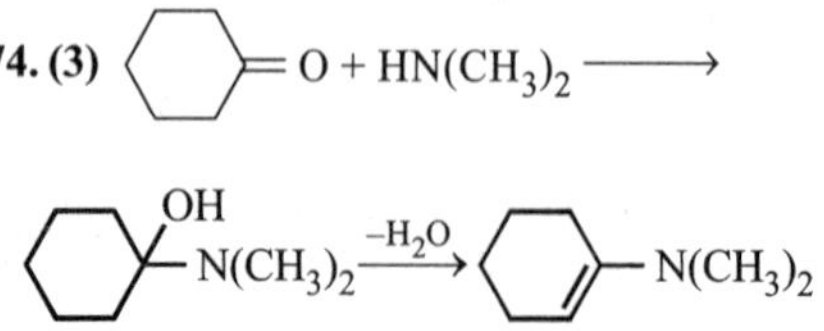

Enamine

175. (3) There is no reaction hence the resultant mixture contains $CH_3COOC_2H_5 + NaCl$.

176. (4) It is Clemmensen's reduction

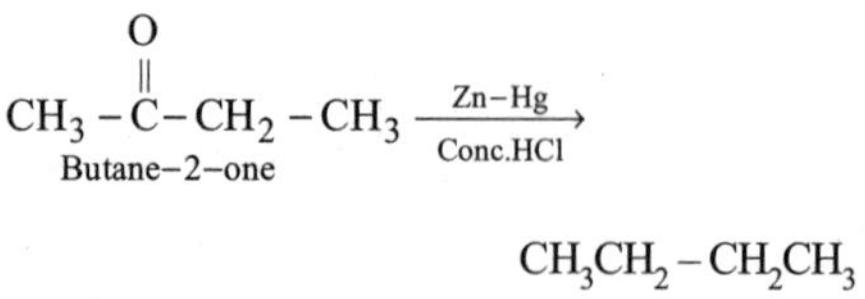

177. (2) This reaction is known as cannizzaro's reaction. In this reaction benzaldehyde in presence of 50%. NaOH undergoes disproportionation reaction and form one mol of benzyl alcohol (reduced product) and one mole of sod. benzoate (oxidation product)

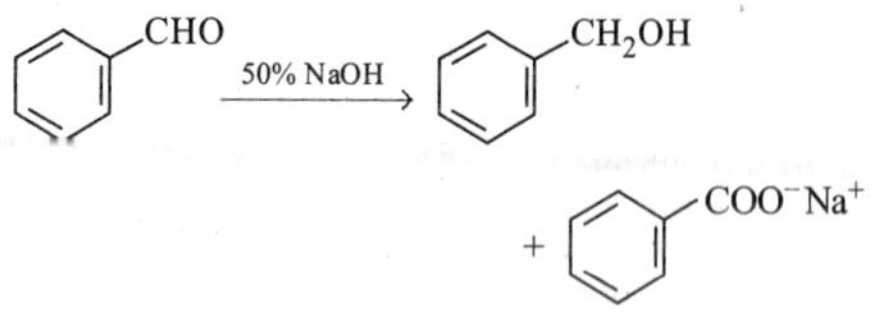

178. (1) 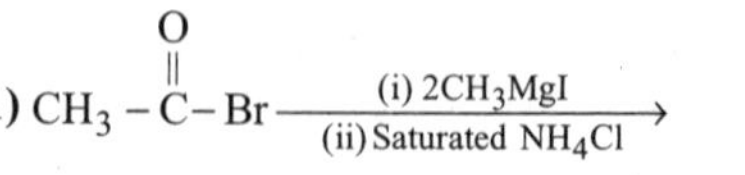

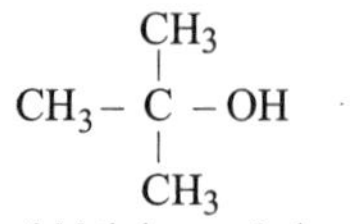

2-Methylpropan-2-ol

179. (2) Double bond is cleaved and oxidised to –COOH, –CHO is also oxidised to –COOH

$$(CH_3)_2C=CHCH_2CHO \xrightarrow{KMnO_4} (CH_3)_2CO + HOOCCH_2COOH$$

180. (16)

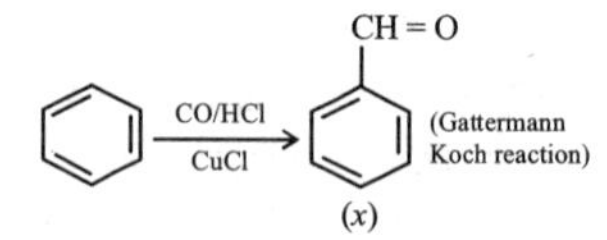

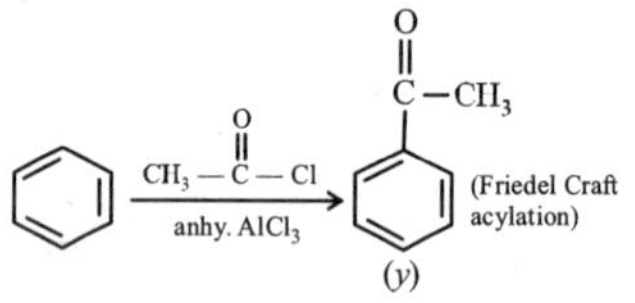

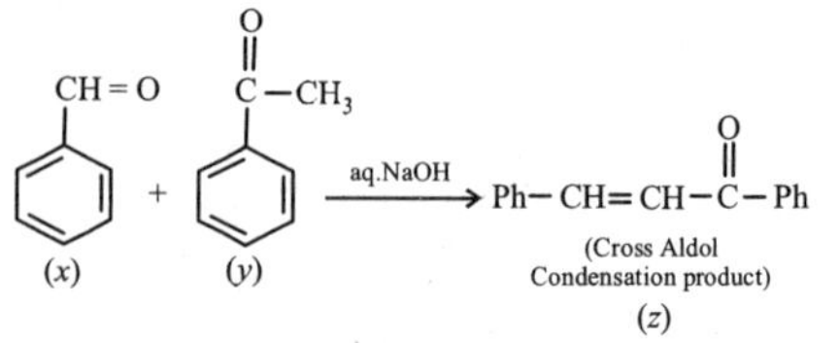

Total numbers of π-bonds in z = 8

Total number of pi (π) electrons in z = 16π electorns

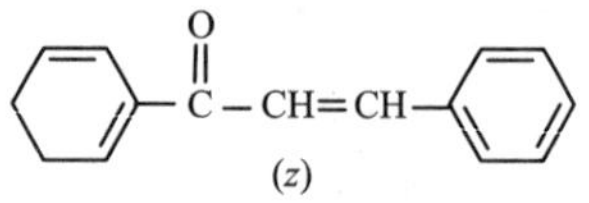

(z)

181. (10.1)

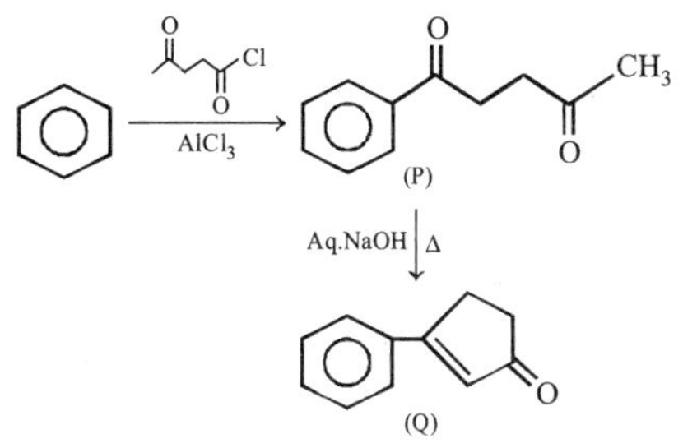

Molecular mass of Q = 157

% of oxygen in product 'Q' = $\frac{16}{157} \times 100 = 10.19\%$

182. (7)

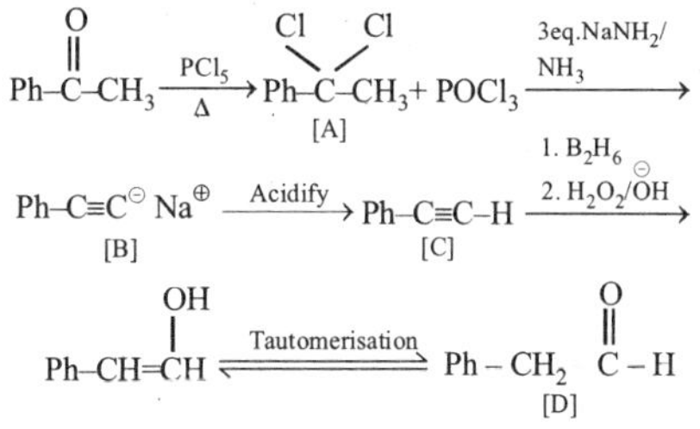

$\Rightarrow$ Number of sp^2 C-atoms in product D = 7

183. (60)

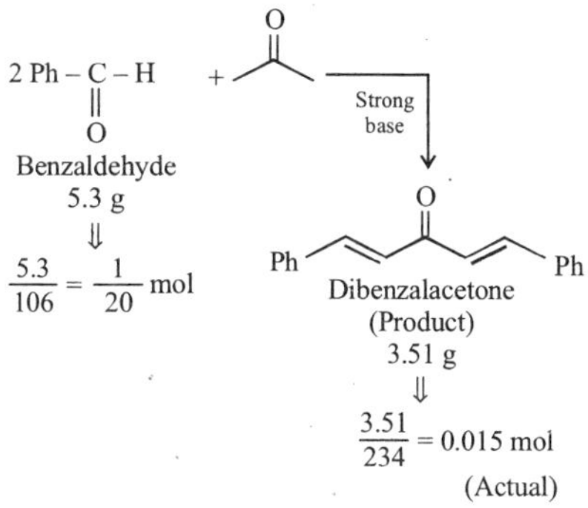

Theroretical yield = $\frac{1}{40}$ mol

$$\% \text{ yield} = \frac{\text{actual yield}}{\text{Theoretical yield}} \times 100$$

$$= \frac{0.015}{1/40} \times 100 = 60\%$$

184. (13)

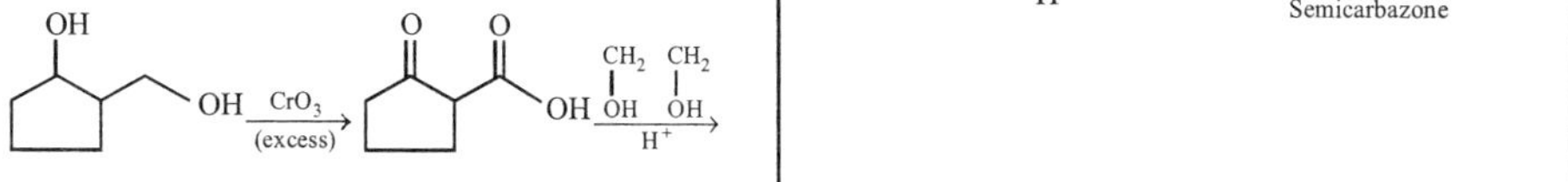

Molar mass of (S) = 130g/mol

Mass of 0.1 mol of (S) = 0.1 × 130 = 13g

185. (318) Mole of $CH_3-\overset{O}{\overset{||}{C}}-CH$ (mol. wt = 58)

$$= \frac{87}{58} = 1.5$$

Mole of dibenzalacetone (mol. wt = 234)

$$= \frac{351}{234} = 1.5 \text{ mol}$$

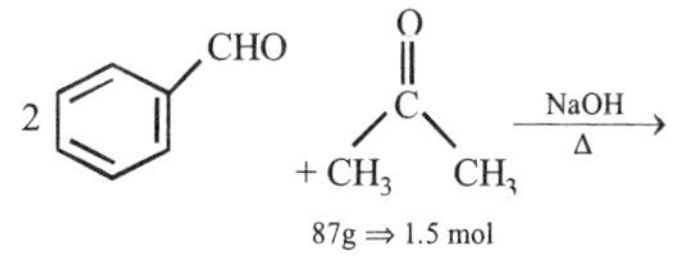

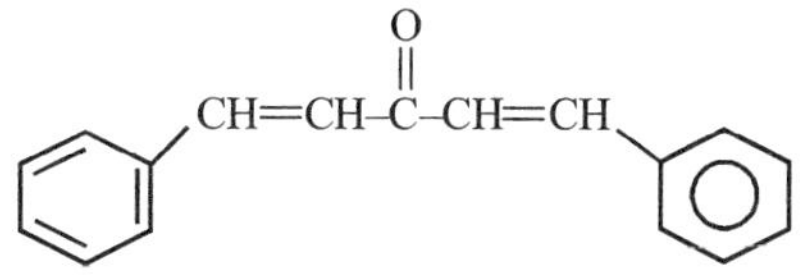

Mol of benzaldehyde required = 1.5 × 2 = 3 mol

Mass of benzaldehyde (mol. wt. = 106) = 3 × 106 = 318 gm.

186. (0) Product benzene has zero hydroxyl group.

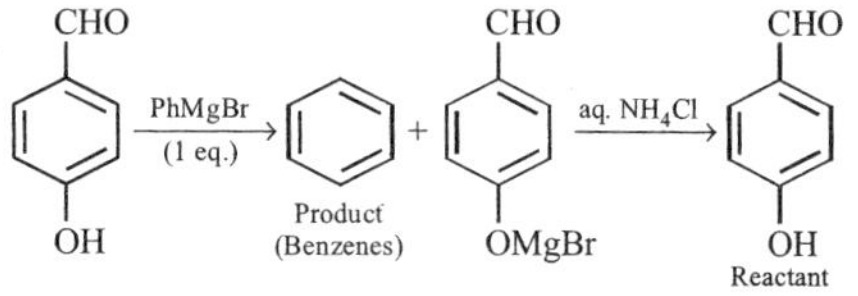

187. (3)

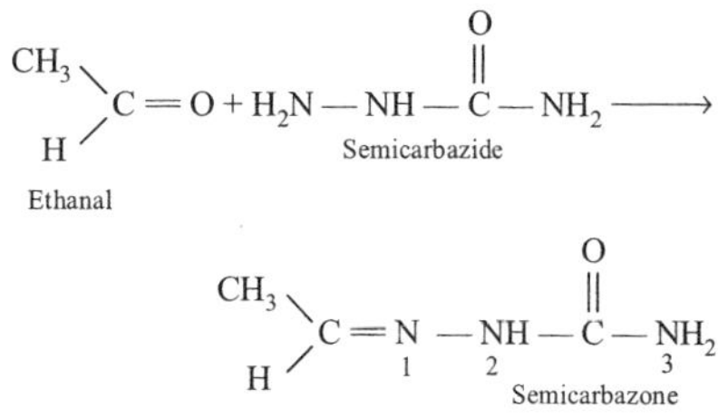

188. (60)

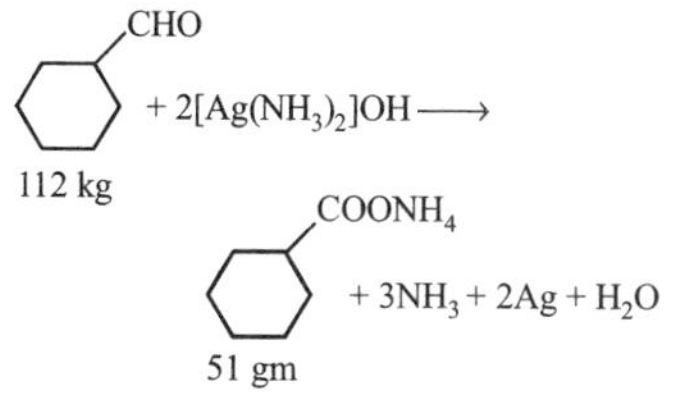

112 g cyclohexanecarbaldehyde produces 51 g NH_3

131.8 kg will produce = W_{NH_3}

$$= \frac{131.8 \times 1000}{112} \times 3 \times 17 = 60 \text{ kg}$$

189. (15)

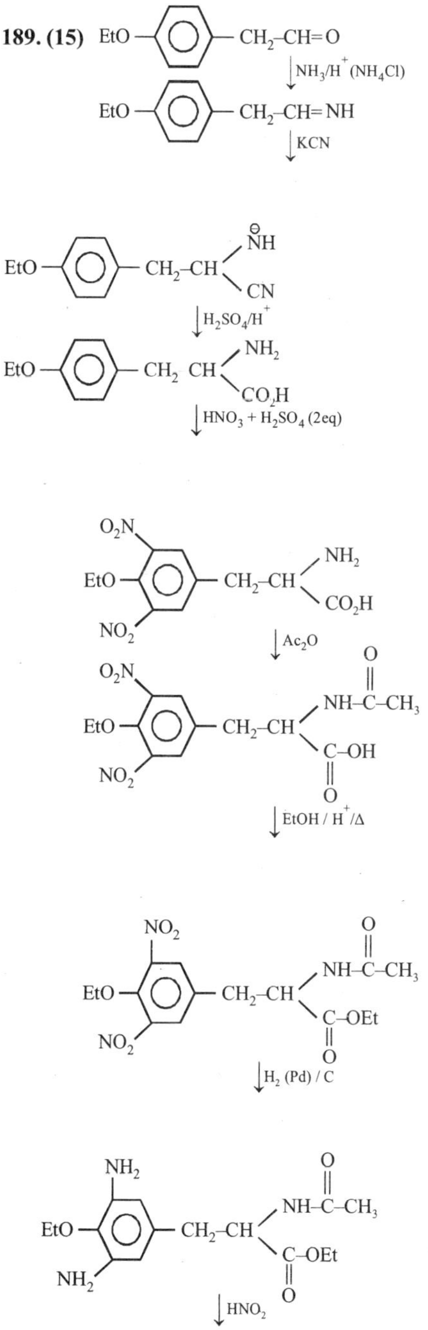

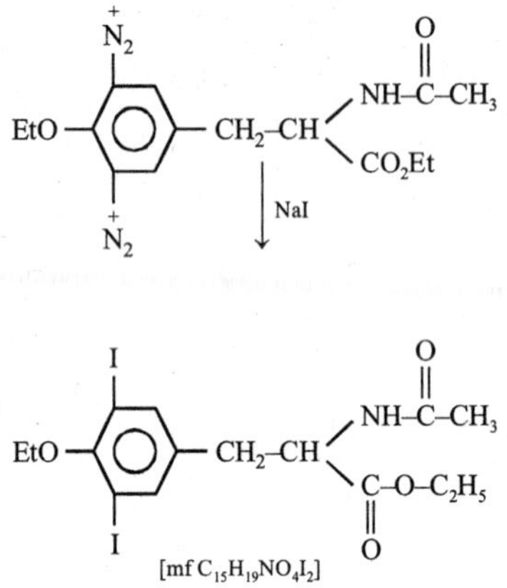

190. (2) 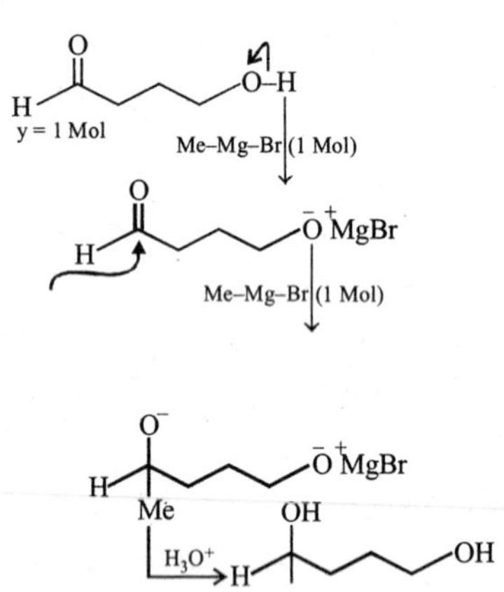

$\therefore x = 2\text{ mol} \Rightarrow \frac{x}{y} = \frac{2}{1} = 2$

191. (4) (a) 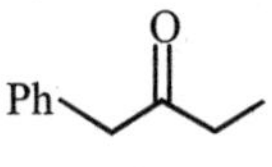 (b) 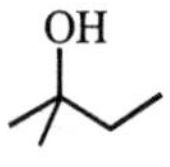

(c) 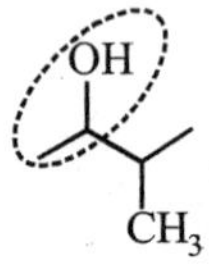 (Positive test for iodoform)

(d) 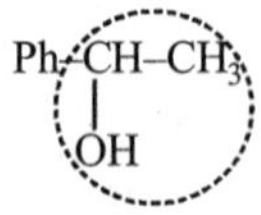 (Positive test for iodoform)

(e) 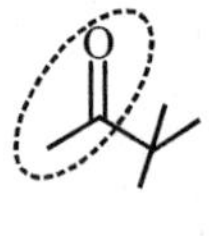 (Positive test for iodoform)

(f) 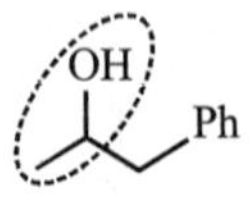 (Positive test for iodoform)

192. (2) As per the conditions given, the compounds do not have acidic or basic component to be dissolved in basic or acidic medium. It has at least one unsaturated bond to be hydrogenated. The molecules do not have >C = O or –CHO functional group.

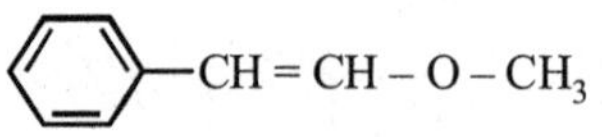

cis-and *trans*-isomers

193. (3) Molecules having

$-\overset{\overset{O}{||}}{C}-CH_3$ and $-\overset{\overset{OH}{|}}{C}H-CH_3$

give positive haloform test.

194. (3) Compound 2nd, 3rd, 7th.

195. (3) 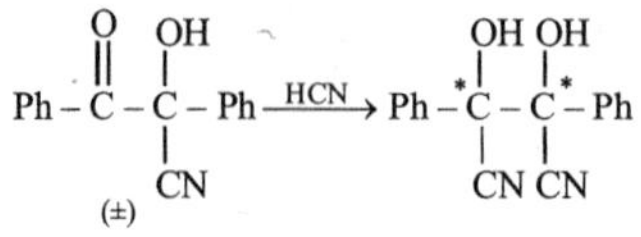

Compound formed has two **similar** chiral carbon atoms, so it has 3 stereoisomers, namely RR, SS, RS.

196. (1) Zymase converts sugar molecule to alcohol and carbon dioxide.

$$C_6H_{12}O_6 \xrightarrow{\text{Zymase}} \underset{(A)}{2CH_3CH_2OH} + CO_2$$

In presense of NaOI, alcohol is oxidised to aldehyde.

$$CH_3-CH_2-OH \xrightarrow[\text{(Oxidation)}]{\text{NaOI}} \underset{(A)}{CH_3-\overset{\overset{O}{||}}{C}-H}$$

$$\underset{\text{(yellow ppt.)}}{CHI_3} + \underset{(B)}{HCOONa} \longleftarrow \underset{(A)}{CH_3-\overset{\overset{O}{||}}{C}-H}$$

197. (8)

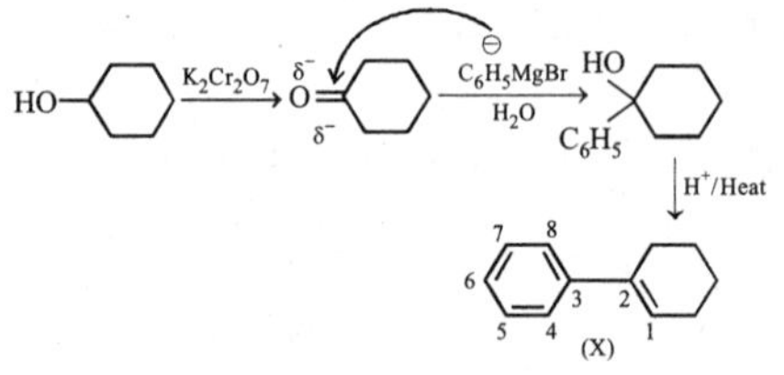

198. (4) 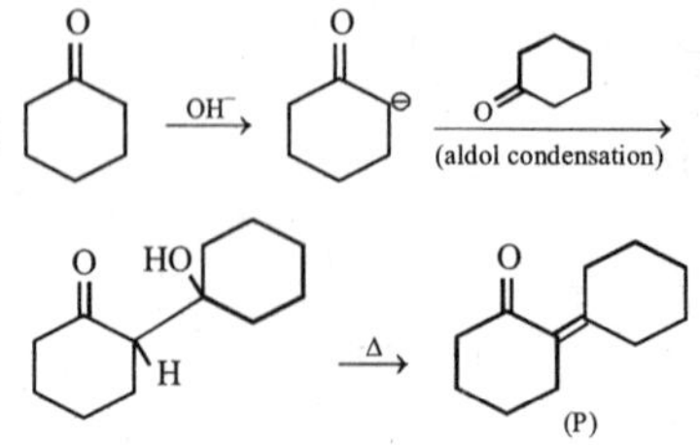

199. (3) 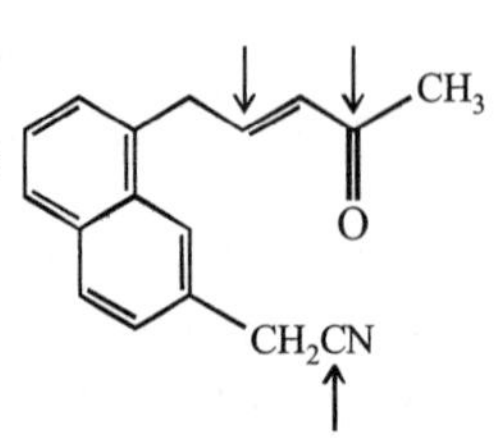

200. (3) 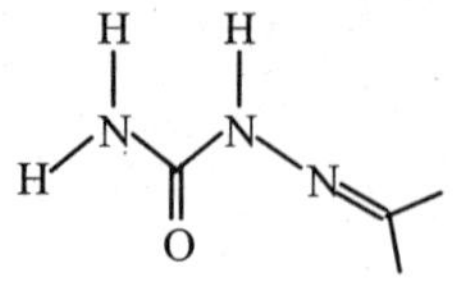

Semicarbazone molecule of acetone

201. (2) Two half reactions are

$$[Ag(NH_3)_2]^+ + e^- \longrightarrow Ag + 2NH_3$$

$$RCHO + 3OH^- \longrightarrow RCO_2^- + 2H_2O + 2e^-$$

Overall reaction is

$$2[Ag(NH_3)_2]^+ + RCHO + 3OH^- \longrightarrow 2Ag + RCO_2^- + 4NH_3 + 2H_2O$$

Therefore, for 1 mol of R–CHO, 2 mol of $[Ag(NH_3)_2]^+$ is reduced.

202. (5) Structure of mesityl oxide is

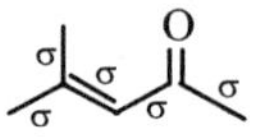

Number of C – C sigma bonds = 5

203. (66.67)

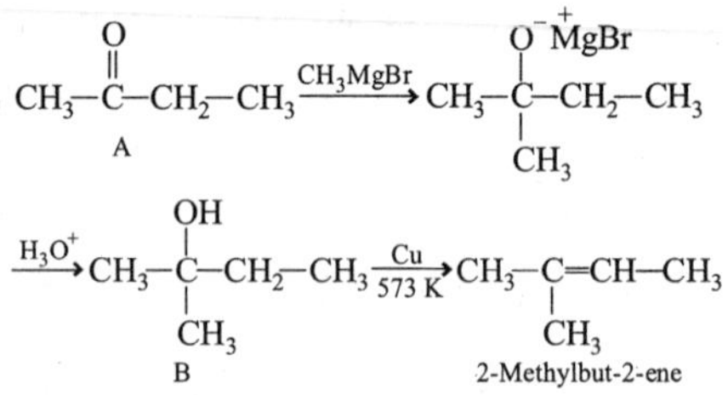

Compound A is $CH_3-\overset{\overset{O}{||}}{C}-CH_2-CH_3$ (C_4H_8O)

Mass percentage of carbon

$$= \left(\frac{12\times4}{48+16+8}\times100\right) = 66.67$$

204. (3) In step 1, when a tertiary butanol is passed over heated copper at 573K it undergoes dehydration to form an alkene.

In step 2, the resulting alkene formed when treated with benzoic acid in acidic medium undergoes electrophilic addition reaction to form an ester.

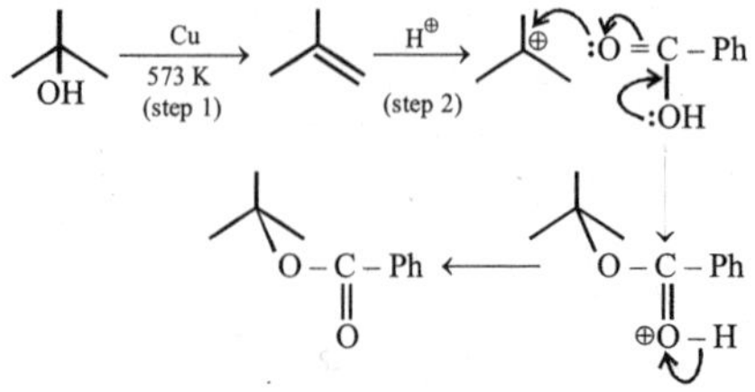

Note: NTA answer is '2' which is not possible due to absence of Lewis acid and benzoic acid have electron withdrawing group which make the ring electron deficient.

205. (4)

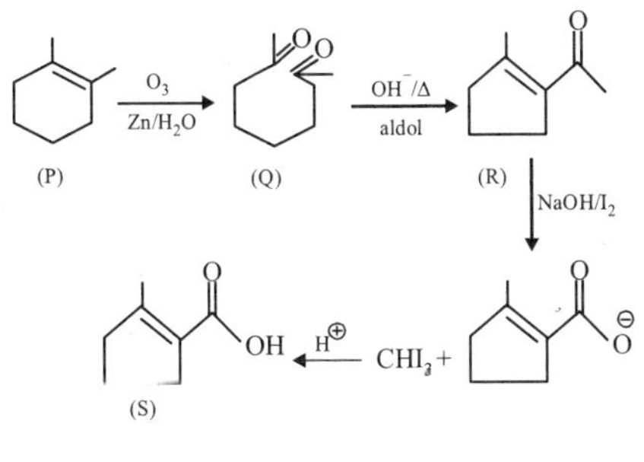

206. (4) Acids are generally more acidic than phenols and alcohols.

Also, electron donating groups decrease (–OCH_3) the acidity and electron withdrawing groups (–NO_2 etc.) increases the acidity.

207. (1) II, III, IV, V, VII

NaOH being a strong base can react with acid. So, compounds which are more acidic than H_2O are soluble in NaOH.

208. (3) If molar ratios are equal, the empirical formula is CHO.

Empirical mass = 12 (C) + 1(H) + 16 (O) = 29 g/mol.

$$n = \frac{\text{Molar mass}}{\text{Empirical Mass}} = \frac{58}{29} = 2.$$

Molecular formula : $C_2H_2O_2$.

Therefore,

$$\underset{\text{(Glyoxal)}}{\begin{matrix} CHO \\ | \\ CHO \end{matrix}} \xrightarrow[\text{Intramolecular cannizaro reaction}]{KOH} \begin{matrix} CH_2OH \\ | \\ COO^{\ominus} \end{matrix} \xrightarrow{H^{\oplus}} \begin{matrix} CH_2OH \\ | \\ COOH \end{matrix}$$

209. (4)

(A) 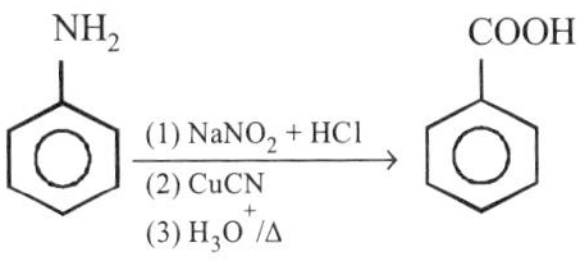

(B) $CH_3CH_2-CHO \xrightarrow[\Delta]{[Ag(NH_3)_2]^+OH^-} CH_3-CH_2-COOH$

(C) $CH_4 + O_2 \xrightarrow[\text{(ii) } Na_2Cr_2O_2 / H^+]{\text{(i) } Mo_2O_3} HCOOH$

(D) $PhCH_2MgBr + CO_2 \xrightarrow[H_3O^+]{\text{Dry ether}} Ph-CH_2-COOH$

Correct order of acidic strength of major product formed in the given reaction is

$HCOOH > C_6H_5COOH > C_6H_5CH_2COOH > CH_3CH_2COOH$

From (C) From (A) From (D) From (B)

210. (3)

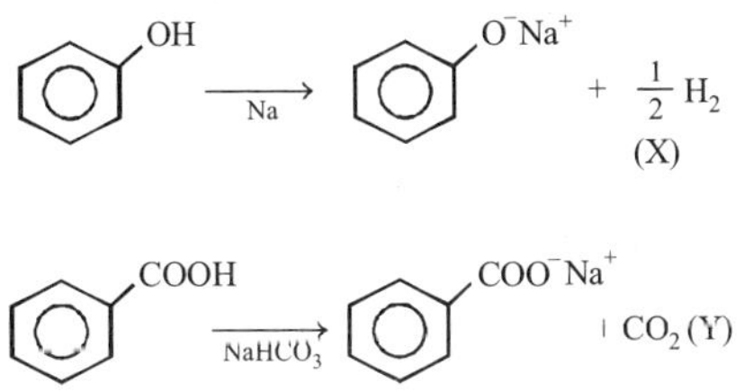

$M_X = 2$ g/mol

$M_Y = 44$ g/mol

211. (3) 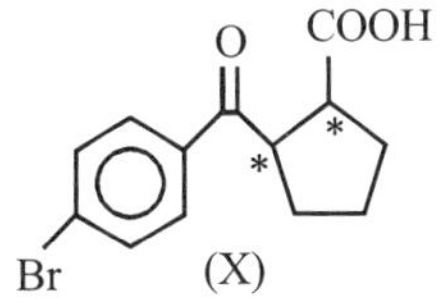

Two chiral centres and due to presence of –COOH group, compound dissolves in $NaHCO_3$.

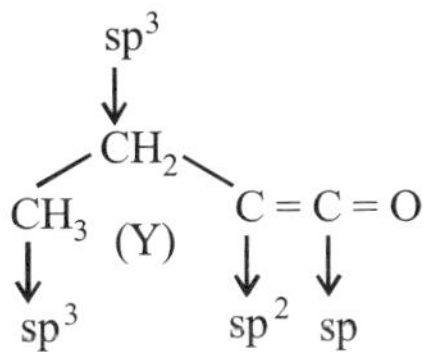

212. (4) 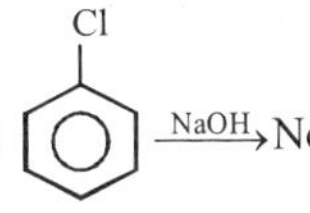 No reaction (due to the absence of an activating group)

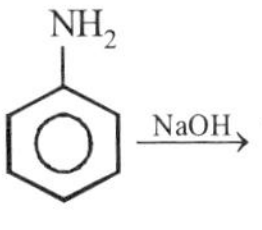 No reaction (due to basic amino group)

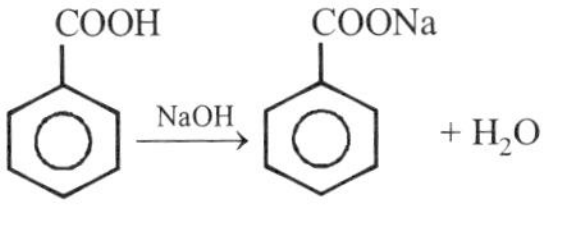

Ethyl acetate layer in funnel are mixture of chloro benene and aniline

213. (4) Those compounds which are more acidic than H_2CO_3 can gives effervescence of CO_2 with aq. $NaHCO_3$. Compound given in option (d) is the weakest acid.

214. (1) $EtO_2C-C\equiv C-H$

Conjugate base of terminal alkyne will be sp hybridised and localised. For other compounds conjugate base will be resonance stablised.

215. (4)

(A) $R-C\equiv N \xrightarrow[\text{mild condition}]{\text{(i)} H^+/H_2O} R-\overset{O}{\overset{\|}{C}}-NH_2$

(B) $R-MgX \xrightarrow[\text{(ii)} H_3O^+]{\text{(i)} CO_2} R-COOH$

(C) $R-C\equiv N \xrightarrow[\text{(ii)} H_3O^+]{\text{(i)} SnCl_2/HCl} R-CHO$

(D) 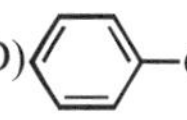 $-COCl \xrightarrow[\text{(ii)} Br_2 \text{ water}]{\text{(i)} H_2/Pd-BaSO_4} Ph-COOH$

216. (4) Rate of hydrolysis ∝ Leaving group ability

$\underset{(p)}{CH_3-CH_2-\overset{O}{\overset{\|}{C}}-Cl} > \underset{(q)}{Et-\overset{O}{\overset{\|}{C}}-O-\overset{O}{\overset{\|}{C}}-CH_3} > \underset{(r)}{Et-\overset{O}{\overset{\|}{C}}-OEt} > \underset{(s)}{Et-\overset{O}{\overset{\|}{C}}-NH_2}$

217. (4) Compounds with at least 1 α–H can undergo oxidation with $KMnO_4$ and resulting in the formation of benzoic acid.

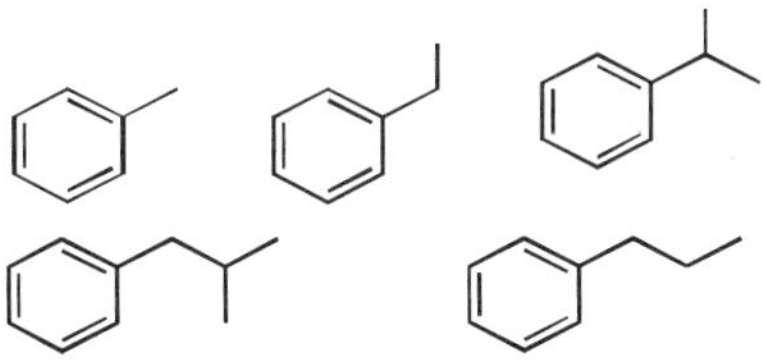

218. (1) This reaction is an example HVZ reaction

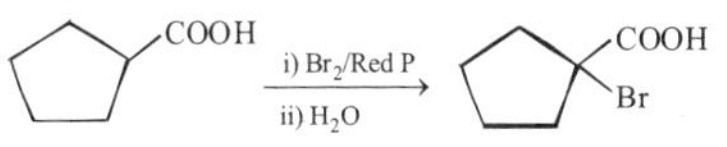

219. (1) +I effect decreases the acidic nature.

Thus the correct order is:

$HCOOH > \underset{\text{Min. +I}}{CH_3COOH} > CH_3CH_2COOH > \underset{\text{Max. +I}}{CH_3CH_2CH_2COOH}$

220. (4) Reaction sequence given below

$CH_3-COOH \xrightarrow{LiAlH_4} CH_3-CH_2OH \xrightarrow{PCC} CH_3-CHO \xrightarrow{HCN/OH^-}$

$CH_3-\overset{OH}{\overset{|}{C}}H-CN \xrightarrow{H_2O/OH^-} \underset{(P)}{CH_3-\overset{OH}{\overset{|}{C}}H-COOH}$

The major product (P) formed in the given reaction is 2-hydroxy propanoic acid.

221. (1) $CH_3-CH_2-OH \xrightarrow[\text{Reagent}]{\text{(i) Jone's}} CH_3-CHO$

$\xrightarrow{\text{(ii) } KMnO_4} CH_3COOH \xrightarrow{\text{(iii) NaOH, CaO, }\Delta} CH_4$ Methane

222. (4)

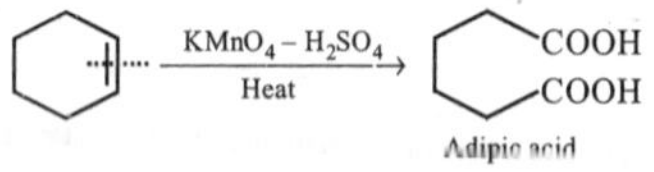

223. (3) Aldehydes and ketones having acidic α-hydrogen show aldol reaction. The acidity of α-H is due to the strong electron withdrawing effect of carbonyl group.

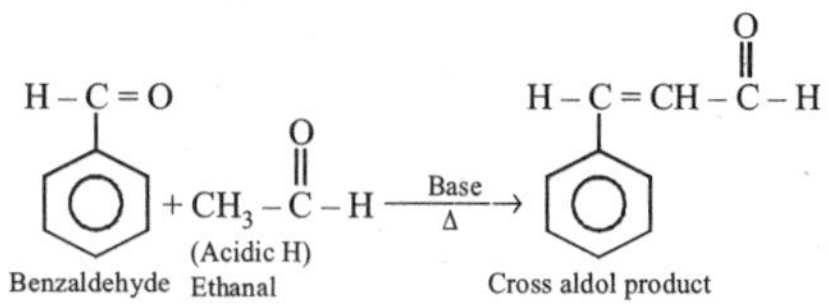

224. (2) (A) – (IV), (B) – (II), (C) – (I), (D) – (III)

(A) DiBAL-H (Di-isobutyl aluminium hydride) is a sterically hindered reducing agent, it reduces ester to aldehyde.

(B) Zn(Hg) with conc. HCl is used in Clemmensen reduction of carbonyl compound to form alkane.

(C) CH_3MgBr is a Grignard reagent, it reacts with carbonyl compound to form alcohol.

(D) $NaBH_4$ is a reducing agent, it reduces carbonyl group.

225. (3)

$C_2H_5Br \xrightarrow{\text{alc. KOH}} H_2C=CH_2 \xrightarrow{Br_2/CCl_4}$ $\underset{Br}{H_2C}-\underset{Br}{CH_2} \xrightarrow{\text{KCN excess}} \underset{CN}{CH_2}-\underset{CN}{CH_2}$

$\xrightarrow{H_3O^+ \text{ excess}} \underset{COOH}{CH_2}-\underset{COOH}{CH_2}$

Succinic acid

226. (2) HCOOH, ⟶ formic acid

CH_3COOH, ⟶ Acetic acid

formula $= C_2H_4O_2$

227. (4)

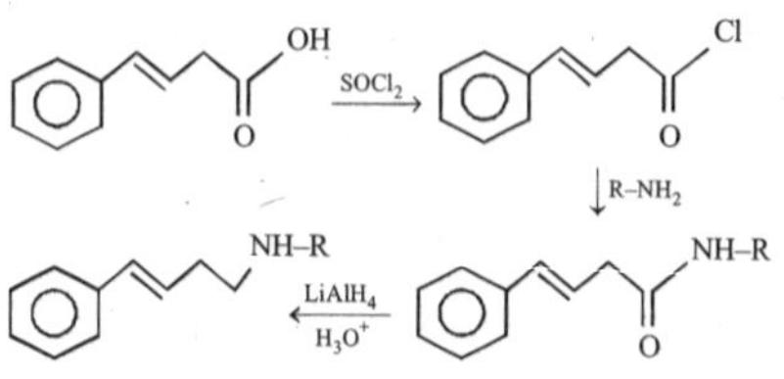

228. (3)

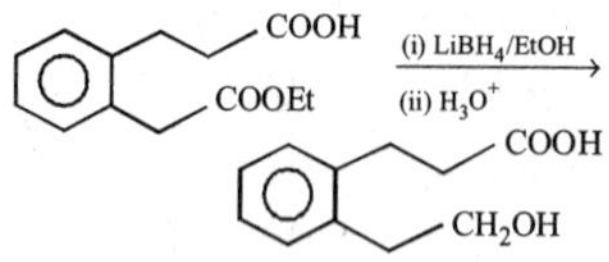

$LiBH_4$ can reduce ester selectively but not carboxylic acids.

Hence correct answer is option (c).

229. (2)

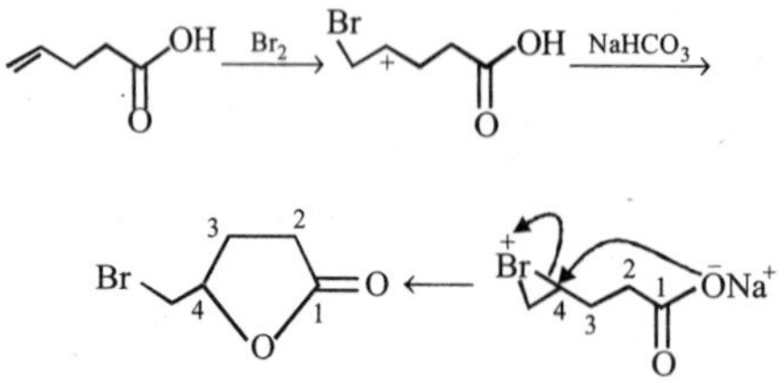

230. (4) Acidic Strength ∝ – I effect

F > Cl > Br : – I effect order

$-CH_3$ gp has + I effect

So option (d) B > D > C > E > A is the correct answer.

231. (1)

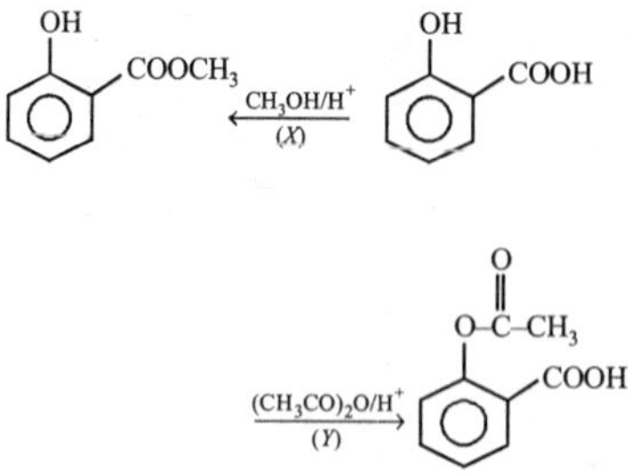

232. (3) The stronger an acid, the greater the ionization, higher the K_a value and lower will be its pK_a value.

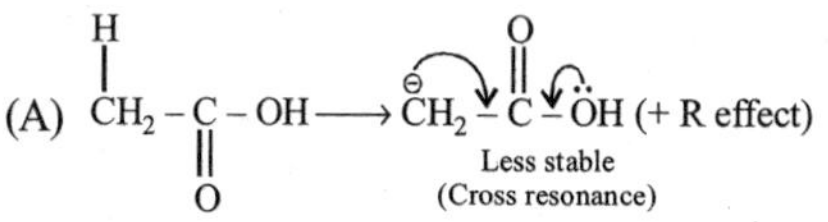

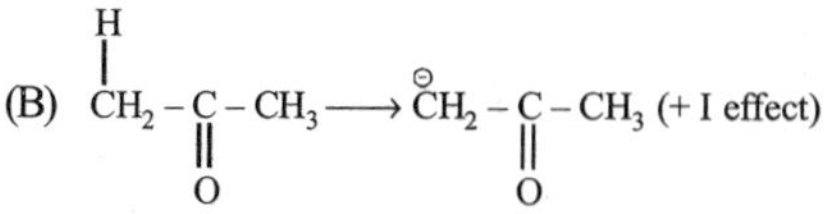

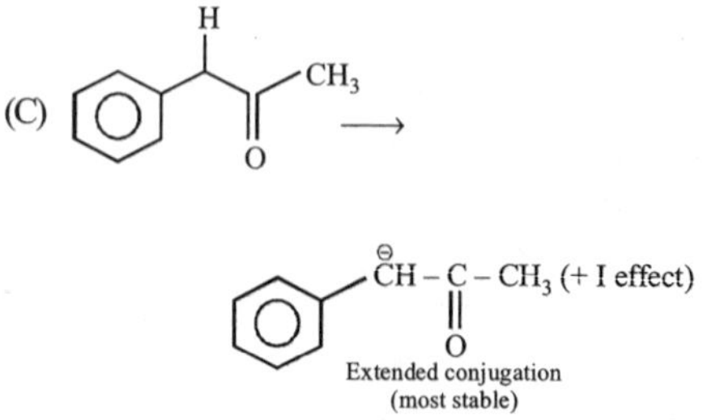

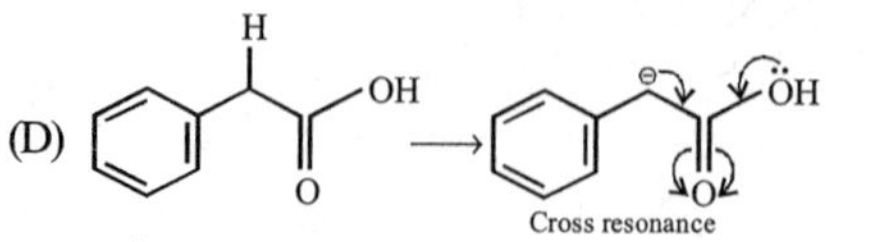

Option 'c' has lowest pK_a value as its anion is resonance stablised through benzene ring.

233. (3)

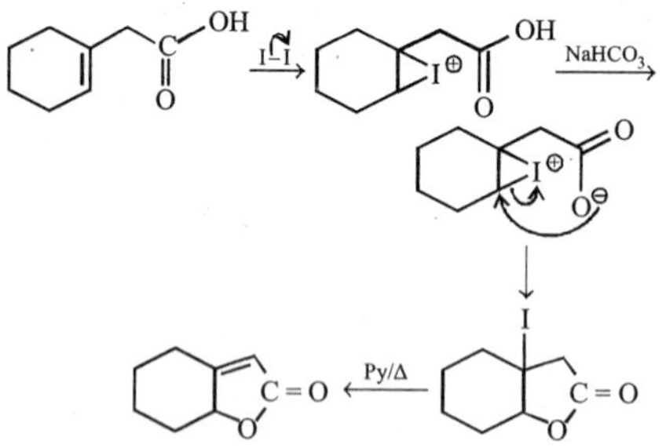

234. (1)

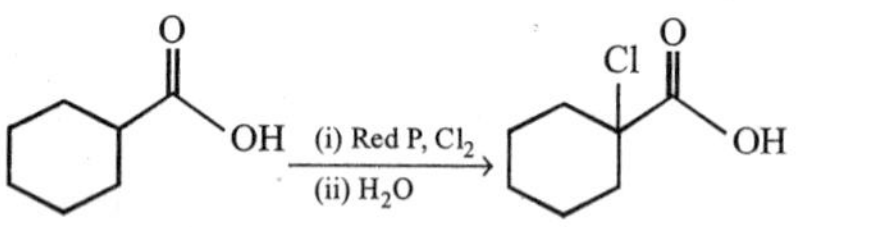

Halogenation takes place at α-position in presence of red phosphorus. It is known as Hell-Volhard-Zelinsky reaction.

235. (2) $(CH_3)_3CLi$ acts as a base or it can give halogen exchange reaction

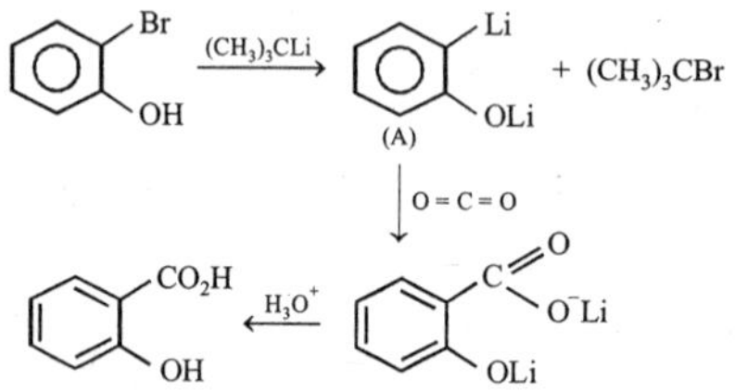

236. (1) In esterification alcohol and carboxylic acid react to form ester

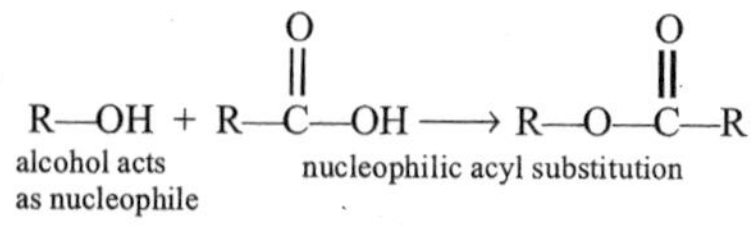

Electron withdrawing groups on carboxylic acid will increase the rate of esterification.

237. (3)

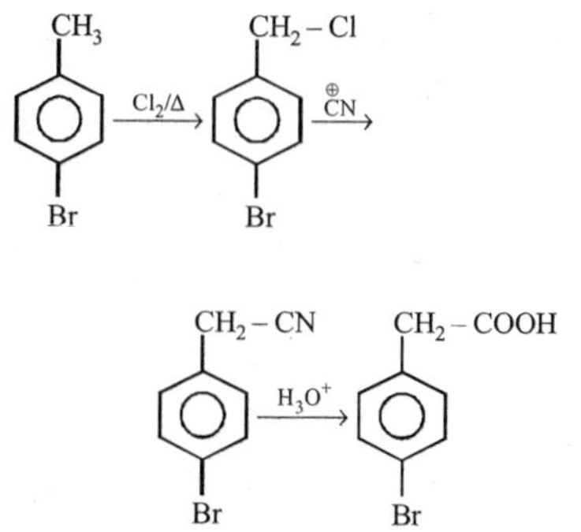

238. (4) Six possible isomers of $C_6H_3(NH_2)_2COOH$ are

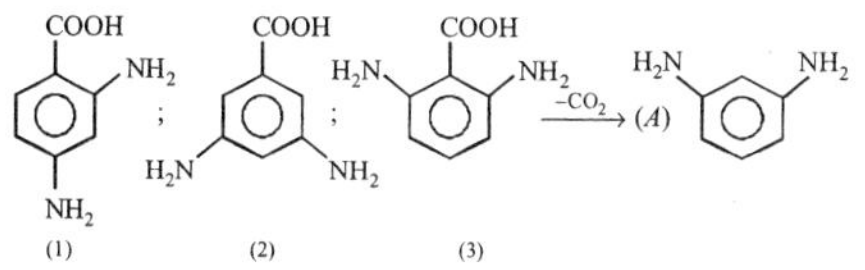

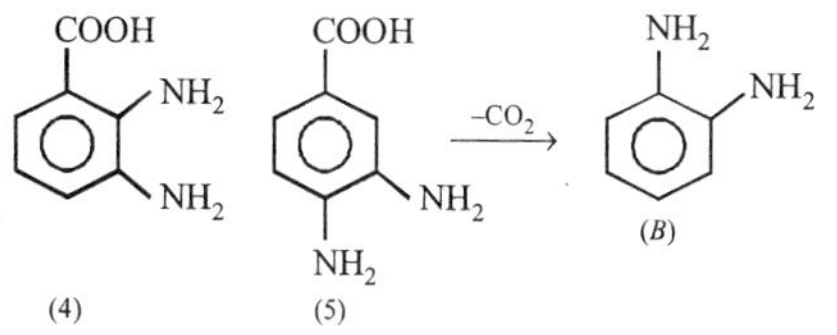

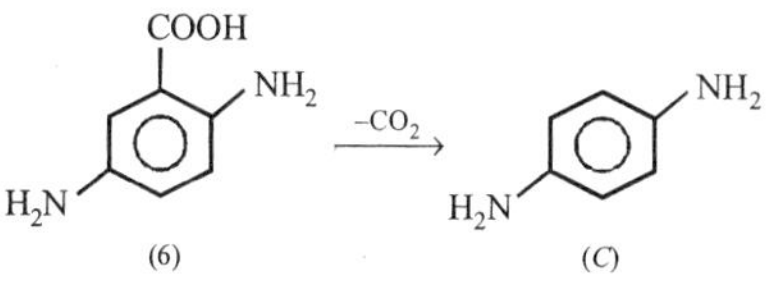

Product (*C*) should have higher melting point (i.e., 142°C) than products (*A*) and (*B*) due to its symmetry.

239. (1) Statement I: Better packing efficiency of monocarboxylic acids with even number of carbon atoms results in higher M.P. Hence statement I is correct.

Statement II: As molar mass increases, size of hydrophobic part increases, hence solubility decreases.

240. (3)

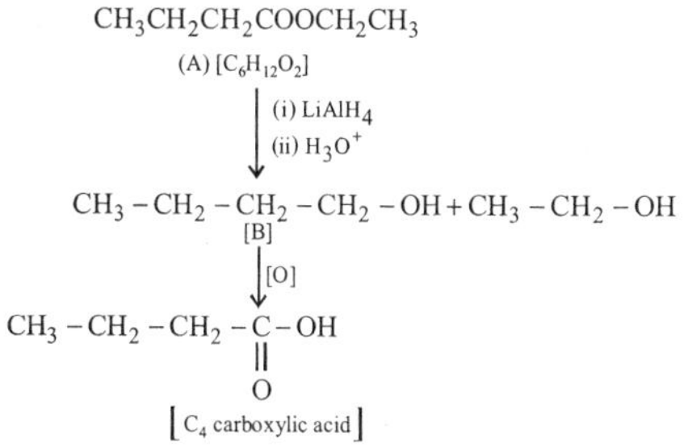

241. (3) In this reaction one mole of ethyl pent-4-yn-oate utilizes three moles of CH_3MgBr :

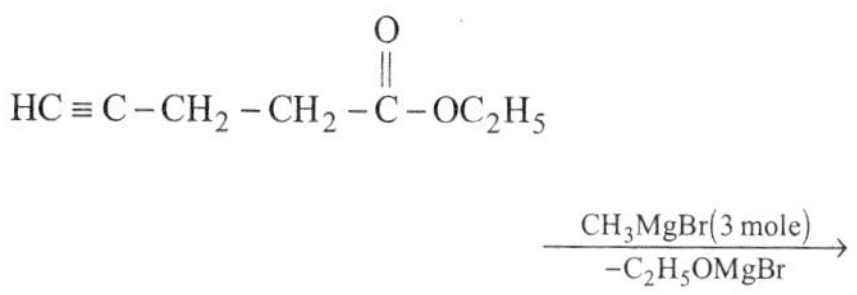

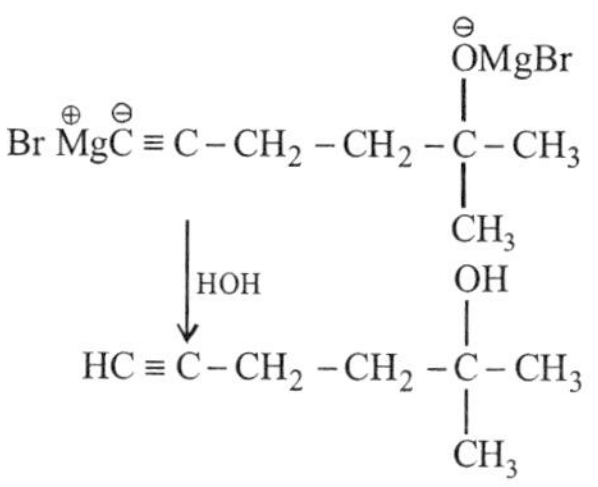

242. (3) Ethylene glycol in presence of H^+ will protect ketonic group as cyclic ketal and the ester group remains intact.

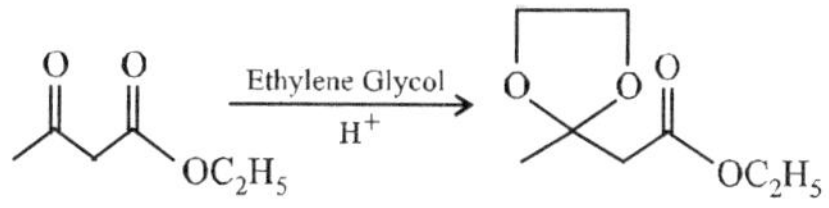

243. (1) Esters are hydrolysed in basic medium (saponification), so it is unstable in aqueous base.

244. (1) Acidity ∝ Stability of conjugate base

Stability order of conjugate bases

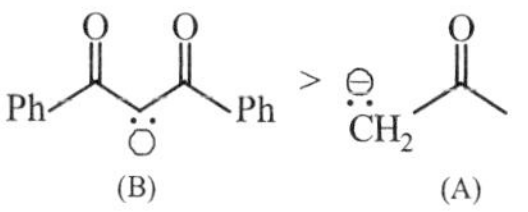

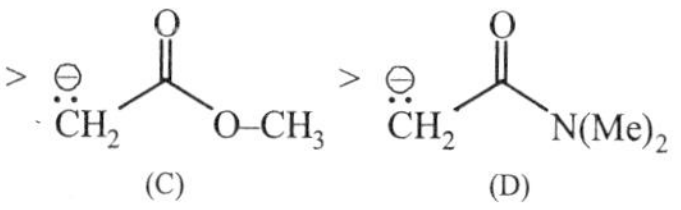

Thus increasing order of acidity is D < C < A < B.

245. (2)

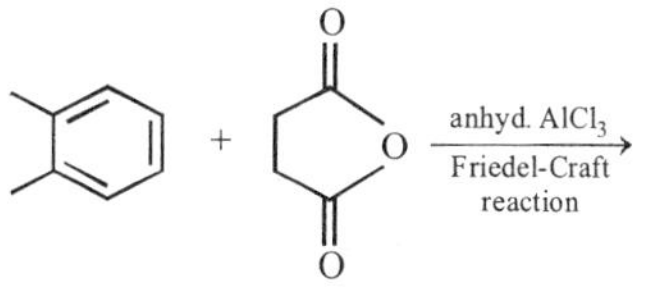

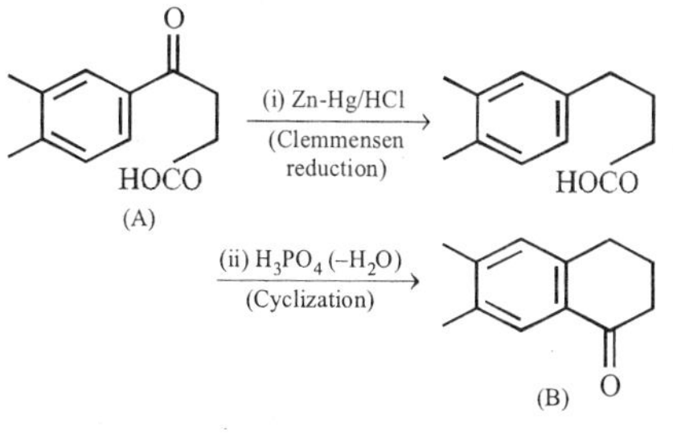

246. (1)

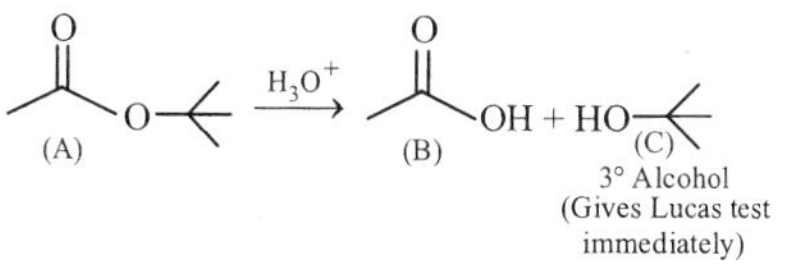

247. (1)

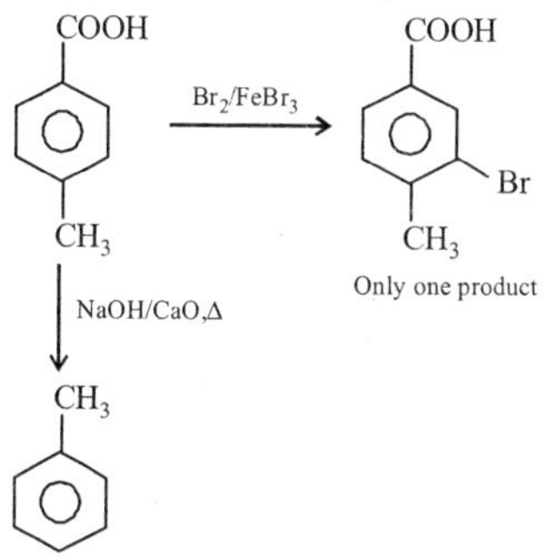

248. (1)

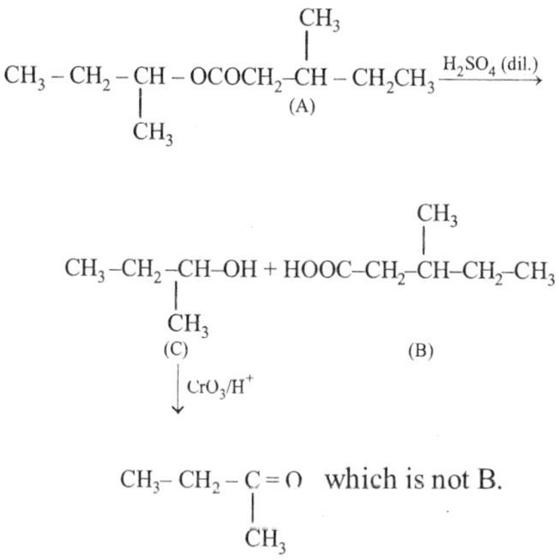

249. (4) Molecule (A) shows intramolecular H-bonding while molecule (B) shows intermolecular H-bonding. Due to presence of intermolecular H-bonding (B) has more b. pt. than molecule (A). Molecule (B) also shows intermolecule H-bonding with water which makes it more soluble than A. Since (B) is more soluble than (A) in water, so it should be less crystalline than (A).

250. (3) Acidic strength ∝ Stability of conjugate base

General order of acidic strength is

R–COOH > Ph–OH > R–C ≡ CH

In between (iii) and (iv), (iii) is more acidic due to –M effect of $-NO_2$.

Thus, decreasing order of acidity is (ii) > (iii) > (iv) > (i).

251. (1) B_2H_6 is a very selective reducing agent and usually used to reduce acid to alcohol.

252. (3)

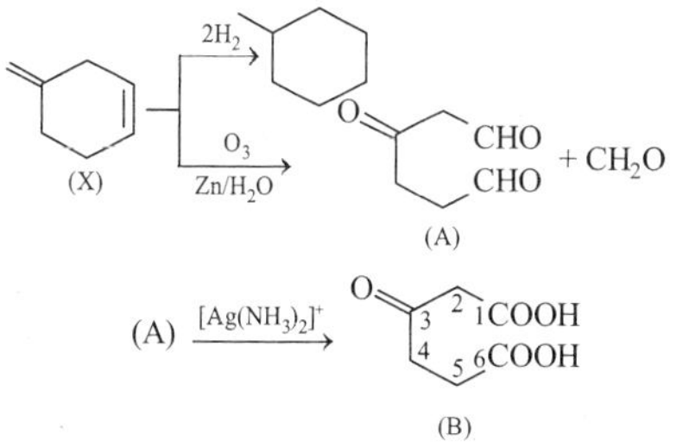

3-Oxo-hexanedicarboxylic acid

253. (4)

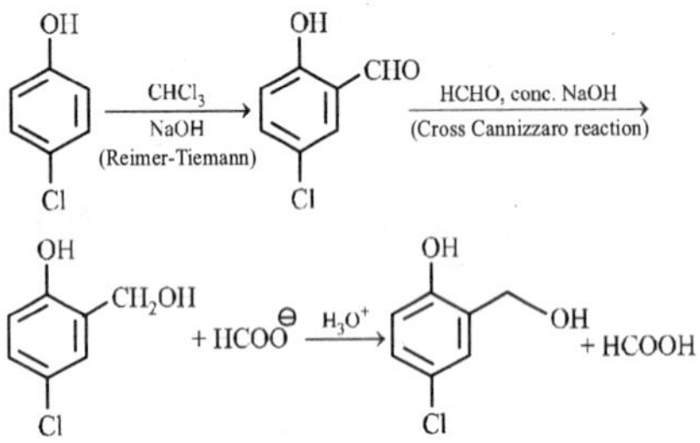

254. (2) $LiAlH_4$ reduces esters to alcohols but does not reduce C = C.

$$CH_3CH=CH-CO_2Me \xrightarrow{LiAlH_4} CH_3CH=CHCH_2OH$$

255. (1) Alkaline $KMnO_4$ converts 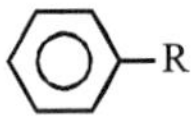 with a benzylic hydrogen into benzoic acid.

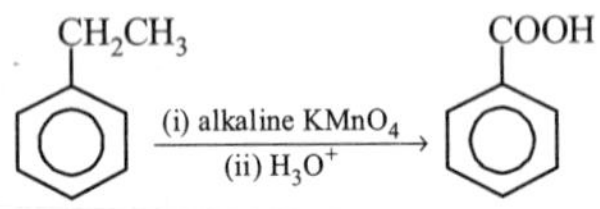

256. (1)

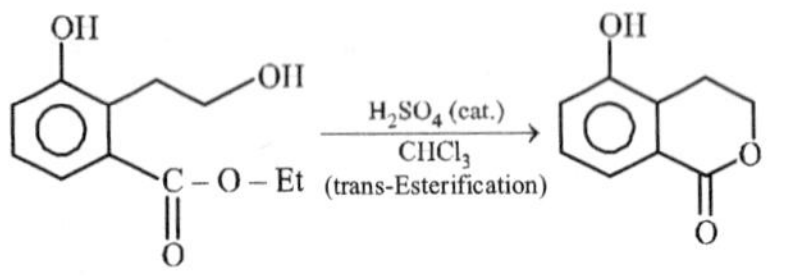

Mechanism:

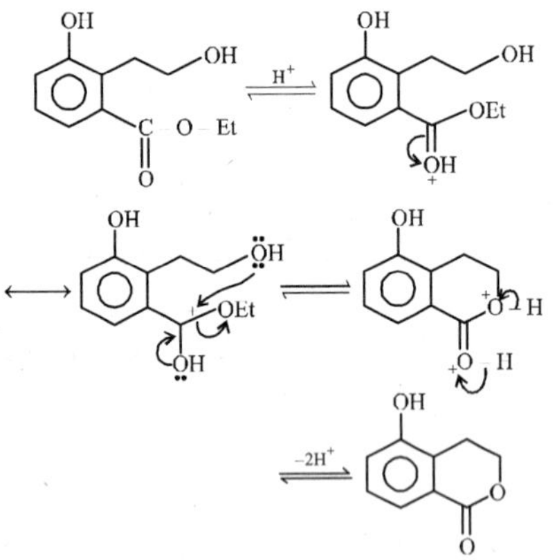

257. (1)

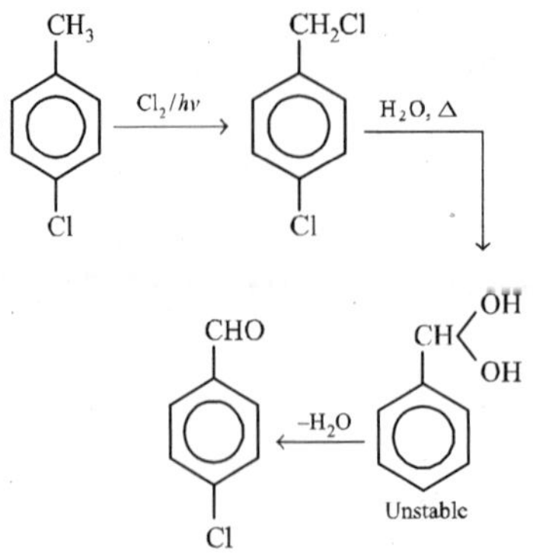

258. (3)

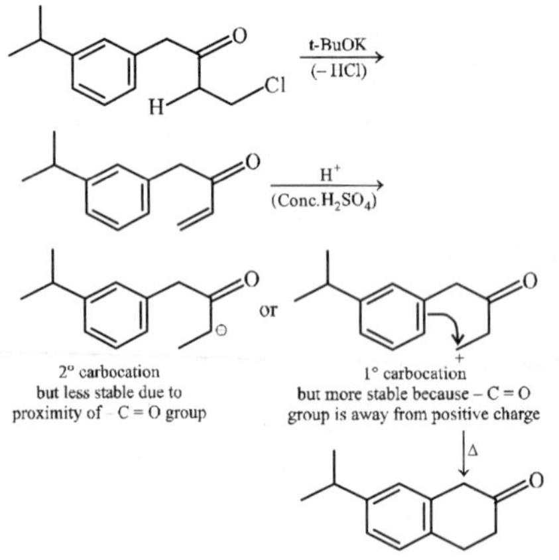

259. (3) Dissobutyl aluminium hydride, commonly abbreviated as DiBAL-H is a reducing agent for some specific functional groups. It reduces $-C\equiv N$ to $-CH=NH$ (an imine) which is easily hydrolysed to – CHO. It also reduces lactones to aldehydes.

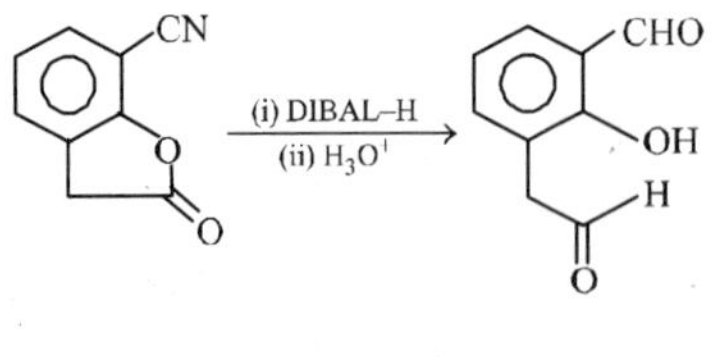

260. (4) The force of attraction between the molecules affects the melting point of a compound. Polarity increases the intermolecular force of attraction and as a result increases the melting point.

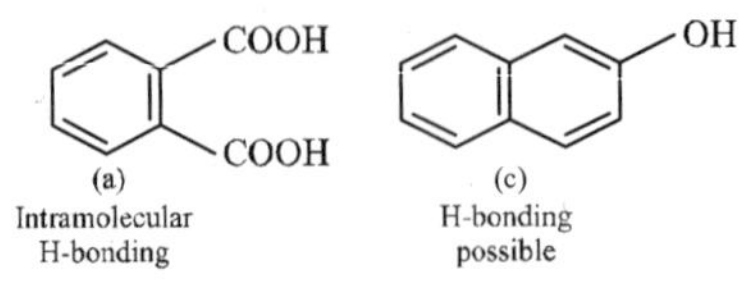

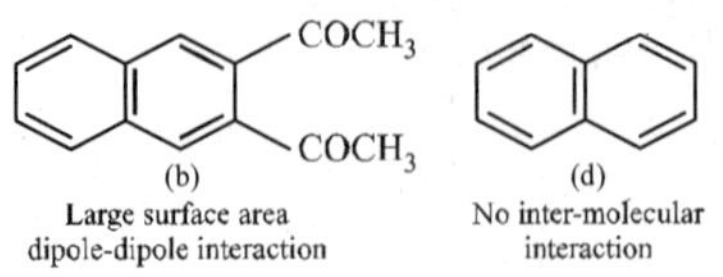

Also, the molecular mass of (d) is lowest.

261. (2) Heating of the given compound in presence of strong base is favoured for elimination reaction resulting in more stable alkene.

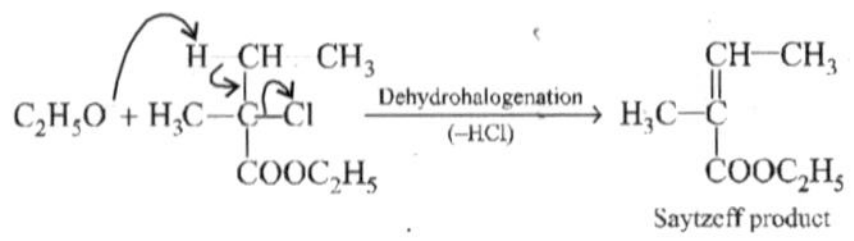

262. (4)

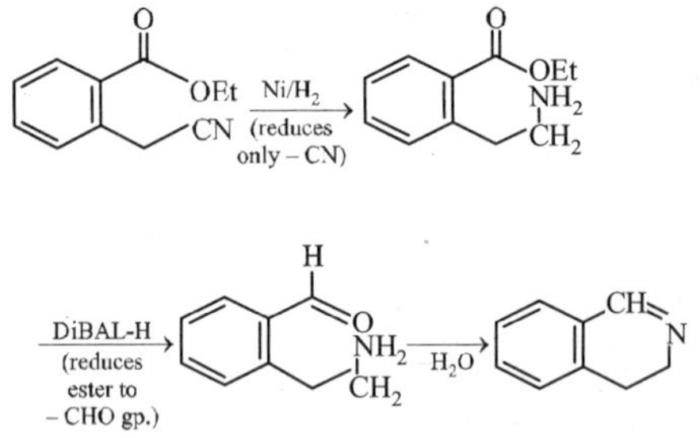

263. (3) Alkaline $KMnO_4$ is a strong oxidising agent and oxidises $-CH_3$ as well as – CO group to – COOH.

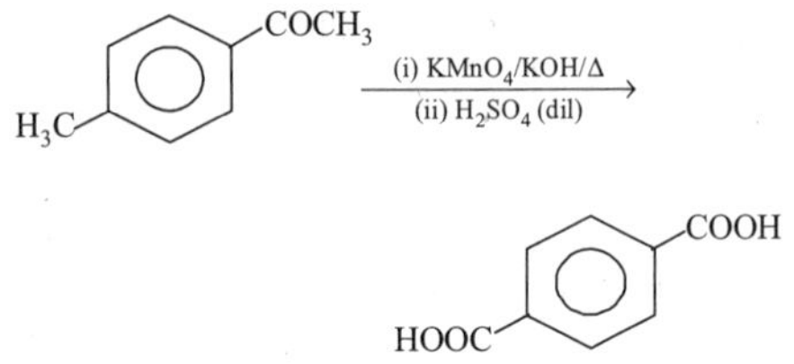

264. (3)

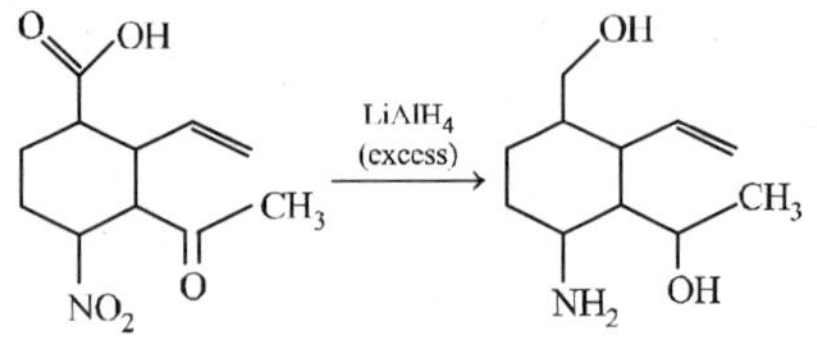

$LiAlH_4$ is a nucleophilic reducing agent, hence it will reduce – COOH to CH_2OH, – CO to – CHOH and – NO_2 to NH_2 but does not reduce C = C linkage.

265. (4)

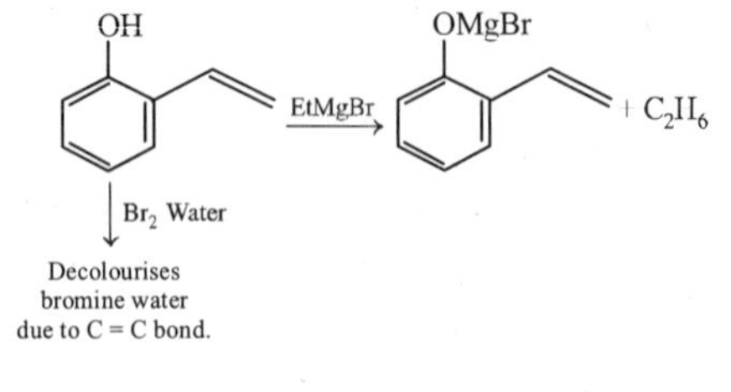

266. (1)

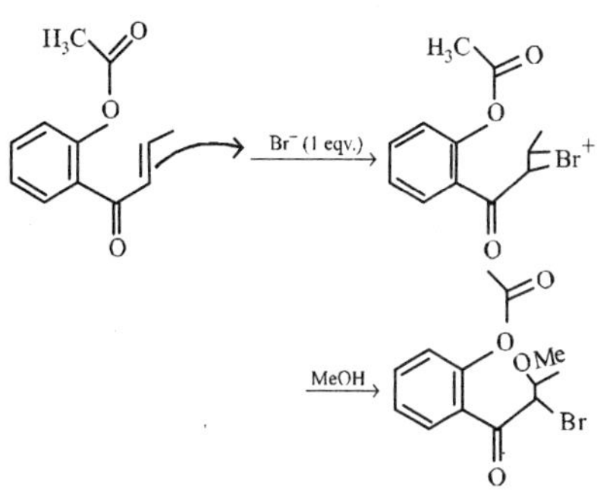

267. (1)

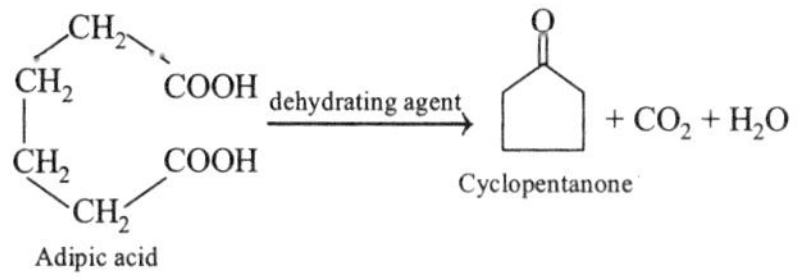

Adipic acid does not form anhydride.

268. (2) Rate of reaction ∝ Electrophilicity of carbonyl carbons, so E.W.G. increases the rate, while E.R.G. decreases the rate.

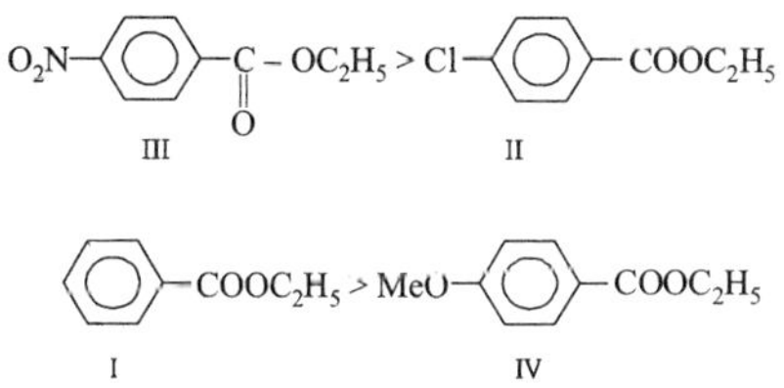

269. (1)

$$\underset{(C_7H_6O_2)}{A} \xrightarrow[\text{(ii) heat}]{\text{(i) aq. } NH_3} B \xrightarrow{Br_2/KOH} \underset{(C_6H_7N)}{C}$$

Going backward, compound C is obtained from B by Br_2 and KOH (Hoffmann bromamide reaction), so B must be an amide ($-CONH_2$) and C an amine ($-NH_2$) or $C_6H_5-NH_2$. Thus A should be benzoic acid, C_6H_5-COOH or $C_7H_6O_2$.

Reaction involved:

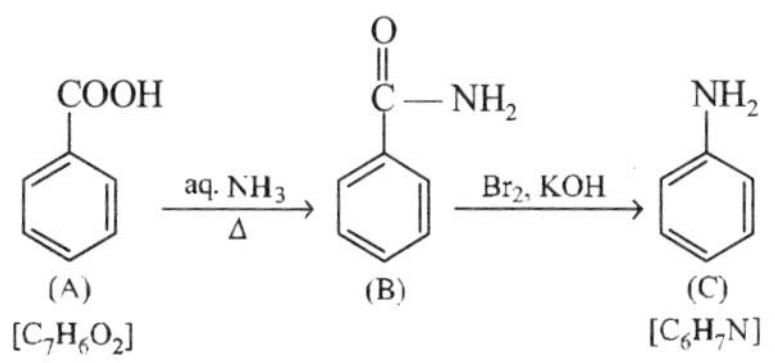

270. (4) Reaction involved:

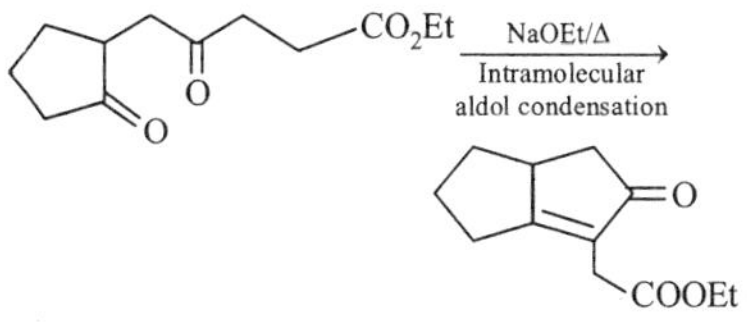

271. (4) The acidic strength of a compound or an acid depends on the inductive effect (–I). Higher the (–I) effect of a substituent, higher will be acidic strength. Now, the decreasing order of (–I) effect of the given substituents is : $NO_2 > CN > F > Cl$.

∴ The correct decreasing order of acidic strength amongst the given carboxylic acids is:

$$NO_2CH_2COOH > CNCH_2COOH > FCH_2COOH > ClCH_2COOH$$

272. (2) Acid chloride is more reactive than aldehyde. Hence, phenolic – OH will react with – COCl group first to form ester. This is followed by cyclisation in presence of conc. sulfuric acid.

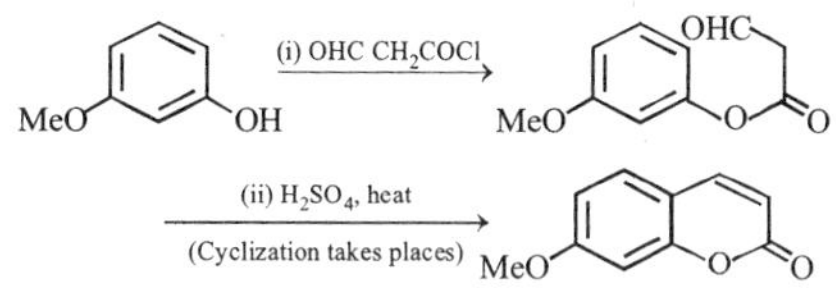

273. (2) PCC oxidizes primary alcohols to aldehydes and secondary alcohols to ketones.

In the above reaction, $-OCOCH_3$ group is hydrolyzed to secondary alcohol which is then oxidised (with PCC) to ketone.

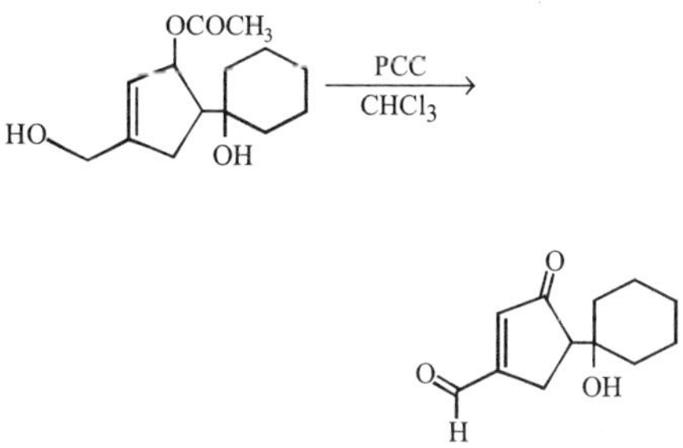

274. (1) The increasing order of the acidity of the carboxylic acids is III < II < IV < I. In aromatic acids, electron withdrawing groups like –Cl, –CN, $-NO_2$ increases the acidity, whereas electron releasing groups like $-CH_3$, –OH, $-OCH_3$, $-NH_2$ decreases the acidity.

275. (3) Since acid chloride is more reactive than alkyl halide, so COCl group will react first.

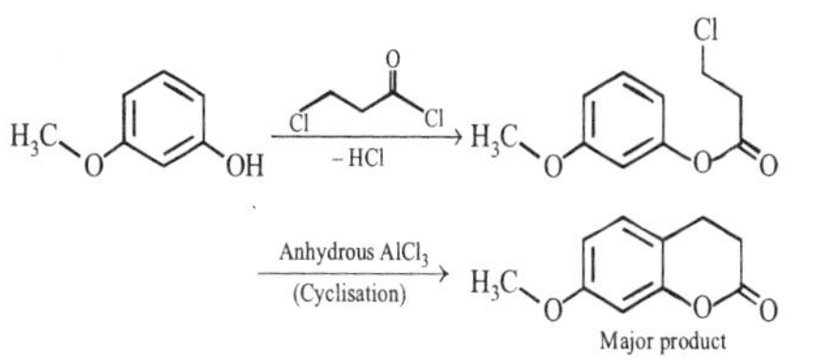

276. (4)

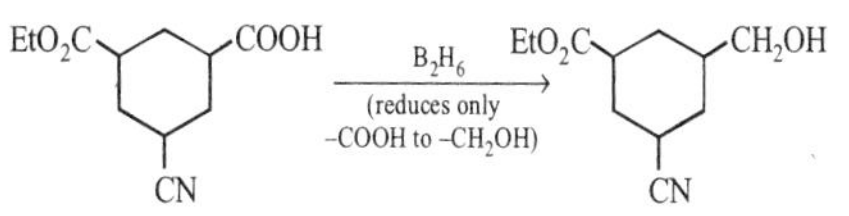

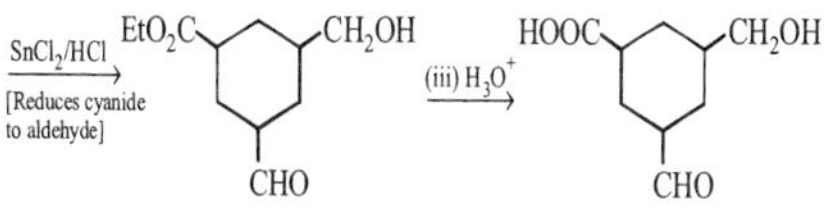

277. (1) Sodium borohydride reduces ketonic group to alcohol, but not the amide group and C = C double bond.

278. (4)

$$\underset{'X'}{Na_2C_2O_4} + \underset{\text{(conc.)}}{H_2SO_4} \rightarrow Na_2SO_4 + CO\uparrow + CO_2\uparrow + H_2O$$

$$\underset{'X'}{Na_2C_2O_4} + CaCl_2 \rightarrow \underset{\text{(white ppt.)}}{CaC_2O_4\downarrow} + 2NaCl$$

$$5CaC_2O_4\downarrow + \underset{\text{(purple)}}{2KMnO_4} + 8H_2SO_4 \rightarrow$$

$$K_2SO_4 + 5CaSO_4 + 2MnSO_4 + 10CO_2 + 8H_2O \quad \text{(colourless)}$$

279. (2) DiBAL-H is a reducing agent. It reduces both ester and carboxylic group into an aldehyde at low temperature.

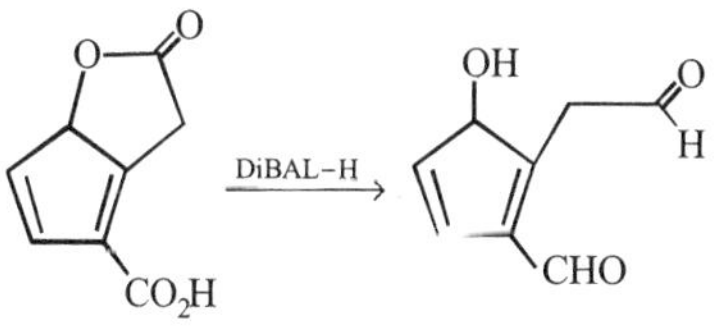

280. (3)

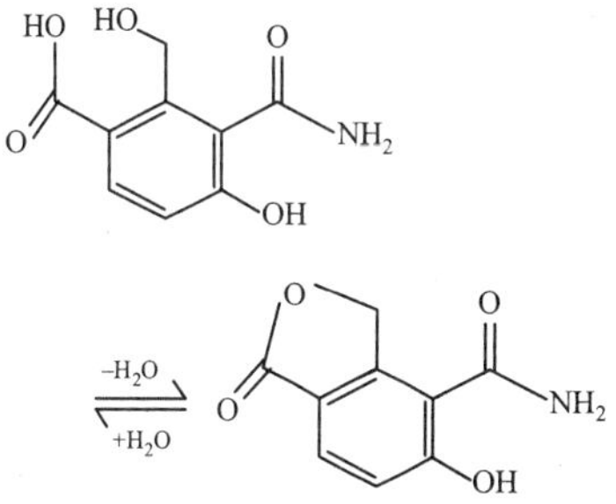

281. (3) The Bouveault reduction is a reaction in which an ester is reduced to primary alcohol using absolute ethanol and sodium.

$$R-\overset{O}{\overset{\|}{C}}-OR \xrightarrow{Na/C_2H_5OH} RCH_2OH + ROH$$

282. (2)

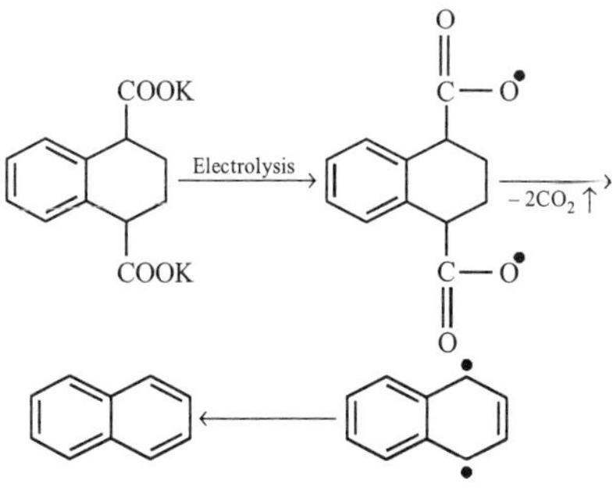

Stable due to aromatic character

283. (1) Drugs which relieve pain are called analgesics drugs. Analgesics are of two types (i) Narcotics and (ii) Non-narcotics. Aspirin (acetylsalicylic acid) is a non-narcotic analgesic.

284. (4) α-Substitution occurs when a carboxylic acid having α-hydrogens is treated with chlorine or bromine in presence of small amount of red phosphorus. This reaction is commonly known as HVZ reaction.

$$R-CH_2COOH + X_2 \xrightarrow{P} R\underset{|}{\overset{X}{C}}H-COOH + HX (X = Cl, Br)$$

285. (3)

$$CH_3COOH \xrightarrow{LiAlH_4} \underset{(A)}{CH_3CH_2OH} \xrightarrow{PCl_5} \underset{(B)}{CH_3CH_2Cl} \xrightarrow{Alc.\ KOH} \underset{(C)}{CH_2 = CH_2}$$

286. (4)

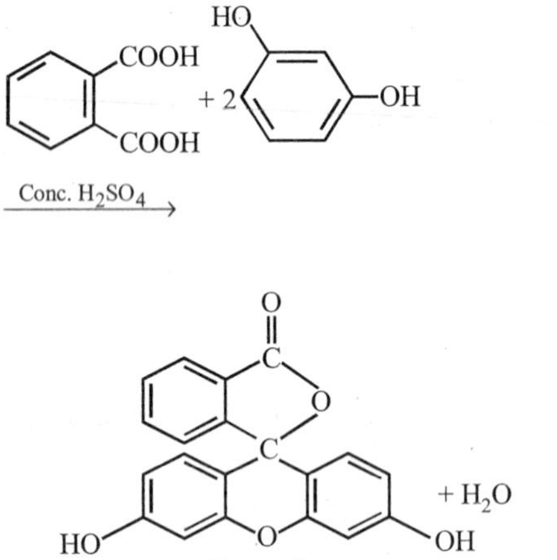

Fluorescein

287. (3) Mono-carboxylic acids are functional isomers of esters, e.g.,

$\underset{\text{Acetic acid}}{CH_3COOH}$ $\underset{\text{Methyl formate}}{HCOOCH_3}$

288. (4) Maleic acid and fumaric acids are geometrical isomers.

289. (1) The correct order of increasing acid strength is

$$(Me)_2CHCOOH < CH_3COOH < MeOCH_2COOH < CF_3COOH$$

Note: Electron withdrawing groups increase the acid strength whereas electron donating groups decrease the acid strength.]

290. (3) $pK_a = -\log K_a$; HCOOH is the strongest acid and hence it has the highest K_a or lowest pK_a value.

291. (1) $LiAlH_4$ can reduce COOH group but not the double bond.

$$CH_2 = CH-COOH \xrightarrow{LiAlH_4} CH_2 = CH-CH_2OH$$

292. (3) $CH_3CH_2COOH \xrightarrow[\text{red P}]{Cl_2} CH_3CHClCOOH \xrightarrow[-HCl]{alc.KOH} \underset{\text{Acrylic acid}}{CH_2 = CHCOOH}$

293. (6) $\underset{(P)}{RMgBr} + H_2O \rightarrow RH\uparrow + \underset{(Q)}{Mg(OH)Br}$

$1.4\ dm^3$ (or 1.4 mL) occupied by 1 gm of Q

$\therefore$ Molecular weight of $Q = \dfrac{22.4}{1.4} = 16$

$\therefore$ Q is CH_4 gas and Grignard reagent is CH_3MgBr

$$CH_3MgBr \xrightarrow[(ii)\ H_3O^+]{(i)\ CO_2\ (dry\ ice)} \underset{\text{(Molecular weight 60)}}{CH_3COOH\ (Z)}$$

$\therefore$ Weight of 0.1 mole of $CH_3COOH = 6$.

294. (5) The group –"COOH" in benzoic acid is electron withdrawl, this ist is meta-directing.

Product (B) forms as:

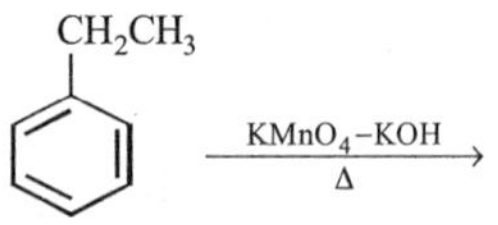

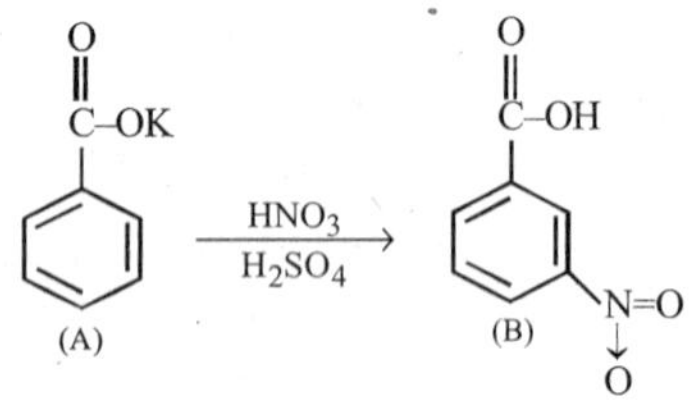

Total number of π bonds in B is 5.

295. (4)

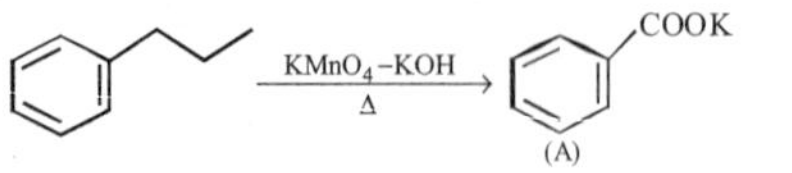

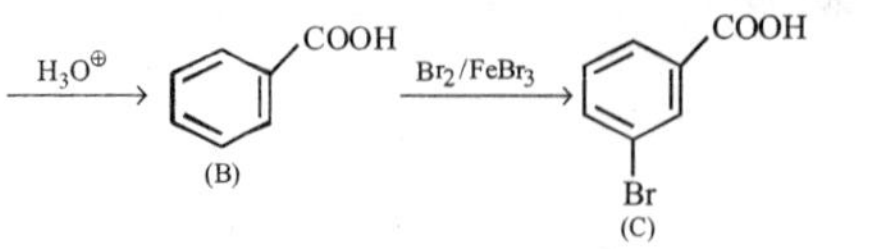

Number of π bonds in (C) = 4

296. (45)

$$\underset{\text{Mol. wt. 45g}}{CH_3CH_2NH_2} \xrightarrow{NaNO_2 + HCl} CH_3CH_2N_2^+Cl^- \xrightarrow{H_2O} CH_3CH_2-OH + \underset{14g}{N_2}$$

Given : Volume of N_2 evolved is 2.24 L i.e. 0.1 mole. i.e. $CH_3CH_2NH_2$ (ethylamine) will be 0.1 mole or 4.5 g

Hence amount (X) of ethylamine $= 45 \times 10^{-1}$ g

297. (18) $HO-\overset{\overset{O}{\|}}{C}-\underset{\underset{O}{\|}}{C}-CH_2-C\equiv C-CH_3$ (π bonds marked: C=O, C=O, two in C≡C; carbons numbered 1–6)

2 – Oxohex – 4 – ynoic acid

Number of σ-bonds = 14; Number of π-bonds = 4

$\therefore$ Total = 18.

298. (2)

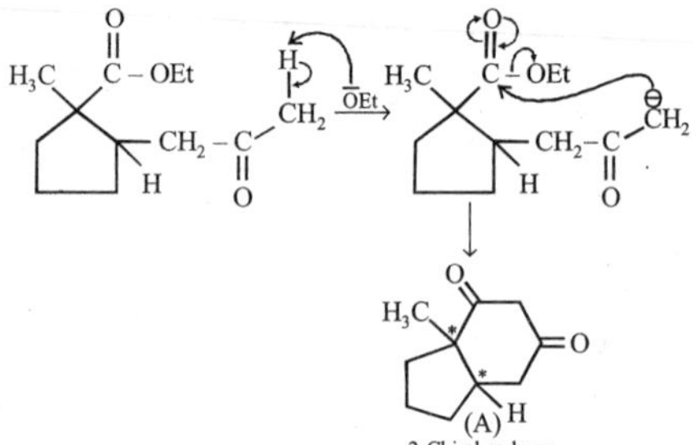

2-Chiral carbons

299. (78)

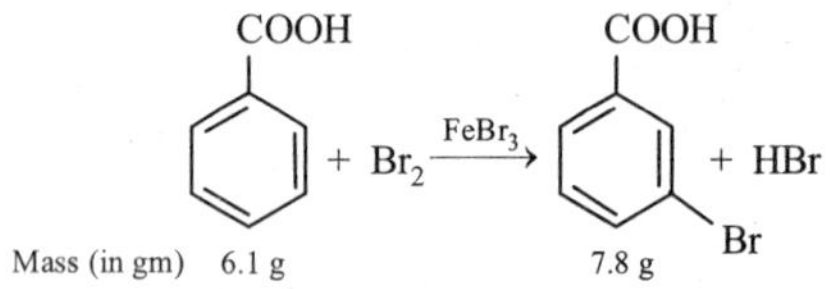

$$\text{Moles} = \frac{6.1}{122} = 0.05$$

$$\text{Moles} = \frac{7.8}{201} = 0.039$$

$$\text{Percentage yield} = \frac{0.039}{0.05} \times 100 = 77.61\%$$

$\approx 78\%$.

300. (1) Only aspirin has –COOH group.

Part-B (JEE Advanced)

1. **(12)**

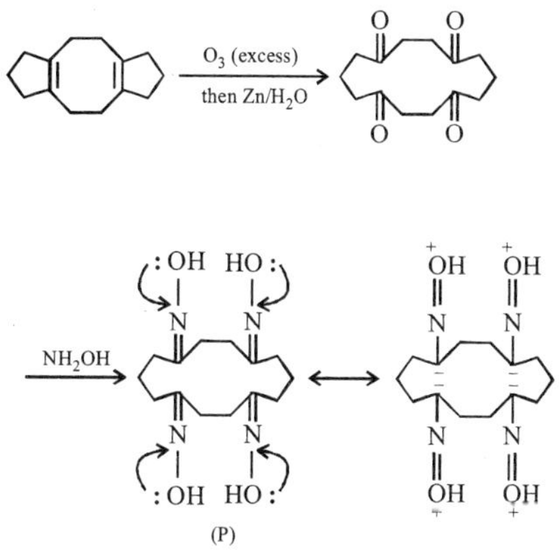

So, total 12 atoms are sp^2 hybridised (4 C atoms, 4 N atoms and 4 O atoms)

2. **(4)**

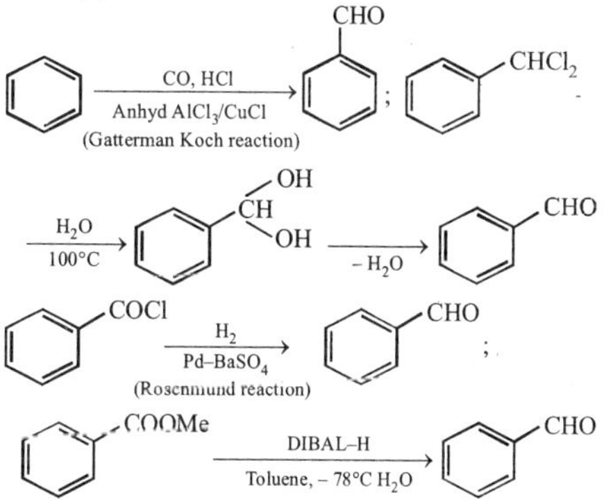

3. **(1)**

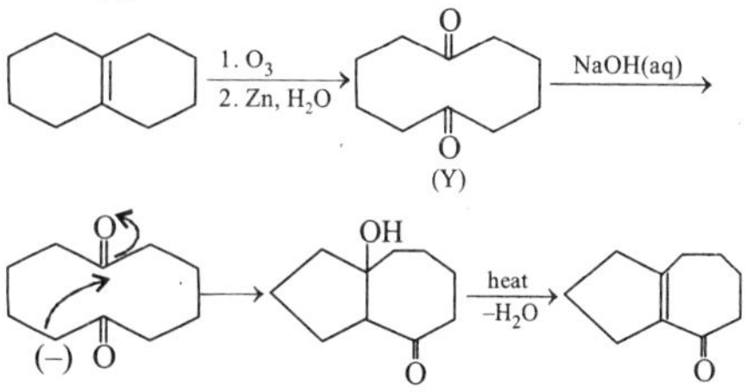

The number of intra molecular aldol condensation products (α, β –unsaturated carbonyl compound formed from **Y** is 1.

4. **(c)**

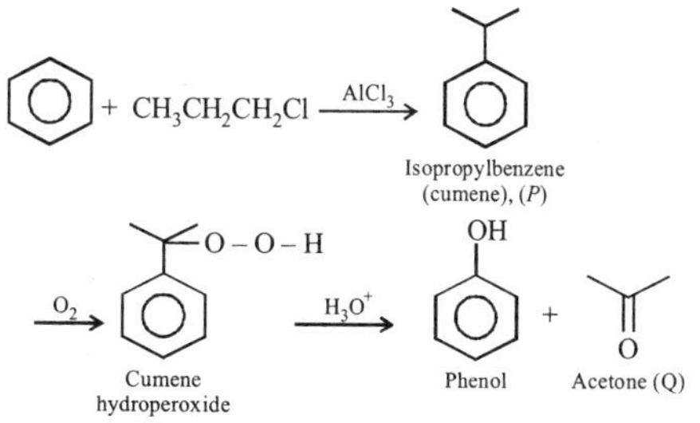

5. **$C_6H_5CH(OCOCH_3)_2$ benzylidene acetate;**

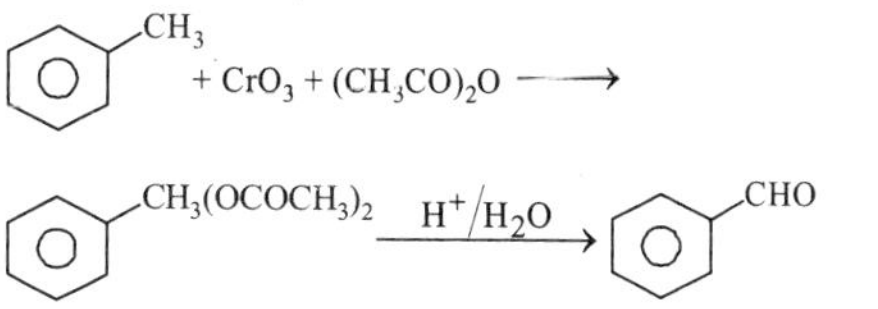

6. **(a, b, d)**

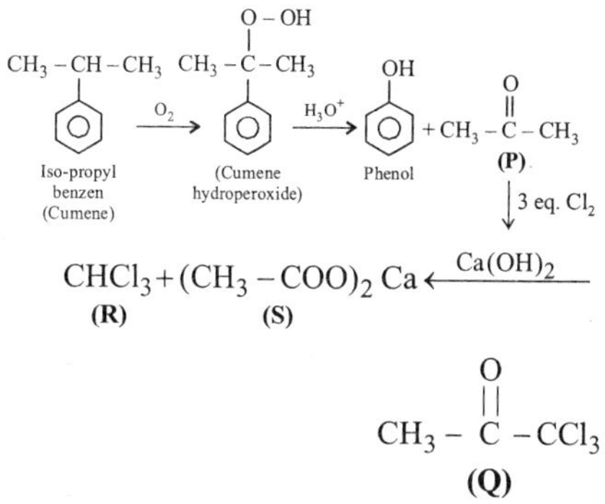

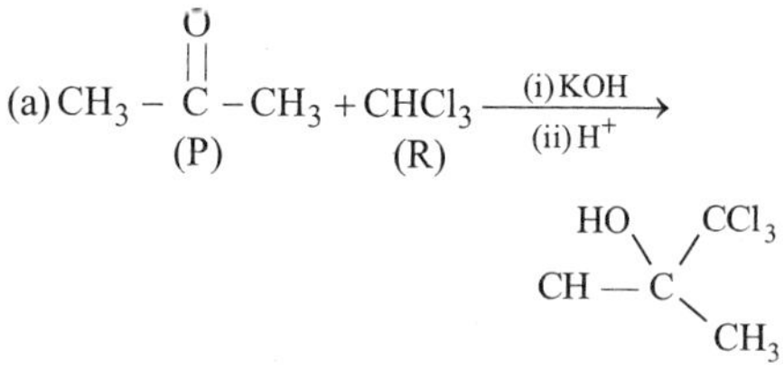

(b) $CHCl_3 + O_2 \xrightarrow{h\nu} COCl_2$
Phosgene gas

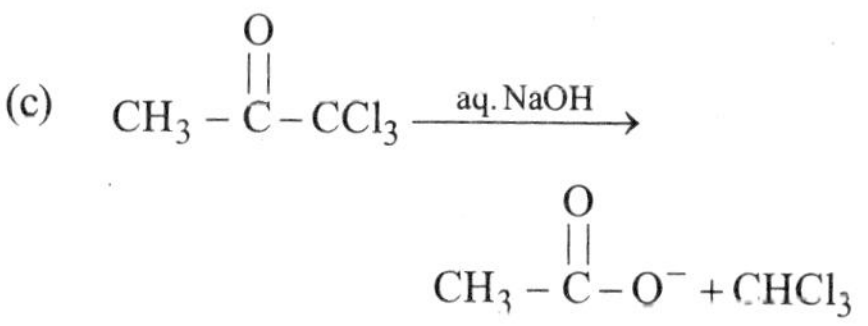

(d) $(CH_3-COO)_2Ca \xrightarrow{\Delta} CH_3-\overset{O}{\overset{||}{C}}-CH_3$

7. **(c, d)**

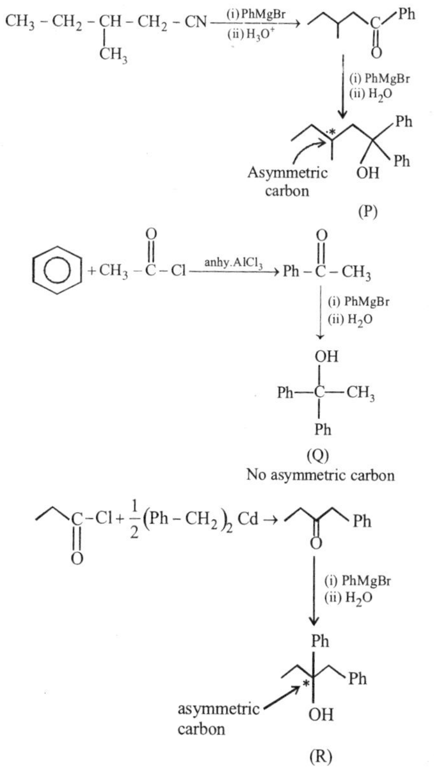

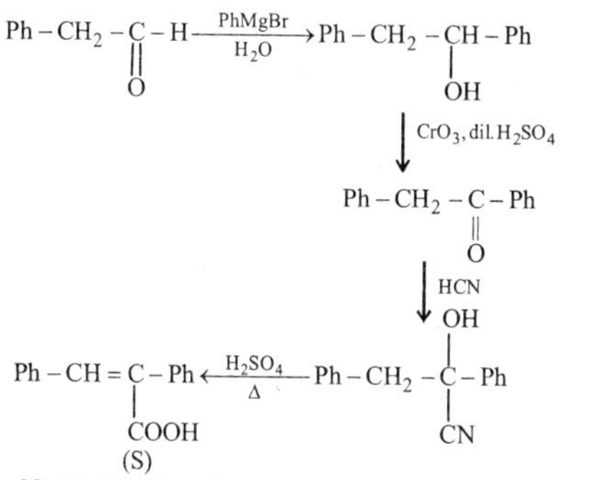

No asymmetric carbon

8. **(a, c)**

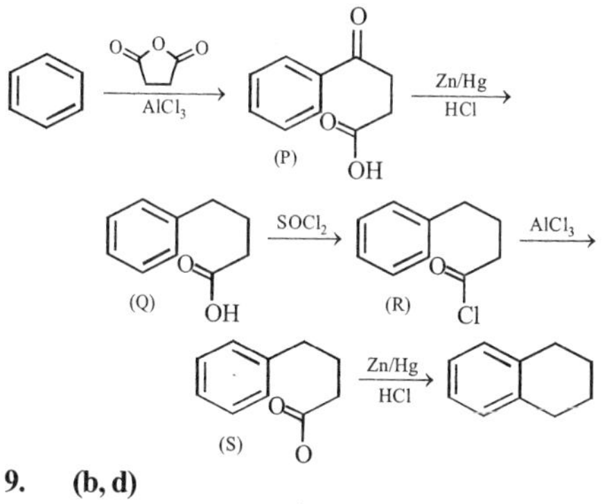

9. **(b, d)**

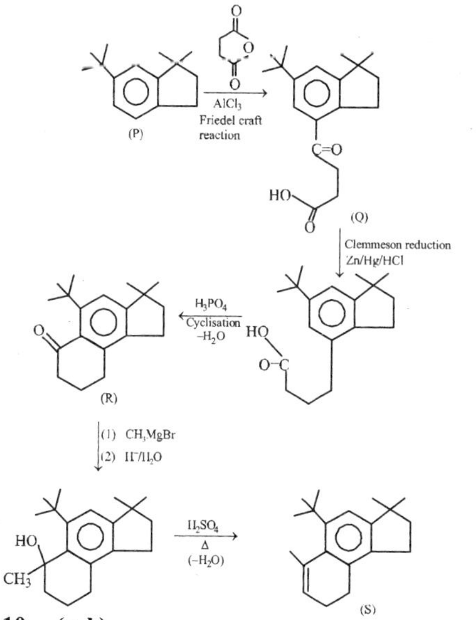

10. **(a, b)**

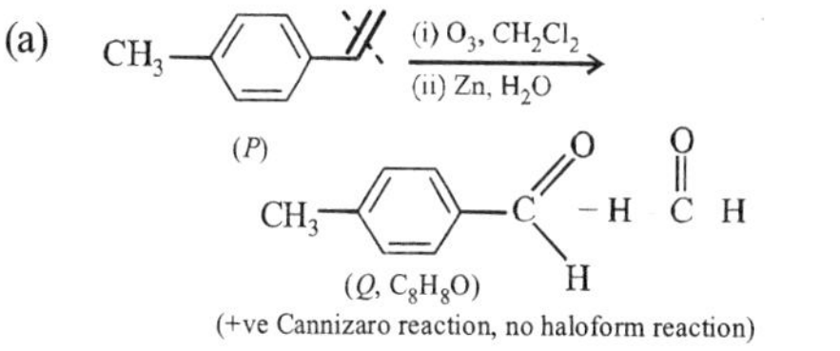

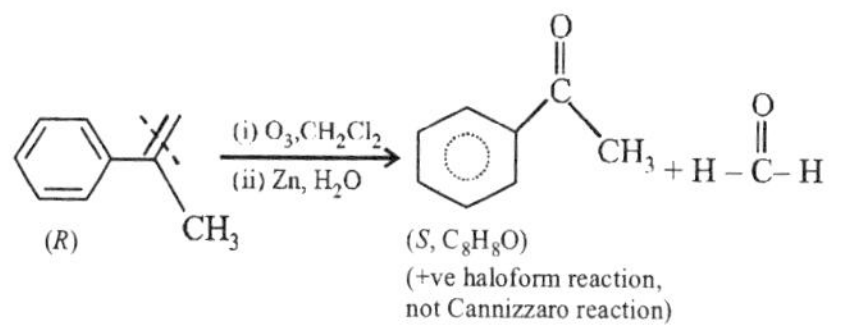

(b)

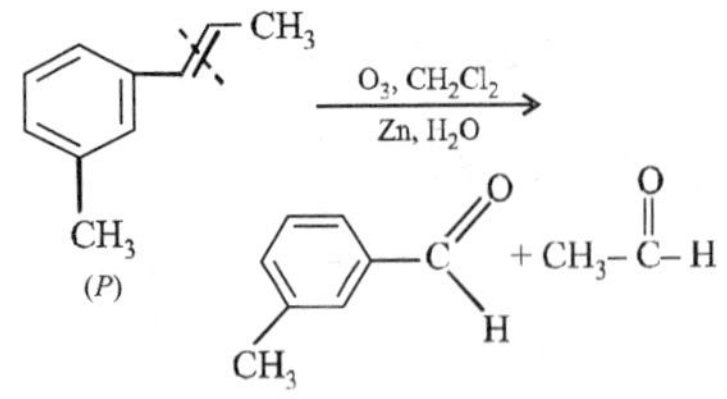

(Q) (+ve Cannizaro reaction, not haloform reaction)

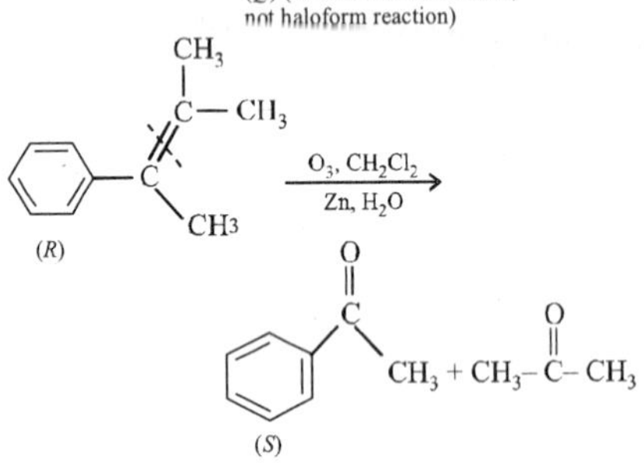

(S) (+ve haloform reaction, not Cannizzaro reaction)

11. (b, c)

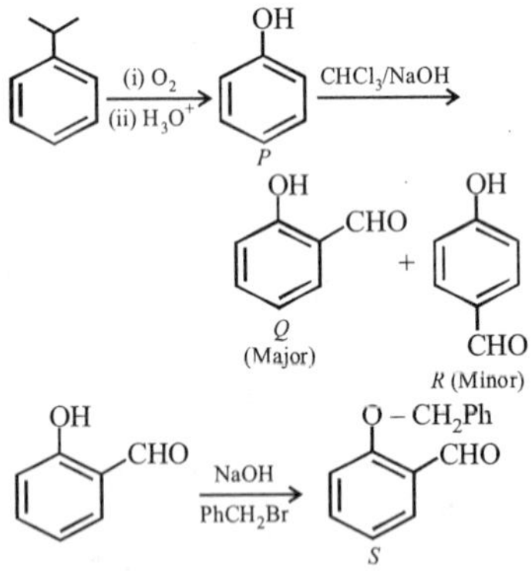

Q is steam volatile, but not R.

Q and R show positive test with 1% aqueous $FeCl_3$ solution but not S.

Q, R, S give yellow precipitate with 2, 4-dinitrophenyl hydrazine.

12. (b, d)

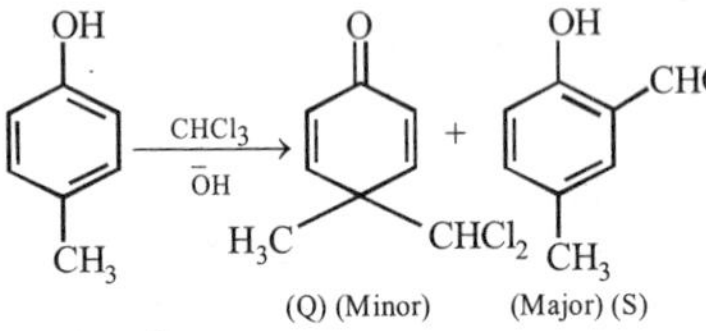

$CHCl_3 + \bar{O}H \longrightarrow :CCl_2 + H_2O + Cl^-$

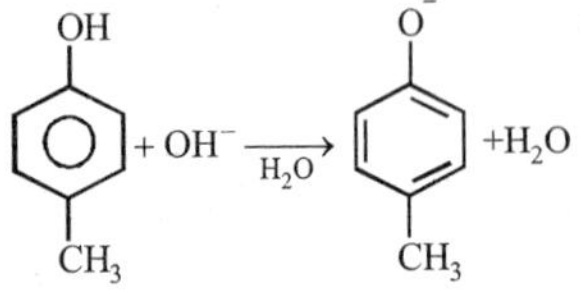

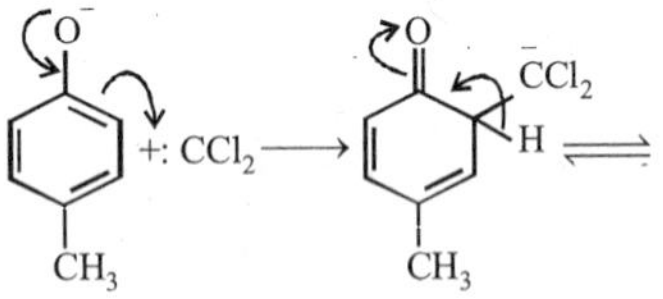

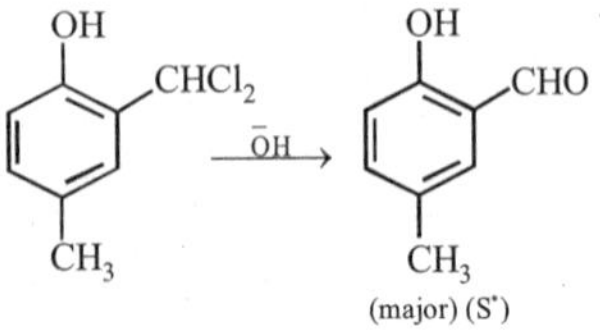

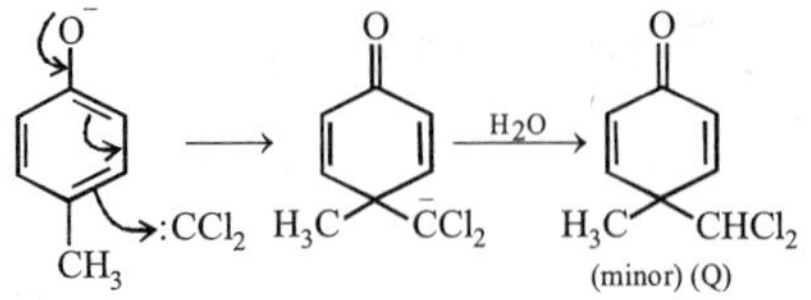

For 13-15 (Passage-IV)

Before answering these question let us complete the sequence of reactions given in data.

The given compound (M) i.e.

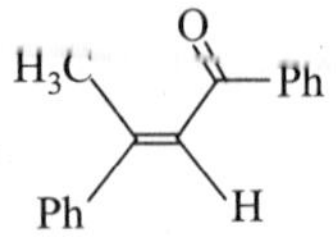

is the only product formed by the action of KOH on compound K.

The compound K is $Ph-\underset{\displaystyle CH_3}{\underset{|}{C}}=O$

$$2Ph-\underset{\displaystyle CH_3}{\underset{|}{C}}=O \xrightarrow[\text{(Aldol conden.)}]{KOH} Ph-\underset{\displaystyle CH_3}{\underset{|}{C}}=CH-\overset{\displaystyle O}{\overset{||}{C}}-Ph$$

Compound K (i.e. $Ph-\underset{\displaystyle CH_3}{\underset{|}{C}}=O$) is one of the products of ozonolysis of compound I. Therefore the compound I may be

$$\underset{(I)}{>C=\underset{\displaystyle CH_3}{\underset{|}{C}}-Ph} \xrightarrow{\text{Ozonolysis}} \underset{(J)}{>CO} + \underset{(K)}{Ph-\underset{\displaystyle CH_3}{\underset{|}{C}}=O}$$

$$\underset{(J)}{>C=O} \xrightarrow{KOH} \underset{\text{Benzylalcohol}}{Ph-CH_2-OH}$$

Thus J seems to be C_6H_5CHO and hence I is

$$\underset{I}{PhCH=\underset{\displaystyle CH_3}{\underset{|}{C}}-Ph}$$

Now we will try to answer the questions.

13. (b) As can be seen from above reaction sequence compound (I) is $Ph-CH=\underset{\displaystyle CH_3}{\underset{|}{C}}-Ph$ and it is formed by catalytic dehydration (acid catalysed) of a tertiary alcohol (compound H). Therefore, compound H is

$$Ph-CH_2-\underset{\displaystyle CH_3}{\overset{\displaystyle OH}{C}}-Ph$$

$$\underset{(H)}{Ph-CH_2-\underset{\displaystyle CH_3}{\overset{\displaystyle OH}{C}}-Ph} \xrightarrow[-H_2O]{H^+} \underset{(I)}{Ph-CH=\underset{\displaystyle CH_3}{\underset{|}{C}}-Ph}$$

(H) can be formed by the action of

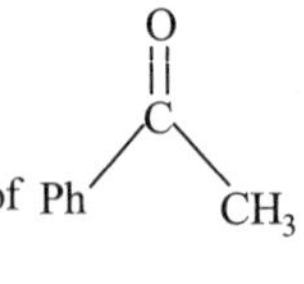

with $PhCH_2MgBr$ as follows

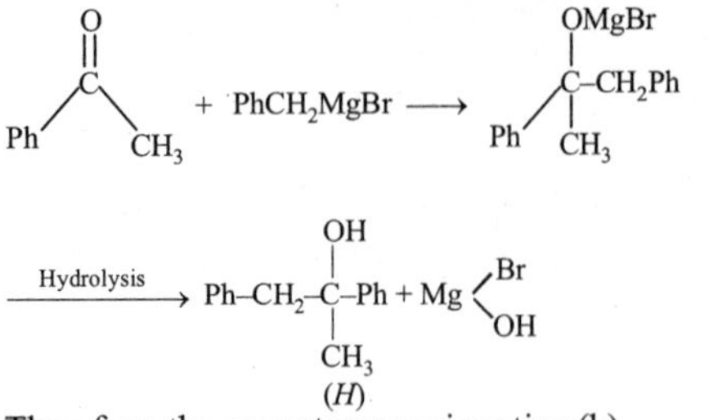

Therefore, the correct answer is option (b)

14. (a) As can be seen form the above sequence of reactions the structure of compound (I) is

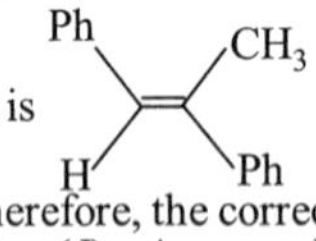

Therefore, the correct answer is option (a).

15. (d) As can be seen from the above sequence of reactions the structures of compounds J, K and L respectively are

$$\underset{(J)}{Ph-CHO}, \underset{(K)}{Ph-\underset{\displaystyle CH_3}{\underset{|}{C}}=O} \text{ and } \underset{(L)}{PhCOO^-K^+}$$

Thus, the correct answer is option (d),

16. (c) (P)

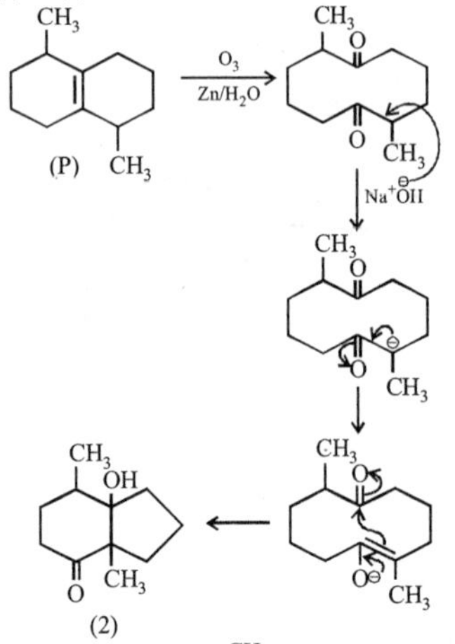

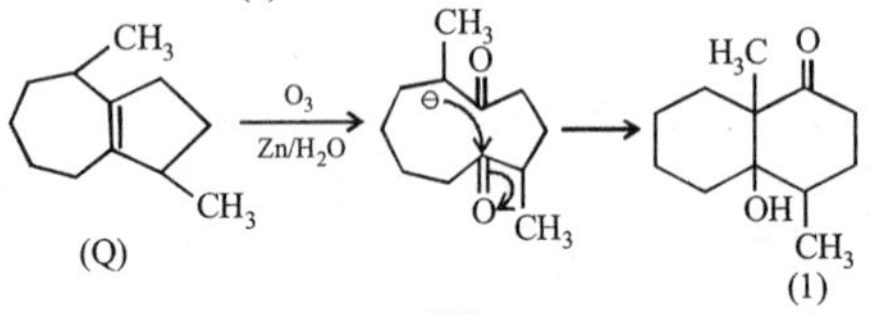

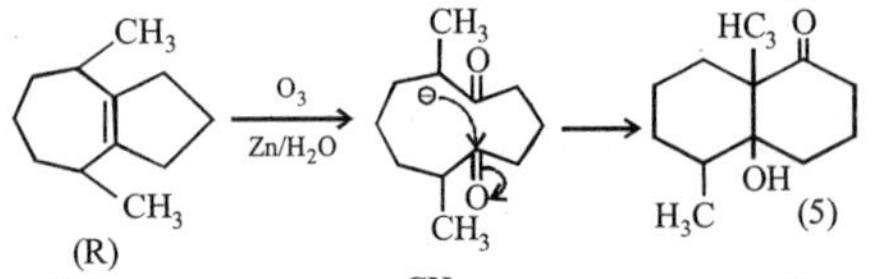

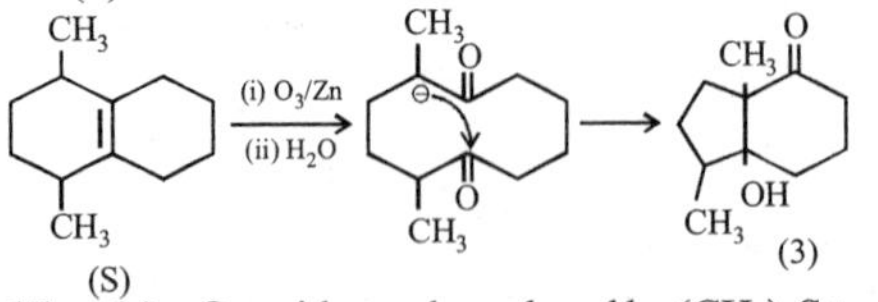

17. (a) Ozonide can be reduced by $(CH_3)_2S$ to give carbonyl compounds and dimethyl sulphoxide.

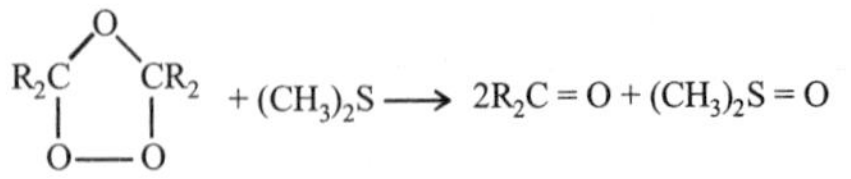

18. $H_2C=O + O=\overset{CH_3}{\overset{|}{C}}-CH=O+O=CH_2$

$\uparrow O_3$

$H_2C=\overset{CH_3}{\overset{|}{C}}-CH=CH_2$ (Monomer)

Thus, the possible polymer should be

$nCH_2=\overset{CH_3}{\overset{|}{C}}-CH=CH_2 \rightarrow \left[CH_2-\overset{CH_3}{\overset{|}{C}}=CH-CH_2\right]_n$

Structure of all *cis* configuration of the polymer.

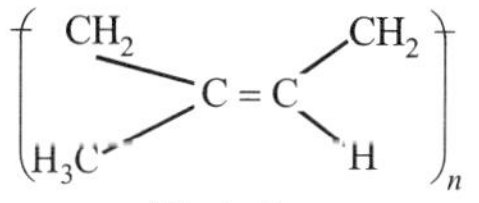

All *cis* form

19.

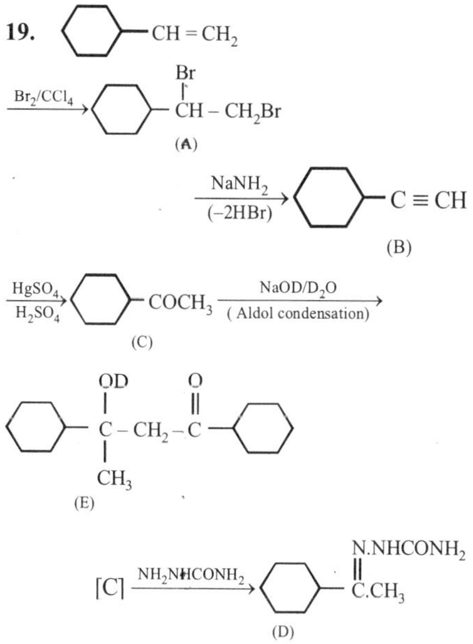

20.

$C_6H_5-\overset{OH}{\underset{H}{C}}-\overset{\delta-}{C}\equiv\overset{\delta+}{C}H \xrightarrow{\overset{\delta+}{H}-\overset{\delta-}{OH}} C_6H_5-\overset{OH}{\underset{H}{C}}-\overset{OH}{\underset{H}{C}}=CH$

$\xrightarrow{\text{tautomerises}} C_6H_5-\overset{OH}{CH}-\overset{H}{CH}-\overset{O}{\overset{||}{CH}} \xrightarrow{-H_2O}$

$C_6H_5-CH=CH-CHO$

21.

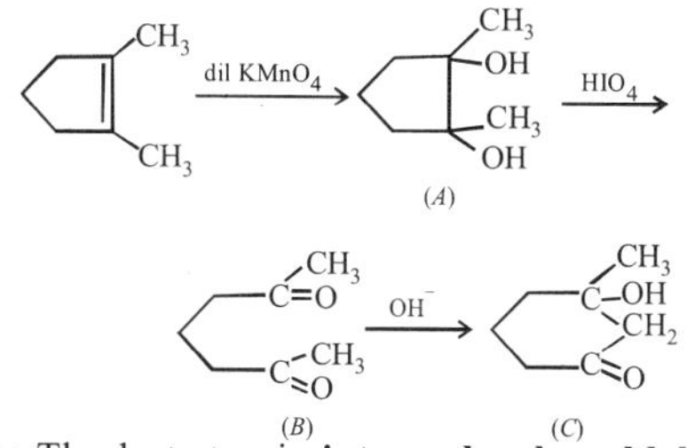

Note: The last step is **intramolecular aldol condensation**.

Function of HIO_4 is the cleavage of 1, 2–dial.

22.

$\underset{(A)}{C_8H_6} \xrightarrow[HgSO_4]{dil.\ H_2SO_4} B \xleftarrow{anhydrous.\ AlCl_3} RCOCl + C_6H_6$

$B \xrightarrow{I_2,\ aq.\ KOH} C + \text{Yellow compound (D)}$

(i) Formation of (B) from benzene and acid chloride in presence of anhydrous $AlCl_3$ (Friedel-Craft reaction) indicates that it is a ketone, C_6H_5COR.

(ii) Further the ketone (B) reacts with alkaline iodine forming yellow compound (D) (haloform reaction). This indicates that one of the alkyl groups in ketone (B) is $-CH_3$. Hence, (B) should be $C_6H_5.CO.CH_3$.

(iii) Since ketone (B) is also formed from the hydrocarbon C_8H_6 (A) by reaction with dil. H_2SO_4 and $HgSO_4$, the hydrocabon (A) must have an acetylenic hydrogen atom, *i.e.* $\equiv C-H$ grouping. Hence, (A) must be $C_6H_5C\equiv CH$.

Thus, compounds (A) to (D) are

$\underset{(A)}{C_6H_5.C\equiv CH}$ $\underset{(B)}{C_6H_5.CO.CH_3}$ $\underset{(C)}{C_6H_5COOH}$ $\underset{(D)}{CHI_3}$

Formation of (B) from (A)

$\underset{\text{Phenylacetylene (A)}}{C_6H_5.C\equiv CH} \xrightarrow[HgSO_4]{dil.H_2SO_4} [C_6H_5-\overset{OH}{\overset{|}{C}}=CH_2]$

$\xrightarrow{\text{rearranges}} \underset{\text{Acetophenone (B)}}{C_6H_5.CO.CH_3}$

23. (*i*) $CH_3CH_2-C\equiv CH + Na^+NH_2^-$

$\longrightarrow CH_3CH_2-C\equiv C^-:Na^+$

$\xrightarrow{CH_3I} CH_3CH_2-C\equiv C-CH_3$

$\xrightarrow[H_2SO_4]{HgSO_4} CH_3CH_2CH_2-\overset{O}{\overset{||}{C}}-CH_3$

(*ii*) $\underset{\text{Acetylene}}{CH\equiv CH} \xrightarrow[Hg^{2+}]{H_2SO_4} CH_3CHO$

$\xrightarrow[(Oxi.)]{KMnO_4} CH_3COOH \xrightarrow{Ca(OH)_2}$

$(CH_3COO)_2Ca \xrightarrow{distil} \underset{\text{Acetone}}{(CH_3)_2CO}$

(*iii*) $\underset{\text{Methanal}}{HCHO} \xrightarrow{[O]} HCOOH \xrightarrow{Ca(OH)_2}$

$(HCOO)_2Ca \xrightarrow[heat]{(CH_3COO)_2Ca} \underset{\text{Ethanal}}{2\,CH_3CHO}$

24. **(c)**

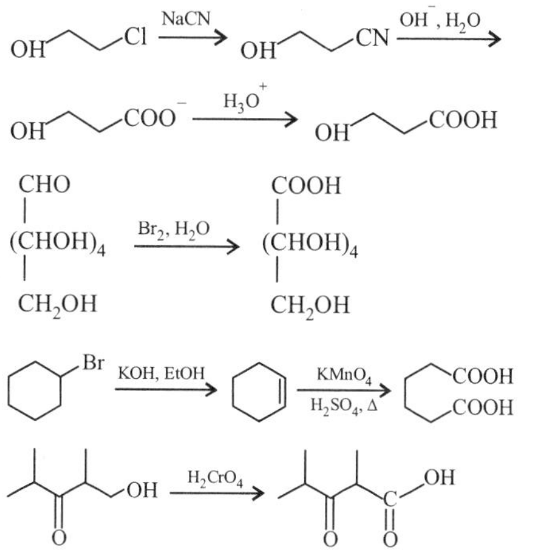

25. **(a)**

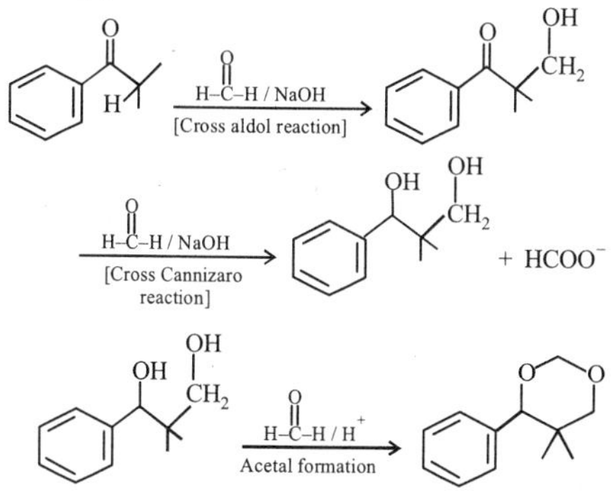

26. **(a)**

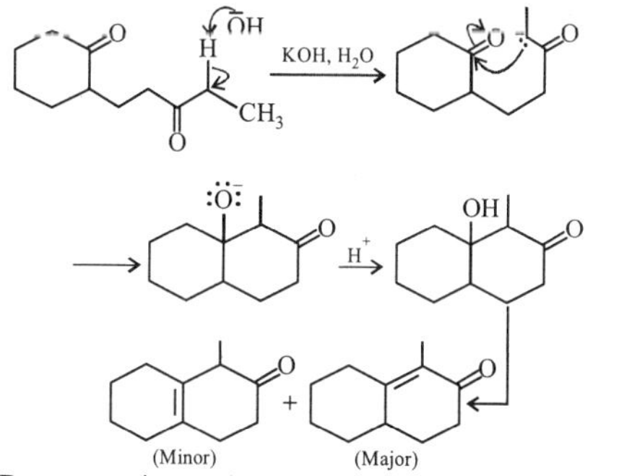

(Due to conjugated double bond)

27. **(d)**

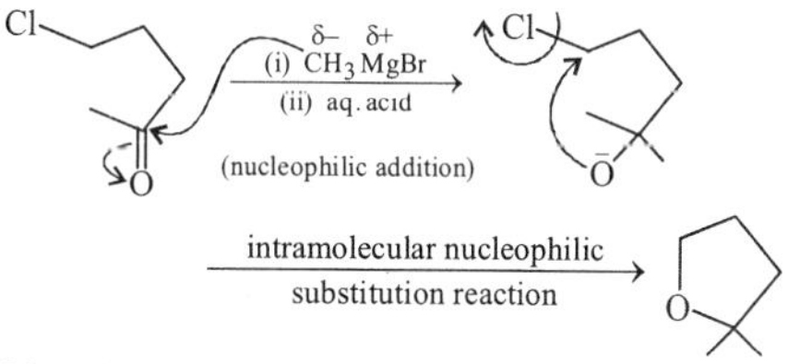

28. **(c)** **Reaction I :**

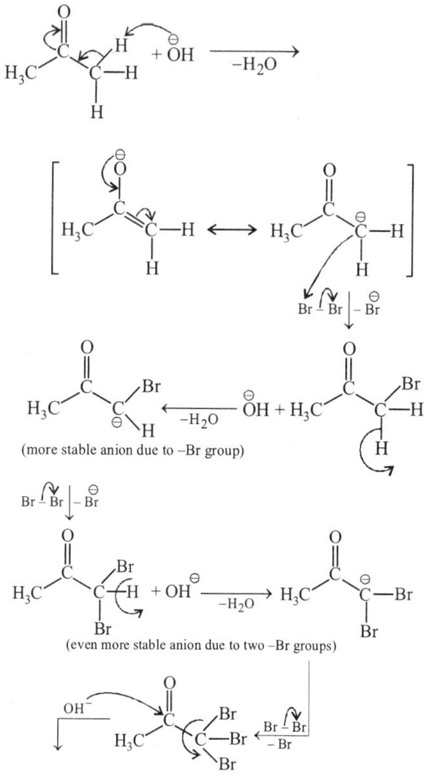

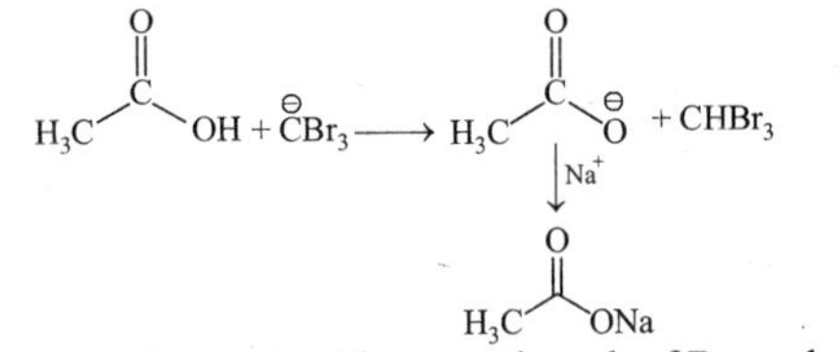

The reaction starts with one mole each of Br_2 and acetone. In this reaction mechanism, we can see that consumption of bromine is more than that of acetone. Thus, acetone will be left after completion of reaction.

The reaction will not stop after forming *P* or *Q*, but continue till the product *I* is obtained.

Only option (c) matches with the result.

Reaction II :

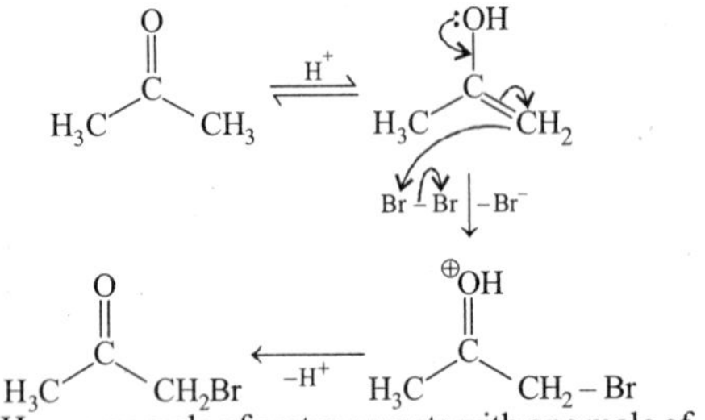

Here, one mole of acetone reacts with one mole of Br_2. This matches with the compound given in *P*.

29. (a)

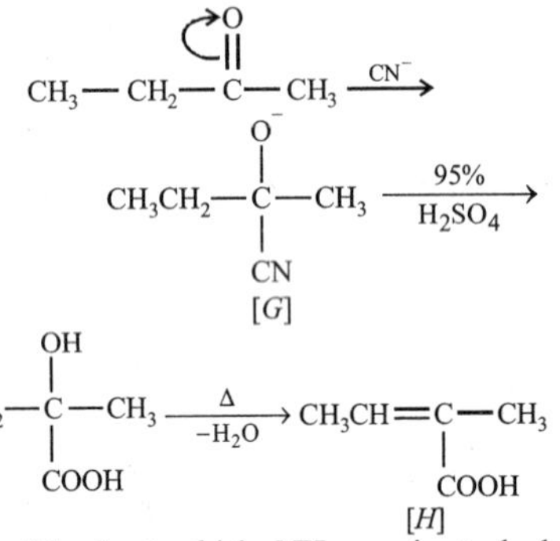

30. (b) The ring to which –NH group is attached is activated due to the lone pairs on N (+M and +E effects); while the ring to which –C = O is attached is deactivated. Hence, the electrophile NO_2^+ would go to the *para*-position of the activated ring.

31. (c) This reaction is an example of "Perkin reaction".

The compound *X* should be $(CH_3CO)_2O$.

In this step the carbanion is obtained by removal of an α–H atom from a molecule of an acid anhydride, the anion of the corresponding acid acting as a necessary base.

32. (c)

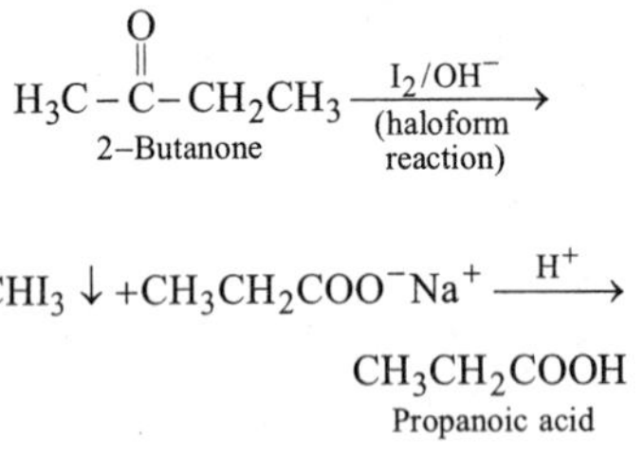

33. (c) Reaction of PhMgBr with carbonyl compounds is an example of nucleophilic addition on carbonyl group which increases with the increase in electron-deficiency of carbonyl carbon.

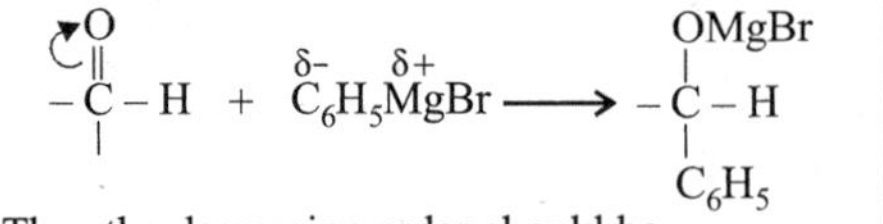

Thus the decreasing order should be

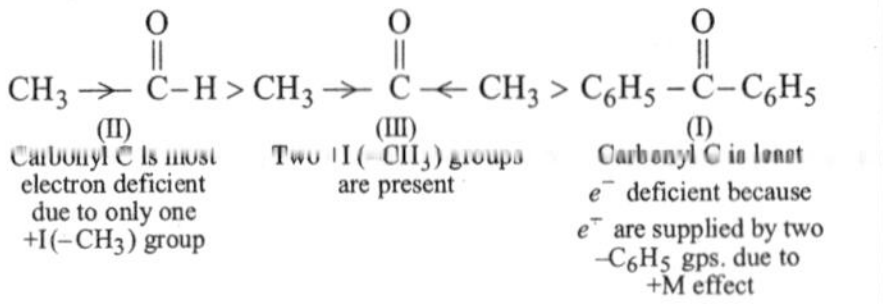

34. (b)

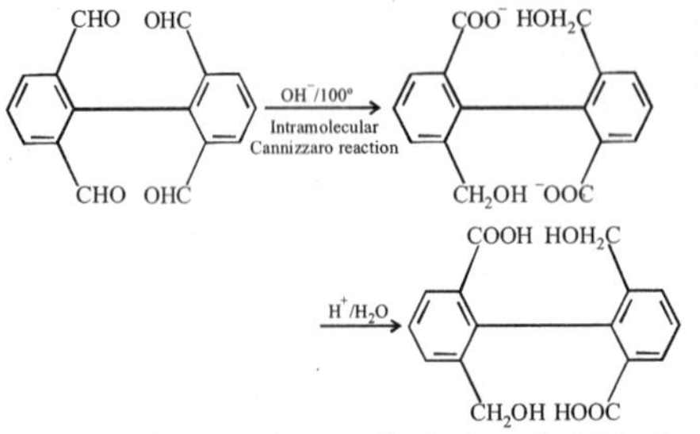

35. (c) Ketone (non-reducing) and aldehyde (reducing) can be distinguished by Fehling solution.

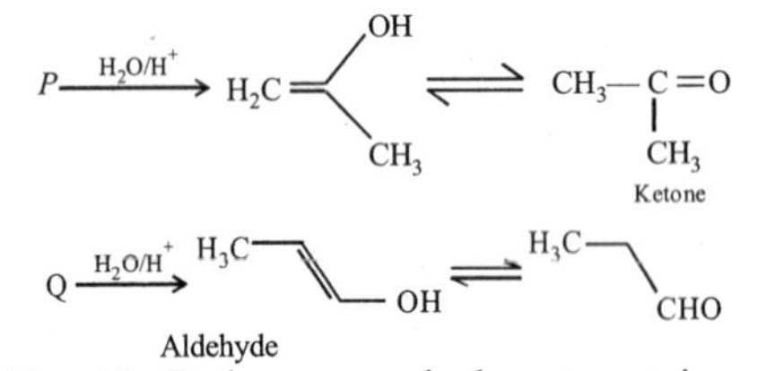

36. (a) Both compounds do not contain α-hydrogen hence undergo Crossed Cannizzaro reaction.

Initially OH^- attacks at the carbonyl carbon of HCHO than that of PhCHO because carbonyl carbon of HCHO is

(i) more electrophilic than that of C_6H_5CHO

(ii) less sterically hindered

to give hydroxymethoxide which acts as hydride donor in next step to give sodium formate.

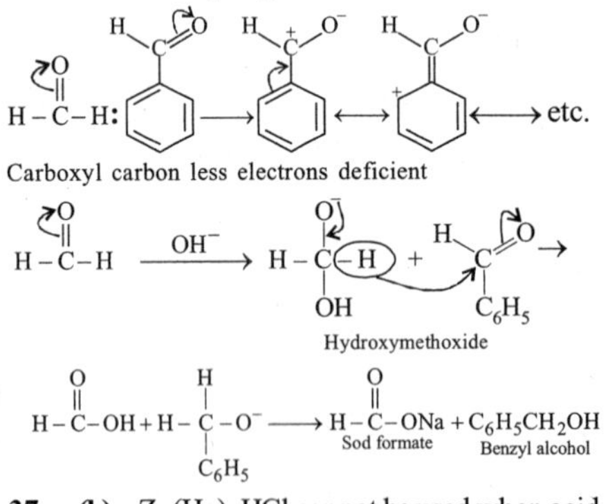

37. (b) Zn(Hg), HCl cannot be used when acid sensitive group like –OH is present, but NH_2NH_2, OH^- can be used.

38. (a)

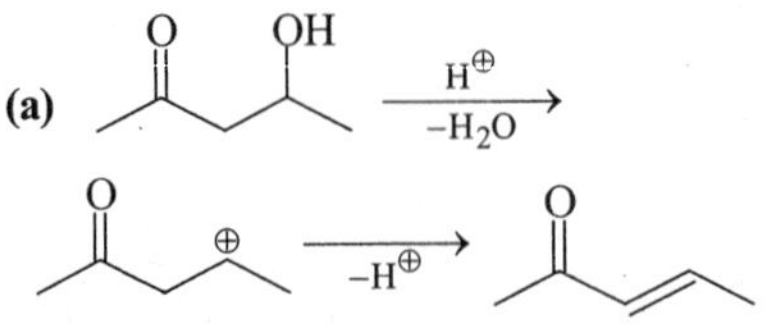

Due to π – σ – π conjugation, this dehydrated product will be most stable.

39. (b) This is a type of deuterium exchange reaction.

Mechanism

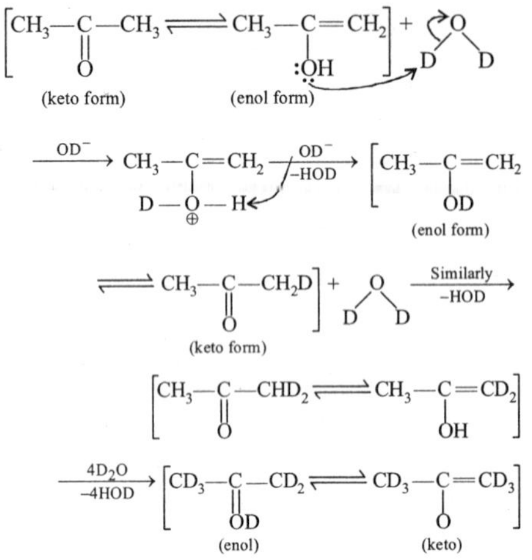

This reaction is fast in basic conditions where all α – H will replace by D.

40. (b) The possible mechanism is

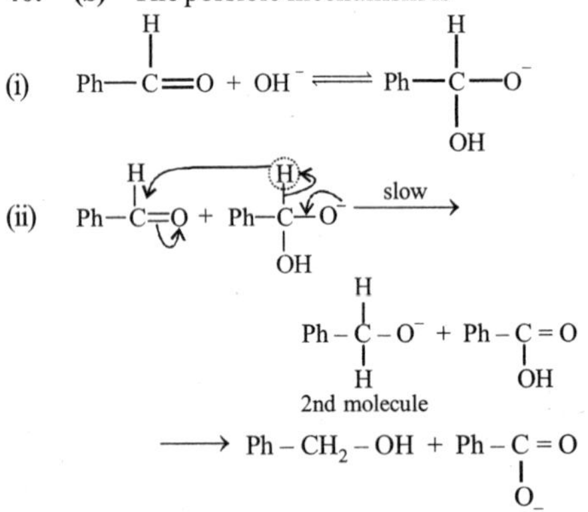

The slowest step is the transfer of hydride to the carbonyl group as shown in step (ii).

41. (b) The addition is initiated by the attack of CN^- group which is a nucleophile.

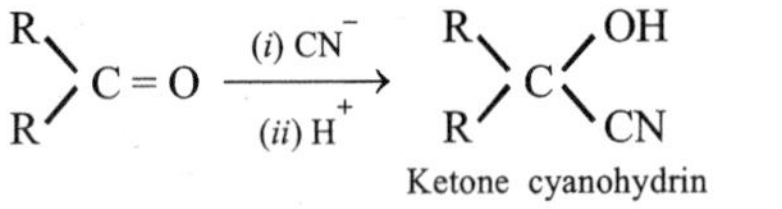

42. (d) *m*-Chlobenzaldehyde does not contains α-H atom and thus undergoes conc. KOH. Cannizzaro reaction with

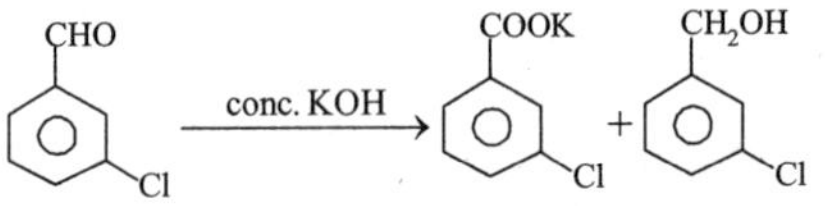

43. (a)

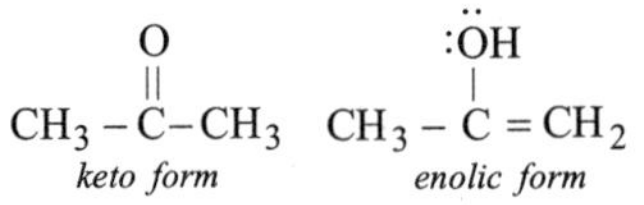

No. of σ bonds in enolic form : 3 + 1 + 1 + 1 + 1 + 2 = 9

No. of π bonds in enolic form : 1

No. of lone pairs of electrons in enolic form = 2

44. (d) –CHO withdraws electrons from the double bond or from a conjugated system towards itself.

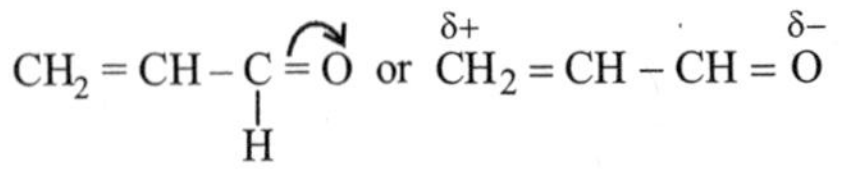

45. (c) Iodoform test is given by compounds having $-COCH_3$, $CHOHCH_3$ and CH_3CH_2OH. Thus diethyl ketone does not give this test.

46. (b) The compound containing α-H atom (CH_3CHO) does not undergo Cannizzaro's reaction. The other three reaction α-hydrogen.

47. (c) $\underset{\text{Fehling solution}}{CH_3CHO + 2Cu^{2+} + OH^-} \rightarrow CH_3COOH + \underset{\text{(red)}}{Cu_2O\downarrow}$

48. (b) Compounds having $-\underset{\underset{O}{\|}}{C}-CH_3$ groups, show positive iodoform test.

Hence, $CH_3-CH_2-CH_2-\underset{\underset{O}{\|}}{C}-CH_3$ (pentanone-2) gives this test.

49. (b) Fehling solution, Schiff's reagent & Tollen's reagent react only with aldehydes, but Grignard reagents react both with aldehydes and ketones.

50. (5) General molecular formula for ketones is $C_nH_{2n}O$

$\therefore \quad C_nH_{2n}O = 100$ or $12n + 2n + 16 = 100, n = 6$

Possible isomeric ketones with 6 carbon atoms are

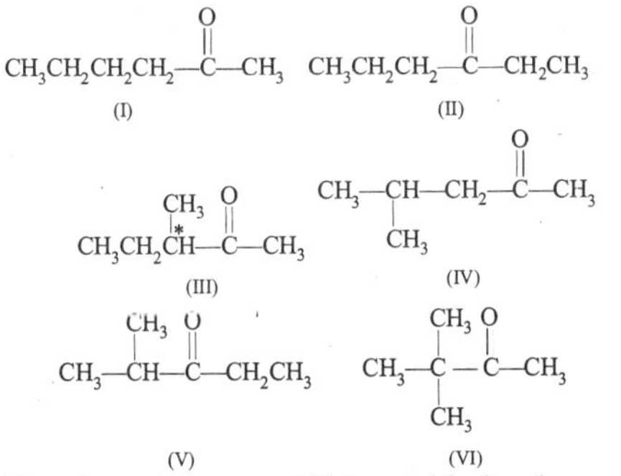

Note that only isomer (III) has a chiral carbon so on reduction with $NaBH_4$ it will give diastereomeric alcohols, while all other five isomers will give racemic mixture.

For example :

$$\underset{\text{(I)}}{CH_3-CH_2-CH_2-CH_2-\overset{\overset{O}{\|}}{C}-CH_3} \xrightarrow{NaBH_4} \underset{(\pm\text{ racemic mixture})}{CH_3-CH_2-CH_2-CH_2-\overset{*}{\underset{\underset{H}{|}}{\overset{\overset{OH}{|}}{C}}}-CH_3}$$

Due to presence of one chiral carbon, the alcohol can be on both optically active forms and the compound will be racemic mixture.

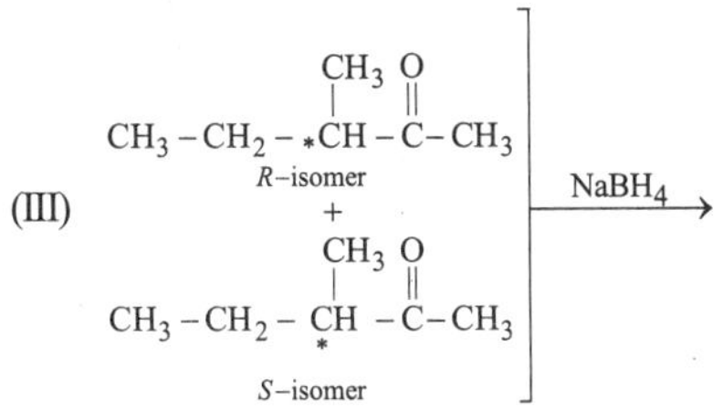

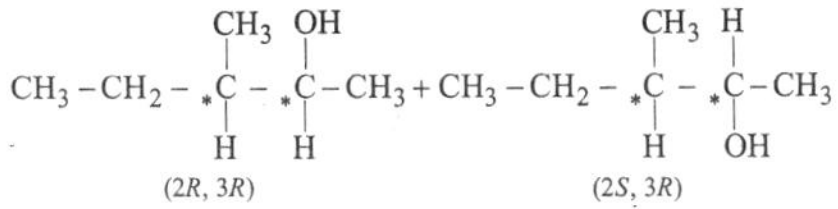

A pair of diastereomers from *R*-isomer of ketone

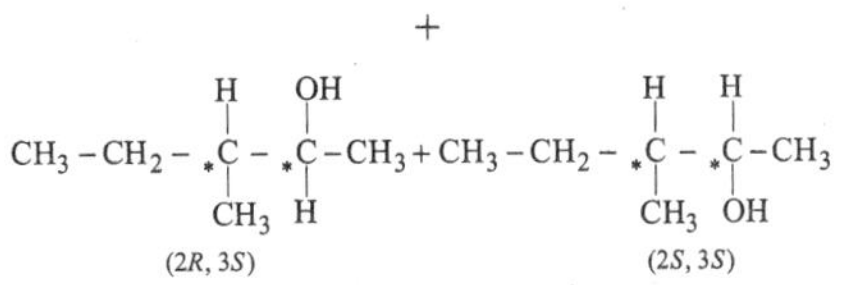

A pair of diastereomers from *S*-isomer of ketone

Note : All stereoisomers are reacted separately.

51. (12) P gives $FeCl_3$ test and there are no intramolecular H-bonds.

Thus, P wil be a trihydric phenol.

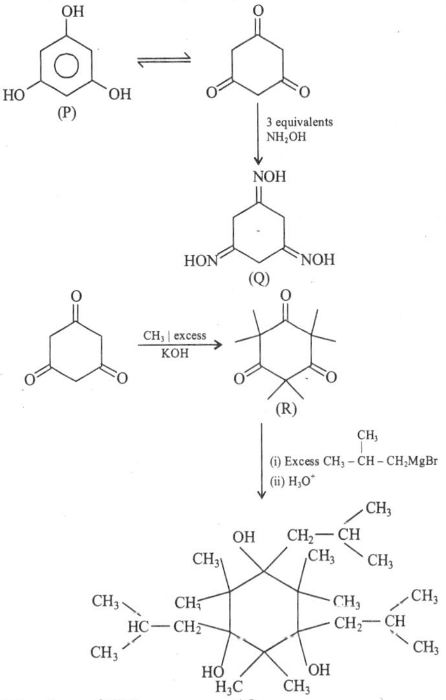

Number of CH_3 groups = 12

52. (18) Reactions taking place as:

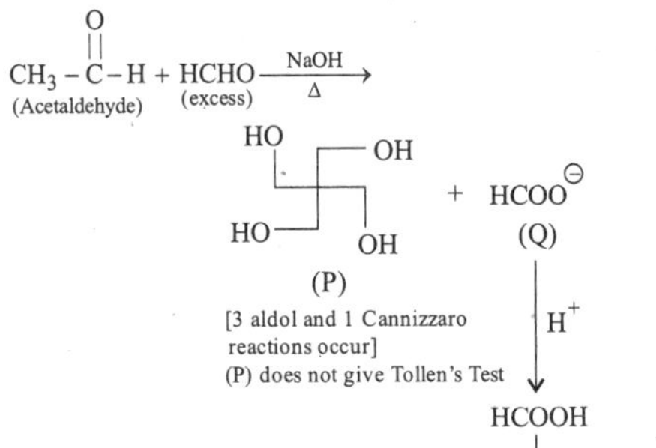

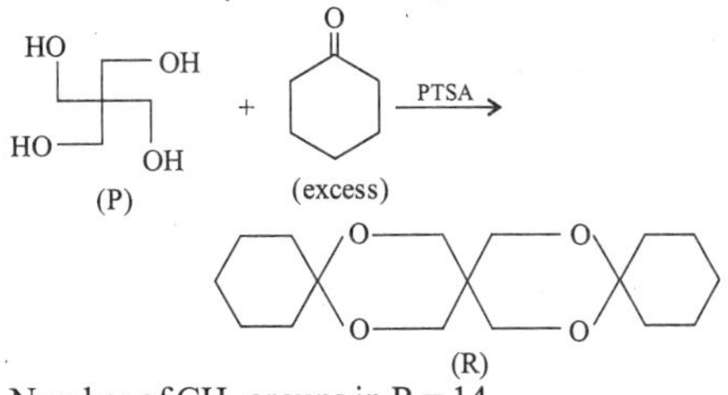

Number of CH_2 groups in R = 14

Number of O-atoms = 4

Required Answer = 14 + 4 = 18

53. 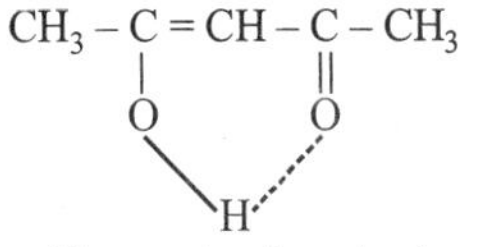

54. sodium potassium tartarate.

55. (*i*) Free radical substitution
(*ii*) Thermite process
(*iii*) Nucleophilic addition
(*iv*) Cyanamide process;
(*v*) Nucleophilic subsitution
(*vi*) Bayer's process
(*vii*) Homologous pair
(*viii*) Electrophilic substitution
(*ix*) Ostwald's process
(*x*) Homolytic addition.

56. False. Grignard reagents react with ketones to form *ter*-alcohols; hence here, *ter*-butanol will be formed.

57. True : Aldehydes (from primary alchols) may further be oxidised easily to acids as compared to ketones (from secondary alcohols).

58. False : Benzaldehyde has no α-hydrogen atom, hence, it does not undergo aldol condensation but undergoes Cannizzaro reaction.

59. (a, b, c)

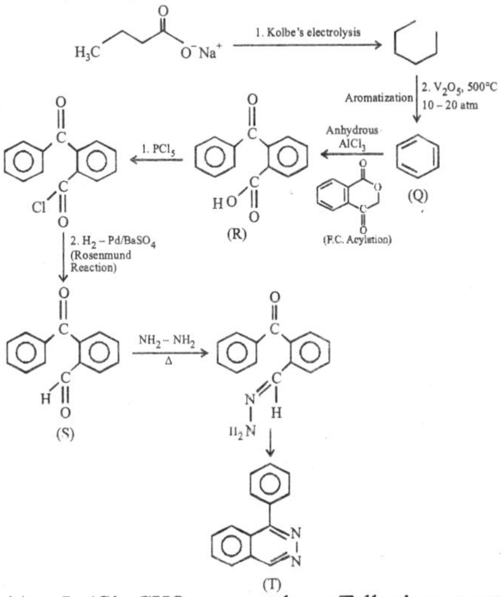

(a) In 'S' - CHO group reduces Tollen's reagent and give silver mirror precipitate.

(b) 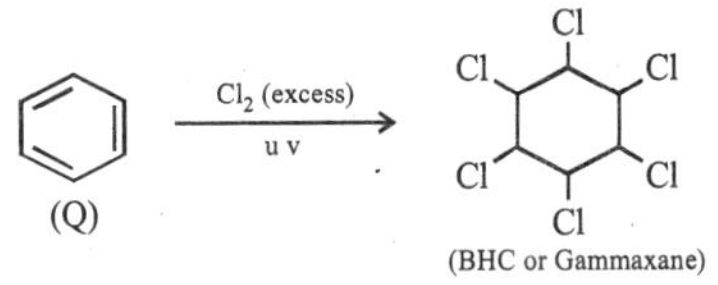

(c) 'T' is a Hetrocyclic compound.

(d) 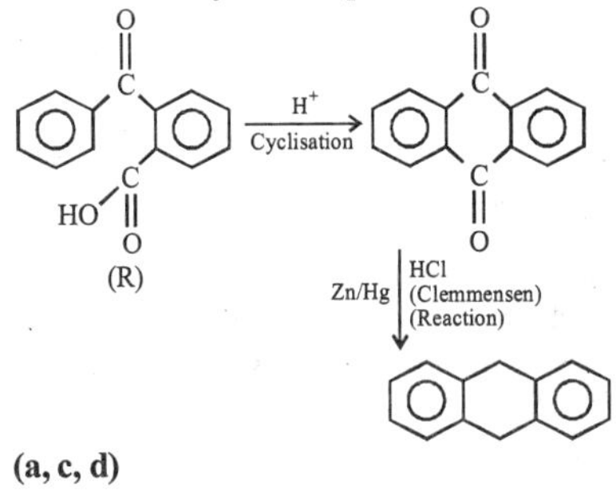

60. (a, c, d)

(a)

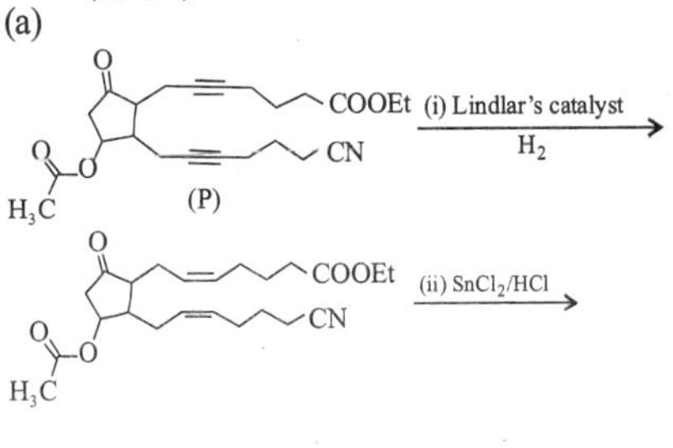

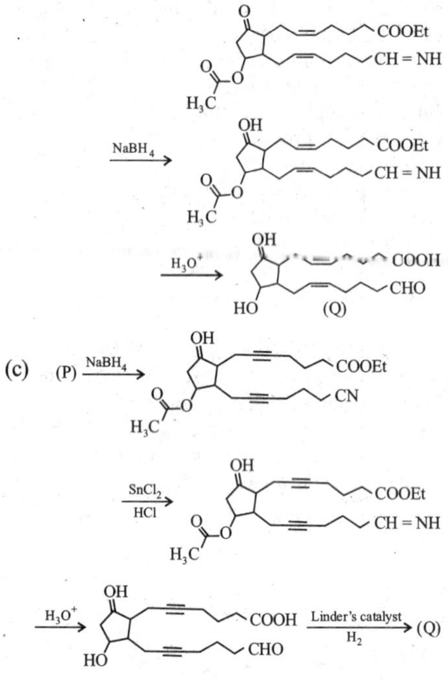

(d) Similarly, (P) $\xrightarrow[\text{(iii) } SnCl_2/HCl, \text{ (iv) } H_3O^+]{\text{(i) Lindlar's catalyst, } H_2, \text{ (ii) } NaBH_4,}$ (Q)

Hydrolysis of – CN gives – COOH and thus option (b) is not correct.

61. (a, b, c) Aldehydes and α-Hydroxyketones show positive Tollen's test.

62. (b, c)

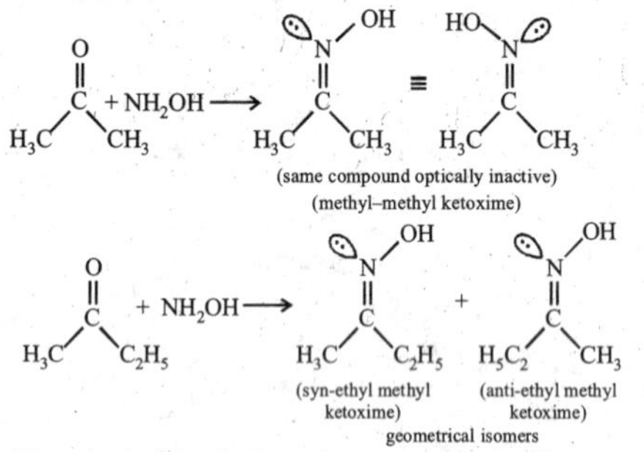

63. (a, b, d) Carbonyl compounds having α – H or α - D undergo aldol condensation.

(a) $\overset{\alpha}{C}H_3CHO$ (b) $CH_3\overset{\alpha}{C}H_2CHO$

(c) 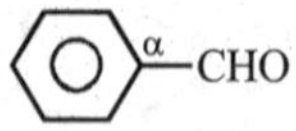 (No α – H) (d) $\overset{\alpha}{C}D_3CHO$ (α–D present)

64. (b) Three Cl of chloral makes its carbonyl carbon highly electron deficient, hence H_2O, a nucleophile easily adds on it forming chloral hydrate, $CCl_3CH(OH)_2$, which is quite stable due to intramolecular H–bonding between two –OH groups.

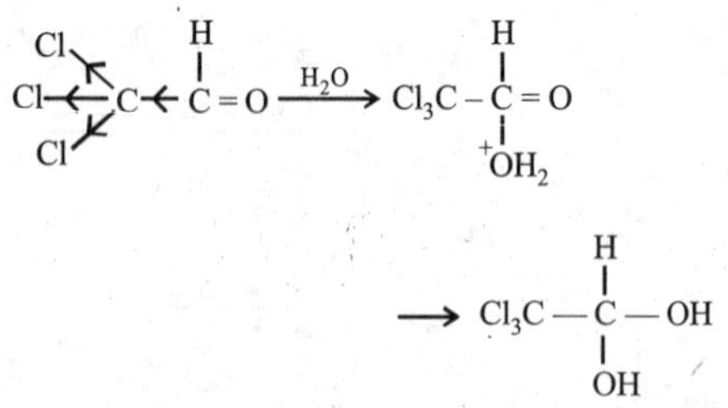

65. (b, d)

(a) $HCHO \xrightarrow{OH^-} CH_3OH + HCOO^-$

(No new C – C bond is formed)

(b) $C_6H_6 + CH_3Cl \xrightarrow{AlCl_3} C_6H_5 - CH_3$

(New C – C bond is formed)

(c) $>C=O \xrightarrow[HCl]{Na/Hg} >CH_2$

(No new C – C bond is formed)

(d) 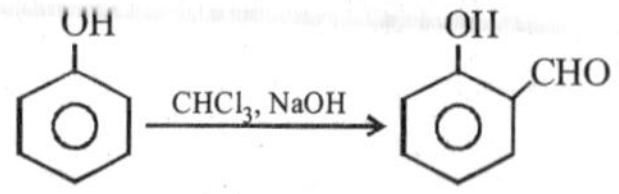

(New C – C bond is formed)

66. (a, b) Aldehydes and ketones containing α-hydrogen atom undergo aldol condensation.

67. (b, d) Keto-enol tautomerism is shown in compounds having α-hydrogen on the C adjacent to the CO group.

68. (a, c) Aldehydes having at least one α-hydrogen atom undergo aldol condensation.

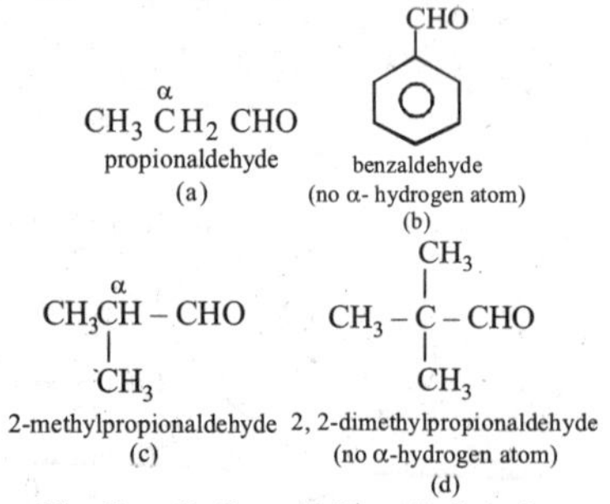

69. (d) P → 3, Q → 4, R → 5, S → 2

(1) $Ph-\overset{O}{\overset{\|}{C}}-CH_3 \xrightarrow{Zn-Hg/HCl} Ph-CH_2-CH_3$
Acetophenone

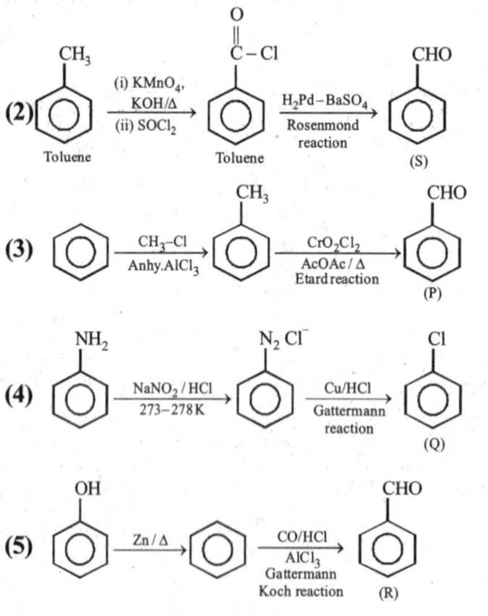

70. (A) - r, s; (B) - p, q; (C) - p, q, r; (D) - p, s

(A) $H_2N-\overset{\oplus}{N}H_3\overset{\ominus}{Cl}$

(p) does not have carbon thus, will not respond to (p).

(r) because of presence of $Cl^{\ominus}$ ion, it will give white precipitate of AgCl with $AgNO_3$.

(s) because of $-NH_2$ group it will react with aldehydes to form hydrazone derivative.

(B) 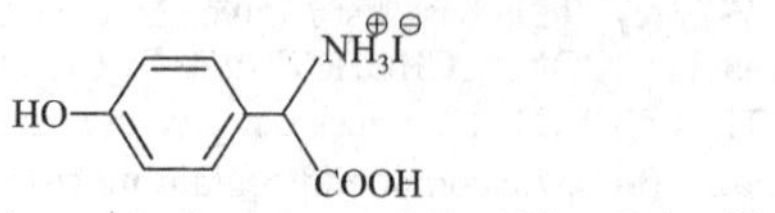

(p) because of presence of C – N bond, it will form prussian blue colour with $FeSO_4$.

(q) presence of –OH group will give positive $FeCl_3$ test.

(r) It will give yellow precipitate of AgI.

(s) because of absence of $-NH_2$ group, it will not give hydrazone derivative.

(C) 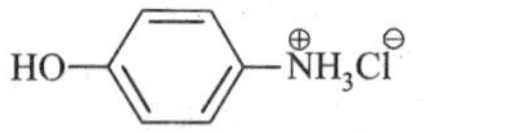

In the similar way, it will respond to p, q and r.

(D) 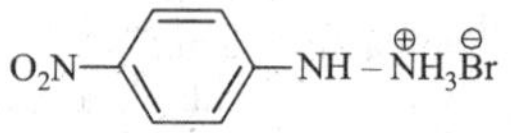

(p) gives positive test with (p)

(r) gives yellow precipitate of AgBr

(s) it has a derivative of NH_2-NH_2, thus it will react with aldehyde to form hydrazone derivative.

71. (2)

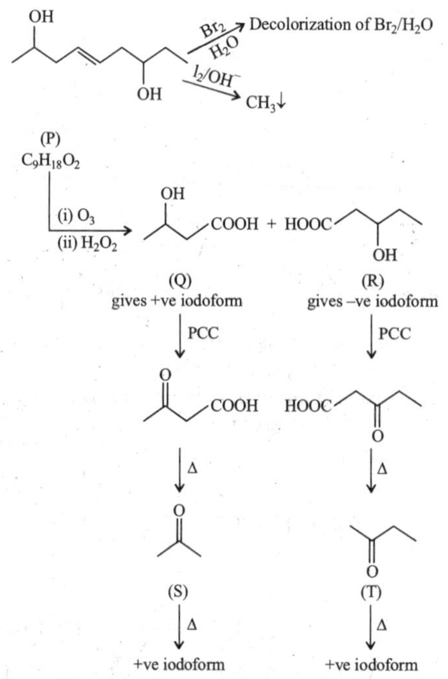

Sum of number of O-atoms in S and T = 1 + 1 = 2

72. (102018)

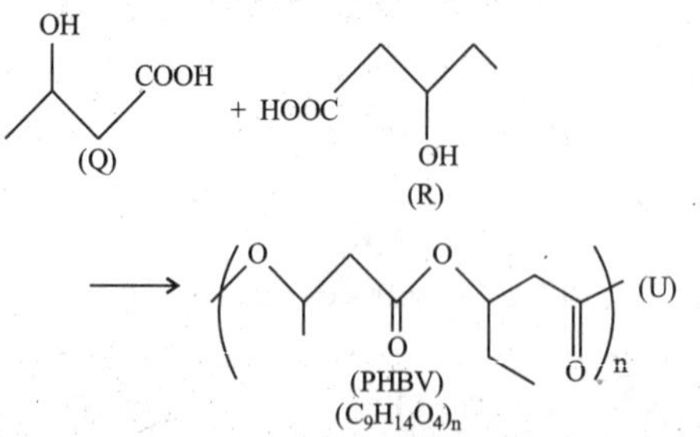

Mol. Wt. of polymer = [(104 × 500) + (118 × 500)] – (18 × 499) (no. of water molecules released)

= (52000 + 59000) – 8982 = 102018 g

For 72-74. (Passage-II)

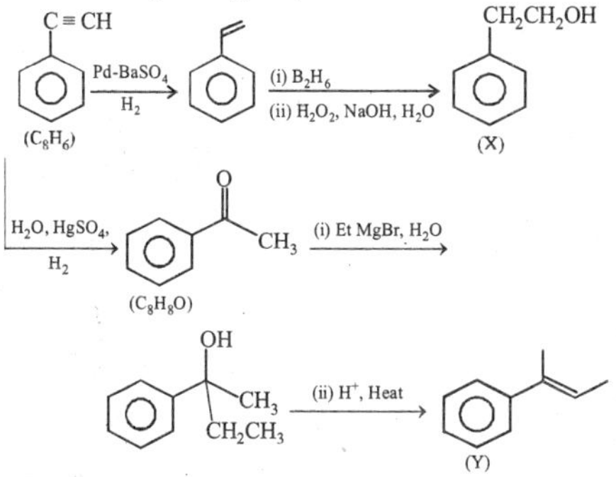

73. (c)
74. (d)

For 75-77. (Passage-III)

Let us summarize the given facts of the problem.

$$P + Q \xrightarrow{aq.K_2CO_3} R \xrightarrow{HCN} S$$
(2 aldehydes)

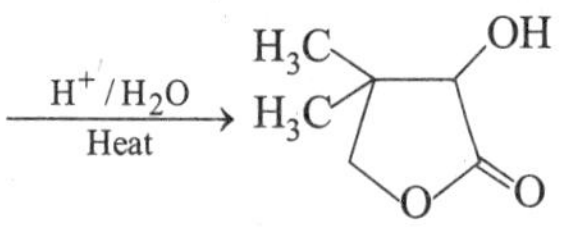

Structures of *P*, *Q*, *R* and *S* can be established on going backward from the known final product.

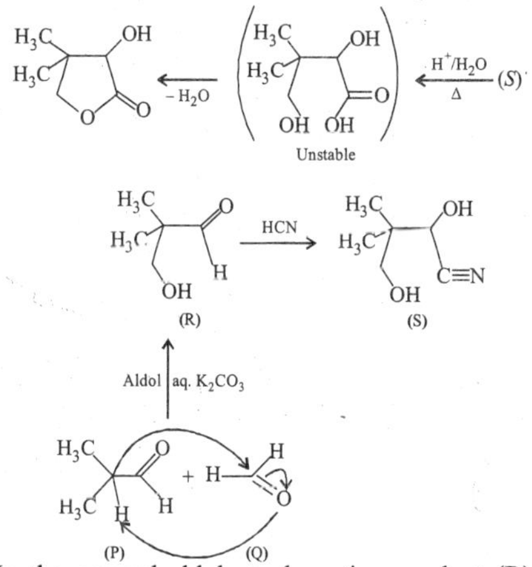

In the crossed aldol condensation product (R) from aldehydes (P) and (Q), dehydration will not occur readily because of formation of non-conjugated system.

75. (b)
76. (a)
77. (d)

For 78-80 (Passage-IV)

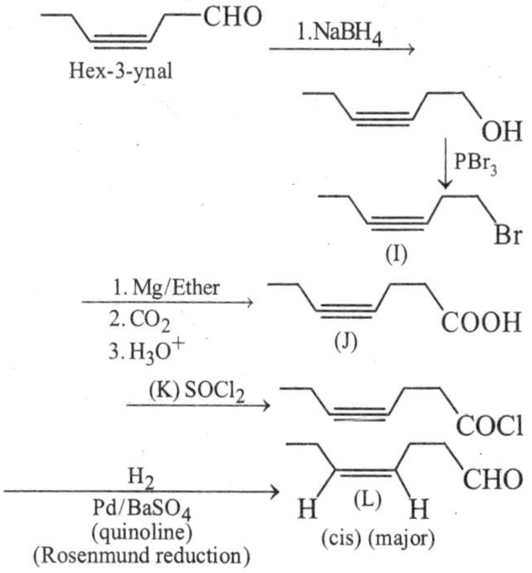

78. (d)
79. (a)
80. (c)

81. (*i*) Formation of HCOONa and a primary alcohol due to **Cannizzaro reaction** of *F* and *G* indicate that either *F* or *G* should be HCHO. Thus, the alkene *A* should have CH_2 = grouping. The remaining 5 C's of *A* should have grouping = HCC_4H_9.

(*ii*) Formation of only *E* by the ozonolysis of *D* indicates that *D* (isomer of A, C_6H_{12}) should have following structure

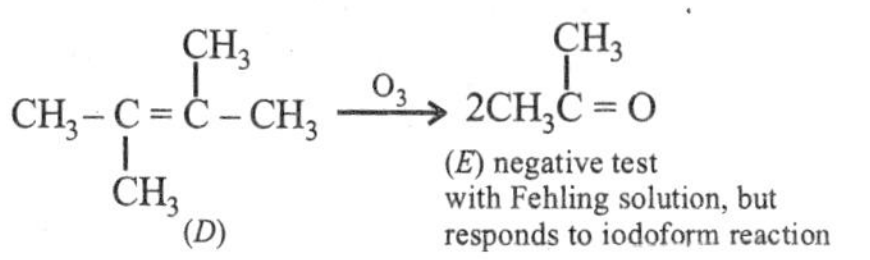

Fehling's test is given by aldehydes and not ketones.

(*iii*) Since *A* is isomer of *D*, former should have following structure.

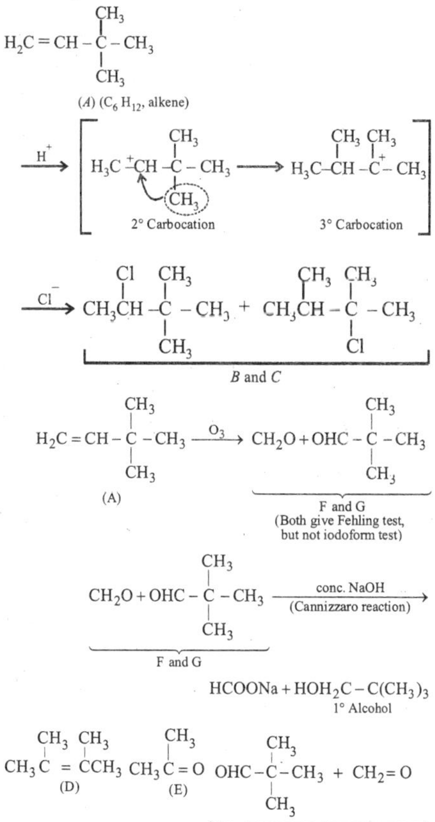

82. 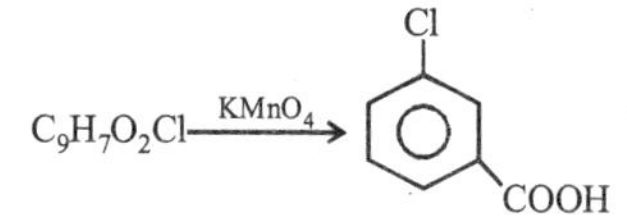

Hence, compound should have following part structure

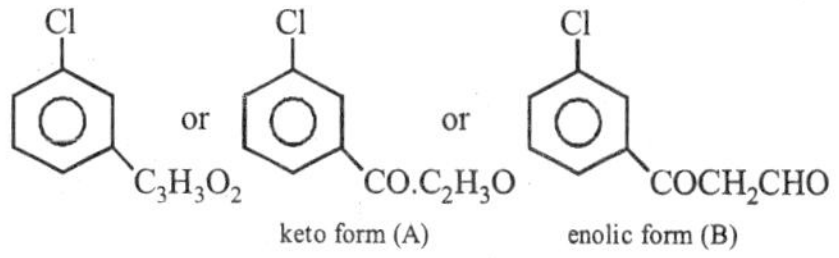

The enolic form of the compound is more stable than the keto form due to hydrogen bonding.

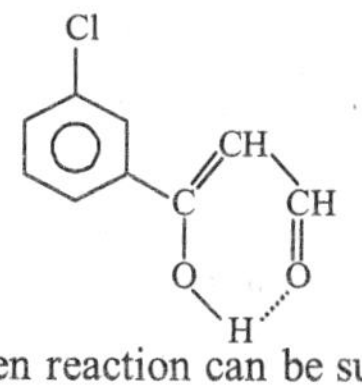

83. The given reaction can be summarised as below :

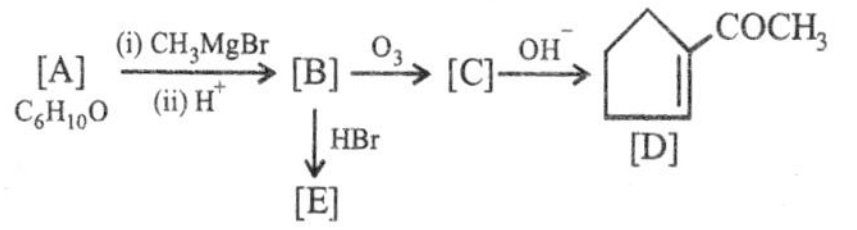

Conclusions from the set of reactions

(*i*) Carbon-hydrogen ratio in *A* indicates that it is a cyclic compound.

(*ii*) Reaction of *A* with CH_3MgBr indicates that it should have a ketonic group.

(*iii*) As *B* undergoes ozonolysis to form *C*, It must have a double bond, and *C* must have two carbonyl groups.

(*iv*) Reaction of *C* (a dicarbonyl compound) with a base gives a cyclic compound, it indicates that intramolecular condensation have occurred during this conversion. Thus, *A* is cyclohexanone which explains all the given reactions.

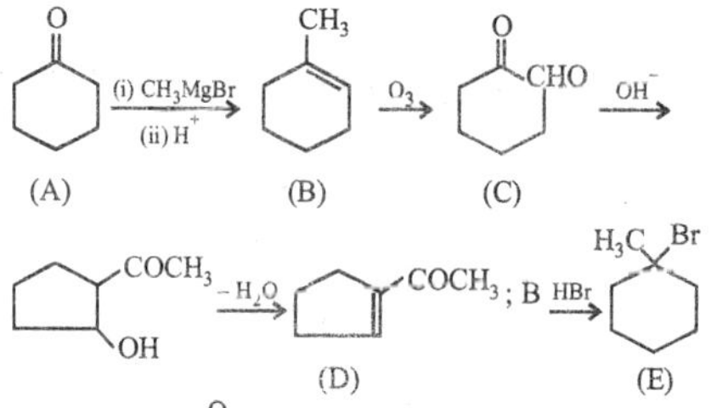

84. (*i*) 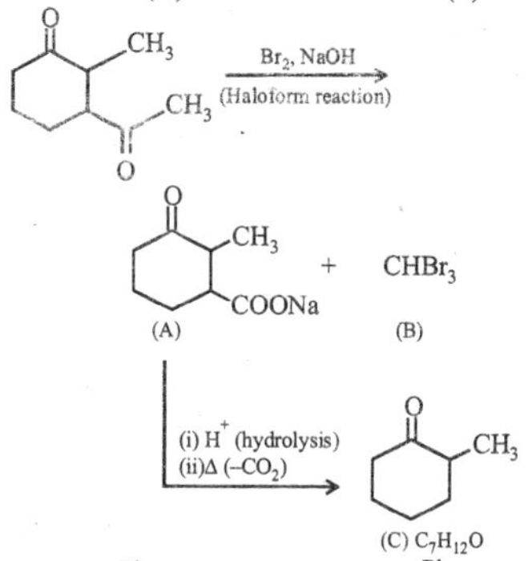

(*ii*) 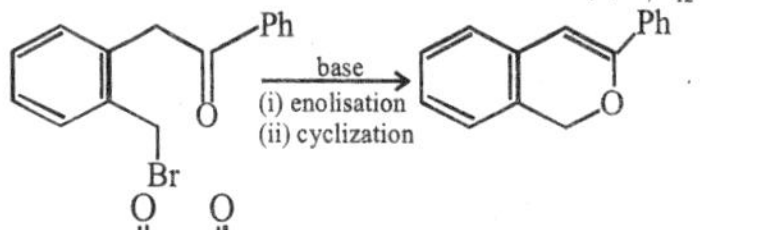

(*iii*) 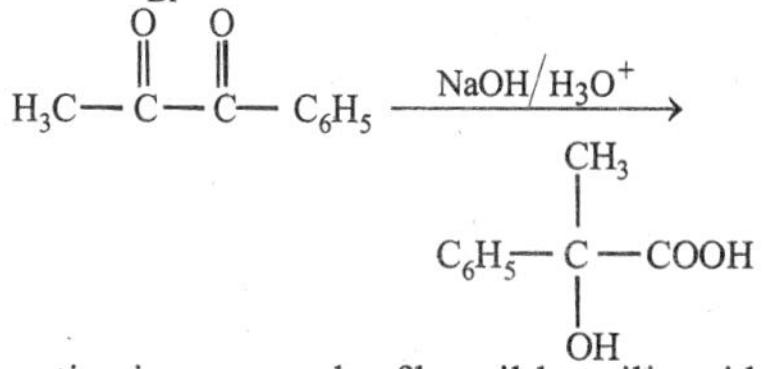

The reaction is an example of benzil-benzilic acid type rearrangement.

(*iv*) 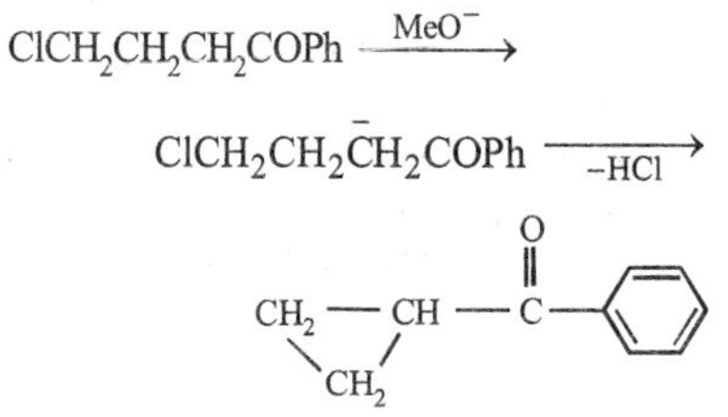

(v) 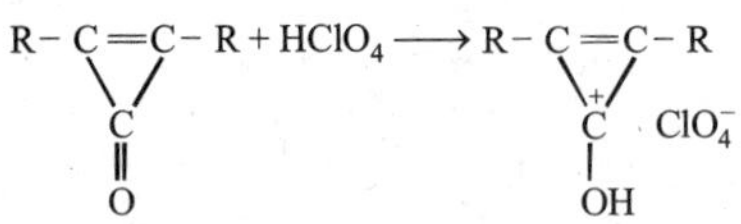

(vi) 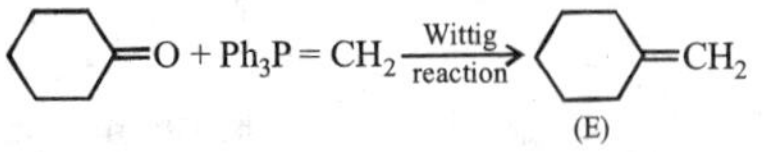

(vii) $KCN + H_2SO_4 \longrightarrow KHSO_4 + HCN$

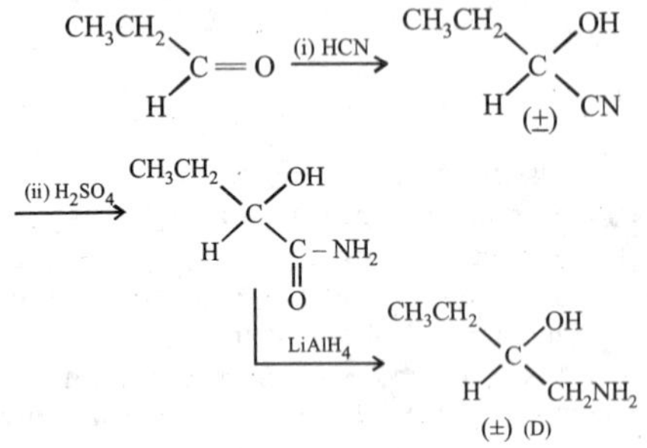

(viii) 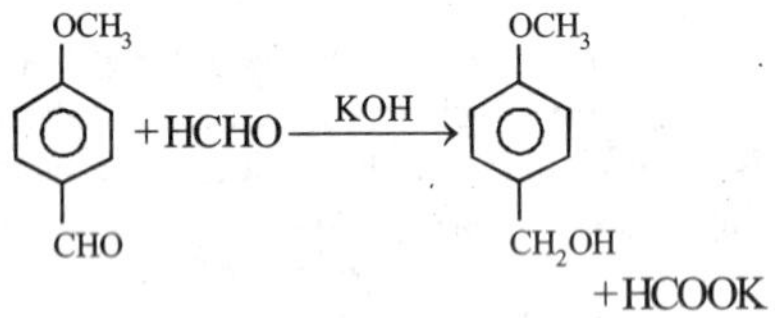

(ix) 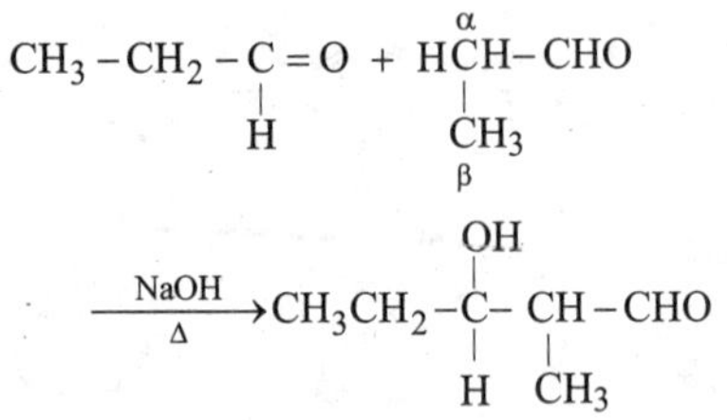

(x) 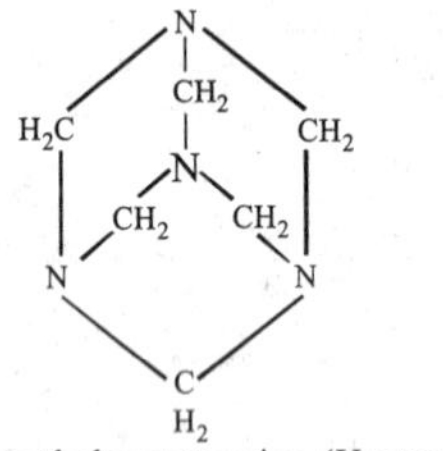

Hexamethylenetetramine (Urotropine)

85.

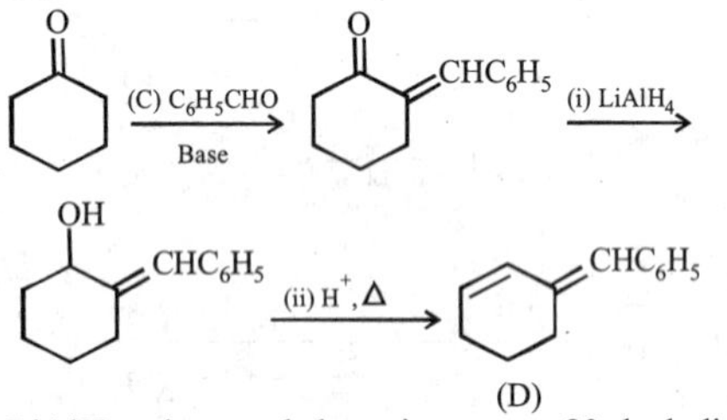

$LiAlH_4$ reduces only ketonic group to 2° alcoholic group without affecting double bond. This dehydration occurs in acidic conditions.

86. Following informations are provided by the problem.

(i) Since aldehyde *A* ($C_{11}H_8O$) gives C_6H_5CHO on ozonolysis, it must have a benzene nucleus and a side chain. The side chain should have five carbon (C_{11}–C_6=C_5), three hydrogen (H_8-H_5=H_3) and one oxygen atom, *i.e.*, it should be C_5H_3O. Further the compound *A* has an aldehydic group, so the side chain can be written as C_4H_2CHO.

(ii) Formation of two moles of *B* from one mole of *A* by ozonolysis indicates that the side chain must possess two unsaturated linkages, one of which must be alkyne type, suggested by very low number of hydrogen atoms.

(iii) Further, since the aldehyde *A* does not undergo aldol condensation, α-hydrogen is absent and hence triple bond should be present between C_2 and C_3.

(iv) Thus the side chain C_4H_2CHO of *A* can be written as $-CH=CH-C\equiv C-CHO$.

(v) Thus compound *A* should possess following structure which explains all the given reactions.

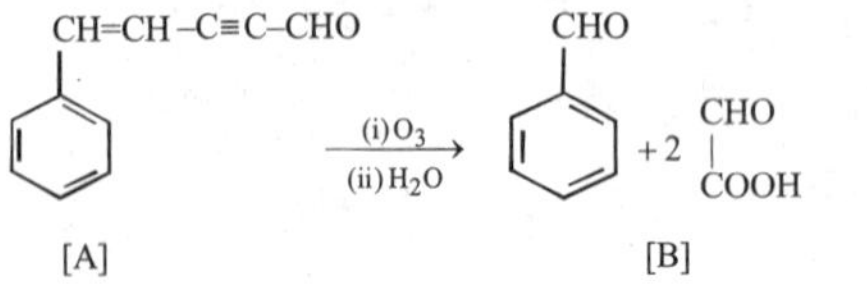

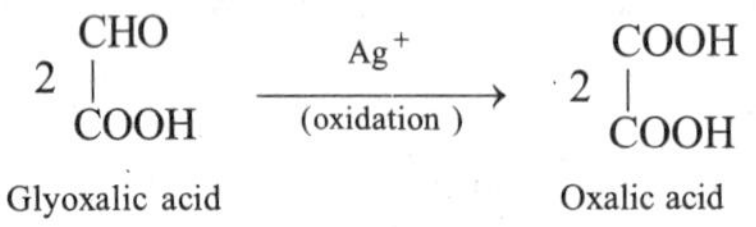

87. 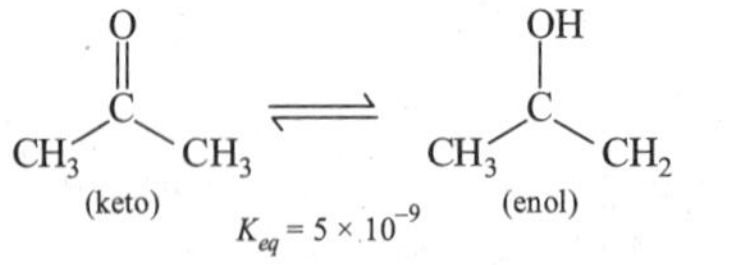

The methyl group stabilizes the C = C in the enol form due to hyperconjugation.

However, the methyl group also stabilizes the carbonyl double bond in the keto form.

In fact, in simple carbonyl compounds, the methyl group stabilizes the carbonyl double bond much more than it stabilizes the enolic C = C bond.

Thus, a carbonyl in a ketone is slightly stronger than that in an aldehyde.

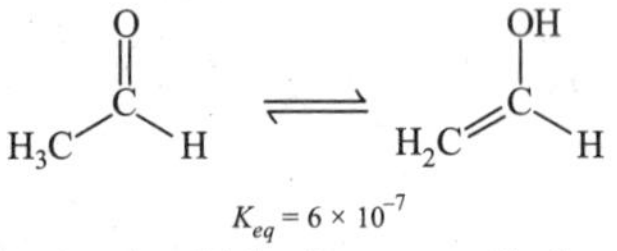

As a result, simple aldehydes generally have a higher enol content than simple ketones.

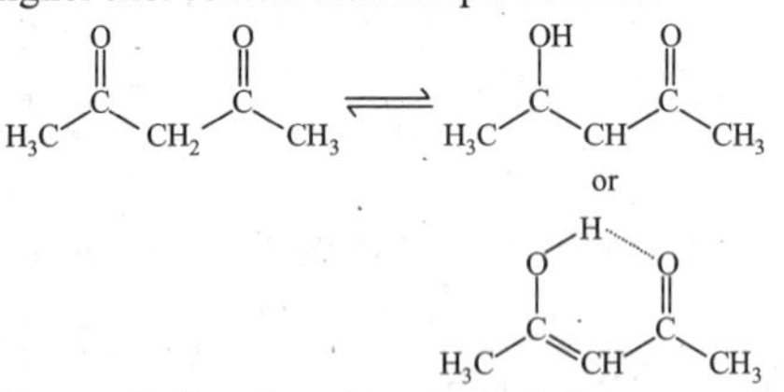

The two β-di carbonyl compounds have a much higher enol content than the two monocarbonyl compounds because of hydrogen bonding, conjugation and 6 membered cyclic form.

In fact, for these compounds, the enol form will be major form at equilibrium. Hence, the correct order of enol content is :

$CH_3COCH_3 < CH_3CHO << CH_3COCH_2COCH_3 < CH_3COCH_2CHO$

88. (*i*) 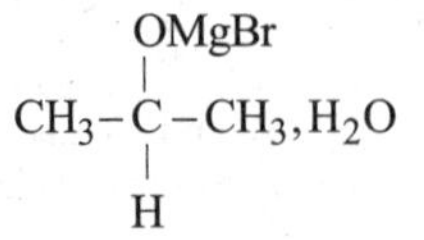

(*ii*) 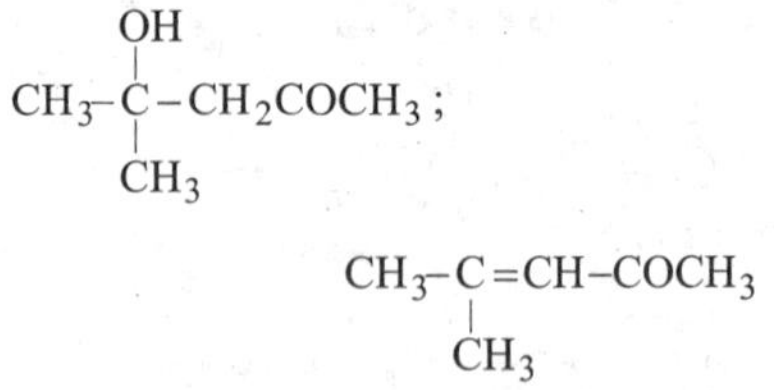

(*iii*)

(a) fuse with NaOH (b) H^+ ; 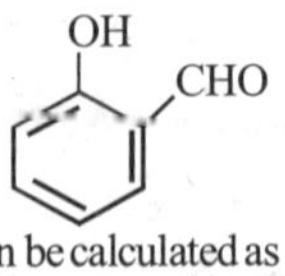

89. (*i*) Empirical formula can be calculated as

Element	Percentage	Relative no. of atoms	Simplest ratio
C	69.77	5.81	5
H	11.63	11.63	10
O	18.60	1.16	1

∴ Empirical formula of compound is $C_5H_{10}O$ and empirical formula wt. = 86.

Also molecular wt. = 86.

∴ Molecular formula of compound is $C_5H_{10}O$.

(*ii*) Compound forms bisulphite addition compound and thus has carbonyl gp, *i.e.* aldehyde or ketone.

(*iii*) It does not reduce Fehling solution and thus it is not an aldehyde but a ketone.

(*iv*) It gives positive iodoform test and thus it has

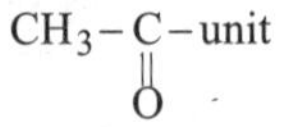

(*v*) Above facts reveal that the compound is $CH_3CH_2CH_2COCH_3$ (Pentan-2-one) or $(CH_3)_2CHCOCH_3$ (3-Methylbutan-2-one)

90. Let us summarise the given facts.

White ppt. $\xleftarrow{AgNO_3}$ (A) $C_6H_{13}Cl$ $\xrightarrow[KOH]{\text{hot alc.}}$ (B) + (C) (Isomeric alkenes, C_6H_{12}) $\xrightarrow{O_3}$

CH_3CHO (*i*) + C_2H_5CHO (*ii*) + CH_3COCH_3 (*iii*) + $(CH_3)_2CHCHO$ (*iv*)

With the help of structures of the four carbonyl compounds, (*i*) to (*iv*), we may write the structures of the two isomeric olefins (B) and (C). The two carbonyl compounds should be joined in such a way that the parent olefin has 6 carbon atoms. Two such possibilities are the combination of carbonyl compounds having (2+4) carbon atoms [*i.e.* (*i*) + (*iv*)] and (3+3) carbon atom [*i.e.* (*ii*) + (*iii*)].

CH_3CHO (i) + $OHC.CH(CH_3)_2$ (iv) $\xleftarrow{O_3}$ $CH_3CH=CH.CH(CH_3)_2$ Olefin B (C_6H_{12})

C_2H_5CHO (ii) + $OC(CH_3)_2$ (iii) $\xleftarrow{O_3}$ $C_2H_5CH=C(CH_3)_2$ Olefin (C_6H_{12})

Thus, the compound (A) should be a chloride that can eliminate a molecule of HCl to give B as well as C.

$CH_3CH_2CH(Cl).CH(CH_3).CH_3 \xrightarrow[\text{alc. KOH}]{\text{hot}}$
2-Methyl-3-chloropentane
(A)

$CH_3CH = CH.CH(CH_3).CH_3 + CH_3.CH_2.CH = C(CH_3).CH_3$
4-Methylpentene -2 (B) 2-Methylpentene–2 (C)

91. Acetaldehyde can be distinguished from acetone by using Tollen's reagent or Fehling solution or Schiff's reagent. Only acetaldehyde responds to all these tests.

$CH_3CHO + [Ag(NH_3)_2]^+ OH^- \rightarrow CH_3COOH + Ag\downarrow$
Tollen's reagent Silver mirror

92. It is a case of nucleophilic addition reaction. More the electron deficiency of the carbonyl carbon, greater will be its reactivity towards nucleophilic addition.

93. (*i*) *o*-Hydroxybenzaldehyde has intramolecular H-bonding while the *p*-isomer has intermolecular H-bonding.

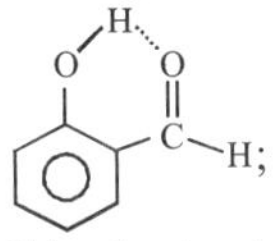

Intramolecular H-bonding in *o*-isomer, low b.p.

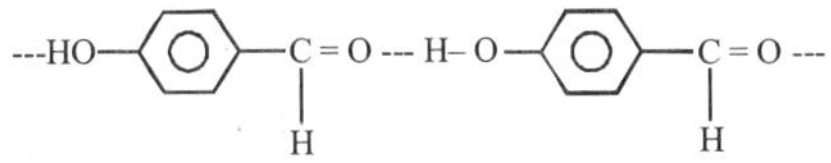

Intermolecular H-bonding in the *p*-isomer (association), high b.p.

(*ii*) In the acylium ion ($R-C\equiv O^+$), each and every element has a complete octet, while in carbonium ion ($R-C^+=O$), carbon bearing positive charge has incomplete octet which makes it more reactive than the acylium ion.

$R:C\vdots\vdots O^+: \longleftrightarrow R:C^+=\ddot{O}:$

Acylium ion
Carbonium ion
(Carbon has only six electrons)

(*iii*) Hypoiodite (^-OI) is a strong base than iodide ion (I^-). (O is more electronegative thus easily accommodates negative charge than I in I^-).
Haloform reaction is base-promoted reaction (the first step involves removal of acidic hydrogen atom of acetone by base). Hypoiodite ion being strong base than iodide ion, can easily remove acidic hydrogen atom.

$CH_3.CO.CH_3 + {^-OI} \rightarrow CH_3.CO.\bar{C}H_2 + HOI$

(*iv*) In weakly acidic medium, carbonyl group is protonated to form conjugate acid.

$>C=O + H^+ \longrightarrow >\overset{+}{C}-OH$

Carbon becomes more electron deficient

In strongly acidic medium (pH < 3.5), the unshared pair of electrons of N of the reagent is protonated, with the result nucleophile (NH_2NH_2) is converted to an electrophile ($NH_2NH_3^+$) which cannot react.

$H_2N-\ddot{N}H_2 \xrightarrow{H^+} H_2N + \overset{+}{N}H_3$

Hence, in highly acidic medium, there is no protonation of the carbonyl group.

(*v*) Alcohols show hydrogen bonding.
The boiling points of carbonyl compounds are lower than the corresponding alcohols since former do not show intermolecular H-bonding like alcohols.

$C_4H_9 - O - H$
$\vdots$
$H - O - C_4H_9$

(H-bonding between two butanol molecules)

Solubility of butanal in water is due to hydrogen bonding between butanal and water molecules (similarity to butanol).

94. (c)

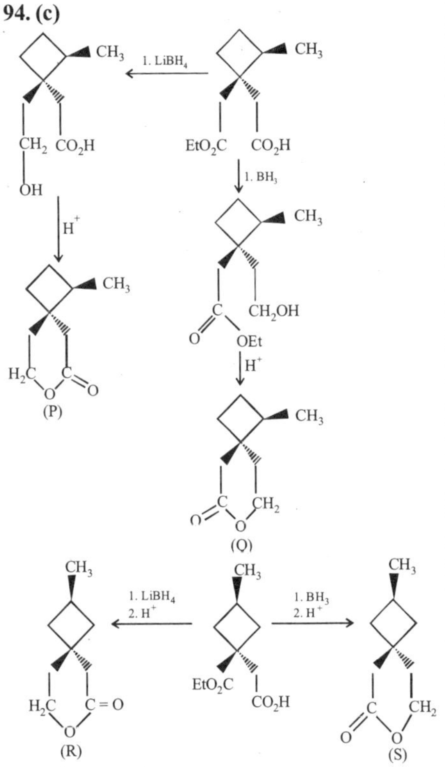

P and Q are diastereomers and R and S are diastereomers.

95. (d)

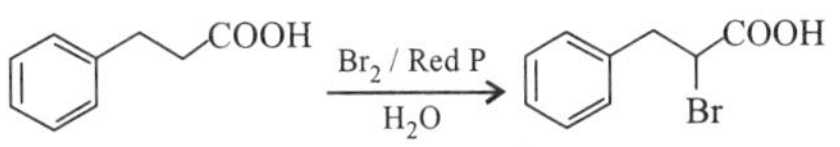

(P)

Number of unsaturated carbons (n) = 7

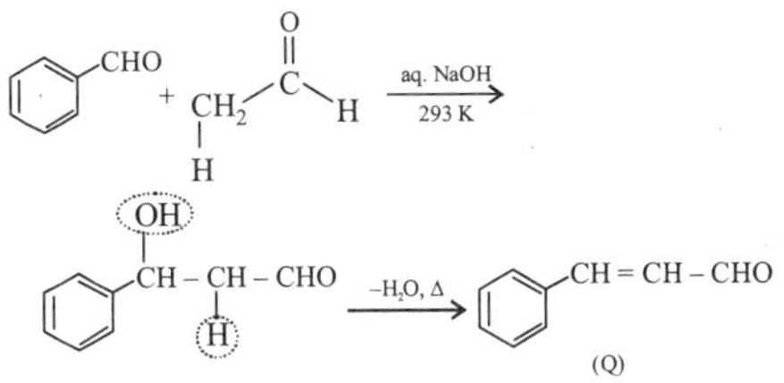

(Q)
$n = 9$

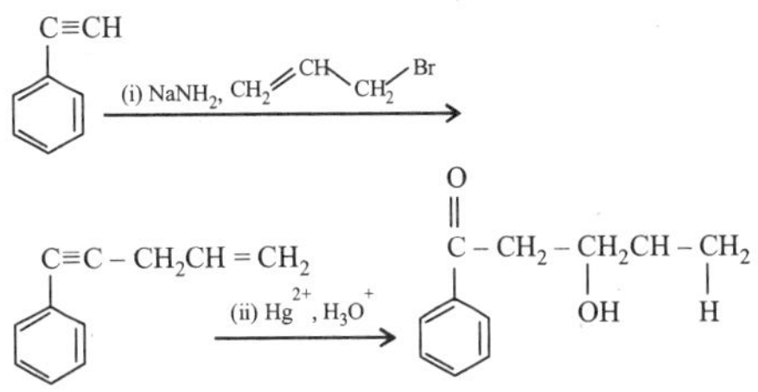

(R)
$n = 7$

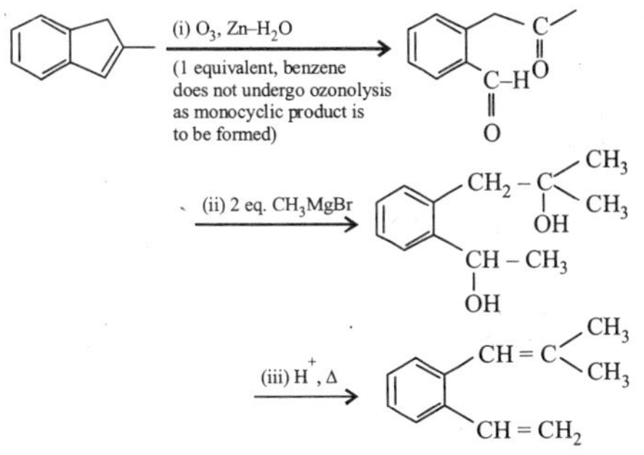

(S)
$n = 10$ (highest)

96. (a) Acidic strength depends upon the stability of anion formed higher the stability of the anion, higher will be acidic strength of the parent acid. Thus anion :

$HC\equiv CCOO^- > H_2C=CHCOO^- > p\text{-}C_6H_4(OMe)COO^- > H_3CCH_2COO^-$
sp sp^2 sp^3

Therefore,

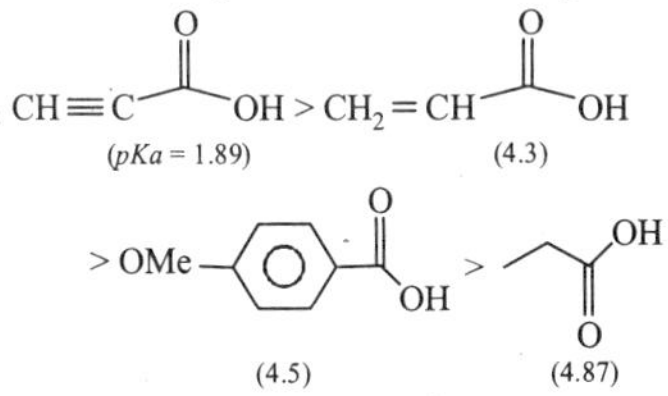

97. (a)

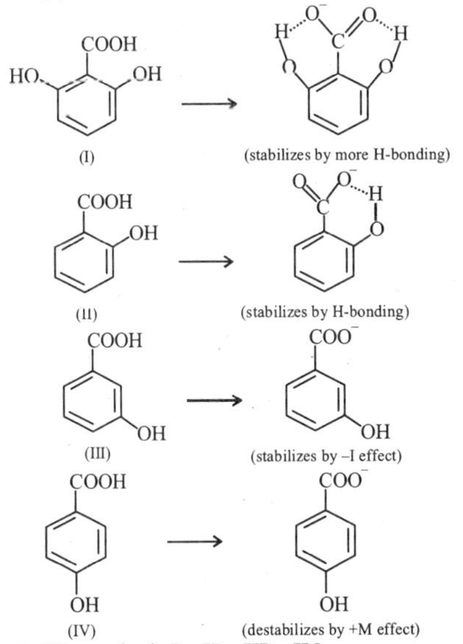

$\therefore$ Acidity order is I > II > III > IV

98. (d) Carbolic acid (Phenol) is weaker acid than carbonic acid and hence does not liberate CO_2 on treatment with aq. $NaHCO_3$ solution.

99. (b) β-Ketoacids undergo decarboxylation easily. It is due to the stability of transition state formed after decarboxylation of β-keto acids. This is not possible in case of other acids.

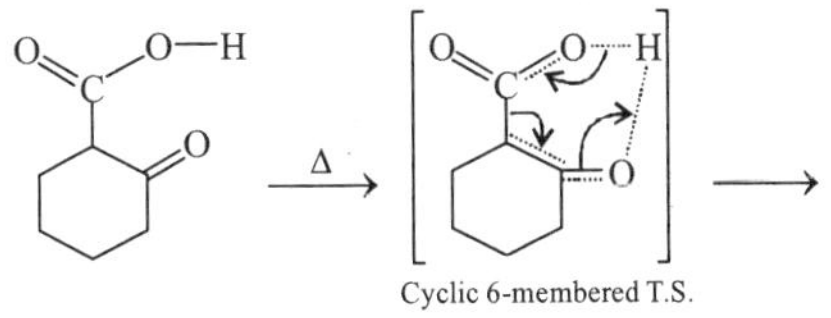

Cyclic 6-membered T.S.

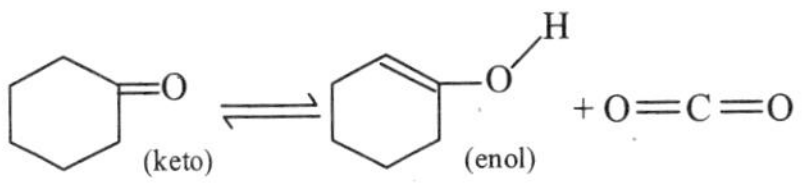

(keto) (enol)

100. (d)

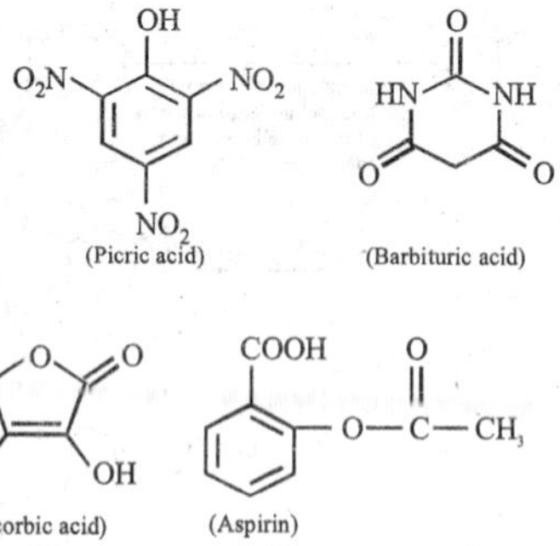

101. (b)

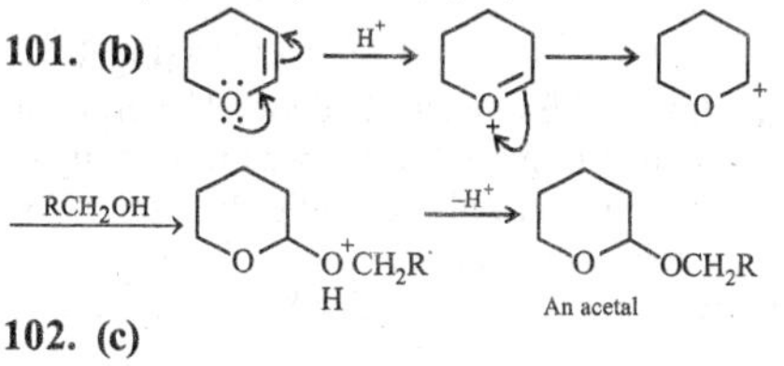

102. (c)

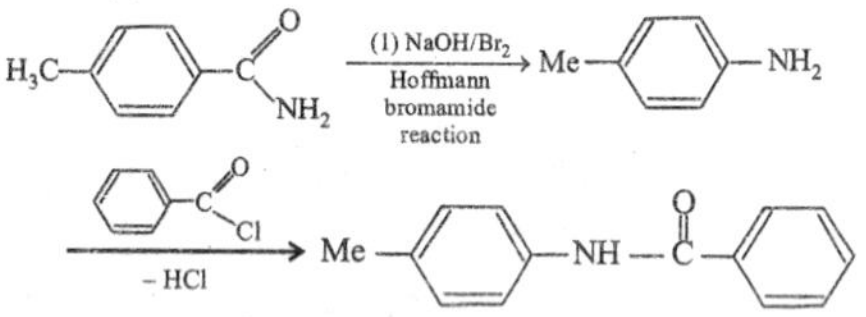

103. (c)

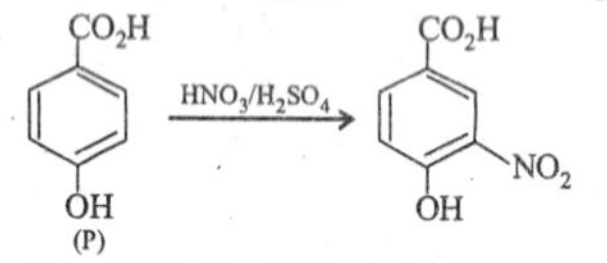

(OH group is strong activating and COOH group is weak deactivating group.)

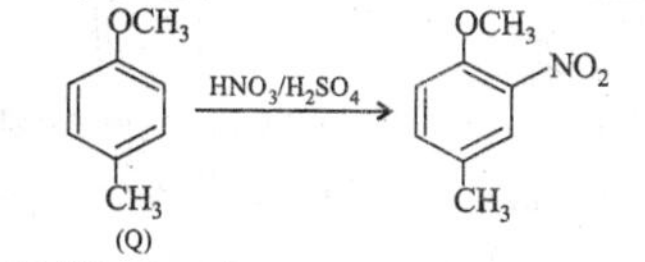

(OCH$_3$ group is more activating)

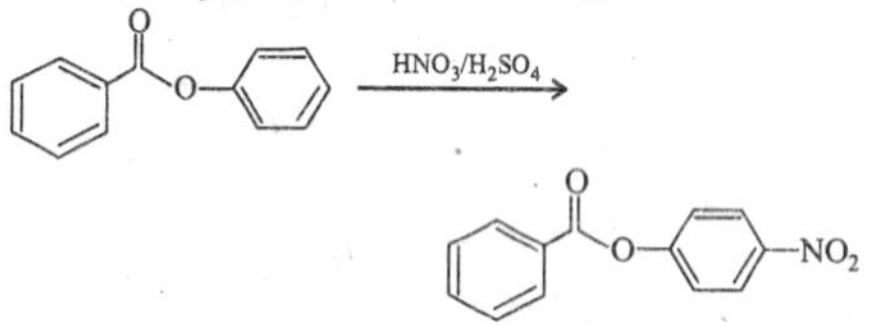

(Benzene ring having –O– is activated, benzene ring having –CO is deactivated)

104. (a) Carboxylic acid is stronger acid than phenol. The presence of electron withdrawing group (*e.g.* Cl) increases acidic strength, while presence of electron donating group (*e.g.* CH_3) decreases acidic strength.

105. (c)

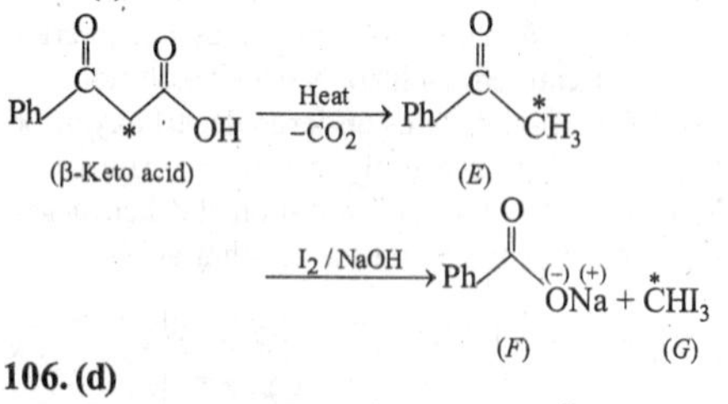

106. (d)

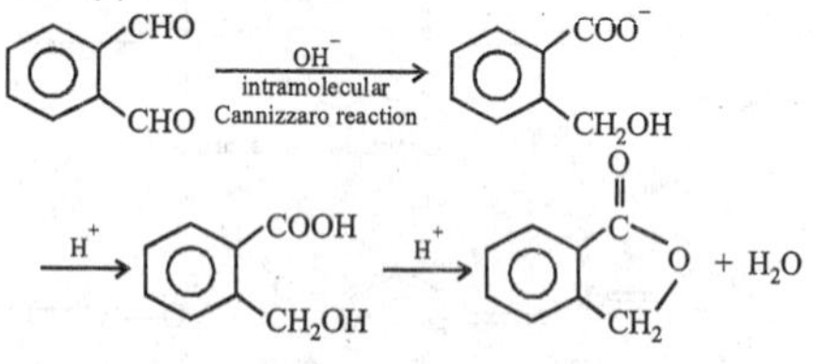

107. (a) This is simply a displacement reaction.

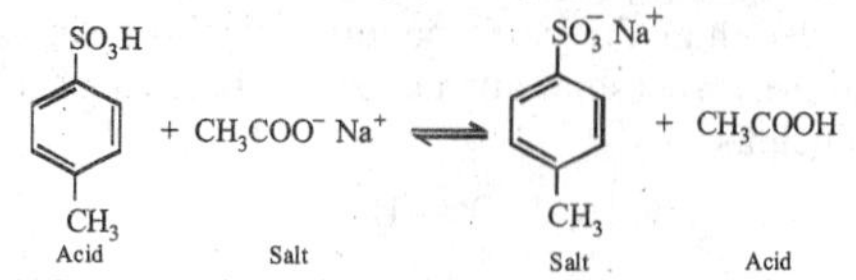

108. (a) The optically active acid will react with *d* and *l* forms of alcohol present in the racemic mixture at different rates to form two diastereomers in unequal amounts leading to optical activity of the product.

109. (a) Recall that, esters react with excess of Grignard reagents to form 3° alcohols having at least two identical alkyl groups corresponding to Grignard reagent.

$$R-\overset{O}{\overset{||}{C}}-OC_2H_5 + R''MgX \longrightarrow R-\underset{R''}{\overset{OH}{C}}-R''$$

Since, here Grignard reagent is CH_3MgBr, the 3° alcohol should have at least two methyl groups. Hence, (a) is the correct choice.

110. (c) $C_6H_5COOH + SOCl_2 \longrightarrow C_6H_5COCl + SO_2 + HCl$

111. (d) $C_2H_5\overset{14}{C}OOH + NaHCO_3 \rightarrow C_2H_5COONa + H_2O + \overset{14}{C}O_2$

112. (a) $LiAlH_4$ is a reducing agent, it reduces –COOH group to –CH_2OH group.

$$C_6H_5COOH \xrightarrow{LiAlH_4} C_6H_5CH_2OH$$

113. (2)

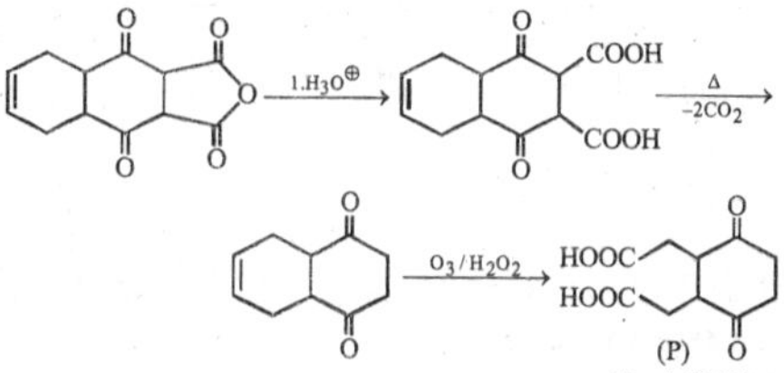

No. of –COOH group is '2'.

114. (4) All carboxylic acids and phenols are soluble in aqueous NaOH. Thus, four compounds are soluble in aqueous NaOH.

115. (4.00)

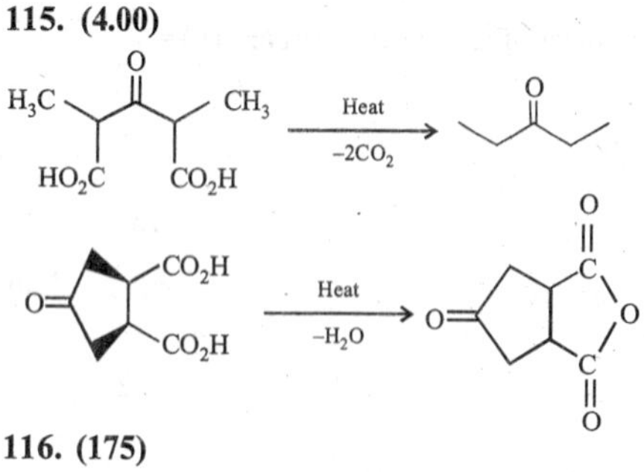

116. (175)

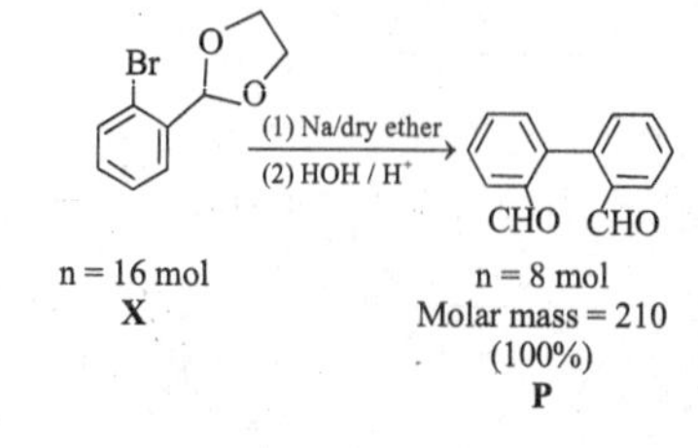

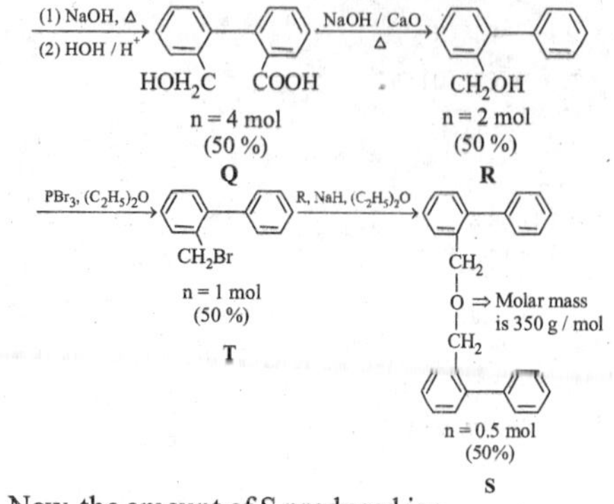

Now, the amount of S produced is :-

$S = 0.5 \times 350 = 175$ g

117. (909) Reactions involves are:

$$H-C\equiv C-(CH_2)_{15}-CO-Et \xrightarrow[H_3O^+]{(i)\,Hg^{2+}} CH_3-\overset{O}{\overset{||}{C}}-(CH_2)_{15}-\overset{O}{\overset{||}{C}}-OEt \xrightarrow{(ii)\,Zn\text{-}Hg/HCl}$$

$$CH_3-(CH_2)-(CH_2)_{15}-\overset{O}{\overset{||}{C}}-OEt \xrightarrow[\Delta]{H_3O^+} \underset{(P)}{CH_3-(CH_2)_{16}-COOH}$$

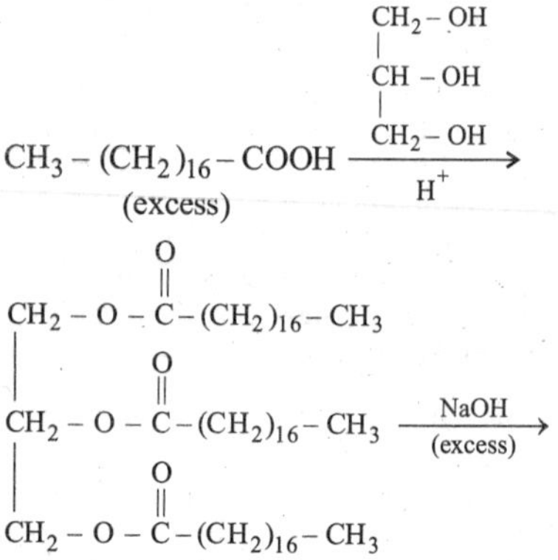

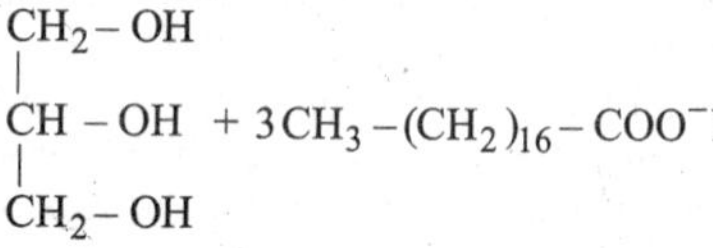

$$\xrightarrow{CaCl_2} \frac{3}{2}(CH_3-(CH_2)_{16}-COO)_2Ca \quad (R)$$

1 mole of Q will 1.5 mole of R.

So, mass of R produced = 606 g × 1.5 = 909 g

118. (18.00)

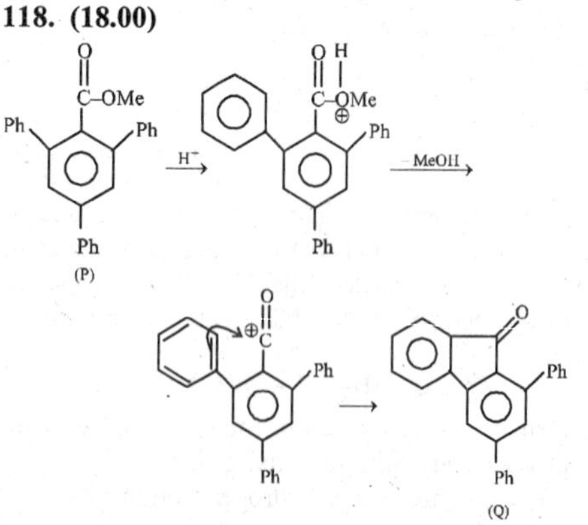

119. CO; $HCOOH \xrightarrow[\text{heat}]{\text{Conc.}H_2SO_4} CO + H_2O$

120. False : H-Bonding in propionic acid is stronger (carboxylic acids can form dimers) than that in butanol.

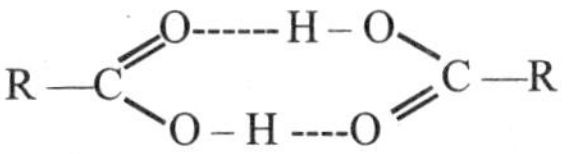

121. False : Saponification is alkaline hydrolysis of esters.

122. (b, c)

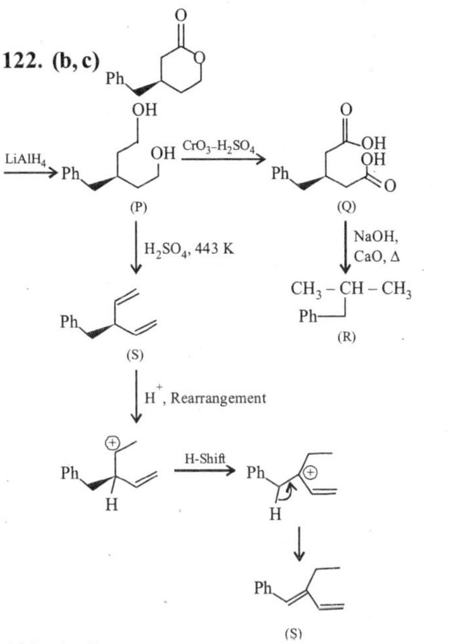

123. (c, d)

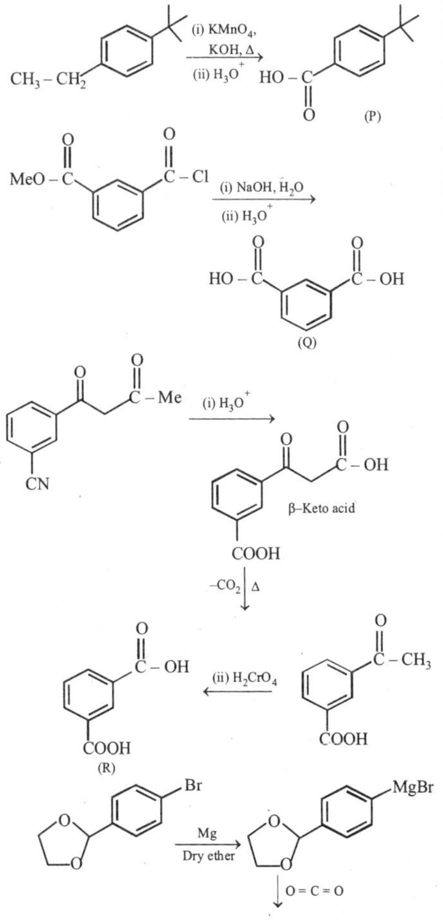

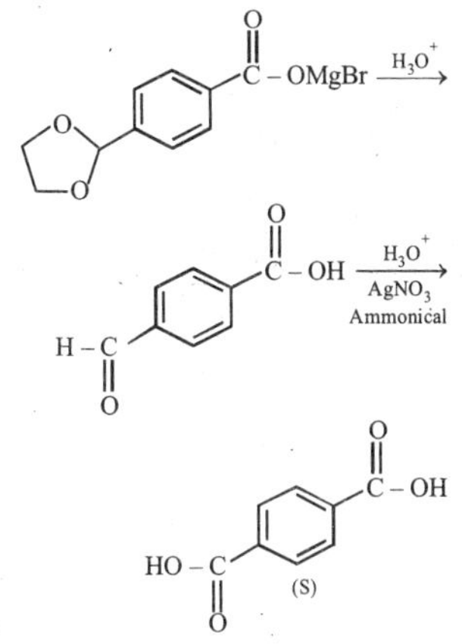

124. (b)

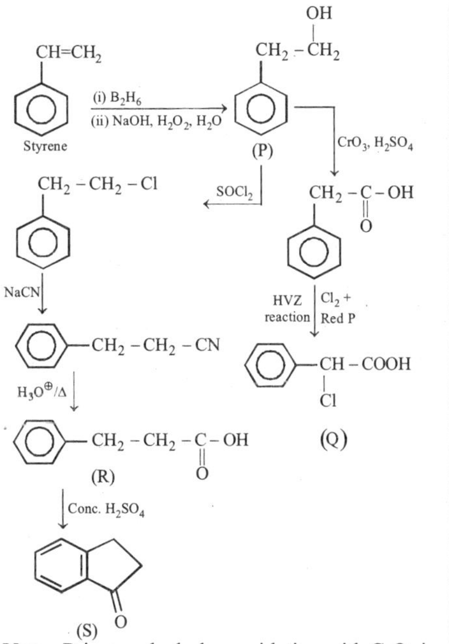

Note : Primary alcohol on oxidation with CrO_3 in anhydrous medium give aldehyde. While with CrO_3, H_2SO_4, carboxylic acid is produced.

125. (a, d)

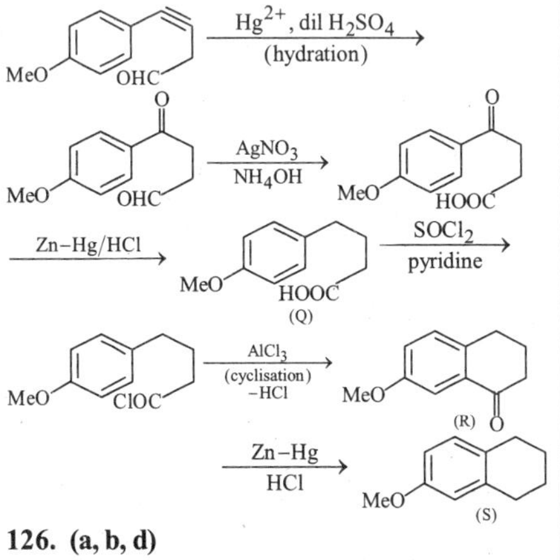

126. (a, b, d)

(a) 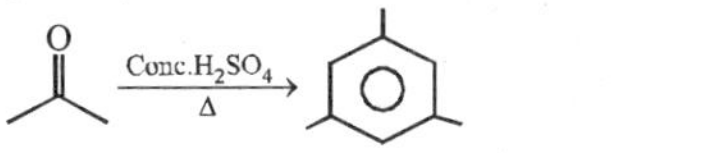

(b) 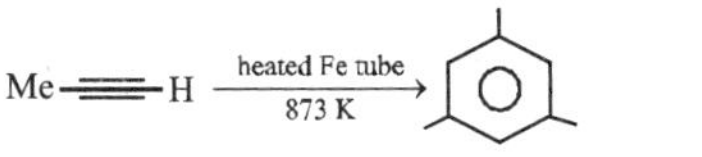

(c) 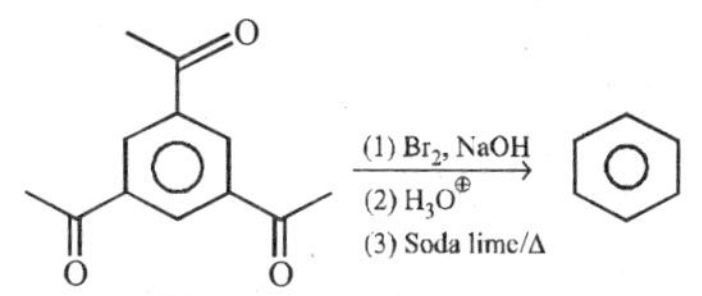

(d) 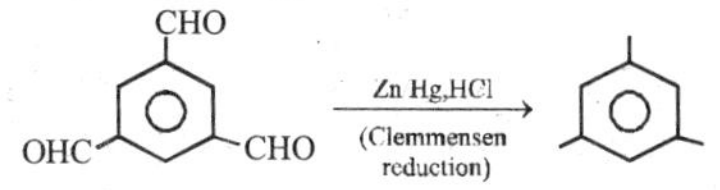

127. (c, d)

$LiAlH_4/(C_2H_5)_2O$: Reduces esters, carboxylic acids, epoxides and aldehydes and ketones.

BH_3 in T.H.F : Reduces –COOH and aldehydes to alcohols but does not reduce esters and epoxides.

$NaBH_4$ in C_2H_5OH : Reduces only aldehydes and ketones to alcohols.

Raney Ni in T.H.F. : Does not reduce –COOH, –COOR and epoxides but it can reduce aldehydes to alcohols.

128. (a, c, d)

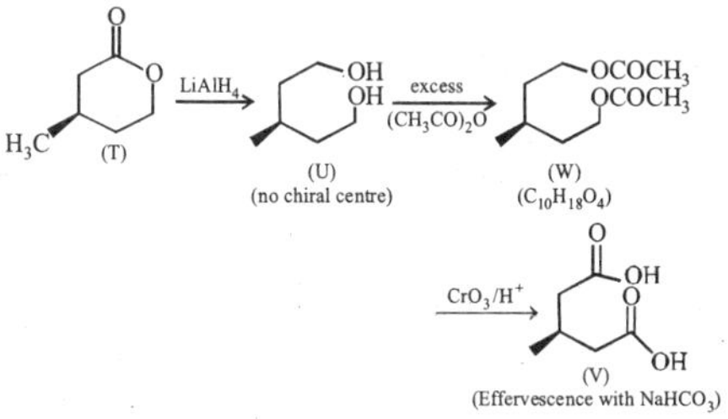

129. (b, d)

(a) Both are soluble in NaOH. Hence, **inseparable**.

(b) Only benzoic acid (C_6H_5COOH) is soluble in NaOH and $NaHCO_3$, while benzyl alcohol ($C_6H_5CH_2OH$) is not. Hence, **separable**.

(c) Although NaOH can enable separation between benzyl alcohol ($C_6H_5CH_2OH$) and phenol (C_6H_5OH) as only the latter is soluble in NaOH. However, in $NaHCO_3$, both are insoluble. Hence, **inseparable**.

(d) α-Phenylacetic acid ($C_6H_5CH_2COOH$) is soluble in NaOH and $NaHCO_3$, while benzyl alcohol ($C_6H_5CH_2OH$) is not. Hence, **separable**.

130. (a, b, d) Ethyl chloride and acetyl chloride react with alc. KCN by nucleophilic substitution reaction while benzaldehyde undergoes benzoin condensation :

$$C_2H_5Cl \xrightarrow{KCN\,(alc.)} C_2H_5CN + KCl$$

$$CH_3COCl \xrightarrow{KCN\,(alc.)} CH_3COCN + KCl$$

$$2\,C_6H_5CHO \xrightarrow{KCN\,(alc.)} C_6H_5CHOHCOC_6H_5$$

131. (a, b) Iodoform reaction is given by the compounds containing $-COCH_3$, $-CH(OH)CH_3$ group and also CH_3CH_2OH and CH_3CHO.

2-Hydroxypropane ($CH_3CHOHCH_3$) contains the grouping CH_3CHOH— and acetophenone ($C_6H_5COCH_3$) contains the grouping CH_3CO–.

132. (b)

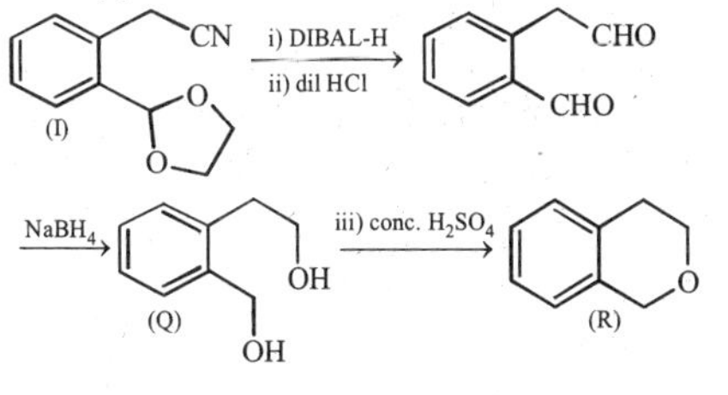

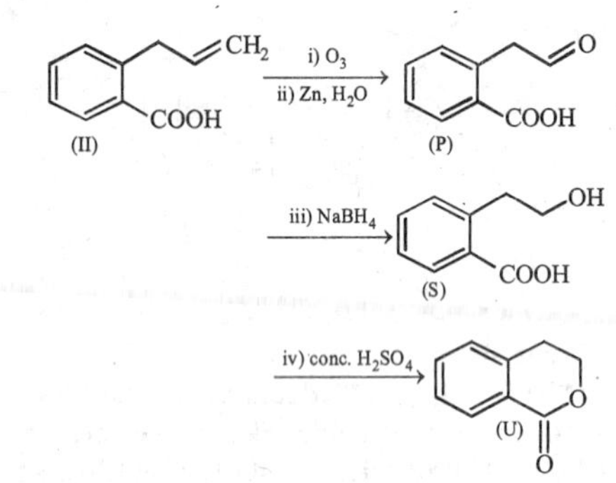

133. (a)

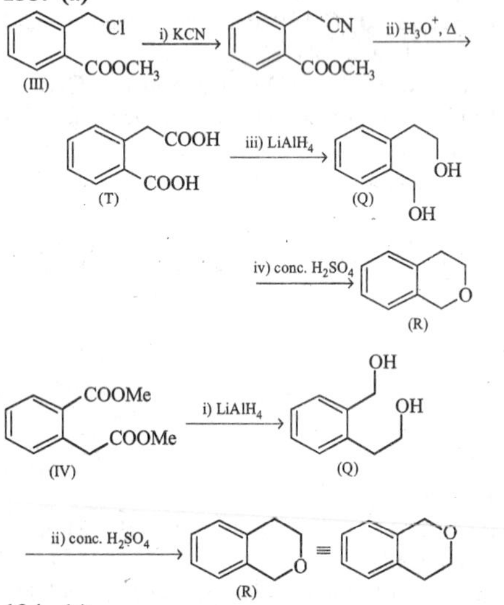

134. (a)

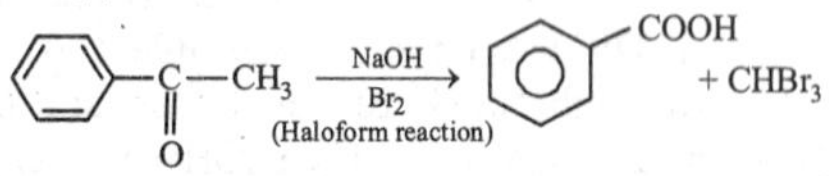

135. (c) Perkin condensation of benzaldehyde with $(CH_3CO)_2O/CH_3COOK$ yields cis and trans form of cinnamic acid.

$$C_6H_5CHO + (CH_3CO)_2O \xrightarrow[(ii)\ H^+]{(i)\ CH_3COOK} C_6H_5CH = CHCOOH$$

Benzaldehyde — Acetic anhydride — Cinnamic acid *cis* and *trans*

136. (d) Alkylbenzenes when treated with Br_2 at high temperature, in the presence of sunlight and absence of halogen carrier undergo **halogenation in the side chain** via free radical formation.

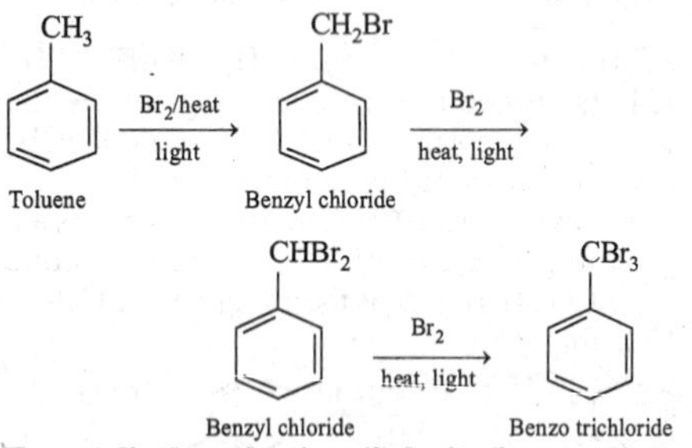

Free radical mechanism (Substitution reaction)

137. (a)

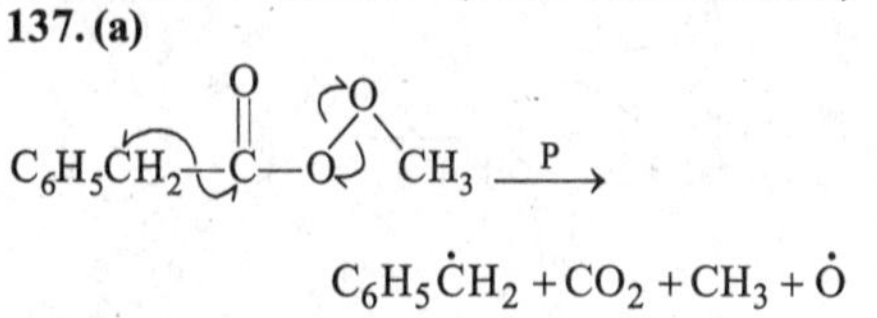

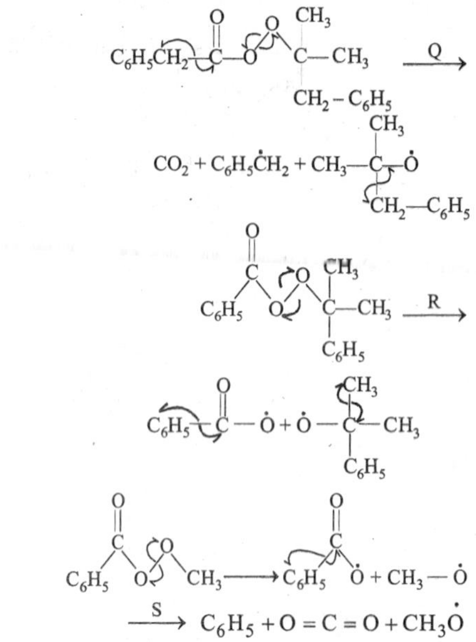

138. A – r, s, t ; B – p, s ;C - r, s; D – q, r

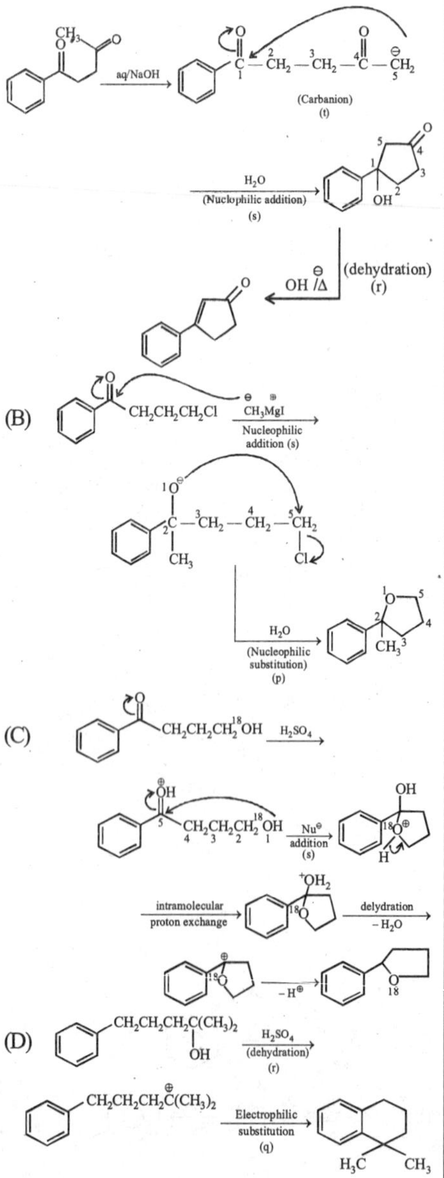

139. (A) - p, q, t ; (B) - p, s, t ; (C) – r, s ; (D) – p

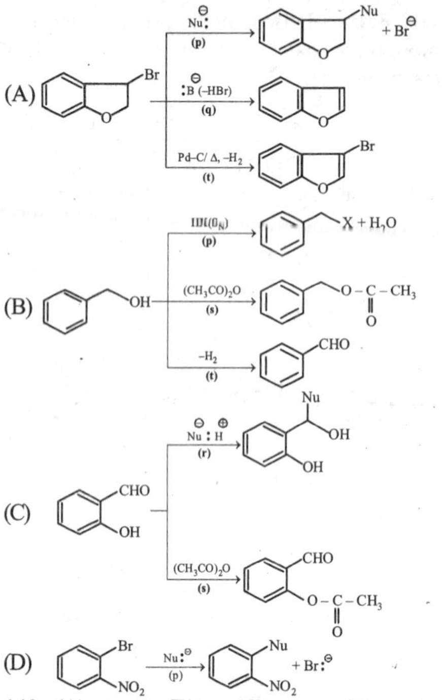

140. (A) - p, q, s; (B) - q; (C) - q, r, s; (D) - q, r

(A) C_6H_5CHO forms ppt. of 2, 4 dibromophenylhydrazone **(p)**, forms silver mirror with ammonical silver nitrate – Tollen's reagent **(q)**, forms cyanohydrin with CN^- **(s)**.

(B) $CH_3C \equiv CH$ gives ppt. with $AgNO_3$ **(q)**

(C) CN^- reacts with $AgNO_3$ to form ppt. of AgCN **(q)**, it is a nucleophile **(r)** and forms cyanohydrin **(s)**

(D) I^- gives ppt. of AgI with $AgNO_3$ **(q)**, and it is a nucleophile **(r)**

For 141-142 (Passage-I)

(*i*) The oxidation product (a dibasic acid) of the compound P reacts with ethylene glycol and gives polymer dacron (a very well known reaction) points out that the oxidation product i.e. the dibasic acid is terephthalic acid.

$$HOOC-C_6H_4-COOH + n\,HOCH_2.CH_2OH \rightarrow Dacron$$

(*ii*) Now since terephthalic acid is obtained by the oxidation of a carboxylic acid $P(C_{11}H_{12}O_2)$, P must have following partial structure ($C_{11}H_{12}O_2 - COOH - C_6H_4 = C_4H_7$)

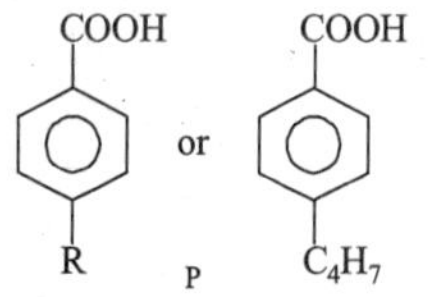

(*iii*) The compound P undergoes reduction by means of H_2/Pd and also undergoes ozonolysis, it must have an unsaturated linkage. Further ozonolysis of P to form an aliphatic ketone indicates = $C(R_2)$ type of linkage for which P should have following final structure,

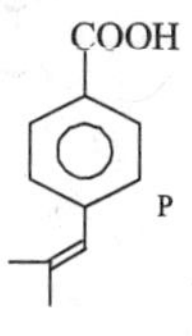

which explains all the given reactions.

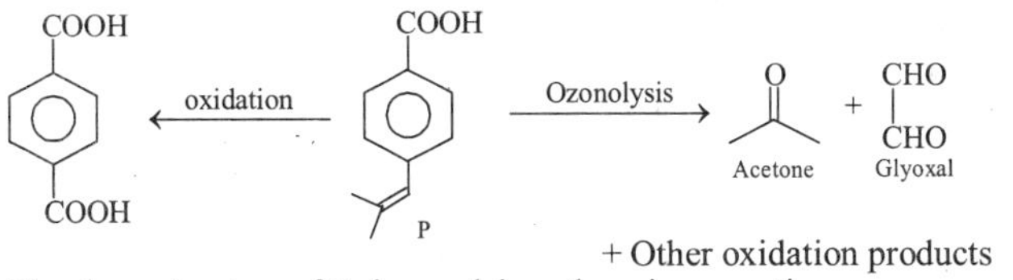

+ Other oxidation products

The above structure of P also explains other given reactions.

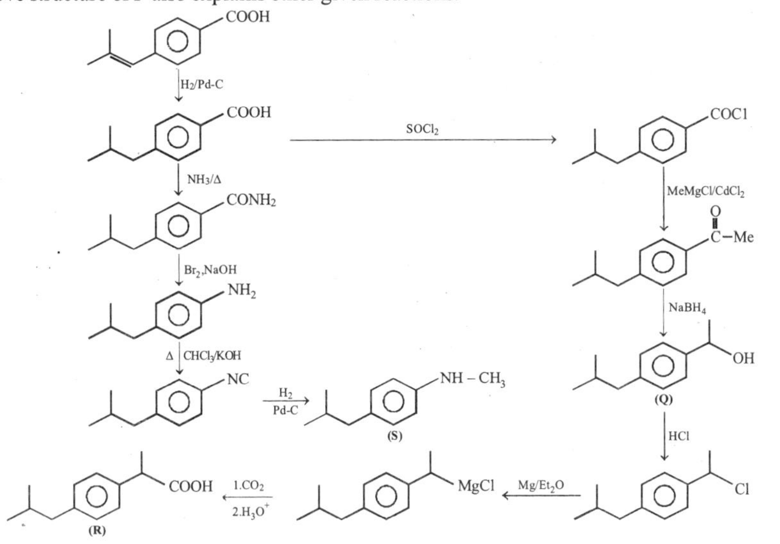

141. (a)

142. (b)

For 143-144 (Passage-II)

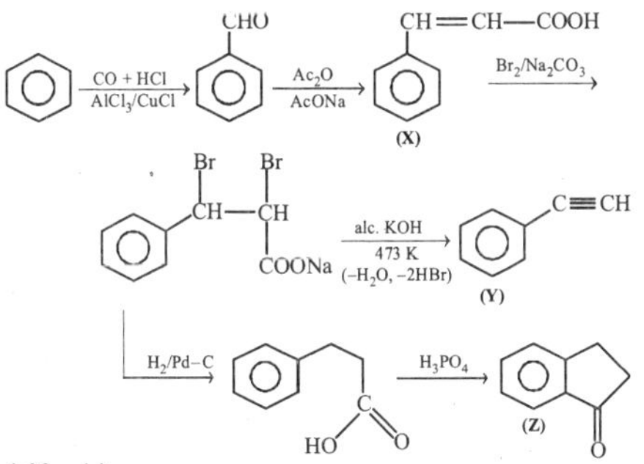

143. (c)

144. (a)

For 145-146 (Passage-III)

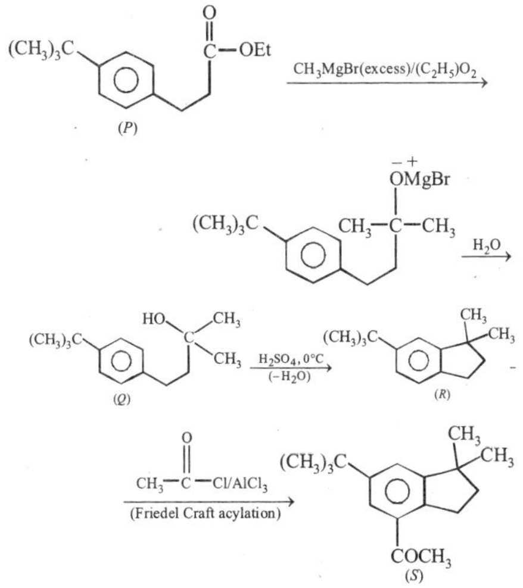

145. (a)

146. (a)

For 147-148. (Passage-IV)

147. (b)

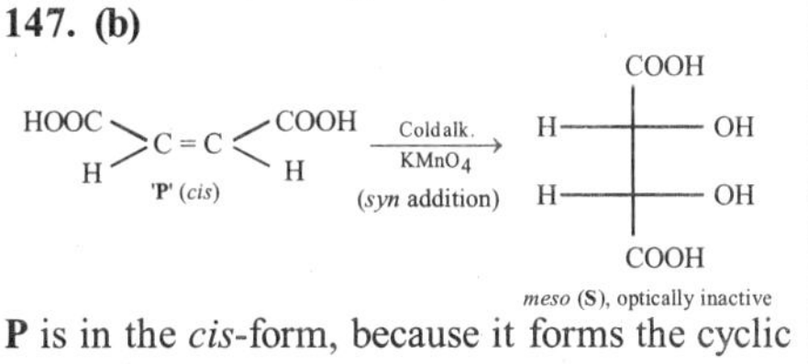

P is in the *cis*-form, because it forms the cyclic anhydride.

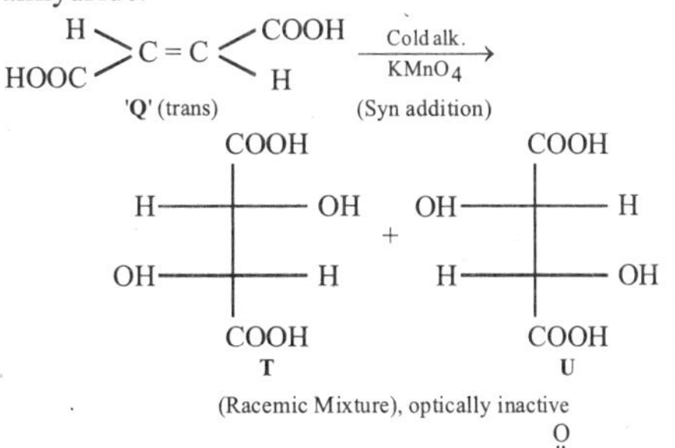

148. (a)

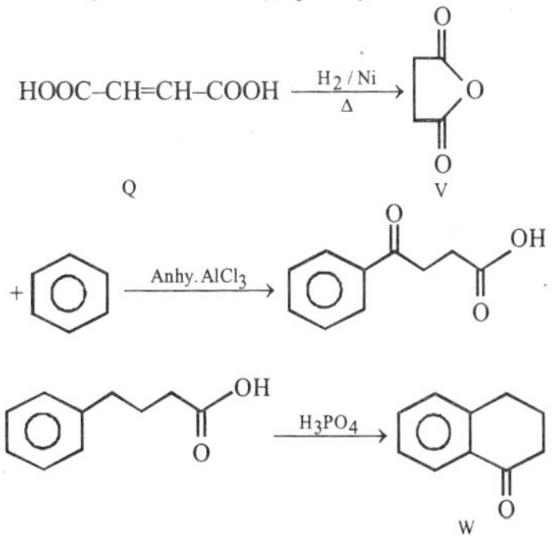

For 149- 150 (Passage-V)

Reactions of compound *J* indicates that it has C = C linkage and – COOH group. Thus, *J* can be written as $C_6H_5CH = CH\ COOH$. Since, *J* is unsaturated carboxylic acid and it is formed by the reaction of compound *I* with $(CH_3CO)_2O$ and CH_3COONa, compound *I* should be an aldehyde (recall Perkin reaction). Thus, the whole series of reactions can be written as below.

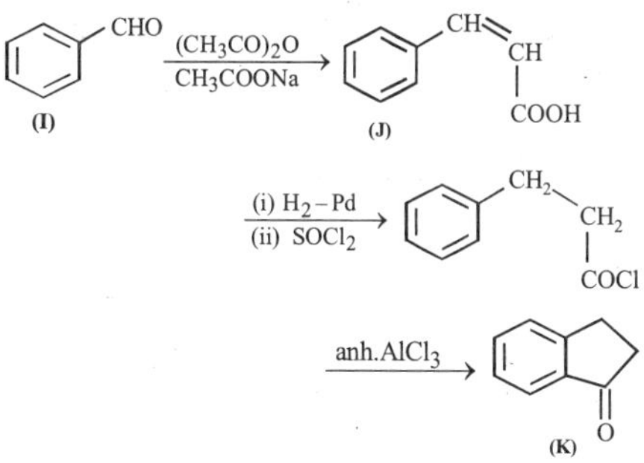

149. (a)

150. (b)

151. (d) *p*-Hydroxybenzoic acid has higher boiling point than *o*-hydroxybenzoic acid due to intermolecular hydrogen bonding. Thus, statement-1 is false. *o*-Hydroxybezoic acid shows intramolecular H-bonding thus, statement-2 is true.

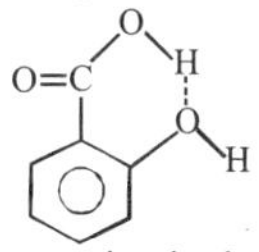

152. (c) Haloform reaction is shown by ketones having $R - CO - CH_3$ structure in which – CO – group is electron deficient, while

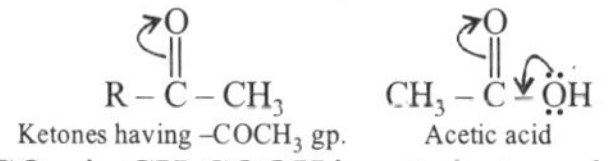

The – CO – in CH_3COOH is not electron deficient due to supply of electrons by the –ÖH part- Acetic has three α–H atoms.

153. (d) Acetate ion is reasonance stabilized while methoxide ion is not.

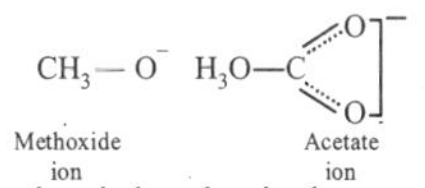

Hence, acetate ion is less basic than methoxide ion.

154. Higher the K_a value, more stronger is the acid. Correct order of acidic strength of the given acids is

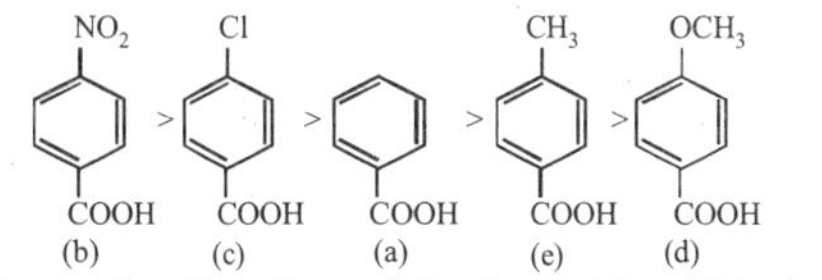

Hence, the K_a values of the five acids will be in the order.

	(b)	(c)	(a)	(e)	(d)
K_a value	30.6×10^{-5}	10.2×10^{5}	6.4×10^{5}	4.2×10^{5}	3.3×10^{5}

155. The two enantiomers of 2-phenylpropanoic acid in the racemic mixture react with (+)– alcohol to form two diastereomers.

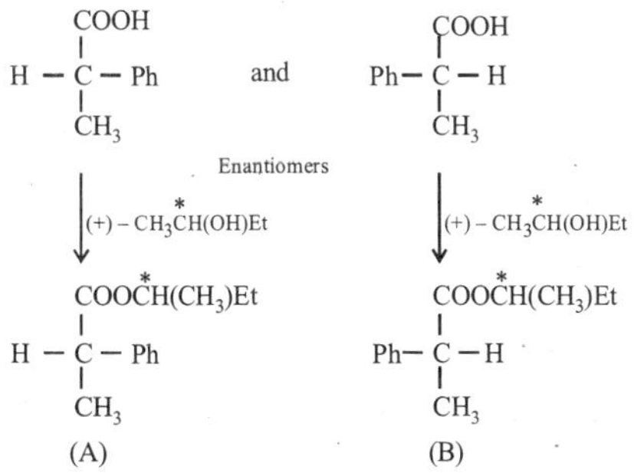

(A) and (B) are diasteromers.

The bonds attached to the chiral carbon in both the molecules are not broken during the esterification reaction. (+) – Acid reacts with (+) – alcohol to give an (+)–(+)–ester, while (–) acid reacts with (+) alcohol to give (–)–(+)–ester. These two esters are diastereoisomers.

(*iv*) Presence of electron withdrawing group increases the acidic character of the –COOH due to –I effect, Further the –I group (–Cl) from –COOH group, weaker is the acid, thus III is weaker than I while presence of electron-donating group (alkyl groups) decreases the acidic character due to +I effect. Thus,

$$IV < II < V < III < I$$

156. $Ba\overset{*}{C}O_3 + H_2SO_4 \longrightarrow \underset{(X)}{\overset{*}{C}O_2} \uparrow$

$$CH_2 = CHBr \xrightarrow[\text{ether}]{\text{(i)Mg}} CH_2 = CHMgBr$$

$$\xrightarrow[\text{(ii)}H_3O^+]{\text{(ii)}\overset{*}{C}O_2(X)} \underset{(Y)}{CH_2 = CH\overset{*}{C}OOH}$$

$$\xrightarrow{LiAlH_4} \underset{(Z)}{CH_2 = CH.\overset{*}{C}H_2OH}$$

Formation of $\overset{*}{C}H_2O$ from (Z)

$$CH_2 = CH.\overset{*}{C}H_2OH \xrightarrow{H^+}$$

$$[CH_2 = CH.\overset{*}{C}H_2^{\oplus} \longleftrightarrow \overset{\oplus}{C}H_2 - CH = \overset{*}{C}H_2]$$

$$\xrightarrow{Br^-} BrCH_2.CH = \overset{*}{C}H_2 \xrightarrow{O_3} \overset{*}{C}H_2 = O$$

157.

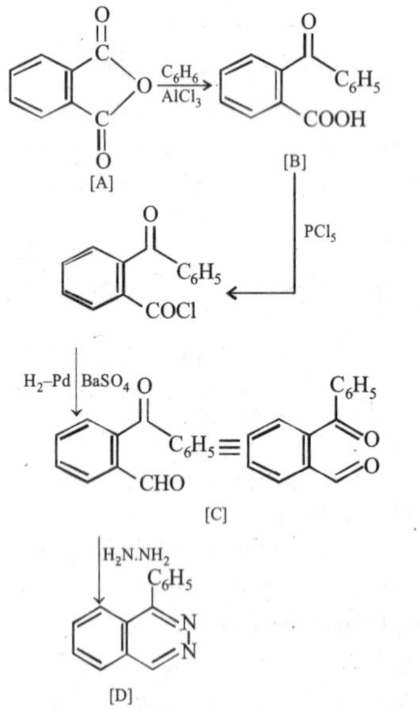

158. (*i*) $CH_3-\overset{O}{\overset{\|}{C}}-\overset{18}{O}C_2H_5 \xrightarrow[H_2O]{H^+}$

$$\underset{A}{CH_3-\overset{O}{\overset{\|}{C}}-OH} + \underset{B}{C_2H_5\overset{18}{O}H}$$

(Ester hydrolysis involves acyl-oxygen fission)

(*ii*)

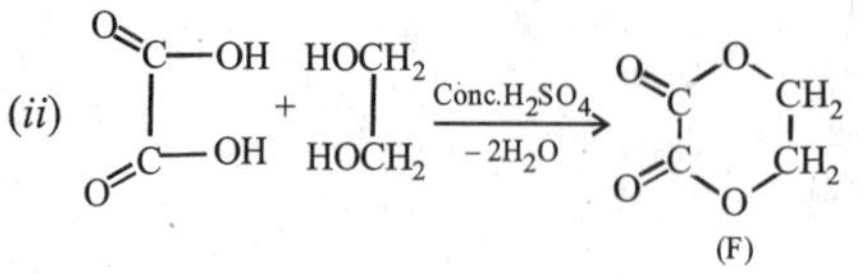

(*iii*)

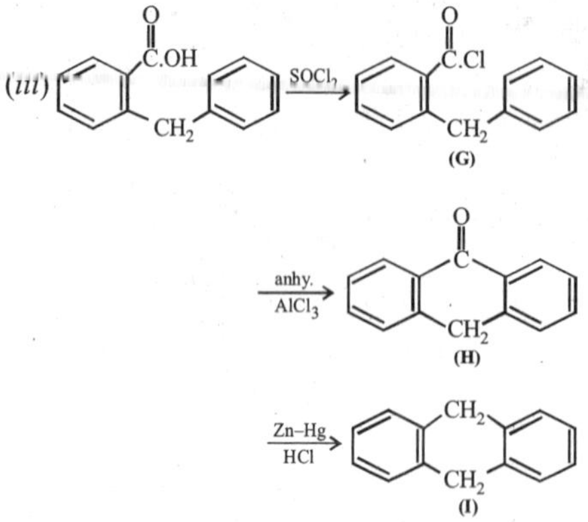

(*iv*) $C_6H_5CHO + CH_3COOC_2H_5$

$$\xrightarrow[\text{absolute alcohol, heat}]{NaOC_2H_5 \text{ in}} \underset{(D)}{C_6H_5CH = CHCOOC_2H_5}$$

(Claisen condensation)

(*v*) $CH_3CH_2COOH \xrightarrow[\text{HVZ reaction}]{\text{P and } Br_2}$

$$\underset{(A)}{CH_3.\overset{Br}{\overset{|}{C}}HCOOH} \xrightarrow[\text{(ii) } H^+]{\text{(i) Alc. KOH (excess)}} \underset{(B)}{CH_2 = CHCOOH}$$

(*vi*) Esters react with excess of RMgX to form 3° alcohols having two alkyl groups corresponding to R of RMgX. Thus,

$$C_6H_5CH_2COOCH_3 \xrightarrow[\text{(ii) } H^+]{\text{(i) } 2CH_3MgBr}$$

$$C_6H_5CH_2\underset{OH}{\underset{|}{C}}(CH_3)_2$$

(*vii*) $C_6H_5COOH + CH_3MgI \longrightarrow$

$$CH_4 + C_6H_5COO\,MgI$$

(*viii*)

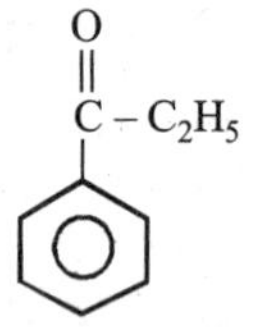

Phenyl ethyl ketone

(*ix*) $\underset{\text{Ethyl acetate}}{CH_3-\overset{O}{\overset{\|}{C}}-OC_2H_5} + 2C_2H_5MgBr \longrightarrow$

$$CH_3-\underset{C_2H_5}{\underset{|}{\overset{OMgBr}{\overset{|}{C}}}}-C_2H_5 + Mg(Br)OC_2H_5 \xrightarrow{H^+}$$

$$\underset{\text{Ethylbutanol-2}}{CH_3-\underset{C_2H_5}{\underset{|}{\overset{OH}{\overset{|}{C}}}}-C_2H_5}$$

159.

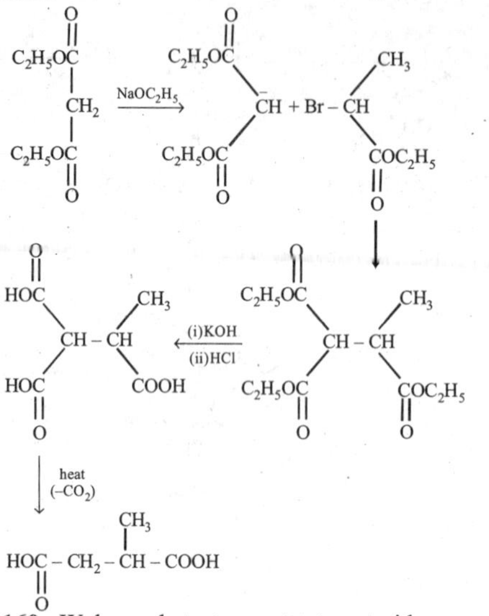

160. We know that esters on treatment with excess of methyl magnesium chloride either give secondary alcohols (from alkyl formates) or tertiary alcohols (from esters other than formates). However, tertiary alcohols are not easily oxidised, hence the alcohol should be secondary alcohol and thus ester is alkyl formate. Hence, ester *A* ($C_4H_8O_2$) should be $HCOOC_3H_7$. Thus the various reactions and nature of compound *B* can be established as below.

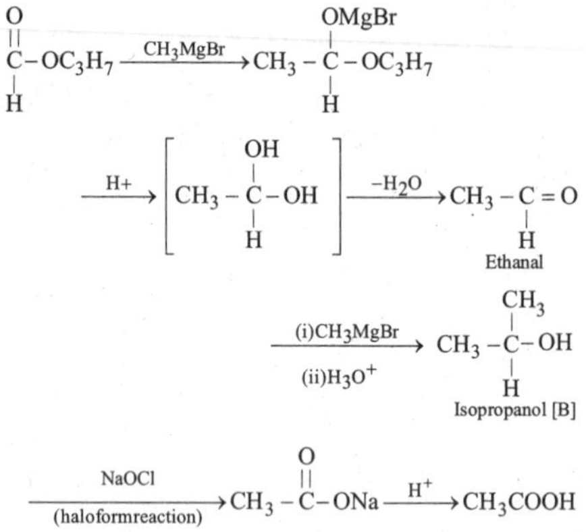

161. (i) Both of the resonating structures of benzoate ion are equivalent, while it is not so in phenoxide ion.

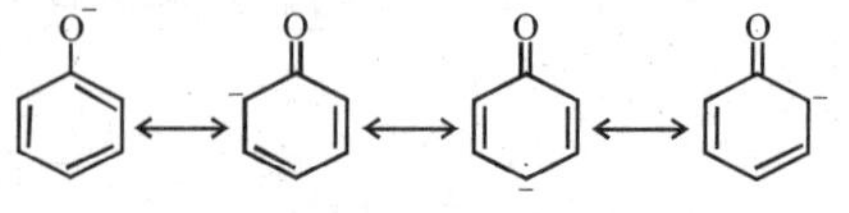

Resonating structures of phenoxide ion

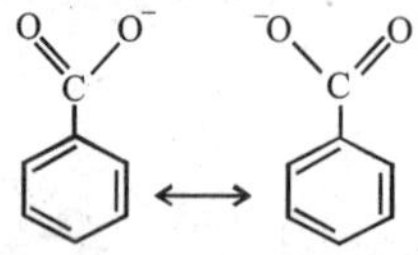

Resonating structures of benzoate ion

The benzoate ion is more stabilized because the negative charge on both structures is on the more electronegative oxygen atom, whereas in phenoxide ion, it is on the less electronegative carbon atoms.

(ii) $H-\underset{O}{\underset{\|}{C}}-OH \rightleftharpoons H-\underset{O}{\underset{\|}{C}}-O^- + H^+$;

$$CH_3-\underset{\underset{O}{\|}}{C}-OH \rightleftharpoons CH_3-\underset{\underset{O}{\|}}{C}-O^- + H^+$$

(iii) Presence of CH_3 gp in acetate ion shows +I effect and thereby intensifies charge on O^- of acetate ion which is thus destabilized. Thus formate ion is more stable than acetate ion or HCOOH loses proton more easily than CH_3COOH.

Formic acid has no alkyl group i.e no $\alpha-$H atom, hence it does not undergo halogenation, while acetic acid has a methyl group (i.e three α – H atoms) on which halogenation takes place.

162. (*i*) β-Keto acids are unstable and undergo decarboxylation most readily. This is due to formation of 6-membered cyclic transition state.

163.

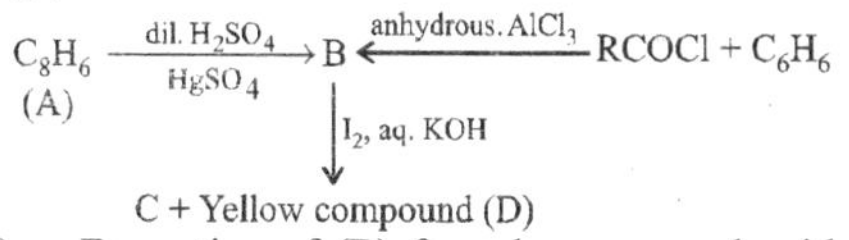

(i) Formation of (B) from benzene and acid chloride in presence of anhydrous $AlCl_3$ (Friedel-Craft reaction) indicates that it is a ketone, C_6H_5COR.

(ii) Further the ketone (B) reacts with alkaline iodine forming yellow compound (D) (haloform reaction). This indicates that one of the alkyl groups in ketone (B) is $-CH_3$. Hence, (B) should be $C_6H_5.CO.CH_3$.

(iii) Since ketone (B) is also formed from the hydrocarbon C_8H_6 (A) by reaction with dil. H_2SO_4 and $HgSO_4$, the hydrocabon (A) must have an acetylenic hydrogen atom, *i.e.* $\equiv C-H$ grouping. Hence, (A) must be $C_6H_5C\equiv CH$.

Thus, compounds (A) to (D) are

$\underset{(A)}{C_6H_5.C\equiv CH}$ $\underset{(B)}{C_6H_5.CO.CH_3}$ $\underset{(C)}{C_6H_5COOH}$ $\underset{(D)}{CHI_3}$

Formation of (B) from (A)

$$\underset{\text{Phenylacetylene (A)}}{C_6H_5.C\equiv CH} \xrightarrow[HgSO_4]{dil.H_2SO_4} [C_6H_5-\overset{OH}{\overset{|}{C}}=CH_2]$$

$$\xrightarrow{\text{rearranges}} \underset{\text{Acetophenone (B)}}{C_6H_5.CO.CH_3}$$

164. $PCl_5 + SO_2 \longrightarrow \underset{(A)}{SOCl_2} + \underset{(B)}{POCl_3}$

$\underset{(A)}{SOCl_2} + CH_3COOH \longrightarrow$

$$\underset{(C)}{CH_3COCl} + SO_2 + HCl$$

$\underset{(C)}{2CH_3COCl} + (CH_3)_2Cd \longrightarrow$

$$\underset{(D)}{2CH_3-CO-CH_3} + CdCl_2$$

165.

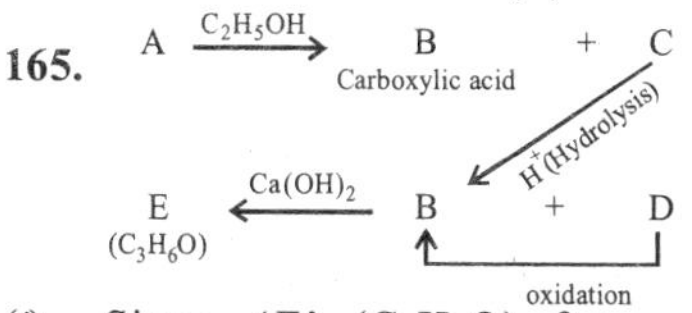

(*i*) Since, 'E' (C_3H_6O) forms a 2, 4-dinitrophenylhydrazone but does not reduce Tollen's reagent and Fehling solution, it must be a ketone, $CH_3.CO.CH_3$.

(*ii*) The compound 'E' (established as ketone) is obtained by heating compound 'B' with $Ca(OH)_2$, 'B' must be CH_3COOH.

(*iii*) Compound 'B' is obtained by the oxidation of 'D', the latter must be ethyl alcohol, C_2H_5OH and hence 'C' must be ethyl acetate, $CH_3COOC_2H_5$.

(*iv*) Since, compound 'A' when treated with ethyl alcohol gives acetic acid 'B' and ethyl acetate 'C', it must be acetic anhydride.

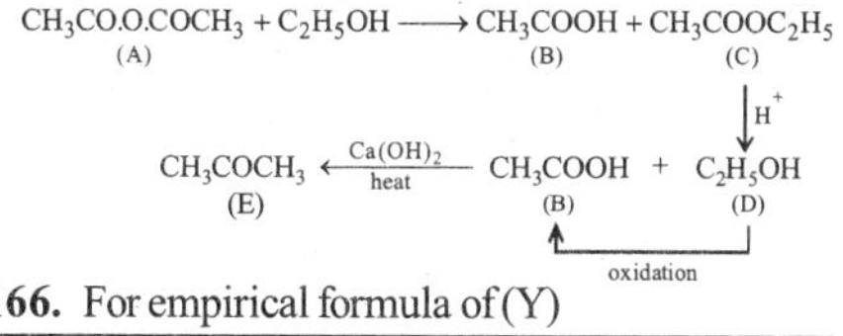

166. For empirical formula of (Y)

Element	%	Relative no. of atoms	Simplest ratio
C	49.31	4.10	3
H	9.59	9.59	7
N	19.18	1.37	1
O	21.92	1.37	1

∴ Empirical formula of (Y) is C_3H_7NO.

(Y) reacts with Br_2 and NaOH to give (Z) and (Z) reacts with HNO_2 to give ethanol and thus (Y) seems to have —$CONH_2$ group.

$$\underset{(Y)}{CH_3CH_2CONH_2} + Br_2 + 4NaOH \longrightarrow$$

$$\underset{(Z)\text{ basic}}{CH_3CH_2NH_2} \xrightarrow{HNO_2} \underset{\text{Ethanol}}{CH_3CH_2OH}$$

(Y) is formed from (X) having Cl on treatment with NH_3 and So, (X) is CH_3CH_2COCl *i.e.*, propanoyl chloride.

$$\underset{(X)}{CH_3CH_2COCl} + NH_3 \rightarrow \underset{(Y)}{CH_3CH_2CONH_2}$$

167. Presence of electron withdrawing group increases the acidic character of the –COOH due to –I effect, Further the –I group (–Cl) from –COOH group, weaker is the acid, thus III is weaker than I while presence of electron-donating group (alkyl groups) decreases the acidic character due to +I effect. Thus, IV < II < V < III < I

168. (*i*) $C_6H_5COOH \xrightarrow{PCl_5} \underset{(C)}{C_6H_5COCl}$

$$\xrightarrow{NH_3} \underset{(D)}{C_6H_5CONH_2} \xrightarrow{P_2O_5} C_6H_5CN$$

$$\xrightarrow{H_2/Ni} \underset{(E)}{C_6H_5CH_2NH_2}$$

(*ii*) $H_3CCH=CHCHO \xrightarrow{NaBH_4}$

$$\underset{(F)}{CH_3CH=CH-CH_2OH} \xrightarrow[ZnCl_2]{HCl}$$

$$\underset{(G)}{CH_3CH=CHCH_2Cl} \xrightarrow[H^+]{KCN}$$

$$\underset{(H)}{CH_3CH=CHCH_2COOH}$$

169.

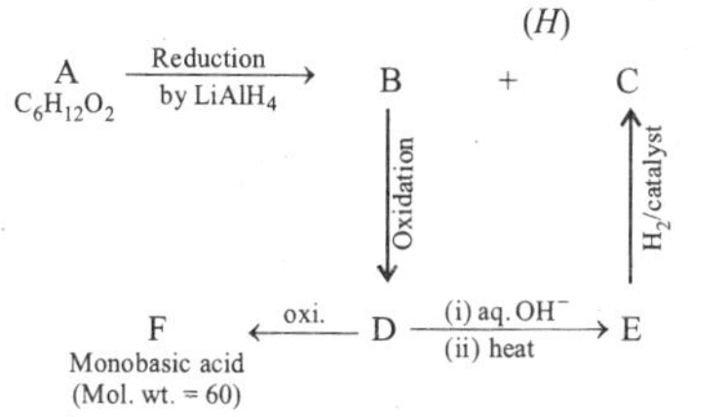

Molecular weight of the monobasic acid (RCOOH) indicates that the R– should be CH_3– *i.e.*, acid *F* should be acetic acid (CH_3COOH, mol. wt. 60). Thus, compound *D* must be acetaldehyde CH_3CHO, and compound *B* which on oxidation gives CH_3CHO must be ethanol, CH_3CH_2OH. Acetaldehyde (*D*) on treating with aqueous alkali will undergo aldol condensation.

$$\underset{(D)}{2CH_3CHO} \xrightarrow{OH^-} CH_3\overset{OH}{\overset{|}{C}}HCH_2CHO \xrightarrow{heat}$$

$$\underset{\text{Crotonaldehyde, }(E)}{CH_3CH=CHCHO}$$

$$\xrightarrow[\text{catalyst}]{H_2/} \underset{\text{Butanol-1, }(C)}{CH_3CH_2CH_2CH_2OH}$$

Nature of A. Thus it is evident that reduction of *A* with $LiAlH_4$ gives two alcohols; *B* (ethanol) and *C* (butanol). Hence, *A* must be an ester *i.e.*, ethyl butanoate ($CH_3CH_2CH_2COOC_2H_5$).

$$CH_3CH_2CH_2COOC_2H_5 \xrightarrow{LiAlH_4}$$

$$CH_3CH_2CH_2CH_2OH + C_2H_5OH$$

Ethyl butanoate, *A* Butanol, *C* Ethanol, *B*

170. The given set of reactions can be represented as below :

$$\text{Aq. NaOH} \xrightarrow[\text{High temp., pressure}]{\text{Gas B}} \text{Sod. salt of acid } A$$

$$\xrightarrow[\text{(ii) } H_2SO_4]{\text{(i) NaOH (heat)}} \text{Dibasic acid, } C$$

Calculation of molecular formula of C

$$\%\text{ of H} = \frac{2}{18}\times\frac{0.08}{0.40}\times 100 = 2.22\%$$

$$\%\text{ of C} = \frac{12}{44}\times\frac{0.39}{0.40}\times 100 = 27.30\%$$

$$\%\text{ of O} = 100-(2.22+27.30) = 71.48\%$$

By usual method, empirical formula of acid *C* = CHO_2

$$\text{Eq. wt. of acid } C = \left(\frac{1.0}{0.71}\times 108\right) - 107 = 45$$

Mol. wt. of acid $C = 45\times 2 = 90$

∴ Mol. formula of $C = C_2H_2O_4$

Since it is dicarboxylic acid, it must have two –COOH groups. Hence *C* is $\underset{\text{Oxalic acid}}{\overset{COOH}{\underset{COOH}{|}}}$

Going back, compound *C* must be produced from sodium oxalate which in turn is produced from sodium formate. Hence *A* is formic acid and *B* is CO_2. Thus the complete series of reactions can be written as below.

$$2NaOH(aq.) \xrightarrow{2CO_2(B)} \underset{(A)}{HCOONa + HCOONa}$$

$$\xrightarrow[\text{heat}]{NaOH} \overset{COONa}{\underset{COONa}{|}} \xrightarrow{H_2SO_4} \underset{(C)}{\overset{COOH}{\underset{COOH}{|}}}$$

171. (*i*) $\underset{\text{Ethanoic acid}}{CH_3COOH} \xrightarrow{PCl_5} CH_3COCl$

$$\xrightarrow{C_6H_5MgBr} C_6H_5COCH_3$$

$$\xrightarrow{C_6H_5MgBr} CH_3-\overset{C_6H_5}{\underset{C_6H_5}{\overset{|}{\underset{|}{C}}}}-OMgBr$$

$\xrightarrow{HOH} (C_6H_5)_2\underset{\displaystyle OH}{\underset{|}{C}}CH_3 \xrightarrow[-H_2O]{heat}$

$(C_6H_5)_2C=CH_2$

$\xrightarrow{CrO_3} C_6H_5COC_6H_5 + \underset{\text{Methanoic acid}}{HCOOH}$

(*ii*)

$\underset{\text{Ethanal}}{CH_3.CHO} \xrightarrow[\text{(cross aldol condens)}]{HCHO/NaOH} \underset{\displaystyle OH}{\underset{|}{CH_2}}CH_2CHO$

$\xrightarrow[\text{(heat)}]{\text{Dehydration}} \underset{\text{propenal}}{CH_2=CHCHO}$

$\xrightarrow{HCN} CH_2=CH-\underset{\displaystyle OH}{\underset{|}{CH}}-CN$

Propenal cyanohydrn

$\xrightarrow[H^+]{\text{hydrolysis}} CH_2=CH-\underset{\displaystyle OH}{\underset{|}{CH}}-COOH$

2–Hydroxy–3–butenoic acid

(*iii*)

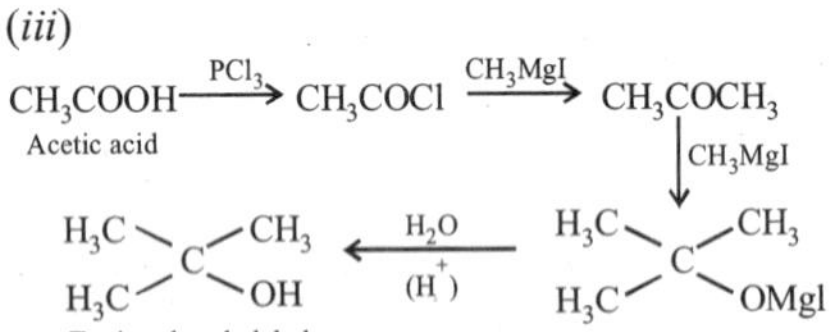

Tertiary-butyl alchol

172. $P/Cl_2, NH_2CH_2COONH_4$

173.

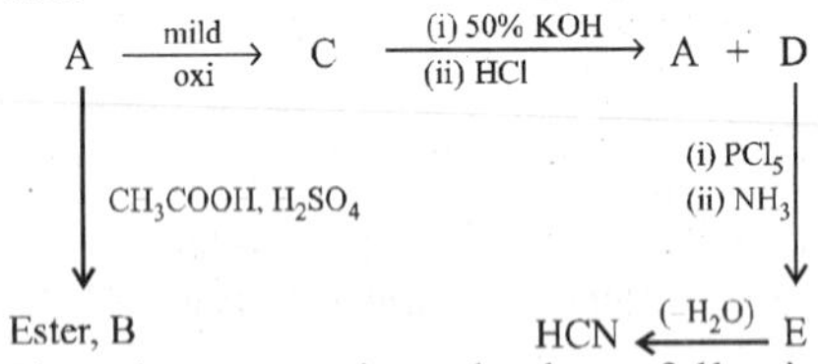

The above reactions lead to following conclusions.

(i) Reaction of (A) with CH_3COOH in presence of H_2SO_4 to form ester (B) indicates that (A) is an alcohol.

(ii) Reaction of (C) with 50% KOH followed by acidification to give alcohol (A) and the compound (D) seems to be the Cannizzaro reaction. Hence, (C) must be an aldehyde and (D) must be an acid. The nature of (C) as aldehyde is again in consistent with the fact that it is obtained by the mild oxidation of (A) which has been established as an alcohol.

(iii) Structure of acid (D) is established by its given facts.

$\underset{(D)}{-COOH} \xrightarrow{PCl_5} -COCl \xrightarrow{NH_3} \underset{(E)}{-CONH_2}$

$\xrightarrow{-H_2O} HCN$

Formation of HCN by the dehydration of (E) establishes that (E) is $HCONH_2$ and hence (D) is HCOOH.

(iv) Thus the alcohol (A) produced along with HCOOH during Cannizzaro reaciton of (C) must be CH_3OH and hence, (C) must be HCHO.

Thus, the various compounds are as below :

(A)	(B)	(C)
CH_3OH	CH_3COOCH_3	HCHO
Methyl alcohol	Methyl acetate	Formaldehyde

(D)	(E)
HCOOH	$HCONH_2$
Formic acid	Formamide

174. (*i*) $CH_3\underset{\displaystyle O}{\underset{||}{C}}-OC_2H_5$

(*ii*) 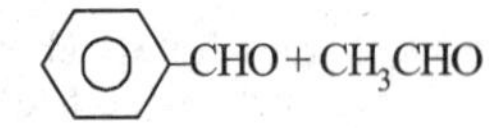

175. (X) is hydrolysed to give a carboxylic acid (Y) and an alcohol (Z) and thus (X) is an ester;

$R-\underset{\displaystyle O}{\underset{||}{C}}-OR'$.

$R-\underset{\displaystyle O}{\underset{||}{C}}-OR' \xrightarrow[H^+]{HOH} \underset{(Y)}{RCOOH} + \underset{(Z)}{R'OH}$

Oxidation of alcohol (Z) gives acid (Y).

$\underset{(Z)}{R'OH} \xrightarrow{\text{Oxidation}} RCOOH$

or $RCH_2OH \rightarrow RCOOH$ $(\therefore R' \text{ is } R—CH_2)$

Hence (X), (Y) and (Z) are

$\underset{\text{(Propyl propanoate) (X)}}{CH_3CH_2\underset{\displaystyle O}{\underset{||}{C}}-OCH_2CH_2CH_3}$, $\underset{\text{(Propanoic acid) (Y)}}{CH_3CH_2COOH}$,

$\underset{\text{(Propan–1–ol) (Z)}}{CH_3CH_2CH_2OH}$

176. The weaker a base better is its leavability. This is an example of nucleophilic substitution where the group X (Cl, NH_2, OC_2H_5, $OCOCH_3$) is replaced by OH. The decreasing basic character of the four concerned groups is:

$NH_2^- > OR^- > OCOR^- > Cl^-$

Hence, Cl^- (the weakest base) will be lost most easily, while NH_2^- (the strongest base) will be lost with most difficulty. Thus, the order of hydrolysis becomes:

$CH_3CONH_2 < CH_3COOC_2H_5 < (CH_3CO)_2O < CH_3COCl$.

177. $CH_3CH_2CHO + [O] \xrightarrow[\text{(Tollen's reagent)}]{AgNO_3/NH_4OH} CH_3CH_2COOH + 2Ag$

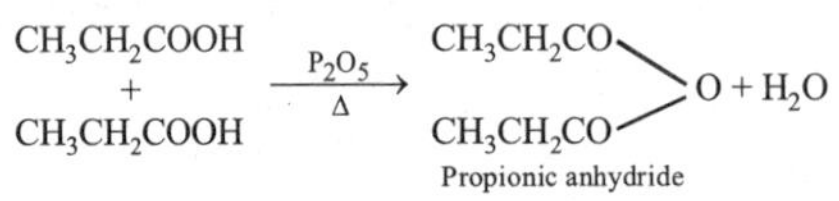

178. (*i*) $C_2H_2 \xrightarrow[Hg^{2+}]{H_2SO_4\text{(dil.)}} CH_3CHO$

$\xrightarrow[\text{reduction}]{H_2/\text{cat.}} CH_3CH_2OH$

(*ii*) $CH_3I + KCN(alc) \xrightarrow{-KI} CH_3CN$

$\xrightarrow{H^+/H_2O} \underset{\text{Acetic acid}}{CH_3COOH}$

179. Ozonolysis of (*A*) to acetone and an aldehyde indicates the following part structure of alkene (*A*) :

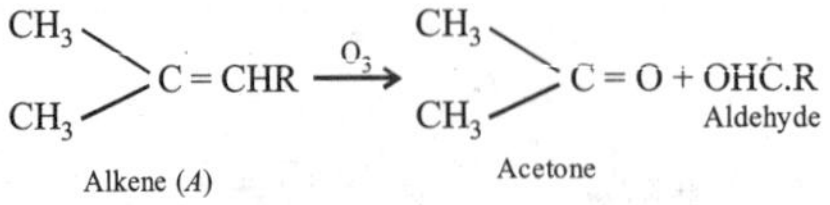

As per problem :

$RCHO \xrightarrow{[O]} RCOOH[B] \xrightarrow{P/Br_2}$

Bromo compound $[C] \xrightarrow{H_2O}$ Hydroxy acid $[D]$

Structure of (*D*) is determined by the reaction :

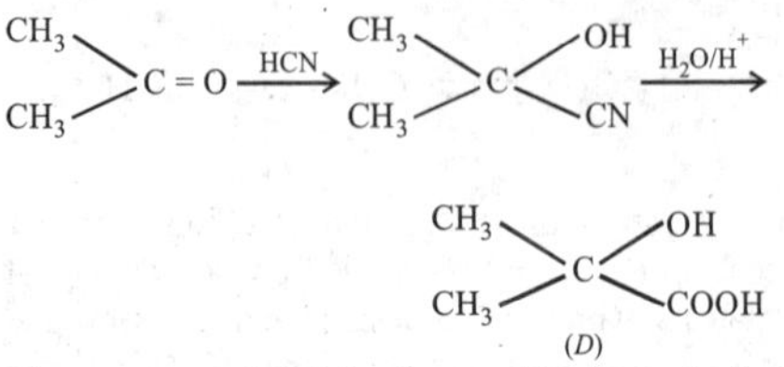

The compound (*D*) is obtained by hydrolysis of (*C*) with aqueous alkali. Since (*C*) is a bromo compound, therefore it has a bromo group whereas the compound (*D*) has a hydroxyl group. Therefore, structure of *C* is

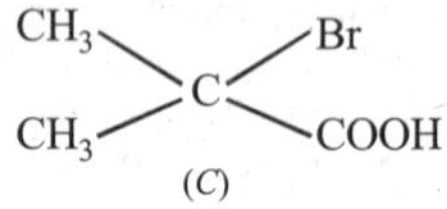

The compound (*C*) is formed by bromination of compound (*B*), therefore, the compound (*B*) is

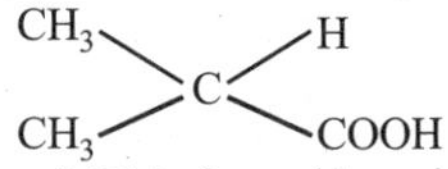

The compound (*B*) is formed by oxidation of an aldehyde therefore, the structure of the aldehyde is

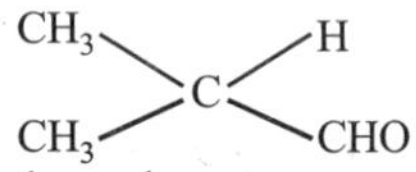

The aldehyde and acetone are formed by ozonolysis of alkene. Therefore, the double bond in alkene should be between the carbon atoms of the two carbonyl compounds (the aldehyde and acetone). Therefore, the compounds and the reactions are identified as

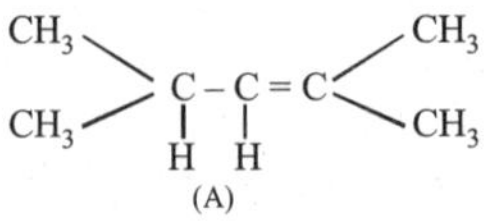

180. $CH_3COCl + AlCl_3 \rightarrow CH_3C^+=O + AlCl_4^-$

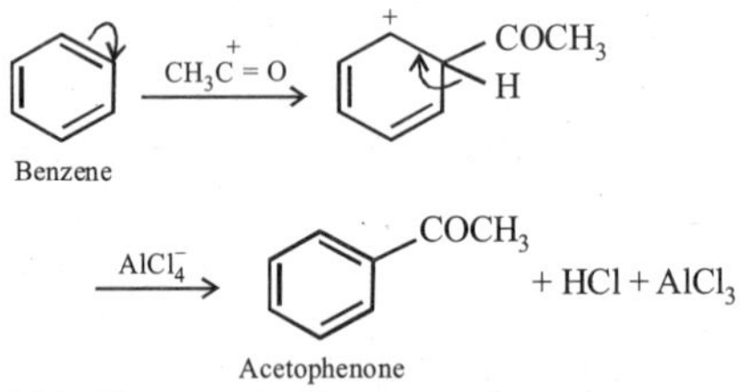

181. First step is Claisen condensation.

$CH_3-\underset{\displaystyle O}{\underset{||}{C}}-O-C_2H_5 + H-CH_2-\underset{\displaystyle O}{\underset{||}{C}}-O-C_2H_5$

$\xrightarrow[C_2H_5OH]{C_2H_5ONa}$

$CH_3-\underset{\displaystyle O}{\underset{||}{C}}-CH_2-\underset{\displaystyle O}{\underset{||}{C}}-O-C_2H_5 + C_2H_5OH$ (β-ketoester)

$\downarrow$ HOH, H^+

$CH_3-\underset{\displaystyle O}{\underset{||}{C}}-CH_2-\underset{\displaystyle O}{\underset{||}{C}}-OH + C_2H_5OH$

(β-keto acid)

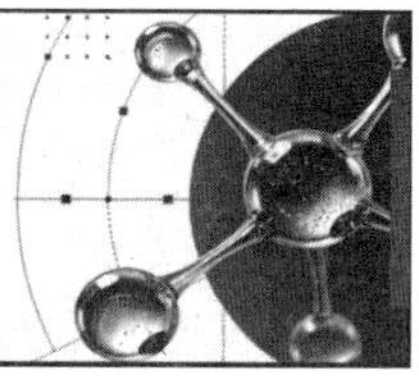

Chapter-13

Amines

Part-A (JEE Main)

1. **(3)** Hoffmann bromamide degradation reaction:

$$\underset{1^\circ \text{ Amide}}{R-\overset{O}{\overset{||}{C}}-NH_2} + 4NaOH + Br_2 \longrightarrow \underset{1^\circ \text{ Amine}}{R-NH_2} + Na_2CO_3 + 2NaBr + 2H_2O$$

From above reaction, we can conclude it requires 4 moles

of NaOH and 1 mole of Br_2 and it is specifically used for

primary amides which can be either alky or aryl amides.

Thus, Statement (A) – False

Statement (B) – False

Statement (C) – Truc

Statement (D) – False

Statement (E) – True

2. **(3)** (ii), (iii), (vi)

(i) $Ph-\overset{O}{\overset{||}{C}}-NH_2 \xrightarrow[\Delta]{Br_2+KOH} Ph-NH_2$

Aniline cannot be prepared by Gabriel phthalimide synthesis.

(ii) $Ph-CH_2-\overset{O}{\overset{||}{C}}-NH_2 \xrightarrow[\Delta]{Br_2+KOH} Ph-CH_2-NH_2$

Also prepared by Gabriel phthalimide synthesis.

(iii) $CH_3-\overset{O}{\overset{||}{C}}-NH_2 \xrightarrow[\Delta]{Br_2+KOH} CH_3-NH_2$

Also prepared by Gabriel phthalimide synthesis.

(iv) 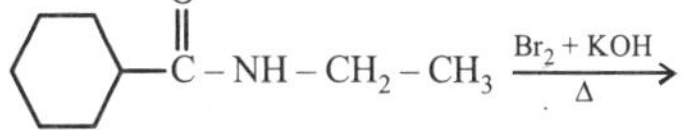

No 1° amine is formed

(v) 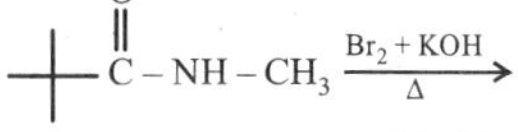

No 1° amine is formed

(vi) 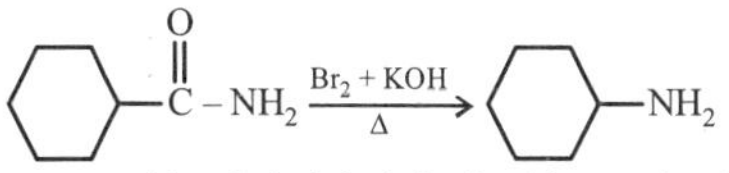

Also prepared by Gabriel phthalimide synthesis.

3. **(3)** Boiling point ∝ Intermolecular attraction ∝ Number of H-bonding.

Compounds with O–H or N–H bonds can form hydrogen bonds which are significantly stronger than alkanes like n–C_4H_{10} which possess weak London dispersion forces, giving them the lowest boiling point.

4. **(3)** **Statement-I**

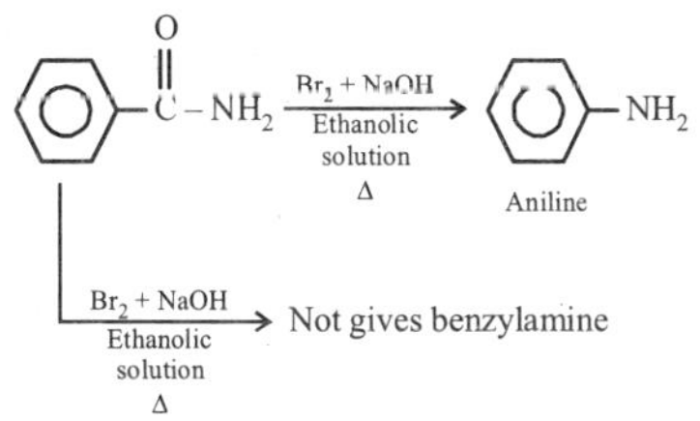

Hence Statement-I is false.

Statement-II

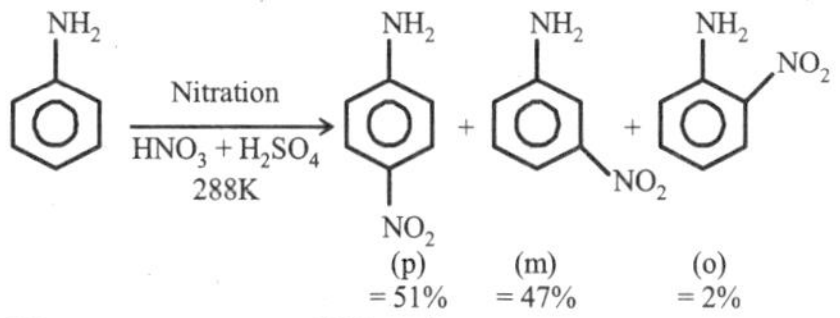

Hence statement-II is true.

5. **(1)**

$$CH_3-CH_2-C\equiv N \xrightarrow[\text{(hydrolysis of cyanide)}]{OH^-/H_2O} CH_3-CH_2COO^- \xrightarrow{H_3O^+} CH_3-CH_2-COOH$$

$$CH_3-CH_2-COOH \xrightarrow[\text{(HVZ Reaction)}]{\text{(i) Red P + } Cl_2 \text{ (ii) } H_2O} \underset{\text{(2-Chloropropanoic acid)}}{CH_3-\underset{Cl}{\underset{|}{C}H}-COOH}$$

6. **(3)**

(A) 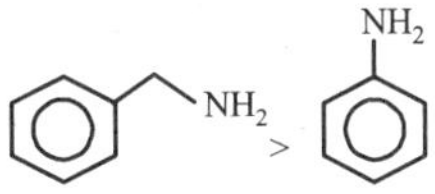

(Basic strength order)

(B) 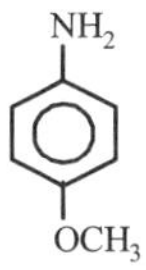

Cannot be synthesized by Gabriel phthalamide synthesis

(C) 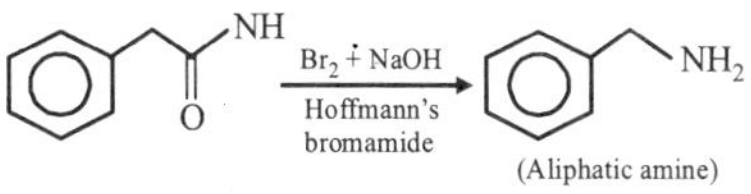

(D) 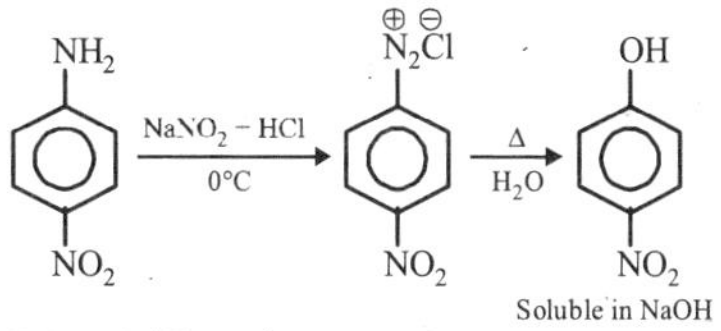

Only (B) and (C) are incorrect.

7. **(3)** Weaker the base, stronger is its conjugate acid and vice-versa.

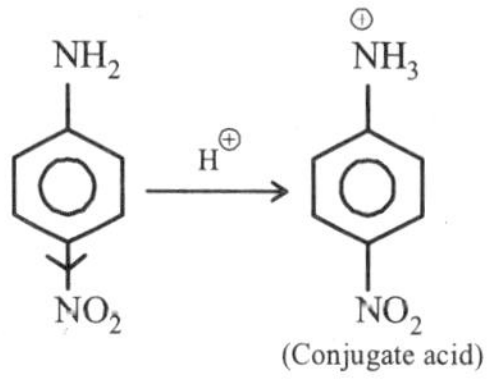

–NO_2 being strong electron withdrawing group decreases the basicity of compound.

8. **(1)** Rate of ESR ∝ nucleophilicity of benzene ring.

In molecule,

P → No substituents ortho to –NMe_2 group

Q → One methyl group at ortho position creating SIR

R → Two methyl groups at ortho position. Resonance is nearly destroyed because when bulky groups are placed ortho to each other, forcing them out of planarity.

9. **(1)** Only sulphonamides formed by the reaction of 2° amines with Hinsberg reagent are solid (insoluble) sulphonamides. Among the given amines only following four are 2° amines.

$Ph-NH-CH_3$, $Me-NH-Me$,

$Ph-NH-CH_2-CH_2-CH_3$, $Ph-NH-Ph$

10. **(3)**

Element	Percentage	Atomic Weight	Atomic ratio	Simple ratio
C	61.01	12.01	5.08	3
H	15.25	1.008	15.14	9
N	23.74	14.01	1.69	1

So, the empirical formula is approximately C_3H_9N

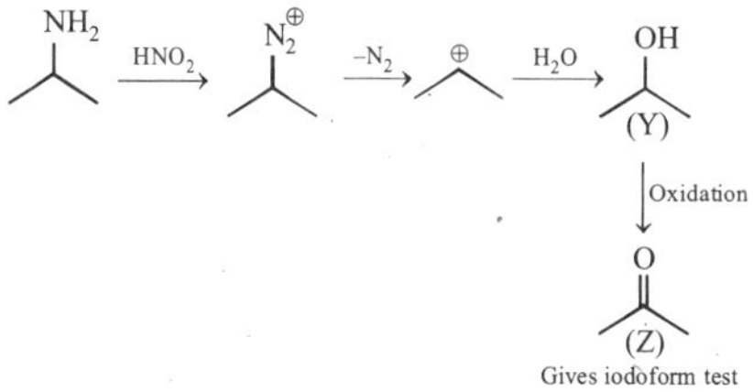

Hence the compound is (C), $(CH_3)_2CHNH_2$.

11. **(4)**

(1) 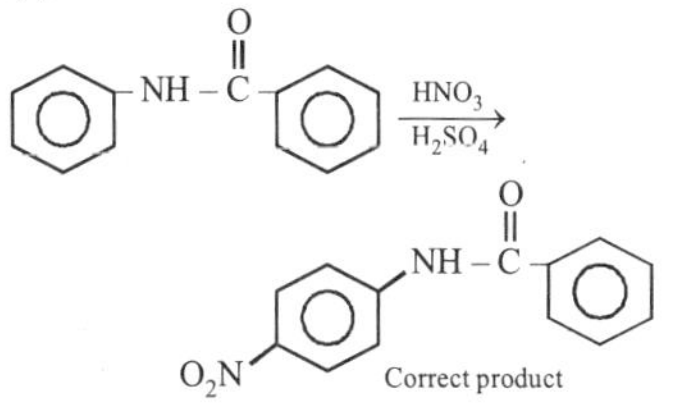

(2) 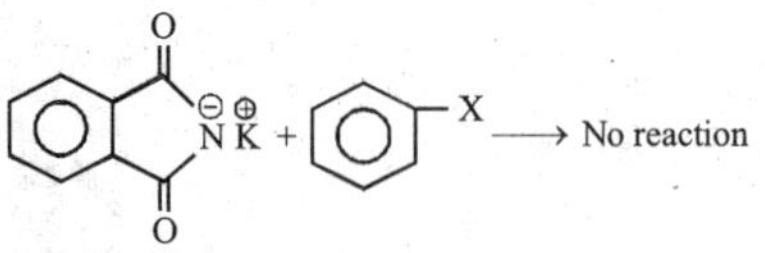

Aromatic halide does not give Gabriel phthalimide reaction

(3) 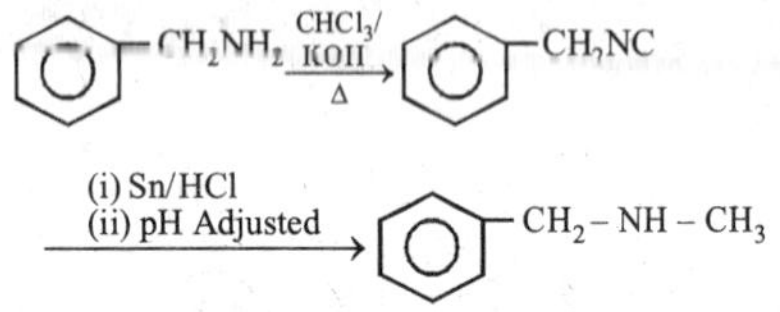

(4) Hoffmann bromamide degradation

$$CH_3-CH_2-\overset{O}{\overset{||}{C}}-NH_2 \xrightarrow[4KOH]{Br_2} CH_2-CH_2-NH_2 + 2KBr + K_2CO_3 + 2H_2O$$

12. (1) 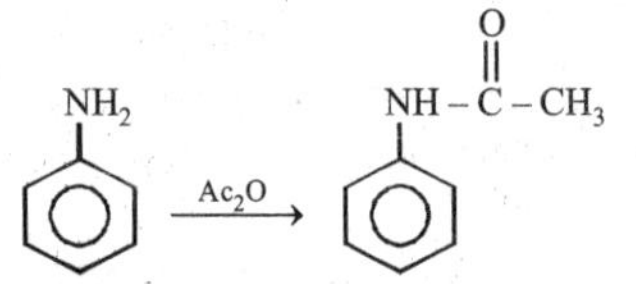

9.3 gm 11 gm

MW = 93 MW = 135

$n = \frac{9.3}{93} = 0.1$ $n = \frac{11}{135} = 0.08148$

$\%\ yield = \frac{0.08148}{0.1} \times 100 = 81.5\%$

13. (4) 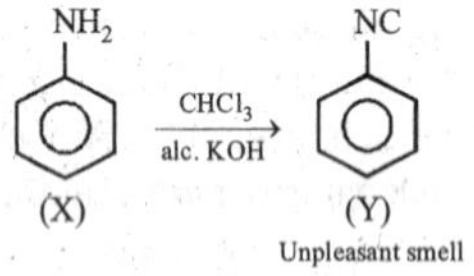

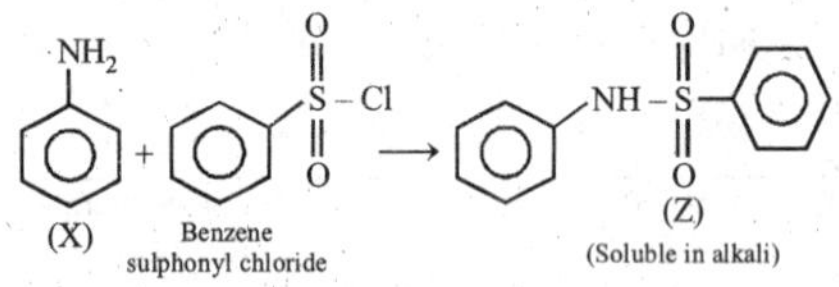

Number of different H atom in 'Z' is 7.

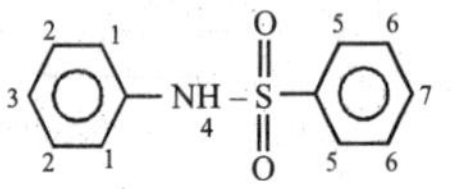

14. (3) 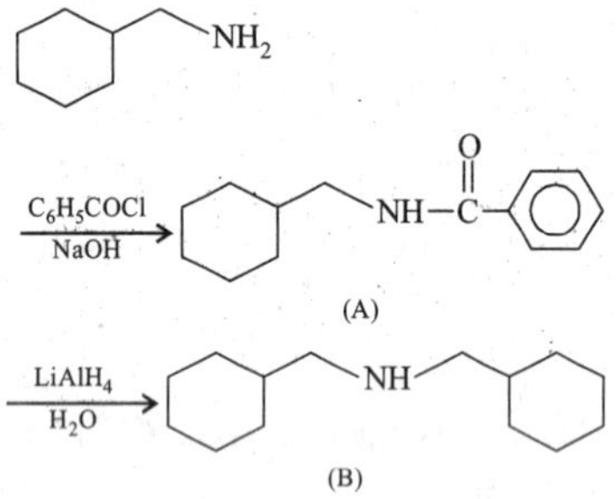

15. (1)

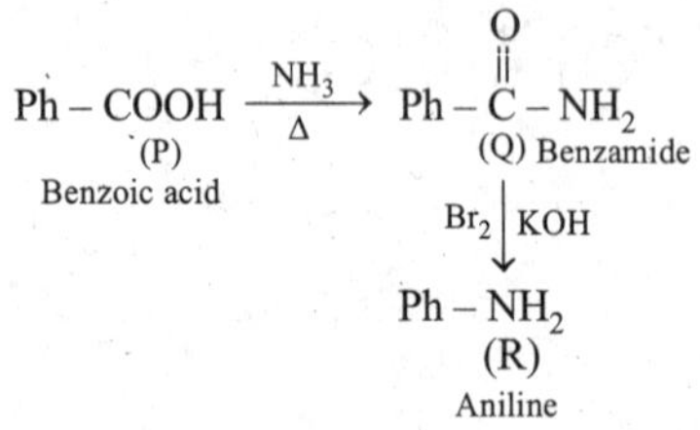

16. (1)

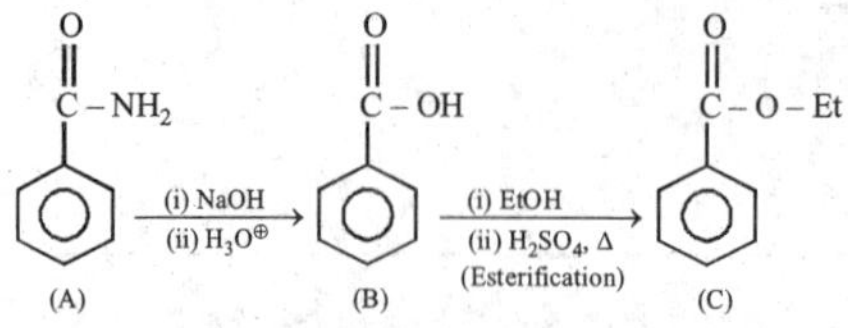

Molar mass = 121

Benzamide (A) Shows positive Lassaigne test for nitrogen and on hydrolysis benzamide gives benzoic acid (B) which gives effervescence with aq. $NaHCO_3$.

Benzoic acid with alcohol in acidic medium gives ester (C) which gives fruity smell.

17. (2) E > D > B > A > C

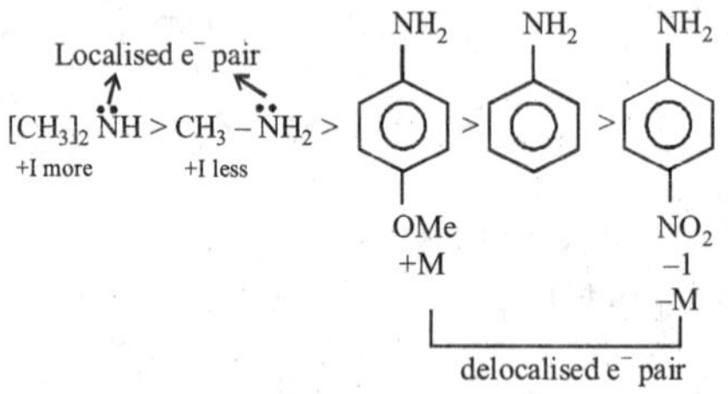

18. (3) Only primary amines gives positive carbylamines test.

Option (A) and (C) are primary amine.

19. (4)

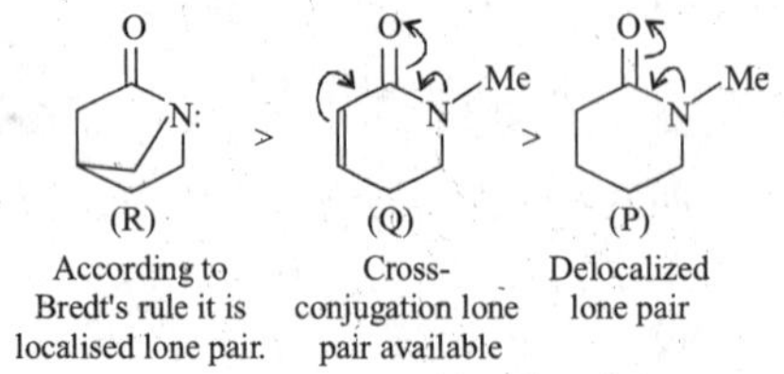

20. (3)

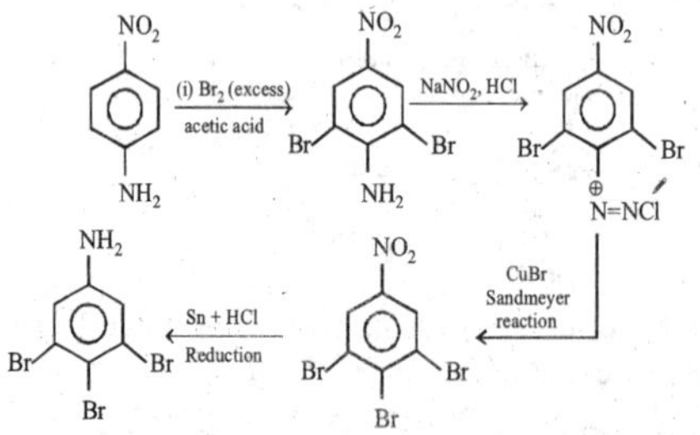

21. (4) Basic strength of amines depends on the cumulative effects of hydrogen bonding and inductive effect.

For ethylamines, order is 2° > 3° > 1°. NH_3 is stronger base than NH_2-NH_2 due to –I effect of $-NH_2$ groups. Thus, order will be,

$NH(Et)_2 > N(Et)_3 > NH_2Et >$

$\ddot{N}H_3 > NH_2-NH_2$

22. (1)

(A) $R-NH_2 \xrightarrow[HCl]{NaNO_2} R-N_2^{\oplus}Cl^{\ominus}$

(B) 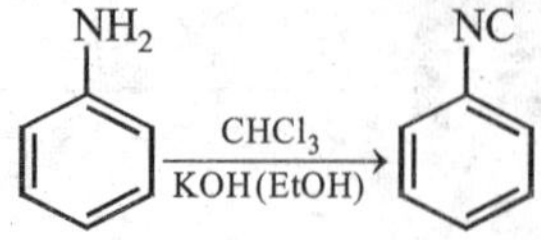

$R-NH_2 \xrightarrow[KOH(EtOH)]{CHCl_3} R-NC$

(C) Only primary amine gives carbylamine test

(D) $Ph-SO_2Cl \longrightarrow$

Hinsberg reagent, Benzene sulphonyl chloride

(E) Tertiary amine do not react with $Ph-SO_2Cl$

So correct options are (B) and (D) only

23. (2)

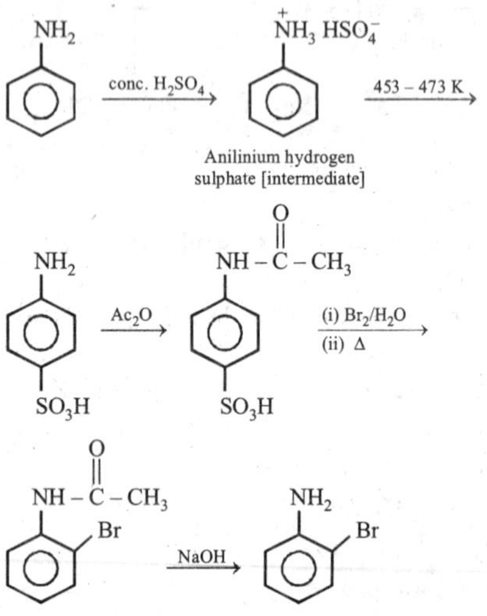

24. (4) B and D are 3° amine which does not have replaceable hydrogen on nitrogen. So they does not react. Primary and secondary amines can reacts with Hinsberg reagent.

25. (4)

(A) $R-\overset{O}{\overset{||}{C}}-NH_2 \xrightarrow[H_2O]{LiAlH_4} R-CH_2-NH_2$

(B) 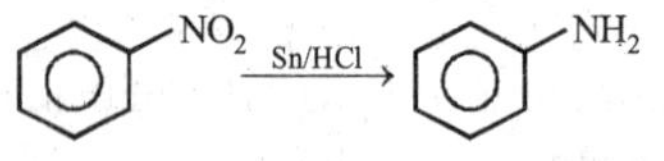

(C) $R-C\equiv N \xrightarrow{H_2/Ni} R-CH_2NH_2$

(D) 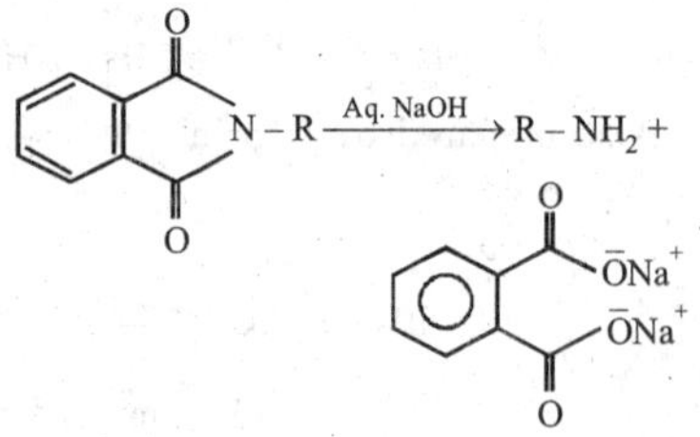

26. (1) Stronger base has its lone pair more easily available. Thus, order of basic strength is: N(sp^3, localized lone pair) in piperidine > N(sp^2, localized lone pair) in pyridine > N(sp^2, delocalized lone pair, aromatic) in pyrrole.

∴ Piperidine > Pyridine > Pyrrole

27. (1) Solubility depends on H-atom since product of 2° amine has no H-atom thus not soluble.

1° and 2° amines give Hinsberg test but not 3°

1° and 2° amines give Hinsberg test but not 3° due to absence of H-atom on nitrogen.

28. (2) Aminobenzene and aniline are same organic compounds.

29. (1) $-NH_2$ is an ortho and para directing group, it activates benzene ring towards electrophilic substitution reaction.

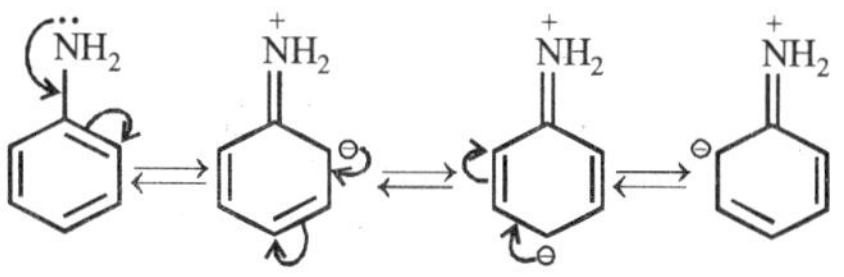

Aniline does not undergo Friedel-Crafts reaction due to salt formation with $AlCl_3$ which deactivate ring towards electrophilic substitution reaction.

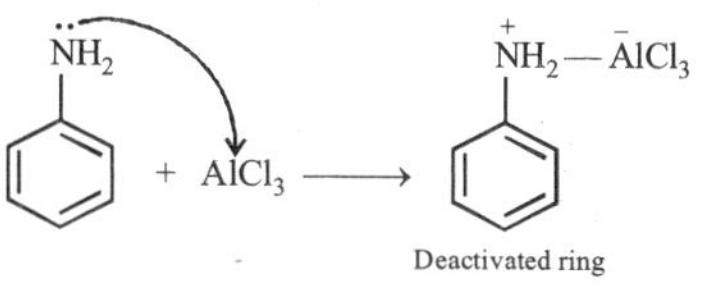

30. (2)

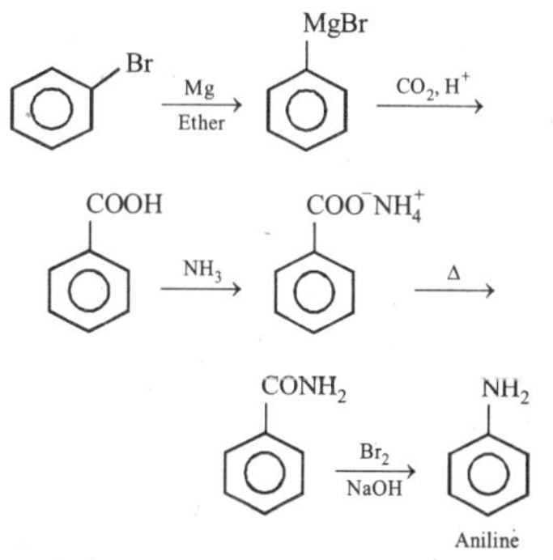

31. (4)

(1)

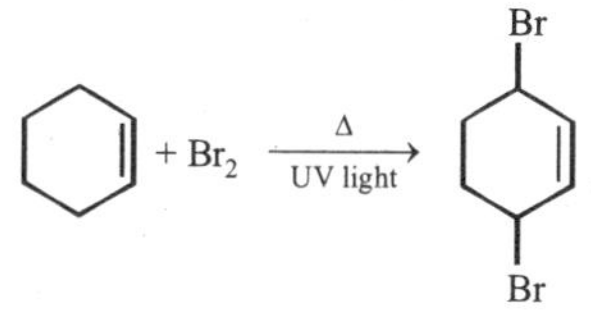

This reaction follows free radical substitution pathway.

(2) This represents Hoffmann – Bromamide reaction which gives a primary amine with one carbon less than in the amide and therefore the given reaction has the same number of carbon atoms in the amine is incorrect.

(3) $CH_3CH_2CH_2NH_2 \xrightarrow[H_2O]{HNO_2,\ 0°C} CH_3CH_2OH + N_2 + HCl$

This is the diazotization of aliphatic amines that gives an unstable diazonium salt that decomposes to give alcohol and N_2 gas but here one C is less in alcohol which is incorrect.

(4)

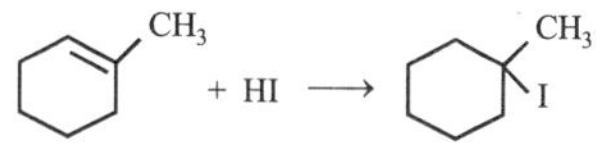

This is hydroiodination reaction that follows Markovnikoff's rule and therefore it is correct.

$C_2H_5CONH_2 + Br_2 + NaOH \longrightarrow C_2H_5CH_2NH_2 + Na_2CO_3 + NaBr + H_2O$

32. (3) Due to strong electron donating effect of $-NH_2$ group, the electron density increases at the *o*-, *p*-positions. Thus, when aniline is treated with Br_2, then Br^+ attacks the benzene ring at *o*-, *p*-positions to form carbocation intermediates (1, 2 and 4, but not 3) which are stabilized by usual resonance of the benzene ring as well as by the $-NH_2$ groups as shown below: Hence *o, p*-products are formed.

o-Bromination:

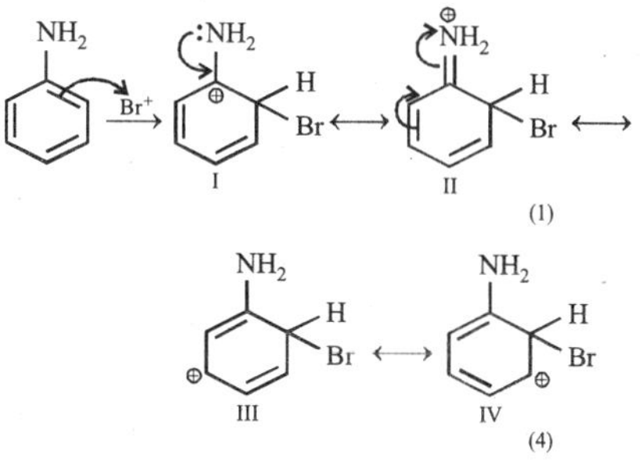

p-Bromination:

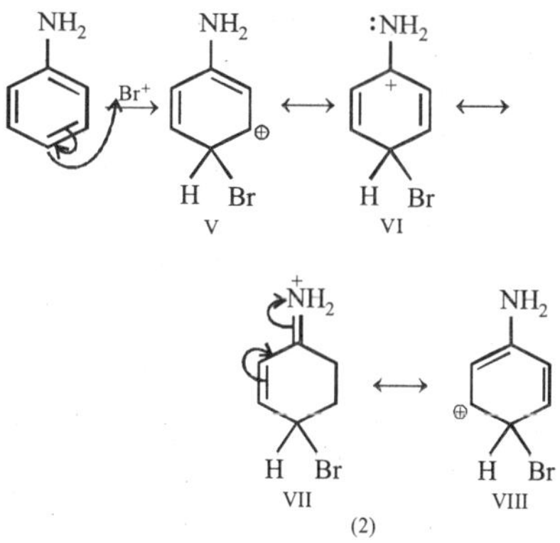

33. (2)

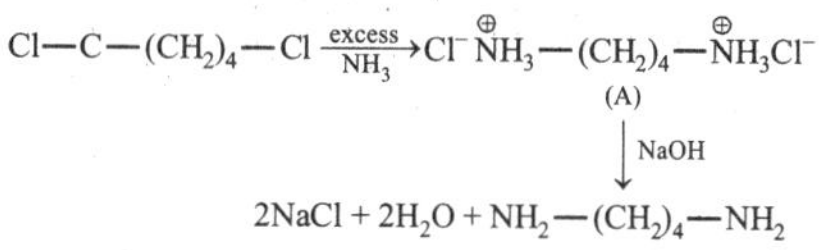

34. (4)

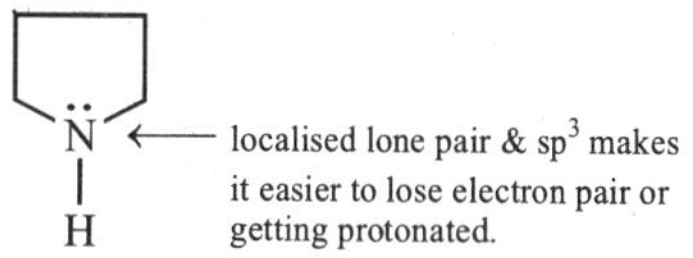

35. (1)

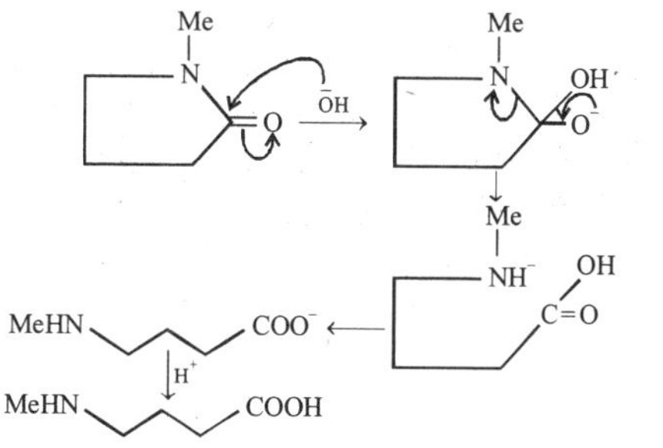

36. (1)

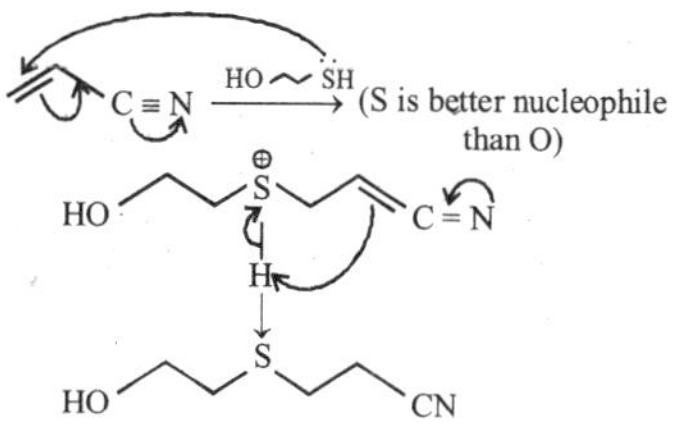

37. (3)

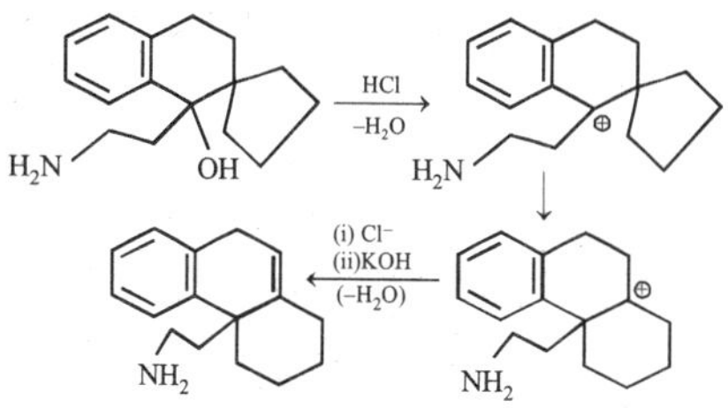

38. (1)

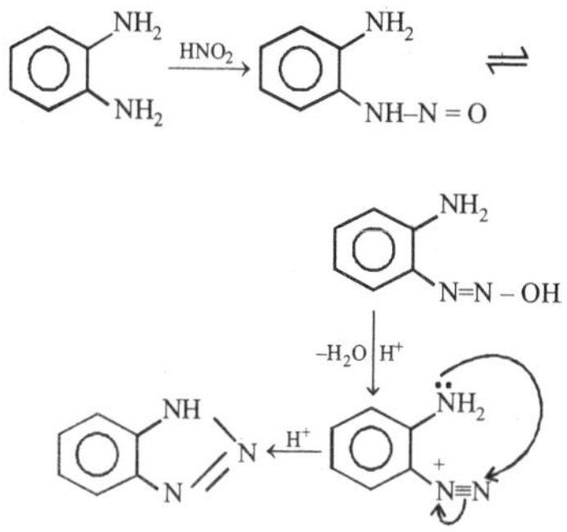

39. (1) (P) Gabriel phthalimide synthesis is used for the preparation of aliphatic primary amines. Aromatic primary amines cannot be prepared by this method.

(Q) 2°–Amines react with Hinsberg's reagent to give solid which is insoluble in NaOH

(R) Aromatic primary amines react with nitrous acid at low temperature (273 – 298 K) to form diazonium salts, which form red dye with β–naphthol

40. (2)

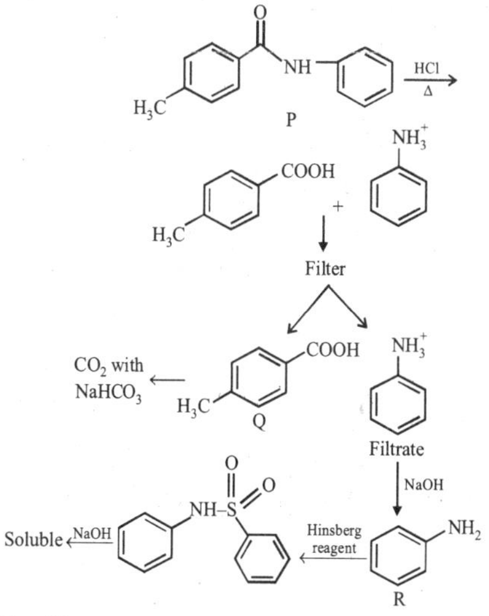

41. (4)

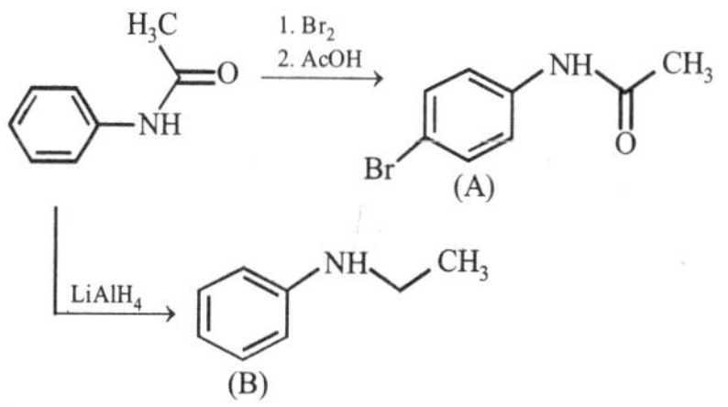

42. (3)

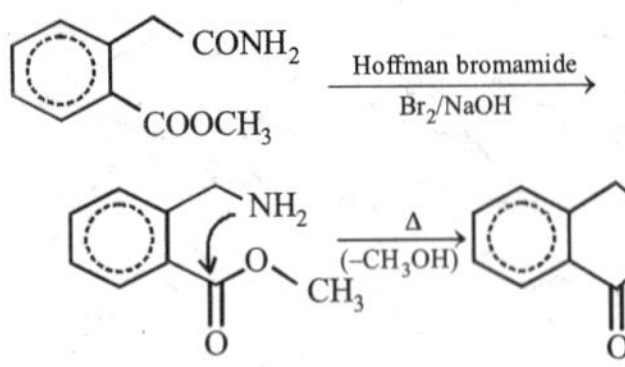

43. (3)

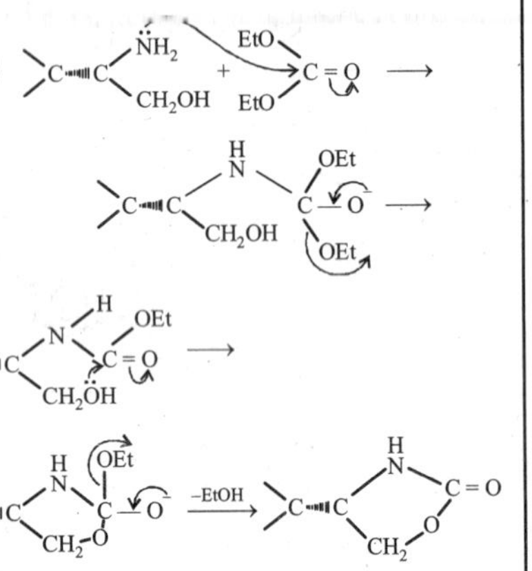

44. (1)

$$CH_3\underset{|}{CH}COOH \xrightarrow{dil.NH_3} CH_3\underset{|}{CH}COOH$$
Cl $\qquad\qquad$ NH_2

α-Amino carboxylic acid is stabilised by zwitter ion formation. While in case of amines obtained by the reaction of aq.NH_3 and alkyl halide, no such stabilisation is possible.

$$CH_3CHCOO^- (\overset{+}{NH_3}) \rightleftharpoons CH_3CHCOOH (NH_2)$$

45. (2)

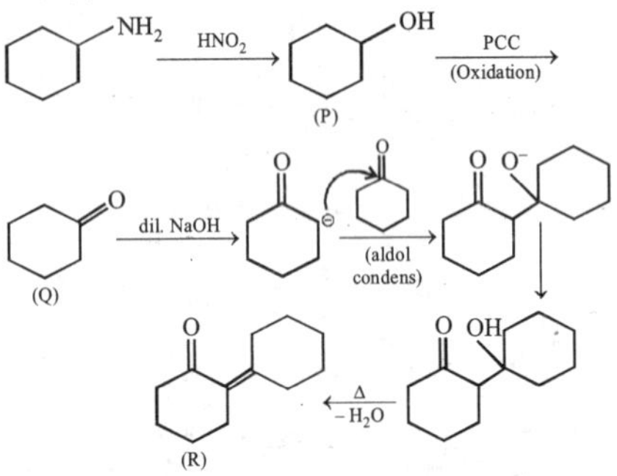

46. (4)

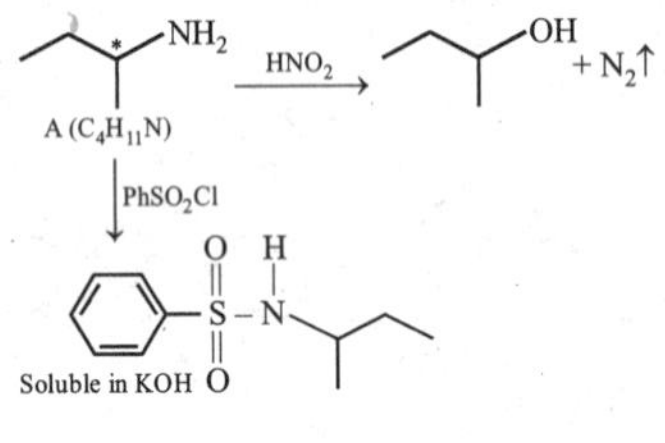

47. (3)

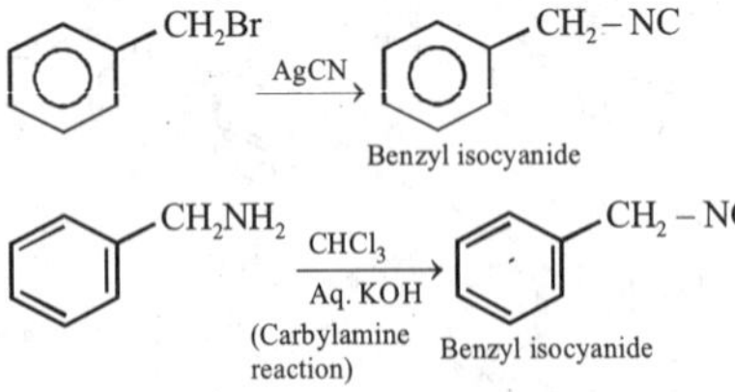

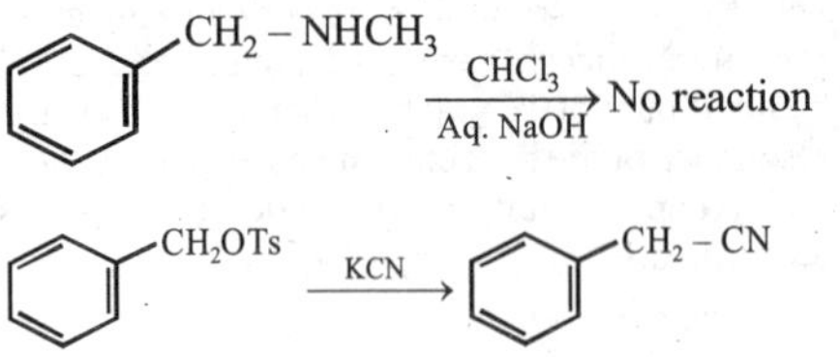

48. (4)

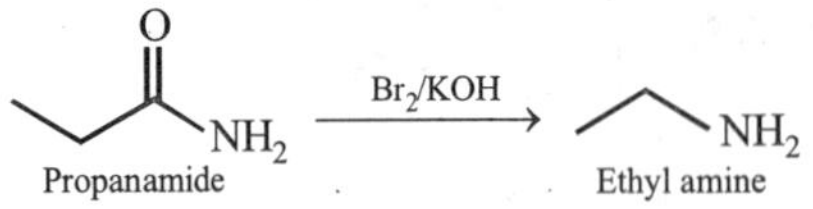

[Hoffmann-Bromamide reaction]

49. (1)

(i) In aqueous medium basic strength is dependent on donating ability of electron density on 'N'

(ii) As well as solvation of cation formed after accepting H^+. After considering all these factors overall basic strength order is

$Me_2NH > MeNH_2 > Me_3N > NH_3$

50. (4)

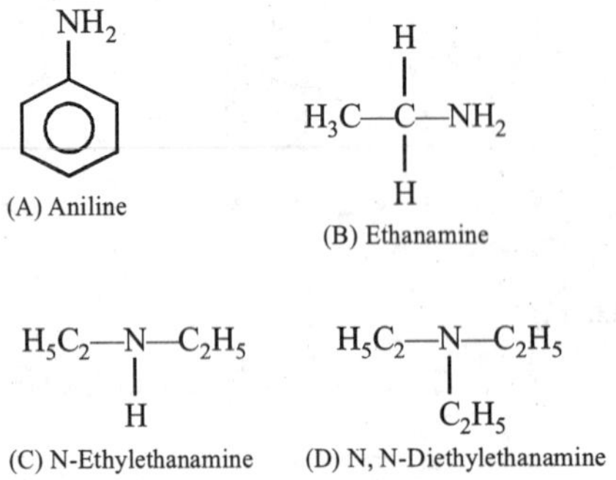

Basic Strength $\alpha \frac{1}{pK_b}$; The order of basicity: A < B < D < C

Due to steric hindrance, lone pair of electrons on N is relatively lower available in (D) than in (C)

51. (3) Pure aniline and arylamines are colourless but get coloured on storage due to atmospheric oxidation as they react readily in air and gets oxidised.

52. (4)

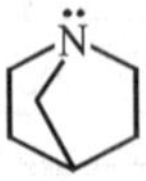

It is most basic due to no amine inversion at N atom.

53. (2)

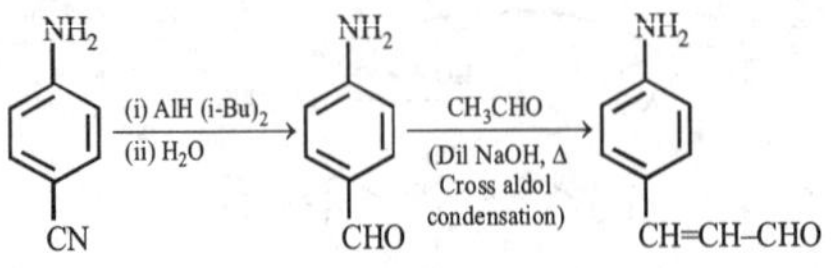

54. (3) Start from the product.

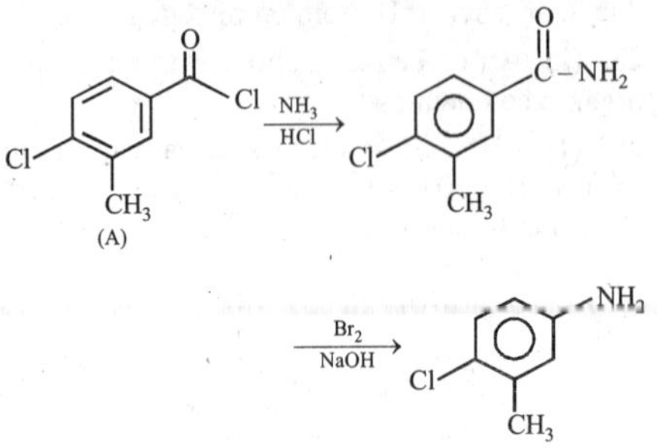

55. (1)

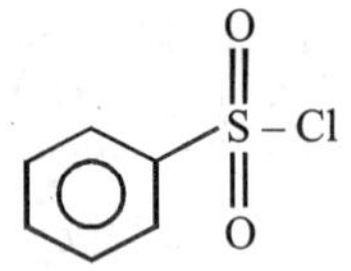

Benene sulphonyl chloride is known's Hinsberg reagent.

56. (3) In Saytzeff's product, proton is removed from the carbon atom which is less substituted.

$$CH_3CH_2CH_2\underset{\overset{|}{\overset{+}{N}(CH_3)_3}}{CH}-CH_3 \xrightarrow[EtOH]{EtO^-} \underset{\text{(major)}\quad (A)}{CH_3CH_2CH_2CH=CH_2}$$

Hofmann product

+

$$\underset{\text{(minor)}\quad (B)}{CH_3CH_2CH=CH-CH_3}$$

Saytzeff product

57. (1) In acidic medium; anilinium ion is formed due to which meta product is also formed in significant amount.

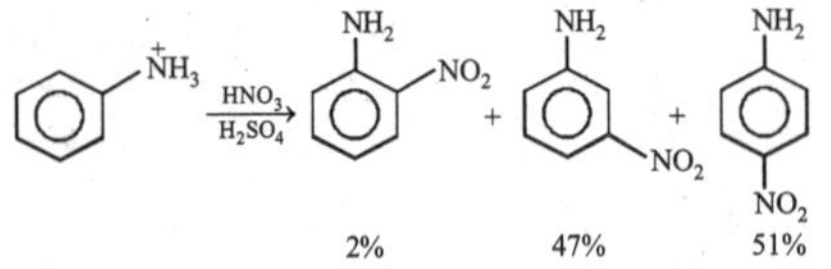

58. (3) A – III, B – IV, C – I, D – II

59. (3) Aniline acts as Lewis base and reacts with Lewis acid $AlCl_3$ to form anilinium hydrochloride. This results in positive charge on the nigrogen atom. Hence, it will not react further. So no *o–m–*or *p*-methyl aniline formed as a product.

60. (1)

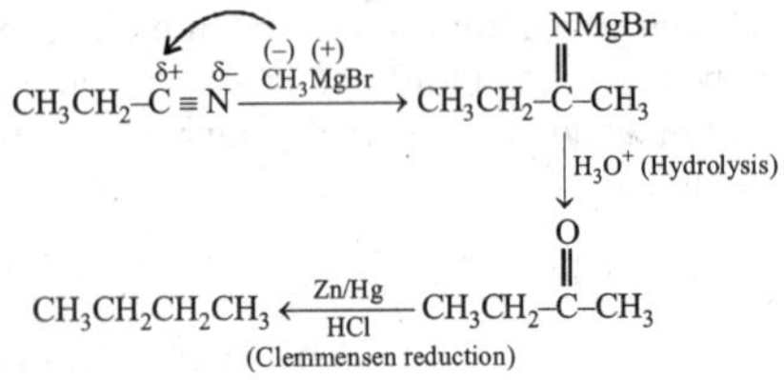

61. (3)

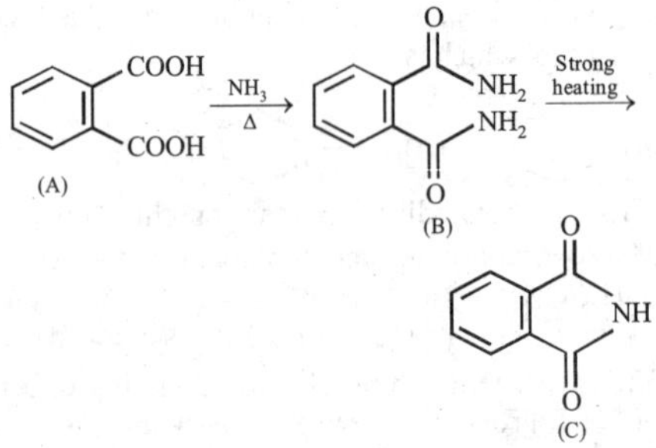

Gabriel phthalimide synthesis :

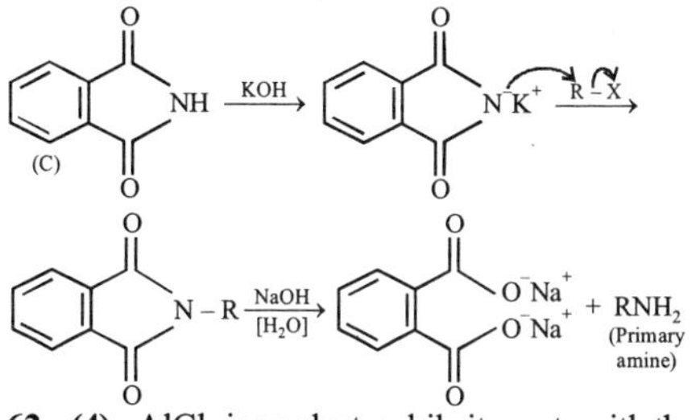

62. (4) $AlCl_3$ is an electrophile it reacts with the aniline to form an adduct

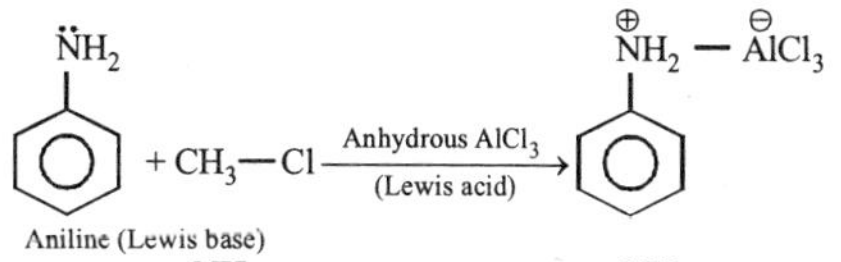

63. (3)

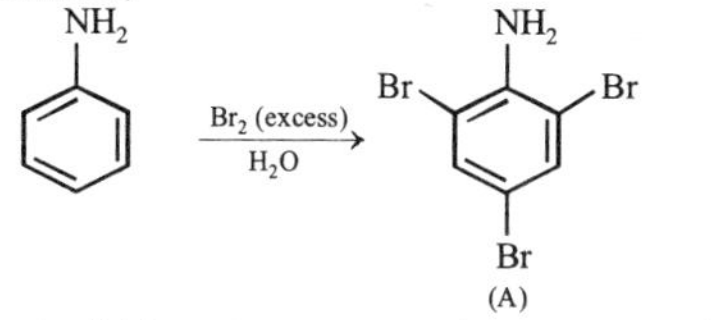

Acetic anhydride acts as protecting agent and decreases the electron donating tendency of $-NH_2$ group.

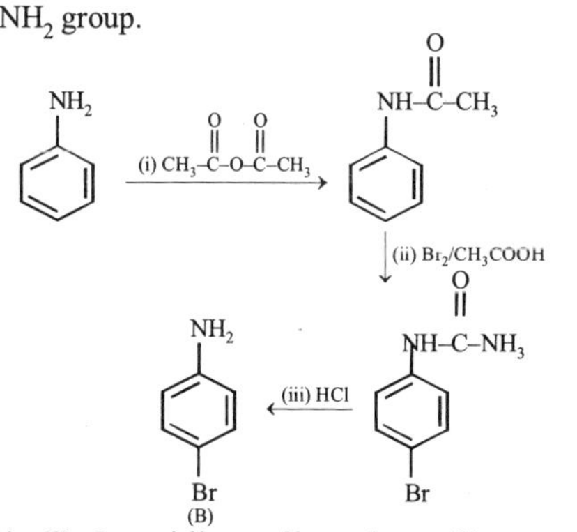

64. (3) In acidic medium the aniline gets protonated to form anilinium ion and hence substantial amount of *m*-product is also formed.

$$\underset{\text{Base}}{HNO_3} + \underset{\text{Acid}}{H_2SO_4} \longrightarrow NO_2^+$$

Sulphuric acid is stronger acid than nitric acid.

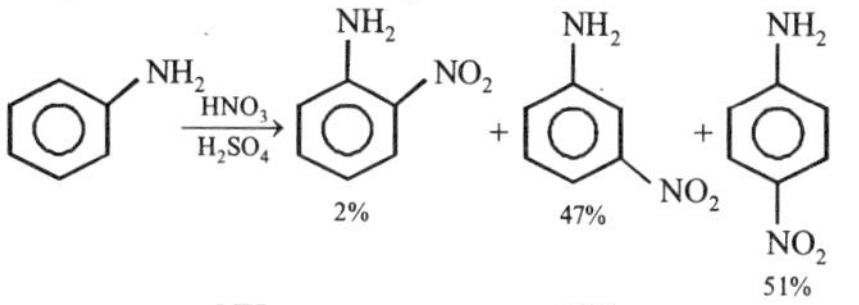

65. (3)

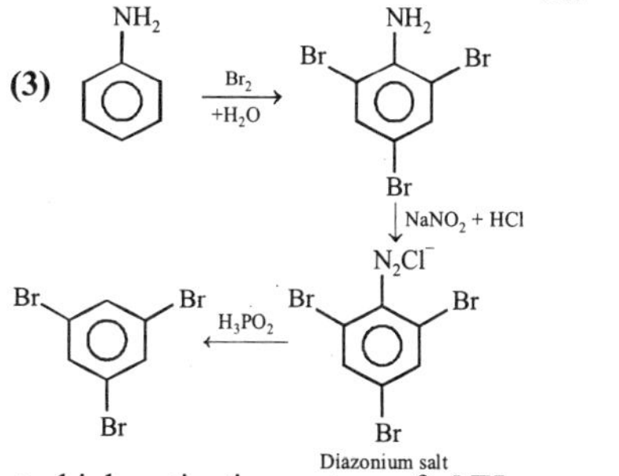

Due to high activating nature of $-NH_2$ group, aniline on reaction with bromine water results in 2, 4, 6-tribromoaniline.

66. (4) In Hofmann degradation reaction, the migration of aryl group is also possible, thus, statement-I is incorrect.

67. (3) *p*-Toluene sulphonyl chloride is derivative of benzene sulphonyl chloride which is commonly known as Hinsberg's reagent.

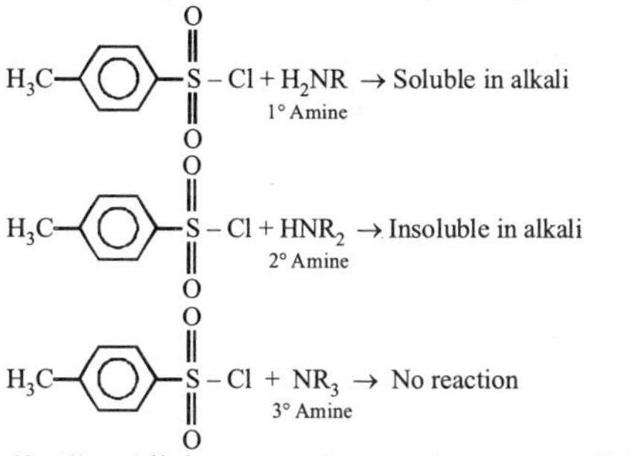

68. (3) All these enamines are interconvertible through their resonating structures. Thus, most stable form is due to least steric factor between – COOH and alkene substituent.

69. (1, 3) The given reaction is Hoffmann-Bromamide degradation method.

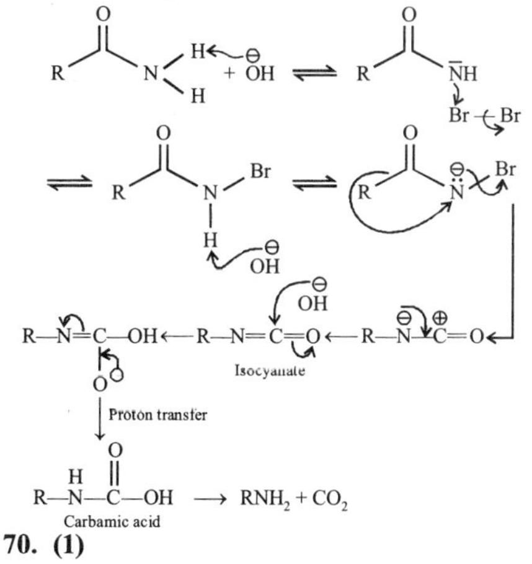

70. (1)

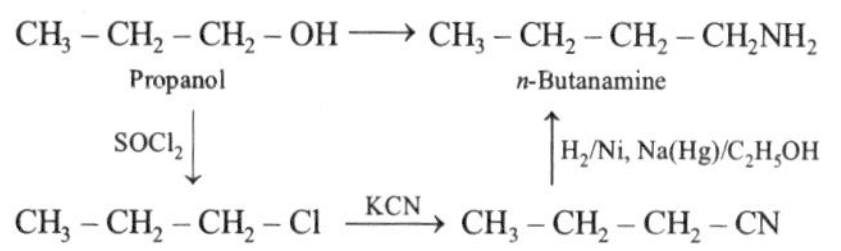

71. (4)

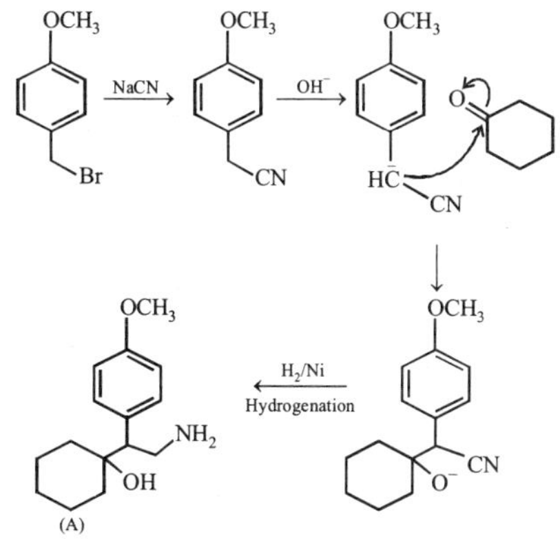

72. (1) The intermolecular association is more prominent in case of primary amines as compared to secondary amines, due to the availability of two hydrogen atoms bounded to N-atom.

73. (3) $R-C\equiv N \xrightarrow[(2)\ H_2O]{(1)\ DiBAL-H} R-\underset{\underset{O}{\|}}{C}-H$

Here, Y is $-\underset{\underset{O}{\|}}{C}-H$ (Aldehyde).

74. (2) Aniline is more basic than acetamide because in acetamide, lone pair of nitrogen is delocalised to more electronegative element oxygen, whereas in aniline,lone pair of nitrogen is delocalised over benzene ring.

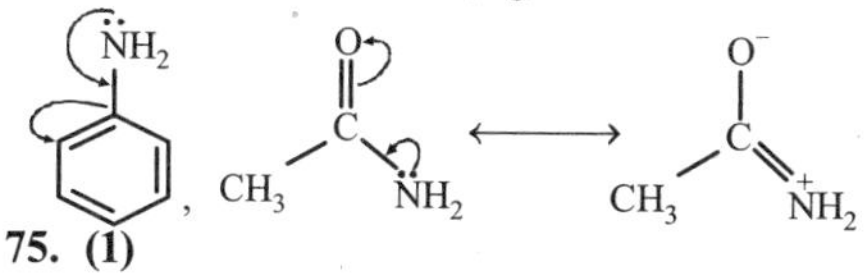

75. (1)

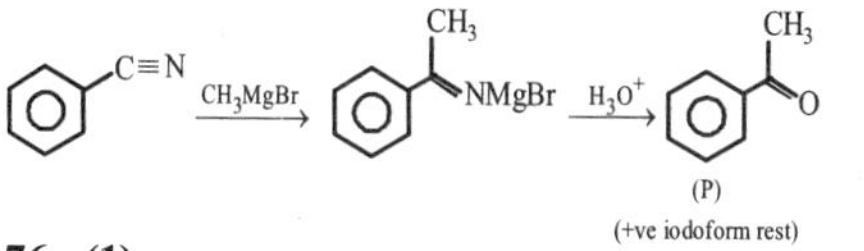

76. (1)

$$(CH_3)_3\overset{\oplus}{N}HCl^{\ominus} + NaHCO_3 \longrightarrow H_2CO_3 + (CH_3)_3N + NaCl$$

$$H_2CO_3 \longrightarrow CO_2 + H_2O$$

77. (1)

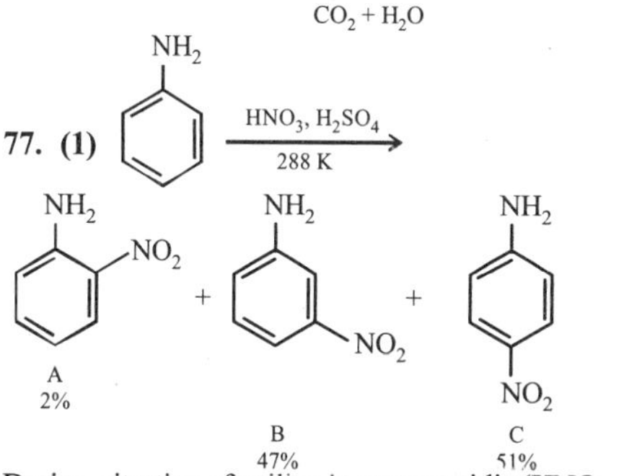

During nitration of aniline in strong acidic (HNO_3, H_2SO_4) medium, aniline changes to anilinium ion that can withdraw electron density. Its effect is felt maximum at ortho followed by meta and then para position. Consequently, very little of ortho nitrated product is formed.

78. (3)

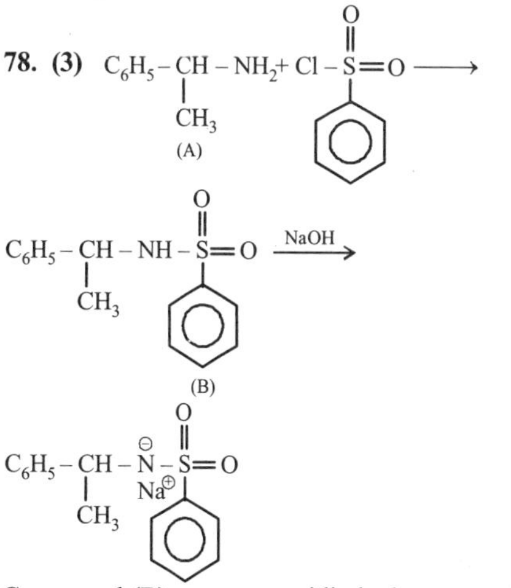

Compound (B) possesses acidic hydrogen, so it is soluble in NaOH.

79. (2)

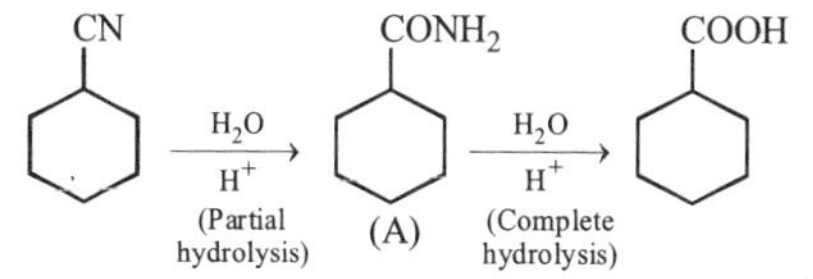

80. (1) Hinsberg reagent is benzene sulphonyl chloride. It is used for the separation of primary, secondary and tertiary amines in a given sample. Use of *p*-toluene sulphonyl-chloride is more effective than Hinsberg reagent.

81. (2) Basic strength ∝ Availability of lone pair of e^-.

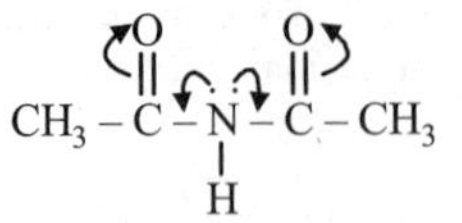

In this case, lone pair of electrons on nitrogen is delocalised due to conjugation with both $>C=O$ group, hence least basic.

82. (1) The amine reacts with benzene sulphonyl chloride and forms a product which is insoluble in alkali. So it should be secondary amine.

$R-NH(R) \xrightarrow{C_6H_5SO_2Cl} R-N(R)SO_2C_6H_5 \xrightarrow{\text{alkali}} \text{Insoluble}$

Since it is prepared by ammonolysis of C_2H_5Cl, it must contain an ethyl group, therefore, the given amine is $CH_3CH_2CH_2NH-CH_2CH_3$.

$CH_3CH_2Cl \xrightarrow[(-HCl)]{NH_3} CH_3CH_2NH_2 \xrightarrow{ClCH_2CH_2CH_3} CH_3CH_2NHCH_2CH_2CH_3$

83. (4) 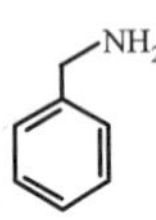 (A) Phenyl methanamine :

(Electron pair on N not involved in resonance)

$pK_b = 4.7$

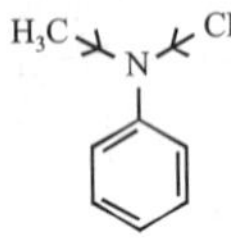 (B) N, N-Dimethylaniline :

$pK_b = 8.92$

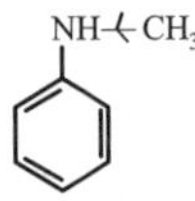 (C) N-Methyl aniline : $pK_b = 9.3$

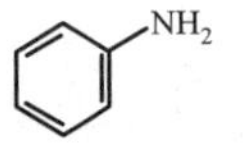 (D) Benzenamine : $pK_b = 9.38$

Electron pair on N in (*A*) is localised, while in other three (*B*, *C* and *D*) electron pair on N is delocalised in the ring. Further, alkyl groups are electron donating, hence increase the basicity of the concerned amine(s). Thus basic order of the amines is A > B > C > D

84. (4)

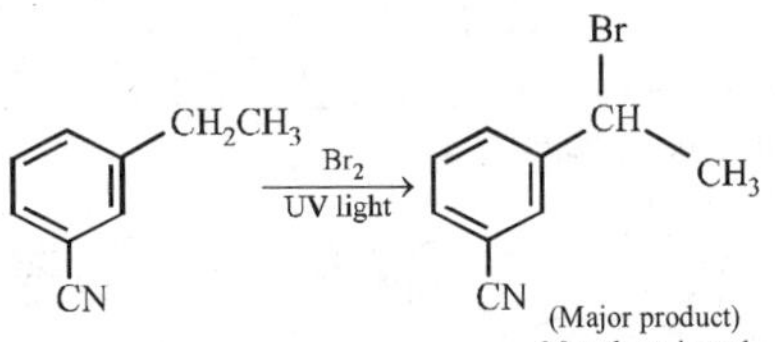

(Major product) Monobrominated

85. (3) $CH_3CH_2CN \xrightarrow[\Delta]{H_3O^+} CH_3CH_2COOH \xrightarrow{SOCl_2} CH_3CH_2\overset{O}{\overset{||}{C}}Cl \xrightarrow[\text{(Rosenmund reduction)}]{Pd/BaSO_4,\ H_2} CH_3CH_2\overset{O}{\overset{||}{C}}-H$

86. (4)

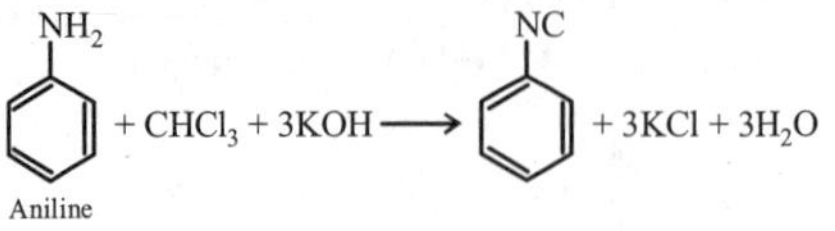

87. (3)

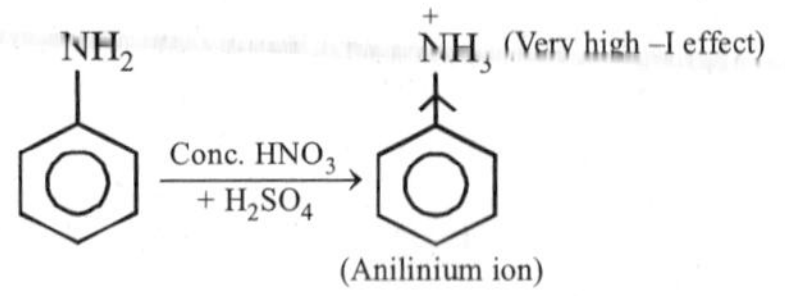

The positive anilinium nitrogen deactivates the benzene ring due to –I effect. –I effect is more on -*o* and -*m* position and less on *p*- position.

Therefore, -*m* product is also formed along with -*o* and -*p* products.

88. (1) Gabriel phthalimide synthesis gives 1° amine in good yield.

89. (2)

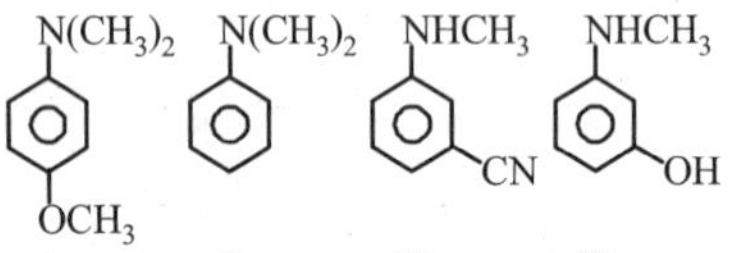

I	II	III	IV
+R effect and –I effect of $-OCH_3$		(–I effect of –CN group)	(–I effect of –OH group)

$-OCH_3$ group increases electron density of ring at *o*- and *p*- position making (I) most basic. (III) is least basic due to –I effect of –CN group at meta position.

Since, –I effect of –CN > –I effect of –OH group.

Hence, correct basic strength will follow the order I > II > IV > III

$\text{Basic strength} \propto \dfrac{1}{pK_b \text{ value}}$

Order of pK_b value is: I < II < IV < III.

90. (4)

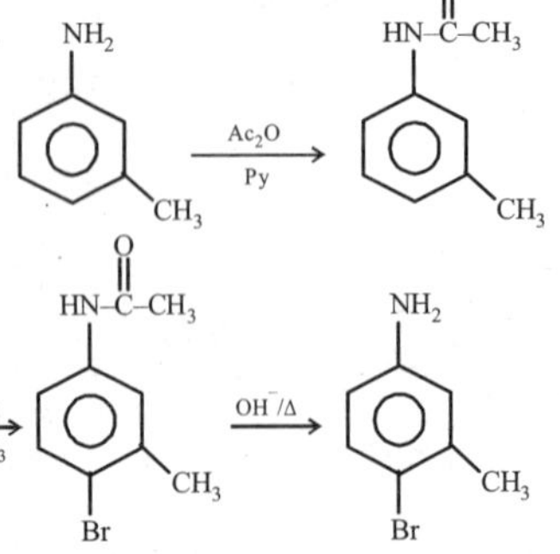

91. (3)

$CH_3CH_2-C\equiv N \xrightarrow{LiAlH_4} CH_3CH_3-CH_2-NH_2$

$NaBH_4$ does not reduce R–CN.

92. (2) (A)  (B) 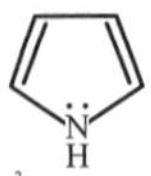

(A) sp^2 hyb. N atom (localised L.P.) (B) sp^2 hyb. N atom (delocalised L.P.)

(C) 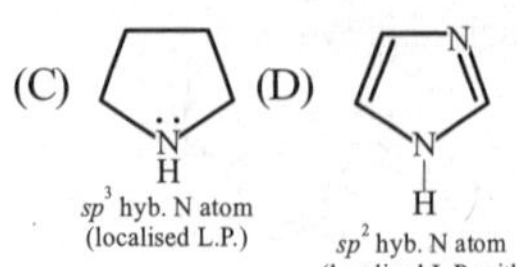

(C) sp^3 hyb. N atom (localised L.P.) (D) sp^2 hyb. N atom (localised L.P. with partial negative ch.)

Thus, increasing order of basicity is

(B) < (A) < (D) < (C).

93. (4) Basic strength of amines depends upon availability of lone pair of electrons. Aliphatic amines are more basic than aromatic amines.

94. (3) Aniline and phenol form complex with Lewis acid. Chlorobenzene produces highest yield in Friedel Craft reaction among the given options.

95. (1)

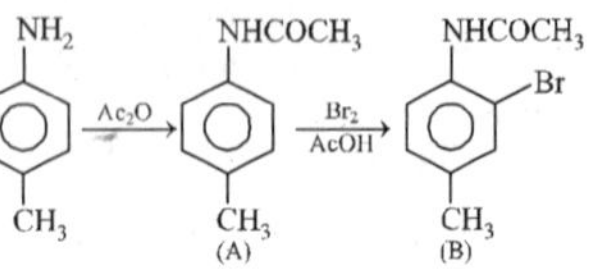

96. (4) Lower the pK_b value of a base (amine), higher will be K_b value of the base. Further basic nature of bases can best be compared by their conjugate acids. A stronger base has weak conjugate acid. The conjugate acids of the three bases are

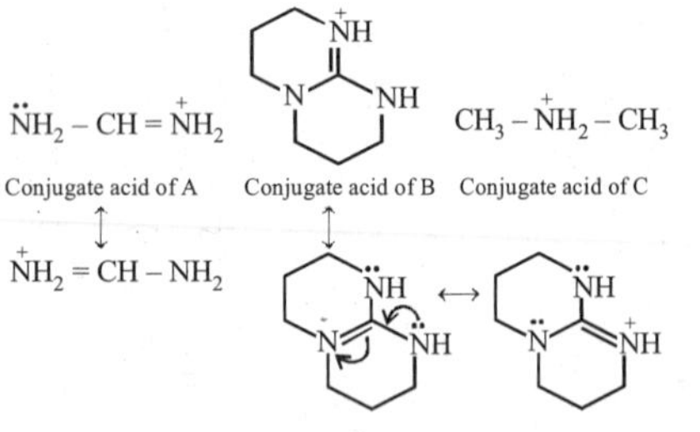

Thus conjugate acid of (B) is more stable than that of (A) which in turn more stable than C. Thus the basic character of the three is B > A > C and hence pK_b value follows the following order (B) < (A) < (C).

97. (3) Electron withdrawing group attached to benzene ring will reduce the basic strength and increase pK_b, while electron donating group decreases pK_b.

Therefore the correct increasing order of pK_b is

(B) < (D) < (A) < (C)

98. (1)

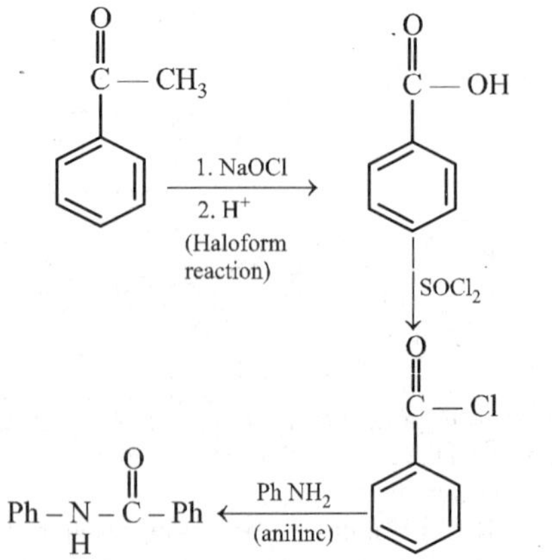

99. (4)

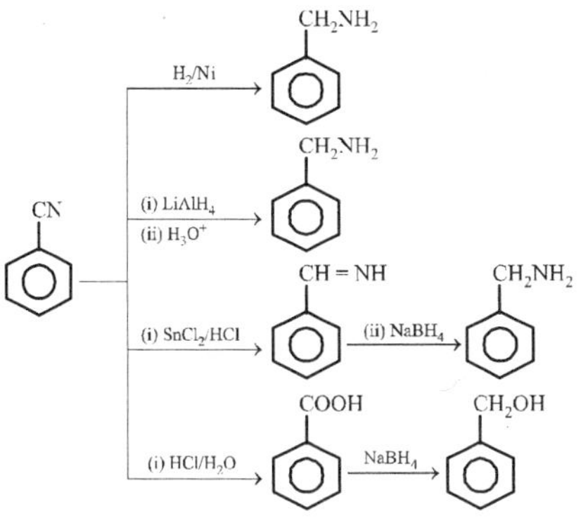

100. (1) N-Ethyl phthalimide on treatment with $NH_2–NH_2$ gives ethylamine (hydrazinolysis).

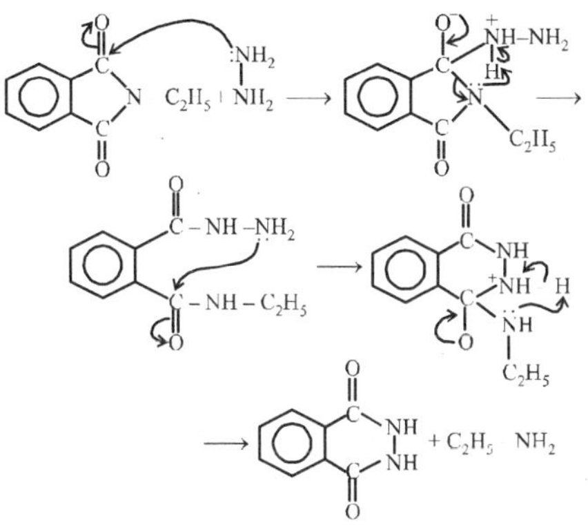

In place of NH_2NH_2, we can also use H_2O in presence of H^+ or OH^- as a catalyst, but the acidic or basic hydrolysis is slow, hence hydrozinolysis is preferred.

101. (2)

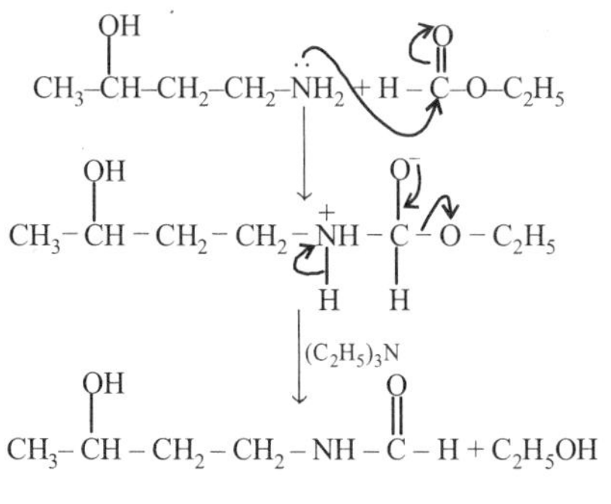

102. (3) Hint: H_2Pd hydrogenate both ketone and cyanide group.

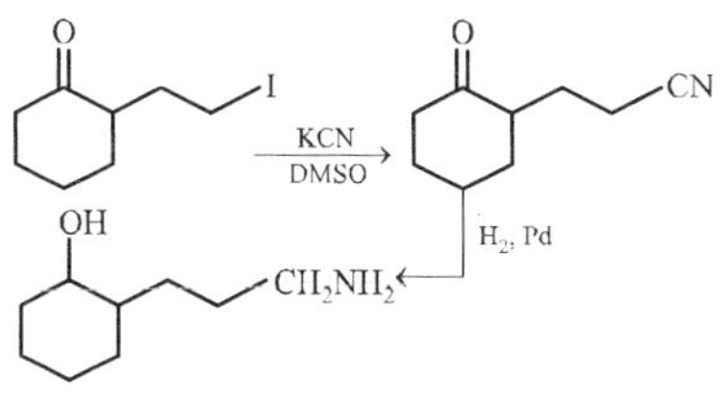

103. (3) Hinsberg's reagent is benzenesulphonyl chloride ($C_6H_5SO_2Cl$). It is used for detection of primary, secondary and tertiary amines.

104. (1) Polysubstitution is a major drawback in Friedel Craft's alkylation because the introduced alkyl group is activating and gives polyalkylated product.

105. (4)

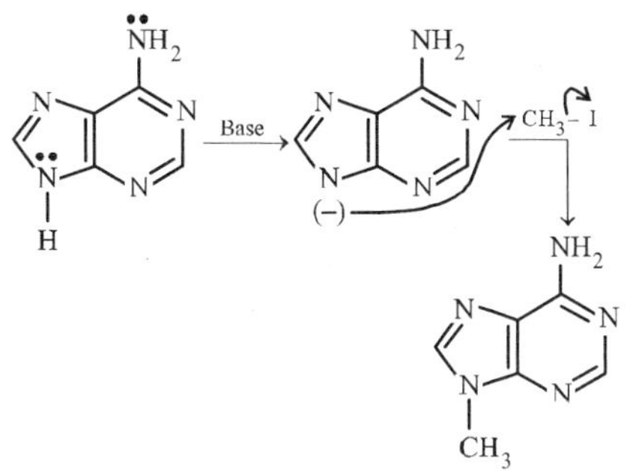

106. (4)

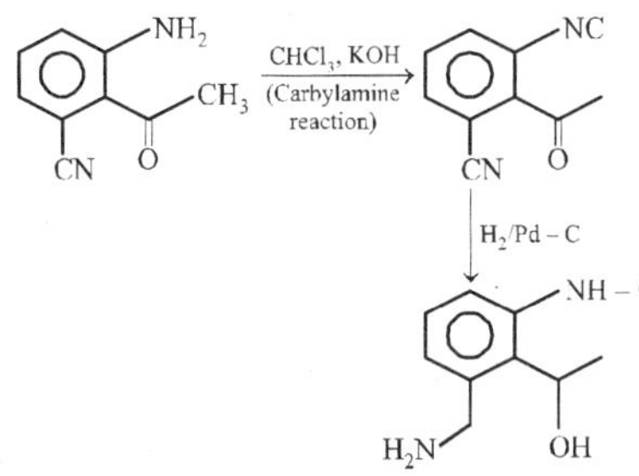

107. (1) Primary amines are prepared by Gabriel phthalimide synthesis

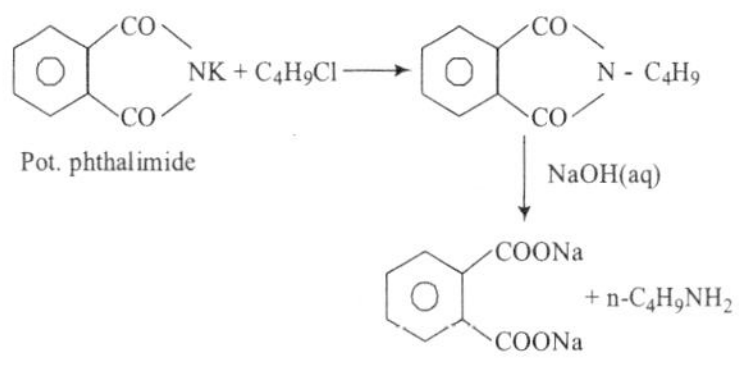

108. (3) $(C_2H_5)_2NH > C_2H_5NH_2 > NH_4$

Two +I gp. one + I gp. No + I gp.

109. (1)

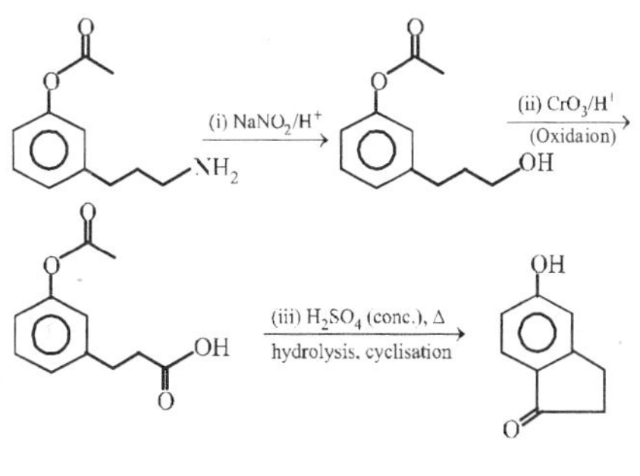

110. (3) The reactivity order of carboxylic acid derivatives depends on the electrophilicity of the carbonyl carbon and leaving tendency of the leaving group.

Leaving ability of the four groups is

$$-Cl > -COOC_2H_5 > -OCH_3 > -NH_2.$$

Therefore, reactivity order towards $LiAlH_4$ is:

$$(A) < (B) < (D) < (C).$$

111. (1) Reactivity order of following nucleophiles.

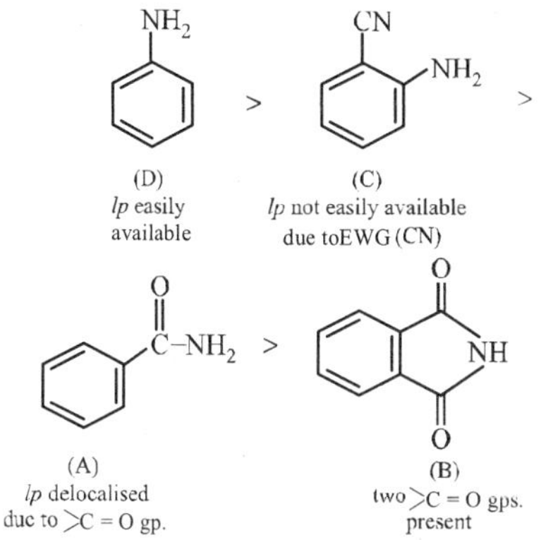

112. (2) Higher the electronegativity ($sp^2 > sp^3$) of an atom, more will be its capacity to take proton. Thus N labelled as (b), (c) and (d) are better protonated than the N at (a) and (e).

113. (3)

$$\underset{(X)}{CH_3CH_2CH_2CONH_2} \xrightarrow[\text{(Hoffmann rearrangement)}]{Br_2/NaOH} \underset{(C_3H_9N)}{CH_3CH_2CH_2NH_2}$$

$$\xrightarrow[\text{(Carbylamine reaction)}]{CHCl_3/KOH} CH_3CH_2CH_2NC$$

114. (4) In the given nitration reaction, major product will be formed as per the activating group, –NH part of $-NH-\overset{O}{\overset{\|}{C}}-$ is activating, while $-\overset{O}{\overset{\|}{C}}-$ part is deactivating group.

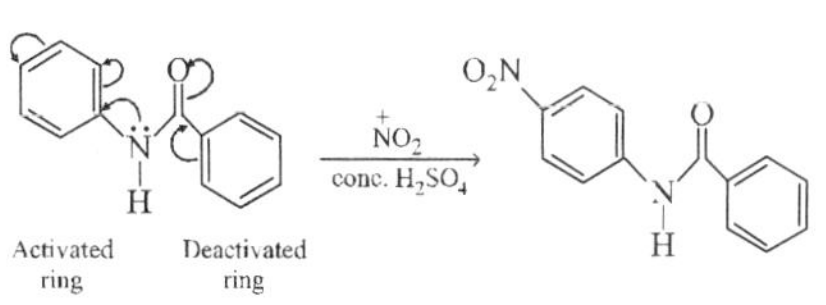

115. (3)

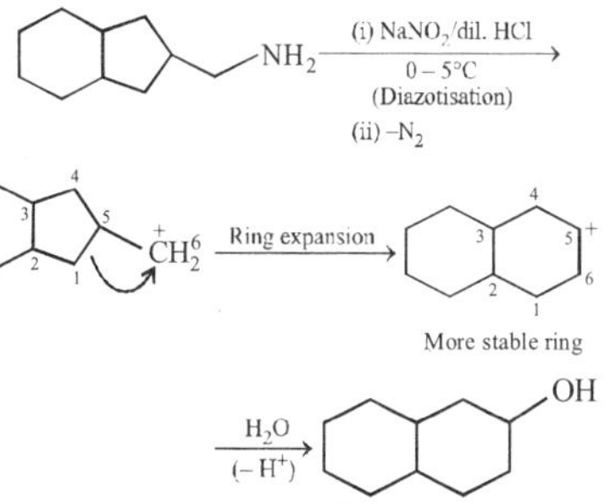

116. (4) $-NH_2$ is a stronger base than $-OH$, hence 1 equation of acetic anhydride will acetylate only $-NH_2$ group.

Reaction involved:

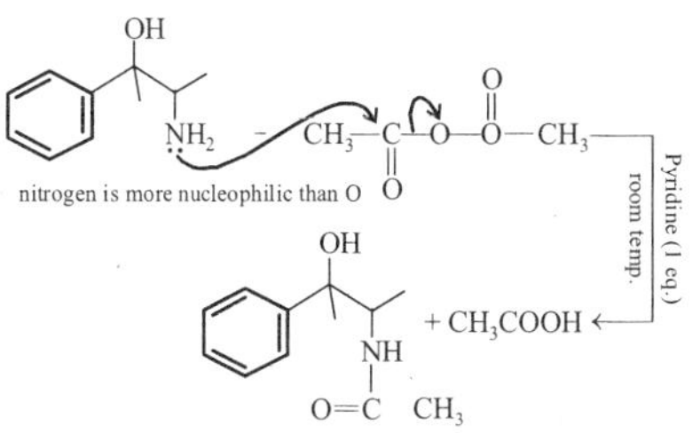

117. (2)

	pK_b
(A) $EtNH_2$	3.29
(B) $(Et)_2NH$	3.00
(C) Me_3N	4.22
(D) Ph – NH – Me	4.7

So, order of basic strength is:

$C_2H_5 \rightarrow NH \leftarrow C_2H_5$ (2C_2H_5 groups) (B) > $C_2H_5NH_2$ (1C_2H_5 group) (A) >

Me_3N (steric factor) (C) > PhNH–Me (e^- pair delocalised over Ph - ring) (D)

(B) > (A) > (C) > (D)

118. (2) Compound, III 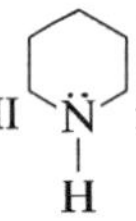 is most basic as the lone pair of nitrogen is easily available for the donation.

In case of compound (I) 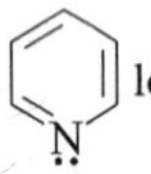 lone pair is not involved in resonance but nitrogen atom is sp^2 hybridsed, whereas in compound II, the lone pair of nitrogen is involved in aromaticity which makes it least basic.

119. (3) Reaction involved for the given reaction:

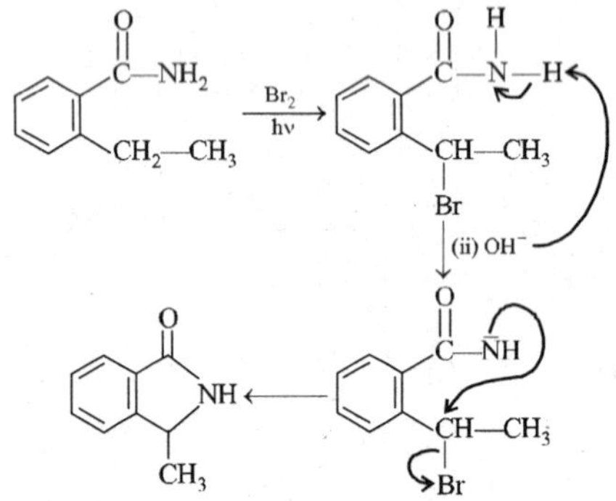

120. (4)

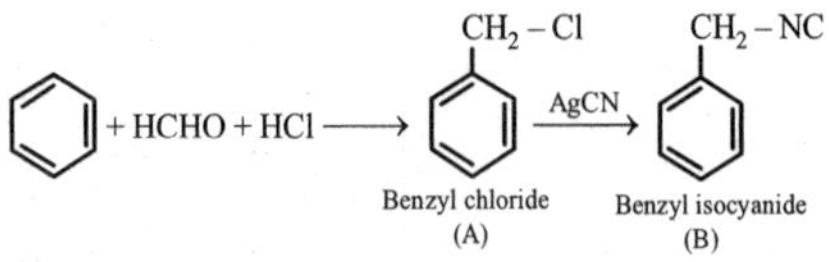

121. (1)

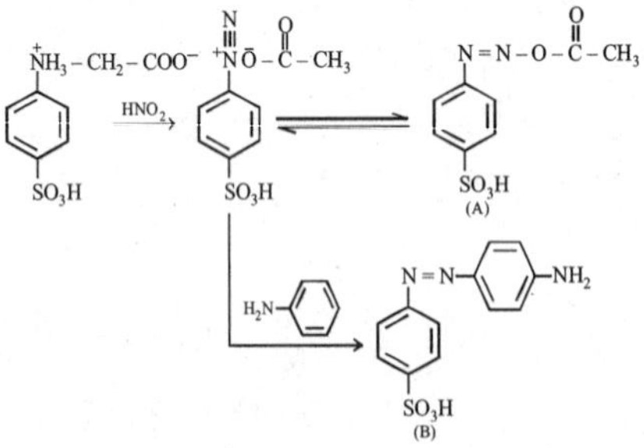

122. (4) Aromatic diazonium salts are more stable than aliphatic diazonium salts. The higher stability of aryl diazonium salts is due to resonance. Electron donating substituents on benzene ring increase electron density. Hereby they increase the stability of diazonium salts. Electron withdrawing substituents decrease electron density on benzene ring. Hereby they decrease the stability of diazonium salts. –$COCH_3$ group is electron withdrawing and hence, diazonium salts from (D) is less stable than that from (B). Although –O–$COCH_3$ is electron donating substituent, but it is present in meta position. Hence, it will not have significant effect on stability.

Therefore, the increasing order of diazotisation is (A) < (D) < (C) < (B).

123. (3) Basic strength of amines depends upon availability of lone pair of electrons. Aliphatic amines are more basic than armoatic amines.

(I) 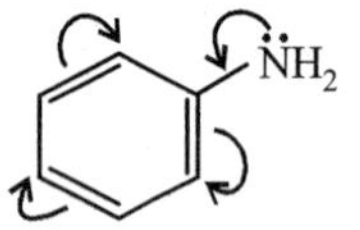 + R effect

the lone pair of electron is less easily available for protonation.

(II) 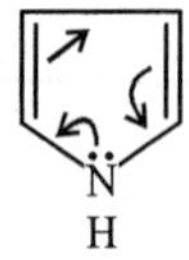 + R effect

the lone pair of electrons is part of aromatic sextet, less available for proton.

(III) 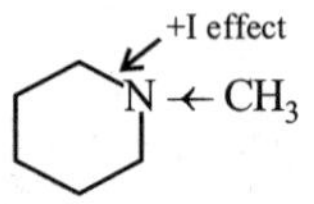

(IV) 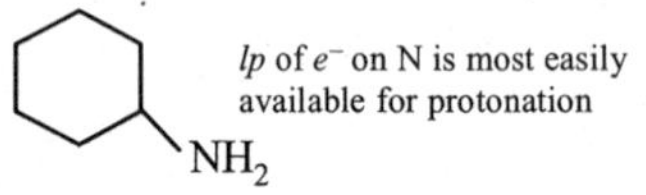

Thus, the correct order is: (II)<(I)<(IV)<(III)

124. (3) Nitration takes place in presence of conc. HNO_3 + conc. H_2SO_4.

In strongly acidic medium, amine is converted into anilinium ion (– NH_3^+); substitution is thus controlled not by –NH_2 group but by –NH_3^+ group which, because of its positive charge, directs the entering group to the meta-position instead of ortho- and para.

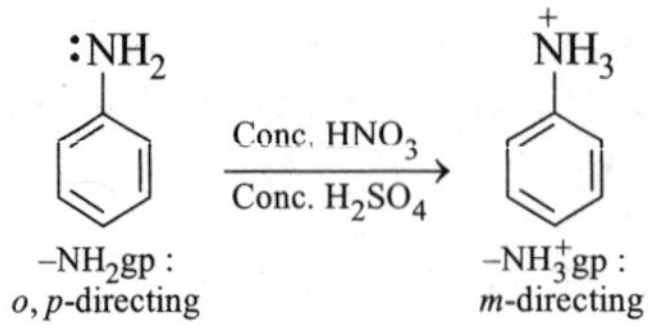

125. (2) $RCONH_2 + Br_2 + 4NaOH \longrightarrow R\text{–}NH_2 + K_2CO_3 + 2NaBr + 2H_2O$

126. (4) Primary, secondary and tertiary amines are distinguished by Hinsberg reagent test. The Hinsberg reagent is an alternative name for benzene sulphonyl chloride ($C_6H_5SO_2Cl$).

127. (1) Lone pair of electrons on N is involved in aromaticity.

128. (3) 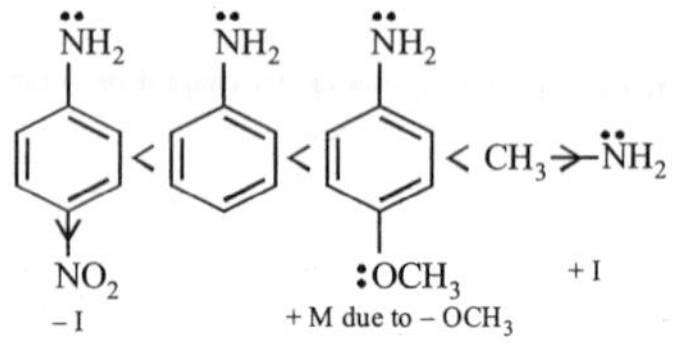

129. (4) $R-CH_2-NH_2 + CHCl_3 + 3KOH\,(alc) \xrightarrow{\text{(Carbylamine reaction)}} R-CH_2-NC + 3KCl + 3H_2O$

Alkyl isocyanide

130. (1) Relative basicities of amines in gas phase is due to +I effect of alkyl group(s).

$(CH_3)_3N > (CH_3)_2NH > CH_3NH_2 > NH_3$

However, in aqueous solution it follows the following order.

$(CH_3)_2NH > CH_3NH_2 > (CH_3)_3N > NH_3$

This is due to solvation effect. Protonated amine forms H-bonding with H_2O and stabilises the concerned conjugate acid.

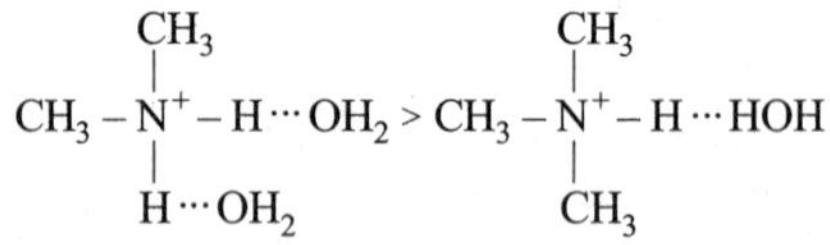

Conjugate acid of 2° amine more stable — Conjugate acid of 3° amine forms only one H-bond

Hence the correct order is

$(CH_3)_2NH > CH_3NH_2 > (CH_3)_3N\ C_6H_5NH_2$

131. (2) $CH_3NH_2 + HNO_2 \longrightarrow CH_3OH + N_2 + H_2O$

132. (2)

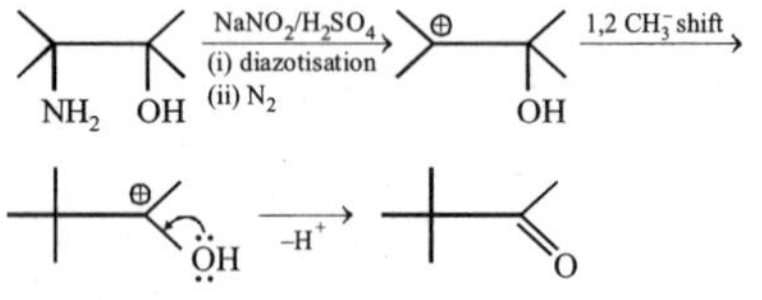

Positive charge better stabilised due to +M effect

133. (2)

$\underbrace{R-NH_2}_{\text{Mol.Mass=16}} + CH_3-\overset{O}{\overset{\|}{C}}-Cl \xrightarrow[-HCl]{} \underbrace{R-NH-\overset{O}{\overset{\|}{C}}-CH_3}_{\text{Mol.mass=58}}$

Now since the molecular mass increases by 42 unit as a result of the reaction of one mole of CH_3COCl with one-NH_2 group and the given increase in mass is (390 – 180) 210. Hence, the number of – NH_2 groups = 210/42 = 5.

134. (3) The correct order of relative basicity of amines in the gas phase is $3° > 2° > 1° > NH_3$

The alkyl group releases electron and thus, tends to disperse the positive charge of the alkyl ammonium ion and therefore, stabilises it. Since, NH_4^+ (from NH_3) has no such alkyl group, it is not stabilised to such an extent as alkylammonium ion.

135. (2) $RNH_2 + CHCl_3 + 3KOH \longrightarrow RNC + 3KCl + 3H_2O$

Mechanism

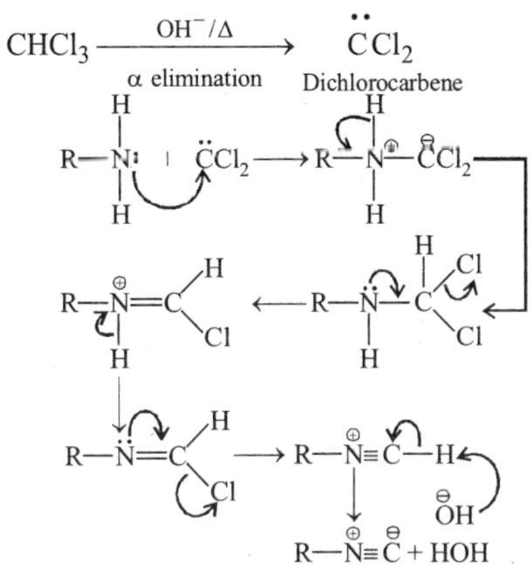

136. (2) 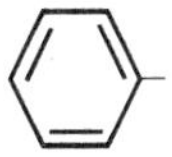 $-CH_2-NH_2$ is most basic due to presence of localised lone pair of electrons on nitrogen atom while in other compounds lone pair of electrons is delocalised.

137. (2) I. 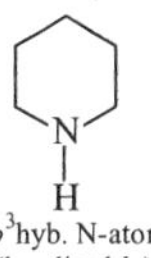
sp^3hyb. N-atom (localised *lp*)

II. 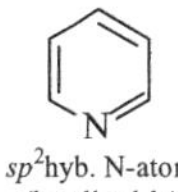
sp^2hyb. N-atom (localised *lp*)

III. 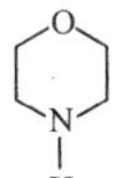
sp^3hyb. N-atom (localised *lp*) (Due to -I effect of O it is less basic than I)

IV. 
sp^3hyb. N-atom (localised *lp*) (*lp* on N contributes towards aromaticity)

Thus, order of basicity: I > III > II > IV

138. (4) Note: Aromatic amines (e.g. aniline) are less basic than aliphatic amines due to delocalisation of electrons. Among aliphatic amines the order of basicity in aq. solution is $2° > 1° > 3°$. The electron density is decreased in 3° amine due to crowding of alkyl group over N atom which makes the approach and bonding by a proton relatively difficult. Therefore the basicity decreases.

∴ The correct order of basic strength is

Dimethylamine > Methyl amine > Trimethyl amine > Aniline.

139. (1) This is carbylamine reaction.

$$CH_3CH_2NH_2 + CHCl_3 + 3KOH \longrightarrow C_2H_5NC + 3KCl + 3H_2O$$

140. (2) Wurtz reaction is used for the preparation of hydrocarbons from alkyl halides.

$$RX + 2Na + XR \longrightarrow R-R + 2NaX$$

141. (2) The alkyl groups are electron releasing (+ I) groups, thus increase the electron density around the nitrogen thereby increasing the availability of the lone pair of electrons to proton or Lewis acid and making the amine more basic. Hence more the number of alkyl groups, more basic is the amine. Therefore, the correct order is $NH_3 < CH_3NH_2 < (CH_3)_2NH$

142. (3) 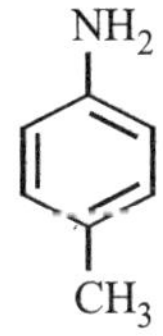 $+ CHCl_3 + 3KOH \longrightarrow$

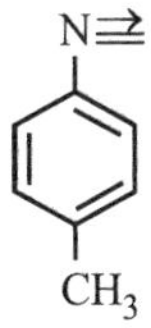 $+ 3KCl + 3H_2O$

143. (4) Ethyl isocyanide on hydrolysis form primary amines.

$$CH_3CH_2N \equiv C + H_2O \xrightarrow{H^+} CH_3CH_2NH_2 + HCOOH$$

144. (1) $C_2H_5NH_2 + CHCl_3 + 3KOH \longrightarrow C_2H_5N \equiv C + 3KCl + 3HCl$

(Ethyl isocyanide)

145. (21) In first step, the – NH – group gets protonated which weakens the adjacent N – N bond, leading to cleavage with the formation of aniline and diazonium salt.

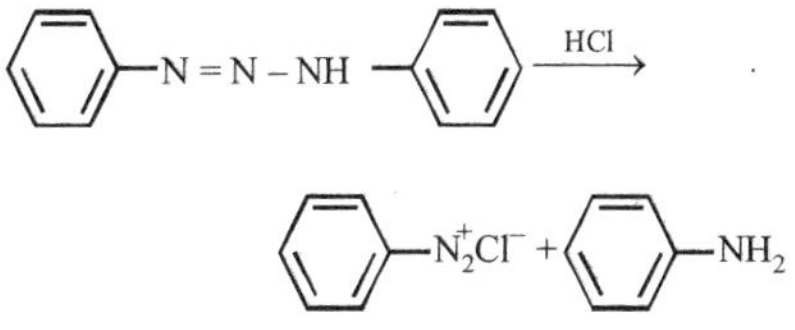

In subsequent steps, there is azo coupling reaction where:

- aniline acts as a solvent and reactant
- The mixture is gently heated to facilitate rearrangement.
- After the reaction is complete, the mixture is cooled and treated with acetic acid to neutralise excess acid and precipitate the final product that is para amino benzene.

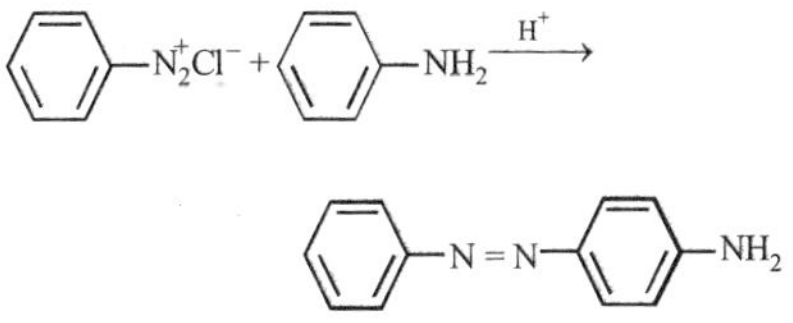

Para amino benzene (yellow dye)

Hence, mass % of nitrogen in the final product is:

$$\frac{42}{197} \times 100 = 21\%$$

146. (1)

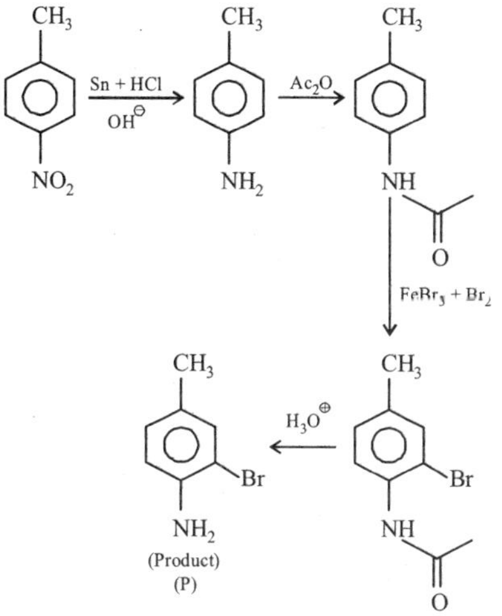

Molar mass of P $(M_P) = (7 \times 12) + (8 \times 1) + 80 + 14 = 186$ g/mol.

Molar mass of AgBr = 108 + 80 = 188 g/mol.

According to Carius method,

1 mole of P → 1 Mole of AgBr

Therefore, 1 g of P → $\frac{188}{186}$ g of AgBr ≈ 1g.

147. (3) 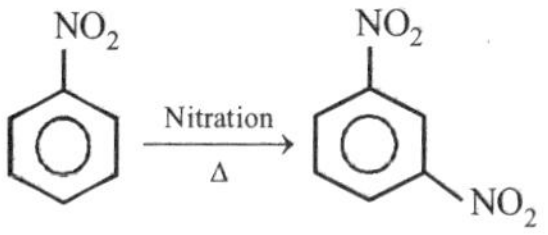

$C_6H_5NO_2$	Molecular formula = $C_6H_5N_2O_4$
Molecular	Molecular mass = 168
mass = 123	Product formed = 4.2g

∴ Number of moles of product formed

$$= \frac{4.2}{168} = 0.025 \text{ mol}$$

Required mass of nitrobenzene = 123 × 0.025 = 3.075 g

∴ Nearest integer is 3.

148. (340) Benzyl Amine is most basic due to localised lone pair.

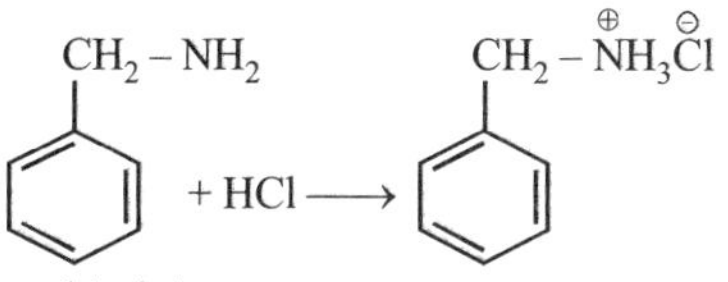

(Benzyl Amine)

Molar mass of Benzylamine = 107g

Mole of benzyl Amine $= \frac{1}{107} = 0.00934$ mol

1 Mol of Benzyl amine consumed 1 mol of HCl

So, Mol of HCl consumed = 0.00934 mol

Mass of HCl consumed = 0.00934 × molar mass of HCl = 0.00934 × 36.5 = 0.34 g = 340 mg

149. (93)

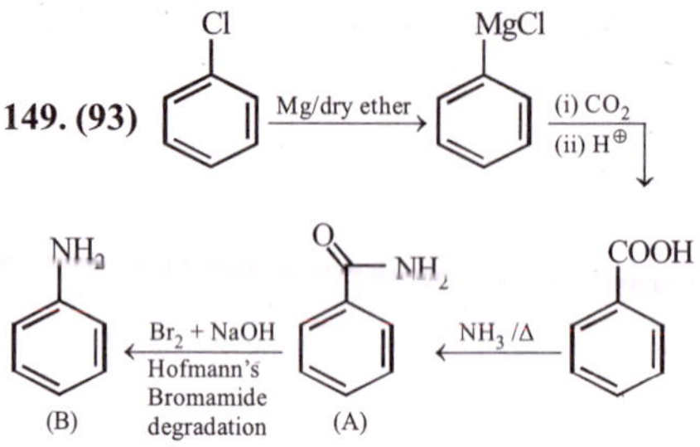

No. of moles of 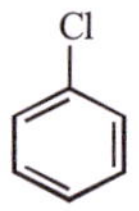 = No. of moles of 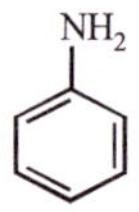

$$\frac{11.25\times10^{-3}}{112.5}=\frac{x\times10^{-1}\times10^{-3}}{93}$$

$$0.1\times10^{-3}=\frac{x\times10^{-3}}{93}$$

$\Rightarrow x = 9.3\times10^{-3}g = 93\times10^{-1}$ mg

150. (4) Compounds which can not undergo Friedel-crafts reaction are

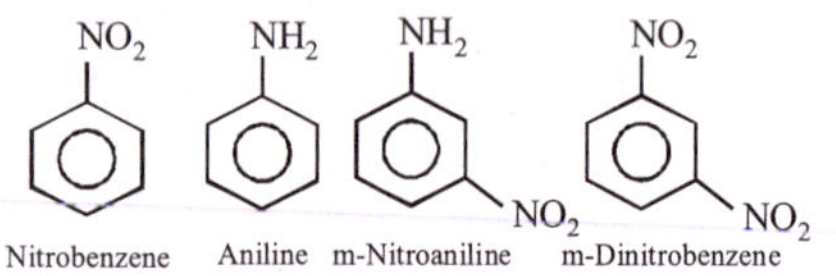

Nitrobenzene Aniline m-Nitroaniline m-Dinitrobenzene

This is due to strongly deactivating groups and– NH_2 group reacts with the acid ($AlCl_3$).

151. (5) Hinsberg's reagent is benzen sulphonyl chloride it reacts with 1° amine to yield N-ethyl benzene sulphonyl amide. The hydrogen attached to nitrogen in sulphonamide is acidic. Hence it is soluble in alkali.

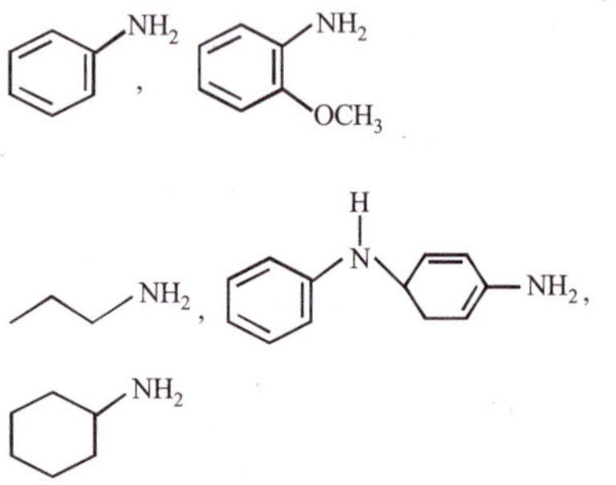

152. (287)

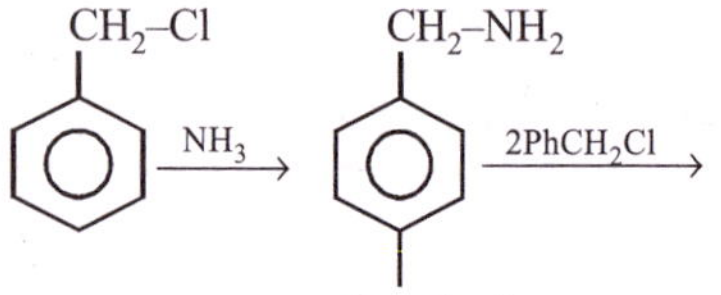

Benzyl amine

$$Ph-CH_2-\underset{\underset{CH_2-Ph}{|}}{N}-CH_2-Ph$$
(X)

Molar mass of amine = 287 g/mol.

153. (80) Reaction is represented as :

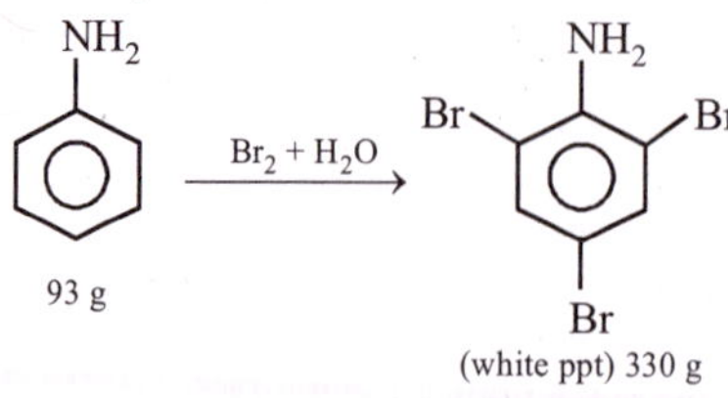

93 g of aniline produces 330 g of 2, 4, 6-tribromoaniline. Hence 9.3 g of aniline should produce 33g of 2, 4, 6-tribromoaniline.

$$\text{Hence percentage yield} = \frac{26.4\times100}{33} = 80\%$$

154. (8)

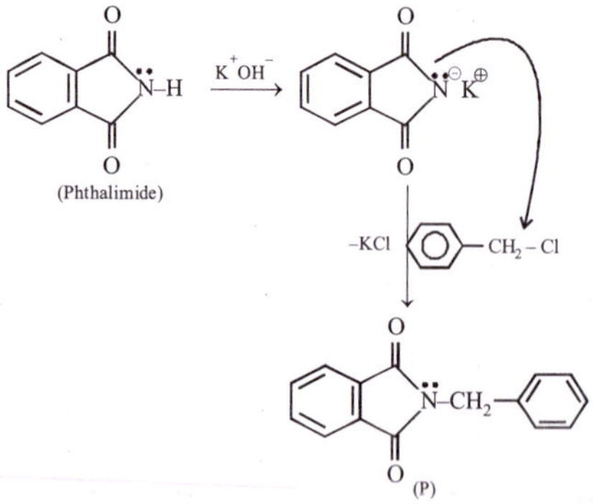

Total number of π-bonds present in product P is 8.

155. (1) Only C is correct.

156. (5)

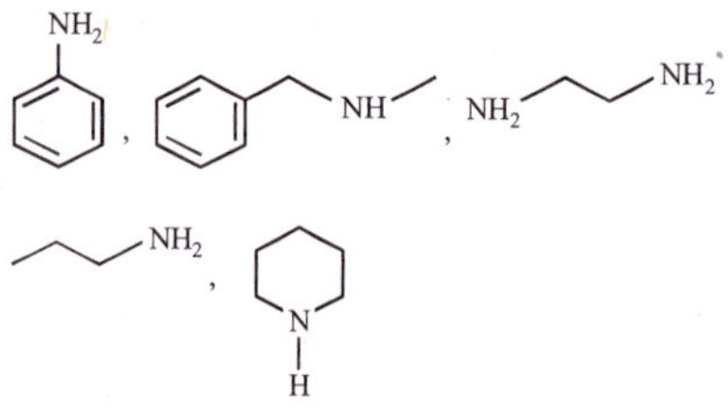

157. (2)

$$\underset{(X)}{R-NH_2} + CH_3-\overset{\overset{O}{||}}{C}-Cl \longrightarrow R-NH-\overset{\overset{O}{||}}{C}-CH_3$$

gain in molecular weight after acetylation with one $-NH_2$ group is 42.

Total increase in molecular weight = 192 – 108 = 84

$$\therefore \text{Number of amino group in (X)} = \frac{84}{42} = 2$$

158. (6) Aryl substituted aliphatic amines which are also considered as aromatic amines can be prepared by Gabriel-pthalimide synthesies.

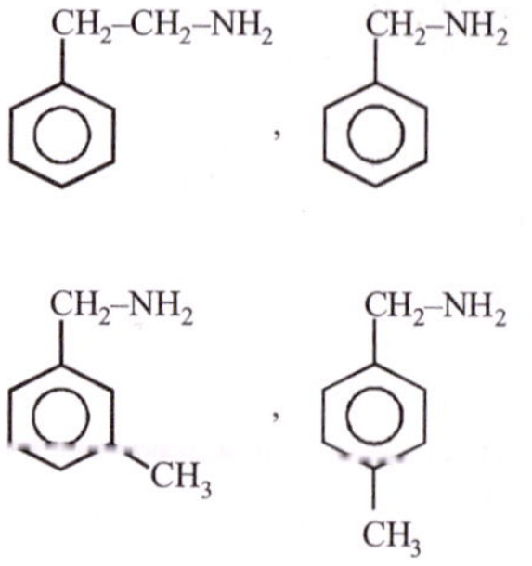

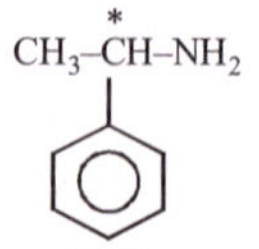

(d, *l*)

159. (3) No reaction will be observed as in Gabriel phthalimide synthesis (B) because 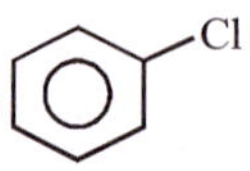 is poor substrate for S_N2.

∴ Aromatic amines will be formed in A, C & D.

160. (5) 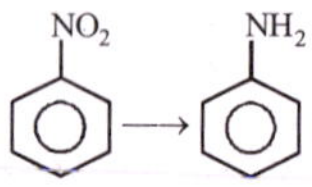 Reagents used can be (i) Sn + HCl (ii) Fe + HCl (iii) Zn + HCl (iv) H_2 – Pd (v) H_2 (Raney Ni)

161. (108°) In gaseous triethyl amine, the "–C–N–C–" bond angle is 108 degree.

162. (77)

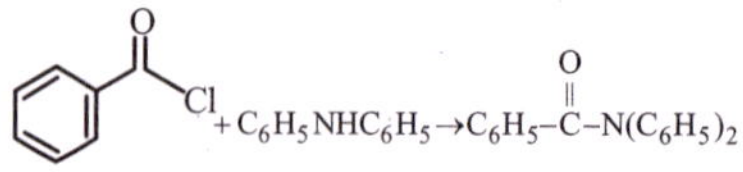

Mass (given)	0.140 g	0.388 g	0.210 g (Observed yield)
Molar mass	140.5 g/mol	169 g/mol	273 g/mol
Moles	$\approx 10^{-3}$ (Limiting Reagent)	2.3×10^{-3}	Theoretical yield $= 10^{-3}\times273$ = 0.273 g

$$\therefore \%\text{ yield} = \frac{\text{Observed yield}}{\text{Theoretical yield}}$$

$$= \frac{0.210}{0.273}\times100 \approx 77\%.$$

Hence amine which can synthesised by Gabriel phthalimide synthesis method is :

(A) $Me_2CH-CH_2-NH_2$

(B) $CH_3CH_2NH_2$

(C) $Ph-CH_2-NH_2$

163. (3) Gabriel phthalimide synthesis is used to prepare 1° aliphatic/alicyclic amine.

164. (1) $CH_3-CH-COOH \xrightarrow[\Delta]{NH_3}$

$CH_3-CH_2-\overset{O}{\overset{||}{C}}-NH_2 \xrightarrow[\text{Hoffmann Bromamide}]{NaOH+Br_2}$

$CH_3-CH_2-NH_2 \xrightarrow[H_2O/0°C]{HNO_2}$

$CH_3-CH_2OH \xrightarrow[0°C]{Ph\overset{+}{N_2}\overset{-}{Cl}}$

$+ CH_3-\overset{O}{\overset{||}{C}}-H + N_2 + HCl$

Major product

165. (2)

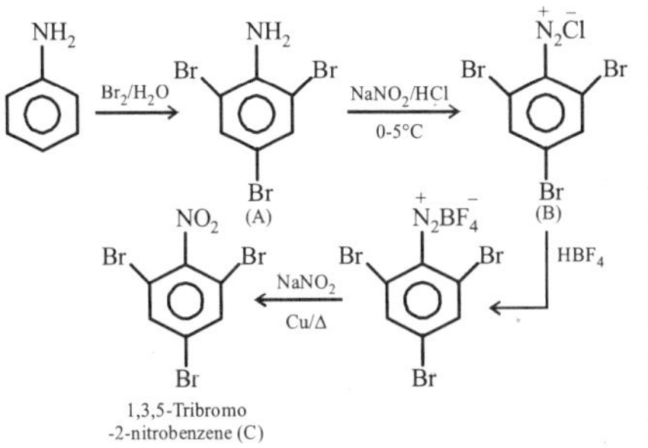

166. (2)

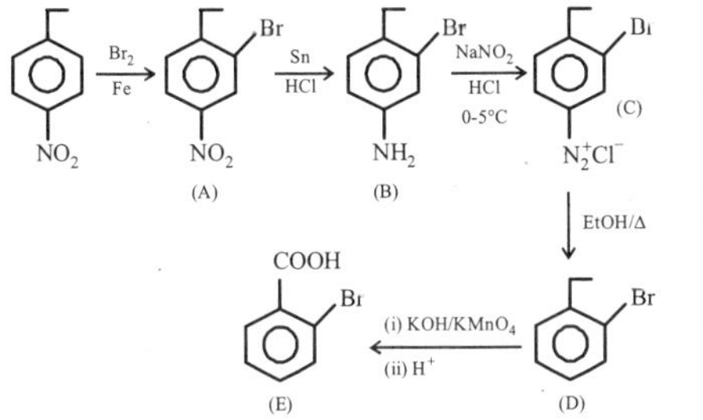

167. (4) Statement I : False

The dipole moment of R – NC is greater than RCN.

$R-NC \xrightarrow{H_3O^+} R-NH_2 + HCOOH$

Statement II : True

$R-CN \xrightarrow{H_3O^+} RCOOH \xrightarrow{SOCl_2}$

$RCOCl \xrightarrow{NH_3} RCONH_2 \xrightarrow{NaOCl/OH^-}$

$RNH_2 \xrightarrow{CHCl_3 + KOH} R-NC$

168. (3) Statement-I:

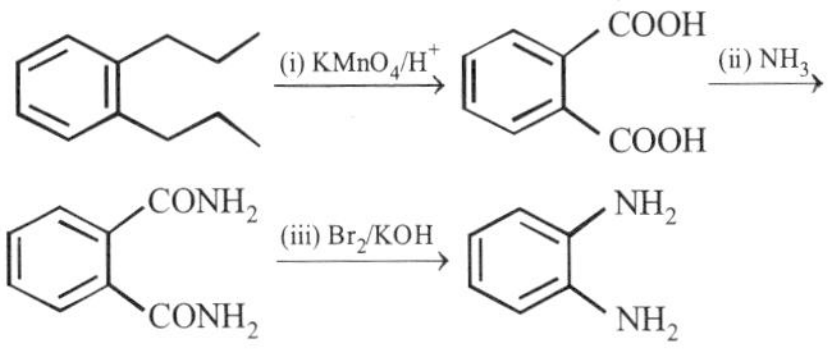

Statement-II:

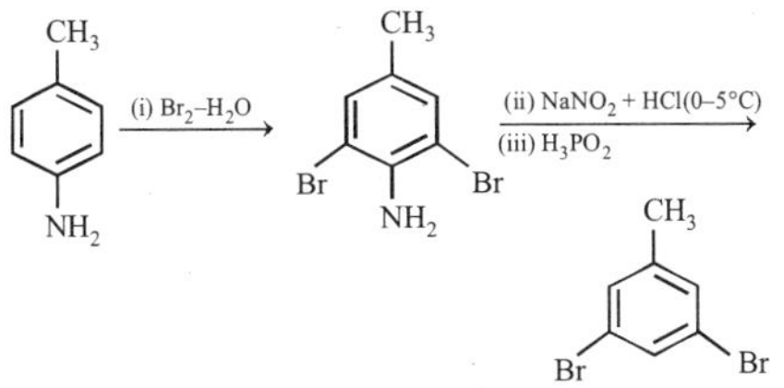

Both statements (I) and (II) are true.

169. (2)

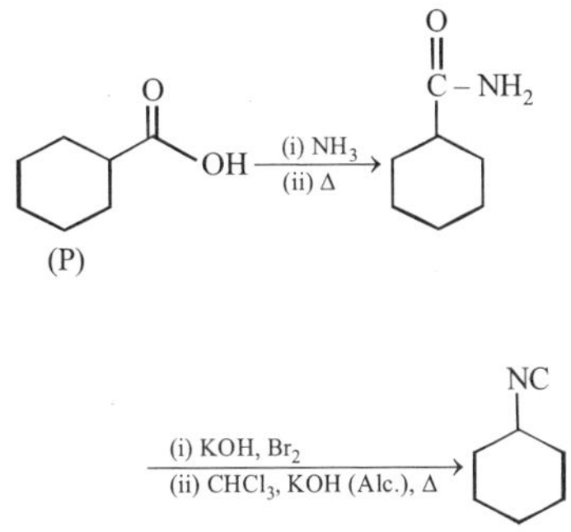

170. (3) Electron donating group (like $-OCH_3$) increases stability of the diazonim salt due to +M effect. On the other hand, electron withdrawing group (like $-NO_2$ and $-CN$) decreases stability of the salt due to –M effect.

Thus the order of stability is

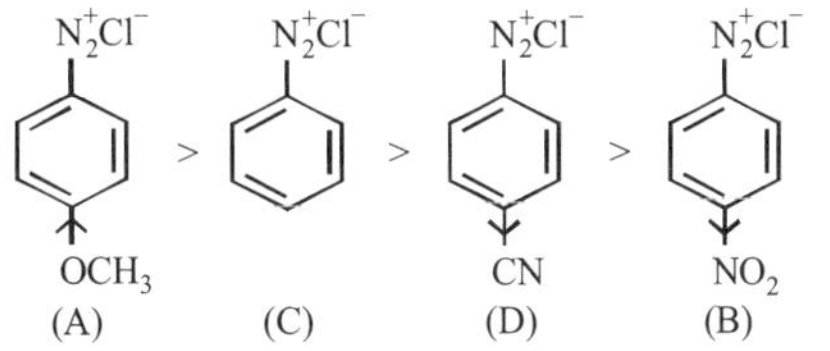

171. (3)

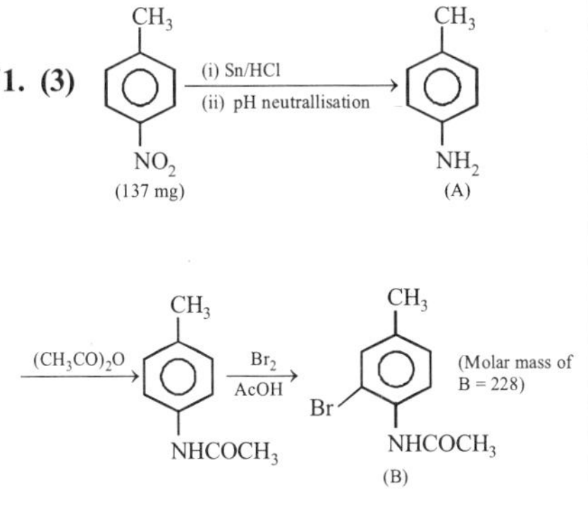

$$\text{Mole} = \frac{137 \times 10^{-3}}{137} = 0.001 \text{ mole}$$

Mole of product = 0.001 mole

Mass of product = 0.001 × 228 gm

= 0.228 gm = 228 mg

172. (3)

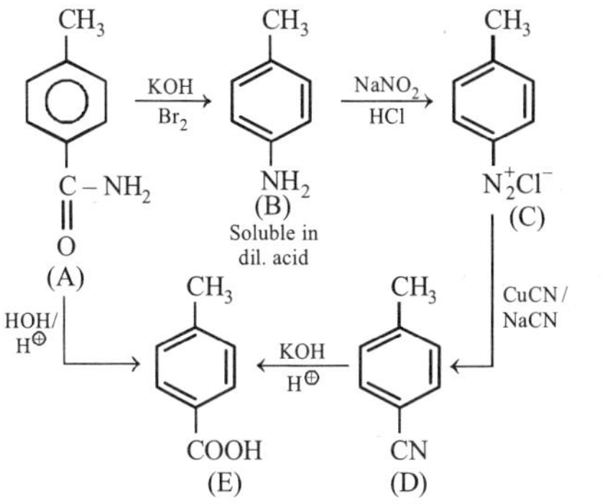

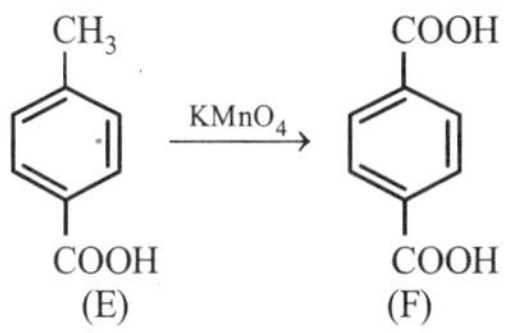

(i) All the 4 H's are identical

(ii) Both H's of –COOH are identical

173. (3)

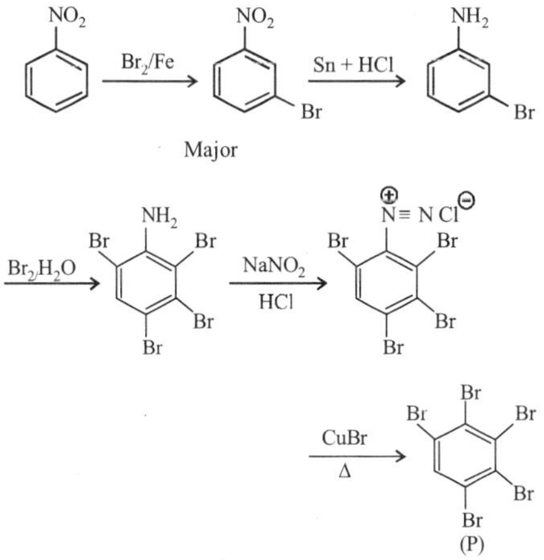

Number of Br atoms in major product (P) = 5

174. (3)

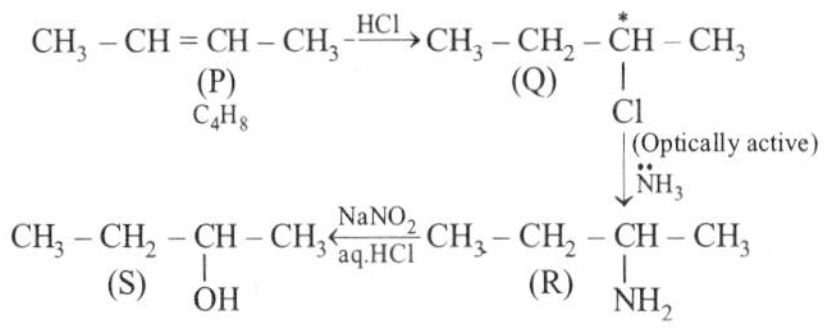

175. (1)

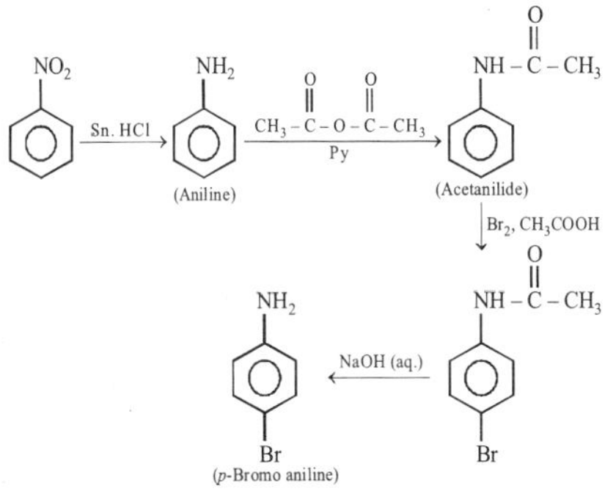

176. (1) 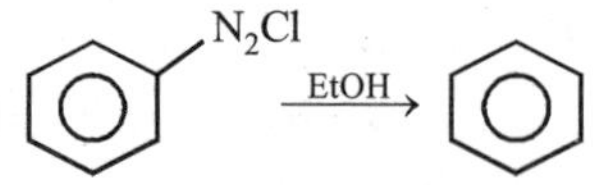

177. (3) 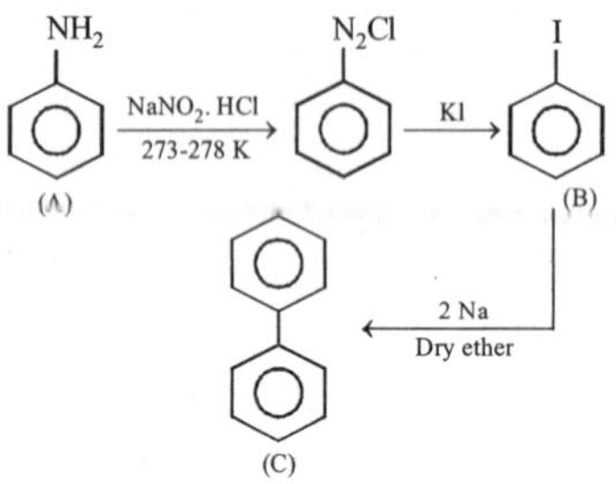

178. (3)

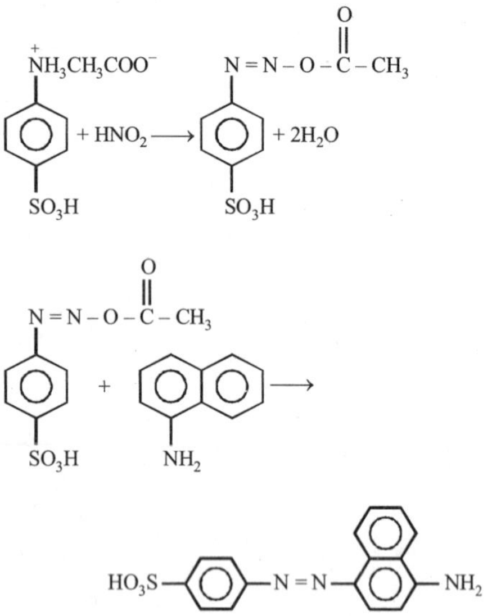

Molar mass of product formed

$= (16 \times 12) + (14 \times 3) + (16 \times 3) + 32 + 13$

$= 327$ g

$\therefore$ Mass of 0.1 mol product = 32.7 g ≈ 33 g

179. (1)

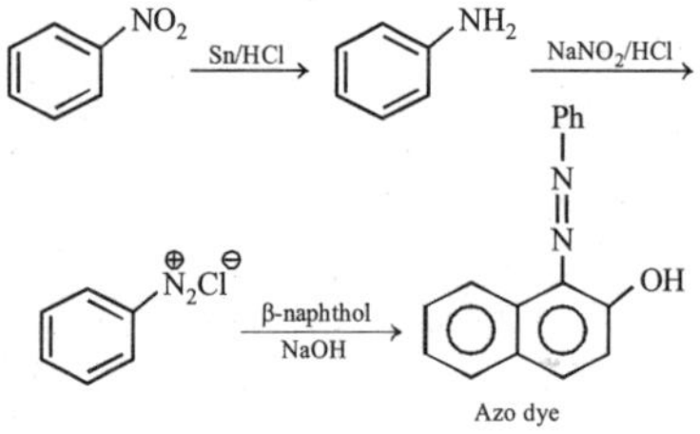

180. (3)

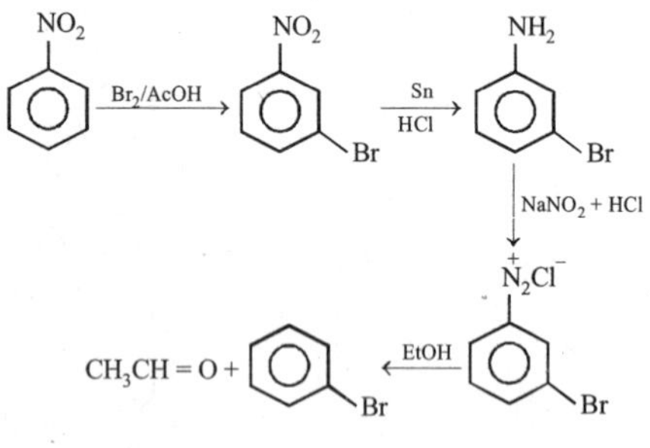

181. (2)

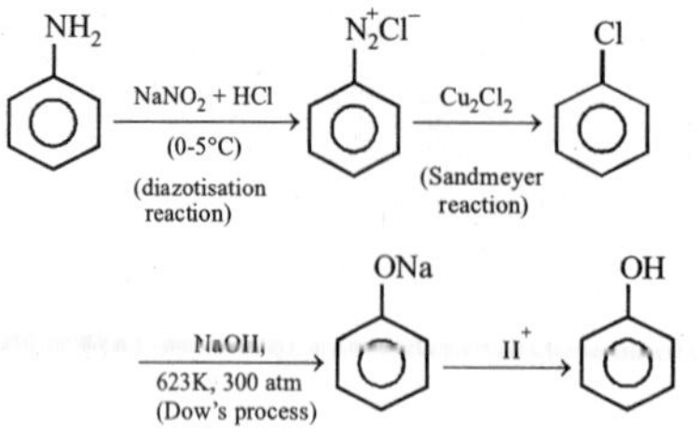

182. (3)

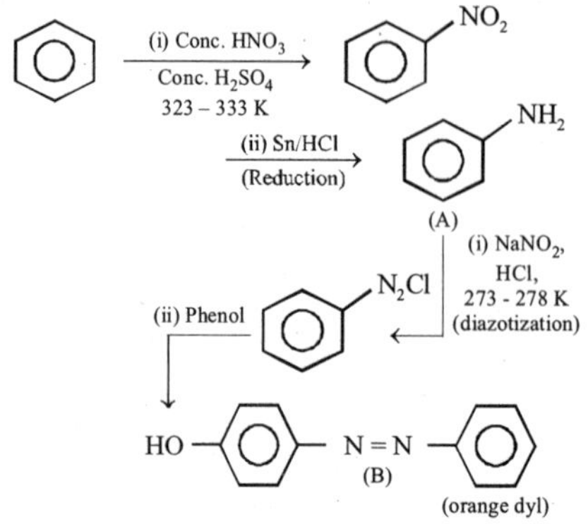

183. (4) 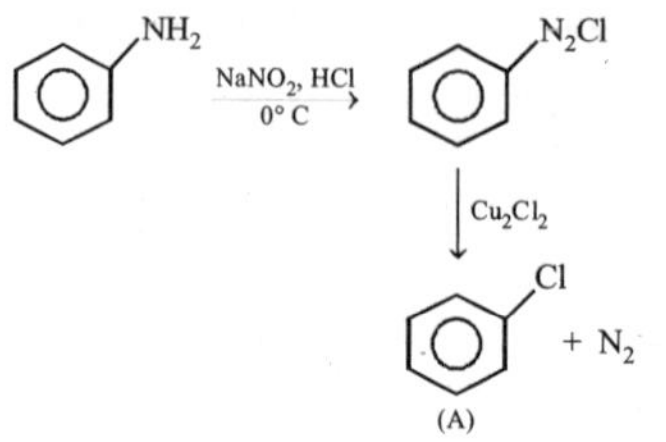

184. (2)

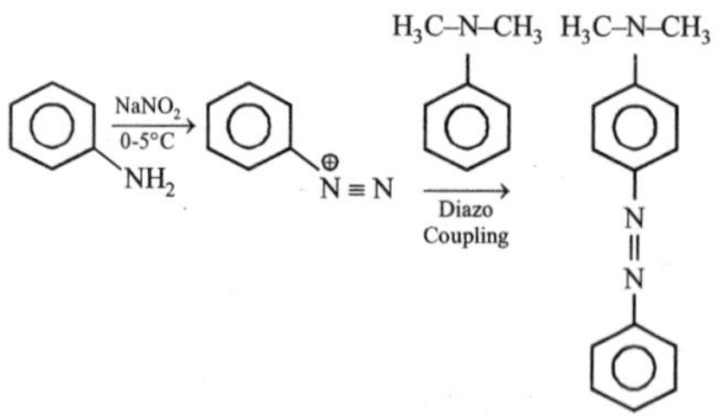

185. (2) $NaNO_2 + HX \rightarrow HNO_2 + NaX$

$$H-O-N=O \xrightarrow[-H_2O]{H^{\oplus}} NO^{\oplus}$$

(Nitrosonium ion)

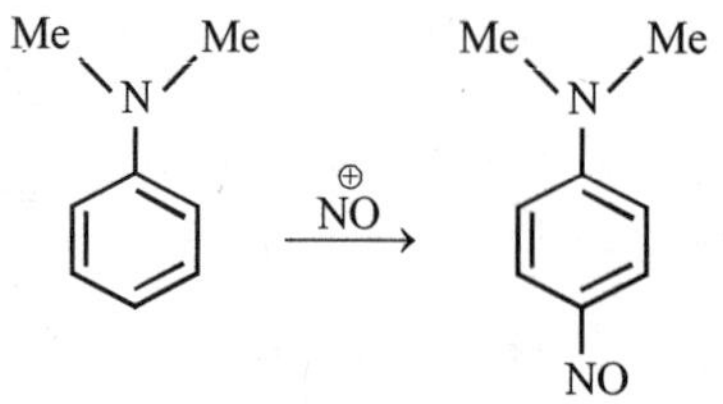

p – Nitroso product

186. (3)

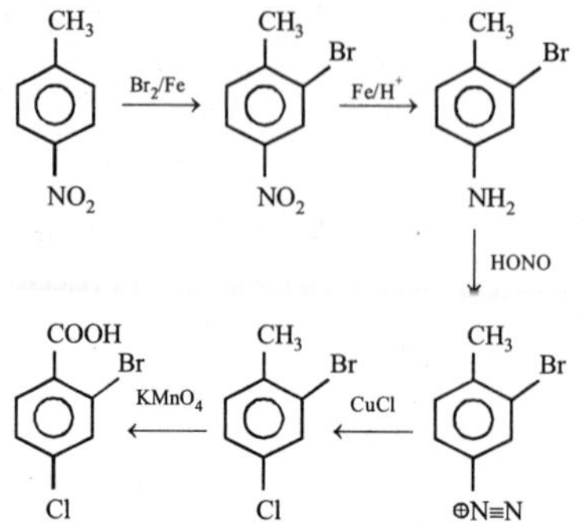

187. (4) 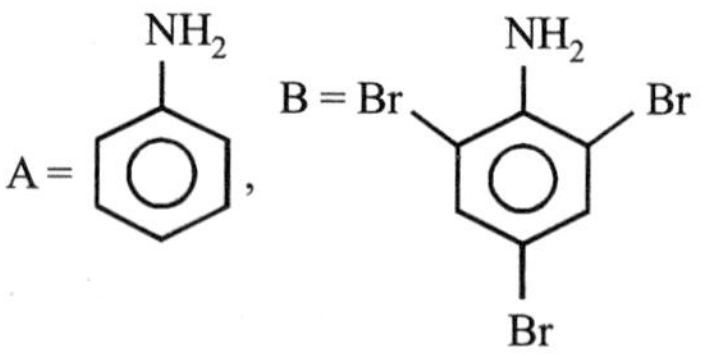

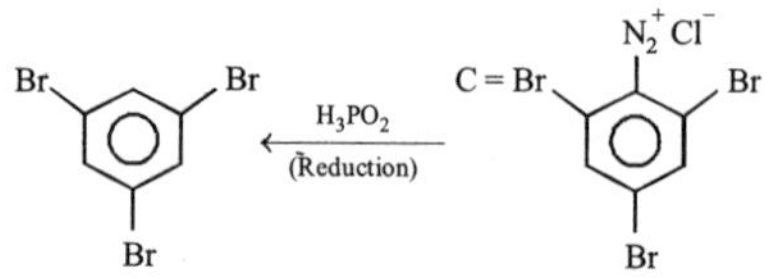

188. (4) A-III, B-I, C-IV, D-II

189. (4)

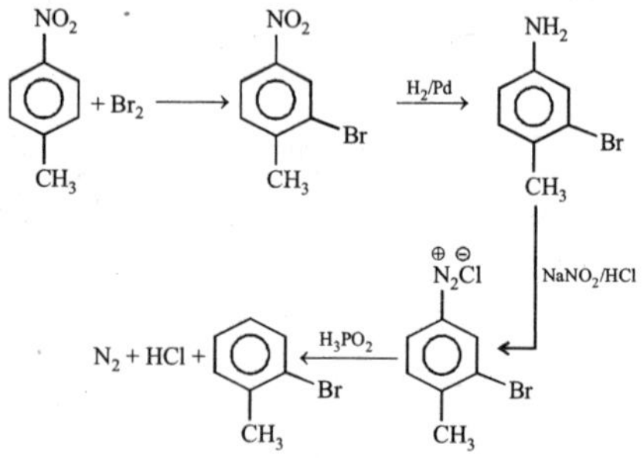

190. (4)

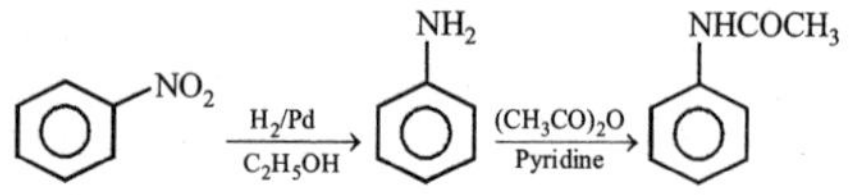

191. (4)

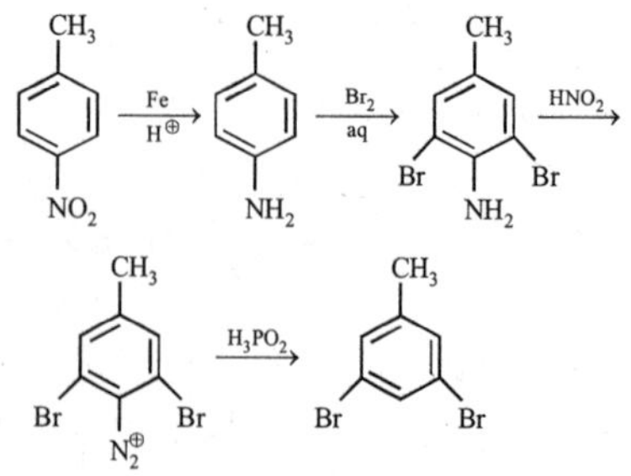

192. (2)

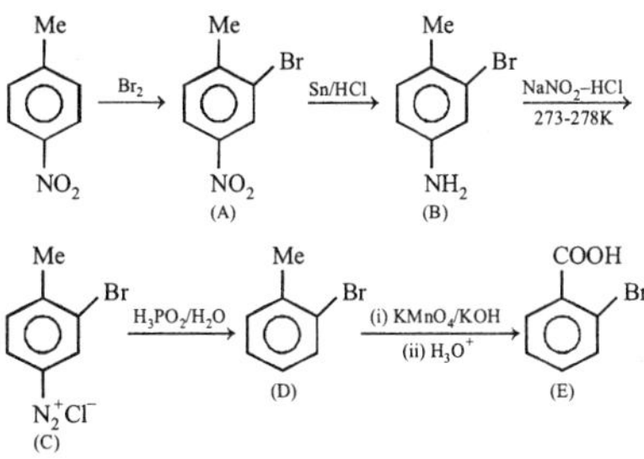

193. (3)

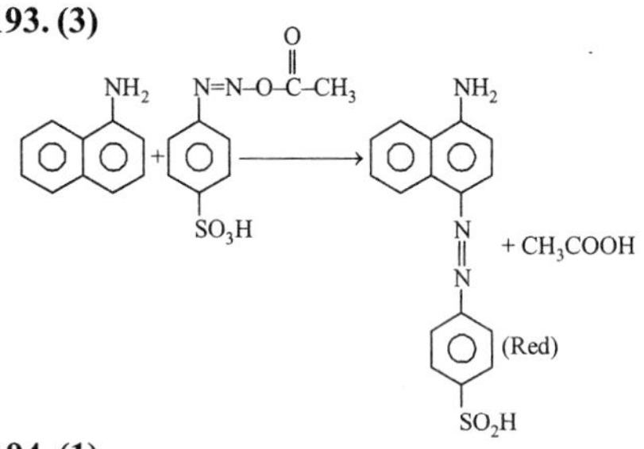

194. (1)

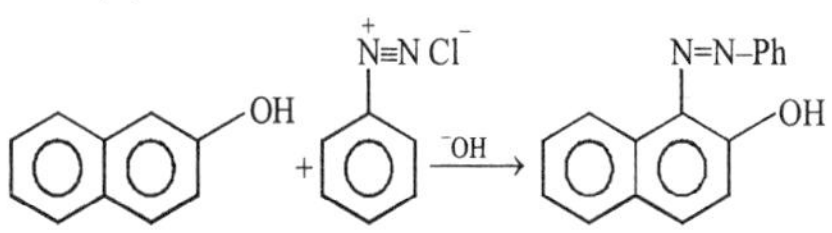

195. (4)

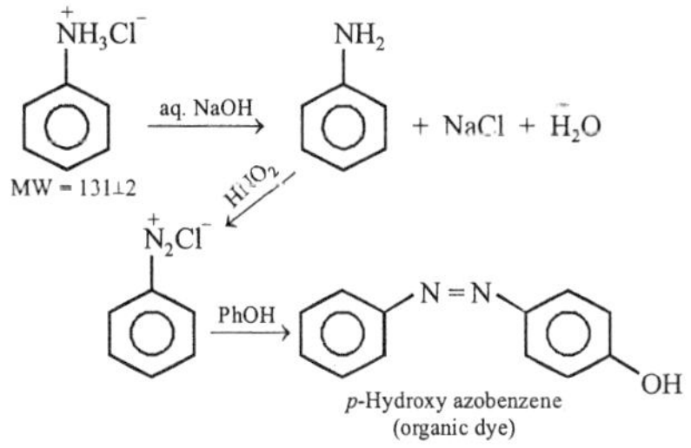

196. (2) Electron donating group increases the electron density and hence stabilises the diazonium salt whereas electron withdrawing group destabilises the diazonium salt.

So the order of stability is

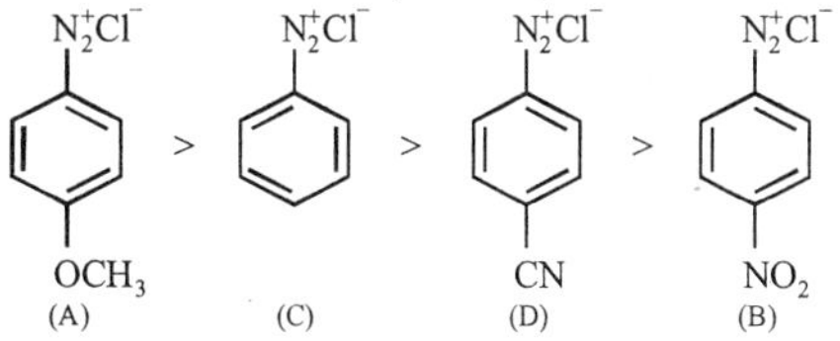

197. (2)

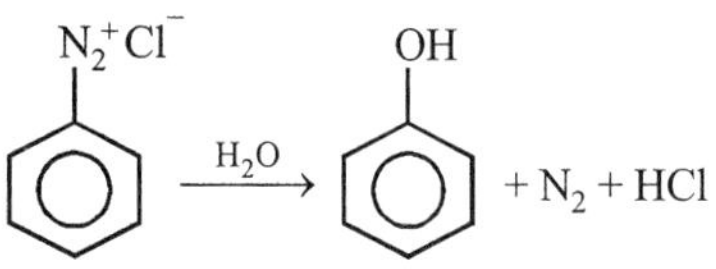

198. (2)

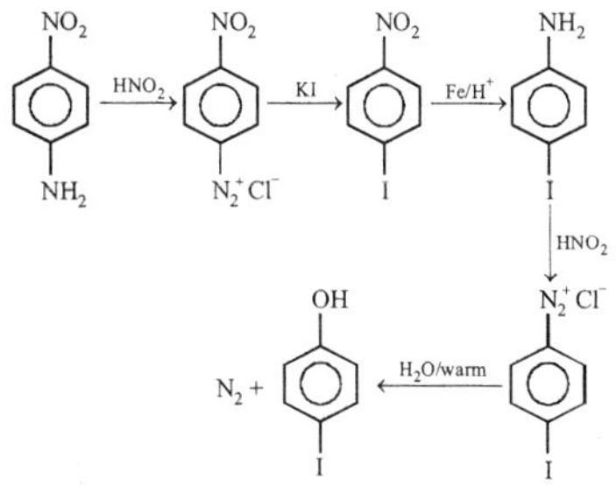

199. (2) Diazonium salts of primary aliphatic amines on heating decompose into carbocation.

$$R-NH_2 \xrightarrow[+HCl]{NaNO_2} R-N_2^+ \rightarrow R^+ \xrightarrow[H_2O]{} \underset{\text{alcohol}}{R-OH}$$

200. (3)

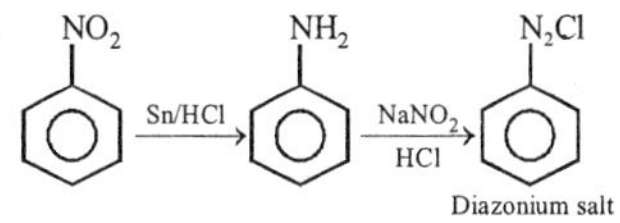

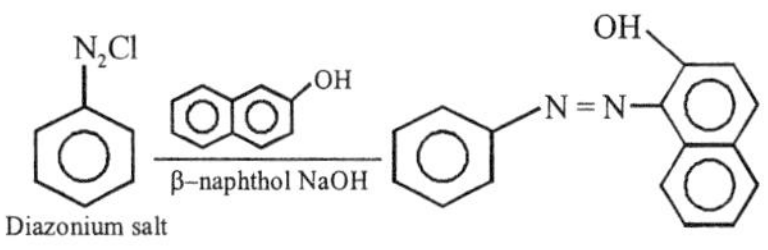

– OH group in the β-napthol increases electron density at ortho, para position.

201. (3)

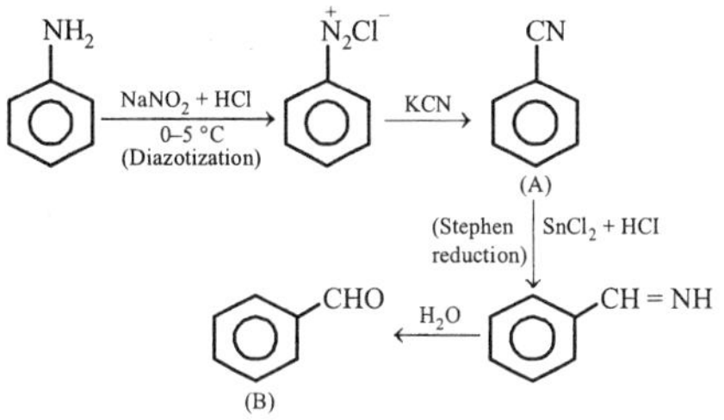

202. (2)

(1)

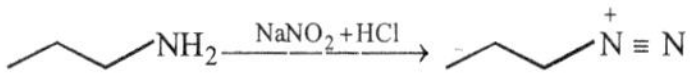

(2)

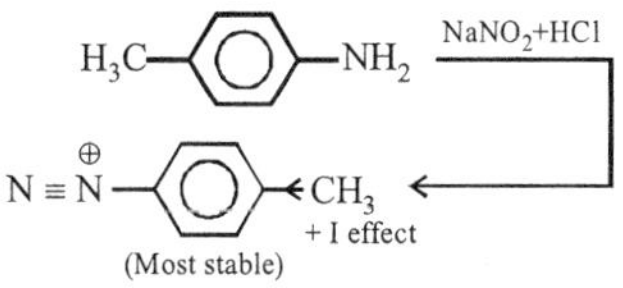

(3) $$CH_3-\underset{CH_3}{\underset{|}{CH}}-NH_2 \xrightarrow{NaNO_2+HCl} H_3C-\underset{CH_3}{\underset{|}{CH}}-\overset{+}{N}\equiv N$$

(4) It is not a primary aromatic amine.

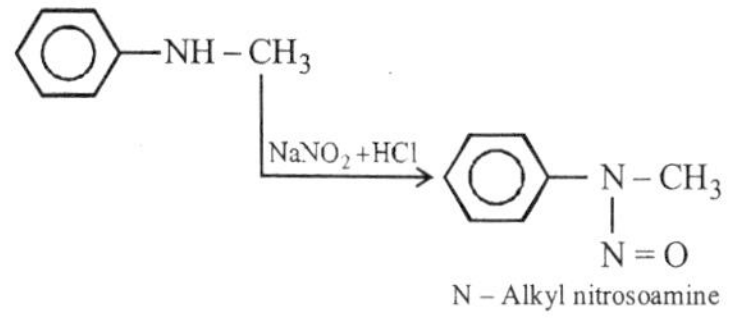

203. (2) – NO_2 group is *m*-directing.

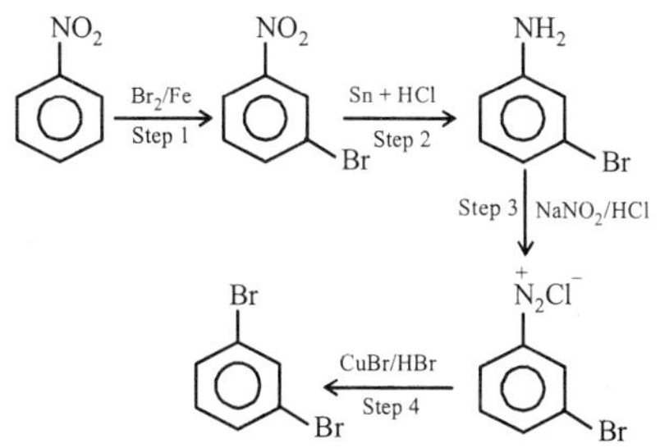

204. (1)

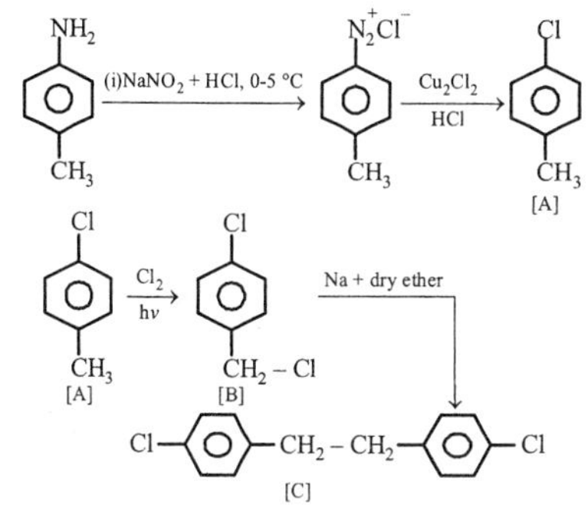

205. (2)

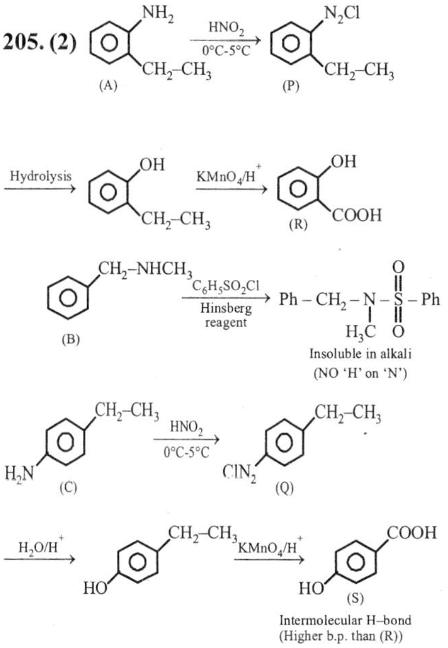

206. (2)

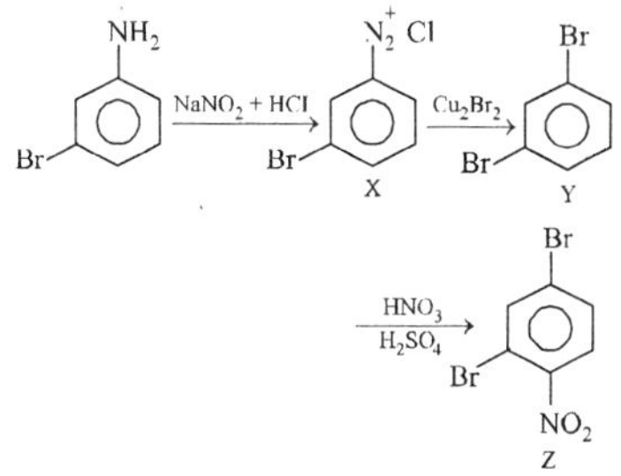

207. (3)

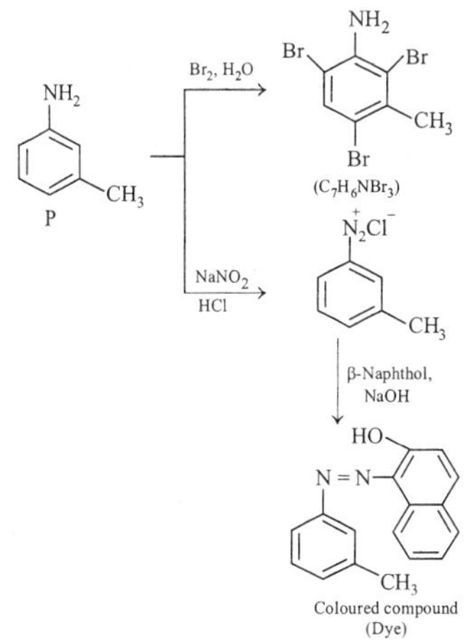

208. (3)

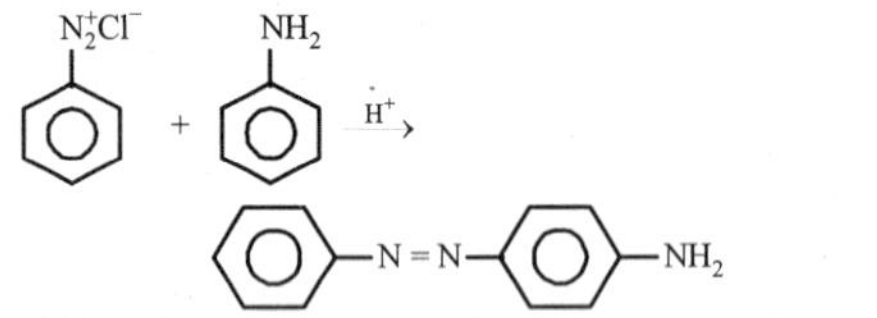

209. (3) In acidic medium, benzene diazonium chloride reacts with aniline (a basic compound) forming a dye (3).

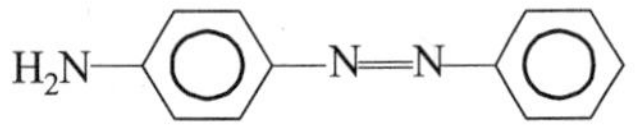

(3)

It will not react with phenol in acidic medium.

210. (3) The substitution will take place at para-position.

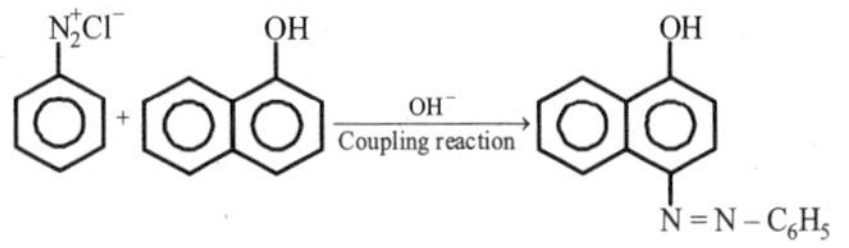

211. (1)

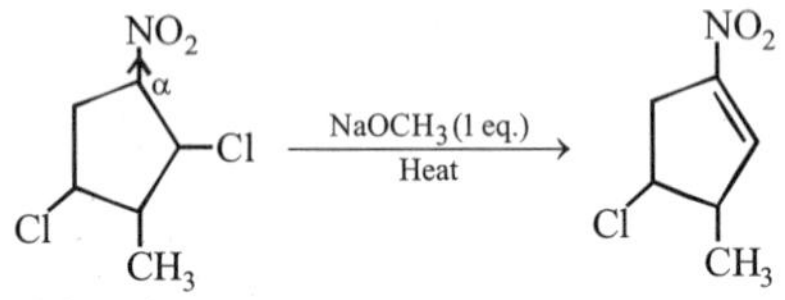

– NO_2 gp. increases acidity of α – H.

212. (1)

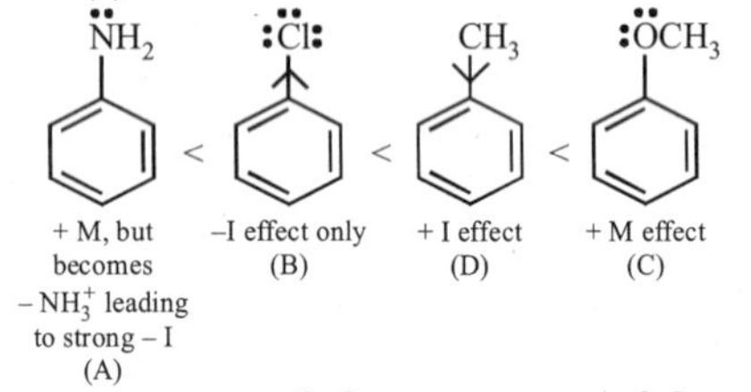

213. (3)

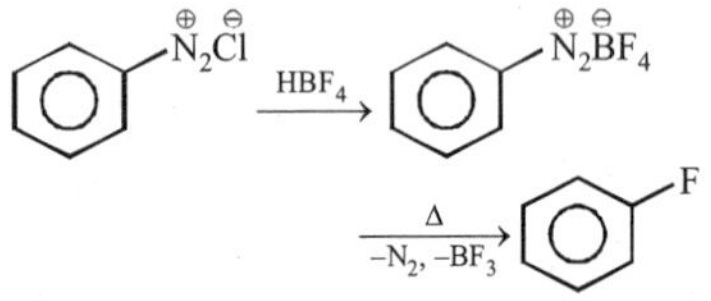

214. (1)

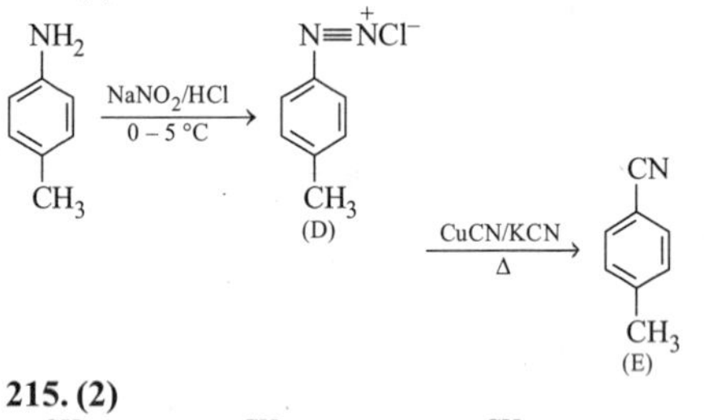

215. (2)

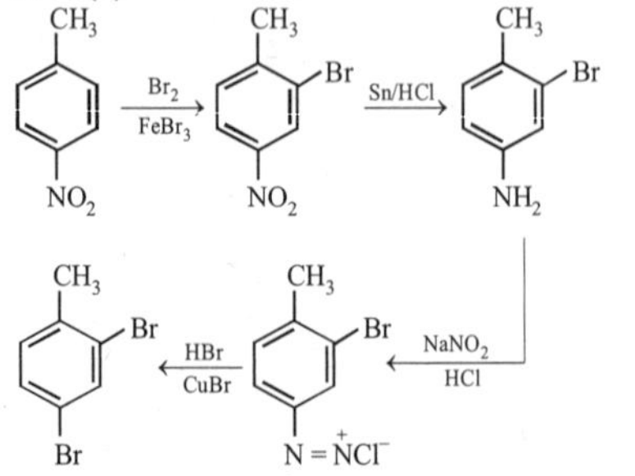

216. (1) On reduction with Zn and HCl, $C_6H_5N_2Cl$ forms aniline as the main product.

217. (3)

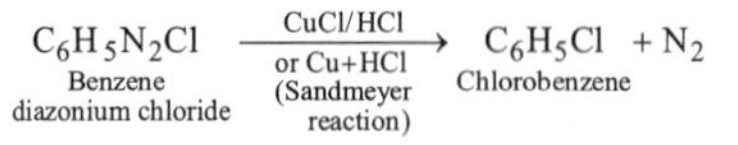

218. (3) Diazonium salts react with copper powder and hydrogen bromide to form bromobenzene. This reaction is known as **Gattermann reaction.**

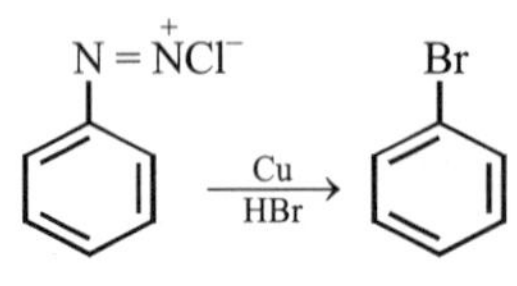

219. (1)

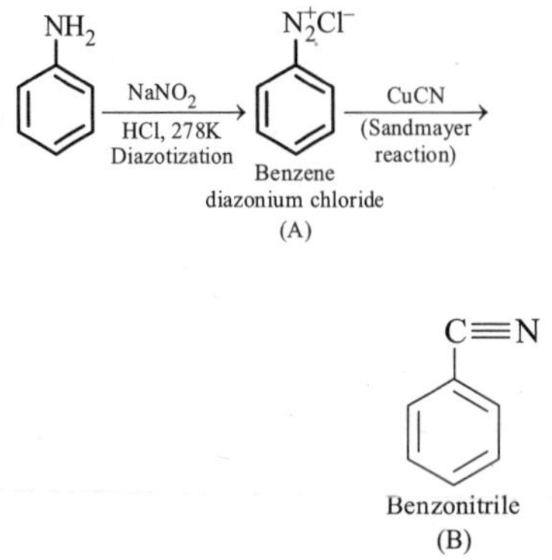

220. (3) $ArNH_2 + NaNO_2 + 2HX \xrightarrow{\text{cold}}$

1° Aromatic amine

$Ar—N\equiv N^+X^- + NaX + 2H_2O$

Arene diazonium salt

The diazonium group can be replaced by fluorine by treating the diazonium salt with fluoroboric acid (HBF_4). The precipitated diazonium fluoroborate is isolated, dried and heated until decomposition occurs to yield the aryl fluoride. This reaction is known as **Balz-Schiemann reaction.**

$$Ar—N_2^+X^- \xrightarrow[(-HX)]{HBF_4} Ar—N_2^+BF_4^- \downarrow$$

$$\xrightarrow{\text{heat}} Ar—F + BF_3 + N_2$$

221. (4)

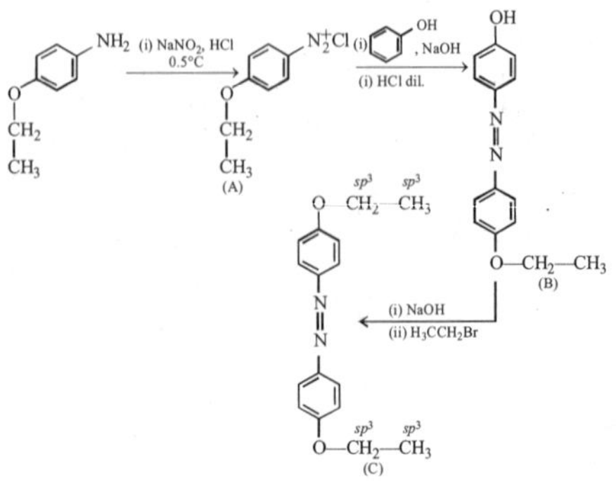

222. (171)

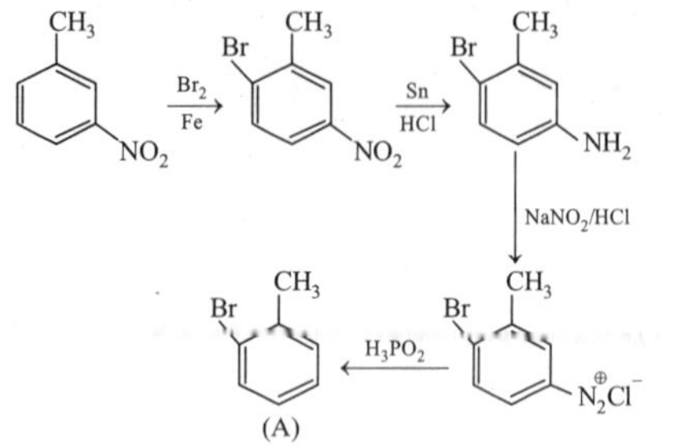

Molar mass of product (C_7H_7Br) (A)

$= 12 \times 7 + 1 \times 7 + 80 = 171$ g/mol

223. (154)

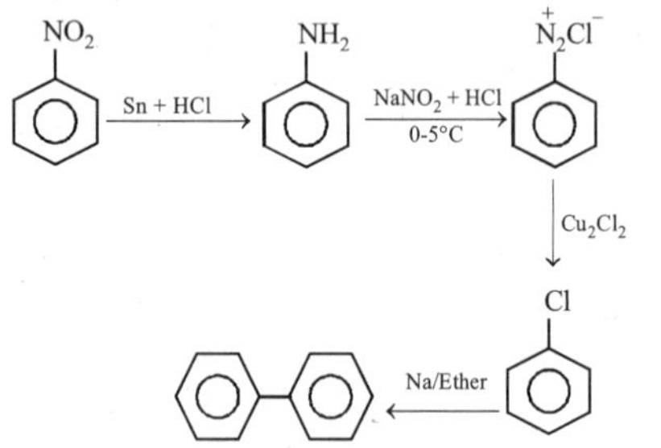

224. (591)

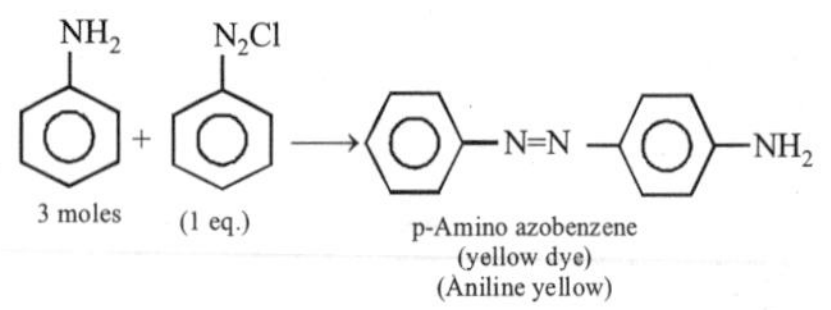

Moles of aniline $= \frac{279}{93} = 3$ moles

As benzenediazonium chloride (1 eq.) is limiting reagent here.

Mole of aniline yellow formed = 3 mole

Molarmass of aniline yellow = 197

Amount of aniline yellow formed = 197 g × 3 = 591 g

225. (20)

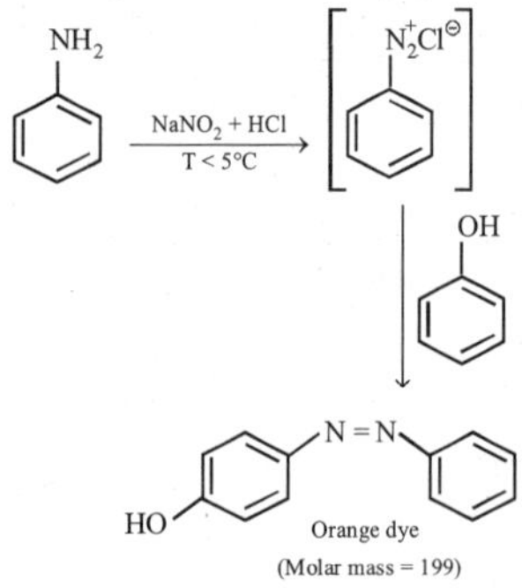

1 mole of aniline give 1 mole of orange dye,

So $(\text{mol})_{\text{aniline}} = (\text{mol})_{\text{orange dye}}$

$$\frac{9.3\text{g}}{93\text{g mol}^{-1}} = \frac{\text{mass of orange dye}}{199\text{g mol}^{-1}}$$

Mass of orange dye = 19.9 g ≃ 20 g.

226. (9) $HCl + NaOH \rightarrow NaCl + H_2O$

mol of HCl = 0.2 mol

$$CH_3CH_2NH_2 \xrightarrow[+HCl]{NaNO_2} CH_3CH_2N_2^+Cl \xrightarrow{H_2O}$$

$$CH_3CH_2OH + N_2 + HCl$$

0.2 mol HCl would be generated by 0.2 mol $C_2H_5NH_2$

X = 0.2 × 45 = 9g.

Part-B (JEE Advanced)

1. **(c)** Higher the stability of the conjugate acid, more is the basic character of the parent amine.

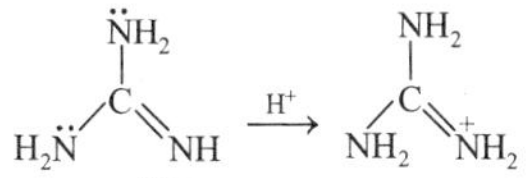

(IV)

The conjugate acid is stabilized by resonance with two different $-NH_2$ groups.

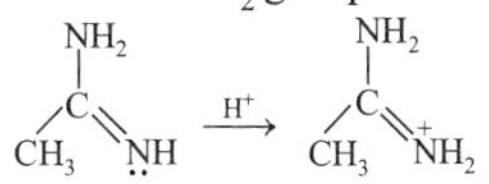

(I)

The conjugate acid is stabilized by resonance with one $-NH_2$ group. Hence, as compared to IV it is less basic.

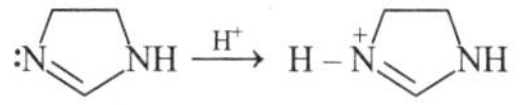

(II)

Lone pair is not involvd in aromaticity. Hence, more available than III.

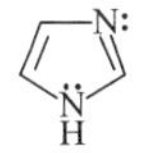

(III)

Lone pair is involved in aromatic sextet. Hence, not available.

Hence, the correct order of basic strength is

IV > I > II > III

2. **(a)**

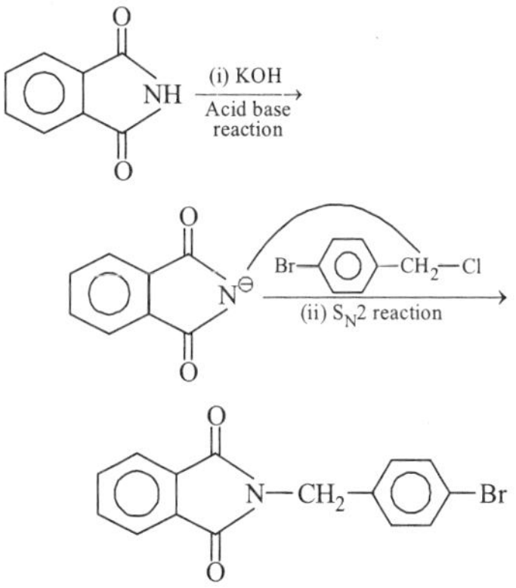

–Br is present on ring, hence less reactive.

3. **(b)** (i) Position (X) is most acidic due to – COOH group.

(ii) $-NH_3^+$ group at position Y is more acidic than at Z because of presence of electron withdrawing – COOH group in close proximity. Hence $-NH_3^+$ group at position Z is least acidic.

4. **(b)** NH is more activating than $-CH_3$ group, whereas C = O group is a deactivating group. Hence, electrophilic substitution at para-position will be governed by the ring having \NH group.

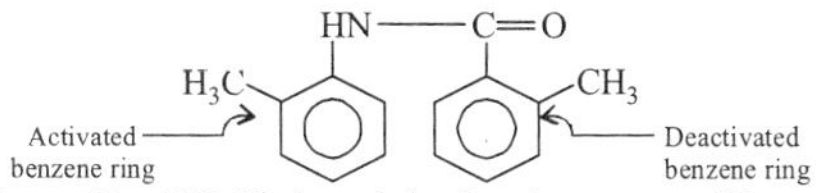

5. **(d)** $POCl_3$ is a dehydrating agent. Hence

$$C_6H_5CONH_2 \xrightarrow{POCl_3} C_6H_5CN + H_2O$$

6. **(a)**

$$\underset{(A)}{C_3H_8O} \xrightarrow{K_2Cr_2O_7/H^+} \underset{(B)}{C_3H_6O} \xrightarrow{\text{amm. } AgNO_3} \text{Silver mirror}$$

$$(B) \xrightarrow{H_2NCONHNH_2.HCl} (C)$$

Reaction of (B) indicates that it is an aldehyde which thus should be C_2H_5CHO or CH_3CH_2CHO, hence (C) should be $CH_3CH_2CH = NNHCONH_2$

7. **(b)**

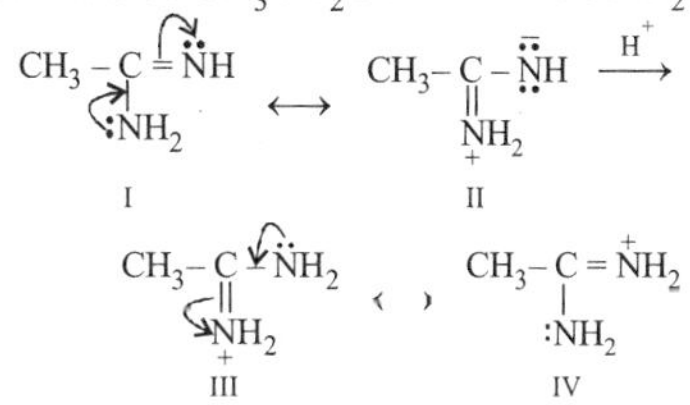

The protonated form of II would be III which is more stable because here the contributing structures (III) and (IV) are equivalent.

In $CH_3 - \ddot{N}H - CH_3$, the availability of electron pair increases due to the +I effect of two CH_3 groups while in $CH_3CH_2NH_2$, +I effect of only one ethyl group is operative. In $CH_3 - CO - NH_2$, the electron availability on nitrogen decreases due to resonance as shown below

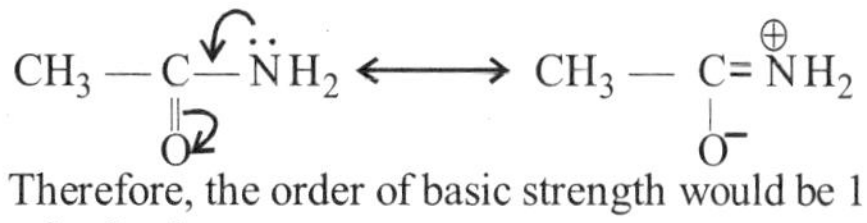

Therefore, the order of basic strength would be 1 > 3 > 2 > 4.

8. **(d)** Aliphatic amines are more basic than aromatic amines because in aliphatic amines electron pair on nitrogen is not involved in resonance.

9. **(b)**

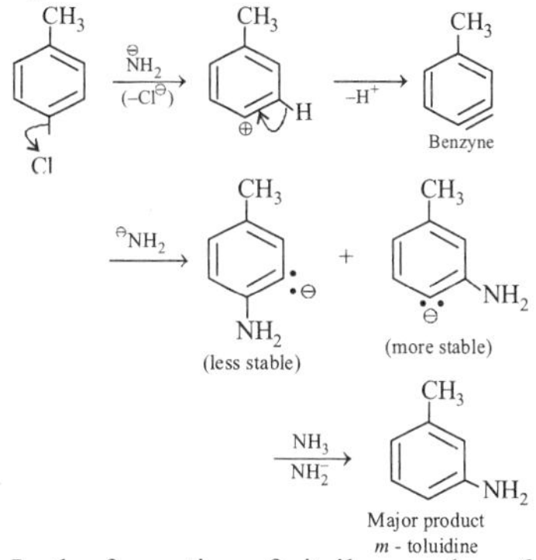

10. **(d)** In the formation of nitrile, number of carbon atoms in parent chain increases by one.

$$CH_3CH_2CH_2Cl + alc. KCN \xrightarrow{100°C} \underset{\text{Propyl chloride}}{CH_3CH_2CH_2CN}$$

11. **(a)** In benzylamine, electron pair on nitrogen is not delocalised due to lack of conjugation; while in all other compounds it is delocalised and hence, lesser available for protonation.

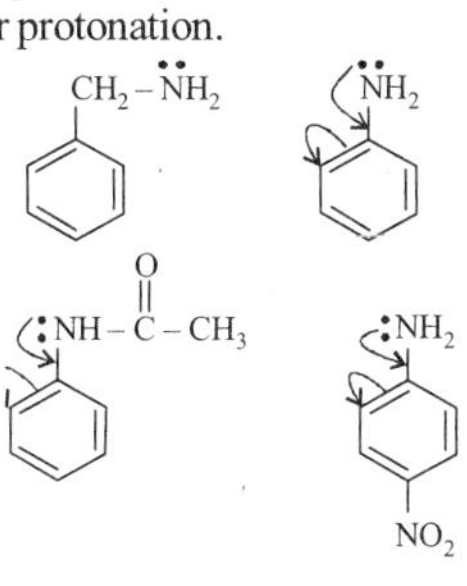

12. **(a)** Toluene has electron-donating methyl group. Hence, reacts fastest while others have either electron withdrawing groups (*i.e.* – COOH or – NO_2 *etc.*) or no substituent.

13. **(b)** Only primary aliphatic and aromatic amines give this test.

$CHX_3 + RNH_2 + 3KOH \rightarrow RNC + 3KX + 3H_2O$

14. **(b)**

$$\underset{\text{Acetamide}}{CH_3CONH_2} \xrightarrow{NaOH + Br_2} \underset{\text{Methylamine}}{CH_3NH_2}$$

This is Hofmann bromomide degradation reaction.

15. **(c)** *Secondary amines* (aliphatic as well as aromatic) *react with nitrous acid to form* **N-nitrosoamines.**

$$(C_2H_5)_2NH + HONO \longrightarrow \underset{\text{N-Nitrosodiethylamine}}{(C_2H_5)_2N-N=O} + H_2O$$

16. **(28)**

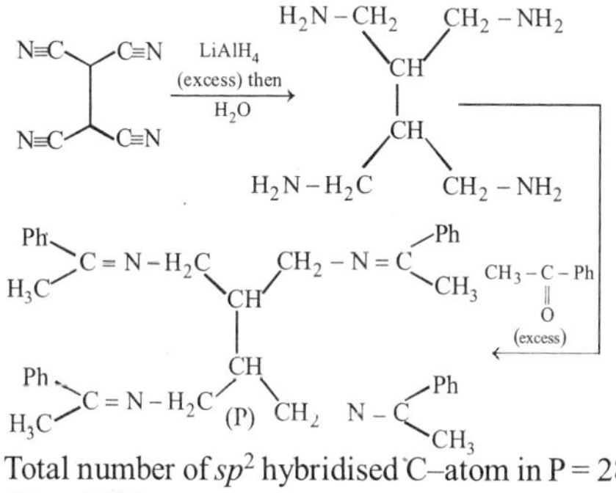

Total number of sp^2 hybridised C–atom in P = 28

17. **(495)**

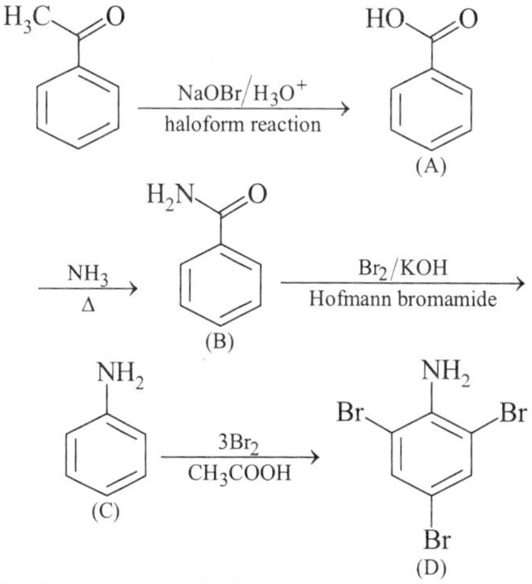

Molecular mass of (D)

$= (6 \times 12) + (4 \times 1) + (1 \times 14) + (3 \times 80)$

$= 330$ g/mol

Moles of (D) formed

$= 10 \times 0.6 \times 0.5 \times 0.5 \times 1 = 1.5$

Mass of (D) formed $= 1.5 \times 330 = 495$ g

18. **Dipolar ion.** (Salt like)

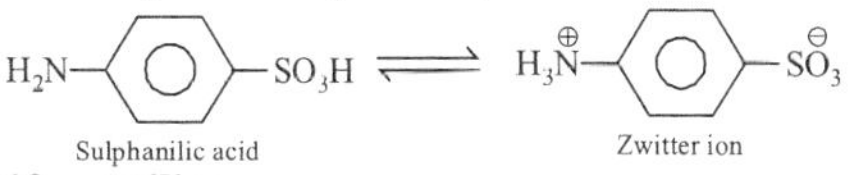

19. **aniline.**

20. **(a, c)**

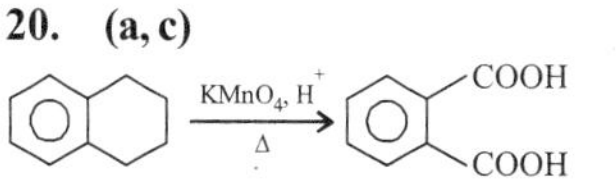

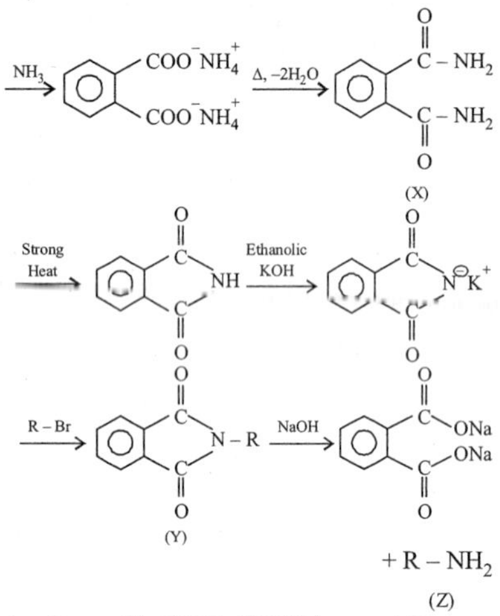

Y on heating with $CHCl_3$/KOH does not form an isocyanide as it is not an amine.

Z reacts with Hinsberg reagent as it is an amine (1° amine). Z is an aliphatic primary amine, not aromatic. Thus, option (a) and (c) are correct.

21. (b, c, d)

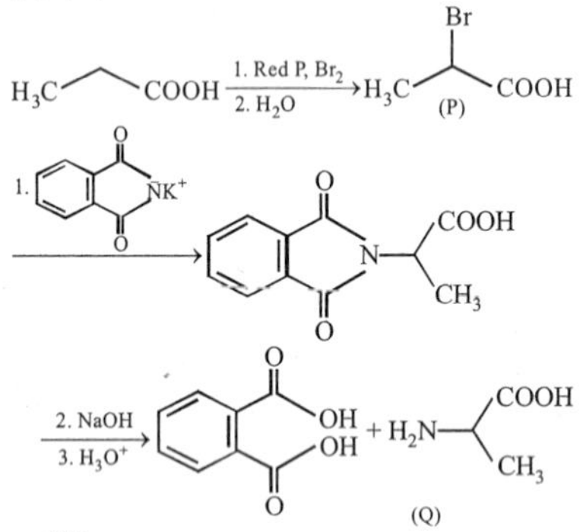

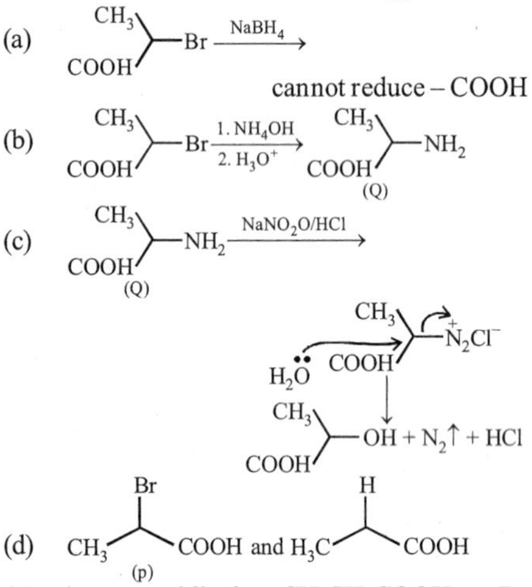

(P) is more acidic than CH_3CH_2COOH as –Br has (–I) effect.

22. (c, d)

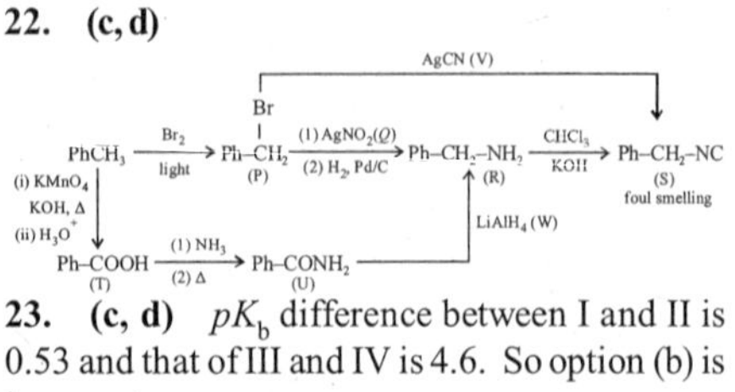

23. (c, d) pK_b difference between I and II is 0.53 and that of III and IV is 4.6. So option (b) is incorrect

Correct statement are (c) and (d)

The most basic compound in the given option is (II) and least basic compound is (III), so option (a) is also incorrect.

In 2,4,6-trinitroaniline (III) due to strong –R effect of the three $–NO_2$ groups, the $\ell.p.$ of electrons on $–\ddot{N}H_2$ is more involved with benzene ring hence it has least basic strength.

Whereas (IV) N,N-dimethyl – 2,4,6-trinitroaniline, due to steric inhibition to resonance (SIR) effect; the lone pair of nitrogen is not in the plane of benzene, hence makes it ($\ell.p.$) more free to protonate

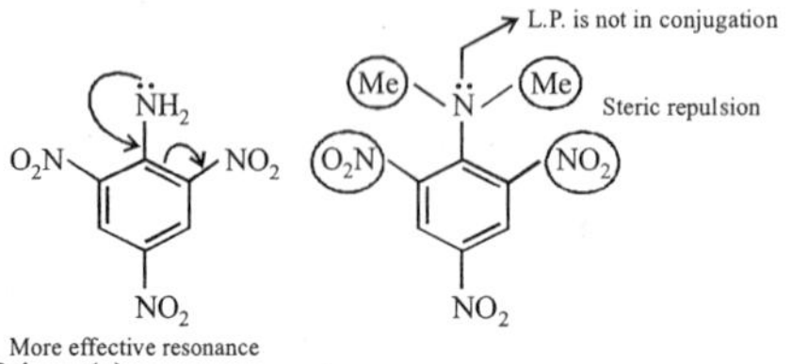

More effective resonance

24. (a)

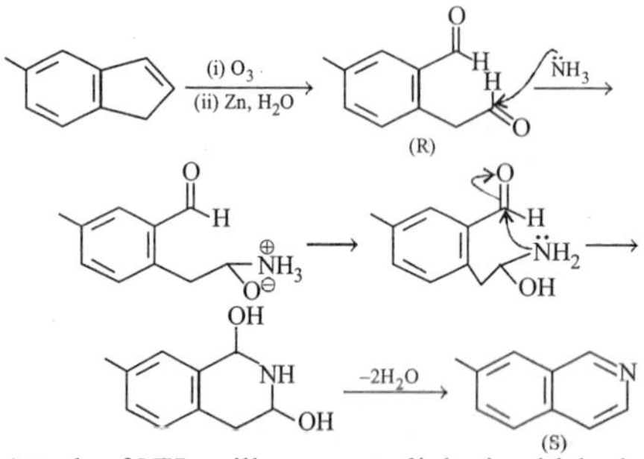

Attack of NH_3 will occur at aliphatic aldehyde group because the aromatic –CHO is in conjugation with benzene ring.

25. (a) $—\ddot{N}H_2$ group is acetylated by acetic anhydride in methylene chloride (solvent). Note that —$CONH_2$ group does not undergo acetylation because here lone pair of electrons is delocalised.

$$CH_2—\ddot{N}H_2 + (CH_3CO)_2O \xrightarrow{CH_2Cl_2} —CH_2—\underset{|}{N}H(COCH_3) + CH_3COOH$$

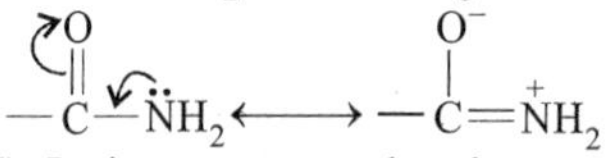

26. (a, b, d) In ice, water molecules are excessively H-bonded giving a cage-like structure which is lighter than water. Primary amines are more basic than tertiary amines, because the protonated 1° amines are extensively H-bonded and hence more stable than the corresponding protonated 3° amines.

$$R—\ddot{N}H_2 \xrightarrow{H^+} R—\overset{+}{N}H_3;$$

$$R_3\ddot{N} \xrightarrow{H^+} R_3\overset{+}{N}H$$

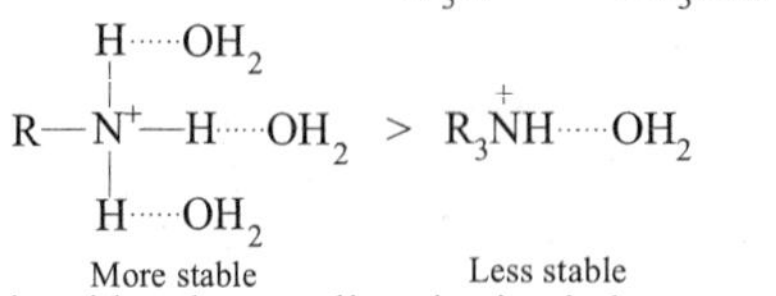

More stable Less stable

Acetic acid undergoes dimerisation in benzene.

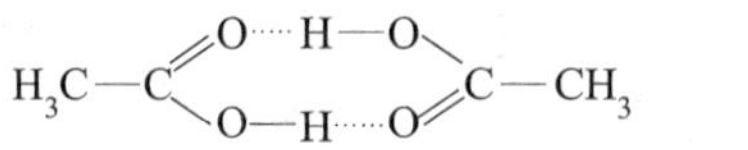

27. (a,b,c) Lower amines like NH_3, CH_3NH_2 and $(CH_3)_2NH$ break diborane molecule unsymmetrically, while larger amines like $(CH_3)_3N$, $(C_2H_5)_3N$ break diborane in symmetrical manner.

$$B_2H_6 + 2NH_3 \longrightarrow [H_2B(NH_3)_2]^+[BH_4]^-$$

$$B_2H_6 + 2(CH_3)_3N \longrightarrow 2H_3B \leftarrow N(CH_3)_3$$

28. (b,d) Only primary amines give carbylamine test.

Hence 2,4-dimethylaniline and P-methyl–benzylamine both give this test.

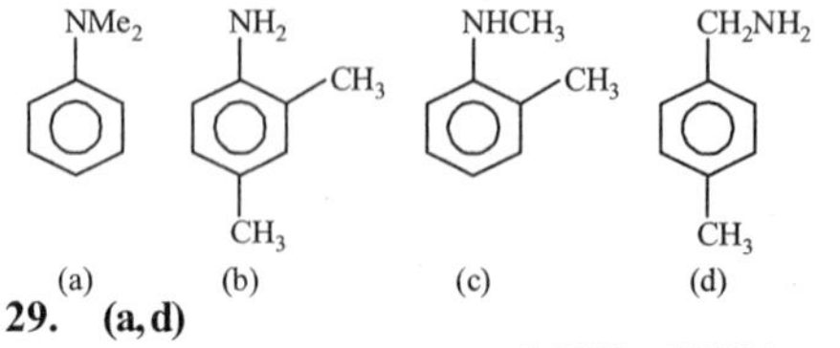

29. (a, d)

$$C_6H_5NH_2 + O{=}C(CH_3)_2 \xrightarrow[-H_2O]{} C_6H_5N = C(CH_3)_2$$

$$C_6H_5NHNH_2 + O = C(CH_3)_2 \xrightarrow[-H_2O]{} C_6H_5NHN = C(CH_3)_2$$

30. (c) Anilinium hydrochloride has ionisable chlorine, whereas p-chloro-aniline has non ionizable chlorine. Thus, anilinium hydrochloride gives white precipitate of AgCl with $AgNO_3$.

$$C_6H_5NH_3^+Cl^- + AgNO_3 \rightarrow C_6H_5NH_3^+NO_3^- + AgCl\downarrow$$

In chloroaniline, –Cl is directly attached to benzene ring, hence it is non-reactive.

31. (c) In I[st] structure N has complete octet, whereas in IInd structure N has 10 electrons in its valence shell, which is not possible.

32. (a, c) This is an example of Hofmann degradation of amides.

$$R-\overset{O}{\overset{||}{C}}-NH_2 + Br_2 \rightarrow R-\overset{O}{\overset{||}{C}}-NHBr + HBr$$

$$R-\overset{O}{\overset{||}{C}}.NHBr + KOH \rightarrow R-\overset{O}{\overset{||}{C}}.N^-Br + H_2O + KBr$$

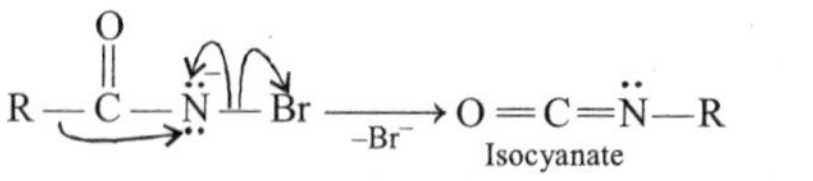

$$\xrightarrow{H_2O} R—NH_2$$

33. (b)

(1) 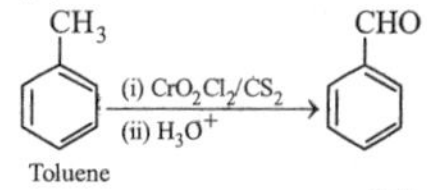

(S) Cannizzaro reaction

(2) 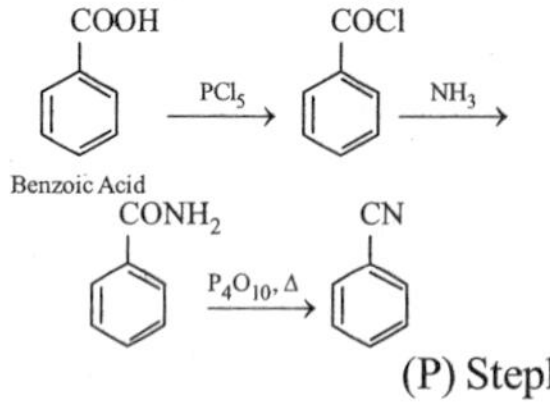

(P) Stephen reaction

(3) 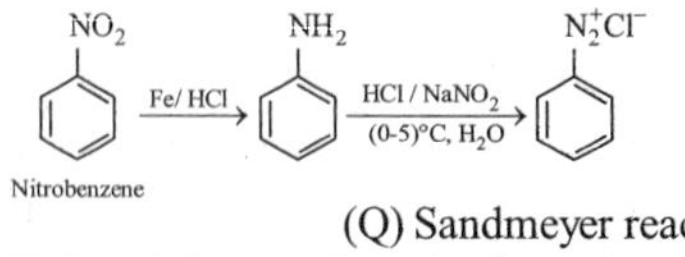

(Q) Sandmeyer reaction

∵ H_2O is being used as a solvent here and not as a reagent.

(4) 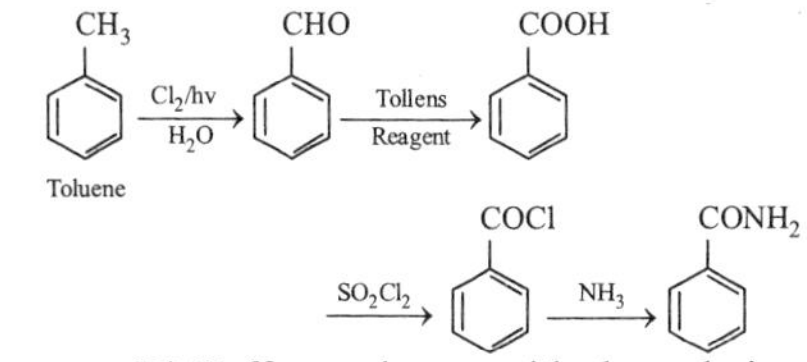

(R) Hoffmann bromamide degradation reaction

(5) 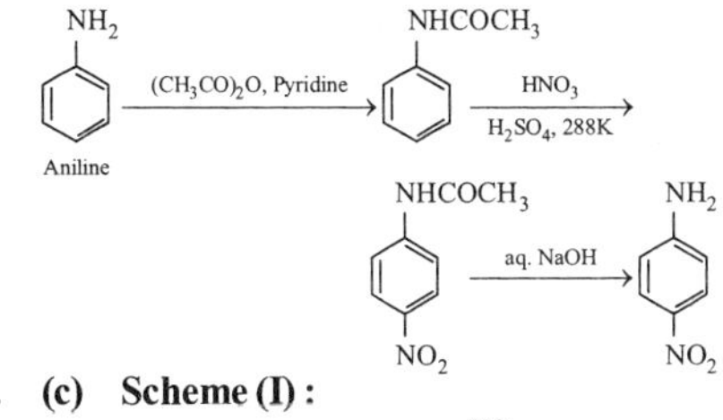

34. (c) Scheme (I) :

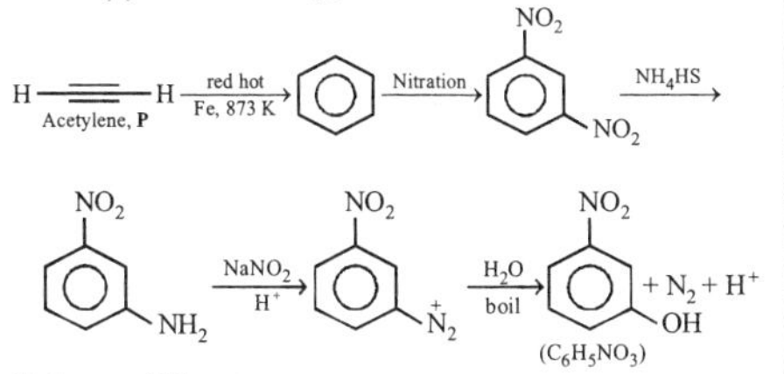

Scheme (II) :

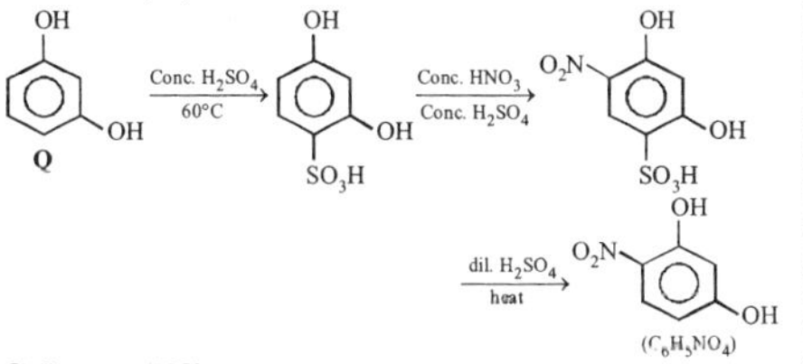

Scheme (III) :

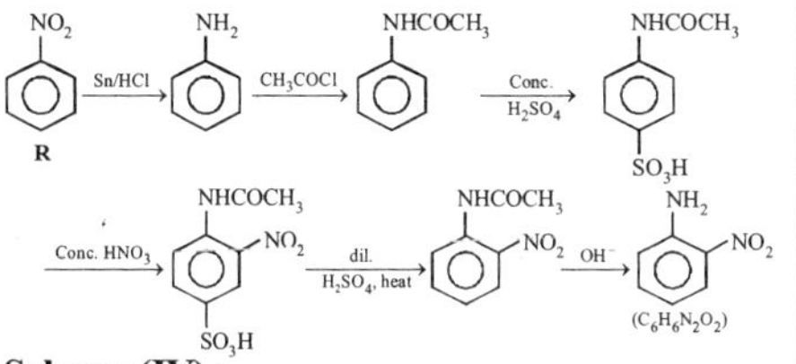

Scheme (IV) :

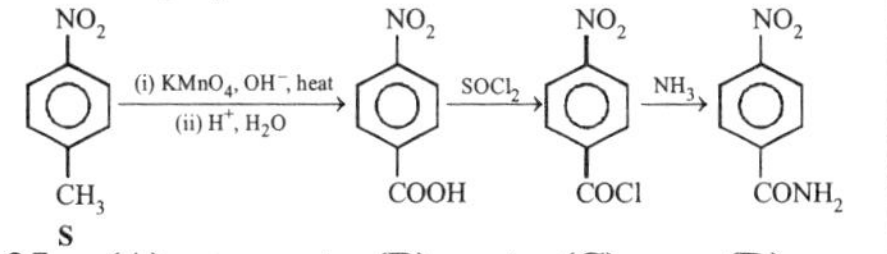

35. (A)– p, q, s, t; (B) – s, t; (C) – p; (D) – r

(A) $CH_3CH_2CH_2CN \xrightarrow[\text{DIBAL-H}]{\text{Pd-C/H}_2\text{ or}} CH_3CH_2CH_2CH_2NH_2$

$CH_3CH_2CH_2CN \xrightarrow[\text{(Stephen reduction)}]{SnCl_2/HCl} CH_3CH_2CH_2CHO$

$CH_3CH_2CH_2CN \xrightarrow[H_2O]{OH^-} CH_3CH_2CH_2COO^-$

(B) $CH_3CH_2OCOCH_3 \xrightarrow{\text{DIBAL-H}} CH_3CH_2OH$

$CH_3CH_2OCOCH_3 \xrightarrow[H_2O]{OH^-} CH_3CH_2OH + CH_3COO^-$

(C) $CH_3CH = CHCH_2OH \xrightarrow{Pd-C/H_2} CH_3CH_2CH_2CH_2OH$

(D) $CH_3CH_2CH_2CH_2NH_2 \xrightarrow[KOH]{CHCl_3} CH_3CH_2CH_2CH_2NC$

Isonitrile (foul smelling)

36. (a)

37. (b)

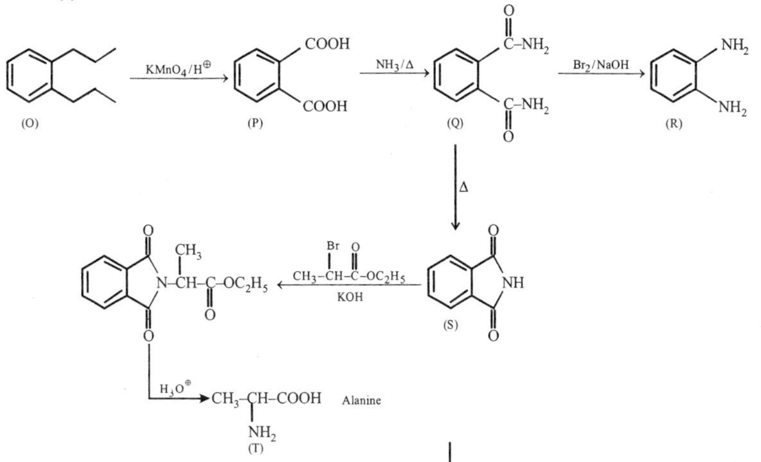

38. (d) The reagent used in Hoffmann bromamide reaction is alkaline halogen (NaOH or KOH + X_2).

39. (d) Conversion of (iii) to (iv) involving rearrangement is the slowest step. Species (iii) is electron deficient (N has only 6 electrons), hence it has a tendency to get its octet completed by migration of alkyl group or aryl (Rearrangement).

40. (b) Since the reaction is intramolecular, no cross product will be formed.

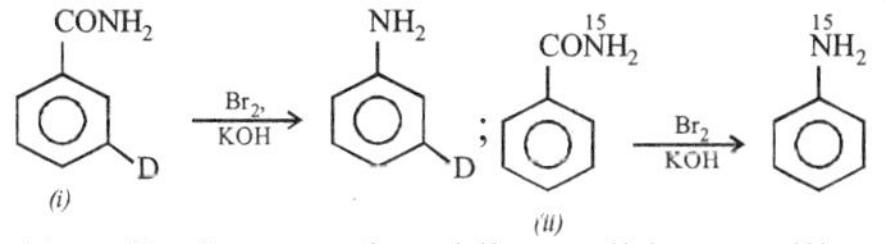

41. (d) In strongly acidic conditions, aniline becomes protonated with the result lone pair of electrons is not available to produce +E and +M effects. Thus, here aniline becomes less reactive towards electrophilic substitution. Thus statement is false. Statement-2 is true.

42. (i)

$$(CH_3)_2CH-{}^*CH(NH_2)-CH_3 \xrightarrow[HCl]{aq.\,NaNO_2} (CH_3)_2C(OH)-CH_2CH_3 + NaCl + N_2 + H_2O$$

[X], Optically active; [Y], 3° Alcohol

(ii) [Y], a 3° alcohol is optically inactive.

(iii) Formation of [Y] from [X].

$$(CH_3)_2CH-CH(NH_2)-CH_3 \xrightarrow[HCl]{NaNO_2}$$

[X]

$$(CH_3)_2CH-CH(N_2^+)-CH_3 \rightarrow (CH_3)_2C(H)-\overset{+}{C}H-CH_3$$

2° Carbocation

$$\xrightarrow{H^-\ \text{Shift}} (CH_3)_2\overset{+}{C}-CH_2CH_3 \xrightarrow{H_2O} (CH_3)_2C(OH)-CH_2CH_3$$

3° carbocation; [Y]

43. (*i*) (a) $-\ddot{N}=O$ group is electron releasing, hence *o*-, *p*-directing

(b) $-NO_2$ group is electron withdrawing, hence *m*-directing

(*ii*) $-NO_2$ group is electron withdrawing, hence *m*-directing, whereas $-CH_2NO_2$ is not.

(a) Given compound is an aryl fluoride having electron-withdrawing $-NO_2$ group at *para* position of fluoride atom which activates the fluoride due to – M and – E effects for nucleophilic substitution (S_NAr), hence reaction with NaOH will liberate F^- as NaF.

(b) The given compound is an aryl fluoride having $-CH_2NO_2$ group in the *meta* position which is not capable of activating aryl fluoride (absence of –M and –E effects because NO_2 group is not conjugated to benzene ring) for nucleophilic substitution, hence aq. NaOH will not displace fluorine here, *i.e.* no F^- will be formed.

(*iii*) The nitro group in nitrobenzene strongly deactivates the benzene ring due to – I and – M effects. This decreases the reactivity of benzene ring towards Friedel - Crafts alkylation.

(*iv*) *o*-Nitrophenol shows intramolecular H-bonding and exists as a single molecule, while *p*-nitrophenol shows intermolecular H- bonding and thus its several molecules associate with each other. The *ortho* isomer goes with the steam due to its low boiling point.

44. 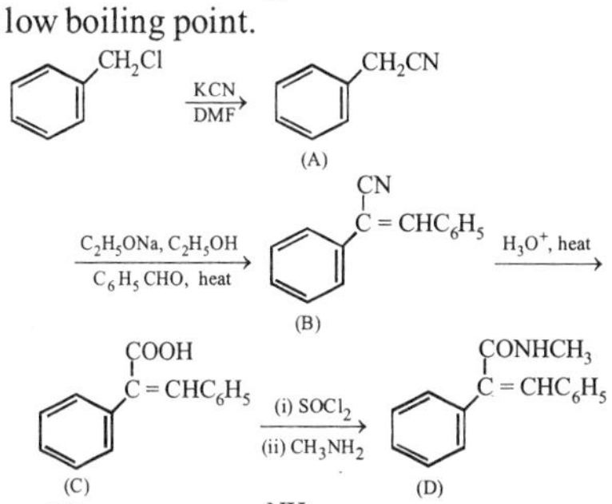

45. 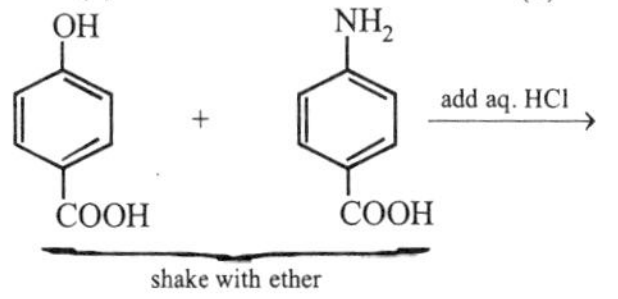

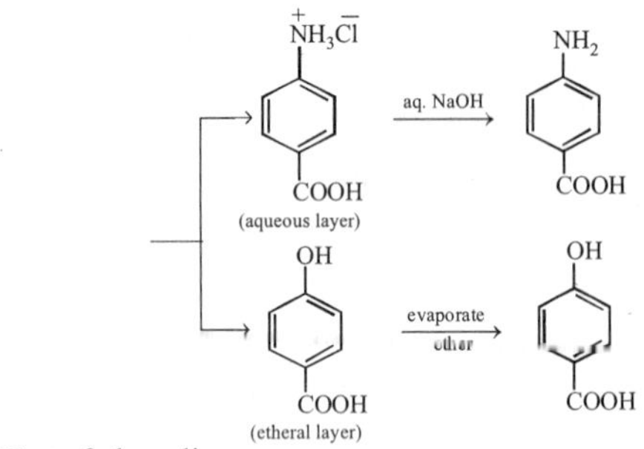

Test of phenolic group :

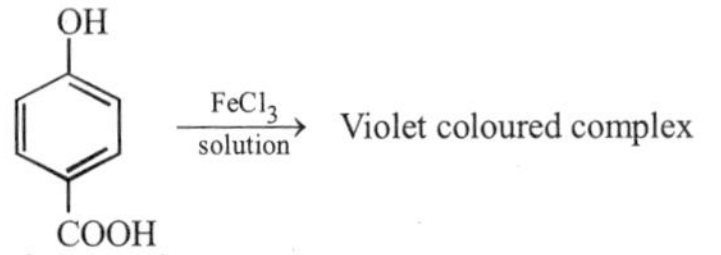

Test of 1° amino group :

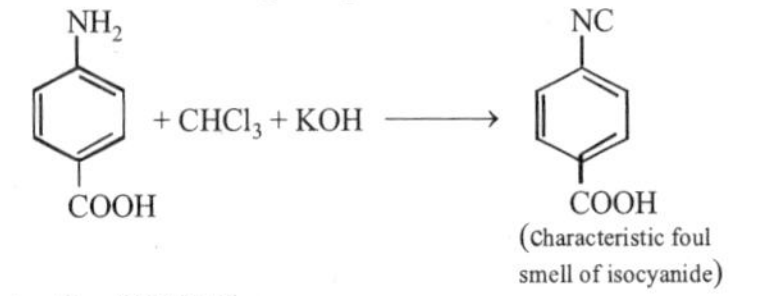

Test of – COOH group :

$$-COOH \xrightarrow{NaHCO_3} CO_2 \uparrow$$
With effervescences

$\xrightarrow{\text{lime water}}$ Milky solution

46.

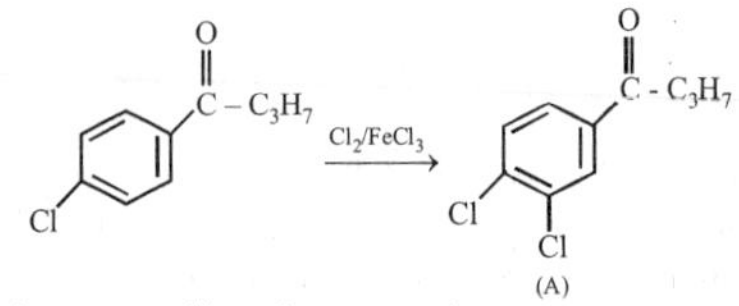

(–COR is a meta directing group)

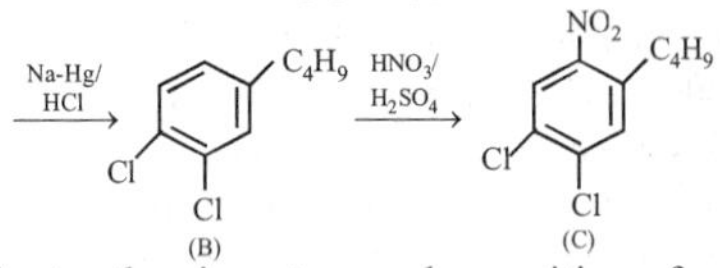

(–R activates the ring at *o*- and *p*-positions for electrophilic aromatic substitution)

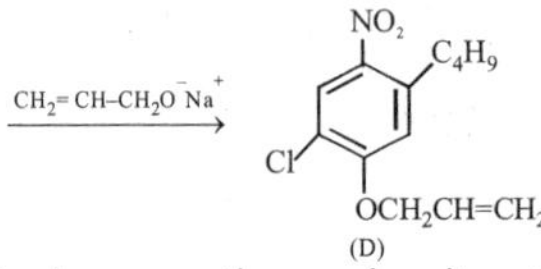

(–NO_2 decrease the e^- density at *o*- and *p*-positions for $Nu^\ominus$ aromatic substitution)

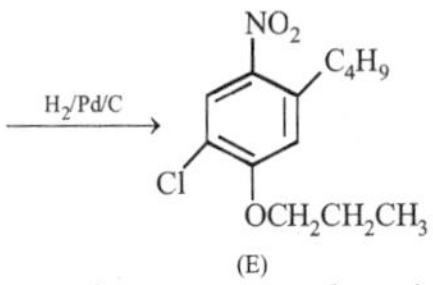

47. The reaction proceeds via benzyne formation

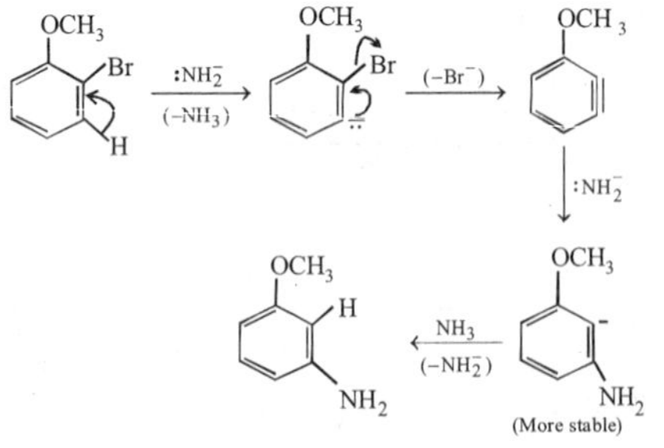

48. Summary of the given facts.

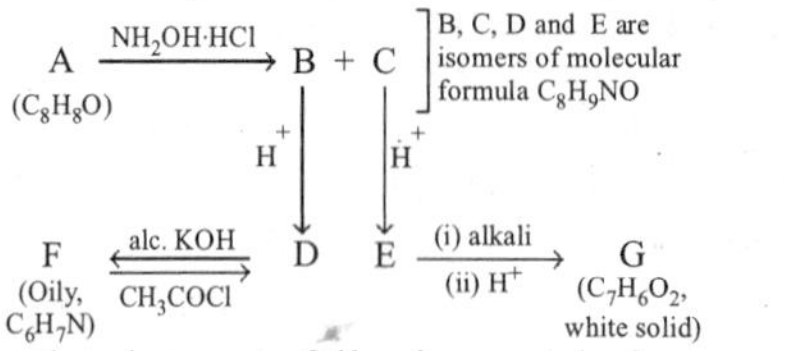

From the above set, following conclusions can be drawn.

(i) Since the oily compound *F* (C_6H_7N) reacts with acetyl chloride, it must have - NH_2 or > NH group. Thus, (*F*) can be written as $C_6H_5NH_2$ and hence, *D* is $C_6H_5NHCOCH_3$.

$$\underset{(F)}{C_6H_5NH_2} \underset{\text{alc. KOH}}{\overset{CH_3COCl}{\rightleftharpoons}} \underset{(D)}{C_6H_5NHCOCH_3}$$

(ii) Compound (*E*) on treatment with alkali followed by acidification gives a white solid compound (*G*), $C_7H_6O_2$. Thus, (*G*) seems to be an acid, hence it is C_6H_5COOH.

(iii) Since (*D*) and (*E*) are isomers of the formula C_8H_9NO, and give $C_6H_5NH_2$ and C_6H_5COOH respectively, both should be amides having different alkyl or aryl group. Thus, (*D*) should be $C_6H_5NHCOCH_3$, and (*E*) must be $CH_3NHCOC_6H_5$.

(iv) Since compounds (*D*) and (*E*) are formed by the rearrangement of compounds (*B*) and (*C*) respectively. Compounds (*B*) and (*C*) should be oximes > C = NOH (recall that oximes rearrange to amides - **Beckmann rearrangement**). Further oximes having different alkyl (aryl) groups show geometrical isomerism (*syn* and *anti*), compounds (*B*) and (*C*) must have following structures.

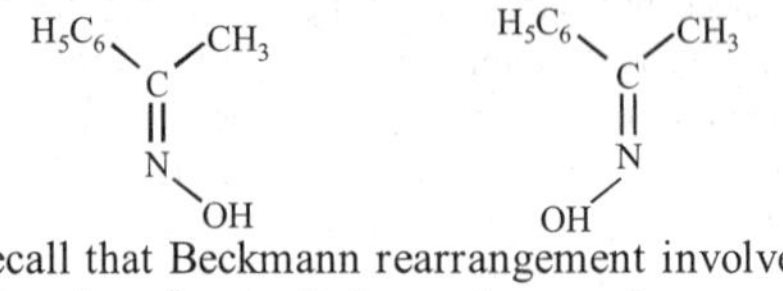

Recall that Beckmann rearrangement involves migration of *anti* -alkyl or aryl group, *i.e.*,

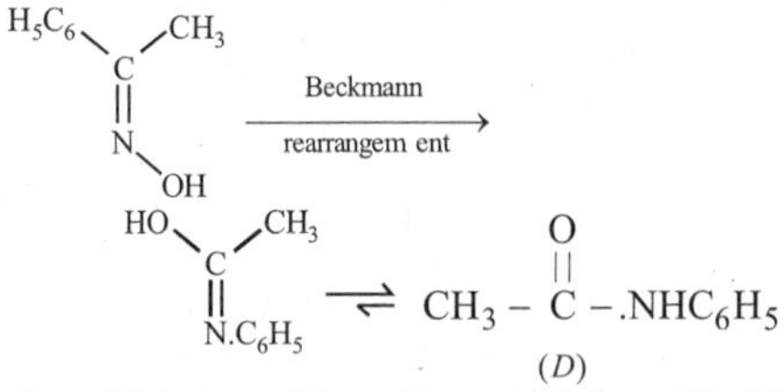

Since (*D*) is formed from (*B*), and (*E*) from (*C*), (*B*) and (*C*) should have following structures.

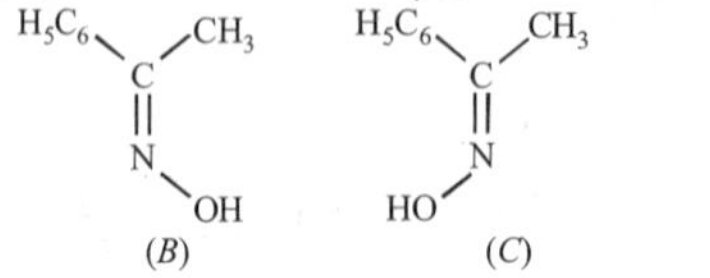

(v) Lastly, oximes (*B*) and (*C*) are formed from (*A*), the latter should be a ketone of the formula $C_6H_5COCH_3$,

$$\underset{(A)}{C_6H_5\overset{O}{\overset{||}{C}}CH_3} \xrightarrow{NH_2OH.HCl} \underset{(B)\ \&\ (C)}{C_6H_5\overset{NOH}{\overset{||}{C}}CH_3}$$
syn- and *anti*-isomors

49. (i) $CH_3-CH_2-NH_2 \xrightarrow{(CH_3CO)_2O} CH_3-CH_2-NH-COCH_3 + CH_3COOH$

(ii)

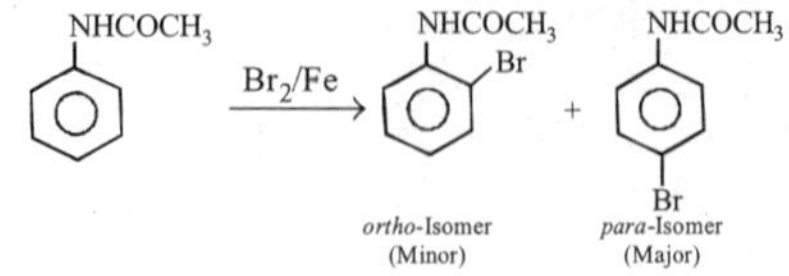

(iii)

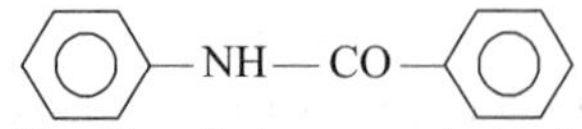

50. (i) Two factors operate in deciding the basicity of alkyl amines.

(a) Inductive effect

(b) Solvation effect

(a) **Inductive effect.** The alkyl group being electron releasing, increases the charge density on nitrogen. This in turn increases the basicity of amines. The expected order of basicity is

$$R_3N > R_2NH > RNH_2 > NH_3$$

(b) **Solvation effect.** Because of the positive charge carried by the conjugate acid of an amine, it is stabilised by the hydrogen bonding with the solvent water. The larger the number of hydrogens attached to the nitrogen in the conjugate acid, the larger is its stability and thus larger is the basicity of the corresponding base. The expected order of basicity of the alkylamines will be

$$NH_3 > RNH_2 > R_2NH > R_3N$$

The inductive and solvent effects predict the opposite trend in the basicity of alkyl amines. In going from R_2NH to R_3N the solvation effect plays a more dominating role as compared to the inductive effect making R_2NH more basic than R_3N.

(ii) Aniline is weak base than cyclohexylamine because of resonance while there is no resonance in cyclohexylamine.

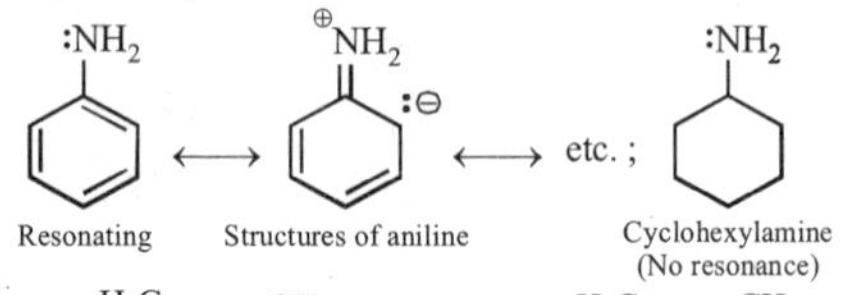

51.

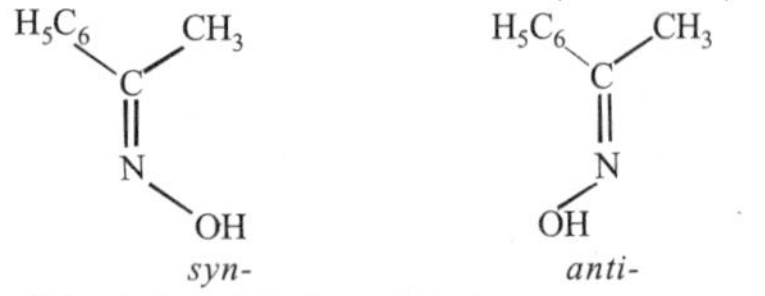

52. $C_6H_5NH_2 + CHCl_3 + 3KOH \longrightarrow C_6H_5N \equiv C + 3KCl + 3H_2O$
Phenyl isonitrile (foul smelling)

53. *A* (C_3H_9N) is a 2° amine $CH_3 - NH - C_2H_5$ (ethylmethylamine) because the product sulphonamide is solid and insoluble in alkali.

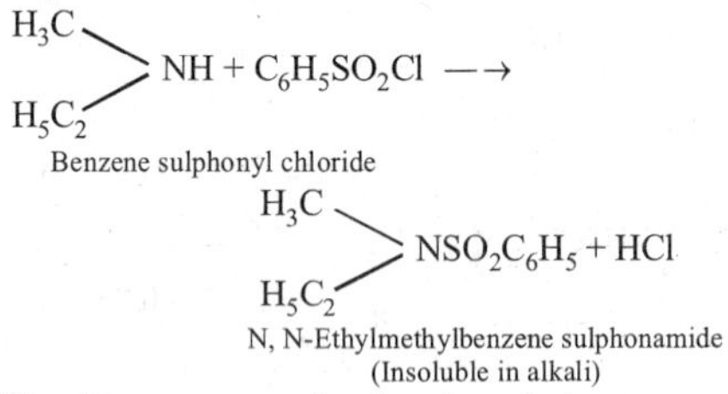

54. Let us summarise the given facts.

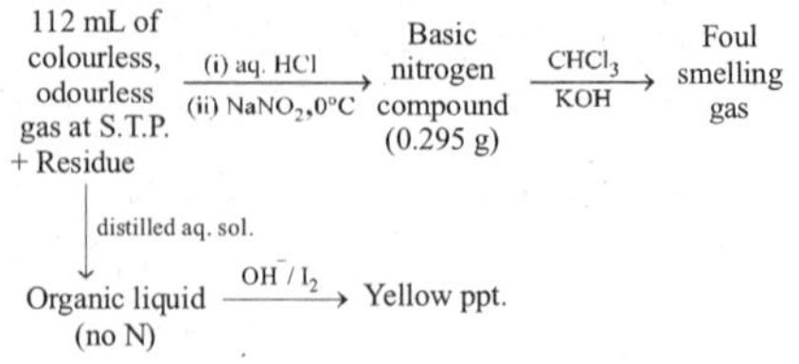

Reaction of the original compound with alcoholic potash and chloroform to give foul smelling gas indicates that it contains a primary $-NH_2$ group.

$R-NH_2 + CHCl_3 + KOH \longrightarrow R-NC\uparrow$
(Basic compound) Carbylamine (foul smelling)

Determination of mol. weight of the amine :
112 mL. of gas is evolved at S.T.P. by 0.295 g of amine
22400 mL. of gas is evolved by

$$= \frac{0.295}{112} \times 22400 = 59\ g$$

Hence, the mol. wt. of the amine = 59
$\therefore$ Mol. wt. of the alkyl group = 59 – 16 = 43
Nature of alkyl gp. of mol. wt. 43 = C_3H_7-
Thus, the amine may be either

$CH_3CH_2CH_2NH_2$ or $(CH_3)_2CHNH_2$

The reaction of amine with $NaNO_2$ at 0ºC and all other reactions may thus be written as below.

$$\underset{n-\text{Propylamine}}{CH_3CH_2CH_2NH_2} \xrightarrow[\text{(ii) } NaNO_2/0°C]{\text{(i) HCl}}$$

$$CH_3CH_2CH_2OH + N_2\uparrow \xrightarrow{\text{aq. sol. distill.}}$$

$$CH_3CH_2CH_2OH \xrightarrow{OH^-, I_2} \text{No yellow ppt.}$$

$$\underset{\text{Isopropylamine}}{(CH_3)_2CHNH_2} \xrightarrow[\text{(ii) } NaNO_2/0°C]{\text{(i) HCl}}$$

$$(CH_3)_2CHOH + N_2\uparrow \xrightarrow{\text{aq. sol. distill.}}$$

$$(CH_3)_2CHOH \xrightarrow[\text{(Haloform reaction)}]{OH^-, I_2} \underset{\text{(yellow)}}{CHI_3\downarrow}$$

Since the given reactions correspond to isopropylamine, the original compound is **isopropylamine**, $(CH_3)_2CHNH_2$.

Alternate Solution:
Starting compound is a primary amine.
$R-NH_2 + CHCl_3 + KOH \rightarrow R-NC$ (Foul Smell)
Also, $R-NH_2 + HNO_2 \rightarrow R-OH + N_2$.

$$\text{Moles of } N_2 = \frac{112}{22500} = 5\times10^{-3}$$

$\because$ one Mole of N_2 is obtained from 1 mol of $R-NH_2$, Mole of $R-NH_2 = 5\times10^{-3}$
$\Rightarrow 5\times10^{-3}$ Mol, $R-NH_2$ weight = 0.295g.

$$1 \text{ Mole of } R-NH_2 \text{ will weigh} = \frac{0.295}{5\times10^{-3}} = 59g$$

In, $R-NH_2$, $-NH_2$ has molar mass = 16 = R- has molar mass = 43
Also, the alcohol (R-OH) given Iodoform test, it must has the structure

$R'-\underset{|\atop OH}{CH}-CH_3$

$\Rightarrow$ R' – has molar mass = 15 (CH_3)
$\Rightarrow$ Original compound is $CH_3-\underset{|\atop CH_3}{CH}-NH_2$

55. Solution of compound *A* in chloroform when treated with alcoholic KOH yields compound *C* (C_7H_5N) having an unpleasant odour which may be due to isocyanide. Hence, the above reaction may be an example of carbylamine reaction. Therefore, compound *A* must be aniline $(C_6H_5NH_2)$ and *C* must be phenylisocyanide (C_6H_5NC).

$$\underset{\text{Aniline, } A}{C_6H_5NH_2} + CHCl_3 + 3KOH \rightarrow \underset{\text{Phenyl isocyanide, } C}{C_6H_5NC} + 3KCl + 3H_2O$$

Alkaline aqueous layer of *B* when heated with chloroform and then acidified gives *D* and *E* which are isomeric with each other and have molecular formula $C_7H_6O_2$. This indicates the possibility that Reimer-Tiemann reaction would have taken place. Hence, compound *B* must be phenol and compounds *D* and *E* are *o*- and *p*-hydroxybenzaldehydes.

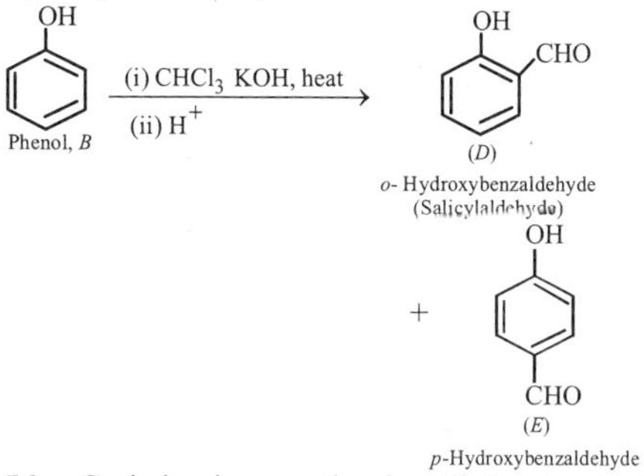

56. Carbylamine reaction is a distinction test for primary amines. All primary amines (alphatic or aromatic) on heating with alcoholic KOH and $CHCl_3$ give unpleasant or foul smell of isocyanide which is easily detected.

$$C_2H_5NH_2 + CHCl_3 + 3KOH(alc.) \longrightarrow \underset{\text{(foul smell) Ethyl isocyanide}}{C_2H_5NC} + 3KCl + 3H_2O$$

Diethylamine, a 2° amine, does not respond this test.

57. Determination of empirical formula :

	C	H	N	O
% age	49.32	9.59	19.18	21.91
mole	$\frac{49.32}{12}$ =4.11	$\frac{9.59}{1}$ =9.59	$\frac{19.18}{14}$ =1.37	$\frac{21.91}{16}$ =1.37
(Dividing by 1.37)				
Simplest ratio	3	7	1	1

$\therefore$ Empirical formula of A will be C_3H_7NO and empirical formula wt = 73

Calculation of molecular wt. of (B)
We know, meq. of Ag = meq. of Ag salt

$$\text{or } \frac{\text{Wt. of Ag}}{\text{Eq. wt. of Ag}} = \frac{\text{Wt. of salt}}{\text{Eq. wt. of salt}}$$

$$\text{or } \frac{59.67}{108} = \frac{100}{E_{salt}} \text{ or } E_{salt} = 181$$

$E_{salt} = E_{Ag} + E_{Anion} \Rightarrow E_{anion} = 181 - 108 = 73$
For monobasic acid (B), adding one H,
Mol. wt. = Eq. wt. + 1 = 73 + 1 = 74
B being monobasic can be represented as $C_nH_{2n+1}COOH$
$\Rightarrow 74 = 12n + (2n + 1).1 + 12 + 16 + 16 + 1$
$\Rightarrow n = 2$
$\therefore B = C_2H_5COOH$

Nature of A : Since B is obtained by the action of A with NaOH followed by hydrolysis, so A is an amide, $CH_3CH_2CONH_2$.
Reaction :

$$\underset{\text{(A) (Propanamide)}}{CH_3CH_2CONH_2} \xrightarrow{NaOH} CH_3CH_2COONa \xrightarrow{HOH} \underset{\text{(B) Ethanoic acid}}{CH_3CH_2COOH}$$

58. (ii)

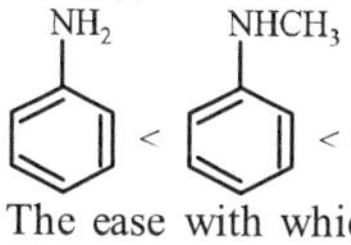

$< CH_3NH_2 < (CH_3)_2NH$

The ease with which the lone pair of electron (unshared) on the N-atom co-ordinates with a proton determines the relative basic strength of amines.
(i) Presence of +I / +M group increases the basicity whereas presence of –I / –M group decreases the basicity.

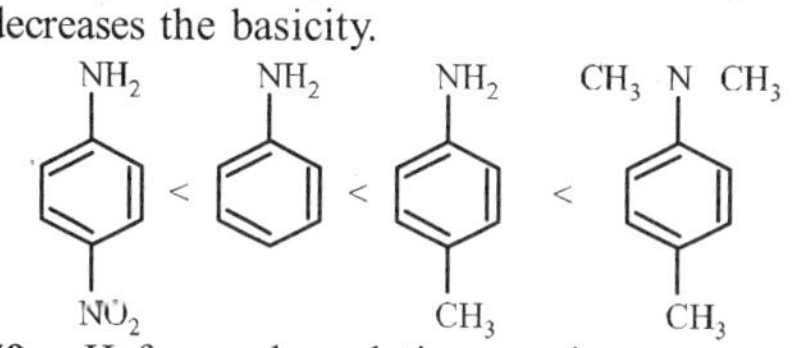

59. Hofmann degradation reaction.

$$\underset{\text{Acetamide}}{CH_3CONH_2} + Br_2 + 4KOH \rightarrow \underset{\text{Methyl amine}}{CH_3NH_2} + K_2CO_3 + 2KBr + 2H_2O$$

60. (i)

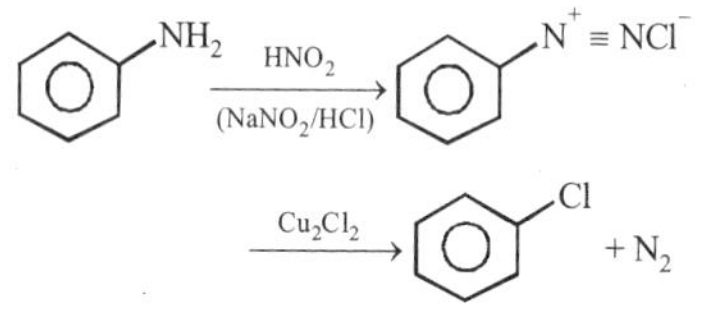

(ii)

$$CH_3CHO \xrightarrow[H_2SO_4]{K_2Cr_2O_7} CH_3COOH \xrightarrow[\text{distill}]{Ca(OH)_2}$$

$$CH_3COCH_3 \xrightarrow{NH_2OH.HCl} \underset{\text{Acetoxime}}{(CH_3)_2C{=}NOH}$$

(iii) $C_6H_6 \xrightarrow[HNO_3\text{ (conc.)}]{H_2SO_4\text{(conc.)}} C_6H_5NO_2 \xrightarrow{Sn/HCl} C_6H_5NH_2$

(iv) $CH_3CH_2Cl \xrightarrow{\text{alc. KCN}}$
$CH_3CH_2.CN \xrightarrow[\text{reduction}]{H_2|Ni} CH_3CH_2CH_2NH_2$

61. **(c)**

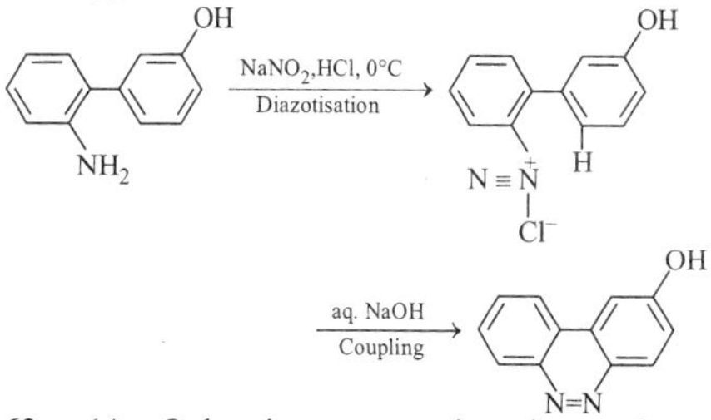

62. **(c)** Only primary aromatic amines undergo diazotisation followed by coupling.

63. **(b)** $H_2C = N^+ = N^- > H_2C^- - N^+ \equiv N$
I: Octet complete, 6 covalent bonds, –ve charge on N
III: Octet complete, 6 covalent bonds, –ve charge on C

$> H_2C^+ - N = N^- > H_2C^- - N = N^+$
II: Octet incomplete, 5 covalent bonds, –ve charge on N
IV: Octet incomplete, 5 covalent bonds, –ve charge on C

64. **(c)** $CHCl_3 + RNH_2 + KOH \xrightarrow[\text{reaction}]{\text{Carbylamine}} R-N^+\equiv C^- + 3KCl + 3H_2O$

65. **(c)** Benzene sulphonic acid, being stronger acid than carbonic acid, would liberate CO_2 when treated with sodium bicarbonate, also *p*-nitrophenol, being more acidic than carbonic acid, will liberate CO_2.

66. (a) Two moles of NH_2^- ions will abstract two moles of most acidic hydrogen out of the four moles of acidic hydrogen present per mole of the compound. The acidic strength is in the order:

–COOH > –OH (phenolic having NO_2 group) > –OH (phenolic) > alkynic H.

67. (a)

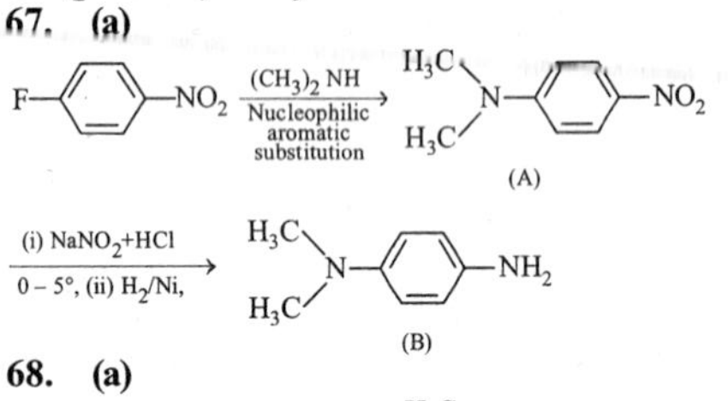

68. (a)

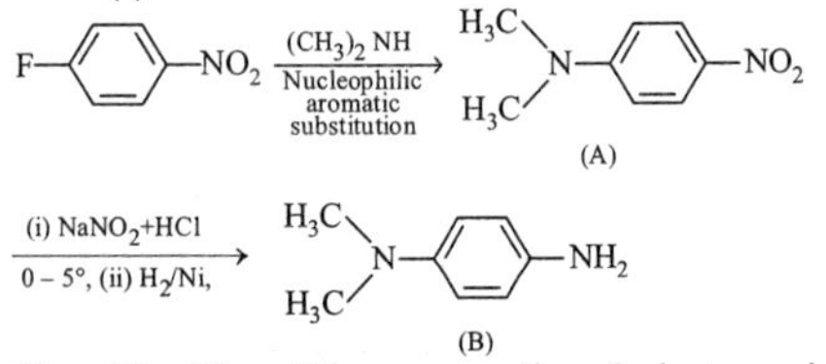

69. (c) N can't have more than 8 electrons in its valence shell as it does not have any *d* orbital. In (c), N has 10 electrons.

70. (4.00)

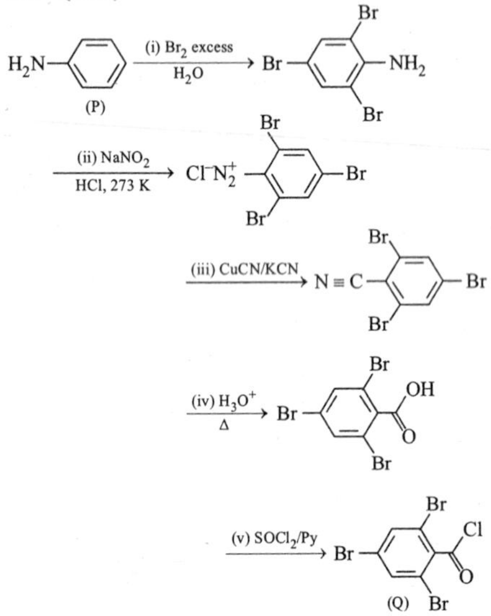

Scheme – II

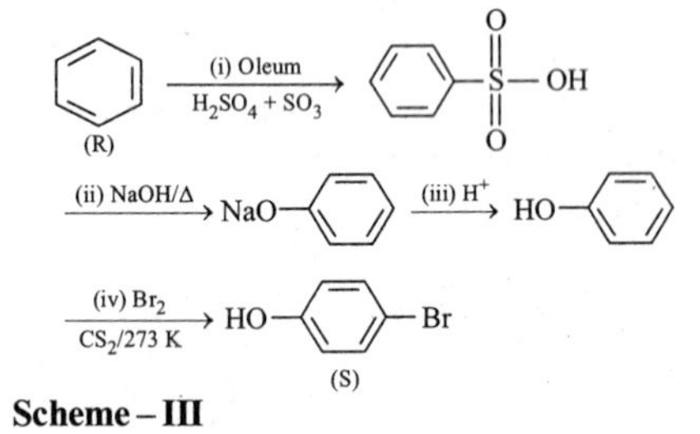

Scheme – III

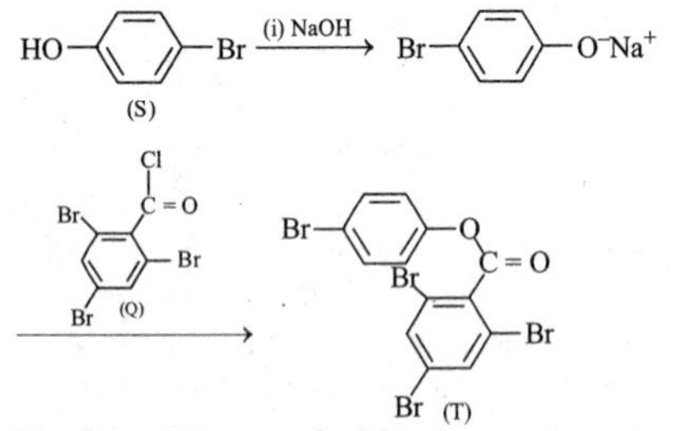

Number of Br atoms in (T) = 4

71. (18.60)

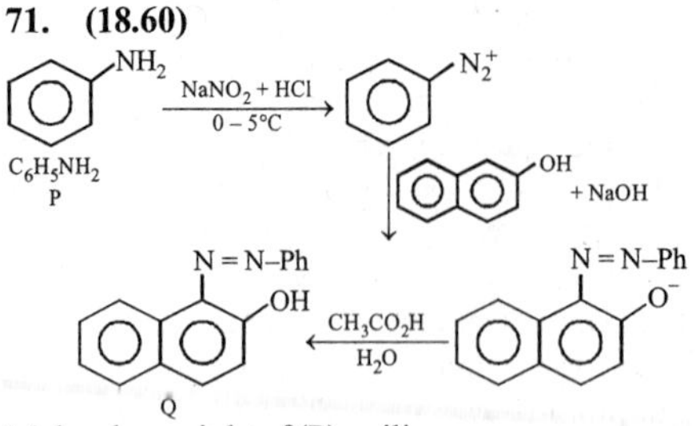

Molecular weight of (P) aniline

$= \text{M.wt. of } C_6H_5NH_2 = 72 + 7 + 14 = 93$

Density of P = 1 g mL^{-1}

$\therefore \quad 9.3 \text{ mL of P} = 9.3 \text{ g P} = \frac{9.3}{9.3} = 0.1 \text{ mole P}$

The mole ratio of $PhNH_2 : PhN_2^+ :$

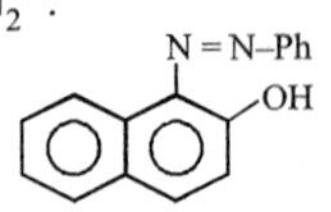

$= 1 : 1 : 1$

So the mole of Q formed will be 0.1 mole and extent of reaction is 100% but if it is 75% yield.

$\text{Then amount of Q} = 0.1 \times \frac{75}{100} = 0.075 \text{ mol}$

The molecular formula of $Q = C_{16}H_{12}ON_2$

So M.wt. of $Q = 16 \times 12 + 12 \times 1 + 16 + 2 \times 14$

$= 192 + 12 + 16 + 28 = 248$ g

So amount of $Q = 248 \times 0.075 = 18.6$ g

72. (51)

Common solution for Q.no. 72 and 73

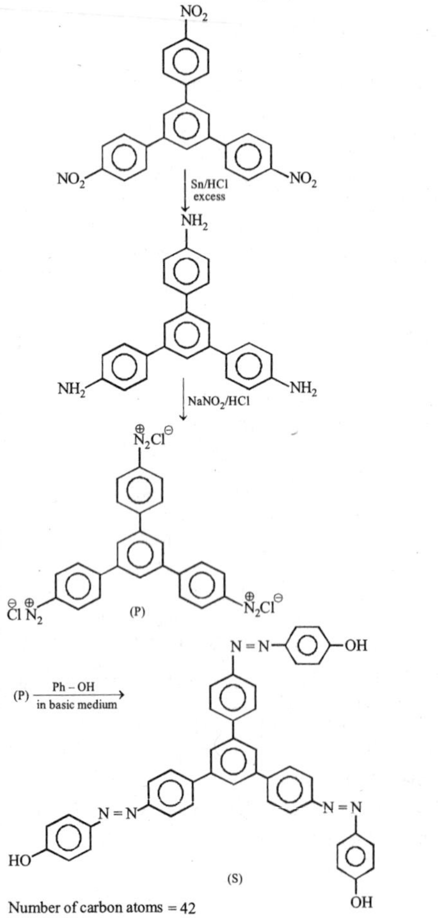

Number of carbon atoms = 42

Number of hetero atoms = 09

Total = 51

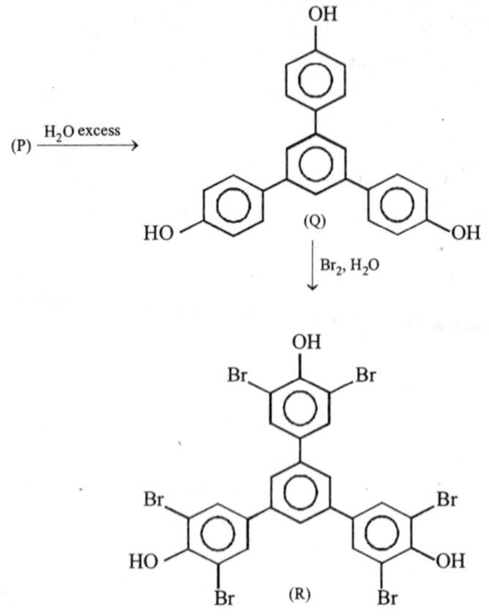

Number of hetero atoms in R is 9.

73. (9)

74. Meta isomer

75. (b,d)

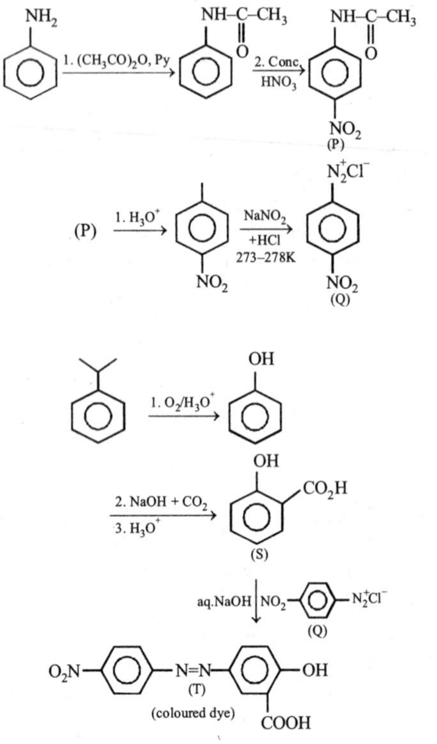

(a) 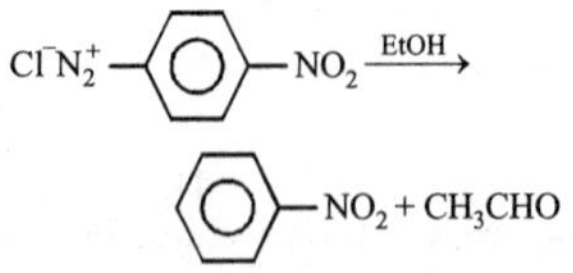 (Incorrect)

(b) S gives positive phthalein dye test because it is phenol derivative. (correct)

(c) P contains only single NO_2 group. (incorrect)

(d) T is coloured dye compound, (correct)

76. (a, b, c)

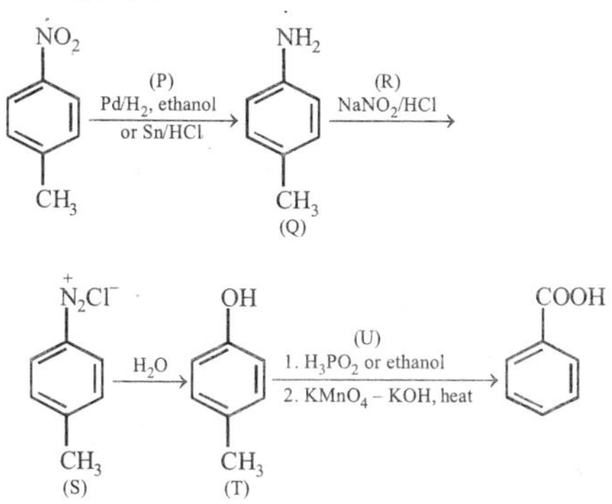

77. (d)

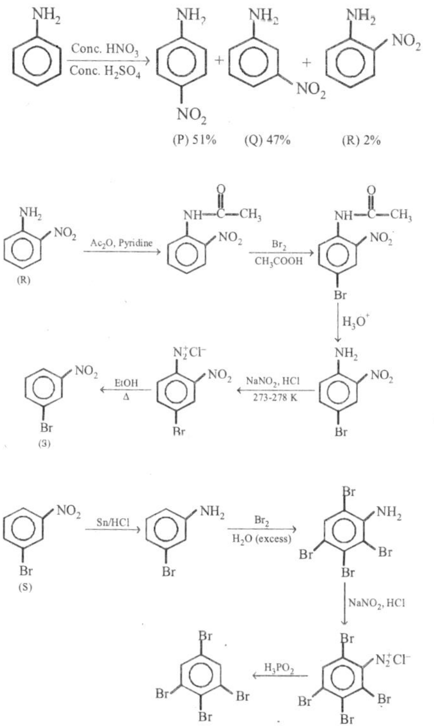

78. (b)

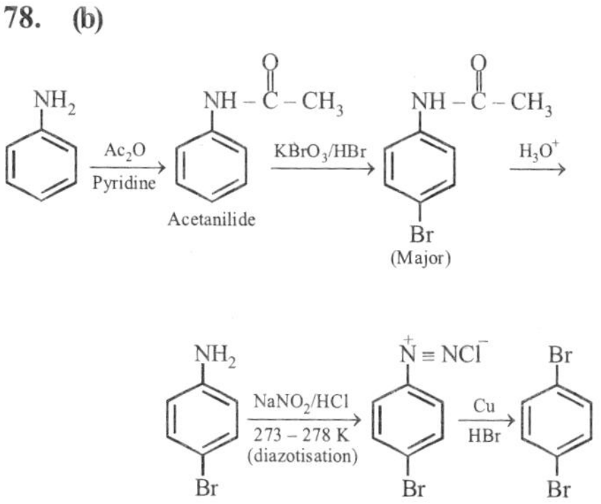

79. (a)

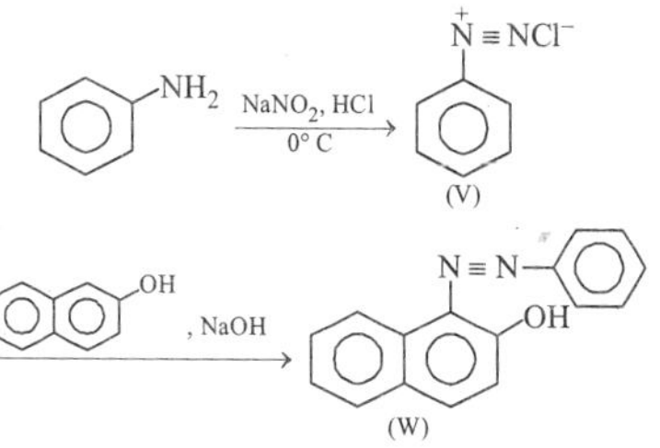

Coupling reaction will occur at ortho position of activated ring.

80. (c)

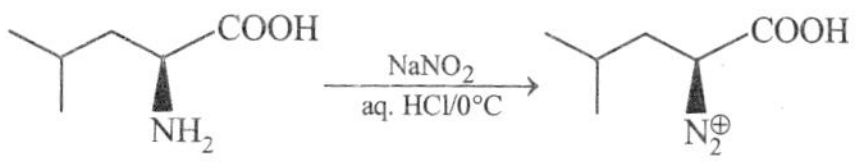

$N_2^{\oplus}$ is a good leaving group but the formed carbocation will not be stable due to –M effect of –COOH group.

Thus, intramolecular attack of oxygen atom from rear side will take place at the carbocation.

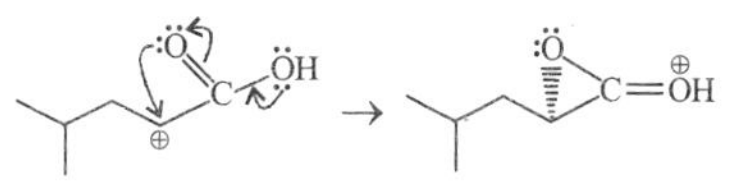

The formed 3 membered ring is not a stable compound. Now, the reaction will proceed with the attack of water molecule.

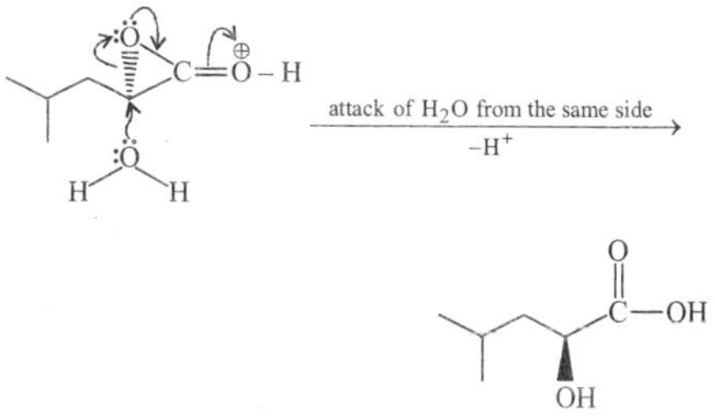

81. (b) $C_6H_5N_2Cl +$

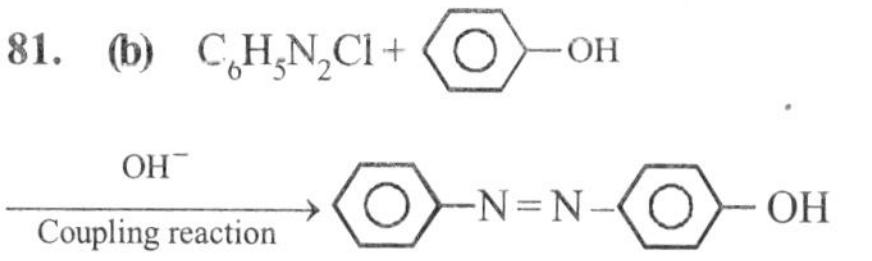

p-Hydroxyazobenzene

82. (a, b, d)

(a) Nitro group decreases the electron density at the *o*- and *p*-positions in comparison to *m*-position due to –M effect.

(b) The below intermediate is a resonance hybrid of three equivalent structures.

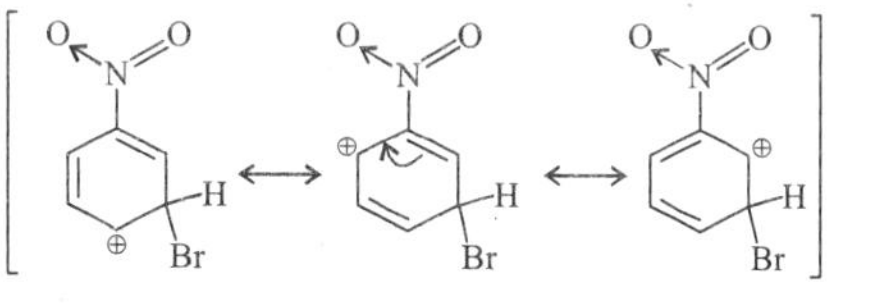

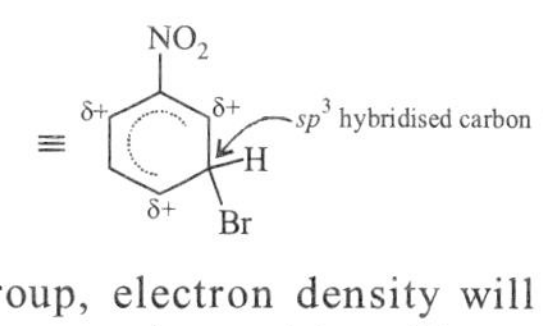

Due to $-NO_2$ group, electron density will decrease more from *o*- and *p*- positions. Hence, the attack of electrophile Br^+ will be favourable at meta position.

In another words, after attack of Br^+ at meta position, intermediate will be least destabilised.

(c) There will be loss of aromaticity when Br^+ attacks at any position because of formation of sp^3 hybridised carbon.

Thus, option (c) is incorrect.

(d) Due to more electron density at meta position, heterolytic fission of C – H bond and removal of H^+ will be easy from this position.

Thus, option (d) is correct.

83. (c, e) Silver nitrite, being a salt of nitrous acid, occurs in two tautomeric forms.

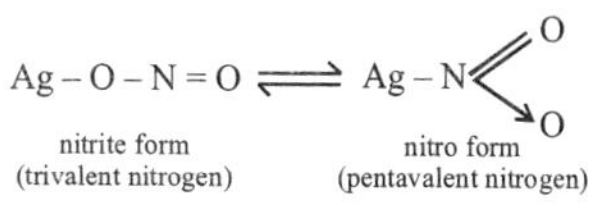

NO_2^- ion from $AgNO_2$ may exist in two tautomeric forms, –O – N = O (nitrite ion) forming **alkyl nitrites**, and $-N\langle^{O}_{O}$ (nitro group) forming **nitroalkanes**.

$2\,C_2H_5Br + 2AgNO_2 \rightarrow C_2H_5NO_2 + C_2H_5ONO + 2\,AgBr$

84. (c)

(P)

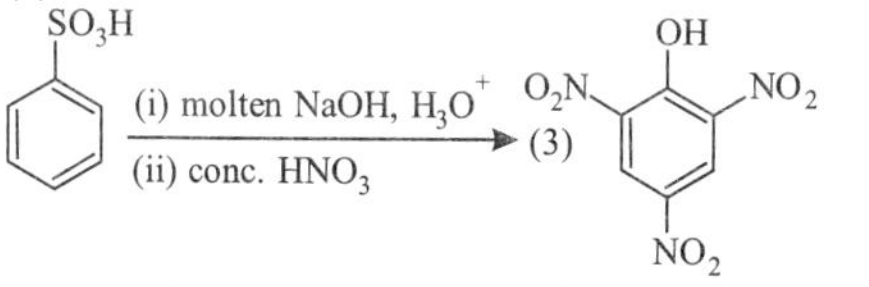

(Q)

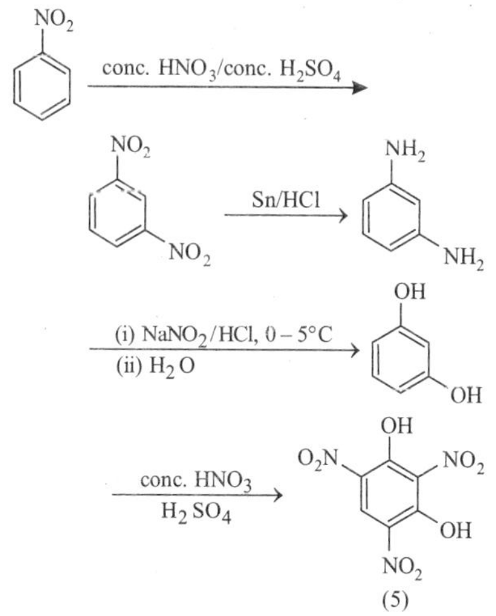

(R)

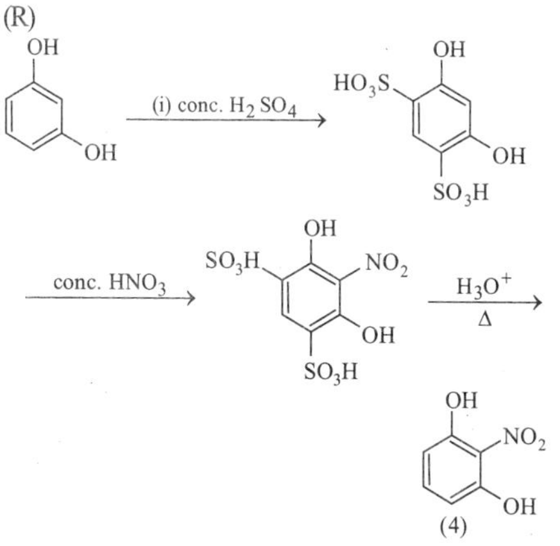

(S)

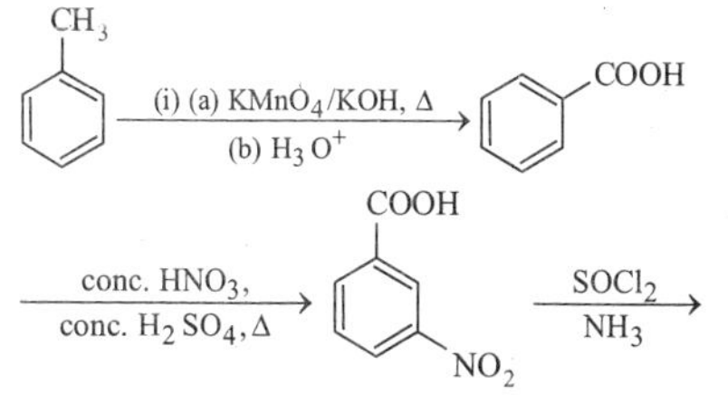

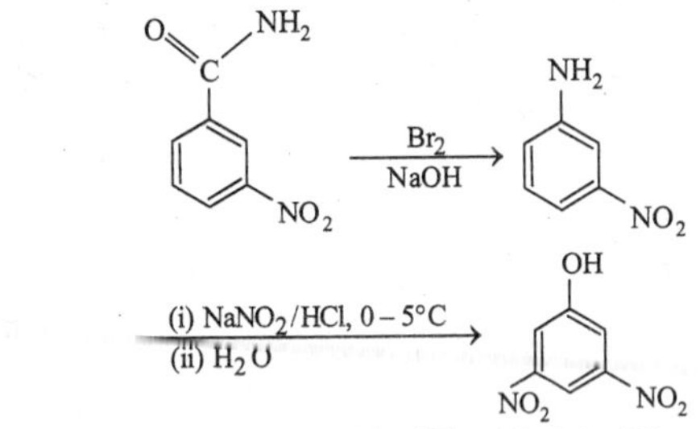

85. (A) – (r), (s) ; (B) – (t) ; (C) – (p), (q) ; (D) – (r)

86. **(d)** Chlorobenzene is resonance stabilized.

Thus aryl halides (chlorobenzene) do not undergo nucleophilic substitution. Reason is correct.

87. **(a)** In *o*-nitrophenol, intramolecular H-bonding is possible because OH and NO_2 groups are close to each other. This makes the *ortho* isomer less acidic as its capacity to donate a proton (H-atom) decreases. There is no such intramolecular H–bonding in the *p*-isomer.

88. (i) Going backward, we can easily solve the problem.

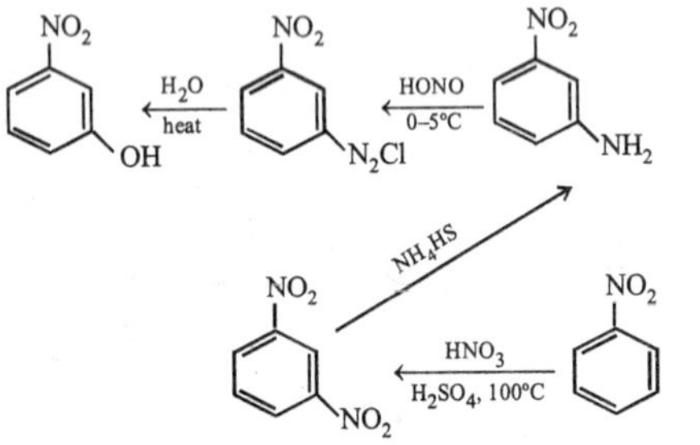

(ii)

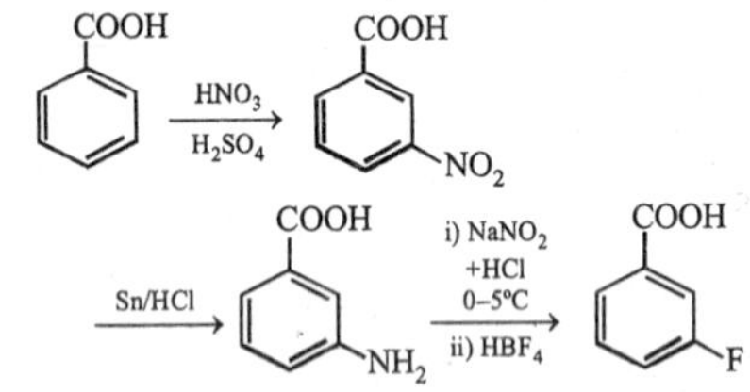

(iii)

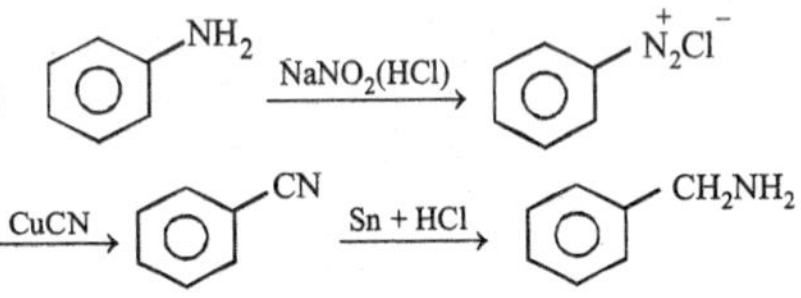

(iv)

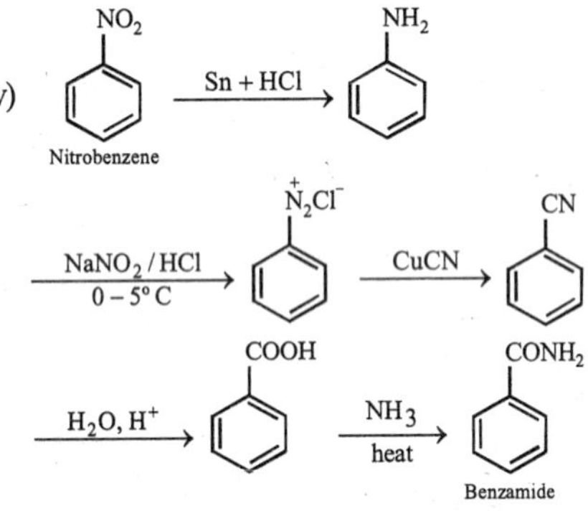

(v)

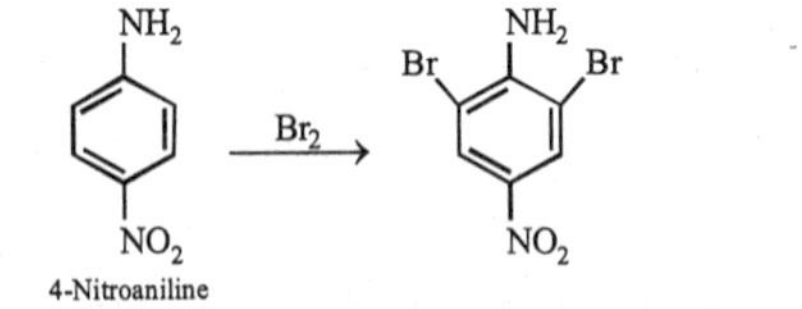

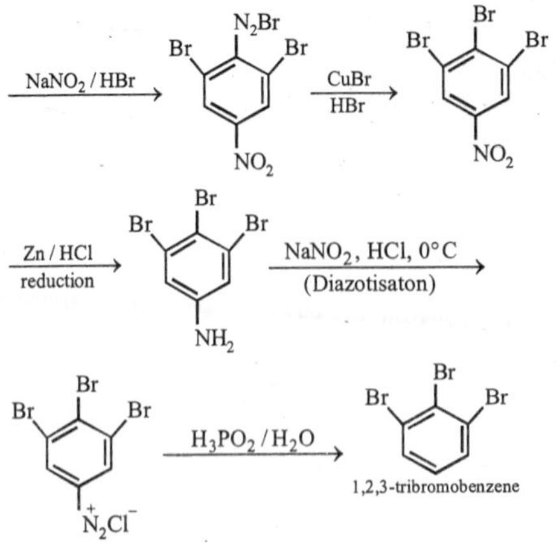

(vi) $C_6H_5CHO \xrightarrow{\text{Oxi.}} C_6H_5COOH$

$\xrightarrow[\text{(ii) } \Delta]{\text{(i) } NH_3} C_6H_5CONH_2 \xrightarrow[KOH]{Br_2}$

$C_6H_5NH_2 \xrightarrow[0°C]{HONO/Cl^-} C_6H_5N_2Cl$

$\xrightarrow{CuCN} C_6H_5CN$
Cyanobenzene

89.

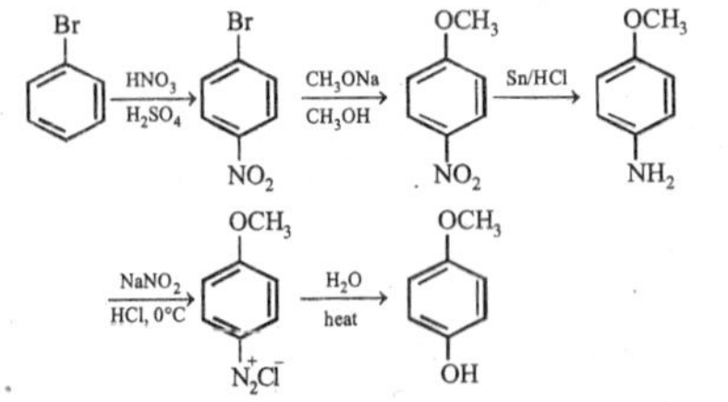

90. (i)

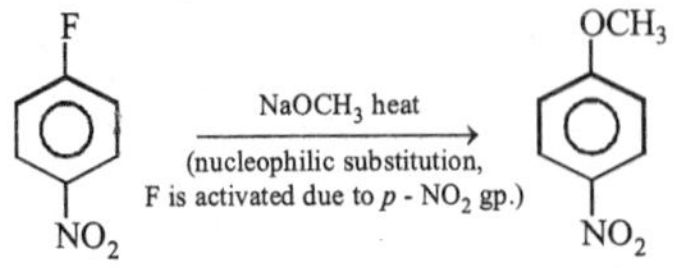

NO_2 group deactivates the ring and thus, decreases the electron density at *o*- and *p*-positions. This makes it for nucleophilic attack at these positions.

(ii)

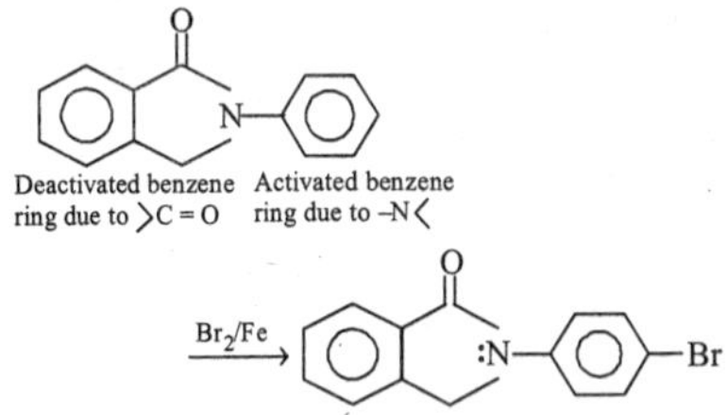

(iii)

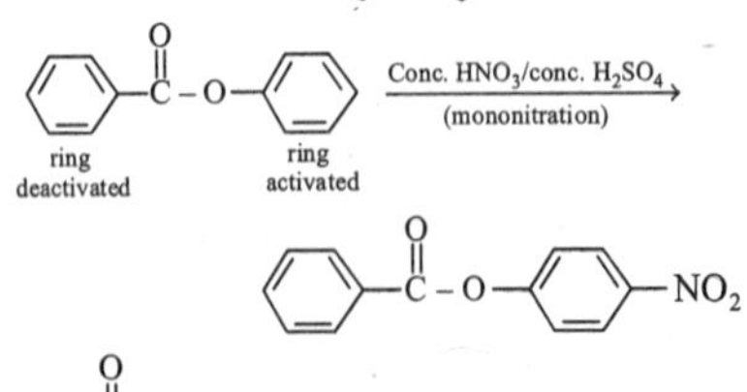

91.

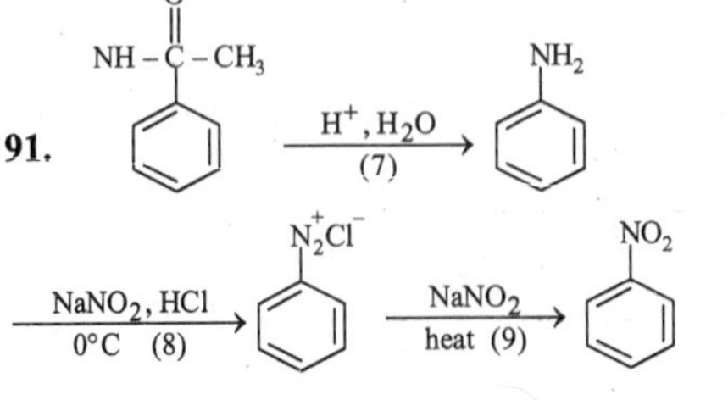

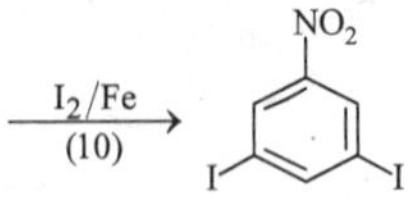

92. (i)

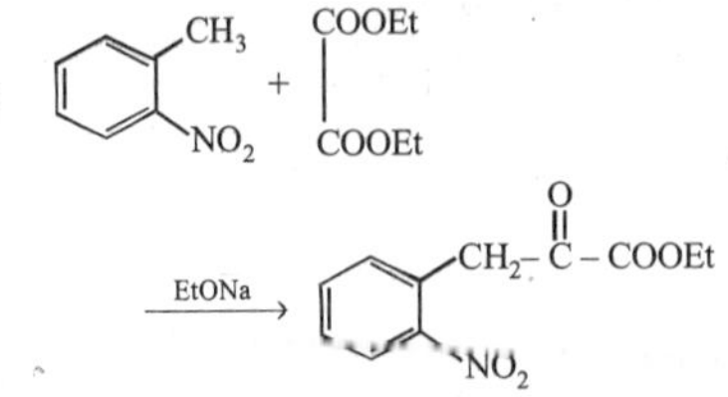

(ii)

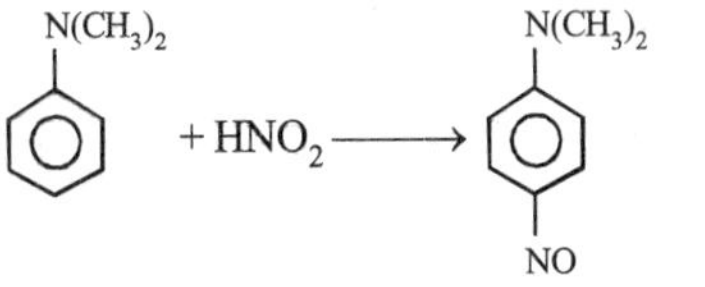

(iii)

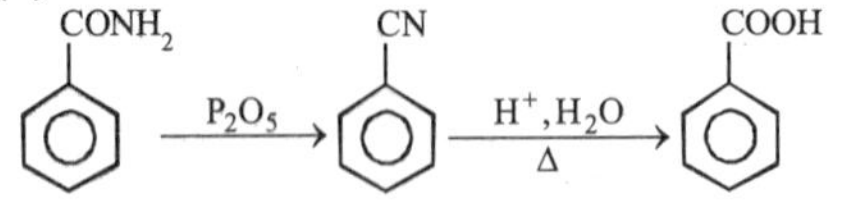

93.

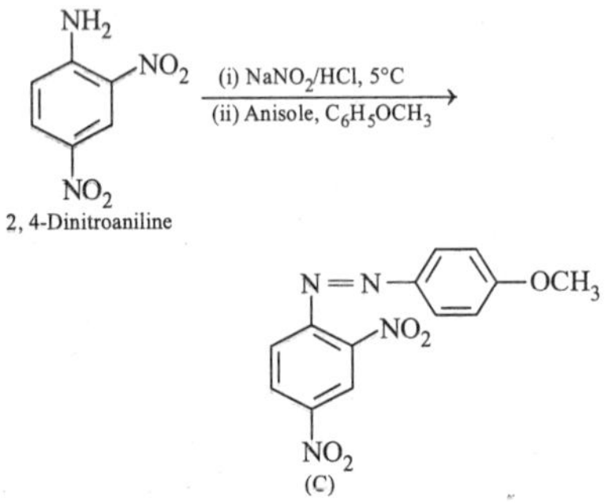

94. (i)

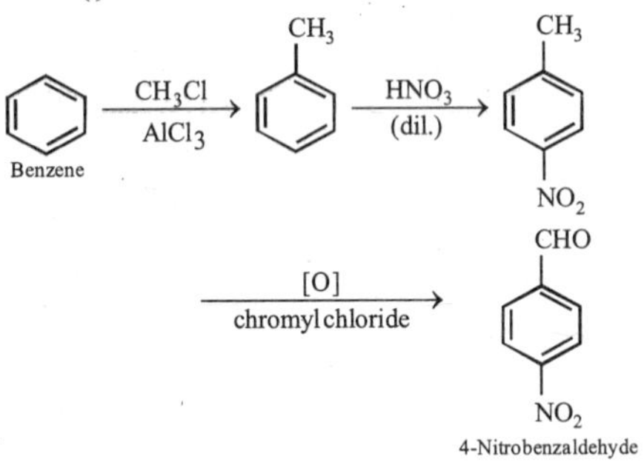

(ii)

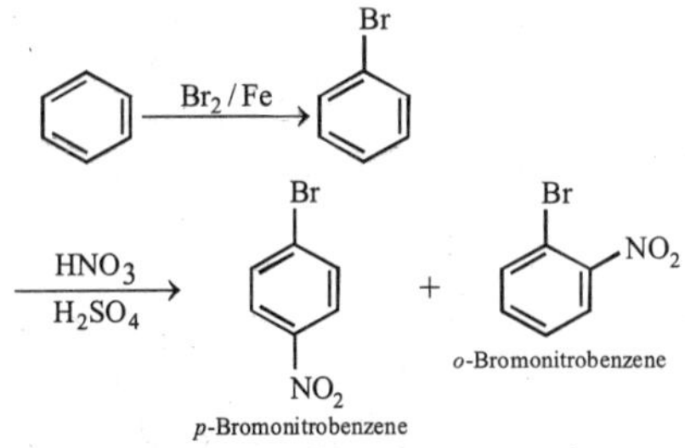

On fractional crystallization, para isomer crystallizes first.

(iii)

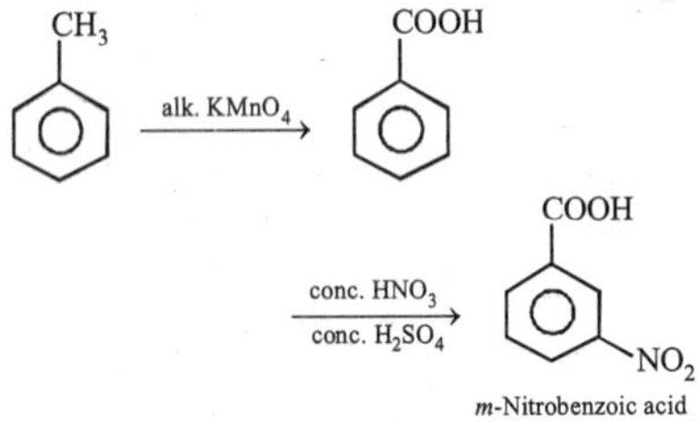

Biomolecules

Part-A (JEE Main)

1. **(2)** The monosaccharide units obtained from hydrolysis
of oligosaccharides may be or may not be same.
Example:

$$\text{Sucrose} \xrightarrow{H_3O^{\oplus}} \underbrace{\alpha\text{-D-Glucose} + \beta\text{-D-Fructose}}_{\text{different monosaccharide units}}$$

$$\text{Maltose} \xrightarrow{H_3O^{\oplus}} \underbrace{\text{Glucose} + \text{Glucose}}_{\text{same monosaccharide units}}$$

Thus statement (2) is wrong.

2. **(4)**

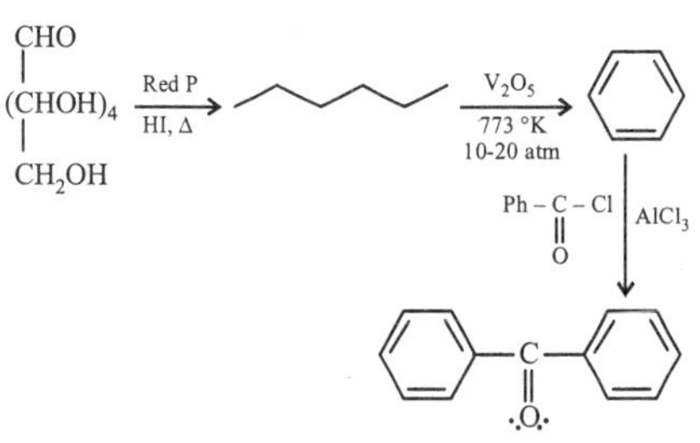

Total number of π electrons and lone pair electrons in final product = 18.

3. **(1)** α-D-Glucose and β-D-Glucose are anomer.

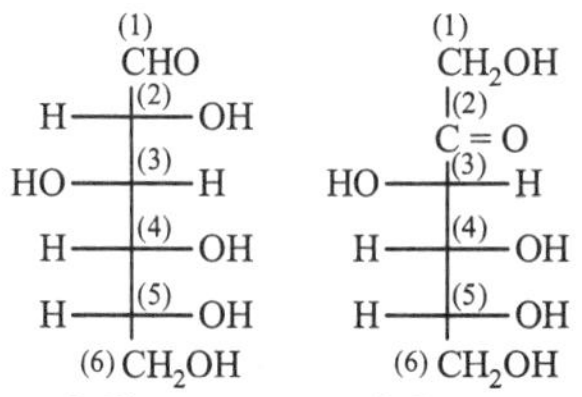

D-Glucose D-Fructose

4. **(4)** α-anomer melting point is 419 K while β-anomer melting point is 423 K. β-anomer is more stable. The specific rotation of α-D Glucose = + 112° and β-D Glucose = + 18.7°

5. **(3)**

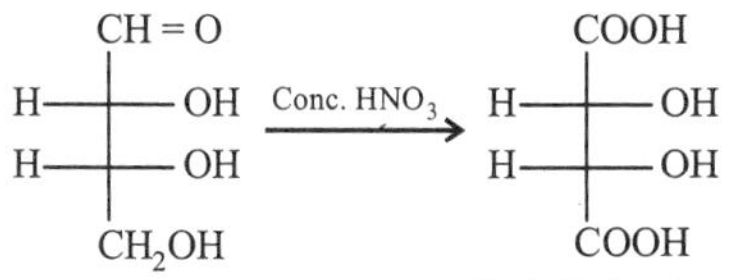

D-Aldotetrose Optically inactive dicarboxylic acid

In D-configuration, –OH group lies in right side of second last carbon.

6. **(4)** Maltose and lactose both have hemiacetal linkage and free aldehyde group can be produced at C-1 carbon of glucose. So, both are reducing sugars.

7. **(1)** Given structure (1) is of sucrose which is non- reducing.
In disaccharides, if the reducing groups of monosaccharides.i.e., aldehydic or ketonic groups are bonded, these are non-reducing sugar.
Alternatively, non-reducing sugars have acetal (not hemi- acetal) linkage,

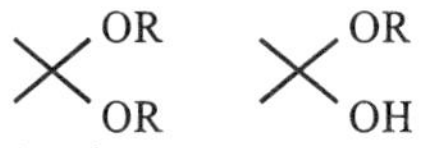

Acetal Hemiacetal

8. **(4)** Compounds in options (B) and (C) are acetals (i.e., not having anomeric –OH). Hence they do not give Tollen's test.
While compounds (A) and (D) are hemiacetals and can reduce Tollen's reagent.

9. **(2)**

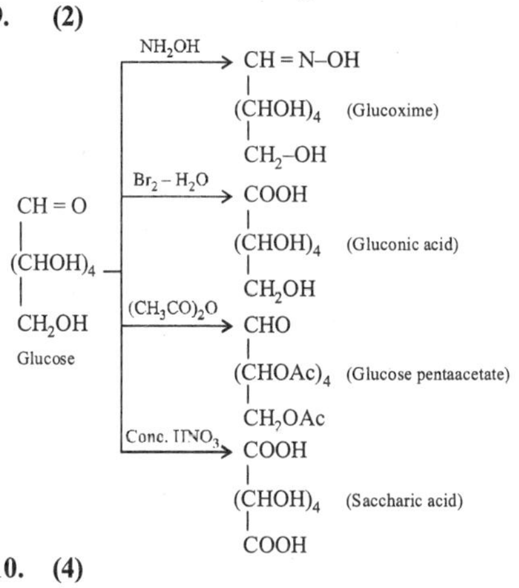

10. **(4)**

$$C_{12}H_{22}O_{11} + H_2O \xrightarrow{HCl} C_6H_{12}O_6 + C_6H_{12}O_6$$

D-Glucose + D Fructose

$[\alpha]_{D\text{-sucrose}} = +66.5°$
$[\alpha]_{D\text{-Glucose}} = +52.5°$
$[\alpha]_{D\text{-Fructose}} = -92.4°$

The laevorotation of fructose (–92.4°) is more than dextrorotation of glucose (+52.5°), the mixture is laevorotatory.
Sucrose is dextrorotatory and hydrolysed product is laevorotatory.

11. **(2)** **Statement-I :** Sucrose is formed by condensation reaction between D-(+)-glucose and D-(–)-fructose not D-(+)-fructose.
Statement-II : Invert sugar is a 1 : 1 molar mixture of D-(+)-glucose and D-(–)-fructose.

12. **(2)** Cellulose derivative have more number of hydroxy groups, so stronger H-bonding is present with water in cellulose derivative of cotton cloths.

13. **(3)**

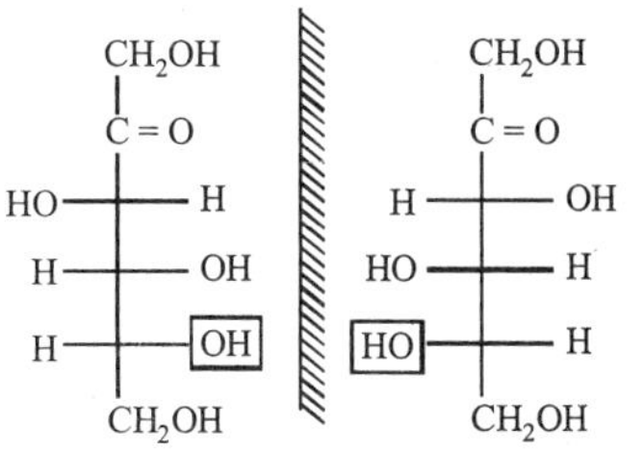

D-Fructose L-Fructose

14. **(2)** (A)-(II), (B)-(III), (C)-(I), (D)-(IV)

15. **(3)**

(A) Starch $\xrightarrow{H^+/H_2O}$ Glucose

(B) Cane sugar (Sucrose) $\xrightarrow{H^+/H_2O}$ glucose + fructose (50% 50%)

(C) Milk sugar $\xrightarrow{H^+/H_2O}$ glucose + galactose (Lactose)

(D) Amylopectin $\xrightarrow{H^+/H_2O}$ Glucose

(E) Amylose $\xrightarrow{H^+/H_2O}$ Glucose

So, correct options are B, C and E only

16. **(1)**
(A) Sucrose $\rightarrow \alpha_1 - \beta_2$ Glycosidic linkage
(B) Maltose $\rightarrow \alpha$ 1– 4 Glycosidic linkage
(C) Lactose $\rightarrow \beta$ 1– 4 Glycosidic linkage
(D) Amylopectin $\rightarrow \alpha$ 1– 4 and α 1 – 6 Glycosidic linkage

17. **(2)** D-glucose pentaacetate does not have a free aldehyde group which is required for this reaction so does not react with 2,4-dinitrophenylhydrazine.

18. **(1)** In Ribose, carbohydrate present in DNA is β-2-Deoxy-D-Ribose whose structure is

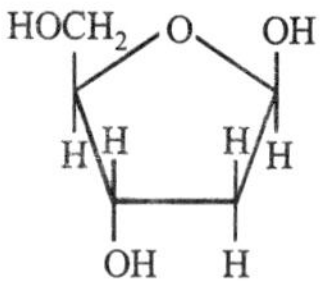

which is a reducing D-sugar in β anomeric form & it is a pentose sugar.

19. **(2)**

20. **(1)**

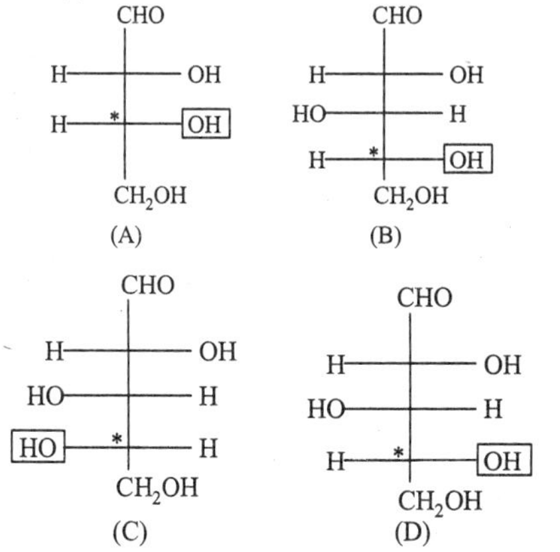

In structure A, B, D, – OH group present in right hand (After 180° rotation) side so D configuration is assigned.

21. **(1)** Glucose is soluble in water due to presence of alcohol functional group and extensive hydrogen bonding.

22. **(1)** Glucose $\xrightarrow[\Delta]{Br_2}$ Gluconic acid

23. **(1)** In L isomer –OH group is on left hand side of lowest asymmetric carbon.

24. **(3)** α- Glucose and α- Galactose differ in configuration of only one asymmetric carbon. Thus, they are epimers.

α and β-glucose differ at anomeric carbon. Thus they are anomers.

α- Glucose and α-fructose have different functional groups. Thus they are functional isomers.

α- Glucose and α-Ribose are homologues.

25. **(3)** A-II, B-IV, C-III, D-I

26. **(1)** Reducing sugar reacts with Fehling's reagent to give reddish brown precipitate of Cu_2O. Non-reducing sugar does not reacts with Fehling's reagent. Lactose, glucose and maltose are reducing sugars but sucrose is a non-reducing sugar.

27. **(2)**

28. **(2)**

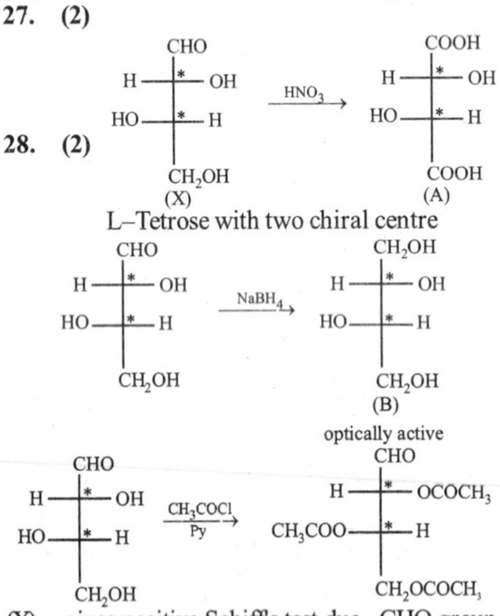

(X) gives positive Schiff's test due –CHO group

(X) is L–tetrose in the open chain structure.

29. **(1)** The given six membered cyclic structure of sugar is represented by Haworth structure (1).

30. **(2)** Seliwanoff's test is a differentiating test for ketoses and aldoses. Seliwanoff's reagent is a mixture of conc. HCl and resorcinol. This test relies on the principle that the ketohexoses are more rapidly dehydrated to form 5-hydroxy- methyl furfural when heated in acidic medium which on condensation with resorcinol gives cherry red or brown red coloured complex rapidly indicating a positive test. Aldose sugars react with the Seliwanoff's reagent to give faint pink colour which intensify with time.

31. **(4)**

32. **(4)** Pyranose ring is due to hemiacetal formation between C_1 and C_5 carbon atoms. For converting Fischer structure to Haworth structure, – OH group lying on left side in Fischer structure will point upwards in Haworth structure and thus –OH group on right will point downwards. Thus for the given cyclic hemiacetal (X) the correct pyranose structure is (4).

33. **(1)**

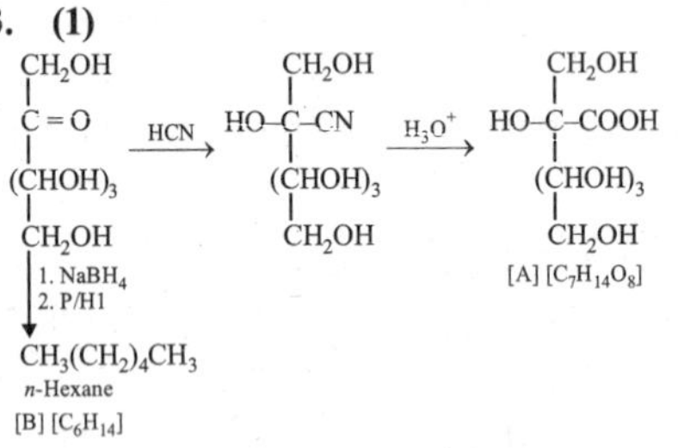

$NaBH_4$ reduces carbonyl group to alcohol which upon further reduction with HI in presence of P is reduced to alkane.

34. **(1)**

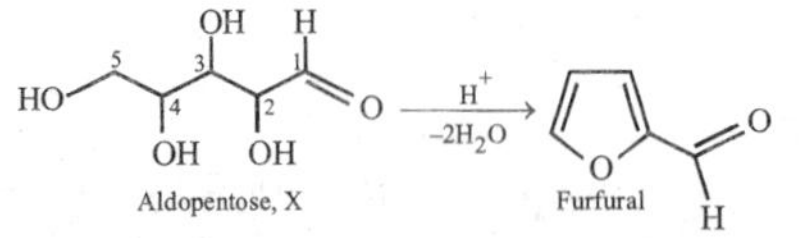

Since furfural is a five membered cyclic compound having an aldedydic group, so the parent sugar (X) should be an aldopentose.

Reaction of aldose/ketose with resorcinol gives coloured compound **(Seliwanoff test)**. Ketoses respond this test immediately while aldoses after sometime.

35. **(1)** (A) - (IV), (B) - (I), (C) - (II), (D) - (III)

36. **(1)** Those carbohydrates which reduce Fehling's solution and Tollen's reagent are referred as reducing sugars. They should have a free ketonic or aldehydic group. In case of cyclic structure those have free OH at anomeric carbon,

i.e., hemiacetal group 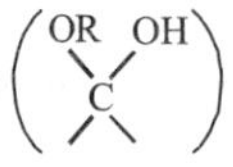 are reducing sugars.

37. **(2)**

- Maltose is composed of two glucose units.
- Sucrose has glycosidic linkage between C_1 of α-glucose and C_2 of β-fructose.
- Lactose is composed of galactose and glucose
- Amylose has glycosidic linkage between C_1 of α-glucose and C_4 of α-glucose.

38. **(2)** Laevorotatory rotates plane polarized light towards left.

$$\underset{\substack{[\alpha]=66.6^\circ\\ \text{Sucrose}}}{C_{12}H_{22}O_{11}} + H_2O \xrightarrow{H^+} \underset{\substack{\text{D-Glucose}\\ [\alpha]=+52.7^\circ\\ (A)}}{C_6H_{12}O_6} + \underset{\substack{\text{D-Fructose}\\ [\alpha]=-92.2^\circ\\ (B)}}{C_6H_{12}O_6}$$

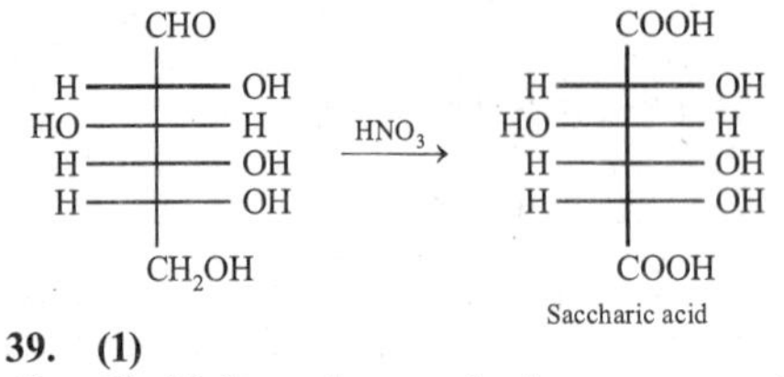

39. **(1)**

40. **(3)** Maltose is a reducing sugar and glycosidic linkage is present between C_1 and C_4 carbon.

41. **(1)** In lactose, β Linkage is between C_1 of galactose and C_4 of glucose.

42. **(3)** Sucrose is an example of disaccharide and non-reducing sugar because it does not have free hemiacetal linkage.

It involves glycoside linkage between C_1 of α-D-glucose C_2 of β-D-fructose.

43. **(4)** Lactose : It is a disaccharide of β–D–galactose and β-D-glucose with C_1 of galactose and C_4 of glucose link.

44. **(4)** Fructose is a six carbon monosaccharide containing ketonic group, so it is a ketohexose.

45. **(1)** Lactose contains β-glycosidic linkage between C_1 of galactose and C_4 of glucose.

46. **(3)**

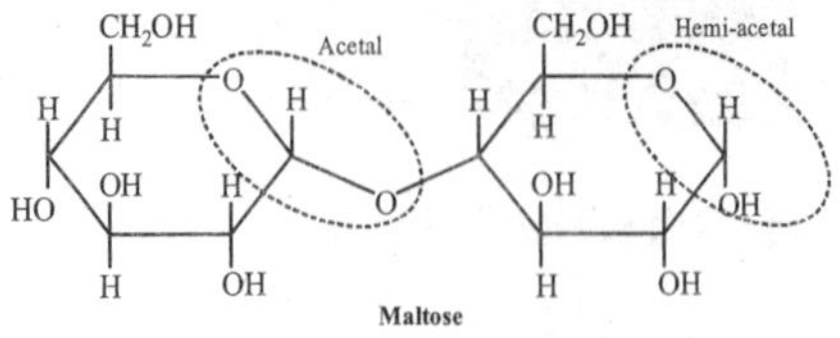

One acetal and one hemiacetal groups are present in maltose.

47. **(2)**

(i)

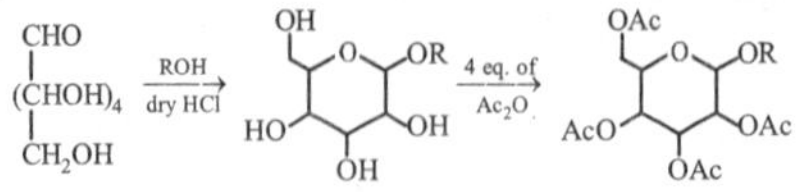

(ii) $CHO-(CHOH)_4-CH_2OH \xrightarrow{H_2/Ni} CH_2OH-(CHOH)_4-CH_2OH \xrightarrow[Ac_2O]{6\text{ eq. of}} CH_2OAc-(CH-OAc)_4-CH_2OAc$

(iii) $CHO-(CHOH)_4-CH_2OH \xrightarrow[Ac_2O]{5\text{ eq. of}} CHO-(CH-OAc)_4-CH_2OAc$

Thus $x = 4$, $y = 6$ and $z = 5$

48. **(3)** Seliwanoff reagent → [Resorcinol + Conc. HCl]

It is used to distinguish aldoses and ketoses. Ketoses show red colour immediately whereas aldoses show light pink colour after some time with Seliwanoff reagent.

49. **(2)** Glucose exists in cyclic form in which aldehyde group is not free, therefore it does not give Schiff's test.

50. **(4)** Maltose on hydrolysis gives two moles of α-D-glucose.

51. **(3)** Gluconic acid is obtained by partial oxidation of glucose by mild oxidising agent, e.g., Tollen's reagent, Fehling solution, Br_2 water.

$$\underset{\text{Glucose}}{CHO-(CHOH)_4-CH_2OH} \xrightarrow{Br_2\text{ water}} \underset{\text{Gluconic acid}}{COOH-(CHOH)_4-CH_2OH}$$

Gluconic acid can not form hemiacetal or acetal.

52. **(2)** Galactose and glucose are C_4 epimers.

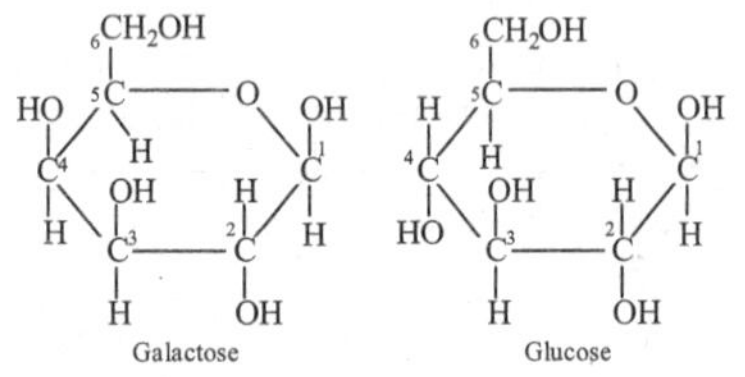

53. **(4)**

54. **(2)** Sucrose contains glycosidic link between C_1 of α-D-glucose and C_2 of β-D-fructose.

$C_{12}H_{22}O_{11} + H_2O \longrightarrow \text{Glucose} + \text{Fructose}$

55. **(4)** Hydrolysis of maltose gives glucose as maltose is composed of two α-D glucose units.

56. **(4)**

(1) Benedict reagent : $CuSO_4$ + Sod. citrate + Na_2CO_3 also gives red ppt. of Cu_2O with both.

(2) Fehling solution : $CuSO_4$ + Sod.–pot. tartarate (Rochelle salt) gives red ppt. of Cu_2O with glucose and fructose both.

(3) Barfoed reagent : 7%$(CH_3COO)_2Cu$ + 1% CH_3COOH + 92% H_2O gives red ppt. of Cu_2O with both.

(4) Seliwanoff's reagent : Resorcinol in conc. HCl gives red colour with both, but a ketose (fructose) reacts more quickly than an aldose (glucose).

57. (1) $\underset{CH_2OH}{\overset{CHO}{(CHOH)_4}} \xrightarrow{HI, \Delta} \underset{n\text{-Hexane}}{CH_3CH_2CH_2CH_2CH_2CH_3}$

58. (4) In amylose, 1, 4-α-glycosidic linkage is present.

59. (1)

60. (2)

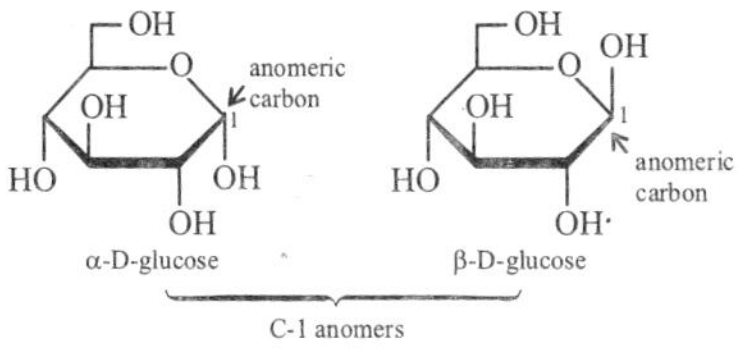

61. (4)

$\underset{\text{Stored in the form of glycogen}}{\text{Glucose}} \xrightarrow[\text{(absence of oxygen; needs only enzymes)}]{\text{enzyme}} \text{Pyruvic acid} \xrightarrow{O_2} CO_2 + H_2O$

$\downarrow$ Lactic acid

During vigorous exercise sufficient oxygen is not available to meet the energy demand, so energy is derived through conversion of pyruvic acid to lactic acid.

62. (4) Sucrose does not contain a free aldehydic or ketonic group, hence it does not show mutarotation.

63. (2) Glycosidic linkage is actually an ether bond as the linkage forming the rings in an oligosaccharide or polysaccharide is not just one bond, but the two bonds sharing an oxygen atom, e.g., sucrose.

64. (1) $6CO_2 + 12NADPH + 18ATP \longrightarrow C_6H_{12}O_6 + 12NADP + 18ADP$

65. (3) Fischer gave the prefix "D" to compounds whose bottom chiral has its OH to the right. So natural glucose is called D-glucose or dextrose.

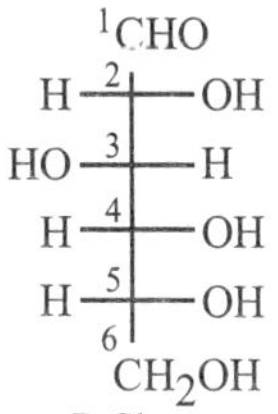

D-Glucose

66. (3) Glycogen is called animal starch and is found in all animal cells. It constitutes the reserve food material.

67. (2) Molisch's test : This is a general test for carbohydrates. One or two drops of alcoholic solution of α-naphthol is added to 2 mL glucose solution. 1 mL of conc. H_2SO_4 solution is added carefully along the sides of the test-tube. The formation of a violet ring at the junction of two liquids confirms the presence of a carbohydrate or sugar.

68. (3) Sucrose is a non-reducing sugar as it does not reduce Fehling reagent and Tollen's reagent.

69. (3) When either of the two forms of glucose is dissolved in water, there is change in rotation till the equilibrium value of + 52.5°. This is known as mutarotation.

$\underset{+111.5°}{\alpha\text{–D(+) Glucose}} \rightleftharpoons \underset{+52.5°}{\text{Equilibrium mixture}} \rightleftharpoons \underset{+19.5°}{\beta\text{–D–(+)Glucose}}$

70. (3) Glucose is considered as a typical carbohydrate which contains –CHO and –OH groups.

71. (3)

72. (2) Cyclization of the open chain structure of D-(+)-glucose has created a new stereocenter at C_1 which explains the existence of two cyclic forms of D-(+)-glucose, namely α– and β–. These two cyclic forms are *diasteromers, such diastereomers which differ only in the configuration of chiral carbon developed on hemiacetal formation* (it is C_1 in glucose and C_2 in fructose) *are called* **anomers** and the hemiacetal carbon (C_1 or C_2) is called the **anomeric carbon.**

73. (2) Both Amylose and Sucrose do not give Benedict's test as they are non-reducing sugars.

74. (9)

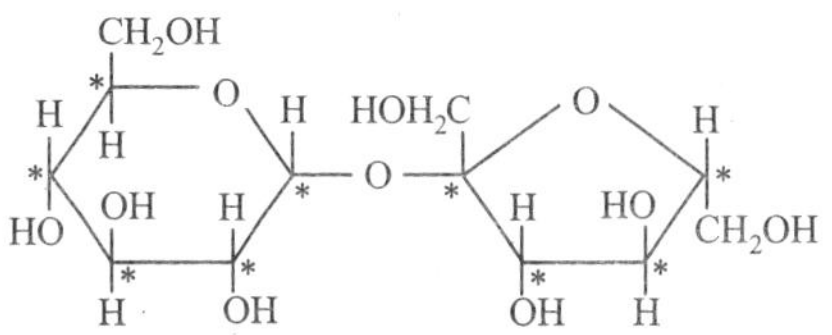

Sucrose

No. of chiral centres = 9.

75. (3)

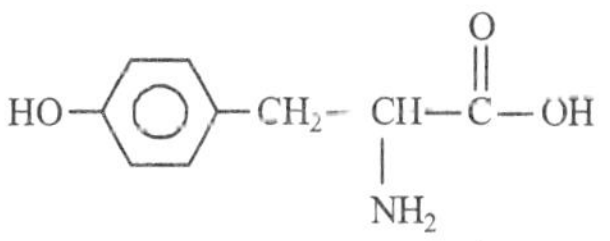

(Tyrosine)

Tyrosine amino acid contains phenolic group, hence it will give violet coloured complex with neutral $FeCl_3$ solution as phenolic group give positive test for $FeCl_3$ test.

76. (2) In the structure of Glutamine,

$\left[H_2N-\overset{O}{\overset{\|}{C}}-CH_2-CH_2-\overset{NH_2}{\overset{|}{CH}}-COOH \right]$

it has amide functional group in the chain. So it gives Hoffmann bromamide reaction.

In the structure of Lysine,

$H_2N-[CH_2]_4-\overset{NH_2}{\overset{|}{CH}}-COOH$ it has primary amino group. So it gives positive Hinsberg test.

In the structure of Tyrosine,

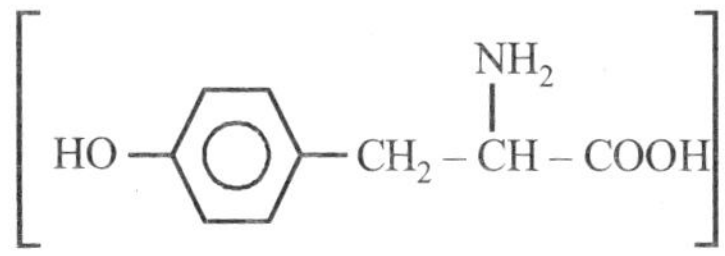

it has phenolic OH. So it gives positive test with neutral $FeCl_3$.

In the structure of Serine,

$HO-CH_2-\overset{NH_2}{\overset{|}{CH}}-COOH$ it has alcoholic –OH group. So it gives positive test with cerric ammonium nitrate.

77. (3) Changing the pH disrupts the tertiary structure of protein but the amino acid remain intact only the peptide bonds are affected.

78. (1) Arginine (R) and Lysine (K) are essential amino acids. Aspartic acid (D) and Glutamic acid (E) are non-essential amino acids.

79. (3) Single letter symbol of Tyrosine amino acid is Y, which is derivative of Thyroxine.

80. (3)

81. (1) gly ala val, gly val ala
val gly ala, val ala gly
ala val gly, ala gly val

∴ Total tripeptides = 6

82. (4)

(B) Histidine does contain heterocyclic ring. Its side chain contains an imidazole ring.

(C) Proline is a five membered cyclic ring amino acid.

(E) Cysteine has characteristic feature of side chain as $-CH_2CH_2-SMe$.

83. (1)

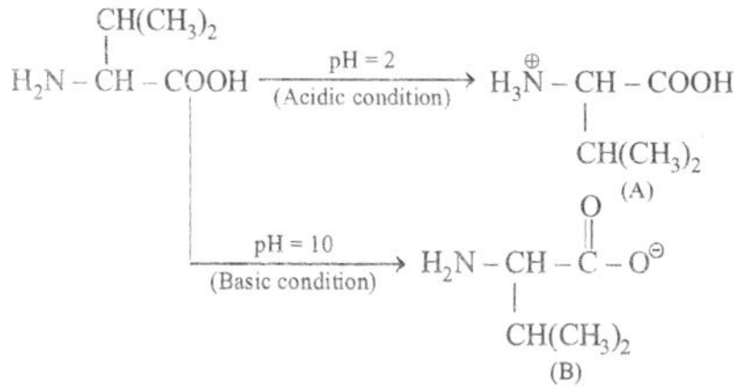

84. (3)

Statement - I: Protein do not yield β-amino acid upon hydrolysis, they produce α-amino acids.

Statement - II: Fibrous protein are generally insoluble, even in their native state.

So both statements are wrong.

85. (2)

86. (1)

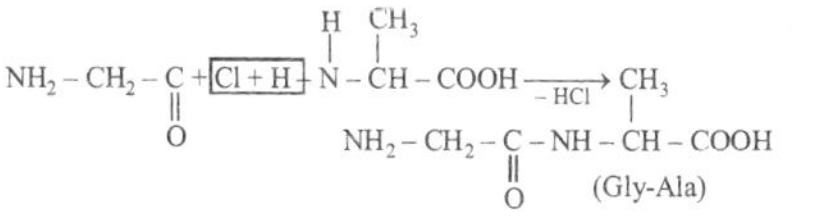

87. (4)

Let A = Gly, B = Ala, C = Val, D = Leu

The tetrapeptides formed including all the four amino acids are given below.

ABCD	BACD	CABD	DABC
ABDC	BADC	CADB	DACB
ACBD	BDAC	CBAD	DBAC
ACDB	BDCA	CBDA	DBCA
ADBC	BCAD	CDAB	DCAB
ADCB	BCDA	CDBA	DCBA

Total 24

88. (4) Isoleucine has 2 chiral centre, thus (1) is incorrect.

Glycine is optically inactive, thus (2) is incorrect.

Aspartic acid also contain COOH group at the side chain. Thus (3) is also incorrect.

Cystine easily dimmerise due to free SH group to form cystine.

89. (1) Valine, Lysine and Threonine are essential amino acids. Proline and Tyrosine non-essential amino acid.

90. (3) α-helix and β-pleated sheet belongs to secondary structure of protein, which have hydrogen bonds.

91. (1) On heating denaturation of protein occurs which result in coagulation of egg.
92. (4) All proteins are the polymers of α-amino acids. So, only α-amino acids are obtained on hydrolysis of proteins.
93. (1) Boiling an egg causes denaturation of its protein resulting in loss of its quarternary, tertiary and secondary structures.

94. (4) 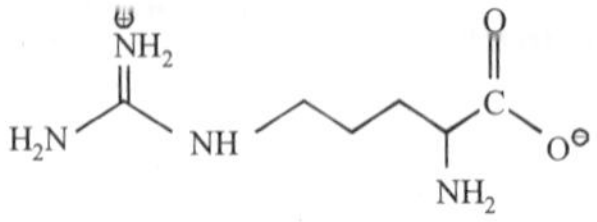

Arginine exist is zwitterion, so solid in nature and soluble in polar solvent.
95. (3) Asparagine has only one basic functional group ($-NH_2$) in its chemical structure.

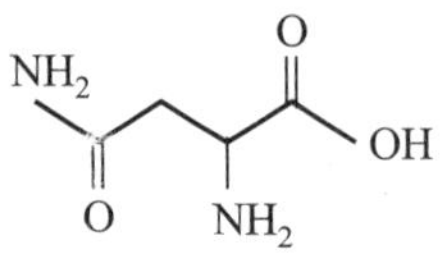

Others are basic amino acids with more than one basic functional group.
96. (3) 2° and 3° structures of proteins are stabilized by hydrogen bonding, disulphide linkages, van der Waals force of attraction and electrostatic force of attraction.
97. (2) Sulphur containing amino acids

cysteine 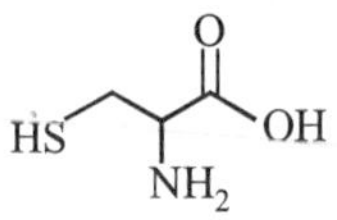 and

methionine

$$CH_3-S-CH_2-CH_2CH(NH_2)-\overset{O}{\overset{\|}{C}}-OH.$$

98. (3) A–III, B–I, C–IV, D–II
99. (4) (A) – IV, (B) – I, (C) – II, (D) – III
100. (2) Only option (2) is an α-Amino acid all other options are not α-amino acids.
101. (4) Two amino acids are

$$\underset{(A)}{H_2N-\underset{R_1}{CH}-COOH},\quad \underset{(B)}{H_2N-\underset{R_2}{CH}-COOH}$$

Tripeptides are formed from three amino acids. Therefore, possibilities are as follows.

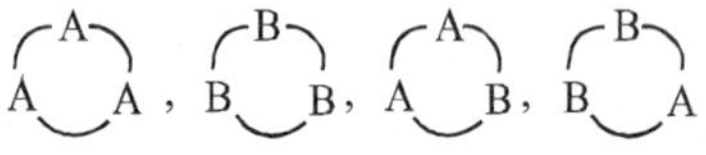

102. (2)
103. (1) Primary structure deals with the number, nature and sequence of amino acids linked together to form polypeptide chain; secondary structure deals with the shape of the peptide chain (linear, cyclic, branched or arranged in the form of helix and forces with which the peptide chains are held together). Tertiary structure of a protein refers to its three - dimensional structure i.e. coiling (folding) of the peptide chains. After denaturation, hydrogen bonds are broken, hence secondary and tertiary structures are destroyed and only primary structure remains intact.

104. (3) The primary structure of protein is associated with the linear sequence of amino acids in the protein.
Primary structure of protein is unaffected by physical 'or' chemical changes.
105. (2) H-bonding between carbonyl oxygen and N of the amino group of every third residue in the helical turn provides stability to the α-helix structure.
106. (2) Keratin, collagen and myosin are example of fibrous protein. They contain hydrogen and disulphide linkages and are insoluble in water. Insulin and albumin are the examples of globular proteins which are water soluble.
107. (2) Albumin is a water soluble globular protein produced in liver.
108. (3) Seliwanoff test is used to distinguish ketoses from aldoses. The reagent of this test consists of resorcinol and conc. HCl. On treatment with a concentrated acid, ketones are dehydrated more rapidly to give furfural derivative and on condensation with resorcinol give cherry red complex.
Xanthoproteic test is used to detect amino acids containing aromatic nucleus. On heating with conc. HNO_3 it gives yellow coloured nitro derivatives.
109. (2) Hydrogen bonding is responsible for stabilisation of α-helix structure of proteins.
110. (1) Tyrosine is a non-essential amino acid.
111. (3) Compounds having 1° amino group give carbylamine reaction with $CHCl_3$ and alc. KOH.
112. (2)

$$Asn \Rightarrow NH_2-\underset{O}{\underset{\|}{C}}-CH_2-\overset{H}{\underset{NH_2}{C}}-COOH$$

$$Ser \Rightarrow HO-CH_2-\overset{H}{\underset{NH_2}{C}}-COOH$$

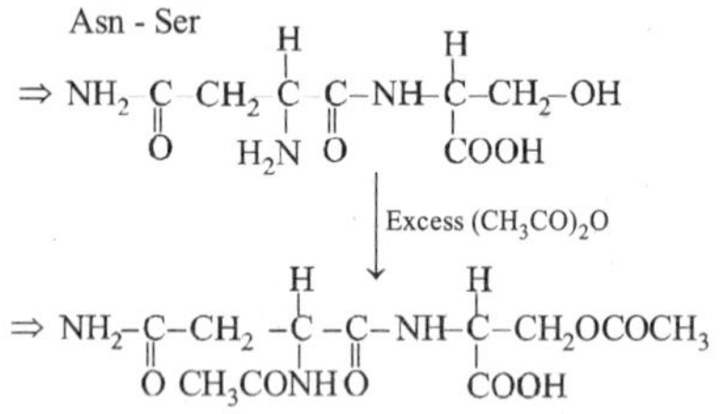

113. (1) Ceric ammonium nitrate test is used for detecting alcohols, while carbylamine test is for primary amines. Among the given peptides, only serine (Ser) has alcoholic group.

$$\underset{\text{Serine}}{HO-CH_2-\underset{NH_2}{CH}}-CO-NH-(CH_2)_4-\underset{\text{Lysine}}{\underset{NH_2}{CH}-COOH}$$

114. (2) Lysine,

$$H_2N-\overset{\alpha}{\underset{COOH}{CH}}-\overset{\beta}{CH_2}-\overset{\gamma}{CH_2}-\overset{\delta}{CH_2}-\overset{\epsilon}{CH_2}-NH_2$$

is most basic due to $\epsilon-NH_2$ group.

115. (3)

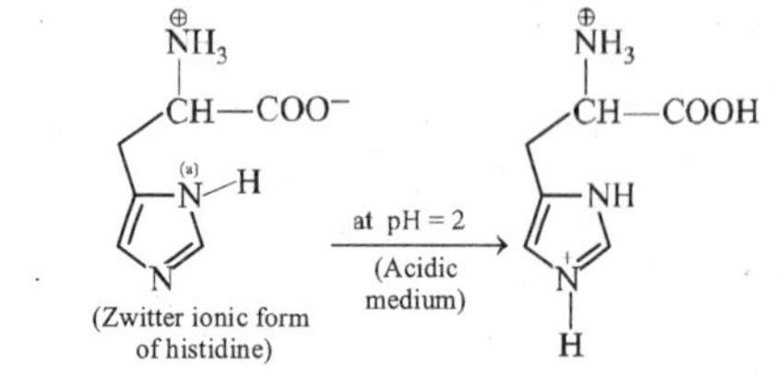

Note that *lp* of electrons on N labelled as N[(a)] are involved in delocalisation, hence not available for protonation.
116. (3)
117. (1)

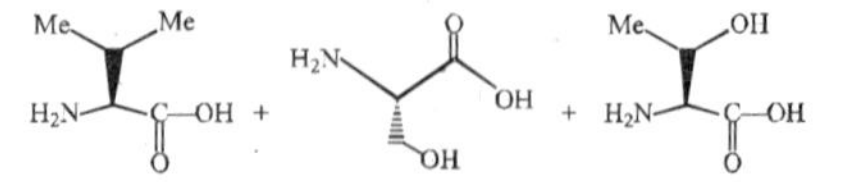

VALINE (Val) SERINE (Ser) THREONINE (Thr)

118. (2) The N atom of amide is not basic, so it will not exist in zwitter ionic form at pH 7.0.
119. (2) Those amino acids that cannot be synthesized in our body and must be supplied in diet is called essential amino acid for ex. valine, histidine, isoluecine etc.
120. (3)
121. (3) With the exception of glycine all other amino acids have different functional groups (atom) on the central tetrahedral alpha carbon.

$$H-\overset{H}{\underset{NH_2}{C}}-COOH$$

Glycine

122. (4) Statement (4) is not correct. Some proteins also contain S, along with C, H, O and N.
123. (1) Biuret test produces violet colour on addition of dilute $CuSO_4$ to alkaline solution of a compound containing peptide linkage. Polypeptides, proteins and urea have $-\underset{O}{\underset{\|}{C}}-NH-$ (peptide) linkage, while carbohydrates have glycosidic linkages. So, test of carbohydtrates should be different from that of other three.
124. (2) The secondary structure of a protein refers to the shape in which a long peptide chain can exist. There are two different conformations of the peptide linkage present in protein, these are α-helix and β-conformation. The α-helix always has a right handed arrangement.
125. (4) Essential amino acids are:-
Arginine, Phenylalanine, Histidine and Valine
126. (9) Tyrosine is a nonessential amino acid the body makes from another amino acid called phenylalanine.
Number of carbon atoms = 9
127. (6) A tripeptide contains 3 amino acids linked by two peptide. If A, B, C are 3 different amino acids, then 6 tripeptides bonds will be formed.
Three different amino acids can be arranged by 3!
In short, $3 \times 2 = 6$

128. (10)

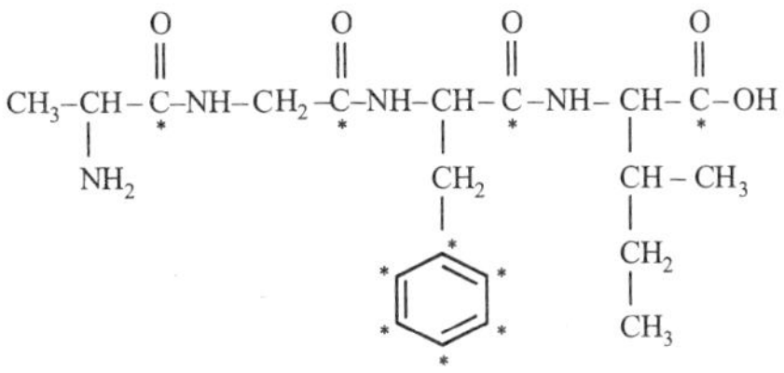

129. (6) Number of peptide linkage
= (No. of α-amino acids – 1) = 7 – 1 = 6

130. (8) No. of possible tripeptide = x^n
where x = No. of α-acids
and n = nature of peptide
Thus, here, amino acids = 2
Hence, No. of tripeptides = $2^3 = 8$
(1) val – val – val (2) pro – pro – pro
(3) val – pro – pr (4) pro – val – pro
(5) val – val – pr (6) val – pro – val
(7) pro – pro – val (8) pro – val – val

131. (1) A tetrapeptide means, there are three peptide bonds, hence number of amino acids should be four. So
No. of amino acids – No. of peptide bonds = 4 – 3 = 1

132. (2) Biuret test is given by compounds having a peptide linkage. So tripeptide and Biuret having peptide bonds, will give biuret test.

133. (4) Ala-Gly-Leu-Ala-Val
Four peptide linkages to join five amino acids.

134. (3) No. of peptide linkage = n – 1
n = No. of reacted amino acids
Now, n = 4 (given)
No. of peptide linkage = 4 – 1 = 3.

135. (4)

136. (4)

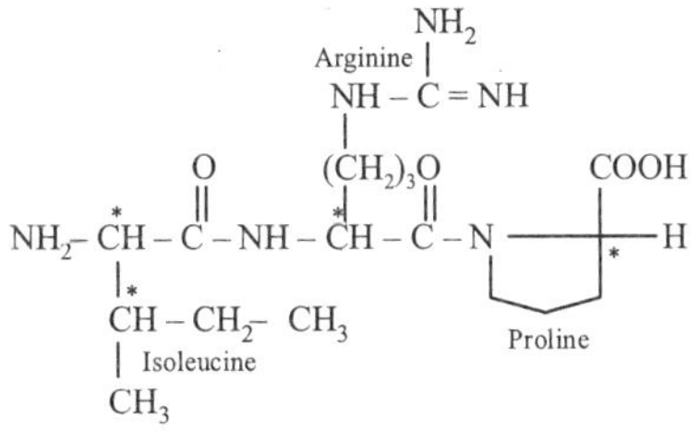

* Represents chiral centre

137. (2)

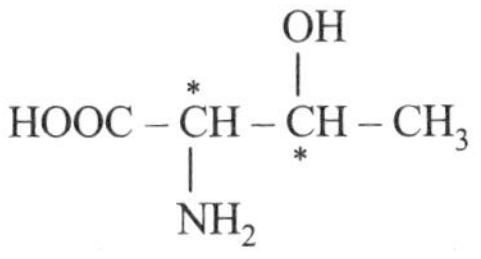

Threonine

No. of chiral centres = 2

138. (5) Asp – Glu – Lys tripeptide is :

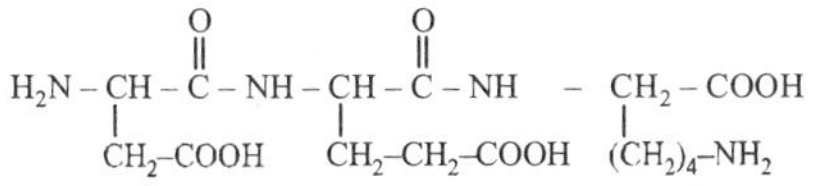

No. of CO groups = 5

139. (37.84)
Molecular formula of histamine is $C_5H_9N_3$
Molecular mass of histamine
$= 5 \times 12 + 9 \times 1 + 3 \times 14 = 111$
Mass percentage of nitrogen in histamine
$$= \frac{42}{111} \times 100 = 37.84\%$$

140. (3) DNA and RNA are chiral molecules due to presence of chiral 2-deoxy-ribose and ribose sugar unit respectively.

141. (3)
- Activitation energy for enzyme catalysed hydrolysis of sucrose is lower than that of acid catalysed hydrolysis.
- During denaturation secondary and tertiary structures of a protein are destroyed but primary structure remains intact.
- Nucleotides are joint together by phosphodiester linkage.
- The unique three dimensonal shape of a fully folded single polypeptide is known as tertiary structure. The quaternary structure refers to the three dimensional structure formed by multiple polypeptide chains.

142. (1) (A)-(III), (B)-(IV), (C)-(II), (D)-(I)

143. (3) DNA contain four bases, which are adenine, guanine, cytosine and thymine. Option (1) is adenine, option (2) is thymine option (4) is cytosine. Option (3) is not the correct structure of guanine. Correct structure of guanine is

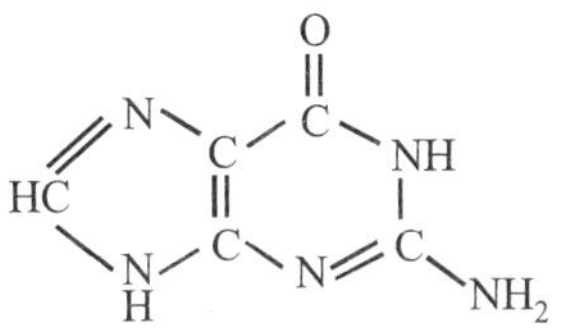

144. (3) Enzymes are biocatalyst and are very specific for a particular reaction and for a particular substrate. Almost all enzymes are globular proteins. Enzyme maltase hydrolyses maltose into glucose.

$$\text{Maltose} \xrightarrow{\text{Maltase}} \text{Glucose}$$

145. (2) In DNA G pairs with C and A pairs with T only.

A T G C T T C A
| | | | | | | |
| | | | | | | |
T A C G A A G T

146. (1) A-II, B-III, C-I, D-IV

147. (1) Phosphodiester linkage joins two nucleotides.

148. (1) DNA mainly consists of a nitrogeneous base, phosphate group and a pentose sugar.
So, hydrolysis of DNA produces a pentose sugar.

149. (2) DNA contains
⇒ β–D–2–deoxyribose
RNA contains ⇒ β–D–ribose

150. (4)

151. (3)

152. (2)

$$\underset{\text{Sucrose}}{C_{12}H_{22}O_{11}} + H_2O \xrightarrow{\text{Invertase}} \underset{\text{Glucose}}{C_6H_{12}O_6} + \underset{\text{Fructose}}{C_6H_{12}O_6}$$

$$\underset{\text{Glu cos e}}{C_6H_{12}O_6} \xrightarrow{\text{Zymase}} 2C_2H_5OH + 2CO_2$$

Invertase enzyme catalyses the hydrolysis of sucrose and give mixture of glucose and fructose which is also known as inverted sugar.
While zymase enzyme catalyses the fermentation of sugar into ethanol and CO_2.
∴ Enzyme *A* is invertase.
Enzyme *B* is Zymase.

153. (2) RNA has a single helix structure, whereas, DNA has a double helix structure.

154. (2) RNA contains adenine (A), guanine (G), cytosine (C) and uracil (U).
(1) is cytosine, found in RNA and DNA, both.
(2) is uracil, found in RNA only.
(3) is thymine, found in DNA only.
(4) is not a pyrimidine, i.e., not a constituent of RNA and DNA.

155. (1)

156. (1) Among 20 naturally occurring amino acids "cysteine" has '– SH' or thiol functional group.
⇒ General formula of amino acid

$$H_2N - \underset{\underset{R}{|}}{CH} - COOH$$

⇒ Value of R = $-CH_2-SH$ in cysteine.

157. (4) The two polynucleotide chains of DNA molecules are twisted around a common axis but run in opposite directions to form a right handed helix. The two chains are joined together by specific hydrogen bonds. Adenine-Thymine (two hydrogen bonds) and cytosine-guanine (three hydrogen bonds)

158. (1) DNA contains ATGC bases
Quinoline is not present in DNA or RNA.

159. (3) Butyric acid, also known as butanoic acid, is found in milk, and butter and is a product of anaerobic fermentation. It has an unpleasant smell and acrid taste.

160. (1) Maltose is obtained by partial hydrolysis of starch by the enzyme diastase present in malt.

$$(C_6H_{10}O_5)_n + nH_2O \xrightarrow{\text{Diastase}} nC_6H_{12}O_6$$
(maltose)

161. (4) The correct structure of thymine is

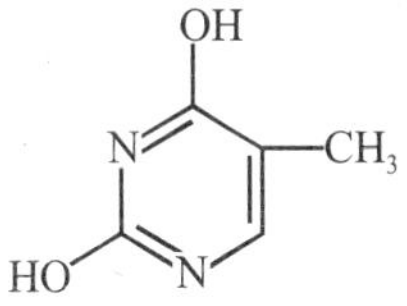

Thymine (T)

162. (1) RNA controls the synthesis of proteins.

163. (1) The pyrimidine bases present in DNA are cytosine and thymine.

164. (2) In DNA and RNA, heterocyclic base and phosphate ester are at C_1' and C_5' respectively of the sugar molecule.

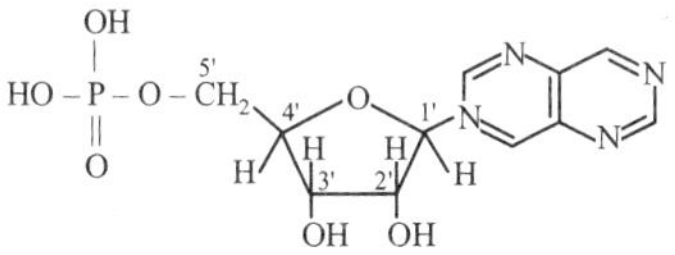

165. (3) RNA contains cytosine and uracil as pyrimidine bases while DNA has cytosine and thymine.
Both have the same purine bases i.e., guanine and adenine.

166. (2) In RNA, the sugar is D–ribose and base is uracil, whereas in DNA, the sugar is D-2 deoxyribose and the nitrogenous base is thymine.

167. (33) Adenine form two hydrogen bonds with thymine and guanine form three hydrogen bond with cytosine.
5' G–G–C–A–A–A–T–C–G–G–C–T–A–3'

In given DNA strand total seven guanine and cytosine bases are present which form total 21 H-bonds and six adenine and thymine bases are present which will form total 12 H-bonds with other DNA strand.

$\therefore$ Total no. of H bonds $= 7 \times 3 + 6 \times 2 = 33$

168. (3) A is false as DNA is regarded as the reserve of genetic information.

B is true as DNA undergoes replication during cell division.

C is false as RNA synthesizes proteins in the cell.

D is true as DNA stores the information about the protein or amino acids to be synthesized in the form of specific sequence.

E is true as DNA undergoes replication and forms strands for the daughter cells that are identical.

169. (25) Mol. wt of uracil, $C_4N_2H_4O_2 = 112$

$$\% N = \frac{28}{112} \times 100 = 25\%$$

170. (9)

171. (3) Deficiency of Vitamin - A leads to Xerophthalmia

Deficiency of Vitamin - C (Ascorbic acid) leads to Scurvy

Deficiency of Vitamin - B_2 (Riboflavin) leads to Cheilosis

Deficiency of Vitamin - B_6 (Pyridoxine) leads to Convulsions

172. (3) A-I, B-III, C-II, D-IV

173. (4) Vitamin D, E, K, A are fat soluble vitamins.

174. (3) Vitamin C is Ascorbic acid.

175. (4) A – III, B – I, C – IV, D – II

176. (1) Noradrenaline is a neurotransmitter and is released by adrenal medulla. Low level of noradrenaline in the human body causes depression.

177. (4) B group vitamins and vitamin C are soluble in water except vitamin B_{12} all are excreted easily.

178. (4) Vitamin B_1 - Thiamin

Vitamin B_6 - Pyridoxine

179. (2) Vitamin K is essential for the formation of coagulation factors as well as several proteins that work together to clot the blood when injuries occur and prevent excessive bleeding. So, deficiency of vitamin K causes increase in blood clotting time.

180. (3) (i) – (c), (ii) – (a), (iii) – (d), (iv) – (b)

181. (3) Water soluble vitamins dissolve in water and are not stored by the body. The water soluble vitamins include the vitamin B-complex group and vitamin C.

182. (3) Deficiency of vitamin D causes rickets.

183. (3) When either of the two forms of glucose is dissolved in water, there is change in rotation till the equllibrium value of $+52.5°$, This is known as mutarotation

$$\underset{+111.5°}{\alpha\text{–D(+) Glucose}} \rightleftharpoons \underset{+52.5°}{\text{Equilibrium mixture}} \rightleftharpoons \underset{+19.5°}{\beta\text{–D–(+)Glucose}}$$

184. (2) Insulin is a biochemically active peptide hormone secreted by pancreas.

185. (5) Vitamins A, D, E, K and B_{12} are stored in our body.

Part-B (JEE Advanced)

1. (a)

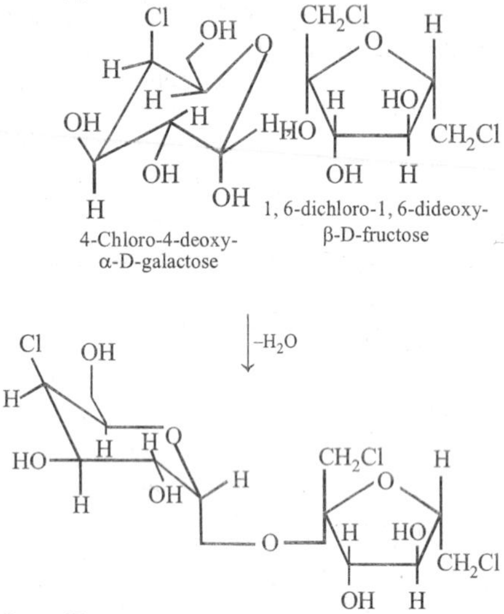

2. (d)

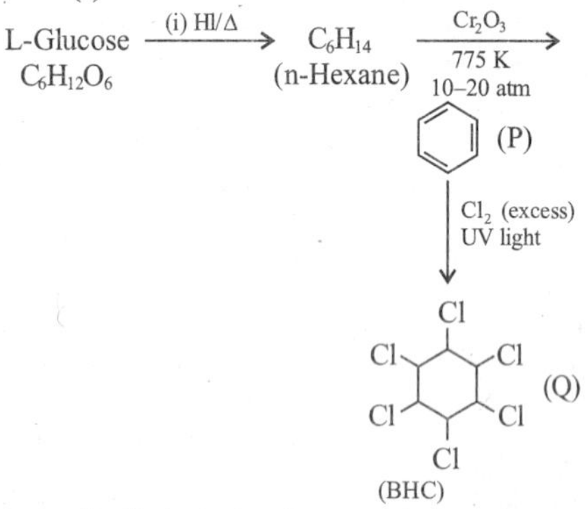

3. (a) Hydrolysis of sucrose brings about a change in the sign of rotation from dextro to laevo.

$$\text{Sucrose} \xrightarrow{H_3O^+} \alpha - D(+)$$

Glucopyranose + β- D(+) Fructofuranose

$$\text{Sucrose} \xrightarrow{Br_2 + H_2O} \text{No reaction}$$

option (b), (c), (d) are reducing sugars, will get oxidized by bromine water.

4. (c)

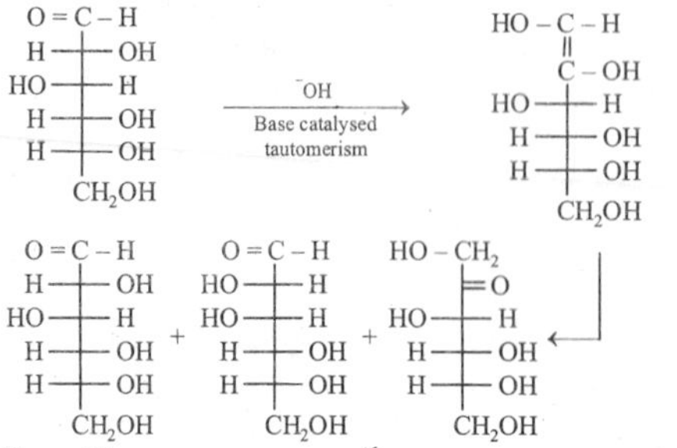

5. (d)

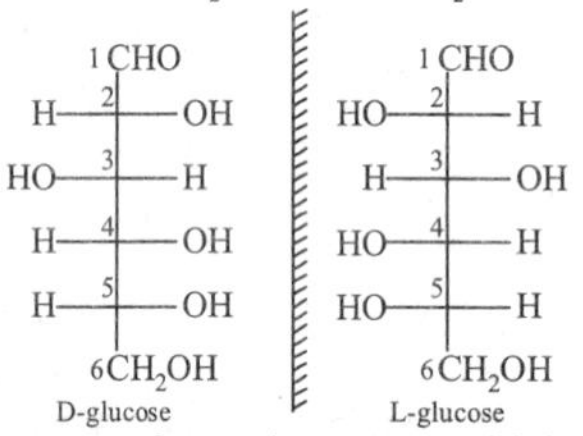

(i) In pyranose form, the groups which are in right side of Fischer projection will be below the plane.

(ii) For β-form, –OH at C_1 and $–CH_2OH$ at C_5 will be on the same side.

(iii) At C2, C3 and C3, C4; –H and –OH are trans to each other.

Hence, the structure of β-L-glucopyranose will be :

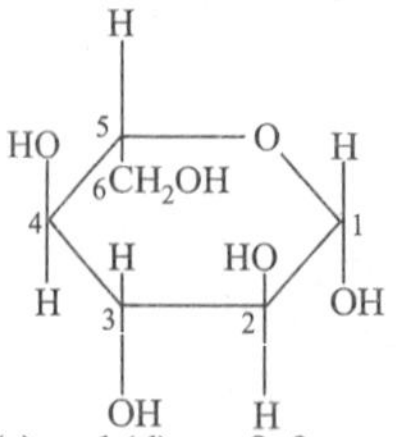

Only option (c) and (d) are β-forms. However, in option (c) at C2 and C3, –OH groups are at the same side. Hence, only option (d) can be the answer.

6. (a)

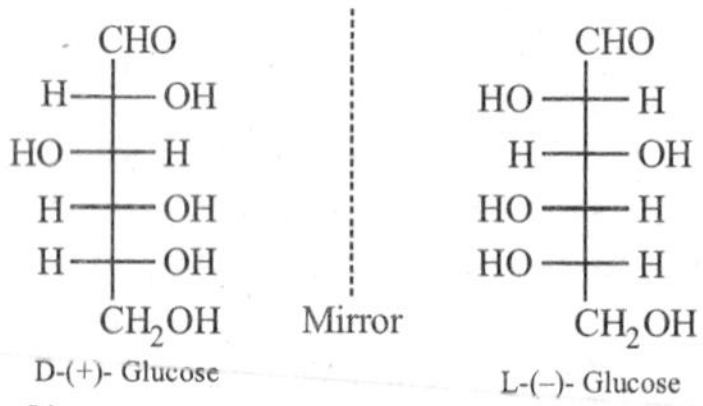

7. (b)

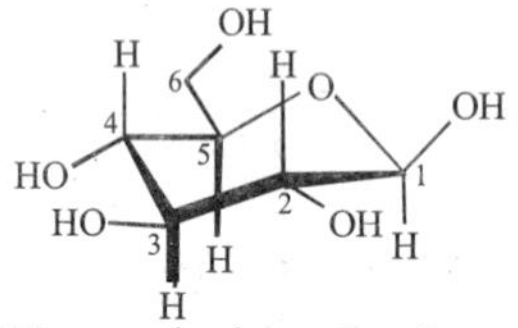

At C_1, –OH group is above the plane, thus, it is β-pyranose.

If hydrogen is present at the anomeric carbon (i.e. C_1), then it is an aldose.

Hence, the given compound is an aldohexose.

8. (a) Ring (A) is pyranose (analogous to pyran) and ring (B) is furanose (analogous to furan). Before forming the disaccharide the monosaccharide (A) has –OH group at C_1, below the plane.

The monosaccharide (B) has –OH group at C_1, above the plane.

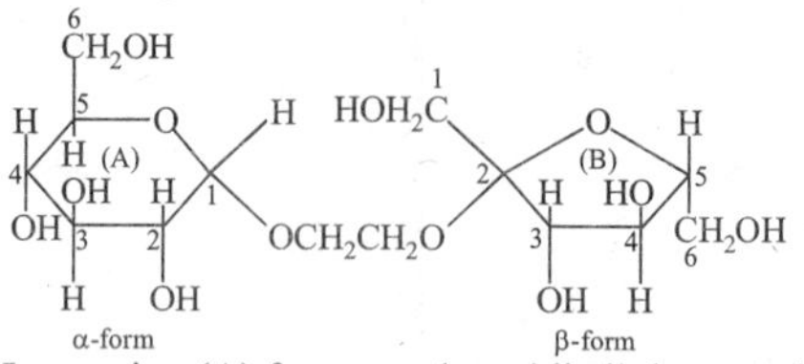

Hence, ring (A) forms α-glycosidic linkage and ring (B) forms β-glycosidic linkage.

9. (a) Cellulose is a polysaccharide composed of only D-glucose units. Every adjacent glucose units are joined by β-glycosidic linkage between C_1 of one glucose and C_4 of the next.

Thus in every glucose units only three –OH groups are free to form triacetate.

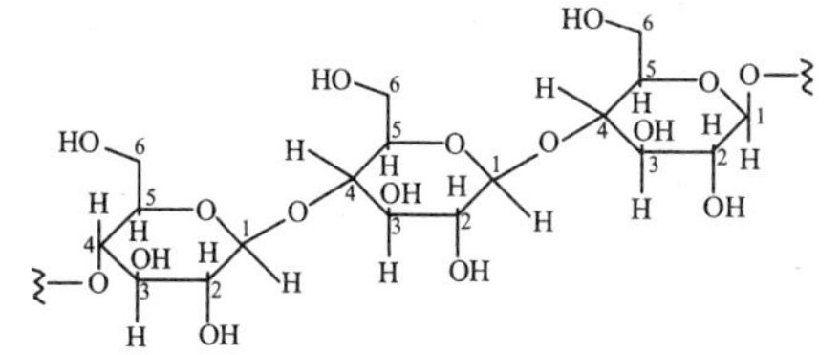

$$(\text{–OH} \rightarrow \text{–OCOCH}_3)$$

$(CH_3CO)_2O \mid H_2SO_4$

Cellulose triacetate

10. (b) The two isomeric forms (α – and β –) of D-glucopyranose differ in configuration only at C–1; hence these are called anomers.

11. (d) Glucose being an aldose responds to Tollen's test while fructose (an α-hydroxy ketone), although a ketose, undergoes rearrangement in presence of basic medium (provided by Tollen's reagent) to form glucose, which then responds to Tollen's test.

12. (8)

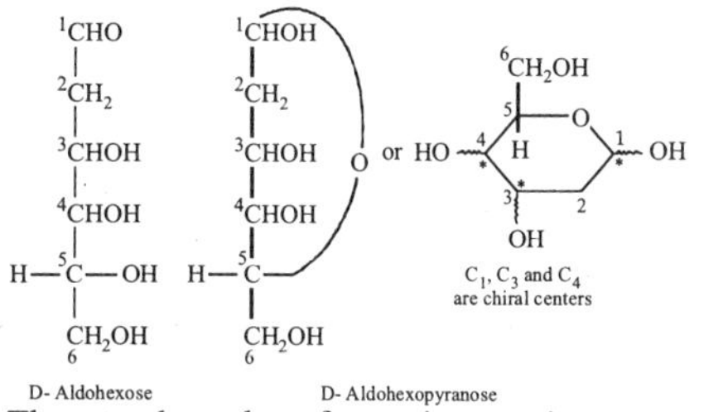

Thus, total number of stereoisomers in pyranose form of D-configuration = $2^3 = 8$

13. (2)

Octasaccharide + 7 H_2O
(molar mass = 1024 g/mol) (molar mass = 126 g/mol)
$\longrightarrow$ Ribose + 2 deoxyribose + glucose
(Total mass = 1024 + 126 = 1150 g/mol)

$$58.26 = \frac{134 \times n}{1150} \times 100$$

$\therefore$ 134n = 669.99 $\Rightarrow$ n = 4.99 $\approx$ 5

$\therefore$ 1150 = (5 × 150) + (x × 150) + (y × 180)

$\Rightarrow$ 1150 = 750 + 150x + 180y

$\therefore$ 750 = 5 unit

$\Rightarrow$ 300 = 150x = 2 unit

180 = 180y = 1 unit

$\therefore$ Ribose unit = 2.

14. (b, c)

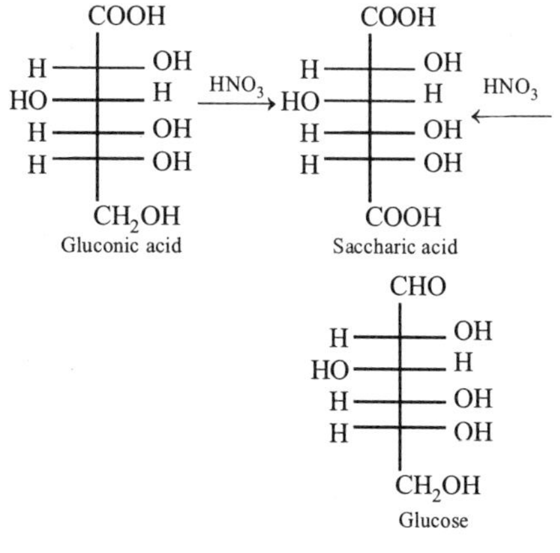

(a) Both gluconic acid and glucose oxidised to saccharic acid. So option (a) is incorrect.

(b) Fructose isomerises to glucose and another aldohexoses in the presence of Fehling's reagent in basic medium via Lobry de Bruyn - van Ekestein transformation.

(c) Invert sugar is equimolar mixture of D-glucose and D-fructose.

(d) Specific rotation of D-fructose $\Rightarrow$ –92.5°
Specific rotation of D-glucose $\Rightarrow$ +52.5°
Specific rotation of invert sugar

$$= -\frac{40}{2} = -20°$$

So option (d) is incorrect.

15. (c, d)

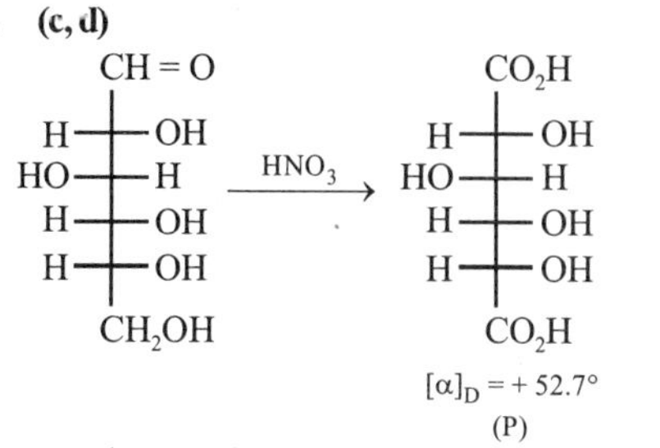

The enantiomer of P has rotation – 52.7° is as follow

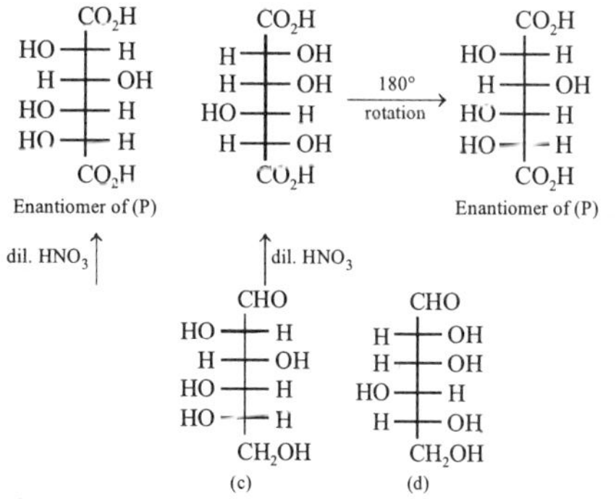

16. (b, c, d)

(a) Glucose $\xrightarrow[H_2O]{Br_2}$ Gluconic acid.

Bromine water oxidises only aldehyde group to carboxylic group. It neither oxidises –OH group nor –CO group.

(b) $\underset{(+)}{\text{Sucrose}} \xrightarrow{H_2O} \underset{(+)}{\text{Glucose}} + \underset{(-)}{\text{Fructose}}$

Sucrose is dextrorotatory (+ 66.6°) in nature but on hydrolysis it gives dextrorotatory glucose (+52.5°) and laevorotatory fructose (–92.4°). Overall solution is laevorotatory since laevorotation is more than dextrorotation.

(c) 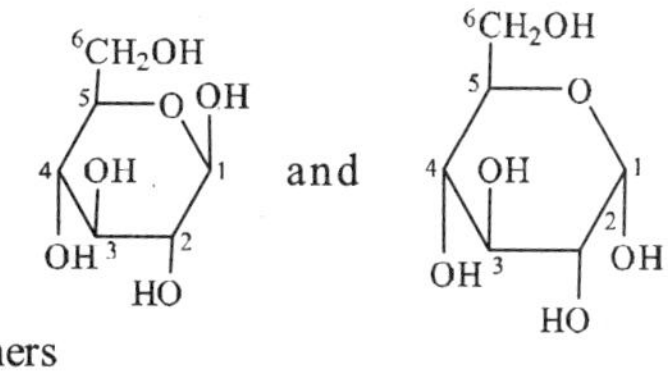 anomers

Because they differ in configuration only around C_1 position.

(d) Monosaccharides do not undergo further hydrolysis due to absence of glycosidic linkages.

17. (b, c) Invert sugar is an equimolar mixture of D-(+) glucose and D-(–) fructose.

$$\underset{(+)\text{ Sucrose}}{C_{12}H_{22}O_{11}} + H_2O \xrightarrow{H^+} \underbrace{\underset{\substack{D(+)-\text{Glucose} \\ (+52°)}}{C_6H_{12}O_6} + \underset{\substack{D(-)-\text{fructose} \\ (-92°)}}{C_6H_{12}O_6}}_{\text{Invert sugar}}$$

Since the specific rotation of the hydrolysis product of (+) sucrose is inverted, i.e., (–), it is known as inverted sugar.

- Specific rotation of invert sugar

$$= \frac{-92° + 52°}{2} = -20°$$

- D-glucose on oxidation with Br_2-water produces gluconic acid and not saccharic acid.

18. (b, c)

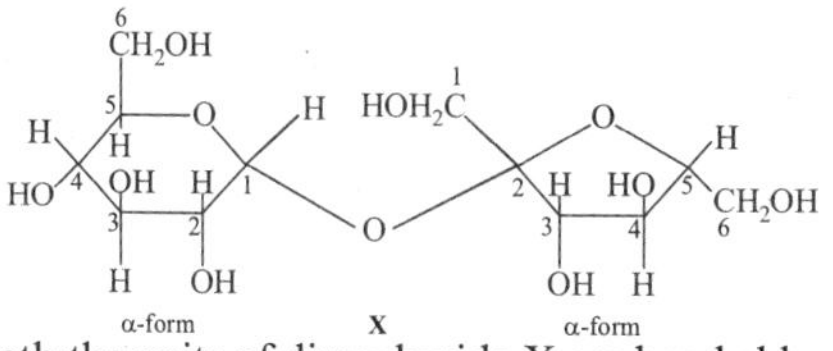

Both the units of disaccharide X are bonded by C1 – C2 linkage. The reducing groups (aldehydic and ketonic respectively) are bonded together. Thus, X is non-reducing sugar.

Also, X has α-glycosidic linkage.

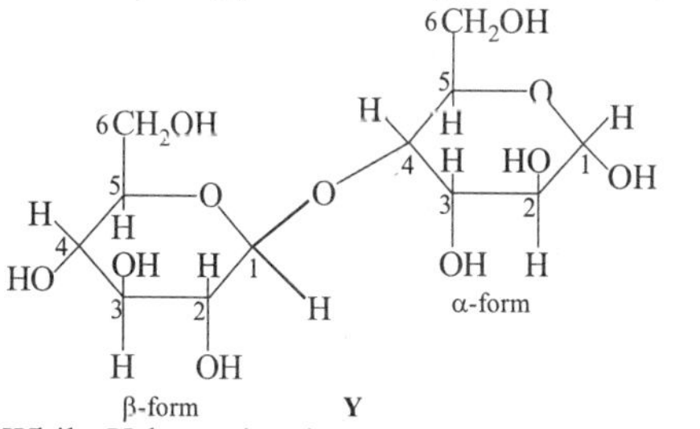

While Y has a hemiacetal group, hence it is reducing sugar. Also Y has β-linkage.

19. (i) – (c), (ii) – (a), (iii) – (d), (iv) – (b)

20. (c) Statement-1 is correct, but statement-2 is incorrect because glucose on reaction with Fehling's solution gives Cu_2O and gluconic acid.

21. In the two disaccharides, structure *A* will be reducing sugar since, both monosaccharides units are not linked through their reducing centers i.e., it has hemiacetal linkage, while in structure *B*, both the monosaccharide units are linked through their reducing centers, i.e., it has acetal linkage, hence it will be non-reducing.

22. L-Glucose is an enantiomer of D-glucose, hence

(i) 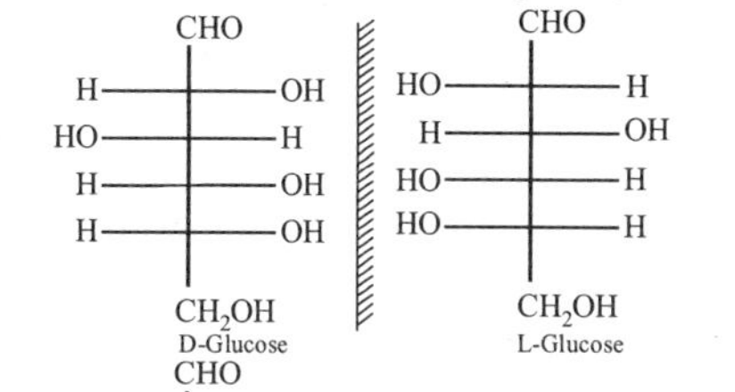

(ii) 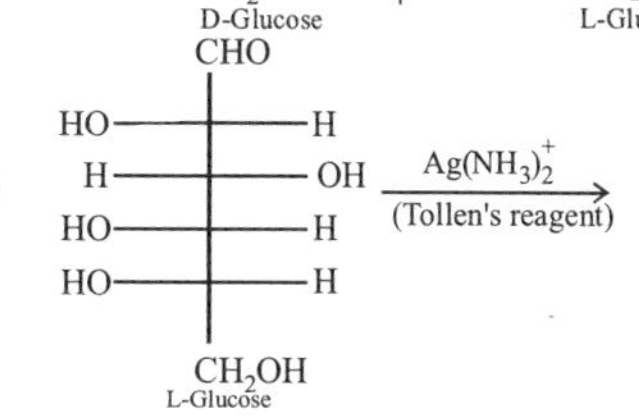

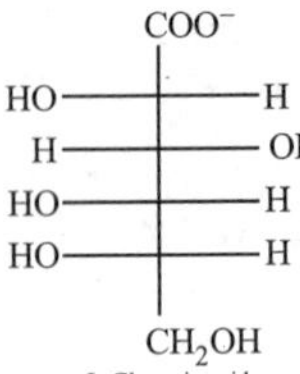

L-Gluconic acid

23. **(5)** $|z_1|+|z_2|+|z_3|=5$

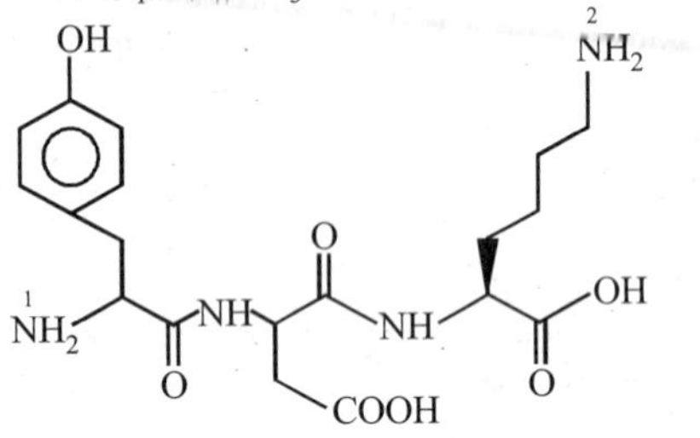

At pH = 2 $\overset{1}{NH_2}$ and $\overset{2}{NH_2}$ of tyrosine and lysine are positively charged (+1 each) $+2\ |z_1|=2$

At pH = 6, NH_2 of Lysine (+1), COOH (–1) of glutamic acid, so because of dipolar ion, $|z_2|=0$

At pH = 11, COOH of glutamic acid (–1)

COOH of lysine (–1)

OH of phenol (–1)

$\therefore\quad |z_3|=3$

24. **(1)** The given structure has peptide bonds which on hydrolysis gives different amino acids, namely A, B, C and D. Out of these only A is naturally occurring which is glycine.

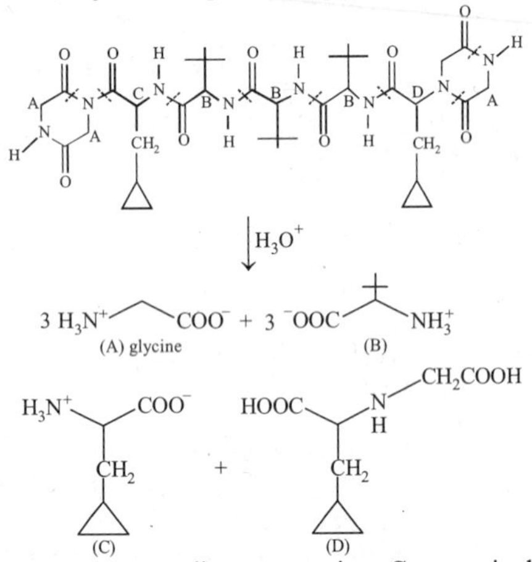

25. **(4)** According to question, C – terminal must be alanine and N – terminal do have chiral carbon means it should not be glycine. So possible sequence is :

Val Phe Gly Ala
Val Gly Phe Ala
Phe Val Gly Ala
Phe Gly Val Ala

The –COOH group of alanine does not participate in forming peptide bond as given in the question.

26. **(4)** Peptides with isoelectric point (pI) more than 7 would exist as cation in neutral solution (pH = 7) which means the given polypeptide is of basic nature, so it must contain two or more amino groups. Hence IV, VI, VIII and IX are correct options.

27. **(6)** Molecular weight of decapeptide = 796 g/mol

Total bonds to be hydrolysed = (10 – 1) = 9 per molecule

Total weight of H_2O added = 9 × 18 = 162 g/mol

Total weight of hydrolysis products = 796 + 162 = 958 g

Total weight % of glycine (given) = 47%

Total weight of glycine in product

$$=\frac{958\times 47}{100}\text{g}=450\text{g}$$

Molecular weight of glycine = 75 g/mol

$$\text{Number of glycine molecules}=\frac{450}{75}=6$$

A dipeptide has one peptide bond. Thus, a decapeptide has 9 peptde bonds.

28. **(2)** The basic groups in the given form of lysine is NH_2 (not $\overset{\oplus}{N}H_3$) and CO_2^-.

29. The structure of the two possible dipeptides are

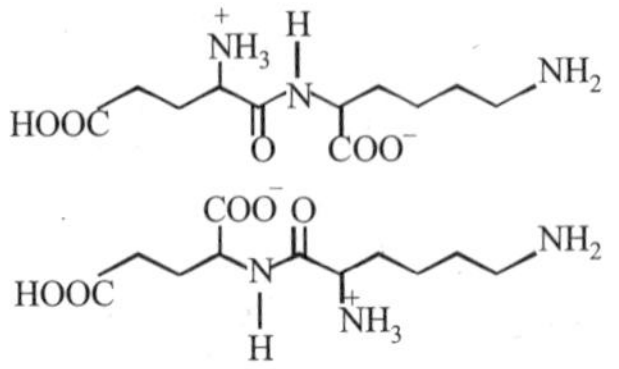

A dipeptide structure has one O=C—N—H linkage.

30.

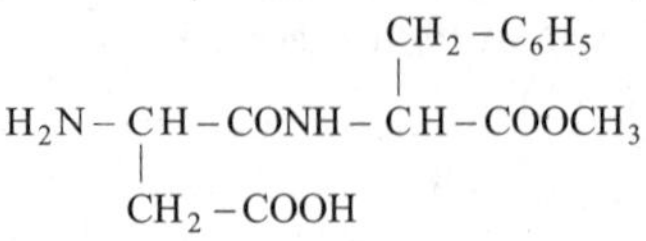

Aspartame

(i) Four functional groups present in it are
(a) $-NH_2$ (Amine)
(b) –COOH (Carboxylic acid)
(c) $-\overset{O}{\overset{||}{C}}-NH-$ (Amide)
(d) $-\overset{O}{\overset{||}{C}}-O-CH_3$ (Ester)

(ii) Zwitterion structure is given as follows :

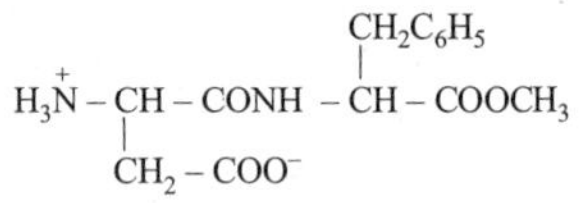

(iii)

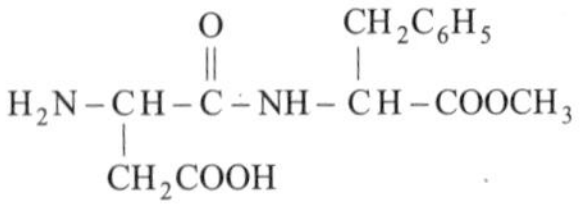

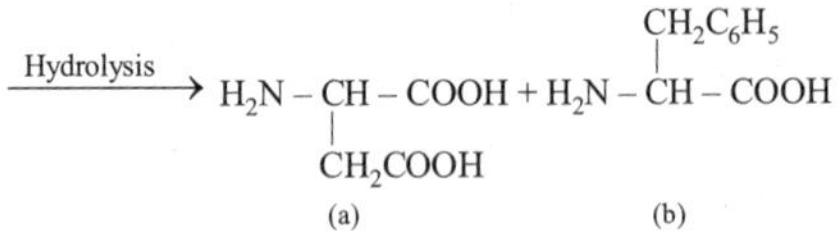

Hence, on hydrolysis two amino acids (a) and (b) are obtained.

(iv) Of the above two amino acids, (b) $NH_2-\underset{CH_2C_6H_5}{\underset{|}{C}}H-COOH$ is more hydrophobic due to presence of non-polar and bulky benzyl group.

31. pH = 2 indicates acidic character whereas pH = 10 indicates basic character.

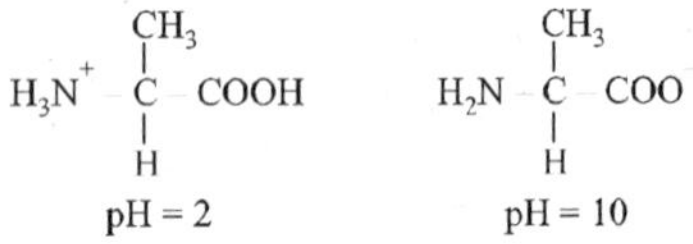

32. **(41)**

A G T C A C G T A A G T C
T C A G T G C A T T C A G

Total energy = [BE of H-bond of A – T × No. of A = T pair × 2] + [BE of H-bond of G – C × No. of G ≡ C pari × 3] = [1 × 7 × 2] + [1.5 × 6 × 3] = 14 + 27 = 41 kcal

Chapter-15

Polymers

Part-A (JEE Main)

1. **(4)** Proteins, starch, rubber are natural polymer. Rayon [cellulose acetate] is semisynthetic polymer.

2. **(2)**

3. **(2)** Copolymer are made up of two or more different monomer species.

Buna-S, PHBV and Butadiene-styrene are copolymer. Only neoprene is homopolymer.

Homopolymer contain one kind of monomer units.

4. **(1)** **Buna - s** is copolymer of butadiene + styrene.

Nylon 6, 6 is condensation polymer of adipic acid and hexanediamine and also a fiber.

Terylene is a fiber not thermosetting polymer.

Buna-N is a copolymer not homopolymer.

5. **(4)** PHBV is an example of polyester.

6. **(3)** Buna-N is an addition copolymer of 1,3-butadiene and acrylonitrile.

$$CH_2 = CH - CH = CH_2 + CH_2 = \underset{\large CN}{\underset{|}{CH}}$$

$$[-CH_3 - CH = CH - CH_2 - CH_2 - \underset{\large CN}{\underset{|}{CH}} -]_n$$

Buna-N

7. **(2)** Buna-N is obtained by addition polymerisation.

$$\underset{\text{1,3-Butadiene}}{nCH_2 = CH - CH = CH_2} + \underset{\text{Acrylonitrile}}{nCH_2 = \overset{\large CN}{\overset{|}{CH}}} \xrightarrow{\text{Copolymerisation}}$$

$$\underset{\text{Buna-N}}{+CH_2 - CH = CH - CH_2 - CH_2 - \overset{\large CN}{\overset{|}{CH}}+_n}$$

Nylon-6, bakelite, nylon-6, 6 are obtained by condensation polymerisation.

8. **(4)**

$$\underset{\textit{iso}\text{-Butylene}}{CH_3 - \underset{\large CH_3}{\underset{|}{C}} = CH_2} \xrightarrow{\text{Polymerisation}} \underset{\text{Polyisobutylene}}{\left[-CH_2 - C(CH_3)_2 - \right]_n}$$

9. **(4)** Nylon –6, 6 is obtained by condensation polymerisation of hexamethylenediamine and adipic acid.

$$NH - (CH_2)_6 - NH \boxed{H + HO} \overset{O}{\overset{\|}{C}} - (CH_2)_4 - \overset{O}{\overset{\|}{C}} \boxed{OH}$$

$$\xrightarrow{-H_2O} \underset{\text{Nylon-6, 6}}{[NH - (CH_2)_6 - NH - CO - (CH_2)_4 - CO]_n}$$

So, nylon-6, 6 is a condensation polymer. Other given polymers are addition polymers.

10. **(3)** Nylon-6 is prepared from caprolactum.

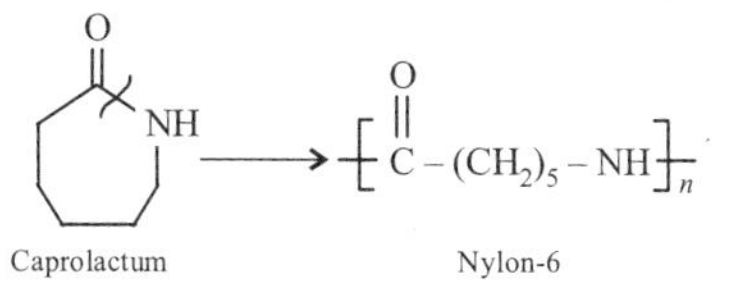

11. **(4)** Teflon or polytetrafluoroethene is an addition polymer and its monomer is alkene derivative ($CF_2 = CF_2$) whereas terylene, melamine and nylon-6, 6 are condensation polymers.

12. **(1)**

13. **(4)** Polymers which change irreversibly into hard and rigid material on heating are known as thermosetting polymers e.g bakelite.

14. **(3)**

15. **(2)** Polystyrene is a chain growth polymer.

16. **(2)** Nylon 6, 6 has amide linkage capable of forming hydrogen bonding.

17. **(3)**

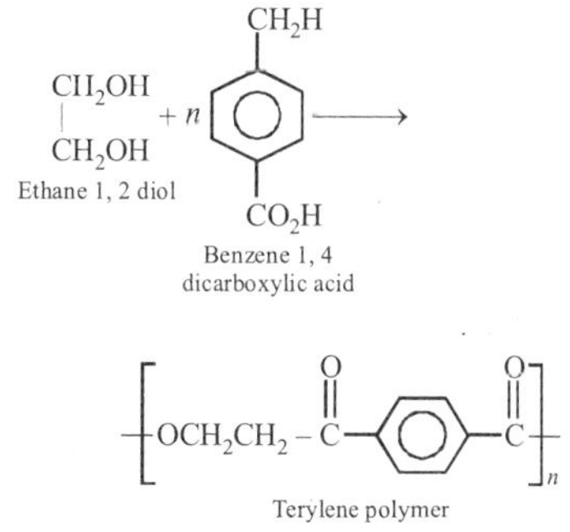

18. **(2)** A – II, B – III, C – I, D – IV

19. **(1)**

20. **(3)** Vulcanization of rubber is carried out by heating a mixture of isoprene & sulphur.

21. **(1)**

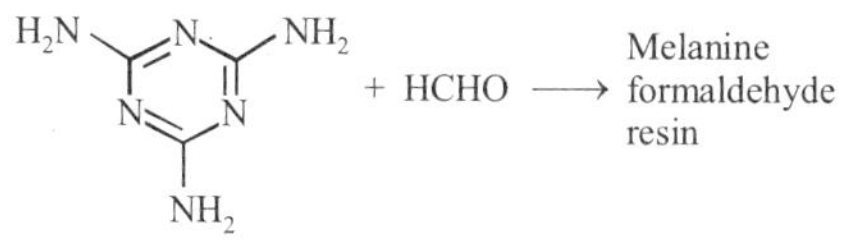

22. **(3)**

$$\underset{\text{Buna-N}}{\left[CH_2 - C = C - CH_2 - CH_2 - \overset{\large CN}{\overset{|}{C}} \right]_n}$$

Buna – N is synthetic rubber which can be stretched and retains its original status on releasing the force.

23. **(1)** 24. **(3)** 25. **(1)** 26. **(3)**

27. **(3)** It is copolymerization of 1, 3-butadiene and acrylonitrile.

$$\underset{\text{1, 3-Butadiene}}{nCH_2 = CH - CH = CH_2} + \underset{\text{Acrylonitrile}}{nCH_2 = \overset{\large CN}{\overset{|}{CH}}}$$

$$\downarrow$$

$$\underset{\text{Buna-N}}{\left[CH_2 - CH = CH - CH_2 - CH_2 - \overset{\large CN}{\overset{|}{CH}} \right]_n}$$

28. **(3)** Monomer of Novolac is *o*-hydroxymethylphenol.

29. **(1)**

$$\text{Novalac} + \text{formaldehyde} \xrightarrow[\text{linkage}]{\text{cross}} \text{Bakelite}$$

30. **(1)** Given structure is a repeating unit of novolac.

$$C_6H_5OH + H - \overset{O}{\overset{\|}{C}} - H \xrightarrow{H^+}$$

$$\underset{\text{(Liner polymer)}}{\text{Novolac}} \xrightarrow{\text{Polymerisation}} \underset{\text{(3D structure)}}{\text{Bakelite}}$$

31. **(3)** Nylon 2-Nylon 6 (polyamide copolymer) is a biodegradable polymer.

Its monomer units are : Glycine + Aminocaproic acid

$$[\underset{\text{Glycine}}{H_2N - CH_2 - COOH} + \underset{\text{Aminocaproic acid}}{NH_2(CH_2)_5 - COOH}]$$

32. **(1)** Orlon fibers are made up of Polyacrylonitrile.

33. **(4)**

34. **(4)** 'S' stand for styrene.

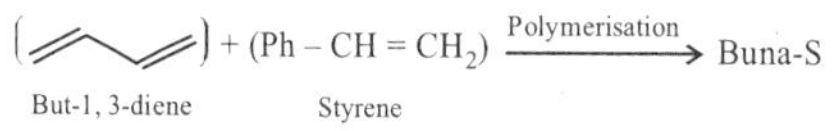

35. **(3)** 36. **(3)**

37. **(4)** $CH_3 - \overset{OH}{\overset{|}{CH}} - CH_2 - COOH$ and

(3-Hydroxybutanoic acid)

$$CH_3 - CH_2 - \overset{OH}{\overset{|}{CH}} - CH_2 - COOH$$

(3-Hydroxypentanoic acid)

Both monomers of PHBV have chiral centre.

38. **(3)** Formation of Bakelite follows electrophilic substitution reaction of phenol with formaldehyde followed by dehydration.

39. **(4)** 40. **(4)**

41. **(2)** The polymer $[NH - \overset{O}{\overset{\|}{C}} - NH - CH_2]_n$ is urea-formaldehyde resin. Its monomers are urea and formaldehyde.

42. (4) As the name of the polymer indicates, it is a copolymer of following two monomers.

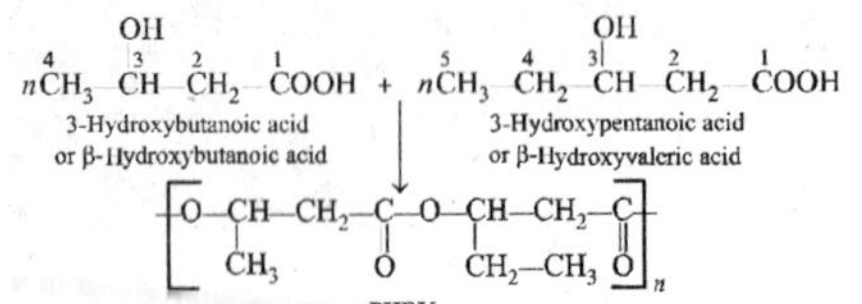

PHBV

43. (1) Adipic acid (having 6C) and hexamethylene diamine (having 6C).

44. (3)

45. (4)

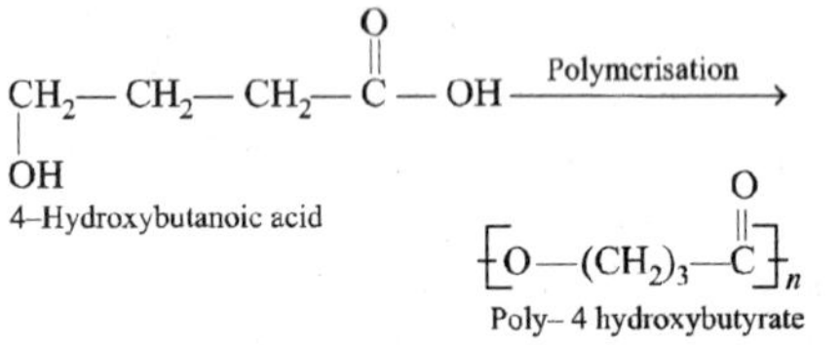

46. (1) **47. (4)** **48. (2)**

49. (2) The statement (b) is not true. Chain growth polymerisation (or addition polymerisation) involves homopolymerisation only. Examples of such polymers include polythene, orlon and teflon.

50. (1) Formation of nylon-6 involves hydrolysis of caprolactam, (its monomer) in initial state.

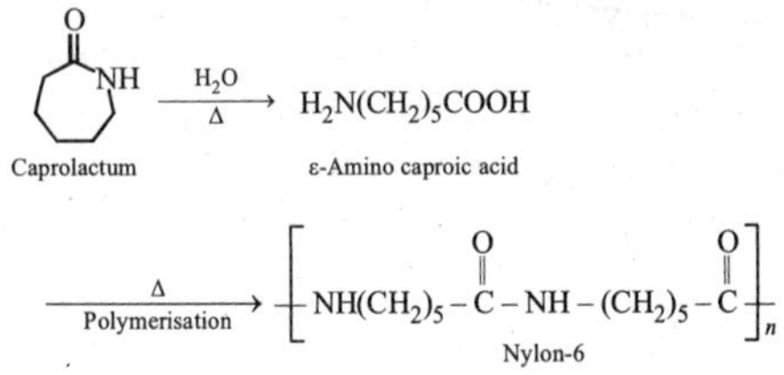

51. (1)

$nH_2N-CH_2-COOH+nH_2N-(CH_2)_5-COOH \rightarrow$

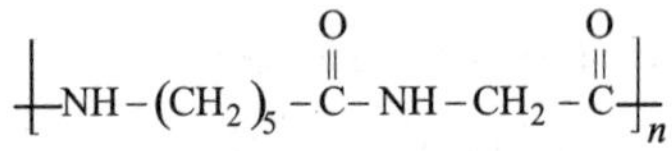
Nylon-2-Nylon-6
biodegradable step-growth copolymer

52. (2) High density polythene is used in the manufacture of housewares like buckets, dustbins, bottles, pipes etc. Low density polythene is used for insulating electric wires and in the manufacture of flexible pipes, toys, coats, bottles etc.

53. (1)

54. (3)

Chloroprene $(CH_2=CH-\overset{Cl}{\overset{|}{C}}=CH_2)$ is a monomer of neoprene.

55. (2)

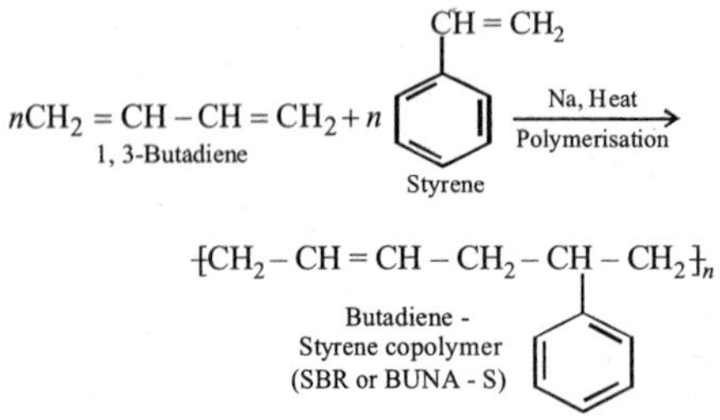

56. (1) Degree of polymerization of I polymer = 100

Degree of polymerization of II polymer = 200

$N_1 = 2; N_2 = 3$

Molar mass of monomer (ethylene) of polythene = 28 g/mol

Now, Degree of polymerization is given as:

$$DP = \frac{\text{Molar mass of monomer}}{\text{Molar mass of polymer}}$$

Thus,

Molar mass of I polymer $(M_1) = 100 \times 28$
$= 2800$ g/mol

Molar mass of II polymer $(M_2) = 200 \times 28$
$= 5600$ g/mol

Now, the weight average molecular weight is given as:

$$M_w = \frac{\Sigma n_i M_i^2}{\Sigma n_i M_i}$$

$$\therefore M_w = \frac{n_1 M_1^2 + n_2 M_2^2}{n_1 M_1 + n_2 M_2}$$

$$= \frac{2(2800)^2 + 3(5600)^2}{2 \times 2800 + 3 \times 5600} = 4900 \text{ g/mol}$$

57. (3)

58. (3) Lewis acids are the most common compounds used for initiation of cationic polymerisation. The more popular Lewis acids are $SnCl_4$, $AlCl_3$, BF_3 and $TiCl_4$.

59. (2) Buna – N is a copolymer of butadiene $(CH_2 = CH - CH = CH_2)$ and acrylonitrile $(CH_2 = CHCN)$.

60. (3) Teflon is a polymer of $CF_2 = CF_2$.

61. (2) Nylon is a polyamide polymer.

62. (1) Polymerisation starts either by condensation or addition reactions between monomers. Condensation polymers are formed by the combination of monomers with the elimination of simple molecules. Whereas addition polymers are formed by the addition of the molecules of the monomer or monomers to form a large molecule without elimination of any thing.

63. (9)

64. (2) Nylon 6,6 is used in making bristles for brushes low density. Polythene is tough but flexible and used in making toys.

High density polythene are more tough and hard. It is used in manufacturing of buckets. Teflon is used in making non-stick utensils.

65. (1) Urea-formaldehyde resin is used in the manufacture of wood laminates.

66. (4)

Polymer	Use
Polystyrene	Manufacture of toys
Glyptal	Paints and lacquers.
P.V.C. Rain	Coats
Bakelite	Computer discs

67. (4) Polymethyl methacrylate is hard, fairly rigid. It is used for optical lenses.

Part-B (JEE Advanced)

1. (i)-(f)-6; (ii)-(c)-2; (iii)-(d)-4; (iv)-(b)-1; (v)-(a)-5; (vi)-(e)-3.

(i) Asbestos was used as an insulator I solid propellant rocket motors.

(ii) Fluorocarbons were used as refrigerant which are carcinogens.

(iii) Lithium metal an easily donate electron and therefore, can be used as a reducing agent.

(iv) Nitric oxide is paramagnetic and an air pollutant.

(v) Zeolites have high porosity due to the presence of cavities of molecular dimensions. They are used as ion exchange for softening of water.

(vi) Zinc oxide is used as a semi-conductor. They have property of fluorescence.

2. (a)

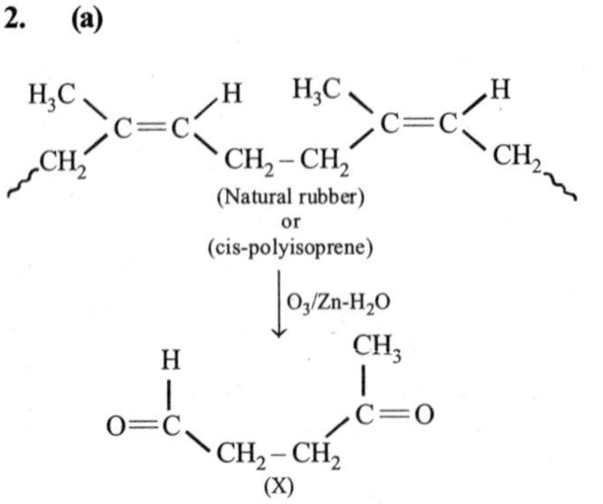

(gives positive iodoform & Tollen's Test)

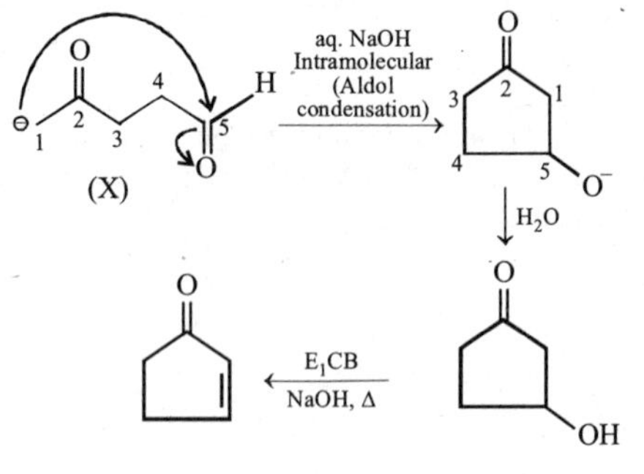

3. **(a)** $CH_2{=}\underset{\substack{|\\ \text{}}}{\overset{CH_3}{\overset{|}{C}}}{-}CH{=}CH_2$ (Isoprene) $\xrightarrow{\text{Polymerisation}}$

$-\!\!\left(CH_2-\overset{CH_3}{\overset{|}{C}}=CH-CH_2\right)\!\!-$ (Natural rubber) $\xrightarrow[\text{catalyst}]{H_2\text{(excess)}}$

$-\!\!\left(CH_2-\overset{CH_3}{\overset{|}{CH}}-CH_2-CH_2\right)\!\!-$ (Completely hydrogenated Natural rubber) $\xleftarrow{\text{Copolymerisation}}$

$CH_2{=}CH_2 + CH_2{=}\overset{CH_3}{\overset{|}{CH}}$

Ethylene Propylene

4. **(d)** Nylon and cellulose, both have intermolecular hydrogen bonding, polyvinyl chloride has dipole-dipole interaction, while natural rubber has van der Waal forces which are weakest.

5. **(85018.00)**

$$CH_3I \xrightarrow{KCN} CH_3CN \xrightarrow[\Delta]{H_3O^+} CH_3COOH \xrightarrow[Br_2]{\text{Red P}}$$

$$\underset{\substack{|\\Br}}{CH_2}-COOH \xrightarrow[\text{excess}]{NH_3} \underset{(X)}{NH_2-CH_2-COOH}$$

(Here 'X' can be $\underset{\substack{|\\NH_2}}{CH}-CO\overset{+}{O}NH_4$ also)

(Caprolactum)

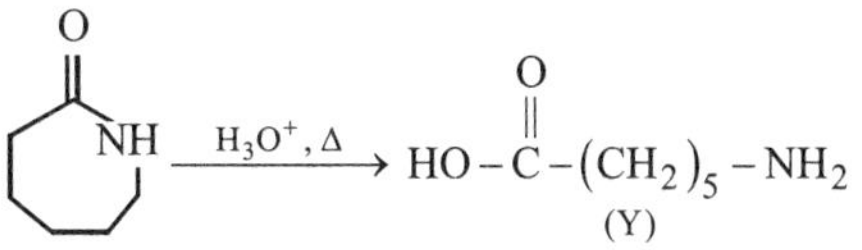

$$\underset{500\text{ mol}}{X} + \underset{500\text{ mol}}{Y} \longrightarrow \text{Nylon -2-nylon-6}$$

Number of water moles lost

$= 500 \times 2 - 1 = 999$

Mass of water lost $= 999 \times 18 = 17982$ g

Mass of Z $= 75 \times 500 + 131 \times 500 - 17982$

$= 37500 + 65500 - 17982$

$= 85{,}018$g

6. **(6)** Structure of melamine is as follows :

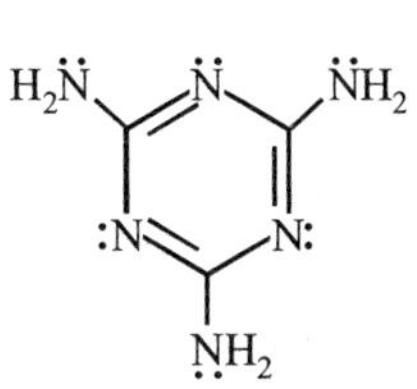

Total no. of lone pairs of electron is '6'.

7. **cross-linking**

8. **(b, c)**

(a) The polymerization of neoprene gives natural rubber.

(d) Ethene in these conditions yield low density polythene.

9. **(a, c)**

(a) Nylon-6. It is obtained by heating caprolactam with water at high temperature and have amide linkage.

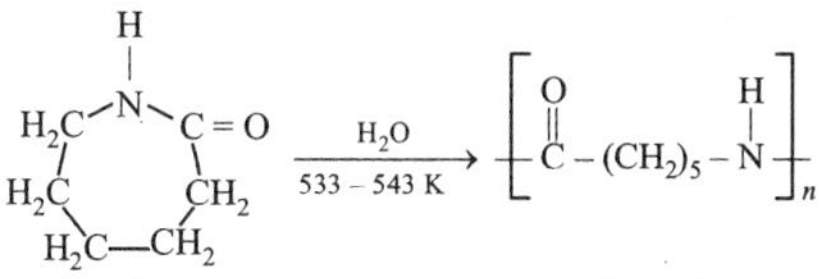

Caprolactum Nylon-6

(b) Cellulose has only β-D-glucose units that are joined by glycosidic linkages between C–1 of one glucose unit and C–4 of the next glucose unit.

(c) Teflon is prepared by heating tetrafluoroethene in presence of a persulphate catalyst at high pressure.

$$\underset{\text{Tetrafluoroethene}}{CF_2{=}CF_2} \xrightarrow[\text{High pressure}]{\text{Catalyst}} \underset{\text{Teflon}}{-\!\!\left[CF_2-CF_2\right]\!\!-_n}$$

(d) Natural rubber is a linear polymer of isoprene (2-methyl-1, 3-butadiene) containing cis alkene units. It is also called cis-1, 4-polyisoprene.

10. **(a,b,c,d)** Condensation polymers are formed by condensation of a diol or diamine with a dicarboxylic acid.

Hence, X may be $-\overset{O}{\overset{\|}{C}}-OR$ or $-\overset{O}{\overset{\|}{C}}-NH_2$

or $-C \equiv N$

(a) $H_3CO-\overset{O}{\overset{\|}{C}}-(CH_2)_4-\overset{O}{\overset{\|}{C}}-OCH_3 \xrightarrow{H_2/Ni(y)} OHH_2C-(CH_2)_4-CH_2OH$

$$\xrightarrow{HO-\overset{O}{\overset{\|}{C}}-(CH_2)_4-\overset{O}{\overset{\|}{C}}-OH} \text{Polyester}$$

(b) $H_2N-\overset{O}{\overset{\|}{C}}-(CH_2)_4-\overset{O}{\overset{\|}{C}}-NH_2 \xrightarrow{H_2/Ni(y)} H_2NCH_2-(CH_2)_4-CH_2NH_2$

$$\xrightarrow{HO-\overset{O}{\overset{\|}{C}}-(CH_2)_4-\overset{O}{\overset{\|}{C}}-OH} \text{Polyamide}$$

$\downarrow$ $Br_2/\overline{OH}, \Delta$ (y) Hofmann bromamide reaction

(c) $H_2N-(CH_2)_4-NH_2$

$$\xrightarrow{HO-\overset{O}{\overset{\|}{C}}-(CH_2)_4-\overset{O}{\overset{\|}{C}}-OH} \text{Polyamide}$$

(d) $N\equiv C-(CH_2)_4-C\equiv N \xrightarrow{H_2/Ni/\Delta(y)} H_2NCH_2-(CH_2)_4-CH_2NH_2$

$$\xrightarrow{HO-\overset{O}{\overset{\|}{C}}-(CH_2)_4-\overset{O}{\overset{\|}{C}}-OH} \text{Polyamide}$$

11. (A) : (p) and (s) Cellulose is a natural polymer and has a $C_1 - C_4$ β-glycosidic linkage.

(B): (q) and (r) Nylon-6, 6 is a synthetic polymer of hexamethylenediamine and adipic acid and has amide linkages.

(C): (p) and (r)Proteins are natural polymers of α amino acids joined by amide linkages (peptide bonds).

(D) : (s)Sucrose is a disaccharide of α-D glucose and β-D-fructose and has an α, β-glycosidic linkage.

12. 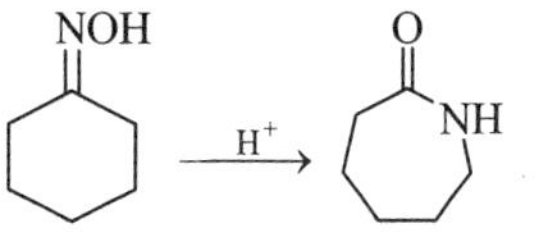

C (Caprolactum)

$$\longrightarrow \underset{\text{Nylon}}{-\!\!\left[CO-NH-\{CH_2\}_5\right]\!\!-_n}$$

Chapter-16

Chemistry in Every Day Life

Part-A (JEE Main)

1. (2) Amytal is a hypnotic drug.
2. (3)
3. (4) **Statement I** Some drugs do not bind to active sites. These bind to different site of enzyme called allosteric sites.
Statement II An inhibitor can form a strong covalent bond with the enzyme.
4. (3)
5. (3)
6. (4)
7. (2) Antagonist
8. (2)
Allosteric site are the region where activator or inhibitor molecules binds to activate or inhibit the enzymes.
9. (3)

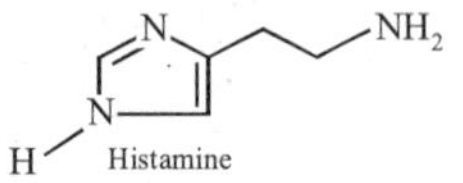
Histamine

Histamine is nitrogenous compound it does not contain sulphur atom.
10. (1)
- Antiseptics is used to kill germ, bacteria, viruses they do not harm the living tissue.
- Disinfectant are also killed bacteria but they can harm living tissue they are used to kill germs on non-living surfaces.

Pesticides → They are used to control pests.
Narcotic analgesics → they are used to provide relief from modesate-to-severe acute or chronic pain.
Antiseptic Dettol is mixture of chloroxylenol and terpineol.
11. (N) Pesticides are chemical substances used to protect crops from pest. All are pesticides.
12. (3) This drug is used to treat occasional heartburn caused by too much release of acid. Tegamet is the brand name of Cimetidine

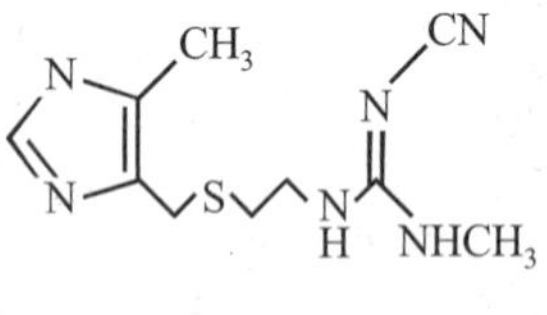

13. (2)
14. (1)
15. (3) Histamine stimulates the secretion of pepsin and HCl in stomach.
16. (2) Synthetic detergents have all the properties of soap but they do not actually contain soap. They can be used in both soft and hard water.

→ is a anionic detergents

$\left[CH_3-(CH_2)_{15}-\overset{CH_3}{\underset{CH_3}{N}}-CH_3\right]Br^{(-)}$ → is a cationic detergents

$CH_3(CH_2)_{16}COO(CH_2CH_2O)_nCH_2CH_2OH$
→ is a non-ionic detergents.

17. (2) Alitame is a second generation dipeptide sweetner that is 200 times sweeter than sucrose.
18. (2) Dettol is mixture of

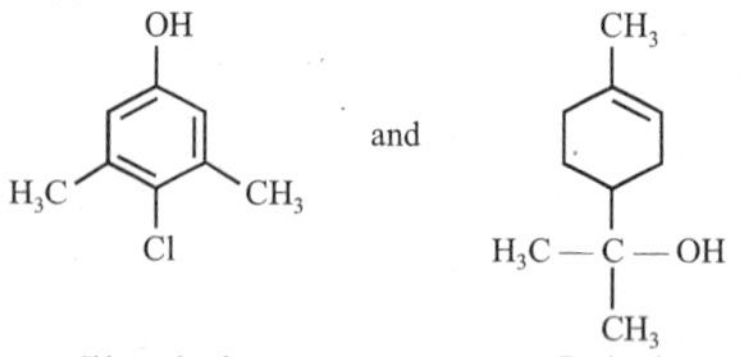

Chloroxylenol (compound A) It has 6πe⁻ and Terpineol (compound B) It has 2πe⁻

Hence compound 'B' is Terpineol.
19. (4) Penicillin G is a narrow spectrum antibiotic.
20. (3)
21. (2) Histamine stimulate the secretion of HCl
22. (2)
23. (1) Forms spherical micelles with hydrophobic party ($CH_3(CH_2)_{16}$ group)
24. (1) CO_2, CCl_4 and $Cl_2C=CCl_2$ are used as dry cleaning agents for clothes while H_2O_2 is used as bleaching agent in laundry.
25. (3)
26. (2) If the level of noradrenaline is low, then the person suffers from depression. In such situations, anti-depressant drugs are required.
27. (4) Synthetic drugs, brompheniramine (dimetapp) acts as antihistamines.
28. (4) Seldane is an anti-histamine drug that has inhibitory action on histamine receptor.
29. (4)
30. (2)
31. (1)
32. (2) Noradrenaline, also called Norepinephrine, is a neurotransmitter. It plays a major role in mood change.
33. (2)
(A) **Norethindrone** is a progesterone derivative, so it is an antifertility drug **(Q)**.
(B) **Ofloxacin** is an antibiotic **(P)**.
(C) **Equanil** is a mild tranquilizer and hence, used in hypertension **(R)**.
34. (1)
35. (4) Structure of histamine

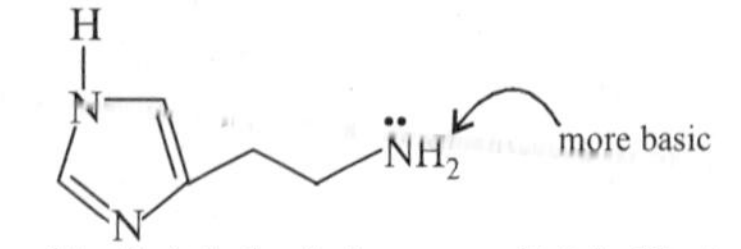

Blood is slightly basic in nature (7.35 pH). At this pH, terminal NH_2 will get protonated due to more basic nature. ∴ Predominant structure of histamine is

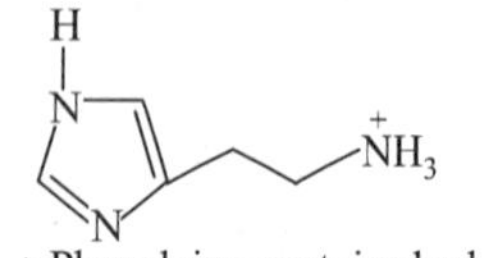

36. (2) → Phenelzine contains hydrazine
→ Chloroxylenol contains phenol
→ Uracil is the pyrimidine base
→ Ranitidine contains furan ring
37. (3)
38. (4) Sodium lauryl sulphate ($C_{11}H_{23}CH_2OSO_3^-Na^+$) is an anionic detergent. Glyceryl oleate is a glyceryl ester of oleic acid. Sodium stearate ($C_{17}H_{35}COO^-Na^+$) is a soap. Cetyltrimethyl ammonium bromide

$\left[CH_3(CH_2)_{15}\overset{+}{N}(CH_3)_3\right]Br^-$ is a cationic detergent.

39. (1) Bactericidal are the drugs that kills bacteria. Ofloxacin works by stopping the growth of bacteria. This antibiotic treats only bacterial infections.
40. (4) Alitame is an artificial sweetner that is 2,000 times as sweet as sugar.
41. (1) Phenelzine is an antidepressant, while others are antacids.
42. (1)

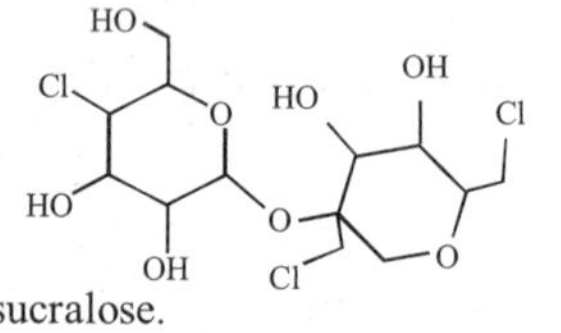

is sucralose.
43. (1) Drugs which relieve pain are called analgesics drugs. Analgesics are of two types (i) Narcotics and (ii) Non-narcotics. Aspirin (acetylsalicylic acid) is a non-narcotic analgesic.
44. (3) Diphenhydramine is used as antihistamine.
45. (1) The aminoglycosides are among the oldest antibiotics.
46. (4) Aspirin is an non-narcotics analgesic.
47. (2) The term "antihistamine" refers only to H_1 antagonists, which is also known as H_1-receptor antagonists and H_1-antihistamine.
48. (3) The antifertility drugs are those chemical compounds which prevent pregnancy in woman. These, drugs controls the menstrual cycle and ovulation and so used as birth control drug or contraceptives.

49. (3)

Morphine $\xrightarrow{\text{acetylation}}$ Morphine diacetate

50. (3) Asprin is prepared by reaction of salicylic acid with acetic anhydride in presence of H_2SO_4.

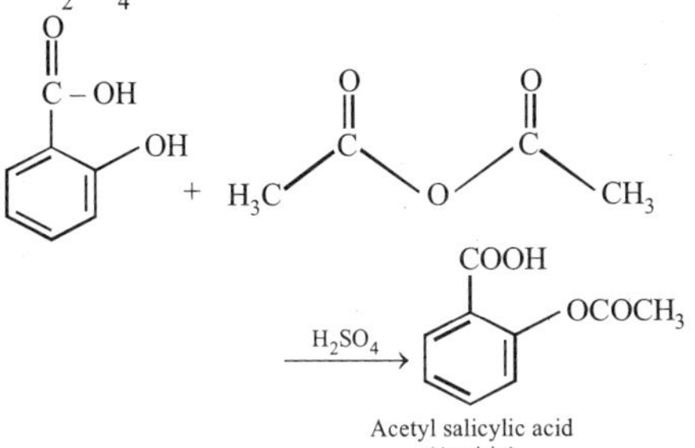

Acetyl salicylic acid (Aspirin)

51. (3) Sulphonamides act as antimicrobials.

52. (3) An antipyretic is a drug which is responsible for lowering the temperature of the feverish body to normal but has no effect on normal temperature states.

53. (2) Liquid hydrogen and liquid oxygen are used as excellent fuel for rockets. $H_2(\ell)$ has low mass and high enthalpy of combustion, whereas oxygen is a strong supporter of combustion.

54. (3) The given compound is aspirin which is antipyretic and analgesic.

55. (4)

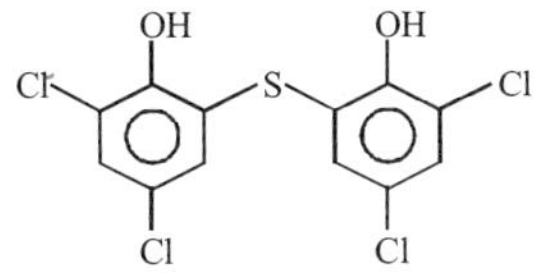

Chlorine atoms = 4.

56. (1) Ofloxacin

57. (3)

Veronal, Luminal and seconal belongs to tranquilzers class.

58. (5)

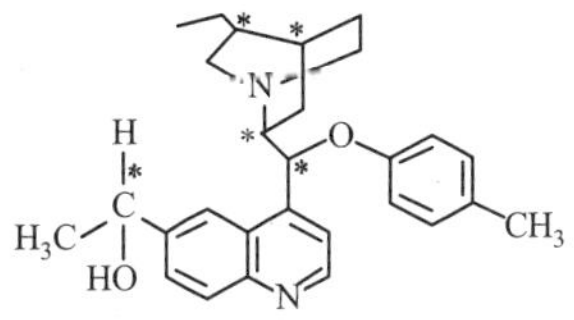

(* = Chiral carbon)

59. (3)

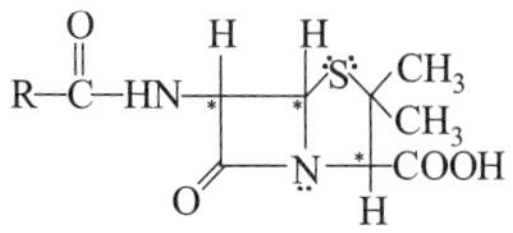

There are three chiral centres in penicillin.

60. (2)

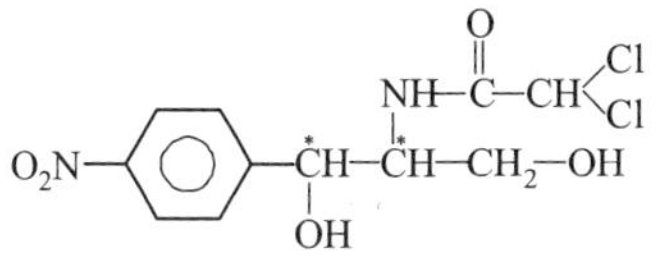

61. (9)

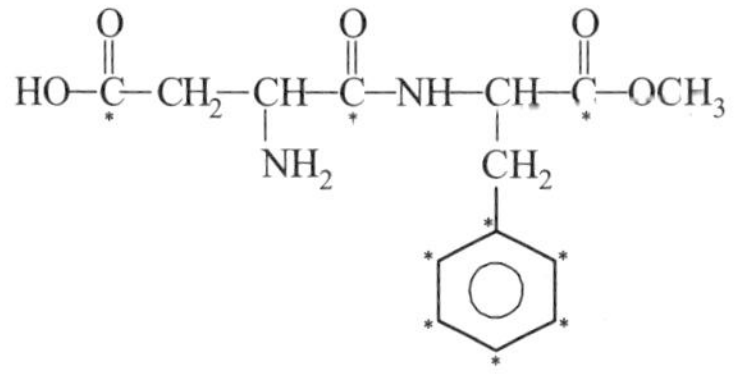

Structure of aspartame is shown above. It is a methyl ester of dipeptide formed from aspartic acid and phenylalanine. sp^2 hybridised carbon atoms are shown by the star mark in the structure.

Part-B (JEE Advanced)

1. (b) Structure of aspartame is :

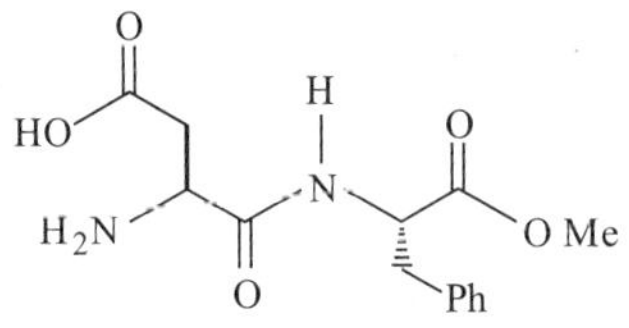

Chapter-17

Principles Related to Practical Chemistry

Part-A (JEE Main)

1. (1) A-III, B-I, C-IV, D-II

2. (4) Phenolphthalein indicator is generally used in acid base titrations and works in pH range of 8.3-10. So it turns colourless in excess of NaOH.

3. (4)

4. (3) (A) – IV, (B) – I, (C) – II, (D) – III

5. (1) A-II, B-III, C-IV, D-I

6. (3) A, C, D produce CO_2 with aqueous $NaHCO_3$ solution. A, C, D acids are stronger acid than H_2CO_3 (Carbonic acid).

7. (4)

8. (3) For Lassaigne's test, sodium is used and not magnesium to convert covalent bond into ionic bond.

9. (3) $CH_3CH{=}O$, $PhCH_2CH{=}O$.

$HOCH_2-\underset{\underset{O}{||}}{C}-(CHOH)_3-CH_2OH$ (Fructose)

All give positive Fehling test

Ketonic group and aromatic – CHO do not give Fehling's test.

10. (2) A-(II), B-(III), C-(IV), D-(I)

11. (2) Apart from aldehyde, Formic acid

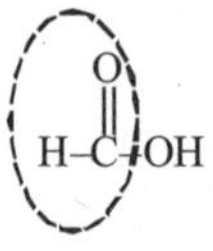

also gives silver mirror test with ammonical silver nitrate.

12. (4)

13. (4) Ninhydrin test is a chemical test useful to identify ammonia, primary and secondary amine, or amino acids. Egg albumin contains protein which is a natural polymer of amino acids which will show positive ninhydrin test.

14. (4) Hydrazine (NH_2–NH_2) have no carbon so does not show Lassaigne's test.

15. (3) Nitrogen, sulphur, phosphorous and halogens Lassaigne's test is used for the detection of nitrogen, sulphur, phosphorous and halogen present in an organic compound.

16. (4)

17. (4) Both statement I and statement II are true.

18. (4)

19. (4) If nitrogen or sulphur is also present in the compound, the sodium fusion extract is boiled with conc. HNO_3 to decompose cyanide or sulphide of sodium formed, these ions would otherwise interfere with silver nitrate test for halogens.

20. (1)

21. (1) Carbylamine Test-Identification of primary amines Lucas Test - Differentiation between 1°, 2° and 3° alcohols

Tollen's Test - Identification of Aldehydes

Phthalein Dye Test - Identification of phenols

22. (4)

23. (2) (A) – (II), (B) – (I), (C) – (III), (D) – (IV)

24. (1) It does not show esterification test. Presence of both sulphur and nitrogen give red colour in Lassigne's test.

25. (4) Aromatic aldehydes do not give Fehling's test. Both nitrogen and sulfur must be present to obtain blood red colour with sodium nitroprusside.

26. (2) $CH_3NH_2 . HCl \xrightarrow[\text{fusion}]{\text{Na}} $ NaCN and NaCl

NaCN gives + ve test for nitrogen and NaCl gives +ve test for halogen.

27. (2) Acetate ions gives deep red colour ppt on reaction with netural ferric chloride solution.

Phenol reacts with freshly prepared ferric chloride solution and gives violet colour complex.

28. (4)

$$\underset{\text{Sodium nitroprusside}}{Na_2S + Na_2[Fe(CN)_5\,NO]} \rightarrow \underset{\substack{\text{Sodium thionitroprusside}\\ \text{(Violet colour)}}}{Na_4[Fe(CN)_5\,NOS]}$$

29. (1) Both statement I and statement II are correct.

30. (2) Sodium fusion extract is boiled with concentrated HNO_3 to remove sodium cyanide and sodium sulphide.

31. (1) By sodium fusion extract we can detect sulphur, nitrogen, phosphorus and halogens, because they are converted into their ionic form with sodium metal.

32. (4) For the test of halogens, it is necessary to remove NaCN and Na_2S from the sodium extract if nitrogen and sulphur are present. This is done by boiling the sodium extract with conc. HNO_3. If NaCN and Na_2S are not decomposed, white or black ppt of AgCN and Ag_2S respectively are formed with $AgNO_3$ solution.

33. (4)

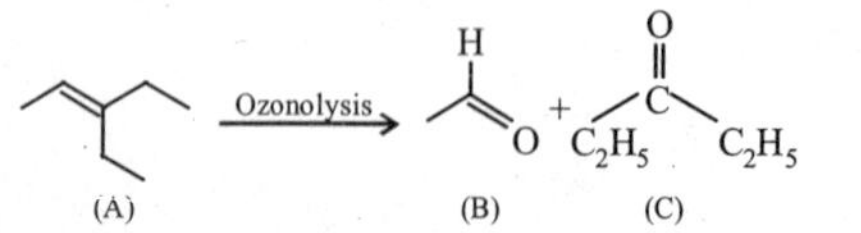

34. (3)

(I) Lucas reagent - Conc. $HCl/ZnCl_2$

(II) Dumas method - CuO/CO_2

(III) Kjeldahl's method - H_2SO_4

(IV) Hinsberg test - $C_6H_5SO_2Cl$/aq. KOH

35. (1)

Test	D	C	Conformation
Iodoform test	–ve	+ve	Presence of $R-\overset{\overset{OH}{|}}{C}H-CH_3$ group
Lucas test	Immediate	after 5-10 Min.	Presence of 3° & 2°, –OH groups respectively
Ceric ammonium nitrate test	+ve	+ve	Presence of –OH group

36. (2) Molisch's test is used to check the presence of carbohydrates while Barfoed test is used for detecting the monosaccharides. Biuret test is used to detect the peptide bonds. Therefore, A and B (monosaccharide) are carbohydrates while C is a protein.

37. (2)

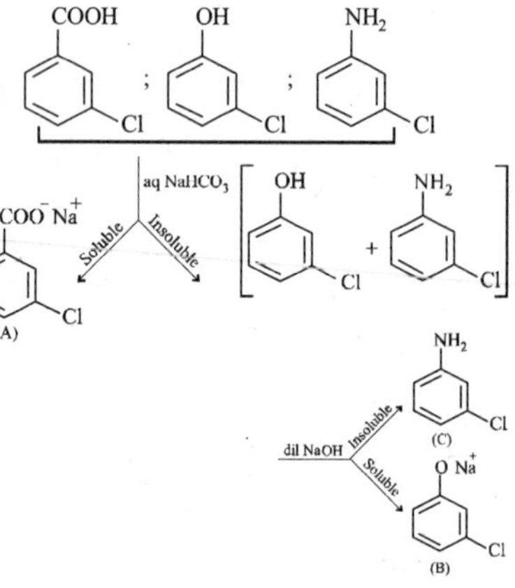

38. (2)

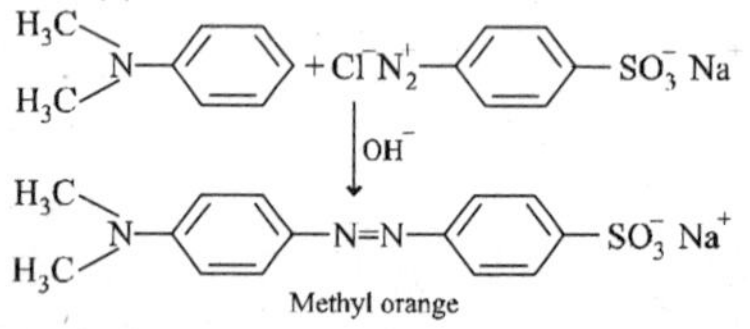

Methyl orange is used as an indicator in acid base titrations.

39. (2) Phosphorus is detected in the form of yellow ppt of ammonium phospho molybdate on reaction with ammonium molybdate.

40. (4) $(CH_3)_3C^+$ from $(CH_3)_3CCl$ is quite stable and hence, easily formed and thus, Cl^- ion forms white precipitate with $AgNO_3$.

41. (1)

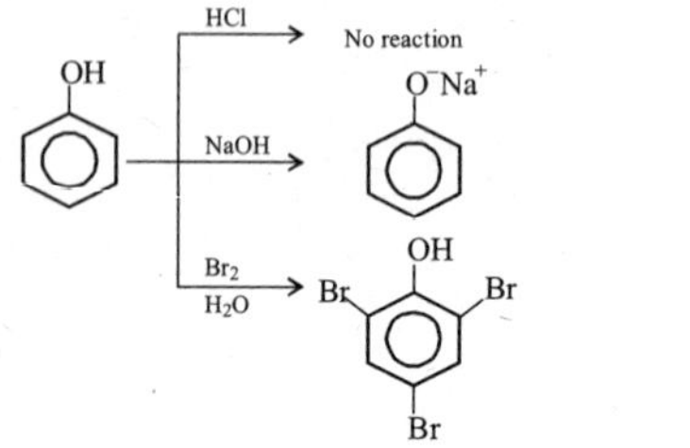

42. (3) Phenols (e.g. *m*-cresol), being weak acid, are soluble in dil NaOH, but insoluble in $NaHCO_3$.

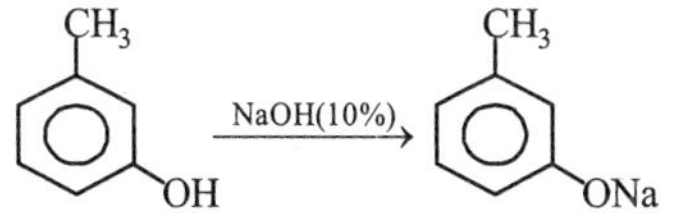

43. (1)

Name	Abrev-ations	Functional group	Test
(P) Tyrosine	Tyr	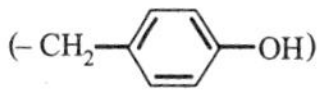	Phthalein dye test (C)
(Q) Aspartic acid	Asp	$(-CH_2COOH)$	Ester test (A)
(R) Serine	Ser	$(-CH_2OH)$	Ester test (A)
(S) Lysine	Lys	$(-(CH_2)_4-NH_2)$	Carbylamine test (B)

44. (2) Only those halides give precipitate with $AgNO_3$ which easily ionises to give halide ion and a stable carbocation. Corresponding carbocation from the four bromides are

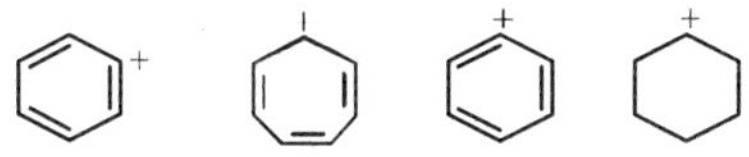

Unstable, (a) Arylcarbocations; Stable, (b) (Aromatic, 6π es); Unstable, (c); Unstable, (d)

45. (1)

46. (2) Barfoed test is used to test the reducing nature of sugar (carbohydrate).

47. (3)

48. (2) Reaction involved in the tests performed is as follows:

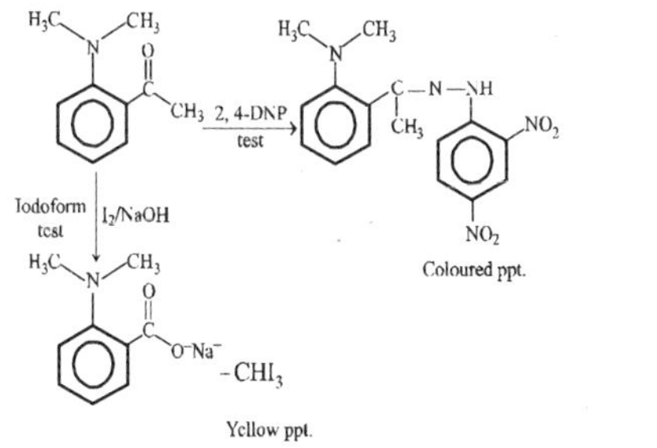

Thus, 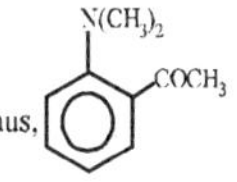 will not give positive azo-dye test due to the presence of an electron-withdrawing group.

49. (2)

50. (2) Beilstein's test : Organic compounds containing halogens when heated over Cu wire loop, give blue or green colour flame due to formation of volatile copper halides.

51. (2) Beilstein's test is used for halogens.

52. (3) Moles of NH_3 absorbed $= 0.5 \times 10^{-3}$

Moles of N in $NH_3 = 0.5 \times 10^{-3}$

Mass of N in organic compound $= 14 \times 0.5 \times 10^{-3}$ g

% N in organic compound

$$= \frac{14 \times 0.5 \times 10^{-3}}{29.5 \times 10^{-3}} \times 100 = 23.7\%$$

53. (2)

Element	%	Relative number of atoms	Simple ratio of atoms
C	20	20/12 = 1.66	1
H	6.67	6.67 / 1 = 6.67	4.16
N	46.7	46.67/14 = 3.33	2.02
O	26.6	26.64 / 16 = 1.66	1

∴ The empirical formula is CH_4N_2O

Empirical weight = 60; Mol. wt. = 60; ⇒

$$\therefore\ n = \frac{60}{60} = 1$$

∴ Molecular formula

$$= CH_4N_2O;\ NH_2 - \overset{\overset{O}{\|}}{C} - NH_2 \text{ (urea)}$$

On heating, urea loses ammonia to give biuret.

$$2NH_2CONH_2 \longrightarrow \underset{\text{Solid residue (Biuret)}}{H_2NCONHCONH_2} + NH_3$$

Biuret with alkaline $CuSO_4$ gives violet colour, i.e., test for –CONH– group.

54. (1) Prussian blue $Fe_4[Fe(CN)_6]_3$ is formed in Lassaigne test for nitrogen.

$$\underset{\text{(From org. comp.)}}{Na + \underbrace{C + N}} \xrightarrow{\text{Fuse}} NaCN$$

$$FeSO_4 + 2NaOH \longrightarrow \underset{\text{Green ppt.}}{Fe(OH)_2\downarrow} + Na_2SO_4$$

$$Fe(OH)_2 + 2NaCN \longrightarrow Fe(CN)_2 + 2NaOH$$

$$Fe(CN)_2 + 4NaCN \longrightarrow \underset{\text{Sod. ferrocyanide}}{Na_4[Fe(CN)_6]}$$

$$3Na_4[Fe(CN)_6] + 4FeCl_3 \longrightarrow \underset{\text{Ferricferrocyanide (Prussian blue)}}{Fe_4[Fe(CN)_6]_3} + 12NaCl$$

55. (95)

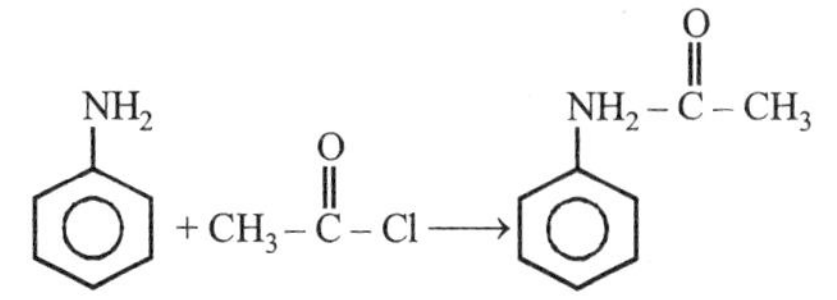

93 g aniline gives = 135 g acetanlide

So, 6.55 g anilne will give $= \frac{135}{93} \times 6.55 = 9.5$ g acetanlide

$= 95 \times 10^{-1}$ acetanlide.

56. (3) Acetaldehyde, methanal and cyclohexane carbaldehyde gives positive Fehling solution test among the given options. Aldehydes gives positive Fehling solution test. Aromatic aldehydes do not respond to this test. Therefore benzaldehyde and substituted benzaldehyde gives negative Fehling solution test.

57. (530) 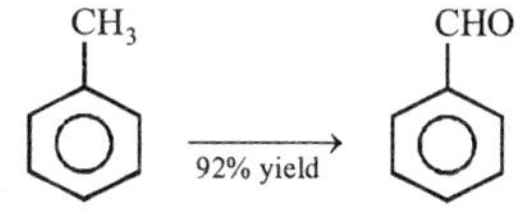

MW of toluene = 92 ;

MW benzaldehyde = 106.

Number of moles toluene $= \frac{5}{92}$

Number of moles of 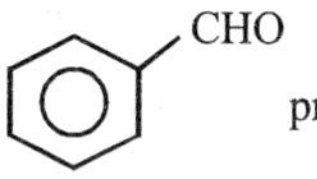 produced

$$= \frac{5}{92} \times \frac{92}{100} = 5 \times 10^{-2}$$

∴ Mass of 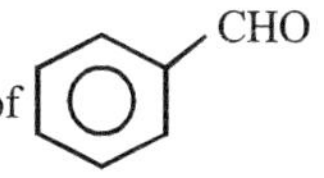

$= 106 \times 5 \times 10^{-2} = 5.3\ g = 530 \times 10^{-2}$ g

58. (2) 233g $BaSO_4$ contains = 32 g sulphur

0.752g $BaSO_4$ contains $= \frac{32}{233} \times 0.752 = 0.1005$g

$$\text{Sulfur\%} = \frac{\text{Weight of Sulphur}}{\text{Weight of Compound}} \times 100$$

$$51.6 = \frac{0.1005}{\text{Weight of compound}} \times 100$$

$$\Rightarrow \text{Weight of compound} = \frac{0.1005}{51.6} \times 100$$

$= 1.9 \times 10^{-1} g \simeq 2 \times 10^{-1}$ g

59. (34) $\underset{0.5g}{O.C} \longrightarrow \underset{0.4g}{AgBr}$

Molar mass at AgBr = 108 + 88 = 188 u

mol of Br = mol of AgBr

$$= \frac{\text{weight of AgBr}}{\text{molar mass of AgBr}} = \frac{0.4}{188} \%Br$$

$$= \frac{\text{weight of Br}}{\text{weight of organic compound}} \times 100$$

$$= \frac{\frac{0.4}{188} \times 80}{0.5} \times 100 = 34.04\%$$

60. (14) Weight of organic compound = 0.2 g

$$\text{Mass of } N_2(g) \text{ evolved} = \frac{22.4 \times 10^{-3}}{22.4} \times 28$$

$= 28 \times 10^{-3}$ g

$$\%\ \text{of N} = \frac{28 \times 10^{-3}}{0.2} \times 100 = 14$$

61. (40) Mass of organic compound = 0.2g

Mass of AgBr = 0.188g

Molar mass of AgBr = 108 + 80 = 188

$$\text{Mass of Br} = \frac{80}{188} \times 0.188 = 0.08g$$

$$Br\% = \frac{0.08}{0.2} \times 100 = 40$$

62. (12) $\% \text{ N} = \dfrac{1.4 \text{ N} \times \text{V}}{\text{W}}$

$\Rightarrow 42 = \dfrac{1.4 \times (1 \times 2) \times \text{V}}{0.8}$

$\Rightarrow$ V = 12 mL

63. (78) 10 g of $C_6H_6 = \dfrac{10}{78}$ moles of C_6H_6

Moles of methyl benzene obtained from

$= \dfrac{10}{78}$ mole of $C_6H_6 = \dfrac{10}{78} \times 92$ g

$\% \text{ yield} = \dfrac{9.2}{10 \times 92} \times 78 \times 100 = 78\%$

64. (68) % of

$\text{Br} = \dfrac{\text{Atomic mass} \times \text{Br} \times m_1}{\text{Molecular mass of AgBr} \times m} \times 100$

m_1 = mass of AgBr obtained

m = mass of organic compound taken

Now, % of Br $= \dfrac{80 \times 0.2397}{188 \times 0.15} \times 100 = 68$

65. (243)

$$\underset{\substack{\text{(Aniline)}\\ \text{(M.M}=93\text{ g)}}}{C_6H_5NH_2} \xrightarrow{90\% \text{ efficiency}} \underset{\substack{\text{(Acetanilide)}\\ \text{(M.M.}=135\text{ g)}}}{C_6H_5-NH-\overset{\overset{\large O}{||}}{C}-CH_3}$$

Given : The amount of aniline reacts to form acetanilide = 1.86 g

Since, 1 mol $C_6H_5NH_2$ gives 1 mol $C_6H_5NH\overset{\overset{O}{||}}{C}CH_3$.

$\therefore$ Moles of $C_6H_5NH_2$ = mole of $C_6H_5NH\overset{\overset{O}{||}}{C}CH_3$

$\Rightarrow \dfrac{1.86}{93} = \dfrac{W_{\text{acetanilide}}}{135}$

$\Rightarrow W_{\text{acetanilide}} = \dfrac{1.86 \times 135}{93} \text{ g} = 2.70 \text{ g}$

But efficiency of reaction is 90% only.

$\therefore$ Mass of acetanilide produced $= 2.70 \times \dfrac{90}{100}$ g

$= 2.43 \text{ g} = 243 \times 10^{-2}$ g.

66. (1)

$$\underset{(X)}{NO + FeSO_4} \xrightarrow{\text{dil. } H_2SO_4} \underset{(Y)}{[Fe(NO)]SO_4}$$

$$[Fe(NO)]SO_4 + H_2O \longrightarrow \underset{\text{Brown ring}}{[Fe(H_2O)_5(NO)]SO_4}$$

67. (4) $X + SO_2 \xrightarrow{H_2SO_4} \underset{\text{Green}}{Cr_2(SO_4)_3}$

SO_2 gas acts as reducing agent.

It turns green due to formation of $Cr_2(SO_4)_3$.

$$K_2Cr_2O_7 + SO_2 \xrightarrow{H_2SO_4} \underset{\text{Green solution}}{Cr_2(SO_4)_3 + SO_4^{2-}}$$

$X \rightarrow K_2Cr_2O_7$.

68. (2) A primary standard must be highly pure, stable, non-hygroscopic $K_2Cr_2O_7$ is used as primary standard indicator, not a base.

$Na_2Cr_2O_7$ is hygroscopic in nature and is not used as primary standard.

Phenolphthalein is weak acid.

69. (4) Group IV reagent : $NH_4OH + H_2S$

Mn^{+2} will be precipitated out and highest possible oxidation state of Mn is +7.

70. (2) MnO_4^- in acidic medium

$$MnO_4^- + 8H^+ + 5e^- \longrightarrow Mn^{+2} + 4H_2O$$

(*n*-factor = 5)

FeC_2O_4 :

$$\left.\begin{array}{l} Fe^{+2} \longrightarrow Fe^{+3} + 1e^- \\ C_2O_4^{2-} \longrightarrow CO_2 + 2e^- \end{array}\right\} (n\text{-factor} = 3)$$

$Fe_2(C_2O_4)_3$: Fe^{+3} (no - oxidation occurs)

$C_2O_4^{2-} \longrightarrow CO_2 + 2e^-$ (*n*-factor = 2)

For $3(C_2O_4^{2-})$; *n*-factor = 2 × 3 = 6

$Fe_2(SO_4)_3$: Fe^{+3} in highest oxidation state, no oxidation occur.

$FeSO_4$: $Fe^{+2} \longrightarrow Fe^{+3} + 1e^-$ (*n*-factor = 1)

M_{eq} of $KMnO_4$ = M_{eq} of FeC_2O_4 + M_{eq} of $Fe_2(C_2O_4)_3$ + M_{eq} of $FeSO_4$

Total mole eq. to be oxidised = 3 + 6 + 1 = 10 eq.

Moles eq. of $KMnO_4$ required $= \dfrac{10}{5} = 2$ eq.

71. (4) Zn^{2+} does not show Borax Bead Test becuase it has a completely filled d-subshell.

Among remaining ions; Fe^{3+} has maximum ionization energy because it contains half filled d-subshell.

72. (2)

(i) Flame test :

Ba^{+2} gives apple green colour while Ca^{+2} and Sr^{+2} gives brick red and crimson red.

(ii) Cation Group Test:

$$Ba^{+2} + CrO_4^{2-} \rightarrow \underset{\text{(Yellow ppt)}}{BaCrO_4}$$

This confirms Ba^{+2}

(iii) Anion Test :

$$PO_4^{3-} + \text{conc. } HNO_3 + (NH_4)_2MoO_4 \rightarrow \underset{\text{(Canary yellow ppt)}}{(NH_4)_3[P(Mo_3O_{10})_4]}$$

73. (2) $Na_2B_4O_7 \xrightarrow{\Delta} \underset{(X)}{2NaBO_2} + \underset{(Y)}{B_2O_3}$

$$CuSO_4 + \underset{(Y)}{B_2O_3} \xrightarrow[\text{Non-luminous flame}]{\Delta} \underset{\substack{\text{(Blue bead)}\\ (Z)}}{\overset{+2}{Cu}(BO_2)_2} + SO_3$$

$$\underset{(Z)}{2Cu(BO_2)_2} + \underset{(X)}{2NaBO_2} + C \xrightarrow[\text{Luminous flame}]{\Delta} \underset{\substack{\text{Colourless}\\ (Q)}}{2\overset{+1}{Cu}BO_2} + Na_2B_4O_7 + CO$$

Thus oxidation state of Cu in (Z) and (Q) are +2 and +1 respectively.

74. (2)

$$NO_2^- + \underset{\text{(Acid)}}{CH_3COOH} \rightarrow HNO_2 + CH_3COO^-$$

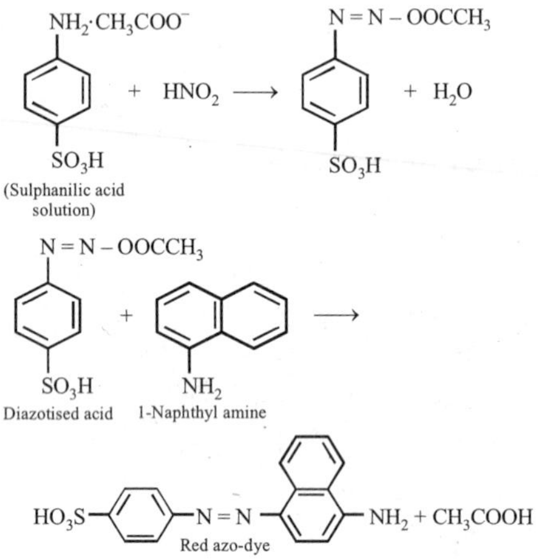

75. (3) Both Ba^{2+} and Ca^{2+} belong to group V.

To identify Ba^{2+} and Ca^{2+}, Reagent $(NH_4)_2CO_3$ + NH_4Cl is used and $BaCO_3$ and $CaCO_3$ are obtained as precipitates.

76. (2) $NH_4^+ + NaOH \longrightarrow H_2O + \underset{(X)}{NH_3} \uparrow$

$$\underset{(Y)}{2[HgI_4]^{2-}} + \underset{(Y)}{3OH^-} + \underset{(X)}{NH_3} \longrightarrow \underset{\text{Brown ppt}}{HgO.Hg\lt^{NH_2}_{I}} + 7I^- + 2H_2O$$

77. (3) Test for sulphide ion

$$S^{2-} + Na_2[Fe(CN)_5NO] \longrightarrow \underset{\text{Deep violet compound}}{[Fe(CN)_5NOS]^{4+}}$$

The formation of this violet complex confirms the presence of sulphide ion in the salt.

78. (3) Mg^{2+} and Ba^{2+} do not give ferrocyanide tests.

79. (2) As_2S_3 and As_2S_5 are yellow colour sulphides, $(NH_4)_2S$ is colourless to yellow. PbS is black, CuS is black in colour

80. (2) Ni^{+2} has d^8 configuration which does not depend on nature of ligand present in octahedral complex.

No significant high spin or low spin effect occurs.

$$Ni^{+2} : t_{2g}^6 e_g^2$$

Ni^{+2} gives violet coloured bead in non-luminous oxidizing flame under hot conditions.

81. (2) $CoS \xrightarrow{\text{Aqua regia}} CoCl_2 \xrightarrow[\text{2.AcOH}]{\text{1. } KNO_2/NH_4OH} \underset{\text{yellow}}{K_3[Co(NO_2)_6]\downarrow}$

82. (2)

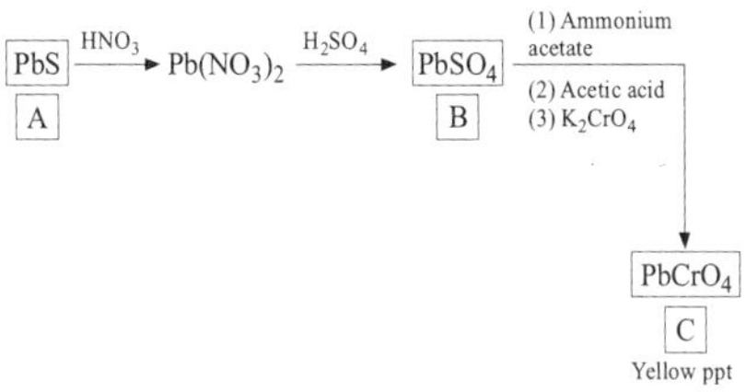

83. (4) $3PbS + 8HNO_3 \rightarrow 3Pb(NO_3)_2 + 2NO + 3S + 4H_2O$

Nitrous oxide (N_2O) is not formed during the reaction.

84. (4)

85. (1) Complexes with their colours are:

$Fe_4[Fe(CN)_6]_3.xH_2O \Rightarrow$ Prussian Blue

$[Fe(CN)_5NOS]^{4-} \Rightarrow$ Violet

$[Fe(SCN)]^{2+} \Rightarrow$ Blood Red

$(NH_4)_3PO_4.12MoO_3 \Rightarrow$ Yellow

86. (3)

$$\underset{\text{Or } HPO_4^-}{PO_4^{3-}} + \underset{\text{Ammonium Molybdate}}{(NH_4)_2MoO_4} \xrightarrow{H^+} \underset{\text{Canary yellow ppt. (Ammonium phospho molybdate)}}{(NH_4)_3PO_4.12MoO_3\downarrow}$$

87. (3)

88. (1) The Br^- ion present in the salt gives pale yellow ppt. of AgBr with $AgNO_3$, insoluble in dil HNO_3 but sparingly soluble in aq. NH_4OH.

$$Br^- + AgNO_3 \rightarrow \underset{\text{Pale yellow}}{AgBr\downarrow} + NO_3^-$$

$$AgBr + 2NH_4OH \rightarrow \underset{\text{Sparingly Soluble}}{\left[Ag(NH_3)_2\right]Br} + 2H_2O$$

89. (1) While preparing crystal of Mohr's salt, dil. H_2SO_4 is added to prevent hydrolysis of ferrous sulphate.

90. (2) $NH_4OH \rightleftharpoons NH_4^+ + OH^-$

$NH_4Cl \rightarrow NH_4^+ + Cl^-$

Due to common ion effect of NH_4^+,

$[OH^-]$ decreases in such extent that only group-III cations can be precipitated, due to their very low K_{sp} in the range of 10^{-38}.

91. (1) Statement I is false but statement II is true Dimethyl glyoxime forms a four membered chelate with $NiCl_2$ solution in presence of NH_4OH

92. (2) Lead sulphate-white

Ammonium sulphide-yellow

Lead iodide-Bright yellow

Ammonium arsinomolybdate-yellow

93. (3) Salts having S^{2-} ion give H_2S on warming with dil. H_2SO_4.

$$H_2S + (CH_3COO)_2Pb \longrightarrow \underset{\substack{\text{Lead Sulphide} \\ \text{Black ppt.}}}{PbS} + 2CH_3COOH$$

94. (1) $CrO_2Cl_2 + 4NaOH \longrightarrow \underset{(A)}{Na_2CrO_4} + 2NaCl + 2H_2O$

$\underset{(A)}{Na_2CrO_4} + 2H_2O_2 + 2HCl \longrightarrow \underset{(B)}{CrO_5} + 3H_2O + 2NaCl$

95. (1) $Ni^{2+} + 2dmg^{2-} \xrightarrow{NH_3} \underset{\text{red ppt.}}{[Ni(dmg)_2]\downarrow}$

dmg = dimethyl glyoxime.

96. (4) $\underset{\text{Nessler's reagent}}{2K_2HgI_4} + 3KOH + \underset{X}{NH_3} \longrightarrow$

$\underset{\text{brown}}{[OHg_2.NH_2]I\downarrow} + 7KI + H_2O$

NH_3 = Zero group basic radical.

97. (1) $4NaCl + 6H_2SO_4 + K_2Cr_2O_7 \rightarrow \underset{\substack{(B) \\ \text{Reddish brown}}}{2CrO_2Cl_2\uparrow} + 2KHSO_4 + 4NaHSO_4 + 3H_2O$

$CrO_2Cl_2 + 4NaOH \rightarrow \underset{\substack{(C) \\ \text{Yellow colour}}}{Na_2CrO_4} + 2NaCl + 2H_2O$

98. (2) In wet testing, $(NH_4)_2CO_3$ is used as group reagent for 5th group cations (Ba^{2+}, Ca^{2+}, Sr^{2+})

$$Ba^{+2} + (NH_4)_2CO_3 \rightarrow \underset{\text{(white precipitate)}}{BaCO_3\downarrow} + NH_4^{\oplus}$$

99. (1)

100. (3) $AgNO_3 + KI \rightarrow AgI\downarrow + KNO_3$

$$AgI \rightarrow \underset{S}{Ag^+} + \underset{S+0.625}{I^-}$$

AgI is a insoluble salt so concentration Ag^+ and I^- will be negligible.

101. (1) Prolonged heating will cause oxidation of Fe^{+2} to Fe^{+3}.

102. (3) $Pb(NO_3)_2$ is a soluble colourless compound so it cannot be used in confirmatory test of Pb^{+2} ion.

103. (4) Nessler's reagent is – $K_2[HgI_4]$

Nitrogen is not present in Nessler's reagent

104. (3) Nessler's Reagent $\rightarrow K_2[HgI_4]$

105. (3) $Fe_4[Fe(CN)_6]_3 \Rightarrow$ prussian blue precipitate

106. (3)

$$Cr_2O_7^{2-} + SO_3^{2-} \xrightarrow{H^+} \underset{\text{Green}}{Cr^{3+}} + SO_4^{2-}$$

107. (1) $Zn^{2+}, Co^{2+}, Ni^{2+}$ = IVth Group,

Fe^{3+} = IIIrd Group

108. (3) Nessler's reagent is K_2HgI_4.

109. (3) Blue green colour is due to formation of $Cu(BO_2)_2$

$CuSO_4 \xrightarrow{\Delta} CuO + SO_3$

$CuO + B_2O_3 \longrightarrow Cu(BO_2)_2$

110. (4) $Ni^{+2} + 2DMG \xrightarrow{NH_3(aq)} \underset{\text{Rosy Red complex}}{[Ni(DMG)_2]}$

111. (4)

112. (2) $Na_2B_4O_7.10H_2O \xrightarrow{\Delta} Na_2B_4O_7 \xrightarrow{\Delta} 2NaBO_2 + B_2O_3$

$$B_2O_3 + CoO \longrightarrow \underset{\text{Blue coloured}}{Co(BO_2)_2}$$

113. (4)

$$4Fe^{3+} + 3[Fe(CN)_6]^{-4} \rightarrow \underset{\text{Prussian Blue}}{Fe_4[Fe(CN)_6]_3}$$

114. (2) Salt is dissolved in water to form a solution. Barium chloride test is performed for the confirmation of sulphite ions.

If the ppt formed after addition of $BaCl_2$ are dissolved in dil. HCl then sulphite ions are present and if the ppt remain undissolved then the ion is sulphate.

$$BaCl_2 + SO_3^{2-} \rightarrow \underset{\text{white}}{BaSO_3\downarrow} \xrightarrow{\text{dil.HCl}} \underset{\text{burning sulphur like smell}}{SO_2\uparrow}$$

115. (4)

116. (2) Ring is formed due to formation of nitrosoferrous sulphate.

117. (1) $Ni^{2+} + 2DMG \rightarrow \underset{\text{(Bright red precipitate)}}{[Ni(DMG)_2]\downarrow}$

118. (1)

	Ion	Colour of the flame
(A)	Cu^{+2}	green flame with blue centre
(B)	Sr^{2+}	Crimson Red
(C)	Ba^{2+}	Apple green
(D)	Ca^{2+}	Brick red

119. (2) (A)-(iii), (B)-(iv), (C)-(ii), (D)-(i)

120. (1) Al^{3+} and Fe^{3+} sulphides hydrolyse in water Ni^{2+} and Zn^{2+} require basic medium with H_2S to form ppt.; Ca^{2+} and Ba^{2+} sulphides are soluble, hence only Cu^{2+} ions will form the ppt of CuS.

121. (3) 1 Naphthyl amine and sulphanilic acid in acetic acid are used for the detection of NO^-_2

$NO^-_2 + CH_3COOH \longrightarrow HNO_2 + CH_3COO^-$

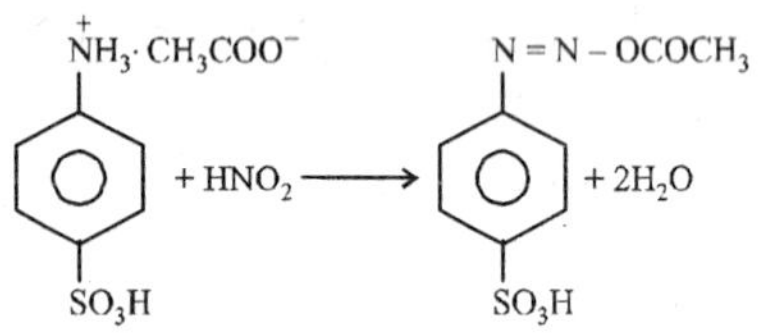

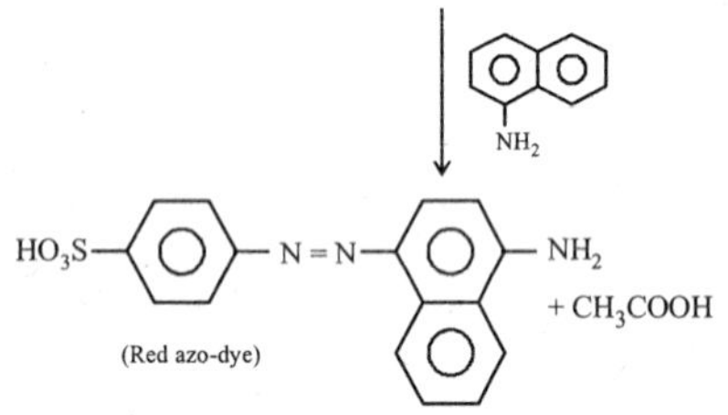

122. (2) $\underset{(X)}{SO_2 + \text{dil. } H_2SO_4 \rightarrow SO_3(g)}$

$SO_3 + K_2Cr_2O_7 \xrightarrow[H_2SO_4]{\text{dil.}} \underset{\text{(Green)}}{Cr_2(SO_4)_3}$

(Y)

123. (2) Due to common ion effect, insufficient concentration of sulphide ion (S^{2-}) is produced. Thus the ionic product of Ni^{2+} and S^{2-} ions is less than the K_{sp} of NiS, so NiS will not precipitate out.

124. (4)

$Cl^- + Na^+ \longrightarrow \underset{\text{(Neutral)}}{NaCl} \xrightarrow{AgNO_3}$

AgCl (white ppt)

(Insoluble in dilute HNO_3)

125. (4)

126. (4) Region 2 (blue flame) will be the hottest region of Bunsen flame as shown in given figure.

127. (2) Fe^{3+}, radical gives blood red with CNS^-

$Fe^{3+} + 3CNS^- \rightarrow Fe(CNS)_3$ Ferric sulphocyanide (Blood red colouration)

Fe^{3+} gives blue colour on treatment with $K_4[Fe(CN)_6]$

$4Fe^{3+} + 3K_4[Fe(CN)_6] \rightarrow 12K^+ + Fe_4[Fe(CN)_6]_3$

Ferricferrocyanide (Prussian blue)

Cl^- radical gives positive chromyl chloride test.

128. (3) Co^{2+} ion is precipitated by H_2S in presence of NH_4OH which is a group reagent of group IV in cationic analysis. Other three cations are palced in group II and are precipitated as sulphides by H_2S in the presence of dil. HCl.

129. (1) Zn^{2+} salts are white. Usually Fe^{2+} salts are rarely pink. Cu^{2+} salts are usually blue in hydrated form Co^{2+} is pink in aqueous solution.

130. (1) The reason for this is the high ionization of Na_2CO_3 as compared to $(NH_4)_2CO_3$, in water into Na^+ and CO_3^{2-}. Now the higher concentration of CO_3^{2-} is available which exceeds the solubility product of group V radicals as well as that of magnesium radicals. Leading to the precipitation of group V radicals, as well as Mg^{2+}.

131. (1) Only Fe^{3+} ions give blood red colouration with SCN^- ions.

$$Fe^{3+} + 3SCN^- \longrightarrow Fe(SCN)_3$$

132. (3) Solid potassium dichromate when heated with concentrated sulphuric acid and a soluble chloride gives orange red vapours of a volatile oily liquid CrO_2Cl_2.

$K_2Cr_2O_7 + 4NaCl + 6H_2SO_4 \longrightarrow 2KHSO_4 + 4NaHSO_4 + \underset{\text{chromyl chloride}}{2CrO_2Cl_2}$

133. (4) $Hg_2Cl_2 + 2NH_4OH \longrightarrow Hg + HgNH_2Cl + NH_4Cl + 2H_2O$

134. (1) Between AgCl and AgI, AgI is less soluble, hence ammonia can dissolve ppt. of AgCl only due to formation of complex as given below:

(1) $AgCl + 2NH_3 \longrightarrow [Ag(NH_3)_2]Cl$

(2) $Fe^{3+} + \overset{(+2)}{[Fe(CN)_6]^{4-}} \longrightarrow \underset{\text{Prussian blue}}{\overset{(+3)\ (+2)}{Fe_4[Fe(CN)_6]_3}}$

(3) Both $KHCO_3$ and $Ca(HCO_3)_2$ are soluble in water and on boiling we get a precipitate of $CaCO_3$.

(4) Mn salts give a colourless borax bead test in the reducing flame.

135. (3) When H_2S is passed through Hg_2S we get a mixture of mercurous sulphide and mercury ($Hg_2S + Hg$).

136. (4) When we add NH_4Cl, it suppresses the ionisation of NH_4OH and prevents the precipitation of higher group hydroxide in gp(III) due to common ion effect.

Note: Further, ferric chloride and chromium chloride form different colour precipitates with NH_4OH.

$FeCl_3 + 3NH_4OH \longrightarrow \underset{\text{Reddish brown}}{Fe(OH)_3\downarrow} + 3NH_4Cl$

$CrCl_3 + 3NH_4OH \longrightarrow \underset{\text{Bluish green}}{Cr(OH)_3\downarrow} + 3NH_4Cl$

Decreasing the concentration of OH^- ions will also do the same thing by shifting the equilibrium towards the supression of ionization of NH_4OH. Thus, both (2) and (3) are correct, and therefore (4) is correct option.

137. (1) Reaction of $KMnO_4$ and KI in basic medium:

$2KMnO_4(aq) + 6KI(aq) + 4H_2O(l) \longrightarrow 2MnO_2 + 3I_2 + 8KOH$

Iodine pass through thiosulphate solution

$$2Na_2S_2O_3 + I_2 \longrightarrow 2NaI + Na_2S_4O_6$$

Gram equivalent of $MnO_4^- = \text{Molarity} \times n_{\text{factor}}$

$= 0.2 \times \frac{500}{1000} \times 3 = 0.3$

Gram equivalent of KI $= 1.5 \times \frac{500}{1000} \times 1 = 0.75$

MnO_4^- act as limiting reagent, thus

Gram equivalent of MnO_4^-

= Gram equivalent of hypo solution ($Na_2S_2O_3$)

$\Rightarrow 0.3 = x \times \frac{300}{1000} \times 1$

Molarity of hypo ($Na_2S_2O_3$) solution = 1 M.

138. (4) NTA Ans. (3)

Zinc ferrocyanide; $Zn_2[Fe(CN)_6]$ = White precipitate

Potassium ferricferrocyanide; $KFe[Fe(CN)_6]$ = Deep blue precipitate (prussian blue)

Cupric ferrocyanide; $Cu_2[Fe(CN)_6]$ = Chocolate brown precipitate

Calcium ferrocyanide; $Ca_2[Fe(CN)_6]$ = White precipitate

139. (0) $Co^{2+} + H_2S \rightarrow \underset{(A)}{CoS\downarrow}$ (Black)

$\underset{(A)}{CoS} + \text{Aqu-regia} \rightarrow \underset{(B)}{Co^{2+}(aq)} + NOCl + S + H_2O$

$Co^{2+}(aq) + KNO_2 + CH_3COOH$

$\downarrow$

$K_3[Co(NO_2)_6] + NO + S + H_2O$

In $K_3[Co(NO_2)_6]$, $Co^{+3}: 3d^6 4s^0$

Co^{3+}: d^2sp^3 Hybridisation

Number of unpaired $e^- = 0$ (NO_2 is a strong Field Ligand)

$\Rightarrow$ Magnetic moment $= \sqrt{n(n+2)} = 0$ B.M

140. (5)

141. (78) CaF_2 does not evolve any gas with concentrated H_2SO_4.

NaBr gives brown coloured gas with conc. H_2SO_4.

$NaNO_3$ gives brown coloured gas with conc. H_2SO_4.

KI gives violet coloured I_2 gas with conc. H_2SO_4.

Molar mass of CaF_2 is 78 mol^{-1}.

142. (4) Flame test is applicable to metals with low ionization energy. Such metals are:

$Sr^{2+}, Ba^{2+}, Ca^{2+}, Cu^{2+}$.

$Zn^{2+}, Co^{2+}, Fe^{2+}$ do not give flame test.

143. (1) $\underset{\text{brown ring}}{[Fe(H_2O)_5NO]^{2+}}$

Oxidation state of Fe = +1. (NO exists as NO^+ in the given complex)

144. (4) PbS, CuS, As_2S_3, CdS are soluble in 50% HNO_3. HgS, Sb_2S_3 are insoluble in 50% HNO_3.

145. (2) Primary standard must be soluble for standard solution formation.

146. (4) The primary standard is a highly pure and stable compound with a known exact composition that can be accurately weighed and dissolved to create a solution of known concentration.

NaOH is hygroscopic and can't be used.

$FeSO_4 \cdot 6H_2O$ is unstable and can be easily oxidised.

$Na_2Cr_2O_7$ is hygroscopic and can't be used.

147. (2) When 10 mL, 2M NaOH solution is added to 20 mL of 1M HCl solution :

$$NaOH + HCl \rightarrow NaCl + H_2O$$

Initial : $MV = 2 \times 0.01 = 0.02$ mol $\quad MV = 1 \times 0.02 = 0.02$ mol

Final : 0 $\quad$ 0

$\therefore$ Resulting solution becomes neutral.

Now, 10 mL of above neutral solution is poured into a flask containing 2 mol HCl and made solution 100 mL with distilled water.

$$\Rightarrow \text{Molarity of HCl} = \frac{2}{100} \times 1000 = 20M$$

148. (2) HCl is strong acid and will be neutralised first.

From the given graph 2 mL NaOH solution is used for neutralisation of HCl and 3 mL NaOH solution is used for neutralisation of CH_3COOH.

This replaces H^+ with Na^+ and conductance decreases first.

$\therefore$ Mol of HCl = Mol of NaOH used

$M \times 40 = 0.1 \times 2$

$M = 0.005$

Mol of CH_3COOH = Mol of NaOH used

$M \times 40 = 0.1 \times 3$

$M = 0.0075.$

149. (2) For the titration: Oxalic acid v/s $KMnO_4$

$$2MnO_4^- + 5(COO)_2^{2-} + 16H^+ \rightarrow 10CO_2 + 2Mn^{2+} + 8H_2O$$

At room temperature this reaction proceeds at slow rate but upon heating around 60° C, the reaction becomes fast. Manganese(II) ions catalyse the reaction; thus, the reaction is autocatalytic; once manganese(II) ions are formed, it becomes progressively faster.

No heating is required for Mohr's salt as the reaction is already fast. Also heating may result into oxidation of Fe^{2+} into Fe^{3+}.

150. (1) Higher the number of milli moles of acid or base reacted higher will be temperature rise.

Option (1) n_{acid} or n_{base} reacted = 30 m mol

Option (2) n_{acid} or n_{base} reacted = 30 m mol

but less energy will be released by neutralisation reaction of weak acid hence option (2) can not be correct.

Option (3) $\Rightarrow$ 20 m mol

Option (4) $\Rightarrow$ 25 m mol

151. (4) $KMnO_4$ & NaOH $\rightarrow$ Secondary standard.

Primary standard should not be Hygroscopic.

152. (2) Enthalpy of neutralization of strong acid with strong base is always –57 kJ / mol because strong monoacidic base gives one mole of OH^- ion and strong monobasic acid gives one mole of H^+ ion which form one mole of water.

153. (1) Potassium hydrogen phthalate is a primary standard for standardisation of sodium hydroxide solution as it is economical and its concentration does not changes with time. Phenolphthalein can acts as indicator in acid-base titration as it shows colour in pH range 8.3 to 10.1.

154. (3) Methyl orange is weak base.

Benzenoid structure $\rightleftharpoons$ Quinonoid structure

(yellow coloured) $\quad$ (Red coloured) (more intense)

155. (4) In redox titration, indicators are sensitive to change in oxidation potential and in acid base titration, indicators are sensitive to change in pH of solution. Both statements are false.

156. (3) $N_1V_1 = N_2V_2$

$\Rightarrow 0.02V_1 = 0.02 \times 10 \Rightarrow V_1 = 10$ mL

157. (3) KOH absorb excess CO_2 from the atmosphere and form $KHCO_3$ which remains soluble in the solution. Therefore concentration of KOH should be checked before use.

158. (1)

Indicator	pH range
Methyl orange	3.2 – 4.5
Phenolpthalein	8.3 – 10.5

Methyl orange may be used for a strong acid vs strong base and strong acid vs weak base titration. Phenolphthalein may be used for a strong acid vs strong base and weak acid vs strong base titration.

159. (1) $I^- + H_2O_2 \longrightarrow \underset{(A)}{I_2} + H_2O$

$$I_2 + \underset{(Indicator)}{Starch} \longrightarrow Blue$$

160. (2)

161. (1) For kinetic study of reaction of iodide ion with H_2O.

- The starch solution should be used freshly prepared.
- The conc. of sodium thiosulphate solution should be less than KI solution.
- The time should be recorded after the appearance of blue colour.

162. (1) $2KMnO_4 + 16HCl \rightarrow 2MnCl_2 + 2KCl + 8H_2O + Cl_2$

HCl gets oxidised by $KMnO_4$ into Cl_2

163. (1) Titration curve of NH_4OH vs HCl

164. (1) In acidic medium the reaction for $KMnO_4$ with oxalic acid is

$$\underset{+7}{MnO_4^-} + 8H^+ + 5e^- \longrightarrow \underset{+2}{Mn^{2+}} + 4H_2O$$

Change in oxidation state is $= 7 - 2 = 5.$

165. (1)

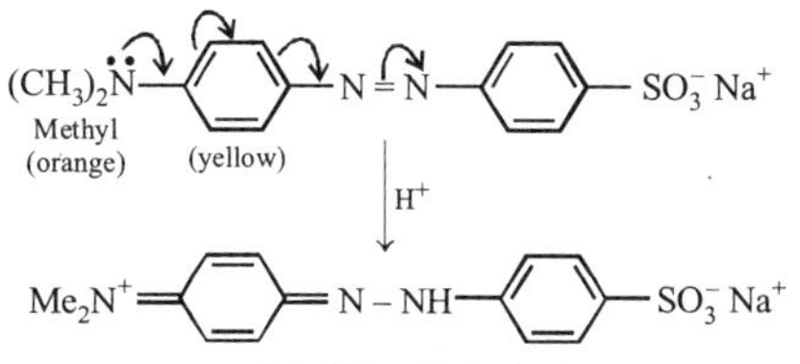

Quinoid form (Red colour)

166. (3) In toilet cleaning liquid, the main constituent is HCl, which can cause skin burn, so it should be treated with $NaHCO_3$, which can easily neutralise the acid.

167. (4) In this acid base titration, bunsen burner and measuring cylinder are of no use while other laboratory equipments will be required, i.e., phenolphthalein, burette and pipette.

168. (2) Oxalic acid is a primary standard solution whereas H_2SO_4 is a secondary standard solution. Fresh oxalic acid is prepared in a conical flask and NaOH is mixed drop by drop from a burette.

169. (3) pH range for methyl orange is

$$\underset{\text{Pinkish red}}{\longleftarrow} 3.9 - 4.5 \underset{\text{Yellow}}{\longrightarrow}$$

Generally, weak bases have pH greater than 7. When methyl orange is added to a weak basic solution, solution becomes yellow. This solution is then titrated against a strong acid, at the end point pH will be less than 3.1.

$\therefore$ Solution becomes pinkish red.

170. (3) Oxalic acid is used as a primary standard for NaOH standardizing.

171. (1) The titration plot of absorbed right (A) vs volume of ligand L (V) will initially remain constant and then increased sharply often the end point reached.

172. (3) The titration of oxalic acid with $KMnO_4$ in presence of HCl gives unsatisfactory result because of the fact that $KMnO_4$ can also oxidise HCl along with oxalic acid. HCl on oxidation gives Cl_2 and HCl reduces $KMnO_4$ to Mn^{2+}, thus the correct answer is (3).

173. (2) Molisch's test : This is a general test for carbohydrates. One or two drops of alcoholic solution of α-naphthol is added to 2 mL glucose solution. 1 mL of conc. H_2SO_4 solution is added carefully along the sides of the test-tube. The formation of a violet ring at the junction of two liquids confirms the presence of a carbohydrate or sugar.

174. (3) $MnO_4^- + I^- \rightarrow MnO_2 + I_2$

$I_2 + S_2O_3^{2-} \rightarrow S_4O_6^{2-} + I^-$

gram eq. of $KMnO_4$ = gram eq. of $Na_2S_2O_3$

$$0.2 \times \frac{500}{1000} \times 3 = 0.1 \times V \times 1$$

$V = 3L$

175. (375)

Mohr's salt is $FeSO_4 \cdot (NH_4)_2SO_4 \cdot 6H_2O$

Only Fe^{2+} is oxidised

$[Fe^{2+} \rightarrow Fe^{3+} + e^-] \times 6$

$Cr_2O_7^{2-} + 14H^+ + 6e^- \rightarrow 2Cr^{3+} + 7H_2O$

$6Fe^{2+} + Cr_2O_7^{2-} + 14H^+ \rightarrow 6Fe^{3+} + 2Cr^{3+} + 7H_2O$

$\therefore$ 1 mole of $K_2Cr_2O_7$ reacts with 6 moles of Mohr's salt

Volume of Mohr's salt = 750 cc = 0.750 L

Moles = 0.6 M × 0.750L = 0.45 moles

$$\text{Moles of } K_2Cr_2O_7 = \frac{0.45}{6} = 0.075 \text{ moles}$$

$$\text{Molarity of } K_2Cr_2O_7 = \frac{0.075}{0.2} = 0.375 \text{ M}$$

Molarity = $X \times 10^{-3}$ M

$0.375 = X \times 10^{-3}$ M

$\therefore X = 375$

176. (1825)

$Ba(OH)_2(aq) + 2HCl(aq) \rightarrow BaCl_2(aq) + 2H_2O(l)$
2.5 m mole   5 m mole

wt of HCl = 5 × 36.5 (milligram)
= 182.5 (milligram)
Hence x = 1825

177. (2500)

Mols of octane = $\frac{1.14}{114} = 0.01$ mol
Heat evolved = C × ΔT = 5 × 5 kJ = 25 kJ
∴ Magnitude of Heat of combustion

$= \frac{\text{Heat}}{\text{Number of moles}} = \frac{25}{0.01} = 2500$ kJ/mol

178. (82)

(1) $CuSO_4(s) + 5H_2O \xrightarrow{-x} CuSO_4.5H_2O$
(2) $CuSO_4.5H_2O + H_2O \xrightarrow{+12} CuSO_4(aq)$

$CuSO_4 + H_2O \xrightarrow{-70} CuSO_4(aq)$

from (1) & (2) –70 = –x + 12 ⇒ x = 82.

179. (2) $H^+ + OH^- \rightarrow H_2O \Rightarrow x$ (for neutralization of HCl)
$2H^+ + 2OH^- \rightarrow 2H_2O \Rightarrow y = 2x$ (for H_2SO_4)
⇒ y/x = 2.

180. (8) Reaction taking place is :

$2MnO_4^- + 16H^+ + 5C_2O_4^{2-} \longrightarrow 2Mn^{2+} + 10CO_2 + 8H_2O$

eq.($KMnO_4$) = eq.($H_2C_2O_4$)
M × 2 × 5 = 2 × 20 × 2
M = 8M

181. (1)

182. (68) By balancing the redox reaction we get

$10[FeSO_4.(NH_4)_2SO_4.6H_2O] + 2KMnO_4 + 8H_2SO_4$
↓
$5Fe_2(SO_4)_3 + 2MnSO_4 + 10(NH_4)_2SO_4 + K_2SO_4 + 68H_2O$

183. (2) $pk_{In} = -\log(4 \times 10^{-10}) = 9.4$
Indicator range
⇒ $pk_{In} \pm 1$, i.e. 8.4 to 10.4
∴ Statement (B) is correct.
In acidic medium, phenolphthalein is in unionized form and is colourless. Hence statement (D) is correct.
It is used as an indicator for the titration of strong acid with strong base and for weak acid with strong base. Phenolphthalein is a weak acid.

184. (24) $\overset{(+6)}{Cr_2}O_7^{2-} \rightarrow Cr^{3+}$; $Fe^{2+} \rightarrow Fe^{3+}$

Equivalents of $K_2Cr_2O_7$ ≡ Eq. of Fe^{2+}
(Molarity × n.f. × volume) of $K_2Cr_2O_7$ ≡
(Molarity × n.f. × volume) of Fe^{2+}
⇒ 0.02 × 6 × 20 = M × 1 × 10
⇒ M = 0.24 = 24 × 10^{-2}

185. (1) No. of equivalents of H_2SO_4
= 100 × 0.1 × 2 = 20
No. of equivalents of NaOH = 50 × 0.1 = 5
No. of equivalents of H_2SO_4 left = 20 – 5 = 15
Total volume = (100 + 50) = 150 mL
⇒ 150 × x = 15

$x = \frac{1}{10} = 0.1N = 1 \times 10^{-1}N$

186. (4) $n_{eq.}$ of I_2 = n_{eq} of $Na_2S_2O_3$
= 20 × 0.002 × 1

2 × n_{mol} of I_2 = 0.4
n_{mol} of I_2 = 0.2 m mol
n_{mol} of Cu^{2+} = 0.2 × 2 × 10^{-3}

$[Cu^{2+}] = \frac{0.4 \times 10^{-3}}{10 \times 10^{-3}} = 0.04 = 4 \times 10^{-2}$

187. (13)

188. (30) The reaction is

$2KMnO_4 + 10FeSO_4(NH_4)SO_4 \cdot 6H_2O + 8H_2SO_4 \longrightarrow K_2SO_4 + 2MnSO_4 + 5Fe_2(SO_4)_3 + 10(NH_4)_2SO_4 + 68H_2O$

Fe in Mohr's salt is oxidised as :

$Fe^{2+} \longrightarrow Fe^{3+} + 1e^-$

$n_{factor} = 1$
Mn in $KMnO_4$ is reduced as :

$MnO_4^- + 8H^+ + 5e^- \longrightarrow Mn^{2+} + H_2O$

$n_{factor} = 5$
at end point; $N_1V_1 = N_2V_2$
⇒ $M_1 \times n_{f_1} \times V_1 = M_2 \times n_{f_2} \times V_2$
0.01 × 5 × V_1 = 0.005 × 1 × 20
V_1 = 20 mL used
∴ Volume left = 50 – 20 = 30 mL

189. (148) Milimoles of acid used = $\frac{30 \times 0.25}{2}$

Milli eq. of NH_3 = 30 × 0.25 = 7.5
Mass% of nitrogen

$= \frac{7.5 \times 10^{-3} \times 14 \times 100}{0.166} = 63\%$

190. (316) Let molarity of $KMnO_4$ = x
$KMnO_4 + FeSO_4 \rightarrow Fe_2(SO_4)_3 + Mn^{2+}$
n = 5 n = 1   (n-factor)
Equivalents of $KMnO_4$ = Equivalents of $FeSO_4$
(5 × x × 10 mL) = 1 × 0.1 × 10 mL ⇒ x = 0.02 M
Molar mass of $KMnO_4$ = 158 g/mol
Strength = Molarity × Molar mass
= 0.02 × 158 = 3.16 g/L

191. (1575) According to law of chemical equivalence, equivalent of $KMnO_4$ = equivalent of oxalic acid
$N_1V_1 = N_2V_2$
$M_1 \times nf \times V_1 = M_2 \times nf \times V_2$ (N = M × nf)
0.05 × 5 × 10 = M_2 × 2 × 10

$M_2 = \frac{0.05 \times 5}{2} = 0.125$

Strength of oxalic acid = Molarity × Molar mass
= 0.125 × 126 = 15.75 = 1575 × 10^{-2} g/L

192. (50)

$Na_2CO_3 + 2HCl \longrightarrow 2NaCl + CO_2 + H_2O$
Most precise value of HCl is 5 mL.
At equivalent point,
$M_{eq.}$ of Na_2CO_3
= $M_{eq.}$ of HCl ⇒ 2 × M × 10 = 1 × 0.2 × 5

⇒ $M = \frac{0.2 \times 5}{20} = 0.05$ M

= 50 × 10^{-3} M = 50 mM

193. (4)

194. (6) Average volume of NaOH solution used at end point

$= \frac{4.5 + 4.5 + 4.4 + 4.4 + 4.4}{5} = 4.44$ mL

At the end point, equivalents of NaOH
= Equivalents of oxalic acid
$N_1V_1 = N_2V_2 \Rightarrow N_1 \times 4.44 = (1.25 \times 2) \times 10$

$\Rightarrow N_1 = \frac{1.25 \times 2 \times 10}{4.44} = 5.63 \simeq 6$

Molarity of NaOH = N × acidity of NaOH
= 6 × 1 = 6 M.

195. (50) M. eq. of $K_2Cr_2O_7$ = M. eq. of FeC_2O_4

$\overset{(+2)}{Fe}\overset{(+3)}{C_2}O_4 + \overset{(+6)}{Cr_2}O_7^{2-} \longrightarrow Fe^{3+} + \overset{(+4)}{C}O_2 + Cr^{3+}$
n-factor of $K_2Cr_2O_7$ = 2(6 – 3) = 6
n-factor of FeC_2O_4 = 1(3 – 2) + 2(4 – 3) = 3
By law of equivalence,
n-factor of $K_2Cr_2O_7$ × no. of moles
= n-factor of FeC_2O_4 × no. of moles

$\Rightarrow \frac{V}{1000} \times 0.02 \times 6 = \frac{0.288}{144} \times 3 \Rightarrow V = 50$ mL

196. (85)

$\overset{-1}{H_2O_2} + \overset{(+7)}{Mn}O_4^- + H^+ \longrightarrow Mn^{2+} + \overset{0}{O_2} + H_2O$

Moles of $KMnO_4$ = $\frac{0.316}{158} = 2 \times 10^{-3}$

Equivalents of H_2O_2 = Equivalents of $KMnO_4$
Equivalents of $KMnO_4$ = 2 × 10^{-3} × 5 = 0.01
Moles of H_2O_2

$= \frac{\text{Equivalents of } H_2O_2}{n\text{-factor of } H_2O_2} = \frac{0.01}{2} = 0.005$

Mass of pure H_2O_2 = 0.005 × 34 = 0.170 g

Percentage of purity = $\frac{0.17}{0.2} \times 100 = 85\%$.

Part-B (JEE Advanced)

1. **(a)** If only S^{2-} is present, the Lassaigne test gives a violet coloured compound.

$S^{2-} + [Fe(CN)_5(NO^+)]^{2-} \rightarrow [Fe(CN)_5(NOS)]^{4-}$ (Violet)

Thus, NO^+ changes to NOS^-.

2. **(1791)**

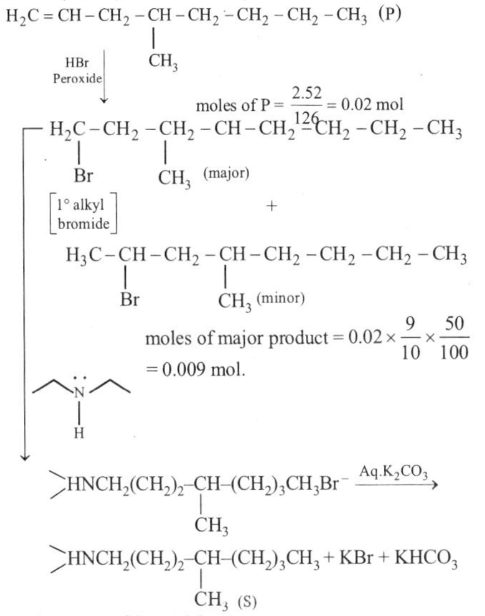

mol. mass of S = 199 g

moles of S = 0.009

mass of S = 0.009 × 199

= 1.7191 g = 1791 mg

3. **(1.62)** $Mg_2C_3 + 4H_2O \rightarrow 2Mg(OH)_2 + CH_3C \equiv CH$

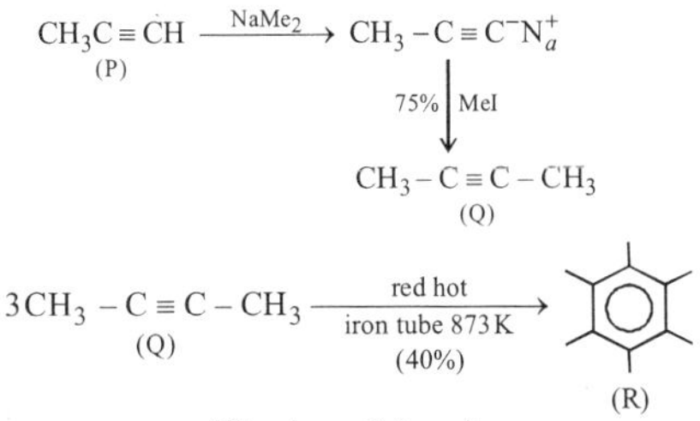

Given, amount of P = 4 g = 0.1 mol

From 0.1 mol of P, 0.01 mol of R will be produced.

So, the value of $x = 162 \times 0.01 = 1.62$ g

4. **(3.2)**

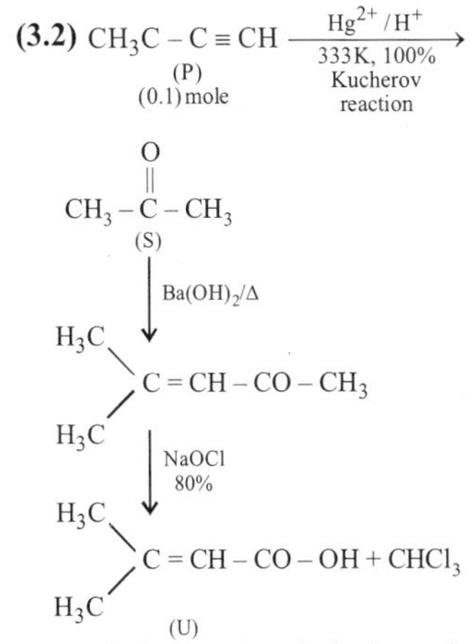

From 0.1 mol of P, 0.032 mol of U is produced.

So, the value of $y = 0.032 \times 100 = 3.2$ g

5. **(3.57)**

6. **(1.23)** The value of x is

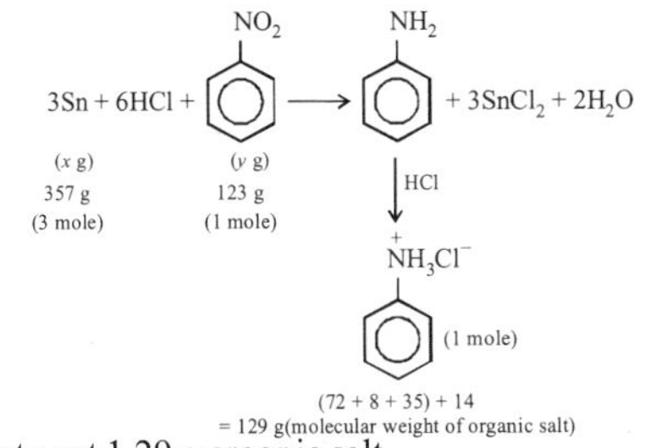

So to get 1.29 g organic salt.

We have to form 0.01 mole salt.

So 0.01 mole nitrobenzene is required or 0.03 mole Sn is required.

So the amount of nitrobenzene = 0.01 × 123 = 1.23 g

the amount of Sn required = 0.01 × 357 = 3.57 g

7. **(a, d)**

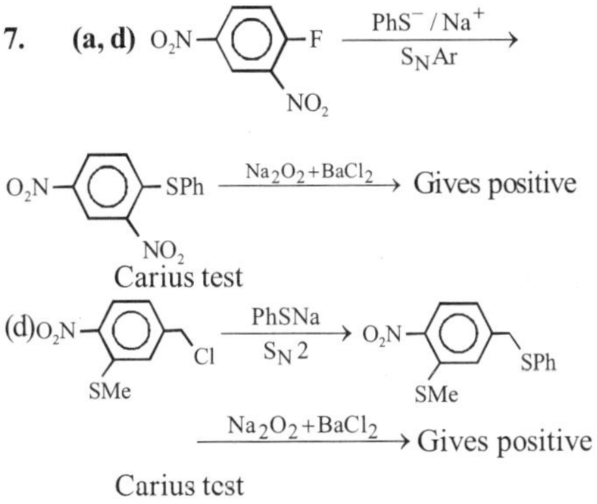

8. **(b)**

(P) 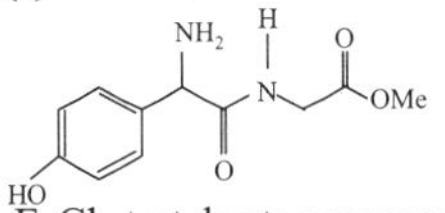

Gives $FeCl_3$ test due to presence of phenolic group and ninhydrin test due to presence of amide group (peptide bond) in the molecule.

(Q) 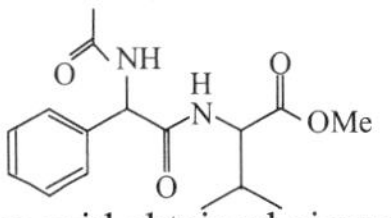

Amino acid obtained gives the ninhydrin test but not phthalein dye test.

(R) 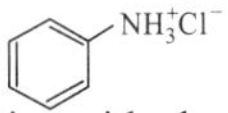

Reaction with phenyl diazonium salt gives yellow dye.

(S) 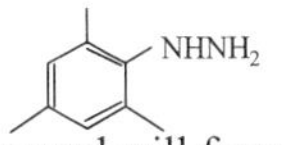

Compound will form hydrazone with aldehydic group of glucose.

9. **(d)**

(I) **Aniline :** 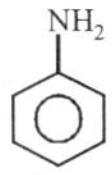 responds in Lassaigne's Test.

Na + (C and N) from aniline $\longrightarrow$ NaCN

$FeSO_4 + 6CN^- \longrightarrow [Fe(CN)_6]^{4-} + SO_4^{2-}$

$3[Fe(CN)_6]^{4-} + 4Fe^{3+} \xrightarrow{xH_2O} Fe_4[Fe(CN)_6]_3 \cdot xH_2O$

Prussian Blue

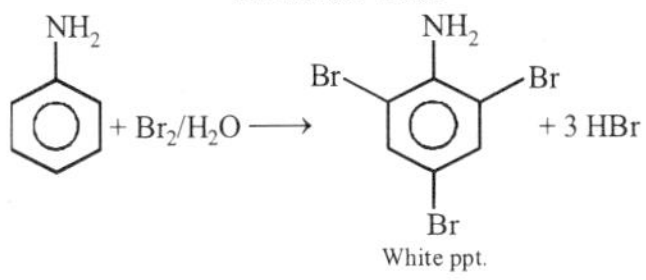

(I) – (P), (S)

(II) 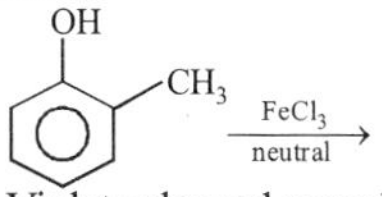

Violet coloured complex. It is the signature test for phenolic –OH

(II) – (T)

(III) $HOOC - CH(NH_2) - CH_2 - SH \xrightarrow[\text{(ii) Na-nitroprusside}]{\text{(i) Na extract}}$ Blood Red colour [NaSCN]

The –COOH group in cysteine produces effervescence with $NaHCO_3$.

(III) → (Q), (R)

(IV) Caprolactum 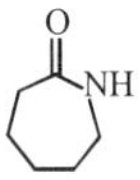

$\xrightarrow[\text{Test}]{\text{Lassaigne}}$ Prussian blue colour

(IV) → (P)

10. **(d)**

(P)

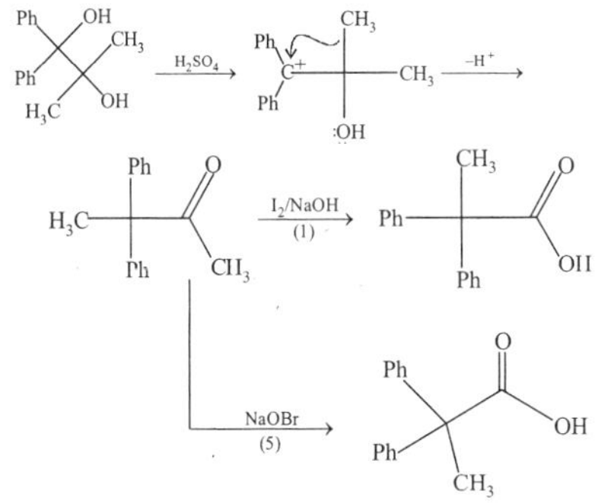

(Q)

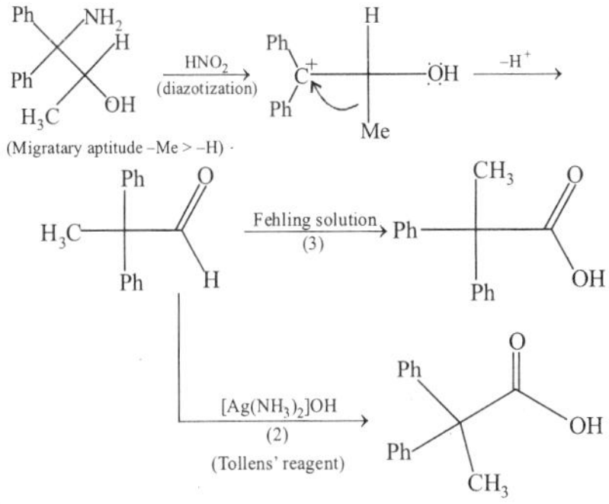

(R)

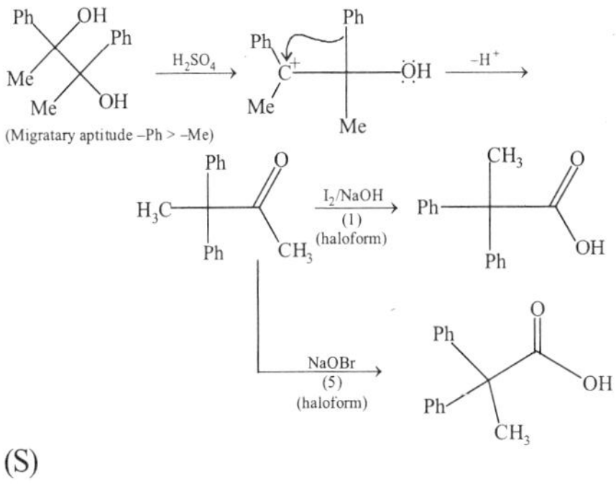

(S)

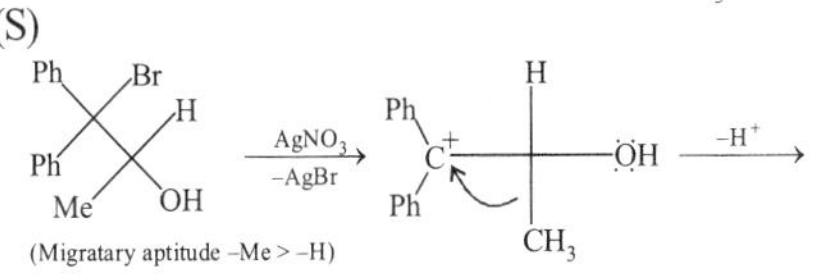

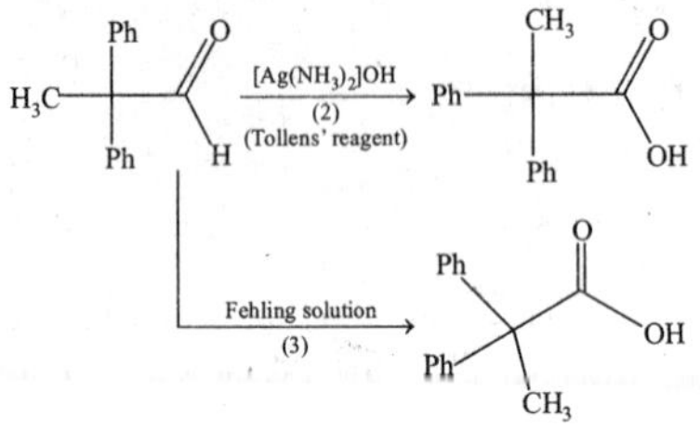

11. (c)

$$Na_2B_4O_7.10H_2O \xrightarrow{\Delta} Na_2B_4O_7 \xrightarrow{\Delta} 2NaBO_2 + B_2O_3$$

$$Cr_2O_3 + 3B_2O_3 \longrightarrow \underset{\text{Chromium meta borate (green colour)}}{2Cr(BO_2)_3}$$

12. (a)

	S^{2-}	SO_4^{2-}
Cu^{2+}	CuS (ppt)	$CuSO_4$ (soluble)
Ba^{2+}	BaS (soluble)	$BaSO_4$ (ppt)
$Pb(OAc)_2$	PbS (ppt)	$PbSO_4$ (ppt)
$Na_2[Fe(CN)_5NO]$	$Na_4[Fe(CN)_5(NOS)]$ Colour (not a ppt)	—

13. (d) The group reagent of fourth group is ammoniacal H_2S by which Zn^{2+} ion will be precipitated as ZnS, whereas Fe^{3+} ion and Al^{3+} ions will be precipitated as hydroxides.

14. (7) All except MnS (buff coloured) and SnS_2 (yellow) are black in colour.

15. (0.32)

$$\underset{\text{MW: 187}}{2Cu(NH_3)_2} + \underset{\text{(excess)}}{5KI} \longrightarrow$$

no. of moles $= \frac{3.74}{187} = 0.02$ mol

$$\underset{\text{(white ppt.)}}{Cu_2I_2\downarrow} + \underset{\text{(Dark Brown Solution)}}{KI_3} + 4KNO_3$$

$$KI_3 + H_2S \longrightarrow \underset{\text{ppt.}}{S\downarrow} + KI + 2HI$$

2 mol of $Cu(NO_3)_2$ produces 1 mole of KI_3 which is further produces – s 1 mol of S precipitate.

$\therefore$ 0.02 mol of $Cu(NO_3)_2$ will produce 0.01 mol

$S = 0.01 \times 32g = 0.32$ g of S

16. (a, b) $Pb(NO_3)_2 \xrightarrow{\text{dil.HCl}} \underset{\text{White ppt.}}{PbCl\downarrow}$

$$Bi(NO_3)_3 \xrightarrow{\text{dil.HCl}} \underset{\text{Water soluble}}{BiCl_3}$$

$$Hg(NO_3)_2 \xrightarrow{\text{dil.HCl}} \underset{\text{Water soluble}}{HgCl_2}$$

$$AgNO_3 \xrightarrow{\text{dil.HCl}} \underset{\text{White ppt}}{AgCl\downarrow}$$

$$Zn(NO_3)_2 \xrightarrow{\text{dil.HCl}} \underset{\text{Water soluble}}{ZnCl_2}$$

$$Pb(NO_3)_2 \xrightarrow{\text{NaOH(dil.)}} \underset{\text{White ppt.}}{Pb(OH)_2\downarrow}$$

$$Zn(NO_3)_2 \xrightarrow{\text{NaOH(dil.)}} \underset{\text{White ppt.}}{Zn(OH)_2\downarrow}$$

$$Bi(NO_3)_2 \xrightarrow{\text{NaOH(dil.)}} \underset{\text{White ppt.}}{Bi(OH)_3\downarrow}$$

$$AgNO_3 \xrightarrow{\text{NaOH(dil.)}} \underset{\text{Brown ppt.}}{Ag_2O\downarrow}$$

$$Hg(NO_3)_2 \xrightarrow{\text{NaOH(dil.)}} \underset{\text{Yellow ppt.}}{HgO\downarrow}$$

17. (b, c, d) Aqua regia is a mixture of conc. HCl and conc. HNO_3 in 3:1 ratio.

When gold dissolves in aqua regia, the species formed in which gold is in +3 oxidation state.

In the absence of air the reaction between gold and aquaregia

$$Au + HNO_3 + 4HCl \rightarrow$$

$$H^+ + [AuCl_4]^- + 2H_2O + NO$$

NO_2 will not be produced. Yellow colour is due its decomposition into NOCl and Cl_2.

$$3HCl + HNO_3 \longrightarrow NOCl + Cl_2 + 2H_2O$$

18. (b,d)

(a) Cu^{2+} shows characteristic green colour whereas Mn^{2+} shows yellowish green colour in flame test. Thus, they cannot be distinguished by flame test.

(b) Only Cu^{2+} can give black precipitate of CuS in acidic medium on passing H_2S. Mn^{2+} forms ppt. in ammonical medium.

(c) Both Cu^{2+} and Mn^{2+} show the formation of precipitate by passing H_2S in faintly basic medium.

(d) $E^{\circ}_{Cu^{2+}/Cu}(+0.34V) > E^{\circ}_{Mn^{2+}/Mn}(-1.18V)$

as per electrochemical series.

19. (c, d) Only group II cations precipitate as sulphide with H_2S in acidic medium that is (Cu^{2+}, Pb^{2+}) and (Hg^{2+}, Bi^{3+})

20. (a)

$$\underset{\text{(Group–4 cation)}}{Mn^{+2}} \xrightarrow{H_2S+NH_4OH} \underset{\text{Pink / buff ppt.}}{MnS\downarrow}$$

$$\underset{\text{(Group–5 cation)}}{Ba^{+2}} \xrightarrow{(NH_4)_2CO_3+NH_4OH} \underset{\text{white ppt.}}{BaCO_3}$$

$$\underset{\text{(Group–3 cation)}}{Al^{+3}} \xrightarrow{NH_4Cl + NH_4OH} \underset{\text{white ppt.}}{Al(OH)_3\downarrow}$$

$$\underset{\text{(Group–2A cation)}}{Cu^{+3}} \xrightarrow{H_2S + \text{dil HCl}} \underset{\text{black ppt.}}{CuS\downarrow}$$

21. (c)

$$K_4[Fe(CN)_6] \xrightarrow[\text{absence of air}]{FeSO_4} \underset{\text{White precipitate (X)}}{K_2Fe[Fe(CN)_6]}$$

$$\downarrow \text{Air}$$

$$\underset{\text{(Prussian Blue)}}{Fe_4[Fe(CN)_6]_3}$$

22. (d)

$$FeSO_4\text{solution} \xrightarrow[\text{slow addition of conc.}H_2SO_4]{NaNO_3} \underset{\text{(Brown ring complex)}}{[Fe(H_2O)_5NO]SO_4}$$

23. (d) $[ZnCl_4]^{2-}$ and $[Zn(CN_4)]^{2-}$ are tetrahedral complexes.

$$\underset{(M2)}{Zn^{2+}} + \underset{(S)}{2OH^-} \longrightarrow \underset{\text{white ppt.}}{Zn(OH)_2\downarrow}$$

$$\xrightarrow[\text{excess}]{OH^-} \underset{\text{soluble}}{[Zn(OH)_4]^{2-}}$$

24. (b) $\underset{(M1)}{Ni^{2+}} + \underset{(Q)}{4HCl} \longrightarrow \underset{\text{Tetrahedral}}{[NiCl_4]^{2-}}$

$$\underset{(M1)}{Ni^{2+}} + \underset{(R)}{4KCN} \longrightarrow \underset{\text{Square planar}}{[Ni(CN)_4]^{2-}}$$

25. (d) The filtrate on treatment with ammoniacal H_2S gives a precipitate which dissolves in aqueous NaOH containing H_2O_2 giving a coloured solution. It contains Cr^{3+} ion.

$$Cr^{+3} + 3NH_4OH \xrightarrow[\text{medium}]{H_2S \text{ in ammoniacal}} \underset{\text{(R)(Green)}}{Cr(OH)_3\downarrow}$$

$$2Cr(OH)_3 + 3H_2O_2 + 4NaOH \rightarrow \underset{\text{(yellow colour)}}{2Na_2CrO_4} + 8H_2$$

26. (a) Lead salts give white precipitate of $PbCl_2$ with dil. HCl which is soluble in hot water.

$Pb^{++} + 2Cl^- \longrightarrow PbCl_2$ (White ppt) soluble in hot water (P)

27. (0.10) No. of eq. of oxalic acid = No. of eq. of NaOH

Using average of the NaOH volume used in five experiments (V_{NaOH})avg. = 10 mL

$$\frac{5.00 \times 0.10}{1000} \times 2 = \frac{10.0 \times M_{NaOH}}{1000} \times 1$$

$\therefore M_{NaOH} = 0.10$ mol/L

28. (126) The balanced equations are

$$2MnCl_2 + 5K_2S_2O_8 + 8H_2O \longrightarrow 2KMnO_4 + 4K_2SO_4 + 6H_2SO_4 + 4HCl \quad ...(1)$$

$$2KMnO_4 + 5H_2C_2O_4 + 3H_2SO_4 \longrightarrow K_2SO_4 + 2MnSO_4 + 8H_2O + 10CO_2 \quad ...(2)$$

Mass of oxalic acid added = 225 mg

Milimoles of oxalic acid added $= \frac{225}{90} = 2.5$

From equation (2)

Milimoles of $KMnO_4$ used to react with oxalic acid = 1

(5 m mole $H_2C_2O_4 \equiv 2$ m mole of $KMnO_4$)

and milimoles of $MnCl_2$ required initially = 1

$\therefore$ Mass of $MnCl_2$ required initially $= 1 \times 126 = 126$ mg

(Molar mass of $MnCl_2 = 126$)